Woldman's Engineering Alloys SIXTH EDITION

Edited by

Robert C. Gibbons, M.S.
Formerly Chief Metallurgist, Utica Division
The Bendix Corp.

American Society for Metals
Metals Park, Ohio 44073

Copyright © 1979
by the
AMERICAN SOCIETY FOR METALS

All rights reserved

No part of this book may be reproduced, stored in a retrieval system, or transmitted, in any form or by any means, electronic, mechanical, photocopying, recording, or otherwise, without the prior written permission of the publisher.

Nothing contained in this book is to be construed as a grant of any right of manufacture, sale, or use in connection with any method, process, apparatus, product or composition, whether or not covered by letters patent or registered trademark, nor as a defense against liability for the infringement of letters patent or registered trademark.

Library of Congress Cataloging in Publication Data

Woldman, Norman Emme, 1899-1969.
  Woldman's Engineering alloys.

  1. Alloys.   I. Gibbons, Robert C.   II. Title.
TA483.W64   1979   620.1'6   79-20379
ISBN 0-87170-086-7

Preface

Because of rapid changes taking place throughout the industrial world, e.g., company mergers, name changes and relocations, and the development of new alloys and the discontinuance of others, an early revision of this book has been found to be in order. One immediately apparent alteration is the new title: *Woldman's Engineering Alloys*. The previous title, *Engineering Alloys,* was revised in order to pay respect to Dr. Norman E. Woldman, who was responsible for developing the various editions of the book into the asset that it is today.

Among industrial changes that have been incorporated into this volume is the use of association numbers, such as those assigned by AISI, CDA and AA in the United States, in recognition of the trend towards eliminating alloy trade names (many of which are registered and protected against unauthorized use). In order to accommodate the estimated 9000 new alloys that have been developed since publication of the Fifth Edition without materially enlarging the book, many obsolete and discontinued alloys have been eliminated from Section I. For reference purposes, these alloys have been included in Section IV, along with some general information on composition and suppliers. Other changes include increased cross indexing of alloy names and more information identifying replaced alloys.

As editor of this edition, I wish to thank the many companies that furnished revised and up-to-date information about their products. I also want to express my gratitude to the Aluminum Association, the American Iron and Steel Association, the Copper Development Association and the Steel Founders' Society for their permission to include their data and charts in Section V. A special thanks is also expressed to Dr. John P. Frick for revising the obsolete alloy section of the Fifth Edition.

Robert C. Gibbons

Contents

Section I: Alloy Data 1

Section II: Alphabetical List of Manufacturers 1609

Section III: Numerical List of Manufacturers 1641

Section IV: Obsolete Alloys 1671

Section V: Association Standards 1763

Abbreviations 1814

This book is divided into five sections, as follows: Section I lists alloys alphabetically and includes data on the manufacturer, chemical composition, typical uses and applications. Section II is an alphabetical arrangement of the manufacturers referred to in Section I. Section III lists these manufacturers in numerical order according to M-number. Section IV is an alphabetical list of obsolete alloys. Section V lists various association standards and gives the alloy designations and compositions covered by each.

Table 1 Chemical Elements and Their Symbols

Name	Symbol	Name	Symbol
Aluminum	Al	Mercury	Hg
Antimony	Sb	Molybdenum	Mo
Argon	A	Neodymium	Nd
Arsenic	As	Neon	Ne
Barium	Ba	Nickel	Ni
Beryllium	Be	Nitrogen	N
Bismuth	Bi	Osmium	Os
Boron	B	Oxygen	O
Bromine	Br	Palladium	Pd
Cadmium	Cd	Phosphorus	P
Calcium	Ca	Platinum	Pt
Cesium	Cs	Potassium	K
Carbon	C	Praseodymium	Pr
Cerium	Ce	Radium	Ra
Chlorine	Cl	Radon	Rn
Chromium	Cr	Rhodium	Rh
Cobalt	Co	Rubidium	Rb
Columbium (a)	Cb	Ruthenium	Ru
Copper	Cu	Samarium	Sa
Dysporsium	Dy	Scandium	Sc
Erbium	Er	Selenium	Se
Europium	Eu	Silicon	Si
Fluorine	F	Silver	Ag
Gadolinium	Gd	Sodium	Na
Gallium	Ga	Strontium	Sr
Germanium	Ge	Sulfur	S
Gold	Au	Tantalum	Ta
Hafnium	Hf	Tellurium	Te
Helium	He	Terbium	Tb
Holmium	Ho	Thalium	Tl
Hydrogen	H	Thorium	Th
Indium	In	Thulium	Tm
Iodine	I	Tin	Sn
Iridium	Ir	Titanium	Ti
Iron	Fe	Tungsten (b)	W
Krypton	Kr	Uranium	U
Lanthanum	La	Vanadium	V
Lead	Pb	Xenon	Xe
Lithium	Li	Ytterbium	Yb
Lutecium	Lu	Yttrium	Y
Magnesium	Mg	Zinc	Zn
Manganese	Mn	Zirconium	Zr

(a) Columbium = Niobium. (b) Tungsten = Wolfram.

Table 2 Physical Constants of the Principal Alloy Forming Elements

Element	Atomic weight	Melting point, °F	Boiling point, °F	Density, Mg/m³	Atomic volume	Linear coefficient of thermal expansion per °C × 10⁻⁶, 0-100 °C	Specific heat, cal/g/°C at room temperature	Thermal conductivity, cal/c³/°C at room temperature	Electrical resistivity, μΩ/c³	Crystallization shrinkage, %	Young's modulus, lb./in.² × 10⁶
Aluminum	27.0	1220	3,272	2.70	10.0	23.1 (25-100 °C)	0.21	0.48	2.83 (20 °C)	6.6	10
Antimony	120.2	1166	2,624	6.69	18.0	10.5	0.05	0.043	38.6 (0 °C)	1.4	11
Arsenic	74.96	1562	680	5.73	13.1	5.0	0.08		35 (0 °C)		
Beryllium	9.1	2336	5,020	1.85	4.9	12.4	0.425	0.385	4.3 (20 °C)		42.7
Bismuth	209.0	519.8	2,606	9.75	21.4	13.2	0.03	0.019	119 (0 °C)	-3.3	4.6
Cadmium	112.4	609.6	1,432	8.64	13.0	31.6	0.055	0.22	7.54 (18 °C)	4.7	10
Calcium	40.07	1490		1.54	26.0		0.16				
Carbon Diamond	12.005		6,512	3.52	3.42						
Graphite				2.25	5.35		0.16	0.04			
Cerium	140.25	1184		6.79	20.6	1.18	0.11				
Chromium	52.0	2939	3,992	6.92	7.5	7.86	0.045		2.6 (0 °C)		30
Cobalt	58.97	2631		8.71	6.8		0.105		9.7 (20 °C)		
Copper	63.57	1981	4,190	8.93	7.15	12.36	0.10	0.92	1.7241 (20 °C)	4.0	17.8
Gold	197.2	1945		19.32	10.2	16.8 (25-100 °C)	0.091	0.70	2.42 (18 °C)	5.2	11.1
Indium	114.8	311		7.28	15.8	13.8 (25-100 °C)	0.032				
Iridium	193.1	4262	4,442	22.42	8.6		0.057	0.14	6.1 (0 °C)		
Iron	55.84	2786	7,88	7.88	7.1	11.7	0.102	0.16	10.6 (25 °C)	3.4	30
Lead	207.20	621	2,777	11.34	18.3	27.09	0.030	0.083	2.04 (0 °C)		
Lithium	6.94	366.8	2,552	0.534	13.0		0.83				
Magnesium	24.32	1202	2,048	1.74	14.0	25.8	0.24	0.38 (0-100 °C)	4.35 (0 °C)	4.2	6.25
Manganese	54.93	2246	3,452	7.42	7.4		0.11		5		
Mercury	200.6	-37.9	675	13.6	14.7		0.033	0.015 (0 °C)	95.74 (20 °C)	3.75	
Molybdenum	96.0	4757	6,512	10.3	9.3	5.32 (20 °C)	0.063	0.35	5.5 (20 °C)		30
Nickel	58.68	2645.6		8.9	6.7	13.2 (25-100 °C)	0.102	0.14	6.93 (20 °C)		
Osmium	190.9	4892		22.5	8.5		0.031				
Palladium	106.7	2822		12.16	8.8	11.76	0.055	0.17	10.2 (0 °C)		
Phosphorus	31.04	111	554	1.83	17.0	125.3 (0-40 °C)	0.18				
Platinum	195.2	3191	7,070	21.37	9.1	8.99	0.030	0.17	10.96 (0 °C)		23.5
Rhodium	102.9	3542		12.44	8.3		0.058				
Ruthenium	101.7	4442		12.06	8.4		0.061				
Silicon	28.1	2588		2.42	11.6	7.63	0.17		58		
Silver	107.88	1761	3,551	10.53	10.2	19.21	0.054	1.00	1.63 (18 °C)	5.0	10.3
Tantalum	181.5	5252		16.6	10.9		0.035	0.13	14.6		
Thorium	232.15	3092		12.16	19.1	22.96	0.028				
Tin	118.7	449	4,118	7.3	16.3		0.053	0.15	13.0 (0 °C)	2.8	5.9
Titanium	48.1	3272		4.5	10.7		0.10		3.2		
Tungsten	184.0	6152	10,526	19.3	9.5	4.44 (27 °C)	0.034	0.48	5.6 (20 °C)		60
Uranium	238.2	3362		18.7	12.7		0.029				
Vanadium	51.0	3128		6.0	8.5		0.11				
Zinc	65.37	787	1,706	7.14	9.1	29.76	0.088	0.26	5.75 (0 °C)	6.5	12.4
Zirconium	90.6	3090		6.4	14.1		0.066				

Table 3 Weights of Alloys and Metals

Alloys and metals	lb./ft.³	lb./in.³
Aluminum	163	0.0943
Aluminum and tin:		
Al 91%, Sn 9%	178	0.103
Aluminum, copper, and tin:		
Al 85%, Cu 7.5%, Sn 7.5%	188	0.1087
Al 6.25%, Cu 87.5%, Sn 6.25%	459	0.2656
Al 5%, Cu 5%, Sn 90%	425	0.2459
Aluminum and magnesium:		
Al 70%, Mg 30%	125	0.0723
Aluminum and zinc:		
Al 91%, Zn 9%	175	0.1012
Antimony	419	0.2424
Babbitt alloy	454	0.2627
Bismuth	611	0.3535
Bismuth, lead, and tin:		
Bi 53%, Pb 40%, Sn 7%	659	0.3813
Wood's metal:		
Bi 50%, Pb 25%, Cd 12.5%, Sn 12.5%	605	0.3501
Brass:		
Cu 90%, Zn 10%	536	0.3101
Cu 70%, Zn 30%	527	0.3049
Cu 60%, Zn 40%	521	0.3015
Cu 50%, Zn 50%	511	0.2957
Bronze:		
Cu 90%, Sn 10%	548	0.3171
Cu 85%, Sn 15%	555	0.3211
Cu 80%, Sn 20%	545	0.3153
Cu 75%, Sn 25%	551	0.3188
Cu 90%, Al 10%	480	0.2777
Cu 95%, Al 5%	522	0.3020
Cu 97%, Al 3%	542	0.3136
Bronze, phosphorus, average	537	0.3107
Bronze, tobin, average	503	0.291
Cadmium and tin:		
Cd 32%, Sn 68%	480	0.2777
Chromium	436	0.2523
Cobalt	533	0.3084
Copper	557	0.3223
Copper and nickel:		
Cu 60%, Ni 40%	554	0.3206
German silver:		
Cu 60%, Zn 20%, Ni 20%	530	0.3607
Cu 52%, Zn 26%, Ni 22%	527	0.3049
Cu 59%, Zn 30%, Ni 11%	520	0.3009
Cu 63%, Zn 30%, Ni 7%	518	0.2997
Gold	1208	0.699
Gold and copper:		
Au 98%, Cu 2%	1176	0.6805
Au 90%, Cu 10%	1071	0.6197
Au 86%, Cu 14%	1027	0.5943
Gun metal, average	544	0.3148
Iridium	1396	0.8078
Iron, cast	450	0.2604
Iron, wrought	480	0.2777
Lead	708	0.4097
Lead and antimony:		
Pb 30%, Sb 70%	450	0.2604
Pb 37%, Sb 63%	460	0.2662
Pb 44%, Sb 56%	475	0.2748
Pb 63%, Sb 37%	514	0.2974

Table 3 Weights of Alloys and Metals (continued)

Alloys and metals	lb./ft.³	lb./in.³
Pb 83%, Sb 17%	596	0.3449
Pb 90%, Sb 10%	658	0.3807
Lead and bismuth:		
Bi 67%, Pb 33%	639	0.3697
Bi 50%, Pb 50%	656	0.3796
Bi 33%, Pb 67%	682	0.3946
Bi 25%, Pb 75%	697	0.4033
Bi 17%, Pb 83%	702	0.4062
Bi 12%, Pb 88%	703	0.4068
Lead and tin:		
Pb 87.5%, Sn 12.5%	661	0.3825
Pb 84%, Sn 16%	644	0.3726
Pb 63.7%, Sn 36.3%	588	0.3402
Pb 46.7%, Sn 53.3%	545	0.3153
Pb 30.5%, Sn 69.5%	514	0.2974
Magnesium	109	0.063
Manganese	499	0.2887
Manganese, copper, and nickel:		
Mn 12%, Cu 84%, Ni 4%	530	0.3067
Mercury	849	0.4913
Nickel	550	0.3182
Osmium	1402	0.8113
Palladium	712	0.412
Platinum	1344	0.7777
Platinum and iridium:		
Pt 90%, Iridium 10%	1348	0.780
Rhodium	755	0.4369
Ruthenium	765	0.4427
Silver	654	0.3784
Steel, cast	490	0.2835
Tin	460	0.2662
Tin and antimony:		
Sn 50%, Sb 50%	424	0.2453
Sn 75%, Sb 25%	442	0.2557
Tin and bismuth:		
Bi 78%, Sn 22%	587	0.3396
Bi 63%, Sn 37%	570	0.3298
Bi 50%, Sn 50%	546	0.3159
Bi 37%, Sn 63%	530	0.3067
Bi 22%, Sn 78%	504	0.2916
Tin and lead:		
Sn 97%, Pb 3%	456	0.2638
Sn 89%, Pb 11%	475	0.2748
Sn 80%, Pb 20%	487	0.2818
Sn 67%, Pb 33%	512	0.2962
Sn 50%, Pb 50%	550	0.3182
Titanium	224	0.1296
Tungsten	1078.7	0.6242
Zinc	437	0.2528

Table 4 Strength – Conversion Table

Kilograms per square millimeter to pounds per square inch
(1 kg/mm² = 1442.34 lb./in.²)

	0	1	2	3	4	5	6	7	8	9
0	...	1422	2845	4267	5689	7112	8534	9956	11379	12801
10	14223	15646	17068	18490	19913	21335	22757	24180	25602	27024
20	28447	29869	31291	32714	34146	35559	36981	38403	39826	41248
30	42670	44093	45515	46937	48360	49782	51204	52627	54049	55471
40	56894	58316	59738	61161	62583	64005	65428	66850	68272	69691
50	71117	72539	73962	75384	76806	78229	79651	81073	82496	83915
60	85340	86763	88185	89607	91030	92452	93874	95297	96719	88148
70	99564	100986	102408	103831	105253	106676	108098	109520	110943	112365
80	113787	115210	116632	118054	119477	120899	122321	123744	125166	126588
90	128011	129433	130855	132278	133700	135122	136545	137967	139389	140812

Thousands of pounds per square inch to kilograms per square millimeter
(1000 lb./in.² = 0.70307 kg/mm²)

	0	1	2	3	4	5	6	7	8	9
0	...	0.70	1.41	2.11	2.81	3.52	4.22	4.92	5.62	6.35
10	7.03	7.73	8.44	9.14	9.84	10.55	11.25	11.95	12.66	13.36
20	14.06	14.76	15.47	16.17	16.87	17.58	18.28	18.98	19.69	20.39
30	21.09	21.80	22.50	23.20	23.90	24.61	25.31	26.01	26.72	27.42
40	28.12	28.83	29.53	30.23	30.93	31.64	32.34	33.04	33.75	34.45
50	35.15	35.86	36.56	37.26	37.97	38.67	39.37	40.07	40.78	41.48
60	42.18	42.89	43.59	44.29	45.00	45.70	46.40	47.11	47.81	48.51
70	49.21	49.92	50.62	51.32	52.03	52.73	53.43	54.14	54.84	55.54
80	56.25	56.95	57.65	58.35	59.06	59.76	60.46	61.17	61.87	62.57
90	63.28	63.98	64.68	65.39	66.09	66.79	67.49	68.20	68.90	69.60

Kilograms per square millimeter to tons per square inch
(1 kg/mm² = 0.635 ton/in.²)

	0	1	2	3	4	5	6	7	8	9
0	...	0.64	1.27	1.91	2.54	3.18	3.81	4.45	5.08	5.72
10	6.35	6.99	7.62	8.26	8.89	9.53	10.16	10.80	11.43	12.07
20	12.70	13.34	13.97	14.61	15.24	15.88	16.51	17.15	17.78	18.41
30	19.05	19.68	20.32	20.95	21.59	22.22	22.86	23.49	24.13	24.76
40	25.40	26.03	26.67	27.30	27.94	28.57	29.21	29.84	30.48	31.11
50	31.75	32.38	33.02	33.65	34.29	34.92	35.56	36.19	36.83	37.46
60	38.10	38.73	39.37	40.00	40.64	41.27	41.91	42.54	43.18	43.81
70	44.45	45.08	45.72	46.35	46.99	47.62	48.26	48.89	49.53	50.16
80	50.80	51.43	52.07	52.70	53.34	53.97	54.61	55.24	55.88	56.51
90	57.15	57.78	58.42	59.05	59.69	60.32	60.96	61.59	62.23	62.86

Tons per square inch to kilograms per square millimeter
(1 ton/in.² = 1.57487 kg/mm²)

	0	1	2	3	4	5	6	7	8	9
0	...	1.57	3.15	4.72	6.30	7.87	9.45	11.02	12.60	14.17
10	15.75	17.32	18.90	20.47	22.05	23.62	25.20	26.77	28.35	29.92
20	31.50	33.07	34.65	36.22	37.80	39.37	40.95	42.52	44.10	45.67
30	47.25	48.82	50.42	51.97	53.54	55.12	56.69	58.27	59.84	61.42
40	62.99	64.57	66.15	67.72	69.29	70.87	72.44	74.02	75.59	77.17
50	78.74	80.32	81.89	83.47	85.04	86.62	88.19	89.77	91.34	92.92
60	94.49	96.07	97.64	99.22	100.79	102.37	103.94	105.52	107.09	108.67
70	110.24	111.82	113.39	114.97	116.54	118.11	119.69	121.26	122.84	124.41
80	125.99	127.56	129.14	130.71	142.29	133.86	133.44	137.01	138.59	140.16
90	141.74	143.31	144.89	146.46	148.04	149.61	151.19	152.76	154.33	155.91

Table 5 Temperature Conversion Table

Degrees Centigrade to Degrees Fahrenheit
(°F = °C x 9/5 + 32)

°C	0	10	20	30	40	50	60	70	80	90
−200	−328	−346	−364	−382	−400	−418	−436	−454	...	...
−100	−148	−166	−184	−202	−220	−238	−256	−274	−292	−310
−0	32	14	−4	−22	−40	−58	−76	−94	−112	−130
0	32	50	68	86	104	122	140	158	176	194
100	212	230	248	266	284	302	320	338	356	374
200	392	410	428	446	464	482	500	518	536	554
300	572	590	608	626	644	662	680	698	716	734
400	752	770	788	806	824	842	860	878	896	914
500	932	950	968	986	1004	1022	1040	1058	1076	1094
600	1112	1130	1148	1166	1184	1202	1220	1238	1256	1274
700	1292	1310	1328	1346	1364	1382	1400	1418	1436	1454
800	1472	1490	1508	1526	1544	1562	1580	1598	1616	1634
900	1652	1670	1688	1706	1724	1742	1760	1778	1796	1814
1000	1832	1850	1868	1886	1904	1922	1940	1958	1976	1994
1100	2012	2030	2048	2066	2084	2102	2120	2138	2156	2174
1200	2192	2210	2228	2246	2264	2282	2300	2318	2336	2354
1300	2372	2390	2408	2426	2444	2462	2480	2498	2516	2534
1400	2552	2570	2588	2606	2624	2642	2660	2678	2696	2714
1500	2732	2750	2768	2786	2804	2822	2840	2858	2876	2894
1600	2912	2930	2948	2966	2984	3002	3020	3038	3056	3074
1700	3092	3110	3128	3146	3164	3182	3200	3218	3236	3254
1800	3272	3290	3308	3326	3344	3362	3380	3398	3416	3434
1900	3452	3470	3488	3506	3524	3542	3560	3578	3596	3614
2000	3632	3650	3668	3686	3704	3722	3740	3758	3776	3794
2100	3812	3830	3848	3866	3884	3902	3920	3938	3956	3974
2200	3992	4010	4028	4046	4064	4082	4100	4118	4136	4154
2300	4172	4190	4208	4226	4244	4262	4280	4298	4316	4334
2400	4352	4370	4388	4406	4424	4442	4460	4478	4496	4514
2500	4532	4550	4568	4586	4604	4622	4640	4658	4676	4694
2600	4712	4730	4748	4766	4784	4802	4820	4838	4856	4874
2700	4892	4910	4928	4946	4964	4982	5000	5018	5036	5054
2800	5072	5090	5108	5126	5144	5162	5180	5198	5216	5234
2900	5252	5270	5288	5306	5324	5342	5360	5378	5396	5414
3000	5432	5450	5468	5486	5504	5522	5540	5558	5576	5594
3100	5612	5630	5648	5666	5684	5702	5720	5738	5756	5774
3200	5792	5810	5828	5846	5864	5882	5900	5918	5936	5954
3300	5972	5990	6008	6026	6044	6062	6080	6098	6116	6134
3400	6152	6170	6188	6206	6224	6242	6260	6278	6296	6314
3500	6332	6350	6368	6386	6404	6422	6440	6458	6476	6494
3600	6512	6530	6548	6566	6584	6602	6620	6638	6656	6674
3700	6692	6710	6728	6746	6764	6782	6800	6818	6836	6854
3800	6872	6890	6908	6926	6944	6962	6980	6998	7016	7034
3900	7052	7070	7088	7106	7124	7142	7160	7178	7196	7214

Table 6 Temperature Conversion Table

Degrees Fahrenheit to Degrees Centigrade
($°C = {}^5/_9 \,[°F - 32]$)

°F	0	10	20	30	40	50	60	70	80	90
−400	−240	−246	−251	−257	−262	−268	−273	...	...	...
−300	−184	−190	−196	−201	−207	−212	−218	−223	−229	−234
−200	−129	−134	−140	−146	−151	−157	−162	−168	−173	−179
−100	−73	−79	−84	−90	−96	−101	−107	−112	−118	−123
−0	−18	−23	−29	−34	−40	−46	−51	−57	−62	−68
0	−18	−12	−7	−1	4	10	16	21	27	32
100	38	43	49	54	60	66	71	77	82	88
200	93	99	104	110	116	121	127	132	138	143
300	149	154	160	166	171	177	182	188	193	199
400	204	210	216	221	227	232	238	243	249	254
500	260	266	271	277	282	288	293	299	304	310
600	316	321	327	332	338	343	349	354	360	366
700	371	377	382	388	393	399	404	410	416	421
800	427	432	438	443	449	454	460	466	471	477
900	482	488	493	499	504	510	516	521	527	532
1000	538	543	549	554	560	566	571	577	582	588
1100	593	599	604	610	616	621	627	632	638	643
1200	649	654	660	666	671	677	682	688	693	699
1300	704	710	716	721	727	732	738	743	749	754
1400	760	766	771	777	782	788	793	799	804	810
1500	816	821	827	832	838	843	849	854	860	866
1600	871	877	882	888	893	899	904	910	916	921
1700	927	932	938	943	949	954	960	966	971	977
1800	982	988	993	999	1004	1010	1016	1021	1027	1032
1900	1038	1043	1049	1054	1060	1066	1071	1077	1082	1088
2000	1093	1099	1104	1110	1116	1121	1127	1132	1138	1143
2100	1149	1154	1160	1166	1171	1177	1182	1188	1193	1190
2200	1204	1210	1216	1221	1227	1232	1238	1243	1249	1254
2300	1260	1266	1271	1277	1282	1288	1293	1299	1304	1310
2400	1316	1321	1327	1332	1338	1343	1349	1354	1360	1366
2500	1371	1377	1382	1388	1393	1399	1404	1410	1416	1421
2600	1427	1432	1438	1443	1448	1454	1460	1466	1471	1477
2700	1482	1488	1493	1499	1508	1510	1516	1521	1527	1532
2800	1538	1543	1549	1554	1560	1566	1571	1577	1582	1588
2900	1593	1599	1604	1610	1616	1621	1627	1632	1638	1643
3000	1649	1654	1660	1666	1671	1677	1682	1688	1693	1699
3100	1704	1710	1716	1721	1727	1732	1738	1743	1749	1754
3200	1760	1766	1771	1777	1782	1788	1793	1799	1804	1810
3300	1816	1821	1827	1832	1838	1843	1849	1854	1860	1866
3400	1871	1877	1882	1888	1893	1899	1904	1910	1916	1921
3500	1927	1932	1938	1943	1949	1954	1960	1966	1971	1977
3600	1982	1988	1993	1999	2004	2010	2016	2021	2027	2032
3700	2038	2043	2049	2054	2060	2066	2071	2077	2082	2088
3800	2093	2099	2104	2110	2116	2121	2127	2132	2138	2143
3900	2149	2154	2160	2166	2171	2177	2182	2188	2193	2199

Table 7 Stress Conversion Table

Multiply \ To Get	MPa	psi	kg-f/mm²	N/mm²	tons/in.²
MPa (megapascals)	1	145	0.102	1	64.7×10^{-3}
psi (pounds/inch²)	6.895×10^{-3}	1	0.703×10^{-3}	6.895×10^{-3}	0.446×10^{-3}
kg-f/mm² (kilograms force/millimeters²)	9.806	1.422×10^{3}	1	9.806	0.634
N/mm² (Newtons/millimeter²)	1	145	0.102	1	64.7×10^{-3}
tons/in.² (British tons/inch²; 2240 pounds/ton)	15.45	2240	1.576	15.45	1

SECTION I

ALLOY DATA

1% NEB-BRONZE.
M-1720; 89-93 Cu, 0.7-1.3 Sn, 0.10 max Pb, 0.05 max Fe, bal Zn.
Good hot and cold working properties; good for welding, brazing, soldering
For jewelry products, flat springs for electrical switchgear.
Copper alloy No. 413.

2 AS.
M-1487.
High speed steel; similar to AISI M2.

2 L 91.
M-106; 4.0-5.0 Cu, 0.25 max Si, 0.25 max Fe, 0.25 max Ti, bal Al.
As sand cast: 220 MPa TS; 165 MPa YS; 7 El; 60 Brin.
BS 1490 LM 11.

2 L 99.
M-106; 0.20-0.45 Mg, 6.5-7.5 Si, 0.20 max Ti, bal Al.
As sand cast: 230 MPa TS; 185 MPa YS; 2 El; 80 Brin.
BS 1490 LM 27.

2V PERMENDUR.
M-115; 49 Fe, 49 Co, 2 V.
Magnetically soft Fe-Co alloy, either airmelted or vacuum arc melted.
Cores for power transformers, pulse transformers, magnetic amplifiers, etc
Maximum permeability: air melt, 3000-4500; vacuum melt; 11,000.

3 A.
M-1083; 0.8-1.2 Ni, 0.16-0.25 P, 0.1-0.15 S, bal Cu.
Bar: 60,000-72,000 TS; 25-34 El; 130-160 Brin.
Forged: 55,000-72,000 TS; 25-32 El; 120-155 Brin.
For electrical switchgear and resistance welding equipment, electrical contacts.
50-60% electrical conductivity.
Was ERM 3A.
BS 4577 ISO 5182. RWMA alloy A/4/1.

03KH16N15M3.
M-USSR; 0.03 max C, 0.8 Mn, 0.6 Si, 15.0-17.0 Cr, 14.0-16.0 Ni, 2 Mo, bal Fe.
Austenitic stainless steel.

03KH16N15M3B.
M-USSR; 0.03 max C, 0.8 Mn, 0.6 Si, 15.0-17.0 Cr, 14.0-16.0 Ni, 2 Mo, 0.25-0.50 Cb, bal Fe.
Austenitic stainless steel.

03KH17N14M2.
M-USSR; 0.03 C, 1.0-2.0 Mn, 0.8 Si, 16.0-18.0 Cr, 13.0-15.0 Ni, 2 Mo, bal Fe.
Austenitic stainless steel.

03KH18N12.
M-USSR; 0.03 max C, 0.4 Mn, 0.4 Si, 17.0-19.0 Cr, 11.5-13.0 Ni, 0 Ti, bal Fe.
Austenitic stainless steel.

03KH18N12T.
M-USSR; 0.03 max C, 2.0 Mn, 0.8 Si, 17.0-19.0 Cr, 11.0-13.0 Ni, 0 (5XC), bal Fe.
Austenitic stainless steel.

03KH18N11.
M-USSR; 0.03 max C, 2.0 Mn, 0.8 Si, 17.0-19.0 Cr, 10.5-12.5 Ni, b
Austenitic stainless steel.
Similar to AISI 304 L.

03KH21N21M4GB.
M-USSR; 0.03 max C, 1.8-2.5 Mn, 0.6 Si, 20.0-22.0 Cr, 20.0-22.0 N 3.4-3.7 Mo, 0.8 Cb (15XC), bal Fe.
Austenitic stainless steel.

3 L 51.
M-106; 0.8-2.0 Cu, 0.05-0.20 Mg, 1.5-2.8 Si, 0.8-1.7 Ni, 0.8-0.14 Fe, Al.
As sand cast: 160 MPa TS; 125 MPa YS; 2 El; 60 Brin.
BS 1490 LM 23.

4-6 CHROME.
M-735; 4-6 Cr, Ti = 4 to 6 x C, 0.10 C, bal Fe.
Annealed: 66,000 TS; 39,000 YS; 45 El; 143 Brin.
For oil stills and refinery tubes; corrosion resistant.

04KH18N10.
M-USSR; 0.04 max C, 2.0 Mn, 0.8 Si, 17.0-19.0 Cr, 9.0-11.0 Ni, ba
Austenitic stainless steel.
Similar to AISI 304L.

4 L 35.
M-106; 3.5-4.5 Cu, 1.2-1.7 Mg, 1.8-2.3 Ni, bal Al. As chill cast: 280 MPa TS; 230 MPA YS; 100 Brin.
BS 1490 LM 14.

5-317 see CARPENTER NO 5-317.

5 CR MO V see HALCOMB 218, AISI H 11.

06KH18N11.
M-USSR; 0.06 max C, 2.0 Mn, 0.8 Si, 17.0-19.0 Cr, 10.0-12.0 Ni, b
Austenitic stainless steel.
Similar to AISI 304.

6 X.
M-US; 0.025 C, 20.25 Cr, 24.25 Ni, 6.25 Mo.
High temperature alloy.

07KH16N6.
M-USSR; 0.05-0.09 C, 0.8 Mn, 0.8 Si, 15.5-17.5 Cr, 5.0-8.0 Ni, ba
Austenitic-martensitic type stainless steel.

07KH21G7AN5.
M-USSR; 0.07 max C, 6.0-7.5 Mn, 0.7 Si, 19.5-21.0 Cr, 5.0-6.0 Ni, 0.15-0.25 N, bal Fe.
Austenitic stainless steel.

08KH10N20T2.
M-USSR; 0.08 max C, 2.0 Mn, 0.8 Si, 10.0-12.0 Cr, 18.0-20.0 Ni, 1 Ti, 1.0 max Al, bal Fe.
Austenitic stainless steel.

08KH13.
M-USSR; 0.08 max C, 0.8 Mn, 0.8 Si, 12.0-14.0 Cr, 1.0-1.8 Al, bal
Ferritic type stainless steel.

08KH15N24V4TR.
M-USSR; 0.08 max C, 0.5-1.0 Mn, 0.6 Si, 14.0-16.0 Cr, 22.0-25.0 N 1.4-1.8 Ti, 4.0-5.0 W, 0.005 max B, 0.025 max Ce, bal Fe.
Austenitic stainless steel.

08KH16N13M2B.
M-USSR; 0.06-0.12 C, 1.0 Mn, 0.8 Si, 15.0-17.0 Cr, 12.5-14.5 Ni, Mo, 0.9-1.3 Cb, bal Fe.
Austenitic stainless steel.

08KH17N5M3.
M-USSR; 0.06-0.10 C, 0.8 Mn, 0.8 Si, 16.0-17.5 Cr, 4.5-5.5 Ni, 3. bal Fe.
Austenitic-martensitic type stainless steel.

08KH17N13M2T.
M-USSR; 0.08 max C, 2.0 Mn, 0.8 Si, 16.0-18.0 Cr, 12.0-14.0 Ni, 2 Mo, 0.7 Ti (5XC), bal Fe.
Austenitic stainless steel.

08KH17N15M3T.
M-USSR; 0.08 max C, 2.0 Mn, 0.8 Si, 16.0-18.0 Cr, 14.0-16.0 Ni, 3 Mo, 0.3-0.6 Ti, bal Fe.
Austenitic stainless steel.

08KH17T.
M-USSR; 0.08 C, 0.8 Mn, 0.8 Si, 16.0-18.0 Cr, 0.80 Ti (5XC), ba
Ferritic type stainless steel.

08KH18G8N2T.
M-USSR; 0.08 max C, 7.0-9.0 Mn, 0.8 Si, 17.0-19.0 Cr, 1.8-2.8 Ni, 0.20-0.50 Ti, bal Fe.
Austenitic-ferritic type stainless steel.

08KH18N10.
M-USSR; 0.08 max C, 2.0 Mn, 0.8 Si, 17.0-19.0 Cr, 9.0-11.0 Ni, ba
Austenitic stainless steel.
Similar to AISI 304.

08KH18N10T.
M-USSR; 0.08 max C, 2.0 Mn, 0.8 Si, 17.0-19.0 Cr, 9.0-11.0 Ni, 0. (5XC), bal Fe.
Austenitic stainless steel.
Similar to AISI 321.

08KH18N12B.
M-USSR; 0.08 max C, 2.0 Mn, 0.8 Si, 17.0-19.0 Cr, 11.0-13.0 Ni, 1 Mo, 1.1 Cb (10XC), bal Fe.
Austenitic stainless steel.
Similar to AISI 347.

08KH20N14S2.
M-USSR; 0.08 max C, 1.5 Mn, 2.0-3.0 Si, 19.0-22.0 Cr, 12.0-15.0 N
Austenitic-ferritic type stainless steel.

08KH21N6M2T.
M-USSR; 0.08 max C, 0.8 Mn, 0.8 Si, 20.0-22.0 Cr, 5.5-6.5 Ni, 1.8 0.20-0.40 Ti, bal Fe.
Austenitic-ferritic type stainless steel.

08KH22N6T.
M-USSR; 0.08 max C, 0.8 Mn, 0.8 Si, 21.0-23.0 Cr, 5.3-6.3 Ni, 0.8 (5XC), bal Fe.
Austenitic-ferritic type stainless steel.

8-N-2.
M-115; 0.72-0.81 C, 1.3-1.8 W, 3.5-4 Cr, 8.0-9.2 Mo, 0.9-1.3 V, bal
For cutting tools, reamers, taps, broaches, drills; high speed steel.

8-N-2 COBALT.
M-115; 0.7 C, 8 Mo, 0.5 V, 4 Cr, 1 W, 5 Co, bal Fe.
For lathe and planer tools, reamers, drills, taps; high speed steel.

9% CR MO.
M-1740; 0.20 max C, 1.0 max Si, 0.30-0.70 Mn, 0.40 max Ni, 8.0-10.0 C 0.9-1.2 Mo, 0.40 max Cu, bal Fe.
For use at temperatures above 400°C.
BS 1463; ASTM A217-65 Gr. C12.

09KH14N16B.
M-USSR; 0.07-0.12 C, 1.0-2.0 Mn, 0.6 Si, 13.0-15.0 Cr, 14.0-17.0 0.9-1.3 Cb, bal Fe.
Austenitic stainless steel.

09KH14N19V2BR.
M-USSR; 0.07-0.12 C, 2.0 Mn, 0.6 Si, 13.0-15.0 Cr, 18.0-20.0 Ni, W, 0.9-1.3 Cb, 0.005 max B, 0.02 max Ce, bal Fe.
Austenitic stainless steel.

09KH14N19VBR1.
M-USSR; 0.07-0.12 C, 2.0 Mn, 0.6 Si, 13.0-15.0 Cr, 18.0-20.0 Ni, W, 0.9-1.3 Cb, 0.025 max B, 0.02 max Ce, bal Fe.
Austenitic stainless steel.

09KH15N8JU.
M-USSR; 0.09 max C, 0.8 Mn, 0.8 Si, 14.0-16.0 Cr, 7.0-9.4 Ni, 0.7 bal Fe.
Austenitic-martensitic type stainless steel.

09KH16N15M3B.
M-USSR; 0.09 max C, 0.8 Mn, 0.8 Si, 15.0-17.0 Cr, 14.0-16.0 Ni, 2 Mo, 0.6-0.9 Cb, bal Fe.
Austenitic stainless steel.

09KH16N4B.
M-USSR; 0.05-0.13 C, 0.5 Mn, 0.6 Si, 15.0-17.0 Cr, 3.5-4.5 Ni, 0. Cb, bal Fe.
Stainless steel.

09KH17N7JU.
M-USSR; 0.09 max C, 0.8 Mn, 0.8 Si, 16.0-17.5 Cr, 7.0-8.0 Ni, 0.5 bal Fe.
Austenitic-martensitic type stainless steel.

09KH17N7JU1.
M-USSR; 0.09 max C, 0.8 Mn, 0.8 Si, 16.5-18.0 Cr, 6.5-7.5 Ni, 0.7 bal Fe.
Austenitic-martensitic type stainless steel.

10KH11N20T3R.
M-USSR; 0.10 max C, 1.0 Mn, 1.0 Si, 10.5-12.5 Cr, 18.0-21.0 Ni, 2 Ti, 0.008-0.020 B, 0.8 max Al, bal Fe.
Austenitic stainless steel.

10KH11N23T3MR.
M-USSR; 0.10 max C, 0.6 Mn, 0.6 Si, 10.0-12.5 Cr, 21.0-25.0 Ni, 1 Mo, 2.6-3.2 Ti, 0.02 max B, 0.8 max Al, bal Fe.
Austenitic stainless steel.

10KH13SJU.
M-USSR; 0.7-0.12 C, 0.8 Mn, 1.2-2.0 Si, 12.0-14.0 Cr, 1.0-1.8 Al,
Ferritic type stainless steel.

10KH14AG15.
M-USSR; 0.10 max C, 14.5-16.5 Mn, 0.8 Si, 13.0-15.0 Cr, 0.15-0.25 Fe.
Austenitic stainless steel.

10KH14G14N3.
M-USSR; 0.09-0.14 C, 13.0-15.0 Mn, 0.7 Si, 12.5-14.0 Cr, 2.8-3.5 Fe.
Austenitic stainless steel.

10KH14G14N4T.
M-USSR; 0.10 max C, 13.0-15.0 Mn, 0.8 Si, 13.0-15.0 Cr, 2.8-4.5 N (5XC), bal Fe.
Austenitic stainless steel.

10KH17N13M2T.
M-USSR; 0.10 max C, 2.0 Mn, 0.8 Si, 16.0-18.0 Cr, 12.0-14.0 Ni, 2 Mo, 0.7 Ti (5XC), bal Fe.
Austenitic stainless steel.

10KH17N13M3T.
M-USSR; 0.10 max C, 2.0 Mn, 0.8 Si, 16.0-18.0 Cr, 12.0-14.0 Ni, 3 Mo, 0.7 Ti (5XC), bal Fe.
Austenitic stainless steel.

10KH23N18.
M-USSR; 0.10 max C, 2.0 Mn, 1.0 Si, 22.0-25.0 Cr, 17.0-20.0 Ni, b
Austenitic stainless steel.

10W-TA Ta-10W.

11KH11N2V2MF.
M-USSR; 0.09-0.13 C, 0.6 Mn, 0.6 Si, 10.5-12.0 Cr, 1.5-1.8 Ni, 0. Mo, 1.6-2.0 W, 0.18-0.30 V, bal Fe.
Martensitic stainless steel; air hardening.

12-2 W, 12-3 W see **GREEK ASCOLOY**.

12KH8VF.
M-USSR; 0.08-0.15 C, 0.5 Mn, 0.6 Si, 7.0-8.5 Cr, 0.6-1.0 W, 0.3-0 bal Fe.
Martensitic stainless steel; air hardening.

12KH13.
M-USSR; 0.09-0.15 C, 0.8 Mn, 0.8 Si, 12.0-14.0 Cr, bal Fe.
Stainless steel; martensitic-ferritic types.
Similar to AISI 410.

12KH17.
M-USSR; 0.12 max C, 0.8 Mn, 0.8 Si, 16.0-18.0 Cr, bal Fe.
Ferritic type stainless steel.
Similar to AISI 430.

12KH17G9AN4.
M-USSR; 0.12 max C, 8.0-10.5 Mn, 0.8 Si, 16.0-18.0 Cr, 3.5-4.5 Ni 0.15-0.25 N, bal Fe.
Austenitic stainless steel.

12KH18N9.
M-USSR; 0.12 max C, 2.0 Mn, 0.8 Si, 17.0-19.0 Cr, 8.0-10.0 Ni, ba
Austenitic stainless steel.
Similar to AISI 302.

12KH18N9T.
M-USSR; 0.12 max C, 2.0 Mn, 0.8 Si, 17.0-19.0 Cr, 8.0-9.5 Ni, 0.8 (5XC), bal Fe.
Austenitic stainless steel.

12KH18N10E.
M-USSR; 0.12 max C, 2.0 Mn, 0.8 Si, 17.0-19.0 Cr, 9.0-11.0 Ni, 0. Se, bal Fe.
Austenitic stainless steel.
Similar to AISI 303 Se.

12KH18N10T.
M-USSR; 0.12 C, 2.0 Mn, 0.8 Si, 17.0-19.0 Cr, 9.0-11.0 Ni, 0.8 T bal Fe.
Austenitic stainless steel.

12KH18N12T.
M-USSR; 0.12 max C, 2.0 Mn, 0.8 Si, 17.0-19.0 Cr, 11.0-13.0 Ni, 0 (5XC), bal Fe.
Austenitic stainless steel.

12KH21N5T.
M-USSR; 0.09-0.14 C, 0.8 Mn, 0.8 Si, 20.0-22.0 Cr, 4.8-5.8 Ni, 0. Ti, 0.08 max Al, bal Fe.
Austenitic-ferritic type stainless steel.

12KH25N16G7AR.
M-USSR; 0.12 max C, 5.0-7.0 Mn, 1.0 Si, 23.0-26.0 Cr, 15.0-18.0 N 0.30-0.45 N, 0.10 max B, bal Fe.
Austenitic stainless steel.

12 MOV.
M-US; 0.25 C, 0.50 Mn, 0.50 Si, 12.0 Cr, 0.5 Ni, 1.0 Mo, 0.30 V,
Martensitic stainless.
Similar to AISI 420.

13KH14N3V2FR.
M-USSR; 0.10-0.16 C, 0.6 Mn, 0.6 Si, 13.0-15.0 Cr, 2.8-3.4 Ni, 0. Ti, 0.18-0.28 V, 1.6-2.2 W, 0.004 max B, bal Fe.
Martensitic stainless steel.

13 MN.
M-1541; 0.96 C, 13.0 Mn, 0.016 P, 0.005 S, 0.53 Si, bal Fe.
Austenitic 13% Mn for high abrasion resistance applications.

14-4 PH.
M-US; 0.07 max C, 13.0-15.0 Cr, 3.0-5.0 Ni, 2.0-3.0 Mo, 2.0-5.0 Cu,
Precipitation hardening steel (Experimental).

01420.
M-USSR; 4-7 Mg, 1.5-2.6 Li, bal Al.
Wrought: 65,000 psi TS; 43,000 psi YS.

14KH17N2.
M-USSR; 0.11-0.17 C, 0.8 Mn, 0.8 Si, 16.0-18.0 Cr, 1.5-2.5 Ni, ba
Stainless steel; martensitic-ferritic type.

15-5 PH see **ARMCO 15-5 PH, REPUBLIC 15-5 PH** and **JOSLYN ST.**

15KH5.
M-USSR; 0.15 max C, 0.50 Mn, 0.50 Si, 4.5-6.0 Cr, bal Fe.
Martensitic stainless steel; air hardening.

15KH5M.
M-USSR; 0.15 max C, 0.50 Mn, 0.50 Si, 4.5-6.0 Cr, 0.45-0.60 Mo, b
Martensitic stainless steel; air hardening.
Similar to AISI 502.

15KH5VF.
M-USSR; 0.15 max C, 0.5 Mn, 0.3-0.6 Si, 4.5-6.0 Cr, 0.4-0.7 W, 0. bal Fe.
Martensitic stainless steel; air hardening.

15KH6SJU.
M-USSR; 0.15 max C, 0.5 Mn, 1.2-1.8 Si, 5.5-7.0 Cr, 0.7-1.1 Al, b
Stainless steel; martensitic-ferritic type.

15KH11MF.
M-USSR; 0.12-0.19 C, 0.7 Mn, 0.5 Si, 10.0-11.5 Cr, 0.6-0.8 Mo, 0. V, bal Fe.
Martensitic stainless steel; air hardening.

15KH12VNMF.
M-USSR; 0.12-0.18 C, 0.5-0.9 Mn, 0.4 Si, 11.0-13.0 Cr, 0.4-0.8 Ni Mo, 0.7-1.1 W, 0.15-0.30 V, bal Fe.
Stainless steel; martensitic-ferritic type.

15KH17AG14.
M-USSR; 0.15 max C, 13.5-15.5 Mn, 0.8 Si, 16.0-18.0 Cr, 0.6 max N 0.25-0.37 N, bal Fe.
Austenitic stainless steel.

15KH18N12S4TJU.
M-USSR; 0.12-0.17 C, 0.5-1.0 Mn, 3.8-4.5 Si, 17.0-19.0 Cr, 11.0-13 0.4-0.7 Ti, 0.13-0.35 Al, bal Fe.
Austenitic-ferritic type stainless steel.

15KH18SJU.
M-USSR; 0.15 max C, 0.8 Mn, 1.0-1.5 Si, 17.0-20.0 Cr, 0.7-1.2 Al,
Ferritic type stainless steel.

15KH25T.
M-USSR; 0.15 max C, 0.8 Mn, 1.0 Si, 24.0-27.0 Cr, 0.90 Ti (5XC)
Ferritic type stainless steel.

15KH28.
M-USSR; 0.15 max C, 0.8 Mn, 1.0 Si, 27.0-30.0 Cr, bal Fe.
Ferritic type stainless steel.

16-6 PH B & W **CROLOY 16-6 PH.**

16-15-6 see **TIMKEN 16-15-6 (DISCONTINUED).**

16-25-6 see **TIMKEN 16-25-6 (DISCONTINUED).**

16KH11N2V2MF.
M-USSR; 0.14-0.18 C, 0.6 Mn, 0.6 Si, 10.5-12.0 Cr, 1.4-1.8 Ni, 0. Mo, 1.6-2.0 W, 0.18-0.30 V, bal Fe.
Martensitic stainless steel; air hardening.

17-4-6 see **WSS 17-4-6.**

17-4 PH see **ARMCO 17-4 PH, JOSLYN STAINLESS TYPE 17-4 P**

17-4 PH, 17-7 PH see **ARMCO, CRUCIBLE REPUBLIC, JOSLYN, ETC., 17-4**

17-7 PH WIRE.
M-1507; 0.09 max C, 1.0 max Mn, 1.0 Max Si, 16.0-18.0 Cr, 6.5-7.75 N 0.75-1.50 Al, bal Fe.
As cold drawn (0.062 inch); 242,000 psi TS.
Cond. "CH" (1 hr 900° (0.062 inch); 297,000-327,000 psi TS.
For stainless springs.

17-10 P see **ARMCO 17-10 P.**

17-14 CU MO.
M-10; 0.12 C, 0.75 Mn, 0.50 Si, 15.9 Cr, 14.1 Ni, 2.5 Mo, 0.45 Cb, Ti, 3.0 Cu, bal Fe.
(Experimental).

17-22 A see **TIMKEN 17-22.**

17KH18N9.
M-USSR; 0.13-0.21 C, 2.0 Mn, 0.8 Si, 17.0-19.0 Cr, 8.0-10.0 Ni, b
Austenitic stainless steel.

17 PS TUNGSTEN.
M-60; 2 ThO$_2$, bal W.
Filament wire.

17 W.
M-US; 0.50 C, 19.0 Ni, 63.0 Fe, 13.0 Cr, 1.0 Mo, 2.0 W.
High temperature alloy.

18-2 MN see **ARMCO 18-2 MN AND ARMCO NITRONIC 32.**

18-5-8 see **USS 18-5-8.**

18-9 LW see **ARMCO 18-9 LW.**

18-9 LW 302 HQ.
M-1280; 0.10 C, 2.0 Mn, 1.0 Si, 17.0-19.0 Cr, 8.0-10.0 Ni, 3.0-4.0 Mo, bal Fe.
Low work hardening austenitic stainless steel.
For cold headed parts.

18-18-2 see **USS 18-18-2.**

18 CR-15 MN see **USS TENELON.**

18KH11MNFB.
M-USSR; 0.15-0.21 C, 0.6-1.0 Mn, 0.6 Si, 10.0-11.5 Cr, 0.5-1.0 Ni 0.8-1.1 Mo, 0.2-0.45 Cb, bal Fe.
Martensitic stainless steel; air hardening.

18KH12VMBFR.
M-USSR; 0.15-0.22 C, 0.5 Mn, 0.5 Si, 11.0-13.0 Cr, 0.4-0.6 Mo, 0.4-0.7 W, 0.15-0.30 V, 0.2-0.4 Cb, 0.003 max B, bal Fe.
Stainless steel; Martensitic-ferritic type.

18 SR see **ARMCO 18 SR.**

19-9 DL; 19-9 DX see **ALTEMP, CARPENTER, UNITEMP 19-9 DL, 19-9 DX.**

20 CB-3 see **CARPENTER STAINLESS NO. 20CB-3.**

20KH12VNMF.
M-USSR; 0.17-0.23 C, 0.5-0.9 Mn, 0.6 Si, 10.5-12.5 Cr, 0.5-0.9 Ni 0.5-0.7 Mo, 0.7-1.1 W, 0.15-0.30 V, bal Fe.
Martensitic stainless steel; air hardening.

20KH13.
M-USSR; 0.16-0.25 C, 0.8 Mn, 0.8 Si, 12.0-14.0 Cr, bal Fe.
Martensitic stainless steel; air hardening.
Similar to AISI 420.

20KH13N4G9.
M-USSR; 0.15-0.30 C, 8.0-10.0 Mn, 0.8 Si, 12.0-14.0 Cr, 3.7-4.7 Ni, bal Fe.
Austenitic-martensitic type stainless steel.

20KH17N2.
M-USSR; 0.17-0.25 C, 0.8 Mn, 0.8 Sr, 16.0-18.0 Cr, 1.5-2.5 Ni, bal Fe.
Martensitic stainless steel; air hardening.

20KH20N14S2.
M-USSR; 0.20 max C, 1.5 Mn, 2.0-3.0 Si, 19.0-22.0 Cr, 12.0-15.0 Ni, bal Fe.
Austenitic-ferritic type stainless steel.

20KH23N13.
M-USSR; 0.20 max C, 2.0 Mn, 1.0 Si, 22.0-25.0 Cr, 12.0-15.0 Ni, bal Fe.
Stainless steel.

20KH23N18.
M-USSR; 0.20 max C, 2.0 Mn, 1.0 Si, 22.0-25.0 Cr, 17.0-20.0 Ni, bal Fe.
Austenitic stainless steel.

20KH25N2052.
M-USSR; 0.20 max C, 1.5 Mn, 2.0-3.0 Si, 24.0-27.0 Cr, 18.0-21.0 Ni, bal Fe.
Austenitic stainless steel.

20 PLUS.
M-1405; 0.35 C, 0.80 Mn, 0.50 Si, 1.7 Cr, 0.40 Mo, bal Fe.
Supplied prehardened at 300 Brin.
Machinable; for plastic moulds, zinc die-casting dies, bolsters.
May be nitrided for increased wear resistance.

20 VM.
M-1290; 0.20 C, 1.3 Mn, bal Fe.
Low alloy cast iron.
SEW 1.1133; GS-20 Mn 5.

21-2N.
M-US; 0.55 C, 8.25 Mn, 0.25 max Si, 20.35 Cr, 2.10 Ni, 0.30 N, 0.0 max S, bal Fe.
For diesel engine exhaust valves.
SAE EV12.

21-4N.
0.53 C, 9.0 Mn, 0.15 Si, 21.0 Cr, 3.75 Ni, 0.07 S, 0.42 N, bal Fe.
1400°F: 62,000 psi TS; 37,000 psi YS; 18 El.
Exhaust valve steel.
SAE No. EV8.

21/4/N see also **FIRTH-BROWN 21/4/N** and **FIRTH-VICKERS 21/4.**

21-6-9.
M-1; 0.04 C, 9.0 Mn, 20.5 Cr, 6.5 Ni, 0.30 N, bal Fe.
112,000 TS; 68,000 YS; 44 El.
Austenitic, corrosion resistant steel.
See also armco 21-6-9 and carpenter stainless 21-6-9.

21-6-9.
M-1772; 21 Cr, 6 Ni, 9 Mn.
Ann: 115,000 psi TS; 70,000 YS; 44 El; 94 Rock B.
30% cold worked: 175,000 psi TS; 150,000 psi YS; 12 El; 32 Rock C.
Welded tubing.

21-12 VALVE STEEL.
M-32; 0.20 C, 1.25 Mn, 0.80 Si, 21.0 Cr, 11.5 Ni, bal Fe.
For diesel and gasoline engine valves.
SAE EV 3.

21-12N VALVE STEEL.
M-32; 0.20 C, 1.25 Mn, 0.50 Si, 21.0 Cr, 11.5 Ni, 0.20 N, bal Fe.
For diesel and gasoline engine valves.
SAE EV-4.

21-55N VALVE STEEL.
M-32; 0.20 C, 5.0 Mn, 0.50 Si, 21.0 Cr, 4.5 Ni, 0.30 N, bal Fe.
Austenitic.
For automotive exhaust valves.
SAE EV-7.

22-4-9 see **ARMCO 22-4-9.**

22-13-5 see **ARMCO 22-13-5 (NOW ARMCO NITRONIC 50).**

25/20 HIGH CARBON STAINLESS STEEL WELD FILLER.
M-108; 0.4 C, 25 Cr, 20 Ni, bal Fe.
For TIG and MIG welding.

25-20-SI see **USS 25-20-SI.**

25KH13N2.
M-USSR; 0.2-0.3 C, 0.8-1.2 Mn, 0.08-0.15 P, 0.15-0.25 S, 0.5 Si, 12.0-14.0 Cr, 1.5-2.0 Ni, bal Fe.
Free-machining, martensitic stainless steel; air hardening.
Similar to AISI 420F.

25 VM.
M-1290; 0.25 C, 1.0 Mn, bal Fe.
Low alloy cast iron.
SEW 1.1136; GS-24 Mn 4.

26-1.
M-38; 0.03 C, 26 Cr, 1 Mo, bal Fe.
High temperature alloy.

26-2.
M-73; 0.005 C, 0.11 Mn, 0.09 Si, 25 Cr, 0.4 Ni, 2.2 Mo, 0.04 V, bal Fe.
Chromium stainless for high temperature equipment.

29-4.
M-US; 0.01 C, 0.010 N, 29 Cr, 4 Mo, bal Fe.
For high temperature equipment.

30KH13.
M-USSR; 0.26-0.35 C, 0.8 Mn, 0.8 Si, 12.0-14.0 Cr, bal Fe.
Martensitic stainless steel; air hardening.
Similar to AISI 420.

30KH13N7S2.
M-USSR; 0.25-0.34 C, 0.8 Mn, 2.0-3.0 Si, 12.0-14.0 Cr, 6.0-7.5 Ni bal Fe.
Stainless steel.

30 VM.
M-1290; 0.30 C, 1.3 Mn, bal Fe.
Low alloy cast iron.
SEW 1.1165; GS-30 Mn 5.

31KH19N9MVBT.
M-USSR; 0.28-0.35 C, 0.8-1.5 Mn, 0.8 Si, 18.0-20.0 Cr, 8.0-10.0 Ni, 0.2-0.5 Ti, 0.2-0.5 Cb, 1.0-1.5 V, bal Fe.
Austenitic stainless steel.

34 SP.
M-1488; 0.32 C, 0.35 Mn, 0.25 Si, 22 Ni, 2.2 Cr, 0.70 Mo, bal Fe.
Oil hardenable to 1220 N/mm^2 min UTS.
For shafts, bolts, levers.
AFNOR 32 CND8.

35 SWC.
M-1290; 0.30-0.40 C, 0.80-1.10 Si, 0.20-0.40 Mn, 0.90-1.20 Cr, 1.80-2.10 W, 0.15-0.20 V, bal Fe.
Hot work tool steel.
W.-Nr. 1.2541.

36KH18N25S2.
M-USSR; 0.32-0.40 C, 1.5 Mn, 2.0-3.0 Si, 17.0-19.0 Cr, 23.0-26.0 Ni, bal Fe.
Austenitic stainless steel.

37KH12N8G8MFB.
M-USSR; 0.34-0.40 C, 7.5-9.5 Mn, 0.3-0.8 Si, 11.5-13.5 Cr, 7.0-9.0 Ni, 1.1-1.4 Mo, 0.25-0.40 Cb, 1.25-1.55 V, bal Fe.
Austenitic stainless steel.

40KH9S2.
 M-USSR; 0.35-0.45 C, 0.8 Mn, 2.0-3.0 Si, 8.0-10.0 Cr, bal Fe.
 Martensitic stainless steel; air hardening.

40KH10S2M.
 M-USSR; 0.35-0.45 C, 0.8 Mn, 1.9-2.6 Si, 9.0-10.5 Cr, 0.7-0.9 Mo, bal Fe.
 Martensitic stainless steel; air hardening.

40KH13.
 M-USSR; 0.36-0.45 C, 0.8 Mn, 0.8 Si, 12.0-14.0 Cr, bal Fe.
 Martensitic stainless steel; air hardening.

40HK15N7G7F2MS.
 M-USSR; 0.38-0.47 C, 6.0-8.0 Mn, 0.9-1.4 Si, 14.0-16.0 Cr, 6.0-8.0 Ni, 0.65-0.95 Mo, 1.5-1.9 V, bal Fe.
 Austenitic stainless steel.

40 VMS.
 M-1290; 0.38 C, 1.0 Mn, 0.5 Si, bal Fe.
 Low alloy steel casting.
 GS 38 Mn Si 4.

42 see **BISHOP 42.**

45KH14N14V2M.
 M-USSR; 0.40-0.50 C, 0.7 Mn, 0.8 Si, 13.0-15.0 Cr, 13.0-15.0 Ni, 0.25-0.40 Mo, 2.0-2.8 W, bal Fe.
 Austenitic stainless steel.

45KH22N4M3.
 M-USSR; 0.40-0.50 C, 0.85-1.25 Mn, 0.7-1.0 Si, 21.0-23.0 Cr, 4.0-5.0 Ni, 2.5-3.0 Mo, bal Fe.
 Austenitic stainless steel.

45 SWC.
 M-1290; 0.40-0.50 C, 0.80-1.10 Si, 0.20-0.40 Mn, 0.90-1.20 Cr, 1.80-2.10 W, 0.15-0.20 V, bal Fe.
 Hot work tool steel.
 W.-Nr. 1.2542.

45 VM.
 M-1290; 0.45 C, 1.0 Mn, bal Fe.
 Low alloy steel casting.
 GS-46 Mn 4.

50 CMV.
 M-1290; 0.47-0.55 C, 0.15-0.35 Si, 0.80-1.10 Mn, 0.90-1.20 Cr, 0.07-0.12 V, bal Fe.
 Cold work tool steel.
 W.-Nr. 1.2241.

52 see **BISHOP 52.**

52 CB BEARING STEEL see **CRUCIBLE 52 CB BEARING STEEL.**

52NI-FE see **WESTINGHOUSE 52NI-FE.**

55KH20G9AN4.
 M-USSR; 0.50-0.60 C, 8.0-10.0 Mn, 0.45 Si, 20.0-22.0 Cr, 3.5-4.5 Ni, 0.30-0.60 N, bal Fe.

57 HW.
 M-24; 0.35 C, 3.25 Cr, 0.3 V, 9 W, bal Fe.
 For dies, trimmers, pliers, punches, shear blades, forming dies; hot work tool steel.

57 SPECIAL see **BETHLEHEM 57 SPECIAL.**

58 CRV 4.
 M-459; 0.58 C, 0.95 Mn, 1.0 Cr, 0.09 V, bal Fe.
 For springs, punches, forging dies; oil hardened, tough.

60 SWC.
 M-1290; 0.55-0.65 C, 0.50-0.70 Si, 0.20-0.40 Mn, 0.90-1.20 Cr, 1.80-2.10 W, 0.15-0.20 V, bal Fe.
 Hot work tool steel.
 W.-Nr. 1.2550.

62 BE-38AL see **LOCKALLOY.**

67 CHISEL.
 M-24; 0.5 C, 2.5 W, 1.15 Cr, 0.75 Si, bal Fe.
 Heat treated: 185,000-337,000 TS; 180,000-255,000 YS; 14-9 El; 47-27 420-670 Brin.
 For chisels, rivet sets, pneumatic tools, upsetters; Type S1; shock resistant.

70-30 ALLOY.
 M-44; 70 Ni, 30 Cr.
 Wire, rod, ribbon for heating elements up to 2200°F.

70/30 COPPER.
 M-108;
 For MIG and TIG welding.

70 LA-2.
 M-1713;
 Weld metal: 0.07 C, 0.90 Mn, 0.50 Si, bal Fe.
 As welded: 79,000 psi TS; 69,500 psi YS; 28 El.
 Lime coated, AC-DC, reverse polarity electrode for general purpose welding and repair.
 AWS Class E7016.

70 WM.
 M-1290; 0.68-0.78 C, 0.20-0.40 Si, 0.40-0.60 Mn, 0.40-0.60 Cr, 0.25-0.40 Mo, 0.40-0.70 W, 0.15-0.30 V, bal Fe.
 Cold work tool steel.
 W.-Nr. 1.2604.

71 ALLOY.
 M-24; 0.60 C, 0.25 Ni, 0.9 Mn, 0.25 max Cr, 2 Si, bal Fe.
 For tools, hand and pneumatic chisels, shear blades, punches; resists shoc

80-20 ALLOY see **CHROMELA NICHROME V.**

85 WCV.
 M-1290; 0.75-0.85 C, 0.15-0.40 Si, 0.30-0.50 Mn, 0.40-0.60 Cr, 0.25-0. V, 0.60-0.80 W, bal Fe.
 Cold work tool steel.
 W.-Nr. 1.2511.

92/8.
M-1256; 8.0 Al, 0.25 Fe, bal Cu.
Weld metal: 25 Tons/sq.in. UTS; 50 El.
Welding wire for automatic processes.

95 KH18.
M-USSR; 0.9-1.0 C, 0.8 Mn, 0.8 Si, 17.0-19.0 Cr, bal Fe.
Martensitic stainless steel; air hardening.
Similar to AISI 440C.

98M2 see HAYNES STELLITE 98M2 ALLOY.

101" VAN PUNCH.
M-1409; 0.5 C, 5 W, bal Fe.
For punches, piercing dies; head, impact and abrasion resistant.

115 W 2 C.
M-1290; 1.05-1.15 C, 0.15-0.30 Si, 0.30-0.40 Mn, 1.10-1.30 Cr, 1.8-2.0 W, 0.15-0.25 V, bal Fe.
Cold work tool steel.
W.-Nr. 1.2521.

134 MO.
M-1290; 0.07 max C, 1.0 max Si, 1.5 max Mn, 12.0-13.5 Cr, 3.5-5.5 Ni, 0.70 max Mo, bal Fe.
Stainless steel casting.
W.-Nr. 1.4313.

151 M.
M-1290; 1.05-1.15 C, 1.0 max Si, 1.0 max Mn, 14.0-16.0 Cr, 0.40-0.60 Mo, 0.10-0.15 V, bal Fe.
Martensitic stainless steel; W.-Nr. 1.4111.

165 CC.
M-1290; 1.55-1.75 C, 0.25-0.40 Si, 0.20-0.40 Mn, 1.2-1.4 Co, 11.0-12.0 0.50-0.60 Mo, bal Fe.
Cold work tool steel.
W.-Nr. 1.2880.

165 CCM.
M-1290; 1.65 C, 0.40 Si, 0.35 Mn, 1.5 Co, 13.5 Cr, 1.2 Mo, bal Fe.
Cold work tool steel.
W.-Nr. 1.2885.

188 E.
M-1290; 0.07 max C, 1.0 max Si, 2.0 max Mn, 16.5-18.5 Cr, 10.5-13.5 N 2.0-2.5 Mo, bal Fe.
Austenitic stainless steel.
W.-Nr. 1.4401.
Similar to AISI 316.

188 ES.
M-1290; 0.10 max C, 1.0 max Si, 2.0 max Mn, 16.5-18.5 Cr, 10.5-13.5 N 2.0-2.5 Mo, Nb=8xC min, bal Fe.
Austenitic stainless steel.
W.-Nr. 1.4580.

188 EST.
M-1290; 0.10 max C, 1.0 max Si, 2.0 max Mn, 16.5-18.5 Cr, 10.5-13.5 Ni, 2.0-2.5 Mo, Ti=5xC min, bal Fe.
Austenitc stainless steel.
W.-Nr. 1.4571.

189 M.
M-1290; 0.85-0.95 C, 1.0 max Si, 1.0 max Mn, 17.0-19.0 Cr, 1.0-1.3 Mo 0.07-0.12 V, bal Fe.
Martensitic stainless steel.
W.-Nr. 1.4112.
Similar to AISI 440C.

203 EZ see ALLEGHENY TYPE 203 EZ.

245 ALLOY.
M-44; 75 Ni, 20 Cr, 1 Si, 5 Al.
For heating elements from 2000-2300°F.

300 M see REPUBLIC 300 M.

304 N see JOSLYN STAINLESS 304 N.

309S CB.
M-1793; 0.08 max C, 2.0 max Mn, 1.0 max Si, 22-24 Cr, 12-15 Ni, Cb = 8xC, bal Fe.
Austenitic stainless steel.

368 SPECIAL.
M-694; 0.40 C, 3.30 Cr, 1.30 Mo, 2.50 W, 1.30 V, bal Fe.
Hot work tool steel for forming cooper, brass and steel.

401" COPPER-NICKEL ALLOY now MONEL ALLOY 401.

476.
M-510; 1.55 C, 12.0 Cr, 0.85 Mo, 0.28 V, bal Fe.
Cold work tool steel.
B.S. 4659 Type BD2; AISI D2.

495 see AMPCOLOY 495.

702 CN.
M-1290; 0.45-0.55 C, 0.15-0.35 Si, 0.40-0.60 Mn, 0.90-1.20 Cr, 3.0-3.5 Ni, bal Fe.
Cold work tool steel.
W.-Nr. 1.2721.

702 W.
M-1290; 0.40-0.50 C, 0.15-0.30 Si, 0.30-0.50 Mn, 1.2-1.5 Cr, 0.15-0.35 Mo, 3.8-4.3 Ni, 0.50 W (optional), bal Fe.
Cold work tool steel.
W.-Nr. 1.2767.

713 C see HAYNES ALLOY NO. 713 C.

718 see INCONEL ALLOY 718 AND ALLVAC 718.

719 CNM.
M-1290; 0.16-0.22 C, 0.15-0.30 Si, 0.30-0.50 Mn, 1.1-1.4 Cr, 0.15-0.25 Mo, 3.8 -4.3 Ni, 0.50 W (optional), bal Fe.
Cold work tool steel.
W.-Nr. 1.2764.

901 see CARPENTER 901 UNITEMP 901 AND UDIMET 901.

930 ALLOY.
M-1256; 6.25 Al, 0.75 Fe, 2.25 Si, bal Cu.
Cast or wrought aluminum bronze.
DGS 8453.

1741 CVM.
M-115; 1.3 C, 17.5 Cr, 4 Mo, 1 V, 0.3 max Mn, bal Fe.
For high temperature bearings, pistons, rings and seals, cryogenic applications.
Good wear and heat resistance in high temperature service.

1845/4.
M-200; 28 Cu, 10 Zn, bal Ag.
Silver braze filler metal.
Melt range: 690-735°C.

1845/5.
M-200; 37 Cu, 19 Zn, bal Ag.
Silver braze filler metal.
Melt range: 700-775°C.

2021.
M-US; 6.3 Cu, 0.3 Mn, 0.06 Ti, 0.1 V, 0.18 Zr, 0.15 Cd, 0.05 Sn, Al.
Precipitation hardenable wrought aluminum alloy.
T 81 Cond: 73,000 TS; 63,000 YS; 9 El.

2155 N.
M-32; 0.20 C, 5.0 Mn, 0.50 Si, 21.0 Cr, 4.5 Ni, 0.30 N, bal Fe.
For exhaust valves.
SAE EV-7.
See also Carpenter 21-55 N.

2300 SUPER ALLOY.
M-95; 0.50 C, 40.0 Ni, 30.0 Cr, 1.5 Si, 1.5 Mn, bal W and Co.
As cast: 61,000 psi TS; 35,000 psi YS; 4.0 El; 220 Brin.
At 2000°F: 12,200 psi TS; 8100 psi YS; 30 El.
Machinable, weldable, non-magnetic.
For high temperature applications.

3693 A.
M-1488; 0.28 C, 1.35 Cr, 0.75 Mo, 0.30 V, B, bal Fe.
Quenched and tempered: 830-930 N/mm^2 UTS; 735 N/mm^2 min YS; El.
For high temperature bolts.
AFNOR 28 CDV 8-08; ASTM A 193 Gr. B 16.

6379.
M-115; 0.82 C, 1.0 Si, 0.3 Mn, 7.75 Cr, 2.5 V, 1.55 Mo, bal Fe.
Heat Treat: 200,000-280,000 TS; 170,000-230,000 YS; 8-6 El; 29-31 R 42-57 Rock.
For shears, slitters, insert dies, hot and cold punches, forging dies, hot and cold shear blades. Wear resistant, tough with good hot hardness.

6542.
M-1405; 0.85 C, 4.0 Cr, 5.0 Mo, 6.25 W, 2.0 V, bal Fe.
Mo-W high speed steel.
For drills, lathe tolls, form tools.
BS 4659 BM 2; AISI M2.

7031 M.
M-1488; 0.60 max C, 0.60 max Mn, 0.50 max Si, 0.60 max Cu, 0.60 max Fe, bal Ni.
Cast, annealed: 350 N/mm^2 min UTS.
Resists potash and soda in all concentrations.
ACI CZ 100.

7152.
M-115; 1.6-1.8 C, 17.0-17.5 Cr, bal Fe.
For pump parts, bearings, pivots; corrosion and wear resistant.

9086 M.
M-1488; 0.12 max C, 15.5-17.5 Cr, 16.0-18.0 Mo, 0.20-0.40 V, 4.0-7.0 Fe, 3.75-5.25 W, bal Ni.
Cast, annealed: 500 N/mm^2 min UTS.
Good resistance to acids and hyperchlorites.
ACI CW 12 M. (Hastelloy C).

9097 M.
M-1488; 0.12 max C, 1.0 max Cr, 26.0-30.0 Mo, 0.20-0.60 V, 4.0-6.0 Fe bal Ni.
Cast, annealed: 500 N/mm^2 min UTS.
Good resistance to acids.
ACI N-12M. (Hastelloy B).

9098 M.
M-1488; 0.10 max C, 8.0-10.0 Si, 2.0-4.0 Cu, 2.0 max Fe, bal Ni.
Cast stainless alloy.
Resists sulphuric acid in all concentrations; also organic acids. (Hastelloy D).

9128 M.
M-1488; 0.30 max C, 1.5 max mn, 1.5 max Si, 26.0-33.0 Cu, bal Ni.
Cast, ann: 450 N/mm^2 min UTS.
For valves, pumps for marine applications.
ACI M 35 (Monel 400).

A

A 5 see **MATTHEY A 5.**

A-6.
M-1769.
Sintered carbide.
240,000 TrS; Density 14.8-15.0; 91.5-92.5 Ra.
Industry code: C-2; ISO K10, 15, 20, M20.

A-7.
M-1769.
Sintered carbide.
200,000 TrS; Density: 14.95-15.2; 91.7-92-7 Ra.
Industry code: C-3-4; ISO K01, 05.

A-7W.
M-373; 2.25 C, 5.25 Cr, 4.75 V, 1.0 W, 1.0 Mo, bal Fe.
Air hardening tool steel; AISI A7.

A 8.
M-Eng; 7.5-9.0 Al, 0.3-1.0 Zn, 0.15-0.4 Mn, bal Mg.
Sand cast: 140 MPa TS; 85 MPa YS; 2 El; 55 Brin.
General purpose sand or chill casting.
BS 2970 MAG 1.

A 8 H.P.
M-Eng; 7.5-9.0 Al, 0.3-1.0 Zn, 0.15-0.7 Mn, bal Mg.
Sand cast: 140 MPa TS; 85 MPa YS; 2 El; 55 Brin.
Higher purity sand or chill casting.
BS 2970 MAG 2.

A9V.
M-78; 8.5 Al, 0.5 Zn, 0.3 Mn, bal Mg.
For castings, light metal parts; heat treatable.

A 25 see **MATTHEY A 25.**

A36CC.
M-173; 0.16 C, 0.50 Mn, 0.04 Si, bal Fe.
36,000 psi min YS.
Good weldability and impact (notch) toughness.
For automotive frames and cross members.
ASTM A-36.

A 70 see **CRUCIBLE A 70.**

A-110.
M-64; 1.2 C, 0.3 Mn, 0.3 Si, bal Fe.
Water hardening tool steel; AISI W1.

A-286.
M-32, M-38, M-114, M-1490 and others; 0.08 max C, 1.4 Mn, 0.4 Si, 15 Cr, 26 Ni, 1.25 Mo, 2.15 Ti, 0.2 Al, 0.003 B, 0.3 V, bal Fe.
For jet engine and supercharger parts, turbine wheels and blades.
Austenitic, heat and corrosion resistant.

A-286.
M-1772; 15 Cr, 24 Ni, 2 Ti, 1.5 Mo, 0.30 V, bal Fe.
As hardened: 135,000-160,000 psi TS; 85,000-115,000 psi YS; 20-30 El; 24-34 Rc.
High strength; corrosion resistant.
Precipitation hardened alloy.
Welded or seamless tubing.

A-286.
M-44; 26 Ni, 15 Cr, 1 Cb, 2 Ti, bal Fe.
Wire and rod for welding wire and fastener stock.

A 300.
M-1290; 0.40-0.50 C, 1.0 max Si, 1.0 max Mn, 12.0-14.0 Cr, bal Fe.
Martensitic stainless steel; W.-Nr. 1.4034.

A 350.
M-1290; 0.17-0.22 C, 1.0 max Si, 1.0 max Mn, 12.0-14.0 Cr, bal Fe.
Martensitic stainless steel.
W.-Nr. 1.4021.

A 400.
M-1290; 0.08 max C, 1.0 max Si, 1.0 max Mn, 13.0-15.0 Cr.
Ferritic type stainless steel.
W.-Nr. 1.4001.

A 500 H.
M-1290; 0.75-0.85 C, 0.30-0.60 Si, 0.20-0.50 Mn, 0.90-1.30 Co, 12.5-14.5 Cr, bal Fe.
Cold work tool steel.
W.-Nr. 1.2883.

A 600.
M-1290; 1.0-1.1 C, 0.15-0.30 Si, 0.80-1.1 Mn, 0.90-1.1 Cr, 1.0-1.3 W, bal Fe.
Cold work tool steel.
W.-Nr. 1.2419.

A 1050.
M-1741; 99.50% Al min.
H12-Temper: 97 MPa TS; 83 MPa YS; 15 El.
H18-Temper: 145 MPa TS; 138 MPa YS; 6 El.
Chemical and process plant and equipment.

A 1100.
M-1741; 99.0% Al min, 0.12 Cu.
O Temper: 90 MPa TS; 34 MPa YS; 35 El.
H18 Temper: 165 MPa TS; 152 MPa YS; 5 El; 44 Brin.
Spinnings, holloware and general sheet metal work.

A 1145.
M-1741; 99.45% Al min.
General foil uses.

A 1199.
M-1741; 99.99% Al min.
Electrical and electronic foil uses.

A 2011.
M-1741; 5.5 Cu, 0.5 Pb, 0.5 Bi, bal Al.
T3 Temper: 379 MPa TS; 296 MPa YS; 15 El: 110 VHN.
T4 Temper: 310 MPa TS; 145 MPa YS; 20 El.
Screw machine products not requiring decorative anodizing.

A 2024.
M-1741; 4.5 Cu, 0.6 Mn, 1.5 Mg, bal Al.
O Temper: 166 MPa TS; 76 MPa YS; 20 El.
T42 Temper: 469 MPa TS; 310 MPa YS; 20 El; 120 Brin.
Aircraft sheeting.

A 3003.
M-1741; 1.2 Mn, 0.12 Cu, bal Al.
O Temper: 110 MPa TS; 41 MPa YS; 30 El.
H18 Temper: 200 MPa TS; 186 MPa YS; 4 El; 55 Brin.
Chemical equipment, sheet metal work, rigid foil containers.

A 3004.
M-1741; 1.2 Mn, 1.0 Mg, bal Al.
O Temper: 179 MPa TS; 69 MPa YS; 20 El.
H38 Temper: 283 MPa TS; 248 MPa YS; 5 El; 70 Brin.
Storage tanks, car bodies, seam welded tubing.

A 3203.
M-1741; 1.2 Mn, bal Al.
H12 Temper: 131 MPa TS; 117 MPa YS; 32 El.
H18 Temper: 200 MPa TS; 172 MPa YS; 4 El.
Sheet metal work, high strength foil.

A 4543.
M-1741; 6.0 Si, 0.3 Mg, bal Al.
T1 Temper: 145 MPa TS; 83 MPa YS; 20 El; 40 VHN.
T8 Temper: 207 MPa TS; 179 MPa YS; 12 El; 75 VHN.
Architectural extrusions.

A 5005.
M-1741; 0.8 Mg, bal Al.
O Temper: 124 MPa TS; 41 MPa YS; 25 El.
H18 Temper: 200 MPa TS; 193 MPa YS; 4 El; 56 VHN.
H38 Temper: 200 MPa TS; 186 MPa YS; 5 El; 51 Brin.
Appliances and utensiles, high strength foil.

A 5052.
M-1741; 2.5 Mg, 0.25 Cr, bal Al.
O Temper: 193 MPa TS; 90 MPa YS; 25 El; 47 Brin.
H38 Temper: 290 MPa TS; 255 MPa YS; 7 El; 77 Brin.
Sheet metal work, marine applications.

A 5056.
M-1741; 5.2 Mg, 0.1 Mn, 0.1 Cr, bal Al.
O Temper: 290 MPa TS; 152 MPa YS; 35 El; 70 VHN.
H38 Temper: 414 MPa TS; 345 MPa YS; 15 El; 100 Brin.
Cable sheathing, rivets, zippers, screen wire.

A 5086.
M-1741; 4.0 Mg, 0.5 Mn, 0.15 Cr, bal Al.
O Temper: 262 MPa TS; 117 MPa YS; 22 El; 60 Brin.
H38 Temper: 359 MPa TS; 303 MPa YS; 7 El.
Unfired pressure vessels, TV towers.

A 5252..
M-1741; 2.5 Mg, bal Al.
H24 Temper: 221 MPa TS; 159 MPa YS; 12 El.
H28 Temper: 283 MPa TS; 241 MPa YS; 5 El; 75 Brin.
High strength automobile trim.

A 5454.
M-1741; 2.7 Mg, 0.8 Mn, 0.10 Cr, bal Al.
O Temper: 248 MPa TS; 117 MPa YS; 22 El; 62 Brin.
H34 Temper: 303 MPa TS; 241 MPa YS; 10 El; 81 Brin.
Welded structures, pressure vessels.

A 5457.
M-1741; 1.0 Mg, 0.2 Mn, 0.1 Cu, bal Al.
O Temper: 131 MPa TS; 48 MPa YS; 22 El; 32 Brin.
H25 Temper: 179 MPa TS; 159 MPa YS; 12 El; 48 Brin.
Automobile trim.

A 5557.
M-1741; 0.6 Mg, 0.2 Mn, 0.1 Cu, bal Al.
O Temper: 110 MPa TS; 41 MPa YS; 25 El.
H25 Temper: 159 MPa TS; 138 MPa YS; 12 El.
Automobile trim.

A 6061.
M-1741; 1.0 Mg, 0.6 Si, 0.25 Cu, 0.2 Cr, bal Al.
O Temper: 124 MPa TS; 55 MPa YS; 25 El; 30 Brin.
T6 Temper: 310 MPa TS; 276 MPa YS; 12 El; 95 Brin.
Structural applications, transport.

A 6262.
M-1741; 1.0 Mg, 0.6 Si, 0.25 Cu, 0.1 Cr, 0.6 Bi, 0.6 Pb, bal Al.
T6 Temper: 310 MPa TS; 276 MPa YS; 12 El; 100 VHN.
T9 Temper: 400 MPa TS; 379 MPa YS; 10 El; 125 VHN.
Screw machine products suitable for decorative anodizing.

AA 7049..
For production of high quality weapon parts

A.A.A.A.
  M-590; Pb, Sb, bal Sn.
  For bearings; Babbitt, antrifriction metal.

"A" ALLOY-1.
  M-311; 77 Al, 20 Zn, 3 Cu.
  Hot rolled: 60,000 TS; 41,000 YS; 21 El; 36 RA.
  For light alloy wrought parts; non-hardenable.

"A" ALLOY-2.
  M-311; 4.7 Cu, 1.34 Mg, 1.85 Ni, 1.5 Pb, 0.5 Fe, bal Al.
  For light alloy wrought parts; age-hardenable.

A-ALLOY GR. A.
  M-1005; 0.4-0.8 C, 66-68 Ni, 19-21 Cr, 0.5-1.0 Mn, bal Fe.
  Cast: 60,000-80,000 TS; 40,000-50,000 YS; 40 El; 30 RA; 217 Brin.
  For heat and corrosion resistant parts; heat and corrosion resistant.

AA NICKEL-COBALT.
  M-580; Ni-Co.
  For anodes; electroplating.

AB2.
  M-England; 8.5-10.5 Al, 3.5-5.5 Fe, 4.6-6.5 Ni, 1.5 max Mn, 0.50 max Zn, bal Cu.
  For impellers and propellers, gears, pumps, shafts, bushings.
  B.S. 1400 equivalent; heat treatable, corrosion resistant.

AB 2.
  M-1653; 0.70 C, 0.40 Mn, bal Fe.
  Water hardening tool steel.
  UC 70 KU; AISI W1 (0.7C).

AB 3.
  M-1653; 0.80 C, 0.40 Mn, bal Fe.
  Water hardening tool steel.
  UC 85 KU; AISI W1 (0.8C).

AB 4S.
  M-1653; 1.0 C, 0.40 Mn, bal Fe.
  Water hardening tool steel.
  UC 90 KU; AISI W1 (1.0C).

A.B.C.
  M-40, M-487; 0.4 C, 0.5 W, 0.25 Cr, 0.3 Si, 0.5 Mn, bal Fe.
  For tools, chisels, hammers, drills, gears, shafts; tough and hard.

A.B.C.
  M-80; high C, alloy, bal Fe.
  For chisels; shock resistant.

ABC-III.
  M-Germany; 0.7 C, 2.5 Mo, 2.5 V, 3 W, bal Fe.
  For cutting tools; oil hardened.

ABEX NO 4K.
  M-1794; 88 Cu, 10 Sn, 2 Zn.
  Sand cast: 40 ksi TS; 18 ksi YS; 20 El; 75 Brin.
  Gan bronze; CDA 905.

ABEX NO 4L.
  M-1794; 88 Cu, 8 Sn, 4 Zn.
  Sand cast: 40 ksi TS; 18 ksi YS; 20 El; 70 Brin.
  Navy "G" bronze; CDA 903.

ABEX NO 6H.
  M-1794; 74 Cu, 7 Sn, 20 Pb.
  Sand cast: 22 ksi TS; 14 ksi YS; 7 El; 50 Brin.
  High leaded bronze; CDA 945.

ABEX NO 6R.
  M-1794; 70 Cu, 10 Sn, 20 Pb.
  Sand cast: 25 ksi TS; 14 ksi YS; 7 El; 50 Brin.
  High leaded tin bronze.

ABEX NO 6X.
  M-1794; 70 Cu, 5 Sn, 25 Pb.
  Cent. cast: 25 ksi TS; 17 ksi YS; 8 El.
  Soft bronze; CDA 943.

ABEX NO 7A.
  M-1794; 84 Cu, 10 Sn, 2.5 Pb, 3.5 Ni.
  Sand cast: 40 ksi TS; 20 ksi YS; 15 El; 75 Brin.
  Nickel bronze; CDA 915.

ABEX NO 9A.
  M-1794; 89 Cu, 10 Al, 1 Fe.
  As cast: 65 ksi TS; 25 ksi YS; 20 El; 120 Brin.
  Ht. Tr: 80 ksi TS; 40 ksi YS; 12 El; 160 Brin.
  Aluminum bronze; CDA 953.

ABEX NO 9AF.
  M-1794; 85 Cu, 11 Al, 4 Fe.
  As cast: 75 ksi TS; 30 ksi YS; 12 El; 150 Brin.
  Ht. Tr.: 90 ksi TS; 45 ksi YS; 6 El; 180 Brin.
  Aluminum bronze; CDA 954.

ABEX NO 9L.
  M-1794; 81 Cu, 11 Al, 4 Fe, 4 Ni.
  As cast: 90 ksi TS; 40 ksi YS; 6 El; 190 Brin.
  Ht. Tr.: 110 ksi TS; 60 ksi YS; 5 El; 200 Brin.
  Nickel aluminum bronze; CDA 955.

ABEX NO 10B.
  M-1794; 64 Cu, 4 Al, 3 Fe, 26 Zn, 3 Mn.
  Cast: 90 ksi TS; 45 ksi YS; 18 El; 180 Brin.
  High tensile-90 Manganese bronze; CDA 862.

ABEX NO 10C.
  M-1794; 58 Cu, 1 Al, 1 Fe, 39 Zn, 1 Mn.
  Cast: 65 ksi TS; 25 ksi YS; 20 El; 130 Brin.
  Manganese bronze: CDA 865.

ABEX NO 10D.
  M-1794; 63 Cu, 6 Al, 3 Fe, 25 Zn, 3 Mn.
  Cast: 110 ksi TS; 60 ksi YS; 12 El; 200 Brin.
  High tensile-110 Manganese bronze; CDA 863.

ABEX NO 15A.
  M-1794; 73 Cu, 5 Sn, 20 Pb, 3 Zn.
  Sand cast: 22 ksi TS; 14 ksi YS; 7 El; 48 Brin.
  Journal bearing bronze; CDA 941.

ABEX NO 20-A.
M-1794; 97 Cu, 2 Sn, 1 Zn.
Sand cast: 25 ksi TS; 8 ksi YS; 35 El; 35 Brin; 28% IACS conductivity.
Electrode copper.

ABEX NO 20-B.
M-1794; 99.7 Cu.
Sand cast: 25 ksi TS; 7 ksi YS; 35 El; 35 Brin; 70% IACS min conductivity.
Blast furnace copper.

ABEX NO 20-C.
M-1794; 99.9 Cu, (Hi Cond. Cu).
Sand cast: 25 ksi TS; 7 ksi YS; 35 El; 35 Brin; 80% IACS min conductivity.

ABEX NO 20-H.
M-1794; 99.2 Cu, 0.8 Cr. (Chrome Cu.).
Cast, HT; 45 ksi TS; 35 ksi YS; 12 El; 95 Brin; 80% IACS min conductivity.
CDA 815.

ABEX NO 20-L.
M-1794; 99.9 Cu (Hi-Cond. Cu).
Sand cast: 25 ksi TS; 7 ksi YS; 35 El; 35 Brin; 90% IACS min conductivity.

ABEX NO 20-Y.
M-1794; 99 Cu, 1 Sn (Blast furnace Cu).
Sand cast: 25 ksi TS; 7 ksi YS; 35 El; 35 Brin; 30% IACS min conductivity.

ABEX NO 41.
M-1794; 85 Cu, 5 Sn, 5 Pb, 5 Zn.
Sand cast: 30 ksi TS; 14 ksi YS; 20 El; 60 Brin.
Red brass; CDA 836.

ABEX NO 61.
M-1794; 75 Cu, 6 Sn, 16 Pb, 3 Zn.
Sand cast: 20 ksi TS; 14 ksi YS; 7 El; 50 Brin.
TIGER" bronze.
Similar to CDA 939.

ABEX NO 63.
M-1794; 88 Cu, 10 Sn, 2 Pb.
Sand cast: 35 ksi TS; 16 ksi YS; 10 El; 67 Brin.
Leaded tin bronze; CDA 927.

ABEX NO 64.
M-1794; 80 Cu, 10 Sn, 10 Pb, P.
Sand cast: 25 ksi TS; 12 ksi YS; 8 El; 60 Brin.
Phosphor bronze; CDA 937.

ABEX NO 65.
M-1794; 89 Cu, 11 Sn, 0.3 max P.
Sand cast: 35 ksi TS; 18 ksi YS; 10 El; 70 Brin.
Gear bronze; CDA 907.

ABEX NO 65-N.
M-1794; 87 Cu, 11.5 Sn, 1.5 Ni.
Sand cast: 35 ksi TS; 17 ksi YS; 10 El; 75 Brin.
Centr. cast: 50 ksi TS; 25 ksi YS; 10 El; 100 Brin.
Nickel gear bronze; CDA 917.

ABEX NO 622.
M-1794; 88 Cu, 6 Sn, 1.5 Pb, 4.5 Zn.
Sand cast: 34 ksi TS; 16 ksi YS; 22 El; 60 Brin.
Navy M" bronze; CDA 922.

ABEX NO 640.
M-1794; 87 Cu, 11 Sn, 1 Pb, 1 Ni.
Sand cast: 35 ksi TS; 10 El; 67 Brin.
Nickel phosphor bronze; CDA 925.

ABEX NO 660.
M-1794; 83 Cu, 7 Sn, 7 Pb, 3 Zn.
Sand cast: 30 ksi TS; 14 ksi YS; 12 El; 60 Brin.
Modified red brass; CDA 932.

ABEX NO 964.
M-1794; 69.1 Cu, 0.9 Fe, 30 Ni, 1.5 max Mn.
Cast: 60 ksi TS; 32 ksi YS; 15 El; 140 Brin.
Copper-nickel bronze; similar to CDA 964.

ABK METAL.
M-1099; 3-3.75 C, 0.3-0.9 Mn, 0.3-1.2 Si, 3.5-6.5 Ni, 1.5-4 Cr, bal Fe.
Cast: 35,000-80,000 TS; 0.10-0.30 El; 500-650 Brin.
For castings, crushing rolls, liners; wear resistant, cast iron.

ABK NI-HARD.
M-9; 4.5 Ni, 2.7-3.6 C, 1.5 Cr, 0.5-1.5 Si, 0.3-0.7 Mn, bal Fe.
Cast: 30,000-40,000 TS; 575-750 Brin.
For pump plungers, roller bearing races, chilled rolls; corrosion resistant, tough.

ABRACORR 25.
M-884; 0.10 C, 0.50 Mn, 0.75 Ni, 4.0 Cr, 0.65 Al, 0.20 Mo, bal Fe.
Weldable and easily formable.
For abrasion and oxidation resisting parts.

ABRACORR 30.
M-884; 0.10 C, 0.80 Mn, 6.0 Cr, 0.8 Al, 1.5 Ni, + Mo + V, bal Fe.
Weldable and easily formable.
For abrasion and oxidation resisting parts.

ABRADUR C.
M-884; 0.90 C, 0.75 Mn, 0.25 Si, bal Fe.
Hardenable to 60 Rc on surface.
For wear and abrasion resisting parts.

ABRADUR C 65 K.
M-884; 0.65 C, 0.80 Mn, 0.25 Si, 0.30 Cr, bal Fe.
Hardenable to Rc 55 on surface.
For wear and abrasion resisting parts.

ABRADUR C 75 K.
M-884; 0.75 C, 0.80 Mn, 0.25 Si, 0.30 Cr, bal Fe.
Hardenable to Rc 58 on surface.
For wear and abrasion resisting parts.

ABRADUR C 90 K.
M-884; 0.90 C, 0.80 Mn, 0.25 Si, 0.35 Cr, bal Fe.
Hardenable to Rc 61 on surface.
For wear and abrasion resisting parts.

ABRADUR PM 25.
 M-884; 0.20 C, 1.30 Mn, 0.75 Cr, bal Fe.
 Hardenable to about Rc 40; weldable.
 For wear and abrasion resisting parts.

ABRADUR PM 35.
 M-884; 0.16 C, 1.60 Mn, 1.50 Cr, 0.20 Mn, 0.10 V, bal Fe.
 Hardenable to about Rc 35; weldable.
 For wear and abrasion resisting parts.

ABRADUR PM 40.
 M-884; 0.28 C, 1.60 Mn, 0.80 Ni, 1.70 Cr, 0.45 Mo, 0.10 V, bal Fe.
 Hardenable to about Rc 45-50; weldable.
 For wear and abrasion resisting parts.

ABRADUR S.
 M-884; 0.45 C, 0.70 Mn, 1.80 Si, bal Fe.
 Heat treatable to 40-55 Rc.
 For wear and abrasion resisting parts.

ABRADUR SK.
 M-884; 0.65 C, 0.85 Mn, 1.65 Si, 0.85 Cr, bal Fe.
 Hardenable to 46-60 Rc.
 For wear and abrasion resisting parts.

ABRASALLOY.
 M-353; 0.35 C, 1 Cr, 0.25 Mo, 0.75 Mn, 0.35 Si, bal Fe.
 For crusher hammers, coal screens, dredge pumps; tough, wear resistant.

ABRASION RESISTING.
 M-67; 0.35-0.50 C, 1.20-1.65 Mn, 0.10-0.30 Si, bal Fe.
 Z10 Brin. (minimum).
 For wear resistant sheet or plate.

ABRASOCRAFT.
 M-175; C, alloy, bal Fe.
 Welded: 350-500 Brin.
 For hard surfacing electrode; abrasion resistant.

ABRASOHARD.
 M-811; C, alloy, bal Fe.
 For welding rods; abrasion resistant.

ABRASOWELD.
 M-578; 2.1 C, 1.1 Mn, 0.75 Si, 0.40 Mo, 6.5 Cr, bal Fe.
 Hard surfacing arc welding electrodes for resistance to abrasion and impact.

ABRATEC N700.
 M-717.
 Electrode for AC-DC metallic arc abrasion resistant coating of steel; Rc 68-72 (double pass).

ABRAZO 60.
 M-1724; 0.60 max C, 0.25 Si, 0.60 Mn, bal Fe.
 Abrasion resistant steel for application, such as liner plates, chutes, buckets, storage bins, and mould boards.

ABROS.
 M-Eng; 10 Cr, 88 Ni, 2 Mn.
 For electrical resistance, heating elements; stainless and corrosion resistant.

ABSCO METAL 25H.
 M-9; high C, Si, Mn, bal Fe.
 Cast: 25,000 TS; 150-210 Brin.
 For ingot molds, pig molds; cast iron, Class 25.

ABSCO METAL 30.
 M-9; high C Si, Mn, bal Fe.
 Cast: 30,000 TS; 20,000 YS; 0.6 El; 170-240 Brin.
 For gears, pulleys, general castings; cast iron, Class 30.

ABSCO METAL 30H.
 M-9; high C, Si, Mn, alloy, bal Fe.
 Cast: 30,000 TS; 220-280 Brin.
 For heat and wear resistant castings; cast iron.

ABSCO METAL 35.
 M-9; high c, S, Mn, bal Fe.
 Cast: 35,000 TS; 200-260 Brin.
 For gears, housings, general castings; cast iron, Class 35.

ABSCO METAL 40.
 M-9; high C, Si, Mn, bal Fe.
 Cast: 40,000 TS; 210-260 Brin.
 For meter bodies, water valves, compressor cylinders; cast iron, Class 40.

ABSCO METAL 40H.
 M-9; high C, Si, Mn, alloy, bal Fe.
 Cast: 40,000 TS; 230-290 Brin.
 For heat and wear resistant castings; cast iron.

ABSCO METAL 50.
 M-9; high C, Si, Mn, bal Fe.
 Cast: 50,000 TS; 220-280 Brin.
 For press parts, cylinder heads, forming dies; cast iron, Class 50.

ABSCO METAL 55.
 M-9; high C, Si, Mn, bal Fe.
 Cast: 55,000 TS; 45,000 YS; 0.7 El; 230-310 Brin.
 For crankshafts, tool shanks, cutter bodies; cast iron, Class 55.

ABSCO METAL 400.
 M-9; high C, Si, Mn, alloy, bal Fe.
 Cast: 400 Brin.
 For ash-sluice pipe chutes, burner nozzles; heat and wear resistant.

ABYSSINIAN GOLD.
 M-eng.; 88 Cu, 11.5 Zn, 0.5 Au.
 For ornaments, jewelry.

ABYSSINIAN GOLD.
 M-eng.; 90-92 Cu, 8-10 Zn, 1 plated Au.
 For ornaments, jewelry; sheets plated with Au.

AC-254.
M-642; 0.24 C, 1.0 Mn, 0.4 Si, 12.0 Cr, 1.0 Ni, 2.5 Mo, 1.0 W, 0.25 V, 0.22 B, bal Fe.
For steam turbine blading, jet engine components.
Stainless, martensitic, creep resistant.

ACCOLOY CN-1.
M-1005; 0.50 max C, 28.0 Cr, 3.0 Ni, bal Fe.
For chemical, mining, paper industries.
Bushings, pump casings, valve bodies, corrosive resistors; ferritic.

ACCOLOY CN-2.
M-1005; 0.40 C, 28 Cr, 10 Ni, bal Fe.
Cast: 95,000 TS; 45,000 YS; 20 El; 200 Brin.
For furnace parts, heat treating equipment, pots; Type HE; austenitic, heat resistant.

ACCOLOY CN-3.
M-1005; 0.40 C, 25 Cr, 20 Ni, bal Fe.
Cast: 75,000 TS; 50,000 YS; 17 El; 170 Brin.
For furnace parts, heat treating equipment, pots; Type HK; austenitic, heat resistant.

ACCOLOY CN-4.
M-1005; 0.30 C, 25 Cr. 12 Ni, bal Fe.
Cast: 80,000 TS; 50,000 YS; 25 El; 185 Brin.
For furnace parts, heat treating equipment, pots; Type HH2; austenitic, heat resistant.

ACCOLOY CN-6.
M-1006; 0.20-0.60 C, 30.0 Cr, 20.0 Ni, bal Fe.
Cast: 82,000 TS; 52,000 YS; 19 El; 192 Bin.
For furnace fixtures, furnace parts, heat treating equipment, radiant tubes. Type HL; heat and corrosion resistant.

ACCOLOY CN-7.
M-1005; 0.20 max C, 18.0 Cr, 8.0 Ni, bal Fe.
Casting; for chemical, petrochemical, food industries. Pumps, return bends, valve bodies, cylinder liners, for heat and corrosion resistance. ACI Type CF-20.

ACCOLOY CN-8.
M-1005; 0.08 max C, 18.0 Cr, 8.0 Ni, bal Fe.
For blast furnace, guide rollers, headers, pumps, for corrosion resistant operations.
ACI Type CF-8. Casting.

ACCOLOY CN-10.
M-1005; 0.08 max C. 18.0 Cr, 9.0 Ni, 2.0-3.0 Mo, bal Fe.
For corrosion and heat resistance. Casting; Austenitic.
ACI Type CF-8M.

ACCOLOY NC-1.
M-1005; 0.35-0.75 C, 68 Ni, 18 Cr, bal Fe.
Cast: 65,000 TS; 36,000 YS; 9 El; 176 Brin.
For furnace parts, radiant tubes, heating elements, heat treating equipment.
ACI Type HX; austenitic.

ACCOLOY NC-2.
M-1005; 0.35-0.75 C, 60 Ni, 12 Cr, bal Fe.
Cast: 68,000 TS; 36,000 YS; 4 El; 185 Brin.
For pots, electric heat elements, retorts, hearth plates, muffles.
ACI Type HW.

ACCOLOY NC-3.
M-1005; 0.35-0.75 C, 38 Ni, 18 Cr, bal Fe.
Cast: 70,000 TS, 40,000 YS; 9 El; 170 Brin.
For heat treating trays, burner tubes, carburizing retorts, conveyer screws, chains, furnace rolls, radiant tubes.
ACI Type HU; austenitic, heat resistant.

ACCOLOY NC-4.
M-1005; 0.35-0.75 C, 35 Ni, 15 Cr, bal Fe.
Cast: 70,000 TS; 40,000 YS; 10 El; 180 Brin.
For carburizing containers, furnace parts, heat treating equipment and fixtures, radiant tubes, roller rails, feed screws.
Type HT; austenitic, heat resistant.

ACCOLOY NC-6.
M-1005; 0.20-0.50 C, 25 Ni, 20 Cr, bal Fe.
Cast: 68,000 TS; 38,000 YS; 13 El; 160 Brin.
For brazing fixtures, chains, furnace beams and parts, pier caps, radiant tubes, nozzles, trays.
ACI Type HN; austenitic, heat resistant.

ACCOLOY NC-7.
M-1005; 0.20-0.40 C, 21 Cr, 9 Ni, bal Fe.
Cast: 92,000 TS; 45,000 YS; 38 El; 165 Brin.
For annealing boxes and trays, baskets, burner tips, conveyer belts, furnace rails, wear plates, hearth plates.
ACI Type HF, austenitic, heat resistant.

ACCOLOY NC-9MO.
M-1005; 0.35-0.75 C, 22 Cr, 46 Ni, bal Fe.
Castings to resist stress corrosion cracking in petrochemical applications and furnace applications over 2000°F.

ACCOLOY NC-10.
M-1005; 0.35-0.75 C, 25 C, 35 Ni, bal Fe.
Cast: 71,000 TS; 40,000 YS; 11.5 El.
For heat treating, petrochemical and petroleum industries. For ethylene heaters, heat treat fixtures, refinery tubes.
Type HP, austenitic, heat resistant.

ACCUMULATOR METAL.
M-U.S.; 90 Pb, 9.2 Sn, 0.8 Sb.
Bearings, battery plates, antrifriction.

ACCURATE METAL.
M-1676; 29-32 Ni, 68 min Cu, 0.4-0.7 Fe.
Annealed: 57,000 TS; 21,000 YS; 43 El; B 44 Rock.
Hard: 82,000 TS; 75,000 YS; 5 El; B 86 Rock.
Soft: 54,000 TS; 16,000 YS; 35 El; B 40 Rock.
For electronic components, marine hardware, ferrules, pump valves.
Corrosion resistant, non-magnetic.

ACCURLOY.
M-1510; 0.51 C, 1.05 Cr, 0.25 Mo, 0.53 Ni, 0.21 V, 0.97 Mn, bal Fe.
Heat treated: 165,000 TS; 150,000 YS; 20 El; 59 RA; 280-300 Brin.
For shafts, pins, boring bars, piston rods; heat treated and stress-relieved bars, shock resistant.

ACEROID.
M-592; Zn, bal Cu.
For castings.

ACEROLD.
M-592; C, alloy, bal Fe.
For machinery castings.

ACHORN 9W HOT WORK.
M-608; 0.30 C, 3.0 Cr, 0.5 V, 9.5 W, bal Fe.
Hot work tool steel; oil hardening; for forging dies, hot forming dies; AISI H21.

ACHORN 11W HOT WORK.
M-608; 0.40 C, 3.0 Cr, 0.45 V, 12.0 W, bal Fe.
Oil hardening hot work tool steel, tungsten type; for forging and hot working dies. AISI H22.

ACHORN 15 W HOT WORK.
M-608; 0.45 C, 3.5 Cr, 0.7 V, 14 W, bal Fe.
Air or oil hardening hot working tool and die steel; AISI H24.

ACHORN 33 HOT WORK.
M-608; 0.33 C, 0.85 Si, 5.0 Cr, 0.2 V, 1.25 W, 1.45 Mo, bal Fe.
Hot work tool steel, chromium type; AISI H12.

ACHORN 33A HOT WORK.
M-608; 0.40 C, 0.3 Mn, 0.9 Si, 5.0 Cr, 0.5 V, 1.3 Mo, bal Fe.
Oil or air hardening tool and die steel; for forging dies and hot forming dies. AISI H11.

ACHORN 33M HOT WORK.
M-D-608; 0.40 C, 1.0 Si, 5.0 Cr, 1.0 V, 1.0 Mo, bal Fe.
Air or oil hardening hot work tool; AISI H13.

ACHORN 100 CHROMIUM.
M-608; 1.0 C, 1.5 Cr, bal Fe.
Low alloy tool steel, oil hardening.
AISI L1.

ACHORN 225 C HIGH PRODUCTION DIE.
M-608; 2.1 C, 12.0 Cr, 0.75 W, bal Fe.
Oil or air hardening cold work tool steel; chromium type; for punch and trim dies, thread rolling tools, gages.
AISI D3.

ACHORN "350" FINISHING.
M-608; 1.3 C, 3.5 W, bal Fe.
For tools, dies; fast finishing steel.

ACHORN 512.
M-608; 0.10 C, 0.48 Mn, 1.5 Cr, 3.5 Ni, bal Fe.
Low carbon steel for cold hubbing and then case hardening for molds.

ACHORN 18-4-1.
M-608; 0.65 C, 3.75 Cr, 18.5 W, 1.2 V, bal Fe.
For tools, cutters; high speed steel.

ACHORN 6-6-4-2.
M-608; 0.85 C, 4 Cr, 2 V, 6.5 W, 5 Mo, bal Fe.
For tools, cutters; high speed steel.

ACHORN AF-33 HOT WORK.
M-608; 0.30-0.35 C, 0.8-1.0 Si, 5 Cr, 1.1 W, 1.5 Mo, bal Fe.
For hot work dies; hot work steel.

ACHORN ALLOY PIVOT.
M-608; 1.10 C, 0.40 Mn, 0.25 Si, 1.35 Cr, 0.40 Mo. bal Fe.
Oil hardening steel for shafts, pivots, lathe centers. AISI L7.

ACHORN BEST CARBON.
M-608; 0.8-1.2 C, 0.25 Si, 0.25 Mn, bal Fe.
Water hardened: 166,000-216,000 TS; 110,000-150,000 YS; 11-15 El; 32-37 RA; 330-600 Brin.
For taps, reamers, drills, punches, stamps, knurls, mandrels, cutters.
Type W1 water hardening.

ACHORN CARBON DRILL ROD.
M-608; 1.0 C, 0.25 Mn, 0.28 Si, bal Fe.
Water hardening tool steel; AISI W1.

ACHORN CM AIR HARDENING.
M-608; 0.70 C, 2.0 Mn, 0.3 Si, 1.0 Cr, 1.35 Mo, bal Fe.
Medium alloy, air hardening cold work tool steel; AISI A6.

ACHORN COLD DRAWN TOOL STEEL.
M-608; 1.05 C, 0.25 Mn, 0.28 Si, bal Fe.
Water hardening tool steel, for drills, arbors, lathe centers, bushings. AISI W1.

ACHORN COLD HEADING.
M-608; 0.8-1.2 C, 0.25 Si, 0.25 Mn, bal Fe.
Water hardened: 166,000-216,000 TS; 110,000-150,000 YS; 11-15 El; 32-37 RA; 330-600 Brin.
For taps, drills, reamers, punches, stamps, knurls, mandrels, cold heading tools.
Type W1 water hardening.

ACHORN COMPOSITE STEEL.
M-608; 1.0 C, 0.40 Cr, 0.40 V, 1.25 W, bal Fe.
Water or oil hardening steel for shafts, arbors, lathe centers, drill bushings.
AISI F1.

ACHORN CRM-50.
M-608; 0.5-0.55 C, 0.6 Mn, 1.0 Cr, 0.20 Mo, bal Fe.
For hot work dies; hot work steel.

ACHORN CVM.
M-1421; 0.95 C, 0.7 Mn, 5 Cr, 0.25 V, 1.2 Mo, bal Fe.
For punches, blanking and forming dies; air hardened, non-deforming.

ACHORN EXTRA BLADE.
M-608; 1.1 C, 0.25 Mn, bal Fe.
For tools, cutters; oil hardened.

ACHORN EXTRA CARBON.
M-608; 1.05 C, 0.25 Mn, 0.28 Si, bal Fe.
Water hardening tool steel ; AISI W1.

ACHORN EXTRA CHISEL.
M-608; 0.8-1.1 C, 0.25 Si, 0.25 Mn, bal Fe.
Water hardened: 166,000-216,000 TS; 110,000-150,000 YS; 11-15 El; 32-37 RA; 330-600 Brin.
For taps, drills, reamers, punches, stamps, knurls, mandrels.
Type W1 water hardening.

ACHORN EXTRA SOLID DRILL.
M-608; 1.15 C, 0.30 Mn, bal Fe.
For tools, drills; water hardened.

ACHORN FAGERSTA BEST.
M-608; 1.05-1.15 C, 0.20-0.25 Mn, 0.25 Si, bal Fe.
For drills, cutters, reamers, taps, broaches; Type W1; water hardened.

ACHORN FAGERSTA CHISEL.
M-608; 0.8-0.9 C, 0.3 Mn, 0.25 Si, bal Fe.
For chisels, screw drivers; water hardened; Type W1.

ACHORN FAGERSTA COLD HEADING.
M-608; 0.95 C, 0.3 Mn, 0.25 Si, bal Fe.
For cold heading dies, form tools; Type W1; water hardened.

ACHORN FAGERSTA ENVELOPE DIE.
M-608; 1.05-1.15 C, 0.3 Mn, 0.25 Si, bal Fe.
For envelope dies, cutters, drills, form tools; Type W1; water hardened.

ACHORN FAGERSTA EXTRA.
M-608; 1.0-1.1 C, 0.3 Mn, 0.25 Si, bal Fe.
For tools, cutters, drills, taps, reamers; Type W1; water hardened.

ACHORN FAGERSTA EXTRA CUTLERY.
M-608; 1.07-1.12 C, 0.25 Mn, 0.3 Si, bal Fe.
For cutlery tools, drills, taps; Type W1; water hardened.

ACHORN FAGERSTA FINISHING.
M-608; 1.15 C, 0.4 Mn, 0.4 Cr, 2.5 W, bal Fe.
For finishing cutters; water hardened, keen cutting edge.

ACHORN FAGERSTA HIGH PRODUCTION.
M-608; M-385; 1.6 C, 0.8 Mn, 12 Cr, 0.2 V, bal Fe.
For blanking and forming dies; air hardened, non-deforming.

ACHORN FAGERSTA HOT DIE.
M-608; 0.55 C, 0.45 Mn, 5.0 W, 3.0 Ni, bal Fe.
For upsetting and forging dies, extrusion rams and liners; hot work steel, oil hardened.

ACHORN FAGERSTA SHOE DIE.
M-608; M-385; 0.55 C, 0.65 Cr, 0.35 Mo, bal Fe.
For shoe dies, cutting dies for leather and rubber; oil hardening.

ACHORN FAGERSTA SILVER DIE.
M-608; 1.0 C, 0.25 Mn, 0.25 Si, bal Fe.
For silver dies; Type W1; water hardened.

ACHORN FAGERSTA SMOOTH BOR HOLLOW.
M-608; 0.80 C, 0.25 Mn, 0.25 Si, bal Fe.
For rock drills; Type W1; water hardened.

ACHORN FAGERSTA SOLID DRILL.
M-608; 0.8-0.9 C, 0.35 Mn, 0.25 Si, bal Fe.
For solid drills; Type W1; water hardened.

ACHORN FAGERSTA SPECIAL ALLOY DIE.
M-608; 0.45-0.55 C, 0.9 Cr, 1.2 W, 0.2 Mo, bal Fe.
For dies, tools; oil hardened.

ACHORN FAGERSTA STANDARD.
M-608; 0.9-1.0 C, 0.3 Mn, 0.25 Si, bal Fe.
For tools, drills, taps, hobs, reamers; Type W1; water hardened.

ACHORN FAGERSTA STANDARD CUTLERY.
M-608; 0.85-0.95 C, 0.35 Mn, 0.25 Si, bal Fe.
For cutlery tools, drills; reamers; Type W1; water hardened.

ACHORN FAGERSTA SUPERIOR OIL.
M-608; 1.0 C, 1.0 Mn, 0.4 Cr, 0.4 W, bal Fe.
For tools, dies, cold headers; water or oil hardened.

ACHORN FAGERSTA UNBREAKABLE CHISEL.
M-608; 0.45 C, 0.9 Cr, 1.2 W, 0.2 Mo, bal Fe.
For chisels, upsetters; oil hardened, tough.

ACHORN FAGERSTA WHITE GOLD.
M-608; 1.15 C, 0.4 Mn, 0.4 Cr, 1.2 W, bal Fe.
For tools, cutters; oil or water hardened.

ACHORN GRAPHITIC OIL.
M-608; 1.45 C, 0.8 Mn, 1.15 Si, 0.20 Cr, 0.25 Mo, bal Fe.
Oil hardening cold work tool steel; AISI o6.

ACHORN HEAT TREATED MOLD STEEL.
M-608; 0.30 C, 0.80 Mn, 0.50 Si, 1.7 Mn, 0.40 Mo, bal Fe.
Hardened, machinable, for molds; AISI P20.

ACHORN HIGH PRODUCTION.
M-608; 1.6 C, 12 Cr, 0.8 Mo, 0.8 V, bal Fe.
For punches, drawing dies, form tools; air hardened, non-deforming.

ACHORN HOLLOW DRILL.
M-608; 0.75 C, 0.3 Mn, bal Fe.
Heat treated: 180,000 TS; 130,000 YS; 12 El; 36 RA; 360 Brin.
For hollow drills, die blocks; water hardened; Type W1.

ACHORN KLOSTER BRILLIANT AX.
M-608; 0.7 C, 4 Cr, 18 W, 1.75 V, bal Fe.
For tools, cutters, reamers, hobs, lathe and planer tools; high speed steel; Type T1.

ACHORN KLOSTER BRILLIANT WKE.
M-608; 0.7 C, 4.5 Cr, 18 W, 1.5 V, 5 Co, bal Fe.
For lathe and planer tools, hobs, reamers, taps, drills; high speed steel; Type T4.

ACHORN KLOSTER PRIOR EXTRA.
M-608; 0.6-0.7 C, 4-5 Cr, 18-19 W, 1.2 V, 1 Co, bal Fe.
For lathe and planer tools, hobs, reamers, taps, drills; high speed steel, oil hardened.

ACHORN KLOSTER REMA.
M-608; 0.02-0.05 C, 0.02-0.07 Mn, bal Fe.
For molding dies for plastics; hobbing steel.

ACHORN M1 HIGH SPEED.
M-608; 0.80 C, 4.0 Cr, 1.0 V, 1.5 W, 8.0 Mo, bal Fe.
High speed steel, molybdenum type; for drills, lathe tools, milling cutters. AISI M1.

ACHORN M-2.
M-608; 0.8 C, 5 Mo, 6.5 W, 2 V, 4 Cr, bal Fe.
For lathe and planer tools, drills, reamers; high speed steel.

ACHORN M2 1/2 (CLASS 1) HIGH SPEED.
M-608; 1.0 C, 4.0 Cr, 2.5 V, 6.25 W, 6.25 Mo, bal Fe.
High speed tool steel, molybdenum-tungsten type; AISI M3 Class 1.

ACHORN M3 (CLASS 2) HIGH SPEED.
M-608; 1.15 C, 4.0 Cr, 3.0 V, 6.0 W, 5.5 Mo, bal Fe.
High speed steel, tungsten-molybdenum-vanadium-chromium; AISI M3 Class 2.

ACHORN M4 HIGH SPEED.
M-608; 1.28 C, 4.5 Cr, 4.0 V, 5.5 W, 4.5 Mo, bal Fe.
High speed steel, tungsten-molybdenum-vanadium-chromium; AISI M4.

ACHORN M10 HIGH SPEED.
M-608; 0.85 C, 4.0 Cr, 2.0 V, 8.0 Mo, bal Fe.
Molybdenum type high speed steel; AISI M10.

ACHORN M34 HIGH SPEED.
M-608; 0.90 C, 4.0 Cr, 2.0 V, 1.75 W, 8.5 Mo, 8 Co, bal Fe.
High speed steel, molybdenum cobalt type, AISI M34.

ACHORN MANGANESE OIL HARDENING.
M-608; 0.90 C, 1.50 Mn, 0.25 Si, 0.30 Mo, bal Fe.
Oil hardening in small sections; AISI O2.

ACHORN MOLDALOY.
M-608; 0.10 C, 0.5 Mn, 0.5 Cr, 1.2 Ni, bal Fe.
For dies, molds; carburizing grade.

ACHORN NI-CRO-MO.
M-608; 0.68 C, 0.6 Mn, 0.25 Si, 0.65 Cr, 1.4 Ni, 0.20 Mo, bal Fe.
Oil hardening, low alloy tool steel for shafts, arbors, lathe ceners, tool holders. AISI L6.

ACHORN OIL HARDENING DRILL ROD.
M-608; 0.90 C, 1.1 Mn, 0.5 Cr. 0.15 V, 0.5 W, bal Fe.
Oil hardenable in small sections; AISI O1.

ACHORN OILWEAR.
M-608; 1.25 C, 0.3 Mn, 0.35 Si, 0.4 Cr, 0.2 V, 1.4 W, bal Fe.
Oil hardening cold work tool steel, wear resistant type; AISI O7.

ACHORN REMA IRON.
M-608; 0.06 C, 0.21 Mn, bal Fe.
For plastic mold dies; water hardened.

ACHORN SOLID DRILL.
M-608; 0.85 C, 0.35 Mn, bal Fe.
Heat treated: 190,000 TS; 145,000 YS; 10 El, 30 RA; 400 Brin.
For drills, punches, taps, reamers, cutters; water hardened; Type W1.

ACHORN SPRING STEEL SHEET.
M-608; 1.0 C, 0.25 Mn, 0.28 Si, bal Fe.
Water hardened sheet steel, spring temper; AISI W1.

ACHORN STANDARD CARBON.
M-608; 1.0-1.1 C, 0.30 Mn, bal Fe.
For cutters, drills, reamers, punches; water hardened; Type W1.

ACHORN SUPERIOR OIL HARDENING.
M-608; C, alloy, ba Fe.
For tools, dies, punches; oil hardened.

ACHORN T2 HIGH SPEED.
M-608; 0.80 C, 4.0 Cr, 2.0 V, 18.5 W, 0.75 Mo, bal Fe.
Tungsten type high speed steel; AISI T2.

ACHORN T4 HIGH SPEED.
M-608; 0.75 C, 4.0 Cr, 1.05 V, 18.5 W, 5.0 Co, bal Fe.
Tungsten-cobalt type high speed steel; AISI T4.

ACHORN T5 HIGH SPEED.
M-608; 0.80 C, 4.25 Cr, 2.0 V, 19.0 W, 1.0 Mo, 8.0 Co, bal Fe.
Tungsten-cobalt type high speed steel; AISI T5.

ACHORN T15 HIGH SPEED.
M-608; 1.5 C, 4.75 Cr, 5.0 V, 12.5 W, 5.0 Co, bal Fe.
High speed steel, tungsten, cobalt, vanadium type; AISI T15.

ACHORN TOOL STEEL SHEET.
M-608; 1.05 C, 0.25 Mn, 0.28 Si, bal Fe.
Water hardenable sheet steel; AISI W1.

ACHORN UBC.
M-608; 0.48 C, 0.9 Cr, 1.1 W, 0.2 Mo, bal Fe.
For hot work dies; hot work steel.

ACHORN USI STEEL.
M-608; 0.60, 0.85 Mn, 2.0 Si, 0.25 Cr. 0.2 V, 0.25 Mo, bal Fe.
Water or oil hardening tool steel, shock resisting type; AISI S5.

ACHORN V85 STEEL.
M-608; 0.90 C, 0.3 Mn, 0.25 Si, 0.25 V, bal Fe.
Water hardening tool steel; AISI W2.

ACHORN VAMO.
M-608; 0.50 C, 0.45 Mn, 1.0 Si, 5.2 Cr, 1.5 Ni, 1.0 V, 1.4 Mo, bal Fe.
Medium alloy, air hardening cold work tool steel; AISI A9.

ACHORN VBC.
M-608; 0.5 C, 0.90 Cr, 1.25 W, 2 Mo, 0.25 Mn, bal Fe.
For punches, rivet sets, upsetters; oil hardened, hot work steel.

ACIBAL.
Al alloy.
For light alloy parts.

ACIBRADE.
M-1724; 0.60-0.75 C, 1.25-1.55 Mn, 0.05 Si, bal Fe.
Resistant to abrasion and tough in the as-rolled condition. For rod mills coke crushers, mineral dressing, chutes, conveyers, and buckets.

ACICULAR.
M-494; M-1485; 2.9 C, 2 Si, 2.5 Ni, 0.9 Mo, bal Fe.
Cast: 280-350 Brin.
For castings, gears, housings; alloy cast iron.

ACID BRONZE-1.
M-Eng; 88 Cu, 10 Sn, 2 Pb.
For bearings, chemical equipment; corrosion resistant.

ACID BRONZE-2.
M-Eng; 82 Cu, 8 Sn, 8 Pb, 2 Zn.
For chemical equipment; corrosion resistant.

ACID BRONZE-3.
M-Eng; 84 Cu, 9.5 Sn, 6.3 Pb. 0.2 P.
For chemical equipment; corrosion resistant.

ACID BRONZE-4.
M-Eng; 74 Cu, 8 Sn, 17 Pb, 1.5 Zn.
For chemical equipment; corrosion resistant.

ACID METAL.
M-U.S; 88 Cu, 10 Sn, 2 Pb.
For chemical equipment; corrosion resisting.

ACID RESISTING-1.
M-Eng; 10 Cr, 30 Ni, 40.8 Fe, 6.2 W, 1.55 Mn, 0.45 Si.
For chemical apparatus, acid resisting vessels and tanks; stainless and corrosion resistant.

ACID RESISTING-2.
M-Eng; 14.5 Cr, 20 Ni, 56 Fe, 5 Co, 4.5 Cu.
For chemical apparatus; resists attack of HNO_3.

ACID RESISTING-3.
M-Eng; 15 Cr, 53 Ni, 23 Fe, 4 W, 1.25 Mn, 3.75 Si.
For chemical apparatus; heat and corrosion resistant.

ACIDUR.
M-323; 16-17 Si, bal Fe.
Cast.
For chemical apparatus; corrosion resistant.

ACIERAL-1.
M-281; 6 Cu, 0.4 Zn, 0.9 Ni, 0.1 Fe, 0.4 Si, bal Al.
Cast 20,000 TS.
For automotive engine parts; castings.

ACIERAL-2.
M-281; 2.3-3.8 Cu, 0.7-1.4 Fe, 1.0-1.5 Mn, bal Al.
Rolled: 22,000 TS; 2 El.
For automotive engine parts; high strength.

ACIPCO 1005.
M-285; 0.08 max C, 0.3-0.6 Mn, bal Fe.
Cast, normalize and temper: 48,000 TS; 25,000 YS; 35 El; 110 Brin.
For high magnetic permeability parts. AISI 1005.

ACIPCO 1015.
M-285; 0.10-0.20 C, 0.3-0.6 Mn, bal Fe.
Cast, normalize and temper: 55,000 TS; 30,000 YS; 30 El; 130 Brin.
Standard mild steel, high ductility.
AISI 1015.

ACIPCO 1025.
M-285; 0.20-0.30 C, 0.3-0.6 Mn, bal Fe.
Cast, normalize and temper: 65,000 TS; 35,000 YS; 25 El; 150 Brin.
Standard structural steel, good weldability.
AISI 1025.

ACIPCO 1045.
M-285; 0.40-0.50 C, 0.5-0.9 Mn, bal Fe.
Cast, normalize and temper: 85,000 TS; 45,000 YS; 15 El; 180 Brin.
For machine parts, and for flame hardening.
AISI 1045.

ACIPCO 1070.
M-285; 0.65-0.75 C, 0.5-09 Mn, bal Fe.
Cast, normalize and temper: 105,000 TS; 50,000 YS; 5 El; 220 Brin.
Particularly for parts requiring flame or induction hardened areas. AISI 1070.

ACIPCO 4130.
M-285; 0.25-0.35 C, 0.4-0.7 Mn, 0.8-1.1 Cr, 0.15-0.25 Mo, bal Fe.
Cast, normalize and temper; 80,000 TS; 45,000 YS; 17 El; 185 Brin.
Cast, water quench and temper; 100,000-160,000 TS; 65,000-145,000 YS; 16-5 El; 250-450 Brin.
Preferred water quenching grade. AISI 4130.

ACIPCO 4140.
M-285; 0.35-0.45 C, 0.6-1.0 Mn, 0.8-1.1 Cr, 0.15-0.25 Mo, bal Fe.
Cast, normalize and temper; 100,000 TS; 50,000 YS; 16 El; 200 Brin.
Cast, oil quench and temper; 120,000-180,000 TS; 85,000-165,000 YS; 16-5 El; 250-450 Brin.
Oil hardening alloy steel.
AISI 4140.

ACIPCO 4330.
M-285; 0.25-0.35 C, 0.4-0.7 Mn, 1.65-2.0 Ni, 0.7-0.9 Cr, 0.2-0.3 Mo, bal Fe.
Cast, normalize and temper; 100,000 TS; 60,000 YS; 16 El; 200 Brin.
Cast, quench and temper; 120,000-180,000 TS; 85,000-165,000 YS; 16-5 El; 250-450 Brin.
Hardenable in heavy sections. AISI 4330.

ACIPCO 8620.
M-285; 0.15-0.25 C, 0.6-1.0 Mn, 0.4-0.7 Ni, 0.4-0.6 Cr, 0.15-0.25 Mo, bal Fe.
Cast, normalize and temper: 70,000 TS; 40,000 YS; 20 El; 150 Brin.
Low alloy carburizing grade.
AISI 8620.

ACIPCO ACICULAR IRONS.
M-285; 2.5-3.0 C, 0.5-1.5 Mn, 0.25 max P, 0.12 max S, 1.7-2.3 Ni, 0.8-1.2 Mo, bal Fe.
Cast: 60,000 TS; 300 Brin.
For highly stressed rolls and cylinders; heat treatable.

ACIPCO CA-15.
M-285; 0.15 max C, 1.0 max Mn, 1.5 max Si, 11.5-14.0 Cr, 1.0 max Ni, bal Fe.
Cast, normalize and temper: 100,000 TS; 65,000 YS; 18 El; 228 Brin.
Heat treatable chromium stainless.
ACI CA 15; AISI 410.

ACIPCO CF-8.
M-285; 0.08 max C, 1.5 max Mn, 2.0 max Si, 18-21 Cr, 8-11 Ni, bal Fe
Cast: 70,000 TS; 30,000 YS; 30 El; 150 Brin.
Austenitic stainless castings. ACI CF-8; AISI 304.

ACIPCO CF-8M.
M-285; 0.08 max C, 1.50 max Mn, 1.50 max Si, 1,-21 Cr, 9-12 Ni, 2-3 Mo, bal Fe.
Cast: 75,000 TS; 40,000 YS; 35 El; 160 Brin.
Austenitic stainless; for special corrosive conditions.
ACI CF-8M; AISI 316.

ACIPCO CI. 30.
M-285; 3.1-3.5 C, 0.4-0.7 Mn, 0.75 max P, 0.12 max S, 1.2-2.5 Si, bal Fe.
Cast: 30,000 TS; 200 Brin.
Castings for engineering parts.

ACIPCO CI. 40.
M-285; 2.8-3.3 C, 0.4-0.7 Mn, 0.40 max P, 0.12 max S, 1.6-2.2 Si, Si inoculated, bal Fe.
Cast: 40,000 TS; 225 Brin.
Standard high strength gray iron.

ACIPCO CI. 50.
M-285; 2.5-3.0 C, 0.4-07 Mn, 0.25 max P, 0.12 max S, 1.6-2.2 Si, Si inoculated, bal Fe.
Cast: 50,000 TS; 269 Brin.
For highly stressed engineering parts.

ACIPCO CR. ALLOY.
M-285; 3.2-3.6 C, 0.4-0.6 Mn, 0.75 max P, 0.12 max S, 0.5-1.0 Cr, bal Fe.
Cast: 240-300 Brin; For parts requiring increased hardness and wear resistance.

ACIPCO CR-CU IRON.
M-285; (C + Si adjusted), 0.4-0.7 Mn, 0.75 max P, 0.12 max S, 0.3-0.7 Cr, 0.8-1.2 Cu, bal Fe.
Cast: 40,000 TS; 235 Brin.
Moderate hardness gray iron for rolls, etc.

ACIPCO DUCTILE IRON.
As per ASTM A339-55,60-45-10.
ASTM A339-55,80-60-03.

ACIPCO HEAT RESISTING IRON.
M-285; 3.2 C, aloy bal Fe.
Cast.
For retorts for magnesium manufacture; heat and pressure resistant.

ACIPCO HF.
M-285; 0.20-0.40 C, 2.0 max Mn, 2.0 max Si, 19-23 Cr, 9-12 Ni, bal Fe.
Cast: 75,000 TS; 38,000 YS; 28 El; 165 Brin.
Austenitic stainless steel.
ACI HF; AISI 302B.

ACIPCO HH.
M-285; 0.20-0.50 C, 2.0 max Mn, 2.0 max Si, 24-28 Cr, 11-14 Ni, bal Fe.
Cast: 75,000 TS; 38,000 YS; 15 El; 170 Brin.
Austenitic stainless; for furnace parts; service to 2000°F.
ACI HH; AISI 309.

ACIPCO HK.
M-285; 0.20-0.60 C, 2.0 max Mn, 2.0 max Si, 24-28 Cr, 18-22 Ni, bal Fe.
Cast: 70,000 TS; 38,000 YS; 15 El; 187Brin.
Completely austenitic stainless; service to 2000°F.
ACI HK; AISI 310.

ACIPCO HT.
M-285; 0.35-0.75 C, 2.0 max Mn, 2.5 max Si, 13-17 Cr, 33-37 Ni, bal Fe.
Cast: 68,000 TS; 35,000 YS; 5 El; 170 Brin.
Good thermal shock resistance.
ACI HT; AISI 330.

ACIPCO NI-CR-MO IRON.
M-285; (C + Si adjusted), 0.4-0.7 Mn, 0.75 max P, 0.12 max S, 1.3-1.8 Ni, 0.2-0.4 Cr, 0.4-0.6 Mo, bal Fe.
Cast: 270 Brin.
Wear and heat resistant alloy for die parts; heat treatable.

ACIPCO REGULAR.
M-285; 3.2-3.6 C, 0.4-0.7 Mn, 0.75 max P, 0.12 max S, 1.20-2.50 Si, bal Fe.
Cast: 25,000 TS.
General gray iron castings.

ACME 18-8 STAINLESS 302.
M-568; 0.08-0.20 C, 18 Cr, 8 Ni, bal Fe.
Annealed: 80,000 TS; 35,000 YS; 45 El; 65 RA; 150 Brin.
For chemical plant equipment; Type 302; stainless, austenitic.

ACME 18-8 STAINLESS 304.
M-568; 0.11 max C, 17-19 Cr, 7-9 Ni, bal Fe.
Annealed: 75,000 TS; 32,000 YS; 45 El; 60 RA; 150 Brin.
For chemical plant equipment; Type 304; stainless, austenitic.

ACME COLORSTRIP.
M-568; 0.2 C, bal Fe.
For ornamental architecture; cold rolled strip steel in colors.

ACMEITE.
M-1461; C, alloy, bal Fe.
For die blocks; oil hardened.

ACMELOY.
M-654; 2.75-3.25 C, 1.00-1.75 Si, 0.65-1.75 Mn, bal Fe.
Cast: 30,000-60,000 TS; 175-275 Brin.
Heat treated: 75,000-85,000 TS; 400-600 Brin.
For cast iron castings, gears, housings, shafts; high strength corrosion, heat and wear resistant.

ACMELOY.
M-1102; Cu alloy.
For resistance and spot welding electrodes.

ACME M.
M-1142; 0.2 C, 18 Cr, 8 Ni, bal Fe.
For heat resistant castings; corrosion and heat resistant.

ACME NICKEL STEEL.
M-Eng; 30.5 Ni, 14.2 Cr, 0.3 C, bal Fe.
For tanks and vessels to resist corrosion; stainless, corrosion resistant.

ACME SATIN STRIP.
M-568; 0.8 C, 18 Cr, 8 Ni, bal Fe.
For ornamental architecture; cold rolled stainless strip steel.

ACME STAINLESS 410.
M-568; 0.12 max C, 12-14 Cr, bal Fe.
Annealed: 75,000 TS; 40,000 YS; 35 El; 70 RA; 155 Brin.
Cold drawn: 100,000 TS; 85,000 YS; 17 El; 60 RA; 205 Brin.
For springs, hardware, tableware, turbine blades, pistons; Type 410; corrosion resistant.

ACME STAINLESS 425.
M-568; 0.12 max C, 14-16 Cr, bal Fe.
For chemical plant equipment; Type 425; corrosion resistant.

ACME STAINLESS 430.
M-568; 0.12 max C, 16-18 Cr, bal Fe.
Annealed: 70,000 TS; 40,000 YS; 30 El; 55 RA; 140 Brin.
Cold drawn: 130,000 TS; 125,000 YS; 2 El; 185 Brin.
For automotive trim, kitchen sinks, fasteners, bolts; Type 430; corrosion resistant.

ACME STAINLESS STEEL TYPE 301X.
M-568; 0.10 C, 18 Cr, 8 Ni, bal Fe.
For stainless parts; corrosion resistant.

ACMONITAL.
M-Italy; C, Cr, Ni, bal Fe.
For coins; stainless.

ACN3.
M-1176; 2.9-3.2 Cu, 1.5 Fe, 0.7 Si, 0.6 Mg, 0.2 Ti, 0.6 Ni, bal Al.
Heat treated: 47,000-60,000 TS; 43,000-51,000 YS; 5-1 El; 115-150 Brin.
For light alloy parts; age-hardened.

ACN10.
M-1176; 9.5-10.5 Cu, 0.8-1.2 Si, 0.3 Mg, 0.2 Ti, 1.3-1.7 Ni, bal Al.
Heat treated: 45,000-54,000 TS; 37,000-48,000 YS; 1.0-0.5 El; 115-140 Brin.
For engine cylinder heads, pistons; heat resistant.

A-COPPER.
M-789; 99 Cu, 1 Ni.
Hard: 58,000-67,000 TS.
For electrical equipment and instruments.
Resistance alloy. Max working temperature to 200°C.

ACORN BRAND.
M-214; 3.7 Cu, 7.4 Sb, special hardened, bal Sn.
Cast: 12,500 TS; 22.5 El; 24 Brin.
For gas and diesel engine bearings; Babbitt, low coefficient of friction.

ACRON.
M-249; 1 Si, 4 Cu, bal Al.
Annealed: 31,000-35,000 TS; 16,000-19,000 YS; 50-60 Brin.
Heat treated: 42,000-60,000 TS; 28,000-40,000 YS; 90-100 Brin.
For automotive engine parts; age-hardenable, high strength.

ACT.
M-1653; 0.15 max C, 1.0 max Si, 2.0 max Mn, 16.0-18.0 Cr, 6.0-8.0 Ni, bal Fe.
Austenitic stainless steel; work hardens rapidly.
W.-Nr. 1.4310; x 12 CrNi 17 07; A151 301.

A.C.T.
M-353; 0.65 C, 0.15 Mn, 18 W, 4.5 Cr, 2 V, 0.75 Mo, 10 Co, bal Fe.
For tools, drills, cutters, reamers; high speed steel.

ACT 2.
M-1086.
TiC coated sintered carbide.
Excellend abrasion, heat, crater and deformation resistance; medium toughness.
For machining cast iron, stainless steel and non-ferrous alloys.
C-2 general purpose and C-3 finishing cuts.

ACT 5.
M-1086.
TiC coated sintered carbide.
Good abrasion, heat and crater resistance. Tough.
For rough machining carbon, leaded, alloy and tool steels, and 400 series stainless.
C-5 roughing and some C-6 general purpose.

ACT 7.
M-1086.
TiC coated sintered carbide.
Excellent abrasion, heat and crater resistance.
Good toughness and shock resistance.
For general purpose and finishing carbon, leaded, and alloy steels.

ACT CARBIDE.
M-353; WC + Co.
For tools, cutters; sintered carbide.

A-D-70.
M-355; 0.7 C, 0.35 Mn, 0.98 W, 0.3 Cr, 0.12 V, bal Fe.
For dies, crimpers, punches; oil hardening.

ADABRAZE.
M-1086.
Sintered carbide coated with thin layer of pure cobalt.
To improve brazing of carbide to base metal shanks or supports.

ADALLOY.
M-1756; 40% Cr + Mn + Si + C, bal Fe.
Cast alloy hard facing rod; coated for AC-DC; bare for oxy-acetylene; 58-60 Rock C.
Wear resistant; non-machinable; for plows, drag chains, cultivators.

ADALLOY 'A'.
M-1756; 9% Cr + Ni + Mn + C + Si, bal Fe.
Hardfacing electrode; coated for AC-DC; bare for oxy-acetylene; 29-33 Rock C.
Shock and impact resistant; machinable after slow cool; for gears, sprockets, latch pins.

ADALLOY 'B'.
M-1756; 13% Cr + Mn + Si + C, bal Fe.
Hardfacing electrode; coated for AC-DC; bare for oxy-acetylene; 45-50 Rock C.
Forgeable at red heat; machinable.
For general preventive maintence on shovel buckets and lips, crawler running gear.

ADALLOY 'C'.
M-1756; 10% Cr + Mn + Si + B + C, bal Fe.
Hardfacing electrode; coated for AC-DC; bare for oxy-acetylene; 55-60 Rock C.
Forgeable, not readily machinable.
For hardfacing heavy equipment subject to impact and abrasion, as pulverizers.

ADAMANTINE.
M-447; 0.7 C, 0.7 Cr, 0.7 Mn, bal Fe.
Heat treated: 145,000 TS; 77,000 YS; 6.5 El; 16.6 RA; 260-300 Brin
For heat treat steel balls for grinding mills; wear resistant.

ADAMANT SUPER-GENUINE BABBITT.
M-178; Cu, Sb, bal Sn.
12,850 TS; 8,600 YS; 22.8 Brin.
For marine, airplane, diesel and other internal combustion engine bearings; Sp. Gr. 7.34; tough.

ADAMAS 6X.
M-1086; WC, TaC, TiC, Co.
Sintered: 230,000 TrS; A 91.5 Rock.
For general purpose machining steel.
BHMA 6-5-5; ISO P20.

ADAMAS 7X.
M-1086; WC, TaC, TiC, Co.
Sintered: 200,000 TrS; A 92.4 Rock.
For finish machining steel.
BHMA 9-3-5; ISO P10.

ADAMAS 434.
M-1086; WC, TaC, TiC, Co.
Sintered: 270,000 TrS; A 91.0 Rock.
For rough machining steel.
BHMA 5-7-5; ISO P40.

ADAMAS 474.
M-1086; WC, Co, TaC.
Sintered: 300,000 TRS; A 90.0 Rock.
For rough machining steel.
BHMA 3-8-2; ISO P40.

ADAMAS 490.
M-1086; WC, TaC, TiC, 4.5 Co.
Sintered: 10.7 density; 92.9 Rock A; transverse rupture strength, 200, psi.
Very high hardness, very good wear resistance. For finishing operations.
ISO P 10, M 10.

ADAMAS 495.
M-1086; WC, TaC, TiC, 5.0 Co.
Sintered: 12.0 density; 92.4 Rock A; transverse rupture strength, 250, psi.
Carbide grade with extra wear resistance for difficult machining operations.
ISO P 15.

ADAMAS 499.
M-1086; WC, TaC, TiC, 9 Co.
Sintered: 12.15 density; 91.3 Rock A; transverse rupture strength, 325,000 psi.
Tough carbide grade with good shock resistance, wear and heat resistance; for medium to heavy cuts.
ISO P 25, P 30.

ADAMAS 502.
M-1086; 88 WC, 12 Co.
Sintered: 390,000 TrS; A 88.0 Rock.

ADAMAS 548.
M-1086; WC, TaC, TiC, Co.
Sintered: 230,000 TrS; A 92.0 Rock.
For finish machining steel.
BHMA 8-5-4; ISO P10.

ADAMAS 569.
M-1086; 90 WC, 10 Co.
Sintered: 370,000 TrS; A 88.5 Rock.

ADAMAS 783.
M-1086; 89 WC, 11 Co.
Sintered: 385,000 TrS; A 88.4 Rock.

ADAMAS 815.
M-1086; 91 WC, 9 Co.
Sintered: 340,000 TrS; A 89.2 Rock.

ADAMAS A.
M-1086; 94 WC, 6 Co.
Sintered: 300,000 TrS; A 91.7 Rock.
For general purpose machining ferrous and non-ferrous metals. BHMA. 7-6-0; ISO K 20.

ADAMAS AA.
M-1086; 96 WC, 4 Co.
Sintered: 240,000 TrS; A 92.0 Rock.
For finish machining ferrous and non-ferrous metals. BHMA 8-5-8; ISO K10, M10.

ADAMAS AAA.
M-1086; 97 WC, 3 Co.
Sintered: 200,000 TrS; A 92.6 Rock.
For finish boring ferrous and non-ferrous alloys.
BHMA 9-3-0; ISO KO1.

ADAMAS ACT 2, ETC see ACT 2 ETC.

ADAMAS AM.
M-1086; WC, Co, TaC.
Sintered: 240,000 TrS; A 91.9 Rock.
For general purpose machining steel and cast iron.
BHMA 7-5-1; ISO M20.

ADAMAS B.
M-1086; 91 WC, 9 Co.
Sintered: 300,000 TrS; A 90.5 Rock.
For rough machining ferrous and non-ferrous alloys.
BHMA 4-8-0; ISO K30.

ADAMAS BB.
M-1086; 87 WC, 13 Co.
Sintered: 360,000 TrS; A 89.4 Rock.
For rough machining ferrous and non-ferrous metals.
BHMA 2-9-0; ISO K40.

ADAMAS C.
M-1086; 76 WC, 15 TiC; 9 Co.
Sintered: 210,000 TrS; A 92.0 Rock.
For finish machining steel.
BHMA 8-4-6; ISO P10.

ADAMAS GG.
M-1086; WC, Co, TaC.
Sintered: 290,000 TrS; A 89.2 Rock.
For rough machining steel.
BHMA 2-8-5; ISO P50.

ADAMAS GU-1.
  M-1086; WC base sintered carbide.
  Density, 13.84; 88.5 Rock A; transverse rupture strength, 500,000 psi.
  Very high strength carbide; for stamping, cold heading, and saw tips.

ADAMAS HD-15.
  M-1086; 85 WC, 15 Co.
  Sintered: 400,000 TrS; A 87.4 Rock.
  For lamination punches and dies and wear parts.

ADAMAS HD 20.
  M-1086; 80 WC, 20 Co.
  Sintered: 385,000 TrS; A 85.5 Rock.
  For heading dies, swaging dies, nail gripper dies.

ADAMAS HD-20T.
  M-1086; 75 WC, 5 TaC, 20 Co.
  Sintered: 380,000 TrS; A 85.3 Rock.
  For small swaging dies and gripper dies where anti-galling properties are important.

ADAMAS HD-25.
  M-1086; 75 WC, 25 Co.
  Sintered: 360,000 TrS; A 83.5 Rock.
  For dies where high impact strength coupled with high abrasion resistance are necessary.

ADAMAS HD-25T.
  M-1086; 70 WC, 5 TaC, 25 Co.
  Sintered: 380,000 TrS; A 83.5 Rock.
  For dies requiring maximum impact strength and anti-galling properties.

ADAMAS PWX.
  M-1086; WC, Co, TaC.
  Sintered: 230,000 TrS; A 92.3 Rock.
  For finish machining steel.
  BHMA 9-5-1; ISO M10, K05.

ADAMAS TITAN see **ALSO TITAN.**

ADAMAS TITAN 60.
  M-1086; TiC, Mo$_2$C, Ni.
  Sintered: 235,000 TrS; A 91.9 Rock.
  For rough machining ferrous and non-ferrous alloys, including superalloys.

ADAMAS TITAN 80.
  M-1086; TiC, Mo$_2$C, Ni.
  Sintered: 200,000 TrS; A 93.0 Rock.
  For finish machining and boring ferrous and non-ferrous alloys, including superalloys.
  BHMA 9-2-9; ISO PO1, P10.

ADAMITE.
  M-229; high C, 1 Cr, 0.75 Ni, bal Fe.
  For press dies, rolls, mill guides, furnace parts, castings; cast iron-pearlitic, wear resistant.

ADAPTALOY.
  M-815; 1.8 Cu, 1.0 Si, 1.0 Mg, bal Zn.
  Cast: 22,000-25,000 TS; 8000-10,000 YS; 11-7 El; 48 Brin.
  For ornamental grills, valve handles, structural castings; high elongation and impact strength.

ADIC.
  M-1112; 0.37 C, 1.1 Si, 5 Cr. 1 V, 1.3 Mo, bal Fe.
  Annealed: 230 Brin.
  For die casting dies, hot forging dies; hot work steel; air hardened.

ADIC C.T.U.
  M-1112; 0.37 C, 1.1 Si, 5 Cr, 1.75 W, 1 V, 1.3 Mo, bal Fe.
  Annealed: 230 Brin.
  For die casting dies, shear blades, slitters; hot work steel, air hardened.

ADLERSTAHL 03.
  M-1330; 1.0 C, 1.1 Cr, 0.25 Si, 0.07 Mn, bal Fe.
  For bearings, races, liners, sleeves; water hardened, wear resistant.

ADLERSTAHL 1W.
  M-1330; 1.0 C, 0.25 Si, 0.07 Mn, 1.1 Cr, bal Fe.
  For bearings, races, liners, sleeves: water hardened, wear resistant.

ADLERSTAHL 3 MH.
  M-1330; 1.1 C. 0.25 Si, 0.25 Mn, bal Fe.
  Annealed: 105,000 TS; 55,000 YS; 20 El; 40 RA; 205 Brin.
  For drills, hobs, reamers, dies, cutters; Type W1; water hardened.

ADLERSTAHL 4 ZH.
  M-1330; 1.0 C, 0.25 max Si, 0.25 max Mn, bal Fe.
  Annealed: 100,000 TS; 53,000 YS; 21 El; 42 RA; 200 Brin.
  For drills, taps, lathe, and planer tools, cutters; Type W1; water hardened.

ADLERSTAHL 5 ZAH.
  M-1330; 0.85 C, 0.25 Si, 0.25 Mn, bal Fe.
  Heat treated: 190,000 TS; 145,000 YS; 12 El; 35 RA; 400 Brin.
  For drills, taps, tools, springs, cutters; Type W1; water hardened.

ADLERSTAH AZ1.
  M-1330; 1.05 C, 1.0 Cr, 1.15 W, 0.9 Mn, bal Fe.
  For bearings liners, cutters, forming dies; water hardened.

ADLERSTAHL B11.
  M-1330; 1.0 C, 0.2 Si, 0.25 Mn, 0.1 V, bal Fe.
  For drills, taps, punches, reamers, hobs; Type W2; water hardened.

ADLERSTAHL C1.
M-1330; 1.45 C, 1.4 Cr, 0.6 Mn, bal Fe.
For bearings, sleeves, liners; water hardened, wear resistant.

ADLERSTAHL C13.
M-1330; 2.1 C, 11.5 Cr, 0.3 Mn, bal Fe.
For blanking and forming dies, punches; oil hardened, nondeforming.

ADLERSTAHL C13TU.
M-1330; 2.1 C, 11.5 Cr, 0.3 Mn, bal Fe.
For blanking and forming dies, punches; oil hardened, nondeforming.

ADLERSTAHL C13W.
M-1330; 1.65 C, 11.5 Cr, 0.1 V, bal Fe.
For blanking and forming dies, punches; oil hardened, nondeforming.

ADLERSTAHL EC2.
M-1330; 1.3 C, 0.25 max Si, 0.25 max Mn, bal Fe.
For reamers, taps, form cutters, drills, hobs; Type W1; water hardened.

ADLERSTAHL EC3.
M-1330; 1.15 C, 0.25 max Si, 0.25 max Mn, bal Fe.
Annealed: 110,000 TS; 55,000 YS; 18 El; 38 RA; 210 Brin.
For reamers, drills, taps, cutters, broaches; Type W1; water hardened.

ADLERSTAHL EC4.
M-1330; 1.0 C, 0.25 max Si, 0.25 max Mn, bal Fe.
Annealed: 100,000 TS; 53,000 YS; 21 El; 42 RA; 200 Brin.
For springs, tools, cutters, drills, reamers; Type W1; water hardened.

ADLERSTAHL EC5.
M-1330; 0.85 C, 0.25 Si, 0.25 max Mn, bal Fe.
Heat treated: 190,000 TS; 145,000 YS; 10 El; 30 RA; 400 Brin.
For springs, tools, drills, cutters, taps, hammers; Type W1; water hardened.

ADLERSTAHL EC6.
M-1330; 0.70 C, 0.25 max Si, 0.25 max Mn, bal Fe.
Heat treated: 175,000 TS; 128,000 YS; 12 El; 37 RA; 355 Brin.
For springs, tools, rails, axes; Type W1; water hardened.

ADLERSTAHL EGH1.
M-1330; 0.35 C, 0.9 Si, 0.3 Mn, 1.05 Cr, 0.18 V, 1.85 W, bal Fe.
For header dies, upsetters, crimpers, punches, oil hardened, tough.

ADLERSTAHL EGH3.
M-1330; 0.55 C, 0.9 Si, 1.05 Cr, 0.18 V, 1.85 W, bal Fe.
For header dies, upsetters, shears crimpers; oil hardened, tough.

ADLERSTAHL EXTRA.
M-1330; 0.86 C, 4 Cr, 2.5 V, 0.85 Mo, 12 W, bal Fe.
For reamers, taps, drills, hobs, broaches; high-speed steel.

ADLERSTAHL EXTRA P12.
M-1330; 0.76 C, 10 Co, 4.2 Cr, 0.8 Mo, 1.8 V, 18 W, bal Fe.
For lathe and planer tools, reamers, hobs, taps; high-speed steel.

ADLERSTAHL GCN1.
M-1330; 0.55 C, 0.7 Cr, 0.18 Mo, 1.65 Ni, 0.1 V, bal Fe.
For gears, shafts, bolts, studs, crankshafts; oil hardening, shock resistant.

ADLERSTAHL GCN2.
M-1330; 0.56 C, Ni, Cr, Mo, V, bal Fe.
For gears, fasteners, bolts, studs; oil hardened, shock resistant.

ADLERSTAHL GDM.
M-1330; 0.30 C, 1.0 Si, 1.1 Cr, 0.18 V, 3.75 W, bal Fe.
For header dies, upsetters, crimpers; oil hardened, tough.

ADLERSTAHL KJ.
M-1330; 0.74 C, 4.1 Cr, 1.1 V, 18.5 W, bal Fe.
For lathe and planer tools, reamers, broaches; high-speed steel.

ADLERSTAHL KT.
M-1330; 0.86 C, 2.8 Co, 4.3 Cr, 0.85 Mo, 2 V, 12 W, bal Fe.
For lathe and planer tools, drills, taps, reamers; high speed steel.

ADLERSTAHL NL5H.
M-1330; 0.50 C, 1.05 Cr, 3.25 Ni, 0.50 Mn, bal Fe.
For gears, bolts, crankshafts, dies, punches; oil hardened, shock resistant.

ADLERSTAHL NWO.
M-1330; 0.45 C, Ni, Cr, bal Fe.
For gears, bolts, crankshafts; oil hardened, shock resistant.

ADLERSTAHL SICR.
M-1330; 1.1 C, 0.40 Cr. 0.30 Mn, bal Fe.
Heat treated: 190,000 TS; 120,000 YS; 10 El; 30 RA; 375 Brin.
For springs, taps, cutters, drills, reamers; Type W1; water hardened.

ADLERSTAHL SPEZIAL M25.
M-1330; 0.95 C, W, Mo, bal Fe.
For cutters, punches, dies; oil hardened.

ADLERSTAHL WKS.
M-1330; 0.55 C, 1.05 Cr, 0.18 V, 1.85 W. 0.9 Si, bal Fe.
For upsetting and heading dies, punches; oil hardened tough.

ADLERSTAHL WPS.
M-1330; 0.30 C, 2.65 Cr, 0.35 V, 8.5 W, bal Fe.
For extrusion rams and dies, hot punches; oil hardened, hot work steel.

ADLERSTAHL WPS-EXTRA.
M-1330; 0.30 C, 2 Co, 2.4 Cr, 0.25 V, 8.5 W, bal Fe.
For extrusion rams and dies, hot punches; oil hardened, hot work steel.

ADLERSTAHL V.
M-1330; 0.8 C, 4.7 Co. 4.3 Cr, 0.75 Mo. 1.5 V, 18 W, bal Fe.
For lathe and planer tools, taps, reamers; high-speed steel.

ADLERSTAHL P15.
M-1330; 1.35 C, Cr, V, W, Co, bal Fe.
For reamers, broaches, cutters, taps; high-speed steel.

ADLERSTAHL P45.
M-1330; 1.3 C, 4.3 Cr, 0.85 Mo, 3.8 V, 12 W, bal Fe.
For reamers, broaches, cutters, taps; high-speed steel.

ADLERSTAHL PM35.
M-1330; 0.38 C, Si, Cr, V, bal Fe.
For gears, springs, punches, bolts, shafts; oil hardened, tough.

ADLERSTAHL PM45.
M-1330; 0.45 C, Si, Cr, V, bal Fe.
For springs, bolts, crankshafts, gears; oil hardened, shock resistant.

ADLERSTAHL PM50.
M-1330; 0.61 C, 1.18 Cr, 0.10 V, 0.75 Mn, 0.85 Si, bal Fe.
For upsetters, heading and forging dies; oil hardened, tough.

ADLERSTAHL PROPHET EXTRA.
M-1330; 0.86 C, 2.8 Co, 4.3 Cr, 0.85 Mo, 2 V, 12 W, bal Fe.
For lathe and planer tools, cutters, reamers; high-speed steel.

ADLERSTAHL PZM.
M-1330; 1.4 C, 0.1 V, 0.30 Cr, 0.30 Mn, bal Fe.
For bearings, forming and blanking dies; water or oil hardened, wear resistant.

ADLERSTAHL RO.
M-1330; 1.42 C, W, V, bal Fe.
For blanking and forming dies, bearings; water or oil hardened, wear resistant.

ADMIRALTY BRASS.
M-8; 62 Cu, 37 Zn, 1 Sn.
Cast: 56,000 TS; 27,000 YS; 15 El; 25 RA.
Rolled hard: 71,680 TS; 55,500 YS; 20 El; 30 RA.
For pump rods, sheets.

ADMIRALTY BRASS.
M-286, M-1560; 70 Cu, 1 Sn, 0.04 As, bal Zn.
Annealed: 40.300-49,300 TS; 18,000-22,400 YS; 60-75 El; 54-73 Brin.
Hard: 71,700-87,400 TS; 60,000-76,000 YS; 8-15 El; 145-175 Brin.
Good resistance to sea air; for fixtures on marine equipment.

ADMIRALTY BRASS, INHIBITED WITH ANTIMONY, ARSENIC OR PHOSPHORUS.
M-33; 71 Cu, 28 Zn, 1 Sn.
Ann: 53,000 TS; 22,000 YS; 65 EL.
Drawn 35%: 85,000 TS; 65,000 YS; 15 EL.
Corrosion resistant; for condenser and heat exchanger tubes, ferrules.

ADMIRALTY BRONZE MODIFIED.
M-126; 86 Cu, 8 Sn, 2 Pb, 4 Zn.
For valves, fittings; pressure tight.

ADMIRALTY CAG.
M-657; 87 Cu, 9 Sn, 4 Zn or 88 Cu, 10 Sn, 2 Zn.
Cast: 36,000 TS; 18,000 YS; 20 El.
For marine castings; resists marine corrosion.

ADMIRALTY GUN METAL.
M-711; 87.7 Cu, 10.3 Sn, 1.4 Zn, 0.4 As, 0.08 Pb, 0.05 Fe.
Cast: 37,300 TS; 21,300 YS; 9 El; 100 Brin.
For valves, pump parts; corrosion resistant.

ADMIRALTY NICKEL.
M-126; 70 Cu, 29 Zn, 1 Ni.
For condenser tubes; corrosion resistant.

ADMIRALTY WHITE METAL.
M-Eng; 8-9 Sb, 2-7 Cu, bal Sn.
For bearings.

ADMIRO.
M-2; 48 Cu, 10 Ni, 35 Zn, 2 Al, 3 Mn, 2 Fe.
Cast: 57,000-70,000 TS; 40-20 El; 110-130 Brin.
Hot pressed: 63,000-76,000 TS; 40-25 El; 150-170 Brin.
For turbine bearings, tubes; corrosion resistant.

ADMIRO V.
M-2; 43 Cu, 15 Ni, 35 Zn, 2 Al, 3 Mn, 2 Fe.
Cast: 18-5 El; 160 Brin.
Pressed: 20-8 El; 185 Brin.
For turbine bearings; corrosion resistant.

ADMOS.
M-2; 40-55 Cu, 3-15 Ni, 1-3 Mn, 1-2 Fe, 0.5-3.0 Al, bal Zn.
Cast: 56,000-71,680 TS; 35-20 El; 80-90 Brin.
Hot pressed: 69,440-85,120 TS; 35-20 El; 90-120 Brin.
For bearing bushes, worm wheels, condenser tubes, turbine blading, gear wheels, valves; resists corrosion and erosion of superheated steam.

ADRIATICAL.
M-658; 4-5 Cu, bal Al.
For light alloy parts; similar to Duralumin.

A D S.
M-1749; 0.35 C, 0.40 Mn, 1.0 Si, 5.0 Cr, 1.5 Mo, 0.54 V, bal Fe.
Aluminum die-casting steel; air or oil hardenable.

ADSPEC 1076 D.
M-1256; 8.5 Al, 2.5 Fe, bal Cu.
34 Tons/sq in WTS; 30 El.
Wrought aluminum bronze.

ADVALOY.
M-1462; Al alloy.
For castings; permanent mold.

ADVANCE.
M-585; Cu, Sb, bal Sn.
For bearings; Babbitt metal.

ADVANCED DUTY OILITE BRONZE.
M-211; 87.5-90.5 Cu, 1.0 max Fe, 1.75 max C, 9.5-10.5 Sn.
Sintered: 16,000-18,000 TS; 15,000-20,000 YS; 6.8-7.2 density.
Oilite bearing material.
ASTM B255-61 Type 11; SAE 842; PMPMA BT-0010-S.

A.E.B. STAINLESS.
M-111a; 0.95 C, 1 Mn, 13.5 Cr, bal Fe.
For razor blades, surgical instruments; hardenable, corrosion resistant.

AECMA FE PL 52 S (VASCOJET 90).
M-1723; 0.12-0.18 C, 0.20 max Si, 0.80-1.10 Mn, 1.25-1.50 Cr, 0.80-1.0 Mo, 0.20-0.30 V, bal Fe.
Air or oil hardenable.
Hardened: 140,000-184,000 TS; 12,000 min YS; 10 min El.
High strength, weldable, low carbon-low alloy steel; can be case hardened for improved wear.
AFNOR 15 CDV 6.

AELENER ZINC.
3 Sb, 3.5 Cu, 21.6 Pb, 20.5 Sn, 51.4 Zn.
For solder, bearings.

AEM 90.
M-1246.
For permanent magnets; sintered.

AEM 120.
M-1246.
For permanent magnets; sintered.

AEONITE.
M-453; Ni, bal Cu.
For sanitary appliances; white metal.

AERAL.
M-623; 3.5-4.5 Cu, 0.4-1.0 Mn, 0.3-0.75 Mg, bal Al.
56,000-75,000 TS; 30,000-55,000 YS; 20-10 El.
For light alloys, airplane and dirigible parts; "Duralumin" type alloy; age-hardening.

AERAL A.
M-623; 2.0-4.5 Cu, 0.2-1.5 Mg, 2.5 max Si, 0.8 max Fe, 0.4 max Ti, 0.5-2.5 Cd, 0.5 max Mn, bal al.
Solution treated: 28,000-34,000 TS; 22,000-24,000 YS; 5-3 El.
For sand and chill castings; heat treatable.

AERALLOY.
M-237; Pt alloy.
For electrical contacts; heat resistant.

AERALLOY NO. 1.
M-237; Ir, bal Pt.
For electrical contacts, vibrators, temperature gages; resists arc erosion

AERALLOY NO. 2.
M-237; Ir, bal Pt.
For electrical contacts, vibrators, thermostats, resists arc erosion.

AERISWELD.
M-578; 0.05 P, 7.0 Sn, 0.1 Al, 0.02 Pb, bal Cu.
Bronze arc welding electrode.
AWS Class E CuSn-C.

AEROLITE.
M-English; 97 Al, 1.2 Cu, 1.0 Fe, 0.5 Si, 0.4 Mg.
For pistons, auto parts.

AERO METAL (CAST).
M-287; 4.2 Cu, 27.8 Zn, 0.5 Fe, 0.5 Si, 67 Al.
Cast: 25,000 TS.
For automotive engine parts; cast.

AERO METAL (SHEET).
M-287; 0.2-0.6 Cu, 2.1-2.9 Mg, 96 Al, 0.3-1.3 Fe.
Rolled: 20,000 TS.
For automotive engine parts; sheet.

AEROMIN.
M-Germ; 92.7 Al, 6.2 Mn, 0.8 Fe, 0.3 Si.
For light alloy parts.

AERON.
M-320, M-299; 1.8 Cu, 0.8 Mn, 1.0 Si, bal Al.
Heat treated: 54,000-60,000 TS; 26,000-34,000 YS; 18-25 El; 90-120 Brin.
For light alloy parts; heat treatable.

AERON.
 M-584; 4.5 Cu, 0.75 Si, 0.75 Mn, bal Al.
 Annealed: 23,000-35,000 TS; 7,000-12,000 YS; 12-20 El; 45-55 Brin.
 Quenched: 45,000-53,000 TS; 15,000-30,000 YS; 15-22 El; 68-85 Brin.
 For light anti-corrosive alloy parts; light anti-corrosion alloy. Similar to Alcoa No. 25S.

AEROSIL 8F.
 M-1323; 0.12 max C, 0.8 Al, 6.5 Cr, bal Fe.
 For oil refining equipment; creep and heat resistant.

AEROSIL 9F.
 M-1323; 0.12 max C, 1.0 Al, 13 Cr, bal Fe.
 For oil refinery and chemical plant equipment; corrosion and heat resistant.

AEROSIL 10A.
 M-1323; 0.15 C, 19.5 Cr, 9.5 Ni, bal Fe.
 Annealed: 80,000 TS; 35,000 YS; 55 El; 75 RA; 150 Brin.
 For chemical plant equipment; Type 302; stainless, austenitic.

AEROSIL 10F.
 M-1323; 0.12 max C, 1.0 Al, 18 Cr, bal Fe.
 Annealed: 80,000 TS; 50,000 YS; 25 El; 50 RA; 160 Brin.
 For oil refinery equipment; corrosion and heat resistant.

AEROSIL 11FO.
 M-1323; 0.2 C, 25 Cr, 4 Ni, bal Fe.
 For furnace parts, heat treating boxes; heat resistant.

AEROSIL 12A.
 M-1323; 0.15 C, 24 Cr, 19 Ni, bal Fe.
 For furnace equipment heat treating boxes, pots; Type 310; stainless, austenitic.

AEROSIL 12F.
 M-1323; 0.12 max C, 1.5 Al, 24 Cr, bal Fe.
 For furnace parts, conveyors, boxes, fixtures; heat and creep resistant.

AF 2-1DA.
 M-114, M-1800; 0.35 C, 12.0 Cr, 6.0 W, 3.0 Mo, 10.0 Co, 3.0 Ti, 4.6 Al, 0.015 B, 0.10 Zr, 1.5 Ta, bal Ni.
 Nickel base superalloy.

AF-1410.
 M-US; 0.16 C, 10 Ni, 2 Cr, 14 Co, 1 Mo, bal Fe.
 Heat treatable to 230,000 psi YS.
 Weldable and corrosion resistant.

AF 1753.
 M-U.S.; 0.24 C, 16.25 Cr, 7.2 Co, 1.6 Mo, 8.4 W, 3.1 Ti, 1.9 Al, 0.06 Zr, 0.008 B, 9.5 Fe, bal Ni.
 At 70°F: 187,000 TS; 121,200 SS.
 At -320°F: 217,100 TS.
 At -423°F: 219,400 TS, 149,200 SS.
 For cryogenic fasteners.
 Good tensile-impact properties.

AFC 77 see **CRUCIBLE AFC 77.**

AFFIMET AL-CU 5 TI.
 M-871; 4.0-5.0 Cu, 0.25 max Si, 0.25 max Fe, 0.05-0.30 Ti, bal Al.
 Cast aluminium alloy.
 BS 1490 LM11.

AFFIMET AL-CU 7 SI 3 ZN 3.
 M-871; 6.0-8.0 Cu, 2.0-4.0 Si, 2.0-4.0 Zn, 1.0 max Fe, 0.6 max Mn, 0. max Ni, others 0.3 max each, bal Al.
 Cast aluminium alloy.
 AFNOR A-U 8 SZ; BS 1490 LM1.

AFFIMET AL-CU 10 MG.
 M-871; 9.0-11.0 Cu, 2.5 max Si, 0.8 max Zn, 1.0 max Fe, 0.6 max Mn, 0.5 max Ni, 0.2-0.4 Mg, bal Al.
 Cast aluminium alloy.
 AFNOR A-U 10 G; BS 1490 LM12.

AFFIMET AL-MG 5.
 M-871; 0.3 max Si, 0.6 max Fe, 3.0-6.0 Mg, 0.3-0.7 Mn, bal Al.
 Cast aluminium alloy.
 AFNOR A-G3T; BS 1490 LM5.

AFFIMET AL-MG 10.
 M-871; 0.25 max Si, 0.35 max Fe, 9.5-11.0 Mg, bal Al.
 Cast aluminium alloy.
 AFNOR A-G 10; BS 1490 LM10.

AFFIMET AL-SI 2.
 M-871; 0.4 max Cu, 10.0-13.0 Si, 0.2 max Zn, 1.0 max Fe, 0.5 max Mn, bal Al.
 Cast aluminium alloy.
 AFNOR A-S 12; BS 1490 LM20.

AFFIMET AL-SI 2 CU.
 M-871; 1.0-2.5 Cu, 1.5-3.5 Si, 0.3-1.4 Fe, 0.05-2.0 Mg, 0.5-1.7 Ni, 0.05-0.30 Ti, bal Al.
 Cast aluminium alloy.
 BS 1490 LM7.

AFFIMET AL-SI 2 CU 2 MG NI.
 M-871; 1.3-3.0 Cu, 0.6-2.0 Si, 0.8-1.4 Fe, 0.5-2.0 Ni, 0.5-1.7 Mg, 0.05-0.30 Ti, bal Al.
 Cast aluminium alloy.
 BS 1490 LM15.

AFFIMET AL-SI 4 CU 4 ZN.
M-871; 3.0-5.0 Cu, 5.0-7.0 Si, 2.0 max Zn, 1.0 max Fe, 0.2-0.6 Mn, 0. max Ni, 0.1-0.3 Mg, bal Al.
Cast aluminium alloy.
AFNOR A-S 5 UZ; BS 1490 LM21.

AFFIMET AL-SI 4 MG.
M-871; 3.5-6.0 Si, 0.6 max Fe, 0.5 max Mn, 0.3-0.8 Mg, bal Al.
Cast aluminium alloy.
AFNOR A-S 4 G; BS 1490 LM8.

AFFIMET AL-CU 4 NI.
M-871; 3.5-4.5 Cu, 0.6 max Si, 0.6 max Fe, 0.6 max Mn, 1.8-2.3 Ni, 1.2-1.7 Mg, bal Al.
Cast aluminium alloy.
AFNOR A-U 4NT; BS 1490 LM14.

AFFIMET AL-SI 5.
M-871; 4.5-6.0 Si, 0.6 max Fe, 0.5 max Mn, bal Al.
Cast aluminium alloy.
BS 1490 LM18.

AFFIMET AL-SI 5 CU.
M-871; 2.8-3.8 Cu, 4.0-6.0 Si, 0.15 max Zn, 0.6 max Fe, 0.2-0.6 Mn, 0.15 max Ni, bal Al.
Cast aluminium alloy.
AFNOR A-S 5 U; BS 1490 LM22.

AFFIMET AL-SI 5 CU 3.
M-871; 2.0-4.0 Cu, 4.0-6.0 Si, 0.5 max Zn, 0.8 max Fe, 0.2-0.6 Mn, 0. max Ni, bal Al.
Cast aluminium alloy.
AFNOR A-S 5 U; BS 1490 LM4.

AFFIMET AL-SI 5 CU MG.
M-871; 1.0-1.5 Cu, 4.5-5.5 Si, 0.40-0.60 Mg, 0.12-0.20 Ti, bal Al.
Cast aluminium alloy.
AFNOR A-S 5 U G.

AFFIMET AL-SI 5 MG CU.
M-871; 1.0-1.5 Cu, 4.5-5.5 Si, 0.6 max Fe, 0.5 max Mn, 0.25 max Ni, 0.4-0.6 Mg, bal Al.
Cast aluminium alloy.
AFNOR A-S 5 U1; BS 1490 LM16.

AFFMET AL-SI 7 CU 2.
M-871; 1.5-2.5 Cu, 6.0-8.0 Si, 1.0 max Zn, 0.8 max Fe, 0.2-0.6 Mn, 0. max Ni, 0.3 max Mg, bal Al.
Cast aluminium alloy.
AFNOR A-S 7 U 3; BS 1490 LM27.

AFFIMET AL-SI 7 MG.
M-871; 6.5-7.5 Si, 0.5 max Fe, 0.3 max MN, 0.20-0.45 Mg, bal Al.
Cast aluminium alloy.
AFNOR A-S 7 G; BS 1490 LM25.

AFFIMET AL-SI 8 CU 3 FE/A.
M-871; 3.0-4.0 Cu, 7.5-9.5 Si, 1.5 max Zn, 1.3 max Fe, 0.5 max Mn, 0. max Ni, 0.3 max Mg, 0.3 max Pb, 0.2 max Sn, 0.2 max Ti, bal Al.
Cast aluminium alloy.
AFNOR A-S 9U 3/A; BS 1490 LM24 type A.

AFFIMET AL-SI 8 CU 3 FE/B.
M-871; 3.0-4.0 Cu, 7.5-9.5 Si, 3.0 max Zn, 1.3 max Fe, 0.5 max Mn, 0. max Ni, 0.3 max Mg, 0.3 max Pb, 0.2 max Sn, 0.2 max Ti, bal Al.
Cast aluminium alloy.
AFNOR A-S 9U 3/B; BS 1490 LM24 type B.

AFFIMET AL-SI 10 CU.
M-871; 0.7-2.5 Cu, 9.0-11.5 Si, 2.0 max Zn, 1.0 max Fe, 0.5 max Mn, 0.5 max Ni, others 0.3 max each, bal Al.
Cast aluminium alloy.
BS 1490 LM 2.

AFFIMET AL-SI 10 CU 3 MG.
M-1490; 2.0-4.0 Cu, 8.5-10.5 Si, 1.0 max Zn, 1.2 max Fe, 0.5 max Mn, 1.0 max Ni, 0.5-1.5 Mg, bal Al.
Cast aluminium alloy.
BS 1490 LM26.

AFFIMET AL-SI 10 MG.
M-871; 10.0-13.0 Si, 0.6 max Fe, 0.3-0.7 Mn, 0.2-0.6 Mg, bal Al.
Cast aluminium alloy.
AFNOR A-S 9 G; BS 1490 LM9.

AFFIMET AL-SI 12 CU NI.
M-871; 0.7-1.5 Cu, 10.0-12.0 Si, 0.5 max Zn, 1.0 max Fe, 0.5 max Mn, 1.5 max Ni, 0.8-1.5 Mg, bal Al.
Cast aluminium alloy.
AFNOR A-S 1i UN; BS 1490 LM13.

AFFIMET AL-SI 13.
M-871; 10.0-13.0 Si, 0.6 max Fe, 0.5 max Mn, bal Al.
Cast aluminium alloy.
AFNOR A-S 13; BS 1490 LM6.

AFFIMET AL-SI 17 CU 4 MG.
M-871; 4.0-5.0 Cu, 16-18 Si, 1.1 max Fe, 0.3 max Mn, 0.4-0.7 Mg, bal Al.
Cast aluminium alloy.
BS 1490 LM30.

AFFIMET AL-SI 19 CU NI CO.
M-871; 1.3-1.8 Cu, 17-20 Si, 0.7 max Fe, 0.6 max Mn, 0.8-1.5 Ni, 0.8-1.5 Mg, 0.6 max Cr, 0.5 max Co, bal Al.
Cast aluminium alloy.
AFNOR A-S 20 U; BS 1490 LM28.

AFFIMET AL-SI 24 CU NI CO.
M-871; 0.8-1.3 Cu, 22-25 Si, 0.7 max Fe, 0.6 max Mn, 0.8-1.3 Ni, 0.8-1.3 Mg, 0.6 Cr, 0.5 Co, bal Al.
Cast aluminium alloy.
AFNOR A-S 22 UNK; BS 1490 LM29.

AFFIMET AL-Z 10 CU 3.
M-871; 2.5-4.5 Cu, 1.3 max Si, 9-13 Zn, 1.0 max Fe, 0.5 max Mn, 0.5 max Ni, 0.10 max Mg, others 0.3 max each, bal Al.
Cast aluminium alloy.
BS 1490 LM3.

A-G5.
M-French; 5 Mg, 0.5 Mn, bal Al.
For automobile chassis; light weight construction.

AG 15.
M-1361; 1.0 Mn, bal Al.
295 N/mm^2 TS; 100 N/mm^2 YS; 17 El.
Good corrosion resistance, weldability and formability. Type AlMn1.

AG/CU EUTETIC.
M-200; 29 Cu, bal Ag.
Melt point 778°C. Silver braze alloy.

AGGERLIT A.
M-614; 0.3 C, 2.2 Si, 6 Cr, bal Fe.
For oil refinery equipment; heat resistant.

AGGERLIT B.
M-614; 0.5 C, 1.5 Si, 17 Cr, bal Fe.
For heat treating boxes, furnace parts, fixtures; corrosion and heat resistant.

AGGERLIT C.
M-614; 0.6 C, 1.5 Si, 22 Cr, bal Fe.
For furnace parts and equipment, fixtures; corrosion and heat resistant.

AGGERLIT D.
M-614; 0.6 C, 1.5 Si, 22 Cr, bal Fe.
For furnace parts and equipment; fixtures; corrosion and heat resistant.

AGGERLIT E.
M-614; 0.6 C, 1.5 Si, 29 Cr, bal Fe.
For furnace parts and equipment, fixtures; heat resistant.

AGGERSTAHL.
M-614; 1.45 C, 1.4 Cr, 0.25 Si, 0.6 Mn, bal Fe.
For bearings, sleeves, liners; water hardened, wear resistant.

AGILE ACTARC.
M-661; Pure Fe or low alloy steel.
For welding electrodes; for sheet metal.

AGILE NO. 200.
M-661; 15 Sn, bal Cu.
For welding and brazing rods; coated, M.P. 1596°F.

AGILE SILVER A-C ELECTRODE.
M-661; 0.9 C, 0.7 Mn, 0.7 Cr, 1.5 W, bal Fe.
400-650 Brin.
For welding electrodes; deposit should be heat treated.

AGILE SILVER BLACK.
M-661; 0.6 C, 12 Cr, 8 Mo, 8 W, 1.5 B, bal Fe.
For welding high speed steel cutters; high speed steel.

AGILE SILVER CHISEL.
M-661; 0.8-1.0 C, 1.0 Si, 1.5 W, 0.3-0.5 B, bal Fe.
400-620 Brin.
For welding electrodes for chisels and shock resisting tools; tough, shock resisting.

AGILE SILVER D-C-ELECTRODE.
M-661; 1.8-2.2 C, 12-14 Cr, 0.5 V, 0.8 Mo, bal Fe.
400-640 Brin.
For welding electrodes for punches and cold working dies; must be heat treated.

AGILE SILVER ROD.
M-661; 0.75 C, 4 Mo, 6 W, 4 Cr, 0.8 B, bal Fe.
For welding high speed steel tools; high speed steel.

AGILE YELLOW ROD.
M-661; 67 Ni, 31 Cu, 0.2 B.
190 Brin.
For welding, electrode for cast iron; machinable.

AGMOLITE.
M-237; Ag-Mo.
For electrical contracts for circuit breakers; sintered.

AGNILITE.
M-237; Ag, Ni.
For electric contacts for circuit breakers; sintered.

AGPU 11.
M-884; C, S, bal Fe.
For screw machine products; free-cutting.

AGRAPHITE.
M-237; Ag-graphite.
For electrical contacts for circuit breakers; sintered.

AGRICOLA.
M-221; 70 Cu, 30 Pb.
For diesel engine bearings, seals; heavy duty.

AGRICOLA BRONZE.
M-699; 70 Cu, 1 Ag, 29 Pb.
For bearings, bushings.

AGRILITE BEARING METAL.
M-128; 70 Cu, 5 Sn, 25 Pb.
25,000 TS; 16,350 YS; 50 Brin.
For bearings for rolling mill, thrust bearings on steam boats, bearing for washing and ironing machines; self lubricating bearing bronze.

AGRILITE NO. 5.
M-128; 70.49 Cu, 5.39 Sn, 24.0 Pb, 0.09 Ni, 0.005 P.
For connecting rod bearings for aero engines, rolling mill bearings; self lubricating bearing bronze.

A.H. CHROME DIE.
M-1406; 1.0 C, 5 Cr, 1 Mo, 0.3 V, bal Fe.
For dies, gages, mandrels; cold work steel, air hardened, non-deforming.

AH-5 see BETHLEHEM AH-5.

AH NO. 2.
M-926; 30 Ag, 52.45 Cu, 9.75 Zn, 7.8 Ni.
For brazing; M.P. 1400-1735°F.

A-HT.
M-24; 1.0 C, 3.0 Cr, 1.1 Mo, 0.25 V, 1.05 W, 1.00 Ti, bal Fe.
Air hardening tool steel.
For dies, punches, shear blades, blanking tools.

AICH METAL.
M-Eng; 59.37 Cu, 39.68 Zn; 0.95 Fe.
Cast: 60,000 TS; 21,000 YS; 44 El; 45 RA; 90 Brin.
Forged: 64,500 TS; 31,000 YS; 44 El; 64 RA; 107 Brin.
For hydraulic cylinders and forgings; resists oxidation; similar to "Sterro."

AIDION.
M-587; Sb, Pb, bal Sn.
For bearings; antifriction metal.

AIKEN METAL.
M-U.S.; 50 Cu, 50 Pb.
For plastic metal parts.

AIP.
M-1653; 0.10 max C, 1.0 max Si, 2.0 max Mn, 17.0-19.0 Cr, 11.0-13.0 N bal Fe.
Austenitic stainless steel.
W.-Nr. 1.4312; x 8 CrNi 18 12; AISI 305.

AIR-4.
M-24; 0.95 C, 2.0 Mn, 2.2 Cr, 1.1 Mo, 0.15-0.35 Pb, bal Fe.
General purpose tools, cold forming dies, blanking dies, forming rolls, knurling tools.
Free machining type.
Type A4 air hardening tool steel.

AIRALOY see REPUBLIC A-4.

AIR-CHROM.
M-335; 1.0 C, 0.6 Mn, 5.25 Cr, 0.25 V, 1.1 Mo, bal Fe.
For tools, dies.
AISI A2 air hardening tool steel.

AIRCO 35-NI 15-CR TITANIA.
M-663; 0.07 C, 1.8 Mn, 16 Cr, 35 Ni, 0.9 Mo, bal Fe.
For welding rod; heat and corrosion resistant.

AIRCO 43.
M-663; 5% Si Aluminum base electrode, coated; DC reverse polarity.
For arc welding aluminum.
AWS-ASTM Al-43.

AIRCO 46.
M-663; Coated electrode; DC reverse polarity.
Deposit analysis 0.6 C, 0.2 Mn, 0.1 Si, bal Fe. Hardness 20-25 Rock C.
For build-up to resist abrasion and impact; for excavator bucket fingers, locomotive tire flanges, rails and switches.

AIRCO 57.
M-663; 4.5-6.0 Si, 0.8 Fe, 0.3 Cu, 0.1 Zn, 0.2 Ti, bal Al.
General purpose aluminum electrode, coated, DC reverse polarity.
As welded: 14,000 TS min.
For welding aluminum alloys 6061, 6063, 5052, 5154, 5454, 3003, and 2024.
AWS-ASTM Al-43.

AIRCO 70.
M-663; 4.25 Sn, 0.5 Fe, 0.15 Zn, 0.10 P, 0.10 Si, bal Cu.
Phosphor bronze brazing rod, coated: DC reverse polarity.
As welded: 40,000 TS; 20,000 YS; 15 El; 75 Brin.
For joining brass, bronze, copper, steel, cast iron, malleable iron.
AWS-ASTM E Cu Sn-A.

AIRCO 77.
M-663; 0.10 C, 0.22 Mn, 0.17 Si, bal Fe.
AC-DC reverse polarity coated electrode.
For arc welding cast iron cylinder blocks and heads, bearing blocks, machine parts, large frames.
AWS-ASTM E St.

AIRCO 100.
M-663; 9.0-11.0 Al, 1.5 max Fe, bal Cu.
Aluminum bronze brazing rod, coated; DC reverse polarity.
As welded: 77,000 TS; 35,000 YS; 27 El; 120 Brin.
For welding and build-up on brass, manganese bronze, aluminum bronze and dissimilar metals.
AWS-ASTM E Cu Al-A2.

AIRCO 116.
M-663; 11.0-12.0 Al, 3.0-4.25 Fe, bal Cu.
Aluminum bronze brazing rod, coated; DC reverse polarity.
As welded: 89,000 TS; 47,000 YS; 15 El; 160 Brin.
For overlays on bearing surfaces that are subjected to shock or impact loading, and for joining many dissimilar metals.
AWS-ASTM E Cu Al-B.

AIRCO 120.
M-663; 12.0-13.0 Al, 3.0-5.0 Fe, bal Cu.
Aluminum bronze brazing rod, coated; DC reverse polarity.
As welded: 90,000 TS; 46,000 YS; 4 El; 200 Brin.
For surfacing and overlays for good wear and strength for severe service.
AWS-ASTM E Cu Al-C.

AIRCO 125.
M-663; 13.0-14.0 Al, 3.0-5.0 Fe, bal Cu.
Aluminum bronze brazing rod, coated: DC reverse polarity.
As welded: 77,000 TS; 54,000 YS; 1.0 El; 250 Brin.
For surfacing and overlays for rebuilding aluminum bronze dies or finish layer on ferrous dies to reduce galling.
AWS-ASTM E Cu Al-D.

AIRCO 130.
M-663; 14.0-15.0 Al, 3.0-5.0 Fe, bal Cu.
Aluminum bronze brazing rod, coated; DC reverse polarity.
As welded: 80,000 TS; 69,000 YS; 0 El; 300 Brin.
For overlaying dies, shafts, bearings to minimize scratching and galling.
AWS-ASTM E Cu Al-E.

AIRCO 308 DC.
M-663; 0.05 C, 19 Cr, 10 Ni, 1.8 Mn, 0.3 Si, bal Fe.
DC reverse polarity lime coated electrode.
For welding austenitic 18-8 type stainless steels.
AWS-ASTM E 308-15; AISI 308.

AIRCO 308 ELC.
M-663; Low carbon grade of 308.
For welding AISI 304L, 321 and 347 stainless steels.
AWS ER 308 L; AISI 308 L.

AIRCO 309.
M-663; 25 Cr, 12 Ni stainless welding rod; bare wire, cut lengths.
For welding AISI 309 Stainless steel.
AWS ER 309; AISI 309.

AIRCO 309 CB AC-DC.
M-663; 0.09 C, 24 Cr, 13 Ni, 2.1 Mn, 0.55 Si, 0.85 Cb, bal Fe.
AC-DC reverse polarity titania coated electrode.
As welded: 90,000 TS; 70,000 YS; 35 El.
AWS-ASTM E 309 Cb-16; AISI 309 Cb.

AIRCO 309 DC.
M-663; 0.07 C, 23 Cr, 12.75 Ni, 1.9 Mn, 0.7 Si, bal Fe.
DC reverse polarity lime coated electrode.
As welded: 89,000 TS; 63,500 YS; 42 El.
For welding austentic type 309 stainless.
AWS-ASTM E 309-15; AISI 309.

AIRCO 310.
M-663; 25 Cr, 20 Ni, stainless welding rod; bare wire, cut lengths.
For welding AISI 310 stainless by TIG or oxyacetylene (with flux) methods; for elevated temperature operations.
AWS ER 310; AISI 310.

AIRCO 310 CB AC-DC.
M-663; 0.11 C, 26 Cr, 21 Ni, 1.9 Mn, 0.6 Si, 0.85 Cb, bal Fe.
AC-DC reverse polarity titania coated electrode.
As welded: 93,500 TS; 69,500 YS; 42 El.
For welding austenitic stainless steels and some dissimilar metals.
AWS-ASTM E 310 Cb-16; AISI 310 Cb.

AIRCO 310 DC.
M-663; 0.10 C, 26 Cr, 21 Ni, 1.8 Mn, 0.6 Si, bal Fe.
DC reverse polarity lime coated electrode.
As welded: 91,000 TS; 61,000 YS; 36 El.
For welding austenitic types 310 and 314 stainless steels.
AWS-ASTM E 310-15; AISI 310.

AIRCO 310 MO AC-DC.
M-663; 0.10 C, 26 Cr, 21 Ni, 1.9 Mn, 0.55 Si, 2.5 Mo, bal Fe.
AC-DC reverse polarity titania coated electrode.
As welded: 89,000 TS; 62,500 YS; 37 El.
For welding type 310 and similar steels and for overlay mild steel for paper pulp digestors and similar applications.
AWS-ASTM E 310 Mo-16; AISI 310 Mo.

AIRCO 312 AC-DC.
M-663; 0.12 C, 28 Cr, 9.25 Ni, 2.0 Mn, 0.7 Si, bal Fe.
AC-DC reverse polarity titania coated electrode.
As welded: 120,000 TS; 80,000 YS; 36 El.
For welding high strength type 312 base materials particularly for jet aircraft industry, and for dissimilar metals.
AWS-ASTM E312-16; AISI 312.

AIRCO 316.
M-663; 19 Cr, 12 Ni, 2.5 Mo stainless welding rod; bare wire, cut lengths.
For welding AISI 316 stainless steel.
AWS ER 316; AISI 316.

AIRCO 316 CB AC-DC.
M-663; 0.07 C, 18.5 Cr, 11.5 Ni, 1.9 Mn, 0.55 Si, 2.25 Mo, 0.5 Cb, bal Fe.
AC-DC reverse polarity titania coated electrode.
As welded: 90,000 TS; 63,000 YS; 36 El.
For welding type 316 and other austenitic stainless steels for best corrosion resistant joints.

AIRCO 316 ELC.
M-663; Low carbon grade of 316
For welding AISI 316L stainless steel.
AWS ER 316L; AISI 316L.

AIRCO 317 AC-DC.
M-663; 0.07 C, 19 Cr, 13 Ni, 2 Mn, 0.55 Si, 3.5 Mo, bal Fe.
AC-DC reverse polarity titania coated electrode.
As welded: 102,000 TS; 87,000 YS; 31 El.
For welding type 317 stainless for severe corrosion resistance.
AWS-ASTM E317-16; AISI 317.

AIRCO 330 AC-DC.
M-663; 0.20 C, 15 Cr, 34.5 Ni, 1.75 Mn, 0.6 Si, bal Fe.
AC-DC reverse polarity titania coated electrode.
As welded: 89,500 TS; 70,000 YS; 35.5 El.
For welding high nickel alloy castings for high temperature operation.
AWS-ASTM E330-16; AISI 330.

AIRCO 347.
M-663; 19 Cr, 9 Ni, Cb stabilized stainless steel welding rod; bare wire, cut lengths.
For welding AISI 304L, 321 and 347 stainless steels.
AWS ER 347; AISI 347.

AIRCO 361.
M-663; stainless core rod, composite coating.
Deposit comp: 0.50 C, 4.75 Mn, 19.0 Cr, 9.5 Ni, 0.75 Si, bal Fe.
For joining and hardfacing of railroad frogs and switches and similar applications. Impact resistant.

AIRCO 363.
M-663; Coated electrode.
AC or DC, either polarity.
Deposit analysis: 0.20 C, 1.0 Mn, 0.65 Si, 0.5 Cr, bal Fe. Hardness 20-28 Rock C.
For surfacing or intermediate layer with some abrasion and impact resistance; for brake drums, gear teeth, sprockets and bushings.

AIRCO 375.
M-663; 1.2 C, 1.0 Mn, 0.45 Si, 2.75 Fe, bal Ni.
AC-DC reverse polarity.
For welding cast iron.
AWS-ASTM E NiCl.

AIRCO 376.
M-663; Weld deposit: 1.1 C, 0.38 Mn, 0.007 P, 0.004 S, 0.04 Si, 54.5 bal Fe.
To produce machinable welds in cast iron using either AC or DC reverse polarity.
AWS Class E NiFe-CI.

AIRCO 410 DC.
M-663; 0.09 C, 12.5 Cr, 0.6 Ni, 0.8 Mn, 0.6 Si, bal Fe.
DC reverse polarity lime type coated electrode.
As welded: 150,000 TS; 5 El.
Heat treated: 71,500 TS; 53,000 YS; 20 El.
For welding types 403 and 410 chrome corrosion resistant steels and for stainless overlay on carbon steels.
AWS-ASTM E410-15; AISI 410.

AIRCO 502 DC.
M-663; 0.08 C, 5.0 Cr, 0.4 Ni, 0.6 Mn, 0.65 Si, 0.55 Mo, bal Fe.
DC reverse polarity lime type coated electrode.
As welded: 145,000 TS; 2.8 El.
Heat treated: 80,000 TS; 67,500 YS; 38 El.
For welding types 501 and 502 stainless and some alloy steels as 4130.
AWS-ASTM E502-15; AISI 502.

AIRCO 630.
M-663; steel core, composite type coating
AC-DC straight polarity.
Deposit analysis: 4.5 C, 30 Cr, 2 Si, 1 Mn, bal Fe.
For build-ups having resistance to severe abrasion with moderate impact. Hardness 56-61 Rock C. For dredging parts, dipper teeth, sand muller parts.

AIRCO 660.
M-663; Low alloy steel rod, composite coating.
AC-DC either polarity.
Deposit analysis: 1.8 C, 4.0 Cr, 0.5 Mn, 0.5 Si, bal Fe. Hardness 45-55 Rock C.
For surfacing or build-up for resistance to abrasion and impact; for crusher rolls, hammermill hammers, impact breaker bars.

AIRCO 718.
M-663; 12 Si, bal Al.
Bare aluminum brazing rod. For braze-welding 1060, 1100, 3003, 5005, 6061 6063, and cast alloys A612 and C612.
AWS B AlSi-4.

AIRCO 4043.
M-663; 4.5-6.0 Si, 0.3 max Cu, 0.8 max Fe, 0.1 max Zn, 0.2 max Ti, ba Al.
Bare aluminum alloy wire for gas or TIG welding of 3003, 3004, 5052, 6061, 6063, and casting alloys 43, 355, 356 and 214. Flows at 1155°F.
AWS ER 4043.

AIRCO 5183.
M-663; 4.3-5.2 Mg, 0.5-1.0 Mn, 0.4 max Si, 0.4 max Fe, 0.25 max Zn, 0.15 max Ti, 0.05-0.25 Cr, 0.10 max Cu, bal Al.
Bare aluminum alloy welding rod for gas or TIG welding of 5052, 5083, 5356, 5454 and 5456 alloys; flows at 1180°F.
AWS ER 5183.

AIRCO 5356.
M-663; 4.5-5.5 Mg. 0.5 max Si + Fe, bal Al.
Bare aluminum alloy, general purpose, for gas or TIG welding of 5050, 5052, 5083, 5086, 5356, 5454, and 5456. Flows at 1180°F.
AWS ER 5356.

AIRCO 5556.
 M-663; 4.7-5.5 Mg, 0.5-1.0 Mn, 0.25 max Zn, 0.05-0.2 Cr, bal Al.
 Bare aluminum wire for gas or TIG welding 5052, 5083, 5086, 5356, 5454, and 5456 for maximum weld strength. Flows freely at 1180°F.
 AWS ER 5556.

AIRCO 6010.
 M-663; 0.07-0.12 C, 0.2-0.4 Mn, 0.1-0.3 Si, bal Fe.
 DC reverse polarity coated electrode.
 Weld metal: 68,500 TS; 58,250 YS; 28.5 El.
 AWS-ASTM Class E6010.

AIRCO 6010 ARCRODS (001).
 M-663; Mild steel coated electrode; DC reverse polarity.
 AWS-ASTM Class E6010.

AIRCO 6011.
 M-663; 0.08 C, 0.35 Mn, 0.11 Si, bal Fe.
 AC-DC reverse polarity coated electrode.
 Weld metal: 70,000 TS; 62,000 YS; 30 El.
 AWS-ASTM Class E6011.

AIRCO 6011C.
 M-663; 0.10 C, 0.58 Mn, 0.19 Si, bal Fe.
 AC-DC reverse polarity coated electrode.
 Weld metal: 72,000 TS; 59,000 YS; 29.5 El.
 AWS-ASTM Class E6011.

AIRCO 6011LOC.
 M-663; 0.11 C, 0.6 Mn, 0.25 Si, bal Fe.
 AC-DC reverse polarity coated electrode.
 Weld metal: 71,000 TS; 59,000 YS; 22 El.
 AWS-ASTM Class E6011.

AIRCO 6012.
 M-663; 0.08 C, 0.45 Mn, 0.30 Si, bal Fe.
 AC-DC straight polarity coated electrode.
 Weld metal: 83,000 TS; 74,000 YS; 20 El.
 AWS-ASTM E-6012.

AIRCO 6012 C.
 M-663; 0.08 C, 0.38 Mn, 0.25 Si, bal Fe.
 AC-DC straight polarity coated electrode.
 Weld metal: 70,000 TS; 60,000 YS; 28 El.
 AWS-ASTM E6012.

AIRCO 6013.
 M-663; 0.10 C, 0.35 Mn, 0.035 max S, 0.035 max P, 0.40 Si, bal Fe.
 AC-DC all position, either polarity; weld metal: 70,000 psi TS; 62, YS; 20 El.
 AWS Class E6013.

AIRCO 6013 C.
 M-663; 0.09 C, 0.52 Mn, 0.39 Si, bal Fe.
 AC-DC either polarity coated electrode.
 Weld metal: 74,500 TS; 63,500 YS; 27 El.
 AWS-ASTM E6013.

AIRCO 6020.
 M-663; 0.12 C, 0.35 Mn, 0.10 Si, bal Fe.
 AC-DC either polarity coated electrode.
 Weld metal: 72,000 TS; 62,000 YS; 28 El.
 AWS-ASTM E6020.

AIRCO 6020 D.
 M-663; Mild steel coated electrode.
 AC or DC either polarity.
 AWS-ASTM E6020.

AIRCO 6020 F.
 M-663; Mild steel coated electrode.
 AC or DC either polarity.
 AWS-ASTM E6020.

AIRCO 7010-A1.
 M-663; 0.09 C, 0.43 Mn, 0.14 Si, 0.48 Mo, bal Fe.
 DC reverse polarity coated electrode.
 Weld, stress relieved: 80,000 TS; 67,000 YS; 29 El.
 For welding long distance pipe lines.
 AWS-ASTM E7010-A1.

AIRCO 7016.
 M-663; 0.09 C, 0.55 Mn, 0.50 Si, bal Fe.
 AC-DC reverse polarity coated electrode.
 Weld metal: 74,000 TS; 62,000 YS; 30 El.
 Low hydrogen, for assemblies to be enameled.
 AWS-ASTM E7016.

AIRCO 7016 M.
 M-663; 0.07 C, 0.90 Mn, 0.50 Si, 0.35 Mo, 0.25 Ni, 0.15 Cr, bal Fe.
 AC-DC reverse polarity coated electrode.
 As welded: 82,000 TS; 72,000 YS; 29 El.
 Good impact strength after stress anneal.
 AWS-ASTM E7016.

AIRCO 7020-A1.
 M-663; 0.09 C, 0.29 Mn, 0.12 Si, 0.52 Mo, bal Fe.
 AC-DC either polarity coated electrode.
 Weld, stress relieved: 73,000 TS; 63,500 YS; 29 El.
 AWS-ASTM E7020-A1.

AIRCO 8016-B2.
 M-663; 0.07 C, 0.6 Mn, 0.42 Si, 0.45 Mo, 1.30 Cr, bal Fe.
 AC-DC reverse polarity coated electrode.
 Weld, stress relieved: 84,000 TS; 68,000 YS; 28 El.
 AWS-ASTM 8016-B2.

AIRCO 8016-C1.
 M-663; 0.05 C, 0.82 Mn, 0.4 Si, 2.4 Ni, bal Fe.
 AC-DC reverse polarity coated electrode.
 Weld, stress relieved: 82,000 TS; 69,000 YS; 31 El; Charpy V notch, ft.lbs at -75°F: 108.
 AWS-ASTM E8016-C1.

AIRCO 8016-C3.
M-663; Low carbon, low alloy coated electrode.
AC-DC reverse polarity.
For welding low allow high tensile steel.
AWS-ASTM E8016-C3.

AIRCO 9016-B3.
M-663; 0.09 C, 0.57 Mn, 0.42 Si, 0.98 Mo, 2.26 Cr, bal Fe.
AC-DC reverse polarity coated electrode.
Weld, stress relieved: 99,000 TS; 87,000 YS; 22 El.
For welding equipment for steam power and other assemblies for elevated temperature.
AWS-ASTM E9016-B3.

AIRCO 10013-G.
M-663; Low carbon, low alloy coated electrode.
AC-DC straight polarity.
For welding low alloy high tensile steels.
AWS-ASTM E10013-G.

AIRCO 10016-D2.
M-663; 0.09 C, 1.8 Mn, 0.58 Si, 0.4 Mo, bal Fe.
AC-DC reverse polarity coated electrode.
Weld, stress relieved: 106,000 TS; 92,000 YS; 25 El.
For welding higher strength steels.
AWS-ASTM E10016-D2.

AIRCO 10016-G.
M-663; 0.06 C, 0.8 Mn, 0.34 Si, 0.3 Mo, 1.63 Ni, 0.13 V, bal Fe.
AC-DC reverse polarity coated electrode.
Weld, stress relieved: 109,500 TS; 102,000 YS; 25 El.
For high strength assemblies.
AWS-ASTM E10016-G.

AIRCO 12015-G.
M-663; Low hydrogen electrode; reverse polarity only; for welding high strength and armor steels.
AWS-ASTM E12015-G.

AIRCO AIRCOLOY NO 1.
M-663; Bare or coated cobalt base electrode.
Deposit analysis: 2.5 C, 50.0 Co, 31.0 Cr, 13 W, 1.5 Si.
For wear resistant and corrosion resistant build-up and surfacing. Hardness of deposit is 50-58 Rock C.
AWS ASTM E Co Cr-C and R Co Cr-C.

AIRCO AIRCOLOY NO 6.
M-663; Bare or coated cobalt base electrode.
Deposit analysis: 1.0 C, 60.0 Co, 28.0 Cr, 5.0 W, 1.5 Si; Hardness 38-45 Rock C.
For build-ups resistant to corrosion and abrasion at high temperatures.
AWS-ASTM E Co Cr-A and R Co Cr-A.

AIRCO AIR-MANG "HC".
M-663; Steel core rod, composite coating.
Deposit analysis: 0.7 C, 14.0 Mn, 4 Cr, 3.5 Ni, bal Fe.
For joining or build-up of dipper teeth, coal crusher segments, rail frogs and switches. Impact resistant.

AIRCO AIROD.
M-663; Steel tube containing tungsten carbide particles used as a bare filler rod for gas welding for hardfacing plow shares, cane knives, conveyor screws. Abrasion resistant.

AIRCO AX-140.
M-663; 0.07-0.11 C, 1.7-2.0 Mn, 0.25-0.45 S, 2.0-2.5 Ni, 0.85-1.2 Cr, 0.5-0.6 Mo, Ti, bal Fe.
Weld: 140,000 YS; 157,000 TS; 16 El; 57 RA.
For welding high strength steel plates.
Tough, strong and crack free welding wire.

AIRCO BULK TUNGSTEN CARBIDE.
M-663; Hard carbide; to be applied with a binder and subsequently fused into the surface for a hard abrasion resistant skin. Abrasion resistant.

AICRO CODE-ARC 308-MR.
M-663; 0.07 C, 20 Cr. 10 Ni, 1.7 Mn, 0.4 Si, bal Fe.
AC-DC reverse polarity titania coated electrode.
For welding austenitic 18-8 type stainless steels.
ASW-ASTM E 308-16; AISI 308.

AIRCO CODE-ARC 308 ELC-MR.
M-663; 0.03 C, 20 Cr, 10 Ni, 1.9 Mn, 0.4 Si, bal Fe.
AC-DC reverse polarity titania coated electrode.
For welding 304 L, 321, 347 and other 18-8 stainless steels.
AWS-ASTM E 308 L-16; AISI 308 ELC.

AIRCO CODE-ARC 309 MR.
M-663; 0.10 C, 24 Cr, 13 Ni, 1.8 Mn, 0.55 Si, bal Fe.
AC-DC reverse polarity titania coated electrode.
As welded: 89,000 TS; 65,000 YS; 42 El.
For welding austenitic type 309 stainless.
AWS-ASTM E 309-16; AISI 309.

AIRCO CODE-ARC 310 MR.
M-663; 0.11 C, 26.5 Cr, 21.0 Ni, 1.8 Mn, 0.5 Si, bal Fe.
AC-DC reverse polarity titania coated electrode.
As welded: 91,000 TS; 61,000 YS; 36.5 El.
For welding austenitic types 310 and 314 stainless steels.
AWS-ASTM E 310-16; AISI 310.

AIRCO CODE-ARC 316 MR.
M-663; 0.06 C, 18.5 Cr, 13.0 Ni, 1.9 Mn, 0.6 Si, 2.25 Mo, bal Fe.
AC-DC reverse polarity titania coated electrode.
As welded: 93,000 TS; 75,000 YS; 35 El.
For welding type 316 for corrosion resistant or elevated temperature applications.
AWS-ASTM E316-16; AISI 316.

AIRCO CODE-ARC 316 ELC-MR.
M-663; 0.03 C, 18.5 Cr, 13.0 Ni, 1.9 Mn, 0.6 Si, 2.25 Mo, bal Fe.
AC-DC reverse polarity titania coated electrode.
As welded: 88,700 TS; 74,350 YS; 33.5 El.
For welding extra low carbon molybdenum bearing austenitic stainless.
AWS-ASTM E 316L-16; AISI 316 ELC

AIRCO CODE-ARC 347 MR.
M-663; 0.06 C, 19.5 Cr, 10.0 Ni, 1.5 Mn, 0.6 Si, 0.70 Cb, bal Fe.
AC-DC reverse polarity titania coated electrode.
As welded: 109,000 TS; 79,500 YS; 34.5 El.
For welding types 321 and 347 stainless.
AWS-ASTM E347-16; AISI 347.

AIRCO CODE-ARC 7018 MR.
M-663; 0.04 C, 0.65 Mn, 0.55 Si, bal Fe.
AC-DC reverse polarity coated electrode.
As welded: 78,000 TS; 67,000 YS; 30 El.
AWS-ASTM E7018.

AIRCO CODE-ARC 7018-A1 MR.
M-663; 0.04 C, 0.6 Mn, 0.6 Si, 0.48 Mo, bal Fe.
AC-DC reverse polarity coated electrode.
Weld, stress relieved: 81,000 TS; 71,000 YS; 30 El.
AWS-ASTM E 7018-A1.

AIRCO CODE-ARC 8018-C3 MR.
M-663; 0.04 C, 0.9 Mn, 0.4 Si, 1.0 Ni, 0.2 Mo, bal Fe.
AC-DC reverse polarity coated electrode.
As welded: 83,000 TS; 72,000 YS; 31 El.
Good impact properties.
AWS-ASTM E8018-C3.

AIRCO CODE-ARC 11018 MR.
M-663; 0.04 C, 1.6 Mn, 0.4 Si, 0.42 Mo, 0.25 Cr. 1.8 Ni, bal Fe.
AC-DC reverse polarity coated electrode.
Weld, stress relieved: 113,000 TS; 105,000 YS; 23 El.
To weld low alloy high strength steels.
AWS-ASTM E11018-M.

AIRCO CODE-ARC 12018.
M-663; 0.04 C, 1.8 Mn, 0.4 Si, 0.48 Mo, 0.45 Cr, 2.0 Ni, bal Fe.
AC-DC reverse polarity coated electrode.
Weld, stress relieved: 122,000 TS; 113,000 YS; 22 El.
To weld low alloy high strength steels.
AWS-ASTM E12018-M.

AIRCO COMMERCIALLY PURE ALUMINUM A1100.
M-663; 100% Al.
For machine welding of 1100 and 3003 aluminum for window frames, heat exchangers, food containers, electrical bus bars, racks.
AWS ER 1100.

AIRCO EASYARC 308 AC-DC.
M-663.
Weld metal: 0.07 C, 19.6 Cr, 9.5 Ni, 1.65 Mo, 0.68 Si, bal Fe.
As welded: 87,900 TS; 57,400 YS; 42 El.
AC-DC reverse polarity; coating: titania type with metal powders; for welding 18-8 type.
AWS-ASTM E308-16; AISI 308.

AIRCO EASYARC 316 AC-DC.
M-663.
Weld metal: 0.06 C, 18.5 Cr, 12.3 Ni, 1.8 Mn, 0.31 Si, 2.28 Mo, bal Fe.
As welded: 80,400 TS; 56,400 YS; 44 El.
AC-DC reverse polarity; coating: titania type with metal powders; for welding type 316 stainless steel.
AWS-ASTM E316-16; AISI 316.

AIRCO EASYARC 620.
M-663; Alloy steel core rod, chrome iron composite coating; AC-DC either polarity.
Deposit analysis: 6.0 C, 22.0 Cr, 7.0 Mo, 5.0 W, 1.5 Si, 0.5 Mn, bal Fe.
For build-ups resistant to severe abrasion, moderate impact and hot wear up to 1100°F. For crusher parts, coke pusher shoes, hot slag dipper teeth. Hardness 60-65 Rock C.

AIRCO EASYARC 6027.
M-663; 0.07 C, 0.90 Mn, 0.14 Si, bal Fe.
AC-DC either polarity coated electrode.
Weld metal: 68,000 TS; 57,000 YS; 30 El.
AWS-ASTM E6027.

AIRCO EASYARC 7014.
M-663; 0.08 C, 0.45 Mn, 0.30 Si, bal Fe.
AC-DC either polarity coated electrode.
Weld metal: 81,000 TS; 73,000 YS; 26 El.
AWS-ASTM E7014.

AIRCO EASYARC 7018 C.
M-663; 0.07 C, 0.90 Mn, 0.50 Si, bal Fe.
AC-DC reverse polarity coated electrode.
As welded: 84,000 TS; 73,000 YS; 30 El.
AWS-ASTM E7018.

AIRCO EASYARC 7018 MR.
M-663; 0.06 C, 0.65 Mn, 0.50 Si, bal Fe.
AC-DC reverse polarity coated electrode.
As welded: 78,000 TS; 66,000 YS; 28 El.
AWS-ASTM E7018.

AIRCO EASYARC 7024.
M-663; 0.07 C, 0.75 Mn, 0.50 Si, bal Fe.
DC reverse polarity coated electrode.
Weld metal: 76,000 TS; 68,000 YS; 20 El.
AWS-ASTM E7024.

AIRCO EASYARC 7028.
M-663; 0.06 C, 0.70 Mn, 0.35 Si, bal Fe.
AC-DC reverse polarity coated electrode.
As welded: 80,000 TS; 72,000 YS; 30 El.
AWS-ASTM E7028.

AIRCOMATIC A5556 ALUMINUM ALLOY.
M-663.
Wire for welding 5083, 5086, 5456 high tensile aluminum alloys for truck frames, diesel engine bases, storage tanks.
AWS ER 5556.

AIRCO NICKEL MANGANESE C.
M-663; Steel core rod, composite coating.
Weld metal: 0.60 C, 13.0 Mn, 3.0 Ni, 0.4 Cr, 0.75 Si, bal Fe.
For joining and build-up of Hadfield's manganese steel dipper teeth, shovel tracks, crusher pads; impact resistant.
AWS-ASTM E Fe Mn-A.

AIRCO NO. 1.
M-663; 0.15 max C, 0.45 Mn, 0.2 Si, 1.0-1.5 Ni. 0.3 max Cr, bal Fe.
Low alloy steel welding rod, for gas or tungsten inert gas welding of carbon and low alloy steels.
As welded: 62,000 TS; 22 El.
AWS RG 60.

AIRCO NO. 4.
M-663; 0.08-0.20 C, 0.9-1.2 Mn, 0.15-0.30 Si, bal Fe.
Low alloy steel welding rod, for gas or tungsten inert gas welding of carbon and low alloy steels.
As welded: 62,000-67,000 TS; 20-25 El.

AIRCO NO. 7.
M-663; 0.06 max C, 0.25 max Mn, bal Fe.
Low carbon steel welding rod, copper coated, for gas welding of low carbon steel sheet, plate and pipe; and for auto body repair.
As welded: 52,000 TS; 23 El.
AWS RG 45.

AIRCO NO. 9.
M-663; 3.25 C, 0.6 Mn, 3.0 Si, bal Fe.
Square cast iron gas welding rod,-for general gas welding of grey cast iron.
AWS-ASTM RCl.

AIRCO NO. 10.
M-663; 3.3 C, 0.6 Mn, 2.25 Si, 1.4 Ni, 0.35 Mo, bal Fe.
Square low alloy cast iron, for welding low alloy grey irons.
AWS-ASTM RC1-A.

AIRCO NO. 20 BRONZE ROD.
M-663; 0.5-1.0 Sn, 60 Cu, 0.06 max Fe, bal Zn.
Bare or coated rod for braze-welding copper, bronze and nickel alloys. Fuses readily at 1625°F; strength: 40,000 TS min.
AWS RB CuZn-A.

AIRCO NO. 21 NICKEL SILVER ROD.
M-663; 46-50 Cu, 9-11 Ni, bal Zn.
Bare or coated rod for joining dissimilar metals as mild steel, tool steel, cast iron, stainless steel, monel, inconel, bronze, copper, malleable iron. 70,000 min. TS.
AWS RB Cu Zn-D.

AIRCO NO. 22 BRONZE ROD.
M-663; 59 Cu, 0.4 Fe, 0.23 Mn, 1.0 Sn, 0.6 Ni, bal Zn.
Bare or coated rod for braze-welding steel, cast iron, malleable iron, copper base alloys and to build up wearing surfaces and bearings. 45,000 min. TS. AWS R Cu Zn-B.

AIRCO NO. 27 LOW FUMING BRONZE ROD.
M-663; 56-59 Cu, 0.4-0.8 Fe, 0.1 Si, 0.3 Mn, 1.0 Sn, bal Zn.
Bare or coated rod, low fuming; for general purpose braze-welding of steel cast iron, brass and bronze. 50,000 TS min.
AWS R Cu Zn-C.

AIRCO NO. 92 SILICON BRONZE ROD.
M-663; 95.8 Cu, 3.0 Si, 1.0 Mn, 1.0 Zn.
Tin coated rod for brazing galvanized iron by carbon arc process or TIG
AWS R Cu Si-A.

AIRCO NO. 1010 SILICON BRONZE ROD.
M-663; 95.8 Cu, 3.1 Si, 1.1 Mn, 1.0 Zn.
Bare rod for gas welding of copper, copper-silicon and copper-zinc base alloys to themselves and to plain and galvanized steel. 50,000 min TS. Everdue grade. AWS R Cu Si-A.

AIRCO PHOS-COPPER ROD.
M-663; 7.25-7.35 P, bal Cu.
Bare rod for torch or furnace braze of transformers, radiators, motors, water coolers, plumbing installations, heat exchangers.
AWS B Cu P-2.

AIRCO PHOS-SILVER 2 ROD.
M-663; 6.75-7.25 P, 1.75-2.25 Ag, bal Cu.
Bare rod for torch or furnace brazing components of copper, brass and bronze.
Melt range 1435-1455°F.

AIRCO RAILROAD ROD.
M-633; 0.2-0.4 C, 0.85-1.15 Mn, 0.1-0.3 V, 0.9-1.25 Cr, bal Fe.
Welded: 100,000 TS.
For welding rod; for rail ends.

AIRCO SELF HARDENING.
M-663; steel core wire, composite coating. Deposit analysis: 0.85 C, 5.0 Cr, 0.9 Mn, 0.55 Si, 0.35 V, bal Fe; hardness 50-55 Rock C.
For hard surfacing and build-up for resistance to abrasion and moderate to heavy impact.
For dump truck bodies, brake shoe hangers, bulldozer trunnions.

AIRCOMATIC A100 ALUMINUM BRONZE.
M-663; Al, Fe, bal Cu.
Bare rod for inert gas welding of aluminum bronze or bronze surfacing of steel.
AWS E Cu A1-A2.

AIRCOMATIC A102 PHOS BRONZE.
M-663; Sn, P, bal Cu.
Bare rod for inert gas welding of brass, bronze or for overlays on ferrous metals.
AWS E Cu Sn-C.

AIRCOMATIC A104 NI-AL-MN BRONZE.
M-663; Al, Ni, Mn, bal Cu.
Bare rod for inert gas welding of ship propellers and turbine runners.

AIRCOMATIC A110 SILICON BRONZE.
M-663; Si, bal Cu.
Bare rod for inert gas welding of silicon bronze or of many dissimilar metals.
AWS E Cu Si.

AIRCOMATIC A145 COPPER.
M-663; Cu.
Bare rod or wire for inert gas welding of copper and copper base alloys and for overlay on steels.
AWS E Cu.

AIRCOMATIC A158 ALUMINUM BRONZE.
M-663; Al, bal Cu.
Iron free bare wire for welding aluminum bronzes, manganese bronze, copper, cast iron, mild steel and alloy steel, and for surfacing; designed for inert gas processes.
AWS E Cu Al-Al.

AIRCOMATIC A160 ALUMINUM BRONZE.
M-663; Al, bal Cu.
Aluminum bronze bare wire largely for overlay work to produce corrosion resistant and wear resistant surface. Hardness of deposit. 160-210 Brin.
AWS E Cu Al-B.

AIRCOMATIC A308.
M-663.
Wire for welding AISI types 301, 305, and 308 stainless steels, and for depositing stainless overlays on mild steel.
AWS ER 308.

AIRCOMATIC A308 ELC.
M-663; Low carbon grade of A308.
Wire for welding AISI 304 ELC and 308 ELC, also AISI 321 and 347.
AWS-ER 308 L.

AIRCOMATIC A308 L SI.
M-663; Higher silicon version of A308 for improved arc stability.

AIRCOMATIC A309.
M-663.
Wire for welding AISI 309 stainless for high temperature operation; also for welding some straight chromium stainless grades.
AWS-ER 309.

AIRCOMATIC A310.
M-663.
Wire for welding AISI 310 stainless steel for high temperatures or corrosion resistant applications.
AWS ER 310.

AIRCOMATIC A312.
M-663.
Wire designed for welding stainless steels to carbon steels.
AWS ER312.

AIRCOMATIC A316.
M-663.
Wire for welding AISI 316 stainless.
AWS ER 316.

AIRCOMATIC A316 ELC.
M-663; Low carbon grade of A316.
For welding AISI 316 L stainless steel.
AWS ER 316 L.

AIRCOMATIC A316 L SI.
M-663; Higher silicon version of a316, for improved arc stability.

AIRCOMATIC A347.
M-663.
Wire for welding AISI types 321, 347 and 304 L, particularly for elevated temperature operation or corrosive conditions.
AWS ER 347.

AIRCOMATIC A4043 ALUMINUM ALLOY.
M-663; AA 4043 composition (5.2 Si).
For welding 2014, 4043, 6061, 6062, 6063 alloys for truck bodies, pressure vessels, structural members.
AWS ER 4043.

AIRCOMATIC A5183 ALUMINUM ALLOY.
M-663; Mg, bal Al.
Wire for welding liquid oxygen and liquid nitrogen containers. AWS ER 5183.

AIRCOMATIC A5356 ALUMINUM ALLOY.
M-663; 5.0 Mg. 0.12 Cr, 0.12 Mn, bal Al.
Wire for welding 5056, 5083, 5086, 5154, 5356 aluminum alloys for truck frames, diesel engine bases, cargo tanks, gun mount bases. AWS ER 5356.

AIRCOMATIC A5554 ALUMINUM ALLOY.
M-663.
Wire for welding 5052 and 5454 aluminum alloys for truck tankers, dump bodies, railroad tank cars. AWS-ER 5554.

AIRCOSIL 3.
M-663; 50 Ag, 3 Ni, 16 Cd, 15.5 Cu, bal Zn.
For silver brazing of carbide tools; M.P. 1170-1270°F.
AWS A5.8-62 and ASTM B260-62; BAg-3.

AIRCOSIL 5.
M-663; 5 Ag, 88.75 Cu, 6.25 P.
For brazing alloy; M.P. 1190-1480°F.
AWS,ASTM BCuP-3.

AIRCOSIL 15.
M-663; 15 Ag, 80 Cu, 5 P.
For silver brazing of Cu alloys; M.P. 1185-1500°F; self-fluxing.
AWS,ASTM BCuP-5.

AIRCOSIL 35.
M-663; 35 Ag, 18 Cd, 26 Cu, 21 Zn.
For silver brazing, general purpose; M.P. 1125-1295°F; high plastic range.
AWS,ASTM BAg-2.

AIRCOSIL 45.
M-663; 45 Ag, 24 Cd, 15 Cu, 16 Zn.
For silver brazing; M.P. 1125-1145°F, free-flowing.
AWS,ASTM BAg-1.

AIRCOSIL 50.
M-663; 50 Ag, 18.0 Cd, 15.5 Cu, 16.5 Zn.
For silver brazing for all metals; M.P. 1160-1175°F; high strength.
AWS,ASTM BAg-la.

AIRCOSIL 60.
M-663; 60 Ag, 30 Cu, 10 Sn.
For brazing alloy; M.P. 1115-1325°F.
AWS,ASTM BAg-18.

AIRCOSIL A.
M-663; 9 Ag, 53 Cu, 38 Zn.
For brazing alloy; M.P. 1410-1565°F.

AIRCOSIL AE-100.
M-663; 92.5 Ag, 7.3 Cu. 0.2 Li.
For brazing alloy; M.P. 1435-1635°F.
AWS-ASTM BAg-19.

AIRCOSIL B.
M-663; 20 Ag, 45 Cu, 35 Zn.
For brazing alloy; M.P. 1315-1500°F.

AIRCOSIL C.
M-663; 20 Ag, 45 Cu, 30 Zn, 5 Cd.
For brazing alloy; M.P. 1140-1500°F.

AIRCOSIL D.
M-663; 30 Ag, 38 Cu, 32 Zn.
For brazing alloy; M.P. 1370-1410°F.

AIRCOSIL E.
M-663; 40 Ag, 30 Cu, 28 Zn, 2 Ni.
For brazing alloy for carbides, M.P. 1240-1435°F.
AWS,ASTM BAg-4.

AIRCOSIL EASY.
M-663; 65 Ag, 20 Cu, 15 Zn.
For brazing alloy for sterling silver; M.P. 1240-1325°F.
AWS,ASTM BAg-9.

AIRCOSIL F.
M-663; 40 Ag, 36 Cu, 24 Zn.
For brazing alloy; M.P. 1325-1415°F.

AIRCOSIL G.
M-663; 45 Ag, 30 Cu, 25 Zn.
For brazing alloy in food industry; M.P. 1250-1370°F; AWS-ASTM BAg-5.

AIRCOSIL H.
M-663; 50 Ag, 34 Cu, 16 Zn.
For brazing alloy in food industry; M.P. 1270-1425°F; AWS-ASTM BAg-6.

AIRCOSIL HARD.
M-663; 75 Ag, 22 Cu, 3 Zn.
For brazing alloy for sterling silver; M.P. 1365-1450°F.
AWS-ASTM BAg-11.

AIRCOSIL J.
M-663; 56 Ag, 22 Cu, 17 Zn, 5 Sn.
For brazing alloy on stainless steels; M.P. 1145-1205°F.
AWS,ASTM BAg-7.

AIRCOSIL K.
M-663; 60 Ag, 25 Cu, 15 Zn.
For brazing alloy; M.P. 1245-1325°F.

AIRCOSIL L.
M-663; 54 Ag, 40 Cu, 5 Zn, 1 Ni.
For brazing alloy; M.P. 1325-1575°F.

AIRCOSIL M.
M-663; 72 Ag, 28 Cu.
For brazing alloy in electronic industry; M.P. 1435°F; eutectic alloy.
AWS,ASTM BAg-8.

AIRCOSIL MEDIUM.
M-663; 70 Ag, 20 Cu, 10 Zn.
For brazing alloy for sterling silver; M.P. 1275-1360°F.
AWS,AWTM BAg-10.

AIRCOSIL N.
M-663; 80 ag, 16 Cu, 4 Zn.
For brazing alloy; M.P. 1345-1490°F.

AIRCOSIL P.
M-663; 85 Ag, 15 Mn.
For brazing alloy for high temperature applications; M.P. 1760-1778°F.
AWS,ASTM BAg-Mn.

AIRCOSIL R.
M-663; 40 Ag, 30 Cu, 25 Zn, 5 Ni.
For brazing alloy; M.P. 1240-1560°F.

AIRCOSIL S.
M-663; 25 Ag, 52.5 Cu, 22.5 Zn.
For brazing alloy; M.P. 1250-1575°F.

AIRCO SPECIAL "B" TUNGTUBE.
M-663; Steel tube containing tungsten carbide particles, used as bare filler rod for build-up on work core and rock drill bits and post hole digger blades. Abrasion and impact resistant.

AIRCO TRACKWEAR.
M-663; Steel core rod, composite coating.
Deposit comp: 0.85 C, 17.0 Mn, 0.45 V, bal Fe.
For joining and build-up of railroad frogs, switches and cross-overs. Impact resistant.

AIRCO TUBE AIRCOLITE.
M-663; Deposit analysis: 4 C, 30 Cr, 2 Si, 1.5 Mn, bal Fe; hardness 60 Rock C.
Bare tubular wire for gas welding and build-up for resistance to severe abrasion with moderate impact. For plow shares, cultivator spades, corn picker runners.

AIRCO TUNGTUBE.
M-663; Steel tube containing tungsten carbide used as a bare filler rod for oxyacetylene flame for hard facing. Abrasion resistant.

AIRCRAFT.
M-Eng.; 10-11.5 Al, 4-6 Fe, 4-6 Ni, bal Cu.
For corrosion resisting parts, aircraft; corrosion resistant.

AIRCRAT.
M-1149; 1.0 C, 0.5 Mn, 1.1 Mo, 0.2 V, 5.0 Cr, bal Fe.
For tools, dies, punches, crimpers; air hardened, ground flat stock.

AIRDI 150.
M-38; 1.55 C, 11.5 Cr, 0.9 V, 0.8 Mo, bal Fe.
Hardened: C 58-63 Rock.
Air hardening, somewhat corrosion resistant.
For blanking and coining dies, gauges, punches, thread rolling dies, shafts.
AISI Type D2 Cold work tool steel.

AIRDI 150-S.
M-38; 1.5 C, 12 Cr, 0.9 V, 0.8 Mo, 0.15 S, bal Fe.
For blanking and forming dies, punches; air hardened, non-deforming.

AIR-EEZ.
M-335; 0.95 C, 2.0 Mn, 0.35 Si, 2.2 Cr, 1.1 Mo, 0.30 Pb, bal Fe.
Free machining grade.
AISI A4 low temperature tool steel.

AIRESIST 13.
M-1491; 21 Cr, 11 W, 0.45 C, 2 Cb, 3.5 Al, 0.5 Y, bal Co.
For jet engine and gas turbine components.
Corrosion resistant. Good high temperature strength. Resists sulphidation and oxidation.

AIRESIST 213.
M-1491; 3.5 Al, 0.10 Y, 0.2 C, 20 Cr, 0.5 Ni, 4.5 W, 6.5 Ta, 3.5 Al, 0.5 Fe, bal Co.
Sheet: 160,000 TS; 100,000 YS; 10 El.
At 1800°F: 35,000 TS; 22,000 YS; 70 El.
For jet engine and gas turbine components.
Resists sulphur attack up to 2200°F.
Corrosion and oxidation resistant.

AIRESIST 215.
M-1491; 19 Cr, 4.5 W, 4.3 Al, 0.35 C, 0.5 Ni, 0.10 Y, 7.5 Ta, 0.5 Fe, 0.1 Zr, bal Co.
Resists oxidation and sulphidation.

ISC AIR HARD.
M-1482; 1.5 C, 0.8 Mo, 0.5 V, 0.7 Mn, 12 Cr, bal Fe.
For dies, cams, punches, blanking and forming tools; air hardened, non-deforming.

AIR-HARD.
M-115; 1.0 C, 0.3 Si, 0.7 Mn, 5.25 Cr, 1.1 Mo, 0.25 V, bal Fe.
Annealed: 55,000 YS; 18 El; 38 RA; 223 Brin.
Hardened: 184,000 YS; 5 El; 14 RA; C 46-54 Rock.
For trimming and blanking dies, cold shears, gauges, punches.
Air hardening, non-deforming, wear resistant.

AIR HARDENING.
M-1424; 1.0 C, 0.7 Mn, 0.3 Si, 5.25 Cr, 0.3 V, 1.15 Mo, bal Fe.
AISI Type A2 air hardening tool steel.

AIRITE NO. 21.
M-Eng.; Ni alloy.
For compressor blades, aircraft rotor discs; corrosion and heat resistant.

AIRKOOL.
M-38, M-1214; 1 C, 5 Cr, 0.4 V, 1.1 Mo, bal Fe.
For dies, gauges, punches, shears, broaches; air hardened, non-deforming.

AIRKOOL-S.
M-38; 1 C, 5.2 Cr, 0.3 V, 1.1 Mo, 0.15 S, bal Fe.
For blanking and forming dies, punches; air hardened, non-deforming. AISI A2.

AIRKOOL-V.
M-38; 2.25 C, 5.25 Cr, 4.75 V, 1 Mo, 1 W, bal Fe.
Heat treated: 650 Brin.
For blanking dies, brick mold liners; oil hardened, non-deforming.

AIRMO.
M-57; 1.0 C, 2 Mn, 1 Cr, 1 Mo, bal Fe.
For cold work dies, crimpers; cold work steel, air hardened.

AIRMOLD.
M-38; 0.1 C, 5 Cr, 0.5 Mo, bal Fe.
Hardened: 157,000 TS; 110,000 YS; 17 El; 60 RA; 321 Brin.
For cold hubbed plastic molds; deep hardening.

AIROLITE.
M-1463; C, alloy, bal Fe.
For high temperature applications; heat resistant.

AIRPLANE BABBITT.
M-665; 16 Sn, 10 Sb, 72 Pb, 2 Cu, Bi.
For bearings; M.P. 600-700°F.

AIRQUE.
M-140; 1 C, 0.7 Mn, 5.25 Cr, 0.25 V, 1.15 Mo, bal Fe.
For gauges, punches, shear blades, dies; air hardened, non-deforming.

AIRQUE-V.
M-140; 1.25 C, 5.25 Cr, 1.15 Mo, 1.0 V, bal Fe.
For thread rolling dies, blanking and forming dies; air hardened, non- deforming.

AIR SERVICE ALUMINUM ALLOY NO. 1.
M-Eng.; 92 Al, 8 Cu.
For light alloy parts, airplane construction; non-hardenable.

AIR SERVICE ALUMINUM ALLOY NO. 2.
M-Eng.; 88.5 Al, 10 Cu, 0.25 Mg, 1.25 Fe.
For aircraft parts, light alloy parts; non-hardenable.

AIR SERVICE ALUMINUM ALLOY NO. 3.
M-Eng.; 95.75 Al, 2.5 Cu, 0.5 Mg, 1.25 Fe.
For aircraft parts; heat treatable.

AIR SERVICE ALUMINUM ALLOY NO. 4.
M-Eng.; 94 Al, 5 Cu, 1 Si.
For aircraft parts; age-hardenable.

AIR SERVICE ALUMINUM ALLOY NO. 5.
M-Eng.; 93 Al, 4 Cu, 3 Si.
For aircraft parts; age-hardenable.

AIR SERVICE ALUMINUM ALLOY NO. 6.
M-Eng.; 4 Cu, 1-5 Mg, 2 Ni, bal Al.
For aircraft parts; age-hardenable.

AIR SHOCK.
M-341; 0.50 C, 0.7 Mn, 3.25 Cr, 1.4 Mo, bal Fe.
Air or oil hardening.
For rivet sets, hot gripper dies, punches, chisels, plastic mold, die casting dies.
Type 57 shock resisting tool steel.

AIRTEM.
M-688; 1.3 C, 0.5 Mn, 5 Cr, 0.3 V, 1.25 Mo, 1.0 Si, bal Fe.
For dies, punches, knives; air hardened.

AIRTREAT A.H.
M-1423; C, alloy, bal Fe.
For dies; air hardened.

AIR-TRUE.
M-1015; 1.0 C, 3.0 Mn, 1.0 Cr, 1.0 Mo, bal Fe.
For dies, templates, gauges, punch and die facings; air hardening, non- deforming.

AIR-TRUE.
M-1703; 0.95-1.05 C, 0.65-0.75 Mn, 0.25-0.35 Si, 5.25 Cr, 0.20-0.25 V 1.0-1.2 M, bal Fe.
Air hardening, non-distorting steel; particularly for drill rod.
AISI A2.

AIRTRUE see ALSO SIMONDS AIRTRUE.

AIRTRUE LC.
M-373; 0.55 C, 5.0 Cr, 1.5 Mo, 1.25 W, 1.0 Si, bal Fe.
Cold work tool steel; shock resisting type.
AISI A8.

AIRVAN.
M-57; 1 C, 5.25 Cr, 1.15 Mo, 0.25 V, bal Fe.
Air hardened: 615 Brin.
For tools, cutters, cold work dies, shear blades, gauges, punches; air hardening, non-shrinking.

AIS.
M-1653; 0.06 max C, 1.0 max Si, 2.0 max Mn, 17.0-19.0 Cr, 8.0-11.0 Ni Fe.
Austenitic stainless steel.
W.-Nr. 1.4301; x 5 CrNi 18 10; AISI 304.

AISC.
M-1653; 0.08 max C, 1.0 max Si, 2.0 max Mn, 17.0-19.0 Cr, 9.0-12.0 Ni 10 x C min, bal Fe.
Stabilized, austenitic stainless.
W.-Nr. 1.4543; AISI 347.

AISL.
M-1653; 0.03 max C, 1.0 max Si, 2.0 max Mn, 17.0-19.0 Cr, 9.0-12.0 Ni Fe.
Low carbon austenitic stainless steel.
W.-Nr. 1.4306; x 2 CrNi 18 11; AISI 304L.

AIST.
M-1653; 0.08 max C, 1.0 max Si, 2.0 max Mn, 17.0-19.0 Cr, 9.0-12.0 Ni 5 x C min, bal Fe.
Stabilized austenitic stainless steel.
W.-Nr. 1.4541; AISI 321.

AJAX.
M-English; 70-30 Fe, 25-50 Ni, 5-20 Cu.
For bearings.

AJAX 1.
M-114; 1 C, 4 Cr, 0.5 V, 0.4 Mo, bal Fe.
For tools, dies, air hardened.

AJAX 2.
M-114; 1 C, 4 Cr, 0.5 V, 0.4 Mo, bal Fe.
For tools, forging mandrels; hot work steel.

AJAX DRILL ROD.
M-339; 1.1 C, bal Fe.
For drills, taps; water hardening.

AJAX DRILL ROD.
M-1703; 0.95-1.05 C, 0.25-0.35 Mn, 0.15-0.3 Si, bal Fe.
Water hardening tool steel; AISI W1.

AJAX T-M.
M-1673; 0.40 C, 0.5 Mn, 1.0 Si, 3.3 Cr, 2.5 Mo, 0.35 V, bal Fe.
For extrusion and forging dies, hot punches and shears.
Hot work die steel, tough, shock resistant.

AK 13.
M-1361; 4.0 Cu, 0.80 Mg, 0.90 Mn, bal Al.
375-410 N/mm^2 TS; 215-275 N/mm^2 YS; 8-14 El.
High strength heat treatable alloy.
Type AlCuMg 1.
Similar to AA 2017.

AK 15.
M-1361; 3.80 Cu, 0.70 Mg, 0.50 Mn, bal Al.
375-410 N/mm^2 TS; 215-275 N/mm^2 YS; 8-14 El.
Special quality for hard drawn tube, rods, bars and wire.
Type AlCuMg 1.
Similar to AA 2017.

AK 24.
M-1361; 4.20 Cu, 1.4 Mg, 0.80 Mn. bal Al.
410-470 N/mm^2 TS; 245-335 N/mm^2 YS; 6-12 El.
High strength heat treatable alloy.
Type AlCuMg 2.
Similar to AA 2024.

AK 34.
M-1361; 4.30 Cu, 0.60 Mg, 0.80 Si, 0.90 Mn, bal Al.
400-460 N/mm^2 TS; 335-385 N/mm^2 YS; 6-8 El.
High strength heat treatable alloy.
Type AlCuSiMn.
Similar to AA 2014.

AK 39.
M-1361; 4.30 Cu, 0.60 Mg, 0.80 Si, 0.60 Mn, bal Al.
400-430 N/mm^2 TS; 335-385 N/mm^2 YS; 6-8 El.
High strength heat treatable alloy, particularly suitable for hand forging.
Type AlCuSiMn.
Similar to AA 2014.

AKRIT.
M-U.S.; 38 Co, 30 Cr, 16 W, 10 Ni, 4 Mo, 25 C.
For tools, high speed cutting tips on lathe tools; corrosion and heat resistant.

AKRON.
M-Eng.; 63 Cu, 1 Sn, 36 Zn.
For fittings, hardware; corrosion resistant.

AL 6X.
M-1; 20.0 Cr, 1.5 Mn, 6.0 Mo, 24 Ni, bal Fe.
Alloy for service in chloride and other pitting environments.

AL-7.
M-1779; 0.50 C, 0.80 Mn, 0.25 Si, 3.25 Cr, 1.45 Mo, bal Fe.
General purpose air hardening die steel with toughness and resistance to softening by heat.
For blanking and forming dies, plastic mold dies, shear blades and punches usual hardness 54-57 Rc.
AISI 5-7.

A L 46.
M-1779; 1.24 C, 4.0 Cr, 3.15 V, 2.0 W, 8.25 Mo, 8.25 Co, bal Fe.
High speed steel, molybdenum-cobalt type; AISI M-46.

A L 75.
M-106; 2.8-3.8 Cu, 4-6 Cr, 0.3-0.6 Mn, bal Al.
Chill cast: 36,000-40,000 TS; 14,500-16,800 YS; 8-10 El; 70-80 Brin; 3.5 Izod.
Gravity die aluminum castings requiring strength and ductility.
Gen. Eng. BS 1490:LM 22-W.

A L 173.
M-1779; 0.40 C, 0.55 Mn, 1.0 Si, 3.3 Cr, 0.5 V, 2.5 Mn, bal Fe.
A hot work die steel capable of exceptionally good resistance to softening at elevated temperatures.
Air or oil hardening. AISI H10.

A L 600.
M-1; 0.15 max C, 14.0 Cr, 6.0 min Fe, 72.0 Ni.
Strengthened by cold work it provides excellent corrosion resistance at elevated temperatures.
UNS No. 6600.

A L 601.
M-1; 0.15 max C, 14.0 Cr, 72 Ni, Cb.
Solid solution with excellent high temperature properties.
UNS No. 6601.

A L-609.
  M-1779; 0.60 C, 0.8 Mn, 2.0 Si, 0.25 Cr, 0.2 V, 0.25 Mo, bal Fe.
  AISI Type S5 shock resisting tool steel.

A L 625.
  M-1; 0.10 max C, 20.0 Cr, 8.0 min Mo, 60 Ni.
  High strength and toughness from cryogenic temperatures to 200°F.
  UNS No. 6625.

A L 718.
  M-1; 0.08 max C, 17.0 Cr, 2.8 min Mo, 4.75 min Cb + Ta, 50.0 Ni, bal
  Age hardenable, high strength, corrosion resistant alloy.
  Similar to Inconel Alloy 718.
  UNS No. 7718.

A L 800.
  M-1, M-1779; 0.10 C, 19 Cr, 30 Ni, 0.6 max Al, 0.6 max Ti, bal Fe.
  Resistant to oxidation and carburization at elevated temperatures.
  UNS NO8800.

A L 825.
  M-1, M-1779; 0.05 max C, 19.5 Cr, 38.0 Ni, 3.5 max Mo, bal Fe.
  Excellent corrosion resistance.
  UNS NO8825.

AL-840 see ALLEGHENY 334.

ALACRITE NO. 52N.
  M-1115; 55 Co, 26 Cr, 7 W, 10 Ni, 0.5 C.
  For hard facing electrode; corrosion and wear resistant.

ALACRITE NO 165 ND.
  M-1115; 64 Co, 28 Cr, 5 Mo, 2 Ni, 0.3 max C, 2 max Fe.
  For hard facing electrode; corrosion and wear resistant.

ALACRITE NO 502.
  M-1115; 55 Co, 30 Cr, 13 W, 1.0 C, 2 max Fe.
  For hard facing electrode; corrosion and wear resistant.

ALACRITE NO 554.
  M-1115; 60 Co, 30 Cr, 8 W, 1.8 Fe, 2.0 C.
  For hard facing electrode; corrosion and wear resistant.

ALACRITE NO 602.
  M-1115; 66 Co, 27 Cr, 5 W, 1.0 C, 2 max Fe.
  For hard facing electrode; corrosion and wear resistant.

ALADAR.
  M-Eng.; 0.13 Cu, 0.40 Fe, 12.5 Si, 0.09 Mn, bal Al.
  For light alloys, castings; same as "Alpax", "Alcoa No. 47", and "Silumen."

ALADDIN.
  M-666; Al, Cu, bal Zn.
  For welding rod; for white metal die castings.

ALAIS 4N2.
  M-678; 0.07-0.13 C, 1.3-2.3 Ni, 0.5 Mn, bal Fe.
  Annealed: 64,000 TS; 47,000 YS; 25 El.
  Hardened: 121,000 TS; 99,500 YS; 12 El.
  For exhaust valves, gears, shafts; case-hardening steel.

ALAIS 5C1.
  M-678; 0.09-0.15 C, 0.6-1.0 Cr, 0.6-0.9 Mn, bal Fe.
  Annealed: 71,100 TS; 42,600 YS; 20 El.
  Hardened: 114,000 TS; 78,000 YS; 12 El.
  For shafts, axles; case-hardening steel.

ALAIS 5N2.
  M-678; 0.18-0.25 C, 1.3-2.3 Ni, 0.5 Mn, bal Fe.
  Annealed: 78,000 TS; 60,000 YS; 20 El.
  Hardened: 160,000 TS; 128,000 YS; 8 El.
  For gears, shafts; case-hardening steel.

ALAIS 5NC3.
  M-678; 0.08-0.13 C, 0.6-0.9 Cr, 2.5-3.0 Ni, 0.35-0.65 Mn, bal Fe.
  Annealed: 71,000 TS; 57,000 YS; 25 El.
  Hardened: 140,000 TS; 102,000 YS; 12 El.
  For shafts, gears; tough, case-hardening.

ALAIS 5 1/2C1.
  M-678; 0.25-0.30 C, 0.8-1.2 Cr, 0.6-0.9 Mn, bal Fe.
  Annealed: 78,000 TS; 48,400 YS; 16 El.
  Hardened: 128,000 TS; 85,300 YS; 10 El.
  For tools, shafts, gears; case-hardening steel.

ALAIS 5 1/2NCD1.
  M-678; 0.07-0.13 C, 0.4-0.7 Cr, 1-1.3 Ni, 0.1-0.2 Mo, 0.5-0.9 Mn, bal Fe.
  Annealed: 78,000 TS.
  Hardened: 140,000 TS; 100,000 YS; 10 El.
  For shafts, gears; tough, case-hardening.

ALAIS 6CD1.
  M-678; 0.08-0.16 C, 0.8-1.2 Cr, 0.15-0.30 Mo, 0.6-0.9 Mn, bal Fe.
  Annealed: 78,000 TS; 50,000 YS; 25 El.
  Hardened: 170,000 TS; 114,000 YS; 8 El.
  For shafts, gears; case-hardening steel.

ALAIS 6 1/2.
  M-678; 0.38 C, bal Fe.
  Rolled: 85,000-121,000 TS; 57,000-83,000 YS; 16-12 El.
  For shafts, tools; water hardening.

ALAIS AE4.
  M-678; 0.1 C, bal Fe.
  Rolled: 50,000-60,000 TS; 35,000 YS; 30 El.
  For forgings, shafts; weldable.

ALAIS AE4 1/2.
M-678; 0.12 C, bal Fe.
Rolled: 57,000-72,000 TS; 40,000 YS; 22 El.
For forgings, shafts; weldable.

ALAIS AE5.
M-678; 0.18 C, bal Fe.
Rolled: 64,000-100,000 TS; 37,000-60,000 YS; 23-15 El.
For forgings, axles, shafts; case hardening.

ALAIS AE6.
M-678; 0.32 C, bal Fe.
Rolled: 78,000-114,000 TS; 50,000-79,000 YS; 20-13 El.
For shafts, axles; water hardening.

ALAIS AES7.
M-678; 0.4-0.5 C, 1.6-2.1 Si, 0.4-0.8 Mn, bal Fe.
Annealed: 99,500 TS; 60,000 YS; 20 El.
Hardened: 210,000 TS; 160,000 YS; 5 El.
For tools, dies; shock resistant.

ALAIS AES8.
M-678; 0.52-0.58 C, 1.5-1.9 Si, 0.5-0.9 Mn, bal Fe.
Annealed: 120,000 TS; 64,000 YS; 15 El.
hardened: 200,000 TS; 160,000 YS; 6 El.
For tools, springs, dies; shock resistant.

ALAIS HC3.
M-678; 0.10 C, bal Fe.
Rolled: 47,000-72,000 TS; 34,000-43,000 YS; 34-30 El.
For case hardened parts; carburized.

ALAIS HC4.
M-678; 0.12 C, bal Fe.
Rolled: 57,000-85,000 TS; 28,000-47,000 YS; 30-22 El.
For case hardened parts; carburized.

ALAIS HCC1.
M-678; 0.09-0.15 C, 0.6-1.0 Cr, 0.6-0.9 Mn, bal Fe.
Annealed: 74,000 TS; 50,000 YS; 22 El.
Hardened: 93,000 TS; 78,000 YS; 20 El.
For gears, shafts; for carburized parts.

ALAIS HCCD1.
M-678; 0.08-0.16 C, 0.8-1.2 Cr, 0.15-0.30 Mo, 0.6-0.9 Mn, bal Fe.
Annealed: 78,000 TS; 50,000 YS; 25 El.
Hardened: 170,000 TS; 114,000 YS; 8 El.
For gears, shafts; case-hardening steel.

ALAIS HCN2.
M-678; 0.07-0.13 C, 1.3-2.3 Ni, bal Fe.
Annealed: 60,000 TS; 45,500 YS; 30 El.
Hardened: 99,500 TS; 71,200 YS; 26 El.
For gears, shafts; case-hardening steel.

ALAIS HCN5.
M-678; 0.07-0.13 C, Ni, bal Fe.
Annealed: 72,000 TS; 57,000 YS; 23 El.
Hardened: 160,000 TS; 128,000 YS; 10 El.
For gears, shafts; case-hardening steel.

ALAIS HCNC1 1/2.
M-678; 0.11-0.18 C, 0.8-1.2 Cr, 1.2-1.6 Ni, bal Fe.
Hardened: 170,000-200,000 TS; 140,000 YS; 6 El.
For gears, shafts; case-hardening steel.

ALAIS HCNCD1.
M-678; 0.07-0.13 C, 0.20 Cr, 0.8-1.2 Ni, 0.1-0.2 Mo, 0.4 Mn, bal Fe.
Annealed: 78,000 TS.
Hardened: 155,000 TS; 99,500 YS; 10 El.
For gears, shafts; case-hardening steel.

ALAR 00.12.
M-137, M-1272; 0.4 C, 0.15 Mg, 10-13 Si, 0.7 Fe, 0.5 Mn, 0.1 Ni, 0.2 Zn, 0.2 Ti, bal Al.
Sand cast: 23,400 TS; 8000 YS; 3.5 El; 55 Brin.
Permanent mold: 27,000 TS; 9000 YS; 5 El; 60 Brin.
For general purpose light castings; corrosion resistant, good castability.

ALAR 00.5.
M-137, M-1272; 0.1 Cu, 0.1 Mg, 4.5-6.0 Si, 0.6 Fe, 0.5 Mn, 0.1 Ni, bal Al.
Sand cast: 17,000 TS; 8,000 YS; 3 El; 40 Brin.
Permanent mold: 20,200 TS; 9000 YS; 4 El; 50 Brin.
For general purpose light castings; good castability, corrosion resistant.

ALAR 21.
M-1272; 2-4 Cu, 4-6 Si, 0.8 Fe, 0.4 Mn, 0.3 Zn, 0.2 Ti, bal Al.
Sand cast: 20,200 TS; 89,000 YS; 2 El; 65 Brin.
Permanent mold: 22,400 TS; 11,200 YS; 2 El; 70 Brin.
For crankcases, gear cases, covers, housings; age-hardenable, good castability.

ALARGAN.
M-Eng.; Ag-Al.

ALASKA WHITE BRASS.
M-U.S.; Zn, Ni, bal Cu.
For bearings, bushings; heavy duty.

ALA-ZE 63.
M-1639; 5-6 Zn, 2-3 rare earth metals, 0.4-1.0 Zr, bal Mg.
Sand cast: 45,000 TS; 25,000 YS; 8 El.
For general purpose castings, oil pans, cases, housings.
Hydrogen heat treated, high strength and good ductility, pressure tight.

ALBA.
M-296; 30 Pd, 60 Ag, 5 Au, and other elements.
For fountain pen points, dental industry; corrosion resistant, heat treatable.

ALBALOY.
M-580; 55 Cu, 30 Sn, 15 Zn.
For bright alloy plating; corrosion resistant.

ALBATRA METAL.
M-U.S.; 57.5 Cu, 22.5 Zn, 18.75 Ni, 1.25 Pb.
For automobile trimmings, hardware, fittings, (Albata).

ALBIDUR-ALUMINUM.
M-Eng.; Al alloy.
For light alloy parts; non-hardenable.

ALBIN.
M-Eng.; Zn, Ni, bal Cu.
For fixtures, hardware, fittings; white brass.

ALBION HIGH SPEED.
M-340; 0.4 C, 0.05 Mn, 2.3 Cr, 15.54 W, bal Fe.
For tools, high speed cutters; high speed steel.

ALBION SPECIAL.
M-340; 0.7 C, 18 W, 4 Cr, 1 V, bal Fe.
For high speed cutting tools; high speed steel.

ALBION TOOL.
M-340; 1.27 C, 0.33 Mn, bal Fe.
For tools, drills, reamers; water hardening.

ALBONDUR.
M-116; 1.5 max Mn, 1.0 max Mg, 1.2 max Si, bal Al.
For tanks, furniture, marine parts; good formability and weldability.

ALBONDUR.
M-Germany; 3.5-5.5 Cu, 0.3-0.5 Si, 0.3-1.0 Mn, 0.2-0.7 Mg, bal Al.
Heat treated: 74,000-80,000 TS; 57,000-63,000 YS; 8-12 El; 130-150 Brin.
For general purpose applications.
Age hardenable, high strength.

ALBOR.
M-261; 0.9 C, 1.2 Cr, bal Fe.
For dies for cold stamping white gold and all hard metals; oil hardening.

ALBOR DIE.
M-261; 0.85 C, 0.7 Cr, 0.2 Mo, bal Fe.
For coining and medal dies, cutlery dies; deep hardening.

ALBRO.
M-593; 10 Al, 1 Fe, bal Cu.
For acid resisting castings; corrosion resistant.

ALCALOY.
M-Eng.; 9-11 Al, bal Cu.
Cast: 85,000 TS; 41,000 YS; 20 El; Rockwell B50-80.
For safety tools, chains, fittings, hand tools; corrosion resistant, Al-bronze.

ALCALOY.
M-1276; 87-89 Cu, 9-11 Al, 1-2 Fe, 0.1-0.5 Be.
Cast: 85,000 TS; 41,000 YS; 20 El; 83-150 Brin.
For safety tools, chains, chain fittings, hand tools; non-sparking, corrosion resistant.

ALCAN 1S.
M-219; 99.5 Al, 0.02 max Cu, 0.15 max Si, 0.15 max Fe, 0.03 max Mn, 0.06 max Zn, others 0.2 max.
Wrought: O Temper: 7500-13,500 TS; 29-35 El; 18 Brin.
H4 Temper: 13,500-17,000 TS; 5-8 El; 33 Brin.
H8 Temper: 18,000 TS; 3-5 El; 44 Brin.
For food and chemical plant equipment, food containers, atomic energy, compression accessories for electrical conductors. Available as sheet and forgings. Not heat treatable.

ALCAN 2S.
M-219; 99.0 Al, 0.10 max Cu, 0.5 max Si, 0.7 max Fe, 0.1 max Mn, 0.1 max Zn. others 1.0 max.
Sheet, plate extrusions, tube; not heat treatable.
O Temper: 10,000-15,000 TS; 20-30 El; 21 Brin.
H4 Temper: 15,000-20,000 TS; 3-5 El; 36 Brin.
H6 Temper: 17,800-21,300 TS; 2-4 El; 42 Brin.
H8 Temper: 20,000 TS; 2-4 El; 46 Brin.
For paneling and mouldings, lightly stressed and decorative assemblies, food and brewing equipment.

ALCAN 3S.
M-219; 0.8-1.5 Mn, 0.1 max Cu, 0.1 max Mg, 0.7 max Fe, 0.2 max Zn, ba Al.
Sheet: 0: 12,800-18,500 TS; 20-24 El; 27 Brin.
H4: 20,000-25,000 TS; 3-6 El; 46 Brin.
H6: 22,800-27,700 TS; 2-4 El; 50 Brin.
H8: 25,000 TS; 2-4 El; 57 Brin.
For building sheet, vehicle paneling and sheet metal work, packaging, hollow-ware, not heat treatable.

ALCAN 16S.
M-219; 2.5 Cu, 0.7 max Fe, 0.3 Mg, 0.7 max Si, bal Al.
S-T Temper: 43,000 TS.
For aircraft and structural rivets; heat treatable.

ALCAN 17S.
M-219; 3.8-4.5 Cu, 0.5-0.8 Mg, 0.5-0.7 Mn, 0.5 max Fe, bal Al.
Forgings: TB Cond; 67,000 TS; 31,000 YS; 14 El; 31,000 SS; 120 Brin.
For stressed parts in aircraft and other structures; heat treatable.

ALCAN 18S.
M-219; 4 Cu, 0.6 Mg, 0.9 max Si, bal Al.
S-T Temper: 55,000 TS; 40,000 YS; 10 El; 100 Brin.
For forgings for light temp service; heat treatable.

ALCAN 24S.
M-219, M-643; 3.8-4.9 Cu, 1.2-1.8 Mg, 0.5 max Si, 0.5 max Fe, 0.3-0.9 Mn, 0.2 max Zn, bal Al.
Sheet and plate, heat treatable.
Sheet: O: 31,000 max TS.
TB: 61,000 TS; 40,000 YS; 120 Brin.
For all types of stressed components in aircraft and general engineering.

ALCAN 24S ALCLAD.
M-219, M-643; 3.8-4.9 Cu, 1.2-1.8 Mg, 0.5 max Si, 0.5 max Fe, 0.3-0.9 Mn, 0.2 max Zn, bal Al.
Sheet and plate, heat treatable.
Sheet: 0: 31,000 max TS.
TB: 61,000 TS; 40,000 YS; 15 El.
For all types of stressed components in aircraft and general engineering.

ALCAN 26S.
M-219; 3.9-5.0 Cu, 0.2-0.8 Mg, 0.5-1.0 Si, 0.7 max Fe, 0.4-1.2 Mn, 0. max Zn, 0.1 max Cr, bal Al.
Plate, extrusions and forgings, heat treatable.
Plate: TB: 54,700 TS; 34,800 YS; 10-14 El.
Ht. Tr. to TF: 62,500 TS; 54,000 YS; 6-9 El.
For stressed components of all types in aircraft and general engineering.

ALCAN 28S.
M-219; 5.6 Cu, 0.2-0.6 Pb, 0.2-0.6 Bi, 0.7 max Fe, 0.4 max Si, 0.3 ma Zn, bal Al.
Extrusions: 41,000-44,000 TS; 27,000-35,000 YS; 6 El; 25,000-26,000 SS. (as heat treated).
For repetition machined parts; free machining screw machined parts; heat treatable.

ALCAN 50S.
M-219; 0.10 max Cu, 0.4-0.9 Mg, 0.3-0.7 Si, 0.40 max Fe, 0.2 max Zn, bal Al.
Extrusions, tube, forgings, heat treatable.
Extrusions: Ht. Tr. to TF cond: 26,000 TS; 22,700 YS; 7-8 El; 75 Brin.
Architectural members as window frames, window-screen sections, road transport.

ALCAN 55S.
M-219; 1.2 Mg, 0.3 Cr, bal Al.
For rivets for aircraft; S.S. 200000.

ALCAN 56S.
M-219; 5 Mg, 0.1 Mn, 0.1 Cr, bal Al.
S-O Temper: 42,000 TS; 20,000 YS; 35 El.
S-H Temper: 58,000 TS; 48,000 YS; 15 El.
For rivets, wire products; resists sea water corrosion.

ALCAN 61S.
M-219; 0.3 max Cu, 0.6 Mg, 1 Si, 0.3 Cr, bal Al.
S-T temper: 44,000 TS; 34,000 YS; 14 El; 90 Brin.
For automotove and aircraft forgings; heat treatable.

ALCAN 65S.
M-219; 0.3 Cu, 1 Mg, 0.6 Si, 0.3 max Cr, bal Al.
S-O Temper: 18,000 TS; 8000 YS; 22 El; 30 Brin.
S-W Temper: 35,000 TS; 21,000 YS; 22 El; 65 Brin.
S-T Temper: 45,000 TS; 40,000 YS; 12 El; 95 Brin.
For structures; heat treatable, corrosion resistant.

ALCAN 75S.
M-219; 1.5 Cu, 2.5 Mg, 5.6 Zn, 0.3 Cr, 0.2 max Ti, bal Al.
S-O Temper: 33,000 TS; 15,000 YS; 17 El.
S-T Temper: 82,000 TS; 72,000 YS; 11 El; 150 Brin.
For aircraft, structures; heat treatable.

ALCAN 75S ALCLAD.
M-219; 1.5 Cu, 2.5 Mg, 5.6 Zn, 0.3 Cr, 0.2 max Ti, bal Al.
S-O Temper: 32,000 TS; 14,000 YS; 17 El.
S-T Temper: 76,000 TS; 67,000 YS; 11 El.
For aircraft, structures; heat treatable corrosion resistant.

ALCAN 99.8.
M-219; 99.8 Al, total all others 0.2 max.
Sheet: 0: 13,000 TS max; 16 Brin.
H4: 13,500-17,000 TS; 28 Brin.
H8: 18,000 TS; 40 Brin.
For fully supported roofing and flashing, collapsible tubes, atomic energy equipment; ann. material is very ductile, malleable and durable. Not heat treatable.

ALCAN 100.
M-219; 99.5 Al, total all others 0.50 max.
Casting: 8500 TS; 24 Brin.
For unstressed castings in food and chemical industries and electrical equipment.

ALCAN 117.
M-219; 3 Cu, 0.1 Mg, 0.5 Mn, 4.5 Si, bal Al.
Cast: 25,000 TS; 16,000 YS; 3-5 El; 50 Brin.
Cast: 29,000 TS; 17,000 YS; 4 El; 65 Brin.
For sand and permanent mold castings; for medium stresses.

ALCAN 123.
M-219; 5 Si, 0.2 max Ti, bal Al.
Cast: 19,000 TS; 9000 YS; 6 El; 40 Brin.
Cast: 24,000 TS; 9000 YS; 10 El; 40 Brin.
For sand and permanent mold castings; for thin sections.

ALCAN 125.
M-219; 2.0-4.0 Cu, 4.0-6.0 Si, 0.8 max Fe, 0.3-0.7 Mn, bal Al.
Cast: 23,000 TS; 15,000 YS; 2 El; 85 Brin.
Cast and aged 40,000 TS; 115 Brin.
General purpose castings for light weight and low cost; gear boxes, clutch housings, valve bodies, radiators; non-solution heat treatable.

ALCAN 160.
M-219; 0.1 Cu, 0.1 max Mg, 10.0-13.0 Si, 0.6 max Fe, 0.5 max Mn, 0.1 Ni, 0.1 Zn, 0.2 Ti, bal Al.
Cast: 26,000 TS; 10,500 YS; 6 El; 60 Brin.
Not heat treatable; for large castings, general purpose, marine and electrical, valve bodies, radiators, grids and grills.

ALCAN 160N.
M-219; 12 Si, bal Al.
Sand cast: 25,000 TS; 12,000 YS; 8 El; 50 Brin.
For castings; high fluidity.

ALCAN 160X.
M-219; 12 Si, bal Al.
Die cast: 37,000 TS; 18,000 YS; 2 El.
For die castings, instrument cases; for thin wall sections.

ALCAN 162.
M-219; 1 Cu, 1 Mg, 12 Si, bal Al.
Permanent mold: 36,000 TS; 28,000 YS; 0.5 El; 105 Brin.
For permanent mold castings, pistons; low coef. expansion.

ALCAN 225.
M-219; 4.5 Cu, 1.5 max Si, 0.2 max Ti, bal Al.
T-22 Temper: 39,000 TS; 28,000 YS; 5 El; 80 Brin.
For sand castings; stressed parts, heat treatable.

ALCAN 236.
M-219; 0.65-0.80 Cu, 0.95-1.4 Fe, 1-3 Si, 0.2 max Ti, bal Al.
Cast: 22,000 TS; 14,000 YS; 2 El; 65 Brin.
For sand castings, good machinability.

ALCAN 250.
M-219; 10 Cu, 1.5 max Fe, 0.2 Mg, 0.6 max Si, 0.2 max Ti, bal Al.
A62 Temper: 25,000 TS; 20,000 YS; 1 El; 75 Brin.
T25 Temper: 36,000 TS; 30,000 YS; 1 El; 100 Brin.
For engine parts; heat treatable.

ALCAN 340.
M-219; 8 Mg, bal Al.
Die cast: 42,000 TS; 23,000 YS; 7 El.
For die castings; corrosion resistant.

ALCAN 350.
M-219; 10 Mg, bal Al.
Cast T4: 47,000 TS; 26,000 YS; 17 El; 75 Brin.
For sand castings; heat treatable, corrosion resistant.

ALCAN 1050A.
M-643; 99.50 min Al.
Same as ALCAN GB-1S.
Similar to AA 1050.

ALCAN 1080A.
M-643; 99.80 min Al.
Similar to AA 1080.

ALCAN 2017A.
M-643; 0.5 Si, 0.7 Fe, 4.0 Cu, 0.7 Mn, 0.6 Mg, bal Al.
Wrought, heat treatable, high strength.
Similar to AA 2017.

ALCAN 2618A.
M-643; 2.5 Cu, 1.5 Mg, 1.2 Ni, 0.2 Si, 1.0 Fe, bal Al.
Same as ALCAN GB-D19S.
Similar to AA 2618.

ALCAN 3103.
M-643; 1.2 Mn, 0.5 Fe, bal Al.
Same as ALCAN GB-3S.

ALCAN 5083.
M-643; 4.3 Mg, 0.6 Mn, 0.13 Cr, bal Al.
Same as ALCAN BG-D54S.

ALCAN 5251.
M-643; 2.1 Mg, 0.3 Mn, bal Al.
Same as ALCAN GB-M57S.

ALCAN 6082.
M-643; 0.8 Mg, 1.0 Si, 0.7 Mg, bal Al.
Same as ALCAN GB-B51S.

ALCAN A56S.
M-219; 5 Mg, 0.3 Mn, bal Al.
F-temper: 38,500 TS; 23,000 YS; 28 El; 70 Brin.
For shipbuilding, structural members; extrusions, good formability.

ALCAN A111.
M-219; 1.2 Cu, 1 Fe, 1 Ni, 2.5 Si, bal Al.

ALCAN A-143.
M-219; 8.5-10.5 Si, 2.0-4.0 Cu, 0.5-1.5 Mg, 0.5-1.5 Ni, 1.0 max Fe, 0.5 max Mn, 0.5 max Zn, 0.2 max Ti, bal Al.
Cast: heat treated: 30,000 TS; 23,000 YS; 105 Brin.
Low thermal expansion, good strength and frictional properties; mainly for automotive pistons.

ALCAN A 320.
M-219; 4 Mg, 0.5 Si, 0.2 max Ti, bal Al.
Cast: 25,000 TS; 12,000 YS; 9 El; 50 Brin.
For architectural usage, ornamental and marine parts; sand casting, corrosion resistant.

ALCAN B26S.
M-219; 3.9-5.0 Cu, 0.2-0.8 Mg, 0.5-1.0 Si, 0.7 max Fe, 0.4-1.2 Mn, 0.2 max Zn, bal Al.
Sheet: 0: 25,500 TS; 18-20 El.
Ht. Tr. to TF cond: 61,000 TS; 53,000 YS; 6-7 El; 145 Brin.
For stressed components of all types in aircraft and general engineering. Heat treatable.

ALCAN B26S ALCLAD.
M-219, M-643; 3.9-5.0 Cu, 0.2-0.8 Mg, 0.5-1.0 Si, 0.7 max Fe, 0.4-1.2 Mn, 0.2 Max Zn, 0.1 max Cr, bal Al.
Sheet: 0: 24,000 TS; 18-20 El (est).
Ht. Tr to TF Cond: 57,000 TS; 46,000 YS; 7-8 El.
For parts requiring good strength plus good corrosion resistance; heat treatable.

ALCAN B51S.
M-219; 0.10 Cu, 0.5-1.2 Mg, 0.7-1.3 Si, 0.5 max Fe, 0.4-1.0 Mn, 0.2 max Zn, 0.25 max Cr, bal Al.
Sheet, plate, extrusions, tube, forgings.
Plate: Ht. Tr. to TF Cond: 42,000 TS; 34,000 YS; 8 El; 95 Brin.
Medium strength, good properties, heat treatable.
For structural applications as bridges, cranes, transport equipment, roof trusses.

ALCAN B53S.
M-219; 0.10 Cu, 3.1-3.9 Mg, 0.5 max Si, 0.5 max Fe, 0.5 max Mn, 0.2 max Zn, 0.25 max Cr, bal Al.
Sheet, plate, extrusions, tube, forgings.
Plate, rolled: 30,000 TS approx; 75 Brin.
Not heat treatable, used in as-rolled or as-forged for road and rail transport, pressure vessels; good weldability and corrosion resistance.

ALCAN B54S.
M-219; 4.4 Mg, 0.3 Mn, bal Al.
F-temper: 43,000 TS; 31,000 YS; 16-20 El, 70 Brin.
For shipbuilding, structural members; good formability.

ALCAN C1S.
M-219, M-643; 99.5 Al, 0.04 max Cu, 0.1 max Si, 0.4 max Fe.
Extrusions: Not heat treatable.
M Temper: 8500 TS; 25 El.
H2 Temper: 12,000 TS; 15 El.
For electrical conductors; bus bar, overhead lines, windings, insulated cable.

ALCAN C50S.
M-219; 0.45-0.85 Mg, 0.2-0.6 Si, 0.05-0.2 Cu, 0.15 max Fe, bal Al.
Extrusions: Sol. Tr. TB Cond: 17,800 TS; 16 El.
Ht. Tr. to TF Cond: 26,000 TS; 22,700 YS; 10 El.
Heat treatable.
Excellent formability, takes good finish.
For motor car trim and other applications requiring bright finish even with anodizing.

ALCAN C75S.
M-219; 5.2-6.2 Zn, 2.2-3.2 mg, 0.3-0.7 Cu, 0.3-0.7 Mn 0.5 max Si, 0.5 max Fe, 0.3 max Ti, 0.1 max Ni, bal Al.
Forging: TF: 70,000 TS; 64,000 YS; 6 El; 160 Brin.
Heat treatable; for aircraftr and other structures where strength/weight ratio is of prime importance.

ALCAN D19S.
M-219; 2.5 Cu, 1.5 Mg, 1.3 Ni, 1.0 Fe, bal Al.
Plate and forgings.
Plate: Ht. Tr. to TF Cond: 61,000 TS; 54,000 YS; 5 El.
Heat treatable.
Aircraft structural parts operating up to 200°C; good strength at 200°C.

ALCAN D50S.
M-219; 0.04 Cu, 0.4-0.9 Mg, 0.3-0.7 Si, 0.5 Fe, bal Al.
Extrusions: Ht. Tr. to TF Cond: 26,000 TS; 22,700 YS; 7-8 El.
Best combination of electrical conductor and mechanical prop; 55% conductivity. Heat treatable. For bus bars, electrical conductors and fittings.

ALCAN D54S.
M-219; 4.0-4.9 Mg, 0.4 max Si, 0.4 max Fe, 0.5-1.0 Mn, 0.2 max Zn, 0.25 max Cr, bal Al.
Sheet, plate, extrusions, forgings.
Rolled plate: 40,000 TS approx; 10-12 El; 65-70 Brin.
Not heat treatable, used in road transport equipment and in ship building; weldable, good resistance to marine atmospheres.

ALCAN D135.
M-219; 0.1 max Cu, 0.20-0.45 Mg, 6.5-7.5 Si, 0.5 max Fe, 0.1 max Ni, 0.3 max Mn, bal Al.
Cast: 22,700 TS; 12,000 YS; 3 El; 60 Brin.
Ht. Tr. to TF; 40,000 TS; 35,000 YS; 2 El; 95 Brin.
Castings for engine components, cylinder blocks, chemical and food equipment; heat treatable.

ALCAN GB-D135.
M-643; Al-Si7 Mg.
Aluminum alloy casting.
As cast, (M): 23,000 TS; 12,000 YS; 60 Brin.
Heat treated(TF): 40,000 TS; 36,000 YS; 115 Brin.
For engine cylinder blocks, chemical and food equipment, many uses.

ALCAN GB-1S.
M-643; 99.5 Al.
Higher purity aluminum.
For deep drawn containers or spun components.

ALCAN GB-2S.
M-643; 99.0 Al.
Commercially pure aluminum.
For panelling, cooking utensils.

ALCAN GB-3S.
M-643; 1.2 Mn, 0.5 Fe, bal Al.
O temper: 90 MN/m^2 TS; 20-25 El.
For cooking utensils, containers, roofing sheet.

ALCAN GB-24S.
M-643; 4.2 Cu, 1.5 Mg, 0.6 Mn, bal Al.
TB (Sol. Tr.): 440 MN/m^2 TS; 305 MN/m^2 YS; 12 El.
High strength alloy with good elongation, used mainly for aircraft structures.

ALCAN GB-50S.
M-643; 0.6 Mg, 0.4 Si, bal Al.
Extruded, heat treated: 150-185 MN/m^2 TS; 130-160 MN/m^2 Y El.
For window frames, architectural sections.

ALCAN GB-51S.
M-643; 0.8 Mg, 1.0 Si, bal Al.
Extruded, TF(STA): 300 MN/m^2 TS; 240 MN/m^2 YS; 7-8.
Medium strength, good corrosion resistance, good surface finish.

ALCAN GB-54S.
M-643; 3.5 Mg, 0.5 Mn, bal Al.
H2: 245 MN/m^2 TS; 165 MN/m^2 YS; 5-8 El.
Suitable for marine and low temperature applications.

ALCAN GB-65S.
M-643; 1.0 Mg, 0.6 Si, 0.25 Cu, 0.25 Cr, bal Al.
H4: 160 MN/m^2 TS; 185 MN/m^2 YS; 5 El.
HF(STA): 295 MN/m^2 TS; 225 MN/m^2 YS; 7-9 El.
Values listed are for drawn tubes, since use is mainly for tubular furniture.

ALCAN GB-B26S.
M-643; 4.6 Cu, 0.7 Mg, 0.7 Si, 0.7 Mn, bal Al.
TF(STA): 280-310 MN/m^2 TS; 240-270 MN/m^2 YS; 5-8 El.
Strongest general utility alloy; for aircraft and structural applications.

ALCAN GB-B51S.
M-643; 0.8 Mg, 1.0 Si, 0.7 Mg, bal Al.
TF(STA); 295 MN/m^2 TS; 240 MN/m^2 YS; 8 El.
Good strength for structural purposes, as bridges, vehicle structures, cranes.

ALCAN GB-B53S.
M-643; 2.7 Mg, 0.75 Mn, 0.12 Cr, bal Al.
H4: 280 MN/m^2 TS; 205 MN/m^2 YS; 6 El.
Good corrosion resistance; for chemical and transport industries.

ALCAN GB-B57S.
M-643; 0.8 Mg, 0.6 Fe, bal Al.
H4: 160 MN/m^2 TS; 140 MN/m^2 YS; 8 El.
Medium strength for general metalworking.

ALCAN GB-B79S.
M-643; 3.2 Mg, 0.6 Cu, 4.2 Zn, bal Al.
Very high strength alloy.
Similar to AA 7079.

ALCAN GB-D19S.
M-643; 2.5 Cu, 1.5 Mg, 1.2 Ni, 0.20 Si, 1.0 Fe, bal Al.
TF(STA): 402-417 MN/m^2 TS; 309-324 MN/m^2 YS; 6-9 El
High strength alloy for use up to 300°C.

ALCAN GB-D50S.
M-643; 0.5 Mg, 0.5 Si, bal Al.
For bus bar, electrical conductors.
Similar to AA 6101.

ALCAN GB-D54S.
M-643; 4.3 Mg, 0.60 Mn, 0.13 Cr, bal Al.
H4: 345 MN/m^2 TS; 270 MN/m^2 YS; 4-8 El.
Strongest of common non-heat treatable alloys with good corrosion resistance.
For ship building, low temperatures. Weldable.

ALCAN GB-D75S.
M-643; 0.6 Cu, 2.9 Mg, 5.8 Zn, 0.4 Mn, bal Al.
TF(STA): 494-540 MN/m^2 TS; 417-463 MN/m^2 YS; 5-7 El
Very high strength alloy; for aircraft and structural applications.

ALCAN GB-E74S.
M-643; 2.6 Mg, 3.9 Zn, bal Al.
TF(STA): 390 MN/m^2 TS; 320 MN/m^2 YS; 10 El.
Weldable medium strength alloy.

ALCAN GB-L57S.
M-643; 0.80 Mg, bal Al.
For motor car trim and hollow ware.
Similar to AA 5657.

ALCAN GB-M57S.
M-643; 2.1 Mg, 0.3 Mn, bal Al.
H6: 225 MN/m^2 TS; 175 MN/m^2 YS; 3-5 El.
Medium strength alloy; weldable; for panelling on marine and road vehicles, and containers.

ALCAN L57S.
M-219; 99.8 A, 0.8 Mg.
Sheet: 0: 22,000 max TS.
H4: 22,000-29,000 TS.
H6: 26,000-33,000 TS.
H8: 30,000 TS.
Excellent formability; takes good finish; For motor car trim and other applications requiring a bright finish; hollow ware; not heat treatable.

ALCAN M57S.
M-219; 0.1 Cu, 1.7-2.4 Mg, 0.5 max Si, 0.5 max Fe, 0.5 max Mn, 0.2 max Zn, 0.25 max Cr, bal Al.
Sheet, plate.
Plate: 0: 22,700-28,500 TS; 8500 YS; 18-20 El.
H4: 32,000-39,000 TS; 25,000 YS; 3-5 El.
Not heat treatable, good weldability and corrosion resistance, used in panelling and structures for boats, road transport, and aircraft.

ALCAN M75S.
M-219; 5.3-6.5 Zn, 2.2-3.2 Mg, 0.3-1.5 Cu, 0.3 Mn, 0.1 Ni, 0.5 Si, 0.5 max Fe, 0.3 max Ti, bal Al.
Sheet and plate, heat treatable.
Sheet: 0: 27,000 TS; 18-20 El; 15,6000 SS.
Ht. Tr. to TF: 70,000 TS; 61,000 YS; 8 El; 40,000 SS.
Very high strength; for aircraft and other structures where strength/ weight ratio is of prime importance.

ALCAN M75S ALCLAD.
M-219; 5.3-6.5 Zn, 2.2-3.2 Mg, 0.3-1.5 Cu, 0.3 Mn, 0.1 Ni, 0.5 Si, 0.5 max Fe, 0.3 max Ti, bal Al.
Sheet and plate, heat treatable, coated with corrosion resistant mat.

ALCET.
M-1698.
High temperature refractory material, sintered; for aluminum vapor deposition crucibles and other refractory applications.

ALCHROME 3.
M-61; 14 Cr, 3 Al, bal Fe.
For resistors, heating elements; operating below 1500°F.

ALCHROME 750.
M-61; 15.0 Cr, 4.5 Al, bal Fe.
For resistors, low cost heating elements.
Max operating temperature 1900°F.

ALCHROME-D.
M-61; 79.5 Fe, 15 Cr, 5.5 Al.
Wire: Specific resistivity: 137 micro ohm/cm; temperature coefficient: $\pm 20 \times 10^{-6}$ ohm/ohm/°C.
For resistance elements; high specific electrical resistance.

ALCHROME DK.
M-61; 20.0 Cr, 4.5 Al, 0.5 Co, bal Fe.
For resistors, heating elements.
Max operating temperature 2350°F.

ALCO.
M-114; 0.5 C, 1 Si, 1.7 Cr, 2.2 W, 0.2 V, 0.5 Mo, bal Fe.
For tools, chisels, punches, pneumatic drills; for hot and cold work.

ALCO.
M-594; Zn, bal Cu.
For metalware.

ALCOA 2EC now **AA 6101.**

ALCOA 13 now **AA 413.0.**

ALCOA 43 now **AA 443.0.**

ALCOA 050.
M-1715; 99.5 min Al.
Ann(0): 13,700 TS max; 22-32 El.
Max cold work (H8): 19,500 TS min; 3-4 El.
Good corrosion resistance and ductility.
For formed parts for chemical plant, dairy, brewery and bakery equipment.
B.S. 1470, 1471, 1472, 1474.

ALCOA 051.
M-1715; 99.5 min Al, Cu + Si + Fe 0.05 max.
Extruded: 9500 TS; 4500 YS; 25 El, H2 Cond; 12,200 min TS; 15 El
Electrical conductivity: 60% I.A.C.S. min. For electric purposes, bus bars. B.S. 2898: E1E.

ALCOA 102.
M-1715; 99.0 min Al.
Ann: 13,000 TS; 5,000 YS; 45 El.
Good corrosion resistance.
For equipment for chemical, petroleum, pharmaceutical, brewery and food industry; panelling and fully supported roof cladding.
B.S. 1470, 1471, 1474.

ALCOA 108 now **AA 208.0.**

ALCOA 122 now **AA 222.**

ALCOA 138 now **AA 238.0.**

ALCOA 142 now **AA 242.0.**

ALCOA 190.
M-1715; 0.6 max Si, 0.7 max Fe, 0.8-1.5 Mn, 0.1 max Cu, 0.1 max Mg, 0.2 max Zn, bal Al.
Cond O: 13,000-19,000 TS; 20-25 El.
Cond H8: 25,400 min TS; 2-4 El.
For general purpose non-heat treatable parts and assemblies.
Good corrosion resistance. B.S. 1470: NS3.

ALCOA 195 now **AA 295.0.**

ALCOA 214 now **AA 514.0.**

ALCOA 218 now **AA 518.0.**

ALCOA 220 now **AA 520.0.**

ALCOA 319 now **AA 319.0.**

ALCOA 333 now **AA 333.0.**

ALCOA 355 now **AA 355.0.**

ALCOA 356 now **AA 356.0.**

ALCOA 364 now **AA 364.0.**

ALCOA 380 now **AA 380.0.**

ALCOA 384 now **AA 384.0.**

ALCOA 505.
 M-1715; 0.2 max Cu, 0.5-1.2 Mg, 0.4 max Si, 0.7 max Fe, 0.5 max Mn, 0.2 max Zn, bal Al.
 Cond O: 14,000-22,000 TS; 18-22 El.
 Cond H8: 27,000 min TS; 24,000 min YS; 1-3 El.
 Medium strength alloy, particularly suitable for anodizing.
 B.S. 4 300/7: NS41 (Al-Mg-1).

ALCOA 510.
 M-1715; 0.1 max Cu, 1.7-2.4 Mg, 0.5 max Si, 0.5 max Fe, 0.5 max Mn + Cr; 0.2 max Zn, bal Al.
 Cond O: 31,000-40,000 TS; 19,000 YS; 12-18 El.
 Cond H6: 33,000-40,000 TS; 25,000 min YS; 3-5 El.
 Medium strength alloy for architectural and marine applications; panelling containers, domestic appliances, welded assemblies.
 B.S. 1470: NS4; B.S. 1472: NF4, etc.

ALCOA 520.
 M-1715; 0.10 max Cu, 3.1-3.9 Mg, 0.5 max Si, 0.5 max Fe, 0.5 max Mn + Cr, 0.2 max Zn, bal Al.
 Cond O: 31,000-40,000 TS; 12,000 min YS; 12-18 El.
 Cond H4: 40,000-48,000 TS; 33,000 min YS; 4-6 El.
 Strong non-heat treatable alloy for general fabrication and for welded structures.
 B.S. 1470: NS5; B.S. 1472: NF5, etc.

ALCOA 540.
 M-1715; 0.15 max Cu, 0.7-1.2 Mg, 0.10 max Si, 0.10 max Fe, 0.30 max Mn, bal Al.
 Cond O: 22,500 max TS.
 Cond H8: 31,000 min TS.
 Alloy developed specifically for car and other bright trim.

ALCOA 550.
 M-1715; 0.15 max Cu, 0.3-0.8 Mg, 0.10 max Si, 0.10 max Fe, 0.2 max Mn bal Fe.
 Cond O: 18,000 max TS.
 Cond H4: 18,000-27,000 TS.
 Cond H8: 27,000 min TS.
 Aluminum alloy more ductile than Alcoa 540; for car and other bright trim.

ALCOA 750 now **AA 850.0.**

ALCOA 800.
 M-1715; 5.0-6.0 Cu, 0.4 max Si, 0.7 max Fe, 0.2-0.7 Pb, 0.2-0.7 Bi, bal Al.
 Cond TD: 53,000 TS; 45,000 YS; 11 El.
 Cond TF: 56,000 TS; 43,000 YS; 13 El.
 A free machining, heat-treatable aluminum alloy for high speed automatic work.
 Nearest U.S. Spec AA 2011.

ALCOA 910.
 M-1715; 0.10 max Cu, 0.4-0.9 Mg, 0.3-0.7 Si, 0.4 max Fe, bal Al.
 Cond O: 14,500 TS; 30-32 El.
 Cond TF: 37,000 TS; 32,000 YS; 13-14 El.
 A general purpose weldable, heat-treatable alloy for glazing bars, window frames.
 B.S. 1471: HT9 etc; nearest U.S. Spec AA 6063.

ALCOA 920.
 M-1715; 0.10 max Cu, 0.5-1.2 Mg, 0.7-1.3 Si, 0.5 max Fe, 0.4-1.0 Mn, 0.25 max Cr, bal Al.
 Cond O: 18,000 TS; 9,400 YS; 26-28 El.
 Cond TF: 49,000 TS; 43,000 YS; 10-12 El.
 Heat treatable, forgeable, weldable alloy for structural sections subject to stresses and shock loading such as lorry body components.
 B.S. 1470: HS 30 etc.

ALCOA 945.
 M-1715; 0.15-0.40 Cu, 0.8-1.2 Mg, 0.4-0.8 Si, 0.5 max Fe, 0.1 max Mn, 0.1 max Cr, bal Al.
 Cond TB: 31,000 TS; 19,000 YS; 19 El.
 Cond TF: 45,000 TS; 40,000 YS; 12 El.
 For structural road transport sections; forgeable, weldable, heat- treatable.

ALCOA 946.
 M-1715; 0.15-0.40 Cu, 0.8-1.2 Mg, 0.4-0.8 Si, 0.5 max Fe, 0.4-0.7 Pb, 0.4-0.7 Bi, bal Al.
 Free machining version of Alcoa 945.

ALCOA A13 now **AA A413.0.**

ALCOA A-108 now **AA A208.0.**

ALCOA-A-132 now **AA A 332.0.**

ALCOA A 142 now **AA A242.0.**

ALCOA A356 now **AA A356.0.**

ALCOA A 360 now **AA A360.0.**

ALCOA A380 now **AA A380.0.**

ALCOA A 612 now **AA A712.0.**

ALCOA A750 now **AA A850.0.**

ALCOA B-195 now **AA-B295.0.**

ALCOA B214 now **AA B214.0.**

ALCOA B750 now **AA B 850.0.**

ALCOA C-113 now **AA213.0.**

ALCOA C355 now **AA C355.0.**

ALCOA C 612 now **AA C712.0.**

ALCOA F132 now **AA F332.0.**

ALCOA F 214 now **AA F214.0.**

ALCOA EC now **AA 1350.**

ALCOA NO. 360 now **AA 360.0.**

ALCOA NO. 716.
M-4; 10 Si, 4 Cu, bal Al.
For brazing alloy; wire.

ALCO BRONZE.
M-404; 10 Sn, 5 Ni, bal Cu.
For ships forgings, propellers, skates, surgical instruments; tough, hard, corrosion resistant.

ALCODIE.
M-35; 0.35 C, 5 Cr, 0.4 V, 1.5 W, 1.5 Mo, bal Fe.
Oil or air hardening tool steel for die casting dies, hot blanking and forging dies, shear blades; AISI H12.

ALCOLOY 688.
M-279; 73.5 Cu, 22.7 Zn, 3.4 Al, 0.4 Co.
Ann: 77,000-87,000 TS; 47,000-62,000 YS; 30-40 El.
Cold rolled: 87,000-140,000 TS; 63,000-120,000 YS; 2-29 El.
Excellent corrosion resistance, high strength.
For relay springs, electric terminals, shells, connectors.

ALCO M AND ALCO S replaced by **CYCLOPS SL.**

ALCOMAX-II.
M-Eng; 8.1 Al, 11.9 Ni, 21 Co, 4 Cu, bal Fe.
For magnetic and electrical equipment. Permanent magnet, high coercive force.

ALCOMAX III.
M-Eng; 13.5 Ni, 25.5 Co, 3 Co, 7.5 Al, 0.9 Cb, 0.17 Si, bal Fe.
For magnets for motors, loud speakers; permanent magnet.

ALCOMAX IV.
M-Eng; 7.4 Al, 13.3 Ni, 24.5 Co, 3 Cu, 2.5 Cb, bal Fe.
For magnets for cycle dynamos; permanent magnet.

ALCONIT-0.
M-Germ.; Ni, Al, bal Fe.
For permanent magnets.

ALCONIT-10.
M-Germ.; Ni, Al, bal Fe.
For permanent magnets.

ALCONIT 16.
M-Germ.; Ni, Al, bal Fe.
For permanent magnets.

ALCONITE.
M-1464; C, alloy, bal Fe.
For castings.

ALCOP BRONZE.
M-508; 62 Cu, 2.5 Al, 35.5 Zn, 0-15 Pb.
Cast: 73,000 TS; 37,000 YS; 35 El; 150 Brin.
For centrifugal castings, bushings, liners, bearing cages; heavy duty.

ALCOP NO. 3.
M-1168; 61.5-63.5 Cu, 35-38 Zn, 0.75-1.25 Pb, 0.9-1.3 Al.
Cast: 63,000 TS; 35,000 YS; 30 El; 102 Brin.
For hardware, die castings; leaded bronze, free-cutting.

ALCRES.
M-U.S.; 12 Cr, 5 Al, 83 Fe.
For electrical resistances; heat resisting.

ALCRESS.
M-Eng.; 70 Fe, 20 Cr, 5 Al.
For electrical resistors; heat resistant.

ALCUFONT 4.5.
M-1176; 4.5 Cu, 0.8 max Fe, 1 max Si, 0.2 max Ti, bal Al.
Heat treated: 36,000-43,000 TS; 28,000-33,000 YS; 4-2 El; 75-95 Brin
For light alloy parts; age-hardenable.

ALCUFONT 8.
M-1176; 7.0-8.5 Cu, 0.2 max Ti, 0.5 max Si, 0.6 max Fe, bal Al.
Heat treated: 20,000-26,000 TS; 11,000-13,500 YS; 5-2 El; 55-70 Brin
For pistons, light alloy parts; age-hardened.

ALCUFONT 10.
M-1176; 9.3-10.7 Cu, 0.5-1.3 Fe, 0.2-0.4 Mg, bal Al.
Heat treated: 40,000-51,200 TS; 34,000-37,000 YS; 0.5-0.2 El; 125-150 Brin.
For pistons; high strength.

ALCUFONT 12.
M-1776; 11.0-12.5 Cu, 0.2 Ti, 1 Fe, bal Al.
Heat treated: 38,400-42,700 TS; 22,800-25,600 YS; 1.5-1.0 El; 110-125 Brin.
For pistons; high strength.

ALCUMAN.
M-Germany; 4.8-6.0 Cu, 0.2 max Si, 0.5 Mn, 1.5 Mg, bal Al.
Heat treated: 60,000-71,000 TS; 50,000-57,000 YS; 5-10 El; 120-140 Brin.
For screw machine products, construction and transportation equipment. Age-hardenable. High strength.

ALCUPLATE.
M-926; Cu clad to 2S aluminum alloy.
Soft: 17,000 TS; 7800 YS; 33 El.
Hard: 30,000 TS; 25,000 YS; 5 El.
For electrical contacts; Al alloys plated by lamination with copper, 80% Al-20% Cu.

ALDA.
M-667; 0.1-0.2 C, bal Fe.
For welding rods; copper coated.

ALDAL.
M-767; 3.5-4.5 Cu, 0.4-1.0 Mn, 0.3-0.75 Mg, bal Fe.
60,000 TS; 22 El; 120 Brin.
For light alloy parts, aircraft construction; Duralumin type.

ALDREI.
Italy; Si, Mg, bal Al.
42,500-51,000 TS; 38,000-44,000 YS; 5-9 El.
For light alloy parts.

ALDREY.
M-249; 0.2-1.0 Si, 0.4 Mg, 0.3 Fe, bal Al.
Heat treated: 44,000-49,000 TS; 38,000-43,000 YS; 9-5 El; 90-100 Brin.
For auto engine parts, electric transmission wires and cables; age-hardenable.

ALDREY.
M-493; 0.5-0.6 Si, 0.4-0.5 Mg, bal Al.
Heat treated: 43,000-50,000 TS; 38,400-44,000 YS; 5-9 El; 70-80 Brin
For general structural members, aircraft and aerospace components.
Age hardenable, corrosion resistant.

ALDREY 14.
M-624; M-1634, M-541; 0.4-0.6 Si, 0.4-0.6 Mg, 0.3 max Fe, 0.3 max Ti, bal Al.
Annealed: 11,400-17,000 TS; 4300-7100 YS; 20-3 El.
THA16 Temper: 42,700-50,000 TS; 38,400-44,100 YS; 4-9 El.
For marine structures and hardware, architectural applications. Heat treatable.

ALDUR, ALDUR G, ALDUR D, ALDUR DG.
Varying quality and use grades of numbered ALDUR steels.

ALDUR 35.
M-1735; 0.09 C, 0.25 Si, 0.6 Mn, bal Fe.
Steel plates.

ALDUR 41.
M-1735; 0.15 C, 0.25 Si, 0.7 Mn, bal Fe.
Steel plates.

ALDUR 44.
M-1735; 0.16 C, 0.30 Si, 1.0 Mn, bal Fe.
Steel plates.

ALDUR 45/60.
M-1735; 0.18 C, 0.45 Si, 1.4 Mn, bal Fe.
Steel plates.

ALDUR 47.
M-1735; 0.17 C, 0.35 Si, 1.2 Mn, bal Fe.
Steel plates.

ALDUR 50.
M-1735; 0.18 C, 0.40 Si, 1.3 Mn, bal Fe.
Steel plates.

ALDUR 50/65.
M-1735; 0.20 C, 0.45 Si, 1.5 Mn, bal Fe.
Steel plates.

ALDUR 55.
M-1735; 0.19 C, 0.45 Si, 1.5 Mn, bal Fe.
Steel plates.

ALDUR 55/68.
M-1735; 0.20 C, 0.50 Si, 1.6 Mn, bal Fe.
Steel plates.

ALDUR 58.
M-1735; 0.20 C, 0.50 Si, 1.6 Mn, bal Fe.
Steel plates.

ALDUR 58/72.
M-1735; 0.20 C, 0.50 Si, 1.7 Mn, bal Fe.
Steel plates.

ALDURBRA.
M-282; M-150; 76 Cu, 22 Zn, 2 Al.
80,000 TS; 44,000 YS; 16 El; 130 Brin.
For condenser tubes; corrosion resistant.

ALEMITE.
M-595; 10 Si, bal Al.
For light alloy parts; die casting.

ALFENIDE.
59.1 Cu, 30.12 Zn, 9.7 Ni, 1.0 Fe.
For ornaments, hardware, fittings; nickel silver.

ALFENOL 12.
M-1084; 12 Al, bal Fe.
Residual inductance, 4,000; maximum permeability, 20,000; coercive force (oersted), .100.
Wear resistant, high permeability magnetically soft alloy.

ALFENOL 16.
M-1084; 16 Al, bal Fe.
Residual inductance, 3,000; maximum permeability, 110,000; coercive force (oersted), .015.
Wear resistant, high permeability magnetically soft alloy.

ALFER.
M-926; low C-steel clad with Al alloy.
For election tubes.

ALFER.
Japan; 12-14 Al, bal Fe.
For permanent magnets; high permeability,

ALFERON.
M-44; 23 Cr, 5 Al, bal Fe.
Rolled: 110,000 TS; 60,000 YS; 20 El.
For electric resistances; magnetic.

ALGIERS (ALGERS) METAL.
Eng; 94.5 Sn, 0.5 Sb, 5 Cu.
For bearings; anti-friction.

ALGIERS METAL-1.
Eng; 90 Sn, 10 Sb.
For bearings; corrosion resistant.

ALGIERS METAL-2.
Eng; 75 Sn, 25 Sb.
For bearings; corrosion resistant.

ALGOFORM 45.
M-1253; 0.06 max C, 0.35 Mn, 0.04 Al, 0.010 Cb, bal Fe.
0.100 in. plate: 55,000 psi (379 MPa) min TS; 45,000 psi (310 MPa) mi 40 min El.
High strength - low alloy steel plate.

ALGOFORM 50.
M-1253; 0.06 max C, 0.35 Mn, 0.04 Al, 0.015 Cb, bal Fe.
0.100 in. plate: 60,000 psi (414 MPa) min TS; 50,000 psi (345 MPa) mi 36 min El.
High strength - low alloy steel plate.

ALGOFORM 60.
M-1253; 0.06 max C, 0.35 Mn, 0.04 Al, 0.035 Cb, bal Fe.
0.100 in. plate: 70,000 psi (483 MPa) min TS; 60,000 psi (414 MPa) mi 32 min El.
High strength - low alloy steel plate.

ALGOFORM 70.
M-1253; 0.06 max C, 0.55 Mn, 0.04 Al, 0.065 Cb, bal Fe.
0.100 in. plate: 80,000 psi (552 MPa) min TS; 70,000 (483 MPa) min Y 28 min El.
High strength - low alloy steel plate.

ALGOFORM 80.
M-1253; 0.06 max C, 0.75 Mn, 0.04 Al, 0.10 Cb, bal Fe.
0.100 in. plate: 90,000 psi (620 MPa) min TS; 80,000 psi (552 MPa) mi 24 min El.
High strength - low alloy steel plate.

ALGOMA AR.
M-1253; 0.30-0.45 C, 1.3-1.65 Mn, 0.15-0.35 Si, bal Fe.
100,000 psi TS.
For applications where abrasive resistance is required.

ALGONQUIN.
M-435; 1.0 C, 0.25 Mn, 0.2 Si, bal Fe.
For dies; water hardened.

ALGO-TUF 50.
M-1253; 0.10-0.18 C, 0.90-1.25 Mn, 0.15-0.30 Si, 0.40-0.60 Cr, 0.25-0. Ni, 0.20-0.40 Cu, 0.01-0.10 V, bal Fe.
50,000 psi min YS.
For welded, bolted or riveted construction.
HSLA steel.

ALGO-TUF 70.
M-1253; 0.20 max C, 1.60 max Mn, 0.15-0.35 Si, 0.20-0.50 Cu, 0.25-0.50 0.03-0.10 V, 0.06 max Cb, bal Fe.
70,000 psi min YS in plates.
For welded, bolted or riveted construction.
HSLA steel.

AL HX.
M-1; 22.0 Cr, 9.0 Mo, 1.5 Co, 48 Ni, bal Fe.
Austenitic nickel-based alloy with exceptional strength at elevated temperatures.
UNS NO6002.

ALIDIE.
M-275; 1.5 C, 12 Cr, 0.2 V, 1 Mo, bal Fe.
For blanking and forming dies, punches; Type D2; air hardened, non-deforming.

ALJEP.
France; 3.8 Cu, 1.3 Mg, 1.5 Ni, 0.5 V, bal Al.

ALKADUR 300.
M-324; 3.5-4.5 Cu, 0.2-1.0 Si, 0.3-1.2 Mn, 0.4-1.4 Mg, bal Al.
Heat treated: 62,700-71,100 TS; 48,000-57,000 YS; 5-15 El; 110-150 Brin.
For aircraft components, fittings, fasteners, rivets.
Heat treatable. High strength and fatigue.

ALKALI-RESISTING METAL.
Eng; 95 Fe, 5 Ni.
For corrosion resisting parts, chemical apparatus; corrosion resistant.

ALKRONID.
M-1331; 0.34 C, 1.1 Al, 1.4 Cr, bal Fe.
For oil refinery equipment; creep resistant.

ALKROTHAL.
M-70, M-1150; 15 Cr, 4.5 Al, bal Fe.
For electrical resistances, for service up to 1050°C; wire, ribbon, strip, foil.

ALLADIN.
Eng; Zn alloy.
For welding rod; for white metal die castings.

ALLAN NO 2 BRONZE.
U.S.; 66 Cu, 9 Sn, 25 Pb.
For bearings, hardware; tough.

ALLEN RED BRONZE.
U.S.; 70-55 Cu, 20-40 Pb, 10-5 Sn, 1 S.
For bearings, hardware; tough.

ALLAN RED BRONZE.
U.S.; 62.5 Cu, 30 Pb, 7.5 Sn.
For bearings, hardware; tough.

ALLAN RED METAL.
M-124; 50 Pb, 50 Cu.
13.5 Brin.
For bearings, crank-pins for facing pistons; not a true alloy but a mechanical mixture.

ALLAN RED METAL NO. 2.
M-124; 60 Cu, 40 Pb.
For high speed bearings, piston wearing rings, operating conditions not exceeding 600°F.

ALLAUTAL.
M-Germany; surface of pure Al on base of Laurel."
Annealed: 20,000 TS; 25 El.
Normalized: 60,000 TS; 18 El.
For light alloy parts; similar to Alclad."

ALLAUTAL.
M-1548; 4.5-5.5 Cu, 0.2-0.5 Si, bal Al.
Heat treated: 60,000-71,000 TS; 42,000-57,000 YS; 2-10 El; 120-140 Brin.
For aircraft components, fasteners, fittings, rivets.
Age-hardenable, high tensile and fatigue strength.

ALLCAST.
M-906; 2.5-3.5 Cu, 4.5-6.0 Si, 0.15 max Mg, 1.0 max Fe, 0.2 max Ni, 0.5 max Mn, bal Al.
Cast: 24,000 TS: 12,000 YS; 2 El; 60 Brin.
Heat treated: 30,000 TS; 18,000 YS; 2.5 El; 75 Brin.
For sand and permanent mold castings; age-hardenable.

ALLCAST 46.
M-130; 0.6 max Cu, 6.5-7.5 Si, 0.2-0.4 Mg, 0.5 max Zn, 0.75 max Fe, 0.1-0.2 Ti, bal Al.
Cast: 23,000 TS; 12,000 YS; 3.5 El; 55 Brin.
T6 Temper: 33,000 TS; 24,000 YS; 4 El; 70 Brin.
For instrument cases, gear cases, fittings, crankcases, oil pans.
Heat treatable. Similar to Aluminum 356.
Corrosion resistant.

ALLCAST 46A.
M-130; 0.20 Cu, 6.5-7.5 Si, 0.25-0.40 Mg, 0.3 max Zn, 0.5 max Fe, 0.2 max Ti, bal Al
Permanent Mold Cast: 25,000 TS; 13,000 YS; 7 El; 65 Brin.
T6 Temper: 40,000 TS; 27,000 YS; 5 El; 90 Brinell.
For permanent mold castings, oil pans, instrument cases, crankcases.
Heat treatable. Similar to Aluminum 356.

ALLCAST 50.
M-130; 3.5-4.5 Cu, 2.5-3.5 Si, 0.10 max Mg, 1.0 max Zn, 1.0 max Fe, 0.2 max Ti, bal Al.
Permanent Mold Cast: 27,000 TS; 16,000 YS; 3 El; 65 Brin.
T6 Temper: 40,000 TS; 26,000 YS; 4 El; 85 Brin.
For castings, housings, oil pans, manifold and valve bodies, machine frames, piano plates and frames.
Heat treatable, permanent mold. Similar to Aluminum 108.

ALLCAST 51.
M-130; 3.5-4.5 Cu, 5.5-6.5 Si, 0.1 max Mg, 1.0 max Zn, 1.0 max Fe; 0.2 max Ti, bal Al.
Cast: 28,000 TS; 16,000 YS; 2 El; 70 Brin.
T6: 45,000 TS; 28,000 YS; 4 El; 105 Brin.
For permanent mold castings, housings, oil pans, casings, fittings,
Heat treatable. Similar to Aluminum A108.

ALLCAST 52.
M-130; 4-5 Cu, 1.5 max Si, 0.5 max Zn, bal Al.
T6 Temper: 37,000 TS; 24,000 YS; 5 El; 75 Brin.
At 600°F: 4000 TS; 3000 YS; 75 El.
For flywheel housings, aircraft wheels, fittings, crankcases, instrument cases.
Heat treatable. Similar to Aluminum 195.
Strong and rugged.

ALLCAST 52A.
M-130; 4-5 Cu, 1.5 max Si, 0.35 max Zn, 0.8 max Fe, 0.25 max Ti, bal Al.
T4 Temper: 32,000 TS; 16,000 YS; 8 El; 60 Brin.
T62 Temper: 41,000 TS; 32,000 YS; 2 El; 95 Brin.
For flywheel housings, crankcases, aircraft and bus wheels.
Heat treatable. Strong and rugged.
Similar to Aluminum 195.

ALLCAST 53.
M-130; 4.0-5.0 Cu, 2.0-3.0 Si, 0.5 max Zn, 0.35 max Ni, 0.2 max Ti, 1.0 max Fe, bal Al.
Cast: 30,000 TS; 14,000 YS; 3.5 El; 70 Brin.
T6: 40,000 TS; 26,000 YS; 5 El; 90 Brin.
For aircraft fittings and wheels, fuel pump bodies, seat frames.
Heat treatable, permanent mold. Similar to Aluminum B195.

ALLCAST 59A.
M-130; 3.5-4.5 Cu, 0.6 max Si, 1.3-1.8 Mg, 0.25 max Cr, 1.7-2.3 Ni, 0.25 max Ti, bal Al.
Cast: 34,000 TS; 24,000 YS; 1 El; 105 Brin.
T6: 54,000 TS; 51,000 YS; 0.8 El; 120 Brin.
For air cooled cylinder heads, diesel engine pistons, generator housings.
Heat treatable, permanent mold. Similar to Aluminum 142.

ALLCAST 60.
M-130; 6 Si, 3.5 Cu, bal Al.
F-temper: 27,000 TS; 16,000 YS; 2.5 El; 65 Brin.
T6-temper: 35,000 TS; 24,000 YS; 3.5 El; 85 Brin.
For pressure tight castings; heat treatable.

ALLCAST 70.
M-130; 6 Si, 3.5 Cu, bal Al.
F-temper: 26,000 TS; 16,000 YS; 2.0 El; 70 Brin.
T6-temper: 36,000 TS; 24,000 YS; 2.0 El; 80 Brin.
For pressure tight castings; heat treatable.

ALLCAST 78.
M-130; 3.0-4.0 Cu, 7.5-8.5 Si, 1.0 max Zn, 0.6 max Fe, 0.25 max Ti, bal Al.
Die cast: 43,000 TS; 26,000 YS; 2 El.
At 300°F: 37,000 TS; 24,000 YS; 4 El.
For motor frames, hand truck wheels, oil pans, ornamental parts.
Good high temperature strength.

ALLCAST 125.
M-130; 9.7-10.5 Mg, 0.25 max Ti, bal Al.
T4 Temper: 57,000 TS; 29,000 YS; 30 El; 80 Brin.
For aircraft fittings, car frames, marine parts, lever brackets.
Age hardenable, corrosion resistant, shock resistant, tough.

ALLCAST 400.
M-130; 3.6-4.2 Mg, 0.5 max Si, 0.5 max Zn, 0.6 max Mn, 0.5 max Fe, 0.25 max Ti, bal Al.
Cast (Sand): 25,000 TS; 12,000 YS; 9 El; 50 Brin.
At 300°F: 22,000 TS; 12,000 YS; 7 El.
For food handling equipment, cooking utensils, chemical and textile plant equipment, marine hardware. Similar to Aluminum 214, good corrosion resistance. Not heat treatable.

ALLCAST 405.
M-130; 4.0-5.0 Zn, 0.5 max Si, 1.0 max Fe, 0.2 max Ni, 0.15 max Mn, 0.9-1.4 Sb, bal Al.
Die cast: 22,000 TS; 11,000 YS; 10 El.
For oil pans, instrument cases, housings, ornaments.
Good corrosion resistant.

ALLCAST 417.
M-130; 6.5-7.5 Mg, 0.12 max Si, 0.10-0.25 Mn, 0.10-0.25 Ti, bal Al.
Sand cast: 36,000 TS; 19,000 YS; 10 El; 70 Brin.
For automotive and marine castings, cooking utensils, gasoline fuel gages, textile machinery.
Similar to Aluminum 2185P and Almag 35.
High strength and shock resistant.

ALLCAST 418.
M-130; 7.5-8.5 Mg, 0.3 max Si, 0.9-1.3 Fe, 0.30 max Mn, 0.10 max Ni, bal Al.
Die cast: 43,000 TS; 25,000 YS; 5 El; 80 Brin.
For cylinder blocks, marine fittings, hardware, brake shoes, wheel flanges Similar to Aluminum 218. Good strength and corrosion resistance.

ALLCAST 420.
M-130; 9.7-10.5 Mg, 0.15 max Si, 0.20 max Fe, 0.25 max Ti, bal Al.
T4 Temper: 46,000 TS; 25,000 YS; 16 El; 75 Brin.
T5 Temper: 35,000 TS; 110 Brin.
For aircraft fittings, marine parts, railroad car frames.
Sand castings. Similar to Aluminum 220
High strength and corrosion resistance.

ALLCAST 440.
M-130; 0.5-0.65 Mg, 5.2-6.0 Zn, 0.4-0.6 Cr, 0.15-0.25 Ti, 0.4 max Fe, bal Al.
Cast: 28,000 TS; 16,000 YS; 12 El; 55 Brin.
Aged: 36,000 TS; 25,000 YS; 6 El; 75 Brin.
For manifolds, aircraft components, transmission housings, tight castings.
Self-aging sand casting. Natural aging.
Similar to Aluminum 40 E.

ALLCAST 470.
M-130; 0.4-1.0 Cu, 0.25-0.50 Mg, 7.0-8.0 Zn, 0.8 max Fe, 0.25 max Ti, bal Al.
Sand Cast: 28,000 TS; 15,000 YS; 8 El; 55 Brin.
Aged: 34,000 TS; 23,000 YS; 5 El; 75 Brin.
For housings, machinery parts, fittings, hardware, automobile parts. Similar to Tenzalloy. Ages at room temperature.

ALLCAST 700.
M-130; 2-3 Cu, 0.7 max Si, 0.9-1.5 Ni, 0.20 max Ti, 6.0-7.5 Sn, bal Al.
T5 or T533 Temper: 20,000 TS; 16,000 YS; 7 El; 55 Brin.
For bearings, bushings, heat treatable.

ALLCAST 700A.
M-130; 0.7-1.3 Cu, 5.5-7.0 Sn, 0.7 max Si, 0.2 max Ti, 0.7-1.3 Ni, bal Al.
T5 or T533 Temper: 20,000 TS; 8500 YS; 10 El; 55 Brin.
For bearings. Similar to Aluminum 750. Good imbeddability.

ALLCAST 46-10.
M-130; 0.10 max Cu, 6.5-7.5 Si, 0.28-0.40 Mg, 0.1 max Zn, 0.15 max Fe 0.2 max Ti, bal Al.
Permanent Mold Cast: 27,000 TS; 14,000 YS; 8 El, 55 Brin.
T61 Temper: 43,000 TS; 28,000 YS; 12 El; 90 Brin.
For gear housings, fittings, oil pans, crankcases, instrument housings.
Heat treatable, permanent mold. Similar to Aluminum A356.

ALLCAST D400.
M-130; 1.6-2.0 Si, 3.6-4.5 Mg, 0.5 max Fe, 0.6 max Mn, 0.25 max Ti, bal Al.
Sand Cast: 20,000 TS; 13,000 YS; 2 El; 50 Brin.
Permanent Mold: 22,000 TS; 13,000 YS; 2 El; 50 Brin.
For cooking utensils, marine fittings.
Similar to Aluminum B214. Good corrosion resistance. Not heat treatabl

ALLCAST P400.
M-130; 3.6-4.5 Mg, 1.5-2.0 Zn, 0.5 max Si, 0.5 max Fe, 0.6 max Mn, 0.25 max Ti, bal Al.
Permanent Mold; 27,000 TS; 16,000 YS; 7 El; 60 brin. Die Cast: 40,000 TS; 22,000 YS; 10 El; 65 Brin.
For food and dairy handling equipment, chemical and textile plant equipment, marine hardware.
Similar to Aluminum A214. Not heat treatable, corrosion resistant.

ALLCAST SC8.
M-130; 2.5-4.0 Cu, 5.0-6.5 Si, 0.15 max Mg, bal Al.
Permanent mold cast: 28,000-36,000 TS; 15,000-20,000 YS; 2-3 El; 70-8 Brin.
For permanent mold castings; light alloy parts.

ALLCAST SC8.
M-906; 2.5-3.5 Cu, 4.5-6.0 Si, bal Al.
Cast: 24,000 TS; 12,000 YS; 2 El; 60 Brin.
T6-Temper: 30,000 T; 18,000 YS; 2.5 El, 75 Brin.
For light alloy castings; sand and permanent mold, hardenable.

ALLCAST Z50.
M-130; 3.0-4.0 Cu, 2.5-3.5 Si, 0.1 max Mg, 2.0-3.0 Zn, 1.0 max Fe, 0.2 max Ti, bal Al.
Permanent Mold Cast: 27,000 TS; 16,000 YS; 3 El; 65 Brin.
T6 Temper: 40,000 TS; 26,000 YS; 4 El; 85 Brin.
For casings, housings, fittings, crankcases, axle housings.
Heat treatable, permanent mold, corrosion resistant.

ALLCAST Z51.
M-130; 3.0-4.0 Cu, 5.5-6.5 Si, 0.1 max Mg, 2-3 Zn, 1.0 max Fe, 0.2 ma Ti, bal Al.
Cast: 39,000 TS; 22,000 YS; 5.5 El; 90 Brin.
T6: 45,000 TS; 28,000 YS; 4 El; 105 Brin.
For permanent mold castings, casings, housings, fittings, oil pans. Heat treatable.

ALLEGHENY 19-9 DL.
M-1; 0.30 C, 1.0 Mn, 0.5 Si, 19.0 Cr, 9.0 Ni, 1.4 Mo, 1.3 W, 0.4 C + Ta, 0.3 Ti, bal Fe.
Heat treated: 148,200 TS; 118,000 YS; 19% El.
Stress-Relieved: 118,500 TS; 69,000 YS; 58% %el.
At 1500°F: 32,500 TS; 29,500 YS; 53% El.
Uses: Gas turbine and turbo supercharger component parts as rotors, buckets, fasteners, tail cones.
Stainless. High temperature properties. Heat resistant to 1200°F. Not hardenable.

ALLEGHENY 216.
M-1; 0.08 max C, 19.75 Cr, 8.25 Mn, 6.0 Ni, 2.5 Mo, 0.37 N, bal Fe.
Annealed: 100,000 TWs; 55,000 YS; 45 El; B 92 Rock.
At 1000°F: 75,400 TS; 36,300 YS; 43.3 El, 67.8 RA.
For marine engineering components, space hardware, woven wire cloth, bushings, bolts, shafting. Springs, valve seats. Stainless, improved high temperature strength and endurance limit.
Good weldability.

ALLEGHENY 309 S CB.
M-1; 0.08 max C, 23.0 Cr, 13.5 Ni, Cb, bal Fe.
Austenitic stainless steel, modified AISI, with low carbon, and Colum added to improve welding characteristics.

ALLEGHENY 332.
M-1; 0.08 max C, 20.0 Cr, 32.0 Ni, 0.5 Ti, 0.5 Al, bal Fe.
Modified AISI 330 for improved oxidation resistance and high temperature resistance.

ALLEGHENY 334.
M-1; 0.05 max C, 20.0 Cr, 20.0 Ni, 0.5 Ti, 0.5 Al, bal Fe.
Austenitic stainless steel with good oxidation resistance at elevated temperature.

ALLEGHENY 404.
M-1; 13.5 Cr, C, bal Fe.
Low residual stainless steel.

ALLEGHENY 410 HC.
M-1; 0.21 C, 12.5 Cr, bal Fe.
Modified AISI 410 stainless with higher carbon for higher mechanical properties.
Similar to AISI 420.

ALLEGHENY 410 KB.
M-1; 0.30 C, 12.5 Cr, bal Fe.
Higher carbon than AISI 410 to produce higher mechanical properties af heat treating.
AISI 420.

ALLEGHENY 410 S.
M-1; 0.05 C, 12.5 Cr, bal Fe.
Low carbon grade of AISI 410 for welding purposes.
UNS 41008.

ALLEGHENY 434.
M-1; 0.12 max C, 17 Cr, 1.0 Mo, bal Fe.
Modified AISI 430 for improved corrosion resistance.
UNS S 43400.

ALLEGHENY 436.
M-1; 0.12 max C, 17.0 Cr, 1.0 Mo, Cb, bal Fe.
Modified type 434 with Cb added.
UNS S436000.

ALLEGHENY 439.
M-1; 0.07 max C, 18.0 Cr, Ti, bal Fe.
Modified type 430 stainlesss for improved corrosion resistance after welding.
UNS S43035.

ALLEGHENY A-286.
M-1, M-1214; 0.08 max C, 1-2 Mn, 13.5-16.0 Cr, 24-28 Ni, 1.3 Mo, 2.0 Ti, 0.3 V, 0.2 Al, bal Fe.
Rolled: 135,000-160,000 TS; 85,000-115,000 YS; 30-20 El; 55-30 RA.
For jet engine and supercharger parts, afterburners; age-hardenable, austenitic, heat resistant.

ALLEGHENY AM-363.
M-1; 0.05 max C, 0.30 max Mn, 0.15 max Si, 11.0-12.0 Cr, 4.0-5.0 Ni, 0.3-0.6 Ti, bal Fe.
Hot rolled: 138,000 TS; 131,000 YS; 11 El.
Annealed: 124,000 TS; 105,000 YS; 8 El.
Aged: 124,000 TS; 118,000 YS; 11.5 El.
For automobile mufflers, jet engine and gas turbine components.
Precipitation-hardenable.
Ferromagnetic, stainless, maraging.

ALLEGHENY METAL AM-350 see AM-350.

ALLEGHENY METAL AM-355 see AM-355.

ALLEGHENY METAL H-17 see AISI 440C.

ALLEGHENY METAL S-590.
M-1; 0.38-0.47 C, 1.0-2.0 Mn, 19-21 Cr, 19-21 Ni, 19-21 Co, 3.5-4.5 Mo, 3.5-4.5 W, 3.5-4.5 Cb, bal Fe.
Heat resistance to 1100-1400°F. Corrosion resistant.

ALLEGHENY METAL S-816.
M-1; 0.4 C, 1.2 Mn, 20.0 Cr, 20.0 Ni, 43.0 Co, 4.0 W, 3.0 Fe, 4.0 Cb.
For turbine blades, rotors, high temperature hardware, jet propulsion engine parts.
High heat and corrosion resistant.
For high temperature service to 1520°F.

ALLEGHENY TYPE 201 see AISI 201.

ALLEGHENY METAL (TYPE 347).
M-1; 0.15 max C, 17-19 Cr, 8-12 Ni, (Cb over 10 times the C), bal Fe
Annealed: 75,000 TS; 50 El: 60 RA; 130 Brin.
For corrosion resistant parts; immune to intergranular corrosion.

ALLEGHENY METAL (TYPE 348).
M-1; 0.15 max C, 19-20 Cr, 8-12 Ni, (Cb over 10 times the C), bal Fe
Annealed: 75,000 TS; 50 El; 60 RA; 130 Brin.
For corrosion resistant parts; immune to intergranular corrosion.

ALLEN "B" METAL.
M-210; 0.22-0.27 C, 0.20-0.35 Si, 0.60-0.80 Mn, bal Fe.
For machinery parts, gears; castings.

ALLEN CAST IRON SOLDER.
M-242.
For soldering cast iron.

ALLEN C. CU.
M-210; 0.30-0.35 C, 0.20-0.35 Si, 0.60-0.80 Mn, 0.80-1.1 Cu, bal Fe.
For machinery parts; castings.

ALLEN CR. MO. 329.
M-210; 0.20-0.35 C, 0.20-0.35 Si, 0.40-0.60 Mn, 3.0-3.25 Cr, 0.45-0.55 Mo, bal Fe.
For oil refinery equipment; castings.

ALLEN CR. MO./F. MO.
M-210; 0.50-0.55 C, 0.20-0.35 Si, 0.50-0.80 Mn, 0.7-0.9 Cr, 0.25-0.35 Mo, bal Fe.
For gears, shafts, axles, bolts, crankshafts; castings.

ALLEN CR. MO. "P".
M-210; 0.43-0.48 C, 0.20-0.35 Si, 0.50-0.80 Mn, 0.7-0.9 Cr, 0.25-0.35 Mo, bal Fe.
For gears, shafts, crankshafts, bolts, housings; castings.

ALLEN "C" STEEL.
M-210; 0.30-0.35 C, 0.20-0.35 Si, 0.60-0.80 Mn, bal Fe.
For gears, pinions, machinery parts; castings.

ALLEN "D" STEEL.
M-210; 0.40-0.50 C, 0.20-0.35 Si, 0.70-1.0 Mn, bal Fe.
For gears, machinery parts, housings; castings.

ALLEN HI LUSTER.
M-242.
For solder for stainless steel; bright finish.

ALLEN MAGNESIUM SOLDER.
M-242.
For solder for magnesium.

ALLEN MT. MO. "P".
M-210; 0.18-0.25 C, 0.20-0.35 Si, 1.2-1.6 Mn, 0.2-0.3 Mo, bal Fe.
For machinery parts, gears, camshafts; castings.

ALLEN MT. MO. "R".
M-210; 0.25-0.33 C, 0.20-0.35 Si, 1.2-1.6 Mn, 0.20-0.30 Mo, bal Fe.
For gears, shafts, carnkshafts, castings.

ALLEN MT. "P".
M-210; 0.22-0.27 C, 0.20-0.35 Si, 1.2-1.6 Mn, bal Fe.
For machinery parts, gears, camshafts; castings.

ALLENITE A.S.
M-210; WC.
For cutting tools; sintered carbide.

ALLENITE C.H.
M-210; WC.
For cutting tools for chilled iron rolls; sintered carbide.

ALLENITE D.N.
M-210; WC.
For drawing dies, wood working tools; sintered carbide.

ALLENITE E.S.
M-210; WC.
For cutting tools for finish machining; sintered carbide.

ALLENITE L. C. G.
M-210; WC.
For percussion drilling tools; sintered carbide.

ALLENITE N.
M-210; WC.
For cutting tools for cast iron; sintered carbide.

ALLENITE T.N.
M-210; WC.
For cutting tools for intermittent cuts; sintered carbide.

ALLENITE T.S.
M-210; WC.
For cutting tools for rough maching; sintered carbide.

ALLEN MT. "R".
M-210; 0.25-0.33 C, 0.20-0.35 Si, 1.2-1.6 Mn, bal Fe.
For machinery parts, camshafts, gears; castings.

ALLEN PRESTO SOLDER.
M-242; Sn, Pb, Ag.
For solder.

ALLEN "T" STEEL.
M-210; 0.14-0.16 C, 0.20-0.35 Si, 0.40-0.50 Mn, bal Fe.
For gears, shafts, housings; castings.

ALLENOY.
M-668; 0.35-0.40 C, 0.9 Mn, 0.2-0.3 Mo, bal Fe.
Hardened: 170,000 TS; 150,000 YS; 12 El; 35 RA; 400 Brin.
For set and cap screws; water hardening.

ALLENOY 1038.
M-668; 0,35-0.42 C, 0.6-0.9 Mn, bal Fe.
Heat treated: 145,000-165,000 TS; 1158000-135,000 YS; 11 El; 35 RA; 350-40 Brin.
For pipe plugs; similar to AISI 1038.

ALLENOY 4137.
M-668; 0.35-0.40 C, 0.70-0.90 Mn, 0.80-1.1 Cr, 0.15-0.25 Mo, bal Fe.
Heat treated: 170,000-225,000 TS; 150,000-180,000 YS; 12-8 El; 35-25 RA; 400-500 Brin.
For nuts, cap screws, set screws; similar to AISI 4137.

ALLENOY 8650.
M-668; 0.48-0.53 C, 0.75-1,0 Mn, 0.40-0.70 Ni, 0.40-0.60 Cr, 0.15-0.25 Mo, bal Fe.
For wrenches; similar to AISI 8650.

ALLENS IMPERIAL SPECIAL.
M-U.S.; 0.68 C, 0.29 Mn, 1.64 Cr, 19.1 W, 1.37 V, bal Fe.
For tools, cutters, dies, punches; high speed steel.

ALLENS OIL HARD.
M-U.S.; 0.80 C, 1.64 Mn, bal Fe.
For tools, dies; non-deforming.

ALLIAGE-30.
M-897; 2 Ni, bal Cu.
For heating elements; useful operating temperature to 300°C.

ALLIAGE-60.
M-897; 5-7 Ni, bal Cu.
For heating elements; useful operating temperature to 300°C.

ALLIAGE-90.
M-897; 11-13, Ni, bal Cu.
For heating elements; useful operating temperature to 400°C.

ALLIAGE-180.
M-897; 21-23 Ni, bal Cu.
For heating elements; useful operating temperature to 950°F.

ALLIS-CHALMERS X-1.
M-642; C, alloy, bal Fe.
600 Brin.
For hard facing electrodes; for abrasion and mild impact.

ALLIS-CHALMERS X-2.
M-642; C, Cr, Mo, bal Fe.
For hard surfacing arc-welding electrodes; resists abrasion, erosion and mild impact.

ALLIS-CHALMERS X-3.
M-642; C, Cr, Mo, bal Fe.
For hard surfacing arc-welding electrodes; resists severe abrasion, erosion and impact.

ALLISITE.
M-642; 3 C, bal Fe.
For machinery castings; high strength cast iron.

ALLITE.
M-1227; Zn, alloy.
For forming and stamping dies.

ALLOY 3-C DIE.
M-389; 0.9 C, 1.2 Mn, bal Fe.
Rockwell C 55-62.
For ball mills, grinding balls, wear shoes, liners, chutes; abrasion resistant, martensitic white iron.

ALLOY 10.
M-140; 0.8 C, bal Fe.
For springs, punches, drills, taps; Type S5.

ALLOY-10.
M-926 and M-1562; 36 Ni, 64 Fe.
Annealed: 70,000 TS; 24,000 YP; 36 El; 143 Brin.
For instruments, geodetic equipment, textile machinery parts.
Controlled expansion, corrosion resistant.

ALLOY 12-2.
M-151; 1.1-1.4 C, 12-14 Mn, 1.9-2.1 Mo, 0.4-1.0 Si, bal Fe.
Heat treated: 130,000 TS; 70,000 YS; 30 El; 30 RA; 215 Brin.
For wear and abrasion resistant parts; wear and abrasion resistant.

ALLOY 15-3.
M-151; 3-4 C, 1 max Si, 0.5-0.9 Mn, 12-18 Cr, 2.4 Mo, bal Fe.
Cast: 550 Brin.
Heat treated: 620 Brin.
For sand pumps, chute liners, brick mold liners; tough and wear resistant cast iron.

ALLOY 17-W.
M-114; 0.5 C, 12 Cr, 19 Ni, 2.2 W, 1 Mo, bal Fe.
For supercharger wheels for jet engines; heat resistant.

ALLOY 21VFS.
M-951; 2.5 Cu, 1.0-2.5 Ni, 0.5-2.0 Mg, Si, Fe, V, bal Al.
For pistons, cylinder heads; age-hardened.

ALLOY 22-3.
M-1626; 22.0 Ni, 3.1 Cr, 0.10 C, 0.50 Mn, 0.25 Si, bal Fe.
Ann: 49 kg/mm^2 TS; 28 kg/mm^2 YS; 35 El; 74 Rb.
Mean coefficient of thermal expansion, 30-450°C: 19.9 cm/cm/°C x 10^{-6}.

ALLOY 29-17.
M-1626; 29.0 Ni, 0.02 C, 0.40 Mn, 0.05 Si, 17.0 Co, bal Fe.
Ann: 53 kg/mm^2 TS; 35 kg/mm^2 YS; 30 El; 68 Rb.
Mean coefficient of thermal expansion, 30-450°C: 5.20 cm/cm/°C x 10^{-6}.
For sealing hard glasses 7040, hard pyrex glasses, ceramics, transistors, etc., to metal.

ALLOY-30.
M-789; 97 Cu, 3 Ni.
Hard: 58,000-67,000 TS.
For electrical equipment and instruments.
Resistance alloy. Max working temperature to 250°C.

ALLOY 35.
M-118; 2.5 Ag, 0.25 Cu, bal Pb.
For solder; M.P. 304°C.

ALLOY 36.
M-1626; 36 Ni, 0.12 C, 0.35 Mn, 0.30 Si, bal Fe.
Ann: 46 kg/mm^2 TS; 28 kg/mm^2 YS; 35 El; 70 Rb.
Mean coefficient of thermal expansion, -18 to +175°C: 1.63 cm/cm/° 10^{-6}.

ALLOY 39.
M-1626; 39.0 Ni, 0.08 C, 0.40 Mn, 0.25 Si, bal Fe.
Ann: 53 kg/mm^2 TS; 27 kg/mm^2 YS; 30 El; 76 Rb.
Mean coefficient of thermal expansion, 25-200°C: 2.9 cm/cm/°C x 10^{-6}.

ALLOY 40N.
M-U.S.; 54 TiC, 40 Ni, 6 Cr_3C_2.
For cutting tools, wear parts; sintered carbides.

ALLOY 41SM.
M-French; Al alloy.
For ornaments; corrosion resistant.

ALLOY 42.
M-1626; 41.0 Ni, 0.02 C, 0.40 Mn, 0.15 Si, bal Fe.
Ann: 58 kg/mm^2 TS; 28 kg/mm^2 YS; 30 El; 76 Rb.
Mean coefficient of thermal expansion, 30 to 450°C: 6.91 cm/cm/°C 10^{-6}.
For glass-to-metal and ceramic-to-metal seals.

ALLOY 42-6.
M-1626; 42.5 Ni, 5.75 Cr, 0.02 C, 0.40 Mn, 0.15 Si, bal Fe.
Ann: 56 kg/mm^2 TS; 28 kg/mm^2 YS; 30 El; 80 Rb.
Mean coefficient of thermal expansion, 30 to 425°C: 10.1 cm/cm/°C 10^{-6}.
For sealing 0120 glass to metal.

ALLOY 42B.
Mg, Si, Be, bal Al.
Sand cast: 46,000 TS; 36,000 YS; 5 El.
Permanent mold: 50,000 TS; 40,000 YS; 6 El.
For gear cases and housings, fittings; heat treatable, high strength.

ALLOY 42-C.
M-1131; 0.5 C, 0.6 Mn, 0.4 Si, 35 Ni, 15 Cr, 5 Mo, 5 W, 25 Co, bal Fe.
For jet engine and turbine blades; high heat resistance.

ALLOY 45-6.
M-1626; 45.0 Ni, 6.0 Cr, 0.02 C, 0.40 Mn, 0.15 Si, bal Fe.
Ann: 56 kg/mm^2 TS; 28 kg/mm^2 YS; 30 El; 80 Rb.
Mean coefficient of thermal expansion, 30-425°C: 10.5 cm/cm/°C x 10^{-6}.
For sealing 0120 glass to metal.

ALLOY 46.
M-1626; 46.0 Ni, 0.02 C, 0.40 Mn, 0.15 Si, bal Fe.
Ann: 58 kg/mm² TS; 24 kg/mm² YS; 27 El; 76 Rb.
Mean coefficient of thermal expansion, 30-500°C: 8.5 cm/cm/°C x 10^{-6}.
For sealing soft glasses to metal.

ALLOY 48.
M-1626; 48 Ni, bal Fe.
Soft magnetic alloy strip. High saturation and high initial and maximum permeability, developed by high temperature hydrogen anneal.
For shielding, medium value.

ALLOY 48S5.
M-U.S.; 0.15 C, 1.09 Mn, 0.09 V, 0.009 Ti, bal Fe.
For pressure vessels; impact resistant.

ALLOY 52.
M-1626; 51.0 Ni, 0.02 C, 0.40 Mn, 0.15 Si, bal Fe.
Ann: 56 kg/mm² TS; 28 kg/mm² YS; 35 El; 83 Rb.
Mean coefficient of thermal expansion, 30 to 550°C: 10.4 cm/cm/°C 10^{-6}.
For sealing 0120 glass, soft glasses and ceramics to metal.

ALLOY 60 CASE.
M-1400; 0.60 C, bal Fe.
Heat treated: 600 Brin.
For shafts, rolls, piston rods, axles.

ALLOY 73J.
M-England; 0.73 C, 1.0 Mn, 23 Cr, 6 Ni, 6 Mo, 2 Ta, bal Co.
For valves, pump parts; age-hardenable, corrosion resistant.

ALLOY-90.
M-789; 89 Cu, 11 Ni.
Hard: 58,000-67,000 TS.
For electrical equipment and instruments.
Resistance alloy. Max working temperature to 250°C.

ALLOY 95-M-255.
M-U.S.; 67 Ni, 1 Fe, 25 Mo, 6 Al.
Cast.
For high temperature applications; high strength, heat resistant.

ALLOY 100NT-2.
M-England; 1.0 C, 1.5 Mn, 20 Cr, 30 Ni, 3 Mo, 2.2 W, 2 Ta, 20 Co, bal Fe.
Heat treated: 100,000 TS; 1.5 El; 1.0 RA.
For high temperature abrasion resistant parts; abrasion and corrosion resistant.

ALLOY 110VT-2.
M-England; 1.1 C, 20 Ni, 23 Cr, 6 Mo, 2 Ta, bal Co.
For high temperature applications; heat and corrosion resistant.

ALLOY 19-9 DL.
M-114; 0.26-0.36 C, 8-10 Ni, 18-22 Cr, 1.0-1.5 Mo, 1.0-1.5 W, 0.2-0.8 Cb, 0.2-0.6 Ti.
For jet and gas engine parts, blades, buckets, wells, rockets; heat resistant.

ALLOY 19-9 WMO.
M-114; 0.08-0.12 C, 8-10 Ni, 18-22 Cr, 0.2-0.5 Mo, 1-1.5 W, 0.2-0.6 Cb, 0.2-0.6 Ti, bal Fe.
For jet engines, gas turbines, buckets, wheels, welding rod; heat resistan cast alloy.

ALLOY 223.
M-1197; 81 Cr, 16 Mo, 3 Si.
For high temperature applications; oxidation and creep resistant.

ALLOY 234A5.
0.1 C, 19 Cr, 5 Ni, Mo, W, Cb, Ti, 4 Mn, bal Fe.
For jet and turbine parts; high heat resistance.

ALLOY 234-A-5.
0.38 C, 4.2 Mn, 18.5 Cr, 4.5 Ni, 1.3 Mo, 1.3 W, 0.57 Cb, bal Fe.
For high temperature applications; heat resistant.

ALLOY 300M.
M-24, M-114; 0.40-0.47 C, 1.4-1.8 Si, 1.6-2.0 Ni, 0.7-0.95 Cr, 0.30-0.45 Mo, 0.05 min V, bal Fe.
Heat treated: 290,000 TS; 250,000 YS; 20 El; 10 RA; 540 Brin.
For aircraft landing gears, bolts, fittings; fatigue and creep resistant.

ALLOY 3-2-1.
M-151; 3.3-3.6 C, 2.75-3.25 Ni, 1.5-2.0 Cr, 0.7-1.1 Mo, 0.5-0.8 Mn, 0.3-0.6 Si, bal Fe.
Rockwell C55-62.
For ball mills, grinding balls, wear shoes, liners, chutes; abrasion resistant, martensitic white iron.
(M-151 for information only.).

ALLOY 325M.
M-114; C, alloy, bal Fe.
For aircraft landing gears, bolts; fatigue and creep resistance.

ALLOY 348.
M-Germ.; 46 Fe, 38 Ni, 16 Al.
For permanent magnet.

ALLOY 348.
M-England; 46 Fe, 38 Ni, 16 Al.
For permanent magnets; high permeability.

ALLOY 349.
M-England; 78 Fe, 14 Ni, 8 Al.
For permanent magnets; high permeability.

ALLOY 479.
M-1295; 92 Pt, 8 W.
Rolled: 300,000 TS.
For potentiometers; available bare or enamel insulated.

ALLOY 600 see INCONEL ALLOY 600.

ALLOY 644.
M-8; 4.6 Ni, 1.1 Si, 4.0 Al, bal Cu.
Precipitation hardenable alloy; can be hardened and cold worked to about 150,000 psi YS; elec. cond: 10-13% IACS.
For springs, switches, electrical connections.

ALLOY 694C.
0.4 C, 24 Ni, 15 Cr, bal Fe.
For furnace equipment, heat treat boxes; heat and corrosion resistant.

ALLOY 713 C.
M-937, M-84, M-1491; 0.08-0.2 C, 12-14 Cr, 3.8-5.2 Mo, 0.5-1.0 Ti, 5.5-6.5 Al, 1.0-3.0 Cb, bal Ni.
Cast: 123,000 TS; 107,000 YS; 8 El; C 30-42 Rock.
At 1500°F: 120,000 TS; 95,000 YS; 6 El.
For gas turbine blades in jet engines, high temperature bolting.
Cast alloy with high rupture strength to 1700°F. High thermal fatigue resistance.

ALLOY 713LC.
M-937, M-84, M-1491; 0.05 C, 0.01 B, 0.1 Zr, 12 Cr, 4.5 Mo, 0.7 Ti, 6 Al, 2 Cb, bal Ni.
Cast: 130,000 TS; 109,000 YS; 15 El; 21 RA.
At 1500°F: 123,000 TS; 99,000 YS; 15 El; 17 RA.
For turbine wheels high temperature parts, turbine blading. casting alloy. Good thermal fatigue.
Combines good low temperature ductility and high temperature strength.

ALLOY 718 see INCONEL ALLOY 718.

ALLOY 1751.
M-18; 23.2 Co, 76.8 Pt.
Rolled: H_c = 4300, Br = 6450, (BH) max = 9.5×10^6.
For permanent magnets in watches, instruments.
High coercive force and high energy product.

ALLOY 2112.
M-U.S.; 21 Cr, 12 Ni, 0.8 Si, 1.4 Mn, 2 C, bal Fe.
For exhaust valves; heat resistant, austenitic.

ALLOY 4002.
M-1625; 3.5-4.5 Si, 0.35 Fe, 0.05-0.15 Cu, 0.03 Mn, 0.05-0.15 Mg, 0.1 0.8-1.4 Cd, 0.02 Ti, bal Al.
Coiled sheet, slab or plate for connecting rod bearings, main bearings, flange bearings, crank and camshaft bearings, for automotive and diesel engines.
Conforms to AA 4002.

ALLOY 8081.
M-1625; 0.7 Si, 0.7 Fe, 0.7-1.3 Cu, 0.10 Mn, 0.05 Zn, 18.0-22.0 Sn, Ti, bal Al.
Coiled sheet, slab or plate for connecting rod bearings, main bearings, flange bearings, crank and camshaft bearings, for automotive and diesel engines.
Conforms to AA 8081.

ALLOY 8280.
M-1625; 1.0-2.0 Si, 0.7 Fe, 0.7-1.3 Cu, 0.10 Mn, 0.2-0.7 Ni, 0.05 Zn 5.5-7.0 Sn, 0.10 Ti, bal Al.
Coiled sheet, slab or plate for connecting rod bearings, main bearings, flange bearings, crank and camshaft bearings, for automotive and diesel engines.
Conforms to AA 8280.

ALLOY 9257.
M-1405; 0.93 C, 0.23 Mn, 0.13 Cr, 0.2 Si, 0.2 V, bal Fe.
For cutters, punches, dies; oil or water hardened.

ALLOY 4-22-19.
M-U.S.; 0.35-0.50 C, 23-29 Cr, 13-17 Ni, 5.7 Mo, 2 max Fe, bal Co.
For turbo supercharger and jet engine parts; corrosion and heat resistant.

ALLOY A226.
M-England; C, alloy, bal Fe.
Heat treated: 19,000 TS.
For jet engine components; age hardenable, high heat resistance.

ALLOY AP.
M-England; 3.45 C, 1.6 Si, 1.0 Mn, 0.15 P, 1.2 Ni, 0.9 Cr, bal Fe.
Cast.
For gears, shafts, machine tool housings; cast iron.

ALLOY B.
M-114; 1.05 C, 0.35 Mn, 1.35 Cr, bal Fe.
For forming rolls, deep drawing dies; oil hardening.

ALLOY B5.
M-1270; 0.07 max C, 29 Ni, 20 Cr, 2-3 Mo, 4 Cu, bal Fe.
Cast: 65,000-75,000 TS; 28,000-38,000 YS; 50-35 El; 50-40 RA; 120-150 Brin.
For chemical plant equipment, tanks; resists mixed acids, austenitic.

ALLOY B-1900.
M-114 and M-115; 0.10 C, 8 Cr, 10 Co, 6 Mo, 6 Al, 4.3 Ta, 1.0 Ti, 0.015 B, 0.07 Zr, bal Ni.
Hardened: 125,000 TS; 110,000 YS; 5 min El.
At 1600°F: 97,000 TS; 83,000 YS; 1.5 min El.
For turbine blades and other high temperature applications.
Precipitation hardened. Vacuum-melted and vacuum-cast.

ALLOY "B" FINISHING.
M-261; 1.4 C, 4.5 W, 0.8 Cr, bal Fe.
For finishing tools, draw dies; wear resistant.

ALLOY BSZ.
M-Italy.
For solder; soft.

ALLOY "C".
M-261; 2.2 C, 13 Cr, bal Fe.
For tools, dies, gauges; non-deforming.

ALLOY C-422.
M-U.S.; 0.2 C, 13 Cr, 1 Mo, 0.8 W, 0.25 V, bal Fe.
For turbine wheels; corrosion and heat resistant.

ALLOY CB-7.
M-48; 0.75 Zr, 7.5 Ti, bal Cb.
Rolled: 60,800 YS; 32 El.
For high temperature applications; high temperature strength to 2000°F

ALLOY CB-16.
M-48; 67 Cb, 10 Ti, 20 W, 3 V.
For high temperature applications; oxidation resistant to 2000°F.

ALLOY CB-22.
M-48; 94 Cb, 3 Al, 3 V.
For high temperature applications; oxidation resistant to 2000°F.

ALLOY CB-74.
M-48; 85 Cb, 10 W, 5 Zr.
At 2200°F: 44,800 TS; 42,400 YS.
For high temperature applications; oxidation resistant to 2000°F.

ALLOY CM1.
M-U.S.; 65 TiC, 11 CbC, 11 Fe, 5.7 Cr, 5.6 Ni, 0.7 TaC.
For cutting tools, wear parts, sintered carbides.

ALLOY CSA-39.
M-Eng.; 0.10 C, 27 Ni, 40 Fe, 19 Cr, 9 Mo, 3 W.
For jet engine components; high strength, heat resistant.

ALLOY CW.
0.10 C, 1.5 Mn, 0.5 Si, 21.5 Cr, 20.5 Ni, 20 Co, 3 Mo, 2 W, 1 Cb, 0.14 N, bal Fe.
At 70°F: 118,000 TS; 58,000 YS; 49 El.
At 1200°F: 73,500 TS; 37,600 YS; 28 El.
For jet engine components; oxidation and heat resistant.

ALLOY D-31.
M-U.S.; 10 Ti, 10 Mo, 80 Cb.
At 70°F: 100,000 TS.
At 2000°F: 34,500 TS.
For high temperature applications; high oxidation resistance.

ALLOY EL 437B.
M-Russia; 0.05 C, 0.23 Mn, 20.5 Cr, bal Ni.
For high temperature applications; high heat and oxidation resistance.

ALLOY FDP.
M-England; C, Cr, Ni, bal Fe.
For aircraft structures; austenitic, stainless.

ALLOY FINISHING.
M-140; 1.4 C, 3 W, bal Fe.
For cutting and finishing tools; non-deforming.

ALLOY H40.
M-English; 0.25 C, 3 Cr, 0.8 V, 0.5 W, 0.5 Mo, 5 Ni, bal Fe.
For high temperature bolts, turbine wheels; ferritic, heat resistant.

ALLOY I-336.
M-U.S.; 0.20 C, 50 Co, 15 Ni, 1 Fe, 20 Cr, 12 W, 1 Cb.
For high temperature applications; high strength, heat resistant.

ALLOY I-1360.
M-U.S.; 0.1 C, 70 Ni, 6 Fe, 10 Cr, 5 Mo, 2 Cb, 6 Al, 0.5 Be.
Cast.
For high temperature applications; high strength, heat resistant.

ALLOY IIIVT2-2.
M-England; 1.11 C, 23 Cr, 6 Mo, 2 Ta, bal Co.
Cast: 120,000 TS; 4 El; 1 RA.
For valves, pump parts; corrosion, abrasion and heat resistant.

ALLOY L-251.
M-U.S.; 0.40 C, 54 Co, 10 Ni, 1 Fe, 19 Cr, 14 W.
Cast.
For high temperature applications; high strength, heat resistant.

ALLOY L605.
M-114; 0.05 C, 10 Ni, 53 Co, 1 Fe, 20 Cr, 15 W.
Rolled: 160,000 TS; 85,000 YS; 55 El.
For gas turbine buckets, afterburners; heat resistant.

ALLOY LM-5.
40 Mo, 40 Si, 8 Cr, 2 B, 10 Al.
Uses: Coating for columbium alloys for high temperature applications.
High oxidation resistance.

ALLOY-M.
M-926 and M-1562; .08 C, 18 Cr, 8 Ni, bal Fe.
Annealed: 80,000 TS; 35,000 YS; B 80 Rock.
For chemical plant equipment, mixers, digesters, tanks, agitators.
Stainless, austenitic.

ALLOY M22 VC.
M-86; 0.08-0.16 C, 5.0-6.5 Cr, 5.9-6.6 Al, 1.5-2.5 Mo, 10.5-11.5 W, 2.6-3.4 Ta, 0.4-0.8 Zr, 1.0 max Fe, bal Ni.
High temperature applications.
Vacuum cast.

ALLOY M-252.
M-1214, M-114; 0.2 C, 9-11 Co, 3 Fe, 18-20 Cr, 9-11 Mo, 2.5 Ti, 1 Al, bal Ni.
Annealed: 175,000 TS; 20 El.
At 1000°F: 165,000 TS; 93,000 YS.
At 1600°F: 70,000 TS; 60,000 YS; 35 El.
For jet engine buckets; high heat and corrosion resistant.

ALLOY M-255.
M-259; 0.5 Al_2O_3, bal Al.
At 70°F: 22,600 TS; 17,600 YS; 22 El.
At 300°F: 15,400 TS; 13,600 YS; 22 El.
For high temperature applications; powder metallurgy.

ALLOY M-257.
M-249; 7.8 Al_2O_3, bal Al.
Sintered: 40,000 TS; 26,000 YS; 14 El.
For high temperature applications; powder metallurgy.

ALLOY M-276.
M-249; 16.5 Al_2O_3, bal Al.
Sintered: 54,000 TS; 35,000 YS; 3 El.
For high temperature applications; powder metallurgy.

ALLOY M-308.
M-114; 0.08 C, 14 Cr, 33 Ni, 4 Mo, 0.25 Al, 0.25 Zr, 2 Ti, 6.5 W, bal Fe.
For jet engine parts; heat resistant.

ALLOY MC-102.
M-86; 0.02-0.06 C, 0.1-0.4 Si, 0.1-0.5 Mn, 4.0 max Fe, 19.0-20.5 Cr, 5.0 max Co, 5.5-6.5 Mo, 6.2-6.7 Cb + Ta, 2.0-3.0 W, bal Ni.
For gas turbine stator blades, turbine rotors, pre-combustion chambers of diesel engines.
Good oxidation resistance and strength to 900°C. Age-hardenable.

ALLOY MCS.
M-114; 0.70 C, 3.5 Cr, 5.5 Mo, 1 Si, bal Fe.
For high temperature bearings; heat resistant.

ALLOY MN5.
M-Russia; Ni, bal Cu.
For high temperature applications; heat resistant.

ALLOY N 153.
M-114; 0.1 C, 1.5 Mn, 0.5 Si, 17 Cr, 15 Ni, 13 Co, 3 Mo, 2 W, 1Cb 0.14 N, bal Fe.
For jet engine components; oxidation and heat resistant.

ALLOY N 155.
M-114; 0.1 C, 1.5 Mn, 0.5 Si, 21.5 Cr, 20.5 Ni, 20 Co, 3 Mo, 2 W, 1 Cb, 0.14 N, bal Fe.
At 70°F: 118,000 TS; 58,000 YS; 49 El.
At 1200°F: 73,500 TS; 37,600 YS; 28 El.
At 1600°F: 39,000 TS; 30,000 YS; 15 El.
For jet engine components, heat and oxidation resistant.

ALLOY NO 4.
M-England; 42-47 Ni, 4-6 Cr, bal Fe.
For glass to metal seals; controlled expansion.

ALLOY NO. 188 see HAYNES ALLOY NO. 188.

ALLOY S 844.
M-U.S.; 0.30 C, 44 Co, 20 Ni, 2 Fe, 25 Cr, 3 Mo, 2 W, 2 Cb.
For high temperature applications; high strength, heat resistant.

ALLOY SSMP.
M-England; 2.98 C, 1.14 Si, 1.1 Mn, 0.18 P, bal Fe.
Cast.
For gears, shafts, machine tool housings; cast iron.

ALLOY V814.
M-German; 45 CC, 45 TiC, 7 Ni, 3 Co.
For cutting tools bits, dies; sintered carbides.

ALLOY VCM.
M-England; 0.28 C, 0.7 Mn, 1.0 Cr, 1.0 Mo, 0.7 Ni, bal Fe.
For gears, shafts, cams; nitriding steel.

ALLOY W.
M-Germany; 0.5-0.9 Cu, 0.28-0.36 Ni, 0.25 Co, 0.16 Mg, V, W, Cr, Be, bal Al.
For anodized die castings, housings.

ALLOY Z 132 now AA 332.

ALLOYMET 1005.
M-1795; 4.5-5.5 Cr, bal Cu.
For Cr additions to Cu alloys; master alloy.

ALLOYMET 1010.
M-1795; 9.5-10.5 Cr, bal Cu.
For Cr additions to Cu alloys; master alloy.

ALLOYMET 1110.
M-1795; 9-11 Fe, bal Cu.
For Cu additions to steel and cast iron; master alloy.

ALLOYMET 1115.
M-1795; 4.5-5.5 Fe, bal Cu.
For Cu additions to steel and cast iron; master alloy.

ALLOYMET 1130.
M-1795; 28-32 Fe, bal Cu.
For Cu additions to steel and cast iron; master alloy.

ALLOYMET 1230A.
M-1795; 28-32 Mn, bal Cu.
For Mn additions for Ni alloys; hardener, deoxidizer.

ALLOYMET 1230B.
M-1795; 23.5-26.5 Mn, 4-6 Fe, bal Cu.
For Mn additions to bronzes; hardener, deoxidizer.

ALLOYMET 1315.
M-1795; 15 Ni, bal Cu.
For alloying additions for bronzes.

ALLOYMET 1330.
M-1795; 30 Ni, bal Cu.
For alloying additions.

ALLOYMET 1350.
M-1795; 50 Ni, bal Cu.
For alloying additions.

ALLOYMET 1410.
M-1795; 9-11 Si, bal Cu.
For high conductivity copper; master alloy.

ALLOYMET 1420.
M-1795; 18-22 Si, bal Cu.
For high conductivity copper; master alloy.

ALLOYMET 1430.
M-1795; 28-32 Si, bal Cu.
For high conductivity copper; hardener, master alloy.

ALLOYMET 2000.
M-1795; 99.8 Ni.
For anodes for plating and additions to iron and steel.

ALLOYMET 2023.
M-1795; 60 Ni, 25 Cu, bal Mn, Si, Fe and C.
For Ni and Cu additions to steel and cast iron; alloying master alloy.

ALLOYMET 2030.
M-1795; 30 Cu, 70 Ni.
For Ni and Cu additions to steel and cast iron; alloying master alloy.

ALLOYMET 2030M.
M-1795; 65 Ni, 30 Cu, 2 max Fe, 1 max Mn, 0.15 max C.
For Ni and Cu additions to steel and cast iron; alloying master alloys.

ALLOYMET 2041.
M-1795; 59 Ni, 29 Si, bal Fe.
For alloying additions of nickel to ferrous and some non-ferrous alloys.

ALLOYMET 2115.
M-1795; 77 min Ni, 14-17 Cr, 6-10 Fe, 1 Mn, 0.5 max Cu.
For alloying additions to stainless steels; for high temperature castings.

ALLOYMET 2115 WH.
M-1795; 33-37 Ni, 14-17 Cr, 0.5-2.0 Mn, 0.8-2.5 Si, bal Fe.
For alloying additions to stainless steels; for high temperature castings.

ALLOYMET 2120.
M-1795; 77-79 Ni, 19-20 Cr, 1 Fe, bal Mn, C, Si.
For alloying additions to stainless steels; alloying master alloy.

ALLOYMET 2140.
M-1795; 57 min Ni, 14-18 Cr, 3 max Mn, 1.5 max Si, bal Fe.
For alloying additions to stainless steels; alloying master alloy.

ALLOYMET 2165.
M-1795; 35 Ni, 10 Cr, bal Fe.
For alloying additions to Ni Hard Iron and to Stainless Steel.

ALLOYMET 2200.
M-1795; 56 Ni, 23 Cu, 7.5 Cr, bal Fe.
For alloying additions for Ni-Resist Iron.

ALLOYMET 2350.
M-1795; 47-50 Ni, bal Fe.
For alloying for high permeability; soft magnets.

ALLOYMET 2364.
M-1795; 36 Ni, bal Fe.
For alloying for low expansion applications; glass seals.

ALLOYMET 2470.
M-1795; 30 Ni, 15 Cu, 0.2 Mn, bal Fe.
For glass to metal seals; low expansion.

ALLOYMET AM1.
M-1795; 30 Ni Ferro Nickel.
For alloying additions to alloy iron.

ALLOYMET AM2.
M-1795; 40 Ni Ferro Nickel.
For alloying additions to stainless steel.

ALLOYMET MG.
M-1795; 76 Ni, 14 Cu, bal Fe.
For high permeability alloys; low field strength.

ALLOYMET NI CR MO 1.
M-1795; 35 Ni, 10 Cr, 5 Mo, bal Fe.
For alloying additions to alloy steel.

ALLOYMET NI CU 1.
M-1795; 50 Ni, 25 Cu, bal Fe.
For alloying additions to low alloy steels.

ALLOYMET NI CU 2.
M-1795; 45 Ni, 15 Cu, bal Fe.
For alloying additions to ferrous and some non-ferrous alloys.

ALLOYMET NI MO 1.
M-1795; 35 Ni, 5 Mo, bal Fe.
For alloy additions to ferrous alloys.

ALLOYMET PERMAG.
M-1795; 20 Ni, 60 Cu, bal Fe.
For permanent magnets.

ALLOY Z-FM.
M-353; 0.4 C, 1 Cr, 0.75 Ni, 0.35 Mo, 0.35 Pb, bal Fe.
Heat treated: 143,000 TS; 122,500 YS; 16 El; 52 RA; 300 Brin.
For machinery parts, gears, pins, shafting; oil hardened, shock resistant.

ALL-PO.
M-507; 0.90 C, 0.4 Si, 0.4 Mn, 9.5 Cr, 0.6 Mo, bal Fe.
Coated, hard facing electrode.
As arc welded: 185,000 psi, 52-55 Rc.
For overlay or buildup on scrapers, loaders, dozer bits, blades, shovels, buckets. Good abrasion resistance.

ALL-STATE AH, HSS, HW, OH AND WH see **TOOL-ARC AIR HARDENING, HIGH SPEED, HOT WORK, OIL HARDENING AND WATER HARDENING.**

ALL-STATE CHAMFER ROD.
M-1713.
Electrode for AC-DC straight polarity for chamfering and gouging.

ALL-STATE CUTTING ELECTRODE.
M-1713.
Electode for AC-DC straight polarity cutting without the use of oxygen or compressed air; for nickel, copper, brass, bronze, aluminum, stainless and alloy steels.

ALL-STATE GALVOVER.
M-1713.
Electrode to restore galvanized surface on welded or damaged galvanized parts.
Working temperature: 600°F.

ALL-STATE LOW FUMING BRONZE.
M-1713.
Bronze rod for torch brazing steel, cast iron, galvanized malleable and copper base alloys.
Working temperature: 1600°F; strengths up to 60,000 psi.

ALL-STATE MANGANESE BRONZE.
M-1713.
Alloy for torch brazing various metals; resistant to salt water.

ALL-STATE MONOWELD.
M-1713.
Electrode for AC-DC, reverse polarity, all-position welding of mild and low alloy steels.
As welded: up to 62,000 psi YS; 20-22 El.

ALL-STATE NICKEL BRONZE.
M-1713.
Ni-bronze alloy for torch brazing various alloys.
Working range: 1200-1750°F; strengths up to 85,000 psi.

ALL-STATE SEALCOR.
M-1713.
Flux cored rod for torch brazing of weldable grades of aluminum.
Working temperature; 1100°F; strengths up to 30,000 psi.

ALL-STATE SILFLO "0".
M-1713; 7 P, bal Cu.
Self-fluxing brazing alloy; very fluid.
Working temperature: 1350-1550°F.

ALL-STATE SILFLO "5".
M-1713; 5 Ag, 6 P, bal Cu.
Self-fluxing brazing alloy; for wider gaps.
Working temperature: 1300-1500°F.

ALL-STATE SILFLO "15".
M-1713; 15 Ag, 5 P, bal Cu.
Self-fluxing brazing alloy; for ductile joints.
Working range: 1300-1500°F.

ALL-STATE SILICON BRONZE.
M-1713; Si, bal Cu.
Electrode for tungsten inert gas or carbon arc brazing or welding silicone bronze, cooper alloys and some iron base metals.
Strengths up to 58,000 Psi.

ALL-STATE STEELARC.
M-1713.
Electrode for AC-DC, reverse polarity, all-position welding mild steel particularly dirty, rusty sheet metal.
As welded; up to 70,000 psi Ts.

ALL-STATE STEELARC PLUS.
M-1713.
Improved grade of ALL-STATE STEELARC.

ALL-STATE NO 3.
M-1713.
Copper coated cast iron alloy for torch brazing manifolds, cylinder heads and other cast iron. Machinable.
175-225 Brin; tensile up to 45,000 psi; working temperature; 1500°F.

ALL-STATE NO 4.
M-1713.
175-225 Brin.
Electrode for AC-DC, reverse polarity welding thin sections of cast iron.

Section I: Alloy Data

ALL-STATE NO 4-60.
M-1713.
200-300 Brin.
Electrode for AC-DC reverse polarity, all position welding heavy sections of cast iron.

ALL-STATE NO 4 IMP.
M-1713.
Very machinable; strength to 50,000 psi.
Electrode fo AC-DC, reverse polarity welding of cast iron.

ALL-STATE NO 6 IMP.
M-1713.
Electrode for AC-DC, reverse polarity welding of cast iron.
Strength to 60,000 psi; not readily machinable.

ALL-STATE NO 7.
M-1713.
Solder for sealing cracks in cast iron, steel, copper, brass and bronze.
Working temperature: 450-600°F.

ALL-STATE NO 8.
M-1713.
175-225 Brin.
Electrode for AC-DC, reverse polarity, all position welding thin sections of cast iron.

ALL-STATE NO 8-60.
M-1713.
200-300 Brin; strength up to 80,000 psi.
Electrode for AC-DC, reverse polarity, all position welding of heavy sections of cast iron and ductile iron.

ALL-STATE NO 11.
M-1713; 47 Cu 43 Zn, 10 Ni.
For brazing; M.P. 1665-1715°F.

ALL-STATE NO 13.
M-1713; 48 Cu, 9.6 Ni, 0.29 Si, 0.18 Fe, 0.16 Mn, bal Zn.
For brazing alloy; nickel silver.

ALL-STATE NO 13FC.
M-1713.
Flux coated grade of ALL-STATE NO 13.

ALL-STATE NO 16.
M-1713.
125-150 Brin.
Electrode for DC reverse polarity, all position welding of Ni-Cu alloys.

ALL-STATE NO 18.
M-1713.
Electrode for DC reverse polarity welding of Ni-Cr-Fe alloys includin 9% Ni steels for cryogenic purposes.

ALL-STATE NO 20.
M-1713.
120-150 Brin; strengths up to 91,000 psi.
Aluminum bronze electrode for DC reverse polarity welding and bulid-up o ferrous and non-ferrous metals.

ALL-STATE NO 21.
M-1713.
Working temperature: 1445-1460°F.
Phosphor-copper brazing rod for torch brazing copper tubing and pipe.

ALL-STATE NO 23.
M-1713; Ag, P, bal Cu.
Working teimperature: 1435-1450°F; strength up to 45,000 psi.
Brazing alloy for torch brazing copper, brass and bronze.

ALL-STATE NO. 24.
M-1713; 7-9 Sn, 0.03-0.35 P, bal Cu.
Welded: 45,000 TS; 80-160 Brin.
For P-bronze welding electrodes; for welding steel, cast iron and copper alloys.

ALL-STATE NO 31.
M-1713.
Working temperature 1075°F; strengths up to 30,000 psi.
Alloy for brazing thin sheet aluminum ; (not recommended for heat-treatabl alloys).

ALL-STATE NO 33.
M-1713.
Working temperature: 1050°F.
Alloy for torch brazing aluminum castings.

ALL-STATE NO 34.
M-1713.
Aluminum base electrode for DC reverse polarity, all-position welding an repair of aluminum housings.
(Also called SMOOTHCOTE NO 34.)

ALL-STATE NO. 39.
M-1713; 59 Sn, 41 Zn.

ALL-STATE NO. 41.
M-1713; 58 Cu, 1 Sn, 1 Fe, bal Zn.
For brazing rod; for gas welding.

ALL-STATE NO 41FC.
M-1713.
Working temperature: 1400-1650°F; strengths up to 60,000 psi.
Flux coated bronze for torch brazing; general maintenance and repair steel cast iron, brass bronze.

ALL-STATE NO. 53.
M-1713; 4 Pb, 3 Al, 2 Cu, bal Zn.
For Zn base welding rod for Zn alloys; for gas welding.

ALL-STATE NO 55 RUBBON.
M-1713.
Working temperature; 705-720°F; strength up to 24,000 psi.
Lead free, self-fluxing aluminum solder.

ALL-STATE NO. 100.
M-1713; 40 Ag, 30 Cu, 28 Zn, 2 Ni.
For brazing, silver solder; M.P. 1220-1435°F.

ALL-STATE NO 101.
M-1713; 45 Ag, Cu Zn, Cd.
Silver solder; melt range: 1125-1145°F.
For joining wide range of metals; strength up to 50,000 psi.

ALL-STATE NO 101FC.
M-1713.
Flux coated grade of ALL-STATE NO 101.
(Also known as TRUCOTE NO 101FC.)

ALL-STATE NO 107.
M-1713.
Silver bearing soft solder for torch, furnace or induction soldering of ferrous and non-ferrous alloys.
Melt range: 480-600°F; strength up to 20,000 psi.

ALL-STATE NO 111.
M-1713; 41% Ag, brazing alloy.
Melt range: 1100-1150°F (lowest melting); TS to 50,000 psi.

ALL-STATE NO 155.
M-1713.
Silver solder (Cadmium free).
For brazing stainless steel food handling equipment to hot water systems.
Melt range: 1150-1200°F; TS to 50,000 psi.

ALL-STATE NO 155FC.
M-1713.
Flux coated grade of ALL-STATE NO 155.
(Also known as TRUCOTE NO 155FC.)

ALL-STATE NO 164.
M-1713.
Silver solder for torch or furnace brazing of "hard-to-solder" applications, particularly carbide tips to steel tools.
Working temperature 1500°F.

ALL-STATE NO 252.
M-1713.
Electrode for AC-DC, reverse polarity, all-position welding of 310, 31 stainless and other high temperature steels. 85 Rb; YS up to 65,000 psi (room temp.).

ALL-STATE NO.275.
M-1713.
Electrode for AC-DC reverse polarity, all-position welding of high strength steels.
As welded: up to 120,000 psi TS; work-hardenable to 180,000 psi.

ALL-STATE NO. 321.
M-1713; 30 Sn, 30 Zn, 40 Pb, plus flux.
powder for galvanizing.
Working temperature: 450°F.

ALL-STATE NO. 430.
M-1713.
Low temperature silver bearing solder; cadmium free.
Melt temp: 430°F; strength up to 15,000 psi.
For food handling equipment.

ALL-STATE NO 4-60.
M-1713.
200-300 Brin.
Electrode for AC-DC reverse polarity, all position welding heavy sections of cast iron.

ALL-STATE NO. 509 STRONGSET.
M-1713; Cd-Zn.
Solder for joining aluminum and other metals (except magnesium).
Melt temp: 509°F: strength up to 29,000 psi.

ALL-STATE NO. 616.
M-1713.
Electrode for AC-DC, reverse polarity, all-position welding, particula high tensile and sulphur bearing steels.
As welded: up to 80,000 psi TS.

ALLVAC 3-2.5.
M-1490; 3.0 Al, 2.5V, bal Ti.
Titanium alloy.
Density: 0.162 lb/cu in.

ALLVAC 5-2.5.
M-1490; 5.0 Al, 2.5 Sn, bal Ti.
Titanium alloy.
Density: 0.161 lb/cu in.

ALLVAC 6-2-4-2.
M-1490; 6.0 Al, 2.0 V, 4.0 Zr, 2.0 Mo, bal Ti.
Titanium alloy.
Density: 0.164 lb/cu in.

ALLVAC 6-2-4-6.
M-1490; 6.0 Al, 2.0 V, 4.0 Zr, 6.0 Mo, bal Ti.
Titanium alloy.
Density 0.168 lb/cu in.

ALLVAC 6-4.
M-1490; 6.0 Al, 4.0 V, bal Ti.
Titanium alloy.
Density: 0.160 lb/cu in.

ALLVAC 6-6-2.
M-1490; 6.0 Al, 6.0 V, 2.0 Sn, bal Ti.
Titanium alloy.
Density 0.164 lb/cu in.

ALLVAC 8-1-1.
M-1490; 8.0 Al, 1.0 V, 1.0 Mo, bal Ti.
Titanium alloy.
Denstiy: 0.156 lb/cu in.

ALLVAC 30.
M-1490; 0.10 Fe, 0.10 O, bal Ti.
Titanium alloy; CP Grade I.
Density: 0.163 lb /cu in.

ALLVAC 40.
M-1490; 0.10 Fe, 0.15 O, bal Ti. (Also with 0.15 Pd).
Titanium alloy; CP Grade II.
Density: 0.163 lb/cu in.

ALLVAC 50.
M-1490; 0.20 Fe, 0.20 O bal Ti.
Titanium alloy; CP Grade III.

ALLVAC 70.
M-1490; 0.35 Fe, 0.35 O bal Ti.
Titanium alloy; CP Grade IV.

ALLVAC 316 SS.
M-1490; 0.05 C, 17.4 Cr, 2.7 Mo, 13.8 Ni, 1.75 Mn, 0.60 Si, bal Fe.
Corrosion resistant alloy; nuclear quality only.

ALLVAC 500 ZB.
M-1490; 0.08 C, 19 Cr, 18.8 Co, 4 Mo, 54 Ni, 3 Ti, 3 Al, 0.006 B, 0.06 Zr.
At R.T.: 176,000 TS; 110,000 YS; 16 El.
At 1400°F: 151,000 Ts; 106,000 YS; 21 El.
For missiles, space equipment, jet engine and gas turbine parts.
High corrosion and oxidation resistant.
Good creep resistance and notch toughness.

ALLVAC 520.
M-1490; 0.04 C, 19.0 Cr, 13.0 Co, 6.3 Mo, 3.0 Ti, 2.0 Al, 0.005 B, 1.0 W, bal Ni.
Corrosion resisting and high temperature alloy.

ALLVAC 700.
M-1490; 0.13 C, 15.3 Cr, 18.5 Co, 5.0 Mo, 3.5 Ti, 4.4 Al, 0.014 B, bal Ni.
For jet engine components and space vehicles.

ALLVAC 718.
M-1490; 0.03-0.01 C, 17-21 Cr, 2.8-3.3 Mo, 50-55 Ni, 5.0-5.5 Cb, 0.65-1.15 Ti, 0.4-0.8 Al, bal Fe.
Heat treated: 190-000-215,000 Ts; 160,000-185,00 YS; 15-25 El, 20-30 RA, C 40-44 Rock.
For heat resistant parts to 1300°F.
Age hardening, high strength.
Corrosion and oxidation resistant.

ALLVAC ASTROLOY.
M-1490; 0.06 C, 15 Cr, 17 Co, 5 Mo, 3.5 Cb, 3.5 Ti, 4 Al, 0.03 B, 0.06 Zr, bal Ni.
Heat treated: 205,000 TS; 152,000 YS; 12 El; 14 RA.
At 1400°F: 160,000 TS; 132,000 YS; 15 El; 17 RA.
For turbine wheels and blades, jet engine components, nozzles, compressor discs.
Heat and corrosion resistant.
High creep and stress-rupture strength.

ALLVAC I-700.
M-1490; 0.13 C, 15 Cr, 29 Co, 3.75 Mo, 46 Ni, 2.55 Ti, 3.25 Al.
At R.T.: 170,000 TS; 103,000 YS 25 El.
At 1400°F: 127,000 TS; 8900 YS; 11 El.
For jet engine and gas turbine components, space vehicles and missiles.
Oxidation and corrosion resistant.
Good creep resistance and notch toughness.

ALLVAC M-252.
M-1490; 0.10-0.15 C, 0.5 max Mn, 18-20 Cr, 9-11 Co, 9-11 Mo, 2.25-2.75 Ti, 0.5-1.25 Al, 0.006 B, 5 max Fe, bal Ni.
Bar: 180,000 TS; 20 El; C 40 Rock; 122,000 YS.
At 1200°F: 168,000 TS; 108,000 YS; 11 El; C 33 Rock.
For heavy duty jet engine and gas turbine components.
High corrosion and heat resistance.
Age hardenable. Good creep strength.

ALLVAC N-105.
M-1490; 0.15 C 15.0 Cr, 20.0 Co, 5.0 Mo, 1.3 Ti, 4.5 Al, 0.007 B, bal Ni.
Corrosion resisting and high temperature alloy.

ALLVAC N-115.
M-1490; 0.16 C, 15.0 Cr, 15.0 Co, 4.0 Mo, 4.0 Ti, 5.0 Al, 0.017 B, bal Ni.
Corrosion resisting and high temperature alloy.

ALLVAC R-235.
M-1490; 0.12 C, 15.5 Cr, 5.5 Mo, 65 Ni, 2.5 Ti, 2 Al, 10 Fe
At R.T.: 160,000 TS; 100,000 YS; 30 El.
At 1500°F: 105,000 TS; 85,000 YS; 18 El.
For high temperature components.
Precipitation hardening. Good oxidation and corrosion resistance.

ALLVAC REN AE 41.
M-1490; 0.09 C, 19 Cr, 11 Co, 10 Mo, 3 Ti, 1.5 Al, 0.006 B, bal Ni.
At 1400°F: 126,000 TS; 88,000 YS; 11 El.
At R.T.: 17,000 TS; 103,000 YS; 25 El.
For jet engine and gas turbine components, after-burners, high temperature bolts.
High temperature, high strength properties.
Corrosion and oxidation resistant.

ALLVAC WASPALOY.
M-1490; 0.10 max C, 12-15 Co, 3.5-5.0 Mo, 2.6-3.2 Ti, 1.0-1.6 Al, 0.0 Zr, 0.008 B, 18-21 Cr, bal Ni.
Annealed: 121,000 TS; 70,000 YS; 54 El; B 90 Rock.
At 1400°F: 120,000 TS; 99000 YS; 28 El; 42 RA.
For turbine engines, airframe assemblies, missile components, fasteners.
High stress-rupture strength and oxidation resistant.

ALMADUR MP6.
M-German; Al alloy.
For bearings.

ALMADUR MZ3.
M-German; Al alloy.
For bearings.

ALMAG.
M-871; 3.5-4.5 Cu, 0.4-1.0 Mn, 0.3-0.75 Mag, bal Al.
Rolled: 54,000 TS; 22 El; 100 Brin.
For light alloy parts, aircraft construction; Duralumin type.

ALMAG 35.
M-1022; 0.1 Ti, 0.1 Mn, 0.05 Be, 6.5-7.5 Mg, bal Al.
Cast: 38,000-42,000 TS; 18,000-21,000 YS; 10-15 El; 70 Brin.
For castings, ornaments, machine tool parts; not hardenable, corrosion resistant.

ALMASILIUM.
M-767; 2 Si, 1 Mg, bal Al.
50,000 TS; 6 El; 90 Brin.
For light alloy parts; corrosion resistant.

ALMECO.
M-1784.
Plated Al alloy wire.
For insulated building wire.

ALMELEC.
M-1784; 0.5-0.6 Mg, 0.5-0.6 Si, bal Al.
Heat treated to 34.5 Kg/mm$_2$ TS; 4 El.
For overhead line conductors.

ALMELEC.
M-979; 0.4-1.2 Mg 0.5-1.2 Si, 0-0.3 Fe, bal Al.
44,600 TS; 6-8 El; 90 Brin.
For electrical transmission lines; heat treated wrought French alloy.

ALMET 302.
M-248; 0.20 max C, 17-19 Cr, 8-10 Ni, bal Fe.
Soft: 105,000 TS; 70,000 YS; 35 El.
Hard: 330,000 TS; 310,000 YS; 2 El.
For springs, trim, wire parts; stainless, austenitic; Type 302.

ALMET 303.
M-248; 0.15 C, 17-19 Cr, 8-10 Ni, 0.07 min S or Se, bal Fe.
Soft: 100,000 TS; 45,000 YS; 25 El; 159 Brin.
Hard: 150,000 TS; 130,000 YS; 15 El; 240 Brin.
For screw machine products; stainless, austenitic, free-cutting.

ALMET 304.
M-248; 0.08 max C, 18-20 Cr, 8-10 Ni, bal Fe.
Soft: 90,000 TS; 35,000 YS; 55 El; 144 Brin.
Hard: 230,000 TS; 220,000 YS; 2 El; 430 Brin.
For woven wire, cold headed parts; stainless, austenitic; Type 304.

ALMET 305.
M-248; 0.12 max C, 18 Cr, 12 Ni, bal Fe.
Annealed: 85,000 TS; 47,000 YS; 60 El; 77 RA; 144 Brin.
Soft: 100,000 TS; 54,000 YS; 58 El; 74 RA; 156 Brin.
For spun, cold drawn and cold headed parts; austenitic, stainless.

ALMET 308.
M-249; 0.08 max C, 19-21 Cr, 10-12 Ni, bal Fe.
Soft: 85,000-105,000 TS; 35,000-40,000 YS; 25-55 El; 150 Brin.
Hard: 160,000-190,000 TS; 150-000-185,000 YS; 4-8 El; 380-430 Brin.
For electrical equipment; austenitic, stainless.

ALMET 309.
M-248; 0.20 max C, 22-24 Cr, 12-15 Ni, bal Fe.
Soft: 85,000 TS; 35,000 YS; 55 El; 159 Brin.
Hard 190,000 TS; 185,000 YS; 4 El; 430 Brin.
For heat resistant parts; corrosion and heat resistant.

ALMET 310.
M-248; 0.25 max C, 24-26 Cr, 19-22 Ni, bal Fe.
Soft: 85,000 TS; 35,000 YS; 25 El; 139 Brin.
Hard: 190,000 TS; 185,000 YS; 4 El; 430 Brin.
For heat resistant parts; corrosion and heat resistant.

ALMET 314.
M-248; 0.25 max C, 23-26 Cr, 19-22 Ni, 2-3 Si, bal Fe.
Soft: 85,000 TS; 35,000 YS; 25 El; 139 Brin.
Hard: 190,000 TS; 185,000 YS; 4 El; 430 Brin.
For heat resistant parts; corrosion and heat resistant.

ALMET 316.
M-248; 0.10 max C, 18 Cr, 10-14 Ni, 2-3 Mo, bal Fe.
Soft: 85,000 TS; 35,000 YS; 25 El; 139 Brin.
Hard: 190,000 TS; 185,000 YS; 4 El; 430 Brin.
For acid and chemical equipment; stainless, austenitic.

ALMET 317.
M-248; 0.10 C, 18 Cr, 12 Ni, 3-4 Mo, bal Fe.
Soft: 85,000 TS; 35,000 YS; 25 El; 139 Brin.
Hard: 190,000 TS; 185,000 YS; 4 El; 430 Brin.
For welded stainless structures; stainless, austenitic.

ALMET 321.
M-248; 0.08 max C, 17-19 Cr, 8-11 Ni, Ti = 5 x C, bal Fe.
Soft: 90,000 TS; 35,000 YS; 25 El; 137 Brin.
Hard: 230,000 TS; 220,000 YS; 2 El; 430 brin.
For welded stainless parts; austenitic, stainless, stabilized.

ALMET 330.
M-248; 0.20 max C, 15-17 Cr, 34-36 Ni, bal Fe.
Soft: 85,000 TS; 35,000 YS; 25 El; 139 Brin.
Hard: 190,000 TS; 185,000 YS; 4 El; 430 Brin.
For carburizing and heat treating equipment; heat resistant, austenitic.

ALMET 347.
M-248; 0.08 max C, 17-19 Cr, 9-12 Ni, Cb = 10 x C, bal Fe.
Soft: 85,000 TS; 35,000 YS; 25 El; 139 Brin.
Hard: 190,000 TS; 185,000 YS; 4 El; 430 Brin.
For welded stainless parts; stainless, austenitic.

ALMET 410.
M-248; 0.15 max C, 11.5-13.5 Cr, bal Fe.
Soft: 75,000 TS; 35,000 YS; 15 El; 150 Brin.
Hard: 190,000 TS; 145,000 YS; 8 El; 300 Brin.
For valve trim, cold headed parts; corrosion resistant.

ALMET 416.
M-248; 0.15 max C, 12-14 Cr, 0.07 min S or Se, bal Fe.
Soft: 75,000 TS; 35,000 YS; 14 El; 150 Brin.
1/2 H-temper: 115,000 TS; 80,000 YS; 16 El; 185 Brin.
For valve trim, cold headed parts; corrosion resistant, free-cutting.

ALMET 430.
M-248; 0.12 max C, 14-18 Cr, bal Fe.
Soft: 75,000 TS; 40,000 YS; 15 El; 156 Brin.
Hard: 140,000 TS; 130,000 YS; 4 El; 300 Brin.
For screws, rivets, formed parts; corrosion resistant, good formability.

ALMET C-20.
M-248; 0.07 max C, 29 Ni, 20 Cr, 3 min Cu, bal Fe.
Rolled: 85,000 TS ; 35,000 YS; 35-50 El; 50-70 RA; 150-180 Brin.
For pump and valve parts; corrosion and heat resistant.

ALMINAL A2.
M-1273, M-1274; 0.02 Cu, 0.15 Si, 0.15 Fe, 99.8 min Al.
M-temper: 8000 TS; 30 El.
For structural members; corrosion resistant.

ALMINAL A3.
M-1273, M-1274; 0.05 Cu, 0.3 Si, 0.4 Fe, 0.05 Mn, bal Al.
M-temper: 9000 TS; 25 El.
For structual members; corrosion resistant.

ALMINAL A4.
M-1273, M-1274; 0.1 Cu, 0.5 Si, 0.7 Fe, 0.1 Mn, bal Al.
M-temper: 9000 TS; 20 El.
For structural members; corrosion resistant.

ALMINAL C2.
M-1273, M-1274; 0.7-2.5 Cu, 9-11 Si, 0.3 Mg, 1 Fe, 0.5 Mn, 1 Ni, 1. Zn, 0.2 Ti, bal Al.
Sand cast: 18,000 TS.
Permanent mold: 20,000 TS.
For pressure tight castings; corrosion resistant.

ALMINAL C4.
M-1273, M-1274; 2-4 Cu, 4-6 Si, 0.15 Mg, 0.8 Fe, 0.6 Mn, 0.3 Ni, 0. Zn, 0.2 Ti, bal Al.
Sand cast: 20,000 TS; 2 El.
Permanent mold: 22,400 TS; 2 El.
For general castings; corrosion resistant.

ALMINAL C5.
M-1273, M-1274; 0.1 Cu, 3-6 Mg, 0.3 Si, 0.6 Fe, 0.5 Mn, 0.1 Ni, 0.1 Zn, bal Al.
Sand cast: 20,000 TS; 3 El.
Permanent mold: 24,600 TS; 5 El.
For structural members and castings; corrosion resistant.

ALMINAL C6.
M-1273, M-1274; 0.1 Cu, 0.1 Mg, 10-13 Si, 0.5 Mn, 0.1 Ni, 0.1 Zn, bal Al.
Sand cast: 23,500 TS; 5 El.
Permanent mold: 27,000 TS; 7 El.
For pressure tight casting; corrosion resistant.

ALMINAL C7.
M-1273, M-1274; 10.0-2.5 Cu, 0.2 Mg, 1.5-3.5 Si, 0.3-1.4 Fe, 0.5-1.7 Ni, 0.1 Mn, bal Al.
Sand cast: 20,000 TS; 2 El.
Permanent mold: 22,400 TS; 2 El.
For structural members and castings; pressure tight castings.

ALMINAL C8.
M-1273, M-1274; 0.1 Cu, 0.4 Mg, 0.6 Fe, 4.5-6.0 Si, 0.4 Mn, 0.1 Ni, 0.1 Zn, bal Al.
Sand cast: 24,600 TS; 2.5 El.
WP-temper: 38,000 TS; 2.5 El.
For pressure tight castings; corrosion resistant, good castability.

ALMINAL C9.
M-1273, m-1274; 0.1 Cu, 0.4 Mg, 0.6 Fe, 10-13 Si, 0.4 Mn, 0.1 Ni, 0.1 Zn, bal Al.
Sand Cast: 24,600 TS; 1.5 El.
WP-temper: 42,500 TS.
For pressure tight castings; corrosion resistant, good castability.

ALMINAL C10.
M-1273, M-1274; 0.1 Cu, 9.5-11 Mg, 0.35 Si, 0.35 Fe, 0.2 Ti, bal Al.
Sand cast: 40,000 TS; 8 El.
Permanent mold: 44,800 TS; 12 El.
For structural castings; corrosion resistant.

ALMINAL C11.
M-1273, M-1274; 4-5 Cu, 0.25 Si, 0.25 Fe, 0.2 Ti, bal Al.
Sand Cast: 40,000 TS; 7 El.
Permanent mold: 44,800 TS; 19 El.
For aircraft engine castings; age-hardened.

ALMINAL C12.
M-1273, M-1274; 9.0-10.5 Cu, 0.4 Mg, 2 Si, 0.5-1.5 Fe, 0.6 Mn, 0.5 Ni, 0.1 Zn, bal Al.
Sand cast: 100 Brin.
Permanent mold: 150 Brin.
For structural members and castings; age-hardened.

ALMINAL C13.
M-1273, M-1274; 0.5-1.3 Cu, 0.8-1.5 Mg, 11-13 Si, 2-3 Ni, 0.8 Fe, 0.5 Mn, bal Al.
Sand cast: 24,500 TS; 100 Brin.
Permanent mold: 36,000 TS; 150 Brin.
For pressure tight castings; age-hardened, corrosion resistant.

ALMINAL C15.
M-1273, M-1274; 1.3-3.0 Cu, 0.5-1.7 Mg, 0.6-2.0 Si, 0.8-1.4 Fe, 0.5-2.0 Ni, 0.2 Ti, bal Al.
Sand cast: 40,400 TS; 100 Brin.
Permanent mold: 47,100 TS; 150 Brin.
For structual members; age-hardened, high temperature uses.

ALMINAL W4.
M-1273, M-1274; 0.15 Cu, 1.75-2.25 Mg, 0.6 Si, 0.7 Fe, 0.5 Mn, 0.2 Ti, bal Al.
M-temper: 24,600 TS; 18 El.
For structural members; corrosion resistant.

ALMINAL W5.
M-1273, M-1274; 3-4 Mg, 0.15 Cu, 0.6 Si, 0.7 Fe, 1 Mn, 0.2 Ti, bal Al.
M-temper; 31,400 TS; 12,000 YS; 18 El.
For structural members; corrosion resistant.

ALMINAL W6.
M-1273, M-1274; 0.15 Cu, 4.5-5.5 Mg, 0.6 Si, 0.7 Fe, 1 Mn, 02 Ti, bal Al.
M-temper: 36,000 TS; 17,000 YS; 18 El.
For structural members; corrosion resistant.

ALMINAL W7.
M-1273, M-1274; 0.15 Cu, 6.5-7.5 Mg, 0.6 Si, 0.7 Fe, 1 Mn, 0.2 Ti, bal Al.
M-temp: 44,800 TS; 19,000 YS; 18 El.
For structural members; corrosion resistant.

ALMINAL W9.
M-1273, M-1274; 0.15 Cu, 0.4-0.9 Mg, 0.3-0.07 Si, 0.6 Fe, 0.2 Ti, bal Al.
M-temper: 15,700 TS; 10,000 YS; 15 El.
W-temper: 19,500 TS; 11,000 YS; 18 El.
WP-temper: 27,000 TS; 22,400 YS; 12 El.
For structural members; age-hardened.

ALMINAL W10.
M-1273, M-1274; 0.15 Cu, 0.4-1.5 Mg, 0.75-1.3 Si, 0.6 Fe, 1 Mn, 0.2 Ti, bal Al.
W-temper: 27,000 TS; 15,700 YS; 18 El.
Wp-temper: 40,400 TS; 33,600 YS; 10 El.
For structural members; age-hardened.

ALMINAL W11.
M-1273, M-1274; 1-2 Cu, 0.5-1.25 Mg, 0.75-1.25 Si, 0.75 Fe, 1 Mn, 0. Ti, bal Al.
W-temper: 38,100 TS; 22,400 YS; 15 El.
Wp-temper: 56,000 TS; 42,600 YS; 6 El.
For structural members; age-hardened.

ALMINAL W12.
M-1273, M-1274; 1.8-2.5 Cu, 0.65-1.2 Mg, 0.55-1.2 Si, 0.6-1.2 Fe, 0.6-1.4 Ni, 0.2 Ti, bal Al.
Wp-temper: 60,500 TS; 47,100 YS; 10 El.
For aircraft engine components; age-hardened, high strength.

ALMINAL W14.
M-1273, M-1274; 3.5-5.0 Cu, 0.4-1.2 Mg, 0.7 Si, 0.7 Fe, 0.4-1.2 Mn, 0.3 Ti, bal Al.
Heat treated: 53,500-56,000 TS; 31,000-33,600 YS; 15-10 El.
For structural members; age-hardened.

ALMINAL W15.
M-1273, M-1274; 3.5-4.8 Cu, 0.6 Mn, 1.5 Si, 1 Fe, 1.2 Mn, 0.3 Ti, bal Al.
W-temper: 53,500-56,000 TS; 31,400-33,600 YS; 15-10 El.
WP-temper: 62,700-67,200 TS: 53,800-58,000 YS; 8-6 El.
For structual members; age-hardened.

ALMINAL W16.
M-1273, M-1274; 3 Cu, 4 Mg, 0.6 Si, 0.6 Fe, 1 Mn, 0.2 Ti, 4.0-8.5, Zn, bal Al.
WP-temper: 78,000-85,000 TS; 67,000-74,000 YS; 5-4 El.
For structural members; age-hardened, high strength.

ALMINAL W-17.
M-1273, M-1274; 3.5-4.5 Cu, 1.2 Mg, 0.6 Fe 0.6 Si, 0.3 Ti, 1.8-2.3 Ni, bal Al.
WP-temper: 49,300 TS; 8 El.
For aircraft engine components; age-hardened, high temperature uses.

ALMINAL W18.
M-1273, M-1274; 1.8-2.5 Cu, 1.2-1.8 Mg, 0.5-1.3 Si, 0.6-1.2 Fe, 0.6-1.4 Ni, bal Al.
W-temper: 53,800 TS; 6 El.
For aircraft engine components; age-hardened, high temperature uses.

ALMINAL W150.
M-1273, M-1274; 4.8 Cu, 0.85 Mg, 0.9 Si, 0.3 Ti, 1.2 Mn, bal Al.
W-temper: 49,000-56,000 TS; 27,000-33,600 YS; 15-12 El.
For structural members; age-hardened.

ALMINAL W151.
M-1274; 0.4 Si, 0.7 Fe, 5-6 Cu, 0.3 Zn, 0.4 Bi, 0.4 Pb, bal Al.
Heat Treated: 57,000 TS; 47,000 YS; 17 El; 97 Brin.
For screw machine products, fasteners, screws, clips.
Free-machining. Heat treatable.

ALMINAL W160.
M-1273, M-1274; 1.5 Cu, 2.0-3.5 Mg, 0.5 Si, 0.5 Fe, 4.5-6.5 Zn, 0.3 Ti, 0.8 Mn, bal Al.
WP-temper: 72,000-78,500 TS; 60,000-67,000 YS; 7-5 El.
For structural members; age-hardened.

ALMOLD-20.
M-1779; 0.30 C, 0.70 Mn, 0.50 Si, 1.70 Cr, 0.40 Mo, bal Fe.
For zinc die casting dies and plastic molds.
Type P20 tool steel, mold quality.

ALMOS.
M-678; 0.08-0.16 C, 0.8-1.2 Cr, 0.15-0.30 Mo, 0.6-0.9 Mn, bal Fe.
Annealed: 78,000 TS; 50,000 YS; 25 El.
Hardened: 170,000 TS; 114,000 YS 8 El.
For shafts, gears; case-hardening steel.

ALNEON.
M-283; 70-90 Al, 2-3 Cu, 7-22 Zn, 0.4-10 other elements.
28,000-48,500 TS; 10,700-17,000 YS; 1-3 El; 100-150 Brin.
For light alloy parts; age-hardening, high endurance limit.

ALNESIUM.
M-U.S.; 3 Mg, bal Al.
For cases, containers, corrosion resistant.

ALNI.
M-Russia; 0.10 max C, 24 Ni, 13 Al, 3.5 Cu, bal Fe.
For permanent magnets; high permeability.

ALNICO.
M-Russia; 0.10 max C, 17 Ni, 10 Al, 12.5 Co, 6 Cu, bal Fe.
For permanent magnets; high permeability.

ALNICO 1.
M-670, M-1754; 12 Al, 21 Ni, 5 Co, 3 Cu, bal Fe.
Cast permanent magnet material.
Cast: 7200 Residual Flux Density Br. (gauss) 470 Coercive Force, Hc, (oersted).
1.40 Max. Energy Product, (BH) max (MGO) (MG.Oe) -Megagauss-oersteds.

ALNICO 1.
M-1076; 12 Al, 21 Ni, 5 Co, 2 Cu, bal Fe.
Cast permanent magnet, heat treated.
Peak energy product, (B_dH_d) max X 10^{-6}, 1.4
Residual Induction, Br, (kilogauss) 7.2
Coercive force Hc (oersteds), 470.

ALNICO 1.
M-373; 12 Al, 22 Ni, 5 Co, 0.35 Ti, bal Fe.
Maximum energy product, BH max x 10^{-6}: 1.38; residual flux density 6800; coercive force Hc (oersted), 530.
Cast, furnished heat treated.

ALNICO 2.
M-373; 10 Al, 17 Ni, 12.5 Co, 6 Cu, 0.45 Ti, bal Fe.
Maximum energy product, BH max x 10^{-6}, 1.65; residual flux density, 7500; coercive force Hc (oersted), 550.
Cast, furnished heat treated.

ALNICO 2.
M-670; 10 Al, 19 Ni, 13 Co, 3 Cu, bal Fe.
Cast permanent magnet material.
Cast: 7500 Br.; 560 Hc.; (BH) max (MGO).

ALNICO 2.
M-1076; 10 Al, 19 Ni, 13 Co, 3 Cu, bal Fe.
Cast permanent magent, heat treated.
Peak energy product (B_dH_d) max X 10^6, 1.7
Residual Induction, Br (kilogauss) 7.5
Coercive force Hc (oersteds) 560.

ALNICO 2C.
M-373; 10 Al, 21 Ni, 12.5 Co, 6 Cu, 0.30 Ti, bal Fe.
Maximum energy product, BH max x 10^{-6}, 1.60; residual flux density ., 7100; coercive force Hc (oersted), 580.
Cast, furnished heat treated.

ALNICO 2 SINTERED.
M-1076; 10 Al, 19 Ni, 13 Co, 3 Cu, bal Fe.
Peak energy product (B_dH_d) max X 10^{-6}, 1.5
Residual Induction, Br, (kilogauss) 7.1
Coercive force Hc (oersteds) 550.
Permanet magnet.

ALNICO 3.
M-670; 12 Al, 25 Ni, 3 Cu, bal Fe.
Cast permanent magnet material.
Cast: 7000 Br.; 480 Hc.; 1.35 (BH) max (MGO).

ALNICO 3.
M-1076; 12 Al, 25 Ni, 2 Cu, bal Fe.
Cast permanent magnet, heat treated.
Peak energy product (B_dH_d) max X 10^6, 1.35
Residual Induction, Br (kilogauss) 7.0
Coercive force Hc (oersteds) 480.

ALNICO 3.
M-373; 12 Al, 26 Ni, bal Fe.
Maximum energy products, BH max x 10^{-6}, 1.20; residual flux density Br, 6600; coercive force Hc (oersted), 450.
Cast, furnished heat treated.

ALNICO 4.
M-670; 12 Al, 27 Ni, 5 Co, bal Fe.
Cast permanent magnet material.
Cast: 5600 Br.; 720 Hc.; 1.35 (BH) max (MGO).

ALNICO 4.
M-1076; 12 Al, 27 Ni, 5 Co, 2 Cu, bal Fe.
Cast permanent magnet, heat treated.
Peak energy product (B_dH_d) max X 10^6, 1.5
Residual Induction, Br (kilogauss) 5.6.
Coercive force Hc (oersteds) 720.

ALNICO 4.
M-373; 12 Al, 25 Ni, 5 Co, 0.40 Ti, bal Fe.
Maximum energy product, BH max x 10^{-6}, 1.40; residual flux density Br, 6300; coercive force Hc (oersted), 630.
Cast, furnished heat treated.

ALNICO 5.
M-670; 8 Al, 14 Ni, 24 Co, 3 Cu, bal Fe.
Cast permanent magnet material.
Cast: 12,800 Br.; 640 Hc.; 5.50 (BH) max (MGO).

ALNICO 5.
M-1076; 8 Al, 14 Ni, 24 Co, 3 Cu, bal Fe.
Cast permanent magnet, cooled in magnetic field.
Peak energy product (B_dH_d) max X 10^6, 5.5
Residual Induction, Br (kilogauss) 12.5
Coercive force Hc (oersteds) 640.

ALNICO 5.
M-373; 8 Al, 14.5 Ni, 24 Co, 3 Cu, bal Fe.
Maximum energy product, BH max x 10^{-6}, 5.25; residual flux density Br., 12700; coercive force Hc (oersted), 640.
Cast, directional magnetic properties; furnished heat treated.

ALNICO 5-7.
M-1076; 8 Al, 14 Ni, 24 Co, 3 Cu, bal Fe.
Cast permanent magnet, cooled in magnetic field.
Peak energy product (B_dH_d) max X 10^{-6}, 7.50
Residual Induction Br (kilogauss) 13.4
Coercive force Hc (oersteds) 730.

ALNICO 5 COL.
M-670; 8 Al, 14 Ni, 24 Co, 3 Cu, bal Fe.
Cast permanent magnet material.
Cast: 13,500 Br.; 740 Hc.; 7.55 (BH) max (MGO).

ALNICO 5 DG.
M-670; 8 Al, 14 Ni, 24 Co, 3 Cu, bal Fe.
Cast permanent magnet material.
Cast: 13,300 Br.; 670 Hc.; 6.50 (BH) max (MGO).

ALNICO 5 DG.
M-1754; 8Al, 14 Ni, 24 Co, 3 Cu, bal Fe.
Permanent magnet.
Residual fulx, Br. (gauss), 13,300; coercive force, Hc (oersteds), 670.

ALNICO 5 E.
M-373; 8 Al, 15 Ni, 22.5 Co, 3.6 Cu, 0.45 Ti, bal Fe.
Maximum energy product, BH max x 10^{-6}, 4.60
Residual flux density, Br, 11800; coercive force, Hc (oersted), 700.
Cast, directional magnetic properties; furnished heat treated.

ALNICO 5 SINTERED.
M-1076; 8 Al, 14 Ni, 24 Co, 3 Cu, bal Fe.
Pressed from powder and sintered.
Peak energy product (B_dH_d) max X 10^6, 3.9
Residual Induction, Br, (kilogauss) 10.9
Coercive force Hc (oersteds) 620.
Cooled in magnetic field; permanent magnet.

ALNICO 6.
M-670; 8 Al, 16 Ni, 24 Co, 3 Cu,1 Ti, bal Fe.
Cast permanent magnet material.
Cast: 10,500 Br.; 780 Hc.; 3.90 (BH) max (MGO).

ALNICO 6.
M-1076; 8 Al, 16 Ni, 24 Co, 3 Cu,1 Ti, bal Fe.
Cast permanent magnet, heat treated.
Peak energy product (B_dH_d) max X 10^6, 3.5
Residual Induction Br (kilogauss) 10.1
Coercive force Hc (oersteds) 750.

ALNICO 6.
M-373; 8 Al, 15 Ni, 24 Co, 3.4 Cu, 1.25 Ti, bal Fe.
Maximum energy product BH max x 10^{-6}, 3.65; residual flux density, Br., 10400; coercive force Hc (Oersted), 700.
Cast, directional magnetic properties; furnished heat treated.

ALNICO 6 SINTERED.
M-1076; 8 Al, 16 Ni, 24 Co, 3 Cu, 1 Ti, bal Fe.
Pressed from powder and sintered.
Peak energy product (B_dH_d) max X 10^6
Residual Induction, Br, (kilogauss) 8.8
Coercive force Hc (oersteds) 800.
Cooled in magnetic field; permanent magnet.

ALNICO 7.
M-670; 8.5 Al, 18 Ni, 24 Co, 3.25 Cu, 5 Ti, bal Fe.
Cast permanent magnet material.
Cast: 7700 Br.; 1050 Hc.; 2.85 (BH) max (MGO).

ALNICO 8.
M-670; 35 Co, 15 Ni, 4 Cu, 5 Ti, 7 Al, bal Fe.
Cast permanent magnet material.
Cast: 8200 Br.; 1650 Hc.; 5.3 (BH) max (MGO).

ALNICO 8B.
M-1076; 7 Al , 15 Ni, 35 Co, 3 Cu, 5 Ti, bal Fe.

ALNICO 8H.
M-1076; 8 Al 14 Ni 38 Co, 3 Cu, 8 Ti, bal Fe.
Cast permanent magnet, cooled in magnetic field.
Peak energy product (B_dH_d) max X 10^6, 5.0
Residual induction, Br, (kilogauss) 7.0
Coercive force Hc (oersteds) 1900.

ALNICO 8H SINTERED.
M-1076; 7 Al, 14 Ni, 38 Co, 3 Cu, 8 Ti, bal Fe.
Permanent magnet.
Pressed from powder and sintered.
Peak energy product (B_4H_4) max x 10^6, 4.5; residual induction Br (kilogauss), 6.5; coercive force Hc (oersteds), 1800.

ALNICO 8HC.
M-1192.
Cast Alnico magnet.
Peak energy (B_dH_d) max x 10^6, 6.5; residual induction (Gausses), 7700; coercive force Hc (Oersteds) 2200; incremental permeability, 2.8.

ALNICO 8HC.
M-1754; 8 Al, 14 Ni, 38 Co, 3 Cu, 8 Ti, bal Fe.
Permanent Magnet.
Residual flux, Br (gauss), 7200; coercive force, Hc (oersteds), 1900.

ALNICO 8HE.
M-1192.
Cast Alnico magnet.
Peak energy (B_dH_d) max x 10^6, 6.75 Residual induction (gausses), 8800; coercive force Hc (oersteds) 1800; incremental permeability, 2.5.

ALNICO 8HE.
M-1076; 7 Al, 15 Ni, 35 Co, 4 Cu, 5 Ti, bal Fe.
Cast permanent magnet, cooled in magnetic field.
Peak energy product (B4H) max X 10^6, 6.0
Residual Induction, Br (kilogauss) 9.25
Coercive force Hc (oersteds) 1550.

ALNICO 8HE SINTERED.
M-1076; 7Al, 15 Ni, 35 Co, 4 Cu, 5 Ti, bal Fe.
Permanent magnet.
Pressed from powder and sintered.
Peak energy product (B_4H_4) max x 10^6, 5.5; Residual induction Br (kilogauss), 8.6; coercive force Hc (oersteds), 1500.

ALNICO 9.
M-670; 35 Co, 15 Ni, 4 Cu, 5 Ti, 7 Al, bal Fe.
Cast permanent magnet material.
Cast: 10,500 Br.; 1500 Hc.; 9.0 (BH) max (MGO) (tentative properties).

ALNICO 9.
M-1076; 35 Co, 15 Ni, 4 Cu, 5 Ti, 7 Al, bal Fe.
Cast permanent magnet, cooled in magnetic field.
Peak energy product (B_dH_d) max X 10^6, 10.0
Residual Induction, Br, (kilogauss) 10.5
Coercive force Hc (oersteds) 1600.

ALNICO 9NB.
M-1192.
Cast Alnico magnet.
Peak energy (B_dH_d) max x 10^6, 11.8; residual induction Br (Gausses), 11,400; coercive force Hc (oersteds), 1640; incremental permeability, 2.5.

ALNICO 160.
M-1246; Al, Ni, Co.
For permanent magnets; sintered.

ALNICO 350.
M-Germany; 7.8 Al, 15 Ni, 34 Co, 3.5 Cu, 5 Ti, bal Fe.
For electrical and magnetic equipment.
Permanent magnet; high permeability.

ALNICO 400.
M-Poland; 8 Al, 14 Ni, 24 Co, 3 Cu, 0.5 Ti, bal Fe.
For electrical and magnetic equipment.
Permanent magnet, high permeability.

ALNICUS GR. USM65.
M-1534; 8 Al, 14 Ni, 24 Co, 3 Cu, bal Fe.
Cast: 5500 TS; 500 Brin.
For magnetic chucks, meters, starting devices, electrocardiographs; permanent magnets, oriented.

ALNICUS GR. USM75.
M-1534; 8 Al, 14 Ni, 24 Co, 3 Cu, bal Fe.
Cast: 5500 TS; 500 Brin.
For magnetic chucks, meters, starting devices, electrocardiographs; permanent magnets, oriented.

ALNIFER.
M-926; low C-steel clad with Al on one side and Ni on other side.
For electron tubes.

ALOYCO 18-8S.
M-1023; 0.07 max C, 18-21 Cr, 8-11 Ni, bal Fe.
Cast: 73,000 TS; 33,000 YS; 50 El; 160 Brin.
For valves, fitting pumps.
Corrosion resistant.
ACI CF8.

ALOYCO 18-8 SCB.
M-1023; 0.07 max C, 18-21 Cr, 9-12 Ni, 1.0 Cb + Ta, bal Fe,
Cast: 75,000 TS; 35,000 YS; 40 El; 160 Brin.
For valves, fittings, pumps.
Corrosion resistant.
ACI CF8C.

ALOYCO 18-8S ELC.
M-1023; 0.03 max C, 17-21 Cr, 8-11 Ni, bal Fe.
Cast: 74,000 TS; 34,000 YS; 50 El, 160 Brin.
For valves, fittings, pumps.
Corrosion resistant.
ACI CF3.

ALOYCO 18-8S MO.
M-1023; 0.07 max C, 18-21 Cr, 9-12 Ni, 2-3 Mo, bal Fe.
Cast: 75,000 TS; 35,000 YS; 45 El; 160 Brin.
For valves, fittings pumps.
Corrosion resistant.
ACI CF8M.

ALOYCO 18-8S MO ELEC.
M-1023; 0.03 max C, 17-21 Cr, 9-13 Ni, 2-3 Mo, bal Fe.
Cast; 75,000 psi TS; 34,000 YS; 45 El, 160 Brin.
For valves, fittings, pumps.
Corrosion resistant.
ACI CF3M.

ALOYCO 20.
M-1023; 0.07 max C, 19-22 Cr, 27.5-30.5 Ni, 2-3 Mo, 3-4 Cu, bal Fe.
Cast: 67,000 TS; 28,000 YS; 45 El; 155 Brin.
For valves, fittings, pumps.
Corrosion resistant to sulphuric acid, seawater, nitric acid, organic acid.
ACI CN7M.

ALOYCO-25-20.
M-1023; 0.20 max C, 22-26 Cr, 19-22 Ni, bal Fe.
Cast: 70,000 TS; 32,000 YS; 30 El; 160 Brin.
For valves, fittings, pumps.
Corrosion resistant.
ACI CK20.

ALOYCO-CA6NM.
M-1023; 0.06 max Co 1.0 max Mo, 1.0 max Si, 3.5-4.5 Ni, 0.4-1.0 Mo, 11.5-14.0 Cr, bal Fe.
Cast: 120,000 TS; 87,000 YS; 15 El; 260Brin.
For valves, fittings, pumps.
Corrosion and cavitation resistant.

ALOYCO-INCONEL.
M-1023; 0.40 max C, 1.5 max Mn, 3.0 max Si, 14-17 Cr, 11.0 max Fe, bal Ni.
Cast: 72,000 TS; 30,000 YS; 30 El; 160 Brin.
For valves, fittings, pumps.
ACI CY40.

ALOYCO MONEL.
M-1023; 0.35 max C, 1.5 max Mn, 2.0 max Si, 26-33 Cu, 3.5 max Fe, bal Ni.
Cast: 67,500 TS; 30,000 YS; 25 El; 160 Brin.
For valves, fittings, pumps.
ACI M35.

ALOYCO-N-2.
M-1023; 0.12 max C, 26-30 Mo, 1.0 max Cr,0.2-0.6 V, 4-6 Fe, bal Ni.
Cast: 80,000 TS; 48,000 YS; 6 El; 170 Brin.
For valves, fittings, pumps.
Corrosion resistant.
ACI Ni2M-1.

ALOYCO-N-3.
M-1023; 0.12 max C, 16-18 Mo, 15.5-17.95 Cr, 3.75-5.25 W, 0.2-0.4 V, 4.5-7.5 Fe, bal Ni.
Cast: 75,000 TS; 48,000 YS; 4 El; 170 Bin.
For valves, fittings, pumps.
Corrosion resistant.
ACI CW12-M-1.

ALOYCO-NICKEL.
M-1023; 1.0 max C, 1.5 max Mn, 2.0 max Si, bal Ni.
Cast: 55,000 TS; 20,000 YS; 10 El; 140 Brin.
For valves, fittings, pumps.
ACI CZ-100.

ALOY-NUM.
M-549; 4 Si, 2 Cu, bal Al.
For pistons, cylinder heads; age-hardenable.

ALPACCA.
M-Eng.; 65.9 Cu, 19.25 Zn, 14.6 Ni, 0.5 Fe.
For base for plated table ware; corrosion resistant.

ALPAKKA (ALPACA).
M-Eng.; 60 Cu, 19 Zn, 15 Ni, 2 Ag.
For base metal for silver plated tableware; resists corrosion.

ALPAX.
M-75; 10-13 Si, 0.2-0.6 Mg, 0.3-0.7 Mn, bal Al.
Sand cast: 24,000 TS; 9,000 YS; 6-8 El; 55 Brin.
Chill cast: 30,500 TS; 12,500 YS; 10-12 El; 65 Brin.
For light corrosion resistant castings; corrosion resistant.

ALPAX.
M-106; 10-13, bal Al.
Sand cast; 23,500-26,800 TS; 6,700-7,800 YS; 5-8 El; 45-50 Brin; 4.5 Izod.
For castings of thin sections, hydraulic parts.
Gen. Eng. BS 1490; LM 6-M; BS 4L 33.

ALPAX.
M-137; 10-13 Si, 0.2 max Ti, 0.6 max Fe, 0.5 max Mn, bal Al.
Sand cast: 23,100 TS; 10,000 YS; 5-10 El; 50 RA; 60 Brin.
Permanent mold: 28,000 TS; 12,000 YS; 8-14 El; 60 RA; 65 Brin.
For water pumps, engine parts; corrosion resistant, good castability.

ALPAX.
M-620; 4.5 Cu, 0.5 Mg, 0.4 Ti, bal Al.
36,000-43,000 TS; 29,000-36,000 YS; 2-1 El.
For light alloy parts; age-hardenable.

ALPAX ALPHA.
M-75; 10-13 Si, 0.2-0.6 Mg, 0.3-0.7 Mn, bal Al.
Sand cast: 22,500 TS; 9,000 YS; 2-4 El; 64 Brin.
Chill cast: 3-5 El; 70 Brin.
For light corrosion resistant castings; corrosion resistant.

ALPAX ALPHA.
M-137; 10-13 Si, 0.6 Mg, bal Al.
Sand cast: 22,000 TS; 2-4 El; 64 Brin.
Permanent mold: 30,000 TS; 3-5 El; 70 Brin.
For gear cases, instrument housings; corrosion resistant, good castability

ALPAX "B" (ALPAX BETA).
M-106; 0.2-0.6 Mg, 10-13 Si, 0.3-0.7 Mn, bal Al.
Sand cast: 24,600-29,000 TS; 14,500-16,000 YS; 1.5-3 El; 60-70 Brin; 1.5 Izod.
General purpose sand or chill cast alumium.
Gen. Eng. BS 1490: LM 9-P.

ALPAX BETA.
M-75; 10-13 Si, 0.2-0.6 Mg, 0.3-0.7 Mn, bal Al.
Sand cast: 24,000 TS: 16,000 YS; 2-3 El; 75 Brin.
Chill cast: 34,000 TS; 20,000 YS; 2-4 El; 85 Brin.
For light corrosion resistant castings; corrosion resistant.

ALPAX BETA.
M-137; 10-13 Si, 0.2-0.6 Mg, 0.6 Fe, 0.2 max Ti, bal Al.
Sand cast: 26,000 TS; 16,000 YS; 2-3 El; 75 Brin.
Permanent mold: 34,000 TS; 18,000 YS; 2-4 El; 85 Brin.
For gear boxes, housings, general castings; corrosion resistant, good castability.

ALPAX "G" (ALPAX GAMMA).
M-106; 0.2-0.6 Mg, 10-13 Si, 0.3-0.7 Mn, bal Al.
Sand cast: 35,000-37,000 TS; 29,000-31,000 YS; 0-2 El; 90-100 Brin; 0.9 Izod.
General purpose sand or chill cast aluminum.
Gen. Eng, BS 1490: LM 9 WP; BS L 75.

ALPAX GAMMA.
M-75; 10-13 Si, 0.2-0.6 Mg, 0.3-0.7 Mn, bal Al.
Sand cast: 38,000 TS; 32,000 YS; 1-2 El; 95 Brin.
Chill cast: 36,000 TS; 1-3 El; 100 Brin.
For light corrosion resistant castings; corrosion resistant.

ALPAX GAMMA.
M-137; 10-13 Si, 0.6 Mg, bal Al.
Aged: 36,000 TS; 30,000 YS 1-2 El; 95 Brin.
For engine components, general castings; corrosion resistant, good castability.

ALPAX HAU COBALT.
M-75; 9.3 Si, 0.45 Co, 0.2 Mg, bal Al.
Cast:48,000 TS; 26,000 YS; 4 El.
For light weight housings, casings, cover plates. Corrosion resistant.

ALPERM.
M-Japan; 14-16 Al, bal Fe.
For magnetic laminations; high permeability.

ALPHA.
M-1; High C, bal Fe.
For drills.

ALPHA GAMMA.
M-137; 10-13 Si, Mg, bal Al.
For gears, bolts, springs, crankshafts; oil hardened, shock resistant.

ALPINE AHK now **VEW K455.**

ALPINE AKL now **VEW K244.**

ALPINE ECR15 now **VEW E410.**

ALPINE ECR20 now **VEW E400.**

ALPINE ED now **VEW E900.**

ALPINE EFW now **VEW E920.**

ALPINE ERM15 now **VEW E304.**

ALPINE ERM20 now **VEW E300.**

ALPINE EXTRA MH now **VEW K992.**

ALPINE EXTRA ZAH now **VEW K976.**

ALPINE EXTRA ZH I now **VEW K988.**

ALPINE EXTRA ZH II now **VEW K984.**

ALPINE FAM now **VEW F180.**

ALPINE FRS now **VEW F204.**

ALPINE HIRA 45 now **VEW V204.**

ALPINE HMV15 now **VEW V930.**

ALPINE HMV35 now **VEW V762.**

ALPINE HMV42 now **VEW V742.**

ALPINE HMV50 now **VEW F550.**

ALPINE HR10 now **VEW V500.**

ALPINE HRIM15 now **VEW V155.**

ALPINE HRIM20 now **VEW V145.**

ALPINE HRM15 now **VEW V340.**

ALPINE HRM35 now **VEW V330.**

ALPINE HRM40 now **VEW V320.**

ALPINE KL3 now **VEW R100.**

ALPINE MA3 now **VEW V960.**

ALPINE MA4 now **VEW V945.**

ALPINE MA5 now **VEW V935.**

ALPINE MA6 now **VEW V920.**

ALPINE MN13 now **VEW K701.**

ALPINE NMV now **VEW V304.**

ALPINE PRG now **VEW K204.**

ALPINE RAV now **VEW K510.**

ALPINE SF now **VEW F112.**

ALPINE WAM now **VEW W106.**

ALPINE WAMV now **VEW W105.**

ALPINE WKL3 now **VEW K200.**

ALPINE WMC now **VEW W501.**

ALPRO NO. 11.
M-126.
For deoxidizer for Cu.

ALPRO NO. 12.
M-126; 10 P, 90 Cu.
For Cu deoxidizer.

ALPRO NO. 15.
M-126.
For Ni degasifying alloy.

ALPRO NO. 19 AX.
M-126; 30 Mn, 70 Cu, C-free.
For adding Mn to alloys.

ALPRO NO. 19 AZ.
M-126; 30 Mn, 70 Cu, C-free.
For adding Mn to alloys.

ALPRO NO. 19 BX.
M-127; 30 Mn, 0.15 max C, 2.5 Fe, bal Cu.
For adding Mn to alloys.

ALPRO NO. 19 CX.
M-127; 30 Mn. 0.25 max C, 4.5 max Fe, bal Cu.
For adding Mn to alloys.

ALPRO NO. 30A.
M-126; 50 Cu, 50 Al.
For copper hardener for Al alloys.

ALPRO NO. 31.
M-126; 75 Al, 25 Mn.
For adding Mn to alloys.

ALPRO NO. 34.
M-126; 20 Ni, 80 Al.
For hardeners for Ni-Al alloys.

ALPRO NO. 35.
M-126; 50 Si, 50 Al.
For Si hardener for Al alloys.

ALPRO NO. 39.
M-126; 10 Fe, 90 Al.
For adding Fe in Al alloys.

ALPRO NO. 40.
M-126; 25 Ni, 25 Cu, 50 Al.
For hardeners for Ni-Al alloys.

ALPRO NO. 41.
M-126; 25 Ni, 25 Mn, 50 Al.
For hardeners for Ni-Al alloys.

ALPRO NO. 42.
M-126; 25 Ni, 25 Sn, 50 Al.
For hardeners for Ni-Al alloys.

ALPRO NO. 44.
M-126; 25 Cu, 50 Al, 25 Mn.
For adding Mn to alloys.

ALPRO NO. 46.
M-126; 25 Fe, 50 Al, 25 Mn.
For adding Mn to alloys.

ALPRO NO. 50 B.
M-126; 50 Ni, 50 Cu.
For addition of Ni of bronzes.

ALPRO NO. 55.
M-126; Ni-Cu.
For deoxidizer of Cu alloys; red composition densifying alloy.

ALPRO NO. 112.
M-126; 15 P, 85 Cu.
For Cu deoxidizer; for copper alloys.

ALPRO NO. 126 A.
M-126; Ni-Cu.
For hardener of Al bronzes.

ALPRO NO. 126 C.
M-126; Ni-Cu.
For hardener of Al bronzes.

ALPRO NO. 126 D.
M-126; Ni-Cu.
For hardener of Al bronzes.

ALPRO NO. 139.
M-126; 20 Fe, 80 Al.
For adding Fe to Al alloys.

ALPRO NO. 140.
M-126; 20 Ni, 40 Cu, 40 Al.

ALPRO NO. 143.
M-126; 40 Ni, 20 Cu, 40 Al.
For hardener for Ni-Al alloys.

ALPRO NO. 146.
M-126; Ni Cr, Fe.
For gray iron improver; hardener.

ALPRO NO. 157.
M-126; 50 Ni, 50 Fe.
For gray iron improver; hardener.

ALPRO NO. 158.
M-126; 50 Ni, 50 Mn.
For adding Mn to alloys.

ALPRO NO. 159.
M-126; 75 Ni, 2 Cr.
For gray iron improver; hardener.

ALPRO NO. 220.
M-126; 10 Al, 90 Cu.
For gears, cams, valves; resists abrasion, wear and shock.

ALPRO NO. 221.
M-126; 10 Al, 89 Cu, 1 Fe.
For ball bearings; resists abrasion wear and shock.

ALPRO NO. 223.
M-126; 10 Al, 87 Cu, 3 Fe.
For bearings.

ALPRO NO. 224.
M-126; 10 Al, 86 Cu, 4 Fe.
For bearings.

ALPRO NO. 300.
M-126; 8 Cu, 92 Al.
For general light alloy castings.

ALPRO NO. 301.
M-126; Al alloy.
For castings to be forged or die pressed.

ALPRO NO. 302.
M-126; 10 Cu, 90 Al.
For die castings; non-hardenable.

ALPRO NO. 304.
M-126; 2 Si, 8 Cu, bal Al.
For castings.

ALPRO NO. 306.
M-126; 4 Si, 6 Cu, bal Al.
For castings.

ALPRO NO. 307.
M-126; Al alloy.
For core boxes.

ALPRO NO. 308.
M-126; Al alloy.
For cast cooking utensils.

ALPRO NO. 309.
M-126; Al alloy.
For pistons.

ALPRO NO. 310.
M-126; 7.5-8.5 Cu, 1.5 max Fe + Si, bal Al.
For castings; non-hardenable.

ALPRO NO. 311.
M-126; 9.5-10.5 Cu, 1.5 max Fe + Si, bal Al.
For die castings; non-hardenable.

ALPRO NO. 312.
M-126; 2 Si, 8 Cu, 2 Zn, bal Al.
For castings; non-hardenable.

ALPRO NO. 314.
M-126; 4 Si, 6 Cu, bal Al.
For castings.

ALPRO NO. 315.
M-126; 7.5-8.5 Cu, 1.75-2.25 Zn, 1.75 max Fe + Si, bal Al.
For castings; non-hardenable.

ALPRO NO. 316.
M-126; 4 Si, 4 Cu, bal Al.
For castings.

ALPRO NO. 317.
M-126; 2 Si, 6 Cu, bal Al.
For castings.

ALPRO NO. 320.
M-126; Zn, bal Al.
For light alloy castings.

ALPRO NO. 330.
M-126; Al alloy.
For match plates; low degree of top shrinkage.

ALPRO NO. 331.
M-126; Si, bal Al.
For marine castings; resists sea-water corrosion.

ALPRO NO. 332.
M-126; Si, bal Al.
For general utility castings; resists sea-water corrosion.

ALPRO NO. 333.
M-126; Al alloy.
For high pressure castings; tough, dense, strong.

ALPRO NO. 335.
M-126; 4.5-5.5 Si, 0.75 max, Fe, bal Al. 19,000-21,000 TS; 4-3 El; 37 Brin.
For marine parts, castings; corrosion resistant.

ALPRO NO. 338.
M-126; 11.0 -13.5 Si, 0.8 max Fe, bal Al. 25,000-27,500 TS; 7-6 El; 45 Brin.
For die castings; corrosion resistant.

ALPRO NO. 351.
M-126; 5 Si, bal Al.
For castings; sand or die cast.

ALPRO NO. 352.
M-126; 12 Si, bal Al.
For castings; die cast.

ALPRO NO. 360.
M-126; 2 Ni, 6 Cu, bal Al.
For castings.

ALPRO NO. 361.
M-126; 2 Ni, 6 Cu, 2 Si, bal Al.
For castings.

ALPRO NO. 362.
M-126; 4 Ni, 4 Cu, 92 Al.
For pistons, light alloy parts.

ALPRO NO. 363.
M-126; 2 Ni, 5 Cu, 2 Si, 91 Al.
For castings.

ALPRO NO. 364.
M-126; 2 Ni, 4 Cu, 4 Si, 90 Al.
For castings; age-hardened.

ALPRO NO. 365.
M-126; 3 Ni, 5 Cu, 2 Si, 90 Al.
For castings; age-hardened.

ALPRO NO. 370.
M-126; 4 Cu, 2 Ni, 1.5 Mg, 92.5 Al.
At 65°F, 29,000 TS. At 482°F, 25,500 TS.
For aeronautical parts; similar to "Y" alloy.

ALPRO NO. 371.
M-126; 2 Cu, 1.5 Ni, 1 Mg, 0.6 Si, 94.9 Al.
For pistons; similar to "Magnalite".

ALPRO NO. 380.
M-126; Cu, Mn, bal Al.
For light alloy parts; modified duralumin type.

ALPRO NO. 385.
M-126; 6.5 Cu, 1.5 Ni, 0.1 Cr, 0.5 Mg, 0.5 Mn, 90.9 Al.
For sand castings; modified Duralumin.

ALPROPAT.
M-1265; Cu, Si, bal Al.
Cast: 27,000 TS; 15,000 YS; 4 El; 60 Brin.
For match plates and patterns; low shrinkage.

ALRAY A 8.
M-248; 20 Cr, 80 Ni.
For electrical resistance; heat resistant.

ALSEX.
M-621; 0.5 Si, 0.5 Mg, bal Al.
Heat treated: 29,00-42,500 TS; 25,000-35,000 YS; 7-5 El; 90 Brin.
For light alloy parts, forgings; conductivity 52% of copper.

ALSIA.
M-672; 20 Si 1 Cu, 0.7 Fe, bal Al.
Cast: 20,000 TS; 18,000 YS; 1-2 El; 80-90 Brin.
For pistons; same as "Alusil".

ALSIFONT 5C.
M-1176; 4 Cu, 5.5 Si, bal Al.
Heat treated: 28,000-32,000 TS; 20,000-23,000 YS; 4-1.5 El; 55-75 Brin.
For light alloy parts; heat treatable.

ALSIPLATE.
M-926; Ag clad to 2S aluminum alloy.
For electrical contacts; Al alloys plated by lamination with silver.

ALSOCO M-1001.
M-1158; Zn-Sn-Pb.
For aluminum solder; M.P. 728°F.

ALSOCO M-1002.
M-1158; Zn-Sn-Pb.
For aluminum solder; M.P. 680°F.

ALSOCO M-1003.
M-1158.
For aluminum solder.

ALTAM.
M-109; 68-88 Ti, 11 Al, 1.25 Fe, 1.18 Si, bal TiO_2.
For making ductile titanium.

ALT BLITZ.
M-1312; 0.79 C, 4.7 Co, 4.3 Cr, 0.7 Mo, 1.5 V, 18 W, bal Fe.
For lathe and planer tools, hobs, reamers, taps; high speed steel.

AL TECH 6X.
M-1779; 21.0 Cr, 25.0 Ni, 6.5 Mo, 0.03 max C, bal Fe.
2150°F water quench anneal.
Superior corrosion resistance, especially crevice.

AL TECH 21-2.
M-1779; 0.55 C, 8.3 Mn, 21.0 Cr, 2.2 Ni, 0.30 N, bal Fe.
For valves, valve seats.
SAE EV12.

AL TECH 21-4.
M-1779; 0.53 C, 9.0 Mn, 21.0 Cr, 3.75 Ni, 0.42 N, bal Fe.
For valves and valve seats.
SAE EV8.

AL TECH 21-6-9.
M-1779; 21.0 Cr, 6.0 Ni, 9.0 Mn, 0.04 max C, 0.35 N, bal Fe.
Solution strengthened austenitic alloy.
High temperature, high strength, corrosion resistant.
Used where non-magnetic, high strength needed; aircraft hydraulic tubing.

AL TECH 21-12N.
M-1779; 0.2 C, 21.0 Cr, 12.0 Ni, 0.21 N, bal Fe.
For valves, valve seats for high temperature applications.
SAE EV4.

AL TECH 26-1S.
M-1779; 26.0 Cr, 1.0 Mo, 0.03 max C, Ti or Cb stabilized, bal Fe.
Superior corrosion resistant ferritic stainless.
For fasteners, screens.

AL TECH 415.
M-1779; 12.0 Cr, 2.5 Ni, 1.5 Mo, 0.12 C, 0.04 N, bal Fe.
For turbine parts; martensitic stainless steel; heat and corrosion resistant.

AL TECH 88X.
M-1774; 0.2-0.3 C, 10-12.5 Mn, 7-8.5 Ni, bal Fe.
For large electrical transformers, switchboard parts; non-magnetic, high electric resistivity.

AL TECH A-8.
M-1779; 0.55 C, 0.30 Mn, 0.90 Si, 5.0 Cr, 1.25 W, 1.30 Mo, bal Fe.
Usual hardness 57-60 Rc.
Punch and die steel for hot and cold work applications, For shear blades, chipper knives, forming dies, forging dies, back up rolls, and plastic molds.
AISI A-8.

AL TECH ALMAR 18.
M-1779; 0.03 C, 17.0-19.0 Ni, 7.0-9.5 Co, 3.0-5.2 Mo, 0.15-0.80 Ti, 0.05-0.15 Al, 0.02 Zr, 0.003 B, bal Fe.
Annealed: 148,000 TS; 105,000 YS; 16 El; 68 RA; C 30 Rock.
Aged: 286,000 TS; 280,000 YS; 11 El: 55 RA; C 54 Rock.
For rocket motor cases, airframes, airborne equipment, large machinery parts, bolts, cryogenic equipment, high temperature equipment.
Maraging steel, shock resistant.

AL TECH ALMAR 18 (200).
M-1779; 0.03 max C, 17-19 Ni, 8-9 Co, 3.0-3.5 Mo, 0.15-0.25 Ti, 0.02 Zr, 0.003 B, 0.05-0.15 Al, bal Fe.
Maraged: 210,000 TS; 200,000 YS; 15 El; 67 Ra.
For fuel rocket cases, aircraft landing gears, missiles and aircraft parts
Maraging steel, high strength and toughness and ductility.

AL TECH ALMAR 18 (250).
M-1779; 0.03 max C, 17-19 Ni, 7.0-8.5 Co, 4.6-5.2 Mo, 0.3-0.5 Ti, 0.0 Zr, 0.003 B, 0.05-0-15 Al, bal Fe.
Maraged: 243,000 TS; 232,000 YS; 12 El; 57 Ra; C 48 Rock.
For fuel rocket cases, aircraft landing gears, missile and aircraft components.
Maraging steel, high strength, ductility and toughness.

AL TECH ALMAR 18 (300).
M-1779; 0.03 max C, 18-19 Ni, 8.5-9.5 Co, 4.6-5.2 Mo, 0.5-08 Ti, 0.02 Zr, 0.003 B, 0.05-0.15 Al, bal Fe.
Maraged: 295,000 TS; 292,000 YS; 11.5 El; 59 RA, C 48-50 Rock.
For rocket cases, aircraft landing gears, missile and aircraft components.
Maraging steel, high strength, ductility and toughness.

AL TECH ALMAR 20.
M-1779; 0.03 max C, 19.0-20.0 Ni, 1.3-1.6 Ti, 0.15-0.30 Al, 0.5 Cb, 0.02 Zr, 0.003 B, bal Fe.
Heat treated: 253,000-265,500 TS; 244,000-260,000 YS; 10-13 El; 52-59 RA; C 49-53 Rock.
For rocket motor cases, airframes, airborne equipment, large machinery parts, bolts, cryogenic and high temperature equipment.
Maraging steel, shock resistant.

AL TECH ALMAR 25.
M-1779; 0.03 C, 25.0-26.0 Ni, 1.3-1.6 Ti, 0.15-0.30 Al, 0.02 Zr, 0.00 0.50 Cb, bal Fe.
Heat treated: 240,000-265,000 TS; 212,000-246,000 YS; 3-5 El.
For rocket motor cases, airframes, airborne equipment, large machinery parts, bolts, cryogenic equipment, high temperature equipment. Maraging steel, shock resistant.

AL TECH ALMAR 362.
M-1779; 0.05 max C, 14.0-14.5 Cr, 6.25-7.0 Ni, 0.6-09 Ti, bal Fe.
Annealed: 125,000 TS; 108,000 YS; 16 El; C 25 Rock.
Age Hardened: 188,000 TS; 182,000 YS; 13 El, C 41 Rock.
For jet engine, aircraft and missile components, hydraulic and pneumatic equipment, valve and pump parts.
Age-hardenable, martensitic, corrosion resistant. UNS S36200.

AL TECH ALMAR 363.
M-1779; 0.04 C, 11.5 Cr, 4 Ni, 0.4 Ti, bal Fe.
Strip Annealed: 123,000 TS; 106,000 YS; 12 El; C 28 Rock.
At 600°F: 97,700 TS; 89,800 YS; 5 El.
For jet engine, missile and space components.
Maraging steel, good fracture toughness.
Corrosion resistant. Good fabricability.

ALTECH ALTEMP 19-19DL.
M-1779; 0.28-0.35 C, 18-20 Cr, 8-11 Ni, 1.0-1.7 Mo, 1.0-1.7 W, 0.4 Cb, 0.5 max Ti, bal Fe.
Hot rolled: 118,000 TS; 69,000 YS; 58 El; 55 RA; 216 Brin.
Heat treated: 103,000 TS; 43,000 YS; 54 El; 57 RA; 189 Brin.
For jet engine and gas turbine components; austenitic, heat resistant.

ALTECH ALTEMP 19-19DX.
M-1179; 0.28-0.35 C, 18-20 Cr, 8-11 Ni, 1.2-2.0 Mo, 1.0-1.7 W, 0.6 Ti, 0.5 max Cu, bal Fe.
Hot rolled: 118,500 TS; 69,000 YS; 58 El; 55 RA; 216 Brin.
Heat treated: 103,000 TS; 43,000 YS; 54 El; 57 RA; 189 Brin.
For jet engine and gas turbine components; austenitic, heat resistant.

ALTECH ALTEMP A-286.
M-1179; 0.08 max C, 15 Cr, 26 Ni, 1.2 Mo, 0.3 V, 2 Ti, 0.35 max Al, bal Fe.
Bar: 135,000-160,000 TS; 85,000-115,000 YS; 20-3.0 El; 30-55 RA.
At 1000°F: 131,000 TS; 87,500 YS; 18.5 El; 31.2 RA.
At 1500°F: 36,500 TS; 68.5 El; 37.5 RA.
For turbine blades, bolting, jet and supercharger parts; austenitic, heat resistant.

AL TECH AM.
M-1779; 0.10 max C, 16.5-17.5 Cr 4.0-4.5 Ni, 2.5-3.0 Mo, bal Fe.
Heat treated: 161,000-200,000 TS; 45,000-142,000 YS; 22-8 El; 200-390 Brin.
For valves, springs, knife blades; for applications up to 1000°F.

AL TECH AM 355.
M-1779; 0.13 C, 15.5 Cr, 4.35 Ni, 2.75 Mo, 1 N, bal Fe.
SCT-heat treatment: 190,000 TS; 165,000 YS; 10 El; 20 RA.
For jet engine and missile components; high heat resistance.

ALTECH CHECKWAITE 8.
M-1779; 0.10 C, 1.0-2.0 Mn, 0.04 max P, 0.03 max S, 1.0 max Si, 18.5- 21.5 Cr, 23.0-26.0 Ni, 1.75-2.75 Mo, bal Fe.
For balance weights. Vacuum consumable electrode melted. Stainless, austenitic, non-magnetic.

AL TECH D-5.
M-1779; 1.40 C, 0.30 Mn, 0.50 Si, 12.50 Cr, 0.80 Mo, 3.30 Co, 0.50 Ni, bal Fe.
Usual hardness 57-62 Rc.
Air hardenable high carbon-high chromium with Co die steel with improved abrasion resistance and red hardness.
For long run blanking and forming dies, hot punches and shear blades.
AISI D-5.

AL TECH E-BRITE 26-1.
M-1779; 26.0 Cr, 1.0 Mo, 0.01 max C, 0.015 N, bal Fe.
High purity ferritic alloy with superior corrosion resistance, especially to stress corrosion cracking.
For instrument parts, screening.

ALTECH NITRALLOY 135 MODIFIED.
M-1779, M-24; 0.40 C, 0.5 Mn, 1.0 Al, 1.75 Cr, 0.4 Mo, bal Fe.
Heat treated: 158,000 TS; 141,000 YS; 17 El; 56 RA; 320 Brin.
For cylinder barrels, nitrided parts; nitriding steel.

ALTECH NITRALLOY 135 TYPE G.
M-1779; 0.36 C, 0.4-0.7 Mn, 0.2-0.4 Si, 0.85-1.2 Al, 1.49 Cr, 0.18 M bal Fe.
Heat treated: 224,000-104,000 TS; 180,000-85,000 YS; 11-18 El; 36-59 445-200 Brin.
For nitrided gears, shafts, cams, clutches, rollers; nitriding steel.

AL TECH RELAY 2.
M-1779; 1.0-1.5 Si, bal Fe.
For armatures, relays, solenoid switches; high permeability.

AL TECH RELAY 2SS.
M-1779; 1.0-1.5 Si, bal Fe.
For armatures, relays, solenoid switches; high permeability.

AL TECH RELAY 5.
M-1779; 2.5 Si, bal Fe.
For armatures, switches, solenoid switches; high permeability.

AL TECH RELAY 5SS.
M-1779; 0.5 C, 0.7 Cr, 0.5 Mo, bal Fe.
For tools; oil hardening, shock resisting.

AL-TECH SILCHROME 10.
M-1779; 0.37 Cu, 19 Cr, 8 Ni, 3 Si, 1 Mn, 0.1 N, 0.3 Mo, 0.3 Cu, b
For exhaust valves.
Austenitic, stainless, heat resistant.

AL TECH SILCROME 25-12.
M-1779; 0.20 C, 25 Cr, 12 Ni, bal Fe.
Annealed: 95,000 TS; 48,000 YS; 42 El; 50 RA; 167 Brin.
For corrosion and heat resistant parts; heat and corrosion resistant.

AL TECH SILCROME 25-20.
M-1779; C, 25 Cr, 20 Ni, bal Fe.
For heat and corrosion resistant parts; heat and corrosion resistant.

AL TECH SILCROME TPA.
M-1779; 0.4-0.5 C, 0.3-0.8 Si, 13-15 Ni, 13-15 Cr, 2-3 W, bal Fe.
170-210 Brin.
For airplane valves; heat resistant.

AL TECH SILCROME XB.
M-1779; 0.75-0.85 C, 1.5-2.5 Si, 19-23 Cr, 1-2 Ni, bal Fe.
270-530 Brin.
For valves and valve seats; heat resistant.

ALTEN NO. 11.
M-1040; 1.7 Si, 0.5 Mn, 3.5 T.C., 2.7 Gr., 0.07 Cr, 0.03 Mo, bal Fe.
For castings, gears, housings; cast iron.

ALTEN NO. 15.
M-1040; 1.8 Si, 0.6 Mn, 3.5 T.C., 3.1 Gr., 1.4 Cr, 0.16 Ni, 0.6 Mo, 0.08 V, 0.04 Cu, bal Fe.
For castings, gears, housings; cast iron.

ALTEN NO. 18.
M-1040; 1.8 Si, 0.5 Mn, 3.7 T.C., 3 Gr., 1 Cr, 0.18 Ni, 0.04 Mo, 0.09 V, bal Fe.
For castings, stokers, grate bars; cast iron.

ALTEN TYPE 30.
M-1040; 3.3 C, 2.2 Si, 0.8 Mn, 0.2 max Ni, 0.1 max Cr, 0.1 max Mo, bal Fe.
Cast: 30,000 TS; 170-223 Brin.
For valve components, machinery castings; cart iron; Tr.S. 2200; SAE 111.

ALTEN TYPE 35 HR.
M-1040; 3.4 C, 2 Si, 0.8 Mn, 1 Ni, 0.10 max Cr, 0.4 Mo, bal Fe.
Cast: 35,000 TS; 190-210 Brin.
For glass molds, stop cocks, heat resistant parts; cast iron, heat resistant.

ALTEN TYPE 45 HR.
M-1040; 3.4 C, 1.9 Si, 0.8 Mn, 1.2 Cr, 0.3 Mo, 0.2 max Ni, bal Fe.
Cast: 50,000 TS; 300-330 Brin.
For furnace parts, heat resistant parts; cast iron, heat resistant.

ALTEN TYPE 45 WR.
M-1040; 3.1 C, 2 Si, 0.7 Mn, 0.5 min Ni, 0.3 min Cr, 0.25 min Mo, bal Fe.
Cast: 45,000 TS; 205-230 Brin.
For valves, pressure castings, wear resistant parts; cast iron, wear resistant; SAE 122.

ALTEN TYPE 50WR.
M-1040; 3.2 T.C., 0.7 C.C., 1.8 Si, 0.7 Mn, 0.2 Ni, 0.5 Cr, 0.5 Mo, 0.09 Cu, bal Fe.
Cast: 50,000 TS; 215 Brin.
For valves, wear plates, superheaters; wear and corrosion resistant; C.I.

ALTIOR BRAND ALUMINUM SOLDER.
M-577; Pb, bal Sn.
Al solder.

ALTIOR BRAND ANTIFRICTION METAL.
M-577; Pb, Sb, bal Sn.
For bearings; white antifriction metal.

ALTMAG.
M-Russian; 5-8 Mg, 0.5-1.0 Mn, 0.1-0.5 Ti, bal Al.
For light alloy parts; age hardenable.

ALTO.
M-15; C, alloy, bal Fe.
Rolled: 151,000 TS; 98,000 YS; 51 El; 50 RA; 220 Brin.
Heat treated: 217,000 TS; 220,000 YS; 16 El; 49 RA; 555 Brin.
For axes, axles, drills, chisels, crowbars; non-tempering, shock resistant, water hardened.

ALTOLOY A.
M-15; C, alloy, bal Fe.
For pneumatic tools, chisels; shock resistant; non-tempering, water hardening.

ALTOLOY B.
M-15; C, alloy, bal Fe.
For pneumatic tools, chisels; tough, oil hardening.

ALUCABLE.
M-673; 0.55 Si, 0.4 Mg, bal Al.
For light alloy parts; non-hardenable.

ALUCHROM O.
M-297; 30 Cr, 5 Al, bal Fe.
Annealed: 99,500-106,000 TS; 30-25 El.
For heat resistant parts; max operating temp. 1250°C.

ALUCHROM 1.
M-297; 22 Cr, 5 Al, bal Fe.
Annealed: 92,400-99,500 TS; 30-25 El.
For heat resistant parts; max operating temp. 1250°C.

ALUCHROM II.
M-297; 8 Cr, 6 Al, bal Fe.
Annealed: 85,200-88,500 TS; 30-25 El.
For heat resistant parts; max operating temp. 1100°C.

ALUDUR.
M-284; 0.5 Mg, 0.7 Si, 0.45 Fe, bal Al.
Heat treated: 50,000 TS; 6 El; 90 Brin.
For overhead transmission lines; self-aging alloy; high corrosion resistant.

ALUDUR 275.
M-1376; 1.5-3.0 Mg, 0.5-1.5 Mn, 0.3 max Cr, bal Al.
For aircraft structural parts.

ALUDUR 300.
M-1376; 2-4 Mg, 0.4 max Mn, 0.4 max Cr, bal Al.
For aircraft structural parts.

ALUDUR 500.
M-1376; 4-5 Mg, 0.8 max Mn, 0.3 max Cr, bal Al.
Soft: 42,000 TS; 22,000 YS; 35 El; 65 Brin.
Hard: 60,000 TS; 50,000 YS; 15 El; 100 Brin.
For aircraft structures, light alloy parts; corrosion resistant.

ALUDUR 533.
M-Germany; 0.3-1.0 Si, 0.3-0.8 Mn, 0.5-1.3 Mg, bal Al.
Soft: 16,000-18,500 TS; 8,500-11,30 YS; 20-27 El; 30-40 Brin.
For structural fittings, hardware, aircraft components. Age-hardenable.

ALUDUR NO. 533D.
M-284; 0.8 Mg, 0.7 Si, bal Al.
Heat treated: 45,000-50,000 TS; 10-8 El; 90-100 Brin.
For overhead transmission lines; corrosion resistant.

ALUDUR 570.
M-Germany; 2.5-5.0 Cu, 0.2-1.8 Mg, 0.3-1.5 Mn, Bal Al.
For light alloy parts; age hardenable.

ALUDUR-570.
M-1548; 3.5-5.5 Cu, 0.3-0.6 Si, 0.3-1.0 Mn, 0.3-0.7 Mg, bal Al.
Annealed: 26,000 TS; 10,000 YS; 22 El; 45 Brin.
Hardened: 60,000 TS; 40,000 YS; 20 El; 105 Brin.
For construction and transportation equipment, fittings, screw machine products.
High strength, age-hardenable.

ALUDUR NO. 570D.
M-284; 0.8 Mn, 1.0 Si, bal Al.
Heat treated: 60,000 TS; 17 El; 90-100Brin.
For overhead transmission lines; corrosion resistant; self-aging alloy.

ALUDUR-580.
M-Germany; 3.5-5.5 Cu, 0.3-0.6 Si, 0.3-1.0 Mn, 0.3-0.7 Mg, bal Al.
Annealed: 26,000 TS; 10,000 YS; 22 El; 45 Brin.
Hardened: 62,000 TS; 40,000 YS; 20 El; 105 Brin.
For construction and transportation equipment, fittings, hardware, screw machine products, fasteners.
High strength, age-hardenable.

ALUDUR 580.
M-1376; 2.5-5.0 Cu, 0.2-1.8 Mg, 0.3-1.5 Mn, bal Al.
For light alloy parts; age hardenable.

ALUDUR-630.
M-Germany; 3.5-5.5 Cu, 0.3-0.6 Si, 0.3-1.0 Mn, 0.3-0.7 Mg, bal Al.
Annealed: 26,000 TS; 10,000 YS; 22 El; 45 Brin.
Age-hardened: 60,000 TS; 40,000 YS; 20 El; 105 Brin.
For fittings, hardware, screw machined products, fasteners.
Age-hardenable, high strength.

ALUDUR 630/OM.
M-1376; 2.5-5.0 Cu, 0.2-1.8 Mg, 0.3-1.5 Mn, bal Al.
For light alloy parts; age hardenable.

ALUDUR 700.
M-1376; 5.5-7.5 Mg, 0.8 max Mn, 0.3 max Cr, bal Al.
For light alloy parts; corrosion resistant.

ALUFER.
M-German; Al alloy.
For light alloy parts; non-hardenable.

ALUFLEX.
M-1444; 0.75 Mg, bal Al.
For electric conductors, electrical wires, and cable for braiding.
High electrical conductivity.

ALUFONT II.
M-249; 4 Cu, 2.0 Si, 0.6 Mn, 0.2 Mg, 0.15 Ti, bal Al.
Heat treated: 33,000-55,000 TS; 27,000-45,000 YS.
For castings where high strength, but no high corrosion resistance is required; high fluidity.

ALUFONT-II.
M-Germany; 3.7-4.3 C, 2.0 Si, 0.5-0.7 Mn, 0.2 Mg, bal Al.
Cast: 21,000 TS; 14,000 YS; 2.5 El; 55 Brin.
For manifold and valve bodies, machine frames, oil pans.
Good castability and machinability.

ALUFONT 3.
M-249; M-1605, M-1634; 4-5 Cu, 0.2-0.5 Ti, bal Al.
Sand cast: 41,000 TS; 28,000 YS; 6 El; 85 Brin.
Chill cast: 53,000 TS; 37,000 YS; 6 El; 100 Brin.
For auto and aircraft construction; heat treatable.

ALUFONT "H".
M-249; 0.4 Si, 1.4 Fe, 0.15 Mg, 2.0 Cu, 12 Zn, bal Al.
Sand cast: 43,000 TS; 36,000 YS; 0.5 El.
Chill cast: 44,000 TS; 42,000 YS; 1.5 El.
For large castings for general purposes; corrosion resisting.

ALUFONT "W".
M-249; 2.0 Si, 0.6 Fe, 0.5 Mn, 2.0 Cu, 12 Zn, bal Al.
Sand cast: 35,000 TS; 20,000 YS; 3.5 El.
Chill cast: 40,000 TS; 35,00 YS; 3.5 El.
For large castings, light alloy parts.

ALUFRAN.
M-1784.
Trade name for several aluminum grades of ingots, billets, casting alloys.

ALUGIR.
M-674; 3 Cu, 0.8 Mg, 1.2 Ni, bal Al.
57,000-64,000 TS; 6-22 El.
For light alloy parts; heat tretable.

ALUMA.
M-622; 1.5 Mn, bal Al.
14,500-18,500 TS; 7,000-9,000 YS; 40-30 El.
For light alloy parts; corrosion resistant.

ALUMAG 15.
M-770; 1.5 Mg, 0.4 Mn, bal Al.
Annealed: 23,000 TS; 10,000 YS; 20 El; 32 Brin.
Hardened: 40,000 TS; 35,000 YS; 55 Brin.
For light alloy parts; corrosion resistant.

ALUMAG 25.
M-770; 2.5 Mg, 0.4 Mn, bal Al.
Soft: 43,000 TS; 14,000 YS; 22 El; 40 Brin.
Hard: 60,000 TS; 39,000 YS; 67 Brin.
For light alloy parts; corrosion resistant.

ALUMAG 35.
M-770; 3.5 Mg, 0.4 Mn, bal Al.
Soft: 36,00 TS 17,000 YS; 25 El; 47 Brin.
Hard: 55,000 TS; 46,000 YS; 78 Brin.
For light alloy parts; corrosion resistant.

ALUMAG 50.
M-770; 5 Mg, 0.4 Mn, bal Al.
Soft: 43,000 TS, 22,000 YS; 25 El; 53 Brin.
Hard: 64,000 TS; 54,000 YS; 95 Brin.
For light alloy parts; corrosion resistant.

ALUMAG 65.
M-770; 6.5 Mg, 0.4 Mn, bal Al.
Soft: 49,000 TS, 27,000 YS; 27 El.
Hard: 69,000 TS; 57,000 YS.
For high temperature applications; high temp. properties.

ALUMAG GRADE 1.
M-675; 4-5 Mg, bal Al.
For light alloy parts; high resistance to corrosion.

ALUMAG GRADE 2.
M-675; 8-10 Mg, bal Al.
For light alloy parts; high resistance to corrosion.

ALUMALUN.
M-250; Al, Si alloy plus abrasive grains imbedded in wearing surface.
For floor plates, stair treads, car steps; wear resistant; anti-slip.

ALUMAN.
M-249; M-624, M-493; 1.5 Mn, 0.25 max Si, 0.25 max Fe, 0.1 max Cu, bal Al.
Soft: 15,000 TS; 7000 YS; 30 El; 30 Brin.
Hard: 36,000 TS; 29,000 YS; 7 El; 50 Brin.
Extruded: 17,000 TS; 10,000 YS; 24 El; 35 Brin.
For panelling, roofing, containers, welded structures; corrosion resistant, non-hardenable.

ALUMAN-30.
M-624, M-541; 1.0-1.5 Mn, 0.1 max Cu, 0.5 max Fe, 0.5 max Si, 0.5 max bal Al.
Annealed: 11,400-18,500 TS; 5,700-10,000 YS; 25-45 El; 25-35 Brin.
Hard: 17,000-35,600 TS; 24,200-35,600 YS; 3-8 El; 50-65 Brin.
For cooking utensils, heat exchangers, storage tanks, plumbing fixtures.

ALUMAN - 100.
M-493; 1.0 Mn, bal Al.
Ann: 90 MPa TS; 28 El.
Hard: 200 MPaTS; 3 El.
For containers.

ALUMAN (AW15).
M-493; 1.4-1.6 Mn, bal Al.
Soft: 17,000 TS; 7,000 YS; 35 El; 30 Brin.
Hard: 32,000 TS; 30,000 YS; 4 El; 60 Brin.
For heat exchangers, truck panels, tanks reflectors, ducts.
Hardened by cold work only.

ALUMANINGOT see SPECIALLOY 5813
ALUMANINGOT TM.

ALUMAR 4043 ETC (15 ALLOYS).
M-677; Aluminum Association Type 4043 and 15 other AA Type compositions.
Bar aluminum welding wire; for welding similar aluminum alloys.

ALUMASOD.
M-676; 76 Sn, 23 Zn, 1 Cd.
For Al solder.

ALUMAWELD.
M-393; Sn, Zn, Pb, and chemical catalyzer.
12,000 TS.
For soldering of all metals, especially Al and white metals.

ALUMAWELD ALL METAL SOLDER.
M-393; Sn, Zn, Pb.
For soldering Al and Mg alloys: M.P. 550°F, no flux required for Mg.

ALUMAWELD H-3.
M-393.
For solder for Al, Mg, foil capacitors; M.P. 510-650°F.

ALUMAWELD H-6.
M-393.
For solder for Al and other metals; M.P. 775-785°F; hard and strong.

ALUMAWELD H-8.
M-393.
For solder for Al and other metals; M.P. 715-740°F; good for friction soldering.

ALUMAWELD NO. 1.
M-393.
For solder for Al and Mg alloys: M.P. 364-620°F.

ALUMAWELD NO. 2.
M-393.
For solder for Al and Mg foil capacitors; M.P. 393-563°F.

ALUMAWELD NO. 7.
M-393.
For solder for Al-bronze, Mg, and cast iron; M.P. 333-620°F.

ALUMAWELD NO. 22.
M-393.
For solder for Al, Mg, foil capacitors; M.P. 393-635°F.

ALUMAWELD NO. 23.
M-393.
For solder for Al, Mg, foil capacitors; M.P. 336-414°F; good wetting power.

ALUMAWELD SPECIAL.
M-393; Sn, Zn, Pb.
For soldering Al and Mg alloys; M.P. 550°F.

ALUMBRO.
M-188, M-282; 76 Cu, 22 Zn, 2 Al.
Drawn: 88,000 TS; 67,000 YS; 10 El; 175 Brin.
Annealed: 54,000 TS; 50 El; 65 Brin.
For condenser tubing to resist sea water corrosion, plates; high resistance to corrosion.

ALUMEL.
M-65; 95 Ni, 2 Mn, 2 Al (plus 8 minor constituents).
Ann: 85,000 TS.
EMF: Negative.
Thermocouple wire,-used with Chromel P for temperature measurement to 2400°F.
Corrosion and heat resistant.

ALUMEND 1100.
M-677; 1.0 max Si + Fe, 0.2 max Cu, 99.0 min Al.
For coated aluminum welding rod.

ALUMEND 4043.
M-677; 4.5-6.0 Si, 0.8 max Fe, 0.3 max Cu, 0.2 max Ti, 0.1 max Zn, 0.05 max Mn, 0.05 max Mg, bal Al.
For coated aluminum welding rod.

ALUMETAL.
M-597; 5 Si, bal Al.
For Al solder; corrosion resistant.

ALUMINARK.
M-118; 5 Si bal Al.
For Al welding electrodes; coated.

ALUMINITE.
M-U.S.; 73 Al, 23 Zn, 2.7 Cu, 0.4 Fe, 0.2 Si.
For light alloy castings; non-hardenable.

ALUMINUM.
M-1755; Al.
Purities: zone refined: 99.9999%, 99.999%, 99.99%; 99.9%.
Forms; ingots, rods, powders, wires, sheets, foils, single crystals.

ALUMINUM 3L-8.
M-86, M-28; 11-13 Si, bal Al.
Die cast: 24,500 TS; 15,500 YS; 1 El; 80 Brin.
Sand cast: 20,000 TS; 1 El; 75 Brin.
For housings, cases, general engineering castings; corrosion resistant, good castability.

ALUMINUM-100.
M-1565; 0.1 Cu, Fe = 1.5 x Si, min, bal Al.
Electric motor rotors.
High purity; corrosion resistant.

ALUMINUM 108-Z.
M-1279; 3-4 Cu, 2.5-3.5 Si, 2-3 Zn, 1 max Fe, 0.5 max Ni, 0.25 max Ti, 0.5 max Mn, bal Al.
For valve bodies, manifolds, truck wheels, piano frames, oil pans.
Non-heat treatable.

ALUMINUM 333.
M-Eng; Cu, Si, bal Al.
Cast: 34,000 TS; 19,000 YS; 2 El; 90 Brin.
For castings, pistons, sole plates; permanent molds.

ALUMINUM 354 now AA 354.0.

ALUMINUM 380B.
3.5 Cu, 9.0 Si, 1.0 Fe, 0.5 Mn, 0.10 Mg, 2.0 Zn, 0.5 Ni, bal Al.
Cast: 40,000 TS, 17,00 YS; 2.5 El.
For general purpose die castings.
Corrosion resistant, high strength.

ALUMINUM 380Z.
3.5 Cu, 9.0 Si, 1.0 Fe, 0.5 Mn, 0.10 Mg, 3.0 Zn, 0.5 Ni, bal Al.
Cast: 40,000 TS 17,000 YS; 2.5 El.
For general purpose die castings.
Corrosion resistant, pressure tight.

ALUMINUM 384 now AA 384.0.

ALUMINUM 392 now AA 392.0.

ALUMINUM 1060 now AA 1060.

ALUMINUM 5040 now AA 5040.

ALUMINUM 5056 now AA 5056.

ALUMINUM 5083 now AA 5083.

ALUMINUM 5086 now AA 5086.

ALUMINUM 5154 now AA 5154.

ALUMINUM 5183.
M-1147; 0.4 Si, 0.4 Fe, 0.1 Cu, 0.3 -1.0 Mn, 4.5-5.2 Mg, 0.05-0.25 Cr, 0.15 max Ti, 0.0005 max Be, bal Al.
For structural and marine applications, truck and trailer bodies, power shovels; good formability and weldability, welding electrodes.

ALUMINUM 5252 now AA 5252.

ALUMINUM 5357 now AA 5357.

ALUMINUM 5657.
M-671; 0.6-1.0 Mg, 0.08 max Si, 0.10 max Fe, 0.10 max Cu, 0.03 max Zn, bal Al.
H25 Temper: 23,000 TS; 20,000 YS; 12 El; 40 Brin.
H38 Temper: 28,000 TS; 24,000 YS; 7 El; 50 Brin.
For deep drawn parts and stampings, interior and exterior trim for automoblies and appliances.
Bright, corrosion resistant.

ALUMINUM 6011 now **AA 6011.**

ALUMINUM 6063 now **AA 6063.**

ALUMINUM 6066.
M-1235; 0.7-1.2 Cu, 0.9-1.8 Si, 0.8-1.4 Mg, 0.6-1.1 Mn, 0.2 Ti, 0.4 Cr, bal Al.
O-temper: 22,000 TS; 12,000 YS; 18 El; 43 Brin.
T-6 temper: 62,000 TS; 55,000 YS 12 El; 120 Brin.
For truck and trailer bodies, aircraft construction; age hardened, good forming and welding properties.

ALUMINUM 6070 now **AA 6070.**

ALUMINUM 6101 now **AA 6101.**

ALUMINUM 6151 now **AA 6151.**

ALUMINUM 6201 now **AA 6201.**

ALUMINUM 6262 now **AA 6262.**

ALUMINUM 6351 now **AA 6351.**

ALUMINUM 6463 now **AA 6463.**

ALUMINUM 7001.
M-1235; 0.35 Si, 0.4 Fe, 1.2-2.6 Cu, 2.6-3.4 Mg, 0.18-0.40 Cr, 6.8-8.0 Zn, bal Al.
T6-Temper: 98,000 TS; 88,000 YS; 10 El.
For missile bodies, trailers, airframes, curtain walls; high strength.

ALUMINUM 7004 now **AA 7004.**

ALUMINUM 7005 now **AA 7005.**

ALUMINUM 7039 now **AA 7039.**

ALUMINUM 7076 now **AA 7076.**

ALUMINUM 7079 now **AA 7079.**

ALUMINUM 7175 now **AA 7175.**

ALUMINUM 7178 now **AA 7178.**

ALUMINUM 7179 now **AA 7179.**

ALUMINUM 7277 now **AA 7277.**

ALUMINUM 8001 now **AA 8001.**

ALUMINUM 8081.
M-1625; 18-22 Sn, 0.7 max Si, 0.10 max Mn, 0.7 max Fe, 0.7-1.3 Cu, 0.10 max Ti, 0.15 max others, bal Al.
H25 Temper: 24,000 TS; 21,500 YS; 13 El.
H112 Temper: 28,000 TS; 25,000 YS; 10 El.
For bearings. Steel backed automotive bearings, bearings for light duty diesel engines, gas turbines, farm tractors; improved corrosion resistance.

ALUMINUM 8280.
M-1625; 0.7 -1.3 Cu, 0.2-0.7 Ni, 1.0-2.0 Si, 5.5-7.0 Sn, bal Al.
Annealed: 17,000 TS; 70,000 YS; 28 El; H 65 Rock.
H16 Temper: 27,000 TS; 25,000 YS; 5 El; H 92 Rock.
For bearings made from continuous coils for automotive engines.
High load carrying capacity, excellent fatigue strength.

ALUMINUM-A13.
M-671, M-4; 1.0 Cu, 12 Si, bal Al.

ALUMINUM A-43.
M-815; 5.3 Si, bal Al.
Die cast: 33,000 TS; 14,000 YS; 9 El; 50 Brin; 19,000 SS, 17,000 Fatigue St.
For marine castings, water jackets, meter housings, carburetor bodies.
High fluidity and castability, Corrosion resistant. Die casting alloy.

ALUMINUM A-57G.
M-France; 6.5-7.0 Si, bal Al.
Cast: 44,000 TS; 7 El; 100 Brin.
Corrosion resistant.

ALUMINUM-A140 now **AA A2400.**

ALUMINUM-A344 now **AA 440.0.**

ALUMINUM-A356 now **AA A356.0.**

ALUMINUM-A612 now **AA A7120.**

ALUMINUM ALLOY B.E.S.A. NO. 361.
M-Eng; 6-8 Cu, 0.1 Zn, 1.0 max Fe, 1.0 max Si, bal Al.
Chill cast: 20,160 YS; 3 El; 60 Brin.
Sand cast: 18,000 TS; 1-2 El.
For general castings, cylinder heads; good machinability.

ALUMINUM ALLOY B.E.S.A. NO. 362.
M-Eng.; 11-13 Cu, 0.1 Zn, 1.0 max Fe, 1.0 max Si, bal Al.
Chill cast: 20,160 TS; 80 Brin.
For permanent mold castings, pistons; non-hardenable.

ALUMINUM ALLOY B.E.S.A. NO. 363.
M-Eng.; 2.5-3.0 Cu, 12.5-14.5 Zn, 1.0 max Fe, 1.0 max Si, bal Al.
Sand cast: 25,000-30,000 TS; 1 El; 70 Brin.
For sand castings, crankcases; non-hardenable.

ALUMINUM ALLOY BRITISH L-5.
M-92; 85 Al, 13-15 Zn, 2.5-3.0 Cu, traces Fe, Si.
Sand cast: 24,500 TS; 3 El; 80 Brin.
For light alloy, parts, regular and intricate castings; sand castings.

ALUMINUM ALLOY BRITISH L-8.
M-92; 89 Al, 10 Cu, 1 Sn.
20,000 TS.
For light alloys, parts; non-hardenable.

ALUMINUM AS-5.
M-1219, M-1220, M-541; 0.6-0.9 Cu, 4.5-5.5 Si, 0.5 Mg, bal Al.
Cast: 35,600-45,600 TS; 25,600-29,900 YS; 4-5 El; 90-100 Brin.
For instrument housings, oil pans, crankcases; high fluidity and corrosion resistant.

ALUMINUM AS-10.
M-1219, M-1220, M-541; 2.2 Cu, 10 Si, 1.4 Mg, 1 Ni, bal Al.
Cast: 32,700-35,600 TS; 27,800-32,700 YS; 0.5-0.7 El; 100-120 Brin.
For pistons; high temperature resistance.

ALUMINUM-AS62.
M-541, M-1219, M-1220; 6 Si, 2 Cu, 0.3 Mg, bal Al.
Cast: 27,000 TS; 18,000 YS; 70 Brin.
Ht.Tr.: 36,000 TS; 24,000 YS; 80 Brin.
For automobile cylinder heads, crankcases, housings.
Permanent molds, age-hardening.

ALUMINUM-AS132.
M-541, M-1219, M-1220; 12 Si, 2 Cu, 0.3 Mn, 0.7 Fe, bal Al.
For crankcases, oil pump bodies, die castings for automobile components.
Die cast alloy.

ALUMINUM AW-15.
M-493; 1.5 Mn, bal Al.
Hard worked: 37,000 TS; 3 El; 59 Brin.
Annealed: 17,000 TS; 27 El; 32 Brin.
For deep drawing, stamping and pressing; weldable, corrosion resistant.

ALUMINUM A.W. 160.
M-493; 1.0 Si, 0.7 Mn, 0.65 Mg, bal Al.
For light alloy parts; same as Anticorodal.

ALUMINUM B-95,.
M-USSR; 5-7 Zn, 1.8-2.8 Mg, 1.4-2.0 Cu, 0.2-0.6 Mn, 0.1-0.25 Cr, bal Al.
For high strength castings; age-hardenable.

ALUMINUM-B133.
M-297; Mg, Zn, bal Al.
For light alloy parts; corrosion resistant.

ALUMINUM B218 now AA B535.

ALUMINUM BA355 now AA 2014 CLAD..
M-28; 3.5-4.8 Cu, 0.8 Mg, 0.9 Si, 1 Fe, 1.2 Mn, 0.3 Ti, bal Al.

ALUMINUM BRASS.
M-33; 77.5 Cu, 20.5 Zn, 2 Al, 0.055 As.
Ann: 60,000 TS; 27,000 YS; 55 El. Drawn 35%: 85,000 TS; 65,000 Y 15 El.
Corrosion resistant; for condenser and heat exchanger tubes, ferrules.

ALUMINUM BRONZE.
M-129; 87.5 Cu, 4.1 Al, 1.7 Fe, 6.6 Zn.
Hot rolled: 63,9000 TS; 35,000 YS; 6.5 El.
Cold drawn: 98,000 TS; 90,500 YS; 13.0 El; 14.0 RA.
For bearings, gears; heavy duty.

ALUMINUM BRONZE.
M-508; 86-90 Cu, 1-4 Fe, 8-12 Al, 1-4 Mn.
Cast: 60,000-90,000 TS; 20-4 El; 150-250 Brin.
For centrifugal castings, valve seat rings; heat treatable.

ALUMINUM BRONZE 5%.
M-8; 95 Cu, 5 Al.
Hard: 105,00 TS; 5 El.
Soft: 52,000 TS; 70 El.
For diaphragms to withstand pressure, condenser tubes; M.P. 1060°C.

ALUMINUM BRONZE 135.
M-177; 10 Al, bal Cu.
For hardware; Al-bronze.

ALUMINUM BRONZE 136.
M-177; 10 Al, 1 Fe, bal Cu.
For propellers; Al-bronze.

ALUMINUM BRONZE 707.
M-141; 7.15 Al, 2 Si, bal Cu.
Soft: 90,000 TS; 50,000 YS; 30 El; 165 Brin.
For boat shafting, pump rods, gears; Al-bronze, corrosion resistant.

ALUMINUM BRONZE 712.
M-141; 95,5 Cu, 1.0 Si, 3.5 Al.
Spring: 114,000 TS; 80,000 YS; 3 El; B97 Brin.
Soft: 65,000 TS; 24,000 YS; 55 El; B60 Brin.
For spring contacts, diaphragms, bellows; spring properties, tough; 12% conductivity.

ALUMINUM BRONZE 715.
M-141; 2.9 Al, 0.35 Si, bal Cu.
Hard: 75,000 TS; 15 El.
Soft: 40,000 TS; 55 El.
For bolts, nuts, screws; aluminum bronze, corrosion resistant.

ALUMINUM BRONZE E-707.
  M-1191; 6.2-8.0 Al, 1.6-2.2 Si, bal Cu.
  Forged: 85,000-90,000 TS; 30,00-40,000 YS; 25-20 El; 120-142 Brin.
  For marine hardware, propellers, gears, fasteners, bolts; high strength; corrosion resistant.

ALUMINUM BRONZE NO. 1.
  M-1518; Al, Fe, bal C.
  Cast: 71,700 TS; 20 El; 90-140 Brin.
  For high strength castings; corrosion resistant, good strength at high temperature.

ALUMINUM BRONZE NO. 2.
  M-1518; Al, Fe, bal Cu.
  Cast: 89,600 TS; 12 El; 156-187 Brin.
  For high strength castings; corrosion resistant, retains strength at high temperature.

ALUMINUM C46.
  M-1219; M-541; 9.5-10.5 Cu, 0.8-1.2 Si, 0.3 Mg, 0.2 Ti, 1.5 Ni, bal Al.
  Cast: 51,200-54,100 TS; 42,700-48,400 YS; 0.5-1.0 El; 125-140 Brin.
  For light alloy castings; general purpose.

ALUMINUM C-113 now **AA 2130.**

ALUMINUM-C355 now **AA C355.0.**

ALUMINUM CD1S now **AA 1160.**

ALUMINUM-CH91.
  M-4; 6-8 Cu, 1-3 Si, 2.2 max Zn, 1.4 max Fe, bal Al.
  Die Cast: 44,000 TS; 22,000 YS; 4 El; 28,000 SS.
  For commercial die castings, large pressure tight engine covers, grills.
  High strength. Similar to Aluminum 13.

ALUMINUM-D16T.
  M-Russia; 4.4-4.52 Cu, 1.6 Mg, 0.7 Mn, 0.38 Fe, 0.26 Si, 0.01 Ni, 0.2 Zn 0.03 Ti bal Al.
  Ht.Tr.: 70,000 TS; 56,000 YS; 130 Brin.
  For aircraft structures, fittings, hardware, fasteners.
  Heat treatable, high strength.

ALUMINUM DTD-25.
  M-England; 5 Si, bal Al.
  For instrument housings, gear cases, general castings; corrosion resistant.

ALUMINUM-EC now **AA 1350.**

ALUMINUM-F132 now **AA F332.0.**

ALUMINUM-G8A.
  M-158, M-4, M-671; 0.44 Si, 0.14 Cu, 1.72 Fe, 8.80 Mg, 0.02 Mn, 0.01 Ni, bal Al.
  As Cast: 47,900 TS; 28,500 YS; 4.2 El.
  For die castings, instrument cases, housings. Corrosion resistant.

ALUMINUM GM-3889M.
  M-165; 4 Si, 1 Cd, bal Al.
  For bearings; requires steel backing.

ALUMINUM H10.
  M-England; 0.07 Cu, 0.20 Fe, 0.58 Mn, 0.61 Mn, 0.94 Si, Ti, bal Al.

ALUMINUM HD-11.
  M-1389; Al alloy.
  For impact extrusions; ductile.

ALUMINUM-HS30.
  M-England; 0.1 Cu, 0.4-1.5 Mg, 0.6-1.3 Si, 0.6 Fe, 0.1 Zn, 0.4-1.0 Mn, 0.5 Cr, 0.2 Ti, bal Al.
  W.P. Temper: 46,000 TS; 42,800 YS; 11 El.
  At 320°F: 61,600 TS; 51,100 YS; 17.8 El.
  For aircraft and space vehicle structures and components.
  Heat treatable. Good weldability.

ALUMINUM HZM-100.
  M-1235; 7.3 Zn, 3 Mg, 2 Cu, 0.2 Cr, bal Al.
  T-6 temper: 100,000 TS; 92,000 YS; 9 El; 160 Brin.
  O-temper: 40,000 TS; 26,000 YS; 10 El.
  For aircraft parts and structures; age hardenable.

ALUMINUM IRON BRONZE "B".
  85.16 Cu, 9.43 Al, 4.74 Fe, 0.38 Pb, 0.09 P.
  For strong corrosion resistant structural parts; corrosion resistant.

ALUMINUM IRON BRONZE "H".
  89.43 Cu, 6.97 Al, 3.41 Fe.
  For strong corrosion resistant structural parts; corrosion resistant.

ALUMINUM IRON BRONZE "R".
  85.16 Cu, 6.6 Al, 7.52 Fe, 0.5 Mn.
  For strong corrosion resistant structrural parts; corrosion resistant.

ALUMINUM-K.
  M-297; 11.5-13.5 Si, 0.4-1.5 Cu, 0.8-1.5 Mg, 1-2 Ni, 0.5 max Mn, bal Al.
  At 75°F: 55,000 TS; 46,000 YS; 9 El; 120 Brin.
  At 700°F: 35,000 TS; 2000 YS; 90 El.
  For forged pistons; for elevated temperature use.

ALUMINUM KO-1.
  4.5-5.2 Cu, 0.4-1.0 Ag, 0.18-0.35 Mg, 0.15-0.35 Ti, 0.4 max Zn, bal Al.
  Cast: 62,000-66,000 YS; 72,300-73,000 TS; 10-11 El.
  At 350°F: 60,000 TS; 57,000 YS; 11 El.
  At 500°F: 46,000 TS; 44,000 YS; 13 El.
  Heat treated: 65,000 TS; 55,000 YS; 3 El.
  For impellers, throttle levers, fins. Age hardenable.

ALUMINUM KS280.
  M-Germany; 21-22 Si, 1.5 Cu, 1.5 Ni, 0.6 Mn, 0.5 Mg, 1.2 Co, bal Al.
  For bearings.

ALUMINUM KSS.
M-624; 1.7-2.3 Mg, 1.0-1.4 Mn, bal Al.
For light alloy parts for marine service; corrosion resistant.

ALUMINUM L8N.
M-1219, M-541; 11.0-12.5 Cu, 0.7 Fe, 0.5 Ni, bal Al.
Cast: 38,400-42,700 TS; 22,800-25,600 YS; 1-2 El; 110-125 Brin.
For light alloy castings; general purpose.

ALUMINUM L8T.
M-1219, M-541; 11.0-12.5 Cu, 1 Fe, 0.5 Si, 0.2 Ti, bal Al.
Cast: 38,400-42,700 TS; 22,800-25,600 YS; 1-1.5 El; 110-125 Brin.
For light alloy castings; general purpose.

ALUMINIUM L-15.
M-Eng.; 2-4 Mg, 1.2-1.5 Mn, 0-0.2 Sb or Ti, bal Al.
For light alloy castings; resists sea water corrosion.

ALUMINUM LM11.
M-England; 4.2-5.0 Cu, 0.15-0.35 Mg, 0.5 max Fe, 0.3 max Si, 0.3 max Ti, 0.1 max Zn, bal Al.
Chill cast: 40,000 TS; 9 El.
For pressure die castings, cases, housings. Heat treatable.

ALUMINUM-NICKEL NIAL-14TI.
M-297; 4-5 Al, 0.5 Ti, bal Ni.
For heating elements; heat and corrosion resistant.

ALUMINUM-S12A.
M-158; 11.76 Si, 0.53 Cu, 0.81 Fe, 0.01 Mg, 0.19 C, Mn 0.04 Ni, bal Al.

ALUMINUM P35.
M-624; 3.2-3.8 Mg, 0.5 Mn, 0.3 max Si, 0.4 max Fe, bal Al.
For light alloy parts for marine service; corrosion resistant.

ALUMINUM V-95.
M-Russian; Mg, Zn, Cu, bal Al.
For aircraft parts; age-hardenable.

ALUMINUM-X149 now AA 249.0.

ALUMINUM XB-805.
M-England; 6.5 Sn, 1 Ni, 1 Cu, bal Al.
For bearings; requires steel backing.

ALUMINUM XZM-100.
M-1235; Al alloy.
Extruded: 92,00 TS; 11 El.
For light alloy parts.

ALUMINUM S6063.
M-671; 0.5 Si, 0.8 Mg, 0.35 max Fe, bal Al.
T5: 27,000 TS; 21,000 YS; 12 El; 60 Brin.
T42: 25,000 TS; 13,000 YS; 22 El.
For architecutural applications, portable irrigation systems, moldings and trim.
Heat treatable extrusion alloy.

ALUMINUM SS6063.
M-671; 0.5 Si, 0.8 Mg, 0.35 max Fe, bal Al.
T5: 27,000 TS; 21,000 YS; 12 El; 60 Brin.
For architectural applications, portable irrigation systems, moldings and trim.
Heat treatable extrusion alloy.

ALUMINUM T6063.
M-671; 0.5 Si, 0.8 Mg, 0.35 max Fe, bal Al.
T5: 27,000 TS; 21,000 YS; 12 El; 60 Brin.
For architecutural applications, portable irrigation systems, molding and trim.
Heat treatable extrusion alloy.

ALUMINUM OILITE.
M-211; Al alloy.
For bearings; self-lubricating, porous.

ALUMINUM OILITE 201.
M-211; 4.5 Cu, 0.8 Si, 0.5 Mg, 94 Al.
Sintered: 40,000 TS; 32,000 YS; 2.0 El; 90 RH; density, 2.5-2.6.
Aluminum base powder metal.

ALUMINUM OILITE 601.
M-211; 0.25 Cu, 1.0 Mg, 98 Al.
Sintered: 35,000 TS; 30,000 YS; 2.0 El; 80 RH; density, 2.5-2.6.
Aluminum base powder metal.

ALUMINUM-OY.
M-Eng.; 4 Cu, bal Al.
For light alloy parts; age-hardened.

ALUMINUM SOLDER (A).
30 Zn, 65 Sn, 5 Bi.
For Al solder.

ALUMINUM SOLDER (B).
50 Zn, 1.5 Cu, 33 Sn, 2 Sb, 12.5 Pb.
For Al solder.

ALUMINUM SOLDER (C).
57 Zn, 43 Cd.
For Al solder.

ALUMINUM SOLDER (D).
M-Eng.; 67.5 Sn, 1 Cu, 16.5 Zn, bal Al.
7,100 TS.
For Al solder.

ALUMINUM SOLDER (E).
M-Eng.; 53 Sn, 40 Zn, 7 Al.
6,700 TS.
For Al solder.

ALUMINUM SOLDER, FRISMUTH (A).
47.5 Zn, 5.5 Cu, 31.5 Sn, 10.5 Al, 5.5 Ag.
For Al solder.

Section I: Alloy Data / 91

ALUMINUM SOLDER, FRISMUTH (B).
47.4 Zn, 5.3 Cu, 36.8 Sn, 10.5 Al.
For Al solder.

ALUMINUM SOLDER, FRISMUTH (C).
67 Sn, 27 Pb, 3 Al.
For Al solder.

ALUMINUM SOLDER, GRIMM'S (A).
69.1 Sn, 28,8 Pb, 1.44 Zn, 0.72 Ag.
For Al solder.

ALUMINUM SOLDER, GRIMM'S (B).
50 Sn, 25 Pb, 25 Zn.
For Al solder.

ALUMINUM SOLDER RICHARDS.
25 Zn, 71.5 Sn, 3.5 Al.
For Al solder.

ALUMINUM TITANIUM BRONZE.
M-U.S.; 90-89 Cu, 9-10 Al, 1 Fe, Ti.
For propellers; strong, tough.

ALUMINUM UNION 17S.
M-644; 3.5 -4.5 Cu, 0.4-0.7 Mn, 0.4-0.7 Mg bal Al.
For light alloy parts, aircraft; heat treatable.

ALUMINUM UNION 22S.
M-644; 3.5-4.7 Cu, 0.7-1.5 Si, 0.4-1.0 Mg, 0.4-1.5 Mn, bal Al.
Heat treated: 56,000 TS; 44,000 YS; 8 El.
For light alloy parts, aircraft; heat treatable.

ALUMINUM UNION 24S.
M-644; 3.5-4.8 Cu, 0.8-1.8 Mg, 0.3-1.5 Mn, bal Al.
Heat treated: 56,000 TS; 35,000 YS; 15 El.
For light alloy parts, aircraft; heat treatable.

ALUMINUM UNION 27S.
M-644; 4.5 Cu, 0.8 Mn, 0.8 Si, 0.05 Sn, bal Al.
Heat teated: 58,000 TS; 30,000 YS.
For forged aircraft parts; heat treatable.

ALUMINUM UNION 35S.
M-644; 10-13 Si, bal Al.
For light alloy paneling; non heat treatable.

ALUMINUM UNION 51S.
M-644; 0.5-1.5 Si. 0.5-1.5 Mg, bal Al.
Heat treated: 40,000 TS; 31,000 YS; 8 El.
For light alloy architectural structure; heat treatable.

ALUMINUM UNION 57S.
M-644; 1.5-3.0 Mg, 0.1-0.35 Cr, bal Al.
Heat treated: 32,000 TS; 28,000 YS; 5 El.
For light alloy paneling, aircraft; heat treatable.

ALUMINUM UNION 123.
M-644; 5 Si, bal Al.
Cast: 16,000 TS; 4 El; 40 Brin.
For light alloy castings, aircraft; corrosion resistant.

ALUMINUM UNION 125.
M-644; 4.5-5.5 Si, 1.0-1.5 Cu, 0.4-0.6 Mg, bal Al.
Heat treated: 30,000 TS; 22,000 YS.
For cylinder heads, valve bodies; pressure proof castings.

ALUMINUM UNION 126.
M-644; 1.4 Cu, 0.5 Mg, 0.75 Mn, 0.75 Ni, 5.0 Si, bal Al.
Heat treated: 26,000-32,000 TS; 3-1 El.
For aircooled cylinder heads; heat treatable.

ALUMINUM UNION 135.
M-644; 0.3 Mg, 7.0 Si, bal Al.
Heat treated: 26,000-32,000 TS; 6-12 El.
For light alloy castings; corrosion resistant.

ALUMINUM UNION 160.
M-644; 12 Si, bal Al.
For light alloy castings; high fluidity.

ALUMINUM UNION 162.
M-664; 12-15 Si, 1.5-2.5 Ni, 0.75-1.25 Mg, 0.5-1.0 Cu, bal Al.
For pistons; low thermal expansion.

ALUMINUM UNION 218.
M-664; 4.0 Cu, 1.5 Mg, 2.0 Ni, bal Al.
For pistons, cylinder heads; Y" alloy.

ALUMINUM UNION 225.
M-664; 4 Cu, bal Al.
Heat treated: 36,000 TS; 4 El.
For aircraft light alloy castings; age-hardenable.

ALUMINUM UNION 350.
M-664; 8.5-11.5 Mg, bal Al.
Cast: 36,000 TS; 22,000 YS; 14 El.
Heat treated: 44,000 TS; 28,000 YS; 12 El.
For light alloy castings; highest strength, ductility and impact.

ALUMINUM UNION A51S.
M-644; 0.6 Mg, 1.0 Si, 0.25 Cr, bal Al.
Heat treated: 40,000 TS; 15 El.
For light alloy forgings; non-hardenable.

ALUMINUM UNION NO. 3S.
M-644; 1.5 max Mn, bal Al.
Hard: 25,000 TS; 22,000 YS.
Soft: 14,000 TS; 5,000 YS.
For light alloy body paneling and molding; corrosion resistant.

ALUMINUM UNION NO. 14S.
M-644; 4.4 Cu, 0.35 Mg, 0.75 Mn, 0.8 Si, bal Al.
Heat treated: 62,000 TS; 18 El; 130 Brin.
For forged aircraft light alloy parts; heat treatable.

ALUMINUM Y TI.
M-1219, m-1220, m-549; 3.8-4.2 Cu, 1.3-1.7 Mg, 0.2 Ti, 1.8-2.3 Ni, bal Fe.
For pistons, cylinder heads; age hardenable, heat resistant.

ALUMINWELD.
M-578; 4.5 Si, 0.8 Fe, 0.3 Cu, 0.5 Mg, 0.10 Zn, 0.20 Ti, bal Al.
Aluminum alloy arc welding electrodes.
AWS Class Al-43.

ALUNEON.
Al alloy.
For light alloy parts.

ALUSIL.
M-Germany; 1.0-2.0 Cu, 20-21 Si, 0.7 max Ni 0.7 max Fe, bal Al.
For engine pistons, cylinder liners, pumps.
Low density and low expansivity.

ALUSIN N. 280.
M-Germany; 22 Si, 1 Cu, 1 Ni, 1 Co, 1 Mg, 0.5 Cr, bal Al.
Cast: 25,000-35,000 TS; 100-140 Brin.
For pistons in engines, pumps, cylinder liners.
Low density and low expansivity.

ALUVAC.
M-679; 4 Cu, bal Al.
For light alloy parts; heat treatable.

ALUWANGAN.
M-1372; 0.5-1.5 Mn, 0.3 max Cr, bal Al.
Soft: 16,000 TS; 6000 YS; 40 El.
Hard: 29,000 TS; 27,000 YS; 10 El.
For cooking utensils, heat exchangers, tanks, furniture; good weldng and forming properties.

ALUWE-1.
M-1373; Al alloy.
For light alloy parts.

ALUWE-4.
M-1373; Al alloy.
For light alloy parts.

ALUWE-6.
M-1373; Al alloy.
For light alloy parts.

ALUWE-8.
M-1373; Al alloy.
For light alloy parts.

ALUWE-22.
M-1373; Al alloy.
For light alloy parts.

ALUWE-53.
M-1373; Al alloy.
For light alloy parts.

ALUWE-55.
M-1373; Al alloy.
For light alloy parts.

ALUWE-57.
M-1373; Al alloy.
For light alloy parts.

ALVA 36.
M-German; 3 Pb, 3 Sb, 2 Cu, Mn, bal Al.
Wrought: 30-80 Brin.
For bearings.

ALVA EXTRA.
M-38; 1.0 C, 0.25 V, bal Fe.
For blanking, threading, forming dies, taps; 1150°C max; water hardening. W2.

ALW.
M-1083; 0.60-1.0 Te, 0.06-1.0 Ni, bal Cu.
Bar: 37,000-61,000 TS; 14-20 El; 80-100 Brin.
Forged: 37,000-43,000 TS; 14-20 El; 80-95 Brin.
For spot and seam welding of aluminum alloys; high conductivity (85-90%) resistance welding electrode. Was ERM ALW.
BS 4577 and ISO 5182 alloy A/1/2.

ALX CAST ALLOY.
M-1; 2 C, 17 W, 33 Cr, 40 Co, B, bal Fe.
For cutting tool bits; fast cutting, centrifugally cast.

ALZEN.
M-French; 66.6 Al, 33.4 Zn.
For strong light alloy parts; not hardenable.

ALZEN 305.
M-French; Zn, Al, bal Cu.
For bearings; bronze.

ALZIN.
French; 20 Zn, bal Al.
Cast: 18,000-28,000 TS; 2-3 El; 62-65 Brin.
For light alloy parts; non-hardenable.

ALZINC.
M-French; 20 Zn, 80 Al.
21,000 TS; 2 El; 80 Brin.
For light castings; Sibley casting alloy.

AM.
M-1740; 1.0-1.2 C, 1.0 max Si, 0.006 max S, 0.07 max P, 11.0-14.0 Mn, bal Fe.
High resistance to impact wear.
B. S. 1457; ASTM A128-64.

AM 1.
M-1653; 0.75-0.90 C, 3.5-4.5 Cr, 8.0-9.5 Mo, 1.0-2.0 W, 0.80-1.30 V, bal Fe.
Molybdenum high speed tool steel.
W.-Nr. 1.3346; AISI M1.

A.M. 1 SPECIAL.
M-210; 0.35 C, 1.0 Si, 1.35 W, 5.0 Cr, 1.5 Mo, 0.45 V, bal Fe.
For extrusion dies, heading dies, mandrels; hot work steel, resists heat checking.

A M 3 DIE STEEL.
M-210; 0.40 C, 1.05 Si, 5.0 Cr, 1.35 Mo, 1.1 V, bal Fe.
For die casting dies, extrusion dies, upsetters; air hardened, resists heat checking.

AM 05.
M-1361; 0.5 Mg, bal Cu.
Extruded, wire, bars, tube for ornamental moldings.

AM 6.
M-German; 6 Ce, 2 Mn, bal Mg.
For light alloy parts for high temperature use; ASTM-EM62; cast and wrought.

AM 10.
M-German; 10 Ce, bal Mg.
For pistons, supercharger impeller; ASTM-E10.

AM 10.
M-1361; 1.0 Mg, bal Al.
100-155 N/mm^2 TS; 40-135 N/mm^2 YS, 4-22 El.
Good formability, weldability, anodizing.
Extrusions, wire, bars, tube.
Type AlMg 1.
Similar to AA 5005.

AM 11.
M-1361; 1.0 Mg, bal Al.
Extrusions, wire, bars, tube for ornamental purposes.

AM 18.
M-1361; 1.9 Mg, bal Al.
145-205 N/mm^2 TS; 60-155 N/mm^2 YS; 4-17 El.
Good formability, weldability, corrosion resistance.
Type AlMg 2.
Similar to AA 5051.

AM 18.10.
M-1290; 0.12 C, 17.5 Cr, 10.5 Ni, bal Fe.
Sol. ann. & age: 64,000 min TS; 28,000 min YS; 20 min El.
Non-magnetic steel casting.
DIN G-X12 CrNi 1811; W.-Nr. 1.3955.

AM 18 11 S.
M-1290; 0.07 max C, 1.0 max S, 2.0 max Mn, 16.0-18.0 Cr, 10.0-12.0 Ni 0.10 max N, bal Fe.
Nonmagnetic cast steel.
SEW 1.3944; G-x 5 CrNi 18 11.

AM 18 15 MO.
M-1290; 0.07 max C, 18 Cr, 15 Ni, 0.3 Mo, bal Fe.
Nonmagnetic cast steel.
SEW 1.3950; G-X 5 CrNi Mo 18 15.

AM 21.
M-1361; 2.0 Mg, 0.8 Mn, bal Al.
175-255 N/mm^2 TS; 80-175 N/mm^2 YS; 4-17 El.
Good formability, weldability, corrosion resistance.
Type AlMg Mn.

AM 30.
M-1361; 2.9 Mg, 0.3 Mn, bal Al.
175-255 N/mm^2 TS; 80-175 N/mm^2 YS; 4-17 El.
Good corrosion resistance, weldability and formability.
Type AlMg 3.

AM 36.
M-1361; 2.9 Mg, bal Al.
175-255 N/mm^2 TS; 80-175 N/mm^2 YS; 4-17 El.
Modification of AM 30 for decorative purposes.

AM 40.
M-1361; 4.5 Mg, 0.8 Mn, Cr, bal Al.
255-275 N/mm^2 TS; 110-155 N/mm^2 YS; 12 El.
Good corrosion resistance, weldability and formability.
Type AlMg 4.5 Mn.
Similar to AA 5083.

AM 54.
M-1361; 4.5 Mg, 0.2 Mn, bal Al.
235-325 N/mm^2 TS; 110-235 N/mm^2 YS; 4-18 El.
Good corrosion resistance and weldability.
Type AlMg 5.
Similar to AA 5182.

AM 58.
M-1361; 5.2 Mg, 0.5 Mn, bal Al.
235-325 N/mm^2 TS; 110-235 N/mm^2 YS; 4-18 El.
Good corrosion resistance and weldability.
For shipbuilding applications.
Type AlMg 5.
Similar to AA 5056.

AM 60A (MG INGOT).
M-43; 5.7-6.3 Al, 0.15 min Mg, 0.20 max Zn, 0.20 max Si, 0.25 max Cu, 0.01 max Ni, bal Mg (others 0.30 max).
For use in die castings.

AM 100A.
M-Various foundries; 10.0 Al, 0.10 min Mn, bal Mg.
F-Temper: 20,000-22,000 TS; 10,000-12,000 YS; 0-2 El.
T4-Temper: 34,000-40,000 TS; 10,000-13,000 YS; 6-10 El.
T6-Temper: 34,000-40,000 TS; 17,000-22,000 YS; 0-1 El.
Magnesium permanent mold casting with good pressure tightness and strength and good weldability.
For housings for motors and power tools.
ASTM B199-68; AMS 4483; QQ-M-55; SAE 502.

AM 100A (MG INGOT).
M-43; 9.4-10.6 Al, 0.13 min Mg, 0.20 max Zn, 0.20 max Si, 0.04 max Cu 0.01 max Ni, bal Mg. (others 0.30 max).
ASTM B93-66.

AM 350.
M-1, M-1491; M-1772; M-38; 0.10 C, 1.0 Mn, 0.4 Si, 16.5 Cr, 4.25 Ni, 2.75 Mo, 0.1 N, bal Fe.
Precipitation hardening stainless steel with 10% Delta ferrite.
AMS 5546; AISI 633; uNS S35000.

AM 350 see ALSO CARPENTER PYROMET 350 UNITEMP 350.

AM 355.
M-1, M-1491; M-38; 0.15 C, 1.0 Mn, 0.4 Si, 15.5 Cr, 4.25 Ni, 2.75 Mo, 0.1 N, bal Fe.
Precipitation hardenable to 216,000 psi TS.
Good strength, wear and corrosion resistance.
AMS 5359; AISI 634; UNS S35500.

AM-362 see ALMAR 362 AND ALLEGHENY AM 362.

AM 503.
M-325; 1.5 Mn, bal Mg.
Plate, sheet, strip, extrusions.
24,000-30,000 TS; 8,000-14,000 YS(est); 4-5 El; 35-55 VDH.
Low strength alloy with very good welding characteristics; for fuel and oil tanks.

AM 537.
M-German; Al, Zn, bal Mg.
For light alloy parts; heat treatable.

AM 858.
M-1290; 0.27 C, 8.5 Mo, 7.5 Cr, 5.5 Ni, bal Fe.
Sol. ann.: 71,000 min TS; 31,000 min YS; 30 min El.
Non-magnetic steel casting.
DIN G-X25 MnCrNi 886; W.-Nr. 1.3966.

AMALLOY.
M-1638; 7 Mg, 0.1 Mn, 0.15 Fe, 0.14 Ti, 0.01 Cu, bal Al.
As Cast: 34,000-38,000 TS; 19,000 YS; 7-10 El; 70 Brin.
For dairy, agricultural, cookware, hardware, aircraft, transportation and oil field equipment.
Corrosion and impact resistant.
Sand, permanent mold and die castings.

AMALOY.
M-132; 0.60-0.70 C, 0.70-0.90 Mn, Mo, bal Fe.
For tools, dies; water hardened.

AMALOY-1.
M-524; 97.5 Pb, 2.5 Sn.
For protective coating against corrosion; applied to ferrous and copper alloys.

AMALOY-2.
M-524; 90-97.5 Pb, 2.5-10 Sn.
For protective coating against corrosion; applied to ferrous and copper alloys.

AMALOY-3.
M-U.S.; Ni, Cr, W.
For corrosion and heat resistant parts; corrosion and heat resistant.

AMAX LP.
M-1778; 0.005-0.012 P, bal Cu.
(Oxygen free).
Elec. cond: 90% (min) of IACS.
Rods, tubes, flat products; for plumbing, refrigerator units, steam lines, commutators.
CDA 108.

AMAX METAL.
M-U.S.; 81 Cu, 11 Sn, 7.4 Pb, 0.3 P.
For bearings; heavy duty.

AMAX MZ COOPER.
M-1778; 0.03 Mg, 0.03 Zr, Cu + Ag + Mg + Zr = 99.95 min.
Cold Worked 40%: 365 MPa (53,000 psi) TS; 358 MPa (52,000 psi) YS 3.5 El.
Elect. cond.: 93% IACS at 68°F.
For heat sinks, high temperature wire, lead frames, semiconductor bases.

AMAX MZC COOPER.
M-1778; 0.03-0.06 Mg, 0.06-0.15 Zr, 0.40-0.80 Cr, Cu + Ag + Mg + Zr + Cr = 99.95% min.
Cold Orked 40%, aged: 496 MPa (72,000 psi) TS; 455 MPa (66,000 psi) YS; 10 El.
Elect. cond: 80% IACS at 68°F.
Electrical and electronic components, contacts, heat sinks, molds for continuous casting.

AMAX XLP.
M-1778; 0.001-0.005 P, bal Cu (Oxygen free).
Elec. cond: 98.16% IACS.
Rods, tubes, flat products for electrical conductors and terminals, waveguide tubing, thermostatic control tubing.
CDA 103.

AMBEROID.
M-135; 15 Ni, 23 Zn, bal Cu.
Annealed: 50,000 TS; 20,000 YS; 60 El; 63 Brin.
Hard: 114,000 TS; 110,000 YS; 2 El; 228 Brin.
For costume jewelry, hollowware, optical equipment; nickel silver.

AMBO BW O.
M-1315; 0.85-0.95 C, 0.15-0.30 Si, 0.20-0.40 Mn, 0.70-090 Cr, bal Fe.
Cold work took steel, for mandrels, punches.
W.-Nr. 1.2036.

AMBO BW1.
M-1315; 1.05 C, 1.0 Cr, 0.3 Mn, bal Fe.
For bearings, sleeves, liners; water hardened, wear resistant.

AMBO BWC.
M-1315; 1.0 C, 1.55 Cr, 0.35 Mn, bal Fe.
For bearings, sleeves, liners; water hardened, wear resistant.

Section I: Alloy Data / 95

AMBO BWW.
M-1315; 1.45 C, 1.4 Cr, 0.6 Mn, bal Fe.
For forming and blanking dies, bearings; water hardened, wear resistant.

AMBO C15W3.
M-1315; 0.15 C 0.25 Si, 0.37 Mn, bal Fe.
Annealed: 70,000 TS; 40,000 YS; 25 El; 60 RA; 145 Brin.
For gears, pinions, camshafts, cams; case hardening steel.

AMBO C22W3.
M-1315; 0.22 C, 0.25 Si, 0.45 Mn, bal Fe.
Annealed: 75,000 TS; 43,000 YS; 20 El; 55 RA; 150 Brin.
For gears, pinions, camshafts, cams; case hardening steel.

AMBO C35W3.
M-1315; 0.35 C, 0.25 Si, 0.55 Mn, bal Fe.
Hot rolled: 85,000 TS; 54,000 YS; 30 El; 53 RA; 185 Brin.
For gears, pinions, shafts, bolts; water hardened.

AMBO C45W3.
M-1315; 0.45 C, 0.25 Si, 0.65 Mn, bal Fe.
Hot rolled: 98,000 TS; 59,000 YS; 24 El; 45 RA; 212 Brin.
For gears, pinions, shafts, bolts, fasteners; water hardened.

AMBO C53W3.
M-1315; 0.53 C, 0.38 Si, 0.55 Mn, bal Fe.
Normalized: 100,000 TS; 55,000 YS; 18 El; 26 RA; 200 Brin.
For gears, pinions, shafts, axles, bolts, fasteners; water hardened.

AMBO C60W3.
M-1315; 0.60 C, 0.25 Si, 0.65 Mn, bal Fe.
Heat treated: 160,000 TS; 115,000 YS; 12 El; 40 RA; 325 Brin.
For springs, rails, machine tool parts, axes; water hardened.

AMBO CMO.
M-1315; 0.20 C, 1.25 Mn, 1.25 Cr, bal Fe.
For gears, cams, camshafts; case hardened, tough.

AMBO CNM6.
M-1315; 0.56 C, Ni, Cr, Mo, V, bal Fe.
For gears, bolts, crankshafts; oil hardened, shock resistant.

AMBO CNMO 6.
M-1315; 0.30-0.38 C, 0.40 max Si, 0.40-0.70 Mn, 1.4-1.7 Cr, 0.15-0.30 Mo, 1.40-1.70 Ni, bal Fe.
Steel for cold extrusion.
W.-Nr. 1.6582.

AMBO CNMO 8.
M-1315; 0.26-0.33 C, 0.15-0.40 Si, 0.30-0.60 Mn, 1.8-2.2 Cr, 0.30-0.50 Mo, 1.8-2.-2 Ni, bal Fe.
Steel for cold extrusion.
W.-Nr. 1.6580.

AMBO CNMV.
M-1315; 0.55 C, 1.65 Ni, 0.9 Cr, 0.18 Mo, 0.1 V, bal Fe.
For gears, bolts, crankshafts; oil hardened, shock resistant.

AMBO CR13.
M-1315; 1.65 C, 11.5 Cr, 0.10 V, bal Fe.
For blanking and forming dies, punches; air hardening, non-deforming.

AMBO CR13CO.
M-1315; 1.65 C, 11.5 Cr, Co, bal Fe.
For blanking and forming dies, punches; air hardened, non-deforming.

AMBO CR13MW.
M-1315; 1.65 C, 11.5 Cr, Mo, V, bal Fe.
For blanking and forming dies, punches; air hardening, non-deforming.

AMBO CR 105.
M-1315; 1.0-1.1 C, 0.15-0.30 Si, 0.25-0.40 Mn, 0.40-0.60 Cr, bal Fe.
Water hardening tool steel.

AMBO CRT.
M-1315; 0.90 C, 19 Mn, 0.1 V, bal Fe.
For punches, dies, upsetters, crimpers; oil hardened, non-deforming.

AMBO CRV50.
M-1315; 0.50 C, 0.95 Mn, 1.05 Cr, 0.1 V, bal Fe.
For gears, springs, crankshafts, bolts, studs; oil hardened, shock resistant.

AMBO CRZ.
M-1315; 2.1 C 11.5 Cr, 0.3 Mn, bal Fe.
For blanking and forming dies, punches; oil hardened, non-deforming.

AMBO CRZW.
M-1315; 2.1 C, 11.5 Cr, 0.7 W, bal Fe.
For blanking and forming dies, punches; oil hardened, non-deforming.

AMBO DCM.
M-1315; 0.40 C, Cr, Mo, Mn, bal Fe.
For gears, bolts, crankshafts, fasteners; oil hardened, tough.

AMBO DCN3.
M-1315; 0.50 C, 1.05 Cr, 3.25 Ni, 0.50 Mn, bal Fe.
For gears, bolts, crankshafts, studs; oil hardened, shock resistant.

AMBO DCN4.
M-1315; 0.31 C, 0.7 Cr, 3.5 Ni, 0.6 Mn, bal Fe.
For gears, bolts, machine tool parts; oil hardened, shock resistant.

AMBO DCN4E.
  M-1315; 0.4 C, 0.7 Cr, 3.5 Ni, 0.6 Mn, bal Fe.
  For gears, bolts, crankshafts, fasteners; oil hardened, shock resistant.

AMBO DCN EXTRA.
  M-1315; 0.32-038 C, 0.15-0.30 Si, 0.40-0.60 Mn, 1.2-1.5 Cr, 0.20-0.40 Mo, 3.8-4.3 Ni, bal Fe.
  Hot work tool steel.
  W.-Nr. 1.2766.

AMBO DR5.
  M-1315; 1.3 C, 4.75 W, bal Fe.
  For cutters, form tools; fast finishing steel, water hardened.

AMBO EM 2 E E.
  M-1315; 1.2-1.35 C, 0.10-0.25 Si, 0.10-0.25 Mn, bal Fe.
  Water hardening tool steel.
  W.-Nr. 1560.

AMBO EM 2 EXTRA EXTRA.
  M-1315; 1.20-1.35 C, 0.10-0.25 Si, 0.10-0.25 Mn, bal Fe.
  Carbon tool steel.
  W.-Nr 1.1560.

AMBO EM3CV.
  M-1315; 1.0 C, 0.07 Mn, 1.1 Cr, bal Fe.
  For bearings, liners, bushings; water hardened, wear resistant.

AMBO EM3 EXTRA.
  M-1315; 1.15 C, 0.25 max Si, 0.25 max Mn, bal Fe.
  Annealed: 110,000 TS; 56,000 YS; 20 El; 40 RA; 210 Brin.
  For springs, cutters, reamers, taps, drills, hobs; water hardened; Type W1.

AMBO EM3 EXTRA EXTRA.
  M-1315; 1.1 C, 0.25 max Si, 0.25 max Mn, bal Fe.
  Annealed: 110,000 TS; 56,000 YS; 20 El; 40 RA; 210 Brin.
  For springs, cutters, reamers, taps, drills, hobs; Type W1; water hardened.

AMBO EM3 PRIMA.
  M-1315; 0.75 C, 0.25-0.50 Si, 0.3-0.8 Mn, bal Fe.
  Heat treated: 180,000 TS; 135,000 YS; 12 El; 35 RA; 375 Brin.
  For springs, rails, punches, upsetters; Type Wl; water hardened.

AMBO EM3W.
  M-1315; 1.2 C, 0.1 V, 0.2 Cr, 1.0 W, bal Fe.
  For bearings, cutters, tools, dies; water or oil hardened, wear resistant.

AMBO EM4 EXTRA.
  M-1315; 1.0 C, 0.25 Si, 0.25 Mn, bal Fe.
  Annealed: 100,000 TS; 53,000 YS; 21 El, 42 RA; 200 Brin.
  For drills, taps, springs, reamers, hobs; Type Wl; water hardened.

AMBO EM4 EXTRA EXTRA.
  M-1315; 1.0 C, 0.25 Si, 0.25 Mn, bal Fe.
  Annealed: 100,000 TS; 53,000 YS; 21 El; 42 RA; 200 Brin.
  For springs, drills, reamers, taps; Type Wl; water hardened.

AMBO EM4 PRIMA.
  M-1315; 0.60 C, 0.25-0.50 Si, 0.3-0.8 Mn, bal Fe.
  Heat treated: 160,000 TS; 113,000 YS; 12 El; 40 RA; 325 Brin.
  For gears, bolts, crankshafts, rails, springs; water hardened.

AMBO EM5 EXTRA.
  M-1315; 0.85 C, 0.25 Si, 0.25 Mn, bal Fe.
  Heat treated: 190,000 TS; 145,000 YS; 10 El; 30 RA; 400 Brin.
  For springs, tools, drills, dies, cutters; Type Wl; water hardened.

AMBO EM5 EXTRA EXTRA.
  M-1315; 0.85 C, 0.25 Si, 0.25 Mn, bal Fe.
  Heat treated: 190,000 TS; 145,000 YS; 10 El; 30 RA; 400 Brin.
  For springs, tools, drills, dies, cutters; Type Wl; water hardened.

AMBO EM5 PRIMA.
  M-1315; 0.45 C, 0.25-0.50 Si, 0.3-0.8 Mn, bal Fe.
  Hot rolled: 98,000 TS; 59,000 YS; 24 El; 45 RA; 215 Brin.
  For gears, pinions, bolts, fasteners, shafts; water hardened.

AMBO EM6 EXTRA.
  M-1315; 0.70 C, 0.25 max Si, 0.25 max Mn, bal Fe.
  Heat treated: 175,000 TS; 128,000 YS; 12 El; 37 RA; 355 Brin.
  For springs, rails, hammers, machine tool parts; Type Wl; water hardened.

AMBO EM6 EXTRA EXTRA.
  M-1315; 0.70 C, 0.25 max Si, 0.25 max Mn, bal Fe.
  Heat treated: 175,000 TS; 128,000 YS; 12 El; 37 RA; 355 Brin.
  For springs, rails, hammers, cutters, punches; Type Wl; water hardened.

AMBO EM6 PRIMA.
  M-1315; 0.35 C, 0.25-0.50 Si, 0.3-0.8 Mn, bal Fe.
  Hot rolled: 85,000 TS; 54,000 YS; 30 El; 53 RA; 185 Brin.
  For gears, botls, fasteners, machine tool parts; water hardened.

AMBO EM 1820.
M-1315; 0.50-0.58 C, 0.30-0.50 Mn, bal Fe.
Carbon tool steel.
W.-Nr. 1.1820.

AMBO EMV.
M-1315; 1.0 C, 0.20 Si, 0.25 Mn, 0.1 V, bal Fe.
For drills, taps, reamers, broaches; Type W2; water hardened.

AMBO EN 15.
M-1315; 0.18 max C, 0.10-0.35 Si, 0.30-0.40 Mo, 1.3-1.6 Cr, bal Fe.
For structural equipment to operate down to -100°C.
W.-Nr. 1.5622.

AMBO FAH.
M-1315; 0.55-0.62 C, 0.15-0.40 Si, 0.70-1.1 Mn, 0.90-1.20 Cr, 0.10-0.2 V, bal Fe.
For heavy sections to be flame or induction hardened.
W.-Nr 1.8161.

AMBO FAW.
M-1315; 0.47-0.55 C, 0.15-0.40 Si, 0.70-1.10 Mn, 0.90-1.20 Cr, 0.10-0.20 V, bal Fe.
For heavy sections to be flame or induction hardened.
W.-Nr. 1.8159.

AMBO FS 7.
M-1315; 0.60-0.70 C, 1.50-1.80 Si, 0.70-1.0 Mn, bal Fe.
For springs.
W.-Nr. 1.5028.

AMBO FSE.
M-1315; 0.68-0.75 C, 1.50-1.80 Si, 0.60-0.80 Mn, bal Fe.
For springs.
W.-Nr. 1.5029.

AMBO HHCA8.
M-1315; 0.12 max C, 0.8 Al, 6.5 Cr, bal Fe.

AMBO HHCA10.
M-1315; 0.12 man C, 1 Si, 1 Al, 18 Cr, bal Fe.

AMBO HHCA12.
M-1315; 0.12 max C, 1.5 Si, 1.5 Al, 24 Cr, bal Fe.

AMBO HHCN100.
M-1315; 0.15 C, 19.5 Cr, 9.5 Ni, bal Fe.

AMBO HHCN120.
M-1315; 0.15 C, 24 Cr, 19 Ni, bal Fe.

AMBO HHCN120S.
M-1315; 0.2 C, 25 Cr, 4 Ni, 1.2 Si, bal Fe.

AMBO KB 1.
M-1315; 0.12-0.20 C, 0.10-0.35 Si, 0.40-0.80 Mn, 0.30 max Cr, 0.25-0.3 Mo, bal Fe.
For high temperature equipment, to 530°C.

AMBO KB2.
M-1315; 0.10-0.18 C, 0.10-0.35 Si, 0.40-0.70 Mn, 0.80-1.15 Cr, 0.45-0.65 Mo, bal Fe.
For high temperature piping to 530°C.
W.-Nr. 1.7335.

AMBO KB 3.
M-1315; 0.13-0.20 C, 0.15-0.35 Si, 0.50-0.80 Mn, 0.90-1.20 Cr, 0.40-0.50 Mo, 0.40 max Ni, bal Fe.
For high temperature forgings, to 530°C.
W.-Nr. 1.7337.

AMBO MN 2 S.
M-1315; 0.24 max C, 0.60 max Si, 1.60 max Mn, 0.5-1.20 Cr, bal Fe.
Structural steel.

AMBO MN 12.
M-1315; 1.2 C, 12.5 Mn, bal Fe.
For wear plates, dipper teeth, rail frogs; wear and abrasion resistant.

AMBO MS 1.
M-1315; 0.58-0.65 C, 0.80-1.0 Si, 0.80-1.2 Mn, bal Fe.
Hot work tool steel.
W.-Nr. 1.2826.

AMBO MS 2.
M-1315; 0.42-0.50 C, 1.50-1.80 Si, 0.50-0.80 Mn, bal Fe.
For springs.
W.-Nr. 1.0902.

AMBO MS 3.
M-1315; 0.47-0.55 C, 1.50-1.80 Si, 0.50-0.80 Mn, bal Fe.
For springs.
W.-Nr. 1.0903.

AMBO MS 4.
M-1315; 0.53 C 0.90 Si, 0.90 Mn, bal Fe.
For punches, chisels, pneumatic tools; oil hardened, tough.

AMBO MS 5.
M-1315; 0.55 C, 1.7 Si, 0.70 Mn, bal Fe.
For springs, chisels, punches, pneumatic tools; oil hardened, shock resistant.

AMBO MS 6.
M-1315; 0.60-0.68 C, 1.50-1.80 Si, 0.70-1.0 Mn, bal Fe.
For springs.
W.-Nr. 1.0906.

AMBO MS 70.
M-1315; 0.65-0.75 C, 1.5-1.8 Si, 0.60-0.80 Mn, bal Fe.
Cold work tool steel.
W.-Nr 1.2823.

AMBO MS 90.
  M-1315; 0.85-0.95 C, 1.05-1.25 Si, 0.60-0.80 Mn, 1.10-1.30 Cr, bal Fe
  Cold work tool steel as shear blades, etc.
  W.-Nr 1.2108.

AMBO MS 125.
  M-1315; 1.2-1.3 C, 1.05-1.25 Si, 0.60-0.80 Mn, 1.10-1.30 Cr, bal Fe
  Cold work tool steel, taps, threading tools.
  W.-Nr. 1.2109.

AMBO MSJ.
  M-1315; 0.53 C, 0.90 Si, 0.90 Mn, bal Fe.
  Annealed: 96,000 TS: 52,000 YS; 16 El; 23 RA; 170 Brim.
  For axles, gears, bolts, tie-rods, bushings; water hardened.

AMBO NR 2 AF.
  M-1315; 0.12 max C, 1.0 max Si, 2.0 max Mn, 16.0-18.0 Cr, 7.0-9.0 Ni, bal Fe.
  Austenitic stainless steel, work hardens rapidly; for cold worked springs.
  W.-Nr. 1.4310.
  Similar to AISI 301.

AMBO NR 2 AZ.
  M-1315; 0.15 max C, 1.0 max Si, 2.0 max Mn, 17.0-19.0 Cr, 8.0-10.0 Ni 0.15-0.35 S, bal Fe.
  Free-machining austenitic stainless steel, for screw machined parts.
  W.-Nr. 1.4305.
  AISI 303.

AMBO NR 4 AW.
  M-1315; 0.03 max C, 1.0 max Si, 1.0 max Mn, 16.5-18.5 Cr, 11.0-14.0 Ni, 2.0-2.5 Mo, bal Fe.
  Austenitic stainless steel; for chemical industry equipment.
  W.-Nr. 1.4404.
  AISI 316L.

AMBO PD 1.
  M-1315; 0.22-0.30 C, 0.30-0.50 Si, 0.20-0.40 Mn, 0.60-0.90 Cr, 0.20-0.40 Mo, 1.30-1.60 Ni, 0.15-0.20 V, bal Fe.
  Hot work tool steel.
  W.-Nr.1.2726.

AMBO PD4.
  M-1315; 0.28 C, Ni, Mo, bal Fe.
  For gears, bolts, fasteners, shafts, oil hardened, tough.

AMBO PDZ.
  M-1315; 0.28 C, Ni, Cr, Mo, V, bal Fe.
  For gears, bolts, shafts, crankshafts; oil hardened, shock resistant.

AMBO SP1V.
  M-1315; 0.38 C, Si, Cr, V, bal Fe.
  For springs, gears, crankshafts; oil hardened, shock resistant.

AMBO SP2V.
  M-1315; 0.45 C, Si, Cr, V, bal Fe.
  For springs, gears, crankshafts; oil hardened, shock resistant.

AMBO SPW3.
  M-1315; 1.42 C, W, V, bal Fe.
  For blanking and forming dies, engravers' tools; water or oil hardened, wear resistant.

AMBO SPW4.
  M-1315; 0.45 C, 1.4 Cr, 0.70 Mo, 0.30 V, 0.7 Mn, bal Fe.
  For forging and heading dies, upsetters, punches; oil hardened, tough.

AMBO SPW5.
  M-1315; 0.30 C, 1.1 Cr, 0.18 V, 3.75 W, bal Fe.
  For chisels, punches, pneumatic tools; oil hardened, shock resistant.

AMBO SPW6.
  M-1315; 0.30 C, 2.35 Cr, 0.6 V, 4.25 W, 0.3 Mn, bal Fe.
  For pneumatic tools, extrusion rams and dies, upsetters; oil hardened, shock resistant.

AMBO SPWD.
  M-1315; 0.35 C, 0.90 Si, 1.05 Cr, 0.18 V, 1.85 W, bal Fe.
  For pneumatic tools, punches, upsetters; oil hardened, shock resistant.

AMBO SPWH.
  M-1315; 0.55 C, 0.9 Si, 1.05 Cr, 0.18 V, 1.85 W, bal Fe.
  For pneumatic tools, upsetters, rivet sets; oil hardened, shock resistant.

AMBO V25.
  M-1315; 0.82 C, 4.1 Cr, 0.85 Mo, 1.6 V, 8.7 W, bal Fe.
  For lathe and planner tools, reamers; high speed steel.

AMBO V35.
  M-1315; 0.86 C, 4.1 Cr, 0.85 Mo, 2.5 V, 12 W, bal Fe.
  For lathe and planer tools, reamers, broaches; high speed steel.

AMBO V50.
  M-1315; 1.3 C, 4.3 Cr, 0.85 Mo, 3.8 V, 12 W, bal Fe.
  For blanking and forming dies, engravers' tools; high speed steel.

AMBO V66.
  M-1315; 0.85 C, W, Mo, Cr, V, bal Fe.
  For lathe and planer tools, reamers, taps, drills; high speed steel.

AMBO V111.
  M-1315; 0.95 C, 4 Cr, V, W, Mo, bal Fe.
  For lathe and planer tools, reamers, broaches; high speed steel.

AMBO VANADIUM.
M-1315; 0.74 C, 4.1 Cr, 1.1 V, 18.5 W, bal Fe.
For lathe and planer tools, drills, taps, hobs; high speed steel.

AMBO VC 135.
M-1315; 0.30-0.37 C, 0.15-0.40 Si, 0.60-0.90 Mn, 0.90-1.20 Cr, bal Fe.
Heat treatable steel for axles, shafts.
W.-Nr. 1.7033.

AMBO WCM1.
M-1315; 1.05 C, 1.0 Cr, 0.90 Mn, 1.15 W, bal Fe.
For cold work tools, cutters, form dies; oil hardened, tough.

AMBO WF 1000.
M-1315; 0.08-0.15 C, 0.15-0.50 Si, 0.40-0.70 Mn, 2.0-2.5 Cr, 0.90-1.20 Mo, bal Fe.
For high temperature equipment to 530°C.
W.-Nr. 1.7380.

AMBO WMOV.
M-1315; 0.45 C, 0.45 Mo, 1.35 Cr, 0.8 V, 0.45 W, bal Fe.
For forging and heading dies; oil hardened, tough.

AMBO WOF.
M-1315; C, Cr, Mo, bal Fe.
For machine tool parts; oil hardened, tough.

AMBRAC 850.
M-8; 20 Ni, 4.4 Zn, 0.6 Mn, bal Cu.
Soft: 50,000 TS; 18,000 YS; 35 El.
Hard: 70,000 TS; 60,000 YS; 20 El.
For plumbing, hardware, corrosion resistant.

AMBRALOY 606.
M-8; 5 Al, 95 Cu.
Soft: 55,000 TS; 22,000 YS; 65 El.
Hard: 92,000 TS; 65,000 YS; 7 El.
For condenser tubes, forged parts; corrosion resistant.

AMBRALOY 612.
M-8; 8 Al, 92 Cu.
Soft: 65,000 TS; 25,000 YS; 65 El.
Hard: 80,000 TS; 50,000 YS; 30 El.
For screw machine parts, wire, rod; corrosion resistant.

AMBRALOY 614.
M-8; 7 Al, 2.75 Fe, bal Cu.
Plate: 70,000 TS; 30,000 YS; 35 El.
For condenser tubes, heat exchangers; high corrosion resistance. ABRALOY 930.

AMBRALOY 630.
M-8; 82 Cu, 9.5 Al, 1.0 Mn, 5 Ni, 2.5 Fe.
Soft: 90,000 TS; 12 El.
Hard: 105,000 TS; 12 El.
For forgings; heat treatable.

AMBRALOY-687.
M-8; 22 Zn, 2 Al, 0.04 As, bal Cu.
Soft: 52,000 TS; 20,000 YS; 65 El.
Hard: 85,000 TS; 60,000 YS; 10 El.
For condenser tubes, ferrules; corrosion resistant.

AMBRAZE 35.
M-1483; 35 Ag, 26 Cu, 21 Zn, 18 Cd.
For brazing alloy for torch brazing; BAg-2.

AMBRAZE 45.
M-1483; 45 Ag, 15 Cu, 16 Zn, 24 Cd.
For brazing alloy for dissimilar metals; BAg-1, M.P. 1125°F- 1145°F.

AMBRAZE 50.
M-1483; 50 Ag, 15.5 Cu, 16.5 Zn, 18 Cd.
For general purpose brazing alloy for ferrous and non-ferrous alloys; BAg-1A, M.P. 1160°F-1175°F.

AMBRAZE 111.
M-1483; Ni, Zn, bal Cu.
Cast: 80,000 TS.
For brazing alloy; nickel silver, M.P. 1700°F, corrosion resistant.

AMBRAZE 111FC.
M-1483; Ni, Zn, bal Cu.
For brazing alloy; nickel silver.

AMBRAZE 131.
M-1483; Ni, Zn, bal Cu.
For brazing alloy; nickel silver.

AMBRAZE 131 FC.
M-1483; Ni, Zn, bal Cu.
For brazing alloy; nickel silver.

AMBRAZE 181.
M-1483; 5 P, 15 Ag, bal Cu.
For brazing alloy; self-fluxing.

AMBRAZE-211.
M-1483; 7 P, bal Cu.
For brazing copper.
Phos-copper, self-fluxing.

AMBRAZE-231.
M-1483; 7.3 P, 2 Ag, bal Cu.
For brazing alloy.
Self-fluxing.

AMBRAZE-311.
M-1483; 12 Si, 4 Cu, 1 Fe, bal Al.
For brazing aluminum alloys, filler metal.
Corrosion resistant.

AMBRAZE-331.
M-1483; 10 Si, 4 Cu, 0.5 Fe, bal Al.
For brazing cast aluminum alloy.

AMBRAZE-411.
M-1483; 58 Cu, 1 Sn, 1 Fe, bal Zn.
For brazing bronzes, filler metal.
Corrosion resistant.

AMBRAZE-411FC.
M-1483; 58 Cu, 1 Sn, 1 Fe, bal Zn.
For brazing bronzes.
Flux coated, corrosion resistant.

AMBRAZE 430.
M-1483; Ag alloy.
For silver solder for stainless steel; M.P. 425°F.

AMBRAZE-611.
M-1483; 9 Al, 0.15 Mn, 2 Zn, bal Mg.
for brazing magnesium alloys, filler metal.

AMBRAZE-1001.
M-1483; 40 Ag, 30 Cu, 28 Zn, 2 Ni.
For brazing alloy, filler metal.
Corrosion resistant.

AMBRAZE-1010.
M-1483; 45 Ag, 18 Cu, 18 Zn, 19 Cd.
For brazing, filler metal.
Corrosion resistant.

AMBRAZE-1110.
M-1483; 41 Ag, 18 Cu, 15 Zn, 27 Cd.
For brazing, filler metal.
Corrosion resistant.

AMBRAZE-1201.
M-1483; 20 Ag, 45 Cu, 30 Zn, 5 Cd.
For brazing, filler metal.
Corrosion resistant.

AMBRAZE-1551.
M-1483; 56 Ag, 22 Cu, 17 Zn, 5 Sn.
For brazing, filler metal.
Corrosion resistant.

AMBRO.
2 Al, 76 Cu, 22 Zn.
For condenser tubes; corrosion resistant.

AMBRONZE 405.
M-8; 4 Zn, 1 Sn, 0.03 P, bal Cu.
Hard: 60,000 TS; 50,000 YS; 6 El; 125 Brin.
Soft: 40,000 TS; 15,000 YS; 40 El; 55 Brin.
For electrical springs; corrosion resistant.

AMBRONZE 413.
M-8; 7 Zn, 1 Sn, bal Cu.
Hard: 63,000 TS; 55,000 YS; 7 El; 137 Brin.
Soft: 42,000 TS; 16,000 YS; 40 El; 57 Brin.
For strips, tubes; corrosion resistant.

AMBRONZE 430.
M-8; 11 Zn, 2 Pb, bal Cu.
Hard: 73,000 TS; 65,000 YS; 10 El; 160 Brin.
Soft: 46,000 TS; 18,000 YS; 50 El; 59 Brin.
For springs; corrosion resistant.

AMBRONZE-4222.
M-8; 11 Zn, 1 Sn, bal Cu.
Soft: 43,000 TS; 16,000 YS; 47 El.
Hard: 72,000 TS; 60,000 YS; 8 El.
For weather strip; corrosion resistant.

AMBRONZE 4301.
M-8; 10 Zn, 2 Pb, bal Cu.
Soft: 20,000-46,000 TS; 55 El; 59 Brin.
For condenser and heat exchanger tubes; corrosion resistant to mine waters.

A. M. C.
M-365; 0.7 C, 18 W, 4 Cr, 1 V, bal Fe.
For tools, cutters, broaches; high speed steel.

AMCARB D-5.
M-1564; 3.5 Co, 96.5 WC.
Sintered: A 92.5 Rock, 170,000 Tr. S.
For extrusion and drawing dies, cutting tools.
Sintered carbide, wear and abrasion resistant.

AMCARB D-10.
M-1564; 95.5 WC, 4.5 Co.
Sintered: Rockwell A 92; 180,000 Tr. S.
For extrusion press nibs, drawing dies, cutting tools.
Sintered carbide, wear and abrasion resistant.

AMCARB D-15.
M-1564; 6 Co, 94 WC.
Sintered: Rock 91.5 A, 225,000 Tr. S.
For drawing and extrusion dies, cutting tools.
Sintered carbides. Wear and abrasion resistant.

AMCARB D-20.
M-1564; 6.0 Co, 94.0 WC.
Sintered: A 90.8 Rock, 275,000 Tr. S.
For extrusion and drawing dies, cutting tools.
Sintered carbide, wear and abrasion resistant.

AMCARB D-30.
M-1564; 9.0 Co, 91.0 WC.
Sintered: A 89.5 Rock, 325,000 Tr. S.
For extrusion and drawing dies, header punches, cutting tools.
Sintered carbide, wear and abrasion resistant.

AMCARB D-35.
M-1564; 9.0 Co, 81.0 WC, 10.0 TaC.
Sintered: A 89.0 Rock, 325,000 Tr. S.
For extrusion and drawing dies, header punches, and inserts, cutters.
Sintered carbide, wear and abrasion resistant.

AMCARB D-40.
M-1564; 13 Co, 87 WC.
Sintered: A 88.5 Rock, 400,000 Tr. S.
For extrusion, drawing and header dies, cutters.
Sintered carbide, wear and abrasion resistant.

AMCARB D-43.
M-1564; 13 Co, 27 TaC, 60 WC.
Sintered: A 88 Rock, 350,000 Tr. S.
For extrusion and drawing dies.
Wear and abrasion resistant. Sintered carbide.

AMCARB D-50.
M-1564; 15 Co, 85 WC.
Sintered: A 87 Rick, 425,000 Tr. S.
For extrusion, drawing and header dies, cutters.
Sintered carbide, wear and abrasion Resistant.

AMCOAB D-55.
M-1564; 16 Co, 57 WC, 27 TaC.
Sintered: A 86 Rock, 360,000 Tr. S.
For extrusion, drawing and header dies, punches.
Sintered carbides, wear and abrasion resistant.

AMCARB D-57.
M-1564; 15.0 Co, 78.0 WC, 7.0 TaC.
Sintered: A 86.5 Rock, 350,000 Tr. S.
For extrusion, drawing and header dies.
Sintered carbides, wear and abrasion resistant.

AMCARB D-60.
M-1564; 20 Co, 5 TaC, 75 WC.
Sintered: A 85 Rock, 390,000 Tr. S.
For extrusion, drawing and header dies-light impact.
Sintered carbides, wear and abrasion resistant.

AMCARB D-65.
M-1564; 20 Co, 80 WC.
Sintered: A 85 Rock, 400,000 Tr, S.
For extrusion, drawing and header dies-light impact.
Sintered carbides, wear and abrasion resistant.

AMCARB D-70.
M-1564; 25 Co, 5 TaC, 70 WC.
Sintered: A 83 Rock, 380,000 Tr. S.
For extrusion, drawing and header dies-normal impact.
Sintered carbides, wear and arasion resistant.

AMCARB D-75.
M-1564; 25 Co, 75 WC.
Sintered: A 83 Rock, 390,000 Tr. S.
For extrusion, drawing and header dies.
Sintered carbides, wear and abrasion resistant.

AMCARB D-80.
M-1564; 27 Co, 5 TaC, 68 WC.
Sintered: A 82 Rock, 375,000 Tr. S.
For extrusion, drawing and header dies-heavy impact.
Sintered carbides, wear and abrasion resistant.

AMCARB D-85.
M-1564; 30 Co, 5 TaC, 65 WC.
Sintered: A 81 Rock, 360,000 Tr. S.
For extrusion, drawing and header dies-very heavy impact.
Sintered carbides, wear and abrasion resistant.

AMCARB T-38.
M-1564; 71 WC, 12 TaC, 8 TiC, 9 Co.
Sintered: Rockwell A 92.5.
For extrusion press nibs, drawing and header dies.
Sintered carbide, wear and abrasion resistant.

AMCO COPPER.
M-1778; Electrolytic tough pitch copper.
For electrical applications and general usage in sheet, strip and rod applications.

AMCOH.
M-365; 0.90 C, 1.15 Mn, 0.5 Cr, 0.5 W, bal Fe.
Oil hardening tool steel; AISI 01.

AMCOH EXTRA SPECIAL.
M-365; 0.7 C, 18 W, 4 Cr, 1 V, bal Fe.
For tools, cutters, hobs; high speed steel.

AMCOH HOLLOW DIE.
M-365; 0.8 C, 0.9 Cr, bal Fe.
For hollow dies; oil hardening.

AMCOH OIL HARDENING.
M-365; 0.9 C, 0.5 Cr, 0.5 W, 1.0-1.25 Mn, bal Fe.
For tools, dies, punches; oil hardening.

A.M.C.O.H. SPECIAL.
M-365; 0.9 C, 1.2 Mn, 0.5 Cr, 0.5 W, bal Fe.
For die, punches; non-deforming.

AMCOLOY.
M-365; 0.75 C, 0.75 Mn, 0.90 Cr, 1.75 Ni, 0.35 Mo, bal Fe.
Oil hardening steel for shafts, arbors, lathe centers; AISI L6.

AMCOLOY 70.
M-365; 0.75 C, 1.25 Ni, 0.80 Cr, 0.25 Mo, 0.15 V. bal Fe.
Heat Treated: 121,000-208,000 TS; 90,000-180,000 PL, C 29-45 Rock.
For brake dies, bushings, cams, shear blades, gears, set screws, forming and swaging dies, punches.
Type L 6, tough, oil hardening.

AMCR.
M-1488; 0.30 C, 18 Mn, 0.60 Si, 10 Cr, 1 Ni, bal Fe.
Water quenched: 700 n/mm², min UTS.
Nonmagnetic alloy.
AFNOR Z 30 MCN 18-10.

AM CR.
M-1740; 1.0-1.2 C, 1.0 max Si, 0.06 max S, 0.07 max P, 11.0 min Mn, 1.5-2.0 Cr, bal Fe.
High resistance to impact wear.
ASTM A128-64 Gr.C.

AMCROM.
M-1778; 0.4-1.2 Cr, bal Cu.
Aged; 80,000 TS.
For electrical contacts, rocket nozzles, resistance welding tips, current carrying components.
80-90% elect., cond., age-hardenable, good resistance to softening at high temperatures.

A.M.D.
M-822; 0.34 C, 0.4 Mn, 4.75 Cr, 1.10 W, 1.45 Mo, bal Fe.
For hot work dies; hot work steel.

AMDRY 721.
M-1734; 0.15 C, 6.0 Cr, 1.25 B, 3.25 Si, 3.4 Fe, bal Ni.
Powder for flame spraying.
Approximate fusing temp: 2050°F.

AMDRY 722.
M-1734; 0.05 C, 1.25 B, 3.25 Si, 1.0 Fe bal Ni.
Powder for flame spraying.
Approximate fusing temp: 1925°F.

AMDRY 723.
M-1734; 0.05 C, 1.0 B, 2.5 Si, 1.0 Fe, bal Ni.
Powder for flame spraying.
Approximate fusing temp: 1950°F.

AMDRY 724.
M-1734; 0.05 C, 1.0 B, 2.5 Si, 1.0 Fe, bal Ni.
Powder for flame spraying.
Approximate fusing temp: 1950°F.

AMDRY 754.
M-1734; 0.45 C, 10.0 Cr, 1.25 B, 3.5 Si, 3.65 Fe, bal Ni.
Hard surfacing powder for flame spraying.
Approximate fusing temp: 1980°F.

AMDRY 755.
M-1734; 0.55 C, 12.0 Cr, 2.4 B, 4.0 Si, 4.15 Fe, bal Ni.
Hard surfacing powder for flame spraying.
Approximate fusing temp: 1930°F.

AMDRY 756.
M-1734; 0.60 C, 13.0 Cr, 2.75 B, 4.25 Si, 4.4 Fe, bal Ni.
Hard surfacing powder for flame spraying.
Approximate fusing temp: 1910°F.

AMDRY 761.
M-1734; 0.65 C, 14.0 Cr, 3.15 B, 4.4 Si, 4.5 Fe, bal Ni.
Hard surfacing powder for flame spraying.
Approximate fusing temp: 1890°F.

AMDRY 769.
M-1734; 0.65 C, 14.5 Cr, 3.15 B, 4.4 Si, 4.5 Fe, 2.3 Cu, 2.3 Mo, bal Ni.
hard surfacing powder for flame spraying.
Approximate fusing temp: 1900°F.

AMDRY 850.
M-1734; 0.70 C, 19.0 Cr, 3.25 B, 3.5 Si, 3.65 Fe, 18.0 Ni, 10.0 W, bal Co.
Hard facing powder for flame spraying.
Approximate fusing temp: 2040°F.

AMDRY ALLOYS see **ALSO AMI ALLOYS**.

AMERA-MAG.
M-1057; 0.08-0.15 c, 1.2 Mn, 0.12 P,0.05 Al, 0.1 max Ni, 0.1 max Mo, bal Fe.
Rolled: 70,000 TS; 50,000 YS; 34 El; 53 RA; 147 Brin.
For tanks, bodies; high tensile.

AMERICAN ALLOY.
M-U.S.; 95 Al, 3 Cu, 1 Mg, 1 Mn.
For light alloy parts; heat treated.

AMERICAN SILVER-1.
M-U.S.; 49.4 Cu, 20.7 Zn, 24,2 Ni, 1.3 Fe, 3.8 Mn, 0.5 Sn.
For ornaments, hardware; corrosion resistant.

AMERICAN SILVER-2.
M-U.S.; 59 Cu, 23 Zn, 11 Ni, 3 Pb, 1.5 Al, 5 P + Sn.
For ornaments, hardware; corrosion resistant.

AMERICAN SILVER-3.
M-U.S.; 58-49 Cu, 24-24 Zn, 15-24 Ni, <4 Mn + Sn + Fe + Al + Pb.
for ornaments, hardware; corrosion resistant.

AMERVAN.
U.S.; 63.5 Fe, 35 V, 1.5 Si.
For alloy steel making; vanadium additions.

A METAL.
Eng.; 5-7 Cu, 44 Ni, bal Fe.
For sound transmitting devices.

A.M.I.
M-210; C, alloy, bal Fe.
For dies; tough and abrasion resistant.

AMI ALLOY 74.
M-1734; 0.75 C, 4.5 Si, 14.5 Cr, 3.5 B, 4.5 Fe, bal Ni.
Plasma spray and hard facing alloy powder.
AMS 4775.

AMI ALLOY 75.
M-1734; 0.90 C, 4.0 Si, 16.5 Cr, 3.75 B, 4.0 Fe, bal Ni.
Plasma spray and hard facing alloy powder.
AMS 4775.

AMI ALLOY 100.
M-1734; 0.03 C, 19.0 Cr, 10.0 Si, bal Ni.
High temperature brazing alloy.
Brazing temp: 2075-2200°F.
AMS 4782; B14Y3; B50TF81.

AMI ALLOY 101.
M-1734; 11.5 Cr, 6.0 Si, bal Ni.
For honeycomb, wide gap and general purpose high temperature brazing.
Brazing temp: 2200-2250°F.

AMI ALLOY 102.
M-1734; 15.2 Cr, 8.0 Si, bal Ni.
For honeycomb, wide gap and general purpose high temperature brazing.
Brazing temp: 2150-2200°F.
B50T51A; B50T1403.

AMI ALLOY 103.
M-1734; 17.1 Cr, 9.2 Si, 0.08 B, bal Ni.
For honeycomb, wide gap and gerneral purpose high temperature brazing.
Brazing temp: 2100-2150°F.
B50T51B.

Section I: Alloy Data / 103

AMI ALLOY 104.
M-1734; 11.4 Cr, 6.8 Si, 0.30 B, bal Ni.
For honeycomb, wide gap and general purpose high temperature brazing.
Brazing temp: 2100-2150°F.
P50T9K.

AMI ALLOY 105.
M-1734; 13.3 Cr, 7.6 Si, 0.23 B, bal Ni.
For honey comb, wide gap and general purpose high temperature brazing.
Brazing temp: 2100-2150°F.
P50T9L.

AMI ALLOY 131.
M-1734; 16.0 Cr, 2.0 Ni, bal Fe.
Plasma spray powder.
Similar to AISI 431 stainless.

AMI ALLOY 134.
M-1734; 18.0 Cr, 10.0 Ni, bal Fe.
Plasma spray powder.
Similar to AISI 304 stainless.

AMI ALLOY 136.
M-1734; 17.0 Cr, 2.5 Mo, 12.0 Ni, bal Fe.
Plasma spray powder.
Similar to AISI 316 stainless.

AMI ALLOY 201.
M-1734; 1.3 Fe, 2.9 Cr, 2.8 Si, 1.9 B, bal Ni.
For honeycomb, wide gap and general purpose high temperature brazing.
Brazing temp: 2000-2110°F.
P50T9A.

AMI ALLOY 202.
M-1734; 1.0 Fe, 2.2 Cr, 2.6 Si, 1.6 B, bal Ni.
For honeycomb, wide gap and general purpose high temperature brazing.
Brazing temp.: 2000-2110°F.
P50T9B.

AMI ALLOY 205.
M-1734; 1.1 Fe, 2.7 Cr, 2.8 Si, 1.9 B, bal Ni.
For honeycomb, wide gap and general purpose high temperature brazing.
Brazing temp.: 2000-2120°F.
P50T9F.

AMI ALLOY 207.
M-1734; 2.4 Fe, 5.6 Cr, 3.2 Si, 2.8 B, bal Ni.
For honeycomb, wide gap and general purpose high temperature brazing.
Brazing temp: 2000-2075°F.
P50T9N.

AMI ALLOY 300.
M-1734; 0.03 C, 19.5 Cr, 9.5 Si, 9.5 Mn, bal Ni.
High temperature brazing alloy.
Brazing temp: 2025-2125°F.
B50T50.

AMI ALLOY 301.
M-1734; 3.9 C, 12.0 Co, bal W.
Plasma spray and hard facing powder.
PWA 1301.

AMI ALLOY 302.
M-1734; 3.9 C, 12.0 Co, bal W.
Plasma spray and hard facing powder.
PWA 1302.

AMI ALLOY 304.
M-1734; 12.5 C, 87.0 Cr.
Plasma spray and hard facing powder.
PWA 1304.

AMI ALLOY 305.
M-1734; 5.0 Cr, 20.0 Ni, 75.0 Cr C.
Plasma spray and hard facing powder.
PWA 1305.

AMI ALLOY 306.
M-1734; 13.0 C, 86.0 Cr.
Plasma spray and hard facing powder.
PWA 1306; B50TF39.

AMI ALLOY 307.
M-1734; 5.0 Cr, 20.0 Ni, 75.0 CrC.
Plasma spray and hard facing powder.
PWA 1307.

AMI ALLOY 308.
M-1734; 3.0 Cr, 12.0 Ni, 85.0 CrC.
Plasma spray and hard facing powder.
PWA 1308.

AMI ALLOY 313.
M-1734; 99.5 Mo.
Plasma spray and hard facing powder.
PWA 1313.

AMI ALLOY 315.
M-1734; 20.0 Cr, bal Ni.
Plasma spray powder.
PWA 1315. B50TF40A.

AMI ALLOY 316.
M-1734; 0.50 C, 25.5 Cr, 10.5 Ni, 7.5 W, bal Co.
Plasma spray and hard facing powder.
PWA 1316.

AMI ALLOY 317.
M-1734; 20.0 Cr. bal Ni.
Plasma spray powder.
PWA 1317; B50TF40B.

AMI ALLOY 318.
M-1734; 0.50 C, 25.5 Cr, 10.5 Ni, 7.5 W, bal Co.
Plasma srray and hard facing powder.
PWA 1318.

AMI ALLOY 319.
M-1734; 20.0 Cr, bal Ni.
Plasma spray powder.
PWA 1319.

AMI ALLOY 336.
M-1734; 5.0 Cr, 5.5 Fe, 24.5 Mo, bal Ni.
Plasma spray and hard facing powder.
PWA 1336.

AMI ALLOY 338.
M-1734; 99.5 Mo.
Plasma spray and hard facing powder.
PWA 1338.

AMI ALLOY 358.
M-1734; 30.0 Co, 0.90 Y, bal Al.
Plasma spray powder.
PWA 1358.

AMI ALLOY 400.
M-1734; 0.40 C, 19.0 Cr, 8.0 Si, 0.80 B, 4.0 W, 16.5 Ni, bal Co.
High temperature brazing alloy.
Brazing temp: 2150-2200°F.
B50T56; PWA 713.

AMI ALLOY 500.
M-1734; 36.5 Ni, 5.0 In, bal Cu.
Plasma spray powder.
B50TF72.

AMI ALLOY 716.
M-1734; 0.03 max C, 37.5 Mn, 9.5 Ni, bal Cu.
High temperature brazing alloy.
Brazing temp: 1750-1850°F.

AMI ALLOY 717.
M-1734; 0.03 C, 23.5 Mn, 9.0 Ni, bal Cu.
High temperature brazing alloy.
Brazing temp: 1800-1850°F.
B50TF80.

AMI ALLOY 750.
M-1734; 0.90 C, 16.5 Cr, 4.0 Si, 3.75 B, 4.0 Fe, bal Ni.
High temperature brazing alloy.
Brazing temp: 1950-2200°F.
AMS 4775.

AMI ALLOY 760.
M-1734; 0.03 C, 16.5 Cr, 4.0 Si, 3.75 B, 4.0 Fe, bal Ni.
High temperature brazing alloy.
Brazing temp: 1975-2200°F.
AMS 4776.

AMI ALLOY 770.
M-1734; 0.03 C, 7.0 Cr, 4.5 Si, 2.75 B, 3.0 Fe, bal Ni.
High temperature brazing alloy.
Brazing temp: 1950-2150°F.
AMS 4777.

AMI ALLOY 780.
M-1734; 0.03 C, 4.5 Si, 3.0 B, 1.0 Fe, bal Ni.
High temperature brazing alloy.
Brazing temp: 1925-2150°F.
AMS 4778.

AMI ALLOY 780 B.
M-1734; 4.5 Si, 3.0 B, bal Ni.
Plasma spray powder.
B50TF84.

AMI ALLOY 790.
M-1734; 0.03 C, 3.5 Si, 2.0 B, 1.0 Fe, bal Ni.
High temperature brazing alloy.
Brazing temp: 1975-2150°F.
AMS 4779.

AMI ALLOY 912.
M-1734; 2.5 Si, 10.0 Cr, 2.5 B, 2.5 Fe, bal Ni.
Plasma spray and hard facing powder.

AMI ALLOY 914.
M-1734; 0.03 max C, 4.5 Si, 3.0 B, 20.0 Co, bal Ni.
High temperature brazing alloy.
Brazing temp: 2050-2175°F.
BTS1205.

AMI ALLOY 915.
M-1734; 0.02 max C, 13.0 Cr, 4.0 Si, 2.75 B, 4.0 Fe, bal Ni.
High temperature brazing alloy.
Brazing temp: 2050-2150°F.
BMS7-141.

AMI ALLOY 916.
M-1734; 0.60 C, 13.0 Cr, 4.0 Si, 3.0 B, 4.0 Fe, bal Ni.
Plasma spray and hard facing alloy.
High temperature brazing alloy.
Brazing temp: 1950-2150°F.
AMS 4775.

AMI ALLOY 930.
M-1734; 0.01 C, 7.0 Si, 5.0 Cu, 22.5 Mn, bal Ni.
High temperature brazing alloy.
Brazing temp: 1870-2000°F.
BMS7-141.

AMI ALLOY 931.
M-1734; 0.01 C, 8.0 Si, 17.0 Mn, bal Ni.
High temperature brazing alloy.
Brazing temp: 1900-2000°F.

AMI ALLOY 942.
M-1734; 38.0 Ni, bal Cu.
Plasma spray powder.
B50TF42.

AMI ALLOY 955.
M-1734; 18.5 Al, 81.5 Ni.
Plasma spray powder.
PWA 1339.

AMI ALLOY 956.
M-1734; 5.0 Al, 95.0 Ni.
Plasma spray powder.

AMINO 3.
M-72; 0.38-0.43 C, 12.5-13.5 Cr, 0.6 Ni, bal Fe.
For tools, cutlery, knives; corrosion resistant, hardenable.

AMNIC 10.
M-1778; 10 Ni, bal Cu. (others very low).
Ann: 221 MPa (38,000 psi) TS; 131 MPa (19,000 psi) YS; 41 El.
50% Cold worked: 476 MPa (69,000 psi) TS; 428 MPa (62,000 psi) YS; 4 El.
High purity copper-nickel alloy for electronic and cryogenic applications.

AMNIC 30.
M-1778; 30 Ni, bal Cu. (others very low).
Ann: 372 MPa (54,000 psi) TS; 152 MPa (22,000 psi) YS; 41 El.
50% Cold worked: 552 MPa (80,000 psi) TS; 524 MPa (76,000 psi) YS; 4 El.
High purity copper-nickel alloy for electronic and cryogenic applications.

AMOL.
M-1049; 0.8-0.9 C, bal Fe.
For tools, cutters, water hardening.

AMOLA now **AISI 40XX STEELS.**

AMOTUN.
M-353; 0.85 C, 4 Cr, 6 Co, 1.75 V, 1.5 W, 8 Mo, bal Fe.
For cutting tools; high speed steel.

AMOUTUN.
M-353; 0.80 C, 4 Cr, 8 Mo, 1.5 W, 1 V, bal Fe.
Hardened: C 64-66 Rock.
For lathe and planer cutters, drills, reamers, broaches, hobs, form cutters, taps.
Type M 1 high speed steel, high red-hardness.

AMPCO 8.
M-13; 6-8 Al, 1.5-3.0 Fe, bal Cu.
Plate: 72,000 TS; 35,000 YS; 34 El; 130 Brin.
Extruded: 82,000 TS; 55,000 YS; 35 El; 149 Brin.
For wear plate service, slides, gibs, bushings, bearings, bolts, fittings.
Alpha phase, corrosion resistant.
High impact and fatigue strength.

AMPCO 12.
M-13; 8.0-9.5 Al, 2.25-3.25 Fe, bal Cu.
Cast: 60,000 TS; 30,000 YS; 40 El; 130 Brin.
For bushings, adjusting nuts, fittings, inlet hoppers, machinery parts.
Corrosion resistant.

AMPCO 15.
M-13; 8.5-10.0 Al, 2.5-3.75 Fe, 0.5 max others, bal Cu.
Wrought: 92,000 TS; 46,000 YS; 25 El; 25 RA; 174 Brin.
For valve stems, washers, worm gears, cam rollers, bushings, shear pins, studs.
Good bearing qualities, wear and corrosion resistant.

AMPCO 16.
M-13; 9.3-10.3 Al, 2.75-4.0 Fe, bal Cu.
Cast: 90,000 TS; 32,000 YS; 22 El; 155 Brin.
For bushings, gears, worm wheels, machinery parts, trolley shoes, trolley wheels.
Corrosion and shock resistant.

AMPCO 18.
M-13; 10.0-11.2 Al, 3.0-4.2 Fe, 0.5 max others, bal Cu.
Cast: 77,000-90,000 TS; 37,000-42,000 YS; 10-14 El; 6-12 RA, 165-18 Brin.
For acid equipment, gears, worm wheels, trolley shoes, piston rods, valve seats.
Corrosion and shock resistant.

AMPCO 18-22.
M-13; 10.0-11.2 Al, 3.0-4.2 Fe, bal Cu.
Heat treated: 90,000-100,000 TS; 45,000-55,000 YS; 3-7 El; 3-7 RA; 202-223 Brin.
For bearings; heat treatable, wear resistant, impact resistant.

AMPCO 18-23.
M-13; 10.0-11.2 Al; 3.0-4.2 Fe, bal Cu.
Heat treated: 100,000 TS; 50,000 YS; 14 El; 14 RA; 179-207 Brin.
For aircraft, propellers, motors, heavy duty worm gears; castings.

AMPCO 20.
M-13; 11.0-12.2 Al, 3.2-4.5 Fe, bal Cu.
Cast: 83,000-90,000 TS; 40,000-43,000 YS; 4-6 El; 3.5 RA, 212-241 Brin
For cams, rollers, safety tools, welding jaws, worm gears, valve bodies.
Corrosion and wear resistant.

AMPCO 20-13.
M-13; 11.0-12.2 Al, 3.2-4.5 Fe, bal Cu.
Heat treated: 85,000-102,000 TS; 35,000-50,000 YS; 6-12 El; 192-207 Brin.
For bearings, bushings, corrosion and wear resistant; non-seizing.

AMPCO 21.
M-13; 12.5-13.6 Al, 3.5-5.0 Fe, 0.5 max others, bal Cu.
Cast: 75,000-95,000 TS; 55,000-60,000 YS; 1.5 El; 0.5 RA; 285-302 Brin.
For valves, bushings, forming dies, slides, roller bushings, gears, worm gears.
Corrosion and wear resistant.

AMPCO 22.
M-13; 13.6-14.6 Al, 4.0-5.2 Fe, 0.5 max others, bal Cu.
Cast: 85,000 TS; 70,000 YS; 0.5 El; 321-341 Brin.
For forming and drawing dies.
Corrosion and wear resistant.
High compressive strength.

AMPCO 24.
M-13.
Cooper base castings.
351-364 Brin.
For draining and forming operations.

AMPCO 25.
M-13.
Copper base castings.
364-375 Brin.
Tough, strong die material.

AMPCOLOY 43.
M-13; 89 Cu, 10 Al, 1 Fe.
62,000-72,000 TS; 35,000-40,000 YS; 8-15 El; 9-16 RA; 126-156 Brin.
For propellers, nuts; corrosion resistant.

AMPCOLOY 45.
M-13; 9.7-10.9 Al, 2.0-3.5 Fe, 4.5-5.5 Ni, 1.5 max Mn, bal Cu, 0.6 max others.
Extruded: 118,000 TS; 75,000 YS; 15 El; 15 RA; 220 Brin.
For bearings, sleeves, valve guides, shafts; heat treatable, corrosion resistant.

AMPCOLOY 45-HT.
M-13; 9.7-10.9 Al, 2.0-3.5 Fe, 4.5-5.5 Ni, 1.5 max Mn, 0.6 max others, bal Cu.
Wrought: 115,000 TS; 66,000 YS; 15 El; 15 RA; 223 Brin.
For plunger tips, valve guides, shafts.
Corrosion resistant.

AMPCOLOY 46.
M-13; 2-6 Fe, 8-12 Al, 2-6 Ni, bal Cu.
Ht. Tr.: 90,000-96,000 TS; 43,000-55,000 YS; 3-12 El; 10 RA; 180-235 Brin.
For valve bodies, gears, sleeves, liners.
Heat treatable castings.

AMPCOLOY 46-HT.
M-13; 8-12 Al; 2-6 Fe, 2-6 Ni, 0-2 Mn, bal Cu.
Cast: 95,000 TS; 43,000 YS; 16 El; 16 RA; 234 Brin.
For gears, shafts, valve bodies, shaft sleeves, liners; heat treatable; corrosion resistant.

AMPCOLOY 50.
M-13; 84 Cu, 10 Sn, 3.5 Ni, 2.5 Pb.
Cast: 47,000 TS; 24,000 YS; 10 El; 20 RA; 80 Brin.
For worm gears, bushings, nuts, gears, elevating screw units.

AMPCOLOY 53.
M-13; 5 Ni, 5 Sn, 2 Pb, bal Cu.

AMPCOLOY 54.
M-13; 10-11 Sn, 1.2 Ni, bal Cu.
Cast: 47,000 TS; 24,000 YS; 15 El; 12 RA; 93 Brin.
For rims.

AMPCOLOY 62.
M-13; 38-42 Zn, 1.5 Mn, 1.5 max Al, 1.0 Sn, 0.4-2.0 Fe, bal Cu, 1.4 max others.
Cast: 65,000-75,000 TS; 28,000-38,000 YS; 20-35 El; 20-35 RA; 110-135 Brin.
For castings, gears, cams; low tensile bronze.

AMPCOLOY 64.
M-13; 60-68 Cu, 3-7 Al, 2-4 Fe, 5 max Mn, bal Zn.
Cast: 95,000 TS, 50,000 YS; 25 El, 192 Brin.
For cast gears, gear cases, scrapers, housings.
Medium strength bronze.

AMPCOLOY 66.
M-13; 3-7 Al, 1-4 Fe, 5 max Mn, 20-30 Zn, 60-68 Cu.
Cast: 110,000-120,000 TS; 60,000-70,000 YS; 10-15 El; 8-14 RA; 217-235 Brin.
For castings; gears; high tensile bronze, nuts.

AMPCOLOY 72.
M-13; 7.5-9 Sn, 3-5 Zn, bal Cu.
Cast: 46,000 TS; 24,000 YS; 26 El; 24 RA; 66 Brin.
For castings; venturi meter bodies, slip rings.

AMPCOLOY 74.
M-13; 84-86 Cu, 4-6 Sn, 4-6 Pb, 4-6 Zn, 1.5 max Ni.
Cast: 38,000 TS; 20,000 YS; 26 El; 24 RA; 65 Brin.
For valves, pipe elbows, hydraulic parts; "Ounce Metal"; for high pressure service.

AMPCOLOY 79.
M-13; 86-89 Cu, 9-11 Sn, 1-3 Zn.
Cast: 46,000 TS; 22,000 YS; 24 El; 22 RA; 105 Brin.
For gears, shafts, bearings, sleeves.
Hard bronze, wear and corrosion resistant.

AMPCOLOY 90.
M-13; 0.25 max others, bal Cu.
Cast: 25,000 TS; 9,000 YS; 40 El; 40 RA; 45 Brin.
For transformer secondaries; cable attachments, bus bars; high conductivity.

AMPCOLOY 92.
M-13; 11.6-12.2 Al, 0.35 max others, bal Cu.
Cast: 70,000 TS; 34,000 YS; 6 El; 150 Brin.
For resistance welding electrodes, hinges, bushings, shafts.
10-15% elect. cond., corrosion resistant.

AMPCOLOY 92-H.
M-13; Al, bal Cu.
Cast: 75,000 TS; 2 El; 240 Brin.
For resistance welding electrode; 20% conductivity.

AMPCOLOY 92-S.
M-13; Al, bal Cu.
Cast: 75,000 TS; 25 El; 137 Brin.
For resistance welding electrode; 20% conductivity.

AMPCOLOY 94.
M-13; 1.5-2.5 Ni, 0.5 Si, 0.75 Al, bal Cu.
Cast: 65,000 TS; 45,000 YS; 15 El; 146 Brin.
For resistance welding parts, shafts, bushings, plunger tips.
Medium conductivity, wear resistant.

AMPCOLOY 97.
M-13; 99.5 Cu, 4 Cr, 1 Be.
Cast: 40,000-50,000 TS; 35,000-40,000 YS; 10-15 El; 140-150 Brin.
For welding electrodes; same as Ampco Trodaloy 7.

AMPCOLOY 99.
M-13; Cu alloy.
Bars: 60,000-50,000 TS; 20,000-15,000 YS; 20-25 El; 100-115 Brin.
For resistance welding electrode; for terneplate and galvanized stock.

AMPCOLOY 483.
M-13; 9.0 Al, 4.5 Fe, 4.5 Ni, 1.0 Mn, bal Cu.
Cast: 100,000 TS; 43,000 YS; 25 El; 163 Brin.
For marine propellers, worm wheels, marine hardware.
Tough, strong, corrosion resistant.

AMPCOLOY 495.
M-13; 8 Al, 2.5 Fe, 2 Ni, 12.5 Mn, bal Cu.
Bar: 100,000 TS; 50,000 YS; 30 El; 187 Brin.
For pump and valve parts, propellers, hydraulic equipment.
Corrosion resistant, tough, shock resistant.

AMPCOLOY 521.
M-13; 30 Ni, 1 Mn, 0.5 Fe, 0.5 Si, bal Cu.
Cast: 63,000 TS; 32,000 YS; 35 El; 120 Brin.
For valve bodies and trim, valve stems for pipe fittings, propeller sleeves, pump impellers.
Good corrosion resistance to 700°F.

AMPCOLOY 522.
M-13; 0.5 Fe, 30 Ni, 0.5 Mn, bal Cu.
Forged: 55,000 TS; 23,000 YS; 40 El; 85 Brin.
For valve bodies and trim, valve stems for pipe fittings, propeller sleeves, pump impellers.
Corrosion resistant to 700°F.

AMPCOLOY 525.
M-13; 1.5 Fe, 10 Ni, 1 Mn, 0.1 Si, bal Cu.
Cast: 42,000 TS; 19,000 YS; 35 El; 100 Brin.
For oil refinery equipment, propeller sleeves, pump impellers, marine hardware.
Corrosion resistant to 600°F.

AMPCOLOY 526.
M-13; 1.5 Fe, 10 Ni, bal Cu.
Forged: 40,000 TS; 18,000 YS; 45 El; B 35 Rock.
For oil refinery equipment, propeller sleeves, pump impellers, marine hardware.
Corrosion resistant to 600°F.

AMPCOLOY 551.
M-13; 29 Cu, 3 Al, 1 Mn, bal Ni.
Rolled: 150,000 TS; 105,000 YS; 25 El; 280 Brin.
For pump and valve parts, scrapers.
High strength, wear and corrosion resistant. (K-Monel).

AMPCOLOY 552.
M-13; 30 Cu, 1.4 Fe, 1.0 Mn, bal Ni.
Forged: 80,000 TS; 45,000 YS; 35 El; 140 Brin.
For propeller shafts, valve seats and stems, pump and compressor parts.
Resists corrosion and elevated temperature. (Mondel).

AMPCOLOY 553.
M-13; 30 Cu, 2 Fe, 0.75 Mn, 4 Si, bal Ni.
Cast: 125,000 TS; 100,000 YS; 2 El; 300 Brin.
For pump wear rings, valve trim, steam nozzles.
Non-magnetic, non-galling, corrosion resisant.
Similar to S-Monel.

AMPCOLOY 557.
M-13; 31 Cu, 2 Fe, 0.75 Mn, 3 Si, bal Ni.
Cast: 100,000 TS; 65,000 YS; 8 El; 200 Brin.
For pump wear rings, valve trim, steam nozzles.
Corrosion resistant, non-magnetic.
Similar to H-Monel.

AMPCOLOY 558.
M-13; 33 Cu, 1.2 Fe, 0.8 Mn, 1.7 Si, bal Ni.
Cast: 75,000 TS; 35,000 YS; 30 El; 130 Brin.
For pump impellers, valve bodies and trim, bubble caps, food machinery, steam ejectors.
Weldable; corrosion resistant.

AMPCOLOY 559.
M-13; 33 Cu, 1.2 Fe, 0.8 Mn, 1.7 Si, bal Ni.
Cast: 75,000 TS; 35,000 YS; 30 El; 130 Brin.
For pump impellers, valve bodies, bubble caps, food machinery, steam ejectors.
Corrosion resistant, strong and tough.

AMPCOLOY 570.
M-13; 10 Al, 0.7 Fe, 15 Ni, 1.5 Co, bal Cu.
Cast: 94,500 TS; 55,000 YS; 5 El; 202 Brin.
For glass mold service.
Heat and corrosion resistant.

AMPCOLOY 666.
M-13; 1.45-1.75 Al, 2.8 max Mn, 57-60 Cu, bal Zn, 1.7 max others.
Extruded 82,000-88,000 TS; 42,000-55,000 YS; 18-12 El; 170-187 Brin.
For bushings, gears, bearings, cams, valve stems, connecting rods; high strength bearing bronze.

AMPCOLOY 668.
M-13; 61 Cu, 2.5 Pb, 2.5 Mn, 1.0 Si, bal Zn.
Extruded: 72,000-75,000 TS; 58,000-60,000 YS; 18-20 El; 162-144 Brin.
For bushings, bearings, gears, cams, valve stems, lead screw nuts; free-cutting, leaded manganese bearing bronze.

AMPCOLOY 711.
M-13; 11 Sn, 0.2 P, bal Cu.
Cast: 55,000 TS; 30,000 YS; 16 El; 102 Brin.
For gears, worm wheels.
Gear bronze. Tough.

AMPCOLOY 712.
M-13; 1.5 Ni, 11.75 Sn, 0.2 P, bal Cu.
Cast: 60,000 TS; 32,000 YS; 16 El; 106 Brin.
For gears, worm wheels.
Gear bronze, tough.

AMPCOLOY 715.
M-13; 12 Sn, 0.2 P, bal Cu.
Cast: 60,000 TS; 32,000 YS; 16 El; 106 Brin.
For gears, worm wheels.
Gear bronze, tough, wear resistant.

AMPCOLOY 742.
M-13; 86-89 Cu, 9-11 Sn, 1-3 Zn, 0.3 max Pb.
Cast: 40,000 TS; 18,000 YS; 14 El; R80F Brin.
For gears, shafts; continuous cast rod.

AMPCOLOY 900.
M-13; 99.98 Cu.
Extruded: 30,000 TS; 10,000 YS; 40 El; 40 Brin.
For electric motor components, conductors.
Electrical cond. 100 min.
OFHC Copper.

AMPCOLOY 901.
M-13; 99.9 Cu, 30-oz./ton Ag.
Cast: 26,000 TS; 8,000 YS; 40 El; 44 Brin.
Forged: 31,000 TS; 10,000 YS; 45 El; 47 Brin.
For high temperature (600°F) electrical and acid resistant applications, heat exchangers, computor segments.
Electrical conductivity 95 min.

AMPCOLOY 910.
M-13; 0.12 Zr, 99.88 Cu.
Forged: 55,000 TS; 47,000 YS; 14 El; 100 Brin.
Extruded: 65,000 TS; 60,000 YS; 16 El; 119 Brin.
For rotor wedges, commutators, collector rings, switch gears, soldering tips, welding electrodes.
90% electrical conductivity.

AMPCOLOY A-1.
M-13; 8.5 Al, 3.0 Fe, bal Cu.
Cast: 80,000 TS; 29,000 YS; 30 El; 135 Brin.
For bushings, bearings, gears, wear plates, connecting rods.
Medium strength, light loads.

AMPCOLOY A-2.
M-13; 7.0-10.0 Al, 0.5-2.0 Fe, 0.6 max others, bal Cu.
Cast: 62,000 TS; 23,000 YS; 30 El; 30 RA; 101 Brin.
For pickling hooks, bushings, marine hardware.
Corrosion resistant.

AMPCOLOY B-2.
M-13; 10 Al, 1 Fe, bal Cu.
Bar: 85,000 TS; 42,000 YS; 23 El; 170 Brin.
Cast: 77,000 TS; 29,000 YS; 30 El; 140 Brin.
For bearings, liners, pump parts, pump impellers, bushings. Heat treatable, high strength.
Aluminum bronze. For use with medium loads and average speeds. Corrosion resistant.

AMPCOLOY C-3.
M-13; 10.5 Al, 3.0 Fe, bal Cu.
Bar: 100,000 TS; 48,000 YS; 12 El; 192 Brin.
Cast: 85,000 TS; 35,000 YS; 12 El; 183 Brin.
For bearings, bushings, gears, valve guides and seats, pump and hydraulic valves.
Aluminum bronze. Tough, corrosion and wear resistant.

AMPCOLOY D-4.
M-13; 10 Al, 5 Fe, 5 Ni, bal Cu.
Bar: 105,000 TS; 46,000 YS; 18 El; 195 Brin.
Ht. Tr.: 120,000 TS; 70,000 YS; 18 El; 240 Brin.
For piston guides, valve seats, glands, pump fluid ends.
High strength, tough Al-bronze, heat treatable.

AMPCOLOY E-1.
M-13; 9.5-11 Al, 1.5 max Fe, bal Cu.
Cast: 70,000-80,000 TS; 40,000-45,000 YS; 12-20 El; 14-20 Ra; 131-14 Brin.
For castings, power shovels, worm wheels, pistons, bushings, corrosion resistant.

AMPCOLOY E-5.
M-13; 9.0 Al, 3.0 Fe, bal Cu.
Bar: 90,000 TS; 48,000 YS; 18 El; 174 Brin.
For valve seats, guides, stems, gears, bearings.
Wear and corrosion resistant aluminum bronze.
Tough.

AMPCOLOY F-6.
M-13; 10 Al, 3 Fe, bal Cu.
Vast: 85,000 TS; 31,000 YS; 20 El; 170 Brin.
For bushings, bearings, gears, worm wheels, shifters.
Resists squashing under load.

AMPCOLOY M-100.
M-13; 100 Mo.
Sintered: 80,000 TS; 30 El; B 90 Rock.
For facings on spot welder tips.
High heat resistance.

AMPCOLOY W-100.
M-13.
Bars: 50,000-200,000 TS; A 76 Rock.
For resistance welding electrodes for Red Brass weldings.
RWMA Class B-13.
30% elect. cond., corrosion and wear resistance.

AMPCO-TRODE 7.
M-13; 6.0-6.9 Al, bal Cu (others 0.50 max).
Weld metal: 68 ksi TS; 28 ksi YS; 47 El; 125 Brin.
For overlay deposits; not recommended for joining.

AMPCO-TRODE 10.
M-13; 9.0-11.0 Al, 1.5 max Fe, bal Cu.
Cast: 60,000-77,000 TS; 25,000-35,000 YS; 25-30 El; 28-30 RA; 110-110 Brin.
For bronze welding rod: aluminum bronze.

AMPCO-TRODE 40.
M-13; 7.0-8.0 Al, 2.0-3.0 Fe, 2.0-3.0 Ni, 11.0-12.0 Mn, 0.6 max others bal Cu.
Wrought: 95,500 TS; 56,000 YS; 27 El; 38 RA; 185 Brin.
For manganese bronze welding rod.
Wear resistant.

AMPCO-TRODE 46.
M-13; 8.5-9.5 Al, 4.0-5.5 Ni, 3.0-5.0 Fe, 0.6-3.5 Mn, bal Cu. (others 0.50 max).
Weldmetal: 99 ksi TS; 58 ksi YS; 25 El; 187 Brin.
For welding ship propellers, ship fittings, pump housings.
AWS-A5.6 Class CuNiAl.

AMPCO-TRODE 150.
M-13; 10.0-11.0 Al, 3.0-5.0 Fe, bal Cu. (others 0.50 max).
Weld metal: 90 ksi TS; 40 ksi YS; 20 El; 166 Brin.
For welding AMPCO 18 Alloy.

AMPCO-TRODE 160.
M-13; 11-12 Al, 3.0-4.25 Fe, bal Cu, 0.6 max others.
Cast: 89,000 TS; 47,000 YS; 15 El; 17 RA; 177 Brin.
For welding electrodes, bearing overlay; resists salt water and commercial acids.

AMPCO-TRODE 200.
M-13; 12-13 Al, 3-5 Fe, bal Cu, 0.6 max others.
Cast: 90,500 TS; 46,000 YS; 4 El; 5 RA; 212 Brin.
For welding electrodes, surfacing, overlays; corrosion and acid resistant.

AMPCO-TRODE 250.
M-13; 13-14 Al, 3-5 Fe, bal Cu, 0.6 max others.
Cast: 77,000 TS; 54,000 YS; 1 El; 260 Brin.
For welding electrodes, bushings, bearings, gears, slides; free from galling and scuffing.

AMPCO-TRODE 300.
M-13; 14-15 Al, 3-5 Fe, bal Cu, 0.6 max others.
Cast: 80,000 TS; 69,000 YS; 0 El; 316 Brin.
For welding electrodes, for overlaying, drawing and forming dies; wear resistant; good bearing qualities.

AMPHOS.
M-1778; 0.03 P, bal Cu.
For condenser tubes, heat exchangers, piping; good hot and cold workability.

AMPHOS-40.
M-1778; Oxygen free copper with added phosphorus.
For plating anodes.

AMS 5700 see ALSO CARPENTER AMS 5700.

AMSCO 40.
M-9; 3.5 C, 2 Cr, 5 Mo, bal Fe.
Welded: 540 Brin.
For hard facing electrode; abrasion resistant.

AMSCO 53.
M-9; 3 C, 15 Cr, Mo, bal Fe.
Welded: 540 Brin.
For hard facing electrode; abrasion and impact resistant.

AMSCO 77.
M-9; 2.5 C, 1 Mn, 19 Cr, 1 Mo, bal Fe.
Weld hardness 46 Rc.
Hardfacing electrode, abrasion and impact resistant.

AMSCO AW-72.
M-9; 0.27 C, 5,8 Cr, 0.8 Mo, 1.5 W, bal Fe.
Weld hardness 52 Rc.
Hardfacing wire for submerged melt welding of carbon and low alloy steel parts; resistant to abrasion, impact, thermal shock and compressive loading.

AMSCO AW-79.
M-9; 0.15 C, 2.2 Mn, 0.7 Si, 3.5 Cr, 0.8 Mo, bal Fe.
Weld hardness 44 Rc.
Hardfacing wire for submerged melt welding of carbon and low alloy steel parts, resistant to impact, abrasion, and metal-to-metal wear.

AMSCO AW-84.
M-9; 0.13 C, 2.2 Mn, 0.8 Si, 0.6 Mo, bal Fe.
Weld hardness: 30 Rc.
Build up wire for submerged melt welding of carbon and low alloy steel parts; high compressive strength and impact resistance.

AMSCO AW-87.
M-9; 0.14 C, 2.2 Mn, 0.8 Si, 2.6 Cr, 0.6 Mo, bal Fe.
Weld hardness: 41 Rc.
Hardfacing and buildup wire for submerged melt welding of carbon and low alloy steel parts; resistant to medium impact and abrasion.

AMSCO AW-420.
M-9; 0.16 C, 1.3 Mn, 0.75 Si, 12 Cr, bal Fe.
Weld hardness 45 Rc.
Hardfacing wire for submerged melt welding; resistant to heat, abrasion, metal-to-metal wear and thermal shock.

AMSCO AW-CMS.
M-9; 0.3 C, 15 Mn, 15 Cr, 1 Ni, bal Fe.
Weld hardness: 200 Brin, work hardens.
Wire for submerged melt welding; impact and abrasion resistant.

AMSCO AW-NICROMANG.
M-9; 0.8 C, 15 Mn, 4 Cr, 3.5 Ni, bal Fe.
As welded: 120,00 TS; 70,000 YS; 40 El; 35 RA.
Weld hardness: 200 Brin, work hardens to 500 Brin.
Manganese steel wire for submerged metl welding; impact and abrasion resistant.

AMSCO AW-THERMALLOY 400.
M-9; 0.6 C, 3 Mn, 22 Cr, 8 Ni, 0.5 Mo, bal Fe.
Deposit hardness: 22 Rc, work hardens.
Hardfacing wire for submerged melt welding; resistant to abrasion and impact at high temperature.

AMSCO CHROMANAL.
M-9; 1-1.4 C, 10-14 Mn, 0.2-1.0 Si, 1.3 Cr, bal Fe.
Heat treated: 95,000-145,000 TS; 53,000-65,000 YS; 63-27 El; 40-30 RA; 195-220 Brin.
For crushers, hammer mills, liners, castings; work hardens up to 550 Brin.

AMSCO CHROME MOLY BALL MILL GRATE STEEL.
M-9; 0.5-0.7 C, 0.5-1.5 Mn, 0.2-0.8 Si, 1.8-2.8 Cr, 0.2-0.5 Mo, bal Fe.
Heat treated: 145,000-175,000 TS; 90,000-130,000 YS; 11-8 El; 20-12 RA; 300-350 Brin.
For ball mill grates; abrasion resistant.

AMSCO CHROME MOLY BALL MILL LINER STEEL.
M-9; 0.7-1.0 C, 0.5-1.5 Mn, 0.2-0.8 Si, 1.8-2.8 Cr, 0.2-0.5 Mo, bal Fe.
Heat treated: 150,000-185,000 TS; 95,000-140,000 YS; 8-4 El; 10-4 RA; 300-350 Brin.
For ball mill liners; abrasion resistant.

AMSCO CM.
M-9; 0.5-1.0 C, 0.5-1.5 Mn, 0.2-1.0 Si, 1.8-2.8 Cr, 0.2-0.5 Mo, bal Fe.
Heat treated: 125,000-185,000 TS; 80,000-150,000 YS; 13-2 El; 25-2 RA; 300-400 Brin.
For rod and ball mill grates liners, crushers; high abrasion and moderate impact resistance.

AMSCO-CMH.
M-9; C, Cr, Mo, bal Fe.
Heat treated: 155,000 TS; 130,000 YS; 6 El; 7 RA; 300-400 Brin.
For castings, ball and rod mill shell liners; air hardened, wear resistant.

AMSCO-CML.
M-9; C, Cr, Mo, bal Fe.
Heat treated: 155,000 TS; 130,000 YS; 10 El; 15 RA; 275-375 Brin.
For castings, trunnion liners, feeder scoops; air hardened, abrasion resistant.

AMSCO CONOMANG.
M-9; 0.95 C, 16 Mn, 1.8 Cr, bal Fe.
As welded: 124,000 TS; 80,000 YS; 28 El; 25 Ra.
Weld hardness 230 Brin, work hardens to 500 Brin.
Manganese steel electrode for rebuilding manganese steel parts. Impact resistant.

AMSCO CS.
M-9; 0.2-0.35 C, 0.7-1.5 Cr, 0.8-1.5 Mn, 0.2-0.8 Si, 0.3-0.8 Ni, 0.2-0.6 Mo, bal Fe.
Heat treated: 200,000-240,000 TS; 170,000-210,000 YS; 10-6 El; 30-10 RA; 300-500 Brin.
For castings, dipper teeth; abrasion and wear resistant.

AMSCO F-1.
M-9; 15-18 Cr, 34-37 Ni, bal Fe.
For electric furnace grids, hearth plates, rails, furnace chains; heat resistant, max. operating temperature 2000°F.

AMSCO F-1H.
M-9; 0.4-0.6 C, 16 Cr, 35 Ni, bal Fe.
For furnace parts; heat resisting.

AMSCO F-1HH.
M-9; 0.60-0.85 C, 16 Cr, 35 Ni, bal Fe.
For furnace parts; heat resisting.

AMSCO F-1-N.
M-9; C, 15-18 Cr, 38-40 Ni, bal Fe.
For retorts, muffles, furnace castings, hearth plates; heat and corrosion and shock resistant.

AMSCO F-2.
M-9; 21-24 Cr, 3-6 Ni, bal Fe.
For furnace lead pots; heat resisting, max. operating temperature 1800°F.

AMSCO F-3.
M-9; 26-29 Cr, 0-3 Ni, bal Fe.
For rabble arms, rabble blades; corrosion and heat resisting.

AMSCO F-3H.
M-9; 0.4-0.6 C, 28 Cr, 2 Ni, bal Fe.
For furnace parts; heat resisting.

AMSCO F-3HH.
M-9; 0.6-0.85 C, 28 Cr, 2 Ni, bal Fe.
For furnace parts; heat resisting.

AMSCO F-4.
M-9; 6-9 Cr, 18-21 Ni, bal Fe.
For high strength castings; corrosion resisting.

AMSCO F-5.
M-9; 16-19 Cr, 65-70 Ni, bal Fe.
For carburizing boxes. oil burner parts; heat resisting, very tough.

AMSCO F-6.
M-9; C 12-15 Cr, 58-64 Ni, bal Fe.
For furnace parts, heat treating boxes, retorts; corrosion and heat resistant.

AMSCO F-7H.
M-9; 0.5 C, 25 Cr, 12 Ni, bal Fe.
For furnace parts; heat resistant.

AMSCO F-8.
M-9; 17-22 Cr, 7-10 Ni, bal Fe.
For pots in chemical plants, acid pumps; corrosion resisting.

AMSCO F-10.
M-9; 0.4 C, 28 Cr, 12 Ni, bal Fe.
For furnace parts, dampers, valves, beams; heat resistant.

AMSCO F-10H.
M-9; 0.4-0.6 C, 28 Cr, 12 Ni, bal Fe.
For furnace parts; heat resistant.

AMSCO F-10HH.
M-9; 0.6-0.85 C, 28 Cr, 12 Ni, bal Fe.
For furnace parts; heat resistant.

AMSCO F-12.
M-9; C, 27-30 Cr, 7-10 Ni, bal Fe.
For heat treating furnaces, valves; creep resisting.

AMSCO F-13.
M-9; C, 25-28 Cr, 34-37 Ni, bal Fe.
For furnace parts, retorts, muffles; resists heat and S atmosphere to 2100°F.

AMSCO F-14.
M-9; C, 24-27 Cr, 19-22 Ni, bal Fe.
Cast.
For furnace parts; heat resistant.

AMSCO HC.
M-9; C, Cr, Si, Mn, bal Fe.
Heat treated: 60,000 TS; 7000 YS; 400-600 Brin.
For castings, chute plates, liners; cast iron, abrasion and wear resistant

AMSCO HC-250.
M-9; 2.0-3.0 C, 0.5-1.8 Mn, 0.2-1.0 Si, 22-30 Cr, 0.2-0.6 Mo, bal Fe
Heat treated: 60,000-100,000 TS; 50,000-75,000 YS; 450-550 Brin.
For pump parts, liners; erosion and abrasion resistant.

AMSCO-M.
M-9; 1.0-1.4 C, 10-14 Mn, 0.2-1.0 Si, bal Fe.
Heat treated: 100,000-145,000 TS; 50,000-57,000 YS; 60-30 El; 40-30 RA; 195 Brin.
For dipper buckets, crawler shoes, mill liners, screens; austenitic, work hardens, wear resistant.

AMSCO MANGANESE STEEL.
M-9; **M-231**; 10-14 Mn, 1-1.4 C, 0.2-1.0 Si, bal Fe.
Heat treated: 100,000-145,000 TS; 50,000-57,000 YS; 60-30 El; 40-30 RA; 155-200 Brin.
For steam shovel teeth rolls, ball mill liners, wheels, gears and pinions; tough and shock resisting; resists abrasion.

AMSCO-MM.
M-9; 0.70-1.4 C, 10-14 Mn, 0.2-1.0 Si, Mo, bal Fe.
Heat treated: 90,000-160,000 TS; 45,000-75,000 YS; 65-30 El; 40-30 RA; 200 Brin.
For mining and construction equipment; abrasion and impact resistance with toughness, work hardens.

AMSCO-MMH.
M-9; C, Mo, Mn, bal Fe.
Heat treated: 120,000 TS; 65,000 YS; 20 El; 18 RA.
For castings, rolls, jaws, hammers, dipper teeth; wear resistant, austenitic.

AMSCO-MML.
M-9; C, Mo, Mn, bal Fe.
Heat treated: 120,000 TS; 52,000 YS; 50 El; 40 RA.
For castings, railroad frogs and crossings; austenitic, wear resistant.

AMSCO-MNI.
M-9; C, Mn, Ni, bal Fe.
Heat treated: 110,000 TS; 48,000 YS; 55 El; 40 RA.
For railroad frogs and crossings; austenitic, wear resistant.

AMSCO-MY.
M-9; 1.0-1.4 C, 2.0 Cr, 10-14 Mn, bal Fe.
Cast: 120,000 TS; 56,000 YS; 45 El; 30 RA; 200 Brin.
Heat treated: 150,000 TS; 65,000 YS; 30 El; 30 RA.
For castings, mill liners, screens, dipper teeth jaws, hammers; austenitic wear resistant.

AMSCO NICROMANG.
M-9; 0.8 C, 14.5 Mn, 4 Cr, 3.5 Ni, bal Fe.
As welded: 120,000 TS; 75,000 YS; 42 El; 35 RA.
Weld hardness 200 Brin, work hardens to 500 Brin.
Manganese steel electrode for rebuilding manganese steel and joining manganese steel and other steels.
Impact resistant.

AMSCO NO 1.
M-9; 2.5 C, 31 Cr, 13 W, 50 Co.
Weld hardness 52 Rc.
Hardfacing rod and electrode; heat and corrosion resistant to 1200°F.
For valves and valve seats, dies.

AMSCO NO 6.
M-9; 1.05 C, 28 Cr, 5 W, 60 Cr.
Weld hardness 41 Rc.
Hardfacing rod and electrode; heat and corrosion resistant to 1200°F.
For build-up and repair valves and dies.

AMSCO NO 12.
M-9; 1.5 C, 30 Cr, 9 W, 55 Co.
Weld hardness 45 Rc.
Hardfacing rod and electrode; heat and corrosion resistant to 1200°F.
For build-up and repain valves and dies.

AMSCO SA-53.
M-9; 3.0 C, 16 Cr, 1 Mo, bal Fe.
Weld hardness: 52 Rc.
Hardfacing open arc wire, abrasion and impact resistant.

AMSCO SA-CMO.
M-9; 0.3 C, 16 Mn, 16 Cr, 1 Ni, bal Fe.
As welded: 124,000 TS; 82,000 YS; 38 El; 34 RA.
Weld hardness: 230 Brin, work hardens.
Wire for open arc rebuilding and joining of manganese, carbon and low alloy steel parts.
Impact resistant.

AMSCO SA-MANGANESE.
M-9; 0.9 C, 17.5 Mn, 0.5 V, bal Fe.
As welded: 128,000 TS; 85,000 YS; 25 El; 25 RA.
Weld hardness: 235 Brin, work hardens to 500 Brin.
Manganese steel wire for open arc rebuilding of manganese and other steel parts. Impact resistant with minimun flow.

AMSCO SA-NICROMANG.
M-9; 0.8 C, 15 Mn, 4 Cr, 3.5 Ni, bal Fe.
As welded: 120,000 TS; 70,000 YS; 40 El; 35 RA.
Weld hardness: 200 Brin, work hardens to 500 Brin.
Manganese steel wire for open arc rebuilding and joining of manganese parts and to other steels.
Impact resistant.

AMSCO SA-ROLL BUILD.
M-9; 0.9 C, 17.5 Mn, 0.5 V, bal Fe.
As welded: 230 Brin, work hardens up to 500 Brin.
Manganese steel wire for open arc automatic rebuilding of manganese steel parts.
Impact resistant.

AMSCO SA-ROLL FACE.
M-9; 3.0 C, 16 Cr, 1 Mo, bal Fe.
Weld hardness: 52 Rc.
Hardfacing wire for open arc automatic applications; abrasion and impact resistant.

AMSCO SA-T40.
M-9; 0.6 C, 3.5 Mn, 22 Cr, 9 Ni, 0.5 Mo, bal Fe.
Weld hardness 22 Rc, work hardens.
Hardfacing open arc wire, resistant to impact, heat, metal-to-metal wear and thermal shock.

AMSCO SA-TOUGHWEAR.
M-9; 2.0 C, 1.8 Si, 5 Cr, bal Fe.
Weld hardness 48 Rc.
Hardfacing open arc wire; impact and abrasion resistant; for multiple layer buildup.

AMSCO SA-TUNGSITE.
M-9.
60% WC particles in a steel matrix.
Hardfacing open arc wire, extremely abrasion resistant.

AMSCO SUPER 20.
M-9; 6.0 C, 22 Cr, 7 Mo, 5 W, bal Fe.
Weld hardness 63 Rc.
Hardfacing electrode, extremely resistant to abrasion at normal and elevated temperatures.

AMSCO SUPERCHROME.
M-9; 4.5 C, 2 Si, 3 Cr, bal Fe.
Weld hardness 59 Rc.
Hardfacing electrode, abrasion resistant at normal and elevated temperatures.

AMSCO THERMALLOY 4, now AMSCO THERMALLOY 400.

AMSCO THERMALLOY 400.
M-9; 0.6 C, 22 Cr, 8 Ni, 0.5 Mo, bal Fe.
Weld hardness 22 Rc; work hardens.
Hard facing electrode, resistant to heat, impact, metal-to-metal wear and thermal shock.

AMSCO TRACKWEAR.
M-9; 0.9 C, 17.5 Mn, 0.5 V, bal Fe.
As welded: 128,000 TS; 85,000 YS; 25 El; 25 RA.
Weld hardness 235 Brin, work hardens to 500 Brin.
Manganese steel electrode for rebuilding manganese steel parts. Impact resistant with minimum flow.

AMSCO TUBE CHROMEFACE.
M-9; 4.0 C, 1.5 Mn, 2 Si, 30 Cr, bal Fe.
weld hardness 60 Rc.
Hardfacing rod, extremely resistant to sliding abrasion.

AMSCO TUBE TUNGSITE.
M-9.
60% WC particles in a steel or iron matrix.
Hardfacing rod and electrode; extremely abrasion resistant. Several particle sizes available.

AMSCO TUBE TUNGSITE 30 DOWN.
M-9.
50% WC particles in a steel matrix.
Hardfacing rod; extremely abrasion resistant.

AMSCO TUBE TUNGSITE SRB.
M-9.
38% WC particles in a steel matrix.
Hardfacing rod; abrasion resistant with greater impact resistance.

AMSCO TUNGCHROME.
M-9.
Tungsten carbide particles in a high chromium iron matrix. Hardfacing rod and electrode, abrasion resistant.

AMSCO TUNGROD.
M-9.
60% WC (smaller mesh size) particles in an iron or steel matrix.
Hardfacing rod and electrode; extremely abrasion resistant.

AMSCO X-53.
M-9; 3.5 C, 16 Cr, 1 Mo, bal Fe.
Weld hardness 52 Rc.
Hardfacing electrode, abrasion and impact resistant.

AMSIL.
M-1778; 0.05 Ag, bal Cu.
Elect. cond. 101%.
Wire: 48,000 TS; 40,000 YP; 15 El; B 50 Rock.
For rotor and stator windings in high speed generators, commutators.
High creep resistance, freedom from embrittlement.

AMSULF.
M-1778; 0.3 S, bal Cu.
Extruded: 32,000 TS; 47 El; 60 RA.
Cold drawn: 57,000 TS; 6 El; 23 RA.
For screw machine products, clamps, connectors, clips; high electrical conductivity, free-cutting.

AMTEL.
M-1778; 0.5 Te, bal Cu.
Bar: 50,000 TS; 40,000 YP; 12 El.
For screw machine products, hardware, fasteners, welding tips and nozzles.
Free machining, immune to hydrogen embrittlement.

AMUTIT, NOW VEW K465.
M-27; 1.05 C, 0.9 Mn, 1.5 W, 1 Cr, bal Fe.
For stay bolt taps, screw taps, gages, broaches, reamers, milling cutters; oil hardened. AISI 07.

AMUTIT S, NOW VEW K460.
M-27, M-1182; 1C, 1 Mn, 0.5 Cr, 0.6 W, 0.1 V, bal Fe.
For cold forming, blanking and bending dies; non-deforming, oil hardening. AISI 01.

AMZIRC.
M-1778; 0.7 Zr, bal Cu.
At 70°F; 72,000 TS; 62,000 YS; 12 El.
At 750°F; 51,600 TS; 15 El.
At 900°F: 39,400 TS; 19 El.
For rectifier bases, rotor wedges, resistance welding wheels and tips; hig conductivity, good high temperature strength.

AMZIRC-150.
M-8; 0.17 Zr, 99.83 Cu.
At RT: 71,000 TS; 62,000 YS.
At 752°F: 52,000 TS; 45,000 YS; 8 El.
At 932°F: 35,200 TS; 25,800 YS; 9 El.
For resistance welding wheels and tips, commutators, contacts, switch blades.
92% elect. cond. annealed.
Precipitation-hardening.

AN 40.
M-1361; 2.5 Cu, 1.5 Mg, Fe, Ni, bal Al.
390-430 N/mm² TS; 305-375 N/mm² YS; 3-8 El
Heat treatatable alloy with good elevated-temperature properties.
Type AlCuMgNi.
Similar to AA 2618.

AN 50.
M-1361; 2.0 Cu, 1.5 Mg, 0.8 Si, Fe, Ni, bal Fe.
355-390 n/mm² TS; 245-275 n/mm² YS; 4-6 El.
Heat treatable alloy with good elevated-temperature properties.
Type AlCuMgNiSi.

ANACONDA 61.
M-8; 63 Cu, 37 Zn.
Hard: 65,000 TS; 50,000 YS; 20 El; B75 Brin.
Soft: 46,000 TS; 17,000 YS; 60 El; B20 Brin.
For screw machine parts; yellow brass.

ANACONDA 66.
M-8; 60 Cu, 40 Zn.
Soft: 54,000 TS; 20,000 YS; 40 El; B45 Brin.
For hardware, condenser tubes; high strength.

ANACONDA 271.
M-8; 62 Cu, 35 Zn, 3 Pb.
Hard: 62,000 TS; 50,000 YS; 20 El; 77 RB.
Soft: 47,000 TS; 32,000 YS; 60 El; 16 RB.
For machined parts, hardware; free cutting.

ANACONDA 293.
M-8; 39 Zn, 1 Pb, bal Cu.
Hard: 80,000 TS; 60,000 YS; 6 El; 142 Brin.
Soft: 54,000 TS; 20,000 YS; 40 El; 79 Brin.
For hardware, architectural uses; leaded Muntz metal.

ANACONDA 605.
M-8; 38.55 Zn, 0.75 Sn, 0.70 Pb, bal Cu.
Hard: 63,000 TS; 35,000 YS; 28 El; 102 Brin.
Soft: 56,000 TS; 22,000 YS; 38 El; 83 Brin.
For hardware, screw machine products; Leaded Naval Brass.

ANACONDA 612.
M-8; 60 Cu, 29.5 Zn, 2 Pb, 0.75 Sn.
Hard: 63,000 TS; 35,000 YS; 25 El; B65 Brin.
Soft: 56,000 TS; 22,000 YS; 35 El; B50 Brin.
For screw machine parts; free cutting.

ANACONDA 624.
M-8; 60 Cu, 1 Sn, 1 Pb, 0.1 Al, 0.15 Si, bal Zn.
Die cast: 55,000 TS; 35,000 YS; 8 El
For hardware, fixtures; free-cutting.

ANACONDA 681.
M-8; 57.8 Cu, 40.27 Zn, 0.95 Sn, 0.85 Fe, 0.10 Si, 0.03 Mn.
For oxyacetylene braze welding of steel, cast iron and copper alloys, and for bearing surfaces. Welding rods.
Low fuming bronze. Melting point 870°C, corrosion resistant.

ANACONDA 831.
M-8; 1 Pb, 1 Ni, 0.2 bal Cu.
Bar: 85,000 TS; 75,000 YS; 5 El.
For fasteners, screw machine products, electrical contacts, connectors; free-cutting, corrosion resistant.

ANACONDA 997.
M-8; 40.3 Zn, 0.95 Sn, 0.85 Fe, bal Cu.
For welding rod; low fuming.

ANACONDA 1026.
M-8; 81.5 Cu, 4.25 Sn, 0.15 Mn, bal Zn.
Die cast: 85,000 TS; 50,000 YS; 8 El.
For hardware, fixtures; corrosion resistant.

ANACONDA COPPER 189.
M-8; 98.80 Cu, 0.75 Sn, 0.25 Si, 0.20 Mn.
For inert-gas and oxyacetylene welding of copper for ductile and strong welds, welding rods.
Melting range 1075°C, corrosion resistant.

ANACONDA COPPER 372.
M-8; 98.8 Cu, 0.75 Sn, 0.25 Si, 0.20 Mn.
For welding rod; oxy-acetylene and inert gas arc welding of copper; M.P. 1967°F.

ANATOMICAL ALLOY.
M-U.S.; 54 Bi, 19 Sn, 17 Pb, 11 Hg, Cd.
For fusible alloy for anatomical impressions and casts; fusible.

ANCHOR ALLOY NO. 95/5.
M-1468; A refrigeration solder, contains Sn and Sb; used with copper sweat fittings; M.P.; 450°F.

ANCHOR ALLOY NO. 250.
M-1468; Aluminum solder; M.P.; 390°F.
For joining aluminum to aluminum.

ANCHOR ALLOY NO. 260.
M-1468; Aluminum solder; M.P.; 650°F.
For joining aluminum to aluminum or to dissimilar metals, including castings, die cast parts and aluminum foil.

ANCHOR ALLOY NO. 270.
M-1468; Solder; M.P.: 507°F.
All purpose solder for all metals except magnesium; including aluminum, zinc die casting, copper, etc.

ANCHOR ALLOY NO. 304.
M-1468; Solder, contains Ag, no Cd.
M.P. 430°F; 14,000-20,000 TS.
For joining stainless steels to dissimilar metals.

ANCHOR ALLOY NO. 520.
M-1468; Solder for repairing, sealing and filling cracks and blow holes in cast iron, M.P.: 500°F.

ANCHOR ALLOY NO. 530.
M-1468; Solder for repairing zinc die castings as radiator grills, door handles. M.P.: 700°F.

ANCHOR GALVANIZED BAR.
M-1468; Galvanizing repair solder; M.P.: 550°F.
For repairing damaged or burned galvanized surface; no flux required.

ANCHOR REDIMIX SOLDER PASTE.
M-1468; Solder paste for metals except aluminum; contains solder and flux M.P.: 370°F.

ANCHOR-TITE.
M-598; WC.
For tools, dies; cemented WC.

ANCOLOY.
M-1526; 1.75 Ni, 0.5 Mo, 0.02 C, 1.5 Cu, bal Fe.
Sintered: 54,400 TS; 41,500 YS; 1-2 El.
Heat treated: 115,000 TS; 104,000 YS; 0.3 El.
For general machinery parts.
Heat treatable. Prealloyed powder, high strength.

ANCOLOY SPONGE IRON POWDER.
M-1781; 95 Fe, 0.02 C, 1.55 Cu, 1.75 Ni, 0.60 Mo, (0.45 H_2- loss).
Compressibility @ 30 tsi - 6.2 g/cm^3.
For medium density P/M parts requiring heat treatment.

ANCOR 303 STAINLESS STEEL POWDER.
M-1781; 0.03 max C, 17.5 Cr, 12.0 Ni, 0.9 Si, 0.25 S, bal Fe.
Compressibility @ 30 tsi - 6.05 g/cm^3.
For austenitic stainless P/M parts requiring machinability.

ANCOR 304L STAINLESS STEEL POWDER.
M-1781; 0.03 max C, 18.5 Cr, 10.0 Ni, 0.9 Si, bal Fe.
Compressibility @ 30 tsi - 6.1 g/cm^3.
Non-magnetic P/M parts.

ANCOR 316L STAINLESS STEEL POWDER.
M-1781; 0.03 max C, 17.5 Cr, 13.0 Ni, 0.9 Si, 2.2 Mo, bal Fe.
Compressibility @ 30 tsi - 6.25 g/cm^3.
For corrosion resistant high strength P/M parts and filters.

ANCOR 410L STAINLESS STEEL POWDER.
M-1781; 0.015 max C, 13.2 Cr, 0.09 Si, bal Fe.
Compressibility @ 30 tsi - 5.9 g/cm^3.
If blended to build up carbon content, the P/M parts can have good hardness and strength as well as corrosion resistance.

ANCOR MH-100 SPONGE IRON POWDER.
M-1781; 99 Fe, 0.02 C, 0.045 H_2 (loss when heated).
Compressibility @ 30 tsi - 6.3 g/cm^3.
For low and medium density P/M parts.

ANCOR MH-1024 M-1S SPONGE IRON.
M-1781; 98 Fe, 0.02 C, (0.45 H_2 - loss), 1 S.
Compressibility @ 30 tsi - 6.2 g/cm^3.
Improved machining P/M parts because of sulphur.

ANCORMET 101 SPONGE IRON POWDER.
M-1781; 99.0 Fe, 0.18 C, 0.40 H_2 (loss when heated).
Compressibility @ 30 tsi - 6.2 g/cm^3.
For medium density, high strength P/M parts.

ANCORSPRAY 120.
M-1781; 0.03 C, 1.5 B, 2.5 Si, 1.5 Fe, bal Ni.
Self-fluxing atomized powder for spraying.
Fusion temp: 2000°F; hardness: 13-15 Rc.
For soft coatings that may be hand finished.

ANCORSPRAY 125.
M-1781; 0.06 C, 1.5 B, 3.5 Si, 1.5 Fe, bal Ni.
Self-fluxing atomized powder for spraying.
Fusion Temp: 1975°F; hardness: 25-28 Rc.
For rebuilding damaged articles; may be refinished by machining.
AMS 4779.

ANCORSPRAY 135.
M-1781; 0.05 C, 10.5 Cr, 2.0 B, 3.25 Si, Fe bal Ni.
Self-fluxing atomized powder for spraying.
Fusion temp: 1925°F; hardness: 33-38 Rc.
For build-up; either final coat or as an undercoat; soft, machinable.

ANCORSPRAY 140.
M-1781; 0.30 C, 7.5 Cr, 1.35 B, 4.0 Si, 1.5 Fe, bal Ni.
Self-fluxing atomized powder for spraying.
Fusion temp: 1925°F; hardness: 35-40 Rc.
For build-up; may be machined with carbide tools.

ANCORSPRAY 150.
M-1781; 0.65 C, 14.0 Cr, 2.8 B, 3.8 Si, 4.2 Fe, bal Ni.
Fusion temp: 1900°F; hardness: 50-55 Rc.
For build-up; strong, hard, but still somewhat flexible.
AMS 4775.

ANCORSPRAY 155.
M-1781; 0.50 C, 16.0 Cr, 4.0 B, 4.25 Si, 4.0 Fe, 2.5 Cu, 2.5 Mo, 2.5 W, bal Ni.
Fusion temp: 1900°F; hardness: 55-59 Rc.
For build-up of heavy coating, and for final coat. Hard, wear resistant.

ANCORSPRAY 160.
M-1781; 0.90 C, 16.5 Cr, 3.25 B, 4.25 Si, 4.5 Fe, bal Ni.
Fusion temp; 1900°F; hardness: 60-64 Rc.
Hardest and most wear resistant grade; finished by grinding.
AMS 4775.

ANCORSPRAY 250.
M-1781; 0.10 C, 18.5 Cr, 3.2 B, 3.3 Si, 2.0 Fe, 27.0 Ni, 5.5 Mo, bal Co.
Fusion temp: 2050°F; hardness 47-53 Rc.
For special build-up jobs; good hardness and wear resistance, corrosion resistant and resistant to thermal cracking.

ANCORSTEEL 45P ATOMIZED STEEL POWDER.
M-1781; 98.8 Fe, 0.02 C, 0.45 P, (0.13 H_2 - loss).
Compressibility @ 30 tsi - 6.8 g/gm^3.
Phosphorus speeds sintering and enhances properties: strength/ductility and electromagnetic.

ANCORSTEEL 1000 ATOMIZED IRON POWDER.
M-1781; 99.2 Fe, 0.015 C, (0.25 H_2 - loss).
Compressibility @ 30 tsi - 6.7 g/cm^3.
For high density P/M parts.

ANCORSTEEL 1000B ATOMIZED IRON POWDER.
M-1781; 99.4 Fe, 0.01c, (0.12 h_2 - loss).
Compressibility @ 30 tsi - 6.8 g/cm^3.
For high density and electromagnetic P/M parts.

ANCORSTEEL 1015 ATOMIZED STEEL POWDER.
M-1781; 99.1 Fe, 0.15 C (0.30 H_2 - loss).
Compressibility @ 30 tsi - 6.5 g/cm^3.
Fast sintering of high density P/M parts.

ANCORSTEEL 2000 ATOMIZED STEEL POWDER.
M-1781; 98.2 Fe, 0.6 Mo, 0.45 Ni, 0.3 Mn, 0.2 C (0.13 H_2 - loss).
Compressibility @ 30 tsi - 6.6 g/cm^3.
Low cost P/M parts requiring heat treatment.

ANCORSTEEL 4600V ATOMIZED STEEL POWDER.
M-1781; 96.9 Fe, 1.8 Ni, 0.5 Mo, 0.25 Mn, 0.02 C, (0.13 H_2 - loss).
Compressibility @ 30 tsi - 6.5 g/cm^3.
Good hardenability parts required high impact strength.

ANCOR W-100 SPONGE IRON POWDER.
M-1781; 97 Fe, 0.08 C, 0.02 max S, (0.80 H_2 - loss).
100 mesh powder for flux core wire and small diameter stick electrodes.

ANCOR W-423A SPONGE IRON POWDER.
M-1781; 97.0 Fe, 0.08 C, 0.02 max S, (0.8 H_2 - loss).
40 mesh powder for low yield stick electrode coatings.

ANCOR W-428 SPONGE IRON POWDER.
M-1781; 97.0 Fe, 0.08 C, 0.02 max S, (0.8 H_2 - loss).
40 mesh powder for medium yeild stick electrode coatings.

ANDARD.
M-358; 1.1 C, bal Fe.
For tools, drills, water hardening.

ANGEL.
M-485; C, alloy, bal Fe.
For high speed tools, high speed steel.

ANGSBURG.
71.9 Cu, 27.6 Zn.
For tubes, fittings, brass.

"A" NICKEL ELECTRON GRADE now NICKEL 205.

ANKO-2.
M-USSR; 9 Al, 20 N; 15 Co, 4 Cu, bal Fe.
Heat treated magnetic alloy.
For electrical and magnetic equipment.

ANKO-3.
M-USSR; 10 Al, 19 Ni, 18 Co, 3 Cu, bal Fe.
Permanent magnet alloy.

ANKO-4.
M-USSR; 8 Al, 14 N:, 24 Co, 3 Cu bal Fe.
Permanent magnet alloy.
Similar to ALNICO 5.

ANODE METAL-1.
U.S.; 94 Pb, 6 Sb.
For batteries, plates; hard.

ANODE METAL-2.
U.S.; 98.8 Pb, 1 Ag, 0.2 As.
For lead storage battery parts; hard.

ANOREFRACT 50.
M-1645; 0.25 C, 25.0 Cr, 12.0 Ni, bal Fe.
Heat resisting steel.

ANOREFRACT 50 NB.
M-1645; 0.25 C, 25.0 Cr, 12.0 Ni, 1.5 Nb, bal Fe.
Heat resisting steel.

ANORESIST 55.
M-1645; 0.50 C, 18.0 Cr, 35.0 Ni, bal Fe.
Heat resisting steel; for furnace parts.

ANOREXACT 5B.
M-1645; 0.65 C, 0.30 Si, 0.40 Mn, 3.75 Ni, 0.80 Cr, bal Fe.
Cold work tool steel.
AFNOR 65 NC 11.

ANOREXACT 344 B.
M-1645; 0.55 C, 1.0 Si, 0.30 Mn, 1.0 Cr, 2.0 W, bal Fe.
Tool steel, shock resisting type.
AFNOR 55 WC 20.

ANOREXACT D 11.
M-1645; 0.90 C, 0.30 Si, 2.0 Mn, 0.20 V, bal Fe.
Oil or water hardening cold work tool steel.
AFNOR 90MV8.

ANOREXACT SB6.
M-1645; 0.38 C, 1.0 Si, 0.30 Mn, 5.0 Cr, 1.25 Mo, 0.50 V, 1.25 W, bal Fe.
Hot work tool steel.
AFNOR Z 55 CDVW 5.

ANOREXACT SB 7.
M-1645; 0.55 C, 1.0 Si, 0.30 Mn, 5.0 Cr, 1.25 Mo, 0.50 V, bal Fe.
Hot work tool steel.
AFNOR Z 55 CDV 5.

ANOREXACT SB 11.
M-1645; 0.60 C, 8.0 Cr, N, Mo, V, W bal Fe.
Tool Steel.
AFNOR Z 60 CNDVW 8.

ANOREXACT SB 55.
M-1645; 1.0 C, 0.30 S:, 0.30 Mn, 5.0 Cr, 1.0 Mo, 0.30 V, bal Fe.
Air hardening cold work tool steel.
AFNOR Z 100 CDV 5.

ANOREXACT SD 3.
M-1645; 0.38 C, 1.0 Si, 0.30 Mn, 5.0 Cr, 1.25 Mo, 0.50 V, bal Fe.
Hot work tool steel.
AFNOR Z 38 CDV 5.

ANOREXACT SD 9.
M-1645; 0.55 C, 1.7 Si, 0.60 MN, 0.60 Cr, 0.20 Mo, bal Fe.
Tool Steel, shock resisting type.
AFNOR Y 55 SCD 7.

ANOREXACT SD 13.
M-1645; 1.60 C, 0.30 Si, 0.30 Mn, 12.0 Cr, 0.80 Mo, 0.40 V, bal Fe.
High chromium cold work too steel.
AFNOR Z 160 CDV 12.

ANOREXACT SD 13 V.
M-1645; 1.60 C, 0.30 Si, 0.30 Mn, 12.0 Cr, 0.80 Mo, V, bal Fe.
High chromium cold work tool steel.
AFNOR Z 160 CDV 12-01.

ANOREXACT SD 15.
M-1645; 2.0 C, 0.30 Si, 0.30 Mn, 12.0 Cr, bal Fe.
High chromium cold work tool steel.
AFNOR Z 200 C 12.

ANOREXACT W.
M-1645; 0.42 C, 0.30 Si, 0.80 Mn, 1.0 Cr, 0.20 Mo, bal Fe.
Low alloy structural or tool steel.
AFNOR Y 42 CD 4.

ANORINOX 35 C.
M-1645; 0.25 - 0.34 C, 1.0 max Si, 1.0 max Mn, 12.0-14.0 Cr, 1.0 max Ni, bal Fe.
Martensitic type stainless steel.
AFNOR Z 30 C 13.

ANORINOX 70.
M-1645; 0.12 max C, 1.0 max Si, 2.0 max Mn, 17.0 - 19.0 Cr 8.0-9.0 Ni, bal Fe.
Austenitic 18/8 type stainless steel.
AFNOR Z 10 CN 18-08.

ANORINOX 80.
M-1645; 0.12 max C, 1.0 max Si, 2.0 max Mn, 17.0-19.0 Cr, 8.0-9.0 Ni, 2.0-3.0 Mo, bal Fe.
Austenitic 18/8 plus Mo type stainless steel.
AFNOR Z 10 CND 18-08; Similar to AISI 316.

ANOXIN-2P.
M-1421; 0.08 max C, 18-20 Cr, 8-12 Ni, bal Fe.
Annealed: 85,000 TS; 35,000 YS; 60 El; 150 Brin.
Cold Drawn: 100,000-180,000 TS; 50,000-125,000 YS; 180-130 Brin.
For architectural trim, chemical and pharmaceutical plant equipment.
Type 304 stainless steel, austenitic. Good formability and weldability.

ANOXIN-4P.
M-1421; 0.08 max C, 16-18 Cr, 10-14 Ni, 2-3 Mo, bal Fe.
Annealed: 80,000 TS; 30,000 YS; 60 El; B 78 Rock.
Cold rolled: 150,000 TS; 135,000 YS; 6 El; C 32 Rock.
For chemical and texile plant equipment, agitators, kettles, valve trim, acid tanks and vessels.
Type 316 stainless steel, austenitic, acid resistant.

ANSCOL 45.
M-568; 0.16 C, 0.70 Mn, 0.015 min Cb, bal Fe.
Plate: 60,000 TS; 45,000 YS; 30 El.
For railroad and mine cars, bridges, booms, pressure vessels, derricks.
Good fabricability and weldability.

ANSCOL 50.
M-568; 0.18 C, 0.70 Mn, 0.04 Si, 0.015 min Cb, bal Fe.
Plate: 63,000 TS; 50,000 YS; 30 El; 140 Brin.
For railroad and mine cars, bridges, booms, pressure vessels, derricks.
Good fabricability and weldability.

ANTACIRON.
M-639; 14.5 Si, 84.5 Fe.
Cast: 15,000-18,000 TS; 15,000-8,000 YS; 0 El; 0 RA; 450-475 Brin.
For equipment to resist corrosion acids carrying abrasive solids; finished by grinding.

ANTICORODAL.
M-493; 0.6-1.4 Mg, 0.6-1.6 Si, 0.6-1.0 Mn, bal Al.
Annealed: 21,000 TS; 8000 YS; 24 El.
For window frames, fan blades, gutters, boats; good forming and welding properties.

ANTICORODAL-5 SI.
M-249, M-1605, M-1634; 4-6 Si, 0.4-1.0 Mg, 0.5-1.0 Mn, bal Al.
Sand cast: 30,000 TS; 23,000 YS; 3.5 El; 75 Brin.
Chilled cast: 40,000 TS; 37,000 YS; 2 El; 95 Brin.
For architecture, ship and auto construction, chemical plants; heat treatable.

ANTICORODAL-11.
M-541; M-624; 0.6-1.2 Si, 0.5-0.85 Mg, 0.25-0.70 Mn, 0.45 max Fe, bal Al.
TA16 Temper: 42,000-50,000 TS; 36,000-46,000 YS; 12-20 El; 90-120 Brin.
For decorative parts, auto trim, metal furniture, marine structures.
Corrosion resistant.

ANTICORODAL 15.
M-249; 0.5-1.5 Si, 0.2-1.0 Mn, 0.5-1.0 Mg, 0.5-1.5 Cd, bal Al.
Heat treated: 45,000-61,000 TS; 38,000-58,000 YS; 12-20 El; 90-120 Brin.
For aircraft structures, highly stressed parts; corrosion resistant, for forgings and extrusions.

ANTICORODAL 15.
M-1220, M-541; 4.2-5.2 Si, 0.6 Mg, 0.7 Mn, 0.15 Ti, bal Al.
Cast: 35,000-45,500 TS; 28,500-38,400 YS; 1-2 El; 90-105 Brin.
For instrument housings, oil pans, crankcases; high fluidity and corrosion resistant.

ANTICORODAL 15.
M-1776; 4.5 Si, 0.65 Mg, 0.7 Mn, bal Al.
Heat treated: 36,000-46,000 TS; 26,000-36,000 YS; 2-1 El; 90-105 Brin.
For light alloy parts; corrosion resistant.

ANTICORODAL-63.
M-624, M-541; 0.2-0.6 Si, 0.45-0.85 Mg, 0.35 max Fe, 0.1 max Ti, bal Al.
Annealed: 11,400-17,000 TS; 5700-10,000 YS; 25-40 El; 25-35 Brin.
TA16H20 Temper: 37,000-42,700 TS; 32,700-40,000 YS; 2-6 El; 80-90 Brin.
For marine structures and hardware. Corrosion resistant.

ANTICORODAL-045.
M-493; 0.5 Mg, 0.5 Si bal Al.
Solution HT: 215 MPa TS; 12 El.
Bars, rods ,tubing for structural work.

ANTICORODAL - 112.
M-493; 1.0 mg, 1.0 Si, 0.7 Mn, bal Al.
Ann: 110 MPa TS; 25 El.
Sol. HT: 310 MPa TS; 10 El.
Bars, tubing for ship equipment.

ANTICORODAL-CASTING.
M-249; 2.25 Si, 0.45 Fe, 0.70 Mn, 0.70 Mg, 0.15 Ti, bal Al.
Chill cast: 20,000-28,000 TS; 16,000-22,000 YS; 1.5-5.0 El.
Heat treated: 28,000-40,000 TS; 22,000-30,000 YS; 1.0-5.0 El.
For window frames, chemical industry, architecture; high corrosion resistance.

ANTICORODAL G.
M-1176; 2 Si, 0.65 Mg, 0.7 Mn, bal Al.
Heat treated: 36,000-42,000 TS; 29,000-39,000 YS; 3-1 El; 90-105 Brin.
For light alloy parts.

ANTICORODAL G.
M-1220, M-541; 1.8-2.3 Si, 0.7 Mg, 0.7 Mn, bal Al.
Cast: 35,600-42,700 TS; 31,300-39,900 YS; 1-3 El; 90-105 Brin.
For light alloy castings; general purpose.

ANTICORODAL SPECIAL.
M-249; 2 Mg, 1.5 Mn, bal Al.
Soft: 36,000 TS; 21,000 YS; 20 El.
For punchings; same as Peraluman.

ANTICORODAL-WROUGHT.
M-249; M-624; 0.5-1.5 Si, 0.2-1.0 Mn, 0.5-1 Mg, 0.3 Fe, bal Al.
Soft: 17,000 TS; 9000 YS; 30 El; 35 Brin.
Hard: 55,000 TS; 45,000 YS; 15 El; 105 Brin.
For architecture, window frames, chemical plants; heat treatable.

ANTIFRICTION METAL-A.
M-561; 8 Cu, 83 Sn, 9 Sb.
For special admiralty bearings for heavy loads; Babbitt.

ANTIFRICTION METAL-B.
M-561; 1.4 Cu, 69 Sn, 29.6 Sb.
For special admiralty bearings for under water; Babbitt.

ANTIKLOR.
Russia; 0.5-0.7 C, 13-16 Si, bal Fe.
For cast pumps and conduits; acid resistant, cast.

ANTIKORRO.
M-1322; 0.9 C, 18 Cr, 1.1 Mo, 1 V, bal Fe.
Annealed: 107,000 TS; 62,000 YS; 18 El; 35 RA; 220 Brin.
Heat treated: 280,000 TS; 270,000 YS; 3 El; 15 RA; 555 Brin.
For ball bearings, racers, liners, sleeves; stainless, wear resistant.

ANTIKORRO AK1B.
M-1322; 0.1 max C, 17.5 Ti = 7 x C, bal Fe.
Annealed: 80,000 TS; 50,000 YS; 25 El; 50 RA; 150 Brin.
For welded structures in chemical plants and oil refineries; corrosion and heat resistant.

ANTIKORRO AK1W.
M-1322; 0.12 max C, 13 Cr, 0.4 Si, bal Fe.
Annealed: 75,000 TS; 40,000 YS; 35 El; 70 RA; 155 Brin.
Cold drawn: 100,000 TS; 85,000 YS; 17 El; 60 RA; 205 Brin.
For turbine blades, chemical plant equipment, cutlery; Type 410; corrosion resistant.

ANTIKORRO AK2.
M-1322; 0.22 C, 17 Cr, 1.5 Ni, bal Fe.
Annealed: 125,000 TS; 95,000 YS; 20 El; 55 RA; 260 Brin.
For oil refinery equipment; Type 431; corrosion resistant.

Section I: Alloy Data / 119

ANTIKORRO AK2S.
M-1322; 0.2 C, 0.4 Si, 13 Cr, bal Fe.
Annealed: 95,000 TS; 50,000 YS; 25 El; 55 RA; 196 Brin.
Cold drawn: 105,000 TS; 85,000 YS; 17 El; 50 RA; 215 Brin.
For turbine blades, cutlery, gauges, instruments; Type 420; corrosion resistant.

ANTIKORRO AK4.
M-1322; 0.4 C, 0.4 Si, 13 Cr, bal Fe.
Annealed: 95,000 TS; 50,000 YS; 25 El; 55 RA; 200 Brin.
For cutlery, valves, surgical and dental instruments; Type 420; corrosion resistant.

ANTIKORRO AK5.
M-1322; 0.4 C, 0.4 Si, 13 Cr, bal Fe.
Annealed: 95,000 TS; 50,000 YS; 25 El; 55 RA; 200 Brin.
For cutlery, valves, surgical instruments; Type 420; stainless.

ANTIKORRO AKC.
M-1322; 0.15 C, 24 Cr, 19 Ni, bal Fe.
Annealed: 100,000 TS; 45,000 YS; 50 El; 65 RA; 185 Brin.
For furnace parts, valves, pumps, turbine parts; Type 310; stainless, austenitic.

ANTIKORRO AKC2.
M-1322; 0.15 C, 19.5 Cr, 9.5 Ni, bal Fe.
Annealed: 80,000 TS; 35,000 YS; 55 El; 75 RA; 150 Brin.
For chemical plant equipment, tanks, vessels; Type 302; stainless, austenitic.

ANTIKORRO AKH.
M-1322; 0.85 C Cr, V, bal Fe.
For cutting tools, bearings, dies, liners; water or oil hardened.

ANTIKORRO AKL.
M-1322; 0.1 max C, 12.5 Cr, 12 Ni, bal Fe.
For valves, oil refinery equipment; corrosion resistant.

ANTIKORRO AKSZ.
M-1322; 0.05 C, 18 Cr, 10 Ni, Cu, bal Fe.
Annealed: 85,000 TS; 35,000 YS; 60 El; 70 RA; 150 Brin.
For chemical plant equipment; stainless, austenitic.

ANTIKORRO AKV.
M-1322; 0.15 max C, 18 Cr, 8.5 Ni, bal Fe.
Annealed: 80,000 TS; 35,000 YS; 55 El; 75 RA; 160 Brin.
For chemical plant equipment, tanks, mixers; Type 302; stainless, austenitic.

ANTIKORRO AKV EXTRA.
M-1322; 0.12 max C, 18 Cr, 2 Mo, 10.5 Ni, Ti = 4 x C, bal Fe.
Annealed: 85,000 TS; 35,000 YS; 50 El; 65 RA; 160 Brin.
Cold drawn: 150,000 TS; 135,000 YS; 6 El; 300 Brin.
For welded chemical plant equipment; Type 316 Ti; stainless, austenitic.

ANTIKORRO AKVM.
M-1322; 0.15 max C, 18 Cr, 8.5 Ni, bal Fe.
Annealed: 80,000 TS; 35,000 YS; 55 El; 75 RA; 160 Brin.
For chemical plant equipment, tanks; Type 302; stainless, austenitic..

ANTIKORRO AKVS.
M-1322; 0.12 max C, 18 Cr, 9.5 Ni, Cb = 8 x C, bal Fe.
Annealed: 90,000 TS; 45,000 YS; 56 El; 65 RA; 160 Brin.
Cold drawn: 100,000 TS; 65,000 YS; 45 El; 60 RA; 205 Brin.
For welded chemical plant equipment, tanks, fermenters; Type 347; stainless, austenitic.

ANTIKORRO AKX10.
M-1322; 0.12 max C, 18 Cr, 1 Al, bal Fe.
Annealed: 80,000 TS; 50,000 YS; 25 El; 50 RA; 150 Brin.
For oil refinery and chemical plant equipment; corrosion resistant.

ANTIKORRO AKX12.
M-1322; 0.12 max C, 1.5 Al, 24 Cr, bal Fe.
Annealed: 90,000 TS; 50,000 YS; 30 El; 55 RA; 180 Brin.
For furnace parts, heat treating boxes; corrosion and heat resistant.

ANTIKORRO AKX SPEZIAL.
M-1322; 0.2 C, 25 Cr, 4 Ni, bal Fe.
Cast: 90,000 TS; 65,000 YS; 2 El; 212 Brin.
For furnace parts, grids, conveyors, heat treating boxes; corrosion and heat resistant.

ANTIMONY.
M-1755; Sb.
Purities: zone refined: 99.9999%, 99.999%, 99.99%.
Forms: Ingot, shot, powders, granules, lump, single crystals, foil.

ANTIMONIAL LEAD.
M-88; 0-25 Sb, 75-100 Pb.
For storage battery plates, bullets, type metal, pipes.

ANTINIT AS2 now **VEW A505.**

ANTINIT AS2F now **VEW A520.**

ANTINIT AS2G now **VEW A550G.**

ANTINIT AS2W now **VEW A500.**

ANTINIT AS2Z now VEW A506.

ANTINIT AS3W now VEW A122.

ANTINIT AS4M now VEW A100.

ANTINIT AS4W now VEW A120.

ANTINIT AS5W now VEW A102.

ANTINIT EAS2 now VEW A600.

ANTINIT EAS4 now VEW A200.

ANTINIT EAS4M now VEW A205.

ANTINIT KW8 now VEW N104.

ANTINIT KW-10 now VEW N100.

ANTINIT KW10A1 now VEW N110.

ANTINIT KW10M now VEW N132.

ANTINIT KW15 now VEW N315.

ANTINIT KW15Z now VEW N316.

ANTINIT KW20 now VEW N320.

ANTINIT KW20M now VEW N330.

ANTINIT KW30 now VEW N530.

ANTINIT KW35M now VEW N335.

ANTINIT KW40 now VEW N540.

ANTINIT KW80 now VEW N685.

ANTINIT KW100 now VEW N690.

ANTINIT KWA now VEW N200.

ANTINIT KWB now VEW N350.

ANTINIT KWZA now VEW N310.

ANTINIT NG now VEW V622.

ANTINIT RAM now VEW A900.

ANTINIT SAS2 now VEW A750.

ANTINIT SAS4 now VEW A350.

ANTINIT SAS4M now VEW A354.

ANTINIT SAS8 now VEW A960.

ANTINIT SAS10 now VEW A955.

ANTINIT SAST2 now VEW A700.

ANTINIT SAST4 now VEW A300.

ANTINIT SAST4M now VEW A305.

ANTINIT SKWL now VEW N238.

ANTITHERM F now VEW A700.

ANTITHERM FA now VEW H300.

ANTITHERM FB8 now VEW H160.

ANTITHERM FB8G now VEW H160G.

ANTITHERM FB9 now VEW H140.

ANTITHERM FB10 now VEW H120.

ANTITHERM FB10G now VEW H120G.

ANTITHERM FB12 now VEW H100.

ANTITHERM FBS90.
 M-27; 0.12 max C, 2.3 Si 6 Cr, bal Fe.

ANTITHERM FBS95.
 M-27; 0.12 max C, 2.2 Si, 13 Cr, bal Fe.

ANTITHERM FBS105.
 M-27; 0.06 C, 17 Cr, 10 Ni, bal Fe.

ANTITHERM FF now VEW H550.

ANTITHERM FFB now VEW H525.

ANTITHERM FFB 400 now VEW H520.

ANTITHERM FFBG now VEW H537.

ANTITHERM FFG now VEW H551.

ANTOXYD.
 M-96; 0.20 C, 21 Cr, 38.5 Ni, bal Fe.
 For heat and corrosion resistant parts; corrosion and heat resistant.

ANV-300.
 M-USSR; 0.10 max C, 15.5 Cr, 1.7 Ti, 5.0 Al, 8.5 W, 0.1 max B, 5 max Fe, bal Ni.
 Cast nickel-base supervolly.
 For automotive turbine blades.

ANVIL BRASS.
 M-U.S.; 62.5 Cu, 37.5 Zn.
 For hardware, tubes, yellow brass.

ANVIL HEADING DIE.
 M-387; 0.6 C, 0.8 Cr, 0.2 Mo, bal Fe.
 For tools, heading dies; oil hardened.

ANVILOY 1100.
 M-625; 90 W, 4 N:, 4 Mo 2 Fe.
 125,000 psi UTS, 30 Rc.
 At 800°C: 49 RA.
 For high temperature tooling.

ANVILOY 1150.
 M-265; 90 w, 4 Ni, 4 Mo, 2 Fe.
 140,000 psi UTS; 34 Rc.
 At 800°c: 75,000 psi UTS.
 For die casting, hot extrusion dies and high strength application.

ANVILOY 1200.
 M-265; 90 w, 4 Ni, 2 Fe, 4 Mo.

ANVILOY WELD ROD.
 M-265; W, Ni, Fe.
 For welding high density (Tungsten base) materials.

AO 20.
M-365; 0.50 C, 0.3 Mn, 0.75 Si, 1.15 Cr, 0.2 V, 2.5 W, bal Fe.
Shock resisting tool steel; S1.

AO 3900.
M-690; 16.0-18.0 Si, 1.3 max Fe, 4.0-5.0 Cu, 0.10 max Mn, 0.45-0.65 Mg, 0.10 max Zn, 0.2 max Ti bal Al.
Cast aluminum alloy. 390.

"AO" QUALITY.
M-179; Cu, bal Zn.

A.P. 33.
M-620; 4.5 Cu, 0.5 Mg, 0.4 Ti, bal Al.
36,000-43,000 TS; 29,000-36,000 YS; 2-1 El.
For light alloy parts; age-hardenable.

AP33M.
M-French; 4-6 Cu, 0.5 Mg, 0.4 Ti, bal Al.
Heat treated: 47,000 TS.
For light metal parts, gear cases; heat treatable.

AP201.
M-1741; 4.0-5.0 Cu, 0.20 max Si, 0.25 max Fe, 0.05 max Mn, 0.05 max Mg, 0.05 max Ni, 0.10 max Zn, 0.20 max Ti, bal Al.
Sand cast, T4: 216 MPa min TS; 138 MPa YS; 8 El; 80 Brin.
Per. mold cast, T6: 324 MPa TS; 193 MPa YS; 10 El; 105 Brin.
ADC H49-9; BS LM11.

AP301.
M-1741; 5.5-6.5 Cu, 5.5-6.5 Si, 0.40 max Fe, 0.05 max Mn, 0.25-0.50 Mg, 0.05 max Ni, 0.10 max Zn, 0.05 max Ti, 0.05 max Sn, 0.05 max Cr, bal Al.
Per. mold cast, T5: 241 MPa TS; 193 MPa YS; 4 El; 95 Brin.

AP303.
M-1741; 2.0-4.0 Cu, 5.0-6.5 Si, 0.6 max Fe, 0.7 max Mn, 0.10 max Mg, 0.10 max Ni, 0.10 max Zn, 0.20 max Ti, bal Al.
Sand cast, T6: 248 MPa TS; 165 MP YS; 2 El; 80 Brin.
Per. mold cast, T6: 310 MPa TS; 165 MPa YS; 2 El; 110 Brin.
ADC H49-3; BS LM4.

AP309.
M-1741; 1.0-1.5 Cu, 4.5-5.5 Si, 0.14-0.25 Fe, 0.05 max Mn, 0.5-0.6 Mg, 0.05 max Zn, 0.20 max Ti, bal Al.
Sand cast, HT T6; 241 MPa TS; 172 MPa YS; 3 El; 80 Brin.
Per. mold cast, T62: 310 MPa TS; 276 MPa YS; 2 El; 105 Brin.
ADC H49-13; BS LM16.

AP311.
M-1741; 1.0-1.5 Cu, 4.0-6.0 Si, 0.15 max Fe, 0.05 max Mn, 0.05 max Mg, 0.10 max Zn, 0.20 max Ti, bal Al.
Per. mold cast, F1: 124 MPa TS; 48 MPa YS; 7 El, 40 Brin.

AP315.
M-1741; 3.0-4.5 Cu, 10.5-12.0 Si, 0.6-1.0 Fe, 0.10 max Mn, 0.10 max Mg, 0.10 max Ni, 0.10 max Zn, 0.10 max Sn, bal Al.
Pressure cast, F1: 331 MPa TS; 165 MPa YS; 2.5 El.
SAE 303.

AP403.
M-1741; 0.10 max Cu, 4.5-6.0 Si, 0.6 max Fe, 0.1 max Mn, 0.05 max Mg, 0.1 max Zn, 0.2 max Ti, bal Al.
Sand cast, F1: 131 MPa TS; 55 MP YS; 8 El, 40 Brin.
Per. mold cast, F1; 165 MPa TS; 62 MPa YS; 9 El; 45 Brin.
ADC H49-14; BS LM18.

AP501.
M-1741; 0.10 max Cu, 0.3 max Si, 0.3 max Fe, 0.5 max Mn, 3.6-4.5 Mg. 0.10 max Zn, 0.2 max Ti, bal Al.
Sand cast, F1: 172 MPa TS; 83 MPa YS; 9 El, 50 Brin.
ADC H49-4; BS LM5.

AP601.
M-1741; 0.05 max Cu, 6.5-7.5 Si, 0.12-0.20 Fe, 0.05 max Mn, 0.30-0.40 Mg, 0.05 max Zn, 0.2 max Ti, bal Al.
Sand cast, T6: 255 MPa TS; 166 MPa YS; 5 El; 70 Brin.
Per. mold cast, T61: 296 MPa TS; 207 MPa YS; 8 El, 105 Brin.
ADC H49-18; BS LM25.

AP701.
M-1741; 0.10 max Cr, 0.25 max Si, 0.4 max Fe, 0.1 max Mn, 0.50-0.75 Mg, 0.1 max Ni, 4.8-5.7, Zn, 0.15-0.25 Ti, 0.05 max Sn, 0.4-0.6 Cr, bal Al.
Per. mold, T1: 255 MPa TS; 179 MPa YS; 5 El; 80 Brin.
ADC H49-17.

AP703.
M-1741; 0.10 max Cu, 0.15 max Si, 0.15 max Fe, 0.10 max Mn, 0.65-0.85 Mg, 6.5-7.5 Zn, 0.10-0.25 Ti, 0.06-0.15 Cr, bal Al.
Sand cast, T6: 345 MPa TS; 276 MPa YS; 9 El; 93 Brin.

APACHE.
M-1779; 0.70 C, 2.0 Mn, 0.3 Si, 1.0 Cr, 1.35 Mo, bal Fe.
Heat treated: 293,000 TS; 265,000 YS; 1 El; Usual 56/59 Rc.
For dies and punches, blanking and forming dies, gages, shear blades, bending tools, stripper plates, master hubs, air hardening.
AISI A6 tool steel, wear resistant. Outstanding size stability in heat treatment.

APEX.
M-688; 0.9 C, bal Fe.
For tools, drills, punches; drill rod, water hardened.

APEX 12.
M-130; 6.5 Cu, 3 Si, 2 Zn bal Al.
Sand cast: 17,000-23,000 TS; 14,000-15,000 YS; 3-1 El; 65-70 Brin.
For general castings, sand and permanent mold castings; non-hardenable.

APEX 12.
M-564; 85 Cu, 5 Sn, 5 Zn, 5 Pb.
24,000 TS; 12,000 YS; 16 El; 48 Brin.
For pump bodies valves, pressure or steam parts; "Ounce Metal".

APEX 12 PM.
M-130; 6.5 Cu, 3 Si, 2 Zn, bal Al.
Sand cast: 23,000 TS; 15,000 YS; 1.5 El; 70 Brin.
Permanent mold: 29,000 TS; 22,000 YS; 1.0 El; 70 Brin.
For sand and permanent mold castings; good machinability.

APEX 13.
M-130; 11-13 Si, bal Al.
Die cast: 45,000 TS; 25,000 YS; 4 El; 70 Brin.
For die castings, housings; corrosion resistant.

APEX 14.
M-564; 85 Cu, 15 Zn.
For flanges which are brazed onto copper pipe; "Brazing Brass."

APEX 16.
M-564; 56 Cu, 41 Zn, 1 Fe, 1 Al, 0.5 Mn, 0.5 Sn.
65,000 TS; 32,000 YS; 25 El; 110 Brin.
For valve stems, gears, ship propellers; "Manganese Bronze;" tough.

APEX 22.
M-564; 86 Cu, 8 Sn 2 Zn 4 Pb.
28,000 TS; 15,000 YS; 17 El.
For bearings for industrial equipment; heavy duty.

APEX 24 (COMPOSITION G).
M-564; 88 Cu, 10 Sn, 2 Zn.
40,000 TS; 22,000 YS; 24 El.
For superheated steam or hydraulic casting, gears, heavy pressure bearings pressure tight.

APEX 28.
M-130; 1.7-2.3 Cu, 10.0 Si, bal Al.
For light alloy castings; high fluidity.

APEX 31.
M-130; 9-11 Cu, 1.5 Fe, 3.5-4.5 Si, 0.15-0.35 Mg, 1.5 max Zn, 0.25 max Ti, 1.0 max Ni, bal Al.
For permanent mold castings.

APEX 32.
M-130; 1 Cu, 12 Si, 1.1 Mg, bal Al.
PM-F temper: 27,000 TS; 19,000 YS; 1.0 El; 88 Brin.
PM-T 65 temper: 47,000 TS; 43,000 YS; 0.5 El; 125 Brin.
PM-T 551 temper: 36,000 TS; 28,000 YS: 0.5 El; 105 Brin.
For pistons, high temperature uses; permanent mold, heat treatable.

APEX 34.
M-130; 3 Cu, 9 Si, 1 Mg, 1 Ni, bal Al.
PM-F temper: 35,000 TS; 26,000 YS; 1 El; 85 Brin.
PM-T 5 temper: 36,000 TS; 28,000 YS; 1 El; 100 Brin
For pistons; permanent mold.

APEX 35.
M-130; 10 Cu, 0.3 Mg, bal Al.
SC-F temper : 26,000 TS; 21,000 YS; 0.5 El; 85 Brin. SC-T 6 temper : 40,000 TS; 30,000 YS; 0.5 El; 115 Brin. PM-T 65 temper: 48,000 TS; 36,000 YS 0.5 El; 140 Brin
For sand and permanent mold castings; wear resistant.

APEX 36.
M-130; 0.6 max Cu, 9-10 Si, 0.4-0.6 Mg, bal Al.
Cast: 46,000 TS; 27,000 YS; 3.5 El; 75 Brin.
For light alloy castings; high fluidity.

APEX 37.
M-130; 4.5 Cu, 4.5 Si, bal Al.
SC-F temper: 25,000 TS; 15,000 YS; 2 El; 65 Brin.
PM-F temper: 27,000 TS; 13,000 YS; 2.5 El; 70 Brin.
For castings; sand and permanent mold.

APEX 38.
M-130; 10 Cu, 4 Si, 0.25 Mg, bal Al.
PM-F temper : 32,000 TS; 24,000 YS; 1.5 El; 100 Brin.
For sole plates, permanent mold castings; retains hardness at elevated temperatures.

APEX 39.
M-130; 3.5 Cu, 9 Si, bal Al.
DC-F temper: 47,000 TS; 27,000 YS; 3.5 El; 80 Brin.
For die castings; good machinability.

APEX 39-11.
M-130; 3-4 Cu, 10.5-11.5 Si, 1 max Zn, 1 max Fe, 0.5 max Ni, bal Al. Die cast: 44,000-46,000 TS; 23,000-27,000 YS; 2 El; 85 Brin.
For general die castings; SS 28,000; Charpy 2.2.

APEX 39A.
M-130; 3-4 Cu, 8.5-9.5 Si, bal Al.
Cast: 44,000 TS; 26,000 YS; 3 El; 80 Brin.
For light alloy castings; high fluidity.

APEX 39B.
M-130; 3.5-4.5 Cu, 8.9-9.5 Si, 1.5-2.5 Zn, 1.2 max Fe, 0.1 max Mg, ba bal Al.
Die cast: 43,000 TS; 22,000 YS; 3 El; 80 Brin.
For general die castings; corrosion resistant.

APEX 39P.
M-130; 4 Cu, 8 Si, bal Al.
SC-F temper: 26,000 TS; 15,000 YS; 2 El; 65 Brin.
PM-F temper: 32,000 TS; 20,000 YS; 2.5 El; 75 Brin.
PM-T6 temper: 40,000 TS; 27,000 YS; 3.5 El; 95 Brin.
For sand and permanent mold castings; for thin sections.

APEX 41.
M-130; 1.5 Cu, 5 Si, 0.5 Mg, bal Al.
SC-F temper: 25,000 TS; 12,000 YS; 2.5 El; 65 Brin.
SC-T 6 temper: 35,000 TS; 25,000 YS; 2.5 El; 80 Brin.
PM-T 6 temper: 43,000 TS; 27,000 YS; 4 El; 90 Brin.
For sand and permanent mold castings; heat treatable.

APEX 41-10.
M-130; 1.0-1.5 Cu, 4.5-5.5 Si, 0.45-0.6 Mg, 0.1 Zn, 0.1 Mn, bal Al.
Sand cast: 27,000 TS; 18,000 YS; 2.5 El; 65 Brin.
T6 temper: 42,000 TS; 30,000 YS; 4.0 El; 90 Brin.
Good pressure tightness, good resistance to hot cracking.
SAE 335; AMS 4215; ASTM B26-68 SC51B; ASTM B108-68 SC51B.

APEX 41A.
M-130; 1-1.5 Cu, 4.5-5.5 si, 0.4-0.6 Mg, 0.25 max Cr, bal Al.
T4-Temper: 25,000 TS; 14,000 YS; 3.5 El; 75 Brin.
T6-Temper: 35,000 TS; 25,000 YS; 2.5 El; 80 Brin.
For light alloy castings; age-hardenable.

APEX 43.
M-130; 5 Si, bal Al.
SC-F temper: 21,000 TS; 10,000 YS; 5 El; 45 Brin.
PM-F temper: 25,000 TS; 9000 YS; 5 El; 45 Brin.
Dc-F temper: 30,000 TS: 14,000 YS; 7 El; 60 Brin.
For castings; pressure tight, corrosion resistant.

APEX 43A.
M-130; 0.10 max Cu, 4.5-6 Si, 0.3 max Zn, bal Al.
Sand cast: 19,000 TS; 9000 YS; 7 El; 40 Brin.
Permanent mold: 24,000 TS; 9000 YS; 9 El; 45 Brin.
For castings; corrosion resistant.

APEX 43C.
M-130; 0.3 max Cu, 4.5-6 Si, 0.3 max Zn, bal Al.
Sand cast: 19,000 TS; 9000 YS; 7 El; 40 Brin.
Permanent mold: 24,000 TS; 9000 YS; 5 El; 65 Brin.
For castings, high fluidity.

APEX 44.
M-564; 70 Cu, 4 Sn, 26 Pb.
For self-lubricating bearing bronze; heavy duty.

APEX 45 ALLOY TERNALLOY.
M-130; 0.3 Cr, 1.5-2.5 Mg, 3.5-4.5 Zn, bal Al.
For light alloy parts.

APEX 46-10.
M-130; 0.1 Cu, 6.5-7.5 Si, 0.28-0.4 Mg, 0.1 Zn, 0.1 Mn, bal Al.
Sand cast: 24,000 TS; 14,000 YS; 4 El; 55 Brin.
T6 temper: 40,000 TS; 30,000 YS; 5 El; 85 Brin.
Good pressure tightness, good resistance to hot cracking.
SAE 336; ASTM B26-68 SG70B; ASTM B108-68 SG70B.

APEX 46 A.
M-130; 7 Si, 0.3 Mg, bal Al.
SC-F temper: 23,000 TS; 12,000 YS; 3.5 El; 55 Brin.
SC-T 6 temper: 33,000 TS; 24,000 YS; 4 El; 70 Brin. PM-T 6 temper: 40,000 TS; 27,000 YS; 5 El; 90 Brin.
For sand and permanent mold castings; pressure tight.

APEX 50.
M-130; 3.7-4.7 Cu, 2.5-3.5 Si, bal Al.
SC-F-temper: 24,000 TS; 14,000 YS; 2.5 El; 60 Brin.
SC-T6-temper: 36,000 TS; 26,000 YS; 1.5 El; 85 Brin.
PM-T6 temper: 40,000 TS; 26,000 YS; 4 El; 85 Brin.
For light castings: sand and permanent mold.

APEX 50-50 COPPER ALLOY.
M-130; 50 Al, 50 Cu.
For rich hardener; for alloying.

APEX 51.
M-130; 4.0 Cu, 6.9 Si, bal Al.
SC-F temper: 23,000 TS; 13,000 YS; 2 El; 55 Brin.
PM-F temper: 28,000 TS; 16,000 YS; 2 El; 70 Brin.
PM-T 6 temper: 45,000 TS; 28,000 YS; 4 El; 105 Brin.
For castings; sand and permanent mold.

APEX 52.
M-130; 4.4 Cu, bal Al.
SC-F temper: 22,000 TS; 10,000 YS; 3 El; 50 Brin.
SC-T6 temper: 36,000 TS; 24,000 YS; 5 El; 75 Brin.
SC-T6Z temper: 40,000 TS; 30,000 YS; 2 El; 95 Brin.
For sand castings; age-hardenable.

APEX 52 A.
M-130; 4-4.8 Cu, 1.5 max Si, 0.2 max Ti, bal Al.
T4-temper: 32,000 TS; 16,000 YS; 8.5 El; 60 Brin.
T6-temper: 36,000 TS; 24,000 YS; 5 El; 75 Brin.
For castings; age hardenable.

APEX 53.
M-130; 4.5 Cu, 2.5 Si, bal Al.
PM-F temper: 30,000 TS; 14,000 YS; 3.5 El; 70 Brin.
PM-T6 temper: 40,000 TS; 26,000 YS; 5 El; 90 Brin.
For permanent mold castings; age-hardenable.

APEX 56.
M-130; 4 Cu, 5 Si, bal Al.
DC-F temper: 38,000 TS; 22,000 YS; 3 El; 70 Brin.
For die castings; good machinability.

APEX 56A.
M-130; 3-4 Cu, 4.5-5.5 Si, bal Al.
For castings; age-hardenable.

APEX 59.
M-130; 4 Cu, 2 Ni, 1.5 Mg, bal Al.
F-temper: 28,000 TS; 24,000 YS; 1 El; 80 Brin.
T6-temper: 54,000 TS; 51,000 YS; 0.8 El; 120 Brin.
For high temperature castings; heat treatable.

APEX 59A.
M-130; 3.5-4.5 Cu, 0.6 max Si, 1.3-1.8 Mg, 1.7-2.3 Ni, 0.2 max Ti, ba Al.
T4-Temper: 37,000 TS; 30,000 YS; 1 El; 95 Brin.
T6-temper: 38,000 TS; 100 Brin.
For castings; age-hardenable.

APEX 60.
M-130; 3.0-4.0 Cu, 5.5-6.5 Si, 0.15 Mg, 1.0 Zn, 0.5 Mn, 0.3 Ni, bal Al.
Sand cast; T6: 32,000-38,000 TS; 20,000-28,000 YS; 2.5-3.0 El; 90 Brin.
Per. mold cast; T61: 40,000-50,000 TS; 24,000-42,000 YS; 2 El; 100 Brin.
ASTM B 108-68 SC64D; ASTM B 26-68 SC64D.

APEX 60-40 COPPER ALLOY.
M-130; 60 Al, 40 Cu.
For rich hardeners for Zn base die cast alloys.

APEX 62.
M-130; 11 Zn, 2.7 Cu, 0.3 Mg, bal Al.
Sand cast: 29,000 TS; 17,000 YS; 2.5 El; 70 Brin.
Permanent mold: 35,000 TS; 30,000 YS; 1.0 El; 105 Brin.
For castings, light alloy parts.

APEX 66.
M-130; 5.5-5.7 Cu, 5.0-6.0 Si, 0.25-0.6 Mg, 0.8 Zn, 0.8 Mn, bal Al.
Per. mold cast; T5: 32,000-35,000 TS; 25,000-30,000 YS; 95 Brin.
Good fluidity, resistance to hot cracking.
SAE 300; ASTM B108-68 CS66A.

APEX 66.
M-564; 86 Cu, 10 Al, 4 Fe.
85,000 TS; 44,000 YS; 18 El.
For gears, bearings, non-sparking tools; strong, tough.

APEX 70.
M-130; 3.0-4.0 Cu, 5.5-6.5 Si, 0.1 Mg, 1.0 Zn, 0.5 Mn, 0.35 Ni, bal Al.
Sand cast; T6: 36,000 TS; 24,000 YS; 2.0 El; 80 Brin; 22,000 SS.
Per. mold cast; T6: 40,000 TS; 27,000 YS; 3.0 El; 99 Brin.
Good, strong, general purpose aluminum casting.
SAE 326, 329; ASTM B108-68 SC64D.
ASTM B26-68 SC64D.

APEX 75.
M-130; 7 Cu, 5.5 Si, 0.5 Mg, bal Al.
PM-T 6 temper: 35,000 TS; 27,000 YS; 90 Brin.
PM-T 51 temper: 40,000 TS; 31,000 YS; 1.0 El; 100 Brin.
For pistons; permanent mold.

APEX 75-25 SILICON HARDENER.
M-130; 75 Al, 25 Si.
For rich hardener for alloying.

APEX 77.
M-130; 7 Cu, 2.5 Si, bal Al.
SC-F temper: 23,000 TS; 15,000 YS; 1.5 El; 70 Brin.
For patterns; sand cast.

APEX 78.
M-130; 4 Cu, 8 Si, bal Al.
For match plates; plaster mold casting.

APEX 80/20 COPPER HARDENER.
M-130; 20 Cu, bal Al.
For alloying hardener.

APEX 97 DC.
M-130; 1 Mg, 97 min Al.
For alloying to zinc; shot.

APEX 95-5 ALUMINUM SILICON ALLOY.
M-130; 4.5-6.0 Si, 1.0 Fe-Cu, 0.2 max Zn, bal Al.
17,000-21,000 TS; 9,000 YS; 4-2 El; 40 Brin.
For thin, intricate dense castings; easy to cast.

APEX 97 DCZ.
M-130.
Zinc hardener.
ASTM ZG71A.

APEX 110.
M-130; 10 Cu, bal Al.
SC-F temper: 23,000 TS; 16,000 YS; 1 El; 85 Brin.
SC-T 61 temper: 36,000 TS; 32,000 YS; 0.5 El; 110 Brin.
PM-F temper: 27,000 TS; 19,000 YS; 1.5 El; 90 Brin.
For bearings; sand and permanent mold.

APEX 125.
M-130; 9.7-10.5 Mg, 0.10 max Si, 0.25 max Ti, bal Al.
T4 temper: 57,000 TS; 29,000 YS; 30 El; 80 Brin.
For sand castings, instrument cases, housing, oil pans.
Age-hardenable, corrosion resistant.

APEX 400.
M-130; 0.1 max Cu, 3.6-4.2 Mg, bal Al.
24,000-26,000 TS; 10,000 YS; 8-6 El; 50-60 Brin.
For light alloy parts, cooking utensils, marine castings; similar to Permbrite."

APEX 405.
M-130; 5 Zn, 0.4 Mg. 0.2 Mn, bal Al.
For die castings.

APEX 417.
M-130; 0.1 max Cu, 0.15 max Si, 6.0-7.5 Mg, 0.2 Ti, bal Al.
Cast: 40,000 TS; 21,000 YS; 13 El; 70 Brin.
For agricultural and aircraft castings, marine parts; corrosion and impact resistant.

APEX 418.
M-130; 8 Mg, bal Al.
DC-F temper: 43,000 TS; 27,000 YS; 8 El; 80 Brin.
For die castings; corrosion resistant.

APEX 420.
M-130; 10 Mg, bal Al.
SC-T4 temper: 46,000 TS; 25,000 YS; 14 El; 75 Brin.
For sand castings; high strength, corrosion resistant.

APEX 440.
M-130; 0.20 max Cu, 0.50-0.65 Mg, 5.2-6.0 Zn, 0.4-0.6 Cr, 0.15-0.25 Ti, bal Al.
Cast: 28,000 TS; 16,000 YS; 12 El; 55 Brin.
Aged: 36,000 TS; 25,000 YS; 6 El; 75 Brin.
For sand castings, housings, oil pans, instrument cases, gear cases. Self aging.

APEX 461.
M-130; 5.5 Zn 1.4 Cu, 0.2 Mg, bal Al.
For general purpose castings; room temperature aging.

APEX 470.
M-130; 0.7 Cu 0.4 Mg, 7.5 Zn, bal Al.
For general purpose castings; room temperature aging.

APEX 630.
M-130; 6 Al, 3 Zn, bal Mg.
SC-F temper: 29,000 TS; 14,000 YS; 6 El; 50 Brin.
SC-T4 temper: 40,000 TS; 14,000 YS; 12 El; 55 Brin.
SC-T6 temper: 40,000 TS; 19,000 YS; 5 El; 70 Brin.
For sand castings; heat treatable.

APEX 700.
M-130; 2.5 Cu, 1.2 Ni, 7.5 Si, bal Al.
PM-T5 temper: 20,000 TS; 85,000 YS; 10 El; 45 Brin.
For bearings; permanent mold.

APEX 700A.
M-130; 6.2 Sn, 1.0 Cu, 1.0 Ni, bal Al.
T5-temper: 20,000 TS; 8500 YS; 10 El; 55 Brin.
For bearings; age-hardened.

APEX 750.
M-130; 0.7-1.3 Cu, 0.7 max Si, 0.10 max Mn, 0.7-1.3 Ni, 5.5-7.0 Sn, bal Al.
Sand or permanent mold cast.
T5 cond: 23,000 TS; 11,000 YS; 12 El; 45 Brin; 15,000 SS.
AA No. 850.0 T5.

APEX 910.
M-130; 9 Al, 0.6 Zn, bal Mg.
DC-F temper: 33,000 TS; 22,000 YS; 3 El; 60 Brin.
For die castings.

APEX 920.
M-130; 9 Al; 2 Zn, bal Mg.
SC-F temper; 24,000 TS; 14,000 YS; 2 El; 65 Brin.
SC-T4 temper: 40,000 TS; 16,000 YS; 10 El; 65 Brin.
SC-T6 temper: 40,000 TS; 23,000 YS; 2 El; 85 Brin.
For castings; heat treatable.

APEX A 750.
M-130; 0.7-1.3 Cu, 2.0-3.0 Si, 0.10 max Mn, 0.3-0.7 Ni, 5.5-7.0 Sn, bal Al.
Sand or permanent mold cast.
T5 Cond: 20,000 TS; 11,000 YS; 5.0 El; 45 Brin; 14,000 SS.
AA No. 850.OT5.

APEX ALLOY NO. 2.
M-130; 3.9-4.3 Al, 2.5-2.9 Cu, 0.02-0.05 Mg, 0.075 max Fe, 0.004 max Pb, 0.003 max Cd, bal Zn.
Die cast: 52,000 TS; 93,000 Com.S; 7 El; 100 Brin; 46,000 SS.
General purpose die casting, motor frames, household hardware, novelties.

APEX ALLOY NO. 3.
M-130; 3.9-4.3 Al, 0.10 Cu, 0.025-0.05 Mg, 0.075 max Fe, 0.004 max Pb, 0.003 max Sn, bal Zn.
Die cast: 41,000 TS; 60,000 Com S; 10 El; 76 Brin; 31,000 SS.
For housings, cases, general die castings, motor frames, household hardware.
ASTM: B240-64, AG40A.

APEX ALLOY NO 5.
M-130; 3.9-4.3 Al, 0.75-1.25 Cu, 0.03-0.06 Mg, 0.075 max Fe, 0.004 max Pb, 0.003 max Cd, bal Zn.
Die cast: 47,000 TS; 87,000 Com S; 7 El; 88 Brin; 38,000 SS.
For housings, cases, general die castings, motor frames, radio parts, hardware.
ASTM: B240-64, AC41.

APEX ALLOY NO 7.
M-130; 3.9-4.3 Al, 0.10 max Cu, 0.10-0.20 Mg, 0.075 max Fe, 0.004 max Pb, 0.003 max Cd, 0.005-0.020 Ni, bal Zn.
Die cast: 41,000 TS; 87,000 Com.S; 14 El; 76 Brin; 31,000 SS.
General die casting, automotive.
SAE 903.

APEX AS-1.
M-130; 95 min Al.
For deoxidizing steel.

APEX AS-2.
M-130; 92 min Al.
For deoxidizing steel.

APEX AS-3.
M-130; 90 min Al.
For deoxidizing steel.

APEX AS-4.
M-130; 85 min Al.
For deoxidizing steel.

APEX B750.
M-130; 1.7-2.3 Cu, 0.4 max Si, 0.6-0.9 Mg, 0.10 max Mn, 0.9-1.5 Ni, 5.5-7.0 Sn, bal Al.
Sand or permanent mold cast.
T5 Cond: 32,000 TS; 23,000 YS; 5.0 El; 70 Brin; 21,000 SS. 11,000 Limit.
AA No. B850.OT5.

APEX BRONZE.
M-545; 86 Cu, 9.5 Al, bal Fe.
For marine parts, hardware; Sillman Bronze.

APEX D 400.
M-130; 4 Mg, 1.8 Si, bal Al.
DC-F temper: 41,000 TS; 20,000 YS; 10 El.
For die casting; corrosion resistant.

APEX DRILL RODS.
M-688; 1.0 C Mn, bal Fe.
Type AISI W1 water hardening tool steel.

APEX NO. 6.
M-130; 4.1 Al, 1 Cu, bal Zn.
DC-F temper: 43,500 TS; 7.3 El; 90 Brin.
For die castings.

APEX NO 7 ZINC BASE ALLOY.
M-130; 4.1 Al, 2.7 Cu, bal Zn.
Cast: 47,000 TS; 8 El; 83 Brin.
For die castings housings; Iz-15.

APEX NO. 405 ZINC BASE ALLOY.
M-130; 4.1 Al; 0.03 Mg, bal Zn.
Cast: 36,000 YS; 4 El; 62 Brin.
For die castings, housing; Iz-18.

APEX NO. 810.
M-130; 7.5 Al, 0.7 Zn, bal Mg.
For light alloy castings; heat treatable.

APEX NO. 4100 ZINC BASE ALLOY.
M-130; 4.1 Al, 1.25 Cu, bal Zn.
Cast: 39,000 TS; 10 El; 71 Brin.
For die castings, housing; Iz-18.

APEX NO. 4102 ZINC BASE ALLOY.
M-130; 4.1 Al, 1.0 Cu, bal Zn.
Cast: 41,000 TS; 4 El; 73 Brin.
For die castings, housing; Iz-17.

APEX P 400.
M-130; 4 Mg, 1.8 Zn, bal Al.
PM-F temper: 27,000 TS; 16,000 YS; 7 El; 60 Brin.
For castings; permanent mold.

APEX PATTERN ALUMINUM ALLOY.
M-130; 4-6 Cu, 2-3.5 Si, 1.5 Zn + Fe, bal Al.
16,000-22,000 TS; 16,000 YS; 3-1 El; 68 Brin.
For matchplate and patterns; age-hardenable.

APEX PISTON ALLOY.
M-130; 9.25-10.75 Cu, 0.2 max Zn, 0.15-0.35 Mg, 1.0-1.25 Fe, bal Al.
Chill cast: 34,000 TS; 1 El; 95-125 Brin.
For pistons; heat treated.

APEX PRESSURE ALUMINUM ALLOY.
M-130; 3-5 Cu, 2-4 Si, 1.2 Fe + Zn, bal Al.
17,000-21,000 TS; 17,000 YS; 2 El; 60 Brin.
For castings requiring extreme density; age-hardenable.

APEX "T" SERIES ZINC BASE ALLOY.
M-130; Al, Cu, bal Zn.
For castings; die casting.

APEX Z28.
M-130; 1.7-2.3 Cu, 9.5-10.5 Si, 0.10 Mg 2.0-3.0 Zn, 0.50 Mn, 0.50 Ni bal Al.
Aluminum die casting.
As cast; 44,000 TS; 23,000 YS; 4.0 El; 75 Brin. 26,000 SS.
For housings, motor frames, radio parts, hardware.

APEX Z-33.
M-130; 4 Si, 3 Cu, 3 Zn, bal Al.
F-temper: 24,000 TS; 13,000 YS; 3.5 El; 55 Brin.
T6-temper: 30,000 TS: 16,000 YS; 3.0 El; 65 Brin.
For sand and permanent mold castings: good castability.

APEX Z-39.
M-130; 3-4 Cu, 8.5-9.5 Si, 2-3 Zn, 1 max Fe, bal Al.
Die cast: 46,000 TS; 25,000 YS; 3 El; 80 Brin.
For general die castings; SS 29,000.

APEX Z-39-11.
M-130; 3-4 Cu, 10.5-11.5 Si, 2-3 Zn, 1 max Fe, bal Al. Die cast: 46,000 TS; 27,000 YS; 2 El; 85 Brin.
For general die castings; SS 28,000.

APEX Z-50.
M-130; 3.5-4.5 Cu, 2.5-3.5 Si, 2.5 Zn, bal Al.
F-temper: 24,000 TS; 14,000 YS; 2.5 El; 60 Brin.
T4-temper: 30,000 TS; 18,000 YS; 4 El; 65 Brin.
T6-temper: 36,000 TS; 26,000 YS; 1.5 El; 85 Brin.
For sand and permanent mold castings; heat treatable.

APEX Z-51.
M-130; 4-5 Cu, 5-6 Si, 1.5-2.5 Zn, 1 max Fe, 0.6 max Mn, bal Al.
Cast: 23,000 TS; 13,000 YS; 2 El; 55 Brin.
Aged: 24,000 TS; 16,000 YS; 1 El; 75 Brin.
For gear cases, housings, machine tool castings; age hardenable.

APFI.
M-1653; 0.20 max C, 1.5 Si, 2.0 max Mn, 24.0-26.0 Cr, 19.0-22.0 Ni, bal Fe.
Austenitic stainless steel.
W.-Nr 1.4841; AISI 310.

APFI-S.
M-1653; 0.08 max C, 1.5 max Si, 2.0 max Mn, 24.0-26.0 Cr, 19.0-22.0 Ni, bal Fe.
Austenitic stainless steel.
W.-Nr. 1.4335; x 8 CrNi 25 20; AISI 310S.

APFR.
M-1653; 0.20 max C, 1.0 max Si, 1.0 max Mn, 22.0-24.0 Cr, 12.0-15.0 Ni, bal Fe.
Austenitic stainless steel.
W.-Nr. 1.4828; x 16 CrNi 23 14; AISI 309.

APFR-S.
M-1653; 0.08 max C, 1.0 max Si, 2.0 max Mn, 22.0-22.0 Cr, 12.0-15.0 Ni.
Austenitic stainless.
X6 CrNi 23 14.

APHIT.
70 Cu, 20 Ni, 5.5 Zn, 4.5 Cd.
For condenser tubes, chemical equipment; corrosion resistant.

APHTIT.
M-Eng; 75-70 Cu, 20-21 Ni, 2.4-5.5 Zn, 1.8-4.5 Cd.
For corrosion resistant parts; (Aphtit); corrosion resistant.

APIS.
M-501; 1.65 C, 12 Cr, Co, bal Fe.
For blanking and forming dies, punches, cutters; air hardened, non-deforming.

APM.
M-1653; 0.08 max C, 1.0 max Si, 2.0 max Mn, 16.0-18.5 Cr, 10.5-13.5 Ni, 2.0-2.5 Mo, bal Fe.
Austenitic stainless steel.
W.-Nr. 1.4401; X 5 CrNiMo 17 12; AISI 316.

APMC.
M-1653; 0.08 max C, 1.0 max Si, 2.0 Max Mn, 16.0-18.5 Cr, 10.5-13.5 Ni, 2.0-2.5 Mo, Nb= 10 X C min, bal Fe.
Stablized austenitic stainless steel.
W.-Nr. 1.4580; X 8 CrNiMoNb 17 12; modified AISI 316.

APM-L.
M-1653; 0.03 C, 1.0 max Si, 2.0 max Mn, 16.0-18.5 Cr, 11.0-14.0 Ni, 2.0-3.0 Mo, bal Fe.
Austenitic stainless steel.
W.-Nr. 1.4404; X 2 CrNiMo 17 12; AISI 316L.

APML/SPEC.
M-1653; 0.03 max C, 1.0 max Si, 2.0 max Mn, 16.0-18.5 Cr, 11.5-14.5 Ni, 2.5-3.0 Mo, bal Fe.
Austenitic stainless steel.
W.-Nr. 1.4435: AISI 316 L.

APM/SPEC.
M-1653; 0.04-0.10 C, 0.75 max Si, 2.0 max Mn, 16.0-18.0 Cr, 11.0-14.0 Ni, 2.5-3.0 Mo, bal Fe.
Austenitic stainless steel.
W.-Nr. 1.4436; Similar to AISI 316.

APMT.
M-1653; 0.08 max C, 1.0 max Si, 2.0 max Mn, 16.0-18.5 Cr, 10.5-13.5 Ni, 2.0-2.5 Mo, Ti = 5 X C min, bal Fe.
Stabilized austenitic stainless steel.
W.-Nr. 1.4571; modified AISI 316.

APOLLO.
M-114; 0.45 C, 2.0 Cr, 0.20 V, bal Fe.
For die casting dies; flying shears; oil hardening.

APOLLO.
M-1705; 2.5 C, 12.0 Cr, 4.0 V, 0.8 Mo, bal Fe.
High carbon, high chromium tool and die steel; air or oil hardening. AISI D7.

APOLLO CHROMSTEEL.
M-687; Ni-Cr plated steel.
For heat resistant parts; heat resistant to 800°F.

APOLLO CROM.
M-Eng; Cr plated sheet Zn.
For corrosion resistant parts; stainless and corrosion resistant.

APOLLOY.
M-131; 0.08 C, 0.25 Cu, bal Fe.
For building construction; corrosion resisting to weather.

APS 10, ETC., see **POMPEY APS 10, ETC.**

APS-10 M 4.
M-884; 0.12 C, 2 Cr, 0.3 Mo, 0.3 Al, bal Fe.
Tube: 100,000 min TS; 80,000 min YS.
For gas well piping.
Resists H_2S corrosion and stress cracking.

APW NO. 129.
M-686; 60 Pd, 40 Ni.
For high temperature brazing; M.P 2260°F; good oxidation resistance.

APW NO. 133.
M-686; 80 Pd, 14.5 Ag, 5.5 Al.
For high temperature brazing; M.P. 1925-2050°F; good oxidation resistance.

APW NO. 200.
M-686; 100 Au.
For brazing filler metal; M.P. 1945°F.

APW NO. 238.
M-686; 80 Au, 20 Cu.
For brazing filler metal; M.P. 1620-1630°F.

APW NO. 241.
M-686; 50 Au, 50 Cu.
For brazing filler metal; M.P. 1697-1733°F.

APW NO. 242.
M-686; 37.5 Au, 62.5 Cu.
For brazing filler metal; M.P. 1755-1815°F.

APW NO. 243.
M-686; 35 Au, 62 Cu, 3 Ni.
For brazing filler metal; M.P. 1815-1875°F.

APW NO. 253.
M-686; 20 Au, 80 Cu.
For brazing filler metal; M.P. 1850-1880°F.

APW NO. 255.
M-686; 82 Au, 18 Ni.
For brazing filler metal; M.P. 1742°F.

APW NO. 259.
M-686; 94 Au, 6 Cu.
For brazing filler metal; M.P. 1742-1796°F.

APW NO. 260.
M-686; 35 Au, 65 Cu.
For brazing filler metal; M.P. 1778-1850°F.

APW NO. 261.
M-686; 75 Au, 20 Cu, 5 Ag.
For brazing filler metal; M.P. 1640-1650°F.

APW NO. 265.
M-686; 72 Au, 22 Ni, 6 Cr.
For brazing filler metal; M.P. 1785-1835°F.

APW-G-355.
M-686; 56 Ag, 22 Cu, 17 Zn, 5 Sn.
For silver solder, brazing; M.P. 1145-1205°F.

APW 431.
M-686; 90 Ag, 10 Pd.
For high temperature brazing for honeycomb assemblies; M.P. 1850-1975°F.
High oxidation resistance.

APW NO.440.
M-686; 75 Ag, 20 Pd, 5 Mn.
For high temperature brazing; M.P. 1960-2050°F; high strength.

APW NO. 441.
M-686; 64 Ag, 33 Pd, 3 Mn.
For high temperature brazing; M.P. 2100-2250°F; high strength.

AQUILA.
M-1322; 0.15 C, Cr, Mo, bal Fe.
For cams, gears, camshafts; case hardening steel, tough.

AR.
M-604; 0.40 C 1.75 Mn, 0.25 Si.
As rolled: 225 Brin.
Mild forming.
Resistant to sliding abrasion.
For chutes, conveyer troughs.

AR-213.
M-US; 0.12 C, 20.3 Cr, 4.5 W, 4.8 Al, 2.3 Ta, 0.6 Fe, 0.34 Zr, bal Co.
Age-hardened: C 47 Rock.
For gas turbine components.
Age-hardenable, corrosion and heat resistant.
Sulphidation and oxidation resistant.

AR 235.
M-24; 0.35-0.50 C, 1.4-2.0 Mn, bal Fe.
Rolled: 110,000 TS; 70,000 YS; 16 El; 220 Brin.
Ht. Tr.: 155,000 TS; 130,000 YS; 10 El; 330 Brin.
For shovels, crushers, scraper blades, hoppers.
Wear resistant. Was Bethlehem Abrasion Resisting.

AR-235.
M-24; 0.35-0.50 C, 1.40-2.00 Mn, 0.15-0.30 Si, bal Fe.
As rolled: 235 Brin. min.
Abrasion resisting steel plate, 3/16 to 15 inch thick.

AR-300.
M-604; 0.30 C, 1.50 Mn, 0.30 Si, 0.001 B, bal Fe.
W.Q + Temp: 315 Brin.
Mild forming and welding.
Resistant to sliding and light impact abrasion.
For chutes, hoppers.

AR-350.
M-604; 0.30 C, 1.50 Mn, 0.30 Si, 0.001 B, bal Fe.
W.Q. + T.: 360 Brin.
Mild forming and welding.
Resistant to sliding and light impact abrasion.
For frames, conveyer troughs.

AR-360.
M-604; 0.30 C, 1.40 Mn, 0.25 Si, 0.55 Cr, 0.11 Mo, 0.001 B, bal Fe.
W.Q. + T.: 400 Brin.
Mild forming and welding.
Resistant to sliding and moderate impact abrasion.
For deck plates, hoppers, dump-bodies.

AR-360.
M-240; 0.31 C, 1.4 Mn, 0.24 Si, 0.50 Cr, 0.11 Mo, 0.004 B, bal Fe.
Heat treated: 360-400 Brin.
Wear resistant plate.

A.R. ALLOY.
M-2a; 10 Sn, 2 Pb, bal Cu.
10-5 El; 50-60 Brin.
For bearings for cold rolling mills, landing gears for airplanes; tough, strong.

ARBOGA.
M-Eng; Tungsten carbide.
For tools, hard cutting tools; sintered alloy.

ARC 1628.
M-1488; 0.03 C, 65 Ni, 26 Mo. 0.4 V, 6 Fe.
Corrosion resistant alloy.
For valves, pumps for hot acids and other chemicals.
AFNOR ND 27 Fe V; W.Nr. 2.4600; similar to HASTELLOY B.

ARC 6015.
M-1488; 0.04 C, 59 Ni, 15.5 Cr, 17 Mo, 4.5 W, 0.35 V, 5 Fe.
Corrosion resistant alloy; for use with sulphuric and phosphoric acids, chlorine.
AFNOR NC 17 DWY; similar to HASTELLOY C.

ARC M1628 see ARC 1628.

ARC WEL.
M-683; Zn, bal Cu.
For welding rod.

ARC ZCR.
M-104; 0.10 max C, 24 Cr, 5 Ni, Mo, bal Fe.
For furnace parts and equipment; heat resistant.

ARCALOY 8N12.
M-1713.
Weld metal: 0.03 C, 14.0 Cr, 5.5 Mn, 9.0 Fe, 0.75 Si, 1.9 Cb, 0.60 Ti, bal Ni.
As welded: 96,000 psi TS; 62,000 psi YS; 44 El.
Covered electrode for welding many dissimilar alloys: Monel, Inconel, carbon steels, cupronickels.
AWS Class ENiCrFe-3.

ARCALOY 9N10.
M-1713.
Weld metal: 0.03 C, 29.0 Cu, 0.9 Si, 0.74 Fe, 0.80 Ti, 64.3 Ni, 3.7 Mn, 0.25 Al.
As Welded: 79,500 psi TS; 51,000 psi YS; 33 El.
Covered electrode for welding Monel and similar alloys.
AWS Class ENiCu-2.

ARCALOY 308.
M-1713.
Weld metal: 0.07 max C, 19.0 Cr, 9.5 Ni, 1.6 Mn, 0.50 Si, bal Fe.
As welded: 85,000-95,000 psi TS; 40-50 El.
Covered electrode for welding 18-8 type austenitic stainless steels.
AWS E308-15,16; ASME F-5, A-8.

ARCALOY 308ELC.
M-1713.
Weld metal: 0.04 max C, 19.0 Cr, 9.5 Ni, 1.0 Mn, 0.30 Si, bal Fe.
As welded: 80,000-90,000 psi TS; 40-50 El.
For welding low carbon austenitic stainless steels.
AWS Class E308-15,16; ASME F-5, A-8.

ARCALOY 309.
M-1713.
Weld metal: 0.10 max C, 23.0 Cr, 13.0 Ni, 1.6 Mn, 0.50 Si, bal Fe.
As welded: 85,000-95,000 psi TS; 35-45 El.
Covered electrode for welding type 309 stainless steel.
AWS E309-15,16; ASME F5, A-8.

ARCALOY 309 CB.
M-1713.
Weld metal: 0.10 max C, 23.0 Cr, 13.0 Ni, 0.80 Cb, 1.6 Mn, 0.60 Si, bal Fe.
As welded: 85,000-95,000 psi TS; 30-40 El.
Covered electrode to inhibit carbide precipitation when welding type 309 and other stainless steels.
AWS E309Cb-15,16; ASM F-5, A-8.

ARCALOY 309 MO.
M-1713.
Weld metal: 0.10 max C, 23.0 Cr, 13.0 Ni, 2.2 Mo, 1.7 Mn, 0.50 Si, bal Fe.
As welded: 85,000-95,000 psi TS; 35-45 El.
Covered electrode, often used for welding type 316 clad steels.
AWS E309Mo-15; ASME F-5, A-8.

ARCALOY 310.
M-1713.
Weld metal: 0.20 max C, 26.0 Cr, 21.0 Ni, 1.8 Mn, 0.40 Si, bal Fe.
As welded: 85,000-95,000 psi TS; 35-45 El.
Covered electrode for welding type 310 stainless steel.
AWS E310-15,16; ASME F-5, A-9.

ARCALOY 310CB.
M-1713.
Weld metal: 0.12 max C, 26.0 Cr, 21.0 Ni, 0.80 Cb, 1.8 Mn, 0.40 Si, bal Fe.
As welded: 85,000-95,000 psi TS; 30-40 El.
Covered electrode for welding types 310,321 and 347 stainless steels.
AWS E310Cb-15,16; ASM F-5, A-9.

ARCALOY 310MO.
M-1713.
Weld metal: 0.12 max C, 26.0 Cr, 21.0 Ni, 2.0 Mo, 1.8 Mn, 0.40 Si bal Fe.
As welded: 85,000-95,000 psi TS; 35-45 El.
Covered electrode for welding type 316 clad steels; also for lining digesters in the paper industry.
AWS E310Mo-15,16: ASME F-5, A-9.

ARCALOY 312.
M-1713.
Weld metal: 0.15 max C, 29.0 Cr, 9.5 Ni, 1.9 Mn, 0.50 Si, bal Fe.
As welded: 110,000-120,000 psi TS; 22-25 El.
Covered electrode for welds requiring high strength and stainless properties.
AWS E312-15,16; ASME F-5.

ARCALOY 316.
M-1713.
Weld Metal: 0.07 max C, 18.0 Cr, 13.0 Ni, 2.25 Mo, 1.7 Mn, 0.40 Si, bal Fe
As welded: 85,000-95,000 psi TS; 35-45 El.
Covered electrode for welding type 316 steel.
AWS E316-15,16; ASME F-5,A-8.

ARCALOY 316CB.
M-1713.
Weld metal: 0.07 max C, 18.0 Cr, 12.0 Ni, 2.25 Mo, 0.80 Cb, 1.6 Mn, 0.60 Si, bal Fe.
As welded: 85,000-95,000 psi TS; 30-40 El.
Covered electrode for welding types 316 and 316 ELC stainless.

ARCALOY 316ELC.
M-1713.
Weld metal: 0.04 max C, 18.0 Cr, 13.0 Ni, 2.25 Mo, 1.0 Mn, 0.30 Si, bal Fe.
As welded: 80,000-90,000 psi TS; 35-45 El.
Covered electrode for welding types 316 and 316 ELC stainless steels.
AWS E316L-15,16; ASME F-5,A-8.

ARCALOY 317.
M-1713.
Weld metal: 0.07 max C, 19.0 Cr, 13.0 Ni, 3.5 Mo, 1.7 Mn, 0.50 Si, bal Fe.
As welded: 85,000-95,000 psi TS; 35-45 El.
Covered electrode for welding 317 stainless.
AWS E317-15,16; ASME F-5, A-8.

ARCALOY 317ELC.
M-1713.
Weld metal: 0.04 max C, 19.0 Cr, 13.0 Ni, 3.5 Mo, 1.7 Mn, 0.50 Si, bal Fe.
As welded: 80,000-90,000 psi TS; 35-45 El.
Covered electrode for welding type 317 stainless.
ASME F-5, A-8.

ARCALOY 318.
M-1713.
Weld metal: 0.07 max C, 18.0 Cr, 12.0 Ni, 2.25 Mo, 0.80 Cb, 1.6 Mn, 0.60 Si, bal Fe.
As welded: 85,000-95,000 psi TS; 30-40 El.
Covered electrode for welding types 316, 316 ELC stainless steels.
AWS E318-15,16; ASME F-5, A-8.

ARCALOY 320.
M-1713.
Weld metal: 0.07 max C, 20.0 Cr, 29.0 Ni, 3.0 Cu, 2.0 Mo, 0.5 Cb, 1.5 Mn, 0.40 Si, bal Fe.
As welded: 75,000-85,000 psi TS; 35-40 El.
Covered electrode for welding Carpenter 320 and Durimet 20.
AWS E320-15.

ARCALOY 330.
M-1713.
Weld metal: 0.25 max C, 15.0 Cr, 35.0 Ni, 1.6 Mn, 0.30 Si, bal Fe.
As welded: 75,000-85,000 psi TS; 25-35 El.
Covered electrode for welding type 330 stainless steel and castings of similar alloy.
AWS E330-15,16; ASME F-5.

ARCALOY 330HC.
M-1713.
Weld metal: 0.80 C, 15.0 Cr, 35.0 Ni, 2.0 Mn, 0.80 Si, bal Fe.
As welded: 75,000-85,000 psi TS; 25-35 El.
Covered electrode for welding high carbon grades of castings.
Similar to ACI HT.

ARCALOY 347.
M-1713.
Weld metal: 0.07 max C, 19.0 Cr, 9.5 Ni, 0.80 Cb, 1.6 Mn, 0.60 Si, bal Fe.
As welded: 85,000-95,000 psi TS; 35-45 El.
Covered electrode for welding types 347 and 321 stainless steels.
AWS E347-15,16; ASME F-5,A-8.

ARCALOY 410.
M-1713.
Weld metal: 0.10 max C, 12.5 Cr, 0.60 Mn, 0.40 Si, bal Fe.
Welded annealed: 80,000-90,000 psi TS; 30-35 El.
Covered electrode for welding type 410 stainless.
AWS E410-15,16; ASME F4, A-6.

ARCALOY 430.
M-1713.
Weld metal: 0.10 max C, 16.0 Cr, 0.60 Si, 0.60 Mn, bal Fe.
Welded annealed: 75,000-80,000 psi TS; 30-35 El.
Covered electrode for welding type 430 stainless.
AWS E430-15,16; ASME F-5, A-7.

ARCALOY 502.
M-1713.
Weld metal: 0.05 max C, 5.10 Cr, 0.56 Mo, 0.55 Mn, 0.40 Si, bal Fe.
Welded annealed: 79,000 psi TS; 22-35 El.
Covered electrode for welding type 502 steel.
AWS E502-15; ASME F-4, A-4.

ARCALOY ER 308.
M-1713.
Weld Metal: 0.07 max C, 20.5 Cr, 9.75 Ni, 1.75 Mn, 0.40 Si bal Fe.
Bare wire or rod for welding 18-8 stainless steel.

ARCALOY ER 308ELC.
M-1713.
Weld metal: 0.03 max C, 20.5 Cr, 9.75 Ni, 1.75 Mn, 0.40 Si.
Bare wire or rod for welding low carbon 18-8 stainless steels.

ARCALOY ER 309.
M-1713.
Weld metal: 0.12 C, 24.25 Cr, 13.50 Ni, 1.75 Mn, 0.40 Si, bal Fe.
Bare wire or rod for welding type 309 stainless steel.

ARCALOY ER 310.
M-1713.
Weld metal: 0.08-0.15 C, 26.5 Cr, 21.5 Ni, 1.75 Mn, 0.40 Si, bal Fe.
Bare wire or rod for welding type 310 and many dissimilar steels.

ARCALOY ER 312.
M-1713.
Weld metal: 0.08-0.15 C, 30.0 Cr, 8.75 Ni, 1.75 Mn, 0.40 Si, bal Fe.
Bare wire or rod for welding stainless steels to mild steels and for welding high strength steels.

ARCALOY ER 316ELC.
M-1713.
Weld metal: 0.03 max C, 19.0 Cr, 13.25 Ni, 2.2 Mo, 1.75 Mn, 0.40 Si, bal Fe.
Bare wire or rod for welding type 316elc.

ARCALOY ER 347.
M-1713.
Weld metal: 0.07 max C, 20.0 Cr, 9.75 Mi, 0.80 Cb+Ta, 1.75 Mn, 0.40 Si, bal Fe.
Bare wire or rod for welding types 347 and 321 stainless steels.

ARCAST.
M-685; 3 C, 2 Si, bal Fe.
For welding electrode; for cast iron.

ARCHANGEL.
M-485; 485; C, alloy, bal Fe.
For high speed tools; high speed steel.

ARCHITECTURAL BRONZE-385.
M-8; 56.0 Cu, 41.5 Zn, 2.5 Pb.
Soft rod: 60,000 TS; 20,000 YS; 25 El; B 65 rock.
For ornamental hardware, architectural applications, hinges, lock bodies, hardware.
Free-cutting, corrosion resistant.

ARCHITECTURAL BRONZE-385.
M-33; 58.5 Cu, 38.25 Zn, 3.25 Pb.
As extruded: 60,000 TS; 20,000 YS; 30 El.
For architectural parts and fittings.

ARCM 30 MO.
M-1488; 0.90-1.20 C, 1.5 max Mn, 1.5 Max Si, 28.0-30.0 Cr, 1.5-2.5 Mo.
Full ann: 270-340 Brin.
Stainless casting; extreme resistance to both corrosion and abrasion.
AFNOR Z100 CD29.2M.

ARCM 098.
M-1488; 0.10 max C, 1.5 max Mn, 1.5 max Si, 18.0-21.0 Ni, 24.0-27.0 Cr, 1.5-2.5 Mo, bal Fe.
Cast, ann: 135-165 Brin.
Austenitic stainless; corrosion resistant to nitric acid and mixed supho- nitrics.
AFNOR Z 8 CND25.19.2M.

ARCM 990.
M-1488; 0.50-0.70 C, 1.0 max Mn, 1.0 max Si, 4.0-6.0 Ni, 21.5-23.5 Cr, 2.0-3.0 Cu, bal Fe.
Quenched, tempered: 240-280 Brin.
Stainless; for both abrasion and corrosion resistance.
AFNOR Z60 CNW 22.5 M.

ARCM 1628.
M-1488; 0.12 max C, 1.0 max Cr, 26.0-30.0 Mo, 0.20-0.60 V, 4.0-6.0 Fe, bal Ni.
Cast, ann: 500 n/mm² min UTS.
Good resistance to acids.
ACE N-12M (Hastelloy B).

ARCM 2233 C8.
M-1488; 0.08 max C, 1.5 max Mn, 1.5 max Si, 10.5-12.5 Ni, 17.0-19.5 Cr, 2.0-2.5 Mo, Nb = 8 X C, bal Fe.
Cast, ann: 490 n/mm² min UTS.
Austenic stainless, weldable, for pumps, valves, faucets; improved resistance to corrosion.
AFNOR Z4CNDNb18.12M; W.Nr.1.4581.

ARCM 2233 C10.
M-1488; 0.10 max C, 1.5 max Mn, 1.5 max Si, 12.0-15.0 Ni, 17.0-19.0 Cr, 2.0-2.5 Mo, Nb = 10 X C, bal Fe.
Cast, ann: 490 n/mm² min UTS.
Austenitic stainless, weldable, for pumps, valves, faucets; improved resistance to corrosion.

ARCM 2266 A3.
M-1488; 0.03 max C, 1.5 max Mn, 1.5 max Si, 9.0-13.0 Ni, 17.0-21.0 Cr, 2.0-3.0 Mo, bal Fe.
Cast, ann: 490 N/mm² min UTS.
Austenitic stainless, weldable, for pumps, valves, faucets; improved resistance to corrosion.
AFNOR Z3CND20.10M; ACI CF-3M.

ARCM 2266 A5.
M-1488; 0.05 max C, 1.5 max Mn, 1.5 max Si, 12.0-15.0 Ni, 17.5-20.5 Cr, 3.0-3.5 Mo, bal Fe.
Cast, ann: 490 n/mm² min UTS.
Austenitic stainless for pumps, valves, faucets; improved resistance to corrosion.
AFNOR Z4CND19.13M; similar to AISI 317.

ARCUM 2266 A8.
M-1488; 0.08 max C, 1.5 max Mn, 1.5 max Si, 9.0-12.0 Ni, 18.0-21.0 Cr, 2.0-3.0 Mo, bal Fe.
Cast, ann: 490 N/mm² min UTS.
Austenitic stainless for pumps, valves, faucets; improved resistance to corrosion.
AFNOR Z5CND20.10M; similar to ACI CF-8M.

ARCM 2266 A10.
M-1488; 0.10 max C, 1.5 max Mn, 1.5 max Si, 9.0-11.0 Ni, 17.0-20.0 Cr, 3.0-3.5 Mo, bal Fe.
Cast, ann: 490 N/mm² min UTS.
Austenitic stainless for pumps, valves, faucets; improved resistance to corrosion.
AFNOR Z8CND 18.10.3M.

ARCM 2266 CU.
M-1488; 0.06 max C, 1.0 max Mn, 1.0 max Si, 7.0-9.0 Ni, 23.5-25.5 Cr, 3.0-4.0 Mo, 1.0-2.0 Cu, bal Fe.
Cast ann: 600 N/mm² min UTS.
Austenitic Stainless; resistant to stress corrosion.
AFNOR Z4CNUD 25.8M.

ARCM 2702 A3.
M-1488; 0.03 max C, 1.5 max Mn, 2.0 max Si, 8.0-12.0 Ni, 18.0-21.0 Cr, bal Fe.
Cast, ann: 490 N/mm² min UTS.
Austenitic stainless steel, weldable for faucets, valves, pumps.
AFNOR Z3 CN19.9M; ACI CF-3.

ARCM 2702 A8.
M-1488; 0.08 C, 1.5 max MN, 2.0 max Si, 8.0-11.0 Ni, 18.0-21.0 Cr, bal Fe.
Cast ann: 490 N/mm² min UTS.
Austenitic stainless for pumps, valves, parts requiring corrision resistance.
AFNOR Z6CN19.9N; ACI-CF8.

ARCM 2702 A12.
M-1488; 0.12 max C, 1.5 max Mn, 2.0 max Si, 8.0-10.0 Ni, 17.0-19.5 Cr, bal Fe.
Cast, ann: 490 N/mm² min UTS.
Austenitic stainless for faucets, valves, pump parts; weldable.
AFNOR Z10CN18.9; similar to ACI CF20.

ARCM 2702 C8.
M-1448; 0.08 max C, 1.5 max Mn, 2.0 max Si, 9.0-12.0 Ni, 18.0-21.0 Cr, Nb = 8 X C, bal Fe.
Cast, ann: 490 N/mm² min UTS.
Stabilized austenitic stainless, preferred for welded pumps, valves, faucets.
AFNOR Z4CNNb19.10M; ACI CF-8C.

ARCM 6015.
M-1488; 0.12 max C, 15.5-17.5 Cr, 16.0-18.0 Mo, 0.20-0.40 V, 4.0-7.0 Fe, 3.75-5.25 W, bal Ni.
Cast, ann: 500 N/mm² min UTS.
Good resistance to acids and hyperchlorites.
ACI CW 12 M (Hastelloy C).

ARCM 8510.
M-1488; 0.10 max C, 8.0-10.0 Si, 2.0-4.0 Cu, 2.0 max Fe, bal Ni.
Cast stainless alloy.
Resists sulphuric acid in all concentrations; also organic acids. (Hastelloy D).

ARCOL 360.
M-1724; 0.13 max C, 1.35 Mn, 0.80 Cr, 0.40 Mo, bal Fe.
Abrasion resisting steel for applications such as liner plates, chutes, buckets, storage bins, and mould boards.

ARCOSARC 70.
M-677; mild steel 70,000 psi.
For flux-cored welding wire, class E70T-2.

ARCOSARC 70X.
M-677; mild steel 70,000 psi.
For flux-cored welding wire, multi-pass deposition without slag removal, class E70T-2.

ARCOSARC 1CM.
M-677; 1.25 Cr, 0.5 Mo steel.
For flux-cored welding wire.

ARCOSARC 2CM.
M-677; 2.25 Cr, 1.0 Mo steel.
For flux-cored welding wire.

ARCOSARC 5M.
M-677; 5 Cr, 0.5 Mo steel.
For flux-cored welding wire.

ARCOSARC 72.
M-677; mild steel 70,000 psi, high impact strength.
For flux-cored welding wire, class E70T-1.

ARCOSARC 80.
M-677; 1.0 Ni steel, 80,000 psi, high impact strength.
For flux-cored welding wire.

ARCOSARC 100.
M-677; 1.5 Mn, 0.5 Mo steel, 100,000 psi.
For flux-cored welding wire.

ARCOSARC 110T.
M-677; 2.0 Ni, 1.0 Cr, 0.5 Mo steel, 110,000 psi.
For flux-cored welding wire.

ARCOSARC 410 NI.
M-677; 0.04 C, 0.6 Mn, 0.4 Si, 12 Cr, 4 Ni, 0.8 Mo, bal Fe.
For flux-cored welding wire; welding CA6NM castings.

ARCOSARC CI.
M-677; cast iron.
For flux-cored welding wire.

ARCOSARC SS-1.
M-677; type 308 stainless steel.
For flux-cored welding wire.

ARCOSARC SS-2.
M-677; type 309 stainless steel.
For flux-cored welding wire.

ARCOSARC SS-3.
M-677; type 316 stainless steel.
For flux-cored welding wire.

ARTIC see CRANE ARCTIC.

ARCTIC D.
M-1724; 0.08-0.14 C, 1.5 max Mn, 0.5 max Si, 0.10 Nb, 0.009 max N, bal Fe.
High yield stress/ultimate TS ratio; weldable; with high notch tough properties at low temperatures.
For liquid gas ship steel.

ARCTIC GRADE X70.
M-558; 0.05 C, 1.60 Mn, 0.05 Si, 0.25 Mo, 0.06 Cb, bal Fe.
For arctic pipe lines.

ARDAL METAL.
M-581; 2 Cu, 1.7 Fe, 0.6 Ni, bal Al.
31,000 TS; 17,000 YS; 15 El; 60 Brin.
For bearings, pistons, wire, tubes; cannot be hardened by heat treatment.

ARDHO NO 2.
M-1744; 0.47 C, 0.47 Mn, 0.90 Si, 1.1 Ni, 0.6 Cr, bal Fe.
1% Nickel-chromium tool steel, for smiths' tools, cold chisels, punches.

ARDOLOY 1A.
M-1806; Sintered carbide tool material.
Cutting tool for semi-finishing and finishing cast iron and non-ferrous materials; suitable for milling. For turning and reaming all aluminum alloys.

ARDOLOY 2A.
M-1806; Sintered carbide tool material.
Shock resistant grade of cutting tool used largely for planing cast iron; and for general wear resistant application.
ISO K 30.

ARDOLOY AD.
M-1806; Sintered carbide tool material.
General purpose cutting tool for machining cast iron and non-ferrous materials at medium feeds and speeds.
ISO M10, M20.

ARDOLOY AF.
M-1806; Sintered carbide tool material.
Very hard grade of straight tungsten carbide; particularly for finishing cast iron, non-ferrous, plastics and abrasive non-metallics.
ISO K05.

ARDOLOY AK.
M-1806; Sintered carbide tool material.
Very tough cutting tool for heavy roughing cuts at slow speeds; good on interrupted cuts.
ISO P40, P50.

ARDOLOY CNO1(B).
M-1806; Sintered carbide tool material.
Very hard and wear resistant straight tungsten grade due to low cobalt content and fine grain structure; for finishing and boring cast iron and non-ferrous.
ISO K01.

ARDOLOY CN10.
M-1806; Sintered carbide tool material.
Hard and wear resistant grade of straight tungsten carbide; for finish turning and boring of cast iron and non-ferrous metals.
ISO K10.

ARDOLOY CN20(K).
M-1806; Sintered carbide tool material.
General pupose cutting tool for use on cast iron, non-ferrous metals and o stainless steel.
ISO K20.

ARDOLOY CN30(M).
M-1806; Sintered carbide tool material.
Tough, shock-resisting grade, used mainly for planing cast iron.
ISO K30.

ARDOLOY CR10.
M-1806; Sintered carbide tool material.
High crater and wear resistant grades for medium and high speed cutting speeds, and finishing steel.
ISO P10.

ARDOLOY CR 20.
M-1806; Sintered carbide tool material.
For general purpose, semi-finishing of steels; good crater and wear resistance.
ISO P20.

ARDOLOY CR25(T).
M-1806; Sintered carbide tool material.
General purpose cutting grade for all types of steel at medium speeds and feeds.
ISO P20.

ARDOLOY CR30.
M-1806; Sintered carbide tool material.
General purpose grade for roughing and semi-finishing all types of steel a medium feeds and speeds.
ISO P30.

ARDOLOY CR40(R).
M-1806; Sintered carbide tool material.
Tough grade of cutting tool for roughing cuts at slow speeds and heavy loads.
ISO P40.

ARDOLOY ICC.
M-1806; Sintered carbide tool material.
Grade of cutting tool for semi-finishing and finishing of cast iron and non-ferrous material; used on all aluminum alloys.
ISO K10.

ARDOLOY S3.
M-1806; Sintered carbide tool material.
General purpose cutting tool for all types of steel at medium speeds and feeds.
ISO P30.

ARDOLOY S48.
M-1806; Sintered carbide tool material.
Tough cutting tool for roughing operations, particularly on steel.
ISO P40.

ARDOLOY S200.
M-1806; Sintered carbide tool material.
General purpose cutting tool particularly for lathe turning most metals. Resistant to wear and cratering.
ISO P20.

ARDOLOY SK2.
M-1806; Sintered carbide tool material.
Titanium carbide grade for finish turning and boring steel; resists flank and crater wear.
ISO P01.

ARDORIT 6.
M-1338; 0.12 max C, 0.8 Si, 0.8 Al, 6.5 Cr, bal Fe.
Bars: 70,000 TS; 30,000 YS; 28 El; 65 RA; 160 Brin.
For oil refinery equipment; creep and heat resistant.

ARDORIT 13A1.
M-1338; 0.12 max C, 1.2 Si, 1.0 Al, 13 Cr, bal Fe.
Annealed: 75,000 TS; 40,000 YS; 30 El; 70 RA; 160 Brin.
For oil refinery equipment; Type 410 Al; stainless, creep resistant.

ARDORIT 18A1.
M-1338; 0.12 max C, 1 Si, 1 Al, 18 Cr, bal Fe.
Annealed: 80,000 TS; 50,000 YS; 22 El; 48 RA; 150 Brin.
For oil refinery equipment; Type 430 Al; heat and creep resistant.

ARDORIT 20/10.
M-1338; 0.15 C, 19.5 Cr, 9.5 Ni, bal Fe.
Annealed: 80,000 TS; 35,000 YS; 55 El; 75 RA; 150 Brin.
Cold drawn: 180,000 TS; 150,000 YS; 10 El; 250 Brin.
For chemical plant equipment, tanks, mixers, agitators; Type 302; stainless, austenitic.

ARDORIT 24/20.
M-1338; 0.15 C, 2 Si, 24 Cr, 19 Ni, bal Fe.
Annealed: 100,000 TS; 45,000 YS; 50 El; 65 RA; 185 Brin.
For furnace parts, pumps, valves, turbine parts; Type 310; stainless, austenitic.

ARDORIT 25N.
M-1338; 0.2 C, 1.2 Si, 25 Cr, 4 Ni, bal Fe.
Aged: 115,000 TS; 80,000 YS; 15 El.
Cast: 70,000 TS; 65,000 YS; 2 El; 190 Brin.
For furnace parts, heat treat boxes, baffles; heat and corrosion resistant

AREMITE.
M-195; Synthetic cast iron.
30,000-50,000 TS; 165 Brin.
For pipes, fittings, hardware.

ARGAL.
M-1362; 2-4 Mg, 0.4 max Mn, 0.3 max Cr, bal Al.
Sof: 28,000 TS; 13,000 YS; 30 El; 47 Brin.
Hard: 40,000 Ts; 35,000 YS; 10 El; 73 Brin.
For aircraft tanks and fittings, marine parts, fuel lines; resists sea water corrosion.

ARGALIUM.
M-1367; 1.5-3.0 Mg; 0.6 1.3 Mn; 1.3 max Si, 0.3 Cr, 0.2 Ti, bal Al.
Soft: 26,000 TS; 10,000 YS; 20 El; 45 Brin.
Hard: 41,000 TS; 36,000 YS; 5 El; 77 Brin.
For roofing, hydraulic tubing, architectural trim; good forming and welding properties.

ARGELITE.
M-Eng; 90 Al, 6 Cu, 2 Si, 2 Bi.
For light alloy parts; non-hardenable.

ARGENT.
M-547; 2 Ag, bal Sn.
6775 TS; 3.7 El.
Al-solder; Sc-14.

ARGENT FRANCAIS.
M-France; Cu-Zn-Ni.
For jewelry, ornaments; Moussets Silver.

ARGENTAL.
M-Eng; 60-75 Al, 15-16 Ag, 7-20 Zn, 3-5 Cu.
For jewelry; corrosion resistant.

ARGENTAL.
10 Sn, 5 Co, bal Cu.
For jewelry; corrosion resistant.

ARGENTALIUM.
M-Eng; 5 Ag, 0.1-1 Mg, bal Al.
For light alloy parts; non-hardenable.

ARGENTAN.
M-Eng; 48.35 Cu, 34.05 Zn, 17.60 Ni.
For table cutlery, ornaments; high resistance to corrosion.

ARGENTAN.
M-US; 50-90 Cu, 3-40 Ni, 0-10 Al, 0-40 Zn.
For cutlery; corrosion resistant.

ARGENTAN.
M-547; 2 Al, 0.5 Ag, bal Cu.
Rolled: 23,660 TS; 2.3 El.
For contacts.

ARGENTAN, BERLIN.
M-Ger; 55.5 Cu, 29.1 Zn, 15.5 Ni.
For electrical resistances; corrosion resistant.

ARGENTAN, CHINESE.
M-China; 40.4 Cu, 25.4 Zn, 31.6 Ni, 2.6 Fe.
For ornaments, jewelry, hardware; corrosion resistant.

ARGENTAN, ENGLISH.
M-Eng; 63.36 Cu, 17 Zn, 19.13 Ni, 0.38 Fe.
For domestic ware; corrosion resistant.

ARGENTAN, FRENCH.
M-French; 50.32 Cu, 30.94 Zn, 18.4 Ni.
For electrical resistances; corrosion resistant.

ARGENTAN (NEUSILBER).
M-Ger; 26 Ni, 56 Cu, 18 Zn.
For chemical equipment construction; substitute for Ag, a nickel silver.

ARGENTAN, RUSSIAN.
M-USSR; 63.88 Cu, 17.58 Zn, 17.58 Ni, 0.33 Fe, 0.32 Pb.
For ornaments, architectural purposes; corrosion resistant.

ARGENTAN RUSSIAN (CAST).
M-USSR; 57.52 Cu, 18.94 Zn, 20.35 Ni, 3.15 Fe.
For ornaments, hardware; corrosion resistant.

ARGENTAN SHEET.
M-USSR; 40-65 Cu, 17-32 Zn, 15-30 Ni.
For ornaments, jewelry, hardware; corrosion resistant.

ARGENTAN, SOLDER.
M-USSR; 35 Cu, 57 Zn, 8 Ni.
For solder; corrosion resistant.

ARGENTAN, VIENNA.
M-Austria; 55.6 Cu, 21.8 Zn, 22.2 Ni, 0.38 Fe.
For ornaments, jewelry, hardware; corrosion resistant.

ARGENTIN.
M-Eng; 85 Sn, 14.5 Sb, 0.5 Cu.
For bearings; antifriction.

ARGENTINE METAL.
M-US; 85.5 Sn, 14.5 Sb.
For statuettes and small ornaments; expands on cooling.

ARGESTE 3956 AM.
M-1333; 0.12 max C, 16.5-18.5 Cr, 11.0-13.0 Ni bal Fe.
Austenitic, ductile, corrosion resistant, less tendency to work harden.
For cold formed or stamped stainless parts.
W. Nr. 3956; AISI Type 305 stainless Was 80 AM.

ARGESTE 3962 AM.
M-1333; 0.05-0.20 C, 10.5-12.5 Cr, 9.0-11.0 Ni, 5.5-6.5 Mn, bal Fe.
Austenitic, corrosion resistant. Was 80 AB.
W. Nr. 3962.

ARGESTE 4001 IW.
M-1333; 0.08 max C, 13-15 Cr, 1.0 max Si, 1.0 max Mn, bal Fe.
Ann: 50-65 kg/mm^2 TS; 30 kg/mm^2 YS; 20 El.
Magnetic type stainless.
For structural parts, building fittings.
W. Nr. 4001. Was 13W.

ARGESTE 4006 IH.
M-1333; 0.08-0.12 C, 12-14 Cr, 1.0 max Mn, 1.0 max Si, bal Fe.
Martensitic type stainless, hardenable.
For structural parts in water and steam.
W. Nr. 4006, AISI type stainless. Was 13H.

ARGESTE 4016 IM.
M-1335; 0.10 max C, 15.5-17.5 Cr, bal Fe.
Ann: 45-60 kg/mm^2 TS; 30 kg/mm^2 YS; 20 El.
Ferritic, non-hardenable stainless.
For structural parts, armatures, building fittings.
W. Nr. 4016, AISI Type 430. Was 17.

ARGESTE 4021 YB.
M-1333; 0.17-0.22 C, 12.0-14.0 Cr, 1.0 max Mn, 1.0 max Si, bal Fe.
Magnetic, hardenable, corrosion resistant.
For structural parts requiring higher strength.
W. Nr. 4021; Nearest AISI 420. Was K 20.

ARGESTE 4024 YA.
M-1333; 0.12-01.17 C, 12.0-14.0 Cr, bal Fe.
Magnetic, hardenable, corrosion resistant.
For structural parts and turbine blades.
W. Nr. 4024. Was K 15.

ARGESTE 4034 YD.
M-1333; 0.40-0.50 C, 12.0-14.0 Cr, bal Fe.
Magnetic, hardenable, corrosion resistant.
For high temperature parts, arbors, spindles, bolts, shafts. Was K 40.
W. Nr. 4034.

ARGESTE 4057 YN.
M-1333; 0.17-0.25 C, 16.0-18.0 Cr, 1.0-2.5 Ni, bal Fe.
Maganetic, hardenable, corrosion resistant.
Heat treated parts have good strenght to 500°C.
W. Nr. 4057; AISI Type 431. Was K 20N.

ARGESTE 4104 IU.
M-1333; 0.10-0.17 C, 15.5-17.5 Cr, 0.2-0.3 Mo, 0.15-0.25 Si, bal Fe.
Magnetic, free machining, non-hardenable stainless.
For stainless parts made on automatic screw machines; bolts, studs, nuts. Was 17 U.
W. Nr. 4104.

ARGESTE 4112 YL.
M-1333; 0.85-0.95 C, 17.0-19.0 Cr, 1.0-1.3 Mo, bal Fe.
Magnetic, hardenable, corrosion resistant.
For cutlery, kitchen knives, harden shafts operating at elevated temperature. Was K 90L.
W. Nr. 4112; AISI Type 440B stainless.

ARGESTE 4113 IL.
M-1333; 0.10 max C, 15.5-17.5 Cr, 0.9-1.3 Mo, bal Fe.
Ann: 45-65 Kg/mm^2 TS; 30 Kg/mm^2 YS; 30 El.
Magnetic, stainless.
For parts for motor cars requiring higher resistance to corrosion.
W. Nr. 4113. Was 17L.

ARGESTE 4120 YL.
M-1333; 0.17-0.22 C, 12.0-14.0 Cr, 1.0-1.3 Mo, bal Fe.
Magnetic, hardenable, corrosion resistant.
For high quality structural parts as turbine blades.
W. Nr. 4120. Was K 20L.

ARGESTE 4122 YL.
M-1333; 0.33-0.43 C. 15.5-17.5 Cr. 1.0-1.3 Mo, bal Fe.
Magnetic, hardenable, corrosion resistant.
For arbors, splindles, bolts operating at elevated temperatures. Was K 35L.
W. Nr. 4122.

ARGESTE 4301 PA.
M-1333; 0.07 max, 17-20 Cr, 9-11.5 Ni, bal Fe.
Austenitic, hardenable by cold work only, corrosion resistant, weldable.
For food processing equipment, dairy equipment.
W. Nr, 4301; AISI Type 304 stainless, Was 80 P.

ARGESTE 4305 UA.
M-1333; 0.15 max C, 17-19 Cr, 8-10 Ni, 0.1-0.2 S, bal Fe.
Ann: 50-70 Kg/mm^2 TS; 22 Kg/mm^2 YS; 50 El.
Austenic, free machining, corrosion resistant.
For stainless bolts, nuts, studs, shafts made on automatic screw machines. Was 80 U.
W. Nr. 4305; AISI Type 303 stainless.

ARGESTE 4306 LA.
M-1333; 0.03 max C, 17-20 Cr, 10-12.5 Ni, bal Fe.
Austenitic, weldable, corrosion resistant.
For welded assembles for food, dairy and chemical industries. Was 80 LC
W. Nr. 4306; AISI Type 304L.

ARGESTE 4310 FA.
M-1333; 0.15 max C, 16-18 Cr, 7-9 Ni, bal Fe.
Austenitic, corrosion resistant; hardens readily by cold work.
For cold worked wire and strip for springs for operation to 350°C. Was 80 FH.
W. Nr. 4310; AISI Type 301 stainless.

ARGESTE 4310 IT.
M-1333; 0.10 max C, 16-18 Cr, Ti = 7 x %C, bal Fe.
Ann: 45-60 Kg/mm^2 TS; 30 Kg/mm^2 YS; 20 El.
Magnetic, non-hardenable stainless, weldable.
For welded assemblies of stainless.
W. Nr. 4510. Was 17T.

ARGESTE 4401 PA.
 M-1333; 0.07 max C, 16.5-18.5 Cr, 2.0-2.5 Mo, 10.5-13.5 Ni, bal Fe.
 Austenitic, very good resistance to corrosion.
 For parts and apparatus for texile industry.
 W. Nr. 4401; AISI Type 316 stainless. Was 82 P.

ARGESTE 4404 LA.
 M-1333; 0.03 max C, 16.5-18.5 Cr, 2.0-2.5 Mo, 11.0-14.0 Ni, bal Fe.
 Austenitic, very good resistance to corrosion.
 For parts and apparatus of chemical and textile industry. Was 82 LC.
 W. Nr. 4404; AISI Type 316L stainless.

ARGESTE 4435 LA.
 M-1333; 0.03 max C, 16.5-18.5 Cr, 2.5-3.0 Mo, 12.5-15.0 Ni, bal Fe.
 Austenitic, weldable, very good resistance to corrosion.
 For welded assemblies requiring good corrosion resistance. Was 83 LC.
 W. Nr. 4435.

ARGESTE 4436 PA.
 M-1333; 0.07 max C, 16.5-18.5 Cr, 2.5-3.0 Mo, 12.0-14.5 Ni, bal Fe.
 Austenitic, weldable, very good resistance to corrosion.
 For welded assemblies requiring corrosion resistance.
 W. Nr. 4436. Was 83 P.

ARGESTE 4449 PA.
 M-1333; 0.07 max C, 16.0-18.0 Cr, 4.0-5.0 Mo, 12.5-14.5 Ni, bal Fe.
 Austentic, good corrosion resistance.
 For parts and apparatus with high pitting stability.
 W. Nr. 4449. Was 135 P.

ARGESTE 4505 BA.
 M-1333; 0.07 max C, 16.5-18.5 Cr, 2.0-2.5 Mo, 19.0-21.0 Ni, Cu, Nb, bal Fe.
 Austenitic, ductile, corrosion resistant.
 For parts in chemical industry in contact with sulphuric acid. Was 182 Nb.
 W. Nr. 4505.

ARGESTE 4506 TA.
 M-1333; 0.07 max C, 16.5-18.5 Cr, 2.0-2.5 Mo, 19.0-21.0 Ni, Cu, Ti, bal Fe.
 Austenitic, ductile, corrosion resistant.
 For parts in chemical industry in contact with sulphuric acid. Was 182 RT.
 W. Nr. 4506.

ARGESTE 4541 TA.
 M-1333; 0.10 max C, 17-19 Cr, 9-11.5 Ni, Ti = 5 x %C, bal Fe.
 Austenitic, weldable, corrosion resistant.
 For welded assemblies for food, chemicals; and for operation at elevated temperatures. Was 80 T.
 W. Nr. 4541; AISI Type 321 stainless.

ARGESTE 4550 BA.
 M-1333; 0.10 max C, 17-19 Cr, 9-11.5 Ni, Nb = %C, bal Fe.
 Austenitic, weldable corrosion resistant.
 For welded assemblies for food, chemicals and for operation at elevated temperature. Was 80 Nb.
 W. Nr. 4550; AISI Type 347 stainless.

ARGESTE 4568 GA.
 M-1333; 0.09 max C, 16.0-18.0 Cr, 6.5-7.75 Ni, plus special additions, bal Fe.
 Austenitic, corrosion resistant. Was 80 SG.
 W. Nr. 4568.

ARGESTE 4571 TB.
 M-1333; 0.10 max c, 16.5-18.5 Cr, 2.0-2.5 Mo, 10.5-13.5 Ni, Ti = 5 x %C min, bal Fe.
 austentic, weldable, very good resistance to corrosion.
 For welded assemblies for chemical, texile, and cellulose industries. Was 82 T.
 W. Nr, 4571.

ARGESTE 4580 BA.
 M-1333; 0.10 max C, 16.5-18.5 Cr, 2.0-2.5 Mo, 10.5-13.5 Ni, Nb = 8 x %C, min, bal Fe.
 Austenitic, weldable, very good resistance to corrosion.
 For welded assemblies in chemical industry.
 W. Nr 4580. Was 82 Nb.

ARGILITE.
 M-French; 90 Al, 6 Cu, 2 Si, 2 Bi.
 For automotive engine parts.

ARGO-BRAZE 50.
 M-200; 50 Ag, plus Cu, Zn, Cd, Ni, Mn.
 Silver brazing alloy; melt range 639-668°C.

ARGO-BRAZE 56.
 M-200; 56 Ag, plus Cu, Zn, Ni.
 Silver brazing alloy; melt range 600-711°C.

ARGOFIL.
 M-1805; 0.25 Mn, 0.25 Si, bal Cu.
 For filter rod for argon-arc and inert gas shielded metal arc welding of copper.
 M.P. 1083°C.

ARGO-FLO.
 M-200; 39 Ag, plus Cd, Zn.
 Silver brazing alloy; melt range 605-651°C.

ARGO HIGH SPEED.
M-Eng.; 0.82 C, 0.17 Mn, 1.74 Cr, 4.87 W, 0.10 V, bal Fe.
For tools, cutters, dies, reamers, gages, punches; high speed steel.

ARGOID METAL.
M-Eng.; 52.6 Cu, 25.8 Zn, 21.6 Ni.
For ornamental parts; Nickel Silver.

ARGOZOIL.
M-U.S.; 54 Cu, 28-20 Zn, 14 Ni, 2 Sn, 2-10 Pb.
For ornamental parts; corrosion resistant.

ARGUZOID.
M-Ger.; 48-56 Cu, 13-21 Ni, 23-31 Zn, 0-4 Sn, 0-4 Pb.
For cutlery, ornaments, utensils; German Silver.

ARGYROID (ARGIROIDE).
M-Eng.; Cu, Ni, Zn.
For resistances, ornaments; Nickel Silver.

ARGYROLITH.
M-Eng.; 50-70 Cu, 10-20 Ni, 5-30 Zn.
For resistances, ornaments; Nickel Silver.

ARGYROPHAN.
M-Eng.; Cu, Ni, Zn.
For resistances, ornaments; Nickel Silver.

ARISTOLOY.
M-18; 67-70 Ag, 25-29 Sn, 5-3 Cu, 0-1 Zn.
For dental amalgams and dies; U.S.P. 1963085.

ARK.
M-261; 0.7 C, 3.7 Cr, 0.6 Mo, 14 W, 1 V, bal Fe.
For tools, cutters, punches; high speed steel.

ARK HIGH SPEED.
M-261; 0.84 C, 1.53 Mn, W, Cr, bal Fe.
For tools, high speed cutters; high speed steel.

ARK SUPERIOR.
M-261; 0.76 C, 4.25 Cr, 0.6 Mo, 1.4 V, 18 W, bal Fe.
For drills, broaches, reamers, hobs, gear cutters; high speed steel.

ARK SUPERIOR EXTRA.
M-261; 0.78 C, 4.25 Cr, 0.6 Mo, 1.4 V, 22 W, bal Fe.
For form cutters, tools for chilled roll turning; high speed steel.

ARK SUPERLATIVE.
M-261; 0.8 C, 4.8 Cr, 0.6 Mo, 1.6 V, 18.5 W, 5.7 Co, bal Fe.
For tools, cutters, milling cutters; high speed steel.

ARK SUPREME.
M-261; 0.8 C, 22 W, 5 Cr, 17 Co, 1.7 V, 0.6 Mo, bal Fe.
For tools, cutters, planing and boring tools; high speed steel.

ARK TRIUMPH.
M-261, M-486; 1.2 C, 4.35 Cr, 4.5 V, 14 W, bal Fe.
For form tools, broaches, reamers; high speed steel.

ARKIT.
M-French; 38 Co, 30 Cr, 16 W, 10 Ni, 4 Mo, 2.5 C.
For high speed cutting tips of lathe tools; high heat resistance.

ARKO.
M-U.S.; 80 Cu, 20 Zn.
For tubes, fittings, hardware; drawn and spun.

ARKTURUS.
M-1306; 0.55 C, 1.05 Cr, 0.18 V, 1.85 W, bal Fe.
For cold work tools, headers, upsetters; oil hardened, tough.

ARM 4.
M-Russia; 0.08-0.14 C, 0.3-1.0 Mn, 0.6 max Si, 14-16 Cr, 14-16 Ni, 1.8-2.2 Mo, 0.8-1.2 W, 2.5-3.0 Co, 0.3 Ti, bal Fe.
Annealed: 90,000 TS; 40,000 YS; 40 El; B 82 Rock.
For high temperature applications; welded structures, agitators.
Heat and corrosion resistant.

ARM 6.
M-Russia; C, 16 Cr, 14 Ni, 2.5 Mo, 6 W, 1 Cb, 0.4 V, bal Fe.
Annealed: 85,000 TS; 35,000 YS; 50 El, B 80 Rock.
For high temperature applications, chemical plant equipment, fasteners, mixers.
Heat and corrosion resistant.

ARM 12.
M-Russia; 0.20 C, 11.5 Cr, 0.5 max Ni, 0.5 Mo, 0.25 V, 2.5 W, 2.2 Co, 0.01 B, bal Fe.
For high temperature applications, furnace parts.
Heat resistant to 600°C., corrosion resistant.

ARMACAST.
M-112; 0.25-0.30 C, 2.65-3.15 Cr, 0.45-0.55 Mo, bal Fe.
Cast: 110,000 TS; 90,000 YS; 20 El; 45 RA; 200-225 Brin.
For road construction and railway equipment castings; tough.

ARMALOY.
M-132; C, alloy, bal Fe.
For cutting tools, bits, blades, ratchets, wrenches; cast alloy.

ARMASTEEL.
M-690; 2.7 C, 1.2 Si, 0.15 S, bal Fe.
Cast: 55,000-108,000 TS; 42,000-95,000 YS; 8-1.5 El; 143-285 Brin.
For gears, rocker arms, universal joints, camshafts; pearlitic malleable iron.

ARMASTEEL GM 84 M.
M-690; 2.55 C, 1.4 Si, 0.45 Mn, 0.12 S, 0.05 P, bal Fe. Pearlitic malleable cast iron.
100,000 TS; 80,000 YS; 2 El; 241-269 Brin.
For parts requiring good strength; e.g. connecting rods.
SAE Grade 70002; ASTM Grade 80002.

ARMASTEEL GM 85 M.
M-690; 2.55 C, 1.4 Si, 0.45 Mn, 0.12 S, 0.05 P, bal Fe. Pearlitic malleable cast iron.
80,000 TS; 60,000 YS; 3 El; 197-241 Brin.
This grade preferred for additional local rehardening by flame or induction; as for planet carriers.
SAE & ASTM Grade 60003.

ARMASTEEL GM 85 M MODIFIED.
M-690; 2.55 C, 1.4 Si, 0.45 Mn, 0.12 S, 0.05 P, bal Fe. Pearlitic malleable cast iron.
90,000 TS; 60,000 YS; 3 El; 217-269 Brin.
For automotive crankshafts, transmission gears.
SAE & AST Grade 60003.

ARMASTEEL GM 86 M.
M-609; 2.55 C, 1.4 Si, 0.45 Mn, 0.12 S, 0.05 P, bal Fe. Pearlitic malleable cast iron.
70,000 TS; 48,000 YS; 4 El; 163-207 Brin.
For lightly stressed automotive type parts; e.g. certain compressor crankshafts.
SAE Grade 48005; ASTM Grade 48004.

ARMASTEEL GM 88 M.
M-690; 2.55 C, 1.4 Si, 0.45 Mn, 0.12 S, 0.05 P, bal Fe. Pearlitic malleable cast iron.
105,000 TS; 85,000 YS; 2 El; 269-302 Brin.
Compressive strength: 250,000 psi ultimate.
For parts requiring good strength and wear resistance; gears. ASTM Grade 80002.

ARMCO 9% NICKEL A353.
M-10; 0.13 max C, 0.90 max Mn, 0.15-0.30 Si, 8.5-9.5 Ni, bal Fe.
2" max thickness: 100,000-120,000 TS; 75,000 YS; 20 El.
Heat treated nickel alloy steel, weldable; for storage and process vessels for liquefied gases to -320°F. Double normalized and tempered.

ARMCO 9% NICKEL A553 GRADE A.
M-10; 0.13 max C, 0.90 max Mn, 0.15-0.30 Si, 8.5-9.5 Ni, bal Fe.
2" max thickness: 100,000-120,000 TS; 100,000 YS; 20 El.
Quenched and tempered: nickel alloy steel, weldable; for storage and process vessels, for liquified gases to -320°F.

ARMCO 15-5 PH.
M-10; 0.07 C, 14.0-15.5 Cr, 3.5-5.5 Ni, 2.5-4.5 Cu, bal Fe.
Martensitic stainless; precipitation hardenable.
Cond H 900: 204,000 TS; 182,000 YS; 14 El.
For springs, strong stampings; better transverse ductility and toughness than 17-7 PH.

ARMCO 17-4 PH.
M-10; 0.05 C, 17 Cr, 4 Ni, 4 Cu, bal Fe.
Age-hardened: 210,000 TS; 200,000 YS; 6-15 El; 450 Brin.
For gears, cams, chains, valves, pump parts; stainless, age-hardenable.

ARMCO 17-7 PH.
M-10; 0.09 C, 16.0-18.0 Cr, 6.5-7.75 Ni, 0.75-1.25 Al, bal Fe.
Semi-austenitic; precipitation hardened.
Annealed: 130,000 TS; 40,000 YS; 35 El.
Hardened: 200,000-265,000 TS; 185,000-260,000 YS; 2-9 El; (wire can be stronger).
For springs, instrument stampings as gears, levers, cams, fasteners; stainless.

ARMCO 17-10P.
M-10; 0.15 max C, 16.0-18.0 Cr, 9.5-12.0 Ni, 0.2-0.4 P, bal Fe.
Austenitic stainless; precipitation hardenable.
Sol. Tr: 89,000 TS; 38,000 YS; 70 El; 143 Brin.
Double aged: 144,000 TS; 99,000 YS; 20 El; 31 Rock C; 70,000 psi endurance limit.

ARMCO 18 SR.
M-10; 0.05 C, 0.5 Mn, 1.0 Si, 18.0 Cr, 0.5 Ni, 2.0 Al, 0.4 Ti, bal Fe.
Ferritic stainless sheet and strip, weldable.
Room temp: 80,000-90,000 TS; 60,000-70,000 YS; 25-30 El; 90 Rock B.
1000°F: 48,700 TS; 33,900 YS; 30 El.
For industrial ovens, heaters, furnace tubes, annealing boxes, baffle plates, pyrometer tubes; resists scaling at high temperature.

ARMCO 18-2 MN now ARMCO NITRONIC 32.

ARMCO 18-9 LW.
M-10; 0.10 max C, 2.0 max Mn, 1.0 max Si, 17-19 Cr, 8-10 Ni, 3-4 Cu, bal Fe.
Annealed: 78,000 TS.
Cold drawn: 152,000 TS.
For cold formed parts, fasteners, ferrules; good formability, stainless, austenitic.

ARMCO 20-10 MN TYPE 307.
M-10; 0.08-0.20 C, 19-22 Cr, 9-12 Ni, bal Fe.
Annealed: 85,000 TS; 35,000 YS; 55 El; 70 RA; 150 Brin.
For chemical plant equipment; Type 304; stainless.

ARMCO 20-45-5.
M-10, M-1491; 0.05 C, 5.0 Mn, 0.40 Si, 20.0 Cr, 45.0 Ni, 2.25 Mo, 0.15 Cb, bal Fe.
For heat exchanger and condenser tubing.

ARMCO 21-6-9 now ARMCO NITRONIC 40.

ARMCO 22-4-9.
M-10; 0.45-0.60 C, 7-10 Mn, 20-23 Cr, 3-5 Ni, 0.4 N, 0.12 max C, bal Fe.
Heat treated: 162,000 TS; 102,000 YS; 9.0 El; 9.0 RA; 344 Brin.
At 900°F: 114,000 TS; 58,000 YS; 20.0 El; 21.0 RA; 236 Brin.
At 1600°F: 38,000 TS; 25,000 YS; 27.0 El; 39.0 RA; 177 Brin.
For high temperature steam valves and gas turbines; age hardenable, corrosion, wear and heat resistant.

ARMCO 22-13-5 now **ARMCO NITRONIC 50.**

ARMCO 304.
M-10; 0.08 max C, 18.0-20.0 Cr, 8.0-10.5 Ni, bal Fe.
Austenitic stainless steel; non-hardenable by heat treatment.
Annealed: 85,000 TS; 35,000 YS; 60 El; 150 Brin.
Good grade of 18-8 type stainless steel; for food, beverage, meat handling equipment, AISI 304.

ARMCO 304 L.
M-10; 0.03 max C, 18.0-20.0 Cr, 8.0-10.0 Ni, bal Fe.
Austenitic stainless steel; non-hardenable by heat treatment.
Annealed: 80,000 TS; 30,000 YS; 60 El; 140 Brin.
For welded assemblies without subsequent heat treatment; for food, beverage, meat processing equipment. AISI 304L.

ARMCO 305.
M-10; 0.12 max C, 17.0-19.0 Cr, 10.5-13.5 Ni, bal Fe.
Austenitic stainless steel; non-hardenable by heat treatment.
Annealed: 85,000 TS; 35,000 YS; 60 El; 140 Brin.
Low work hardening rate; for spinning, deep drawing and cold heading work. AISI 305.

ARMCO 316 L.
M-10; 0.03 max C, 16-18 Cr, 10-14 Ni, 2-3 Mo, bal Fe.
Annealed: 80,000 TS; 30,000 YS; 60 El; 70 RA; 140 Brin.
Cold drawn: 90,000 TS; 60,000 YS; 45 El; 60 RA; 190 Brin.
For chemical and textile plant equipment, welded structures; Type 316L; stainless, austenitic.

ARMCO 400.
M-10; 0.05 max C, 1.0 max Mn, 1.0 max Si, 12.0-13.0 Cr, 0.5 max Ni, 0.5 max Al, bal Fe.
Ferritic stainless, non-hardenable, weldable.
Room temp: 61,500 TS; 29,500 YS; 35 El.
1200°F: 15,000 TS; 10,000 YS; 63 El.
For office furniture, dryer drums, air conditioning panels, sink strainers; resists scaling up to 1300°F.

ARMCO 409.
M-10; 0.05 max C, 10.0-12.0 Cr, 0.50 Ti, bal Fe.
Ferritic, non hardenable stainless.
Sheet: 68,000 TS; 40,000 YS.
Low cost stainless sheet; formable.

ARMCO 410 CB.
M-10; 0.15 max C, 11.5-13.5 Cr, 0.25 max Cb, bal Fe.
Martensitic stainless steel, magnetic, hardenable.
Hardened, tempered 500°F: 195,000 TS; 161,000 YS; 16 El; 43 Rock C; 65 Ft.lbs. Charpy impact.
Tempered 1100°F: 137,000 TS; 121,000 YS; 19 El; 277 Brin; 73,000 psi endurance limit.

ARMCO A-286.
M-10; 0.05 C, 1.45 Mn, 0.5 Si, 14.75 Cr, 25.25 Ni, 1.3 Mo, 2.15 Ti, 0.15 Al, 0.3 V, 0.005 B, bal Fe.
Austenitic, precipitation hardenable, stainless.
Annealed: 90,000 TS; 35,000 YS; 45 El; 83 Rock B.
Sol.tr.& aged: 150,000 TS; 100,000 YS; 25 El; 34 Rock C.
Very good high temperature properties as creep strength, fatigue, scaling, impact up to 1300°F; for high temperature fasteners, etc. AMS 5731, 5732 etc.

ARMCO ABRASION RESISTING STEEL.
M-10; 0.43 max C, 1.4-2.0 Mn, 0.15-0.35 Si, bal Fe.
As rolled: 225-285 Brinell; good abrasion resistance for chutes, hoppers, spouts, truck bodies, ore cars; weldable with special precautions.

ARMCO ALUMINIZED STEEL TYPES 1 AND 2.
M-10; sheet steel coated with Al.
Rolled: 45,000 TS; 30,000 YS; 35 El; 55 Rb.
For structural uses, mufflers; corrosion resistant.

ARMCO C 42 ETC see **ARMCO HIGH STRENGTH C 42, ETC.**

ARMCO COLD ROLLED PAINTGRIP.
M-10.
Mill applied phosphate coating.
For caskets, interior painted applications.

AMCO CRYONIC 5.
M-10; 0.13 max C, 0.3-0.6 Mn, 4.75-5.25 Ni, 0.20-0.35 Mo, 0.05-0.12 Al, 0.02 max Ni, bal Fe.
For cryogenic tanks.

ARMCO CT N.
M-10; 0.18 C, 1.15-1.6 Mn, 0.15-0.30 Si, 0.02-0.05 Cb, bal Fe.
Normalized: 70,000-90,000 TS; 50,000 min. YS (up to 2").
HSLA steel; for structural purposes.
ASTM A633 Gr. C. (Struc.).

ARMCO CT QT.
Quenched and tempered condition of Armco CT: YS 60,000 psi min.

ARMCO DI MAX M 15.
M-10.
Cold rolled, fully processed silicon sheet steel; non-oriented; low core loss.
For high efficiency distribution transformers and large power transformers

ARMCO DI MAX M 19.
M-10.
Cold rolled, fully processed silicon sheet steel; non-oriented; low core loss.
For 2, 4 or many pole large rotating machine laminations.

ARMCO DI MAX M 22.
M-10.
Cold rolled fully and semi-processed silicon sheet steel; non-oriented; low core loss.
For transformers, radio chokes, large rotating machines.

ARMCO DI MAX M 27.
M-10.
Cold rolled fully and semi-processed silicon sheet steel; non-oriented; low core loss.
For transformers, radio chokes, domestic appliance motors.

ARMCO DI MAX M 36.
M-10.
Cold rolled fully and semi-processed silicon sheet steel; non-oriented; low core loss.
For transformers, magnetos, motor laminations.

ARMCO DI MAX M 43.
M-10.
Cold rolled fully and semi-processed silicon sheet steel; non-oriented; low core loss.
For generators and motor laminations.

ARMCO DI MAX M 45.
M-10.
Cold rolled fully and semi-processed silicon sheet steel; non-oriented; low core loss.
For generators, motor laminations.

ARMCO DI MAX M 47.
M-10.
Cold rolled semi-processed silicon sheet steel; non-oriented; low core loss.
For motor laminations.

ARMCO ENAMELING IRON.
M-10; 0.1 max C, bal Fe.
For enameling purposes; sheets.

ARMCO FORMABLE 50.
M-10; 0.03 C 0.60 Mn, 0.03 Cb, bal Fe.
Hot rolled: 68,000 TS; 58,000 YS; 30 El; 75 Rock B.
For automotive structurals.

RMCO FORMABLE 70.
M-10; 0.04 C, 0.10 Cb, bal Fe.
Hot rolled: 88,000 TS; 77,000 YS; 23 El; 92 Rock B.
Automotive structurals.

ARMCO GAINEX (35 40 45 50 SA).
M-10; 0.1-0.2 C, 0.6-0.9 Mn, 0.005-0.010 Cb, 0.008-0.015 N, bal Fe.
Hot rolled: 58-75,000 TS; 40-55,000 YS; 33-29 El; 60-80 Rock B.
For LPG tanks, automotive structurals, shelving.

ARMCO HIGH STRENGTH A.
M-10; 0.12 max C, 0.6 max Mn, 0.25-0.7 Si, 0.50-1.0 Cr, 0.8 max Ni, 0.1 max Mo, 0.3-0.5 Cu, 0.07 max Ti, bal Fe.
1/2" max thickness: 70,000 TS; 50,000 YS; 21 El.
(thicker metal has higher C, Mn, and lower strength)
For bridges, buildings, railroad cars, agricultural equipment, vent stacks.
ASTM Standards A242, A588.

ARMCO HIGH STRENGTH A588.
M-10; 0.20 C, 1.0 Mn, 0.3 Cu, 0.55 Cr, 0.40 Ni, V, bal Fe.
70 Ksi TS; 50 Ksi YS; 18 El.
HSLA steel with good atmosphere corrosion resistance.

ARMCO HIGH STRENGTH B.
M-10; 0.22 max C, 0.85-1.25 Mn, 0.30 max Si, 0.2 min Cu, 0.02 min V, bal Fe.
3/4" max thickness: 70,000 TS; 50,000 YS.
(thicker metal has lower strength).
For bridges, buildings, railroad cars, dam components, tractors, trailers.
ASTM Standard A441.

ARMCO HIGH STRENGTH C-42.
M-10; 0.21 max C, 1.35 max Mn, 0.3 max Si, 0.005-0.05 Cb or 0.01-0.10 V, bal Fe.
60,000 TS; 42,000 YS; 24 El.
For truck trailers, auto parts, utility poles.
ASTM Standard A572.

ARMCO HIGH STRENGTH C-45.
M-10; 0.22 max C, 1.35 max Mn, 0.30 max Si, 0.005-0.05 Cb or 0.01-0.1 V, bal Fe.
60,000 TS; 45,000 Y S; 19 EL (8 in. gage).
For miscellaneous structural work on bridges, buildings, railroad cars.
ASTM Standard A572.

ARMCO HIGH STRENGTH C-50.
M-10; 0.23 max C, 1.35 max Mn, 0.30 max Si, 0.005-0.05 Cb, and/or 0.01 -0.1 V, bal Fe.
65,000 TS; 50,000 YS; 18 El(8 in. gage).
For miscellaneous structural work on bridges, buildings, truck and trailer bodies.
ASTM Standard A572.

ARMCO HIGH STRENGTH C-55.
M-10; 0.25 max C, 1.35 max Mn, 0.30 max Si, 0.005-0.05 Cb and/or 0.01- V, bal Fe.
70,000 TS; 55,000 YS; 17 El (8 in. gage).
For miscellaneous structural work on bridges, buildings, railroad cars, trucks and trailers.
ASTM Standard A572.

ARMCO HIGH STRENGTH C-60.
M-10; 0.26 max C, 1.35 max Mn, 0.30 max Si, 0.005-0.05 Cb and/or 0.01-0.10 V, bbal Fe.
75,000-TS; 60,000 YS; 15 El (8 in. gage).
For miscellaneous structural work on bridges, buildings, railroad cars, tanks, trucks.
ASTM Standard A 572.

ARMCO HIGH STRENGTH C-65.
M-10; 0.26 max C, 1.35 max Mn, 0.30 max Si, 0.005-0.05 Cb, and/or 0.01-0.10 V, bal Fe.
80,000 TS; 65,000 YS; 15 El (8 in. gage).
For miscellaneous structural work on bridges, transmission towers, utility poles, buildings.
ASTM Standard A 572.

ARMCO HIGH STRENGTH C-70.
M-10; 0.26 max C, 1.35 max Mn, 0.30 max Si, 0.005-0.05 Cb, and/or 0.01-0.10 V, bal Fe.
85,000 TS; 70,000 YS.
For miscellaneous structural work on bridges, transmission towers, buildings, railroad cars, truck and trailer bodies, tanks.

ARMCO HIGH STRENGTH D.
M-10; 0.28 max C, 1.1-1.6 Mn, 0.3 max Si, 0.20 min Cu, bal Fe.
3/4" max thickness: 70,000 TS; 50,000 YS; 18 El.
For bolted or riveted bridges, buildings, mining applications.
ASTM Standard A440.

ARMCO HY-80.
M-10; 0.18 max C, 0.10-0.40 Mn, 0.15-0.35 Si, 2.0-3.25 Ni, 1.0-1.8 Cr 0.2-0.6 Mo, bal Fe..
5/8" max thickness: 80,000-100,000 YS; 19 El.
For submarine hulls, deck and hatch sections, diving bells, helicopter decks, other navy and marine applications; weldable.
MIL-S-16216 G (ships), (plate).

ARMCO HY-100.
M-10; 0.20 C, 0.10-0.40 Mn, 0.15-0.35 Si, 2.25-3.50 Ni, 1.0-1.8 Cr, 0.2-0.06 Mo, bal Fe.
5/8" max thickness: 100,000-120,000 YS; 18 El.
For submarine hulls, marine diving bells, deck sections; weldable.
MIL-S-16216 G (Ships), (plate).

ARMCO IF STEEL (HR OR CR).
M-10; 0.006 C, 0.25 Mn, 0.06 Cb, 0.06 Ti, bal Fe.
Annealed: 45,000 TS; 22,000 YS; 45 El; 45 Rock B.
For deep drawing, automotive and appliances.

ARMCO LONG TERNES.
M-10.
Pb-Sn coated steel.
Electronic chassis and fuel tanks.
ASTM Standard A440.

ARMCO LO-TEMP, CLASS 1.
M-10; 0.20 max C, 0.70-1.35 Mn, 0.15-0.50 Si, 0.25 max Cr, 0.25 max N 0.08 max Mo, 0.35 max Cu, bal Fe.
1 1/4" max thickness: 70,000-90,000 TS; 50,000 YS; 18 El (8 in. gage).
Weldable, fine-grain, good toughness to - 75°F.
For off shore platforms, ships, process and pressure tanks, storage tanks.
ASTM A537, Cl. 1.

ARMCO LTM.
M-10; 0.14 max C, 0.09-1.35 Mn, 0.15-0.30 Si, bal Fe.
3/4" max thickness: 62,000-82,000 TS; 42,000 YS; 20 El (8 in. gage).
For pressure vessels; fine grain; weldable.
Good toughness to - 75°F. for pipes, tanks, refrigerated containers.

ARMCO LTM-N.
Normalized condition of ARMCO LTM: 42,000 psi min YS.
ASME SA 662.
ARMCO LTM QT.
Quenched and tempered condition of ARMCO LTM: 50,000 psi min YS. (102").
ASTM A 678 Gr A (Struc.).

ARMCO NI-COP.
M-10; 0.07 max C, 0.4-0.7 Mn, 0.7-1.0 Ni, 0.6-0.9 Cr, 1.0-1.3 Cu, 0.15-0.25 Mo, 0.02 Cb, bal Fe.
General structural applications, usually when low ambient temperatures are encountered.

ARMCO NITRONIC 32 STAINLESS STEEL.
M-1746, M-10; 0.15 max C, 1.0 max Si, 11.0-14.0 Mn, 0.06 max P, 0.03 max S, 16.5-19.0 Cr, 0.05-2.50 Ni, 0.20-0.45 N, bal Fe.
Ann: 120,000 (827 MPa) TS; 65,000 (448 MPa) YS; 55 El; 70 RA; 9 Rb.
Non-magnetic, austenitic stainless for pole line hardware, springs, cold-headed parts.

ARMCO NITRONIC 33 STAINLESS STEEL.
M-1746, M-10; 0.08 max C, 1.0 max Si, 11.5-14.5 Mn, 0.060 max P, 0.03 max S, 17.0-19.0 Cr, 2.25 -3.75 Ni, 0.2-0.4 N, bal Fe.
Ann: 120,000 psi (827 MPa) TS; 65,000 psi (448 MPa) YS; 55 El; 70 RA; 96 Rb.
Austenitic stainless for heat exchangers, chemical equipment.
ASTM A242 XM-29.

ARMCO NITRONIC 40 STAINLESS STEEL.
M-1746,M-10; 0.08 max C, 1.0 max Si, 8.0-10.0 Mn, 0.06 max P, 0.030 max S, 19.0-21.0 Cr, 5.5-5.7 Ni, 0.15-0.40 N, bal Fe.
Ann: 112,000 psi (772 MPa) TS; 68,000 psi (469 MPa) YS; 44 El; 96 Rb.
High tensile: 145,000 psi(1000 MPa) TS; 130,000 psi (896 MPa) YS; 20 El; 34 Rc.
Aircraft exhaust systems; cryogenic applications.
Austenitic stainless.
ASM 5656; 5595.

ARMCO NITRONIC 50 STAINLESS STEEL.
M-1746, M-10; 0.06 max C, 1.0 max Si, 4.0-6.0 Mn, 0.040 max P, 0.030 max S, 20.5-23.5 Cr, 11.5-13.5 Ni, 1.5-3.0 Mo, 0.20-0.40 N, 0.10-0.30 Cb, 0.10-0.30 V, bal Fe.
Ann: 120,000 psi (862 MPa) TS; 65,000 psi (448 MPa) YS; 45 El; 65 Ra; 23 Rc.
Austenitic stainless for petrochemical, pulp and paper, food processing industries.
ASM 5861, 5764; ASTM A240 XM-19.

ARMCO NITRONIC 60 STAINLESS STEEL.
M-1746, M-10; 0.10 max C, 3.5-4.5 Si, 7.0-9.0 Mn, 16.0-18.0 Cr, 8.0-9.0 Ni, 0.08-0.18 N, bal Fe.
Austenitic stainless steel with good galling resistance.

ARMCO ORIENTED M-3.
M-10.
Iron-silicon electrical steel sheet with oriented properties; low core loss; thickness 9 mils.

ARMCO ORIENTED M-4.
M-10.
Iron-silicon electrical steel sheet with oriented properties; low core loss; thickness 11 mils.

ARMCO ORIENTED M-5.
M-10.
Iron-silicon electrical steel sheet with oriented properties; low core loss; thickness 11, 12 mils.

ARMCO ORIENTED M-6.
M-10.
Iron-silicon electrical steel sheet with oriented properties; low core loss; thickness 12, 14 mils.

ARMCO PH 12 9 MO.
M-10; 0.035 C, 0.002 Mn, 0.03 Si, 11.85 Cr, 8.65 Ni, 1.5 Mo, 1.65 Al, 0.005 N, bal Fe.
Precipitation hardening steel (Experimental).

ARMCO PH 13-8 MO.
M-10; 0.05 max C, 12.25-13.25 Cr, 7.5-8.5 Ni, 2.0-2.5 Mo, 0.9-1.35 Al, bal Fe.
Martensitic stainless; precipitation hardenable.
Cond H 950: 225,000 TS; 205,000 YS; 12 El.
For springs, stampings, fasteners; better transverse properties than 17-7 PH.

ARMCO PH 14-8 MO.
M-10; 0.05 C, 13.5-15.5 Cr, 7.5-9.5 Ni, 2.0-3.0 Mo, 0.75-1.25 Al, bal Fe.
Semi-austenitic; precipitation hardened.
Cond SRH 900: 230,000 TS; 215,000 YS; 6 El.
For springs, stampings, watch and instrument parts, fasteners; stainless.

ARMCO PH 15-7 MO.
M-10; 0.09 max C, 14.0-16.0 Cr, 6.5-7.75 Ni, 2.0-3.0 Mo, 0.75-1.50 Al bal Fe.
Semi-austenitic; precipitation hardened.
Annealed: 130,000 TS; 55,000 YS; 35 El.
Hardened: 220,000-265,000 TS; 210,000-260,000 YS; 7-2 El; 45-50 Rock C.
For springs, watch and instrument parts as gears, cams, levers, fasteners; stainless.

ARMCO QTC.
M-10; 0.20 max C, 1.15-150 Mn, 0.20-0.50 Si, 0.20 Max Cr, 0.15 max Ni, 0.08 max Mo, 0.20 min Cu, 0.0015 min B, bal Fe.
3/4" max thickness: 100,000-120,000 TS; 80,000 YS; 18 El.
Quenched and tempered, fine-grain, weldable; for off-shore platforms, ships, railroad cars.

ARMCO QTC.
M-10; 0.20 C, 1.0-1.6 Mn, 0.20-0.50 Si, bal Fe.
Q+T: 95,000 TS; 75,000 YS; 19 El.
HSLA Steel.

ARMCO SSS 100 A514 A517 GRADE E.
M-10; 0.20 max C, 0.40-0.70 Mn, 0.20-0.35 Si, 1.4-2.0 Cr, 0.4-0.6 Mo, 0.2-0.4 Cu, 0.04-0.1 Ti or V, 0.0015-0.005 B, bal Fe.
2 1/2" max thickness: 115,000-135,000 TS; 100,000 YS; 18-16 El.
Quenched and tempered steel for bridges, buildings, crane booms, ships, earth moving equipment.
ASTM Standard A514, a517.

ARMCO SSS 100A A 514 A517 GRADE D.
M-10; 0.20 max C, 0.40-0.70 Mn, 0.20-0.35 Si, 0.85-1.2 Cr, 0.15-0.25 Ni, 0.2-0.4 Cu, 0.04-0.1 Ti or V, 0.0015-0.005 B, bal Fe.
1 1/4" max thickness: 115,000-135,000 TS; 100,000 YS; 18-16 El.
Quenched and tempered steel for bridges, buildings, crane booms, ships, earth moving equipment.
ASTM Standard A514, A517.

ARCO SSS 100B, A514 A517 GRADE L.
M-10; 0.20 max C, 0.40-0.70 Mn, 0.20-0.35 Si, 1.15-1.65 Cr, 0.25-0.40 Mo, 0.20-0.40 Cu, 0.04-0.10 Ti or V, 0.0015-0.005 B, bal Fe.
2" max thickness: 115,000-135,000 TS; 100,000 YS; 18-16 El.
Quenched and tempered steel for buildings, bridges, ordnance, ships, earth moving equipment, crane booms.
ASTM Standard a514, a517.

ARMCO SUPER-LO-TEMP, CLASS 2.
M-10; 0.20 max C, 0.70-1.35 Mn, 0.15-0.50 Si, 0.25 max Cr, 0.25 max Ni, 0.08 max Mo, 0.35 max Cu, bal Fe.
1 1/4" max thickness: 80,000-100,000 TS; 60,000 YS; 22 El.
Quinched and tempered, fine-grain, excellent low temperature properties. For refrigerated storage tanks.
ASTM A537.

ARMCO TRAN COR H 2 (0 012").
M-10.
Iron silicon electrical steel sheet with oriented properties; very low core loss and high permeability.
For power and distribution transformers.

ARMCO TRAN COR H 3 (0 014").
M-10.
Iron silicon electrical steel sheet with oriented properties; very low core loss and high permeability.
For power and distribution transformers.

ARMCO TYPE 301.
M-10; 0.10-0.15 C, 16-18 Cr, 6-8 Ni, bal Fe.
Annealed: 90,000 TS; 45,000 YS; 55 El; 65 RA; 143 Brin.
For chemical plant equipment; Type 301; stainless, austenitic.

ARMCO TYPE 302 B.
M-10; 0.08-0.15 C, 17-19 Cr, 8-10 Ni, 2-3 Si, bal Fe.
For heat resistant parts; Type 302 B; heat and corrosion resistant.

ARMCO TYPE 303.
M-10; 0.15 max C, 17-19 Cr, 8-10 Ni, 0.07 max S, P, Se, 0.6 max Zr, Mo, bal Fe.
Cold drawn: 100,000 TS; 60,000 YS; 40 El; 53 RA; 212 Brin.
Annealed: 90,000 TS; 35,000 YS; 50 El; 55 RA; 160 Brin.
For screw machine products, hardware, bolts, nuts, screws; Type 303; stainless, free-cutting.

ARMCO TYPE 303 SE.
M-10; 0.15 max C, 17-19 Cr, 8-10 Ni, 0.15 min Se, bal Fe.
Annealed: 80,000 TS; 35,000 YS; 50 El; 70 RA; 150 Brin.
For stainless parts, screw machine products; Type 303 Se; stainless, free-cutting.

ARMCO TYPE 308.
M-10; 0.08 max C, 19-21 Cr, 10-12 Ni, bal Fe.
Annealed: 85,000 TS; 35,000 YS; 50 El; 65 RA; 165 Brin.
For chemical and plastic plant equipment; stainless, austenitic.

ARMCO TYPE 309.
M-10; 0.20 max C, 22-24 Cr, 12-15 Ni, bal Fe.
Annealed: 90,000 TS; 40,000 YS; 45 El; 55 RA; 150 Brin.
For furnace parts, heat treating boxes; Type 309; heat and corrosion resistant.

ARMCO TYPE 310.
M-10; 0.25 max C, 24-26 Cr, 19-22 Ni, bal Fe.
Annealed: 95,000 TS; 40,000 YS; 45 El; 65 RA; 185 Brin.
Cold drawn: 125,000 TS; 90,000 YS; 20 El; 60 RA; 220 Brin.
For furnace equipment, chemical plant equipment; Type 310; stainless, austenitic.

ARMCO TYPE 316.
M-10; 0.10 max C, 16-18 Cr, 10-14 Ni, 2-3 Mo, bal Fe.
Annealed: 80,000 TS; 30,000 YS; 60 El; 70 RA; 150 Brin.
Cold drawn: 90,000 TS; 60,000 YS; 45 El; 60 RA; 180 Brin.
For chemical and textile plant equipment; Type 316; stainless, austenitic

ARMCO TYPE 321.
M-10; 0.08 max C, 2 max Mn, 0.75 max Si, 17-19 Cr, 8-11 Ni, Ti = 5 x C, bal Fe.
Annealed: 90,000 TS; 35,000 YS; 55 El; 60 RA; 160 Brin.
For high temperature service and welded corrosion resistant; stabilized, austenitic, stainless.

ARMCO TYPE 348.
M-10; 0.10 max C, 1.5 max Mn, 17-19 Cr, 9-12 Ni, Cb = 10 x C, bal Fe.
Annealed: 85,000 TS; 40,000 YS; 45 El; 65 RA; 140 Brin.
Cold drawn: 110,000 TS; 65,000 YS; 40 El; 60 RA; 220 Brin.
For welded structures, chemical plant equipment; stainless, austenitic.

ARMCO TYPE 405.
M-10; 0.08 max C, 11.5-13.5 Cr, 0.2 Al, bal Fe.
Annealed: 75,000 TS; 45,000 YS; 30 El; 65 RA; 160 Brin.
Cold drawn: 85,000 TS; 70,000 YS; 20 El; 60 RA; 185 Brin.
For pumps, valves, tanks; Type 405; corrosion resistant.

ARMCO TYPE 410.
M-10; 0.15 max C, 11.5-13.5 Cr, bal Fe.
Annealed: 75,000 TS; 41,000 YS; 35 El; 70 RA; 155 Brin.
Cold drawn: 110,000 TS; 85,000 YS; 23 El; 65 RA; 230 Brin.
For pumps, valves, pressure vessels, motor shafts; Type 410; corrosion resistant.

ARMCO TYPE 414.
M-10; 0.15 max C, 11.5-13.5 Cr, 1.25-2.5 Ni, bal Fe.
Annealed: 120,000 TS; 95,000 YS; 20 El; 60 RA; 250 Brin.
Cold drawn: 130,000 TS; 115,000 YS; 15 El; 58 RA; 270 Brin.
For valves, propeller shafts, exhaust port inserts; Type 414; corrosion resistant.

ARMCO TYPE 416.
M-10; 0.15 max C, 12-14 Cr, 0.07 max P, S, Se, Mo, 0.6 max Zr, bal Fe.
Cold drawn; 100,000 TS; 85,000 YS; 13 El; 40 RA; 205 Brin.
Annealed: 75,000 TS; 40,000 YS; 30 El; 60 RA; 105 Brin.
For screw machine products, screws, shafts, gears; Type 416; corrosion resistant, free-cutting.

ARMCO TYPE 420.
M-10; 0.15 min C, 12-14 Cr, bal Fe.
Annealed: 95,000 TS; 50,000 YS; 25 El; 55 RA; 195 Brin.
Cold drawn: 105,000 TS; 85,000 YS; 17 El; 50 RA; 215 Brin.
For cutlery, ball bearings, surgical instruments; Type 420; corrosion resistant.

ARMCO 420 F.
M-10; 0.15-0.45 C, 12-14 Cr, 0.18-0.35 Se, bal Fe.
Annealed: 95,000 TS; 50,000 YS; 20 El; 50 RA; 195 Brin.
For cutlery, ball bearings, dental and surgical instruments; Type 420 F; free-cutting, corrosion resistant.

ARMCO TYPE 430.
M-10; 0.12 max C, 14-18 Cr, bal Fe.
Annealed: 75,000 TS; 50,000 YS; 30 El; 65 RA; 155 Brin.
Cold drawn: 85,000 TS; 70,000 YS; 20 El; 60 RA; 185 Brin.
For trim, hardware, fixtures; industrial equipment; Type 430; corrosion and heat resistant.

ARMCO TYPE 431.
M-10; 0.2 max C, 15-17 Cr, 1.5-2.5 Ni, bal Fe.
Annealed: 125,000 TS; 95,000 YS; 20 El; 55 RA; 250 Brin.
Cold drawn: 130,000 TS; 110,000 YS; 15 El; 35 RA; 270 Brin.
For aircraft and marine parts, bolts; type 431; corrosion resistant.

ARMCO TYPE 440 A.
M-10; 0.60-0.75 C, 16-18 Cr, 0.75 max Mo, bal Fe.
Annealed: 105,000 TS; 60,000 YS; 20 El; 45 RA; 215 Brin.
For cutlery, ball bearings, pump shafts, valves; Type 440 A; corrosion resistant, hardenable.

ARMCO TYPE 440 B.
M-10; 0.75-0.95 C, 16-18 Cr, 0.75 max Mo, bal Fe,
Annealed: 107,000 TS; 62,000 YS; 18 El; 35 RA; 220 Brin.
For cutlery, ball bearings, pump shafts, valves; Type 440 B; corrosion resistant, hardenable.

ARMCO TYPE 440 C.
M-10; 0.95-1.2 C, 16.18 Cr, 0.75 max Mo, bal Fe.
Annealed: 110,000 TS; 65,000 YS; 14 El; 25 RA; 230 Brin.
For cutlery, ball bearings, pump shafts, valves; Type 440 C; corrosion resistant, hardenable.

ARMCO TYPE 440 F.
M-10; 0.95-1.2 C, 16-18 Cr, 0.75 max Mo, 0.2 S, bal Fe.
Annealed: 110,000 TS; 65,000 YS; 12 El; 20 RA; 230 Brin.
For cutlery, bearings, pump shafts, valves; Type 440 F; corrosion resistant, hardenable.

ARMCO TYPE 440F SE.
M-10; 0.95-1.2 C, 16-18 Cr, 0.60 max Mo or Zr, bal Fe.
For bearings, races; Type 440 FSe; corrosion resistant.

ARMCO TYPE 446.
M-10; 0.35 max C, 1.0 Si, 23-27 Cr, bal Fe.
Annealed: 80,000 TS; 50,000 YS; 23 El; 55 RA; 175 Brin.
For furnace parts, heat treating boxes; Type 446; heat resistant.

ARMCO UNIVIT.
M-10; 0.008 max C, 0.20 Mn, bal Fe.
Ann: 40,000 TS; 25,000 YS; 45 El; 35 Rock B.
Decarburized enameling steel.
For enameling drawn parts, appliances and plumbing.

ARMCO VNT.
M-10; 0.22 max C, 1.15-1.50 Mn, 0.15-0.50 Si, 0.04-0.11 V, 0.01-0.03 N, bal Fe.
80,000-100,000 TS; 60,000 YS; 23 El.
For pressure vessels, truck bodies, heavy machinery, construction, farm equipment. Weldable, good shock resistance to -75°F.
ASTM A 633.

ARMCO VNT N.
M-10; 0.22 max C, 1.15-1.50 Mn, 0.15-0.50 Si, 0.04-0.11 V, 0.01-0.03 N, 0.35 max Cu, 0.25 max Cr, 0.25 max Ni, 0.08 max Mo, bal Fe.
Normalized: 80,000-100,000 TS; 60,000 min YS; 23 El (2").
For pressure vessels, truck bodies, farm equipment. Weldable; nil-ductility transition temperature -50°F.
ASTM A633 Gr E (Struc.).

ARMCO ZGPG ZINCGRIP-PAINTGRIP.
M-10; Galvanized sheet steel phosphated.
For structures, sheet metal work; mill bonderized.

ARMCO ZINC GRIP.
M-10; 0.03 max C, bal Fe.
For roofing, culverts; automotive; galvanized sheets.

ARMCO ZINCGRIP OS (ONE SIDE).
M-10.
Galvanized steel.
Auto body components.

ARMCO ZINCROMETAL.
M-10.
Zn rich painted steel - one side.
Auto body components.

ARMELEC.
M-691; 0.5 Si, bal Fe.
For armatures, motors; good permeability.

ARMIDE.
M-132; Carbide alloy TaC.
250,000 TS.
For tipped cutters and tools; hard, tough and wear resisting.

ARMIDE (78).
M-132; Sintered carbide tool material.
For cutting tools, finishing and light roughing cuts on steel.

ARMIDE (78B).
M-132; Sintered carbide tool material.
For cutting tools, general purpose machining of steel.

ARMIDE (350).
M-132; Sintered carbide tool material.
For cutting tools; light roughing and general finishing cuts on steel.

ARMIDE (370).
M-132; Sintered carbide tool material.
Cutting tools for heavy roughing cuts on steel.

ARMIDE (883).
M-132; Sintered carbide tool material.
For cutting tools, general purpose cutting of cast iron and non-ferrous metals.

ARMIDE GRAY.
M-132; carbide.
For tipped tools; for machining cast iron and brass.

ARMIDE RED.
M-132; carbide.
For tipped tools; for machining steel.

ARMIN 90.
M-1307; 1.3 C, 4.3 Cr, 0.85 Mo, 3.8 V, 12 W, bal Fe.
For engravers' tools, blanking and forming dies; high speed steel.

ARMORLOY see MCKAY ARMORLOY.

ARMORLOY C.
M-849; 0.09-0.15 C, 25 min Cr, 20 min Ni, bal Fe.
Welded: 87,000 TS; 60,000 YS; 45 El.
For welding electrodes for armor; shielded arc, work hardenable.

ARMSTEEL 88M.
M-690; 3 C, Si, Mn, bal Fe.
Cast: 105,000 TS; 85,000 YS; 2 El; 302 Brin.
For transmissions gears and shafts, universal joints; pearlitic malleable iron.

ARMSTRONG.
M-U.S.; 12 Cr, 5 Si, 0.5 C, bal Fe.
For heat resisting steel parts; heat resisting stainless steel.

ARMSTRONG-1.
M-U.S.; 3-50 Cr, 0.5-8 Si, C, bal Fe.
For heat and corrosion resisting parts; heat resisting.

ARMSTRONG-2.
M-U.S.; 21.6 Cr, 3.4 Si, 2.35 C, 2.25 Mn, bal Fe.
For heat and corrosion resisting parts; heat resistant cast iron.

ARMSTRONG-3.
M-U.S.; 12 Cr, 5 Si, 0.45 C, bal Fe.
For heat and corrosion resisting parts; heat and corrosion resistant.

ARMSTRONG HIGH SPEED.
M-132; 0.7 C, 18 W, 4 Cr, 1 V, bal Fe.
For tools, cutters; high speed steel.

ARMSTRONG METAL.
M-Eng.; 0.10 C, 4-6 Mn, 17 Cr, 8 Ni, 3 Cu, bal Fe.
Annealed: 70,000 TS; 52 El; 71 RA; 126 Brin.
For heat resisting parts; heat and corrosion resistant.

ARMSTRONG SELF-HARDENING.
M-132; C, alloy, bal Fe.
For tool shanks; oil hardening.

ARNE.
M-111; 0.9 C, 1.2 Mn, 0.5 Cr, 0.5 W, 0.1 V, bal Fe.
For dies; for cold work. AISI O1.

ARNGRIM 2.
M-111; 0.68 C, 0.25 Cr, 6 W, bal fe.
For permanent magnets.

ARNGRIM 3.
M-111; 0.72 C, 0.65 Cr, 6 W, bal Fe.
For permanent magnets.

ARNOX-I.
M-670; Barium ferrite + iron oxide.
Hc = 1800; Hm = 10,000; Br = 2200.
(BdHd) max = 1,000,000; Bd = 1150.
For magnets in door latches, holding assemblies, D.C. motors, focusing devices.
Permanent magnet, non-conducting. Non-oriented.

ARNOX-III.
M-670; Oxides.
Hm = 8000; Br = 1500.
Hc = 1100; Bd = 800.
(BdHd) max = 400,000.
For permanent magnets in toys and novelties, timing motors. Non-oriented.
Molded type magnet. High coercive force and low residual induction and energy product.

ARNOX-V.
M-670; Barium ferrite.
Hm = 10,000; Br = 3950; Hc = 2200.
(BdHd) max = 3,500,000; Bd = 2000.
For permanent magnets in loudspeakers, motors, generators, magnetic separation applications.
Ceramic magnet. High coercive force. Magnetized in direction of orientation.

ARNOX-VA.
M-670; Barium ferrite.
Max energy product 3,500,000.
Remanence, Br, gauss 3850.
Coercive force, Hc, oersteds 2000.
For permanent magnets (non-conducting).
Hard brittle ceramic, high coercive force; low flux density.

ARO.
M-1705; 0.50 C, 0.7 Mn, 0.7 Si, 0.2 V, 0.45 Mo, bal Fe.
Oil or water hardening shock resisting tool steel. AISI S2.

AROSTIT C13.
M-1325; 0.12 max C, 0.4 Si, 13 Cr, bal Fe.
Annealed: 75,000 TS; 40,000 YS; 35 El; 70 RA; 180 Brin.
For turbine blades, chemical plant equipment; Type 410; corrosion resistant.

AROSTIT C13S.
M-1325; 0.12 max C, 16.5 Cr, 0.25 Mo, 0.2 S, bal Fe.
Annealed: 70,000 TS; 40,000 YS; 30 El; 55 RA; 150 Brin.
Cold drawn: 130,000 TS; 120,000 YS; 2 El; 270 Brin.
For chemical plant equipment, bolts, oil refinery parts; Type 430F; corrosion and heat resistant.

AROSTIT C15.
M-1325; 0.25 C, 14.5 Cr, 1 max Ni, bal Fe.
Annealed: 95,000 TS; 50,000 YS; 25 El; 55 RA; 195 Brin.
Cold drawn: 105,000 TS; 85,000 YS; 17 El; 50 RA; 215 Brin.
For cutlery, valves, surgical instruments; Type 420; corrosion resistant.

AROSTIT C18.
M-1325; 0.25 C, 17 Cr, 1 max Ni, bal Fe.
Annealed: 75,000 TS; 42,000 YS; 28 El; 53 RA; 160 Brin.
For chemical plant equipment, bolts, shafts; Type 430; corrosion and heat resistant.

AROSTIT C20.
M-1325; 0.2 C, 13 Cr, bal Fe.
Annealed: 95,000 TS; 50,000 YS; 25 El; 55 RA; 195 Brin.
Cold drawn: 105,000 TS; 85,000 YS; 17 El; 50 RA; 215 Brin.
For turbine blades, cutlery, surgical instruments; Type 420; corrosion resistant.

AROSTIT C20N.
M-1325; 0.20 C, 13 Cr, 1.15 Mo, bal Fe.
Annealed: 100,000 TS; 55,000 YS; 20 El; 50 RA; 205 Brin.
For turbine blades, cutlery, valves; corrosion resistant.

AROSTIT C22K.
M-1325; 0.22 C, 17 Cr, 1.5 Ni, bal Fe.
Annealed: 125,000 TS; 95,000 YS; 20 El; 55 RA; 260 Brin.
Cold drawn: 130,000 TS; 110,000 YS; 15 El; 35 RA; 270 Brin.
For pump shafts, marine hardware, valve trim; Type 431; corrosion and heat resistant.

AROSTIT C22L.
M-1325; 0.35 C, 16.5 Cr, 1.15 Mo, bal Fe.
Annealed: 130,000 TS; 100,000 YS; 18 El; 52 RA; 275 Brin.
For pump shafts, marine hardware, valve trim; corrosion resistant.

AROSTIT C30.
M-1325; 0.4 C, 29 Cr, 1.3 Si, bal Fe.
For furnace parts and equipment, heat treat boxes; heat resistant, ferrititic.

AROSTIT C30H.
M-1325; 1.2 C, 1.3 Si, 29 Cr, bal Fe.
For wear plates, rolls, crushers; corrosion and heat resistant.

AROSTIT C30N.
M-1325; 1.3 C, 1.3 Si, 29 Cr, 2 Mo, bal Fe.
For wear plates, rolls, crushers; corrosion and heat resistant.

AROSTIT C40.
M-1325; 0.4 C, 13 Cr, bal Fe.
Annealed: 100,000 TS; 55,000 YS; 22 El; 52 RA; 200 Brin.
Cold drawn: 110,000 TS; 90,000 YS; 15 El; 45 RA; 225 Brin.
For turbine blades, gages, valve trim, cutlery; Type 420; corrosion resistant.

AROSTIT C90N.
M-1325; 0.9 C, 18 Cr, 1.5 Mo, 1 V, bal Fe.
Annealed: 107,000 TS; 62,000 YS; 18 El; 35 RA; 220 Brin.
For cutlery, valve parts, bearings, instruments; Type 440B; corrosion resistant, hardenable.

AROSTIT CK4.
M-1325; 0.4 C, 1.3 Si, 27 Cr, 4 Ni, bal Fe.
For furnace equipment, heat treat boxes; heat resistant.

AROSTIT CK8.
M-1325; 0.15 C, 1.5 Si, 18 Cr, 8.5 Ni, bal Fe.
Annealed: 80,000 TS; 35,000 YS; 50 El; 65 RA; 160 Brin.
For chemical plant equipment, tanks, mixers; Type 302; stainless, austenitic.

AROSTIT CK8 EXTRA.
M-1325; 0.12 max C, 18 Cr, 9.5 Ni, Ti = 4 x C, bal Fe.
Annealed: 85,000 TS; 35,000 YS; 55 El; 65 Ra; 150 Brin.
For welded structures, chemical plant equipment; Type 321; stainless, austenitic.

AROSTIT CK8SI.
M-1325; 0.1 max C, 18 Cr, 8.5 Ni, bal Fe.
Annealed: 85,000 TS; 35,000 YS; 60 El; 70 RA; 140 Brin.
For welded structures, chemical plant equipment; Type 302; Stainless, austenitic.

AROSTIT CK9N.
M-1325; 0.15 C, 2 Si, 2 Mo, 9.5 Ni, 18 Cr, bal Fe.
Annealed: 80,000 TS; 30,000 YS; 60 El; 80 RA; 135 Brin.
Cold drawn: 150,000 TS; 135,000 YS; 6 El; 300 Brin.
for acid resistant equipment, mixers, filters, agitators; Type 316; stainless, austenitic.

AROSTIT CK9N EXTRA.
M-1325; 0.12 max C, 18 Cr, 10.5 Ni, 2 Mo, Ti = 4 x C, bal Fe.
Annealed: 85,000 TS; 35,000 YS; 50 El; 75 RA; 160 Brin.
Cold drawn: 150,000 TS; 135,000 YS; 6 El; 30 Brin.
For welded structures, chemical plant equipment; Type 316 Ti; Stainless austenitic.

AROSTIT CK12.
M-1325; 0.07 max C, 17 Cr, 4.7 Mo, 13 Ni, bal Fe.
Annealed: 90,000 TS; 40,000 YS; 45 El; 65 RA; 170 Brin.
For acid resistant equipment, mixers, agitators; Type 317; stainless, austenitic.

ARPOCALLOY.
M-1012; C Ni, Mo, bal Fe.
For tool shanks; cast iron.

ARROW HIGH SPEED.
M-355; 0.7 C, 18 W, 4 Cr, 2 V, bal Fe.
For tools, dies, broaches, drills; high speed steel.

ARROW NO 16.
0.95 C, 1.2 Mn, 0.5 Cr, 0.5 W, 0.1 V, bal Fe.
For tools and dies; non-deforming oil hardening.

ARROW OIL HARDENING.
M-336; 0.6 C, 1.5 W, 0.9 Mn, 3 Cr, 1 V, bal Fe.
88,800 TS; 55,100 YS; 30 El; 189 Brin.
For tools, dies, perforated dies, taps, intricate shapes; non-shrinking.

ARROW OIL HARDENING.
M-387; 0.7 C, 2 W, bal Fe.
For tools, dies; oil hardened.

ARROW OIL HARDENING SWEDISH STEEL.
M-336; 0.95 C, 1.25 Mn, 0.5 Cr, 0.5 W, 0.1 V, bal Fe.
For thread gages, taps, master tools, dies; non-shrinking.

ARROW SPECIAL.
M-335; 1.1 C, bal Fe.
For tools, dies.

ARS.
M-373; 2.35 C, 12.0 Cr, 4.0 V, 1.0 Mo, bal Fe.
Air or oil hardening cold work tool steel, chromium type; AISI D7.

ARSENIC.
M-1755; AS, (crystalline & amorphous).
Purities: 99.9999%, 99.999%, 99.99%, commercial grade.
Forms: lump, powder.

ARSENIC BRONZE.
M-U.S.; 80 Cu, 10 Sn, 9.2 Pb, 0.8 As.
30,000 TS; 60 Brin.
For bearings; heavy duty.

ARSENICAL ADMIRALTY 30.
M-141; 71 Cu, 1.0 Sn, 0.03 As, bal Zn.
Annealed: 53,000 TS; 65 El.
Drawn: 85,000 TS; 10 El.
For condenser and heat exchanger; embrittlement-free.

ARSENICAL ADMIRALTY-443.
M-8; 71.0 Cu, 27.96 Zn, 1.0 Sn, 0.04 As.
Soft: 48,000 TS; 18,000 YS; 65 El; B 25 Rock.
For condenser tubes, heat exchangers.
Resists dezincification, corrosion resistant.

ARSENICAL ALUMINUM BRONZE 53.
M-141; 5.5 Al, 0.25 As, bal Cu.
Annealed: 60,000 TS; 22,000 YS; 60 El.
For condenser tubes; resists sea-water corrosion.

ARSENICAL ALUMINUM BRASS 54.
M-141; 76 Cu, 2 Al, 0.03 As, bal Zn
Annealed: 60,000 TS; 27,000 YS; 55 El.
For condensers and heat exchangers; free of embrittlement.

ARZADE.
M-14; 3 C, 1.5 Si, bal Fe.
For castings; malleable cast iron.

ARZITE.
M-14; 3 C, Si, Mn, bal Fe.
For gears, fittings, housings; malleable iron.

ARZITE METAL.
M-14; 3.2 C, 1.5 Si, 1.5 Ni, 0.8 Cr, bal Fe.
For castings; corrosion resistant malleable iron.

ARZON METAL.
M-14; 0.90 Si, 0.85 Mn, 2.30 C, 0.09 S, 0.16 P, bal Fe.
Cast: 90,000 TS; 80,000 YS; 5 El.
For malleable iron castings, fittings, plumbing; close grain, white fracture.

AS 05.
M-1361; 0.5 Mg, 0.5 Si, bal Al.
130-245 N/mm^2 TS; 70-195 N/mm^2 YS; 10-15 El.
Heat treatable, corrosion resistance, weldable and good anodizing characteristics.
Type Al Mg Si 0.5.
Similar to AA 6063.

AS 08.
M-1361; 0.80 Mg, 0.80 Si, bal Al.
195-275 N/mm^2 TS; 100-195 N/mm^2 YS; 12-16 El.
Heat treatable, corrosion resistant, weldable and good anodizing characteristics.
Type Al Mg Si 0.8.

AS 5.
M-1176; 1-3 Cu, 5 Si, 0.5 Mg, bal Al.
Heat treated: 50,000-57,000 TS; 40,000-46,000 YS; 5-2 El; 110-140 Brin.
For light alloy parts; age hardenable.

AS 9.
M-1361; 0.40 Mg, 10.0 Si, bal Al.
For welding rod.

AS 10.
M-1176; 2.0-2.5 Cu, 9.5-10.5 Si, 1.4 Mg, 0.8-1.2 Ni, bal Al.
Cast: 36,000-50,000 TS; 34,000-42,000 YS; 0.5-0.3 El; 95-125 Brin.
For light alloy parts; corrosion resistant.

AS 10.
M-1361; 0.9 Mg, 1.0 Si, 0.6 Mn, Cr, bal Al.
195-313 N/mm^2 TS; 100-225 N/mm^2 YS; 6-18 El.
Heat treatable, corrosion resistant, weldable and good anodizing characteristics.
Type Al Mg Si 1.

AS 15.
M-1361; 0.8 Mg, 1.0 Si, 0.5 Mn, bal Al.
195-315 N/mm^2 TS; 100-255 N/mm^2 YS; 6-18 El.
Heat traetment, corrosion resistant, weldable and good anodizing characteristics.
Type AL Mg Si 1.

AS 17.
M-1361; 0.80 Mg, 1.0 Si, 0.6 Mn, Cr, Be, bal Al.
195-315 N/mm^2 TS; 100-255 N/mm^2 YS; 6-18 El.
Special mining quality.
Type Al Mg Si 1 Be.

AS 20.
M-1361; 0.25 Cu, 0.9 Mg, 0.7 Si, Cr, bal Al.
193-315 N/mm^2 TS; 100-255 N/mm^2 YS; 6-18 El.
Heat treatable, corrosion resistant, weldable and good anodizing characteristics.
Type Al Mg Si 1 Cu.
Similar to AA 6061.

AS 41A (DIE CASTING).
M-Various foundries; 3.5-5.0 Al, 0.20-0.50 Mn, 0.12 max Zn, 0.50-1.5 Si, 0.06 max Cu, 0.03 max Ni, bal Mg. (other 0.30 max).
As cast: 32,000 psi TS; 22,000 psi YS; 4 El.

AS 41 A (INGOT).
M-43; 3.7-4.8 Al, 0.22-0.48 Mn, 0.10 max Zn, 0.60-1.4 Si, 0.04 max Cu 0.01 max Ni, bal Mg (other 0.30 max).
Mg ingots to be used for die castings.

AS 55.
M-31, M-60; 5 W, 1 Zr, 0.05 Cr, 0.06 C, Y, bal Cb.
For jet engines, missile components.

AS 303.
M-1741; 2.0-4.0 Cu, 4.0-6.0 Si, 0.08 max Fe, 0.07 max Mn, 0.15 max Mg 0.30 max Ni, 0.50 max Zn, 0.20 max Ti, 0.05 max Sn, bal Al.
Sandcast, T6: 248 MPa TS; 138 MPa YS; 1 El; 100 Brin.
ADC H49-3; BS LM4.

AS 305.
M-1741; 2.0-4.0 Cu, 8.5-10.5 Si, 0.9 max Fe, 0.5 max Mn, 0.6-1.5 Mg, 0.5 max Ni, 1.0 max Zn, 0.25 max Ti, bal Al.
Per. Mold cast, T6: 248 MPa TS; 193 MPa YS; 1 El; 105 Brin.

AS 307.
M-1741; 0.7-2.5 Cu, 9.0-11.5 Si, 1.0 max Fe, 0.5 max Mn, 0.3 max Mg, 1.0 max Ni, 1.2 max Zn, 0.2 max Ti, 0.2 max Sn, bal Al.
Pressure cast, F1: 248 MPa TS; 3 El; 80 Brin.
ADC H49-2; BS LM2.

AS 313.
M-1741; 3.0-4.0 Cu, 7.5-9.5 Si, 1.3 max Fe, 0.5 max Mn, 0.3 max Mg, 0.5 max Ni, 3.0 max Zn. 0.2 max Ti, 0.2 max Sa, bal Al.
Pressure cast, F1: 269 MPa TS; 159 MPa YS; 2 El, 85 Brin.
ADC H49-16; BS LM24.

AS 315.
M-1741; 3.0-4.5 Cu, 10.5-12.0 Si, 1.3 max Fe, 0.5 max Mn, 0.1 max Mg, 0.5 max Ni, 1.0 max Zn, 0.35 max Ti, 0.35 max Sn, bal Al.
Pressure cast, F1; 331 MPa YS; 165 MPa YS; 2.5 El.
SAE 303.

AS 317.
M-1741; 1.5-2.5 Cu, 6.0-8.0 Si, 0.08 max Fe, 0.2-0.6 Mn, 0.3 max Mg, 0.5 max Ni, 1.0 max Zn, 0.2 max Ti, 0.1 max Sn, bal Al.
Sand cast, HT-T6: 280 MPa min TS; 250 MPa YS; 0.5 min El; 110 Brin.
BS LM27.

AS 401.
M-1741; 0.6 max Cu, 11.0-13.0 Si, 1.0 max Fe, 0.5 max Mn, 0.15 max Mg 0.5 max Ni, 0.4 max Zn, 0.2 max Ti, 0.15 max Sn, bal Al.
Pressure cast, F1; 262 MPa TS; 131 MPa YS; 2 El; 65 Brin.
BS LM20.

AS 601.
M-1741; 0.25 Cu, 6.5-7.5 Si, 0.5 max Fe, 0.35 max Mn, 0.30-0.50 Mg, 0.35 max Zn, 0.25 max Ti, bal Al.
Sandcast, T6: 228 MPa TS; 152 MPa YS; 4 El; 75 Brin.
ADC H49-18. BS LM25.

AS 607.
M-1741; 0.10 max Cu, 10.0-13.0 Si, 0.6 max Fe, 0.3-0.7 Mn, 0.20-0.6 Mg, 0.10 max Ni, 0.10 max Zn, 0.2 max Ti, 0.05 max Sn, bal Al.
Sand cast, T6: 240 MPa min TS; 230 MPa YS; 0.1 El, 95 Brin.
BS LM9.

ASARCO ZDC NO. 7.
M-815, m-314; 3.5-4.3 Al, 0.25 max Cu, 0.005-0.020 Mg, 0.005-0.020 Ni, 0.003 max Cd, bal Zn.
Die Cast: 41,000 TS; 10 El; 82 Brin.
For hardware, housings, instument cases. Die castings.

ASARCOLO 117.
M-815; 23 Pb, 5 Cd, 45 Bi, 8 Sn, 19 In.
Cast: 5400 TS; 12 Brin.
For hermetic seals, heat transfer medium, fusible cores; M.P. 117.

ASARCOLO 158.
M-815; 27 Pb, 10 Cd, 50 Bi, 13 Sn.
6000 TS; 9 Brin.
For foundry cores, pipe bending, for anchoring; M.P. 158°F; fusible alloy.

ASARCOLO 158-162.
M-815; 50 Bi, 25 Pb, 12.5 Cd, 12.5 Sn.
For fire protection devices; fusible alloy; M.P. 158-162°F.

ASARCOLO 158-190.
M-815; 42.5 Bi, 37.7 Pb, 11.3 Sn, 8.5 Cd.
5500 TS; 9 Brin.
For foundry patterns, coating core boxes, spotting fixtures; M.P. 158-190°F; fusible alloy.

ASARCOLO 205.
M-815; 52 Bi, 32 Pb, 16 Sn.
For fire protection devices; fusible alloy; M.P. 205°F.

ASARCOLO 205-215.
M-815; 50 Bi, 31 Pb, 19 Sn.
For fire protection devices; fusible alloy; M.P. 205-215°F.

ASARCOLO 217-440.
M-815; 28.5 Pb, 14.5 Sn, 9 Sb, 48 Bi.
13,000 TS; 19 El.
For anchoring punches in stamping dies, making chuck jaws; M.P. 217-440°F; fusible alloy.

ASARCOLO 243-260.
M-815; indium alloy.
For soldering and sealing; M.P. 243-260°F; fusible alloy.

ASARCOLO 255.
M-815; 45 Pb, 55 Bi.
6400 TS; 10 Brin.
For foundry molds dies, proof casting cavities; M.P. 255°F; fusible alloy.

ASARCOLO 255-300.
M-815; 50 Bi, 50 Pb.
For fire protection devices; fusible alloy; M.P. 255-300°F.

ASARCOLO 281.
M-815; 58 Bi, 42 Sn.
8000 TS; 22 Brin.
For molds and patterns; M.P. 281°F; fusible alloy.

ASARCOLO 281-338.
M-815; 40 Bi, 60 Sn.
8000 TS; 22 Brin.
For soldering and sealing; M.P. 281-338°F; fusible alloy.

ASARCOLO 362.
M-815; 38 Pb, 62 Sn.
For fire protection devices; fusible alloy; M.P. 362°F.

ASARCOLOY NO. 7.
M-314; 0.75-3.0 Ni, bal Cd.
Cast: 164,000 TS; 117,000 YS; 19 El; 43 RA; 33 Brin.
For bearings; for severe service.

ASARCON 3.
M-815, m-314; 4.15 Al, 0.5 Mg, bal Zn.
Die Cast: 40,000 TS; 63,000 YS; 8 El; 80 Brin.
For die castings, hardware, door knobs, instrument cases, coin chutes, ornaments. High fluidity and castability.

ASARCON 7.
M-815, M-314; 3.5-4.3 Al, 0.25 max Cu, 0.010-0.020 Mg, 0.005-0.020 Ni 0.003 max Cd, bal Zn.
Die Cast: 41,000 TS; 14 El; 76 Brin.
For die castings, hardware, instrument cases, ornaments, coin chutes. High fluidity and castability.

ASARCON 50N.
M-314; 88 Cu, 5 Sn, 2 Zn, 5 Ni.
H: 80,000 TS; 55,000 YS; 10 El; 180 BHN.

ASARCON 55.
M-314, m-815; 5 Sn, 5 Pb, 5 Zn, 85 Cu.
Cast: 45,000 TS; 21,400 YS; 28 El; 72 Brin.
For water pump impellers, bushings, fittings; SAE 40; leaded red brass.

ASARCON 59.
M-314, m-815; 5 Sn, 9 Pb, 1 Zn, bal Cu.
Cast: 38,000 TS; 21,000 YS; 20 El; 66 Brin.
For bearings, bushings, sleeves; SAE 66; leaded bronze.

ASARCON 61.
M-314, m-815; 6 Sn, 1.5 Pb, 4.5 Zn, 88 Cu.
Cast: 45,500 TS; 23,000 YS; 35 El; 76 Brin.
For gears, bearings, bushings; SAE 622; continuous cast.

ASARCON 77.
M-314, m-815; 7 Sn, 7 Pb, 3 Zn, 83 Cu.
Cast; 40,000 TS; 27,000 YS; 16 El; 72 Brin.
For bearings, spring bushings, thrust washers; SAE 660; leaded bronze.

ASARCON 80.
M-314, m-815; 8 Sn. 4 Zn. 88 Cu.
Cast: 49,000 TS; 23,000 YS; 18 El; 77 Brin.
For gears, bearings, sleeves; SAE 620; hard bronze.

ASARCON 100.
M-314, m-815; 10 Sn, 2 Zn, 88 Cu.
Cast: 51,000 TS; 28,000 YS; 18 El; 92 Brin.
For gears, bearings, liners, sleeves; SAE 62; hard bronze.

ASARCON 102.
M-314, m-815; 10 Sn, 2 Pb, 88 Cu.
Cast: 49,000 TS; 25,000 YS; 18 El; 86 Brin.
For bushings, bearings, sleeves; SAE 63; leaded gun metal.

ASARCON 102N.
M-314; 84 Cu, 10 Sn, 2.5 Pb, 3.5 Ni, bal Zn.
Cast 53,000 TS; 31,000 YS; 15 El; 92 Brin.
For hardware, marine parts, liners; free-cutting, leaded bronze.

ASARCON 110.
M-314, m-815; 11 Sn, 89 Cu.
Cast: 51,000 TS; 29,000 YS; 18 El; 100 Brin.
For gears, worms, shafts; Al-bronze; continuous cast.

ASARCON 210.
M-314, m-815; 2.5 Sn, 10 Pb, 7.5 Zn, 80 Cu.
Cast: 34,000 TS; 18,000 YS; 22 El; 62 Brin.
For bearings, bushings, liners; leaded bronze.

ASARCON 230.
M-815, m-314; 1.5-2.5 Sn, 27.0-35.0 Pb, 0.25-0.75 Ni, bal Cu.
Cast: 16,300 TS; 9000 YS; 10 El; 30 Brin.
For self-lubricating bearings, bushings and seals, jet engine fuel pumps.
Leaded bronze. Free-machining.

ASARCON 310 N.
M-314, m-815; 77 Cu, 3 Sn, 10 Pb, 8 Zn, 2 Ni.
Cast: 48,000 TS; 28,000 YS; 8 El; 80 Brin.
For bearings. Continuous cast.

ASARCON 520.
M-314, m-815; 5 Sn, 20 Pb, 75 Cu.
Cast: 28,700 TS; 22,800 YS; 8 El; 57 Brin.
For bearings, bushings, liners; leaded bronze.

ASARCON 616.
M-314; 78 Cu, 6 Sn, 16 Pb.
34,200 TS; 23,000 YS; 12 El; 62 BHN.

ASARCON 773.
M-314, m-815; 7 Sn, 7 Pb, 3 Zn, 83 Cu.
Cast: 40,000 TS; 27,000 YS; 16 El; 72 Brin.
For bearings, bushings, liners; SAE 660; leaded bronze.

ASARCON 1010.
M-314, m-815; 10 Sn, 10 Pb, 80 Cu.
Cast: 41,000 TS; 26,000 YS; 10 El; 80 Brin.
For bearings, bushings, liners; SAE 64 leaded bronze.

ASARCON 1010B.
M-815; 10 Sn, 10 Pb, bal Cu.
Cast: 45,000 TS; 29,000 YS; 5 El; 93 Brin.
For gears, bearings, bushings, shafts; continous cast.

ASC 1.
M-569; 0.25-0.30 C, 0.70 Mn, 0.40 Si, bal Fe.
Ann: 70,000 TS; 40,000 YS; 24 El; 150 Brin.
Uses: castings for construction, railroad.

ASC 2.
M-569; 0.40-0.50 C, 0.70 Mn, 0.40 Si, bal Fe.
Ann: 80,000 TS; 40,000 YS; 18 El; 170 Brin.
Uses: castings for dies.
Flames hardenable, wear resistant.

ASC 3.
M-569; 0.60-0.80 C, 0.70 Mn, 0.40 Si, bal Fe.
Norm. & Temper: 110,000 TS; 60,000 YS; 10 El; 190 Brin.
Uses: castings for dies.
Flame hardenable, wear resistant.

ASC 5.
M-569; 0.20 C, 0.70 Mn, 0.50 Si, 1.5 Cr, 1.1 Mo, bal Fe.
Norm. & Temper: 70,000 TS; 40,000 YS; 20 El; 160 Brin.
Uses: Turbine casings and supports.
High temp. steel castings.

ASC 6.
M-569; 0.18 C, 0.60 Mn, 2.5 Cr, 1.1 Mo, bal Fe.
Norm. & Temper: 70,000 TS; 40,000 YS; 20 El; 160 Brin.
Uses: Turbine casings and supports.
High temp. steel castings.

ASC 7.
M-569; 0.25 C, 0.60 Mn, 0.50 Si, 0.60 Mo, bal Fe.
Norm. & Temper; 70,000 TS; 45,000 YS; 22 El; 160 Brin.
Uses; Turbine casings and supports.
High temp. steel castings.

ASC 8.
M-569; 0.30 C, 1.30 Mn, 0.50 Si, 0.40 Mo, bal Fe.
Norm. & Temper: 90,000 TS; 60,000 YS; 180 Brin.
Quenched: 200,000 TS; 150,000 YS; 400 Brin.
High strength castings - dies.
Weldable, water hardenable.

ASC 11.
M-569; 0.40 C, 1.15 Mn, 0.40 Si, 1.15 Cr, 0.40 Mo, bal Fe.
Physicals depend on heat treatment.
High strength castings.
Similar to wrought AISI 4140.

ASC 15.
M-569; 0.40 C, 1.15 Mn, 0.40 Si, 1.15 Cr, 1.75 Ni, 0.40 Mo, bal Fe.
Physicals depend on heat treatment.
High strength casting - dies.
Similar to wrought AISI 4340.

ASC-B10.
M-1248; 0.4 Be, 2.6 Co, bal Cu.
Annealed: 50,000 TS; 25,000 YS; 30 El; 10 RA; 240 Brin.
For current carrying springs, switch parts; heat treatable, corrosion and wear resistant.

ASC-B25.
M-1248; 1.9 Be, 0.3 Co, bal Cu.
Annealed: 75,000 TS; 30,000 YS; 50 El; 70 RA; 380 Brin.
Aged: 205,000 TS; 170,000 YS; 1 El; 3 RA; 380 Brin.
For electrical contacts, springs, diaphragms; age hardenable, corrosion resistant.

ASC-B165.
M-1248; 1.7 Be, 0.2-0.35 Co, bal Cu.
Annealed: 60,000 TS; 28,000 YS; 60 El; 79 Brin.
Aged: 200,000 TS; 185,000 YS; 1 El; 410 Brin.
For electrical contacts, springs, slips, diaphragms; heat treatable, corrosion and wear resistant.

ASCO.
M-351; 1.7-1.2 C, bal Fe.
For general tools; water hardening.

ASGM.
M-1785; 0.7-1.3 Si, 0.50 max Fe, 0.40-1.0 Mn, 0.6-1.2 Mg, bal Al.
Wrought aluminum alloy, AA 6082.
For production of high pressure gas containers.

ASH-21.
M-1249; 3.0-3.6 C, 4.0-4.75 Ni, 1.4-3.5 Cr, 0.4-0.7 Si, bal Fe.
Sand cast: 45,000 TS; 600 Brin.
Permanent mold: 55,000 TS; 675 Brin.
For cams, dies, rollers, bearing races; white cast iron, corrosion resistant, hard.

ASHBERRY METAL-A.
M-Eng.; 80 Sn, 1 Zn, 14 Sb, 2 Cu, 2 Ni, Al.
For tableware and utensils; same as "Brittania."

ASHBERRY METAL-B.
M-Eng.; 78-80 Sn, 0-2.8 Zn, 14-19 Sb, 0-3 Cu.
For utensils, bearings; Babbitt.

ASHBERRY METAL-C.
M-Eng.; 80 Sn, 14 Sb, 2 Cu, 1 Zn, 3 Ni.
For utensils, bearings; Babbitt.

ASHBERRY METAL-D.
M-Eng.; 79 Sn, 15 Sb, 3 Cu, 2 Zn, 1 Ni.
For utensils, bearings; Babbitt.

ASM-122.
M-1646; 0.25 max C, 13 Cr, 0.5 Mo, bal Fe.
Annealed: 95,000 TS; 50,000 YS; 25 El; B 92 Rock.
For cutlery, surgical instruments, gauges, needle valves, bearings.
Corrosion resistant, hardenable, heat treatable.

ASM-123.
M-1646; 0.30 max C, 13 Cr, 0.25 Mo, 1.25 Ni, bal Fe.
Annealed: 98,000 TS; 52,000 YS; 22 El; B 95 Rock.
For bearings, cutlery, gauges, needle valves, surgical instruments.
Corrosion resistant, hardenable.

A.S. NO. 5.
M-694; 1.0 C, 0.7 Mn, 5 Cr, 0.25 V, 1.1 Mo, bal Fe.
For tools, dies; air hardening. AISI A2.

A.S. NO. 7.
M-694; 0.45 C, 0.9 Si, 1.15 Cr, 0.2 V, 2.5 W, bal Fe.
For master hobs, shear blades, punches; Type Sl; shock resisting steel.

ASP 11.
M-1743; 0.04 max C, 0.80 max Si, 1.5 max Mn, 0.30 max Cu, 5.3-6.9 Ni, 23.5-25.0 Cr, 1.45-1.95 Mo, 0.30-0.50 Nb, 0.0010-0.0030 B, bal Fe.
Water quenched: 49.6 kg/mm^2 YS; 68.4 kg/mm^2 TS; 36.6 Elon: 70.5 RA; 229 BHN; 34 Charpy.
Magnetic, ferrite-austenite stainless with excellent corrosion resistance: hardware for chemical equipment.

ASP 23 H.S.S.
M-694; 1.27 C, 4.2 Cr, 5.0 Mo, 6.4 W, 3.1 V, bal Fe.
High speed steel for cutting tools, broaches, cold work applications.
High tensile, high toughness, excellent grindability.

ASP 30 H.S.S.
M-694; 1.27 C, 4.2 Cr, 5.0 Mo, 6.4 W, 3.1 V, 8.5 Co, bal Fe.
High speed steel; cutting tools for hard to machine materials.
High red hardness and temper resistance, good grindability.

ASP 60 H.S.S.
M-694; 2.30 C, 4.0 Cr, 7.0 Mo, 6.5 W, 10.5 Co, 6.5 V, bal Fe.
High carbon, high alloy, high speed steel.
Hardness to 69 Rc; used on the most demanding cutting tools. Good grindability.

ASR 1.
M-French; 0.36 C, 18 Cr, 9 Ni, 10 W, 2 Ti, 12 Co, bal Fe.
For gas turbine components; heat and creep resistant.

ASR 2.
M-French; C, 18 Cr, 8 Ni, 10 W + Co, bal Fe.
For gas turbines; heat creep and oxidation resistant.

ASTAR 811-C.
M-118; 8 W, 1 Re, 1 Hf, 0.025 C, bal Ta.
For space power systems utilizing liquid alkali metals as coolants.
High creep resistant, good ductility and weldability.

ASTRA.
M-1405; 0.55 C, 2 W, 0.5 Si, 1.0 Cr, bal Fe.
For cold punches, shears, piercers; oil hardened, tough.

ASTRALLOY.
M-97; 0.25 C, 3.5 Ni, 1.4 Cr, 0.28 Mo, bal Fr.
Normalized Plate: 241,000 TS; 157,000 YS; 11.7 El; 39 RA; C 47 R
Quenched: 250,000 TS; 191,000 YS; 30 El; 39 RA.
For shafts, abrasion-resistant parts, coal mine chutes and hoppers.
High impact strength, tough wear resistant, good weldability.

ASTRALLOY GR. 1.
M-176; 0.23-0.27 C, 0.7-1.0 Mn, 0.02 max P & S, 3.25-3.75 Ni, 1.25-1.75 Cr, 0.2-0.3 Mo, bal Fe.
Normalized: 241,000 TS; 157,000 YS; 39 RA; 477 Brin.
Quenched: 250,000 TS; 190,000 YS; 39 RA.
For paper and mining industry equipment, coal mine chutes and hoppers.
High toughness and ductility. Good weldability. Abrasion resistant.

ASTROLOY.
M-1490, M-1491; 0.05 C, 15 Cr, 1m Co, 5 Mo, 3.5 Ti, 4 Al, 0.06 Zr, bal Ni.
At R.T. 190,000 TS; 130,000 YS; 15 El.
For turbine wheels and blades, jet engine components, nozzles, compressor discs.
High heat and corrosion resistance.
High creep and stress-rupture strength.

ASTROLOY see CARPENTER ASTROLOY.

A.S.V.
M-57; 0.7 C bal Fe.
For tools, cutters; water hardened.

A.T. 2.
M-140; 2.2 C, 12 Cr, 0.25 V, 0.9 Mo, bal Fe.
Hardened: C 63-65 Rock.
For blanking and forming dies, gauges, trimmer and drawing dies, lathe centers, brick mold liners.
Cold work Type D4, oil or air hardening.
High abrasion resistance.

A.T. 2 DIE.
M-140; 2.2 C, 12 Cr, 0.25 V, 0.90 Mo, bal Fe.
For blanking and forming dies, gauges; oil hardened, non-deforming.

AT-3.
M-USSR; 3.2 Al, 0.84 Cr, 0.40 Fe, 0.34 Si, 0.01 B, bal Ti.
At 70°F: 126,000 TS; 122,000 YS; 16 El; 52 RA.
At 750°F: 84,000 TS; 80,000 YS; 16 El; 67 RA.
At 1110°F: 62,000 TS; 60,000 YS; 21 El; 86 RA.
For aircraft structures, high temperature fasteners.
Pseudoalpha alloy. Corrosion resistant.

AT-4.
M-USSR; 4.4 Al, 0.79 Cr, 0.60 Fe, 0.40 Si, 0.01 B, bal Ti.
At 70°F: 147,000 TS; 146,000 YS; 12 El; 41 RA.
At 750°F: 100,000 TS; 92,000 YS; 15 El; 55 RA.
At 1110°F: 58,000 TS; 55,000 YS; 28 El; 84 RA.
For high temperature fasteners, aircraft structures.
Pseuodoalpha alloy. Corrosion resistant.

AT-6.
M-USSR; 5.8 Al, 0.64 Cr, 0.38 Fe, 0.32 Si, 0.01 B, bal Ti.
At 70°F: 162,000 TS; 157,000 YS; 13 El; 36 RA.
At 750°F: 102,000 TS; 91,000 YS; 12 El; 58 RA.
At 1110°F 90,000 TS; 84,000 YS; 20 El; 78 RA.
For high temperature fasteners, aircraft structures.
Corrosion resistant.

AT-8.
M-USSR; 6.5 Al, 0.73 Cr, 0.39 Fe, 0.4 Si, 0.01 B, bal Ti.
At 750°F: 162,000 YS; 156,000 YS; 15 El; 38 RA.
At 750°F: 119,000 TS; 109,000 YS; 10 El; 47 RA.
1110°F: 82,000 TS; 74,000 YS; 16 El, 67 RA.
For airplane structures, high temperature fasteners.
Corrosion resistant.

AT 250 see LESCALLOY MARVAC 250.

ATG 33.
M-1488; 0.06 C, 45.5 Ni, 25.5 Cr, 3.25 Co, 3.25 Mo, 3.25 W, bal Fe.
High temperature alloy; for radiant tubes, furnace parts; heat treat fixtures.
AFNOR Z 6 NCKDW 45.
Similar to RA 333.

ATG C 1.
M-1488; 0.04 C, 52 Ni, 19 Cr, 3 Mo, 0.8 Ti, 0.5 Al, 5.25 Nb, bal Fe.
Good high temperature properties.
For rocket motors, pump bodies, jet engines.
AFNOR NC19FeNb; INCONEL 718.

ATG E.
M-1488; 0.10 C, 22 Cr, 1.5 Co, 9 Mo, 0.6 W, 18.5 Fe, bal Ni.
Good strength at elevated temperature.
For aircraft and jet engine parts; weldable.
AFNOR NC22FeD; HASTELLOY X.

ATG E 2.
M-1488; 0.07 C, 21.5 Cr, 9 Mo, 3.65 Nb, 2.0 Fe, bal Ni.
Good strength at elevated temperature.
For fuel nozzles, after burners.
AFNOR NC22FeDNb; INCoNEL 625.

ATG F.
M-1488; 0.05 C, 15 Cr, 2.5 Ti, 0.7 Al, 7.0 Fe, bal Ni.
Good strength at elevated temperature.
For gas turbine parts, furnace equipment.
AFNOR NC15TNbA; INCONEL X750.

ATG H.
M-1488; 0.10 C, 10 Ni, 20 Cr, 15 W, 3 max Fe, bal Co.
Good strength at elevated temperature.
For turbine blades and discs.
AFNOR KC20WN; HAYNES ALLOY No. 25.

ATG M 2.
M-1488; 0.18 C, 10 Cr, 15 Co, 3 Mo, 4.7 Ti, 5.5 Al, 1.0 V, bal Ni.
Good strength at elevated temperature.
For jet engine parts, turbine blades.
AFNOR NK15CAT; IN 100.

ATG R.
M-1488; 0.06 C, 19.5 Cr, 5 max Co, 0.4 Ti, 5 max Fe, bal Ni.
Good strength at elevated temperatures.
For combustion chambers, turbines.
AFNOR NC20T; similar to NIMONIC 75.

ATG S 3.
M-1488; 0.07 C 19 Cr, 2.5 Ti, 1.5 Al, 1.0 max Fe, bal Ni.
Good strength at elevated temperature.
For gas turbine blades.
AFNOR NC20TA; NIMONIC 80A.

ATG S 4.
M-1488; 0.07 C, 19 Cr, 19 Co, 2.5 Ti, 1.5 Al, 1.0 max Fe, bal Ni.
Good strength at elevated temperatures. For jet engine parts.
AFNOR NC20KTA; NIMONIC 90.

ATG S 8.
M-1488; 0.12 C, 15 Cr, 27 Co, 3 Mo, 2.10 Ti, 3.0 Al, 4 max Fe, bal
Good strength at elevated temperatures.
For gas turbine parts.
AFNOR NK27CADT; (INCoNEL 700).

ATG S 9.
M-1488; 0.12 C, 13 Cr, 1.0 Co, 4.5 Mo, 0.7 Ti, 6.0 Al. 2.0 Nb, 2.0 max Fe, bal Ni.
Good strength at elevated temperature.
For jet engine components.
AFNOR NC13AD; INCONEL 713 C.

ATG W 0.
M-1488; 0.06 C, 20 Cr, 20 Co, 5.9 Mo, 2.15 Ti, 0.45 Al, bal Ni.
High strength at elevated temperature.
For gas turbine components.
AFNOR NCK20D; WIGGIN C-263.

ATG W 1.
M-1488; 0.06 C, 20 Cr, 13 Co, 4.0 Mo, 3.0 Ti, 1.25 Al, 2.0 max Fe, bal Ni.
High strength at elevated temperature.
For jet engine components.
AFNOR NC20K14; WASPALOY.

ATG W 2.
M-1488; 0.10 C, 18 Cr, 18 Co, 4 Mo, 3 Ti, 3 Al, 4 max Fe, bal Ni.
High strength at elevated temperatures.
For engine components.
AFNOR NC20KDTA; UDIMET 500.

ATG W 3.
M-1488; 0.010 C, 15 Cr, 18 Co, 5 Mo, 3 Ti, 4 Al, 4 max Fe, bal Ni.
High strength at elevated temperatures.
For combustion chambers, turbine blades.
AFNOR NK18CDAT.
Similar to UDIMET 700.

ATG W 4.
M-1488; 0.07 C, 18 Cr, 15 Co, 3 Mo, 1.5 W, 5 Ti, 2.5 Al, bal Ni.
High strength at elevated temperatures.
For land-based gas turbine blades.
AFNOR NCK18TDA; UDIMET 710.

ATG X.
M-1488; 0.12 C, 20 Ni, 21 Cr, 20 Co, 3 Mo, 2.5 W, 1.0 Nb, bal Fe.
High temperature alloy; for gas turbine blades, jet engine parts.
AFNOR Z12CNKDW 20.
Similar to N-155.

ATG XX.
M-1488; 0.40 C, 20 Ni, 20 Cr, 20 Co, 4 Mo, 4 W, 4 Nb, bal Fe.
High temperature alloy.
For jet engines, turbine wheels and buckets.
AFNOR Z42CKNDW 20.
Similar to S590.

ATHA PNEU.
M-38; 0.50 C, 2.75 W, 1.25 Cr, 0.25 V, bal Fe.
For pneumatic chisels, rivet busters; hot work steel.
AISI S1.

ATHOS.
M-Eng; 22 Ni, 8 Cr, 1.75 Si, 1.0 Cu, 0.7 Mn, $\leq$ 0.5 C, bal Fe.
For stainless parts; heat and corrosion resistant.

ATLAN.
M-353; 0.90 C, 0.50 Cr, 1.15 Mn, 0.5 W, bal Fe.
For tools, cutters, drawing dies; non-deforming.
AISI 01.

ATLAN H.C.C. DIE.
M-353; 1.5 C, 12 Cr, 1 V, 0.8 Mo, bal Fe.
For tools, blanking dies; non-deforming. AISI D2.

ATLAN NON-SHRINK DIE.
M-353; 0.9 C, 1.5 Mn, bal Fe.
For tools dies; non-deforming.

ATLANTALOY NO. 8A MANGANESE BRONZE.
M-695; 58.0 Cu, 0.5 Sn, 1.0 Fe, 1.0 Al, 39.5 Zn.
Cast: 71,000 psi TS; 28,000 psi YS; 30 El; 130 Brin.
For housings, hardware, geared impellers.
CDA 865.

ATLANTALOY NO. 10 YELLOW BRASS.
M-695; 58 Cu, 1.0 Sn, 1.0 Pb, 0.3 Al, bal Zn.
Cast: 50,000 psi TS; 20,000 psi YS; 25 El; 70 Brin.
For hardware castings, brush broxes; free cutting.
CDA 857.

ATLANTALOY NO. 13B SILICON BRONZE.
M-695; 82.5 Cu, 4.0 Si, 13.5 Zn.
Cast: 68,000 psi TS; 30,000 psi YS; 17 El; 120 Brin.
Hardware castings, levers.
CDA 875.

ATLANTALOY NO. 20C BERYLLIUM COPPER.
M-695; 2.0 Be, 0.50 Co, 0.25 Si, bal Cu.
As cast: 133 Brin.
Hardened and aged: 160,000 psi TS; 150,000 psi YS; 1 El; 400 Brin.
For brush holders, electrical contacts.
AMS 4890.

ATLANTALOY NO. 25 ALUMINUM BRONZE.
M-695; 81.5 Cu, 4.0 Ni, 4.0 Fe, 10.5 Al.
As cast: 98,000 psi TS; 45,000 psi YS; 14 El; 190 Brin.
Heat treated: 117,000 psi TS; 70,000 psi YS; 240 Brin.
For gears, levers, non-sparking tools.
CDA 955.

ATLANTALOY NO. 31 MANGANESE BRONZE.
M-695; 64 Cu, 3.0 Fe, 6.0 Al, 3.5 Mn, bal Zn.
Cast: 115,000 psi TS; 68,000 psi YS; 18 El; 210 Brin.
Manganese bronze casting for gears, marine hardware.
CDA 863; AMS 4862.

ATLANTIC NO. 33.
M-353; 0.33 C, 0.75 Cr, 0.75 Mo, 0.75 Cu, bal Fe.
Heat treated: 135,000 TS; 116,000 YS; 18 El; 54 RA; 277 Brin.
For chisels, dies, blacksmith tools, hot work tools; water hardened, non-tempering.

ATLANTALOY NO. 40.
M-695; 0.8 Cu, 0.4 Mg, 8 Zn, bal Al.
Cast: 38,000 psi TS; 25,000 psi YS; 4 El; 70 Brin.
For light castings, housings.

ATLANTALOY NO. 70C BERYLLIUM COPPER.
M-695; 0.05 Be, 0.80 Cr, bal Cu.
Cast, HT: 53,000 psi TS; 36,000 psi YS; 11 El.
Electrical contacts; good electrical conductivity.

ATLANTIC C.
M-353; 0.65 C, 4 Cr, 4 Co, 18 W, 1 Mo, bal Fe.
For shaping tools, millers, cutters; high speed steel.

ATLANTIC DIE.
M-353; 0.70 C, 0.75 Cr, 1.50 Ni, 0.25 Mo, bal Fe.
Special purpose die steel.
AISI L 6.

ATLANTIC H.S.
M-353; 0.75 C, 18.0 W, 4.0 Cr, 1.0 V, bal Fe.
High speed steel.
AISI T 1.

ATLANTIC N.T.
M-353; 0.40 C, 0.4 Mn, 0.65 Si, 1.0 Cr, 0.5 Ni, 0.75 Mo, 0.5 Cu, bal Fe.
Shock resisting tool steel.

ATLANTIC STANDARD.
M-353; 0.8-1.2 C, 0.25 Si, 0.25 Mn, bal Fe.
Water hardened: 166-000-216,000 TS; 110,000-150,000 YS; 11-15 El; 32-37 RA; 330-600 Brin.
For taps, drills, reamers, punches, stamps, knurls, mandrels.
Type W1 water hardening.

ATLANTIC "V".
M-353; 0.8 C, 4.25 Cr, 2 V, 18.5 W, 0.65 Mo, bal Fe.
For cutting tools; high steel.

ATLAS 4-79.
M-435; 0.06 max C, 78-80 Ni, 0.85 Co, 0.3-0.8 Mn, bal Fe.
For communication and telephone equipment; high permeability.

ATLAS 12-12.
M-435; 0.12 max C, 12-14 Cr, 12-14 Ni, bal Fe.
Annealed: 80,000 TS; 35,000 YS; 50 El; 60 RA; 135 Brin.
For cutlery, valves; low rate of work hardening.

ATLAS 20.
M-435; 0.08 max C, 2 Mn, 19-21 Cr, 2-3 Mo, 27-30 Ni, 3-4 Cu, bal Fe.
For chemical and paper industries; high resistance to H_2SO_4.

ATLAS 40.
M-435; 0.24 C, 1.2 Cr, 0.26 Mo, 3.5 Ni, 1.2 Al, bal Fe.
For plastic molds, die casting dies; precipitation hardening.

ATLAS 45% NI ALLOY.
M-435; 0.06 max C, 0.4-1.0 Mn, 43.5-46.5 Ni, 0.6 max Co, bal Fe.
1/2 Hard: 140,000 TS; 128,000 YS; 1.5 El; 331 Brin.
For telephone equipment; high permeability and high induction.

ATLAS 78 1/2 NI ALLOY.
M-435; 0.06 max C, 77.5-79.5 Ni, 0.6 max Co, bal Fe.
Annealed: 86,000 TS; 35,000 YS; 41 El; 130 Brin.
For communication circuits; high permeability.

ATLAS 93.
M-1779; 0.55 C, 0.55 Mn, 0.2 Si, 0.7 Cr, 0.4 Mo, bal Fe.
Shock resisting tool steel.

ATLAS 301.
M-435; 0.15 max C, 16-18 Cr, 6-8 Ni, bal Fe.
Annealed: 110,000 TS; 40,000 YS; 60 El; 70 RA; 165 Brin.
For doctor blades and springs, auto and furniture trim; high strengh and wear resistance, stainless.

ATLAS 302.
M-435; 0.08-0.20 C, 17-19 Cr, bal Fe.
Annealed: 80,000 TS; 35,000 YS; 50 El; 70 RA; 150 Brin.
For chemical plant equipment; Type 302; stainless, austenitic.

ATLAS 303.
M-435; 0.15 max C, 17-19 Cr, 8-10 Ni, 0.07 min Se, 0.09-0.17 P, bal Fe.
Annealed: 90,000 TS; 45,000 YS; 40 El; 60 RA; 170 Brin.
For screw machine products, shafts, fasteners; Type 303; stainless, free-cutting.

ATLAS 303 MX.
M-435; 0.15 max C, 18 Cr, 9 Ni, 1.5 Mn, 0.15 min S, bal Fe.
Free-machining austenitic stainless for automatic screw machines.

ATLAS 304.
M-435; 0.08 max C, 2 max Mn, 18-20 Cr, 8-10 Ni, bal Fe.
Annealed: 85,000 TS; 35,000 YS; 60 El; 70 RA; 160 Brin.
Cold drawn: 180,000 TS; 125,000 YS; 10 El; 330 Brin.
For architectural molding and trim, kitchen equipment, chemical plant equipment; Type 304; stainless, austenitic.

ATLAS 305.
M-435; 0.12 max C, 16-19 Cr, 8-11 Ni, bal Fe.
Annealed: 85,000 TS; 38,000 YS; 50 El; 60 RA; 140 Brin.
For spinning, cold heading and drawing operations; low rate of work hardening, stainless.

ATLAS 308.
M-435; 0.08 max C, 19-21 Cr, 10-12 Ni, bal Fe.
Annealed: 85,000 TS; 35,000 YS; 50 El; 60 RA; 150 Brin.
For welding rods; heat and corrosion resistant.

ATLAS 310.
M-435; 0.25 max C, 2 max Mn, 24-26 Cr, 19-22 Ni, bal Fe.
Annealed: 100,000 TS; 45,000 YS; 50 El; 65 RA; 185 Brin.
For furnace parts and equipment, heat treat boxes, baffles; Type 310; heat resistant, austenitic.

ATLAS 316.
M-435; 0.10 max C, 16-18 Cr, 10-14 Ni, 2-3 Mo, bal Fe.
Annealed: 80,000 TS; 30,000 YS; 60 El; 80 RA; 150 Brin.
Cold drawn: 150,000 TS; 135,000 YS; 6 El; 300 Brin.
For chemical plant equipment, tanks, evaporators, valve trim; Type 316; stainless, austenitic.

ATLAS 317.
M-435; 0.10 max C, 18-20 Cr, 11-14 Ni, 3-4 Mo, bal Fe.
Annealed: 90,000 TS; 40,000 YS; 45 El; 160 Brin.
For chemical plant equipment, tanks; high corrosion resistance.

ATLAS 321.
M-435; 0.08 max C, 17-19 Cr, 8-11 Ni, Ti = 5 x C, bal Fe.
Annealed: 85,000 TS; 33,000 YS; 58 El; 75 RA; 180 Brin.
Cold drawn: 95,000 TS; 60,000 YS 40 El; 60 RA; 190 Brin.
For welded structures, chemical plant equipment, tanks; Type 321; stainless, stabilized.

ATLAS 330.
M-435; 0.25 max C, 14-16 Cr, 33-36 Ni, bal Fe.
Annealed: 80,000 TS; 40,000 YS; 30 El; 30 RA; 185 Brin.
At 1800°F: 147,000 TS; 43 El; 40 RA.
For furnace parts and equipment, cracking units, fixtures; Type 330; heat resistant, austenitic.

ATLAS 331.
M-435; 0.08 max C, 19-21 Cr, 30-34 Ni, bal Fe.
Annealed: 95,000 TS; 45,000 YS; 45 El; 60 RA; 160 Brin.
For carburizing boxes, heater tubing, nitriding fixtures; good resistance to thermal shock.

ATLAS 347.
M-435; 0.08 max C, 17-19 Cr, 9-12 Ni, Cb = 10 x C, bal Fe.
Annealed: 90,000 TS; 35,000 YS; 50 El; 65 RA; 160 Brin.
Cold drawn: 100,000 TS; 65,000 YS; 40 El; 60 RA; 212 Brin.
For welded structures, chemical plant equipment; Type 347; stainless, stabilized.

ATLAS 403.
M-435; 0.15 max C, 11.5-13.5 Cr, bal Fe.
Annealed: 75,000 TS; 40,000 YS; 34 El; 72 RA; 155 Brin.
Heat treated: 215,000 TS; 140,000 YS; 14 El; 54 RA; 415 Brin.
For dental and surgical instruments, valves, cutlery; Type 403; corrosion resistant, hardenable.

ATLAS 405.
M-435; 0.08 max C, 11.5-13.5 Cr, 0.10-0.30 Al, bal Fe.
Sub-annealed: 65,000 TS; 40,000 YS; 20 El; 50 RA; 165 Brin.
For mufflers, oil refinery equipment, annealing boxes; creep resistant.

ATLAS 410.
M-435; 0.15 max C, 11.5-13.5 Cr, bal Fe.
Annealed: 75,000 TS; 35 El; 70 RA; 155 Brin.
Cold drawn: 100,000 TS; 85,000 YS; 60 El; 205 Brin.
For flat springs, tableware, valve parts, turbine blades; Type 410; corrosion resistant, hardenable.

ATLAS 416.
M-435; 0.15 max C, 12-14 Cr, 0.18-0.35 S bal Fe.
Annealed: 75,000 TS; 40,000 YS; 30 El; 60 RA; 155 Brin.
Heat treated: 110,000 TS; 85,000 YS; 18 El; 55 RA; 230 Brin.
For screw machine products, shafts, valve trim; Type 416; corrosion resistant, free-cutting.

ATLAS 420.
M-435; 0.15 min C, 12-14 Cr, bal Fe.
Annealed: 95,000 TS; 50,000 YS; 25 El; 55 RA; 196 Brin.
Cold drawn: 105,000 TS; 85,000 YS 17 El; 50 RA; 215 Brin.
For cutlery, surgical instruments, valve trim; Type 420; corrosion resistant, hardenable.

ATLAS 420F.
M-435; 0.15 min C, 12-14 Cr 0.18-0.35 S, bal Fe.
Annealed: 95,000 TS; 50,000 YS; 20 El; 50 RA; 196 Brin.
Cold drawn: 105,000 TS; 85,000 YS; 15 El; 45 RA; 215 Brin.
For cutlery, surgical instruments, screw machine products; Type 420 F; corrosion resistant, free-cutting.

ATLAS 430.
M-435; 0.12 max C, 14-18 Cr, bal Fe.
Annealed: 70,000 TS; 40,000 YS; 30 El; 55 RA; 160 Brin.
Cold drawn: 130,000 TS; 120,000 YS; 2 El; 190 Brin.
For oil refinery and chemical plant equipment, hardware, bolts; Type 430; corrosion resistant.

ATLAS 431.
M-435; 0.2 max C, 15-17 Cr, 1.2-2.5 Ni, bal Fe.
Annealed: 95,000 TA; 50,000 YS; 20 El; 50 RA; 195 Brin.
For oil refinery and chemical plant equipment; Type 431; corrosion resistant.

ATLAS 440A.
M-435; 0.6-0.75 C, 16-18 Cr, 0.75 max Mo, bal Fe.
Annealed: 95,000 TS; 55,000 YS; 20 El; 240 Brin.
Heat treated: 275,000 TS; 240,000 YS; 2 El; 555 Brin.
For bearings, cutlery, surgical instruments, pivots; Type 440A; corrosion resistant, hardenable.

ATLAS 440B.
M-435; 0.75-0.95 C, 16-18 Cr, 0.75 max Mo, bal Fe.
Annealed: 107,000 TS; 62,000 YS; 18 El; 35 RA; 220 Brin.
Cold drawn: 120,000 TS; 95,000 YS; 9 El; 20 RA; 250 Brin.
For bearings, valves, surgical instruments; cutlery; Type 440B; corrosion resistant, hardenable.

ATLAS 440C.
M-435; 0.95-1.2 C, 16-18 Cr, 0.75 max Mo, bal Fe.
Annealed: 100,000 TS; 60,000 YS; 18 El; 200 Brin.
Heat treated: 280,000 TS; 250,000 YS; 2 El; 575 Brin.
For bearings, cutlery, valves, pivots, surgical instruments; Type 440C; corrosion resistant, hardenable.

ATLAS 442.
M-435; 0.20 max C, 23-27 Cr, bal Fe.
Sub-annealed: 80,000 TS; 50,000 YS; 20 El; 50 RA; 165 Brin.
For oil burner furnace and boiler parts; high corrosion and oxidation resistance.

ATLAS 446.
M-435; 0.35 max C, 23-27 Cr, bal Fe.
Annealed; 75,000 TS; 45,000 YS; 35 El; 65 RA; 160 Brin.
Cold drawn: 175,000 TS; 155,000 YS; 2 El; 25 RA; 250 Brin.
For oil burner parts, heat treat boxes, furnace equipment; Type 446; corrosion resistant.

ATLAS 501.
M-435; 0.10 min C, 1 max Mn, 5 Cr, bal Fe.
Annealed: 70,000 TS; 30,000 YS; 28 El; 65 RA; 160 Brin.
Heat treated: 175,000 TS; 135,000 YS; 15 El; 50 RA; 370 Brin.
For oil refinery equipment, valve trim, furnace parts; good to 1200°F, creep resistant.

ATLAS 502.
M-435; 0.10 max C, 1 max Mn, 4-6 Cr, 0.5 Mo, bal Fe.
Annealed: 75,000 TS; 35,000 YS; 26 El; 62 RA; 170 Brin.
Heat treated: 175,000 TS; 135,000 YS; 15 El; 50 RA; 370 Brin.
For oil refinery equipment, valve trim, furnace part; good to 1200°F, creep resistant.

ATLAS A.
M-1779; 0.30 C, 3.5 Cr, 9.0 W, 0.45 V, bal Fe.
W-C hot work tool steel for heavy duty tooling requirements; punches, mandrels, dies.
AISI H-21.

ATLAS ACX.
M-435; 0.83 C, 18.5 W, 4 Cr, 1.7 V, 1 Mo, 10.5 Co, bal Fe.
For hogging cutters, tools; high speed steel.

ATLAS AHT-28.
M-435, M 1636; 0.30 C, 0.50 Mn, 1.4 Cr, 4 Ni, 2 Mo, bal Fe.
Heat treated: 241,000 TS; 204,000 YS; 12.5 El; 477 Brin.
For gears, rolls, camshafts, transmission components, carbide bit holders; oil or air hardenable.

ATLAS ALPHA.
M-435; 0.8 C, bal Fe.
For punches, dies, blacksmith tools; water hardened.

ATLAS ALPHA-8.
M-435; 0.8 C, 0.3 Mn, bal Fe.
For hammer dies, picks, set screws, wedges; water hardening, shock resistant.

ATLAS BRAKE DIE.
M-435; 0.50 C, 1.1 Mn, 0.65 Cr, 0.15 Mo, bal Fe.
Heat treated: 130,000 TS; 118,000 YS; 20 El; 262 Brin.
For brake press dies, gears, drive shafts; high toughness and wear resistance.

ATLAS-CM.
M-435, M 1636; 0.40 C, 1.2 Mn, 0.15 Mo, 0.6 Cr, bal Fe.
Normalized: 100,000 TS; 65,000 YS; 24 El; 53 RA; 217 Brin.
For machinery parts, bolts axles, jigs, fasteners, gears, shafts.
Shock resistant, tough, water hardening.

ATLAS CW26.
M-435; 2.4 C, 0.40 Mn, 8.15 1.1 Mo, 5 V, bal Fe.
For drawing and stamping dies; high abrasion and wear resistance.

ATLAS D319.
M-435; 0.07 max C, 2 Mn, 17.5-19.5 Cr, 2.25-3.0 Mo, 13-15 Ni, bal Fe
For pulp and paper industry; high corrosion resistance.

ATLAS DIE CASTING STEEL.
M-435; 0.4 C, 0.7 Mn, 0.6 Cr, 0.15 Mo, 125 Ni, bal Fe.
For die casting dies; oil hardened.

ATLAS EXTRA.
M-435; 1.05 C, 1.25 Mn, bal Fe.
For tools, shear blades, cutters; oil hardening.

ATLAS FNS.
M-435; 1.5 C, 12 Cr, 0.8 Mo, bal Fe.
For tools, lamination dies; hot work steel.

ATLAS HOBBING IRON.
M-435; 0.05 C, 0.2 Mn, 0.15 Si, bal Fe.
For plastic mold dies; hubbed cavity.

ATLAS HOT DIE.
M-253, M 1; 0.5 C, 4 Cr, 18 W, 1 V, bal Fe.
For hot dies; punches, shears; hot work steel.

ATLAS HW-7.
M-435; 0.45 C, 1 Si, 5 Cr, 3.75 W, 1 Mo, 0.5 V, 0.5 Co, bal Fe.
At 1000°F: 217,000 TS; 8.5 El; 40 RA; 477 Brin.
For hot punches, shears, forging dies, cold punches; high hot strength and shock resistance.

ATLAS H W 24.
M-435; 0.35 C, 0.55 Mn, 1.3 Si, 3.5 Cr, 4.25 Mo, 0.85 V, bal Fe.
Hot work tool steel.

ATLAS-KK.
M-435; 1.1 C, 1.7 Cr, 0.4 Mo, 1.4 V, bal Fe.
For bearings, machinery parts; oil hardening.

ATLAS M 3.
M-435; 1.03 C, 0.25 Mn, 0.3 Si, 4.0 Cr, 6.0 Mo, 6.25 W, 2.5 V, bal Fe.
High speed steel.

ATLAS M-4.
M-435; 1.25 C, 6 W, 4.25 Cr, 4.75 Mo, bal Fe.
For reamers, tool bits, form rolls, drill; high abrasion and wear resistance.

ATLAS M-34.
M-435; 0.90 C, 1.45 W, 3.75 Cr, 8.7 Mo, 8.25 Co, 2.05 V, bal Fe.
For tool bits, reamers, form tools; high red hardness.

ATLAS NN.
M-435; 2.2 C, 12 Cr, bal Fe.
For tools dies, blanking dies; non-deforming.

ATLAS NO. 50A.
M-1; 0.3 C, 3.5 Cr, 9 W, 0.4 V, bal Fe.
For hot work tools, punches; hot work tool steel.

ATLAS NO. 50B.
M-1; 0.4 C, 3 Cr, 11 W, 0.5 V, bal Fe.
For hot work tools, punches; hot work tool steel.

ATLAS NO. 57.
M-696; 82 Cu, 11 Sn, 3.5 Zn, 0.5 Pb, 3.0 Ni-Sn hardener.
For gears, bearings; wear resistant.

ATLAS NO. 83A.
M-696; 85 Cu, 2 Sn, 5 Zn, 8 Pb.
For gears, worms, nuts, bearings.

ATLAS NON-MAG.
M-435; 0.4 C, 11.5-12.5 Mn, 7.0-8.5 Ni, bal Fe.
For transformer parts, retainer rings; low magnetic permeability.

ATLAS "Q".
M-435; 1.2 C, 0.5 Cr, bal Fe.
For tools, taps, reamers; water harding.

ATLAS ROLL.
M-1636; 1.05 C, 0.3 Mn, 0.015 S, 0.02 P, 1.45 Cr, bal Fe.
Annealed: 95,000 TS; 54,000 YS; 26 El; 55 RA; 192 Brin.
Hardened: 189,000 TS; 131,000 YS; 14 El, 35 RA; 331 Brin.
For rolls, feed and pinch rolls, bearings, flaring tools.
Water hardening; wear and fatigue resistant.

ATLAS SPECIAL ALLOY 8.
M-435; 0.8 C, 0.2 Mn, 0.2 V, bal Fe.
For impact tools, rivet sets; water hardened, shock resistant.

ATLAS SPECIAL ALLOY 10.
M-435; 1.05 C, bal Fe.
For tools, striking dies; water hardening.

ATLAS-SPS.
M-1636, M-435; 0.40 C, 0.75 Mn, 0.60 Cr, 1.25 Ni, 0.15 Mo, bal Fe.
Annealed: 105,000 TS; 70,000 YS; 25 El; 212 Brin.
Hardened: 145,000 TS; 132,000 YS; 20 El; 293 Brin.
For heavy duty shafts, gears, axles, spindles, tool holders, feed screws.
Oil hardening, shock resistance.

ATLAS STAINLESS NO. 20.
M-435; 0.07 C, 29 Ni, 20 Cr, 2-3 Mo, 4 Cu, bal Fe.
Cast: 65,000-75,000 TS; 28,000-38,000 YS; 50-35 El; 50-40 RA; 120-150 Brin.
For chemical plant equipment; resists mixed acids, austenitic.

ATLAS SUPERIOR.
M-1474; Sn, bal Cu.
For springs, bearings; corrosion resistant.

ATLAS SUPERIOR.
M-1636; 0.40 C, 1.1 Mn, 0.08 S, 0.02 P, 0.15 Mo, bal Fe.
Rolled: 108,000 TS; 84,000 YS; 18 El; 46 RA; 229 Brin.
For arbors, bolts, camshafts, collets, compressor shafts, guide bars, mandrels.
Free-machining, water hardening.

ATLAS TRIPLE EXTRA.
M-435; 1.3 C, 3.5 W, bal Fe.
For fast finishing tools, drills; water hardening.

ATLAS REFINED-8.
M-435; 0.8 C, 0.3 Mn. bal Fe.
For flaring tools, vise jaws, sledges, drill; water hardening.

ATLAS REFINED-10.
M-435; 1.0 C, 0.25 Mn, bal Fe.
For stamping and blanking dies, arbors, lathe centers, drills; water hardening.

ATLAS X-10.
M-435; 1.05 C, 0.25 Mn, bal Fe.
For shear blades, dies and cutting tools; water hardening.

ATLAS X-12.
M-435; 1.2 C, 0.25 Mn, bal Fe.
For drills, taps, reamers, files, cold forming dies; water hardening.

ATLAS XLO.
M-435; 0.55 C, 0.7 Mn, 0.8 V, 2 Cr, 0.55 Mo, bal Fe.
For drop forging die blocks; oil hardening.

ATLAS XX-95.
M-435; 0.95 C, bal Fe.
For cold beading dies; water hardened.

ATLAS XXX.
M-435; 1.35 C, 3.75 W, bal Fe.
For tools, forming rolls; maximum wear resistance.

ATLOY D.D.
M-353; C, alloy, bal Fe.
Heat treated: 310,000 TS.
For bushings, gears, clutches, gages, spindles; tough, non-deforming.

ATLOY FM.
M-353; C, B, Pb, alloy, bal Fe.
For gears, arbors, worms, collets, motor shafts; free-cutting, fatigue and wear resistant.

ATLOY H.T.
M-353; C, Cr, Mn, bal Fe.
Heat treated: 260,000-112,000 TS; 230,000-95,000 YS; 9.5-21.5 El; 26-62.5 RA; 514-241 Brin.
For arbors, axles, bolts, cams, gears, jaws; fatigue and wear resistant.

ATLOY Z.
M-353; C, B, alloy, bal Fe.
Heat treated: 280,000 TS.
For gears, axles, connecting rods, propeller shafts; high strength.

ATMODIE.
M-35; 1.5 C, 12 Cr, 0.8 Mo, 0.85 V, bal Fe.
For blanking and forming dies, punches, shear blades; tough and wear resistant; air hardening; AISI Type D2.

ATMODIE 4.
M-35; 2.22 C, 0.4 Mn, 0.45 Si, 0.8 Mo, 11.65 Cr, 0.20 V, bal Fe.
Air hardening cold work tool steel.
For coining dies, slitters, forming mandrels. AISI D4.

ATMODIE 5.
M-35; 1.55 C, 0.4 Mn, 0.4 Si, 0.8 Mo, 11.9 Cr, 0.5 V, 3.25 Co, 0.4 Ni, bal Fe.
High hardenability, high wear resistance; air hardening cold work tool steel.
For draw dies, trim dies, extrusion mandrels. AISI D5.

ATMODIE SMOOTHCUT.
M-35; 1.5 C, 12 Cr, 0.9 Mo, 1.0 V, bal Fe.
For drawing and blanking dies, punches, mandrels, gages; air or oil hardening; free machining, wear and corrosion resistant. AISI D2.

ATMOS.
M-1422; C, alloy, bal Fe.
For dies, punches; air hardened.

AT-NICKEL.
M-121; 99.0 min Ni.
For chemical plant equipment.
Welding grade. Good resistance to alkaline salts and organic acids.

AT NICKEL see NICKEL 205.

ATOM ARC 502.
M-1713; weld metal: 0.05 C, 0.62 Mn, 0.60 Si, 5.72 Cr, 0.62 Mo, bal Fe.
Welded, stress annealed 1025°F: 96,000 psi TS; 78,000 psi YS; 22
Iron powder, low hydrogen electrode for welding 4-6 Cr steels as AISI 501,502.
AWS Class E501-16(-18).

ATOM ARC 4130.
M-1713; weld metal: 0.18 C, 1.25 Mn, 0.40 Si, 0.50 Cr, 1.28 Ni, 0.20 Mo, bal Fe.
Welded, Quench + Temper 1100°F: 138,000 psi TS; 121,000 psi YS; El.
All-position, iron-powder, low-hydrogen electrode to weld AISI 8630, 4130 and similar alloys.

ATOM ARC 4140.
M-1713; weld metal: 0.38 C, 0.95 Mn, 0.62 Si, 0.80 Cr, 0.33 Mo, bal Fe.
All-position, iron-powder, low-hydrogen electrode to weld AISI 4140 and similar steels, including castings.
Properties after heat treatment are similar to AISI 4140 base metal.

ATOM ARC 4340.
M-1713; weld metal: 0,38 C, 0.94 Mn, 0.70 Si, 0.80 Cr, 0.40 Mo, 1.30 Ni, bal Fe.
All-position, iron-powder, low-hydrogen electrode to weld AISI 4340 an similar steels, including castings.
Properties after heat treatment are similar to AISI 4340 base metal.

ATOM ARC 7018.
M-1713; weld metal: 0.06 C, 1.10 Mn, 0.50 Si, bal Fe.
As welded: 75,000 psi TS; 68,000 psi YS; 34 El.
Iron-powder, low hydrogen electrode for welding wide variety of carbon steels.
AWS class E7018.

ATOM ARC 7018 MO.
M-1713; weld metal: 0.05 C, 0.75 Mn, 0.56 Si, 0.53 Mo, bal Fe.
As welded: 79,000 psi TS; 68,000 psi YS; 31 El.
Iron-powder, low hydrogen electrode for welding low alloy, high tensile steels of 50,000 psi YS, and also the 0.5 Mo steels.
AWS class E7018-A1.

ATOM ARC 8018.
M-1713; weld metal: 0.05 C, 1.06 Mn, 1.04 Ni, bal Fe.
As welded: 84,000 psi TS; 73500 psi YS; 30 El.
Iron-powder, low hydrogen electrode for welding high strength steels in the 70,000-80,000 psi tensile strength range and low temperature operation.
AWS Class E 8018-C3.

ATOM ARC 8018 C1.
M-1713; weld metal: 0.04 C, 1.06 Mn, 0.31 Si, 2.37 Ni, bal Fe.
As welded: 88,500 psi TS; 73,800 psi YS; 28 El.
Iron-powder, low hydrogen electrode for welding 2 1/3% Ni steels for low temperature operation.
AWS Class E8018-C1.

ATOM ARC 8018 CM.
M-1713; weld metal: 0.05 C, 0.68 Mn, 0.60 Si, 1.24 Cr, 0.4 Mo, bal Fe.
As welded: 92,000 psi TS; 83,000 psi YS; 27 El.
Iron-powder, low hydrogen electrode for welding low Cr-Mo steels in power piping and boiler work.
AWS Class E 8018-B2.

ATOM ARC 8018 N.
M-1713; weld metal: 0.05 C, 0.84 Mn, 0.37 Si, 0.30 Ni, bal Fe.
As welded: 94,000 psi TS; 83,000 psi YS; 25 El.
Iron-powder, low hydrogen electrode for welding 2-4% Ni steels; for low temperature equipment; and for welds subject to impact.
AWS Class E 8018-C2.

ATOM ARC 8018 NM.
M-1713; weld metal: 0.06 C 1.10 Mn, 0.40 Si, 1.00 Ni, 0.50 Mo, bal Fe.
As welded: 93,000 psi TS; 82,000 psi YS; 25 El.
Iron-powder, low hydrogen electrode for welding quenched and tempered Mn-Ni-Mo steels, especially for pressure vessels.
AWS Class E 8018-G.

ATOM ARC 9018.
M-1713; weld metal: 0.05 C, 1.11 Mn, 0.32 Si, 1.72 Ni, 0.28 Mo, bal Fe.
As welded: 94,400 psi TS; 84,600 psi YS; 27 El.
Iron-powder, low hydrogen electrode for welding HY-80 and other high tensile, quenched and tempered steels.
AWS Class E 9018-M.

ATOM ARC 9018CM.
M-1713; weld metal: 0.05 C, 0.75 Mn, 0.60 Si, 2.20 Cr, 1.05 Mo, bal Fe.
As welded: 96,000 psi TS; 83,000 psi YS; 25 El.
Iron-powder, low hydrogen electrode for welding 2% Cr steels.
AWS Class E 9018-B3.

ATOM ARC 9018HT.
M-1713; weld metal: 0.14 C, 0.80 Mn, 0.65 Si, 2.30 Cr, 1.0 Mo, bal Fe.
Welded, normalized + tempered 1275°F: 99,000 psi TS; 82,000 psi YS; 24 El.
For welding Cr-Mo steel castings in all-position, with iron powder, low, hydrogen electrode.

ATOM ARC 10018.
M-1713; weld metal: 0.05 C, 1.20 Mn, 0.54 Si, 1.73 Ni, 0.33 Mo, bal Fe.
As welded: 103,000 psi TS; 96,000 psi YS; 24 El.
Iron powder, low hydrogen electrode for welding where joints must equal 100,000 psi tensile strength.
AWS Class E 10018-M.

TOM ARC 10018MM.
M-1713; weld metal: 0.06 C, 1.77 Mn, 0.68 Si, 0.44 Mo, bal Fe.
As welded: 106,000 psi TS; 101,000 psi YS; 22 El.
Iron powder, low hydrogen electrode for welding low-alloy, high-tensile steels requiring weld strengths of 100,000 psi TS.
AWS Class E 10018-D2.

ATOM ARC 12018.
M-1713; weld metal: 0.05 C, 1.9 Mn, 0.25 Si, 0.85 Cr, 2.0 Ni, 0.50 Mo, bal Fe.
As welded: 132,000 psi TS; 120,000 psi YS; 20 El.
Iron powder, low hydrogen electrode for welding low-alloy, high-tensile steels requiring weld strengths of 120,000 psi minimum tensile.
AWS Class E 12018-M.

ATOM ARC "T".
M-1713; weld metal: 0.06 C, 1.53 Mn, 0.27 Si, 0.31 Cr, 1.88 Ni, 0.42 Mo, bal Fe.
As welded: 115,000 psi TS; 103,000 psi YS; 22 El.
Iron powder, low hydrogen electrode for high strength welds usable also at low temperature.
AWS Class E 11018-M.

ATOMINPHY B2.
M-1488; 0.10 max C, 2.0 max Mn, 1.5 max Si, 13.0-15.0 Ni, 17.0-20.0 Cr, 0.20 max Co, 1.5-2.0 Bo, bal Fe.
Stainless casting: 220-250 Brin.
Protective screens, nuclear energy.
AFNOR Z8CNB19.14M.

ATR ALLOY.
M-83, M-1637; 0.5 Cu, 0.5 Mo, 1.5 Sn, 0.12 Fe, 0.10 Cr, 0.05 Ni, bal Zr.
For nuclear reactors.
Resists CO_2 corrosion.

ATRAX A 6 ETC see A 6.

ATRIX 100.
M-912; 0.17 C, 0.30 Si, 1.10 Mn, bal Fe.
Ht Tr: 410-590 N/mm² TS.
Weldable, forgeable, for boiler flanges.
W.-Nr. 0481.

ATRIX 101.
M-912; 0.20 C, 0.50 Si, 1.15 Mn, bal Fe.
Ht Tr: 490-640 N/mm² TS.
Weldable, forgeable, for boiler flanges.
W-Nr. 0482.

ATRIX 110.
M-912; 0.19 C, 0.12 Si, 1.35 Mn, 0.55 Mo, 0.55 Ni, bal Fe.
Ht Tr: 550-700 N/mm² TS.
Weldable forgeable, for boiler flanges.
W.-Nr. 6310.

ATRIX 120.
M-912; 0.17 C, 0.25 Si, 0.60 Mn, 0.30 Mo, bal Fe.
Ht Tr.: 440-580 N/mm² TS.
Weldable, forgeable; resistant to temperature to 530°C.
W.-Nr. 5415.

ATRIX 122.
M-912; 0.15 C, 0.25 Si, 0.55 Mn, 0.50 Cr, 0.60 Mo, 0.28 V, bal Fe.
Ht Tr: 490-690 N/mm² TS.
Weldable, forgeable; resistant to temperatures up to 400°C.
W.-Nr. 7715.

ATRIX 140 CU.
M-912; 0.15 C, 0.35 Si, 1.0 Mn, 0.35 Mo, 1.10 Ni, 0.60 Cu, 0.02 Nb, bal Fe.
Ht Tr: 610-760 N/mm² TS.
Weldable, forgeable; for boiler flanges.
W.-Nr. 6368.

ATRIX 200.
M-912; 0.15 C, 0.25 Si, 0.60 Mn, 0.90 Cr, 0.50 Mo, bal Fe.
Ht Tr: 440-590 N/mm² TS.
For boiler tubes and parts, for temperatures up to 530°C.
W.-Nr. 7335.

ATRIX 203.
M-912; 0.24 C, 0.25 Si, 0.70 Mn, 1.10 Cr, 0.25 Mo, bal Fe.
Ht Tr: 590-735 N/mm² TS.
For bolts and nuts resistant up to 530°C.
W.-Nr. 7258.

ATRIX 204.
M-912; 0.12 Cr, 0.25 Si, 0.50 Mn, 2.15 Cr, 1.0 Mo, bal Fe.
Ht Tr: 440-590 N/mm² TS.
Boiler water tubes and parts up to 530°C.
W.-Nr. 7380.

ATRIX 231.
M-912; 0.21 C, 0.50 Si, 0.40 Mn, 1.35 Cr, 1.10 Mo, 0.30 V, bal Fe.
Ht Tr: 690-835 N/mm² TS.
For bolts and nuts up to 530°C.
W.-Nr. 8070.

ATRIX 232.
M-912; 0.24 C 0.25 Si, 0.45 Mn, 1.35 Cr, 0.55 Mo, 0.20 V, bal Fe.
Heat treat: 685-835 N/mm² TS.
For bolts and nuts resistant to elevated temperatures up to 530°C.
W.-Nr. 7733.

ATRIX 234.
M-912; 0.21 C, 0.25 Si, 0.60 Mn, 1.35 Cr, 0.75 Mo, 0.30 V, bal Fe.
Ht Tr: 700-850 N/mm² TS.
For nuts and bolts resistant to elevated temperatures.
W.-Nr. 7709.

ATRIX 300.
M-912; 0.20 C 0.25 Si, 0.65 Mn, 0.40 Cr, 0.65 Mo, 0.75 Ni, bal Fe.
Heat treated: 560-710 N/mm² TS.
Weldable, forgeable parts for boilers.
W.-Nr. 6751.

ATRIX 321.
M-912; 0.22 C, 0.25 Si, 0.70 Mn, 1.90 Cr, 0.75 Mo, 1.05 Ni, bal Fe.
Heat treated: 750-900 N/mm² TS.
For steam turbine equipment.
W.-Nr. 6740.

ATRIX 380.
M-912; 0.19 C, 0.20 Si, 0.70 Mn, 1.30 Cr, 1.0 Mo, 0.70 Ni, 0.30 V, bal Fe.
Ht Tr: 700-850 N/mm² TS.
For steam turbine parts.

ATRIX 381.
M-912; 0.29 C, 0.20 Si, 0.70 Mn, 1.30 Cr, 0.95 Mo, 0.70 Ni, 0.30 V, bal Fe.
Ht Tr: 700-850 N/mm² TS.
For steam turbine parts.
W.-Nr. 6983.

ATRIX 382.
M-912; 0.30 C, 0.20 Si, 0.70 Mn, 1.30 Cr, 1.10 Mo, 0.70 Ni, 0.30 V, bal Fe.
Ht Tr: 700-850 N/mm² TS.
For steam turbine parts to 450°C.
W.-Nr. 6946.

ATRIX N 3.
M-912; 0.22 C, 0.20 Si, 0.65 Mn, 2.50 Cr, 0.50 Mo, bal Fe.
Ht Tr: 540-690 N/mm² TS.
For parts operating to 450°C.
W.-Nr. 7273.

ATRIX N 9.
M-912; 0.20 C, 0.25 Si, 0.40 Mn, 3.15 Cr, 0.55 Mo, 0.50 V, bal Fe.
Ht Tr: 640-785 N/mm² TS.
For part operating up to 450°C.
W.-Nr. 7779.

ATSCO.
M-353; C, alloy, bal Fe.
For tools, water hardening. AISI W1.

ATSCO AR.
M-353; 0.9 C, 0.7 Mn, 0.3 Mo, 1.1 Cr, bal Fe.
For shovel teeth, bucket lips; water hardened, abrasion resistant.

ATSCO EXTRA.
M-353; 0.8-1.1 C, 0.25 Si, 0.25 Mn, bal Fe.
Water hardened: 166,000-216,000 TS; 110,000-150,000 YS; 11-15 El; 32-37 RA; 330-600 Brin.
For taps, drills, reamers, punches, stamps, knurls, mandrels.
Type W1 water hardening.

ATSCO SPECIAL.
M-353; 0.5-1.0 C, 0.5-1.0 Mn, 0.3 Mo, bal Fe.
For rock drills, concrete wreckers, demolition tools; water hardening.

ATSIL.
M-353; 0.5 C, 0.6 Mn, 0.5 W, 0.3 Mo, 1.3 Si, bal Fe.
For shear blades, rivet sets, cutting tools; shock resisting. AISI S5.

A.T. STEEL.
M-32; 0.35 C, 0.5 Mn, 0.3 Si, 1.5 Cr, 3.5 Ni, bal Fe.
Air hardening tool steel.

ATVS 2.
M-1488; 0.04 C, 26 Ni, 13.5 Cr, 1.0 Co, 2.75 Mo, 1.8 Ti, 0.3 Al, bal Fe.
Good strength at elevated temperatures.
For turbine discs, extrusion dies, high temperature bolts.
AFNOR Z3NCT25; DISCALOY.

ATVS 7.
M-1488; 0.10 C, 30 Ni, 18 Cr, 20 Co, 2.0 Ti, 0.8 Al, bal Fe.
Good strength at elevated temperature.
AFNOR Z10NKC30.

ATVS 7 MO.
M-1488; 0.06 C 37 Ni, 18 Cr, 20 Co, 3.0 Mo, 2.75 Ti, bal Fe.
Good strength at elevated temperature.
For turbine blades and parts.
AFNOR Z6NKCDT38; REFRACTOLOY 26.

ATVS MO.
M-1488; 0.05 max C, 15.0 Cr, 26.0 Ni, 1.2 Mo, 2.0 Ti, 0.3 V 0.2 N, bal Fe.
Wrought, ann, aged: 930 N/mm^2 min UTS.
Austenitic, non-magnetic stainless, good strength; resistant to high temperature oxidation.
AFNOR Z6NCTDV25.15.2; AISI 660.

ATW 432 ATOMIZED IRON POWDER.
M-1781; 97.0 Fe, 0.08 C, 0.03 max S, (0.80 H$_2$ - loss).
40 mesh powder for high yield stick electrode coatings.

A-U4G1.
M-USSR; 4.36 Cu, 1.5 Mg, 0.6 Mn, 0.5 Si, bal Al.
Ht. Tr.: 75,000 TS; 58,000 YS; 135 Brin.
For aircraft structures, fittings, hardware, rivets, fastenings.
Similar to Aluminum 2024, age-hardenable, high strength.

A-U4SG.
M-USSR; 4.35 Cu, 0.85 Si, 0.8 Mn, 0.5 Mg, bal Al.
Ht. Tr.: 70,000 TS; 60,000 YS; 12 El; 135 Brin.
For hydraulic fittings, structures, hardware, engine components.
Similar to Aluminum 2014, age-hardenable, high strength.

AU4SG.
M-1785; 0.5-1.2 Si, 0.7 max Fe, 3.5-5.0 Cu, 0.4-1.2 Mn, 0.2-0.8 Mg, bal Al.
Wrought, heat treatable aluminum alloy.
AA 2014. For production of high quality weapon parts.

AU 18/8.
M-1653; 0.12 max C, 1.0 max Si, 2.0 max Mn, 17.0-19.0 Cr, 8.0-10.0 Ni bal Fe.
Austenitic stainless steel.
W.-Nr. 1.4300; X 10 CrNi 18 08; AISI 302.

AU 18/8 Z.
M-1653; 0.12 max C, 1.0 max Si, 2.0 max Mn, 17.0-19.0 Cr, 8.0-10.0 Ni 0.60 max Mo, 0.15-0.25 S, bal Fe.
Free machining austenitic stainless steel.
W.-Nr. 1.4305; AISI 303.

AUBERT & DUVAL-56R.
M-1115; 0.08 C, 12 Cr, 1 Mo, 0.25 V, 1.0 W, bal Fe.
Annealed: 75,000 TS; 35,000 YS; 30 El; 70 RA; B 82 Rock.
Cold drawn: 95,000 TS; 80,000 YS; 15 El; 60 RA; B 92 Rock.
For springs, table flatware, knives, oil refinery and chemical plant equipment.
Corrosion resistant.

AUBERT & DUVAL-56T5.
M-1115; 0.20 C, 11 Cr, Mo, V, bal Fe.
Annealed: 95,000 TS; 40,000 YS; 25 El; B 92 Rock.
Hardened: 240,000 TS; 205,000 YS; 8 El; C 50 Rock.
For valves, bearings, cutlery, surgical instruments, gears. Corrosion resistant, hardenable.

AUBERT & DUVAL 56 T G.
M-1115; 0.20 C, 0.50 Mn, 0.50 Si, 1.30 Cr, 0.80 Mo, 0.25 V, bal Fe.
For high temperature equipment.
AFNOR 20CDV6.

AUBERT & DUVAL 819B.
M-1115; 0.3-0.4 C, 3.65-4.0 Ni, 1.5-2.0 Cr, 0.15-0.35 Mo, bal Fe.
For gears, shafts, machinery parts; oil hardened, shock resistant.

AUBERT & DUVAL 897D.
M-1115; 0.25-0.35 C, 3.0-3.5 Ni, 0.5-1.0 Cr, 0.8 Mn, bal Fe.
For crankshafts, axles, connecting rods; AISI 3335; oil hardened, shock resistant.

AUBERT & DUVAL AD3.
M-1115; 0.12 max C, 3.0-3.5 Ni, 0.5-1.0 Cr, 0.5 Mn, bal Fe.
For shafts, cams, camshafts, machinery parts; AISI 3310; case hardened shock resistant.

Section I: Alloy Data / 165

AUBERT & DUVAL APX.
M-1115; 0.10-0.20 C, 1.5-3.0 Ni, 16-19 Cr, bal Fe.
Annealed: 107,000 TS; 62,000 YS; 18 El; 35 RA; 220 Brin.
For cutlery, valves, surgical instruments, ball bearings; Type 431; corrosion resistant.

AUBERT & DUVAL APZ2.
M-1115; 0.6-0.8 C, 1-2 Ni, 18-21 Cr, bal Fe.
Heat treated: 143,000 TS.
For exhaust valves; corrosion and heat resistant.

AUBERT & DUVAL BM 3.
M-1115; 0.15 max C, 0.50 Mn, 2.2 Cr, 1.0 mo, bal Fe.
Steel for high temperature operation.
AFNOR 12CD9 10.

AUBERT & DUVAL BM 6.
M-1115; 0.15 C, 0.30 Si, 0.45 Mn, 4.0-6.0 Cr, 0.50 Mo, bal Fe.
For stainless requirements or high temperature operation.
AFNOR Z15CD5.05.

AUBERT & DUVAL B M V 4 STEEL.
M-1115; 0.40 C, 5.0 Cr, 1.3 Mo, 0.45 V bal Fe.
Deep hardening steel for highly stressed heavy equipment; or hot work tool steel.
Similar to AFNOR 40CDV20; AISI H 11 tool steel.

AUBERT & DUVAL B X O.
M-1115; 0.19 C, 1.5 Ni, 0.9 Cr, 0.2 Mo, bal Fe.
Carburizing steel; core hardenable to 140 Kg/mm^2 TS; 110 Kg/mm^2 YS; 12 El.
AFNOR 18NCD6; similar to AISI 4320.

AUBERT & DUVAL B X 1.
M-1115; 0.14 C, 1.5 Ni, 0.9 Cr, 0.15 Mo, bal Fe.
Carburizing steel; core hardenable to 135 Kg/mm^2 TS; 105 Kg/mm^2 YS; 13 El.
Similar to AFNOR 16NCD6.

AUBERT & DUVAL B X 3.
M-1115; 0.10 C 0.75 Mn, 1.4 Ni, 1.0 Cr, 0.20 Mo, bal Fe.
Alloy carburizing steel.
AFNOR 10NCD6.

AUBERT & DUVAL CNS.
M-1115; 0.3-0.4 C, 0.9-1.4 Ni, 0.5-1.0 Cr, 0.8 Mn, 0.5 Si, bal Fe.
For crankshafts, axles, bolts, gears; AISI 3135; oil hardened, shock resistant.

AUBERT & DUVAL F65.
M-1115; 0.34-0.40 C, 1.0-1.5 Cr, 0.10-0.30 Mo, 0.8 Mn, bal Fe.
For axles, gears, machinery parts; AISI 4137; oil hardened.

AUBERT & DUVAL F66J.
M-1115; 0.28-0.34 C, 1.0-1.5 Cr, 0.10-0.30 Mo, 0.8 Mn, bal Fe.
For gears, shafts, machinery parts; AISI 4130; oil hardened.

AUBERT & DUVAL F66S.
M-1115; 0.21-0.28 C, 1.1-1.5 Cr, 0.10-0.30 Mo, 0.8 Mn, bal Fe.
For gears, shafts, machinery parts; AISI 4125; tough.

AUBERT & DUVAL FAD.
M-1115; 0.17 max C, 3.0-3.5 Ni, 0.80-1.3 Cr, bal Fe.
For gears, cams, crankshafts; AISI 9315; case hardened, shock resistant.

AUBERT & DUVAL F A D H STEEL.
M-1115; 0.16 C, 3.0 Ni, 1.0 Cr, 0.30 Mo, bal Fe.
Deep hardening carburizing steel; core hardenable to 140 Kg/mm^2 TS 110 Kg/mm^2 YS; 13 El.
For highly stressed heavy sections in aircraft and automotive industry.

AUBERT & DUVAL F A D S STEEL.
M-1115; 0.16 C, 4.3 Ni, 1.2 Cr, 0.20 Mo, bal Fe.
Deep hardening carburizing steel; core hardenable to 145 Kg/mm^2 TS 120 Kg/mm^2 YS; 12 El.
For highly stressed heavy sections in aircraft and automotive industry.
AFNOR 16NCD17.

AUBERT & DUVAL FDMA.
M-1115; 0.25-0.35 C, 3.2-3.7 Ni, 1.0-1.5 Cr, 0.3-0.6 Mo, bal Fe.
For crankshafts, gears, connecting rod; oil hardened, shock resistant.

AUBERT & DUVAL GK3.
M-1115; 0.25-0.35 C, 2.75-3.3 Cr, 0.20-0.50 Mo, 0.4 max Al, bal Fe.
For gears, shafts, axles, machinery parts; nitriding steel.

AUBERT & DUVAL GK5.
M-1115; 0.15-0.25 C, 2.7-3.3 Cr, 0.2-0.5 Mo, 0.8 Mn, bal Fe.
For gears, shafts, machinery parts; nitriding steel.

AUBERT & DUVAL GKH.
M-1115; 0.32 C, 0.70 max Mn, 0.50 max Si, 3.0 Cr, 1.0 Mo, 0.25 V bal Fe.
For elevated temperature operation.
AFNOR 32 CDV 12.

AUBERT & DUVAL JD 19.
M-1115; 0.20 C, 0.80 Mn, 0.55 Ni, 0.50 Cr, 0.20 Mo, 0.35 max Cu, bal Fe.
Carburizing steel.
AFNOR 20NCD2.

AUBERT & DUVAL LK3.
  M-1115; 0.40-0.50 C, 1.5-2.0 Cr, 0.5-1.1 Al, 0.10-0.40 Mo, bal Fe.
  For gears, axles, machinery parts; nitriding steel.

AUBERT & DUVAL LK5.
  M-1115; 0.25-0.35 C, 1.5-2.0 Cr, 0.10-0.40 Mo, 0.5-1.1 Al, bal Fe.
  For gears, shafts, machine tool parts; nitriding steel.

AUBERT & DUVAL MARVAL 18 STEEL.
  M-1115; 0.03 max C, 18.0 Ni, 8.0 Co, 5.0 Mo, 0.50 Ti, bal Fe.
  Maraged: 170 kg/mm^2 TS; 155 kg/mm^2 YS; 8 El.
  Maraging steel.
  AFNOR Z 2 NKD 18.

AUBERT & DUVAL MEP.
  M-1115; 0.4-0.5 C, 1.0-2.0 Cr, 2.0-3.0 W, bal Fe.
  For impact tools, rivet sets, chisels, dies; cold work steel, oil hardened

AUBERT & DUVAL MES.
  M-1115; 0.3-0.4 C, 2.5-3.0 Cr, 2.5-3.0 W, bal Fe.
  For rivet sets, extrusion rams and dies; hot work steel, oil or air hardened.

AUBERT & DUVAL MMR0.
  M-1115; 0.20 max C, 1.1-1.6 Ni, 0.5-1.0 Cr, 0.8 Mn, bal Fe.
  For shafts, cams, gears, axles, machine tool parts; AISI 3120; case hardened, shock resist.

AUBERT & DUVAL MMR1.
  M-1115; 0.17 max C, 1.1-1.6 Ni, 0.5-1.0 Cr, bal Fe.
  For gears, cams, camshafts; AISI 3115; case hardened.

AUBERT & DUVAL MMR3.
  M-1115; 0.13 max C, 1.1-1.6 Ni, 0.5-1.0 Cr, bal Fe.
  For gears, cams, camshafts; AISI 3115; case hardened.

AUBERT & DUVAL MOC2.
  M-1115; 0.4-0.5 C, 1.0-1.5 Cr, 0.10-0.30 Mo, bal Fe.
  For gears, axles, shafts, bolts, spindles; AISI 4145; oil hardened.

AUBERT & DUVAL MUV.
  M-1115; 0.9-1.1 C, 0.1-0.4 V, bal Fe.
  For punches, dies, threading taps, drawing dies; water hardened; Type W2.

AUBERT & DUVAL N.C. 40 M STEEL.
  M-1115; 0.40 C, 1.8 Ni, 0.80 Cr, 0.25 Mo, bal Fe.
  Deep hardening structural steel; for axles, gears, torsion bars.
  AFNOR 30 NCD 10; AISI 4340.

AUBERT & DUVAL NCAV.
  M-1115; 0.14 max C, 3.0-3.5 Ni, 0.5-1.0 Cr, 0.5 Mn, bal Fe.
  For cams, gears, camshafts, machinery parts; AISI 3310; case hardened, shock resistant.

AUBERT & DUVAL NCAV2.
  M-1115; 0.17 max C, 3.0-3.5 Ni, 0.5-1.0 Cr, bal Fe.
  For shafts, gears, camshafts, machinery parts; AISI 3316; case hardened, shock resistant.

AUBERT & DUVAL NCAV4.
  M-1115; 0.20 max C, 3.0-3.5 Ni, 0.5-1.0 Cr, 0.5 Mn, bal Fe.
  For camshafts, cams, gears, machinery parts; AISI 3316; case hardened, shock resistant.

AUBERT & DUVAL NCAV5.
  M-1115; 0.22 max C, 3.0-3.5 Ni, 0.5-1.0 Cr, bal Fe.
  For camshafts, cams, gears, machinery parts; AISI 3320; case hardened, shock resistant.

AUBERT & DUVAL P.E.R. 2 U ALLOY.
  M-1115; 0.12 max C, 20.0 Cr, 20.0 Co, 2.0 max Ti, 1.0 max Ai, 5.0 max Fe, bal Ni.
  High temperature alloy for blades and wheels for jet or turbo-prop engines.
  AFNOR NCK 20 TA.

AUBERT & DUVAL P.E.R. 3 ALLOY.
  M-1115; 19 Cr, 14 Co, 4 Mo, 3 Ti, Al, bal Ni.
  High temperature alloy; good high temperature creep strength.
  Blades and wheels for gas turbines.

AUBERT P DUVAL PYRAD 49 D.
  M-1115; 20.0-23.0 Cr, 17.0-20.0 Fe, 8.0-10.0 Mo, bal Ni.
  HT, air cool: 79 kg/mm^2 TS; 39 kg/mm^2; 54 El.
  Corrosion resistant and good high temperature strength for jet engine parts.

AUBERT & DUVAL R2C.
  M-1115; 0.90-1.1 C, 1.8-2.2 Cr, bal Fe.
  For stamping tools, dies, jigs, templates; oil hardened.

AUBERT & DUVAL RAD.
  M-1115; 0.9-1.2 C, 1.4-1.9 Cr, 0.5 Si, 0.8 Mn, bal Fe.
  For roller bearings, thrust bearings; AISI 52100; water hardened, wear resistant.

AUBERT & DUVAL SCV.
  M-1115; 0.15 C, 0.30 max Si, 1.0 max Mn, 1.3 Cr, 0.90 Mo, 0.25 V, bal Fe.
  For elevated temperature operation.
  AFNOR 15 CDV 6.

AUBERT & DUVAL SM1.
  M-1115; 0.6 C, 2.25-3.30 Ni, 0.6-1.0 Cr, 0.40 Mo, bal Fe.
  For die blocks; hot work steel, air hardened.

AUBERT & DUVAL SMV.
M-1115; 0.3-0.4 C, 4.5-5.5 Cr, 1-2 Mo, W, V, bal Fe.
For forging dies, punches, mandrels, hot shear blades; air hardened, hot work steel.

AUBERT & DUVAL SMV 3.
M-1115; 0.38 C, 1.0 Si, 5.0 Cr, 1.25 Mo, 0.50 V, 1.25 W.
Hot work tool steel.
AFNOR Z 38 CDWV 5; similar to AISI H12.

AUBERT & DUVAL S.O.S. 3 STEEL.
M-1115; 0.40 C, 8.5 Cr, 3.25 Si, Mo, bal Fe.
For motor car inlet valves, also for supercharged diesel engines.
Similar to: EN 52; AFNOR Z 45 CS 09 03.

AUBERT & DUVAL SYCOB.
M-1115; 2.0-2.5 C, 11-13 Cr, V, Co, bal Fe.
For cutting, pressing and stamping tools, drawing dies; oil or air hardened, non-deforming.

AUBERT & DUVAL TA 3.
M-1115; 0.32 C, 4.5 Ni, 1.2 Mo, 0.5 Cr, bal Fe.
Hot work tool steel.
AFNOR 32 NDC 18.12.

AUBERT & DUVAL V300.
M-1115; 0.4-0.5 C, 1.3-1.8 Si, 0.5-1.0 Cr, 0.1-0.3 Mo, bal Fe.
For suspension springs, torsion bars; AISI 9245; tough.

AUBERT & DUVAL X 13.
M-1115; 0.25-0.35 C, 12-14 Cr, 0.8 Mn, bal Fe.
For valves, cutlery, turbine blades, surgical instruments; Type 420; corrosion resistant.

AUBERT & DUVAL X 13 D STEEL.
M-1115; 0.20 C, 13.0 Cr, bal Fe.
Martensitic stainless steel; hardenable to 140 kg/mm^2 TS; 115 kg/mm^2 YS; 15 El.
For steam turbine blades, valves, cocks, fittings.
AFNOR Z 20 C 13; G.B 25.62, En 56 C; similar to AISI 420.

AUBERT & DUVAL X 13 E STEEL.
M-1115; 0.13 C, 13.0 Cr, 1.0 max Ni, bal Fe.
Martensitic stainless steel; hardenable to 135 kg/mm^2 TS.
For compressor blades and wheels; pump and valve parts.
AFNOR Z 12 C 13; G.B. En 56 B.

AUBERT & DUVAL X 13 M.
M-1113; 0.15 max C, 12-14 Cr, 0.8 Mn, bal Fe.
For valves, cutlery, surgical instruments; Type 410; corrosion resistant.

AUBERT & DUVAL X 15 D 2.
M-1115; 0.09 C, 1.0 max Si, 1.0 max Mn, 15 Cr, 7 Ni, 2.5 Mo, 1.0 Al, bal Fe.
Precipitation hardening stainless.
AFNOR Z 8 CND 15 07.

AUBERT & DUVAL X 16 D 3 STEEL.
M-1115; 0.20 max C, 16.0 Cr, 4.0 Ni, 3.0 Mo, bal Fe.
Ht; 150 kg/mm^2 TS; 115 kg/mm^2 YS; 16 El.
Good strength and corrosion resistance.
AFNOR Z 15 CND 16 04 03; USA AM 355.

AUBERT & DUVAL X 17 T STEEL.
M-1115; 13 Ni, 17 Cr, W, Ti, bal Fe.
High temperature alloy; used up to 750°C.
For jet engine and gas turbine parts.

AUBERT & DUVAL X 17 U4.
M-1115; 0.07 max C, 16.5 Cr, 4.0 Ni, 2.5-4.5 Cu, bal Fe.
Precipitation hardening stainless.
Z 8 CNU 17 04; similar to 17-4 PH.

AUBERT & DUVAL X 18.
M-115; 0.12 max C, 8-11 Ni, 17-20 Cr, bal Fe.
Annealed: 80,000 TS; 35,000 YS; 55 El; 75 RA; 150 Brin.
For chemical plant equipment, tanks, mixers; Type 302; stainless, austenitic.

AUBERT & DUVAL X 18 BC.
M-1115; 0.05 max C, 10-13 Ni, 17-22 Cr, bal Fe.
Annealed: 85,000 TS; 35,000 YS; 60 El; 70 RA; 150 Brin.
For chemical plant equipment, tanks, mixers; Type 304; austenitic, stainless.

AUBERT & DUVAL X 18 J.
M-1115; 0.07 C, 1.0 max Si, 2.0 max Mn, 18.0 Cr, 9.0 Ni, bal Fe.
Austenitic stainless steel.
AFNOR Z 6 CN 18 09; AISI 304.

AUBERT & DUVAL X 18 MBC.
M-1115; 0.05 max C, 10-13 Ni, 16-19 Cr, 2-3 Mo, bal Fe.
Annealed: 85,000 TS; 35,000 YS; 50 El; 65 RA; 160 Brin.
For acid resistant chemical plant equipment; Type 316; stainless, austenitic.

AUBERT & DUVAL X 18 MNB.
M-1115; 0.08 max C, 1.0 max Si, 2.0 max Mn, 16.0-18.0 Cr, 11.0-13.0 Ni, 2.0-2.5 Mo, (Nb+Ta) = 10 x C to 1.0, bal Fe.
Stabilized austenitic stainless steel for welded chemical equipment.
AFNOR Z6 CNDNb 17-12; modified AISI 316.

AUBERT & DUVAL X 18 MP.
M-1115; 0.10 max C, 1.0 max Si, 2.0 max Mn, 16.0-18.0 Cr, 11.0-13.0 Ni, 2.0-2.5 Mo, Ti = 5 x C to 0.60, bal Fe.
Stabilized austenitic stainless steel for welded chemical equipment.
AFNOR Z 8 CNDT 17-12; modified AISI 316.

AUBERT & DUVAL X 18 NB.
M-1115; 0.08 max C, 1.0 max Si, 2.0 max Mn, 17.0-19.0 Cr, 10.0-12.0 Cr, (Nb+Ta) = 10 x C to 1.0, bal Fe.
Stabilized austenitic stainless steel.
AFNOR Z 6 CNNb 18-11; AISI 347.

AUBERT & DUVAL X 18 P.
M-1115; 0.12 max C, 9-12 Ni, 17-20 Cr, Ti = 5 x C, bal Fe.
Annealed: 85,000 TS; 35,000 YS; 55 El; 65 RA; 150 Brin.
For welded chemical plant equipment, mixers, tanks; Type 321; stainless, austenitic.

AUBERT & DUVAL X 20 T.
M-1115; 0.2-0.3 C, 7-11 Ni, 19-23 Cr, 1.75-3.75 W, bal Fe.
Heat treated: 114,000 TS; 65,000 YS; 40 El.
For valves for diesel engines; austenitic, corrosion and heat resistant.

AUBERT & DUVAL X 25.
M-1115; 0.15 max C, 18-20 Ni, 24-26 Cr, bal Fe.
Annealed: 100,000 TS; 45,000 YS; 50 El; 65 RA; 185 Brin.
For furnace parts, valve parts, jet engine parts; Type 310; stainless, austenitic.

AUBERT & DUVAL X.D.B.. STEEL.
M-1115; 1.0 C, 17.0 Cr, 0.50 max Mo, bal Fe.
Martensitic stainless steel; hardenable to HV 700; for valve seats and guide rings.
Similar specs: AFNOR Z 100 C 17; AISI 440 C.

AUBERT & DUVAL X.M. 114 STEEL.
M-1115; 0.50 C, 0.25 max Si, 9.0 Mn, 21.0 Cr, 4.0 Ni, bal Fe.
Sol. Tr and aged: 105 kg/mm² TS; 65 kg/mm² YS; 11 El; 330.
For exhaust valves.
AFNOR Z 50 CMN 22.

AUBERT & DUVAL XN 26 TW.
M-1115; 0.06 C, 1.5 Mn, 25 Ni, 15 Cr, 1.25 Mo, 2.0 Ti, bal Fe.
For high temperature equipment.

AUBURN CAST.
M-697; 0.7-1.0 C, bal Fe.
For tools, drills, taps; water hardened.

AUBURN EXTRA.
M-697; 0.8-1.2 C, bal Fe.
For tools, drills, taps; water hardened.

AUBURN HOT DIE.
M-697; 0.5 C, 4 W, bal Fe.
For tools, hot shears; hot work steel.

AUBURN PERFECTION.
M-697; C, alloy, bal Fe.
For tools, punches; water hardened.

AUBURN SPECIAL.
M-697; 0.7-1.4 C, bal Fe.
For tools, drills, taps; water hardened.

AUBURN STANDARD.
M-697; 0.7-0.9 C, bal Fe.
For tools, drills, taps; water hardened.

AUBURN TOOL MAKER.
M-697; 0.9 C, 1.2 Mn, bal Fe.
For tools, dies, punches; non-deforming.

AUDEN WIRE.
M-200; 75 Au, plus Pt.
Wire for dental purposes.

AUDIO 52.
4.5-4.75 Si, bal Fe.
For laminations for electrical equipment; high permeability.

AUDIO 58.
3.5-3.75 Si, bal Fe.
For laminations for electrical equipment; high permeability.

AUDIO 65.
3.25-3.75 Si, bal Fe.
For laminations for electrical equipment; high permeability.

AUDIO 72.
3.25-3.75 Si, bal Fe.
For laminations for electrical equipment; high permeability.

AUDIO 82.
2.25-2.75 Si, bal Fe.
For laminations for electrical equipment; high permeability.

AUDIO 101.
2.25-2.75 Si, bal Fe.
For laminations for electrical equipment; high permeability.

AUDIO 117.
1-1.5 Si, bal Fe.
For laminations for electrical equipment; high permeability.

AUDIO 130.
0.5-0.75 Si, bal Fe.
For laminations for electrical equipment; high permeability.

AUDIO 145.
0.25-0.5 Si, bal Fe.
For laminations for electrical equipment; high permeability.

AUDIO 165.
M-Under 0.25 Si, bal Fe.
For laminations for electrical equipment; high permeability.

AUER METAL-A.
M-Ger.; 35 Fe, 65 Ce.
For gas and cigarette lighters; pyrophoric.

AUER METAL-B.
M-Ger.; 53 Fe, 30 Mn, 10 Sb, 7 Misch metal.
For gas and cigarette lighters; pyrophoric, sparking alloy.

AUER METAL-C.
M-Ger.; 35 Fe, 35 Ce, 29 Misch metal.
For gas and cigarette lighters; pyrophoric.

AUER METAL-D.
M-Ger.; 35 Fe, 35 Ce, 24 La, 3 Yb, 2 Er.
For gas and cigarette lighters; pyrophoric.

AUGER.
M-289; C, bal Fe.
For machine tool parts; water hardened.

AUGER SECTION A.
M-289; 0.75 C, bal Fe.
For tools, mining drills; water hardened.

AUGER SECTION B.
M-289; 0.75 C, bal Fe.
For tools, mining drills; water hardened.

AUMAT.
M-200; 83.3% Au, plus Pt.
Wire for dental purposes.

AURIGA MARK I.
M-479; 0.25 C, 0.7 Mn, bal Fe.
Annealed: 67,200-78,500 TS; 34,000 YS; 20 El.
For structural castings; water hardening.

AURIGA MARK III.
M-476; 0.4 C, 0.7 Mn, bal Fe.
Annealed: 83,000-101,000 TS; 42,000 YS; 15 El.
For gears and wear resistant castings; water hardening.

AURIGA MARK VIII.
M-476; 0.20 C, Mo, bal Fe.
Normalized: 67,200 TS; 40,300 YS; 20 El.
For high temperature castings to 1020°F; heat resistant.

AURIGA MARK VIII A.
M-476; 0.15 C, Cr, Mo, bal Fe.
Normalized: 78,000 TS; 44,800 YS; 15 El.
For castings; high creep strength to 1020°F.

AURIGA MARK IX.
M-476; 0.3 C, Cr, Mo, bal Fe.
Hardened: 102,000-141,000 TS; 78,000-101,000 YS; 15-12 El.
For high strength castings; oil hardened.

AURIGA MARK X.
M-479; C, 11-14 Mn, bal Fe.
For wear resistant castings; Mn steel.

AURIGA MARK XV.
M-476; 0.3 C, 3 Cr, Mo, bal Fe.
Hardened: 102,000-141,000 TS; 78,000-101,000 YS; 15-12 El.
For high strength castings; oil hardened.

AURIGA MARK XVI.
M-476; 0.3 C, Cr, Ni, Mo, bal Fe.
Hardened: 102,000-141,600 TS; 78,000-100,000 YS; 15-12 El.
For high strength castings; oil hardened.

AURIGA MARK XVII.
M-476; 0.15 C, 18 Cr, 8 Ni, 1.5 W, bal Fe.
Annealed.
For heat resistant castings; corrosion and heat resistant.

AURIGA MARK XXI.
M-476; 0.2 C, 5 Cr, Mo, bal Fe.
Hardened: 90,000 TS; 60,000 YS; 18 El.
For castings for high temperature service; heat resistant.

AURIGA MARK XXII.
M-476; 0.2 C, 9 Cr, 1.2 Mo, bal Fe.
Hardened: 90,000 TS; 60,000 YS; 18 El.
For castings for high temperature service; heat resistant.

AURIGA MARK XXIII.
M-476; 0.2 C, 13 Cr, bal Fe.
Hardened: 78,500 TS; 49,600 YS; 20 El.
For castings; corrosion resistant.

AURIGA MARK XXIV B.
M-476; 0.15 C, 18 Cr, 8 Ni, Cb, bal Fe.
Annealed: 68,500 TS; 30,300 YS; 20 El.
For castings; stainless, austenitic.

AURIGA V.
M-476; 0.55 C, 0.4 Si, 0.7 Mn, bal Fe.
Steel casting.

AUR-O-MET 11B.
M-603; 10.5 Al, 1.0 Fe, bal Cu.
Sand Cast: 75,000-85,000 TS; 32,000-42,000 YS; 20-35 El; 71-81 RA.
Heat treatable aluminum bronze, corrosion resistant.
For gears, worm wheels, valve seats, valve guides, shafts.
ASTM B148-52-9B; SAE 68 B; CDA 953.

AUR-O-MET 56.
M-603; 10.5 Al, 4.75 Ni, 4.25 Fe, bal Cu.
Sand Cast: 90,000-100,000 TS; 50,000-60,000 YS; 5-10 El; 89-93 Rb.
For gears, machine and structural parts, impellers; corrosion and wear resistant.
ASTM B148-52-9D; CDA 955.

AUR-O-MET 123.
M-603; 3 Sn, 7 Pb, 9 Zn, bal Cu.
Sand cast: 29,000-39,000 TS; 13,000-17,000 YS; 18-30 El; 22 Rb (approx).
Semi-red brass for general purpose plumbing and machine parts; ductile and corrosion resistant, free machining.
ASTM B145-52 5A; CDA 844.

AUR-O-MET 145.
M-603; 14 Zn, 5 Si, bal Cu.
Die cast: 70,000-92,000 TS; 40,000-60,000 YS;
20-15 El; 20 RA; 128-160 Brin.
For gears, water pumps, impellers, marine parts;
corrosion and wear resistant. CDA 875.

AUR-O-MET 150.
M-603; 60 Cu, 39.5 Zn, 0.5 Al.
Cast: 40,000-45,000 TS; 11,000-14,000 YS; 23-28
El; B 40-50 Rock.
For fittings, hardware, permanent and shell
mold castings.
Yellow brass, corrosion resistant. CDA 858.

AUR O MET 245.
M-603; 88 Cu, 6 Sn, 1.5 Pb, 4.5 Zn.
40,000 psi TS; 20,000 psi YS; 30 El; 14.3% Elec.
Cond. IACS.
CDA 922; SAE 622; QQ-C-390 Alloy D4.

AUR-O-MET 420.
M-603; 0.75 Sn, 0.75 Pb, 37.0 Zn, 1.25 Fe, 0.75
Al, 0.5 Mn, bal Cu.
Cast: 60,000-78,000 TS; 20,000-40,000 YS; 15-30
El; B 40-51 Rock.
For propellers, marine hardware, shafts, gears,
fittings. Good corrosion resistance.
Manganese bronze, high strength.
ASTM-B147-52 7A; CDA 864.

AUR-O-MET 421.
M-603; 39.25 Zn, 1.25 Fe, 1.25 Al, 0.25 Mn, bal
Cu.
Cast: 65,000-80,000 TS; 25,000-40,000 YS; 20-40
El; B 54-68 Rock.
For gear shift forks, spiders, fittings, brackets,
landing gears.
Good corrosion resistance, high strength.
Manganese bronze, SAE 43; CDA 865.

AUSCO 80.
M-1152; 1.3-1.4 C, 0.7 Si, 0.9 Mn, 0.10 max Cr,
0.10 max Cu, 0.5 Ni, 0.15 Mo, bal Fe.
Heat treated: 110,000 TS; 80,000 YS; 7 El; 270
Brin.
For crankshafts; castings.

AUSMAN 12.
M-1766; 1.20 C, 12.5 Mn, 0.50 Si, bal Fe.
Austenitic manganese steel for resistance to
abrasion; for excavator bucket teeth, wear
plates.
IHA F-642; DIN x 120 MN 12.

AUSTALON.
M-937; 0.08 C, 18 Cr, 8 Ni, 2 Mo, bal Fe.
For orthodontic and denture applications;
stainless, austenitic.

AUSTENITE.
M-1749; 0.65 C, 4.0 Cr, 14.0 W, 0.50 V bal Fe.
14% Tungsten high speed tool steel.
For drills, turning tools, cutters.

AUSTINOX.
M-884; 0.06 C, 18 Cr, 10 Ni, bal Fe.
Annealed: 85,000 TS; 35,000 YS; 60 El; 70 RA;
150 Brin.
For chemical plant equipment, tanks, vessels,
filters; Type 304; stainle austenitic.

AUSTINOX B.
M-884; 0.06 C, 18 Cr, 12 Ni, 2-3 Mo, bal Fe.
Annealed: 85,000 TS; 35,000 YS; 50 El; 65 RA;
160 Brin.
For acid resistant chemical plant equipment;
Type 316; stainless, austenitic.

AUSTINOX F.
M-884; 0.10 C, 18.0 Cr, 9.0 Ni, 0.20 S, + Mo,
bal Fe.
Free machining austenitic steeless steel.
AISI 303.

AUSTINOX S.
M-884; 0.08 C, 18 Cr, 10 Ni, Ti = 5 x C, bal
Fe.
Annealed: 85,000 TS; 35,000 YS; 55 El; 65 RA;
150 Brin.
For welded chemical plant equipment, mixers,
tanks; Type 321; stainless, austenitic.

AUSTINOX SB.
M-884; 0.08 C, 18 Cr, 12 Ni, 2-3 Mo, Ti = 5 x
C, bal Fe.
Annealed: 85,000 TS; 35,000 YS; 50 El; 65 RA;
160 Brin.
For welded acid resistant chemical plant
equipment; Type 316 Ti; stainless, austenitic.

AUSTRIAN ALLOY (SPANDAU).
M-Austria; 4-6 Cu, 2-3.5 Al, bal Zn.
For die castings; Spandau Alloy.

AUSTRIAN (GERSDORF).
M-Austria; 60-50 Cu, 20-25 Zn, 20-25 Ni.
For costume jewelry; German silver.

AUSTRIAN JOURNAL BOX.
M-Austria; 92.5 Cu, 7.5 Zn.
For bearing purposes; gilding metal.

AUTO.
M-1322; 0.41 C, 1.1 Cr, 0.7 Mn, 0.25 Si, bal Fe.
For gears, bolts, shafts, axles, bolts; oil or water
hardened, shock resistant.

AUTOCHROM CV4.
M-1322; 0.5 C, 0.85 Mn, 1 Cr, 0.09 V, bal Fe.
For gears, bolts, springs, crankshafts; oil
hardened, shock resistant.

AUTO EXTRA PA now VEW V204.

AUTOGRIP.
M-1787.
WC, W, others.
Antiskid and wear-resistant alloy for anti-skid
studs for automotive tires snow and ice chains
with studded straps.

AUTO LITE A1-5.
M-851; 4.5-6.0 Si, 2 max Fe, bal Al.
Cast: 30,000 TS; 5 El.
For die castings; high fluidity.

AUTO LITE A1-6.
M-851; 11-13 Si, 2 max Fe, bal Al.
Cast: 37,000 TS; 1.8 El.
For die castings; high fluidity.

AUTO-LITE A1-9.
M-851; 7.5-9.5 Si, 3-4 Cu, 2 max Fe, bal Al.
Cast: 43,000 TS; 2 El.
For die castings; high fluidity.

AUTO-LITE Z-3.
M-851; 3.5-4.2 Al, 0.03-0.08 Mg, bal Zn.
Cast: 43,000 TS; 10 El; 82 Brin.
For die castings.

AUTOMANG.
M-1100; 20% total Mn, Cr, Mo, Ni, Cu, Si, bal Fe.
Self-shielded, flux-cored wire for joining and build-up of high manganese and other steels.
As deposited: 20-24 Rc; work hardened: 50-54 Rc.

AUTOMANG 2.
M-1100; 31% total Mn, Cr, Si, Mo, Ni, bal Fe.
Cr-Mn austenitic manganese welding wire, for semi-automatic (DC reverse) build-up on high manganese and other steels.
As deposited: 15-20 Rc: work hardened: 47-57 Rc.

AUTOMANG 3.
M-1100; 36% total C, Mn, Cr, Si, bal Fe.
Cr-Mn austenitic manganese welding wire for semi-automatic (DC reverse) build-up and joining of high manganese and other steels.
As deposited: 15-20 Rc; work hardened: 47-57 Rc.

AUTOMATIC AW-79.
M-9; 0.3 C, 2 Mn, 4 Cr, 1 Mo, bal Fe.
Welded: 450 Brin.
For tractor rolls and idlers, steel car wheels; hard facing electrode.

AUTOPAN.
M-297; 0.6-1.0 Mg, 0.6-1.2 Si, 0.3 max Cr, 0.5-2.5 Pb, Sn, Cd, Bi, bal Al.
Annealed: 21,000 TS; 8000 YS; 20 El.
For screw machine products, fasteners; free-cutting.

AUTOTHERMIC.
M-Eng.; steel coated with Al.
For fire walls; heat resistant.

AV 1.
M-38; 0.50 C, 7.5 Mn, 0.17 Si, 21.5 Cr, 3.0 Ni, 2.0 W, 1.0 Cb, 0.3 V, 0.50 N, bal Fe.
For diesel engine valves.

AVC.
M-U.S; 70 Mo, 30 W.
For rocket nozzle components subject to molten zinc.

AVESTA 248 SV.
M-16; 0.05 max C, 16 Cr, 5 Ni, 1 Mo, bal Fe.
Heat treated: 128,000 TS; 100,000 YS; 20 El; 280 Brin.
Weldable, hardenable stainless steel. As rolled for pressure vessels and as cast for propellers, turbines, valves etc. Corrosion resistant, austenitic-martensitic-ferritic.

AVESTA 249 MV.
M-16; 0.05 C, 17 Cr, bal Fe.
Annealed: 74,000 TS; 52,000 YS; 30 El; 150 Brin.
Pressed and welded articles subject to slight corrosion attack. Oil refinery equipment, oil burners.
Corrosion resistant, type 430, ferritic.

AVESTA 252 M.
M-16; 0.07 C, 23 Cr, 14.5 Ni, bal Fe.
Annealed: 100,000 TS; 50,000 YS; 40 El; 170 Brin.
For furnace parts, heat treating boxes, oil burners; Type 309S, corrosion and heat resistant; austenitic.

AVESTA 252 S.
M-16; 0.03 max C, 21.5 Cr, 15 Ni, 2.7 Mo, 0.2 N, bal Fe.
Annealed: 106,000 TS; 50,000 YS; 50 El; 190 Brin.
Pipes, fittings, valves, etc.; mainly for salt water application. Corrosion resistant, austenitic.

AVESTA 254.
M-16; 0.10 C, 19.5 Cr, 20.5 Ni, bal Fe.
Annealed: 85,000 TS; 36,000 YS; 54 El; 160 Brin.
For furnace parts. Low sensibility to sigma phase formation. Corrosion and heat resistant, austenitic.

AVESTA 254 EM.
M-16; 0.08 max C; 25 Cr, 21 Ni, bal Fe.
Annealed: 93,000 TS; 43,000 YS; 50 El; 170 Brin.
For annealing boxes, tanks, recuperators; Type 310 S, corrosion and heat resistant, austenitic.

AVESTA 254 SLX.
M-16; 0.02 max C, 20 Cr, 25 Ni, 4.5 Mo, 1.5 Cu, bal Fe.
Annealed: 85,000 TS; 36,000 YS; 50 El; 155 Brin.
For pumps, chemical plant equipment, resists H_2SO_4, corrosion resistant, austenitic.

AVESTA 393.
M-16; 0.12 C, 13.5 Cr, bal Fe.
Annealed: 78,000 TS; 50,000 YS; 30 El; 190 Brin.
Heat treated: 185,000 TS; 157,000 YS; 17 El; 390 Brin.
For press and caul plates in hardboard, chipboard and plastic laminating ind.
Type 410, hardenable, corrosion resistant.

AVESTA 393 M.
M-16; 0.07 max C, 14 Cr, bal Fe.
Annealed: 74,000 TS; 47,000 YS; 26 El; 150 Brin.
For table flatware, wood pulp conveyor screws, steam and water turbine parts. Recuperators. Non-hardenable, corrosion resistant.

AVESTA 453 E.
M-16; 0.09 C, 26 Cr, 4.5 Ni, bal Fe.
Annealed: 100,000 TS; 70,000 YS; 25 El; 210 Brin.
For furnace plates, lead and salt bath crucibles. Ferritic-austenitic, heat resistant.

AVESTA 453 S.
M-16; 0.08 C, 26 Cr, 5 Ni, 1.5 Mo, bal Fe.
Annealed: 100,000 TS; 70,000 YS; 23 El; 220 Brin.
For valves, pumps, shafts and equipment in the cellulose, dyeing, chemical and dairy industries. Corrosion resistant, ferritic-austenitic (Type 329)

AVESTA 453 SG.
M-16; 0.04 max C, 22 Cr, 6.0 Ni, 1.7 Mo, bal Fe.
Annealed: 90,000 TS; 56,000 YS; 20 El; 220 Brin.
Cast material, for ship propellers. Corrosion resistant, ferritic- austenitic.

AVESTA 664 M.
M-16; 0.07 max C, 17 Cr, 5.5 Ni, 6.5 Mn, bal Fe.
Annealed: 100,000 TS; 50,000 YS; 70 El; 190 Brin.
For sinks, food processing equipment, in the architecture.
For Type 202, corrosion resistant, austenitic.

AVESTA 664 MV.
M-16; 0.05 max C, 18 Cr, 5 Ni, 8 Mn, bal Fe.
Annealed: 106,000 TS; 54,000 YS; 60 El; 190 Brin.
For nitric acid and brewery equipment. Type 204, corrosion resistant, austenitic.

AVESTA 739 G.
M-16; 0.12 C, 13 Cr, 1 Ni, bal Fe.
Annealed: 100,000 TS; 64,000 YS; 20 El; 210 Brin.
For castings, water turbine parts, vanes, bottom plates, corrosion resistant.

AVESTA 739 S.
M-16; 0.14 C, 13.5 Cr, 1 Mo, bal Fe.
Annealed: 93,000 TS; 64,000 YS; 18 El; 190 Brin.
Hardened: 192,000 TS; 128,000 YS; 10 El; 385 Brin.
For paper mill knives, ship propellers, corrosion resistant, hardenable.

AVESTA 739 SG.
M-16; 0.12 C, 13 Cr, 1 Ni, 1 Mo, bal Fe.
Annealed: 92,000 TS; 58,000 YS; 16 El; 50 RA; 190 Brin.
For castings, ship propellers; corrosion resistant.

AVESTA 831.
M-16; 0.2 C, 1.0 Si, 25 Cr, 0.8 Ni, bal Fe.
Annealed: 83,000 TS; 57,000 YS; 30 El; 175 Brin.
For furnace equipment, Type 446, heat resistant, non-hardenable.

AVESTA 832 H.
M-16; 0.11 C, 17 Cr, 8 Ni, bal Fe.
Annealed: 107,000 TS; 50,000 YS; 65 El; 200 Brin.
For press and caul plates in hardboard, chipboard and plastic laminating ind. and chain type bottle conveyors. Type 302, corrosion resistant, austenitic.

AVESTA 832 M.
M-16; 0.05 max C, 18 Cr, 9 Ni, bal Fe.
Annealed: 85,000 TS; 40,000 YS; 70 El; 155 Brin.
Cold strip rolled. For household articles and in the architectural industry. For equipment in the food-processing industry. Type 304, corrosion resistant, austenitic.

AVESTA 832 MV.
M-16; 0.05 max C, 18 Cr, 9 Ni, bal Fe.
Annealed: 83,000 TS; 37,000 YS; 62 El; 150 Brin.
Hot rolled. For welded structures, acid vessels and other equipment in food processing and chemical ind. Type 304, corrosion resistant, austenitic.

AVESTA 832 MVN.
M-16; 0.05 max C, 18.5 Cr, 8.5 Ni, 0.20 N, bal Fe.
Annealed: 90,000 TS; 48,000 YS; 50 El; 160 Brin.
High proof stress stainless steel. Cryogenic applications. Corrosion resistant, austenitic.

AVESTA 832 MVNB.
M-16; 0.08 max C, 17.5 Cr, 9.5 Ni, Cb, bal Fe.
Annealed: 86,000 TS; 40,000 YS; 56 El; 160 Brin.
For nitric acid ind., aircraft internal combustion and jet engine parts. Furnace equipment; Type 347, corrosion resistant, austenitic.

AVESTA 832 MVR.
M-16; 0.03 max C, 18.5 Cr, 10 Ni, bal Fe.
Annealed: 77,000 TS; 36,000 YS; 60 El; 150 Brin.
For nitric acid equipment, nuclear ind. Type 304 L, corrosion resistant, austenitic.

AVESTA 832 MVRN.
M-16; 0.03 max C, 18.5 Cr, 10.5 Ni, 0.20 N, bal Fe.
Annealed: 88,000 TS; 40,000 YS; 50 El; 150 Brin.
High proof stress stainless steel. Equipment in the chemical ind. Corrosion resistant, austenitic.

AVESTA 832 MVT.
M-16; 0.08 max C, 17.5 Cr, 10 Ni, Ti, bal Fe.
Annealed: 86,000 TS; 40,000 YS; 55 El; 160 Brin.
For nitric acid ind.; aircraft internal combustion and jet engine parts. Furnace equipment. Type 321, corrosion resistant, austenitic.

AVESTA 832 SF.
M-16; 0.05 max C, 17 Cr, 11 Ni, 2.3 Mo, bal Fe.
Annealed: 85,000 TS; 42,000 YS; 58 El; 150 Brin.
Equipment in the chemical, cellulose and food processing ind. Type 316, corrosion resistant, austenitic.

AVESTA 832 SFR.
M-16; 0.03 max C, 17 Cr, 11.5 Ni, 2.3 Mo, bal Fe.
Annealed: 83,000 TS; 42,000 YS; 58 El; 150 Brin.
Equipment in the chemical and food processing ind. Type 316 L, corrosion resistant, austenitic.

AVESTA 832 SFT.
M-16; 0.08 max C, 17 Cr, 12 Ni, 2.3 Mo, Ti, bal Fe.
Annealed: 85,000 TS; 42,000 YS; 55 El; 160 Brin.
Equipment in the chemical ind., Type 316 T, corrosion resistant, austenitic.

AVESTA 832 SI.
M-16; 0.02 max C, 4 Si, 17.5 Cr, 14.5 Ni, bal Fe.
Annealed: 96,000 TS; 42,000 YS; 55 El; 160 Brin.
Vessels, pipes, valves, etc. Especially good resistance against highly concentrated nitric acid. Austenitic.

AVESTA 832 SK.
M-16; 0.05 max C, 17 Cr, 11.5 Ni, 2.7 Mo, bal Fe.
Annealed: 85,000 TS; 42,000 YS; 58 El; 150 Brin.
Main type for the chemical and cellulose ind. Type 316, corrosion resistant, austenitic.

AVESTA 832 SKER (832 SKR-5).
M-16; 0.03 max C, 17.5 Cr, 13.5 Ni, 2.7 Mo, bal Fe.
Annealed: 85,000 TS; 40,000 YS; 58 El; 150 Brin.
Low ferrite grade mainly for the urea and acetic acid ind. (Type 316) corrosion resistant, austenitic.

AVESTA 832 SKNB.
M-16; 0.06 max C, 17 Cr, 13 Ni, 2.7 Mo, Cb, bal Fe.
Annealed: 85,000 TS; 42,000 YS; 55 El; 160 Brin.
For equipment in the chemical ind. Type 316 Cb, corrosion resistant, austenitic.

AVESTA 832 SKR.
M-16; 0.03 max C, 17.5 Cr, 13 Ni, 2.7 Mo, bal Fe.
Annealed: 85,000 TS; 42,000 YS; 57 El; 150 Brin.
For equipment in the chemical ind. Type 316 L, corrosion resistant, austenitic.

AVESTA 832 SKRN.
M-16; 0.03 max C, 17.5 Cr, 13 Ni, 2.7 Mo, 0.2 N, bal Fe.
Annealed: 90,000 TS; 50,000 YS; 50 El; 160 Brin.
High proof stress stainless steel. Equipment in the chemical and cellulos ind. For chemical tankers. Corrosion resistant, austenitic.

AVESTA 832 SKT.
M-16; 0.08 max C, 17 Cr, 13 Ni, 2.7 Mo, bal Fe.
Annealed: 85,000 TS; 42,000 YS; 55 El; 160 Brin.
For equipment in the chemical ind. Type 316 Ti, corrosion resistant, austenitic.

AVESTA 832 SL.
M-16; 0.05 max C, 16.5 Cr, 15 Ni, 4.3 Mo, bal Fe.
Annealed: 88,000 TS; 46,000 YS; 51 El; 150 Brin.
For equipment in the chemical ind. Highly resistant to acids, austenitic.

AVESTA 832 SN.
M-16; 0.5 max C, 18.5 Cr, 14.5 Ni, 3.3 Mo, bal Fe.
Annealed: 85,000 TS; 43,000 YS; 50 El; 150 Brin.
For sulfite digesters and equipment in the chemical ind. Type 317, corrosion resistant, austenitic.

AVESTA 832 SNR.
M-16; 0.03 max C, 18.5 Cr, 14.5 Ni, 3.3 Mo, bal Fe.
Annealed: 85,000 TS; 43,000 YS; 50 El; 150 Brin.
For sulfite digesters and equipment in the chemical ind. Type 317 L, corrosion resistant.

AVESTA 832 SV.
M-16; 0.05 max C, 17.5 Cr, 9.5 Ni, 1.5 Mo, bal Fe.
Annealed: 84,000 TS; 40,000 YS; 61 El; 150 Brin.
For equipment in the cellulose and food processing industry. Window frames. Corrosion resistant, austenitic.

AVESTA 832T/NB.
M-16; 0.10 C, 18 Cr, 9.5 Ni, Cb, bal Fe.
Annealed: 90,000 TS; 45,000 YS; 56 El; 65 RA; 150 Brin.
For welded chemical plant equipment, tanks, mixers; Type 347; corrosion resistant.

AVIAL.
M-698; 0.5 Si, 2.5 Cu, 0.6 Mg, 1.0 Ni, 0.7 Cr, bal Al.
Heat treated: 57,000 TS; 26 El.
For light alloy parts; hardenable.

AVIALITE-915.
M-8; 89.25 Cu, 9.25 Al, 0.4 Sn, 0.5 Ni, 0.6 Fe.
Soft: 80,000 TS; 40,000 YS; 22 El.
Hard: 95,000 TS; 55,000 YS; 16 El.
For valve seats, spark-plug bushings in aircraft engines; same as "Millard Metal."

AVIOL.
M-USSR.; 0.7 Si, 0.6 Mg, bal Al.
Annealed: 17,000 TS; 7000 YS; 27 El; 40 Brin.
For aircraft parts.

AVIONAL.
M-493, M-541; 2.5-5.0 Cu, 0.2-1.8 Mg, 0.3-1.5 Mn, bal Al.
Annealed: 27,000 TS; 11,000 YS; 22 El; 47 Brin.
Heat Treated: 72,000 TS; 57,000 YS; 130 Brin.
For aircraft structures and fittings; age-hardenable.

AVIONAL.
M-624; 3-4 Cu, 0.6 Mg, 0.6 Mn, 0.3 max Si, 0.3 max Fe, bal Al.
For aircraft parts; age hardenable.

AVIONAL-14.
M-624, M-249, M-541; 4.4 Cu, 0.8 Si, 0.4 Mg, 0.8 Mn, bal Al.
Annealed: 27,000-30,000 TS; 11,400-17,800 YS; 12-18 El; 45-55 Brin.
TA-Temper: 67,500-75,400 TS; 58,300-64,000 YS; 6-9 El; 125-140 Brin.
For aircraft structures, general engineering components.
Age hardenable; high strength.

AVIONAL-20.
M-249; Cu, Mg, bal Al.
For clad sheets for orthopedics; age-hardened, malleable as heat treated.

AVIONAL-21.
M-624, M-541; 2.5 Cu, 0.3 Si, 0.3 Mg, bal Al.
Annealed: 14,200-21,400 TS; 5700-10,000 YS; 25-40 El; 35-45 Brin.
TN-Temper: 38,400-50,000 TS; 21,500-28,400 YS; 16-23 El; 80-90 Brin.
For aircraft components, marine and transportation equipment.

AVIONAL-22.
M-249, M-541; 3.5-5.0 Cu, 1 max Si, 0.2-1.5 Mn, 0.2-1.5 Mg, bal Al.
Annealed: 22,750 TS; 8500 YS; 30 El; 45 Brin.
Heat Treated: 62,500 TS; 42,500 YS; 20 El; 120 Brin.
For aircraft and bus construction, machine tool parts; age hardened, high fatigue strength.

AVONAL-23.
M-249; 3.5-5.0 Cu, 0.2-1.5 Mg, 0.2-1.5 Mn, 0-1 Si, bal Al.
Annealed: 27,000 TS; 11,500 YS; 20 El; 50 Brin.
Heat treated: 71,000 TS; 48,250 YS; 20 El; 125 Brin.
For aircraft construction, machine tool parts; age-hardened, high fatigue strength.

AVIONAL-24.
M-249, M-541; 3.5-5.0 Cu, 0.1 Si, 0.2-1.5 Mn, 0.2-1.5 Mg, bal Al.
Annealed: 28,500 TS; 12,750 YS; 22 El; 55 Brin.
Heat Treated: 74,000 TS; 51,000 YS; 15 El; 150 Brin.
For aircraft and bus construction, machine tool parts; age-hardened, high fatigue strength.

AVIONAL-25.
M-249; Cu, Mg, bal Al.
Annealed: 22,750 TS; 8500 YS; 30 El; 50 Brin.
Heat treated: 74,000 TS; 68,250 YS; 10 El; 150 Brin.
For aircraft and bus body construction; age-hardened, corrosion resistant.

AVIONAL 102.
M-493; 4.0 Cu, 1.5 Mg, bal Al.
Sol. HT: 380 MPa TS; 10 El.
Bars, forgings, for vehicle and machine construction.

AVIONAL 411.
M-249; 1.0 Si, 4.8 Cu, 0.5 Mg, 0.8 Mn, bal Al.
Heat treated: 64,000-74,000 TS; 54,000-63,000 YS; 8-12 El; 115-130 Brin.
For light alloy parts; hardenable.

AVIONAL D.
M-249, M-624; 3.5-5.0 Cu, 0.2-1.5 Mg, 0.2-1.5 Mn, 0-1 Si, bal Al.
Soft: 25,000 TS; 11,000 YS; 22 El; 50 Brin.
Heat treated: 58,000-87,000 TS; 38,000 YS; 21-2 El; 105 Brin.
For aviation and automobile parts; corrosion resisting; heat treatable.

AVIONAL D TI.
M-249; 0.2 Si, 3-8 Cu, 0.5 Mn, 0.5 Mg, 0.2 Fe, 0.16 Ti, bal Al.
For aircraft parts; age-hardenable.

AVIONAL M.
M-249; 0.7 Si, 4.2 Cu, 0.65 Mg, 0.7 Mn, bal Al.
Heat treated: 57,000-65,000 TS; 40,000-44,000 YS; 16-20 El; 105-120 Brin.
For aircraft parts; age hardened.

AVIONALPLAT.
M-249; Al alloy.
For light alloy parts; Avional coated with Al.

AVIONAL S.
M-249; 3.5-5 Cu, 0.2-1.5 Mg, 0.2-1.5 Mn, 0.2 Fe, 0.1-1.0 Si, bal Al.
Extruded: 71,000 TS; 54,000 YS; 15 El; 125 Brin.
For aircraft parts, auto construction; heat treatable.

AVIONAL SK.
M-249; 3.5-5.0 Cu, 0.5-0.8 Mn, 0.5-0.8 Mg, bal Al.
Sheets: 65,000-72,000 TS; 48,000-56,000 YS; 18-14 El.
For light alloy parts; age-hardenable.

AVIONAL Z.
M-249; 0.4 Si, 0.3 Fe, 0.75 Mn, 0.80 Mg, 4.5 Cu, bal Al.
Wrought: 62,700-71,500 TS; 37,000-46,000 YS; 15-12 El.
For aviation and automobile parts; similar to "Duralumin."

AVONMOUTH.
M-699; Al alloy.
For light alloy parts.

AVROCAN M7-12.
M-435; 0.40 C, 1.0 Si, 5.0 Cr, 1.4 Mo, 0.5 V bal Fe.
Heat Treated: 225,000-305,000 TS; 195,000-255,000 YS; 8-15 El; 30-50 RA; C 45-56 Rock.
For extrusion dies, forging dies and inserts, hot punches, header and gripper dies.
Tough, hot work tool steel, Type H11.

AVS.
M-1653; 0.75-0.85 C, 1.75-2.50 Si, 0.80 max Mn, 19.0-21.0 Cr, 1.0-1.70 Ni, bal Fe.
Stainless steel for valves.
W.-Nr. 1.4747.

AVS METAL.
M-700; 0.2-0.4 C, bal Fe.
For structural parts; water hardened.

AVW.
M-1653; 0.40-0.50 C, 2.0-3.0 Si, 0.80-1.50 Mn, 17.0-20.0 Cr, 8.0-10.0 Ni, 0.80-1.20 W, bal Fe.
Stainless steel for valves.
W.-Nr. 1.4873.

AWARNITE (NATURAL).
M-Eng.; 75 Ni, 25 Fe.
For heat and corrosion resistant parts; heat and corrosion resistant.

AWCO 1050 A.
M-982; 99.5 min Al.
Soft: 75 N/mm^2 TS.
Hard: 140-195 N/mm^2 TS.
General mechanical wire, welding wire, rivet stock.

AWCO 1080 A.
M-982; 99.8 min Al.
Soft: 70 N/mm^2 TS.
Hard: 130-160 N/mm^2 TS.
General mechanical wire, welding wire.

AWCO 1350.
M-982; 99.5 min Al.
Soft: 75 N/mm^2 TS.
Hard: 160-200 N/mm^2 TS.
For electrical conductors.

AWCO 3103.
M-982; 1.25 Mn, bal Al.
Soft: 115 N/mm^2 TS.
Hard: 205-245 N/mm^2 TS.
General mechanical wire, welding wire.

AWCO 4043 A.
M-982; 5 Si, bal Al.
Welding and brazing wire.

AWCO 4047 A.
M-982; 11.5 Si, bal Al.
Brazing wire.

AWCO 4145 A.
M-982; 10 Si, 4 Cu, bal Al.
Brazing wire.

AWCO 5056 A.
M-982; 0.15 Mn, 5.0 Mg, 0.10 Cr, bal Al.
Soft: 300 N/mm^2 TS.
Hard: 400-450 N/mm^2 TS.
General mechanical wire, rivet and bolt stock, welding wire, zip fastening wire.

AWCO 5154 A.
M-982; 0.3 Mn, 3.5 Mg, bal Al.
Soft: 250 N/mm^2 TS.
Hard: 355 N/mm^2 TS.
General mechanical wire, rivet stock, welding wire.

AWCO 5251.
M-982; 0.3 Mn, 2.2 Mg, bal Al.
Soft: 200 N/mm^2 TS.
Hard: 280-310 N/mm^2 TS.
General mechanical wire.

AWCO 5556 A.
M-982; 0.70 Mn, 5.1 Mg, 0.10 Cr, 0.10 Ti, bal Al.
Welding wire.

AWCO 6101 A.
M-982; 0.50 Si, 0.60 Mg, bal Al.
TH: 300 N/mm² TS.
Electrical conductors.

AWCO OTHER NUMBERS see ALUMINUM ASSOCIATION GRADES.

AXALOY.
M-702; 0.4 C, bal Fe.
For truck axles; water hardened.

AXITE.
M-703; Co-Cr-W.
For valve seats, hard facing electrodes; wear resistant.

AXLOY 20.
M-703; 3.6 C, 2.5 Si, 0.6 Mn, bal Fe.
Cast: 20,000 TS; 163 Brin.
For glass molds; soft cast iron.

AXLOY 30.
M-703; 3.4 C, 1.8 Si, 0.7 Mn, 0.25-1.0 Ni, 0.4 Cr, 0.4 Mo, bal Fe.
Normalized: 30,000-35,000 TS; 180 Brin.
For machinery castings, gears; cast iron, wear resistant.

AXLOY 30 CN.
M-703; 3.2 C, 2.5 Si, 1.5 Ni, bal Fe.
For hydraulic cylinders, valves; cast iron.

AXLOY 30 CNM.
M-703; 3.3 C, 0.25-1.00 Ni, 0.25-1.00 Cr, 0.25-1.00 Mo, bal Fe.
Cast: 35,000 TS; 207 Brin.
For liners, cams, cylinder blocks, gears; cast iron.

AXLOY 30 CRN.
M-703; 3.2 C, 1.5 Ni, 0.8 Cr, 2.5 Si, bal Fe.
Cast: 30,0000 TS.
For oil field valves, pump liners; cast iron, corrosion resistant.

AXLOY 35.
M-703; 3.2 C, 1.9 Si, 0.85 Mn, bal Fe.
Normalized: 35,000-45,000 TS; 220 Brin.
For machinery castings, gears; cast iron, wear resistant.

AXLOY 35 CN.
M-703; 3 C, 2 Si, 2 Ni, 1 Cr, bal Fe.
Cast: 35,000 TS.
For brake drums, gears, pump liners; abrasion and wear resistant.

AXLOY 50.
M-703; 3.0 C, 2.2 Si, 1 Mn, bal Fe.
Normalized: 50,000-60,000 TS; 240 Brin.
For gears, machinery castings; cast iron.

AXLOY 50 CN.
M-703; 3.2 C, 2.2 Si, 2 Ni, 1 Cr, bal Fe.
Cast: 50,000 TS.
For hydraulic cylinders, valves, steam chests; cast iron.

AXLOY 50 CNM.
M-703; 3.2 C, 2 Si, 2 Ni, 1 Cr, bal Fe.
Cast: 50,000 TS.
For cylinder heads, pressure gates; cast iron.

AXLOY CNM.
M-703; 3 C, 2 Si, 1 Ni, bal Fe.
Cast: 35,000 TS.
For dies, pistons, liners, cams; wear resistant.

AZ.
M-699; Al alloy.
For light alloy parts.

AZ5G.
M-French; 5 Zn, 0.5 Mg, 0.2 Ti, 0.3 Cr, bal Al.
Aged: 31,300-35,600 TS; 18,500-22,800 YS; 5-9 El; 70 Brin.
For aircraft components, high strength castings; age-hardenable, shock resistant.

AZ8GU.
M-French; 7.0-8.5 Zn, 1.7-3.0 Mg, 1-2 Cu, 0.2 Cr, 0.4 Mn, bal Al.
Hardened: 95,000 TS; 85,000 YS; 8 El.
For light alloy parts; age-hardenable.

AZ8GU.
M-1785; 0.25 max Si, 0.35 max Fe, 1.2-1.9 Cu, 0.2-2.9 Mg, 0.1-0.22 Cr, 7.2-8.2 Zn, bal Al.
Wrought, heat treatable aluminum alloy.
AA 7049. For production of high quality weapon parts.

AZ 14.
M-1361; 1.2 Mg, 0.10 Mn, 4.7 Zn, Cr, Zr, bal Al.
315-390 N/mm² TS; 275-295 N/mm² YS; 6-10 El.
High strength heat treatable alloy with good corrosion resistance, weldability and formability.
Type Al Zn Mg 1.
Similar to AA 7005.

AZ 24.
M-1361; 0.10 Cu, 2.0 Mg, 0.20 Si, 0.25 Mn, 4.0 Zn, Zr, bal Al.
370-385 N/mm² TS; 310-330 N/mm² YS; 3-8 El.
Heat treatable alloy, - special quality.
Type AL Zn Mg 2.

Section I: Alloy Data / 177

AZ31B.
  M-43; 3.0 Al, 1.0 Zn, bal Mg.
  O-temper: 32,000-37,000 TS; 18,000-22,000 YS; 21-12 El; 56 Brin.
  F-temper: 32,000-36,000 TS; 16,000-24,000 YS; 6-18 El; 46 Brin.
  H24-temper: 37,000-42,000 TS; 24,000-32,000 YS; 6-19 El; 73 Brin.
  Sheet, extrusions; formable, weldable; for use to 300°F.
  ASTM B107-69; QQ-M-31b.

AZ 31C.
  M-43; 9 Al, 1 Zn, bal Mg.
  Extruded: 37,000 TS; 28,000 YS; 12 El; 49 Brin.
  O-temper: 37,000 TS; 22,000 YS; 21 El; 56 Brin.
  For truck bodies, dock boards, aircraft frames; formerly Dowmetal FS; good formability.

AZ 40.
  M-1361; 0.70 Cu, 3.2 Mg, 0.1 Mn, 4.5 Zn, Cr, bal Al.
  460-500 N/mm^2 TS; 385-430 N/mm^2 YS; 5-8 El.
  High strength heat treatable alloy, special extrusion quality.
  Type Al Sn Mg Cu 0.5.
  Similar to AA 7079.

AZ 54.
  M-1361; 0.70 Cu, 3.4 Mg, 0.1 Mn, 4.6 Zn, Cr, bal Al.
  High strength heat treatable alloy.
  Type Al Zn Mg Cu 0.5.
  Similar to AA 7079.

AZ61A.
  M-43; 6.5 Al, 1.0 Zn, bal Mg.
  F-temper: 36,000-45,000 TS; 16,000-33,000 YS; 7-16 El; 50-60 Brin.
  General purpose alloy for use up to 300°F.
  Used as welding rod and wire, extruded bars and plate and tube, and forgings.
  ASTM B107-69, B91-68, A5.19-69.

AZ 63.
  M-1361; 1.4 Cu, 2.3 Mg, 5.7 Zn, Cr, bal Al.
  Special extrusion quality, hardenable to: 420-530 N/mm^2 TS; 355-46 N/mm^2 YS; 5-8 El.
  Type Al Zn Mg Cu 1.5.
  Similar to AA 7075.

AZ63A (MG INGOT).
  M-43; 5.5-6.5 Al, 0.18 min Mn, 2.7-3.3 Zn, 0.04 max Cu, 0.01 max Ni, 0.20 max Si, 0.30 max other, bal Mg.
  Ingots for remelting. ASTM B93-66.

AZ63A.
  M-Various foundries; 6.0 Al, 3.0 Zn, 0.15 min Mn, bal Mg.
  F-temper: 24,000-29,000 TS; 10,000-14,000 YS; 4-6 El.
  T6-temper: 34,000-40,000 TS; 16,000-19,000 YS; 3-5 El.
  Sand and permanent mold castings; are hardenable; good corrosion reistance pressure tight.
  For airplane wheel and brake castings, gear housings.
  ASTM B93-66; AMS 4420, 4422; QQ-M-55b; SAE 50.

AZ 64.
  M-1361; 1.4 Cu, 2.5 Mg, 5.7 Zn, Cr, bal Al.
  420-530 N/mm^2 TS; 355-460 N/mm^2 YS; 5-8 El.
  High strength heat treatable alloy.
  Type Al Zn Mg Cu 1.5.
  Similar to AA 7075.

AZ 67.
  M-1361; 1.0 Cu, 2.3 Mg, 5.7 Zn, Cr, Be, bal Al.
  420-530 N/mm^2 TS; 355-460 N/mm^2 YS; 5-8 El.
  Special mining quality.
  Type Al Zn Mg Cu 1 Be.

AZ 74.
  M-1361; 1.0 Cu, 2.4 Mg, 5.8 Zn, Cr, Ag, bal Al.
  450-550 N/mm^2 TS; 390-490 N/mm^2 YS; 5-9 El.
  High strength alloy with resistance to stress-corrosion cracking.
  Type Al Zn Mg Cu Al.

AZ 75.
  M-1361; 1.0 Cu, 2.5 Mg, 6.0 Zn, Cr, Zr, Ag, bal Al.
  450-550 N/mm^2 TS; 390-490 N/mm^2 YS; 5-10 El.
  High strength alloy, mainly for die forgings having low level of internal stresses.
  Type Al Zn Mg Cu Ag Zr.

AZ 79.
  M-1361; 1.0 Cu, 2.5 Mg, 6.0 Zn, Cr, Ag, bal Al.
  450-550 N/mm^2 TS; 390-490 N/mm^2 YS; 5-10 El.
  Improved hardenability and fracture toughness.
  Type Al Zn Mg Cu Ag.

AZ80A.
  M-43; 8.5 Al, 0.5 Zn, bal Mg.
  T5-temper: 48,000-55,000 TS; 33,000-40,000 YS; 4-7 El; 82 Brin.
  Extruded rods and forgings; good strength.
  ASTM B107-69, B91-68; AMS 4360.

AZ81A.
M-Various foundries; 7.6 Al, 0.7 Zn, 0.13 min Mn, bal Mg.
T4-temper: 34,000-40,000 TS; 10,000-12,000 YS; 7-15 El.
Sand and permanent mold castings; age hardenable, good corrosion resistance.
For aircraft equipment where good elongation and toughness are required.
ASTM B80-69; QQ-M-56; SAE 505.

AZ91A (MG INGOT).
M-43; 8.5-9.5 Al, 0.0003-0.001 Be, 0.15 max Mn, 0.45-0.9 Zn, 0.08 max Cu, 0.08 max Ni, 0.20 max Si, 0.30 other, bal Mg.
Ingots for remelting.
ASTM B93-66.

AZ91A.
M-Various foundries; 9.0 Al, 0.7 Zn, 0.13 min Mn, bal Mg.
F-temper: 34,000 TS; 23,000 YS; 3 El.
Magnesium die cast alloy with good pressure tightness and corrosion resistance.
For housings and small precision parts.
ASTM B94-57; QQ-M-38; AMS 4490; SAE 501.

AZ91B.
Similar to AZ91A.

AZ91B (MG INGOT).
M-43; 8.5-9.5 Al, 0.0003-0.001 Be, 0.15 min Mn, 0.50-0.90 Zn, 0.25 max Cu, 0.01 max Ni, 0.20 max Si, 0.30 max other, bal Mg.
Ingots for remelting for die casting and permanent mold casting use only.
ASTM B93-66.

AZ91C (MG INGOT).
M-43; 8.3-9.2 Al, 0.15 min Mn, 0.45-0.90 Zn, 0.04 max Cu, 0.01 max Ni 0.20 max Si, 0.30 max other, bal Mg.
Ingot for remelting.
ASTM B93-66.

AZ91C.
M-various foundries; 8.7 Al, 0.7 Zn, 0.13 min Mn bal Mg.
F-temper: 18,000-24,000 TS; 10,000-14,000 YS; 0.2 El.
T6-temper: 34,000-40,000 TS; 16,/00-19,000 YS;3-5 El.
Sand and permanent mold castings,-good pressure tightness; age hardenable.
For light weight housings for motors and various accessories.
ASTM B80-69; QQ-M-56; QQ-M-55; SAE 504.

AZ92A (MG INGOT).
M-43; 8.5-9.5 Al, 0.13 min Mn, 1.7-2.3 Zn,0.4 max Cu, 0.01 max Ni, 0.20 max Si, 0.30 max other, bal Mg.
Ingot for remelting. ASTM B93-66.

AZ 92 A (WELD ROD).
M-43; 8.3-9.7 Al, 0.15 min Mn, 0.4-1.5 Zn, bal Mg, (other 0.30 max).
Melting point: 1110°F.
For welding rod and electrode.

AZ 100 A replaced by AM 100 A.

AZA.
M-1638; 0.45-0.85 Cu, 0.5 Fe, 0.15 Si, 7.1-7.8 Zn, 0.2-0.5 Mg, 0.1-0.2 Ti, 0.3 Mn, 0.2 Cr, 0.0003 Be, 0.05 Ni, bal Al.
Aged: 38,00 TS; 26,000 YS; 5 El; 77 Brin.
For hardware, aircraft and oil field equipment, dairy and agriculture equipment.
Pressure tight castings.

A.Z. ALLOY.
M-591; 33 Zn, 1 Pb, bal Cu.
For plates for photoengraving; free-cutting.

AZM.
M-325; 1.0 Zn, 0.3 Mn, 6.0 Al, bal Mg.
Extrusions anf forgings.
30,000-36,000 TS; 20,000-22,000 proof stress; 7-10 El; 55-70 VDH.
Medium strength alloy, machines readily, difficult to cold form and to weld; may need protection against corrosion.

AZZALON.
M-96; 0.75-0.85 C, 0.05 Si, bal Fe.
For welding rod.

B

B-4.
M-1769.
Sintered carbide.
350,000 TS; Density: 14.0-14.25.
Hardness: 87.5-88.5 RA.
Industry code: C-12-13-14.

B9-HIGH SPEED.
M-114; 8.84 C, 4.2 Cr, 18.5 W, 2 V, bal Fe.
For tools, taps, reamers; high speed steel.

B-10 HIGH SPEED STEEL.
M-114; 0.8 C, 4.5 Cr, 18.5 W, 2 V, 9 Co, bal Fe.
For lathe and form tools, reamers, broaches; high speed steel for heavy cuts.

B 15 LC2.
M-1488; 1.60 C, 12.0 Cr, 0.90 Mo, 0.90 V, bal Fe.
Cold work tool steel; air hardening.
For stamping dies, thread rollers, shears.
AFNOR Z160CDV 12; AISI D2.

B 15 M.
M-1488; 2.0 C, 12.0 Cr, 0.60 Mo, bal Fe.
Cold work tool steel; air hardening.
For stamping and trimming dies, shears.
AFNOR Z190CD12; similar to AISI D3.

B24 Q-TEMP.
M-604; 0.20-0.25 C, 0.80-1.10 Mn, 0.15-0.30 Si, 0.0005 B, bal Fe.
120-200 ksi TS; 92-180 ksi YS; 375 Brin, as Ht Tr.
For high strength parts.

B33 see **WESTINGHOUSE B-33.**

B-47 HOT WORK.
M-1779; 0.40 C, 0.35 Mn, 4.2 Cr, 4.2 W, 4.2 Co, 2.2 V, 0.4 Mo, bal Fe.
At 800°F: 178,000 TS; 9 El; 36 RA.
At 1200°F: 110,000 TS; 16.5 El; 37 RA.
W-Cr-Co hotwork tool steel for heavy duty applications; forging die inserts, hotwork punches, brass extrusion tooling.
AISI H-19.

B-50.
M-1683; 0.50 C, 0.85 Mn, 1.1 Cr, 0.15 V, 0.25 Mo, bal Fe.
Oil hardening tool steel for shafts, arbors, lathe centers.

B-66 see **WESTINGHOUSE B-66.**

B-76 see **HEPPENSTALL B76.**

B80.
M-1739.
Bronze, sintered.
Density: 6.8-7.2 gms/cc; UTS: 14 kg/mm^2.
Elon: 5 min; 26-18% porosity.
Medium duty bronze.
B.S.S. A110; MPIE BT-0010-S.

B80/0 1-138.
M-1739.
Bronze, sintered.
Density: 5.8-8.0 gms/cc as ordered.
UTS: 9.5-31.5 kg/mm^2 as ordered.
Bearings, piston rings, structural parts.

B80/1.
M-1739.
Bronze-Graphite, sintered.
Density: 6.8-7.2 gms/cc UTS: 14 kg/mm^2.
Elon: 6 min; 22-16 porosity.
For bearings.
SAE 842; B.S.S. A113; MPIE BT-0010-S.

B85.
M-1739.
Bronze, sintered.
Density: 8.0-8.4 gms/cc; UTS: 35 kg/mm^2.
Elon: 8 min; 9-5 porosity.
Heavy duty bronze.
MPIE BT-0100-W.

B88 see **COLUMBIUM XB-88; WESTINGHOUSE XB-88.**

B99.
22 W, 2 Hf, 0.07 C, bal Cb.
High temperature, refractory alloy.

B-110.
M-1776; 85.0 Cu, 15 Zn.
Bronze powder, 60 mesh.
For fabrication of compacted-sintered mechanical P/M articles.

B-120VCA see **CRUCIBLE B 120 VAC.**

B-129.
M-1776; 78.5 Cu, 1.5 Pb, bal Zn.
Bronze powder, 60 mesh.
For fabrication of compacted-sintered mechanical P/M articles.

B-155.
M-1776; 70.0 Cu, 1.5 Pb, bal Zn.
Bronze powder, 60 mesh.
For fabrication of compacted-sintered mechanical P/M articles.

B-161.
M-1776; 89.0 Cu, 1.5 Pb, bal Zn.
Bronze powder, 60 mesh.
For fabrication of compacted-sintered mechanical P/M articles.

B-174.
M-1776; 63.5 Cu, 1.5 Pb, bal Zn.
Bronze powder, 60 mesh.
For fabrication of compacted-sintered mechanical P/M articles.

B-412.
M-1776.
Bronze powder, 60 mesh.
For fabrication of compacted-sintered mechanical P/M articles.

B1080.
M-1741; 99.80% Al min.
O Temper: 62 MPa TS; 28 MPa YS; 45 El.
H18 Temper: 138 MPa TS; 131 MPa YS; 6 El.
Chemical and process plant and equipment.

B1200.
M-1741; 99.0% Al min.
O-Temper: 90 MPa TS; 34 MPa YS; 35 El.
H18 Temper: 165 MPa TS; 152 MPa YS; 5 El; 44 Brin.
Spinnings, holloware and general sheet metal work.

B-1900 see also **ALLOY B-1900**.

B 1900.
M-1491; 0.10 C, 8 Cr, 10 Co, 6 Mo, 1 Ti, 6 Al, 0.015 B, 0.10 Zr, 4 Ta, bal Ni.
At R.T.: 141,000 TS; 120,000 YS; 8 El.
At 1400°F: 138,000 TS; 117,000 YS; 4 El.
For high temperature applications, jet engines, turbines.
Heat and corrosion resistant.
Casting alloy. Good thermal fatigue and creep resistance.

B-1910.
M-1491; 0.10 C, 10.0 Cr, 10.0 Co, 3.0 Mo, 1.0 Ti, 6.0 Al, 0.015 B, 0.10 Zr, 7.0 Ta, bal Ni.
High temperature alloy; for jet engine blades.

B1914.
M-1788; 10 Cr, 10 Co, 3 Mo, 5.5 Al, 5.3 Ti, 0.10 B, bal Ni, (C, Zr as low as possible).
Cast, ann: 2100°F, 1 hr, air cool; ann 1650°F, 10 hr, air cool: Room Temp: 145,000 psi TS; 112 ksi YS; 12 El. 1400°F: 150,000 psi TS; 131 ksi YS; 20 El.
Good high temperature creep strength.
For turbine blades, wheels.

B1925.
M-1788; 12 Cr, 8.5 Co, 1.8 Mo, 4.5 W, 4 Ta, 3.5 Al, 4 Ti, 0.10 B, bal Ni, (C, Zr as low as possible).
Cast, ann: 160-174 ksi TS; 140-145 ksi YS; 4-8 El at room temp, and up to 1400°F.
Good high temperature creep strength.
For turbine blades, wheels.

B1964.
M-1788; 8.8 Cr, 10 Co, 1 Mo, 8.5 W, 2.5 Ta, 3.5 Al, 5.3 Ti, 0.11 B 0.02 C, 0.02 Zr, bal Ni.
Cast, ann: 1975°F, 4 hrs, AC; 1650°F, 10 hrs, AC. Room Temp: 176,000 psi TS; 144 ksi YS; 7 El. 1400°F: 168,000 psi TS; 135 ks YS; 7 El.
Good high temperature creep strength.
For turbine blades, wheels.

B1981.
M-1788; 16 Cr, 8.5 Co, 1.8 Mo, 2.6 W, 1.8 Ta, 0.9 Cb, 3.4 Al, 3.4 Ti, 0.10 B, bal Ni, (C, Zr low as possible).
Cast, ann: 2050°F, 2 hrs, AC; + 1550°F, 24 hrs, AC. Room Temp: 160,000 psi TS; 140 ksi YS; 4.5 El. 1400°F: 163,000 psi TS; 128 ksi YS; 12 El.
Good high temperature creep strength.

B2014.
M-1741; 4.4 Cu, 0.8 Si, 0.8 Mn, 0.6 Mg, bal Al.
T4 Temper: 448 MPa TS; 290 MPa YS; 20 El.
Aircraft structures, forgings, heavy duty structural applications.

B5083.
M-1741; 4.5 Mg, 0.7 Mn, 0.15 Cr, bal Al.
O Temper: 290 MPa TS; 148 MPa YS; 22 El; 70 VHN.
H321 Temper: 331 MPa TS; 228 MPa YS; 16 El; 82 Brin.
Cryogenics, marine aircraft, drilling rigs.

B6063.
M-1741; 0.7 Mg, 0.4 Si, bal Al.
O Temper: 90 MPa TS; 48 MPa YS; 30 El; 30 VHN.
T6 Temper: 241 MPa TS; 214 MPa YS; 12 El; 73 Brin.
Furniture, general purpose extrusions.

B6101.
M-1741; 0.6 Mg, 0.5 Si, bal Al.
T5 Temper: 207 MPa TS; 179 MPa YS; 12 El.
Electrical conductors.

B6351.
M-1741; 0.6 Mg, 1.0 Si, 0.6 Mn, bal Al.
T4 Temper: 241 MPa TS; 165 MPa YS; 20 El; 70 VHN.
T6 Temper: 331 MPa TS; 310 MPa YS; 11 El; 103 VHN.
Transport applications.

BA. 21 now **AA 5251**.

B.A. 24 now **AA 6063**.

B.A. 25 now **AA 6082**.

BA. 27 now **AA 5154A**.

BA. 28 now **AA 5056 A**.

BA. 60 now **AA 3103**.

B.A. 212.
 M-28; 0.6 Mg, bal Al.
 Alloy sheet and strip for bright anodizing.

B.A. 226.
 M-28; 1.0 Mg, bal Al.
 Alloy aluminium tubing.

BA. 703.
 M-28; 1 Cu, 2.6 Mg, 0.25 Mn, 5.7 Zn, 0.1 Cr, bal Al.
 WP-temper: 78,000 TS; 67,000 YS; 7 El.
 For light alloy parts; age-hardened.

B.A. 705.
 M-28; 0.5 Cu, 2.5 Mg, 0.5 Mn, 5.7 Zn, bal Al.
 Aluminium alloy; solution treated and aged.

BA 2014 ETC same as **AA 2014 ETC.**

BABBIT.
 Lead base antifriction metals.

BABBIT see **BABBITT GRAPHO BABBITT.** see also **NBD ARMATURE BABBITT ETC.**

BACHITE.
 M-704; 0.2 C, 18 Cr, 8 Ni, bal Fe.
 For stainless parts, chemical equipment; austenitic.

BACK CASE METAL.
 M-Eng.; 62 Cu, 20 Zn, 18 Ni.
 For heat and corrosion resistant parts.

BADALL.
 M-1535; 0.25 C, 0.028 S, 0.65 Si, 0.15 Mo, 0.85 Mn, 0.60 max Cr, 0. max B, bal Fe
 Plate: 245,000 TS; 210,000 YS; 11 El; 44 RA; 500 Brin.
 Plate: 190,000 TS; 175,000 YS; 16 El; 55 RA; 360-400 Brin.
 For severe service applications, pressure vessels, wear plates.
 Heat treated, wear resistant plates.

BADGER.
 M-73; 0.95 C, 0.3 Si, 1.2 Mn, 0.5 W, 0.5 Cr, bal Fe.
 Heat Treat: 250,000 TS; 225,000 YS; 8 El; 16 RA; 50 Rock C.
 For blanking dies, punches, reamers, threading dies, non-deforming; oil hardening tool steel; AISI 01.

BADGER.
 M-435; 1.2 C, 1.5 W, 0.4 Cr, 0.2 V, bal Fe.
 For tools, cutters, broaches; hot work tools; non-deforming.

BADGER DRILL ROD.
 M-1702; 0.94 C, 1.2 Mn, 0.3 Si, 0.5 Cr, 0.5 W, bal Fe.
 Oil hardening tool steel, AISI 01.

BADGER FLAT GROUND.
 M-1702; 0.94 C, 1.2 Mn, 0.3 Si, 0.5 Cr, 0.5 W, bal Fe.
 Oil hardening tool steel, flat ground condition, AISI 01.

BADGER STEEL.
 M-253; 0.9-1.2 C, bal Fe.
 For taps, reamers, threading dies.

BAHN ALUMINIUM.
 M-621; 6 Cu, bal Al.
 Extruded: 29,000 TS; 13,000 YS; 20 El; 50 Brin.
 Cold drawn: 36,000 TS; 17,000 YS; 10 El; 70 Brin.
 For light alloy parts, electric pantographs.

BAILY'S METAL.
 M-Eng.; 82.1 Cu, 12.8 Sn, 5.1 Zn.
 For bearings, corrosion resistant castings; corrosion resistant.

BAKER 1729.
 M-18; 40.5 Pd, 10 Fe, bal Au.
 Wire: 200,000-220,000 TS.
 For resistors; high electrical resistance.

BAKER 1780.
 M-18; 40 Au, 10 Fe, bal Pd.
 Wire: 175,000-210,000 TS.
 For resistors; high electrical resistance.

BAKER ALLOY 1757.
 M-18; 79 Pt, 15 Rh, 6 Ru.
 Wire: 278,000 TS; 1.8 El.
 For electrical contact equipment.
 Corrosion resistant.

BAKER ALLOY NO. 934.
 M-18; Pt alloy.
 For electrical resistances; resistance wire.

BAKER ALLOY NO. 1534.
 M-18; 8.5 Pt, 72.5 Au, 4 Ag, 14 Cu, 1 Zn.
 Wire: 182,500 TS; 1.3 El.
 For electrical resistances; resistance wire.

BAKER ALLOY NO. 1765.
 M-18; 92 Pt, 8 W.
 300,000 TS; 2 El.
 For electrical resistances; resistance wire.

BAKER CONTACT ALLOY NO. 846.
 M-18; 50 Pt, 26 Pd, 20 Au, 4 Ag.
 Annealed: 107,000 TS; 85,000 YS; 25 El.
 Hard: 171,000 TS; 163,000 YS; 3 El.
 For electrical contacts; heat resistant.

BAKER NO. 175L.
 M-18; 60 Ag, 20 Cu, 20 Zn.
 Annealed: 78,600 TS; 18 El.
 For coined parts, bearings, Silver solder; M.P. 680°C.

BALCO.
M-61; 30 Fe, bal Ni.
Rolled: 70,000 TS.
For voltage resistors.
Magnetic, high permeability.

BALCO.
M-897; 29-31 Fe, bal Ni.
Annealed: 80,000 TS; 55,000 YS; 25 El.
For thermometer bulbs, ballast tubes; heat resistant to 590°C.

BALDER.
M-111; 0.05 C, 0.1 Si, 0.1 Mn, bal Fe.
For electrical equipment; soft magnet iron.

BALDWIN NO. 1.
M-1510; 0.75 C, 0.91 Cr, 0.40 Mo, 0.81 Mn, 1.78 Ni, bal Fe.
For dies, punches, shear blades, knives, rolls; oil hardened, shock resistant.

BALDWIN NO. 711.
M-1510; 0.35 C, 0.86 Cr, 0.81 Mn, 0.35 Mo, 0.48 Si, 0.34 Cu.
Heat treated: 267,000-275,000 TS; 209,000-221,000 YS; 13-15 El; 37-51 RA; 560-580 Brin.
For chisels, punches, hard tools, furnace bars, dies, shears; non-tempering, water hardened.

BALDWIN AH.
M-1510; 1.1 C, 0.75 Mn, 5.5 Cr, 0.3 V, 1.2 Mo, bal Fe.
For dies, gages, forming rolls, master hobs, cutters; non-deforming, air hardened.

BALFORS ULTRA-CAPITAL + 1.
M-Eng.; C, alloy, bal Fe.
For cutters, tools; high speed steel.

BALFOSTEEL.
M-359; 0.7-1.2 C, bal Fe.
For punches, dies, reamers, saws; water hardening.

BALFOUR 4.T.S.S.
M-1112; 1.25 C, 1.2 Cr, 4.5 W, 0.25 V, bal Fe.
For reamers, rolls, punches, drawing dies; oil hardened, abrasion resistant.

BALFOUR 227.
M-1112; 0.28 C, 9 W, 3 Cr, 0.3 V, bal Fe.
For bolt headers, extrusion dies and liners; hot work steel, tough.

BALFOUR 293.
M-1112; 0.36 C, 3 Cr, 9 W, 0.3 V, bal Fe.
Annealed: 230 Brin. For extrusion and hot piercing dies, rams, liners; hot work steel, oil hardened.

BALFOUR 351.
M-1112; 0.5 C, 1 Cr, 2 W, 0.3 V, bal Fe.
For punches, forging dies, shear blades; hot work steel, oil hardened.

BALFOUR A.B.75.
M-1112; 0.6 C, 2 Si, 0.85 Mn, 0.25 Cr, bal Fe.
For pneumatic chisels, shear blades, punches; oil hardened, tough.

BALFOUR A.G.S.
M-1112; 1.05 C, 1.3 Cr, 0.35 Mn, 0.25 Si, bal Fe.
Annealed: 210 Brin.
For forming rolls, cams, jigs, trimming dies; oil hardened, tough.

BALFOUR BLUE LABEL.
M-1112; 0.7-0.8 C, 0.3 Mn, bal Fe.
For drills, taps, punches, crimpers; Type W1; water hardened.

BALFOUR DARWINS AR1.
M-1725.
Corrosion resistant alloy useful in handling sulphuric acid.

BALFOUR DARWINS AR2.
M-1725.
Nickel-chromium-copper-molybdenum alloy for resistance to hot concentrated sulphuric acid.

BALFOUR DARWINS AR3.
M-1725.
Nickel-chromium-copper-iron alloy for resistance to sulphuric and phosphoric acids; used in picking processes and in drainage disposal plants.

BALFOUR DARWINS AR4A.
M-1725.
Nickel-molybdenum-iron alloy for use with hydrochloric and sulphuric acids.

BALFOUR DARWINS AR5B.
M-1725.
Nickel-molybdenum-iron alloy; higher molybdenum than AR4A for greater corrosion resistance.

BALFOUR DARWINS AR6C.
M-1725.
Nickel-molybdenum-chromium-iron alloy for handling oxidizing acids as nitric acid and mixed acids and salts containing free chlorine.

BALFOUR DARWINS AR7.
M-1725.
A duplex, age-hardenable, corrosion-resistant stainless steel with good strength, weldability, resistance to corrosion and galling; used particularly in wood pulp processing.

BALFOUR DSW.
M-1112; 1.3 C, 4.5 W, bal Fe.
For dies, cutters; water or oil hardened.

BALFOUR E.X.D.I.
M-1112; 0.33 C, 1.5 Cr, 5.5 W, 3.75 Ni, bal Fe.
For extrusion dies, cr, 5.5 upsetting dies, mandrels; hot work steel, oil hardened.

BALFOUR NSS3.
M-1112; 0.95 C, 2.0 Mn, 0.3 Si, bal Fe.
For punches, reamers, taps, dies, cams, pawls; oil hardened, non-deforming.

BALFOUR OO.
M-1112; 0.6 C, 1 Cr, 2 W, 0.3 V, bal Fe.
For dies, punches, trimmers; oil hardened, non-deforming.

BALFOUR P.R.N.2.
M-1112; 1.05 C, 1.1 Mn, 1.5 Cr, bal Fe.
For blanking and forming dies, master hobs; oil hardened, abrasion resistant.

BALFOUR R. 9030.
M-1112; 0.30 C, 0.75 Cr, 0.5 Mo, 2.75 Ni, bal Fe.
For plastic mold dies, piercing punches; hot work steel, oil hardened.

BALFOUR R9030.
M-1112; 0.95 C, 1.25 Mn, 0.5 Cr, 0.5 W, bal Fe.
For tools, dies, punches, forming dies, crimpers, upsetters; oil hardened, non-deforming.

BALFOUR RSD.
M-1112; 0.95 C, 1.25 Mn, 0.5 W, 0.5 Cr, bal Fe.
For tools, dies, punches, headers, upsetters; non-deforming, oil hardened.

BALFOUR SC13.
M-1112; 2.15 C, 11.5 Cr, 0.75 Mo, bal Fe.
For-blanking and forming dies, engravers rolls, hobs; non-deforming, air hardened.

BALFOUR S.C. 25.
M-1112; 1.5 C, 0.25 V, 0.75 Mo, 12 Cr, bal Fe.
For blanking and forming dies, shear blades; air hardened, non-deforming.

BALFOUR S.C. 26.
M-1112; 1.8 C, 12 Cr, 0.3 Mn, 0.2 Si, bal Fe.
For rolling dies, shear blades, rim rolls; air hardened, non-deforming.

BALFOUR SCP.
M-1112; 0.9 C, 0.15 V, bal Fe.
For chisels, drills, taps, punches, reamers, hobs; Type W2; water hardened.

BALFOUR S.D. 20.
M-1112; 0.35 C, 0.60 Mn, 1.2 Cr, 1.75 Ni, bal Fe.
Annealed: 200 Brin.
For plastic mold dies, punches, extrusion dies; oil hardened, shock resistant.

BALFOURS SHOE DIE.
M-350; 0.5 C, 0.6 Mn, 0.3 Mo, bal Fe.
For tools, leather cutters; oil hardened.

BALFOUR S.I.W.
M-1112; 0.65 C, 1.1 Si, 0.70 Mn, 0.2 Cr, 1.3 W, bal Fe.
Annealed: 220 Brin.
For pneumatic chisels, rivet snaps, shear blades; oil hardened, fatigue resistant.

BALFOUR TIO. H.
M-1112; 0.9 C, 1.2 Mn, 0.4 Cr, 0.5 W, bal Fe.
For blanking and forming dies, cams, pauls; oil hardened, non-deforming.

BALFOUR TSS.
M-1112; 1.3 C, 4.5 W, bal Fe.
For dies, cutters; water or oil hardened.

BALLAST.
M-897; 99.8 Ni.
Soft: 60,000 TS; 18,000 YS; 50 El.
For cathodes and filaments for electronic tubes; heat resistant.

BALLAST NICKEL.
M-61; 99.6 Ni.
Annealed: 56,000 TS.
For ballast wire for voltage control electrical equipment; resistance 52 ohms/mil. ft.

BALL BEARING STAINLESS STEEL.
M-521; 0.4 C, 0.25 Si, 0.3 Mn, 11.5 Cr, bal Fe.
Annealed: 110,000 TS; 72,000 YS; 20 El; 240 Brin.
For ball bearings; corrosion resistant.

"B" ALLOY.
M-U.S.; 25 Zn, 3 Cu, bal Al.
For light alloy parts; non-hardenable.

BA. LM8.
M-28; 0.3-0.8 Mg, 3.5-6.0 Si, 0.5 max Mn, bal Al.
Sand Cast: 17,920 min TS; 11,200 PS; 2 min El.
WP-Sand Cast: 33,600 min TS; 31,400 PS; 5 min El.
WP-Chill Cast: 40,320 min TS; 31,400 PS; 2 min El.
For cylinder heads, propeller gear boxes, gear housings, crankcases.
Pressure-tight leak-proof castings.

BALTIC C.D.V.1.
M-1406; 2.25 C, 11.5 Cr, 0.5 Mo, 0.3 V, bal Fe.
For blanking and extrusion dies; oil hardened, non-deforming.

BALTIC C.D.V.2.
M-1406; 1.6 C, 12 Cr, 0.8 Mo, 0.5 V, bal Fe.
For blanking and extruding dies, gages, shear blades; air hardened, non-deforming.

BALTIC C.L.15.
M-1406; 0.38-0.42 C, 3.0-3.5 Ni, 0.60 max Mn, bal Fe.
For chisels, snaps, tools, punches; tough and shock resistant.

BALTIC C.L.40T.
M-1406; 0.55 C, 2 Si, 0.90 Mn, 0.30 Cr, 0.20 V, bal Fe.
For punches, shear blades, concrete breakers; tough, shock resistant.

BALTIC C.L.45.
M-1406; 0.50 C, 1.2 Cr, 0.7 Mn, 0.2 V, bal Fe.
For chisels, pneumatic tools, pistons; tough, shock resistant.

BALTIC C.L. 60.
M-1406; 0.60 C, 0.60 Cr, 0.70 Mn, bal Fe.
For vice grips, general tools; water hardened.

BALTIC C.L. 222.
M-1406; 0.35 C, 0.80 Cr, 0.70 Mo, 0.50 W, bal Fe. For punches, snaps, shear blades, pneumatic tools; non-tempering, water hardened.

BALTIC C.L. 224.
M-1406; 0.55 C, 1.5 Ni, 0.75 Cr, 0.3 Mo, bal Fe.
For drop forging dies; tough and water resistant.

BALTIC C.L. 225.
M-1406; 0.50 C, 0.50 Mn, 1.25 Ni, 0.50 Cr, 0.25 Mo, bal Fe.
For drop forging dies; tough and wear resistant.

BALTIC C.L.400.
M-1406; 0.9 C, 3.75 Cr, 0.5 Mo, 0.5 V, bal Fe.
For forging and extrusion dies, hot shears and punches; hot die steel, air or oil hardened.

BALTIC C.L.444.
M-1406; 0.35 C, 5 Cr, 0.9 Si, 1.5 Mo, 1.25 W, 0.3 V, bal Fe.
For forging and extrusion dies, hot shears and punches; shock resistant, hot die steel.

BALTIC C.L.444W.
M-1406; 0.35 C, 0.90 Si, 5 Cr, 5 W, 0.30 V, bal Fe.
For hot heading and gripping dies; oil hardened, hot die steel.

BALTIC C.L.666.
M-1406; 0.60-0.70 C, 0.40 Mn, 6 W, 0.6 Cr, bal Fe.
For coal cutters, picks; air or oil hardened.

BALTIC L.C.H.D.
M-1406; 0.30 C, 9-10 W, 2.75 Cr, 0.4 V, bal Fe.
For dies, inserts and liners; oil hardened, hot die steel.

BALTIC P.C.S.K.
M-1406; 0.45 C, 2 W, 1.5 Cr, 0.25 V, bal Fe.
For chisels, snaps, pneumatic tools; tough and shock resistant.

BARAL.
M-511; up to 50 Ba, bal Al.
For getters in electrical discharge devises and vacuum tubes; good stability in air.

BARE-BRITE A632.
M-663; Bare wire for inert gas welding of T1 steel, HY-80 and similar grades.
MIL-E-19822A, Type B88.

BARE-BRITE AX-90.
M-663; Low alloy bare wire for welding high strength low alloy steel for 100,000 psi TS in weld area. Used for ships, military vessels and equipment, earth moving equipment.
MIL-E-23765/2 Type 100 S-1.

BARE-BRITE AX-110.
M-663; Low alloy bare wire for welding high strength low alloy steel for 120,000 psi TS in weld area.
MIL-E-23765/2 Type 120S-1.

BARE-BRITE AX-140.
M-663; 0.10 C, 1.8 Mn, 0.35 Si, 2.25 Ni, 1.0 Cr, 0.55 Mo, bal Fe.
Low alloy bare wire for welding high strength low alloy steel for 140,000 psi TS in weld area.
MIL-E-24355 Type 140S.

BARIO.
M-Eng.; 21.4 Cr, 57.4 Ni, 1 Fe, 15.4 W, 0.3 C.
For resistor elements; stainless and corrosion resisting.

BARIO-HARD.
30 Co, 30 Cr, 25 W, 10 Mn, 5 Ti.
For tools, high temperature applications; corrosion and heat resistant.

BARIO, SHEET.
90 Ni, 4.3 Cr, 1.2 W, 0.3 Si, traces Co, Cu, Fe.
For tools, high corrosion resistant parts; stainless and corrosion resistant.

BARIO SOFT.
60 Co, 20 Cr, 20 W.
For tools, heat and corrosion resistant parts; corrosion and heat resistant.

BARIUM.
M-1775; Ba.
Purities: 99.9+%, 99.5%-99.7%, 98+%.
Forms: Sticks, rod, billets, powder, lump, wire.

BARIUM 13 CR.
M-705; 0.12 C, 12-15 Cr, bal Fe.
Annealed: 65,000-75,000 TS; 40,000-45,000 YS; 30 El; 60 RA.
For stainless parts; corrosion resistant.

BARIUM 17 CR.
M-705; 0.10 C, 16-23 Cr, bal Fe.
Annealed: 70,000-80,000 TS; 40,000-45,000 YS; 28 El; 55 RA.
For stainless trim; corrosion resistant.

BARIUM 18-8.
M-705; 0.05-0.20 C, 16-23 Cr, 7-11 Ni, bal Fe.
Annealed: 80,000-89,000 TS; 35,000-45,000 YS; 50-60 El; 55-65 RA.
For stainless utensils; heat and corrosion resistant.

Section I: Alloy Data / 185

BARIUM DIE STEEL.
M-705; 0.4-0.6 C, 1-2 Ni, 0.5-1.0 Cr, 0.2-0.3 Mo, bal Fe.
For forging dies; in 3 grades of varying hardness.

BARMAG.
M-511; 35 Ba, bal Mg.
For getters in electrical discharge devices and vacuum tubes.

BARNITE.
M-Eng.; 5.5 Cu, Mg, Cr, Si, Ti, bal Al.
For light alloy parts; age-hardenable.

BARR ALLOY NO. 00C.
M H ALLOY NO. 00C.
M-212; 84 Cu, 16 Sn.
Cast: 25,000-35,000 TS; 20,000-28,000 YS; 1-2 El; 80-100 Brin.
For bearings for turn tables and movable bridges; compression of 0.1%-18000.

BARR ALLOY NO. 1.
M H ALLOY NO. 1.
M-212; 88 Cu, 8-10 Sn, 2-4 Zn.
Cast: 40,000-50,000 TS; 18,000-22,000 YS; 20-40 El; 20-30 RA; 65-74
For steam and hydraulic castings, air valves, small gears; U.S.N. "G" Bronze; "Gun Metal"; "SAE No. 62."

BARR ALLOY NO. 4.
M H ALLOY NO. 4.
M-212; 80 Cu, 10 Sn, 10 Pb.
Cast: 30,000-50,000 TS; 19,000-21,000 YS; 20-10 El; 20-10 RA; 55-65 Brin.
For high speed bearings; for high pressures.

BARR ALLOY NO. 5.
M H ALLOY NO. 5.
M-212; 85 Cu, 5 Sn, 5 Pb, 5 Zn.
Cast: 30,000-38,000 TS; 15,000-19,000 YS; 15-20 El; 15-20 RA; 50-59 Brin.
For bearings, pumps, steam valves; carburetors; "Ounce Metal."

BARR ALLOY NO. 9.
M H ALLOY NO. 9.
M-212; 58 Cu, 40 Zn, 1.5 Fe, 0.5-1.5 Al, 0.0-0.25 Mn.
Grade A: 70,000-80,000 TS; 30,000-34,000 YS; 25-40 El; 100 Brin.
Grade B: 80,000-90,000 TS; 40,000-45,000 YS; 15-25 El; 120 Brin.
For propeller blades and hubs, valve stems, engine framing; compression of 0.1%-20000.

BARR ALLOY NO. 11.
M H ALLOY NO. 11.
M-212; 89 Cu, 10 Al, 1 Fe.
Cast: 65,000-80,000 TS; 23,000-27,000 YS; 20-27 El; 20-27 RA; 93-100 Brin.
Heat treated: 75,000-95,000 TS; 55,000-65,000 YS; 3-15 El; 140-200 Brin.
For gears, feed nuts, bearings; resists wear, repeated shock and corrosion.

BARR ALLOY NO. 14.
M H ALLOY NO. 14.
M-212; 90 Cu, 6.5 Sn, 1.5 Pb, 2.0 Zn.
Cast: 34,000-40,000 TS; 16,000-19,000 YS; 23-35 El; 23-53 RA; 50-60
For plain bearings, backs of Babbit-lined shells; low coefficient of friction and expansion.

BARR ALLOY NO. 15.
M H ALLOY NO. 15.
M-212; 88.5 Cu, 11 Sn, 0.25 Pb, 0.25 P.
Sand cast: 33,000-40,000 TS; 21,000-24,000 YS; 10-15 El; 8-15 RA; 7
80 Brin.
Chill cast: 50,000 TS; 30,000 YS; 6 El; 100 Brin.
For worm gears mating with hardened worms; "SAE-65"; compression of 0.1%-17000.

BARR ALLOY NO. 20.
M H ALLOY NO. 20.
M-212; 65 Cu, 23 Zn, 2 Fe, 3 Mn, 7 Al.
Cast: 90,000-100,000 TS; 50,000-60,000 YS; 20-5 El; 180-200 Brin.
For general castings; compression of 0.1%-65000.

BARR ALLOY NO. 25 B.
M H ALLOY NO. 25 B.
M-212; 69 Cu, 5 Sn, 25 Pb, 1 Ni.
20,000-24,000 TS; 14,000-16,000 YS; 16-12 El.
For general castings, bearings; acid resisting.

BARR ALLOY NO. 48.
M H ALLOY NO. 48.
M-212; 84 Cu, 10 Sn, 2.5 Pb, 3.5 Ni.
Sand cast: 40,000-50,000 TS; 25,000-28,000 YS; 25-15 El; 20-10 RA; 80-93 Brin.
For gears; compression of 0.1%-20000-24000.

BARRONIA METAL.
M-20; 83 Cu, 0.5 Pb, 4 Sn, bal Zn.
Cast: 42,400 TS; 20,400 YS; 38 El; 33 RA; 72 Brin.
Drawn: 84,000 TS; 74,000 YS; 18 El; 46 RA; 171 Brin.
For condenser tubes, heat exchangers, evaporators, fittings; for superheated steam work.

BARROW B.H. BRAND PIG IRON.
M-619; approx 4.0 T.C., 1.5 -4.0 Si, 0.025 max S, 0.03 max P, 0.25-1. Mn, bal Fe.
For metallurgical applications; for making steel and cast iron.

BARROW B.H.R. BRAND (CUPOLA GRADE).
M-619; 3.25-3.45 T.C., 0.5-1.1 Si, 0.12-0.18 S, 0.045 P, 0.25 Mn, bal Fe.
For metallurgical applications, for malleable foundries; various grades.

BARROW B.H.S. BRAND BESSEMER HAEMATITE PIG IRON.
M-619; approx 4.0 T.C., 1.5-4.5 Si, 0.03 max S, 0.35 max P, 0.25-1.2 Mn, bal Fe.
For metallurgical applications; for making steel.

BARROW B H S BRAND FOUNDRY HAEMATITE PIG IRON.
M-619; approx 3.85 T.C., 2.0-3.5 Si, 0.035 max S, 0.03 max P, 0.6-1.0 Mn, bal Fe.
For metallurgical applications; for general foundry work.

BARROW BRAND FOUNDRY HAEMATITE PIG IRON.
M-619; approx 4.0 T.C., 1.5-3.0 Si, 0.02 max S, 0.03 P, 0.90-1.25 Mn, bal Fe.
For metallurgical applications, for heat resisting castings; for making alloy steels.

BARROW B.X. BRAND PIG IRON.
M-619; approx 4.0 T.C., 1.5-4.0 Si, 0.02 max S, 0.025 max P, 0.25-1.5 Mn, bal Fe.
For metallurgical applications; for making steel and cast iron.

BARROW B.X.X. BRAND PIG IRON.
M-619; approx 4.0 T.C., 1.5-3.0 Si, 0.02 max S, 0.02 max P, 0.25-1.5 Mn, bal Fe.
For metallurgical applications; for making steel and cast iron.

BARROW B.X.X. BRAND PIG IRON "SWEDISH".
M-619; approx 4.0 T.C., 1.0-2.5 Si, 0.012 max S, 0.02 max P, 0.25 max Mn, bal Fe.
For metallurgical applications; for making steel and cast iron.

BARROW S.P. BRAND FOUNDRY PIG IRON.
M-619; 3.5-3.75 Tc, 1.5-4 Si, 0.2-0.03 S, o.1-0.3 P, 1.0-1.5 Mn, bal Fe.
For foundry iron for cylinders, valves, pumps, steam chests; for making steel and cast iron.

BARTO.
M-32; 0.50 C, 0.50 Mn, 0.25 Si, 1.0 Cr, 1.75 Ni, bal Fe.
Heat treated: 310,000 TS; 24,000 YS; 5 El; C 56 Rock.
For expander punches, feeder rolls, clutch parts, vice jaws, chuck jaws.
Tough, shock resistant.

BARTO see **CARPENTER BARTO.**

BARWORTH B.C.M.
M-1407; 1.0 C, 1.5 Mn, 0.5 Cr, 0.4 Si, bal Fe.
For press tools, gages, taps, dies; non-deforming, oil hardened.

BARWORTH B.M.S.
M-1407; 0.30-0.35 C, 3 Ni, 1 Cr, 0.4 Mo, bal Fe.
For plastic mold dies; oil hardened.

BARWORTH B.S.C.C.
M-1407; 1.5 C, 1 Mn, 1 Cr, 0.5 W, bal Fe.
For press tools, taps, dies; oil hardened, wear resistant.

BARWORTH B.S.W. 5CO.
M-1407; 0.8 C, 20 W, 4.5 Cr, 1.5 V, 1 Mo, 5 Co, bal Fe.
For cutters, drills, reamers, hobs, broaches; high speed steel.

BARWORTH B.S.W. 6/6/2.
M-1407; 0.80-0.87 C, 4 Cr, 6.5 W, 2 V, 5 Mo, bal Fe.
For lathe and planer tools, reamers, drills, hobs; high speed steel.

BARWORTH B.S.W. 10CO.
M-1407; 0.8 C, 22 W, 4.5 Cr, 1.5 V, 1 Mo, 10 Co, bal Fe.
For cutting tools, reamers, hobs, broaches; high speed steel.

BARWORTH B.S.W. 14.
M-1407; 0.65 C, 14 W, 3.75 Cr, 0.5 V, bal Fe.
For drills, turning tools, hot punches, die cores; high speed steel.

BARWORTH B.S.W. 16.
M-1407; 0.75 C, 16 W, 4.5 Cr, 1 V, 1.5 Mo, bal Fe.
For lathe and planer tools, drills, punches, taps; high speed steel.

BARWORTH B.S.W. 18.
M-1407; 0.75 C, 18 W, 4.5 Cr, 1 V, 0.5 Mo, bal Fe.
For lathe and planer tools, hot punches; high speed steel.

BARWORTH B.S.W. 18V2.
M-1407; 0.75-0.80 C, 18 W, 4.5 Cr, 2 V, 0.5 Mo, bal Fe.
For lathe and planer tools, drills, reamers, taps; high speed steel.

BARWORTH B.S.W. 22.
M-1407; 0.76 C, 22 W, 4 Cr, 1.25 V, bal Fe.
For drills, turning tools, cutters, hot punches; high speed steel.

BARWORTH C.D.S.H.
M-1407; 2 C, 13 Cr, bal Fe.
For forging and blanking dies, punches; oil or air hardened, abrasion resistant.

BARWORTH C.R.V.
M-1407; 0.45-0.50 C, 2 Cr, 0.2 V, bal Fe.
For die casting dies; oil hardened, for short runs.

BARWORTH C.T.S.
M-1407; 0.50-0.55 C, 1 Cr, 2 W, 1 Si, 0.6 Mn, bal Fe.
For hot stamping and pressing dies, pneumatic tools; cold work steel.

BARWORTH H.D.3.
M-1407; 0.35 C, 3.5 Cr, 3 W, 0.3 V, bal Fe.
For hot and cold stamping dies, extrusion dies; hot work steel, oil hardened.

BARWORTH J.H.D.
M-1407; 0.3 C, 3 Cr, 10 W, 0.2 V, bal Fe.
For stamping, pressing and extrusion dies; hot work steel, oil hardened.

BARWORTH J.H.D.N.
M-1407; 0.3 C, 3 Cr, 10 W, 0.2 V, 2 Ni, bal Fe.
For stamping, pressing and extrusion dies; hot work steel, oil hardened.

BARWORTH N.T. TYPE A.
M-1407; 0.4 C, 0.3 Mn, 0.6 Cr, bal Fe.
For chisels, shock tools; water hardened, tough.

BARWORTH N.T. TYPE B.
M-1407; 0.4 C, 0.7 Mn, 1.0 Cr, bal Fe.
For chisels, pneumatic tools; oil hardened, shock resistant.

BARWORTH S.C.R.
M-1407; 1.0 C, 1.0 Mn, 0.8 Cr, bal Fe.
For press tools, taps, dies; oil hardened, wear resistant.

BARWORTH T.C.
M-1407; 0.40 C, 5 Cr, 2 W, 1 Mo, bal Fe.
For die casting dies; oil hardened, for long runs.

BASCH METAL.
M-92; 91 Al, 7 Cu, 0.8 Mg, 0.2 Mn, 1.4 Ni.
Chill cast: 23,000 TS; 2 El; 90 Brin.
Heat treated chill castings; light weight.

BATALBRA.
M-1551; 76 Cu, 22 Zn, 2 Al, 0.03 As.
Inhibited brass for condenser and ships' piping carrying sea water, chemical plant equipment, heat exchangers.

BATH METAL-1.
M-Eng.; 83 Cu, 17 Zn.
For plumbing, pipe; red brass.

BATH METAL-2.
M-Eng.; 55 Cu, 45 Zn.
For bath fixtures; yellow brass.

BATNAVAL.
M-1551; 62 Cu, 37 Zn, 1 Sn.
For heat exchangers, marine condensers, power station condensers.

BATNICKON.
M-1551; 5 Ni, 1.2 Fe, 0.5 Mn, bal Cu.
For ships' piping.

BATTERIUM.
M-21; 89 Cu, 9 Al, 1 Ni, and 1 other metal.
78,000-101,000 TS; 35-48 El; 158-168 Brin.
For chemical apparatus, valves, plug cocks, plate terminals; non-corrodible, acid resistant.

BATTERY COPPER.
M-U.S.; 94 Cu, 6 Zn.
For batteries, gilding metal.

BATTERY PLATES.
M-U.S.; 94 Pb, 6 Sb.
For storage battery plates; hard.

BATURNAL.
M-1551; 2 Pb, 35 Zn, bal Cu.
Free-machining brass, particularly for tubing.

BAUDRINS METAL.
M-Eng.; 72 Cu, 7.10 Zn, 17 Ni, 2.5 Fe, 1.8 Co, 0-0.5 Al.
For ornamental and corrosion resistant parts; corrosion resistant.

BAUDRINS METAL NO. 1.
72 Cu, 16.6 Ni, 2.5 Sn, 7.1 Zn, 1.8 Co.
For ornamental and corrosion resistant parts; corrosion resistant.

BAUDRINS METAL NO. 2.
0.5 Al, 75 Cu, 1.5 Fe, 16 Ni, 2.75 Sn, 2.25 Zn, 2 Co.
For ornamental and corrosion resistant parts; corrosion resistant.

BAZAR.
M-135; 10 Ni, Zn, bal Cu.
For domestic utensils, ornaments; nickel silver.

B.B.
M-261; 0.6 C, 9.5 W, 2.5 Cr, 0.1 V, bal Fe.
For tools, dies; hot die steel.

B & B.
30 Co, 15 Cr, 5 Mo, 2.5 Ti, 3 Al, 0.5 B, bal Ni.
For jet engine components, buckets; high heat resistance.

BBB WELDING ELECTRODE.
M-706; 0.2 C, bal Fe.
For welding electrode.

B.B.D.C. STANDARD ALLOY.
M-Eng.; 88.5 Cu, 0.25 Pb, 1 Ni, 10 Sn, 0.25 P.
For gears, bearings; tough.

B.B. HOT DIE.
M-261; 0.4 C, 10 W, 3.5 Cr, 0.5 V, bal Fe.
For hot dies, punches; hot die steel.

BC-35.
M-1766; 0.37 C, 0.75 Mn, 0.25 Si, 1.0 Cr, bal Fe.
Chromium structural steel.
AFNOR 38C4; DIN 37Cr4; AISI 5135.

BC-40.
M-1766; 0.41 C, 0.75 Mn, 0.25 Si, 1.0 Cr, bal Fe.
Chromium structural steel.
AFNOR 45C4; AISI 5140; BS EN 18-D.

BCV-42.
M-176; 0.25 max C, 1.50 max Mn, 0.15-0.30 Si, 0.40-0.65 Cr, 0.25-0.40 Cu, 0.02-0.10 V, bal Fe.
Plate, normalized: 63,000 psi TS; 42,000 psi YS; 19 El.
For structural purposes, particularly transmission tower uprights.

BCV-46.
M-176; 0.25 max C, 1.50 max Mn, 0.15-0.30 Si, 0.40-0.65 Cr, 0.25-0.40 Cu, 0.02-0.10 V, bal Fe.
Plate, normalized: 67,000 psi TS; 46,000 psi YS; 19 El.
For structural purposes, particularly transmission tower uprights.

BCV-50.
M-176; 0.25 max C, 1.50 max Mn, 0.15-0.30 Si, 0.40-0.65 Cr, 0.25-0.40 Cu, 0.02-0.10 V, bal Fe.
Plate, normalized: 70,000 psi TS; 50,000 psi YS; 19 El.
For structural purposes, particularly transmission tower uprights.

BCV-55.
M-176; 0.25 max C, 1.50 max Mn, 0.15-0.30 Si, 0.40-0.65 Cr, 0.25-0.40 Cu, 0.02-0.10 V, bal Fe.
Plate, normalized: 70,000 psi TS; 55,000 psi YS; 19 El.
For structural purposes, particularly transmission tower uprights.

BCV-60.
M-176; 0.25 max C, 1.50 max Mn, 0.15-0.30 Si, 0.40-0.65 Cr, 0.25-0.40 Cu, 0.02-0.10 V, bal Fe.
Plate, Q+T: 70,000 psi TS; 60,000 psi YS; 19 El.
For structural purposes, particularly transmission tower uprights.

BCV-70.
M-176; 0.25 max C, 1.50 max Mn, 0.15-0.30 Si, 0.40-0.65 Cr, 0.25-0.40 Cu, 0.02-0.10 V, bal Fe.
Plate, Q+T: 80,000 psi TS; 70,000 psi YS; 16 El.
For structural purposes, particularly transmission tower uprights.

BCV-X-50.
M-176; 0.24 max C, 1.0-1.6 Mn, 0.15-0.50 Si, 0.35 max Cu, 0.25 max Ni 0.25 max Cr, 0.08 max Mo, bal Fe.
Plate: 70,000-90,000 psi TS; 50,000 psi min YS; 20 min El.
For structural purposes; transmission lines requiring improved low temperature impact.

BCV-X-52.
M-176; 0.24 max C, 1.0-1.6 Mn, 0.15-0.50 Si, 0.35 max Cu, 0.25 max Ni 0.25 max Cr, 0.08 max Mo, bal Fe.
Plate: 73,000-93,000 psi TS; 52,000 psi min YS; 20 min El.
For structural purposes; transmission lines requiring improved low temperature impact.

BCV-X-56.
M-176; 0.24 max C, 1.0-1.6 Mn, 0.15-0.50 Si, 0.35 max Cu, 0.25 max Ni 0.25 max Cr, 0.08 max Mo, bal Fe.
Plate: 75,000-95,000 psi TS; 56,000 psi min YS; 20 min El.
For structural purposos; transmission lines requiring improved low temperature impact.

BCV-X-60.
M-176; 0.24 max C, 1.0-1.6 Mn, 0.15-0.50 Si, 0.35 max Cu, 0.25 max Ni 0.25 max Cr, 0.08 max Mo, bal Fe.
Plate: 80,000-100,000 psi TS; 60,000 psi min YS; 22 min El.
For structural purposes; transmission lines requiring improved low temperature impact.

BD. 30.
M-341; 0.51 C, 0.95 Cr, 0.87 Mn, 0.20 Mo, bal Fe (typical).
Prehardened to 248-293 Brin.
For dies for mechanically and hand operated press brakes for forming operations.

BE 5.
M-1687; 5 Sn, bal Cu.
Tin bronze, wrought; 70-100 Brin.

BE 12.
M-1687; 11.5 Sn, 1 Zn, 0.1 min P, bal Cu.
Phosphor bronze; castor wrought; 80-95 Brin.
Afnor : UE 12 Z1; UE 12 P; ASTM B103-74.

BE 20.
M-1687; 18 Sn, bal Cu.
Bronze; wrought; 170 Brin.

BEACON.
M-433; 0.8-1.0 C, bal Fe.

BEARCAT.
M-24; 0.5 C, 0.7 Mn, 0.3 Si, 3.25 Cr, 1.4 Mo, bal Fe.
Heat treated: 342,000-145,000 TS; 205,000-130,000 YS; 4-20 El; 7-60 RA; 590-310 Brin.
For rivet sets, punches, chisels, hobs; shock resistant, air or oil hardened. AISI S7.

BEARDSHAW H.M.1.
M-1406; 0.95 C, 1.6 Mn, 0.25 Mo, 0.25 V, bal Fe.
For punches, gages, shear blades, reamers; oil hardened, non-deforming.

BEARING METAL NO. 600.
M-Eng.; 59.7 Cu, 35 Zn, 0.2 Sn, 0.33 Pb, 2.6 Mn, 1.38 Al, 0.7 Si.
For bearings, bushings; tough.

BEARINGOY.
M-208; Pb, Cu, Sb, bal Sn.
For machine bearings; long life under poor lubrication.

BEARING STANDARD.
M-24; 1.0 C, 0.35 Mn, 0.23 Si, 1.35 Cr, bal Fe.
For tools, bearings; water hardening.

BEARITE.
M-214; 80.1 Pb, 0.37 Cu, 0.13 Bi, 16.75 Sb, 2.65 hardener.
At 70°F: 9750 TS; 1.5 El; 29.1 Brin.
At 212°F: 24.5 Brin.
For bearings not subject to vibration or pounding; high compressive value, low coef. fric.

BEARIUM B-4.
M-431; 70 Cu, 26 Pb, bal Sn.
Cast: 21,650 TS; 9750 YS; 16 El; 40 Brin.
For bearings, bushings; poor lubrication.

BEARIUM B-8.
M-431; 70 Cu, 22 Pb, 8 Sn.
Cast: 24,500 TS; 11,500 YS; 12 El; 50 Brin.
For bearings, bushings, thrust washers; for heavy duty requirements.

BEARIUM B-10.
M-431; 20 Pb, 70 Cu, bal Sn.
Cast: 25,500 TS; 12,000 YS; 10 El; 55 Brin.
For bearings, bushings; non-scoring; severe applications.

BEARIUM B-11.
M-431; 17.5 Pb, 70 Cu, bal Sn.
Cast: 30,000 TS; 8 El; 195 Brin.
For bearings, bushings; severe applications.

BEAUCALLOY.
M-1183; 1.4 C, Ni, Cr, bal Fe.
Hardened: 175,000 TS; 150,000 YS; 400 Brin.
For conveyor chains; casting.

BEAUTYWELD.
M-717.
Electrode for AC-DC metallic arc welding all low carbon steel sheet, forms, plates; 75,000 psi TS.

BEAVER.
M-435; 0.68 C, 1 Si, 8.25 Cr, 1 V, 1.4 Mo, 1.5 Ni, bal Fe.
For slitter knives, blanking dies, cold work tools; high wear resistance.

BEAVER PRECISION GROUND FLATS & SQUARES.
M-370; 1.55 C, 0.38 Si, 0.25 Mn, 12.0 Cr, 0.80 Mo, 0.80 V, bal Fe.
AISI D2. Coldwork tool & die steel.

BECKET ALLOY.
M-Eng.; 25-30 Cr, 3 Si, 1.5-3.0 C, bal Fe.
For stainless castings; corrosion resistant.

BEDCO ALLOY.
M-1429; 0.55 C, 1 Mn, 2 Si, 0.4 Mo. 0.35 V, bal Fe.
Heat treated: 350,000 TS; 295,000 YS; 6 El; 12 RA; 600 Brin.
For chisels, punches, pneumatic and beading tools; shock resistant, oil hardened. AISI S5.

BEDCO M-2.
M-1429; 0.65 C, 4 Cr, 2 V, 6.5 W, 5 Mo, bal Fe.
Hardened: C58-60 Rock.
For hot extrusion dies, punches, shear blades, forging mandrels.
High speed steel for hot working; shock resisting.

BEDCO M-2 HIGH SPEED.
M-1429; 0.82 C, 5 Mo, 6.5 W, 4 Cr, 2 V, bal Fe.
Oil hardened: C63-65 Rock.
For lathe and planer cutters, drills, taps, chasers, drawing dies, punches hobs.
Type M2 high speed steel, high red-hardness.

BEDCO T1.
M-1429; 0.75 C, 0.3 Mn, 1.3 Si, 4.0 Cr, 1.15 V, 18.0 W, bal Fe.
Standard grade of tungsten high speed steel for cutting tools; AISI T1.

BEDEL 1 COURONNES 0.
M-455; C, bal Fe.
For wood tools without hardening; crucible steel.

BEDEL 1 COURONNES 1.
M-455; C, bal Fe.
For saws, lathe tools; crucible steel.

BEDEL 1 COURONNES 2.
M-455; C, bal Fe.
For cutlery, taps, chisels, mandrels, razors, metal saws; crucible steel.

BEDEL 1 COURONNES 3.
M-455; C, bal Fe.
For drills, borers, mining bars, chisels, engraving tools, stone tools, needle dies; crucible steel.

BEDEL 1 COURONNES 4.
M-455; C, bal Fe.
For rams, shear blades, dies, hammers; crucible steel.

BEDEL 1 COURONNES 5.
M-455; C, bal Fe.
For surgical instruments, forge tools; crucible steel.

BEDEL 1 TREFLE 5.
M-455; C, Cr, 4.5 Ni, Mo, bal Fe.
Heat treated: 157,000 TS; 135,000 YS; 15 El.
For valves operating at high temperatures; heat resistant.

BEDEL 2 COURONNES 0.
M-455; C, bal Fe.
For wood tools used without hardening; crucible steel.

BEDEL 2 COURONNES 1.
M-455; C, bal Fe.
For saws for hard wood, lathe tools; crucible steel.

BEDEL 2 COURONNES 2.
M-455; C, bal Fe.
For general cutlery, chisels, mandrels, razors, metal saws; crucible steel.

BEDEL 2 COURONNES 3.
M-455; C, bal Fe.
For drills, borers, mining bars, chisels, punches, stone tools; crucible steel.

BEDEL 2 COURONNES 4.
M-455; C, bal Fe.
For shear blades, cutting and stamping dies, hammers; crucible steel.

BEDEL 2 COURONNES 5.
M-455; C, bal Fe.
For surgical instruments, forge tools; crucible steel.

BEDEL 3 COURONNES 0.
M-455; C, bal Fe.
For wood tools used without hardening; crucible steel.

BEDEL 3 COURONNES 1.
M-455; 0.9-1.1 C, bal Fe.
For saws for hard wood, lathe tools; crucible steel.

BEDEL 3 COURONNES 2.
M-455; 1.0-1.1 C, bal Fe.
For general cutlery, taps, chisels, mandrels, razors, metal saws; crucible steel.

BEDEL 3 COURONNES 3.
M-455; 1.2-1.3 C, bal Fe.
For drills, borers, chisels, engraving tools, needle dies, stone tools; crucible steel.

BEDEL 3 COURONNES 4.
M-455; 0.9 C, bal Fe.
For hot and cold shear blades, hammers, cutting and stamping dies; crucible steel.

BEDEL 3 COURONNES 5.
M-455; C, bal Fe.
For forge tools, surgical instruments; crucible steel.

BEDEL 405.
M-455; 1.6 C, 4.5 Cr, 12 W, 0.4 Mo, 0.5 V, 5 Co, bal Fe.
For special tools, engravers tools; textile needles. High carbon, high-speed steel, wear and abrasion resistant.

BEDEL BCN.
M-455; 0.4 C, Cr, 2-3 Ni, bal Fe.
Heat treated: 242,000-135,000 TS; 213,000-121,000 YS; 10-15 El.
For general engineering construction; oil hardening.

BEDEL BCN 3 S 2.
M-455; high C, Cr, 3 Ni, bal Fe.
Heat treated: 270,000 TS; 242,000 YS; 6 El.
For general engineering construction; oil hardening.

BEDEL BCN 4 S.
M-455; C, Cr, 4 Ni, bal Fe.
Air hardened: 270,000-260,000 TS; 235,000-220,000 YS; 6-8 El.
For general engineering construction; air hardening.

BEDEL BK 3 B.
M-455; low C, 3.5 Ni, Cr, bal Fe.
Annealed: 78,000 TS; 58,000 YS; 30 El.
Heat treated: 135,000 TS; 128,000 YS; 11 El.
For crankshafts, general case hardened parts for severe service; case hardening steel.

BEDEL BN 23.
M-455; C, 25 Ni, Cr, bal Fe.
Annealed: 100,000 TS; 45 El.
For chemical plant equipment; non-magnetic; corrosion and acid resistant.

BEDEL BN 33.
M-455; C, 33 Ni, Cr, bal Fe.
Annealed: 115,000 TS; 30 El.
For chemical plant equipment; high acid and corrosion resistant.

BEDEL BNAV 0.
M-455; C, Cr, 3 Ni, Mo, bal Fe.
Heat treated: 228,000 TS; 199,000 YS; 9 El.
For general engineering construction; oil hardening.

BEDEL BNAV 1.
M-455; C, Cr, 4 Ni, Mo, bal Fe.
Heat treated: 315,000 TS; 260,000 YS; 6 El.
For general engineering construction; air hardening.

BEDEL BNAV 2.
M-455; medium C, Cr, 4 Ni, Mo, bal Fe.
Heat treated: 242,000 TS; 228,000 YS; 9 El.
For aircraft motor crankshafts; oil hardening.

BEDEL BNAV 3.
M-455; low C, 4 Ni, Cr, Mo, bal Fe.
Annealed: 107,000 TS.
Heat treated: 143,000 TS; 135,000 YS; 14 El.
For crankshafts, general case hardened parts for severe service; case hardening steel.

BEDEL C.3.
M-455; low C, 2.5 Ni, bal Fe.
Annealed: 85,000 TS; 78,000 YS; 20 El.
For aviation and marine case hardened parts; case-hardening steel.

BEDEL C.5.
M-455; low C, 5 Ni, bal Fe.
Annealed: 90,000 TS; 78,000 YS; 20 El.
For aviation and marine case hardening parts; case hardening steel.

BEDEL CHROME 4.
M-455; C, 2-3 Cr, bal Fe.
For cold stamping dies and punches; non-deforming.

BEDEL CKN.
M-455; low C, 3 Ni, Cr, bal Fe.
Annealed: 85,000 TS.
Heat treated: 164,000 TS; 143,000 YS; 9 El.
For crankshafts, general case hardened parts for severe service; case hardening steel.

BEDEL "DIAMONT BEDEL".
M-455; 4 W, 4 Cr, V, bal Fe.
For cutting tools; oil hardened.

BEDEL DOUBLE TREFLE.
M-455; C, 14 W, Cr, bal Fe.
For general usage, medium speed cutters; high speed steel.

BEDEL DOUBLE TREFLE E.
M-455; C, 15-16 W, bal Fe.
For tools, dies; oil hardening.

BEDEL INO 2.
M-455; medium C, 13 Cr, bal Fe.
Heat treated: 170,000 TS; 157,000 YS; 5 El.
For marine valves; corrosion resistant.

BEDEL INO 3.
M-455; C, 13 Cr, V, bal Fe.
Heat treated: 135,000 TS; 110,000 YS; 10 El.
For automobile valves; corrosion resistant.

BEDEL INO 6.
M-455; C, 17 Cr, V, bal Fe.
For cutlery, surgical instruments; high heat and corrosion resistant.

BEDEL INO 11.
M-455; C, 6 Ni, 25 Cr, bal Fe.
For rolled or forged parts, pump and furnace parts, valves; heat and corrosion resistant; non-hardenable.

BEDEL INO 12.
M-455; C, 20 Cr, 8 Ni, Mo, bal Fe.
For valves operating up to 900°C; austenitic, heat resistant.

BEDEL INO 13.
M-455; C, 8 Ni, 20 Cr, bal Fe.
For carburizing boxes, recuperator tubes, salt bath pots; heat and corrosion resistant; non-hardenable.

BEDEL INO 14.
M-455; C, 6 Ni, 25 Cr, bal Fe.
For carburizing boxes, recuperator tubes, salt bath pots; heat and corrosion resistant; non-hardenable.

BEDEL INO 15.
M-455; C, 6 Ni, 25 Cr, bal Fe.
For carburizing boxes, recuperator tubes, salt bath pots; heat and corrosion resistant; non-hardenable.

BEDEL INO 16.
M-455; C, 6 Ni, 25 Cr, Si, bal Fe.
For carburizing boxes, recuperator tubes, salt bath pots; heat and corrosion resistant; non-hardenable.

BEDEL INO 20.
M-455; C, 25 Cr, 20 Ni, bal Fe.
For pump, furnace and valve parts, carburizing boxes; heat and corrosion resistant; non-hardenable.

BEDEL INO 21.
M-455; C, high Ni, high Cr, bal Fe.
For pump, furnace and valve parts, carburizing boxes; heat and corrosion resistant; non-hardenable.

BEDEL INO 22.
M-455; C, 25 Cr, 20 Ni, bal Fe.
For pump, furnace and valve parts, carburizing boxes; heat and corrosion resistant; non-hardenable.

BEDEL INO 23.
M-455; C, 20 Cr, 20 Ni, Si, bal Fe.
For pump, furnace and valve parts, carburizing boxes; heat and corrosion resistant; non-hardenable.

BEDEL INO 31.
M-455; C, 14 Cr, 15 Ni, W, Mo, bal Fe.
For valves operating up to 900°C; austenitic, heat resistant.

BEDEL INO 32.
M-455; C, 15 Cr, 20 Ni, Mo, bal Fe.
For valves operating up to 900°C; austenitic, heat resistant.

BEDEL INO 33.
M-455; C, 25 Ni, 10 Cr, W, bal Fe.
For valves operating up to 900°C; austenitic, heat resistant.

BEDEL INO 40.
M-455; C, 35 Ni, 15 Cr, bal Fe.
For turbine blading, marine parts; maximum heat and corrosion resistant.

BEDEL INO 41.
M-455; C, 40 Ni, 10 Cr, bal Fe.
For marine parts; maximum heat and corrosion resistant.

BEDEL INO 50.
M-455; C, 15 Cr, 50 Ni, bal Fe.
For carburizing boxes, furnace parts; maximum heat and corrosion resistant.

BEDEL INO 51.
M-455; C, 50 Ni, 15 Cr, bal Fe.
For carburizing boxes, furnace parts; maximum heat and corrosion resistant.

BEDEL M.G.B.A.
M-455; C, Co, bal Fe.
For magnets; magnet steel.

BEDEL M.G.B.C.
M-455; C, 35 Co, 8 W, bal Fe.
For magnets; magnet steel.

BEDEL M.G.B.S.
M-455; C, 5-6 W, Cr, bal Fe.
For magnets; magnet steel.

BEDEL NO. 600 C.
M-455; low C, 1.5 Ni, Cr, Mo, V, bal Fe.
Heat treated: 150,000-177,000 TS; 143,000-157,000 YS; 12 El.
For crankshafts, general case hardened parts for severe service; case- hardening steel.

BEDEL NO. 644.
M-455; C, W, Cr, V, bal Fe.
For high speed cutters; high speed steel.

BEDEL QUATRE TREFLES.
M-455; C, W, Cr, V, Co, bal Fe.
For high speed cutters; high speed steel for extra hard pieces at high speed.

BEDEL QUATRE TREFLES C.
M-455; C, 10 Co, Mo, V, bal Fe.
For tools, dies; oil hardening.

BEDEL R.B. 6.
M-455; C, 2.5 Cr, 0.5 Ni, bal Fe.
For bearings, dies, punches; not shock resistant.

BEDEL R.B. 7.
M-455; high C, 1.5 Cr, bal Fe.
For bearings, dies, punches; not shock resistant.

BEDEL SOC.
M-455; C, 13 Cr, Mo, Co, bal Fe.
Heat treated: 145,000 TS; 130,000 YS; 6 El.
For valves operating at temperatures up to 800°C; very oxidation resistant.

BEDEL SOC 2.
M-455; C, 20 Cr, high Co, bal Fe.
Heat treated: 135,000 TS; 115,000 YS; 7.5 El.
For valves operating at temperatures up to 800°C; very oxidation resistant.

BEDEL SPECIAL F.
M-455; C, 4 W, Cr, Mo, bal Fe.
For wire drawing dies, cold drawn wire dies; non-deforming.

BEDEL SPECIAL FA.
M-455; C, 13 Cr, Ni, Mo, V, bal Fe.
For dies and punches; non-deforming.

BEDEL SPECIAL FB.
M-455; C, 13 Cr, Mo, V, bal Fe.
For dies and punches; non-deforming.

BEDEL SPECIAL FK.
M-455; C, 13 Cr, Mo, V, bal Fe.
For dies and punches; non-deforming.

BEDEL SPECIAL FL.
M-455; C, 13 Cr, bal Fe.
For wire drawing, dies for cold drawn hard steel; non-deforming.

BEDEL SPECIAL 000.
M-455; C, 13 Cr, bal Fe.
For dies, punches, rolling mill rolls; non-deforming.

BEDEL SPECIAL OZ.
M-455; C, 10 W, Cr, V, bal Fe.
For hot forging tools, stamping and punching, hot shear blades; hot work steel.

BEDEL SPECIAL X.
M-455; C, Cr, V, bal Fe.
For calipers, gages, dies, punches; non-deforming.

BEDEL SPECIAL Y.2.
M-455; C, 3 Ni, 0.7 Cr, bal Fe.
For pneumatic or hydraulic rivet sets; tough.

BEDEL SPECIAL Y.5.
M-455; C, 2 W, 1 V, bal Fe.
For pneumatic chisels, punches, shear blades; tough.

BEDEL SPECIAL Z3.
M-455; C, 10 W, Cr, Ni, V, bal Fe.
For tools, dies; oil hardening.

BEDEL SPECIAL Z4.
M-455; C, 10 W, Cr, V, Co, bal Fe.
For tools, dies; oil hardening.

BEDEL TRIPLE TREFLE.
M-455; C, 18 W, Cr, bal Fe.
For heavy roughing and milling cutters; high speed steel.

BEDEL TRIPLE TREFLE-E.
M-455; 0.75 C, 18 W, 5 Cr, 1 Mo, 1 V, bal Fe.
For tools, cutters, drills, reamers, broaches, lathe and planer tools.
High-speed steel, high red hardness.

BEDEL TRIPLE TREFLE E.S.
M-455; C, W, Cr, V, 6 Co, bal Fe.
For high speed cutters; super high speed steel.

BEDEL TRIPLE TREFLE ESD.
M-455; C, 5 Co, Mo, V, bal Fe.
For tools, dies; oil hardening.

BEDEL TRIPLE TREFLE F.R.
M-455; C, high W, high Cr, bal Fe.
For taps, forming tools, bores, finishing tools; high speed steel.

BEDEL UN TREFLE 1.
M-455; C, 2-3 W, Cr, V, bal Fe.
For hot and cold forging punches, dies; hot work steel.

BEDEL UN TREFLE 2.
M-455; C, 4 Ni, Cr, Mo, bal Fe.
For hot stamping dies and punches, pipe mandrel, shear blades; shock resistant.

BEDEL UN TREFLE 3.
M-455; C, Cr, Ni, bal Fe.
For pneumatic chisels, hot slicing; shock resistant.

BEDEL UN TREFLE COURONNE.
M-455; C, 5-6 F, Cr, bal Fe.
For hot punches and dies for screw machines; hot work steel.

BELAIS WHITE GOLD-1.
M-237; 75-85 Au, 8-18 Ni, 4-14 Zn.
For jewelry, ornaments; corrosion resistant.

BELAIS WHITE GOLD-2.
M-237; 75-85 Au, 10-18 Ni, 2-9 Zn, 0.05-0.5 Pt or 0.5-2.0 Mn.
For jewelry, ornaments; corrosion resistant.

BELGIAN ALUMINUM PISTON ALLOY.
M-Belgium; 90.5 Al, 7 Cu, 2.5 Zn.
For pistons; non-hardenable.

BELL BRAND.
M-368; 0.7-1.0 C, bal Fe.
For tools; water hardened.

BELL BRASS.
M-Eng.; 64 Cu, 35 Zn, 0.85 Sn.
For bells.

BELL BRONZE.
M-657; 80 Cu, 20 Sn.
For bells.

BELLER NO. 4.
M-1117; 0.75 C, bal Fe.
For tools, taps, springs, punches; water hardened.

BELL METAL-1.
M-Eng.; 83 Al, 10 Mn, 7 Cd.
For bells, chimes, whistles, ornaments.

BELL METAL-2.
M-Eng.; 85-60 Cu, 15-40 Sn.
For bells, chimes; hard bronze.

BELL METAL, JAPANESE KARAKANE-1.
M-Japan; 72-61 Cu, 14-25 Sn, 0.14 Pb, 0-9.4 Zn, 0-3 Fe.
For bells; corrosion resistant.

BELL METAL, JAPANESE KARAKANE-2.
M-Japan; 71.42 Cu, 14.2 Sn, 14.3 Pb.
For bells; free-cutting.

BELL METAL, JAPANESE KARAKANE-3.
M-Japan; 70 Cu, 19 Sn, 3 Zn, 8 Pb.
For bells; free-cutting.

BELL METAL, JAPANESE KARAKANE-4.
M-Japan; 65.95 Cu, 17.25 Sn, 3.45 Zn, 10.35 Pb.
For bells; free-cutting.

BELL METAL, JAPANESE KARAKANE-5.
M-Japan; 64 Cu, 24 Sn, 9 Zn, 3 Fe.
For bells; corrosion resistant.

BELL METAL, JAPANESE KARAKANE-6.
M-Japan; 61 Cu, 18 Sn, 6 Zn, 3 Fe, 12 Pb.
For bells; free-cutting.

BELL METAL, MUSICAL.
M-U.S.; 84 Cu, 16 Sn.
For bells; bronze.

BELL SPECIAL.
M-368; 1.2 C, 0.2 V, bal Fe.
For tools, fixtures, jigs; water hardened.

BELMONT.
M-1705; 0.90 C, 1.5 Mn, 0.25 Si, 0.3 Mo, bal Fe.
Oil hardening tool steel. AISI 02.

BELMONT NO. 40.
M-1255.
For aluminum solder; M.P. 400°F.

BELMONT NO. 60.
M-1255.
For aluminum solder; M.P. 650°F.

BELMONT HOLTITE.
M-1255; 50 Cu, 50 Zn, granular.
For brazing all metals except aluminum.

BELMONT MAB.
M-1255.
For aluminum brazing, M.P. 720°F.

BELMONT SUPERDIE.
M-1255; Zinc base, plus Al, Cu, and Ti.
For cast stamping dies, blow molds.

BELT LACE.
M-U.S.; 62 Cu, 38 Zn.
For belt lacing, light alloy parts; high impact strength.

BEMIT.
M-Ger.; 96.91 Al, 2.0 Cu, 0.45 Mn, 0.27 W.
For light alloy welded, drawn or stamped parts; also called "Benit;" German alloy.

BEN2.
M-1687; 11 Sn, 2 Ni, bal Cu.
Bronze, as cast: 80 Brin.

BEN5.
M-1687; 11 Sn, 5 Ni, bal Cu.
Nickel bronze; as cast: 80 Brin.
ASTM B584.

BENDIX O1.
M-1694; 0.90 C, 1.2 Mn, 0.35 Si, 0.5 Cr, 0.2 V, 0.5 W, bal Fe.
Oil hardening tool steel. AISI O1.

BENDIX O1 DRILL ROD.
M-1694; 0.90 C, 1.2 Mn, 0.5 Cr, 0.2 V, 0.5 W, bal Fe.
Oil hardening tool steel. AISI O1.

BENDIX O1 FLAT GROUND STOCK.
M-1694; 0.90 C, 1.2 Mn, 0.35 Si, 0.5 Cr, 0.2 V, 0.5 W, bal Fe.
Oil hardening tool steel. AISI O1.

BENDIX A2 FLAT GROUND STOCK.
M-1694; 1.0 C, 0.5 Mn, 5.0 Cr, 0.3 V, 1.25 Mo, bal Fe.
Air hardening tool steel. AISI A2.

BENDIX COMMERCIAL DRILL ROD.
M-1694; 1.0 C, 0.3 Mn, 0.2 Si, bal Fe.
Water hardening tool steel. AISI W1.

BENDIX D2.
M-1694; 1.5 C, 0.4 Mn, 0.3 Si, 12.0 Cr, 0.9 V, 0.8 Mo, bal Fe.
High carbon, high chromium tool steel. AISI D2.

BENDIX D2 FLAT GROUND STOCK.
M-1694; 1.55 C, 0.25 Mn, 0.38 Si, 12.0 Cr, 0.8 V, 0.8 Mo, bal Fe.
High carbon, high chromium tool steel. AISI D2.

BENECKE.
M-823; 0.7-1.0 C, alloy, bal Fe.
For tools, dies; oil hardening.

BENEDICT PLATE.
M-Eng.; 57 Cu, 28 Zn, 15 Ni.
For white metal for flat work; corrosion resistant.

B E NO 4 ALLOY.
M-U.S.; 4 Cu, 0.25 Mg, $\leq$ 0.5 Fe, $\leq$ 0.1 Si, bal Al.
For boxes, covers, face-plates; U.S.N. Bureau of Engineering Alloy.

BENSON.
M-959; 10 Sn, 15 Sb, bal Pb.
For bearings, bushings; Babbitt. SAE 14.

BENUM.
M-1405; 0.3 C, 0.5 Mn, 1.3 Cr, 0.3 Si, 0.3 Mo, 4.2 Ni, bal Fe.
For molds, dies; air hardened, tough.

BENZ AVIATEK.
M-Eng.; 80 Al, 6 Cu, 12 Zn, 1.5 Fe.
For pistons; non-hardenable.

BERA AUTO A.
M-1261; 88.7 Sn, 7.8 Sb, 3.5 Cu.
Cast: 12,800 TS; 26 Brin.
For engine bearings; M.P. 440-620°F; shock resistant.

BERA AUTO SPECIAL.
M-1261; 90.3 Sn, 6.5 Sb, 3 Cu, 0.2 Ni.
Cast: 12,800 TS; 25 Brin.
For engine bearings; M.P. 440-610°F; shock resistant.

BERACO.
M-1260, M-1261; 10 Sn, 13.5 Sb, 0.5 Cu, 76 Pb.
Cast: 12,900 TS; 30 Brin.
For transmission bearings; M.P. 470-705°F.

BERA COMMON A.
M-1261; 55 Sn, 10 Sb, 2.5 Cu, 32.5 Pb.
Cast: 11,400 TS; 22 Brin.
For refrigerator and generator bearings; M.P. 355-620°F; good castability.

BERA COMMON B.
M-1261; 24.5 Sn, 13 Sb, 0.5 Cu, 62 Pb.
Cast: 12,800 TS; 26 Brin.
For refrigerator and generator bearings; M.P. 355-535°F; good castability.

BERA DIESEL.
M-1261; 83.5 Sn, 8 Sb, 6.5 Cu, 2 Pb.
Cast: 17,100 TS; 31 Brin.
For diesel engine bearings; M.P. 355-710°F; wear resistant.

BERALITE 35.
M-1011; 35 Be, bal Al.
Drawn: 62,800 TS; 58,400 YS; 1.5 El; 94 Brin.
Annealed: 41,400 TS; 23,100 YS; 14.5 El; 68 Brin.
For instruments; corrosion resistant.

BERALOY A.
M-897; 2 Be, 0.25 Co, bal Cu.
Heat treated: 180,000 TS; 95,000 YS; 1 El.
For springs; corrosion resistant, hardenable.

BERA MARINE.
M-1261; 85 Sn, 10 Sb, 5 Cu.
Cast: 19,900 TS; 28 Brin.
For engine bearings; M.P. 440-630°F; shock resistant.

BERA STEAM A.
M-1261; 80 Sn, 11.5 Sb, 5.5 Cu, 3 Pb.
Cast: 17,800 TS; 34 Brin.
For steam turbine and generator bearings; M.P. 355-675°F; wear resistant.

BERA STEAM B.
M-1261; 74 Sn, 9 Sb, 4 Cu, 13 Pb.
Cast: 14,200 TS; 28 Brin.
For steam turbine bearings; M.P. 355-625°F; wear resistant.

BERDO ALLOY NO. 1.
M-20; Cu, Sn, Pb, Ni.
30,000 TS; 20,000 YS; 1.5 El; 2.6 RA; 99 Brin.
For bearings; heavy service; C.Y.P.-38,000.

BERDO ALLOY NO. 2.
M-20; Cu, Sn, Pb, Ni.
75 Brin.
For bearings; general service; C.Y.P.-28,000.

BERDO ALLOY NO. 3.
M-20; Cu, Sn, Pb, Ni.
44 Brin.
For bearings; light service; C.Y.P.-14,000.

BERDO NO. 6.
M-20; Cu-Sn-Ni-Pb-P.
For mill bearings, rock drill parts, gears; wear resistant.

BERDO NO. 7.
M-20; Cu-Sn-Ni-Pb-P.
For mill bearings, rock drill parts, gears; wear resistant.

BERGAL.
M-708; 94.1 Al, 4.0 Cu, 0.8 Mg, 0.7 Mn, 0.4 Si.
For light alloy parts; age-hardenable.

BERGISCHE-CMVW11.
M-1290; 0.20 C, 12.5 Cr, 1.1 Mo, 0.6 Ni, 0.3 V, 0.5 W, bal Fe.

BERGISCHE-CMW11.
M-1290; 0.20 C, 12.5 Cr, 1.1 Mo, 0.6 Ni, 0.3 V, bal Fe.

BERGIT 1.
M-1290; 0.10 C, Ni, Cr, Mo, Cu, bal Fe.

BERGIT A.
M-1290; 0.10 C, Ni, Mo, bal Fe.

BERGSOE F.
M-1260; 88.7 Sn, 7.8 Sb, 3.5 Cu.
Cast: 17,000 TS; 25 Brin.
For engine bearings; M.P. 440-620°F; shock resistant.

BERGSOE G.
M-1260, M-1261; 88 Sn, 8 Sb, 4 Cu.
Cast: 18,500 TS; 27 Brin.
For engine bearings; M.P. 440-625°F; shock resistant.

BERGSOE M.
M-1260, M-1261; 6 Sn, 16 Sb, 79 Pb.
Cast: 29 Brin.
For transmission bearings; M.P. 470-520°F.

BERGSOE P.
M-1260, M-1261; 16 Sb, 84 Pb.
Cast: 19 Brin.
For transmission bearings; M.P. 475-520°F.

BERGSOE WM.
M-1260, M-1261; 5 Sn, 12 Sb, 83 Pb.
Cast: 25 Brin.
For transmission bearings; M.P. 470-490°F.

BERGSOL MN. CU.
M-1260, M-1261; 8 Sn, 14 Sb, 0.5 Cu, 77.5 Pb.
Cast: 29 Brin.
For transmission bearings; M.P. 470-715°F.

BERGSTROM TYPE S.
M-1128; C, Mn, Ni, bal Fe.
300 Brin.
For hard surfacing electrodes; work hardening rod.

BERKSHIRE.
M-32; 1.2 C, 1.34 W, bal Fe.
For taps, cutters, engravers tools; fast finishing steel, water hardened.

BERKSHIRE, HIGH MANGANESE.
M-32; 1.2 C, 1.0 Mn, 0.4 Si, 0.5 Cr, 0.2 V, 1.4 W, bal Fe.
Water or oil hardening high carbon tool steel, for shafts, arbors, drill bushings, lathe centers.

BERLIN NO. 1.
M-Germany; 56 Cu, 29 Zn, 16 Ni.
For electrical resistances, ornaments; corrosion resistant.

BERLIN NO. 2.
M-Germany; 48 Cu, 24 Zn, 24 Ni, 3.6 Fe.
For ornamental parts; spinning, drawing, stamping.

BERMAX.
M-709; 9-11 Sb, 0.5 max Cu, bal Pb.
Cast: 11,000 TS; 4.5 El; 20 Brin.
For bearings; anti-friction, non-magnetic.

BERRY METAL.
M-1281; 10 Pb, 8 Ni, 2 Sb, bal Cu.
Cast: 20,000 TS; 13,000 YS; 6 El.
For bearings; wear resistant.

BERSCH METAL.
M-Germ.; 93 Al, 7 Ni.
For bearings.

BERTHIER'S ALLOY-1.
M-Eng.; 68 Cu, 32 Ni.
For evaporators, stills; corrosion resistant.

BERTHIER'S ALLOY-2.
M-Eng.; 72 Cu, 25 Zn, 2 Pb, 1.2 Sn.
For screw machine products; free-cutting.

BERUDA ALLOY.
M-Eng.; Cu-Zn.
For strong corrosion resistant parts; high tensile brass, corrosion resistant.

BERYLCO 10.
M-646; 0.4 Be, 2.6 Co, bal Cu.
Annealed: 50,000 TS; 25,000 YS; 30 El; 50 RA; 70 Brin.
Heat treated: 125,000 TS; 105,000 YS; 8 El; 10 RA; 240 Brin.
For springs, resistance welding electrodes; age-hardenable.

BERYLCO 10 C.
M-646; 0.6 Be, 2.6 Co, bal Cu.
Cast: 60,000 TS; 30,000 YS; 20 El; 35 RA; 90 Brin.
Annealed: 40,000 TS; 20,000 YS; 30 El; 50 RA; 70 Brin.
Hardened: 90,000 TS; 75,000 YS; 10 El; 15 RA; 185 Brin.
For resistance welding parts, switches, contacts; age-hardenable, corrosion resistant.

BERYLCO 20C.
M-646; 2.0-2.25 Be, 0.35-0.60 Co, bal Cu.
Heat treated: 110,000-160,000 TS; 95,000-140,000 YS; 8-2 El; 10-3 RA; 250-440 Brin.
For plastic molds, safety tools, cams, pump parts; wear and impact resistant, age-hardenable.

BERYLCO 20CR.
M-646; 2.0-2.25 Be, 0.35-0.65 Co, bal Cu.
Cast: 70,000 TS; 40,000 YS; 15 El; 138 Brin.
Heat treated: 155,000 TS; 115,000 YS; 0 El; 380 Brin.
For plastic molds, safety tools, gears, bushings, bearings; age hardenable.

BERYLCO 25.
M-646; 2 Be, 0.3 Co, bal Cu.
Heat treated: 200,000 TS; 170,000 YS; 2 El; 3 RA; 400 Brin.
For springs, diaphrams; age-hardenable.

BERYLCO 25S.
M-646; 2 Be, 0.3 Co, bal Cu.
Annealed: 165,000 TS; 135,000 YS; 5 El; 8 RA; 350 Brin.
4 No. Hard: 190,000 TS; 160,000 YS; 2 El; 3 RA; 400 Brin.
For springs, diaphragms, clips, bellows; corrosion and heat resistant.

BERYLCO 40.
M-646; 2 Be, 0.25 Co, bal Cu.
Rolled: 98,000 TS; 2 El.
Heat treated: 180,000 TS; 95,000 YS; 1 El.
For springs; age-hardenable.

BERYLCO 50.
M-646; 0.3-0.55 Be, 1.4-1.7 Co, 1.0 Ag, bal Cu.
Heat treated: 110,000-95,000 TS; 90,000-65,000 YS; 5-10 El; 5-10 RA 195-240 Brin.
For springs, resistance welding electrodes and dies; age-hardenable.

BERYLCO 70.
M-646; 0.1 Be, 0.5 Cr, bal Cu.
Heat treated: 80,000 TS; 60,000 YS; 15 El; 25 RA; 150 Brin.
For springs, current carrying parts; age-hardenable.

BERYLCO 165.
M-646; 1.6-1.8 Be, 0.2-0.35 Co, bal Cu.
Annealed: 70,000 TS; 30,000 YS; 40 El; 45 RA; 95 Brin.
Hardened: 185,000 TS; 155,000 YS; 2 El; 3 RA; 355 Brin.
For springs, terminal plugs, jewelry, clips; heat treatable, corrosion resistant.

BERYLCO 185.
M-646; 1.5-2.0 Be, 0.5 max Ni or Co, 0.25 max Fe, bal Cu.
Annealed: 68,000 TS; 49 El.
Heat treated: 195,000 TS; 3 El.
For springs; age-hardenable.

BERYLCO 200C.
M-646; 2.05-2.25 Be, 0.35-0.65 Co, bal Ni.
For bearings, gears, cams, plungers, drawing dies; age-hardenable, corrosion and wear resistant.

BERYLCO 250-C.
M-646; 2.6-2.8 Be, 0.35-0.65 Co, bal Cu.
For bearings, gears, cams, valves, drawing dies; age-hardenable, corrosion and wear resistant.

BERYLCO 275CR.
M-646; 2.6-2.8 Be, 0.35-0.65 Co, bal Cu.
Heat treated: 125,000-175,000 TS; 110,000-145,000 YS; 5-2 El; 8-3 R ; 270-415 Brin.
For plastic mold dies, safety tools; age-hardenable, corrosion resistant.

BERYLCO-717.
M-646; 0.5 Be, 0.7 Fe, 30 Ni, bal Cu.
Bar: 110,000-140,000 TS; 10-15 El.
Age-hardened: 145,000 min TS; 125,000 min YS; 10 min El; C 29 Rock.
For marine hardware, piping, high velocity heat exchangers, hydrophone parts.
High sea water corrosion resistant. Age-hardenable.

BERYLCO-717 C.
M-646; 0.50 Be, 1.0 Fe, 0.95 Mn, 0.06 C, 29 Ni, bal Cu.
Cast: 76,000 TS; 44,700 YS; 14 El; 15.3 RA; B 82 Rock.
Heat Treated: 121,500 TS; 84,400 YS; 14.5 El; C 20 Rock.
For marine hardware, heat exchangers.
Age hardenable. High sea water corrosion resistant.

BERYLCO CR-1.
M-646; 2.75 Be, 0.5 max C, bal Ni.
Heat treated: 125,000-195,000 TS; 200,000-180,000 YS; 2-0 El; 300-520 Brin.
Cast: 115,000 TS; 60,000 YS; 5 El; 240 Brin.
For aircraft fuel pumps, impellers, core drill bits; age hardened, corrosion resistant castings.

BERYLCO CR-2.
M-646; 2.75 Be, 0.75-1.1 C, bal Ni.
For aircraft fuel pumps, impellers, core drill bits; age hardened, corrosion and wear resistant.

BERYLCO HPA.
M-646; 98.0 Be, 1.8 BeO, 0.18 Fe, 0.15 Al, 0.08 Mg, 0.02 Mn, 0.02 Cu, 0.10 Si, 0.05 Zn, 0.06 Ni.
Vacuum hot pressed: 40,000 TS; 30,000 YS; 1 El; Rock B 80.
Hot Extruded and Annealed: 70,000 TS; 45,000 YS; 10 El; Rock B 90
For nuclear space applications.

BERYLCO NICKEL 440.
M-646; 1.95 Be, 0.50 Ti, bal Ni.
Annealed: 105,000 TS; 45,000 YS; 40 El; B 70 Rock.
Hard Drawn: 170,000 TS; 165,000 YS; 2 El; C 35 Rock.
HT-Temper: 270,000 TS; 230,000 YS; 8 El; C 51 Rock.
For heat resistant springs, and switches, diaphragms, bellows, retainer clips, electrical shunts, contact springs.
High strength and corrosion resistant. Age-hardenable.

BERYLDUR.
M-646; 0.8-1.2 Be, 0.2 min Ni, 4.0 others, bal Cu.
Annealed: 55,000-65,000 TS; 18,000-30,000 YS; 40-60 El.
Aged: 125,000-140,000 TS; 105,000-125,000 YS; 12-20 El; 230-300 Brin.
For springs, clips, connectors; age-hardenable, high strength, corrosion resistant.

BERYLLIUM.
M-1755; Be.
Purities: distilled 99.99+ %, Grade AA 99.96+%, Grade A 99.87%, Nuclear grade 99.5+%.
Forms: Flake, powders, plates, sheets, foils, wires, rods, single crystals.

BERYLLIUM-ALUMINUM.
M-646; 5.0 Be, 1.0 Mg, bal Al.
For master alloy for remelting; pigs.

BERYLLIUM BRONZE 1.0% BE.
M-198, M-296; 1.0 Be, bal Cu.
Untempered: 67,000 TS; 62,000 YS; 7.5 El.
For corrosion resistant parts; corrosion resistant.

BERYLLIUM BRONZE 1.5% BE.
M-198, M-296; 1.5 Be, bal Cu.
Untempered: 68,000 TS; 45 El.
Tempered: 94,000 TS; 8 El.
For corrosion resistant parts; corrosion resistant.

BERYLLIUM BRONZE 2.0% BE.
M-198, M-296; 2.0 Be, bal Cu.
Untempered: 72,000 TS; 15,500 YS; 40 El.
Tempered: 121,000 TS; 100,000 YS; 1 El.
For corrosion resistant parts; corrosion resistant.

BERYLLIUM BRONZE 2.5% BE.
M-198, M-199, M-296; 97.5 Cu, 2.5 Be.
Soft: 62,000 TS; 20,000 YS; 52 El; 66 RA; 98 Brin.
Hard: 176,000 TS; 160,000 YS; 0.6 El; 396 Brin.
For springs and parts subjected to frictional wear; wear resistant.

BERYLLIUM-COBALT.
M-646; 50 Be, bal Co.
For master alloy for remelting.

BERYLLIUM-COPPER.
M-646; 4.0 Be, bal Cu.
Used to introduce Be in Cu alloys; master alloy.

BERYLLIUM COPPER NO. 175.
M-960; 2.15 Be, 0.35 Ni, bal Cu.
Heat treated: 160,000-185,000 TS; 140,000-160,000 YS; 5-2 El; 360-450 Brin.
For springs, clips, diaphragms; age hardenable, high fatigue strength.

BERYLLIUM I-400.
M-1037; 92 Be, 4.5 BeO, 0.2 max Al, 0.5 max C, 0.3 max Fe, 0.1 max Mg, 0.15 max Si.
Unnotched: 129,000 TS; 86,400 YS; 10.6 El.
Notched: 127,100 TS.
For super-critical gyro applications.
For aerospace structures, pressure vessels, jet engine turbine wheels and blades. High stiffness-weight ratio.

BERYLLIUM-IRON.
M-646; 50 Be, bal Fe.
Used to introduce Be in Cu-Fe alloys; master alloy.

BERYLLIUM-MAGNESIUM-ALUMINUM.
M-646; 5.0 Be, 5.0 Mg, bal Al.
For master alloy for remelting.

BERYLLIUM MALLEABLE.
99.8 Be, 0.2 Ti.
For X-ray windows, camera shutters; very malleable.

BERYLLIUM MANGANESE BRONZE 1.0% BE.
M-198, M-296; 1.0 Be, 10 Mn, bal Cu.
Untempered: 74,000 TS; 38,000 YS; 22 El.
Tempered: 146,000 TS; 127,700 YS; 2.3 El.
For corrosion resistant parts; corrosion resistant.

BERYLLIUM MANGANESE BRONZE 1.5% BE.
M-198, M-296; 1.5 Be, 3 Mn, bal Cu.
Untempered: 78,000 TS; 38,000 YS; 25 El.
Tempered: 150,000 TS; 135,000 YS; 4.5 El.
For corrosion resistant parts; corrosion resistant.

BERYLLIUM-NICKEL.
M-646; 50 Be, bal Ni.
Used to introduce Be in Cu-Ni alloys; master alloy.

BERYLLIUM S-200-C.
M-1037; 98 Be, 2 BeO, 0.16 max Al, 0.15 max C, 0.18 max Fe, 0.08 max Mg, 0.08 max Si.
Hot Pressed: 44,800 TS; 33,800 YS; 1.6 El.
For aerospace structures, pressure vessels, jet engine components. High stiffness-weight ratio.

BERYLLIUM-SILVER.
M-646; 50 Be, bal Ag.
For master alloy for remelting.

BERYVAC 60.
M-1631; 0.4-0.6 Be, 2.5 Co, bal Cu.
For springs, resistance welding electrodes; can be age-hardened.

BERYVAC 170.
M-1631; 1.6-1.8 Be, Ni and/or Co 0.2 max, bal Cu.
For springs, plugs, connectors; can be heat treated; corrosion resistant.

BERYVAC 200.
M-1631; 1.8-2.05 Be, Ni and/or Co 0.2 max, bal Cu.
For springs, diaphragms; can be age-hardened.

BERYVAC 520.
M-1631; 2 Be, additions, bal Ni.
Springs, diaphragms, plug connections; can be age hardened.

BERYVAC M 25.
M-1631; 1.8-2.05 Be, Ni and/or Co 0.2 max, Pb, bal Cu.
Free cutting grade (round bar only).
Can be age hardened.

BESCOLOY.
M-424; C, bal Fe.
For piercer points.

BESPLATE.
M-710; Ni.
For anodes.

BESPLATE.
M-710; 0.2-0.6 C, 0.2-0.8 Si, bal Ni.
Nickel anodes for electroplating.

BEST.
M-336; 1.0 C, 0.3 Mn, bal Fe.
78,000 TS; 41,600 YS; 28 El; 167 Brin.
For tools, taps, reamers, dies; "Soderfors Best;" non-deforming.

BEST.
M-364; 0.7-1.1 C, bal Fe.
For general tools, taps, drills; water hardening.

BEST.
M-24; 0.7-1.1 C, 0.2 V, bal Fe.
For tools, punches, dies, broaches, reamers, shear blades; water hardened.

BEST 2.
M-1287; 1.0 C, 0.25 max Si, 0.25 max Mn, bal Fe.
Annealed: 100,000 TS; 53,000 YS; 21 El; 42 RA; 200 Brin.
For springs, drills, taps, reamers, broaches; Type W1; water hardened.

BEST 3K.
M-1287; 0.85 C, 0.25 max Si, 0.25 max Mn, bal Fe.
Heat treated: 190,000 TS; 145,000 YS; 10 El; 30 RA; 400 Brin.
For springs, drills, taps, punches, tools; Type W1; water hardened.

BEST 4W.
M-1287; 0.70 C, 0.25 max Si, 0.25 max Mn, bal Fe.
Heat treated: 174,000 TS; 128,000 YS; 12 El; 37 RA; 355 Brin.
For springs, rails, hammers, crimpers; Type W1; water hardened.

BEST BRONZE.
M-Eng.; 90 Cu, 10 Zn.
For jewelry trade as base for fire enameling, primers, bullet shells; gilding brass.

BEST TYPE METAL.
M-Eng.; 50 Pb, 25 Sn, 25 Sb.
For type metal.

BETA III see CRUCIBLE BETA III.

BETHADUR 420.
M-24; 0.3 C, 13 Cr, bal Fe.
Heat treated: 250,000 TS; 214,000 YS; 10 El; 22 RA; 514 Brin.
For surgery and dental tools, table cutlery; corrosion resisting, hardenable.

BETHADUR 501.
M-24; 0.16-0.20 C, 4-6 Cr, 0.5 Mo, bal Fe.
Annealed: 79,000 TS; 33,000 YS; 33 El; 75 RA; 143 Brin.
For oil refinery equipment; corrosion resistant.

BETHADUR 502.
M-24; 0.08 C, 4-6 Cr, bal Fe.
For oil refinery equipment; corrosion resistant.

BETHALLOY.
M-24; 0.75 C, 0.75 Mn, 0.3 Si, 0.9 Cr, 0.37 Mo, 1.8 Ni, bal Fe.
For tools, broaches, taps, slitting cutters, reamers; oil hardened, tough.

BETHCO MACHINE SCREW.
M-24; 0.3 C, bal Fe.
For machine screws; special cold heading steel, special process.

BETHCO WOOD SCREW.
M-24; 0.25 C, bal Fe.
For wood screws; special cold heading steel, special process.

BETH-CU-LOY.
M-24; 0.3 Cu, 0.2 C, bal Fe.
For sheet metal construction, roofing, siding; resists atmospheric corrosion.

BETH-LED.
M-24; C, Mn, S, Pb, bal Fe.
For screw machine products; free-cutting.

BETHLEHEM 01 DRILL ROD.
M-24; 0.90 C, 1.2 Mn, 0.5 Cr, 0.2 V, 0.5 W, bal Fe.
For pin header, special punches, precision gages, dies, drills, plugs.
AISI Type 01 oil hardening tool steel.

BETHLEHEM 66 HS now **M-2 HIGH SPEED STEEL.**

BETHLEHEM AH-5.
M-24; 1.0 C, 0.6 Mn, 5.25 Cr, 1.1 Mo, 0.25 V, bal Fe.
For cutters, tools, dies, punches; air hardening, tough wear resistant. AISI A2.

BETHLEHEM COMMERCIAL WATER HARDENING DRILL ROD.
M-24; 1.0 C, 0.30 Mn, 0.20 Si, bal Fe.
For automatic screw machine parts, dowel pins, mandrels, punches, small tools, dies.
AISI Type W1 water hardening tool steel.

BETHLEHEM CR-MO-W see **CROMO W.**

BETHLEHEM HM now **M-1 HIGH SPEED STEEL.**

BETHLEHEM M-10 now **M-10 HIGH SPEED STEEL.**

BETHLEHEM SILICO-MANGANESE SPRING STEEL.
M-24; 0.7 C, 0.9 Mn, 0.2 Si, bal Fe.
Heat treated: 210,000-235,000 TS; 180,000-195,000 YS; 12-9 El; 30-35 RA; 415 Brin.
For springs; oil hardened.

BETHLEHEM SOLID DRILL.
M-24; 0.8 C, bal Fe.
For drills, bits, blacksmith tools; shock resistant.

BETHLEHEM SPECIAL HS now **T-1 HIGH SPEED STEEL.**

BETHLEHEM STANDARD BEARINGS.
M-24; 1.0 C, 1.35 Cr, bal Fe.
For bearings, rollers, master gauges; bearing steel.

BETHLEHEM T1 HIGH SPEED (B.S.H.S.).
M-24; 0.73 C, 4.0 Cr, 1.1 V, 18.0 W, bal Fe.
General purpose high speed cutting tool, all types of cutting operations.
AISI Type T1 high speed tool steel.

BETHLEHEM X.
M-24; 0.7-1.1 C, bal Fe.
For general tools, quarrying tools, crow bars, soft rock drills, chisels; water hardening.

BETHNAMEL.
M-24; 0.1 C, bal Fe.
Rolled: 26,000 TS; 43,000 YS; 38 El; 77 Brin.
For porcelain enameled articles; good drawability.

BF 954.
M-1154; 0.3 C, 13 Cr, bal Fe.
For knives, crusher jaws, guide plates, blasting nozzles; wear resistant.

BFD.
M-373; 1.20 C, 0.35 Mn, 0.25 Si, 0.6 Cr, 0.2 V, 1.5 W, bal Fe.
AISI Type 07 oil hardening tool steel.

"B-F" HIGH SPEED see **REPUBLIC T1.**

BG42 see **LESCALLOY BG42.**

BHT 80.
M-1732; 0.06 C, 0.67 Si, 2.52 Mn, 0.015 P, 0.007 S, 0.30 Cu, 0.30 Ni, 0.04 Nb, bal Fe.
4.5 mm thick plate, Q+T: 122,000 psi TS; 105,000 psi YS; 20 El.
High strength steel, good formability for welded structures as earth moving equipment and crane booms.

B.H.T.A. METAL.
2 Cu, 1.25-1.75 Ni, 1.75-2.0 Fe, 0.8 Mg, 0.5 max Si, bal Al.
Heat treated: 60,000 TS; 47,000 YS; 10 El.
For light alloy parts, forgings; Duralumin type, age-hardening.

BICALOY.
M-711; Cu alloy.
For resistance welding electrodes; wear and deformation resistant.

BICOP OXYGEN FREE COPPER.
Cu.
For electrical motors, generators; high conductivity.

BIDDERY, HEINIES.
84.3 Zn, 11.4 Cu, 2.9 Pb, 1.4 Sn.
For bearings; Babbitt.

BIDERY (BIDDERY).
M-U.S.; 90.2 Zn, 6.3 Cu, 2.6 Pb, 0.8 Sn.
For buttons, ornaments; free-cutting.

BIDERY BUTTONS "A".
48.5 Cu, 33.3 Zn, 6.06 Sn, 12.15 Pb.
For buttons, ornamental parts; free-cutting.

BIDERY BUTTONS "B".
48.5 Cu, 33.32 Zn, 6.06 Sn, 12.15 Pb.
For buttons; free-cutting.

BIERMAN TUNGSTEN BRONZE.
M-Germany; 95 Cu, 3.4 Sn, 1.6 W.
For strong corrosion resistant parts; corrosion resistant.

BIG J.
M-477; 0.7 C, 18 W, 4 Cr, 1 V, bal Fe.
For high speed tools; high speed steel.

BILGEN BRONZE.
M-Germany; 97 Cu, 1.9 Sn, 0.5 Fe, 0.2 Pb.
For electrical bronze; corrosion resistant.

BINAL.
M-1469; Sinter cast boron carbide.
(Made to customers, requirement).

BINDING BRASS.
M-U.S.; 63.25 Cu, 35 Zn, 1.75 Pb.
For automatic screw machine products; free-cutting.

BIRDSBORO-20A6.
M-136; 0.25 max C, 0.90 Mn, 0.50 Cr, 0.50 Ni, 0.50 Mo, bal Fe.
Cast: 80,000-120,000 TS; 50,000-90,000 YP; 150-210 Brin; 14-20 El.
For general and structural applications, gears, shafts, housings. Good weldability.
AISI 8620 modified.

BIRDSBORO-20 CN.
M-136; 0.25 max C, 0.70 Mn, bal Fe.
Cast: 60,000-75,000 TS; 30,000-45,000 YP; 115-140 Brin; 22-30 El.
For general and structural applications, gears, housings. Excellent weldability.
AISI 1020.

BIRDSBORO-22 CMN.
M-136; 0.25 max C, 1.0 Mn, bal Fe.
Cast: 70,000-80,000 TS; 40,000-55,000 YP; 22-30 El; 30-50 RA.
For general and structural applications, gears, housings.
ASTM A216, WCC.

BIRDSBORO-25A6.
M-136; 0.30 max C, 0.90 Mn, 0.50 Cr, 0.50 Ni, 0.50 Mo, bal Fe.
Cast: 105,000-145,000 TS; 85,000-120,000 YP; 14-17 El; 30-35 RA.
For general and structural applications, gears, housings, shafting.
AISI 8625 modified.

BIRDSBORO-25 CN.
M-136; 0.30 max C, 0.70 Mn, bal Fe.
Cast: 65,000-80,000 TS; 35,000-55,000 YP; 125-160 Brin; 24-30 El.
For general and structural applications, gears, housings.
Excellent weldability. AISI 1025.

BIRDSBORO-30A1.
M-136; 0.35 max C, 0.90 Mn, 0.20 Mo, 1.0 Cu, bal Fe.
Cast: 80,000-105,000 TS; 50,000-90,000 YS; 150-210 Brin; 15-30 El; 30-40 RA.
For general and structural applications, gears, housings. Good weldability.

BIRDSBORO-30A2.
M-136; 0.35 max C, 0.80 Mn, 0.95 Cr, 0.20 Mo, bal Fe.
Cast: 90,000-120,000 TS; 65,000-100,000 YP; 14-20 El; 30-40 RA; 180-260 Brin.
For general and structural castings, shafts, gears, housings. AISI 413 shock resistant.

BIRDSBORO-30A3.
M-136; 0.35 max C, 0.75 Mn, 0.85 Cr, 2.0 Ni, 0.3 Mo, bal Fe.
Heat Treated: 120,000-145,000 TS; 95,000-120,000 YP; 14-20 El; 30-40 RA; 200-250 Brin.
For general and structural castings, gears, shafting, housing.
AISI 4330. Shock resistant.

BIRDSBORO-30A4.
M-136; 0.35 max C, 1.5 Mn, 0.5 Cr, 0.5 Ni, 0.5 Mo, bal Fe.
Cast: 120,000-150,000 TS; 95,000-125,000 YP; 9-14 El; 22-30 RA; 240-310 Brin.
For general and structural castings, shafts, housings, gears.
Tough, shock resistant.

BIRDSBORO-30A6.
M-136; 0.35 max C, 0.90 Mn, 0.50 Cr, 0.50 Ni, 0.50 Mo, bal Fe.
Cast: 120,000-145,000 TS; 95,000-120,000 YP; 14-20 El; 30-40 RA.
For general and structural applications, gears, shafts, housings.
AISI 8630 modified.

BIRDSBORO-30 CN.
M-136; 0.35 max C, 0.70 Mn, bal Fe.
Cast: 65,000-80,000 TS; 35,000-55,000 YP; 125-160 Brin; 22-30 El; 30-50 RA.
For general and structural applications, gears, housings.
Excellent weldability. AISI 1030.

BIRDSBORO-40A2.
M-136; 0.45 max C, 0.80 Mn, 0.95 Cr, 0.20 Mo, bal Fe.
Cast: 90,000-165,000 TS; 60,000-140,000 YP; 9-20 El; 22-40 RA; 240- 310 Brin.
For general and structural castings, gears, shafts, housings. AISI 414 , tough.

BIRDSBORO-40A3.
M-136; 0.45 C, 0.75 Mn, 0.85 Cr, 2.0 Ni, 0.30 Mo, bal Fe.
Ht.Tr.: 180,000-210,000 TS; 150,000-180,000 YP; 20-25 RA; 330-400 Brin.
For general and structural castings, shafts, gears, housings. AISI 4340. Shock resistant.

BIRDSBORO-40A5.
M-136; 0.45 max C, 0.95 Mn, 0.5 Cr, 0.5 Ni, 0.5 Mo, bal Fe.
Cast: 120,000-150,000 TS; 95,000-125,000 YP; 9-14 El; 22-30 RA; 290-350 Brin.
For general and structural castings, shafts, gears, housings.
High strength, shock resistant.
AISI 8640 Modified.

BIRDSBORO-40 CN.
M-136; 0.50 max C, 0.80 Mn, bal Fe.
Cast: 80,000-95,000 TS; 40,000-60,000 YP; 140-180 Brin, 30-40 RA, 18-28 El.
For general and structural applications, gears, housings. Weldable with caution. AISI 1040.

BIRDSBORO-45A2.
M-136; 0.50 max C, 0.80 Mn, 0.95 Cr, 0.20 Mo, bal Fe.
Cast: 120,000-165,000 TS; 95,000-140,000 YP; 14-9 El; 22-30 RA; 300 Brin.
For general and structural castings; gears, shafts, housings. AISI 4145, tough.

BIRDSBORO-45A7.
M-136; 0.50 max C, 0.80 Mn, 2.0 Cr, 0.5 Mo, bal Fe.
Ht.Tr.: 175,000-210,000 TS; 145,000-180,000 YP; 6-12 El; 12-30 RA; 290-350 Brin.
For general and structural castings, gears, shafts, housings. Tough, shock resistant.

BIRDSBORO-45 CN.
M-136; 0.50 C, 0.80 Mn, bal Fe.
Cast: 80,000-95,000 TS; 40,000-60,000 YP; 140-180 Brin; 18-28 El; 30-40 RA.
For general and structural applications, gears, housings. Weldable with caution.

BIRDSBORO-50 CN.
M-136; 0.55 max C, 0.80 Mn, bal Fe.
Cast: 80,000-95,000 TS; 50,000-65,000 YP; 22-30 El; 25-40 RA.
For general and structural applications, gears, housings. AISI 1050.

BIRDSBORO CA15.
M-136; 0.15 max C, 12.5 Cr, bal Fe.
Cast: 90,000-110,000 TS; 65,000-80,000 YS; 18-25 El.
Mildly corrosion resistant.

BIRDSBORO-CMO.
M-136; 0.25 max C, 1.0 Mn, 0.25 Mo, bal Fe.
Cast: 65,000-80,000 TS; 35,000-50,000 YP; 24-30 El; 35-50 RA.
For general and structural applications, gears, shafts, housings.
AISI 4024 modified, tough.

BIRDSBORO-CNI.
M-136; 0.45 max C, 0.8 Mn, 3.5 Ni, bal Fe.
Ht.Tr.: 120,000-145,000 TS; 95,000-120,000 YP; 14-20 El; 30-40 RA.
For general and structural castings, shafts, gears, housings.
AISI 2340, tough, shock resistant.

BIRDSBORO-CRMO1.
M-136; 0.25 max C, 0.5 Cr, 0.5 Mo, bal Fe.
Cast: 70,000-80,000 TS; 45,000-55,000 YS; 22-30 El; 35-50 RA.
For general and structural castings, shafts, gears, housings.
ASTM A356, Alloy 5.

BIRDSBORO-CRMO2.
M-136; 0.20 max C, 0.65 Mn, 1.25 Cr, 0.5 Mo, bal Fe.
Cast: 70,000-80,000 TS; 45,000-55,000 YP; 22-30 El; 35 RA.
For general and structural castings, shafts, gears, housings.
ASTM A356 Alloy A.

BIRDSBORO-CRMO2 + V.
M-136; 0.18 max C, 0.55 Mn, 1.25 Cr, 0.50 Mo, 0.20 V, bal Fe.
Cast: 70,000-80,000 TS; 40,000-55,000 YP; 20-30 El; 35-50 RA.
For general and structural castings, shafting, gears, housings. Shock resistant.

BIRDSBORO-CRMO3.
M-136; 0.18 max C, 0.55 Mn, 2.3 Cr, 1.0 Mo, bal Fe.
Cast: 70,000-105,000 TS; 40,000-85,000 YP; 17-30 El; 30-45 RA.
For general and structural castings, gears, shafts, housings. Tough, shock resistant.
ASTM A217, Alloy WC9.

BIRDSBORO-CRMO4.
M-136; 0.20 max C, 0.55 Mn, 5.25 Cr, 0.55 Mo, bal Fe.
Cast: 90,000-120,000 TS; 60,000-95,000 YP; 18-25 El; 35-45 RA.
For general and structural castings, gears, shafts, housings.
ASTM A217 Alloy C5.

BIRDSBORO FOR COMPOSITION OF OTHER BIRDSBORO CASTINGS see AMERICAN CASTING INSTITUTE GRADES IN APPENDIX.

BIRDSBORO HY 80.
M-136; 0.20 max C, 1.50 Cr, 2.85 Ni, 0.45 Mo, 0.65 Mn, bal Fe.
80,000-95,000 psi YS; 20-30 El.
High yield, good notch toughness.

BIRDSBORO-LC2.
M-136; 0.25 max C, 0.65 Mn, 2.5 Ni, bal Fe.
Cast: 80,000 TS; 50,000 YP; 24 El; 35 RA.
For general and structural applications, gears, shafts, housings. Shock resistant.

BIRDSBORO-LC3.
M-136; 0.15 max C, 0.65 Mn, 3.5 Ni, bal Fe.
Cast: 60,000-80,000 TS; 40,000-55,000 YP; 24-30 El; 35-50 RA.
For general and structural castings, gears, housings, shafts.
ASTM A352, LC3; AISI 4812 Modified.
Tough, shock resistant.

BIRDSBORO-MM.
M-136; 0.32 max C, 1.5 Mn, 0.50 Mo, bal Fe.
Cast: 80,000-120,000 TS; 50,000-90,000 YS; 14-22 El; 150-210 Brin; 30-35 RA.
For general and structural applications, gears, housings. Good weldabilit ASTM A148.

BIRDSBORO NO. 30.
M-136; 0.30 C, 0.20 Mo, 1.0 Cu, bal Fe.
85,000 TS; 55,000 YS; 22 El; 40 RA; 180 Brin.
For engineering construction; corrosion and fatigue resisting.

BIRMABRIGHT.
M-106; 3-6 Mg, 0.3-0.7 Mn, bal Al.
Sand cast: 20,000-22,400 TS; 20,000-22,400 YS; 3-6 El; 55-60 Brin; 5.8 Izod.
Good corrosion resistance; takes good finish, for decorative parts.
Gen. Eng. Bs 1490:LM 5-M.

BIRMABRIGHT.
M-137; 3-6 Mg, 0.25-0.75 Mn, bal Al.
Sand cast: 20,000-22,000 TS; 11,000-15,000 YS; 5.5-3.0 El; 54-60 Brin.
Drawn: 34,000-43,000 TS; 20,000-27,000 YS; 25.0-15.0 El; 70-80 Brin.
For tubes, rivets, bolts, nuts, screws, architectural metal work; corrosion resistant to salt water.

BIRMABRIGHT 5454.
M-325; 0.7 Mn, 2.8 Mg, bal Al.
Sheet and strip, cond. O: 215-285 MPa TS; 80 min MPa YS; 12-8 El.
AA 5454.

BIRMABRIGHT BB1.
M-325; 0.5 Mn, 1.0 Mg, bal Al.
Drawn tube, cond. O: 162 MPa TS; 70 MPa YS; 25 El.

BIRMABRIGHT BB2.
M-325; 0.25 Mn, 2.0 Mg, bal Al.
Sheet and strip, cond H6: 230 MPa TS; 190 MPa YS; 8 El.
AA 5251.

BIRMABRIGHT BB3.
M-325; 0.25 Mn, 3.5 Mg, bal Al.
Sheet and strip, cond H4; 290 MPa TS; 260 MPa YS; 7 El.
AA 5154A.

BIRMABRIGHT BB4.
M-325; 0.7 Mn, 4.7 Mg, bal Al.
Sheet and strip, cond H4; 370 MPa TS; 290 MPa YS; 8 El.
AA 5083.

BIRMABRIGHT BB 5.
M-325; 0.3 Mn, 5.0 Mg, bal Al.
Sheet and strip, cond. O: 278 MPa TS; 147 MPa YS; 25 El.
AA 5056A.

BIRMABRIGHT BB 17.
M-325; 0.6 Mg, bal Al.
Sheet and strip, ann.: 110 MPa TS; 20 El.

BIRMAL.
M-137; 2-6 Mg, 0.25-0.75 Mn, bal Al.
Cast: 22,000 TS; 15,000 YS; 3.0 El; 58 Brin.
For light alloy parts; heat treatable.

BIRMALITE.
M-137; 9-11 Cu, 0.1-0.5 Mg, 1.0 Fe, bal Al.
Cast: 19,000 TS; 17,000 YS; 0.5 El; 80 Brin.
Heat treated: 40,000 TS; 0.5 El; 150 Brin.
For pistons; heat treatable.

BIRMAL L4.
M-137; 2-4 Cu, 0.5-1.5 Mg, 8.5-10.5 Si, 0.5-1.5 Ni, 0.2 Ti, bal Al.
Permanemt mold: 32,500 TS; 0 El; 120 Brin.
For pistons; age-hardenable.

BIRMAL MB7.
M-137; 0.7-1.3 Cu, 0.75-1.2 Mg, 0.35-0.85 Si, 1.6 Ni, 7.0 Sn, 0.2 Ti, bal Al.
Cast.
For bearings; age-hardenable.

BIRMAL P83.
M-137; 3-4 Cu, 7.5-9.5 Si, 1.3 max Fe, 1.0 max Zn, bal Al.
Die cast: 40,400 TS; 15,700 YS; 3 El; 85 Brin.
For general purpose die castings; SAE 306.

BIRMASIL.
M-137; 11-14 Si, bal Fe.
For light alloy castings; die cast alloy.

BIRMASIL SPECIAL.
M-137, 10-13 Si, 2.5-3.5 Ni, 0.2 Ti, 0.6 Mg, bal Al.
Chill cast: 42,000-36,000 TS; 22,000-16,000 YS; 3-6 El; 70-80 Brin.
Sand cast: 28,000-29,000 TS; 14,000-16,000 YS; 4-2 El; 52-60 Brin.
For automotive engine water cooled cylinder blocks, cylinder heads, fire brick molds, motor car fittings; British Patent 342, 152; good corrosion resistance.

BIRMETAL-AZM.
M-321; 0.3 Mn, 6.0 Al, bal Mg.

BIRMETAL B.B. 016.
M-325; 1 Mg, 0.6 Si, 0.25 Cu, bal Al.

BIRMETAL B.B. 019.
M-325; 1.25 Mn, 0.9 Si, 0.15 Cu, 0.2 Ti, bal Al.

BIRMETAL BBZ 36.
M-321; 0.3 Cu, 3.5 Mg, 0.3 Mn, 6.5 Zn, 0.3 Cr, bal Al.

BIRMETAL BMB 055.
M-321; 0.5 Mg, 0.4 Si, bal Al.
Extruded, ann: 110 MPa TS; 75 MPa YS; 30 El.
AA 6063.

BIRMETAL BMB 065.
M-321.
(Same as BMB 055 but more suitable for bright anodizing).

BIRMETAL BMB 071.
M-321; 0.06 Mn, 0.7 Mg, 1.0 Si, bal Al.
Sheet and strip: 120 MPa TS; 20 El: hardness 35 H.V. Ann.
AA 6082.

BIRMETAL BMB 240.
M-321; 2.0 Cu, 0.5 Mg, bal Al.
Sheet and strip, TB cond.: 290 MPa TS; 140 MPa YS; 24 El.
AA 2117.

BIRMETAL BMB 473.
M-321; 4 Cu, 0.7 Mg, 0.7 Mn, 0.3 Si, bal Al.

BIRMETAL BMB 478.
M-321; 4.3 Cu, 0.8 Mn, 0.7 Mg, 0.8 Si, bal Al.
Sheet and strip: 440 MPa TS; 310 MPa YS; 18 El; hardness 120 H.V. (TB cond.).
AA 2014.

BIRMETAL BMB 551.
M-321; 4.5 Cu, 0.5 Mg, 0.7 Mn, 1.0 Si, 0.75 Fe, bal Al.

BIRMETAL BMB 1306.
M-321; 0.75 Cu, 2.75 Mg, 0.5 Mn, 5.5 Zn, bal Al.

BIRMETAL BMB 2024.
M-321; 4.4 Cu, 0.6 Mn, 1.5 Mg, bal Al.
Clad sheet and strip: 430 MPa TS; 285 MPa YS; 18 El. (TB cond.)
AA 2024.

BIRMETAL BMB 2308.
M-321; 1.75 Cu, 2 Mg, 0.2 Mn, 6.5 Zn, 0.15 Cr, bal Al.

BIRMETAL BMB Z12.
M-321; 1.6 Cu, 2.5 Mg, 0.16 Cr, 5.8 Zn.
Extruded, cond. TF: 600 MPa TS; 560 MPa YS; 8 El; 190 H.V. hardness.
AA 7075.

BIRMETAL-ZW1.
M-321; 1.2 Zn, 0.4-0,8 Zr, bal Mg.

BIRMETAL-ZW3.
M-321; 3.0 Zn, 0.4-0.8 Zr, bal Mg.

BIRMID 21.
M-137; 2-4 Cu, 0.3-0.7 Mn, 0.2 max Zn, 0.2 max Ti, bal Al.
Sand cast: 21,300 TS; 9000 YS; 1.7 El; 60 Brin.
Permanent mold: 27,000 TS; 10,000 YS; 2.0 El; 70 Brin.
For moderately stressed parts; not heat treatable.

BIRMID 112.
M-137; 0.7-2.5 Cu, 9-11.5 Si, 0.2 max Ti, bal Al.
Cast: 36,000 TS; 13,500 YS; 1.5 El; 70 Brin.
For instrument housings; die castings.

BIRMID 298/304.
M-137; 4-5 Cu, 0.05-0.30 Ti, 0.25 max Si, bal Al.
Sand cast: 40,400 TS; 4 El.
Permanent mold: 44,800 TS; 9 El.
Heat treated: 42,600 TS; 27,000 YS; 5 El; 100 Brin.
For high strength, shock resistant castings; age-hardenable.

BIRMID 300.
M-137; 9.5-11 Mg, 0.2 Ti, 0.25 Si, 0.1 Mn, bal Al.
Sand cast: 42,600 TS; 22,400 YS; 10 El; 76 Brin.
Permanent mold; 49,300 TS; 24,600 YS; 18 El; 80 Brin.
For high strength castings; age-hardenable.

BIRMID 428.
M-137; 6-8 Cu, 2-4 Si, 2-4 Zn, 0.6 max Mn, 1 max Fe, bal Al.
Cast: 24,600 TS; 14,600 YS; 1 El; 80 Brin.
For housings, gear cases; light stressed parts.

BIRMIDAL.
M-106; 0.3-0.8 Mg, 3.5-6 Si, bal Al.
For grades of castings of same compositions; 18,000-43,000 TS range.
General purpose castings; properties vary with varying heat treatments.
Gen. Eng. BS 1490: LM 8-M, 8-P, 8-W 8-WP.

BIRMIDAL.
M-137; 0.5 Mg, 3.5-6.0 Si, 0.2 max Ti, bal Al.
Cast: 25,000 TS; 15,000 YS; 2-10 El; 60 Brin.
Heat treated: 44,000 TS; 32,000 YS; 1-5 El; 95 Brin.
For gear cases, instrument housings; age-hardenable.

BIRMID D7.
M-137; 2-4 Cu, 4-6 Si, 0.3-0.7 Mn, 0.25 Ni, 0.3 Zn, 0.2 Ti, bal Al.
Sand cast: 20,200 TS; 2 El; 70 Brin.
Permanent mold: 22,400 TS; 2 El; 75 Brin.
For heavy duty castings; age-hardenable, good castability.

BIRMID D8.
M-137; 2-4 Cu, 4-6 Si, 0.3-0.7 Mn, 0.35 Ni, 0.3 Zn, 0.2 Ti, bal Al.
Sand cast: 20,200 TS; 89,600 YS; 2 El; 60 Brin.
Heat treated: 47,000 TS; 40,000 YS; 1 El; 110 Brin.
For gears, cases, hardware, instrument housings; age-hardenable, good castability.

BIRMIDIUM "Y".
M-137; 3.5-4.5 Cu, 1.8-2.3 Ni, 1.2-1.7 Mg, bal Al.
Sand cast: 25,000 TS; 20,000 YS; 0.5 El; 85 Brin.
Aged: 32,000 TS; 29,000 YS; 1.0 El; 100 Brin.
For pistons, cylinder heads, crankcases, oil pans; age-hardenable, heat resistant.

BIRMIDIUM "Y" ALLOY.
M-137; 3.5-4.5 Cu, 1.8-2.3 Ni, 1.2-1.7 Mg, bal Al.
Heat treated: 35,000-42,000 TS; 2-3 El.
For automobile pistons, piston heads, crank cases; corrosion resistant.

BIRMINGHAM.
M-Eng.; 62-50 Cu, 32-20 Zn, 12-30 Ni.
For German silver; corrosion resistant.

BIRMINGHAM 21.
M-137; 2-4 Cu, 4-6 Si, 2 max Zn, 0.2 max Ti, 0.15 max Mg, bal Al.
Sand cast: 21,300 TS; 9000 YS; 1.7 El; 60 Brin.
Permanent mold: 27,000 TS; 11,000 YS; 2 El; 70 Brin.
For cast parts subject to moderate stress; sand and permanent mold castings.

BIRMINGHAM 298/304.
M-137; 4-5 Cu, 0.10 max Mg, 0.25 max Si, 0.05-0.30 Ti, bal Al.
T4-temper: 34,800 TS; 14,000 YS; 12 El; 65 Brin.
T6-temper: 42,000 TS; 27,000 YS; 5 El; 100 Brin.
For shock resistant castings; sand and permanent mold cast, age-hardenable.

BIRMINGHAM 300.
M-137; 0.1 max Cu, 9.5-11.0 Mg, 0.2 max Ti, 0.25 max Si, bal Al.
Sand cast: 43,000 TS; 22,400 YS; 10 El; 76 Brin.
Permanent mold: 49,000 TS; 25,000 YS; 18 El; 80 Brin.
For highly stressed castings; solution heat treated only.

BIRMINGHAM 428.
M-137; 6-8 Cu, 2-4 Si, 2-4 Zn, 0.2 max Ti, bal Al.
Permanent mold: 25,000 TS; 12,500 YS; 1 El; 80 Brin.
For castings for light stress; permanent mold casting.

BIRMINGHAM L.4.
M-137; 2-4 Cu, 8.5-5-10.5 Si, 1.2 max Fe, 0.5-1.5 Mg, 0.2 max Ti, bal Al.
Permanent mold: 32,000 TS; 0 El; 90-120 Brin.
For pistons; permanent mold castings.

BIRMINGHAM P.83.
M-137; 3-4 Cu, 7.5-9.5 Si, 1.3 max Fe, 0.1 max Mg, 1 max Zn, 0.3 max Sn, bal Al.
Die cast: 40,400 TS; 16,000 YS; 3 El; 85 Brin.
For instruments, casings; die castings.

BIRSO.
M-577; C, bal Fe.
For gears, housings, castings.

BISBO ARMA.
M-1286; 2.5 T.C., 0.7 C.C., 1.4 Si, 0.4 Mn, 0.025 Bi, bal Fe.
Ht. Tr.: 70,000-100,000 TS; 48,000-80,000 YS; 2-4 El; 163-269 Brin.
For crankshafts, rocker arms, universal joint yokes, gears. Pearlitic. Malleable iron.

BISCO.
M-334; 0.40-0.45 C, 0.85 Mn, 1.0 Cr, 0.20 Mo, bal Fe.
Heat treated: 285-341 Brin.
Bars 3/4 Rd to 6" Rd; for shafts.

BISHOP 29-17.
M-1772; 29 Ni, 17 Co, 53 Fe.
1/2 hard welded or seamless tubing: 85,000-100,000 psi TS; 60,000-80,000 psi YS; 10-15 El. Coef. thermal expan: 6×10^6 in/in/°C, 20-500°C.
For glass to metal seals.

BISHOP 42.
M-1772; 42 Ni, 57 Fe.
1/2 hard seamless tubing: 90,000-110,000 psi TS; 45,000-65,000 psi YS; El.
Coef. thermal expan: 5.3×10^6 in/in/°C, 20-500°C.
For glass to metal seals.

BISHOP 52.
M-1772; 52 Ni, 47 Fe.
1/2 hard welded or seamless tubing:
85,000-100,000 psi TS; 45,000-65,000 psi YS;
10-25 El.
Coef. thermal expan: 9.8×10^6 in/in/°C, 20-500°C.
For glass to metal seals.

BISMUTH.
M-1755; Bi.
Purities: zone refined 99.9999%, 99.9995%, 99.999%, 99.99%.
Forms: Rods, shot, needles, powders, ingot, foil, single crystals.

BISMUTH BRASS-1.
M-U.S.; 52 Cu, 30 Ni, 12 Zn, 5 Pb, 1 Bi.
For ornaments, hardware; corrosion resistant.

BISMUTH BRASS-2.
M-U.S.; 47 Cu, 31 Ni, 21 Zn, 1 Bi, 1Sn.
For ornaments, hardware; corrosion resistant.

BISMUTH BRONZE.
M-U.S.; 45-53 Cu, 33-10 Ni, 20-22 Zn, 16-15 Sn, 1 Bi, 0-0.1 Al.
For ornaments, hardware; corrosion resistant.

BISON.
M-1705; 0.5 C, 0.75 Si, 1.15 Cr, 0.2 V, 2.5 W, bal Fe.
Oil hardening shock resisting tool steel.
AISI S1.

BKE.
M-USSR; 3-4 Cd, bal Cu.
For electrical contacts; high conductivity.

B K SPECIAL.
M-1260; 90.3 Sn, 6.5 Sb, 3 Cu, 0.2 Ni.
Cast: 12,800 TS; 25 Brin.
For engine bearings; M.P. 440-610°F; shock resistant.

BLA-CALOY.
M-1254; 3.3 C, 0.7 Mn, 2 Si, 0.05 Mg, bal Fe.
For gears, shafts, cams, housings; ductile cast iron.

BLACKALLOY see **BLACKALLOY TX90.**

BLACKALLOY-525.
M-1095; 44 Co, 24 Cr, 20 W, bal Fe.
Cast: C 63-65 Rock.
For cutting and boring tools.
Hard, abrasion and heat resistant.

BLACKALLOY-TX90.
M-1095; 42 Co, 24 Cr, 22 W, bal Fe.
Cast: C 66-67 Rock.
For tools, cutters, form tools, reamers, drills.
Centrifugally cast. High heat and abrasion resistant.

BLACK BEAUTY.
M-691; 0.2 C, bal Fe.
For construction steel; black oxide sheet.

BLACK DEVIL.
M-712; 0.09 C, 0.3 Mn, bal Fe.
For welding rod; E 6020.

BLACK DEVIL E B.
M-712; 0.07 C, 0.39 Mn, 0.21 Si, bal Fe.
Welded: 66,000 TS; 56,000 YS; 26 El; 45 RA.
For welding electrodes; arc, E-6030.

BLACK DEVIL NO. 75.
M-712; 0.09 C, 0.37 Mn, 0.2 Si, 0.56 Mo, bal Fe.
Welded: 70,000-74,000 TS; 57,000-62,000 YS;
34-27 El; 65-50 RA.
For welding electrodes; E-7020.

BLACK DIAMOND.
M-38; 1.0 C, bal Fe.
For tools, cutters, drillers, dies, hammers; water hardening. AISI W1.

"BLACK HEART" FERRITIC 22/14/14.
M-1762; 2.3-2.6 C, 1.3-1.55 Si, 0.4-0.57 Mn, 0.15-0.25 S, 0.08 max P, bal Fe.
Ferritic malleable iron castings.
49,300 TS; 31,400 YS; 14 El; 146 Brin.
For automotive, truck and tractor industries.

BLACK LABEL.
M-341; 1.1 C, 0.5 Cr, 0.2 V, bal Fe.
For tools and dies; water hardening.

BLACK LABEL CAST (EXTRA).
M-261; C, bal Fe.
For tools and dies; water hardening.

BLACKOR.
M-26; 87 W, 9 C.
For oilwell core bits, cutters, tools; W_2 C hard facing compound.

BLACKSKIN ADMIRALTY.
M-319; 70 Cu, 29 Zn, 1 Sn.

BLACK STREAK ALNICO NO. 4.
M-373; 12 Al, 28 Ni, 5 Co, bal Fe.
For cast permanent magnets; high coercive force.

BLANCO BA.
M-1288; 0.10 C, 18 Cr, 8.5 Ni, bal Fe.
Annealed: 80,000 TS; 35,000 YS; 55 El; 75 RA; 150 Brin.
Cold drawn: 180,000 TS; 150,000 YS; 10 El; 250 Brin.
For chemical plant equipment, tanks, fermenters; Type 302; stainless, austenitic.

BLANCO BA88.
M-1288; 0.12 C, 18 Cr, 8 Ni, 0.25 Mo, 0.2 S, bal Fe.
Annealed: 80,000 TS; 35,000 YS; 40 El; 60 RA; 150 Brin.
For screw machine products, fasteners; Type 303; stainless, free-cutting.

BLANCO B SPEZIAL.
M-1288; 0.1 max C, 18 Cr, 8.5 Ni, bal Fe.
Annealed: 80,000 TS; 35,000 YS; 55 El; 75 RA; 150 Brin.
Cold drawn: 180,000 TS; 150,000 YS; 10 El; 250 Brin.
For chemical plant equipment, tanks, mixers, vessels; Type 302; stainless, austenitic.

BLANCO B SUPER.
M-1288; 0.1 max C, 18 Cr, 2 Mo, 9.5 Ni, bal Fe.
Annealed: 85,000 TS; 35,000 YS; 50 El; 65 RA; 160 Brin.
Cold drawn: 150,000 TS; 135,000 YS; 6 El; 300 Brin.
For acid resistant chemical plant equipment, tanks; Type 316; stainless, austenitic.

BLANCO CM2.
M-1288; 0.35 C, 16.5 Cr, 1.15 Mo, bal Fe.
Annealed: 90,000 TS; 55,000 YS; 20 El; 45 RA; 180 Brin.
For oil refinery equipment, furnace parts; corrosion and heat resistant.

BLANCO CN12.
M-1288; 0.1 max C, 12.5 Cr, 12 Ni, bal Fe.
For valves, pumps; corrosion and heat resistant.

BLANCO CNM.
M-1288; 0.07 max C, 17 Cr, 13 Ni, 4.7 Mo, bal Fe.
Annealed: 90,000 TS; 40,000 YS; 45 El; 60 RA; 180 Brin.
For acid resistant chemical plant equipment; Type 317; stainless, austenitic.

BLANCO CT.
M-1288; 0.12 max C, 18 Cr, 19.5 Ni, Ti = 4 x C, bal Fe.
Annealed: 85,000 TS; 35,000 YS; 55 El; 65 RA; 150 Brin.
For welded structures, chemical plant equipment; Type 321; stainless, austenitic.

BLANCO CT2N.
M-1288; 0.12 max C, 18 Cr, 9.5 Ni, Cb = 8 x C, bal Fe..
Annealed: 90,000 TS; 45,000 YS; 56 El; 65 RA; 160 Brin.
Cold drawn: 100,000 TS; 65,000 YS; 40 El; 60 RA; 202 Brin.
For welded structures, chemical plant equipment; Type 347; stainless, austenitic.

BLANCO CU.
M-1288; 0.07 max C, 17.5 Cr, 2 Mo, 17.5 Ni, 2 Cu, Ti = 7 x C, bal Fe.
For valves, pumps, chemical plant equipment; corrosion and heat resistant.

BLANCO G.
M-1288; 0.20 C, 13 Cr, bal Fe.
Annealed: 95,000 TS; 50,000 YS; 25 El; 55 RA; 196 Brin.
Cold drawn: 105,000 TS; 85,000 YS; 17 El; 50 RA; 215 Brin.
For turbine blades, cutlery, valves, instruments; Type 420; stainless, hardenable.

BLANCO H now **VEW N540.**

BLANCO K(H).
M-1288; 0.4 C, 13 Cr, bal Fe.
Annealed: 100,000 TS; TS; 55,000 YS; 20 El; 50 RA; 200 Brin.
For valves, cutlery, surgical and dental instruments; Type 420; stainless, hardenable.

BLANCO L1.
M-1288; 0.12 max C, 13 Cr, bal Fe.
Annealed: 75,000 TS; 40,000 YS; 35 El; 70 RA; 155 Brin.
Cold drawn: 100,000 TS; 85,000 YS; 17 El; 60 RA; 205 Brin.
For turbine blades, surgical instruments; Type 410; stainless.

BLANCO L2.
M-1288; 0.12 max C, 13 Cr, bal Fe.
Annealed: 75,000 TS; 40,000 YS; 35 El; 70 RA; 155 Brin.
Cold drawn: 100,000 TS; 85,000 YS; 17 El; 60 RA; 205 Brin.
For turbine blades, cutlery, surgical and dental instruments; Type 410; stainless.

BLANCO L3.
M-1288; 0.12 max C, 13 Cr, 0.4 Si, bal Fe.
Annealed: 75,000 TS; 40,000 YS; 35 El; 70 RA; 155 Brin.
Cold drawn: 100,000 TS; 85,000 YS; 17 El; 60 RA; 205 Brin.
For cutlery, turbine blades, instruments; Type 410; stainless.

BLANCO M now **VEW N320.**

BLANCO M1.
M-1288; 0.2 C, 0.4 Si, 13 Cr, bal Fe.
Annealed: 95,000 TS; 50,000 YS; 25 El; 55 RA; 196 Brin.
Cold drawn: 105,000 TS; 85,000 YS; 17 El; 50 RA; 215 Brin.
For cutlery, valves, turbine blades, knives; Type 420; stainless, hardenable.

BLANCO M2.
M-1288; 0.22 C, 0.4 Si, 17 Cr, 1.5 Ni, bal Fe.
Annealed: 125,000 TS; 95,000 YS; 20 El; 55 RA; 260 Brin.
Cold drawn: 130,000 TS; 110,000 YS; 15 El; 35 RA; 270 Brin.
For pumps, marine hardware, valve trim, shafts; Type 431; stainless.

BLANCO M3.
M-1288; 0.4 C, 0.4 Si, 13 Cr, bal Fe.
Annealed: 95,000 TS; 50,000 YS; 25 El; 55 RA; 195 Brin.
Cold drawn: 105,000 TS; 85,000 YS; 17 El; 50 RA; 215 Brin.
For valves, pumps, turbine blades, cutlery, knives; Type 420; stainless, hardenable.

BLANCO M3E.
M-1288; 0.9 C, 0.4 Si, 18 Cr, 1.15 Mo, 1 V, bal Fe.
For valves, bearings, instrument pivots, rollers; corrosion and wear resistant, hardenable.

BLANCO M4.
M-1288; 0.12 C, 16.5 Cr, 0.25 Mo, 0.20 S, bal Fe.
Annealed: 80,000 TS; 50,000 YS; 25 El; 50 RA; 160 Brin.
For gears, shafts, screw machine products; Type 430F; stainless, free-cutting.

BLANCO RCM.
M-1288; 0.20 C, 13 Cr, 1.15 Mo, bal Fe.
Annealed: 95,000 TS; 50,000 YS; 25 El; 55 RA; 196 Brin.
For valves, cutlery, valve trim, turbine blades; Type 420 Mo; stainless, hardenable.

BLANCO SPEZIAL.
M-1288; 0.07 C, 18 Cr, 9.5 Ni, bal Fe.
Annealed: 85,000 TS; 35,000 YS; 60 El; 70 RA; 150 Brin.
Cold drawn: 180,000 TS; 125,000 YS; 10 El; 330 Brin.
For welded structures, chemical plant equipment, tanks; Type 304; stainless, austenitic.

BLANCO SPEZIAL now **VEW A550.**

BLANCO SUPER.
M-1288; 0.07 max C, 18 Cr, 2 Mo, 10.5 Ni, bal Fe.
Annealed: 85,000 TS; 35,000 YS; 50 El; 65 RA; 160 Brin.
Cold drawn: 150,000 TS; 135,000 YS; 6 El; 300 Brin.
For acid resistant chemical plant equipment, tanks; Type 316; stainless, austenitic.

BLANCO SUPER EXTRA.
M-1288; 0.12 max C, 18 Cr, 2 Mo, 10.5 Ni, Ti = 4 X C, bal Fe.
Annealed: 85,000 TS; 35,000 YS; 50 El; 65 RA; 160 Brin.
Cold drawn: 150,000 TS; 135,000 YS; 6 El; 300 Brin.
For welded structures, chemical plant equipment, tanks: Type 316 Ti; stainless, austenitic.

BLANCO SUPER EZN.
M-1288; 0.12 max C, 18 Cr, 2 Mo, 10.5 Ni, Cb = 8 x C, bal Fe.
Annealed: 85,000 TS; 35,000 YS; 50 El; 65 RA; 160 Brin.
For welded structures, chemical plant equipment, tanks; Type 316 Cb; stainless, austenitic.

BLANCO T70.
M-1288; 0.1 max C, 17.5 Cr, Ti = 7 x C, bal Fe.
Annealed: 80,000 TS; 50,000 YS; 25 El; 50 RA; 150 Brin.
For heat treating boxes, furnace parts, rabble arms; Type 430 Ti; stainless, ferritic.

BLANCO W.
M-1288; 0.12 max C, 0.4 Si, 13 Cr, bal Fe.
Annealed: 75,000 TS; 40,000 YS; 35 El; 70 RA; 155 Brin.
Cold drawn: 100,000 TS; 85,000 YS; 17 El; 60 RA; 205 Brin.
For valve trim, turbine blades, cutlery; Type 410; stainless.

BLANKO-BLECH.
M-1288; 80 Cu, 20 Ni.
For turbine blades, condenser tubes; corrosion resistant.

BLATT (LEAF) GOLD.
M-Eng.; 77 Cu, 23 Zn.
For gold leaf substitute, signs; corrosion resistant.

BLATT (LEAF) SILVER.
M-Eng.; 91 Sn, 8.3 Zn, 0.4 Pb, 0.2 Fe.
For bearings, tin foil for wrappers; Babbitt.

BLAUPUNKT.
M-1331; 0.82 C, 4.1 Cr, 0.85 Mo, 1.6 V, 8.7 W, bal Fe.
For lathe and planer tools, drills, reamers, taps; high speed steel.

BLAW-KNOX C-7.
M-112, M-1222; 0.3-0.4 C, 0.8-1.0 Mn, 0.3-0.45 Si, 0.8 Cr, 1.5 Ni, 0.4 Mo, bal Fe.
Cast: 125,000 TS; 100,000 YP; 15 El; 35 RA; 220-365 Brin.
For gears, shafts, housings. Shock resistant, tough.

BLAW-KNOX C-8.
M-112, M-1222; 0.25-0.35 C, 0.7-0.9 Mn, 0.30-0.45 Si, 0.50 Cr, 0.50 Ni, 0.4 Mo, bal Fe.
Cast: 115,000 TS; 90,000 YP; 22 El; 50 RA; 200-325 Brin.
For gears, shafts, housings.
Shock resistant. Water hardening.

BLEIZINNBRONZE 7.
M-1420; Zn, Sn, bal Cu.
Cast: 25,000 TS; 8 El; 100 Brin.
For hardware; bronze.

BLEIZINNBRONZE 8.
M-1420; Zn, Sn, bal Cu.
Cast: 22,000 TS; 8 El; 60 Brin.
For hardware; bronze.

BLEIZINNBRONZE 10.
M-1420; Zn, Sn, bal Cu.
Cast: 26,000 TS; 15 El; 70 Brin.
For hardware; bronze.

BLENDALLOY 22-1000 HIGH PURITY NICKEL.
M-1626; 0.005 max C, 0.005 max Fe, 99.98 Ni.
For plater bars, tube cathodes, fluorescent lamp components, and resistanc thermometers.

BLENDALLOY 22-9604.
M-1626; 0.005 max C, 0.005 max Fe, 96 Ni, 4 W. (High purity product).
For specialized electronic and magnetic applications.

BLENDALLOY 22-9800.
M-1626; 0.005 max C, 0.005 max Fe, 98 Ni, 0.50 Ti, 0.25 Mg. (High purity product).
For springs requiring high electrical conductivity and parts requiring good thermal conductivity.

BLENDALLOY 25-4200 ("42-ALLOY").
M-1626; 0.005 max C, 42 Ni, 58 Fe. (High purity product).
For electronic, magnetic and electrical applications. Glass sealing.

BLENDALLOY 25-4206.
M-1626; 0.005 max C, 43 Ni, 48 Fe, 2.5 Ti, 5 Cr, 0.5 Al. (High purity product).
For springs, mechanical filters, tuning forks.

BLENDALLOY 25-4400.
M-1626; 44.0 Ni, bal Fe. Others 0.02 max. (Free from non-metallic inclusions).
For mechanical filters.

BLENDALLOY 25-4601 FM.
M-1626; 0.005 max C, 46 Ni, 53 Fe, 0.5 Mn, 0.10 Se. (High purity product).
For glass sealing applications.

BLENDALLOY 25-4803.
M-1626; 0.005 max C, 48 Ni, 49 Fe, 3 Mo. (High purity product).
High magnetic permeability with minimum energy loss. For communication and electronic equipment.

BLENDALLOY 25-5000 (ORTHONOL).
M-1626; 0.005 max C, 50 Ni, 50 Fe. (High purity product).
Soft magnetic alloy with very high squareness and high core gain. For saturable reactors, magnetic amplifiers, switching devices, and power inverter-converter applications.

BLENDALLOY 25-5025.
M-1626; 50.0 Ni, 0.25 Mn, bal Fe. (Others 0.02 max).
Soft magnetic iron strip. Magnetic properties vary with thickness and hardness.

BLENDALLOY 25-5200 ("52" ALLOY).
M-1626; 0.005 max C, 52 Ni, bal Fe. (High purity product).
Precise thermal expansion, uniformly fine-grained microstructure, excellent electrical and thermal conductivity.
For glass sealing in dry reed switches and mercury wetted relays.

BLENDALLOY 25-7904 (PERMALLOY).
M-1626; 0.005 max C, 80 Ni, 15 Fe, 5 Mo. (High purity product).
Soft magnetic alloy with high initial permeability. For use in transformers, reed relays, amplifiers, vacuum tubes, microphones, loud speakers, cables.

BLENDALLOY 25-8000.
M-1626; 0.005 max C, 80 Ni, bal Fe. (High purity product).
For electronic and magnetic applications.

BLENDALLOY 25-8004.
M-1626; 80 Ni, 4.5 Mo, bal Fe. (Others 0.02 max).
Soft magnetic iron strip; square hysteresis loop. Magnetic properties var with thickness and hardness.

BLENDALLOY 25-8004 (PERMALLOY).
M-1626; 0.005 max C, 80 Ni, 15 Fe, 5 Mo. (High purity product).
Soft magnetic alloy with high squareness and high core gain, low coercive force.
For preamplifiers and modifiers, converters.

BLENDALLOY 25-8300.
M-1626; 0.005 max C, 83 Ni, bal Fe. (High purity product).
For electronic and magnetic applications.

BLENDALLOY 26-5446.
M-1626; 46 Cu, 0.002 C, bal Ni.
For electronic tube and glass-to-metal sealing applications.

BLENDALLOY 32-9010.
M-1626; 0.005 max C, 0.005 max Fe, 90 Co, 10 Ni.
For electronic, magnetic, magnetostrictive, and plating applications.

BLENDALLOY 33-1000, GRADE 1.
M-1626; 99.5 Co, 10 ppm Cu, 900 ppm Fe, 800 ppm Ni, 150 ppm S.
For use in electroplating, alloying.

BLENDALLOY 33-9505.
M-1626; 5 Fe, 10 ppm Cu, 1000 ppm Ni, 300 ppm S, bal Co.
More ductile bar and strip.

BLITZ FEE.
M-1312; 0.86 C, 4.3 Cr, 0.85 Mo, 2.1 V, 12 W, bal Fe.
For lathe and planer tools, drills, taps, hobs; high speed steel.

BLOCK BRASS.
M-U.S; 66.5 Cu, 32 Zn, 1.5 Pb.
For free machining brass parts.

BLOCK'S ALLOY.
M-U.S; 54 Co, 45 Ni, 0.9 Si.
For tools; corrosion and heat resistant.

BLOMBIT.
M-621; 2-7 Ag, 93-98 Cu.
Forged: 72,000-78,000 TS; 8-5 El; 120-140 Brin.
For electrodes and tips for spot welding; 83-90% conductivity.

BLUE CHIP.
M-57; 0.70 C, 0.25 Mn, 4 Cr, 18 W, 1 V, bal Fe.
Annealed: 105,100 TS; 75,000 YS; 14 El; 17 RA; 230 Brin.
For tools, cutters, reamers, drills, punches, dies, shears, taps; high speed steel.

BLUEDAC.
M-712; 0.08 C, 0.57 Mn, 0.21 Si, bal Fe.
Cast: 63,000 TS; 55,000 YS; 29 El; 55 RA.
Welded: 73,000 TS; 64,000 YS; 23 El; 35 RA.
For welding electrodes; arc welding all positions; E-6011.

BLUE DEVIL YOLOY.
M-712; 0.2 C, 0.5 Cu, bal Fe.
80,000 TS; 75,000 YS; 25 El; 53 RA.
For welding rod; flux coated.

BLUE DIAMOND.
M-542; C, alloy, bal Fe.
For road machinery cutting edges; wear resistant.

BLUE DOT SPECIAL.
M-750; C, alloy, bal Fe.
For maintenance and repair operations; oil hardening.

BLUE EDGE.
M-275; 0.5 C, 1 Cr, 1 W, bal Fe.
For pneumatic chisels, shear blades, cold rivet sets; shock resisting. AISI S3.

BLUE GOLD.
M-U.S.; 75 Au, 25 Fe.
For ornaments, jewelry; corrosion resistant.

BLUE LABEL.
M-341; 1.0-1.1 C, 0.15-0.25 V, Mn, Si, bal Fe.
For cold swaging dies, forming rolls, knives, punches, collets, pipe cutters, drill bushings. AISI W2 water hardening tool steel.

BLUE LABEL.
M-343; 0.8-1.1 C, bal Fe.
For general tools; "Heller Special Tool."

BLUE LABEL.
M-350; C, bal Fe.
For tools, drills, taps; water hardened.

BLUE LABEL.
M-694; 1.0 C, 0.3 Mn, 0.25 Si, bal Fe.
Type W1 water hardening tool steel.

BLUE LABEL EXTRA.
M-373; 0.6-1.4 C, 0.25 Si, 0.25 Mn, bal Fe.
For drills, taps, cutters, punches; Type W1; water hardened.

BLUE LABEL SF-2.
M-341; same as blue label but surface finished on two sides.

BLUE RIBAND VICTORY NO. 7.
M-713; 0.7 C, 18 W, 4 Cr, 1 V, 5 Co, bal Fe.
For cutters, tools; high speed steel.

BLUE STREAK 18-4-1.
M-822; 0.72 C, 4 Cr, 18 W, 1 V, bal Fe.
For cutting tools; high speed steel.

BLUE STREAK COBALT.
M-822; 1.25 C, 4.0 Cr, 3.0 V, 9.0 W, 3.0 Mo, 9.0 Co, bal Fe.
High speed steel, extra wear and red hardness properties; tungsten-cobalt type.

BLUE STREAK MOLY.
M-822; 0.8 C, 4 Cr, 5.75 W, 4.5 Mo, 1.6 V, bal Fe.
For cutting tools; high speed steel.

BLUE TIP NAVAL BRASS.
M-266; 60 Cu, 0.75 Sn, bal Zn.
Hard: 68,000 TS; 45,000 YS; 35 El.
Annealed: 63,000 TS; 40,000 YS; 40 El.
For bolts, nuts, marine hardware, valve stems; corrosion resistant.

BM 2.
M-1653; 0.75-0.90 C, 3.5-4.5 Cr, 4.5-5.5 Mo, 5.5-7.0 W, 1.5-2.2 V, bal Fe.
Mo-W high speed tool steel.
W.-Nr. 1.3343; AISI M2.

BM 78.
M-1739; 9 Cu, 1 C, bal Fe, sintered.
Density: 5.9-6.2 gms/cc; UTS: 34.5 kg/mm^2; Elon: 1%; Hardness: VPN-5-200 min.
Machinable.

B-METAL (BAHN-METALL).
M-315, M-316, M-402; 0.03-0.05 Li, 0.68-0.76 Ca, 0.62-0.72 Na, 0.02-0.04 K, 0.2 Al, bal Pb.
For locomotive bearings and journal bearings in railroad cars; retains high hardness at high temperature.

"B.M.S".
M-344; 0.7 C, 5 Cr, 1.5 W, bal Fe.
For tools, chisels, hard punches, shear blades; air hardening.

"BNC" STEEL.
M-Eng; C, Cr, Ni, bal Fe.
Hardened: 194,000 TS; 15 El; 50 RA; 400 Brin.
For gears, pinions; case hardening.

BND.
M-1177; 0.7 C, Cr, bal Fe.
For tools, gripper dies; tough.

B.N.F. COPPER-NICKEL-IRON ALLOY.
M-396; 5-10 Ni, 1-2 Fe, 0.3-0.8 Mn, bal Cu.
For marine parts and hardware; resists sea water corrosion.

B.N.F. LEAD ALLOY NO. 1.
M-396; 0.25 Cd, 0.5 Sb, bal Pb.
For lead sheathing of electric cables.

B.N.F. LEAD ALLOY NO. 2.
M-296; 0.25 Cd, 1.5 Sn, bal Pb.
For water pipes, lead sheathing of electric cables; corrosion resistant.

B.N.F. LEAD ALLOY NO. 3.
M-396; 0.15 Cd, 0.4 Sn, bal Pb.
For sheathing of ship cables.

B NO. 4.
M-114; 0.7 C, 18 W, 4 Cr, 1 V, bal Fe.
For drills, reamers, cutters, hobs, taps, broaches; high speed steel, oil hardened.

B NO. 9.
M-114; 0.7 C. 9.5 Mo, 1.5 W, 4 Cr, 1 V, bal Fe.
For drills, reamers, hobs, broaches, taps, cutters; high speed steel, oil hardened.

BOBIERRE'S METAL.
M-France; 66-58 Cu, 34-42 Zn.
For bolts, nuts, sheathing; yellow brass.

BOCHLET FCHD.
M-1435; 0.14 max C, 17-19 Mn, 10-13 Cr, 0.5-2.0 Si, 0.6 Ti, bal Fe.
For turbine blades, jet engine components; austenitic, heat resistant.

BOCHUM 85C7.
M-1331; 0.85 C, 1.75 Cr, 0.35 Mn, bal Fe.
For bearings, liners, forming dies; water hardened, wear resistant.

BOCHUM BC13V.
M-1331; 1.15 C, 0.65 Cr, 0.1 bal Fe.
For blanking and forming dies; water hardened, wear resistant.

BOCHUM BCMH.
M-1331; 1.0 C, 1.0 Mn, 0.37 Si, bal Fe.
Annealed: 100,000 TS; 53,000 YS; 21 El; 42 RA; 200 Brin.
For drills taps, reamers, cutters; Type W1; water hardened.

BOCHUM BCMW.
M-1331; 0.9 C, 0.37 Si, 1.0 Mn, bal Fe.
Heat treated: 190,000 TS; 145,000 YS; 10 El; 30 RA; 400 Brin.
For drills, taps, reamers, broaches, hobs; Type W1; water hardened.

BOCHUM BCR.
M-1331; 0.85 C, Cr, bal Fe.
For bearings, liners, sleeves; water hardened, wear resistant.

BOCHUM BCVH.
M-1331; 0.90 C, Cr, V, bal Fe.
For bearings, liners, sleeves; water hardened, wear resistant.

BOCHUM BCVW.
M-1331; 0.80 C, Cr, V, bal Fe.
For bearings, liners, sleeves; water hardened, wear resistant.

BOCHUM BFK6.
M-1331; 0.60 C, 0.25-0.50 Si, 0.3-0.8 Mn, bal Fe.
Heat treated: 160,000 TS; 113,000 YS; 12 El; 40 RA; 325 Brin.
For gears, rails, punches, hammers; water hardened.

BOCHUM BFK7.
M-1331; 0.75 C, 0.25-0.50 Si, 0.3-0.8 Mn, bal Fe.
Heat treated: 185,000 TS; 140,000 YS; 15 El; 40 RA; 400 Brin.
For springs, tools, punches, crimpers; water hardened.

BOCHUM BFK8.
M-1331; 0.90 C, 0.25-0.50 Si, 0.3-0.8 Mn, bal Fe.
Heat treated: 190,000 TS; 145,000 YS; 10 El; 30 RA; 400 Brin.
For springs, tools, cutters, drills; Type W1; water hardened.

BOCHUM BGK9.
M-1331; 0.85 C, 0.1-0.4 Si, 0.5-0.7 Mn, bal Fe.
For springs, tools, drills, taps, reamers; water hardened; Type W1.

BOCHUM BSW10.
M-1331; 1.2 C, W, Cr, bal Fe.
For cutters, bearings; water hardened, wear resistant.

BOCHUM BSW18.
M-1331; 1.2 C, W, Cr, bal Fe.
For cutters, bearings; water hardened, wear resistant.

BOCHUM C58G.
M-1331; 0.25 C, 14.5 Cr, 0.1 max Ni, bal Fe.
For cutlery, valves, surgical and dental instruments; stainless, hardenable.

BOCHUM C68G.
M-1331; 0.25 C, 17 Cr, 1.8 max Ni, bal Fe.
Annealed: 125,000 TS; 95,000 YS; 20 El; 55 RA; 260 Brin.
For pumps, marine hardware, cutlery; corrosion and heat resistant.

BOCHUM CSG.
M-1331; 0.60 C, Cr, V, bal Fe.
For springs, gears, crankshafts; oil hardened, shock resistant.

BOCHUM D7.
M-1331; 0.70 C, 0.25 max Si, 0.25 max Mn, bal Fe.
Heat treated: 175,000 TS; 128,000 YS; 12 El; 37 RA; 355 Brin.
For rails, axes, hammers, crimpers; water hardened; Type W1.

BOCHUM D8.
M-1331; 0.85 C, 0.25 max Si, 0.25 max Mn, bal Fe.
Heat treated: 190,000 TS; 145,000 YS; 10 El; 30 RA; 400 Brin.
For drills, taps, springs, reamers; Type W1; water hardened.

BOCHUM D10.
M-1331; 1.0 C, 0.25 max Si, 0.25 max Mn, bal Fe.
Annealed: 100,000 TS; 53,000 YS; 20 El; 42 RA; 200 Brin.
For springs, cutters, hobs, drills, taps; Type W1; water hardened.

BOCHUM D11.
M-1331; 1.15 C, 0.25 max Si, 0.25 max Mn, bal Fe.
Annealed: 110,000 TS; 56,000 YS; 18 El; 40 RA; 210 Brin.
For springs, cutters, taps, drills, hobs; Type W1; water hardened.

BOCHUM D13.
M-1331; 1.3 C, 0.25 max Si, 0.25 max Mn, bal Fe.
For engravers' tools, blanking and forming dies; Type W1; water hardened.

BOCHUM EK10.
M-1331; 0.10 C, 0.25 Si, 0.37 Mn, bal Fe.
Cold drawn: 72,000 TS; 60,000 YS; 22 El; 58 RA; 145 Brin.
For gears, shafts, fasteners; case hardening steel.

BOCHUM EK15.
M-1331; 0.15 C, 0.25 Si, 0.37 Mn, bal Fe.
Annealed: 70,000 TS; 40,000 YS; 25 El; 60 RA; 145 Brin.
For gears, shafts, fasteners; case hardening steel.

BOCHUMER BC1.
M-1343; 1.0 C, 1.55 Cr, bal Fe.
For bearings, sleeves, liners, blanking dies; water hardened, wear resistant.

BOCHUMER BC2.
M-1343; 0.90 C, 0.8 Cr, 0.30 Mn, bal Fe.
For springs, tools, punches, dies, hammers; Type W1; water hardened.

BOCHUMER BCE SPEZIAL.
M-1343; 1.45 C, 1.4 Cr, 0.6 Mn, bal Fe.
For bearings, liners, sleeves; water hardened, abrasion resistant.

BOCHUMER BCL200.
M-1343; 2.1 C, 11.5 Cr, 0.3 Mn, bal Fe.
For blanking and forming dies, punches; oil or air hardened, nondeforming.

BOCHUMER BCOL155.
M-1343; 1.65 C, 11.5 Cr, 0.3 Mn, bal Fe.
For blanking and forming dies, punches; air hardened, nondeforming.

BOCHUMER BCOV40.
M-1343; 0.45 C, 1.4 Cr, 0.7 Mo, 0.3 V, 0.7 Mn, bal Fe.
For gears, bolts, fasteners, machine tool parts; oil hardened, tough.

BOCHUMER BCS120.
M-1343; 1.25 C, 1.15 Si, 1.2 Cr, 0.7 Mn, bal Fe.
For bearings, sleeves, liners; water or oil hardened, wear resistant.

BOCHUMER BC SPEZIAL.
M-1343; 1.45 C, 0.6 Mn, 1.4 Cr, bal Fe.
For bearings, liners, sleeves; water hardened, abrasion resistant.

BOCHUMER BCSV/G.
M-1343; 0.61 C, 1.18 Cr, 0.1 V, 0.7 Mn, bal Fe.
For springs, gears, crankshafts; oil hardened, shock resistant.

BOCHUMER BCSV/K.
M-1343; 0.61 C, 1.18 Cr, 0.1 V, 0.75 Mn, bal Fe.
For springs, gears, crankshafts; oil hardened, shock resistant.

BOCHUMER BCV115.
M-1343; 1.15 C, 0.65 Cr, 0.1 V, 0.3 Mn, bal Fe.
For cutters, dies, forming dies; water hardened, wear resistant.

BOCHUMER BCVOW.
M-1343; 0.45 C, 0.45 Mo, 1.35 Cr, 0.8 V, 0.45 W, bal Fe.
For cold work tools, upsetters, headers; oil hardened, tough.

BOCHUMER BCWL200.
M-1343; 2.1 C, 11.5 Cr, 0.7 W, bal Fe.
For blanking and forming dies, punches; oil hardened, nondeforming.

BOCHUMER BFS5 1/2.
M-1343; 0.46 C, 1.7 Si, 0.65 Mn, bal Fe.
For springs, punches, crimpers; oil hardened, tough.

BOCHUMER BFS5 1/2+.
M-1343; 0.51 C, 1.7 Si, 0.65 Mn, bal Fe.
For springs, punches, crimpers; oil hardened, tough.

BOCHUMER BFS6.
M-1343; 0.6 C, Si, Mn, bal Fe.
Heat treated: 160,000 TS; 115,000 YS; 12 El; 40 RA; 325 Brin.
For springs, punches, crimpers; oil hardened, tough.

BOCHUMER BMS1.
M-1343; 0.53 C, 0.9 Si, 0.9 Mn, bal Fe.
Heat treated: 160,000 TS; 113,000 YS; 12 El; 40 RA; 320 Brin.
For gears, springs, shafts; water or oil hardened.

BOCHUMER BMV85.
M-1343; 0.9 C, 1.9 Mn, 0.1 V, bal Fe.
For punches, forming dies, crimpers, upsetters; oil hardened, shock resistant.

BOCHUMER BNCO/G.
M-1343; 0.35 C, 0.25 Mo, 1.35 Cr, 3.9 Ni, bal Fe.
For gears, bolts, crankshafts, studs; oil hardened, shock resistant.

BOCHUMER BNCO/K.
M-1343; 0.19 C, 1.25 Cr, 0.2 Mo, 3.75 Ni, bal Fe.
For gears, bolts, cams, studs, camshafts; case hardening steel, shock resistant.

BOCHUMER BVA3.
M-1343; 0.2 C, 0.55 Mo, 1.0 Mn, 0.25 Si, bal Fe.
For gears, bolts, cams, studs, camshafts; case hardening steel, tough.

BOCHUMER BVT40.
M-1343; 0.16 C, 1.05 Cr, 0.45 Mo, 0.65 Mn, bal Fe.
For gears, pinions, machine tool parts; case hardening steel, tough.

BOCHUMER BVT40N.
M-1343; 0.16 C, Cr, Mo, V, bal Fe.
For gears, machine tool parts; case hardening steel, tough.

BOCHUMER BVT50.
M-1343; 0.24 C, 1.25 Cr, 0.45 Mo, bal Fe.
For gears, pinions, bolts, fasteners, shafts; oil hardened, tough.

BOCHUMER BVT50N.
M-1343; 0.22 C, Cr, Mo, V, bal Fe.
For bolts, cams, camshafts, fasteners; case hardening steel, tough.

BOCHUMER BVT60.
M-1343; 0.24 C, 1.35 Cr, 0.55 Mo, 0.2 V, bal Fe.
For bolts, machine tool parts, shafts, studs; oil hardened, tough.

BOCHUMER BVT60N.
M-1343; 0.22 C, 1.15 Cr, 0.25 Mo, 0.2 V, bal Fe.
For bolts, machine tool parts, shafts, studs; oil hardened, tough.

BOCHUMER BVT90.
M-1343; 0.18 C, Cr, Mo, V, bal Fe.
For gears, bolts, machine tool parts; case hardening steel, tough.

BOCHUMER BVT125 EXTRA.
M-1343; 1.3 C, 4.75 W, 0.20 max Cr, bal Fe
For cutters, form cutters, engravers' tools; water hardened.

BOCHUMER BVT130.
M-1343; 0.20 C, 1.25 Cr, 0.22 Mo, 0.55 Mn, bal Fe.
For gears, bolts, machine tool parts; case hardening steel, tough.

BOCHUMER BVT130V.
M-1343; 0.22 C, 12 Cr, 1 Mo, 0.4 Ni, 0.3 V, bal Fe.
Annealed: 95,000 TS; 40,000 YS; 25 El; 55 RA; B 92 Rock.
Heat Treated: 240,000 TS; 20,000 YS; 10 El; 25 RA; C 50 Rock.
For cutlery, surgical instruments, knives, shafts, gears, scissors.
Corrosion resistant, hardenable.

BOCHUMER BVT130VSO.
M-1343; 0.20 C, 12 Cr, 1 Mo, 0.3 V, 0.5 W, bal Fe.
Annealed: 95,000 TS; 40,000 YS; 25 El; 55 RA; B 92 Rock.
Heat Treated: 240,000 TS; 200,000 YS; 10 El; 25 RA; C 50 Rock.
For gears, shafts, cutlery, hardware, surgical instruments.
Corrosion resistant, hardenable.

BOCHUMER BWC100.
M-1343; 1.0 C, 0.9 Mn, 1 Cr, 1.15 W, bal Fe.
For cutters, dies, shears, upsetters; oil hardened, tough.

BOCHUMER BWCO SUPRA.
M-1343; 0.65 C, 3.75 Cr, 0.85 Mo, 0.7 V, 8.5 W, bal Fe.
For lathe and planer tools, reamers, taps, drills; high speed steel.

BOCHUMER BWCV25 EXTRA.
M-1343; 0.30 C, Cr, V, W, bal Fe.
For upsetters, crimmers, header dies; oil hardened, tough.

BOCHUMER BWCV25 SPEZIAL.
M-1343; 0.30 C, 1.0 Si, 1.1 Cr, 0.18 V, 3.75 W, bal Fe.
For upsetters, riveters, header dies; oil hardened, tough.

BOCHUMER BWCV25 SUPRA.
M-1343; 0.30 C, 2.65 Cr, 0.35 V, 8.5 W, bal Fe.
For extrusion press dies and rams, upsetters; hot work steel, oil hardened.

BOCHUMER BWCV30 SPEZIAL.
M-1343; 0.35 C, 1.05 Cr, 0.18 V, 1.85 W, bal Fe.
For header dies, upsetters, crimpers; oil hardened, tough.

BOCHUMER BWCV40 SPEZIAL.
M-1343; 0.45 C, 1.05 Cr, 0.2 V, 1.85 W, bal Fe.
For header dies, upsetters, crimpers; oil hardened, tough.

BOCHUMER BWCV50 SPEZIAL.
M-1343; 0.55C 1.05 Cr, 0.18 V, 1.85 W, bal Fe.
For header dies, upsetters, crimpers; oil hardened, tough.

BOCHUMER BWV115.
M-1343; 1.2 C, 0.2 Cr, 0.1 V, 1.0 W, bal Fe.
For cutters, bearings, crimpers; water hardened, wear resistant.

BOCHUMER CBV.
M-1343; 0.15 C, 0.65 Cr, 0.50 Mn, 0.25 Si, bal Fe.
For gears, bolts, machine tool parts; case hardened.

BOCHUMER CBV1.
M-1343; 0.33 C, 1.0 Cr, 0.65 Mn, bal Fe.
For gears, bolts, fasteners; oil hardened, tough.

BOCHUMER CBV2.
M-1343; 0.37 C, 0.65 Mn, 1.0 Cr, bal Fe.
For gears, bolts, fasteners; oil hardened, tough.

BOCHUMER CBV/H.
M-1343; 0.36 C, 0.65 Mn, 1.55 Cr, bal Fe.
For gears, bolts, fasteners; oil hardened, tough.

BOCHUMER CBVO/D.
M-1343; 0.30 C, 0.55 Mn, 1.5 Cr, 0.18 Mo, bal Fe.
For gears, bolts, machine tool parts; oil hardened, tough.

BOCHUMER CBVO/G.
M-1343; 0.40 C, Cr, Mn Mo, V, bal Fe.
For gears, bolts, machine tool parts; oil hardened, tough.

BOCHUMER CBVO/H.
M-1343; 0.50 C, 0.65 Mn, 1.0 Cr, 0.2 Mo, bal Fe.
For gears, bolts, machine tool parts; oil hardened, tough.

BOCHUMER CBVO/M.
M-1343; 0.25 C, 0.65 Mn, 1.0 Cr, 0.2 Mo, bal Fe.
For gears, bolts, machine tool parts; oil hardened, tough.

BOCHUMER CBVO/MD.
M-1343; 0.21 C, 0.65 Mn, 0.85 Cr, 0.25 Mo, bal Fe.
For gears, bolts, machine tool parts; oil hardened, tough.

BOCHUMER CBVO/TM.
M-1343; 0.24 C, 0.55 Mn, 1.15 Cr, 0.25 Mo, bal Fe.
For gears, bolts, machine tool parts; oil hardened, tough.

BOCHUMER CBVO/V.
M-1343; 0.30 C, 0.55 Mn, 2.5 Cr, 0.2 Mo, 0.15 V, bal Fe.
For gears, bolts, machine tool parts; oil hardened, tough.

BOCHUMER CBVO/W.
M-1343; 0.16 C, 0.65 Mn, 1.05 Cr, 0.20 Mo, bal Fe.
For gears, cams, camshafts; case hardened.

BOCHUMER CBVO/Z.
M-1343; 0.42 C, 0.65 Mn, 1.05 Cr, 0.20 Mo, bal Fe.
For gears, bolts, crankshafts, fasteners; oil hardened, tough.

BOCHUMER CBV/V.
M-1343; 0.22 C, Cr, V, bal Fe.
For gears, bolts, machine tool parts; oil hardened, tough.

BOCHUMER CBVZ.
M-1343; 0.41 C, 1.0 Cr, 0.65 Mn, 0.25 Si, bal Fe.
For gears, bolts, machine tool parts; oil hardened, tough.

BOCHUMER CMBV/H.
M-1343; 0.20 C, 1.25 Mn, 1.15 Cr, bal Fe.
For gears, cams, camshafts, fasteners; case hardened, tough.

BOCHUMER CMBV/W.
M-1343; 0.16 C, 1.15 Mn, 0.95 Cr, bal Fe.
For gears, cams, camshafts, fasteners; case hardened, tough.

BOCHUMER CNBV1W.
M-1343; 0.15 C, 1.55 Cr, 1.55 Ni, bal Fe.
For gears, bolts, machine tool parts; case hardened, tough.

BOCHUMER CNBV2W.
M-1343; 0.18 C, 2 Cr, 2 Ni, 0.5 Mn, bal Fe.
For gears, bolts, machine tool parts; case hardened, tough.

BOCHUMER CNBVO/M.
M-1343; 0.25 C, 0.50 Mn, 1.0 Cr, 0.18 Mo, 1.5 Ni, bal Fe.
For gears, bolts, fasteners; oil hardened, shock resistant.

BOCHUMER CRVO.
M-1343; 0.33 C, 0.65 Mn, 1.0 Cr, 0.2 Mo, bal Fe.
For gears, bolts, machine tool parts; oil hardened, tough.

BOCHUMER CRW.
M-1343; 0.85 C, 1.75 Cr, 0.35 Mn, bal Fe.
For bearings, bushings, liners, sleeves; water or oil hardened, wear resistant.

BOCHUMER CV23.
M-1343; 0.22 C, 0.65 Mn, 1.1 Cr, 0.2 V, bal Fe.
For gears, pinions, bolts, shafts; oil hardened, tough.

BOCHUMER CV40/H.
M-1343; 0.42 C, Cr, V, bal Fe.
For gears, bolts, springs, crankshafts; oil hardened, tough.

BOCHUMER CV48.
M-1343; 0.50 C, 1.0 Cr, 0.09 V, bal Fe.
For gears, bolts, springs, studs; oil hardened, shock resistant.

BOCHUMER CV58.
M-1343; 0.50 C, 0.85 Mn, 1.0 Cr, 0.1 V, bal Fe.
For gears, bolts, springs, crankshafts; oil hardened, shock resistant.

BOCHUMER EMC.
M-1331; 0.20 C, 1.25 Mn, 1.15 Cr, bal Fe.
For camshafts, cams, gears; case hardened.

BOCHUMER ES5.
M-1343; 0.85 C, 0.25 Si, 0.25 Mn, bal Fe.
Heat treated; 188,000 TS; 143,000 YS; 12 El; 35 RA; 390 Brin.
For springs, drills, taps, reamers; Type W1; water hardened.

BOCHUMER ES6.
M-1343; 0.70 C, 0.25 Si, 0.25 Mn, bal Fe.
Heat treated: 175,000 TS; 128,000 YS; 12 El; 37 RA; 355 Brin.
For springs, rails, punches, hammers; Type W1; water hardened.

BOCHUMER EVL.
M-1343; C, Cr, Ni, bal Fe.
For machine tool parts; oil hardened, shock resistant.

BOCHUMER FCK.
M-1343; C, Cr, bal Fe.
For machine tool parts; oil hardened.

BOCHUMER F EXTRA.
M-1343; C, alloy, bal Fe.
For machine tool parts; oil hardened.

BOCHUMER FRS.
M-1343; 0.67 C, 1.3 Si, 0.5 Cr, 0.5 Mn, bal Fe.
For punches, crimpers, upsetters; oil or water hardened.

BOCHUMER GKS.
M-1343; 0.15 C, Si, 1.1 Mn, bal Fe.
For machine tool parts; case hardened.

BOCHUMER HMFB EXTRA.
M-1343; 0.55 C, Ni, Cr, Mo, V, bal Fe.
For forging dies, punches, upsetters, shears; oil hardened, shock resistant.

BOCHUMER HMFB SPEZIAL.
M-1343; 0.55 C, 0.70 Cr, 0.18 Mo, 1.65 Ni, 0.1 V, bal Fe.
For forging and header dies, punches; oil hardened, shock resistant.

BOCHUMER HMS 35.
M-1343; 0.10 C, 0.25 Si, 0.37 Mn, bal Fe.
Annealed: 64,000 TS; 48,000 YS; 28 El; 65 RA; 135 Brin.
For plastic mold dies, rivets, screws, fan blades; case hardened.

BOCHUMER HMS 40.
M-1343; 0.15 C, 0.25 Si, 0.37 Mn, bal Fe.
Annealed: 70,000 TS; 55,000 YS; 25 El; 60 RA; 145 Brin.
For screws, bolts, fan blades, bushings, gears; case hardened.

BOCHUMER HMS 45.
M-1343; 0.22 C, 0.25 Si, 0.45 Mn, bal Fe.
Annealed: 73,000 TS; 61,000 YS; 22 El; 58 RA; 150 Brin.
For gears, bolts, fan blades, camshafts; water hardened.

BOCHUMER HMS 55.
M-1343; 0.35 C, 0.25 Si, 0.55 Mn, bal Fe.
Hot rolled: 85,000 TS; 54,000 YS; 30 El; 53 RA; 185 Brin.
For gears, shafts, axles, bolts, screws; water hardened.

BOCHUMER HMS 65.
M-1343; 0.45 C, 0.25 Si, 0.65 Mn, bal Fe.
Hot rolled: 98,000 TS; 59,000 YS; 24 El; 45 Ra; 212 Brin.
For axles, gears, bolts, tie rods, bushings; water hardened.

BOCHUMER HMS 80.
M-1343; 0.60 C, 0.25 Si, 0.65 Mn, bal Fe.
Heat treated: 160,000-115,000 TS; 113,000-77,000 YS; 12-23 El; 40-54 RA; 321-229 Brin.
For wheels, die blocks, girders, shafts; water hardened.

BOCHUMER HMS 85.
M-1343; 0.60 C, 0.25 Si, 0.65 Mn, bal Fe.
Heat treated: 115,000-160,000 TS; 77,000-113,000 YS; 23-12 El; 54-40 RA; 229-331 Brin.
For wheels, die blocks, rails, girders; water hardened.

BOCHUMER HSS.
M-1343; C, alloy, bal Fe.
For machine tool parts, gears; oil hardened, tough.

BOCHUMER HSS EXTRA.
M-1343; C, alloy bal Fe.
For machine tool parts, gears; oil hardened, tough.

BOCHUMER HSS SPEZIAL.
M-1343; C, alloy, bal Fe.
For machine tool parts, gears; oil hardened, tough.

BOCHUMER KMC 17E.
M-1343; 0.20 C, 1.15 Cr, 1.25 Mn, bal Fe.
For camshafts, cams, gears; case hardened.

BOCHUMER KSS/CV.
M-1343; C, Cr, V, bal Fe.
For saws, saw blades; oil hardened.

BOCHUMER LFS.
M-1343; C, Mn, Si, bal Fe.
For springs; oil hardened.

BOCHUMER LKLS.
M-1343; C, Si, Mn, bal Fe.
For machine tool parts.

BOCHUMER LKRS.
M-1343; C, Cr, bal Fe.
For machine tool parts.

BOCHUMER M14.
M-1343; 0.17 C, 0.30 Si, 1.05 Mn, bal Fe.
For gears, cams, camshafts; case hardened.

BOCHUMER M17.
M-1343; 0.19 C, 0.50 Si, 1.15 Mn, bal Fe.
For gears, cams, camshafts; case hardened.

BOCHUMER MBV.
M-1343; 0.4 C, 0.37 Si, 0.95 Mn, bal Fe.
Hot rolled: 90,000 TS; 58,000 YS; 17 El; 50 RA; 200 Brin.
For gears, shafts, machine tool parts; water hardened.

BOCHUMER MBV/M.
M-1343; 0.30 C, 0.25 Si, 1.35 Mn, bal Fe.
For gears, machine tool parts, shafts; water hardened.

BOCHUMER MCV 24.
M-1343; 0.27 C, Mn, Cr, V, bal Fe.
For gears, bolts, machine tool parts; oil hardened, shock resistant.

BOCHUMER MINIMUM R.
M-1343; C, alloy, bal Fe.
For machine tool parts; oil hardened.

BOCHUMER MNFS.
M-1343; 0.46 C, Mn, Si, bal Fe.
Hot rolled: 98,000 TS; 60,000 YS; 24 El; 45 RA; 215 Brin.
For gears, shafts, machine tool parts; water hardened.

BOCHUMER MS.
M-1343; 0.55 C, 1.05 Cr, 0.18 V, 1.85 W, bal Fe.
For header dies, upsetters, dies, crimpers; oil hardened, tough.

BOCHUMER MSBV.
M-1343; 0.37 C, 1.25 Si, 1.25 Mn, bal Fe.
For gears, shafts, fasteners; oil or water hardened.

BOCHUMER MSS.
M-1343; C, alloy, bal Fe.
For machine tool parts; oil hardened.

BOCHUMER MSS/CV.
M-1343; C, Cr, V, bal Fe.
For saws; oil hardened.

BOCHUMER MSS EXTRA.
M-1343; C, 2 W, bal Fe.
For saws; oil hardened.

BOCHUMER MSS SPEZIAL.
M-1343; C, 1 W, bal Fe.
For saws; oil hardened.

BOCHUMER MV38.
M-1343; 0.42 C, 0.25 Si, 1.75 Mn, 0.1 V, bal Fe.
For gears, bolts, crankshafts, dies; oil hardened, tough.

BOCHUMER NFS.
M-1343; C, alloy, bal Fe.
For files; oil hardened.

BOCHUMER NZF S/H.
M-1343; 0.65 C, 1.7 Si, 0.70 Mn, bal Fe.
For springs, upsetters, punches; oil hardened, tough.

BOCHUMER NZF S/M.
M-1343; 0.51 C, 1.7 Si, 0.65 Mn, bal Fe.
For springs, upsetters, punches; oil hardened, tough.

BOCHUMER NZF S/O.
M-1343; 0.55 C, 1.7 Si, 0.70 Mn, bal Fe.
For springs, punches, upsetters; oil hardened, tough.

BOCHUMER NZF S/W.
M-1343; 0.46 C, 1.7 Si, 0.65 Mn, bal Fe.
For springs, punches, rivet sets; oil hardened, tough.

BOCHUMER P400.
M-1343; C, alloy, bal Fe.
For machine tool parts; oil hardened.

BOCHUMER P600.
M-1343; C, alloy, bal Fe.
For machine tool parts; oil hardened.

BOCHUMER PFS.
M-1343; C, alloy, bal Fe.
For machine tool parts; oil hardened.

BOCHUMER PFS/1.
M-1343; C, alloy, bal Fe.
For machine tool parts; oil hardened.

BOCHUMER PKL/V-KPV.
M-1343; 1.0 C, 0.25 Mn, 0.10 V, bal Fe.
For header dies, forming rolls; Type W2; water hardened.

BOCHUMER PNC45.
M-1343; 0.45 C, 1.05 Cr, 3.25 Ni, bal Fe.
For gears, bolts, crankshafts, forging dies; oil hardened, shock resistant

BOCHUMER RKS.
M-1343; C, alloy, bal Fe.
For machine tool parts; oil hardened, tough.

BOCHUMER RKS SPEZIAL.
M-1343; C, alloy, bal Fe.
For machine tool parts; oil hardened, tough.

BOCHUMER SAFS.
M-1343; 0.70 C, 1.7 Si, 0.70 Mn, bal Fe.
Heat treated: 340,000 TS; 280,000 YS; 5 El; 20 RA; 600 Brin.
For springs; oil hardened, shock resistant.

BOCHUMER SC 60.
M-1343; 0.67 C, 1.3 Si, 0.5 Cr, 0.5 Mn, bal Fe.
Heat treated: 320,000 TS; 250,000 YS; 8 El; 25 RA; 550 Brin.
For springs; oil hardened, shock resistant.

BOCHUMER SFS60.
M-1343; 0.6 C, bal Fe.
Heat treated: 160,000 TS; 113,000 YS; 12 El; 40 RA; 320 Brin.
For wheels, die blocks, rails, springs; water hardened.

BOCHUMER SFS65.
M-1343; 0.69 C, bal Fe.
Heat treated: 175,000 TS; 128,000 YS; 12 El; 37 RA; 350 Brin.
For springs, clutch discs, girders, rails; water hardened.

BOCHUMER SFS70.
M-1343; 0.72 C, bal Fe.
Heat treated: 175,000 TS; 128,000 YS; 12 El; 37 RA; 350 Brin.
For springs, clutch discs, girders, rails; Type W1; water hardened.

BOCHUMER SMFS/O.
M-1343; 0.60 C, Si, Mn, bal Fe.
Heat treated: 160,000 TS; 113,000 YS; 12 El; 40 RA; 320 Brin.
For wheels, die blocks, rails, springs; water hardened.

BOCHUMER SPEZIAL REZISTANCE STAHL.
M-1343; 0.30 C, 1.0 Si, 1.1 Cr, 0.18 V, 3.75 W, bal Fe.
For extrusion rams, dies; hot work steel, oil hardened.

BOCHUMER SPEZIAL W.
M-1343; 0.30 C, 2.65 Cr, 0.35 V, 8.5 W, bal Fe.
For extrusion rams, dies; hot work steel, oil hardened.

BOCHUMER SSCV.
M-1343; C, V, bal Fe.
For machine tool parts; water hardened.

BOCHUMER SSS.
M-1343; C, bal Fe.
For machine tool parts; water hardened.

BOCHUMER SSS1.
M-1343; 0.70 C, 1.7 Si, 0.70 Mn, bal Fe.
For springs, punches, upsetters, chisels; oil hardened, shock resistant.

BOCHUMER SSS1 EXTRA.
M-1343; C, W, V, bal Fe.
For heading dies, cutters, liners; oil or water hardened.

BOCHUMER SSSK.
M-1343; 0.50 C, 1.0 Cr, 0.09 V, 0.85 Mn, bal Fe
For springs, gears, fasteners, crankshafts; oil hardened, shock resistant.

BOCHUMER TSE.
M-1343; C, Cr, Mn, bal Fe.
For machine tool parts; water hardened.

BOCHUMER TSV.
M-1343; C, bal Fe.
For machine tool parts; water hardened.

BOCHUMER UFS.
M-1343; C, bal Fe.
For machine tool parts; water hardened.

BOCHUMER UGS.
M-1343; C, bal Fe.
For machine tool parts; water hardened.

BOCHUMER UMS45.
M-1343; 0.22 C, 0.45 Mn, 0.25 Si, bal Fe.
Annealed: 73,000 TS; 41,000 YS; 22 El; 58 RA; 140 Brin.
For screws, bolts, gears, shafts, rivets; case hardened.

BOCHUMER WFS SPEZIAL.
M-1343; 0.50 C, Cr, Mo, V, bal Fe.
For header and forming dies, punches; oil hardened, tough.

BOCHUMER WFS SUPRA.
M-1343; 0.30 C, 2.35 Cr, 0.6 V, 4.25 W, bal Fe.
For extrusion dies, rams, liners, punches; oil hardened, tough.

BOCHUMER WK45.
M-1343; 0.45 C, 0.25-0.50 Si, 0.30-0.80 Mn, bal Fe.
Hot rolled: 98,000 TS; 59,000 YS; 24 El; 45 RA; 212 Brin.
For axles, gears, bolts, fasteners, crankshafts; water hardened.

BOCHUMER WK60.
M-1343; 0.60 C, 0.25-0.50 Si, 0.30-0.80 Mn, bal Fe.
Heat treated: 160,000 TS; 113,000 YS; 12 El; 40 RA; 320 Brin.
For wheels, die blocks, springs, girders, rails; water hardened.

BOCHUMER ZFS.
M-1343; 0.70 C, 1.7 Si, 0.70 Mn, bal Fe.
For springs, punches, upsetters; oil hardened, shock resistant.

BOCHUMER ZMCV.
M-1343; 1.4 C, 0.30 Mn, 0.30 Cr, 0.1 V, bal Fe.
For engravers tools, cutters, bearings; water hardened, wear resistant.

BOCHUM EU.
M-1331; 0.15 C, 0.25-0.50 Si, 0.30-0.80 Mn, bal Fe.
Annealed: 70,000 TS; 40,000 YS; 25 El; 60 RA; 145 Brin.
For machine tool parts, gears, fasteners; case hardened.

BOCHUM K55 SPEZIAL.
M-1331; 0.55 C, 0.1-0.4 Si, 0.5-0.7 Mn, bal Fe.
Annealed: 100,000 TS; 55,000 YS; 15 El; 20 RA; 180 Brin.
For axles, gears, bolts, shafts; water hardened.

BOCHUM K65.
M-1331; 0.60 C, 0.10-0.40 Si, 0.50-0.70 Mn, bal Fe.
Heat treated: 160,000-115,000 TS; 113,000-77,000 YS; 12-23 El; 40-54 320-230 Brin.
For wheels, die blocks, girders, clutch discs; water hardened.

BOCHUM KPN.
M-1331; 0.50 C, 1.05 Cr, 3.25 Ni, 0.5 Mn, bal Fe.
For gears, bolts, crankshafts, axles; oil hardened, shock resistant.

BOCHUM M80.
M-1331; 0.16 C, 0.95 Cr, 1.15 Mn, 0.25 Si, bal Fe.
For gears, bolts, camshafts, cams; case hardened, tough.

BOCHUM M90.
M-1331; 0.20 C, 1.25 Mn, 1.15 Cr, bal Fe.
For gears, bolts, camshafts, cams, case hardened, tough.

BOCHUM M100.
M-1331; 0.22 C, 1.25 Cr, 1.30 Mn, bal Fe.
For gears, bolts, camshafts, cams; water or oil hardened.

BOCHUM MMB3.
M-1331; C, alloy, bal Fe.
For machine tool parts; oil hardened.

BOCHUM MN18G.
M-1331; 1.3 C, Mn, Cr, bal Fe.
For cutters, wear plates; oil or water hardened.

BOCHUM MNH(G).
M-1331; 1.2 C, 12.5 Mn, 0.4 Si, bal Fe.
For tracks, frogs, dipper teeth shovels; wear and abrasion resistant.

BOCHUM MNHW.
M-1331; 1.2 C, 0.4 Si, 12.5 Mn, bal Fe.
For wear plates, dipper teeth, shovels; wear and abrasion resistant.

BOCHUM PKS10H.
M-1307; 1.1 C, 0.25 max Si, 0.25 max Mn, bal Fe.
Heat treated: 200,000 TS; 130,000 YS; 8 El; 25 RA; 400 Brin.
For springs, taps, drills, reamers, cutters, hobs; Type W1; water hardened.

BOCHUM PKS EXTRA.
M-1307; 1.0 C, 0.1 V, 0.25 Mn, bal Fe.
Heat treated: 200,000 TS; 130,000 YS; 8 El; 25 RA; 400 Brin.
For reamers, drills, drawing and stamping dies; oil or water hardened, wear resistant.

BOCHUM PKS EXTRA SPEZIAL.
M-1331; 1.0 C, Cr, V, bal Fe.
Heat treated: 200,000 TS; 130,000 YS; 8 El; 25 RA; 400 Brin.
For reamers, drills, drawing dies; oil or water hardened.

BOCHUM QKS10W.
M-1307; 1.0 C, 0.25 max Si, 0.25 max Mn, bal Fe.
Heat treated: 200,000 TS; 130,000 YS; 8 El; 25 RA; 400 Brin.
For springs, taps, drills, reamers, cutters, hobs; Type W1; water hardened.

BOCHUM RR8.17.
M-1331; 0.08 C, 17 Cr, 0.4 Si, bal Fe.
Annealed: 80,000 TS; 50,000 YS; 25 El; 50 RA; 150 Brin.
For oil refinery equipment, sinks, soot blowers; Type 430; stainless.

BOCHUM RR8.17E.
M-1331; 0.1 max C, 17.5 Cr, Ti = 7 X C, bal Fe.
Annealed: 80,000 TS; 50,000 YS; 25 El; 50 RA; 150 Brin.
For welded oil refinery equipment; Type 430 Ti; stainless.

BOCHUM RR10.13.
M-1331; 0.12 max C, 0.4 Si, 13 Cr, bal Fe.
Annealed: 75,000 TS; 40,000 YS; 35 El; 70 RA; 155 Brin.
For turbine blades, cutlery, valves, surgical instruments; Type 410; stainless.

BOCHUM RR10.13MO.
M-1331; 0.12 max C, 13 Cr, 0.2 Mo, bal Fe.
Annealed: 75,000 TS; 40,000 YS; 35 El; 70 RA; 155 Brin.
For turbine blades, cutlery, valves, surgical instruments; Type 410; stainless.

BOCHUM RR10.13T.
M-1331; 0.12 max C, 13 Cr, 0.4 Si, bal Fe.
Annealed: 75,000 TS; 40,000 YS; 35 El; 70 RA; 155 Brin.
For turbine blades, cutlery, valves, knives; Type 410; stainless.

BOCHUM RR10.17MOS.
M-1331; 0.12 C, 16.5 Cr, 0.25 Mo, 0.2 S, bal Fe.
Annealed: 80,000 TS; 50,000 YS; 20 El; 45 RA; 150 Brin.
For screw machine products; Type 430F; stainless, free-cutting.

BOCHUM RR10.17S.
M-1331; 0.10 C, 20 Ni, 25 Cr, Mo, Si, Cu, bal Fe.
For chemical plant equipment, tanks; stainless, austenitic.

BOCHUM RR15.13.
M-1331; 0.15 max C, 0.4 Si, 13 Cr, bal Fe.
Annealed: 75,000 TS; 40,000 YS; 35 El; 70 RA; 155 Brin.
For turbine blades, cutlery, valves, surgical instruments; Type 410; stainless.

BOCHUM RR20.13.
M-1331; 0.20 C, 0.4 Si, 13 Cr, bal Fe.
Annealed: 95,000 TS; 50,000 YS; 25 El; 55 RA; 195 Brin.
For turbine blades, cutlery, valves; Type 420; stainless.

BOCHUM RR20.13MO.
M-1331; 0.20 C, 13 Cr, 1.15 Mo, bal Fe.
Annealed 95,000 TS; 50,000 YS; 25 El; 55 RA; 195 Brin.
For turbine blades, valves, cutlery; Type 420 Mo; stainless.

BOCHUM RR20.13NI.
M-1331; 0.20 C, 13 Cr, Ni, bal Fe.
For valves, cutlery, surgical and dental instruments; corrosion resistant.

BOCHUM RR22.17NI.
M-1331; 0.22 C, 17 Cr, 1.5 Ni, 0.4 Si, bal Fe.
Annealed: 125,000 TS; 95,000 YS; 20 El; 55 RA; 260 Brin.
For pumps, marine hardware, valves; Type 431; heat and corrosion resistant.

BOCHUM RR35.17MO.
M-1331; 0.35 C, 16.5 Cr, 1.15 Mo, bal Fe.
For chemical plant equipment; corrosion resistant.

BOCHUM RR40.13.
M-1331; 0.4 C, 0.4 Si, 13 Cr, bal Fe.
Annealed: 100,000 TS; 55,000 YS; 20 El; 50 RA; 200 Brin.
For valves, cutlery, pump parts; Type 420; stainless.

BOCHUM RR45.13MO.
M-1331; C, alloy, bal Fe.
For machine tool parts; oil hardened, tough.

BOCHUM RR90.17MOV.
M-1331; 0.90 C, 18 Cr, 1.15 Mo, 1.0 V, bal Fe.
For bearings, cutlery, valves; oil hardened, wear and corrosion resistant.

BOCHUM U35.
M-1331; 0.35 C, 0.40 Si, 0.60 Mn, bal Fe.
Hot rolled: 85,000 TS; 54,000 YS; 30 El; 53 RA; 185 Brin.
For gears, bolts, axles, shafts, fasteners; water hardened.

BOCHUM U45.
M-1331; 0.45 C, 0.40 Si, 0.60 Mn, bal Fe.
Hot rolled: 98,000 TS; 59,000 YS; 24 El; 45 RA; 212 Brin.
For axles, gears, bolts, crankshafts; water hardened.

BOCHUM U60.
M-1331; 0.60 C, 0.40 Si, 0.60 Mn, bal Fe.
Heat treated: 160,000 TS; 113,000 YS; 12 El; 40 RA; 320 Brin.
For wheels, die blocks, girders, springs, rails; water hardened.

BOCHUM U75.
M-1331; 0.75 C, 0.40 Si, 0.60 Mn, bal Fe.
Heat treated: 180,000 TS; 135,000 YS; 12 El; 36 RA; 375 Brin.
For springs, tools, hammers, rails, girders, Type W1; water hardened.

BOCHUM U90.
M-1331; 0.90 C, 0.40 Si, 0.60 Mn, bal Fe.
Heat treated: 190,000 TS; 145,000 YS; 10 El; 30 RA; 400 Brin.
For springs, taps, reamers, drills; Type W1; water hardened.

BOCHUM VK22.
M-1331; 0.22 C, 0.25 Si, 0.45 Mn, bal Fe.
Annealed: 78,000 TS; 40,000 YS; 22 El; 58 RA; 140 Brin.
For screws, bolts, gears; water hardened.

BOCHUM VK35.
M-1331; 0.35 C, 0.25 Si, 0.45 Mn, bal Fe.
Hot rolled: 85,000 TS; 54,000 YS; 30 El; 53 RA; 185 Brin.
For gears, shafts, axles, bolts; water hardened.

BOCHUM VK45.
M-1331; 0.45 C, 0.25 Si, 0.65 Mn, bal Fe.
Hot rolled: 98,000 TS; 58,000 YS; 24 El; 45 RA; 212 Brin.
For gears, bolts, axles, shafts, crankpins; water hardened.

BOCHUM VK53.
M-1331; 0.53 C, 0.25 Si, 0.65 Mn, bal Fe.
Annealed: 96,000 TS; 52,000 YS; 16 El; 23 RA; 170 Brin.
For gears, axles, bolts, shafts; water hardened.

BOCHUM VK60.
M-1331; 0.61 C, 0.25 Si, 0.65 Mn, bal Fe.
Heat treated: 160,000 TS; 115,000 YS; 12 El; 40 RA; 325 Brin.
For wheels, die blocks, rails, girders; water hardened.

BOCHUM W3G.
M-1331; 0.90 C, Cr, bal Fe.
For bearings, liners, cutters; oil or water hardened, wear resistant.

BODVAR 1.
M-111; 1.0 C, 3.6 Cr, 0.5 W, 2 Co, bal Fe.
For permanent magnets.

BODVAR 2.
M-111; 0.95 C, 5.5 Cr, 0.5 W, 3.25 Co, bal Fe.
For permanent magnets.

BODYRITE SOLDER.
M-815; Pb, bal Sn.
For solder; for auto bodies.

BOFORS 2R27.
M-1184; 0.10 max C, 13.5 Cr, 0.3 max Ni, bal Fe.
Annealed: 71,100 TS; 42,700 YS; 160 Brin.
For corrosion resistant parts; ferritic, corrosion resistant.

BOFORS 2R29.
M-1184; 0.09 C, 17 Cr, bal Fe.
Annealed: 80,000 TS; 50,000 YS; 25 El; 60 Brin.
For household articles, meat hooks, furnace grates, furnace parts.
Type 430 stainless, non-hardenable.

BOFORS 2R37.
M-1184; 0.3 C, 13 Cr, bal Fe.
Annealed: 75,000 TS; 48,000 YS; 26 El; 150 Brin.
Hardened: 250,000 TS; 215,000 YS; 8 El; C 52 Rock.
For chemical and oil refinery equipment, surgical instruments, marine hardware.
Corrosion resistant steel, Type 420, stainless, hardenable.

BOFORS 2R47.
M-1184; 0.20 C, 13 Cr, 0.3 max Ni, bal Fe.
Hardened: 106,700 TS; 92,500 YS; 240 Brin.
For corrosion resistant parts; hardenable martensitic.

BOFORS 2R57.
M-1184; 0.30 C, 13 Cr, 0.4 Ni, bal Fe.
Annealed: 78,000 TS; 50,000 YS; 25 El; 155 Brin.
Hardened: 145,000 TS; 117,000 YS; 25 El; 280 Brin.
For chemical and oil refinery equipment, surgical instruments, gears, shafts.
Corrosion and heat resistant. Martensitic.

BOFORS 2R77.
M-1184; 0.35 C, 13.5 Cr, 0.3 max Ni, bal Fe.
Hardened: 135,000 TS; 107,000 YS; 300 Brin.
For corrosion resistant parts; hardenable, martensitic.

BOFORS 2R107.
M-1184; 0.55 C, 14 Cr, bal Fe.
Annealed: 110,000 TS; 72,000 YS; 20 El; 240 Brin.
Hardened: 156,000 TS; 128,000 YS; 10 El; 325 Brin.
For edge tools, instruments, springs, intake valves, cutlery, surgical instruments.
AISI Type 420, hardenable, corrosion resistant.

BOFORS 2RA27.
M-1184; 0.06 C, 11.5-14.5 Cr, 0.1-0.3 Al, bal Fe.
Annealed: 70,000 TS; 35,000 YS; 30 El; 60 RA; 160 Brin.
Cold Drawn: 85,000 TS; 70,000 YS; 20 El; 60 RA; 185 Brin.
For heat treating equipment, oil refinery equipment.
Type 405 corrosion resistant steel.

BOFORS 2RL2.
M-1184; 0.10 C, 13 Cr, 0.2 S, bal Fe.
Annealed: 72,000 TS; 35,000 YS; 30 El; 155 Brin.
Hardened: 100,000 TS; 78,000 YS; 25 El; 210 Brin.
For screw, bolts, fasteners, shafts, screw machine products.
Type 416 stainless, free-cutting.

BOFORS 2RM2.
M-1184; 0.08 C, 13 Cr, 6 Ni, bal Fe.
Heat Treated Casting: 106,000-142,000 TS; 74,000 min YS; 10 min El; 230-290 Brin.
For water turbine castings, turbine blades and wheels, ship propellers, feed screws for the cellulose industry. Hardenable, stainless cast alloy.

BOFORS 2RO26.
M-1184; 0.10 C, 12 Cr, 0.5 Ni, 0.5 Mo, bal Fe.
Annealed: 78,000 TS; 50,000 YS; 27 El; 165 Brin.
Hardened: 107,000 TS; 85,000 YS; 22 El; 220 Brin.
For tableware, knives, turbine parts, machinery components.
Resistant to stress corrosion cracking.
Martensitic, hardenable.

BOFORS 2RO27.
M-1184; 0.10 C, 13.5 Cr, 1.2 Mo, bal Fe.
Annealed: 78,000 TS; 50,000 YS; 25 El; 155 Brin.
Hardened: 135,000 TS; 107,000 YS; 25 El; 280 Brin.
For chemical and oil refinery equipment.
Martensitic, corrosion and heat resistant.

BOFORS 2RO46.
M-1184; 0.20 C, 12 Cr, 1.2 Mo, bal Fe.
Hardened: 142,000 TS; 121,000 YS; 17 El; 300 Brin.
For steam turbine blading, compressor parts, gas turbine components, propeller shafts.
Corrosion and creep resistant, hardenable.

BOFORS 2RO-189.
M-1184; 0.85 C, 17 Cr, 0.3 max Ni, 0.6 Mo, bal Fe.
Hardened: 590 Brin.
For corrosion resistant edge tools; martensitic, corrosion resistant, hardenable.

BOFORS A286.
M-1184; 0.06 C, 15 Cr, 25 Ni, 1.3 Mo, 0.3 V, 2.0 Ti, bal Fe.
Rolled: 150,000 TS; 105,000 YS; 22 El; 290 Brin.
For gas turbine components, jet engine parts, afterburners.
High heat and creep resistant, austenitic.

BOFORS ARO-75.
M-1184; 0.3 C, 1.0 Cr, 0.25 Mo, 1.1 Al, bal Fe.
Heat treated: 121,000 TS; 99,200 YS; 245-290 Brin.
For fuel pump pistons, valve parts; nitriding steel.

BOFORS B4.
M-1184; 0.25 C, bal Fe.
For machine parts; construction steel.

BOFORS B4V.
M-1184; 0.15 C, 0.65 N, bal Fe.
Heat treated: 71,000-142,000 TS; 36,000-42,000 YS; 150-300 Brin.
For gears, shafts; case hardening.

BOFORS B7.
M-1184; 0.35 C, bal Fe.
For machinery parts, construction steel.

BOFORS B10.
M-1184; 0.45 C, bal Fe.
For machine parts, gears; construction steel.

BOFORS B12.
M-1184; 0.6 C, bal Fe.
For machinery parts; construction steel.

BOFORS B14.
M-1184; 0.70 C, 0.2 Si, 0.6 Mn, bal Fe.
Stress relieved: 107,000-121,000 TS; 57,000 YS; 13 El; 220-250 Brin.
Heat treated: 175,000 TS; 130,000 YS; 12 El; 350 Brin.
For machinery components, gears, shafts, control rods, axes, hammers, springs.
AISI 1070 steel.
Wear resistant, water hardening.

BOFORS B15T.
M-1184; 0.75 C, 0.2 Si, 0.6 Mn, bal Fe.
Stress relieved: 110,000 TS; 60,000 YS; 12 El; 240 Brin.
Heat treated: 180,000 TS; 135,000 YS; 10 El; 370 Brin.
For rock drills, chisels, reamers, shafts, axes, hammers, springs, hand tools, pliers.
Water hardening, AISI 1075 steel. Wear resistant.

BOFORS B15 V.
M-1184; 0.75 C, bal Fe.
For tools, fixtures; water hardening.

BOFORS B20V.
M-1184; 1.0 C, 0.20 Si, 0.30 Mn, bal Fe.
For tools, drills, reamers; water hardened; Type W1.

BOFORS B24V.
M-1184; 1.2 C, 0.2 Si, 0.3 Mn, bal Fe.
For reamers, drills, taps, hobs, cutters; water hardened; Type W1.

BOFORS B28V.
M-1184; 1.4 C, bal Fe.
For knives, trimming tools; water hardening.

BOFORS CR83.
M-1184; 0.4 C, 0.8 Cr, 1.25 Ni, bal Fe.
114,500 TS; 85,100 YS; 240 Brin.
For machine parts, shafts, crankshafts; water or oil hardened.

BOFORS CRO861.
M-1184; 0.35 C, 0.7 Mn, 1.4 Cr, 1.4 Ni, 0.2 Mo, bal Fe.
Hardened: 156,000-178,000 TS; 128,000 YS; 12 El; 45 RA; 330-370 Brin.
For shafts, gears, connecting rods, bolts, crankshafts.
AISI 4337, oil hardening, tough, shock resistant.

BOFORS DR34.
M-1184; 0.15 C, 0.7 Mn, 0.8 Cr, 1.5 Ni, bal Fe.
Hardened: 100,000-206,000 TS; 57,000-100,000 YS; 10-14 El; 210-430 Brin.
For gears, bolts, camshafts, shafting worms, chains.
Case hardening, tough.

BOFORS DR44.
M-1184; 0.20 C, 0.7 Mn, 0.8 Cr, 1.5 Ni, bal Fe.
Hardened: 114,000-228,000 TS; 71,000-114,000 YS; 10-12 El; 240-480 Brin.
For gears, bolts, camshafts, worms.
Case hardening, tough.

BOFORS DRO-1133.
M-64; 0.55 C, 0.6 Cr, 1.6 Ni, 0.3 Mo, bal Fe.
For drop forging dies; oil hardened.

BOFORS DRS16.
M-1184; 0.80 C, 21 Cr, 1.5 Ni, 2 Si, bal Fe.
Hardened: 137,000 TS; 111,000 YS; 15 El; 290 Brin.
For outlet valves for internal combustion engines.
Martensitic, heat and corrosion resistant. Resists leaded fuels.

BOFORS FR-86.
M-1184; 0.35 C, 1.2 Cr, 2.6 Ni, bal Fe.
Rolled: 128,000 TS; 107,000 YS; 270-300 Brin.
For crankshafts, shafts; oil hardened.

BOFORS HR-19.
M-1184; 0.55 C, 1.5 Cr, 3 Ni, bal Fe.
For cold and hot work dies, bakelite dies; oil hardening.

BOFORS HR33.
M-1184; 0.15 C, 0.75 Cr, 3.0 Ni, bal Fe.
Heat treated: 99,500-185,000 TS;, 71,100-113,800 YS; 210-390 Brin.
For gears, shafts; case hardened.

BOFORS HR 44.
M-1184; 0.20 C, 3.0 Ni, 0.7 Cr, bal Fe.
Heat treated: 142,200-213,000 TS; 92,500-142,200 YS; 300-430 Brin.
For gears, cams, shafts; case hardened.

BOFORS HRD-1243.
M-1184; 0.55 C, 1 Cr, 3 Ni, 0.3 Mo, bal Fe.
For cold and hot work dies, forging dies; oil hardening.

BOFORS HRP 1152.
M-1184; 0.50 C, 0.85 Mn, 1.05 Cr, 0.15 V, bal Fe.
Annealed: 105,000 TS; 75,000 YP; 27 El; C 16 Rock.
Hardened: 300,000 TS; 263,000 YP; 4 El; 570 Brin.
For gears, springs, shafts, machinery parts, axles, clutches, small tools.
Oil hardened, shock resistant.

BOFORS IR 34.
M-1184; 0.13 C, 0.4 Mn, 0.7 Cr, 3.5 Ni, bal Fe.
Hardened: 120,000-192,000 TS; 78,000-114,000 YS; 10-12 El; 250-400 Brin.
For gears, worms, camshafts, bolts, fasteners.
AISI 3310 steel, case hardening, tough and shock resistant.

BOFORS IR74.
M-1184; 0.30 C, 0.60 Mn, 0.75 Cr, 3.5 Ni, bal Fe.
For gears, pinions, shafts; oil hardening, tough.

BOFORS KR35.
M-1184; 0.12 C, 1.25 Cr, 4.5 Ni, bal Fe.
For camshafts, crankshafts, gears; case hardening, tough.

BOFORS KR75.
M-1184; 0.3 C, 1.3 Cr, 4.3 Ni, bal Fe.
Heat treated: 156,500 TS; 128,000 YS; 330-370 Brin.
For shafts, gears, wear parts; air hardening.

BOFORS N82.
M-1184; 0.4 C, 1.25 Mn, bal Fe.
99,500 TS; 71,000 YS; 240 Brin.
For machinery parts; water or oil hardened.

BOFORS N91.
M-1184; 0.45 C, 0.8 Mn, bal Fe.
Normalized: 85,000-107,000 TS; 46,000 YS; 18 El; 180-230 Brin.
For induction and flame hardened machinery components, gears, pinions, shafting.
AISI 1045 steel. Water hardening.

BOFORS NON-SHRINKING STEEL RT-1733.
M-267; 0.85-0.95 C, 1.0-1.25 Mn, 0.5-0.6 Cr, 0.5-0.6 W, bal Fe.
For blanking dies, broaches, cutters, tools, drills, plugs, gages; non-shrinking tool steel.

BOFORS P02.
M-1184; 85 C, 4.3 Cr, 6.4 W, 5 Mo, 2 V, bal Fe.
For cutters, twist drills; high speed steel.

BOFORS P03.
M-1184; 0.85 C, 7 W, 4 Cr, 3-5 Mo, 2 V, bal Fe.
Heat treated: C 64-67 Rock.
For lathe and planer tools, milling cutters, drills, broaches, reamers.
High speed steel. High red hardness, wear and abrasion resistant.

BOFORS P10.
M-1184; 0.75 C, 4 Cr, 18 W, 1.2 V, bal Fe.
For dies, shears, cutters; high speed steel.

BOFORS P15.
M-1184; 0.8 C, 18 W, 4 Cr, 1.5 V, 2 Co, 1 Mo, bal Fe.
For cutting tools; high speed steel.

BOFORS P121.
M-1184; 0.60-1.05 C, 0.10 V, bal Fe.
Annealed: 120,000 TS; 70,000 YS; 15 El; 220 Brin.
Hardened: 185,000 TS; 135,000 YS; 13 El; 400 Brin.
For axes, cutting tools, drills, springs, punches.
Water hardening, wear resistant. Type W2.

BOFORS P171.
M-1184; 0.8 C, 0.1 V, bal Fe.
For pneumatic tools; water hardened.

BOFORS P181.
M-1184; 0.9 C, 0.1 V, bal Fe.
For tools, dies, punches; water hardening.

BOFORS P211.
M-1184; 1 C, 0.1 V, bal Fe.
For cold working dies, pneumatic pistons; water hardening.

BOFORS Q5.
M-1184; 0.8 C, 4 Cr, 18 W, 1.7 V, 5.5 Co, 1 Mo, bal Fe.
For cutters, taps, hobs, reamers; high speed steel.

BOFORS Q10.
M-1184; 0.8 C, 4 Cr, 18 W, 1.7 V, 10.5 Co, 1 Mo, bal Fe.
For cutters, millers, tool bits, hobs; high speed steel.

BOFORS QRO45.
M-1184; 0.30 C, 2.8 Cr, 2.8 Mo, 0.5 V, 2.8 Co, bal Fe.
Hardened: C 45-55 Rock.
For die casting and forging dies, extrusion dies, mandrels, punches.
Oil or air hardening.
Hot work steel. Develops high red hardness and toughness to 1300°F.

BOFORS QX2.
M-1184; 1.0 C, 7 W, 4 Cr, 3-5 Mo, 2 V, 9.5 Co, bal Fe.
Hardened: C 64-67 Rock.
For lathe and planer tools, milling cutters, drills, broaches, reamers, shears.
High speed steel. High red-hardness. Wear and abrasion resistant.

BOFORS R10-214.
M-1184; 0.10 max C, 27 Cr, 5 Ni, 1.5 Mo, bal Fe.
Rolled: 85,330 TS; 64,000 YS; 210 Brin.
For sulphite-cellulose industries; high corrosion resistant.

BOFORS R309.
M-1184; 1.5 C, 2 Cr, bal Fe.
For files, trimmer tools; oil hardened.

BOFORS R10214.
M-1184; 0.20 C, 23-28 Cr, 2.5-5.0 Ni, 1-2 Mo, bal Fe.
Normalized: 103,000 TS; 78,000 YS; 18 El; 45 RA; 235 Brin.
Annealed: 95,000 TS; 41,000 YS; 29 El; 60 RA; 225 Brin.
For valves, pumps, furnace parts; Type 327; heat resistant.

BOFORS RCK3.
M-1184; 0.15 C, 25 Cr, 13 Ni, 0.2 N, bal Fe.
Annealed: 94,000 TS; 50,000 YS; 54 El; 170 Brin.
For feeding plates in clinker coolers in cement industry.
Austenitic, stainless, high oxidation resistance. Type 309S.

BOFORS RCK4.
M-1184; 0.20 C, 21 Cr, 12 Ni, 0.2 N, bal Fe.
Annealed: 117,000 TS; 67,000 YS; 41 El; 240 Brin.
For outlet valve heads in internal combustion engines.
Austenitic, corrosion resistant, oxidation resistant.

BOFORS RCT.
M-1184; 0.40 C, 12 Ni, 14.5 Cr, 2.3 Si, bal Fe.
Hardened: 106,700 TS; 64,000 YS; 240 Brin.
For engine valves; heat resistant.

Section I: Alloy Data / 223

BOFORS RCT3.
M-1184; 0.20 C, 21 Cr, 13 Ni, 1.2 Si, 3.0 W, bal Fe.
Annealed: 104,000 TS; 47,000 YS; 27 El; 220 Brin.
For stator blades, turbolators, rings in jet engines, gas and steam turbines, diesel engine valves.
Austenitic, high heat resistant.

BOFORS RD-653.
M-1184; 0.25 C, 1.1 Cr, 0.2 Mo, bal Fe.
Rolled: 128,000 TS; 99,500 YS; 270-310 Brin.
For shafts, cams, gears; oil hardened.

BOFORS RE39.
M-1184; 0.20 C, 17.5 Cr, 2.1 Ni, bal Fe.
Heat treated: 121,000-135,000 TS; 92,500-112,000 YS; 20-15 El; 60-45 RA; 260-288 Brin.
For pumps, spindles, propeller shafts; Type 431; corrosion resistant.

BOFORS REB-210.
M-1184; 0.15 C, 22.5 Cr, 24 Ni, bal Fe.
71,000 TS; 28,500 YS; 180 Brin.
For heat resistant parts; resists S and SO_2.

BOFORS RES210.
M-1184; 0.2 C, 22-24 Cr, 12-15 Ni, 2 max Mn, 1 max Si, bal Fe.
Annealed: 85,000-95,000 TS; 40,000-50,000 YS; 45-55 El; 150-185 Brin.
For heat treating boxes, furnace parts, chemical plant equipment; Type 309; austenitic, heat resistant.

BOFORS RIM21.
M-1184; 0.08 C, 18 Cr, 11 Ni, 1.5 Mo, bal Fe.
Annealed: 85,000 TS; 36,000 YS; 60 El; 160 Brin.
For acid resisting shafts, bolts, chemical plant equipment.
Austenitic, stainless, non-hardening.

BOFORS RIM-29.
M-1184; 0.08 max C, 18 Cr, 18 Ni, bal Fe.
Annealed: 82,500 TS; 28,500 YS; 60 El; 70 RA; 165 Brin.
Cold drawn: 125,000 TS; 95,000 YS; 25 El; 55 RA; 277 Brin.
For chemical and oil refinery equipment, tanks, vessels; Type 302; stainless, austenitic.

BOFORS RIM-92.
M-1184; 0.08 max C, 18-20 Cr, 8-11 Ni, 2 max Mn, bal Fe.
Annealed: 90,000 TS; 45,000 YS; 60 El; 135 Brin.
Cold drawn: 180,000 TS; 150,000 YS; 10 El; 330 Brin.
For chemical plant equipment, welded structures; Type 304; stainless, austenitic.

BOFORS RIM 200.
M-1184; 0.04 C, 18 Cr, 9 Ni, bal Fe.
Annealed: 85,000 TS; 35,000 YS; 55 El; Rock B 80.
For architectural trim, kitchen equipment, chemical and textile plant equipment.
Type 304 stainless, austenitic.

BOFORS RIM-210.
M-1184; 0.07 max C, 19 Cr, 10 Ni, 1.5 Mo, bal Fe.
Annealed: 78,400 TS; 28,500 YS; 165 Brin.
For sulfite, cellulose industries; high corrosion resistance.

BOFORS RIM 213.
M-1184; 0.07 C, 18 Cr, 11 Ni, 2.3 Mo, 0.4 Ti, bal Fe.
Annealed: 85,000 TS; 38,000 YS; 50 El; 165 Brin.
For chemical plant equipment and structural parts operating at 900-1150°F.
Stabilized, austenitic, stainless, weldable grade.

BOFORS RIM-215.
M-1184; 0.08 max C, 18 Cr, 11 Ni, 2.7 Mo, bal Fe.
Annealed: 82,500 TS; 165 Brin.
For chemical industries; corrosion resistant; austenitic.

BOFORS RIM 290.
M-1184; 0.03 max C, 18 Cr, 10 Ni, bal Fe.
Annealed: 76,000 TS; 32,000 YS; 60 El; 140 Brin.
For welded structures, chemical and textile plant equipment, food and drug equipment.
Type 304L stainless, austenitic steel.

BOFORS RIM-291.
M-1184; 0.08 max C, 18-20 Cr, 8-11 Ni, 2 max Mn, bal Fe.
Annealed: 90,000 TS; 45,000 YS; 60 El; 135 Brin.
Cold drawn: 180,000 TS; 150,000 YS; 10 El; 330 Brin.
For chemical plant equipment, welded structures; Type 304; stainless, austenitic.

BOFORS RIM-294.
M-1184; 0.08 max C, 18 Cr, 10 Ni, 0.8 Mn, 0.5 Ti, bal Fe.
Annealed: 82,500 YS; 165 Brin.
For chemical industries; heat and corrosion resistant.

BOFORS RIM295.
M-1184; 0.08 max C, 17-19 Cr, 9-12 Ni, Cb = 10 x C, bal Fe.
Annealed: 85,000-95,000 TS; 35,000-45,000 YS; 55-50 El; 175 Brin.
For welded structures, chemical plant equipment; Type 347; corrosion and heat resistant.

BOFORS RK 214.
M-1184; 0.10 max C, 28 Cr, 4.5 Ni, bal Fe.
Annealed: 85,300 TS; 64,000 YS; 210 Brin.
For furnace parts, nitric acid contact; heat and corrosion resistant.

BOFORS RLH 2.
M-1184; 0.10 C, 18 Cr, 8 Ni, 0.25 S, bal Fe.
Annealed: 85,000 TS; 28,000 YS; 60 El; 160 Brin.
For screws, bolts, fasteners, shafts, screw machine products.
Type 303 stainless, free-cutting, austenitic, non-hardenable.

BOFORS RNK 29.
M-1184; 0.09 C, 18 Cr, 5 Ni, 8.5 Mn, 0.2 N, bal Fe.
Annealed: 104,000 TS; 56,000 YS; 59 El; 200 Brin.
For shafts, valve parts, bolts, fittings, foodstuff and chemical equipment
Type 202 stainless, austenitic.

BOFORS RO211.
M-1184; 0.12 C, 2.5 Cr, 1.1 Mo, bal Fe.
Normalized: 88,000 TS; 70,000 YS; 26 El; 190 Brin.
For steam and gas turbine parts, operating up to 1100°F.
Martensitic, corrosion resistant.

BOFORS RO-346.
M-1184; 0.15 C, 0.8 Cr, 0.6 Mo, bal Fe.
For camshafts, crankshafts, gears; case hardened, tough.

BOFORS RO-752.
M-1184; 0.35 C, 1.05 Cr, 0.20 Mo, bal Fe.
For shafts, gears; tough.

BOFORS RO-952.
M-1184; 0.40 C, 1.10 Cr, 0.25 Mo, bal Fe.
For shafts, gears; oil hardened, tough.

BOFORS RO4154.
M-1184; 0.25 C, 3 Cr, 0.5 Mo, bal Fe.
Hardened: 124,000 TS; 102,000 YS; 20 El; 260 Brin.
For construction elements subjected to stresses at temperatures up to 1100°F.
Martensitic. Good impact strength and resistance to temper brittleness.

BOFORS RO-7155.
M-1184; 0.30 C, 2.7 Cr, 0.5 Mo, bal Fe.
For shafts, gears; oil hardened.

BOFORS ROP10.
M-1184; 0.08 C, 3.0 Cr, 0.8 Mo, 0.2 V, bal Fe.
For hobbed molds for pressing and injection molding of plastics, hobbed dies for zinc die castings. Case-hardening, shock resistant.

BOFORS ROP18.
M-1184; 0.2 C, 1.3 Cr, 0.5 Mo, 0.2 V, bal Fe.
For hobbed hot work dies, die casting and drop forging dies.
Hot work steel, oil or air hardening.

BOFORS ROP19.
M-1184; 0.40 C, 5.3 Cr, 1.4 Mo, 1.0 V, 1.0 Si, bal Fe.
Annealed: 98,000 TS; 74,000 YS; 28 El; 210 Brin.
Heat treated: 290,000 TS; 228,000 YS; 3 El; C 55 Rock.
For die casting and hot pressing dies, extrusion dies, mandrels, punches.
Type H13 hot work steel. Tough and red hard.

BOFORS ROP21.
M-1184; 1 C, 5.3 Cr, 0.2 V, 1.1 Mo, bal Fe.
For cold work dies, shears, press tools; air hardened.

BOFORS ROP43.
M-1184; 0.23 C, 12 Cr, 1.2 Mo, 0.6 Ni, 0.3 V, bal Fe.
Annealed: 95,000 TS; 40,000 YS; 25 El; B 92 Rock.
Hardened: 240,000 TS; 205,000 YS; 9 El; C 48 Rock.
For cutlery, bearings, valves, surgical instruments. Corrosion resistant, hardenable.

BOFORS ROP46.
M-1184; 0.20 C, 12 Cr, 0.5 Mo, 0.35 V, bal Fe.
Annealed: 90,000 TS; 38,000 YS; 26 El; B 90 Rock.
Hardened: 235,000 TS; 200,000 YS; 9 El; C 50 Rock.
For valves, cutlery, bearings, surgical instruments. Corrosion resistant, hardenable.

BOFORS ROP57.
M-1184; 1.5 C, 12 Cr, 0.2 V, 0.8 Mo, bal Fe.
For dies, stamping and shearing tools; oil or air hardened.

BOFORS ROP-63.
M-1184; 0.30 C, 0.3 Mn, 2.5 Cr, 0.2 V, 0.5 Ni, 0.3 Mo, bal Fe.
Heat treated: 142,200 TS; 121,000 YS; 295-330 Brin.
For gears, shafts, nitrided parts; nitriding steel.

BOFORS ROP5462.
M-1184; 0.24 C, 1.3 Cr, 0.5 Mo, 0.2 V, bal Fe.
Hardened: 121,000 TS; 107,000 YS; 20 El; 260 Brin.
For steam turbine rings, bolts, screws, nuts at temperatures to 1025°F.
Martensitic, high temperature steel.

BOFORS ROP9653.
M-1184; 0.45 C, 1.5 Cr, 0.7 Mo, 0.3 V, val Fe.
For die casting and hot pressing dies, extrusion dies and mandrels, punches, forging dies.
Hot work, oil hardening steel.

BOFORS ROPT.
M-1184; 0.20 C, 3.0 Cr, 0.6 Mo, 0.5 W, 0.8 V, bal Fe.
Hardened: 128,000 TS; 107,000 YS; 17 El; 275 Brin.
For turbine discs for jet engines, gas and steam turbines, shafts, rotors, bolts.
Martensitic, high creep resistance.

BOFORS RP1152 see **BOFORS HRP 1152.**

BOFORS RR27.
M-1184; 0,08 max C, 11.5-13 Cr, 0.1-0.3 Al, bal Fe.
Annealed: 71,000 TS; 42,600 YS; 22 El; 70 RA; 150 Brin.
Heat treated: 175,000 TS; 145,000 YS; 21 El; 64 RA; 352 Brin.
For oil refinery and chemical plant equipment; Type 405; corrosion resistant.

BOFORS RR77.
M-1184; 0.15 min C, 12-14 Cr, bal Fe.
Annealed: 88,000 TS; 40,000 YS; 32 El; 68 RA; 170 Brin.
For cutlery, valve trim, turbine blades; Type 420; stainless, hardenable.

BOFORS RR212.
M-1184; 0.35 C, 23-27 Cr, bal Fe.
Annealed: 90,000 TS; 60,000 YS; 20 El; 45 RA; 180 Brin.
For furnace parts, heat treating boxes; Type 446; heat resistant.

BOFORS RR412.
M-1184; 0.35 C, 23-27 Cr, bal Fe.
Annealed: 90,000 TS; 60,000 YS; 20 El; 45 RA; 180 Brin.
For furnace parts, heat treating boxes; Type 446; heat resistant.

BOFORS RT27.
M-1184; 0.3 C, 1 Cr, 5.5 W, 3.5 Ni, bal Fe.
For extrusion dies, mandrels; oil hardened, hot work steel.

BOFORS RT45.
M-1184; 0.3 C, 3 Cr, 10 W, 2 Ni, 0.3 V, bal Fe.
For die casting dies, hot work dies; hot work steel, oil hardened.

BOFORS RT46.
M-1184; 0.30 C, 12 Cr, 12 W, 0.5 V, bal Fe.
For die casting and pressing dies, extrusion dies and mandrels, punches. Hot work steel Type H23. Develops red hardness and toughness up to 1300°F.

BOFORS RT60.
M-1184; 2 C, 13 Cr, 1 W, bal Fe.
For stamping and shearing tools, dies; oil hardening.

BOFORS RT 225.
M-1184; 1.3 C, 5 W, bal Fe.
For impact extrusion dies; water hardened.

BOFORS RT-1733.
M-1184; 0.9 C, 0.6 Cr, 0.6 W, 1.2 Mn, bal Fe.
For cutting, threading tools; oil hardening.

BOFORS RTO 712.
M-1184; 0.4 C, 1.2 Cr, 2.5 W, 0.3 Mo, 0.15 V, bal Fe.
For chisels, pneumatic tools; tough.

BOFORS RTO-912.
M-1184; 0.45 C, 1 Cr, 2.5 W, 0.25 Mo, 0.15 V, bal Fe.
For chisels, shear blades; hot work steel.

BOFORS S-145.
M-1184; 0.55 C, 1.75 Si, 0.75 Mn, bal Fe.
Hardened: 185,000 TS; 164,000 YS; 380-440 Brin.
For springs; tough, oil hardened.

BOFORS SIR.
M-1184; 0.20 C, 25 Cr, bal Fe.
Annealed: 71,000 TS; 43,000 YS; 170 Brin.
For furnace parts, heat treating boxes, quenching baskets.
Corrosion and heat resistant to 2000°F.

BOFORS SRO.
M-1184; 0.45 C, 9 Cr, 0.3 Mo, 2.8 Si, bal Fe.
Hardened: 144,000 TS; 121,000 YS; 30 El; 320 Brin.
For valves, oil refinery equipment.
Martensitic, heat and corrosion resistant.

BOFORS SRO2.
M-1184; 0.45 C, 13 Cr, 1.0 Ni, 2.8 Si, bal Fe.
Heat Treated: 142,000 TS; 121,000 YS; 16 El; 300 Brin.
Uses: Intake valves, outlet valves.
Martensitic steel, corrosion and oxidation resistant.

BOHLER 2M now **VEW F180.**

BOHLER 5 NW now **VEW P602.**

BOHLER 751 now **VEW K980.**

BOHLER ACE now **VEW V810.**

BOHLER ACN now **VEW V820.**

BOHLER ACV now **VEW V350.**

BOHLER AS-2 now **VEW A505.**

BOHLER AS2G now **VEW A505G.**

BOHLER BLITZ now **VEW V960.**

BOHLER BTH now **VEW K960.**

BOHLER CC now **VEW S300.**

BOHLER CC55N now **VEW S308.**

BOHLER CC55 SPEZIAL now VEW S307.

BOHLER CMO now VEW V340.

BOHLER CRV now VEW F550.

BOHLER CSF now VEW K243.

BOHLER DCM42 now VEW V340.

BOHLER DCM910 now VEW D320.

BOHLER DCMV55 now VEW D240.

BOHLER DCMV511 now VEW D220.

BOHLER DMO3 now VEW D500.

BOHLER DMV83 now VEW D404.

BOHLER EB60 now VEW E525.

BOHLER EBM now VEW M100.

BOHLER ECL 80 now VEW E304.

BOHLER ECL 100 now VEW E300.

BOHLER ECN150 now VEW E230.

BOHLER ECN 200 now VEW E220.

BOHLER EL now VEW M150.

BOHLER EPB EXTRA W now VEW M130.

BOHLER EPB SPEZIAL now VEW M120.

BOHLER ESC now VEW K240.

BOHLER ES PRIMA now VEW E216.

BOHLER ES SPECIAL now VEW E200.

BOHLER EWH now VEW V920.

BOHLER EXTRA FM now VEW K505.

BOHLER EXTRA K now VEW K310.

BOHLER EXTRA S now VEW K761.

BOHLER EXTRA SC now VEW K240.

BOHLER EXTRA ZAH now VEW K980.

BOHLER EXTRA ZAHNHART now VEW K990.

BOHLER EXTRA ZH100 now VEW K990.

BOHLER F130 now VEW K995.

BOHLER FB8 now VEW H160.

BOHLER FB8G now VEW H160G.

BOHLER FB10 now VEW H120.

BOHLER FC now VEW K205.

BOHLER FM now VEW K505.

BOHLER GNM now VEW W501.

BOHLER GNME now VEW W500.

BOHLER HM now VEW V930.

BOHLER HMA now VEW V940.

BOHLER IN 5 now VEW D328.

BOHLER IN 9 now VEW D204.

BOHLER INC5 now VEW D310.

BOHLER K150 now VEW K200.

BOHLER KHS now VEW F126.

BOHLER KHSW now VEW F100.

BOHLER KL now VEW K463.

BOHLER KMCW now VEW V510.

BOHLER KW20 now VEW N320.

BOHLER KW30 now VEW N530.

BOHLER KW40 now VEW N540.

BOHLER KWA now VEW N200.

BOHLER MO RAPID EXTRA 3 now VEW S610.

BOHLER MO RAPID EXTRA 9 now VEW S401.

BOHLER MO RAPID EXTRA 500 now VEW S705.

BOHLER MO RAPID EXTRA 800 now VEW S500.

BOHLER MO RAPID EXTRA 1200 now VEW S700.

BOHLER MO RAPID EXTRA V30 now VEW S607.

BOHLER MPD now VEW W322.

BOHLER MST now VEW K720.

BOHLER MYA now VEW K247.

BOHLER MYAH now VEW K507.

BOHLER MY EXTRA now VEW K450.

BOHLER NBS now VEW K605.

BOHLER NBSN now VEW W500.

BOHLER NMH now VEW V130.

BOHLER NO. 2 SI now VEW F105.

BOHLER NWM now VEW K600.

BOHLER PPA now VEW E204.

BOHLER PRIMA MITTELHART 100 now VEW K990.

BOHLER PRIMA WEICH now VEW K971.

BOHLER PRIMA ZAH now VEW K980.

BOHLER SAS2 now VEW A750.

BOHLER SAS4 now VEW A350.

BOHLER SAS8 now VEW A960.

BOHLER SIC20 now **VEW H730.**

BOHLER SPECIAL K now **VEW K100.**

BOHLER SPECIAL K5 now **VEW K305.**

BOHLER SPECIAL K8 now **VEW K300.**

BOHLER SPECIAL KMV now **VEW K110.**

BOHLER SPECIAL KN now **VEW K116.**

BOHLER SPECIAL KNL now **VEW K105.**

BOHLER SPECIAL KR now **VEW K107.**

BOHLER SPECIAL VERY HARD now **VEW K400.**

BOHLER SPEZIAL KR now **VEW K107.**

BOHLER SPN now **VEW F180.**

BOHLER SSC now **VEW K510.**

BOHLER SSWV now **VEW K405.**

BOHLER SUPER RAPID EXTRA now **VEW S200.**

BOHLER SUPER RAPID EXTRA 500 now **VEW S305.**

BOHLER SUPER RAPID EXTRA MO now **VEW S600.**

BOHLER THM now **VEW K451.**

BOHLER TWR now **VEW K459.**

BOHLER TWVW now **VEW K403.**

BOHLER UM8 now **VEW P550.**

BOHLER US SPEZIAL now **VEW W326.**

BOHLER US ULTRA now **VEW W300.**

BOHLER US ULTRA 2 now **VEW W302.**

BOHLER US ULTRA 4 now **VEW W304.**

BOHLER V6 N now **VEW K630.**

BOHLER VB 135 now **VEW V510.**

BOHLER VB200 now **VEW V554.**

BOHLER VC2 now **VEW B400.**

BOHLER VCL125 now **VEW V340.**

BOHLER VCL135 now **VEW V330.**

BOHLER VCL 140 now **VEW V320.**

BOHLER VCL150 now **VEW V310.**

BOHLER VCL230 now **VEW V350.**

BOHLER VCN 100 now **VEW V165.**

BOHLER VCN150 now **VEW V155.**

BOHLER VCN 200 now **VEW V145.**

BOHLER VSW now **VEW H800.**

BOHLER WKV now **VEW W327.**

BOHLER WKW4 now **VEW M310.**

BOHLER WKZ now **VEW W 100.**

BOHLER WKZ50 now **VEW W 105.**

BOHLER WM now **VEW W 106.**

BOHLER WMD now **VEW W 320.**

BOHLER WPN now **VEW W 103.**

BOHNALITE 62S.
M-138; 0.8-1.2 Mg, 0.4-0.8 Si, 0.15-0.40 Cu, 0.15 max Ti, bal Al.
Annealed: 17,500 TS; 6500 YS; 30 El; 28 Brin.
T4-temper: 35,000 TS; 21,000 YS; 25 El; 65 Brin.
T6-temper: 45,000 TS; 40,000 YS; 17 El; 95 Brin.
For doors, window frames, trim; age hardenable.

BOHNALITE A.
M-138; 5-7 Sn, 0.75-1.25 Cu, 1.5-2.2 Si, 0.75-1.25 Ni, bal Al.
Cast.
For bearings; similar to Alcoa X750.

BOHNALITE X-10.
M-138; 9-11 Al, 0.1 Mn, bal Mg.
Cast: 22,000 TS; 12,000 YS; 1-3 El; 53 Brin.
Heat treated: 36,000 TS; 17,000 YS; 1-4 El; 65 Brin.
For castings; permanent mold, sand or die cast.

BOHNALLOY MB6.
M-138; 0.80-1.40 Cd, 3.5-4.5 Si, 0.05-0.15 Cu, 0.25 Fe, 0.03 Mn, 0.03 Ti, bal Al.
Bearings and bushings.
SAE 781.

BOHNALLOY MB7.
M-138; 6.5-7.5 Sn, 1.5-1.8 Ni, 0.70-1.3 Cu, 0.75-1.25 Mg, 0.50-1.25 Si, 0.60 Fe, 0.10 Ti, 0.25 Mn, 0.15 Zn, bal Al.
Solid permanent mold cast bearings and bushings.
Similar to SAE 770.

BOHNALLOY MB8.
M-138; 5.5-7.0 Sn, 0.20-0.70 Ni, 0.70-1.30 Cu, 1.0-2.0 Si, 0.70 Fe, 0.10 Mn, 0.10 Ti, bal Al.
Bearings and bushings.
SAE 780.

BOHNALLOY NO. 3.
M-138; 7.0-8.0 Sb, 0.50 Pb, 3.0-4.0 Cu, 0.08 Fe, 0.10 As, 0.08 Bi, 0.005 Zn, 0.005 Al, bal Sn.
Tin base babbitt.
Main and rod bearings and bushings - high corrosion resistance.
SAE 12; ASTM B23 Alloy No. 2.

BOHNALLOY NO. 9.
M-138; 5.0-7.0 Sn, 9.0-11.0 Sb, 0.50 Cu, 0.25 As, 0.10 Bi, 0.005 Zn, 0.005 Al, 0.05 Cd, bal Pb.
High lead babbitt.
Main and rod bearings and bushings, moderate loads.
SAE 13, ASTM B23 Alloy No. 13.

BOHNALLOY NO. 15.
M-138; 9.2-10.7 Sn, 14.0-16.0 Sb, 0.50 Cu, 0.6 As, 0.10 Bi, 0.005 Zn 0.005 Al, 0.05 Cd, bal Pb.
High lead babbitt.
Main and rod bearings and bushings, - moderate loads.
SAE 14, ASTM B23 Alloy 7.

BOHNALLOY NO. 18.
M-138; 0.9-1.3 Sn, 14.0-15.5 Sb, 0.50 Cu, 0.8-1.2 As, 0.10 Bi, 0.005 Zn, 0.005 Al, 0.02 Cd, bal Pb.
High lead babbitt.
Main and rod bearings and bushings, - moderate loads.
SAE 15, ASTM B23 Alloy 15.

BOHNALLOY PMB20.
M-138; 77.0-81.0 Cu, 9.0-11.0 Sn, 8.0-11.0 Pb, 0.75 max Zn, 0.20 max Sb, 0.50 max Ni, 0.15 max Fe, 0.03 max P.
Bushings and wear plates. Hard shaft desirable.
SAE 797.

BOHNALLOY PMB20A.
M-138; 77.0-81.0 Cu, 9.0-11.0 Sn, 8.0-11.0 Pb, 0.75 max Zn, 0.20 max Sb, 0.50 max Ni, 0.15 max Fe, 0.03 max P.
Bushings. Hard shaft desirable. Softer than PMB20.
SAE 797.

BOHNALLOY PMB21.
M-138; 70.0-75.0 Cu, 3.0-4.0 Sn, 21.0-25.0 Pb, 0.50 max Zn, 0.20 max Sb, 0.50 max Ni, 0.15 max Fe, 0.03 max P.
Rocker arm, transmission and camshaft bushings; intermediate load applications.
SAE 794 or 799.

BOHNALLOY PMB22.
M-138; 86.0-89.0 Cu, 3.5-4.5 Sn, 7.0-9.0 Pb, 0.20 max Zn, 0.50 max Sb 0.50 max Ni, 0.15 max Fe, 0.03 max P.
Transmission and chassis bushings, - medium to high load applications.
SAE 798.

BOHNALLOY PMB24.
M-138; 23.0-27.0 Pb, 0.6-1.0 Sn, 0.25 max Fe, 0.10 max Ni, 0.10 max Sb, 0.10 max Zn, 0.03 max P, bal Cu.
Heavy duty rod and main bearings, - hard or soft shaft.
SAE 49.

BOHNALLOY PMB30.
M-138; 29.0-33.0 Pb, 0.50 max Sn, 0.25 max Fe, 0.10 max Ni, 0.10 max Sb, 0.10 max Zn, 0.03 max P, bal Cu.
Rod and main bearings, - medium to heavy loads, - hard or soft shaft without overplate.
SAE 48.

BOHNALLOY PMB42.
M-138; 54.0-58.0 Cu, 40.0-44.0 Pb, 1.50-3.50 Sn, 0.25 max Fe, 0.20 max Sb, 0.10 max Ni, 0.10 max Zn, 0.03 max P.
Intermediate bearing material usually without overlay. May be used on soft shaft.
SAE 484.

BOHNALLOY PMB224.
M-138; 22.0-26.0 Pb, 2.0-2.8 Sn, 0.30 max Ni, 0.30 max Zn, 0.15 max Fe, 0.10 max Sb, 0.03 max P, 0.002 max S, 0.005 max Si, 0.005 max Al, bal Cu.
Engine bearings and bushings, - high load capacity, - good resistance to erosion.
SAE 49.

BOHNALLOY R55.
M-138; 56.0-58.0 Cu, 2.25-3.50 Pb, 0.35 max Fe, bal Zn.
Soft: 60,000 psi TS; 20,000 psi YS; 30 El; 55 Rock B.
For wrought architectural parts, casements, lock bodies, extrusions.
Architectural bronze.

BOHNALLOY R-56.
M-138; 56-59 Cu, 0.7-1.2 Al, 0.8-1.3 Fe, 0.50 max Mn, bal Zn.
Hard: 80,000 TS; 50,000 YS; 12 El; 85 Rock B.
Soft: 70,000 TS; 25,000 YS; 20 El; 70 Rock B.
For wrought architectural shapes, marine hardware; wrought alloy Mn bronze.

BOHNALLOY R58NE.
M-138; 58.0-62.0 Cu, 1.6-2.5 Pb, 0.30 max Fe, bal Zn.
Soft: 58,000 psi TS; 23,000 psi YS; 40 El; 40 Rock B.
For forgings and good machinability.
Forging brass.

BOHNALLOY R60L.
M-138; 59.0-62.0 Cu, 0.40-1.0 Pb, 0.5-1.0 Sn, 0.10 max Fe, bal Zn.
Half hard: 62,000 TS; 40,000 YS; 18 El; 65 Rock B.
Soft: 54,000 TS; 22,000 YS; 32 El; 50 Rock B.
Low leaded naval brass; for marine parts and screw machine products.

BOHNALLOY R60N.
M-138; 59.0-62.0 Cu, 0.5-1.0 Sn, 0.20 max Pb, 0.10 max Fe, bal Zn.
Half hard 62,000 TS; 40,000 YS; 22 El; 65 Rock B.
Soft: 54,000 TS; 22,000 YS; 40 El; 50 Rock B.
For water pump parts, shafting, bushings and turnbuckle bands.
Naval brass, uninhibited.

BOHNALLOY R61.
M-138; 60.0-63 Cu, 2.5-3.7 Pb, 0.25 max Fe, bal Zn.
Half hard: 60,000 TS; 45,000 YS; 18 El; 75 Rock B.
Soft: 48,000 TS; 18,000 YS; 45 El; 20 Rock B.
For high speed automatic screw machine parts.
Free cutting brass.

BOHNALLOY R62.
M-138; 59.0-64.5 Cu (60.0 min for rod), 2.0-3.0 Pb, 0.17 max Fe, bal Zn.
Soft: 52,000 TS; 25,000 YS; 25 El; 55 Rock B.
For moderate thread rolling, slight cold working.
Extra high leaded brass.

BOHNALLOY R63.
M-138; 59.0-64.5 Cu (61.0 min for rod), 1.3-2.3 Pb, 0.10 max Fe, bal Zn.
Extruded: 55,000 TS; 40,000 YS; 30 El; 65 Rock B.
For thread rolling, staking & bending.
High leaded brass.

BOHNALLOY R64.
M-138; 63.0-66.0 Cu, 0.50-1.5 Pb, 0.15 max Fe, bal Zn.
Extruded: 55,000 TS; 40,000 YS; 30 El; 65 Rock B.
For heading, upsetting, roll threading and riveting.
Low leaded brass.

BOHNALLOY R67.
M-138; 58-63 Cu, 0.40-3.0 Pb, 0.50 max Fe, 0.30 max Sn, 0.25 max Ni, 0.25 max Al, 2.0-3.5 Mn, 0.50-1.5 Si, bal Zn.
Half hard: 87,000 TS; 65,000 YS; 15 El; 85 Rock B.
For bearings and bushings.
Free cutting bearing bronze.

BOHNALLOY R71.
M-138; 57.0-60.0 Cu, 0.35 max Pb, 0.25 max Fe, 0.50-2.0 Al, 2.0-3.5 Mn, 0.50-1.5 Si, bal Zn.
Half hard: 89,000 TS; 60,000 YS; 15 El; 85 Rock B.
Soft: 68,000 TS; 40,000 YS; 20 El; 75 Rock B.
For bushings, connecting rods, marine hardware and gears.
Forgeable bearing alloy.

BOHNALLOY R72N.
M-138; 55.0-58.0 Cu, 0.25-1.50 Pb, 0.35 max Fe, 1.50-2.50 Ni, 0.25 max Al, 2.0-3.50 Mn, 0.50-1.50 Si, bal Zn.
1/2 hard 1 in. rod: 75,000 TS; 55,000 YS; 19 El; 81 Rock B.
For gear blanks and hydraulic cylinder barrels.
Free cutting bearing bronze.

BOLONEY.
M-714; Sn, Pb, bal Cu.
For bearings; heavy duty.

BOLSTER SILVER.
M-Eng.; 65.5 Cu, 16 Zn, 18 Ni, 0.5 P.
For ornaments, hardware; spun and drawn.

BOLTOMET 103.
M-575; 99.99 Cu.
Oxygen free H.C. copper, electronic grade.
B.S. No. C 110.

BOLTOMET 104.
M-575; 99.95 Cu.
Oxygen free H.C. copper.
B.S. No. C 103.

BOLTOMET 105.
M-575; 99.90 Cu.
Electrolytic tough pitch H.C. copper.
B.S. No. C 101.

BOLTOMET 107.
M-575; 99.90 Cu.
Fire refined tough pitch H.C. copper.
B.S. No. C 102.

BOLTOMET 112.
M-575; 0.09 Ag, 99.95 Cu+Ag.
Oxygen free H.C. copper strip.

BOLTOMET 113.
M-575; 0.07 Ag, 99.90 Cu+Ag.
Tough pitch H.C. copper.
For commutator bars.

BOLTOMET 115.
M-575; 0.15 Ag, 99.90 Cu+Ag.
Tough pitch H.C. copper.

BOLTOMET 117.
M-575; 0.09 Ag, 99.90 Cu+Ag.
Low oxygen H.C. copper strip.

BOLTOMET 123.
M-575; 0.03 P, 99.85 Cu.
Phosphor deoxidized non-arsenical copper.
B.S. No. C 106.

BOLTOMET 160.
M-575; 0.06 P, 99.85 Cu.
High phosphorus deoxidized non-arsenical copper wire.

BOLTOMET 162.
M-575; 99.85 Cu, 0.04 P.
Phosphorus deoxidized copper.
Nominal resistivity: 0.023 microhm metre.

BOLTOMET 210.
M-575; 0.9 Cd, 0.01 P, bal Cu.
1% Cadmium copper.
B.S. No. C 108.

BOLTOMET 302.
M-575; 0.4 Sn, 0.05 P, bal Cu.
0.4% Tin bronze wire.

BOLTOMET 304.
M-575; 1.0 Sn, 0.02 P, bal Cu.
1% Tin bronze; nominal resistivity: 0.035 microhm metre.

BOLTOMET 305.
M-575; 1.0 Sn, 0.08 P, bal Cu.
1% Tin bronze, in strip form.

BOLTOMET 307.
M-575; 1.25 Sn, 0.025 P, bal Cu.
1.25% Tin bronze; nominal resistivity: 0.041 microhm metre.

BOLTOMET 309.
M-575; 2.5 Sn, 0.045 P, bal Cu.
2.5 Tin bronze; nominal resistivity: 0.061 microhm metre.

BOLTOMET 317.
M-575; 5.0 Sn, 0.13 P, bal Cu.
5% Tin bronze.
B.S. No. PB 102.

BOLTOMET 320.
M-575; 6.15 Sn, 0.24 P, bal Cu.
6% Tin bronze wire.
B.S. No. PB 103.

BOLTOMET 338.
M-575; 7.7 Sn, 0.30 P, bal Cu.
8% Tin bronze strip and wire.

BOLTOMET 510.
M-575; 10 Zn, 90 Cu.
90/10 Brass strip.
B.S. No. CZ 101.

BOLTOMET 514.
M-575; 15 Zn, 85 Cu.
85/15 Brass strip.
B.S. No. CZ 102.

BOLTOMET 516.
M-575; 20 Zn, 80 Cu.
80/20 Brass strip and wire.
B.S. No. CZ 103.

BOLTOMET 518.
M-575; 30 Zn, 70 Cu.
70/30 Brass strip.
B.S. No. CZ 106.

BOLTOMET 520.
M-575; 35 Zn, 65 Cu.
65/35 Brass wire.
B.S. No. CZ 107.

BOLTOMET 522.
M-575; 37 Zn, 63 Cu.
63/37 Brass strip, wire and rod.
B.S. No. CZ 108.

BOLTOMET 611.
M-575; 59 Cu, 1.8 Pb, bal Zn.
Leaded brass rod, strip and wire; free machining.
B.S. No. CZ 120/122.

BOLTOMET 613.
M-575; 58 Cu, 2.5 Pb, bal Zn.
Leaded brass rod; free machining.
B.S. No. CZ 121.

BOLTOMET 710.
M-575; 62.5 Cu, 1.2 Sn, bal Zn.
Naval brass.
Nominal resistivity: 0.071 microhm metre.
B.S. No. CZ 112.

BOLTOMET 807.
M-575; 10 Al, 5 Fe, 5 Ni, bal Cu.
Aluminium bronze rod.
B.S. No. CA 104; DTD 197.

BOLTOMET 814.
M-575; 0.7 Cr, bal Cu.
Copper - Chromium.

BOLTOMET 818.
M-575; 0.7 Cr, 0.05 Mg, bal Cu.
Copper - Chromium - Magnesium.

BOLTOMET 917.
M-575; 0.4 S, 0.005 P, bal Cu.
Sulphurized copper.
B.S. No. C111.

BONDED CARBIDE.
M-140; 0.7 C, 0.2 Mn, 4.5 Cr, 1.5 V, 18 W, 0.7 Mo, 10 Co, bal Fe.
For tools, dies, cutters, drills, taps, reamers; high speed steel.

BONDUR.
M-116; 3.5-5.5 Cu, 0.3-0.5 Si, 0.25-1.0 Mn, 0.2-0.7 Mg, bal Al.
Annealed: 31,000 TS; 15 El.
Heat treated: 65,000 TS; 43,000 YS; 16 El; 125 Brin.
For light alloy parts, aircraft parts; age-hardening. Duralumin type alloy.

BONDUR.
M-1548; 3.5-5.5 Cu, 0.4 Si, 0.3-1.0 Mn, 0.2-0.7 Mg, bal Al.
Heat treated: 74,000-80,000 TS; 57,000-63,000 YS; 8-12 El; 130-150 Brin.
For structural components, transportation equipment, fittings, fasteners. Age-hardenable, high strength.

BONDURPLATE.
M-116; 2.5-5.0 Cu, 0.2-1.8 Mg, 0.3-1.5 Mn, bal Al, clad with 0-1.5 Mn 0-1.0 Mg, 0-1.2 Si, bal Al.
Aged: 55,000-60,000 TS; 35,000-45,000 YS; 20-15 El.
For aircraft construction, wings, fuselage; age-hardenable clad alloy.

BONDWICH.
M-926; shim material with silver brazing alloy clad on both sides.
For sandwich brazing of carbide-tipped cutting tools; shim absorbs stresses during brazing and cutting.

BONGRIP.
M-1787.
WC plus alloy.
For anti-skid tire studs and studded straps for snow and ice chains.

BOOTH MG1 replaced by **ALCAN GB-B57S.**

BOOTH MG2 replaced by **ALCAN GB-M57S.**

BOOTH MG3 replaced by **ALCAN GB-54S.**

BOOTH MG3C replaced by **ALCAN GB-B53S.**

BOOTH MG5S replaced by **ALCAN GB-D54S.**

BORAL.
M-1395.
Composite of Boron carbide and Aluminum. High neutron absorbing properties.
For shielding and spent fuel storage containers.

BORAWIRE.
M-200; 61 Au, plus Pt.
Wire for dental purposes.

BORCHER ALLOY.
M-Eng.; 24 Ni, 32.5 Cr, 0.5 Ag, 1.8 Mo, bal Fe.
For chemical apparatus, crucibles, pyrometer tubes; heat and corrosion resistant.

BORCHER ALLOY 1.
M-U.S.; 30 Cr, 35 Co, 35 Ni.
For chemical apparatus; heat and corrosion resistant.

BORCHER ALLOY 2.
M-U.S.; 36 Cr, 60 Fe, 4 Mo.
For heat treating and annealing pots; heat and corrosion resistant.

BORCHER ALLOY 3.
M-U.S.; 65 Cr, 35 Fe.
For pyrometer tubes, crucibles; heat and corrosion resistant.

BORCHER ALLOY 4.
M-U.S.; 34 Co, 34 Ni, 30 Cr, 2 Ag.
For chemical apparatus; heat and corrosion resistant.

BORCHER ALLOY 5.
M-U.S.; 35 Co, 35 Ni, 30 Cr, 0.5-5.0 Mo.
For heat resisting parts, annealing pots; heat and corrosion resistant.

BORCHER'S "A".
65-68 Ni, 30 Cr, 0.5-5.0 Au, 0.25-1.5 Ag.
For resistances, heat and corrosion resistant parts; heat and corrosion resistant.

BORCHER'S "B".
52-68 Ni, 30 cr, 0,15-1.5 Ag, bal Au.
For heat and corrosion resistant parts; heat and corrosion resistant.

BORE 2.
M-111; 1.1 C, 0.3 Cr, 1 W, 0.1 V, bal Fe.
For twist drills, finishing cutters; water hardened.

BORE 2 LEDLOY.
M-111; 1.1 C, 0.3 Cr, 1 W, 0.1 V, 0.2 Pb, bal Fe.
For twist drills, punches; water hardened.

BORITE.
M-333; 1-1.2 C, bal Fe.
For tools, drills; water hardening.

BORITE HOLLOW.
M-333; 0.70 C, 0.20 Si, 0.20 Mn, bal Fe.
Heat treated: 175,000 TS; 128,000 YS; 12 El; 37 RA; 360 Brin.
For drills, tools, springs, hammers; Type W1; water hardened.

BORITE SOLID DRILL.
M-333; 0.75 C, 0.20 Si, 0.20 Mn, bal Fe.
For moil points; blacksmith tools; shock resistant.

BORIUM.
M-202; WC + W_2C.
For welding rod for hard facing, hard inserts; hard and abrasion resisting.

BOROBEST 2D.
M-1343; 0.20 C, 13 Cr, bal Fe.
Annealed: 95,000 TS; 50,000 YS; 25 El; 55 RA; 195 Brin.
For cutlery, turbine blades, surgical instruments; Type 420; stainless, hardenable.

BOROBEST 4K.
M-1343; 0.40 C, 13 Cr, bal Fe.
Annealed: 95,000 TS; 50,000 YS; 20 El; 50 RA; 200 Brin.
For cutlery, valves, surgical instruments; Type 420; stainless, hardenable.

BOROD.
M-202; 62 WC+ 38 steel.
For welding rod for hard facing; hard and abrasion resisting.

BOROFIL see IMI 161.

BOROLITE.
M-1284; Boron carbide.
For cutting tools; sintered carbide.

BORON.
M-1755; B, (crystalline and amorphous).
Standard purities: 99.9999%, 99.999%, 99.9%, 99.5%, 95-97%, 85-92% (Commercial grade).
Forms: Zone refined rod, lump, powders, single crystals, Isotopes B_{10}, B_{11} (Various enrichments).

BORON DEOXIDIZED COPPER 1170.
M-8; 0.01 B, bal Cu.
Hard: 48,000 TS; 40,000 YS; 6 El; B 50 Rock.
Soft: 33,000 TS; 10,000 YS; 45 El; F 45 Rock.
For electrical and electronic parts, magnetrons, synchrotons, vacuum switchgear.
Resists oxide penetration and thermal stress cracking. Resists grain growth.

BOROSIL.
M-398; 3-4 B, 38-42 Si, bal Fe.
For boron additions to steel.

BOROTAL Z7.
M-Ger.; 3-4 Cu, 2 max Fe, 3 max Pb, 3 max Zn, 0.1 graphite, bal Al.
For bearings.

BOROTEC 10009.
M-717.
Powder for spray coating that resists metal-to-metal friction. RC 55-62.

BOROTO B.K.
M-Germany; graphitic white metal.
31 Brin.
For bearings; for high stresses.

BOROTO B.L.
M-Ger.; graphitic white metal.
35 Brin.
For bearings; high friction and heavy torque.

BOROTO B.N.
M-Ger.; graphitic white metal.
18 Brin.
For bearings; for high speeds.

BOROTO B.R.
M-Ger.; graphitic white metal.
27 Brin.
For bearings; high friction and speed.

BOSCH AL 1.
M-304; 2.5-5.0 Cu, 0.2-1.8 Mg, 0.3-1.5 Mn, bal Al.
For aircraft structures; age hardenable.

BOSCH (AL-5).
M-304; 6 Cu, 3 Si, bal Al.
Sand cast: 21,200 TS; 1-2 El.
Chill cast: 24,400 TS; 1-2 El; 80 Brin.
Die cast: 28,000 TS; 1-2 El.
For Al castings; very good machinability.

BOSCH (AL-7).
M-304; 89 Al, bal Zn + modifying agent.
Annealed: 36,000 TS; 20 El.
Soft drawn: 42,500 TS; 12 El; 30 RA; 85 Brin.
For parts for machinery and apparatus, pressed, forged and drawn parts; ages at room temperature.

BOSCH (AL-9).
M-304; 4 Cu, 2 Si, 1.5 Ni, bal Al.
Die cast: 28,000 TS; 2 El; 80 Brin.
For ordinary die cast parts; good fluidity.

BOSCH AL 12.
M-304; 99.5 Al.
Annealed: 13,000 TS; 5000 YS; 45 El; 23 Brin.
Hard: 24,000 TS; 22,000 YS; 15 El; 44 Brin.
For structures, machine tool parts; corrosion resistant.

BOSCH (AL-20).
M-304; 88 El, 11 Zn, 0.5 Mg.
Annealed: 46,000 TS; 36,000 YS; 18 El; 25 RA.
Pressed: 50,000 TS; 18 El; 20 RA; 100 Brin.
For high quality parts for fine mechanism; good hot working properties.

BOSCH AL 24.
M-304; 0.6-1.4 Mg, 0.6-1.6 Si, 0.6-1.0 Mn, 0.3 max Cr, bal Al.
Annealed: 21,000 TS; 8000 YS; 24 El; 36 Brin.
Hard: 32,000 TS; 29,000 YS; 6 El.
For structural members.

BOSCH AL 25.
M-304; 4.0-5.5 Mg, 0.8 max Mn, 0.3 max Cr, bal Al.
Annealed: 42,000 TS; 22,000 YS; 35 El; 65 Brin.
For light alloy parts, marine structures; resists gasoline corrosion.

BOSCH AL 27.
M-304; 2.5-5.0 Cu, 0.2-1.8 Mg, 0.3-1.5 Mn, Pb, Sn, Cd, Bi, bal Al.
For screw machine products; free-cutting.

BOSCH AL 29.
M-304; 6-10 Mg, 0.2-0.7 Mn, 1.5 max Fe, bal Al.
For gasoline meters, aircraft parts; corrosion resistant.

BOSCH (AL-34).
M-304; 89 Al, 11 Cu.
Chill cast: 21,000 TS; 1-2 El; 80 Brin.
Sand cast: 20,000 TS; 1-2 El; 75 Brin.
For automobile pistons, sand and permanent mold parts; good machinability.

BOSCH (AL-36).
M-304; 89 Al, 6.5 Cu, 3.5 Ni.
Annealed; 19,000 TS; 1-2 El.
Die cast: 27,000 TS; 2 El; 80 Brin.
For die castings; good fluidity and machinability.

BOSCH AL 39.
M-304; 1.0-1.5 Mn, 0.3 max Cr, bal Al.
Annealed: 16,000 TS; 6000 YS; 40 El; 28 Brin.
Hard: 29,000 TS; 27,000 YS; 10 El; 55 Brin.
For light alloy tanks, formed parts; good formability and corrosion resistance.

BOSCH AL 42.
M-304; 5.0-6.5 Si, 2-3 Cu, 0.2-0.6 Mn, 1.5 max Fe, bal Al.
For light alloy parts; high corrosion resistance.

BOSCH AM-14.
M-304; 90 Cu, 10 Al.
Pressed: 85,000 TS; 25 El; 150 Brin.
For parts exposed to H_2SO_4 and HNO_3; acid resistant.

BOSCH AM-158.
M-304; 1.5 Mn, bal Al.
Pressed: 17,800 TS; 20 El; 50 RA; 35 Brin.
Drawn: 21,300 TS; 6 El; 30 RA; 50 Brin.
For rods, tubes, pressings; corrosion resistant.

BOSCH (AN-4).
M-304; 4 Zn, 3 Cu, balance Al plus modifying agent.
Die cast: 50,100 TS; 2 El; 110 Brin.
For die castings; high tensile and bending strength.

BOSCH (BR-6).
M-304; 66 Cu, 6 Sn, bal Zn.
Pressed: 50,000 TS; 0.5 El; 140 Brin.
For bearings subject to heavy loads and wear; hard, dense bronze.

BOSCH (BR-14).
M-304; 86 Cu, 10 Sn, 2 Zn, 2 Pb.
Sand cast: 36,000 TS; 6 El; 80 Brin.
For bearings, phosphor bronze castings; corrosion and high wear resistant, tough, hard.

BOSCH (BR-32).
M-304; 76 Cu, Al, Ni, Fe, Mn.
Pressed: 120,000 TS; 72,000 YS; 0.5 El; 20 RA; 250 Brin. For gears, forged parts; special bronze; wear resistant.

BOSCH (CU-28).
M-304; 99.9 Cu.
Annealed: 40,300 TS; 40 El.
Pressed: 33,000 TS; 40 El; 60 RA; 50 Brin.
For bars, shapes, tubes, pressed parts; electrolytic copper; corrosion resistant.

BOSCH (ME-1).
M-304; 58 Cu, 1.8 Pb, bal Zn.
Annealed: 67,000 TS; 27 El; 30 RA; 115 Brin.
Pressed: 62,500 TS; 18 El; 20 RA; 105 Brin.
For hot pressed parts for machines, apparatus and equipment; easily workable.

BOSCH (ME-2).
M-304; 58 Cu, 2.2 Pb, bal Zn.
Annealed: 67,000 TS; 30 El; 35 RA; 110 Brin.
Soft drawn: 72,000 TS; 20 El; 40 RA; 130 Brin.
For automatic parts, spring boxes, ornamental and hot pressed parts; easil workable.

BOSCH (ME-3).
M-304; 59 Cu, 2.5 Pb, bal Zn.
Pressed: 65,000 TS; 30 El; 38 RA.
Soft drawn: 72,000 TS; 20 El; 25 RA; 120 Brin.
For screws and turned parts for fine work, clock manufacture; somewhat cold workable.

BOSCH (ME-4).
M-304; 59 Cu, 1 Pb, bal Zn.
Pressed: 65,000 TS; 30 El; 45 RA.
Soft drawn: 72,000 TS; 20 El; 25 RA; 100 Brin.
For wire, hot pressed and forged parts; can be cold drawn.

BOSCH (ME-5) RIVET BRASS.
M-304; 62 Cu, 0.5 Pb, bal Zn.
Pressed: 58,000 TS; 40 El; 60 RA.
Soft drawn: 65,000 TS; 30 El.
For clock manufacture and fine mechanical work, rivets, tubes; cold workable.

BOSCH (ME-6) (BOSCH SPECIAL).
M-304; 57 Cu, some Ni and Mn, bal Zn.
Pressed: 72,000 TS; 30 El; 40 RA.
Soft drawn: 78,500 TS; 20 El.
For piston rods, screws, valve spindles, condenser shells; corrosion resistant to sea water.

BOSCH ME-15.
M-304; 78 Cu, 4 Pb, 3 Sn, bal Zn.
Sand cast: 32,000 TS; 15 El; 70 Brin.
For housings, fittings; free cutting.

BOSCH (ME-16).
M-304; 61.5 Cu, 2 Pb, bal Zn.
Pressed: 50,000 TS; 35 El; 40 RA; 80 Brin.
Soft drawn: 58,000 TS; 30 El; 40 RA; 90 Brin.
For rods, bars, and shapes handled by automatic machinery; cold workable.

BOSCH (ME-17).
M-304; 63 Cu, bal Zn.
Pressed: 46,000 TS; >40 El; 55 RA; 75 Brin.
Soft drawn: 55,000 TS; >35 El; 50 RA; 85 Brin.
For cold formed rods and shapes, rivets; excellent cold forming.

BOSCH (ME-25) (BOSCH METAL).
M-304; 57 Cu, Mn, Al, Ni, bal Zn.
Pressed: 86,000 TS; 15 El; 20 RA; 135 Brin.
Soft drawn: 94,000 TS; 12 El; 15 RA; 155 Brin.
For bushings, bars, shapes; resists sea water corrosion and wear.

BOSCH NS-1.
M-304; Cu, Ni, Fe, Mn.
Pressed: 85,000 TS; 30 El; 30 RA; 150 Brin.
Hard: 100,000 TS; 15 El; 20 RA; 180 Brin.
For plates of nickel color; corrosion resistant.

BOSCH PB-7.
M-304; 74 Pb, 10 Sn, 15 Sb, 1 Cu.
Cast: 8000 TS; 30 Brin.
For bearing shells; anti-friction.

BO-STAN.
M-1469; 98.7 Type 304 stainless, 1.2 B, bal Fe.
Annealed: 104,500 TS; 7.5 El; 5.5 RA; 220 Brin.
Cold worked: 420 Brin.
For pressurized water reactors and controls; sintered, corrosion and heat resistant.

BOURBOUNES.
M-Eng.; 51 Sn, 49 Al, 0.3 Fe, 0.3 Cu.

BOWCO.
M-336; C, W, Mn, Cr, V, Mo, Co, bal Fe.
For reamers, punches, dies, stamping and cutting tools; high speed steel.

BOWCO 7720.
M-336; C, Cr, Mn, Mo, bal Fe.
For blanking dies; oil hardening.

BOWCO COBALT.
M-336; C, W, Co, Mn, Cr, V, Mo, bal Fe.
For tools for trimming and heavy cutting on hard and abrasive material; high speed steel.

BOWCO FAST FINISHING.
M-336; 1.3 C, 1.8 Cr, 4 W, 0.2 V, bal Fe.
For dies, cutting tools; oil hardening.

BOWCO NONSHRINK.
M-336; 0.9 C, 1.2 Mn, 0.5 Cr, 0.5 W, 0.35 Si, bal Fe.
For tools, dies; oil hardening, non-deforming.

BOWCO OIL HARDENING.
M-336; 0.9 C, 1.3 Mn, 0.5 Cr, 0.5 W, 0.1 V, bal Fe.
For thread gages, taps, dies; non-shrinking.

BOWCO ONE STAR.
M-336; 0.7 C, 14 W, 4 Cr, 2 V, bal Fe.
For tools, cutters for intermittent cutting; high speed steel.

BOWCO WATER HARDENING TUBING.
M-336; 1.0-1.1 C, bal Fe.
For forming and blanking dies; water-hardened.

BOWER 315.
M-709, M-97; 0.10-0.15 C, 0.5 Mn, 0.3 Si, 2.8 Ni, 1.5 Cr, 5.0 Mo, bal Fe.
For bearings for service up to 600°F, case hardened.

BOWSTEEL FC.
M-1434; 1.0 C, 0.25 V, bal Fe.
For tools, cutters, taps, reamers; Type W2; water hardened.

BOWSTEEL FCCR.
M-1434; 1.5 C, 12 Cr, 1 Mo, 3 Co, bal Fe.
For blanking and forming tools and dies; Type D5; air hardened, non-deforming.

BOWSTEEL FND.
M-1434; 0.90 C, 1 Mn, 0.5 Cr, 0.5 W, bal Fe.
For dies, punches, broaches, taps, hobs; Type 01; oil hardened, non-deforming.

BOW WIRE.
M-Eng.; 93 Cu, 2 Zn, 5 Sn.
For corrosion resistant wire; corrosion resistant.

BOYD.
M-1705; 0.75 C, 0.75 Mn, 0.9 Cr, 1.75 Ni, 0.35 Mo, bal Fe.
Tool steel, miscellaneous applications as dies, wood working tools, shears. AISI L6.

BOYD-WAGNER B-W POINT 5.
M-336; 0.02 C, 0.18 Mn, trace Si, 0.012 S, 0.004 P, bal Fe.
For plastic molds, white metal, die casting dies; water hardening.

BOYD-WAGNER CHROME MAGNET.
M-336; C, Cr, bal Fe.
For magnets; magnet steel.

BOYD-WAGNER EZ9W.
M-336; 0.35 C, 5 W, bal Fe.
For punches, shear tools, gripper dies; hot work steel, oil hardened.

BOYD-WAGNER NO. 4 SWEDISH.
M-336; 1.05 C, 0.5 Cr, 0.1 V, bal Fe.
For wood and machine screw cold header dies; water hardening.

BOYD-WAGNER NO. 9 SWEDISH.
M-336; 0.9 C, 0.5 Cr, bal Fe.
For large cold header dies; water hardening.

BOYD-WAGNER NO. 41 SWEDISH.
M-336; 0.7 C, 0.6 Mn, bal Fe.
For chipping chisels; water hardening.

BOYD-WAGNER TUNGSTEN MAGNET.
M-336; C, W, bal Fe.
For magnets; magnet steel.

BP 24.
M-912; 0.14 C, 0.55 Mn, 0.75 Cr, 3.45 Ni, bal Fe.
Case hardening steel; for plastic moulds.
W.-Nr. 2735.

BP 24 EXTRA.
M-912; 0.19 C, 0.40 Mn, 1.30 Cr, 0.20 Mo, 4.0 Ni, bal Fe.
Air hardenable, case hardening steel with low distortion; good for plastic moulding; may be nitrided and highly polished.
W.-Nr. 2764.

BP 28.
M-912; 0.20 C, 1.25 Mn, 1.15 Cr, bal Fe.
Case hardenable steel for plastic molds.
W.-Nr. 2162.

BP 41.
M-912; 0.23 C, 0.45 Mn, 13.50 Cr, 0.90 Ni, bal Fe.
Corrosion resistant steel for use with corrosive plastics.

BP 42.
M-912; 0.40 C, 0.70 Mn, 16.0 Cr, 1.1 Mo, bal Fe.
Corrosion resistant steel for use with corrosive plastics.
W.-Nr. 2316.

BP313.
M-1741; 3.0-4.0 Cu, 7.5-9.5 Si, 0.6 max Fe, 0.10 max Mn, 0.10 max Mg, 0.10 max Ni, 0.10 max Zn, bal Al.
Pressure cast F1: 324 MPa TS; 159 MPa YS; 4 El, 80 Brin.
ADC H49-16; BS LM24.

BP401.
M-1741; 0.10 max Cu, 11.0-13.0 Si, 0.40 max Fe, 0.1 max Mn, 0.05 max Mg, 0.05 max Ni, 0.1 max Zn, 0.2 max Ti, bal Al.
Per. mold cast, F1: 207 MPa TS; 90 MPa YS; 9 El, 60 Brin.
ADC H49-5; BS LM6.

BP601.
M-1741; 0.05 max Cu, 6.5-7.5 Si, 0.11 max Fe, 0.05 max Mn, 0.30-0.40 Mg, 0.05 max Zn, 0.2 max Ti, bal Al.
Per. mold cast, T6: 276 MPa TS; 165 MPa YS; 17 El, 85 Brin.
BS LM25.

BP605.
M-1741; 0.10 max Cu, 9.0-10.0 Si, 0.6 max Fe, 0.05 max Mn, 0.45-0.60 Mg, 0.05 max Zn, bal Al.
Per. mold cast: 303 MPa TS; 248 MPa YS; 5 El; 100 Brin.
ADC H49-23.

BPS.
M-1655; 0.21 C, 1.20 Mn, 0.30 Si, 1.0 Cr, bal Fe.
For plastic molds, measuring instruments.
W.-Nr. 1.2162.

BPS 2.
M-1655; 0.40 C, 0.30 Mn, 0.40 Si, 13.0 Cr, bal Fe.
Dies and molds for corrosive plastics.
W.-Nr. 1.2083.

BPS 3.
M-1655; 0.21 C, 1.20 Mn, 0.30 Si, 1.0 Cr, bal Fe.
For plastic molds, measuring instruments.
W.-Nr. 1.2082.

BR-2 DIE STEEL.
M-73; 2.5 C, 0.3 Si, 0.75 Mn, 5.25 Cr, 1.1 Mo, 4.5 V, bal Fe.
Hardened: C 55-66 Rock.
For dies, brick and tile mold liners and sleeves requiring extreme abrasion resistance; air hardenable, good temperature resistance.

BR-3 see LATROBE BR3.

BR-3 & BR-4 see LATROBE BR-3 AND BR-4.

BR-4 FM see LATROBE BR-4FM.

BRAEBURN M-33.
M-140; 0.90 C, 1.5 W, 9.5 Mo, 4.0 Cr, 1.15 V, 8.0 Co, bal Fe.
Cobalt-Moly high speed steel; for extra cutting ability.
AISI M-33.

BRAEBURN W-5.
M-140; 1.1 C, 0.50 Cr, bal Fe.
Water hardening tool steel, for extra wear properties.
AISI W-5.

BRAECUT.
M-140; 1.15 C, 4.25 Cr, 2.25 V, 5.25 W, 12 Co, 6.25 Mo, bal Fe.
For cutting tools for high temperature alloys: high speed steel.

BRAEFOUR.
M-140; 1.25 C, 4.5 Cr, 4 V, 5.5 W, 4.5 Mo, bal Fe.
Hardened: C 64-66 Rock.
For cutting, finishing and form tools, broaches, end mills, hobs, gauges, lathe and planer tools, reamers.
Type M-4 high speed steel, excellent abrasion resistance and high red-hardness.

BRAEMAX.
 M-140; 1.1 C, 1.5 W, 9.5 Mo, 3.75 Cr, 1.15 V, 8 Co, bal Fe.
 Hardened: C 66-70 Rock.
 For broaches, form cutters, hobs, end mills, milling cutters, gear shaper cutters, taps.
 High speed steel. High red-hardness and abrasion resistance.

BRAEMOW M-2.
 M-140; 0.8 C, 0.2 Mn, 4.2 Cr, 2 V, 6.5 W, 5.0 Mo, bal Fe.
 For tools, cutters; high speed steel.

BRAETWIST.
 M-140; 0.90 C, 0.30 Mn, 0.30 Si, 9.5 Mo, 1.6 W, 3.75 Cr, 1.15 V, 8.0 Co, bal Fe.
 Cobalt-moly high speed steel for drills, taps, reamers, broaches, milling cutters and end mills.
 AISI M-33.

BRAEVAN.
 M-140; 1.02 C, 5.7 Mo, 6.25 W, 4 Cr, 2.5 V, bal Fe.
 Heat treated: 630-660 Brin.
 For broaches, reamers, form tools, drills; high speed steel.

BRAEVAN 2.
 M-140; 1.15 C, 5.5 Mo, 5.6 W, 4 Cr, 3.3 V, bal Fe.
 For broaches, reamers, end mills, form tools; high speed steel.

BRAEVAN-M-3.
 M-334; 1 C, 6.2 W, 5.6 Mo, 2.5 V, 4.0 Cr, bal Fe.
 For lathe and planer tools, reamers, broaches; high speed steel.

BRAGE 1.
 M-111; 1.05 C, 0.5 Cr, 0.1 V, bal Fe.
 For threading tools; water hardening.

BRAGE 2.
 M-111; 1.1 C. 0.25 Cr, 1 W, 0.1 V, bal Fe.
 For threading tools, twist drills; water hardening.

BRAKE DIE.
 M-24; 0.51 C, 0.95 Cr, 0.87 Mn, 0.2 Mo, bal Fe.
 For dies, press brakes; pre-heat treated.

BRAKE DIE.
 M-341; 0.51 C, 0.95 Cr, 0.87 Mn, 0.2 Mo, bal Fe.
 Hardened: 140,000 TS; 293 Brin.
 For press brake dies; oil hardening.

BRAND 110.
 M-1409; 0.55.-0.65 C, bal Fe.
 For drop forging die blocks; water hardened.

BRAND 113.
 M-1409; 0.5 C, 1 Ni, bal Fe.
 For drop forging die blocks; water hardened.

BRAND 123.
 M-1409; C, Ni, Cr, bal Fe.
 For cold chisels and stakes; oil hardened, shock resistant.

BRAND 124.
 M-1409; 0.5 C, 3 Ni, 0.75 Cr, bal Fe.
 For drop forging dies; oil hardened, shock resistant.

BRAND 125.
 M-1409; 0.5 C, 1.5 Ni, 1 Cr, 0.25 Mo, bal Fe.
 For drop forging dies; oil hardened, shock resistant.

BRAND 131.
 M-1409; C, Ni, Cr, bal Fe.
 For cold chisels and stakes; oil hardened, shock resistant.

BRAND 133.
 M-1409; 0.15 C, bal Fe.
 Annealed: 70,000 TS; 40,000 YS; 25 El; 60 RA; 143 Brin.
 For gears, pinions, shafts; case hardened.

BRAND 134.
 M-1409; 0.15 C, 3 Ni, bal Fe.
 For gears, pinions, camshafts, cams; case hardened, tough.

BRAND 135.
 M-1409; 0.15 C, 2 Ni, Mo, bal Fe.
 For gears, pinions, camshafts, cams, shafts; case hardened, tough.

BRAND 138.
 M-1409; 0.15 C, 5 Ni, bal Fe.
 For gears, pinions, camshafts, cams; case hardened, tough.

BRAND 140.
 M-1409; 0.15 C, Ni, Cr, bal Fe.
 For gears, pinions, camshafts, fasteners; case hardened, tough.

BRAND F.T. 20-30.
 M-1409; 0.12 C, 25 Cr, 20 Ni, bal Fe.
 For furnace parts, glass molds; heat and scale resistant to 1250°F.

BRAND F.T. 1000.
 M-1409; 0.2-0.3 C, 18-20 Cr, 0.4 Ni, 1.2 Mo, bal Fe.
 For furnace parts, glass molds; heat and scale resistant to 1150°C.

BRASS, DRAWING.
 M-8; 67-70 Cu, 30-33 Zn.
 Sand cast: 40,000 TS; 23,000 YS; 35 El; 35 RA; 45 Brin.
 Hard rolled: 67,200 TS; 67,000 YS; 15 El; 50 RA; 145 Brin.
 For seamless tubes; deep drawing.

BRASS E-133.
M-1191; 58.5-60.5 Cu, 1.5-2.0 Pb, bal Zn.
Forged: 55,000-60,000 TS; 20,000-25,000 YS; 40-30 El; 80-100 Brin.
For hardware, machinery parts; leaded brass, free-cutting.

BRASS, ESCUTCHEON.
M-U.S.; 64.5 Cu, 35.07 Zn, 0.43 Pb.
For escutcheon pins; yellow brass.

BRASS, HIGH.
M-8; 65 Cu, 35 Zn.
Cold rolled: 47,000-75,000 TS; 20,000-60,000 YS; 5-60 El; 5-75 RA; 45-180 Brin.
Used for drawing, forming and spinning parts; high strength.

BRASS, KRUPP NICKEL.
M-72; 48.5 Cu, 24.3 Zn, 24.3 Ni, 2.9 Fe.
For ornamental and corrosion resistant parts; corrosion resistant.

BRASS, LANCASHIRE.
M-Eng.; 73 Cu, 25 Zn, 2 Pb.
For parts to be brazed or soldered; free-cutting.

BRASS, LEADED HIGH.
M-8; 65 Cu, 0.5-1.5 Pb, bal Zn.
For cupped, formed or drawn parts; does not foul cutting tools.

BRASS, LEADED LOW.
M-U.S.; 78 Cu, 20 Zn, 1.7 Pb, traces Fe.
Hard: 80,000 TS; 5 El.
Soft: 40,000 TS; 35 El.
For rivets, pins, wire; free-cutting.

BRASS, LEADED SCREEN WIRE.
M-U.S.; 69 Cu, 30 Zn, 1.0 Pb, traces Fe.
For screens, hardware; high ductility.

BRASS LK80-3L.
M-USSR; 79.28 Cu, 0.28 Pb, 17.35 Zn, 0.24 Fe, 2.8 Si.
Cast: 63,000 TS; 24,000 YS; 36 El; 44 RA.
For hardware, plumbing; red brass.

BRASS LMTS58-2.
M-USSR; 57.2 Cu, 40.15 Zn, 0.08 Fe, 1.54 Mn, 0.01 P.
Rolled: 72,000 TS; 48,000 YS; 26 El; 60 RA.
For marine parts; strong and corrosion resistant.

BRASS LS59-1.
M-USSR; 55.32 Cu, 1.42 Pb, 40.25 Zn.
Rolled: 75,000 TS; 55,000 YS; 18 El; 31 RA.

BRASS, MANGANESE.
M-U.S.; 85-54 Cu, 2-40 Zn, 1-25 Mn, 0-2.4 Fe, 0-2.5 Ni, Al, Pb.
For propellers, marine parts, bolts, fittings; corrosion resistant.

BRASS, MANGANESE NICKEL.
M-U.S.; 66-53 Cu, 5-40 Zn, 2-18 Ni, 1.5-20 Mn, Al, Fe, Sn, Pb.
For condenser tubes, pump parts, corrosion resisting fittings; corrosion resistant.

BRASS, PIN WIRE.
M-U.S.; 61 Cu, 39 Zn.
Wire: 51,000 TS; 20 El.
For brass pins, condenser tubes, water pipes; corrosion resistant.

BRASS STEEL.
M-538; brass coated steel.
For fabricated parts; easily formed, stamped, drawn.

BRAZE 053 (TEC).
M-63; 5 Ag, 95 Cd.
Melt point 640°F (340°C); Flow point 740°F (395°C).
High temperature solder for medium strength joints.

BRAZE 056 (TEC Z).
M-63; 4.5-5.5 Ag, 15.6-17.6 Zn, 77.9-78.9 Cd, others 0.15 max.
Melt point 480°F (250°C); Flow point 600°F (315°C).
High temperature solder for medium strength joints.

BRAZE 071.
M-63; 7 Ag, 85 Cu, 8 Sn.
For brazing alloy; M.P. 1225-1805°F.

BRAZE 090 (TL).
M-63; 9 Ag, 53 Cu, 38 Zn.
Melt point 1410°F. Flow point 1565°F.
For brazing copper base alloys such as band instruments.

BRAZE 200.
M-63; 20 Ag, 45 Cu, 30 Zn, 5 Cd.
Melt point 1140°F. Flow point 1500°F.
For brazing ferrous and non-ferrous alloys, good color match for yellow brass.

BRAZE 202.
M-63; 20 Ag, 45 Cu, 35 Zn.
For brazing steel; M.P. 1315-1500°F.

BRAZE 250 (NE).
M-63; 25 Ag, 52.5 Cu, 22.5 Zn.
Melt point: 1250°F; Flow point 1575°F.
For brazing ferrous and non-ferrous metals that are not damaged by 1600°F temperature.

BRAZE 285.
M-63; 28.5 Ag, 32.3 Cu, 34.2 Zn, 5 Mn.
Melt point 1305°F (705°C); Flow point 1385°F (750°C).
Economical narrow-melt-range filler metal for ferrous and non-ferrous alloys.

BRAZE 300.
M-63; 30 Ag, 38 Cu, 32 Zn.
Melt point 1250°F (675°C); Flow point 1410°F (765°C).
For brazing steel and non-ferrous alloys melting above 1450°F (790°C), as nickel-silver knife handles, electrical equipment.
AWS BAg-20.

BRAZE 400 (DT).
M-63; 40 Ag, 36 Cu, 24 Zn.
M.P. 1235°F; Flow P. 1415°F.
For brazing copper base alloys, monel, mild steel; can braze wide joints.

BRAZE 401.
M-63; 40 Ag, 30 Cu, 30 Zn.
Melt point 1245°F (675°C); Flow point 1340°F (725°C).
For brazing copper alloys, mild steel, nickel and Monel, and for wide gap joints.

BRAZE 403 (SS).
M-63; 40 Ag, 30 Cu, 28 Zn, 2 Ni.
Melt point 1220°F; Flow Point 1435°F.
For brazing tungsten carbide tool tips, and for stainless food handling equipment.
AWS BAg-4.

BRAZE 404 (SS-5).
M-63; 40 Ag, 30 Cu, 25 Zn, 5 Ni.
Melt Point 1220°F; Flow Point 1580°F.
For brazing tungsten carbide tips, and for stainless steel.

BRAZE 450 (DE).
M-63; 45 Ag, 30 Cu, 25 Zn.
Melt Point 1225°F; Flow Point 1370°F.
For brazing ships' piping, band instruments, aircraft engine oil coolers, brass lamps. AWS BAg-5.

BRAZE 495.
M-63; 49 Ag, 16 Cu, 23 Zn, 7.5 Mn, 4.5 Ni.
Melt point 1160°F (625°C); Flow point 1300°F (705°C).
For low temperature brazing of tungsten carbides and stainless steels.

BRAZE 501 (ETX).
M-63; 50 Ag, 34 Cu, 16 Zn.
Melt Point 1250°F; Flow Point 1425°F.
For brazing steam turbine blading and heavily galvanized or tinned steel.
AWS BAg-6.

BRAZE 505.
M-63; 50 Ag, 20 Cu, 28 Zn, 2 Ni.
Melting range: 1220-1305°F.
For brazing 300 series stainless food handling equipment.

BRAZE 541.
M-63; 54 Ag, 40 Cu, 5 Zn, 1 Ni.
Melt Point 1340°F; Flow Point 1575°F.
For atmosphere furnace brazing steel and stainless; for applications up to 700°F.
AMS 4772 B; AWS BAg-13.

BRAZE 559.
M-63; 56 Ag, 42 Cu, 2 Ni.
Melting range: 1420-1640°F.
For furnace brazing of steels and high temp. alloys where zinc fumes are undesirable.
AMS-4765; AWS BAg-13a.

BRAZE 560.
M-63; 56 Ag, 22 Cu, 17 Zn, 5 Sn.
Melt point 1145°F; Flow Point 1205°F.
For brazing food handling equipment requiring a low melting cadmium free alloy.
AWS BAg-7.

BRAZE 580.
M-63; 57.5 Ag, 32.5 Cu, 7 Sn, 3 Mn.
Melt point 1120°F (605°C); Flow point 1345°F (730°C).
For brazing tungsten and chrome carbides, and vacuum brazing of high manganese stainless steels.

BRAZE 600 (RT).
M-63; 60 Ag, 25 Cu, 15 Zn.
Melt Point 1245°F; Flow Point 1325°F.
For brazing Monel and other nickel base alloys and for silverware.

BRAZE 603.
M-63; 60 Ag, 30 Cu, 10 Sn.
Melt Point 1115°F; Flow Point 1325°F.
For brazing marine heat exchangers, ferrous and non-ferrous alloys, vacuum tube seals.
AMS 4773A; AWS BAg-18.

BRAZE 630.
M-63; 63 Ag, 28.5 Cu, 6 Sn, 2.5 Ni.
Melt Point 1275°F; Flow Point 1475°F.
For brazing 400 series stainless, alloy steels, food handling equipment. Good for combined furnace brazing and hardening for low alloy steels.
AMS 4774A.

BRAZE 650 (EASY).
M-63; 65 Ag, 20 Cu, 15 Zn.
Melt Point 1240°F; Flow Point 1325°F.
For brazing silverware, iron and nickel alloys.
ASME BAg-9.

BRAZE 655.
M-63; 65 Ag, 28 Cu, 5 Mn, 2 Ni.
For jet engine components, brazing alloy; M.P. 1385-1560°F.

BRAZE 700 (MEDIUM).
M-63; 70 Ag, 20 Cu, 10 Zn.
Melt Point 1275°F; Flow Point 1360°F.
For brazing silverware when subsequent joints are made with Braze 650.
ASME BAg-10.

BRAZE 720 (BT).
M-63; 72 Ag, 28 Cu.
Melt Point and Flow Point 1435°F. (Eutectic alloy)
For brazing electric components requiring highest electrical and thermal conductivity; low volatile materials and impurities; for furnace brazing.
AWS BAg-8.

BRAZE 750 (HARD).
M-63; 75 Ag, 22 Cu, 3 Zn.
Melt Point 1365°F; Flow Point 1450°F.
For brazing silverware, for step brazing or subsequent enameling; iron and nickel base alloys.
ASME BAg-11.

BRAZE 800 (IT).
M-63; 80 Ag, 16 Cu, 4 Zn.
Melt Point 1340°F; Flow Point 1490°F.
For brazing silver, iron and nickel base alloys.

BRAZE 852.
M-63; 85 Ag, 15 Mn.
Melt Point 1760°F; Flow Point 1780°F.
For brazing stainless steels, Stellite and Inconel.

BRAZE 999.
M-63; 99.9 Ag.
Melt point 1761°F (960°C).
For metallizing ceramics to be used as conductors.

BRAZE ATT see **BRAZE 200 (ATT)**.

BRAZE BT see **BRAZE 720 (BT)**.

BRAZE DE see **BRAZE 450 DE**.

BRAZE DT see **BRAZE 400 DT**.

BRAZE EASY see **BRAZE 650 (EASY)**.

BRAZE ETX see **BRAZE 501 ETX**.

BRAZE HARD see **BRAZE 750 (HARD)**.

BRAZE IT see **BRAZE 800 (IT)**.

BRAZE MEDIUM see **BRAZE 700 (MEDIUM)**.

BRAZE NE see **BRAZE 250 NE**.

BRAZE RT see **BRAZE 600 (RT)**.

BRAZE SS see **BRAZE 403 (SS)**.

BRAZE TL see **BRAZE 090(TL)**.

BRAZINAL.
M-1231; Si, bal Al.
Cast: 45,000 TS; 80 Brin.
For brazing alloy for Al.

BRAZING BRASS.
M-8; 80-75 Cu, 20-25 Zn.
For brazing.

BRAZING BRASS 282.
M-279; 59 Cu, 38.8 Zn, 0.2 P.
Brazing sheets, annealed.
Low melting temperature.

BRAZING METAL.
M-8; 84-86 Cu, 14-16 Zn.
Cast: 95,000-50,000 TS, 7 El; 180 Brin.
For brazed joints on steel parts; tough and ductile.

BRAZING METAL "F".
M-Eng.; 85 Cu, 15 Zn.
For brazing metal, water pipe, architectural purposes; low m.p.

BREARLY PATENT.
M-U.S.; 9-16 Cr, 0.7 max C, bal Fe.
For cutlery, surgical instruments; stainless and corrosion resisting.

BREDA CM10.
M-1449; 0.09-0.12 C, 13 Cr, 0.5 max Si, bal Fe.
Annealed: 75,000 TS; 40,000 YS; 35 El; 70 RA; 155 Brin.
For valves, cutlery, surgical and dental instruments; Type 403; corrosion resistant.

BREDA CM13.
M-1449; 0.26-0.37 C, 13 Cr, bal Fe.
Annealed: 95,000 TS; 50,000 YS; 25 El; 55 RA; 195 Brin.
For valves, cutlery, surgical and dental instruments; Type 420; corrosion resistant.

BREDA CMC.
M-1449; 0.38-0.45 C, 13 Cr, bal Fe.
Annealed: 100,000 TS; 55,000 YS; 23 El; 52 RA; 210 Brin.
For valves, cutlery, surgical and dental instruments; Type 420; corrosion resistant.

BREDA LYS.
M-1449; 0.07-0.12 C, 17 Cr, bal Fe.
Annealed: 80,000 TS; 50,000 YS; 25 El; 50 RA; 150 Brin.
For oil refining equipment, oil burners and heaters; Type 430; corrosion resistant.

BREDA NK1.
M-1449; 0.25 max C, 17 Cr, 2 Ni, bal Fe.
Annealed: 125,000 TS; 95,000 YS; 20 El; 55 RA; 260 Brin.
For pumps, marine hardware, valves; Type 431; corrosion and heat resistant.

BREDA-NK2.
M-1449; 0.20 max C, 24 Cr, 12 Ni, bal Fe.
Annealed: 90,000 TS; 40,000 YS; 50 El; 65 RA; 170 Brin.
For furnace parts, pumps, oil burners, heat treat boxes; Type 309; stainless, austenitic.

BREDA-NK3.
M-1449; 0.25 max C, 25 Cr, 20 Ni, 1.5 max Si, bal Fe.
Annealed: 100,000 TS; 45,000 YS; 50 El; 65 RA; 185 Brin.
For furnace parts, pumps, valves, turbine and jet parts; Type 310; stainless, austenitic.

BREDA RAS.
M-1449; 0.40 C, 10 Cr, 1 Mo, 2.5 Si, bal Fe.
For oil refinery equipment; heat resistant.

BREDA RF302.
M-1449; 0.35 C, 25 Cr, bal Fe.
Annealed: 85,000 TS; 50,000 YS; 30 El; 55 RA; 180 Brin.
For furnace parts and equipment; Type 446; heat resistant.

BRESCIANA AK.
M-1451; 0.09-0.12 C, 13 Cr, 0.5 max Si, bal Fe.
Annealed: 75,000 TS; 40,000 YS; 35 El; 70 RA; 155 Brin.
For turbine blades, valves, cutlery knives; Type 403 and 410; corrosion resistant.

BRESCIANA ATK.
M-1451; 0.07-0.12 C, 17 Cr, bal Fe.
Annealed 80,000 TS; 50,000 YS; 25 El; 50 RA; 150 Brin.
For oil refinery equipment, oil burners and heaters; Type 430; corrosion resistant.

BRESCIANA AU.
M-1451; 0.17-0.25 C, 18 Cr, 8 Ni, bal Fe.
Annealed: 80,000 TS; 35,000 YS; 55 El; 75 RA; 150 Brin.
For chemical plant equipment, tanks, mixers; Type 302; stainless, austenitic.

BRESCIANA AUS.
M-1451; 0.11-0.16 C, 18 Cr, 8 Ni, bal Fe.
Annealed: 80,000 TS; 35,000 YS; 55 El; 75 RA; 150 Brin.
For chemical plant equipment, tanks, mixers; Type 301 and 302; stainless, austenitic.

BRESCIANA AUT.
M-1451; 0.09-0.15 C, 18 Cr, 9 Ni, bal Fe.
Annealed: 85,000 TS; 35,000 YS; 55 El; 65 RA; 150 Brin.
For welded chemical plant equipment, tanks; Type 321; stainless, austenitic.

BRESCIANA CNV.
M-1451; 0.20 max C, 24 Cr, 12 Ni, bal Fe.
Annealed: 90,000 TS; 40,000 YS; 50 El; 65 RA; 170 Brin.
For furnace parts, heat treating boxes, pumps, oil burners; Type 309; corrosion and heat resistant.

BRESCIANA INC.
M-1451; 0.26-0.37 C, 13 Cr, bal Fe.
Annealed: 100,000 TS; 55,000 YS; 22 El; 52 RA; 200 Brin.
For valves, cutlery, surgical instruments; Type 420; corrosion resistant.

BRESCIANA KK.
M-1451; 0.35 max C, 25 Cr, bal Fe.
Annealed: 85,000 TS; 50,000 YS; 30 El; 55 RA; 180 Brin.
For furnace parts and equipment; Type 446; heat resistant.

BRESCIANA MIC.
M-1451; 0.38-0.45 C, 13 Cr, bal Fe.
Annealed: 110,000 TS; 60,000 YS; 20 El; 50 RA; 220 Brin.
For valves, cutlery, surgical instruments; Type 420; corrosion resistant.

BRESCIANA VV.
M-1451; 0.25 max C, 25 Cr, 20 Ni, 1.5 max Si, bal Fe.
Annealed: 100,000 TS; 45,000 YS; 50 El; 65 RA; 185 Brin.
For furnace parts, valves, pumps, turbine components; Type 310; corrosion and heat resistant.

BRICROME.
M-715; 1.75 T.C., 1.5 Si, 0.8 Mn, 30 Cr, bal Fe.
Cast: 72,000 TS; 310 Brin.
For piston rings, cylinder liners; wear resistant.

BRIDGE BRONZE A.
M-Eng.; 80 Cu, 20 Sn, 0.1 P.
3 El; 1.8 RA; 80 Brin.
For tubes; P-Bronze.

BRIDGE BRONZE B.
M-Eng.; 85 Cu, 15 Sn, 0.1 P.
50 Brin.
For springs, electrical parts; P-Bronze.

BRIDGE BRONZE C.
M-Eng.; 80 Cu, 10 Sn, 0.7-1.0 P, 10 Pb.
15 El; 74 Brin.
For bearings; heavy duty P-Bronze.

BRIDGE BRONZE D.
M-Eng.; 88 Cu, 10 Sn, 0.3 P, 2 Zn.
For bearings, gears, worm wheels; P-Bronze.

BRIDGEPORT NO. 1.
M-141; 66 Cu, 34 Zn.
Hard: 74,000 TS; 60,000 YS; 8 El.
Soft: 47,000 TS; 15,000 YS; 62 El.
For hardware, general drawing and forming; yellow brass.

BRIDGEPORT NO. 2.
M-141; 1.8 Pb, 63.25 Cu, bal Zn.
Hard: 55,000 TS; 42,000 YS; 30 El; B 75 Rock.
Soft: 45,000 TS; 15,000 YS; 55 El; F 65 Rock.
For screw machine parts requiring some cold working such as roll threading, knurling, etc., binding posts, toggle switch parts.
High leaded brass, free-cutting.

BRIDGEPORT NO. 3.
M-141; 65.5 Cu, 33.9 Zn, 0.60 Pb.
Hard: 74,000 TS; 60,000 YS; 8 El.
Soft: 49,000 TS; 17,000 YS; 57 El.
For hardware, bolts, studs; low leaded brass.

BRIDGEPORT NO. 5.
M-141; 80 Cu, 20 Zn.
Hard: 74,000 TS; 59,000 YS; 7 El.
Soft: 44,000 TS; 14,000 YS; 50 El.
For tubing, hardware, diaphragms; low brass.

BRIDGEPORT NO. 6.
M-141; 35.35 Zn, 3.4 Pb, bal Cu.
Hard: 58,000 TS; 45,000 YS; 25 El; 144 Brin.
Soft: 49,000 TS; 20,000 YS; 40 El; 61 Brin.
For screw machine products; free-cutting brass.

BRIDGEPORT NO. 14.
M-141; 40.25 Zn, bal Cu.
Hard: 72,000 TS; 50,000 YS; 25 El; B 78 Rock.
Soft: 54,000 TS; 21,000 YS; 50 El; F 80 Rock.
For architectural trim, fasteners, condenser plates, heat exchangers.
Good hot working qualities, high corrosion resistance.

BRIDGEPORT NO. 16.
M-141; 65 Cu, bal Zn.
For wood and machine screws; yellow brass.

BRIDGEPORT NO. 18.
M-141; 67 Cu, 0.5 Pb, bal Zn.
Hard: 75,000 TS; 60,000 YS; 7 El.
For garden sprayers; free-cutting.

BRIDGEPORT NO. 19.
M-141; 0.3 Mn, 0.7 Sn, 1 Fe, 58.5 Cu, bal Zn.
Hard: 83,000 TS; 55,000 YS; 25 El; 165 Brin.
Soft: 72,000 TS; 30,000 YS; 45 El; 100 Brin.
For bolts, valve parts, tie rods; manganese bronze.

BRIDGEPORT NO. 24.
M-141; 60 Cu, 0.12 Pb, 0.65 Sn, bal Zn.
Hard: 75,000 TS; 53,000 YS; 20 El; 156 Brin.
Soft: 57,000 TS; 25,000 YS; 47 El; 100 Brin.
For nuts, bolts, marine hardware; Naval Brass.

BRIDGEPORT NO. 25.
M-141; 90 Cu, 10 Zn.
Soft: 38,000 TS; 12,000 YS; 45 El.
For threaded fasteners; resists season cracking.

BRIDGEPORT NO. 26.
M-141; 95 Cu, 5 Zn.
Hard: 56,000 TS; 50,000 YS; 5 El; 114 Brin.
Soft: 35,000 TS; 11,000 YS; 45 El; 50 Brin.
For jewelry; gilding metal.

BRIDGEPORT NO. 28.
M-141; 60 Cu, 0.6 Pb, 0.65 Sn, bal Zn.
Hard: 75,000 TS; 53,000 YS; 20 El; 156 Brin.
Soft: 57,000 TS; 25,000 YS; 47 El; 100 Brin.
For marine hardware; Leaded Naval Brass.

BRIDGEPORT NO. 29.
M-141; 60 Cu, 0.65 Sn, 1.75 Pb, bal Zn.
Hard: 75,000 TS; 53,000 YS; 15 El; 162 Brin.
For hardware; Leaded Naval Brass.

BRIDGEPORT NO. 30.
M-141; 71 Cu, 0.03 As, 1.05 Sn, bal Zn.
Soft: 53,000 TS; 22,000 YS; 65 El; 68 Brin.
For condenser tubes; Arsenical Admiralty Metal.

BRIDGEPORT NO. 32.
M-141; 98.53 Cu, 0.07 P, 1.4 Sn.
Hard: 65,000 TS; 50,000 YS; 8 El; 137 Brin.
Soft: 40,000 TS; 14,000 YS; 48 El; 56 Brin.
For flexible hose; P-bronze, corrosion resistant.

BRIDGEPORT NO. 34.
M-141; 89.85 Cu, 0.15 P, 10.0 Sn.
For springs; P-bronze, corrosion resistant.

BRIDGEPORT NO. 35.
M-141; 8 Sn, 0.1 P, bal Cu.
Hard: 93,000 TS; 72,000 YS; 10 El; B 93 Rock.
Soft: 60,000 TS; 24,000 YS; 63 El; F 82 Rock.
For springs, fuse clips, contacts, meterparts, diaphragms, bellows, snap switches, cutter pins, terminals.
Phosphor bronze-13% elect. cond.

BRIDGEPORT NO. 36.
M-141; 5.5 Sn, 0.15 P, bal Cu.
Annealed: 47,000 TS; 19,000 YS; 64 El; B 26 Rock.
Hard: 81,000 TS; 75,000 YS; 10 El; B 87 Rock.
Spring: 100,000 TS; 80,000 YS; 4 El; B 95 Rock.
For springs, fuse clips, contacts, relay parts, snap switches, terminals, diaphragms, bellows, cotter pins.
Phosphor bronze-18% elect. cond.

BRIDGEPORT NO. 37.
M-141; 69.5 Cu, 30.5 Zn.
Hard: 76,000 TS; 63,000 YS; 8 El.
Soft: 47,000 TS; 15,000 YS; 62 El.
For cartridge cases, deep drawn parts; cartridge brass.

BRIDGEPORT NO. 41.
M-141; 0.75 Pb, 60 Cu, bal Zn.
Annealed: 54,000 TS; 21,000 YS; 50 El; F 80 Rock.
1/4 hard: 72,000 TS; 50,000 YS; 25 El; B 78 Rock.
For screw machine products, bolts, fasteners, hardware, valve stems.
Low leaded brass, free cutting, corrosion resistant.

BRIDGEPORT NO. 42.
M-141; 38.65 Zn, 1.10 Pb, bal Cu.
Hard: 80,000 TS; 60,000 YS; 6 El; B 85 Rock.
Soft: 54,000 TS; 20,000 YS; F 80 Rock.
For screw machine products, bolts, fasteners, hardware. Corrosion resistant.
Free-cutting Muntz Metal.

BRIDGEPORT NO. 45.
M-141; 61.0 Cu, 38.35 Zn, 0.65 Sn.
Hard: 75,000 TS; 50,000 YS; 20 El; B 80 Rock.
Soft: 54,000 TS; 25,000 YS; 50 El; B 55 Rock.
For bolts, fasteners, propellers, valve stems, rivets, marine hardware.
Good resistance to salt water corrosion.
Cold heading Naval Brass. Corrosion resistant.

BRIDGEPORT NO. 53.
M-141; 5 Al, 0.25 As, bal Cu.
Soft: 60,000 TS; 22,000 YS; 60 El.
For propellers, hardware; aluminum bronze.

BRIDGEPORT NO. 54.
M-141; 20.87 Zn, 2.1 Al, 0.03 As, bal Cu.
Soft: 60,000 TS; 27,000 YS; 55 El.
For hardware; corrosion resistant; Al-Bronze.

BRIDGEPORT NO. 62.
M-141; 62.25 Cu, 35.75 Zn, 2 Pb.
Hard: 74,000 TS; 60,000 YS; 7 El; 150 Brin.
Soft: 49,000 TS; 17,000 YS; 52 El.
For hardware, bolts, nuts; free-cutting.

BRIDGEPORT NO. 63.
M-141; 65.5 Cu, 33.4 Zn, 1.1 Pb.
Hard: 74,000 TS; 60,000 YS; 7 El.
Soft: 49,000 TS; 17,000 YS; 54 El.
For hardware, bolts, nuts; free-cutting.

BRIDGEPORT NO. 64.
M-141; 67 Cu, 31.25 Zn, 1.75 Pb.
Hard: 75,000 TS; 60,000 YS; 7 El; 150 Brin.
Soft: 52,000 TS; 20,000 YS; 50 El; 70 Brin.
For screw machine products; leaded brass.

BRIDGEPORT NO. 69.
M-141; 70 Cu, 30 Zn.
Hard: 65,000 TS; 25 El.
Soft: 50,000 TS; 60 El.
For hollow rivets, fasteners; good formability, ductile.

BRIDGEPORT NO. 77.
M-141; 30 Zn, 0.05 Hg, bal Cu.
Hard: 78,000 TS; 64,000 YS; 8 El; 110 brin.
Soft: 47,000 TS; 15,000 YS; 65 El; 58 brin.
For heat exchangers, condensers; resists biofouling and dezincification.

BRIDGEPORT NO. 85.
M-141; 85 Cu, 15 Zn.
Hard: 70,000 TS; 57,000 YS; 5 El.
Soft: 40,000 TS; 12,000 YS; 47 El.
For flexible hose, deep drawn parts; red brass.

BRIDGEPORT NO. 87.
M-141; 88 Cu, 12 Zn.
For costume jewelry; resembles 14K gold.

BRIDGEPORT NO. 89.
M-141; 89 Cu, 2 Pb, 8.5 Zn.
Hard: 52,000 TS; 45,000 YS; 18 El; B 58 Brin.
Soft: 37,000 TS; 12,000 YS; 45 El; F 55 Brin.
For hardware; free-cutting.

BRIDGEPORT NO. 90.
M-141; 88.5 Cu, 2 Pb, 1 Ni, 0.07 P, bal Zn.
Hard: 70,000 TS; 60,000 YS; 12 El; 144 Brin.
For screw machine products, hardware; free-cutting, corrosion resistant.

BRIDGEPORT NO. 92.
M-141; 89.0 Cu, 9.1 Zn, 1.9 Sn.
Hard: 90,000 TS; 72,000 YS; 3 El; B 90 Rock.
For springs, spring contacts, slide contacts.
Substitute for phosphor bronze. Corrosion resistant.

BRIDGEPORT NO. 100.
M-141; Ag, bal Cu, (20 oz. Troy/ton Avoir.)
Soft: 32,000 TS; 10,000 YS; 45 El; F 40 Rock.
For radio, television, radar and computer parts, bus conductors.
Silver bearing copper.

BRIDGEPORT NO. 101.
M-141; Ag, bal Cu (10 oz. Troy/ton Avoir.)
Soft: 32,000 TS; 10,000 YS; 45 El; F 40 Rock.
For radio, television, radar and computer parts, wave guides, bus conductors.
Does not soften as readily as pure copper during soldering. Silver bearing copper.

BRIDGEPORT NO. 102.
M-141; 99.90 min Cu.
Hard: 50,000 TS; 45,000 YS; 4 El; 100 Brin.
Soft: 34,000 TS; 11,000 YS; 45 El; 45 Brin.
For electrical uses and hollowware; tough pitch copper.

BRIDGEPORT NO. 103.
M-141; 99.92 min Cu.
Hard: 53,000 TS; 51,000 YS; 3 El; B 56 Rock.
Soft: 32,000 TS; 8000 YS; 43 El; F 33 Rock.
For electronic and radar parts, bus conductors, wave guides, transistor and rectifier bases, heat sinks.
OFHC copper - 102% elect. cond.

BRIDGEPORT NO. 104.
M-141; Cu.
For electronic parts, vacuum tube parts; oxygen free copper.

BRIDGEPORT NO. 105.
M-141; 99.9 min Cu, 0.02 P.
Hard: 55,000 TS; 50,000 YS; 8 El; 107 Brin.
Soft: 32,000 TS; 10,000 YS; 45 El; 40 Brin.
For deep drawing and water tubes; deoxidized copper.

BRIDGEPORT NO. 106.
M-141; 99.98 min Cu.
Hard: 55,000 TS; 50,000 YS; 8 El; B 60 Rock.
Soft: 32,000 TS; 10,000 YS; 45 El; F 40 Rock.
For tubular bus bars.
Deoxidized copper.

BRIDGEPORT NO. 108.
M-141; 0.02 P. 0.3 As, 99.4 + min Cu.
Hard: 40,000 TS; 32,000 YS; 25 El; F 77 Rock.
For condensers, heat exchangers.
Inhibited.
Arsenical deoxidized copper.

BRIDGEPORT NO. 110.
M-141; 0.02 P, 99.91 + Cu.
Hard: 50,000 TS; 45,000 YS; 14 El; B 40 Rock.
Soft: 32,000 TS; 10,000 YS; 45 El; F 40 Rock.
For heat sinks.
85% elect. cond., corrosion resistant.

BRIDGEPORT NO. 112.
M-141; 99.50 Cu, 0.50 Te.
1/2 H-temper: 44,000 TS; 42,000 YS; 20 El; 78 Brin.
For screw machine products; high conductivity, free-cutting.

BRIDGEPORT NO. 120.
M-141; 99.7 Cu, 0.3 S.
1/2 H-temper: 44,000 TS; 42,000 YS; 20 El; 78 Brin.
For screw machine products, fasteners; high conductivity, free-cutting.

BRIDGEPORT NO. 133.
M-141; 59 Cu, 1.75 Pb, bal Zn.
Hard: 72,000 TS; 50,000 YS; 15 El; 144 Brin.
Soft: 52,000 TS; 20,000 YS; 45 El; 72 Brin.
For hardware, machine tool parts; BPT forging rod.

BRIDGEPORT NO. 134.
M-141; 2.25 Pb, 38 Zn, bal Cu.
Rolled: 75,000 TS; 52,000 YS; 22 El; B 80 Rock.
For hot forged and machined parts, hardware, fasteners, bolts.
27% elect. cond. Forging brass.
Corrosion resistant, free-cutting.

BRIDGEPORT NO. 141.
M-141; 62.25 Cu, 37.75 Zn.
Annealed: 54,000 TS; 21,000 YS; 45 El; F 80 Rock.
Half Hard: 70,000 TS; 50,000 YS; 10 El; B 75 Rock.
For architectural trim, fasteners, hardware, heat exchangers, brazing rod, condenser tubes.
Muntz Metal. Good corrosion resistance.

BRIDGEPORT NO. 142.
M-141; 62.0 Cu, 37.9 Zn, 0.1 As.
Hard:56,000 TS; 23,000 YS; 50 El; F 82 Rock.
For condensers, heat exchangers.
Free from dezincification.
Arsenical Muntz Metal. Inhibited.

BRIDGEPORT NO. 192.
M-141; 58 Cu, 0.9 Sn, 0.6 Fe, bal Zn.
For welding rod, brazing; non-fuming.

BRIDGEPORT NO. 285.
M-141; 81 Cu, 4.25 Si, 0.15 Pb, bal Zn.
Annealed: 95,000 TS; 50,000 YS; 30 El; 185 Brin.
For valve stems, pump parts; corrosion and wear resistant.

BRIDGEPORT NO. 511.
M-141; 88.85 Cu, 10 Ni, 1.15 Fe.
Hard: 50,000 TS; 45,000 YS; 18 El.
Soft: 45,000 TS; 15,000 YS; 40 El.
For condensers and heat exchangers; cupronickel, corrosion resistant.

BRIDGEPORT NO. 520.
M-141; 78.85 Cu, 20 Ni, 0.4 Fe, 0.75 Mn.
Soft: 49,000 TS; 14,000 YS; 40 El.
For condensers and heat exchangers; cupronickel, corrosion resistant.

BRIDGEPORT NO. 521.
M-141; 20 Ni, bal Cu.
Hard: 103,000 TS; 101,000 YS; 2 El.
Spring: 85,000 TS; 81,300 YS; 2.5 El.
For wave guides, radar equipment, relay springs.
Resists stress corrosion cracking.

BRIDGEPORT NO. 531.
M-141; 67.75 Cu, 31 Ni, 0.5 Fe, 0.75 Mn.
Soft: 60,000 TS; 25,000 YS; 45 El; 75 Brin.
For condensers and heat exchangers; cupronickel, corrosion resistant.

BRIDGEPORT NO. 548.
M-141; 48 Cu, 9.5 Ni, bal Zn.
For nickel silver welding rod; corrosion resistant.

BRIDGEPORT NO. 555.
M-141; 55 Cu, 18 Ni, bal Zn.
Drawn: 100,000 TS; 3 El.
Spring: 115,000 TS; 90,000 YS; 2.5 El; B99 Brin.
For telephone switch parts, springs; corrosion resistant.

BRIDGEPORT NO. 558.
M-141; 12 Ni, 56.5 Cu, bal Zn.
Annealed: 55,000 TS; 22,000 YS; 50 El; B 40 Rock.
Hard: 100,000 TS; 90,000 YS; 5 El; B 93 Rock.
For spring parts and contacts for telephone boards, radios, controls, springs, resistance wire, diaphragms. Good workability and weldability.
Corrosion resistant nickel silver.

BRIDGEPORT NO. 565.
M-141; 65 Cu, 18 Ni, bal Zn.
Hard: 85,000 TS; 75,000 YS; 3 El; 172 Brin.
Soft: 60,000 TS; 30,000 YS; 35 El; 64 Brin.
For hollowware, zippers, tableware, optical goods; nickel silver, corrosion resistant.

BRIDGEPORT NO. 566.
M-141; 65 Cu, 12 Ni, 0.15 Mn, bal Zn.
For jewelry, hardware; nickel silver, corrosion resistant.

BRIDGEPORT NO. 567.
M-141; 65.25 Cu, 9.75 Ni, 0.15 Mn, bal Zn.
Hard: 86,000 TS; 75,000 YS; 4 El; 180 Brin.
Soft: 60,000 TS; 28,000 YS; 36 El; 100 Brin.
For hardware, hollowware, jewelry; nickel silver, corrosion resistant.

BRIDGEPORT NO. 606.
M-141; 96.05 Cu, 3 Si, 0.95 Mn.
Hard: 94,000 TS; 58,000 YS; 8 El.
Soft: 60,000 TS; 25,000 YS; 60 El.
For pole line hardware, stampings; silicon bronze, corrosion resistant.

BRIDGEPORT NO. 609.
M-141; 98 Cu, 2 Si.
Annealed: 45,000 TS; 15,000 YS; 50 El.
Drawn: 80,000 TS; 50,000 YS; 12 El.
For transmission lines, marine hardware, bolts, welding rod; corrosion resistant.

BRIDGEPORT NO. 632.
M-141; 0.10 Fe, 2.95 Si, bal Cu.
Hard: 94,000 TS; 58,000 YS; 8 El; B 93 Rock.
Soft: 60,000 TS; 25,000 YS; 60 El; F 85 Rock.
For bolts, fasteners, springs, marine fittings, hardware. High fatigue strength.
High silicon bronze, corrosion resistant.

BRIDGEPORT NO. 635.
M-141; 97.5 Cu, 1.9 Ni, 0.6 Si.
Annealed: 40,000 TS; 12,000 YS; 50 El; 56 Brin.
Aged: 103,000 TS; 97,000 YS; 17 El; 216 Brin.
For cold headed bolts, fasteners, switch gear, springs, contacts; age-hardenable, corrosion resistant.

BRIDGEPORT NO. 707.
M-141; 7 Al, 2.0 Si, bal Cu.
Soft: 90,000 TS; 50,000 YS; 30 El; B 85 Rock.
For pump rods, gears, valve stems, pole-line hardware, oil burner nozzles.
High strength and corrosion resistance.
Duronze III.

BRIDGEPORT NO. 708.
M-141; 91.5 Cu, 1.75 Si, 6.75 Al.
Annealed: 85,000 TS; 40,000 YS; 30 El; 165 Brin.
For valve stems, bolts, fasteners; corrosion and wear resistant, hot forgeable.

BRIDGEPORT NO. 712.
M-141; 1.0 Si, 3.5 Al, bal Cu.
Aluminum silicon bronze; for bolts, cold headed nuts, pole line hardware.
CDA C63600.

BRIDGEPORT NO. 715.
M-141; 96.75 Cu, 2.9 Al, 0.35 Si.
Hard: 75,000 TS; 15 El.
Soft: 40,000 TS; 55 El.
For bolts, nuts, screws; Al-bronze.

BRIDGEPORT NO. 819.
M-141; 4 Zn, 4 Sn, 4 Pb, bal Cu.
For bearings, bushings, thrust washers; free machining bronze.
CDA C83600.

BRIDGEPORT NO. 820.
M-141; 95.35 Cu, 0.15 P, 4.5 Sn.
Hard: 81,000 TS; 75,000 YS; 10 El; 172 Brin.
Soft: 50,000 TS; 21,000 YS; 52 El; 71 Brin.
For springs, clutch discs, diaphragms; P-bronze.

BRIDGEPORT NO. 828.
M-141; 92 Cu, 1.9 Sn, 6.1 Zn.
Hard: 72,000 TS; 8 El.
1/2 H-temper: 58,000 TS; 20 El.
Spring: 90,000 TS; 2 El; 172 Brin.
For electrical switches, springs, contacts; substitute for p-bronze, 37% electrical conductivity.

BRIDGEPORT NO. 835.
M-141; 98.85 Cu, 0.75 Sn, 0.25 Si, 0.15 Mn.
For copper welding rods.

BRIDGEPORT NO. 840.
M-141; 98.6 Cu, 1.4 Sn.
Hard: 65,000 TS; 10 El.
Soft: 40,000 TS; 40 El.
For pole line hardware, marine parts; corrosion resistant.

BRIDGEPORT NO. 980.
M-141; 99 Cu, 1 Cd.
Hard: 55,000 TS; 45,000 YS; 6 El; B 65 Brin.
Soft: 35,000 TS; 15,000 YS; 60 El; F 35 Brin.
For electrical applications; high electrical conductivity.

Section I: Alloy Data / 245

BRIDGEPORT NO. 985.
M-141; 99.1 Cu, 0.9 Cd.
Hard: 55,000 TS; 48,000 YS; 6 El; 116 Brin.
Soft: 37,000 TS; 12,000 YS; 50 El; 45 Brin.
For trolley wire, marine hardware; cadmium copper.

BRIDGEPORT NO. 992.
M-141; 0.13 Zr, bal Cu.
Annealed: 30,000 TS; 10,000 YS; 45 El; 40 Brin.
Aged: 60,000 TS; 55,000 YS; 15 El; 116 Brin.
For resistance welding electrodes, grid wires; heat treatable, high conductivity.

BRIDGEPORT NO. 1232.
M-141; 3 Si, bal Cu.
Welded: 55,000 TS; 55 El.
For welding rod; bronze.

BRIDGEPORT NO. 1426.
M-141; 95.5 Cu, 3.5 Al, 1.0 Si.
For tubing; corrosion resistant.

BRIDGEPORT NO. 1552.
M-141; 1.9 Sn, 6.1 Zn, bal Cu.
For fuse clips, springs, diaphragms; good electrical conductivity.

BRIDGEPORT BRONZE.
M-141; 60 Cu, 0.75 Sn, bal Zn.
Drawn: 75,000 TS; 15 El.
For condenser tubes, water pipes, nuts, bolts; corrosion resistant.

BRIDGEPORT F-37.
M-141; 30 Zn, bal Cu.
For bearing retainer cages; cartridge brass.

BRIDGEPORT F-3034.
M-141; 0.10 S, 0.25 P, bal Cu.
For hardware; P-bronze, corrosion resistant.

BRIDGEPORT FORGING ROD.
M-141; 59.5 Cu, 1.75 Pb, 0.2 Sn, bal Zn.
Annealed: 52,000 TS; 20,000 YS; 45 El.
For forgings, hardware; free cutting.

BRIDGEPORT JEWELRY BRONZE NO. 92.
M-141; 89 Cu, 1.9 Sn, bal Zn.
Hard: 90,000 TS; 72,000 YS; 3 El; B 90 Rockwell.
For costume jewelry, electrical contacts; resembles red gold.

BRIDGEPORT MANGANESE BRONZE 19.
M-141; 58.5 Cu, 1.0 Fe, 0.75 Sn, 0.3 Mn, bal Zn.
Annealed: 60,000 TS; 15 El.
Drawn: 85,000 TS; 35 El.
For welding rod, forgings, brazing; corrosion resistant.

BRIDGEPORT PHOSPHOR 36 BRONZE GRADE A.
M-141; 5.0 Sn, 0.15 P, bal Cu.
Annealed: 50,000 TS; 52 El.
Drawn: 81,000 TS; 10 El.
For springs, diaphragms, clutch discs; corrosion and fatigue resistant.

BRIDGEPORT PHOSPHOR BRONZE GRADE C.
M-141; 8 Sn, 0.10 P, bal Cu.
Annealed: 60,000 TS; 65 El.
Drawn: 112,000 TS; 3 El.
For springs, diaphragms; corrosion and fatigue resistant.

BRIDGEPORT PLUMRITE.
M-141; 61 Cu, 38.6 Zn, 0.4 Pb.
For brass pipe for fresh water service and chemical plants; high strength.

BRIDGEPORT SULPH-COPPER.
M-141; 0.3 S, 99.7 Cu.
1/2 H-temper: 42,000 TS; 39,000 YS; 20-35 El; 77 Brin.
H-temper: 48,000 TS; 46,000 YS; 18-20 El; 81 Brin.
For nozzles, welding tips, motor parts; free-cutting, high electrical and thermal conductivity.

BRIGHT ALLOY.
M-580.
For anodes; electroplating.

BRIGHT CAP GILDING.
M-Eng.; 90 Cu, 9.9 Zn, 0.4-0.1 Pb.
For gilding, jewelry; corrosion resistant.

BRIGHT E-3 see **CRUCIBLE BRIGHT E-3.**

BRIGHTNER ALLOY.
M-91; 10 Al, bal Zn.
To brighten galvanized metal.

BRIGHTRAY C.
M-121; 80 Ni, 19 Cr, 0.5 max Fe, 1.5 Si.
Cold drawn: 106,000 TS; 47 El; 63 RA; 196 Brin.
For electrical resistances, heating elements.
Heat resistant to 1150°C.

BRIGHTRAY F.
M-121; 37 Ni, 18 Cr, 2 Si, bal Fe.
For electrical resistances, heating elements, electric furnace elements.
Heat resistant to 1000°C.

BRIGHTRAY S.
M-121; 80 Ni, 20 Cr.
Annealed: 107,000 TS; 50,000 YS; 47 El; 63 RA; 187 Brin.
For heavy industrial electrical and resistance heating elements; rod and strip.

BRIGHTWAY ALLOY 35.
M-121; 0.05 C, 35.0 Ni, 42.0 Fe, 20.0 Cr, 2.0 Si.
Elec. res.: 102 microhm/cm at 20°C.
Electrical resistance alloy for heating elements up to 1050°C.
ASTM B344.

BRIGHTWAY B.
  M-121; 15-17 Cr, 58-60 Ni, 20 Fe.
  Annealed: 99,000 TS; 41,000 YS; 44 El; 66 RA; 190 Brin.
  For heating elements, electrical resistances; heat resistant to 950°C.

BRILLALUMAG 3.
  M-770; 3 Mg, bal Al.
  Annealed: 39,000 TS; 14,000 YS; 30 El; 35 Brin.
  Hardened: 53,000 TS; 50,000 YS; 4 El; 65 Brin.
  For decorative, light alloy parts; takes high polish.

BRILLALUMAG 5.
  M-770; 5 Mg, bal Al.
  Annealed: 47,000 TS; 31,000 YS; 28 El; 48 Brin.
  Hardened: 69,000 TS; 50,000 YS; 3 El; 90 Brin.
  For decorative, light alloy parts; takes high polish.

BRILLIANT.
  M-333; 0.70 C, 4 Cr, 1 V, 18 W, bal Fe.
  For twist drills, reamers, taps, milling cutters; high speed steel.

BRILLIANT MM.
  M-333; 0.7 C, 5 Mo, 6 W, 4 Cr, 4 V, bal Fe.
  For cutters, dies; Type M4; high speed steel.

BRILLIANT WW.
  M-333; 0.7 C, 18 W, 4 Cr, 1 V, bal Fe.
  For drills, taps, cutters, reamers; Type T1; high speed steel.

BRILLUM.
  M-469; 1.5 Cu, 2 Ni, bal Al.
  For pistons; light alloy.

BRILYBDENUM.
  M-715; 3.2 C, 2 Si, 0.4 Ni, 0.2 Cr, bal Fe.
  For piston rings; heat resistant.

BRIMCO.
  M-1579; 87-88 Cu, 2.36-3.25 Sn, 0.016-0.06 P, 0.21-0.23 Mn, bal Zn.
  For contacts; corrosion resistant.

BRIMCOLLOY-200.
  M-1579; 86.0 Cu, 1.0 Sn, 13.0 Zn.
  Hard: 77,500 TS; 65,200 YS; 5 El; B 82 Rock.
  Spring: 84,000 TS; 76,000 YS; 2 El; B 88 Rock.
  For electrical contacts, springs, fuse clips, heat exchangers, condensers.
  Corrosion resistant, readily fabricated.

BRIMOL.
  M-715; 2.8 C, 2 Si, 14 Ni, 3 Cr, 7 Cu, bal Fe.
  For cylinder liners, rings; austenitic.

BRINALLOY (NS-8).
  M-264; 70 Ni, 16 Cr, 10 Si.
  Cast: 120,000 TS; 600 Brin.
  For valve seats and discs; resists wear, galling and corrosion.

BRINALLOY (NS-9).
  M-264; 73 Ni, 15 Cr, 8 Si.
  Cast: 332 Brin.
  For valve seats and discs.

BRISTAHL.
  M-716; 0.2 C, 18 Cr, 8 Ni, bal Fe.
  For stainless steel parts; stainless.

BRISTOL.
  M-Eng.; 60.8-75.7 Cu, 24.3-39.2 Zn.
  For hardware, clocks.

BRISTOL.
  M-European; 0.30 C, 0.90 Cr, 0.65 Ni, 1.0 Mo, 0.6 Al, bal Fe.
  Heat treated: 150,000-200,000 TS; 135,000-175,000 YS; 18-16 El; 55-50 RA; 300-400 Brin.
  For nitrided parts, gears, cams; nitriding steel.

BRISTOL ALLOY BUTTONS.
  M-Eng.; 57-61 Cu, 36-37 Zn, 2.7-5.3 Sn.
  For buttons, fixtures; corrosion resistant.

BRISTOL BRASS.
  M-Eng.; 76-61 Cu, 24-39 Zn.
  For condenser tubes, pipes, ornamental purposes; same as Princes Metal."

BRITEST.
  M-715; 3.1 T.C., 2.0 Si, 0.8 Mn, 1 Cr, bal Fe.
  Cast: 58,000 TS; 280 Brin.
  For piston rings, cylinder liners; wear resistant.

BRITISH ALUMINUM PISTON ALLOY.
  M-968; 88 Al, 12 Cu.
  For pistons, castings; non-hardenable.

BRITISH ALUMINUM PISTON ALLOY.
  M-968; 85 Al, 14 Cu, 1 Mn.
  For pistons, castings; non-hardenable.

BRITISH ALUMINUM PISTON ALLOY.
  M-968; 94.5 Al, 5.5 Ni.
  For pistons, castings; non-hardenable.

BRITISH I STEEL.
  0.45 C, 0.62 Mn, 2.8 Cr, 0.43 Ni, 0.9 Mo, 0.20 V, bal Fe.
  At 70°F: 167,000 TS; 3.7 El; 7.0 RA.
  At 1000°F: 88,000 TS; 9.2 El; 17.5 RA.
  At 1200°F: 48,000 TS; 25 El; 55.4 RA.
  For ordnance mortar tubes.
  Good fatigue resistance.

BRITISH NAVY ANTIFRICTION METAL.
  M-561; 5 Cu, 85 Sn, 10 Sb.
  For admiralty lining, plastic bearings; Babbitt.

BRITOR.
  W, C.
  For welding; welding to steel.

BRITTANIA, ENGLISH.
  M-Eng.; 90-85 Sn, 0-3 Zn, 1.3 Cu, 5-10 Sb.
  For bearings, table ware; corrosion resistant.

BRITTANIA, GERMAN.
M-Ger.; 70-94 Sn, 0-5 Zn, 1.8-5 Cu, 3.7-5 Sb, 0-9 Pb.
For bearings; Babbitt.

BRIX.
M-Eng.; 75-60 Ni, 15-20 Cr, 5 Cu, 4 Si, 3 Ti, 2 Al, 1-4 W, 1 B
For heating elements, heat and corrosion resisting parts; heat and corrosion resistant.

BRM.
M-289; 0.75-0.85 C, 4.2 Cr, 2.1 V, 19 W, 0.6 Mo, bal Fe.
For tools and cutters; oil hardening, high speed steel.

BRM.
M-783; 0.8 C, 4.15 Cr, 2.1 V, 19 W, 0.6 V, bal Fe.
For cutting tools; oil hardening.

BR-NICKEL.
M-303; 99.6 Ni.
For chemical equipment.

BROCKHOUSE.
M-961; 0.2 C, 18 Cr, 8 Ni, bal Fe.
For case hardening boxes, steam boiler parts; heat resisting.

BROLUNICK.
M-Eng.; 82 Cu, 7 Al, 5.5 Ni, 4 Fe, 2 Mn.
For corrosion resisting parts; Al-Bronze; tough.

BROLUNICK BRONZE.
M-French; 7 Al, 5.5 Ni, 4 Fe, 2 Mn, bal Cu.
For gears, housings, propellers, marine hardware; tough, corrosion resistant.

BROMET.
M-Australia; W, C.
For dies; sintered.

BRONCO.
M-926; copper clad with P-bronze.
For springs; high current carrying capacity.

BRONWITE.
M-815; 20 Mn, 20 Zn, 1 Al, bal Cu.
Die cast: 75,000 TS; 40,000 YS; 20 El.
Sand cast: 60,000-70,000 TS; 30,000-35,000 YS; 25-40 El.
For general die castings, hardware, fixtures.
For die casting copper alloys.
Die castable bronze, corrosion resistant.

BRONZALUN.
M-250; a bronze with abrasive grains cast in the metal 85 Cu, 15 alumina or carborundum.
For floor plates, stair treads, car steps and door saddles; wear resistant and "anti-slip."

BRONZARK.
M-118; Sn, bal Cu.
For welding electrodes; for cast iron or copper.

BRONZE AU-CADMIUM.
M-897; 1 Cd, bal Cu.
For heating elements; useful operating temperature to 350°C.

BRONZE DEVIL.
M-712; 8.5 Sn, 0.25 P, bal Cu.
For P-bronze welding rods; arc welding.

BRONZE FILTER POWDER.
M-1775; 90 Cu, 10 Sn.
Powder for compressing and sintering to make bronze filters. (many mesh sizes).

BRONZE OTSSN3-7-5-1.
M-Russian; 85.08 Cu, 3.58 Pb, 7.35 Zn, 0.01 P, 3.4 Sn, 0.3 Ni.
Cast: 27,000 TS; 13,000 YS; 17 El; 26 RA.
For hardware, plumbing; free-cutting.

BRONZE, PHOSPHOR GRADE "A".
M-8; 96 Cu, 3.75 Sn, 0.25 P
Cold rolled: 90,000 TS; 45,000 YS; 10 El.
Wire: 150,000 TS; 0.5 El.
For springs, electrical switches, diaphragms; high strength.

BRONZE, PHOSPHOR GRADE "C".
M-8; 92 Cu, 8 Sn.
Plates: 50,000 TS; 20,000 YS; 73 El; 60 Brin.
Wire: 150,000 TS.
For springs, switches, fittings; tough.

BRONZE, PHOSPHOR GRADE "D".
M-8; 89.5 Cu, 10.5 Sn.
Cold rolled: 92,000 TS; 83,000 YS; 29 El; 218 Brin.
For worm wheels, pumps; tough.

BRONZE, STEAM FITTING.
M-U.S.; 88 Cu, 8 Sn, 2 Zn, 2 Pb.
For steam fittings; pressure tight.

BRONZE, STEAM VALVE.
M-U.S.; 88 Cu, 10 Sn, 2 Zn.
For steam valves.

BRONZE, STEEL STAHL.
M-U.S.; 52-59 Cu, 36-43 Zn, 1 Fe, 2.5-3.0 Mn, 1 Al.
For propellers, marine parts, hardware; same as "Uchatius Bronze."

BRONZE WABBLER.
M-U.S.; Sn, bal Cu.
For rolls; tough and hard.

BRONZE, WEATHERSTRIP.
M-U.S.; 89 Cu, 9.5 Zn, 1.5 Sn.
For weather strips.

BRONZE, WHITE.
M-U.S.; 54 Cu, 42 Zn, 4 Ni, 0.3 Fe + Al.
For ornamental and architectural parts; corrosion resistant.

BRONZE, WIRE.
M-U.S.; 98.75 Cu, 1.2 Sn, 0.05 Pb.
For electrical purposes, roofing, gutters.

BRONZOCHROM 10185.
M-717; Nickel-base alloy powder for overlays on ferrous and nickel base alloys.

BRONZ-ROD NO. 61.
M-1096; 97 Cu, 2.85 Si, 0.15 Fe.
Welded: 55,000 TS; 22,000 YS; 8 El; Rockwell F85.
For brazing rod for steels and cast iron; high tensile brazing.

BROTERNAL.
M-282; 92 Cu, 1 Mn, 7 Al.
Drawn: 112,000 TS; 85,000 YS; 6 El; 200 Brin.
For periscope tubes, paper mill and dye works equipment; Al bronze, corrosion resistant.

BROWN & SHARPE PLAIN.
M-1172; 0.90 C, 1.0 Mn, 0.5 Cr, 0.5 W, bal Fe.
Oil hardening flat ground stock.
AISI 01.

BROWNIE EXTRA.
M-1182; 0.80 C, bal Fe.
For drills, taps, reamers, cutters; Type W1; water hardened.

BROWN LABEL.
M-341; 0.5 C, 0.2 Mn, 1.15 Cr, 2.5 W, 0.2 V, 0.75 Si, bal Fe.
For coining and swaging dies, chisels; oil hardening, tough; AISI S1, shock resistant tool steel.

BROWN LABEL.
M-343; 0.85-1.0 C, 3-4 Cr, bal Fe.
For tools, hot work tools and dies, hot shear blades, bull dies; Hellers Hot Die Steel."

BRT see BETHLEHEM BRT.

BRUSH-10.
M-1037; 0.40-0.70 Be, 2.35-2.70 Co, bal Cu.
Cast: 45,000-50,000 TS; 5000-10,000 PL; 20-30 El; 72-81 Brin.
Cast and aged: 100,000 TS; 50,000 PL; 3-15 El; 191-241 Brin.
Wrought and aged: 110,000-125,000 TS; 100,000-120,000 YS; 70,000-95, PL; 5-8 El; 79-82 Rock 30 T.
For springs, switches, instrument parts, clutch rings, resistance welding electrodes.
Tough, strong, corrosion resistant, heat treatable. (CA 175).

BRUSH 10-C.
M-1252, M-1037; 0.45-0.65 Be, 2.4-2.6 Co, bal Cu.
Heat treated: 90,000-120,000 TS; 70,000-90,000 YS; 5-12 El; 5-18 RA 195-262 Brin.
For slip rings, contact arms, switch parts; age-hardenable, good conductivity.

BRUSH 10-X.
M-1037; 0.4-0.7 Be, 2.35-2.7 Co, bal Cu.
Ht-Temper: 110,000-130,000 TS; 100,000-120,000 YS; 8-20 El; B 95-102 Rock.
For wind tunnel components, springs, switch gears, resistance welding electrodes.
High strength up to 800°F. Corrosion resistant. Heat treatable (CA 175).

BRUSH 17.
M-1037; 31 Ni, 0.70 Fe, 1.0 max Mn, 0.3-0.7 Be, 0.15 max Si, bal Cu.
Cast: 76,000 TS; 45,000 YS; 14 El; B 82 Rock.
Age Hardened: 120,000 TS; 85,000 YS; 14 El; C 20 Rock.
For marine hardware, sonar cases, heat exchangers, desalinization equipment.
Corrosion resistant. Age-hardenable.

BRUSH 18.
M-1037; 30 Ni, 0.62 Fe, 1.5 max Mn, 0.4-0.6 Be, 0.7-0.9 Si, bal Cu.
Cast: 100,000 TS; 70,000 YS; 12 El; B 94 Rock.
Age Hardened: 120,000 TS; 80,000 YS; 13 El; C 30 Rock.
For marine hardware, desalinization equipment, heat exchangers, valve bodies, sonar cases.
Corrosion resistant. Age-hardenable.

BRUSH 20-C.
M-1252, M-1037; 1.90-2.15 Be, 0.35-0.65 Co, bal Cu.
Heat treated: 150,000-175,000 TS; 125,000-160,000 YS; 1-4 El; 1-5 RA; 352-426 Brin.
For bearings, gears, valve and pump parts; age-hardenable, wear and corrosion resistant.

BRUSH 25.
M-1037; 1.8-2.0 Be, 0.2-0.3 Co, bal Cu.
For corrosion and wear resistant parts; corrosion and wear resistant.

BRUSH 35-C.
M-1252, M-1037; 0.25-0.50 Be, 1.4-1.6 Ni, bal Cu.
Heat treated: 70,000-90,000 TS; 50,000-70,000 YS; 5-17 El; 5-25 RA; 180-210 Brin.
For resistance welding dies and jaws; age-hardenable, high conductivity.

BRUSH 50.
M-1037; 0.25-0.50 Be, 1.4-1.7 Co, 0.9-1.1 Ag, bal Cu.
Annealed: 35,000-55,000 TS; 25,000 YS; 20-35 El; B 35 Rock.
HT-Temper: 120,000 TS; 110,000 YS; 8-20 El; B 95-102 Rock.
For resistance welding applications, electrodes.
Heat treatable, corrosion resistant.

BRUSH 50 C.
M-1037; 0.40-0.65 Be, 1.4-1.7 Co, 1.0-1.15 Ag, bal Cu.
Cast: 45,000-60,000 TS; 15,000-35,000 YS; B 50-65 Rock.
Heat treated: 100,000 TS; 75,000 YS; 3-15 El; B 92-100 Rock.
For resistance welding electrodes; dies and holders.
RWMA Class III, heat treatable, corrosion resistant.

BRUSH 55 C.
M-1037; 0.45-0.65 Be, 2.4-2.6 Co, bal Cu.
Cast: 90,000-120,000 TS; 70,000-90,000 YS; 5-12 El; 5-18 RA; B 90-103 Rock.
For resistance welding electrodes, circuit breaker and switch parts, slip rings, contact arms, welder bearings.
Good conductivity. Resists elevated temperatures to 700°F. RWMA Class 3.

BRUSH 125.
M-1037; 1.8-2.0 Be, 0.12-0.18 Co + Ni, bal Cu.
Magnetic mass susceptibility -0.5 X 10^{-1}.
Magnetic permeability 0.999997.
Heat treated: 200,000 TS; 175,000 YS; 2-5 El, C 42 Rock.
For strain members in nonmagnetic cables, wire forms, springs.
High strength, nonmagnetic, heat treatable, corrosion resistant.

BRUSH 165.
M-1037; 1.6-1.8 Be, 0.2-0.3 Co, bal Cu.
For corrosion and wear resistant parts; corrosion and wear resistant.

BRUSH 165 C.
M-1037; 1.6-1.85 Be, 0.2-0.65 Co, bal Cu.
Cast: 73,000 TS; 38,000 YS; 23 El; B 78 Rock.
Heat treated: 150,000 TS; 140,000 YS; 2-4 El; C 37 Rock.
For plastic tooling and pressure containers.
RWMA Class IV, corrosion resistant, heat treatable, tough.

BRUSH 190.
M-1037; 1.80-2.05 Be, 0.2-0.3 Co, 0.20 min Co + Ni, bal Cu.
Condition AM: 105,000 TS; 95,000 YS; 20 El; Rock 40 (30N).
Condition XHMS: 190,000 TS; 180,000 YS; 3 El; Rock 61 (30N).
For springs, shafts, fasteners.
(CA 172) Corrosion resistant, heat treatable.

BRUSH 200 C.
M-1037; 2 Be, 4 max additives, bal Ni.
Annealed: 130,000 TS; 57,000 YS; 30 El; B 97 Rock.
Aged: 240,000 TS; 190,000 YS; 5 El; C 54 Rock.
For safety tools, plastic molds and cores, fuel pump impellers, mechanical seals.
Good structural stability at elevated temperatures.

BRUSH 220-C.
M-1037; 2.0-2.3 Be, 0.4 max C, bal Ni.
For corrosion and wear resistant parts; corrosion and wear resistant.

BRUSH 220-CC.
M-1037; 2.0-2.3 Be, 0.8 max Cr, 0.4 max C, bal Ni.
For corrosion and wear resistant parts; corrosion and wear resistant.

BRUSH 221-C.
M-1037; 2.0-2.3 Be, 0.5-1.0 C, bal Ni.
For corrosion and wear resistant parts; corrosion and wear resistant.

BRUSH 221-CC.
M-1037; 2.0-2.3 Be, 0.8 max Cr, 0.5-1.0 C, bal Ni.
For corrosion and wear resistant parts; corrosion and wear resistant.

BRUSH 245-C.
M-1037; 2.3-2.55 Be, 0.35-0.65 Co, bal Cu.
Cast: 83,000 TS; 45,000 YS; 23 El; B 84 Rock.
Heat treated: 175,000 TS; 165,000 YS; 1-2 El; C 40-45 Rock.
For plastic tooling, pressure containers.
Heat treatable, corrosion resistant, impact resistant.

BRUSH 250-C.
M-1037; 2.50-2.75 Be, 0.35-0.65 Co, bal Cu.
Cast: 140,000-165,000 TS; 110,000-130,000 YS; 0-2 El; 0-2 RA; C 42-48 Rock.
For plastic molds, deep drawing dies, zinc die casting dies.
High wear resistance.

BRUSH 260-C.
M-1252, M-1037; 2.55-2.80 Be, 0.4 max C, bal Ni.
Cast: 115,000-125,000 TS; 60,000-70,000 YS; 7-12 El; 5-10 RA; 250-283 Brin.
Heat treated: 200,000-220,000 TS; 190,000-210,000 YS; 0-2 El; 0-1 RA; 500-560 Brin.
For aircraft parts, molds for plastics; age-hardenable, heat resistant to 800°F.

BRUSH 260-CC.
M-1037; 2.55-2.8 Be, 0.8 max Cr, 0.4 max C, bal Ni.
For springs, bellows; age hardenable.

BRUSH-261 C.
M-1252, M-1037; 2.55-2.80 Be, 0.5-1.0 C, bal Ni.
For aircraft parts, molds for plastics; age-hardenable, heat resistant to 800°F.

BRUSH 261-CC.
M-1037; 2.55-2.8 Be, 0.8 max Cr, 0.5-1.0 C, bal Ni.
For springs, bellows; age hardenable.

BRUSH 275-C.
M-1252, M-1037; 2.50-2.75 Be, 0.35-0.65 Co, bal Cu.
Heat treated: 140,000-165,000 TS; 110,000-130,000 YS; 0-2 El; 0-2 RA; 393-460 Brin.
For plastic molds, deep drawing dies; age-hardenable, wear resistant.

BRUSH BERYLLIUM NO. 6.
M-1037; 2.4-2.6 Be, 1.2-1.3 Ni, bal Cu.
Heat treated: 76,000-193,000 TS; 68,000-85,000 YS; 34-0 El; 33-2 RA 135-387 Brin.
For springs, diaphragms, instrument parts; age-hardenable; now Brush 240-C.

BRUSH M25.
M-1037; 1.8-2.05 Be, 0.20-0.30 Co, 0.2 Pb, bal Cu.
IIT Temper: 175,000-215,000 TS; 150,000-200,000 YS; 2-5 El; C 39-45 Rock.
Annealed: 65,000-85,000 TS; 25,000 YS; 50 El; B 65 Rock.
For gears, shafts, screw machine products, fasteners.
Free-cutting (CA 172), heat treatable, corrosion resistant.

BRUSH M220C.
M-1037; 2.0-2.3 Be, 0.5-0.75 C, bal Ni.
Heat treated: 170,000-190,000 TS; 160,000-180,000 YS; 1-2 El; C 48-5 Rock.
For molds, plungers, forming tools and plungers in the glass industry.
Age-hardenable. High strength, corrosion and oxidation resistant.

BRUSH M260C.
M-1037; 2.55-2.80 Be, 0.4 max C, bal Ni.
Cast: 115,000-125,000 TS; 60,000-70,000 YS; 7-12 El; 5-10 RA; C 24-30 Rock.
Aged: 210,000-230,000 TS; 200,000-210,000 YS; 1-2 El; 0-1 RA; C 50-54 Rock.
For turbine wheels, glass molds, valve bodies, fuel injection tips.
Age-hardenable, corrosion resistant.
High temperature resistant.

BRUSH QMV.
M-1037; Be, bal Cu.
For corrosion and wear resistant parts; age-hardenable.

BRUSH WIRE.
M-U.S.; 64.25 Cu, 35 Zn, 0.75 Sn.
For wire.

BRUSH ZRII-5% BE.
M-1037; 5 Be, bal Zr.
For brazing Zircalloy alloys.
High temperature brazing alloy.

BRYAN.
M-1705; 1.10 C, 0.30 Mn, 0.50 Si, 0.25 Cr, bal Fe.
Water hardening carbon tool steel, good hardness and wear; AISI W4.

BRYIRON.
M-718; 2.96 T.C., 0.73 Mn, 1.2 Si, 1.5 Ni, bal Fe.
51,000 TS; 221 Brin.
For liners, pistons; Tr. S. 4160; Def. 0.207".

BRYMILL BRM-1.
M-478; 0.10-0.18 C, 0.5-1.1 Mn, bal Fe.
Annealed: 60,000 TS; 36,000 YS; 38 El; B 70 Rock.
For gears, fasteners, bolts, machine tool parts. Case hardening.

BRYMILL BRM-5.
M-478; 0.25-0.35 C, bal Fe.
Drawn: 89,600 TS; 15 El; 40 RA.
For bolts, studs, jack spindles; water hardened.

BRYMILL BRM-6.
M-478; 0.35-0.45 C, bal Fe.
Drawn: 103,000 TS; 10 El; 35 RA.
For axles, shafts, racks, clutch plates, chain links; water hardened.

BRYMILL CH-2.
M-478; 0.12-0.18 C, 0.90-1.5 Mn, 0.08-0.15 S, bal Fe.
Heat treated: 98,500 TS; 20 El; 50 RA.
For screw machine products, gears, bolts, fasteners; case hardened, free-cutting.

BRYMILL CH-8.
M-478; 0.08-0.55 C, 2.75-3.5 Ni, bal Fe.
Heat treated: 107,500 TS; 30 El; 50 RA.
For gears, bolts, crankshafts, camshafts; case hardened or thru hardened, tough.

BRYMILL DH-11.
M-478; 0.4 C, 1 Ni, bal Fe.
For shafts, axles, gears, machinery parts; case hardening.

BRYMILL DH12.
M-478; 0.30-0.40 C, 1.3-1.7 Mn, bal Fe.
For gears, bolts, machine tool parts; oil hardened.

BRYMILL DH13.
M-478; 0.25-0.35 C, 0.35-0.75 Mn, 2.75-3.5 Ni, bal Fe.
For gears, bolts, crankshafts; oil hardened, shock resistant.

B.S.2.
M-1116; 0.57 C, 0.3 Cr, 0.85 Mn, 1.9 Si, 0.2 V, 0.3 Mo, bal Fe.
Oil or water hardening tool steel; for stamps, cold working dies, pneumatic tools, cold cutting tools for thick metal, shock resisting tools.
AFNOR: 60 SMD 08.03; AISI 35.

BS401.
M-1741; 0.14 max Cu, 11.0-13.0 Si, 0.6 max Fe, 0.5 max Mn, 0.1 max Mg 0.1 max Ni, 0.14 max Zn, 0.2 max Ti, 0.05 max Sn, bal Al.
Per. Mold, cast F1; 207 MPa TS; 62 MPa YS; 9 El; 60 Brin.
ADC H49-5; BS LM6.

BSI V12.
M-1290; 0.29 C, 1.35 Cr, 0.4 Mo, 0.1 V, bal Fe.
Heat treated: 128,000 min TS; 100,000 min YS; 10 min El.
Low alloy steel casting. (formerly PZA)
Werkstoff Nr. 1.7725.

BSI V22.
M-1290; 0.35 C, 2.45 Cr, 0.4 Mo, 0.1 V, bal Fe.
Heat treated: 100,000-120,000 TS; 78,000-92,000 YS; 12-11 El.
Low alloy steel casting. (formerly PZB)
Werkstoff Nr. 1.7755.

B.S.SEEWASSER.
M-116; 5-10 Mg, bal Al.
Soft: 47,000 TS; 22,000 YS; 25 El.
Hard: 78,000 TS; 65,000 YS; 3 El.
For light alloy parts, ship and seaplane parts; corrosion resistant to sea water.

B.S.SEEWASSER.
M-196; 7.5 Mg, 0.3 Mn, bal Al.
Heat treated: 30,000 TS; 20,600 YS; 4 El; 10 RA; 80 Brin.
Rolled: 64,000 TS; 45,000 YS; 8 El; 50 RA; 120 Brin.
For furniture, interior light fixtures, wire, castings; resists sea water corrosion.

BS-SEEWASSER 05.
M-116; 4.0-5.5 Mg, 0.8 max Mn, 0.3 max Cr, bal Al.
Soft: 42,000 TS; 22,000 YS; 35 El; 65 Brin.
Hard: 60,000 TS; 50,000 YS; 10 El; 105 Brin.
For aircraft and marine parts; good corrosion resistance.

BS-SEEWASSER 07.
M-116; 5.5-7.5 Mg, 0.8 max Mn, 0.3 max Cr, bal Al.
For aircraft and marine parts; good corrosion resistance.

BS-SEEWASSER 63/03.
M-116; 2-4 Mg, 0.4 max Mn, bal Al.
Soft: 28,000 TS; 13,000 YS; 30 El; 47 Brin.
Hard: 40,000 TS; 35,000 YS; 10 El; 73 Brin.
For aircraft tanks, and fittings, fuel lines, marine parts; resists sea water corrosion.

BT-O.
M-1776; 0.10 C, 0.90 Mn, 0.10 max Si, bal Fe.
For low temperature operation down to -70°C.
ASTM A333 Gr.1. (similar spec).

BT-3.
M-1766; 0.15 C, 0.50 Mn, 0.25 Si, 3.5 Ni, bal Fe.
For low temperature operation, down to -100°C.

BT-4.
M-USSR; 4 Al, 1.5 Mn, bal Ti.
Titanium alloy.

BT-9.
M-1766; 0.10 C, 0.70 Mn, 0.25 Si, 9.0 Ni, bal Fe.
For low temperature operation, down to -200°C.

BTF ALLOY.
M-1429; 0.55 C, 1 Mn, 0.35 Mo, 0.30 Cr, 2 Si, bal Fe.
For punches, rivet sets, pneumatic tools; oil hardened, shock resistant.

BTG.
M-104; 60 Ni, 12 Cr, 2 W, bal Fe.
For ammonia synthesis tubes; heat and NH_3 resistant.

B.T.G. STEEL.
Eng.; 60 Ni, 25 Fe, 12 Cr, 2 Mn, 0.5 C, 3 W.
Sand cast: 50,000-70,000 TS; 40,000-70,000 YS; 2-10 El; 130-180 Brin.
Rolled: 90,000-109,000 TS; 50,000-70,000 YS; 25-45 El; 50-70 RA; 180-200 Brin.
For heating elements; heat and corrosion resistant.

BTH ALLOY NO. 12.
M-306; Ni-Cr-Fe.
For seals to lead glass; metal to glass seal.

BTR.
M-24; 0.7 C, 1.2 Mn, 0.5 Cr, 0.2 V, 0.5 W, bal Fe.
For tools, master tools and dies; taps, hobs, reamers, broaches; non- deforming; oil hardened.

BTR see **BETHLEHEM BTR.**

BUCKEYE BRONZE.
U.S.; 41.6 Sn, 0.6 Cu, 15.6 Pb, 37.8 Zn, 4.3 Al.
For bearings; Babbitt.

BUCKEYE M1.
M-1705; 0.78 C, 3.9 Cr, 1.05 V, 1.6 W, 8.5 Mo, bal Fe.
High speed tool steel, molybdenum type.
AISI M1.

BUCKEYE M2.
M-1705; 0.80 C, 4.0 Cr, 1.75 V, 6.0 W, 5.0 Mo, bal Fe.
High speed tool steel, Mo-W type.
AISI M2.

BUCKEYE M3 CLASS 1.
M-1705; 1.0 C, 4.0 Cr, 2.5 V, 6.25 W, 6.25 Mo, bal Fe.
High speed tool steel, Mo-W type.
AISI M3 (Class 1).

BUCKEYE M3 CLASS 2.
M-1705; 1.15 C, 4.0 Cr, 3.0 V, 6.0 W, 5.5 Mo, bal Fe.
High speed tool steel, Mo-W type.
AISI M3 (Class 2).

BUCKEYE M4.
M-1705; 1.3 C, 4.0 Cr, 4.0 V, 5.5 W, 4.5 Mo, bal Fe.
High speed steel for cutting tools, high hardness and good wear; AISI M4.

BUCKEYE M7.
M-1705; 1.0 C, 3.75 Cr, 2.0 V, 1.75 W, 8.75 Mo, bal Fe.
High speed tool steel, molybdenum type.
AISI M7.

BUCKEYE M10.
M-1705; 0.86 C, 4.0 Cr, 2.0 V, 8.5 Mo, bal Fe.
High speed tool steel, molybdenum type.
AISI M10.

BUCKEYE M42.
M-1705; 1.10 C, 3.75 Cr, 1.15 V, 1.70 W, 9.5 Mo, 8.0 Co, bal Fe.
High speed tool steel, Mo-Co type.
AISI M42.

BUCKEYE M43.
M-1705; 1.2 C, 3.75 Cr, 1.6 V, 2.7 W, 8.0 Mo, 8.0 Co, bal Fe.
High speed tool steel, Mo-Co type.
AISI M43.

BUCKEYE T1.
M-1705; 0.75 C, 4.0 Cr, 1 V, 18.0 W, bal Fe.
High speed tool steel, tungsten type.
AISI T1.

BUCKEYE T2.
M-1705; 0.83 C, 4.25 Cr, 2.1 V, 18.5 W, 0.85 Mo, bal Fe.
High speed tool steel, tungsten type.
AISI T2.

BUCKEYE T4.
M-1705; 0.75 C, 4.5 Cr, 1.0 V, 19.0 W, 5.0 Co, bal Fe.
High speed tool steel, tungsten-cobalt type.
AISI T4.

BUCKEYE T5.
M-1705; 0.80 C, 4.0 Cr, 2.0 V, 18.0 W, 8.0 Co, bal Fe.
High speed tool steel, tungsten-cobalt type.
AISI T5.

BUCKEYE T15.
M-1705; 1.5 C, 4.75 Cr, 5.0 V, 12.5 W, 5.0 Co, bal Fe.
High speed tool steel W-Co-V type.
AISI T15.

BUCKLE BRASS-1.
M-Eng.; 90 Cu, 9 Zn, 1 Pb.
For bullets, shells, tubes, cartridges; free-cutting.

BUCKLE BRASS-2.
M-Eng.; 65 Cu, 34 Zn, 1 Pb.
For bullets; free-cutting.

BUDERUS RCC now **RCC.**

BUDERUS RT6 now **RT6.**

BUDERUS RT11 now **RB11.**

BUFF ALOY.
M-962; high C, bal Fe.
For wire cloth; abrasion resisting.

BUFLOKAST.
M-143; 3.2 C, 2.4 Si, 0.7 Mn, bal Fe.
For drums, dryers, kettles, chemical engineering equipment; cast iron for chromium plating.

BUILDTEC 10225.
M-717; Nickel base alloy powder for spraying on cast iron.

BUILDTEC 25685.
M-717.
Alloy powder for metal spraying one-step machinable coating with good wear resistance.

BULL ALLOY.
M-Eng.; 18.3 Sb, 0.6 Fe, 80.9 Pb.
For bearings; antifriction.

BULLDOG.
M-1433; 0.50 C, 1.2 Cr, 0.25 V, 2.5 W, bal Fe.
For chisels, upsetters, crimpers, punches; Type S1; shock resistant.

BULLET BRASS.
M-U.S.; 90 Cu, 9-10 Zn, 0-1 Pb.
For bullets, shells, cartridges.

BULL'S METAL.
M-719; 0.5-1.5 Al, 0.5-1.0 Mn, 0.5-1.0 Sn, 0.8-1.2 Fe, 57-60 Cu, bal Zn.
Cast: 65,000 TS; 30,000 YS; 25 El; 30 RA; 420 Brin.
For marine hardware, propellers; corrosion resistant.

BULL'S WHITE METAL B.
M-719; 80 Sn, 10 Sb, 6 Pb, 4 Cu.
For bearings, linings for marine; white metal, wear resistant.

BUNDYWELD STEEL TUBING.
M-401; 0.2 C, bal Fe.
For tubing, gasoline and oil lines, refrigerator coils; hydrogen welded, copper coated, rolled steel tubing.

BU-NITE.
M-Eng.; Ni-Al, bal Cu.
For pistons; retain strength up to 370°C.

BUNTING 100.
M-720; 87.5-90.5 Cu, 1.0 max Fe, 1.75 max C, 9.5-10.5 Sn.
Sintered: density 5.8-6.2.
For bearings.
SAE 840; ASTM B438 Gr.1 Type I.

BUNTING 101.
M-720; 87.5-90.5 Cu, 1.0 max Fe, 1.75 max C, 9.5-10.5 Sn.
Sintered: density 6.4-6.8.
For bearings.
SAE 841; ASTM B 438 Gr.1 Type II.

BUNTING 104.
M-720; 82.6-88.5 Cu, 1.0 max Fe, 2.0-4.0 Pb, 1.75 max C, 9.5-10.5 Sn.
Sintered: density 6.5-6.9.
For bearings.
SAE 843; ASTM B438 Gr2 Type I.

BUNTING 105.
M-720; 87.5-90.5 Cu, 1.0 max Fe. 1.75 max C, 9.5-10.5 Sn.
Sintered: density 6.8-7.2. 16,000-20,000 TS; 15,000-20,000 YS (comp).
For mechanical components.
SAE 842; ASTM B255 Type II.

BUNTING 108.
M-720; 77-80 Cu, 0.1 max Sn, 1.0-2.0 Pb, 0.25 max Fe, 0.1 max Ni, bal Zn.
Sintered: 7.2-7.6 density; 7% porosity, 9-10 El; 24,000 psi TS.
For mechanical components.
MPIF CZP-0218-T; SAE 890; ASTM B282 Type I.

BUNTING 109.
M-720; 77-80 Cu, 0.1 max Sn, 1.0-2.0 Pb, 0.25 max Fe, 0.1 max Ni, bal Zn.
Sintered: 7.6-8.0 density; 10.0-13.0 El; 28,000 psi TS; 14,000 psi Comp. St.
For mechanical components.
MPIE CZP-0218-U; SAE 891; ASTM B282 Type II.

BUNTING 201.
M-720; 96.25 min Fe, 0.25 max C, 3.0 max other.
Sintered: density 5.7-6.1; porosity 18.
For bearings.
SAE 850; ASTM B439 Gr.1.

BUNTING 203.
M-720; 18.0-22.0 Cu, bal Fe.
Sintered: 5.8-6.2 density.
For bearings.
SAE 863; ASTM B439 Gr.4.

BUNTING 204.
M-720; 7.0-11.0 Cu, bal Fe.
Sintered: 5.8-6.2 density; porosity 18.
For bearings.
SAE 862; ASTM B439 Grade 3.

BUNTING 205-1.
M-720; 1.0-2.5 Cu, 94.5 min Fe, 0.6-1.0 C.
Sintered: 6.0 max density; Rb 20.
Hardened: 40,000 psi TS; Rc 28.
For mechanical components.
MPIF FC-0208-N; SAE 864-A; ASTM B426 Gr.1, Type I.

BUNTING 205-2.
M-720; 1.0-2.5 Cu, 94.5 min Fe, 0.6-1.0 C.
Sintered: 6.0-6.4 density; Rb 50.
Hardened: 49,000 psi TS; Rc 36.
For mechanical components.
MPIF FC-0208-P; SAE 864-B; ASTM B426 Gr.1, Type II.

BUNTING 205-3.
M-720; 1.0-2.5 Cu, 94.5 min Fe, 0.6-1.0 C.
Sintered: 6.4-6.8 density; Rb 65.
Hardened: 89,000 psi TS; Rc 40.
For mechanical components.
MPIF FC-0208-R; ASTM B426 Gr.1, Type III.

BUNTING 205-4.
M-720; 1.0-2.5 Cu, 94.5 min Fe, 0.6-1.0 C.
Sintered: 6.8 min density; Rb 80.
Hardened: 124,000 psi TS; Rc 44.
For mechanical components.
MPIF FC-0208-S; ASTM B426 Gr.1, Type IV.

BUNTING 205-5.
M-720; 2.5-6.0 Cu, 91 min Fe, 0.6-1.0 C.
Sintered: 6.0 max density; Rb 30.
Hardened: 45,000 psi TS; Rc 28.
For mechanical components.
MPIF FC-0508-N; ASTM B426 Gr.2 Type I; SAE 865-A.

BUNTING 205-6.
M-720; 2.5-6.0 Cu, 91 min Fe, 0.6-1.0 C.
Sintered: 6.0-6.4 density; Rb 60.
Hardened: 54,000 psi TS; Rc 40.
For mechanical components.
MPIF FC-0508-P; ASTM B426 Gr.2 Type II, SAE 865-B.

BUNTING 205-7.
M-720; 2.5-6.0 Cu, 91 min Fe, 0.6-1.0 C.
Sintered: 6.4-6.8 density; Rb 75.
Hardened: 92,000 psi TS; Rc 47.
For mechanical components.
MPIF FC-0508-R; ASTM B426 Gr.2 Type III.

BUNTING 205-8.
M-720; 2.5-6.0 Cu, 91 min Fe, 0.6-1.0 C.
Sintered: 6.8 min density.
For mechanical components.
MPIF FC-0508-S; ASTM B426 Gr.2 Type IV.

BUNTING 205-9.
M-720; 6.0-11.0 Cu, 86.0 min Fe, 0.6-1.0 C.
Sintered: 6.0 max density; Rb 30.
Hardened: 45,000 psi TS; Rc 26.
For mechanical components.
MPIF FC-0808-N; ASTM B426 Gr.3 Type I, SAE 866-A.

BUNTING 205-10.
M-720; 6.0-11.0 Cu, 86.0 min Fe, 0.6-1.0 C.
Sintered: 6.0-6.4 density; Rb 60.
Hardened: 56,000 psi TS; Rc 33.
For mechanical components.
MPIF FC-0808-P; ASTM B426 Gr.3 Type II, SAE 866-B.

BUNTING 205-11.
M-720; 6.0-11.0 Cu, 86.0 min Fe, 0.6-1.0 C.
Sintered: 6.4-6.8 density.
For mechanical components.
MPIF FC-0808-R; ASTM B426 Gr.3 Type III.

BUNTING 205-12.
M-720; 6.0-11.0 Cu, 86.0 min Fe, 0.6-1.0 c.
Sintered: 6.8 min density.
For mechanical components.
MPIF FC-0808-S; ASTM B426 Gr.3 Type IV.

BUNTING 206-1.
M-720; 2.5 max Cu, 91.6 min Fe, 1.0-3.0 Ni, 0.6-0.9 C.
Sintered: 6.4-6.8 density; Rb 62.
Hardened: 100,000 psi TS; Rc 34.
For mechanical components.
MPIF FN-0208-R; ASTM B484 Gr.1, Type I, Class C.

BUNTING 206-2.
M-720; 2 max Cu, 89.6 min Fe, 3.0-5.5 Ni, 0.6-0.9 C.
Sintered: 6.8-7.2 density; Rb 79.
Hardened: 135,000 psi TS; Rc 45.
For mechanical components.
MPIF FN-0208-S; ASTM B484 Gr.1, Type II, Class G.

BUNTING 207-IA.
M-720; 97.75 min Fe, 0.25 max C.
Sintered: 5.7-6.1 density; 16,000 psi TS; 2.0 El.
For mechanical components.
MPIF F-0000-N; ASTM B310 Type I Class A. SAE 850.

BUNTING 207-IB.
M-720; 97.4 min Fe, 0.26-0.6 C.
Sintered: 5.7-6.1 density; 20,000 psi TS; 1.5 El.
Hardened: 30,000 psi TS.
For mechanical components.
MPIF F-0005-N; ASTM B310 Type I Class B. SAE 851.

BUNTING 207-IC.
M-720; 97.0 min Fe, 0.61-1.0 C.
Sintered: 5.7-6.1 density; 18% porosity; Rb 35.
Hardened: 40,000 psi TS.
For mechanical components.
MPIF F-0008-N; ASTM B310 Type I Class C; SAE 852.

BUNTING 207-IIA.
M-720; 97.75 min Fe, 0.25 max C.
Sintered: 6.1-6.5 density; 20,000 psi TS; 3.0 El.
For mechanical components.
MPIF F-0000-P; ASTM B310 Type II Class A; SAE 853.

BUNTING 207-IIB.
M-720; 97.40 min Fe, 0.26-0.6 C.
Sintered: 6.1-6.5 density; 26,000 psi TS; 5.0 El.
Hardened: 40,000 psi TS.
For mechanical components.
MPIF F-0005-P; ASTM B310 Type II Class B.

BUNTING 207-IIC.
M-720; 97.0 min Fe, 0.61-1.0 C.
Sintered: 6.1-6.5 density; 34,000 psi TS; Rb 50.
Hardened: 50,000 psi TS.
For mechanical components.
MPIF F-0008-P; ASTM B310 Type II Class C. SAE 855.

BUNTING 207-IIIA.
M-720; 97.75 min Fe, 0.25 max C.
Sintered: 6.5-6.9 density; 26,000 psi TS; 5.0 El.
For mechanical components.
MPIF F-0000-R; ASTM B310 Type III Class A.

BUNTING 207-IIIB.
M-720; 97.40 min, 0.26-0.6 C.
Sintered: 6.5-6.9 density; 34,000 psi TS; 3.0 El.
Hardened: 50,000 psi TS.
For mechanical components.
MPIF F-0005-R; ASTM B310 Type III Class B.

BUNTING 207-IIIC.
M-720; 97.0 min Fe, 0.61-1.0 C.
Sintered: 6.5-6.9 density; 44,000 psi TS; 1.0 El.
Hardened: 64,000 psi TS.
For mechanical components.
MPIF F-0008-R; ASTM B310 Type III Class C.

BUNTING 207-IVA.
M-720; 97.75 min Fe, 0.25 max C.
Sintered: 6.9-7.3 density; 30,000 psi TS; Rf 60.
For mechanical components.
MPIF F-0000-S; ASTM B310 Type IV Class A.

BUNTING 207-IVB.
M-720; 97.40 min Fe, 0.26-0.6 C.
Sintered: 6.9-7.3 density; 50,000 psi TS; 4.0 El.
Hardened: 70,000 psi TS.
For mechanical components.
MPIF F-0005-S; ASTM B310 Type IV Class B.

BUNTING 207-IVC.
M-720; 97.0 min Fe, 0.61-1.0 C.
Sintered: 6.9-7.3 density; 60,000 psi TS; 2.0 El.
Hardened: 80,000 psi TS.
For mechanical components.
MPIF F-0008-S; ASTM B310 Type IV Class C.

BUNTING 208.
M-720; 15.0-25.0 Cu, 69-84 Fe, 0.61-1.0 C.
Sintered: density: 7.1 min. 85,000 TS; 120,000 YS (Comp).
For mechanical components.
SAE 872; ASTM B303 Class C.

BUNTING NO. 327.
M-720; 10 Sn, 2.5 Pb, 1 Zn, 0.1 P, 0.2 Fe, 0.2 Sb, bal Cu.
Cast: 40,000 TS; 18,000 YS; 20 El; 65 Brin.
For bearings; C.S. 22,000.

BUNTING NO. 905.
M-720; 88 Cu, 10 Sn, 2 Zn, 0.06 Fe, 0.20 Sb, 0.1 P.
Cast: 40,000 TS; 18,000 YS; 20 El; 67 Brin.
For bearings for aircraft; C.S. 23,000.

BUNTING NO. 907.
M-720; 89 Cu, 11 Sn, 0.2 P.
Cast: 40,000 TS; 18,000 YS; 20 El; 67 Brin.
For gears; C.S. 23,000.

BUNTING NO. 923.
M-720; 88 Cu, 8 Sn, 1 Pb, 3 Zn, 0.2 Fe, 0.2 Sb, 0.1 P.
Cast: 35,000 TS; 17,000 YS; 15 El; 65 Brin.
For bearings: C.S. 17,000.

BUNTING NO. 925.
M-720; 86.5 Cu, 11 Sn, 1.2 Pb, 0.3 Fe, 0.2 Sb, 1.0 Ni, 0.25 P.
Cast: 40,000 TS; 18,000 YS; 12 El; 75 Brin.
For bearings; C.S. 23,000.

BUNTING NO. 932.
M-720; 83 Cu, 7 Sn, 7 Pb, 3 Zn, 0.07 P, 0.2 Fe, 0.2 Sb.
Cast: 34,000 TS; 16,000 YS; 20 El; 58 Brin.
For bearings; C.S. 20,000.

BUNTING NO. 935.
M-720; 85 Cu, 5 Sn, 9 Pb, 1 Zn, 0.2 Fe, 0.2 Sb, 0.1 P.
Cast: 28,000 TS; 14,000 YS; 15 El; 50 Brin.
For bearings; C.S. 17,000; low friction.

BUNTING NO. 937.
M-720; 78.5-81.5 Cu, 9-11 Sn, 9-11 Pb.
Cast: 35,000 TS; 17,000 YS; 20 El; 60 Brin.
For bearings for machinery and automobiles; C.S. 21,000.

BUNTING NO. 938.
M-720; 8 Sn, 15 Pb, 1 Zn, 0.2 Fe, 0.2 Sb, 0.05 P, bal Cu.
Cast: 30,000 TS; 14,000 YS; 15 El; 52 Brin.
For bearings; C.S. 18,000.

BUNTING NO. 941.
M-720; 4.5 Sn, 20.5 Pb, 1 Zn, 0.2 Fe, 0.2 Sb, 0.05 P, bal Cu.
Cast: 22,000 TS; 11,000 YS; 12 El; 40 Brin.
For bearings for electric motors; C.S. 14,000.

BUNTING 954.
M-720; 85 Cu, 4 Fe, 11 Al.
Cast: 102,000 TS; 45,000 YS; 18 El; 170 Brin.
Bearings for heavy loads.
SAE 954; AMS 4870B.

BURDOX 9.
M-1467; 56-60 Cu, 1 max Sn, 0.25-1.0 Mn, bal Zn.
For welding rod; M.P. 1400-1600°F.

BURDOX 91.
M-1467; 57-61 Cu, 0.25-1.0 Sn, 0.05 max Pb, 0.01 max Al, bal Zn.
For welding rod; M.P. 1630-1650°F.

BURDOX 92.
M-1467; 58-60 Cu, 0.6-0.9 Sn, 0.35-0.50 Fe, 0.15-0.30 Mn, 0.25-0.40 Ni, bal Zn.
For welding rod.

BURGESS ALUMINUM SOLDER.
M-Eng.; 76 Sn, 21 Zn, 3 Al.
For aluminum solder.

BURLOY A.
M-570; 0.12 max C, 16-23 Cr, 7-11 Ni, bal Fe.
For stainless and corrosion resistant parts; stainless, heat and corrosion resistant.

BURLOY B.
M-570; 0.21-0.35 C, 12-15 Cr, bal Fe.
For corrosion resistant parts; corrosion resistant.

BURLOY C.
M-570; 0.12 max C, 24-30 Cr, 12-15 Ni, bal Fe.
For heat and corrosion resistant parts; heat and corrosion resistant.

BURLOY C-1.
M-570; 0.36-0.50 C, Si, 16-23 Cr, 7-11 Ni, bal Fe.
For heat and corrosion resistant parts; heat and corrosion resistant.

BURNDY NO. 60.
M-722.
Ductile cast iron.
60,000 psi min TS; 45,000 psi min YS; 10 min El.

BURNDY NO. 86.
M-722.
Gilding metal, hard.
57,000 psi min TS; 54,000 psi YS, typ.; 6 El.

BURNDY NO. 87.
M-722.
Gilding metal, hard.
50,000 psi min TS; 46,000 psi YS, typ.; 5 El.

BURNDY NO. 101.
M-722; Cu.
Cast: 20,000 TS; 9,000 YS; 50 El; 35 Brin.
For castings.

BURNDY NO. 102 AND 103.
M-722.
Cast alloys from BURNDY NO. 101.

BURNDY NO. 115.
M-722.
BURNDY cast alloy.
30,000 psi min TS; 14,200 psi min YS; 20 El.

BURNDY NO. 215.
M-722.
Leaded commercial bronze, half hard.
50,000 psi min TS; 30,000 psi min YS; 14 El; 80% machinability.

BURNDY NO. 329.
M-722.
Everdur 1014 soft.
88,000 psi TS; 44,000 psi YS; 25 El; 60% machinability.

BURNDY NO. 330.
M-722.
Aluminum bronze.
55,000 psi min TS; 21,000 psi min YS; 6 El.

BURNDY NO. 335.
M-335.
Spec. Aluminum-Nickel bronze.
90,000 psi min TS; 40,000 psi min YS; 6 El; 35% machinability.

BURNDY NO. 339.
M-722.
Aluminium bronze.
85,000 psi min TS; 28,000 psi min YS; 10 El; 35% machinability.

BURNDY NO. 701.
M-722.
Everdur 1015 soft.
40,000 psi TS; 15,000 psi YS; 50 El; 30% machinability.

BURNDY NO. 715.
M-722.
DURIUM VI Hard.
75,000 psi TS; 15 El; 30% machinability.

BURNDY NO. 801.
M-722.
Phosphor bronze, hard.
75,000 psi TS; 63,000 psi YS; 15 El; 80% machinability.

BURNDY NO. 802.
M-722.
Phosphor bronze, hard.
94,000 psi TS; 13 El; 20% machinability.

BURNDY NO. 851.
M-722.
High brass, half hard.
57,000 psi TS; 25,000 psi YS; 18 El; 100% machinability.

BURR-A.
M-Eng.; 62 Cu, 38 Zn.
For pipes, tubes, hardware.

BURR-B.
M-Eng.; 90 Cu, 10 Zn.
For window screen wire, radiators.

BURYS C.16.B.
M-1415; 0.5 C, 1 Si, 1.5 Cr, 2 W, 0.2 V, bal Fe.
For pneumatic tool pistons, chisels, riveters; tough and impact resistant.

BURYS H.18.S.
M-1415; 1.0 C, 1.5 Cr, bal Fe.
For ball and roller bearings, forming rolls; oil or water hardened.

BUSHINGS.
M-Eng.; 85-86.2 Cu, 10.2-11.0 Sn, 3.6-4.0 Zn.
For bushings, bearings; high strength.

BUSTER ALLOY.
M-35; 0.50-0.58 C, 2.2 W, 1.25 Cr, 0.25 Mn, 1 Si, 0.25 V, bal Fe.
Ht. Tr.: 283,000 TS; 246,000 YS; 10 El; C 53 Rock.
Annealed: 190 Brin.
For dies, chisels, shear blades, punches, forging dies.
Type S1 hot work tool steel, shock resistant.

BUSTER ALLOY 50.
M-35; 0.50 C, 0.80 Si, 2.25 W, 1.35 Cr, 0.25 V, bal Fe.
Oil hardening shock resisting tool steel.
For rivet busters, pneumatic chisels, punches.
AISI S1.

BUSTER ALLOY 60.
M-35; 0.58 C, 0.80 Si, 2.25 W, 1.35 Cr, 0.25 V, bal Fe.
Oil hardening shock resisting tool steel.
For rivet sets, chisels, heading dies.
AISI S1.

BUTTON.
M-U.S.; 80 Zn, 20 Cu.
For die castings, ornaments; red brass.

BUTTON ALLOY-1.
M-Eng.; 60-50 Cu, 30-45 Zn, 0-10 Sn.
For buttons, ornaments; corrosion resistant.

BUTTON ALLOY-2.
M-Eng.; 43 Cu, 57 Zn.
For buttons, ornaments; low strength.

BUTTON BRASS.
M-Eng.; 90 Cu, 10 Zn, 0.5 Sn.
For buttons, ornaments.

BVR 30.
M-72; 0.34 C, 1.5 Cr, 1.5 Ni, 0.25 Mo, bal Fe.
Ann: 690 N/mm² max UTS.
For cold forming. Heat treatable steel.

BVT 30.
M-72; 0.22 C, 0.55 Mn, 0.30 max Cr, 0.35 Mo, bal Fe.
For thick-walled high pressure tubes to operate up to 530°C.
W. Nr. 1.5419.

BVT 35.
M-72; 0.14 C, 0.55 Mn, 1.0 Cr, 0.55 Mo, bal Fe.
For water boilers and superheater tubes operating below 530°C.
W. Nr. 1.7335.

BVT 40.
M-72; 0.17 C, 1.0 Cr, 0.45 Mo, bal Fe.
For steam turbine forgings.
W. Nr. 1.7337.

BVT 50.
M-72; 0.22 C, 1.05 Cr, 0.45 Mo, 0.60 max Ni, bal Fe.
Forgings for steam turbines.

BVT 60.
M-72; 0.24 C, 1.35 Cr, 0.55 Mo, 0.60 max Ni, 0.20 V, bal Fe.
Oil or water hardenable bolts and nuts, for operation up to 530°C.
W. Nr. 1.7733.

BVT 90.
M-German; 0.3 C, 13 Cr, 2 Mo, 1 V, bal Fe.
For turbine rotors; high heat resistant.

BVT-130.
M-1343; 0.20 C, 12 Cr, 1.2 Mo, bal Fe.
Annealed: 82,000 TS; 38,000 YS; B 82 Rock.
For chemical plant and oil refinery equipment, gears, shafts, marine hardware.
Corrosion resistant, hardenable.

BVT 130V.
M-72; 0.20 C, 1.0 max Mn, 11.0 Cr, 1.0 Mo, 0.5 Ni, 0.3 V, bal Fe.
Corrosion resistant steel for use at elevated temperature.
W. Nr. 1.4922.

"B-W" STANDARDIZED.
M-213; 0.15-0.25 C, 0.30-0.60 Mn, bal Fe.
Rolled: 55,000-65,000 TS; 35,000-40,000 YS; 30 El; 55 RA.
Heat treated: 65,000-75,000 TS; 40,000-45,000 YS; 20 El; 60 RA; 158 Brin.
For general use, boiler plates.

B&W 0117.
M-17; 0.22 max C, 0.85-1.25 Mn, 0.20 min Cu, 0.02 min V, bal Fe.
Hot finished or normalized: 70,000 TS; 50,000 YS; 25 El.
Shock resistant; good resistance to atmospheric corrosion.

B&W 1118FM.
M-17; 0.14-0.20 C, 1.3-1.6 Mn, 0.08-0.13 S, 0.040 max P, bal Fe.
Hot finished: 70,000-75,000 TS; 40,000-45,000 YS; Rb 75-85.
Cold drawn: 90,000-105,000 TS; 75,000-90,000 YS; Rb 92-100.
For machining applications.

B&W 52100.
M-17; 0.95-1.10 C, 0.25-0.45 Mn, 1.3-1.6 Cr, bal Fe.
Ann: 90,000 TS; 51,000 YS; 31 El; BHN 180.
Normalized: 161,000 TS; 106,000 YS; 7 El; BHN 373.
Hardened: Oil Q+T: tensile strengths in excess of 300,000 psi are obtainable.
For bearings, bushings, tools.

B&W CARBON MOLY (T/P-1).
M-17; 0.10-0.20 C, 0.30-0.80 Mn, 0.10-0.50 Si, 0.44-0.65 Mo, bal Fe.
Ann: 55,000 min TS; 30,000 min YS; 30 min El, Rb 80 max.
For service conditions requiring higher creep strength than carbon steel with no increase in corrosion or oxidation resistance.
Oxidation resistance to 1050°F in air.

B&W CARBON STEEL.
M-17; 0.06-0.18 C, 0.27-0.63 Mn, bal Fe.
Ann: 47,000 min TS; 26,000 min YS; 35 min El; Rb 77 max.
For tubular heat exchangers, condensers, and similar heat transfer apparatus.

B&W CROLOY 1/2 (T/P-2).
M-17; 0.10-0.20 C, 0.30-0.61 Mn, 0.10-0.30 Si, 0.50-0.81 Cr, 0.44-0.65 Mo, bal Fe.
Ann: 60,000 min TS; 30,000 min YS; 30 min El; Rb 85 max.
Superior to CARBON MOLY for creep strength and graphitization; for high temperature steam piping. Oxidation resistance to 1075°F in air.

B&W CROLOY 1 1/4 (T/P-11).
M-17; 0.15 max C, 0.30-0.60 Mn, 0.50-1.0 Si, 1.0-1.5 Cr, 0.44-0.65 Mo bal Fe.
Ann: 60,000 min TS; 30,000 min YS; 30 min El; Rb 85 max.
Good creep strength properties; more corrosion resistant than Cr free grades. Oxidation resistance to 1100°F in air.

B&W CROLOY 2 (T/P-3B).
M-17; 0.15 max C, 0.30-0.60 Mn, 0.50 max Si, 1.65-2.35 Cr, 0.44-0.65 Mo, bal Fe.
Ann: 60,000 min TS; 30,000 min YS; 30 min El; Rb 85 max.
For resisting both oxidation and corrosion, with excellent high temperature strength.
Oxidation resistance to 1150°F in air.

B&W CROLOY 2 Al.
M-17; 0.15 max C, 0.50 max Mn, 1.0-1.4 Si, 1.75-2.25 Cr, 0.45-0.65 Mo bal Fe.
Ann: 60,000 min TS; 30,000 min YS; 30 min El; Rb 85 max.
Good corrosion resistance to high temperature (vapor phase) acids. Slightly greater resistance to oxidation.

B&W CROLOY 2 1/4 (T/P 22).
M-17; 0.15 max C, 0.30-0.60 Mn, 0.50 max Si, 1.9-2.6 Cr, 0.87-1.13 Mo bal Fe.
Ann: 60,000 min TS; 30,000 min YS; 30 min El; Rb 85 max.
Exceptionally high creep strength for polymerization and high pressure cracking. Oxidation resistance to 1175°F in air.

B&W CROLOY 3M (T/P 21).
M-17; 0.15 max C, 0.30-0.60 Mn, 0.50 max Si, 2.65-3.35 Cr, 0.80-1.06 Mo, bal Fe.
Ann: 60,000 min TS; 30,000 min YS; 30 min El; Rb 85 max.
Better creep properties and corrosion and oxidation resistance than CROLOY 2. Oxidation resistance to 1175°F in air.

B&W CROLOY 5 (T/P 5).
M-17; 0.15 max C, 0.30-0.60 Mn, 0.50 max Si, 4.0-6.0 Cr, 0.45-0.65 Mo bal Fe.
Ann: 60,000 min TS; 30,000 min YS; 30 min El; Rb 85 max.
Normalized: 160,000 TS; 120,000 YS; 16 El; Rc 34.
Very good corrosion resistance, creep strength. Oxidation resistance to 1200°F in air.

B&W CROLOY 5 SI (T/P 5B).
M-17; 0.15 C, 0.30-0.60 Mn, 1.0-2.0 Si, 4.0-6.0 Cr, 0.45-0.65 Mo, bal Fe.
Ann: 60,000 min TS; 30,000 min YS; 30 min El; Rb 89 max.
For operating conditions where oxidation is a primary requirement; resistant to scaling by oxidation. Oxidation resistance to 1300°F in air.

B&W CROLOY 7 (T/P 7).
M-17; 0.15 max C, 0.30-0.60 Mn, 0.50-1.0 Si, 6.0-8.0 Cr, 0.45-0.65 Mo, bal Fe.
Ann: 60,000 min TS; 30,000 min YS; 30 min El; Rb 89 max.
Corrosion resistance is intermediate between CROLOYS 5 and 9. Oxidation resistance to 1250°F in air.

B&W CROLOY 9M (T/P 9).
M-17; 0.15 max C, 0.30-0.60 Mn, 0.25-1.0 Si, 8.0-10.0 Cr, 0.90-1.10 Mo, bal Fe.
Ann: 60,000 min TS; 30,000 min YS; 30 min El; Rb 89 max.
Normalized: 185,000 TS; 150,000 YS; 17 El; Rc 42.
For severe operating conditions where high corrosion and oxidation resistance are essential. Oxidation resistance to 1300°F in air.

B&W CROLOY 12 (T/P 410).
M-17; 0.15 max C, 1.0 max Mn, 0.75 max Si, 11.5-13.5 Cr, bal Fe.
Ann: 60,000 min TS; 30,000 min YS; 20 min El; Rb 95 max.
Normalized and tempered: 106,000-180,000 Ts; 85,000 175,000 YS, 15-25 El; 21-39 Rc.
For use where mechanical properties plus corrosion resistance are important. Oxidation resistance to 1300°F in air.

B&W CROLOY 12A1 (T/P 405).
M-17; 0.08 max C, 1.0 max Mn, 0.75 max Si, 11.5-13.5 Cr, 0.10-0.30 Al bal Fe.
Ann: 60,000 min TS; 30,000 min YS; 20 min El; Rb 95 max.
Non-hardenable by heat treating; weldable. Good oxidation resistance.

B&W CROLOY 12-3W.
M-17; 0.15-0.20 C, 0.50 max Mn, 12.0-14.0 Cr, 1.8-2.2 Ni, 2.5-3.5 W, bal Fe.
Martensitic stainless that can be heat treated to high hardness.

B&W CROLOY 16-1.
M-17; 0.03 max C, 1.0 max Mn, 14.0-16.0 Cr, 1.0-1.5 Ni, bal Fe.
Ann: 60,000 min TS; 30,000 min YS; 20 Min El; Rb 95 max.
In tubular operations requiring increased resistance to chloride stress corrosion cracking.

B&W CROLOY 16-6 PH.
M-17; 0.025-0.045 C, 0.70-0.90 Mn, 15.0-16.0 Cr, 7.0-8.0 Ni, 0.25-0.45 Al, 0.30-0.50 Ti, bal Fe.
Solution ann: 134,000 TS; 110,000 YS; 16 El; Rc 28.
Sol. ann. and aged: 190,000 TS; 185,000 YS; 15 El; Rc 40.
Good combination of hardness, strength and corrosion resistance.

B&W CROLOY 18 (T/P 430).
M-17; 0.12 max C, 1.0 max Mn, 16.0-18.0 Cr, 0.50 max Ni, bal Fe.
Ann: 60,000 min TS; 35,000 min YS; 20 min El; Rb 90 max.
Ferritic, magnetic, non-hardenable steel with good corrosion resistance; good for nitration work and nitric acid manufacture.

B&W CROLOY 27 (T/P 446).
M-17; 0.20 max C, 1.5 max Mn, 23.0-30.0 Cr, 0.50 max Ni, 0.10-0.25 N_2, bal Fe.
Ann: 70,000 min TS; 40,000 min YS; 18 min El; Rb 95 max.
Good resistance to oxidation at 1500-2100°F, and resists attack by sla ang flue dust.

B&W CROLOY 299.
M-17; 0.12-0.25 C, 14.0-15.5 Mn, 16.5-18.5 Cr, 1.15-1.75 Ni, 0.32-0.40 N_2, bal Fe.
Ann: 115,000 TS; 68,000 YS; 72 El; Rc 18.
Cold worked: 173,000 TS; 160,000 YS; 26 El; Rc 39.
Austenitic stainless steel; work-hardens to high strength without becoming magnetic.

B&W CROLOY 304.
M-17; 0.08 max C, 2.0 max Mn, 18.0-20.0 Cr, 8.0-11.0 Ni, bal Fe.
Ann: 75,000 min TS; 30,000 min YS; 28 min El; Rb 90 max.
Resistant to corrosion and heat. (Available also as 304H, 304L, 321, 347, 348; 1/8, 1/4, 1/2 hard).

B&W CROLOY 310.
M-17; 0.15 max C, 2.0 max Mn, 24.0-26.0 Cr, 19.0-22.0 Ni, bal Fe.
Ann: 75,000 min TS; 35,000 min YS; 25 min El; Rb 90 max.
For extreme resistance to oxidation and corrosion, for high pressure, high temperature applications.

B&W CROLOY 316.
M-17; 0.08 max C, 2.0 max Mn, 16.0-18.0 Cr, 11.0-14.0 Ni, 2.0-3.0 Mo, bal Fe.
Ann: 75,000 min TS; 30,000 min YS; 28 min El; Rb 90 max.
Superior to Type 304 in resistance to creep and corrosion. (Also available as 316H and 316L).

B&W CROLOY 600.
M-17; 0.15 max C, 1.0 max Mn, 14.0-17.0 Cr, 72.0 min Ni, 6.0-10.0 Fe.
Ann: 80,000 min TS; 35,000 min YS; 30 min El; Rb 92 max.
Excellent high temperature strength and useful oxidation resistance up to 2100°F.

B&W CROLOY 800.
M-17; 0.10 max C, 1.50 max Mn, 19.0-23.0 Cr, 30.0-35.0 Ni, 0.15-0.60 Al, 0.15-0.60 Ti, bal Fe.
Ann: 75,000 min TS; 30,000 min YS; 30 min El; Rb 95 max.
Resistant to elevated temperature oxidation and carburization.

B&W IRON.
M-17; 0.05 max C, 0.25-0.40 Mn, 0.10 max Cu, bal Fe.
Hot finished: 45,000-50,000 TS; 30,000-41,000 YS; 49-70 El; Rb 53-57
Cold drawn: 68,000-86,000 TS; 60,000-80,000 YS; 25-42 El; Rb 76-88.
For electric motor and generator housings; good magnetic characteristics.

B&W NICOLOY 3 1/2.
M-17; 0.06-0.12 C, 0.31-0.64 Mn, 0.18-0.37 Si, 3.18-3.82 Ni, bal Fe.
Normalized: 65,000 min TS; 35,000 min YS; 30 min El; Rb 90 max.
Combination of high strength and resistance to brittle fracture at temperatures down to -150°F.

B&W NICOLOY 9.
M-17; 0.13 max C, 0.90 max Mn, 0.13-0.32 Si, 8.4-9.6 Ni, bal Fe.
Double normalized and tempered: 100,000 min TS; 75,000 min YS; 22 min El.
Combination of high strength and resistance to brittle fracture at temperatures down to -320°F.

B&W NO. 441.
M-447; 0.20 C, 5.0 Cr, 0.55 Mo, bal Fe.
Ann: 250 max Brin.
For oil still headers, castings, pumps, pipes; moderate corrosion resistance.

B&W NO. 600.
M-447; 0.20 C, 19.5 Cr, 9.5 Ni, bal Fe.
Cast: 130-160 Brin.
For corrosion resisting parts; chemical industries; austenitic alloy; high corrosion resistance.

B&W NO. 642.
M-447; 0.08 C, 19.5 Cr, 10.5 Ni, 2.5 Mo, bal Fe.
Cast: 140-165 Brin.
For corrosion resisting parts; paper mill sulfite; resists mineral acids; austenitic; corrosion resistant.

B&W NO. 661.
M-447; 0.35 C, 26.0 Cr, 12.5 Ni, 2.0 Si, bal Fe.
Cast: 75,000 TS; 190 Brin.
For oil still tube supports, furnace parts; heat and corrosion resistant to 1800°F.

B&W NO. 800.
M-447; 1.25 C, 25.0 Cr, 11.0 Ni, bal Fe.
Ht. Tr.: 300 Brin.
For heat resisting parts, wire guides; heat resistant.

B&W NO. 850.
M-447; 1.55 C, 19.0 Cr, 7.0 Ni, bal Fe.
Ht. Tr.: 320-430 Brin.
For rolling mill plugs, wire guides; wear and heat resistant.

B&W NO. 1101.
M-447; 0.40 C, 26.0 Cr, 20.0 Ni, bal Fe.
Cast: 190 Brin.
For heat resisting castings, furnace parts; heat resistant to 2000°F.

B&W NO. 1440.
M-447; 0.45 C, 1.5 Cr, 7.5 Mo, bal Fe.
Ann: 250 Brin.
For tube mill piercer points; wear resistant.

B&W NO. 1501.
M-447; 0.60 C, 27.0 Cr, 9.5 Ni, bal Fe.
Cast: 200 Brin.
For oil burner parts; heat and corrosion resistant.

B&W NO. 5150.
M-447; 1.0 C, 1.4 Cr, bal Fe.
Ht. Tr.: 300-500 Brin.
For grinding units, wear plates; abrasion resistant.

B&W STROLOY 1.
M-17; 0.14-0.20 C, 0.60-1.0 Mn, 0.15-0.35 Si, 0.40-0.65 Cr, 0.70-1.0 Ni, 0.40-0.60 Mo, 0.03-0.08 V, 0.001 B, bal Fe.
Q+T; 115,000-145,000 TS; 100,000 min YS; 15 min El. (up to 1.5 inch thick).
Good strength, toughness and weldability; for structural applications.

B&W STROLOY 2A.
M-17; 0.15-0.21 C, 0.70-1.0 Mn, 0.20-0.35 Si, 0.80-1.1 Cr, 0.40-0.70 Ni, 0.20-0.30 Mo, 0.002-0.004 B, bal Fe.
Q+T; 110,000-145,000 TS; 100,000 min YS; 15 min El. (up to 0.750 inc thick).
Good strength, toughness and weldability; for structural applications.

B&W STROLOY 5C.
M-17; 0.15-0.21 C, 0.70-1.0 Mn, 0.20-0.35 Si, 0.75-1.1 Cr, 0.15-0.25 Mo, 0.001-0.005 B, bal Fe.
Q+T; 110,000-145,000 TS; 100,000 min YS; 15 min El. (up to 0.375 inc thick).
Good strength, toughness and weldability; for structural applications.

BX-3 see JESSOP BX-3.

B-XX.
M-424; 0.3-0.4 C, bal Fe.
For gears, shafts; water hardening.

BYRWILL "B".
M-723; 91-93 Zn, 2-3 Pb, 6-7 Cu.
28,000 TS; 67 Brin.
For bearings; antifriction.

C

C-4 HOT DIE STEEL.
M-432; 0.95 C, 4 Cr, bal Fe.
For hot work dies, gripper dies, bolt and rivet header dies; hot work steel.

CA6NM.
M-U.S.; 0.06 max C, 1.0 max Mn, 0.65 max Si, 11.5-14.0 Cr, 3.0-5.0 Ni, 1.0 max Mo, bal Fe.
Corrosion resistant casting alloy.
ASTM A296-77 CA-6NM.

CA-10 ETC see also CARMET CA-10 ETC.

C 18.
M-1290; 0.15-0.23 C, 1.0 max Si, 1.0 max Mn, 16.0-18.0 Cr, 1.5-2.5 Ni, bal Fe.
Martensitic stainless steel.
W.-Nr. 1.4057, similar to AISI 431.

C 18 8 S.
M-1290; 0.07 max C, 2.0 max Si, 2.0 max Mn, 17.5-20.0 Cr, 9.0-11.0 Ni 0.70 max Mo, bal Fe.
Austenitic stainless cast steel for low temperature service.
SEW 6902; G-X 6 CrNi 18 10.

C 18 8 SS.
M-1290; 0.10 max C, 1.0 max Si, 2.0 max Mn, 17.0-19.0 Cr, 10.0-12.0 Ni, 0.50 max Mo, 1.0 Nb, bal Fe.
Austenitic stainless cast steel for low temperature service.
SEW 1.6905; G-X 7 CrNiNb 18 10.

C 18 E.
M-1290; 0.33-0.43 C, 1.0 max Si, 1.0 max Mn, 15.5-17.5 Cr, 1.0-1.3 Mo, 1.0 max Ni, bal Fe.
Martensitic stainless steel.
W.-Nr. 1.4122.

C 18 W.
M-1290; 0.10 max C, 1.0 max Si, 1.0 max Mn, 15.5-17.5 Cr, bal Fe.
Ferritic stainless steel.
W.-Nr. 1.4016.

C20CB3 see CARPENTER STAINLESS NO. 20 CB-3.

C 40.
M-Germany; 60 Al_2O_3, 40 mixed carbides.
For hard cutters to machine steel and cast iron. Sintered Alloy.

C 46.
M-Italian; 10 Cu, 1 Si, 0.25 Mg, 0.15 Ti, 1.5 Ni, bal Al.
For castings; heat resistant.

C-67.
M-Switzerland; 5 Sn, 7 Al, bal Cu.
Non-staining, gold color. For watch cases.

C-74 ARMOR.
M-1222, M-112; 0.25-0.30 C, 1.0-1.6 Mn, 0.3-0.4 Si, 0.5-1.0 Cr, 0.75-1.45 Ni, 0.4-0.6 Mo, bal Fe.
Cast: 120,000 TS; 95,000 YP; 18 El; 40 RA; 225-270 Brin.
For gears, countershafts, axles, crankshafts.
Oil hardening, tough, shock resistant.

C80.
M-1739.
Pure copper sintered.
Porosity: 11-8%.
Good electrical properties.

C-103 see COLUMBIUM C-103.

C120AV see CRUCIBLE C-120AV.

C129 see COLUMBIUM C-129.

C196.
M-8; 98.7 Cu, 1.0 Fe, 0.3 P.
Ann: 45,000 psi TS; 21,000 psi YS; 36 El.
Hard: 62,000 psi TS; 59,000 psi YS; 5 El.
Elec. cond: 74-76% IACS.
Good strength, good conductivity, retains fine grain to 850°C; for current conducting springs.

C-242.
M-239; 0.3 C, 10 Co, 20 Cr, 10 Mo, 1 max Fe, bal Ni.
For high temperature applications, jet engine and gas turbine components.
Heat and oxidation resistant.

C263 see WIGGIN ALLOY 263.

C-276 see HASTELLOY ALLOY C-276 and NICKELVAC H-C276.

C-501.
M-724; 2.1 Mg, 1.5 Mn, bal Al.
For light alloy parts; similar to "Pearlman."

C-502.
M-724; 11-14 Si, bal Al.
For light alloy parts; similar to "Silumin."

C-503.
M-724; 4 Cu, 0.4 Ti, bal Al.
Heat treated: 40,000-43,000 TS; 26,000-27,000 YS; 4-5 El; 80-90 Brin
For light alloy parts; age-hardenable.

C 1023.
M-48; 0.15 C, 15.5 Cr, 9.7 Co, 8.4 Mo, 3.6 Ti, 4.2 Al, bal Ni.
For nozzle vanes.

C5152.
M-1741; 2.0 Mg, bal Al.
O Temper: 186 MPa TS; 76 MPa YS; 24 El; 48 VHN.
H38 Temper: 283 MPa TS; 241 MPa YS; 5 El; 92 VHN.
Sheet metal work, hydraulic tube.

C5154.
M-1741; 3.5 Mg, 0.25 Cr, bal Al.
O Temper: 241 MPa TS; 117 MPa YS; 27 El; 59 VHN.
H32 Temper: 269 MPa TS; 207 MPa YS; 15 El; 92 VHN.
Storage tanks, welded structures.

CADMET.
M-1091; WC + Co.
For cutting tool; sintered carbides.

CADMIUM.
M-1755; Cd.
Purities: Zone refined 99.9999%, 99.9995%, 99.999%, 99.9 + % (Nuclear grade).
Forms: Bars, rods, sheets, foils, shot, powders, wires, single crystals.

CADMIUM AMALGAM.
M-Eng.; Cd, bal Hg.
For modeling purposes, fusible plugs; softens when moderately heated, soft as wax.

CADMIUM COPPER.
M-33; 99 Cu, 1 Cd.
Ann: 36,000 TS; 12,000 YS; 57 El.
Drawn hard: 73,000 TS; 69,000 YS; 9 El.
High conductivity; for trolley wire, connectors, switch gear.

CADMIUM COPPER, 0.1% CADMIUM.
M-1789; 99.9 Cu, 0.1 Cd.
Ann: 32,000 TS; 45 El.
Rolled hard: 51,000 TS; 6 El.
High conductivity, resistance to softening.
Cooling fins for radiators and air conditioners.

CADMIUM COPPER, 0.2% CADMIUM.
M-1789; 99.8 Cu, 0.2 Cd.
Ann: 32,000 TS; 45 El.
Rolled hard: 51,000 TS; 6 El.
Good conductivity, resistance to softening.
For electrical contacts, terminals, springs.

CADMIUM COPPER 985.
M-141; 99.1 Cu, 0.9 Cd.
Hard: 55,000 TS; 48,000 YS; 6 El; 116 Brin.
Soft: 37,000 TS; 12,000 YS; 50 El; 45 Brin.
For trolley wire, marine hardware; corrosion resistant.

CADMIUM SILVER.
M-Eng.; Cd, bal Ag.
For corrosion resistant parts.

CAF 78.
M-1587; 0.5 max C, 1.0 max Mn, 2.0 max Si, 26.0-30.0 Cr, 4.0 max Ni, 0.5 max Mo, bal Fe.
Corrosion resistant steel casting.
ACI HC.

CAF 303.
M-1587; 0.16 max C, 1.5 max Mn, 2.0 max Si, 18-21 Cr, 9-12 Ni, 1.5 max Mo, 0.20-0.35 Se, bal Fe.
Free-machining, corrosion resistant steel casting.
ACI CF-16F; ASTM A-296.

CAF 304.
M-1587; 0.08 max C, 1.5 max Mn, 18-21 Cr, 8-11 Ni, bal Fe.
Corrosion resistant steel casting.
ACI CF-8; ASTM A-296, A-351 CF-8.

CAF 310.
M-1587; 0.20 max C, 2.0 max Mn, 2.0 max Si, 23-27 Cr, 19-22 Ni, bal Fe.
Corrosion resisting steel casting; for high temperature operation.
ACI CK-20; ASTM A-351 CK-20.

CAF 312.
M-1587; 0.30 max C, 1.5 max Mn, 2.0 max Si, 26-30 Cr, 8-11 Ni, bal Fe.
Corrosion resistant steel casting.
ACI CE-30; ASTM A-296.

CAF 317.
M-1587; 0.08 max C, 1.5 max Mn, 1.5 max Si, 18-21 Cr, 9-13 Ni, 3-4 Mo, bal Fe.
Corrosion resistant steel casting.
ACI CG-8M; ASTM A-296.

CAF 318.
M-1587; 0.08 max C, 1.5 max Mn, 2.0 max Si, 18-21 Cr, 9-12 Ni, 2-3 Mo, 0.70 Cb, bal Fe.
Corrosion resistant steel casting; weldable.

CAF 320.
M-1587; 0.08 max C, 1.5 max Mn, 1.0 max Si, 18-21 Cr, 9-12 N, 2-3 Mo, bal Fe.
Corrosion resistant steel casting.
ACI CF-8M; ASTM A-351 CF8M.

CAF 327.
M-1587; 0.5 max C, 1.5 max Mn, 2.0 max Si, 26-30 Cr, 4-7 Ni, 0.5 max Mo, bal Fe.
Corrosion resistant steel casting.
ACI-HD.

CAF 329H.
M-1587; 0.20 max C, 1.5 max Mn, 2.0 max Si, 22-26 Cr, 12-15 Ni, bal Fe.
Corrosion resisting steel casting.
ACI CF-20; AMS 5358.

CAF 379.
M-1587; 0.2-0.5 C, 2.0 max Mn, 2.0 max Si, 26-30 Cr, 8-11 Ni, 0.5 max Mo, bal Fe.
Corrosion resistant steel casting.
ACI HE; ASTM A-297.

CAF 404.
M-1587; 0.12 max C, 1.0 max Mn, 2.0 max Si, 11.0-13.0 Cr, 1.0 max Ni, 1.5-2.0 Cu, bal Fe.
Cast stainless steel.

CAF 420.
M-1587; 0.20-0.40 C, 1.0 max Mn, 1.5 max Si, 11.5-14.0 Cr, 1.0 max Ni, 0.5 max Mo, bal Fe.
Cast stainless steel.
ACI CA-40; ASTM A-296.

CAF 509.
M-1587; 0.50 max C, 2 max Mn, 2 max Si, 24-28 Cr, 11-14 Ni, 0.5 max Mo, bal Fe.
Corrosion resistant steel casting.
ACI HH.

CAF 602.
M-1587; 0.20-0.40 C, 2.0 max Mn, 2.0 max Si, 18-23 Cr, 8-12 Ni, 0.5 max Mo, bal Fe.
Corrosion resistant steel casting.
ACI HF; ASTM A-297.

CAF 620.
M-1587; 0.07 max C, 1.5 max Mn, 1.5 max Si, 19-22 Cr, 27.5-30.5 Ni, 2.0-3.0 Mo, 3.0-4.0 Cu, bal Fe.
Corrosion resistant steel casting.
ACI CN-7M; ASTM A-351 CN-7M.

CAF 720.
M-1587; 0.07 max C, 1.5 max Mn, 1.5 max Si, 19-22 Cr, 27.5-30.5 Ni, 2.0-3.0 Mo, 3.0-4.0 Cu, 0.6-1.0 Cb + Ta, bal Fe.
Corrosion resistant steel casting.

CAF 820.
M-1587; 0.07 max C, 1.5 max Mn, 1.5 max Si, 19-22 Cr, 35-38 Ni, 2-3 Mo, 3-4 Cu, 1.0 Cb, bal Fe.
Corrosion resisting steel casting.

CALCIUM.
M-1755; Ca.
Purities: Distilled 99.9%, 99.5%, 99%, 98%.
Forms: Granules, ingot, extruded forms, sheet, powder.

CALCLAD L NICKEL.
M-1724.
Layer of nickel on mild steel backing plate.

CALCLAD MONEL.
M-1724.
Layer of Monel on mild steel backing plate.

CALITE 25-10.
M-30; 0.25-0.45 C, 23-28 Cr, 10-14 Ni, 0.25-1.00 Si, 0.25-1.00 Mn, bal Fe.
Cast: 81,000 TS; 45,000 YS; 16 El; 180 Brin.
Annealed: 87,000 TS; 51,000 YS; 5 El; 180 Brin.
For furnace parts, beams, tube supports; heat resistant to 1950°F.

CALITE A.
M-30; 0.4 C, 15-18 Cr, 35-37 Ni, 1-1.7 Si, 1.5 Mn, bal Fe.
At 1600°F: 18,000 TS; 26 El.
Cast: 68,000 TS; 36,200 YS; 8 El; 156 Brin.
For carburizing boxes, hearth plates, retorts; corrosion and heat resisting to 2000°F.

CALITE A-15.
M-30; 0.15 max C, 15-18 Cr, 35-37 Ni, 1.00-1.75 Si, 1.00-1.75 Mn, bal Fe.
Cast: 71,000 TS; 34,000 YS; 19 El; 112 Brin.
Annealed: 90,650 TS; 44,200 YS; 13 El.
For retorts, fan blades; heat resistant to 2000°F.

CALITE B-18.
M-30; 0.29-0.36 C, 18-20 Cr, 8-10 Ni, 1 Mo, bal Fe.
Cast: 80,000 TS; 35,000 YS; 9 El; 3 RA; 170 Brin.
Annealed: 105,000 TS; 48,100 YS; 17 El.
For beams supporting loads at elevated temperatures, oil still tube supports; heat resisting up to 1650°F; resists sulphur; formerly Calite B.

CALITE B-18 LC.
M-30; 0.08 max C, 18-20 Cr, 8-10 Ni, 0.75-1.75 Si, 1.00-1.75 Mn, 0.9-1.1 Mo, 0.1 N_2, bal Fe.
Cast: 156 Brin.
For welded stainless equipment; heat resistant to 1650°F.

CALITE B-18 MO.
M-30; 0.08 max C, 18-20 Cr, 12.5-14.00 Ni, 2.9-3.1 Mo, bal Fe.
Cast: 78,600 TS; 34,100 YS; 44 El; 143 Brin.
Annealed: 77,200 TS; 33,300 YS; 37 El; 149 Brin.
For paper and pulp equipment; corrosion resistant.

CALITE B 28.
M-30; 0.35 C, 25-28 Cr, 10-12 Ni, 1 Mo, 0.1 N.
Cast: 85,000 TS; 57,000 YS; 24 El; 10 RA; 187 Brin.
For furnace parts, beams, tube supports; to sustain loads at high temperatures.

CALITE B 28 N.
M-30; 0.30-0.35 C, 24-26 Cr, 10-12 Ni, 0.75-1.75 Si, 0.75-1.75 Mn, 0.1 N_2, bal Fe.
Cast: 84,100 TS; 47,400 YS; 10 El; 217 Brin.
Annealed: 86,000 TS; 46,500 YS; 7 El.
For furnace parts, hangers, supports; heat resistant.

CALITE B 29.
M-30; 0.29-0.34 C, 28-30 Cr, 15-17 Ni, 1.00-1.75 Si, 0.7-1.7 Mn, 0.9- Mo, 0.1 N_2, bal Fe.
Cast: 64,600 TS; 45,000 YS; 8 El.
Annealed: 85,500 TS; 50,650 YS; 6 El.
For furnace parts; heat resistant to 2100°F.

CALITE E.
M-30; 7.8 Ni, 17.28 Cr, 73.62 Fe, 0.45 Mn, 0.14 C, 0.18 Al, 0.14 Cu.
Rolled, annealed: 85,000 TS; 50 El; 60 RA; 165 Brin.
For stainless parts, bolts, nuts, forged and rolled parts; resists oxidation up to 1800°F.

CALITE E 28.
M-30; 0.2 max C, 22-26 Cr, 12-14 Ni, bal Fe.
Rolled: 95,000 TS; 40 El; 50 RA; 156 Brin.
For malleable bolts, nuts, rivets, welding wire, furnace parts; heat resistant to 1950°F.

CALITE N.
M-30; 0.55 C, 17-20 Cr, 65-68 Ni, 1-1.7 Si, 1-1.7 Mn, bal Fe.
At 100°F: 50,000 TS; 2 El; 1 RA; 190 Brin.
At 2100°F: 8,000 TS; 12 El; 13 RA.
For furnace parts, heat treatment equipment, carburizing boxes; resists heat up to 2100°F; poor S resistance.

CALLOY.
M-1808; 1.1 C, 5.5 Mn, 1.5 Cr, 1.5 Mo, 1.1 Si, 1.5 Co, bal Fe.
Ann: 120,000 psi TS (825 MPa).
Hardened: 58-64 Rc.
Tough and abrasion resistant.

CALLOY CADMIUM BARIUM.
M-511; Cd, Ba.
For alloys for bearings.

CALLOY CADMIUM CALCIUM.
M-511; Cd, Ca.
For alloys for bearings.

CALLOY CADMIUM STRONTIUM.
M-511; Cd, Sr.
For alloys for bearings.

CALLOY (CALAL).
M-511; 8-26 Ca, bal Al.
For the deoxidation of steel.

CALLOY LEAD BARIUM.
M-511; Ba, bal Pb.
For alloys for bearings.

CALLOY LEAD CALCIUM.
M-511; Ca, bal Pb.
For alloys for bearings.

CALLOY LEAD STRONTIUM.
M-511; Sr, bal Pb.
For alloys for bearings.

CALLOY STRONTIUM ALUMINUM.
M-511; up to 50 Sr, bal Al.
For getters in electrical discharge devices and vacuum tubes; good stability in air.

CALMAR 18-8 CB.
M-277; 17-20 Cr, 7-10 Ni, 0.9-1.25 Cb, C, bal Fe.
Annealed: 80,000-85,000 TS; 40,000-45,00 YS; 40-45 El.
For castings; abrasion and corrosion resistant.

CALMAR 19-9.
M-277; 0.15 max C, 18-21 Cr, 8-10 Ni, bal Fe.
Heat treated: 85,000 TS; 40,000 YS; 64 El.
For stainless castings; stainless.

CALMAR 19-9M.
M-277; 0.15 max C, 18-21 Cr, 8-10 Ni, bal Fe.
Heat treated: 90,000 TS; 45,000 YS; 50 El.
For stainless castings; stainless.

CALMAR 19-9S.
M-277; 0.07 max C, 18-21 Cr, 8-10 Ni, bal Fe.
Heat treated: 75,000 TS; 37,000 YS; 55 El.
For stainless castings; stainless.

CALMAX.
M-111; 0.28 C, 12 Cr, 7 W, 9 Co, 0.4 V, bal Fe.
Hardened: C 54-55 Rock.
For cores in die casting dies, extrusion dies, forging dies, mandrels.
Hot work tool steel, high hot hardness, air or oil hardening.

CALMET.
M-726; Ni-Cr.
For thermocouples, resistance wire; heat resistant to 1920°F.

CALO FERRO (30).
M-275; 0.3 C, 9 W, 3.5 Cr, 0.2 V, bal Fe.
For tools, dies, hot piercing punches; hot working dies, mild high speed steel.

CALO FERRO (45).
M-275; 0.45 C, 13 W, 3 Cr, 0.25 V, bal Fe.
For tool, dies, hot piercing punches; hot work tool steel.

CALO FERRO (50).
M-275; 0.50 C, 3 Cr, 0.50 V, 15 W, bal Fe.
For hot extrusion dies and nadrels, upsetters; Type H24; oil hardened.

CALOMIC.
M-108; 60 Ni, 16 Cr, bal Fe.
Electrical resistance alloy for up to 900°C. for heating element.

CALOR 5.
M-1318; 0.30 C, W, Cr, V, bal Fe.
For upsetters, riveters, punches; hot work steel, oil hardened.

CALOR 10.
M-1318; 0.30 C, 2.6 Cr, 0.35 V, 8.5 W, bal Fe.
For extrusion rams and liners, punches; dies; hot work steel, oil hardened.

CALOR 10 CO.
M-1318; 0.30 C, 2 Co, 2.4 Cr, 0.25 V, 8.5 W, bal Fe.
For extrusion rams and liners, dies, punches; hot work steel, oil hardened.

CALOR 304.
M-1318; 0.45 C, 1,4 Cr, 0.7 Mo, 0.3 V, bal Fe.
For gears, punches, crimpers, crankshafts; oil hardened, tough.

CALORIZED 4-6 CR MO.
M-30; 0.15 C, 0.5 Mo, 4-6 Cr, bal Fe.
For parts to resist erosion and hydrogen sulfide; 35% Al-1/32" case.

CALOR LR.
M-1318; 0.30 C, 1.1 Cr, 0.18 V, 3.75 W, bal Fe.
For cold water tools, upsetters, punches; oil hardened, tough.

CALOXO 25 12.
M-277; 24-27 Cr, 11-14 Ni, 0.25 max C bal Fe.
Annealed: 75,000-85,000 TS; 40,000-45,000 YS; 30-35 El.
For stainless castings; stainless.

CALOXO 26.
M-277; 25-30 Cr, 0.25 min C, bal Fe.
For corrosion resisting parts; corrosion resistant.

CALOXO 26-4.
M-277; 25-30 Cr, 2.5-4.0 Ni, 0.25 min C, bal Fe.
For heat and corrosion resisting parts; corrosion and heat resistant.

CALOXO 28.
M-277; 0.25 max C, 26-30 Cr, bal Fe.
For corrosion and heat resisting castings; corrosion and heat resistant.

CALOXO 28-10.
M-277; 27-30 Cr, 8-12 Ni, 0.25 max C, bal Fe.
For heat and corrosion resisting parts; heat and corrosion resistant.

CALOXO 28-10 M.
M-277; 27-30 Cr, 8-12 Ni, 3-4.5 Mo, 0.25 min C, bal Fe.
For heat and corrosion resisting parts; heat and corrosion resistant.

CALOXO 35-15.
M-277; 0.35 max C, 15-18 Cr, 35-38 Ni, bal Fe.
For heat and corrosion resisting castings; heat and corrosion resistant.

CALSIFER 75.
M-1038; 74-79 Si, 0.75-1.25 Al, 0.50-1.0 Ca, bal Fe.
Ferrosilicon for inoculation of gray and ductile iron; also a source of silicon in iron and steel.

CALSUN BRONZE 951.
M-8; 95.5 Cu, 2 Sn, 2.5 Al.
Soft: 52,000 TS; 40 RA.
Hard: 135,000 TS; 1 RA.
For electrical conductors; corrosion resistant.

CALUMET.
M-1431; 1.0 C, 2 Mn, 0.9 Cr, 0.9 Mo, bal Fe.
For punches, blanking and forming dies; air hardened.

CALUMETAL B.
M-1166; 0.26-0.33 C, 1.2 Mn, 0.25 Mo, bal Fe.
Heat treated: 130,000 TS; 93,000 YS; 16 El; 16 RA; 260 Brin.
Cast: 83,000 TS; 59,000 YS; 23 El; 37 RA; 194 Brin.
For guides for rolling mills; wear resistant, tough.

CALUMETAL C-5.
M-1166; 0.20 max C, 4,0-6,5 Cr, 0.45-0.65 Mo, bal Fe.
Annealed: 90,000 TS; 60,000 YS; 18 El; 35 RA; 230 Brin.
For castings for high temperature service to 1400°F; corrosion and heat resistant.

CALUMETAL L.
M-1166; 0.38-0.43 C, 0.7 Mn, 0.4 Si, 1.5-2.0 Ni, 0.7 Cr, 0.3 Mo, bal Fe.
Normalized: 100,000-130,000 TS; 70,000-95,000 YS; 10-20 El; 25-40 R 210-250 Brin.
Heat treated: 120,000-250,000 TS; 100,000-225,000 YS; 3-18 El; 7-40 RA; 250-500 Brin.
For machine tool parts, gears, shafts, castings; shock resistant, oil hardened.

CALUMETAL MC.
M-1166; 0.20 max C, 0.7 Mn, 0.4 Si, 0.8-1.1 Cr, 0.15-0.25 Mo, bal Fe
Heat treated: 100,000-200,000 TS; 90,000-180,000 YS; 6-18 El; 35-60 200-390 Brin.
For gears, pinions, shafts, castings; tough. carburizing steel.

CALUMETAL WCI.
M-1166; 0,25 max C, 0.7 Mn, 0.6 max Si, 0.45-0.65 Mo, bal Fe.
Annealed: 65,000 TS; 35,000 YS; 24 El; 35 RA; 200 Brin.
For castings, gears, bolts, housings; for high temperature service.

CALUMETAL WC6.
M-1166; 0.20 max C, 0.7 Mn, 0.4 Si, 1.0-1.5 Cr, 0.4-0.6 Mo, bal Fe.
Annealed: 70,000 TS; 40,000 YS; 20 El; 35 RA; 200 Brin.
For pressure tight castings for high temperature uses; tough, oil hardened.

CALUMETAL WC9.
M-1166; 0.18 max C, 0.6 Mn, 0.4 Si, 2.0-2.75 Cr, 0.9-1.1 Mo, bal Fe.
Annealed: 90,000 TS; 40,000 YS; 20 El; 35 RA; 200 Brin.
For pressure tight castings; to operate up to 1000°F.

CALUMETAL X-1.
M-1166; 0.08 max C, 1.5 max Mn, 1.5 max Si, 8-11 Ni, 18-21 Cr, bal Fe.
Annealed: 85,000 TS; 35,000 YS; 55 El; 60 RA; 120 Brin.
For food and chemical plant equipment, castings; stainless, austenitic.

CALUMETAL X-2.
M-1166; 0.08 max C 1.5 max Mn, 1.5 max Si, 8-11 Ni, 18-21 Cr, 0,07-0. S, bal Fe.
Annealed: 85,000 TS; 35,000 YS; 40 El; 50 RA; 120 Brin.
For screw machine products; stainless, austenitic, free-cutting.

CALUMETAL X-4.
M-1166; 0.08 max C, 8-11 Ni, 18-21 Cr, 2-3 Mo, bal Fe.
Annealed: 85,000 TS; 40,000 YS; 50 El; 55 RA; 160 Brin.
For castings to handle hot chlorides and acids, austenitic, stainless; Type 316.

CALUMETAL X-5.
M-1166; 0.30 max C, 1.5 max Mn, 11-13 Cr, 23-27 Ni, bal Fe.
Cast: 90,000 TS; 40,000 YS; 20 El; 30 RA; 170 Brin.
For furnace parts, heat treating boxes; austenitic, stainless.

CALUMETAL X-8.
M-1166; 0,15 max C, 11.5-13.5 Cr, bal Fe.
Heat treated: 97,000-150,000 TS; 70,000-100,000 YS; 10-20 El; 40-55 RA; 190-350 Brin.
For chemical plant equipment, stainless castings; corrosion resistant; ACI-CK20.

CALUMETAL X-15.
M-1166; 0.35 max C, 33-37 Ni, 14-17 Cr, bal Fe.
Cast: 65,000 TS; 40,000 YS; 6 El; 10 RA; 170 Brin.
At 1600°F: 18,500 TS; 14,000 YS; 22 El; 53 RA; 55 Brin.
For conveyors, furnace parts, heat treating boxes; resists oxidation and carburization.

CALUMET E.
M-1166; 0.36-0.43 C, 1.2 Mn, 0.25 Mo, bal Fe.
Cast: 83,000 TS; 59,000 YS; 23 El; 37 RA; 194 Brin
Heat treated: 125,000 TS; 114,000 YS; 17 El; 40 RA; 286 Brin.
For guides for rolling mills; wear resistant, tough.

CALUMET G.
M-116; 0.32-0.39 C, 0.7-1.3 Ni, 0.6 Mo, bal Fe.
Cast: 100,000 TS; 65,000 YS; 20 El; 35 RA; 200 Brin.
Heat treated: 200,000 TS; 173,000 YS; 6 El; 10 RA; 600 Brin.
For gears, connecting rods, cams, dies; tough, wear resistant.

CALUMET H.
M-1166; 0.25-0.35 C, 1.6 Mn, 0.45 Si, bal Fe.
Cast: 77,000 TS; 48,000 YS; 30 El; 56 RA; 168 Brin.
Heat treated: 127,000 TS; 115,000 YS; 17 El; 47 RA; 286 Brin.
For sprockets, gears, housings; abrasion resistant.

CALUMET M.
M-1166; 0.32-0.39 C, 0.7 Mn, 1 Cr, 0.2 Mo, bal Fe.
Cast: 96,000 TS; 63,000 YS; 22 El; 42 RA; 178 Brin.
Heat treated: 126,000 TS; 106,000 YS; 12 El; 29 RA; 330 Brin.
For gears, shafts, sprockets, housings; abrasion resistant.

CALYPSO 25 A.
M-1793; 0.20 Fe, 0.15 Si, 4.5-5.5 Cu, 1.7-2.3 Zn, 0.20-0.35 Mg, 0.05 Mn, 0.05 Ni, 0.15-0.25 Ti, bal Al.
Cast, Sol. Tr + Age: 450 MPa TS; 390 MPa YS; 7 El.
High strength, for gun breeches.

CALYPSO 41 R.
M-1793; 0.15.-0.30 Fe, 10.2-11.8 Si, 0.05 Cu, 0.10 Zn, 0.05 Mg, 0.10 Mn, 0.05 Ni, 0.05 Ti, bal Al.
Per. mold, as cast: 170 MPa TS; 75 MPa YS; 17 El.
For high voltage electric line equipment.

CALYPSO 43 P.
M-1793; 0.45-0.61 Fe, 12.5-14.0 Si, 0.03 Cu, 0.05 Zn, 0.03 Mg, 0.10 Mn, 0.03 Ni, 0.035 Ti, 0.10 Co, 0.03 Cr, bal Al.
Per. mold, as cast: 150 MPa TS; 80 MPa YS; 4 El.
For thin section permanent mold castings, as tool boxes, lawn mower parts, letter boxes.

CALYPSO 61 S.
M-1793; 0.15 Fe, 10.0-12.0 Si, 0.10 Cu, 0.10 Zn, 0.10-0.25 Mg. 0.30 Mn, 0.10 Ni, 0.05 Pb, 0.05 Sn, 0.20 Ti, bal Al.
Per. mold, as cast: 185 MPa TS; 80 MPa YS; 15 El.
Low pressure wheel castings.

CALYPSO 67 B.
M-1793; 0.15 Fe, 6.5-7.5 Si, 0.05 Cu, 0.10 Zn, 0.25-0.40 Mg, 0.10 Mn, 0.05 Ni, 0.08-0.16 Ti, bal Al.
Per. mold cast, Sol Tr. Age: 290 MPa TS; 200 MPa YS; 18 El.
Suspension arms for automotive.

CALYPSO 67 B1.
M-1793; 0.15 Fe, 6.5-7.5 Si, 0.05 Cu, 0.10 Zn, 0.45-0.60 Mg, 0.10 Mn 0.05 Ni, 0.08-0.16 Ti, bal Al.
Per. mold, Sol Tr. Aged: 340 MPa TS; 285 MPa YS; 10 El.
For high quality aeronautical components.

CALYPSO 67 N.
M-1793; 0.15 Fe, 6.5-7.5 Si, 0.05 Cu, 0.10 Zn, 0.25-0.40 Mg, 0.10 Mn, 0.05 Ni, 0.08-0.16 Ti, bal Al.
Per. mold cast, Sol. Tr. Aged: 290 MPa TS; 200 MPa YS; 18 El.
For disc brake fittings, pivots and suspension arms.

CALYPSO 67 N1.
M-1793; 0.15 Fe, 6.5-7.5 Si, 0.05 Cu, 0.10 Zn, 0.45-0.60 Mg, 0.10 Mn, 0.05 Ni, 0.08-0.16 Ti, bal Al.
Per. mold cast, Sol. Tr. Aged: 340 MPa TS; 285 MPa YS; 10 El.
High strength castings.

CALYPSO 67 R.
M-1793; 0.15 Fe, 6.5-7.5 Si, 0.05 Cu, 0.10 Zn, 0.25-0.40 Mg, 0.05 Mn, 0.05 Ni, 0.08-0.16 Ti, 0.05 Cr, bal Al.
Per. mold cast: 195 MPa TS; 95 MPa YS; 15 El.
For auotmotive wheels.

CALYPSO 67 R1.
M-1793; 0.15 Fe, 6.5-7.5 Si, 0.05 Cu, 0.10 Zn, 0.45-0.65 Mg, 0.05 Mn, 0.05 Ni, 0.08-0.16 Ti, 0.05 Cr, bal Al.
Per. mold, Sol. Tr. Age: 350 MPa TS; 295 MPa YS; 10 El.
For mechanical parts and electronic assemblies.

CALYPSO 73 A.
M-1793; 0.30 Fe, 3.5-4.5 Si, 0.05 Cu, 12.0-14.0 Zn, 0.15-0.25 Mg, 0.05 Mn, 0.10 Ti, bal Al.
Per. mold, as cast: 310 MPa TS; 230 MPa YS; 5 El.
High strength as cast.
One use: spirit levels.

CALYPSO 82 N.
M-1793; 0.60 Fe, 10.5-12.0 Si, 0.80-1.20 Cu, 0.05 Zn, 1.0-1.4 Mg, 0.10 Mn, 0.80-1.20 Ni, 0.10 Ti, bal Al.
Per. mold cast; St. rel.: 240 MPa TS; 220 MPa YS.
For petrol engine pistons.

CALYPSO 82 P.
M-1793; 0.30 Fe, 11.8-13.2 Si, 0.80-1.20 Cu, 0.05 Zn, 1.1-1.6 Mg, 0.10 Mn, 0.80-1.20 Ni, 0.10 Ti, bal Al.
Per. mold cast, stress, rel.: 240 MPa TS; 220 MPa YS.
Engine pistons and cylinder heads for air-cooled automotive engines.

CALYPSO 85 R.
M-1793; 0.15 Fe, 4.5-5.5 Si, 2.5-3.5 Cu, 0.20 Zn, 0.25-0.40 Mg, 0.10 Mn, 0.05 Ni, 0.05 Pb, 0.05 Sn, 0.05-0.10 Ti, 0.05 Cr, bal Al.
Per. mold cast, Sol. Tr. Age: 400 MPa TS; 300 MPa YS; 2 El.
Good strength, machinability, decorative anodizing; for textile beams, automotive.

CALYPSO 92 A.
M-1793; 1.2-1.6 Fe, 0.15 Si, 0.10 Zn, 0.05 Mg, 0.05 Mn, 0.05 Ti, 1.4-1.8 Co, bal Al.
Die cast: 115 MPa TS; 45 MPa YS; 17 El.
Good strength at elevated temperatures.

CAMBRILOY 1.
M-727; 0.11 C, 1.3 Cr, 0.8 Si, 0.55 Mo, bal Fe.
For woven wire conveyor belts, wire cloth and slings.

CAMBRILOY 3.
M-727; 0.15 max C, 3.0 Cr, 1.2 Si, 0.55 Mo, bal Fe.
For woven wire conveyor belts, wire cloth and slings.

CAMBRILOY 35-19.
M-727; 0.15 max C, 20 Cr, 35 Ni, 2 Si, bal Fe.
For woven wire conveyor belts, wire cloth and slings.

CAMBRILOY 35-19 CB.
M-727; 0.15 max C, 20 Cr, 35 Ni, 2 Si, 1.2 Cb, bal Fe.
For woven wire conveyor belts, wire cloth and slings; high temperature operation.

CAMBRILOY 80-20.
M-727; 0.15 max C, 1.2 Si, 1.0 Fe, 20 Cr, bal Ni.
For woven wire conveyor belts, wire cloth and slings; high temperature operation.

CAMBRILOY 80-20 CB.
M-727; 0.15 max C, 1.2 Si, 1.2 Cb, 20 Cr, bal Ni.
For woven wire conveyor belts, wire cloth and slings.

CAMBRILOY AL.
M-727; aluminum coated carbon steel for woven wire conveyor belts, wire cloth and slings.

CAMELIA METAL.
M-Eng.; 70 Cu, 15 Pb, 10 Zn, 4.2 Sn, 0.5 Fe.
For hardware; free-cutting.

CAMVAC 200 see **NICKEL 200.**

CAMVAC 400 see **MONEL 400.**

CAMVAC 600 see **INCONEL 600.**

CAMVAC 800 see **INCOLOY 800.**

CAMVAC B see **HASTELLOY B.**

CAN METALS.
M-Ger.; 95 Pb, 1.75 Ca, 1.35 Cu, 1 Sr, 1 Ba.
For bearings.

CANNON.
M-40; 0.70 C, 16 W, 3.5 Cr, 1.0 V, bal Fe.
For high speed tools, planing and lathe tools, twist drills, punches; high speed steel.

CANNONITE.
M-347; 2.7 C, 1.5 Si, bal Fe.
50,000 TS; 0 El; 250 Brin.
For diesel engine and refrigeration parts, crankshafts, pistons, cylinders brake drums; high test cast iron; Tr. S. 8300.

CANNON-MUSKEGON NO. 6.
M-1491; 28 Cr, 4 W, 1 C, bal Co.
For hard facing bushings, pressure valves, tong bits, hot trimming dies.
Heat and abrasion resistant.

CANNON-MUSKEGON NO. 12.
M-1491; 29 Cr, 8 W, 1.5 C, 1.0 Si, 3 max Fe, bal Co.
For hardfacing bushings, saw teeth, pressure bars, entry guides.
Wear and abrasion resistant.

CANNON-MUSKEGON NO. 40.
M-1491; 15 Cr, 4.5 Fe, 3.25 B, 4.25 Si, 0.75 C, bal Ni.
For hardfacing gear teeth, plump plungers, valve plugs.
Wear, oxidation and abrasion resistant.

CANNON-MUSKEGON NO. 41.
M-1491; 13 Cr, 4 Fe, 3 B, 4 Si, 0.6 C, bal Ni.
For hardfacing and welding, oil expeller screws, scraper blades.
Heat and wear resistant.

CANNON SPECIAL.
M-40; 0.70-0.75 C, 4 Cr, 18-20 W, 0.50 Mo, 2.0-2.25 V, bal Fe.
For tools, cutters.

CANNON VANADIUM.
M-40; 1.0 C, 4 Cr, 18-20 W, 0.5-0.8 Mo, 3.5 V, bal Fe.
For tools, cutters; high speed steel.

CAP GILDING.
M-U.S.; 90 Cu, 10 Zn.
For ornamental window screens; commercial bronze.

CAPI 205-WI.
M-1748; 0.40 C, 21 Cr, 11 W, 2 Cb, 2 Fe, bal Co.
As cast: 120,000 TS; 90,000 YS; 5 El; 10 RA; 341 Brin.
High temperature strength, oxidation resistance. For jet engine vanes.

CAPI 224.
M-1748; 0.25 C, 28 Cr, 6 Mo, bal Co.
121,000 TS; 79,000 YS; 12 El; 14 RA; 269 Brin.
High yield strength; resists body fluids. For hip joints, prosthetics-dental.

CAPI 309X.
M-1748; 0.10 C, 22 Cr, 9 Mo, 1 W, 1 Co, 18 Fe, bal Ni.
As cast: 85,000 TS; 48,000 YS; 15 El; 20 RA; 179 Brin.
Good strength and oxidation resistance to 2200°F; for furnace hardware burner parts, retorts, vanes.

CAPI 320.
M-1748; 0.05 C, 21 Cr, 9 Mo, 4 Cb, bal Ni.
As cast: 80,000 TS; 45,000 YS; 30 El; 30 RA; 170 Brin.
Easily fabricated; good corrosion resistance. For burner hardware, cryogenic and chemical equipment.

CAPI 462.
M-1709, M-1748; 0.10 max C, 0.30 max Mn, 1.0 Si, 1.0 max Fe, 0.50 max Ti, 0.25 max Al, 48.0-52.0 Cr, bal Ni.
Cast: 80,000 ps min TS; 50,000 Min YS; 5 min El; Rc 30 max.
For parts to resist corrosion at elevated temperatures.

CAPI 864 CB.
M-1748; 0.05 C, 19 Cr, 10.5 Ni, 1 Cb, bal Fe.
Solution ann: 78,000 TS; 34,000 YS; 54 El; 57 RA.
Columbium stabilized stainless; for general corrosion resistant and heat resistant parts.

CAPI 960.
M-1709, M-1748; 0.35-0.45 C, 1.5 max Mn, 1.5 max Si, 23.0-25.0 Cr, 21.0-23.0 Ni, 0.50 max Mo, 1.25-2.0 Cb, bal Fe. (Pb-100 ppm max).
Cast alloy similar to HK40 but stabilized with Cb to resist thermal cycling in petrochemical environments.

CAPITAL-562.
M-1112; 0.80 C, 4.2 Cr, 6 W, 2 V, 5 Mo, bal Fe.
For lathe and planer tools, milling cutters; high speed steel.

CAPITO-VK5M.
M-1654; 0.20 C, 13 Cr, 1.2 Mo, bal Fe.
Annealed: 95,000 TS; 40,000 YS; 25 El; 55 RA; B 92 Rock.
Heat treated: 240,000 TS; 200,000 YS; 8 El; 25 RA; C 50 Rock.
For gears, shafts, surgical instruments, knives, scissors, cutlery.
Corrosion resistant, hardenable.

CAPIVAC IV.
M-1709, M-1748; 0.55-0.65 C, 23.0-26.0 Cr, 9.0-11.0 Ni, 1.5 max Fe, 3.5-4.5 Ta, 6.5-7.5 W, 0.35-0.60 Al, 0.20-0.50 Ti, 0.50 max Cu, bal Co.
Cast: 105,000 psi min TS; 70,000 psi min YS; 3.0 min El.
Stress rupture: 23 hr min at 9000 psi at 2000°F.
For high temperature operation.

CAPIVAC 326.
M-1748; 0.05 C, 19 Cr, 3 Mo, 5 Cb, 18 Fe, 0.5 Al, 1.0 Ti, bal Ni.
Aged: 150,000 TS; 115,000 YS; 20 El; 20 RA; 363 Brin.
For structural parts up to 1400°F; cryogenic gear.

CAPIVAC 371.
M-1748; 0.10 C, 13 Cr, 5 Mo, 2 Cb, 6 Al, 0.8 Ti, bal Ni.
As cast: 120,000 TS; 110,000 YS; 5 El; 8 RA; 331 Brin.
Good high temperature strength; for gas turbine blades, wheels.

CAPSULE METAL.
M-U.S.; 92 Pb, 8 Sn.
For bearing.

CARAS.
M-1405; alloy, bal Fe.
For molds; case hardened, oil hardened.

CABRIDE G1.
M-Germany; 94 WC, 6 Co.
For cutting tools; sintered carbide.

CARBIDE H1.
M-Germany; 94 WC, 6 Co.
For cutting tools; sintered carbide.

CARBIDE H2.
M-Germany; 91.5 WC, 1.5 TiC, 7 Co.
For cutting tools; sintered carbide.

CARBIDE L2.
M-Germany; 77-78 WC, 14-15 TiC, 8 Co.
For cutting tools; sintered carbide.

CARBIDE LO.
M-Germany; 88 WC, 5 TiC, 7 Co.
For cutting tools; sintered carbide.

CARBIDE T5K10.
M-USSR; 85 WC, 6 TiC, 9 Co.
For cutting tools; sintered carbide.

CARBIDE T15K6.
M-USSR; 79 WC, 15 TiC, 6 Co.
For cutting tools; sintered carbide.

CARBIDE WK3.
M-USSR; WC, 3 Co.
For cutting tools; sintered carbide.

CARBIDE WK6.
M-USSR; 94 WC, 6 Co.
For cutting tools; sintered carbide.

CARBIDE WK8.
M-USSR; 92 WC, 8 Co.
For cutting tools; sintered carbide.

CARBIDIE CD-18.
M-1695; Sintered carbide. 12% Co. 89-90 RA. Special wear resistant die grade for extra abrasive stamping applications; gall resistant. Stamping and lamination tooling, drawing and forming; can tooling.

CARBIDIE CD-20.
M-1695; Sintered carbide. 3% Co. 91.8-92.8 RA. Extreme wear applications where no shock is involved; highest compression strength. Glass and plastic cut-off tooling; gages.

CARBIDIE CD-24X.
M-1695; Sintered carbide. 5% Co. 91.5-92.5 RA. Wear applications in no-shock or light shock tooling, such as ceramic compacting; will hold a highly polished, wear resistant finish. Valve balls and seats, bearings and seals; corrosion resistant uses, pyrochemical liners and plungers, gages.

CARBIDIE CD-30.
M-1695; Sintered carbide. 6% Co. 90.5-91.5 RA. Wear and die applications. Light shock metal powder tooling, drawing, blanking, boring bars and tooling.

CARBIDIE CD-35.
M-1695; Sintered carbide. 9% Co. 89.8-90.8 RA. Dies and wear applications. Metal powder tooling wear resistance is important; blanking; wire, bar and tube drawing.

CARBIDIE CD-35F.
M-1695; Sintered carbide. 9% Co. 90.2-91.2 RA. Wear and gall resistant; will hold high polish. Aluminum, stainless and brass draw applications, wire straightening, bearings and seals, valve ball and seats.

CARBIDIE CD-36 1/4.
M-1695; Sintered carbide. 10% Co. 89.5-90.5 RA. Wear and die grade; moderate shock resistance. Normal metal powder tooling; blanking and drawing; slitter knives.

CARBIDIE CD-38 3/4.
M-1695; Sintered carbide. 12% Co. 88.8-89.8 RA. Forming and blanking grade; wear resistant; will resist light shock. Structural parts, wire bar and tube drawing.

CARBIDIE CD-40.
M-1695; Sintered carbide. 13% Co. 88.5-89.5 RA. Die and blanking grade, non-ferrous material. Wear resistant blanking, forming, drawing; general purpose metal powder tooling; slitter knives for paper and steel.

CARBIDIE CD-45.
M-1695; Sintered carbide. 14% Co. 88.0-90.0 RA. Wear resistant ferrous blanking grade. Very light impact applications, forming dies, drawing, general can tooling.

CARBIDIE CD-50.
M-1695; Sintered carbide. 15% Co. 87.5-88.5 RA. General all purpose die, stamping, drawing. Will take moderate shock with good die resistance.

CARBIDIE CD-53.
M-1695; Sintered carbide. 16% Co. 87.0-88.0 RA. Shock resistant, medium impact-shock applications, spring tooling.

CARBIDIE CD-60.
M-1695; Sintered carbide. 20% Co. 83.0-84.5 RA. Moderate shock, cold heading grade. Nail grippers, and tooling, spring tooling.

CARBIDIE CD-70.
M-1695; Sintered carbide. 25% Co. 81.5-83.0 RA. Severe shock, cold heading grade. Heavy bolt making and sizing. Nail tooling.

CARBIDIE CD-337.
M-1695; Sintered carbide. 11% Co. 88.0-89.0 RA. High strength, wear resistance. For cold forming, structural-wear-impact applications; swaging hammers, mandrels, rotary percussion applications.

CARBIDIE CD-637.
M-1695.
Sintered carbide, 10% Co; hardness 90.5-91.5 RA.
Sub-micron material for use in the most abrasive stamping and wear applications. High strength and hardness, and medium shock resistance.

CARBIDIE CD-650.
M-1695; Sintered carbide. 15% Co. 89.0-90.5 RA. Sub-micron material for use in highly abrasive and severe stamping and wear applications. Extreme strength, hardness, and shock resistance.

CARBIUM.
M-729; 4-5 Cu, bal Al.
For light alloy parts; heat treatable.

CARBO.
M-English; WC, Co.
For cutting tools; cemented.

CARBOBRONZE.
M-Eng; 92 Cu, 8 Sn, 0.3 P.
For tubing; resists acid and alkalis.

CARBOLOY O-30.
M-31; Sintered cemented oxide.
90,000 TrS; A 94.0 Rock.
For high speed machining of cast iron and steel; high heat and wear resistance.

CARBOLOY 44A.
M-31; Sintered cemented carbide; 94 WC, 6 Co.
320,000 TrS; A 91.0 Rock.
For semi-roughing of cast iron, non-ferrous and non-metallics; medium high shock resistance.

CARBOLOY 55A.
M-31; Sintered cemented carbide; 87 WC, 13 Co.
390,000 TrS; A 88.2 Rock.
For heavy duty roughing of cast iron, non-ferrous and non-metallics; variety of die applications; high shock resistance.

CARBOLOY 55B.
M-31; Sintered cemented carbide; 84 WC, 16 Co.
420,000 TrS; A 86.8 Rock.
For medium and large dies; cold work, piercing, blanking and drawing; extremely high shock resistance.

CARBOLOY 77B.
M-31; Sintered cemented carbide; 57 WC, 27 TaC, 16 Co.
390,000 TrS: A 85.0 Rock.
For cutting hot flash, extrusion of aluminum wire bar and tube.

CARBOLOY 78.
M-31; Sintered cemented carbide; 76 WC, 4 TaC, 12 TiC, 8 Co.
250,000 TrS; A 92.0 Rock.
For light-roughing and finishing of steels; high abrasion resistance.

CARBOLOY 78B.
M-31; Sintered cemented carbide; 79 WC, 4 TaC, 8 TiC, 9 Co.
260,00 TrS; A 91.2 Rock.
For medium roughing of steel; die applications.

CARBOLOY 90.
M-31; Sintered cemented carbide; 90 WC, 10 Co.
350,000 TrS; A 89.0 Rock.
For rock drilling; excellent wear resistance.

CARBOLOY 115.
M-31; Sintered cemented carbide; 88.5 WC, 11.5 Co.
400,00 TrS; A 88.5 Rock.
For percussive mining bits and large size impact punches: good shock and wear resistance.

CARBOLOY 120.
M-31; Sintered cemented carbide; 88 WC, 12 Co.
410,00 TrS; A 86.5 Rock.
For rotary and percussive mining tools and small impact extrusion punches; impact and wear resistant.

CARBOLOY 190.
M-31; Sintered cemented carbide; 75 WC, 25 Co.
450,000 TrS; A 84.0 Rock.
For heavy impact die and punch applications; high toughness.

CARBOLOY 210.
M-31; Sintered cemented carbide; 28 WC, 2 TaC, 64 TiC, 2 Cr_3C_2, 4 Co.
100,000 TrS; A 94.5 Rock.
For finish machining of steel and cast iron; high wear resistance at elevated temperatures.

CARBOLOY 231.
M-31; Sintered cemented carbide; 90 WC, 10 Co.
400,00 TrS; A 87.8 Rock.
For mining applications; shock resistant.

CARBOLOY 241.
M-31; Sintered cemented carbide; 90 WC, 10 Co.
400,000 TrS; A 88.4 Rock.
For mining applications; shock and wear resistant.

CARBOLOY 248.
M-31; Sintered cemented carbide; 89 WC, 11 Co.
450,000 TrS; A 89.5 Rock.
For rock drilling and interrupted metal cutting; excellent wear resistance relative to shock resistance.

CARBOLOY 258.
M-31; Sintered cemented carbide; 87 WC, 13 Co.
470,000 TrS; A 88.5 Rock.
For mining and for cold extrusion; shock resistant.

CARBOLOY 268.
M-31; Sintered cemented carbide; 84 WC, 16 Co.
490,000 TrS; A 87.0 Rock.
For mining and for impact extrusion; high shock resistance.

CARBOLOY 320.
M-31; Sintered cemented carbide; 64 WC, 4.5 TaC, 25.5 TiC, 6 Co.
200,000 TrS; A 93.0 Rock.
For high speed finishing of steel and high tensile cast irons; excellent wear resistance.

CARBOLOY 350.
M-31; sintered cemented carbide; 71 WC, 12 TaC, 12.5 TiC, 4.5 Co.
200,000 TrS; A 92.4 Rock.
For light-roughing and semi-finishing of steel, ferrous castings, stainless steel and some high temperature alloys; excellent wear resistance and toughness.

CARBOLOY 370.
M-31; Sintered cemented carbide; 72 WC, 11,5 TaC, 8 TiC, 8.5 Co.
250,000 TrS; A 91.2 Rock.
For heavy duty roughing of steels, ferrous castings, stainless steel and some high temperature alloys; excellent shock and wear resistance.

CARBOLOY 390.
M-31.
Sintered complex Tungsten carbide, uncoated.
For heavy duty and extremely heavy duty roughing of carbon, alloy and chromium stainless steels, at relatively low speeds.

CARBOLOY 514.
M-31; Sintered TiC coated cemented carbide.
For finishing and precision finishing of steel and cast iron.

CARBOLOY 516.
M-31; Sintered TiC coated cemented carbide.
For finishing and precision finishing of steel and cast iron.

CARBOLOY 518.
M-31.
Sintered complex Tungsten carbide, TiC coated.
For light to heavy roughing of alloy steels, tool steels and stainless steels.

CARBOLOY 519.
M-31.
Sintered complex Tungsten carbide, TiC coated.
For machining.

CARBOLOY 523.
M-31; Sintered TiC coated cemented carbide.
For general purpose and finish machining of cast iron and non-ferrous materials.

CARBOLOY 545.
M-31.
Sintered complex Tungsten carbide, Al_2O_3 coated.
For high speed finishing cast iron and steel.

CARBOLOY 570.
M-31.
Sintered complex Tungsten carbide, Al_2O_3 coated.
For machining.

CARBOLOY 616.
M-31; Sintered cemented carbide; 82 WC, 16 Ni, 2 Cr.
370,000 TrS; A 87.8 Rock.
For components rather than cutting tool.

CARBOLOY 779.
M-31; Sintered cemented carbide; 91 WC, 9 Co.
340,000 TrS; A 89.5 Rock.
For medium-size wire drawing dies, bar and tube dies, shape dies, cupping and compression dies; mining tool and rock-drilling applications.

CARBOLOY 820.
M-31; Sintered cemented carbide; 90 WC, 10 Co.
450,000 TrS; A 51.0 Rock.
For finish and semi-finish machining.

CARBOLOY 860.
M-31; Sintered cemented carbide; 91 WC, 4 TaC, 5 Co.
270,000 TrS; A 92.0 Rock.
For machining high tensile cast iron; good wear resistance.

CARBOLOY 883.
M-31; Sintered cemented carbide; 94 WC, 6 Co.
290,000 TrS; A 92.0 Rock.
For general purpose machining of non-steel work materials; also used for small compacting dies, burnishing rings and nozzles; high wear resistance.

CARBOLOY 895.
M-31; Sintered cemented carbide; 94 WC, 6 Co.
250,000 TrS; A 92.9 Rock.
For light machining of cast iron and non-ferrous materials.

CARBOLOY 905.
M-31; Sintered cemented carbide; 92 WC, 4 TaC, 4 Co.
240,000 TrS; A 92.2 Rock.
For light finishing of cast iron, non-ferrous and non-metallic materials.

CARBOLOY 907.
M-31; Sintered cemented carbide; 74 WC, 20 TaC, 6 Co.
270,000 TrS; A 91.3 Rock.
For finish machining of fine-grained cast iron, malleable iron, aluminum and magnesium alloys; crater and abrasive resistant.

CARBOLOY 999.
M-31; Sintered cemented carbide; 97 WC, 3 Co.
230,000 TrS; A 92.9 Rock.
For machining cast iron, non-ferrous metals, and non-metallics; also for fine wire dies and small nozzles; excellent abrasion resistance.

CARBOLOY HM-1 HEVIMET.
M-31; 90 W, 7.5 Ni, 2.5 Cu.
Sintered alloy.
300,000 TrS; C 25 Rock.
For weights, counter balances, radiation shielding, vibration damping; high density tungsten alloy.

CARBOLOY HM-3 HEVIMET.
M-31; 90 W, 7.5 Ni, 2.5 Cu.
Sintered alloy.
For governor counter weights, radiation shielding, vibration damping; high density tungsten alloy.

CARBOMANG.
M-157; 0.9-1.0 C, 1.0-1.25 Mn, 0.45-0.60 Cr, 0.4-0.6 W, bal Fe.
For tools and dies, machine parts; castings; oil hardenable; cast to shape

CARBON + GRAPHITE.
M-1755; C.
Purities: 99.9999%, 99.999%, 99.99%.
Forms: Rods, powder, electrodes, single crystals.

CARBON BRONZE.
M-England; 15 Pb, 10 Sn, bal Cu.
Cast: 28,000 TS; 6 El.
For bearings; plastic bronze.

CARBON COLD HEADER.
M-24; 0.95 C, 0.40 Mn, 0.40 Si, bal Fe.
For header dies, gripper dies, swaging and forming dies.

CARBON COLD HEADER.
M-73; 0.9 C, bal Fe.
For cold heading dies; water hardening.

CARBONDALE SILVER.
M-U.S.; 66 Cu, 18 Ni, 16 Zn.
Hard: 96,000 TS; 2 El; 158 Brin.
Soft: 58,000 TS; 33 El; 77 Brin.
For spinning and drawing, flatware, spoons, forks, knives, etc., to be plated; German silver.

CARBON DRILL ROD.
M-114; 1.2 C, bal Fe.
For tools; water hardened.

CARBONIZED NICKEL.
M-44; 99.5 Ni.
Strip for electron tube plates.

CARBON STEEL.
M-373; 0.7-1.4 C, bal Fe.
For tools, drills, taps; water hardened.

CARBON-VANADIUM DRILL ROD.
M-342, M-115; 1.0 C, 0.2 V, bal Fe.
For drill rod, punches; water hardening.

CARBON VANADIUM STEEL.
M-1409; 0.8-1.3 C, V, bal Fe.
For shear blade, punches, chisels, heading and swaging dies; water hardened; Type W2.

CARBORTAM.
M-109; 16-17 Ti, 2.5-3.0 Si, 6.5-7.5 C, 1.25-1.0 B, bal Fe.
For steel making conditions; adds deep hardening properties.

CARBURITE.
M-U.S.; 47-48 C, 28 Fe, 0.3 S, 0.2 P.
For recarburizer in steel; for steel making.

CARB-X.
M-U.S.; C, bal Fe.
Used in manufacturing cast iron molds; ferro-alloy, exothermic.

CARDINAL RAPID.
M-459; 0.7 C, 18 W, 4 Cr, 1 V, bal Fe.
For high speed cutters; high speed steel.

CARECO.
M-U.S.; Pb-Sn-Sb.
For bearings; white bearing alloy.

CARILLOY FC.
M-604; 0.50 C, 1.10 Mn, 0.25 Si, 0.75 Cr, 0.20 Mo, 0.08 S, bal Fe.
HT bar and plate: 130 ksi TS; 110 ksi YS; 20 El.
Free machining, high strength steel for gears, shafts, structural plates.

CARIRON.
M-718; 2.96 T.C., 0.75 Mn, 1.2 Si, 1.55 Ni, bal Fe.
50,600 TS, 221 Brin.
For liners, pistons; Tr.S. 4160; Def. 0.207.

CARLOY.
M-1146; C, Mo, bal Fe.
For tools; water hardened.

CARLSON 600.
M-1793; 0.08 C, 0.5 Mn, 0.008 S, 0.25 Si, 15.5 Cr, 8.0 Fe, 0.25 Cu, 76 Ni.
Nickel base super-alloy.
ACI CY-40; INCONEL 600.

CARLSON 825.
M-1793; 0.03 max C, 0.50 Mn, 0.25 Si, 21.5 Cr, 42.0 Ni, 3.0 Mo, 0.9 Ti, 0.10 Al, 2.25 Cu, 30.0 Fe.
Corrosion and heat resistant alloy.
Reference: INCOLOY 825.

CARLTON.
M-1705; 0.95 C, 2.0 Mn, 2.2 Cr, 1.1 Mo, bal Fe. (contains lead)
Free machining grade of air hardening tool steel.
AISI A4.

CARMELIA BRONZE.
Pb, Zn, bal Cu.
For machinery bearings; heavy duty.

CARMET.
M-1; WC.
For tools; cemented carbides.

CARMET CA-1.
M-1; WC + Co.
For tipped tools; for steel.

CARMET CA-2.
M-1; WC + Co.
For tipped tools; for finishing cuts.

CARMET CA-3.
M-1; WC + Co.
For tipped tools; for roughing cuts.

CARMET CA-3.
M-1696; Sintered carbide.
For rough cutting operations on cast iron, non-ferrous metals and non-metallics. For gages.

CARMET CA-4.
M-1; WC + Co.
For tipped tools; for soft metals.

CARMET CA-4.
M-1696; Sintered carbide.
Cutting tools for finish cutting cast iron, non-ferrous metals and non-metallics. Also for gages.

CARMET CA-5.
M-1; WC + Co.
For cutting tool bits for slow speeds and heavy cuts on steel; sintered; Tr.S. 200,000.

CARMET CA-6.
M-1; WC + Co.
For cutting tool bits, steel precision boring; sintered, for light precision cuts, Tr.S. 150,000.

CARMET CA-7.
M-1; WC, Co.
For cutting tool bits for hard cast iron; sintered, wear resistant;
Tr.S. 190,000.

CARMET CA-7.
M-1696; Sintered carbide.
Cutting tools for finish cutting of non-ferrous metals and non-metallics.

CARMET CA-8.
M-1; WC, Co.
For cutting tool bits for light precision cuts on cast iron; sintered, high resistance to wear;
Tr.S. 190,000.

CARMET CA-8.
M-1696; Sintered carbide.
For cutting tools for finish cutting of non-ferrous metals and non-metallics.

CARMET CA-9.
M-1; WC, Co.
For cutting tool bits for soft plastics, gauge blanks; sintered; Tr.S. 220,000.

CARMET CA-10.
M-1; WC Co.
For cutting tool bits, dies; sintered, for heavy cuts; Tr.S. 320,000.

CARMET CA-10.
M-1696; Sintered carbide
Cold work dies for blanking dies.

CARMET CA-11.
M-1; WC.
86 Ra.
For cold header dies; sintered; Tr.S. 350,000.

CARMET CA-11.
M-1696; Sintered carbide.
Cold header dies; cold work dies for blanking.

CARMET CA-12.
M-1; WC.
89 Ra.
For cold forming dies; sintered; Tr.S. 300,000.

CARMET CA-12.
M-1696; Sintered carbide.
For cold forming dies; rough machining non-ferrous metals.

CARMET CA-20.
M-1; WC.
84 Ra.
For cold header dies; sintered; Tr.S. 375,000.

CARMET CA-21.
M-1; WC.
86 Ra.
For blanking and cold work dies; sintered; Tr.S. 275,000.

CARMET C.A. 51.
M-1; WC.
90 Ra.
For high speed planer tools, heavy cuts on steel; sintered; Tr.S. 225,000.

CARMET CA-51.
M-1696; Sintered carbide.
For cutting tools, heavy roughing cuts on cast iron and on steel.

CARMET CA-310.
M-1696.
Sintered carbide.
For cutting tools.

CARMET CA-315.
M-1696.
Sintered carbide.
For cutting tools.

CARMET CA-425.
M-1696; Sintered carbide.
For cold header dies.

CARMET CA-604.
M-1; WC, TaC, TiC.
For cutting tools for high speed machines; sintered carbides.

CARMET CA-606.
M-1; WC.
For cutting tools, dies; sintered carbides.

CARMET CA-606.
M-1696; Sintered carbide.
For cutting tools for finish cutting of steel.

CARMET CA-608.
M-1; TiC.
For cutting tools; sintered carbides.

CARMET CA-609.
M-1; WC.
For cutting tools, dies; sintered carbides.

CARMET CA-610.
M-1; TiC.
For cutting tools; sintered carbides.

CARMET CA-610.
M-1696; Sintered carbide.
For cutting tools, for roughing and general purpose cutting of steel.

CARMET CA-704.
M-1696; Sintered carbide.
For cutting tools, precision finishing and boring on steel.

CARMET CA-711.
M-1696; Sintered carbide.
For cutting tools, finish cutting of steel.

CARMET CA-720.
M-1696; Sintered carbide.
For cutting tools, for roughing and general purpose cutting of steel.

CARMET CA-740.
M-1696; Sintered carbide.
For cutting tools, heavy roughing interrupted cuts.

CARMET CA-815.
M-1; WC.
For hot extrusion dies, precision gauges; sintered carbides.

CARMET CA-815.
M-1696; Sintered chromium carbide.
Used for gages.

CARMET CA-9443.
M-1696.
Coated sintered CA-443.

CARMET CA-9721.
M-1696.
Coated sintered CA-721.

CARMET CA-9740.
M-1696.
Coated sintered CA-740.

CARMET CA-B.
M-1696.
Al_2O_3 + TiC hot-pressed ceramic.
For rough and finished machining of hard cast irons and hardened steel.

CARMET CA-W.
M-1696.
Al_2O_3 base sintered ceramic.
For machining cast irons, heat-treated steels.

CARNOLIA.
M-405; Pb-Sn-Sb.
For antifriction alloy for bearings; Babbitt.

CARO.
91.6 Cu, 8.5 Sn, 0.19 P.
For tubes; corrosion resistant.

CAROBRONZE.
M-251; 91.2 Cu, 8.5 Sn, 0.3 P.
60,000-95,000 TS; 50-20 El; 90-190 Brin.
For solid cold drawn tubes, bushes for highly stressed bearings; resists corrosion.

CARO BRONZE.
M-730; 91.2 Cu, 8.5 Sn, 0.3 P.
Annealed: 63,000 TS; 60 El.
Hard: 99,000 TS; 12 El.
For bearings and wearing parts; wrought.

CARPENTER 19-9DL.
M-32; 0.28-3.35 C, 18-21 Cr, 9-11 Ni, 1.5 Mo, 1.5 W, 0.5 Cb, 0.3 Ti, bal Fe.
Annealed: 2,000 TS; 42,00 YS; 53 El; 58 RA; 188 Brin.
Hot rolled: 114,000 TS; 68,000 YS; 35 El; 50 RA; 241 Brin.
For turbine blades, supercharger wheels, blades; austenitic, stainless.
AISI 651; Uns K 63198.

CARPENTER 19-9DL.
M-32; 0.28-0.35 C, 18-21 Cr, 9-11 Ni, 1.5 Mo, 1.5 W, 0.5 Cb, 0.3 Ti, bal Fe.
Annealed: 102,000 TS; 42,000 YS; 53 El; 58 RA; 188 Brin.
Hot rolled: 114,000 TS; 68,000 YS; 35 El; 50 RA; 241 Brin.
For turbine blades, supercharger wheels, blades; austenitic, stainless.
AISI 651; UNS K63198.

CARPENTER 19-9DX.
M-32; 0.28-0.35 C, 18-21 Cr, 8-11 Ni, 1.7 Mo, 1.5 W, 0.5 Ti, bal Fe.
Annealed: 102,000 TS; 42,000 YS; 53 El; 188 Brin.
Hot rolled: 114,000 TS; 68,000 YS; 35 El; 241 Brin.
For turbine wheels, buckets, blades.
Austenitic stainless, non-hardenable.
AISI 652; UNS K63199.

CARPENTER 20CB-3 see **CARPENTER STAINLESS 20 CB-3.**

CARPENTER 80-20.
M-32; 0.15-0.30 C, 0.6-1.0 Mn, 1.0 max Fe, 0.3 Si, 19-21 Cr, bal Ni.
Annealed: 121,000 TS; 60,000 YS; 35 El; 50 RA; 205 Brin.
For aircraft valves, aircraft engine exhaust systems.
Resists corrosion of leaded gasoline exhausts.

CARPENTER 430 FR SOLENOID QUALITY.
M-32; 0.06 C, 0.50 Mn, 1.25 Si, 0.02 P, 0.30 S, 17.5 Cr, 0.30 Mo, bal Fe.
Corrosion resistant, magnetically soft metal for solenoid valves.

CARPENTER 430F SOLENOID QUALITY.
M-32; 0.06 C, 0.50 Mn, 0.03 P, 0.30 S, 0.50 Si, 17.5 Cr, 0.30 Mo, bal Fe.
Corrosion resistant soft magnetic steel for solenoids.
Properties vary with annealing treatment.

CARPENTER 434 HS.
M-32; 0.12 C, 1.0 Mn, 1.0 Si, 16.0-18.0 Cr, 0.75-1.25 Mo, bal Fe.
Closely controlled chemical balance of Type 434; for dependable response to heat treatment.
AISI 434.

CARPENTER 636 ALLOY (TYPE 422).
M-32; 0.20-0.25 C, 1.0 max Mn, 1.0 max Si, 12.0-14.0 Cr, 0.50-1.0 Ni, 0.75-1.25 Mo, 0.75-1.25 W, 0.2-0.5 V, bal Fe.
Air or oil hardenable corrosion resistant steel.
Hardened, tempered 1200°F, tested at: Room temp: 149,000 TS; 125, YS; 1100°F: 76,000 TS; 72,000 YS.
For buckets and blades in steam turbines, high temperature bolting, compressor parts.
AISI 422.

CARPENTER 709 TYPE 2.
M-32; 0.45 C, 0.55 Mn, 0.25 Si, 1.0 Cr, 0.5 Mo, 0.3 V, bal Fe.
Heat treated: 142,000-260,000 TS; 130,000-230,000 YS; 20-10 El; 60-41 RA; 340-510 Brin.
For steam turbine valves, pressure vessels; operating temperature to 1200°F.

CARPENTER A-8.
M-32; 0.55 C, 0.30 Mn, 0.90 Si, 5.0 Cr, 1.25 Mo, 1.20 W, bal Fe.
Medium carbon, air hardening tool steel used for slitting cutters, shear blades, forming dies and blanking dies.

CARPENTER ALLOY 182-FM see CARPENTER PROJECT 70 182-FM.

CARPENTER AMS 5616.
M-32; 0.15-0.20 C, 0.50 Mn, 0.50 Si, 12.0-14.0 Cr, 1.8-2.2 Ni, 0.50 max Mo, 2.5-3.5 W, bal Fe.
Air or oil hardenable corrosion resistant steel.
Max. heat treated strength: about 217,000 TS.
Tested 1100°F: 135,000 TS; 120,000 YS; 18 El.
For steam turbine buckets and blades, gas turbine parts, high temperature bolts.
AMS 5616; Greek Ascoloy.

CARPENTER AMS 5700.
M-32; 0.45 C, 14 Cr, 14 Ni, 2.5 W, 0.50 max Mo, bal Fe.
Annealed: 110,000 TS; 48,000 YS; 40 El; 45 RA; 216 Brin.
At 1500°F: 37,000 TS; 25,000 YS; 55 El; 58 RA; 93 Brin.
For aircraft valves, valve seat inserts, steam turbine bolts; austenitic, high heat resistance.
UNS K66009.

CARPENTER ASTROLOY.
M-32; 0.06 C, 15.0 Cr, 15.0 Co, 3.5 Ti, 4.4 Al, 5.25 Mo, 0.03 B, 0.06 max Zr, bal Ni.
Precipitation hardening nickel base super-alloy.
Used for turbine discs in aircraft gas turbine engines.

CARPENTER BARTO.
M-32; 0.50 C, 0.5 Mn, 0.25 Si, 1.0 Cr, 1.75 Ni, bal Fe.
Oil hardenable alloy steel; for expander punches, feeder rolls, clutch parts, vise jaws and chuck jaws.

CARPENTER CHROME-NICKEL STAINLESS WITH BORON.
M-32; 0.08 C, 2.0 Mn, 1.0 Si, 18.0-20.0 Cr, 12.0-15.0 Ni, 2.0 max (3 ranges), bal Fe.
Modified 304 stainless,-three boron ranges for thermal neutron absorption.

CARPENTER CONSUMET CORE IRON.
M-32; 0.06 max C, 0.20 max Mn, 0.25 max Si, 0.1 max V, bal Fe.
For solenoid switches, armatures; high magnetic permeability.

CARPENTER CONSUMET M-50 BEARING STEEL see CARPENTER CONSUMET M-50 HIGH SPEED STEEL.

CARPENTER CONSUMET M-50 HIGH SPEED STEEL.
M-32; 0.80 C, 0.25 Mn, 0.25 Si, 4.0 Cr, 0.10 Ni, 1.0 V, 4.5 Mo, bal Fe.
High speed steel; used largely for bearings in aircraft and gas turbine engines operating at temperatures up to 800°F.
AISI M-50.

CARPENTER CUSTOM 450.
M-32; 0.05 max C, 0.5 max Mn, 0.5 max Si, 14.5-16.5 Cr, 5.5-7.0 Ni, 0.5-1.0 Mo, 1.25-1.75 Cu, Cb = 8 x C min, bal Fe.
Martensitic age hardenable stainless steel.
Annealed: 144,000 TS; 117,000 YS; 14 El.
Hardened: 196,000 TS; 184,000 YS; 14 El.
UNS S45000.

CARPENTER CUSTOM 450.
M-32; 0.05 max C, 1.0 max Mn, 1.0 max Si, 14.0-16.0 Cr, 5.0-7.0 Ni, 0.50-1.0 Mo, 1.25-1.75 Cu, Cb 8 x C min, bal Fe.
Martensitic age-hardenable stainless steel.
Sol. ann: TS: 143 ksi (986 MPa); YS: 118 ksi (814 MPa).
Aged (900°F): TS: 195 ksi (1344 MPa); YS: 186 ksi (1282 MPa).
AMS 5763 etc; (545000).

CARPENTER CUSTOM 455.
M-32; 0.05 max C, 0.50 max Mn, 0.50 max Si, 11.0-12.5 Cr, 7.5-9.5 Ni, 0.80-1.4 Ti, 0.10-0.50 Cb + Ta, 1.5-2.5 Cu, 0.50 max Mo, bal Fe.
High strength precipitation-hardenable stainless.
Hardenable to 140-210 ksi (1000-1447 MPa).
S45500.

CARPENTER CUSTOM 630 (17 CR-4 NI).
M-32; 0.07 C, 15.5-17.5 Cr, 3.0-5.0 Ni, 3.0-5.0 Cu, 0.15-0.45 Cb + Ta, 1.0 max Mn, 1.0 max Si, bal Fe.
Martensitic, precipitation/age hardening stainless steel.
Aged: 125,000-200,000 TS; 85,000-185,000 YS; (depending on aging treatment).
For oil field valve parts, aircraft fittings, paper mill equipment, jet engine parts.
AISI 630; UNS S17400.

CARPENTER C-XB VALVE STEEL.
M-32; 0.80 C, 0.4 Mn, 2.25 Si, 20.0 Cr, 1.5 Ni, bal Fe.
Hardenable silicon-chrome alloy.
At 1200°F: 39,500 TS; 29,500 YS; 33.5 El.
For exhaust valves, valve seat inserts, intake valves in internal combustion engines.
UNS K65006.

CARPENTER C-XB VALVE STEEL.
M-32; 0.80 C, 0.40 Mn, 2.25 Si, 20.0 Cr, 1.5 Ni, bal Fe.
For exhaust valves in gasoline engines.
K-65006.

CARPENTER D-6A.
M-32; 0.48 C, 0.75 Mn, 0.25 Si, 1.1 Cr, 0.55 Ni, 1.0 Mo, 0.10 V, bal Fe.
A tough alloy tool steel.

CARPENTER DOUBLE VACUUM MELTED (VIN-VAR) M-50 HIGH SPEED STEEL same as **CARPENTER CONSUMET M-50 HIGH SPEED STEEL.**

CARPENTER ELECTRICAL IRON.
M-32; 0.02 max C, 0.12 Mn, 0.12 Si, 0.010 P, 0.010 S, 0.10 Cr, 0.08 Ni, 0.03 Al, bal Fe.
Soft magnetic iron for solenoids, magnetic pole pieces, magnetic circuit core members.

CARPENTER FIBRALLOY 460.
M-32; 0.02 C, 0.04 Mn, 0.04 Si, 11.25 Cr, 9.0 Ni, 2.25 Cu, 0.80 Al, 0.45 Cb, bal Fe.
Martensitic precipitation hardening stainless steel; hardness Rc 45; touch, ductile, retains hardness and oxidation resistance at elevated temperature.
Designed for use in the synthetic fiber textile industry.

CARPENTER FOUR STAR.
M-32; 1.30 C, 0.3 Mn, 0.3 Si, 4.5 Cr, 4.5 Mo, 5.5 W, 4.0 V, bal Fe.
High carbon, high speed steel for extra wear resistance; for reamers, drills, lathe tools, thread chasers, broaches.
AISI M4; UNS T11304.

CARPENTER FREE CUT INVAR 36.
M-32; 0.12 max C, 0.35 Mn, 36 Ni + Co, 0.20 Se, bal Fe.
For low expansion alloy; free cutting.

CARPENTER GLASS SEALING "27".
M-32; 0.15 max C, 28 Cr, bal Fe.
Annealed: 85,000 TS; 55,000 YS; 25 El; 185 Brin.
For glass sealing, metal to glass.

CARPENTER GLASS SEALING "42".
M-32; 0.10 max C, 0.50 Mn, 0.25 Si, 41.5 Ni, bal Fe.
Cold drawn: 120,000 TS; 3 El; 240 Brin.
Annealed: 82,000 TS; 30 El; 140 Brin.
For glass to metal seals; for hard and soft glass.

CARPENTER GLASS SEALING "42-6".
M-32; 0.10 max C, 0.50 Mn, 0.25 Si, 42.5 Ni, 5.75 Cr, bal Fe.
Strip and wire with thermal expansion matching characteristics of 0120 glass; for sealing into glass.

CARPENTER GLASS SEALING "45-6".
M-32; 0.10 max C, 0.3 Mn, 0.3 Si, 45 Ni, 6 Cr, bal Fe.
Annealed: 80,000 TS; 40,000 YS; 30 El; B 80 Rock.
For metal to glass sealing (0120 and 9010 glass).
Controlled coef. expansion.
Vacuum melted, low gas content.

CARPENTER GLASS SEALING 46.
M-32; 46 Ni, bal Fe.
For metal to glass seals; controlled expansion.

CARPENTER GLASS SEALING 52.
M-32; 0.10 max C, 0.50 Mn, 0.25 Si, 51 Ni, bal Fe.
Annealed: 80,000 TS; 40,000 YS; 35 El; Rockwell B 83.
For glass to metal seals; for soft glass and ceramics.

CARPENTER GLASS SEALING "52" PHOTO-ETCH QUALITY.
M-32; 0.10 C, 0.50 Mn, 0.25 Si, 50.50 Ni, bal Fe.
Strip designed to produce sharp, square edges in photo-etching and to match expansion of soft glasses and some ceramics.

CARPENTER GREEN LABEL DRILL ROD.
M-32; 1.2 C, 0.2 Mn, 0.2 Si, bal Fe.
AISI Type W1 water hardening tool steel.

CARPENTER H-9 DOUBLE HEADER DIE STEEL.
M-32; 0.90 C, 0.4 Mn, 0.4 Si, bal Fe.
Water hardening tool steel; for solid and gripper cold heading dies, inserts, coining dies, knurls.
AISI W1; UNS G10900.

CARPENTER H-46.
M-32; 0.15-0.20 C, 10-14 Cr, 0.3-0.6 Ni, 0.5-0.8 Mo, 0.2-0.4 V, 0.2-0.6 Cb, 0.06-0.10 N, bal Fe.
At 70°F: 151,000 TS; 124,000 YS; 20 El; 56 RA; 302 Brin.
At 1200°F: 60,500 TS; 56,200 YS; 30 El; 76 RA.
For jet aircraft engine compressor blades and rotor discs; good strength and ductility to 1200°F.

CARPENTER HAMPDEN.
M-32; 2.1 C, 0.35 Mn, 0.3 Si, 12.0 Cr, 0.5 Ni, bal Fe.
Oil hardening high carbon, high chromium die steel with extra wear resistance; for slitting cutters, lamination dies, cold rolls, blanking and forming dies.
AISI D3.

CARPENTER HIGH PERMEABILITY 49.
M-32; 0.02 C, 0.50 Mn, 0.35 Si, 48.0 Ni, bal Fe.
3 grades with varying permeability depending on treatment and size or shape.
For rotors, transformers, solenoid cores and magnetic shields.

CARPENTER HIGH PERMEABILITY "49".
M-32; 0.05 C, 0.05 Mn, 0.35 Si, 48.0 Ni, bal Fe.
Annealed: 70,000 TS; 22,000 YS; 45 El.
Flux density: 15,000 gausses; low hysteresis loss.
Available in 3 grades; for magnet cores, solenoids, laminations, transformers.

CARPENTER HI MAG PERM.
M-1425; 0.03-0.05 C, 0.005-0.009 P, 0.01-0.02 Si, 0.03-0.07 Cr, 006-0.01 Al, 0.04-0.07 Mo, bal Fe.
Annealed: 40,000 TS; 20,000 YS; 40 El; 78 RA; 69 Brin.
For electrical applications, magnetic control devices, magnetic clutches and chucks.

CARPENTER HIPERCO 27.
M-32; 0.015 C, 0.25 Mn, 0.25 Si, 0.60 Cr, 27.0 Co, bal Fe.
A high magnetic saturation alloy for use in magnetic flux carrying members magnetic pole caps, and laminations for aircraft motors and generators.

CARPENTER HIPERCO 50.
M-32; 0.01 C, 0.05 Mn, 0.05 Si, 48.75 Co, 1.9 V, bal Fe.
A high magnetic saturation alloy for use as magnetic pole caps, lamination for aircraft motors and generators, transformer laminations, and tape toroids.

CARPENTER HIPERCO 50-FM.
M-32; 0.04 C, 0.80 Mn, 0.40 Si, 49.0 Co, 1.9 V, 0.20 Se, bal Fe.
Ann: Bm: 22,000 gausses.
For sonar applications, ultrasonic transducers.

CARPENTER HIPERNOM.
M-32; 4.2 Mo, 80.0 Ni, 0.35 Si, 0.02 C, 0.50 Mn, bal Fe.
Soft magnetic alloy capable of high permeability.
For shielding applications.

CARPENTER HI-SHOCK 60.
M-32; 0.68 C, 0.5 Mn, 0.3 Si, 1.0 Cr, 0.5 Ni, 1.0 Mo, 2.5 Cu, 0.15 V, bal Fe.
Air hardened: 363,000-226,000 TS; 316,000-206,000 YS; 1.6-10.2 El; 3.1-30.9 RA; 580-500 Brin.
For hobs, punches, mandrels, tools and dies; air hardened, shock resistant

CARPENTER HI SHOCK 60.
M-32; 0.68 C, 0.35 Mn, 0.3 Si, 1.0 Cr, 0.5 Ni, 1.0 Mo, 2.5 Cu, 0.15 V, bal Fe.
Air or oil hardening tool steel; for large blanking and forming dies, trimming dies, shear blades, mandrels, forming tools.

CARPENTER HY MU "80".
M-32; 0.15 C, 0.50 Mn, 0.35 Si, 80.0 Ni, 4.2 Mo, bal Fe.
Unoriented alloy with high initial permeability and maximum permeability with minimum hysteresis loss.
For transformer cores, tape wound toroids, and laminations.

CARPENTER HYMU 80 MARK II.
M-32; 0.015 C, 0.50 Mn, 0.30 Si, 80.0 Ni, 4.6 Mo, bal Fe.
Unoriented high initial permeability alloy; maximum permeability with minimum hysteresis loss; slightly better than Hy Mu 80.
For transformer lamps, tape toroids, and magnetic pick-up head laminations.

CARPENTER HYMU 800.
M-32; 0.03 C, 0.50 Mn, 0.35 Si, 80.0 Ni, 4.0 Mo, bal Fe.
Vacuum melted soft magnetic alloy for tape wound toroids and laminations.
Minimum permeability of 60,000 at 40 gauss.

CARPENTER HYMU 800.
M-32; 0.01 C, 0.50 Mn, 0.15 Si, 80.0 Ni, 5.0 Mo, bal Fe.
Soft magnetic material.
For laminations or toroids.

CARPENTER HY MU "800" PHOTO CHEM QUALITY.
M-32; 0.01 C, 0.50 Mn, 0.15 Si, 80.0 Ni, 5.0 Mo, bal Fe.
Vacuum melted soft magnetic sheet for use with photochemical techniques for lamination or toroids. Minimum permeability of 60,000 at 40 gauss.

CARPENTER HY-RA 49.
M-32; 49 Ni, bal Fe.
For electronic devices, magnetic amplifiers; high permeability.

CARPENTER HY-RA 80.
M-32; 0.015 C, 0.5 Mn, 0.15 Si, 79 Ni, 4.2 Mo, bal Fe.
For electrical equipment, electronic devices; square hysteresis loop properties, high permeability.

CARPENTER HY-RA 80.
M-32; 79 Ni, 0.05 C, 4 Mo, 0.5 Mn, 0.15 Si, bal Fe.
For electronic devices, magnetic amplifiers; high permeability.

CARPENTER INVAR 36.
M-32; 0.12 max C, 0.35 Mn, 36 Ni + Co, bal Fe.
Rolled: 90,000 TS; 70,000 YS; 20 El; 60 RA; B 90 Brin.
For precision instruments, radio parts, bi-metal; low coefficient of expansion.

CARPENTER INVAROD.
M-32; 0.10 max C, 2.50 Mn, 0.20 max Si, 36.0 Ni, 0.75 Ti, bal Fe.
Welding alloy for joining Invar "36" to itself or other metals without preheating or postheating.

CARPENTER JY ROLL STEEL.
M-32; 0.85 C, 0.30 Mn, 0.30 Si, 2.0 Cr, 0.2 V, bal Fe.
High carbon, low alloy steel used for cold rolls.

CARPENTER KOVAR.
M-32; 0.02 max C, 0.30 Mn, 0.20 Si, 29.0 Ni, 17.0 Co, bal Fe.
Vacuum-melted low-expansion alloy for making hermetic seals.
Coef. of Expan: 5.06×10^{-6} in/in/°F, 77-1292°F.
(Was NICOSEAL).

CARPENTER KOVAR PHOTO-ETCH QUALITY.
M-32; 0.02 max C, 0.30 Mn, 0.20 Si, 29.0 Ni, 17.0 Co, bal Fe.
Designed to secure square edges in photoetching of "flatpak" lead preforms for integrated circuits. Corrosion resistant; controlled thermal expansion (Was NICOSEAL).

CARPENTER K W.
M-32; 1.3 C, 3.5 W, 0.3 Si, 0.3 Mn, bal Fe.
For draw dies, punches, mandrels, gauges; water hardened.
AISI F2; UNS T60202.

CARPENTER L-605.
M-32; 0.12 C, 1.55 Mn, 20 Cr, 12 Ni, 15 W, bal Co.
For jet engine components, after burners; heat and corrosion resistant.

CARPENTER LAPELLOY.
M-32; 0.25-0.35 C, 0.95-1.25 Mn, 0.50 max Si, 11.0-12.0 Cr, 0.50 max Ni, 2.5-3.0 Mo, 0.2-0.3 V, bal Fe.
Air or oil hardenable corrosion resistant steel.
Hardened; tested 1200°F: 60,000 TS; 50,000 YS.
For steam turbine buckets and blades, compressor blades, valve stems.

CARPENTER LAPELLOY "C".
M-32; 0.20-0.25 C, 0.65-1.00 Mn, 11-12 Cr, 0.50 max Ni, 2.50-3.00 Mo, 1.75-2.25 Cu, 0.06-0.10 Ni, bal Fe.
Heat treated: 135,000-203,000 TS; 105,000-170,000 YS; 17-18 El; 47-55 RA.
For compressor wheels, turbine shafts, compressor buckets, blades, bolts; for high stressed parts operating to 1200°F.

CARPENTER LOW EXPANSION 39.
M-32; 39 Ni, bal Fe.
For instruments; low expansion alloy.

CARPENTER LOW EXPANSION 42.
M-32; 42 Ni, bal Fe.
Annealed: 82,000 TS; 30 El; 140 Brin.
Cold drawn: 120,000 TS; 3 El; 240 Brin.
For low expansion and high temperature applications; coefficient of expansion 3×10^{-6}/°F at 70-650°F.

CARPENTER MANIFLEX 11.
M-32; 0.08 C, 1.25 Mn, 0.80 Si, 0.03 S, 21.0 Cr, 11.5 Ni, bal Fe.
Strip room: 593 MPa TS; 269 MPa YS; 59 El 1600°F (871°C): 139 MPa TS; 96 MPa YS; 73 El.
For butterfly valves in emission control devices. UNS K63016.

CARPENTER MANIFLEX-FM.
M-32; 0.20 C, 1.25 Mn, 0.80 Si, 0.02 P, 0.20 S, 21.0 Cr, 11.5 Ni, bal Fe.
Free machining Austenitic stainless with excellent high temperature strength and hardness.
For shafts and bushings in manifold exhaust heat control valves and emission control devices.

CARPENTER MARINALOY 17.
M-32; 0.07 C, 1.0 Mn, 1.0 Si, 15.5-17.5 Cr, 3.0-5.0 Ni, 3.0-5.0 Cu, 0.15-0.45 Cb + Ta, bal Fe.
Martensitic age hardening stainless steel with high strength and excellent corrosion resistance. Minimum tensile strength: 135,000 psi; for boat shafting.

CARPENTER MARINALOY HN.
M-32; 0.08 C, 2.0 Mn, 1.0 Si, 18.0-20.0 Cr, 8.0-10.5 Ni, 0.16-0.30 N, bal Fe.
Nitrogen strengthened austenitic stainless steel with unique combination of strength, toughness and corrosion resistance.
For boat shafting.

CARPENTER MEL-TROL SPEED STAR see **SPEED STAR.**

CARPENTER MEL-TROL STAR-ZENITH see **STAR ZENITH.**

CARPENTER MIRROMOLD.
M-32; 0.10 C, 0.20 Mn, 0.10 V, bal Fe.
Low carbon steel for easy cold hubbing and subsequent case hardening for plastic molds.
AISI P1.

CARPENTER MOLY ASCOLOY.
M-32; 0.08-0.15 C, 0.50-0.90 Mn, 0.35 max Si, 11.0-12.5 Cr, 2.0-3.0 Ni, 1.5-2.0 Mo, 0.25-0.40 V, bal Fe.
Hardenable martensitic stainless steel.
Hardened, tested at 1200°F: 121,000 psi (834 MPa) TS; 97,000 psi (669 MPa) YS; 18 El.
For steam turbine buckets, high temperature bolts.
UNS K64152.

CARPENTER NIMARK 200.
M-32; 0.008 C, 18.5 Ni, 4.25 Mo, 7.5 Co, 0.20 Ti, 0.10 Al, bal Fe.
Low carbon, high nickel maraging steel.

CARPENTER NIMARK I see **NIMARK 250.**

CARPENTER NIMARK 300.
M-32; 0.03 max C, 18.0-19.0 Ni, 4.7-5.1 Mo, 8.0-9.5 Co, 0.50-0.80 Ti, 0.05-0.15 Al, 0.03 max Zr, 0.005 max B, 0.05 max Ca, bal Fe.
Maraging type alloy; as hardened: 294,000 TS; 290,000 YS; 11 El; 52 Rock C.
Weldable; for high strength parts and assemblies.

CARPENTER NITREX I + II see **NITREX I + II.**

CARPENTER NO. 1-JR.
M-32; 0.15 max C, 12-14 Cr, 3.25-4.5 Al, bal Fe.
At 70°F: 86,000 TS; 25 El; 57 RA.
At 1400°F: 13,500 TS; 77 El; 93 RA.
For magnetic cores, resistors; corrosion and oxidation resistant.

CARPENTER NO 5-317.
M-32; 0.50 C, 0.50 Mn, 0.25 Si, 1.0 Cr, 1.75 Ni, bal Fe.
Nickel chrome alloy steel.
Oil hardenable to above 55 Rock C.
For shafts, pinions, vise jaws.

CARPENTER NO. 5-BG.
M-32; 0.15 max C, 1.25 max Mn, 0.06 max P, 0.15 min S, 1.0 max Si, 12.0-14.0 Cr, bal Fe.
Martensitic free-machining stainless.
AISI Type 416; S41600.

CARPENTER 5-F see **CARPENTER STAINLESS NO. 5F.**

CARPENTER NO. 11 SPECIAL.
M-32; 1.05 C, 0.20 Mn, 0.20 Si, bal Fe.
Water hardening tool steel for drills, taps, reamers, punches, blanking dies, bushings.
AISI W1. UNS G10910.

CARPENTER NO. 11 SPECIAL.
M-32; 1.05 C, 0.20 N, 0.20 Si, bal Fe.
Water hardening tool steel for drills, taps, reamers, punches, blanking dies, bushings.
AISI W1. UNS G10910.

CARPENTER NO. 158.
M-32; 0.10 C, 0.50 Mn, 0.30 Si, 1.50 Cr, 3.50 Ni, bal Fe.
Deep hardening carburizing steel for heavy duty gears, power tool cams, clutch levers.

CARPENTER NO. 158 PLASTIC MOLD STEEL.
M-32; 0.10 C, 0.5 Mn, 0.3 Si, 1.5 Cr, 3.5 Ni, bal Fe.
Oil hardenable case hardening steel; designed for molds and cavities requiring high case hardness and good core strength. Also used for case hardened gears. AISI P6; SAE 3312; UNS T51606.

CARPENTER NO. 345 HOT WORK DIE STEEL.
M-32; 0.35 C, 0.30 Mn, 1.0 Si, 5.0 Cr, 1.25 W, 1.5 Mo, bal Fe.
Hot work tool steel; for aluminum extrusion dies, dummy blocks, forging dies, die casting dies.
AISI H 12; UNS T20812.

CARPENTER NO. 404.
M-32; 0.05 max C, 11-12 Cr, 1.25-2.00 Ni, 0.03 max N, 1.0 max Mn, 0.5 max Si, bal Fe.
Heat treated: 108,000-163,000 TS; 95,000-125,000 YS; 23-10 El; 70-40 RA; 180-350 Brin.
For steam turbine buckets, blades and bucket covers; for highly stressed parts at temperatures up to 1050°F.

CARPENTER NO. 408 PUNCH STEEL.
M-32; 0.50 C, 0.50 Mn, 0.25 Si, 0.75 Cr, 3.0 Ni, bal Fe.
Heat treated: 275,000 TS; 250,000 YS; 8 El; C 56 Rock.
For punches, clutch parts, vise jaws, chuck jaws, feeder rolls.
Tough, wear resistant.

CARPENTER NO. 481.
M-32; 0.55 C, 0.8 Mn, 1.9 Si, 0.25 Cr, 0.4 Mo, bal Fe.
Oil hardening shock resistant steel.
For pneumatic tools, chipping chisels, rivet sets, shear blades, collets.

CARPENTER NO. 484.
M-32; 1.0 C, 0.80 Mn, 0.30 Si, 5.25 Cr, 1.1 Mo, 0.20 V, bal Fe.
Air-hardening tool steel for blanking dies, thread rollers, coining dies, gages.
AISI A2; UNS T30102.

CARPENTER NO. 610.
M-32; 1.50 C, 0.50 Mn, 0.30 Si, 12.0 Cr, 0.80 Mo, 0.90 V, bal Fe.
Air hardening, high carbon, high chromium steel with high wear resistance.
For blanking and coining dies and rolls.
AISI Type D2; T30402.

CARPENTER NO. 610-FM.
M-32; 1.50 C, 0.5 Mn, 0.3 Si, 12.0 Cr, 0.8 Mo, 0.9 V, plus alloy sulphides, bal Fe.
Air hardening, high carbon, high chromium tool steel, free-machining grade; for blanking and forming dies, drawing dies, edging rolls, extrusion dies. AISI D2.

CARPENTER NO. 882.
M-32; 0.40 C, 0.35 Mn, 1.0 Si, 5.0 Cr, 0.4 V, 1.35 Mo, bal Fe.
Hot work tool steel; for forging dies, die casting dies, hot heading dies, aluminum extrusion dies.
AISI H 11; UNS T20811.

CARPENTER NO. 883.
M-32; 0.37 C, 0.35 Mn, 1.0 Si, 5.25 Cr, 1.0 V, 1.3 Mo, bal Fe.
Chrome hot work tool steel, air or oil hardenable; for forging dies, die casting dies, extrusion dies.
AISI H 13.

CARPENTER O-1.
M-32; 0.90 C, 1.2 Mn, 0.5 Cr, 0.5 W, 0.2 V, bal Fe.
Oil hardening tool steel; for blanking and forming dies, broaches, collets, thread gages, spindles.
AISI O1, UNS T31501.

CARPENTER PH 13-8 MO.
M-32; 0.05 max C, 0.10 max Mn, 0.010 max P, 0.008 max S, 0.10 max Si, 12.25-13.25 Cr, 7.5-8.5 Ni, 0.90-1.35 Al, 2.0-2.5 Mo, 0.01 max N, bal Fe.
Martensitic precipitation/age-hardening stainless steel. For aircraft components.
S13800.

CARPENTER PROJECT 70 182-FM.
M-32; 0.08 max C, 1.25-1.50 Mn, 1.0 max Si, 0.04 max P, 0.15-0.40 S, 17.5-19.5 Cr, 1.5-2.5 Mo, bal Fe.
Free machining 18-2 type stainless.
Ferritic type.
UNS S18200.

CARPENTER PYROMET 31.
M-32; 0.04 C, 22.7 Cr, 55.5 Ni, 2.0 Mo, 2.3 Ti, 1.3 Cb, 0.85 Cb, 0.005 B, bal Fe.
Precipitation hardenable superalloy.
At 1500°F: 100,000-115,000 psi (687-798 MPa) TS; 93,000 psi (640 MPa) YS; 16-21 El.
For truck and locomotive diesel valves; gas and oil well equipment.

CARPENTER PYROMET 41.
M-32; 0.60-0.12 C, 18-20 Cr, 9-10.5 Mo, 10-12 Co, 3.0-3.3 Ti, 1.4-1.6 Al, 0.003-0.010 B, 5.0 max Fe, bal Ni.
Aged: 206,000 TS; 154,000 YS; 14 El.
For jet engine and aircraft components, high temperature bolts; age-hardenable, useful to 1800°F.

CARPENTER PYROMET 80A.
M-32; 0.06 C, 20 Cr, 0.007 S, 2.35 Ti, 1.25 Al, 0.05 Cu, 1.0 Co, ba Ni.
Aged Hardened: 145,000 TS; 90,000 YS; 39 El.
At 1200°F: 115,000 TS; 80,000 YS; 21 El.
For high temperature applications.
Heat treatable. High creep and fatigue resistance.
Oxidation and corrosion resistance.

CARPENTER PYROMET 88.
M-32; 0.03 C, 2.2 Mn, 16.4 Cr, 6.7 Fe, 3.05 Ti, bal Ni.
High strength, high temperature nickel base alloy, age hardenable; useful to 1500°F; good corrosion resistance.

CARPENTER PYROMET 90.
M-32; 0.07 C, 19.5 Cr, 18.0 Co, 2.4 Ti, 1.4 Al, bal Ni.
Age hardenable nickel base alloy.
Room Temp: 179,000 TS; 117,00 YS; 33 El.
1600°F: 48,000 TS; 38,000 YS.
For high temperature applications; good resistance to scaling and corrosion.
UNS N07090.

CARPENTER PYROMET 95.
M-32; 0.15 C, 0.15 max Mn, 0.20 max Si, 14.0 Cr, 8.0 Co, 2.5 Ti, 3.5 Al, 3.5 W, 3.5 Cb, bal Ni.
Precipitation hardening nickel base superalloy. Used for turbine and compressor discs, shafts and seals in aircraft gas turbine engines.

CARPENTER PYROMET 102.
M-32; 0.10 max C, 14.0-16.0 Cr, 2.75-3.75 Nb, 2.75-3.75 Mo, 2.75-3.75 W, 5.0-9.0 Fe, 0.30-0.70 Al, 0.40-0.70 Ti, 0.003-0.008 B, bal Ni.
Non-magnetic, Ni-Cr base alloy; corrosion resistant and high temperature strength.
80°F: 137,000-149,000 TS; 72,000-85,000 YS; 41-47 El.
1500°F: 48,000-52,000 TS; 45,000-51,000 YS; 88-112 El.
For heat shields, piping for steam and gas turbine and rocket engines, furnace hardware.
UNS N06102.

CARPENTER PYROMET 350.
M-32; 0.07-0.13 C, 0.50-1.25 Mn, 0.50 max Si, 16.0-17.0 Cr, 4.0-5.0 Ni, 2.5-3.25 Mo, 0.07-0.13 N, bal Fe.
Martensitic or precipitation hardenable stainless. Hardenable to 160-198 ksi; 112-139 kg/mm^2.
For gas turbine compressor components.
AISI 633; AMS 5745; S35000.

CARPENTER PYROMET 355.
M-32; 0.10-0.15 C, 0.50-1.25 Mn, 0.5 max Si, 15.0-16.0 Cr, 4.0-5.0 Ni 2.5-3.25 Mo, 0.07-0.13 N, bal Fe.
Precipitation hardenable stainless steel; wide range of mechanical properties after various heat treatments;-to 220,000 TS max; weldable; machinable in solution treated condition. AISI 634.
UNS S21904.

CARPENTER PYROMET 538.
M-32; 0.03 C, 8.0-10.0 Mn, 1.0 max Si, 18.0-21.5 Cr, 5.5-7.5 Ni, 0.15-0.40 N, bal Fe.
Strengthened austenitic stainless.
Room T.: 112,000 TS; 65,000 YS; 42 El.
1000°F: 71,000 TS; 29,000 YS; 35 El.
For steam and autoclave parts, airframe and aircraft engine, chemical processing equipment.

CARPENTER PYROMET 600.
M-32; 0.10 max C, 1.0 max Mn, 0.50 max Si, 14.0-17.0 Cr, 72.0 Ni, 0.50 max Cu, 6.0-10.0 Fe.
Non-magnetic nickel base alloy, corrosion resistant and temperature resistant.
1000°F: 84,000 TS; 28,500 YS; 47 El.
1600°F: 15,000 TS; 9,000 YS; 80 El.
AMS 5665; 5540, 5580.
UNS N06600.

CARPENTER PYROMET 625.
M-32; 0.10 max C, 20.0-23.0 Cr, 8.0-10.0 Mo, 5.0 max Fe, 0.40 max Ti, 1.0 max Co, 3.15-4.15 Cb + Ta, 0.40 max Al, bal Ni.
Non-magnetic, corrosion resistant and high temperature alloy.
400°F: 116,000 TS; 43,000 YS; 57 El.
1800°F: 18,000 TS; 16,000 YS; 120 El.
For heat shields, furnace hardware, gas turbine engine ducting, chemical plant hardware.
AMS 5666; 5599; UNS NO 6625.

CARPENTER PYROMET 680.
M-32; 0.05-0.15C, 20.5-23.0 Cr, 0.5-2.5 Co, 8.0-10.0 Mo, 0.2-1.0 W, 1 -20.0 Fe, bal Ni.
Non-magnetic heat and corrosion resistant nickel base alloy. Room T.: 104,000 TS; 49,000 YS; 40 El.
1000°F: 94,000 TS; 41,500 YS; 45 El.
1800°F: 21,000 TS; 17,000 YS; 43 El.
For gas turbine rotors, afterburner parts, hardware for furnaces and chemical.
AMS 5536 etc. (Reference Hastelloy X).
UNS No 6002.

CARPENTER PYROMET 718.
M-32; 0.10 max C, 17.0-21.0 Cr, 50.0-55.0 Ni + Co, 1.0 max Co, 2.8-3. Mo, 4.75-5.50 Cb + Ta, 0.65-1.15 Ti, 0.35-0.85 Al, 0.001-0.006 B, 0.15 max Cu, bal Fe.
Precipitation hardenable nickel base alloy.
Hardened, Room T.: 210,000 TS; 175,000 YS; 22 El.
1400°F: 110,000 TS; 110,000 YS; 27 El.
For jet engine parts as buckets, spacers, wheels, high temperature fasteners.
UNS No 7718.

CARPENTER PYROMET 751.
M-32; 0.04 C, 0.70 Mn, 0.30 Si, 15.0 Cr, 0.007 S, 1.0 Cb, 6.75 Fe, 2.5 Ti, 1.2 Al, 0.05 Cu, bal Ni.
Sol. Tr. and Aged: 1207 MPa TS; 758 MPa YS; 20 El.
Age hardenable alloy with good high-temperature stress rupture properties. (Formerly X-751).

CARPENTER PYROMET 800 PIPE & TUBING.
M-32; 0.10 C, 1.5 Mn, 1.0 Si, 19.0-23.0 Cr, 30.0-35.0 Ni, 0.15-0.60 Al, 0.15-0.60 Ti, 0.75 max Cu, bal Fe.
Non-magnetic Ni-Cr-Fe Alloy. At 1000°F 72,000 TS; 31,700 YS; 38.5 El. At 1500°F: 24,800 TS; 14,200 YS; 91 El.
Weldable, corrosion resistant; for high temperature piping.

CARPENTER PYROMET 860.
M-32; 0.10 max C, 1.0 max Mn, 1.0 max Si, 12.0-16.0 Cr, 40.0-45.0 Ni, 5.0-7.0 Mo, 3.5-4.5 Co, 2.75-3.75 Ti, 0.75-1.50 Al, 0.0008-0.012 B, bal Fe.
Austenitic Fe-Ni base precipitation hardening alloy. Hardened: Room T.: 180,000 TS,; 115,000 YS; 21 El. 1200°F: 159,000 TS; 125,000 YS; 19 El. 1500°F: 106,000 TS; 103,000 YS; 16 El.
For corrosion and scale resisting parts at high temperatures; turbine engine parts, steam turbine bolting.

CARPENTER PYROMET 882.
M-32; 0.40 C, 1.0 Si, 5.0 Cr, 1.5 Mo, 0.40 V, bal Fe.
Air hardened, tempered 1000°F (double). At 500°F: 260,000 TS; 220,000 YS. At 1000°F: 215,000 TS; 182,000 YS.
For forging dies, aluminum extrusion dies, hot piercing and forming punches AISI 610; (similar to AISI H 11); UNS T20811.

CARPENTER PYROMET 901.
M-32; 0.10 max C, 1.0 max Mn, 11-14 Cr, 40-45 Ni, 5-7 Mo, 2.35-3.10 Ti, 0.5 max Cu, 0.35 max Al, 0.010-0.020 B, bal Fe.
At 70°F: 175,000 TS; 125,000 YS; 15 El; 19 RA.
At 1000°F: 156,000 TS; 113,000 YS; 17 El; 29 RA.
At 1500°F: 81,000 TS; 79,000 YS; 13 El; 20 RA.
For aircraft, gas turbines, rotors, compressor discs, hubs, shafts; precipitation hardened, heat and corrosion resistant to 1400°F.

CARPENTER PYROMET A-286.
M-32; 0.08 max C, 2.0 max Mn, 1.0 max Si, 13.5-16.0 Cr, 24.0-27.0 Ni, 1.0-1.75 Mo, 1.9-2.3 Ti, 0.10-0.50 V, 0.35 max Al, 0.003-0.010 B, bal Fe.
At 1200°F: 103,500 TS; 88,000 YS; 13 El.
At 1500°F: 36,500 TS; 33,000 YS; 68 El.
For turbine blades, jet engine components; age-hardened for service to 1300°F.
AISI 660; AMS 5731 etc; UNS K66286.

CARPENTER PYROMET CTX-1.
M-32; 0.03 C, 0.20 max Mn, 0.20 max Si, 0.20 max Cr, 0.20 max Mo, 0.50 max Cu, 37.7 Ni, 3.0 Cb (+Ta), 1.75 Ti, 1.0 Al, 0.0075 B, 16.0 Co, bal Fe.
Precipitation hardening alloy; good stress rupture properties to 1200°F. Low coefficient of expansion.
For ordnance hardware, turbine blades, springs gauge blocks, die casting dies.

CARPENTER PYROMET M-252.
M-32; 0.10-0.20 C, 18-20 Cr, 9-10.5 Mo, 9-11 Co, 0.75-1.25 Al, 2.25-2.75 Ti, 0.02-0.15 Zr, 0.001-0.01 B, 5 max Fe, bal Ni.
Heat treated: 175,000 TS; 110,000 YS; 25 El.
At 1500°F: 91,000 TS; 84,000 YS; 24 El.
For jet engine and gas turbine buckets.
Resists heat to 1600°F. Precipitation hardening.

CARPENTER PYROMET N-155.
M-32; 0.08-0.16 C, 1.0-2.0 Mn, 1.0 max Si, 20.0-22.5 Cr, 19.0-21.0 Ni, 18.5-21.0 Co, 2.5-3.5 Mo, 2.0-3.0 W, 0.75-1.25 Cb + Ta, 0.50 max Cu, 0.10-0.20 N, bal Fe.
Hardened: Room T.: 118,000 TS; 59,000 YS; 40 El.
1800°F: 19,000 TS (131 MPa).
For aircraft tail cones and tail pipes, exhaust manifolds, combustion chambers, afterburners.

CARPENTER PYROMET X-15.
M-32; 0.03 max C, 15.0 Cr, 20.0 Co, 2.9 Mo, 0.20 max Ni, bal Fe.
Precipitation hardenable martensitic alloy.
Hardened: Room T.: 235,000 TS; 215,000 YS; 17 El.
1000°F: 190,000 TS; 170,000 YS; 18 El.
For highly stressed parts operating at temperatures up to 1050°F.

CARPENTER PYROMET X-23.
M-32; 0.02 max C, 10.0 Cr, 10.0 Co, 5.5 Mo, 7.0 Ni, bal Fe.
Low carbon martensitic alloy; high strength from cryogenic temperatures to 1000°C (538°C).

CARPENTER PYROMET X-750.
M-32; 0.05 max C, 14.0-17.0 Cr, 70.0 min Ni + Co, 1.0 max Co, 2.25- 2.70 Ti, 0.4-1.0 Al, 0.70-1.20 Cb + Ta, 0.5 max Cu, 5.0-9.0 Fe.
Precipitation hardenable nickel base alloy. Room T.: 161,000 TS; 92,000 YS; 22 El.
1400°F: 80,000 TS; 68,000 YS; 10 El.
For high temperature springs, parts for gas turbine, jet engines, extrusion dies. AISI 688; AMS 5667, 5668; (Reference: Inconel X); UNS No 7750.

CARPENTER PYROMET X-751 see **CARPENTER PYROMET 75.**

CARPENTER PYROTOOL A.
M-32; 0.04 C, 1.20 Mn, 0.50 Si, 14.5 Cr, 25.0 Ni, 1.5 Mo, 2.2 Ti, bal Fe.
Austenitic precipitation hardening iron base alloy.
Room temp: 155,000 psi (1069 MPa) TS; 100,000 psi (689 MPa) YS; 30-35 Rc.
Corrosion resistant; heat resistant.
UNS K66286.

CARPENTER PYROTOOL 7.
M-32; 0.05 C, 19.0 Cr, 52.5 Ni + Co, 3.0 Mo. 5.25 Cb + Ta, 1.0 Ti, 0.6 Al, bal Fe.
Austenitic precipitation hardenable nickel base alloy for high strength up to 1300°F.
Hardened; at 1300°F: 120,000 TS; 115,000 YS.
For high temperature tooling, extrusion dies, forging dies, rams, liners. UNS no 7718.

CARPENTER PYROTOOL 15.
M-32; 0.03 C, 15.0 Cr, 20.0 Co, 2.9 Mo, 0.20 Ni, bal Fe.
Low carbon martensitic alloy having good strength up to 1050°F. Solution treatable.
Hardened; at 1100°F: 145,000 TS; 125,000 YS.
For arbors, cams, collets, dies, fixtures.

CARPENTER PYROTOOL EX.
M-32; 0.05 C, 0.2 Mn, 0.2 Si, 14.0 Cr, 42.5 Ni, 6.0 Mo, 4.0 Co, 3.0 Ti, 1.2 Al, bal Fe.
Austenitic precipitation hardenable nickel base alloy for high strength in 1000-1500°F range.
Hardened: at 1500°F: 106,000 TS; 103,000 YS.
For high temperature tooling, extrusion dies, forging dies, rams, liners.

CARPENTER PYROTOOL M.
M-32; 0.12 C, 0.2 Mn, 0.2 Si, 19.0 Cr, 10.0 Mo, 10.0 Co, 1.0 Al, 2.5 Ti, 2.0 Fe, bal Ni.
Austenitic precipitation hardenable nickel base alloy for operation up to 1500°F.
Hardened; at 1500; 91,000 TS; 84,000 YS.
For high temperature tooling, extrusion dies, forging dies, mandrels, dummy blocks.

CARPENTER PYROTOOL V.
M-32; 0.04 C, 0.25 Mn, 0.25 Si, 14.5 Cr, 27.0 Ni, 1.25 Mo, 3.0 Ti, 0.2 V, bal Fe.
Austenitic precipitation hardenable iron base alloy. Hardened; at 1500°F: 60,000 TS; 49,000 YS.
For high temperature tooling, liners, extrusion dies, forging dies, dummy blocks.

CARPENTER PYROTOOL W.
M-32; 0.05 C, 0.2 Mn, 0.2 Si, 19.5 Cr, 4.25 Mo, 13.0 Co, 3.10 Ti, 1.20 Al, 1.0 Fe, bal Ni.
Austenitic hardenable nickel base alloy, high strength and hardness up to 1500°F.
Hardened; at 1600°F: 76,000 TS; 75,000 YS.
For high temperature tooling, extrusion dies, forging dies. UNS No 07001.

CARPENTER R.D.S.
M-32; 0.70 C, 0.35 Mn, 0.25 Si, 1.0 Cr, 1.75 Ni, bal Fe.
Oil hardening tool steel with extra toughness; for collets, thread rolling dies, punches, knuckle pins, slitting shears. AISI L 6. UNS T6 1206.

CARPENTER S-7.
M-32; 0.50 C, 0.70 Mn, 0.30 Si, 3.25 Cr, 1.40 Mo, bal Fe.
Air hardening tool steel having high impact and shock resistance.
For blanking dies, rivet sets, pneumatic tools. AISI S7.

CARPENTER SAMSON EXTRA PLASTIC MOLD STEEL.
M-32; 0.08 C, 0.40 Mn, 0.30 Si, 2.3 Cr, bal Fe.
Low carbon steel for easy cold hubbing and subsequent case hardening for plastic molds. AISI P5; UNS T51605.

CARPENTER SEAT RING DIE STEEL.
M-32; 0.65 C, 3.0 Cr, 3.0 Mo, 1.0 V, 3.5 W, bal Fe.
Oil or air hardenable modified high speed tool steel; designed for hot forming of automotive valves; good wear resistance at elevated temperatures.

CARPENTER SEVEN STAR.
M-32; 1.00 C, 0.25 Mn, 0.25 Si, 4.0 Cr, 8.75 Mo, 1.75 W, 2.0 V, bal Fe.
Molybdenum high speed steel for cutting tools, blanking dies, drills.
AISI M7; UNS T11307.

CARPENTER SIL. NO. 1.
M-32; 0.45 C, 0.50 Mn, 3.25 Si, 8.5 Cr, bal Fe.
At 77°F: 133,000 TS; 82,500 YS; 22.5 El; 49.0 RA; 269 Brin. At 1000°F: 67,500 TS; 53,000 YS; 40.5 El; 71.0 RA; 125 Brin.
For intake valves, exhaust valve stems; hardenable, heat resistant.

CARPENTER SILICON CORE IRON A.
M-32; 0.05 C, 0.15 Mn, 1.0 Si, bal Fe.
Saturation (B) 21,000; Res. Ind. (Br) 6500; Hc 0.90; Max permeability 4500.
For motor armatures, pole pieces, solenoid switches, relays.

CARPENTER SILICON CORE IRON "A-FM".
M-32; 0.05 max C, 0.15 max Mn, 1.0 max Si, 0.18 P, bal Fe.
Free machining silicon core iron. DC U max 4500 at H = 1.36 Oersted. Hc from 10,000 gausses 0.70/0.80 Oersted. Br from 10,000 gausses 6000 gausses. Saturation at H = 200 21,000 gausses.
For machined magnet cores.

CARPENTER SILICON CORE IRON "B".
M-32; 0.05 C, 0.15 Mn, 2.50 Si, bal Fe.
Saturation (Bs)-Gausses 20,600. Residual Induction (Br) from 10,000 Gausses: 6,000. Coercive force (Hc) from 10,000 Gausses: 0.70. Max permeability: 7000.
For solenoid switches, armatures, pole pieces.

CARPENTER SILICON CORE IRON "B-FM".
M-32; 0.05 C, 0.40 Mn, 2.50 Si, 0.12 P, bal Fe.
Free machining grade of Carpenter Silicon Core Iron "b."
Properties and uses similar.

CARPENTER SILICON CORE IRON C.
M-32; 0.03 C, 0.15 Mn, 4.0 Si, bal Fe.
Saturation (B) 20,000; Res. Ind. (Br) 4000; Hc 0.60; Max permeability 4000.
For motor armatures, pole pieces, solenoid switches, relays.

CARPENTER SOLAR.
M-32; 0.50 C, 0.40 Mn, 1.0 Si, 0.50 Mo, bal Fe.
Water hardening tool steel with more than normal toughness; for chisels, pneumatic tools, rivet busters, screw drivers. AISI S2.

CARPENTER SPECIAL WATER-HARDENING TOOL STEEL.
M-32; 0.70-1.25 C, 0.25 Mn, 0.25 Si, bal Fe.
Water hardening for punches, drills, stamps, jigs, bushings.
AISI W1; UNS G10900 and G10950.

CARPENTER SPEED STAR.
M-32; 0.82 C, 0.3 Mn, 0.25 Si, 4.25 Cr, 5.0 Mo, 6.25 W, 1.8 V, bal Fe.
High speed steel; for roughing tools as lathe tools, milling cutters, end mills, form tools, drills and reamers.
AISI M2; UNS T11302.

CARPENTER STAINLESS 15CR-5NI.
M-32; 0.07 max C, 1.0 max Mn, 1.0 max Si, 0.04 max P, 0.03 max S, 14.0-15.5 Cr, 3.5-5.5 Ni, 2.5-4.5 Cu, 0.045 max N, 0.15-0.45 Cb + Ta, bal Fe.
Martensitic precipitation hardening stainless.
ASM 5658; ASTM A564 (Grade XM-12).
UNS S15500.

CARPENTER STAINLESS 18CR-2NI-12MN.
M-32; 0.15 max C, 11.0-14.0 Mn, 1.0 max Si, 16.5-19.0 Cr, 0.5-2.5 Ni, 0.20-0.45 N, bal Fe.
Ann: 122,000 psi (841 MPa) TS; 68,000 psi (469 MPa) YS; 58 El; 76 RA.
Can be cold worked (40%) to 206,000 psi TS.
Austenitic; for springs, fasteners, pump shafts.

CARPENTER STAINLESS 18-18 PLUS.
M-32; 0.15 C, 17.0-19.0 Mn, 1.0 max Si, 17.0-19.0 Cr, 0.75-1.25 Mo, 0.75-1.25 Cu, 0.4-0.6 N, bal Fe.
Nickel-free high strength austenitic stainless.
Ann, R.T.; 120 ksi (827 MPa) TS; 69 ksi (476 MPa) YS; 65 El; 75 RA.
Good high temperature strength; good corrosion resistance.

CARPENTER STAINLESS 21CR-6NI-9MN.
M-32; 0.03 C, 8.0-10.0 Mn, 1.0 Si, 18.0-21.5 Cr, 5.5-7.5 Ni, 0.15-0.4 N, bal Fe.
Nitrogen strengthened austenitic stainless steel; readily weldable, and non-magnetic after severe cold work.
UNS S21904.

CARPENTER STAINLESS 22CR-13NI-5MN.
M-32; 0.06 max C, 4.0-6.0 Mn, 1.0 max Si, 20.5-23.5 Cr, 11.5-13.5 Ni, 1.5-3.0 Mo, 0.10-0.30 Cb, 0.10-0.30 V, 0.20-0.40 N, bal Fe.
Ann: 120,000 psi (827 MPa) TS; 65,000 psi (448 MPa) YS; 45 El; 65 RA.
Good strength and corrosion resistance; austenitic.
For marine hardware, petrochemical equipment.
UNS S20910.

CARPENTER STAINLESS CUSTOM FLO 302 HQ.
M-32; 0.08 C, 2.0 max Mn, 1.0 max Si, 17.0-19.0 Cr, 8.0-10.0 Ni, 3.0- Cu, bal Fe.
Austenitic stainless steel with low work-hardening tendencies. Annealed: 73,000 TS; 27,000 YS; 65 El.
For upset head bolts, nuts, screws and other cold formed parts. UNS S30430.

CARPENTER STAINLESS NO.5 (TYPE 416).
M-32; 0.15 max C, 1.25 max Mn, 0.15 min S, 1.0 max Si, 12.0-14.0 Cr, bal Fe.
Free machining, air or oil hardenable corrosion resistant steel. As hardened: 160,000-195,000 TS.
For machined parts as studs, bolts, shafts, lead screws, axles for resistance to rust. AISI 416; UNS 541610.

CARPENTER STAINLESS NO. 5-F.
M-32; 0.10 max C, 1.0 max Mn, 0.06 max P, 0.30 Min S, 1.0 max Si, 13.0-14.0 Cr, 0.50 max Ni, 0.40-0.60 Mo, bal Fe.
Corrosion resistant steel, very free machining, not designed for heat treating.
As rolled: 90,000 psi, 63 kg/mm^2 TS; 75,000 psi, 53 kg/mm^2 YS; 15 El.
For valve trim, fuel nozzles, studs.
S41600.

CARPENTER STAINLESS NO 7-MO (TYPE 329).
M-32; 0.20 max C, 23-28 Cr, 2.5-5.0 Ni, 1.0-2.0 Mo, bal Fe. Welded stainless steel tubing, ferrite matrix with some austenite. Annealed: 105.600 TS; 80,000 YS; 25 El.
For heat exchangers, paper & pulp, chemical industries. Good corrosion resistance.
UNS S32900.

CARPENTER STAINLESS NO 10 (TYPE 384).
M-32; 0.08 max C, 15-17 Cr, 17-19 Ni, bal Fe. Annealed: 75,000 TS; 35,000 YS; 55 El; 72 RA; 145 Brin.
For bolts, fasteners, cold headed and upset parts; austenitic stainless, low work hardening.
UNS 538400.

CARPENTER STAINLESS NO. 20 CB-3.
M-32; 0.06 max C, 2.0 max Mn, 1.0 max Si, 19.0-21.0 Cr, 32.5-35.0 Ni, 2.0-3.0 Mo, 3.0-4.0 Cu, Cb + Ta = 8 x C min 1.0 max, bal Fe. Austenitic stainless steel with superior resistance to 10-40% sulphuric acid and other chemicals.
Weldable; for mixing tanks, heat exchangers, process piping, pump shafts and rods.
UNS NO 8020.

CARPENTER STAINLESS NO 309.
M-32; 0.20 max C, 2 max Mn, 22-24 Cr, 12-15 Ni, bal Fe. At 70°F; 80,000 TS; 58,000 YS; 52 El; 71 RA. At 1200°F: 52,000 TS; 31,00 YS; 36 El; 44 RA.
For furnace parts, fire box sheets, high temperature containers, weld wire Heat, oxidation and corrosion resistance.
UNS 30900.

CARPENTER STAINLESS NO 309 S.
M-32; 0.08 max C, 2 max Mn, 22-24 Cr, 12-15 Ni, bal Fe. Annealed: 95,000 TS; 40,000 YS; 45 El; 65 RA; 160 Brin.
For furnace parts, fire box sheets, high temperature containers, weld wire
Heat, oxidation and corrosion resistance.
UNS 30903.

CARPENTER STAINLESS NO 310.
M-32; 0.20 max C, 2 max Mn, 24-26 Cr, 19-22 Ni, bal Fe. Annealed: 95,000 TS; 45,000 YS; 50 El; 65 RA; 185 Brin. At 1200°F: 58,000 TS; 22,000 YS; 32 El; 45 RA.
For furnace parts, heat treating boxes. High oxidation and heat resistance for service up to 2100°F.
UNS S31000.

CARPENTER STAINLESS NO 310 S.
M-32; 0.08 max C, 2 max Mn, 24-26 Cr, 19-22 Ni, bal Fe.
Annealed: 95,000 TS; 45,000 YS; 50 El; 65 RA; 185 Brin.
At 1200°F 58,000 TS; 22,000 YS; 32 El; 45 RA.
For furnace parts, heat treating boxes. High oxidation and heat resistance for service up to 2100°F.
UNS S31008.

CARPENTER STAINLESS NO 317.
M-32; 0.10 max C, 18-20 Cr, 11-15 Ni, 3-4 Mo, bal Fe. Annealed: 85,000 TS; 150 Brin.
For chemical plant equipment; stainless, austenitic.
UNS S31700.

CARPENTER STAINLESS NO 347 (TYPE 347).
M-32; 0.08 max C, 2.0 max Mn, 1.0 max Si, 17.0-10.0 Cr, 9.0-13.0 Ni, Cb - Ta = 10 x C min bal Fe (0.10 max Ta, 0.20 max Cb).
Austenitic, 18-8 stainless steel, weldable and elevated temperature grade.
At 1500°F: 25,000 TS; 17,000 YS; 40 El.
For welded assemblies in food, dairy, beverage, textile and chemical industries, and in elevated temperature operations; weld rods.
AISI 347; UNS S34700.

CARPENTER STAINLESS NO. 348 (TYPE 348).
M-32; 0.08 max C, 2.0 max Mn, 1.0 max Si, 17.0-19.0 Cr, 9.0-13.0 Ni, Cb + Ta = 10 x C min, 0.10 max Ta, 0.20 max Co, bal Fe.
Stabilized austenitic 18-8 type stainless steel; mainly for nuclear applications.
AISI 348; UNS S34800.

CARPENTER STAINLESS NO 404.
M-32; 0.05 max C, 1.0 max Mn, 0.5 Si, 11.0-12.5 Cr, 1.25-2.0 Ni, 0.03 N bal Fe.
Hardenable to 163,000 TS; weldable, tough, corrosion resistant; for stressed parts to 1050°F.

CARPENTER STAINLESS NO 431.
M-32; 0.12-0.17 C, 1.0 max Mn, 1.0 max Si, 1.5-2.5 Ni, 15.5-17.0 Cr, bal
Air or oil hardenable, corrosion resistant steel.
As hardened: Room Temp.: 200,000 TS; 165,000 YS.
As hardened: at 1000°F: 105,000 TS; 20 El.
Designed for highly stressed and elevated temperature bolts, bomb rack fasteners, pump shafts, valve stems. AISI 431; UNS S43100.

CARPENTER STAINLESS NO 440-B (TYPE 440-B).
M-32; 0.75-0.95 C, 1.0 max Mn, 1.0 max Si, 16.0-18.0 Cr, 0.75 max Mo, Fe.
Air or oil hardenable corrosion resistant steel.
Hardenable to 53-58 Rock C; 280,000 TS approx.
For cutlery, ball bearings, scissors.
AISI 440 B; UNS S44003.

CARPENTER STAINLESS NO 440 C.

M-32; 0.95-1.20 C, 1.0 max Mn, 1.0 max Si, 16.0-18.0 Cr, 0.75 max Mo, bal Fe.
Air or oil hardenable corrosion resistant steel. Hardenable to 55-61 Rock C; 285,000 TS approx.
For bushings, cutlery, needle valves, ball and roller bearings and races, ball chuck valves. Resists softening and scaling to about 900°F.
AISI 440 C; UNS S44004.

CARPENTER STAINLESS NO N-1 (TYPE 414).

M-32; 0.15 max C, 1.0 max Mn, 1.0 max Si, 11.5-13.5 Cr, 1.25-2.50 Ni, bal Fe. Air or oil hardening, corrosion resistant steel. As quenched: 210,000 TS; 155,000 YS; 15 El.
For high strength bolts, nuts, studs and other hardware operating up to about 900°F. AISI 414.

CARPENTER STAINLESS TYPE 20 MOLD STEEL.

M-32; 0.33 C, 0.40 C, 0.50 Si, 13.5 Cr, bal Fe.
Air or oil hardenable corrosion resistant steel; hardenable to 48-54 Rock C max; for plastic molds.
(Replaces Carpenter Stainless Type 2 Mold Steel).
AISI 420; UNS S42000.

CARPENTER STAINLESS TYPE 302.

M-32; 0.15 max C, 2.0 max Mn, 1.0 Si, 17.0-19.0 Cr, 8.0-10.0 Ni, bal Fe.
Austenitic 18-8 stainless steel.
Annealed: 85,000 TS; 35,000 YS; 60 El.
For salt water fishing tackle, dairy equipment, ice cream molds, camera parts.
AISI 302; UNS S30200.

CARPENTER STAINLESS TYPE 303.

M-32; 0.12 max C, 2.0 max Mn, 0.03 max P, 0.18-0.35 S, 1.0 max Si, 17 19.0 Cr, 8.0-10.0 Ni, bal Fe.
Free machining austenitic stainless steel.
Annealed: 90,000 TS; 35,000 YS; 50 El; 160 Brin.
For machined parts as shafts, valves, valve bodies, studs, fittings; not hardenable by heat treating.
AISI 303; UNS S30300.

CARPENTER STAINLESS TYPE 303 SE.

M-32; Same as Carpenter Stainless Type 303 except 0.15-0.35 Se replaces the sulphur as free-machining agent.
UNS S30323.

CARPENTER STAINLESS TYPE 304 HN.

M-32; 0.08 max C, 2.0 max Mn, 1.0 max Si, 18.0-20.0 Cr, 8.0-10.5 Ni, 0.16-0.30 N, bal Fe.
Ann: 758 MPa TS; 448 YS; 48 El.
Austenitic stainless; work hardens readily; good temperature resistance.
For aircraft and aerospace, marine shafting.
UNS S20452.

CARPENTER STAINLESS TYPE 304 L.

M-32; 0.03 max C, 2.0 max Mn, 1.0 max Si, 18.0-20.0 Cr, 8.0-12.0 Ni, bal Fe.
Low carbon, austenitic, 18-8 stainless steel, for welding without subsequent heat treat.
Annealed: 75,000 TS; 28,000 YS; 60 El; 150 Brin.
For welded assemblies in laundry, dairy, food and beverage industries. AISI 304 L; UNS S30403.

CARPENTER STAINLESS TYPE 304 PROJECT 70.

M-32; 0.08 max C, 2.0 max Mn, 1.0 max Si, 18.0-20.0 Cr, 8.0-12.0 Ni, bal Fe.
Austenitic 18-8 stainless steel. Annealed: 85,000 TS; 35,000 YS; 60 El.
For parts in food, dairy and beverage industries. AISI 304; UNS S30400.

CARPENTER STAINLESS TYPE 305.

M-32; 0.12 max C, 2.0 max Mn, 1.0 max Si, 1m.0-19.0 Cr, 10.5-13.0 Ni, Fe.
Austenitic stainless steel; work-hardens less than 18-8 type. Annealed: 80,000 TS; 36,000 YS; 56 El; 156 Brin.
Used for severe deep drawing and spinning operations, also cold headed bolt and screws.
UNS 30500.

CARPENTER STAINLESS TYPE 316.

M-32; 0.08 max C, 2.0 max Mn, 1.0 max Si, 16.0-18.0 Cr, 10.0-14.0 Ni, 2.0-3.0 Mo, bal Fe.
Austenitic stainless steel with extra corrosion resistance and better machinability. Project 70.
AISI 316; S31600.

CARPENTER STAINLESS TYPE 316 L.

M-32; 0.03 max C, 2.0 max Mn, 1.0 max Si, 16.0-18.0 Cr, 10.0-14.0 Ni. 2.0-3.0 Mo. bal Fe.
Austenitic stainless steel, low carbon for improved weldability, extra corrosion resistance. Useful to 1600°F.
For pulp handling equipment, photographic chemicals, inks, rayon, rubber, welded assemblies. AISI 316 L; UNS S31603.

CARPENTER STAINLESS TYPE 316N.

M-32; 0.08 C, 2.0 Mn, 1.0 Si, 16.0-18.0 Cr, 10.0-14.0 Ni, 2.0-3.0 Mo, 0.10-0.16 N, bal Fe.
Nitrogen strengthened Type 316 to increase strength with minimum effect on ductility and corrosion resistance.
AISI 316.

CARPENTER STAINLESS TYPE 321.
M-32; 0.08 max C, 2.0 max Mn, 1.0 max Si, 17.0-19.0 Cr, 9.0-12.0 Ni, Ti = 5 x C min, bal Fe.
Austenitic, 18-8 type stainless steel, weldable and elevated temperature grade.
At 1400°F: 30,000 TS; 21,000 YS; 33 El.
For aircraft collector rings, exhaust manifolds, expansion joints, hot chemical process equipment.
AISI 321; UNS S32100.

CARPENTER STAINLESS TYPE 330.
M-32; 0.08 C, 2.0 Mn, 0.75-1.50 Si, 17.0-20.0 Cr, 34.0-37.0 Ni, bal Fe.
Austenitic, non-hardenable, heat and corrosion resistant alloy; weldable and machinable.
For high temperature applications.
AISI 330.

CARPENTER STAINLESS TYPE 347F SE.
M-32; 0.08 C, 2.0 Mn, 1.0 Si, 17.0-19.0 Cr, 9.0-13.0 Ni, Cb + Ta = 10 x C, 0.15 Se, bal Fe.
Improved machinability, good high temperature scale resistance; not recommended for welding.

CARPENTER STAINLESS TYPE 405.
M-32; 0.08 max C, 1.0 max Si, 11.5-14.5 Cr, 0.10-0.30 Al, bal Fe.
Non-hardenable chromium corrosion resistant steel; weldable without hard weld metal; for steam nozzles, partitions, fabricated and welded assemblies for mining equipment and chemical plants.
AISI 405; UNS S40500.

CARPERNTER STAINLESS TYPE 410.
M-32; 0.15 max C, 1.0 max Mn, 10 max Si, 11.5-13.0 Cr, bal Fe.
Air or oil hardenable corrosion resistant steel; as quenched hardness 35-44 Rock C.
For steam turbine buckets, gas turbine compressor blades, nuclear reactor rod mechanisms; weldable.
AISI 410; AMS 5613, 5504; UNS S41000.

CARPENTER STAINLESS TYPE 414.
M-32; 0.15 max C, 1.0 max Mn, 1.0 max Si, 11.5-13.5 Cr, 1.25-1.50 Ni, bal Fe.
Martensitic, hardenable stainless.
For high strength bolts and nuts.
AISI 414; UNS S41400.

CARPENTER STAINLESS TYPE 416.
M-32; 0.15 max C, 1.25 max Mn, 1.0 max Si, 0.15-0.35 S, 12.0-13.5 Cr, bal Fe.
Free machining, air or oil hardenable corrosion resistant steel.
As hardened: 160,000-190,000 TS.
For machined hardware as studs, bolts, gears to resist rust and scaling up to about 800°F.
AISI 416; AMS 5610; UNS S41600.

CARPENTER STAINLESS TYPE 420.
M-32; 0.15 min C, 1.0 max Mn, 1.0 max Si, 12.0-14.0 Cr, bal Fe. Air or oil hardenable corrosion resistant steel. As hardened: 190,000-260,000 TS; 140,000-220,000 YS.
At 1200°F: 48,000 TS; 42,000 YS.
For springs, valve parts, valves, gears, cams, pivots, bearings, plastic molds, magnets.
AISI 420; UNS S42000.

CARPENTER STAINLESS TYPE 420 F.
M-32; 0.15 min C, 12.0-14.0 Cr, 0.60 max Mo, 0.15 min S or Se, bal F
Free machining, air or oil hardenable, corrosion resistant steel.
Properties similar to Carpenter Stainless Type 420 except not recommended for vessels containing gases or liquids under pressure, nor for welded assemblies. Good machinability.
AISI 420 F; UNS S42020.

CARPENTER STAINLESS TYPE 420 MOLD STEEL.
M-32; 0.33 C, 0.40 Mn, 0.50 Si, 13.5 Cr, bal Fe.
Hardenable to 48-52 Rc; resists softening to 900°F. (482°C); corrosion and oxidation resistant.
For plastic molds.
UNS S42000.

CARPENTER STAINLESS TYPE 430.
M-32; 0.12 max C, 1.0 max Mn, 1.0 max Si, 16.0-18.0 Cr, bal Fe.
Non-hardenable chromium corrosion resistant steel; for hub caps, body molding, radiator grills, gas tank caps, hardware. AISI 430; UNS S43000.

CARPENTER STAINLESS TYPE 430 F.
M-32; 0.12 max C, 1.25 max Mn, 0.06 max P, 0.15 min S or Se, 1.0 max Si, 14.0-18.0 Cr, 0.60 max Mo, bal Fe.
Free machining, non-hardenable corrosion resistant steel.
Cold drawn: 90,000 TS; 65,000 YS; 15 El.
For screw machined parts as studs, cap screws, shafts, that are non-rusting.
AISI 430 F; UNS S43020.

CARPENTER STAINLESS TYPE 440A.
M-32; 0.60-0.75 C, 1.0 max Mn, 1.0 max Si, 16.0-18.0 Cr, 0.75 max Mo, bal Fe.
Air or oil hardenable corrosion resistant steel.
Hardenable to 52-56 Rock C; 220,000 min TS.
For pivot pins, dental and surgical instruments, cutlery, valve parts, permanent magnets; resists scaling and softening to about 900°F.
AISI 440A; UNS S44002.

CARPENTER STAINLESS TYPE 440 F.
M-32; 0.95-12.0 C, 16.0-18.0 Cr, 0.75 max Mo, 0.15 S or Se, bal Fe.
Free machining grade of Carpenter Stainless No. 440 C; properties and uses similar.
AISI 440 F; UNS S44020.

CARPENTER STAINLESS TYPE 443.
M-32; 0.20 max C, 18.0-23.0 C, 0.90-1.25 Cu, bal Fe.
Heat resisting, non-hardenable, corrosion resisting steel.
Annealed: 90,000 TS; 50,000 YS; 22 El.
At 1300°F: 26,000 TS; 11,000 YS; 41 El.
For furnace parts, miscellaneous hardware, magnetic cores.

CARPENTER STAINLESS TYPE 446.
M-32; 0.15 max C, 24.0-30.0 Cr, 0.50 max Ni, bal Fe.
Chromium corrosion resistant steel.

CARPENTER STAR MAX.
M-32; 0.81 C, 0.30 Mn, 0.30 Si, 4.0 Cr, 8.5 Mo, 1.1 V, 1.5 W, bal Fe.
General purpose high speed steel for drills, cutting tools.
AISI M1. T11301.

CARPENTER STAR MAX FM.
M-32.
Free machining grade of CARPENTER STAR MAX.

CARPENTER STAR-ZENITH.
M-32; 0.72 C, 0.25 Mn, 0.20 Si, 4.0 Cr, 18.25 W, 1.15 V, bal Fe.
Tungsten high speed steel; for lathe and planer tools, form cutters, milling cutters, broaches, thread chasers, taps.
AISI T1; UNS T12001.

CARPENTER STAR ZENITH LOW CARBON.
M-32; 0.5 C, 18 W, 4 Cr, 1 V, bal Fe.
For cutting tools, drills, punches; high speed steel.

CARPENTER STENTOR.
M-32; 0.90 C, 1.60 Mn, 0.25 Si, bal Fe.
General purpose oil hardening tool steel; for blanking and forming dies, broaches, collets, spindles, thread gages.
AISI 02; UNS T31502.

CARPENTER SUPER SAMSON.
M-32; 0.10 C, 0.3 Mn, 0.2 Si, 5.0 Cr, 0.9 Mo, 0.25 V, bal Fe.
Low carbon, chromium air hardening steel to be case hardened for plastic molds or for die casting dies. AISI P4.

CARPENTER SUPERSTAR.
M-32; 1.08 C, 0.25 Si, 0.25 Mn, 3.75 Cr, 9.5 Mo, 1.5 W, 1.15 V, 8.0 Co, bal Fe.
High carbon Moly-Cobalt high speed steel.
For broaches, taps, endmills, gear cutters.
AISI Type M42.

CARPENTER TEMPERATURE COMPENSATOR "30" (TYPE 1).
M-32; 32% nickel-iron alloy whose magnetic permeability decreses at a controlled rate with increse in temperature,-to compensate for variations in ambient temperature.

CARPENTER TEMPERATURE CAMPENSATOR "30" (TYPES 2, 3, 4 AND 5).
M-32; four types of 30% nickel-iron alloys whose magnetic permeabilities decrease at controlled rates with increase in temperature to compensate for temperature variation by using as shunts in watt-hour meters, speedometer, tachometers, voltage regulators.

CARPENTER TEN STAR.
M-32; 0.85 C, 0.2 Mn, 0.3 Si, 4.0 Cr, 8.0 Mo, 2.0 V, bal Fe.
Molybdenum high speed steel for cutting tools, blanking dies.
AISI M10. T11310.

CARPENTER TGS.
M-32; 0.20 C, 1.30 Mn, 0.20 Si, bal Fe.
Oil hardening carburizing grade steel.
For case hardened gears, guides, cams, and gages.

CARPENTER T-K.
M-32; 0.35 C, 0.3 Mn, 0.3 Si, 3.5 Cr, 9.0 W, 0.4 V, bal Fe.
Air or oil hardenable hot work tool steel; for hot shear blades, hot gripper dies, die casting dies, hot punches, hot extrusion dies.
AISI H21.

CARPENTER V 57.
M-32; 0.05 C, 14 Cr, 26.5 Ni, 1.5 Mo, 3 Ti, 0.25 Al, 0.007 B, bal Fe.
Rolled: 175,000 TS; 125,000 YS; 21 El; 35 RA; 331 Brin.
For jet engine blades and buckets, turbine wheels, torque rings; vacuum melted superalloy, high heat resistance up to 1300°F.
AISI No. 663.

CARPENTER VACUMET KOVAR (NICOSEAL).
M-32; 0.02 max Cx, 0.30 Mn, 0.20 Si, 29.0 Ni, 17.0 Co, bal Fe.
Vacuum melted low-expansion alloy for making hermetic seals with the harder Pyrex glasses and ceramic materials.
70-392°F: 2.89 in/in/°F x 10^{-6}.

CARPENTER VACUMET KOVAR (NICOSEAL) PHOTO-ETCH QUALITY.
M-32; 0.02 max Cx, 0.30 Mn, 0.20 Si, 29.00 Ni, 17.0 Co, bal Fe.
Photo-etching quality strip grade of Carpenter Vacumet Nicoseal.

CARPENTER VACUMET NICOSEAL see KOVAR.

CARPENTER VACUMET NICOSEAL PHOTO-ETCH QUALITY see **KOVAR PHOTO-ETCH QUALITY.**

CARPENTER VACUUM MELTED 52100.
M-32; 1.0 C, 0.3 Mn, 0.25 Si, 1.4 Cr, bal Fe.
Vacuum melted high-carbon, chromium-bearing steel for ball and roller bearings.
UNS G12516.

CARPENTER VEGA.
M-32; 0.70 C, 2.0 Mn, 0.3 Si, 1.0 Cr, 1.35 Mo, bal Fe.
Non-deforming, air-hardening die steel for coining dies, precision tools, gages.
AISI A6; UNS T30106.

CARPENTER VEGA-FM.
M-32; 0.70 C, 2.25 Mn, 0.3 Si, 1.0 Cr, 1.35 Mo, plus alloy sulfides, bal Fe.
Air hardening, non deforming die steel, free-machining grade; for blanking forming and trimming dies. mandrels, punches, precision tools. AISI A 6.

CARPENTER V S M.
M-32; 0.70 C, 1.1 Si, 3 Cr, 5.25 Mo, 0.5 Mn, bal Fe.
Heat treated: 550-640 Brin.
For valve seat inserts and stems, guides, spindles; resists heat checking and wear.

CARPENTER WASPALLOY.
M-32; 0.03-0.10 C, 18-21 Cr, 3.5-5.0 Mo, 12-15 Co, 2.6-3.25 Ti, 1.0-1.5 Al, 0.02-0.12 Zr, 0.003-0.008 B, 2.0 max Fe, bal Ni.
At 70°F: 190,000 TS: 130,000 YS: 22 El; 25 RA.
At 1600°F: 74,000 TS; 73,000 YS; 40 El; 55 RA.
For gas turbine parts, jet engine components; age-hardenable, heat resistant.
UNS N07001; AISI 685.

CARRIAGE WHEEL BEARING.
M-Eng.; 84 Cu, 16 Sn.
For bearings; corrosion eresistant.

CARR'S QUALITY 06S.
M-1410; 0.95 C, 0.25 V, bal Fe.
Carbon-vanadium tool steel.
AISI W2.

CARR'S QUALITY 08S.
M-1410; 1.45 C, 12.5 Cr, 1.0 Mo, 3.0 Co, bal Fe.
High Chrome-Cobalt cold work tool steel.
AISI D5.

CARR'S QUALITY 09B.
M-1410; 0.95 C, 1.25 Mn, 0.50 W, 0.50 Cr, 0.30 V, bal Fe.
Tungsten-Chrome general purpose oil-hardening tool steel.
AISI 01.

CARR'S QUALITY 12S.
M-1410; 0.30 C, 3.25 Cr, 9.5 W, 0.40 V, bal Fe.
Tungsten-Chrome hot work tool steel.
AISI H21.

CARR'S QUALITY 14S.
M-1410; 2.2 C, 12.0 Cr, 0.80 Mo, 0.40 V, bal Fe.
High Carbon-Chrome hot work tool steel.
AISI D4.

CARR'S QUALITY 23S.
M-1410; 2.2 C, 12.0 Cr, bal Fe.
High Carbon-High Chrome cold work tool steel.
AISI D3.

CARR'S QUALITY 24S.
M-1410; 0.24 C, 2.5 Ni, 3.0 Cr, 8.5 W, bal Fe.
Tungsten-Chrome-Nickel hot work tool steel.

CARR'S QUALITY 28F.
M-1410; 0.10 C, 0.30 Mn, 5.0 Cr, 1.0 Mo, 0.30 V, bal Fe.
5% Chrome carburizing steel.

CARR'S QUALITY 32S.
M-1410; 1.0 C, 5.0 Cr, 1.0 Mo, 0.30 V, bal Fe.
5% Chrome-air hardening cold work tool steel.
AISI A2.

CARR'S QUALITY 53S.
M-1410; 0.40 C, 1.0 Si, 5.25 Cr, 1.4 Mo, 1.0 V, bal Fe.
5% Chrome hot work tool steel.
AISI H13.

CARR'S QUALITY 58S.
M-1410; 0.35 C, 1.0 Si, 5.25 Cr, 1.5 Mo, 1.25 W, 0.30 V, bal Fe.
Special 5% Chrome hot work tool steel.
AISI H12.

CARR'S QUALITY 65S.
M-1410; 0.40 C, 1.0 Si, 5.25 Cr, 1.40 Mo, 0.40 V, bal Fe.
5% Chrome hot work tool steel; AISI H11.

CARR'S QUALITY 67S.
M-1410; 1.35 C, 3.75 W, 0.75 C, bal Fe.
Carbon-Tungsten tool steel.
AISI F3.

CARR'S QUALITY 69S.
M-1410; 1.50 C, 12.0 Cr, 0.80 Mo, 0.90 V, bal Fe.
High Carbon-Chrome cold work tool steel.
AISI D2.

CARR'S QUALITY 74S.
M-1410; 0.30 C, 2.3 Cr, 3.0 Mo, 4.25 W, 0.60 V, bal Fe.
4 1/2 Tungsten-Chrome hot work tool steel.

CARR'S QUALITY 82S.
M-1410; 0.33 C, 1.0 Si, 3.0 Cr, 2.8 Mo, 1.0 W, 0.90 V, 3.0 Co, bal Fe.
Chrome-Molybdenum-Cobalt hot work tool steel.

CARR'S QUALITY BCC.
M-1410; 0.50 C, 1.0 Si, 2.25 W, 1.2 Cr, 0.25 V, bal Fe.
Heavy duty general purpose tool steel.
AISI S1.

CARR'S QUALITY BCD 37.
M-1410; 0.37 C, 2.0 W, 1.6 Cr, 0.30 V, bal Fe.
Non-tempering chisel steel.

CARR'S QUALITY BLUE LABEL.
M-1410; 0.70 C, 0.30 Mn, bal Fe
Water hardening plain carbon tool steel.
AISI W1.

CARR'S QUALITY 'O' BRAND.
M-1410; 0.35 C, 1.50 Mn, bal Fe.
Carbon-Manganese steel.
Similar to AISI 1335.

CARR'S QUALITY P 20.
M-1410; 0.32 C, 0.80 Mn, 1.6 Cr, 0.40 Mo, bal Fe.
Prehardened Chrome-Molybdenum plastic mould steel.
AISI P20.

CARR'S QUALITY P 151.
M-1410; 0.15 C, 0.80 Mn, bal Fe.
Low carbon steel for case hardening.

CARR'S QUALITY P 153.
M-1410; 0.17 C, 0.45 Mn, 2.0 Ni, 0.25 Mo, bal Fe.
2% Nickel-Molybdenum steel for case-hardening intermediate size sections.

CARR'S QUALITY P 155.
M-1410; 0.15 C, 0.45 Mn, 3.4 Ni, 1.0 Cr, bal Fe.
3% Nickel-Chrome carburizing steel; for heavy sections.

CARR'S QUALITY P 158.
M-1410; 0.15 C, 0.40 Mn, 4.25 Ni, 1.2 Cr, 0.30 Mo, bal Fe.
4 1/2% Ni-Cr-Mo carburizing steel; for very heavy sections.

CARR'S QUALITY P 256.
M-1410; 0.55 C, 0.70 Mn, bal Fe.
Bolster steel.

CARR'S QUALITY P 280.
M-1410; 0.60 C, 0.60 Mn, 0.60 Cr, bal Fe.
Carbon-Chrome steel.

CARR'S QUALITY P 552.
M-1410; 0.30 C, 0.50 Mn, 4.25 Ni, 1.25 Cr, 0.30 Mo, bal Fe.
Nickel-Chrome-Molybdenum structural steel; deep hardening; plastic moulds.

CARR'S QUALITY P 553.
M-1410; 0.40 C, 0.60 Mn, 1.5 Ni, 1.1 Cr, 0.30 Mo, bal Fe.
Nickel-Chrome-Molybdenum steel.
Similar to AISI 4340.

CARR'S QUALITY P 558.
M-1410; 0.30 C, 0.60 Mn, 2.5 Ni, 0.70 Cr, 0.50 Mo, bal Fe.
Nickel-Chrome-Molybdenum structural steel; deep hardening; tough.

CARR'S QUALITY P 564.
M-1410; 0.30 C, 0.60 Mn, 3.0 Ni, 0.75 Cr, bal Fe.
Nickel-Chrome structural steel.

CARR'S QUALITY P 602.
M-1410; 0.40 C, 0.65 Mn, 1.1 Cr, 0.30 Mo, bal Fe.
Chrome-Molybdenum hot work tool steel.

CARR'S QUALITY P 609.
M-1410; 0.40 C, 0.80 Mn, 1.0 Cr, bal Fe.
1% Chrome hot work tool steel.

CARR'S QUALITY P 618.
M-1410; 0.40 C, 0.30 Si, 0.50 Mn, 0.30 Ni, 3.0 Cr, 1.0 Mo, 0.25 V, bal Fe.
Chrome-Molybdenum-Vanadium hot work tool steel.

CARR'S QUALITY P 704.
M-1410; 0.95 C, 1.65 Mn, 0.20 Cr, bal Fe.
Manganese oil-hardening general purpose tool steel.
AISI O2.

CARR'S QUALITY P 720.
M-1410; 1.0 C, 0.40 Mn, 1.40 Cr, bal Fe.
Carbon-Chrome oil-hardening tool steel.
AISI L1.

CARR'S QUALITY P 1000.
M-1410; 0.10 C, 0.40 Mn, 0.40 Ni, 13.0 Cr, bal Fe.
Martensitic Chrome stainless.
AISI 410.

CARR'S QUALITY P 1001.
M-1410; 0.16 C, 0.40 Mn, 0.40 Ni, 13.0 Cr, bal Fe.
Martensitic Chrome stainless steel.

CARR'S QUALITY P 1002.
M-1410; 0.21 C, 0.40 Mn, 0.40 Ni, 13.0 C, bal Fe.
Martensitic Chrome stainless.
AISI 420.

CARR'S QUALITY P 1003.
M-1410; 0.30 C, 0.40 Mn, 0.40 Ni, 13.0 Cr, bal Fe.
Martensitic Chrome stainless steel.
Similar to AISI 420.

CARR'S QUALITY P 1008.
M-1410; 0.35 C, 0.40 Mn, 1.0 Ni, 13.0 Cr, bal Fe.
Stainless mould steel.

CARR'S QUALITY P 1009.
M-1410; 0.18 C, 0.60 Mn, 2.1 Ni, 17.0 Cr, bal Fe.
Martensitic stainless steel.
AISI 431.

CARR'S QUALITY P 1010.
M-1410; 0.16 max C, 18.0 Cr, 8.0 Ni, bal Fe.
Austenitic stainless.
AISI 302.

CARR'S QUALITY P 1011.
M-1410; 0.15 max C, 18.0 Cr, 8.0 Ni + Ti, bal Fe.
Stabilized austenitic stainelss.
AISI 321.

CARR'S QUALITY P 1012.
M-1410; 0.15 max C, 18.5 Cr, 10.0 Ni + Ti, bal Fe.
Stabilized austenitic stainless.
AISI 321.

CARR'S QUALITY P 1013.
M-1410; 0.15 max C, 13.0 Cr, 13.0 Ni, bal Fe.
Austenitic stainless steel.

CARR'S QUALITY P 1014.
M-1410; 0.08 max C, 19.0 Cr, 9.5 Ni, bal Fe.
Austenitic stainless steel.
AISI 304.

CARR'S QUALITY P 1015.
M-1410; 0.15 max C, 18.0 Cr, 8.0 Ni, + Nb, bal Fe.
Stabilized austenitic stainless.
AISI 347.

CARR'S QUALITY P 1016.
M-1410; 0.15 max C, 18.5 Cr, 10.0 Ni, + Nb, bal Fe.
Stabilized austenitic stainless.
AISI 347.

CARR'S QUALITY P 1017.
M-1410; 0.12 max C; 18.5 Cr, 10.0 Ni, 2.0 Mo, + Ti or Nb, bal Fe.
Stabilized austenitic stainless.

CARR'S QUALITY P 1018.
M-1410; 0.12 max C, 18.5 Cr, 10.0 Ni, 3.0 Mo, + Ti or Nb, bal Fe.
Stabilized austenitic stainless.
AISI 316.

CARR'S QUALITY RED LABEL.
M-1410; 0.90 C, 0.30 Mn, bal Fe.
Water hardening plain carbon tool steel.
AISI W1.

CARR'S QUALITY ST BRAND.
M-1410; 0.70 C, 0.30 Mn, bal Fe.
Water hardening plain carbon tool steel.

CARR'S QUALITY YELLOW LABEL.
M-1410; 1.15 C, 0.30 Mn, bal Fe.
Water hardening plain carbon tool steel.
AISI W1.

CARTER WHITE GOLD.
M-Eng.; 83.3 Au, 16.7 Ni.
For jewelry, ornaments; corrosion resistant.

CARTOS.
M-64; 0.93 C, 0.4 Mn, 0.6 Si, 0.09 V, bal Fe.
Annealed: 250 Brin.
Ht. Tr.: 620 Brin.
For shear knives, trimmers.
Oil hardening, wear resistant.

CARTOS 2V72.
M-64; 0.8 C, 0.4 Mn, 0.6 Si, 0.1 V, bal Fe.
Water hardening tool steel; AISI W 2.

CARTOS 2V90.
M-64; 0.9 C, 0.4 Mn, 0.6 Si, 0.1 V, bal Fe.
Water hardening tool steel; AISI W 2.

CARTRIDGE BRASS.
M-33, M-1789; 70 Cu, 30 Zn.
Annealed: 47,000 TS; 15,000 YS; 62 El.
Rolled: 76,000 TS; 64,000 YS; 8 El.
For cartridge cases, primer cups, springs; deep drawing.

CARTRIDGE BRASS 42.
M-8; 70 Cu, 30 Zn.
1/4 H-temper: 54,000 TS; 40,000 YS; 43 El; 63 Brin.
1/2 H-temper: 62,000 TS; 52,000 YS; 23 El; 120 Brin.
H-temper: 76,000 TS; 63,000 YS; 8 El; 156 Brin.
For cartridges, primers, shot shells, tanks, fasteners, deep drawing, high ductility.

CARTRIDGE BRASS 260.
M-279, M-8; 70 Cu, 30 Zn.
Ann: 45,000-65,000 TS; 10,000-34,000 YS; 35-67 El.
All purpose deep drawing brass. For all deep drawn brass parts.
Cold rolled: 49,000-110,000 TS; 20,000-100,000 YS; 1-56 El.
For less drastic forming, spring clips.

CARTRIDGE GILDING.
M-U.S.; 93 Cu, 7 Zn.
For cartridge shells, ornaments, base for fire enameling; gliding metal.

CARTUN.
M-432; 1.35 C, 2.75 W, bal Fe.
For fast finishing tools, cutters; water or oil hardened.

CAR-VAN.
M-750; 0.75-1.1 C, 0.2 V, bal Fe.
For tools, dies, shear blades, punches; water hardening.

CAR-VAN SPECIAL.
M-750; 0.80-1.35 C, 0.2 V, bal Fe.
For tools, dies, drills, cutters; water hardening.

CASAR.
M-1312; 1.3 C, 0.85 Mo, 3.8 V, 12 W, 4.3 Cr, bal Fe.
For lathe and planer tools, reamers, broaches, taps; high speed steel.

CASAR C.
M-1312; 1.35 C, 4.2 Cr, Mo, W, V, bal Fe.
For blanking and forming dies, engravers' tools; high speed steel.

CASCADE.
M-73; 0.13 C, 0.12 Cr, 4 Ni, 0.5 V, bal Fe.
207-350 Brin.
For plastic mold die zinc die casting dies; can be nitrided.

CASE DIE.
M-373; 0.45 C, 1.0 W, 0.20 Mo, 0.90 Cr, bal Fe.
Shock resisting tool steel.
AISI S3.

CASINO.
0.7 C, 18 W, 4 Cr, 1 V, bal Fe.
For shapers, cutters; high speed steel.

CAST 14-14.
1.0 C, 0.8 Mn, 3.0 Si, 14.5 Cr, 14.5 Ni, bal Fe.
Exhaust valve steel.
SAE EV 10.

CASTALOY.
M-157; 1.5-1.6 C, 12-14 Cr, 0.7-0.8 Mo, 0.45-0.55 Mn, bal Fe.
For dies; easily machined Rockwell "C"-26-30.

CASTALOY.
M-378; 4.1 Al, 0.04 Mg, bal Zn.
Die cast: 40,300 TS; 5 El; 74 Brin.
For chemical apparatus, clamps, holders; SAE903.

CAST COMPOSITE.
M-210; C, bal Fe.
For tools, blanking dies; water hardening.

CAST COMPOSITE.
M-365; 1.0-1.05 C, bal Fe.
For tools, drills, taps; water hardening.

CASTEC.
M-717.
For AC-DC all-position welding of cast iron; machinable deposits; 53,000 psi TS.

CASTEEL 15.
M-1166; 0.13-0.18 C, 0.6 Mn, 0.4 Si, bal Fe.
Normalized: 57,000-65,000 TS; 32,000-37,000 YS; 30-35 El; 50-55 RA; 110-125 Brin.
For gears, pinions, shafts; case hardening.

CASTEEL 22.
M-1166; 0.18-0.25 C, 0.6 Mn, 0.4 Si, bal Fe.
Normalized: 65,000-75,000 TS; 38,000-45,000 YS; 25-30 El; 45-50 RA; 130-145 Brin.
For gears, pinions, housings, castings; case hardening.

CASTEEL 28.
M-1166; 0.26-0.32 C, 0.7 Mn, 0.4 Si, bal Fe.
Normalized: 75,000-85,000 TS; 45,000-55,000 YS; 20-25 El; 36-45 RA; 140-160 Brin.
For structural and machinery parts, castings; water hardened.

CASTEEL 40.
M-1166; 0.37-0.43 C, 0.7 Mn, 0.4 Si, bal Fe.
Normalized: 85,000-95,000 TS; 55,000-65,000 YS; 15-20 El; 25-35 RA; 165-180 Brin.
Heat treated: 90,000-110,000 TS; 62,000-70,000 YS; 18-23 El; 40-50 RA; 180-220 Brin.
For gears, pinions, shafts, housings, castings; water hardened.

CASTINGWELD.
M-118; C, bal Fe.
For welding electrodes; non-machinable, welds on cast iron.

CAST-I-NICKEL.
M-712; Ni.
For cast iron welding rod; flux coated, machinable.

CAST-I-STEEL.
M-712; 0.10-0.14 C, 0.4-0.6 Mn, 0.02 max Si, bal Fe.
For nonmachinable welds on cast iron; coated arc electrode.

CASTOR 3.
M-111; 0.72 C, 4.5 Cr, 18 W, 1.2 V, bal Fe.
For tools, cutters; high speed steel.

CASTOR 7.
M-111; 0.8 C, 4.5 Cr, 1.25 Mo, 18.5 W, 2.5 Co, 1.6 V, bal Fe.
For tools, cutters; high speed steel.

CASTOR 8.
M-111; 0.8 C, 4.5 Cr, 1.25 Mo, 18.5 W, 5.5 Co, 1.6 V, bal Fe.
For tools, cutters; high speed steel.

CASTOR 9.
M-111; 0.8 C, 4.5 Cr, 1 Mo, 18.5 W, 10.5 Co, 1.6 V, bal Fe.
For tools, cutters; high speed steel.

CASTOR 1939.
M-111; 0.86 C, 2.8 Co, 4.3 Cr, 0.85 Mo, 2 V, 12 W, bal Fe.
For lathe and planer tools, reamers, broaches; high speed steel.

CAT 497 STARRETT PRECISION GROUND DIE STOCK see **STARRETT NO. 497.**

CATALOY.
M-481; Pb-Cu-Sn.
For catalyst; for non-ferrous metals.

CATALOY NO. 1.
M-481; 20 Pb, 72 Cu, 4 Sn, 4 other elements.
26,750 TS; 18 El; 25 RA; 50 Brin.
For bearings; heavy duty.

CATALOY NO. 2.
M-481; 25 Pb, 68 Cu, 3 Sn, 4 others.
21,350 TS; 13 El; 25 RA; 50 Brin.
For bearings; heavy duty.

CATALOY NO. 3.
M-481; 30 Pb, 63 Cu, 3 Sn, 4 others.
Cast: 19,550 TS; 11.7 El; 19.5 RA; 39 Brin.
For bearings; heavy duty.

CATALOY NO. 4.
M-481; 25 Pb, 70 Cu, 1 Sn, 4 others.
Cast: 13,700 TS; 10 El; 9.7 RA; 34 Brin.
For bearings; heavy duty.

CATALOY NO. 5.
M-481; 40 Pb, 52 Cu, 8 others.
Cast: 9,910 TS; 5.5 El; 9.8 RA; 27 Brin.
For bearings; heavy duty.

CATALOY NO. 6.
M-481; 40 Pb, 51 Cu, 1 Sn, 8 others.
Cast: 10,500 TS; 5.5 El; 19.5 RA; 30 Brin.
For bearings; heavy duty.

CATALOY NO. 7.
M-481; 50 Pb, 42 Cu, 8 others.
Cast: 6,400 TS; 10 El; 8.6 RA; 13 Brin.
For bearings; heavy duty.

CATALOY NO. 8S.
M-481; 13 Pb, 80 Cu, 5 Sn, 2 others.
Cast: 30,700 TS; 18.8 El; 24.9 RA; 50 Brin.
For bearings; heavy duty.

CATALOY NO. 9N.
M-481; 11 Pb, 80 Cu, 5 Sn, 4 others.
Cast: 31,000 TS; 14.1 El; 8.2 RA; 53 Brin.
For bearings; heavy duty.

CATARACT METAL.
M-92; 19-21 Ni, 16-20 Zn, 2.5 max Fe, 2.5 max Pb, bal Cu.
Bar: 45,000-55,000 TS; 24,000 YS; 22 El; 28 RA.
For machinery parts, chemical apparatus, valves on superheated steam lines. Corrosion resistant.

CAT BRAND 14% W.
M-1409; 0.65 C, 14.5 W, 3.75 Cr, 0.5 V, bal Fe.
For drills, punches, taps, piercing and blanking dies; high speed steel.

CAT BRAND S.H.X.
M-1409; 0.65 C, 15.5 W, 3.9 Cr, 0.5 V, bal Fe.
For slitting saws; high speed steel.

CATHALOY A-31.
M-945; 3.7-4.2 W, 0.10 max Cu, 0.01-0.06 Mg, 0.02-0.06 Si, bal Ni.
Tempered: 85,000-140,000 TS; 15,000-125,000 YS; 50-4 El; 150-350 Brin.
For electronic valves; cathode nickel.

CATHALOY P-50.
M-945; 0.04 max Cu, 0.05 max Fe, 0.05 max C, 0.01 max Ti, bal Ni.
Tempered: 75,000-120,000 TS; 15,000-110,000 YS; 50-4 El; 125-250 Brin.
For cathodes in electron tubes; ASTM Alloy 22.

CATHALOY P-51.
M-945; 3.75-4.25 W, 0.04 max Cu, 0.05 max Fe, bal Ni.
Tempered: 85,000-140,000 TS; 65,000-125,000 YS; 50-4 El; 150-350 Brin.
For electronic tubes, cathodes; passive cathode material.

CATHODE NICKEL.
M-61; 96-99 Ni, selected additives.
Annealed: 60,000 TS.
For cathodes in electron tubes; several types.

CATHOLOY A-33.
M-945; 1.75-2.25 W, 0.05-0.10 Mn, 0.05-0.10 C, 0.04-0.8 Zr, bal Ni + Co.
Tubing for electron tubes; all purpose cathode. Active.

CATHOLOY A-34.
M-945; 3.75-4.25 W, 0.01-0.06 Mg, bal Ni + Co; low Mn, Si, C.
Tubing for electron tubes subject to shock, vibration or use above rated heater voltage. Active.

CATHOLOY P-52.
M-945; 0.40-0.70 Cb, 1.5-2.5 Mo, bal Ni + Co.
Tubing as non-emitting alloy for disc cathode shanks in electron tubes. Passive.

CATHOLOY P-53.
M-945; Pure Ni + Co.
Tubing for low sublimation disc cathode shanks. Passive.

CAV1.
M-495; 0.6 C, 0.9 Si, 0.9 Mn, 1.35 Cr, 0.15 V, bal Fe.
For tools, dies; shock resistant.

CAV2.
M-459; 0.5 C, 1.5 Si, 0.75 Mn, 1.2 Cr, 0.15 V, bal Fe.
For tools, dies; shock resistant.

CAV3.
M-459; 0.35 C, 1.5 Si, 0.7 Mn, 1.1 Cr, 0.15 V, 1.5-7 W, bal Fe.
For tools, dies; shock resistant.

CAZIN.
 Eng.; 82.6 Cd, 17.4 Zn.
 For solder for steel; M.P. 263°C.

CB-1.
 M-US; 30 W, 1 Zr, 0.06 C, 0.032 N, bal Cb.
 Similar to XB-88.

CB-1ZR same as **WAH CHANG WC-1ZR.**

CB-10HF-1TI-0.5 ZR.
 M-1596, M-1537; 9-11 Hf, 0.7-1.3 Ti, 0.7 max Zr, bal Cb.
 Cold rolled: 105,000 TS; 96,500 YS; 5 El.
 Recrystallized: 59,000 TS; 45,400 YS; 26 El.
 For high temperature applications, space vehicles.
 Good TIG weldability.

CB-10W-1ZR-0.1C see **COLUMBIUM D-43.**

CB 10W-2.5ZR see **WAH CHANG 10W-2.5ZR.**

CB 10W-2.5ZR see **COLUMBIUM 10W-2.5ZR.**

CB-10W-10HF see **COLUMBIUM C-129.**

CB 12 see **FANSTEEL 80.**

CB-15W-5MO-1ZR see **COLUMBIUM 15W-5MO-1ZR.**

CB-20W-10TI-6MO see **COLUMBIUM D-41.**

CB-28 TA-10W-1ZR see **FANSTEEL FS-85.**

CB-28W-2HF-0.07C see **COLUMBIUM XB-88.**

CB = 33TA-1ZR see **FANSTEEL 82 (NOW DISCONTINUED).**

CB 45 see **STELCO CB 45-60.**

CB 74.
 M-48; 5 Zr, 10 W, bal Cb.
 At 70°F: 88,000 TS; 70,000 YS; 26% El.
 At 2200°F: 39,000 TS; 33,000 YS; 25% El.
 (Recrystallized 1 hr. a 2700°F).
 For space vehicles, nuclear reactors. Good combination of density, strength and oxidation resistance at high temperatures.

CB132 see **COLUMBIUM CB-132M.**

CB-751 see **FANSTEEL 80.**

CB-752 see **COLUMBIUM CB-752.**

C.B.F. CHROMIUM BRONZE.
 M-452; 1.5-3.5 Sn, 0.5-2.0 Cr, 0.5-3 Fe, 0.5 max Mn, bal Cu.
 47,000 TS; 21,000 YS; 30-40 El; 90-130 Brin.
 For engine valve guides, gears, rock drill twist nuts; heavy duty.

C-BRITE 29-4.
 M-663; 29 Cr, 4 Mo, balance Fe.

CBS 600 see **TIMKEN CBS 600.**

CB-TZM.
 M-U.S.; 1.5 Cb, 0.5 Ti, 0.3 Zr, 0.03 C, bal Mo.

CB/V 42.
 M-1124; 0.21 C, 1.35 Mn, 0.30 Si, 0.005 min Cb, 0.01 min V, bal Fe.
 Wrought: 60,000 psi TS; 42,000 psi YS; 24 El.
 HSLA steel; ASTM A572.

CB/V 45.
 M-1253, M-1124; 0.22 max C, 1.35 max Mn, 0.005 max Cb, 0.01 max V, bal Fe.
 45,000 psi min YS, HSLA steel.

CB/V 50.
 M-1253, M-1124; 0.23 max C, 1.35 max Mn, 0.005 max Cb, 0.01 max V, bal Fe.
 50,000 psi min YS. HSLA steel.

CB/V 55.
 M-1124; 0.25 C, 1.35 Mn, 0.30 Si, 005 min Cb, 0.01 min V, bal Fe.
 wrought: 70,000 psi TS; 55,000 psi YS; 20 El.
 HSLA; ASTM A572.

CB/V 60.
 M-1253, M-1124; 0.26 max C, 1.35 max Mn, 0.005 max Cb, 0.01 max V, bal Fe.
 60,000 psi min YS. HSLA steel.

CB/V 65.
 M-1124; 0.26 C, 1.35 Mn, 0.30 Si, 0.005 min Cb, 0.01 min V, bal Fe.
 Wrought: 80,000 psi TS; 65,000 psi YS; 15 El.
 HSLA steel; ASTM A572.

CC/1 BIS.
 M-1488; 0.15 max C, 1.0 max Mn, 1.0 max Si, 11.5-14.0 Cr, bal Fe.
 Stainless casting; Ann: 180-220 Brin.
 For petroleum and food industries.
 AFNOR Z12C13M; ACI CA 15.

CC/2.
 M-1488; 0.25-0.35 C, 1.0 Max Mn, 1.0 max Si, 13.0-15.0 Cr, bal Fe.
 Stainless casting; Ann: 250-000 Brin.
 Faucets and pumps for paper industries.
 AFNOR Z30C13M; similar to ACI CA-40.

CC/2C.
 M-1488; 0.35-0.45 C, 1.0 max Mn, 1.0 max Si, 13.0-15.0 Cr, bal Fe.
 Stainless casting; Ann: 260-300 Brin.
 For pump parts in paper industries.
 AFNOR Z38C13M; W.Nr. 1.4034.

CC/4 BIS.
 M-1488; 0.15-0.25 C, 1.0 max Mn, 1.0 max Si, 1.5-3.0 Ni, 15.0-18.0 Cr, bal Fe.
 Stainless casting; Ann: 275-310 Brin.
 For parts subject to seawater or to dilute organic acids.
 AFNOR Z 20 Cn 17.2 M; similar to AISI 431.

CC 4000.
 M-1488; 0.08-0.15 C, 1.0 max Mn, 1.0 max Si, 0.50-1.50 Ni.
 Stainless casting; annealed: 170-290 Brin.
 For pumps and faucets handling cold dilute organic acids in food industries.
 AFNOR Z12CN13M; W.Nr. 1.4008.

CC 4027.
 M-1488; 0.18-0.25 C, 1.0 max Mn, 1.0 max Si, 12.5-14.5 Cr, bal Fe.
 Stainless casting; ann 200-250 Brin.
 Pumps and faucets for food industries.
 AFNOR Z20C13M; similar to ACI CA-40.

C C M see **SIMONDS CCM.**

CCM 5.
 M-1290; 0.90-1.05 C, 0.20-0.40 Si, 0.40-0.70 Mn, 4.8-5.5 Cr, 0.90-1.20 Mo, 0.10-0.30 V, bal Fe.
 Cold work tool steel; W.-Nr. 1.2363.

CCN.
 M-1290; 1.90-2.20 C, 0.20-0.40 Si, 0.20-0.40 Mn, 11.0-12.0 Cr, bal Fe
 Cold work tool steel.
 W.-Nr. 1.2080.

CCNW.
 M-1290; 2.0-2.25 C, 0.25-0.40 Si, 0.20-0.40 Mn, 11.0-12.0 Cr, 0.60-0.8 W, bal Fe.
 Cold work tool steel.
 1.2436.

CCR.
 M-772; 2.25 C, 13 Cr, bal Fe.
 For tools, dies; non-deforming.

CCS.
 M-1083; 0.5-1.2 Cr, 0.01-0.1 Si, 0.02-0.05 S, bal Cu.
 Bar: 54,000-77,000 TS; 15-30 El; 110-165 Brin.
 Forged: 52,000-67,000 TS; 17-32 El; 105-135 Brin.
 For switchgear and resistance welding equipment, rotor bars, 78-85% electrical conductivity.
 BS 4577, ISO 5182 and RWMA alloy A/2/1.

CCS/Z.
 M-1083; 0.5-1.2 Cr, 0.1-0.2 Zr, 0.01-0.1 Si, bal Cu.
 Bar: 54,000-77,000 TS; 15-30 El; 110-165 Brin.
 Forged: 52,000-67,000 TS; 17-32 El; 105-135 Brin.
 For resistance welding equipment, rotor bars, 78-85% electrical conductivity.
 BS 4577, ISO 5182 and RWMA alloy A/2/2.

CCZ.
 M-1405; 0.35 C, 3.0 Cr, 2.75 Mo, 0.60 V, 3.0 Co, bal Fe.
 Hot work tool steel; for die-casting dies and hot extrusion dies.

CD01.
 M-1806.
 Sintered carbide tool material.
 For die insert requiring extreme wear resistance, as wire drawing die or pressing abrasive material.

CD-4M CU see **COOPER ALLOY CD4M CU** and **OHIDLOY CD4M CU.**

CD 18 see **CARBIDIE CD 18.**

CD20 AND CD25.
 M-1806.
 Sintered carbide tool material.
 For general purpose wire drawing dies and for powder metal pressing dies.

CD30.
 M-1806.
 Sintered carbide tool material.
 Hardest cutanit grade recommended for lamination die work.

CD40.
 M-1806.
 Sintered carbide tool material.
 General purpose insert material for bar and tube drawing dies, and for press tool dies and punches.

CD50.
 M-1806.
 Sintered carbide tool material.
 For swaging dies, dies for cold heading, bar and tube dies, notching dies.

CD55 F.
 M-1806.
 Sintered carbide tool material.
 For lamination dies and punches.

CD55, CD60 AND CD65.
 M-1806.
 Sintered carbede tool materials.
 For cold heading and extrusion dies in nut, bolt and fastener manufacture. CD65 is the toughest; CD55 the most wear resistant.

CDA see **COOPER DEVELOPMENT NO. IN INDEX.**

CDC MANGANESE NO. 715.
 M-1187; Mn, Ni, bal Cu.
 For springs; corrosion resistant.

C.D.C. MANGANESE ALLOY NO. 762.
 M-1187; 62-65 Mn, bal Cu.
 Quenched and aged: 78,000 TS; 38,000 YS; 35 El; 70-75 Rock B.
 Damping capacity at torsile stress of 4000 psi-30.
 High electrical resistivity and damping capacity.

C.D.C. MANGANESE ALLOY NO. 780.
M-1187; 80 Mn, 20 Cu.
Quenched: 68,000 TS; 24,000 YS; 35 El.
Cold rolled: 130,000 TS; 115,000 YS; 9 El.
High electrical resistivity and high damping properties.

C.D.C. NO. 720.
M-1187; 60 Cu, 20 Ni, 20 Mn,
Hardenable by cold work and/or heat treatment.
Ann: 98,000 TS; 80,000 YS; 30 El; 140 VHN.
Hardened: 200,000-220,000 TS; 475-515 VHN.
For springs, diaphragms.

C.D.C. NO. 772.
M-1187; 72 Mn, 18 Cu, 10 Ni.
Wrought: 115,000 TS; 95,000 YS; 6.5 El; 220 VHN.
Elec. res.: 1050 ohms per cir. mil. ft.
Low temp. coefficient of resistance.
For low temp. resistance application, rheostats, circuit breaker parts, misc. elec. appliance controls.

CDV.
M-1744; 0.39 C, 1.0 Si, 1.5 Mo, 5.0 Cr, 1.0 V, bal Fe.
Chromium-Vanadium hot work steel, for forging dies for non-ferrous metals

C.D.W.
M-1116; 0.90 C, 4.25 Cr, 1.5 W, 4.25 Mo, 1.30 V, bal Fe.
Air or oil harden to 62-65 Rc.
For drills, reamers, broaches, threading tools.
AFNDR: Z 90 CDWV 04.04.

CEC IMPACT.
M-35; 0.58 C, 0.85 Mn, 1.95 Si, 0.30 Cr, 0.25 V, bal Fe.
Oil or water hardening shock resisting tool steel.
For pneumatic tools, chisels, rivet sets.
AISI S4.

CECO.
M-Eng.; 62.5 Cu, 32 Pb, 4.6 Sn, 0.9 Ni.
18,000 TS; 3 El; 0.35 RA; 52 Brin.
For bearings, bushings; heavy duty.

CECOLLOY A.
M-474; 3.0 TC., 0.5 Mo, 0.60 Ni, bal Fe.
40,000-60,000 TS.
For anvils, frames for forges, forming dies, steam cylinder liners, large valves, series of synthetic alloys; made in air furnaces.

CECOLLOY B.
M-474; 2.8 TC., 0.5 Mo, 0.35 Cr, bal Fe.
For liners, frames, castings, valves; shock resistant.

CEKAS.
M-U.S.; 11.2 Cr, 59.7 Ni, 2 Mn, bal Fe.
For heat and corrosion resisting parts; stainless and corrosion resistant.

CEKAS.
M-1327; 0.16 C, 1.05 Cr, 0.25 Mo, 0.65 Mn, bal Fe.
For gears, bolts, machine tool parts; case hardened, tough.

CEKAS EXTRA 2.
M-1327; 0.10 max C, 1.3 max Si, 1.0 max Mn, 1.2 max Co, 19-22 Cr, 5.5 max Al, bal Fe.
For high temperature equipment.
W.-Nr. 1.4767.

CEKAS EXTRA 3.
M-1327; 0.10 max C, 1.0 max Si, 0.6 max Mn, 22-25 Cr, 0.6 max Al, 1.0 max Ti, bal Fe.
For high temperature equipment.
W.-Nr. 1.4765.

CEKAS M151.
M-1327; 0.15 C, 13 Cr, 1 Mo, bal Fe.
Annealed: 75,000 TS; 40,000 YS; 30 El; B 82 Rock.
Ht. Tr.: 135,000 TS; 105,000 YS; 10 El; C 30 Rock.
For flat springs, hardware, fittings, tableware, chemical plant and oil refining equipment.
Corrosion resistant. Hardenable.

CEKAS M152.
M-1327; 0.20 C, 13 Cr, 1 Mo, bal Fe.
Annealed: 95,000 TS; 50,000 YS; 25 El; B 92 Rock.
Oil hardened: 250,000 TS; 215,000 YS; 8 El; C 50 Rock.
For surgical instruments, cutlery, pivots, ball bearings, gears.
Corrosion resistant, hardenable.

CELEBRATED 101.
M-1409; 0.75 C, 18 W, 4.2 Cr, 1.5 V, bal Fe.
For swaging dies, hobs, lathe and planer tools; high speed steel.

CELSIT N now VEW L208.

CELSIT P now VEW L216.

CELSIT V now VEW L219.

CELTO.
M-731; 0.2 C, bal Fe.
For welding electrodes; shielded arc.

CEMENTED TUNGSTEN CARBIDE.
M-57; 5.5 C, 86.5 W, 8 Co.
For hard cutting tools, tool tips; WC + Co; M.H. 9.8.

CENTANIN.
M-789; 67 Cu, 5 Ni, 27 Mn.
Annealed: 72,000-78,000 TS.
Temp. coef. resistance per °C ± 0.00002.
For electrical equipment and instruments.
Resistance alloy. Max working temperature 300°C.

Section I: Alloy Data / 297

CENTAUR.
M-261; C, bal Fe.
For tools, drills, taps; water hardened.

CANTRAL A.
M-732; 12 Si, 2.5 Cu, 1.25 Mg, 1.25 Mn, 2.5 Ni, 0.25 Ti, bal Al.
Cast: 24,000 TS; 1 El; 110 Brin.
Heat treated: 33,000 TS; 1 El; 140 Brin.
For light alloy parts; heat treatable.

CENTRAL V.
M-732; 12 Si, 1.25 Mg, 2.5 Mn, 0.25 Ti, bal Al.
Cast: 23,000 TS; 1 El; 100 Brin.
Heat treated: 31,000 TS; 1 El; 130 Brin.
For light alloy parts; heat treatable.

CENTRALLOY.
M-649; 0.15 C, 0.8 Ni, 0.2 Cr, bal Fe.
For bus and railway bodies; high strength.

CENTRARD.
M-494, M-1485; 2.7 C, 1.5-1.75 Cr, 1.5-1.75 Al, bal Fe.
Annealed: 45,000-56,000 TS; 250 Brin.
For cylinder liners, general parts to resist severe abrasion; nitrogen hardened alloy cast iron.

CENTREX HT.
M-1422; C, alloy, bal Fe.
Pre-heat-treated to 150,000 TS; 115,000 YS; 24 El; 302 Brin.
For gears, bolts, shafts; abrasion resistant.

CENTREX P & G.
M-1422; C, alloy, bal Fe.
Pre-heat-treated to 155,000 TS; 135,000 YS; 24 El; 302 Brin.
For gears, bolts, shafts; shock and fatigue resistant.

CENTREX PBD.
M-1422; C, alloy, bal Fe.
Heat Treated: 248-293 Brin.
For brake dies.
Resists severe wear and high impact forces.

CENTREX PRESS BRAKE see **CENTREX-PBD.**

CENTREX RHINO-CHEK.
M-1422; C, alloy, bal Fe.
Heat Treated: 370 Brin; 175,000 TS; 150,000 YS.
For fan blades; lining chutes.
Wear and corrosion resistant.

CENTREX RHINO-TUF.
M-1422; C, alloy, bal Fe.
Heat Treated: 175,000 TS; 150,000 YS; 360-400 Brin; (can be furnished harder).
For anchors, coal chutes, conveyers, furnace liners, grader blades, mixers.
Special heat treated wear and corrosion resistant plate.

CENTRICAST MARK 2.
M-494, M-1485; 3.3 C, 2.5 Si, 0.6 P, 0.3 Cr, bal Fe.
Cast: 36,000 TS; 241-302 Brin.
For cylinder liners, general castings; cast iron.

CENTRICAST MARK 3.
M-494, M-1485; 3.2 C, 1.9 Si, 0.2 P, 1.4 Ni, 0.4 Cr, bal Fe.
Cast: 40,000 TS; 217-285 Brin.
For gears, cylinder liners; heat treated castings.

CENTRICAST MARK 4.
M-494, M-1485; 3.2 C, 2.1 Si, 0.8 Mn, 0.2 P, 2 Ni, bal Fe.
Cast: 38,000 TS; 195-255 Brin.
For gears, cylinder liners; alloy cast iron.

CENTRICAST MARK 5.
M-494, M-1485; 3.3 C, 1.8 Si, 0.2 P, bal Fe.
Cast: 36,000 TS; 179-241 Brin.
For piston rings, liners; cast iron.

CENTRICAST MARK 9.
M-494, M-1485; 3.1 C, 2 Si, 0.4 P, 0.7 Cr, bal Fe.
Cast: 58,000 TS; 269-302 Brin.
For piston rings, automobile castings; cast iron.

CENTRICAST MARK 10-HI-TEN.
M-494, M-1485; 3.3 C, 2.1 Si, 0.1 P, 1.6 Ni, 0.75 Al, bal Fe.
Cast: 63,000-90,000 TS; 10-3 El.
For rollers, shafts; nitriding cast iron.

CENTRICAST MARK 11.
M-494, M-1485; 3.2 C, 2.3 Si, 0.7 Mn, 0.1 P, 0.2 Cr, 0.6 Ni, 0.15 Mo, bal Fe.
Cast: 52,000 TS; 207-255 Brin.
For cylinder liners; cast iron, wear resistant.

CENTRICAST MARK 12-4K6.
M-494, M-1485; 3.4 C, 2.3 Si, 0.6 P, 0.3 Cr, bal Fe.
Cast: 36,000 TS; 229-293 Brin.
For automobile piston rings; cast iron, wear resistant.

CENTRICAST MARK 14.
M-494, M-1485; 3.2 C, 2.4 Si, 0.6 P, 0.4 Cr, 0.6 Mo, bal Fe.
Cast: 45,000 TS; 255-293 Brin.
For aircraft piston rings; cast iron, wear resistant.

CENTRICAST MARK 15.
M-494, M-1485; 2.9 C, 2.4 Si, 0.4 P, 0.9 Cr, 0.9 Mo, bal Fe.
Cast: 58,000 TS; 269-302 Brin.
For aircraft compression and piston rings; cast iron, wear resistant.

CENTRICAST MARK 16.
M-494, M-1485; 1.9 C, 1.9 Si, 0.06 P, 16 Cr, bal Fe.
Cast: 78,000 TS; 269-321 Brin.
For piston rings, cylinder liners; heat and abrasion resistant.

CENTRICAST MARK 17.
M-494, M-1485; 1.2 C, 2 Si, 33 Cr, bal Fe.
Cast: 78,000 TS; 285-341 Brin.
For liners and pump cylinders; heat and abrasion resistant.

CENTRICAST MARK 18.
M-494, M-1485; 2.7 C, 2 Si, 14 Ni, 1.75 Cr, 7 Cu, bal Fe.
Cast: 31,500 TS; 175-235 Brin.
For cylinder liners; austenitic, cast iron, corrosion resistant.

CENTRICAST MARK 20.
M-494, M-1485; 3.0 C, 2.5 Si, 0.1 P, 1.3 Cr, bal Fe.
Cast: 58,000 TS; 269-311 Brin.
For valve seats; heat resistant, cast iron.

CENTRICAST MARK 21.
M-494, M-1485; 2.8 C, 2.7 Si, 0.2 P, 1.5 Cr, 1.0 Al, bal Fe.
Cast: 54,000 TS; 269-311 Brin.
For cylinder liners; nitriding cast iron.

CENTRICAST MARK 23.
M-494, M-1485; 3.3 T.C., 2.1 Si, 0.3 Mn, 0.1 P, 1.6 Ni, bal Fe.
Normalized: 90,000 TS; 260 Brin.
Annealed: 60,000 TS; 223 Brin.
For rollers, cylinder liners; spheroidal cast iron.

CENTRICAST MARK 24-LODED.
M-1485; 3.2 C, 2.9 Si, 0.8 Mn, 0.7 P, 0.7 Cr, bal Fe.
Cast iron, harder and more wear resistant than ordinary gray iron.
As cast: 241-302 Brin.
Liners for internal combustion engines.

CENTRICAST MARK 27.
M-1489; 3.3 C, 2.2 Si, 0.8 Mn, 0.10 P, 0.15 Cr, 0.25 V, bal Fe.
As cast: 212-293 Brin.
For cylinder liners.

CENTRICAST MARK 28.
M-1485; 3.2 C, 1.8 Si, 0.7 Mn, 0.25 P, 0.8 Cu, 0.4 V, 0.04 Ti, bal Fe.
As cast: 210-262 Brin.
For large diesel engine liners, particularly in the marine industry.

CENTRICAST MARK 29.
M-1485; 3.2 C, 2.5 Si, 0.8 Mn, 0.5 P, 0.5 Cr, bal Fe.
As cast: 229-302 Brin.
General purpose cast iron for motor vehicle industry.

CENTURY 1050.
M-97.
105,000 psi YS.
High strength steel.

CENTURY 1141.
M-97; 0.37-0.45 C, 1.35-1.65 Mn, 0.08-0.13 S, 0.045 max P, bal Fe.
Hot Rolled: 98,000 TS; 60,000 YS; 25 El; 52 RA; 200 Brin.
Hardened: 185,000 TS; 168,000 YS; 12 El; 40 RA; 365 Brin.
For axles, shafts, screws, bolts, arbors, bushings, gears, lead screws.
Free-machining. Water hardenable.

CENTURY 1144.
M-97; 0.40-0.48 C, 1.35-1.65 Mn, 0.24-0.33 S, 0.045 max P, bal Fe.
Annealed: 94,000 TS; 54,000 YS; 24 El; 183 Brin.
Water Hardened: 175,000-295,000 TS; 155,000-188,000 YS; 4-11 El; 11-38 RA; 363-555 Brin.
For axles, shafts, tie-rods, arbors, ball joints, gears, pinions, clutches.
Free machining, water or oil hardening.

CEOT.
M-912; 0.45 C, 1.65 Cr, 0.70 Mo, 0.30 V, bal Fe.
Hot work tool steel; for extrusion press tools.
W.-Nr. 2323; AFNOR 45 CDV 7.

CERALLOY 75.
M-1497; 25 Fe, bal Misch Metal.
For additives for ductile cast iron; desulphurizer and graphite nodularizer.

CERALLOY 75M2.
M-1497; 23 Fe, 2 Mg, bal Misch Metal.
For additives for ductile cast iron; desulphurizer and graphite nodularizer.

CERALLOY 80A.
M-1497; 20 Al, bal Misch Metal.
For steel deoxidizer and desulphurizer, scavenger.

CERALLOY 90M.
M-1497; 10 Mg, bal Misch Metal.
For additives for ductile cast iron; desulphurizer and graphite nodularizer.

CERALLOY 400.
M-1497; 80 Th. bal Ce + Al.
For vacuum tube getter; continuous getter.

CERALLOY 420.
M-1497; 20 Fe, 2 Mg, bal Misch Metal.
For vacuum tube getter, anode in voltage regular gas tubes; flash getter.

CERALLOY MISCHMETAL 95 M.
M-1497; 5 Mg, bal mixed rare earth metals.
Ignition alloy.

Section I: Alloy Data / 299

CERALLOY MISCHMETAL 100X.
M-1497; 99.9 rare earth metals, typically 50 Ce + 27 La + 16 Nd + 5 Pr bal other rare earth metals.
Additive for non-ferrous alloys.

CERALLOY MISCHMETAL FG.
M-1497; 4% Fe, bal mixed rare earth metals.
Additive for iron and steel; desulfurizer and graphite nodularizer.

CERALLOY MISCHMETAL M.
M-1497; 2.5% Mg, bal mixed rare earth metals.
Additive for magnesium alloys.

CERALUMIN 21.
M-452; 3.5-4.5 Cu, 0.5-1.0 Co, 1.2-2.5 Mg, bal Al.
Cast: 54,000-58,000 TS; 20-22 El.
For sand and die castings for elevated temperature service.
High strength and hardness at elevated temperatures.

CERALUMIN 22.
M-452; Si, Cu, Ce, bal Al.
For light alloy parts; forgings.

CERALUMIN A.
M-452; 1.25 Si, 2.5 Cu, 0.8 Mg, 1.2 Fe, 1.5 Ni, 0.15 Ce, bal Al.
Heat treated: 51,000-60,000 TS; 45,000-54,000 YS; 1 El; 130-140 Brin.
For light alloy parts; heat treatable.

CERALUMIN "ASM".
M-452; 1-2 Cu, 0.25-1 Ni, 0.5 max Fe, 0.7-1.3 Si, 0.4-0.8 Mg, 0.03-0. Cb, bal Al.
Sand cast: 38,000 TS; 4-7 El; 95 Brin.
Chill cast: 47,000 TS; 14-21 El; 95 Brin.
For castings: good corrosion resistance.

CERALUMIN B.
M-452; 1.5 Cu, 1.5 Ni, 0.2 Mg, 0.7 Fe, 1.5 Si, 0.05-0.3 Ce, bal Al.
Sand cast: 26,000 TS; 20,000 YS; 2-3 El; 80 Brin.
Permanent mold: 31,000 TS; 20,000 YS; 4-9 El; 82 Brin.
For pistons, cylinder heads, gear boxes, impellers; good castability.

CERALUMIN C.
M-452; 2-3 Cu, 1-2 Ni, 1-1.4 Fe, 1-1.4 Si, 0.05-0.20 Ce, 0.5-1.0 Mg, 0.05-0.3 Cb, bal Al.
Aged: 63,000 TS; 55,000 YS; 9 El; 145 Brin.
Permanent mold: 59,000 TS; 49,000 YS; 1-4 El; 135 Brin.
For castings for high temperature use; age-hardenable.

CERALUMIN D.
M-452; 2-3 Cu, 1-2 Ni, 0.5-1.0 Mg, 1.2 Si, 1.2 Fe, 0.05-0.20 Ce, bal Al.
Heat treated: 28,000-40,000 TS; 25,000 YS; 5-1 El; 90-100 Brin.
For cylinder heads, general castings; age-hardenable.

CERALUMIN F.
M-452; 1.3-30 Cu, 1.5-2.0 Ni, 0.05-0.3 Ce, 0.5-1.25 Mg, 0.03-0.03 Cr, 1. Fe, 0.9 Si, bal Al.
For light alloy castings; heat treatable.

CERAMVAR.
M-61; 48 Fe, 27 Ni, 25 Co.
For ceramic-to-metal seals.
Expansion characteristics match aluminum.

CERIUM.
M-1755; Ce, Metallothermic + Electrolytic.
Purities: 99.9%, 99.6%.
Forms: Ingot, turnings, powder, wire, sheet, rod, lump, foil.

CERIUM METAL.
M-1497.
Comparatively pure Ce; for alloying.

CERROBASE.
M-216, M-122; 55.5 Bi, 44.5 Pb.
Cast: 6200 TS; 4000 YS; 64 El; 10 Brin.
For molds, tube filler, proof casting forging dies; M.P. 123.5°C; formerly "Basaloy."

CERROBEND.
M-216; 50 Bi, 26.7 Pb, 13.3 Sn, 10 Cd.
Cast: 5990 TS; 200 El; 9.2 Brin.
For filler for bending tube, low melting solder. sealing glass to metal; formerly called Bendalloy, M.P. 158°F.

CERROCAST.
M-216; 40 Bi, 60 Sn.
Cast: 8000 TS; 200 El; 22 Brin.
For wax pattern molds; M.P. 281-338°F.

CERRODENT.
M-216; 38.1 Bi, 26.4 Pb, 31.7 Sn, 2.65 Cd, 0.06 Cu. 15 Brin.
For dental models; M.P. 75-118°C.

CERROLOW 105.
M-216; 7.97 Sn, 42.91 Bi, 21.7 Pb, 5.09 Cd, 18-33 In, 4 Hg.
For fusible alloy; M.P. 100-110°F.

CERROLOW 117.
M-216; 44.7 Bi, 22.6 Pb, 8.3 Sn, 5.3 Cd, 19.1 In.
Cast: 5400 TS; 1.5 El; 9-12 Brin.
For low melting alloys, solders, fuses; eutectic alloy; M.P. 117°F.

CERROLOW 117 B.
M-216; 11.3 Sn, 44.7 Bi, 22.6 Pb, 5.3 Cd, 16.1 In.
For fusible alloy; M.P. 117-126°F.

CERROLOW 136.
M-216; 49 Bi, 18 Pb, 12 Sn, 21 In.
Cast: 6300 TS; 50 El; 14 Brin.
For low melting alloys, solder, fuses; M.P. 136°F; eutectic alloy.

CERROLOW 136 B.
M-216; 15 Sn, 49 Bi, 18 Pb, 18 In.
For fusible alloy; M.P. 136-156°F.

CERROLOW 140.
M-216; 12.6 Sn, 47.5 Bi, 25.4 Pb, 9.5 Cd, 5 In.
For fusible alloy; M.P. 134-149°F.

CERROLOW 147.
M-216; 12.77 Sn, 48 Bi, 25.63 Pb, 9.6 Cd, 4 In.
Cast: 4950 TS; 13.5 El; 11 Brin.
For thermal safety controls; M.P. 142-149°F.

CERROLOW 174.
M-216; 57 Bi, 17 Sn, 26 In.
For safety devices, fusible elements, solder; M.P. 174°F.

CERROMATRIX.
M-216; 48 Bi, 14.5 Sn, 28.5 Pb, 9 Sb.
Cast: 13,000 TS; 1 El; 19 Brin.
For dental alloys, fusible alloy for die amounting; expands on cooling; M.P. 218-4407.

CERROSAFE.
M-216; 42.5 Bi, 37.7 Pb, 11.3 Sn, 8.5 Cd.
Cast: 5400 TS; 200 El; 9 Brin.
For toy casting sets; formerly "Saffalloy;" M.P. 158-194°F.

CERROSEAL 25.
M-216; 50 In. 50 Cd.
In glass to glass or glass to metal seals; M.P. 227-462°F.

CERROSEL 35.
M-216; 50 Sn, 50 In.
For soldering glass to metal; M.P. 250-260°F. useful in vacuum and low vapor pressure.

CERROTRU.
M-216; 58 Bi, 42 Sn.
Cast: 8000 TS; 200 El; 22 Brin.
For castings, molds, patterns; zero volume change from molten to solid state, M.P. 280°F.

CERTANIUM 250.
M-1600; Composite of an austenitic matrix plus TiC and CrC and Silicides.
For hard surfacing.
Resists high stress abrasion. Hard and tough electrode for AC and DC welding.

CESCO DIAMOND.
M-356; 0.5 C, 0.85-1.0 Mn, 2.2 Si, 0.5 Mo, 0.25-0.35 V, bal Fe.
Heat treated: 320,500-231,600 TS; 259,400-185,700 YS; 2-5 El; 5-27 RA; 659 Brin.
For hollow and solid drills, concrete busters, hammer pistons, springs; shock resistant.

CESCO DIAMOND SPECIAL.
M-356; 0.50-0.65 C, 2.0 Si, 1.0 Mn, 0.6-0.7 Mo, 0.34-0.36 V, bal Fe.
Heat treated: 350,000 TS; 290,000 YS; 6 El; 11 RA; 615 Brin.
For chisels, concrete busters, drills, caulking and beading tools; shock and abrasion resistant.

CESCO SPECIAL.
M-356; 0.55 C, 1.0 Mn, 0.4 Mo, 0.35 V, 2.0 Si, bal Fe.
For tools; shock resistant.

CESIUM.
M-1755; Cs.
Purities: 99.99%, 99.9+%, 99.5 %.
Packaging: Glass ampules, steel containers (under vacuum, inert gas, oil).

C.E.S. NO. 2.
M-440; 0.36-0.50 C, 16-23 Ni, 7-11 Cr, bal Fe.
For heat and corrosion resistant parts; heat and corrosion resistant.

CETAL.
M-299; 6.5 Si, 3 Cu, 10 Zn, bal Al.
Cast: 26,000-31,000 TS; 0.5-3.0 El; 65-90 Brin.
For light alloy parts; non-hardenable.

CETO.
M-1549; 80 Th, 15 Al, 5 Mischmetal.
For getters.
Absorbs gases.

CF-43.
M-U.S.; 0.5 C, 25 Cr, 10 Ni, 7.5 W, 1.5 Fe, bal Co.
For marine gas turbines, jet engine components. Similar to X-40.

C.F.S.
M-73; 0.9-1.3 C, 0.5 Cr, bal Fe.
For dies, draw dies; water hardened.
AISI W5.

CG 27 see **CRUCIBLE CG27.**

CG ALLOY.
M-1038; 4.0-5.0 Mg, 8.5-10.5 Ti, 0.20-0.35 Ce, 4.0-5.5 Ca, 1.0-1.5 Al 48.0-52.0 Si, bal Fe.
Ferroalloy for production of compacted graphite iron.

CHACE NO. 772.
M-582; 18 Cu, 10 Ni, bal Mn.
Treatment: 50% Red.
105,000 TS; 90,000 YS; 8.5 El; 220 Brin.
Uses: high expansion, high resistivity; Max. temp: 600°F.

CHACE NO. 1050.
M-582; 36 Ni, bal Fe/100 Cu/22 Ni, 3 Cr, bal Fe.
Thermostatic bimetal; Low electrical resistivity.

CHACE NO 2400.
M-582; 36 Ni, bal Fe/22 Ni, 3 Cr, bal Fe.
Thermostatic bimetal; range 0-300°F.

CHACE NO 2500.
M-582; 50 Ni, bal Fe/25 Ni, 3 Cr, bal Fe.
Thermostatic bimetal; max. temp. 1000°F.

CHACE NO 3700.
M-582; 40 Ni, bal Fe/22 Ni, 3 Cr, bal Fe.
For thermostatic bimetal; range 100-550°F.

CHACE NO 4700.
M-582; 38 Ni, 7 Cr, bal Fe/19 Ni, 7 Cr, bal Fe.
For thermostatic bimetal; corrosion resistant.

CHACE NO 6125.
M-582; 36 Ni, bal Fe/100 Ni/22 Ni, 3 Cr, bal Fe.
For thermostatic bimetal; range 0-300°F.

CHACE NO 6150.
M-582; 36 Ni, bal Fe/100 Ni/22 Ni, 3 Cr, bal Fe.
Thermostatic bimetal; range 0-300°F.

CHACE NO 6650.
M-582; 36 Ni, bal Fe/18 Cu, 10 Ni, bal Mn.
For thermostatic bimetal; high deflection.

CHACE NO 6850.
M-582; 36 Ni, bal Fe/18 Cu, 10 Ni, bal Mn.
For thermostatic bimetal; high electrical resistivity.

CHAIN BRONZE.
M-U.S.; 4.9 Sn, 0.1 P, bal Cu.
Rolled: 28,000-69,000 TS; 15,000-40,000 YS; 30-6 El.
For chains, springs, diaphragms; high strength, corrosion resistant.

CHAIN IRON.
M-Eng.; 0.10 max Mn, Si, bal Fe.
For chains for cranes, slings, hoists, steam shovels, marine uses; wrought iron.

CHAMAX.
M-1116; 0.82 C, 4.5 Cr, 18 W, 0.8 Mo, 1.75 V, 5 Co, bal Fe.
Air or oil harden to 63-66 Rc.
For lathe tools, form tools, gear cutters, counter bores, planing tools.
AFNOR: Z 85 WK 18.05.
US-AISI T4.
Germany: S 18.1.2.5 (E 18 Co 5); W. Nr. 1.3255.

CHAMAX 00.
M-1116; 0.95, 4.5 Cr, 18 W, 0.7 Mo, 1.75 V, 5 Co, bal Fe.
Air or oil harden to 62-67.5 Rc.
For lathe tools, milling cutters, gear cutters.
AFNOR: Z 95 WK 18.05.

CHAMAX 11.
M-1116; 0.82 C, 4.5 Cr, 18 W, 0.8 Mo, 1.75 V, 11 Co, bal Fe.
Air or oil harden to 63-67 Rc.
For lathe tools and drills for special jobs - higher speeds, higher temp., special alloys.
AFNOR: Z 85 WK 18.10.
US-AISI T5 or T6.
Germanay: S 18.1.2.10 (E 18 Co 10): W. Nr. 1.3265.

CHAMAX. CV.
M-1116; 1.55 C, 4.3 Cr, 12.5 W, 0.8 Mo, 5 V, 5 Co, bal Fe.
Air or oil harden to 62-68 Rc.
For tools for machining stainless steels; for punches for silicon sheets.
AFNOR: Z 150 WKV 12.05.05.
US-AISI T15.
Germany: S 12.1.1.5 (EV 4 Co); W. Nr. 1.3202.

CHAMAX. CV. 11.
M-1116; 1.65 C, 4.4 Cr, 12 W, 0.8 Mo, 5 V, 10 Co, bal Fe.
Air or oil harden to 62-69 Rc.
For lathe tools for finish cuts at high speeds; also for machining special materials.
AFNOR: Z 160 WKV 12.10.05.

CHAMET LEADED BRONZE.
M-U.S.; 38.5 Zn, 0.75 Pb, 0.75 Sn, bal Cu.
For fittings, pipes, hardware; free-cutting, high strength.

CHAMFERTRODE.
M-717.
For chamfering and grooving electrodes, (AC-DC) on ferrous or non-ferrous.

CHAMPALOY.
M-38; 0.75 C, 1.5 Ni, 0.75 Cr, 0.3 Mo, bal Fe.
Heat treated: 208,000 TS; 180,000 YS; 450 Brin.
For tools, collets, jigs, shear blades, rolls; oil hardening.
AISI L6.

CHAMPION 2.
M-712; 0.10 max C, 2 Cr, 0.5 Mo, bal Fe.
Welded: 70,000-75,000 TS; 35-30 El.
For welding rods for low alloy steels; subject to air hardening.

CHAMPION 9.
M-712; 0.1 max C, 9 Cr, 0.5 Mo, bal Fe.
Welded: 90,000-95,000.
For welding rods for low Cr alloys; high creep resistance.

CHAMPION 255.
M-1683; 0.6 C, 0.8 Mn, 2.0 Si, 0.2 Cr, 0.2 Mo, 0.2 V, bal Fe.
Annealed: 107,000 TS; 64,000 YP; 27 El; 212 Brin.
Oil hardened: 145,000-340,000 TS; 127,000-283,000 YP; 5-24 El; 20-44 Ra; 293-611 Brin.
For shear blades, pneumatic tools, punches, caulking tools.
Type S5 shock resisting tool steel. Oil or water hardening.

CHAMPION 308.
M-712; 0.07 max C, 18.5-20.0 Cr, 10.5-11.25 Ni, bal Fe.
Welded: 85,000-95,000 TS; 50-40 El.
For welding rods for stainless steel; stainless, austenitic; Type 308.

CHAMPION 309.
M-712; 0.10 max C, 22.5-24.0 Cr, 13-14 Ni, bal Fe.
Welded: 85,000-95,000 TS; 45-35 El.
For welding rods for stainless steel; resists oxidation to 2000°F; Type 309.

CHAMPION 309 CB.
M-712; 0.10 max C, 22.5-24.0 Cr, 13-14 Ni, 0.8-1.0 Cb, bal Fe.
Welded: 85,000-95,000 TS; 40-30 El.
For welding rods for stainless steel; prevents carbide ppt., austenitic.

CHAMPION 310.
M-712; 0.09-0.17 C, 25.0-26.5 Cr, 20-21 Ni, bal Fe.
Welded: 85,000-95,000 TS; 45-35 El.
For welding rods for 5% Cr steel; will not air harden, austenitic.

CHAMPION 310 CB.
M-712; 0.10 max C, 25.0-26.5 Cr, 20-21 Ni, 0.8-1.0 Cb, bal Fe.
Welded: 85,000-95,000 TS; 40-30 El.
For welding rods for stainless steel; resists carbide ppt, stabilized.

CHAMPION 310 MO.
M-712; 0.10 max C, 25.0-26.5 Cr, 20-21 Ni, 2.0-2.5 Mo, bal Fe.
Welded: 85,000-95,000 TS; 45-35 El.
For welding rods for stainless steel; acid resistant, stainless, austenitic.

CHAMPION 316.
M-712; 0.07 max C, 17.5-19.0 Cr, 12.5-13.5 Ni, 2.0-2.5 Mo, bal Fe.
Welded: 85,000-95,000 TS; 50-40 El.
For welding rods for stainless steel; acid resistant, austenitic.

CHAMPION 316 CB.
M-712; 0.07 max C, 17.5-19.0 Cr, 12.5-13.5 Ni, 2.0-2.5 Mo, 0.5-0.75 Cb, bal Fe.
Welded: 85,000-95,000 TS; 45-35 El.
For welding rods for stainless steel; stabilized, stainless.

CHAMPION 317.
M-712; 0.07 max C, 17.5-19.0 Cr, 12.5-13.5 Ni, 3.5-4.0 Mo, bal Fe.
Welded: 85,000-95,000 TS; 50-40 El.
For welding rods for stainless steel; stainless, austenitic.

CHAMPION 330.
M-712; 0.15 max C, 14.5-16.0 Cr, 34-35 Ni, bal Fe.
Welded: 65,000-70,000 TS; 30-25 El.
For welding rods for furnace construction; heat resistant.

CHAMPION 347.
M-712; 0.07 max C, 18.5-20.0 Cr, 10.5-11.25 Ni, 0.75-1.00 Cb, bal Fe.
Welded: 85,000-95,000 TS; 45-35 El.
For welding rods for stainless steel; stabilizer, stainless; Type 347.

CHAMPION 410.
M-712; 0.10 max C, 12.0-13.5 Cr, 0.40-0.65 Mo, bal Fe.
Welded: 85,000-95,000 TS; 35-30 El.
For welding rods for oil refinery vessels; corrosion and heat resistant.

CHAMPION 430.
M-712; 0.10 max C, 15.25-16.75 Cr, bal Fe.
Welded: 75,000-80,000 TS; 35-30 El.
For welding rods to resist HNO_3; corrosion resistant.

CHAMPION 442.
M-712; 0.10 max C, 17.5-19 Cr, bal Fe.
For welding rods for high temperature applications; corrosion and heat resistant.

CHAMPION 446.
M-712; 0.13 max C, 27-28.5 Cr, bal Fe.
For welding rods for high temperature applications; resists oxidation at high temperature.

CHAMPION 502.
M-712; 0.10 max C, 4.5-6 Cr, 0.4-0.65 Mo, bal Fe.
For welding rods for oil refinery equipment; corrosion resistant.

CHAMPION C-30.
M-712; 0.22-0.26 C, 0.45 Mn, 0.4-0.6 Mo, 0.10-0.15 V, bal Fe.
Welded: 215 Brin.
For welding electrodes; for worn surfaces, wear resistant.

CHAMPION DOUBLE SPECIAL.
M-US; 1.22 C, 0.25 Mn, 7.53 W, bal Fe.
For tools, dies; oil hardening.

CHAMPION EXTRA.
M-38; 1.2 C, 1.25 W, 0.3 Cr, bal Fe.
For cutters, broaches, drills; oil hardening.

CHAMPION GRAY DEVIL NO. 2.
M-712; 0.09 C, 0.35 Mn, bal Fe.
Welded: 78,000 TS; 67,000 YS; 24 El; 45 RA.
For welding rod; flux coated, E 6012.

CHAMPION HW1.
M-1683; 0.35 C, 5.0 Cr, 0.4 V, 1.5 W, 1.5 Mo, bal Fe.
Hot work tool steel, air or oil hardening, for forging and hot forming die AISI H12.

CHAMPION HW3.
M-1683; 0.35 C, 1 Si, 5 Cr, 1 V, 1.5 Mo, bal Fe.
Annealed: 98,000 TS; 74,000 YS; 28 El; 210 Brin.
Hardened: 135,000-290,000 TS; 100,000-228,000 YS; 3-16 El; 7-48 RA; C 27-55 Rock.
For forging and heading dies, compression tools, die casting dies, piercing and forming punches, bolt and swaging dies.
Type H13 hot work steel, red-tough, shock and impact resistant.

CHAMPION HW21.
M-1683; 0.35 C, 3.5 Cr, 9.0 W, bal Fe.
Hot work tool steel, air or oil hardening, tungsten type; for hot forming operations. AISI H21.

CHAMPION NO. 59.
M-712; 3 C, 4 Cr, 4 Mo, bal Fe.
550 Brin.
For hard surfacing electrodes; for high carbon and manganese steel, abrasion resistant.

CHAMPION S1.
M-1683; 0.50 C, 1.5 Cr, 2.5 W, bal Fe.
Oil hardening tool steel, shock resisting type.
AISI Type S1 shock resisting tool steel.

CHAMPION TOOL.
M-US; 0.77 C, 0.18 Mn, bal Fe.
For tools, punches; water hardening.

CHAMPLAIN.
M-207; 0.90 C, 0.50 W, 0.50 Cr, 1.2 Mn, 0.2 V, bal Fe.
Heat treated: 170,000-280,000 TS; 155,000-272,000 YS; 2-4 El; 2-18 RA; 335-535 Brin.
For dies, knives, guides, shear blades, vise jaws, forming rolls.
Oil hardening, non-deforming, tough.

CHARDON-S.
M-1546; 0.60-1.10 C, bal Fe.
Annealed: 100,000 TS; 52,000 YS; 20 El; 200 Brin.
Water hardened: 185,000-220,000 TS; 125,000-155,000 YS; 32-37 RA; 11-13 El; 330-600 Brin.
Uses: tools, cutters, hammers, chisels, punches.
Water hardening, Type W1 tool steel.

CHARLES LEONARD.
M-261; 0.7-1.0 C, 0.3 Mn, 0.3 Si, bal Fe.
For tools, cutters, drills; water hardened; Type W1.

CHAR-PAC.
M-604; 0.17 C, 1.25 Mn, 0.35 Si, 0.25 Cu, 0.15 Ni, 0.12 Cr, 0.04 Mo bal Fe.
Heat treated: 60,000 min YS; 80,000-100,000 TS; 15 min Charpy.
Normalized: 70,000-95,000 TS; 50,000 min YS.
For storage tanks, penstocks, bridges and buildings, pressure vessels.
Heat treated carbon steel plates.

CHARPY ALLOY.
M-Eng; 5.6 Cu, 83.3 Sn, 11.1 Sb.
For bearings; Babbitt.

CHARPY PHOSPHOR BRONZE.
M-US; 12.2-13.4 Sn, 0.4 P, bal Cu.
For bearings, gears, bushings; heavy duty.

CHASE 192 HC.
M-1789; 98.97 Cu, 1 Fe, 0.03 P.
Annt 1/4 hard: 50,000 TS; 37,000 YS; 25 El.
Rolled hard: 65,000 TS; 60,000 YS; 7 El.
Good conductivity, resistance to softening.
For electrical contacts, terminals, springs, lead frames, cable wrap.

CHASE 690.
M-1789; 73.3 Cu, 22.7 Zn, 3.4 Al, 0.6 Ni.
Ann: 82,000 TS; 52,000 YS; 35 El.
Rolled hard: 113,000 TS; 101,000 YS; 4 El.
For springs, bellows, diaphragms, fuse clips, switch parts.

CHATEAUGAY LOW PHOSPHORUS PIG IRON.
M-257; 4.0 C, 0.75-4.05 Si, 0.1-2.0 Mn, 0.035 max. S and P, bal Fe.
For casting rolls, gears, cylinders, steel and iron; produced form New York State ores.

CHATILLON 709.
M-1486; 0.21 max C, 0.85-1.2 Cr, 0.9-1.1 Mo, 0.15-0.30 V, bal Fe.
Heat treated: 105,000-178,000 TS; 92,400-106,000 YS; 17-8 El.
For steam turbine rotors and stators, high temperature bolts; creep resistant, structural steel.

CHD.
M-1744; 0.33 C, 0.35 Mn, 1.0 Si, 5.0 Cr, 1.5 W, 0.3 V, 1.5 Mo, bal Fe.
Chromium-Tungsten hot work steel, for die casting dies for light alloys, extrusion dies, punches.

CHECO.
M-US; 89 Ag, 10 Sn, 1 Pt.
For dentures.

CHEMALLOY.
M-76; Zn, Pb, plus additives.
Homogenized alloy for fluxless soldering of aluminum, copper base alloys, galvanized iron. Melt point 500°F. (higher melting grades are available).

CHEMBRITE A608.
M-663; 0.10 C, 1.9 Mn, 0.68 Si, 0.06 Ni, 0.5 Mo, bal Fe.
Low alloy bare for machine welding to give welds of about 100,000 psi TS.
AWS E 70 S-1B.

CHEMBRITE A675.
M-663; 0.12 C, 1.19 Mn, 0.59 Si, bal Fe.
Bare mild steel welding wire for CO_2 and inert gas shielded arc welding.
AWS E 70 S-3.

CHEMBRITE A681.
M-663; 0.11 C, 1.64 Mn, 0.86 Si, bal Fe.
Mild steel bare wire for CO_2 and inert gas welding of steel.
AWS E 70 S-6.

CHEMICAL LEAD.
M-88; 99.93 Pb, 0.07 Cu, 0.005 Ag.
Cast: 1,750 TS; 1,000 YS; 40 El; 90 RA; 5.5 Brin.
Rolled: 2,000 TS; 1,000 YS; 54 El; 5.5 Brin.
For tank linings, valves, pipes, fuses; M.E. 2,000,000.

CHENITE.
M-Eng.; Ti, V, Cr, Hf.
For cutting tools, dies, air hardening.

CHINA BRASS.
M-China; 56.6 Cu, 27-37 Zn, 0.2-1.0 Sn, 0-0.8 Pb.
For sheet, hardware, fittings; good workability.

CHINESE BRONZE.
M-China; 10-20 Pb, 1-14 Zn, 1-13 Sn, bal Cu.
For bearings; heavy duty.

CHINESE GERMAN SILVER.
M-China; 40.4-41 Cu, 25.4-26.5 Zn, 30.8-31.6 Ni, 2.6-2.7 Fe.
For electrical resistances, ornaments; corrosion resistant.

CHINESE SILVER.
M-China; 58 Cu, 17.5 Zn, 11.5 Ni, 2 Ag.
For ornaments, electrical parts; nickel silver.

CHINESE SPECULUM-1.
M-Eng.; 81 Cu, 11 Sn, 8.5 Sb.
For mirrors, optical grading.

CHINESE SPECULUM-2.
M-Eng.; 81 Cu, 9 Pb, 8 Sb, 2 Sn.
For mirrors; (Elsner's).

CHIPPEWA.
M-435; 0.45 C, 0.6 Mn, 0.25 Si, 0.4 V, 3 Cr, 0.25 Mo, bal Fe.
For mining bit shanks; hollow shank steel.

CHISEL 3581.
M-40; 0.55 C, 0.7 Mn, 0.2 V, 0.35 Mo, 1.6 Si, bal Fe.
For pneumatic tools, chisels; oil hardened, shock resistant.

CHITONAL.
M-624; 1.5 Si, 5 Cu, bal Al.
For light alloy parts, ship parts, bi-metal, coated "Avional."

CHITONAL-14.
M-624, M-541; 3.9-5.0 Cu, 0.5-1.2 Si, 0.2-0.8 Mg, 0.4-1.2 Mn, bal Al
Annealed: 27,000-30,000 TS; 11,400-14,200 YS; 16-20 El.
TA-Temper: 66,800-75,400 TS; 58,300-66,800 YS; 6-12 El.
For general engineering components. Age hardenable, high strength.

CHITONAL-24.
M-624, M-541; 3.8-4.9 Cu, 1.2-1.8 Mg, 0.3-0.9 Mn, bal Al.
Annealed: 27,000 TS; 14,000 YS; 14-22 El;
TH06N-Temper: 74,000 TS; 62,000 YS; 10 El.
For aircraft structures, general engineering components.
Age hardenable, high strength.

CHIZ-ALAIR.
M-335; 0.50 C, 0.7 Mn, 0.25 Si, 3.25 Cr, 1.4 Mo, bal Fe.
Air hardening; for tools, dies.
AISI S7 shock resisting tool steel.

CHIZ-ALLOY.
M-335; 0.5 C, 1.15 Cr, 2.5 W, 0.2 V, 0.75 Si, bal Fe.
For hand and pneumatic chisels and punches; shock resisting; oil hardening octagons and hexagons; AISI S1 tool steel.

CHLORIMET 2.
M-46; 63 Ni, 30-33 Mo, 3.0 max Fe, 1.0 max Si, 1.0 max Mn, 0.07 max
Cast, Heat treated: 76,000 TS; 46,000 YS; 20 El; 200 Brin.
For corrosion resistant pumps and valves; resistant to HCl, boiling H_2SO_4, boiling H_3PO_4, brine.

CHLORIMET 3.
M-46; 60 Ni, 17-20 Cr, 17-20 Mo, 3.0 max Fe, 1.0 max Si, 1.0 max Mn, 0.07 max C.
Cast, Heat treated: 72,000 TS; 46,000 YS; 25 El; 200 Brin.
For corrosion resistant pumps and valves; resistant to HCl, hot H_2SO_4, hot H_3PO_4, wet chlorine, ferric chloride, bleach.

CHOYCE-77.
M-1683; 0.90 C, 1.2 Mn, 0.2 V, 0.5 W, 0.5 Cr, bal Fe.
Annealed: 84,000 TS; 60,000 YS; 26 El; 185 Brin.
Heat treated; 145,000-280,000 TS; 125,000-272,000 YS; 2-8 El; 2-31 RA; 290-535 Brin.
For blanking and bending dies, master tools, knurling tools, punches, cutters, master dies, gauges.
Type 01, oil hardening; cold work steel, shock resisting.

C.H.Q. DIE STEEL.
M-57; C, alloy, bal Fe.
For cold header dies; cold work steel.

CHRISTOFLE METAL.
M-Eng; silver plated German Silver (2.0 Ag).
For cutlery, hardware; corrosion resistant.

CHROGO "U-42".
40 Au, 45 Cu, 14 Ni, 1 Cr, traces Pt.
For dental alloy; corrosion resistant.

CHROMADOR STEEL.
M-470; 0.30 max C, 0.7-1.0 Mn, 0.7-1.0 Cr, 0.25-0.50 Cu, bal Fe.
Rolled: 83,000-98,500 TS; 52,000 YS; 17 El; 40 RA; 160-200 Brin.
For structural work, tanks, rivets; high tensile steel; resists corrosion.

CHROMAL.
M-Swedish; 2-4 Cr, Ni, Mn, bal Al.
Rolled: 55,000-60,000 TS.
For airplane parts, propellers, cooking utensils, milk and oil separators; non-hardenable.

CHROMALOY V.
M-408; 80 Ni, 20 Cr.
Annealed: 95,000 TS; 25-35 El; 185 Brin.
For heating and resistance elements; heat resistant to 1150°F.

CHROMAL STEEL.
M-Eng; 0.8 Cr, 0.8 Mn, 0.8 Mo, 0.3 C, bal Fe.
For gears, pinions, shafts; water hardening.

CHROMAX.
M-44; 35 Ni, 20 Cr, 45 Fe, 1 Si.
Wore, rod or ribbon for heating elements up to 1700°F.

CHROMCARB N-6006.
M-717.
Electrode for hard but tough overlays on steel: AC-DC, for crusher jaws, clam shell bucket lips, dipper teeth. Hardness C 57-60 Rock.

CHROMDIE.
M-627; 1.5 C, 12 Cr, 0.4 V, 1.0 Mo, bal Fe.
Hardened: 278,000 TS; 214,000 YS; 1 El; 567 Brin.
For blanking and drawing dies; wire drawing and stamping dies, broaches, hobs, punches, gauges, thread rolling dies.
Type D2 air hardening, nondeforming tool steel, tough.

CHROME 9A.
M-1488; 1.0 C, 5.0 Cr, 1.2 Mo, 0.40 V, bal Fe.
Cold work tool steel; air hardening.
AFNOR Z 100 CCV 5; AISI A2.

CHROME ALLOY PRODUCTS KA-2.
M-146, M-436; 0.13 C, 18.5 Cr, 9 Ni, bal Fe.
For stainless parts; corrosion and heat resistant.

CHROME ALLOY PRODUCTS KA-2 MO.
M-146, M-436; 0.14 C, 20 Cr, 9 Ni, 3 Mo, bal Fe.
For stainless parts; corrosion and heat resistant.

CHROME ALLOY PRODUCTS KA-2S.
M-146, M-436; 0.06 C, 20 Cr, 9 Ni, bal Fe.
For stainless parts; corrosion and heat resistant.

CHROME ALLOY PRODUCTS KNC-3.
M-146, M-436; 0.17 C, 25 Cr, 19 Ni, 1.75 Si, bal Fe.
For furnace parts; corrosion and heat resistant.

CHROME ALLOY PRODUCTS STAINLESS NO. 7.
M-146, M-436; 0.25 C, 20.5 Cr, 1 Cu, bal Fe.
For stainless parts; stainless.

CHROME B.
M-344; C, Cr, bal Fe.
For cutting dies; non-deforming.

CHROME "B-15".
M-344; C, 15 Cr, bal Fe.
For tools, cold and threading dies; punch stamps; non-deforming.

CHROME BRASS.
M-538; Cr coated brass
For fabricated parts; easily formed, stamped, drawn.

CHROME CAST.
M-569; 1.1 min C, 16-23 Cr, bal Fe.
For heat, wear and corrosion resistant parts; heat, wear and corrosion resistant.

CHROME-COPPER.
M-538; Cr coated copper.
For stampings, display cases, refrigerators, kitchen equipment; pure Cu base bonded with Ni then Cr on surface.

CHROME DIE.
M-15; 1.5 C, 12.0 Cr, 1.0 Mo, 1.0 V, bal Fe.
Cold work tool Steel; used with manganese rod in "two-tone" arc method.
AISI D2.

CHROME DIE.
M-253, M-1; 0.5 C, 5 Cr, bal Fe.
For hot dies, punches, shears; hot work steel.

CHROME HOT WORK.
M-365; 0.9 C, 3.75 Cr, bal Fe.
For tools, dies, mandrels; hot work steel.

CHROME IRON.
M-Eng; 13-14 Si, Cr, bal Fe.
For pumps, valves, corrosion resisting vessels; similar to "Duriron."

CHROMEL 1.
M-65; 37 Ni, 21 Cr, 2 Si, bal Fe.
Drawn wire: 115,000 psi TS.
For heating elements; corrosion and heat resistant 1900°F.

CHROMEL 70/30.
M-65; 70 Ni, 30 Cr.
Drawn wire: 128,000 psi TS.
For heating elements to 2150°F; corrosion and heat resistant.

CHROMEL A.
M-65; 20.0 Cr, 0.5 Fe, 1.0 Si, bal Ni.
21 gauge wire: 120,000 TS.
Elec. resistance: 650 ohms/circular mil ft.
For restance wire for heating elements up to 2150°F; corrosion and heat resistant.
Toasters, electric furnaces, electric ranges.

CHROMEL AA.
M-65; 20.0 Cr, 8.3 Fe. 2.0 Si, bal Ni.
21 gauge wire: 130,000 TS.
Elec. resistance: 700 ohms/circular mil ft.
Resistance wire for furnaces operating in controlled atmosphere to 2250°F; more resistant to reducing atmospheres.

CHROMEL C.
M-65; 16.0 Cr, 25.5 Fe, 1.3 Si, bal Ni.
21 gauge wire: 110,000 TS.
Elec. resistance: 675 ohms/circular mil ft.
Resistance wire for heating elements up to 1850°F; corrosion and heat resistant. Flat irons, dryers, waffle and sandwich grills.

CHROMEL D.
M-65; 18.5 Cr, 44.0 Fe, 1.5 Si, bal Ni.
21 gauge wire: 105,000 TS.
Elec. resistance: 600 ohms/circular mil ft.
Resistance wire for heating elements to 1800°F.
Industrial furnaces operating to 1800°F.
Good hot strength.

CHROMEL P.
M-65; 90 Ni, 10 Cr, (plus 9 minor constituents).
Ann: 95,000 TS.
EMF: positive.
Thermocouple wire-matched with Alumel for temperature measurement to 2400°F; corrosion and heat resistant.

CHROME-MOLY.
M-157; 2.75-3.0 C, 2-2.6 Si, 0.8-1.0 Mn, 0.4 Cr, 0.4-0.6 Mo, bal Fe.
Cast: 35,000-45,000 TS; 230-270 Brin.
For heavy dies; tough, close grained.

CHROMENAR 308 ETC.
M-677; one of 36 bare stainless steel wires to 24 AISI standard stainless steel grades plus 12 modified grades. For welding or build-up on stainless steels.

CHROMEND 1 MA.
M-677; 0.5 C, 0.75 Mn, 0.6 Si, 0.8-1.1 Cr, 0.45-0.65 Mo, bal Fe.
Welded: 85,000-100,000 TS; 68,000-88,000 YS; 23-20 El; 65-50 RA.
For welding rod; repairing 1% Cr steel castings and pipe.

CHROMEND 2M.
M-677; 0.10 C, 2 Cr, 0.9 Mo, bal Fe.
Welded: 150,000 TS; 120,000 YS; 4 El; 10 RA; 325 Brin.
For welding rod; for low Cr steels.

CHROMEND 2MA.
M-677; 0.03 C, 0.7 Mn, 0.6 Si, 2.0-2.5 Cr, 0.9-1.2 Mo, bal Fe.
Welded: 113,000 TS; 99,000 YS; 18 El; 48 RA; stress relieved (1350°F): 92,000 TS; 85,000 YS; 25 El; 72 RA.
For welding rod, type E9018-B3L; welding 21/4 Cr, 1 Mo steel.

CHROMEND 5M.
M-677; 0.10 C, 4-6 Cr, 0.5 Mo, bal Fe.
For welding rod.

CHROMEND 14/75.
M-677; 75 Ni, 14 Cr, bal Fe.
For welding electrodes; "Inconel," heat and corrosion resistant.

CHROMEND 16-8-2.
M-677; 0.05 C, 2.2 Mn, 0.3 Si, 16.0 Cr, 8.5 Ni, 1.5 Mo, bal Fe.
For welding electrodes; stainless steel for high temperature service free from sigma-phase embrittlement.

CHROMEND 307.
M-677; 0.07-0.17 C, 3.3-4.75 Mn, 0.25 max Mo, 18-20.5 Cr, 9-10.5 Ni, bal Fe.
Welded: 84,000-96,000 TS; 56,000-62,000 YS; 46-37 El; 49-35 RA.
For welding electrodes; for armor, coated.

CHROMEND 308.
M-677; 0.07 C, 19 Cr, 9 Ni, bal Fe.
For welding rod; stainless.

CHROMEND 308HC.
M-677; 0.14 C, 1.7 Mn, 0.4 Si, 20.0 Cr, 9.7 Ni, bal Fe.
For welding electrodes; specially coated for welding austenitic Cr-Ni stainless steel.

CHROMEND 308L.
M-677; 0.035 C, 1.8 Mn, 0.4 Si, 20.0 Cr, 9.8 Ni, bal Fe.
For welding electrodes; stainless steel type 308l.

CHROMEND 309.
M-677; 0.10 C, 25 Cr, 12 Ni, bal Fe.
For weld metal; stainless.

CHROMEND 309 CB.
M-677; 0.10 C, 25 Cr, 12 Ni, 0.90-1.1 Cb, bal Fe.
For welding rod; corrosion and heat resistant.

CHROMEND 309MO.
M-677; 0.07 C, 1.8 Mn, 0.4 Si, 23.0 Cr, 12.8 Ni, 2.5 Mo, bal Fe.
For welding electrodes; Mo-bearing stainless steel for root pass welding of 316,316L, 319 Clad steel.

CHROMEND 310.
M-677; 0.10 C, 25 Cr, 20 Ni, bal Fe.
For welding rod; heat and corrosion resistant.

CHROMEND 310 CB.
M-677; C, 25 Cr, 20 Ni, Cb, bal Fe.
For welding electrodes; stainless, austenitic.

CHROMEND 310HC.
M-677; 0.22 C (also 0.30 C, also 0.40 c) 1.8 Mn, 0.4 Si, 26.5 Cr, 21.0 Ni, bal Fe.
For welding electrodes; welding ACI types CK-20, HK, HL, HN castings; supplied in three carbon ranges.

CHROMEND 310 MO.
M-677; C, 25 Cr, 20 Ni, Mo, bal Fe.
For welding electrodes; stainless, austenitic.

CHROMEND 312.
M-677; C, 29 Cr, 9 Ni, bal Fe.
For welding rod; heat resistant.

CHROMEND 316.
M-677; 0.08 C, 2 Mo, 18 Cr, 13 Ni, bal Fe.
For welding electrodes; stainless.

CHROMEND 316L.
M-677; 0.035 C, 1.8 Mn, 0.4 Si, 19.5 Cr, 12.5 Ni, 2.2 Mo, bal Fe.
For welding electrodes; stainless steel type 316L.

CHROMEND 317.
M-677; 0.07 C, 18 Cr, 12 Ni, 3 Mo, bal Fe.
For welding rod; corrosion resistant.

CHROMEND 318.
M-677; C, 18 Cr, 12 Ni, Mo, Cb, bal Fe.
For welding electrodes; stainless, austenitic.

CHROMEND 320.
M-677; 0.05 C, 1.1 Mn, 0.4 Si, 20.0 Cr, 34.0 Ni, 2.5 Mo, 0.6 Cb, 3.4 Cu, bal Fe.
For welding electrodes; Carpenter 20 Cb-3.

CHROMEND 330.
M-677; C, 15 Cr, 35 Ni, bal Fe.
For welding rod; heat resistant.

CHROMEND 330HC.
M-677; 0.30 C (also 0.40 C), 1.6 Mn, 0.6 Si, 16.0 Cr, 34.0 Ni, bal Fe.
For welding electrodes; welding ACI types HT and HU castings; supplied in two carbon ranges.

CHROMEND 347.
M-677; 0.07 C, 19 Cr, 9 Ni, 0.8-1.0 Cb, bal Fe.
For welding rod; corrosion resistant.

CHROMEND 349.
M-677; C, 19 Cr, 9 Ni, W, Mo, bal Fe.
For welding electrodes; stainless, austenitic.

CHROMEND 410.
M-677; 0.10 C, 12 Cr, bal Fe.
For welding rod; corrosion resistant.

CHROMEND 410NI.
M-677; 0.04 C, 0.6 Mn, 0.4 Si, 12 Cr, 4 Ni, 0.8 Mo, bal Fe.
For welding electrodes; welding CA6NM casting.

CHROMEND 430.
M-677; 0.10 C, 16 Cr, bal Fe.
For welding rod; corrosion resistant.

CHROMEND 442.
M-677; 0.10 C, 18 Cr, bal Fe.
For welding rod; corrosion resistant.

CHROMEND 446.
M-677; C, 28 Cr, bal Fe.
For welding rod; heat resistant.

CHROMEND 505.
M-677; 0.2 C, 8-10 Cr, 1.5 Mo, bal Fe.
For welding electrodes; for high temperatures, corrosion resistant.

CHROMEND CMV.
M-677; Timken 17/22A(S) steel, 0.3 C, 1.25 Cr, 0.5 Mo, 0.25 V, bal
Heat treated: 140,000-240,000 psi.
For welding electrodes.

CHROME NICKEL.
M-297; 20 Cr, 77 Ni, 2 Mn.
Annealed: 121,000 TS; 23 El.
Hard: 195,000 TS; 1 El.
For heat resistances, electrical heating elements, rheostats; max operating temperature 1150°C.

CHROME NICKEL IRON.
M-297; 20 Cr, 70 Ni, 8 Fe, 2 Mn.
Annealed: 114,000 TS; 24 El.
Hard drawn: 182,000 TS; 1 El.
For electrical heating elements, rheostats, electrical measuring instruments; max operating temperature 1150°C.

CHROME NICKEL SILVER.
M-538; Ni, Zn, bal Cu.
For automatic fabrication, sheets and strips; nickel silver sheet bonded with chromium, rust proof.

CHROME-NICKEL STEEL ARMOR PLATE.
M-Eng; 4 Ni, 2 Cr, 0.33 C, 0.32 Mn, 0.06 Si, 0.03 S, 0.14 P, bal
For armor plate; oil hardening.

CHROME ROLL.
M-85; 0.9 C, 1 Cr, bal Fe.
For tools, cutters, dies; water hardening.

CHROMESCO 1.
M-1495; 0.14 C, 0.5 Cr, 0.5 Mo, bal Fe.
Good creep rupture properties up to 550°C.
For boilers, steampipes, superheaters.

CHROMESCO 3.
M-1495; 0.15 max C, 1.9-2.6 Cr, 0.80-1.2 Mo, bal Fe.
Annealed: 60,000 TS; 30,000 YS; 30 El; 163 Brin.
For oil refinery tubes and equipment; ASTM-A-200.

CHROME SILICON SPRING.
M-12; 0.5-0.6 C, 0.5-0.8 Mn, 1.2-1.6 Si, 0.5-0.8 Cr, bal Fe.

CHROME STEEL.
M-538; Cr bonded to steel.
For floor plates, reflectors, trim and moulding, toasters; resists heat to 1050°F; Cr bonded to steel.

CHROMET.
M-German; 10 Si, bal Al.
For bearings.

CHROME TIN.
M-538; Cr coated Sn.

CHROMETOUGH.
M-1724; 0.55 C, 0.55 Mn, 0.50 Cr, 0.20 V, bal Fe.
Oil hardening general purpose tool steel; for boring bars, chisels, rivet snaps.

CHROME WEAR.
M-115; 2.3 C, 0.7 Mn, 0.4 Si, 5.25 Cr, 4.75 V, 1.1 W, 1.1 Mo, bal Fe.
Heat treatable to high hardness and wear resistance; for tube manufacturing rolls, brick mold liners.
Type A7 air hardening tool steel.

CHROMEWEAR 300.
M-115; 2.70 C, 0.40 Si, 0.70 Mn, 8.25 Cr, 4.5 V, 1.12 Mo, bal Fe.
Wear resistant tool steel; air hardenable to 66 Rc.
For deep drawing dies, brick mold liners.
Liners for sand slingers, extrusion punches for ceramics.

CHROMEX 2.
M-USSR; 1.2-1.7 C, 32-35 Cr, 1.2-2.0 Si, bal Fe.
For furnace equipment; heat resistant, cast iron.

CHROMEX 3.
M-USSR; 1.8-2.8 C, 32-32 Cr, 1.5-2.5 Si, bal Fe.
For furnace equipment; heat resistant, cast iron.

CHROMIDIUM-1.
M-185; 3.2 T.C., 0.7 C.C., 2.2 Si, 0.25-0.45 Cr, 0.8 Mn, bal Fe.
Cast: 35,000-45,000 TS; 35,000-45,000 YS; 0 El; 0 RA; 200-250 Brin.
For auto engine blocks, cylinders, brake drums; wear resistant, cast iron.

CHROMIDIUM 2.
M-185; 3.1-3.5 C, 2.0-2.4 Si, 0.75-1.0 Mn, 0.2-0.4 Cr, bal Fe.
Cast: 37,000 TS; 200-240 Brin.
For brake drums, brake discs; Tr,S. 72,000; cast iron.

CHROMIN.
M-Ger.
For anodes for Cr-plating.

CHROMIUM.
M-1755; Cr.
Purties: 99.999%, 99.99%, 99.9%, 99.6+% (commercial grade).
Forms: Crystal bar (ductile), degassed and regular powders, pellets, foil rods (hot pressed), tubules, flake, lumps, composites, single crystals, discs (arc melted).

CHROMIUM BORIDE B401.
M-U.S.; CrB.
Sintered.
For high heat applications; operating temperature range above 2000°F.

CHROMIUM COPPER-182.
M-8; 99.14 Cu, 0.85 Cr, 0.01 Si.
Hard Rod: 70,000 TS; 60,000 YS; 20 El; B 82 Rock.
Soft Rod: 35,000 TS; 15,000 YS; 40 El.
For cable connectors, electronic devices, resistance welding electrode tips, grid supports, circuit breakers. Age hardenable. Corrosion resistant.
Hardened elect. cond. 80%.

CHROMIUM-COPPER 999.
M-8; 99.05 Cu, 0.85 Cr, 0.10 Si.
Soft: 35,000 TS; 15,000 YS; 40 El; 877 Brin.
Hard: 72,000 TS; 61,000 YS; 20 El.
For electrical apparatus; high conductivity.

CROMNICKEL.
M-297; 20 Cr, 77 Ni, 2 Mn.
For heating resistances up to 1100°C.

CHROMNICKEL 1.
M-297; 78 Ni, 20 Cr, 2 Mn.
For furnace parts, electric heating elements; heat and corrosion resistant.

CHROMNICKEL II.
M-297; 70 Ni, 20 Cr, 2 Mn, 8 Fe.
For electrical heating elements; heat and acid resistant.

CHROMODUR-22.
M-72; 0.20 C, 12 Cr, 1 Mo, 0.4 Ni, 0.3 V, bal Fe.
Annealed: 90,000 TS; 40,000 YS; 22 El; B 95 Rock.
Heat treated: 240,000 TS; 200,000 YS; 8 El; 24 RA; C 50 Rock.
For gears, knives, shafts, hardware, surgical instruments.
Corrosion resistant, hardenable.

CHROMODUR-33.
M-72; 0.20 C, 12 Cr, 1 Mo, 0.4 Ni, 0.3 V, 0.5 W, bal Fe.
Annealed: 95,000 TS; 40,000 YS; 25 El; B 92 Rock.
Heat treated: 240,000 TS; 205,000 YS; 10 El; C 48 Rock.
For cutlery, hardware, knives, oil refinery and chemical plant equipment.
Corrosion resistant, hardenable.

CHROMOLD-VM.
M-115; 0.06-0.12 C, 0.15-0.25 Si, 0.10-0.40 Mn, 0.010 max S, 0.010 max P, 2.15-2.45 Cr, bal Fe.
Heat treated: 110,000-120,000 TS; 70,000-90,000 YS; C 20-25 Rock in core and C 64 in case.
Plastic molds cavities.
Vacuum melted. Case hardening steel. Easy hubbing for mold cavities.

CHROMO-LOY.
M-507; 3.0 C, 2.0 Si, 0.43 Mn, 21.6 Cr, 4.2 Mo, 3.4 Ni, bal Fe.
Coated, hardfacing electrode.
As arc welded: 70,000 psi, approx., 56 Rc.
For overlay or buildup on roll crushers, shovel teeth, bucket lips and oil field tool joints; tough and wear resistant.

CHROMO-N.
M-24; 0.23 C, 1.25 Si, 0.75 Ni, 10 Cr, 1.2 Mo, 1.0 V, bal Fe.
For extrusion mandrels, hot work tools and dies; hot work steel.

CHRO-MOW.
M-38; 0.35 C, 1.05 Si, 5.0 Cr, 1.25 W, 0.35 V, 1.35 Mo, bal Fe.
Air harden to 51-53 Rc.
For hot work punches, shell piercing tools, mandrels for Al & Mg extrusion.
Type H12 Hot work tool steel.

CHROMSOL.
M-48; 62 Cr, 5 Mn, 1.5 Si, 5.25 C bal Fe.
For chromium additions to steel melts.
Steel alloying agent.

CHROM SPECIAL now **VEW K100.**

CHROM SPECIAL EXTRA now **VEW K107.**

CHROM SPECIAL SUPRA now **VEW K103.**

CHROM-SUPER SERVICE.
M-736; Cr, Sb, Pb, bal Sn.
Cast.
For Babbitt; antifriction.

CHROMTEC 10680.
M-717.
Nickel base alloy powder for steel, stainless, nickel alloys.

CHROM-TEC 19222.
M-717.
Alloy powder for metal spraying corrosion resistant dense coating; machinable.

CHROM-X.
M-1059; ferrochrome, C, Si.
For chromium addition to steel; for alloy steels.

CHRONIFER F 14 now **WESTIG 4006.**

CHRONIFER F 17 now **WESTIG 4016.**

CHRONIFER SPEZIAL 4 SUPRA now **WESTIG 4401.**

CHRONIFER SPEZIAL D now **WESTIG 4305.**

CHRONIFER SPEZIAL EXTRA now **WESTIG 4541.**

CHRONIFER SPEZIAL SUPRA now **WESTIG 4301.**

CHRONIFER V-13 now **WESTIG 4021.**

CHRONIKA 1565.
M-1292; 0.2 Cr, 57 Ni, 15 Cr, bal Fe.
For high temperature applications;
X15NiCr5715; heat resistant.

CHRONIKA 2035.
M-1292; 0.15 C, 30 Ni, 21 Cr, bal Fe.
For high temperature applications;
X15NiCr3021; heat resistant.

CHRONIKA 2080.
M-1292; 0.15 C, 77 Ni, 18 Cr, bal Fe.
For high temperature applications;
X15NiCr7718; heat resistant.

CHRONIKA 2520.
M-1292; 0.15 C, 2 Si, 24 Cr, 19 Ni, bal Fe.
For furnace parts, heat treat boxes;
X15CrNiSi2419; corrosion and heat resistant.

CHRONIN.
M-Eng.; 83.7 Ni, 14.7 Cr.
For resistance alloy.

CHRONIN "85".
M-303; 13 Cr, 85 Ni, bal Cu, Fe, Mn, Al impurities.
Annealed: 100,800 TS; 30 El.
For electric resistances, resistors, thermocouple element against Ni; heat resistance to 1200°C.

CHRONIN "100".
M-303; 18 Cr, 80 Ni, bal Cu, Fe, Mn, Al.
Annealed: 117,000 TS; 30 El.
For electric resistance, resistors, thermocouple element; heat resistant to 1200°C.

CHRONIN 110.
M-303; 20 Cr, 80 Ni.
For electrical equipment, heating elments, resistors; heat and electrical resistance.

CHRONIT 14.
M-1309; 0.20 C, 13.0 Cr, bal Fe.
Annealed: 95,000 TS; 50,000 YS; 25 El; B 92 Rock.
Hardened: 250,000 TS; 215,000 YS; 8 El; C 52 Rock.
For cutlery, surgical instruments, rules, gears, valves, springs, pivots.
Corrosion resistant. Hardenable.

CHRONIT 14 M.
M-1309; 0.18 C, 13.0 Cr, 0.6 Mo, 0.7 Ni, bal Fe.
Annealed: 100,000 TS; 55,000 YS; 20 El; B 95 Rock.
Hardened: 240,000 TS; 210,000 YS; 8 El; C 50 Rock.
For cutlery, surgical instruments, gears, springs, valve parts.
Corrosion resistant. Hardenable.

CHRONIT 14 MS.
M-1309; 0.18 C, 13.0 Cr, 0.6 Mo, 0.7 Ni, others, bal Fe.
Annealed: 100,000 TS; 55,000 YS; 20 El; B 95 Rock.
Hardened: 240,000 TS; 210,000 YS; 8 El; C 50 Rock.
For cutlery, surgical instruments, gears, springs, valve parts.
Corrosion resistant. Hardenable.

CHRONIT 14 N.
M-1309; 0.17 C, 13 Cr, 1.1 Ni, bal Fe.
Annealed: 95,000 TS; 52,000 YS; 22 El; B 92 Rock.
Hardened: 220,000 TS; 180,000 YS; 12 El; B 48 Rock.
For cutlery, surgical instruments, gears, springs, valve trim.
Corrosion resistant. Hardenable.

CHRONIT 14 P.
M-1309; 0.10 C, 12.3 Cr, 0.6 Mo, 0.5 Ni, bal Fe.
Annealed: 85,000 TS; 45,000 YS; 30 El; B 85 Rock.
For chemical and oil refinery equipment, valve trim.
Corrosion resistant.

CHRONIT 18 M.
M-1309; 0.20 C, 16.5 Mo, 1.2 Ni, bal Fe.
Annealed: 80,000 TS; 45,000 YS; 26 El; 140 Brin.
For hardware, burner parts, oil refinery equipment, fasteners, furnace parts, storage tanks.
Corrosion resistant.

CHRONIT 18 N.
M-1309; 0.20 C, 16.5 Cr, 1.2 Ni, bal Fe.
Annealed: 82,000 TS; 43,000 YS; 28 El; 140 Brin.
For hardware, burner parts, oil refinery equipment, furnace parts.
Corrosion resistant.

CHRONIT 30.
M-1309; 1.0 C, 28.0 Cr, bal Fe.
Cast, ann: 260-330 Brin.
Ferrite-carbide structure; stainless and acid resistant cast steel.
DIN G-X120 Cr 29.

CHRONIT 30 C.
M-1309; 1.50 C, 28.0 Cr, bal Fe.
Cast, ann: 260-330 Brin.
Ferrite-carbide structure; stainless and acid resistant cast steel.

CHRONIT 30 M.
M-1309; 1.0 C, 28.0 Cr, 2.25 Mo, 1.25 Cu, bal Fe.
Cast, ann: 260-330 Brin.
Ferrite-carbide structure: stainless and acid resistant cast steel.
DIN G-X 120 Cr Mo 292.

CHRONIT 30 MW.
M-1309; 0.70 C, 28.0 Cr, 2.25 Mo, bal Fe.
Cast, ann: 219-280 Brin.
Ferrite-carbide structure; stainless and acid resistant cast steel.
DIN G-X CrMo292.

CHRONIT 30 W.
M-1309; 0.70 C, 28.0 Cr, bal Fe.
Cast, ann: 210-280 Brin.
Ferrite-carbide structure; stainless and acid resistant cast steel.
DIN G-X 70 Cr29.

CHRONIT 165 M.
M-1309; 0.07 max C, 16.0 Cr, 2.0 Mo, 6.0 Ni, others, bal Fe.
Annealed: 85,000 TS; 35,000 YS; 50 El; B 80 Rock.
Precipitation hardenable to about 400 Brin.
For chemical and pharmaceutical plant equipment digesters, valve trim, acid containers, tanks.
Corrosion resistant.

CHRONIT 218.
M-1309; 0.15 max C, 18.0 Cr, 9.0 Ni, bal Fe.
Annealed: 90,000 TS; 40,000 YS; 50 El; B 85 Rock.
Cold rolled: 150,000 TS; 100,000 YS; 25 Eli; B 100 Rock.
For chemical and pharmaceutical equipment, valve trim, fasteners, molding, acid tanks, digesters.
Stainless, austentic, non-magnetic.

CHRONIT 218 E.
M-1309; 0.10 max C, 18.5 Cr, 10.0 Ni, Cb = 8 x C, bal Fe.
Annealed: 85,000 TS; 35,000 YS; 50 El; B 82 Rock.
Cold rolled: 130,000 TS; 90,000 YS; 30 El; B 100 Rock.
For welded structures, chemical plant equipment, vessels, agitators. Austenitic, stabilized.
Welding grade, stainless, non-magnetic.

CHRONIT 218 S.
M-1309; 0.07 max C, 18.5 Cr, 10.0 Ni, bal Fe.
Annealed: 85,000 TS; 35,000 YS; 50 El; B 82 Rock.
For chemical plant equipment, tanks, evaporators, agitators, kettles.
Stainless, austenitic, non-magnetic.

CHRONIT 274 R.
M-1309; 0.04 C, 27.0 Cr, 4.0 Ni, bal Fe.
Cast, ann: 95,000-120,000 TS; 4 El; 210-260 Brin.
Ferrite-austenite-carbide structure.
Stainless and acid resistant cast steel.
DIN G-X CrNi274.

CHRONIT 274 RM.
M-1309; 0.40 C, 27.0 Cr, 2.25 Mo, 4.0 Ni, bal Fe.
Cast, quenched: 210-260 Brin; 95,000-120,000 TS; 4 El.
Ferrite-austenite-carbide; stainless and acid resistant cast steel.

CHRONITE 275 RM.
M-1309; 0.10 C, 27.0 Cr, 1.5 Mo, 5.0 Ni, bal Fe.
Cast, quenched: 85,000-110,000 TS; 50,000 YS; 8 El; approx. 240 Brin
Ferrite-austenite structure; stainless and acid resistant cast steel.
DIN G-X10 CrNiMo 275.

CHRONIT 418.
M-1309; 0.15 max C, 18.0 Cr, 2.25 Mo, 10.0 Ni, bal Fe.
Annealed: 90,000 TS; 40,000 YS; 40 El; B 85 Rock.
For chemical plant and oil refinery equipment digesters, valve trim, evaporators.
Stainless, austenitic, non-magnetic.

CHRONIT 418 E.
M-1309; 0.10 max C, 17.5 Cr, 2.25 Mo, 11.5 Ni, Cb = 9 x C, bal Fe.
Annealed: 90,000 TS; 40,000 YS; 40 El; B 85 Rock.
For welded structures, tanks, vessels, chemical plant and oil refinery equipment austenitic.
Welding grade, stainless, non-magnetic.

CHRONITE 418 ES.
M-1309; 0.10 max C, 17.5 Cr, 2.25 Mo, 11.5 Ni, Cb = 9 x C, bal Fe.
Annealed: 85,000 TS; 35,000 YS; 50 El; B 82 Rock.
For acid tanks, evaporators, kettles, chemical plant equipment. Austenitic, stabilized.
Welding grade, stainless, non-magnetic.

CHRONIT 418 S.
M-1309; 0.07 max C, 18.0 Cr, 2.25 Mo, 11.0 Ni, bal Fe.
Annealed: 85,000 TS; 35,000 YS; 50 El; B 82 Rock.
For chemical plant equipment, kettles, agitators, evaporators.
Stainless, austenitic, non-magnetic.

CHRONIT 1417.
 M-1309; 0.07 max C, 17.0 Cr, 4.5 Mo, 13.5 Ni, Cb = 8 x C, bal Fe.
 Annealed: 90,000 TS; 45,000 YS; 35 El; 160 Brin.
 For chemical and oil refinery equipment, digesters, acid tanks, agitators.
 Austenitic, non-hardenable.
 Welding grade, stainless, non-magnetic.

CHRONIT 1618.
 M-1309; 0.08 max C, 17.5 Cr, 2.25 Mo, 20.0 Ni, 2.0 Cu, Nb = 8 x C, bal Fe.
 Cast, quenched; 65,000-92,000 TS; 25,000 YS; 15 El; 130-180 Brin.
 Acid resistant chrome-nickel austenitic cast steel. DIN G-X8CrNiMoCu1818.

CHRONIT 2025.
 M-1309; 0.08 max C, 20.0 Cr, 2.75 Mo, 25.0 Ni, 2.0 Cu, Cb = 8 x C, bal Fe.
 Annealed: 95,000 TS; 45,000 YS; 45 El; B 90 Rock.
 For furnace parts, heat treating boxes, carburizing boxes, valves, pumps, jet engine parts.
 Non-hardenable. Welding grade, heat resistant, austenitic.

CHRONIT CNM100 now **VEW K618.**

CHRONIT VHS now **VEW F126.**

CHRONIT VM now **VEW K724.**

CHRONIT VMG.
 M-27; 1.3 C, 25 Cr, 4 W, 67 Co, 2 max Fe.
 Cast: Rock C 42-45.
 For hardfacing. Wear and corrosion resistant.

CHRONIT VMH.
 M-27; 0.9 C, 1.8 Mn, bal Fe.
 Heat treated: 171,000 TS; 350 Brin.
 For wear resistant parts, punches; oil hardened.

CHRONIT ZII.
 M-27; 2 C, 2 Cr, bal Fe.
 Heat treated: 185,000 TS; 370 Brin.
 For wear resistant parts; oil hardened.

CHRONOS now **VEW K700.**

CHRYSIODE.
 92 Ag, 8 Al.

CHRYSITE.
 M-Eng.; 63 Cu, 37 Zn, 0.24 Pb.
 For dental alloy.

CHRYSOKALK-1.
 M-Eng.; 95 Cu, 4.5 Zn, 0.5 Pb.
 For cheap jewelry; tarnishes easily.

CHRYSOKALK-2.
 M-Eng.; 59 Cu, 40 Zn, 1 Pb.
 For tubes.

CHRYSOKALK-3.
 M-Eng.; 91 Cu, 7.9 Zn, 1-6 Pb.
 For cheap jewelry.

CHRYSORIN.
 M-Eng.; 72-63 Cu, 28-37 Zn.
 40,000-56,000 TS; 20,000-35,000 YS; 78-35 El; 78-63 RA.
 For tubes, pipes, plumbing; high strength.

CHUGAL.
 M-Russia; 2.5-3.2 C, 1.6-2.3 Si, 5.5-20.0 Al, bal Fe.
 For furnace equipment; cast iron.

C.H.W.
 M-73; 0.5 C, 2.75 Cr, 0.5 V, 15.0 W, bal Fe.
 For dies, punches, shear blades, trimmers; hot work steel.

CICRON - 1.
 M-1766; 0.65 C, 1.2 max Si, 0.80 Cr, 0.30 Mo, bal Fe.
 Hot work tool steel.

CICRON-2.
 M-1766; 0.60 C, 1.2 max Si, 0.80 Cr, 0.30 Mo, bal Fe.
 Hot work tool steel.

C.I.G. NO. 400.
 M-734; 10 Sn, bal Cu.
 For bronze welding rod; fluxed.

C.I.G. NO. 401.
 M-734; Mn, Sn, bal Cu.
 For bronze welding rod; fluxed.

CINDAL.
 M-737; 0.8 Zn, 0.3 Mg, 0.3 Cr, bal Al.
 27,000-58,000 TS; 5-20 El.
 For light alloy parts; heat treatable.

CINDAL ALLOY "J-551".
 M-298; 5 Si, bal Al.
 22,000-26,000 TS; 11,000-15,000 YS; 17-15 El.
 Used to resist acetic acid, edible oils, acids, beer, cider, fruit juices; stainless and non-corrosive in sea water.

CINDALL 50 "A".
 M-298; Al alloy.
 36,000 TS; 22,500 YS; 1-1.5 El; 140 Brin.
 Domestic castings, light gear wheels, light pulleys, bearings and brushings; to replace cast iron where lightness and corrosion resistance are required.

CINDALL ALLOY "J-51".
 M-298; Al alloy.
 20,000-24,000 TS; 9,000-11,000 YS; 28-24 El.
 For parts to resist sea water, fruit acids, beer and cider vats; non-corrosive to sea water.

Section I: Alloy Data / 313

CINDALL E-11 "A".
M-298; Al alloy.
31,500 TS; 14,000 YS; 5-8 El; 66 Brin.
For water pipes, carburetor parts; sand or chill castings subjected to atmospheric, sea water, or other corrosive influences; boron modified.

CINDALL E-11 "B".
M-298; Al alloy.
36,000 TS; 18,500 YS; 4-6 El; 72 Brin.
For propellers, water cooled cylinders, cylinder heads, pump impellers, stern tubes, water pump bodies; sand castings, boron modified.

CINDALL "J-12".
M-298; Al alloy.
15,500-21,200 TS; 11,000-13,000 YS; 11-9 El.
For resistance to industrial waters and alkaline solutions, radiators, water meters; non-hardenable.

CINDALL "L-316".
M-298; Al alloy.
29,000 TS; 16,000 YS; 2-4 El; 68 Brin.
For exhaust manifolds, automobile parts; sand and die castings.

CINIDUR.
England; 0.25 C, 19 Cr, 24 Ni, 2 Mo, 1 W, 2.25 Ti, 1.0 Al, bal Fe.
For high temperature applications; heat and corrosive resistant.

CINIDUR.
U.S.; 0.25 C, 19 Cr, 24 Ni, 2 Mo, 1 W, 2.25 Ti, 1 Al, bal Fe.
For oil refinery equipment, furnace parts; heat resistant.

CINSEAL.
M-269; 29 Ni, 53 Fe, 17.5 Co, 0.5 Mn.
For telecommunications, hermetic seals; metal-to-glass seal.

CIRCLE "C".
M-155; 0.77 C, 9 Co, 18.5 W, 4.5 Cr, 2 V, 1.0 Mo, bal Fe.
For tools, cutters, drills, tools for milling, slotting, forming; super high speed steel; tough heavy duty.

CIRCLE L 3 replaced by **LEBANON 4140.**

CIRCLE L 5 replaced by **LEBANON 4330.**

CIRCLE L 8 replaced by **LEBANON WC9.**

CIRCLE L 9 replaced by **LEBANON WC1.**

CIRCLE L 10 replaced by **LEBANON C5.**

CIRCLE L 11 replaced by **LEBANON 442.**

CIRCLE L 12 replaced by **LEBANON CA15.**

CIRCLE L 12M replaced by **LEBANON CA15M.**

CIRCLE L 13 replaced by **LEBANON CA40.**

CIRCLE L 15 replaced by **LEBANON CC50.**

CIRCLE L 17 replaced by **LEBANON 17-4.**

CIRCLE L 19 replaced by **LEBANON LC2.**

CIRCLE L 22 replaced by **LEBANON CF8.**

CIRCLE L 22 AGXM replaced by **LEBANON CG8M.**

CIRCLE L 22L replaced by **LEBANON CF3.**

CIRCLE L 22M replaced by **LEBANON CF8C.**

CIRCLE L 22XM replaced by **LEBANON CF8M.**

CIRCLE L 22 XML replaced by **LEBANON CF3M.**

CIRCLE L 23 replaced by **LEBANON CF20.**

CIRCLE L 30H replaced by **LEBANON HH.**

CIRCLE L 31 replaced by **LEBANON CE30.**

CIRCLE L 31H replaced by **LEBANON HE.**

CIRCLE L 32 replaced by **LEBANON HT.**

CIRCLE L 33 replaced by **LEBANON 33.**

CIRCLE L 34 replaced by **LEBANON CN7M.**

CIRCLE L 41 replaced by **LEBANON HX.**

CIRCLE L 46 replaced by **LEBANON CK20.**

CIRCLE L 46H replaced by **LEBANON HK.**

CIRCLE L 91 replaced by **LEBANON C-12.**

CIRCLEF L 109 replaced by **LEBANON WC6A.**

CIRCLE L 130 replaced by **LEBANON CHW.**

CIRCLE L 205A replaced by **LEBANON 8630.**

CIRCLE L 205AL replaced by **LEBANON 8630-1.**

CIRCLE L 205B replaced by **LEBANON 8630-2.**

CIRCLE L 205C replaced by **LEBANON 8630-3.**

CIRCLE L 205D replaced by **LEBANON 8630-4.**

CIRCLE L 206 replaced by **LEBANON 8613.**

CIRCLE L 209 replaced by **LEBANON WC6.**

CIRCLE L 219 replaced by **LEBANON LC3.**

CIRCLE L 431 replaced by **LEBANON 431.**

CIRCLE LA replaced by **LEBANON 1040.**

CIRCLE LB replaced by **LEBANON WCA.**

CIRCLE LB20 replaced by **LEBANON WCB.**

CIRCLE LCD4MCU replaced by **LEBANON CD.**

CIRCLE LHB replaced by **LEBANON HAB.**

CIRCLE LHC replaced by **LEBANON HAC.**

CIRCLE LIN replaced by **LEBANON INC.**

CIRCLE LM replaced by **LEBANON ME.**

CIRCLE M.
M-57; 0.85 C, 6 W, 5 Mo, 4 Cr, 2 V, 8 Co, bal Fe.
For drills, hobs, reamers, taps, broaches, cutters; high speed steel, Type T5.

CITROEN.
M-English; 12 Cu, 0.1 Mg, 0.5 Mn, 0.5 Fe, 0.5 Si, bal Al.
For pistons; cast.

CLAMERS ALLOY.
25-5 Ni, 5-25 Co, bal Fe.
For electrical machinery; heat and corrosion resistant.

CLARITE.
M-35; 0.7 C, 18 W, 4 Cr, 1.1 V, 0.25 Mn, 0.6 Mo, bal Fe.
Ht. Tr.: C 64 Rock; 400,000 bend strength, 300,000 torsion strength.
For lathe and planer tools, drills, punches, dies, cutters, springs.
Type T 1 high-speed steel, high red-hardness.

CLARITE HW26.
M-35; 0.53 C, 18.0 W, 4.0 Cr, 1.0 V, bal Fe.
For punches, piercing tools, shear blades, forming rolls.
AISI Type H 26 hot work tool steel.

CLARITE HW 60.
M-35; 0.63 C, 0.3 Mn, 0.3 Si, 4.0 Cr, 18.0 W, 1.1 V, bal Fe.
Hot work tool steel.

CLARK'S ALLOY.
M-Eng.; 75 Cu, 7.2 Zn, 14 Ni, 1.9 Co.
For chemical equipment; corrosion resistant.

CLARK'S PATENT.
M-Eng.; 75 Cu, 14 Ni, 7.2 Zn, 1.9 Sn, 1.9 Co.
For corrosion resisting parts; corrosion resistant.

CLARUS METAL.
M-407; 1.5 Cu, 4 Si, bal Al.
For light alloy parts; will not oxidize; 60% stronger than Al.

CLASS "P".
M-210; 0.7-0.9 C, bal Fe.
For tools, punches, dies, engraved dies, trimming dies; water hardening.

CLAY-LOY.
M-1529; 0.22 C, 1.25 Mn, 0.35 Si, 0.50 Cu, 0.20 V, bal Fe.
Rolled: 70,000 TS; 50,000 YS.
For railroad and bus bodies; high strength, low alloy construction steel.

CLEBRIUM-1.
M-Eng.; 2 Ni, 13.1 Cr, 0.75 Mn, 3.6 Mo, 1.5 Si, 2.6 C, bal Fe.
For heat and corrosion resisting cast iron castings; high heat resistant.

CLEBRIUM-2.
M-Eng.; 4.6 Ni, 18.3 Cr, 2.8 Mn, 2.0 C, 2 Cu, bal Fe.
For heat and corrosion resisting cast iron castings; heat and corrosion resistant.

CLEREMONT DRILL ROD.
M-38; 1.05 C, bal Fe.
For tools, pivots; drill rod.

CLEVITE 77 see **CLEVITE F-77.**

CLEVITE 100 see **CLEVITE F-100.**

CLEVITE 112 see **CLEVITE F-112.**

CLEVITE 153 see **CLEVITE F-153.**

CLEVITE 250 see **CLEVITE F-250.**

CLEVITE 500.
M-739; 44-58 Pb, 0.5-1.5 Sn, bal Cu, sintercast with steel backing.
For bearings.

CLEVITE F-1.
M-739; 3.0-3.5 Cu, 7.25-7.75 Sb, 0.1-0.14 Te, bal Sn.
For engine bearings; steel backed Babbitt.

CLEVITE F-4.
M-739; 7-9 Pb, 3.5-4.5 Sn, 4 Zn, bal Cu.
For wrist pin and transmission bearings; with steel backing. Was Clevite No. 8.

CLEVITE F-5.
M-739; 9-11 Pb, 9-11 Sn, 0.5 max Zn, bal Cu.
For wrist pin and transmission bearings; with steel backing. Was Clevite No. 10.

CLEVITE F-7.
M-739; 21-25 Pb, 3-4 Sn, 3 Zn, bal Cu.
For steering knuckle bushings; steel backing. Was Clevite No 25.

CLEVITE F-17.
M-739; 8-12 Pb, 8-12 Sn, bal Cu, overlay 3-5 Cu, 7-8 Sb, bal Sn.
For heavy duty bearings; with steel backing. Was F-17 Trimetal.

CLEVITE F-23.
M-739; 0.90-1.25 Sn, 14.75-15.50 Sb, 0.8-1.1 As, 0.6 max Cu, bal Pb.
For bearings.
SAE 15. Was F-23.

CLEVITE F-66.
M-739; 3.5-4.5 Si, 0.5-1.0 Cu, 7.5-9.5 pb, 1.25-1.75 Sn, bal Al.
Bimetal/steel backed. For intermediate range plain bearings, bushings and washers.

CLEVITE F-77.
M-739; 22-26 Pb, 0.15-0.50 Sn, bal Cu, overlay 8-12 Sn, 2.3 Cu, bal Pb.
For bearings; with steel backing. Was Clevite 77.

CLEVITE F-100.
M-739; 9-11 Pb, 9-11 Sn, bal Cu.
For transmission bushings and washers; sintered on steel back. Was Clevite 100.

CLEVITE F-112.
M-739; 21-27 Pb, 1.75-2.75 Sn, bal Cu, overlay 8-12 Sn, 2-3 Cu, bal Pb.
For bearings; with steel backing. Was Clevite 112.

CLEVITE F-153.
M-739; 3.5-4.5 Sn, 0.75-1.4 Cd, bal Al, overlay 8-12 Sn, 2-3 Cu, bal Pb.
For bearings, aluminum clad to steel. Was Clevite 153.

CLEVITE F-154.
M-739; 3.5-4.5 Si, 0.75-1.4 Cd, 0.05-0.15 Cu, 0.10-0.20 Mg, bal Al.
Overlay: 87.5 Pb, 10 Sn, 2.5 Cu.
Trimetal/steel backed. For heavy-duty automotive, truck and diesel engine bearings.

CLEVITE F-250.
M-739; 21-25 Pb, 3-4 Sn, bal Cu.
For transmission bushings and washers; sintered on steel back. Was Clevite 250.

CLEVITE NO. 8 see CLEVITE F-4.

CLEVITE NO 10 see CLEVITE F-5.

CLEVITE NO 25 see CLEVITE F-7.

CLEVITE S-56.
M-739; 80-86 Cu, 3.5-5.0 Sn, 3.5-4.5 Pb, 4 Zn.
For bushings, bearings; SAE 791.
Was No. 444 Alloy.

CLICHIER METAL.
M-Eng; 5 Pb, 80 Sn, 15 Bi.
For bearings, fuses; antifriction.

CLICHIER METAL.
M-Eng.; 33 Pb, 48 Sn, 9 Bi, 11 Sb.
For bearings, fuses; antifriction.

CLICHIER METAL.
M-Eng.; 50 Pb, 36 Sn, 14 Cd.
For bearings, fuses; antifriction.

CLIMAX 6-1.
M-151; 1.2-1.35 C, 5.5-6.75 Mn, 0.40-0.70 Si, 0.90-1.10 Mo, 0.05 max P, 0.50 max Cr, bal Fe.
Austenitic in water quenched condition.
Good abrasion resistance casting but less tough than Hatfield type manganese steel.
For ball-mill liners, ball mill discharge grates, grizzly screens, drag chain and crusher liners, scoop lips.

CLIMAX 6-2-1.
M-151; 1.05-1.20 C, 5.25-6.50 Mn, 0.30-0.70 Si, 1.5-2.0 Cr, 0.90-1.10 Mo, 0.05 max P, bal Fe.
Lean alloy casting, austenitic after quench.
Good abrasion resistance,-work hardenable.
For ball-mill liners and discharge grates, scoop lips, drag chain and crusher liners subject to rapid wear and moderate impact.

CLIMAX 12.
M-151; 3.0-3.5 C, 0.5-0.8 Mn, 0.5-0.8 Si, 11.0-14.0 Cr, 0.5-1.0 Mo, 1.0 max Cu, bal Fe.
Hardened: C 60-67 Rock.
For grinding balls, classifier wear shoes, shot blast wheel blades.
Martensitic casting.

CLIMAX 12-2 HCS.
M-151; 1.30-1.45 C, 12.0-14.0 Mn, 0.30-0.75 Si, 1.8-2.1 Mo, 0.05 max P, 0.50 max Cr, bal Fe.
Alloy casting, austenitic as quenched, work hardenable, high abrasion resistance and good toughness.
For crusher liners, scraper blades, heavy-duty grizzly screens.

CLIMAX 12-2 LCS.
M-151; 1.15-1.30 C, 12.0-14.0 Mn, 0.30-0.75 Si, 1.8-2.1 Mo, 0.05 max P, 0.50 max Cr, bal Fe.
Alloy casting, austenitic as quenched, work hardenable, high abrasion resistance and good toughness.
For crusher liners, scraper blades, heavy-duty grizzly screens.

CLIMAX 15-2-1.
M-151; 2.8-3.5 C, 0.6-0.9 Mn, 0.4-0.8 Si, 14.0-16.0 Cr, 1.9-2.2 Mo, 0.8-1.2 Cu, bal Fe.
Casting, hardened: C 60-67 Rock.
For rod and ball mill liners, tires and grinding rings for roller mill pulverizers.

CLIMAX 15-3 HC.
M-151; 3.2-3.6 C, 0.70-1.0 Mn, 0.30-0.80 Si, 14.0-16.0 Cr, 2.5-3.0 Mo bal Fe.
High carbon grade martensitic white cast iron; hard and abrasion resistant.
For shot blast impeller blades and liners, and sand pump impellers, jaw plates in small jaw crushers, garbage disposal wearing parts.

CLIMAX 15-3 LC.
M-151; 2.4-2.8 C, 0.5-0.8 Mn, 0.30-0.80 Si, 14.0-16.0 Cr, 2.4-2.8 Mo, bal Fe.
Lower carbon grade martensitic white iron; hard and abrasion resistant.
For heavier section castings for rod and ball mill liners, tires for rolle mill pulverizers and heavy pulverized hammers.

CLIMAX 15-3 MC.
M-151; 2.8-3.2 C, 0.6-0.9 Mn, 0.30-0.80 Si, 14.0-16.0 Cr, 2.5-3.0 Mo, bal Fe.
Medium carbon grade martensitic white cast iron; hard and abrasion resistant.
For die bushings in clay product molds, impact pulverizer blow bars, chute liners.

CLIMAX 15-3 XHC.
M-151; 3.60-4.30 C, 0.70-1.0 Mn, 0.30-0.80 Si, 14.0-16.0 Cr, 2.5-3.0 Mo, bal Fe.
Extra high carbon grade martensitic white cast iron; hard, brittle, abrasion resistant.
For unstressed liners and parts handling abrasive slurries without impact, and fluidized solids up to about 1200°F.

CLIMAX 18-2 (18 CR-2 MO).
M-151; 0.04 max C, 18-20 Cr, 1.75-2.25 Mo, 1.0 max Mn, 1.0 max Si, 0. max Cu+Ni, Ti = 5 x C + N (0.25 mim); Cb = 9 x C + N min, bal Fe. Ann: 467 N/mm² TS; 302 N/mm² YS; 37 El.
60% cold rolled: 829 N/mm² TS; 823 N/mm² YS; 5.5 El.
Ferritic type stainless steel with very good resistance to corrosion, stress-corrosion cracking, and good formability. Weldable.

CLIMAX 20-2-1.
M-151; 2.6-2.9 C, 0.6-0.9 Mn, 0.4-0.9 Si, 18.0-21.0 Cr, 1.4-2.0 Mo, 0.8-1.2 Cu, bal Fe.
Ann: C 38-43 Rock.
Hardened: C 60-67 Rock.
Alloy steel casting, martensitic as quenched.
For rod and ball mill liners, sand and dredge-pump parts, clay working machine parts, pulverizer impactor and blow bars.

CLIMAX 321.
M-151; 3.3-3.6 C, 1.75-2.25 Cr, 0.70-1.1 Mo, 2.75-3.25 Ni, bal Fe.
For ball mill parts, grinding balls, pug mill knives; martensitic, cast iron, abrasion resistant.

CLIMAX ALLOY 42 replaced by **CLIMAX 15-3.**

CLIMAX CR13MO.
M-151; 0.25-0.40 C, 0.4-0.6 Mn, 0.4-0.8 Si, 13-14 Cr, 0.6-0.75 Mo, bal Fe.
Annealed: 180-220 Brin.
Hardened: 500-550 Brin.
For grinding mill liners, sand and dredge pump castings.
Abrasion and corrosion resistant.
For information only.

CLIMAX FERROMOLYBDENUM.
M-151; 58-64 Mo, 1.0 max Si, 0.10 max C, bal Fe.
For metallurgical applications in steel and cast iron; Mo-additions.

CLIMAX POWDER METALLURGY MOLYBDENUM.
M-151; made from powder 99.95 min Mo.
Sintered parts and wrought products from sintered parts.
For glass melting electrodes, glass furnace hardware, electrical and electronic parts, furnace heat shields, heating elements and supports.

CLIMELT LOW-CARBON MOLYBDENUM.
M-151; 99.97 min Mo, 0.005 max C.
Vacuum arc melted.
For space power generator parts, grinding quills, chemical handling equipment, machined components.

CLIMELT MO-30W.
M-151; 30 W, bal Mo.
At 72°F: 121,500 TS; 106,900 YS; 26 El; 40 Ra; 198 Brin.
At 1800°F: 65,700 TS; 25 EL; 77 RA; 80 Brin.
For high temperature applications; high heat resistance; has excellent resistance to attack by molten zinc and certain other liquid metals.

CLIMELT MOLYBDENUM.
M-151; 99.94 min Mo, 0.03 max C.
Vacuum arc melted.
For rocket nozzle parts, electrical and electronic parts, spot welding tips, brazing contacts, boring bars, furnace heating elements and supports.

CLIMELT TZM.
M-151; 0.01-0.04 C, 0.40-0.55 Ti, 0.06-0.09 Zr, bal Mo.
At 72°F: 144,000 TS; 129,000 YS; 21 El; 46 RA; 172 Brin.
At 1600°F: 88,000 TS; 48,000 YS; 21 El; 76 RA; 82 Brin.
For high temperature applications, heat engines, heat exchangers; heat and corrosion resistant, high strength and hardness at elevated temperatures.

CLINCHING SCREW WIRE.
M-Eng.; 69 Cu, 29.5 Zn, 1.5 Pb.
For brass screws.

CLIPPER.
M-335; 0.73 C, 18.0 W, 4.0 Cr, 1.1 V, bal Fe.
For cutting tools.
AISI T1. High speed steel.

CLOCK BRASS-243.
M-8; 37 Zn, 2 Pb, 62 Cu.
Hard: 75,000 TS; 60,000 YS; 10 El.
Soft: 45,000 TS; 17,000 YS; 50 El.
For clock gears and frame, meter parts; free milling.

CLOMO.
M-111; 0.97 C, 1.15 Cr, 0.32 Mo, bal Fe.
For hollow mine drills; oil hardening.

CLOVERLEAF.
M-741; Sn, Pb, bal Cu.
For bearings, bushings.

CLUNISE.
M-Eng.; 40 Cu, 32 Ni, 25 Zn, 2.6 Fe.
For decorative parts.

C.L.W.
M-73; 0.30 C, 3.5 Cr, 0.5 V, 9 W, bal Fe.
For extrusion and trimming dies, punches, shears; hot work steel.
AISI H22.

CLYDE ALLOY.
M-742; 0.3-0.6 C, bal Fe.
For machinery parts; water hardening.

CM.
M-365; 0.38 C, 1.0 Si, 5.0 Cr, 0.45 V, 1.25 Mo, bal Fe.
Air or oil hardening hot work tool steel for forging dies, and hot forming tools. AISI H11.

CM-1.
M-1769.
Sintered carbide.
325,000 TrS; Density: 14.3-14.5; Hardness: 88.3-89.3 Ra.
Industry code: C-16.

CM 1.
M-1290; 0.25 C, 1.0 Cr, 0.25 Mo, bal Fe.
Heat treated; 85,000-170,000 TS; 64,000-120,000 YS; 16-5 El.
Low alloy steel casting. Also useful to -110°C.
Werkstoff Nr. 1.7218.

CM 1 H.
M-1290; 0.42 C, 1.0 Cr, 0.20 Mo, bal Fe.
Heat treated: 106,000-184,000 TS; 78,000-142,000 YS; 12-4 El.
Low alloy steel casting.
Werkstoff Nr. 1.7225.

CM 1 K.
M-1290; 0.34 C, 1.0 Cr, 0.25 Mo, bal Fe.
Heat treated: 78,000-150,000 TS; 42,500-106,000 YS; 20-9 El.
Low alloy steel casting.
Werkstoff Nr. 1.7220.

CM1KM.
M-1290; 0.35 C, 0.8 Mn, 1.0 Cr, 1.0 Mo, bal Fe.
Low alloy cast steel; GS-34 CrMoMn 44.

CM-2.
M-1769.
Sintered carbide.
325,000 TrS; Density: 14.4-14.6; 89.0-90.0 Ra.
Industry code: C-16.

CM 2 H.
M-1290; 0.50 C, 0.70 Mn, 1.0 Cr, 0.25 Mo, bal Fe.
Lowalloy cast steel; GS-50 CrMo 4.

CM 3.
M-1653; 1.0-1.1 C, 3.5-4.5 Cr, 4.5-5.5 Mo, 5.5-7.0 W, 2.2-2.6 V, bal Fe.
Mo-W high speed tool steel.
AISI M3 Class 1.

CM-3.
M-1769.
Sintered carbide.
350,000 TrS; Density: 14.1-14.35; 88.3-89.3 Ra.
Industry code: C-16.

CM25.
M-Ger.; 0.3-0.4 C, Cr, 0.2 V, bal Fe.
For turbine blades for jet engines; water hardening.

CM 44.
M-1740; 0.40-0.48 C, 0.80-1.0 Si, 0.50-0.70 Mn, 2.8-3.1 Cr, 0.30-0.40 Mo, bal Fe.
350-500 BHN.

CM 60.
M-1724; 0.13-0.18 C, 0.15-0.35 Si, 0.80-1.10 Mn, 0.80 max Cr+Mo+Ni, 0.0005 min B, bal Fe.
Low alloy carburizing grade for small parts.

CM 70.
M-1724; 0.13-0.19 C, 0.15-0.35 Si, 1.1-1.4 Mn, 0.80 max Cr+Mo+Ni, 0.0005 min B, bal Fe.
Similar to CM 60, but higher strength.

CM 80.
M-1724; 0.17-0.23 C, 0.15-0.35 Si, 1.2-1.5 Mn, 0.80 max Cr+Mo+Ni, 0.0005 min B, bal Fe.
Similar to CM 70, but higher strength.

CM 90.
M-1724; 0.20-0.25 C, 0.15-0.35 Si, 1.3-1.6 Mn, 0.80 max Cr+Mo+Ni, 0.0005 min B, bal Fe.
Similar to CM 80, but higher strength.

CM 469.
M-1491; 0.03 C, 60 Cr, 25 Mo, 14 Fe.
For jet engine parts.
High heat and oxidation resistant.

CM 718.
M-1491; 0.08 C, 20 Cr, 52 Ni, 0.4 Co, 5.25 Cb, 3 Mo, 0.83 Ti, 0.55 Al, 0.004 B, 0.02 Cu, bal Fe.
Annealed: 120,200 TS; 53,400 YS; 0.45 EL.
Aged: 191,000 TS; 172,000 YS; 22 EL; C 45 Rock.
For aircraft structural parts operating at 800-1400°F.
High oxidation and corrosion resistance.

C.M.A.1.
M-452; 10-15 Mn, 7-8 Al, 2.5-3.5 Fe, bal Cu.
Cast: 94,000-105,000 TS; 40,000-49,000 YS; 20-35 El; 165-200 Brin.
For propellers, marine hardware.
Aluminum bronze.

C.M.A. BEARING METAL.
M-Eng.; Pb, alkali earth metals.
For bearings.

CMCW.
M-1290; 2.0-2.25 C, 0.20-0.40 Si, 0.20-0.40 Mn, 0.80-1.10 Co, 11.5-12.5 Cr, 0.3-0.5 Mo, 0.6-0.8 W, bal Fe.
Cold work tool steel; W.-Nr. 1.2884.

CMH.
M-1653; 0.27-0.34 C, 0.40-0.70 Mn, 2.7-3.3 Cr, 0.30-0.40 Mo, bal Fe.
Cr-Mo structural steel. 30 CrMo 12.

CM L-605.
M-1491; 0.12 C, 1.65 Mn, 19.8 Cr, 9.9 Ni, 15.2 W, 1.6 Fe, bal Co.
At R.T.: 144,300 TS; 65,400 YS; 55 El.
At 1400°F: 73,000 TS; 49,000 YS; 20 El; C 25 Rock.
For jet engine afterburners, exhaust cone assemblies, nozzle diaphragm valves, high temperature springs, turbine buckets. Good oxidation and corrosion resistance.

CM N-155.
M-1491; 0.10 C, 1.5 Mn, 20.75 Cr, 19.85 Ni, 19.5 Co, 2.95 Mo, 2.35 W, 1.15 Cb, 0.2 Cu, bal Fe.
At R.T.: 112,000 TS; 57,000 YS; 41 El; B 77 Rock.
At 1500°F: 39,000 TS; 25,000 YS; 35 El.
For turbine rotors and blading, afterburner rings, rocket chambers. High oxidation and corrosion resistance.

C.M.P.
M-740; 0.2 C, 0.7 Ni, bal Fe.
For machinery parts; precision cold rolled strip steel.

CM-R41.
M-1491; 0.09 C, 19 Cr, 11 Co, 10 Mo, 3 Ti, 1.5 Al, bal Ni.
At 70°F: 206,000 TS; 154,000 YS; 14 El.
At 1500°F: 126,000 TS; 118,000 YS; 14 El.
For jet engine components and high speed airframes. Afterburners, turbine castings, combustion liners, fasteners.
Precipitation hardening. High temperature alloy. Vacuum melted.

C.M.S.
M-743; 0.10-0.20 C, 1.2-1.5 Mn, 0.075-0.15 S, bal Fe.
Rolled: 80,000 TS; 75,000 YS; 22 El, 49 RA; 192 Brin.
For machinery parts; free machining.

CMSZ.
M-48; 40-56 Cr, 4-6 Mn, 13-21 Si, 1.25-1.75 Zr, 3-5 C, bal Fe.
For ladle additions of Cr to cast iron; for harder and stronger iron.

CMSZ-4 MIXTURE.
M-48; 45-49 Cr, 4-6 Mn, 18-21 Si, 1.25-1.75 Zr, 3-4.5 C, bal Fe.
For ladle additions of Cr to cast iron; no increase in chill.

CMSZ-5 MIXTURE.
M-48; 50-56 Cr, 4-6 Mn, 13.5-16 Si, 0.75-1.25 Zr, bal Fe.
For ladle additions of Cr to cast iron; no increase in chill.

C.M.V.
M-510; 0.38 C, 1.05 Si, 0.35 Mn, 5.25 Cr, 1.35 Mo, 1.0 V, bal Fe.
Hot work tool steel.
B.S. 4659 Type BH13.
AISI H13.

CMVA.
M-1740; 0.24-0.28 C, 0.30-0.50 Si, 0.70-0.80 Mn, 0.5 max Ni, 0.95-1.15 Cr, 0.45-0.55 Mo, 0.03-0.07 V, bal Fe.
Yield: 38 tons/sq. in. min, W.T.S. 47 tons/sq. in. min.

CMW.
M-365, M-618; 0.35 C, 5 Cr, 1.3 W, 1.7 Mo, 1 Si, bal Fe.
For hot work tools, die casting dies; hot work steel.
AISI H12.

CMW 3.
M-265; Cu, Cr.
Elec. cond.: 80% IACS (Ht Tr).
Sand cast: 50,000 UTS, 46,000 YS, 20 El; 110 Brin.
Bars: 75,000 UTS; 70,000 YS; 15 El; 150 Brin.
For resistance welding electrodes and high electrical conductivity, high strength applications (RWMA Class 2).

CMW 8CC.
M-265; Cu, Cd, Si.
Cast: 50,000 UTS; 2 El; 50% IACS elec. cond.
For electrical contracts.

CMW 28.
M-265; Cu, Zr.
Forgings: 54,000 UTS, 48,000 YS; 90% IACS elec. cond.; 65 Rb (Ht Tr).
Rod: 70,000 UTS, 63,000 YS; 90% IACS elec. cond.; 75 Rb (Ht Tr).
For resistance welding electrodes.

CMW 44 STRIP.
M-265; Cu, Cr, Cd.
Elec. cond.: 78% IACS.
70,000 psi WTS; 75 Rb (Ht Tr).
For high electrical conductivity; high strength applications.

CMW 53.
M-265; Cu, Ni, Si, Mn.
Cast: 60,000 UTS, 180 Brin (Ht Tr).
Elec. cond.: 45% IACS.
For flash welding dies, bearings and bushings.

CMVB.
M-1740; 0.30-0.40 C, 0.40-0.60 Si, 0.50-0.70 Mn, 2.5-3.0 Cr, 0.70-1.0 Mo, 0.10-0.20 Y, bal Fe.
300 BHN max.

CMW 53 B.
M-265; Cu, Ni, Be.
Cast: 80,000 UTS, 90 Rb (Ht Tr).
Elec. cond.: 45% IACS.
For flash welding dies, bearings, bushings.

CMW 73.
M-265; Cu, Be, Co.
Cast: 110,000 UTS; 38 Rc (Ht Tr).
Worught: 170,000 UTS; 38 Rc (Ht Tr).
Elec. cond.: 20% IACS.
Good abrasion resistance for current carrying shafts, collets, and bearings.
RWMA Class 4 welding electrodes.

CMW 100.
M-265; Cu, Co, Be.
Rolled: 110,000 UTS; 10 El; 100 Rb (Ht Tr).
Cast: 95,000 UTS; 6 El; 95 Rb (Ht Tr).
Elec. cond.; 48% IACS.
For resistance welding electrodes, conducting arms, springs.
RWMA Class 3.

CMW 150.
M-265; Cu, Ag, Be.
110,000 UTS; 95 Rb (Ht Tr).
Elec. cond.: 52% IACS.
For resistance welding dies and fixtures.

CMW 1000.
M-265; 6 Ni, 4 Cu, bal W.
110,000 UTS; 80,000 YS; 4 El, 25 Rc.
Sintered density: 17.00 gm/cc.
For weights, radiation shielding and boring bars.
cmw 2000.
M-265; 3.5 Ni, 1.5 Cu, bal W.
110,00 UTS; 85,000 YS; 3 El, 27 Rc.
Sintered density: 18.00 gm/cc.
For weights, radiation shielding.

CMW 3000.
M-265; 7 Ni, 3 Fe, bal W.
130,000 UTS, 85,000 YS; 18 El; 25 Rc.
Sintered density: 17.00 gm/cc.
For armor piercing projectiles, high strength weights.

CMW 3950.
M-265; 3.5 Ni, 1.5 Fe, bal Fe.
130,000 WTS; 90,000 YS; 15 EL; 28 Rc.
Sintered denstity: 18.00 gm/cc.
For armor piercing projectiles, high strength weights.

CMW 3970.
M-265; 2.1 Ni, 0.9 Fe, bal W.
130,000 UTS; 90,000 YS; 10 El; 30 Rc.
Sintered density: 18.50 gm/cc.
For armor piercing projectiles, high strength weights.

CM-WASPALLOY.
M-1491; 0.06 C, 0.08 Mn, 0.01 S, 0.004 P, 0.1 Si, 20.5 Cr, 14.2 Co, 4.2 Mo, 3.0 Ti, 1.5 Al, 0.03 Zr, 0.003 B, 0.05 Fe, 0.05 Cu, bal Ni.
Annealed: 121,400 TS; 69,700 YS; 54 El; B 90 Rock.
Aged: 186,000 TS; 138,000 YS; 26 El; C 39 Rock.
For jet engine turbine buckets and discs, high temperature bolts, missile systems.
Precipitation hardened. High temperature strength.

CMW D50.
M-265; 15 Ni, bal Ag.
Elec. cond.: 65% IACS; 50 Rf ann hardness.
Density: 10.0 gm/cc.
For electrical contacts; resistant to mechanical wear.

CMW D50F.
M-265; 15 Ni, bal Ag.
27,000 UTS; 40 Rf (ann hardness).
89 Rf (worked); Elec. cond.: 80% IACS.
For electrical contacts; resistant to mechanical wear.

CMW D54.
M-265; 10 CdO, bal Ag.
Elec. cond.: 75% IACS, 40 Rf (ann hardness).
Density 9.80 gm/cc.
For electrical contacts; especially resistant to welding and high surge currents.

CMW D54F.
M-265; 10 CdO, bal Ag.
Ann: 39,000 UTS; 71 Rf.
Worked: 46,000 UTS; 90 Rf.
Elec. cond.: 82% IACS.
For electrical contacts; resistant to welding and high surge currents.

CMW D54X.
M-265; 10 CdO, bal Ag.
Ann: 27,000 UTS, 45 Rf.
Elec. cond.: 75% IACS.
For electrical contacts; resistant to welding and high surge currents.

CMW D55.
M-265; 15 CdO, bal Ag.
Elec. cond.: 55% IACS; 25 Rf ann hardness.
Density: 9.6 gm/cc.
For electrical contacts; resistant to welding and high surge currents.

CMW D55F.
M-265; 15 CdO, bal Ag.
Ann: 40,000 UTS; 70 Rf.
Worked: 48,000 UTS; 90 Rf.
Elec. cond.: 72% IACS.
For electrical contacts; resistant to welding and to high surge currents.

CMW D55X.
  M-265; 15 CdO, bal Ag.
  Ann: 30,000 UTS, 50 Rf.
  Elec cond.: 65% IACS.
  For electrical contacts, resistant to welding and to high surge current.

CMW D56.
  M-265; 30 Ni, bal Ag.
  Elec. cond.: 55% IACS; 45 RF (Ann).
  Density: 9.7 gm/cc.
  For electrical contacts; resistant to mechanical wear.

CMW D57F.
  M-265; 10 Fe, bal Ag.
  Ann : 31,000 WTS; 48 Rf.
  Worked: 39,000 WTS; 81 Rf.
  Elec. cond.: 90% IACS.
  For electrical contacts; AC resistance loads.

CMW D58.
  M-265; 5 C, bal Ag.
  Density: 8.65 gm/cc; Elec. cond.: 55%; 25 Rf.
  For electrical contacts; resistant to sliding wear.

CMW D63X.
  M-265; 0.41 MgO, 0.25 NiO, bal Ag.
  Heat treated: 70,000 UTS; 97 Rf.
  Elec. cond.: 70% IACS.
  For electrical contacts; high hardness; resists annealing.

CMW D64F.
  M-265; 0.3 CaO, bal Ag.
  Ann: 24,000 UTS; 14 Rf.
  Worked: 40,000 UTS; 82 Rf.
  Elec. cond.: 101% IACS.
  For electrical contacts, voltage regulators; low voltage DC.

CMW D154F.
  M-265; 13.3 CdO, bal Ag.
  Ann: 40,000 UTS; 70 Rf.
  Worked: 47,000 UTS; 90 Rf.
  Elec. cond.: 75% IACS.
  For electrical contacts; resistant to welding and high surge currents.

CMW D155F.
  M-265; 17 CdO, bal Ag.
  Ann: 40,000 UTS; 70 Rf.
  Worked: 50,000 UTS; 90 Rf.
  Elec. cond.: 70% IACS.
  For electrical contacts; resistant to welding and high surge currents.

CMW D158F.
  M-265; 1 C, bal Ag.
  Ann: 23,000 UTS; 36 Rf.
  Worked: 35,000 UTS; 69 Rf.
  Elec. cond: 99% IACS.
  For electrical contacts; reduced sticking; low friction.

CMW D355F.
  M-265; 20 CdO, bal Ag.
  Ann: 40,000 WTS; 70 RF.
  Worked: 51,000 WTS; 90 Rf.
  Elec. cond.: 68% IACS.
  For electrical contacts; resistant to welding and high surge currents.

CMW D505F.
  M-265; 5 Ni, bal Ag.
  Ann: 24,000 UTS; 32 Rf.
  Worked: 84 Rf.
  Elec. cond.: 95% IASC.
  For electrical contacts.

CMW D510F.
  M-265; 10 Ni, bal Ag.
  Ann: 25,000 WTS; 35 Rf.
  Worked: 89 Rf.
  Elec. cond.: 87% IASC.
  For electrical contacts.

CMW D581F.
  M-265; 0.5 C, bal Ag.
  Ann.: 24,000 WTS; 44 Rf.
  Worked: 36,000 WTS; 72 RF.
  Elec. cond.: 102% IASC.
  For electrical contacts for reduced sticking, low friction.

CMW D582F.
  M-265; 0.25 C, bal Ag.
  Ann: 25,000 UTS; 45 Rf.
  Worked: 37,000 UTS; 73 Rf.
  Elec. cond.: 103% IACS.
  For electrical contacts for reduced sticking, low voltage DC.

CMW D583F.
  M-265; 0.75 C, bal Ag.
  Ann.: 24,000 UTS; 39 Rf.
  Worked: 35,000 UTS; 70 Rf.
  Elec. cond.: 100% IACS.
  For electrical contacts for reduced sticking, low friction.

CMW D1058.
  M-265; 10 C, bal Ag.
  Elec. cond.: 35% IACS; hardness 3 Rf.
  Density: 6.4 gm/cc.
  For electrical contacts, resistant to sliding wear.

CMWK-TUNGSTEN.
  M-265; W.
  Wrought: 300,000 TS.
  For contacts for ignition system; 32% conductivity.

CMX-A.
  M-1653; 0.60-0.75 C, 1.0 max Si, 1.0 max Mn, 13.0-15.0 Cr, 0.50-0.60 Mo, bal Fe.
  Martensitic stainless steel; W.-Nr. 1.4109.
  Similar to AISI 440A.

CNDC 1.
M-1488; 0.18 C, 0.75 Mn, 1.35 Ni, 0.85 Cr, 0.20 Mo, bal Fe.
Alloy carburizing steel; deep hardening.
AFNOR 18 NCD 6.

CN20.
M-1806. Sintered carbide tool material.
Hardness: 1550-1650 VHN.
For general purpose machining cast iron, non-ferrous and non-metallics.
ISO K20.

CMX-B.
M-1653; 0.75-0.95 C, 1.0 max Si, 1.0 max Mn, 16.0-18.0 Cr, 0.75 max Mo, bal Fe.
Martensitic stainless steel.
AISI 440B.

CMX-C.
M-1653; 0.95-1.20 C, 1.0 max Si, 1.0 max Mn, 16.0-18.0 Cr, 0.75 max Mo, bal Fe.
Martensitic stainless steel. W.-Nr. 1.4125.
AISI 440C.

CN01.
M-1806.
Sintered carbide tool material.
Hardness: over 1750 VHN.
For finish machining cast iron and non-ferrous at high speeds, light cuts.
ISO K01.

CN-7MCB3.
M-1262; 0.05 C, 30 Cr, 33 Ni, 2.5 Mo, 0.5 Cb, bal Fe.
Cast 70,000 TS; 32,000 YS; 30 El; 163 BHN.
For chemical plant equipment, valves, agitators, pumps and tanks; resists mixed acids.

CN10.
M-1806.
Sintered carbide tool material.
Hardness: over 1750 VHN.
For high speed machining of nodular iron and semi-steels.
ISO K10 and M10.

CN15.
M-1806.
Sintered carbide tool material.
Hardness: 1625-1725 VHN.
For finish machining with light feeds on cast iron, non-ferrous and non-metallics.
ISO K10.

CN20.
M-1290; 0.20 max C, 1.8-2.3 Si, 2.0 max Mn, 24.0-26.0 Cr, 19.0-21.0 Ni, bal Fe.
Heat resisting steel.
W.-Nr. 1.4841: similar to AISI 310.

CN25.
M-1806.
Sintered carbide tool material.
Hardness; 1475-1575 VHN.
For machining cast iron, non-ferrous and stainless even under vibration and some shock. Tough.
ISO K20-K30.

CN30.
M-1806.
Sintered carbide tool material.
Hardness: 1425-1500 VHN.
For machining cast iron and non-ferrous on old machines or at slow speeds; also for interrupted cuts.
ISO K30.

CN40.
M-1806.
Sintered carbide tool material.
Hardness: 1275-1375 VHN.
Tough grade, shock resistant; for heavy duty planing and interrupted cuts of cast iron.
ISO K40.

CNC 1.
M-1488; 0.16 C, 0.85 Mn, 0.95 Ni, 0.95 Cr, bal Fe.
Carburizing steel; oil hardening.
AFNOR 16NC6.

CNDB 1.
M-1488; 0.38 C, 0.65 Mn, 0.85 Ni, 0.85 Cr, 0.20 Mo, bal Fe.
Oil hardenable to 980-1130 N/mm² UTS.
For structural parts.

CNDB1-SV75.
M-1488; 0.38 C, 0.85 Ni, 0.85 Cr, 0.20 Mo, bal Fe.
Bars, treated: 980-1130 n/mm² WTS; 835 N/mm² min YS; 12 E 293 min Brin.
For structural purposes.

CNS-2.
M-69; 2.15 C, 12 Cr, 0.25 Co, 0.18 V, bal Fe.
For blanking and forming dies, punches; oil hardened, non-deforming.
AISI D3.

CNS NO. 1 (AISI D2).
M-69; 1.55 C, 12.5 Cr, 8.8 V, 0.8 Mo, bal Fe.
For blanking and forming dies, punches; air hardened, non-deforming. AISI D2.

CO 6 see **HAYNES STELLITE ALLOY NO. 6B.**

CO 12 see **HAYNES STELLITE ALLOY NO. 12.**

CO 19 see **HAYNES STELLITE ALLOY NO. 19.**

CO 21 see **HAYNES STELLITE ALLOY NO. 21.**

CO 31 see **HAYNES STELLITE ALLOY NO. 31.**

COANAILIUM.
M-218; Al alloy.
For marine parts; corrosion resistant.

COAST CM-119.
M-1060.
For hard facing, hot friction guides; resists molten Cu.

COAST METALS NP.
M-1060; 50 Ni, 30 Fe, 12 Si, 4 P, 4 Mo.
For brazing stainless steel atomic fuel elements for service in 565°F. pressurized water. Corrosion resistant.

COAST NO. 1.
M-1060; 4.0 C, 16 Cr, 6 Ni, bal Fe.
Cast: 420-550 Brin.
For hard facing welding rod; austenitic, wear and heat resistant.

COAST NO. 4.
M-1060; 4.0 C, 16 Cr, 6 Ni, 5 Si, bal Fe.
Cast: 600-630 Brin.
For hard facing welding rod: austenitic, wear and heat resistant.

COAST NO. 7.
M-1060; 3.0 C, 25 Cr, 12 Ni, 8 Mo, bal Fe.
Cast: 300-450 Brin.
For hard facing welding rod; heat and abrasion resistant.

COAST NO. 8.
M-1060; 3.0 C, 16 Cr, 6 Ni, 20 Co, bal Fe.
Cast: 400-530 Brin.
For hard facing welding rod; heat and abrasion resistant.

COAST NO. 9.
M-1060; 1.0 C, 29 Cr, bal Fe.
Cast: 450-550 Brin.
For hard facing welding rod; heat resistant; discontinued.

COAST NO. 10.
M-1060; 4 C, 16 Cr, 2 Ni, 8 Mo, bal Fe.
Cast: 480-600 Brin.
For hard facing welding rod; non-magnetic, hot abrasion resistant.

COAST NO. 15.
M-1060; 4 C, 16 Cr, 4 Ni, 6.5 Mo, 20 Co, bal Fe.
Welded: 530-560 Brin.
For hard facing rod for acetylene welding; corrosion and wear resistant, tough, hard.

COAST NO. 17.
M-1060; 3 C, 40 Cr, 8 Mo, 30 Co, 1 Si, bal Fe.
Welded: 530-600 Brin.
For acetylene welding rod for chemical equipment; corrosion and abrasion resistant.

COAST NO. 18.
M-1060; 3.0 C, 25 Cr, 15 Ni, 8 Mo, 30 Co, 1 Si, bal Fe.
Cast: 400-470 Brin.
For hard-facing electrode; heat and abrasion resistant.

COAST NO. 19.
M-1060; 1 C, 35 Cr, 2 Cu, bal Fe.
Welded: 300-400 Brin.
For acetylene welding rod for rolling mills; heat and abrasion resistant.

COAST NO. 40.
M-1060; 3 C, 30 Cr, 40 Ni, 14 W, bal Fe.
Welded: 380-420 Brin.
For acetylene welding rod for valves; corrosion, heat and wear resistant.

COAST NO. 50.
M-1060; 93.25 Ni, 3.5 Si, 2.25 B, 1 others.
Cast: 390-440 Brin.
High temperature nickel brazing alloy, M.P. 1825°F. For thin walled joints, wide or close tolerance joints; oxidation resistant.

COAST NO. 50B.
M-1060; 2.5 Si, 1.5 B, bal Ni.
Metal powder for repair brazing of castings; overlays on dies and molds; machinable.

COAST NO. 50C.
M-1060; 3.0 Si, 1.8 B, bal Ni.
Brazing alloy for casting repairs; overlays on dies and molds; machinable. Metal powders.

COAST NO. 52.
M-1060; 91.25 Ni, 4.5 Si, 4.5 B, 1 others.
Cast: 570-620 Brin.
For brazing alloy; M.P. 1825°F; oxidation resistant. High strength joints.

COAST NO. 52 SPECIAL.
M-1060; 4.5 Si, 0.15 C, 20.0 Co, 2.9 B, bal Ni.
For brazing high temperature alloys.
Brazing temperature 2150-2175°F.

COAST NO. 53.
M-1060; 82.1 Ni, 4.5 Si, 2.9 B, 7 Cr, 3 Fe, 0.5 others.
Cast: 570-620 Brin.
For brazing alloy; M.P. 1825°F; oxidation resistant.

COAST NO. 54.
M-1060; 0.35 C, 9.0 Cr, 2.5 Si, 2.1 Fe, 1.66 B, bal Ni.
Metal powder for repair or build-up on valves, valve seats; average hardness 38 Rock C; machinable.

Section I: Alloy Data

COAST NO. 55.
M-1060; 0.35 C, 10.0 Cr, 4.35 Si, 4.0 Fe, 2.1 B, bal Ni.
Metal powder for build-up or repair on pump sleeves, wear rings; average hardness 45 Rock C.

COAST NO. 56.
M-1060; 72.5 Ni, 4 Si, 3.75 B, 16 Cr, 4 Fe, 1 others.
Cast: 570-620 Brin.
For brazing alloy; M.P. 1880°F; oxidation resistant.

COAST NO. 60.
M-1060; Ni-Cr-Si.
Alloy powder for nuclear brazing.

COAST NO. 62.
M-1060; Mn-Ni-Co.
Alloy powder for brazing wide gaps and for high temperature strengths with good corrosion resistance, on such alloys as Rene 41, 15-7 Mo, A-286, tungsten. AMS 4780.

COAST NO. 80A.
M-1060; Mo, Cr, W, bal Ni.
For hard facing electrode; corrosion and abrasion resistant.

COAST NO. 81.
M-1060; 30.0 Mo, 5.0 Fe, 2.0 B, bal Ni.
Metal powder for repair or build-up of chemical and petro-chemical plant equipment; excellent corrosion resistance.

COAST NO. 90.
M-1060; 3 C, 5 Cr, 3 Ni, bal Fe.
For hard surfacing electrode; resists abrasion and impact, acetylene welded.

COAST NO. 91A.
M-1060; C, Cr, bal Fe.
For hard facing electrode; abrasion and impact resistant.

COAST NO. 92A.
M-1060; C, Cr, Si, Mn, bal Fe.
For hard facing electrode; impact and abrasion resistant.

COAST NO. 96A.
M-1060; C, Cr, Mn, Si, bal Fe.
For hard facing electrode; work hardens, wear resistant.

COAST NO. 98A.
M-1060; Cr, Mn, Si, bal Fe.
For hard facing electrode; corrosion and impact resistant.

COAST NO. 100X.
M-1060; 1 C, 3 Cr, 3 Ni, 7 Mo, bal Fe.
For hard facing electrodes; arc welding, can be forged.

COAST NO. 100X.
M-1060; 0.7 C, 18 W, 4 Cr, 1 V, bal Fe.
Cast: 450-530 Brin.
For hard facing welding rod; high speed steel.

COAST NO. 101.
M-1060; 4.0 C, 16 Cr, 6 Ni, bal Fe.
Cast: 420-550 Brin.
For hard facing welding rod; austenitic, wear and heat resistant.

COAST NO. 104.
M-1060; 4.0 C, 16 Cr, 6 Ni, 5 Si, bal Fe.
Cast: 530-600 Brin.
For hard facing welding rod; austenitic, wear and heat resistant.

COAST NO. 107.
M-1060; 3.0 C, 25 Cr, 12 Ni, 8 Mo, bal Fe.
Cast: 300-430 Brin.
For hard facing welding rod; heat and abrasion resistant.

COAST NO. 108.
M-1060; 3.0 C, 16 Cr, 6 Ni, 20 Co, bal Fe.
Cast: 400-530 Brin.
For hard facing welding rod; heat and abrasion resistant.

COAST NO. 110.
M-1060; 4 C, 16 Cr, 2 Ni, 8 Mo, bal Fe.
Cast: 480-600 Brin.
For hard facing welding rod; non-magnetic, hot abrasion resistant.

COAST NO. 111.
M-1060; C, Cr, Ni, bal Fe.
Cast: 450-560 Brin.
For hard facing electrode; corrosion resistant; discontinued.

COAST NO. 112.
M-1060; 4.0 C, 16 Cr, 6 Ni, 8 Mg, bal Fe.
Cast: 520-590 Brin.
For hard facing electrode; resists heat and dry abrasion.

COAST NO. 115.
M-1060; 4 C, 16 Cr, 4 Ni, 6.5 Mo, 20 Co, bal Fe.
For hard facing rod for arc welding; corrosion and wear resistant.

COAST NO. 117.
M-1060; 3 C, 40 Cr, 8 Mo, 30 Co, 1 Si, bal Fe.
Welded: 530-600 Brin.
For arc welding rod for chemical equipment; corrosion and abrasion resistant.

COAST NO. 118.
M-1060; 3.0 C, 25 Cr, 15 Ni, 8 Mo, 30 Co, 1 Si, bal Fe.
Cast: 400-470 Brin.
For hard facing welding rod; heat and abrasion resistant.

COAST NO. 119.
M-1060; 1 C, 35 Cr, 2 Cu, bal Fe.
Welded: 300-400 Brin.
For arc welding rod for rolling mills; heat and abrasion resistant.

COAST NO. 140.
M-1060; 3 C, 30 Cr, 40 Ni, 14 W, bal Fe.
For arc welding rod for valves; corrosion, heat and wear resistant.

COAST NO. 190.
M-1060; 3 C, 5 Cr, 3 Ni, bal Fe.
For hard surfacing electrodes; resists abrasion and impact, arc welded.

COAST NO. 1600-N.
M-1060; Mn-Cu-Ni.
Wire, powder or foil for brazing and welding 300 and 400 stainless steels, AM 350, H 11, and SAE 4130 steels.

COAST NO. X.
M-1060; 0.7 C, 18 W, 4 Cr, 1 V, bal Fe.
Cast: 450-530 Brin.
For hard facing welding rod; high speed steel.

COAST NO. X.
M-1060; 1 C, 3 Cr, 3 Ni, 7 Mo, bal Fe.
For acetylene welding rod for hot shears; forgeable, tough.

COATING METAL 560.
M-88; 95 Pb, 2.5 Sn, 2.5 Zn.
For coated iron for roofing, pipes.

COBAL 1.
M-111; 1 C, 6.5 Cr, 1 Mo, 7.5 Co, bal Fe.
For permanent magnets.

COBAL 3.
M-111; 0.95 C, 0.7 Mn, 5.7 Cr, 4.5 W, 15.5 Co, bal Fe.
For permanent magnets.

COBAL 4.
M-111; 0.85 C, 0.7 Mn, 5.5 Cr, 4.5 W, 35 Co, bal Fe.
For permanent magnets.

COBALT.
M-1755; Co.
Purities: 99.999%, 99.99%, 99.9%, 99.7%. (Commercial grade).
Forms: sponge, powders, rod, wires, sheets, foils, platelets, cakes, single crystals.

COBALT.
M-140; 0.74 C, 18 W, 4 Cr, 1 V, 5 Co, 0.8 Mo, bal Fe.
For lathe and planer tools, reamers, hobs, broaches, drills; high speed steel, oil hardened.

COBALT.
M-344; 0.7 C, 18 W, 4 Cr, 1 V, 5 Co, bal Fe.
For tools, cutters; high speed steel.

COBALT 1.
M-677; 2-3 C, 0.4-2.0 Si, 26-35 Cr, 11-14 W, 3 max Ni, 1 max Mo, bal Co.
For bare (1B) and coated (1C) hard-facing electrodes; grades R Co Cr- C and E Co Cr-C.

COBALT 2.
M-677; 2-3 C, 1 max Si, 29-35 Cr, 16-19 W, 2.5 max Ni, bal Co.
For bare hard-facing electrodes; abrasion resistant at high temperatures.

COBALT 3%.
M-1076; 3.25 Co, 4 Cr, 1 C, bal Fe.
Peak energy product (Bd Hd) max x 10^6 0.38; residual induction Br (Gauss) 9700; coercive force Hc (Oersted) 80.
Permanent magnet.

COBALT 6.
M-677; 0.9-1.4 C, 0.4-2.0 Si, 26-32 Cr, 3-6 W, 3 max Ni, 1 max Mo, bal Co.
For bare (6B) and coated (6C) hard-facing electrodes, grades RCoCr- A and ECoCr-A.

COBALT 7.
M-677; 0.2-0.4 C, 25-29 Cr, 2-4 Ni, 5-6 Mo, bal Co.
For bare (7B) and coated (7C) hard-facing electrodes; impact resistant, crack resistant.

COBALT 12.
M-677; 1.2-1.7 C, 26-32 Cr, 7-9.5 W, 3 max Ni, 1 max Mo, bal Co.
For bare (12B) and coated (12C) hardfacing electrodes; grades RCoCr-B and ECoCr-B.

COBALT ASCOLOY.
M-U.S.; 0.20 C, 12.25 Cr, 5.0 Co, 3.0 W, 0.25 V, bal Fe.
High temperature alloy; heat and corrosion resistant.

COBALTCROM.
M-40; C, 3.7 Co, 13.6 Cr, bal Fe.
For dies, broaches, cutters, valves; nonscaling.

COBALT CHROME.
M-73; 1.35 C, 12.5 Cr, 0.3 Ni, 3.0 Co, 0.8 Mo, bal Fe.
For broaches, burnishing tools, valves, punches; high abrasion resistance, non-deforming.

COBALTCROM KXK STAINLESS.
M-40; 1.0 C, 18 Cr, 1.1 Co, 0.2 V, 1.1 Mo, bal Fe.
For dies, press tools, pump valves; corrosion and abrasion resistant.

COBALTCROM PRK-HT.
M-40; C, Co, Cr, bal Fe.
For welding rod; coated.

COBALTCROM STAINLESS.
M-40; 3.3 Co, 18 Cr, 0.75 Mo, 1.10 C, 0.25 V, 0.30 Mn, bal Fe.
For special cutlery and corrosion resisting parts; air hardening steel; non-deforming.

COBALTLOY.
M-950; 1.5 C, 0.9 Mo, 12 Cr, 0.5 V, 3.25 Co, bal Fe.
For hot forming dies; abrasion resistant.

COBALT MAGNET.
M-373; C, alloy, bal Fe.
For magnet.

COBANIC.
M-61; 55 Ni, 45 Co, 0.1 C, bal Fe.
Annealed: 65,000-80,000 TS; 38 El.
For vacuum tube filament wire; heat and corrosion resistant.

COBANIC.
M-897; 54.5 Ni, 44.5 Co, 1 Fe.
Annealed: 80,000 TS; 60,000 YS; 25 El.
For cathodes and filaments in electronic tubes; heat resistant.

COBAR 6.
M-677; Weld metal: 1.0 C, 30.0 Cr, 4.5 W, 2.0 Ni, 2.0 Fe, bal Co.
Cored wire hardfacing electrode.

COBEND 1.
M-677; 2.0-3.0 C, 26-33 Cr, 11-14 W, 3 Max Ni, 1 max Mo, bal Co.
Welded: 45-48 Rc.
For hardfacing electrodes; Grade ECoCr-C; resists metal to metal wear.

COBEND 6.
M-677; 0.9-1.4 C, 26-32 Cr, 3-6 W, 3 max Ni, 1 max Mo, bal Co.
Welded: 37-40 Rc; work hardened: 47-51 Rc.
For hardfacing electrodes; grade ECoCr-A; resists abrasion, corrosion, galling, oxidation, erosion in 1200-1800°F range.

COBEND 7.
M-677; 0.2-0.4 C, 25-29 Cr, 2-4 Ni, 5-6 Mo, bal Co.
Welded: 34-37 Rc; work hardened: 47-40 Rc.
For hardfacing electrodes; resistant to cracking, erosion, hot abrasion.

COBEND 7W.
M-677; 0.25-0.45 C, 24-26 Cr, 4.5-6 W, 2-4 Ni, 4.5-6 Mo, bal Co.
Welded: 35-37 Rc.
For hardfacing electrodes; for cutting edge of hot forming tools.

COBEND 12.
M-677; 1.2-1.7 C, 26-32 Cr, 7-9.5 W, 3 max Ni, 1 max Mo, bal Co.
Welded: 35-39 Rc.
For hardfacing electrodes; grade ECoCr-B.

COBITE.
M-35; 0.81 C, 18.4 W, 4.25 Cr, 2.05 V, 8.75 Co, bal Fe.
W-Co high speed tool steel; very high red hardness.
For lathe tools, boring tools, cutoff tools.
AISI T5.

COBRA.
M-1422; 0.7 C, 18 W, 4 Cr, 1 V, bal Fe.
Hardened: C 64-66 Rock.
For cutters, lathe and planer tools, reamers, hobs, form cutters.
High-speed steel, Type T1, high red hardness.

COBRAZE 6991.
M-279; 30 Mn, 4 Zn, 1.2 Fe, 0.5 Al, bal Cu.
Ann: 85,000 max TS.
Hard: 100,000 min TS.
Brazing filler metal; brazing temp 885-900°C.

COBRON 664.
M-279; 11 Zn, 1.5 Fe, 0.5 Co, bal Cu.
Ann: 63,000 TS; 45,000 YS; 25 El.
Spring temper: 100,000 TS; 94,000 YS; 3 El.
For fuse clips, electrical terminals, connectors, springs.
Excellent formability and strength.
Electrical conductivity 30%.

COCHROME.
M-English; 60 Co, 24 Fe, 12 Cr, 2 Mn.
For heating elements, filaments; high heat resistance.

COCHROME.
M-503; 0.3 C, 1.2 Cr, 0.3 Mo, bal Fe.
For valve and valve fittings, castings for 250 lb steam pressure; high strength.

COCK BRONZE.
M-England; 8-10 Sn, 2-6 Zn, bal Cu.
For cocks, fittings, hardware; corrosion resistant.

CODE-ARC see AIRCO CODE-ARC.

CODE-ROD NO. 120.
M-1096; 0.08-0.13 C, 0.45-0.65 Mn, 0.15-0.25 Si, 0.03 S, 0.03 P, bal Fe.
Welded: 70,000-75,000 TS; 60,000-65,000 YS; 22-27 El; 50-55 RA.
For all-purpose welding rod; AWS-E6012.

COE BRONZE.
M-8; 89.5 Cu, 10.5 Sn.
Hard sheet; 115,000 TS; 95,000 YS; 5 El; 190 Brin.
Soft Sheet: 60,000 TS; 40,000 YS; 65 El; 74 Brin.
For gears; tough.

CO-ELINVAR.
M-U.S.; 57-60 Co, 25-35 Fe, 8-15 Cr.
For instruments; constant modulus.

COFLEX.
M-237; bimetal.
For thermostatic bimetal; corrosion resistant.

COG BRONZE.
M-Eng.; 11 Sn, 4 Zn, bal Cu.
For cogs, worms; tough, corrosion resistant.

COGNE CX.
M-1450; 0.38-0.45 C, 13 Cr, bal Fe.
Annealed: 100,000 TS; 55,000 YS; 20 El; 50 RA; 210 Brin.
For bearings, valves, cutlery, surgical instruments; Type 420; corrosion resistant.

COGNE FEOX.
M-1450; 0.07-0.12 C, 17 Cr, bal Fe.
Annealed: 80,000 TS; 50,000 YS; 25 El; 50 RA; 150 Brin.
For oil refinery equipment, oil burners and heaters, dairy and food equipment; Type 430; corrosion resistant.

COGNE FNOX.
M-1450; 0.25 max C, 17 Cr, 2 Ni, bal Fe.
Annealed: 125,000 TS; 95,000 YS; 20 El; 55 RA; 260 Brin.
For pumps, marine hardware, valves; Type 431; corrosion and heat resistant.

COGNE IOX1.
M-1450; 0.13-0.18 C, 13 Cr, bal Fe.
Annealed: 85,000 TS; 45,000 YS; 25 El; 55 RA; 200 Brin.
For valves, cutlery, surgical instruments, turbine blades; Type 410 and 420; corrosion resistant.

COGNE IOX3.
M-1450; 0.26-0.37 C, 13 Cr, bal Fe.
Annealed: 85,000 TS; 50,000 YS; 25 El; 55 RA; 195 Brin.
For valves, cutlery, turbine blades, surgical instruments; Type 420; corrosion resistant.

COGNE IOXA.
M-1450; 0.06 C, 13 Cr, 0.2 Al, bal Fe.
Annealed: 70,000 TS; 42,000 YS; 25 El; 160 Brin.
Cold Drawn: 85,000 TS; 70,000 YS; 20 El; 185 Brin.
For annealing boxes, quenching racks, oil refinery equipment; not hardenable.
Type 405 stainless steel, magnetic.

COGNE-IOXKM.
M-1450; 0.11 C, 12.75 Cr, 0.5 Mo, bal Fe.
Annealed: 75,000 TS; 35,000 YS; 30 El; B 92 Rock.
For springs, table flatware, oil refinery equipment.
Corrosion resistant, hardenable.

COGNE IOXO.
M-1450; 0.09-0.12 C, 13 Cr, 0.5 max C, bal Fe.
Annealed: 75,000 TS; 40,000 YS; 35 El; 70 RA; 155 Brin.
For turbine blades, valves, cutlery, surgical instruments; Type 403; corrosion resistant.

COGNE IOXOO.
M-1450; 0.08 max C, 13 Cr, bal Fe.
Annealed: 75,000 TS; 40,000 YS; 35 El; 70 RA; 155 Brin.
For turbine blades, valves, cutlery, surgical instruments; Type 403 and 410; corrosion resistant.

COGNE KXO2.
M-1450; 0.15 max C, 13 Cr, 2 Ni, bal Fe.
Annealed: 80,000 TS; 40,000 YS; 32 El; 68 RA; 160 Brin.
For oil refinery equipment; Type 414; corrosion resistant.

COGNE LIOX.
M-1450; 0.08-0.16 C, 12 Cr, 12 Ni, bal Fe.
For valves, pump parts; corrosion resistant.

COGNE NIOX.
M-1450; 0.10 C, 18 Cr, 8 Ni, bal Fe.
Annealed: 80,000 TS; 35,000 YS; 55 El; 75 TA; 155 Brin.
For chemical plant equipment, mixers, tanks, filters; Type 301 and 302; austenitic, stainless.

COGNE NIOX-C.
M-1450; 0.06 C, 18 Cr, 12 Ni, 0.5 Cb, bal Fe.
Annealed: 90,000 TS; 40,000 TS; 55 El; B 82 Rock.
Cold Drawn: 100,000 TS; 65,000 YS; 40 El; B 95 Rock.
For exhaust manifolds, steam pipes, radiant superheaters, welded structures.
Type 347 stainless steel, stabilized, austenitic.

COGNE NIOX-D.
M-1450; 0.11-0.16 C, 18 Cr, 8 Ni, bal Fe.
Annealed: 80,000 TS; 35,000 YS; 55 El; 75 RA; 155 Brin.
For chemical plant equipment, tanks, mixers, filters; Type 302; austenitic, stainless.

COGNE NIOX-L.
M-1450; 0.03 max C, 18-20 Cr, 8-12 Ni, bal Fe.
Annealed: 77,000 TS; 30,000 YS; 60 El; 110 Brin.
Cold Drawn: 100,000-180,000 TS; 50,000-125,000 YS; 10-50 El; 180-330 Brin.
For kitchen utensils, architectural trim, welded components, chemical and textile plant equipment.
Type 304L stainless steel.

COGNE NIOX-M.
M-1450; 0.10 max C, 19 Cr, 12 Ni, 3.5 Mo, bal Fe.
Annealed: 90,000 TS; 40,000 YS; 48 El; 62 RA; 170 Brin.
For acid resistant chemical plant equipment; Type 317; stainless, austenitic.

COGNE NIOX-MT.
M-1450; 0.12 max C, 18 Cr, 8 Ni, 2.5 Mo, Ti, bal Fe.
Annealed: 85,000 TS; 35,000 YS; 50 El; 65 RA; 160 Brin.
For welded acid resistant chemical plant equipment; Type 316 Ti; stainless, austenitic.

COGNE NIOX-S.
M-1450; 0.08 max C, 18-20 Cr, 8-12 Ni, bal Fe.
Annealed: 85,000 TS; 35,000 YS; 60 El; 150 Brin.
Cold Drawn: 100,000-180,000 TS; 50,000-125,000 YS; 10-50 El; 180-330 Brin.
For kitchen equipment, architectural trim, welded components, chemical and textile plant equipment. Type 304 stainless steel.

COGNE NIOX-T.
M-1450; 0.08 max C, 18 Cr, 8 Ni, Ti, bal Fe.
Annealed: 85,000 TS; 35,000 YS; 55 El; 65 RA; 150 Brin.
For welded chemical plant equipment, tanks, mixers; Type 321; stainless, austenitic.

COGNE RIOX.
M-1450; 0.25 max C, 25 Cr, 20 Ni, 1.5 max Si, bal Fe.
Annealed: 100,000 TS; 45,000 YS; 50 El; 65 RA; 185 Brin.
For furnace parts, valves, pumps, turbine and jet parts; Type 310 and 314 stainless, austenitic.

COGNE SIOX.
M-1450; 0.04 C, 10 Cr, 1 Mo, 2.5 Si, bal Fe.
For oil refinery equipment; creep and heat resistant.

COGNE VIOX.
M-1450; 0.20 max C, 24 Cr, 12 Ni, bal Fe.
Annealed: 90,000 TS; 40,000 YS; 50 El; 65 RA; 170 Brin.
For furnace parts, heat treat boxes, pumps, oil burners; Type 309; corrosion and heat resistant.

COG WHEEL BRAND NO. II.
M-1742; 11% Sn content Phosphor Bronze.
Valves, bearing, bushes, pumps and conn. rods.

COG WHEEL BRAND NO. VII.
M-1742; 12% Sn content Phosphor Bronze.
Worm and gear wheels, piston rings, wear plates, gib keys.

COG WHEEL BRAND NO. VIII.
M-1742; 14% Sn content Phosphor Bronze.
Heavy duty/high speed bearings.

COG WHEEL BRAND NO. XI.
M-1742; 10% Lead content Phosphor Bronze.
Rolling mills, crushing equipment.

COINAGE BRONZE.
M-U.S.; 3-4 Sn, 1-2 Zn, bal Cu.
For coins; corrosion resistant.

COIN BRONZE see also IMI 341.

COIN SILVER.
M-53; 90 Ag, 10 Cu.
Ann: 68 R-15T.
For electrical contacts.

CO J see HAYNES STELLITE STAR J METAL.

COLALLOY.
M-744; 2 Mg, 1 Mn, 1 Si, bal Al.
Heat treated: 12,000-61,000 TS; 20-110 Brin.
For chemical and food handling eqiupment; corrosion resistant.

COLCLAD 13CR AL.
M-1724.
Layer of stainless.
AISI 405 steel on mild steel backing plate.

COLCLAD 18/8 ELC.
M-1724.
Layer of stainless.
AISI 304L steel on mild steel backing plate.

COLCLAD 18/8 TI.
M-1724.
Layer of stainless.
AISI 321 steel on mild steel backing plate.

COLCLAD 18/10/2 ELC.
M-1724.
Layer of stainless.
AISI 316L steel on mild steel backing plate.

COLCLAD 18/10/2 TI.
M-1724.
Layer of titanium stabilized AISI 316 stainless steel on mild steel backing plate.

COLCLAD INCONEL.
M-1724.
Layer of Inconel on mild steel backing plate.

COLD HEADER NO 4.
M-336; 1.0 C, 0.3 Mn, 0.2 Si, bal Fe.
For cold heading dies; water hardened.

COLD HEADER VANADIUM.
M-69; 1.0 C, 0.25 Mn, 0.2 Si, 0.20 V, bal Fe.
Water hardenable to 60 Rock C.
Can be tempered to desired hardness from 200°F to 800°F.
For taps, dies, punches and pneumatic tools.
AISI Type W 2 water hardening tool steel.

COLD HEADING 4.
M-122; 66.5 Cu, bal Zn.
60,000 TS; 35,000 YS; 42 El; 75 RA; 112 Brin.
For cold upsetting work, rivets, thread rolled products; for cold deformation.

COLDIE.
M-140; 0.90 C, 0.20 V, 0.25 Si, 0.25 Mn, bal Fe.
For blanking and forming dies, rivet sets; water hardened; Type W 2.

COLESCO.
77 Pb, 8 Sn, 14 Sb, 1 Cu.
For bearings.

COLFORM "E.T.D." 1527.
M-669; 0.22-0.29 C, 1.20-1.50 Mn, 0.15-0.30 Si, bal Fe.
Cold finished: 120,000 min TS; 92,000 min YS; 14 El; 25-34 min Rc.
For cold forming or cold heading without subsequent heat treatment for strong parts.

COLFORM "E.T.D." 1541 A.
M-669; 0.40-0.45 C, 1.30-1.65 Mn, 0.15-0.30 Si, bal Fe.
Cold finished: 150,000 min TS; 130,000 min YS; 120,000 min proof strength; 32 min Rc.
For cold forming or cold heading without subsequent heat treatment for strong parts.

COLFORM "E.T.D." 1541 B.
M-669; 0.36-0.44 C, 1.35-1.65 Mn, 0.15-0.30 Si, bal Fe.
Cold finished: 140,000 min TS; 125,000 min YS; 10 El; 28 min Rc.
For cold forming or cold heading without subsequent heat treatment for strong parts.

COL-GRAPH.
M-35; 1.45 C, 0.8 Mn, 1.2 Si, 0.2 Cr, 0.25 Mo, bal Fe.
For blanking dies, gages, machine parts, taps, wear plates.
AISI Type 06 oil hardening tool steel.

COLLET BRASS.
M-U.S.; 61 Cu, 37 Zn, 2.5 Pb.
For hot forged parts, collets; free-cutting.

COLMO 900 GRADE 29.
M-1724; 0.18-0.25 C, 0.10-0.30 Si, 0.50-0.90 Mn, 0.45-0.65 Mo, bal Fe
For boiler and pressure vessels; good creep resistance to 480°C.

COLMO 900 GRADE 31.
M-1724; 0.20-0.27 C, 0.10-0.30 Si, 0.50-0.90 Mn, 0.45-0.65 Mo, bal Fe.
For boiler and pressure vessels; good creep resistance to 480°C.

COLMO 950.
M-1724; 0.17 max C, 0.10-0.30 Si, 0.40-0.70 Mn, 0.70-1.1 Cr, 0.45-0.65 Mo, bal Fe.
For boiler and pressure vessels; good creep resistance to 520°C.

COLMO 1000.
M-1724; 0.13 max C, 0.10-0.30 Si, 0.40-0.70 Mn, 0.25-0.50 Cr, 0.50-0.7 Mo, 0.22-0.30 V, bal Fe.
For boiler and pressure vessels; good creep resistance up to 575°C.

COLMONOY NO. 1.
M-963; 0.9 C, 11 Cr, 2.5 B, 1.75 Si, 0.4 Mn, bal Fe.
Cast: 58-63 Rock C.
For hard facing alloy; abrasion resistant.

COLMONOY NO. 4.
M-963; 0.45 C, 10 Cr, 2 B, 2.25 Si, 2.5 Fe, bal Ni.
Cast: 35-40 Rock C.
For hard facing alloy; corrosion and heat resistant.

COLMONOY NO. 4SA.
M-963.
Same as COLMONOY NO. 4 but as atomized powder.

COLMONOY NO. 5.
M-963; 0.65 C, 11.5 Cr, 2.5 B, 3.75 Si, 4.25 Fe, bal Ni.
Cast: 45-50 Rock C.
For hard facing alloy; corrosion and heat resistant.

COLMONY NO. 5SA.
M-963.
Same as COLMONOY NO. 5 but as atomized powder.

COLMONOY NO. 6.
M-963; 0.75 C, 13.5 Cr, 3 B, 4.25 Si, 4.75 Fe, bal Ni.
Cast: 56-61 Rock C.
For hard facing alloy; corrosion and heat resistant.

COLMONOY NO. 8.
M-963; 0.95 C, 26.0 Cr, 3.3 B, 4.0 Si, 1.0 Fe, bal Ni.
Weld: C 53-58 Rock.
For hard facing by spray welder.
Resists oxidation and corrosion. High abrasion resistance.

COLMONOY NO. 15.
M-963; 93.0 Cu, 7.0 P.
Hardness: B 78-82 Rock; M.P. 1460°F (approx).
Powder for buildup or braze welding of copper and copper alloys.

COLMONOY NO. 20.
M-963; 0.25 C, 5 Cr, 1 B, 3 Si, 3.5 Fe, bal Ni.
Cast: 15-20 Rock C.
For glass mold repairs; corrosion and heat resistant.

COLMONOY NO. 21.
M-963; 0.25 C, 5.0 Cr, 3.25 Si, 1.25 B, 1.0 Fe, bal Ni.
As cast: C 26-31 Rock; M.P. 2050°F (approx).
For medium hard facing for glass container molds; heat, impact and corrosion resistant.

COLMONOY NO. 22.
M-963; 0.1 C, 3.15 Si, 0.75 Fe, 1.25 B, bal Ni.
Cast: C 28-33 Rock; M.P. 1925°F (approx).
Atomized powder for machinable buildup on shafts and worn or undersize parts; good impact resistance.

COLMONOY NO. 23A.
M-963; 0.1 C, 1.0 Fe, 2.3 Si, 1.25 B, bal Ni.
Cast: C 14-19 Rock; M.P. 1950°F (approx).
Atomized powder for machinable buildup on shafts and worn or undersize parts; good impact resistance.

COLMONOY NO. 24.
M-963; 0.1 C, 2.3 Si, 1.0 Fe, 1.25 B, bal Ni.
Cast: C 14-19 Rock; M.P. 1950°F (approx).
Atomized powder for machinable buildup on shafts and worn or undersize parts; good impact resistance.

COLMONOY NO. 43.
M-963; 0.45 C, 10.0 Cr, 2.75 Fe, 2.75 Si, 2.0 B, bal Ni.
Cast: C 35-40 Rock; M.P. 2025°F (approx).
For building up and surface finishing shafts and wear surfaces; machinable with carbide tools; impact resistance better than Colmonoy No. 53.

COLMONOY NO. 53.
M-963; 0.65 C, 11.5 Cr, 4.25 Fe, 3.75 Si, 2.5 B, bal Ni.
Cast: C 45-50 Rock; M.P. 1950°F (approx).
For hard facing shafts, wear surface; good red hardness, corrosion resistance, and improved shock resistance.

COLMONOY NO. 56.
M-963; 0.7 C, 12.5 Cr, 2.75 B, 4 Si, 4.5 Fe, bal Ni.
Cast: 50-55 Rock C.
For hard facing alloy; abrasion, galling and corrosion resistance.

COLMONOY NO. 62.
M-963; 0.75 C, 14.0 Cr, 4.5 Si, 4.75 Fe, 3.0 B, bal Ni.
Cast: C 58-63 Rock; M.P. 1875°F (approx).
For hard surfacing shafts, sleeves, valve trim; good resistance to abrasion, corrosion and galling.

COLMONOY NO. 62SA.
M-963.
Same as COLMONOY NO. 62 but as atomized powder.

COLMONOY NO. 63.
M-963; 0.70 C, 14.0 Cr, 3.75 Fe, 3.75 Si, 3.5 B, bal Ni.
Cast: C 58-63 Rock; M.P. 1875°F (approx).
For hard surfacing by Fusewelder Torch, on shafts, sleeves, valve trim; good resistance to abrasion, corrosion and galling.

COLMONOY NO. 69.
M-963; 0.70 C, 14.0 Cr, 4.0 Si, 3.0 B, 1.75 Cu, 1.75 Mo, 4.35 Fe, bal Ni.
Cast: C 58-63 Rock; M.P. 1900°F (approx).
For hard surfacing, usually by Spraywelder; for shafts, sleeves, valve trim; good resistance to heat, abrasion, corrosion and galling.

COLMONOY NO. 70.
M-963; 0.4-0.7 C, 10-13 Cr, 1.7-3.2 B, 15-17 W, 2.5-4.0 Si, 2.5-4.0 Fe, bal Ni.
As cast: 50-55 Rock C.
For hard facing; wear, heat and corrosion resistant.

COLMONOY NO. 72.
M-963; 0.75 C, 13.0 Cr, 12.0 W, 3.6 Si, 3.5 Fe, 2.7 B, bal Ni.
Cast: C 58-63 Rock; M.P. 1940°F (approx).
For hard surfacing; excellent resistance to fretting corrosion and abrasion.

COLMONOY NO. 75.
M-963; 0.70 C, 13.5 Cr, 3 B, 4.2 Si, 4.75 Fe, WC, bal Ni.
Rockwell C 54-59.
For hard facing; resists galling and abrasion.

COLMONOY NO. 705.
M-963; Colmonox No. 63 with tungsten carbide particles added.
Cast: C 58-63 Rock; M.P. 1900°F (approx).
For hard facing surfaces requiring extreme abrasion resistance as buffing fixtures.

COLMONOY NO. 805.
M-963.
Same as COLMONOY NO. 6 with Chromium Boride added.

COLMONOY NO. C-290.
M-963; 0.45 C, 2.5 Si, 37.0 Ni, 1.5 B, 13.25 Cr, bal Fe.
For hard facing by Spraywelder, building up shaft bearing surfaces.

COLMONOY NO. C-395.
M-963; 0.50 C, 2.5 Si, 13.25 Cr, 37.0 Ni, 1.5 B, bal Fe.
Weld metal: 35-45 Rc, melting point, 2700°F approx.
Metallizing powder for spray welder for build-up to resist metal-to-metal wear in rotary type engines.

COLMONOY NO. HC240.
M-963; 5.0 C, 1.5 Mn, 1.5 Si, 30.0 Cr, bal Fe.
Weld metal: 60 Rc approx.
Semi-automatic tube wire for build-up of high abrasion resistant and hot wear coating.

COLMONOY SPECIAL NO. 1.
M-963; 1.0 C, 13.0 Cr, 3.0 Si, 3.0 B, 0.7 Mn, bal Fe.
Coated rods; as cast 60-65 Rc; melting point 2450°F, approx.
For abrasion resistant coatings as coal chutes, dredge cutters.

COLMONOY SWEAT ON PASTE.
M-963; 82 Cr, 18 B.
Rockwell C 70.
For welding hard facing alloy paste, wear and abrasion resisting.

COLONIAL HIGH SERVICE.
M-Eng; 0.90 Cu, 0.20 Mn, 0.14 V, bal Fe.
For tools, drills, taps; water hardening.

COLONIAL HIGH SPEED.
M-34; 0.52 c, 0.28 Mn, 3.8 Cr, 18.25 W, 0.24 V, bal Fe.
For high speed cutting tools; high speed steel.

COLONIAL NO. 6.
M-34, M-115; 0.95 C 1.3 Mn, 0.5 Cr, 0.15 V, 0.5 W, bal Fe.
Hardened: 140,000 TS; 120,000 YS; 8 El; 30 RA; 302 Brin.
For blanking dies, chasers, gages, hobs, master tools; non-shrinking, oil hardening.

COLONIAL NO. 7.
M-115; 0.6-1.4 C, 0.18 V, bal Fe.
For tools, chisels, pneumatic riveters, cutting tools, thread cutting dies tough, shock and vibration resistant.

COLONIAL NO. 14.
M-115; 0.6-1.2 C, bal Fe.
For tools; water hardening.

COLONIAL NO. 101 (CDA 842).
M-1393; 80 Cu, 5 Sn, 2.5 Pb, 12.5 Zn.
For pipe fittings, elbows, bushings.

COLONIAL NO. 115 (CDA 836).
M-1393; 85 Cu, 5 Sn, 5 Pb, 5 Zn.
For plumbing, hardware, impellers, valves.

COLONIAL NO. 120 (CDA 838).
M-1393; 83 Cu, 4 Sn, 6 Pb, 7 Zn.
For low pressure valves, plumbing, hardware.

COLONIAL NO. 123 (CDA 844).
M-1393; 81 Cu, 3 Sn, 7 Pb, 9 Zn.
For low pressure valves, plumbing, hardware.

COLONIAL NO. 125 (CDA 845).
M-1393; 78 Cu, 3 Sn, 7 Pb, 12 Zn.
For low pressure valves, plumbing, air and gas fittings.

COLONIAL NO. 130 (CDA 848).
M-1393; 76 Cu, 3 Sn, 6 Pb, 15 Zn.
For low pressure valves, plumbing, air and gas fittings.

COLONIAL NO. 131 (CDA 833).
M-1393; 93 Cu, 1.5 Sn, 1.5 Pb, 4 Zn.
For terminal ends for electrical cables.

COLONIAL NO. 132.
M-1393; 87 Cu, 2 Sn, 2 Pb, 9 Zn.
For electrical fittings.

COLONIAL NO. 194 (CDA 913).
M-1393; 81 Cu, 19 Sn.
For bearings, bridge plates, bells.

COLONIAL NO. 196.
M-1393; 14-16 Sn, bal Cu.
For bearings.

COLONIAL NO. 197 (CDA 910).
M-1393; 85 Cu, 14 Sn, 1 Zn.
For bearings, piston rings.

COLONIAL NO. 199.
M-1393; 12-14 Sn, bal Cu.
For bearings.

COLONIAL NO. 200 (CDA 925).
M-1393; 87 Cu, 11 Sn, 1 Pb, 1 Ni.
For gears, automotive rings.

COLONIAL NO. 205. (CDA 907).
M-1393; 9.75-12.0 Sn, bal Cu.
For bearings, bushings, gears.

COLONIAL NO. 206 (CDA 927).
M-1393; 88 Cu, 10 Sn, 2 Pb.
For bearings, impellers, steam fittings.

COLONIAL NO. 206A.
M-1393.
Same as no. 206 with Ni added.
For bearings, better machining than 210.

COLONIAL NO. 210 (CDA 905).
M-1393; 88 Cu, 10 Sn, 2 Zn.
For bearings, gears, steam fittings.

COLONIAL NO. 215 (CDA 926).
M-1393; 87 Cu, 10 Sn, 1 Pb, 2 Zn.
For bearings, gears, steam fittings.

COLONIAL NO. 225 (CDA 903).
M-1393; 88 Cu, 8 Sn, 4 Zn.
For bearings, gears, steam fittings.

Section I: Alloy Data

COLONIAL NO. 230 (CDA 923).
M-1393; 87 Cu, 8 Sn, 4 Zn.
For valves, high pressure steam castings.

COLONIAL NO. 235.
M-1393; 87 Cu, 8 Sn, 4 Zn, & Pb.
For bearings.

COLONIAL NO. 242. (CDA 902).
M-1393; 6-8 Sn, 0.5 max Zn, bal Cu.
For bearings.

COLONIAL NO. 245 (CDA 922).
M-1393; 88 Cu, 6 Sn, 1.5 Pb, 4.5 Zn.
For valves, fittings, for use up to 550°F.

COLONIAL NO. 255.
M-1393; 90 Cu, 5 Sn, 2.50 Pb, 2.50 Zn.
For plaques.

COLONIAL NO. 295 (CDA 928).
M-1393; 76 Cu, 16 Sn, 5 Pb.
For piston rings.

COLONIAL NO. 296 (CDA 940).
M-1393; 72 Cu, 13 Sn, 15 Pb.
For bearings.

COLONIAL NO. 305 (CDA 937).
M-1393; 80 Cu, 10 Sn, 10 Pb.
For high speed and heavy pressure bearings.

COLONIAL NO. 310 (CDA 934).
M-1393; 84 Cu, 8 Sn, 8 Pb.
For bearings and bushings.

COLONIAL NO. 311.
M-1393; 84 Cu, 8 Sn, 8 Pb.
For bearings.

COLONIAL NO. 312 (CDA 944).
M-1393; 81 Cu, 8 Sn, 11 Pb.
For general utility bushings and bearings.

COLONIAL NO. 319 (CDA 938).
M-1393; 78 Cu, 7 Sn, 15 Pb.
For bearings, mine water impellers.

COLONIAL NO. 321 (CDA 945).
M-1393; 73 Cu, 7 Sn, 20 Pb.
For high speed, low load bearings.

COLONIAL NO. 322 (CDA 943).
M-1393; 70 Cu, 5 Sn, 25 Pb.
For high speed bearings for light loads.

COLONIAL NO. 325 (CDA 941).
M-1393; 70 Cu, 5.5 Sn, 18.5 Pb, 3 Zn.
For high speed, low load bearings.

COLONIAL NO. 326 (CDA 935).
M-1393; 85 Cu, 5 Sn, 9 Pb.
For small bearings and bushings.

COLONIAL NO. 351 (CDA 932).
M-1393; 83 Cu, 7 Sn, 7 Pb, 3 Zn.
For general utility bearings & bushings.

COLONIAL NO. 400 (CDA 852).
M-1393; 72 Cu, 1 Sn, 3 Pb, 24 Zn.
For plumbing fittings, ornamental, and hardware.

COLONIAL NO. 403 (CDA 854).
M-1393; 67 Cu, 1 Sn, 3 Pb, 29 Zn.
For general purpose yellow brass.

COLONIAL NO. 405.
M-1393; 66 max Cu, 1.0 Sn, 3.0 Pb, bal Zn.
Yellow brass, for plumbing, hardware.

COLONIAL NO. 405.1 (CDA 858).
M-1393; 58 Cu, 1 Sn, 1 Pb, 40 Zn.
For general purpose yellow brass die castings.

COLONIAL NO. 405.2 (CDA 857).
M-1393; 63 Cu, 1 Sn, 1 Pb, 35 Zn.
For ornamental and hardware fittings.

COLONIAL NO. 406.
M-1393; 66 max Cu, 1.0 Sn, 1.0 Pb, bal Zn.
For hardware and fittings.

COLONIAL NO. 407 (CDA 853).
M-1393; 70 Cu, 30 Zn.
For yellow brass hardware.

COLONIAL NO. 407.5 (CDA 834).
M-1393; 90 Cu, 10 Zn.
For moderate conductivity, rotating bends.

COLONIAL NO. 408.
M-1393; 85 Cu, 15 Zn.
For brazing applications.

COLONIAL NO. 410 (CDA 973).
M-1393; 56 Cu, 2 Sn, 10 Pb, 12 Ni, 20 Zn.
For ornamental and hardware castings.

COLONIAL NO. 411 (CDA 974).
M-1393; 59 Cu, 3 Sn, 5 Pb, 17 Ni, 16 Zn.
For valves, hardware, fittings.

COLONIAL NO. 412 (CDA 976).
M-1393; 64 Cu, 4 Sn, 4 Pb, 20 Ni, 8 Zn.
For marine hardware, sanitary fittings.

COLONIAL NO. 413.
M-1393; 5.0 max Sn, 10.0 max Pb, 22-27 Ni, bal Cu+Zn.
For valves and sanitary fittings.

COLONIAL NO. 413B (CDA 978).
M-1393; 66 Cu, 5 Sn, 2 Pb, 25 Ni, 2 Zn.
For valves, sanitary fittings, musical instruments.

COLONIAL NO. 415A (CDA 952).
M-1393; 88 Cu, 3 Fe, 9 Al.
For general alumimum bronze castings.

COLONIAL NO. 415B (CDA 953).
M-1393; 89 Cu, 1 Fe, 10 Al.
For marine equipment, nuts, gears.

COLONIAL NO. 415C (CDA 954).
M-1393; 85 Cu, 4 Fe, 11 Al.
For gears, bushings, valve seats.

COLONIAL NO. 415D (CDA 955).
M-1393; 81 Cu, 4 Ni, 4 Fe, 11 Al.
For corrosion resistant parts, ship propellers.

COLONIAL NO. 420 (CDA 864).
M-1393.
60,000 psi tensile Mn bronze.
Free machining manganese bronze.

COLONIAL NO. 421 (CDA 865).
M-1393.
65,000 psi tensile Mn bronze.
For valve stems, gears, machinery.

COLONIAL NO. 422 (CDA 867).
M-1393.
80,000 psi tensile Mn bronze.
For high strength free machining parts.

COLONIAL NO. 423 (CDA 861).
M-1393.
90,000 psi tensile Mn bronze.
For marine castings, bushings, and bearings.

COLONIAL NO. 424 (CDA 863).
M-1393.
110,000 psi tensile Mn bronze.
For high strength manganese bronze.

COLONIAL NO. 500A (CDA 875).
M-1393.
Silicon bronze 12A
For bearings, bells, impellers, pumps.

COLONIAL NO. 500B (CDA 874).
M-1393; 83 Cu, 14 Zn, 3 Si. .d Silicon brass 13A.
For small boat propellers, valve stems.

COLONIAL NO. 500C (CDA 875).
M-1393; 82 Cu, 14 Zn, 4 Si.
Silicon brass 13B.
For small boat propellers, valve stems.

COLONIAL NO. 500D (CDA 876).
M-1393; 90 Cu, 5.5 Zn, 4.5 Si.
Silicon bronze 13C.
For valve stems.

COLONIAL NO. 500E (CDA 878).
M-1393; 82 Cu, 14 Zn, 4 Si.
Silicon brass ZS144A.
For thin wall die castings, brush holders.

COLONIAL NO. 500F (CDA 879).
M-1393; 65 Cu, 34 Zn, 1 Si.
Silicon brass ZS331A.
For general purpose die casting.
ASTM 176-70 Z5-331A.

COLONIAL NO. 600A (CDA 957).
M-1393; 75 Cu, 2 Ni, 3 Fe, 8 Al, 12 Mn.
For propellers, impellers, pumps.

COLONIAL NO. 600B (CDA 958).
M-1393; 81 Cu, 5 Ni, 4 Fe, 9 Al, 1 Mn.
For propeller hubs, blades for salt water.

COLORADO METAL.
M-US; 57 Cu, 25 Ni, 18 Zn.
For electrical resistances, corrosion resisting parts; corrosion resistant.

COLOSSO.
M-344; 0.3 Cr, 0.5 Ni, 0.3 W, 0.1 V, 0.6 Mo, bal Fe.
For tools, flogging and pneumatic tools, rivet sets; no drawing of temper.

COLUMAX.
M-1010; Eng.; 7.8 Al; 13.5 Ni, 25 Co, 3 Cu, 0.8 Cb, bal Fe.
13,600 Br; 750 Hc; 7.5×10^6 BH max.
Heat treatable, for permanent magnets.

COLUMAX-5.
M-1192; 8 Al, 14 Ni, 24 Co, 3 Cu, bal Fe.
Peak energy product 7,500,000.
Residual induction 13,500, Hc 750, 5000 TS; 8000 Tr.M.R., C 50 Roc
For permanent magnets in electrical and magnetic equipment and instruments.
Similar to Alnico V-7, cast magnet, hardenable. Hard and brittle.

COLUMBIA 16.
M-35; 0.98 C, 0.35 Mn, 1.38 Si, 4.25 Cr, 2.5 Mo, 1.1 V, 0.40 W, bal Fe.
Cold work tool steel, for punches, forming rolls, thread rolling dies.

COLUMBIA A 8.
M-35; 0.56 C, 5.0 Cr, 1.25 W, 1.25 Mo, 1.0 Si, bal Fe.
For shear blades, forming rolls, heavy blanking dies, punches.
Type A 8 air hardening tool steel.

COLUMBIA EXTRA.
M-35; 1.06 C, 0.25 Mn, 0.25 Si, bal Fe.
Water hardening tool steel.
For threading dies, shear blades, hand stamps.
AISI W1-2.

COLUMBIA EXTRA.
M-35; 0.7-1.2 C, 0.25 Si, 0.25 Mn, bal Fe.
For cold chisels, shear blades, punches, drills; Type W1; water hardened.

COLUMBIA EXTRA HEADERDIE.
M-35; 0.95 C, 0.35 Mn, 0.25 Si, bal Fe.
Water hardening tool steel.
For stamping dies, punches, heading dies.
AISI W1-2H.

COLUMBIA SPECIAL.
M-35; 1.06 C, 0.30 Mn, 0.25 Si, 0.20 Cr, 0.05 V, bal Fe.
Water hardening tool steel.
For taps, reamers, paper knives, gages.
AISI W1-1.

COLUMBIA STANDARD.
M-35; 1.06 C, 0.30 Mn, 0.25 Si, bal Fe.
Water hardening tool steel.
For blacksmith tools, dowel pins, drift pins.
AISI TYPE W1-3.

COLUMBIUM 10W-2.5 ZR.
M-1537; 9-11 W, 2-3 Zr, bal Cb.
Bar: 85,000-89,000 TS; 66,000-71,000 YS; 26-33 El; 64-70 Brin.
Sheet at 2000°F: 28,000 TS; 22,000 YS; 25 min El.
Bar at 2000°F: 36,500 TS; 27,500 YS; 32 El.
For space vehicles and nuclear reactors.
High strength at elevated temperatures.
Insensitive to notch stress concentration.

COLUMBIUM-10W-10 HF.
M-1537; 9-11 W, 9-11 Hf, bal Cb.
Cold rolled: 130,800 TS; 122,200 YS; 4 El.
Stress-relieved: 112,300 TS; 100,500 YS; 16 El.
Recrystallized: 88,500 TS; 71,500 YS; 26 El.
For space vehicles, missiles, high temperature components.
Good oxidation resistance and high temperature properties.

COLUMBIUM-15W-5 MO-1 ZR.
M-1537; 15 W, 5 Mo, 1 Zr, bal Cb.
Annealed: 115,000 TS; 95,000 YS; 5 El, 8 RA.
At 2200°F: 43,500 TS; 39,000 YS; 22 El; 40 RA.
For missile and space vehicle components, high temperature fasteners.
High heat, oxidation and corrosion resistant.

COLUMBIUM C-103.
M-1596; 9.0-11.0 Hf, 0.7-1.3 Ti, 0.7 max Zr, bal Cb.
Cold rolled: 105,000 TS; 96,600 YS; 4.5 El.
Recrystallized: 58,800 TS; 45,400 YS; 26 El.
For high temperature applications, space vehicles.
Good TIG weldability.

COLUMBIUM C-129.
M-1596 M-1537; 10 W, 10 Hf, bal Cb.
Rolled: 131,000 TS; 123,000 YS; 5 El.
Stress relieved: 112,000 TS; 100,000 YS; 16 El.
Recrystallized: 85,000 TS; 71,000 YS; 30 El.
For high temperature service, space vehicles, reactors, missiles.
Recrystallized at 1900-2200°F. Heat resistant.

COLUMBIUM C-129-Y.
M-1596, M-1537; 9-11 w, 9-11 Hf, 0.05-0.3 Y, bal Cb.
Recrystallized: 89,800 TS; 75,700 YS; 25 El.
At 2000°F: 39,600 TS; 29,500 YS; 44 El.
At 3000°F: 11,000 TS; 9900 YS; 75 El.
For high temperature applications, space vehicles, missiles.
Good elevated temperature properties combined with heat and oxidation resistance.

COLUMBIUM CB 132 M.
M-114, M-1537, M-1596; 0.001 C, 1.5 Zr, 15 W, 5 Mo, 20 Ta, 58.5 Cb.
Extruded: 131,800 TS; 122,100 YS; 4 El.
At 2400°F: 56,700 TS; 49,200 YS; 37 El (extruded).
Heat treated: 89,200 TS; 0.5 El.
At 2400°F: 58,300 TS; 49,900 YS; 20 El; (heat treated).
For jet engine and gas turbine blades.
High strength, high temperature alloy.

COLUMBIUM CB-752.
M-48, M-1537; 9-11 W, 2-3 Zr, bal Cb.
Annealed sheet: 78,000 TS; 58,000 YS; 20 min El.
At 2200°F: 28,000 TS; 22,000 YS; 25 min El.
For high temperature applications.
High heat and corrosion resistance.

COLUMBIUM D-10.
M-1563; 0.0175 O_2, 0.0020 H_2, 0.01 N_2, 0.01 C, bal C.
Recrystallized sheet: 30,000 min TS; 15,000 min YS; 15 min El.
Recrystallized bar: 25,000 min TS; 12,000 min YS; 25 min El.
For high temperature applications. High heat resistance.

COLUMBIUM D-11.
1563; 0.75-1.25 Zr, bal Cb.
Bar (Recryst.) 30,000 min TS; 16,000 min YS; 25 min El.
Extruded: 40,000 TS; 20,000 YS; 25 El.
At 1800°F: 27,000 TS; 16,000 YS; 23 El.
Sheet (Recryst): 35,000 min TS; 20,000 min YS; 12 min El.
For space vehicles and nuclear reactors.
High heat resistant.

COLUMBIUM D-14.
M-1563; 4.4-5.6 Zr, bal Cb.
At 78°F: 68,000 TS; 56,000 YS; 15 El.
At 2000°F: 32,000 TS; 22,000 YS; 40 El.
Stress relieved sheet: 55,000 min TS; 45,000 min YS; 12 min El.
For hot surfaces on space re-entry vehicles.
Moderate strength alloy for elevated temperature service.

COLUMBIUM D-36.
M-1563; 9.0-11.0 Ti, 4.0-6.0 Zr, bal Cb.
Sheet recrystallized: 80,000 TS; 71,000 YS; 20 El.
At 800°F: 65,000 TS; 45,000 YS; 18 El.
At 2000°F: 22,000 TS; 20,000 YS; 50 El.
Bar Recryst: 65,000 min TS; 58,000 min YS; 20 min El.
For space re-entry vehicles.
High heat resistance.

COLUMBIUM D-40.
M-1563; 15 W, Mo, 1 Zr, 0.06 C, bal Cb.
Cold rolled: 125,000 TS; 85,000 YS; 25 El.
At 2200°F: 50,000 TS; 30,000 YS; 22 El.
For space missiles and rocket components.
Heat resistant. Retains usable strength to 2500°F.

COLUMBIUM D-41.
M-1563; 19.0-21.0 W, 9.0-11.0 Ti, 5.5-6.5 Mo, bal Cb.
Extruded: 130,000 TS; 127,000 YS; 5.3 El.
At 2000°F: 56,000 TS 53,000 YS; 26 El.
At 2400°F: 34,000 TS; 29,000 YS; 33 El.
For space vehicles and nuclear reactors. High heat resistant.

COLUMBIUM D-43.
M-1563; 10 W, 1 Zr, 0.1 C, bal Cb.
Sheet: 92,000 TS; 78,000 YS; 22 El.
At 2000°F: 50,000 TS; 45,000 YS; 15 El.
For missiles and space craft.
High creep and rupture strength.

COLUMBIUM (NIOBIUM).
M-1755; Cb (Nb).
Purities; Zone refined 99.99+%, dendritic 99.99+%, nuclear grade 99.8%, 99.5%.
Forms: Powders, crystal bar, sintered bar, wires, sheets, ingots, etc., single crystals.

COLUMBIUM XB-88.
M-118; 28 W, 2 Hf, 0.067 C, bal Cb.
For gas turbine buckets.
Heat resistant. High rupture strength.

COMALLOY.
M-England; 17 Mo, 12 Co, bal Fe.
For permanent magnets, electrical and magnetic equipment.
Precipitation hardened, high permeability.

COMANCHE.
M-1432; 0.9 C, 1.2 Mn, 0.5 Cr, 0.2 V, 0.5 W, bal Fe.
Oil hardening tool steel; AISI 01.

COMET.
M-32; 1.05 C, 0.25 Mn, bal Fe.
For taps, reamers, dies for blanking, trimming and forming; water hardening, "Carpenter No. 11."

COMINCO NO. 2570.
M-1679; Zn alloy.
Forged: 60,000-72,000 TS; 45,000-60,000 YS; 14 El; 120 Brin.
For industrial applications, hardware, fire fighting equipment, automotive parts, plumbing fixtures, solenoid valves.

COMINCO NO. 2573.
M-1679; Zn alloy.
Forged: 62,000-75,000 TS; 45,000-60,000 YS; 13 El; 125 Brin.
For industrial applications, hardware, fire fighting equipment, automotive parts, plumbing fixtures, solenoid valves.

COMINCO NO. 3130.
M-1679; Zn alloy.
Forged: 35,000 TS; 23,000 YS; 30 El; 65 Brin.
For structural and pressure components, plumbing fixtures, solenoid valves
Good creep resistance.

COMINCO NO. 3330.
M-1679; Zn alloy.
Forged: 48,000 TS; 35,000 YS; 20 El; 90 Brin.
For structural and pressure components, plumbing fixtures, valves, regulators. Good creep resistance.

COMMANDO.
M-435; 1.2 C, 0.25 Mn, 0.2 Si, bal Fe.
For dies; drill rod.

COMMANDO.
M-373; 0.45 C, 1.4 Cr, 0.4 Si, 2 W, bal Fe.
For hot punches, tools; hot work steel, shock resistant.
AISI S1.

COMMELL SPECIAL.
M-Eng; 1.20 C, 0.21 Mn, 2.92 W, bal Fe.
For tools, dies, cutters; fast finishing tool steel.

COMMERCE H.
M-1248; Cu alloy.
Cast: 65,000-75,000 TS; 12,000-16,000 YS; 10-2 El; 116-165 Brin.
For resistance welding electrode; 10-15% electrical conductivity.

COMMERCIAL BRASS.
M-8; 65 Cu, 35 Zn.
Hard: 76,000 TS; 4 El; 153 Brin.
Soft: 45,000 TS, 60 El; 52 Brin.
For fixtures, radiators, ornaments; must be annealed.

COMMERCIAL BRONZE.
M-33, M-1789; 90 Cu, 10 Zn.
Sheets: 32,000 TS; 10,000 YS; 40 El; 50 Brin.
Wire: 100,000 TS; 50,000 YS; 1 El.
For window screen wire, automobile radiators, hardware, ornaments.

COMMERCIAL BRONZE 25.
M-141; 90 Cu, 10 Zn.
Hard: 61,000 TS; 54,000 YS; 5 El; 125 Brin.
Soft: 38,000 TS; 12,000 YS; 45 El; 53 Brin.
For weather stripping; corrosion resistant.

COMMERCIAL BRONZE 220.
M-8; 10 Zn, bal Cu.
Sheet: 32,000 TS; 10,000 YS; 40 El; 50 Brin.
Wire: 100,000 TS; 50,000 YS; 1 El.
For window screen wire, auto radiators, hardware; corrosion resistant.

COMMERCIAL BRONZE 226.
M-8; 12.5 Zn, bal Cu.
Hard: 65,000 TS; 52,000 YS; 7 El; 125 Brin.
Soft: 39,000 TS; 13,000 YS; 45 El; 72 Brin.
For costume jewelry, slide fasteners; base for gold plate.

COMMERCIAL CARBON DRILL ROD.
M-1702; 0.95-1.05 C, 0.3-0.5 Mn, 0.15-0.3 Si, bal Fe.
AISI Type W 1 water hardening tool steel.

COMMERCIAL CASTINGS B. C.
M-Eng; 62 Cu, 30 Zn, 6 Sn, 2 Pb.
For name plates, oil cups, instrument cases.

COMMON FORMULA.
M-Eng; 55 Cu, 25 Zn, 20 Ni.
For electrical resistance; German silver.

COMMON TOMBAC.
M-French; 28 Zn, 72 Cu.
Rolled: 42,000 TS; 25,000 YS; 35 El; 30 RA; 45 Brin.
For cartridge cases, condenser tubes, brazing; corrosion resistant; ductile.

COMMON TYPE METAL "A".
M-U.S.; 60 Pb, 10 Sn, 30 Sb.
For type metal.

COMMON TYPE METAL "B".
M-U.S.; 55.5 Pb, 40 Sn, 4.5 Sb.
For type metal.

COMMONWEALTH.
M-1034; 0.3 C, 2 Ni, bal Fe.
80,000 TS.
For trunk frames; tough.

COMO.
M-140; 0.7 C, 9 Mo, 1.5 W, 4 Cr, 1 V, 5 Co, bal Fe.
For lathe and planer tools, drills, hobs, reamers; high speed steel. AISI M-30.

COMPAX.
M-111, M-111a; 0.13 C, 1.4 Cr, 3.8 Ni, 0.5 Mn, bal Fe.
Annealed: 100,000 TS; 200-220 Brin.
Core Hardened: 145,000 TS; 112,000 YP; 290 Brin.
For plastic molds, large cavity dies for compression molding.
Type P 6 tool steel, case hardening.

COMPENSATOR ALLOY.
M-England; Fe-Ni.
For compensating shunts for electrical equipment; temperature sensitive, magnetic.

COMPLEX ENGLISH METAL.
M-Eng; 87 Sn, 6 Sb, 2 Ni, 2 Cu, 1.5 W, 1 Zn, 0.5 Bi.
For bearings.

COMPO 61-A ETC see **GKN 61-A ETC.**

COMPOSITE DIE.
M-261; C, bal Fe.
For tools, dies; water hardening.

COMPOSITION BRASS.
84-85 Cu, 4-6 Zn, 4-6 Sn, 4-6 Pb.
26,000 TS; 12,000 YS; 15 El.
For bearings, screws, hardware; (ounce metal).

COMPOSITION NO. 1.
M-508; 6 Sn, 3 Pb, 4 Zn, bal Cu.
34,000 TS; 23,000 YS; 20 El; 75 Brin.
For centrifugal castings; general purpose.

COMSOL.
M-200; Ag, Sn, Pb.
Cast: 56,000 TS; 40 El.
For soft solder; M. P. 296°C; retains strength at high temperature.

CONDENSER FOIL.
M-Eng.; 90 Pb, 9 Sn, 1 Sb.
For accumulators; accumulator metal.

CONDOR SPECIAL.
M-253; 0.7-1.4 C, Mn, Si, bal Fe.
For drill rods, tools; see Pompton Special.

CONDUCTAL.
M-1784.
Name of several Al alloys.
For busbars and insulated conductors.

CONDULOY.
M-1037; 0.23-0.32 Be, 1.4-1.5 Ni, bal Cu.
Cast: 54,000 TS; 21,000 YS; 15 El; 14 RA; 110 Brin.
Heat treated: 84,000 TS; 64,000 YS; 4 El; 7 RA; 189 Brin.
For diaphragms, springs, instrument parts; age-hardenable; now Brush 35- alloy.

CONEL.
M-942; 38.5 Ni, 3.3 Mn, 3.3 Si, 4.6 Cr, bal Fe.
Cold worked and annealed: 185,000 TS; 145,000 YS; 4 El.
For springs; low thermoelastic coefficient.

CO-NETIC A-152.
M-1567; Fe, bal Ni.
Sheet: 21,000 YP; B 43-47 Rock.
For magnetic shielding.

CO-NETIC AA.
M-1567; 20 Fe, bal Ni.
Sheet: 57,000 TS; 28,500 YP; B 47-53 Rock.
For electrostatic shielding.
High permeability.

CO-NETIC B.
M-1567; Fe, bal Ni.
Sheet: 58,000 TS; 19,000 YP.
For magnetic shielding.

CONFLEX 216.
M-1562; 10% copper layer on AISI 1065 steel.
Heat treated: 215,000 TS; 185,000 YS; 5 El.
For current carrying springs.
Magnetic, 16% conductivity.

CONFLEX 316.
M-1562; Carbon spring steel clad with Cu both sides, 90% steel, 10% Cu.
Annealed: 74,000 TS; 65,000 YS; 25 El.
Rolled: 130,000 TS; 110,000 YS; 2 El.
Heat treated: 213,000 TS; 185,000 YS; 5 El.
For current carrying springs. 16% conductivity.
High elasticity, low contact resistance.

CONFLEX 326.
M-1562; Carbon spring steel clad with Cu both sides. 80% steel, 20% Cu.
Annealed: 70,000 TS; 61,000 YS; 25 El.
Rolled: 120,000 TS; 110,000 YS; 2 El.
Heat treated: 204,000 TS; 177,000 YS; 6 El.
For current carrying springs.
High elasticity, low contact resistance.

CONFLEX 335S.
M-1562; Core of age-hardened 17-7 PH double clad with copper.
Aged: 155,000 TS; 140,000 YS; 2.5 El.
For current carrying springs.
30-40% conductivity.

CONFLEX 545.
M-1562; AISI 6150 steel double clad on a copper base.
Heat Treated: 145,000 TS; 133,000 YS; 4-5 El.
For flat or cantilever type springs. Two thin copper layers clad on the outer surfaces to provide resistance to corrosion.
Five layers bonded together. Not brazed together. Copper is 45% of cross-section.

CONFLEX 720.
M-1562; 20 Mn, 20 Ni, bal Cu.
Hardened: 102,650-175,600 TS; 49,800-149,300 YS; 0.5-15 El; 200-425 VHN.
For springs, diaphragms, clips, watch cases, pencaps.
Nonmagnetic, corrosion resistant. Age-hardenable.

CONGO.
M-140; 0.8 C, 4 Cr, 1.5 V, 4 W, 5 Mo, 12 Co, bal Fe.
For form tools, cut-off knives, reamer blades; high speed steel, oil hardened.

CONICO.
50 Cu, 29 Co, bal Ni.
For magnets for electrical equipment; magnetically soft.

CONLY.
M-746; 4-5 Cu, bal Al.
For bicycles; tubes.

CON-PAC 80.
M-604; 0.18 C, 1.25 Mn, 0.30 Si, 0.15 Ni, 0.15 Cr, 0.04 Mo, 0.001 B, bal Fe.
W.Q.+T.; 100 ksi TS; 80 ksi YS; 20 El.
For heavy duty equipment, trucks, mobile machinery.
ASTM A678.

CON-PAC 90.
M-604; 0.18 C, 1.25 Mn, 0.30 Si, 0.15 Ni, 0.15 Cr, 0.04 Mo, 0.001 B, bal Fe.
W.Q+T.: 110 ksi TS; 90 ksi YS; 20 El.
For heavy duty equipment, trucks, mobile machinery.
ASTM A678.

CON-PAC 100.
M-604; 0.18 C, 1.25 Mn, 0.30 Si, 0.15 Ni, 0.15 Cr, 0.04 Mo, 0.001 B, bal Fe.
W.Q.+T.: 120 ksi TS; 100 ksi YS; 20 El.
For heavy duty equipment, trucks, mobile machinery.
ASTM A678.

CON-PAC M.
M-604; 0.18 C, 1.30 Mn, 0.40 Si, bal Fe.
W.Q.+T.: 90 ksi TS; 75 ksi YS; 22 El.
Readily formed and welded.
Construction equipment, off-shore platforms, ship hulls. Good toughness.
ASTM A678.

CONPERNIK.
M-118; 50 Fe, 50 Ni, trace Mn.
Annealed: 55,000 TS; 20,000 YS; 33 El.
For laminations for transformers, choke coils; high magnetic permeability.

CONQUEROR.
M-688; 0.9 C, 0.4 Mn, 0.3 Si, bal Fe.
For stone drills and tools, punches, chisels; tough and wear resistant.

CONQUEROR.
M-1406; 0.70-0-0.85 C, bal Fe.
Heat treated: 188,000 TS; 143,000 YS; 12 El; 35 RA; 385 Brin.
For press tools, chisels, springs; Type W1; water hardened.

CONQUEROR 14%.
M-1406; 0.65 C, 14 W, 4 Cr, 0.75 V, bal Fe.
For shear blades, punches, drills, taps; high speed steel.

CONQUEROR AA.
M-1406; 0.70 C, 17 W, 4 Cr, 1 V, 0.5 Mo, bal Fe.
For cutters, taps, reamers, hobs, drills; high speed steel.

CONQUEROR HOLLOW DRILL.
M-688; 0.75 C, 0.3 Mn, 0.2 Si, bal Fe.
For hollow stone drills, pneumatic tools; abrasion and shock resistant.

CONQUEROR L.C.
M-1406; 0.45 C, 14 W, 2.7 Cr, 0.4 V, bal Fe.
For forging and extrusion dies, mandrels; oil hardened, hot die steel.

CONQUEROR O.H.D.
M-1406; 1.0 C, 0.25 Mn, 1.6 Cr, 0.5 W, bal Fe.
For cutters, gages, drawing dies, rolls, reamers; cold work steel, oil hardened.

CONSIL 852.
M-63; 85 Ag, 15CdO.
For electric contacts.
Internally oxidized.

CONSIL 866.
M-63; 86.6 Ag, 13.4 Cd.
For electric contacts.

CONSIL 880.
M-63; 88 Ag, 12 Cd.
For electric contacts.

CONSIL 900.
M-63; 90 Ag, 10 CdO.
For electric contacts.
Internally oxidized.

CONSIL 910.
M-63; 91 Ag, 9 Cd.
For electric contacts.

CONSIL 995.
M-63; 0.25 max Mg, 0.25 max Ni, 99.4 min Ag.
Annealed: 28,000-34,000 TS; 30-40 El; 60-65 Rock 15 T.
Cold Worked: 55,000 TS; 6-8 El; 50-60 Rock 30 T.
Hardened: 68,000 TS; 56,000 YS 5-15 El; 68 Rock 30 T.
For electrical contacts, cable connectors, spring clips, relay springs.
Air hardened. High elect. conductivity.

CONSTAHL.
M-716; 0.1 C, 19 Cr, 9 Ni, bal Fe.
For stainless parts; stainless.

CONSTANTAN.
M-44, M-61; 60-45 Cu, 40-55 Ni, 0-1.4 Mn.
Annealed: 60,000-70,000 TS; 20,000-30,000 YS; 40-60 El; 50-70 RA; 100-120 Brin.
Rolled: 140,000 TS; 125,000 YS; 1 El; 5 RA; 300 Brin.
For electrical resistances, thermocouples; high heat resistance.

CONSTANTAN.
M-65; 46 Ni; 54 Cu, 0.2 Fe, 0.7 Mn.
For resistance alloy, rheostats, thermocouples; heat resistant.

CONSTANT SPECIAL now **VEW K720.**

CONSTRUCTAL 1.
M-116; Cu, Zn, Mg, bal Al.
For light alloy parts.

CONSTRUCTAL 2.
M-116; 1.2 Cu, 0.9 Mg, 0.6 Si, 0.5 Ti, bal Al.
Aged: 54,000-60,000 TS; 18-25 El; 95-115 Brin.
For aircraft and heavy duty forgings; age-hardenable.

CONSTRUCTAL 8.
M-116; 7 Zn, 2.5 Mg, 0.2 Si, 1 Mn, bal Al.
Heat treated: 67,000-74,000 TS; 15-18 El; 120-140 Brin.
For light alloy structural parts; heat treatable.

CONSTRUCTAL 20/42.
M-116; 4.5 Zn, 2.5 Mg, 0.4 Si, 0.6 Mn, 0.1 Cu, bal Al.
Aged: 65,000-71,000 TS; 51,000-60,000 YS; 10-12 El.
For light alloy parts; high strength, age hardenable.

CONSTRUCTAL 20/53.
M-116; 5.0 Zn, 3.0 Mg, 0.4 Si, 0.3 max Cr, 0.1 Cu, bal Al.
Aged: 71,000-77,000 TS; 64,000-68,000 YS; 8 El.
For light alloy parts; high strength, age-hardenable.

CONSTRUCTAL 87.
M-116; 0.28 Fe, 1.24 Mn, 6.87 Zn, 1.62 Mg, 0.75 Si, bal Al.
For light alloy structural parts; heat treatable.

CONSUMET see **CARPENTER CONSUMET.**

CONSUMET 882.
M-32; 0.35 C, 5.0 Cr, 1.5 Mo, 0.4 V, bal Fe.
Ht. Tr.: 300,000 TS; 250,000 YS; 6 El; C 55 Rock.
For die casting dies, extrusion and forging dies, hot punches, aircraft structural fittings.
Vacuum melted Type H-11 tool steel.

CONTACT BRONZE.
M-141; 8.95 Zn, 1.9 Sn, 0.15 P, bal Cu.
Annealed: 48,000 TS; 54 El; 67 Brin.
Hard: 78,000 TS; 7.5 El; 165 Brin.
Spring: 97,500 TS; 3 El; 190 Brin.
For electrical contacts, switches, springs; good spring properties.

CONTACT BRONZE 92.
M-141; 89 Cu, 0.15 P, 1.9 Sn, bal Zn.
Hard: 90,000 TS; 72,000 YS; 3 El; 185 Brin.
For electrical contacts; corrosion resistant.

CONTRALOY.
M-184; C, 28 Cr, 15 Ni, bal Fe.
For heat and corrosion resistant parts; now Michiana No. 63.

COO 75.
M-1488; 0.48 C, 0.40 Mn, 0.25 Si, bal Fe.
For plastic molds or die-casting molds for aluminum or zinc.
AFNOR $Y_2 48$.

COOK A.L.Z.
M-1405; 0.4 C, 5 Cr, 0.1 W, 1 Si, 1.5 Mo, 0.6 V, bal Fe.
For punches, pneumatic tools; oil hardened, tough.

COOK B.K.V.
M-1405; 1.9 C, 0.35 Mn, 13.5 Cr, 0.6 Si, bal Fe.
For punches, plastic mold dies, form tools; air or oil hardened, non-deforming.

COOK CCV.
M-1405; 0.50 C, 0.75 Mn, 1.0 Cr, 0.15 V, bal Fe.
Oil hardening tool steel.
For chuck jaws, hammer pistons, axes.
Similar to AISI 6150.

COOK C.K.K.
M-1405; 1.0 C, 1.4 Cr, 0.3 Mn, 0.25 Si, bal Fe.
For bearings, bushings, blanking and forming dies; water hardened, wear and abrasion resistant.

COOK C.M.C.
M-1405; 0.9 C, 1.3 Mn, 0.95 Cr, 0.3 Si, bal Fe.
For blanking and forming tools, taps, punches; oil hardened, non-deforming.

COOK C.R.P.
M-1405; 0.9 C, 1.3 Mn, 1.0 Cr, 0.25 Si, bal Fe.
For press tools, guages, master hobs, rollers; oil hardened, non-deforming.

COOK CRP 01.
M-1405; 0.95 C, 1.25 Mn, 0.50 Cr, 0.20 V, 0.50 W, bal Fe.
Oil hardening tool and die steel.
For jigs, fixtures, slitting cutters, blanking tools.
BS 4659 B01; AISI 01.

COOK E.T.A.
M-1405; 0.9 C, 0.35 Mn, 0.2 Si, 0.1 V, bal Fe.
For drawing, pressing and forming dies; cold work steel, water hardened.

COOK ETH. (RED LABEL).
M-1405; 1.14 C, 0.40 Mn, 0.15 V, bal Fe.
Water hardening tool steel.
For mandrels, knurling tools.

COOK M.K.Z.
M-1405; 0.38 C, 3 Cr, 9 W, 0.4 Mo, 0.2 V, bal Fe.
For hot forging and extrusion dies and tools; hot work steel, oil hardened.

COOK'S ALLOY.
M-Eng; 56-69 Sb, 32-44 Zn.

COOK SILVER LABEL.
M-1405; 1.4 C, 0.40 Mn, 0.60 Cr, bal Fe.
Water hardening tool steel.
Punches and dies, lathe and planer tools.

COOPER.
M-Eng.; 40-50 Au, 50-60 Pd.
For dental alloy; corrosion resistant.

COOPER ALLOY 13-4.
M-36; 0.06 C, 1.0 Si, 1.0 Mn, 11.5-14.0 Cr, 3.5-4.5 Ni, 0.40-1.0 Mo, bal Fe.
Cast: 758 MPa TS; 552 MPa YS; 15 El, 35 Ra.
Corrosion resistant casting.
ACI CA6NM.

COOPER ALLOY 14 H.
M-36; 0.25-0.35 C, 1 max. Mn, 11-14 Cr, 0.75 max. Ni, bal Fe.
Annealed: 99,000 TS; 55,000 YS; 27 El; 48 RA; 190 Brin.
Hardened: 250,000 TS; 225,000 YS; 0 El; 0 RA; 550 Brin.
For grinding parts, castings; wear and abrasion resistant.

COOPER ALLOY 14 I.
M-36; 0.12-0.20 C, 1 max. Mn, 11-14 Cr, 0.75 max. Ni, bal Fe.
Cast: 115,000-120,000 TS; 90,000-100,000 YS; 25-20 El; 60-55 RA; 200-240 Brin.
For valve trim, fittings, pump parts; corrosion resistant.

COOPER ALLOY 14S.
M-36; 0.10 C, 12.5 Cr, 0.5 Ni, 0.75 Si, bal Fe.
Heat treated: 197,000-116,000 TS; 139,000-96,000 YS; 16-23 El; 42-60 RA; 372-230 Brin.
For valves, valve trim, pump parts; corrosion resistant.

COOPER ALLOY 14SM.
M-36; 0.15 C, 0.65 Si, 1.0 Mn, 11.5-14.0 Cr, 1.0 Ni, 0.15-1.0 Mo, bal Fe.
Cast: 621 MPa TS; 448 MPa YS; 18 El; 30 RA.
Corrosion resistant casting.
ACI CA-15M.

COOPER ALLOY 15B.
M-36; 1.4-1.6 C, 0.30-0.60 Si, 0.20-0.40 Mn, 11.0-12.5 Cr, 0.25 Ni, 0.70-1.0 Mo, 0.75-1.25 V, bal Fe.
Abrasion resistant cast alloy.

COOPER ALLOY 16.
M-36; 0.12-0.30 C, 16-20 Cr, bal Fe.
Annealed: 70,000-100,000 TS; 50,000-75,000 YS; 12-5 El; 15-5 RA; 200-225 Brin.
For valve trim for high pressure steam, pump valves; stainless steel; formerly "Sweetaloy No. 16."

COOPER ALLOY 16A.
M-36; 0.12-0.20 C, 1.0 Si, 1.0 Mn, 15.0-17.0 Cr, 1.25-2.5 Ni, bal Fe.
Abrasion resistant cast alloy.
AISI 431.

COOPER ALLOY 17.
M-36; 0.20 max C, 18-20 Cr, 8-10 Ni, bal Fe.
Annealed: 75,000-85,000 TS; 40,000-50,000 YS; 40-50 El; 40-55 RA; 180-160 Brin.
For oil refineries, canneries, dairies, pump valves, cocks, fittings; best general acid resisting base iron; "Sweetaloy No. 17."

COOPER ALLOY 17-4 PH.
M-36; 0.05 C, 16.5 Cr, 4.5 Ni, 0.65 Si, 2.8 Cu, bal Fe.
Heat treated: 165,000 TS; 144,000 YS; 19 El; 49 RA; 401 Brin.
For high strength, corrosion resistant parts and castings; heat treatable, stainless.

COOPER ALLOY 17ELC.
M-36; 0.025 C, 20 Cr, 8.5 Ni, 1.25 Si, bal Fe.
Annealed: 85,000 TS; 43,000 YS; 54 El; 59 RA; 145 Brin.
For chemical plant equipment, stainless castings; stainless, austenitic.

COOPER ALLOY 17GM.
M-36; 0.06-0.08 C, 17-19 Cr, 12-14 Ni, 0.5 max Mo, bal Fe.
Annealed: 78,000 TS; 38,000 YS; 55 El; 140 Brin.
For chemical plant equipment, agitators, mixers, vessels, valve bodies, pumps.
Corrosion resistant, austenitic.

COOPER ALLOY 17 LL.
M-36; 0.2 max. C, 2 max. Si, 18-20 Cr, 8-10 Ni, 0.2-0.3 Se, bal Fe.
W.Q.: 75,000-85,000 TS; 40,000-50,000 YS; 55-40 El; 55-40 RA; 130-160 Brin.
For fittings, valves, pumps; corrosion resistant, freecutting.

COOPER ALLOY 17-S.
M-36; 0.10 C, 18 Cr, 8-10 Ni, bal Fe.
Cast: 84,500 TS; 45,000 YS; 50 El; 50 RA; 165 Brin.
For stainless parts, valves, pumps, fittings; stainless, corrosion resistant.

COOPER ALLOY 17SCB.
M-36; 0.8 C, 18-20 Cr, 8-10 Ni, Cb = 10 x C, bal Fe.
Cast: 80,000 YS; 40,000 YS; 47 El; 50 RA; 150 Brin.
Annealed: 90,000 TS; 55,000 YS; 55 El; 55 RA; 140 Brin.
For stainless steel castings, pumps, oil refining; stainless, resists acids, stabilized tough welding.

COOPER ALLOY 17SELC.
M-36; 0.03 max C, 17-21 Cr, 8-12 Ni, bal Fe.
Annealed: 77,000 TS; 37,000 YS; 55 El; 140 Brin.
For autoclaves, filter press plates, pasteurizers, mixing kettles, pumps, pump sleeves, spray nozzles.
ACI-CF3; AISI 304, austenitic, stainless.

COOPER ALLOY 17 SLL.
M-36; 0.10 max. C, 2 Max. Si, 18-20 Cr, 8-10 Ni, 0.2-0.3 Se, bal Fe
W.Q.: 70,000-80,000 TS: 35,000-45,000 YS; 60-45 El; 60-45 RA; 130-160 Brin.
For fittings, valves, pumps; corrosion resistant, freecutting.

COOPER ALLOY 17SM.
M-36; 0.08 max C, 19 Cr, 10 Ni, 1.25 Si, 2.5 Mo, bal Fe.
Water quenched: 90,000 TS; 55,000 YS; 50 El; 55 RA; 155 Brin.
For paper mill equipment; corrosion resistant, austenitic.

COOPER ALLOY 17SM ELC.
M-36; 0.3 max C, 17-21 C, 9-13 Ni, 2-3 Mo, bal Fe.
Annealed: 80,000 TS; 42,000 YS; 50 El; 155-170 Brin.
For chemical plant equipment, acid mixers, filter presses, acid pumps, spray nozzles.
ACI-CF3M; AISI 316 L.
Corrosion resistant, austenitic.

COOPER ALLOY 17XM.
M-36; 0.08 max C, 18-21 Cr, 9-13 Ni, 3-4 Mo, bal Fe.
Annealed: 80,000 TS; 42,000 YS; 50 El; 155-170 Brin.
For chemical plant equipment, acid mixers and pumps, valves, filter presses, spray nozzles.
ACI-CG8M; AISI 317.
Corrosion resistant, austenitic.

COOPER ALLOY 17XM ELC.
M-36; 0.03 C, 1.5 Si, 1.5 Mn, 17.0-21.0 Cr, 9.0-13.0 Ni, 3.0-4.0 Mo, bal Fe.
Corrosion resistant casting.
ACI CG3M.

COOPER ALLOY 19A.
M-36; 0.35 max C, 26-30 Cr, 2-3 Ni, bal Fe.
Cast: 75,000-100,000 TS; 50,000-65,000 YS; 25-15 El; 25-15 RA; 190-210 Brin.
For furnace parts, stainless castings; heat and corrosion resistant.

COOPER ALLOY 19BH.
M-36; 2.5-2.8 C, 0.30-0.70 Si, 1.0 Mn, 27.0-30.0 Cr, 1.5-2.5 Ni, bal Fe.
Abrasion resistant cast alloy.

COOPER ALLOY 20.
M-36; 0.3-0.5 C, 34-37 Ni, 13-18 Cr, bal Fe.
Cast: 60,000-70,000 TS; 35,000-40,000 YS; 6-12 El; 6-12 RA; 165-185 Brin.
For furnace parts, enameling racks, tube supports, retorts; resists heat and flue gases at high temperature; "Sweetaloy No. 20."

COOPER ALLOY 21.
M-36; 0.3-0.5 C, 62-68 Ni, 12-18 Cr, bal Fe.
Cast: 60,000-75,000 TS; 30,000-50,000 YS; 4-10 El; 4-10 RA; 170-185 Brin.
For carburizing boxes, pots, retorts; heat resisting alloy; "Sweetaloy No. 21."

COOPER ALLOY 22.
M-36; 0.3-0.4 C, 24-28 Cr, 11-14 Ni, bal Fe.
Cast: 80,000-100,000 TS; 40,000-55,000 YS; 20-30 El; 20-30 RA; 165-
For sulfite pulp mills, digester fittings, pump valves, furnace parts; resists sulfurous acids and ammonia; "Sweetaloy No. 22."

COOPER ALLOY 22 P.
M-36; 0.3 max. C, 2 max. Si, 27-30 Cr, 1.5 max. Mn, 8-11 Ni, bal Fe.
Cast: 80,000-100,000 TS; 40,000-55,000 YS; 30-10 El; 30-10 RA; 170-210 Brin.
Furnace parts, sulfite fittings; heat and corrosion resistant, austenitic.

COOPER ALLOY 22 S.
M-36; 0.2 max. C, 1.5 max Mn, 2 max. Si, 24-28 Cr, 11-14 Ni, bal Fe.
W.Q.: 80,000-90,000 TS; 35,000-55,000 YS; 45-25 El; 55-25 RA; 135-190 Brin.
For fittings, valves; corrosion and heat resistant to hot HNO_3.

COOPER ALLOY 23.
M-36; 0.35-1.0 C, 15-19 Cr, 64-68 Ni, 2.5 max Si, 2.0 max Mn, 0.5 max Mo, bal Fe.
Cast: 65,000 TS; 36,000 YS; 9 El; 176 Brin.
Aged: 73,000 TS; 44,000 YS; 9 El; 185 Brin.
For autoclaves, carburizing boxes, cyanide pots, roller hearths, salt pots.
ACI-HX, heat and corrosion resistant.

COOPER ALLOY 24.
M-36; 0.35-0.75 C, 17-21 Cr, 37-41 Ni, 2.5 max Si, 2 max Mn, 0.5 max Mo, bal Fe.
As cast: 70,000 TS; 40,000 YS; 9 El; 170 Brin.
For carburizing retorts, cyanide pots, dipping baskets, muffles, lead pots.
ACI-HU, heat and corrosion resistant.

COOPER ALLOY 26.
M-36; 0.2-0.6 C, 28-32 Cr, 18-22 Ni, 2 max Si, 2 max Mn, 0.5 max Mo, bal Fe.
Cast: 82,000 TS; 52,000 YS; 19 El; 192 Brin.
For enameling furnace parts, furnace skids, stack dampers, radiant tubes.
ACI-HL, heat and corrosion resistant.

COOPER ALLOY 27.
M-36; 0.2-0.5 C, 26-30 Cr, 14-18 Ni, 2 max Si, 2 max Mn, 0.5 max Mo, bal Fe.
Cast: 80,000 TS; 45,000 YS; 12 El; 180 Brin.
For billet skids, brazing fixtures, conveyor rollers, furnace rails, lead pots, retorts.
ACI-HI, heat and corrosion resistant.

COOPER ALLOY 28.
M-36; 0.2-0.4 C, 19-23 Cr, 9-12 Ni, 2.0 max Si, 2 max Mn, 0.5 max Mo, bal Fe.
Cast: 85,000 TS; 45,000 YS; 35 El; 165 Brin.
For arc furnace electrodes, annealing boxes, trays and baskets, burner tips, conveyor belts and chains.
ACI-HF; AISI 302 B, corrosion resistant, austenitic.

COOPER ALLOY 29.
M-36; 0.2-0.5 C, 26-30 Cr, 8-11 Ni, 2.0 max Si, 2 max Mn, 0.5 max Mo bal Fe.
Cast: 95,000 TS; 45,000 YS; 20 El; 200 Brin.
For billet skids, furnace chains and conveyors, oil burner parts, rabble arms and blades, recuperators.
ACI-HE, heat and corrosion resistant, austenitic.

COOPER ALLOY 30.
M-36; 0.5 max C, 26-30 Cr, 4-7 Ni, 2 max Si, 1.5 max Mn, 0.5 max Mo, bal Fe.
Cast: 85,000 TS; 48,000 YS; 16 El; 190 Brin.
For brazing furnace parts, furnace blowers, rabble shoes, salt pots, recuperators, gas burners.
ACI-HD, heat and corrosion resistant.

COOPER ALLOY 50.
M-36; 0.12 C, 1.0 Si, 1.0 Mn, 15.5-17.5 Cr, 16.0-18.0 Mo, 0.20-0.40 V, 3.75-5.25 W, 4.5-7.5 Fe, bal Ni.
Cast: 496 MPa TS; 317 MPa YS; 4 El.
Corrosion resistant casting.
ACI CW-12M-1; ASTM A-296/A-494.

COOPER ALLOY 51.
M-36; 0.12 C, 1.0 Si, 1.0 Mn, 1.0 Cr, 26.0-30.0 Mo, 0.20-0.60 V, 4.0-6.0 Fe, bal Ni.
Cast: 503 MPa TS; 317 MPa YS; 6.0 El.
Corrosion resistant casting.
ACI N-12M-1.
ASTM A-296/A-494.

COOPER ALLOY 52.
M-36; 0.12 C, 7.5-8.5 Si, 0.5-1.25 Mn, 1.0 Cr, 2.0-4.0 Cu, 2.0 Fe, bal Ni.
Corrosion resistant casting.

COOPER ALLOY 60.
M-36; 0.10 C, 1.0 Si, 0.30 Mn, 58.0-62.0 Cr, 1.0 Fe, bal Ni.
Cast: 760 MPa TS; 590 MPa YS.
Corrosion resistant casting.
ASTM A-560.

COOPER ALLOY CD4MCU.
M-36; 0.04 C, 25-27 Cr, 4.75-6.0 Ni, 1.75-2.25 Mo, 2.75-3.25 Cu, bal Fe.
Annealed: 112,700 TS; 81,000 YS; 28 El; 248 Brin.
Aged: 141,000 TS; 100,000 YS; 25 El; 302 Brin.
For pumps, valves, impellers, high pressure components.
Precipitation hardening, corrosion and heat resistant.

COOPER ALLOY CF10MC.
M-36; 0.10 C, 1.5 Si, 1.5 Mn, 15.0-18.0 Cr, 13.0-16.0 Ni, 1.75-2.25 Mo, Cb = 10 x C min, 1.20 max, bal Fe.
Cast: 483 MPa TS; 207 MPa YS; 20 E.
Heat resistant alloy.
ASTM A-351 CF10MC.

COOPER ALLOY FA 20.
M-36; 0.07 max. C, 1.5 max. Si, 3.5 Mo, 19-21 Cr, 2.5 max. Mn, 28-30 Ni, 4-5 Cu, bal Fe.
W.Q.: 65,000-85,000 TS; 40,000-50,000 YS; 45-35 El; 50-40 RA; 140-150 Brin.
For fittings, valves, pumps; corrosion resistant; austenitic.

COOPER ALLOY FA20CB.
M-36; 0.07 max C, 19-22 Cr, 27-5.30.5 Ni, 2-3 Mo, 3-4 Cu, 1 max Cb, bal Fe.
Annealed: 75,000 TS; 45,000 YS; 40 El; 145 Brin.
For fittings, valves, pumps.
ACI-CN7MCb, corrosion resistant, austenitic.

COOPER ALLOY FA-20CB3.
M-36; 0.07 C, 1.5 Si, 1.5 Mn, 19.0-22.0 Cr, 32.5-35.0 Ni, 2.0-3.0 Mo 3.0-4.0 Cu, Cb = 8 x C min-1.0 max, bal Fe.
Corrosion resistant casting.

COOPER ALLOY HK30.
M-36; 0.25-0.35 C, 1.75 Si, 1.50 Mn, 23.0-27.0 Cr, 19.0-22.0 Ni, bal Fe.
Cast: 448 MPa TS; 241 MPa YS; 10 El.
Heat resistant alloy.
ASTM A-351 HK30.

COOPER ALLOY HK40.
M-36; 0.35-0.45 C, 1.75 Si, 1.5 Mn, 23.0-27.0 Cr, 19.0-22.0 Ni, bal Fe.
Cast: 431 MPa TS; 241 MPa YS; 10 El.
Heat resistant alloy.
ASTM A-351 HK40.

COOPER ALLOY HT30.
M-36; 0.25-0.35 C, 2.5 Si, 2.0 Mn, 13.0-17.0 Cr, 33.0-37.0 Ni, 0.50 Mo, bal Fe.
Cast: 448 MPa TS; 193 MPa YS; 15 El.
Heat resistant alloy.
ASTM A-351 HT30.

COOPER ALLOY: INCONEL.
M-36; 0.40 C, 3.0 Si, 1.5 Mn, 14.0-17.0 Cr, 11.0 Fe, bal Ni.
Cast: 483 MPa TS; 193 MPa YS; 30 El.
Corrosion resistant casting.
ACI CY-40; ASTM A-296 CY-40.

COOPER ALLOY KNC-3.
M-36; 0.2-0.4 C, 24-28 Cr, 19-22 Ni, 1-3 Si, 1-2 Mn, bal Fe.
Cast: 65,000-85,000 TS; 35,000-50,000 YS; 10-30 El; 10-30 RA; 170-195 Brin.
Heat and corrosion resistant parts; furnace parts; heat and corrosion resistant.

COOPER ALLOY MM.
M-36; 0.25 C, 63 Ni, 1.5 Si, 2.5 Fe, bal Cu.
Cast; 65,000 TS; 33,000 YS; 25 El; 28 RA; 150 Brin.
For chemical plant equipment; corrosion resistant.

COOPER ALLOY MMCB.
M-36; 0.30 C, 1.0-2.0 Si, 1.5 Mn, 26.0-33.0 Cu, 3.5 Fe, 1.0-3.0 Cb, bal Ni.
Corrosion resistant casting.
FED QQ-N-288-E.

COOPER ALLOY MM-H.
M-36; 0.25 C, 63 Ni, 3 Si, 2 Fe, bal Cu.
Cast: 80,000 TS; 50,000 YS; 10 El; 14 RA; 225 Brin.
For valve trim, valve parts, pump and turbine parts; corrosion and abrasion resistant.

COOPER ALLOY MM-S.
M-36; 0.25 C, 63 Ni, 4 Si, 2 Fe, bal Cu.
Aged hardened: 130,000 TS; 100,000 YS; 2 El; 2 RA; 350 Brin.
For valve trim, pump and turbine parts; age-hardenable, corrosion resistant.

COOPER ALLOY: NICKEL.
M-36; 1.0 C, 2.0 Si, 1.5 Mn, 3.0 Fe, 1.25 Cu, bal Ni.
Cast: 345 MPa TS; 124 MPa YS; 10 El.
Corrosion resistant casting.
ACI CZ-100; ASTM A-296 CZ100.

COOPER ALLOY PH 55A.
M-36; 0.08 max C, 3.0-3.75 Si, 20 Cr, 9 Ni, 4 Mo, bal Fe.
For corrosion resistant castings; austenitic, stainless, high strength.

COOPER ALLOY PH 55B.
M-36; 0.08 max C, 20 Cr, 9 Ni, 5 Mo, 4 Cu, bal Fe.
For corrosion resistant castings; corrosion and shock resistant.

COOPER ALLOY P.H.55C.
M-36; 0.08 max C, 3.5 Si, 20 Cr, 9 Ni, 4 Mo, 3 Cu, bal Fe.
For corrosion resistant castings; stainless, austenitic.

COOPER ALLOY PH-55D.
M-36; 0.05 max C, 18-21 Cr, 9-12 Ni, 3-5 Si, 3.75-4.25 Mo, 0.75-1.25 Cb, 0.10 N, bal Fe.
For pump parts; precipitation hardening, corrosion and abrasion resistant.

COOPERITE.
M-Eng.; 80 Ni, 14 W, 6 Zr.
For cutting tools; extreme hardness.

COOPER'S SPECULUM.
M-Eng.; 57.8 Cu, 27.3 Sn, 3.6 Zn, 1.2 As, 10 Pt.
For mirrors, reflectors.

COPALOY.
M-545; Sn, Sb, Pb.
For bearings; antifriction.

COPAN.
M-US.; 80-90 Sn, 10-15 Sb, 2-5 Cu, 0.2 Pb.
For bearings; Babbitt.

COPEL.
M-65; 55 Cu, 45 Ni.
60,000 TS; 40 El.
For electrical instruments, rheostats, hot resistances; resistance alloy up to 800°F.

COPELMET-D.
M-1552; W + Cu.
For electrical contacts, welding electrodes.

COPELMET-P.
M-1552; W + Cu.
For electrical contacts, resistance welding electrodes.

COPELMET-PW.
M-1552; W + Cu.
For electrical contacts, resistance welding electrodes.

COPELMET-PW4.
M-1552; WC + Cu.
For facing of electrical contacts, spot, projection and butt welding electrodes, hot riveting dies. Sintered carbide.

COPERNICK.
M-205; 50 Fe, 50 Ni.
For transformer cores; resistance alloy.

COPPER.
M-1755; Cu.
Purities: Zone refined 99.9999%, 99.9997%, 99.999%, 99.99+% (OFHC).
Forms: Rods, powders, wires, sheet, foil, ingots, shot, single crystals.

COPPER 102 OF.
M-279; 99.95 min Cu.
Ann: 26,000-36,000 TS; 6000-12,000 YS; 20-44 El.
Cold Rolled: 34,000-60,000 TS; 30,000-58,000 YS; 1-33 El.
For electrical assemblies and severely formed parts; resists hydrogen embrittlement.

COPPER 104 OFS.
M-279; 99.95 min Cu + Ag, 0.027 min Ag.
Ann: 26,000-36,000 TS; 6000-12,000 YS; 20-44 El.
Cold Rolled: 34,000-60,000 TS; 30,000-58,000 YS; 1-33 El.
For electrical assemblies and severely formed parts; resistant to hydrogen embrittlement and elevated temperature softening.

COPPER 110 ETP.
M-279; 99.9 min Cu.
Ann: 26,000-36,000 TS; 6,000-12,000 YS; 20-44 El.
Cold rolled: 34,000-60,000 TS; 30,000-58,000 YS; 1-33 El.
Drawn or formed electrical shells and parts; General purpose copper for electrical parts.

COPPER 122 DHP.
M-279; 99.90 min Cu, 0.015 min P.
Ann: 26,000-36,000 TS; 6000-12,000 YS; 20-44 El.
For deep drawn copper shells, brazed or welded assemblies.
Cold rolled: 34,000-60,000 TS; 30,000-58,000 YS; 1-33 El.
For formed copper parts.

COPPER 129 FRSTP.
M-279; 99.9 min Cu, 0.054 min Ag.
Ann: 26,000-36,000 TS; 6000-12,000 YS; 20-44 El.
Cold rolled: 34,000-60,000 TS; 30,000-58,000 YS; 1-33 El.
Excellent solderability and high softening temperature.
For radiator air channels, radiator fins.

Section I: Alloy Data / 343

COPPER 1102, LOW OXYGEN ETP.
M-279; 99.90 min Cu, 0.02 max oxygen.
Ann: 26,000-36,000 TS; 6,000-12,000 YS; 20-44 El.
Cold rolled: 34,000-60,000 TS; 30,000-58,000 YS; 1-33 El.
Formed and deep drawn parts; excellent ductility (soft); good tear resistance.

COPPER 1142 LOW OXYGEN STP.
M-279; 99.90 min Cu, 0.044 min Ag, 0.02 max Oxygen.
Ann: 26,000-36,000 TS; 6000-12,000 YS; 20-44 El.
Cold rolled: 34,000-60,000 TS; 30,000-58,000 YS; 1-33 El.
For commutator and collector rings, severely formed parts; resists softening at elevated temperatures.

COPPER ALLOY Z30A.
M-U.S.; 57 min Cu, 1.5 max Pb, 30 min Zn.
Die cast: 55,000 TS; 30,000 YS; 15 El; 10 Brin.
For die castings, hardware; yellow brass, good machinability.

COPPER ALLOY ZS144A.
80-83 Cu, 3.75-4.25 Si, bal Zn.
Die Cast: 85,000 YS; 50,000 YS; 25 El; 70 Charpy; B 85-90 Rock.
For die castings.

COPPER ALLOY ZS331A.
63-67 Cu, 0.75-1.25 Si, bal Zn.
Die Cast: 70,000 TS; 35,000 YS; 25 El; 50 Charpy; B 68-72 Rock.
For die castings.

COPPER ARC.
M-685; 0.2 C, 0.5 Cu, bal Fe.
For weldng electrodes; for cast iron.

COPPER-BERYLLIUM 50.
M-1037, M-646; 0.38 Be, 1.55 Co, 1.0 Ag, bal Cu.
Annealed: 45,000 TS; 27 El; 67 Brin.
Cold rolled: 70,000 TS; 13 El; 132 Brin.
Aged: 120,000 TS; 14 El; 234 Brin.
For springs, contacts, diaphragms; corrosion and wear resistant.

COPPERCLAD.
M-725; Ni steel core with pure Cu sleeve.
For sealing in vacuum tubes; 42% Ni in steel.

COPPER, DEOXIDIZED, DHP.
M-1789; 99.9 min cu, 0.02 P.
ann: 32,000 TS; 10,000 YS; 45 El.
Rolled hard: 50,000 TS; 45,000 YS; 6 El.
DHP means Deoxidized, High Phosphorus.
Readily brazed and welded.
For heat exchangers.

COPPER, DEOXIDIZED, DLP.
M-1789; 99.9 min Cu, 0.007 P.
Ann: 32,000 TS; 10,000 YS; 45 El.
Rolled hard: 50,000 TS; 45,000 YS; 6 El.
DLP means Deoxidized, Low Phosphorus.
For welded or brazed components.

COPPER, DEOXIDIZED, XLP.
M-1789; 99.95 min Cu, 0.003 P.
Ann: 32,000 TS; 10,000 YS; 45 El.
Rolled hard: 50,000 TS; 45,000 YS; 6 El.
XLP means Extra Low Phosphorus.
For high conductivity components.

COPPER, ELECTROLYTIC TOUGH PITCH.
M-33, M-1789; 99.9 min Cu, 0.04 O.
Ann: 32,000 TS; 10,000 YS; 45 El.
Rolled hard: 50,000 TS; 45,000 YS; 6 El.
Electrical conductors, roofing, switches.

COPPERIOR.
M-613; 0.1 C, Cu, bal Fe.
For roofing, culverts; corrosion resistant.

COPPER, LOW OXYGEN STP 1162.
M-279; 99.90 min Cu, 0.085 min Ag(25 oz/ton) 0.02 O.
Ann: 26,000-30,000 TS; 6000-12,000 YS; 20-44 El.
Cold rolled: 34,000-60,000 TS; 30,000-58,000 YS; 1-3 El.
For lead frames, heat sinks, electronic components.

COPPER M3.
M-Russian; 99.86 Cu, 0.01 Pb, 0.03 Zn, 0.01 Fe, 0.02 Sn.
Rolled: 32,000 TS; 43 El; 69 RA.

COPPER NICKEL, 10%.
M-33; 88.6 Cu, 10.0 Ni, 1.4 Fe.
Ann: 47,000 TS; 18,000 YS; 48 El.
Drawn 35%: 70,000 TS; 65,000 YS; 10 El.
Condenser and heat exchanger tubes.

COPPER NICKEL 30%.
M-33; 69.5 Cu, 30 Ni, 0.5 Fe.
Ann: 54,000 TS; 20,00 YS; 55 El.
Drawn 35%: 85,000 TS; 74,000 YS; 5 El.
Condenser and heat exchanger tubes, ferrules.

COPPER-NICKEL 30%.
M-297; 30-33 Ni, 70-67 Cu.
Annealed: 60,500 TS; 34 El.
For electrical resistors, condensers, corrosion resisting apparatus; corrosion resistant.

COPPER-NICKEL 40%.
M-297; 40-45 Ni, 60-55 Cu.
Annealed: 70,000 TS; 38 El.
For electrical resistors, thermocouple element; heat resistant.

COPPER-NICKEL 67%.
M-297; 67 Ni, 33 Cu.
Annealed: 71,500 TS; 35-40 El.
For machine parts, in locomotives, ships, turbines, pumps, corrosion resisting containers; corrosion resistant.

COPPER-NICKEL 401 now MONEL ALLOY 401.

COPPER-NICKEL-TIN.
M-1426; 8.5-10.5 Ni, 2.0 Sn, bal Cu.
UNS C 725 00.

COPPER-NICKEL-TIN ALLOY 725.
M-279; 88.2 Cu, 9.5 Ni, 2.3 Sn.
Ann: 55,000 TS; 22,000 YS; 35 El.
Spring temper: 91,000 TS; 90,000 YS; 1 El.
For springs in connectors, relays and switches.

COPPER-NICKEL-TITANIUM ALLOY.
M-1778; 5 Ni, 2.5 Ti, bal Cu.
Solution Ht, 90% cold work, aged: 745 MPa (108,000 psi) TS; 634 MPa (92,000 psi) YS; 10 El.
Elec. cond.; 53% IACS at 68°F.
For electronic applications at elevated temperatures, current carrying springs, heat sinks, switch parts, radar components for high temperature service.

COPPEROID.
M-531; C, 0.2 min Cu, bal Fe.
For sheets for construction purposes, automobile furniture; resists rust and atmospheric corrosion.

COPPEROID.
M-U.S.; Cu alloy.
For bearings; copper bearing alloy.

COPPER, OXYGEN FREE HIGH CONDUCTIVITY.
M-33, M-1789; 99.99 Min Cu.
Ann: 32,000 TS; 10,000 YS; 45 El.
Rolled hard: 50,000 TS; 45,000 YS; 6 El.
Electrical conductors.
Certified or uncertified.

COPPER, PHOSPHORUS.
M-33; 99.9 min Cu, 0.003, 0.007 or 0.02 P.
Ann: 32,000 TS; 10,000 YS; 45 El.
Rolled hard: 50,000 TS; 45,000 YS; 6 El.
Electrical Conductors, heat exchangers, kettles, tanks.

COPPER'S GOLD.
M-Eng.; 81-67 Cu, 19-30 Pt, 0-4 Zn.
For jewelry, ornaments; corrosion resistant.

COPPER-SILICON ALUMINUM 8-2.
M-137; 8 Cu, 2.5 Si, 0.8 Fe, bal Al.
Cast: 17,000 TS; 1 El; 70 Brin.
For gear boxes, radiator tanks; non-hardenable.

COPPER, SILVER BEARING.
M-33; 99.9 min Cu, 0.027-0.085 Ag.
Ann: 33,000 TS; 10,000 YS; 50 El.
Drawn hard; 55,000 TS; 50,000 YS; 10 El.
Motor commutators, switches.
Higher softening temperature.

COPPER'S MIRROR.
M-Eng.; 58 Cu, 28 Sn, 9.5 Pt, 3.5 Zn, 1.5 As.
For reflectors, mirrors; corrosion resistnt.

COPPER'S PEN METAL-1.
M-Eng.; 50 Cu, 25 Au, 25 Ag.
For pen points; corrosion resistant.

COPPER'S PEN METAL-2.
M-Eng.; 12 Cu, 38 Ag, 50 Pt.
For pen points; corrosion resistant.

COPPER STEEL.
M-538; copper coated steel.
For fabricated parts; easily stamped, formed, drawn.

COPPER STEEL.
M-U.S.; 0.37 C, 0.17 Mn, 4 Cu, 0.22 Si, bal Fe.
Rolled: 138,000 TS; 116,000 YS; 11 El; 23 RA; 302 Brin.
Annealed: 97,000 TS; 80,500 YS; 16 El; 42 RA; 212 Brin.
For structural parts; corrosion resistant.

COPPER TIN.
M-538; copper coated Sn.
For fabricated parts; easily stamped, formed, drawn.

COPPER TITANIUM ALLOY.
M-1778; 4.3 Ti, bal Cu.
Sol. Ht, 40% cold work, aged: 1076 MPa (156,000 psi) TS; 965 MPa (140,000 psi) YS; 5 El. Elec. cond.: 10% IACS at 68°F.
For non-sparking tools, springs, electrical contacts, diaphragms.

COPPERWELD WIRE.
M-711; Cu on steel core; for cable; clad.

COPREX.
M-824; Cu alloy.
For bearings; sintered.

COPR-TRODE.
M-13; 1.0 max Sn, 0.5 max Mn, 0.5 max Si, 0.15 max P, 98.0 min Cu. (others 0.50 max).
Weld metal: 29 ksi TS; 8 ksi YS; 29 El; 54 Brin.
For welding deoxidized copper castings with gas metal-arc and gas-tungsten arc.

COP-SIL-LOY.
M-1217; Cu-Ag-Pb.
For reducing wear on friction surfaces; low coefficient of friction.

COP-SIL-LOY.
M-1217; 47-53 Pb, bal Cu.
For special parts; sintered.

CORBIN.
M-Eng.; 87.5 Al, 12.5 Cu.
For pistons; non-hardenable.

CORE IRON.
M-32; 0.06 C, 1 V, bal Fe.
For magnetic and electrical equipment; high permability.

CORINTH.
M-1779, M-1; 0.90 C, 0.30 Mn, bal Fe.
For tools, cutters; water hardening.
AISI W1.

CORMIN BRONZE.
44 Cu, 37 Ni, 11 Sn, 8 Pb.
For hardware; corrosion resistant.

CORNISH BRONZE.
M-England; 9.6 Sn, 16.5 Pb, 0.8 P, bal Cu.
For bearings; heavy duty.

CORNIX-2.
M-1307; 0.20 C, 0.5 Mn, 12 Cr, 1.15 Mo, 0.5 Ni, 0.3 V, bal Fe.
Annealed: 95,000 TS; 40,000 YS; 25 El; 50 RA; B 92 Rock.
Heat treated: 240,000 TS; 205,000 YS; 9 El; 26 RA; C 50 Rock.
For surgical instruments, hardware, oil refinery and chemical plant equipment.
Corrosion resistant, hardenable.

CORODENT.
M-200; 71 Au, plus Pt.
For dental purposes.

COROMANT C1.
M-101, M-101A; WC, Co.
For cutting tools; sintered carbide.

COROMANT C3.
M-101, M-101a; WC, Co.
For finishing cutting tools; sintered carbide.

COROMANT C5.
M-101, M-101a; TaC, WC, Co.
For cutting tools for roughing; sintered carbide.

COROMANT C6.
M-101, M-101a; WC, TiC, Co.
For general purpose cutting tools; sintered carbide.

COROMANT C7.
M-101, M-101a; WC, TiC, Co.
For cutting tools, dies; sintered carbide.

COROMANT C8.
M-101, M-101a; WC, TiC.
For finishing cutters and tools; sintered carbide.

COROMANT F0A.
M-101, M-101a; Sintered carbide tool material.
For finish cuts and precision machining of steel; crater resistant grade.

COROMANT H05.
M-101, M-101a; Sintered carbide tool material.
Cutting tool for precision finishing of cast iron, hard irons and steels, and heavy metals.

COROMANT H1P.
M-101, M-101a; Sintered carbide tool material.
For cutting tools for machining cast iron; Good wear resistance and toughness.

COROMANT H13.
M-101, M-101a; Sintered carbide tool material.
Cutting tool for general purpose machining of cast iron, steels, hard and tough metals.

COROMANT H20.
M-101, M-101a; Sintered carbide tool material.
Cutting tool for general purpose rough machining cast iron and non-ferrous metals.

COROMANT R1P.
M-101, M-101a; Sintered carbide tool material.
Cutting tool for finish cuts on nickel base, high strength, heat resistant alloys.

COROMANT S1P.
M-101, M-101a; Sintered carbide tool material.
For light roughing and finishing cuts on steel; high wear resistance.

COROMANT S2.
M-101, M-101a; Sintered carbide tool material.
Cutting tool for general purpose machining of steel; tough.

COROMANT S4.
M-101, M-101a; Sintered carbide tool material.
Cutting tool for rough machining of steel, good toughness and wear resistance.

COROMANT S6.
M-101, M-101a; WC, TaC, Co.
For cutting tools for roughing; sintered carbide.

COROMANT SH.
M-101, M-101a; WC, TaC, Co.
For cutting tools for general machining; sintered carbide.

CORONA EXTRA M.
M-1310; 0.85 C, 9 W, 1 Mo, 4 Cr, bal Fe.
For milling cutters, lathe and planer tools, hobs; high speed steel.

CORONA EXTRA V.
M-1310; 0.86 C, 4.1 Cr, 0.9 Mo, 2.5 V, 12 W, bal Fe.
For lathe and planer tools, drills, hobs; high speed steel.

CORONA KOBALT 3.
M-1310; 0.86 C, 2.8 Co, 4.3 Cr, 0.85 Mo, 2 V, 12 W, bal Fe.
For lathe and planer tools, drills, hobs, reamers; high speed steel.

CORONA KOBALT 5.
M-1310; 1.35 C, W, Cr, V, Co, bal Fe.
For fast finishing cutters; high-speed steel.

CORONA KOBALT 5W.
M-1310; 0.8 C, 4.7 Co, 4.3 Cr, 0.75 Mo, 1.5 V, 18 W, bal Fe.
For lathe and planer tools, reamers, broaches, taps, hobs, drills; high speed steel.

CORONA KOBALT 10.
M-1310; 0.76 C, 10 Co, 4.2 Cr, 0.8 Mo, 1.8 V, 18 W, bal Fe.
For lathe and planer tools, milling cutters, hobs; high speed steel.

CORONA PRIMA M.
M-1310; 0.95 C, W, Mo, Cr, V, bal Fe.
For cutters, drills, reamers, broaches; high speed steel.

CORONA PRIMA V.
M-1310; 0.82 C, 4.1 Cr, 0.8 Mo, 1.6 V, 8.7 W, bal Fe.
For reamers, drills, broaches, cutters; high speed steel.

CORONA PRIMA W.
M-1310; 0.74 C, 4 Cr, 1 V, 18.5 W, bal Fe.
For lathe and planer tools, reamers, broaches; high speed steel.

CORONA VIERBRENZ V.
M-1310; 1.3 C, 4.3 Cr, 0.8 Mo, 3.8 V, 12 W, bal Fe.
For cutters, fast finishing tools; high speed steel.

CORONEL.
M-Eng.; 65-70 Ni, 30-35 Cu.
Cast: 64,000 TS; 32,000 YS; 34 EL; 32 RA.
Rolled: 78,000 TS; 48,000 YS; 42 El; 35 RA.
For valves, pumps, turbine blades; similar to "Monel Metal."

CORONZE 638.
M-279; 95.0 Cu, 2.8 Al, 1.8 Si, 0.4 Co.
Ann: 77,000-87,000 TS; 41,000-67,000 YS; 29-42 El.
Cold rolled: 90,000-130,000 TS; 70,000-114,000 YS; 2-30 El.
Soft-for-cans, bases, electronic parts; cold rolled for springs, terminals circuit frames.
Stress corrosion resistant, oxidation resistant, good strength.

CORORESIST 19300.
M-717.
Low carbon stainless alloy powder for metal spraying.

CORRESIST 17/12 NB.
M-1342; 0.07 max C, 17.0 Cr, 12.0 Ni, Nb = 8 x C, bal Fe.
As quenched: 50-75 kp/mm² TS; 23 kp/mm² min YS; 40 min El.
Rolled and wrought stainless steel.
For chemical and petrochemical industries.

CORRESIST 18/8.
M-1342; 0.13 max C, 18.0 Cr, 18.0 Ni, bal Fe.
As quenched: 50-70 kp/mm² TS; 22 kp/mm² min YS; 50 min El.
Rolled and wrought stainless steel.
For chemical and petrochemical industries.

CORRESIST 18/8 NB.
M-1342; 0.07 max C, 18.0 Cr, 10.0 Ni, Nb = 8 x C, bal Fe.
As quenched: 50-75 kp/mm² TS; 21 kp/mm² min YS; 40 min El.
Rolled and wrought stainless steel.
For chemical and petrochemical industries.

CORRESIST 18/8 S.
M-1342; 0.07 max C, 18.0 Cr, 10.0 Ni, bal Fe.
As quenched: 50-70 kp/mm² TS; 19 kp/mm² min YS; 50 min El.
Rolled and wrought stainless steel.
For chemical and petrochemical industries.

CORRESIST 18/8 TI.
M-1342; 0.07 max C, 18.0 Cr, 10.0 Ni, Ti = 5 x C, bal Fe.
As quenched: 50-75 kp/mm² TS; 21 kp/mm² YS; 40 min El.
Rolled and wrought stainless steel.
For chemical and petrochemical industries.

CORRESIST 18/10 MONB.
M-1342; 0.07 max C, 18.0 Cr, 10.0 Ni, 2.2 Mo, Nb = 8 x C, bal Fe.
As quenched: 50-75 kp/mm² TS; 23 kp/mm² min YS; 40 min El.
Rolled and wrought stainless steel.
For chemical and petrochemical industries.

CORRESIST 18/10 MOS.
M-1342; 0.07 max C, 18.0 Cr, 10.0 Ni, 2.2 Mo, bal Fe.
As quenched: 50-70 kp/mm² TS; 21 kp/mm YS; 45 Min El.
Rolled and wrought stainless steel.
For chemical and petrochemical industries.

CORRESIST 18/10 MOTI.
M-1342; 0.07 max C, 18.0 Cr, 10.0 Ni, 2.2 Mo, Ti = 5 x C, bal Fe.
As quenched: 50-75 kp/mm² TS; 23 kp/mm² YS; 40 min El.
Rolled and wrought stainless steel.
For chemical and petrochemical industries.

CORRESIST 18/18.
M-1342; 0.13 max C, 18.0 Cr, 18.0 Ni, bal Fe.
As quenched: 50-70 kp/mm^2 TS; 21 kp/mm^2 min YS; 40 min El.
Rolled and wrought stainless steel.
For chemical and petrochemical industries.

CORRESIST 25/25 MOTI.
M-1342; 0.06 max C, 25.0 Cr, 25.0 Ni, 2.2 Mo, Ti = 5 x C, bal Fe.
As quenched: 50-70 kp/mm^2 TS; 23 kp/mm^2 min YS; 35 min El.
Rolled and wrought stainless steel.
For chemical and petrochemical industries.

CORRESIST G 14.
M-1342; 0.20 C, 13.0 Cr, bal Fe.
Tempered: 60-80 kp/mm^2 TS; 45 kp/mm^2 min YS; 12 min El.
Cast stainless steel.
For chemical and petrochemical industries.

CORRESIST G 17.
M-1342; 0.20 C, 17.0 Cr, bal Fe.
Tempered: 80-100 kp/mm^2 TS; 60 kp/mm^2 min YS; 4 min El.
Cast stainless steel.
For chemical and petrochemical industries.

CORRESIST G 18/8 ELC.
M-1342; 0.03 max C, 18 Cr, 8 Ni, bal Fe.
As quenched: 45-65 kp/mm^2 TS; 16 kp/mm^2 min YS; 25 min El.
Cast stainless steel.
For chemical and petrochemical industries.

CORRESIST G 18/8 NB.
M-1342; 0.07 max C, 18 Cr, 8 Ni, Nb = 8 x C, bal Fe.
As quenched: 45-65 kp/mm^2 TS; 18 kp/mm^2 min YS; 20 min El.
Cast stainless steel.
For chemical and petrochemical industries.

CORRESIST G 18/8 S.
M-1342; 0.07 max C, 18 Cr, 8 Ni, bal Fe.
As quenched: 45-65 kp/mm^2 TS; 18 kp/mm^2 min YS; 20 min El.
Cast stainless steel.
For chemical and petrochemical industries.

CORRESIST G 18/10 MONB.
M-1342; 0.07 max C, 18 Cr, 11 Ni, 2.2 Mo, Nb = 8 x C; bal Fe.
As quenched: 45-65 kp/mm^2 TS; 19 kp/mm^2 min YS; 20 min El.
Cast stainless steel.
For chemical and petrochemical industries.

CORRESIST G 18/10 MOS.
M-1342; 0.07 max C, 18 Cr, 11 Ni, 2.2 Mo, bal Fe.
As quenched: 45-65 kp/mm^2 TS; 19 kp/mm^2 min YS; 20 min El.
Cast stainless steel.
For chemical and petrochemical industries.

CORRESIST G 18/18.
M-1342; 0.13 max C, 18 Cr, 18 Ni, bal Fe.
As quenched: 45-65 kp/mm^2 TS; 18 kg/mm^2 min YS; 20 min El.
Cast stainless steel.
For chemical and petrochemical industries.

CORRESIST G 26/6.
M-1342; 0.08 max C, 26 Cr, 6 Ni, bal Fe.
As quenched: 60-80 kp/mm^2 TS; 45 kp/mm^2 min YS; 15 min El.
Cast stainless steel.
For chemical and petrochemical industries.

CORRESIST G 28.
M-1342; 0.50 C, 28 Cr, bal Fe.
Cast stainless steel.
For chemical and petrochemical industries.

CORRESIST G 28/5.
M-1342; 0.40 C, 28 Cr, 5 Ni, bal Fe.
As quenched: 50-80 kp/mm^2 TS; 3 min El.
Cast stainless steel.
For chemical and petrochemical industries.

CORRESIST G 28/5 MO.
M-1342; 0.40 C, 28 Cr, 5 Ni, 2.2 Mo, bal Fe.
As quenched: 50-80 kp/mm^2 TS; 3 min El.
Cast stainless steel.
For chemical and petrochemical industries.

CORRESIST G 28 H.
M-1342; 1.3 C, 28 Cr, bal Fe.
Cast stainless steel.
For chemical and petrochemical industries.

CORRESIST G 28 HMO.
M-1342; 1.3 C, 28 Cr, 2.2 Mo, bal Fe.
Cast stainless steel.
For chemical and petrochemical industries.

CORRESIST G 28 MO.
M-1342; 0.50 C, 28 Cr, 2.2 Mo, bal Fe.
Cast stainless steel.
For chemical and petrochemical industries.

CORRIX METAL.
M-German; 88.1 Cu, 8.7 Al, 3.2 Fe.
Cast: 78,000-90,000 TS; 30-20 El; 30 RA.
For worm wheels, gears, marine and mining applications; corrosion resistant to acids and alkalis.

CORRODUR 13 4 MO.
M-1290; 0.07 C, 1.0 max Si, 1.5 max Mn, 12.0-13.5 Cr, 0.70 max Mo, 3.5-5.0 Ni, bal Fe.
Stainless steel casting.
W.-Nr. 1.4313; Din G - X 5 CrNi 13 4.

CORRODUR 17.13 ESS.
M-1290; 0.05 C, 17.0 Cr, 13.5 Ni, 4.75 Mo, bal Fe.
Sol. ann. & age: 69,000 min TS; 35,000 min YS; 20 min El.
Corrosion resistant casting.
Werkstoff Nr. 1.4448; DIN G-X 6 CrNiMo 1713.

CORRODUR 18.
M-1290; 0.25 C, 17 Cr, 1.8 max Ni, bal Fe.
For furnace parts, retorts, heat treating boxes; Type CB-30; corrosion and heat resistant.

CORRUDUR 18.8.
M-1290; 0.10 C, 18.0 Cr, 8.5 Ni, bal Fe.
Sol. ann. & age: 64,000-85,000 TS; 30,000 min YS; 25 min El.
Corrosion resistant casting.
Werkstoff Nr. 1.4312; DIN G-X 10 CrNi 188.

CORRODUR 18.8 E.
M-1290; 0.10 C, 18.0 Cr, 9.5 Ni, 2.25 Mo, bal Fe.
Sol. ann. & age: 71,000-92,000 TS; 30,000 min YS; 25 min El.
Corrosion resistant casting.
Werkstoff Nr. 1.4410; DIN G-X CrNiMo 189.

CORRODUR 18.8 E 3 S.
M-1290; 0.07 max C, 2.0 max Si, 2.0 max mn, 16.5-18.5 Cr, 2.5-3.0 Mo, 11.5-13.5 Ni, bal Fe.
Stainless steel casting; W.-Nr. 1.4437; Din G - X 6 CrNiMo 18 12.

CORRODUR 18.8 E 3 SS.
M-1290; 0.07 max C, 2.0 max Si, 2.0 max Mn, 16.5-18.5 Cr, 2.5-3.0 Mo, 11.5-13.5 Ni, Nb, bal Fe.
Stainless steel casting.
Din G - X 7 CrNiMoNb 18 12.

CORRODUR 18.8 ES.
M-1290; 0.07 C, 17.5 Cr, 10.5 Ni, 2.25 Mo, bal Fe.
Sol. ann. & age: 64,000-85,000 TS; 30,000 min YS; 25 min El.
Corrosion resistant casting.
Werkstoff Nr. 1.4408; DIN G-X 6 CrNiMo 1810.

CORRODUR 18.8 ESS.
M-1290; 0.08 C, 17.5 Cr, 11.0 Ni, 2.25 Mo, Nb, bal Fe.
Sol. ann. & age: 64,000-85,000 TS; 30,000 min YS; 20 min El.
Corrosion resistant casting.
Werkstoff Nr. 1.4581; DIN G-X7CrNiMoNb 1810.

CORRUDUR 20 25 ESS.
M-1290; 0.05 C, 25.0 Ni, 20.0 Cr, 3.25 Mo, 1.35 Cu, bal Fe.
Sol. ann. & age: 64,000-78,000 TS; 28,000 min YS; 15 min El.
Corrosion resistant casting.
Werkstoff Nr. 1.4500; DIN G-X7NiCrMoCu2520.

CORRUDUR 28 4.
M-1290; 0.35 C, 27.0 Cr, 4.0 Ni, bal Fe.
As cast: 64,000-85,000 TS; 42,500 min YS; 4 min El.
Corrosion resistant casting.
Werkstoff Nr. 1.4340; DIN G-X40CrNi274.

CORRODUR 28 4 MO.
M-1290; 0.10 max C, 1.0 max Si, 2.0 max Mn, 26,0-28.0 Cr, 1.3-2.0 Mo, 4.0-5.0 Ni, bal Fe.
Corrosion resistant cast steel.
W.-Nr. 1.4460; DIN G - X 8 CrNiMo 27 5.

CORRODUR 30.
M-1290; 0.60 C, 29.0 Cr, bal Fe.
As cast: 57,000-64,000 TS.
Corrosion resistant casting.
Werkstoff Nr. 1.4085; DIN G-X70Cr29.

CORRODUR 30 E.
M-1290; 0.65 C, 28.0 Cr, 2.0 Mo, bal Fe.
As cast: 57,000-64,000 TS.
Corrosion resistant casting.
Werkstoff Nr. 1.4136; DIN G-X70 CrMo292.

CORRODUR 30 EW.
M-1290; 0.15 C, 28 Cr, 2 Mo, bal Fe.
Stainless steel casting.
Din G - X 15 CrMo 29.

CORRODUR 30 W.
M-1290; 0.15 C, 28 Cr, bal Fe.
Stainless steel casting.
Din G X 15 Cr 29.

CORRODUR BERGIT B.
M-1290; 28.5 Mo, bal Ni.
Sol. ann. & age: 78-92 ksi TS; 57 ksi min YS; 8 min El.
Corrosion resistant nickel base coating.
Werkstoff Nr. 2.41810, DIN G-X 8 NiMo 6530.

CORRODUR BERGIT C.
M-1290; 19.0 Mo, 17.0 Cr, bal Ni.
Sol. ann. & age: 71-85 ksi TS; 57 ksi min YS; 5 min El.
Corrosion resistant nickel base coating.
Werkstoff Nr. 2.4472; DIN G-X 8 NiMoCr 6018.

CORRODUR C14.
M-1290; 0.22 C, 14.5 Cr, bal Fe.
Annealed: 78,000-92,000 TS; 50,000 min YS; 3 min El.
Heat treated: 85,000-115,000 TS; 64,000-min YS; 12 min El.
Corrosion resistant casting.
Werkstoff Nr. 1.4027; DIN G-X 22 Cr 14.

CORRODUR C 14 W.
M-1290; 0.12 C, 12.5 Cr, 0.50 Ni, bal Fe.
Annealed: 71,000-92,000 TS; 42,500 min YS; 2 min El.
Heat treated: 85,000-112,000 TS; 57,000 min YS; 4 min El.
Corrosion resistant casting.
Werkstoff Nr. 1.4008; DIN G-X 12 Cr14.
ASTM A296 Gr.CA 15.

CORROFEST 13/1.
M-1293; 0.10 C, 13 Cr, bal Fe.
Annealed: 75,000 TS; 40,000 YS; 35 El; 70 RA; 155 Brin.
Cold drawn: 100,000 TS; 85,000 YS; 17 El; 60 RA; 205 Brin.
For furnace parts, heat treat boxes; X10Cr13; heat resistant.

CORROFEST 13/2.
M-1293; 0.2 C, 0.4 Si, 13 Cr, bal Fe.
Annealed: 95,000 TS; 50,000 YS; 25 El; 50 RA; 195 Brin.
Heat treated: 250,000 TS; 215,000 YS; 8 El; 25 RA; 500 Brin.
For cutlery, knives, valves, surgical instruments; X20Cr13; corrosion resistant, hardenable.

CORROFEST 13/2 MO.
M-1293; 0.2 C, 13 Cr, 1.15 Mo, bal Fe.
Annealed: 100,000 TS; 55,000 YS; 22 El; 48 RA; 200 Brin.
For cutlery, knives, chemical plant equipment; X20CrMo13; corrosion resistant, hardeanble.

CORROFEST 13/4.
M-1293; 0.4 C, 0.4 Si, 13 Cr, bal Fe.

CORROFEST 17/1.
M-1293; 0.1 max C, 0.4 Si, 17.5 Cr, bal Fe.
Annealed: 80,000 TS; 45,000 YS; 30 El; 55 RA; 170 Brin.
For furnace parts, heat treat boxes: X8Cr17; corrosion and heat resistant.

CORROFEST 17/1 MO.
M-1293; 0.1 max C, 0.4 Si, 17.5 Cr, 1 Mo, Ti = 7 x C, bal Fe.
Annealed: 80,000 TS; 45,000 YS; 30 El; 55 RA; 170 Brin.
For chemical plant equipment; X8CrMoTi17; corrosion and heat resistant.

CORROFEST 17/2.
M-1293; 0.22 C, 0.4 Si, 17 Cr, 1.5 Ni, bal Fe.
Annealed: 80,000 TS; 45,000 YS; 30 El; 55 RA; 170 Brin.
For stainless machine tool parts; X22CrNi17; corrosion resistant, heat treatable.

CORROFEST 17/9.
M-1293; 0.9 C, 0.4 Si, 18 Cr, 1.15 Mo, 1 V, bal Fe.
For bearings, bushings, pivots; X90CrMoV18; corrosion resistant, hardenable.

CORROFEST 18/8.
M-1293; 0.15 max C, 0.4 Si, 18 Cr, 8 Ni, bal Fe.
Annealed: 80,000 TS; 35,000 YS; 55 El; 65 RA; 150 Brin.
Cold drawn: 180,000 TS; 125,000 YS; 10 El; 330 Brin.
For chemical plant equipment, tanks, vessels, piping; Type 302; stainless, austenitic.

CORROFESTAL.
M-981; 0.5-2.0 Mg, 0.3-1.5 Si, 0-1.5 Mn, bal Al.
For light alloy parts; corrosion resistant, hardenable.

CORROFOND M3.
M-1445; 3 Mg, 0.3 Mn, bal Al.
Cast: 21,000-29,000 TS; 8,000-12,000 YS; 10-6 El; 45-55 Brin.
For light alloy parts; corrosion resistant.

CORROFOND M5.
M-1445; 5 Mg, 0.4 Mn, bal Al.
Heat treated: 28,000-33,000 TS; 13,000-17,000 YS; 12-8 El; 60-80 Brin.
For light alloy parts; corrosion resistant.

CORROFOND M7.
M-1445; 7 Mg, 0.4 Mn, bal Al.
Heat treated: 37,000-42,000 TS; 19,000-22,000 YS; 11-5 El; 70-80 Brin.
For marine hardware; resists sea water corrosion, age-hardenable.

CORROFOND M-10.
M-1445; 10 Mg, 0.4 Mn, bal Al.
Heat treated: 50,000-56,000 TS; 21,000-26,000 YS; 13-6 El; 75-85 Brin.
For marine and aircraft parts; heat treatable, resists sea water corrosion.

CORROFOND S2.
M-1445; 2 Si, 0.65 Mg, 0.7 Mn, bal Al.
Heat treated: 36,000-42,000 TS; 29,000-39,000 YS; 3-1 El; 90-105 Brin.
For light alloy parts.

CORROFOND S4.
M-1445; 4.2 Si, 0.5 Mn, 0.15 Ti, bal Al.
Cast: 19,000-26,000 TS; 7,000-13,000 YS; 15-9 El; 45-60 Brin.
For light alloy parts; corrosion resistant.

CORROFOND S45.
M-1445; 4.5 Si, 0.65 Mg, 0.7 Mn, bal Al.
Heat treated: 36,000-46,000 TS; 26,000-36,000 YS; 2-1 El; 90-105 Brin.
For light alloy parts; corrosion resistant, age-hardened.

CORRONEL B.
M-897; 66 Ni, 28 Mo, 6 Fe.
Rolled: 134,400 TS; 62,000 YS; 40 El; 40 RA; 250 Brin.
For heat exchangers, pump parts, reaction vessels; corrosion resistant.

CORRONIUM.
M-Eng.; 80 Cu, 15 Zn, 5 Sn.
4,000 TS; 38 El; 37 RA; 90 Brin.
Cast or wrought.

CORRONON-1.
M-1294; 0.12 max C, 0.4 Si, 13 Cr, bal Fe.
Annealed: 90,000 TS; 45,000 YS; 28 El; 55 RA; 180 Brin.
For turbine blades, cutlery, knives; X10Cr13; corrosion resistant.

CORRONON-2.
M-1294; 0.2 C, 13 Cr, bal Fe.
Annealed: 95,000 TS; 50,000 YS; 25 El; 50 RA; 195 Brin.
Heat treated: 250,000 TS; 215,000 YS; 8 El; 25 RA; 500 Brin.
For cutlery, knives, turbine blades; Type 320; corrosion resistant.

CORRONON-8.
M-1294; 0.12 max C, 0.8 Si, 0.8 Al, 6-5 Cr, bal Fe.
For oil refinery equipment; X10CrA17; creep resistant.

CORRONON-9.
M-1294; 0.20 C, 13 Cr, 1 Al, bal Fe.
Annealed: 95,000 TS; 50,000 YS; 25 El; 50 RA; 195 Brin.
Heat treated: 250,000 TS; 215,000 YS; 8 El; 25 RA; 500 Brin.
For cutlery, knives, turbine blades, surgical instruments; X20CrAl13; corrosion resistant.

CORRONON-10.
M-1294; 0.12 max C, 1 Si, 1 Al; 18 Cr, bal Fe.
Annealed: 80,000 TS; 45,000 YS; 30 El; 55 RA; 170 Brin.
For oil refinery and chemical plant equipment; X10CrA118; corrosion and heat resistant.

CORRONON-11.
M-1294; 0.2 C, 1.2 Si, 25 Cr, 4 Ni, bal Fe.
For furnace parts and equipment, heat treat boxes; X20CrNiSi254; heat resistant.

CORRONON-12.
M-1294; 0.12 max C, 1.5 Si, 1.5 Al, 24 Cr, bal Fe.
For furnace parts and equipment, heat treat boxes; X10CrA124; heat resistant.

CORRONON-16.
M-1294; 0.1 max C, 0.4 Si, 17.5 Cr, bal Fe.
Annealed: 80,000 TS; 45,000 YS; 30 El; 55 RA; 170 Brin.
For chemical plant and oil refinery equipment; X8Cr17; corrosion and heat resistant.

CORRONON-16N.
M-1294; 0.22 C, 17 Ni, bal Fe.
For chemical plant equipment; X22Ni17; corrosion resistant.

CORRONON-16S.
M-1294; 0.1 max C, 17.5 Cr, Ti = 7 x C, bal Fe.
Annealed: 80,000 TS; 45,000 YS; 30 El; 55 RA; 170 Brin.
For welded structures; X8CrTi17; heat resistant, stabilized.

CORRONON-18.
M-1294; 0.1 max C, 17.5 Cr, 1 Mo, Ti = 7 x C, bal Fe.
Annealed: 85,000 TS; 48,000 YS; 28 El; 52 RA; 180 Brin.
For chemical plant equipment; X8CrMoTi17; heat and corrosion resistant.

CORRONON-18H.
M-1294; 0.9 C, 18 Cr, 1.15 Mo, 1 V, bal Fe.
For bearings, linings; X90CrMoV18.

CORRONON 20/10.
M-1294; 0.15 C, 2 Si, 19.5 Cr, 9.5 Ni, bal Fe.
Annealed: 85,000 TS; 40,000 YS; 50 El; 60 RA; 180 Brin.
For chemical plant equipment; X15CrNiSi199; corrosion and heat resistant.

CORRONON 23/20.
M-1294; 0.15 C, 2 Si, 24 Cr, 19 Ni, bal Fe.
For furnace parts and equipment, heat treat boxes; corrosion and heat resistant, austenitic.

CORRONON 184E.
M-1294; 0.12 max C, 18 Cr, 2 Mo, 10.5 Ni, bal Fe.
Annealed: 80,000 TS; 30,000 YS; 60 El; 80 RA; 135 Brin.
Cold drawn: 150,000 YS; 135,000 YS; 6 El; 300 Brin.
For chemical plant equipment, mixers, agitators; filters; Type 316; stainless, austenitic.

CORRONON 184S.
M-1294; 0.07 max C, 18 Cr, 2 Mo, 10.5 Ni, bal Fe.
Annealed: 80,000 TS; 30,000 YS; 60 El; 80 RA; 135 Brin.
Cold drawn: 150,000 TS; 135,000 YS; 6 El; 300 Brin.
For chemical plant equipment, mixers, agitators, filters; Type 316L; stainless, austenitic.

CORRONON 188.
M-1294; 0.12 C, 18 Cr, 8 Ni, bal Fe.
Annealed: 80,000 TS; 35,000 YS; 55 51; 65 RA; 150 Brin.
Cold drawn: 180,000 TS; 135,000 YS; 10 El; 330 Brin.
For chemical plant equipment, tanks, vessels, mixers; X12CrNi18-8; austenitic, stainless.

Section I: Alloy Data / 351

CORRONON 188E.
M-1294; 0.12 max C, 18 Cr, 9.5 Ni, Ti = 4 x C, bal Fe.
Annealed: 85,000 TS; 35,000 YS; 55 El; 65 RA; 150 Brin.
Cold drawn: 95,000 TS; 60,000 YS; 40 El; 60 RA; 185 Brin.
For welded structures, chemical plant equipment; Type 321; austenitic, stainless.

CORRONON 188S.
M-1294; 0.07 max C, 18 Cr, 9.5 Ni, bal Fe.
Annealed: 80,000 TS; 35,000 YS; 55 El; 65 RA; 150 Brin.
Cold drawn: 180,000 TS; 135,000 YS; 10 El; 330 Brin.
For kitchen equipment, architectural trim; Type 304; stainless, austenitic.

CORROSALLOY 13.
M-1485; 0.25 max C, 11.5-13.5 Cr, 1 max Ni, bal Fe.
Cast: 101,000 TS; 15 El; 160-260 Brin.
For chemical plant equipment; Brit. BS 1630 B, corrosion resistant.

CORROSALLOY 18.
M-1485; 0.25 max C, 1.0-1.3 Ni, 16-20 Cr, bal Fe.
Cast: 123,200 TS; 10 El; 250 Brin.
For chemical plant and oil refinery equipment; Brit. S 80, corrosion and heat resistant, hardenable.

CORROSALLOY DU.
M-1485; 0.10 max C, 28-30 Ni, 19-21 Cr, 2-3 Mo, 3.5-4.5 Cu, bal Fe.
Cast: 62,800 TS; 30 El; 160 Brin.
For chemical plant equipment; corrosion resistant to H_2SO_4.

CORROSALLOY H.T.
M-1485; 0.05 C, 1.0 max Si, 1.0 max Mn, 4.0 Ni, 17 Cr, 3.5 Cu, bal Fe.
Cast: 140,000 TS; 15 El; 290-330 Brin.
High strength and corrosion resistant alloy for pump and valve applications.

CORROSALLOY HT.
M-1485; 0.07, 1.0 max Si, 1.0 max Mn, 15.0-17.5 Cr, 3.0-5.0 Ni, 3.0-5.0 Cu, bal Fe.
Precipitation hardening stainless steel casting. Hardenable to 300-350 Brin.

CORROSALLOY M.
M-1485; 0.12 max C, 10-12 Ni, 18-20 Cr, 2.5-3.5 Mo, Cb = 10 x C, bal Fe.
Cast: 78,500 TS; 25 El; 180-220 Brin.
For acid resistant chemical plant equipment, welded parts; Brit. BS 163 B, stainless, austenitic.

CORROSALLOY NDP.
M-1485; 0.12 max C, 8-10 Ni, 18-20 Cr, Cb = 10 x C, bal Fe.
Cast: 78,500 TS; 30 El; 170-210 Brin.
For welded chemical plant equipment, mixers; Brit. BS 1631 B, stainless, austenitic.

CORROSALLOY S.
M-1485; 0.08 max C, 2.0 max Si, 1.5 max Mn, 8.0-11.0 Ni, 18.0-21.0 Cr bal Fe.
Austenitic stainless casting for corrosion resisting applications.
ASTM A351 CF8.

CORROSALLOY S.
M-1485; 0.15 max C, 8-10 Ni, 18-20 Cr, bal Fe.
Cast: 78,500 TS; 30 El; 170-210 Brin.
For chemical plant and oil refinery equipment; austenitic, stainless, Brit. 1631 A.

CORROSIL.
M-1709; 0.70 C, 14 Si, bal Fe.
As cast: 514 Brin.
Corrosion and wear resistant iron, corrosion control anodes.

CORROSIRON.
M-192; M-23; 0.8-1.0 C, 13.5-14.5 Si, bal Fe.
Cast: 16,000 TS; 16,000 YS; 0 El; 0 RA; 350 Brin.
For acid pans, drains, pumps, valves, cocks; corrosion resistant, not forgeable.

CORROSIST B.
M-1485; 0.15 max C, 28-32 Mo, 6 max Fe, bal Ni.
Cast: 78,500 TS; 6 El; 250-300 Brin.
For pumps, valves, turbine parts, chemical plant equipment; resists HCl.

CORROSIST C.
M-1485; 0.15 max C, 17-20 Mo, 3-4 W, 8 max Fe, 14-16 Cr, bal Ni.
Cast: 74,000 TS; 10 El; 250-300 Brin.
For drains, pumps, chemical plant equipment; resists HCl and wet Cl_2.

CORROSIST D.
M-1485; 0.15 max C, 2.0-3.5 Cu, 5 max Fe, 9.0-10.5 Si, bal Fe.
Cast: 49,300 TS; 550-600 Brin.
For valves, pumps, drains, chemical plant equipment; resists boiling H_2SO_4, hard and brittle.

CORROSIST IL.
M-1485; 0.15 max C, 22-24 Cr, 3.5-4.5 Mo, 7-8 Cu, 7 max Fe, bal Ni.
Cast: 56,000 TS; 12 El; 190-220 Brin.
For chemical plant equipment; resists H_2SO_4.

CORSAIR see **ELECTRITE CORSAIR.**

CORTEN see **ALSO REPUBLIC CORTEN USS CORTEN.**

COR-TEN A.
M-604, M-1724; 0.10 C, 0.40 Mn, 0.12 P, 0.05 max S, 0.50 Si, 1.0 Cr 0.65 max Ni, 0.40 Cu, 0.02 min Ti, bal Fe.
Plate: 70,000 TS; 50,000 YS; 19 El.
For bridges, booms, derricks, mine cars, bus and truck bodies.
High strength, low alloy steel.
ASTM A 242.

COR-TEN B.
M-604, M-1724, M-67; 0.10-0.19 C, 0.90-1.25 Mn, 0.25-0.40 Cu, 0.65 ma Ni, 0.40-0.65 Cr, 0.02-0.10 V, bal Fe.
Plate: 70,000 TS; 50,000 YS; 19 El.
For structures, derricks, booms, mine cars, truck and bus bodies, bridges.
High strength, low alloy steel.
ASTM A 588.

COR-TEN B-QT.
M-604; 0.16 C, 1.15 Mn, 0.25 Si, 0.65 max Ni, 0.50 Cr, 0.04 V, 0.25-0.40 Cu, bal Fe.
W.Q.&T.: 90 ksi TS; 70 ksi YS; 19 El.
High strength, corrosion resistant steel for bridges, towers, buildings.
ASTM A588.

COR-TEN C.
M-604, M-67; 0.12-0.19 C, 0.90-1.35 Mn, 0.65 max Ni, 0.25-0.40 Cu, 0.40-0.70 Cr, 0.04-0.10 V, bal Fe.
Plate: 80,000 TS; 60,000 YS; 16 El.
For structures, bridges, bus and truck bodies, booms, mine cars, derricks.
High strength-low alloy steel.

COSINT 1000.
M-1526; 20 Co, 16 Cr, 5.5 W, 2.5 Mo, 5.5 Al, 2.5 Ti, 0.02 B, 0.3 Zr, bal Ni.
For high temperature applications; pre-alloyed powder, sintered, heat resistant.

COSMOLOY-F.
M-U.S.; 0.04 max C, 15 Cr, 3.8 Mo, 2.2 W, 3.4 Ti, 4.7 Al, 0.08 B, 0.07 Zr, bal Ni.
For high temperature applications; heat and corrosion resistant.

COTHIAS ALLOY.
M-Eng.; 93 Al, 6.5 Cu, 0.5 Zn.
For light weight castings.

COUGAR.
M-1432; 1.55 C, 0.04 Mn, 11.5 Cr, 0.9 V, 0.8 Mo, bal Fe.
AISI Type D2 Cold work tool steel, chromium type.

COURSIER.
M-1546; 0.60-0.70 C, bal Fe.
Hardened: C 58-60 Rock.
For cutting tools, drills, knives, hammers, chisels.
Water hardening.

COUSSINAL A.
M-French; 3 Sn, 2 Pb, 3 Sb, bal Al.
For bearings; anti-friction.

COUSSINAL C.
M-French; 4 Sn, 4 Cu, 3 Mg, 1 Mn, bal Al.
For bearings; antifriction.

COWLES.
M-Eng.; 1.25-11 Al, bal Cu.
For strong corrosion resistant parts; corrosion resistant.

COWLES "A-1".
M-U.S.; 80 Cu, 10 Al.
For hardware, gears; Al-bronze.

COWLES "A-2".
M-U.S.; 80 Cu, 10 Al.
For hardware, gears; Al-bronze.

COWLES ALUMINUM BRONZE.
M-Eng.; 88.4 Cu, 9.74 Al, 0.43 Fe, 1.36 Si.
For gears, trolley wheels, worm wheels; tough, wear resistant.

COWLES "B".
M-U.S.; 92.5 Cu, 7.5 Al.
For hardware, gears; Al-bronze.

COWLES "C-1".
M-U.S.; 94.5 Cu, 5.5 Al.
For hardware, gears; Al-bronze.

COWLES "C-2".
M-U.S.; 95 Cu, 5 Al.
For hardware, gears; Al-bronze.

COWLES "C-3".
M-U.S.; 95 Cu, 5 Al.
For hardware, gears; Al-bronze.

COWLES "D".
M-U.S.; 97.5 Cu, 2.5 Al.
For hardware, gears; Al-bronze.

COWLES "E".
M-U.S.; 98.75 Cu, 1.25 Al.
For hardware; corrosion resistant.

COWLES HIGH MANGANESE BRASS.
M-U.S.; 80-67 Cu, 15-18 Mn, 5-13 Zn, 0.1 Al, Si.
For strong corrosion resistant parts; corrosion resistant.

COWLES SPECIAL "A".
M-U.S.; 80 Cu, 11 Al.
For hardware, gears; Al-bronze.

CP 401.
M-1741; 0.10 max Cu, 12.0-13.0 Si, 0.4 max Fe, 0.05 max Mn, 0.05 max Mg, 0.05 max Ni, 0.10 max Zn, bal Al.
Per. Mold cast, F1; 207 MPa TS; 90 MPa YS; 9 El, 60 Brin.
ADC H49-5; BS LM6.

CP 601.
M-1741; 0.05 Cu, 6.5-7.5 Si, 0.12-0.20 Fe, 0.05 Max Mn, 0.30-0.40 Mg, 0.05 max Zn, 0.05 max Ti, bal Al.
Sand cast, T6: 255 MPa TS; 186 MPa YS; 5 El; 70 Brin.
Per. Mold cast, T6: 276 MPa TS; 186 MPa YS; 10 El; 100 Brin.
BS LM25.

CPI.
M-233; 0.5 C, 2.25 W, 1.5 Cr, 0.25 V, bal Fe.

CPM REX M2S.
M-38; 1.0 C, 4.15 Cr, 6.4 W, 5.0 Mo, 1.95 V, 0.15 S, bal Fe.
Free-machining Mo-W high speed steel.
AISI M2.

CPM REX M35S.
M-38; 0.85 C, 4.15 Cr, 2.0 V, 6.0 W, 5.0 Mo, 5.0 Co, 0.10 S, bal Fe.
High speed steel, free machining grade.
For form tools, deep hole drills, hobs.
AISI M35.

CPM REX M35-2.
M-38; 1.2 C, 4.0 Cr, 3.0 V, 6.25 W, 6.25 Mo, 0.15 S, bal Fe.
Free-machining high speed steel.
AISI M3 C12.

CPM REX T15S see REX T15S.

CPO.
M-669.
Cold finished, chrome-plated carbon steel bars; as per order.

CPT 64AV.
M-739; 0.02 C, 0.3 Fe, 6 Al, 4 V, bal Ti.
Sintered: 119,000 TS; 107,000 YS; Rock C 23.
Heat Treated: 146,000 TS; 133,000 YS; Rock C 27.
For aircraft bearing housings.
Heat treatable. Corrosion resistant.
Sintered powders.

C Q 2.
M-1697; 93.7 WC, 0.3 TaC, 6.0 Co.
290,000 TrS; A 92.0 Rock; Density 14.95.
Sintered carbide cutting tool; very good shock and wear resistance; for general machining.

C Q 3.
M-1697; 95.5 WC, 0.5 TaC, 4.0 Co.
250,000 TrS; A 92.3 Rock; Density 15.10.
Sintered carbide cutting tool; excellent abrasion resistance, for high speed finishing cuts.

C Q 4.
M-1697; 96.5 WC, 0.5 TaC, 3.0 Co.
240,000 TrS; A 92.7 Rock; Density 15.15.
Sintered carbide cutting tool; very high wear resistance; for high speed finishing.

C Q 12.
M-1697; 90.7 WC, 0.3 TaC, 9.0 Co.
340,000 TrS; A 91.0 Rock; Density 14.60.
Sintered carbide cutting tool; high shock resistance; for heavy cuts or rough work including slow speed and heavy feed conditions.

C Q 13.
M-1697; 85.0 WC, 15.0 Co.
400,000 TrS; A 88.0 Rock; Density 14.0.
Sintered carbide tool material.

C Q 14.
M-1697; 87.0 WC, 13.0 Co.
400,000 TrS; A 89.3 Rock; density 14.20.
Sintered carbide tool material.

C Q 22.
M-1697; 89.5 WC, 0.5 TaC, 10.0 Co,
400,000 TrS; A 91.2 Rock; Density 14.51.
Sintered carbide cutting tool; high load strength, for slow speed operations under heavy load, as in high temperature alloys.

C Q 23.
M-1697; 93.5 WC, 0.5 TaC, 6.0 Co.
270,000 TrS; A92.5 Rock; Density 14.95.
Sintered carbide cutting tool; exceptional abrasion resistance and good shock resistance; for machining high temperature alloys.

CR 05.
M-1806.
Sintered carbide tool material.
Hardness: above 1650 VHN.
For continuous finishing cuts on ferritic steels and steel castings.
ISO P01.

CR1A.
M-1296; 0.13 max C, 0.2-0.8 Si, 1.4 max Mn, 0.25-0.35 Cu, 1.0-1.5 Cr, bal Fe.
50 mm max thick plate, sheet and coil: 59,000-89,000 TS; 31,000 or 49,000 min YP.
Sulfuric acid resisting steel for chimneys, chemical equipment, etc.
Available in 2 strength grades: Cr1A-41, CR1A - 50.

CR 2.
M-1296; 0.20 max C, 0.75 max Si, 1.4 max Mn, 0.2-0.7 Cu, 0.65 max Ni, 0.2-1.2 Cr, 0.2 max Mo, 0.15 max Cb + V, bal Fe.
50 mm max thick plate, sheet and coil: 59,000-104,000 TS; 31,000, 49,000 or 63,000 psi YP.
Atmospheric corrosion resisting steel for structures, bridges.
Available in 4 strength grades: CR2-41, CR2-50, CR2-53, CR2-60.

CR2R-H.
M-1296; 0.12 max C, 0.25-0.75 Si, 0.2-0.5 Mn, 0.07-0.15 p, 0.25-0.55 Cu, 0.45 max Ni, 0.3-1.0 Cr, bal Fe.
19mm max thick plate, sheet or coil: 70,000 max TS; 50,000 max YP.
Atmospheric corrosion resisting steel for vehicles.

CR4A.
M-1296; 0.15 max C, 0.55 max Si, 1.2 max Mn, 0.07-0.15 P, 0.2 min Cu, 0.65 max Ni, 0.3-0.8 Cr, 0.15 max Cb + V , bal Fe.
19mm max thick plate, sheet or coil: 59,000-89,000 TS; 33,000 or 47,000 min YP.
Sea water resisting steel.
Available in 2 strength grades: Cr4A-41, CR4A-50.

CR 4 B.
M-1296; 0.15 max C, 0.55 max Si, 1.5 max Mn, 0.2 min Cu, 0.8-1.5 Cr, 0.15 max Cb + V, bal Fe.
50 mm max thick plate, sheet or coil: 59,000-89,000 TS; 33,000 or 43,000 min YP.
Sea water resisting steel.
Available in 2 strength grades: CR4B-41, CR4B-50.

CR 4 T.
M-1296; 0.08 max C, 0.35-0.75 Si, 1.2 max Mn, 0.4 max Cu, 1.8-2.3 Cr, 0.1-0.3 Mo, bal Fe.
6-100 mm thick plate: 71,000-89,000 TS; 31,000 min YP.
Sea Water resisting steel.

CR.5.
M-1116; 1.0 C, 5.0 Cr, 0.5 V, 1.1 Mo, 0.6 Mn, 0.3 Si, bal Fe.
Air or oil hardening tool steel; for stamping and drawing dies, shears, cold forming tube dies.
AFNOR: Z 100 CDV 05; AISI A2.

CR 5 MO.
M-1763; 1.0 C, 1.0 Mo, 5.0 Cr, bal Fe.
Air hardening type cold work tool steel.
AISI A2.

CR.5.T.M.
M-1116; 0.35 C, 5.0 Cr, 1.4 W, 0.4 V, 1.4 Mo, 0.25 Mn, 1.0 Si, bal Fe.
Air or oil hardening tool and die steel; for hot work forming and forging.
AFNOR: Z 35 CDWVS 05; AISI H12.

CR.5.T.M.S.
M-1116; 0.50 C, 4.7 Cr, 1.1 W, 0.35 V, 1.2 Mo, 0.3 Mn, 0.8 Si, bal Fe.
Air or oil hardening tool steel; for hot and cold working shear blades, dies, ejectors, bending tools, forming tools.
AFNOR: Z 50 CDWVS 05.

CR.5.V.4.
M-1116; 1.15 C, 5.0 Cr, 3.9 V, 1.1 Mo, 0.4 Mn, 1.0 Si, bal Fe.
Air or oil hardening tool steel; for stamping and drawing dies requiring good abrasive wear, for crushers and grinders in stone and gravel work.
AFNOR: Z 110 CVDS 05.04.

CR.5.V.6.
M-1116; 1.42 C, 6.25 Cr, 5.9 V, 1.0 Mo, 0.3 Mn, 1.0 Si, bal Fe.
Air or oil hardening tool and die steel; for stamping silicon laminations, and abrasive materials.
AFNOR: Z 135 CVDS 06.06.

CR.5.V.M.
M-1116; 0.35, 5.5 Cr, 1.0 V, 1.1 Mo, 0.3 Mn, 1.0 Si, bal Fe.
Air or oil hardening tool and die steel; for hot work stamping copper and copper alloys; molds for plastics.
AFNOR: Z 35 CDVS 05; AISI H13.

CR 6 MO.
M-1763; 0.10 min C, 1.0 max Mn, 1.0 max Si, 4.0-6.0 Cr, 0.40-0.65 Mo, bal Fe.
Martensitic stainless steel.
AISI 501.

CR.8.V.3.
M-1116; 0.83 C, 8.0 Cr, 2.5 V, 1.5 Mo, 0.3 Mn, 1.0 Si, bal Fe.
Air hardening tool and die steel; for cutting and shaping tools, shears.
AFNOR: Z 85 CVDS 08.03.

CR 10.
M-1806.
Sintered carbide tool material.
Hardness: 1525-1625 VHN.
For finishing and light to medium roughing on ferritic steel and steel castings.
ISO P10.

CR 12.
M-1116; 2.0 C, 12.0 Cr, 0.30 Mn, 0.30 Si, bal Fe.
Air, oil or salt, hardenable to 65 Rc.
For cold stamping and shearing metal sheet, thread rolling, sheet punches, wood working cutters and tools.
AFNOR Z 200 C 12.
US - Similar to AISI D3.
German: W. Nr. 1.2080.

CR 13 CO 3.
M-1763; 1.5 C, 12-13 Cr, 1.0 Mo, 3.0 Co, bal Fe.
Cold work tool steel, punching and coining dies.
AISI D5.

CR 15.
M-1806.
Sintered carbide tool material.
Hardness: 1525-1625 VHN.
For general purpose machining of ferritic steels, with light and medium cuts.
ISO P10 - P20.

CR 20.
M-1806.
Sintered carbide tool material.
Hardness: 1475-1575 VHN.
For medium to heavy cuts at high speeds on ferritic steels.
ISO P20.

CR 25.
M-1806.
Sintered carbide tool material.
Hardness: 1450 - 1550 VHN.
For general purpose machining with medium feeds and medium speeds, including some interrupted cuts.
ISO P20 - P30.

CR 30.
M-1806.
Sintered carbide tool material.
Hardness: 1425 - 1525 VHN.
For heavy interrupted cuts at medium to low speeds, using throwaway tips.
ISO P30.

CR 40.
M-1740; 0.36-0.43 C, 0.50 max Si, 0.60-0.75 Mn, 0.50-0.70 Cr, bal Fe.
Carbon chromium steel; abrasion resistant.

CR 40.
M-1806.
Sintered carbide tool material.
Hardness: 1400-1500 VHN.
For heavy interrupted turning and planing of steel where high metal removal is required.
ISO P30 - P40 and M30.

CR 50.
M-1740; 0.45-0.55 C, 0.75 max Si, 0.50-1.0 Mn, 0.80-1.2 Cr, bal Fe.
Carbon chromium steel; abrasion resistant.
B.S. 1956 Gr. A & B. ASTM. A148-65 Gr.105-85.

CR 50.
M-1806.
Sintered carbide tool material.
Hardness: 1175 - 1275 VHN.
For heavy cutting at low speeds, interrupted cutting, and under bad conditions; tough.
ISO P50 and M40.

CRAIG GOLD.
80 Cu, 10 Ni, 10 Zn.
For jewelry, ornaments.

CRALCLAD.
M-Japan; Al alloy.
For light alloy parts; clad duralumin.

CRALFER.
M-England; 3 C, Mn, Si, Al, bal Fe.
For machinery castings; cast iron, growth resistant.

CRAMP ALLOY NO. 5.
M-309; 79.3 Cu, 10 Sn, 10 Pb, 0.70 P.
Cast: 28,000-32,000 TS; 18,000-19,000 YS; 4-8 El; 70 Brin.
For bearings for medium speeds and light loads, cross-head and mill bearings; SAE 64.

CRAMP ALLOY NO. 7.
M-309; 75 Cu, 18 Pb, 7 Sn.
Cast: 17,000-32,000 TS; 8,000-12,000 YS; 7-10 El; 45-50 Brin.
For bearings of low compressive load and where lubrication is poor; acid corrosion resistant.

CRAMP ALLOY NO. 12.
M-309; 87.75 Cu, 6.25 Sn, 4 Zn, 1.5 Pb, 0.5 Ni.
Cast: 32,000-38,000 TS; 16,000-18,000 YS; 20-25 El; 40-50 Brin.
For valves, fittings, cocks, pressure castings; similar to U.S. Navy Composition "M."

CRAMP ALLOY NO. 13.
M-309; 83 Cu, 13.5 Sn, 3.5 Zn.
Cast: 23,000-32,000 TS; 13,000-15,000 YS; 4-2 El; 55-65 Brin.
For bearings for heavy duty as for bridges, trunnions, machines, center disks; corrosion resistant to weak sulfurous acid.

CRAMP ALLOY NO. 20.
M-309; 85 Cu, 5 Sn, 5 Zn, 5 Pb.
Cast: 30,000 TS; 25 El.
For bearings for general service; similar to "Ounce" metal.

CRAMP ALLOY NO. 45 (PARSONS MN BRONZE).
M-309; 55-60 Cu, 38-42 Zn, 0.20 max Pb, 1.5 max Al, 3.5 max Mn, 1.5 max Sn, 0.4-2.0 Fe.
Cast: 65,000-75,000 TS; 33,000-38,000 YS; 30-20 El; 100 Brin.
For ship propellers, heavy duty bushings, bearings, spur and bevel gears, valve stems, valve bodies; corrosion resistant, tough.

CRAMP ALLOY NO. 49 (SUPER STRENGTH BRONZE).
M-309; 60-70 Cu, 20-30 Zn, bal special alloy.
Cast: 90,000-115,000 TS; 45,000-68,000 YS; 25-10 El; 185-240 Brin.
For housing nuts, trunnion bearings, expansion plates, worms, worm wheels, spur and bevel gears, valve stems; resists high compressive loads and shock.

CRAMP ALLOY NO. 49A.
M-309; 60-70 Cu, 20-30 Zn, bal Mn, Fe, Al.
Cast: 120,000 TS; 90,000 YS; 10 El; 10 RA; 240 Brin.
For bushings, worms, bearings, gears, valve bodies; Mn-bronze, tough.

CRAMP ALLOY NO. 50.
M-309; 60-70 Cu, 20-30 Zn, bal Mn, Fe, Al.
Cast: 90,000 TS; 45,000 YS; 25 El; 20 RA; 175 Brin.
Forged: 93,000 TS; 50,000 YS; 20 El; 20 RA; 195 Brin.
For bushings, worms, bearings, gears, valve bodies; Mn-bronze, tough.

CRAMP ALLOY NO. 50A.
M-309; 60-70 Cu, 20-30 Zn, bal Mn, Fe, Al.
Cast: 100,000 TS; 50,000 YS; 20 El; 18 RA; 150 Brin.
Forged: 105,000 TS; 60,000 YS; 18 El; 18 RA; 210 Brin.
For bushings, worms, bearings, gears, valve bodies; Mn-bronze, tough.

CRAMP ALLOY NO. 54 (PARSONS WHITE BRASS SA).
M-309; 78-61 Sn, 4.5-5 Cu, 6-11 Sb, 3.5-13 Pb.
Cast: 12,250 TS; 10,750 YS; 2.7 El; 32-38 Brin.
For bearings for heavy duty service, main and cross-head bearings; Babbitt.

CRAMP ALLOY NO. 69 (HYDRAULIC METAL).
M-309; 86.5 Cu, 11 Sn, 2 Zn, 0.50 Pb.
35,000-40,000 TS; 16,500-18,500 YS; 10-8 El; 60-65 Brin.
For valve bodies, sleeves, pump parts, cylinder bodies, bearings; for high hydraulic pressures.

CRAMP ALLOY NO. 71 (CRAMP'S NICKEL BRONZE).
M-309; 87.5 Cu, 11 Sn, 1.5 Ni.
Cast: 45,000-50,000 TS; 25,000-27,000 YS; 20-15 El; 85-95 Brin.
For worms, worm gears, bevel and spur gears, high speed bearings, bushings wearing parts; for high contact speeds and heavy loads.

CRAMP ALLOY NO. 73.
M-309; 10 Sn, 10 Pb, 0.7 P, bal Cu.
Cast: 28,000-32,000 TS; 18,000-19,000 YS; 4-8 El; 70 Brin.
For bearings, bushings, shaft sleeves; SAE 64; bearing bronze.

CRAMP ALLOY NO. 78.
M-309; 85-89 Cu, 8-10 Al, 1.0-2.0 Fe.
Cast: 65,000-95,000 TS; 47,000-57,000 YS; 20-4 El; 130-190 Brin.
For acid resistant containers and parts; resistant to acid corrosion at high temperature.

CRAMP ALLOY NO. 83.
M-309; 87-89 Cu, 9-11 Al, 3-4 Fe.
Heat treated: 75,000-85,000 TS; 35,000-40,000 YS; 8-20 El; 165-185 Brin.
For bearings, worm gears, spur gears; resists acid corrosion.

CRAMP ALLOY NO. 84.
M-309; 78-82 Cu, 9-11 Al, 3-6 Fe, 3-6 Ni.
Heat treated: 90,000-115,000 TS; 40,000-70,000 YS; 5-20 El; 5-15 RA 172-300 Brin.
For structural parts; acid resisting.

CRAMP ALLOY NO. 91.
M-309; 87.5 Cu, 8 Sn, 4 Zn, 0.5 Ni.
Cast: 35,000-45,000 TS; 17,500-21,000 YS; 30-20 El; 60-70 Brin.
For shaft liners, bushings, pump parts, valve fittings, flanges, connections; similar to "Composition G."

CRAMP ALLOY NO. 97 (CRAMP COPPER ALLOY).
M-309; 99-99.8 Cu.
17,000-22,000 TS; 5,000-7,000 YS; 40-30 El.
For electrical installations for high conductivity; elec, cond. 75-85%.

CRAMP ALLOY NO. 99.
M-307; 3.5 Si, 1.5 Fe, 4 Zn, bal Cu.
Cast: 45,000-55,000 TS; 20,000-28,000 YS; 23-15 El 120-130 Brin.
For gears, bushings, cross-head bearings; silicon bronze.

CRAMP ALLOY NO. 101 (PARSONS TURBADIUM).
M-309; 47-52 Cu, 40-46 Zn, 1.0-2.5 Ni.
Cast: 65,000-75,000 TS; 33,000-40,000 YS; 18-15 El; 90 Brin.
For pump impellers, turbine runners, buckets for water wheels; resists erosive action of water.

CRAMP ALLOY NO. 114.
M-309; 67 Ni, 29 Cu, 1.5 Fe, 0.9 Mn, 1.25 Si.
Cast: 65,000-80,000 TS; 32,500-45,000 YS; 25-40 El; 25-40 RA; 125-150 Brin.
For impellers, valve seats; acid resistant; "Monel Metal."

CRAMP ALLOY NO. 115.
M-309; 5 Sn, 5 Ni, 2 Zn, bal Cu.
Heat treated: 50,000-85,000 TS; 22,000-55,000 YS; 40-10 El; 95-150 Brin.
For gears, slippers, bearings, valve bodies; hardenable, heavy duty.

CRAMP ALLOY NO. 118.
M-309; 63 Ni, 30 Cu, 2 Fe, 0.9 Mn, 3-4 Si.
Cast: 100,000-130,000 TS; 75,000-95,000 YS; 1-4 El; 250-325 Brin.
For valve seats; S-Monel; acid resistant.

CRAMP ALLOY NO. 151.
M-309; 78-82 Cu, 8-11 Al, 3-6 Fe, 3-6 Ni.
Heat treated: 80,000-115,000 TS; 35,000-70,000 YS; 20-5 El; 172-300 Brin.
For ship propellers; heat treatable, non-galling, corrosion resistant.

CRAMP ALLOY NO. 174.
M-309; 0.20-0.35 Be, 1.4-1.5 Ni, bal Cu.
Cast: 50,000 TS; 20,000 YS; 16 El; 100 Brin.
Heat treated: 80,000 TS; 55,000 YS; 17 El; 210 Brin.
For welding wheels; heat treatable, 40% electrical conductivity.

CRAMP ALLOY NO. 175.
M-309; 0.30-0.90 Cr, 0.5 max Fe, 0.5 max Si, bal Cu.
Cast: 50,000 TS; 40,000 YS; 15 El; 90 Brin.
For welding wheels; heat treatable, 75% electrical conductivity.

CRAMP ALLOY NO. 176.
M-309; 4-5 Cu, 1 Fe, 1.5 max Si, 0.25 max Ti, bal Al.
Cast: 32,000 TS; 16,000 YS; 8 El; 60 Brin.
For cylinder heads, gear cases; age-hardenable.

CRAMP ALLOY NO. 177.
M-309; 4-5 Zn, 0.40-0.65 Mn, 0.3-0.6 Cr, bal Al.
Cast: 31,000 TS; 7 El.
For gear cases, general castings; age-hardens at room temperature.

CRAMP ALLOY NO. 275.
M-309; 2.6-3.8 T.C., 0.15-1.50 Si, 3-6 Ni, 0.5-2.5 Cr, 0.15-1.50 Mn, bal Fe.
Cast: 50,000-80,000 TS; 500-600 Brin.
For pump plungers, rolls for metal rolling, grinding plates; wear and abrasion resistant; "Ni-Hard."

CRAMP'S SUPER-STRENGTH BRONZE, GRADE "A".
M-309; 57 Cu, 28 Zn, bal hardeners of Mn, Fe, Al.
Cast: 115,000 TS; 68,000 YS; 10 El; 10 RA; 240 Brin.
Wrought: 118,000 TS; 70,000 YS; 8 El; 8 RA.
For busings, bearings, gears, worms, die castings, shafts, valve stems, tube sheets; Mn bronze; C.Y.P. 65000.

CRAMP'S SUPER-STRENGTH BRONZE, GRADE "B".
M-309; 57 Cu, 28 Zn, bal hardeners of Mn, Fe, Al.
Cast: 110,000 TS; 63,000 YS; 15 El; 15 RA; 225 Brin.
Wrought: 115,000 TS; 65,000 YS; 15 El; 15 RA.
For bushings, bearings, gears, worms, die castings, shafts, valve stems, tube sheets, pump impellers; Mn bronze; C.Y.P. 60000.

CRAMP'S SUPER-STRENGTH BRONZE, GRADE "C".
M-309; 57 Cu, 28 Zn, bal hardeners of Mn, Fe, Al.
Cast: 100,00 TS; 55,000 YS; 20 El; 18 RA; 200 Brin.
Wrought: 105,000 TS; 60,000 YS; 18 El; 18 RA.
For bushings, bearings, gears, worms, die castings, pump impellers, valve stems; Mn bronze; C.Y.P. 50000.

CRAMP'S SUPER-STRENGTH BRONZE, GRADE "D".
M-309; 57 Cu, 28 Zn, bal hardeners of Mn, Fe, Al.
Cast: 90,000 TS; 45,000 YS; 25 El; 20 RA; 185 Brin.
Wrought; 93,000 TS; 50,000 YS; 20 El; 20 RA.
For bushings, bearings, valve stems, shafts, gears, worms, die castings; Mn bronze; C.Y.P. 40000.

CRANE.
M-387; 0.7-1.2 C, bal Fe.
78,000 TS; 41,000 YS; 150 Brin.
For tools, cold chisels; water-hardening.

CRANE-1.
M-111; C, alloy, bal Fe.
For tool machinery parts; oil hardened.

CRANE-2.
M-111; 0.50 C, 1.05 Cr, 3.25 Ni, bal Fe.
For gears, bolts, crankshafts; oil hardened, shock resistant.

CRANE "ARCTIC".
M-748; 0.15 max C, 0.50-0.80 Mn, 3.0-4.0 Ni, bal Fe.
Cast: 483-655 MPa TS; 276 MPa YS; 24 El.
Valves, fittings for low temperature service.
ASTM A352, Grade LC3.

CRANE EXCELLOY (CAST).
M-748; 0.15 max C, 1.0 max Mn, 1.0 max Si, 1.0 max Ni, 11.5-14.0 Cr, 0.15-1.0 Mo, bal Fe.
Cast: 621-793 MPa TS; 448 MPa YS; 18 El.
Corrosion resistant cast alloy for valves, fittings.
ASTM A487, Grade CA15M.

CRANE EXELLOY (WROUGHT).
M-748; 0.12 max C, 1.0 max Mn, 1.0 max Si, 0.50 max Ni, 11.5-13.5 Cr, bal Fe.
Forged or Rolled: 690 MPa TS; 552 MPa YS; 15 El.
Corrosion resistant alloy for valves, fittings.
ASTM A182, Grade F6.

CRANE FERROSTELL.
M-748; C, Si, 0.75 max P, 0.15 max S, bal Fe.
Higher test gray iron.
Cast: 214 MPa TS; 3300 lbs Transverse St.
ASTM A 126, Class B.

CRANE HIGH TENSILE.
M-748; C, Si, 0.75 max P, 0.15 max S, bal Fe.
High test cast iron.
Cast: 283 MPa TS; 4000 lbs Transverse St.
ASTM A125, Class C.

CRANELOY 20 STAINLESS STEEL.
M-748; 0.07 max C,1.5 max Mn, 1.5 max Si, 27.5-30.5 Ni, 19.0-22.0 Cr, 2.0-3.0 Mo, 3.0-4.0 Cu, bal Fe.
Cast: 431 MPa TS; 172 MPa YS; 35 El.
Austenitic stainless for chemical equipment.
ASTM A361, Grade CN7M.

CRANELOY 20 STAINLESS STEEL (FORGED-WROUGHT).
M-748; 0.07 C, 2.0 max Mn, 1.0 max Si, 32.0-38.0 Ni, 19.0-21.0 Cr, 2.0-3.0 Mo, bal Fe.
Wrought: 500 MPA TS; 240 MPa YS; 30 El.
Austenitic stainless for chemical equipment.
Similar to ASTM B473.

CRANE NO. 1.
M-1109; 75 Pb, 12.5 Sn, 12.5 Sb.
25.2 Brin.
For bearings; Babbitt.

CRANE NO. 2.
M-748; 0.3 C, 2 Ni, 0.75 Cr, 0.25 Mo, bal Fe.
Cast: 100,000 TS.
For castings, valve fittings; for service at 750-1100°F, high pressure service.

CRANE NO. 2.
M-1109; 75 Pb, 10 Sn, 15 Sb.
25.6 Brin.
For bearings; Babbitt.

CRANE NO. 3.
M-748; 0.25 max C, 0.50-0.80 Mn, 0.60 max Si, 2.0-3.0 Ni, bal Fe.
Cast: 483-655 MPa TS; 276 MPa YS; 24 El.
Cast nickel steel for low temperature operations.
ASTM A352, grade LC2.

CRANE NO. 3.
M-1109; 80 Pb, 5 Sn, 15 Sb.
25.4 Brin.
For bearings; Babbitt.

CRANE NO .5.
M-748; 0.20 max C, 0.40-0.70 Mn, 4.0-6.5 Cr, 0.45-0.65 Mo, bal Fe.
Cast: 621-793 MPa TS; 414 MPa YS; 18 El.
For valves, fittings.
ASTM A217, Grade C5.

CRANE NO. 6.
M-1109; 85 Sn, 0.25 Pb, 10.75 Sb, 4 Cu.
30.8 Brin.
For bearings; Babbitt.

CRANE NO. 7.
M-1109; 81.5 Sn, 4 Pb, 10.5 Sb, 4 Cu.
34.6 Brin.
For bearings; Babbitt.

CRANE NO. 7 CHROME-MOLYBDENUM.
M-748; 0.18 max C, 1.0-1.5 Cr, 0.4-0.6 Mo, 0.75 Mn, bal Fe.
Cast: 78,000 TS; 52,000 YS; 28 El; 60 RA; 165 Brin.
For valves, fittings; for temperature up to 1000°F.
ASTM A217, Gr WC6.

CRANE NO. 8.
M-1109; 40 Sn, 48.5 Pb, 10.5 Sb, 1 Cu.
23.1 Brin.
For bearings; Babbitt.

CRANE NO. 9 CHROME-MOLYBDENUM.
M-748; 0.18 max C , 2-3 Cr, 0.8-1.1 Mo, 0.7 Mn, bal Fe.
Cast: 85,000 TS; 40,000 YS; 20 El; 55 RA; 180 Brin.
For valves, fittings; high temperature steam service.

CRANE NO. 49 (NICKEL ALLOY).
M-748; 30.5-34.0 Cu, 10.5-13.0 Sn, 52.0-56.0 Ni, 0.3-0.75 Mn, 0.4-1.0
60,000 psi TS; 414 MPa TS.
For valves, fittings; corrosion resistant.

CRANE SPECIAL BRONZE.
M-748; 5.5-6.5 Sn, 1.0-2.0 Pb, 1.0 max Ni, 3.0-5.0 Zn, 86.0-90.0 Cu.
Cast: 342 MPa TS; 110 MPa YS; 22 El.
Steam or valve bronze castings (was CRANE SPECIAL BRASS).
ASTM B61, Alloy 922.

CRANE STEAM BRONZE.
M-748; 4.0-6.0 Sn, 4.0-6.0 Pb, 4.0-6.0 Zn, 1.0 max Ni, 84.0-86.0 Cu.
Cast: 207 MPa TS; 97 MPa YS; 20 El.
For valves, fittings (was CRANE STEAM BRASS).
ASTM B62, Alloy 836.

CRASCO ALLOY.
M-337; C, W, Cr bal Fe.
For tools, chisels; shock resistant.

CRASCO BLACK LABEL.
M-337; 1.5 C, 0.4 Mn, 11-13 Cr, 0.8 Mo, 0.2 V, bal Fe.
For tools, dies, hot working dies and long run dies; air hardening, non-shrinking.

CRASCO BROACH STEEL.
M-337; C, alloy, bal Fe.
For broaches, punches, dies; resists heavy pressures.

CRASCO DRILL ROD.
M-337; 1.05 c, 0.25 Mn, 0.25 Si, bal Fe.
For pivots, bearings; drill rod.

CRASCO FINISHING.
M-337; 1.4 C, 0.2 V, 4 Cr, 0.5 Si, 0.2 Mn, bal Fe.
For cutters, engravers' tools; fast finishing tool steel.

CRASCO GREEN LABEL.
M-337; 1.0 C, bal Fe.
For tools, dies; water hardening.

CRASCO HIGH SPEED.
M-337; 0.7 C, 4 Cr, 18 W, 1.1 V, bal Fe.
For tools, cutters, drills, reamers; high speed steel.

CRASCO HIGH SPEED MOLY NO. 2.
M-337; 0.83 C, 4.25 Cr, 6.4 W, 5 Mo, 1.8 V, bal Fe.
For tools, cutters, drills, broaches; high speed steel.

CRASCO NO. 7.
M-337; 1,05 C, 0.65 Cr, 0.4 Mn, 0.15 V, bal Fe.
For tools, dies; water or oil hardening.

CRASCO RED LABEL.
M-337; 0.9 C, 1.2 Mn, 0.5 Cr, 0.5 W, bal Fe.
For tools dies; cutters, punches; oil hardening, non-shrinking.

CRASCO SHANK STEEL.
M-337; 0.95 C, bal Fe.
For tools, tool shanks; water hardening.

CRASCO SPECIAL VANADIUM.
M-337; 0.9- 1.05 C, 0.15-0.25 V, bal Fe.
For tools, dies; water hardening.

CRASCO WHITE LABEL.
M-337; 1.0 C, bal Fe.
For tools, dies; water hardeing.

CRASCO YELLOW LABEL.
M-337; 0.45-0.50 C, 0.25 Mn, 0.95 Cr, 0.2 V, bal Fe.
For gears, shafts; oil hardening.

CRASFLOY.
M-451; C, Ni, Cr, bal Fe.
For rolls; hard alloy.

CR.E.
M-1116; 2.0 C, 12.5 Cr, 1.0 W, 0.3 V, 0.3 Mo, 0.5 Mn, 0.4 Si, bal Fe.
Air or oil hardening tool steel; for Zendzimir rolls, wood working tools, brick making tools.
AFNOR: Z 200 CW 13; AISI D3.

CR.E.4.V.
M-1116; 2.30 C, 12.5 Cr, 4.1 V, 1.1 Mo, 0.7 Mn, 0.4 Si, bal Fe.
Oil or air hardening tool and die steel; for punches, shears, cutting tool for abrasive materials; work in sand and gravel.
AFNOR: Z 230 CVD 13.04; AISI D7.

CR.E.15.
M-1116; 1.55 C, 12.0 Cr, 0.8 V, 0.8 Mo, 0.25 Mn, 0.3 Si, 0.8 Co, bal Fe.
Air or oil hardening tool and die steel; for stamping dies for silicon laminations, swaging dies, gages.

CRESCENT.
M-1744.
Carbon tool steel available in 6 tempers and carbon ranges: No. 1 1.20-1.40 C; No. 2 1.05-1.15 C; No. 3 0.90-1.00 C, No. 4 0.80- C; No. 5 0.70-0.80 C; No. 6 0.60-0.70 C.

CREST A.H.
M-1684; 1.0 C, 0.70 Mn, 5.5 Cr, 0.3 V, 1.15 Mo, bal Fe.
Air hardening tool steel for blanking and trimming dies, shear blades, rolling dies, broaches.
AISI A 2.

CRESTALOY.
M-749; 0.5 C, 1.5 Cr, 0.2 V, bal Fe.
For tools, wrenches; oil hardening.

CREST O.H.
M-1684; 0.90 C, 1.15 Mn, 0.65 Cr, 0.25 V, 0.5 W, bal Fe.
Oil hardening tool steel for blanking and bending dies, punches, knurling tools, gauges.
AISI O 1.

CRESTON NO. 5.
M-750; 0.75-1.1 C, bal Fe.
For cold battering tools; water hardening.

CRESTON NO. 6.
M-750; 0.8-1.2 C, bal Fe.
For tools, dies,; water hardening.

CRESTON NO. 7.
M-750; 0.9-1.3 C, bal Fe.
For tools, dies, cutters; water hardening.

CRESTON 8.
M-750; 0.90-1.40 C, bal Fe.
For tools, dies, cutters; water hardening.

CREUSABRO 32.
M-1488; 0.20 C, 1.2 Mn, 1.3 Cr, 0.25 Mo, bal Fe.
Normalized: 142,000 TS; 106,000 YS; 10 El.
For cement plant equipment, ventilators, screens, coal and mine equipment.
Wear resistant.

CREUSABRO 360.
M-1488; 0.20 C, 1.6 Mn, 1.3 Cr, 0.2 Mo, Cu, B, bal Fe.
360 Brin; Weldable.
For mine construction, public works.

CREUSANBRO 400.
M-1488; 0.28 C, 0.75 Mn, 1.40 Ni, 1.40 Cr, 0.50 Mo, bal Fe.
Treated: 380 Brin, weldable.
For strong welded structures.

CREUSABRO M.
  M-1488; 1.2 C, 13 Mn, 0.6 Si, bal Fe.
  Quenched: 128,000 TS; 50,000 YS; 35 El.
  For shafts, scraper buckets, crushers, grinders, safes, helmets.
  Abrasion and wear resistant.

CREUSABRO ML.
  M-1488; 0.45 C, 0.6 Mn, 2 Si, bal Fe.
  Heat treated: 214,000 TS; 115,000 YS; 5 El.
  For conveyors, wear plates.
  Abrasion and wear resistant.

CREUSELSO 34SS.
  M-1488; 0.17 max C, 1.3 Mn, 0.35 max Si, 0.50 Max Ni, bal Fe.
  Normalized: 490-610 N/mm^2 UTS; 335 N/mm^2 min YS; 24 El.
  Containers for liquified gases (propane etc.).

CREUSELSO 38.
  M-1488; 0.20 max C, 1.2 Mn, 0.4 Si, 0.70 max Ni+Cr+Mo+Cu, bal Fe.
  Normalized: 490-640 N/mm^2 WTS.
  Weldable; for pressure vessels.
  AFNOR E375.

CREUSELSO 38W.
  M-1488; 0.20 C, 0.55 max Si, 1.50 max Mn, 0.70 max, Ni+Cr+Mo+Cu, bal Fe.
  Normalized: 510-640 N/mm^2 TS; 335 N/mm^2 min YS; 18 min El.
  Weldable, for pressure vessels.
  AFNOR E 375 C.

CREUSELSO 38 AK.
  M-1488; 0.20 max C, 1.5 max Mn, 0.55 max Si, bal Fe.
  Normalized: 510-640 N/mm^2 UTS; 335 N/mm^2 min YS; 18 min El.
  Parts to be used at low temperature.
  AFNOR E 375 FP.

CREUSELSO 42.
  M-1488; 0.22 max C, 1.4 Mn, 0.4 Si, 0.70 max Ni+Cr+Mo+Cu, bal Fe.
  Normalized: 540-650 N/mm^2 WTS.
  Weldable; for pressure vessels, construction.
  AFNOR E 420.

CREUSELSO 42C.
  M-1488; 0.22 C, 0.55 max Si, 1.60 max Mn, 0.70 max Ni+Cr+Mo+Cu, bal Fe.
  Normalized: 540-650 N/mm^2 UTS; 370 N/mm^2 min YS; 17 min El.
  Weldable; for pressure vessels.
  AFNOR E 420 C.

CREUSELSO 42 FP.
  M-1488; 0.22 C, 1.6 Mn, 0.55 max Si, bal Fe.
  Normalized: 540-650 N/mm^2 UTS; 370 N/mm^2 min YS; 17 min El.
  Parts to be used at low temperature.
  AFNOR E 420 FP.

CREUSELSO 47.
  M-1488; 0.18 max C, 1.5 Mn, 0.35 Si, 0.4 Ni, 0.07 V, 0.70 max Cr+Mo Cu, bal Fe.
  Normalized: 550-710 N/mm^2 UTS.
  Weldable; for marine platforms, mine structures.
  AFNOR E 460.

CREUSELSO 47C.
  M-1488; 0.18 max C, 1.5 Mn, 0.35 Si, 0.4 Ni, 0.70 max, Cr+Mo+Cu, 0.07 V, bal Fe.
  Normalized: 440 N/mm^2 min YS.
  Weldable; for pressure vessels; structures.
  AFNOR E 460 C.

CREUSELSO 47 FP.
  M-1488; 0.18 max C, 1.5 Mn, 0.35 Si, 0.40 Ni, 0.07 V, bal Fe.
  Normalized: 560-710 N/mm^2 UTS; 440 N/mm^2 min YS; 24 El.
  Parts for cryogenic operation.
  AFNOR E 460 FP.

CREUSEM E.
  M-1488; 0.080 max C, 0.375 max Mn, 0.225 Si, bal Fe.
  Normalized: 310 N/mm^2 min UTS; 155 N/mm^2 min YS; 35 min El.
  Soft magnetic material; for relays.

CREUSEM G.
  M-1488; 0.050 max C, 0.30 max Mn, 0.20 Si, bal Fe.
  Normalized: 290 N/mm^2 min UTS; 145 N/mm^2 min YS.
  Soft magnetic material; for relays.

CREUSOT 0.5 FO.
  M-1488; 0.18 max C, 0.50 Cr, 0.50 Mo, bal Fe.
  Normalized and tempered: 265 N/mm^2 min YS.
  Plates, for pressure vessels: weldable.
  AFNOR 15 CD 2-05.
  ASTM A 387 Gr. 2.

CREUSOT 1.1 FO.
  M-1488; 0.18 max C, 1.0 Cr, 0.5 Mo, bal Fe.
  Normalized and tempered: 295 N/mm^2 min YS.
  Plates for pressure vessels; weldable.
  AFNOR 15 CD 4-05.
  ASTM A387 Gr. 12.

CREUSOT 1.2DF3.
  M-1488; 0.40 C, 0.6 Cr, 1.2 Ni, bal Fe.
  Oil hardened: 114,000-157,000 TS; 100,000 YS; 9 El.
  For gears, clutches, connecting flanges, bolts, fasteners, shafts.
  AISI 3140 steel, tough, shock resistant.

CREUSOT 1.2DF6.
  M-1488; 0.20 C, 0.6 Cr, 1.2 Ni, bal Fe.
  Heat treated: 107,000-142,000 TX; 92,000 YS; 10 El.
  For bolts, levers, shafts, pins, fasteners, gears.
  Case-hardening.
  AISI 3120 steel, tough, shock resistant.

CREUSOT 1.2 MOV.
M-1488; 0.15 C, 0.50 Mo, 1.2 Mn, 0.07 V, bal Fe.
Normalized and tempered: 345 N/mm² min YS.
Plates for pressure vessels; weldable.
AFNOR 15 MDV 4-05.

CREUSOT 1.3FOV.
M-1488; 0.24 C, 1.25 Cr, 0.6 Mo, 0.25 V, bal Fe.
Heated treated: 100,000 TS; 82,000 YS; 16 El; C 25 Rock.
For rotors, hot bolts, hardware, gears, shafts.
Tough, shock resistant.

CREUSOT 1.4DF3.
M-1488; 0.30-0.38 C, 1.2-1.6 Ni, 0.8-1.2 Cr, bal Fe.
Heat treated: 295,000 TS; 200,000 YS; 4 El.
For gears, bolts, machine tool parts; oil hardened, tough.

CREUSOT 1.4DF5.
M-1488; 0.22-0.30 C, 1.2-1.6 Ni, 0.80-1.15 Cr, bal Fe.
Heat treated: 250,000 TS; 166,000 YS; 6 El.
For gears, bolts, machine tool parts; oil hardened, tough.

CREUSOT 1.4DF6.
M-1488; 0.16-0.22 C, 1.2-1.6 Ni, 0.85-1.2 Cr, bal Fe.
Heat treated: 222,000 TS; 143,00 YS; 7 El.
For gears, bolts, machine tool parts; case hardened, oil quenched.

CREUSOT 1.4DF7.
M-1488; 0.11-0.18 C, 1.2-1.6 Ni, 0.85-1.2 Cr, bal Fe.
Heat treated: 213,000 TS; 128,000 YS; 7 El.
For gears, bolts, machine tool parts; case hardened, oil quenched.

CREUSOT 1.4DF8.
M-1488; 0.07-0.12 C, 0.6-0.9 Mn, 1.2-1.6 Ni, 0.85-1.2 Cr, bal Fe.
Heat treated: 192,000 TS; 121,000 YS; 8 El.
For gears, bolts, machine tool parts; case hardened, oil quenched.

CREUSOT 1.4DF06.
M-1488; 0.14-0.22 C, 1.2-1.6 Ni, 0.8-1.2 Cr, 0.1-0.3 Mo, bal Fe.
Heat treated: 229,000 TS; 130,000 YS; 6 El.
For gears, bolts, camshafts; case hardened, oil quenched.

CREUSOT 1.4DF07/8.
M-1488; 0.10-0.16 C, 1.2-1.6 Ni, 0.8-1.2 Cr, 0.1-0.3 Mo, bal Fe.
Heat treated: 221,000 TS; 128,000 YS; 7 El.
For gears, bolts, camshafts, cams; case hardened, oil quenched.

CREUSOT 1.5DF02.
M-1488; 0.28 C, 1.4 Cr, 1.5 Ni, 0.38 Mo, bal Fe.
Oil hardened: 200,000 TS; 170,000 YS; 8 El.
For aircraft engine parts, propeller shafts, gears, linings, fasteners.
Tough, shock resisting.

CREUSOT 1.5D06S.
M-1488; 0.20 max C, 0.6 Cr, 1.5 Ni, 0.25 Mo, bal Fe.
For gears, bolts, shafts, bridge members; resists low temperature.

CREUSOT 1.5DF0V1S.
M-1488; 0.22 max C, 1.3 Cr, 1.5 Ni, 0.27 Mo, 0.17 V, bal Fe.
For welded structures, bridge members; heat resistant to 550°C.

CREUSOT 1.7DFOV.
M-1488; 0.55 C, 1.5 Ni, 1 Cr, 0.3 Mo, 0.1 V, bal Fe.
For hot stamping dies, punches.
Oil hardening hot-work tool steel.

CREUSOT 1.8DFO 2/3.
M-1488; 0.40 C, 0.8 Cr, 1.8 Ni, 0.25 Mo, bal Fe.
Oil hardened: 156,000-214,000 TS; 128,000 YS; 9 El.
For gears, shafts, crankshafts, fasteners.
AISI 4340 steel. Tough and shock resistant.

CREUSOT 1.8DFO 6/7.
M-1488; 0.20 C, 0.8 Cr, 0.25 Mo, 1.8 Ni, bal Fe.
Oil hardened: 156,000-214,000 TS; 128,000 YS; 8 El.
For gears, pinions, cams, crankshafts, cam shafts, pawls.
AISI 4320 steel. Case hardening, tough and shock resistant.

CREUSOT 2.2 FO.
M-1488; 0.15 max C, 2.25 Cr, 1.0 Mo, bal Fe.
Normalized and tempered; 295 N/mm² min YS; 510/610 N/mm² W TS; 21 El.
Plates for pressure vessels, weldable.
AFNOR 10 CD 9-10.
ASTM 387 Gr. 22.

CREUSOT 3DF4/5.
M-1488; 0.26-0.33 C, 2.7-3.3 Ni, 0.6-0.9 Cr, bal Fe.
Heat treated: 150,000 TS; 100,000 YS; 13 El.
For gears, bolts, machine tool parts; oil hardened, tough.

CREUSOT 3DF7/8.
M-1488; 0.11-0.16 C, 2.7-3.3 Ni, 0.6-0.9 Cr, 0.3-0.6 Mn, bal Fe.
Heat treated: 207,000 TS; 120,000 YS; 9 El.
For gears, bolts, machine tool parts; case hardened, tough.

CREUSOT 3DF8.
M-1488; 0.08-0.14 C, 2.7-3.0 Ni, 0.6-0.9 Cr, 0.3-0.6 Mn, bal Fe.
Heat treted: 166,000 TS; 100,000 YS; 12 El.
For gears, bolts, camshafts, cams; case hardened, oil quenched.

CREUSOT 3DFO1.
M-1488; 0.30-0.38 C, 2.5-3.0 Cr, 1.2-1.6 Ni, 0.2-0.4 Mo, bal Fe.
Heat treated: 235,000 TS; 192,000 YS; 4 El.
For gears, bolts, machine tool parts; oil hardened, tough.

CREUSOT 3.5D8.
M-1488; 0.20 C, 3.5 Ni, 0.9 max Mn, 0.35 max Si, bal Fe.
Heat treated: 100,000-128,000 TS; 78,000 YS; 14 El.
For cryogenic vessels, tanks, equipment.
Shock resistant at cryogenic temperatures.
ASTM A 203D.

CREUSOT 3.5 DFO1.
M-1488; 0.3-0.4 C, 3.2-3.8 Ni, 1.2-1.7 Cr, 0.2-0.5 Mo, bal Fe.
Heat treated: 250,000 TS; 208,000 YS; 5 El.
For gears, bolts, machine tool parts; oil hardened, tough.

CREUSOT 4DF2.
M-1488; 0.30-0.38 C, 3.5-4.0 Ni, 1.5-1.8 Cr, bal Fe.
Heat treated: 242,000 TS; 193,000 YS; 5 El.
For gears, bolts, crankshafts; oil hardened, tough.

CREUSOT 4DF6.
M-1488; 0.08-0.15 C, 3.4-3.9 Ni, 0.7-1.1 Cr, bal Fe.
Heat treated; 228,000 TS; 135,000 YS; 8 El.
For gears, bolts, camshafts, cams; case hardened, oil quenched.

CREUSOT 4DF01.
M-1488; 0.30-0.40 C, 3.5-4.5 Ni, 1.5-2.0 Cr, 0.35-0.60 Mo, bal Fe.
Heat treated: 270,000 TS; 207,000 YS; 4 El.
For gears, bolts, machine tool parts; oil hardened, tough.

CREUSOT 4DO6.
M-1488; 0.09-0.15 C, 3.7-4.4 Ni, 0.70-1.2 Mo, bal Fe.
Heat treated: 172,000 TS; 135,000 YS; 8 El.
For gears, bolts, camshafts, cams; case hardened, oil quenched.

CREUSOT 5D6.
M-1488; 0.13-0.20 C, 0.4-0.7 Mn, 4.7-5.4 Ni, bal Fe.
Heat treated: 213,000 TS; 127,000 YS; 6 El.
For gears, bolts, machine tool parts; case hardened, oil quenched.

CREUSOT 5 FO.
M-1488; 0.10 C, 5.0 Cr, 0.50 Mo, bal Fe.
Normalized and tempered: 370 N/mm^2 min YS; 590 N/mm^2 min TS; 15 min El.
Plates for pressure vessels; weldable.
AFNOR Z 10 CD 5.
ASTM A 387.

CREUSOT 9D6.
M-1488; 0.08 max C, 0.8 max Mn, 0.35 max Si, 8.5-10,0 Ni, bal Fe.
Normalized: 93,000 TS; 71,000 YS.
Sheets: 70,000 min TS; 60,000 min YS; 19 min El.
For liquid gas industry, storage or transportation tanks, weldable cryogenic steel, good low temperature ductility.

CREUSOT M.
M-1488; 1.20 C, 12 Mn, 0.6 Si, bal Fe.
Water quenched: 880 N/mm^2 min UTS.
Non-magnetic alloy.
AFNOR Z 120 M 12.

CREUSOT MO 6/7.
M-1488; 0.20 max C, 0.80 Mn, 0.52 Mo, bal Fe.
Normalized: 300 N/mm^2 min YS.
Weldable; plates for pressure vessels.
AFNOR 18 MD 4.05.
ASTM A 204 Gr. C.

CRH 4-17.
M-1647; 0.1 C, 13 Cr, 0.45 Mo, 0.6 max Ni, bal Fe.
Annealed: 75,000 TS; 40,000 YS; 35 El; 70 RA; 155 Brin.
Cold drawn: 100,000 TS; 85,000 YS; 17 El; 60 RA; 205 Brin.
For oil refinery and chemical plant equipment, knives, hardware, table flatware.
Corrosion resistant.

CRILLEY METALS.
M-183; Hg, Sn, bal Cu.
23,000 TS.
For bearings; C.S.-92,000.

CRISIL.
M-1620; 0.55 C, 0.5-0.8 Cr, 1.2-1.6 Si, 0.5-0.8 Mn, bal Fe.
Wire: 280,000-335,000 TS; 45 min RA.
For mechanical springs, operating under high stress and moderately high temperatures.
Hard drawn steel wire.

CRISTITE 1.
M-152; 17 W, 10 Cr, 3.5 C, 2.5 Mo, bal Fe.
For facing dredger, dipper or shovel teeth and bucket lips, agricultural implements, knives, tools, drills, chisels; resists abrasion and high temperature.

CRISTITE 2.
M-152; 17 W, 12 Cr, 3 C, 2 Mo, 0.75 Ti, 0.25 Hf, bal Fe.
For hard facing electrode; wear resistant.

CRISTITE 3.
M-152; 16 Cr, 3 C, 0.80 Ti, 0.28 Hf, 6 Ni, bal Fe.
For hard facing electode; wear resistant.

CRITNIC.
M-U.S; 97 Ni, Ti, 0.2 C.
For heat and corrosion resistant parts; corrosion and heat resistant.

CRM 4.
M-680; 0.02 C, 0.40 Mn, 0.1 Si, 0.5 Cr, 0.5 Ni, 0.2 Cu, (all max), 5.5-6.5 Al, bal Fe.
Low carbon low strength, wrought ferritic alloy; good oxidation resistance to 1200°F.

CRM 6D.
M-680, M-1491; 1.0-1.1 C, 20-22 Cr, 5 Ni, 5 Mn, 1 W, 1 Mo, 1 Cb, 0.55 Si, 0.002-0.008 B, 0.05 max N, bal Fe.
Cast, austenitic; good strength above 1500°F.
For high temperature rotating parts as turbine wheels.

CRM-15D.
M-680, M-1491; 1.0-1.1 C, 20-22 Cr, 5 Ni, 5 Mn, 2 W, 2 Mo, 2 Cb, 0.55 Si, 0.002-0.008 B, 0.15-0.25 N, bal Fe.
Cast, austenitic; good strength above 1500°F.
For high temperature roatating parts as turbine wheels.

CRM-17D.
M-680, M-1491; 0.65-0.75 C, 18-20 Cr, 5 Ni, 5 Mn, 1 W, 1 Mo, 2 Cb, 0.55 Si, 0.002-0.008 B, 0.15-0.25 N, bal Fe.
Cast, austenitic; for operation at 1200-1400°F as nozzle blades and nozzle support struts.

CRM-18D.
M-680, M-1491; 0.70-0.80 C, 22.5-24.5 Cr, 5 Ni, 5 Co, 5 Mn, 1 W, 1 Mo, 2 Cb, 0.55 Si, 0.002-0.008 B, 0.25-0.35 N, bal Fe.
Cast, austenitic; for operation up to 1800°F such as first stage nozzle and burner vortex.

CRMN 55.
M-1541; 0.18 C, 1.28 Mn, 0.016 P, 0.009 S, 0.29 Si, 0.70 Cr, bal Fe.
88,000 psi TS; 61,200 psi YS; 29 El.
Cr-Mn low alloy steel for abrasion resistance applications.

CRMN 60.
M-1541; 0.30 C, 1.27 Mn, 0.021 P, 0.014 S, 0.35 Si, 0.67 Cr, bal Fe
111,000 psi TS; 64,000 psi YS; 20 El.
Cr-Mn low alloy steel for abrasion resistance applications.

CRMO 18.
M-1740; 0.16-0.20 C, 0.30-0.60 Si, 0.50-0.80 Mn, 0.40 max Ni, 1.0-1.5 Cr, 0.45-0.65 Mo, 0.40 max Cu, bal Fe.
B.S. 1504-621.

CRMO 23.
M-1740; 0.20-0.25 C, 0.75 max Si, 0.30-0.70 Mn, 0.40 max Ni, 2.5-3.5 Cr, 0.35-0.60 Mo, 0.40 max Cu, bal Fe.
For service at elevated temperature, hardenable.
B.S. 1461.

CRMO 60.
M-1740; 0.55-0.65 C, 0.75 max Si, 0.50-1.0 Mn, 0.80-1.50 Cr, 0.20- 0.40 Mo, bal Fe.
Hardenable; high abrasion resistant steel.
B.S. 1956 Gr. C.

CRMO A.
M-1740; 0.50-0.60 C, 0.30-0.60 Si, 0.50-0.80 Mn, 0.40-0.80 Ni, 2.3- 3.0 Cr, 0.30-0.50 Mo, 0.40 max Cu, bal Fe.
Hardenable; good abrasion resistance.

CRMO B.
M-1740; 0.55-0.65 C, 0.30-0.60 Si, 0.50-0.80 Mn, 0.40-0.80 Ni, 2.3- 3.0 Cr, 0.30-0.50 Mo, 0.30 Max Cu, bal Fe.
Hardenable; good abrasion resistance.

CRMO D.
M-1740; 0.18 max C, 0.30-0.50 Si, 0.45-0.65 Mn, 0.40 max Ni, 2.0-2.75 Cr, 0.90-1.2 Mo, 0.40 max Cu, bal Fe.
ASTM A217-65 Gr. WC9.

CRMO E.
M-1740; 0.15-0.22 C, 0.35-0.60 Si, 0.60-0.90 Mn, 0.60-0.80 Cr, 0.20- 0.30 Mo, bal Fe.

CRMO F.
M-1740; 0.28-0.33 C, 0.30-0.50 Si, 0.60-0.80 Mn, 2.0-2.5 Cr, 0.20- 0.30 Mo, bal Fe.
Oil hardenable, tough steel.

CRMO G.
M-1740; 0.35-0.45 C, 0.30-0.70 Si, 0.60-0.85 Mn, 0.80-1.10 Cr, 0.25- 0.35 Mo, bal Fe.
Similar to AISI 4140.

CR-MO-W.
M-341; 0.33 C, 1.05 Si, 1.55 W, 1.65 Mo, 5 Cr, bal Fe.
For hot work dies; air hardening.

CR-NI ROTOR FM.
M-32; 0.55 C, 0.6 Mn, 0.3 Si, 1.5 Cr, 4.0 Ni, bal Fe.
Air hardening tool steel.

CR. N.O.
M-1116; 0.38 C, 1.6 Cr, 3.7 Ni, 0.3 Mo, 0.4 Mn, 0.3 Si, bal Fe.
Deep hardening alloy steel, air or oil hardening.
For cold forming tools, cold-heading and rivet sets.
AFNOR: 35 NCD 15.

CRO-13 MO.
M-1648; 0.15 C, 13 Cr, 0.6 max Mo, bal Fe.
Annealed: 75,000 TS; 40,000 YS; 35 El; 70 RA; 155 Brin.
Cold drawn: 100,000 TS; 85,000 YS; 17 El; 60 RA; 205 Brin.
For oil refinery and chemical plant equipment, knives, hardware, table flatware.
Corrosion resistant.

CROBALITE NO. 1.
M-1052; 48 Co, 30 Cr, 14 W.
Welded: 550 Brin.
For hard facing electrode; corrosion and abrasion resistant.

CROBALITE NO. 2.
M-1052; 40 Co, 33 Cr, 18 W, carbides. 620 Brin.
For tools, cutters, dies; cast to shape, wear resistant.

CROBALITE NO. 3.
M-1052; 40 Co, 33 Cr, 20 W, carbides, 630 Brin.
For tools, cutters, dies; cast to shape, wear resistant.

CROBALITE NO. 6.
M-1052; 57 Co, 30 Cr, 5 W. 440 Brin.
For hard facing electode; corrosion and abrasion resistant.

CROBALITE NO. 12.
M-1052; 52 Co, 30 Cr, 10 W, 480 Brin.
For hard facing electrode; corrosion and abrasion resistant.

CROBALT.
M-751; 40-50 Co, 25-30 Cr, 14-20 W.
Cast: 75,000 TS.
For cutters, tools; wear resistant.

CROBALT NO. 1.
M-1052; 48 Co, 30 Cr, 14 W, high C, bal Fe.
Cast: 75,000 TS; 0 El; 0 RA; 580-620 Brin.
For cutting tool bits, milling cutters.
High red-hardness.

CROBALT NO. 2.
M-1052; 40 Co, 33 Cr, 18 W, bal Fe.
Cast: C 61-62 Rock.
For cutters, tools, punches.
Cast alloy, wear resistant.

CROBALT NO. 3.
M-1052; 40 Co, 33 Cr, 20 W, bal Fe.
Cast: C 63-64 Rock.
For cutters, tools, punches.
Cast alloy, wear resistant.

CROCAR.
M-115; 12 Cr, 2.15 C, 0.75 V, 0.5 Co, 0.30 Si, bal Fe.
103,000 TS; 51,000 YS; 14 RA; 212 Brin.
For thread rolling, threading, dies, extrusion dies, gages; heat, abrasion resisting.

CROCEM-18.
M-1766; 0.16 C, 1.10 Mn, 0.25 Si, 1.0 Cr, bal Fe.
Chromium carburizing steel.
DIN 16 Mn Cr 5; AFNOR 16 MC5; UNI 16MC5.

CRODI.
M-435; 0.35 C, 0.5 Mn, 1.2 W, 5 Cr, 0.3 V, 1.4 Mo, bal Fe.
For punches, dies, shear blades, extension dies; hot work steel.

CRODON.
M-980; Cr-Fe.
For anodes in electrolytic copper cells; electroplating alloy.

CRODUR.
M-1331; 2.1 C, 11.5 Cr, bal Fe.
For blanking and forming dies, punches; oil or air hardened, nondeforming.

CRODUR SPEZIAL.
M-1331; 2.1 C, 11.5 Cr, 0.7 W, bal Fe.
For blanking and forming dies, punches; oil or air hardened, nondeforming.

CRODUR V.
M-1331; 2.2 C, 11.5 Cr, 0.1 V, 0.2 Mo, bal Fe.
For blanking and forming dies, punches; oil hardened, nondeforming.

CRODUR ZAH.
M-1331; 1.65 C, 11.5 Cr, 0.1 V, bal Fe.
For blanking and forming dies, punches; air hardened, nondeforming.

CROFER 106.
M-297; 0.12 max C, 6-7 Cr, 0.6-0.9 Si, 0.6-0.9 Al, bal Fe.
For oil refinery equipment; heat and creep resistant.

CROFER 113.
M-297; 0.12 max C, 12-14 Cr, 1.0-1.3 Si, 0.8-1.1 Al, bal Fe.
For oil refinery equipment; heat and creep resistant.

CROFER 118.
M-297; 0.12 max C, 17-19 Cr, 0.8-1.1 Si, 0.8-1.1 Al, bal Fe.
For oil refinery equipment; heat and creep resistant.

CROFER 124.
M-297; 0.12 max C, 23-25 Cr, 1.3-1.6 Al, bal Fe.
For oil refinery equipment; corrosion and heat resistant.

CROFER 218.
M-297; 0.12 max C, 18 Cr, 2 Si, bal Fe.
Annealed: 80,000 TS; 50,000 YS; 25 El; 50 RA; 150 Brin.
For oil refinery and dairy equipment, bolts; Type 430; corrosion resistant.

CROFER 230.
M-297; 0.12 C, 29 Cr, bal Fe.
Annealed: 85,000 TS; 50,000 YS; 30 El; 55 RA; 180 Brin.
For furnace parts, heat treat boxes; heat resistant.

CROFER 1300.
M-297; 0.12 max C, 12-14 Cr, bal Fe.
Annealed: 75,000 TS; 40,000 YS; 35 El; 70 RA; 155 Brin.
For valves, cutlery, turbine blades; corrosion resistant.

CROFER 1300AL.
M-297; 0.06 C, 13 Cr, 0.2 Al, bal Fe.
Annealed: 75,000 TS; 45,000 YS; 20 El; 180 Brin.
For heat treating boxes, furnace parts, oil refinery equipment.
Type 405 stainless steel, ferritic, corrosion and heat resistant.

CROFER 1700.
M-297; 0.0 max C, 15.5-17.5 Cr, bal Fe.
Annealed: 80,000 TS; 50,000 YS; 25 El; 50 RA; 150 Brin.
For oil and chemical plant equipment; Type 430; stainless.

CROFER 1700NB.
M-297; 0.08 C, 17 Cr, 8 Ni, 0.5 Cb, bal Fe.
Annealed: 85,000 TS; 40,000 YS; 50 El; B 82 Rock.
For chemical plant equipment, welded tanks and structures.
Type 17-8 Cb. Welding grade, austenitic, stainless.

CROFER 1700TI.
M-297; 0.10 max C, 16-18 Cr, Ti = 7xC, bal Fe.
Annealed: 80,000 TS; 50,000 YS; 25 El; 50 RA; 150 Brin.
For welded oil and chemical plant equipment; Type 430 Ti; stainless.

CROFER 1702TI.
M-297; 0.10 max C, 16-18 Cr, 1.5-2.0 Mo, Ti = 7xC, bal Fe.
Annealed: 125,000 TS; 95,000 YS; 20 El; 55 RA; 260 Brin.
For pumps, marine hardware, valves; Type 431 Ti; stainless.

CROFER 1919.
M-297; 0.15 C, 19.5 Cr, 9.5 Ni, 2 Si, bal Fe.
Annealed: 80,00 TS; 35,000 YS; 55 El; 75 RA; 150 Brin.
For chemical plant equipment, tanks, mixers, filters; Type 302; stainless, austenitic.

CROFER 2419.
M-297; 0.15 C, 2 Si, 24 Cr, 19 Ni, bal Fe.
Annealed: 100,000 TS; 45,000 YS; 50 El; 65 RA; 185 Brin.
For valves, pumps, turbine and jet parts; Type 310; stainless, austenitic.

CROFER 2420.
M-297; 0.15 C, 24 Cr, 19 Ni, 1.2 Mn, bal Fe.
Annealed: 100,000 TS; 45,000 YS; 50 El; 65 RA; 185 Brin.
For valves, pumps, turbine and jet parts; Type 310; stainless, austenitic.

CROFER 2504.
M-297; 0.2 C, 25 Cr, 4 Ni, bal Fe.
Cast: 90,000 TS; 65,000 YS; 2 El; 212 Brin.
For cylinder liners, valve seats and bodies; Type CC-50; corrosion and heat resistant.

CROFORM.
M-English; 60.0 Co, 30.0 Cr, 5.0 Mo, bal Fe.
Cast: 100,000-119,000 TS; 50,000-95,000 YS; 5-4 El; 390 Brin.
For dental alloy; corrosion resistant.

CROLOY 1/2 see **B&W CROLOY 1/2.**

CROLOY 1 1/4 see **B&W CROLOY 1 1/4.**

CROLOY 2 see **B&W CROLOY 2.**

CROLOY 2 1/4 see **B&W CROLOY 2 1/4.**

CROLOY 3M see **B&W CROLOY 3M.**

CROLOY 5 see **B&W CROLOY 5.**

CROLOY 5 SI see **B&W CROLOY 5 SI.**

CROLOY 7 see **B&W CROLOY 7.**

CROLOY 9M see **B&W CROLOY 9M.**

CROLOY 12 see **B&W CROLOY 12.**

CROLOY 12 AL see **B&W CROLOY 12 A1.**

CROLOY 16-1 see **B&W CROLOY 16-1.**

CROLOY 16-6 PH see **B&W CROLOY 16-6 PH.**

CROLOY 18 see **B&W CROLOY 18.**

CROLOY 18-8 see **B&W CROLOY 304.**

CROLOY 18-13-3 see **B&W CROLOY 316.**

CROLOY 25-20 see **B&W CROLOY 310.**

CROLOY 27 see **B&W CROLOY 27.**

CROMA.
M-688; 0.35 C, 1.0 Cr, 0.8 Mn, bal Fe.
For machinery parts; formerly Lehigh Croma.

CROMADUR.
M-72; 0.15 max C, 18 Mn, 12.5 Cr, 1 V, 0.2 Ni, bal Fe.
For blading for gas turbine on jet engine; heat resistant.

CROMAL.
M-USSR; 1.8 C, 20-22 Cr, 2.2 Si, 3-4 Al, bal Fe.
For furnace equipment; corrosion resistant, cast iron.

CROMALOY.
M-408; 61 Ni, 15 Cr, 4 Mn, 20 Fe.
107,500 TS.
For electric heating devices; heat resistant to 950°C.

CROMALOY IV.
M-408; 80 Ni, 20 Cr.
134,500 TS.
For heating appliances for high temperature work; heat resistant up to 1150°C.

CROMANSIL.
M-48; 0.10 C, 0.4-0.6 Cr, 1.1-1.4 Mn, 0.7-0.8 Si, bal Fe.
Rolled: 65,000 TS; 45,000 YS; 45 El; 70 RA; 130 Brin.
For staybolts, boilers, pressure vessels, ship plates, tanks; ductile.

CROMANSIL.
M-48; 0.15 C, 0.4-0.6 Cr, 1.1-1.4 Mn, 0.7-0.8 Si, bal Fe.
Rolled: 140,000 TS; 101,000 YS; 38 El; 65 RA; 150 Brin.
For staybolts, ship plates, tanks, pressure vessels.

CROMANSIL.
M-48; 0.20 C, 0.4-0.6 Cr, 1.1-1.4 Mn, 0.7-0.8 Si, bal Fe.
Rolled: 90,000 TS; 60,000 YS; 28 El; 62 RA; 200 Brin.
For boilers and pressure vessels; heat treated after welding.

CROMANSIL.
M-48; 0.25 C 0.4-0.6 Cr, 1.1-1.4 Mn, 0.7-0.8 Si, bal Fe.
Rolled: 115,000 TS; 70,000 YS; 25 El; 60 RA; 220 Brin.
For seamless tubing, boilers, pressure vessels; good workability.

CROMANSIL.
M-48; 0.30 C, 0.4-0.6 Cr, 1.1-1.4 Mn, 0.7-0.8 Si, bal Fe.
Rolled: 140,000 TS; 90,000 YS; 20 El; 50 RA; 265 Brin.
Heat treated: 172,000 TS; 131,000 YS; 11 El; 32 RA; 364 Brin.
For heat treated parts, gears, building, bridges; Iz-11.

CROMAR.
M-1331; 1.0 C, Cr, Mo, V, bal Fe.
For tools, dies, cutters; oil hardened.

CROMAR W.
M-1331; 1.0 Cr, Mo, W, V, bal Fe.
For engravers tools, reamers, broaches; high speed steel.

CROMAX F.
M-210; 0.50-0.55 C , 0.20-0.35 Si, 0.50-0.80 Mn, 0.70-0.90 Cr, bal Fe.
For gears, housings, shafts; castings.

CROMAX F6 MO.
M-1488; 0.50-0.70 C, 1.5 max Mn, 1.5 max Si, 28.0-30.0 Cr, 1.5-2.5 Mo, bal Fe.
Full ann: 220-270 Brin.
Stainless steel; for pumps requiring resistance to both corrosion and abrasion.
AFNOR Z 60 CD 29.2 M.

CROMAX F13.
M-1488; 1.20-1.50 C, 1.5 max Mn, 1.5 max Si, 28.0-30.0 Cr, bal Fe.
Full ann: 260-330 Brin. (Casting).
Stainless steel; for pumps requiring corrosion resistance and extreme abrasion resistance.
AFNOR Z 130 C 29 M.

CROMAX H.
M-210; 0.70-0.75 C, 0.20-0.35 Si, 0.50-0.80 Mn, 1.8-2.0 Cr, bal Fe.
For machinery parts, wear plates; castings.

CROMETEX NO. 1.
M-1009; 0.38-0.43 C, 0.7 Mn , 0.3 Si, 0.8 Cr, 0.25 Mo, 1.8 Ni, bal Fe.
For arbors, axles, gears, piston rods; shock resistant.

CROMEX-1.
M-USSR; 0.5-1.0 C, 25-32 Cr, Si, bal Fe.
For furnace parts and equipment; heat resistant.

CROMIN-1.
M-897; 9-11 Ni, 19-21 Cr, bal Fe.
For heating elements; useful operating temperature to 650°F.

CROMINO 0.
M-1315; 0.12 max C, 13 Cr, bal Fe.
Annealed: 75,000 TS; 40,000 YS; 35 El; 70 RA; 155 Brin.
For turbine blades, valves, cutlery, surgical instruments; Type 410; stainless.

CROMINO 1.
M-1315; 0.2 C, 0.4 Si, 13 Cr, bal Fe.
Annealed: 95,000 TS; 50,000 YS; 25 El; 55 RA; 195 Brin.
For turbine blades, valves, cutlery, surgical instruments; Type 420; stainless, hardenable.

CROMINO 1 M.
M-1315; 0.17-0.22 C, 1.0 max Si, 1.0 max Mn, 12.0-14.0 Cr, 0.90-1.30 Mo, 1.0 max Ni, bal Fe.
Martensitic stainless steel; for valve cones, turbine blades.
W.-Nr. 1.4120.

CROMINO-I.
M-1763; 0.15 max C, 1.0 max Mn, 1.0 max Si, 11.5-13.5 Cr, bal Fe.
Martensitic stainless steel.
AISI 410.

CROMINO-IV.
M-1763; 0.20 max C, 1.0 max Mn, 1.0 max Si, 15.0-17.0 Cr, 1.25-2.5 Ni bal Fe.
Martensitic stainless steel.
AISI 431.

CROMINO 8.
M-1315; 0.85-0.95 C, 1.0 max Si, 1.0 max Mn, 17.0-19.0 Cr, 0.90-1.30 Mo, 0.07-0.12 V, bal Fe.
Martensitic stainless steel; for cutlery.
W.-Nr. 1.4112; similar to AISI 440 C.

CROMINO 8M.
M-1315; 0.9 C, 18 Cr, 0.2 Mo, bal Fe.
Annealed: 110,000 TS; 65,000 YS; 18 El; 35 RA; 220 Brin.
Heat treated: 280,000 TS; 270,000 YS; 3 El; 15 RA; 555 Brin.
For cutlery, valves, ball bearings, surgical instruments; corrosion and wear resistant, hardenable.

CROMINO 8MS.
M-1315; 0.12 C, 16.5 Cr, 0.25 Mo, 0.2 S, bal Fe.
Annealed: 80,000 TS; 50,000 YS; 25 El; 50 RA; 150 Brin.
For screw machine products; Type 430F; stainless, free-cutting.

CROMINO 15.
M-1315; 0.12-0.17 C, 1.0 max Si, 1.0 max Mn, 12.0-14.0 Cr, bal Fe.
Martensitic stainless steel, for structural parts.
W.-Nr. 1.4024; similar to AISI 410.

CROMINO 15 M.
M-1315; 0.12-0.17 C, 1.0 max Si, 1.0 max Mn, 12.0-14.0 Cr, 1.0-1.3 Mo, bal Fe.
Martensitic stainless steel, for steam turbine blades.
W.-Nr. 1.4119.

CROMINO 17.
M-1315; 0.10-0.17 C, 1.0 max Si, 1.5 max Mn, 15.5-17.5 Cr, 0.20-0.30 Mo, 0.15-0.35 S, bal Fe.
Ferritic type stainless steel; free cutting for machined parts. W.-Nr. 1.4104.
AISI 430 F.

CROMINO 18.
M-1315; 0.08 C, 0.4 Si, 17 Cr, bal Fe.
Annealed: 80,000 TS; 50,000 YS; 25 El; 50 RA; 150 Brin.
Cold drawn: 130,000 TS; 120,000 YS; 2 El; 185 Brin.
For oil refinery and food processing equipment, sinks; Type 430; stainless, ferritic.

CROMINO 100.
M-1315; 0.85-0.95 C, 1.0 max Si, 1.0 max Mn, 15.5-17.5 Cr, 0.40-0.60 Mo, 0.20-0.30 V, 1.2-1.8 Co, bal Fe.
Martensitic stainless; for cutlery.
W.-Nr. 1.4535.

CROMINO X.
M-1315; 0.08 max C, 1.0 max Si, 1.0 max Mn, 13.0-15.0 Cr, bal Fe.
Ferritic type stainless steel; for structural parts and tableware. W.Nr. 1.4001.
Similar to AISI 405.

CROMO.
M-111; 0.4 C, 0.65 Mn, 1.1 Cr, 0.25 Mo, bal Fe.
For gears, shafts, bolts; oil hardening.

CROMO-CO.
M-24; 0.27 C, 0.6 Mn, 1.25 Si, 11.0 Cr, 0.75 Ni, 10.0 Co, 1.2 Mo, 0.45 W, 1.0 V, 0.10 N, bal Fe.
Martensitic stainless steel.

CROMODI.
M-1432; C, alloy, bal Fe.
Heat treated: 280 Brin.
For brake dies; preheat treated, non deforming.

CROMO HIGH V.
M-24, M-341; 0.40 C, 5.0 Cr, 1.2 Mo, 1.5 V, bal Fe.
Hot work tool steel; for die casting dies, shear blades; oil hardened; AISI H13.

CROMOL C-9.
M-451; C, Cr, Mo, V, bal Fe.
Air hardened: 115,000-181,500 TS; 90,000-127,000 YS; 20-14 El; 63-46 RA; 217-375 Brin.
For rams, sow blocks, dies; alloy steel casting.

CRO-MOL C-10.
M-451; C, Cr, V, bal Fe.
For dies; rams, sow blocks; casting.

CRO-MO-LOY.
M-435; 1.0 C, 1.0 Mn, 5.0 Cr, 0.25 V, 1.0 Mo, 1 Co, bal Fe.
For tools, dies, shear blades; air hardening, non-deforming.

CROMO N.
M-24; 0.25 C, 1.0 Mn, 1.0 Si, 1.0 Ni, 11.0 Cr, 1.0 Mo, 0.5 W, 0.95 W, 0.10 N, bal Fe.
For hot work applications, extrusion mandrels, forging die inserts, pierce punches, die casting dies.

CROMO N.
M-341; 0.23 C, 0.6 Mn, 1.25 Si, 0.75 Ni, 10.0 Cr, 1.2 Mo, 1.0 V, 0. W, 0.10 N, bal Fe.
For die casting dies, extrusion mandrels, forging and extrusion tooling.
Resists heat checking and cracking.
Air hardening hot work tool steel.

CROMONITE.
M-451; C, alloy, bal Fe.
For chill rolls; wear resistant.

CROMOTEX NO. 2.
M-1009; 0.43-0.48 C, 0.8 Mn, 0.8-1.1 Cr, 0.2 Mo, 1.8 Ni, bal Fe.
For arbors, axles, gears, piston rods; shock resistant.

CROMOTUNG.
M-289; C, alloy, bal Fe.
For taps, drills, reamers, dies, punches; oil hardened, tough.

CROMO V.
M-24; 0.35 C, 5 Cr, 0.4 V, 1.5 Mo, bal Fe.
Heat treated: 220,000 TS; 186,000 YS; 13 El; 40 RA.
At 1000°F: 145,000 TS; 105,000 YS; 19 El; 65 RA.
For extrusion dies, gripper and header dies, aluminum and magnesium forging dies.
Type H11 hot work steel.
Resists heat checking.

CROMOVAN.
M-57; 1.4-1.7 C, 12-14 Cr, 0.5-1.0 Mo, 1-1.5 V, bal Fe.
Annealed: 110,000 TS; 210 Brin.
Hardened: 278,000 TS; 214,000 YS; 1 El; 0.2 RA; 570 Brin.
For dies for trimming, blanking extrusion, swaging, pressing, thread rolling, etc, taps, punchers, reamers, broaches, shear blades; heat and wear resistant "Chromovan Triple Die Steel," non-deforming.

CROMOVAN F.M.
M-57; 1.55 C, 12 Cr, 1 V, 1 Mo, 0.12 S, bal Fe.
For blanking and drawing dies, punchers, gages; airhardened, non-deforming.

CROMO W.
M-24; 0.35 C, 1.05 Si, 1.55 W, 1.65 Mo, 5.15 Cr, bal Fe.
Heat treated: 216,000 TS; 185,000 YS; 14 El; 52 RA; Rock C 56.
For trimmer and hot forging dies, die casting dies, hot sheer blades, punches.
Type H12 hot work steel, oil hardening.

CROMO W-55.
M-24; 0.55 C, 0.85-1.10 Si, 1.0-1.5 W, 4.8-5.1 Cr, 1.0-1.5 Mo, bal Fe.
Annealed: 103,000 TS; 65,000 YS; 24 El; C 18 Rock.
Heat treated: 290,000 TS; 235,000 YS; 9 El; 22 RA; C 54 Rock.
For blanking and beading dies, cold forming dies, pneumatic tools.
Type A8 hot work steel, air harden.
Tough and shock resistant.

CRONI 11.
M-1766; 0.14 C, 0.80 Mn, 0.25 Si, 1.0 Ni, 1.0 Cr, bal Fe.
Ni-Cr carburizing steel.
AFNOR 16NC6; DIN 15 Cr Ni 6; BS EN-352.

CRONI-13.
M-1766; 0.13 C, 0.50 Mn, 0.25 Si, 2.9 Ni, 0.70 Cr, bal Fe.
Ni-Cr carburizing steel; deep hardening.
AFNOR 14NC11; UNI 15NC11; BS EN-36A.

CRONI 19.
M-1766; 0.19 C, 0.90 Mn, 0.25 Si, 1.1 Ni, 0.95 Cr, bal Fe.
Ni-Cr carburizing steel.
AFNOR 20 NC6; BS EN 352; WNI 19CN5.

CRONIDUR 4967.
M-72; 0.10 C, 20 Cr, 10 Ni, 15 W, 3.0 Max Fe, bal Co.
For gas turbine and after burners.

CRONIFER II.
M-297; 65 Ni, 15 Cr, bal Fe.
Annealed: 92,400-106,600 TS; 30-25 El; 55-45 RA; 130 Brin.
For heat resistant parts; maximum operating temp. 1075°C.

CRONIFER II EXTRA.
M-297; 65 Ni, 15 Cr, bal Fe.
Annealed: 92,400-106,600 TS; 30-25 El; 55-45 RA; 130 Brin.
For heat resistant parts; maximum operating temp. 1125°C.

CRONIFER III.
M-297; 30 Ni, 20 Cr, bal Fe.
Annealed: 92,400-99,500 TS; 32-30 El; 50 RA; 150 Brin.
For heat resistant parts; maximum operating temperature 1100°C.

CRONIFER III EXTRA.
M-297; 30 Ni, 20 Cr, bal Fe.
Annealed: 92,400-99,500 TS; 32-30 El; 50 RA; 150 Brin.
For heat resistant parts; max operating temp. 1150°C.

CRONIFER IV.
M-297; C, 20 Ni, 24 Cr, bal Fe.
Rolled: 85,000-108,000 TS; 43,000 YS.
For furnace parts, salt pots, heat treating equipment; heat resistant to 1050°C. in continuous operation.

CRONIFER IV-EXTRA.
M-297; C, 20 Ni, 24 Cr, bal Fe.
Rolled: 85,000-108,000 TS; 43,000 YS.
For furnace parts, salt pots, heat treating equipment; heat resistant to 1100°C. in continuous operation.

CRONIFER 1212.
M-297; 0.10 max C, 11.5-13.5 Cr, 12.0-14.0 Ni, bal Fe.
For valves, pumps; corrosion resistant.

CRONIFER 1613NB.
M-297; 0.10 max C, 15-17 Cr, 12-14 Ni, Nb = 10 x C + 0.4, bal Fe.
For chemical plant equipment, stainless.

CRONIFER 1613NBN.
M-297; 0.10 max C, 15.5-17.5 Cr, 12.5-14.5 Ni, 1.1-1.5 Mo, 0.6-0.8 V, 0.1 N, Nb = 10 x C + 0.4, bal Fe.
For chemical plant equipment; stainless.

CRONIFER 1616NB.
M-297; 0.10 max C, 15.5-17.5 Cr, 15.5-17.5 Ni, 1.6-2.0 Mo, Nb = 10 x C + 0.4, bal Fe.
For chemical plant equipment; stainless.

CRONIFER 1704.
M-297; 0.15 max C, 16-18 Cr, 3.5-5.5 Ni, 5.5-7.5 Mn, 0.17-0.25 N, bal Fe.
For chemical plant equipment; stainless, austenitic.

CRONIFER 1707.
M-297; 0.15 max C, 16-18 Cr, 7-8 Ni, bal Fe.
Annealed: 80,000 TS; 35,000 YS; 55 El; 75 RA; 150 Brin.
For chemical plant equipment; Type 301; stainless, austenitic.

CRONIFER 1805.
M-297; 0.15 max C, 17-19 Cr, 4-6 Ni, 7.5-10.0 Mn, 0.17-0.25 N, bal Fe.
For chemical plant equipment; stainless, austenitic.

CRONIFER 1808.
M-297; 0.15 max C, 0.4 Si, 18 Cr, 8 Ni, bal Fe.
Quenched: 78,100-106,000 TS; 50 El; 130-180 Brin.
For textile and food industries; stainless, austenitic.

CRONIFER 1809.
M-297; 0.07 max C, 0.4 Si, 18 Cr, 9 Ni, bal Fe.
Quenched: 78,100-99,500 TS; 50 El; 135-180 Brin.
For textile and food equipment; stainless, austenitic.

CRONIFER 1809 NB.
M-297; 0.12 max C, 18 Cr, 9.5 Ni, Cb = 8 x C, bal Fe.
Annealed: 90,000 TS; 45,000 YS; 56 El; 65 RA; 160 Brin.
Cold drawn: 100,000 TS; 65,000 YS; 40 El; 60 RA; 205 Brin.
For welded chemical plant equipment, tanks, mixers; Type 347; stainless, austenitic.

CRONIFER 1809NC.
M-297; 0.03 C, 17-19 Cr, 9-11 Ni, bal Fe.
Annealed: 85,000 TS; 35,000 YS; 65 El; 75 RA; 140 Brin.
For chemical plant equipment; Type 304L; stainless, austenitic.

CRONIFER 1809 TI.
M-297; 0.12 max C, 18 Cr, 9.5 Ni, Ti = 4 x C, bal Fe.
Annealed: 85,000 TS; 35,000 YS; 55 El; 65 RA; 150 Brin.
Cold drawn: 95,000 TS; 60,000 YS; 40 El; 60 RA; 185 Brin.
For welded chemical plant equipment, tanks, filters; Type 321; stainless, austenitic.

CRONIFER 1810.
M-297; 0.07 max C, 0.4 Si, 18 Cr, 2 Mo, 10 Ni, bal Fe.
Quenched: 78,100-99,500 TS; 45 El; 140-180 Brin.
For chemical equipment; stainless, austenitic.

CRONIFER 1810NB.
M-297; 0.10 max C, 16.5-18.5 Cr, 10.5-12.5 Ni, 2.0.-2.5 Mo, Cb = 8 x C, bal Fe.
Annealed: 85,000 TS; 35,000 YS; 50 El; 65 RA; 160 Brin.
For welded chemical plant equipment; Type 316 Cb; stainless, austenitic.

CRONIFER 1810 TI.
M-297; 0.12 max C, 18 Cr, 2 Mo, 10.5 Ni, = 4 x C, bal Fe.
Annealed: 85,000 TS; 40,000 YS; 50 El; 65 RA; 170 Brin.
For welded acid resistant chemical plant equipment; Type 316 Ti; stainless, austenitic.

CRONIFER 1812.
M-297; 0.07 max C, 16.5-18.5 Cr, 11.5-13.5 Ni, 2.5-3.0 Mo, bal Fe.
Annealed: 85,000 TS; 35,000 YS; 50 El; 65 RA; 160 Brin.
For acid resistant equipment; Type 316; stainless, austenitic.

CRONIFER 1812NB.
M-297; 0.10 max C, 16.5-18.5 Cr, 11.5-13.5 Ni, 2.5-3.0 Mo, Nb = 8 x C, bal Fe.
Annealed: 85,000 TS; 35,000 YS; 50 El; 65 RA; 160 Brin.
For acid resistant equipment; Type 316 Cb; stainless, austenitic.

CRONIFER 1812TI.
M-297; 0.10 max C, 16.5-18.5 Cr, 11.5-13.5 Ni, 2.5-3.0 Mo, Ti = 5 x C, bal Fe.
Annealed; 85,000 TS; 35,000 YS; 50 El; 65 RA; 160 Brin.
For welded chemical plant equipment; Type 316 Ti; stainless, austenitic.

CRONIFER 1813.
M-297; 0.07 max C, 16.5-18.5 Cr, 11.5-13.5 Ni, bal Fe.
Annealed: 85,000 TS; 35,000 YS; 60 El; 70 RA; 150 Brin.
For chemical plant equipment; Type 304; stainless, austenitic.

CRONIFER 1818NB.
M-297; 0.07 max C, 16.5-18.5 Cr, 16.5-18.5 Ni, 2.0-2.5 Mo, 1.8-2.2 Cu Nb = 8 x C, bal Fe.
For acid resistant equipment; stainless, austenitic.

CRONIFER 1818TI.
M-297; 0.07 max C, 16.5-18.5 Cr, 16.5-18.5 Ni, 2.0-2.5 Mo, 1.8-2.2 Cu Ti = 5 x C, bal Fe.
Annealed: 85,000 TS; 35,000 YS; 50 El; 65 RA; 160 Brin.
For acid resistant equipment; Type 316 Ti; stainless, austenitic.

CRONIFER 1910 NB.
M-297; 0.12 max C, 18 Cr, 2 Mo, 10.5 Ni, Cb = 8 x C, bal Fe.
Annealed: 85,000 TS; 40,000 YS; 50 El; 65 RA; 170 Brin.
For welded acid resistant chemical plant equipment; Type 316 Cb; stainless, austenitic.

CRONIFER 2012.
M-297; 0.20 max C, 19-21 Cr, 11-13 Ni, 1.8 -2.3 Si, bal fe.
For high temperature applications; heat resistant.

CRONIFER 2504.
M-297; 0.15-0.25 C, 24-26 Cr, 3.5-4.5 Ni, 0.8-1.3 Si, bal Fe.
For furnace parts, salt pots; heat resistant.

CRONIFER 2520.
M-297; 0.20 max C, 24-26 Cr, 17-21 Ni, 1.8-2.3 Si, bal Fe.
For high temperature applications; heat resistant.

CRONIFER 2520NV.
M-297; 0.20 max C, 23-25 Cr, 19-21 Ni, 1.0-1.5 Si, bal Fe.
For furnace parts, heat treat boxes; austenitic, corrosion and heat resistant.

CRONIFER 2525TI.
M-297; 0.06 max C, 24-26 Ni, 2.0-2.5 Mo, Ti = 5 x C, bal Fe.
For furnace parts, heat resistant equipment; corrosion and heat resistant.

CRONIRO.
M-1493; 72 Au, 22 Ni, 6 Cr.
M.P. 1787-1832°F.
For high temperature brazing.

CRONIT.
M-Eng; 60 Ni, 40 Cr.
For resistance alloys.

CRONITE.
M-37; 63.5-67.0 Ni, 13.5-16.0 Cr, 1 Mn, 0.4 Si, 0.8 Al, bal Fe.
At 20°C: 67,000-78,000 TS; 0.5-1.5 El; 200-250 Brin.
At 1000°C: 25,000 TS.
For carburizing boxes, grids, fire doors, furnace parts; maximum operating temperature 1000°C, heat resistant.

CRONITE 428.
M-37; 12 Ni, 25 Cr, bal Fe.
For high temperature uses; corrosion and heat resistant.

CRONITE ORD.
M-37; 55 Ni, 18 Cr, bal Fe.
For high temperature uses; corrosion and heat resistant .

CRONITE SR.
M-37; 35 Ni, 20 Cr, bal Fe.
For high temperature uses; corrosion and heat resistant.

CRONITE W.X.2.
M-37; Ni, Cr, Fe, W.
70,000-120,000 TS.
For drastic temperature conditions.

CRONIX.
M-297; 80 Ni, 20 Cr.
Annealed: 92,400-106,600 TS; 35-25 El; 55-50 RA; 145 Brin.
For heating elements; maximum operating temp. 1150°C.

CRONIX EXTRA.
M-297; 80 Ni, 20 Cr.
Annealed; 92,400-106,600 TS; TS; 35-25 El; 55-50 RA; 145 Brin.
For heating elements; maximum operating temp. 1200°C.

CROTERITE IV ETC see DELTA COROTERITE IV ETC.

CROUSE FUSIBLE ALLOY.
M-Eng.; 45 Bi, 25 Sn, 25 Pb, 5 Cd.
For fusible alloy, fire extinguishers; M.P. 88°C.

CROVA.
M-1331; 1.15 C, 0.65 Cr, 0.1 V, bal Fe.
For cutters, drills, bearings; water hardened, wear resistant.

CROVA 31.
M-1331; 0.31 C, 0.65 Cr, 0.1 V, bal Fe.
For gears, bolts, fasteners, crankshafts; oil hardened, tough.

CROVA 50.
M-1331; 0.5 C, 0.95 Mn, 1.05 Cr, 0.1 V, bal Fe.
For gears, bolts, springs, crankshafts; oil hardened, shock resistant.

CROVA 115.
M-1331; 1.2 C, 0.65 Cr, 0.1 V, bal Fe.
For bearings, cutters, forming dies; water hardened, wear resistant.

CROVAC.
M-1631; Cr, Co, Fe.
Deformable permanent magnet materials.

CROVAN.
M-435; 0.35 C, 1 Si, 5 Cr, 1.4 Mo, 0.9 V, bal Fe.
At 1000°F: 186,200 TS; 149,100 YS; 18.6 El; 48.2 RA; 477 Brin.
For die casting dies, extrusion tools, plastic molds; resists thermal shock, non-deforming.

CROVANI.
M-1318; 0.58 C, 1.05 Cr, 0.1 V, bal Fe.
For springs, gears, bolts, studs; oil hardened, shock resistant.

CROWN.
M-73; 0.5 C, 0.9 Cr, 0.2 V, bal Fe.
For tools shafting; high fatigue resistance.
AISI L2.

CROWN.
M-261; 0.44 C, 1.3 Cr, 0.15 V, 2.3 W, bal Fe.
For hot or cold trimming dies, punches, chipping chisels, pneumatic tools, mandrels, hot heading dies.
Oil hardening, shock resistant.

CROWN.
M-657; 88 Cu, 10 Sn, 2 Zn, 87 Cu, 9 Sn, 4 Zn; 87 Cu, 8 Sn, 5 Zn.
Cast: 32,000 TS; 16,000 YS; 21 El.
For bearings, pumps, engine casting; heavy duty.

CROWN "W".
M-387; 0.8 C, 2.5 W, bal Fe.
For tools, dies; oil hardened.

CR. T.
M-1116; 0.50 C, 0.7 Cr, 2.0 W, 0.15 V, 0.35 Mo, 0.35 Mn, 0.5 Si, bal Fe.
Oil or water hardening tool steel; for pneumatic tools, riveting hammers, chisels.
AFNOR: 50 WCS 20.03; AISI S1.

CRUCIBLE 25-25.
M-38; 0.05 C, 25.0 Cr, 25.0 Ni, bal Fe.
Ann: 90,000 TS; 40,000 YS; 40 El.
For heating elements, thermal reactors.

CRUCIBLE 26-1.
M-38; 0.04 max C, 0.75 max Mn, 0.75 max Si, 0.30 max Ni, 25.0 Cr, 0.75 Mo, 0.20 max Cb, 0.04 max N, 0.20-1.0 Ti, bal Fe.

CRUCIBLE 35% NICKEL-IRON.
M-38; 35 Ni, bal Fe.
Annealed: 70,000 TS; 24,000 YS; 36 El; 68 RS; 143 Brin.
Cold drawn: 90,000 TS; 70,000 YS; 20 El; 60 RA; 185 Brin.
For instruments, geodetic parts, thermostats; controlled thermal expansion low.

CRUCIBLE 42% NICKEL-IRON.
M-38; 42 Ni, 0.08 max C, bal Fe.
Annealed: 80,000 TS; 30,000 YS; 30 El.
For metal to glass seals, thermostats, temperature controls; low coefficient of expansion (controlled).

CRUCIBLE 52 CB BEARING STEEL.
M-38; 0.75-0.90 C, 0.3-0.45 Mn, 0.6-0.9 Si, 0.8-1.1 Cr, 0.25 Ni, 0.50-0.65 Mo, bal Fe.
Oil hardenable to 62-66 Rc.
For bearings, bearing races, hardened shafts.

CRUCIBLE 120 ZA.
M-38; 12 Zr, 4.5 Al, bal Ti.
For high temperature applications; heat and creep resistant.

CRUCIBLE 218 see HALCOMB 218.

CRUCIBLE 223.
M-38; 0.08 C, 12.0 Mn, 15.5 Si, 0.5 max Ni, 0.5 max Mo, 1.0 Cu, 0.25 N, bal Fe.
Austenitic manganese steel.
For earth moving equipment.

CRUCIBLE 301.
M-38; 0.2 max C, 7 Ni, 17 Cr, bal Fe.
Annealed: 110,000 TS; 40,000 YS; 60 El; 165 Brin.
Rolled: 185,000 TS; 140,000 YS; 8 El; 410 Brin.
For aircraft structural members, trailer bodies, diaphragms, household utensils; Type 301; stainless, austenitic.

CRUCIBLE 302.
M-38; 0.2 max C, 18 Cr, 9 Ni, bal Fe.
Annealed: 90,000 TS; 40,000 YS; 60 El; 70 RA; 165 Brin.
For aircraft structural members, chemical plant equipment; Type 302; stainless, austenitic.

CRUCIBLE 303.
M-38; 0.20 max C, 0.30 S, 9 Ni, 18 Cr, bal Fe.
Annealed: 85,000 TS; 40,000 YS; 45 El; 60 RA; 170 Brin.
For screw machine products, bolts, screws, fasteners; Type 303; free-cutting, stainless.

CRUCIBLE 303 PLUS.
M-38; 0.15 max C, 2.0 max Mn, 1.0 max Si, 0.20 max P, 0.15 min S, 17-19 Cr. 8-10 Ni, 0.6 max Mo, bal Fe.
Ann: 90,000 TS; 35,000 YS; 50 El; 170 Brin.
Good machinability, austenitic, corrosion resistant.
For stainless hardware to be made on automatic screw machines, fasteners, bolts, studs. AMS 5640.

CRUCIBLE 303 PLUS X.
M-38; 0.15 max C, 2.5-4.5 Mn, 1.0 max Si, 0.20 max P, 0.15 min S, 17-19 Cr, 7.0-10.0 Ni, 0.6 max Mo, bal Fe.
Ann: 90,000 TS; 35,000 YS; 50 El; 170 Brin.
Austenitic, good machinability, corrosion resistant.
For stainless screw machine parts, studs, bolts, shafts, fasteners; chemical plants. AMS 5640.

CRUCIBLE 304.
M-38; 0.08 max C, 9.5 Ni, 19 Cr, bal Fe.
Annealed: 85,000 TS; 35,000 YS; 60 El; 70 RA; 160 Brin.
For architectural molding and trim; Type 304; stainless, austenitic.

CRUCIBLE 304L.
M-38; 0.03 max C, 9.5 Ni, 19 Cr, bal Fe.
Annealed: 85,000 TS; 35,000 YS; 60 El; 70 RA; 160 Brin.
For welded structures, chemical plant equipment; Type 304L; austenitic, stainless.

CRUCIBLE 304 PLUS.
M-38; 0.08 C, 2.0 max Mn, 1.0 max Si, 18-10 Cr, 8-12 Ni, bal Fe.
Ann: 85,000 TS; 35,000 YS; 60 El; 170 Brin.
Austenitic, non-magnetic, corrosion resistant; ductile.
For shafts, bar and fountain equipment, food and dairy equipment, valves, marine equipment.
AMS 5639.

CRUCIBLE 305.
M-38; 0.12 max C, 11.5 Ni, 18 Cr, bal Fe.
Annealed: 85,000 TS; 35,000 YS; 55 El; 70 RA; 180 Brin.
For spun parts, cold heading, special drawing; Type 305; austenitic, stainless.

CRUCIBLE 307.
M-38; 0.08 max C, 4 Mn, 9.5 Ni, 20.5 Cr, bal Fe.
For welding electrodes; austenitic, stainless.

CRUCIBLE 308.
M-38; 0.08 max C, 11 Ni, 20 Cr, bal Fe.
For welding electrodes; austenitic, stainless.

CRUCIBLE 309.
M-38; 0.20 max C, 13.5 Ni, 23 Cr, bal Fe.
Annealed: 90,000 TS; 40,000 YS; 50 El; 65 RA; 160 Brin.
For furnace parts, heat treat boxes, boiler baffles; Type 309; austenitic corrosion and heat resistant.

CRUCIBLE 309S.
M-38; 0.08 max C, 13.5 Ni, 23 Cr, bal Fe.
Annealed: 90,000 TS; 40,000 YS; 50 El; 65 RA; 150 Brin.
For furnace parts, heat treat boxes, baffles, tube supports; Type 309S; austenitic, corrosion and heat resistant.

CRUCIBLE 310.
M-38; 0.25 max C, 20.5 Ni, 25 Cr, bal Fe.
Annealed: 100,000 TS; 45,000 YS; 50 El; 65 RA; 180 Brin.
For furnace parts and equipment, heat treat boxes, pots; Type 370; austenitic, heat resistant.

CRUCIBLE 310CB.
M-38; 0.08 max C, 20.5 Ni, 25 Cr, 1 Cb, bal Fe.
Annealed: 100,000 TS; 45,000 YS; 50 El; 65 RA; 185 Brin.
For welded parts, valves, pumps, furnace parts; austenitic, stainless, heat resistant.

CRUCIBLE 310S.
M-38; 0.08 max C, 20.5 Ni, 25 Cr, bal Fe.
Annealed: 100,000 TS; 45,000 YS; 50 El; 65 RA; 180 Brin.
For furnace parts and equipment, heat treat boxes; Type 310S; austenitic heat resistant.

CRUCIBLE 310S.
M-38; 0.08 max C, 20.5 Ni, 25 Cr, bal Fe.
Annealed: 100,000 TS; 45,000 YS; 50 El; 65 RA; 180 Brin.
For furnace parts and equipment heat treat boxes; Type 310S; austenitic, heat resistant.

CRUCIBLE 316.
M-38; 0.10 max C, 12 Ni, 17 Cr, 2.5 Mo, bal Fe.
Annealed: 80,000 TS; 30,000 YS; 60 El; 80 RA; 140 Brin.
Cold drawn: 150,000 TS; 135,000 YS; 6 El; 300 Brin.
For chemical plant equipment, agitators, digesters, kettles; Type 316; stainless, austenitic.

CRUCIBLE 316L.
M-38; 0.03 max C, 12 Ni, 17 Cr, 2.5 Mo, bal Fe.
Annealed: 80,000 TS; 30,000 YS; 60 El; 80 RA; 140 Brin.
Cold drawn: 150,000 TS; 135,000 YS; 6 El; 300 Brin.
For chemical plant equipment, agitators, kettles, digesters; Type 316L; stainless, austenitic.

CRUCIBLE 316 PLUS.
M-38; 0.08 max C, 2.0 max Mn, 1.0 max Si, 16-18 Cr, 10-14 Ni, 2.0-3.0 Mo, bal Fe.
Ann: 85,000 TS; 35,000 YS; 55 El; 170 Brin.
Austenitic, very good corrosion resistance, nonmagnetic, weldable.
For parts for food and beverage equipment, pulp handling, chemical plants.
AMS 5648 AISI Type 316 stainless.

CRUCIBLE 319L.
M-38; 0.025 C, 18 Cr, 13.35 Ni, 2.5 Mo, bal Fe.
Ann: 84,000 TS; 39,000 YS; 52 El.
For corrosion resistant parts in process industries.

CRUCIBLE 321.
M-38; 0.08 max C, 9.5 Ni, 18 Cr, 0.5 Ti, bal Fe.
Annealed: 85,000 TS; 35,000 YS; 55 El; 65 RA; 165 Brin.
Cold drawn: 95,000 TS; 60,000 YS; 40 El; 60 RA; 185 Brin.
For chemical plant equipment, welded construction; stainless; Type 321; austenitic, stabilized.

CRUCIBLE 347.
M-38; 0.08 max C, 10.5 Ni, 18 Cr, 1 Cb, bal Fe.
Annealed: 90,000 TS; 40,000 YS; 50 El; 70 RA; 180 Brin.
For welded structures, chemical plant equipment; Type 347; stainless, stabilized.

CRUCIBLE 347-F-SE.
M-38; 0.08 max C, 17-19 Cr, 9-13 Ni, Cb = 10 X C, P, Se, bal Fe.
For chemical plant equipment; free-cutting, stainless, austenitic.

CRUCIBLE 403.
M-38; 0.15 max C, 12 Cr, bal Fe.
Annealed: 75,000 TS; 40,000 YS; 35 El; 70 RA; 140 Brin.
Cold drawn: 100,000 TS; 85,000 YS; 60 El; 205 Brin.
For flat springs, hardware, cutlery, tableware; Type 403; corrosion resistant.

CRUCIBLE 410.
M-38; 0.15 max C, 12 Cr, bal Fe.
Annealed: 75,000 TS; 40,000 YS; 35 El; 70 RA; 140 Brin.
Cold drawn: 100,000 TS; 85,000 YS; 60 El; 205 Brin.
For flat springs, hardware, cutlery, tableware; Type 410; corrosion resistant. AMS 5613.

CRUCIBLE 414.
M-38; 0.15 max C, 2 Ni, 12 Cr, bal Fe.
Annealed: 115,000-120,000 TS; 90,000-105,000 YS; 15-20 El; 60 RA; 235-255 Brin.
For corrosion resistant parts, hardenable, martensitic, stainless.

CRUCIBLE 416.
M-38; 0.15 max C, 0.30 S, 13 Cr, bal Fe.
Annealed: 75,000 TS; 40,000 YS; 30 El; 60 RA; 155 Brin.
Heat treated: 110,000 TS; 85,000 YS; 18 El; 55 RA; 230 Brin.
For screw machine products, gears, shafts; Type 416; free-cutting, stainless.

CRUCIBLE 416 PLUS.
M-38; 0.15 max C, 1.25 max Mn, 1.0 max Si, 0.06 max P, 0.15 min S, 12-14 Cr, 0.60 max Mo, bal Fe.
Ann: 75,000 TS; 40,000 YS; 30 El; 155 Brin.
Free machining, magnetic, hardenable, corrosion resistant.
For corrosion resistant hardware, screw machine parts, bolts, nuts, shafts.
AISI 416; AMS 5610.

CRUCIBLE 416 PLUS X.
M-38; 0.15 max C, 1.5-2.5 Mn, 13 Cr, 1.0 max Si, 0.06 max P, 0.15 min S, 0.6 max Mo, bal Fe.
Ann: 75,000 TS; 40,000 YS; 30 El; 155 Brin.
Free machining, magnetic, hardenable, corrosion resistant.
Parts made on automatic screw machines, bolts, studs, nuts, shafts.
AISI 416.AMS 5610.

CRUCIBLE 420.
M-38; 0.35 C, 13 Cr, bal Fe.
Annealed: 95,000 TS; 50,000 YS; 25 El; 55 RA; 190 Brin.
Cold drawn: 105,000 TS; 85,000 YS; 17 El; 50 RA; 215 Brin.
For cutlery, dental and surgical instruments, valve trim; Type 420; stainless, hardenable.

CRUCIBLE 420 DENSIFIED.
M-38; 0.3-0.4 C, 1.0 max Mn, 1.0 max Si, 12-14 Cr, bal Fe.
Air hardenable to 250,000 TS.
For plastic molds, glass molds, injection molds.

CRUCIBLE 420F.
M-38; 0.35 C, 0.30 S, 13 Cr, bal Fe.
Annealed: 95,000 TS; 50,000 YS; 25 El; 55 RA; 190 Brin.
Cold drawn: 105,000 TS; 85,000 YS; 17 El; 50 RA; 215 Brin.
For cutlery, dental and surgical instruments, valve trim; Type 420F; free-cutting, stainless.

CRUCIBLE 420-F-SE.
M-38; 0.15 min C, 12-14 Cr, P, Se, bal Fe.
For valves, cutlery, knives, free-cutting, corrosion resistant.

CRUCIBLE 420S.
M-38; 0.20 C, 13 Cr, bal Fe.
Annealed: 95,000 TS; 50,000 YS; 25 El; 55 RA; 196 Brin.
For corrosion resistant parts, cutlery, surgical instruments; hardenable, free-cutting, stainless.

CRUCIBLE 422.
M-38; 0.23 C, 0.9 Mn, 0.14 Si, 0.7 Ni, 13.2 Cr, 0.25 V, 1 W, 1 Mo, bal Fe.
At 80°F: 149,000 TS; 125,000 YS; 18 El; 52 RA.
At 1000°F: 96,000 TS; 82,000 YS; 25 El; 67 RA.
For compressor blades, valves, furnace parts; stainless, ferritic.

CRUCIBLE 430.
M-38; 0.12 max C, 16 Cr, bal Fe.
Annealed: 70,000 TS; 40,000 YS; 30 El; 55 RA; 140 Brin.
Cold drawn: 130,000 TS; 120,000 YS; 2 El.
For kitchen sinks, chemical and dairy equipment, bolts; Type 430; stainless, ferritic.

CRUCIBLE 430F.
M-38; 0.12 max C, 0.30 S, 16 Cr, bal Fe.
Annealed: 70,000 TS; 40,000 YS; 30 El; 55 RA; 140 Brin.
Cold drawn: 130,000 TS; 120,000 YS; 2 El.
For screw machine products, bolts, fasteners, gears; Type 430F; stainless, free-cutting.

CRUCIBLE 430-F-SE.
M-38; 0.12 C, 14-18 Cr, 0.15 min Se, bal Fe.
For stainless parts, shafts, valves; free-cutting, corrosion resistant.

CRUCIBLE 431.
M-38; 0.20 max C, 2 Ni, 16 Cr, bal Fe.
Annealed: 125,000 TS; 95,000 YS; 20 El; 55 RA; 260 Brin.
For corrosion resistant parts, valves, marine hardware; hardenable, martensitic, stainless.

CRUCIBLE 440A.
M-38; 0.65 C, 17 Cr, bal Fe.
Annealed: 95,000 TS; 55,000 YS; 20 El; 250 Brin.
Heat treated: 275,000 TS; 240,000 YS; 2 El; 555 Brin.
For cutlery, valves, dental and surgical instruments, bearings; Type 440A; stainless, hardenable.

CRUCIBLE 440B.
M-38; 0.85 C, 17 Cr, bal Fe.
Annealed: 107,000 TS; 62,000 YS; 18 El; 35 RA; 220 Brin.
Heat treated: 280,000 TS; 270,000 YS; 3 El; 15 RA; 555 Brin.
For cutlery, valves, dental and surgical instruments, bearings; Type 440B; stainless, hardenable.

CRUCIBLE 440BM.
M-38; 0.95 C, 18 Cr, 0.5 Mo, bal Fe.
Annealed: 107,000 TS; 62,000 YS; 18 El; 35 RA; 220 Brin.
For corrosion and wear resistant parts, cutlery, bearings; hardenable, martensitic, stainless.

CRUCIBLE 440BMF.
M-38; 0.95 C, 0.30 Si, 18 Cr, 0.5 Mo, bal Fe.
Annealed: 107,000 TS; 62,000 YS; 18 El; 35 RA; 220 Brin.
For corrosion and wear resistant parts; free-cutting, hardenable, stainless.

CRUCIBLE 440C.
M-38; 1.05 C, 17 Cr, bal Fe.
Annealed: 110,000 TS; 65,000 YS; 15 El; 30 RA; 225 Brin.
Heat treated: 290,000 TS; 275,000 YS; 2 El; 12 RA; 575 Brin.
For cutlery, valves, bearings, pivots; Type 440C; stainless, hardenable.

CRUCIBLE 440-F-SE.
M-38; 0.95-1.2 C, 16.5-18.5 Cr, P, Se, bal Fe.
For bearings, valves, pivots; stainless, free-cutting, hardenable.

CRUCIBLE 442.
M-38; 0.25 max C, 20.5 Cr, bal Fe.
For corrosion and heat resistant parts; non-hardenable, ferritic, stainless.

CRUCIBLE 446.
M-38; 0.20 max C, 1.5 max Mn, 23-27 Cr, bal Fe.
Ann: 80,000 TS; 50,000 YS; 25 El; 86 Rock B.
Magnetic, corrosion resistant, non-hardenable, weldable.
For heat treat equipment, radio tube parts, rotary driers and retorts, tank cars, combustion chambers.
Type 446.

CRUCIBLE 501.
M-38; 0.20 C, 4-6 Cr, bal Fe.
Annealed: 70,000 TS; 30,000 YS; 28 El; 65 RA; 160 Brin.
Heat treated: 175,000 TS; 135,000 YS; 15 El; 50 RA; 370 Brin.
For furnace parts, valve trim, oil refinery equipment; Type 501; creep resistant.

CRUCIBLE 502.
M-38; 0.10 max C, 4-6 Cr, 0.5 Mo, bal Fe.
Annealed: 75,000 TS; 35,000 YS; 26 El; 60 RA; 160 Brin.
For furnace parts, valve trim, oil refinery equipment; Type 502; creep resistant.

CRUCIBLE 3003.
M-38; 0.60 C, 4.5 Mn, 15 Ni, 12.5 Cr, 3 Al, bal Fe.
For corrosion resistant parts; age-hardenable, high strength, stainless.

Section I: Alloy Data / 375

CRUCIBLE 3311.
 M-38; 0.15 C, 23 Ni, 21.5 Cr, 3.25 Al, bal Fe.
 For corrosion resistant parts; age-hardenable, high strength, stainless.

CRUCIBLE 3329.
 M-38; 0.40 C, 6.25 Ni, 24 Cr, 4.25 Mo, bal Fe.
 For corrosion resistance and strength; hardenable by sigma phase, stainless.

CRUCIBLE A-40.
 M-38; 0.10 C, 0.07 N, 0.01 H, bal Ti.
 Rolled: 65,000 TS; 50,000 YS; 28 El; 50 RA.
 For high temperature applications; for maximum ductility and formability.

CRUCIBLE A 55.
 M-38; 0.20 max C, 0.015 max H, 0.08 max N, bal Ti.
 Annealed: 75,000 TS; 65,000 YS; 25 El; 50 RA.
 For non-structural aircraft parts; commercial titanium.

CRUCIBLE A 70.
 M-38; 0.05-0.15 C, 0.07 max N, bal Ti.
 Annealed: 90,000 TS; 80,000 YS; 20 El; 40 RA.
 For aircraft parts; commercial titanium.

CRUCIBLE A286 see **A-286.**

CRUCIBLE AFC 77.
 M-38; 0.15 C, 14.5 Cr, 13 Co, 5 Mo, 0.4 V, bal Fe.
 High strength, high temperature alloy.
 AMS 5748.

CRUCIBLE AM 350 see **AM 350.**

CRUICBLE AM 355 see **AM 355.**

CRUCIBLE BETA III.
 M-38; 11.5 Mo, 6 Zr, 4.5 Sn, bal Ti.
 Ht. tr.: 205,000 TS; 191,000 YS; 7 El; 29 RA.
 Annealed: 122,000 TS: 107,000 YS; 20 El; 32 RA.
 For aircraft and structural fasteners.
 Heat treatable beta alloy.

CRUCIBLE BRIGHT E3.
 M-38; 0.08 max C, 12.0 Cr, 0.5 max Ni, Ti = 5 X C, bal Fe.
 Corrosion resistant steel.
 For mufflers, sink hardware, plumbing hardware.

CRUCIBLE C-120 AV.
 M-38; 6 Al, 4 V, bal Ti.
 Annealed: 140,000 TS; 130,000 YS; 15 El; 40 RA.
 Heat treated: 190,000 TS; 180,000 YS; 10 El; 30 RA.
 For jet engine components, ordnance equipment; high temperature strength, creep resistant.

CRUCIBLE CG27.
 M-38; 0.05 C, 5.75 Mo, 13.0 Cr, 38.0 Ni, 1.6 Al, 2.5 Ti, 0.70 Cb, 0.01 B, bal Fe.
 Austenitic, forgeable high temperature alloy.
 Turbine wheel forgings for use up to 1500°F.

CRUCIBLE CPM REX 76 HIGH SPEED STEEL.
 M-38; 1.50 C, 3.75 Cr, 3.10 V, 10.0 W, 5.25 Mo, 9.0 Co, bal Fe.
 Super high speed steel, hardenable to 70 Rc.
 High abrasion resistance and superior red hardness.
 For difficult machining operations as lathe tools, milling cutters, end mills.

CRUCIBLE CSM 6 DENSIFIED.
 M-38; 0.70 C, 2.0 Mn, 0.30 Si, 1.0 Cr, 1.35 Mo, bal Fe.
 Air hardening tool steel; for injection molds, compression molds, transfer molds, lens molds.
 AISI A6.

CRUCIBLE CSM 420.
 M-38; 0.30-0.40 C, 1.0 max Mn, 1.0 max Si, 12.0-14.0 Cr, bal Fe.
 Martensitic stainless steel, hardenable to about 500 Brin.
 For plastic molds, glass molds, transfer molds.
 AISI Type 420.

CRUCIBLE EZ.
 M-38; 0.08 max C, 10.5-11.75 Cr, 0.50 max Ni, Ti = 5 X C min, bal Fe.
 Corrosion resisting steel; non-hardenable.

CRUCIBLE-F.
 M-1191; 0.4 C, 1.5 Ni, bal Fe.
 Cast: 80,000 TS; 45,000 YS; 180 Brin.
 For crankshafts, gears, machinery castings; oil hardening, shock resistant.

CRUCIBLE HNM.
 M-38; 0.30 C, 3.5 Mn, 0.5 Si, 9.5 Ni, 18.5 Cr, 0.25 P, bal Fe.
 At 80°F: 168,000 TS; 124,000 YS; 19.5 El; 31.5 RA; 380 Brin.
 At 1200°F: 89,000 TS; 80,000 YS; 19.0 El; 38.5 RA.
 At 1800°F: 49,000 TS; 46,000 YS; 4.0 El; 16.5 RA.
 For aircraft and jet engine components, structural members; age hardened, stainless.

CRUCIBLE HOLDER BLOCK STEEL.
 M-38; 0.50 C, 1.25 Mn, 0.08 S, 0.65 Cr, 0.18 Mo, bal Fe.
 Hardenable to 300,000 TS max.
 For backers for forging dies, brake dies, frames for plastic molds.

CRUCIBLE LAPELLOY.
 M-38; 0.30 C, 1.0 Mn, 12.0 Cr, 0.30 Ni, 2.75 Mo, 0.25 V, bal Fe.
 For high temperature bolts and struts.

CRUCIBLE M50 VAR.
M-38; 0.80 C, 4.0 Cr, 1.0 V, 4.25 Mo, bal Fe.
Hardenable to 64 RC.
Used largely for bearings operating at temperatures up to 800°F.

CRUCIBLE NITRIDING 135 MOD.
M-38; 0.35 C, 0.80 Mn, 0.10 S, 1.15 Cr, 0.2 Mo, 1.1 V, bal Fe.
For gears, shafts, camshafts, cams; nitriding steel.

CRUCIBLE RENE 41.
M-38; 0.12 C, 19 Cr, 11.3 Co, 10 Mo, 3 Ti, 1.5 Al, bal Ni.
At 70°F: 206,000 TS; 154,000 YS; 14 El.
At 1500°F: 126,000 TS; 118,000 YS; 14 El.
For jet engine components, after-burner parts; severely stressed high temperature applications.

CRUCIBLE S7.
M-38; 0.50 C, 3.25 Cr, 0.25 Si, 0.70 Mn, 1.50 Mo, bal Fe.
Shock resisting tool steel.
For punches, shears, chisels, rivet sets, plastic molds.

CRUCIBLE SA22.
M-38; 0.22 C, 1.15 Mn, 1 Si, 1.15 Ni, 0.5 Cr, 0.4 Mo, bal Fe.
Heat treated: 156,000 TS; 129,000 YS; 15 El; 51 RA; 340 Brin.
For structures, aircraft sheet components; good weldability, resists softening or tempering.

CRUCIBLE SCB.
M-38; 0.08 max C, 13.5 Ni, 23 Cr, 1 Cb, bal Fe.
For welded parts; stabilized austenitic stainless steel.

CRUCIBLE T303 PLUS X.
M-38; 0.15 max C, 2.5-4.5 Mn, 17-19 Cr, 7-10 Ni, 0.6 max Mo, bal Fe.
Austenitic stainless steel; modified.
AISI 302.

CRUCIBLE WF11.
M-38; 0.15 C, 19-21 Cr, 9-11 Ni, bal Fe.
Heat treated: 160,000 TS; 85,000 YS; 55 El.
For aircraft turbine blades, after burners; corrosion and heat resistant.

CRUCIN.
M-303; 44 Ni, 56 Cu.
For kitchen utensils, kettles; corrosion and wear resistant.

CRUSADER.
M-1702; 1.2 C, 4.1 Cr, 6.0 W, 3.0 V, 5.0 Mo, S, bal Fe.
M3 high speed steel tool bits; good wear resistance. (Electrite Crusader).

CRUSADER XL.
M-73; 1.2 C, 4.1 Cr, 6.0 W, 3.0 V, 5.0 Mo, alloy sulphides, bal Fe.
Hardened: C 65-68 Rock.
For form cutters, roll turning cutters, lathe and planer tools.
Good machineability rating.
Type M-3 Type 2 high-speed steel, high red-hardness and edge toughness.

CRUSCO STEEL.
M-153; 0.7 C, 18 W, 4 Cr, 1 V, 5 Co, bal Fe.
For piercing points, rolling mill plugs, dies, forming rolls; high abrasion resistance at high temperature.

CR-VICALLOY.
M-Japan; 12 (Cr + V), 52 Co, bal Fe.
For electrical and magnetic equipment.
Permanent magnet.

CRW.
M-275; 0.40 C, 0.35 Mn, 1.0 Si, 5.25 Cr, 0.25 Mo, 5.0 W, 0.2 V, bal Fe.
Hot work tool steel; forging dies.

CRYOGENIC TENELON.
M-604; 0.10 C, 15.1 Mn, 0.70 Si, 17.5 Cr, 5.5 Ni, 0.38 N, bal Fe.
High strength at low temperatures.

CRYOPERM.
M-1631.
High Ni content, soft magnetic alloy for cryogenic engineering.

CRYSTALLOY.
M-1664; 3 Si, bal Fe.
For transformers, lamination for motors and generators.
High magnetic permeability.

CR Z.
M-1763; 2.25 C, 12.0 Cr, bal Fe.
Cold work tool steel, punching and coining dies.
AISI D3.

CR Z S.
M-1763.
High chromium cold work tool steel.

CS 2700.
M-801.
Aluminum casting alloy; (0.6 max Cu).
Sand cast: 24,000 TS; 16,000 YS; 2 El; 57 Brin.
T6: 37,000 TS; 27,000 YS; 2 El; 80 Brin.
Improved corrosion resistance.

CS 2700 D.
M-801.
Die cast aluminum alloy CS 2700.
As cast: 41,000 TS; 8.5 El.

CSF-10.
M-1038; 9-11 Ce, 10.5-15.0 total rare earths, 36-40 Si, 0.50 max Al, 0.20 max Ca, bal Fe.
Ferroalloy for inoculant for gray iron and nodularizing supplement for ductile iron.

CSM-2.
M-38, M-1214; 0.30 C, 0.75 Mn, 0.5 Si, 1.65 Cr, 0.40 Mo, bal Fe.
For Zn and Sn die casting dies; oil or water hardened.

C.S.M. STEEL.
M-259; 0.15 C, 0.30-0.70 Cr, bal Fe.
Oil treated: 110,000 TS; 25 El; 68 RA; 219 Brin.
For springs, gears, axles, shafts; tough.

CSN 13-123.
M-Poland; 0.23 C, 1.1 Mn, 0.2 V, bal Fe.
For gears, shafts, hardware.
Cast case hardening and water hardening.

C.T. METAL.
M-815; Pb alloy.
For slush castings, statues; low M.P.

C.T.V. HOT WORK.
M-261; 0.40 C, 3.5 Cr, 2.4 W, 0.5 V, bal Fe.
For bolt and extrusion dies; hot work steel; oil hardening.

CU30.
M-1806.
Sintered carbide tool material.
For mining and quarrying.
National Coal Board Grade H (England).

CU35.
M-1806.
Sintered carbide tool material.
For mining and quarrying.
National Coal Board Grade M.

CU40.
M-1806.
Sintered carbide tool material.
For mining and quarrying.
National Coal Board Grade T.

CU50.
M-1806.
Sintered carbide tool material.
For mining and quarrying.
National Coal Board Grade XT.

CU-BE 50.
M-108; 0.5 Be, 2.5 Co, bal Cu.
Annealed: 50,000 TS; 25,000 YS; 30 El; B 30 Rock.
Ht tr.: 120,000 TS; 105,000 YS; 10 El; B 100 Rock.
For wire and strip springs, clips, connectors.
Age-hardenable. Corrosion resistant.
Non-magnetic and non-sparking.

CU-BE 250.
M-108; 2 Be, 0.25 Co, bal Cu.
Ht. Tr.: 200,000 TS; 180,000 YS; 1-3 El.
For wire and strip springs, clips, connectors.
Non-magnetic, non-sparking.
Age-hardenable. Corrosion resistant.

CU-BE 275.
M-108; 2.3-2.8 Be, 0.3-0.6 Co, bal Cu.
Beryllium copper alloy.

CUBEX.
M-118; 3.2 Si, bal Fe.
For electrical and magnetic equipment, motors, transformers; cores.
Doubly grain oriented.

CUBOND 14L.
M-1728.
One of several copper brazing pastes for use in brazing ferrous components in a reducing atmosphere furnace.

CUFENIUM.
M-US; 72-60 Cu, 22-20.5 Ni, bal Fe.
For tableware; nickel silver.

CUFENLOY-30.
M-319; 29.1 Ni, 0.5 Fe, 0.35 Mn, bal Cu.
At 75°F; 77,000 TS; 61,000 YS; 5, El.
At 700°F; 62,000 TS; 19,000 YS; 62 El.
At 1050°F; 43,000 TS; 16,000 YS; 48 El.
For heat exchanger tubes; high strength, ductility and stress corrosion resistance.

CUFERCO.
M-118; 96 Cu, 2 Fe, 2 Co.
For spot welding tips; hardenable electric conductivity = 70% Cu.

CUIVRELECT.
M-French; Cu alloy.
For welding rods.

CUIVRE POLI.
M-France; 70 Cu, 30 Zn.
For cartridges, shell cases, condenser tubes; maximum ductility.

CU-LEAD-ITE NO. 1.
M-481; 70 Cu, 20 Pb, 10 Sn.
Cast: 31,000 TS; 11 El; 12 RA; 68 Brin.
For nuts, parts for pneumatic tools subjected to severe pounding and abuse tough, hard casting.

CU-LEAD-ITE NO. 2.
M-481; 68 Cu, 25 Pb, 7 Sn.
Cast: 25,000 TS; 16 El; 15 RA; 48 Brin.
For severe service such as bearings in rolling mills, wire mills, diesel engines, ball mills and gyratory crushers; heavy duty.

CU-LEAD-ITE NO. 3.
  M-481; 65 Cu, 30 Pb, 5 Sn.
  Cast: 23,200 TS; 17 El; 18 RA; 43 Brin.
  For all round bearing service in high speed engines, locomotives, mills and railroads; heavy duty.

CU-LEAD-ITE NO. 4.
  M-481; 60 Cu, 38 Pb, 2 Sn.
  Cast: 18,500 TS; 14 El; 15 RA; 38 Brin.
  For bearings for all small machinery, deep well pumps, conveyors, connecting rods, automobiles and trucks; heavy duty.

CU-LEAD-ITE NO. 5.
  M-481; 50 Cu, 50 Pb.
  Cast: 6000 TS; 20 El; 21 RA; 14 Brin.
  For railroad locomotives, piston packers, metallic packing, bibs, etc; withstands superheated steam.

CU-LEAD-ITE NO. 6.
  M-481; 66 Cu, 34 Pb.
  Cast: 8910 TS; 15 El; 16 RA; 28 Brin.
  For small high speed machinery, pumps, electric drills, loose pulleys, small tools; heavy duty.

CULVER NCS.
  M-1195; 0.50 C, 2 W, 1.65 Cr, 0.25 V, bal Fe.
  Annealed: 70,000 TS; 30 El; 64 RA; 165 Brin.
  Heat treated: 245,000 TS; 5 El; 18 RA; 525 Brin.
  For punches, dies, chisels, pneumatic tools; shock resistant, oil hardened.

CUMBERLAND.
  M-213; 0.18-0.23 C, bal Fe.
  Rolled: 60,000-70,000 TS; 35,000-45,000 YS; 35-25 El; 60-50 RA.
  For shafts, gears; case hardening.

CUMLOY see **WEST NO. 9, 10, ETC.**

CUNIC.
  M-602; 45 Ni, 55 Cu.
  Annealed: 62,000 TS; 25 El.
  For rheostats, shunts, thermocouples; low temperature resistance.

CUNICO 1.
  M-1076; 50 Cu, 21 Ni, 29 Co.
  Rolled: 85,000 TS; 210 Brin.
  Annealed: 3400 Br; 710 Hc; 2000 Bo.
  For magnitic and electrical equipment. Permanent magnet, high coercive force.

CUNIFE.
  M-65; 60 Cu, 20 Ni, 20 Fe.
  Ductile permanent magnet alloy.
  Speedmeters, instruments, electronic equipment, and control systems.

CUNIFE 1.
  M-1076; 60 Cu, 20 Ni, 20 Fe.
  Permanent magnet, cold reduced.
  Peak energy product (BdHd) max X 10^6 1.4.
  Residual Induction, Br, (kilogauss) 5.5.
  Coercive force Hc (oersteds) 530.
  Most ductile grade of permanent magnet.

CUNIFER 10.
  M-297; 9-11 Ni, 0.5-1.0 Mn, 1.0-1.8 Fe, bal Cu.
  For chemical industry, power stations, heat exchangers, sea water desalination.

CUNIFER 30.
  M-297; 30-32 Ni, 0.5-1.5 Mn, 0.4-1.0 Fe, bal Cu.
  For chemical engineering, power stations, heat exchangers, sea-water desalination.

CUNILOY.
  M-US; 25 Cu, 3.8 Mn, 1 Pb, bal Ni.
  For pump rods, valve parts; corrosion resistant.

CUNIP.
  M-63; 1.1 Ni, 0.2-0.3 P, bal Cu.
  Wire or strip: 57,000-98,000 TS; 3-5 El.
  For electron tube components, cathode supports, tuning fingers, spring clips. Heat treatable.
  High strength and electrical conductivity.

CUNISIL 647.
  M-8; 0.6 Si, 1.9 Ni, bal Cu.
  Ht. Tr.: 100,000 TS; 85,000 YS; 15 El; Rock B 95.
  For electrical equipment, electrical hardware, machined mechanical fasteners.
  Precipitation hardening. Corrosion resistant. 35% electrical conductivity.

CUNISIL 837.
  M-8; 97.5 Cu, 1.9 Ni, 0.6 Si.
  Precipitation hardened: 90,000 TS; 70,000 YS; 8 El; 90 Rock B.
  For electrical apparatus; corrosion resistant.

CUPA.
  M-204; Cu plated Al.

CUPAL.
  M-127; Cu-clad Al sheet.
  For panels for railroad cars; corrosion resisting.

CUPAL.
  M-US; 80 Al, 20 Cu.
  For copper-clad aluminum; 10 Cu-80 Al-10 Cu.

CUPALLOY.
  M-USSR; 0.5 Cr, bal Cu.
  For electrical equipment, motors; high conductivity.

CUPLAT.
  M-1493; 40 Pt, 60 Cu.
  For brazing cathode structures.
  M.P. 1185-1216°C., corrosion resistant.
  Low vapor pressure.

CUPRALINOX 100.
M-1687; 10.0 Al, bal Cu.
Aluminium bronze; cast or wrought; hardness 140-180 Brin.
Afnor: UA10.

CUPRALINOX 115.
M-1687; 13.5 Al, 4.5 Fe, 3.5 Mn, bal Cu.
Aluminium bronze; cast or wrought; hardness: 290-350 Brin.
Forming dies for stainless steels.

CUPRALINOX C.
M-1687; 8.5 Al, bal Cu.
Aluminium bronze; wrought.
Ann: 90-140 Brin; Hard: 180-230 Brin.
Afnor: UA9.

CUPRALINOX CN.
M-1687; 6.0 Al, 2.0 Ni, bal Fe.
Aluminium bronze plates.
Afnor: UN6 N2.

CUPRALINOX NC2.
M-1687; 9 Al, 5 Ni, 2.5 Fe, 0.5 Mn, bal Cu.
Aluminium-Nickel bronze; wrought or cast.
Hardness 152 Brin min.
Afnor: UA9 NFe; ASTM B171; 628.

CUPRALINOX NC4.
M-1687; 10 Al, 5 Ni, 4 Fe, bal Cu.
Aluminium-Nickel bronze; wrought or cast.
Hardness: 170-210 Brin.
Afnor: UA10N; SAE 701C.

CUPRALINOX NCK.
M-1687.
Bronze, wrought, non-magnetic grade.

CUPRALINOX NCL.
M-1687; 12 Al, 6 Ni, 5 Fe, bal Cu.
Aluminium-Nickel-Iron bronze, wrought.
Hardness: 240-300 Brin.
DIN 17665-CuAl11Ni.

CUPRALINOX NCS.
M-1687; 11 Al, 5 Ni, 5 Fe, bal Cu.
Aluminium-Nickel-Iron bronze; wrought.
Hardness: 185-235 Brin.
Afnor: UA11 N.

CUPRALINOX NCVB.
M-1687.
Bronze, cast; for glass molds.

CUPRALINOX TM.
M-1687; 9 Al, 2 Ni, 2 Fe, 1.5 Mn, bal Cu.
Aluminium bronze; cast or wrought; hardness: 115-180 Brin.
Afnor: UA9 NFe.

CUPRALINOX VE.
M-1687; 8.5 Al, 5 Mn, bal Cu.
Aluminium-Manganese bronze; wrought 1/4 hard: 100-150 Brin; 1/2 hard: 150-195 Brin; hard: 190-230 Brin.

CUPRALINOX VN3.
M-1687; 8.5 Al, 2 Ni, 2 Fe, 6 Mn, bal Cu.
Aluminium-Manganese bronze; wrought.
Hardness: 152 Brin min.
Afnor: UA9 NFe.

CUPRALINOX VNC.
M-1687; 9.5 Al, 2.5 Fe, 7.0 Mn, bal Cu.
Aluminium-Manganese bronze; wrought.
Hardness: 170-210 Brin.
Afnor: UA10 M.

CUPRALIUM.
M-Ger.; 7-8 Cu, bal Al.
For light alloy parts; non-hardenable.

CUPRALIUM 12.
M-Ger.; 11.5-12.5 Cu, bal Al.
For light alloy castings; non-hardenable.

CUPRALUM.
M-1193; lead clad copper.
For chemical plant equipment; acid resistant.

CUPRALUMIN 8.
M-Cu, 0.5 Fe, 1.5 Si, bal Al.
For light alloy parts; similar to Alcoa 12.

CUPRALUMIN 12.
M-10 Cu, bal Al.
Cast: 21,000-26,000 TS; 20,000-24,000 YS; 0.5-1.5 El; 80-90 Brin.
For light alloy parts; non-hardenable.

CUPRANIUM.
M-Eng; Ni, Zn, bal Cu.
For corrosion resistant parts; corrosion resistant.

CUPRO-ALUMINUM.
M-US; 10 Al, Fe, bal Cu.
For worm wheels, gears; aluminum bronze.

CUPROCHROME.
M-61; 1 Cr, bal Cu.
Annealed: 40,000 TS; 27,000 YS; 25 El.
For electron tubes; high conductivity.

CUPRODIE see FINKL CUPRODIE.

CUPROMAGNESIUM.
M-US; 90 Cu, 10 Mg.
For cast iron inoculant; graphite spheroidizer.

CUPROMANGANESE.
M-US; 90 Cu, 10 Mn.
For staybolts, heat resisting parts; heat resistant.

CUPROMANGANESE TUBES.
M-US; 96 Cu, 4 Mn.
For tubes; corrosion resistant.

CUPRON.
M-61; 55 Cu, 45 Ni.
Annealed: 62,000 TS.
For rheostats, voltmeters, shunts, resistances; similar to "Advance," "Ideal," "Constantan" and "Ia-Ia."

CUPRON.
M-897; 45 Ni, 55 Cu.
Annealed: 85,000 TS; 50,000 YS; 50 El.
For strain gauges, rheostats; low temperature coefficient of resistance.

CUPRONAR 900.
M-677; deoxidized copper, 1.0 Sn.
For welding wire; AWS RCu, ECu.

CUPRONAR 910.
M-677; 2.8-4.0 Si, 0.5 Fe, 94 min Cu.
For welding wire; silicon bronze, AWS RCuSi-A, ECuSi.

CUPRONAR 920A.
M-677; 4-6 Sn, 0.1-0.35 P, 93.5 min Cu.
For welding wire; phosphorus bronze, AWS RCuSn-A, ECuSn-A.

CUPRONAR 920C.
M-677; 7-9 Sn, 0.05-0.35 P, bal Cu.
For welding wire; phosphorus bronze, AWS ECuSn-C.

CUPRONAR 950.
M-677; 6-9 Al, 0.1 Si, bal Cu.
For welding wires; aluminum bronze, AWS ECuAl-Al.

CUPRO-NICKEL 5% 704.
M-279; 93 Cu, 5.5 Ni, 1.5 Fe.
Available annealed or cold rolled.
For welded heat exchanger tubes, corrosion resistant.

CUPRO-NICKEL 7% 705.
M-279; 93 Cu, 7 Ni.
Ann: 36,000-42,000 TS; 35-42 El.
Cold rolled: 39,000-70,000 TS; 1-32 El.
For resistance strips for fuses.
Constant resistivity and melting temperature.

CUPRO-NICKEL 9% 725.
M-8; 88.78 Cu, 9.0 Ni, 2.0 Sn, 0.22 Mn.
Sheet, hard: 80,000 TS; 76,000 YS; 3 El; 90 Rb.
Soft: 52,000 TS; 22,000 YS; 40 El; 42 Rb.
For connectors used in telephone, computer and other electrical and electronic systems; tableware, boat hardware.

CUPRO-NICKEL 10%-510.
M-141; 10 Ni, 0.85 Fe, bal Cu.
Annealed: 46,000 TS; 42 El.
Drawn: 75,000 TS; 10 El.
For condensers and heat exchangers; corrosion resistant.

CUPRO-NICKEL 10%-511.
M-141; 88.85 Cu, 10 Ni, 1.15 Fe.
Hard: 50,000 TS; 45,000 YS; 18 El.
Soft: 45,000 TS; 15,000 YS; 40 El.
For condensers and heat exchangers; corrosion resistant.

CUPRO-NICKEL 10% 706.
M-8; 88.35 Cu, 10 Ni, 1.25 Fe, 0.40 Mn.
Hard tube: 60,000 TS; 57,000 YS; 15 El; B 68 Rock.
Soft tube: 44,000 TS; 22,000 YS; 46 El; B 25 Rock.
For condenser tubes, heat exchangers, marine equipment, oil refinery condensers and evaporators.
Resists general corrosion and stress corrosion cracking.

CUPRO-NICKEL 10% 706.
M-279; 88 Cu, 10 Ni, 1.4 Fe. 0.6 Mn.
Ann: 40,000-50,000 TS; 10,000-25,000 YS; 33-40 El.
Cold rolled: 54,000-88,000 TS; 48,000-84,000 YS; 1-30 El.
For heat exchanger tubes and plates.
Excellent salt water corrosion resistance.

CUPRO-NICKEL 10%-755.
M-8; 10 Ni, 0.4 Mn, 1.25 Fe, bal Cu.
Soft: 44,000 TS; 22,000 YS; 46 El.
Hard: 60,000 TS; 57,000 YS; 15 El.
For condenser tubes; resists sea water corrosion.

CUPRO-NICKEL 20%-520.
M-141; 78.85 Cu, 20 Ni, 0.4 Fe, 0.75 Mn.
Soft: 49,000 TS; 14,000 YS; 40 El.
For condensers and heat exchangers; corrosion resistant.

CUPRO-NICKEL 20%-710.
M-8; 78.75 Cu, 20 Ni, 0.75 Fe, 0.5 Mn.
Annealed: 50,000 TS; 22,000 YS; 45 El.
For feed water heaters; corrosion resistant.

CUPRO-NICKEL 25% 713.
M-279; 75 Cu, 25 Ni.
Ann: 48,000-58,000 TS; 14,000-24,000 YS; 35-45 El.
Cold rolled: 54,000-94,000 TS; 16,000-89,000 YS; 2-38 El.
For heat exchanger tubes and plates.
Excellent salt water corrosion resistance.

CUPRO-NICKEL 30%-531.
M-141; 67.75 Cu, 31 Ni, 0.5 Fe, 0.75 Mn.
Soft: 60,000 TS; 25,000 YS; 45 El; 75 Brin.
For condensers and heat exchangers; corrosion resistant.

CUPRO NICKEL 30%-702.
M-8; 68.9 Cu, 30 Ni, 0.6 Mn, 0.5 Fe.
Hard: 77,000 TS; 70,000 YS; 5 El; 162 Brin.
Soft: 55,000 TS; 22,000 YS; 40 El; 71 Brin.
For condenser tubes; corrosion resistant.

CUPRO-NICKEL 30%-707.
M-8; 64.15 Cu, 30 Ni, 5.25 Fe, 0.60 Mn.
Annealed: 74,000 TS; 36,000 YS; 30 El.
For heat exchanger tubes; high strength, corrosion resistant.

CUPRO-NICKEL 30% 715.
M-8; 68.9 Cu, 30 Ni, 0.60 Mn, 0.50 Fe.
Hard tube: 70,000 TS; 60,000 YS; 10 El; B 80 Rock.
Soft tube: 55,000 TS; 22,000 YS; 45 El; B 35 Rock.
For vessel condensers, and salt water tubes, heat exchangers, cold headed fasteners.
Tough and corrosion resistant.

CUPRO-NICKEL 30% 715.
M-279; 68.5 Cu, 31 Ni, 0.5 Fe.
Ann: 54,000-60,000 TS; 20,000-22,000 YS; 40-45 El.
Cold rolled: 58,000-94,000 TS.
For heat exchanger tube and sheets.
Good corrosion resistance to high velocity salt water.

CUPRO NICKEL 30% 716.
M-8; 64.15 Cu, 30 Ni, 5.25 Fe, 0.60 Mn.
Soft tube: 74,000 TS; 36,000 YS; 30 El.
Drawn: 112,000 TS; 90,000 YS; 12 El.
For feedwater heater tubes, heat exchanger tubes, high pressure air and hydraulic lines.
Corrosion resistant, high strength.
Elect. cond: 4.6.

CUPRO NICKEL 826.
M-8; 30 Ni, 0.6 Mn, 0.15 Si, bal Cu.
For welding rod; for steel and copper-nickel.

CUPRO-NICKEL BULLET JACKETS.
M-U.S.; 85; Cu, 15 Ni.
For bullet jackets; ductile.

CUPRO-NICKEL COMMERCIAL.
M-U.S.; 98-60 Cu, 2-40 Ni.
For hardware; corrosion resistant.

CUPRO-NICKEL DRIVING BANDS.
M-U.S.; 97.5-95 Cu, 2.5-5 Ni.
For driving bands; corrosion resistant.

CUPRO-NICKEL ELECTRODES.
M-1713; Weld metal: 0.03 C, 0.07 Si, 31.0 Ni, 1.3 Mn, 0.55 Fe, 68.0 Cu.
As welded: 52,000 psi TS; 35,000 psi YS; 40 El.
For welding 70-30, 90-10 and similar alloys.

CUPRO-NICKEL LOCOMOTIVE TUBES.
M-U.S.; 97 Cu, 3 Ni.
For locomotive tubes; corrosion resistant.

CUPRO-NICKEL NO. 300 ALLOY.
89 Cu, 11 Ni.
For corrosion resistant parts; corrosion resistant.

CUPROR.
M-U.S.; 94 Cu, 5.8 Al.
For pump rods, valve stems; corrosion resistant.

CUPRO-SILICON.
M-U.S.; 55 Si, bal Cu.
For hardener for copper alloys.

CUPROSIL NS5.
M-1687; 2 Ni, 0.5 Si, bal Cu.
Cast or wrought; hardness 160-200 Brin.
AFNOR UN35.

CUPROSIL SI45Z.
M-1687; 3 Si, 2 Fe, 3 Zn, bal Cu.
Silicon bronze; wrought; hardness 90-190 Brin.
AFNOR US3 2Fe; AMS 4616B.

CUPROTEC 10180.
M-717.
Powder for spray joining thin walled copper/copper alloys. Gas and liquid tight joints under pressure. 42,000 psi TS.

CUPROTEC 10180.
M-717.
Copper-base alloy powder for brazing thin walled copper alloy sections.

CUPROTHAL 30.
M-70, M-1150; 2 Ni, bal Cu.
Annealed: 30,000-60,000 TS; 60 max El.
For heating and resistance elements; max operating temperature 600°F, nonmagnetic.

CUPROTHAL 60.
M-70, M-1150; 6 Ni, bal Cu.
Annealed: 35,000-70,000 TS; 55 max El.
For heating and resistance elements; max operating temperature 600°F, nonmagnetic.

CUPROTHAL 90.
M-70, M-1150; 11 Ni, bal Cu.
Annealed: 35,000-75,000 TS; 50 max El.
For heating and resistance elements; max operating temperature 750°F, nonmagnetic.

CUPROTHAL 180.
M-70, M-1150; 22 Ni, bal Cu.
Annealed: 50,000-100,000 TS; 40 max El.
For heating and resistance elements; max operating temperature 1000°F, nonmagnetic.

CUPROTHAL 294.
M-70, M-1150; 45 Ni, bal Cu.
Rolled: 60,000-100,000 TS; 30 El.
For resistors; low electrical resistance, nonmagnetic maximum temperature 1000°F.

CUPROTHERM now **WIELAND K60.**

CUPROVAC E.
M-1214; Cu.
For electrical equipment; gas free, high purity copper.

CUPTEN-G.
M-1612; 0.12 max C, 0.60 max Si, 0.60 max Mn, 0.06-0.12 P, 0.040 max S, 0.20-0.60 Cu, 0.40-1.20 Cr, 0.35 max Mo, 0.10 max V, bal Fe.
Hot rolled, 1.4-13.0 mm thick: 36 kg/mm^2 YS.
Cold rolled, 0.5-2.6 mm thick: 33 kg/mm^2 YS.
Weather resistant steel.

CUPTEN-R.
 M-1612; 0.12 max C, 0.25-0.75 Si, 0.20-0.50 Mn, 0.07-0.15 P, 0.040 max S, 0.25-0.55 Cu, 0.45 max Ni, 0.30-1.0 Cr, bal Fe.
 Hot rolled, 1.4-13.0 mm thick: 35 kg/mm^2 YS. Cold rolled, 0.5-2.6 mm thick: 32 kg/mm^2 YS.
 Weather resistant steel.

CURTISOL.
 M-1502; silver alloy.
 For solder for titanium alloys; M.P. 1300-1400°F.

CURTISS.
 M-Eng; 95.2 Al, 2.5 Cu, 1.5 Mg.
 For pistons; heat treatable.

CUSIL.
 M-1493; 72 Ag, 28 Cu.
 M.P. 1436°F.
 For high temperature brazing.
 Eutectic alloy. Excellent flow.
 High vapor pressure.

CUSILOY.
 M-U.S.; 95 Cu, 3-1 Si, 1-1.5 Sn, 0.7-1 Fe.
 Soft: 50,000-60,000 TS; 15,000-20,000 YS; 50 El.
 For wire; corrosion resistant.

CUSTOM 455.
 M-32; 0.03 C, 11.75 Cr, 8.5 Ni, 1.2 Ti, 0.3 Cb, 2.25 Cu, bal Fe.
 Hardened: 332,000 TS; 244,000 YS; 13 El; C 50 Rock.
 For high temperature bolts, springs, valves.
 Maraging, precipitation hardening, tough.

CUSTOM 630 ETC see **CARPENTER CUSTOM 630 ETC.**

CUTANIT.
 M-261, M-486; Mo, Ti, WC.
 For taps, tools, cutters; cemented carbides.

CUTTRODE 1.
 M-717.
 Electrode for AC-DC to cut, pierce, clean castings, remove flash and risers-all metals.

CUZINAL.
 M-141; 77 Cu, 2.1 Al, 0.03 As, bal Zn.
 Annealed: 50,000 TS; 65 El.
 Drawn: 85,000 TS; 10 El.
 For condenser and heat exchangers; embrittlement free.

CV see **DARWIN CV.**

C.V.M.
 M-1067; 1.0 C, 0.7 Mn, 5 Cr, 1.1 Mo, 0.25 V, bal Fe.
 For dies, blanking, trimming and forming dies; air hardening.

C-W.
 M-1067; 0.9 C, 1.1 Mn, 0.5 Cr, 0.5 W, 0.35 Si, bal Fe.
 For drills, milling cutters, dies; oil hardened.

CW01.
 M-1806.
 Sintered carbide material.
 For wear parts; for optimum wear with complete freedom from shock.

C.W.3.
 M-1116; 1.25 C, 3.0 W, 0.4 max Mn, 0.3 Si, bal Fe.
 Water hardening tool steel; for dies for cartridge cases, threading brass and stamping light alloys and thin steel sheet.
 AFNOR: 125 W30; AISI F2.

CW 11.
 M-1655; 1.20 C, 0.25 Mn, 0.20 Si, 1.0 W, 0.10 V, bal Fe.
 Cold work tool steel for center drills, twist drills, milling cutters.
 W.-Nr. 1.2516.

CW25.
 M-1806.
 Sintered carbide material.
 For wear resisting applications under good conditions with little shock.

CW30.
 M-1806.
 Sintered carbide material.
 For wear resisting applications where some shock is encountered.

CW50.
 M-1806.
 Sintered carbide material.
 For wear resisting applications where heavy shock is encountered.

C.W. CHISEL.
 M-114; 0.7 C, bal Fe.
 For tools, chisels; oil hardening.

C.W. OIL.
 M-908; 0.85-0.95 C, 1.0-1.2 Mn, 0.4-0.6 Cr, 0.4-0.6 W, bal Fe.
 For tools, dies; non-deforming.

CX3.
 M-1697; 59.0 WC, 27.0 TaC, 14.0 Co.
 360,000 TrS; A 86.7 Rock; Denisty 13.91.
 Sintered carbide tool material.

C-XB VALVE STEEL see **CARPENTER C-XB VALVE STEEL.**

CY2.
 M-1697; 76.0 WC, 15.0 TiC, 9.0 Co.
 200,000 TrS; A 92.0 Rock; Density 11.20.
 Sintered carbide cutting tool; excellent wear resistance; for light and general purpose machining.

CY-4.
M-1697; Sintered carbide tool material.

CY5.
M-1697; 82.0 WC, 8.0 TiC, 10.0 Co.
260,000 TrS; A 91.0 Rock; Density 12.45.
Sintered carbide cutting tool; for general purpose and semi-roughing.

CY12.
M-1697; 83.5 WC, 3.5 TiC, 13.0 Co.
300,000 TrS; A 90 Rock; Density 13.30.
Sintered carbide cutting tool; excellent shock resistance; for heavy feed, low speed machining of rough and irregular steel.

CY14.
M-1697; 75.0 WC, 10.0 TaC, 9.0 TiC, 6.0 Co.
250,000 TrS; A 92.6 Rock; Density 12.60.
Sintered carbide cutting tool; excellent wear resistance and resistance to cratering; for general purpose and finishing cuts.

CY16.
M-1697; 72.0 WC, 11.5 TaC, 8.0 TiC, 8.5 Co.
275,000 TrS; A 91.3 Rock; Density 12.6.
Sintered carbide cutting tool; high resistance to cratering and to high temperatures; for general purpose machining.

CY17.
M-1697; 71.0 WC, 11.5 TaC; 8.0 TiC, 9.5 Co.
300,000 TrS; A 91.0 Rock; Density 12.50.
Sintered carbide cutting tool; very high shock and good wear resistance; for milling or operations on rough work.

CY31.
M-1697; 76.0 WC, 12.0 TaC, 8.0 TiC, 4.0 Co.
180,000 TrS; A 93.5 Rock; Density 12.90.
Sintered carbide cutting tool; highest abrasion resistance; for high speed precision boring and turning.

CY/A.
M-752; 2.4-3.0 C, 0.5-1.5 Si, 0.2-0.8 Mn, 2.0 max Cr, 0.15 max P, bal Fe.
White cast iron; 250 min Brin.
Abrasion resistant.

CY/C.
M-752; 2.4-3.2 C, 1.0 max Si, 0.5-1.5 Mn, 22-28 Cr, 1.5 max Mo, 1.0 max Ni, 1.2 max Cu, 0.1 max P, bal Fe.
High chrome abrasion resistant casting; 450 min Brin.

CYCLO.
M-114; 0.7 C, 1.2 Mn, bal Fe.
For tools, dies, punches; oil hardening.

CYCLOPE.
M-1763; 0.35 C, 9.0 W, 3.5 Cr, bal Fe.
Hot work tool steel, for dies.
AISI H21.

CYCLOPS 14 MV replaced by **UNITEMP 14MV.**

CYCLOPS 17A replaced by **UNILOY 325.**

CYCLOPS 67 replaced by **CYCLOPS S 5.**

CYCLOPS 2570.
M-114; 0.47 C, 0.30 Mn, 1.0 Si, 8.5 Cr, 1.15 Mo, 1.15 V, bal Fe.
Shear blades, forging dies.

CYCLOPS B-6.
M-114; 0.75 C, 18 W, 1 V, bal Fe.
For lathe tools, taps, reamers, twist drills, milling cutters, dies, lathe centers; high speed steel; wear and abrasion resistant.

CYCLOPS B 6X replaced by **THERMOLD H 26.**

CYCLOPS B-9.
M-114; 0.84 C, 0.25 Mn, 0.30 Si, 4.5 Cr, 18.5 W, 2.25 V, 0.5 Mo, bal Fe.
For drills, taps, broaches, form tools, milling cutters; high speed steel.

CYCLOPS B-10.
M-114; 0.80 C, 1.0 Mo, 18.5 W, 4.5 Cr, 2.0 V, 9.0 Co, bal Fe.
For cutters, lathe and planer tools, form tools, checking tools, cut-off tools.
High abrasion resistance and red-hardness.
AISI-T 5 high speed steel.

CYCLOPS B-44 replaced by **THERMOLD H22.**

CYCLOPS B44J replaced by **THERMOLD H21.**

CYCLOPS EXTRA replaced by **CYCLOPS W1.**

CYCLOPS K.
M-114; C, Cr, W, bal Fe.
For hot work dies, punches; hot work steel.

CYCLOPS K-L DIE STEEL.
0.35 C, 6 Cr, 6 W, bal Fe.
For coining dies, punches, hot heading dies; oil hardened.

CYCLOPS K-M DIE STEEL.
0.45 C, 6 Cr, 6 W, bal Fe.
For blanking, forming and gripper dies, hot piercers; water hardened.

CYCLOPS L1.
M-114; 1.05 C, 0.35 Mn, 0.30 Si, 1.4 Cr, bal Fe.
Water hardened: 237,000 TS; 226,000 YP; 444 Brin.
Cold Drawn : 107,000 TS; 87,000 YP; 17 El; 55 RA; 229 Brin.
For gauges, knurls, knife edges, taps, dies, arbors, rolls.
AISI Type L1. Oil hardening. Wear resistant.

CYCLOPS L2.
M-114; 0.50 C, 0.70 Mn, 0.25 Si, 1.0 Cr, 0.2 V, bal Fe.
Annealed: 103,000 TS; 74,000 YS; 27 El; 52 RA; 201 Brin.
Hardened: 298,000 TS; 263,000 YP; 1 El; 5 RA; 610 Brin.
For gears, forgings, arbors, crankpins, chuck jaws, die rings, gun barrels, jack screws, rivet sets, shear blades.
Tough, shock resisting.
AISI Type L2 Oil hardening. Fatigue resistant.

CYCLOPS L6.
M-114; 0.75 C, 0.4 Mn, 0.25 Si, 1.0 Cr, 1.5 Ni, bal Fe.
Heat treated: 190,000-305,000 TS; 178,000-280,000 YS; 5-12 El; 17-38 RA; C 43-61 Rock.
For arbors, blanking dies, clutch parts, forming dies, brake dies, punches pinions, shear blades, spindles, swages.
Tough and wear resistant.
AISI Type L6 oil hardening tool steel.

CYCLOPS M4.
M-114; 1.3 C, 4.5 Mo, 5.5 W, 4.5 Cr, 4.0 V, bal Fe.
Hardened: C 64-66 Rock.
For cutters, broaches, reamers, milling cutters, lathe and planer tools, taps, Counterbore tools.
Good abrasion resistance, tough.
AISI Type M4. High speed steel.

CYCLOPS M42.
M-114; 1.10 C, 3.75 Cr, 1.2 V, 1.5 W, 9.5 Mo, 8.0 Co, bal Fe.
Ht. Tr.: 63-67 Rock C.
For finish machining, boring, precision turning, at high speeds; good temperature and wear resistance.
AISI Type M42 High speed tool steel.

CYCLOPS M-T.
0.75 C, 1.0 V, 1.5 W, 7 Mo, bal Fe.
For tools; see Motung.

CYCLOPS N-9 replaced by CYCLOPS L6.

CYCLOPS S1.
M-114; 0.50 C, 0.35 Mn, 0.30 Si, 1.5 Cr, 2.25 W, 0.25 V, 0.30 Mo, bal Fe.
Hardened: 260,000-280,000 TS; C 48-55 Rock.
For bolt header dies, chipping and caulking tools, concrete drills, pneumatic tools, shear blades, track tools. Type S1. Tough and fatigue resistant.

CYCLOPS S2.
M-114; 0.50 C, 0.45 Mn, 1.1 Si, 0.2 V, 0.5 Mo, bal Fe.
Water Hardened: 235,000-323,000 TS; 229,000-300,000 YS; 4-10 El; 12- RA; C 47-58 Rock.
For hand and pneumatic tools, chisels, stamps, spindles, pipe cutters, rivet sets; flaring tools.
Tough and shock resistant.
Type S2. Tough and shock resistant.

CYCLOPS S5.
M-114; 0.55 C, 2.0 Si, 0.9 Mn, 0.4 Mo, bal Fe.
Heat Treated: 338,000 TS; 281,000 YS; 5 El; 600 Brin.
Heat Treated: 220,000 TS; 207,000 YS; 11 El; 455 Brin.
For shear blades, punches, pneumatic tools, rivet sets, chisels.
Shock and impact resistant. Type S5.

CYCLOPS SCK.
M-114; 0.70 C, 0.35 Mn, 1.0 Si, 8.5 Cr, 1.0 V, 1.4 Mo, 1.5 Ni, bal Fe.
Ht. Tr.: 55-61 Rock C.
For cold shears, trimmers, woodworking chippers, punches; air hardening; resists tempering, good wear resistance.

CYCLOPS SPECIAL COLD HEADER replaced by CYCLOPS W2.

CYCLOPS T-15.
M-114; 1.5 C, 12.5 W, 4.75 Cr, 5.0 V, 5.0 Co, bal Fe.
Oil harden: C 64-66 Rock.
For heavy duty cutters, form tools, lathe and planer tools, broaches, milling cutters, blanking dies, punches. High red-hardness and abrasion resist.
AISI-T15 High speed steel.

CYCLOPS W1.
M-114; 0.80-1.25 C, 0.25 Mn, 0.25 Si, bal Fe.
Annealed: 95,000 TS; 50,000 YS; 22 El; 195 Brin.
Water Hardened: 200,000-215,000 TS; 138,000-152,000 YS; 11-12 El; 400-600 Brin.
For punches, beading tools, cold heading dies, axles, drills, reamers, files, woodworking tools.
Wear and abrasion resistance.
Water hardening high carbon steel.

CYCLOPS W2.
M-114; 0.80-1.25 C, 0.25 Mn, 0.25 Si, 0.25 V, bal Fe.
Annealed: 100,000 TS; 55,000 YS; 21 El; 200 Brin.
Water Hardened: 216,000 TS; 152,000 YS; 11 El; 32 RA; 600 Brin.
For punches, beading tools, cold heading dies, axles, drills, reamers, woodworking tools, files, die rings.
Wear and abrasion resistant.
Water hardening high carbon steel.

CY/H.
M-752; 2.4-3.4 C, 0.5-1.5 Si, 0.2-0.8 Mn, 2.0 max Cr, 0.15 max P, bal Fe.
White cast iron; 400 min Brin.
Abrasion resistant.

CYKLOP.
M-1315; 0.30 C, 2.65 Cr, 0.35 V, 8.5 W, bal Fe.
For extrusion press rams and liners, punches; hot work steel, oil hardened.

CYKLOP CO.
M-1315; 0.3 C, 2 Co, 2.4 Cr, 0.25 V, 8.5 W, bal Fe.
For extrusion press rams and liners, punches; hot work steel.

CYKLOP EXTRA.
M-1315; 0.65 C, 3.75 Cr, 0.85 Mo, 0.7 V, 8.5 W, bal Fe.
For lathe and planer tools, reamers, drills, hobs; high speed steel.

CYLINDER IRON.
M-165; 3-3.25 C, 0.40-0.75 Mn, 2-2.25 Si, bal Fe.
For engine cylinders, pistons, piston rings; cast iron.

CYLINDER IRON.
M-165; 3-3.5 C, 0.5-0.8 Mn, 1.75-2.75 Si, 1-1.5 Ni, 0.3-0.4 Cr, bal Fe.
For engine cylinders; cast iron.

CYMBAL METAL.
M-U.S.; 78 Cu, 22 Zn.
For architectural and ornamental parts; red brass.

CYPRUS BRONZE.
65 Cu, 30 Pb, 5 Sn.
17,000 TS; 6-8 El; 41 Brin.
For bearings, hardware; heavy duty.

D

D-2.
M-739; 3.0-3.5 Cu, 7.25-7.75 Sb, 0.1-0.14 Te, bal Sn.
For engine bearings; steel backed Babbitt.

D-2 see CLEVITE F-1.

D-6 AC.
M-1602, M-1527, M-97; 0.46 C, 0.76 Mn, 0.22 Si, 1.10 Cr, 0.50 Ni, 1 Mo, 0.08 V, bal Fe.
Q + T.: 195 ksi min TS; 180 ksi min YS; 8 min El; 25 min RA.
For rocket motor case rings.

D6C see also UNITEMP D6C.

D7 ALLOY.
M-137; 2.2-2.8 Cu, 4.5-5.5 Si, 0.3-0.7 Mn, bal Al.
Cast: 26,000-36,000 TS; 8,000-12,000 YS; 10-5 El; 60-65 Brin.
For heavy duty castings; age-hardenable.

D8 ALLOY.
M-137; 2-4 Cu, 3-6 Si, 0.3-0.7 Mn, 0.35 max Ni, bal Al.
Cast: 22,000-35,000 TS; 10,000-30,000 YS; 3-2 El; 60-100 Brin.
For hydraulic units, gear cases; hardenable.

D8 (D8W; D8 WP).
M-106; 2-4 Cu, 4-6 Si, 0.3-0.7 Mn, bal Al.
Sand cast, Sol. treat (D8W): 22,400-26,800 TS; 12,300 YS; 2-3 El; Brin; 1-3 Izod.
Chill cast, Sol. treat and age (D8WP) 45,000-47,000 TS; 40,000 YS; El; 110 Brin; 0.6 Izod.
Castings for good strength, hydraulic pressure, BS L79.

D9 ALLOY.
M-137; 3.2-4.0 Cu, 4.5-5.5 Si, 0.3-0.7 Mn, bal Al.
36,000-44,000 TS; 15,000-28,000 YS; 8-4 El; 75-105 Brin.
For light alloy parts; hardenable.

D-11.
M-1773; 87 Sn, 6.25 Cu, 6.75 Sb.
Tin based Babbitt; half shell bearings, bushings and thrust washers.
Soft, excellent embeddability.

D-12.
M-1773; 89 Sn, 3.5 Cu, 7.5 Sb.
Tin based Babbitt; half shell bearings, bushings and thrust washers.
Good corrosion resistance.

D12 ALLOY.
M-137; 6.2-7.2 Cu, 0.15-0.40 Mg, 5-6 Si, 0-1.25 Fe, bal Al.
Heat treated: 40,000-49,000 TS; 34,000 YS; 0 El; 130 Brin.
For pistons, cylinder heads; age-hardenable.

D-13.
M-1773; 84 Pb, 6.0 Sn, 10.0 Sb.
Lead based Babbitt; half shell bearings, bushings and thrust washers.
For lightly loaded pump bearings.

D 14 see also COLUMBIUM D-14.

D-14.
M-1773; 75 Pb, 10.0 Sn, 15.0 Sb.
Lead based Babbitt; half shell bearings, bushings and thrust washers.
Good score resistance.

D-15.
M-1773; 83 Pb, 1.0 Sn, 15.0 Sb, 1.0 As.
Lead based Babbitt; half shell bearings, bushings and thrust washers.

D-48.
M-1773; 70 Cu, 29 Pb, 1.0 Sn.
Copper lead alloy for half shell bearings for connecting rod and main bearings. Moderately hard.

D-50.
M-1773; 99.9 min Ag.
When plated with 8 Sn-92 Pb as half shell bearing can be used for extremely heavy duty main and connecting rod bearings.

D-51.
M-1773; 95 Al, 4 Si, 1.0 Cd.
Aluminum based alloy for half shell bearings for intermediate loads.

D-52A.
M-1773; 52 Cu, 44 Pb, 4.0 Sn.
Copper lead alloy for half shell bearings for intermediate loaded main and connecting rod bearings.

D-56.
M-1773; 95 Al, 4 Si, 1.0 Cd.
Aluminum based alloy for half shell bearings. Should be over plated with 8 Sn-92 Pb for highly loaded main and connecting rod bearings.

D-57.
M-1773; 80 Cu, 10 Pb, 10 Sn.
Copper-lead alloy for bushings and thrust washers. Maximum shock and load capacity. For steering knuckles, wear plates.

D-58.
M-1773; 88 Cu, 8 Pb, 4 Sn.
Copper-lead alloy for bushings and thrust washers. General purpose loading.

D-59.
M-1773; 73.5 Cu, 23.0 Pb, 3.5 Sn.
Copper-lead alloy for bushings and thrust washers. For intermediate loads on oscillating and rotating shafts.

D-61.
M-1773; 91 Al, 6.5 Sn, 1.0 Cu, 1.5 Si.
Aluminum based alloy for half shell bearings. Hard, usually plated; for highly loaded main and connecting rod engine bearings.

D-62.
M-1773; 90.5 Al, 6.5 Sn, 1.0 Cu, 1.5 Si.
Aluminum based alloy for bushings and thrust washers. For highly loaded transmission and motor bushings.

D-63.
M-1773; 79 Al, 20 Sn, 1.0 Cu.
Aluminum based alloy for bushings and thrust washers. Intermediate to heavy loaded transmission and camshaft bushings.

D319L.
M-1793; 0.03 max C, 2.0 max Mn, 1.0 max Si, 17.5-19.5 Cr, 11-15 Ni, 2.25-3.0 Mo, bal Fe.
Austenitic stainless steel.

D-406, ETC see **DIAMOND D-406, ETC.**

D421.
M-1744; 0.44 C, 0.40 Mn, 0.70 Si, 1.3 Cr, 2.3 W, 0.3 V, bal Fe.
2 1/4% Tungsten shock resisting steel, for hand and pneumatic chisels.

D979 see **UDIMET D-979.**

D1045.
M-1741; 99.45% Al min.
O Temper: 83 MPa TS; 28 MPa YS; 40 El.
H19 Temper: 186 MPa TS; 165 MPa YS.
Electrical conductors.

D1150.
M-1741; 99.35% Al min.
O Temper: 69 MPa TS; 28 MPa YS; 35 El.
H18 Temper: 145 MPa TS; 138 MPa YS; 5 El.
50 VHN.
Sheet metal components requiring decorative finishing.

D3005.
M-1741; 1.2 Mn, 0.35 Mg, bal Al.
O Temper: 131 MPa TS; 62 MPa YS; 25 El.
H18 Temper: 241 MPa TS; 228 MPa YS; 4 El.
High strength foil, roofing sheet.

D5050.
M-1741; 1.4 Mg, bal Al.
O Temper: 145 MPa TS; 55 MPa YS; 24 El; 36 Brin.
H38 Temper: 221 MPa TS; 200 MPa YS; 6 El; 63 Brin.
Coiled tubes, refrigerator trim.

D-5116.
M-1733; 0.16 C, 0.25 Si, 1.15 Mn, 0.95 Cr, bal Fe.
Alloy carburizing steel.
DIN 16 Mn Cr 5.

D-5119.
M-1733; 0.20 C, 0.25 Si, 1.25 Mn, 1.15 Cr, bal Fe.
Alloy carburizing steel.
DIN 20 Mn Cr 5.

D-6158.
M-1733; 0.58 C, 0.25 Si, 0.95 Mn, 1.05 Cr, 0.10 V, bal Fe.
Spring steel.
DIN 58 Cr V 4.

D6201.
M-1741; 0.7 Mg, 0.6 Si, bal Al.
T8 Temper: 310 MPa TS; 290 MPa YS; 4 El.
Electrical conductors.

D6463.
M-1741; 0.7 Mg, 0.4 Si, bal Al.
T1 Temper: 152 MPa TS; 90 MPa YS; 20 El.
T6 Temper: 241 MPa TS; 214 MPa YS; 12 El; 74 Brin.
Extrusions for trim requiring decorative finishing.

D8011.
M-1741; 0.8 Fe, 0.7 Si, bal Al.
Sheet for bottle closures.

DA-47.
M-1773; 52 Cu, 44 Pb, 4.0 Sn.
Copper-lead alloy for half shell bearings for intermediate loads on connecting rod and main bearings.

DA-49.
M-1773; 75 Cu, 24 Pb, 1.0 Sn.
Copper lead alloy for half shell bearings for highly loaded main and connecting rod bearings.

DA-95.
M-1773; 73.5 Cu, 23.0 Pb, 3.5 Sn.
Copper lead alloy for half shell bearings for heavy duty main and connecting rod bearings.

DAIDO ST-1.
M-1545; 0.13-0.18 C, 1.35-1.85 Mn, 0.3-0.6 Si, 0.35 max Cu, bal Fe.
Rolled: 78,000 min TS; 51,000 min YP; 18% min El.
For buildings, bridges, agricultural equipment, structural members, case hardened parts.
Constructional steel, shock resistant.

DAIDO ST-2.
M-1545; 0.15-0.20 C, 1.2-1.5 Mn, 0.30-0.65 Si, 0.35 max Cu, bal Fe.
Rolled: 78,000 min TS; 51,000 min YP; 18% min El.
For buildings, bridges, structural members, agricultural equipment.
Constructional steel, tough.

DAIMLER BEARING METAL.
M-Eng.; 76 Cu, 3 Sn, 20 Zn, 1 Pb.
For bearings, bushings; corrosion resistant.

DAIRY METAL.
M-252; 68 Cu, 1.5 Sn, 28 Ni, 0.5 Pb, 2.0 flux.
For stainless filters; corrosion resisting.

DAIRYWHITE.
M-U.S.; Zn, Ni, bal Cu.
For dairy equipment; nickel silver, corrosion resistant.

DALTON FUSIBLE ALLOY.
M-Eng.; 60 Bi, 15 Sn, 25 Pb.
For fusible alloy, fire extinguishers; M.P. 92°C.

DAMAR.
M-Eng.; 76 Cu, 13 Pb, 11 Sn.
For bearings; heavy duty.

DAMASCUS.
M-754; Sn, Pb, bal Cu.
For bearings.

DAMASCUS.
M-73; 0.55 C, 0.25 Cr, 0.90 Mn, 0.20 V, 1.95 Si, bal Fe.
For chisels, stamps, cold cutter, punches, shears; oil hardening.

DAMASCUS BRONZE.
M-Eng.; 13 Pb, 10 Sn, bal Cu.
For bearings, ornaments; plastic bronze.

DAMPING.
M-U.S.; 87 Mn, 13 Cu.
For alloy for meters; low temperature resistance coefficient.

DAMSTADT BELL METAL.
M-Ger.; 74-72.5 Cu, 21.7-21.1 Sn, 2.12 Pb, 0.19-0.05 Fe, 2-2.6 Ni.
For bells; corrosion resistant.

DANA AUTO SPECIAL.
M-1260, M-1261; 6.5 Sb, 3 Cu, bal Sn.
Cast: 69 El; 24-25 Brin.
For engine bearings; shock resistant, Babbitt metal.

DANA COMMON.
M-1260, M-1261; 10 Sn, 13.5 Sb, 0.5 Cu, bal Pb.
Cast: 42 El; 28-30 Brin.
For transmission bearings; Babbitt.

DANA DIESEL.
M-1260, M-1261; 2 Pb, 6.5 Cu, 8 Sb, bal Sn.
Cast: 22 El; 30-31 Brin.
For diesel engine bearings; wear resistant, Babbitt metal.

DANA STEAM.
M-1260, M 1261; 11.5 Sb, 5.5 Cu, 3 Pb, bal Sn.
Cast: 19 El; 33-35 Brin.
For steam turbine and generator bearings; wear resistant, Babbitt metal.

DANALLOY-I.
M-1606; Au-Ni-Mg, bal Ag.
Rolled: 82,000 TS; 78,000 YS.
For circuit board retainers, micro circuit back-up plates, relay contacts.
High heat and electrical conductivity. Non-magnetic. Heat treatable.

DANALLOY-II.
M-1606; Au-Ni-Mg, bal Ag.
Rolled: 66,000 TS; 60,000 YS.
For circuit board retainers, micro-circuit back-up plates, relay contacts.
High heat and electrical conductivity. Non-magnetic. Heat treatable.

D&D SPECIAL O.H.
M-755; 0.7 C, 0.8 Cr, 0.2 Mo, bal Fe.
For tools, dies; oil hardening.

DANDELION METAL.
M-Eng.; 72 Pb, 18 Sb, 10 Sn.
For heavy duty machine bearings and locomotive cross-head linings; high strength Babbitt.

D & J ANTIFRICTION METAL-1.
M-Eng.; 10 Sb, 80-85 Zn, 5-8 Sn.
For bearings, bushings; Babbitt.

D & J ANTIFRICTION METAL-2.
M-Eng.; 1.6 Cu, 0.4 Sb, 52 Zn, 46 Sn.
For bearings, bushings; Babbitt.

D & M-7-G.
M-40; C, alloy, bal Fe.
For arc welding electrodes; coated, hard surfacing, air hardening.

D & M-7-0.
M-40; C, alloy, bal Fe.
For arc welding electrodes; coated, self hardening.

D & M-92.
M-40; C, alloy, bal Fe.
For welding rods; shock resistant.

D & M-305.
M-40; C, bal Fe.
For arc welding rods; coated.

D & M-LT.
M-40; C, alloy, bal Fe.
For arc welding rods; coated.

D & M-U2.
M-40; C, alloy, bal Fe.
For welding rods; self hardening.

DANISH MINT.
M-Eng.; 92 Cu, 6 Al, 2 Ni.
For coinage; corrosion resistant.

DANNEMORA.
M-434; 0.7 C, 18 W, 4 Cr, 1 V, bal Fe.
For coining dies; high speed steel.

DANNEMORA AD 95.
M-1067; C, Cr, W, V, bal Fe.
For coining dies; oil hardening.

DANNEMORA BEST.
M-387; 0.9-1.1 C, bal Fe.
For drills, reamers, punches, broaches; Type W1; water hardened.

DANNEMORA DB59.
M-608; 0.7 C, 18 W, 4 Cr, 1 V, 5 Co, bal Fe.
For cutters, hobs, high speed steel.

DANNEMORA EXTRA BEST.
M-350; 0.7-1.2 C, bal Fe.
For tools; water hardened.

DANNEMORA NO. O.
M-350; C, W, bal Fe.
For tools, dies; oil hardened.

DANNEMORA SELF-HARDENING.
M-350; C, W, bal Fe.
For tools, cutters, dies; oil hardened.

DANNEMORA STANDARD.
M-387; 0.9-1.1 C, bal Fe.
For tools, drills, taps; water hardened.

DANNEMORA VERY BEST.
M-387; 1.0-1.2 C, bal Fe.
For tools, cutters, drills; water hardened.

DARCET FUSIBLE ALLOY.
M-Eng.; 50 Bi, 25 Sn, 25 Pb
For boiler safety plugs, fire extinguishers; M.P. 93°C.

DARGRAPH.
M-40; 1.45 C, 1.0 Si, 0.25 Mo, bal Fe.
Annealed: 84,500 TS; 49,500 YS; 25 El; 40 RA; 197 Brin.
Heat treated: 218,000-164,000 TS; 177,000-136,000 YS; 8.5-13 El; 14-2 RA; 388-302 Brin.
For wear plates, cams, cutters, dies, punches; oil hardened, wear resistant graphite steel.

DARK RED GOLD.
M-Eng.; 50 Au, 50 Cu.
For ornaments; corrosion resistant.

DART.
M-73; 0.42 C, 1.05 Si, 0.6 Mn, 3.3 Cr, 2.5 Mo, 0.37 V, bal Fe.
Hardened: 53-59 Rock C.
Hot work die steel with extra high carbon for higher hardness and wear resistance; forging dies, hot forming and press dies.
AISI Type H 10 hot work tool steel.

DARWIN.
M-487, M-40; C, alloy, bal Fe.
For high speed tools, hacksaw blades, tools, cutters; super high speed steel.

DARWIN-1.
M-40; 1.5 C, 12 Cr, 1 Mo, bal Fe.
For drawing and forming dies, punches; Type D2; air hardened, non- deforming.

DARWIN 5V.
M-40; 1.50 C, 5.0 Cr, 5.0 V, 12.5 W, 5.0 Co, bal Fe.
High speed tool steel, Cr-W-V-Co type, for cutting tools for tough metal; good red hardness and good wear properties; AISI T15.

DARWIN 19.
M-40; 0.7 C, 18 W, 4 Cr, 2 V, 8 Co, bal Fe.
For tools, cutters, reamers; high speed steel.

DARWIN 93.
M-40; 0.30 C, 3.5 Cr, 0.30 V, 10 W, 0.4 Si, bal Fe.
For dies; hot work steel.

DARWIN 505 SPECIAL.
M-40; 0.80 C, 4.0 Cr, 2.0 V, 18.0 W, 0.8 Mo, 9.0 Co, bal Fe.
High speed steel cutting tool, tungsten-cobalt type; good red hardness; AISI T5.

DARWIN 3581.
M-40; C, alloy, bal Fe.
For chisels; tough and shock resistant.

DARWIN ACD.
M-40; C, bal Fe.
For dies; water hardening.

DARWIN "ALNI".
M-487; 59 Fe, 4 Cu, 24 Ni, 13 Al.
For permanent magnets; magnetic steel.

DARWIN BEST WARRANTED.
M-40, M-487; High C, bal Fe.
For tools; water hardening.

DARWIN "BKM".
M-40, M-487; C Ni, Cr, bal Fe.
For Bakelite molds; oil hardening.

DARWIN BRAKE DIE STEEL.
M-40; C, Cr, Mo, bal Fe.
For brake dies; oil hardened.

DARWIN BRAND "H".
M-40; 0.95 C, 0.4 Cr, 0.2 V, 1.0 Mn, bal Fe.
For dies, blanking, forming and trimming dies, taps, broaches; oil hardening, non-deforming; resistant to wear.

DARWIN BRAND L-35.
M-40; C, alloy, bal Fe.
For heading dies.

DARWIN C C.
M-40; 0.55-0.65 C, 0.3-0.4 Cr, 0.60-0.75 W, 0.55-0.65 Mo, 0.95-1.1 Si bal Fe.
For tools, punches; oil hardening.

DARWIN C.L. NO.1.
M-40; C, alloy, bal Fe.
For arc welding electrodes, hard surfacing; coated, tough and abrasion resistant.

DARWIN COBALT MAGNET.
M-40; C, 3 Co, bal Fe.
For permanent magnets.

DARWIN COBALT MAGNET.
M-40; C, 6 Co, bal Fe.
For permanent magnets.

DARWIN COBALT MAGNET.
M-40; C, 9 Co, bal Fe.
For permanent magnets.

DARWIN COBALT MAGNET.
M-40; C, 15 Co, bal Fe.
For permanent magnets.

DARWIN COBALT MAGNET.
M-40; C, 35 Co, bal Fe.
For permanent magnets.

DARWIN C V.
M-40; 1.1-1.15 C, 0.2-0.4 Cr, 0.25-0.35 V, bal Fe.
For tools, taps, drills; water hardening.

DARWIN D93.
M-40, M-487; C, alloy, bal Fe.
For hot working dies and tools; hot die steel.

DARWIN "DCI".
M-40, M-487; C, alloy, bal Fe.
For electrical equipment, magnetic parts; magnetic alloy.

DARWIN DUREX.
M-40, M-487; high C, 18 W, 4 Cr, 1 V, bal Fe.
For tools, cutters, reamers, dies, gauges, punches; high speed steel.

DARWIN EE.
M-40; 1.35 C, 0.15 Cr, 4.0 W, 0.35 V, bal Fe.
For cutters, shears, tools; water hardening.

DARWIN EXTRA QUALITY.
M-40; M-487; 1.0 C, bal Fe.
For drills, taps, tools, cutters, reamers; Type W1; water hardened.

DARWIN EXTRA SPECIAL.
M-40; M-487; 1.1 C, bal Fe.
For drills, taps, hobs, reamers; Type W1, water hardened.

DARWIN EXTRA TOUGH.
M-40; C, alloy, bal Fe.
For punches, chisels, pneumatic tools; Type S6; oil hardened, shock resistant.

DARWIN FLAME HRD.
M-40; 0.50 C, 1.2 Mn, 0.5 Si, 1.4 Cr, 0.1 V, 0.4 Mo, bal Fe.
Air hardening tool steel for shafts, arbors.

DARWIN H.A.W.
M-40; 0.35 C, 5 Cr, 0.4 V, 1.5 Mo, bal Fe.
For shears, punches, hot work tools; Type H11; hot work steel.

DARWIN H.W.S.
M-40; 0.35 C, 5 Cr, 0.4 V, 1.5 W, 1.5 Mo, bal Fe.
For dies, punches, hot work tools; Type H12; hot work steel.

DARWIN I W I.
M-40; 0.40 C, 3 Cr, 9-10 W, 0.2-0.3 V, bal Fe.
For tools, dies, punches; hot work steel.

DARWIN LOW AIR.
M-40; 0.75 C, 2.0 Mn, 0.3 Si, 1.0 Cr, 1.35 Mo, bal Fe.
Air hardening tool steel; AISI A6.

DARWIN M3.
M-40; 1.0 C, 4 Cr, 2.7 V, 6 W, 5 Mo, bal Fe.
For reamers, drills, taps, lathe and planer tools; Type M3; high speed steel.

DARWIN MT-6.
M-40; 0.85 C, 6 W, 4 Cr, 1.5 V, 6 Mo, bal Fe.
For cutters, tools; high speed steel.

DARWIN N-32.
M-40; 0.38 C, 3-4 Ni, 0.6 Mn, bal Fe.
Oil hardened.
For chisels; oil hardened.

DARWIN NO. 1 AIR HARDENING.
M-40; 1.45-1.60 C, 11-12 Cr, 0.20-0.35 V, 0.7-0.8 Mo, bal Fe.
For tools, dies; non-deforming.

DARWIN NO. 1 FM.
M-40; 1.5 C, 12 Cr, 0.3 V, 0.8 Mo, 0.2 S, bal Fe.
For dies, tools, punches, crimpers; air or oil hardened, good machinability; resists galling.

DARWIN NO. 505.
M-40; 0.65-0.75 C, 3.75-4.25 Cr, 17-18 W, 1.5-1.75 V, 7-8 Co, 0.7-1.0 Mo, bal Fe.
For tools, cutters; high speed steel.

DARWIN NO. 1366.
M-40, M-487; 0.70 C, 4.5 Cr, 20 W, 2 V, 12 Co, bal Fe.
For lathe and planer tools, form cutters; high speed steel, oil hardened.

DARWIN OHT.
M-40, M-487; 0.9 C, 1.0 Mn, 0.5 Cr, 0.5 W, bal Fe.
For crimpers, punches, jaws, dies, cutters; Type O1; non-deforming, oil hardened.

DARWIN P-20.
M-40; 0.30 C, 0.75 Cr, 0.25 Mo, bal Fe.
Oil or water hardening tool steel for molds; AISI P20.

DARWIN PRK-33.
M-40; 3.7 C, 13.5 Cr, bal Fe.
For welding electrodes; air hardening.

DARWINS 6/5/2.
M-487; 0.85 C, 6 W, 4 Cr, 2 V, 5 Mo, bal Fe.
Heat treated: C 63-67 Rock.
For woodworking machine tools, taps, slitting saws, reamers, boring and planing tools, drawing dies. High speed steel Type M2.

DARWINS 18/8.
M-487; 0.14 max C, 7.5-9.0 Ni, 17.5-19.0 Cr, bal Fe.
Annealed: 78,000 TS; 27,000 YS; 30 El; 130 Brin.
Cold drawn: 180,000 TS; 150,000 YS; 10 El; 250 Brin.
For food, chemical and brewing process equipment; Type 302; stainless, austenitic.

DARWINS 18/8 (316).
M-487; 0.14 max C, 17.5-19 Cr, 10-12 Ni, 2-3 Mo, bal Fe.
Bars: 78,000 TS; 27,000 YS; 30 El.
For equipment handling sulfite liquors; Type 316; stainless, austenitic.

DARWINS 18/8 (316A).
M-487; 0.08 max C, 17.5-19.0 Cr, 8.5-10.5 Ni, 1.25-2.25 Mo, 0.25-0.40 Ti, bal Fe.
Bars: 78,000 TS; 27,000 YS; 30 El.
For acid resistant chemical plant equipment; Type 316Ti; stainless, austenitic.

DARWINS 18/8 CB.
M-487; 0.10 max C, 17.5-19.0 Cr, 10-12 Ni, Cb = 10 X C, bal Fe.
Bars: 78,000 TS; 27,000 YS; 30 El.
For welded chemical plant equipment; Type 347; austenitic, stainless.

DARWINS 18/8 FZ.
M-487; 0.14 max C, 7.9-9.0 Ni, 17.5-19.0 Cr, 0.2 S, 0.5 max Mo, bal
Bars: 78,000 TS; 27,000 YS; 30 El.
For stainless screw machine products; Type 303; free-cutting, stainless.

DARWINS 18/8 LC.
M-487; 0.07 max C, 8.0-10.0 Ni, 17.5-19.0 Cr, bal Fe.
Annealed: 78,000 TS; 27,000 YS; 30 El.
Cold drawn: 180,000 TS; 125,000 YS; 10 El; 330 Brin.
For welded chemical plant and dyeing equipment; Type 304; stainless, austenitic.

DARWINS 18/8 MO.
M-487; 0.14 max C, 17.5-19.0 Cr, 8-10 Ni, 3-3.5 Mo, bal Fe.
Bars: 78,000 TS; 27,000 YS; 30 El.
For acid resistant equipment; Type 317; stainless, austenitic.

DARWINS WDP.
M-487; 0.07 max C, 8-10 Ni, 17.5-19.0 Cr, Ti = 5 X C, bal Fe.
Bars: 78,000 TS; 27,000 YS; 30 El.
For welded structures and chemical plant equipment; Type 321; austenitic, stainless.

DARWINS 26.
M-487; Ni, bal Fe.
For glass-to-metal seals; controlled expansion.

DARWINS 55.
M-487; 0.2 max C, 23 Cr, 55 Ni, 4 Mo, 2 W, 10 max Fe, 5 Cu.
For chemical plant and pickling equipment; resists hot or cold H_2SO_4.

DARWINS 168.
M-487; 0.75 C, 17 Cr, bal Fe.
For knives, bearings, cutlery, valves; Type 440A; stainless.

DARWINS 426.
M-487; Ni. bal Fe.
For glass-to-metal seals; controlled expansion.

DARWINS 654A.
M-487; 0.1 max C, 60 Ni, 20 Mo, bal Fe.
For chemical plant equipment; resists HCl and H_2SO_4.

DARWINS 655B.
M-487; 0.1 max C, 65 Ni, 27 Mo, bal Fe.
For chemical plant equipment; resists HCl and H_2SO_4.

DARWINS 656C.
M-487; 0.1 max C, 14 Cr, 58 Ni, 17 Mo, 5 W, bal Fe.
For chemical plant and high temperature equipment; resists HNO_3 and HCl.

DARWINS 1031.
M-487; Ni, bal Fe.
For thermostats; controlled expansion.

DARWINS 1366.
M-487; 0.73 C, 20 W, 4.5 Cr, 2 V, 12 Co, bal Fe.
Heat Treated: C 64-66 Rock.
For boring and turning tools, broaches, milling cutters.
High speed steel Type T6, high red-hardness, abrasion resistant.

DARWINS COBALT FAST WORK.
M-487; C, Co, Mo, bal Fe.
For press tools, cutters; oil hardened.

DARWINS D.C.C.M.
M-487; C, alloy, bal Fe.
For punches, crimpers, upsetters; cold work steel.

DARWINS DSC.
M-487; 0.30-0.40 C, 13-14 Cr, 1 max Ni, bal Fe.
Heat Treated: 100,000 TS; 20 El; 152-255 Brin
For cutlery, knives, valves; hardenable, stainless; EN56D.

DARWIN'S DUROR.
M-487; C, alloy, bal Fe.
For files; water hardened.

DARWINS EXTRA.
M-487; C, alloy, bal Fe.
For files; water hardened.

DARWINS F.
M-487; Ni, bal Fe.
For glass-to-metal seals; controlled expansion.

DARWINS HS22.
M-487; 0.75 C, 22 W, 5 Cr, 2 V, bal Fe.
For roll turning tools, lathe and planer cutters, hacksaws; high speed steel.

DARWINS H.W.1.
M-487; C. alloy, bal Fe.
For pressure casting dies; hot work steel, resists thermal shock.

DARWINS H.W.2.
M-487; C, alloy, bal Fe.
For pressure casting dies; hot work steel, abrasion resistant.

DARWINS H.W.3.
M-487; C, alloy, bal Fe.
For extrusion and drawing dies, forging tools; hot work steel, oil hardened.

DARWINS H.W.4.
M-487; C, alloy, bal Fe.
For die casting dies; hot work steel, resists thermal shock.

DARWINS H.W.5.
M-487; C, alloy, bal Fe.
For hot piercing and forming punches, bolt dies, forging tools; hot work steel, oil hardened.

DARWINS K-ALLOY.
M-487; 0.15 max C, 23 Cr, 23 Ni, 2 Mo, 4 Cu, bal Fe.
For chemical plant equipment; resists H_2SO_4.

DARWINS N.1932.
M-487; C, alloy, bal Fe.
For chisels, punches, upsetters; shock resistant, oil hardened.

DARWINS S61.
M-487; 0.12 max C, 12-14 Cr, 1 max Ni, bal Fe.
Heat treated: 100,000 TS; 20 El, 155-255 Brin.
For pump rods, valves, cutlery, surgical instruments; Type 410; stainless.

DARWINS S61F.
M-487; 0.12 max C, 12-14 Cr, 0.2-0.3 S, 1 max Ni, bal Fe.
Heat treated: 100,000 TS; 20 El; 155-207 Brin.
For pump rods, valves, cutlery, surgical instruments; Type 410F; stainless, free-cutting.

DARWINS S62.
M-487; 0.18-0.25 C, 12-14 Cr, 1.0 max Ni, bal Fe.
Heat treated: 100,000 TS; 20 El; 155-255 Brin.
For surgical instruments, chemical plant equipment; EN56C-D; stainless hardenable.

DARWINS S62F.
M-487; 0.18-0.25 C, 12-14 Cr, 1 max Ni, 0.2-0.3 S, bal Fe.
Heat treated: 100,000 TS; 20 El; 152-255 Brin.
For screw machine products, bolts, screws; Type 420F; stainless, free-cutting.

DARWINS S80.
M-487; 0.12-0.18 C, 16-18 Cr, 1.6-2.0 Ni, bal Fe.
Heat treated: 123,000 TS; 15 El; 248 Brin.
For pumps, fittings, valves; hardenable, stainless; EN57.

DARWINS S80F.
M-487; 0.15-0.25 C, 16-18 Cr, 1-3 Ni, 0.5 max Mo, 0.25 S, bal Fe.
Heat treated: 123,000 TS; 15 El; 248 Brin.
For pumps, fittings, valves; hardenable, stainless, free-cutting.

DARWIN SPECIAL.
M-40; 1.2 C, bal Fe.
Annealed: 100,000 TS; 53,000 YS; 21 El; 42 RA; 200 Brin.
For drills, taps, hobs, reamers, cutters; Type W1; water hardened.

DARWIN SPECIAL CARBON.
M-40; 0.9 C, bal Fe.
Heat treated: 190,000 TS; 145,000 YS; 10 El; 30 RA; 400 Brin.
For drills, taps, hobs, reamers, lathe cutters; Type W1; water hardened.

DARWIN-SSC.
M-40; C, bal Fe.
For machine tool parts; water hardened.

DARWIN STANDARD.
M-40; 0.8-1.1 C, 0.25 Si, 0.25 Mn, bal Fe.
Water hardened: 166,000-216,000 TS; 110,000-150,000 YS; 11-15 El; 32-37 RA; 330-600 Brin.
For taps, drills, reamers, punches, stamps, knurls, mandrels.
Type W1 water hardening.

DARWIN TEMPER TOUGH.
M-40; 0.7 C, 0.75 Cr, 1.5 Ni, 0.25 Mo, bal Fe.
For brake dies, bushings, jigs, cams, chucks; Type L6; oil hardened.

DARWIN'S TRIPLE LIFE.
M-487; C, alloy, bal Fe.
For files; water hardened.

DARWIN "TTS".
M-40, M-487; C, W, bal Fe.
For tools; oil hardening.

DARWIN'S T.T.S.
M-487; C, 1.5 W, bal Fe.
For surgical instruments, taps, reamers, drills, cutters, hacksaws; water hardened.

DARWINS VANADIA.
M-487; C, W, alloy, bal Fe.
For hacksaw blades: water hardened.

DARWIN WARRANTED.
M-40, M-487; high C, bal Fe.
For tools; water hardening.

DARWIN "W" BRAND.
M-40; 0.45 C, 0.85-0.95 Cr, 1.0 W, bal Fe.
For tools, dies, punches; oil hardening.

DAUPHINOX A3.
M-1649; 0.03 max C, 18.0 Cr, 10.0 Ni, bal Fe.
Austenitic stainless steel, low carbon.
For brewery, dairy, cheese industry equipment.
AFNOR Z 03 CN 18-10.
AISI 304L; W. Nr. 4306.

DAUPHINOX A3I.
M-1649; 0.07 max C, 18.0 Cr, 10.0 Ni, bal Fe.
Austenitic stainless steel.
Navy, railroad, aeronautics equipment.
AFNOR Z 06 CN 18-10.
AISI 304; W.-Nr. 4301.

DAUPHINOX A3II.
M-1649; 0.12 max C, 18.0 Cr, 10.0 Ni, bal Fe.
Austenitic stainless steel.
Equipment for paper mills.
AFNOR Z 10 CN 18-10.
AISI 302; W.-Nr. 4300.

DAUPHINOX A3ML.
M-1649; 0.10 max C, 18.0 Cr, 12.0 Ni, 2.5 Mo, T = 4 x C, bal Fe.
Stabilized Type 316 stainless steel.
For welded assemblies for food and beverage equipment; also navy and aeronautics.
AFNOR Z 08 CNDT 18-12.
AISI 316 T; W.-Nr. 4571.

DAUPHINOX A3M2.
M-1649; 0.03 max C, 18.0 Cr, 13.0 Ni, 2.5 Mo, bal Fe.
Low carbon type 316 stainless.
For chemical industry equipment: organic, acetic, phosphoric acids; fiber industry.
AFNOR Z 03 CND 18-13.
AISI 316L; W.-Nr. 4404.

DAUPHINOX A3MO.
M-1649; 0.10 max C, 18.0 Cr, 12.0 Ni, 2.5 Mo, bal Fe.
Austenitic stainless steel; extra corrosion resistance.
Chemical, photography, cellulose industry equipment.
AFNOR Z 08 CND 18-12.
AISI 316; W.-Nr. 4401.

DAUPHINOX A3T.
M-1649; 0.12 max C, 18.0 Cr, 10.0 Ni, Ti = 4 x C, bal Fe.
Titanium stabilized austenitic steel.
For welded stainless structures.
AFNOR Z 10 CNT 18-10.
AISI 321; W.-Nr. 4541.

DAUPHINOX D1.
M-1649; 0.08 C, 13.0 Cr, bal Fe.
Semi-ferritic type stainless; generally not hardenable.
Fittings, slide calipers, knife handles.
AFNOR Z 08 C 13.
W.-Nr.-4000.

DAUPHINOX D2.
M-1649; 0.08 C, 16.0 Cr, bal Fe.
Ferritic stainless steel; not hardenable.
Corrosion resistant fittings.
AFNOR Z 08 C 16.
AISI 430; W.-Nr. 4016.

DAUPHINOX D2N.
M-1649; 0.15 C, 16.0 Cr, 2.0 Ni, bal Fe.
Martensitic stainless steel, hardenable to 46-29 Rc.
For parts used in sea water and superheated steam.
AFNOR Z 15 CN 16.2.
AISI 431; W.-Nr. 4057.

DAUPHINOX D2S.
M-1649; 0.08 C, 16.0 Cr, 0.20 S, bal Fe.
Free machining ferritic stainless steel; not hardenable.
For screws, bolts, nuts, threaded fittings.
AFNOR Z 08 CF 16.
AISI 430F; W.-Nr. 4104.

DAUPHINOX D3.
M-1649; 0.08 C, 18.0 Cr, bal Fe.
Ferritic stainless steel; not hardenable.
Fittings; good corrosion resistance.
AFNOR Z 08 C 18.
Similar to AISI 430 and W.-Nr.- 4016.

DAUPHINOX T1.
M-1649; 0.40 C, 13.5 Cr, bal Fe.
Martensitic stainless steel; hardenable to 56-53 Rc.
Special cutlery steel; can be highly polished.
AFNOR Z 40 C 14.
AISI 420; W.-Nr. 4034.

DAUPHINOX T1MO.
M-1649; 0.50 C, 13.5 Cr, 0.80 Mo, bal Fe.
Martensitic stainless steel; hardenable to 56-55 Rc.
For cutlery, surgical instruments.
AFNOR Z 50 CD 14.
W.-Nr. 4110.

DAUPHINOX T1ST.
M-1649; 0.70 C, 17.0 Cr, bal Fe.
Martensitic stainless steel; hardenable to 58-55 Rc.
For cutlery, surgical instruments.
AFNOR Z 70 C 17.
AISI 440A.

DAUPHINOX T5.
M-1649; 0.50 C, 14.0 Cr, bal Fe.
Martensitic stainless steel; hardenable to 57-54 Rc.
For surgical instruments.
AFNOR Z 50 Cr 14.

DAUPHINOX T5MO.
M-1649; 0.50 C, 14.0 Cr, 0.25 Mo, bal Fe.
Martensitic stainless steel; hardenable to 55-52 Rc.
For cutlery, surgical instruments.
AFNOR Z 50 CD 14.

DAUPHINOX T7MO.
M-1649; 0.70 C, 14.0 Cr, 0.80 Mo, bal Fe.
Martensitic stainless steel; hardenable to 57-53 Rc.
For cutlery, surgical instruments.
AFNOR Z 70 CD 14.

DAUPHINOX T10MC.
M-1649; 1.0 C, 15.0 Cr, 0.80 Mo, 1.0 Co, bal Fe.
Martensitic stainless steel; hardenable to 59-56 Rc.
For cutlery, surgical instruments.
AFNOR Z 100 CDK 15.
Similar to AISI 440 B or C. W.-Nr. 4535.

DAUPHINOX TP.
M-1649; 0.20 C, 13.5 Cr, bal Fe.
Martensitic stainless steel; hardenable to 49-46 Rc.
For surgical instruments, knife springs, turbine blades and wheels.
AFNOR Z 20 C 13.
AISI 420; W-Nr. 4021.

DAUPHINOX TP1.
M-1649; 0.30 C, 13.5 Cr, bal Fe.
Martensitic stainless steel; hardenable to 52-48 Rc.
For cutlery, scissors, surgical instruments.
AFNOR Z 30 C 13.
AISI 420.

DAUPHINOX TPMO.
M-1649; 0.20 C, 13.5 Cr, 0.80 Mo, bal Fe.
Martensitic stainless steel; hardenable to 49-46 Rc.
Paper stock beater and refiner blades, shears.
AFNOR Z 20 CD 14.
AISI 420; W.-Nr. 4120.

DAUPHINOX TPO.
M-1649; 0.15 C, 13.0 Cr, bal Fe.
Martensitic stainless steel, hardenable to 45-42 Rc.
For plastic moulds, turbine blades and wheels.
AFNOR Z-15 C 13.
AISI 416-410; W.-Nr. 4024.

DAVIGNON.
M-Eng.; 58 Au, 37 Cu, 5 Al.
For jewelry, ornaments; corrosion resistant.

DAVIS METAL.
M-145; 1.5 Mn, 29 Ni, 2 Fe, 67 Cu, 0.5 C, Si.
Cast: 60,000 TS; 35,000 YS; 18 El; 120 Brin.
For valves and fittings, turbine blades, throttle valves; corrosion and heat resisting.

DAVIS METAL.
M-145; 0.2 C, 0.3-1.0 Mn, 25 Ni, 6 Fe, 67 Cu, 0.8 Pb.
For valves, fittings.

DAWSON.
M-1705; 0.65 C, 0.9 Mn, 2.0 Si, bal Fe.
Shock resisting tool steel, water hardening, AISI S4.

DAWSON'S BRONZE.
M-Eng.; 83.9 Cu, 15.9 Sn, 0.10 Pb, 0.05 As.
Cast: 30,000 TS; 146 Brin.
For journal bearings; very fluid.

DBK 50/50.
M-45; 0.10 max C, 0.30 max Mn, 1.0 max Si, 48.0-52.0 Cr, bal Ni.
Cast alloy; corrosion and temperature resistant.
ASTM A560 Gr. 50 Cr-50 Ni.

DBK 50/50 CB.
M-45; 0.10 max C, 0.30 max Mn, 0.50 max Si, 48.0-52.0 Cr, 1.0 Cb, bal Ni.
Cast alloy; corrosion and heat resistant.

DBK CA-15.
M-45; 0.15 max C, 11-14 Cr, bal Fe.
Heat treated: 100,000-203,000 TS; 75,000-173,000 YS; 28-6 El; 62-9 RA; 185-415 Brin.
For pumps, valves, bushings, gears, impellers; ferritic, corrosion resistant; Type 410.

DBK CA40.
M-45; 0.30-0.40 C, 11-14 Cr, bal Fe.
Annealed: 95,000 TS; 60,000 YS; 25 El; 190 Brin.
Heat treated: 220,000 TS; 165,000 YS; 1 El; 470 Brin.
For fan blades, oil pump pistons, chemical plant equipment; heat resistant to 1200°F.

DBK CB-30.
M-45; 0.30 max C, 18-22 Cr, bal Fe.
For valve sleeves, chemical equipment impellers; corrosion resistant.

DBK CB-50.
M-45; 0.50 max C, 26-30 Cr, max Ni, bal Fe.
For lead pots, pump and valve bodies, furnace parts; heat and corrosion resistant.

DBK CF-8.
M-45; 0.08 max C, 1.5 max Mn, 2.0 max Si, 18.0-21.0 Cr, 8.0-11.0 Ni, bal Fe.
Cast austenitic stainless steel; ductile and corrosion resistant.
Similar to AISI 304.

DBK CF-20.
M-45; 0.20 max C, 1.5 max Mn, 2.0 max Si, 18.0-21.0 Cr, 8.0-11.0 Ni, bal Fe.
Cast austenitic stainless steel.
Similar to AISI 302.

DBK CH-20.
M-45; 0.20 max C, 1.5 max Mn, 2.0 max Si, 22.0-26.0 Cr, 12.0-15.0 Ni, bal Fe.
Cast austenitic stainless; resistant to hot dilute H_2SO_4.
Similar to AISI 309.

DBK CK-20.
M-45; 0.20 max C, 2.0 max Mn, 2.0 max Si, 23.0-27.0 Cr, 19.0-22.0 Ni, bal Fe.
Cast austenitic stainless; good corrosion and oxidation resistance at elevated temperatures.
Similar to AISI 310.

DBK HC.
M-45; 0.5 max C, 26-30 Cr, 4 max Ni, 0.5 max Mo, bal Fe.
Cast: 70,000 TS; 65,000 YS; 2 El; 190 Brin.
Aged: 115,000 TS; 80,000 YS; 18 El; 230 Brin.
For roasting furnaces, rabble arms, tuyeres, blowers; heat resistant.

DBK HD.
M-45; C, 26-30 Cr, 3-7 Ni, bal Fe.
Cast.
For oil and gas burner parts; heat and corrosion resistant.

DBK HE.
M-45; 0.20-0.50 C, 2.0 max Mn, 2.0 max Si, 26.0-30.0 Cr, 8.0-11.0 Ni, 0.5 max Mo, bal Fe.
Cast alloy; corrosion and heat resistant.
ASTM A297 Gr. HE.

DBK HF.
M-45; 0.20-0.40 C, 2.0 max Mn, 2.0 max Si, 19.0-23.0 Cr, 9.0-12.0 Ni, 0.5 max Mo, bal Fe.
Cast austenitic stainless; heat and corrosion resistant at elevated temperature.
ASTM A297 Gr. HF.

DBK HH.
M-45; 0.2 C, 9-13 Ni, 23-28 Cr, bal Fe.
Cast: 70,000 TS; 45,000 YS; 50 El.
For furnace parts, soot blowers, kiln parts; resists oxidation up to 2100°F.

DBK HI.
M-45; C, 26-33 Cr, 14-17 Ni, bal Fe.
For retorts for distillation of Mg, heat and corrosion resistant.
Cast alloy.

DBK HK.
M-45; 0.20-0.60 C, 2.0 max Mn, 2.0 max Si, 24,0-28.0 Cr, 18.0-22.0 Ni, 0.5 max Mo, bal Fe.
Cast austenitic stainless; heat and corrosion resistant; similar to AISI 310.
ASTM A297 Gr. HK.

DBK HL.
M-45; 0.20-0.60 C, 2.0 max Mn, 2.0 max Si, 28.0-32.0 Cr, 18.0-22.0 Ni, 0.5 max Mo, bal Fe.
Cast stainless alloy.
ASTM A297 Gr. HL.

DBK HN.
M-45; 0.20-0.50 C, 2.0 max Mn, 2.0 max Si, 19.0-23.0 Cr, 23.0-27.0 Ni, 0.5 max Mo, bal Fe.
Cast stainless alloy.
ASTM A297 Gr. HN.

DBK HOM-3.
M-45; 0.25-0.50 C, 24.0-27.0 Cr, 44.0-47.0 Ni, 1.25 max Mn, 1.5 max Si, 2.5-4.0 W, 2.5-4.0 Mo, 2.5-4.0 Co, bal Fe.
At 1800°F: 16,500 psi TS; 30.5 El.
Cast heat resistant austenitic alloy for use up to 2200°F.

DBK HP.
M-45; 0.35-0.75 C, 2.0 max Mn, 2.0 max Si, 24.0-28.0 Cr, 33.0-37.0 Ni, 0.5 max Mo, bal Fe.
Cast stainless alloy; for high temperature equipment as furnace parts.
ASTM A297 Gr. HP.

DBK HT.
M-45; 0.35-0.75 C, 2.0 max Mn, 2.0 max Si, 15.0-19.0 Cr, 33.0-37.0 Ni, 0.5 max Mo, bal Fe.
Cast stainless alloy; for high temperature.
ASTM A297 Gr. HT.

DBK HT-CB.
M-45; 0.35-0.60 C, 2.0 max Mn, 2.5 max Si, 15.0-19.0 Cr, 33.0-37.0 Ni, 0.5 max Mo, 1.0 max Cb, bal Fe.
Cast stainless alloy; resistant to carburization at elevated temperature.

DBK HU.
M-45; 0.35-0.75 C, 17-21 Cr, 37-41 Ni, bal Fe.
Cast. Stainless alloy.
For salt pots, retorts, annealing boxes; heat and corrosion resistant.
ASTM A297 Gr. HU.

DBK HW.
M-45; 0.35-0.75 C, 2.0 max Mn, 2.5 max Si, 10.0-14.0 Cr, 58.0-62.0 Ni, 0.5 max Mo, bal Fe.
Cast stainless alloy; for high temperature operation.
ASTM A297 Gr. HW.

DBK HX.
M-45; 0.35-0.75 C, 2.0 max Mn, 2.5 max Si, 15.0-19.0 Cr, 64.0-68.0 Ni, 0.5 max Mo, bal Fe.
Cast high temperature stainless alloy; usable to 2100°F.
ASTM A297 Gr. HX.

DBK MO-RE 1.
M-45; 0.40-0.50 C, 25.0-28.0 Cr, 35.0-38.0 Ni, 1.25 max Mn, 1.5 max Si, 1.25-2.0 W, 0.50 max Mo, bal Fe.
At 1800°F: 18,800 psi TS; 34 El.
Cast austenitic alloy for petrochemical and industrial heating applications at 1600-2100°F.

DBK MO-RE 2.
M-45; 0.15-0.25 C, 32.0-34.0 Cr, 48.0-52 Ni, 0.30 max Mn, 0.30 max Si, 15.0-17.0 W, 0.75-1.25 Al, bal Fe.
At 2000°F: 24,000 psi TS.
Cast heat resisting alloy with good high temperature creep resistance and oxidation resistance. For use at 2100-2400°F.

DBK MO-RE 5.
M-45; 0.15 max C, 26.0-30.0 Cr, 1.0 max Mn, 1.0 max Si, 50.0 max Co, bal Ni.
At 1650°F: 18,600 psi TS; 9 El.
Utilized for anti-pickup rolls in steel mill annealing lines.

DBK NA-3.
M-45; 0.20-0.50 C, 21.0-25.0 Cr, 23.0-27.0 Ni, 2.0 max Mn, 2.0 max Si, bal Fe.
At 1800°F: 11,950 psi TS; 51 El.
Cast austenitic stainless; for use as radiant tubes, blow pipes at 1800-2000°F.

DBK NA-6.
M-45; 0.20-0.50 C, 24.0-28.0 Cr, 11.0-14.0 Ni, 2.0 max Mn, 2.0 max Si, bal Fe.
At 1800°F: 9,000 psi TS; 45 El.
Cast austenitic stainless; good high temperature properties to 1800°F; for radiant tubes, furnace parts.

DBK NA-7.
M-45; 0.20-0.60 C, 24.0-28.0 Cr, 18.0-22.0 Ni, 2.0 max Mn, 2.0 max Si, bal Fe.
At 1800°F: 12,400 psi TS; 42 El.
Cast austenitic alloy for use as radiant tubes, carburizing fixtures, cement kiln parts up to 2000°F.

DBK SUPER 22-H.
M-45; 0.40-0.60 C, 26.0-30.0 Cr, 46.0-50.0 Ni, 1.5 max Mn, 1.75 max Si, 4.0-6.0 W, bal Fe.
At 1800°F: 18,000 psi TS; 32 El.
Cast alloy, good strength and oxidation resistance; for use at 1800-2250°F.

DBL-2.
M-1779; 0.8 C, 0.3 Mn, 0.3 Si, 4 Cr, 2 V, 6 W, 5 Mo, bal Fe.
General purpose high speed steel for cutting tools and ultra high strength non-cutting tool applications as punches and die inserts to operate at Rc 62/64.
AISI M-2.

D.B.L. 2 1/2.
M-1779; 1.0 C, 4 Cr, 6.25 W, 2.5 V, 6.25 Mo, bal Fe.
High speed tool steel, W-Mo type for lathe and planer tools, form tools, broaches, drills, reamers, hobs, taps, AISI M3.

DBL 3 see also **LUDLUM DBL 3.**

D.B.L. 3.
M-1779; 1.15 C, 4 Cr, 6 W, 3 V, 5.5 Mo, bal Fe.
High speed tool steel, W-Mo-V type for lathe and planer tools, broaches form tools, hobs, taps, AISI M3.

DC see **FINKL DC.**

D-C-33.
M-335; 0.33 C, 1.05 Si, 1.55 W, 1.65 Mo, 5.15 Cr, bal Fe.
Air hardening; for tools, dies. AISI H12 Hot work steel.

D.C. 33.
M-1116; 0.32 C, 3.0 Cr, 0.5 V, 2.6 Mo, 0.3 Mn, 0.3 Si, bal Fe.
Air or oil hardening tool and die steel; for hot work forging dies, shears and saws.
AFNOR: 32 CDV.12.30; AISI H10.

D-C-33-VA.
M-335; 0.33 C, 1.05 Si, 1.55 W, 1.65 Mo, 5.15 Cr, V, bal Fe.
Air hardening tool & die steel.
AISI H 12 Hot work steel.

DC-66 see **KLOSTER DC-66.**

DCM ALLOY.
M-937, M-31, M-167, M-1317; 0.08 max C, 15 Cr, 3.5 Ti, 0.08 B, 4.6 Al, 4.5-6.0 Mo, 4-6 Fe, bal Ni.
Heat treated: 140,000 TS; 116,000 YS; 5 El; 9 RA.
For turbine blades, jet engine components; age hardenable, high stressrupture strength.

D-DIE.
M-1704; 1.55 C, 11.5 Cr, 0.9 V, 0.8 Mo, bal Fe.
Air or oil hardening cold work tool steel, high carbon-high chromium type.
For punching and trimming dies, gages, broaches.
AISI D2.

DECOBRA.
M-Eng.; 74.4 Cu, 5.4 Zn, 19 Ni, 0.8 Fe, 0.4 Mn.
For ornamental parts, hardware; corrosion resistant.

DEEFIVE.
M-1097; 60-85 Au, bal Ag, Cu, Pt, Zn.
Cast: 112,000 TS; 1 El; 237 Brin.
For dentures, dental inlays; cast, hard.

DEEPEX.
M-526; 0.15 C, 0.35 Mn, bal Fe.
Steel welding rod.

DEEP HARDENING BERKSHIRE.
M-32; 1.23 C, 1.0 Mn, 0.50 Cr, 0.20 V, bal Fe.
Cold work tool steel; for thread chasing dies.

DEGUSSA ALLOY.
M-Eng.; 66 Cu, 34 Zn, traces Fe.
For hardware, lamp fixtures, ornamental parts; yellow brass.

DELAIR.
M-432; 1.0 C, 5.0 Cr, 1.0 Mo, 0.4 V, bal Fe.
Annealed: 105,000 TS; 52,000 YS; 26 El; C 18 Rock.
Heat treated: 253,000 TS; 200,000 YS; 3 El; C 53 Rock.
For blanking and trimming dies, cutters, engravers tools, shear blades, broaches, thread rolling dies.
Type A 2 air hardening tool steel.
Tough, wear resistant.

DELAWARE EXTRA.
M-432; 0.60-1.5 C, bal Fe.
For tools and parts, drills; water hardening.

DELAWARE H.S.
M-432; 0.82 C, 5 Mo, 6.5 W, 4 Cr, 2 V, bal Fe.
Heat treated: Rock C 63-67.
For lathe and planer tools, drills, taps, chasers, reamers, drawing dies.
Type M 2 high-speed steel. High toughness and wear resistant.

DELAWARE S.T.
M-432; C, Mn, Si, V, Mo, bal Fe.
Alloy tool steel, shock resisting type.

DELAWARE STANDARD.
M-432; 0.7-1.2 C, bal Fe.
For tools and parts, taps; water hardening.

DELCAR.
M-1580.
Tube carbide rods and electrodes (3 grades) for hard facing. Cannot be machined or ground.
For ore crusher rolls, grab bucket teeth, coal plow picks.

DELCAR.
M-432; 0.60-1.4 C, bal Fe.
For drills, taps, springs, hobs, reamers, punches; Type W 1; water hardened.

DELCROME 50V.
M-41, M-1580; 27 Cr, 2.75 C, 0.75 V, bal Fe.
Cast: 51-54 Rc.
High wear resistance, heat treatable for machining.
Hardfacing electrode for cold abrasion application.

DELCROME 450.
M-1580; 0.2 C, 11 Cr, 2 Ni, 1 Mo, bal Fe.
Cast: 45 Rock C.
For submerged arc welding; good general properties.

DELCROME 550.
M-1580; 15 Cr, 0.5 C, 0.4 Mo, bal Fe.
Wire for hard surfacing, multi-layer deposits on steel rolls, wear plates operating at temperature below 500°C. Hardness 51 Rc.

DELCROME 600.
M-1580; 27 Cr, 3.0 C, 0.7 Mo, bal Fe.
Wire for hard surfacing; maximum resistance to wear and abrasion. Hardness 54 Rc.

DELCROME-C.
M-1580; 21 Cr, 3.75 C, bal Fe.
Cast: C 54 Rock.
For tappet tips and rocker pads in internal combustion engines.
Wear resistant. Resists mineral abrasion.

DELCROME R.
M-1580; 30 Cr, 3 W, 6 Mn, 61 Fe.
Cast: 53 Rock C.
For hard facing, especially on wearing parts of earth moving equipment.

DELETOTS ALLOY.
80 Cu, 18 Zn, 2 Mn.
For brass solder, cartridge cases; extremely ductile.

DELFER.
M-1580; 16.5 Co, 13.5 Cr, 5.5 W, 9 Mo, 2.5 C, bal Fe.
Cast: 100,000 TS; 600 Brin.
For hard facing electrodes for abrasion resisting castings.
Heat and wear resistant.

DELFER B.
M-1580; 6 Co, 18 Cr, 3.2 C, 16 Mo, 2 V, bal Fe.
Hard facing electrode; for extreme conditions of abrasive wear by hard gritty particles; hardness 62 Rc.

DELHI GRADE A.
M-15; 16-18 Cr, 0.10 max C, 0.5 max Si, 0.5 Mn, bal Fe.
Rolled: 110,000 TS; 10 El; 25 RA; 200 Brin.
Annealed: 70,000 TS; 45,000 YS; 30 El; 60 RA; 160 Brin.
For ornamental work, tanks; max. operating temperature 1600°F; austenitic, corrosion resistant.

DELHI HARD.
M-15; 17.5 Cr, 1.0 C, 0.12 Ni, 1.13 Si, bal Fe.
Annealed: 90,000 TS; 40,000 YS; 25 El; 50 RA; 195 Brin.
Heat treated: 225,000 TS; 185,000 YS; 9 El; 25 RA; 420 Brin.
For structures; see Silcrome 17.

DELHI IRON.
M-15; 16.5-18 Cr, 0.10-0.11 C, 0.75-1.0 Si, bal Fe.
For replacing galvanized iron, for roofing, auto body sheets; see Silcrome 17.

DELHI IRON GRADE "S".
M-15, 11.5-14 Cr, 0.12 max C, 0.5 Si, 0.5 Mn, bal Fe.
Annealed: 45,000 TS; 30,000 YS; 30 El; 60 RA; 130 Brin.
Heat treated: 130,000 TS; 100,000 YS; 21 El; 68 RA; 235 Brin.
For turbine blades, pump rods, valves, cutlery, machine parts; see Silcrome 12.

DELHI SPECIAL.
M-15; 0.14 C, 0.46 Mn, 0.64 Si, 17 Cr, bal Fe.
At 70°F: 65,000 TS; 40,000 YS; 32 El; 63 RA.
At 1500°F: 10,000 TS; 8000 YS; 90 El; 98 RA.
For furnace linings, conveyors, furnace parts; heat and corrosion resistant.

DELHI TOUGH IRON.
M-15; 17 Cr, 1.25 Si, 0.3 Mn, 0.07 C, bal Fe.
For corrosion resisting parts; corrosion resistant; see Silcrome 17.

DELLOY.
M-756; Co-Cr-W.
For hard facing welding electrodes; heat and corrosion resistant.

DELLOY DC-7.
M-756; WC + Co.
For cutters for iron and nonferrous; sintered carbide.

DELLOY DS-8.
M-756; WC + Co.
For cutters for steel; sintered carbide.

DELLOY G-P.
M-756; 0.7 C, 18 W, 4 Cr, 1 V, bal Fe.
For cutters, tools; high speed steel.

DELLOY NO. 4.
M-756; Co-Cr-W.
For cutting tools.

DELLOY NO. 6.
M-756; Co-Cr-W.
For cutting tools.

DELLOY NO. 7.
M-756; Co-Cr-W.
For cutting tools.

DELLOY NO. 9.
M-756; Co-Cr-W.
For cutting tools.

DELORO 40 G.
M-1580; 7.5 Cr, 82 Ni, 5 Fe, 4 Si, 1.2 B.
Melt range 1810-2160°F.
As cast: 386 N/mm^2 TS; 255 N/mm^2 YS; 1 El; 34 Rock C.

DELORO ALLOY 40 G.
M-41; 0.3 C, 7.5 Cr, 4.0 Si, 5.0 Fe, 1.2 B, bal Ni.
As Cast: 741 N/mm² TS; 432 N/mm² YS; 10 El; 29-38 Rc.
Hard facing alloys, good corrosion resistance, machinable.

DELORO ALLOY 45.
M-41, M-1580; 7.5 Cr, 1.5 Fe, 4 Si, 1.5 B, bal Ni.
Cast: 35-42 Rc.
Hardfacing electrode for abrasion and corrosion resistance.

DELORO ALLOY 50.
M-41, M-1580; 10 Cr, 4 Mo, 4 Si, 1.5 B, bal Ni.
Cast: 49-52 Rc.
Hardfacing electrode for abrasion and corrosion resistance.

DELORO ALLOY 60.
M-41, M-1580; 15 Cr, 4.5 Fe, 4.5 Si, 3 B, bal Ni.
Cast: 59-62 Rc.
Hardfacing electrode for abrasion and corrosion resistance.

DELORO ALLOY B.
M-41, M-1580; 0.1 C, 29 Mo, 5 Fe, bal Ni.
Cast: 80,000 TS; 22-28 Rc.
High corrosion and oxidation resistance for chemical industry.

DELORO ALLOY C.
M-41, M-1580; 17 Cr, 5 W, 0.1 C, 17 Mo, 6 Fe, bal Ni.
Cast: 80,000 TS; 23-31 Rc.
For hardfacing drop forging dies, readily machinable.

DELORO ALLOY RT1.
M-1580; 15 Cr, 4 W, 0.4 C, 53 Ni, 15 Mo, 5 Fe, 7 Nb.
Cast: 300 Brin. approx.
For hardfacing and build-up; corrosion resistant and resistant to abrasive wear and impact at high temperatures.

DELORO PW22.
M-1580; 1.2 Cr, 95 Ni, 2.5 Si, 1.3 B.
Cast: 200-250 Brin. approx.
For hardfacing and build-up; good corrosion resistance; machinable.

DELORO SF40.
M-1580; 7.5 Cr, 1.5 Fe, 4 Si, 1.5 B, bal Ni.
Cast: C 39 Rock.
For hard facing; machinable. Abrasion and corrosion resistant.

DELORO SF50.
M-1580; 10 Cr, 4 Fe, 4 Si, 1.5 B, bal Ni.
Cast: C 51 Rock.
For hardfacing. Wear and corrosion resistant.

DELORO SF56.
M-1580; 16 Cr, 2 Cu, 2 W, 0.5 C, 2 Mo, 4 Fe, 4 Si, 4 B, bal Ni.
Hardfacing alloy, powder form; deposited hardness 60 Rock C.

DELORO SF60.
M-1580; 16 Cr, 4.5 Fe, 4.5 Si, 3 B, bal Ni.
Cast: C 60 Rock.
For hardfacing. Abrasion and corrosion resistant

DELORO STELLITE 1.
M-41; 33 Cr, 13 W, 2.5 C, bal Co.
Cast: 80,000 TS; 51-58 Rc.
For hardfacing electrode; heat and wear resistant.

DELORO STELLITE 3.
M-41; 30 Cr, 13 W, 2.4 C, bal Co.
Cast: 80,000 TS; 51-58 Rc.
For castings; high abrasion and corrosion resistant.

DELORO STELLITE 4.
M-41; 31 Cr, 14 W, 1 C, bal Co.
Cast: 130,000 TS; 45-49 Rc.
High temperature strength and wear resistant castings.

DELORO STELLITE 6.
M-41; 26 Cr, 5 W, 1 C, bal Co.
Cast: 116,000 TS; 39-43 Rc.
For hardfacing electrode, heat, shock and wear resistant.

DELORO STELLITE 7.
M-41; 26 Cr, 6 W, 0.4 C, bal Co.
Cast: 120,000 TS; 30-35 Rc.
Corrosion resistant and high temperature strength with good ductility and resistance to thermal shock.

DELORO STELLITE 8.
M-41; 27 Cr, 2 Ni, 0.2 C, 6 Mo, bal Co.
Cast: 120,000 TS; 30-35 Rc.
Thermal shock resistant; high temperature strength; corrosion resistant.

DELORO STELLITE 12.
M-41; 29 Cr, 9 W, 1.8 C, bal Co.
Cast: 108,000 TS; 47-51 Rc.
For hardfacing electrode, heat and wear resistant.

DELORO STELLITE 20.
M-41; 33 Cr, 18 W, 2.5 C, bal Co.
Cast: 80,000 TS; 55-59 Rc.
High abrasion and corrosion resistance.

DELORO STELLITE 100.
M-41; 34 Cr, 19 W, 2.0 C, bal Co.
Cast: 56,000 TS; 61-66 Rc.
Tool bits, milling cutter blades, etc.

DELORO STELLITE 250.
M-41; 0.1 C, 28 Cr, 20 Fe, Nb, bal Co.
As cast: 541 MN/m² TS; 309 MN/m² YS: 8 El; 19-29 Rc.
Excellent corrosion resistance, high temperature strength, ductility, and good resistance to thermal shock; machinable. For turbine blades, brass casting dies, extrusion dies.

DELORO STELLITE X-40.
M-41; 25 Cr, 7 W, 0.3 C, 10 Ni, bal Co.
Cast: 96,000 TS; 30-35 Rc.
Corrosion resistant, high temperature strength and resistance to thermal shock.

DELSTEEL ALLOY.
M-432; C, Mn, S, V, Mo, bal Fe.
Alloy tool steel, shock resisting type. AISI S 5.

DELTA ALUMINUM BRONZE CA1.
M-179, M-156; 10 Al, 5 Fe, 5 Ni, 0.25 Mn, bal Cu.
Light drawn: 110,000 TS; 67,000 YS; 15 El; 220 Brin.
High strength aluminum bronze, corrosion resistant.

DELTA ALUMINUM BRONZE CA2.
M-179, M-156; 10.5 Al, 4.5 Fe, 4.5 Ni, 0.5 Mn, bal Cu.
Light drawn: 115,000 TS; 45,000 YS; 20 El; 160 Brin.
Strong, more ductile aluminum bronze, corrosion resistant.

DELTA ALUMINUM BRONZE CA3.
M-179, M-156; 9.2 Al, 4.2 Fe, 4.2 Ni, 0.25 Mn, bal Cu.
Light drawn: 100,000 TS; 47,000 YS; 25 El; 170 Brin.
For bronze parts subject to shock; 20 ft. lbs. Izod.

DELTA ALUMINUM BRONZE CA4.
M-179, M-156; 11.5 Al, 5.5 Fe, 5.5 Ni, 1.5 Mn, bal Cu.
Light drawn: 130,000 TS; 60,000 YS; 5 El; 230 Brin.
For high strength hot-stampings.

DELTA ALUMINUM BRONZE CA5.
M-179, M-156; 9.6 Al, 2.25 Fe, 1.5 Ni, 0.3 Mn, bal Cu.
Light drawn: 90,000 TS; 47,000 YS; 15 El; 170 Brin.
Wear resistant; for valve guides on internal combustion engines.

DELTA ALUMINUM BRONZE CA6.
M-179, M-156; 8.6 Al, 2 Fe, 0.2 Mn, bal Cu.
Light drawn: 90,000 TS; 43,000 YS; 25 El; 160 Brin.
High impact strength; for pump spindles, valve stems, etc.

DELTA ALUMINUM BRONZE CA7.
M-179, M-156; 9.6 Al, 2 Fe, 0.25 Mn, bal Cu.
Light drawn: 90,000 TS; 43,000 YS; 25 El; 160 Brin.
For corrosion resistant bolts and hardware.

DELTA ALUMINUM BRONZE CA8.
M-179, M-156; 9.4 Al, 0.25 Mn, bal Cu.
Light drawn: 78,000 TS; 37,000 YS; 25 El; 160 Brin.
Resistant to sulphuric acid.

DELTA ALUMINUM BRONZE CA9.
M-179, M-156; 10 Al, 0.25 Mn, 1.5 Pb, bal Cu.
Light drawn: 78,000 TS: 36,000 YS; 10 El; 155 Brin.
For machined parts.

DELTA ALUMINUM BRONZE CA10.
M-156; 6.75 Al, 2 Si, bal Cu.
Light drawn: 80,000 TS; 40,000 YS; 25 El; 170 Brin.
High strength machined parts; highly corrosion resistant.

DELTA ALUMINUM BRONZE CA11.
M-179, M-156; 6.2 Al, 0.9 Fe, 0.3 Mn, 2.2 Si, bal Cu.
Light drawn: 80,000 TS; 40,000 YS; 35 El.
Izod 25 ft. lbs.; for parts subject to shock.

DELTA ALUMINUM BRONZE CA12.
M-179, M-156; 6.2 Al, 0.7 Fe, 0.2 Mn, 2.2 Si, bal Cu.
Light drawn: 80,000 TS; 45,000 YS; 35 El.
For parts subject to shock; Izod 30 ft. lbs.; non-magnetic.

DELTA ALUMINUM BRONZE CA13.
M-156; 7 Al, bal Cu.
Light drawn: 67,000 TS; 25,000 YS; 30 El; 120 Brin.
For cold formed components.

DELTA ALUMINUM BRONZE CA14.
M-179, M-156; 10 Al, 3 Fe, 1.5 Ni, 3 Mn, 0.9 Sn, bal Cu.
Light drawn: 170 Brin.
Wear resistant; for parts subject to stress corrosion.

DELTA ALUMINUM BRONZE CA15.
M-179, M-156; 3 Al, 6.75 Ni, 0.5 Mn, bal Cu.
Rolled: 90,000 TS; 50,000 YS; 20 El; 180 Brin.
For paper mill beater bars, etc.

DELTA ALUMINUM BRONZE CA16.
M-179, M-156; 11 Al, 5.5 Fe, 4.5 Ni, 1.5 Mn, bal Cu.
For high strength stamped parts.

DELTA BRONZE II.
M-103; 55-65 Cu, 30-44.9 Zn, 0.1-5.0 Fe.
Cast: 76,000 TS; 10 El.
Extruded: 84,000 TS; 14 El.
For tubes, ornamental and sanitary fittings, automobile parts; corrosion resistant.

DELTA BRONZE III.
M-103; 40-98 Cu, 1.8-45 Zn, 0.1-5.0 Fe, 0.1-10.0 Sn.
Cast: 49,000 TS; 30 El.
Extruded: 72,000 TS; 29 El.
For solid drawn tubes for hydraulic purposes, condensers, gears, valves, pump parts; corrosion resistant.

DELTA COLD FORMING BRASS CF1.
M-156; 62.25 Cu, bal Zn.
Annealed: 55,000 TS; 17,000 YS; 50 El; 70 Brin.
Hard drawn: 120,000 TS; 70,000 YS; 4 El; 160 Brin.
For screws, rivets, nails cold formed.

DELTA COLD FORMING BRASS CF2.
M-156; 62.25 Cu, bal Zn.
Annealed: 55,000 TS; 17,000 YS; 50 El; 70 Brin.
Hard drawn: 120,000 TS; 70,000 YS; 4 El; 160 Brin.
Mainly for wood screws.

DELTA COLD FORMING BRASS CF3.
M-156; 63 Cu, 0.15 Si, bal Zn.
Annealed: 55,000 TS; 17,000 YS; 50 El; 70 Brin.
Hard drawn: 120,000 TS; 70,000 YS; 4 El; 60 Brin.
Bent and welded into chain links; low fuming, easily welded.

DELTA COLD FORMING BRASS CF4.
M-156; 63.5 Cu, bal Zn.
Annealed: 55,000 TS; 17,000 YS; 50 El; 70 Brin.
Hard drawn: 120,000 TS; 70,000 YS; 4 El; 160 Brin.
Good ductility; for screws, rivets.

DELTA COLD FORMING BRASS CF5.
M-156; 65 Cu, bal Zn.
Annealed: 55,000 TS; 17,000 YS; 50 El; 70 Brin.
Hard drawn: 120,000 TS; 70,000 YS; 4 El; 160 Brin.
For cold forging, cold bending and forming.

DELTA COLD FORMING BRASS CF6.
M-156; 67.5 Cu, bal Zn.
Annealed: 55,000 TS; 17,000 YS; 50 El; 70 Brin.
Hard drawn: 120,000 TS; 70,000 YS; 4 El; 160 Brin.
Wire for metallizing.

DELTA COLD FORMING BRASS CF7.
M-156; 70 Cu, bal Zn.
Annealed: 45,000 TS; 16,000 YS; 60 El; 60 Brin.
Hard drawn: 120,000 TS; 70,000 YS; 4 El; 170 Brin.
For cold extrusion, spinning.

DELTA COLD FORMING BRASS CF8.
M-156; 71 Cu, 0.04 As, bal Zn.
Annealed: 45,000 TS; 16,000 YS; 60 El; 60 Brin.
Hard drawn: 120,000 TS; 70,000 YS; 4 El; 170 Brin.
For redrawing and weaving wire; anti-dezincification.

DELTA COLD FORMING BRASS CF9.
M-156; 80 Cu, bal Zn.
Annealed: 45,000 TS; 15,000 YS; 60 El; 60 Brin.
Hard drawn: 110,000 TS; 70,000 YS; 4 El; 170 Brin.
Excellent cold working properties.

DELTA COLD FORMING BRASS CF10.
M-156; 85 Cu, bal Zn.
Annealed: 45,000 TS; 15,000 YS; 50 El; 60 Brin.
Hard drawn: 95,000 TS; 70,000 YS; 4 El; 150 Brin.
Excellent cold working properties; Red Brass.

DELTA COLD FORMING BRASS CF11.
M-156; 90 Cu, bal Zn.
Annealed: 40,000 TS; 12,000 YS; 45 El; 55 Brin.
Hard drawn: 80,000 TS; 65,000 YS; 4 El; 130 Brin.
Excellent cold working properties; Gilding brass.

DELTA COLD FORMING BRASS CF12.
M-156; 88 Cu, 1 Sn, bal Zn.
Annealed: 40,000 TS; 11,000 YS; 50 El; 50 Brin.
Hard drawn: 80,000 TS; 60,000 YS; 5 El; 155 Brin.
Excellent cold working properties; Fire Bronze.

DELTA COLD FORMING BRASS CF13.
M-156; 80 Cu, 2 Sn, bal Zn.
Annealed: 55,000 TS; 60 El.

DELTA COLD FORMING BRASS CF21.
M-156; 59 Cu, bal Zn.
Light drawn: 57,000 TS; 36,000 YS; 35 El; 110 Brin.
For windshield frame sections, etc; limited cold bending.

DELTA COLD FORMING BRASS CF22.
M-156; 60.5 Cu, bal Zn.
Light drawn: 60,000 TS; 30,000 YS; 35 El; 110 Brin.
For bending, cold forging, hot working; Muntz metal.

DELTA COLD FORMING BRASS CF23.
M-156; 60.5 Cu, 0.5 Pb, bal Zn.
Light drawn: 55,000 TS; 25,000 YS; 35 El; 100 Brin.
For nut blanking, etc.; leaded Muntz metal.

DELTA COLD FORMING BRASS CF24.
M-156; 60.5 Cu, 1 Pb, bal Zn.
Light drawn: 55,000 TS; 27,000 YS; 30 El; 100 Brin.
For machining and cold forming parts.

DELTA COLD FORMING BRASS CF25.
M-156; 61 Cu, 0.3 Pb, bal Zn.
Drawn: 75,000 TS; 5 El.
For bending and some machining.

DELTA COLD FORMING BRASS CF26.
M-156; 63.5 Cu, 0.3 Pb, bal Zn.
Light drawn: 55,000 TS; 40 El.
For cold heading and drilling.

DELTA COLD FORMING BRASS CF27.
M-156; 63 Cu, 1.8 Pb, bal Zn.
Extruded product for subsequent rolling.

DELTA COLD FORMING BRASS CF28.
M-156; 65 Cu, 1.1 Pb, bal Zn.
Extruded product for subsequent rolling.

DELTA COPPER C 1.
M-156; ETP or FRHC extruded and drawn 99.9 Cu.
Annealed: 33,000 TS; 15,000 YS; 50 El; 45 Brin.
Hard drawn: 55,000 TS; 50,000 YS; 5 El; 90 Brin.
Wire for rivets, etc.

DELTA COPPER C 2.
M-156; ETP or FRHC Hot rolled and drawn 99.9 Cu.
Annealed: 33,000 TS; 15,000 YS; 50 El; 45 Brin.
Hard drawn: 55,000 TS; 50,000 YS; 5 El; 90 Brin.
Wire for nails, etc.

DELTA COPPER C 3.
M-156; DHP 99.85 Cu.
Annealed: 34,000 TS; 15,000 YS; 50 El; 45 Brin.
Hard drawn: 54,000 TS; 47,000 YS; 10 El; 95 Brin.
Tubes only; for water, gas, etc.

DELTA CROTERITE IV.
M-156, M-179; Ni, Al, Mn, bal Cu.
Annealed: 50,000 TS; 18,000 YS; 48 El; 95 Brin.
Hardened: 100,000 TS; 20 El.
For paper mill equipment, pump rods, piston rods, turbine blading; tough, corrosion resistant.

DELTA CROTERITE V.
M-156, M-179; Al, bal Cu.
Cast: 65,000 TS; 25,000 YS; 35 El; 25 RA; 100 Brin.
Hard rolled: 90,000 TS; 40,000 YS; 30 El; 50 RA; 140 Brin.
For pump rods, piston rods, valves, impellers, exhaust manifolds, worm wheels, gears; resists high temperatures and corrosion.

DELTA CROTERITE Z.
M-156, M-179; Cu, Al, Mn, 2 Fe, bal Zn.
Extruded: 84,000 TS; 38,000 YS; 28 El; 25 RA; 140 Brin.
Drawn: 84,000 TS; 42,000 YS; 25 El; 25 RA; 150 Brin.
For hardware; corrosion resistant.

DELTA HIGH TENSILE BRASS HT1.
M-179, M-156; 57 Cu, 1 Pb, 0.75 Sn, 0.75 Fe, 1.5 Mn, bal Zn.
Light drawn: 80,000 TS; 40,000 YS; 22 El; 135 Brin.
General purpose high tensile brass; for pump rods, valve spindles, etc.

DELTA HIGH TENSILE BRASS HT2.
M-179, M-156; 57.5 Cu, 1 Pb, 0.5 Sn, 0.75 Fe, 1 Al, 1.5 Mn, bal Zn.
Light drawn: 80,000 TS; 38,000 YS; 20 El; 120 Brin.
For pump rods, valve spindles; not suitable for plating.

DELTA HIGH TENSILE BRASS HT4.
M-179, M-156; 58.25 Cu, 0.9 Sn, 0.9 Fe, 0.6 Al, 0.5 Mn, bal Zn.
Light drawn: 82,000 TS; 40,000 YS; 20 El; 135 Brin.
For valve spindles, general marine work.

DELTA HIGH TENSILE BRASS HT5.
M-179, M-156; 58 Cu, 0.25 Pb, 1 Sn, 0.75 Fe, 1.5 Mn, bal Zn.
Light drawn: 82,000 TS; 43,000 YS; 20 El; 150 Brin.
For general purpose, plating and soldering.

DELTA HIGH TENSILE BRASS HT6.
M-156; 58 Cu, 1.2 Sn, 1.1 Fe, 0.3 Al, 0.2 Mn, bal Zn.
Light drawn: 72,000 TS; 36,000 YS; 25 El; 135 Brin.
For marine shafts, rolled tubes.

DELTA HIGH TENSILE BRASS HT7.
M-179, M-156; 57.5 Cu, 0.75 Sn, 0.75 Fe, 1.25 Mn, bal Zn.
Light drawn: 74,000 TS; 37,000 YS; 25 El; 135 Brin.
For parts subject to shock; Izod 20-25 ft. lbs.

DELTA HIGH TENSILE BRASS HT8.
M-179, M-156; 57 Cu, 1 Sn, 1 Fe, 0.75 Al, 1.5 Mn, bal Zn.
Light drawn: 85,000 TS; 42,000 YS; 25 El; 150 Brin.
For high pressure marine fittings.

DELTA HIGH TENSILE BRASS HT9.
M-179, M-156; 57 Cu, 0.75 Pb, 0.3 Sn, 1 Fe, 0.75 Al, 0.75 Mn, bal Zn.
Light drawn: 75,000 TS; 40,000 YS; 15 El; 120 Brin.
For special hot stampings.

DELTA HIGH TENSILE BRASS HT10.
M-179, M-156; 57 Cu, 0.75 Pb, 0.3 Sn, 1 Fe, 2.5 Al, 1.5 Mn, bal Zn.
For hot stampings; easily machined.

DELTA HIGH TENSILE BRASS HT11.
M-179, M-156; 58 Cu, 0.75 Pb, 0.75 Fe, 1.5 Al, 1 Mn, bal Zn.
Light drawn: 78,000 TS; 40,000 YS; 15 El.
For hot stampings; easily machined.

DELTA HIGH TENSILE BRASS HT12.
M-179, M-156; 57.5 Cu, 0.85 Pb, 0.65 Fe, 0.3 Al, 1.75 Mn, bal Zn.
Light drawn: 160 Brin.
For conveyer belt segments; hard wearing alloy.

DELTA HIGH TENSILE BRASS HT13.
M-179, M-156; 58 Cu, 2.25 Pb, 1 Sn, 0.35 Fe, 0.35 Al, 1.5 Mn, bal Zn.
Light drawn: 72,000 TS; 32,000 YS; 20 El; 115 Brin.
For valve spindles, bearing bushes; free machining.

DELTA HIGH TENSILE BRASS HT14.
M-179, M-156; 58.25 Cu, 1.5 Pb, 0.75 Sn, 0.75 Fe, 0.3 Al, 0.75 Mn, bal Zn.
For high pressure gas fittings; free machining.

DELTA HIGH TENSILE BRASS HT15.
M-156; 62 Cu, 0.4 Pb, 0.25 Sn, 1 Al, 0.4 Mn, bal Zn.
For golf ball moulds.

DELTA HIGH TENSILE BRASS HT16.
M-179, M-156; 65.8 Cu, 0.85 Sn, 1.4 Fe, 3 Al, 0.6 Mn, bal Zn.
Light drawn: 85,000 TS; 42,000 YS; 20 El; 155 Brin.
Wear and corrosion resistant brass.

DELTA HIGH TENSILE BRASS HT17.
M-179, M-156; 67.5 Cu, 1.4 Fe, 4.8 Al 0.5 Mn, bal Zn.
Light drawn: 98,000 TS; 50,000 YS; 15 El; 185 Brin.
For bearing plates, wear strip.

DELTA HIGH TENSILE BRASS HT18.
M-179, M-156; 70 Cu, 2 Fe, 6 Al, 3 Mn, bal Zn.
Light drawn: 110,000 TS; 60,000 YS; 12 El; 220 Brin.
For gears, cams, press slides, etc.

DELTA HIGH TENSILE BRASS HT19.
M-179, M-156; 57.25 Cu, 1 Pb, 0.25 Sn, 0.5 Fe, 0.75 Al, 1 Mn, 2 Ni, bal Zn.
For gas water heating equipment; good hot corrosion resistance.

DELTA HIGH TENSILE BRASS HT20.
M-179, M-156; 57.25 Cu, 0.5 Pb, 1.5 Al, 2.5 Mn, 0.75 Si, bal Zn.
Light drawn: 90,000 TS; 40,000 YS; 15 El; 155 Brin.
For bearings and bushings.

DELTA HIGH TENSILE BRASS HT21.
M-179, M-156; 58 Cu, 1.75 Al, 3 Mn, 1 Si, bal Zn.
Light drawn: 80,000 TS; 40,000 YS; 15 El; 130 Brin.
For gearbox components.

DELTA HIGH TENSILE BRASS HT22.
M-179, M-156; 60 Cu, 3 Pb, 2.5 Mn, bal Zn.
Resistance alloy; resistivity 17.

DELTA HIGH TENSILE BRASS HT23.
M-179, M-156; 56 Cu, 3 Pb, 6.3 Mn, bal Zn.
Resistance alloy; resistivity 20.

DELTA HIGH TENSILE BRASS HT24.
M-156; 58 Cu, 0.75 Sn, 0.5 Fe, 0.2 Mn, bal Zn.
Drawn: 80,000 TS; 60,000 YS; 15 El; 130 Brin.
Metallizing wire.

DELTA IMMADIUM I.
M-156, M-179; Zn, Al, Mn, bal Cu.
Cast: 72,400 TS; 19.5 El; 12 RA; 152 Brin.
Rolled: 70,000 TS; 35,000 YS; 33 El; 125 Brin.
For pump rods, valve spindles, staybolts, propeller shafts; high strength and corrosion resistance.

DELTA IMMADIUM II.
M-156, M-179; Zn, Al, Mn, bal Cu.
Cast: 68,400 TS; 10 El; 9 RA; 149 Brin.
Rolled: 81,000 TS; 40,000 YS; 28 El; 135 Brin.
For pump rods, valve spindles, staybolts, shafts, impellers; high strength and corrosion resistance.

DELTA, KRUPP.
M-72; 54-56 Cu, 41-42 Zn, 0.7-1.8 Pb, 0.86-1.28 Fe, 0.8-1.4 Mn.
For high strength structural castings; tough.

DELTAL.
M-1356; 0.6-1.4 Mg, 0.6-1.6 Si, 0.6-1.0 Mn, 0.3 max Cr, bal Al.
Annealed: 21,000 TS; 80,000 YS; 24 El; 36 Brin.
Hard: 32,000 TS; 29,000 YS; 6 El.
For structural members.

DELTAL.
M-1548; 0.3-1.5 Si, 0.2-1.5 Mn, 0.5-2.0 Mg, bal Al.
Ht. Tr.: 56,000 TS; 52,000 YS; 12 El; 120 Brin.
For general structures, scaffolds, rails, booms, water heater tubes.
Heat treatable, good fabricability.

DELTA MANGANESE BRASS MB1.
M-179, M-156; 57 Cu, 1.25 Mn, 2.25 Pb, bal Zn.
Light drawn: 67,000 TS; 27,000 YS; 30 El; 95 Brin.
Architectural uses; good machinability.

DELTA MANGANESE BRASS MB2.
M-179, M-156; 57 Cu, 1.25 Mn, 1.25 Pb, bal Zn.
Light drawn: 67,000 TS; 27,000 YS; 30 El; 95 Brin.
Architectural uses; "warmer" appearance.

DELTA MANGANESE BRASS MB3.
M-179, M-156; 55.5 Cu, 1.25 Mn, bal Zn.
Architectural uses.

DELTA MANGANESE BRASS MB4.
M-179, M-156; 57.75 Cu, 1.25 Mn, bal Zn.
Architectural uses; good cold bending.

DELTA MANGANESE BRONZE MB5.
M-179, M-156; 58.5 Cu, 0.35 Mn, 1.25 Fe, 1 Sn, bal Zn.
Light drawn: 70,000 TS; 35,000 YS; 15 El; 115 Brin.
American standard CA675.

DELTA-MU.
M-1089; 45 Ni, 5 Cr, bal Fe.
For magnetic cores; high electrical resistivity.

DELTA NAVAL BRASS N 1.
M-156; 61.5 Cu, 1.2 Sn, 0.2 Pb, bal Zn.
Light drawn: 55,000 TS; 24,000 YS; 25 El; 110 Brin.
Limited machining and hot stamping; good corrosion resistance.

DELTA NAVAL BRASS N 2.
M-156; 61.5 Cu, 1.2 Sn, bal Zn.
Light drawn; 55,000 TS; 22,000 YS; 30 El; 90 Brin.
For cold forming, instrument components; non-magnetic.

DELTA NAVAL BRASS N 3.
M-156; 58.5 Cu, 0.8 Sn, bal Zn.
Light drawn: 58,000 TS; 27,000 YS; 20 El; 100 Brin.
Good hot stamping, reasonable machinability.

DELTA NAVAL BRASS N 4.
M-156; 60.75 Cu, 1.2 Sn, 1.5 Pb, bal Zn.
Light drawn: 55,000 TS; 20 El; 100 Brin.
Good machining alloy.

DELTA NAVAL BRASS N 5.
M-156; 60.25 Cu, 0.75 Sn, bal Zn.
Light drawn: 60,000 TS; 27,000 YS; 25 El; 90 Brin.
American standard CA 482.

DELTA NAVAL BRASS N 6.
M-156; 60.25 Cu, 0.75 Sn, 0.6 Pb, bal Zn.
Light drawn: 60,000 TS; 27,000 YS; 20 El; 90 Brin.
American standard CA 482.

DELTA NAVAL BRASS N 7.
M-156; 60.25 Cu, 0.75 Sn, 1.5 Pb, bal Zn.
Light drawn: 60,000 TS; 27,000 YS; 20 El; 90 Brin.
American standard CA 485.

DELTA NAVAL BRASS N 8.
M-156; 58 Cu, 1.2 Sn, 2 Pb, bal Zn.
Light drawn: 60,000 TS; 25,000 YS; 20 El; 100 Brin.
Free machining alloy, also hot-stamping.

DELTA NAVAL BRASS N 9.
M-156; 60 Cu, 1.1 Sn, bal Zn.
Drawn: 80,000 TS; 15 El; 200 Brin.
For electrical switching parts.

DELTA NAVAL BRASS N 10.
M-156; 61 Cu, 1.75 Sn, bal Zn.
Light drawn: 60,000 TS; 20 El; 120 Brin.
For electrical plug pins.

DELTA NICKEL SILVER NS1.
M-156; 45.5 Cu, 9.5 Ni, 2 Pb, 0.3 Mn, bal Zn.
Light drawn: 90,000 TS; 45,000 YS; 20 El; 150 Brin.
For machining and hot stamping decorative and ornamental parts.

DELTA NICKEL SILVER NS2.
M-179, M-156; 40 Cu, 13.25 Ni, 1.9 Pb, 2 Mn, bal Zn.
Light drawn: 90,000 TS; 58,000 YS; 15 El; 150 Brin.
For decorative and ornamental parts; colour of sterling silver.

DELTA NICKEL SILVER NS3.
M-156; 46.5 Cu, 9.25 Ni, bal Zn.
Light drawn: 80,000 TS; 30 El.
For window fittings; warm pink color.

DELTA NICKEL SILVER NS4.
M-156; 51.5 Cu, 5.5 Ni, bal Zn.
For decorative purposes; yellow color.

DELTA NICKEL SILVER NS5.
M-156; 45.5 Cu, 9.5 Ni, 2 Pb, bal Zn.
Light drawn: 89,000 TS; 20 El; 130 Brin.
For instrument components; non-magnetic.

DELTA NICKEL SILVER NS6.
M-179, M-156; 49.5 Cu, 9.8 Ni, 0.35 Mn, bal Zn.
Rolled: 83,000 TS; 42,000 YS; 18 El; 145 Brin.
Rolled bar and sheet; lead free.

DELTA PHOSPHOR BRONZE PB1.
M-156; 5 Sn, 0.2 P, bal Cu.
Annealed: 60,000 TS; 30,000 YS; 50 El; 90 Brin.
Hard drawn: 130,000 TS; 70,000 YS; 4 El; 200 Brin.
Wire for redrawing, etc.

DELTA PHOSPHOR BRONZE PB2.
M-156; 7 Sn, 0.2 P, bal Cu.
Annealed: 60,000 TS; 30,000 YS; 50 El; 90 Brin.
Hard drawn: 130,000 TS; 70,000 YS; 4 El; 220 Brin.
For springs.

DELTA PHOSPHOR BRONZE PB3.
M-156; 8 Sn, 0.3 P, bal Cu.
Annealed: 60,000 TS; 30,000 YS; 50 El; 80 Brin.
Hard drawn: 130,000 TS; 70,000 YS; 4 El; 230 Brin.
For Fourdrinier wire in paper making.

DELTA PHOSPHOR BRONZE PB4.
M-156; 6 Sn, 0.25 P, bal Cu.
Annealed: 60,000 TS; 30,000 YS; 50 El; 90 Brin.
Hard drawn: 130,000 TS; 70,000 YS; 4 El; 210 Brin.
Wire for redrawing, etc.

DELTA STAMPING BRASS S 1.
M-156; 57.25 Cu, 2 Pb, bal Zn.
Light drawn: 60,000 TS; 30,000 YS; 20 El; 120 Brin.
For general purpose hot brass stampings.

DELTA STAMPING BRASS S 2.
M-156; 59 Cu, 1.75 Pb, bal Zn.
Light drawn: 60,000 TS; 25,000 YS; 25 El; 110 Brin.
For hot brass stampings; some cold formability.

DELTA STAMPING BRASS S 3.
M-156; 59.5 Cu, 2.25 Pb, bal Zn.
Light drawn: 60,000 TS; 25,000 YS; 25 El; 110 Brin.
For hot brass stampings; improved machinability.

DELTA STAMPING BRASS S 4.
M-156; 55 Cu, 2.75 Pb, 0.2 Sn, bal Zn.
For gas tap components, etc.

DELTA STAMPING BRASS S 5.
M-156; 59 Cu, 1 Pb, bal Zn.
Light drawn: 62,000 TS; 30,000 YS; 25 El; 110 Brin.
For hot brass stampings, also cold forming.

DELTA STAMPING BRASS S 6.
M-156; 58.5 Cu, 1.75 Pb, bal Zn.
Light drawn: 60,000 TS; 30,000 YS; 20 El; 110 Brin.
For hot brass stampings, non-magnetic.

DELTA STAMPING BRASS S 7.
M-156; 57.5 Cu, 2 Pb, bal Zn.
Light drawn: 60,000 TS; 30,000 YS; 20 El; 120 Brin.
For hot stamped instrument components; non-magnetic.

DELTA STAMPING BRASS S 8.
M-156; 58.25 Cu, 1.75 Pb, bal Zn.
Light drawn: 63,000 TS; 35,000 YS; 15 El; 120 Brin.
For valve bodies, etc.

DELTA STAMPING BRASS S 9.
M-156; 70 Cu, 1.75 Al, 0.03 As, bal Zn.
For hot stamped water fittings; anti-dezincification.

DELTA T II.
M-764; 10 Al, 2 Mn, 2 Fe, bal Cu.
Rolled: 85,000 TS; 42,000 YS; 15 El; 150 Brin.
For paper and chemical industries, pump shafts, valve components.
Aluminum bronze. Wear and corrosion resistant.

DELTA TURNING AND RIVETING BRASS TR1.
M-156; 61.5 Cu, 2.25 Pb, bal Zn.
Light drawn: 55,000 TS; 35,000 YS; 25 El; 100 Brin.
Standard turning and riveting brass; good machinability.

DELTA TURNING AND RIVETING BRASS TR2.
M-156; 62.5 Cu, 2.25 Pb, bal Zn.
Light drawn: 55,000 TS; 30,000 YS; 25 El; 100 Brin.
Turning and riveting brass; machinable and ductile.

DELTA TURNING AND RIVETING BRASS TR3.
M-156; 60.5 Cu, 1.75 Pb, bal Zn.
Light drawn: 55,000 TS; 35,000 YS; 25 El; 100 Brin.
Good turning, riveting, hot stamping.

DELTA TURNING AND RIVETING BRASS TR4.
M-156; 61.5 Cu, 1.25 Pb, bal Zn.
Light drawn: 55,000 TS; 28,000 YS; 25 El; 100 Brin.
Good machining and riveting.

DELTA TURNING AND RIVETING BRASS TR5.
M-156; 60.5 Cu, 2.5 Pb, bal Zn.
Light drawn: 55,000 TS; 35,000 YS; 25 El; 100 Brin.
For machined instrument parts; non-magnetic.

DELTA TURNING AND RIVETING BRASS TR6.
M-156; 62.5 Cu, 1.25 Pb, bal Zn.
Light drawn: 55,000 TS; 28,000 YS; 25 El; 100 Brin.
For cold shaped instrument parts; non-magnetic.

DELTA TURNING AND RIVETING BRASS TR7.
M-156; 63.25 Cu, 1.8 Pb, bal Zn.
Light drawn: 55,000 TS; 30,000 YS; 25 El; 110 Brin.
For machining and riveting; high purity.

DELTA TURNING BRASS T 1.
M-156; 57 Cu, 3 Pb, bal Zn.
Light drawn: 60,000 TS; 25,000 YS; 20 El; 120 Brin.
For general purpose machined components.

DELTA TURNING BRASS T 2.
M-156; 57.7 Cu, 2.75 Pb, bal Zn.
Light drawn: 60,000 TS; 25,000 YS; 20 El; 120 Brin.
For machined components; also hot stamping.

DELTA TURNING BRASS T 3.
M-156; 57.7 Cu, 3.5 Pb, 0.3 Sn, bal Zn.
Light drawn: 60,000 TS; 20 El; 100 Brin.
For high speed machining and intricate slotting and drilling.

DELTA TURNING BRASS T 4.
M-156; 58 Cu, 2 Pb, 0.6 Sn, bal Zn.
For roller bearing cages; good corrosion resistance.

DELTA TURNING BRASS T 5.
M-156; 61.5 Cu, 3.25 Pb, bal Zn.
Light drawn: 55,000 TS; 30,000 YS; 25 El; 110 Brin.
For automatic machining; also for thread rolling.

DELTA TURNING BRASS T 6.
M-156; 65 Cu, 3.25 Pb, bal Zn.
Light drawn: 55,000 TS; 25 El; 110 Brin.
For very fine deep drilling; high purity.

DELTA TURNING BRASS T 7.
M-156; 57.5 Cu, 3 Pb, bal Zn.
Light drawn: 60,000 TS; 25,000 YS; 20 El; 120 Brin.
For machined instrument components; non-magnetic.

DELTA TURNING BRASS T 8.
M-156; 56.5 Cu, 2.25 Pb, bal Zn.
Light drawn: 60,000 TS; 25,000 YS; 20 El; 120 Brin.
For machined components and hot stamping.

DELTA TURNING BRASS T 9.
M-156; 56.25 Cu, 2.75 Pb, bal Zn.
Light drawn: 60,000 TS; 25,000 YS; 20 El; 120 Brin.
For hollow shapes as bolt sections; very good for plating.

DELTA TURNING BRASS T 10.
M-156; 57.5 Cu, 2.5 Pb, 0.5 Al, bal Zn.
Extruded: 55,000 TS; 25,000 YS; 25 El; 100 Brin.
For hinges, lock sections; natural bright finish.

DELTA WELDING ALLOY W 1.
M-156; 60.5 Cu, 0.4 Si, 0.4 Sn, bal Zn.
As welded: 56,000 TS; 40,000 YS; 20 El; 130 Brin.
For brazing copper and mild steel.

DELTA WELDING ALLOY W 2.
M-156; 60.5 Cu, 0.3 Si, bal Zn.
Rod as manufactured: 75,000 TS; 15 El.
For brazing copper and mild steel.

DELTA WELDING ALLOY W 3.
M-156; 59.5 Cu, 0.2 Si, 0.3 Sn, bal Zn.
For brazing copper and mild steel.

DELTA WELDING ALLOY W 4.
M-156; 59.5 Cu, 0.25 Si, 0.4 Sn, 0.4 Fe, 0.2 Mn, bal Zn.
Welded: 56,000 TS.
High tensile brass weld; for brazing copper and cast iron.

DELTA WELDING ALLOY W 5.
M-179, M-156; 48.5 Cu, 9 Ni, 0.25 Si, 0.4 Mn, bal Zn.
Welded: 60,000 TS.
Nickel silver weld metal; for brazing ferrous metals.

DELTA WELDING ALLOY W 6.
M-179, M-156; 48 Cu, 8.5 Ni, 0.25 Si, 0.3 Sn, 0.4 Mn, bal Zn.
Nickel silver weld metal; for brazing ferrous metals.

DELTA WELDING ALLOY W 7.
M-156; 48 Cu, 10 Ni, 0.15 Si, 0.02 P, bal Zn.
Rod as manufactured: 100,000 TS; 56,000 YS; 15 El; 210 Brin.
Nickel silver weld metal; for brazing ferrous metals.

DELTA WELDING ALLOY W 8.
M-156; 58.5 Cu, 0.5 Ni, 0.4 Si, 0.8 Sn, 0.4 Fe, 0.4 Mn, bal Zn.
As welded: 60,000 TS.
High tensile brass weld; for brazing copper and cast iron.

DELTA WELDING ALLOY W 9.
M-156; 57.5 Cu, 0.1 Si, 0.8 Sn, 0.8 Fe, 0.3 Mn, bal Zn.
Rod as manufactured: 80,000 TS; 170 Brin.
High tensile brass weld; for brazing copper and cast iron.

DELTA WELDING ALLOY W 10.
M-156; 7 Al, 0.4 Si, bal Cu.
Welded: 67,000 TS.
For welding aluminum bronzes.

DELTA WELDING ALLOY W 11.
M-156; 9.4 Al; 0.5 Mn, bal Cu.
As welded: 67,000 TS.
For welding aluminum bronze; high impact strength.

DELTA WELDING ALLOY W 12.
M-179, M-156; 9.75 Al, 2.25 Fe, 1.75 Mn, bal Cu.
As welded: 63,000 TS.
For welding aluminum bronzes.

DELTABRA.
M-338; 77 Cu, 2 Al, 0.03 As, bal Zn.
Cold drawn: 82,880 TS; 76,160 YS; 8 El.
For condenser tubes, tanker heater tubes; resists sea water corrosion.

DELTOXAL.
M-1356; 2-4 Mg, 0.4 max Mn, 0.3 max Cr, bal Al.
Annealed: 28,000 TS; 13,000 YS; 30 El; 47 Brin.
Hard: 42,000 TS; 37,000 YS.
For aircraft and missile and marine parts; corrosion resistant to sea water.

DELTUMIN.
M-1548; 3.5-5.5 Cu, 0.2-1.5 Si, 0.1-1.5 Mn, 0.2-2.0 Mg, bal Al.
Ht. Tr.: 70,000 TS; 60,000 YS; 13 El; 135 Brin.
For fittings, hardware, aircraft engine components.
Heat treatable, high tensile and fatigue strength.

DE LUXE-9.
M-1145; 0.9 C, 0.25 Mn, 0.2 Si, bal Fe.
For header dies, reamers.

DE LUXE-10.
M-1145; 1.05 C, 0.25 Mn, 0.2 Si, bal Fe.
For dies, forming tools.

DE LUXE-12.
M-1145; 1.2 C, 0.25 Mn, 0.2 Si, bal Fe.
For cutters; for nonferrous metals.

DEMARK.
M-1078; Ti C, Co.
For thread gauging dies; sintered carbide.

DEM NO. 1.
M-1066; 63.3 Ag, bal Cu.
For silver solder; M.P. 1275°F.

DEM NO. 2.
M-1066; 66.7 Ag, bal Cu.
For silver solder; M.P. 1363°F.

DEM NO. 10.
M-1066; 22.2 Ag, bal Cu.
For silver solder; M.P. 1207°F.

DEM NO. 22.
M-1066; 50 Ag, bal Cu.
For silver solder; M.P. 1255°F.

DEM NO. 23.
M-1066; 60 Ag, bal Cu.
For silver solder; M.P. 1248°F.

DEM NO. 37.
M-1066; 9 Ag, bal Cu.
For silver solder; M.P. 1510°F.

DEM NO. 41.
M-1066; 71.8 Ag, bal Cu.
For silver solder; M.P. 1434°F.

DEM NO. 43.
M-1066; 54 Ag, bal Cu.
For silver solder; M.P. 1223°F.

DEM NO. 48.
M-1066; 17 Ag, bal Cu.
For silver solder; M.P. 1382°F.

DEM NO. 64.
M-1066; 19-20 Ag, bal Cu.
For silver solder; M.P. 1430°F.

DEM NO. 65.
M-1066; 40 Ag, bal Cu.
For silver solder; M.P. 1233°F.

DEM NO. 66.
M-1066; 64-66 Ag, bal Cu.
For silver solder; M.P. 1280°F.

DEM NO. 69.
M-1066; 44-46 Ag, bal Cu.
For silver solder; M.P. 1250°F.

DEM NO. 71.
M-1066; 49-51 Ag, bal Cu.
For silver solder; M.P. 1160°F.

DEM NO. 73.
M-1066; 49-51 Ag, bal Cu.
For silver solder; M.P. 1195°F.

DEM NO. 102.
M-1066; 41.7 Au, bal Cu.
For gold solder; M.P. 1436°F-1508°F.

DEM NO. 104.
M-1066; 41.7 Au, bal Cu.
For gold solder; M.P. 1248°F-1318°F.

DEM NO. 106.
M-1066; 41.7 Au, bal Cu.
For gold solder; M.P. 1378°F-1410°F.

DEM NO. 110.
M-1066; 58.3 Au, bal Cu.
For gold solder; M.P. 1395°F-1310°F.

DEM NO. 112.
M-1066; 50 Au, bal Cu.
For gold solder; M.P. 1420°F-1495°F.

DEM NO. 114.
M-1066; 33.3 Au, bal Cu.
For gold solder; M.P. 1360°F-1392°F.

DEM NO. 115.
M-1066; 33.3 Au, bal Cu.
For gold solder; M.P. 1245°F-1310°F.

DEM NO. 119.
M-1066; 25 Au, bal Cu.
For gold solder; M.P. 1350°F-1141°F.

DEM NO. 120.
M-1066; 41.7 Au, bal Cu.
For gold solder; M.P. 1378°F-1418°F.

DEM NO. 293.
M-1066; 50 Au, bal Cu.
For gold solder; M.P. 1317°F-1255°F.

DEM NO. 426.
M-1066; 33.3 Au, bal Cu.
For gold solder; M.P. 1407°F-1522°F.

Section I: Alloy Data / 407

DEM NO. 429.
M-1066; 50 Au, bal Cu.
For gold solder; M.P. 1392°F-1373°F.

DEM NO. 430.
M-1066; 58.3 Au, bal Cu.
For gold solder; M.P. 1323°F-1248°F.

DEM NO. 431.
M-1066; 58.3 Au, bal Cu.
For gold solder; M.P. 1495°F-1440°F.

DEM NO. SSEI-70.
M-1066; 46 Ag, bal Cu.
For silver solder; M.P. 1171°F.

DEM NO. SSEI-82.
M-1066; 46 Ag, bal Cu.
For silver solder; M.P. 1165°F.

DEMO BRONZE.
61 Cu, 33 Pb, 4-6 Sn, 2-1 Ni.
18,000 TS; 3 El; 0.35 RA; 52 Brin.
For bearings, utensils; C.U.S.-56,000.

DENAVIS.
M-755; 0.7-1.2 C, bal Fe.
For tools, cutters; water hardening.

DENAVIS HIGH SPEED.
M-755; 0.7 C, 18 W, 4 Cr, bal Fe.
For tools, cutters, reamers; high speed steel.

DENINE.
M-435; 1.2 C, 1.5 W, bal Fe.
For taps, drawing dies; water hardened.

DENSALLOY.
M-1388; 90 W, 6 Ni, 4 Cu.
Sintered: 85,000-115,000 TS; 10-2 El; 400 Brin.
For balance weights, centrifugal clutches; sintered; heavy metal.

DENSE STEEL.
M-U.S.; Cr, Si, Mo, C, bal Fe.
For tools.

DENSITE.
M-1423; 1.4 C, 11-13 Cr, 0.8 Mo, 0.2 min V, bal Fe.
For cutting, forming and trimming dies; air hardened.

DENSITE AH.
M-1423; 1.5 C, 12 Cr, 1 Mo, bal Fe.
For forming and drawing dies, punches; Type D 2; air hardened, non-deforming.

DENSO IRON.
M-95; 3 C, Cr, Mo, bal Fe.

DENTAL BURR (1%W).
M-38; 1.15 C, 1.25 W, bal Fe.

DENTAL BURR (3 1/2%W).
M-38; 1.05 C, 0.75 Cr, 3.5 W, bal Fe.

DENTALLOY.
M-216; 38.14 Bi, 26.42 Pb, 31.67 Sn, 2.64 Cd, 1 Sb, 0.06 Cu.
Cast: 15 Brin.
For dental castings, models; see "Cerrodent."

DENTURE CLASP.
M-349.
Gold color clasp and orthodontic wire.
Fusing temp: 1650°F.

DEOXIDIZED COPPER 105.
M-141; 99.9 min Cu, 0.02 P.
Hard: 55,000 TS; 50,000 YS; 8 El; 107 Brin.
Soft: 32,000 TS; 10,000 YS; 45 El; 40 Brin.
For deep drawing, water tubes; deoxidized copper.

DEOXIDIZED LEADED COPPER-129.
M-8; 0.80-1.2 Pb, 99.90 Min Cu + Pb.
Hard: 51,000-60,000 TS; 46,000-55,000 YS; 14-10 El; 83-100 Brin.
For contact pins and inserts, screw machine products; high thermal and electrical conductivity, free-cutting, corrosion resistant.

DEOXOLOY.
M-126.
For general brass and bronze deoxidizer.

DEOXOLOY.
M-1265; Cu alloy.
For deoxidizer for Cu alloys.

DEPAL.
M-757; 2 Cu, 2 Mn, 2 Ni, bal Al.
Cast: 20,000-26,000 TS; 10,000-14,000 YS; 1.5-3.5 El; 55-65 Brin.
For pistons, cylinder heads; age-hardenable.

DEURANCE METAL.
M-Eng.; 33.3 Sn, 44.5 Sb, 22.2 Cu.
For locomotive bearing; Babbitt.

DEUTRO 18/8.
M-1300; 0.80-0.20 C, 2 max Mn, 17-19 Cr, 8-10 Ni, bal Fe.
Annealed: 85,000 TS; 35,000 YS; 60 El; 70 RA; 150 Brin.
Cold drawn: 125,000 TS; 95,000 YS; 25 El; 55 RA; 277 Brin.
For oil refinery and chemical plant equipment; Type 302; stainless, austenitic.

DEUTRO 18/8 MS.
M-1300; 0.10 max C, 16-18 Cr, 10-14 Ni, 2-3 Mo, bal Fe.
Annealed: 85,000-95,000 TS; 35,000-45,000 YS; 60-50 El; 75-60 RA; 150-190 Brin.
For chemical plant equipment, mixers, agitators, filters; Type 316; stainless, austenitic.

DEUTRO 18/8S.
M-1300; 0.08 max C, 18-20 Cr, 8-11 Ni, 2 max Mn, bal Fe.
Annealed: 90,000 TS; 45,000 YS; 60 El; 135 Brin.
Cold drawn: 180,000 TS; 150,000 YS; 10 El; 330 Brin.
For chemical plant equipment, welded structures; Type 304; stainless, austenitic.

DEUTRO 20/15.
M-1300; 0.2 max C, 22-24 Cr, 12-15 Ni, 2 max Mn, 1 max Si, bal Fe.
Annealed: 85,000-95,000 TS; 40,000-50,000 YS; 45-55 El; 150-185 Brin.
For heat treating boxes, oil refinery and chemical plant equipment; Type 309; austenitic, heat resistant.

DEUTRO 23/20.
M-1300; 0.25 max C, 24-26 Cr, 19-22 Ni, bal Fe.
Annealed: 95,000 TS; 45,000 YS; 50 El; 65 Ra; 180 Brin.
At 1200°F: 57,000 TS; 22,000 YS; 32 El; 45 RA.
For furnace parts and equipment, heat treating boxes; Type 310; austenitic, heat resistant.

DEWARD.
M-1779; 0.9 C, 1.5 Mn, 0.2 Si, 0.08 Cr, 0.3 Mo, bal Fe.
Oil hardening, non-deforming tool steel suitable for punches, dies, taps, gages, cutters, rolls and bushings. Usual working hardness 58/62 Rc.
AISI 0-2.

DEWRANCE CM50.
M-1489; 93.25 Ni, 3.5 Si, 2.25 B.
For high temperature brazing; flow point 1055°C, high ductility.

DEWRANCE CM52.
M-1489; 91.25 Ni, 4.5 Si, 2.9 B.
For high temperature brazing; flow point 992°C, resists molten Ni, Hg, and Hg vapor.

DEWRANCE CM53.
M-1489; 82 Ni, 7.1 Cr, 4.5 Si, 2.9 B.
For high temperature brazing; flow point 996°C, corrosion resistant.

DEWRANCE CM56.
M-1489; 71 Ni, 15 Cr, 4.5 Si, 3.5 B.
For high temperature brazing; flow point 1030°C, corrosion resistant.

DEXITE AH.
M-207; 1.15 C, 0.8 Mn, 0.2 Si, 5.45 Cr, 0.35 V, 1.25 Mo, bal Fe.
Air hardening cold work tool steel, high carbon; good wear resistance.

DEXITE AH MODIFIED.
M-207; 0.55 C, 0.9 Mn, 0.25 Si, 3.5 Cr, 0.25 Ni, 0.25 V, 1.4 Mo, bal Fe.
Air hardenable tool steel.

DEXITE NO. 14.
M-207; 0.75 C, 0.8 Mn, 0.94 Cr, 1.8 Ni, 0.4 Mo, bal Fe.
Oil hardening tool steels; for dies, drills, shear blades, swaging dies.

DEXITE NO. 15.
M-207; C, alloy, bal Fe.
For boring tools, cams, rollers, reamers, gauges; deep hardening.

DEXITE NO. 16.
M-207; C, alloy, bal Fe.
For chisels, shovel teeth, wrenches, crushers; abrasion resistant, water hardening, tough.

DEX-TUNG.
M-207; 0.32 C, 0.93 Cr, 0.48 Mo, 0.47 W, bal Fe.
Heat treated: 177,000-265,000 TS; 170,000-208,000 YS; 19-14 El.
For gears, shafting, clutch parts, punches, collets, blacksmith tools; oil hardened.

DFO 95 SUPERVITAC.
M-1488; 0.40 C, 0.90 Ni, 0.80 Cr, 0.25 Mo, bal Fe.
Bars, treated: 930 N/mm^2 UTS; 735 N/mm^2 min YS; 13 El, 2 Brin.
For structural purposes.

DFO 110 SUPERVITAC.
M-1488; 0.30 C, 2.0 Ni, 2.0 Cr, 0.35 Mo, bal Fe.
Bars, treated: 1080 N/mm^2 UTS; 830 N/mm^2 min YS; 11 El; Brin.
For structural purposes.

D.G. ALLOY.
M-2a; 88 Cu, 7.5 Sn, 2.5 Ni, 2 Zn.
25-15 El; 70-80 Brin.
For shafts, staybolts, bearings, stuffing boxes; corrosion resistant.

DH-27.
M-1713; Weld metal: 0.11 Cu, 0.72 Mn, 0.27 Si, bal Fe.
As welded: 73,500 psi TS; 61,000 psi YS; 27 El.
AC-DC electrode for repair of heavy equipment, ship building, pressure vessels.
AWS Class E6027.

D-H 60-OHM ALLOY.
M-44; 5 Ni, bal Cu.
For electrical resistances; 60 ohms/c.m.f.

D-H 90 OHM ALLOY.
M-44; 11 Ni, bal Cu.
For electrical resistance; load banks, power resistors; 90 ohms/c.m.f.

D-H180-OHM ALLOY.
M-44; 23 Ni, bal Cu.
For electrical resistance; 180 ohms/c.m.f.

D-H A-NICKEL.
M-44; Ni.
For magnetostriction, jewelry, mechanical parts.

D-H D-NICKEL.
M-44; 4-5 Mn, bal Ni.
For heating element lead wire, lamp fuse leads.

D-H DURANICKEL.
M-44; 93.5 Ni, 0.35 Fe, 0.3 Mn, 0.5 Si, 0.15 C, 4.5 Al.
Wrought: 90,000-250,000 TS; 30,000-150,000 YS; 50-2 El; 65-15 RA; 140-380 Brin.
For electrical equipment; Z-Nickel Type A.

D-H-E-NICKEL.
M-44; 2 Mn, bal Ni.
For lead wire lamp bulbs, grid wire radio tubes.

D-H NI-HARD.
M-1016; 4.5 Ni, 3.2 C, 1.5 Cr, 1.0 Si, bal Fe.
For ball and rod mill liners, pump casings, impellers; heat resistant castings.

D-H NO. 11.
M-44; Cu.
Lead wire for precious metal thermocouple.

D-H NO. 30.
M-44; 2.0 Ni, bal Cu.
Rolled: 30,000-60,000 TS.
For corrosion resistant parts; corrosion resistant.

D-H NO. 52.
M-44; 50-51 Ni, bal Fe.
Wrought: 150,000-70,000 TS.
For glass to metal seal, grid wire; max operating temperature 500°C.

D-H NO. 99.
M-44; 99.5 Ni.
For electrical equipment, resistance thermocouples; heat and corrosion resistant.

D-H NO. 133.
M-44; 3 Si, bal Ni.

D-H NO. 142.
M-44; 41 Ni, bal Fe.
Rolled: 70,000-150,000 TS.
For metal to glass seals, thermostats; matches 8160 glass expansion.

D-H NO. 146.
M-44; 46 Ni, bal Fe.
Rolled: 70,000-150,000 TS.
For seals with ceramic coated materials; special expansion properties.

D-H NO. 152.
M-44; 51 Ni, bal Fe.
Rolled: 70,000-150,000 TS.
For thermostats, glass seals, high frequency transformers; controlled expansion.

DH NO. 241.
M-44; 19.5 Cr, 1.45 Si, 0.03 C, bal Ni.
For furnace heating elements, moving belts and conveyors, retorts.
High heat and oxidation resistant.

DH NO. 242.
M-44; 0.03 C, 1.2 Si, 19.5 Cr, 1.1 Cb, bal Ni.
For thermocouples, conveyor belts; high heat resistance; stable in reducing atmosphere.

DH NO. 243.
M-44; 19.50 Cr, 1.45 Si, 1.8 Mn, 0.03 C, bal Ni.
For furnace heating elements, furnace belts and conveyors.
High heat and oxidation resistant.

DH NO. 520.
M-44; 35.0 Ni, 21.5 Cr, 2.0 Si, 1.0 Cb, 0.05 C, bal Fe.
For furnace heating elements, furnace belts and conveyors.
High heat and oxidation resistance.

DH NO. 525.
M-44; 35 Ni, 19.5 Cr, 1.45 Si, 0.05 C, bal Fe.
Rolled: 150,000-70,000 TS; 35 El.
For high temperature resistors.
Heat resistant to 800°C.

D-H NO. R. 63.
M-44; 4 Mn, 1 Si, bal Ni.
Wrought: 175,000-70,000 TS.
For spark plug electrodes, resistances; heat resistant to 750°C.

D-H PERMANICKEL.
M-44; 0.2-0.6 Mn, 0.6 max Fe, 0.4 max C, 0.35 max Si, bal Ni.
Age-hardened: 180,000-230,000 TS; 130,000-175,000 YS; 10-2 El.
For springs; Z-Nickel Type B.

D-H PURE NICKEL.
M-44; 99.8 Ni.
For passive and active electron tube cathodes; wire or ribbon.

D-H-S BRONZE NO. 1.
M-758; 62-66 Cu, 4.5-7.0 Al, 2.0-3.5 Fe, 3.0-4.0 Mn, bal Zn.
120,000 TS; 95,000 YS; 12 El; 12 RA; 240 Brin.
For gears, bearings, valve stems; resists acid, wear and shock.

D-H-S BRONZE NO. 2.
M-758; 62-66 Cu, 4.5-7.0 Al, 2.0-3.5 Fe, 3.0-4.0 Mn, bal Zn.
115,000-120,000 TS; 90,000-95,000 YS; 12-16 El; 12-19 RA; 235-275 Brin.
For worm gears, rolling mill housing nuts; resists acid, wear and shock.

D-H-S BRONZE NO. 3.
M-758; 62-66 Cu, 4.5-7.0 Al, 2.0-3.5 Fe, 3.0-4.0 Mn, bal Zn.
105,000 TS; 60,000 YS; 17 El; 17 RA; 200 Brin.
For gears, bearings; resists acid, wear and shock.

D-H-S BRONZE NO. 4.
M-758; 62-66 Cu, 4.5-7.0 Al, 2.0-3.5 Fe, 3.0-4.0 Mn, bal Zn.
90,000 TS; 45,000 YS; 25 El; 25 RA; 185 Brin.
For gears, bearings; resists acid, wear and shock.

D-H T1 (TYPE KP).
M-44; 90 Ni, 10 Cr.
Positive Type K extension and thermocouple wire and ribbon.

DH T2 (TYPE KN).
M-44; 1 Si, 1 Al, 1 Mn, bal Ni.
Negative Type K extension and thermocouple wire and ribbon.

D-H Z-NICKEL.
M-44; 0.3 max C, 4.0-4.75 Al, bal Ni.
For high temperature applications; two types, see Duranickel or Permanickel.

DIALLIST.
M-Eng.; Ni-Al-Co-Fe.
For permanent magnets; high coercive force.

DI-ALLOY.
M-1080; Al, Cu, bal Zn.
For die castings.

DIALLOY.
M-820; C, Mn, Cr, Ni, Mo, bal Fe.
For battering tools, punches; high impact and fatigue resistance.

DIALOX.
M-1246; Al_2O_3.
For cutting tools, wear parts; ceramic aluminum oxide.

DIAMANT 5.
M-1344; 1.3 C, 4.75 W, 0.2 max Cr, bal Fe.
For fast finishing cutters, engravers tools; water hardened.

DIAMANT 12.
M-1344; 2.1 C, 11.5 Cr, bal Fe.
For blanking and forming dies; oil or water hardened.

DIAMANT 28.
M-1655; 2.70 C, 0.20 Mn, 0.30 Si, 26.0 Cr, bal Fe.
For drawing dies.

DIAMANT 44 now VEW K400.

DIAMANTBRONZE.
M-330; 58 Cu, 1 Ni, 1 Mn, 1 Al, 1 Fe, bal Zn. 20 El.
For valve gauge fittings; corrosion resistant.

DIAMANT HK.
M-1318; 1.42 C, W, V, bal Fe.
For fast finishing cutters, engravers tools; water hardened.

DIAMANT R6.
M-1287; 1.42 C, W, V, bal Fe.
For fast finishing cutters, engravers tools; water hardened.

DIAMANT WDSA now VEW K400.

DIAMAX 19112.
M-717.
Alloy powder for metal spraying hard final coat.

DIAMEND 27.
M-677; 0.9-1.1 C, 3.7-4.3 Mn, 0.5-1.0 Si, 11-13 Cr, bal Fe.
Welded: 260-300 Brin.
For hardfacing electrodes; work hardenable, for applications having severe impact.

DIAMEND 63.
M-677; 0.9 C, 1.0 Mn, 0.7 Si, 5 Cr, 8.5 Mo, 2 Co, 1.0 V, 2.5 W, bal Fe.
Welded: 60 Rc.
For hardfacing electrodes; high speed tool steel resistant to abrasion, oxidation, light impact to 1000°F.

DIAMEND 305.
M-677; 0.1 C, 1.2 Mn, 0.6 Si, 2 Cr, 0.5 Mo, bal Fe.
Welded: 37 Rc.
For hardfacing electrodes; pearlitic, machinable steel for build-up and repair of worn machine parts.

DIAMEND 350.
M-677; 0.15 C, 1.0 Mn, 0.6 Si, 2 Cr, 0.4 Mo, bal Fe.
Welded: 37 Rc.
For hardfacing electrodes; pearlitic, machinable steel for repair of worn machine parts.

DIAMEND 500.
M-677; 0.4 C, 1.0 Mn, 0.6 Si, 3.5 Cr, 0.5 Mo, bal Fe.
Welded: 55 Rc.
For hardfacing electrodes; martensitic steel to withstand severe abrasion, heavy impact.

DIAMEND 600.
M-677; 0.6 C, 1.0 Mn, 0.6 Si, 4 Cr, 0.5 Mo, bal Fe.
Welded: 60 Rc.
For hardfacing electrodes; martensitic steel to withstand severe abrasion with mild impact.

DIAMEND 605.
M-677; 0.6 C, 1.0 Mn, 0.9 Si, 4.5 Cr, 0.7 V, bal Fe.
Welded: 57 Rc.
For hardfacing electrodes; martensitic steel to resist severe abrasion especially on alloy steels.

DIAMEND B.
M-677; 2.0 C, 0.4 Mn, 0.5 Si, 4.5 Cr, bal Fe.
Welded: 55 Rc.
For hardfacing electrodes; martensitic cast iron general purpose electrode.

DIAMITE.
M-278; 3-5 Ni, 1-2 Cr, 3-3 1/2 C, 1.1-1.5 Si, 0.6-0.9 Mn, bal Fe.
Cast: 600-725 Brin.
For liners for chutes, pulverizer hammers, sand pump parts, welding rod for hard facing; alloy white iron; resistant to abrasion.

DIAMITE.
M-790; WC.
For cutting tools; sintered carbide.

DIAMOND.
M-1744.
Carbon tool steel plus Vanadium.

DIAMOND A.
M-85; 2.2 C, 0.4 V, 11.5 W, bal Fe.
For dies; air hardening.

DIAMOND A.11.
M-1415; 1.7 C, 7 W, bal Fe.
For turning and planing tools, cold drawing dies; fast finishing steel, wear resistant.

DIAMOND ALLOY.
M-759; 45 Co, 40 Mo, 15 Cr.
For cutters, tools; wear and abrasion resistant.

DIAMOND B.
M-85; 2 C, 12 Cr, 0.8 V, 0.4 Mo, bal Fe.
For cold work tools, blanking and forming dies; abrasion resistant, non-deforming, oil hardened.

DIAMOND BRAND.
M-85; 1.7 C, 0.45 Mn, 18.57 Cr, 0.55 V, bal Fe.
For tools, dies, valves, punches; rust and abrasion resisting.

DIAMOND BRONZE.
M-U.S.; 10 Al, 2 Si, bal Cu.
For pump parts, valves, marine parts, hardware; high strength; corrosion resistant.

DIAMOND D-406.
M-1726; Sintered carbide tool material.
Cutting tools for cast iron; blast nozzles, valve trim.

DIAMOND D-406N.
M-1726; Sintered carbide tool material.
Hard wear resistant material for valve trim; corrosion resistant.

DIAMOND D-409.
M-1726; Sintered carbide tool material.
Cutting tools for machining cast iron.

DIAMOND D-409N.
M-1726; Sintered carbide tool material.
For valve trim; corrosion resistant and erosion resistant.

DIAMOND G BRONZE.
M-741; Pb, Sn, bal Cu.
For bearings, bushings; tough.

DIAMOND M.
M-57; 1.35 C, 0.3 Cr, bal Fe.
For form cutters, tools, dies; water hardened.

DIAMOND NO. 1 B.16.D.
M-1415; 1.25 C, 3 W, 0.15 V, bal Fe.
For drawing and burnishing dies; water hardened, wear resistant.

DIAMOND OO.
M-261, M-486; 1.6 C, 0.6 Cr, 1 V, 5.5 W, bal Fe.
For finishing tools, form tools, rifling tools; for light cuts.

DIAMOND S-1.
M-288; 0.30 C, 18 Cr, bal Fe.
For corrosion resistant parts; corrosion resistant.

DIAMOND T.
M-288; 0.30 C, 12 Cr, 25 Ni, bal Fe.
For heat resistant parts; heat resistant.

DIAMONITE.
M-355; 95.65 W, 3.91 C, bal Fe.
For cutting tools, dies; W_2C + WC.

DIAPHRAGM BRASS.
M-U.S.; 95 Cu, 3 Sn, 2 Zn.
For diaphragms; corrosion resistant.

DIAWELD.
M-72; 4.2-4.4 C, 1.8 Si, 5-6 Mn, 30 Cr, bal Fe.
For hard surfacing electrodes; corrosion resistant.

DICA B.
M-69; 0.33 C, 4.8 Cr, 0.4 Mn, 1.5 Mo, 1.0 Si, 0.2 V, bal Fe.
For Al die casting dies, hot gripper and header dies; hot work steel, air or oil hardened.
AISI H 12.

DICA B MOD.
M-69; 0.37 C, 1 Si, 5 Cr, 0.5 V, 1.35 Mo, bal Fe.
For upsetters, punches, shears, extrusion tools; hot work steel, oil hardened.
AISI H 11.

DICA B-VANADIUM.
M-69; 0.35 C, 5 Cr, 1 V, 1.5 Mo, bal Fe.
For Al and Mg die casting dies; Type H13; oil hardened.
AISI H 13.

DICA (FLAME HARD) CAST TO SHAPE.
M-69; 0.50 C, 0.75 Mn, 0.25 Si, 1.0 Cr, 0.20 Mo, bal Fe.
Hardness normally 250-300 Brin, or can be annealed.
For automobile bumper dies.
Oil hardening tool steel.

DICAL.
M-760; 12 Si, bal Cu.
For light alloy castings; die cast.

DICK'S BEARING BRONZE.
M-Eng.; 80 Cu, 10 Pb, 9.2 Sn, 0.8 P.
For bearings.

DICROME.
M-56; 0.60 C, 0.25 Si, 0.60 Mn, 0.60 Cr, bal Fe.
For punches, heading dies, axes; Brit. EN11.

DIDBY'S CUPRITIC ALLOY.
18 Cu, 18 Cr, 72 Fe.

DIDYMIUM METAL.
M-1497; Ne (Neodymium)-Pr(Praseodymium).
For alloying.

DIECARB.
M-57; 4.6-5.6 C, 65-90 W, 6.25 Co, 0.5-2 Fe, 0-5 Ta.
For blanking and forming dies; shock resistant, sintered.

DIE CAST.
M-336; 0.45 C, 1.0 Si, 1.4 Cr, 2.2 W, 0.2 V, 0.3 Mo, bal Fe.
For aluminum die casting dies; non-warping.

DIE CASTING ALLOY "A".
91-84 Sn, 2-9 Sb, 4.5-8 Cu.
For high grade bearings; anti-friction.

DIE CASTING ALLOY "B".
91-84 Sn, 10-17 Sb, 0-1 Cu.
For light duty bearings; anti-friction.

DIE CASTING ALLOY "C".
80 Sn, 10 Pb, 10 Sb.
For light duty bearings; anti-friction.

DIE CASTING ALLOY "D".
61-5 Sn, 25 Pb, 10.5 Sb, 3 Cu.
For light duty bearings; anti-friction.

DIE CASTING ALLOY "E".
74 Zn, 15 Sn, 6 Al, 5 Cu.
For soft work die castings; anti-friction.

DIE CASTING ALLOY "F".
85 Zn, 8 Sn, 4 Cu, 3 Al.
For standard die castings; anti-friction.

DIE CASTING ALLOY "G".
82-75 Zn, 13.75 Al, 3 Cu.
For hard die castings.

DIE CASTING ALLOY "H".
46 Zn, 31 Sn, 20 Cu, 3 Sb.
For hard die castings.

DIE CASTING ALLOY "I".
86.5 Zn, 10.75 Cu, 2.75 Al.
For die castings.

DIE CASTING ALLOY "J".
90-83 Zn, 5-11 Cu, 2-5 Al, 1-5 Sn.
For die castings.

DIE CAST NO. 1.
M-1423; 0.35-0.40 C, 1.3 Mo, 0.5 V, 1.0 Si, 5.2 Cr, bal Fe.
For blanking and forging dies; hot work steel, tough and wear resistant.

DIECRAT.
M-1149; 0.75 C, 2.0 Mn, 0.3 Si, 1.0 Cr, 1.35 Mo, bal Fe.
Air hardening cold work tool steel; AISI A6.

DIE FLEX.
M-373; 0.40 C, 1.5 Cr, 0.80 Mo, 4.25 Ni, bal Fe.
Hot work die steel; tough.

DIEHL GR. A.
M-822; 1.0 C, 0.3 Mn, 0.25 Si, bal Fe.
For drills, taps, reamers, hobs, broaches, cutters; Type W1; water hardened.

DIE L.
M-343; 1.45-1.55 C, 12.5-14 Cr, 0.65-0.75 Co, 1.1 V, bal Fe.
For dies, broaches, reamers; non-deforming.

DIEMAC.
M-1680; 3.1-3.3 C, 1.7-1.9 Si, 0.6-0.8 Mn, 0.8-1.0 Mo, 1.0-1.2 Ni, 0.05 max P, bal Fe.
Close grained electric furnace cast iron for use in permanent moulds. 180-200 Brin in heavy sections.

DIENETT GERMAN SILVER.
M-Eng.; 51 Cu, 32 Zn, 9.5 Pb, 6.4 Ni, 1.6 Sn.
For ornamental, hardware novelties; free-cutting.

DIESEL BEARINGS.
M-Eng.; 80 Sn, 15 Sb, 5 Cu.
For bearings for Diesel engines; Babbitt.

DIESEL OIL ENGINE BABBITT.
M-88; 9-11 Sb, 4-6 Cu, bal Sn.
34-4 Brin.
For bearings for diesel engines; resists impact loading.

DIETEMPER-1.
M-1431; 0.40 C, 1.5 Cr, 0.25 V, 1 Mo, 0.65 Mn, bal Fe.
For forging and die casting dies; hot work steel, oil hardened.

DIETEMPER-2.
M-1431; 0.30 C, 2.75 Cr, 0.30 Mo, 9.5 W, 1.6 Ni, bal Fe.
For extrusion dies, rams, punches; hot work steel, oil hardened.

DIEWEAR.
M-820; 0.90 C, 1 Mn, 0.5 Cr, 0.5 W, bal Fe.
For dies, punches, blanking and forming dies; Type O1; oil hardened, non-deforming.

DI HARD.
M-1748; 3.5 C, 4.2 Ni, 1.1 B, bal Fe.
As deposited: 578 Brinell.
For lined cylinders; extreme wear resistance, fusible coating.

DI-HARD see XALOY 100.

DI HW.
M-1287; 1.3 C, 0.20 max Cr, 4.75 W, 0.3 Mn, bal Fe.
For bearings, liners, forming and blanking dies; water hardened.

DI-IRON.
M-644; 3.2 C, 2.2 Si, 1.5 Ni, 0.8 Cr, 0.15 Mo, bal Fe.
Cast: 50,000-60,000 TS; 230-290 Brin.
For dies, gears, housings, shafts; cast iron; Tr.S. 4100; Tr.D. 320.

DILATON 36.
M-303; 36 Ni, bal Fe.
For bimetal thermostats; controlled expansion.

DILATON 42.
M-303; 42 Ni, bal Fe.
For bimetal thermostats; controlled expansion.

DILATON 48.
M-303; 48 Ni, bal Fe.
For glass to metal seals; controlled expansion.

DILATON 50.
M-303; 50 Ni, bal Fe.
For glass to metal seals; controlled expansion.

DILATON 60.
M-303; 60 Ni, bal Fe.
For glass to metal seals; controlled expansion.

DI-METAL NO. 2.
M-815; 3.5-4.5 Al, 2.5-3.5 Cu, 0.02-0.10 Mg, 1.0 max Fe, bal Zn.
Die cast: 48,000-52,000 TS; 8-5 El; 100-105 Brin.
For die castings; general purpose.

DI-METAL NO. 3.
M-815; 3.5-4.3 Al, 0.75-1.25 Cu, 0.03-0.08 Mg, 1.0 max Fe, bal Zn.
Die cast: 40,000-41,000 TS; 10-5 El; 75-85 Brin.
For die castings; general purpose.

DI-METAL NO. 5.
M-815; 3.5-4.3 Al, 0.75-1.25 Cu, 0.03-0.08 Mg, 1.0 max Fe, bal Zn.
Die cast: 45,000-48,000 TS; 7-3 El; 95-100 Brin.
For die castings; general purpose.

DI-MOL.
M-357; 0.8 C, 4.0 Cr, 1.0 V, 1.5 W, 9 Mo, bal Fe.

DIMPALLOY.
M-Eng.; Zn, bal Cu.
For brazing of brasses and bronzes.

DIN GERMAN STANDARDS.
DIP METAL.
70 Cu, 0.5 Sn, 0.5 Pb, 0.5 Ni, bal Zn.
Rolled: 35,000 TS; 50 El.
For hardware; free-cutting.

DIPPING BRASS.
M-U.S.; 67 Cu, 33 Zn.
For deep drawn, spun or stamped parts; high ductility.

DIPPIT.
M-565; 94.5 Pb, 5 Sn, 0.5 Cu.

"DI" QUALITY.
M-179; Cu, Mn, bal Zn.

DI R35.
M-1287; 1.42 C, W, V, 0.30 Mn, 0.25 Si, bal Fe.
For cutters, bearings, forming dies; water hardened.

DIRIGOLD.
M-594; 9-10 Al, 1.1-2.0 Ni, 0-1.7 Sn, bal Cu.
For strong corrosion resisting parts, ornamental castings; corrosion resistant.

DIRILYTE METAL.
M-761; 5 Al, bal Cu.
For tableware; color of gold Al bronze.

DIRO 703.
M-1803; 0.45 C, 0.7 Mn, 0.3 Si, 1.5 Cr, 0.3 V, 0.7 Mo, bal Fe.
Hot work tool steel; oil hardening.
For pressure punches.
W-Nr. 1.2323; 48 Cr Mo V 6.7.

DIRO B 30.
M-1803; 0.40 C, 0.3 Mn, 0.4 Si, 13.0 Cr, bal Fe.
Hot or cold work tool steel; air or oil hardening; corrosion resistant.
For tools for plastics and light metals.
W.-Nr. 1.2083; X 42 Cr 13.

DIRO C 10.
M-1803; 0.10 C, 0.40 Mn, 0.25 max Si, bal Fe.
Low carbon structural steel.
W.-Nr. 1.0301; was St.C. 10.61.

DIRO CV 5.
M-1803; 0.50 C, 1.0 Mn, 0.25 Si, 1.0 Cr, 0.1 V, bal Fe.
Cold work tool steel; oil hardening.
For hand tools, screw drivers, shears, shanks for carbide tools.
W.-Nr. 1.2241; 51 Cr V 4.

DIRO CV 6.
M-1803; 0.58 C, 1.0 Mn, 0.25 Si, 1.0 Cr, 0.1 V, bal Fe.
Cold work tool steel; oil hardening.
For punches, chisels, shear blades.
W.-Nr. 1.2242; DIN 59 Cr V 4.

DIRO CW.
M-1803; 1.05 C, 1.0 Mn, 0.25 Si, 1.0 Cr, 1.2 W, bal Fe.
Cold work tool steel; oil hardening.
For shears, broaches, reamers, end mills.
W.-Nr. 1.2419.

DIRO EC 30.
M-1803; 0.13 C, 0.50 Mn, 0.25 Si, 0.40 Cr, bal Fe.
Low alloy carburizing steel.
W.-Nr. 1.7012.

DIRO EXTRA 1.
M-1803; 1.30 C, 0.25 Mn, 0.20 Si, 0.025 P, 0.025 S, bal Fe.
Carbon tool steel; water hardening.
For punches, drills, lathe tools.
W.-Nr. 1.1560; DIN C 125 W 1.

DIRO EXTRA 2.
M-1803; 1.10 C, 0.25 Mn, 0.20 Si, 0.025 P, 0.025 S, bal Fe.
Carbon tool steel; water hardening.
For hand and pneumatic chisels, drills.
W.-Nr. 1.1550; similar to AISI W1.

DIRO EXTRA 3.
M-1803; 1.0 C, 0.25 Mn, 0.20 Si, 0.025 P, 0.025 S, bal Fe.
Carbon tool steel; water hardening.
For hand chisels, shears, punches, hand dies.
W.-Nr. 1.1540; DIN C 100 W1.

DIRO EXTRA 4.
M-1803; 0.85 C, 0.30 Mn, 0.20 Si, 0.025 P, 0.025 S, bal Fe.
Carbon tool steel; water hardening.
For hand tools, rock drills, hammers.
W.-Nr. 1.1530; DIN C 85 W1.

DIRO EXTRA 5.
M-1803; 0.70 C, 0.30 Mn, 0.20 Si, 0.025 P, 0.025 S, bal Fe.
Carbon tool steel; water hardening.
For hand tools, wood working tools.
W.-Nr. 1.1520; DIN C 70 W1.

DIRO FORTIS.
M-1803; 2.0 C, 0.3 Mn, 0.3 Si, 12.0 Cr, bal Fe.
Hot or cold work tool steel; oil hardening.
For hot pressure casting molds; heavy duty cold punching dies and shear blades.
W.-Nr. 1.2080; DIN X 210 Cr 12.

DIRO FORTIS EXTRA.
M-1803; 1.65 C, 0.3 Mn, 0.3 Si, 12.0 Cr, 0.1 V, bal Fe.
Cold work tool steel; air or oil hardening.
For coining dies, broaches, threading tools.
W.-Nr. 1.2201; X 165 Cr V 12.

DIRO FORTIS PRIMA.
M-1803; 1.65 C, 0.3 Mn, 0.3 Si, 12.0 Cr, 0.5 W, 0.1 V, 0.6 Mo, bal Fe.
Cold work tool steel; air or oil hardening.
For heavy duty stamping, drawing and trimming dies; coining dies.
W.-Nr. 1.2601; X Cr Mo V 12.

DIRO FORTIS SPEZIAL.
M-1803; 2.10 C, 0.3 Mn, 0.3 Si, 12.0 Cr, 0.7 W, bal Fe.
Cold work tool steel; oil or air hardening.
For heavy duty punching and trimming dies.
W.-Nr. 1.2436; X 210 Cr W 12.

DIRO HS 4.
M-1803; 0.80 C, 4.0 Cr, 9.0 W, 0.85 V, 1.6 Mo, bal Fe.
High speed tool steel.
For milling cutters, drills, taps.
W.-Nr. 1.3316.

DIRO KH 20.
M-1803; 0.20 C, 1.2 Mn, 0.3 Si, 1.0 Cr, bal Fe.
To be carburized and oil hardened for dies and molds for artifical resin molding press.
W.-Nr. 1.2162; DIN 21 Cr Mn 5.

DIRO KLS.
M-1803; 1.0 C, 0.3 Mn, 0.25 Si, 1.5 Cr, bal Fe.
Cold work tool steel; oil hardening.
For lathe centers, taps, threading tools.
W.-Nr. 1.2067; DIN 100 Cr 6.

DIRO KONSTANT.
M-1803; 1.05 C, 0.25 Mn, 0.3 Si, 1.4 Cr, bal Fe.
Cold work tool steel; oil hardening.
For gauges, small dies and stamps.
W.-Nr. 1.2060; DIN 100 Cr 5.

DIRO KONSTANT 15.
M-1803; 1.45 C, 0.6 Mn, 0.2 Si, 1.4 Cr, bal Fe.
Cold work tool steel; oil or water hardening.
For taps, broaches, milling cutters, counter bores.
W.-Nr. 1.2063; 145 Cr 6.

DIRO KONSTANT M.
M-1803; 1.0 C, 0.7 Mn, 0.25 Si, 1.0 Cr, bal Fe.
Cold work tool steel; oil hardening.
For short run cutting and stamping tools and dies.
W.-Nr. 1.2061; DIN 100 Cr 4.

DIRO KONSTANT S.
M-1803; 1.25 C, 0.7 Mn, 1.2 Si, 1.2 Cr, bal Fe.
Cold work tool steel; oil hardening.
For drills, punches, end mills, broaches.
W.-Nr. 1.2109; DIN 125 Cr Si 5.

DIRO L 18.
M-1803; 0.90 C, 2.0 Mn, 0.2 Si, 0.1 V, bal Fe.
Cold work tool steel; oil hardening.
For small cutting and punching tools, taps.
W.-Nr. 1.2842; DIN 90 Mn V 8.

DIRO MS 53 G.
M-1803; 0.53 C, 1.0 Mn, 1.0 Si, bal Fe.
Hot or cold work tool steel; oil hardening.
For forging or trimming dies; short runs.
W.-Nr. 1.2826; DIN 60 Mn Si 4.

DIRO MV.
M-1803; 0.55 C, 0.7 Mn, 0.3 Si, 1.0 Cr, 1.7 Ni, 0.1 V, 0.5 Mo, bal Fe.
Hot or cold work tool steel; oil or air hardening.
For forging dies, molding dies.
W.-Nr. 1.2714; DIN 56 Ni Cr Mo V 7.

DIRO MW.
M-1803; 0.55 C, 0.6 Mn, 0.3 Si, 0.7 Cr, 1.7 Ni, 0.1 V, 0.2 Mo, bal Fe.
Hot or cold work tool steel; oil hardening.
Forging and extrusion dies.
W.-Nr. 1.2713; DIN 55 Ni Cr Mo V 6.

DIRON.
M-1724.
Mild steel with one coat vitreous enamel.

DIRO NC 3.
M-1803; 0.15 C, 0.4 Mn, 0.3 Si, 0.7 Cr, 3.5 Ni, bal Fe.
To be carburized and hardened for dies for artificial resins.
W.-Nr. 1.2735; 15 Ni Cr 14.

DIRO NC 4.
M-1803; 0.15 C, 0.4 Mn, 0.3 Si, 1.0 Cr, 4.5 Ni, bal Fe.
To be carburized and hardened for dies for artificial resins.
W.-Nr. 1.2745; 14 Ni Cr 18.

DIRO PRIMA 00.
M-1803; 0.55 C, 0.40 max Mn, 0.15 Si, 0.030 P, 0.030 S, bal Fe.
Water hardening carbon tool steel.
For anvils, hammers, knives, axes, and shears.
W.-Nr. 1.1820; DIN C 55 W2.

DIRO PRIMA 1.
M-1803; 1.30 C, 0.30 Mn, 0.25 Si, 0.030 P, 0.030 S, bal Fe.
Water hardening carbon tool steel.
For hand punches, shears, drills.
W.-Nr. 1.1660; DIN C 125 W 2.

DIRO PRIMA 2.
M-1803; 1.15 C, 0.30 Mn, 0.25 Si, 0.030 P, 0.030 S, bal Fe.
Water hardening carbon tool steel.
For hand and pneumatic chisels, punches, threading tools.
W.-Nr. 1.1650; DIN C 110 W2.

DIRO PRIMA 3.
M-1803; 1.0 C, 0.30 Mn, 0.25 Si, 0.030 P, 0.030 S, bal Fe.
Water hardening carbon tool steel.
For hand tools, stamping dies, drills.
W.-Nr. 1.1640; DIN C 100 W2.

DIRO PRIMA 4.
M-1803; 0.85 C, 0.30 Mn, 0.25 Si, 0.030 P, 0.030 S, bal Fe.
Water hardening carbon tool steel.
For hand tools, woodworking tools.
W.-Nr. 1.1630; DIN C 85 W 2.

DIRO PRIMA 5.
M-1803; 0.70 C, 0.30 Mn, 0.25 Si, 0.025 P, 0.025 S, bal Fe.
Water hardening carbon tool steel.
For shears, hand tools, woodworking tools.
W.-Nr. 1.1620; DIN C 70 W 2.

DIRO PS.
M-1803; 0.50 C, 0.5 Mn, 0.25 Si, 1.0 Cr, 3.5 Ni, bal Fe.
Cold work tool steel; air or oil hardening.
For stamping dies, cold forming dies, shear blades.
W.-Nr. 1.2721; 50 Ni Cr 13.

DIRO SC.
M-1803; 0.67 C, 0.5 Mn, 1.3 Si, 0.5 Cr, bal Fe.
Hot or cold work tool steel; oil hardening.
For hand tools, shanks, pressure plates.
W.-Nr. 1.2101; DIN 62 Si Mn Cr 4.

DIRO SD.
M-1803; 0.35 C, 0.3 Mn, 1.0 Si, 1.0 Cr, 2.0 W, 0.2 V, bal Fe.
Hot work tool steel; water hardening; for chisels, shears.
W.-Nr. 1.2541; 35 W Cr V 7.

DIRO SD 45.
M-1803; 0.45 C, 0.3 Mn, 1.0 Si, 1.0 Cr, 2.0 W, 0.2 V, bal Fe.
Hot work tool steel; oil or water hardening.
For hot cutting dies and punches, trimming tools.
W.-Nr. 1.2542.

DIRO SD 50.
  M-1803; 0.55 C, 0.3 Mn, 1.0 Si, 1.0 Cr, 2.0 W, 0.2 V, bal Fe.
  Hot or cold work tool steel; oil hardening.
  For punches, shear blades, chisels.
  W.-Nr. 1.2550; 60 W Cr V 7.

DIRO SEB.
  M-1803; 0.95 C, 4.0 Cr, 3.0 W, 2.8 V, 2.5 Mo, bal Fe.
  High speed tool steel.
  For twist drills, milling cutters, broaches.
  W.-Nr. 1.3333.

DIRO SS.
  M-1803; 0.80 C, 4.5 Cr, 18.0 W, 1.7 V, 1.0 Mo, 10.0 Co, bal Fe.
  Co-W high speed steel.
  For lathe and planer tools.
  W.-Nr. 1.3265; DIN S 18-1-2-10.

DIRO SS 1.
  M-1803; 1.4 C, 4.5 Cr, 12.0 W, 4.0 V, 1.0 Mo, 5.0 Co, bal Fe.
  W-Co-V high speed steel.
  For lathe tools.
  W.-Nr. 1.3202; similar to AISI T15.

DIRO SS 2.
  M-1803; 0.75 C, 4.0 Cr, 18.0 W, 1.0 V, bal Fe.
  Tungsten high speed steel.
  For drills, lathe tools, milling cutters, taps.
  W.-Nr. 1.3355; AISI T1.

DIRO SS 3.
  M-1803; 0.80 C, 4.5 Cr, 12.0 W, 2.0 V, 1.0 Mo, 3.0 Co, bal Fe.
  W-Co high speed steel.
  For lathe tools, milling cutters, taps.
  W.Nr. 1.3211; S 12-1-2-3.

DIRO SS 4.
  M-1803; 0.80 C, 4.5 Cr, 12.0 W, 1.0 V, 1.7 Mo, 5.0 Co, bal Fe.
  Co-W high speed steel.
  For lathe and planing tools.
  W.-Nr. 1.3251.

DIRO SS 5.
  M-1803; 0.80 C, 4.5 Cr, 18.0 W, 1.7 V, 1.0 Mo, 5.0 Co, bal Fe.
  Co-W high speed steel; for lathe tools.
  W.-Nr. 1.3255; DIN S 18-1-2-5.

DIRO SS 45.
  M-1803; 1.3 C, 4.5 Cr, 12.0 W, 4.0 V, 1.0 Mo, bal Fe.
  High speed steel.
  For machining aluminum, bakelite, other plastics.
  W.-Nr. 1.3302.

DIRO SS 183.
  M-1803; 0.90 C, 4.0 Cr, 12.0 W, 0.85 V, 2.5 Mo, bal Fe.
  High speed steel.
  For lathe tools, milling cutters.
  W.-Nr. 1.3318.

DIROSTAHL see **DIRO ALLOYS.**

DIRO SWP.
  M-1803; 0.30 C, 0.3 Mn, 0.2 Si, 2.5 Cr, 9.0 W, 0.4 V, bal Fe.
  Hot work tool steel; oil or air hardening.
  For pressing dies, pressure castings, molds.
  W.-Nr. 1.2581; X 30 W Cr V 9.3.

DIRO WM 4.
  M-1803; 0.30 C, 0.4 Mn, 1.0 Si, 1.0 Cr, 4.0 W, 0.2 V, bal Fe.
  Hot work tool steel; oil or water hardening.
  Pressing mandrels for non-ferrous alloys.
  W.-Nr. 1.2564; DIN X 30 W Cr V 4.1.

DIRO WM 5.
  M-1803; 0.30 C, 0.3 Mn, 0.2 Si, 2.5 Cr, 4.5 W, 0.6 V, bal Fe.
  Hot work tool steel; oil or air hardening.
  For pressure casting molds, forming dies.
  W.-Nr. 1.2567; DIN X 30 W Cr V 5.3.

DIRO WS 15.
  M-1803; 0.15 C, 0.3 Mn, 0.25 Si, 0.035 P, 0.035 S, bal Fe.
  To be carburized and water hardened.
  For punches, lathe jaws, molding presses.
  W.-Nr. 1.1805; DIN C WS.

DIRO WS 35.
  M-1803; 0.35 C, 0.5 Mn, 0.3 Si, 0.035 P, 0.0355 S, bal Fe.
  Water hardening carbon tool steel.
  For wrenches, tongs, pliers.
  W.-Nr. 1.1720; DIN C 35 W 3.

DIRO WS 45.
  M-1803; 0.45 C, 0.7 Mn, 0.3 Si, 0.035 P, 0.035 S, bal Fe.
  Water (or oil) hardening carbon tool steel.
  For hammers, axes, screw drivers, wrenches.
  W.-Nr. 1.1730; DIN C 45 W 3.

DIRO WS 60.
  M-1803; 0.60 C, 0.7 Mn, 0.35 Si, 0.035 P, 0.035 S, bal Fe.
  Water or oil hardening tool steel.
  For lathe tool shanks, hammers, hand tools.
  W.-Nr. 1.1740; DIN C 60 W 3.

DIRO WS 67.
  M-1803; 0.67 C, 0.7 Mn, 0.35 Si, 0.035 P, 0.035 S, bal Fe.
  Oil hardening carbon tool steel.
  For hand saws, woodworking tools.
  W.-Nr. 1.1744.

DISCALOY.
  M-1491; 0.08 max C, 0.35 max Al, 0.6-1.0 Si, 0.6-1.0 Mn, 1.35-1.85 Ti 2.5-3.5 Mo, 12-15 Cr, 24-28 Ni, bal Fe.
  At 70°F: 145,000 TS; 106,000 YS; 19 El; 23 RA.
  At 1200°F: 104,000 TS; 91,000 YS; 19 El; 24 RA.
  At 1350°F: 82,000 TS; 74,000 YS; 14 El; 26 RA.
  For turbine discs, bolts, extrusion dies, non-magnetic gears and pinions; austenitic, heat resistant.

DISCUS.
  M-1724.
  High strength galvanized corrugated sheets; mild steel.

DISSTON D-12-CO.
  M-357; C, alloy, bal Fe.

DISSTON NO. 844.
  M-357; 0.85 C, 0.8 Cr, 0.25 Si, bal Fe.

DISSTON NO. 871.
  M-357; 0.55 C, 0.65 Mn, 0.2 Mo, bal Fe.

DISSTON NO. 872.
  M-357; 0.35 C, 3.25 Cr, 9.25 W, 0.3 V, bal Fe.

DISSTON NO. 873.
  M-357; 0.38 C, 5 Cr, 1.25 W, 1.35 Mo, 0.4 V, 1.0 Si, bal Fe.

DISSTON NO. 5170.
  M-357; 0.7 C, 0.3 Mn, bal Fe.

DISSTON NO. 5190.
  M-357; 0.9 C, 0.3 Mn, bal Fe.

DISSTON NO. 5390.
  M-357; 0.9 C, 0.7 Mn, bal Fe.

DISSTON NO. 51110.
  M-357; 1.1 C, 0.3 Mn, bal Fe.

DISSTON STAINLESS A.
  M-357; 0.30 C, 14 Cr, bal Fe.

DISSTON STAINLESS B.
  M-357; 0.60 C, 17 Cr, bal Fe.

DISSTON STAINLESS D.
  M-357; 0.45 C, 1.0 Si, 8.75 Cr, 1.4 Mo, 2.0 Cu, bal Fe.

DISSTON STAINLESS IRON.
  M-357; 0.10 C, 12-14 Cr, bal Fe.

DIVCO 40/60 NO. 180 SOLDER.
  M-1121; 40 Sn, 60 Pb.
  For solder; rosin core.

DIXIE BRAND.
  M-365; 0.7-1.4 C, bal Fe.
  For tools, dies, drills; water hardening.

DIXI NO. 3.
  M-344; C, Cr, bal Fe.
  For tools and dies.

DIXOILBRONZ.
  M-590; 10 Sn, 2 Zn, bal Cu.
  For bearings, bushings, gears, liners; corrosion resistant.

DM.
  M-268; 19.76 Cu, 55.12 Fe, 0.57 Si, 0.87 Ni, 0.29 C, 15.18 Cr.
  155 Brin.
  For corrosion resistant parts; formerly "Durbin."

DM 7.
  M-1653; 0.95-1.05 C, 3.25-4.25 Cr, 8.0-9.0 Mo, 1.25-2.25 W, 1.8-2.2 V bal Fe.
  Molybdenum high speed tool steel.
  W.-Nr. 1.3348; AISI M 7.

DMAT 13.
  M-1488; 0.55 C, 2.0 Ni, 0.85 Cr, 0.35 Mo, 0.04 V, bal Fe.
  Oil hardening hot work tool steel; for hammer forging and upsetting dies; good resistance to shock.
  AFNOR 55 NCDV 7.

DM; DM2 see **TIMKEN DM, DM-2.**

D-M-E NO. 1.
  M-1568; 0.28-0.34 C, 0.6-0.9 Mn, bal Fe.
  Plate: 165-185 Brin.
  For plastic molds.
  AISI C1030, water hardening.

D-M-E NO. 2.
  M-1079; 0.42-0.49 C, 0.8-1.1 Cr, 0.15 min V, 0.6-0.9 Mn, bal Fe.
  Hardened: 107,500 TS; 88,000 YS; 15.5 El; 33.8 RA; 225-300 Brin.
  For die casting mold bases and molds, cavity plates; for Zn die casting.

D-M-E NO. 2H.
  M-1568, M-1616; 0.55 C, 1.15 Mn, 0.20 Mo, 0.75 Cr, 0.08 max S, 0.035 max P, 0.25 Si, bal Fe.
  Plate: 252-302 Brin; 150,000 TS; 130,000 YS; 20 El; 60 RA.
  For plastic molds and die casting dies, holder blocks.
  Free-cutting, water or oil hardening.

D-M-E NO. 3.
  M-1568, M-1616; 0.36 C, 0.85 Mn, 0.25 Si, 1.0 Cr, 0.2 V, 0.5 Mo, bal Fe.
  Plate: 262-311 Brin.
  Heat Treated: 206,000 TS; 175,000 YP; 12 El; 418 Brin.
  For cavity plates and inserts in die casting dies; zinc die casting dies, aluminum die casting dies.

D-M-E NO. 5.
  M-1568, M-1616; 0.35 C, 5.0 Cr, 1.5 Mo, 1.0 V, bal Fe.
  Annealed: 209 Brin.
  Ht. Tr.: 270,000 TS; 240,000 YS; 10 El; 30 RA; 540 Brin.
  For heavy duty compression tools, die casting dies, bolt dies, aluminum extrusion dies, plastic molds. Vacuum degassed forgings, Type H13 tool steel, ground flat stock.

DMO 5.
  M-1655; 0.85 C, 4.5 Cr, 5.0 Mo, 6.5 W, 1.8 V, bal Fe.
  High speed steel for lathe tools, milling tools, drills.
  W.-Nr. 1.3343; similar to AISI M2.

DMO 5 CO.
  M-1655; 0.90 C, 4.0 Cr, 5.0 Mo, 6.5 W, 1.5 V, 5.0 Co, bal Fe.
  High speed steel for heavy duty lathe tools, milling cutters, drills.
  W.-Nr. 1.3243.

DN-15.
  M-Ger.; 0.4-0.6 C, 11-13 Ni, 4-6 Mn, 3 Cr, 0.5 Mo, bal Fe.
  For aeronautical structures, gas turbines and internal combustion engines.
  Austenitic steel, stainless. Coefficient of expansion similar to that of aluminum.

D.N.V see SIMONDS DNV.

DO-1.
  M-1424; Sintered carbide.
  For rough cutting and chip removal on cast iron and non-ferrous metals.

DO-2.
  M-1424; Sintered carbide.
  General purpose cutting tool for cast iron and non-ferrous metals; good resistance to wear.

DO-3.
  M-1424; Sintered carbide.
  Cutting tool for finish cutting on cast iron and non-ferrous metals.

DO-4.
  M-1424; Sintered carbide.
  Cutting tool for finishing and precision boring of cast iron and non-ferrous metals.

DO-10.
  M-1424; Sintered carbide.
  For wear and shock resistance.

DO-11.
  M-1424; Sintered carbide.
  For light impact duty.

DO-13.
  M-1424; Sintered carbide.
  For medium impact duty.

DO-14.
  M-1424; Sintered carbide.
  For heavy impact duty.

DO-15.
  M-1424; Sintered carbide.
  Cutting tools for heavy roughing and interrupted cuts on steel.

DO-16.
  M-1424; Sintered carbide.
  For general purpose cutting tool on steel.

DO-17.
  M-1424; Sintered carbide.
  Cutting tools for finish cutting steel.

DO-18.
  M-1424; Sintered carbide.
  Cutting tools for finish cutting and precision boring steel.

DO-30.
  M-1424; Sintered carbide.
  Cutting tools for rough machining of cast iron and non-ferrous metal.

DO-34.
  M-1424; Sintered carbide.
  Cutting tools for general purpose machining and finishing of steel.

DO-35.
  M-2424; Sintered carbide.
  Cutting tools for heavy, roughing, interrupted cuts on steel.

DO-36.
  M-1424; Sintered carbide.
  Cutting tools for general purpose machining of steel.

DO 40.
  M-1424.
  Titanium coated carbide, for general purpose machining of steels and metal that tend to alloy or weld to carbides.

DO 80.
  M-1424.
  Ceramic cutting material, for high speed machining of abrasive super alloys, hardened steel and cast iron.

DO ALL AIR HARDENING AISI-A2.
  M-1424; 1.0 C, 0.6 Mn, 5.25 Cr, 1.1 Mo, 0.25 V, 0.12 S, bal Fe.
  Free machining, precision ground.
  For dies, punches, taps, shear blades, reamers, gages.

DO ALL AIR HARDENING AISI-A6.
  M-1424; 0.70 C, 2.1 Mn, 1.0 Cr, 1.3 Mo, 0.3 Si, 0.12 S, bal Fe.
  Free machining, precision ground.
  For blanking and coining dies, shear blades, jigs, and fixtures, punches.

DO ALL LOW CARBON AISI-1018.
 M-1424; 0.18 C, 0.6 Mn, 0.2 Si, bal Fe.
 Precision ground flat stock.
 For patterns, jigs, fixtures, machine parts that do not require heat treatment.

DO ALL M41.
 M-1424; 1.10 C, 4.25 Cr, 2.0 V, 6.75 W, 3.75 Mo, 5.0 Co, bal Fe.
 High speed steel for cutting tools, tungsten-molybdenum-cobalt type; AISI M-41.

DO ALL O.H.
 M-1424; 1.2 C, 0.55 Cr, 0.33 Mo, 0.75 Mn, bal Fe.
 For dies, punches, gages, machine parts; oil hardened.

DO ALL OIL HARDENING AISI-O1.
 M-1424; 0.90 C, 1.2 Mn, 0.3 Si, 0.5 W, 0.5 Cr, 0.2 V, bal Fe.
 Precision ground flat stock.
 For blanking and trimming dies, cutters, punches, special gages.

DO ALLOY.
 M-1424; Co, Cr, W.
 Cast alloy for general purpose machining, high red hardness and shock resistance.

DO ALL T1.
 M-1424; 0.73 C, 4.0 Cr, 1.1 V, 18.0 W, bal Fe.
 High speed steel, general purpose cutting tools, tungsten type; AISI T1.

DO ALL T5.
 M-1424; 0.80 C, 18 W, 4 Cr, 2 V, 8 Co, bal Fe.
 Tungsten-cobalt high speed steel.

DO ALL T8.
 M-1424; 0.75 C, 14 W, 4 Cr, 2 V, 5 Co, bal Fe.
 Tungsten-cobalt high speed steel for hogging of hard, tough material.

DO ALL T15.
 M-1424; 1.5 C, 12 W, 4 Cr, 5 V, 5 Co, bal Fe.
 Tungsten type high speed steel, recommended for extreme abrasion resistance.

DOBLINSCHE ALLOY.
 M-German; 50 max Si, bal Co.
 For chemical construction; corrosion resistant.

DOCO.
 M-912; 1.75 C, 2.50 Co, 12.5 Cr, 0.90 Mo, 0.25 V, bal Fe.
 Cold work tool steel; for heavy duty cutting and stamping dynamo and transformer sheets.
 Similar to W.-Nr. 2880.

DOCTOR METAL.
 M-Eng.; 88 Cu, 9.5 Zn, 2.5 Sn.
 For corrosion resistant brass parts; corrosion resistant.

DODGE.
 M-762.
 All alloys are now to ASTM grades.

DOE RUN COPPERIZED LEAD.
 M-1747; 0.05 Cu, 0.0006 Ag, 0.0003 Cd, bal Pb.
 Cast: 2400-2600 psi TS; 40-50 El; 5 Brin.
 MP: 618°F; Density: 0.41 lbs/in^3.
 For cable sheathing, chemical equipment.

DOE RUN LEAD.
 M-1747; 0.0005 Ag, 0.0003 Cd, 99.99 min Pb.
 Cast: 1700-2000 psi TS; 35-50 El; 4 Brin. Melt point: 621°F; Density 0.41 lbs/in^3.
 For battery oxide, solder, chemical pigments anti-knock compounds.
 Exceeds ASTM B 29-55.

DOFASCO.
 M-1033; 0.3 C, 2-3 Ni, 0.9-1.1 Cr, 0.3-0.4 Mo, bal Fe.
 For valves, fittings, oil crusher roll shells; high pressure service to 1100°F.

DOFASCOLOY 1.
 M-1033; 0.25 max C, 1.25 max Mn, 0.60 max Cu, 0.90 max Ni, 0.005 min Cb, bal Fe.
 Rolled: 70,000 TS; 50,000 YS; 18.0 El.
 For exposed structural members used in bridges, railroad equipment, agricultural equipment.
 Good weathering, welding, impact properties.

DOFASCOLOY 42 W.
 M-1033; 0.21 C, 1.35 Mn, 0.3 Si, 0.005-0.05 Cb, bal Fe.
 60,000 min TS; 42,000 min YS.
 For structural work.
 ASTM Standard A 572.

DOFASCOLOY 45 W.
 M-1033; 0.22 C, 1.35 Mn, 0.3 Si, 0.005-0.10 Cb, bal Fe.
 60,000 min TS; 45,000 min YS; 22 min El.
 For structural work.
 ASTM Standard A 572.

DOFASCOLOY 50 W.
 M-1033; 0.23 C, 1.35 Mn, 0.3 Si, 0.005-0.10 Cb, bal Fe.
 65,000 min TS; 50,000 min YS; 22 min El.
 For structural work.
 ASTM Standard A 572.

DOFASCOLOY 55 W.
 M-1033; 0.25 C, 1.35 Mn, 0.3 Si, 0.005-0.10 Cb, bal Fe.
 70,000 min TS; 55,000 min YS; 20 min El.
 For structural work.
 ASTM Standard A 572.

DOFASCOLOY 60W.
 M-1033; 0.26 C, 1.35 Mn, 0.3 Si, 0.005-0.10 Cb, bal Fe.
 75,000 min TS; 60,000 min YS; 18 min El.
 For structural work.
 ASTM Standard A 572.

DOFASCOLOY 70 W.
M-1033; 0.26 max C, 1.65 max Mn, 0.005-0.10 Cb, bal Fe.
Rolled: 85,000 TS; 70,000 min YS; 16 El.
For automobile and truck components, bridges, and agricultural equipment; weldable.

DOFASCOLOY 80 W.
M-1033; 0.26 max C, 1.65 max Mn, 0.005-0.10 Cb, bal Fe.
Rolled: 95,000 TS; 80,000 min YS; 14 El.
For structural members in bridges, buildings, railroad and automotive equipment.

DOFASCOLOY F.
M-1033; 0.15 max C, 1.65 max Mn, 0.90 max Si, 0.005 min Cb, bal Fe.
Rolled: 55,000-90,000 TS; 45,000-80,000 YS; 18-24 El.
For automotive, railroad and agricultural equipment; weldable, formable.

DOFASCOLOY M.
M-1033; 0.28 max C, 1.6 max Mn, 0.30 min Cu, bal Fe.
Plate: 70,000 min TS; 60,000 min YS; 18 min El.
For tanks, trailers, trucks, auto frames, cranes, booms, buckets, stokers.
High strength low alloy steel.

DOFASCOLOY MV.
M-1033; 0.22 max C, 1.25 max Mn, 0.20 min Cu, 0.02 min V, bal Fe.
Plate: 70,000 min TS; 50,000 min YS; 18 min El.
For tanks, trailers, trucks, auto frames, booms, cranes, buckets, stokers.
High strength low alloy steel.

DOFASCOLOY NO. 2.
M-1033; 0.15 max C, 1.0 max Mn, 0.90 max Ni, 0.60 max Cu, bal Fe.
Plate: 65,000 min TS; 45,000 min YS; 19 min El.
For tanks, trailers, trucks, auto frames, mine cars, buckets, stokers, cranes, booms.
High strength low alloy steel.

DOFASCOLOY P.
M-1033; 0.16 max C, 0.90 max Mn, 0.12 max P, 0.035 max S, 0.15-0.35 Si, 0.90 max Ni, 0.60 max Cr, 0.60 max Cu, bal Fe.
Plate: 70,000 min TS; 50,000 min YS; 18 min El.
For tanks, trailers, trucks, auto frames, mine cars, buckets, stokers, cranes.
High strength low alloy steel.

DOFASCOLOY W.
M-1033; 0.25 max C, 0.50-1.25 Mn, 0.01 min Cb, bal Fe.
Gr. 45 W: 65,000 min TS; 45,000 min YS; 18 min El.
Gr. 50 W: 70,000 min TS; 50,000 min YS; 18 min El.
Gr. 55 W: 70,000 min TS; 55,000 min YS; 17 min El.
Gr. 60 W: 75,000 min TS; 60,000 min YS; 16 min El.
For trucks, tanks, trailers, cranes, booms.
High strength low alloy steel.

D.O.H.
M-1744; 0.9 C, 0.25 Si, 1.7 Mn, bal Fe.
Oil hardening, cold work tool steel, for bushes, collets, lathe centers.

DOHLEN-CSV35.
M-1348; 0.38 C, Cr, V, Si, bal Fe.
For machine tool parts, gears, shafts, fasteners, crankshafts.
Oil hardening, tough, shock resisting.

DOHLEN CSV45.
M-1348; 0.45 C, Cr, V, Si, bal Fe.
For machine tool parts, springs, gears, bolts; oil hardened, tough.

DOHLEN DELTA PEB.
M-1348; 0.55 C. 0.90 Si, 1.05 Cr, 0.18 V, 1.85 W, bal Fe.
For header dies, upsetters, punches; oil hardened, tough.

DOHLEN DP1W.
M-1348; 0.15 C, 0.15-0.35 Si, 0.25-0.50 Mn, bal Fe.
Annealed: 70,000 TS; 40,000 YS; 25 El; 60 RA; 145 Brin.
For gears, cams, camshafts, machine tool parts; case hardening steel.

DOHLEN DP3W.
M-1348; 0.35 C, 0.25-0.50 Si, 0.3-0.8 Mn, bal Fe.
Hot rolled: 85,000 TS; 55,000 YS; 30 El; 53 RA; 185 Brin.
For shears, bolts, machine tool parts, fasteners; water hardened.

DOHLEN DP4W.
M-1348; 0.45 C, 0.25-0.50 Si, 0.3-0.8 Mn, bal Fe.
Hot rolled: 98,000 TS; 60,000 YS; 24 El; 45 RA; 212 Brin.
For gears, bolts, machine tool parts, fasteners; water hardened.

DOHLEN DP6W.
M-1348; 0.60 C, 0.25-0.50 Si, 0.3-0.8 Mn, bal Fe.
Heat treated: 160,000 TS; 115,000 YS; 12 El; 40 RA; 325 Brin.
For gears, bolts, springs, rails, axles, shafts; water hardened.

DOHLEN DP7 EXTRA.
M-1348; 0.70 C, 0.25 max S, 0.25 max Mn, bal Fe.
Heat treated: 175,000 TS; 128,000 YS; 12 El; 37 RA; 355 Brin.
For springs, tools, rails, punches, mandrels; water hardened; Type W 1.

DOHLEN DP7 PRIMA.
M-1348; 0.70 C, 0.25 max Si, 0.25 max Mn, bal Fe.
Heat treated: 175,000 TS; 128,000 YS; 12 El; 37 RA; 355 Brin.
For springs, tools, rails, punches, mandrels; water hardened; Type W 1.

DOHLEN DP7W.
M-1348; 0.70 C, 0.25-0.50 Si, 0.3-0.8 Mn, bal Fe.
Heat treated: 175,000 TS; 128,000 YS; 12 El; 37 RA; 355 Brin.
For springs, tools, punches, mandrels; water hardened; Type W 1.

DOHLEN DP8.
M-1348; 0.85 C, 0.25 max Si. 0.25 max Mn, bal Fe.
Heat treated: 190,000 TS; 145,000 YS; 12 El; 33 RA; 400 Brin.
For tools, springs, drills, taps, cutters; Type W 1; water hardened.

DOHLEN DP8 PRIMA.
M-1348; 0.85 C, 0.25 max Si, 0.25 max Mn, bal Fe.
Heat treated: 190,000 TS; 145,000 YS; 12 El; 33 RA; 400 Brin.
For springs, tools, cutters, taps, drills; water hardened; Type W 1.

DOHLEN DP8W.
M-1348; 0.9 C, 0.25-0.50 Si, 0.3-0.8 Mn, bal Fe.
Heat treated: 195,000 TS; 150,000 YS; 10 El; 30 RA; 410 Brin.
For springs, tools, cutters, dies, drills, taps; Type W 1; water hardened.

DOHLEN DP10 EXTRA.
M-1348; 1.0 C, 0.25 max Si, 0.25 max Mn, bal Fe.
Annealed: 100,000 TS; 53,000 YS; 21 El; 42 RA; 200 Brin.
For springs, tools, cutters, dies, drills, taps; Type W 1; water hardened.

DOHLEN DP10 PRIMA.
M-1348; 1.0 C, 0.25 max Si, 0.25 max Mn, bal Fe.
Annealed: 100,000 TS; 53,000 YS; 21 El; 42 RA; 200 Brin.
For springs, tools, taps, drills, reamers; Type W 1; water hardened.

DOHLEN DP11 EXTRA.
M-1348; 1.1 C, 0.25 max Si, 0.25 max Mn, bal Fe.
Annealed: 110,000 TS; 58,000 YS; 18 El; 40 RA; 210 Brin.
For springs, tools, drills, taps, reamers; Type W 1; water hardened.

DOHLEN DP11 PRIMA.
M-1348; 1.15 C, 0.25 max Si, 0.25 max Mn, bal Fe.
Annealed: 110,000 TS; 58,000 YS; 18 El; 40 RA; 210 Brin.
For springs, tools, reamers, broaches; Type W 1; water hardened.

DOHLEN DP12 PRIMA.
M-1348; 1.3 C, 0.25 max Si, 0.25 max Mn, bal Fe.
For cutters, tools, drills, taps, reamers; Type W 1; water hardened.

DOHLEN DSW1.
M-1348; 0.3 C, 2.65 Cr, 0.35 V, 8.5 W, bal Fe.
For extrusion press liners and rams; oil hardened, hot work steel.

DOHLEN DSW2.
M-1348; 0.30 C, 2.35 Cr, 0.6 V, 4.25 W, bal Fe.
For hot work tools, dies, punches; hot work steel, oil hardened.

DOHLEN GSR.
M-1348; 0.67 C, 1.3 Si, 0.5 Mn, 0.5 Cr, bal Fe.
For punches, dies, shears, upsetters; oil hardened, tough.

DOHLEN KE.
M-1348; 0.20 C, 1.15 Cr, 1.25 Mn, bal Fe.
For camshafts, cams, bolts; case hardened, tough.

DOHLEN P 53.
M-1348; 0.50 C, 1.05 Cr, 3.25 Ni, 0.5 Mn, bal Fe.
For gears, bolts, crankshafts; oil hardened, shock resistant.

DOHLEN SAR.
M-1348; 1.15 C, 0.65 Cr, 0.10 V, 0.30 Mn, bal Fe.
For blanking and header dies; oil hardened, abrasion resistant.

DOHLEN STANDARD.
M-1348; 0.90 C, 1.9 Mn, 0.10 V, bal Fe.
For punches, blanking and forming dies; oil hardened, non-deforming.

DOLLAR BLUE CHIP.
M-Eng.; 0.62 C, 4.33 Cr, 17.82 W, 1.66 V, 3.93 Co, bal Fe.
For tools, cutters, reamers; high speed steel.

DOM.
M-1495; 0.37 C, 0.95 Cr, 0.20 Mo, bal Fe.
High elongation; for mechanical and tool-joints.

DOMAL AM80A.
M-1239; 8 Al, 0.2 Mn, bal Mg.
Cast: 24,000-27,000 TS; 10,000-12,500 YS; 4-7 El; 48-52 Brin.
Forged: 42,000-48,000 TS; 28,000-32,000 YS; 8-12 El; 54-58 Brin.
For aircraft castings, housings, crankcases; age-hardenable, high strength.

DOMAL AM100A.
M-1239; 10 Al, 0.1 Mn, bal Mg.
Cast: 22,000 TS; 12,000 YS; 1-3 El; 55 Brin.
Heat treated: 38,000 TS; 20,000 YS; 1-4 El; 80 Brin.
For high strength castings; age-hardenable.

DOMAL AS100.
M-1239; 10 Al, 0.5 Zn, bal Mg.
For die castings, instrument cases; high strength.

DOMAL AZ21X.
M-1239; 1.2-2.0 Al, 0.4-0.75 Zn, 0.20 min Mn, bal Mg.
Extruded: 32,000 TS; 18,000 YS; 5 El.
Tube: 32,000 TS; 15,000 YS; 5 El.
Forged: 32,000 TS; 17,000 YS; 5 El.
For airframes, cowles, tanks, structures; good weldability and workability.

DOMAL AZ31B.
M-1239; 9 Al, 0.3 Mn, bal Mg.
Extruded: 37,000 TS; 28,000 YS; 12 El; 49 Brin.
For aircraft structures; high ductility.

DOMAL AZ31X.
M-1239; 2.5-3.5 Al, 0.6-1.4 Zn, 0.20 min Mn, bal Mg.
Extruded: 32,000-34,000 TS; 16,000-20,000 YS; 6-8 El.
Plate 1/2 Hard: 33,000-39,000 TS; 16,000-29,000 YS; 10-4 El.
For aircraft parts; good formability.

DOMAL AZ61A.
M-1239; 6.5 Al, 1 Zn, bal Mg.
Extruded: 44,000 TS; 30,000 YS; 14 El; 60 Brin.
For aircraft structures; strong and tough.

DOMAL AZ61X.
M-1239; 5.8-7.2 Al, 0.4-1.5 Zn, 0.2 min Mn, bal Mg.
Extruded: 39,000 TS; 24,000 YS; 9 El.
Forged: 38,000 TS; 22,000 YS; 6 El.
Tube: 36,000 TS; 16,000 YS; 7 El.
For aircraft parts; good weldability.

DOMAL AZ63A.
M-1239; 2.7-3.3 Zn, 0.2 min Mn, bal Mg.
Cast: 27,000 TS; 14,000 YS; 16 El; 50 Brin.
Heat treated: 40,000 TS; 24,000 YS; 10 El; 60 Brin.
For cylinder heads, pistons, high strength castings; age-hardenable.

DOMAL AZ80A.
M-1239; 8.5 Al, 0.5 Zn, bal Mg.
Extruded: 48,000 TS; 32,000 YS; 12 El; 60 Brin.
Heat treated: 52,000 TS; 36,000 YS; 5 El; 82 Brin.
For cylinder heads, valve and pump bodies; age-hardenable.

DOMAL AZ80X.
M-1239; 7.5-9.2 Al, 0.2-0.8 Zn, 0.2 min Mn, bal Mg.
F-temper: 43,000 TS; 28,000 YS; 6 El.
T6-temper: 47,000 TS; 30,000 YS; 2 El.
For aircraft parts; age hardenable.

DOMAL AZ91B.
M-1239; 9 Al, 0.6 Zn, bal Mg.
Die cast: 33,000 TS; 22,000 YS; 3 El; 60 Brin.
For die castings, instrument housings; good strength.

DOMAL AZ91C.
M-1239; 9 Al, 0.5 Zn, bal Mg.
Die cast: 38,000 TS; 19,000 YS; 9 El; 59 Brin.
Heat treated: 40,000 TS; 20,000 YS; 4 El; 66 Brin.
For general high strength castings; age-hardenable.

DOMAL-AZ91X.
M-1239; 8.3-9.7 Al, 0.4-1.0 Zn, 0.2-0.4 Mn, 0.1 max Si, bal Mg.
Die cast: 33,000 TS; 22,000 YS; 3 El; 60 Brin.
Sand cast: 24,000 TS; 14,000 YS; 2 El; 52 Brin.
Ht. Tr.: 40,000 TS; 19,000 YS; 4 El; 66 Brin.
For instrument housings, portable tools, casings. Age-hardenable.

DOMAL AZ92A.
M-1239; 9 Al, 2 Zn, bal Mg.
Cast: 24,000 TS; 14,000 YS; 2 El; 65 Brin.
Heat treated: 40,000 TS; 23,000 YS; 2 El; 84 Brin.
For aircraft engine components; age-hardenable.

DOMAL EZ33.
M-1239; 2.0-3.5 Zn, 2.5-4.0 rare earths, 0.5-1.0 Zr, bal Mg.
T5-temper: 20,000 TS; 13,000 YS; 3 El.
For engine castings, diffuser and compressor casings; heat and creep resistant to 500°F.

DOMAL HZ32.
M-1239; 3.0 Th, 2.3 Zn, 0.7 Mn, bal Mg.
At 20°C: 30,000 TS; 13,000 YS; 7 El.
At 250°C: 14,900 TS; 9000 YS; 50 El.
At 350°C: 10,000 TS; 40 El.
For jet engine and missile components; high creep resistance to 660°F.

DOMAL M1B.
M-1239; 1.2 Mn, bal Mg.
F-temper: 33,000 TS; 27,000 YS; 44 El.
H24-temper: 37,000 TS; 28,000 YS; 7 El; 56 Brin.
For parts subjected to high stresses; good weldability and corrosion resistance.

DOMAL TA54A.
M-1239; 5 Sn, 3 Al, bal Mg.
Forged: 40,000 TS; 28,000 YS; 12 El; 52 Brin.
For pistons; high ductility.

DOMAL ZH62.
M-1239; 5.8 Zn, 1.8 Th, 0.7 Zr, bal Mg.
Cast: 40,000 TS; 24,000 YS; 8.5 El.
Sheet: 48,000 TS; 42,000 YS; 15 El.
Extruded: 49,000 TS; 42,000 YS; 20 El.
For aircraft engine and missile components; good formability and weldability.

DOMAL ZK31.
M-1239; 2.5-3.5 Zn, 0.5-1.0 Zr, bal Mg.
Extruded: 40,000 TS; 28,000 YS; 5 El.
Forged: 38,000 TS; 26,000 YS; 5 El.
For airframe structural parts, landing wheels, gear casings; heavy duty service.

DOMAL ZK60.
M-1239; 4.8-6.5 Al, 0.5 min Zr, 0.003 max Fe, bal Mg.
F-temper: 42,000 TS; 28,000 YS; 5 El.
T5-temper: 44,000 TS; 32,000 YS; 4 El.
Cast: 35,000 TS; 18,000 YS; 8 El.
T6-temper: 42,000 TS; 26,000 YS; 5 El.
For aircraft structural members; age hardenable, shock resistant.

DOMESTIC.
M-389; 0.7-1.2 C, bal Fe.
For tools, punches, taps; water hardening.

DOMINIAL 170.
M-1352; 0.45 C, Cr, Ni, bal Fe.
For gears, bolts, crankshafts, fasteners; oil hardened, shock resistant.

DOMINIAL BA.
M-1352; 1.05 C, 6 Cr, W, bal Fe.
For oil refinery equipment; creep and wear resistant.

DOMINIAL BZM.
M-1352; C, alloy, bal Fe.
For gears, shafts, bolts; oil hardening.

DOMINIAL CH.
M-1352; 2.1 C, 11.5 Cr, bal Fe.
For blanking and forming dies; oil or air hardened, nondeforming.

DOMINIAL CH160W.
M-1352; 1.65 C, 11.5 Cr, 0.2 Mo, 0.1 V, bal Fe.
For blanking and forming dies, punches; air hardened, nondeforming.

DOMINIAL CH165.
M-1352; 1.65 C, 11.5 Cr, 0.1 V, bal Fe.
For blanking and forming dies, punches; air hardened, nondeforming.

DOMINIAL CHW.
M-1352; 2.1 C, 11.5 Cr, 0.7 W, bal Fe.
For blanking and forming dies, punches; oil or air hardened, nondeforming.

DOMINIAL CM167.
M-1352; 0.45 C, 1.4 Cr, 0.7 Mo, 0.3 V, bal Fe.
For gears, bolts, crankshafts, fasteners; oil hardened, tough.

DOMINIAL DAG.
M-1352; 1.45 C, 0.6 Mn, 1.4 Cr, bal Fe.
For bearings, sleeves, blanking dies; water hardened, wear resistant.

DOMINIAL DS.
M-1352; 0.38 C, 6 Cr, 0.9 Si, 0.1 V, bal Fe.
For oil refinery equipment; creep and heat resistant.

DOMINIAL DSW.
M-1352; 0.35 C, 0.9 Si, 1.05 Cr, 0.18 V, 1.85 W, bal Fe.
For cold heading dies, crimpers, punches, upsetters; oil hardened, tough.

DOMINIAL DWK.
M-1352; 1.3 C, 0.2 max Cr, 4.75 W, bal Fe.
For engravers tools, cutters; oil hardened.

DOMINIAL EC4.
M-1352; 0.13 C, 3.5 Ni, 0.7 Cr, bal Fe.
For bolts, fasteners, gears, cams, camshafts; case hardening steel, shock resistant.

DOMINIAL EC5.
M-1352; 0.13 C, 1.1 Cr, 4.5 Ni, bal Fe.
For gears, cams, camshafts; case hardening steel, shock resistant.

DOMINIAL ECNL.
M-1352; 0.19 C, 1.25 Cr, 0.2 Mo, 3.75 Ni, bal Fe.
For gears, bolts, camshafts, fasteners; case hardening steel, shock resistant.

DOMINIAL EXTRA.
M-1352; 1.05 C, 1.2 Cr, 0.5 Mn, bal Fe.
For bearings, liners, sleeves; water hardened, wear resistant.

DOMINIAL FGS.
M-1352; 1.25 C, 1.15 Si, 0.7 Mn, 1.2 Cr, bal Fe.
For bearings, liners, sleeves; oil hardened, wear resistant.

DOMINIAL GH.
M-1352; 2.1 C, 11.5 Cr, bal Fe.
For blanking and forming dies, punches; oil or air hardened, nondeforming.

DOMINIAL KC15.
M-1352; 0.45 C, 1.5 Cr, 1.0 Si, 0.1 V, bal Fe.
For springs, gears, bolts, crankshafts; oil hardened, shock resistant.

DOMINIAL KHP.
M-1352; 0.15 C, 0.65 Cr, 0.5 Mn, bal Fe.
For gears, pinions, cams, camshafts; case hardening steel.

DOMINIAL KL.
M-1352; 0.55 C, 0.9 Si, 0.3 Mn, 1.05 Cr, 0.18 V, 1.85 W, bal Fe.
For cold heading dies, crimpers, punches, upsetters; oil hardened.

DOMINIAL KST.
M-1352; 1.0 C, 0.2 Si, 0.25 Mn, 0.1 V, bal Fe.
For cutters, drills, taps, reamers, broaches; Type W 2; water hardened.

DOMINIAL KTW.
M-1352; 0.40 C, Cr, Mn, Mo, bal Fe.
For gears, shafts, bolts, studs, crankshafts; oil hardened, tough.

DOMINIAL KZR.
M-1352; 1.4 C, 0.3 Cr, 0.1 V, bal Fe.
For engravers tools, cutters, taps, reamers; water hardened.

DOMINIAL MA.
M-1352; 0.3 C, 2.65 Cr, 0.35 V, 8.5 W, bal Fe.
For upsetters, riveters, punches, crimpers; hot work steel, oil hardened.

DOMINIAL MAK.
M-1352; 0.3 C, 2.35 Cr, 0.6 V, 4.25 W, bal Fe.
For upsetters, riveters, punches; hot work steel, oil hardened.

DOMINIAL MA SUPRA.
M-1352; 0.3 C, 2 Co, 2.4 Cr, 0.25 V, 8.5 W, bal Fe.
For extrusion press rams, hot punches, shears; hot work steel, oil hardened.

DOMINIAL MC.
M-1352; 0.3 C, 2.5 Cr, 0.2 Mo, 0.15 V, bal Fe.
For gears, bolts, crankshafts, fasteners; oil hardened, tough.

DOMINIAL MK.
M-1352; 1.05 C, Mn, Cr, bal Fe.
For drills, taps, bearings, races; water hardened, wear resistant.

DOMINIAL MKST.
M-1352; 0.9 C, 0.1 V, 1.9 Mn, 0.25 Si, bal Fe.
For punches, forming dies, shears, crimpers; oil hardened, nondeforming.

DOMINIAL N400.
M-1352; 0.45 C, Ni, Cr, Mo, bal Fe.
For gears, bolts, crankshafts, fasteners; oil hardened, shock resistant.

DOMINIAL PD.
M-1352; 0.65 C, 3.75 Cr, 0.85 Mo, 0.7 V, 8.5 W, bal Fe.
For lathe and planer tools, reamers, broaches, taps; high speed steel.

DOMINIAL PK.
M-1352; 0.45 C, 1.05 Cr, 0.2 V, 1.85 W, bal Fe.
For cold heading tools, punches, mandrels; oil hardened, tough.

DOMINIAL PK17.
M-1352; 0.45 C, W, Cr, V, bal Fe.
For hot shears and punches, upsetters; hot work tools, oil hardened.

DOMINIAL PW15.
M-1352; 0.55 C, 0.7 Cr, 0.18 V, 1.65 Ni, 0.18 Mo, bal Fe.
For gears, bolts, crankshafts; oil hardened, shock resistant.

DOMINIAL PWM.
M-1352; 0.56 C, 0.6 Mn, Cr, Mo, Ni, V, bal Fe.
For gears, bolts, crankshafts, axles; oil hardened, shock resistant.

DOMINIAL R 13F.
M-1352; 0.12 max C, 0.4 Si, 13 Cr, bal Fe.
Annealed: 75,000 TS; 40,000 YS; 35 El; 70 RA; 155 Brin.
Cold drawn: 100,000 TS; 85,000 YS; 17 El; 60 RA; 205 Brin.
For turbine blades, surgical instruments, valves, cutlery; Type 410; stainless.

DOMINIAL RAN.
M-1352; 0.15 max C, 18 Cr, 8.5 Ni, bal Fe.
Annealed: 80,000 TS; 35,000 YS; 55 El; 75 RA; 150 Brin.
Cold drawn: 180,000 TS; 150,000 YS; 10 El; 250 Brin.
For chemical plant equipment, tanks, mixers, filters; Type 302; stainless, austenitic.

DOMINIAL RAS.
M-1352; 0.12 max C, 18 Cr, 9.5 Ni, Ti = 4 x C, bal Fe.
Annealed: 85,000 TS; 35,000 YS; 55 El; 65 RA; 150 Brin.
Cold drawn: 95,000 TS; 60,000 YS; 40 El; 60 RA; 185 Brin.
For welded chemical plant equipment, tanks, mixers; Type 321; stainless, austenitic.

DOMINIAL RAS4.
M-1352; 0.12 max C, 18 Cr, 2 Mo, 10.5 Ni, Ti = 4 x C, bal Fe.
Annealed: 85,000 TS; 35,000 YS; 45 El; 60 RA; 160 Brin.
For welded acid resistant chemical plant equipment; Type 316 Ti; stainless, austenitic.

DOMINIAL RF.
M-1352; 0.4 C, 13 Cr, 0.4 Si, 0.3 Mn, bal Fe.
Annealed: 95,000 TS; 50,000 YS; 25 El; 55 RA; 200 Brin.
For valves, cutlery, surgical and dental instruments; Type 420; stainless hardenable.

DOMINIAL RM13.
M-1352; 0.2 C, 13 Cr, 0.4 Si, bal Fe.
Annealed: 95,000 TS; 50,000 YS; 25 El; 55 RA; 200 Brin.
For valves, cutlery, turbine blades, surgical instruments; Type 420; stainless, hardenable.

DOMINIAL RM13H.
M-1352; 0.4 C, 0.4 Si, 13 Cr, bal Fe.
Annealed: 95,000 TS; 50,000 YS; 25 El; 55 RA; 200 Brin.
For valves, cutlery, surgical and dental instruments; Type 420; stainless hardenable.

DOMINIAL RM13MO.
M-1352; 0.2 C, 0.4 Si, 13 Cr, 1.15 Mo, bal Fe.
Annealed: 95,000 TS; 50,000 YS; 25 El; 55 RA; 200 Brin.
For turbine blades, valves, cutlery; Type 420Mo; stainless.

DOMINIAL RM17.
M-1352; 0.22 C, 17 Cr, 1.5 Ni, 0.4 Si, bal Fe.
Annealed: 125,000 TS; 95,000 YS; 20 El; 55 RA; 260 Brin.
Cold drawn: 130,000 TS; 110,000 YS; 15 El; 35 RA; 270 Brin.
For pumps, valves, marine hardware; Type 431; stainless.

DOMINIAL RM174.
M-1352; 0.35 C, 0.4 Si, 16.5 Cr, 1.15 Mo, bal Fe.
For chemical plant equipment, furnace parts; corrosion and heat resistant.

DOMINIAL RM189.
M-1352; 0.90 C, 18 Cr, 2 Mo, bal Fe.
For cutlery, valves, ball bearings, surgical instruments; Type 440B; corrosion resistant, hardenable.

DOMINIAL SES.
M-1352; 1.15 C, 0.65 Cr, 0.1 V, 0.3 Mn, bal Fe.
For bearings, sleeves, cutters; water hardened, wear resistant.

DOMINIAL SGW.
M-1352; 0.9 C, 0.3 Mn, 0.8 Cr, bal Fe.
For cutters, bearings, sleeves; water hardened, wear resistant.

DOMINIAL SN.
M-1352; 0.5 C, 0.5 Mn, 1.05 Cr, 3.25 Ni, bal Fe.
For gears, bolts, crankshafts, axles, fasteners; oil hardened, shock resistant.

DOMINIAL SP.
M-1352; 0.70 C, 1.7 Si, 0.70 Mn, bal Fe.
For punches, dies, upsetters, springs, crimpers; oil hardened, tough.

DOMINIAL SS50.
M-1352; 0.50 C, 0.85 Mn, 1.0 Cr, 0.09 V, bal Fe.
For gears, springs, bolts, studs, crankshafts; oil hardened, shock resistant.

DOMINIAL VSF.
M-1352; 0.90 C, 1.0 Mn, 0.37 Si, bal Fe.
For drills, taps, reamers, hobs, broaches; water hardened, wear resistant.

DOMINIAL W 44.
M-1352; 0.45 C, 1.35 Cr, 0.45 Mo, 0.8 V, 0.45 W, bal Fe.
For cold heading dies, upsetters, riveters; oil hardened, tough.

DOMINIAL WEH.
M-1352; 0.2 C, 1.25 Mn, 1.15 Cr, bal Fe.
For gears, cams, camshafts, fasteners; case hardening steel.

DOMINIAL WF.
M-1352; 1.2 C, 0.2 Cr, 0.1 V, 1.0 W, bal Fe.
For blanking and forming dies, cutters; water hardened, wear resistant.

DOMINIAL WF2.
M-1352; 1.15 C, W, bal Fe.
For blanking dies, cutters, punches; water hardened, wear resistant.

DOMINIAL WMK.
M-1352; 0.30 C, 1.0 Si, 1.1 Cr, 0.18 V, 3.75 W, bal Fe.
For cold heading dies, punches, upsetters; oil hardened, tough.

DOMINIAL WR.
M-1352; 1.4 C, 1 Si, 0.18 V, 3.75 W, bal Fe.
For bearings, forming dies, cutters; water hardened.

DOMITE 30.
M-763; 3.2 C, 1.5 Ni, 2 Si, bal Fe.
Cast: 30,000 TS; 180 Brin.
For stamping dies, pulleys; cast iron, Tr.S. 2200.

DOMITE 35.
M-763; 3.5 C, 2.5 Si, bal Fe.
Cast: 35,000 TS; 190 Brin.
For pumps, die shoes, evaporators; cast iron; Tr.S. 2400.

DOMITE 45.
M-763; 3.3 C, 2.2 Si, bal Fe.
Cast: 45,000 TS; 220 Brin.
For gears, flywheels, pump impellers; cast iron; Tr.S. 2800.

DOMITE 55.
M-763; 3.1 C, 1.5 Si, bal Fe.
Cast: 55,000 TS; 250 Brin.
For crankshafts, forming dies, gears; cast iron; Tr.S. 3400.

DOMITE NI-HARD.
M-763; 4.2-4.7 Ni, 3-3.6 C, 1.4-2.5 Cr, 0.5 Si, 0.4-6 Mn, bal Fe.
Cast: 50,000-60,000 TS; 600-725 Brin.
For wear resistant castings; wear and abrasion resistant.

DOMITE NODULOY.
M-763; 3.3 T.C., 1.9 Si, bal Fe.
Cast: 115,000 TS; 2 El.
Annealed: 75,000 TS; 15 El.
For high strength castings; high ductility.

DOMITE WEAR RESISTING A.
M-763; high C, alloy, Si, bal Fe.
Cast: 200 Brin.
For wheels, brake shoes, brick dies; pearlitic cast iron.

DOMITE WEAR RESISTING B.
M-763; high C, alloy, Si, bal Fe.
Cast: 300 Brin.
For iron liners, plow points, wheels; wear resistant cast iron.

DOMITE WEAR RESISTING C.
M-763; high C, alloy, Si, bal Fe.
Cast: 475 Brin.
For railway wheels, brake shoes; chilled cast iron.

DOMITE WEAR RESISTING D.
M-763; high C, alloy, Si, bal Fe.
Cast: 600 Brin.
For railway wheels, wear, parts; chilled cast iron.

DONAL.
M-1548; 0.3-0.5 Si, 1.0-2.0 Mn, bal Al.
Soft: 22,000 TS; 35 El; 30 Brin.
Half hard: 26,000 TS; 21,000 YS; 5-15 El; 40-50 Brin.
For general structures, containers, heat exchangers, trim.
Non-hardenable. Corrosion resistant.

DONAL now WEILAND A61.
M-764; 98.5 Al, 1.5 Mn.
For light alloy parts; die castings.

DONEGAL D-1.
M-1223; 0.20 C, 0.65 Mn, bal Fe.
Annealed: 60,000 TS; 30,000 YS; 24 El; 35 RA; 130 Brin.
For machine tool parts, gears, shafts; ASTM Gr. WCA.

DONEGAL D-2.
M-1223; 0.27 C, 0.65 Mn, bal Fe.
Annealed: 70,000 TS; 35,000 YS; 24 El; 35 RA; 140 Brin.
For machine tool parts, gears, housings; ASTM Gr. WCB.

DONEGAL D-3.
M-1223; 0.40 C, 0.65 Mn, bal Fe.
Normalized: 80,000 TS; 40,000 YS; 18 El; 30 RA; 165 Brin.
For gears, housings, shafts, machine tool parts; ASTM Gr. 80-40.

DONEGAL D-4.
M-1223; 0.20 C, 0.65 Mn, 0.90 Ni, 0.65 Cr, 0.55 Mo, bal Fe.
Normalized: 70,000 TS; 40,000 YS; 20 El; 35 RA; 160 Brin.
For gears, bolts, crankshafts, housings; ASTM Gr. WC4; shock resistant.

DONEGAL D-5.
M-1223; 0.20 C, 0.80 Ni, 0.70 Cr, 1.0 Mo, 0.60 Mn, bal Fe.
Normalized: 70,000 TS; 40,000 YS; 20 El; 35 RA; 160 Brin.
For gears, housings, machine tool parts; ASTM Gr. WC5; case hardened, shock resistant.

DONEGAL D-6.
M-1223; 0.20 C, 0.65 Mn, 1.25 Cr, 0.55 Mo, bal Fe.
Normalized: 70,000 TS; 40,000 YS; 20 El; 35 RA; 160 Brin.
For gears, housings, machine tool parts; ASTM Gr. W 6.

DONEGAL D-7.
M-1223; 0.30 C, 1.35 Mn, bal Fe.
Normalized: 90,000 TS; 60,000 YS; 20 El; 40 RA; 190 Brin.
For gears, bolts, crankshafts, machine tool parts; ASTM Gr. 90-60; water or oil hardened.

DONEGAL D-8.
M-1223; 0.30 C, 1.3 Mn, 0.25 Mo, bal Fe.
Normalized: 90,000 TS; 60,000 YS; 20 El; 40 RA; 190 Brin.
For gears, bolts, crankshafts, machine tool parts; tough.

DONEGAL D-9.
M-1223; 0.18 C, 0.5 Cr, 1.0 Mo, 0.60 Mn, bal Fe.
Normalized: 70,000 TS; 40,000 YS; 20 El; 35 RA; 160 Brin.
For machine tool parts; ASTM Gr. WC9.

DONEGAL D-10.
M-1223; 0.30 C, 0.65 Mn, 0.60 Ni, 0.60 Cr, 0.50 Mo, bal Fe.
Normalized: 100,000 TS; 60,000 YS; 15 El; 30 RA; 202 Brin.
Water hardened: 150,000 TS; 125,000 YS; 10 El; 25 RA; 320 Brin.
For machine tool parts.

DONEGAL D-11.
M-1223; 0.20 C, 0.65 Mn, 0.55 Mo, bal Fe.
Normalized: 65,000 TS; 35,000 YS; 24 El; 35 RA; 145 Brin.
For machine tool parts, gears, shafts; ASTM Gr. WC1.

DONEGAL D-12.
M-1223; 0.20 C, 0.50 Mn, 9.0 Cr, 1.0 Mo, bal Fe.
Normalized: 90,000 TS; 60,000 YS; 18 El; 35 RA; 190 Brin.
For oil refinery equipment; ASTM Gr. C 12; creep resistant.

DONEGAL D-13.
M-1223; 0.25 C, 0.65 Mn, 2.5 Ni, bal Fe.
Normalized: 65,000 TS; 40,000 YS; 24 El; 35 RA; 179 Brin.
For gears, machine tool parts; ASTM Gr. CL2; tough.

DONEGAL D-14.
M-1223; 0.15 C, 0.65 Mn, 3.5 Ni, bal Fe.
Normalized: 65,000 TS; 40,000 YS; 25 El; 35 RA; 179 Brin.
For gears, bolts, machine tool parts; ASTM Gr. CL3; shock resistant.

DONEGAL D-15.
M-1223; C, alloy, bal Fe.
For armor; tough.

DONEGAL D-16.
M-1223; 0.20 C, 0.60 Mn, 5.0 Cr, 0.5 Mo, bal Fe.
Normalized: 90,000 TS; 60,000 YS; 18 El; 35 RA; 190 Brin.
For oil refinery equipment; ASTM Gr. C 5, creep resistant.

DONEGAL D-20.
M-1223; 0.07 C, 18-22 Cr, 21-31 Ni, 2.5 Mo, 4.0 Cu, bal Fe.
Cast: 65,000 TS; 30,000 YS; 30 El; 150 Brin.
For chemical plant equipment; ACI Type CN-7M; austenitic, stainless.

DONEGAL D-21.
M-1223; 0.15 max C, 11-14 Cr, 1.0 max Ni, bal Fe.
Cast: 90,000 TS; 65,000 YS; 18 El; 30 RA; 200 Brin.
For chemical plant equipment, valves, cutlery; ACI Type CA15; corrosion resistant.

DONEGAL D-23.
M-1223; 0.20-0.40 C, 11.5-14.0 Cr, 1.0 max Ni, bal Fe.
Cast: 110,000 TS; 75,000 YS; 15 El; 25 RA; 220 Brin.
For valves, cutlery, chemical plant equipment; ACI Type CA40; corrosion resistant.

DONEGAL D-24.
M-1223; 0.30 max C, 18-22 Cr, 2 max Ni, bal Fe.
Cast: 65,000-95,000 TS; 30,000-60,000 YS; 15 El; 170-195 Brin.
For furnace parts, heat treating boxes; ACI Type CB30; corrosion resistant.

DONEGAL D-25.
M-1223; 0.50 max C, 26-30 Cr, 4 max Ni, bal Fe.
Cast: 55,000 TS; 190 Brin.
Heat treated: 97,000 TS; 65,000 YS; 18 El; 210 Brin.
For cylinder liners, bushings, valve seats; ACI Type CC50; corrosion resistant.

DONEGAL D-26.
M-1223; 0.30 max C, 26-30 Cr, 8-11 Ni, bal Fe.
Cast: 80,000 TS; 40,000 YS; 10-20 El; 20 RA; 170 Brin.
For furnace parts, salt pots, heat treating boxes; ACI Type CE30; austenitic, stainless.

DONEGAL D-27.
M-1223; 0.20 max C, 18-21 Cr, 8-11 Ni, bal Fe.
Cast: 70,000 TS; 30,000 YS; 30 El; 160 Brin.
For furnace parts, heat treating boxes, retorts; ACI Type CF20; stainless, austenitic.

DONEGAL D-28.
M-1223; 0.08 max C, 18-21 Cr, 8-11 Ni, bal Fe.
Cast: 70,000 TS; 28,000 YS; 35 El; 150 Brin.
For furnace parts, chemical plant equipment; ACI Type CF8; stainless, austenitic.

DONEGAL D-29.
M-1223; 0.16 max C, 18-21 Cr, 9-10 Ni, 0.20-0.35 Se, bal Fe.
Cast: 70,000 TS; 30,000 YS; 25 El; 150 Brin.
For chemical plant equipment; ACI Type CF16F; free-cutting, stainless.

DONEGAL D-30.
M-1223; 0.08 max C, 18-21 Cr, 9-12 Ni, Cb = 8 x C, bal Fe.
Cast: 70,000 TS; 30,000 YS; 30 El; 150 Brin.
For welded chemical plant equipment; ACI Type CF8C; stainless, austenitic.

DONEGAL D-31.
M-1223; 0.08 max C, 18-21 Cr, 9-12 Ni, 2-3 Mo, bal Fe.
Cast: 70,000 TS; 30,000 YS; 30 El; 150 Brin.
For acid resistant chemical plant equipment; ACI Type CF8M; stainless, austenitic.

DONEGAL D-32.
M-1223; 0.20 max C, 22-26 Cr, 12-15 Ni, bal Fe.
Cast: 70,000 TS; 30,000 YS; 30 El; 150 Brin.
For furnace parts, heat treating boxes, retorts; ACI Type CH20; corrosion and heat resistant.

DONEGAL D-33.
M-1223; 0.20 max C, 23-27 Cr, 19-22 Ni, bal Fe.
Cast: 65,000 TS; 28,000 YS; 30 El; 170 Brin.
For retorts, pots, furnace equipment and parts; ACI Type CK20; corrosion and heat resistant.

DONEGAL D-40.
M-1223; 0.20-0.50 C, 24-28 Cr, 11-14 Ni, bal Fe.
Cast: 75,000 TS; 35,000 YS; 15 El; 200 Brin.
For furnace shafts, beams, rollers, tube supports; ACI Type HH; corrosion and heat resistant.

DONEGAL D-41.
M-1223; 0.20-0.50 C, 26-30 Cr, 8-11 Ni, bal Fe.
Cast: 85,000 TS; 40,000 YS; 9 El; 170 Brin.
For furnace parts, retorts, salt pots; ACI Type HE; corrosion and heat resistant.

DONEGAL D-42.
M-1223; 0.20-0.60 C, 24-28 Cr, 18-22 Ni, bal Fe.
Cast: 75,000 TS; 35,000 YS; 15 El; 170 Brin.
For furnace parts, retorts, skids, stack dampers; ACI Type HK; corrosion and heat resistant.

DONEGAL D-43.
M-1223; 0.35-0.75 C, 13-17 Cr, 33-37 Ni, bal Fe.
Cast: 65,000 TS; 40,000 YS; 4-10 El; 12 RA; 180 Brin.
For salt pots, furnace parts, heat treating boxes; ACI Type HT; corrosion and heat resistant.

DONEGAL DC-40.
M-1223; 0.25 C, 0.75 Mn, 0.75 Si, 1.5 Ni, 19 Cr, bal Fe.
Annealed: 95,000 TS; 60,000 YS; 15 El; 195 Brin.
For chemical and food processing equipment, furnace parts; ACI-CB30; corrosion resistant.

DONEGAL DC-41.
M-1223; 0.10 C, 0.65 Mn, 0.8 Si, 0.5 Ni, 12.5 Cr, bal Fe.
Annealed: 95,000 TS; 78,000 YS; 22 El; 185 Brin.
Heat treated: 163,000-213,000 TS; 75,000-173,000 YS; 28-6 El; 59-9 RA; 185-415 Brin.
For chemical and food processing equipment, ship propellers; ACI-CA15 corrosion resistant.

DONEGAL DC-42.
M-1223; 0.3 C, 0.75 Mn, 0.75 Si, 3 Ni, 28 Cr, bal Fe.
Cast: 95,000 TS; 60,000 YS; 15 El; 193 Brin.
Annealed: 97,000 TS; 65,000 YS; 18 El; 210 Brin.
For bushings, impellers, cylinder liners, valve seats and bodies; ACI-CC50; corrosion resistant.

DONEGAL DC-43.
M-1223; 0.08 C, 0.75 Mn, 1.25 Si, 10.5 Ni, 19.5 Cr, Cb, bal Fe.
Cast: 70,000 TS; 30,000 YS; 30 El; 150 Brin.
For chemical and food processing equipment; ACI-CF8C; corrosion resistant, austenitic.

DONEGAL DC-44.
M-1223; 0.08 max C, 0.75 Mn, 1.25 Si, 9 Ni, 19.5 Cr, bal Fe.
Cast: 70,000 TS; 28,000 YS; 35 El; 150 Brin.
For chemical and food processing equipment; ACI-CF8; corrosion resistant, austenitic.

DONEGAL DC-45.
M-1223; 0.16 max C, 0.75 Mn, 1.25 Si, 9 Ni, 19.5 Cr, Se, bal Fe.
Cast: 70,000 TS; 28,000 YS; 30 El; 150 Brin.
For chemical and food processing equipment; ACI-CF16F; corrosion resistant, austenitic.

DONEGAL DC-46.
M-1223; 0.08 max C, 0.75 Mn, 1.25 Si, 10 Ni, 19.5 Cr, Mo, bal Fe.
Cast: 70,000 TS; 30,000 YS; 30 El; 150 Brin.
For chemical and food processing equipment; ACI-CF8M; corrosion resistant, austenitic.

DONEGAL DC-47.
M-1223; 0.20 max C, 0.75 Mn, 1.25 Si, 9 Ni, 19.5 Cr, bal Fe.
Cast: 70,000 TS; 30,000 YS; 30 El; 150 Brin.
For chemical and food processing equipment; ACI-CF20; corrosion resistant, austenitic.

DONEGAL DC-48.
M-1223; 0.20 max C, 0.75 Mn, 1.25 Si, 13.5 Ni, 24.5 Cr, bal Fe.
Cast: 70,000 TS; 30,000 YS; 30 El; 170 Brin.
For chemical plant equipment, furnace parts; ACI-CH20; corrosion and heat resistant.

DONEGAL DC-49.
M-1223; 0.20 max C, 0.75 Mn, 1.25 Si, 20.5 Ni, 25 Cr, bal Fe.
Cast: 65,000 TS; 28,000 YS; 30 El; 170 Brin.
For chemical plant equipment, furnace parts; ACI-CK20; corrosion and heat resistant.

DONEGAL DC-50.
M-1223; 0.05 C, 16.5 Cr, 4 Ni, 4 Cu, bal Fe.
Heat treated: 180,000-210,000 TS; 165,000-200,000 YS; 15-6 El; 60-30 RA; 375-440 Brin.
For chemical plant equipment; age hardenable, corrosion resistant.

DOPPLOY 30.
M-479; 2.0 Si, 0.8 Mn, 3.0 total C, 18.5 Ni, 2.5 Cr, bal Fe.
Cast: 35,000 TS.
For casting jacketed kettles, agitators, etc., corrosion resistant.

DOQUAT.
M-USSR; Fe-W.
For cutting tools; hard sintered alloy.

DORDENT.
M-200; 72.5 Au, plus Pt.
For dental purposes.

DORIUM "D".
70 Cu, 30 Zn.
For condenser tubes.

DORRENBERG A 50.
M-614; 0.56 C, Ni, Cr, Mo, V, bal Fe.
For crankshafts, punches, dies; oil hardened, shock resistant.

DORRENBERG CN60.
M-614; 0.15 C, Ni, Cr, bal Fe.
For gears, bolts, machine tool parts; case hardened, shock resistant.

DORRENBERG CP10V.
M-614; 2.1 C, 0.35 Si, 0.3 Mn, 11.5 Cr, bal Fe.
For blanking and forming dies, gages, punches; oil hardened, non-deforming.

DORRENBERG CPP.
M-614; 1.65 C, 11.5 Cr, 0.1 V, bal Fe.
For blanking and forming dies, gages, punches; air hardened, non-deforming.

DORRENBERG CPW.
M-614; 2.1 C, 0.25 Si, 11.5 Cr, 0.7 W, bal Fe.
For blanking and forming dies, gages, punches; oil hardened, non-deforming.

DORRENBERG DGS.
M-614; 0.58 C, Mn, Si, bal Fe.
For machine tool parts; oil hardened.

DORRENBERG DM1.
M-614; 0.45 C, 1.4 Cr, 0.7 Mo, 0.3 V, bal Fe.
For gears, machine tool parts; oil hardened, tough.

DORRENBERG DML.
M-614; 0.50 C, Cr, Mo, bal Fe.
For gears, bolts, studs, fasteners; oil hardened, shock resistant.

DORRENBERG ECN4M.
M-614; 0.19 C, 1.25 Cr, 0.2 Mo, 3.75 Ni, bal Fe.
For gears, bolts, camshafts, cams; case hardened.

DORRENBERG EPM1.
M-614; 0.19 C, 1.25 Cr, 0.20 Mo, 3.75 Ni, bal Fe.
For gears, cams, camshafts, fasteners; case hardened, shock resistant.

DORRENBERG EPM2.
M-614; 0.20 C, 1.15 Cr, 1.25 Mn, bal Fe.
For gears, cams, camshafts; case hardened, tough.

DORRENBERG EXTRA 12.
M-614; 6.60 C, 0.25-0.50 Si, 0.30-0.80 Mn, bal Fe.
For springs, tools, rails, hammers; water hardened.

DORRENBERG EXTRA 16.
M-614; 0.55 C, 0.10-0.40 Si, 0.50-0.70 Mn, bal Fe.
Heat treated: 150,000 TS; 110,000 YS; 15 El; 45 RA; 310 Brin.
For tools, punches, hammers; water hardened.

DORRENBERG EXTRA NO. 10.
M-614; 0.60 C, 0.25-0.50 Si, 0.30-0.80 Mn, bal Fe.
Heat treated: 160,000 TS; 113,000 YS; 12 El; 40 RA; 325 Brin.
For tools, punches, springs, axes, hammers; water hardened.

DORRENBERG GWS2.
M-614; 0.90 C, 1.2 Cr, 1.15 Si, 0.70 Mn, bal Fe.
For punches, cutters, bearings, bushings; oil hardened, wear resistant.

DORRENBERG GWS4.
M-614; 1.25 C, 1.2 Cr, 1.15 Si, 0.7 Mn, bal Fe.
For cutters, dies, bearings, bushings; oil hardened, wear resistant.

DORRENBERG H24CN.
M-614; 0.2 C, 1.2 Si, 25 Cr, 4 Ni, bal Fe.
For valve seats and bodies, bushings; corrosion and heat resistant.

DORRENBERG HC15.
M-614; 0.12 max C, 2.2 Si, 13 Cr, bal Fe.
For oil refinery and chemical plant equipment; corrosion and heat resistant.

DORRENBERG HC17M.
M-614; 0.30 C, 16 Cr, Mo, bal Fe.
For oil refinery and chemical plant equipment; corrosion and heat resistant.

DORRENBERG HC22.
M-614; 0.20 C, 18 Cr, 1 Al, bal Fe.
For oil refinery equipment; corrosion and heat resistant.

DORRENBERG HC25.
M-614; 0.12 max C, 1.5 Si, 1.5 Al, 24 Cr, bal Fe.
For heat treating boxes, furnace parts and equipment; heat resistant.

DORRENBERG HC30.
M-614; 0.20 C, 29 Cr, 1 Al, bal Fe.
For heat treating boxes, furnace parts and equipment; heat resistant.

DORRENBERG HC50.
M-614; 0.40 C, 0.40 Si, 0.30 Mn, 13 Cr, bal Fe.
For cutlery, surgical instruments; corrosion resistant.

DORRENBERG HCN.
M-614; 0.15 C, 24 Cr, 19 Ni, bal Fe.
For furnace parts, pots, retorts; Type 314; corrosion resistant.

DORRENBERG HCNN.
M-614; 0.15 C, 19.5 Cr, 9.5 Ni, bal Fe.
Annealed: 80,000 TS; 35,000 YS; 55 El; 75 RA; 150 Brin.
For chemical plant equipment; Type 302; stainless, austenitic.

DORRENBERG MAB.
M-614; 1.1 C, W, Cr, V, bal Fe.
For cutters, dies, tools; oil hardened.

DORRENBERG MCM.
M-614; 0.40 C, Cr, Mn, Mo, bal Fe.
For gears, bolts, fasteners, shafts; oil hardened, tough.

DORRENBERG MS.
M-614; 0.53 C, 0.90 Si, 0.90 Mn, bal Fe.
For gears, bolts, shafts, fasteners; water hardened.

DORRENBERG NC15.
M-614; 0.45 C, Cr, Ni, Fe.
For gears, shafts, machine tool parts; oil hardened, shock resistant.

DORRENBERG NC15A.
M-614; 0.50 C, Ni, Cr, bal Fe.
For gears, shafts, machine tool parts; oil hardened, shock resistant.

DORRENBERG NCM1.
M-614; 0.55 C, 0.6 Mn, 0.70 Cr, 0.18 Mo, 1.65 Ni, 0.1 V, bal Fe.
For forging and heading dies, upsetters; oil hardened, tough.

DORRENBERG P 60.
M-614; 0.61 C, 0.85 Si, 0.75 Mn, 1.18 Cr, 0.10 V, bal Fe.
For dies, punches, gauges, shear blades; oil hardened, tough.

DORRENBERG PC130.
M-614; 0.38 C, Si, Cr, V, bal Fe.
For gears, shafts, machine tool parts; oil hardened, tough.

DORRENBERG PMV.
M-614; 0.58 C, 0.95 Mn, 1.0 Cr, 0.09 V, bal Fe.
For springs, gears, bolts; oil hardened, tough.

DORRENBERG PMVW.
M-614; 0.50 C, 1.0 Cr, 0.09 V, 0.85 Mn, bal Fe.
For springs, gears, bolts; oil hardened, shock resistant.

DORRENBERG PNC EXTRA.
M-614; 0.50 C, 1.05 Cr, 3.25 Ni, 0.5 Mn, bal Fe.
For gears, bolts, crankshafts; oil hardened, shock resistant.

DORRENBERG PV3W.
M-614; 0.85 C, 1.0 Cr, bal Fe.
For bearings, cutters, liners, bushings; water hardened, wear resistant.

DORRENBERG PV4.
M-614; 0.85 C, 1.15 Cr, bal Fe.
For bearings, cutters, liners, bushings; water hardened, wear resistant.

DORRENBERG PV5.
M-614; 1.15 C, 0.65 Cr, 0.1 V, bal Fe.
For bearings, cutters, liners, bushings; water hardened, wear resistant.

DORRENBERG PV15.
M-614; 1.0 C, 1.1 Cr, 0.07 Mn, bal Fe.
For bearings, bushings, liners; water hardened, wear resistant.

DORRENBERG PV35.
M-614; 1.5 C, Cr, Si, bal Fe.
For tools, dies; oil hardened.

DORRENBERG R 15.
M-614; 0.12 max C, 0.4 Si, 13 Cr, bal Fe.
Annealed: 75,000 TS; 40,000 YS; 35 El; 70 RA; 155 Brin.
For turbine blades, cutlery, valves; Type 410; stainless.

DORRENBERG R 17.
M-614; 0.90 C, Si, Cr, Mo, bal Fe.
For bearings, liners, bushings; oil hardened, wear resistant.

DORRENBERG R 18.
M-614; 0.22 C, 17 Cr, 1.5 Ni, bal Fe.
Annealed: 125,000 TS; 95,000 YS; 20 El; 55 RA; 260 Brin.
For pumps, marine hardware, valves; Type 431; corrosion resistant.

DORRENBERG R 25.
M-614; 0.2 C, 13 Cr, 0.4 Si, bal Fe.
Annealed: 95,000 TS; 50,000 YS; 25 El; 55 RA; 195 Brin.
For turbine blades, valves, cutlery; Type 420; stainless.

DORRENBERG R 45.
M-614; 0.40 C, 0.30 Mn, 0.4 Si, 13 Cr, bal Fe.
Annealed: 95,000 TS; 50,000 YS; 25 El; 55 RA; 195 Brin.
For valves, cutlery, turbine blades; Type 420; stainless.

DORRENBERG SA.
M-614; 0.15 max C, 18 Cr, 8.5 Ni, bal Fe.
Annealed: 80,000 TS; 35,000 YS; 55 El; 75 RA; 150 Brin.
For chemical plant equipment, tanks, vessels; Type 302; stainless, austenitic.

DORRENBERG SAT.
M-614; 0.12 max C, 18 Cr, 9.5 Ni, Ti = 4 x C, bal Fe.
Annealed: 85,000 TS; 35,000 YS; 55 El; 65 RA; 150 Brin.
For welded chemical plant equipment, tanks, mixers; Type 321; stainless, austenitic.

DORRENBERG SAW.
M-614; 0.07 max C, 18 Cr, 9.5 Ni, bal Fe.
Annealed: 85,000 TS; 35,000 YS; 60 El; 70 RA; 150 Brin.
For chemical plant equipment, tanks: Type 304; stainless, austenitic.

DORRENBERG SB.
M-614; 0.1 max C, 18 Cr, 9.5 Ni, 2 Mo, bal Fe.
Annealed: 85,000 TS; 35,000 YS; 50 El; 65 RA; 160 Brin.
For acid resistant chemical plant equipment, tanks; Type 316; stainless, austenitic.

DORRENBERG SBT.
M-614; 0.12 max C, 18 Cr, 2 Mo, 10.5 Ni, Ti = 4 x C, bal Fe.
Annealed: 85,000 TS; 35,000 YS; 50 El; 65 RA; 160 Brin.
For welded acid resistant chemical plant equipment; Type 316Ti; austenitic, stainless.

DORRENBERG SBW.
M-614; 0.07 max C, 18 Cr, 10.5 Ni, 2 Mo, bal Fe.
Annealed: 90,000 TS; 40,000 YS; 45 El; 60 RA; 180 Brin.
For acid resistant chemical plant equipment, tanks; Type 317; stainless, austenitic.

DORRENBERG SPEZIAL NR. 2.
M-614; 0.40 C, 0.25-0.50 Si, 0.30-0.80 Mn, bal Fe.
Hot rolled: 91,000 TS; 58,000 YS; 27 El; 50 RA; 200 Brin.
For fishplates, gears, bolts, fasteners; water hardened.

DORRENBERG ST2.
M-614; 0.70 C, 1.7 Si, 0.70 Mn, bal Fe.
For springs, punches, chisels; oil hardened, shock resistant.

DORRENBERG STP.
M-614; 0.67 C, 1.3 Si, 0.50 Mn, 0.5 Cr, bal Fe.
For springs, punches, chisels; oil hardened, shock resistant.

DORRENBERG VNC4.
M-614; 0.40 C, Ni, Cr, Mo, bal Fe.
For gears, bolts, crankshafts; oil hardened, shock resistant.

DORRENBERG W 4.
M-614; 0.30 C, 1.0 Si, 0.4 Mn, 1.1 Cr, 0.18 V, 3.75 W, bal Fe.
For extrusion arms, punches, upsetters; oil hardened, tough.

DORRENBERG W 5.
M-614; 0.30 C, 2.35 Cr, 0.6 V, 4.25 W, bal Fe.
For extrusion arms, punches, upsetters; oil hardened, hot work steel.

DORRENBERG W 9.
M-614; 0.30 C, 2.65 Cr, 0.35 V, 8.5 W, bal Fe.
For extrusion rams and liners, upsetters; oil hardened, hot work steel.

DORRENBERG WC11.
M-614; 0.30 C, 2.4 Cr, 0.25 V, 8.5 W, 2.0 Co, bal Fe.
For extrusion rams and liners, upsetters; oil hardened, hot work steel.

DORRENBERG WD1.
M-614; 1.2 C, 0.2 Cr, 0.1 V, 1.0 W, bal Fe.
For blanking and forming dies; oil hardened, wear resistant.

DORRENBERG Z 1A.
M-614; 1.1 C, Mo, V, bal Fe.
For cutters, blanking and forming dies; oil hardened, wear resistant.

DORRENBERG Z 1B.
M-614; 0.90 C, 1.9 Mn, 0.1 V, bal Fe.
For punches, blanking and forming dies; oil hardened, non-deforming.

DORRENBERG Z 3C.
M-614; C, alloy, bal Fe.
For machine tool parts; oil hardened.

DOUBLE BRONZE.
M-Eng.; 6.5 Al, skin of pure Cu, bal Cu.
For electrical equipment; wire.

DOUBLE CLOCHE 0.1.
M-1546; 1.3 C, 5 W, bal Fe.
Hardened: C 63-65 Rock.
For cold work tools, punches, bending dies, piercing tools, bearings.
Cold working steel, shock and wear resisting. Water or oil hardening.

DOUBLE CLOCHE 2.3.
M-1546; 1.10 C, 0.5 Cr, 2.1 W, bal Fe.
Hardened: C 60-64 Rock.
For cold working tools, punches, hobs, bending dies, drills, cutters.
Oil or water hardening. Cold working steel, shock resisting.

DOUBLE CLOCHE E.
M-1546; 1.45 C, 3.25 W, 0.3 Cr, 0.25 V, bal Fe.
For cold working tools, bending dies, punches, reamers, cutters.
Cold working steel, shock and wear resistant. Oil hardening.

DOUBLE CONQUEROR.
M-1406; 1.1-1.4 C, bal Fe.
For roll turning tools, rock drills, punches; water hardened, wear resistant.

DOUBLE EXTRA BEST.
M-1408; high C, W, bal Fe.
For turning tools; oil or water hardened.

DOUBLE EXTRA LION.
M-1415; 0.75 C, 19 W, 4 Cr, 1.2 V, 2 Co, 0.5 Mo, bal Fe.
For drills, reamers, lathe and planer tools, hobs, broaches; high speed steel.

DOUBLE GEANT.
M-1118; 0.7 C, 18 W, 9 Co, 4 Cr, 2 V, 1 Mo, bal Fe.
For tools, dies, cutters; high speed steel.

DOUBLE GRIFFIN.
M-1112; 0.35 C, 1.75 Cr, 3.5 Ni, 0.3 Mn, bal Fe.
For shears, punches, pneumatic tools; hot work steel, oil hardened, shock resistant.

DOUBLE GRIFFIN SUPER CHISEL.
M-1112; C, alloy, bal Fe.
For tools, chisels; tough.

DOUBLE RAPID.
M-1405; 0.8 C, 4.5 Cr, 18.5 W, 5.5 Co, 0.3 Mo, 1.2 V, bal Fe.
For lathe and planer tools, reamers, drills, hobs; high speed steel.

DOUBLE SEVEN.
M-210; 1.25 C, 12.5 Cr, 2.7 Co.
For shear blades, cold forging dies, blanking dies; air hardened, non-deforming.

DOUBLE SEVEN.
M-365; 1.3 C, 13.3 Cr, 3-4.5 Co, 0.25 Mo, 1 Ni, bal Fe.
For tools, broaches, lamination dies; non-deforming.

DOUBLE-SIX.
M-1702; 0.83 C, 6.5 W, 5.0 Mo, 4.0 Cr, 2.0 V, bal Fe.
M-2 high speed steel tool bits; best shock resistance. (Electrite Double Six M-2).

DOUBLE SIX.
M-210; 1.9 C, 12.5 Cr, 0.8 Mo, 0.25 V, bal Fe.
For dies, plug and ring gages, press tools; oil or air hardened, non-deforming.

DOUBLE SIX.
M-365; 2.25 C, 14 Cr, bal Fe.
For blanking and drawing dies, gages; non-shrinking.

DOUBLE SIX SUPER.
M-210; 1.9 C, 12.5 Cr, 0.8 Mo, 0.25 V, bal Fe.
Ht. Tr.: 275,000 TS; 210,000 YS; 1 El; C 58 Rock.
For blanking, coining and drawing dies. Plug and ring gauges, mandrels, press tools, reamers, broaches. Non-shrinking. Air hardening.

DOUBLE SUPER EXPRESS.
M-1408; 0.7 C, 22 W, 5 Cr, 1 V, bal Fe.
For cutters for asbestos and rubber; high speed steel.

DOUBLE TWELVE.
M-210; 0.35 C, 12.0 Cr, 12.0 W, 1.0 V, bal Fe.
For master hobs for beryllium-copper, mold inserts in the plastic industry, brass die casting molds, extrusion dies. Oil hardening.

DOUBLE YOU CHROME.
M-1113; 2.25 C, 13 Cr, bal Fe.
For tool, dies; non-shrink.

DOUBLE YOU DIE.
M-1113; 0.9 C, 3.75 Cr, bal Fe.
For tools, dies; air hardened.

DOUBLE YOU HOT STUFF.
M-1113; C, 9.5 W, 2.5 Cr, 0.1 V, bal Fe.
For tools, hot dies; hot die steel.

DOUBLE ZEBRA.
M-510; 0.40 C, 0.30 Cr, 3.5 Ni, bal Fe.
Cold work tool steel for chisels; hardenable to about 600 DPN.

DOUX.
M-1495; 0.20 C, 0.25 Si, 0.80 Mn, bal Fe.
Annealed: 47,000 TS; 28,000 YS; 35 El; 137 Brin.
For oil refinery tubes.

DOVER.
M-1705; 1.0 C, 0.25 Mn, 0.25 Si, bal Fe.
Water hardening tool steel. AISI W 1.

DOWMETAL AZ91B replaced by **AZ91B.**

DOWMETAL G replaced by **AM100A.**

DOWMETAL H replaced by **AZ63A.**

DOWMETAL HK31A see **HK31A.**

DOWMETAL J replaced by **AZ61A.**

DOW METAL O1 replaced by **AZ80A.**

DOWMETAL ZK60A see **ZK60A.**

DP3.
M-1296; 0.03 max C, 0.75 max Si, 1.0 max Mn, 0.2-0.8 Cu, 5.5-7.5 Ni, 24.0-26.0 Cr, 2.5-3.5 Mo, 0.1-0.4 W, 0.1 min N, bal Fe.
118,00 psi TS; 88,200 psi YS; 36 El.
Duplex-phase stainless steel resistant to crevice corrosion to sea water at high temperature. Good weldability.

DRACO.
M-114; 0.66-1.15 C, 0.2 V, bal Fe.
For tools, stamping dies, cold heading dies, taps; high strength and toughness.

DRACO DV.
M-114; 1 C, 0.3 Mn, 0.45 V, bal Fe.
For cold header dies, edge tools; shock resistant.

DRAGONITE.
M-1724.
Electro-zinc coated steel; mild steel.

DRAGONZIN.
M-1724.
Hot dipped galvanized steel; mild steel.

DRAWING BRASS.
M-8; 67-70 Cu, 30-33 Zn.
Hard: 77,000 TS; 67,000 YS; 15 El; 50 RA; 145 Brin.
For seamless tubes, condensers and evaporators; deep drawing.

DREWOSTA.
M-1294; 1.42 C, W, V, bal Fe.
For engravers tools, forming and blanking dies; water hardened, wear resistant.

DRILLALLOY.
M-432; 0.65 C, 0.70 Mn, 0.40 Mo, bal Fe.
For mine drills, tools; water hardened.

DRILLEX.
M-1100.
Bulk tungsten sintered carbide screened to -60 + 100 mesh size. This is then alloyed with small amount of Ni, Si + Mn to develop a strong impact resistant matrix for welding.

DRILL ROD.
M-341; 1.2 C, bal Fe.
For drills, tools; water hardened.

DRILL ROD BRASS.
M-U.S.; 62 Cu, 35.5 Zn, 2.5 Pb.
For automatic screw machine parts; drills and turns easily.

DRITTEL SILVER.
M-Ger.; 66.6 Al, 33.3 Ag.
For ornamental trade.

DRIVER 30 ALLOY.
M-61; 98 Cu, 2 Ni.
Annealed: 30,000 TS.
Cold worked: 60,000 TS.
For resistances, precision resistors, rheostats, instruments.
Max. operating temperature 600°F in air.

DRIVER 60 ALLOY.
M-61; 94 Cu, 3.5 Ni.
Annealed: 35,000 TS.
Cold worked: 70,000 TS.
For resistances, resistors, rheostats, instruments.
Max. operating temperature 600°F in air.

DRIVER 90 ALLOY.
M-61; 88 Cu, 12 Ni.
Annealed: 35,000 TS.
Cold worked: 75,000 TS.
For resistances, rheostats, precision resistors, instruments.
Max. operating temperature 800°F in air.

DRIVER 180 ALLOY.
M-61; 78 Cu, 22 Ni.
Annealed: 50,000 TS.
Cold worked: 100,000 TS.
For resistances, precision resistors, rheostats, instruments.
Max. operating temperature 1000°F in air.

DRW COPPER.
M-1778; Electrolytic tough pitch copper.
For electrical applications and general usage in rod and wire applications.

DS (DANISH SPECIFICATION)..
DS-LEAD.
M-1656; 99.9 Pb, 0.1 PbO.
Sheet: 6000 TS; 5800 YS; 18 El; 6500 CS.
At 300 °F: 3250 TS; 20 El.
For chemical construction, storage batteries, counterweights, roofing, gutters.
Dispersion strengthened, fine-grained, corrosion resistant.

DS NICKEL.
M-1573; 98 Ni, 2 ThO$_2$. (dispersion strengthened nickel).
Powder for sintering; and sheet or bar.
Sintered: 8.9 g/cc.
Room T.:71,200 psi TS (491 MPa).
2000°F (1093°C): 17,500 psi TS (121 MPa).
Good high temperature properties.

DS NICR.
M-1573; 20 Cr, 1.7 ThO$_2$, bal Ni.
Powder for sintering.

DSC HIGH C.
M-157; 0.25-0.80 C, bal Fe.
For springs; water hardened.

DTC.
M-1653; 1.8-2.3 C, 12.0-14.0 Cr, (0.15-0.30 V), bal Fe.
Cold work tool steel.
UX 210 Cr 13 KU; AISI D3.

DTC/AR.
M-1653; 1.4-1.6 C, 11.0-13.0 Cr, 0.60-0.90 Mo, bal Fe.
Cold work tool steel.
UX 150 CrMo 12 KU; similar to AISI D2.

DTC W.
M-1653; 1.8-2.2 C, 0.50 Si, 0.50 Mn, 11.5-13.5 Cr, 0.40-0.80 W, 0.15-0.30 V, bal Fe.
Cold work tool steel.
W.-Nr. 1.2436; similar to AISI D3.

DTD.
M-Eng.; 0.40 C, 3 Cr, 0.8 Mo, 0.2 V, bal Fe.
For gas turbine parts; creep resistant.

D.T.D. 49B.
M-Brit.; 0.35-0.45 C, 1.00-1.75 Si, 0.5-1.0 Mn, 12.5-14.5 Cr, 12.5-14.5 Ni, 2-3 W, bal Fe.
For aircraft valves; austenitic.

DTD 424.
M-Brit.; Cu, Si, bal Al.
For castings; age-hardenable.

DTD (SPECIFICATIONS OF DIRECTOR OF MATERIALS RESEARCH AND DEVELOPMENT, ENGLAND)..
DUALOY.
M-358; C, W, bal Fe.
For tools, dies; water hardening.

DUAL SHIELD 78.
M-1713; weld metal; 0.06 C, 1.20 Mn, 0.40 Si, bal Fe.
As welded: 84,000 psi TS; 74,500 psi YS; 28 El.
Flux-cored continuous electrode for single and multipass welding of low and medium carbon steels.
AWS Class E70T-1.

DUAL SHIELD 85-C1.
M-1713; weld metal; 0.05 C, 1.17 Mn, 0.30 Si, 2.75 Ni, bal Fe.
As welded: 95,000 psi TS; 85,100 psi YS; 23 El.
Flux-cored continuous electrode for strong welds of good toughness down to minus 100°F.
AWS Grade E90T.

DUAL SHIELD 85 NM.
M-1713; weld metal; 0.07 C, 1.09 Mn, 0.38 Si, 0.96 Ni, 0.50 Mo, bal Fe.
As welded: 98,500 psi TS; 89,700 psi YS; 23 El.
Used largely for welding nuclear pressure vessels.
AWS Grade E90T.

DUAL SHIELD 88-C3.
M-1713; weld metal; 0.07 C, 1.22 Mn, 0.35 Si, 1.07 Ni, bal Fe.
As welded: 88,600 psi TS; 79,300 psi YS; 24 El.
Flux-cored continuous electrode for welding high tensile steels in the 70,000-80,000 psi tensile strength range.
AWS Grade E80T.

DUAL SHIELD 88 CM.
M-1713; weld metal; 0.06 C, 0.74 Mn, 0.34 Si, 1.13 Cr, 0.50 Mo, bal Fe.
As welded: 106,000 psi TS; 94,000 psi YS; 14 El.
Flux-cored continuous electrode for welding HSLA steel and low alloy steels.

DUAL SHIELD 98.
M-1713; weld metal; 0.07 C, 1.10 Mn, 1.75 Ni, 0.20 Mo, 0.30 Si, bal Fe.
As welded: 94,000 psi TS; 85,000 psi YS; 20 El.
Flux-cored continuous electrode for welding HY-80, HY-90 and T-1 and similar high tensile quenched and tempered steels.
AWS Grade E90T.

DUAL SHIELD 98-CM.
M-1713; weld metal; 0.06 C, 0.70 Mn, 0.40 Si, 2.21 Cr, 1.04 Mo, bal Fe.
As welded: 128,000 psi TS; 109,500 psi YS; 12 El.
(Usually stress annealed to 90,000 min TS).
For welding 2 Cr-1 Mo steels.

DUAL SHIELD 110.
M-1713; weld metal; 0.10 C, 1.29 Mn, 0.82 Si, bal Fe.
As welded: 86,000 psi TS; 81,000 psi YS; 30 El.
Flux-cored continuous electrode for welding low and medium carbon steel in the mill scaled or rusty condition.
AWS Class E70T-2.

DUAL SHIELD 111A.
M-1713; weld metal; 0.08 C, 0.85 Mn, 0.40 Si, bal Fe.
As welded: 74,800 psi TS; 64,200 psi TS; 25 El.
Flux-cored continuous electrode for single and multipass welding of mild steel in CO_2 atmosphere.
AWS Class E70T-1.

DUAL SHIELD 111A-C.
M-1713; weld metal; 0.07 C, 1.45 Mn, 0.48 Si, bal Fe.
As welded: 88,900 psi TS; 77,400 psi YS; 26 El.
Flux-cored continuous electrode for single and multipass welding of mild and medium carbon steels in CO_2 atmosphere.
AWS C E70T-1; AWS A5.20.

DUAL SHIELD 111HD.
M-1713; weld metal; 0.09 C, 1.33 Mn, 0.58 Si, bal Fe.
As welded: 91,000 psi TS; 80,500 psi YS; 26 El.
Flux-cored continuous electrode for single and multipass welding of mild and medium carbon steels in CO_2 atmosphere.
AWS Class E70T-1.

DUAL SHIELD 150.
M-1713; weld metal; 0.06 C, 1.4 Mn, 0.38 Si, 0.62 Mo, bal Fe.
As welded: 108,500 psi TS; 97,000 psi YS; 16 El.
Flux-cored continuous electrode for multipass welding of low alloy-high tensile strength steels.
AWS Grade E100T.

DUAL SHIELD 7000.
M-1713; weld metal; 0.08 C, 1.47 Mn, 0.74 Si, bal Fe.
As welded: 93,000 psi TS; 86,000 psi YS; 25 El.
Flux-cored continuous electrode for welding mild and medium carbon steels in all positions with CO_2 gas.
AWS Class E70T-1.

DUAL SHIELD 7000-A1.
M-1713; weld metal; 0.07 C, 1.13 Mn, 0.57 Si, 0.55 Mo, bal Fe.
As welded: 102,000 psi TS; 97,000 psi YS; 23 El.
Flux-cored continuous electrode designed for welding low alloy-high tensile steels in out-of-position work with 75% A-25% CO_2 gas.
AWS Grade E100T.

DUAL SHIELD 8000-B2.
M-1713; weld metal; 0.05 C, 0.85 Mn, 0.64 Si, 1.41 Cr, 0.55 Mo, bal Fe.
As welded: 107,500 psi TS; 97,000 psi YS; 20 El.
Flux-cored continuous electrode for all-position welding of 1% Cr - 1/2% Mo and similar steels.

DUAL SHIELD 9000-B3.
M-1713; weld metal; 0.06 C, 0.85 Mn, 0.70 Si, 2.3 Cr, 1.10 Mo, bal Fe.
As welded: 124,000 psi TS; 108,000 psi YS; 12 El.
Flux-cored continuous electrode for all-position welding of 2 1/4% Cr-Mo steels.

DUAL SHIELD 9000-C1.
M-1713; weld metal; 0.05 C, 1.3 Mn, 0.60 Si, 2.4 Ni, bal Fe.
As welded: 106,250 psi TS; 101,000 psi YS; 20 El.
Flux-cored continuous electrode for all-position welding of 2-3% Ni steels for low temperature operation.
AWS Grade E100T.

DUAL SHIELD 9000-D1.
M-1713; weld metal; 0.09 C, 1.37 Mn, 0.73 Si, 0.45 Mo, bal Fe.
As welded: 100,500 psi TS; 92,000 psi YS; 23 El.
Flux-cored continuous electrode for all-position welding of low alloy high strength steels requiring 100,000 psi strength.
AWS Grade E100T.

DUAL SHIELD 9000-M.
M-1713; weld metal; 0.06 C, 1.10 Mn, 0.65 Si, 1.80 Ni, 0.25 Mo, bal Fe.
As welded: 103,500 psi TS; 94,000 psi YS; 23 El.
Flux-cored continuous electrode for all-position welding (fillets) of HY-80 and similar steels.
AWS Grade E100T.

DUAL SHIELD SP.
M-1713; weld metal; 0.11 C, 1.05 Mn, 0.61 Si, bal Fe.
As welded: probably 72,000 to 100,000 psi TS.
Flux-cored continuous electrode for welding scaly or rusty steel.
AWS Class E70T-2.

DUAL SHIELD T-8.
M-1713; weld metal; 0.06 C, 1.45 Mn, 0.34 Si, 0.28 Cr, 1.85 Ni, 0.45 Mo, bal Fe.
As welded: 117,000 psi TS; 106,500 psi YS; 20 El.
Flux-cored continuous electrode for welding high strength quenched and tempered steels.
AWS Grade E110T.

DUAL SHIELD T-62.
M-1713; weld metal; 0.08 C, 1.35 Mn, 0.55 Si, bal Fe.
As welded: 82,000 psi TS; 69,000 psi YS; 26 El.
Flux-cored continuous electrode for single and multipass welding of mild and medium carbon steels in CO_2 atmosphere.
AWS Class E70T-1.

DUAL SHIELD T-63.
M-1713; weld metal; 0.08 C, 1.60 Mn, 0.72 Si, bal Fe.
As welded: 92,000 psi TS; 79,000 psi YS; 24 El.
Highly deoxidized, flux-cored electrode for welding steel in spite of oil, scale or rust.
AWS E70T-2.

DUAL SHIELD T-75.
M-1713; weld metal; 0.06 C, 1.40 Mn, 0.45 Si, bal Fe.
As welded: 79,000 psi TS; 67,000 psi YS; 28 El.
Flux-cored continuous electrode producing welds of good impact strength.
AWS Class E70T-5.

DUAL SHIELD T-90C1.
M-1713; weld metal; 0.07 C, 1.0 Mn, 0.40 Si, 2.5 Ni, bal Fe.
As welded: 84,000 psi TS; 70,000 psi YS; 25 El.
Flux-cored electrode for welding 2 1/2-3 3/4% Ni steels for low temperature operations.
AWS Grade E80T.

DUAL SHIELD T-100.
M-1713; weld metal; 0.08 C, 1.4 Mn, 0.50 Si, 0.30 Cr, 2.0 Ni, 0.30 M bal Fe.
As welded: 104,000 psi TS; 91,000 psi YS; 23 El.
Flux-cored electrode for welds requiring 100,000 psi strength.
AWS Grade E100T.

DUAL SHIELD T-115.
M-1713; weld metal; 0.05 C, 2.0 Mn, 0.47 Si, 2.0 Ni, 0.50 Mo, bal Fe.
As welded: 113,000 psi TS; 107,000 psi YS; 24 El.
Flux-cored electrode for welding high strength steels.
AWS Grade E110T.

DUAL SHIELD T-4130.
M-1713; weld metal; 0.20 C, 1.1 Mn, 0.29 Si, 0.35 Cr, 1.25 Ni, 0.22 Fe.
As welded: 110,000 psi TS; 90,000 psi YS; 20 El.
Hardenable (Q + T) to above 170,000 psi YS.
For welding AISI 8630, 4130 and similar low alloy steels, especially for reheat treat.

DUCOL W30 GRADE A.
M-1724; 0.11-0.17 C, 0.40 max Si, 1.0-1.5 Mn, 0.4-0.7 Cr, 0.20-0.28 M 0.70 max Ni, 0.30 max Cu, 0.04-0.12 V, bal Fe.
High strength; weldable.
For pressure vessel applications.

DUCOL W30 GRADE B.
M-1724; 0.09-0.15 C, 0.40 max Si, 0.90-1.30 Mn, 0.40-0.70 Cr, 0.20-0.2 Mo, 0.04-0.12 V, 0.30 max Cu, 0.10 max Nb, 0.015 max N, bal Fe.
High strength; weldable; improved notch ductility.
For pressure vessel applications.

DUCTALLOY.
M-128; 3.0-3.5 T.C., 2.0-2.5 Si, 0.15 Mg, 0.6 Mn, bal Fe.
Cast: 115,000 TS; 1 El; 230 Brin.
Annealed: 60,000 TS; 20 El; 175 Brin.
For machinery castings, gears, shafts; ductile cast iron.

DUCTALUMINUM 356S.
M-1099; 0.20-0.40 Mg, 6.5-7.5 Si, 0.20 max Cu, 0.5 max Fe, 0.2 max Ti, bal Al.
Heat treated: 42,000 TS; 35,000 YS; 3 El.
For aircraft castings; age-hardenable.

DUCTALUMINUM 356T.
M-1099; 0.2-0.4 Mg, 6.5-7.5 Si, 0.20 max Cu, 0.5 max Mn, 0.2 max Ti, bal Al.
Heat treated: 38,000 TS; 28,000 YS; 6 El.
For aircraft castings; age-hardenable.

DUCTILE CECOLLOY.
M-474; 3.3-3.6 T.C., 2.2-2.6 Si, 0.5 Mg, bal Fe.
Cast and annealed: 60,000-80,000 TS; 40,000-60,000 YS; 10-5 El.
For castings, forging hammer components; good machinability, ductile cast iron.

DUCTILE IRON.
M-1267; 3.3 C, 2.5 Si, 0.7 Mn, 0.05 Mg, bal Fe.
Cast: 60,000-90,000 TS; 45,000-65,000 YS; 10-2 El; 163-277 Brin.
For gears, crankshafts, hydraulic castings; high strength ductile iron.

DUCTILE IRON FERRITIC.
M-68; 3-4 T.C., 1-3 Si, 0.10-0.60 Mn, 0.10 max P, 0-2 Ni, bal Fe.
Annealed: 60,000 TS; 45,000 YS; 15 El; 143-207 Brin.
For pressure castings; maximum toughness, ductile cast iron.

DUCTILE IRON, HARDENED.
M-68; 3-4 T.C., 1-3 Si, 0.10-0.60 Mn, 0.10 max P, 0-2 Ni, bal Fe.
Heat treated: 120,000 TS; 9,000 YS; 2 El; 269-388 Brin.
For gears, dies, machine tool parts; ductile cast iron.

DUCTILE IRON, PEARLITIC.
M-68; 3-4 T.C., 1-3 Si, 0.10-0.60 Mn, 0.10 max P, 0-2 Ni, bal Fe.
Normalized: 100,000 TS; 70,000 YS; 3 El; 225-302 Brin.
For gearsa, crankshafts, agricultural equipment; wear resistant, ductile cast iron.

DUCTILE IRON, SEMI-PEARLITIC.
M-68; 3-4 T.C., 1-3 Si, 0.10-0.60 Mn, 0.10 max P, 0-2 Ni, bal Fe.
Cast: 85,000 TS; 55,000 YS; 3 El; 200-280 Brin.
For heavy machinery castings; wear resistant, ductile cast iron.

DUCTILEND 70.
M-677; 0.06 C, 0.6 Mn, 0.4 Si, 0.1 Mo, bal Fe.
For welding electrodes; class E7018.

DUCTILEND 70 MO.
M-677; 0.05 C, 0.08 Mn, 0.6 Si, 0.6 Mo, bal Fe.
For welding electrodes; class E7018-Al carbon-molybenum steel.

DUCTILEND 80.
M-677; 0.6 C, 0.8 Mn, 0.5 Si, 0.9 Ni, 0.15 Mo, bal Fe.
For welding electrodes; class E8018-C3, 1% nickel steel.

DUCTILEND 85.
M-677; 0.05 C, 1.0 Mn, 0.5 Si, 1.5 Ni, 0.3 Mo, bal Fe.
For welding electrodes; class E9018M, for joints in HY80 and steels of 90,000-100,000 psi TS range.

DUCTILEND 90.
M-677; 0.06 C, 1.5 Mn, 0.4 Si, 0.45 Mo, bal Fe.
For welding electrodes; class E9018-D1, manganese-molybdenum steel 90,000 psi tensile strength range.

DUCTILEND 100.
M-677; 0.05 C, 1.4 Mn, 0.5 Si, 1.8 Ni, 0.35 Mo, bal Fe.
For welding electrodes; class E10018-M, for joints in HY80 and steels of 100,000-110,000 psi TS range.

DUCTILEND 110.
M-677; 0.05 C, 1.6 Mn, 0.4 Si, 2.2 Ni, 0.4 Mo, bal Fe.
For welding electrodes; class E11018-M, for joints in HY 80 and "T" steel, etc, of 110,000-120,000 psi TS range.

DUCTILEND 120.
M-677; 0.05 C, 1.9 Mn, 0.55 Si, 0.5 Cr, 2.2 Ni, 0.45 Mo, bal Fe.
For welding electrodes; class E12018-M for joints in HY100 and "T" steel, etc., of 120,000 psi TS.

DUCTILITE.
M-1282; 3.3-3.5 C, 2.2-2.5 Si, 0.7 Mn, 0.05 Mg, bal Fe.
Cast: 70,000-100,000 TS; 15-5 El; 210-250 Brin.
For earth moving equipment; ductile cast iron.

DUCTILOY see NAX HIGH TENSILE.

DUCTIMET 10.
M-1795; 50 Ni, 20 Si, 15 Mg, bal Fe.
Inoculant for production of nodular iron.

DUCTIMET 21.
M-1795; 85 Ni, 15 Mg.
Inoculant for production of nodular iron.

DUCTIMET 80-20.
M-1795; 80 Ni, 20 Mg.
Inoculant for production of nodular iron.

DUCTIMET 90-10.
M-1795; 90 Ni, 10 Mg.
Inoculant for production of nodular iron.

DUCTIMET 95-5.
M-1795; 95 Ni, 5 Mg.
Inoculant for production of nodular iron.

DUCTLIRON TYPE GS.
M-U.S.; 3.3 C, 0.7 Mn, 2.2 Si, 0.05 Mg, bal Fe.
Cast: 75,000 TS; 55,000 YS; 1-4 El; 217-269 Brin.
For gears, shafts, machine tool housings; ductile cast iron.

DUCTLIRON TYPE GSF.
M-U.S.; 3.3 C, 0.7 Mn, 2.2 Si, 0.05 Mg, bal Fe.
Cast: 60,000 TS; 45,000 YS; 5-10 El; 207 Brin.
Annealed: 20 El; 170 Brin.
For gears, shafts, machine tool housings; ductile cast iron.

DUDLEY'S BEARING METALS.
M-Eng.; 98 Sn, 1.6 Cu, 0.3 Pb.
For bearings.

DUDLEY'S BRONZE "B".
M-Eng.; 77 Cu, 8 Sn, 15 Pb.
For bearings, bushings; heavy duty.

DUDLEY'S BRONZE "K".
M-Eng.; 77 Cu, 10.5 Sn, 12.5 Pb.
For bearings, bronze castings; heavy duty.

DUDLEYS PHOSPHOR BRONZE.
M-Eng.; 80 Cu, 10 Sn, 9.6 Pb, 0.8 P.
30,000 TS; 6El; 55 Brin.
For bearings; heavy duty.

DUEX.
M-1546; 0.88 C, 17.5 W, 8 Co, 1.5 V, 1.2 Mo, bal Fe.
Hardened: C 64-66 Rock.
For lathe machining high manganese steels, heavy duty lathe and planer tools, broaches, hobs, boring tools.
High speed steel, oil hardening. High red-hardness.

DUFOUR WHITE GOLD.
M-Eng.; 75 Au, 21.5 Pb, 4.5 Pt.
For ornaments; corrosion resistant.

DUKANE.
M-755; 0.4 C, bal Fe.
For gears, shafts; water hardening.

DUKANE DRILL ROD.
M-370; 0.95-1.05 C, bal Fe.
For dowel pins, mandrels, screw machine parts, shafts, small tools.
AISI Type 1 Water hardening tool steel.

DUKE-KIDD.
M-358; 0.9 C, bal Fe.
For tools, drills, taps; water hardened.

DUKE'S METAL.
M-Eng.; 12 Cr, 1.5 C, 4 Co, 0.6 Si, 0.4 W, 0.2 Mn, bal Fe.
For tools, dies, corrosion and heat resisting parts; non-deforming.

DUKE'S METAL.
M-Eng.; 40 Ni, 30 Cu, 30 Fe.
For corrosion resistant and heat resistant parts; corrosion and heat resistant.

DUKEX.
M-358; C, W, bal Fe.
For cutting tools; water hardening.

DUMET.
M-32, M-118; 46 Ni, 54 Fe, sheath of Cu.
Rolled: 70,000 TS; 18 El.
For glass to metal seals, lead-in wires; thermal expansion similar to soft glass.

DUMORE.
M-289; 0.95-1.05 C, 5.0-5.5 Cr, 0.95-1.25 Mo, bal Fe.
For blanking, forming and drawing dies; air hardened.

DUNNLEVIC & JONES ANTIFRICTION METAL.
M-Eng.; 60 Pb, 20 Sb, 20 Zn.
For bearings, bushings; Babbitt.

DUNNLEVIC & JONES RUSSIAN.
M-USSR; 80 Zn, 8 Cu, 12 Sb.
For bearings; antifriction.

DUPLEX.
M-627; 0.9-1.1 C, 0.2 Si, 0.2 Mn, bal Fe.
Water Hardened: 200,000-216,000 TS;
  138,000-152,000 YS; 11-13 El; 32-35 RA;
  388-600 Brin.
For taps, drills, reamers, punches, stamps,
  knurls, mandrels.
Type W1 water hardening.

DU PONT D-14.
M-522; 5 Zr, bal Cb.
Rolled: 75,000 TS; 60,000 YS.
For hot surfaces on space re-entry vehicles; high
  heat resistance.

DU PONT D-36.
M-522; 10 Ti, 5 Zr, bal Cb.
Rolled: 80,000 TS; 70,000 YS.
For hot surfaces on space re-entry vehicles; high
  heat resistance.

DUPONT FA 22.
M-522; 0.07 C, 29 Ni, 3 Cu, 2 Mo, bal Fe.
For chemical plant equipment; resists H_2SO_4.

DUPONT FA 22.
M-522; 0.07 max C, 2.5 Si, 0.75 Mn, 20.5 Cr,
  28.5 Ni, Cu, bal Fe.
Cast: 72,000 TS; 35,000 YS; 45 El; 150 Brin.
For heat and corrosion resistant casting; heat
  and corrosion resistant.

DUPONT METAL.
M-308; 96 Cu, 3 Si, 1 Mn, 0.1 Fe.
For bolts, screws, turnbuckles for airplanes;
  same as "Everdur"; non-corrosive.

DUQUESNE SPECIAL.
M-451; C, Cr, Mo, bal Fe.
For rolls; subjected to severe service.

DURABIL.
M-358; C, bal Fe.
For tools and dies; water hardening.

DURACAST NO. 7 see **WEST NO. 7, ETC.**

DURACID.
M-323; 16 Si, 3 C, bal Fe. Cast.
For chemical apparatus; corrosion resistant; cast
  iron.

DURAFLEX.
M-8; 1-10 Sn, 0.01-0.30 P, bal Cu.
O-temper: 55,000 TS.
H-temper: 72,000 TS.
Spring temper: 91,000 TS.
For springs, clips, diaphragms; P-Bronze; high
  endurance limit.

DURAL.
M-Eng.; 91.7-93.9 Al, 6-4 Cu, 0.63-0.59 Mn, 0.46
  Mg, 0.8 Fe, 0.2 Si, 0.11 Cr.
For light alloy parts, aircraft structures; age-
  hardenable.

DURAL A.
M-487; C, alloy, bal Fe.
For turning tools.

DURAL B.
M-487; C, alloy, bal Fe.
For coal cutter picks, for welding of hard facing
  material; abrasion resistant.

DURAL C.
M-487; C, alloy, bal Fe.
For valve parts, dry battery manufacture;
  chemical and abrasive resistance.

DURALFA.
M-658; Al alloy.
For light alloy parts.

DURALIMIN.
M-Russia; 4 Cu, bal Al.
For light alloy parts; age-hardenable.

DURALINE.
M-435; 0.35 C, 1.3 Si, 3.5 Cr, 0.85 V, 4.25 Mo,
  bal Fe.
At 1000°F: 216,000 TS; 8.5 El; 28.3 RA; 633
  Brin.
For extrusion tools and dies; high hot strength
  and hardness.

DURALINOX.
M-French; 4.5 Mg, 95.5 Al.
For light alloy parts; formerly "Alumag."

DURALINOX.
M-767; 7 Mg, bal Al.
Rolled: 18-22 El.
For floors, partitions.

DURALINOX 5.
M-767; 5 Mg, 0.5 Mn, bal Al.
For light alloy parts; corrosion resistant.

DURALINOX P.
M-767; 12 Mg, 0.5 Mn, bal Al.
For light alloy parts; corrosion resistant.

DURALIT.
M-French; 3 Cu, 0.5 Ni, 0.5 Mg, 0.7 Si, 1 Ti, bal
  Al.
For forged connecting rods; age-hardenable.

DURALITE.
M-541, M-1219; 3 Cu, 0.7 Si, 0.5 Ni, 0.5 Mg, 1.5
  Fe, 0.2 Ti, bal Al.
Aged: 55,000-67,000 TS; 41,000-51,000 YS; 18-10
  El; 115-140 Brin.
For auto and aircraft parts, pistons; age-
  hardenable.

DURALITE 35.
M-541, M-624; 3.3-3.8 Cu, 1.3-1.6 Fe, 0.5-0.7 Si, 0.55-0.75 Mg, bal
Annealed: 28,000 TS; 18,000 YS; 15 El; 55 Brin.
TA-Temper: 64,000 TS; 50,000 YS; 9 El; 145 Brin.
For general engineering components, structures, fittings, fasteners, bolts.
Age-hardenable, high strength.

DURALLOY.
M-465; 1-3.5 Ni, 1-2 Cr, C, bal Fe.
For castings; cast steel.

DURALOY see also **DBK**.

DURALOY.
M-210; 1.3-1.4 C, 0.5 max Si, 13-15 Mn, 1.5-1.7 Cr, bal Fe.
For wear plates, tread plates; wear and abrasion resistant castings.

DURALOY.
M-1120; 0.25 C, 23-30 Cr, bal Fe.
Annealed: 95,000 TS; 60,000 YS; 25 El; 200 Brin.
For heat treating boxes, furnace parts, fittings, valves, exhaust manifolds.
Type 446 stainless steel, heat and corrosion resistant.

DURALOY CA-15, ETC see **DBK CA-15 ETC.**

DURALPLAT.
M-457; base metal of Cu containing Duralumin coated with a Cu free Duralumin.
For light alloy parts; Brit. Pat. 318999; high corrosion resistance.

DURALUM.
M-Ger.; 79 Al, 11 Mg, 10 Cu, 0.5 P.
For light alloy parts; non-hardened.

DURALUMIN.
M-Eng.; 3.5-4.5 Cu, 0.3-0.6 Mg, 0.0-0.8 Mn, 0.4-1.0 Fe, 0.0-1.0 Si, bal Al.
Annealed: 32,000-36,000 TS; 7,000-10,000 YS; 34 El; 45-55 Brin.
Heat treated: 63,000 TS; 30,000-40,000 YS; 18-25 El.
For airplane and dirigible construction; same as "Alcoa No. 17"; age-hardening alloy.

DURALUMIN 3L-1.
M-Eng.; 4.5 Cu, bal Al.
For housings, cases, structures, screw machine products; age-hardenable, high strength.

DURALUMIN 3L-3.
M-Eng.; 4.5 Cu, bal Al.
For housings, cases, structures, screw machine products; age-hardenable, high strength.

DURALUMIN 681A.
M-457; 4.5 Cu, 0.5 Mg, 0.6 Mn, bal Al.
For light alloy parts; age-hardenable.

DURALUMIN 681B.
M-768, M-457; 4.5 Cu, 0.5 Mg, 0.6 Mn, bal Al.
Heat treated: 65,000 TS; 45,000 YS.
For light alloy parts; heat treatable.

DURALUMIN 681H.
M-768; 2.5 Cu, 0.5 Mg, bal Al.
Heat treated: 51,000 TS; 25 El.
For light alloy parts; heat treatable.

DURALUMIN 681K.
M-768, M-457, M-767; 0.2 Mg, bal Al.
Heat treated: 31,000 TS; 20 El.
For light alloy parts; heat treatable.

DURALUMIN 681ZB.
M-768; 0.6 Si, 4.5 Cu, 0.5 Mg, 0.8 Mn, bal Al.
Heat treated: 62,000-88,000 TS; 14-3 El; 128-175 Brin.
For light alloy parts; heat treatable.

DURALUMIN DM31.
M-768, M-338; 0.6 Si, 4.0 Cu, 1.0 Mg, 1.2 Mn, bal Al.
Heat treated: 71,000-74,000 TS; 57,000-60,000 YS; 10-12 El; 125-140 Brin.
For light alloy parts; heat treatable.

DURALUMIN E.
M-Eng.; 3.0-4.5 Cu, 1.0 max Si, 1.0 max Mg, 1.2 max Mn, bal Al.
60,000 TS; 52,000 YS; 8 El.
For light alloy parts; heat treatable.

DURALUMIN F.
M-338; 1.0 Mg, 0.5 Si, 0.25 Cr, 0.25 Cu, bal Al.
WP-temper: 40,400 TS; 33,600 YS; 10 El.
Annealed: 18,000 TS; 15 El.
For extrusions of intricate sections; age-hardenable.

DURALUMIN F replaced by **ALCAN GB-65S.**

DURALUMIN G replaced by **ALCAN GB-24S.**

DURALUMIN H replaced by **ALCAN GB-B51S.**

DURALUMIN JJ replaced by **ALCAN GB-D19S.**

DURALUMIN IMITATION.
M-Eng.; 79 Al, 11 Mg, 10 Cu, 0.05 P.
For light alloy parts; age-hardened.

DURALUMIN K.
M-457; 0.5-2.0 Mg, 0.3-1.5 Si, 0-1.5 Mn, bal Al.
For light alloy parts; corrosion resistant, hardenable.

DURALUMIN K.
M-1548; 0.3-1.5 Si, 1.5 Mn, 0.5-2.0 Mg, bal Al.
Heat treated: 50,000 TS; 35,000 YS; 10-20 El; 100 Brin.
For hardware, fasteners. Age-hardenable.

DURALUMIN K replaced by **ALCAN GB-D75S.**

DURALUMIN K replaced by **ALCAN GB-B26S.**

DURALUMIN SPECIAL "Y".
M-Eng.; 92.5 Al, 4 Cu, 2 Ni, 1.5 Mg.
For dirigible, airplane and motor car light alloy parts; age hardening.

DURALUMIN SUPER.
M-German; 5.5 Cu, 0.3 Fe, 1.0 Mg, 0.6 Mn, 0.6 Si, bal Al.
71,000-78,000 TS; 10-20 El; 130-140 Brin.
For light alloy parts; heat treatable.

DURALUMIN T.
M-457; 2.1 Cu, 0.5-1.5 Fe, 0.3-1.5 Mg, 2 max Ni, 1 max Mn, bal Al.
Heat treated: 62,000 TS; 41,000 YS; 12 El; 130 Brin.
For aircraft engine forgings; age-hardenable.

DURALUMIN W.
M-78, M-457; 4 Cu, 0.3 Fe, 1.5 Mg, 1 Ni, 0.2 Si, bal Al.
For light alloy parts, aircraft; age-hardenable.

DURALUMIN W.
M-1548; 3.5-4.5 Cu, 1.0-1.8 Mg, 1.8-2.2 Ni, bal Al.
Heat treated: 60,000 TS; 35,000 YS; 10 El; 100-200 Brin.
For fasteners, hardware, engine components.
Heat treatable, high strength.

DURALUMIN X replaced by ALCAN GB-51S.

DURALUMIN Z.
M-Ger.; 4.3-5.0 Cu, 0.3 Fe, 0.9 Mg, 0.6 Mn, 0.6 Si, bal Al.
68,000-71,000 TS; 12-14 El; 130-135 Brin.
For light alloy parts; heat treatable.

DURALUMIN ZC replaced by ALCAN GB-E74S.

DURAMIUM.
M-358; 0.7 C, 18 W, 4 Cr, 1 V, bal Fe.
For reamers, hobs, drills, taps; high speed steel.

DURAMOLD B.
M-24; 0.07 C, 0.25 Ni, 0.95 Cr, 0.25 Mo, bal Fe.
For cold hobbing cavity, molds; oil hardening, medium core hardness. AISI P2.

DURAMOLD N.
M-24; 0.10 C, 1.5 Cr, 3.25 Ni, bal Fe.
Ht. Tr.: 130,000-173,000 TS; 88,000-142,000 YS; 15-20 El; 262-340 Brin.
For plastic molds.
Type P6 case hardening tool steel.

DURANA-1.
M-Ger.; 64-78 Cu, 29.5 Zn, 2.2 Sn, 1.7 Al, 1.5 Fe.
60,000-65,000 TS; 14 El.
For chemical equipment construction; also known as "Duranametal"; corrosion resistant.

DURANA-2.
M-Ger.; 65 Cu, 30 Zn, 1.5 Al, 1.5 Fe, 2 Pb.
For chemical equipment; also known as "Duranametal"; corrosion resistant, high strength.

DURANA-3.
M-Ger.; 59 Cu, 40 Zn, 1 Sn, 0.3 Fe, 0.4 Pb.
For chemical equipment; high strength.

DURANAL.
M-Ger.; 5-10 Mg, 0.6 Mn, 0.2 Si, bal Al.
For light alloy parts.

DURANALIUM.
M-457; 5-10 Mg, bal Al.
For light alloy parts, aircraft castings; corrosion resistant.

DURANALIUM.
M-1548; 0.3-0.6 Mn, 2.5-9.0 Mg, bal Al.
Bars: 26,000-57,000 TS; 8500-34,000 YS; 15-25 El; 45-100 Brin.
For hardware, fittings, fasteners. Age-hardenable.

DURANALIUM 2 S.
M-457; 2.0-2.5 Mg, 1-2 Mn, 0-0.2 Sb, bal Al.
Soft: 26,000 TS; 12,000 YS; 25 El; 45 Brin.
Hard: 45,000 TS; 36,000 YS; 3 El; 75 Brin.
For light alloy parts; corrosion resistant.

DURANALIUM 2S.
M-1548; 1.0-2.0 Mn, 2.0-2.5 Mg, bal Al.
Bar: 26,000-34,000 TS; 8500-17,000 YS; 15-25 El; 50-60 Brin.
For light weight parts.

DURANALIUM 3.
M-457; 2-4 Mg, 0.4 max Mn, 0.3 max Cr, bal Al.
Soft: 28,000 TS; 13,000 YS; 30 El; 47 Brin.
Hard: 42,000 TS; 37,000 YS; 8 El; 77 Brin.
For light alloy parts; resists sea water corrosion.

DURANALIUM 3.5.
M-457; 2-6 Mg, 0.25-0.75 Mn, bal Al.
Cast: 20,000-22,000 TS; 11,000-15,000 YS; 3-5 El; 54-58 Brin.
For rivets, bolts, screws; corrosion resistant.

DURANALIUM 5.
M-457; 4-6 Mg, bal Al.
Soft: 35,000 TS; 18,000 YS; 26 El; 60 Brin.
Hard: 55,000 TS; 45,000 YS; 4 El; 105 Brin.
For light alloy parts; corrosion resistant.

DURANALIUM 7.
M-457; 7 Mg, 0.45 Mn, bal Al.
Rolled: 52,000 TS; 32,000 YS; 12 El; 115 Brin.
For seaplane pontoons, propellers; resists sea water corrosion.

DURANALIUM 9.
M-457; 8-10 Mg, bal Al.
Soft: 50,000 TS; 25,000 YS; 15 El; 80 Brin.
Hard: 65,000 TS; 45,000 YS; 10 El; 110 Brin.
For light alloy parts; corrosion resistant.

DURANCE BEARINGS.
 M-Ger.; 44.5 Sb, 33.3 Sn, 22.2 Cu.
 For bearings; Babbitt.

DURAND'S ALLOY.
 M-Eng.; 67 Al, 33 Zn.
 For strong and tough light weight parts; non-hardenable.

DURANIC.
 M-French; 96 Al, 2 Mn, 2 Ni.
 23,000-26,000 TS; 1.0-2.5 El; 60-70 Brin.
 For light alloy parts; non-hardenable.

DURANICKEL replaced by **DURANICKEL ALLOY 301.**

DURANICKEL ALLOY 301.
 M-1499; 96.5 Ni, 0.15 C, 0.25 Mn, 0.30 Fe, 0.005 S, 0.5 Si, 0.13 Cu 4.38 Al, 0.63 Ti.
 Annealed: 90,000-120,000 TS; 30,000-60,000 YS; 55-35 El; 135-185 Brin.
 Annealed and aged: 150,000-190,000 TS; 110,000-140,000 YS; 30-20 El; 285-360 Brin.
 For extrusion press parts; molds used in glass industry; clips, diaphragms and springs.
 Corrosion resistant, high strength and hardness.

DURANMIUM.
 M-358; C, W, bal Fe.
 For high speed cutting tools; high speed steel.

DURAPLAT.
 M-457; 3.5-5.5 Cu, 0.2-2 Mg, 1 Si, 1 Mn, bal Al.
 For light alloy parts; Al ply alloy, bonded; corrosion resistant.

DURASINT RF7.
 M-1739.
 Non-metallic friction material.
 For oil cooled clutches.

DURASINT GRADE RM 4.
 M-1739.
 Iron, sintered (friction material).
 Moderate duty material; for main drive clutches.

DURASINT GRADES RM5, RM6.
 M-1739.
 Iron, sintered (friction material).
 For brake drums for passenger cars.

DURASINT GRADE S.1.
 M-1739.
 Bronze, sintered (friction material).
 Good general purpose for power shift clutches, disc brake pads.

DURASINT GRADE S.2.
 M-1739.
 Bronze, sintered (friction material).
 For tractor steering and industrial clutches.

DURASINT GRADE S.3.
 M-1739.
 Bronze, sintered (friction material).
 For automatic and power shift transmissions.

DURASINT GRADE S.14.
 M-1739.
 Bronze ceramic, sintered (friction material).
 For heavy duty dry applications as tractor main drive clutches.

DURASINT GRADE S.22.
 M-1739.
 Iron, sintered (friction material).
 High friction levels.
 For main drive clutches.

DURASINT GRADE S.71.
 M-1739.
 Bronze Graphite, sintered (friction material).
 For medium duty dry friction applications.

DURASINT GRADE S.72.
 M-1739.
 Bronze ceramic, sintered (friction material).
 Moderate to high duty, multidisc brake material.

DURASINT GRADE S.73.
 M-1739.
 Bronze Graphite, sintered (friction material).
 Low wear conditions; safety clutches.

DURASINT GRADE S.106.
 M-1739.
 Iron Ceramic, sintered (friction material).
 Low friction level, low wear; aircraft brakes.

DURASINT GRADE S112.
 M-1739.
 Iron Ceramic, sintered (friction material).
 Moderate friction, low wear; aircraft brakes.

DURASINT GRADES S.113, S.114.
 M-1739.
 Iron, sintered (friction material).
 Aircraft brakes.

DURASINT GRADE S.193.
 M-1739.
 Bronze, ceramic (friction material).
 High co-efficient of friction with good wear; for heavy duty main clutches.

DURASINT GRADE S.200.
 M-1739.
 Bronze, sintered (friction material).
 Semi-automatic gear boxes, aircraft actuators.

DURASINT GRADE S.206.
 M-1739.
 Bronze, sintered (friction material).
 Excellent all purpose material.
 Torque limiting clutches on farm machinery.

DURASINT GRADE S.208.
 M-1739.
 Bronze, sintered (friction material).
 For tension and overload clutches.

DURASINT GRADE S.210.
M-1739.
Bronze, sintered (friction material).
For steering clutches and steering band brakes for track laying vehicles.

DURASINT GRADE S.214.
M-1739.
Iron Ceramic, sintered (friction material).
High co-efficient of friction and good wear resistance under high temperature conditions.
For heavy duty main drive clutches.

DURASINT GRADE S.215.
M-1739.
Bronze, sintered (friction material).
High co-efficient of friction.
For automotive and farm equipment clutches.

DURASINT GRADE S.217.
M-1739.
Bronze, sintered (friction material).
Good energy absorption and wear resistance.
For clutches for construction equipment.

DURASINT GRADE S.220.
M-1739.
Iron, sintered (friction material).
Heavy duty brake material.

DURA-STRAND.
M-1598.
High strength, non-magnetic core wire, with copper plus silver plate for coaxial transmission designs.

DURATHERM 200.
M-1631; Co-Ni-Cr, non-magnetic alloy.
Temperature and corrosion resistant spring alloy; can be age-hardened.

DURATHERM 2602.
M-1631; Co-Ni-Cr, non-magnetic alloy.
Temperature and corrosion resistant spring alloy; can be age-hardened.

DUROTHERM 600.
M-1631; Co-Ni-Cr, non-magnetic alloy.
Temperature and corrosion resistant spring alloy; can be age-hardened.

DURAWELD.
M-1246; WC.
For hard facing electrode; sintered.

DURAWELD NO. 1.
M-769.
For hard facing welding rods; wear resistant.

DURAWELD NO. 2.
M-769.
For hard facing welding rods; wear resistant.

DURAWELD NO. 3.
M-769.
For hard facing welding rods; wear resistant.

DURAWELD NO. 4.
M-769.
For hard facing welding rods; wear resistant.

DURAX.
M-333; 0.50 C, 1.5 Cr, 2.5 W, bal Fe.
For tools, dies; Type S1; oil hardened, shock resistant.

DURAX MO3.
M-1496; 0.30 C, 0.30 Mn, 0.30 Si, 3.0 Cr, 2.8 Mo, 0.5 V, bal Fe.
Heat treated: 215,000-230,000 TS.
For extrusion dies, mandrels, die casting dies, hot work tools.
Hot work steel. Thermal shock resistant.

DURAZIT.
M-1246; WC.
For hard facing electrode; sintered.

DURAZIT I.
M-1246; Co-Cr-W.
Cast: 580 Brin.
For chemical equipment; hard facing.

DURAZIT II.
M-1246; Co-Cr-W.
Cast: 460 Brin.
For chemical equipment; hard facing.

DURCILIUM.
M-770; 94.3 Al, 4.0 Cu, 0.7 Mg, 0.5 Mn, 0.5 Si.
Annealed: 34,000 TS; 17,000 YS; 16 El; 45 Brin.
Hardened: 61,000 TS; 37,000 YS; 16 El; 110 Brin.
For light alloy parts, aircraft structures; age-hardenable.

DURCILIUM B.
M-1444; 3.5 Mg, 0.25 Cr, bal Al.
Aluminium wire.
AA 5154.

DURCILIUM C.
M-1444.
Drawn aluminium tubing.

DURCILIUM E.
M-1444; 0.50 Si, 0.6 Mg, bal Al.
Aluminium bar for electrical purposes. AA 6101.

DURCILIUM ET.
M-1444; 0.50 max Cu, bal Al.
1/4 or hard drawn bar for electrical purposes.
BS 2898 E1E; BS 215.

DURCILIUM F.
M-770; 3.4-4.8 Cu, 0.7 Mg, 0.8 Si, 1 Fe. 0.4-1.2 Mn, 0.2 Ni, 0.2 Zn 0.2 Ti, bal Al.
W-temper: 58,200 TS; 42,600 YS; 8-13 El.
For structural members; age-hardened.

DURCILIUM FD.
M-1444 5.2 Si, bal Al.
Aluminium alloy wire; mostly for welding.
AA 4043.

DURCILIUM J.
M-1444; 5.0 Mg, 0.12 Mn, 0.12 Cr, bal Al.
Aluminium wire.
AA 5056.

DURCILIUM K.
M-1444, M-770; 3.5-5.0 Cu, 0.4-1.2 Mg, 0.7 Si, 0.7 Fe, 0.4-1.2 Mn, 0.3 Ti, bal Al.
Heat treated: 53,800-56,000 TS; 31,400-33,600 YS; 15-10 El.
For structural members; age-hardened, corrosion resistant.

DURCILIUM L.
M-770; 3.5-5.8 Cu, 0.6 Mg, 1.5 Si, 1 Fe, 1.2 Mn, 0.3 Ti, bal Al.
W-temper: 53,800-56,000 TS; 31,400-33,600 YS; 15-10 El.
WP-temper: 62,000-67,200 TS; 53,800-58,200 YS; 8-6 El.
For structural members; age-hardened.

DURCILIUM M.
M-770; 1-2 Cu, 0.5-1.2 Mg, 0.7-1.2 Si, 0.7 Fe, 1.0 Mn, 0.2 Ti, bal Al.
W-temper: 42,100 TS; 32,400 YS; 18 El.
WP-temper: 62,000 TS; 49,800 YS; 13 El.
For aircraft and other structural members; age hardenable.

DURCILIUM MX.
M-1444; 2.6 Cu, 0.35 Mg, bal Al.
Aluminium wire.
AA 2117.

DURCILIUM P.
M-1444.
Special aluminium wire to meet DTD 5074 A.

DURCILIUM R.
M-770; 0.15 Cu, 0.4-1.5 Mg, 0.7-1.3 Si, 0.6 Fe, 1 Mn, 0.2 Ti, bal Al.
W-temper: 27,000 TS; 15,700 YS; 18 El.
WP-temper: 40,400 TS; 33,600 YS; 10 El.
For structural members; age-hardened, corrosion resistant.

DURCILIUM S.
M-1444; 0.7-1.3 Si, 0.50 max Fe, 0.6 Mg, 0.6 Mn, bal Al.
Aluminium wire, bar, tubing.
AA 6351.

DURCILIUM T.
M-1444; 0.25 max Si, 0.40 max Fe, 99.5 min Al.
Aluminium wire, bar, tube.
1050 A.

DURCILIUM TA.
M-1444; 0.15 max Si, 0.15 max Fe, 99.8 min Al.
Aluminium wire, bar, tube.
1080 A.

DURCILIUM V.
M-1444; 2.0 Mg, bal Al.
Aluminium wire, bar and tube.
5251.

DURCILIUM W.
M-1444, M-770; 0.15 Cu, 0.4-0.9 Mg, 0.3-0.7 Si, 0.6 Fe, 0.2 Ti, bal Al.
M-temper: 15,700 TS; 11,200 YS; 15 El.
W-temper: 24,600 TS; 15,700 YS; 23 El.
WP-temper: 31,400 TS; 26,900 YS; 14 El.
For structural members; age-hardened.

DURCILIUM WT.
M-1444; 0.40 Si, 0.7 Mg, bal Fe.
Aluminium bar.
AA 6463.

DURCILIUM XT.
M-1444.
Aluminum wire specially processed for use in telephone cable.

DURCO CF-3.
M-46; 17-21 Cr, 8-12 Ni, 0.03 max C, bal Fe.
Heat treated: 65,000 TS; 28,000 YS; 35 El; 160 Brin.
For stainless parts; chemical process and food handling equipment; chemical resistant cast alloy.

DURCO CF-3M.
M-46; 17-21 Cr, 9-13 Ni, 2-3 Mo, 0.03 max C, bal Fe.
Heat treated: 70,000 TS; 30,000 YS; 30 El; 160 Brin.
For stainless parts; chemical process and food handling equipment; chemical resistant cast alloy.

DURCO CF-8.
M-46; 18-21 Cr, 8-11 Ni, 0.08 max C, bal Fe.
Heat treated: 65,000 TS; 28,000 YS; 35 El; 160 Brin.
For stainless parts; chemical process and food handling equipment; chemical resistant cast alloy.

DURCO CF-8M.
M-46; 18-21 Cr, 9-12 Ni, 2-3 Mo, 0.08 max C, bal Fe.
Heat treated: 70,000 TS; 30,000 YS; 30 El; 160 Brin.
For stainless parts; chemical process and food handling equipment; corrosion resistant cast alloy.

DURCO CY-40.
M-46; 14-17 Cr, 11 max Fe, 0.40 max C, bal Ni.
Heat treated: 70,000 TS; 28,000 YS; 30 El; 150 Brin.
For chemical process equipment handling caustic and some salt solutions; cast alloy.

DURCO CZ-100.
M-46; 95 Ni, 3 max Fe, 1.0 max C.
As cast: 50,000 TS; 18,000 YS; 10 El; 130 Brin.
For chemical process equipment handling caustic alkalies, anhydrous HF, and organic acids.

DURCO DC-8.
M-46.
Cobalt base alloy.
For use as shaft sleeve material; resists wear and corrosion.

DURCO M-35.
M-46; 26-33 Cu, 3.5 max Fe, 0.35 max C, bal Ni.
As cast: 65,000 TS; 30,000 YS; 25 El; 150 Brin.
For chemical process equipment handling hydrofluoric acid and alkaline brines.

DURCOMET 100.
M-46; 24.5-26.5 Cr, 4.75-6.0 Ni, 1.75-2.25 Mo, 2.75-3.25 Cu, 0.04 max bal Fe.
Heat treated: 100,000 TS; 70,000 YS; 20 El; 250 Brin.
For equipment handling HNO_3, dilute H_2SO_4, H_3PO_4, fatty acids, and erosion-corrosion services.

DURCO WCB.
M-46; 0.40 max Cr, 0.5 max Ni, 0.25 max Mo, 0.5 max Cu, 0.6 max Si, 1.0 max Mn, 0.3 max C, bal Fe.
Heat treated: 70,000 TS; 36,000 YS; 22 El; 150 Brin.
For process equipment handling weak caustic, many types of water, and a wide variety of organics.

DUREDGE.
M-333; 0.55 C, 1.6 Si, 0.4 Mo, 0.6 V, bal Fe.
For cutting tools, chisels, rivet sets, shear blades; will maintain keen cutting edge, hard and tough.

DUREDGE.
M-336; 0.55 C, 0.75 Mn, 0.35 Mo, 2.0 Si, bal Fe.
For punches, shears, dies; water or oil hardening.

DUREHETE 900.
M-1724; 0.35-0.45 C, 1.0-1.5 Cr, 0.5-0.8 Mo, 0.5 max Si, bal Fe.
Heat treated: 147,000 TS; 136,000 YS; 22 El.
For high temperature bolts, studs, fasteners; high creep resistance.
B.S. 970-En20B; ASTM A193-61T Gr 7A.

DUREHETE 950.
M-1724; 0.30-0.45 C, 1.0-1.5 Cr, 0.5-0.8 Mo, 0.2-0.3 V, bal Fe.
Heat treated: 144,000 TS; 135,000 YS; 25 El; 64 RA.
For high temperature bolts, studs, fasteners; heat resistant to 950°F.
B.S. 1506-661; ASTM A193-61T Gr B16.

DUREHETE 1055.
M-1724; 0.15-0.25 C, 0.9-1.3 Cr, 0.85-1.1 Mo, 0.6-0.8 V, 0.05-0.20 Ti 0.001-0.01 B, bal Fe.
Heat treated: 122,000 min TS; 100,000 min YS; 43 min El.
For bolts required to operate at 500-565°C.
For use in steam power plant.

DURENER MN20.
M-457; 0.5-1.5 Mn, 0.3 max Cr, bal Al.
Soft: 16,000 TS; 6000 YS; 40 El.
Hard: 29,000 TS; 27,000 YS; 10 El.
For cooking utensils, heat exchangers, tanks, furniture; good forming and welding properties.

DURENER Z8.
M-457; Zn, Cu, Mg, bal Al.
For light alloy parts.

DUREX.
M-839; 4.4-4.7 total C, 10 Sn, 83.22 Cu, 0.60 impurities.
30-40 Brin.
For bearings (porous to absorb oil-25% oil by volume); carbon present as graphite.

DUREX.
M-487; 0.7 C, 18.5 W, 4.5 Cr, 1.0 V, bal Fe.
For shaping and planing tools, broaches, hobs; high speed steel.

DUREX BRONZE.
M-839; 95 Cu, 5 Sn.
Cast: 30,000-40,000 TS; 65-100 Brin.
For bearings; sintered.

DUREX BRONZE NO. 46.
M-839; 10 Sn, 1.2 graphite, bal Cu.
Sintered: 13,000-25,000 TS; 2.0 El; 25-40 Brin.
For bearings; sintered; C.S. 79,000.

DUREX IRON 362.
M-839; 20 Cu, 1 C, bal Fe.
75,000 TS; 1 El; 185 Brin.
For gears, cams; copper infiltrated sintered steel.

DUREX IRON NO. 62.
M-839; 1.0 C, bal Fe.
Sintered: 28,000 TS; 0.5 El.
For bearings; sintered, C.S. 160,000.

DUREX IRON NO. 93.
M-839; 7.5 Cu, bal Fe.
Sintered: 15,000-35,000 TS; 1-7 El; 90 Brin.
For bearings; sintered alloy.

DURICHLOR 51.
M-46; 4-5 Cr, 14.2-14.75 Si, 0.70-1.10 C, bal Fe.
For pumps handling H_2SO_4, HNO_3, HCl; and impressed current anodes.

DURIMET 20.
  M-46; 19-22 Cr, 27.5-30.5 Ni, 2-3 Mo, 3-4 Cu, 0.07 max C, bal Fe.
  Cast, Heat treated: 62,500 TS; 25,000 YS; 35 El; 130 Brin.
  For chemical process equipment handling H_2SO_4, HNO_3, mixed acids, and caustic.

DURIRON.
  M-46; 14.2-14.75 Si, 0.70-1.10 C, bal Fe.
  Cast alloy for chemical process equipment handling H_2SO_4 and HNO_3 acids.

DURITAS.
  M-261, M-486; WC.
  For dies, nibs; cemented carbides.

DURMES.
  M-457; 2.5-5.0 Cu, 0.2-1.8 Mg, 0.3-1.5 Mn, 0.5-2.5 Pb, Sn, Cd, Bi, bal Al.
  For screw machine products; free-cutting.

DURO-CHIP.
  M-750; 0.6 C, 0.7 Mn, 1.85 Si, 0.5 Mo, 0.25 V, bal Fe.
  For hand and pneumatic chisels, caulking tools; shock resistant, oil hardening.

DUROCYL.
  M-185; 3 C, 2 Si, 0.8 Mn, 0.3 Cr, bal Fe.
  Cast: 38,000 TS; 240 Brin.
  For cylinder liners; wear plates; centrifugally cast iron.

DURODI.
  M-55; 0.5-0.6 C, 0.90 Mo, 0.55 Mn, 0.8 Cr, 1.6 Ni, bal Fe.
  Heat treated: 225,000-130,000 TS; 200,000-120,000 YS; 15-26 El; 43-51 RA; 460-275 Brin.
  For hot work die steel, punches, upsetting and gripping dies, shear knives; tough steel; air hardening, non-distorting.

DURODIE.
  M-261; 0.4 C, 3.3 Cr, 2.7 W, 0.3 V, bal Fe.
  For bolt dies, hot extrusion dies, swaging dies; hot work steel.

DUROLITH.
  M-Germ.; Zn base alloy.
  For die castings; Br. P. 411557.

DUROMAX 250.
  M-1315; 0.10-0.20 C, 0.15-0.35 Si, 0.80-1.0 Mn, 0.40-0.65 Cr, 0.40-0.60 Mo, 0.70-1.0 Ni, V, Cu, B, bal Fe.
  High strength low alloy steel.

DUROMAX 321.
  M-1315; 0.15 C, 0.25 Si, 0.85 Mn, 0.50 Cr, 0.50 Mo, 0.50 Ni, 0.05 V, B, bal Fe.
  High strength, low alloy steel; 321 Brin.

DURON H.
  M-459; 0.6 C, 0.85 Si, 0.75 Mn, 1.18 Cr, 0.1 V, bal Fe.

DURON W.
  M-459; 0.40 C, Si, Mn, Cr, V, bal Fe.

DURONZE I.
  M-141; 97 Cu, 1 Si, 1.5 Sn.
  Annealed: 45,000 TS.
  Cold drawn: 135,000 TS.
  For general use, bolts, cable, hardware, cap screws; corrosion resisting.

DURONZE II (BRIDGEPORT 632).
  M-141; 3 Si, bal Cu.
  Sand cast: 50,000 TS; 40 El; 50 RA; D 93 Brin.
  Annealed: 70,000 TS; 55 El; 65 RA; F 85 Brin.
  For poleline hardware, marine hardware, water storage tanks, chemical vats, range boilers; strong, hot forging alloy; corrosion resistant.

DURONZE III.
  M-141; 2.5 Si, bal Cu.
  Forged: 90,000 TS; 50,000 YS; 27 El; 35 RA; 165 Brin.
  Annealed: 90,000 TS; 30 El; 35 RA; 160 Brin.
  For boat shafting, pump rods, pinions, gears; corrosion resistant.

DURONZE IV.
  M-141; 5 Al, bal Cu.
  Annealed: 65,000 TS; 45 El.
  For condenser tubes; sea water corrosion resistant.

DURONZE 606.
  M-141; 3.0 Si, 0.95 Mn, bal Cu.
  Hard: 90,000 TS; 45,000 YS; 18 El; B 85 Rock.
  Soft: 55,000 TS; 22,000 YS; 55 El; B 50 Rock.
  For bolts, gears, valve stems, marine hardware, clutch disks. Corrosion resistant, tough, wear resistant.

DURONZE 609.
  M-141; 2 Si, bal Cu.
  Hard: 75,000 TS; 40,000 YS; 10 El; 138 Brin.
  Soft: 50,000 TS; 20,000 YS; 40 El; 63 Brin.
  For bolts, cold headed parts, gears, shafts; formerly Duronze V; corrosion resistant.

DURONZE 632.
  M-141; 2.95 Si, 0.10 Fe, bal Cu.
  Annealed: 63,000 TS; 30,000 YS; 55 El; F 90 Rock.
  Hard: 90,000 TS; 45,000 YS; 18 El; B 85 Rock.
  For poleline hardware, nuts, bolts, wire and cable connectors, cap screws, springs, conduits, rivets. 7% elect. cond.-corrosion resistant.

DURONZE 707.
  M-141; 91 Cu, 2 Si, 7 Al.
  For bolts, gears, valve stems, marine hardware; tough, corrosion resistant.

DURONZE 708.
M-141; 6.75 Al, 1.75 Si, bal Cu.
Annealed: 85,000 TS; 40,000 YS; 30 El; 165 Brin.
For valve parts, bolts, fasteners; corrosion and wear resistant, hot forgeable.

DUROPLAT.
M-German; Al coated Duralumin.
For bimetal parts.

DUROTEC 19910.
M-717.
Alloy powder for metal spraying, makes super hard, grindable final coat.
For bearing areas subject to severe friction.

DURSILIUM.
M-France; 4-5 Cu, bal Al.
For light alloy parts; age-hardenable.

DUTCH BOY 111.
M-88; 50 Pb, 50 Sn.
Cast: 6400 TS; 40 El; 12.7 Brin.
For solder; M.P. 361-414°F; S.S. 5800.

DUTCH BOY 222.
M-88; 55 Pb, 45 Sn.
Cast: 6600 TS; 40 El; 12.7 Brin.
For solder; M.P. 361-424°F; S.S. 5700.

DUTCH BOY 333.
M-88; Pb, Sn.
Cast: 6400 TS; 41 El; 12.6 Brin.
For solder; M.P. 361-437°F; S.S. 5500.

DUTCH BOY 444.
M-88; Pb, Sn.
Cast: 6600 TS; 40 El; 13.3 Brin.
For solder; M.P. 361-437°F; S.S. 5625.

DUTCH BOY 555.
M-88; Pb, Sn.
Cast: 6700 TS; 39 El; 13.3 Brin.
For solder; M.P. 361-448°F; S.S. 5525.

DUTCH BOY 666.
M-88; Pb, Sn.
Cast: 7000 TS; 35 El; 14.0 Brin.
For solder; M.P. 361-460°F; S.S. 5450.

DUTCH BOY 777.
M-88; Pb, Sn.
Cast: 7000 TS; 33 El; 13.9 Brin.
For solder; M.P. 361-460°F; S.S. 5375.

DUTCH BOY 888.
M-88; Pb, Sn.
Cast: 70,000 TS; 30 El; 13.9 Brin.
For solder; M.P. 361-468°F; "Plumbers solder."

DUTCH BOY BEARING METAL.
M-88; Sb, Cu, bal Sn.
Cast: 23.5 Brin.
For bearings for machinery and general mill work; Babbitt, for slow movin machinery.

DUTCH BOY GENUINE BABBITT.
M-88; 88.9 Sn, 7.4 Sb, 3.7 Cu.
Cast: 26.7 Brin.
For bearings; Babbitt.

DUTCH BOY NO. 1 JOURNAL.
M-88; Sb, Cu, bal Sn.
Cast: 32.4 Brin.
For heavy pressure bearings; Babbitt, resists heavy crushing strains.

DUTCH BOY P-A-F.
M-88; Pb-Sn-Sb-Cu.
For bearings; Babbitt.

DUTCH BOY STERLING JOURNAL.
M-88; Sb, Cu, bal Sn.
Cast: 23.3 Brin.
For bearings for stationary gas engines; Babbitt, resists heavy crushing strains.

DUTCH METAL.
76-80 Cu, 24-20 Zn.
Cast: 38,400 TS; 22,750 YS; 33 El; 33 RA; 45 Brin.
Hard rolled: 66,900 TS; 66,900 YS; 12 El; 42 RA; 145 Brin.
For cheap jewelry, gold leaf; red brass.

DUTCH SILVER.
Eng.; 81.5 Sn, 8.8 Sb, 9.6 Cu.
For tableware, novelties, art metal pieces; corrosion resistant.

DUTCH WHITE METAL.
M-Eng.; 81.5 Sn, 8.8 Sb, 9.6 Cu.
For bearings, antifriction purposes; Babbitt.

DUTRAX.
M-358; C, bal Fe.
For machinery parts.

DUTRAX 3.
M-333; 0.45 C, 0.8 Cr, 1.1 W, 0.25 Mo, bal Fe.
For plastic mold dies; oil hardening.

DUTRAX 7.
M-333; 0.45 C, 1.25 Cr, 2.15 W, 0.2 Si, bal Fe.
For plastic mold dies; oil hardening.

DUVAL MU.
M-1115; 1.10 C, 0.3 Mn, 0.25 Si, bal Fe.
For taps, drills, cutters, reamers; Type W1; water hardened.

DUVAL SF1.
M-1115; 0.9-1.1 C, 0.5-1.0 Cr, 3.5-4.5 W, 0.1 V, bal Fe.
For cold work tools, headers, forming dies; cold work steel, oil hardened.

DV2A.
M-38; 0.53 C, 11.5 Mn, 0.28 max Si, 20.50 Cr, 2.0 W, 1.0 Cb, 0.4 V, 0.45 N, bal Fe.
For diesel engine valves.

DYCAST NO. 1.
M-73; 0.40 C, 5 Cr, 0.30 Mn, 0.50 V, 0.80 Mo, 1.0 Si, bal Fe.
For die casting dies and cores; resists heat checking, hot work steel.

DYCRO.
M-1433; 1.5 C, 12 Cr, 0.25 V, 0.8 Mo, bal Fe.
For blanking and forming dies, broaches; air hardened, non-deforming.

DYCRO-1.
M-1431; 1 C, Cr, Mn, Mo, bal Fe.
For tools, dies; air hardened.

DY-KROME.
M-750; 1.5 C, 12 Cr, 0.2 V, 0.8 Mo, bal Fe.
For blanking dies, broaches, cold extrusion dies; air hardening.

DYMAL.
M-750; 0.9 C, 0.5 Cr, 1.15 Mn, 0.5 W, bal Fe.
For Bakelite molding dies, gages, hobs, forming tools; non-deforming.

DYMAL AH.
M-750; 1.65-2.1 C, 0.25 Mn, 11-13 Cr, 0.35 V, 0.90 Mo, bal Fe.
For dies, tools; nondeforming.

DYMAL OH.
M-750; 0.90 C, 1.1 Mn, 0.5 Cr, 0.5 W, 0.2-0.3 V, bal Fe.
For dies, hobs, milling cutters, gauges; oil hardening.

DYNABRAZE 145 FC.
M-717;
Flux coated bronze rod for brazing steel and cast iron. Bonding temp; 1400-1600°F. 65,000 TS.

DYNACUT.
M-1702; 1.2 C, 2.7 W, 3.75 Cr, 1.6 V, 8.0 Mo, 8.2 Co, bal Fe.
M-43 high speed steel tool bits; best red hardness. (Electrite Dynacut).

DYNACUT.
M-73; 1.2 C, 2.7 W, 3.75 Cr, 1.6 V, 8.0 Mo, 8.2 Co, bal Fe.
Hardened: C 65-66 Rock.
For cutters, in heavy duty machining, drills, reamers, broaches.
Type M-43 high-speed steel, wear resistant. High red-hardness.

DYNAFLEX.
M-73; 0.40 C, 0.30 Mn, 0.90 Si, 5.0 Cr, 1.3 Mo, 0.45 V, bal Fe.
Air hardened: 300,000 TS; 250,000 YS; 7 El; 30 RA; 525 Brin.
For punches, upsetters, extrusion dies; hot work steel; AISI Type H 11.

DYNA-KUT DK1211.
M-1774; 0.13 max C, 0.60-0.90 Mn, 0.08-0.15 S, 0.07-0.12 P, bal Fe.
Free machining steel bars.

DYNA-KUT DK1212.
M-1774; 0.13 max C, 0.70-1.0 Mn, 0.07-0.12 P, 0.08-0.15 S, bal Fe.
Free machining steel bars; 47% greater production capability over DK1211.

DYNA-KUT DK1213.
M-1774; 0.09 max C, 0.70-1.0 Mn, 0.07-0.12 P, 0.24-0.36 S, bal Fe.
Free machining steel bars; 87% greater production capability over DK1211.

DYNA-KUT DK1216.
M-1774; 0.15 max C, 0.85-1.15 Mn, 0.06-0.11 P, 0.37-0.46 S, bal Fe.
Free machining steel bars; 131% greater production capability over DK1211.

DYNA-KUT DK12L13.
M-1774; 0.09 max C, 0.70-1.0 Mn, 0.07-0.12 P, 0.24-0.33 S, 0.15-0.35 PB, bal Fe.
Free machining steel bars; 154% greater production capability over DK1211.

DYNA-KUT DK12L16.
M-1774; 0.15 max C, 0.85-1.15 Mn, 0.06-0.11 P, 0.37-0.46 S, 0.15-0.35 Pb, bal Fe.
Free machining steel bars; 200% greater production capability over DK1211.

DYNALLOY 600 now **DYNALLOY 6741 (600).**

DYNALLOY 601 now **DYNALLOY 6680 (601).**

DYNALLOY 602 now **DYNALLOY 6730 (602).**

DYNALLOY 603 now **DYNALLOY 6731 (603).**

DYNALLOY 604 now **DYNALLOY 6732 (604).**

DYNALLOY 605 now **DYNALLOY 6733 (605).**

DYNALLOY 6680 (601).
M-266; 61 Cu, 0.8 Si, 2.5 Mn, bal Zn.
HH Temper: 70,000-85,000 TS; 40,000-65,000 YS; 15-25 El; B 70-87 Rock.
For bearings, worm wheels, bushings.
Mate with soft member. Corrosion resistant.

DYNALLOY 6730 (602).
M-266; 61 Cu, 1.0 Pb, 1.0 Si, 2.5 Mn, bal Zn.
HH Temper: 70,000-85,000 TS; 40,000-65,000 YS; 15-25 El; B 70-87 Rock.
For bearings, bushings, gears.
Leaded forgeable bearing alloy, corrosion resistant.

DYNALLOY 6731 (603).
M-266; 2.75 Mn, 1.0 Pb, 1.0 Si, 60 Cu, bal Zn.
Forged: 50,000 TS; 30,000 YS; 15 El; B 65 Rock.
Heat treated: 75,000 TS; 45,000 YS; 20 El; B 82 Rock.
For gears, bushings, bearings, pump barrels, machinery parts.
Leaded forging brass.

DYNALLOY 6732 (604).
M-266; 61 Cu, 2.5 Pb, 1 Si, 2.5 Mn, bal Zn.
HH Temper: 65,000-80,000 TS; 40,000-60,000 YS; 10-20 El; B 75-86 Rock.
For bearings, pump gears, valve stems.
High leaded bearing alloy, corrosion resistant.

DYNALLOY 6733 (605).
M-266; 62 Cu, 0.6 Pb, 1 Si, 2.5 Mn, bal Zn.
HH Temper: 70,000-75,000 TS; 45,000-55,000 YS; 18-25 El; B 75-82 Rock.
For bearings, pump gears, valve stems.
Low leaded bearing alloy, corrosion resistant.

DYNALLOY 6736 (606).
M-266; 4.0-4.7 Al, 0.50 max Fe, 99.5 min Cu + Al + Fe.
Aluminum bronze forgings.
CDA 606.

DYNALLOY 6741 (600).
M-266; 58 Cu, 0.8 Si, 1.5 Al, 2.5 Mn, bal Zn.
HH Temper: 75,000-100,000 TS; 40,000-65,000 YS; 12-18 El; B 82-88 Rock.
Forged: 68,000 TS; 34,000 YS; 18 El; B 78 Rock.
For bearings, worms and worm wheels, bushings.
Forgeable bearing alloy, corrosion resistant.

DYNA-LOY DL41L43.
M-1774; 0.39-0.45 C, 0.75-1.05 Mn, 0.035 max P, 0.06-0.10 S, 0.15-0.35 Si, 0.80-1.10 Cr, 0.15-0.25 Mo, 0.15-0.35 Pb, bal Fe.
Hardenable to 33-55 Rock C; 150,000-280,000 psi TS.
Free machining alloy steel.

DYNA-LOY DL41L49.
M-1774; 0.46-0.52 C, 0.75-1.05 Mn, 0.035 max P, 0.06-0.10 S, 0.15-0.35 Si, 0.80-1.10 Cr, 0.15-0.25 Mo, 0.15-0.35 Pb, bal Fe.
Hardenable to 43-56 Rc; 175,000-300,000 psi TS.
Free machining alloy steel.

DYNA-LOY HS-32.
M-1774; 0.40 min C, 0.70-1.10 Mn, 0.20-0.35 Si, 0.80-1.20 Cr, 0.15-0.2 Mo, 0.15-0.35 Pb, 0.06-0.10 S, bal Fe.
HT: 150,000 psi min TS; 130,000 psi min YS; 10 El; 37 RA; 32 min Rock C.
Hardened, free machining steel.

DYNAMAX.
65 Ni, 33 Fe, 2 Mo.
For toroidal cores; high permeability.

DYNAMAX see also ELECTRITE DYNAMAX.

DYNAMIC C-1.
M-451; 0.2-0.3 C, 0.5-0.8 Mn, bal Fe.
For cast steel parts; pearlitic.

DYNAMIC C-1-3.
M-451; 0.25-0.35 C, 0.6-0.8 Mn, bal Fe.
For castings; water hardening.

DYNAMIC C-1-4.
M-451; 0.35-0.45 C, 0.65-0.85 Mn, bal Fe.
For gears, pinions, coupling boxes; water hardening.

DYNAMIC C-1-5.
M-451; 0.4-0.5 C, 0.65-0.85 Mn, bal Fe.
For gears, pinions, power shovels; water hardening.

DYNAMIC C-1-6.
M-451; 0.5-0.6 C, 0.65-0.85 Mn, bal Fe.
For cams, bottom plates; wear hardening.

DYNAMIC C-1-7.
M-451; 0.7-0.8 C, 0.65-0.85 Mn, bal Fe.
For castings; abrasion resistant.

DYNAMIC C-2.
M-451; 0.25-0.35 C, 0.9-1.4 Mn, 0.25-0.45 Ni, 0.05-0.15 Mo, bal Fe.
Cast: 85,000 TS; 52,500 YS; 30 El; 57 RA.
For tractor frames, locomotive castings; resists shock and impact; Iz-45.

DYNAMIC C-2-A.
M-451; 0.35-0.45 C, 1.0-1.5 Mn, 0.25-0.45 Ni, 0.05-0.15 Mo, bal Fe.
For crawler shoes, gears, sprockets; wear resistant.

DYNAMIC C-3.
M-451; 0.25-0.35 C, 1.0-1.5 Mn, 0.1-0.2 Mo, bal Fe.
Cast: 96,000 TS; 68,000 YS; 24 El; 48 RA.
For sprockets, spindles, wheel centers, crossheads, gears; wear resistant; resists shock and impact.

DYNAMIC C-3A.
M-451; 0.4 C, 1.2 Mn, 0.15 Mo, bal Fe.
Cast.
For gears, racks, sprockets; wear resistant.

DYNAMIC C-4.
M-451; 0.25 C, 0.9 Mn, 0.8 Cr, 1.9 Ni, 0.5 Mo, bal Fe.
Heat treated: 120,000 TS; 100,000 YS; 15 El; 30 RA; 321 Brin.
For rolling mill pinions; impact resistant.

DYNAMIC C-5.
M-451; 0.3 C, 1 Mn, 0.7 Ni, 0.4 Mo, bal Fe.
Cast.
For rams, saw blocks.

DYNAMIC C-6.
M-451; 0.4-0.7 C, 3 Cr, 0.4 Mo, bal Fe.
Cast.
For sand mills, rock crushers; abrasion resistant.

DYNAMIC C-7.
M-451; 0.4 C, 0.9 Mn, 0.5 Cr, 1.5 Ni, 0.5 Mo, bal Fe.
Cast.
For steel castings, crane wheels; high strength.

DYNAMIC C-8.
M-451; 0.25-0.35 C, 1.0-1.5 Mn, 0.3-0.7 Cr, 0.4-0.8 Ni, 0.2-0.4 V, bal Fe.
For gears, sprockets; tough, high impact resistant.

DYNAMIC C-9.
M-451; C, Mn, V, bal Fe.
Cast.
For steel castings; air hardened.

DYNAMIC C-10.
M-451; C, Cr, Mo, bal Fe.
Cast.
For die blocks, crane wheels; wear resistant.

DYNAMO.
M-200; 0.12 C, 0.2 Si, 0.1 Mn, bal Fe.
For electrical purposes; high magnetic permeability.

DYNAMO.
M-368; 0.7-0.9 C, bal Fe.
For tools, dies, jigs; water hardening.

DYNAMO SHEET STEEL.
M-U.S.; 3-4 Si, <0.1 C, <0.3 Mn, <0.03 P + S, bal Fe.
For laminated sheets for dynamos, electrical equipment; Hadfields silicon steel type.

DYNAMO STEEL.
M-331; 0.09 C, bal Fe.
For dynamo parts; high permeability.

DYNATRODE 666.
M-717;
Electrode for AC-DC welding of mild steel; strength 70,000 psi.

DYNATRODE 777.
M-717;
Mild steel electrode for AC-DC welding of mild steel. 76,000 TS.

DYNAVAN.
M-1702; 1.5 C, 4.5 Cr, 13.5 W, 4.75 V, 5.0 Co, 0.5 Mo, bal Fe.
T15 high speed steel tool bits; best wear resistance. (Electrite Dynavan).

DYNAVAN.
M-73; 1.55 C, 12.5 W, 4.7 Cr, 5 V, 5 Co, bal Fe.
Hardened: C 64-66 Rock.
For cutting tools, blanking dies, broaches, drills, milling cutters.
Type T-15 high-speed steel. Heat and abrasion resistant. High red-hardness.

DYNELEC.
M-691; 2.5 Si, bal Fe.
For armatures, electric motors and generators; high permeability.

DYN-GZ10.
M-1218, M-1218a; 89.25 Cu, 10 Sn, 0.5 Ni, 0.25 P.
Cast: 29,000-50,000 TS; 16,000-27,000 YS; 29-8 El; 70-85 Brin.
For nuts, bearings, valve guides, bushings; phosphor bronze, tough.

DYN-GZ14.
M-1218, M-1218a; 86.15 Cu, 11.5 Sn, 2 Ni, 0.3 P.
Cast: 27,000-42,500 TS; 16,000-25,000 YS; 25-6 El; 90-110 Brin.
For worm wheels, toggle pads, bearings; phosphor bronze, shock resistant.

DYN-RM.
M-1218, M-1218a; 83.2 Cu, 10 Sn, 2.5 Ni, 4 Pb, 0.3 P.
Cast: 25,000-41,000 TS; 16,000-22,500 YS; 20-6 El; 80-100 Brin.
For mill bearings, bushings, liners; phosphor bronze, shock resistant.

D.Y.O.
M-32; 0.3 C, 0.3 Mn, 0.25 Si, 4 Cr, 14.5 W, 0.5 V, bal Fe.
Heat treated: 205,000 TS.
For tools, forging dies; red-hard, hot work steel. Similar to AISI H25.

DYSOID.
M-Eng.; 63 Cu, 18 Pb, 10 Sn, 10 Zn.
For hardware; free-cutting.

DYSPROSIUM.
M-1755; Dy.
Purities: 99.9% (Special distilled grade), 99.5+%.
Forms: Ingot, sponge, wire, sheet, rod, turnings, powder, foil, single crystals.

DZISTALOY 6971.
M-8; 78.0 Cu, 1.0 Pb, 18.0 Zn, 2.85 Si, 0.15 Mn.
70,000 psi TS; 40,000 YS; 25 El; 80 Rock B.
Free machining; for automatic screw machine parts; resistant to corrosion and dezincification.

DZISTALOY DZ6943.
M-8; 81.5 Cu, 0.25 Pb, 14.25 Zn, 4.0 Si.
85,000 psi TS; 50,000 psi YS; 20 El; 90 Rock B.
Forgeable, free machining; for valve stems, valve seats and parts needing wear resistance and resistance to dezincification by fresh water.

D-Z-L MARINE NICKEL GENUINE.
M-178; 88 Sn, bal Sb and Cu.
Cast: 17,0000 TS; 9510 YS; 29.6 Brin.
For bearings subject to shock and heat; pouring temperature 850-925°F.

DZR.
M-815; Cu, Zn (Yellow brass).
Sand cast: 53,000 TS; 22,000 YS; 23 El; 80 Brin.

E

E1 DYNAMO.
M-Russia; 0.10 max C, 1.0 Si, 0.03 max Mn, bal Fe.
For motor stators and rotors; high permeability.

E2 DYNAMO.
M-Russia; 0.10 max C, 0.03 max Mn, 2.0 Si, bal Fe.
For motor stators and rotors; high permeability.

E3 TRANSFORMER.
M-Russia; 0.10 max C, 0.30 max Mn, 3 Si, bal Fe.
For transformers; high permeability.

E4 TRANSFORMER.
M-Russia; 0.10 max C, 0.30 max Mn, 4.0 Si, bal Fe.
For transformers; high permeability.

E 10B46.
M-1736; 0.46 C, 0.22 Si, 1.25 Mn, 0.0017 B, bal Fe.
Boron carbon steel.

E18 CO5.
M-Germany; 0.78 C, 4.2 Cr, 0.6 Mo, 18.8 W, 1.53 V, 5.72 Co, bal Fe.
Hardened: C 64-66 Rock.
For tools and cutters, reamers, taps, broaches, lathe and planer tools.
High speed steel, type T4, high red-hardness.

E18 CO10.
M-Germany; 0.75 C, 18 W, 4 Cr, 1.5 V, 0.7 Mo, 10 Co, bal Fe.
Hardened: C 64-66 Rock.
For lathe and planer tools, drills, reamers, form cutters, broaches, form tools, gear cutters.
For heavy duty work.
High speed steel, Type T5. Red-hard.

E38 MO.
M-1496; 0.38 C, 1.0 Si, 5.3 Cr, 1.5 Mo, 0.45 V, bal Fe.
Annealed: 102,000 TS; 66,000 YS; 28 El; 197 Brin.
Hardened: 300,000 TS; 250,000 YS; 6 El; C 55 Rock.
For Al & Mg die casting dies, components for die casting machines.
Air or oil hardening. Thermal shock resistant, Type H11.

E38 V.
M-1496; 0.4 C, 1.0 Si, 5.3 Cr, 1.4 Mo, 1.0 V, bal Fe.
Annealed: 98,000 TS; 74,000 YS; 28 El; 210 Brin.
Hardened: 220,000 TS; 182,000 YS; 12 El; 430 Brin.
For extrusion press mandrels, aluminum die casting dies.
Air or oil hardening. Thermal shock resistant, Type H13.

E38 W.
M-1496; 0.38 C, 1.2 Si, 5.6 Cr, 1.5 Mo, 0.3 V, 1.5 W, bal Fe.
Ht. Tr.: 216,000 TS; 185,000 YP; 14 El; 53 RA.
For metal extrusion press components, die casting dies, forging dies, hot shear blades.
Air or oil hardening. Thermal shock resistance. Type H12.

E 86B15.
M-1736; 0.15 C, 0.25 Si, 0.80 Mn, 0.60 Cr, 0.50 Mo, 0.85 Ni, 0.06 V 0.004 B, bal Fe.
Boron, low alloy steel, for carburizing.

E-132.
M-690; 11.0-13.0 Si, 1.0 max Fe, 1.8-2.8 Cu, 0.50 max Mn, 0.90-1.30 Mg, 0.10 max Pb, 0.50-1.50 Ni, 1.0 max Zn, 0.25 max Ti, 0.10 max Sn, bal Al.
Cast aluminum alloy.

E 300M.
M-1736; 0.43 C, 1.63 Si, 0.78 Mn, 0.83 Cr, 0.40 Mo, 1.83 Ni, bal Fe.
Cold work tool steel, shock resistant tools.

E 302 ETC (M-1736) see AISI 302 ETC (STAINLESS).

E 1010 ETC (M-1736) see AISI 1010 ETC (STEELS).

E1318.
M-USSR; 0.6 max C, 16-18 Cr, 4.5-6.5 Al, 0.6 max Ni, bal Fe.
For electrical resistance and heating elements; applications to 1100°C.

E1340.
M-USSR; 0.15 max C, 23-27 Cr, 4-7 Al, 0.6 max Ni, bal Fe.
For electrical resistance and heating elements; applications to 1200°C.

E1503.
M-USSR; 0.5-0.6 C, 7.5-9.5 Mn, 0.7 Si, 3.8-4.5 Cr, 8-10 Ni, bal Fe.
For instrument parts; nonmagnetic.

E1919.
M-USSR; 0.83 C, 10.0 Co, 4.0 Cr, 1.8 V, 10.0 W, bal Fe.
Ht. Tr.: Rock C 64-66.
For cutters, lathe and planer tools, form cutters, drills.
High speed steel. High red-hardness.

E1931.
M-USSR; 1.49 C, 4.0 Cr, 4.4 V, 10.6 W, 4.9 Co, bal Fe.
Hardened: C 65-68 Rock.
For cutters, drills, broaches, milling cutters, taps, reamers.
High red-hardness. Abrasion resistant. High speed steel. Type T15.

E1940.
M-USSR; 0.83 C, 4.0 Cr, 2.1 V, 18.0 W, 5.0 Co, bal Fe.
Hardened: C 62-65 Rock.
For cutters, boring tools, lathe and planer tools, gun-barrel drills, roll turning tools.
High speed steel, type T4. Red-hard.

E 2365.
M-1736; 0.35 C, 0.40 Si, 0.40 Mn, 2.95 Cr, 2.85 Mo, 0.50 V, bal Fe.
Hot work tool steel.
W.-Nr. 1.2365.

E 2419.
M-1736; 1.05 C, 0.25 Si, 0.95 Mn, 1.0 Cr, 1.25 W, bal Fe.
Cold work tool steel.
W.-Nr. 1.2419.

E 2542.
M-1736; 0.51 C, 0.95 Si, 0.35 Mn, 1.25 Cr, 0.20 Mo, 0.20 V, 2.35 W, bal Fe.
Shock resisting tool steel.
W.-Nr. 1.2542; similar to AISI S1.

E 2550.
M-1736; 0.60 C, 0.60 Si, 0.30 Mn, 1.1 Cr, 0.20 V, 2.0 W, bal Fe.
Cold work tool steel.
W.-Nr. 1.2550.

E 2601.
M-1736; 1.65 C, 0.30 Si, 0.30 Mn, 12.0 Cr, 0.10 V, 0.60 Mo, 0.50 W, bal Fe.
Cold work tool steel.
W.-Nr. 1.2601.

E 2603.
M-1736; 0.45 C, 0.75 Si, 0.45 Mn, 1.6 Cr, 0.70 Mo, 0.80 V, 0.70 W, bal Fe.
Hot work tool steel.
W.-Nr. 1.2603.

E 2713.
M-1736; 0.55 C, 0.32 Si, 0.60 Mn, 0.70 Cr, 0.35 Mo, 1.50 Ni, bal Fe.
Alloy structural steel, for shafts, gears, springs.
W.-Nr. 1.2713.

E 2714.
M-1736; 0.56 C, 0.25 Si, 0.60 Mn, 1.0 Cr, 0.45 Mo, 1.70 Ni, 0.15 V, bal Fe.
Alloy structural steel, for shafts, gears, springs.
W.-Nr. 1.2714.

E 2721.
M-1736; 0.54 C, 0.30 Si, 0.35 Mn, 0.95 Cr, 0.30 Mo, 3.25 Ni, bal Fe.
Alloy structural steel; deep hardening.
W.-Nr. 1.2721.

E 2764.
M-1736; 0.20 C, 0.30 Si, 0.40 Mn, 1.25 Cr, 0.20 Mo, 3.75 Ni, bal Fe.
Cold work tool steel, or alloy carburizing steel.
W.-Nr. 1.2764.

E 2766.
M-1736; 0.35 C, 0.25 Si, 0.50 Mn, 1.30 Cr, 0.25 Mo, 4.0 Ni, bal Fe.
Alloy structural steel; deep hardening, for heavy sections. Also hot work tool steel.
W.-Nr. 1.2766.

E 2767.
M-1736; 0.41 C, 0.30 Si, 0.40 Mn, 1.20 Cr, 0.45 Mo, 1.0 Ni, 0.45 W, bal Fe.
Hot work tool steel.
W.-Nr. 1.2767.

E 2842.
M-1736; 0.91 C, 0.25 Si, 2.05 Mn, 0.10 min V, bal Fe.
Cold work tool steel.
W.-Nr. 1.2842; similar to AISI O2.

E 7131.
M-1736; 0.16 C, 0.25 Si, 1.15 Mn, 0.95 Cr, bal Fe.
Alloy carburizing steel.
W.-Nr. 1.7131.

E 7147.
M-1736; 0.20 C, 0.25 Si, 1.25 Mn, 1.15 Cr, bal Fe.
Alloy carburizing steel.
W.-Nr. 1.7147.

E 7218.
M-1736; 0.25 C, 0.25 Si, 0.65 Mn, 1.05 Cr, 0.20 Mo, bal Fe.
Chrome-Moly structural steel.
W.-Nr. 1.7218.

E 7228.
M-1736; 0.50 C, 0.25 Si, 0.65 Mn, 1.05 Cr, 0.20 Mo, bal Fe.
Alloy structural steel.
W.-Nr. 1.7228; AISI 4150.

E 8550.
M-1736; 0.35 C, 0.30 Si, 0.55 Mn, 1.70 Cr, 0.20 Mo, 1.0 Ni, 1.05 Al, bal Fe.
Steel for nitriding.
W.-Nr. 1.8550.

E 9335.
M-1736; 0.35 C, 0.30 Si, 0.52 Mn, 1.20 Cr, 0.11 Mo, 3.25 Ni, bal Fe.
Alloy structural steel, deep hardening.

E 9840.
M-1736; 0.40 C, 0.30 Si, 0.80 Mn, 0.80 Cr, 0.25 Mo, 1.0 Ni, bal Fe.
Alloy structural steel.

EAGLE A BABBITT.
M-88; Sn, Sb, bal Pb.
At 70°F: 23.5 Brin.
At 212°F: 11.9 Brin.
For bearings, bushings; Babbitt.

EAGLE MUSIC WIRE.
M-774; 0.9 C, bal Fe.
For wire, springs; water hardened.

"E" ALLOY.
M-311; 20 Zn, 2.5-3 Cu, 0.2-1 Si, 0.5 Mg, 0.5 Mn, Fe, bal Al.
Hot rolled: 67,000 TS; 53,000 YS; 15 El.
For light weight structures; non-hardenable.

EARLUMIN.
M-Spain; 4-5 Cu, bal Al.
For light alloy parts; age-hardenable.

EASTERN 12R.
M-775; 6-8 Cu, 1-4 Si, 2.5 Zn, 1-4 Fe, 0.6 Mg, bal Al.
Cast: 19,000 TS; 14,000 YS; 2 El.
For light alloy parts; Aluminum 113, SAE33.

EASTERN 19-9 DL see 19-9DL. (OTHER GRADES FOLLOW AISI NUMBERING.)

EASTERN 29-9.
M-775; C, 29 Cr, 9 Ni, bal Fe.
Cast: 97,000 TS; 49,000 YS; 28 El; 90-95 Brin.
For valves, pumps, headers, corrosion resistant.

EASTERN 36.
M-775; 0.1 C, 2.3-3.1 Si, 0.03 Mg, 0.2 Zn, 0.6 Fe, 0.2 Mn, bal Al.
Cast: 17,000 TS; 5500 YS; 20 El.
For light alloy parts.

EASTERN 321 SW.
M-1014; 0.08 max C, 2.0 max Mn, 1.0 max Si, 17.0-19.0 Cr, 9.0-12.0 Ni Ti = 6 x C min, bal Fe.
"Slag-washed" grade of AISI 321 stainless steel for improved spinning and forming.

EASTERN 500T.
M-775; 0.4-1.0 Cu, 0.35 Si, 0.2-0.5 Mg, 7-8 Zn, 1 Fe, 0.6 Mn, bal Al.
Aged: 30,000 TS; 22,000 YS; 3 El.
For light alloy parts; Alloy ZC81A.

EASTERN 32510.
M-775; Malleable Iron Casting.
Cast: 50,000 TS; 32,500 YS; 10 El; 110-135 Brin.
Excellent machinability.
ASTM A47-52 Grade 32510.

EASTERN 35018.
M-775; Malleable Iron Casting.
Cast: 53,000 TS; 35,000 YS; 18 El; 110-135 Brin.
Good machinability.
ASTM A47-52 Grade 35018.

EASTERN G-35.
M-775; C, bal Fe.
35,000 TS; 0 El; 215 Brin.
For pressure fittings, cams, castings; gray cast iron.

EASTERN G-40.
M-775; C, alloy, bal Fe.
40,000 TS; 0 El; 240 Brin.
For die blocks, stoker parts, castings; heat resistant cast iron.

EASTERN GR. P1.
M-775; 3 C, Si, Mn, bal Fe.
Cast: 70,000 TS; 48,000 YS; 10 El; 187 Brin.
For gears, shafts, housings; pearlitic malleable iron.

EASTERN GR. P2.
M-775; 3 C, Si, Mn, bal Fe.
Cast: 75,000 TS; 50,000 YS; 8 El; 201 Brin.
For gears, shafts, housings; pearlitic malleable iron.

EASTERN GR. P3.
M-775; 3 C, Si, Mn, bal Fe.
Cast: 80,000 TS; 60,000 YS; 5 El; 217 Brin.
For gears, shafts, housings; pearlitic malleable iron.

EASTERN GR. P4.
M-775; 3 C, Si, Mn, bal Fe.
Cast: 85,000 TS; 55,000 YS; 4 El; 207 Brin.
For gears, shafts, housings; pearlitic malleable iron.

EASTERN GR. P5.
M-775; 3 C, Si, Mn, bal Fe.
Cast: 75,000 TS; 50,000 YS; 4 El; 201 Brin.
For gears, shafts, housings; pearlitic malleable iron.

EASTERN GR. P6.
M-775; 3 C, Si, Mn, bal Fe.
Cast: 65,000 TS; 45,000 YS; 10 El; 187 Brin.
For gears, shafts, housings; pearlitic malleable.

EASTERN H-55.
M-775; 2.2 C, 1.3 Si, alloy, bal Fe.
Cast: 55,000 TS; 0 El; 270 Brin.
For lapping plates, rolls, pressure fittings, castings; gray cast iron Tr S. 3200.

EASTERN H-60.
M-775; 2.2 C, 1 Si, 0.5 Mn, bal Fe.
Cast: 60,000 TS; 0 El; 270 Brin.
For manhole covers, cylinders, rolls, brake drums, die blocks; gray cast iron, wear resistant.

EASTERN H-65.
M-775; 2.2 C, 1 Si, alloy, bal Fe.
Cast: 65,000 TS; 0 El; 300 Brin.
For die blocks, stoker parts, castings; heat resistant cast iron up to 1200°F.

EASTERN N.
M-775; C, 16 Cr, bal Fe.
Annealed: 50,000 TS; 0 El; 195 Brin.
For carburizing and annealing boxes, castings, furnaces; heat resistant cast iron to 1500°F.

EASTERN N-1.
M-775; C, 28 Cr, 12 Ni, bal Fe.
Cast: 75,000 TS; 60,000 YS; 2 El; 260 Brin.
For rabble arms, pump parts, furnace parts; heat resistant cast iron.

EASTERN N-2.
M-775; C, 28 Cr, bal Fe.
Cast: 50,000 TS; 40,000 YS; 1 El; 225 Brin.
For annealing boxes, lead pots, pump and valve bodies; heat resistant cast iron to S atm.

EASTERN N-3.
M-775; C, 23 Cr, bal Fe.
Cast: 65,000 TS; 50,000 YS; 1 El; 260 Brin.
For glass industry, stainless castings, glass molds; stainless, resists scaling to 1600°F.

EASTERN N-4.
M-775; C, 20 Cr, 64 Ni, bal Fe.
Cast: 60,000 TS; 45,000 YS; 2 El; 250 Brin.
For carburizing boxes, glass molds, pyrometer tubes; heat resistant to 1800°F.

EASTERN N-5.
M-775; C, 20 Cr, 38 Ni, bal Fe.
Cast: 65,000 TS; 45,000 YS; 4 El; 235 Brin.
For furnace parts, grids, rollers, hearth plates; heat resistant to 1650°F.

EASTERN N-6.
M-775; C, 8 Cr, 22 Ni, bal Fe.
Cast: 75,000 TS; 40,000 YS; 20 El; 160 Brin.
For ship propellers, pump and valve bodies; stainless, heat resistant to 1800°F.

EASTERN N7.
M-775; 0.95-1.2 C, 1.0 max Mn, 1.0 max Si, 16-18 Cr, 0.75 max Mo, bal Fe.
High carbon martensitic stainless.
AISI 440C.

EASTERN N-8.
M-775; C, 23 Cr, bal Fe.
Cast: 65,000 TS; 55,000 YS; 0 El; 555 Brin.
For heat resistance; abrasion resistant.

EASTERN "Z" METAL.
M-775; 2.0-2.6 C, 1 Si, 1 Mn, bal Fe.
Cast: 75,000 TS; 50,000 YS; 12 El; 180 Brin.
For air drill parts, sprockets, gears, unions; malleable cast iron.

EASTON PM 216.
M-1710; 0.80 C (incl. 0.75 graphite), 0.43 Mn, 0.41 Ni, 0.73 Mo, 2.05 Cu, bal Fe.
140-300 mesh metal powder for sintering.
Produces hardenable steel.

EASTON RZ 365.
M-1710; 98.7 Fe, 0.22 Mn, 0.06 C (typical).
140-300 mesh iron powder for sintering.

EASTON RZ 365 MM.
M-1710; 0.11-0.14 C, 0.55 Mn, 98.6 Fe.
140-300 mesh powder for sintering.

EASTON RZ 365 S.
M-1710; 0.08 C, 0.17 Mn, 0.31 S, 98.5 Fe.
140-300 mesh iron powder for sintering.
For free machining material.

EASTON RZ 4600.
M-1710; 0.08 C, 1.84 Ni, 0.30 Mo, 0.21 Mn, bal Fe.
140-300 mesh metal powder for sintering.
Mixing with varying amounts of graphite before sintering will permit product similar in composition to AISI 46XX.

EASY.
M-686; 65 Ag, 20 Cu, 15 Zn.
For silversmithing solder for sterling; M.P. 1280-1325°F.

EASYARC see also **AIRCO EASYARC.**

EASY-FLO.
M-63; 50 Ag, 15.5 Cu, 16.5 Zn, 18 Cd.
For brazing; M.P. 1175-1200°F; corrosion resistant. AMS 4770C; AWS BAg-la.

EASY-FLO 1.
M-200; 50 Ag, plus Zn, Cd, Cu.
Silver solder; melt range 620-630°C.

EASY-FLO 2.
M-200; 42 Ag, plus Cu, Zn, Cd.
Silver solder; melt range 608-617°C.

EASY-FLO 3.
M-200; 50 Ag, plus Cu, Zn, Cd, Ni.
Silver solder; melt range 634-656°C.
For brazing stainless steels and carbide tips.

EASY-FLO 3.
M-63; 50 Ag, 15.5 Cu, 15.5 Zn, 16 Cd, 3 Ni.
M.P. 1170°F; Flow P. 1270°F.
For brazing stainless steels with corrosion resisting joints; brazing carbide tips.
AMS 4771 A; AWS BAg-3.

EASY-FLO 30.
M-63; 30 Ag, 27 Cu, 23 Zn, 20 Cd.
M.P. 1125°F; Flow P. 1310°F.
60,000 TS expected when joining 1020 steel. General purpose brazing alloys, particularly for leak-tight joints.
AWS BAg-2a.

EASY-FLO 35.
M-63; 35 Ag, 26 Cu, 21 Zn, 16 Cd.
M.P. 1125°F; Flow P. 1295°F.
General purpose low melting brazing alloy for brazing ferrous and non-ferrous alloys.
AMS 4768A; AWS BAg-2.

EASY-FLO 45.
M-63; 45 Ag, 15 Cu, 16 Zn, 24 Cd.
M.P. 1125°F; Flow P. 1145°F.
For brazing ferrous and non-ferrous alloys; lowest flow point of silver braze alloys.
AMS 4769 A; AWS BAg-1.

EATONITE.
M-947; 2.0-2.75 C, 37-41 Ni, 14 W, 27-31 Cr, 9-11 Co, 1 max Si, 8 max Fe.
At 1300°F; 302 Brin.
For valve facing, valve seats and faces; retains hardness at high temperature.

E.B. ALLOY.
M-435; 0.50 C, 4 Cr, 0.4 V, 0.5 Mo, bal Fe.
Heat treated: 217,000-308,000 TS; 176,000-229,000 YS; 6-13 El; 6-34 RA; 44-56 Rock C.
For shears, header dies, hot work steel, air or oil hardening.

"E.B.D." BEARING.
M-Eng.; 90-88 Cu, 10 Sn, 5 P + Sn, 2 Zn.
For bearings; strong.

E-BRITE 26-1.
M-1772; 0.005 max C, 0.40 max Si, 0.40 max Mn, 25-27.5 Cr, 0.75-1.25 Mo, 0.0150 max N, 0.5 max Cu + Ni, bal Fe.
Ann: 70,000 TS; 45,000 YS; 42 El; 86 RA.
(Supplied as ingots for remelting.)
Good resistance to impact, intergranular corrosion, pitting and crevice corrosion, and immune to chloride stress cracking.
For parts for marine and desalting equipment, food processing, petroleum industries.

EC 60.
M-912; 0.16 C, 0.17 Si, 0.50 Mn, 0.65 Cr, bal Fe.
Carburizing steel for roller bearing, camshafts, piston pins, cog wheels.
W.-Nr. 7015.

EC 80.
M-912; 0.16 C, 0.25 Si, 1.15 Mn, 0.95 Cr, 0.20-0.35 S, bal Fe.
Free machining carburizing steel for camshafts, piston pins, cog wheels.
W.-Nr. 7131.

EC 80 B.
M-912; 0.16 C, 0.25 Si, 1.15 Mn, 0.95 Cr, B, bal Fe.
Carburizing steel for piston pins, camshafts, cog wheels.

EC 100.
M-912; 0.20 C, 0.25 Si, 1.25 Mn, 1.45 Cr, 0.20-0.35 S, bal Fe.
Free-machining carburizing steel; for spindles, camshafts, shafts, piston pins, gears.
W.-Nr. 7147.

EC 100 B.
M-912; 0.18 C, 0.25 Si, 1.15 Mn, 1.15 Cr, B, bal Fe.
Carburizing steel; for steering shafts, camshafts, gears.

E.C. 124.
M-Eng.; 12 Si, 1 Cu, 0.5 Mg, 1 N, bal Al.
For light alloy parts; cast or pressed.

ECCENTRIC RING.
M-Eng.; 84 Cu, 14 Sn, 2 Zn.
For bearings, bells.

ECLIPSE 75.
C, alloy, bal Fe.
For cold hobbing tools; water hardened.

E.C.N. ALLOY.
M-107; C, Ni, Cr, bal Fe.
At 20°C: 67,000-78,000 TS.
At 1000°C: 33,000 TS.
For case-hardening boxes, glass molds, retorts, stills, fire boxes, muffles, crucibles, pumps; highly resistant to heat and corrosion.

E CO 3.
M-Germany; 0.8 C, 12 W, 4 Cr, 1.9 V, 0.85 Mo, 3 Co, bal Fe.
Hardened: C 64-67 Rock.
For lathe and planer tools, reamers, broaches, drills, form cutters, taps, hobs.
High red-hardness.
High speed steel, Type T8.

ECONO-2.
M-1683; 0.75 C, 1.25 Ni, 0.80 Cr, 0.25 Mo, 0.15 V, bal Fe.
Heat treated: 121,000-208,000 TS; 90,000-180,000 PL, C 29-45 Rock.
For brake dies, bushings, cams, shear blades, gears, set screws, swaging and forming dies, punches.
Type L6, tough, oil hardening.

ECONO-5.
M-1683; 1.0 C, 5 Cr, 1 Mo, 0.4 V, 0.4 Mn, bal Fe.
Annealed: 103,000 TS; 52,000 YS; 26 El; C 18 Rock.
Hardened: 178,000-253,000 TS; 146,000-200,000 YS; 3-12 El; 7-32 RA; C 41-53 Rock.
For blanking and trimming dies, shears, cutters, gauges, lamination dies.
Type A2, air hardening, nondeforming, cold work tool steel, tough.

ECONO-KROME STAINLESS.
M-15; 0.08 C, 18.0-20.0 Cr, 11.0-12.0 Ni, 2.0 max Mn, 0.80 max Mo, 0.20 max Zn, bal Fe.
Austenitic stainless steel; good resistance to intergranular corrosion.

ECONOMET.
M-1005; 0.35-0.75 C, 8-12 Cr, 28-32 Ni, bal Fe.
For heat treating boxes, furnace parts. Heat and abrasion resistant.

ECONOMET 4.
M-1005; 0.36-0.50 C, 16-23 Cr, 7-11 Ni, bal Fe.
For heat and corrosion resistant parts; stainless, heat and corrosion resistant.

ECONOMY HARDFACE.
M-9; 1.0 C, 5 Cr, 1.7 Mo, bal Fe.
Welded: 450-550 Brin.
For hard facing electrode; abrasion resistant, air hardened.

ECONOMY HARDFACE C.
M-9; 0.5 C, 3 Cr, 1.7 Mo, bal Fe.
Welded: 500 Brin.
For hard facing electrode; wear and abrasion resistant.

ECONO NO. 1-V.
M-1683; 1.05 C, 0.25 V, bal Fe.
Water hardening tool steel; AISI W2.

E C S.
M-604; 0.06 C, 0.75 Mn, 0.75 Si, 11.5 Cr, Ti = 6 x C min - 0.75 max, bal Fe.
Ann: 60 ksi TS; 30 ksi YS; 40 El.
Stainless sheet steel for automotive emission controls.

ED 2.
M-1736; 1.50 C, 0.35 Si, 0.50 Mn, 12.0 Cr, 0.80 Mo, 0.20 V, 0.15 max W, bal Fe.
Cold work tool steel.
W.-Nr. 1.2379; similar to AISI D2.

ED 3.
M-1736; 2.0 C, 0.35 Si, 0.47 Mn, 12.5 Cr, 0.20 V, bal Fe.
Cold work tool steel.
Similar to AISI D3.

ED 6.
M-1736; 2.0 C, 0.57 Si, 0.47 Mn, 12.5 Cr, 0.20 V, 1.0 W, bal Fe.
Cold work tool steel.
W.-Nr. 1.2436.

EDAL.
M-Ger.; 5 Mg, bal Al.
Cast: 40,000-64,000 TS; 21,000-43,000 YS; 5-31 El; 1 RA.
For welding rods, light alloy parts, heat treatable.

EDCO PHOSPHOR BRONZE.
M-1035; 91.5 Cu, 8.25 Sn, 0.25 P.
Cast: 30,000 TS; 7-10 El; 163 Brin.
For welding rod; M.P. 1010°C.

EDELBRONZE.
M-557; Sn, bal Cu.
For chemical plant and textile industry equipment; corrosion resistant.

EDELMESSING.
M-German; Zn, alloy, bal Cu.
For ornamental hardware; high tensile brass.

EDGAR ALLEN 66.
M-210; 2.25 C, 13 Cr, bal Fe.
For dies; non-deforming.

EDGAR ALLEN A.13.
M-210; 0.40 C, 1.15 Cr, 1.5 Ni, 0.3 Mo, 0.6 Mn, bal Fe.
Heat treated: 224,000-251,000 TS; 215,000-224,000 YS; 10-12 El; 477-512 Brin.
For arbors, axles, beaters, bolts, connecting rods; shock resistant, oil hardened.

EDGAR ALLEN A.100.
M-210; 0.32 C, 0.65 Cr, 2.5 Ni, 0.55 Mo, bal Fe.
Heat treated: 160,000-275,000 TS; 138,000-230,000 YS; 11-18 El; 32-58 RA; 300-450 Brin.
For arbors, boring bars, collets, dies, gears; tough, oil hardened.

EDGAR ALLEN AM.1.
M-210; 0.35 C, 1.35 W, 5 Cr, 1.5 Mo, 0.45 V, bal Fe.
For Al and brass extrusion dies, heading tools; hot work steel, resists heat checking.

EDGAR ALLEN AM.3.
M-210; 0.40 C, 5 Cr, 1.35 Mo, 1.1 V, bal Fe.
For Al die casting dies, extrusion and forming tools; hot work steel, resists heat checking.

EDGAR ALLEN DOUBLE 12.
M-210; 0.3 C, 12 Cr, 12 W, bal Fe.
Ht. Tr.: C 38-42 Rock.
For liquid hobbing of beryllium copper cavity molds for plastics, brass and copper extrusion dies. Air or oil hardening.
Hot work steel, Type H23.

EDGAR ALLEN FIVE PLY STEEL.
M-210; C, bal Fe.
For engineering construction; manufactured in plates containing 3 layers of soft steel and 2 layers of hard steel.

EDGAR ALLEN GENUINE DOUBLE SHEAR STEEL.
M-210; C, bal Fe.
For carving knives; water hardening.

EDGAR ALLEN IMPERIAL.
M-210; 1.4 C, 0.75 Cr, 4.5 W, bal Fe.
For cutting tools, drills, reamers, textile needles. Water hardening. Wear resistant.

EDGAR ALLEN K-9.
M-210; 0.95 C, 0.75 Cr, 0.85 Mn, 0.5 W, bal Fe.
For dies, stay taps, broaches, milling cutters, plug gauges; oil hardened, tough.

EDGAR ALLEN N.I.F.E. 30.
M-210; 70 Fe, 30 Ni.
For compensating shunts for electrical equipment; thermo-sensitive, magnetic.

EDGAR ALLEN NO. 5 HOT DIE.
M-210; 0.32 C, 3.25 Cr, 9.5 W, 0.35 V, bal Fe.
For hot shear blades, blades, press and form dies, piercers; hot work steel, oil hardened.

EDGAR ALLEN SPECIAL CHROME.
M-210; C, Cr, bal Fe.
For thread rolling dies; retains hardness at high temperatures.

EDGAR ALLEN STAGPICK.
M-210; C, Cr, V, bal Fe.
For coal cutter picks; tough, hard and shock resisting.

EDGAR ALLEN TOOL CLASS "C".
M-210; 0.6-1.35 C, bal Fe.
For spiral springs, chisels, punches, shear blades, drills, reamers, taps, dies; water hardening.

EDGAR ALLEN TOOL CLASS "V".
M-210; 0.6-1.35 C, bal Fe.
For spiral springs, chisels, punches, shear-blades, drills, reamers, taps; water hardening.

EDGAR ALLEN V.S. 4.
M-210; 1.05 C, 0.30 Mn, 0.25 V, bal Fe.
For cold heading dies; Type W2; water hardened.

EDGAR ALLEN YWA.
M-210; 0.28 C, 0.45 Si, 1.3 Cr, 3.4 Ni, 5.8 W, 0.25 V, bal Fe.
For extension dies, mandrels; oil hardened, shock resistant.

EDM ALLOY "A".
M-57; Cu, bal W.
For electric discharge machining electrodes. Copper infiltrated tungsten.

EDM ALLOY "S".
M-57; 85.91 W, 9.15 Ag.
Sintered: 100,000 TS; B 100 Rock.
For EDM electrodes for electric discharge machining.
Silver infiltrated tungsten.

EDM ALLOY "SA".
M-57; 75 W, 25 Ag.
Sintered: 125,000 TR.S.; B 100 Rock.
For electric discharge machining electrodes. Silver infiltrated tungsten.

EDWARDS SPECULUM.
M-German; 70-63 Cu, 25-32 Sn, 2.4-1.6 As, 2.6-0 Zn.
For reflectors, mirrors; corrosion resistant.

EE see **DARWIN EE.**

EEL BRAND ANTIFRICTION METAL.
M-526; 75 Pb, 6 Sn, 0.9 Cu, 1.5 Cd, 1 Ni, 0.1 P, 0.5 As, Sb.
For bearings, antifriction metal; Babbitt.

EGALITE.
M-161; Al alloy.
For light alloy parts; subject to special process while in molten state.

EH 0000.
M-1655; 1.20 C, 4.5 Cr, 1.0 Mo, 12.0 W, 4.0 V, bal Fe.
High speed steel, usually for finishing tools.
W.-Nr. 1.3302.

EH 0018.
M-1655; 0.75 C, 4.5 Cr, 18.0 W, 1.1 V, bal Fe.
High speed steel; for lathe tools, thread cutters, broaches, hobs.
W.-Nr. 1.3355.

EH-4.
M-Poland; 1.0 C, 1.0 Si, 4.0 Cr, bal Fe.
For magnets.

EH 10.
M-1736; 0.40 C, 1.0 Si, 0.55 Mn, 3.25 Cr, 2.5 Mo, 0.32 V, bal Fe.
Hot work tool steel.
AISI H10; W.-Nr. 1.2365.

EH 11.
M-1736; 0.40 C, 0.85 Si, 0.35 Mn, 5.25 Cr, 1.5 Mo, 0.35 V, bal Fe.
Hot work tool steel.
AISI H11; W.-Nr. 1.2343.

EH 12.
M-1736; 0.30 C, 0.85 Si, 0.35 Mn, 5.25 Cr, 1.5 Mo, 0.25 V, 1.2 W, bal Fe.
Hot work tool steel.
W.-Nr. 1.2606; similar to AISI H12.

EH 13.
M-1736; 0.40 C, 0.95 Si, 0.40 Mn, 5.25 Cr, 1.4 Mo, 0.95 V, bal Fe.
Hot work tool steel.
AISI H13; W.-Nr. 1.2344.

EH 20.
M-1736; 0.30 C, 0.30 Si, 0.40 Mn, 2.75 Cr, 9.1 W, 0.30 V, bal Fe.
Tungsten hot work tool steel.
W.-Nr. 1.2581.

EHRHARDS B.M.
M-England; 84.4 Zn, 0.2 Sn, 10.9 Cu, 1.2 Pb, 2.5 Al.
For bearings; will not resist heat or live steam.

EHRHARDS TYPE.
M-Eng.; 89 Zn, 3 Cu, 6 Sn, 2 Pb.
For type metal.

EHRHARDT'S METAL.
M-German; 89 Zn, 4 Cu, 4 Sn, 3 Pb.
For bearings, bushings; Babbitt.

E.H.W. NO. 1.
M-73; 0.25 C, 15 W, 4 Cr, bal Fe.
For hot work tools and dies; hot work steel.

EI-435.
M-USSR; 0.12 max C, 21 Cr, 0.25 Ti, 0.15 max Al, bal Ni.
Wrought, high temperature, oxidation resistant nickel-base alloy for tailpipes and parts of combustion equipment.

EI 435.
M-USSR; 0.10 C, 20 Cr, 0.2 Ti, bal Ni.
Heat treated: 103,000 TS; 60,000 YS; 50 El; 50 RA.
For gas turbine and jet engine components, combustion chambers.
Heat and oxidation resistant.

EI 437.
M-USSR; C, alloy, bal Fe.
For machine tool parts; oil hardened.

EI-437.
M-USSR; 0.03-0.05 C, 20-21 Cr, 2.5-2.8 Ti, 0.8-1.0 Al, bal Ni.
Heat treated: 147,000 TS; 41 El; 35 RA; C 35 Rock.
For turbine buckets and discs, rotor blades.
High creep strength.
High heat resistant. Similar to Nimonic 80A.

EI-437A.
M-USSR; 0.08 C, 21 Cr, 2.5 Ti, 0.75 Al, bal Ni.
For aircraft rotating turbine blades.

EI-437B.
M-USSR; 0.08 C, 20 Cr, 0.7 Al, 2.6 Ti, 0.005 B, bal Ni.
Heat treated: 132,000 TS; 80,000 YS; 45 El; 36 RA.
At 800°C: 62,000 TS; 53,000 YS; 8 El; 10 RA.
For turbine buckets and discs. High creep strength.
Boron modified Nimonic 80A.
High heat and corrosion resistant.

EI-437R.
M-USSR; 0.08 max C, 18.5 Cr, 2.5 Ti, 0.7 Al, 4.5 W, 4.5 Mo, 0.01 ma B, 0.01 max Ce, 4 max Fe, bal Ni.
For turbine blades.

EI-559A.
M-USSR; 0.10 max C, 16.5 Cr, 3.2 Al, 55-60 Ni, bal Fe.
Wrought, high temperature, oxidation resistant nickel-base alloy for combustion equipment.

EI-598.
M-USSR; 0.12 max C, 17.5 Cr, 2.4 Ti, 1.4 Al, 2.8 W, 5 Mo, 0.01 max B, 0.02 max Ce, 0.9 Cb, bal Ni.
For aircraft turbine blades.

EI-602.
M-USSR; 0.08 max C, 20.5 Cr, 0.55 Ti, 0.55 Al, 2.0 Mo, 1.5 Cb, 3 max Fe, 0.2 max Cu, bal Ni.
Wrought, high temperature, oxidation resistant nickel-base alloy for combustion equipment.

EI-607.
M-USSR; 0.08 max C, 16 Cr, 2.0 Ti, 0.7 Al, 1.3 Cb, 3 max Fe, bal Ni.
For turbine blades.

EI-607A.
M-USSR; 0.08 max C, 16 Cr, 1.6 Ti, 0.7 Al, 1.3 Cb, 3 max Fe, bal Ni.
For turbine blades.

EI-612K.
M-USSR; 0.09 C, 15.0 Cr, 36.5 Ni, 4.1 Co, 3.2 W, 1.5 Ti, 0.12 B
For gas turbine discs, jet engine components.

EI-617.
 M-USSR; 0.12 max C, 15 Cr, 2 Ti, 2 Al, 6 W, 3 Mo, 0.005 B, 0.3 V, bal Ni.
 For turbine blades.

EI-652.
 M-USSR; 0.05 C, 27 Cr, 3 Al, 2 max Fe, bal Ni.
 Wrought, high temperature, oxidation resistant nickel-base alloy for combustion equipment.

EI-747.
 M-USSR; 0.15 C, 11.2 Cr, 0.6 Mo, 0.6 Ni, 0.3 V, bal Fe.
 Corrosion resistant structural steel.

EI-755.
 M-USSR; 0.13 C, 10.85 Cr, 0.75 Mo, 0.3 Ni, 0.09 V, 0.2 W, bal Fe.
 Corrosion resistant structural steel.

EI-756.
 M-USSR; 0.10-0.15 C, 10.5-12.0 Cr, 0.7 Mo, 0.8 max Ni, 0.25 V, 3.7-4.3 W, bal Fe.
 Corrosion resistant structural steel.

EI-765.
 M-USSR; 0.12 max C, 14.5 Cr, 1.2 Ti, 2.0 Al, 5 W, 4 Mo, 0.008 B 3 max Fe, bal Ni.
 For turbine blades.

EI-802.
 M-USSR; 0.11-0.18 C, 11-13 Cr, 0.5 Mo, 0.5-1.0 Ni, 0.15-0.30 V, 0.7-1.0 W, bal Fe.
 Annealed: 85,000 TS; 40,000 YS; 22 El; B 95 Rock.
 For cutlery, valves, bearings, surgical instruments.
 Corrosion resistant, hardenable.

EI-826.
 M-USSR; 0.12 max C, 14.5 Cr, 2.0 Ti, 2.6 Al, 6.0 W, 3.2 Mo, 0.01 max B, 0.02 max Ce, 5 max Fe, 0.03 V, bal Ni.
 For turbine blades.

EI-827.
 M-USSR; 0.47 max C, 10 Cr, 4.3 Al, 5 W, 7.5 Mo, 4 max Fe, bal Ni.
 High temperature alloy.

EI-867.
 M-USSR; 0.10 max C, 9 Cr, 4.5 Al, 5 W, 10.3 Mo, 0.02 max B, 0.02 max Ce, 4 max Fe, 5 Co, bal Ni.
 Wrought nickel-base superalloy; for turbine blades.

EI-868.
 M-USSR; 0.10 max C, 25.5 Cr, 0.5 Ti, 0.5 max Al, 15.0 W, 4 max Fe, bal Ni.
 Wrought, high temperature, oxidation resistant nickel-base alloy for combustion equipment.

EI-869.
 M-USSR; 0.08 max C, 15.5 Cr, 1.7 Ti, 1.2 Al, 0.005 max B, 1.3 Cb, 3 max Fe, 0.005 max Zr, bal Ni.
 Wrought nickel-base superalloy; for turbine blades.

EI-893.
 M-USSR; 0.08 max C, 16.0 Cr, 1.4 Ti, 1.4 Al, 9.0 W, 4.2 Mo, 0.01 max B, 0.025 max Ce, bal Ni.
 Wrought nickel-base superalloy; for turbine blades.

EI-894.
 M-USSR; 0.09 C, 22.5 Cr, 1.1 Ti, 3.1 Al, 5.7 W, 9.7 Fe, bal Ni.
 Wrought nickel-base superalloy.

EI-929.
 M-USSR; 0.12 max C, 10.5 Cr, 1.7 Ti, 4.0 Al, 5.5 W, 5 Mo, 0.02 max B, 5 max Fe, 14 Co, 0.6 V, 0.1 max Ba, bal Ni.
 Wrought nickel-base superalloy; for turbine blades.

EI-961.
 M-USSR; 0.10-0.16 C, 10.5-12.5 Cr, 0.35-0.50 Mo, 1.5-1.8 Ni, 0.18-0.30 V, 1.5-2.0 W, bal Fe.
 Corrosion resistant structural steel.

EI-993.
 M-USSR; 0.20 C, 12 Cr, 0.4 Cb, 0.04 Mo, 0.1 Ni, 0.3 V, 0.6 W, bal Fe.
 Annealed: 95,000 TS; 42,000 YS; 24 El; B 95 Rock.
 Hardened: 240,000 TS; 195,000 YS; 10 El; C 48 Rock.
 For bearings, cutlery, hardware, surgical instruments, valves.
 Corrosion resistant, hardenable.

EIFFEL 32/32T.
 M-884; 0.15 C, 1.10 Mn, 0.15 Si, bal Fe.
 44-54 kg/mm^2 TS; 32 min kg/mm^2 YS; 25 El.
 For beams, frames, chains.

EIFFEL 36/36T.
 M-884; 0.18 C, 1.25 Mn, 0.20 Si, bal Fe.
 52-62 kg/mm^2 TS; 36 min kg/mm^2 TS; 24 El.
 For beams, frames, chains.

EIFFEL 42/42T.
 M-884; 0.16 C, 1.5 Mn, 0.15 Si, 0.10 V, bal Fe.
 50-65 kg/mm^2 TS; 42 min kg/mm^2 YS; 20 El.
 For beams, frames, chains.

EIFFEL 48/48T.
 M-884; 0.17 C, 1.55 Mn, 0.20 Si, 0.12 V, 0.03 Nb, bal Fe.
 60-80 kg/mm^2 TS; 48 min kg/mm^2 YS; 18 El.
 For beams, frames, chains.

EIFFEL 55T.
M-884; 0.17 C, 1.55 Mn, 0.20 Si, 0.30 Mo, 0.10 V, bal Fe.
60-80 kg/mm² TS; 55 min kg/mm² YS; 16 El.
For beams, frames, chains.

EIFFEL 60T.
M-884; 0.18 C, 1.60 Mn, 0.20 Si, 0.40 Mo, 0.10 V, bal Fe.
Plates: 63-85 kg/mm² TS; 58 min kg/mm² YS; 15 El.
For beams, frames, crane booms, girders.

EIGHTEEN PER CENT.
M-Eng.; 65 Cu, 17 Zn, 18 Ni.
For base for silver plated flatware; nickel silver.

EINHEITSMETALL.
M-German; 79 Pb, 14 Sb, 5.3 Sn, 1.5 Cu.
For bearings, bushings; Babbitt.

E.I.S. 49.
M-64; 0.85 C, 5 Cr, 0.25 Mn, bal Fe.
Hardened: 560 Brin.
For trimmers, dies, shears; for heavy shearing.

EIS 77.
M-64; 0.35 C, 4 Cr, 12 W, 0.25 V, bal Fe.
For hot work dies, shears, punches; oil or air hardened.

EIS 96.
M-64; 0.55 C, 0.8 Mn, 1 Cr, 0.45 Mo, 0.08 V, bal Fe.
For plastic molding dies; oil hardened.

EIS 718.
M-64; 0.35 C, 4.95 Cr, 1.25 W, 1.5 Mo, 1.0 Si, bal Fe.
For hot work dies, shears, punches; oil or air hardened.

E.I.S. B-76.
M-64; 1.0 C, 0.6 Cr, 1.2 Mn, 0.20 V, 0.5 W, bal Fe.
For dies, cold blanking and forming dies; non-deforming, oil hardened.

E.I.S. C-57.
M-64; 0.6 C, 1 Cr, 2 Ni, 0.12 V, 0.8 Mo, bal Fe.
For shear blades, knives, punches, upsetters; hot work steel, oil hardened.

EISENBRONZE-1.
M-German; 57.5 Cu, 39.5 Zn, 1 Sn, 0.3 Pb, 1.3 Fe, 0.3 Al.
For marine parts, hardware; iron bronze.

EISENBRONZE-2.
M-German; 82.5 Cu, 4.45 Zn, 8.55 Sn, 3.95 Fe.
For marine parts, hardware; iron bronze.

EISENNICKEL 75.
M-297; 78.5 Ni, bal Fe.
For magnetic and electrical equipment, motors; soft magnet, high permeability.

EIS-H5.
M-64; 0.55 C, 1 Cr, 0.45 Mo, 0.08 V, 0.8 Mn, bal Fe.
For plastic mold dies; oil hardened.

E.I.S. H-41.
M-64; 1.0 C, 4 Cr, 0.15 V, 0.25 Mo, bal Fe.
For hot work dies, trimmers; oil or air hardening.

EIS-H720.
M-64; 0.40 C, 5.25 Cr, 1.15 Mo, 1 V, bal Fe.
For die casting dies and inserts; Type H11; air hardened.

EIS-H-721.
M-64; 0.35 C, 0.35 Mn, 5 Cr, 1.55 Mo, 1.25 W, 1 Si, 0.2 V, bal Fe.
Air hardened: 550 Brin.
For dies, inserts, punches, liners, rams; hot work, press, upsetter, extrusions.

EISLERS BRONZE.
M-U.S.; 6 Zn, bal Cu.
For electrical contacts, springs; good conductivity.

E.I.S. R-43.
M-64; 1.5 C, 12 Cr, bal Fe.
For shear blades, trimmers; air hardened, light shearing.

E.I.S. R-97.
M-64; 0.5 C, 0.4 Cr, 1.0 Mn, 0.4 Mo, 2.1 Si, bal Fe.
For dies, inserts, shear knives; oil hardened, heavy shearing.

EIS-R-718.
M-64; 0.35 C, 5 Cr, 2 Mo, 1 Si, bal Fe.
Air hardened: 550 Brin.
For dies, inserts, punches; hot work, press, upsetter, extrusions.

EIS-R-720.
M-64; 0.4 C, 0.4 Mn, 5.25 Cr, 1.15 Mo, 1 V, bal Fe.
Air hardened: 560 Brin.
For dies and inserts; die casting.

E.I.S. T-51.
M-64; 0.6 C, 0.7 Mn, 1.4 Ni, 0.7 Cr, 2.2 W, 0.2 V, bal Fe.
For trimming and forming dies; oil hardened, hot or cold work steel.

EIS-T-71.
M-64; 0.5 C, 0.3 Mn, 1.2 Cr, 2.4 Ni, 1 Si, bal Fe.
Oil hardened: 560 Brin.
For shear knives, punches; cold heavy duty, cold and semi-hot.

E.I.S. T-73.
M-64; 0.30 C, 3.5 Cr, 0.35 V, 10 W, bal Fe.
For inserts, dies, punches; hot work brass forgings.

EIS-T77.
M-64; 0.33 C, 4 Co, 12 W, 0.25 V, 0.3 Mn, bal Fe.
For hot shearing dies, extrusion liners; oil hardened, hot work steel.

E.I.S. T-79.
M-64; 0.4 C, 5 Cr, 0.2 V, 4.0 W, 0.6 Mo, bal Fe.
For hot work dies, inserts; heavy shearing.

E.I.S. T-717.
M-64; 0.35 C, 5 Cr, 1.35 Mo, 1.2 W, 1 Si, bal Fe.
For shears, knives; oil hardened.

EIS-T721.
M-64; 0.3 C, 5 Cr, 1.5 Mo, 1.25 W, 0.2 V, bal Fe.
For die casting dies; oil hardened, hot work steel.

EIS-V2.
M-64; 0.8 C, 0.4 Mn, V, Si, bal Fe.
Water hardened: 600 Brin.
For shear knives; cold heavy duty.

EIS-V3.
M-64; 0.95 C, 0.4 Mn, Si, V, bal Fe.
Water hardened: 620 Brin.
For shear knives, trimmers; cold light shearing.

EK5.
M-USSR; 0.9-1.0 C, 5.5-6.5 Cr, 5.5-6.5 Co, bal Fe.
For permanent magnet; heat treated.

EK15.
M-USSR; 0.9-1.0 C, 7.5-8.5 Cr, 14-16 Co, 1.5 Mo, bal Fe.
For permanent magnet; heat treated.

EK30A.
M-Various foundries; 3.0 RE (Ce, etc.), 0.3 Zr, bal Mg.
T6-Temper: 20,000-23,000 TS; 14,000-16,000 YS; 2-3 El.
At 600°F: 12,000 TS; 8,000 YS; 70 El.
Magnesium sand casting; good pressure tightness, weldability and corrosion resistance.
For parts operating at 350-500°F. ASTM B80-57T.

EK41A.
M-Various foundries; 4.0 RE (Ce, etc.), 0.6 Zr, bal Mg.
T6-Temper: 22,000-25,000 TS; 16,000-18,000 YS; 1-3 El.
At 600°F: 13,000 TS; 9,000 YS; 53 El.
Magnesium sand casting with good pressure tightness, weldability and corrosion resistance.
For parts operating at 350-600°F.
ASTM B80-57T, B199-57T; ASM 4440, 4441.

EK-81.
M-114; 1.35 C, 12.75 Cr, 0.8 Mo, 3 Co, bal Fe.
For blanking and forming dies, punches; air hardening.

EKATIT 13.
M-1760; 0.08 max C, 1.0 max Si, 1.0 max Mn, 12.0-14.0 Cr, bal Fe.
Ferritic type stainless steel.
W.-Nr. 1.4000.

EKATIT 13 A1.
M-1760; 0.08 max C, 1.0 max Si, 1.0 max Mn, 12.0-14.0 Cr, 0.10-0.30 Al, bal Fe.
Ferritic type stainless steel; non-hardenable.
W.-Nr. 1.4002; similar to AISI 405.

EKATIT 14.
M-1760; 0.08 max C, 1.0 max Si, 1.0 max Mn, 13.0-15.0 Cr, bal Fe.
Ferritic type stainless steel.
W.-Nr. 1.4001.

EKATIT 17.
M-1760; 0.10 max C, 1.0 max Si, 1.0 max Mn, 15.5-17.5 Cr, bal Fe.
Ferritic type stainless steel.
W.-Nr. 1.4016; AISI 430.

EKATIT 17-7.
M-1760; 0.12 max C, 1.0 max Si, 2.0 max Mn, 16.0-18.0 Cr, 7.0-9.0 Ni, bal Fe.
Austenitic stainless steel.
W.-Nr. 1.4310; similar to AISI 301.

EKATIT 17-13 S.
M-1760; 0.07 max C, 1.0 max Si, 2.0 max Mn, 16.0-18.0 Cr, 12.5-14.5 Ni, 4.0-5.0 Mo, bal Fe.
Austenitic stainless steel.
W.-Nr. 1.4449.

EKATIT 17 E.
M-1760; 0.10 max C, 1.0 max Si, 1.0 max Mn, 16.0-18.0 Cr, Ti = 7 x C min, bal Fe.
Ferritic stainless steel.
W.-Nr. 1.4510; Ti stabilized.

EKATIT 17 EMO.
M-1760; 0.10 max C, 1.0 max Si, 1.0 max Mn, 16.5-18.5 Cr, 1.5-2.0 Mo, Ti = 7 x C min, bal Fe.
Ferritic type stainless steel.
W.-Nr. 1.4523; Ti stabilized.

EKATIT 17 N.
M-1760; 0.10 max C, 1.0 max Si, 1.0 max Mn, 16.0-18.0 Cr, Nb = 12 x C, bal Fe.
Ferritic type stainless.
W.-Nr. 1.4511; Nb stabilized.

EKATIT 17 VCO.
M-1760; 0.85-0.95 C, 1.0 max Si, 1.0 max Mn, 15.50-17.50 Cr, 0.40-0.60 Mo, 1.2-1.8 Co, 0.20-0.30 V, bal Fe.
Martensitic stainless steel.
W.-Nr. 1.4535.

EKATIT 18-8 A.
M-1760; 0.15 max C, 1.0 max Si, 2.0 max Mn, 17.0-19.0 Cr, 8.0-10.0 Ni bal Fe.
Austenitic stainless steel.
W.-Nr. 1.4305; AISI 302.

EKATIT 18-8 E.
M-1760; 0.10 max C, 1.0 max Si, 2.0 max Mn, 17.0-18.0 Cr, 9.0-11.5 Ni Ti = 5 x C min, bal Fe.
Titanium stabilized austenitic stainless steel.
W.-Nr. 1.4541; similar to AISI 321.

EKATIT 18-8 N.
M-1760; 0.12 max C, 1.0 max Si, 2.0 max Mn, 17.0-19.0 Cr, 8.0-10.0 Ni bal Fe.
Austenitic stainless steel.
W.-Nr. 1.4300; similar to AISI 302.

EKATIT 18-8 N.
M-1760; 0.07 max C, 1.0 max Si, 2.0 max Mn, 17.0-20.0 Cr, 9.0-11.5 Ni 0.20 max Mo, Nb = 10 x C, bal Fe.
Niobium stabilized austenitic stainless steel.
W.-Nr. 1.4543; similar to AISI 347.

EKATIT 18-8 NB.
M-1760; 0.10 max C, 1.0 max Si, 2.0 max Mn, 17.0-19.0 Cr, 9.0-11.5 Ni Nb = 8 x C min, bal Fe.
Niobium stabilized austenitic stainless steel.
W.-Nr. 1.4550; similar to AISI 347.

EKATIT 18-8 S.
M-1760; 0.07 max C, 1.0 max Si, 2.0 max Mn, 17.0-20.0 Cr, 8.5-10.0 Ni bal Fe.
Austenitic stainless steel.
W.-Nr. 1.4301; similar to AISI 304.

EKATIT 18-8 SW.
M-1760; 0.03 max C, 1.0 max Si, 2.0 max Mn, 17.0-20.0 Cr, 10.0-12.5 Ni, bal Fe.
Austenitic stainless steel.
W.-Nr. 1.4306; similar to AISI 304L.

EKATIT 18-10 E.
M-1760; 0.10 max C, 1.0 max Si, 2.0 max Mn, 16.5-18.5 Cr, 10.5-13.5 Ni, 2.0-2.5 Mo, Ti = 5 x C min, bal Fe.
Stabilized austenitic stainless steel.
W.-Nr. 1.4571.

EKATIT 18-10 NB.
M-1760; 0.10 max C, 1.0 max Si, 2.0 max Mn, 16.5-18.5 Cr, 10.5-13.5 Ni, 2.0-2.5 Mo, Nb = 8 x C min, bal Fe.
Stabilized stainless steel.
W.-Nr. 1.4580; similar to AISI 347.

EKATIT 18-10 S.
M-1760; 0.07 max C, 1.0 max Si, 2.0 max Mn, 16.5-18.5 Cr, 10.5-13.5 Ni, 2.0-2.5 Mo, bal Fe.
Austenitic stainless steel.
W.-Nr. 1.4401; similar to AISI 316.

EKATIT 18-10 SW.
M-1760; 0.03 max C, 1.0 max Si, 2.0 max Mn, 16.5-18.5 Cr, 11.0-14.0 Ni, 2.0-2.5 Mo, bal Fe.
Austenitic stainless steel.
W.-Nr. 1.4404; AISI 316L.

EKATIT 18-12.
M-1760; 0.03 max C, 1.0 max Si, 2.0 max Mn, 16.5-18.5 Cr, 12.5-15.0 Ni, 2.5-3.0 Mo, bal Fe.
Austenitic stainless steel.
W.-Nr. 1.4435.

EKATIT 18-12 E.
M-1760; 0.10 max C, 1.0 max Si, 2.0 max Mn, 16.5-18.5 Cr, 12.0-14.5 Ni, 2.5-3.0 Mo, Ti = 5 x C min, bal Fe.
Stabilized austenitic stainless steel.
W.-Nr. 1.4573.

EKATIT 18-12 S.
M-1760; 0.07 max C, 1.0 max Si, 2.0 max Mn, 16.5-18.5 Cr, 11.5-14.0 Ni, 2.5-3.0 Mo, bal Fe.
Austenitic stainless steel.
W.-Nr. 1.4436.

EKATIT 18-16.
M-1760; 0.03 max C, 0.50 max Si, 0.90 max Mn, 15.5-16.5 Cr, 17.5-18.5 Ni, bal Fe.
Stainless steel.
W.-Nr. 1.4321.

EKATIT 18-16.
M-1760; 0.03 max C, 1.0 max Si, 2.0 max Mn, 17.0-19.0 Cr, 15.0-17.0 Ni, 3.0-4.0 Mo, bal Fe.
Austenitic stainless steel.
W.-Nr. 1.4438.

EKATIT 20-18.
M-1760; 0.07 max C, 1.0 max Si, 2.0 max Mn, 16.5-18.5 Cr, 19.0-21.0 Ni, 2.0-2.5 Mo, 2.2 Cu, Nb = 8 x C min, bal Fe.
Austenitic stainless steel.
W.-Nr. 1.4505.

EKATIT 20-18 T.
M-1760; 0.07 max C, 1.0 max Si, 2.0 max Mn, 16.5-18.5 Cr, 19.0-21.0 Ni, 2.0-2.5 Mo, 2.2 Cu, Ti = 7 x C min, bal Fe.
Austenitic stainless steel.
W.-Nr. 1.4506.

EKATIT 22-18.
M-1760; 0.07 max C, 1.0 max Si, 2.0 max Mn, 16.5-18.5 Cr, 21.5-23.5 Ni, 3.0-3.5 Mo, 2.0 Cu, Nb = 8 x C min, bal Fe.
Stabilized stainless steel.
W.-Nr. 1.4586.

EKATIT 25-7.
M-1760; 0.06 max C, 1.0 max Si, 1.5 max Mn, 24.0-26.0 Cr, 6.5-7.5 Ni, 1.3-2.0 Mo, Nb = 10 x C min, bal Fe.
Stabilized stainless steel.
W.-Nr. 1.4582.

EKATIT 25-12 NB.
M-1760; 0.10 max C, 1.0 max Si, 2.0 max Mn, 16.5-18.5 Cr, 12.0-14.5 Ni, 2.5-3.0 Mo, Nb = 8 x C min, bal Fe.
Stabilized stainless steel.
W.-Nr. 1.4583.

EKATIT 25-25.
M-1760; 0.07 max C, 1.0 max Si, 2.0 max Mn, 24.0-26.0 Cr, 24.0-26.0 Ni, 2.0-2.5 Mo, Ti = 10 x C min, bal Fe.
Stabilized stainless steel.
W.-Nr. 1.4577.

EKATIT 27-5.
M-1760; 0.10 max C, 1.0 max Si, 2.0 max Mn, 26.0-28.0 Cr, 4.0-5.0 Ni, 1.3-2.0 Mo, 0.10 N (op), bal Fe.
Stainless steel.
W.-Nr. 1.4460.

EKATIT 330.
M-1760; 0.17-0.22 C, 1.0 max Si, 1.0 max Mn, 12.0-14.0 Cr, 1.0-1.3 Mo 1.0 max Ni, bal Fe.
Martensitic stainless steel.
W.-Nr. 1.4120.

EKATIT 1013.
M-1760; 0.08-0.12 C, 1.0 max Si, 1.0 max Mn, 12.0-14.0 Cr, bal Fe.
Chromium stainless steel.
W.-Nr. 1.4006.

EKATIT 1217.
M-1760; 0.10-0.17 C, 1.0 max Si, 1.5 max Mn, 15.5-17.5 Cr, 0.2-0.3 Mo bal Fe.
Stainless steel.
W.-Nr. 1.4104.

EKATIT 1513 MO.
M-1760; 0.12-0.17 C, 1.0 max Si, 1.0 max Mn, 12.0-14.0 Cr, 1.0-1.3 Mo bal Fe.
Martensitic stainless steel.
W.-Nr. 1.4119.

EKATIT 2013.
M-1760; 0.17-0.22 C, 1.0 max Si, 1.0 max Mn, 12.0-14.0 Cr, bal Fe.
Martensitic stainless steel.
W.-Nr. 1.4021; similar to AISI 420.

EKATIT 2217.
M-1760; 0.15-0.23 C, 1.0 max Si, 1.0 max Mn, 16.0-18.0 Cr, 1.5-2.5 Ni bal Fe.
Martensitic stainless steel.
W.-Nr. 1.4057; similar to AISI 431.

EKATIT 3517.
M-1760; 0.33-0.43 C, 1.0 max Si, 1.0 max Mn, 15.5-17.5 Cr, 1.0-1.3 Mo 1.0 max Ni, bal Fe.
Martensitic stainless steel.
W.-Nr. 1.4122.

EKATIT 4013.
M-1760; 0.40-0.50 C, 1.0 max Si, 1.0 max Mn, 12.0-14.0 Cr, bal Fe.
Martensitic stainless steel.
W.-Nr. 1.4034.

EKATIT 4024.
M-1760; 0.12-0.17 C, 1.0 max Si, 1.0 max Mn, 12.0-14.0 Cr.
Stainless steel.
W.-Nr. 1.4024.

EKATIT 4108.
M-1760; 1.0-1.1 C, 1.0 max Si, 1.0 max Mn, 12.0-14.0 Cr, 0.40-0.60 Mo bal Fe.
Martensitic stainless.
W.-Nr. 1.4108.

EKATIT 4110.
M-1760; 0.50-0.60 C, 1.0 max Si, 1.0 max Mn, 13.0-15.0 Cr, 0.50-0.60 Mo, bal Fe.
Martensitic stainless steel.
W.-Nr. 1.4110.

EKATIT 4113.
M-1760; 0.07 max C, 1.0 max Si, 1.0 max Mn, 16.0-18.0 Cr, 0.9-1.20 Mo bal Fe.
Ferritic stainless steel.
W.-Nr. 1.4113.

EKATIT 9018.
M-1760; 0.85-0.95 C, 1.0 max Si, 1.0 max Mn, 17.0-19.0 Cr, 1.0-1.3 Mo 0.07-0.12 V, bal Fe.
Martensitic stainless steel.
W.-Nr. 1.4112; similar to AISI 440C.

EKATIT G 1.
M-1760; 0.18-0.25 C, 1.0 max Si, 1.0 max Mn, 12.5-14.5 Cr, bal Fe.
Martensitic stainless steel casting.
W.-Nr. 1.4027.

EKATIT G 2.
M-1760; 0.20-0.27 C, 1.0 max Si, 1.0 max Mn, 16.0-18.0 Cr, 1.0-2.0 Ni bal Fe.
Martensitic stainless steel casting.
W.-Nr. 1.4059.

EKATIT G 3.
M-1760; 0.90-1.30 C, 2.0 max Si, 1.0 max Mn, 27.0-30.0 Cr, bal Fe.
Stainless steel casting.
W.-Nr. 1.4086.

EKATIT G 4.
M-1760; 0.50-0.90 C, 2.0 max Si, 1.0 max Mn, 27.0-30.0 Cr, 2.0-2.5 Mo bal Fe.
Stainless steel casting.
W.-Nr. 1.4136.

EKATIT G 5.
M-1760; 0.90-1.30 C, 2.0 max Si, 1.0 max Mn, 27.0-29.0 Cr, 2.0-2.5 Mo bal Fe.
Stainless steel casting.
W.-Nr. 1.4138.

EKATIT G 6.
M-1760; 0.12 max C, 2.0 max Si, 1.5 max Mn, 17.0-19.5 Cr, 8.0-10.0 Ni bal Fe.
Austenitic stainless steel casting.
W.-Nr. 1.4312.

EKATIT G 7.
M-1760; 0.30-0.50 C, 2.0 max Si, 1.5 max Mn, 26.0-28.0 Cr, 3.5-5.5 Ni bal Fe.
Stainless steel casting.
W.-Nr. 1.4340.

EKATIT G 8.
M-1760; 0.12 max C, 2.0 max Si, 1.5 max Mn, 17.0-19.5 Cr, 9.0-11.0 Ni, 2.0-2.5 Mo, bal Fe.
Austenitic stainless steel casting.
W.-Nr. 1.4410.

EKATIT G 9.
M-1760; 0.08 max C, 1.5 max Si, 2.0 max Mn, 16.5-18.5 Cr, 19.0-21.0 Ni, 2.0-2.5 Mo, 2.4 Cu, 1.0 Nb (op), bal Fe.
Stainless steel casting.
W.-Nr. 1.4585.

EKATIT G 10.
M-1760; 0.08 max C, 1.5 max Si, 2.0 max Mn, 19.0-21.0 Cr, 24.0-26.0 Ni, 2.5-3.5 Mo, 2.5 Cu, 1.0 Nb (op), bal Fe.
Stainless steel casting.
W.-Nr. 1.4500.

EKATIT G 11.
M-1760; 0.50-0.90 C, 2.0 max Si, 1.0 max Mn, 27.0-29.0 Cr, bal Fe.
Stainless steel casting.
W.-Nr. 1.4085.

EKATIT G 17-13 S.
M-1760; 0.07 max C, 1.0 max Si, 2.0 max Mn, 16.0-18.0 Cr, 12.5-14.5 Ni, 4.0-5.0 Mo, bal Fe.
Austenitic stainless steel casting.
W.-Nr. 1.4448.

EKATIT G 18-8 N.
M-1760; 0.08 max C, 1.5 max Si, 1.5 max Mn, 17.5-20.0 Cr, 9.0-11.0 Ni Nb = 8 x C min, bal Fe.
Stabilized austenitic stainless steel casting.
W.-Nr. 1.4552.

EKATIT G 18-8 S.
M-1760; 0.07 max C, 2.0 max Si, 1.5 max Mn, 17.5-20.0 Cr, 9.0-11.0 Ni bal Fe.
Austenitic stainless steel casting.
W.-Nr. 1.4308.

EKATIT G 18-10 N.
M-1760; 0.08 max C, 1.5 max Si, 1.5 max Mn, 17.0-19.5 Cr, 10.5-12.5 Ni, 2.0-2.5 Mo, Nb = 8 x C min, bal Fe.
Stabilized austenitic stainless steel casting.
W-Nr. 1.4581.

EKATIT G 18-10 S.
M-1760; 0.07 max C, 2.0 max Si, 1.5 max Mn, 17.0-19.5 Cr, 10.0-12.0 Ni, 2.0-2.5 Mo, bal Fe.
Austenitic stainless steel casting.
W.-Nr. 1.4408.

EKATIT G 18-12 NB.
M-1760; 0.10 max C, 1.5 max Si, 1.5 max Mn, 16.5-18.5 Cr, 12.0-14.5 N 2.5-3.0 Mo, Nb = 8 x C min, bal Fe.
Stabilized austenitic stainless steel casting.
W.-Nr. 1.4583.

EKH3.
M-Russia; 0.9-1.1 C, 0.4 max Mn, 2.8-3.8 Cr, bal Fe.
For permanent magnet; heat treated.

EKL.
M-1655; 0.60 C, 0.30 Mn, 0.60 Si, 1.1 Cr, 2.0 W, 0.20 V, bal Fe.
Hot forging dies for light metals.
Cold punches for sheet metal.
W.-Nr. 1.2550.

EKL SPEZIAL.
M-1655; 0.60 C, 0.30 Mn, 0.70 Si, 1.1 Cr, 3.0 W, 0.20 V, bal Fe.
Hot or cold work tool steel.
Hot shears, punches, hot stamps.
Cold stamping and forming.

EL 3.
M-1736; 0.85 C, 0.25 Si, 0.30 Mn, 1.7 Cr, 0.12 Mo, 0.10 V, bal Fe.
Special purpose tool steel.
W.-Nr. 1.2237; similar to AISI L3.

ELASTIC.
M-1318; 0.50 C, 1.0 Cr, 0.09 V, bal Fe.
For gears, springs, bolts, studs, countershafts; oil hardened, shock resistant.

ELASTIC NO. 4.
M-349; 81 Au-Pt, bal Cu, Ag, Zn.
Soft: 117,500 TS; 86,500 YS; 15 El.
Heat treated: 173,000 TS; 131,500 YS; 7 El.
For dental wire clasps, orthopedic appliances; M.P. 1925°F.

ELASTIC 12.
M-349.
Platinum color clasp and orthodontic wire.
Fusing temp: 2010°F.

ELASTUF PB.
M-307; 0.50 C, 0.92 Mn, 0.93 Cr, 0.20 Mo, 0.15-0.35 Pb, bal Fe.
Hard: 150,000 TS; 115,000 YS; 17 El; 52 RA; 302 Brin.
For gears, shafting; free-cutting, preheat treated.

ELASTUF PENN MACHINERY STEEL.
M-307, M-213; 0.4-0.48 C, 1.25 Mn, bal Fe.
Hot rolled: 100,000-125,000 TS; 50,000-75,000 YS; 16-15 El; 32-30 RA; 197-217 Brin.
For spindles, lead screws, shafts, chuck and jack screws, bolts, arbors, gears; tough, excellent machinability.

ELASTUF TYPE A2.
M-307; 0.50 C, 0.85 Mn, 1.03 Cr, 0.35 Mo, 0.07 V, bal Fe.
Pre-hardened: 125,000 min YS; Ave. hardness 321 BHN.
For gears, shafts, axles, boring bars, crankshafts.

ELB.
M-1655; 0.50 C, 0.50 Mn, 0.20 Si, 1.0 Cr, 3.5 Ni, bal Fe.
Cold work tool steel, for heading dies.
W.-Nr. 1.2721.

ELB SPEZIAL.
M-1655; 0.45 C, 0.50 Mn, 0.20 Si, 1.3 Cr, 4.0 Ni, 0.20 Mo, bal Fe.
Cold work tool steel, for dies, heading tools, shear blades.
W.-Nr. 1.2767.

ELBEBRONZE LB5.
M-1687; 36 Zn, 0.5 Al, 2 Ni, 0.5 Fe, 2 Mn, 0.5 Pb, bal Cu.
High tensile brass; cast or wrought; hardness 110-200 Brin; free-machining.
AFNOR UZ36 N3.

ELBEBRONZE LBA 3.
M-1687; 36 Zn, 2 Al, 3 Mn, 0.6 Si, 0.5 Pb, bal Cu.
High strength brass; wrought; hardness 165 Brin.
AFNOR UZ36 M3 A2 S.

ELBEBRONZE LBR 1.
M-1687; 34 Zn, 2.5 Al, 0.4 Ni, 0.5 Fe, 3.5 Mn, bal Cu.
High tensile brass; wrought; hardness: 160-195 Brin.

ELBEBRONZE LBU.
M-1687; 39 Zn, 2 Pb, bal Cu.
Free machining brass, wrought; hardness 90-150 Brin.
AFNOR UZ39 Pb2.

ELBEBRONZE NAB.
M-1687; 39 Zn, 1 Sn, bal Cu.
Naval brass; wrought.
SAE 73.

ELCOMET K.
M-263; 22 Ni, 23 Cr, 1.25-1.0 Si, 3.5-4.0 Cu, 2.0 Mo, 0.30 Mn, 0.12 max C, bal Fe.
For spinner heads, valves, pumps; resists abrasion and corrosion.

ELECKTRO.
M-1311; 0.85 C, 0.25 Si, 0.25 Mn, bal Fe.
Heat treated: 190,000 TS; 145,000 YS; 10 El; 30 RA; 400 Brin.
For springs, tools, drills, taps, reamers; Type W1; water hardened.

ELECKTROBRONZE.
M-330; 10 Al, Mn, Fe, bal Cu.
For valve gages, fittings; Al-bronze, corrosion resistant.

ELECTALLOY NO. 2.
M-1208; 2.7-3.2 G.C., 0.4-0.7 C.C., 2.0-2.9 Si, 0.8-1.2 Ni, 0.9-1.3 Mo, 0.4 Cr, bal Fe.
Cast: 60,000 TS.
For piston rings; cast iron, wear resistant.

ELECTALLOY NO. 3.
M-1208; 2.8-3.2 T.C., 0.7-1.15 C.C., 2.1-2.7 Si, 0.5-1.2 Mn, 0.6-1.2 Mo, 0.8-1.2 Cr, bal Fe.
Cast: 70,000 TS.
For piston rings; cast iron, wear resistant.

ELECTALLOY NO. 4.
M-1208; 3.75-3.85 T.C., 2.8 max Si, 0.7 Mn, 1 Ni, 0.9 Mo, 0.3 Cr, bal Fe.
Cast: 50,000 TS.
For piston rings; cast iron, wear resistant.

ELECTRA JAP.
M-114; 1.2 C, 0.6 Mn, 0.55 Cr, bal Fe.
For taps, edge tools; water or oil hardening.

ELECTRA JAP M.
M-114; 1.2 C, 0.85 Mn, 0.5 Cr, 0.6 Mo, bal Fe.
For taps, edge tools; water or oil hardening.

ELECTRAL CB 4.
M-1687; 2.0 Co, 0.5 Be, bal Cu.
Wrought or cast Berylium-Copper alloy.
AFNOR UK2 Be.

ELECTRAL CD 1.
M-1687; 1.0 Cd, bal Cu.
Wrought Cadmium-Copper.
Good electrical conductivity.

ELECTRAL CRM 16M.
M-1687; 0.2-0.10 Zr, 0.4 max Cr, bal Cu.
Copper castings; quenched and tempered.
Good electrical conductivity.
AFNOR UC1 Zr.

ELECTRAL CRM 16N.
M-1687; 0.2-0.10 Zr, 0.4 max Cr, bal Cu.
Wrought copper bars; quenched and tempered.
Good electrical conductivity.
AFNOR UC1 Zr.

ELECTRICAL GOLD ALLOY.
M-Eng.; 70 Au, 25 Ag, 5 Ni.
For contacts.

ELECTRITE NO. 5.
M-73; 0.6 C, 18 W, 4 Cr, 1 V, bal Fe.
For bolt trimmers, header inserts, punches; high speed steel.

ELECTRITE CO-6.
M-73; 0.9 C, 6 W, 4 Cr, 2 V, 5 Mo, 9 Co, bal Fe.
For tools, cutters; high speed steel.

ELECTRITE COBALT.
M-73; 0.7 C, 5 Co, 18 W, 4 Cr, 1 V, bal Fe.
For tools for heavy hogging cuts; high speed steel.

ELECTRITE CORSAIR.
M-73; 1.02 C, 6.1 W, 4 Cr, 2.4 V, 6 Mo, S, bal Fe.
For broaches, form tools; high speed steel.

ELECTRITE CRUSADER.
M-73; 1.2 C, 4.1 Cr, 6.0 W, 3.0 V, 5.0 Mo, alloy sulphides, bal Fe.
Hardened: C 65-68 Rock.
For form cutters, roll turning tools, lathe and planer tools.
Good machinability rating.
Type M3 Type 2 high-speed steel, high red-hardness and edge toughness.

ELECTRITE DOUBLE SIX M-2.
M-73; 0.83 C, 6.5 W, 5 Mo, 4 Cr, 2 V, bal Fe.
For cutting tools, drills, taps, broaches, reamers; high speed steel.

ELECTRITE DOUBLE SIX M-2XL.
M-73; 0.8 C, 4 Cr, 6 W, 6 Mo, 2 V, bal Fe.
For lathe and planer tools, reamers, hobs, taps; high speed steel.

ELECTRITE DYNACUT.
M-73; 1.2 C, 2.7 W, 3.75 Cr, 1.6 V, 8.0 Mo, 8.2 Co, bal Fe.
Hardened: C 65-66 Rock.
For cutters in heavy duty machining, broaches, chasers, drills.
Type M-43 high-speed steel, wear resistant. High red-hardness.

ELECTRITE DYNAMAX.
M-73; 1.08 C, 1.5 W, 3.75 Cr, 1.15 V, 9.5 Mo, 8.0 Co, bal Fe.
Hardened: C 65-70 Rock.
For cutting tools requiring long production runs or heavy duty machining; very high wear resistance; good red hardness.
AISI Type M-42 high speed tool steel.

ELECTRITE DYNAVAN.
M-73; 1.5 C, 4.5 Cr, 13.5 W, 4.75 V, 5 Co, 0.5 Mo, bal Fe.
For inserted blade cutting tools; high speed steel.

ELECTRITE HS-12.
M-73; 0.99 C, 0.72 W, 3.9 Cr, 1.96 V, 8-11 Mo, bal Fe.
Hardened: C 63-66 Rock.
For lathe tools and other cutting tools and dies requiring extra hardness and wear resistance.
AISI Type M-7 (Mod) high speed tool steel.

ELECTRITE HS 29 XL.
M-73; 0.98 C, 4.15 Cr, 6.3 W, 1.85 V, 5.05 Mo, bal Fe, plus alloy sulphides.
Hardened: C 64-67 Rock; free machining.
For lathe tools and other cutting tools requiring extra high hardness and wear resistance.
AISI Type M2 (Mod) high speed tool steel.

ELECTRITE KELVAN.
M-73; 0.88 C, 1.7 W, 3.75 Cr, 1.15 V, 9.6 Mo, 8.4 Co, bal Fe.
Hardened: C 63-67 Rock.
For lathe tools and other machine tools, very good red hardness and wear resistance.
AISI Type M-33 high speed tool steel.

ELECTRITE LACOMO.
M-73; 0.8 C, 4 Cr, 5 Co, 1.25 V, 1.5 W, 8.5 Mo, bal Fe.
For hobs, saw teeth, lathe tools, reamers; high speed steel.
AISI M30.

ELECTRITE MCH.
M-73; 0.50 C, 6.2 Mo, 1 W, 3.7 Cr, 0.75 V, bal Fe.
For forging dies, punches, shear blades; hot work steel, resists heat checking.

ELECTRITE MCL.
M-73; 0.30 C, 6.2 Mo, 1 W, 3.7 Cr, 0.75 V, bal Fe.
For forging dies, dummy blocks, extrusion dies; hot work steel, oil hardened.

ELECTRITE MV-1.
M-73; 0.8 C, 4.1 Cr, 4.25 Mo, 1.1 V, bal Fe.
For cutting tools, small taps, thread rolls; low alloy high speed steel.

ELECTRITE MV-2.
M-73; 0.88 C, 4.1 Cr, 4.25 Mo, 2.0 V, bal Fe.
For cutting tools, pipe taps, thread chasers; low alloy high speed steel.

ELECTRITE NO. 1.
M-73; 0.70 C, 4 Cr, 18 W, 1.0 V, 0.30 Mn, bal Fe.
For twist drills, reamers, milling cutters, tools; high speed steel.

ELECTRITE NO. 7.
M-73; 0.65 C, 6.5 W, 5 Mo, 4 Cr, 2 V.
For extrusion dies, header inserts; high speed steel.

ELECTRITE NO. 19.
M-73; 0.84 C, 4 Cr, 0.25 Mn, 2 V, 18 W, bal Fe.
For tools, cutters, blanking and boring tools, broaches; abrasion resistant, high speed steel.

ELECTRITE STARK.
M-73; 1.33 C, 5.5 W, 4.5 Cr, 4.0 V, 4.5 Mo, bal Fe.
Hardened: C 64-66 Rock.
For drills, reamers, lathe and planer tools, chisels, plastic core pins, chasers.
Type M-4 tool steel. Abrasion resistant and tough.
High speed steel.

ELECTRITE SUPER COBALT.
M-73; 0.8 C, 2 V, 9 Co, 18 W, 4 Cr, bal Fe.
For tools, high speed cutting tools; high speed steel.
AISI T5.

ELECTRITE TATMO.
M-73; 0.8 C, 3.25-4.25 Cr, 1.25-2 W, 0.75-1.5 V, 7.5-9.5 Mo, (Co), bal Fe.
For reamers, lathe tools, drills, broaches, dies, hot work dies, cutting tools; high speed steel; wear resistant.

ELECTRITE TATMO COBALT.
M-73; 0.90 C, 1.50 W, 4.0 Cr, 2.0 V, 8.75 Mo, 8.50 Co, bal Fe.
Desegatized super high-speed steel for heavy duty machining of hard or heat treated material.
Similar to AISI M-34.

ELECTRITE TATMO V.
M-73; 1.0 C, 1.75 W, 3.75 Cr, 2.05 V, 8.75 Mo, bal Fe.
Hardened: C 65-68 Rock.
For cutters, drills, reamers, lathe and planer tools, form tools, broaches.
Type M-7 high speed steel. Good red-hardness and edge toughness.

ELECTRITE T.N.W.
M-73; 0.85 C, 4 Cr, 28 V, 8 Mo, bal Fe.
For form cutters, taps, drills; high speed steel.

ELECTRITE ULTRA COBALT.
M-73; 0.8 C, 2.0 V, 12 Co, 18 W, 4 Cr, bal Fe.
For tools, high speed cutting tools; high speed steel.

ELECTRITE ULTRAVAN.
M-73; 1.5 C, 6.3 W, 4.25 Cr, 4.75 V, 5 Mo, 5 Co, bal Fe.
For inserted blade cutting tools; high speed steel.

ELECTRITE VANADIUM.
M-73; 1.1 C, 18 W, 4 Cr, 3.5 V, bal Fe.
C 67 Brin.
For high speed tools for cutting hard materials; high speed steel.

ELECTRO.
M-388; 0.7 C, 6 W, 6 Mo, bal Fe.
For high speed cutting tools; high speed steel.

ELECTRO-ALLOY H3.
M-England; Ni, Cr.
For high temperature applications; heat and corrosion resistant.

ELECTRO HIGH SPEED.
M-Eng.; 0.76 Cu, 0.45 Mn, 2.95 Cr, 13.24 W, 1.51 V, bal Fe.
For tools, cutters, drills; high speed steel.

ELECTROLOY.
M-778; Cu alloy.
For welding equipment.

ELECTROLOY.
M-1523; 15 Cr, 60 Ni, bal Fe.
Wire. 95,000-175,000 TS.
For resistors; high electrical resistance.

ELECTROLOY 1.
M-778; Cu-W.
Bars: 135,000 TS; 130 Brin.
For resistance welding electrode; 35% electrical conductivity, for stainless steel.

ELECTROLOY 10.
M-778; Cu-W.
Bars: 160,000 TS; 205 Brin.
For resistance welding electrode; 28% electrical conductivity, for inserts and facings.

ELECTROLOY 20.
M-778; Cu-W.
Bars: 170,000 TS; 228 Brin.
For resistance welding electrode; for heavy projection welding.

ELECTROLOY 100.
M-778; Cu-W.
Bars: 200,000 TS; Rockwell A76.
For resistance welding electrode; 30% electrical conductivity, for Red Brass.

ELECTROLOY GRADE B.
M-778; Cu alloy.
175,000 TS; 5 El; 365 Brin.
For facing and inserts on projection welding electrodes; electrical conductivity 40-45%.

ELECTROLOY GRADE BX.
M-778; Cu alloy.
95,000 TS; 4-8 El; 210-228 Brin.
For flash and butt welding dies; electrical conductivity 50%.

ELECTROLOY GRADE C.
M-778; Cu alloy.
60,000 TS; 20 El; 150 Brin.
For spot welding tips; electrical conductivity 55-65%.

ELECTROLOY GRADE XX.
M-778; Cu alloy.
70,000 TS; 20 El; 150 Brin.
For spot welding tips; electrical conductivity 90%.

ELECTROLOY MOLIN 2.
M-778; Cu alloy.
Cast: 65,000-75,000 TS; 12,000-16,000 YS; 10-2 El; 116-165 Brin.
For resistance welding electrode; 10-15% electrical conductivity.

ELECTROLOY TX.
M-778; Cu alloy.
Bars: 100,000 TS; 50,000 YS; 10 El; 185 Brin.
Cast: 85,000 TS; 45,000 YS; 10 El; 180 Brin.
For resistance welding electrode; for stainless steel.

ELECTROLYTIC IRON.
M-205; 0.006 C, 0.004 S, 0.005 Si, 0.015 Cu, 99.965 Fe.
Forged: 55,000 TS; 48,500 YS; 33 El; 83 RA.
For thin seamless tubing, magnetic cores, electrical instruments; brittle as deposited.

ELECTROLYTIC TOUGH PITCH COPPER 100.
M-8; 99.9 + Cu.
Hard: 55,000 TS; 45,000 YS; 5 El.
For electric conductors, pipes, tubes; high conductivity.

ELECTROLYTIC TOUGH PITCH COPPER 110.
M-8; 99.9 + Cu.
Hard: 48,000 TS; 40,000 YS; 15 El; B 45 Rock.
Soft: 32,000 TS; 10,000 YS; 50 El; F 45 Rock.
For roofing, flashing, gutters, architectural shapes, fasteners, plating anodes.
Elect. cond. 101.

ELECTROMATIC.
M-United States; 0.75 C, bal Fe.
For springs; oil tempered wire.

ELECTROMET CALCIUM-ALUMINUM-SILICON.
M-48; 10-14 Ca, 8-12 Al, 50-53 Si.
For deoxidizing and degasifying steel.

ELECTROMET CALCIUM-MANGANESE-SILICON.
M-48; 16-20 Ca, 14-18 Mn, 53-59 Si, bal Fe.
Used as a scavenger for steel; cleanser of oxides and gases.

ELECTROMET CALCIUM-SILICON.
M-48; 30-33 Ca, 60,65 Si, 1.5-3 Fe.
For metallurgical applications, deoxidizer and degasifier; graded according to Ca content.

ELECTROMET CHROMIUM CARBIDE.
M-48; 86.1-86.6 Cr, 12.8-13.3 total C.
Sintered: Rockwell A88.
For extrusion dies, gage blocks; pressed or slip-cast and sintered.

ELECTROMET CHROMIUM-COPPER.
M-48; 8-11 Cr, 88-90 Cu, 1 max Fe, 0.5 max Si.
For metallurgical applications.

ELECTROMET CHROMIUM (HIGH CARBON).
M-48; 9-11 C, 87-90 Cr, bal Fe.
For production of nonferrous Cr alloys, deoxidized; graded according to carbon content.

ELECTROMET CHROMIUM METAL (LOW C).
M-48; 97 min Cr, 1 max Fe, 0.2 or 0.5 max C.
For Cr bearing alloy production; Cr-additions.

ELECTROMET COLUMBIUM METAL.
M-48; 0.27 C, 0.01 N, 0.04 O, 0.1 Ta, 0.5 ppm B, bal Cb.
Cold drawn: 82,500 TS; 72,000 YS; 17 El; 68 RA.
Annealed: 50,000 TS; 36,000 YS; 49 El; 82 RA.
For atomic reactors, chemical processing equipment, high temperature alloys; high heat and corrosion resistance.

ELECTROMET DISTILLED CALCIUM.
M-48; 99.48 Ca, 0.045 Fe, 0.025 Mn, 0.03 Ni, 0.03 C, 0.004 Cr.
Rolled: 16,700 TS; 12,300 YS; 7 El; 35 RA.
Annealed: 6960 TS; 1990 YS; 51 El; 58 RA.
For alloying agent for Al, Pb, Si, copper deoxidizer; getter for electronic tubes.

ELECTROMET FERRO-BORON.
M-48; 17-25 B, 0.5 max Al, 1 max Si, 0.5 max C, bal Fe.
For deoxidizer and degasifier additions to steel; boron; boron additions.

ELECTROMET FERRO-CHROME (FOUNDRY).
M-48; 5-7 C, 62-66 Cr, 7-10 Si, bal Fe.
For metallurgical applications, ladle additions of Cr to cast iron; dissolves readily.

ELECTROMET FERRO-CHROME (HIGH CARBON).
M-48; 4.5-7 C, 1-3 Si, 65-70 Cr, bal Fe.
For production of Cr steels and cast irons; graded according to carbon content.

ELECTROMET FERRO-CHROME (LOW CARBON).
M-48; 67-73 Cr, 0.02-2.0 C, 0.2-1.0 Si, bal Fe.
For stainless steels; graded according to carbon content.

ELECTROMET FERRO-CHROME (NITROGEN).
M-48; approximately 0.75 N, 0.10 max C, 67-71 Cr, 0.3-1 Si, bal Fe.
For metallurgical applications for high Cr steels; graded according to N_2 content.

ELECTROMET FERROCHROME-SILICON.
M-48; 33-57 Cr, 23-48 Si, 0.05 max C, bal Fe.
For production of stainless steel; to reduce metal value in slag.

ELECTROMET FERRO-COLUMBIUM.
M-48; 50-60 Cb, 8.0 max Si, 0.4 max C, bal Fe.
For additions to stainless steel and high temperature alloys of Cb; increases strength at elevated temperatures.

ELECTROMET FERRO-MANGANESE (LOW CARBON).
M-48; 0.07-0.50 C, 85-90 Mn, bal Fe.
For high Mn additions to low C steels, metallurgical applications; grade according to carbon content.

ELECTROMET FERRO-MANGANESE (LOW IRON).
M-48; 85-90 Mn, 2 max Fe, 3 max Si, approx 7.0 C, bal Fe.
For metallurgical applications in non-ferrous alloys; where high Mn, low Fe are needed.

ELECTROMET FERRO-MANGANESE (MEDIUM CARBON).
M-48; 1.5 C, 1.5 max Si, 80-85 Mn, bal Fe.
For high Mn additions to medium C steels, metallurgical applications; fo C-Mn steels.

ELECTROMET FERROMANGANESE SILICON.
M-48; 63-66 Mn, 28-32 Si, 0.10 max C, 0.05 max P, bal Fe.
For slag reducing agent and Mn additions to steel.

ELECTROMET FERRO-MANGANESE (STANDARD).
M-48; 74-76 Mn, 1 max Si, 0.3 max P, approx 7 C, bal Fe.
For metallurgical applications, alloy deoxidizer.

ELECTROMET FERRO-SILICON (15%).
M-48; 14-20 Si, 1.0 max C, 74-84 Fe, 0.05 max P, 0.04 max S.
For addition of Si to steel, metallurgical applications; graded according to Si content.

ELECTROMET FERRO-SILICON (50%).
M-48; 47-51 Si, bal Fe.
For metallurgical applications, graphitizer and deoxidizer; graded according to Si content.

ELECTROMET FERROSILICON (50% LOW AL).
M-48; 47-51 Si, 0.4 max Al, bal Fe.
For additions of Si to steel and cast iron; deoxidizer.

ELECTROMET FERROSILICON (65% LOW AL).
M-48; 64.5-69.5 Si, 0.5 max Al, bal Fe.
For production of electrical sheet steel; deoxidizer; Si-additions.

ELECTROMET FERRO-SILICON (75%).
M-48; 73-78 Si, bal Fe.
For metallurgical applications, silicon additions; graded according to Si content.

ELECTROMET FERROSILICON (75% LOW AL).
M-48; 73-78 Si, 0.5 max Al, bal Fe.
For deoxidizer and alloy for steel, silicon additions; inoculant for cast iron.

ELECTROMET FERRO-SILICON (85%).
M-48; 83-88 Si, bal Fe.
For high silicon alloys, metallurgical applications; inoculant for cast iron.

ELECTROMET FERROSILICON (85% LOW AL).
M-48; 83-88 Si, 0.5 max Al, bal Fe.
For deoxidizer and alloy for steel; inoculant for cast iron.

ELECTROMET FERRO-SILICON (90%).
M-48; 92-95 Si, bal Fe.
For high silicon alloys, metallurgical applications; graded according to Si content.

ELECTROMET FERROSILICON (90% LOW AL).
M-48; 92-95 Si, 0.5 max Al, bal Fe.
For deoxidizer and alloy for steel; inoculant for cast iron.

ELECTROMET FERROSILICON-CHROME.
M-48; 50-54 Cr, 28-32 Si, 1.25 max C, bal Fe.
For ladle additions of Cr and Si to steel; alloying.

ELECTROMET FERRO-TUNGSTEN.
M-48; 70-80 W, 0.6 max C, bal Fe.
For steel alloys, alloying W to steel; graded according to C content.

ELECTROMET FERRO-VANADIUM.
M-48; 55-50 V, 10 max Si, 3 max C, bal Fe.
For metallurgical applications, steel alloys, V additions; graded according to C and Si content.

ELECTROMET LOW PHOSPHORUS FERROMANGANESE.
M-48; 90 min Mn, 0.06 max P, 0.07 max C, bal Fe.
For Mn additions to austenitic stainless steels; alloying.

ELECTROMET MANGANESE.
M-48; 2.5 max Fe, 1.5 max Si, 0.2 max C, 96 min Mn.
For metallurgical applications, alloying, deoxidizer; graded according to and Fe content.

ELECTROMET MANGANESE-BORON.
M-48; 75 Mn, 17.5 min B, 5 max Fe, 1.5 max Si, 3 max C.
For deoxidizer, degasifier and cleanser; for nonferrous alloys.

ELECTROMET MANGANESE-COPPER.
M-48; 28-32 Mn, 65-70 Cu, 4 max Fe, 0.75 max Si.
For non-ferrous metallurgical applications; graded according to Fe content.

ELECTROMET MANGANESE-NICKEL TITANIUM.
M-48; 0.10 max C, 6-8 Mn, 29-31 Ni, 46.5-48.5 Ti, 12.5-14.5 Al, 1 max Fe, 0.2 max Si.
For Ti additions to high temperature alloys.

ELECTROMET NICKEL BORON.
M-48; 15-18 B, 3 max Fe, 0.5 max C, bal Ni.
For addition agent for Ni and Al alloys; for alloying in hard abrasion resistant alloys.

ELECTROMET NICKEL-ZIRCONIUM.
M-48; 43-50 Ni, 25-30 Zr, 2 max Fe, 10 max Si, approx 15 Al.
For deoxidizer and degasifier for Ni and Fe alloys.

ELECTROMET SILICO-MANGANESE.
M-48; 3.0 max C, 65-68 Mn, 12-21 Si, bal Fe.
For metallurgical applications, alloying, and deoxidizing; graded according to carbon content.

ELECTROMET SILICON (REFINED).
M-48; 96 min Si, 2 max Fe.
For metallurgical applications for Cu and Al silicon alloys; graded according to Fe content.

ELECTROMET SILICON-SPIEGEL.
M-48; 25-30 Mn, 5-8 Si, 2-4 C, bal Fe.
For metallurgical applications; furnished in standard and special grades.

ELECTROMET SILICON TITANIUM.
M-48; 40-50 Ti, 45-50 Si, 3 max Fe.
For tool and die steels, also nonferrous alloys; titanium additions.

ELECTROMET SPIEGELEISEN.
M-48; 19-28 Mn, 1.0 max Si, bal Fe, C.
For metallurgical applications; graded according to Mn content.

ELECTROMET TANTALUM METAL.
M-48; 0.015 C, 0.001 H, 0.01 Fe, bal Ta.
Cold drawn: 114,200 TS; 104,500 YS; 19 El; 76 RA; 205 Brin.
Recrystallized: 54,900 TS; 37,700 YS; 54 El; 92 RA; 103 Brin.
For heat exchangers, bayonet heaters, pickling tanks; high corrosion resistance.

ELECTROMET TITANIUM CARBIDE.
M-48; 95 min TiC, 19.2 min combined C, 0.2 max Fe, 0.3 max free C.
For additions to tungsten carbide; pressed or slip-cast and sintered.

ELECTROMET VANADIUM METAL.
M-48; 0.10 C, 0.10 H, 0.10 N, 0.10 O, bal V.
Annealed: 60,000-70,000 TS; 49,100 YS; 30-20 El; 80-70 RA; 140 Brin.
For atomic reactor structures.

ELECTROMET ZIRCONIUM (12-15%).
M-48; 12-15 Zr, 39-43 Si, 0.30 max C, bal Fe.
For ferrous metallurgical applications, steel deoxidizer; nitride and sulfide former.

ELECTROMET ZIRCONIUM (35-40%).
M-48; 35-40 Zr, 47-52 Si, 0.5 max C, bal Fe.
For ferrous metallurgical applications, deoxidizer; nitride and sulfide former.

ELECTRON AZ-31.
M-78; 3.3 Al, 0.3-0.5 Mn, 1.25 Zn, bal Mg.
Cast: 36,000-40,000 TS; 25,000-29,000 YS; 12-8 El; 50 Brin.
For light alloy parts, structures; C.S.-49000-52000.

ELECTRON AZ61.
M-Eng.; 5.5-7.0 Al, 0.1-1.0 Zn, 0.1-0.5 Mn, bal Mg.
Used to hold broken bones together.

ELECTRON AZ-855.
M-78; 8.5-9.0 Al, 0.2-0.6 Zn, 0.1-0.3 Mn, 0.5 max impurities, bal Mg.
Forged: 40,500-49,000 TS; 22,500-31,000 YS; 15-8 El; 15-10 RA; 65-7 Brin.
For forgings, aircraft components; Iz-4-6.

ELECTRON "A.Z.G".
M-78, M-217; 6 Al, 3 Zn, 0.35 Mn, 0.3 Si, bal Mg.
Cast: 23,000-29,000 TS; 14,500-19,000 YS; 6-3 El; 7 RA; 55 Brin.
For automobile and airplane engine crankshafts; high fatigue strength.

ELECTRON "A.Z.M".
M-78, M-217, M-338; 6 Al, 1 Zn, 0.2-0.5 Mn, 0.2 Si, bal Mg.
Rolled: 49,000 TS; 31,000 YS; 14 El; 27 RA; 70 Brin.
Forgings: 45,000 TS; 29,000 YS; 8 El; 70 Brin.
For extruded and forged light alloy parts, airplane propellers, bus body frames; general structural work.

ELECTRON AZMQ.
M-78; 6.5 Al, 1 Zn, 0.3-0.5 Mn, bal Mg.
Cast: 41,000-47,000 TS; 31,000-33,000 YS; 16-10 El; 55 Brin.
Extruded: 46,000 TS; 32,000 YS; 13 El; 27 RA; 55 Brin.
For light alloy parts; now obsolete; C.S.-50000-56000.

ELECTRON ZS-32.
M-78; 5 Al, 1 Zn, bal Mg.
Forged: 7 El; 15-5 RA; 53 Brin.
For forgings, pistons; hardenable.

ELECTROPLATE.
M-Eng.; 50-70 Cu, 10-20 Ni, 5-30 Zn.
Base for silver plate, tableware.

ELECTROTYPE (STANDARD).
M-Eng.; 93 Pb, 4 Sb, 3 Sn.
For electrotype.

ELECTRUM.
M-Eng.; 52 Cu, 26 Ni, 23 Zn.
For corrosion resisting parts; nickel silver.

ELECTRUM.
M-Eng.; 55-58 Au, 15-45 Ag.
For jewelry, ornaments; corrosion resistant.

ELECTRUNITE ENDURO.
M-97; 0.15 C, 18 Cr, 8 Ni, bal Fe.
For welded tubing; corrosion and heat resistant.

ELECTRUNITE STEEL TUBING.
M-97; 0.2 C, 0.8 Ni, 0.4 Cr, bal Fe.
For mechanical tubing; electric welded.

ELEFANT EXTRA ZH.
M-614; 1.0 C, 0.25 max Si, 0.25 max Mn, bal Fe.
Annealed: 100,000 TS; 53,000 YS; 21 El; 42 RA; 200 Brin.
For drills, taps, reamers, hobs, broaches; Type W1; water hardened.

ELEFANT HART.
M-614; 1.3 C, 0.25 max Si, 0.25 max Mn, bal Fe.
for engravers tools, taps, milling cutters, hobs; Type W1; water hardened.

ELEFANT MH.
M-614; 1.15 C, 0.25 max Si, 0.25 max Mn, bal Fe.
Annealed: 110,000 TS; 60,000 YS; 18 El; 40 RA; 210 Brin.
For drills, taps, reamers, hobs, broaches, Type W1; water hardened.

ELEFANT NR 3.
M-614; 1.0 C, 0.25 max Si, 0.25 max Mn, bal Fe.
Annealed: 100,000 TS; 53,000 YS; 21 El; 42 RA; 200 Brin.
For drills, taps, reamers, hobs, cutters, punches; Type W1; water hardened.

ELEFANT NR 4.
M-614; 0.85 C, 0.25 max Si, 0.25 max Mn, bal Fe.
Heat treated: 190,000 TS; 145,000 YS; 12 El; 35 RA; 400 Brin.
For drills, taps, springs, hobs; Type W1; water hardened.

ELEFANT NR 5.
M-614; 0.70 C, 0.25 max Si, 0.25 max Mn, bal Fe.
Heat treated: 175,000 TS; 130,000 YS; 12 El; 36 RA; 360 Brin.
For springs, tools, hammers, rails, axes; Type W1; water hardened.

ELEFANT ZAH.
M-614; 0.70 C, 0.25 max Si, 0.25 max Mn, bal Fe.
Heat treated: 175,000 TS; 130,000 YS; 12 El; 36 RA; 360 Brin.
For springs, tools, rails, axes, hammers; Type W1; water hardened.

ELEFANT ZH.
M-614; 0.85 C, 0.25 max Si, 0.25 max Mn, bal Fe.
Heat treated: 190,000 TS; 145,000 YS; 12 El; 35 RA; 400 Brin.
For springs, cutters, drills, taps, reamers; Type W1; water hardened.

ELEKTROBRONZE.
M-330; 81 Cu, 10 Al, bal Mn, Fe.
20 El; 160-170 Brin.
For valve gauge fittings; corrosion resistant.

ELEKTRON A5.
M-137; 5 Al, bal Mg.
For light alloy parts.

ELEKTRON A8.
M-78; 8 max Al, 0.4 max Zn, 0.4 Mn, bal Mg.
Sand cast: 29,400 TS; 115,000 YS; 4 El; 7 RA; 55 Brin.
Aged: 36,000 TS; 21,000 YS; 13 El; 9 RA; 60 Brin.
For engine components, housings, gear cases; sand and permanent mold castings, age-hardenable.

ELEKTRON A8.
M-137; 7.9-9.0 Al, 0.3-1.0 Zn, 0.15-0.40 Mn, 0.15 max Cu, bal Mg.
Sand cast: 22,800 TS; 11,400 YS; 4 El; 55 Brin.
Permanent mold: 27,000 TS; 10,000 YS; 4.5 El; 60 Brin.
Heat treated: 36,000 TS; 13,000 YS; 12 El; 60 Brin.
For aircraft parts, general purpose castings; age-hardenable.

ELEKTRON A8.
M-784; 8.0 Al, 0.5 Zn, 0.3 Mn, 0.2 Si, bal Mg.
Cast: 22,000 TS; 11,000 YS; 5 El; 50 Brin.
Heat treated: 38,000 TS; 11,000 YS; 9 El; 60 Brin.
For gear and blower casings for aero engines; sand or die castings.

ELEKTRON A8.
M-1130; 7.5 Al, 0.7 Zn, 0.3 Mn, bal Mg.
Cast: T4; 34.0 ksi TS; 11.0 ksi YS; 7 El; 50-60 Brin.
General purpose alloy.

ELEKTRON A8HT.
M-78; 8 Al, 1 Zn, 0.5 Mn, bal Mg.
Heat treated: 26,000-32,000 TS; 6-10 El; 50 Brin.
For light alloy parts; sand and permanent mold castings.

ELEKTRON A9V.
M-137; 8.5 Al, 0.5 Zn, 0.3 Mn, bal Mg.
For light alloy parts; heat treatable.

ELEKTRON AM-503.
M-78; 0.2 Al, 0.2 Zn, 2.5 Mn, 0.2 Cu, 0.4 Si, bal Mg.
Cast: 14,000-18,000 TS; 3 El; 30 Brin.
Rolled: 40,000 TS; 22,000 YS; 5 El; 20 RA.
For light alloy parts, aircraft panelling, cowling, fairing; easy to weld, rolled sheets.

ELEKTRON AM503.
M-137; 1-2 Mn, bal Mg.
Rolled: 29,000-40,000 TS; 13,500-22,500 YS; 14-5 El; 35-45 Brin.
For sheets, shapes, aircraft; for welding.

ELEKTRON AM503.
M-784; 1.8 Mn, 0.2 Si, bal Mg.
Cast: 14,000 TS; 40,000 YS; 3 El; 45 Brin.
For light alloy parts; corrosion resistant.

ELEKTRON AM503.
M-1130; 1.5 Mn, bal Mg.
Extruded: 230 N/mm^2 TS; 130 N/mm^2 Proof stress; 4 El; 45-55 VHN.
Low strength, general purpose; weldable; good corrosion resistance.

ELEKTRON AZ31.
M-1130; 3.0 Al, 1.0 Zn, 0.3 Mn, bal Mg.
Extruded (10mm): 230 N/mm^2 TS; 150 N/mm^2 Proof stress; 8 E 50-65 VHN.
Medium strength; good formability, weldable.
BS: 3373 MAG-E-101M.

ELEKTRON AZ31.
M-137; 2.5-3.5 Al, 0.5-1.5 Zn, 0.2-0.4 Mn, bal Mg.
Cast: 21,000 TS; 11,000 YS; 7 El; 45 Brin.
Chilled: 30,000 TS; 11,000 YS; 15 El; 45 Brin.
For brake shoes; sand or gravity die cast.

ELEKTRON AZ91.
M-1130; 9.0 Al, 0.7 Zn, 0.3 Mn, bal Mg.
Diecast (F): 34.0 ksi TS; 23.0 ksi YS; 3 El; 63 Brin.
General purpose die casting alloy.
AMS 4490E.

ELEKTRON AZ91.
M-78; 9.5 Al, 0.5 Zn, 0.2 Mn, bal Mg.
Cast: 24,000 TS; 13,500 YS; 2 El; 65 Brin.
For die castings, engine parts; thin sections.

ELEKTRON AZ91.
M-137; 9.0-10.5 Al, 0.1-1.0 Zn, 0.2-0.4 Mn, bal Mg.
Sand cast: 20,000 TS; 16,000 YS; 2 El; 60 Brin.
Permanent mold: 28,000 TS; 16,000 YS; 5 El; 70 Brin.
For engine casings, aircraft parts; age-hardenable.

ELEKTRON AZ91.
M-784, M-1130; 9.5 Al, 0.5 Zn, 0.3 Mn, 0.2 Si, bal Mg.
Cast: 20,000 TS; 11,000 YS; 3 El; 50 Brin.
Heat treated: 34,000 TS; 18,000 YS; 1 El; 80 Brin.
For aircraft engine parts, crankcases; sand or die castings.

ELEKTRON AZ92A.
M-1130; 9.0 Al, 2.0 Zn, 0.3 Mn, bal Mg.
Cast: T6; 34.0 ksi TS; 18.0 ksi YS; 1 El.
General purpose alloy.

ELEKTRON AZ855.
M-784, M-1130; 7.8 Al, 0.4 Zn, 0.3 Mn, 0.1 Si, bal Mg.
Forged: 44,000-49,900 TS; 28,000-32,000 YS; 12-8 El; 65-75 Brin.
For propeller blades, forgings; press forgings.

ELEKTRON AZM.
M-1130; 6.0 Al; 1.0 Zn, 0.3 Mn, bal Mg.
Extruded: 270 N/mm^2 TS; 180 N/mm^2 Proof stress; 8 El; 60-7 VHN.
General purpose alloy; gas and arc weldable.
BS: 3373 MAG-E-121M.

ELEKTRON AZM.
M-784; 6.2 Al, 1.0 Zn, 0.3 Mn, 0.2 Si, bal Mg.
Forged: 40,000 TS; 24,000 YS; 10 El; 70 Brin.
Extruded: 49,000 TS; 31,000 YS; 10 El; 70 Brin.
For aircraft wings and fuselage; general construction.

ELEKTRON C.
M-784; 8 Al, 0.4 Zn, 0.3 Mn, bal Mg.
Cast: 24,000 TS; 12,000 YS; 2 El; 60 Brin.
Heat treated: 36,000 TS; 17,000 YS; 2 El; 80 Brin.
For light parts, engine parts; heat treatable.

ELEKTRON C.
M-1130; 7.5-9.5 Al, 0.3-1.5 Zn, 0.15 min Mn, bal Mg.
Cast: 24,100 TS; 11,200 YS; 4 El; 60 Brin.
T4-temper: 35,000 TS; 11,200 YS; 8 El; 60 Brin.
T6-temper: 35,000 TS; 14,000 YS; 2 El; 80 Brin.
For aircraft engine components, rear axle casings; age hardenable.

ELEKTRON MCZ.
M-1130; 0.4-1.0 Zr, 2.5-4.0 Mischmetal, bal Mg.
Annealed: 20,000-25,000 TS; 11,000-13,500 YS; 6-3 El; 45-55 Brin.
For engine components, aircraft structural parts; high creep strength at 500°F.

ELEKTRON MSR/QE22.
M-1130; 2.5 Ag, 2.0 Re, 0.6 Zr, bal Mg.
Cast: T6; 35.0 ksi TS; 25.0 ksi YS; 2 El; 70-90 Brin.
Heat treatable casting alloy.
AMS 4418C.

ELEKTRON MTZ.
M-1130; 3.0 Th, 0.7 Zr, bal Mg.
Cast: T6; 27.0 ksi TS; 13.0 ksi YS; 4 El; 50-60 Brin.
Pressure tight and weldable; creep-resistant up to 650°F for short time applications.
AMS 4445C.

ELEKTRON QH21A.
M-1130; 2.0-3.0 Ag, 0.4-1.6 Th, 0.6-1.5 Re, (75% min Nd), 0.5-1.0 Zr, 1.6-2.2 Re plus Th, bal Mg.
Rare earth casting alloy.
Good mechanical properties to 250°C.

ELEKTRON RZ5.
M-1130; 4.2 Zn, 1.3 Re, 0.7 Zr, bal Mg.
Cast: T5; 29.0 ksi TS; 19.5 ksi YS; 2.5 El; 55-70 Brinell.
Easily cast; weldable, pressure tight; useful strength at elevated temperatures.
AMS 4439.

ELEKTRON RZ5.
M-137, M-452; 0.75-1.75 rare earths, 3.5-5.0 Zn, 0.4-1.0 Zr, bal Mg.
Heat treated: 30,000-32,000 TS; 20,000-22,000 YS; 5-3 El; 65-75 Brin.
For structural and engine parts; age-hardenable, high temperature use to 400°F.

ELEKTRON TZ6.
M-1130; 5.5 Zn, 1.8 Th, 0.7 Zr, bal Mg.
Cast: T5; 35.0 ksi TS; 22.0 ksi YS; 5 El; 65-75 Brin.
Weldable, pressure tight castings.
AMS 4438B.

ELEKTRON Z5Z.
M-1130; 4.5 Zn, 0.7 Zr, bal Mg.
Cast: T5; 34.0 ksi TS; 20.0 ksi YS; 5 El; 65-75 Brin.
General purpose casting with useful properties to about 300°F.
AMS 4443B.

ELEKTRON Z5Z.
M-137; 3.5-5.5 Zn, 0.4-1.0 Zr, 0.15 max Mn, bal Mg.
Cast: 33,600 TS; 17,500 YS; 10 El; 55 Brin.
Heat treated: 38,200 TS; 22,000 YS; 7 El; 65 Brin.
For high strength castings; age-hardenable.

ELEKTRON ZE63.
M-1130; 5.8 Zn, 2.5 Re, 0.7 Zr, bal Mg.
Cast: T6; 39.0 ksi TS; 26.0 ksi YS; 5 El; 60-85 Brin.
Excellent castability; pressure tight and weldable, good in thin wall castings.
AMS 4425.

ELEKTRON ZM21.
M-1130; 2.0 Zn, 1.0 Mn, bal Mg.
Extruded (10mm): 230 N/mm^2 TS; 150 N/mm^2 Proof stress; 8 E 50-65 VHN.
Weldable by Argon arc process; sheet is easily formed.
BS: 3373 MAG-E-131M.

ELEKTRON ZREO.
M-1130; 0.5 Zn, 0.6 Zr, 2.7 rare earths, bal Mg.
Annealed: 20,000-24,500 TS; 12,000-15,000 YS; 6-3 El; 50-60 Brin.
For jet and turbine parts; creep resistant, used to 500°F.

ELEKTRON ZRE1.
M-1130; 2.5 Zn, 3.0 Re, 0.6 Zr, bal Mg.
Cast: T5; 20.0 ksi TS; 14.0 ksi YS; 2 El; 50-60 Brin.
Excellent castability; pressure tight and weldable; creep-resistant to 500°F.
AMS 4442B.

ELEKTRON ZRE1.
M-137; 2.5-4.0 rare earths, 0.8-3.0 Zn, 0.4-1.0 Zr, bal Mg.
At 20°C: 24,000 TS; 14,600 YS; 5 El; 50 Brin.
At 200°C: 20,200 TS; 8960 YS; 30 El.
For engine and turbine castings; high creep and heat resistance.

ELEKTRON ZRE1.
M-784; 2.75 Ce, 0.6 Zr, 2.25 Zn, bal Mg.
Annealed: 22,000 TS; 12,000 YS; 5 El; 55 Brin.
Heat treated: 40,500 TS; 23,500 YS; 5 El; 75 Brin.
For light parts, aircraft engine parts; heat resistant, operating to 500°F.

ELEKTRON ZT1.
M-1130; 3.0 Th, 2.2 Zn, 0.7 Zr, bal Mg.
Cast: T5; 27.0 ksi TS; 13.0 ksi YS; 4 El; 50-60 Brin.
Pressure tight and weldable; creep-resistant up to 650°F.
AMS 4447B.

ELEKTRON ZTX.
M-1130; 2.5 Th, 1.0 Zn, 0.6 Zr, bal Mg.
At 20°C: 33,600 TS; 20,200 YS; 15 El.
At 225°C: 14,600 TS; 9500 YS; 32 El.
For light weight parts, housings, castings.
Extrusion and forging alloy.

ELEKTRON ZTY.
M-1130; 0.8 Th, 0.6 Zn, 0.6 Zr, bal Mg.
Extruded: 230 N/mm^2 TS; 130 N/mm^2 Proof stress; 6 El; 50-6 VHN.
Weldable; creep resistant to 350°C.

ELEKTRON ZW1.
M-1130; 1.3 Zn, 0.6 Zr, bal Mg.
Extruded (10mm): 250 N/mm^2 TS; 170 N/mm^2 Proof stress; 6-8 60-75 VHN.
Can be forged and welded; sheet and plate available.
BS: 3373 MAG-E-141M.

ELEKTRON ZW 2.
M-784; 2 Zn, 0.7 Zr, bal Mg.
Tube: 39,000 TS; 5 El; 75 Brin.
For light parts; non-hardenable.

ELEKTRON ZW 3.
M-784; 3 Zn, 0.7 Zr, bal Mg.
Rolled: 37,000 TS; 24,000 YS; 18 El; 60 Brin.
Forgings; 49,000 TS; 33,000 YS; 8 El; 80 Brin.
For light parts, aircraft structures; heavy duty.

ELEKTRON ZW3.
M-1130; 3.0 Zn, 0.6 Zr, bal Mg.
Extruded (10mm): 280 N/mm^2 TS; 200 N/mm^2 Proof stress; 8 E 65-75 VPN.
Can be forged or welded; sheet, plate available.
BS: 3373 MAG-E-151M.

ELEKTRON ZW6.
M-1130; 5.5 Zn, 0.06 Zr, bal Mg.
Extruded and precipitation hardened: 315 N/mm^2 TS; 230 N/mm^2 Proof stress; 8 El; 60-75 VHN.
High strength alloy; not weldable.
BS: 3373-MAG-E-161TE.

ELEKTRON ZZ.
M-1130; 0.5-1.0 Zr, 1.0-5.0 Zn, 4 max Cd, bal Mg.
For aircraft extrusions; high temperature use.

ELEPHANT BRAND HARDENING NO. 1.
M-103; 4 P, 50 Sn, bal Cu.
For hardener for Cu alloys.

ELEPHANT BRAND HARDENING NO. 2.
M-103; 7 P, 10 Sn, bal Cu.
For hardener for Cu alloys.

ELEPHANT BRAND NO. 2 METAL.
M-103; 97.85 Cu, 2 Sn, 0.15 P.
For flexible hose, welding rod, line wire; Phosphor Bronze Grade E.

ELEPHANT BRAND NO. 3 METAL.
M-103; 96.95 Cu, 3 Sn, 0.05 P.
For electrical contacts, cold heading stock; Phosphor Bronze Grade E.

ELEPHANT BRAND NO. 13 D.
M-103; 4-6 Sn, 4-6 Pb, 0.1 P, bal Cu.
55,000 TS; 30,000 YS; 12 El.
For hardware; free-cutting.

ELEPHANT BRAND NO. 16 METAL.
M-103; 95.7 Cu, 4 Sn, 0.3 P.
For springs, knife switches, diaphragms, elec. contacts; Phosphor Bronze Grade A.

ELEPHANT BRAND NO. 22 METAL.
M-103; 94.7 Cu, 5 Sn, 0.3 P.
For diaphragms, welding rods, springs; Phosphor Bronze Grade A.

ELEPHANT BRAND NO. 25 METAL.
M-103; 94.45 Cu, 4 Sn, 0.25 Pb, 0.3 P.
Rods: 55,000-70,000 TS; 5-25 El.
For bolts, nuts, hardware; Phosphor Bronze Grade B.

ELEPHANT BRAND NO. 28 METAL.
M-103; 9 Sn, 0.3 Zn, 0.01-0.35 P, bal Cu.
Rolled: 70,000-100,000 TS; 55,000-80,000 YS; 15-20 El.
For diaphragms; P-Bronze Gr. D.

ELEPHANT BRAND NO. 113.
M-103; 90 Cu, 10 Al.
42,500 TS; 19,000 YS; 9 El.
For hardware; corrosion resistant.

ELEPHANT BRAND NO. 126R METAL.
M-103; 97.2 Cu, 2.75 Sn, 0.05 P.
For springs, diaphragms, friction plates; Phosphor Bronze Grade E.

ELEPHANT BRAND NO. 133 METAL.
M-103; 93.75 Cu, 5 Sn, 1 Pb, 0.25 P.
Rods: 55,000-70,000 TS; 5-25 El.
For screw machine parts; Phosphor Bronze Grade B 1.

ELEPHANT BRAND NO. 170 METAL.
M-103; 91.85 Cu, 8 Sn, 0.15 P.
For springs, diaphragms, gears, welding rod; Phosphor Bronze Grade C.

ELEPHANT BRAND NO. 175 METAL.
M-103; 93.8 Cu, 6 Sn, 0.2 P.
For fuse clips, welding rod, wire cloth; Phosphor Bronze Grade F.

ELEPHANT BRAND NO. 192 METAL.
M-103; 98.95 Cu, 1 Sn, 0.05 P.
For springs, hot forgings, signal wire; Phosphor Bronze Grade E.

ELEPHANT BRAND PHOSPHOR BRONZE, GRADE A.
M-103; 95.5 Cu, 3.9 Sn, 0.3 P.
Rolled rod: 45,000-80,000 TS; 50-25 El.
For springs, knife switches, diaphragms; very tough.

ELEPHANT BRAND PHOSPHOR BRONZE, GRADE B.
M-103; 96 Cu, 3.25 Sn, 0.25 Pb, 0.5 P.
Rolled: 45,000-80,000 TS; 40-12 El.
For machine castings, pinions, cogwheels, propeller screws, piston rods, hardware; corrosion resistant, tough and hard.

ELEPHANT BRAND PHOSPHOR BRONZE, GRADE B-1.
M-156, M-103; 94 Cu, 5.0 Sn, 1 Pb, 0.5 P.
Rolled rod: 60,000-70,000 TS; 25 Brin.
For rods; rolled or drawn.

ELEPHANT BRAND PHOSPHOR BRONZE, GRADE C.
M-103; 92 Cu, 8 Sn, 0.5 P.
Rolled rod: 75,000-106,000 TS; 52-20 El.
For valves, cocks, cylinder liners; wear and corrosion resistant; hard and durable.

ELEPHANT BRAND PHOSPHOR BRONZE, GRADE D.
M-103; 90 Cu, 10 Sn, 0.5 P.
Rolled rod: 89,000-135,000 TS; 29-16 El.
For gears, valves, pumps, plungers, slides, powder mill tools; hard.

ELEPHANT BRAND PHOSPHOR BRONZE, GRADE E.
M-103; 2 Sn, 0.15 P, bal Cu.
For valves, bearings of heated rolls; very hard.

ELEPHANT BRAND PHOSPHOR BRONZE, GRADE F-2.
M-103; 18.5 Sn, 0.5 P, bal Cu.
Chill cast: 43,000 TS; 0.2 El; 170 Brin.
For slow moving bearings under extreme pressures; hard.

ELEPHANT BRAND PHOSPHOR BRONZE, GRADE G.
M-103; 90 Cu, 5 Sn, 5 Pb, 0.35 P.
For rods, bolts; very tough.

ELEPHANT BRAND PHOSPHOR BRONZE, GRADE H.
M-156, M-103; 97.5 Cu, 2.5 Sn, 0.08 P.
For sheets, wire; rolled or drawn.

ELEPHANT BRAND PHOSPHOR BRONZE, GRADE S.
M-103; 0.7-1.0 P, 9-11 Sn, 8-11 Pb, bal Cu.
Cast: 35,000-40,000 TS; 19,000-31,000 YS; 4-3 El; 57-90 Brin.
For bearings for locomotive, marine and stationary engines, roll neck bearings, piston rings; very hard and durable.

ELEPHANT BRAND PHOSPHOR BRONZE, GRADE V.
M-156, M-103; 87 Cu, 7 Sn, 6 Pb, 0.5 P.
For rods; rolled or drawn.

ELEPHANT BRAND PHOSPHOR BRONZE NO. 64.
M-103; 79.8 Cu, 10 Sn, 10 Pb, 0.2 P.
For castings; pressure tight.

ELEPHANT BRAND PHOSPHOR BRONZE NO. 130.
M-103; 90 Cu, 5 Sn, 5 Pb, P.
Cast: 55,000 TS; 45,000 YS; 20 El.
For bearings; heavy loads.

ELEPHANT BRAND PHOSPHOR BRONZE NO. 187.
M-103; 88 Cu, 8 Sn, 4 Pb, P.
For bearings; heavy duty.

ELEPHANT BRAND PHOSPHOR BRONZE NO. 190.
M-450; 10 Sn, 5 Pb, bal Cu.
For bearings; heavy duty.

ELEPHANT BRAND PHOSPHOR BRONZE NO. 146L.
M-103; 88 Cu, 10 Sn, 2 Pb.
For castings, hardware; free-cutting.

ELEPHANT BRAND PHOSPHOR BRONZE NO. GK.
M-103; 83.5 Cu, 8 Sn, 8 Pb, 0.5 P.
For castings, hardware; free-cutting.

ELEPHANT BRAND PHOSPHOR BRONZE NO. SN 1.
M-103; 78.9 Cu, 10.5 Sn, 9 Pb, 0.75 Ni, 0.85 P.
For bearings; heavy duty.

ELGILOY.
M-1712; 40 Co, 20 Cr, 15 Ni, 7 Mo, 2 Mn, 15 Fe, 0.15 C, 0.05 Be.
Heat treated: 380,000 TS; 280,000 YS; 702 Brin.
Available in wire, rod, strip, cable, tubing for springs, watch parts; non-magnetic, corrosion resistant.

ELGINITE.
M-1087; Cr, Ni, Fe, Mo, Mn, bal Co.
For temperature compensating hair springs; constant flexibility from -35 to 122°F.

ELHANCO.
M-Eng.; 0.7-1.2 C, 0.2 V, bal Fe.
For general tools; water hardening.

ELIANITE I.
M-Italy; 15 Si, 0.6 Mn, bal Fe.
For pumps, valves, drains, chemical plant equipment; acid resistant, brittle.

ELIANITE II.
M-Italy; 15 Si, 0.5 Mn, 2.2 Ni, 0.82 C, bal Fe.
For pumps, valves, chemical plant equipment; acid resistant, brittle.

ELINVAR.
M-373, M-108; 36 Ni, 12 Cr, 1-2 Si, 0.8 C, W, bal Fe.
Rolled: 107,000 TS; 64,500 YS.
For watches, chronometers, hair springs, resistances; slightly magnetic.

ELINVAR EXTRA.
M-1084; 43 Ni, 5 Cr, 2.75 Ti, 0.04 C, 0.30 Al, 0.50 Si, 0.35 Co, bal Fe.
Precipitation hardening alloy; aged 160,000 psi TS; cold worked and aged 200,000 psi TS.
For springs, orthodontic wires, flexures.

ELINVAR EXTRA.
M-68; 41-43 Ni, 2.4 Ti, 5.5 Cr, 0.6 max C, 0.5 Mn, 0.4 Si, bal Fe.
Heat treated: 90,000 TS; 35,000 YS; 40 El; 125 Brin.
Aged: 180,000 TS; 115,000 YS; 18 El; 330 Brin.
Cold drawn: 200,000 TS; 180,000 YS; 7 El; 420 Brin.
For springs, diaphragms, bourdon tubes; constant modulus.

ELITE.
M-1294; 1.35 C, W, Co, bal Fe.
For blanking and forming dies, punches; oil hardening, wear resistant.

ELKALOY A.
M-265; Cu, Cd.
Bar, rod, strip and shapes: 65,000 UTS; 60,000 YS; 15 El; 70 Rb.
Elec. cond.: 90% IACS.
Resistance welding electrodes (RWMA Class 1), electrical conducting parts, and contacts.

ELKALOY D.
M-265; Cu, Al.
Cast: 70,000 UTS; 30,000 YS; 12 El; 75 Rb.
Elec. cond.: 18% IACS.
For corrosion resistant jigs and fixtures, resistance welding electrodes (RWMA Class 5); flash welding dies for steel.

ELKALOY D110.
M-265; Cu, Al, Fe.
Cast: 90,000 UTS; 12 El; 88 Rb.
Elec. cond.: 12% IACS.
For corrosion and wear resistant jigs and fixtures; flash welding dies for steel.

ELKALOY D120.
M-265; Cu, Al, Fe.
Cast: 85,000 UTS; 4 El; 95 Rb.
Elec. cond.: 10% IACS.
For corrosion and wear resistant jigs and fixtures; heavy duty flash welding dies for steel.

ELKALOY D130.
M-265; Al, Fe, bal Cu.
Cast: 85,000 TS; 0.5 El; 350 Brin.
For forming dies and rolls, guide posts; 9% conductivity, Al-bronze.

ELKALOY D140.
M-265; Cu, Al, Fe, Ni.
Cast: 80,000 UTS; 0.1 El; 42 Rc.
Elec. cond.: 5% IACS.
For corrosion and very wear resistant dies, rolls and guide posts where impact is minimal.

ELKONITE 1W3.
M-265; 45 Cu, bal W.
Elec. cond.: 53% IACS; 75 Rb hardness.
Density: 12.5 gm/cc; transverse rupture strength: 110,000 psi.
For arcing and current carrying electrical contacts.
RWMA Class 10 resistance welding electrodes.

ELKONITE 3W3.
M-265; 32 Cu, bal W.
Elec. cond.: 50% IACS, 90 Rb hardness.
Density: 13.93 gm/cc; transverse rupture strength: 130,000 psi.
For arcing and current carrying electrical contacts, power transformer contacts, resistance welding electrodes.

ELKONITE 3W53.
M-265; W, Cu, Ni, Si.
Elec. cond.: 30% IACS; 105 Rb hardness.
Density: 14.00 gm/cc; Ht.Tr.: 120,000 psi TS.
For resistance welding electrodes.

ELKONITE 5W3.
M-265; 30 Cu, bal W.
Elec. cond.: 48% IACS; 95 Rb hardness.
Density: 14.2 gm/cc; TrS: 140,000 psi.
For arcing and current carrying electrical contacts; power transformer contacts.

ELKONITE 10W3.
M-265; 25 Cu, bal W.
Elec. cond.: 45% IACS; 98 Rb hardness.
Density: 14.9 gm/cc; TrS: 150,000 psi.
For electrical contacts resistant to arcing, power transformer switches: RWMA Class 11, resistance welding electrodes.
Semi conductor heat sink; EDM electrodes.

ELKONITE 10W53.
M-265; W, Cu, Ni, Si.
Elec. cond.: 27% IACS; 109 Rb hardness.
Density: 14.75 gm/cc. Ht Tr: 160,000 psi TS.
For resistance welding products.

ELKONITE 20S.
M-265; 26 Ag, 74 W.
Elec. cond.: 50% IACS; 90 Rb hardness.
Density: 15.85 gm/cc. TrS: 130,000 psi.
For electrical contacts resistant to arcing and welding for circuit breakers, etc.

ELKONITE 30S.
M-265; 30 Ag, bal W.
Elec. cond.: 50% IACS; 85 Rb hardness.
Density: 15.25 gm/cc. TrS: 125,000 psi.
For electrical contacts, circuit breakers.
Resistant to arcing and welding.

ELKONITE 30W3.
M-265; 20 Cu, bal W.
Elec. cond.: 40% IACS; 40 Rb hardness.
Density: 15.55 gm/cc; 98,000 psi TS.
For projection welding dies, die facing.
RWMA Class 12 resistance welding electrode; low expansion structural members.

ELKONITE 35S.
M-265; 34 Ag, bal W.
Elec. cond.: 55% IACS; 78 Rb hardness.
Density: 14.85 gm/cc. TrS: 120,000 psi.
For arcing and current carrying electrical contacts; household and power circuit breaker contacts.

ELKONITE 40S.
M-265; 40 Ag, bal W.
Elec. cond.: 58% IACS: 70 Rb hardness.
Density: 14.25 gm/cc.
For arcing and current carrying electrical contacts.

ELKONITE 40W3.
M-265; 13.35 Cu, bal W.
Elec. cond.: 40% IACS; 25 Rc hardness.
Density: 16.6 gm/cc.
For vacuum switch contacts, low expansion structural members.

ELKONITE 45S.
M-265; 45 Ag, bal W.
Elec. cond.: 62% IACS; 62 Rb hardness.
Density: 13.8 gm/cc.
For arcing and current carrying electrical contacts.

ELKONITE 50S.
M-265; 49 Ag, bal W.
Elec. cond.: 65% IACS; 55 Rb hardness.
Density: 13.5 gm/cc. TrS: 80,000 psi.
For arcing and current carrying electrical contacts.

ELKONITE 50W3.
M-265; 10.36 Cu, bal W.
Elec. cond.: 35% IACS; 27 Rc hardness.
Density: 17.15 gm/cc.
For vacuum switch contacts.

ELKONITE 100M.
M-265; Molybdenum (Mo).
Elec. cond.: 31% IACS.
110,000 psi UTS; 90 Rb.
RWMA Class 14 welding electrodes.

ELKONITE 100W.
M-265; Tungsten (W).
100,000 psi UTS; 70 Ra hardness.
RWMA Class 13 welding electrodes and electrical contacts.

ELKONITE 2050C.
M-265; 50 Cu, bal W.
Elec. cond.: 58% IACS; 70 Rb hardness.
Density: 12.0 gm/cc.
For current carrying and arcing electrical contacts.

ELKONITE 2110.
M-265; 90 Ag, bal W.
Wrought, ann: 25 Rb.
Elec. cond.: 92% IACS.
For electrical contacts.

ELKONITE 2125C.
M-265; 75 Cu, 25 W.
For electrical contacts.

ELKONITE 2140C.
M-265; 58 Cu, 2 Ni, bal W.
Density: 10.5 gm/cc.
For structural rotors.

ELKONITE 2150.
M-265; 50 Ag, 50 W.
For electrical contacts.
60% elect. cond.

ELKONITE 3042.
M-265; 58 Ag, bal WC.
Elec. cond.: 52% IACS, 80 Rb hardness.
Density: 11.95 gm/cc; TrS: 80,000 psi.
For arcing and current carrying electrical contacts resistant to welding and sticking.

ELKONITE 3150.
M-265; 50 Ag, bal WC.
Elec. cond.: 45% IACS; ann; 60 Rb hardness.
Density: 11.93 gm/cc; TrS: 62,000 psi.
For arcing and current carrying electrical contacts resistant to welding and sticking.

ELKONITE 3250-C.
M-265; Cu-WC.
190 Brin.
For electrical contacts; elec. cond. 45%.

ELKONITE 4050.
M-265; 50 Ag, bal W.
Elec. cond.: 62% IACS; 65 Rb hardness.
Density: 13.4 gm/cc. TrS: 110,000 psi.
For current carrying and arcing electrical contacts.

ELKONITE 4055.
M-265; 46 Ag, 19.6 WC, bal W.
Elec. cond.: 55% IACS; 85 Rb hardness.
Density: 13.3 gm/cc. TrS: 90,000 psi.
For current carrying and arcing electrical contacts.

ELKONITE G-12.
M-265; 65 Ag, bal WC.
Elec. cond.: 57% IACS; 57 Rb hardness.
Density: 11.55 gm/cc. TrS: 65,000 psi.
For current carrying and arcing electrical contacts resistant to welding and sticking.

ELKONITE G-13.
M-265; 50 Ag, bal WC.
Elec. cond.: 47% IACS; 91 Rb hardness.
Density: 12.35 gm/cc. TrS: 95,000 psi.
For current carrying and arcing electrical contacts resistant to welding and sticking.

ELKONITE G-14.
M-265; 40 Ag, bal WC.
Elec. cond.: 37% IACS; 100 Rb hardness.
Density: 12.9 gm/cc. TrS: 120,000 psi.
For current carrying and arcing electrical contacts resistant to welding and sticking.

ELKONITE G-17.
M-265; 39.5 Ag, 0.7 Cu, bal Mo.
Elec. cond.: 47% IACS; 82 Rb hardness.
Density: 10.2 gm/cc; TrS: 135,000 psi.
For arcing electrical contacts especially where low mass is required.

ELKONITE G-18.
M-265; 49.5 Ag, 0.7 Cu, bal Mo.
Elec. cond.: 55% IACS; 67 Rb hardness.
Density: 10.23 gm/cc. TrS: 110,000 psi.
For current carrying and arcing electrical contacts, especially where low mass is required.

ELKONITE TC-5.
M-265; 50 Cu, 5 W, bal WC.
Elec. cond.: 45% IACS; 94 Rb hardness.
Density: 11.25 gm/cc. TrS: 160,000 psi.
For wiping and arcing electrical contacts and light duty welding tips.

ELKONITE TC-10.
M-265; 44 Cu, 3.3 W, bal WC.
Elec. cond.: 42% IACS; 100 Rb hardness.
Density: 11.65 gm/cc. TrS: 180,000 psi.
For wiping and arcing electrical contacts and production welding tips.

ELKONITE TC-20.
M-265; WC, Cu.
Elec. cond.: 30% IACS; 37 Rc hardness.
Density: 12.65 gm/cc; 85,000 psi TS.
For electroforging and electrical upsetting dies.

ELKONITE TC-53.
M-265; WC, Cu, Ni, Si.
Elec. cond.: 18% IACS; 47 Rc hardness.
Density: 12.65 gm/cc. Ht Tr: 150,000 psi TS.
For electroforging and electrical upsetting dies.

ELKONIUM 1.
M-265; 75.0 Ag, 24.5 Cu, 0.5 Ni.
Ann: 45,000 TS; 32 El; 78 R-15T.
Cold worked: 80,000 TS; 4 El; 85 R-15 T.
Elec. cond. 75% I.A.C.S.; Density 5.27.
For electrical contacts.

ELKONIUM 17.
M-265; 77 Ag, 22.6 Cd, 0.4 Ni.
Annealed: 35,000 TS; 50 El.
Wrought: 76,000 TS; 3 El.
For electrical contacts.
31% elect. cond.

ELKONIUM 18.
M-265; 86.8 Ag, 5.5 Cd, 0.2 Ni, 7.5 Cu.
Wrought: 75,000 TS; 50,000 YS; 3 El.
For current carrying reeds and springs, make and break electrical contacts.
43% elect. cond.

ELKONIUM 22.
M-265; 72 Ag, 28 Cu.
Annealed: 53,000 TS; 20 El; Rock 15 T-79.
Cold worked: 80,000 TS; 5 El; Rock 15 T-85.
Elect. cond. 84.
For electrical contacts.

ELKONIUM 23.
M-265; 1.5 Cu, bal Ag.
Wrought: 53,000 TS; 3 El.
For electrical contacts; 97% cond.

ELKONIUM 28.
M-265; 5 Mn, bal Ag.
Wrought: 33,000 psi UTS; 7 El; 73 R-15T.
Elec. cond.: 10% IACS.
For electrical contacts.

ELKONIUM 30.
M-265; 99.9 Pt.
Annealed: 20,000 TS; 35 El.
Wrought: 35,000 TS; 5 El.
For electrical contacts.
15% elect. cond.

ELKONIUM 33.
M-265; 89 Pt, 11 Ru.
Annealed: 85,000 TS; 12 El.
Wrought: 150,000 TS; 5 El.
For electrical contacts.
3% elect. cond.

ELKONIUM 34.
M-265; 92 Pt, 8 Ru.
Annealed: 70,000 TS; 15 El.
Wrought: 130,000 TS; 5 El.
For electrical contacts.
4% elect. cond.

ELKONIUM 36.
M-265; 3 Pt, bal Ag.
Wrought, worked: 47,000 psi TS; 3 El; 77 R-15T.
Ann: 25,000 psi TS; 37 El; 45 R-15T.
Elec. cond.: 45% IACS.
For electrical contacts.

ELKONIUM 36.
M-265; 97 Ag, 3 Pt.
For electrical contacts for automotive voltage regulators.
45% elect. cond.

ELKONIUM 40.
M-265; 99.9 Pd.
Wrought: 47,000 TS; 5 El.
Annealed: 28,000 TS; 28 El.
For electrical contacts.
16% elect. cond.

ELKONIUM 41.
M-265; 26 Ag, 2 Ni, bal Pd.
Wrought, worked: 100,000 psi UTS; 2 El; 90 R-15T.
Ann: 68,000 psi UTS; 13 El; 82 R-15T.
Elec. cond.: 4% IACS.
For electrical contacts.

ELKONIUM 42.
M-265; 40 Ag, bal Pd.
Wrought, worked: 100,000 UTS; 5 El; 91 R-15T.
Ann: 54,000 UTS; 20 El; 65 R-15T.
Elec. cond.: 4% IACS.
For electrical contacts.

ELKONIUM 43.
M-265; 23 Pd, 60 Ag, 12 Cu, 5 Ni.
Wrought: 110,000 TS; 3 El.
For electrical contacts.
11.5% elect. cond.

ELKONIUM 45.
M-265; 3 Pd, bal Ag.
Wrought, worked: 48,000 UTS; 3 El; 77 R-15T.
Ann: 27,000 UTS; 37 El; 45 R-15T.
Elec. cond.: 58% IACS.
For electrical contacts.

ELKONIUM 46.
M-265; 1 Pd, bal Ag.
Wrought, worked: 47,000 UTS; 3 El; 76 R-15T.
Ann: 26,000 UTS; 42 El; 44 R-15T.
Elec. cond.: 79% IACS.
For electrical contacts.

ELKONIUM 63.
M-265; 99.55 Ag, 0.25 Mg, 0.20 Ni.
Wrought: 50,000 TS; 6 El; 100 Brin.
For electrical contacts.
71% elect. cond.

ELKONIUM 70.
M-265; 72 Au, 2.62 Ag, 1.8 Ni.
Wrought: 50,000 TS; 5 El; 81 Rock (15 T).
For electrical contacts.
14% elect. cond.

ELKONIUM 71.
M-265; 90 Ag, 10 Au.
Annealed: 29,000 TS; 28 El.
Wrought: 46,000 TS; 3 El.
For electrical contacts.
40% elect. cond.

ELKONIUM 72.
M-265; 69 Au, 25 Ag, 6 Pt.
Wrought: 55,000 TS; 4 El; 66 Brin.
For electrical contacts.
11% electrical conductivity.

ELKONIUM 73.
M-265; Cu, Cd, Ag, Au.
Wrought, worked: 65,000 UTS; 2 El; 84 R-15T.
Ann: 37,000 UTS; 40 El; 51 R-15T.
Elec. cond.: 90% IACS.
For electrical contacts.

ELKONIUM 74.
M-265; 50 Au, 50 Ag.
Elec. cond.: 16%.
For electrical contacts.
Corrosion resistance.

ELKONIUM 76.
M-265; 90 Au, 10 Cu.
Annealed: 58,000 TS; 76 Rock 15T.
Cold worked: 102,000 TS; 91 Rock 15T.
Elec. cond.: 16%.
For electrical contacts.

ELKONIUM 217.
M-265; 85 Ag, 15 Cd.
Annealed: 28,000 TS; 55 El.
Wrought: 58,000 TS; 5 El.
For electrical contacts.
35% elect. cond.

ELKONIUM 301.
M-265; 85 Pt, 15 Ir.
Annealed: 75,000 TS; 12 El.
Wrought: 120,000 TS; 5 El.
For electrical contacts.
6% elect. cond.

ELKONIUM 302.
M-265; 73.4 Pt, 18.4 Pd, 8.2 Ru.
Wrought: 125,000 TS; 2 El; 240 Brin.
For electrical contacts.
4% electrical conductivity.

ELKONIUM 305.
M-265; 5 Ru, bal Pt.
Wrought, worked: 115,000 UTS; 5 El; 89 R-15T.
Ann: 60,000 UTS; 18 El; 84 R-15T.
Elec. cond.: 5% IACS.
For electrical contacts.

ELKONIUM 306.
M-265; 6 Ru, bal Pt.
For electrical contacts.

ELKONIUM 405.
M-265; 89 Pd, 11 Ru.
Wrought: 100,000 TS; 2 El.
For electrical contacts.
6% elect. cond.

ELMARID.
M-Eng.; 5.9 C, 83 W, 4.5 Co, 0.4 Fe.
For tips for high speed tools, dies; W_2C + WC.

ELMEDUR.
M-1246; W-alloy.
For resistance welding electrodes; sintered alloy.

ELMET.
M-1246; TiC, Co, Cr, Ni.
For electrical contacts; sintered.

ELMET 2H.
M-780; Mo-Ag.
130 Brin.
For electrical contacts; sintered.

ELMET 3C.
M-780; W-Cu-Ni.
198 Brin.
For electrical contacts; sintered.

ELMET 4.
M-780; W-Cu-Ni.
275 Brin.
For electrical contacts; sintered.

ELMET 5D.
M-780; W-Ag.
180 Brin.
For electrical contacts; sintered.

ELMET 5K.
M-780; W-Ag.
110 Brin.
For electrical contacts; sintered.

ELMET C.
M-780; Cu alloy.
120 Brin.
For welding electrode; 82% conductivity; for welding iron (spot).

ELMET CU.
M-780; Cu alloy.
350 Brin.
For welding electrodes; 36% conductivity; spot welding of copper.

ELMET H-1.
M-780; Cu alloy.
120 Brin.
For welding electrodes; butt welding, 59% conductivity.

ELMET H-2.
M-780; Cu alloy.
180 Brin.
For butt welding electrodes; 45% conductivity.

ELMET H-3.
M-780; W, bal Cu.
For butt welding electrodes; 36% conductivity.

ELMET H-3.
M-1246; W-alloy.
For resistance welding electrodes; sintered alloy.

ELMET HEAVY METAL.
M-1246; W-alloy.
For flywheel, weights; for protection against radioactivity.

ELMET HR.
M-1246; TiC, Ni, Co, Cr.
Sintered.
For dies, tools, jet engine components; high oxidation resistance.

ELMET-ROTUNG.
M-1246; W, Cu.
Sintered.
For electrical contacts.

ELMET S17.
M-1246; W, Cu, Ni.
For balancing weights, isotope containers; heavy metal.

ELMET S18.
M-1246; W, Ni.
For balancing weights, isotope containers; heavy metal.

ELMET-SILVUNG.
M-1246; W, Ag.
For electrical contacts; sintered alloy.

ELMET U.
M-780; Cu alloy.
180 Brin.
For spot weldng electrodes; 68% conductivity.

ELMET-W.
M-1246; Cu, Cr.
For resistance welding electrodes; sintered alloy.

ELNERS GERMAN SILVER.
M-Eng.; 57.4 Cu, 26.6 Zn, 13 Ni, 3 Fe.
For ornaments; nickel silver.

ELOXAL.
M-Ger.; Al alloy.
For light alloy parts; anodically oxidized.

ELPHAL.
M-1724.
Aluminum coated mild steel.

ELVERITE C.
M-447; 3.5 C, 0.60 Si, 1.5 Cr, 4.5 Ni, 0.60 Mo, bal Fe.
Cast: 400-650 Brin.
For coal and cement pulverizers, crushers, mixers; abrasion resistant.

EM 2.
M-1736; 0.88 C, 0.25 Si, 0.25 Mn, 4.15 Cr, 5.15 Mo, 6.3 W, 1.95 V, bal Fe.
High speed steel.
AISI M2; W.-Nr. 1.3343.

EM2.
M-1495; 0.17 max C, 2.13-2.67 Ni, 0.13-0.32 Si, bal Fe.
Annealed: 65,000-76,000 TS; 37,000 YS; 25 El.
For low temperature tubing; Afnor 9N10, minimum operating temperature -60°C.

EM5.
M-1495; 0.15 max C, 6-8 Cr, 0.45-0.65 Mo, bal Fe.
Annealed: 60,000 TS; 24,000 YS; 30 El; 179 Brin.
For oil refinery tubes and equipment; ASTM-A-335; Afnor Z8CD7.

EM6.
M-1495; 0.15 max C, 1.0-2.0 Si, 4-6 Cr, 0.45-0.65 Mo, bal Fe.
Annealed: 60,000 TS; 24,000 YS; 30 El; 179 Brin.
For oil refinery tubes and equipment; ASTM-A-213.

EM7.
M-1495; 0.15 max C, 4-6 Cr, 0.50 max Mn, 0.45-0.65 Mo, bal Fe.
Annealed: 60,000 TS; 24,000 YS; 30 El; 163 Brin.
For oil refinery tubes; ASTM-A-199.

EM8.
M-1495; 0.15 max C, 1.6 max Si, 4-6 Cr, 0.45-0.65 Mo, 1.0 max Al, bal Fe.
Annealed: 70,000-88,000 TS; 33,000 YS; 18 El; 179 Brin.
For oil refinery tubes and equipment; Afnor Z7CSAD5.

EM9.
M-1495; 0.15-0.20 C, 5-7 Cr, 0.5-0.6 Mo, 0.35-0.45 V, bal Fe.
Rolled: 92,000-114,000 TS; 65,000 YS; 18 El.
For oil refinery tubes and equipment; Afnor Z17CDV6.

EM10.
M-1495; 0.15 max C, 0.25-1.0 Si, 8-10 Cr, 0.8-1.2 Mo, bal Fe.
Annealed: 60,000 TS; 30,000 YS; 30 El; 179 Brin.
For oil refinery tubes and equipment; ASTM-A-212; Afnor Z10CD9.

EM11.
M-1495; 0.15 max C, 1.0-1.5 Cr, 0.44-0.65 Mo, bal Fe.
Annealed: 60,000 TS; 30,000 YS; 30 El; 163 Brin.
For oil refinery tubes and equipment; ASTM-A-199.

EM 12.
M-1495; 0.15 C, 9.5 Cr, 2.0 Mo, 0.4 Nb, bal Fe.
Good creep rupture properties up to 650°C.
For boilers, superheaters.

EM 17.
M-1495; 0.34 C, 1.0 Cr, 0.3 Mo, bal Fe.
High tensile strength and high resilience under low temperature.
For aeronautics, automotive, gas cylinders.

EM 18.
M-1495; 0.25 C, 1.0 Cr, 0.22 Mo, bal Fe.
Good strength; for automotive, aeronautics.

EM 24.
M-1495; 0.30 C, 0.75 Cr, 2.75 Ni, bal Fe.
Good tensile strength; for aeronautics.

EM 26.
M-1495; 0.10 C, 1.25 Cr, 0.55 Mo, bal Fe.
Good elongation; for boilers, refineries.

EM 29.
M-1495; 0.40 C, 0.95 Cr, 0.20 Mo, bal Fe.
High tensile strength; for car industry, aeronautics.

EM 35.
M-1736; 0.90 C, 0.30 Si, 0.30 Mn, 4.0 Cr, 5.0 Mo, 6.3 W, 2.0 V, 5.0 Co, bal Fe.
High speed steel.
W.-Nr. 1.3243.

EM35.
M-1495; 0.19 max C, 3.18-3.82 Ni, bal Fe.
Annealed: 65,000 TS; 35,000 YS; 30 El; 190 Brin.
For low temperature tubing; Afnor 10N14, minimum operating temperature -100°C.

EM 36A.
M-1495; 0.05 C, 9.0 Ni, bal Fe.
Good properties at low temperature; for liquified gas lines, low temperature heat exchangers.

EM 44.
M-1736; 1.20 C, 0.25 Si, 0.30 Mn, 4.15 Cr, 5.0 Mo, 6.4 W, 2.75 V, 0.30 max Ni, 10.0 Co, bal Fe.
Cobalt high speed steel.
W.-Nr. 1.3207; similar to AISI M44.

EM 51.
5-6 Ce, 1.2 Mn, bal Mg.
For aircraft engine parts; heat treatable.

EM-62.
M-German; 6 Ce, 2 Mn, bal Mg.
For engine cylinder heads; high temperature use.

EM 75.
M-1495; 0.44 C, 0.15 Cr, 0.20 Mo, bal Fe.
Good tensile strength; for oil industry, special casing.

EM 80.
M-1495; 0.36 C, 0.15 Mo, 0.04 V, bal Fe.
Good tensile strength; for drill pipes, casing and tubing.

EM 110.
M-1495; 0.32 C, 0.50 Cr, 0.30 Mo, 0.15 V, bal Fe.
High tensile strength; for casing.

EME see also **UNITEMP EME**.

EME.
M-85; 0.15 C, 19 Cr, 12 Ni, 3 W, 1 Cb, bal Fe.
Rolled: 135,000 TS; 100,000 YS; 20 El; 45 RA.
Forged: 115,000 TS; 80,000 YS; 25 El; 50 RA.
For jet engine and missile components; stainless and heat resistant, creep resistant.

EMK.
M-Eng.; 71 Fe, 19 Mo, 19 Co, 0.8 Si.
For alloy for sealing in glass; coefficient of expansion 92 x 10^{-7}.

EMPEDUR.
M-1246; WC, Fe.
Sintered.
For hard facing electrode; sintered.

EMPEROR.
M-107; 0.2 C, 20 Cr, 10 Ni, bal Fe.
For hardening boxes, carburizing pans, furnaces, crucibles, glass molds; heat and corrosion resistant.

EMPEROR.
M-Ger.; Sn, Sb, Pb.
For bearings; anti-friction.

EMPEROR BRASS.
M-Eng.; 60 Cu, 20 Al, 20 Zn.
For corrosion resistant parts; corrosion resistant.

EMPEROR CHROME NICKEL ALLOY.
M-107; 0.2 C, 9 Ni, 19 Cr, bal Fe.
For furnace parts; heat resistant, stainless, austenitic.

EMRO.
M-1432; C, alloy, bal Fe.
For gears, shafts, hooks, chains, chisels; shock and fatigue resistant.

EMS-1.
M-947; 0.45 C, 0.40 Mn, 8.50 Cr, 3.25 Si, bal Fe.
Exhaust valve alloy.
Known commercially as SIL-1; SAE HNV-3.

EMS-10.
M-947; 0.52 C, 9.0 Mn, 27.0 Cr, 4.0 Ni, 0.44 N, bal Fe.
Exhaust valve alloy.
Known commercially as 21-4N; SAE EV-5.

EN12KHG.
M-USSR; 0.5-0.6 C, 4.5-5.5 Mn, 3.0-4.5 Cr, 11-13 Ni, bal Fe.
For instrument parts; paramagnetic.

EN25.
M-USSR; 0.3-0.6 C, 2-3 Cr, 22-25 Ni, bal Fe.
For instrument parts; paramagnetic.

EN36.
M-USSR; 0.26 max C, 0.7 max Mn, 0.35 max Si, 35-37 Ni, bal Fe.
For electrical equipment, instruments; low coefficient of expansion.

EN42.
M-USSR; 0.3 max C, 0.8 max Mn, 0.4 max Si, 42-44 Ni, bal Fe.
For calibrating instruments; controlled expansion coefficient.

EN60.
M-USSR; 0.25 max C, 0.7-1.5 Mn, 12-15 Cr, 3.5-5.5 Al, bal Fe.
For electrical resistance and heating elements; applications to 1100°C.

ENDEWRANCE-50.
M-1489; 1.8-5.0 B, 3.0-5.0 Si, 2 max Cr, 3.5 max Fe, bal Ni.
Cast: C 40-62 Rock.
For hard facing electrodes for pump sleeves, steam valves, gauges.
Heat, corrosion and wear resistant.

ENDEWRANCE-51.
M-1489; 1.8-5.0 B, 3.0-5.0 Si, 2 max Cr, 3.5 max Fe, bal Ni.
Cast: C 40-62 Rock.
For hard facing electrodes for pump sleeves, steam valves, gauges.
Heat, corrosion and wear resistant.

ENDEWRANCE-52.
M-1489; 1.8-5.0 B, 3.0-5.0 Si, 2 max Cr, 3.5 max Fe, bal Ni.
Cast: C 40-62 Rock.
For hard facing pump sleeves, steam valves, gauges.
Corrosion, heat and wear resistant.

ENDEWRANCE-53.
M-1489; 1.8-5.0 B, 3.0-5.0 Si, 2 max Cr, 3.5 max Fe, bal Ni.
Cast: C 40-62 Rock.
For hard facing pump sleeves, steam valves, gauges.
Corrosion, heat and wear resistant.

ENDEWRANCE 202.
M-1489; 0.05-0.12 C, 0.18 Ti, 0.6 Cb, 26-30 Cr, 47-52 Co, bal Fe.
Cast: 135,000 TS; 48,000 YS; 10 El; 10 RA; 250 Brin.
Wrought: 132,000 TS; 61,000 YS; 7 El; 6 RA; 350 Brin.
For furnace baffles, burner tips, sintering grates, quench baskets.
Corrosion and heat resistant. Resists thermal shock.

ENDEWRANCE-CM2.
M-1489; 23 Cr, 28 Mo, 10 Co, bal Fe.
Cast: 112,000 TS; Rock C 50-54.
For hard facing welding rods, rollers, steel mill guides, pug mill paddles, coal pulverizing hammers.
Heat and corrosion resistant. High wear resistance.

ENDEWRANCE-CM.18.
M-1489; 25 Cr, 15 Ni, 8 Mo, 25 Co, bal Fe.
Cast: Rock C 40-42; 62,000 TS.
For welding rods, steam valves, punches, pump sleeves, hot trimming dies.
High wear resistance. Heat and corrosion resistant.

ENDEWRANCE-CM.40.
M-1489; 30 Cr, 10 Ni, 10 Co, 14 W, bal Fe.
Cast: 64,000 TS; C 38-42 Rock.
For welding rods, valve and valve seats, gas turbine parts, superchargers.
Heat, corrosion and wear resistant.

ENDURANCE AA.
M-657; 87 Cu, 12 Sn, 1 P.
For aero and motor bronze castings; wear resistant, low coefficient of friction.

ENDURANCE BB.
M-657; 89 Cu, 10.5 Sn, 0.5 P.
For bushings; resists abrasive wear.

ENDURANCE CC.
M-657; 89-65 Cu, 10 Sn, 0.35 P.
For pumps, collars, bearings; corrosion resistant.

ENDURANCE DD.
M-657; 80 Cu, 10 Sn, 9.75 Pb, 0.25 P.
For railway bearings, connecting rod bearings; heavy duty.

ENDURANCE EE.
M-657; 84 Cu, 9.5 Sn, 6.3 Pb, 0.2 P.
For railway bearings, connecting rod bearings; heavy duty.

ENDURO 17-7 now ENDURO 301.

ENDURO 18-8 now ENDURO 302.

ENDURO 18-8 B now ENDURO 302B.

ENDURO 18-8 CB now ENDURO 347.

ENDURO 18-8 FM now ENDURO 303.

ENDURO 18-8 FS now ENDURO 305.

ENDURO 18-8 S now ENDURO 304.

ENDURO 18-8 S-CB now ENDURO 348.

ENDURO 18-8 S-MO now ENDURO 316.

ENDURO 18-8 S-MO-CB now ENDURO 318.

ENDURO 18-8 S-TI now ENDURO 321.

ENDURO 19-9 S-MO now ENDURO 317.

ENDURO 20-10 S now ENDURO 308.

ENDURO 201.
M-97; 0.15 max C, 5.5-7.5 Mn, 16-8 Cr, 3.5-5.5 Ni, 0.25 N, bal Fe.
Rolled: 115,000 TS; 55,000 YS; 55 El; 185 Brin.
For railroad car bodies; Type 201; stainless, austenitic.

ENDURO 202.
M-97; 0.15 max C, 7.5-10.0 Mn, 17-19 Cr, 4-6 Ni, 0.25 max N, bal Fe.
Rolled: 105,000 TS; 55,000 YS; 55 El; 185 Brin.
For kitchen utensils, storage and pasteurizing equipment; Type 202; stainless, austenitic.

ENDURO 301.
M-97; 0.15 max C, 2 max Mn, 1 max Si, 16-18 Cr, 6-8 Ni, bal Fe.
Annealed: 110,000 TS; 40,000 YS; 60 RA; 165 Brin.
1/2 H-temper: 150,000 TS; 110,000 YS; 16 El; 320 Brin.
Hard: 185,000 TS; 140,000 YS; 9 El; 410 Brin.
For transportation equipment and aircraft parts; stainless, austenitic; Type 301.

ENDURO 302.
M-97; 0.15 max C, 2.0 Mn, 17-19 Cr, 8-10 Ni, bal Fe.
Annealed: 85,000-90,000 TS; 35,000-40,000 YS; 50-60 El; 65-70 RA; 1 Brin.
For evaporators, vessels, chemical engineering equipment, sterilizers, marine parts; stainless; resists scaling to 1450°F; Type 302.

ENDURO 302 B.
M-97; 0.15 max C, 2 max Mn, 2-3 Si, 17-19 Cr, 8-10 Ni, bal Fe.
Annealed: 90,000-95,000 TS; 40,000 YS; 55-50 El; 65 RA; 156-160 Br.
For annealing boxes, furnace parts, pump parts, pre-heaters; stainless, resists oxidation to 1700°F; Type 302B.

ENDURO 303.
M-97; 0.15 max C, 17-19 Cr, 8-10 Ni, 0.07 P, S or Se, 0.6 max Mo or bal Fe.
Heat treated: 101,700 TS; 46,900 YS; 55 El; 62 RA; 179 Brin.
Annealed: 90,000 TS; 35,000 YS; 50 El; 55 RA; 160 Brin.
For tanks, vessels, chemical engineering equipment, pump shafts, valves; stainless and heat resistant; free machining.

ENDURO 304.
M-97; 0.08 max C, 2 max Mn, 18-20 Cr, 8-11 Ni, bal Fe.
Annealed: 85,000 TS; 35,000 YS; 55 El; 70 RA; 150 Brin.
For welded stainless structures where severe acid corrosion and high temperatures are encountered; stainless, corrosion resistant; Type 304.

ENDURO 305.
M-97; 0.12 max C, 17-19 Cr, 10-13 Ni, bal Fe.
Annealed: 85,000 TS; 38,000 YS; 50 El; 880 Brin.
For spinners, stainless parts; stainless, austenitic; Type 305.

Section I: Alloy Data / 483

ENDURO 308.
M-97; 0.08 max C, 19-21 Cr, 10-12 Ni, 0.2 max Mn, bal Fe.
Annealed: 85,000-95,000 TS; 40,000-60,000 YS; 45-50 El; 60 RA; 150
For working electrodes and filler rods; heat and corrosion resistant.

ENDURO 309.
M-97; 0.20 max C, 20 max Mn, 1.0 max Si, 22-24 Cr, 12-15 Ni, bal Fe.
Annealed: 90,000 TS; 45,000 YS; 50 El; 60 RA; 170 Brin.
For furnace parts, oil refining and chemical plant equipment, skid rails; stainless, corrosion and heat resistant; resists heat to 1950°F.

ENDURO 309 S.
M-97; 0.08 max C, 2 max Mn, 1 max Si, 22-24 Cr, 12-15 Ni, bal Fe.
Annealed: 90,000 TS; 45,000 YS; 50 El; 60 RA; 170 Brin.
For furnace parts, skid rails, oil refining and chemical equipment; stainless and heat resistant to 1950°F.

ENDURO 310.
M-97; 0.25 max C, 2.0 Mn, 1.5 Si, 24-26 Cr, 19-22 Ni, bal Fe.
Annealed: 95,000 TS; 45,000 YS; 45 El; 65 RA; 183 Brin.
For retorts, valves, furnace parts, pump parts, rails; heat and corrosion resistant to 2000°F.

ENDURO 316.
M-97; 0.10 max C, 2 Mn, 16-18 Cr, 10-14 Ni, 2-3 Mo, 0.1 max Si, bal Fe.
Annealed: 80,000-90,000 TS;; 30,000-40,000 YS; 50-60 El; 65-70 RA; 150-156 Brin.
For textile and dye equipment, chemical apparatus, paper and pulp mill equipment; stainless, corrosion resistant; Type 316.

ENDURO 317.
M-97; 0.10 max C, 2 max Mn, 1 max Si, 18-20 Cr, 11-14 Ni, 3-4 Mo, ba Fe.
Annealed: 90,000 TS; 40,000 YS; 45 El; 165 Brin.
For corrosion and heat resistant parts; corrosion resistant, high temperature strength, austenitic.

ENDURO 318.
M-97; 0.08 max C, 2 max Mn, 1 max Si, 17-19 Cr, 13-15 Ni, 2-3 Mo, C 10 x C min, bal Fe.
Annealed: 90,000 TS; 40,000 YS; 40 El; 172 Brin.
For corrosion and heat resistant parts; corrosion resistant, high temperature strength, austenitic.

ENDURO 321.
M-97; 0.08 C, 2.0 Mn, 1.0 max Si, 17-19 Cr, 7-9 Ni, Ti = 5 x C, bal Fe.
Annealed: 85,000-90,000 TS; 35,000 YS; 55-50 El; 65 RA; 150-156 Brin.
For welded stainless structures where severe acid corrosion and high temperatures are encountered; stainless, corrosion resistant; Type 321.

ENDURO 347.
M-97; 0.15 C, 17-19 Cr, 7.0-9.5 Ni, Cb = 6-10 x C, bal Fe.
Annealed: 80,000-90,000 TS; 35,000-45,000 YS; 60-55 El; 60-55 RA; 135-185 Brin.
For stainless parts, welded structures, tanks; non-magnetic, austenitic.

ENDURO 348.
M-97; 0.08 max C, 17-19 Cr, 9-12 Ni, 2 max Mn, Cb = 10 x C, bal Fe.
Annealed: 80,000-95,000 TS; 35,000-45,000 YS; 60-55 El; 60-55 RA; 160 Brin.
For airplane exhaust stacks, collector rings, heat resistors; stainless, austenitic, corrosion and heat resistant; Type 347.

ENDURO 403.
M-97; 0.15 max C, 11.5-13.0 Cr, bal Fe.
Annealed: 70,000 TS; 40,000 YS; 30 El; 60 RA; 155 Brin.
Hardened: 110,000 TS; 85,000 YS; 20 El; 55 RA; 225 Brin.
For turbine blades and parts; corrosion and heat resistant; Type 410.

ENDURO 405.
M-97; 0.08 max C, 11.5-13.5 Cr, 0.10-0.30 Al, bal Fe.
Annealed: 65,000 TS; 40,000 YS; 25 El; 170 Brin.
Cold drawn: 85,000 TS; 70,000 YS; 20 El; 60 RA; 185 Brin.
For oil refining equipment; creep and heat resistant.

ENDURO 410.
M-97; 0.15 max C, 11.5-13.5 Cr, bal Fe.
Annealed: 65,000 TS; 35,000 YS; 35 El; 70 RA; 155 Brin.
Heat treated: 110,000 TS; 85,000 YS; 23 El; 65 RA; 225 Brin.
For cutlery, pump shafts, steam turbine parts, valve seats, rifle and revolver barrels; stainless and heat resistant; Type 410.

ENDURO 414.
M-97; 0.15 max C, 1 max Mn, 1 max Si, 11.5-13.0 Cr, 1.25-2.50 Ni, bal Fe.
Heat treated: 210,000 TS; 105,000 YS; 15 El; 228 Brin.
For heat treated springs; corrosion resistant; Type 414.

ENDURO 416.
   M-97; 0.15 max C, 1.25 max Mn, 12-14 Cr, 0.07 max P, 1 S, Se, 0.6 max Mo, Zr, bal Fe.
   Annealed: 75,000 TS; 40,000 YS; 30 El; 60 RA; 160 Brin.
   Hardened: 110,000 TS; 85,000 YS; 18 El; 55 RA; 230 Brin.
   For turbine blades, valve stems, chemical engineering equipment, pump shafts, bushings; stainless, corrosion and heat resistant; free machining; Type 416.

ENDURO 420.
   M-97; 0.15 min C, 12-14 Cr, bal Fe.
   Annealed: 95,000 TS; 50,000 YS; 25 El; 55 RA; 195 Brin.
   For corrosion resistant parts, valves, cutlery; corrosion resistant; Type 420.

ENDURO 430.
   M-97; 0.12 max C, 14-18 Cr, 1.0 max Mn, bal Fe.
   Annealed: 80,000 TS; 50,000 YS; 25 El; 60 RA; 163 Brin.
   For valves, heat transfer parts, condenser tubes, nitrogen fixation apparatus, pumps; non-hardening, corrosion resistant; Type 430; heat resistant to 1400°F.

ENDURO 430 F.
   M-97; 0.12 max C, 14-18 Cr, 0.60 max Mn, 0.07 min P, S or Se, 0.60 m or Zr, bal Fe.
   Annealed: 80,000 TS; 55,000 YS; 25 El; 60 RA; 170 Brin.
   Rolled: 90,000 TS; 80,000 YS; 15 El; 55 RA; 190 Brin.
   For pump shafts, trim, oil burner parts, hardware, tanks, valves, screw machine parts; free machining, corrosion resistant.

ENDURO 431.
   M-97; 0.20 max C, 1 max Mn, 1 max Si, 15-17 Cr, 1.25-2.50 Ni, bal Fe.
   Wire: 135,000 TS; 115,000 YS; 10 El; 50 RA; 290 Brin.
   Bars: 125,000 TS; 95,000 YS; 20 El; 55 RA; 260 Brin.
   For spring temper applications; corrosion resistant; Type 431.

ENDURO 440C.
   M-97; 0.9-1.2 C, 16-18 Cr, bal Fe.
   Annealed: 125,000 TS; 90,000 YS; 10 El; 30 RA; 265 Brin.
   Hardened: 240,000 TS; 200,000 YS; 5 El; 10 RA; 550 Brin.
   For corrosion resistant parts, valves, bearings; corrosion and wear resistant; Type 440 C.

ENDURO 446.
   M-97; 0.35 max C, 23-27 Cr, 1.0 max Mn, bal Fe.
   Annealed: 80,000 TS; 45,000 YS; 25 El; 45 RA; 179 Brin.
   For annealing boxes, glass molds, valves, fittings, rabble arms, furnace parts; stainless, corrosion resistant; resists heat to 1900°F; Type 44.

ENDURO "A.A." now **ENDURO 430.**

ENDURO AA-FM now **ENDURO 430F.**

ENDURO AA HIGH CARBON now **ENDURO 440 C.**

ENDURO AA-NI now **ENDURO 431.**

ENDURO "F.C." now **ENDURO 416.**

ENDURO "H C" now **ENDURO 446.**

ENDURO "HCN" now **309.**

ENDURO HCN LOW CARBON enduro **309 S.**

ENDURON.
   M-1485, M-494; 1.9 C, 1.9 Si, 0.06 P, 16 Cr, bal Fe.
   Cast: 78,000 TS; 269-321 Brin.
   For piston rings, cylinder liners; heat and abrasion resistant.

ENDURO NC-3 now **ENDURO 310.**

ENDURO S-1 now **ENDURO 410.**

ENDURO S-1 ALUMINUM now **ENDURO 405.**

ENDURO S-1-NI now **ENDURO 414.**

ENDURO "S" HIGH CARBON now **ENDURO 420.**

ENDURO "S" TURBINE QUALITY now **ENDURO 403.**

ENEBRA BRONZE.
   M-557; Sn, bal Cu.
   For high static and dynamic parts; corrosion resistant.

ENEBRA EDELBRONZE NI.
   M-557; Sn, bal Cu.
   Sand cast: 50,000-65,000 TS; 21,000-36,000 YS; 30-20 El; 110-130 Brin.
   For chemical industries; corrosion resistant.

ENEBRA EDELBRONZE TO.
   M-557; Sn, bal Cu.
   Sand cast: 71,000-85,000 TS; 26,000-36,000 YS; 30-20 El; 120-150 Brin.
   For chemical and textile industries; corrosion resistant.

ENEBRA EDELBRONZE TU-I.
M-557; Sn, bal Cu.
Sand cast: 92,000-107,500 TS; 43,000-56,000 YS; 15-8 El; 150-190 Brin.
For chemical and textile industries; corrosion resistant.

ENEBRA EDELBRONZE TU-II.
M-557; Sn, bal Cu.
Sand cast: 78,500-92,000 TS; 28,000-43,000 YS; 30-20 El; 120-150 Brin.
For chemical and textile industries; corrosion resistant.

ENEBRA LAGERBRONZE B-33.
M-557; Sn, bal Cu.
For machine parts, automobile and diesel engine parts; high strength.

ENEBRA LAGERBRONZE BZ-20.
M-557; Sn, bal Cu.
For machine parts, automobile and diesel engine parts; high strength.

ENEBRA LAGERBRONZE BZ-25.
M-557; Sn, bal Cu.
For machine parts, automobile and diesel engine parts; high strength.

ENEBRA LAGERBRONZE PN.
M-557; Sn, bal Cu.
36,000-51,000 TS; 23,000-28,000 YS; 18-10 El; 70-90 Brin.
For pump parts; corrosion resistant.

ENEBRA SONDERMESSINGE S-IM.
M-557; Zn, bal Cu.
78,400-93,000 TS; 36,000-43,000 YS; 15-10 El; 130-160 Brin.
For general construction.

ENEBRA SONDERMESSINGE S-IIM.
M-557; Zn, bal Cu.
56,000-72,000 TS; 19,000-25,000 YS; 35-25 El; 90-120 Brin.
For general construction.

ENEBRA SPEZIALBRONZE EH-I.
M-557; Sn, bal Cu.
85,000-100,800 TS; 36,000-50,000 YS; 15-10 El; 150-180 Brin.
For electrical industries.

ENEBRA SPEZIALBRONZE EH-II.
M-557; Sn, bal Cu.
72,000-85,000 TS; 28,000-36,000 YS; 25-15 El; 130-150 Brin.
For electrical industries.

ENEBRA SPEZIALBRONZE EH-III.
M-557; Sn, bal Cu.
72,000-78,500 TS; 25,000-31,000 YS; 30-20 El; 110-130 Brin.
For electrical industries.

ENGALOY 129.
M-686; 60 Pd, 40 Ni.
For brazing electronic components.
M.P. 2260°F Eutectic alloy. Good wettability.

ENGALOY 135.
M-686; 21 Pd, 48 Ni, 31 Mn.
Shear strength: 38,000 at 1250°F and 20,000 at 1500°F and 50,000 at R.T.
For brazing electronic components.
Eutectic alloy. Melting point 2050°F. Good wettability and ductility.

ENGALOY 142.
M-686; 60 Pd, 39.8 Ni, 0.05 B, 0.15 Li.
For brazing electronic components.
M.P. 1630°F. Eutectic alloy.
Good wettability and ductility.

ENGALOY 238.
M-686; 80 Au, 20 Cu.
Annealed: 67,200 TS; 17 El.
For brazing electronic components.
M.P. 1630°F. Eutectic alloy. Good wettability and ductility.

ENGALOY 241.
M-686; 50 Au, 50 Cu.
Annealed wire: 77,800 TS; 32 El.
For brazing filler metal for electronic applications.
M.P. 950-975°C.

ENGALOY 242.
M-686; 37,5 Au, 62,5 Cu.
Annealed: 53,400 TS; 35 El.
For brazing electronic components.
M.P. 1814-1859°F.
Good wettability and ductility.

ENGALOY 243.
M-686; 35 Au, 62 Cu, 3 Ni.
Annealed: 93,700 TS; 9 El.
For brazing electronic components.
M.P. 1787-1886°F. Good wettability and ductility.

ENGALOY 255.
M-686; 82 Au, 18 Ni.
Annealed: 100,800 TS; 5 El.
For brazing electronic components.
M.P. 1742°F. Eutectic alloy. Good wettability and ductility.

ENGALOY 259.
M-686; 94 Au, 6 Cu.
Annealed wire: 39,600 TS; 31 El.
For brazing filler metal for electronic applications.
M.P. 965-990°C.

ENGALOY 260.
M-686; 35 Au, 65 Cu.
Annealed: 53,900 TS; 40 El.
For brazing electronic components.
M.P. 1832-1870°F. Good wettability.

ENGALOY 261.
M-686; 75 Au, 5 Ag, 20 Cu.
Annealed wire: 78,900 TS; 36 El.
Uses: Brazing alloy for electronic components.
M.P. 1625-1640°F. Good wettability and ductility.

ENGALOY 265.
M-686; 72 Au, 22 Ni, 6 Cr.
Uses: Brazing alloy for electronic components.
M.P. 1785-1835°F.
Good wettability and ductility.

ENGALOY 269.
M-686; 81.5 Au, 15.5 Cu, 3 Ni.
Annealed: Wire 90,000 TS; 25 El.
For brazing filler metal for electronic applications.
M.P. 900-910°C.

ENGALOY 428.
M-686; 5 Pd, 95 Ag.
Annealed wire: 31,500 TS; 28 El.
Uses: Brazing alloy for electronic components.
M.P. 1780-1850°F.
Good wettability and ductility.

ENGALOY 431.
M-686; 10 Pd, 90 Ag.
Annealed wire: 31,600 TS; 44 El.
For brazing alloy for electronic components.
M.P. 1835-1950°F.
Good wettability and ductility.

ENGALOY 440.
M-686; 20 Pd, 75 Ag, 5 Mn.
Shear strength of 19,000 at 1250°F and 14,000 at 1500°F.
For brazing electronic components.
M.P. 1830-2050°F. Good wettability.

ENGALOY 441.
M-686; 33 Pd, 64 Ag, 3 Mn.
Shear strength of 27,000 at 1250°F and 18,000 at 1500°F.
For brazing electronic components.
M.P. 2100-2190°F. Good wettability.

ENGALOY 447.
M-686; 20 Pd, 80 Ag.
Annealed wire: 39,000 TS; 34 El.
For brazing alloy for electronic components.
M.P. 1960-2150°F.

ENGALOY 478.
M-686; 5 Pd, 68 Ag, 27 Cu.
Annealed wire: 65,100 TS; 23 El.
For brazing alloy for electronic components.
M.P. 1480-1490°F.
Good wettability and ductility.

ENGALOY 485.
M-686; 30 Pd, 70 Ag.
Uses: Brazing alloy for electronic components.
M.P. 2120-2250°F. Good wettability and ductility.

ENGALOY 490.
M-686; 15 Pd, 65 Ag, 20 Cu.
Annealed wire: 68,300 TS; 20 El.
Uses: Brazing alloy for electronic components.
M.P. 1565-1650°F.
Good wettability and ductility.

ENGALOY 491.
M-686; 10 Pd, 58 Ag, 32 Cu.
Annealed wire: 69,500 TS; 22 El.
Uses: Brazing alloy for electronic components.
M.P. 1520-1565°F.
Good wettability and ductility.

ENGALOY 492.
M-686; 25 Pd, 54 Ag, 21 Cu.
Annealed wire: 73,000 TS; 15 El.
Uses: Brazing alloy for electronic components.
M.P. 1650-1740°F.
Good wettability and ductility.

ENGALOY 493.
M-686; 20 Pd, 52 Ag, 28 Cu.
Annealed wire: 71,700 TS; 25 El.
Uses: Brazing alloy for electronic components.
M.P. 1615-1650°F. Good wettability and ductility.

ENGALOY 845.
M-686; 84.6 Ag, 7.5 Cu, 2.2 Pb, 5.5 In, 0.2 Li.
For brazing electronic components.
M.P. 1400-1615°F.
Good wettability and ductility.

ENGALOY El 14597.
M-686; 55 Cu, 20 Pd, 15 Ni, 10 Mn.
Shear strength: 27,000 at 1000°F; 21,000 at 1250°F and 18,000 at 1500°F.
For brazing electronic components.
M.P. 1940-2020°F.
Good wettability and ductility.

ENGELHARD 4556.
M-237, M-1036; 1.7 Mg, bal Ag.
Annealed: 46,000 TS; 15 El; Rock (30T) 55.
Cold drawn: 62,000 TS; 5 El; Rock (30T) 66.
For secondary emitter in television camera tubes.
Elec. cond. 35 annealed, 32 cold rolled.

ENGELHARD 15065.
M-237, M-1036; 0.2 Mg, 0.20 Ni, bal Ag.
Annealed: 32,000-38,000 TS; 12-27 El; 75 Knoop Hard.
Hardened: 68,000-74,000 TS; 2-6 El; 169 Knoop Hard.
For springs, switches, relays.
Oxidation hardened. Good spring properties.
75% elec. cond. as hardened.

ENGELHARD 16527.
M-237, M-1066; 0.15 Mg, 0.20 Ni, bal Ag.
Annealed: 28,000-34,000 TS; 20-38 El; 82 Knoop.
Hardened: 58,000-64,000 TS; 9-16 El; 143 Knoop.
For springs, switches.
Oxidation hardened.
75% elec. cond. as hardened.

ENGELHARD ALLOY 1751.
M-18; 23.2 Co, 76.8 Pt.
Bar: C 26 Rock; 9,500,000 (BH) max, 6450 Br, 4300 Hc, 3770 Bo.
For permanent magnets in watches, electrical and magnetic instruments.
Ductile, high coercive force.

ENGELHARD ALLOY NO. 18.
M-18; 75 Pt, 25 Ir.
For thermocouple wire, magnetos, electrical contacts, thermostats, lab. ware, surgical tools, voltage regulators, potentiometer wire, jewelry.

ENGELHARD ALLOY NO. 26.
M-18; 70 Pt, 30 Ir.
For thermocouple wire, magnetos, electrical contacts, thermostats, lab. ware, surgical tools, voltage regulators, potentiometer wire, jewelry.

ENGELHARD ALLOY NO. 109.
M-18; 90 Pt, 10 Ir.
For thermocouple wire, magnetos, electrical contacts, thermostats, lab. ware, surgical tools, voltage regulators, potentiometer wire, jewelry.

ENGELHARD ALLOY NO. 141.
M-18; 6 Pt, 69 Au, 25 Ag.
For electrical contacts.

ENGELHARD ALLOY NO. 158.
M-18; 85 Pt, 15 Ir.
For thermocouple wire, magnetos, electrical contacts, thermostats, lab. ware, surgical tools, voltage regulators, potentiometer wire, jewelry.

ENGELHARD ALLOY NO. 190.
M-18; 3 Pt, 97 Ag.
For electrical contacts.

ENGELHARD ALLOY NO. 208.
M-18; 80 Pt, 20 Ir.
For thermocouple wire, magnetos, electrical contacts, thermostats, lab. ware, surgical tools, voltage regulators, potentiometer wire, jewelry.

ENGELHARD ALLOY NO. 349.
M-18; 75 Pd, 25 Ag.
For hydrogen purification.

ENGELHARD ALLOY NO. 430.
M-18; 95 Pt, 5 Ru.
For contact material, resistance wire, jewelry.

ENGELHARD ALLOY NO. 661.
M-18; 60 Pd, 40 Cu.
For sliding contacts and slip rings.

ENGELHARD ALLOY NO. 1453.
M-18; 50 Pd, 50 Ag.
Dental material.

ENGELHARD ALLOY NO. 1560.
M-18; 10 Pt, 35 Pd, 10 Au, 30 Ag, 14 Cu, 1 Zn.
For electrical contacts.

ENGELHARD ALLOY NO. 1990.
M-18; 40 Pd, 60 Au.
For thermocouples, resistance wire.

ENGELHARD ALLOY NO. 2006.
M-18; 100% Ag.
For solder, electrical contacts, chemical processing equipment.

ENGELHARD ALLOY NO. 2246.
M-18; 90 Pt, 5 Pd, 5 Rh.
For nitric acid catalyst.

ENGELHARD ALLOY NO. 2576.
M-18; 30 Pt, 70 Au.
For spinnerettes.

ENGELHARD ALLOY NO. 3004.
M-18; 100% Au.
For chemical equipment.

ENGELHARD ALLOY NO. 3947.
M-18; 75 Au, 12 Ag, 13 Cu.
18 K gold jewelry.

ENGELHARD ALLOY NO. 4002.
M-18; 100% Ir.
For crucibles, spark plug electrodes, high temperature apparatus.

ENGELHARD ALLOY NO. 5009.
M-18; 100% Os.
Alloying element for: pen nibs, record player needles, electrical contacts and instrument pivots.

ENGELHARD ALLOY NO. 5033.
M-18; 75 Au, 18.5 Ni, 1 Cu, 5.5 Zn.
18 K white gold jewelry.

ENGELHARD ALLOY NO. 5124.
M-18; 4 Pt, 26 Pd, 25 Au, 28 Ag, 16 Cu, 1 Zn.
Dental material.

ENGELHARD ALLOY NO. 5330.
M-18; 13.95 Pt, 83 Pd, 3.05 Au.
For thermocouple wire.

ENGELHARD ALLOY NO. 5355.
M-18; 31.4 Pt, 55 Pd, 13.6 Au.
For thermocouple wire.

ENGELHARD ALLOY NO. 6007.
M-18; 100% Pd.
For telephone relay contacts.

ENGELHARD ALLOY NO. 6031.
M-18; 92 Pt, 8 W.
For spark plug electrodes, electrical contacts, strain gages.

ENGELHARD ALLOY NO. 6346.
M-18; 56.25 Au, 24.96 Cu, 12.64 Ni, 6.15 Zn.
14 K jewelry alloy.

ENGELHARD ALLOY NO. 6361.
M-18; 56.25 Au, 8.75 Ag, 30.63 Cu, 4.37 Zn.
14 K jewelry alloy.

ENGELHARD ALLOY NO. 6395.
M-18; 97 Pt, 3 Ru.
For contact material, resistance wire, jewelry.

ENGELHARD ALLOY NO. 6429.
M-18; 72.9 Au, 16.1 Ag, 11 Cu.
For electrical contacts.

ENGELHARD ALLOY NO. 6551.
M-18; 85 Pt, 15 Ni.
For cathode emitter wire.

ENGELHARD ALLOY NO. 7005.
M-18; 100% Rh.
For plated reflectors, sliding electrical contact plate, jewelry plate.

ENGELHARD ALLOY NO. 7070.
M-18; 95 Pt, 5 Au.
For electrical ribbon, spinnerettes, brazing wire, lab. ware.

ENGELHARD ALLOY NO. 7146.
M-18; 70 Pt, 30 Rh.
For thermocouples, resistance heaters, spinnerettes, glass manufacturing, nitric acid catalyst.

ENGELHARD ALLOY NO. 7450.
M-18; 95 Pt, 5 Rh.
For thermocouples, resistance heaters, spinnerettes, glass manufacturing, nitric acid catalyst.

ENGELHARD ALLOY NO. 7500, 7526, 7542.
M-18; 90 Pt, 10 Rh.
For thermocouples, resistance heaters, spinnerettes, glass manufacturing, nitric acid catalyst.

ENGELHARD ALLOY NO. 7609; 11,775.
M-18; 80 Pt, 20 Rh.
For thermocouples, resistance heaters, spinnerettes, glass manufacturing, nitric acid catalyst.

ENGELHARD ALLOY NO. 7658.
M-18; 91 Pt, 9 W.
For spark plug electrodes, electrical contacts, strain gages.

ENGELHARD ALLOY NO. 7674.
M-18; 34.95 Pd, 65.05 Au.
For thermocouple wire.

ENGELHARD ALLOY NO. 7872.
M-18; 90 Pt, 10 Ru.
For contact material, resistance wire, jewelry.

ENGELHARD ALLOY NO. 8003.
M-18; 100% Ru.
Hardener for Pt and Pd; alloyed with 90% Pt for aircraft magneto contacts.

ENGELHARD ALLOY NO. 8227.
M-18; 75 Au, 22 Ag, 3 Ni.
For thermostats.

ENGELHARD ALLOY NO. 8318.
M-18; 92 Pd, 8 W.
Potentiometer resistor wire.

ENGELHARD ALLOY NO. 8383.
M-18; 60 Pt, 40 Rh.
For thermocouples, resistance heaters, spinnerettes, glass manufacturing, nitric acid catalyst.

ENGELHARD ALLOY NO. 8391.
M-18; 95.5 Pd, 4.5 Ru.
For electrical contacts, jewelry.

ENGELHARD ALLOY NO. 8904.
M-18; 96 Pt, 4 Ru.
For contact material, resistance wire, jewelry.

ENGELHARD ALLOY NO. 8938.
M-18; 94 Pt, 6 Ru.
For contact material, resistance wire, jewelry.

ENGELHARD ALLOY NO. 8961.
M-18; 87 Pt, 13 Rh.
For thermocouples, resistance heaters, spinnerettes, glass manufacturing, nitric acid catalyst.

ENGELHARD ALLOY NO. 8987.
M-18; 93.5 Pd, 6.5 Ru.
For resistor wire, white jewelry alloy.

ENGELHARD ALLOY NO. 9001.
M-18; 100% Pt.
For electrical contacts, thermocouples, resistance thermometers, metal to glass seals, lab. ware, cathodic protection.

ENGELHARD ALLOY NO. 17632.
M-18; 8.5 Pt, 72.5 Au, 4.75 Ag, 14 Cu, 0.25 Zn.
For electrical contacts.

ENGELHARD ALLOY NO. 19182.
M-18; 77.7 Pt, 22.3 Co.
For magnets.

ENGLISH-1.
M-Eng.; 61.3 Cu, 19.1 Zn, 19.1 Ni.
For knives, forks, etc. for silver plating; nickel silver.

ENGLISH-2.
M-Eng.; 70.3 Cu, 29.3 Zn, 0.17 Sn, 0.26 Pb.
For hardware; free-cutting.

ENGLISH ALLOY.
M-Eng.; 53 Sn, 33 Pb, 11 Sb, 2.4 Cu, 1 Zn.
For bearings, bushings; Babbitt.

ENGLISH B.M.
 M-England; 80 Zn, 14.5 Sn, 5.5 Cu.
 For bearings; will not resist heat or live steam.

ENGLISH BRASS.
 M-Eng.; 70 Cu, 29 Zn, 0.3 Pb, 0.2 Sn.
 For hardware, novelties; free-cutting.

ENGLISH GERMAN SILVER.
 M-Eng.; 59.4 Cu, 25 Zn, 13 Ni, 3 Fe.
 For ornamental parts; corrosion resistant.

ENGLISH LINOTYPE.
 M-Eng.; 83 Pb, 5 Sn, 12 Sb.
 For type metal for linotype machines.

ENGLISH PENNY BRONZE.
 M-England; 3 Sn, 1.5 Zn, bal Cu.
 For springs; corrosion resistant.

ENGLISH PEWTER.
 M-Eng.; 81.2 Sn, 5.7 Sb, 1.6 Cu, 11.5 Pb.
 For bearings, ornaments, household utensils; corrosion resistant.

ENGLISH PHOSPHOR BRONZE.
 M-England; 10 Sn, 9.6 Pb, 1 P, bal Cu.
 For bearings, gear wheels; heavy duty.

ENGLISH SPECULUM.
 M-Eng.; 67 Cu, 33 Zn.
 For deep drawing, spinning, stamping; high ductility.

ENGLISH STEREOTYPE (STANDARD).
 M-Eng.; 82.5 Pb, 4.5 Sn, 13 Sb.
 For type metal.

ENGLISH TYPE METAL (A).
 M-Eng.; 63.2 Pb, 12 Sn, 24 Sb, 0.8 Cu.
 For type metal.

ENGLISH TYPE METAL (B).
 M-Eng.; 60.5 Pb, 14.5 Sn, 24.2 Sb, 0.8 Cu.
 For type metal.

ENGLISH TYPE METAL "D".
 M-Eng.; 58 Pb, 15 Sn, 26 Sb, 1.0 Cu.
 For type metal.

ENGLISH TYPE METAL "E".
 M-Eng.; 77.5 Pb, 6.5 Sn, 16 Sb.
 For type metal.

ENGLISH TYPE METAL OLD "C".
 M-Eng.; 69.2 Pb, 9.1 Sn, 19.5 Sb, 1.7 Cu.
 For type metal.

ENGRAVERS BRASS.
 M-U.S.; 66 Cu, 33 Zn, 1 Pb.
 For engravings; free-cutting.

ENGRAVERS BRASS 63.
 M-141; 65.5 Cu, 33.4 Zn, 1.1 Pb.
 Hard: 74,000 TS; 60,000 YS; 7 El; 150 Brin.
 Soft: 49,000 TS; 17,000 YS; 54 El; 61 Brin.
 For engraved items, dials; brass.

"E" NICKEL now **NICKEL 212.**

EO 1.
 M-1736; 0.95 C, 0.30 Si, 1.25 Mn, 0.52 Cr, 0.10 V, 0.55 W, bal Fe.
 Oil hardening tool steel.
 AISI 01; W.-Nr. 1.2510.

EP 20.
 M-1736; 0.36 C, 0.30 Si, 0.60 Mn, 1.8 Cr, 0.20 Mo, 1.0 V, bal Fe.
 Special purpose tool steel, for molds.
 W.-Nr. 1.2330; similar to AISI P20.

EP 65.
 M-USSR; 0.23 C, 13 Cr, 1 Mo, 1 Ni, 1 V, 1 W, bal Fe.
 Hardenable to 50 Rc min.
 For cutlery, surgical instruments.

EPCO-OIL HARD.
 M-781; 1.0 C, 0.5 Cr, 1.0 Mn, 0.5 W, bal Fe.
 For dies, gauges, master tools; non-deforming.

EPCO-WATER HARD.
 M-781; 1.05 C, 0.25 Mn, bal Fe.
 For blanking, forming, and trimming dies; water hardened.

EPD.
 M-1655; 0.60 C, 0.30 Si, 0.30 Mn, 3.8 Cr, 0.90 Mo, 0.70 V, 9.0 W, bal Fe.
 Hot work tool steel.
 For tube pressing mandrels.
 W.-Nr. 1.2622.

EPD MO.
 M-1655; 0.60 C, 0.30 Si, 0.30 Mn, 3.6 Cr, 8.5 Mo, 1.75 V, 1.50 W, bal Fe.
 Hot work tool steel.
 For hot pressing dies and blocks.

EPM.
 M-1655; 0.45 C, 1.0 Si, 0.30 Mn, 1.0 Cr, 0.20 V, 2.0 W, bal Fe.
 Hot or cold work forming, punching and trimming dies.
 W.-Nr. 1.2542.

EPS.
 M-1655; 0.35 C, 0.20 Si, 0.30 Mn, 2.5 Cr, 0.40 V, 8.5 W, bal Fe.
 Hot work tool steel.
 Pressing and forging dies.
 W.-Nr. 1.2581; similar to AISI H21.

EPS 25.
 M-1655; 0.45 C, 1.0 Si, 0.30 Mn, 1.7 Cr, 0.20 V, 2.0 W, bal Fe.
 Hot work tool steel.
 Pressing dies for lead, zinc and light metals.
 W.-Nr. 1.2547.

EPS 33.
M-1655; 0.30 C. 0.40 Si, 0.40 Mn, 3.0 Cr, 2.8 Mo, 0.50 V, bal Fe.
Hot work tool steel.
Piercing mandrels; pressing dies.
W.-Nr. 1.2365.

EPS 35.
M-1655; 0.30 C. 1.0 Si, 0.40 Mn, 1.0 Cr, 0.20 V, 4.0 W, bal Fe.
Hot work tool steel.
For water cooled forming tools for bolts, screws.
W.-Nr. 1.2564.

EPS 45.
M-1655; 0.30 C. 0.20 Si, 0.30 Mn, 2.2 Cr, 0.60 V, 4.5 W, bal Fe.
Hot work tool steel.
Heading tools for screws, bolts, non-ferrous alloys.
W.-Nr. 1.2567.

EPS 51.
M-1655; 0.38 C, 1.0 Si, 0.40 Mn, 5.5 Cr, 1.3 Mo, 0.30 V, bal Fe.
Hot work tool steel.
Forging dies; casting molds for light metals.
W.-Nr. 1.2343.

EPS 51 V.
M-1655; 0.38 C, 0.40 Mn, 1.1 Si, 5.5 Cr, 1.0 V, 1.5 Mo, bal Fe.
Hot work tool steel.
Forging dies; casting molds; extrusion and piercing mandrels.
W.-Nr. 1.2344.

EPSK 2.
M-1655; 0.30 C, 0.20 Si, 0.30 Mn, 2.0 Co, 2.5 Cr, 0.30 V, 9.0 W, bal Fe.
Hot work tool steel; forging tools.
W.-Nr. 1.2662.

EPSK SPEZIAL.
M-1655; 0.65 C, 0.30 Si, 0.30 Mn, 4.0 Cr, 0.50 V, 15.0 W, 5.0 Co, bal Fe.
Hot work tool steel.
Forging and pressing dies.

EPSNI.
M-1655; 0.45 C, 0.70 Mn, 1.4 Si, 13.5 Cr, 2.5 W, 0.50 V, 13.0 Ni, bal Fe.
Hot work tool steel; pressing dies.
W.-Nr. 1.2731.

EPW.
M-1655; 0.35 C, 1.0 Si, 0.30 Mn, 1.0 Cr, 0.20 V, 2.0 W, bal Fe.
Hot or cold work tool steel.
For water-cooled hot work tools as hot shears; also cold punches.
W.-Nr. 1.2541.

ER 347 ETC see ARCALOY ER 347 ETC.

"ERA" MANGANESE STEEL.
M-62; 1.25 C, 11-14 Mn, bal Fe.
85,000-100,000 TS; 60-80 El.
For crushers, dredging machines, soldiers' helmets; wear resisting, non-magnetic.

ERAYDO.
M-226; 1-3 Cu, 0.03-0.1 Ag, bal Zn.
Rolled: 30,000-50,000 TS; 12,000-22,000 YS; 40-15 El.
For panels, indoor trim, electrical appliances, drawn cases; rustless, non-magnetic.

ERBIUM.
M-1755; Er.
Purities: 99.9% (Special distilled grade), 99.5+%.
Forms: Ingot, lump, wire, sheet, foil, turnings, sponge, powder, single crystals.

ERCO.
M-1425; 0.8 C, bal Fe.
For dies to resist alternate heating and cooling; water hardened.

ERFTAL.
M-Germany; 0.04 Fe, 99.9 Al.
For reflectors; a high gloss mill aluminum.

ERGAL 55.
M-541, M-1220, M-1283; 5.8 Zn, 2.5 Mg, 1.6 Cu, 0.2 Mn, 0.15 Cr, 0.08 Ti, bal Al.
TA-Temper: 90,000 TS; 80,000 YS; 6 El; 160 Brin.
For cast aircraft components, hardware.

ERGAL 60.
M-1220, M-1283; 6.8 Zn, 2.5 Mg, 1.6 Cu, 0.1 Ti, 0.2 Mn, 0.15 Cr, bal Al.
Cast.
For aircraft castings; self-aging.

ERGAL 65.
M-541, M-1220, M-1283; 7.8 Zn, 2.5 Mg, 1.6 Cu, 0.2 Mn, 0.15 Cr, 0.1 Ti, bal Al.
TA-Temper: 82,000-96,000 TS; 72,000-85,000 YS; 4-9 El; 150-180 Brin.
For cast aircraft components.
Self-aging, high strength cast alloy.

ERGALPLAT 65.
M-541; 1.5-1.7 Cu, 2.5 Mg, 0.2 Mn, 8 Zn, bal Al.
R-Temper: 27,000-37,000 TS; 23,000 YS; 8-20 El.
TA-Temper: 80,000-94,000 TS; 71,000-85,000 YS; 4-14 El.
For aircraft constructional parts; self-aging.
Heat and corrosion resistant.
High strength and fatigue resistant.

ERGES 4.
M-1368; 0.6-1.4 Mg, 0-1.6 Si, 0.6-1.0 Mn, 0-0.3 Cr, bal Al.
For chemical equipment; corrosion resistant.

ERGSTE 13 ETC see **ARGEST 13 ETC.**
M-1333; 0.12 max C, 0.4 Si, 13 Cr, bal Fe.
Annealed: 75,000 TS; 40,000 YS; 35 El; 70 RA; 155 Brin.
Cold drawn: 100,000 TS; 85,000 YS; 17 El; 60 RA; 205 Brin.
For turbine blades, valve trim, chemical plant equipment; Type 410; corrosion resistant.

ERHARD BRONZE.
M-German; Zn, Cu, Al.
For ornamental fittings.

ERIE.
M-1705; 1.0 C, 0.35 Mn, 0.2 Si, bal Fe.
Water hardening tool steel. AISI W1.

ERIE NO. 5.
M-1425; 0.7 C, alloy, bal Fe.
For shear blades, punches, blacksmith tools; water hardened.

ERIE AA.
M-1425; 0.7 C, 18 W, 4 Cr, 1 V, bal Fe.
For lathe and planer tools, drills, reamers, taps; high speed steel.

ERIE EXTRA.
M-1705; 1.0 C, 0.35 Mn, 0.2 Si, bal Fe.
Water hardening tool steel. AISI W1.

ERIE RA.
M-1425; 0.7 C, 18 W, 4 Cr, 1 V, 5 Co, bal Fe.
For tools, drills, lathe and planer cutters; high speed steel.

ERIE SPECIAL.
M-1705; 1.0 C, 0.35 Mn, 0.2 Si, bal Fe.
Water hardening tool steel. AISI W1.

ERK ADG.
M-1336; 2.1 C, 11.5 Cr, 0.3 Mn, bal Fe.
For blanking and forming dies, gauges; oil hardened, non-deforming.

ERKENZWEIG BRA.
M-1336; 0.20 C, 0.40 Si, 0.3 Mn, 13 Cr, bal Fe.
Annealed: 95,000 TS; 50,000 YS; 25 El; 55 RA; 196 Brin.
For turbine blades, valves, cutlery, dental instruments; Type 420; stainless.

ERKENZWEIG EX.
M-1336; 1.4 C, 0.30 Cr, 0.10 V, 0.30 Mn, bal Fe.
For forming dies, engraving tools, cutters; water hardened, wear resistant.

ERKENZWEIG EXTRA BEST MH.
M-1336; 1.1 C, 0.25 max Si, 0.25 max Mn, bal Fe.
Annealed: 110,000 TS; 58,000 YS; 20 El; 40 RA; 210 Brin.
For springs, tools, cutters, drills, broaches; Type W1; water hardened.

ERKENZWEIG EXTRA BEST Z.
M-1336; 0.85 C, 0.25 max Si, 0.25 max Mn, bal Fe.
Heat treated: 190,000 TS; 145,000 YS; 10 El; 30 RA; 400 Brin.
For punches, drills, taps, reamers, cutters; Type W1; water hardened.

ERKENZWEIG EXTRA BEST ZH.
M-1336; 1.0 C, 0.25 max Mn, 0.25 max Si, bal Fe.
Annealed: 100,000 TS; 53,000 YS; 21 El; 42 RA; 200 Brin.
For springs, taps, drills, cutters; Type W1; water hardened.

ERKENZWEIG HB.
M-1336; 1.4 C, 0.30 Cr, 0.1 V, bal Fe.
For header dies, cutters, bearings; oil hardened, tough.

ERKENZWEIG KM45.
M-1336; 0.50 C, 1.05 Cr, 3.25 Ni, bal Fe.
For gears, pinions, crankshafts, forging dies; oil hardened, shock resistant.

ERKENZWEIG KWB.
M-1336; 1.05 C, 1.0 Cr, 1.15 W, 0.90 Mn, bal Fe.
For cutters, shears, header dies; oil hardened, tough.

ERKENZWEIG NH.
M-1336; 0.95 C, W, Mo, Cr, bal Fe.
For lathe and planer tools, drills; high speed steel.

ERKENZWEIG NHA.
M-1336; 1.35 C, W, Co, Cr, V, bal Fe.
For engravers' tools, forming and blanking dies; high speed steel.

ERKENZWEIG NHB.
M-1336; 0.79 C, 4.75 Co, 4.3 Cr, 0.75 Mo, 1.5 V, 18 W, bal Fe.
For lathe and planer tools, broaches, reamers, drills; high speed steel.

ERKENZWEIG NHC.
M-1336; 0.76 C, 10 Co, 4.2 Cr, 0.8 Mo, 1.8 V, 18 W, bal Fe.
For lathe and planer tools, drills, taps, hobs; high speed steel.

ERKENZWEIG NHE.
M-1336; 1.3 C, 4.3 Cr, 0.85 Mo, 3.8 V, 12 W, bal Fe.
For engravers' tools, blanking and forming dies; high speed steel.

ERKENZWEIG NHH.
M-1336; 0.82 C, 4.1 Cr, 0.85 Mo, 1.6 V, 8.7 W, bal Fe.
For lathe and planer tools, reamers, broaches; high speed steel.

ERKENZWEIG NHPC.
M-1336; 0.86 C, 4.3 Cr, 0.85 Mo, 2.1 V, 12 W, bal Fe.
For lathe and planer tools, reamers, broaches; high speed steel.

ERKENZWEIG NHUC.
M-1336; 0.86 C, 0.85 Mo, 2.5 V, 12 W, 4.1 Cr, bal Fe.
For lathe and planer tools, drills, taps, hobs; high speed steel.

ERKENZWEIG PKU.
M-1336; 1.0 C, 0.1 V, 0.25 Mn, bal Fe.
Heat treated: 200,000 TS; 130,000 YS; 8 El; 25 RA; 400 Brin.
For reamers, drills, drawing and forming dies; oil or water hardened, wear resistant.

ERKENZWEIG PLDI.
M-1336; 0.55 C, 0.18 V, 1.85 W, 0.9 Si, 1.05 Cr, bal Fe.
For heading and forging dies, upsetters, punches; oil hardened, tough.

ERKENZWEIG PLD III.
M-1336; 0.35 C, 0.9 Si, 1.05 Cr, 0.18 V, 1.85 W, bal Fe.
For heading and forging dies, upsetters, punches; oil hardened, tough.

ERKENZWEIG PSA.
M-1336; 1.45 C, 1.4 Cr, 0.6 Mn, bal Fe.
For blanking and forming dies, cutters, punches; oil hardened, wear resistant.

ERKENZWEIG PSX.
M-1336; 1.4 C, 0.30 Cr, 0.10 V, 0.30 Mn, bal Fe.
For blanking and forming dies, bearings; oil or water hardened.

ERKENZWEIG SPEZIAL K.
M-1336; 2.1 C, 11.5 Cr, 0.30 Mn, bal Fe.
For blanking and forming dies, punches; oil hardened, non-deforming.

ERKENZWEIG SPWC.
M-1336; 0.30 C, 2.0 Co, 2.4 Cr, 0.25 V, 8.5 W, bal Fe.
For extrusion rams, dies, liners; oil hardened, hot work steel.

ERKENZWEIG SWK III.
M-1336; 0.67 C, 1.3 Si, 0.5 Mn, 0.5 Cr, bal Fe.
For springs, shear blades, punches; oil hardened, shock resistant.

ERKENZWEIG SWK V.
M-1336; 0.45 C, Cr, Ni, bal Fe.
For gears, bolts, crankshafts, fasteners; oil hardened, shock resistant.

ERKENZWEIG SWK VI.
M-1336; 2.1 C, 11.5 Cr, 0.70 W, bal Fe.
For blanking and forming dies, punches; oil hardened, non-deforming.

ERKENZWEIG SWK VII.
M-1336; 0.45 C, W, Cr, V, bal Fe.
For shear blades, forging and heading dies; oil hardened, tough.

ERKENZWEIG SWK VIII.
M-1336; 0.30 C, 2.65 Cr, 0.35 V, 8.5 W, bal Fe.
For extrusion rams, dies and liners; hot work steel, oil hardened.

ERKENZWEIG WCR.
M-1336; 1.42 C, W, V, bal Fe.
For blanking and heading dies; oil hardened, wear resistant.

ERKENZWEIG WNC 40.
M-1336; 0.50 C, 1.05 Cr, 3.25 Ni, bal Fe.
For gears, bolts, crankshafts; oil hardened, shock resistant.

ERMAL.
M-51; 2.2-2.4 C, 0.9-1.0 Si, 0.7-0.8 Mn, bal Fe.

ERODUR 15 3.
M-1290; 3.50 C, 15.0 Cr, 2.7 Mo, bal Fe.
Abrasion resistant cast steel; G-X 350 CrMo 15 3.

ERODUR 16 SR.
M-1290; 1.6 C, 12.0 Cr, 0.75 Mo, 0.25 V, bal Fe.
Hardenable to 500-600 Brin.
Castings for wear resistant parts or tools.
DIN G-X160CrMoV12.

ERODUR 16 SRV.
M-1290; 1.60 C, 12.0 Cr, 0.75 Mo, 1.0 V, bal Fe.
Abrasion resistant cast steel.
SEW 1.2379; G-X 155 CrVMo 12 1.

ERODUR 28.
M-1290; 2.7 C, 28.0 Cr, bal Fe.
Hardenable to 500-600 Brin.
Castings for wear resistant parts or tools.
DIN G-X270Cr29.

ERODUR CM 1H.
M-1290; 0.42 C, 1.0 Cr, 0.20 Mo, bal Fe.
Cast, heat treated: 118,000-184,000 TS; 100,000-145,000 YS; 9-4 El.
Castings for wear resistant parts or tools.
DIN GS-42CrMo4; Werkstoff Nr. 1.7225.

ERODUR MNA 1.
M-1290; 1.2 C, 13.0 Cr, bal Fe.
Sol. ann. + age: 84,000 min TS; 42,500 min YS; 25 min El.
Austenitic manganese steel casting; work hardens to hard, wear resistant martensite.
DIN G-X120Mn12; Werkstoff 1.3401.

ERODUR V 2 Z.
M-1290; 0.32-0.39 C, 0.30-0.50 Si, 0.50-0.80 Mn, 2.2-2.7 Cr, 0.30-0.50 Mo, 0.05-0.15 V, bal Fe.
Abrasion resistant cast steel.
GS-35 CrMoV 10 4; SEW 1.1755.

ERW 3.
M-1655; 1.40 C, 0.30 Mn, 0.20 Si, 0.30 Cr, 0.25 V, 3.0 W, bal Fe.
Hardenable to 67 Rc for scrapers, rifflers, and engraving tools.
W.-Nr. 1.2562.

ES 1.
M-1736; 0.45 C, 0.95 Si, 0.30 Mn, 1.0 Cr, 0.20 Mo, 1.95 W, bal Fe.
Shock resisting tool steel.
W.-Nr. 1.2542; similar to AISI S1.

ES-15-35.
M-1014; 0.15 max C, 17-20 Ni, 34-37 Cr, bal Fe.
Annealed: 80,000 TS; 40,000 YS; 40 El; 150 Brin.
For salt pots, furnace equipment, heat treating boxes; heat resistant, austenitic; Type 330.

E.S.A.
M-73; 1.4 C, 4 W, 0.5 Cr, 0.3 V, bal Fe.
For finishing tools, reamers; abrasion resistant.

ESC.
M-1655; 2.20 C, 0.40 Mn, 0.30 Si, 12.0 Cr, 0.12 V, bal Fe.
Hot or cold work tool steel.
For zinc die casting dies, hammer cores.
For cold heavy duty stamping dies.
W.-Nr. 1.2080; similar to AISI D3.

E.S.C. ALLOYS see B.S.C. ALLOYS.

ESCO 1B.
M-438; 0.30 C, 1.0 Mn, 0.40-0.60 Si, bal Fe.
Cast: 79,000 TS; 46,000 YS; 30 El; 47 RA; 163 Brin.
For housings, gears, shafts, machine tool parts; water hardened.

ESCO 5T.
M-438; 0.20 max C, 8-10 Cr, 0.9-1.2 Mo, bal Fe.
Cast: 90,000-105,000 TS; 65,000-85,000 YS; 18-22 El; 40-55 RA; 200-240 Brin.
For oil refinery equipment; Type HA; corrosion resistant.

ESCO 5W.
M-438; 0.20 max C, 4.0-6.5 Cr, 0.45-0.65 Mo, bal Fe.
Cast: 90,000-100,000 TS; 75,000-80,000 YS; 18-24 El; 45-60 RA; 190-248 Brin.
For oil refinery equipment; Type 501; corrosion resistant.

ESCO 5X.
M-438; 0.20 max C, 4.0-6.5 Cr, 0.45-0.65 Mo, bal Fe.
For oil refinery equipment; Type 502; corrosion resistant.

ESCO 6-T.
M-438; 0.15 max C, 3-4 Ni, 0.6 Mn, 0.5 Si, bal Fe.

ESCO 12M.
M-438; 0.25-0.35 C, 0.2-0.3 Mo, 1.0-1.3 Mn, 0.5 Si, bal Fe.
Cast: 102,000 TS; 81,000 YS; 21 El; 225 Brin.
Heat treated: 240,000 TS; 190,000 YS; 5 El; 495 Brin.
For earth moving equipment, dredging and cement equipment; wear, shock and abrasion resistant.

ESCO 14.
M-438; 1.15-1.25 C, 12-14 Mn, 0.4-1.0 Si, bal Fe.
Cast: 120,000 TS; 55,000 YS; 20 El; 200 Brin.
For rock and ore crushers, earth moving equipment; impact, shock and abrasion resistant.

ESCO 16W.
M-438; 0.20 max C, 1.0-1.5 Cr, 0.45-0.65 Mo, 0.5-0.8 Mn, bal Fe.
Cast: 80,000-90,000 TS; 50,000-70,000 YS; 22 El; 210-220 Brin.
For structural and steam turbine components; high temperature use.

ESCO 16Z.
M-438; 0.18 max C, 2.0-2.7 Cr, 0.9-1.2 Mo, bal Fe.
Cast: 70,000-100,000 TS; 40,000-75,000 YS; 21-20 El; 200-220 Brin.
For structural and steam turbine components; high temperature use.

ESCO 18CW.
M-438; 0.65-0.75 C, 1.9-2.4 Cr, 0.4-0.6 Ni, 0.25 max Mo, bal Fe.
Cast: 135,000-180,000 TS; 118,000-130,000 YS; 12-4 El; 290-390 Brin.
For liners for abrasion slurries; wear resistant.

ESCO 20.
M-438; 0.07 max C, 19-22 Cr, 27.5-31 Ni, 1.8-2.5 Mo, 3.0-3.5 Cu, bal Fe.
Cast: 68,000 TS; 30,000 YS; 47 El; 60 RA; 125 Brin.
For acid resistant equipment; Type CN-7M; austenitic, corrosion resistant.

ESCO 20E.
M-438; 0.07 max C, 19-21 Cr, 30-38 Ni, 2.0-3.0 Mo, 3.0-4.0 Cu, bal Fe.
Cast: 68,000 TS; 28,000 YS; 47 El; 60 RA; 160 Brin.
Austenitic; for acid resistant equipment; specific for H_2SO_4.

ESCO 22.
M-438; 0.30 max C, 62-68 Ni, 3 max Fe, bal Cu.
Cast: 75,000 TS; 32,000 YS; 50 El; 50 RA; 120 Brin.
For corrosion resistant parts; cast Monel, corrosion resistant.

ESCO 22H.
M-438; 0.35 max C, 0.5-1.5 Mn, 2.5 max Fe, 1 Si, 62-68 Ni, bal Cu.
Cast: 85,000-120,000 TS; 50,000-80,000 YS; 20-10 El; 175-250 Brin.
For chemical plant equipment; cast H-Monel, corrosion resistant.

ESCO 22S.
M-438; 0.35 max C, 0.5-1.5 Mn, 2.5 max Fe, 1.0-2.2 Si, 62-68 Ni, bal Cu.
Cast: 120,000-145,000 TS; 80,000-130,000 YS; 4-1 El; 275-350 Brin.
For chemical plant equipment; age-hardenable, cast S-Monel, corrosion resistant.

ESCO 23.
M-438; 0.20 max C, 12-15 Cr, 76-79 Ni, 6 max Fe.
Cast: 80,000 TS; 37,000 YS; 16 El; 5 RA; 150 Brin.
At 1200°F: 63,000 TS; 25,000 YS; 39 El.
At 1800°F: 72,000 TS; 3800 YS; 118 El.
For nitriding containers; cast Inconel, corrosion and heat resistant.

ESCO 26.
M-438; 0.12 max C, 26-30 Mo, 4-7 Fe, bal Ni.
Cast: 78,000 TS; 56,000 YS; 12 El; 12 RA; 196 Brin.
For corrosion resistant equipment; Hastelloy B; corrosion resistant.

ESCO 27.
M-438; 0.15 max C, 15.5-17.5 Cr, 16-18 Mo, 3.7-4.7 W, 4.0-7.0 Fe, bal Ni.
Cast: 70,000 TS; 50,000 YS; 15 El; 22 RA; 200 Brin.
For corrosion resistant equipment; Hastelloy C.

ESCO 28.
M-438; 11 max Si, 4 max Cu, 1 Mn, 1 Al, bal Ni.
Cast: 36,000-40,000 TS; 35,000-40,000 YS; 0 El; 0 RA; 484-545 Brin.
For pump valves; Hastelloy D; resists HCl and H_2SO_4.

ESCO 32.
M-438; 0.20-0.40 C, 12 Cr, 1 max Ni, bal Fe.
Heat treated: 200,000 TS; 150,000 YS; 5 El; 7 RA; 450 Brin.
For cutlery, valves, furnace parts; Type CA40; corrosion resistant.

ESCO 32B.
M-438; 0.06-0.15 C, 11.5-13.5 Cr, 0.25 max Ni, bal Fe.
Heat treated: 160,000 TS; 145,000 YS; 12 El; 25 RA; 360 Brin.
For turbine blades, valves, cutlery; Type 403; corrosion and heat resistant.

ESCO 32C.
M-438; 0.06-0.15 C, 11.5-14 Cr, 1 max Ni, 0.5 max Mo, bal Fe.
Heat treated: 200,000-135,000 TS; 150,000-115,000 YS; 7-17 El; 6-43 RA; 390-260 Brin.
Cast: 115,000-100,000 TS; 100,000-75,000 YS; 20-30 El; 52 RA; 225-185 Brin.
For cutlery, valves, pump parts; Type CA15; corrosion resistant.

ESCO 33-A.
M-438; 0.60-0.75 C, 1 Mn, 16-18 Cr, 0.75 Mo, bal Fe.
Heat treated: 400 Brin.
For bearings, races, pivots, valves, cutlery; Type 440A; stainless, hardenable.

ESCO 33-B.
M-438; 0.75-0.95 C, 16-18 Cr, 0.75 max Mo, 1 Mn, 1 Si, bal Fe.
Annealed: 107,000 TS; 62,000 YS; 18 El; 35 RA; 220 Brin.
For bearings, cutlery, valves; Type 440 B; stainless, hardenable.

ESCO 33C.
M-438; 0.95-1.2 C, 16-18 Cr, 0.75 max Mo, bal Fe.
For bearings, valves; Type 440C; corrosion resistant.

ESCO 33 G.
M-438; 0.06 max C, 11.5-14.0 Cr, 3.5-4.5 Ni, 0.5-1.0 Mo, bal Fe.
Ht Tr: 110,000 min TS; 80,000 min YS; 15 El; 35 RA; 363 Brin.
Type CA6NM corrosion resistant, for large propellers, water wheels, impellers, valves, pump parts.

ESCO 35AW.
M-438; 2.3-2.8 C, 1.25 Mn, 24-28 Cr, bal Fe.
Cast: 630 Brin.
Annealed: 370 Brin.
For roasters, impellers; high abrasion resistance.

ESCO 35H.
M-438; 0.50 max C, 26-30 Cr, 4 max Ni, bal Fe.
At 1400°F: 10,500 TS; 8700 YS; 65 El.
At 1800°F: 2500 TS; 2100 YS; 110 El.
at 70°F: 75,000 TS; 66,000 YS.
For furnace rabble arms, sintering bars, grates, dampers; heat resistant to 2000°F in S atmosphere.

ESCO 35-T.
M-438; 0.50 C, 26-30 Cr, 4-7 Ni, 0.5 max Mo, bal Fe.
Cast: 97,000 TS; 65,000 YS; 18 El; 212 Brin.
For cylinder liners, valves, furnace parts; Type CC50; corrosion and heat resistant.

ESCO 36.
M-438; 0.20 max C, 26-30 Cr, 3-4 Ni, 1.5 Mo, bal Fe.
For corrosion and heat resisting parts; acid resistant.

ESCO 36F.
M-438; 0.07 max C, 14.0-15.5 Cr, 3.5-5.5 Ni, 2.5-4.5 Cu, bal Fe.
Ht Tr (age hardenable 925 F. age): 175,000 min TS; 150,000 min YS; 5 Elon; 375 Brin.
High strength cast, corrosion resistant steel. CB 7 Cu-2 15-5 PH.

ESCO 36PH.
M-438; 0.07 C, 16 Cr, 4 Ni, 2.5 Cu, bal Fe.
Heat treated: 160,000 TS; 145,000 YS; 12 El; 25 RA; 360 Brin.
For corrosion resistant castings; age-hardenable, high strength.

ESCO 37 PH.
M-438; 0.04 C, 1 Mn, 1 Si, 25-27 Cr, 4.75-6.0 Ni, 1.75-2.25 Mo, 2.75-3.25 Cu, bal Fe.
Heat treated: 130,000-144,000 TS; 95,000-105,000 YS; 10-22 El; 15-41 RA; 285-321 Brin.
For chemical plant equipment, digesters, autoclaves; high corrosion resistance.

ESCO 40B.
M-438; 0.08 max C, 18-21 Cr, 8-11 Ni, 1.75-2.25 B, bal Fe.
Cast: 80,000 TS; 50,000 YS; 2 El; 302 Brin.
Neutron absorbing stainless steel; for control rods in nuclear reactors.

ESCO 40F.
M-438; 0.16 max C, 18-21 Cr, 9-12 Ni, 1.5 max Mo, 0.20-0.35 Se, bal Fe.
Cast: 72,000 TS; 35,000 YS; 58 El; 60 RA; 140 Brin.
For chemical plant equipment; Type CF-16F; corrosion resistant, free-cutting.

ESCO 40H.
M-438; 0.20-0.40 C, 18-23 Cr, 9-12 Ni, 0.5 max Mo, bal Fe.
At 1200°F: 57,000 TS; 16 El.
At 1600°F: 22,000 TS; 22 El.
At 70°F: 90,000 TS; 50,000 YS; 25 El.
For furnace dampers, oil still supports, annealing furnaces; strength and scale resistant to 1600°F; Type HF.

ESCO 40L.
M-438; 0.03 max C, 18-21 Cr, 8-11 Ni, bal Fe.
Cast: 65,000 TS; 28,000 YS; 55 El; 60 RA; 140 Brin.
For chemical plant equipment, tanks, mixers; Type 304L; austenitic, stainless.

ESCO 40S.
M-438; 0.08 max C, 18-21 Cr, 8-11 Ni, bal Fe.
Cast: 72,000 TS; 32,000 YS; 60 El; 65 RA; 140 Brin.
For chemical plant equipment; Type CF-8; austenitic, corrosion resistant.

ESCO 40T.
M-438; 0.08 max C, 17-20 Cr, 9-12 Ni, bal Fe.
Cast: 65,000 TS; 28,000 YS; 45 El; 50 RA; 125 Brin.
For chemical plant equipment; Type 305; austenitic, stainless.

ESCO 41.
M-438; 0.08 max C, 18-21 Cr, 9-12 Ni, Cb = 8 x C, bal Fe.
Cast: 72,000 TS; 36,000 YS; 35 El; 35 RA; 145 Brin.
For welded chemical plant equipment; Type CF8C; austenitic, stainless.

ESCO 43C.
M-438; 0.20 max C, 22-26 Cr, 12-15 Ni, bal Fe.
Cast: 70,000 TS; 30,000 YS; 35 El; 60 RA; 150 Brin.
For furnace parts, heat treat boxes; Type CH-20; austenitic, heat resistant.

ESCO 43H.
M-438; 0.20-0.50 C, 24-28 Cr, 11-14 Ni, 1.5 max Mo, 0.2 N, bal Fe.
At 1400°F: 35,000 TS; 18,000 YS; 12 El.
At 1800°F: 11,000 TS; 7000 YS; 30 El.
At 70°F: 85,000 TS; 45,000 YS; 15 El.
For furnace shafts, beams and rollers, tube supports; Type HH; scale resistant at 2000°F.

ESCO 44.
M-438; 0.03 max C, 17.5-18.5 Cr, 12.0-13.0 Ni, 2.0-3.0 Mo, 0.2-0.3 Cb 0.5-1.0 Mn, bal Fe.
Cast: 71,500-76,500 TS; 42,750-48,250 YS; 39-47 El; 45-50 RA; 170 Brin.
For heat exchanger tubes; corrosion and heat resistant.

ESCO 45L.
M-438; 0.03 max C, 18-21 Cr, 10-13 Ni, 2-3 Mo, bal Fe.
Cast: 70,000 TS; 35,000 YS; 40 El; 55 RA; 170 Brin.
For acid resistant chemical plant equipment; Type 316L; austenitic, stainless.

ESCO 45M.
M-438; 0.08 max C, 18-21 Cr, 9-13 Ni, 3-4 Mo, bal Fe.
Cast: 75,000 min TS: 35,000 min YS; 25 El; 170 Brin.
For chemical plant equipment, sulfite pulp mills, acid resistant service.
Type CF8M austenitic stainless.

ESCO 45S.
M-438; 0.08 max C, 18-21 Cr, 9-12 Ni, 2-3 Mo, bal Fe.
Cast: 80,000 TS; 42,000 YS; 45 El; 65 RA; 170 Brin.
For acid resistant equipment; Type CF-8M; stainless.

ESCO 45T.
M-438; 0.08 max C, 16-19 Cr, 13-15 Ni, 2-3 Mo, bal Fe.
Cast: 70,000 TS; 30,000 YS; 50 El; 62 RA; 125 Brin.
For acid resistant equipment; Type 316; stainless, austenitic.

ESCO 46.
M-438; 0.2-0.5 C, 26-30 Cr, 14-18 Ni, bal Fe.
For heat resistant parts, furnace equipment; stainless and heat resistant, austenitic.

ESCO 48.
M-438; 0.2-0.6 C, 28-32 Cr, 18-22 Ni, bal Fe.
Cast: 85,000 TS; 50,000 YS; 10 El; 190 Brin.
For furnace parts and equipment, salt pots; Type HL; heat resistant.

ESCO 49.
M-438; 0.05 max C, 8-11 Mn, 14.5-17.5 Cr, 18-27 Ni, 1.75-2.75 Mo, bal Fe.
Cast RT: 65,000 TS; 30,000 YS; 50 El; 12.5 ft-lbs. 423 F.: 125,000 TS; 80,000 YS; 40 El.
Non-magnetic; very low temperature service; liquid hydrogen, liquid nitrogen; excellent weldability.
Kromarc 55.

ESCO 50.
M-438; 0.12 max C, 2.5-3.0 Si, 20-23 Cr, 25-28 Ni, 2.5 Mo, 1.5 Cu, bal Fe.
Cast: 70,000 TS; 35,000 YS; 50 El; 60 RA; 140 Brin.
For salt pots, heat and corrosion resistant parts; Type CN-M-Cu; austenitic, corrosion resistant.

ESCO 51.
M-438; 0.30 max C, 26-30 Cr, 8-11 Ni, 0.5 max Mo, bal Fe.
Cast: 90,000 TS; 57,000 YS; 25 El; 34 RA; 200 Brin.
For furnace parts, heat treat boxes; Type CE-30 austenitic, stainless.

ESCO 51 H.
M-438; 0.20-0.50 C, 26-30 Cr, 8-11 Ni, bal Fe.
Cast: 85,000 min TS; 40,000 min YS; 9 El.
Heat resistant service, good resistance to high sulphur content gases, ore roasting equipment.
ACI Type HE.

ESCO 52C.
M-438; 0.10 C, 14-17 Cr, 33-37 Ni, bal Fe.
Cast: 70,000 TS; 40,000 YS; 10 El; 12 RA; 170 Brin.
For salt pots, furnace pots, heat treat boxes; Type HT; heat resistant.

ESCO 52H.
M-438; 0.35-0.75 C, 13-17 Cr, 33-37 Ni, 0.5 max Mo, bal Fe.
At 1200°F: 42,000 TS; 28,000 YS; 5 El.
At 1800°F: 11,000 TS; 8000 YS; 28 El.
At 70°F: 67,000 TS; 40,000 YS; 10 El.
For retorts, radiant tubes, salt pots, hearth plates, carburizing boxes; resists oxidation and thermal cycling.

ESCO 53C.
M-438; 0.20 max C, 23-27 Cr, 19-22 Ni, bal Fe.
Cast: 75,000 TS; 35,000 YS; 35 El; 42 RA; 140 Brin.
For furnace parts, salt pots, heat treat boxes; Type CK-20; austenitic, heat resistant.

ESCO 53H.
M-438; 0.20-0.60 C, 24-28 Cr, 18-22 Ni, 0.5 max Mo, bal Fe.
At 1600°F: 23,000 TS; 21 El.
At 70°F: 75,000 TS; 40,000 YS; 18 El.
For gas dissociation equipment, fixtures, baskets, calcining tubes; good strength and oxidation resistance to 2100°F; Type HK.

ESCO 54.
M-438; 0.2-0.5 C, 19-23 Cr, 23-27 Ni, bal Fe.
For heat treating boxes, furnace parts; Type HN; heat resistant.

ESCO 55.
M-438; 0.3-0.5 C, 17-21 Cr, 37-41 Ni, bal Fe.
For salt pots, heat treat boxes, furnace parts; Type HU; heat resistant.

ESCO 56.
M-438; 0.35-0.75 C, 10-14 Cr, 58-62 Ni, bal Fe.
At 1400°F: 32,000 TS; 23,000 YS.
At 1800°F: 10,000 TS; 8000 YS; 40 El.
At 70°F: 70,000 TS; 40,000 YS; 6 El.
For carburizing and hardening fixtures, retorts; Type HW; resists thermal cycling.

ESCO 57.
M-438; 0.3-0.7 C, 15-19 Cr, 64-68 Ni, bal Fe.
For heat treat boxes, furnace equipment; heat and corrosion resistant.

ESCO 58.
M-438; 0.40-0.60 C, 24-28 Cr, 33-37 Ni, 4.5-5.5 W, bal Fe.
For heat resistant service at 2000-2200°F.

ESCO 61A.
M-438; 3 max C, 1.7-2.5 Cr, 13.5-17.5 Ni, 5.5-7.5 Cu, bal Fe.
Cast: 25,000 TS; 135 Brin.
For corrosion and wear resistant parts; Ni-Resist Type 1; corrosion resistant cast iron.

ESCO 62A.
M-438; 3.0 max C, 1.7-2.5 Cr, 18-22 Ni, bal Fe.
Cast: 25,000 TS; 135 Brin.
For corrosion and wear resistant parts; Ni-Resist Type 2; corrosion resistant cast iron.

ESCO 63.
M-438; 3 max C, 2 Cr, 20 Ni, bal Fe.
For corrosion and wear resistant parts; Ni-Resist Type 2B; corrosion resistant cast iron.

ESCO 70.
M-438; 0.08-0.16 C, 1-2 Mn, 20.0-22.5 Cr, 19-21 Ni, 2.5-3.5 Mo, 18.5-21.0 Co, 2-3 W, 0.75-1.25 Cb, 0.15 N, bal Fe.
For high temperature applications; Multimet N-155.

ESCO 72.
M-438; 0.1 C, 0.18 Ti, 0.6 Cb, 26-30 Cr, 47-52 Co, bal Fe.
Cast: 135,000 TS; 48,000 YS; 10 El; 10 RA; 250 Brin.
For furnace baffles, burner tips, sintering grates, quench baskets.
Corrosion, heat and thermal shock resistant.

ESCO 625.
M-438; 0.10 max C, 20-23 Cr, 8-10 Mo, 3.15-4.15 Cb, 4.5 max Fe, bal Ni.
Electroslag remelted billet, forging stock and shaped forms.
Inconel 625 Type.

ESCO 718C.
M-438; 0.08 C, 17-21 Cr, 2.0-4.0 Mo, 4.5-5.75 Cb, 0.40-1.30 Ti, 19-21 Fe, bal Ni.
Electroslag remelted billet, forging stock.
Inconel 718C Type.

ESCO 800.
M-438; 0.05 C, 21 Cr, 32.5 Ni, 0.40 Ti, 0.40 Al, bal Fe.
Electroslag remelted billet, forging stock and shaped forms.
Incoloy 800.

ESCO 802.
M-438; 0.35 C, 21 Cr, 32.5 Ni, 0.55 Al, 0.75 Ti, bal Fe.
Electroslag remelted billet, forging stock and shaped forms.
Incoloy 802.

ESCO A286.
M-438; 0.08 max C, 13.5-16.0 Cr, 24-27 Ni, 1.0-1.5 Mo, 1.9-2.35 Ti, bal Fe.
Electroslag remelted billet, forging stock, and shaped forms.
AISI Type 660.

ESCO N155 (ESCO 70 CASTINGS).
M-438; 0.8-0.16 C, 20-22.5 Cr, 19-21 Ni, 18.5-21 Co, 2.5-3.5 Mo, 0.75-1.25 W, bal Fe.
Electroslag remelted billet, forging stock and shaped forms.
AiSI Type 661, N155.

ESCO NO. 21.
M-438; 0.70 max C, 96.5 min Ni, 2 max Si, 0.75 max Mn, bal Fe.
Cast: 60,000 TS; 25,000 YS; 25 El; 30 RA; 120 Brin.
For corrosion and heat resistant castings; cast nickel, corrosion and heat resistant.

ESCO NO. 35C.
M-438; 0.50 C, 28 Cr, 3 max Ni, bal Fe.
For furnace parts, heat resistant parts; Type CC-50.

E.S.D. 31.
M-1754; 20.7 Fe, 11.6 Co, 67.7 Pb.
Permanent magnet.
Residual flux density, Br. (gauss) 5000.
Coercive force, Hc (oersteds) 1000.
Anisotropic.

E.S.D. 32.
M-1754; 18.3 Fe, 10.3 Co, 72.4 Pb.
Permanent magnet.
Residual flux density, Br (gauss) 6800.
Coercive force, Hc (oersteds) 960.
Anisotropic.

E.S.D. 41.
M-1754; 20.7 Fe, 11.6 Co, 67.7 Pb.
Permanent magnet.
Residual flux density, Br (gauss) 3600.
Coercive force, Hc (oersteds) 970.

E.S.D. 42.
M-1754; 18.3 Fe, 10.3 Co, 72.4 Pb.
Residual flux density, Br (gauss) 4800.
Coercive force, Hc (oersteds) 830.

ESMO.
M-1655; 1.65 C, 0.30 Mn, 0.30 Si, 12.0 Cr, 0.60 Mo, 0.50 W, 0.10 V, bal Fe.
For broaches, coining, punching dies for thin sheet.
W.-Nr. 1.2601; similar to AISI D2.

ESMO 2.
M-1655; 1.65 C, 12.0 Cr, 0.80 Mo, 0.90 V, bal Fe.
Air or oil hardenable to 58-63 Rc.
For punching and coining dies for sheet metal.
Similar to AISI D2.

ESS.
M-1655; 2.10 C, 0.30 Mn, 0.30 Si, 12.0 Cr, 0.80 W, 0.10 V, bal Fe.
Cold work tool steel for heavy duty shears, punching and trimming dies.
W.-Nr. 1.2436.

ESS 1.
M-1655; 2.1 C, 0.35 Mn, 0.30 Si, 12.5 Cr, 1.20 Co, 0.50 Mo, 0.70 W, bal Fe.
Cold work tool steel for stamping transformer laminations.
W.-Nr. 1.2884.

ESSHETE 316.
M-1724; 0.04-0.09 C, 0.8 max Si, 1.0-2.0 Mn, 11.0-14.0 Ni, 16.0-18.0 Cr, 2.0-2.75 Mo, B, bal Fe.
Sol. tr.: 74,000 min TS; 30,000 min YS; 30 El.
For superheater tubes, power plant components.
B.S. 3605-855 (1963); ASTM 213-64T TP316H.

ESSHETE 321.
M-1724; 0.04-0.09 C, 1.0 max Si, 2.0 max Mn, 9.0-11.0 Ni, 17.0-19.0 Cr, 0.70 max Ti, bal Fe.
Sol. tr.: 74,000 min TS; 30,000 min YS; 30 El.
For superheater tubes, power plant components.
B.S. 970-En58B; ASTM 213-64T TP321H; AISI 321.

ESSHETE 347.
M-1724; 0.04-0.09 C, 0.8 max Si, 0.5-2.0 Mn, 9.0-13.0 Ni, 17.0-19.0 Cr, Nb = 10 x %C min, bal Fe.
Sol. tr.: 74,000 min TS; 32,600 min YS; 30 El.
For superheater tubes, power plant components.
B.S. 3605-822 Nb(1963): AISI 347; ASTM 213-64T TP347H.

ESSHETE 600.
M-1724; 0.15 max C, 0.50 max Si, 1.0 max Mn, 14.0-17.0 Cr, 6.0-10.0 F 0.50 max Cu, bal Ni (+ Co).
Good high temperature strength and corrosion properties; above 600°C.

ESSHETE 800.
M-1724; 0.10 max C, 1.0 max Si, 1.5 max Mn, 19.0-23.0 Cr, 30.0-35.0 N 0.75 max Cu, 0.15-0.60 Al, 0.15-0.60 Ti, bal Fe.
Good high temperature strength and corrosion properties.

ESSHETE 800L.
M-1724; 0.03 max C, 1.0 max Si, 1.5 max Mn, 20.0-23.0 Cr, 32.0-35.0 N 0.75 max Cu, 0.15-0.60 Al, 0.15-0.60 Ti, 0.03 max N, bal Fe.
Good high temperature strength and improved corrosion properties.

ESSHETE 1250.
M-1724; 0.15 max C, 0.2-1.0 Si, 5.5-7.0 Mn, 9.0-11.0 Ni, 14.0-16.0 Cr, 0.8-1.2 Mo, 0.15-0.4 V, 0.75-1.25 Nb, 0.003-0.009 B, bal Fe.
Sol. tr.: 72,000 min TS; 25,800 min YS; 30 El.
For superheater tubes, pressure vessels, steam piping, parts operating up to 650°C.

ESSHETE CML.
M-1724; 0.12 max C, 0.4-0.7 Mo, 0.75-1.25 Cr, bal Fe.
Normalized: 62,000-78,000 TS; 22,000-25,000 YS; 45-41 El; 70-63 RA.
For superheater tubes, steam piping; resists heat to 1000°F; creep resistant.
BS 1507, 1508; ASTM A199-64T-Gr T11.

ESSHETE CRM1.
M-1724; 0.10-0.15 C, 1.0 Cr, 0.5 Mo, bal Fe.
For applications up to 540°C.

ESSHETE CRM2.
M-1724; 0.15 max C, 2.0-2.5 Cr, 0.9-1.1 Mo, bal Fe.
Normalized: 93,000 TS; 80,000 YS; 30 El.
Annealed: 72,500 TS; 34,500 YS; 40 El.
For steam piping, superheater tubes; creep resistance to 1100°F.
B.S. 1503, 1508; ASTM A200-62T GrT22.

ESSHETE CRM5.
M-1724; 0.12 max C, 4.0-6.0 Cr, 0.45-0.65 Mo, bal Fe.
Annealed: 67,000 TS; 30,000 YS; 40 El.
For heat exchangers, valves, superheaters, steam and oil tubes; resists scaling to 1100°F.

ESSHETE CRM9.
M-1724; 0.15 max C, 0.25-1.0 Si, 0.3-0.6 Mn, 8.0-10.0 Cr, 0.90-1.10 Mo, bal Fe.
.2% proof stress at 540°C: 32,500-39,000 psi.
B.S. 1607-GrP17: ASTM 199 Gr T9; DIN X9CrMo91: Werkstoff Spec 213.

ESSHETE CRM 12.
M-1724; 0.15 max C, 12.0 Cr, 1.0 Mo, 0.25 V, bal Fe.
For long time rupture strength up to about 600°C.
For power plant components, especially superheater tubes.

ESSHETE D4C.
M-1724; 0.15 C, 0.5 Mo, bal Fe.
Normalized: 56,000-68,000 TS.
For superheater tubes, steam piping; resists heat to 950°F: creep resistant.

ESSHETE MV.
M-1724; 0.08-0.15 C, 0.4-0.7 Mn, 0.25-0.50 Cr, 0.5-0.7 Mo, 0.22-0.30 V, bal Fe.
Nor. & Temp.: 67,000 min TS; 42,200 min YS; 17 El.
For steam piping.
B.S. 3604 CD.660(1963).

ESS SPEZIAL.
M-1655; 1.65 C, 0.30 Mn, 0.30 Si, 12.0 Cr, 1.3 Co, 0.50 Mo, bal Fe.
Cold work tool steel for stamping transformer laminations.
W.-Nr. 1.2880.

ESSW.
M-1655; 1.65 C, 0.40 Mn, 0.30 Si, 12.0 Cr, 0.80 W, 0.10 V, bal Fe.
Oil or air hardenable to 60-63 Rc.
For cold punching and forming dies.

ESTHONIA 5 MARK.
M-Eng.; 70 Cu, 20 Zn, 10 Ni.
For coinage; corrosion resistant.

ESV.
M-1655; 2.20 C, 0.40 Si, 0.40 Mn, 12.5 Cr, 1.10 Mo, 4.0 V, bal Fe.
Hardenable to 65-66 Rc.
For cold stamping and coining; for precision rollers.

ESW.
M-1655; 1.65 C, 0.30 Mn, 0.30 Si, 12.0 Cr, 0.10 V, bal Fe.
Cold work tool steel for punching and shearing dies, trimming and coining dies.
W.-Nr. 1.2201; similar to AISI D2.

ET-8.
M-1766; 0.10 max C, 0.35 Mn, 0.06 max Si, bal Fe.
Low carbon steel for cold heading.

ET-21.
M-1766; 0.20 C, 0.40 Mn, 0.08 max Si, bal Fe.
Steel for cold heading and extrusion.

E.T.D see also **COLFORM E.T.D.**

"E.T.D."150.
M-669; 0.40 min C, 0.75-1.0 Mn, 0.20-0.35 Si, 0.80-1.10 Cr, 0.15-0.25 Mo, 0.04-0.05 Te, bal Fe.
Bar: 150,000 min TS; 130,000 min YS; 10 El; 32 min Rc; 75% machinability.
For shafts, studs, bolts, gears; high strength without heat treatment.

"E.T.D."180.
M-669; 0.40 min C, 0.75-1.0 Mn, 0.20-0.35 Si, 0.80-1.10 Cr, 0.15-0.25 Mo, bal Fe.
Bar: 180,000 min TS; 165,000 min YS; 5-10 El; 38 min Rc; 56% machinability.
For shafts, gears, studs, bolts; high strength without heat treatment.

ETERNOS 10 now **ETERNOS NI10.**

ETERNOS 23 see **ETERNOS A1 24.**

ETERNOS A1 7.
M-1339; 0.12 max C, 0.5-1.0 Si, 1.0 max Mn, 6.0-8.0 Cr, 0.5-1.0 Al, bal Fe.
For annealing boxes; elevated temperature.
W.-Nr. 1.4713.

ETERNOS A1 13.
M-1339; 0.12 max C, 0.7-1.2 Si, 1.0 max Mn, 12.0-14.0 Cr, 0.7-1.2 Al, bal Fe.
Furnace equipment; elevated temperature.
W.-Nr. 1.4724.

ETERNOS A1 18.
M-1339; 0.12 max C, 1.0-1.5 Si, 1.0 max Mn, 17.0-19.0 Cr, 0.7-1.2 Al, bal Fe.
For furnace parts; elevated temperature.
W.-Nr. 1.4742.

ETERNOS A1 24.
M-1339; 0.12 max C, 1.0-1.5 Si, 1.0 max Mn, 24.0 Cr, 1.2-1.7 Al, bal Fe.
For high temperature equipment.
W.-Nr. 1.4762.

ETERNOS CS.
M-1339; 0.10 max C, 1.5-1.8 Si, 1.0 max Mn, 1.5-2.0 Cr, bal Fe.
Non-highly stressed parts for elevated temperature.
W.-Nr. 1.4700.

ETERNOS NAT 32.
M-1339; 0.10 max C, 1.0 max Si, 1.5 max Mn, 19-23 Cr, 30-34 Ni, 0.4 Al, 0.4 Ti, bal Fe.
Parts for furnace and steam boiler equipment.
W.-Nr. 1.4876.

ETERNOS NI 4.
M-1339; 0.10-0.20 C, 1.15 Si, 2.0 max Mn, 25 Cr, 3.5-5.5 Ni, bal Fe.
For high temperature equipment.
W.-Nr. 1.4821.

ETERNOS NI 10.
M-1339; 0.12 max C, 17.0-19.0 Cr, 9.0-11.5 Ni, Ti = 4 x C min, bal Fe.
For furnace and boiler parts; high temperature.
W.-Nr. 1.4878; similar to AISI 321.

ETERNOS NI 12.
M-1339; 0.20 max C, 2.0 Si, 2.0 max Mn, 20 Cr, 12 Ni, bal Fe.
For annealing boxes and bells.
W.-Nr. 1.4828; similar to AISI 309.

ETERNOS NI 20.
M-1339; 0.20 max C, 2.0 Si, 2.0 max Mn, 25 Cr, 20 Ni, bal Fe.
For annealing and carburizing equipment.
W.-Nr. 1.4841; AISI 310.

ETERNOS NI 35.
M-1339; 0.15 max C, 1.0-2.0 Si, 2.0 max Mn, 15-17 Cr, 34-37 Ni, bal Fe.
For furnace parts for high temperature.
W.-Nr. 1.4864.

ETERNOS SI 6.
M-1339; 0.12 max C, 2.0-2.5 Si, 1.0 max Mn, 5.5-6.5 Cr, bal Fe.
For carburizing boxes.
W.-Nr. 1.4712.

ETERNOS SI 13.
M-1339; 0.12 max C, 1.9-2.4 Si, 1.0 max Mn, 12.0-14.0 Cr, bal Fe.
Furnace rails, grates; thermocouple tubes.
W.-Nr. 1.4722.

ETERNOS SI 18.
M-1339; 0.12 max C, 1.9-2.4 Si, 1.0 max Mn, 17.0-19.0 Cr, bal Fe.
For furnace parts; elevated temperature.
W.-Nr. 1.4741.

EUGENE VADERS.
M-Eng.; 57.5 Cu, 0.3 Fe, 2.5 Mn, 1.5 Al, 0.5 Si, bal Zn.
For propellers, gears, high strength castings; corrosion resistant.

EUREKA.
M-U.S.; 60 Cu, 40 Ni.
Hot rolled: 60,500-65,000 TS; 20,000-22,400 YS; 51-45 El; 67-65 RA.
For resistance materials and thermocouples; similar to "Constantan" and "Ferry Metals."

EUREKA.
Pb, Sb, bal Sn.
For Babbitt bearings; Babbitt metal.

EUREKA 1 ELECTRODE.
M-765; 2.50 C, Cr, W, high cobalt.
For surfacing parts subject to severe wear, heat and abrasion.

EUREKA 2 HIGH SPEED BARE ROD.
M-765; 0.75 C, 4.0 Cr, 4.0 Mo, W, plus high cobalt.
For welding high speed cutting tools all types by G.T.A. method.

EUREKA 4 HIGH SPEED BARE ROD.
M-765; 1.55 C, 12.0 W, 4.85 V, Cr, Co, bal Fe.
For welding high speed tool where a 63-65 Rock C hardness is required.

EUREKA 6 ELECTRODE.
M-765; 1.45 C, 28.5 Cr, 3.9 W, Ni, Si, high cobalt.
For surfacing punches, dies, shear blades subject to severe heat and impact.

EUREKA 12 ELECTRODE.
M-765; 1.40-1.45 C, Cr, Ni, W, Mo and high cobalt.
For surfacing parts subject to severe wear, heat and abrasion.

EUREKA 31-A.
M-765; same composition as AISI H-13 steel, W-free.
For welding die cast dies, hot work steel in general.

EUREKA 35 H. W. ELECTRODE.
M-765; 0.07 C, 4.35 Cr, 0.85 Ni, Mo, Mn, bal Fe.
For welding hot work die units, forging and upsetter dies.

EUREKA 45-A H. W. BARE ROD.
M-765; 0.08 C, Cr, Ni, Mo, Si, bal Fe.
Very low carbon hot work for welding hot work units that must be machinable, forging and die cast dies.

EUREKA 45 H. W. ELECTRODE.
M-765; 0.10 C, 4.25 Cr, 2.0 W, Mn, Mo, Ni, V, bal Fe.
For welding hot work die units, forging and upsetter dies.

EUREKA 45-N H.W. ELECTRODE.
M-765; 0.10 C, 4.25 Cr, 2.0 W, 25-30 Ni, Mn, Mo, V, bal Fe.
Tougher weld metal for welding hot work die units, forging and upsetter dies.

EUREKA 52 ELECTRODE.
M-765; Pure deoxidized copper.

EUREKA 60 NICKEL ELECTRODE.
M-765; 60 Ni, bal Fe.
For strong welds on all cast iron.

EUREKA 70-A O. H. BARE ROD.
M-765; 0.90 C, Cr, V, W, Mo, Mn, bal Fe.
For welding oil hard tool steel by G.T.A. process.

EUREKA 70-W O. H. ELECTRODE.
M-765; 0.65-0.75 C, 1.4 Cr, 2.15 W, Mn, V, Si, bal Fe.
For welding "W" type oil hardening tool steels.

EUREKA 71-M O. H. ELECTRODE.
M-765; 0.65-0.75 C, 1.4 Cr, 1.15 Mo, Mn, V, bal Fe.
For welding oil hardening tool and die steels.

EUREKA 72-A H. W. BARE ROD.
M-765; 0.36 C, 5.0 Cr, W, Mo, Si, bal Fe.
For welding hot work tool steel punches, shear blades, etc.

EUREKA 72 H. W. ELECTRODE.
M-765; 0.25-0.30 C, 5.0 Cr, 2.75 W, 2.25 Mo, Mn, V, bal Fe.
For welding hot work die steels.

EUREKA 73-A H. W. BARE ROD.
M-765; 0.50 C, 7.5 Cr, 1.48 V, Mo, Si, bal Fe.
For welding hot work shear blades and punches subject to high heat.

EUREKA 73 H. W. ELECTRODE.
M-765; 0.40-0.50 C, 4.25 Cr, 9.5 W, 1.4 Mo, bal Fe.
For welding extreme hot work die steels.

EUREKA 75-XA W. H. BARE ROD.
M-765; 1.15 C, Si, Mn, V, bal Fe.
For welding water hard tool steel by G.T.A. process.

EUREKA 75-X H. W. ELECTRODE.
M-765; 0.75-1.0 C, 0.80 Cr, 0.40 V, Mo, Mn, bal Fe.
For welding water hardening tools and dies.

EUREKA 78-A H. W. BARE ROD.
M-765; 0.50 C, 7.5 Cr, 1.48 V, Mo, Si, bal Fe.
For welding high C, high Cr and similar alloys where a high Rockwell C hardness is required.

EUREKA 88-A H. W. BARE ROD.
M-765; 0.39 C, 4.25 Cr, 4.1 W, V, Co, bal Fe.
For welding hot work dies where extreme heat and impact is a problem.

EUREKA 88 H. W. ELECTRODE.
M-765; 0.40-0.50 C, 5.0 Cr, 5.0 W, 3.50 Co, Mn, Mo, V, bal Fe.
For welding extreme hot work die steels.

EUREKA 99 NICKEL ELECTRODE.
M-765; 97.0 Ni, bal Fe.

EUREKA 100-A PURE NICKEL BARE ROD.
M-765; 99.0 Ni.

EUREKA 100 ELECTRODE.
M-765; 99 Ni, bal Fe.

EUREKA 130-A BARE ROD.
M-765; 0.32 C, Mo, Cr, Mn, Si, bal Fe.
Low alloy for welding flame hardened dies and SAE 4130 steel.

EUREKA 130 ALLOY ELECTRODE.
M-765; 0.25-0.35 C, 1.15 Cr, Mo, Mn, bal Fe.
For welding SAE 4130 and flame hardening units.

EUREKA 145-A BARE ROD.
M-765; 0.45-0.50 C, Cr, Mo, Mn, Si, bal Fe.
Medium alloy for welding flame hardened dies and 6145 steel.

EUREKA 145 ALLOY ELECTRODE.
M-765; 0.40-0.50 C, 1.0 Cr, Mn, Mo, V, bal Fe.
For welding SAE 6145 and also flame hardening dies.

EUREKA 200 HARD SURFACING ELECTRODE.
M-765; 0.06-0.12 C, Cr, Ni, Mo, Mn, Cb, bal Fe.
Work hardening; 200 BHN + for surfacing parts subject to wear, heat and impact; austenitic.

EUREKA 240 DRAWALLOY HARD SURFACING ELECTRODE.
M-765; 0.10-0.20 C, Cr, Ni, Mo, Mn, bal Fe.
Work hardening, 240 BHN + for surfacing parts subject to wear, heat, an impact; austenitic.

EUREKA 340 DRAWALLOY HARD SURFACING ELECTRODE.
M-765; 0.60-0.75 C, Mn, Cr, W, Mo, Cb, bal Fe.
Work hardening; 340 BHN +, for surfacing parts subject to wear, heat, and impact; austenitic.

EUREKA 350-A ALLOY BARE ROD.
M-765; 0.06 C, 12.5 Cr, 1.85 Ni, bal Fe.
For joining or repairing cracked die sections.

EUREKA 400 BUILD UP ELECTRODE.
M-765; 0.55-0.65 C, Mn, Cr, Si, bal Fe.
For build up on water hardening tool and die steels.

EUREKA 440 DRAWALLOY HARD SURFACING ELECTRODE.
M-765; 0.75-1.0 C, Mn, Cr, Ni, W, Mo, Cb, bal Fe.
Work hardening; 440 BHN + for surfacing parts subject to wear, heat and impact; austenitic.

EUREKA 500 HIGH ALLOY ELECTRODE now **EUREKA 590 HIGH ALLOY ELECTRODE.**

EUREKA 590 HIGH ALLOY ELECTRODE.
M-765; low C, Cr, Ni, Mn, Mo, Si, bal Fe.
For welding high carbon steel and joining hard to weld steels.

EUREKA 1215-A A. H. BARE ROD.
M-765; 1.0 C, 5.0 Cr, Mo, W, Mn, bal Fe.
For welding air hard tool steel by G.T.A. process.

EUREKA 1215 A. H. ELECTRODE.
M-765; 0.75-1.0 C, 4.75 Cr, 1.05 W, 1.85 Mo, V, Mn, bal Fe.
For welding air hardening tool and die steels.

EUREKA 1216-A HIGH SPEED BARE ROD.
M-765; 0.93 C, 8.5 Mo, 3.5 Cr, V, W, bal Fe.
High Mo high speed for welding tools of the M1-M2 type.

EUREKA 1216 H. S. ELECTRODE.
M-765; 0.60-0.75 C, 4.60 Cr, 8.75 Mo, W, V, Mn, bal Fe.
For welding high speed steel and other hot work units.

EUREKA 1220-A A. H. BARE ROD.
M-765; 2.15 C, 11.5 Cr, Mo, V, Mn, bal Fe.
For welding high carbon, high chromium tool steel by G. T. A. method.

EUREKA 5545 ELECTRODE.
M-765; 45 Ni, 55 Cu.

EUREKA 8510 W. H. ELECTRODE.
M-765; 0.50-0.75 C, 1.15 Cr, 0.45 Mo, V, Mn, bal Fe.
For welding water hardening tools and dies.

EUREKA EXP-10 SURFACING ELECTRODE.
M-765.
For overlaying sheet metal, drawing and forming dies.

EUREKA EXP-20 SURFACING ELECTRODE.
M-765.
For overlaying sheet metal, drawing and forming dies. Harder than EXP-10.

EUREKA "FORGEWELD" LOW ALLOY ELECTRODE.
M-765; Low C, Mo, Si, Mn, bal Fe.
For repairing forge shop components, rams and sow blocks.

EUREKA "HAMMERWELD" ELECTRODE.
M-765; Low C, Cr, Ni, Mo, Si, Mn, bal Fe.
For welding and repairing forging dies.

EUREKALLOY "C" ELECTRODE.
M-765; 0.05-0.10 C, 55.0 Ni, 16.25 Cr, 17.5 Mo, W, Mn, V, bal Fe.
For surfacing parts subject to heat, impact, oxidation and certain chemicals.

EUREKALLOY "X" ELECTRODE.
M-765; 0.30-0.35 C, Ni, Mo, high cobalt.
For surfacing units subject to pressure, impact, heat and galling, hot extrusion and forge dies.

EUREKA MARWELD BARE ROD.
M-765; Maraging steel for repairing dies of maraging steel or overlaying hot work dies.

EUREKAMATIC 2 CORED WIRE.
M-765; Low C, med. alloy, bal Fe.
For repairing rams and sow blocks.

EUREKAMATIC 5 CORED WIRE.
M-765; Low C, med. alloy, bal Fe.
For forge die impression and shanks.

EUREKAMATIC 45 CORED WIRE.
M-765; Low C, Cr, Ni, Mn, Mo, W, bal Fe.
For filling forge die impressions.

EUREKAMATIC 45 SOLID WIRE.
M-765; 0.08 C, med. Cr, Mo, bal Fe.
Semi hot work grade, for repairing upsetter and forge dies.

EUREKAMATIC 72 SOLID WIRE.
M-765; Low C, Cr, Mo, W, bal Fe.
For repairing shear blades, forge dies, and building composite units, hot work.

EUREKAMATIC 73 SOLID WIRE.
M-765; Low C, med Cr, high W, bal Fe.
For repairing shear blades and other hot work steel subject to severe heat.

EUREKAMATIC 78 SOLID WIRE.
M-765; Med C, high Cr, (W free) bal Fe.
Hot work grade for repairing hot or cold die units.

EUREKAMATIC 130 SOLID WIRE.
M-765; 0.32 C, Cr, Mo, Mn, Si, bal Fe.
For overlay with mild alloy; meets SAE 4130 Spec.

EUREKAMATIC 145 SOLID WIRE.
M-765; 0.45-0.50 C, Cr, V, bal Fe.
Medium carbon, meets SAE 6145 Spec.

EUREKAMATIC 350 ALLOY SOLID WIRE.
M-765; 0.06 C, Mn, Mo, high Cr, bal Fe.
For joining tool steel, repairing crane wheels and forge dies.

EUREKAMATIC 1216 SOLID WIRE.
M-765; C, Cr, high Mo, bal Fe.
Type M1 and M2.

EUREKAMATIC "BUILD-UP" SOLID WIRE.
M-765; Low C, alloy, bal Fe.
For preliminary build up on die sections before facing with harder material.

EUREKAMATIC FORGEWELD CORED WIRE.
M-765.
Low alloy wire for repairing forge shop components as rams, sow blocks, die holders.

EUREKAMATIC FORGEWELD CORED WIRE (GASLESS).
M-765.
Low alloy wire for repairing forge shop components, - can be applied without covering gas.

EUREKAMATIC 3 CORED WIRE.
M-765.
Analysis similar to typical forge block.
For repair and flooding forge die blocks; 38-40 Rc.

EUREKAMATIC 72 CORED WIRE.
M-765; composition similar to AISI H-12.
For repair welding dies, shear blades; 51-54 Rc.

EUREKAMATIC 130 CORED WIRE.
M-765; composition similar to AISI 4130.
For welding and build-up of AISI 4130 and 4140 steel.

EUREKAMATIC 350 CORED WIRE.
M-765; very low C, alloy, bal Fe.
For joining tool steels and high alloys that must be heat treated, 38/40 Rc.

EUREKAMATIC 450.
M-765; low carbon, alloy, bal Fe.
For repairing forge die blocks, 44/46 Rc.

EUREKAMATIC 550.
M-765; low carbon, alloy, bal Fe.
For repairing forge die blocks, 50 Rc approx.

EUREKAMOLD P-6 BARE ROD.
M-765; similar composition to AISI P-6.
For repairing plastic molds where extreme high lustre is required.

EUREKAMOLD P-20 BARE ROD.
M-765; same composition as AISI P-20.
For welding molds in generel, those that must be grained or etched.

EUREKA NO. 1 HARD FACING BARE RODS.
M-765; 2.50 C, Cr, W, high cobalt.
For surfacing parts subject to severe wear, heat and abrasion.

EUREKA NO. 6 HARD FACING BARE ROD.
M-765; 1.45 C, 28.5 Cr, 3.9 W, Ni, Si, high cobalt.
For surfacing punches, dies, shear blades subject to severe heat and impact.

EUREKA NO. 12 HARD FACING BARE ROD.
M-765; 1.40 C, Ni, W, Mo, high cobalt.
For surfacing parts subject to severe wear, heat and abrasion.

EUREKA NO. 5545-A CUPRO-NICKEL BARE ROD.
M-765; 45 Ni, 55 Cu.
For repairing cast iron patterns.

EUREKA T.G.A. LOW AIR ELECTRODE.
M-765; 0.70-0.75 C, 0.75 Cr, 1.70 Ni, 1.65 Mo, 1.15 Mn, bal Fe.
For welding low air tool steels.

EUROPEAN "REAMUR".
M-Eng.; 2.8-3.5 C, graphite, 0.6-0.8 Si, bal Fe.
For castings; malleable cast iron.

EUROPIUM.
M-1755; Eu.
Purities: 99.9% (special distilled grade) 99.5+%.
Forms: Ingot, lump, wire, sheet, foil.

EUTALLITE UNIVERSAL 10092.
M-717.
Powder for spray welding of steel. Resists scaling and softening at elevated temperatures. Rc 45-50.

EUTECROD 14 FC.
M-717; Electrode for torch brazing or building up cast iron, as machine bases, motor and gear housings.
Bonding temperature 1400-1600°F; Hard: 200 Brin.

EUTECROD 15.
M-717; Electrode for low temperature torch build up and sealing of cast iron, malleable, steel, copper and nickel alloys, such as crankcases, water jackets.
Bonding temperature 450-600°F: pressure strength 1500 psi.

EUTECROD 16.
M-717; Copper base with high nickel content. Rod for torch brazing ferrous and non-ferrous metals.
Bonding temperature 1400-1600°F; 100,000 TS.

EUTECROD 16 FC.
M-717; Flux coated modification of EUTECROD 16.

EUTECROD 18.
M-717; Zn, bal Cu.
For gas welding rod; for bronze and brass.

EUTECROD 18 FC.
M-717; Flux coated grade of EUTECROD 18.

EUTECROD 21.
M-717.
For torch welding fillet and bead joints on sheet, tubular, extruded and cast aluminum.
Bonding temp. 1090°F; 33,000 psi TS.

EUTECROD 21-FC-E.
M-717.
For torch joining of aluminum; Melts 1090°F; 33,000 TS.

EUTECROD 141.
M-717.
Rod for torch brazing cast iron; easily machinable; bonding temp: 1400-1600°F; 40,000 psi TS.

EUTECROD 146.
M-717.
Electrode for torch brazing of cast iron, steels and copper base alloys at low heat; Bonding temperature 1400-1600°F; Strength 65,000 psi.

EUTECROD 146 FC.
M-717; Flux coated modification of EUTECROD 146.

EUTECROD 157.
M-717; 95.5 Sn, 3.46 Ag.
Solder type alloy free from lead, zinc, antimony and cadium. Melts 425°F; 15,000 TS.

EUTECROD 157-B.
M-717.
Flux cored solder type alloy free from lead, zinc, antimony and cadium; melts 425°F; 15,000 TS.

EUTECROD 180.
M-717.
Electrode for torch brazing copper base alloys; melts 1290°F; 42,000 TS.

EUTECROD 185.
M-717.
Bronze electrode for torch build-up; machinable and wear resistant; bonding temp. 1400-1600°F; 130 Brin, work hardens to 200 Brin.

EUTECROD 185 FC.
M-717.
Machinable overlays on cast iron, steel, copper and nickel alloys.
Flux coated; for torch, bonding temperature 1400-1600°F; hardness 200 Brin.

EUTECROD 190.
M-717.
For torch welding aluminum; Melts 1070°F; 34,000 TS.

EUTECROD 196.
M-717.
For torch build-up and joining zinc die castings; M. P. 700°F; 28,000 TS.

EUTECROD 1600 SUPER.
M-717.
Cadmium free silver solder type for high strength braze joints on steels, copper, brass, dissimilar metals. Bonding temp: 1375°F; 60,000 psi TS.

EUTECROD 1601.
M-717.
Rod for torch, furnace or induction brazing ferrous and non-ferrous.
Temperature 1225°F; 60,000 TS.

EUTECROD 1702.
M-717.
Electrode for torch brazing steel, stainless steel, carbide tips; Melts 1200°F; 85,000 TS.

EUTECROD 1800.
M-717; Ag, bal Cu.
For gas welding rod; for steel and copper alloys.

EUTECROD 1801.
M-717; 51 Ag, 22 Cu, 21 Zn, 5 Cd, 1 Sn.
For brazing ferrous and non-ferrous; M.P. 1120°F; 90,000 TS.

EUTECROD 1804.
M-717.
For torch brazing copper and copper alloys; Melts 1185°F; 50,000 TS

EUTECROD 1810.
M-717.
Silver solder type for joining copper, brass, steels and dissimilar metals. Bonding temp: 1195°F; 78,000 psi TS.

EUTECROD 1900.
M-717.
For torch welding magnesium; M.P. 1100°F; 30,000 TS.

EUTECSIL 1020 FC.
M-717.
Flux coated alloy for torch brazing of ferrous and non-ferrous alloys; Melts 1050°F; 85,000 TS.

EUTECSIL 1030 FC.
M-717.
Flux coated rod, silver braze type, for joining ferrous and non-ferrous metals; Temperature 1150°F; 60,000 TS.

EUTECSIL 1801 FC.
M-717.
Silver solder type alloy for joining ferrous and non-ferrous metals and dissimilar metals. Flux coated; bonding temp.: 1120°F; 90,000 psi TS.

EUTECSILVERWELD ECON 1.
M-717.
General purpose silver alloy and flux in unitized container; bonding temp: 1295°F; 58,000 psi TS.

EUTECSILVERWELD ECON 2.
M-717.
Cadmium-free silver alloy and flux in unitized container; bonding temp: 1130°F; 85,000 psi TS.

EUTEC-SILWELD 1618.
M-717.
For ferrous and non-ferrous joining; M.P. 1125°F; 85,000 TS.

EUTECTAL.
M-620; 1.5 Cu, 0.8 Mn, 1.58 Mg, 0.35 Ti, 0.25 Si, bal Al.
Heat treated: 38,000 TS; 4 El; 98 Brin.
For light alloy parts; age hardenable.

EUTECTIC ALLOY-1.
M-U.S.; 50 Bi, Pb, 13 Sn, 10 Cd.
For fusible alloys, fuses.

EUTECTIC ALLOY-2.
M-U.S.; 52 Bi, 40 Pb, 8 Cd.
For fusible alloys, fuses.

EUTECTIC ALLOY-3.
M-U.S.; 53 Bi, 32 Pb, 15 Sn.
For fusible alloys, fuses.

EUTECTIC ALLOY-4.
M-U.S.; 54 Bi, 26 Sn, 20 Cd.
For fusible alloys, fuses.

EUTECTIC ALLOY-5.
50 Sn, 18 Cd, 32 Pb.

EUTECTRODE 4.
M-717.
Electrode for AC-DC metallic arc tough coating on manganese and alloy steels; good shock resistance; Rb 90, work hardens to Rc 45.

EUTECTRODE 27.
M-717.
For repair welding cast iron; AC-DC; for machine bases, frames, supports. Strength 60,000 psi.

EUTECTRODE 40.
M-717.
Electrode for AC-DC metallic arc build-up and join manganese steels; deposits can be flame cut; Rb 80-90, work hardens to Rc 45-50.

EUTECTRODE 53.
M-717.
For AC-DC metallic arc welding 316 stainless steel; Moly bearing; 85,000 psi TS.

EUTECTRODE 53-L.
M-717.
For AC-DC metallic arc welding of 316 L; moly bearing with low carbon 85,000 psi TS.

EUTECTRODE 54.
M-717.
For AC-DC metallic arc joining of 301, 302, 304, 305 and 308 stainless steels; 80,000 psi TS.

EUTECTRODE 54-L.
M-717.
308L-16 stainless; low carbon content; for joining types 304, 304L, 308, 347; 75,000 psi TS.

EUTECTRODE 57.
M-717.
309-16 stainless steel for AC-DC welding of stainless steels; good oxidation resistance at elevated temperatures; 85,000 psi TS.

EUTECTRODE 66.
M-717.
Electrode for AC-DC welding of mild steel, beams, channel iron, pipes; 80,000 TS; machinable.

EUTECTRODE 71.
M-717.
Electrode for AC-DC metallic arc welding thin sections of 4130, 4140 and 8620 steels; similar response to heat treatment; 100,000 psi TS.

EUTECTRODE 232.
M-717.
For AC-DC metallic arc machinable welds on alloy cast irons including ductile and high phosphorus types; 55,000 psi TS.

EUTECTRODE 240.
M-717.
For AC-DC metallic arc build up and fill grey cast iron with machinable deposits. 53,000 psi TS.

EUTECTRODE 280.
M-717.
Electrode for reverse DC welding copper alloys to ferrous metals; 60,000 TS.

EUTECTRODE 501.
M-717.
Electrode for AC-DC metallic arc welding of mild steel sheet, plate, angle iron and pipe; 68,000 psi TS.

EUTECTRODE 518.
M-717.
Electrode for AC-DC metallic arc joining of low alloy and medium carbon steels in all positions; 80,000 psi TS.

EUTECTRODE 526.
M-717.
Electrode for AC-DC metallic arc welding of most steels including dissimilar combinations; 120,000 psi TS.

EUTECTRODE 554-L.
M-717.
All purpose extra low carbon stainless for AC-DC welding of dissimilar combinations; 75,000 psi TS.

EUTECTRODE 670.
M-717; Cr, Ni steel.
Electrode for AC-DC welding of low alloy steel and some stainless steel; 95,000 TS.

EUTECTRODE 680.
M-717.
Electrode for AC-DC welding of alloy steels, leaf and coil springs, some stainless grades; 120,000 TS.

EUTECTRODE 1851 (DC).
M-717.
For arc welding copper base alloys.

EUTECTRODE 2101.
M-717.
For DC reverse welding of aluminum alloys; 34,000 TS.

EUTECTRODE 6800.
M-717.
Electrode for AC-DC metallic arc welding and build-up of ferrous and nickel alloys. Heat and corrosion resistant; 85,000 psi TS.

EUTECTRODE N 2.
M-717.
For hard, wear resistant overlays on steels and cast iron; AC-DC; for plow shares, cement grinder rings, mixers, excavator teeth; hardness C 50-55 Rock.

EUTECTRODE N 90.
M-717.
Hard overlay for ferrous metals; AC-DC; deposits retain high strength and hardness at elevated temperatures; for ash remover impellers, exhaust valves, hot punches.

EUTECTRODE N5005.
M-717.
Electrode for AC-DC metallic arc coating on carbon, alloy and manganes steel for combined hardness and toughness; Rc 57-60.

EUTECTRODE SUPER 110.
M-717.
Electrode for AC-DC welding steel, angle iron, I-beams, channel iron; 110,000 TS.

EUTHERM A11.
M-1340; 0.15 C, 19.5 Cr, 9.5 Ni, bal Fe.
Annealed: 80,000 TS; 35,000 YS; 55 El; 75 RA; 150 Brin.
For chemical plant equipment, tanks, mixers, filters; Type 302; stainless austenitic.

EUTHERM A22.
M-1340; 0.15 C, 24 Cr, 19 Ni, bal Fe.
Annealed: 100,000 TS; 45,000 YS; 50 El; 65 RA; 185 Brin.
For furnace parts, pumps, valves, turbine parts; Type 310; stainless, austenitic.

EUTHERM F8.
M-1340; 0.12 max C, 2.3 Si, 6 Cr, bal Fe.
Annealed: 70,000 TS; 30,000 YS; 28 El; 65 RA; 160 Brin.
For oil refinery equipment; heat and creep resistant.

EUTHERM F10.
M-1340; 0.12 max C, 2 Si, 18 Cr, bal Fe.
Annealed: 80,000 TS; 50,000 YS; 25 El; 50 RA; 150 Brin.
For shafts, gears, chemical plant equipment, sinks; Type 430; stainless.

EUTHERM F30.
M-1340; 0.12 max C, 1.5 Al, 24 Cr, bal Fe.
Annealed: 85,000 TS; 50,000 YS; 30 El; 55 RA; 180 Brin.
For oil refinery and chemical plant equipment; heat resistant.

EUTHERM FA25.
M-1340; 0.2 C, 25 Cr, 4 Ni, bal Fe.
Cast: 80,000 TS; 55,000 YS; 200 Brin.
For furnace parts, heat treating boxes, valves, pumps; Type CC; corrosion and heat resistant.

EUT-O-MAT 3010A.
M-717.
Continuous electrode for AC-DC build-up on low and medium carbon steel for joining and filling. Rc 25.

EUT-O-MAT 3205.
M-717.
Continuous electrode for AC-DC build-up and coating of manganese steel; Rc 25 but work hardens to Rc 50-55.

EUT-O-MAT 3220A.
M-717.
Continuous electrode for joining and coating manganese steel; base for harder coating. AC-DC; Rc 20, work hardens to Rc 50.

EUT-O-MAT 4601A.
M-717.
Continuous electrode for AC-DC coating on all steels. Rc 55-60; maintains hardness at moderate temperatures.

EUT-O-MAT 4625A.
M-717.
Continuous electrode for AC-DC build-up on all steels; resists abrasion, impact, compression. Rc 40-45.

EUT-O-MAT AN690.
M-717.
Continuous electrode for high strength steel joining and tough, wear resistant cladding. DC; 90,000 psi TS.

EUZONIT.
M-1335; 55 Ni, 20 Mo, 20 Fe.
For corrosion and heat resistant equipment; corrosion and heat resistant.

EUZONIT see also **MARKER EUZONIT.**

EUZONIT 60.
M-1335; 0.10 C, 17 Cr, 60 Ni, 20 Mo, bal Fe.
For corrosion and heat resistant parts; corrosion and heat resistant. Resists sulphuric acid.

EUZONIT 70.
M-1335; 67 Ni, 30 Mo, bal Fe.
For high temperature applications; corrosion and heat resistant. Resists sulphuric acid.

EUZONIT 85.
M-1335; 85 Ni, 9 Si, 4 Cu.
For high temperature applications; corrosion and heat resistant.

EV4CO.
M-Germany; 1.3 C, 5 Co, 1 Mo, 4 V, 12 W, 4 Cr, bal Fe.
Hardened: C 64-66 Rock.
For cast cutting tools, reamers, drills, broaches, engravers tools.
High speed steel, Type T15, high hot-hardness and wear resistance.

EV5.
M-Russia; 0.7-0.8 C, 0.4 max Mn, 0.3-0.5 Cr, 5.0-6.5 W, bal Fe.
For permanent magnet; heat treated.

EV. 12.
M-752; 0.2-0.5 C, 1.75 max Si, 23-26 Cr, 11-14 Ni, 1.5-2.0 W, bal Fe.
Stainless alloy casting; good hot strength, and resistance to oxidizing and scaling.
BS 1648/1957 Gr.E; ASTM A297 Gr. HH.

EV. 20.
M-752; 0.2-0.5 C, 2.25-3.0 Si, 24.0-26.0 Cr, 17.0-20.0 Ni, bal Fe.
Austenitic stainless casting; good creep resistance at elevated temperatures.
BS 1648/1957 Grade F; ASTM A297 Gr. HK.

EV. 25.
M-752; 0.5 max C, 3.0 max Si, 17-23 Cr, 23-28 Ni, bal Fe.
Alloy casting; for elevated temperature work.
BS. 1648/1957 Gr. G; ASTM A297 Gr. HN.

EV. 37.
M-752; 0.5 max C, 2.5 max Si, 17.0-21.0 Cr, 37.0-41.0 Ni, bal Fe.
Austenitic stainless casting; for high temperature equipment, as furnace parts.
BS 1648/1957 Gr.H; ASTM A297 Gr.HU.

EV. 60.
M-752; 0.35-0.75 C, 2.5 max Si, 15.0-20.0 Cr, 58.0-62.0 Ni, bal Fe.
Austenitic stainless casting; good hot strength and scale resistance to 1100°C.
BS. 1648/1957 Gr. K; ASTM A297 Gr. HW.

EV. 300.
M-752; 0.5-1.0 C, 2. 0 max Si, 27.0-30.0 Cr, bal Fe.
Ferritic type alloy casting; good resistance to scaling and sulphurous atmospheres to 1150°C.
BS. 1648/1957 Gr. B; ASTM A297 Gr. HC.

EV. 500.
M-752; 0.05-0.12 C, 0.5-1.0 Si, 26.0-30.0 Cr, 47.0-52.0 Co, bal Fe.
Alloy casting; resistant to oxidization and sulphurous atmospheres to 1200°C. Good thermal shock and abrasion resisting properties.
Equivalent to UMCO 50.

E.V. ALLOY.
M-752; C, 10-12 Ni, 22-24 Cr, bal Fe.
Annealed: 65,000-75,000 TS; 5 El.
For carburizing boxes, furnace parts; resists heat to 1175°C.

EVANOHM.
M-61; 20 Cr, 75 Ni, 2.75 Al, 2.75 Cu.
Annealed: 150,000 TS; 117,000 YS; 20 El.
For precision resistors; high electrical resistance, low resistance change with temperature.

EVANOHM.
M-897; 20 Cr, 3 Al, 2 Cu, bal Ni.
Annealed: 90,000 TS; 65,000 YS; 40 El.
For precision resistors; heat resistant to 300°C.

EVANS PEERLESS.
M-259; 0.9-1.2 C, bal Fe.
For tools, dies, jigs; water hardened.

EVANSTEEL, GRADE 1.
M-146; 1.5-2.0 Ni, 0.75-1.0 Cr, 0.3 C, bal Fe.
Annealed: 95,000-100,000 TS; 50,000-60,000 YS; 27-22 El; 40-30 RA; 200 Brin.
For sprockets, gears, high pressure valves; resists abrasion and wear; resists temperature to 1000°F.

EVANSTEEL, GRADE II.
M-146; 0.3 C, 1.5-2.0 Ni, 0.75-1.00 Cr, bal Fe.
Annealed: 115,000-125,000 TS; 65,000-75,000 YS; 20-15 El; 30-25 RA; 275 Brin.
For tractor shoes, bucket lips, dipper teeth, sheaves; abrasion resistant; heat resistant to 1000°F.

EVANSTEEL, GRADE III.
M-146; 0.3 C, 1.5-2.0 Ni, 0.75-1.00 Cr, bal Fe.
Annealed: 145,000-160,000 TS; 120,000 YS; 12 El; 10 RA; 400-500 Brin.
For dipper teeth, liner plates, guard rails, pulverizer hammers; abrasion resistant; heat resistant to 1000°F.

EVERBRITE.
M-642; 29.5-31.5 Ni, 59-65 Cu, 5-8 Fe, 0.6 max Mn, 0.25 max C, 0.6 max Si.
Cast: 75,000 TS; 50,000 YS; 14 El; 20 RA; 200 Brin.
For valve disks, valve seat rings, steam turbine nozzle blocks and valves; machinable; resists action of steam.

EVERBRITE.
M-129; 60 Cu, 30 Ni, 3 Fe, 3 Si, 3 Cr.
Cast: 75,000 TS; 45,000 YS; 14 El; 170 Brin.
For valves, chemical plants; corrosion resistant.

EVERCLAD.
M-1724.
Galvanized plastic-coated roofing and cladding steel; mild steel base.

EVERDUR see also **IMI 705.**

EVERDUR 637.
M-8; 2 Si, 7.25 Al, bal Cu.
Annealed: 88,000 TS; 44,000 YS; 25 El; Rock B 85.
For screw machine products and forgings, studs, bolts, switchgear, electrical hardware, valve stems.
Corrosion resistant, nonmagnetic.

EVERDUR 651.
M-8; 1.5 Si, 0.25 Mn, bal Cu.
Soft: 40,000 TS; 15,000 YS; 50 El; Rock F 55.
Hard: 70,000 TS; 55,000 YS; 15 El; Rock B 80.
For cold headed and roll threaded bolts, hardware marine fittings, pole line hardware, rigid conduit, welding rod. Corrosion resistant.
Good for severe cold working.

EVERDUR 655.
M-8; 95.8 Cu, 3.1 Si, 1.1 Mn.
Annealed Rods: 55,000 TS; 20,000 YS; 65 El; B 35 Rock.
Hard: 90,000 TS; 55,000 YS; 18 El; B 90 Rock.
For bearing plates, bolts, screws, forgings, hardware, springs, tanks, marine fittings, oil storage tanks.
High corrosion resistance and strength.

EVERDUR 656.
M-8; 95.8 Cu, 3.1 Si, 1.1 Mn.
For welding rods to weld copper and copper alloys, also plain and galvanized steel.
Good for carbon-arc and acetylene welding and inert gas welding.

EVERDUR 661.
M-8; 3 Si, 1 Mn, 0.40 Pb, bal Cu.
Soft: 58,000 TS; 22,000 YS; 70 El.
Hard: 90,000 TS; 60,000 YS; 18 El.
For screw machine parts; free cutting, wear and corrosion resistant.

EVERDUR 1000 see **EVERDUR 6552.**

EVERDUR 1010 see **EVERDUR 655.**

EVERDUR 1012 see **EVERDUR 661.**

EVERDUR 1014 see **EVERDUR 637.**

EVERDUR 1015 see **EVERDUR 651.**

EVERDUR 6552.
M-8; 4 Si, 1.1 Mn, bal Cu.
Cast: 50,000 TS; 20,000 YS; 25 El.
For castings, pipe fittings, valves; M.P. 1000°C.

EVERDUR A see **EVERDUR 655.**

E.V.M.
M-115; 0.82 C, 0.35 Si, 0.25 Mn, 18.5 W, 4.3 Cr, 2.1 V, 0.65 Mo, bal Fe.
For drills, chasers, lathe tools, forming tools; high speed steel.

E.V. STEEL.
M-752; Ni-Cr heat resisting steel.
Six alloys ranging from 12 to 60% Ni.
For heat resistant parts as furnace parts, heat treat boxes, retorts and trays, burner nozzles.

EW 1.
M-1736; 0.70 C, 0.25 Si, 0.27 Mn, bal Fe.
Carbon tool steel, water hardening.
W.-Nr. 1.1620; AISI W1.

EW 2.
M-1736; 1.0 C, 0.25 Si, 0.27 Mn, 0.20 V, bal Fe.
Carbon tool steel, water hardening.
W.-Nr. 1.1640; AISI W2.

EW 108.
M-1736; 0.85 C, 0.25 Si, 0.27 Mn, bal Fe.
Carbon tool steel, water hardening.
W.-Nr. 1.1620; AISI W1.

EW 110.
M-1736; 1.0 C, 0.25 Si, 0.27 Mn, 0.20 max Cr, bal Fe.
Carbon tool steel, water hardening.
W.-Nr. 1.2005; AISI W1.

EW 711.
M-1731; 0.13 max C, 0.9 Mn, 0.10 max P, 0.18-0.25 S, bal Fe.
Free machining steel for automatic screw machines.
W.-Nr. 1.0711.

EW AERO.
M-1731; 1.05 C, 1.0 Mn, 1.0 Cr, 1.2 W, bal Fe.
For punching and trimming dies.
W.-Nr. 1.2419.

EW APM.
M-1731; 0.90 C, 2.0 Mn, 0.35 Cr, 0.10 V, bal Fe.
For shear blades, reamers.
W.-Nr. 1.2842.

EW METEORIT I.
M-1731; 0.60 C. 0.60 Si, 1.05 Cr, 2.0 W, 0.20 V, bal Fe.
Hot or cold work punches, shears.
W.-Nr. 1.2550.

EW METEORIT II.
M-1731; 0.45 C, 1.5 Si, 1.5 Cr, 0.10 V, bal Fe.
Hot or cold punches, shears.
W.-Nr. 1.2249.

EW METEORIT III.
M-1731; 0.45 C, 1.0 Si, 1.1 Cr, 2.0 W, 0.20 V, bal Fe.
For punches, trimming dies, pneumatic tools.
W.-Nr. 1.2542.

EW ORION III.
M-1731; 1.45 C, 0.20 Si, 0.6 Mn, 1.5 Cr, bal Fe.
Oil hardening cold work tool steel.
Werkstoff Nr. 1.2063.

EW ORION SPEZIAL.
M-1731; 0.90 C, 1.15 Si, 0.70 Mn, 1.2 Cr, bal Fe.
Cold work tool steel; for shear blades, milling cutters.
W.-Nr. 1.2108.

EW SATURN.
M-1731; 0.30 C, 3.0 Cr, 2.8 Mo, 0.5 V, bal Fe.
Hot work tool steel; for pressure casting molds.
W.-Nr. 1.2365.

EW SATURN W.
M-1731; 0.30 C, 2.7 Cr, 8.5 W, 0.35 V, bal Fe.
Hot work tool steel; for extrusion dies, pressure casting molds, forging dies.
W.-Nr. 1.2581.

EW URUS.
M-1731; 0.12 max C, 0.10 P, 0.27 S, 0.70 Mn, 0.20 Pb, bal Fe.
Leaded free machining steel.
Werkstoff Nr. 1.0716; DIN 9SPb23.

EW URUS MN.
M-1731; 0.14 max C, 1.1 Mn, 0.10 max P, 0.24-0.32 S, 0.15-0.30 Pb, bal Fe.
Free machining steel for automatic screw machines.

EW ZIRKON 1.
M-1731; 2.0 C, 11.5 Cr, bal Fe.
Pressure casting molds; cold work sheet punching dies.
W.-Nr. 1.2080.

EW ZIRKON W.
M-1731; 2.1 C, 11.5 Cr, 0.7 W, bal Fe.
Heavy duty sheet punching dies.
W.-Nr. 1.2436.

EX-2.
0.65-0.75 C, 0.25-0.45 Mn, 0.20-0.35 Si, 0.70-1.0 Ni, 0.15-0.30 Cr, 0.08-0.15 Mo, bal Fe.
Hardened: Rock C 60 min.
Annealed: 200 Brin max.
For spindles, rolls, bearings, plungers, valves, meters, pins, cylinders. Wear resistant, deep hardening.

EX. "B" METAL.
M-Eng.; 77 Cu, 15 Pb, 8 Sn, P.
For bearings, bushings; heavy duty.

EXCELITE.
M-965; 32 Cr, 18 W, 45 Co, 2 C, 0.2 B, bal Fe.
For cutting tools; cast.

EXCELLO.
M-32; 0.50 C, 0.30 Mn, 0.30 Si, 1.5 Cr, 2.6 W, 0.25 V, bal Fe.
Shock resisting tool steel used for heading dies, punches, chisels and impact tooling.

EX-CELL-O 6A.
M-976; 7.0 Co, 7.0 TiC, 10.0 TaC, 76.0 WC.
Rock A 91.9.
General purpose medium roughing grade for steels and alloy cast irons.

EX-CELL-O 6AX.
M-976; 5.0 Co, 12.5 TiC, 11.0 TaC, 71.5 WC.
Rock A 92.6.
For light roughing and finishing of steels and alloy cast irons.

EX-CELLO-8A.
M-976; 8.5 Co, 7.5 TiC, 12.0 TaC, 72.0 WC.
Rock A 91.2.
General purpose heavy duty roughing grade for steels and alloy cast irons.

EX-CELL-O 10A.
M-976; 11.0 Co, 7.5 TiC, 8.0 TaC, 73.5 WC.
Rock A 90.8.
For heavy cuts at slow speeds, interrupted cuts, and severe milling of steels and alloy cast irons.

EX-CELL-O 509.
M-976; 8.0 Co, 16 TiC, 76.0 WC.
Rock A 92.5.
For precision boring and turning of steel and alloy cast iron at high speed and low feeds.

EX-CELL-O 606.
M-976; 8.0 Co, 10.0 TiC, 4.0 TaC, 78.0 WC.
Rock A 91.8.
For medium roughing cuts on steels and alloy cast irons.

EX-CELL-O E3.
M-976; 3.0 Co, 0.5 TaC, 96.5 WC.
Rock A 93.3.
For precision turning and boring of cast iron and non-ferrous materials.

EX-CELL-O E5.
M-976; 5.0 Co. 0.5 TaC, 94.5 WC.
Rock A 92.8.
For general purpose semi-finishing of cast iron and non-ferrous materials.

EX-CELL-O E6.
M-976; 6.5 Co, 0.5 TaC, 93.0 WC.
Rock A 92.2.
For wear application with little shock, and general purpose roughing of cast iron and non-ferrous materials.

EX-CELL-O E8.
M-976; 6.0 Co, 94.0 WC.
Rock A 91.0.
For heavy duty roughing cuts of cast iron and non-ferrous materials.

EX-CELL-O E9.
M-976; 9.0 Co, 91.0 WC.
Rock A 90.7.
General purpose wear grade with medium shock, and roughing cuts on high temperature alloys.

EX-CELL-O E16.
M-976; 16.0 Co, 84.0 WC.
Rock A 87.5.
For moderate impact use where good wear resistance is required.

EX-CELL-O E20.
M-976; 20.0 Co, 80.0 WC.
Rock A 86.0.
For high impact applications.

EX-CELL-O E25.
M-976; 25.0 Co, 75.0 WC.
Rock A 85.0.
For applications requiring maximum shock resistance.

EX-CELL-O W12C.
M-976; 12.0 Co, 88.0 WC.
Rock A 89.0.
For wear applications where heavy shock is present.

EX-CELL-O XL85.
M-976; 21.0 Ni, 9.0 Mo, 70.0 TiC.
Rock A 91.0.
For roughing use at moderate speeds on steels.

EX-CELL-O XL86.
M-976; 18.0 Ni, 9.0 Mo, 73.0 TiC.
Rock A 91.5.
For general purpose machining of steels.

EX-CELL-O XL88.
M-976; 12.5 Ni, 11.0 Mo, 76.5 TiC.
Rock A 93.0.
For finish and semi-finish machining of steels and alloy cast irons at high speeds.

EX-CELL-O XL620.
M-976; 6.0 Co, 20.0 TaC, 74.0 WC.
Rock A 91.4.
Special purpose grade where good hot hardness and high temperature lubricity are required.

EX-CELL-O XL028.
M-976; 6.0 Co, 3.0 TaC, 91.0 WC.
Rock A 92.8.
For milling, broaching, reamer applications on cast iron and non-ferrous materials.

EX-CELL-O XL061.
M-976; 8.0 Co, 6.0 TiC, 6.0 TaC, 80.0 WC.
Rock A 92.0.
General purpose milling grade for steels and alloy cast irons.

EXELLOY see CRANE EXELLOY.

EX. "K" METAL.
M-Eng.; 77 Cu, 12.5 Pb, 10.5 Sn, P.
For bearings, bushings; heavy duty.

EXL-DIE.
M-35; 0.95 C, 1.3 Mn, 0.5 Cr, 0.5 W, 0.1 V, bal Fe.
Oil hardening cold work tool steel.
For taps, reamers, broaches, extrusion and engraving dies, form tools; non-deforming.
AISI 01.

EXOCUT.
M-1779; 1.09 C, 3.75 Cr, 8.0 Co, 9.5 Mo, 1.6 W, 1.15 V, 0.2 Mo, 0.3 Si, bal Fe.
Heavy duty high speed steel designed to machine high temperature alloys, titanium, and prehardened alloy steels. Usual heat-treated hardness: Rc 67/70.
AISI M-42.

EXOTEC 29904.
M-717.
Alloy powder for metal spraying bond coat, - on all base metals except copper.

EXPANDAL see LESCALLOY EXPANDAL.

EXPANDED METAL.
M-U.S.; 0.15-0.25 C, bal Fe.
For reinforced concrete or plaster walls; low carbon steel mesh.

EXPANDING ALLOY.
M-Eng.; 67 Pb, 25 Sn, 8.3 Bi.
For type metal, mounting purposes.

EXPANSIVE METAL.
M-216; 75 Pb, 16.7 Sb, 8.3 Bi.
For plugging castings; expands on freezing.

EXPORT.
M-1260, M-1261; 92 Sn, 3.9 Sb, 3.9 Cu, 0.2 Ni.
Cast: 7000 TS; 19 Brin.
For engine bearings and bushings; M.P. 440-600°F; shock resistant.

EXPRESS.
M-1408; 0.7 C, 14 W, 4 Cr, 1 V, bal Fe.
For lathe and planer tools, reamers, drills, broaches; high speed steel.

EXPRESS COBALT FIVE.
M-1408; 0.7 C, 4 Cr, 18 W, 1 V, 5 Co, bal Fe.
For lathe and planer tools, reamers, broaches; high speed steel.

EXPRESS COBALT TEN.
M-1408; 0.7 C, W, 4 Cr, Mo, V, 10-12 Co, bal Fe.
For cutters for heat treated steels; high speed steel.

EXPRESS E.Z.
M-1546; 0.72 C, 10.5 W, 1.3 V, 0.5 Mo, 0.25 Co, bal Fe.
Hardened: C 64-66 Rock.
For lathe and planer tools, hobs, broaches, end mills, form cutters.
High-speed steel, oil hardening, high red-hardness.

EX-TEN 42.
M-604; 0.21 max C, 0.02 min V, 0.01 Cb, 1.35 max Mn, 0.30 max Si, bal Fe.
Plate: 42,000 min YS; 63,000 min TS; 24 El; 42,000 CYS.
For structural applications, automotive and truck parts, storage tanks, pipe lines, cargo containers and construction machinery.
Good formability and weldability. Tough.

EX-TEN 45.
M-604; 0.22 max C, 1.35 max Mn, 0.10 Si, 0.02 Cb, 0.02 min V, bal Fe.
Rolled: 60,000 TS; 45,000 YS; 25 El.
For railroad car and bus bodies, cargo containers.
High strength low alloy construction steel.

EX-TEN 50.
M-604; 0.23 max C, 1.35 max Mn, 0.01 min Cb, 0.02 min V, 0.10 Si, bal Fe.
Rolled: 65,000 TS; 50,000 YS; 22 El.
For railroad car and bus bodies, cargo containers.
High strength low alloy construction steel.

EX-TEN 55.
M-604; 0.25 max C, 1.35 max Mn, 0.10 Si, 0.02 Cb, 0.02 min V, bal Fe.
Rolled: 70,000 TS; 55,000 YS; 20 El.
For automotive and truck bodies, cargo containers, tote boxes. High strength low alloy steel.

EX-TEN 60.
M-604; 0.26 max C, 1.35 max Mn, 0.02 min Cb, 0.02 min V, 0.012 max N, bal Fe.
Bar: 75,000 min TS; 60,000 min YP; 18 El.
For automotive and truck parts, cargo containers, tote boxes, gas cylinders, construction machinery.
Low-alloy, high strength steel.

EX-TEN 65.
M-604; 0.26 max C, 1.35 max Mn, 0.01 min Cb, 0.02 min V, 0.012 max N, bal Fe.
Bar: 80,000 min TS; 65,000 min YP; 16 El.
For automotive and truck parts, cargo containers, tote boxes, gas cylinders, construction machinery.
Low-alloy, high strength steel.

EX-TEN 70.
M-604; 0.26 max C, 1.35 max Mn, 0.40 max Si, C, V, bal Fe.
590 MPa(85 ksi) TS; 485 MPa(70 ksi) YS; 14 El.
For railroad freight cars, bridges, towers.

EXTRA.
M-140; 0.8-1.20 C, 0.25 Mn, 0.25 Si, bal Fe.
Water hardening tool steel; AISI W1.

EXTRA.
M-783; 1.15 C, 0.3 Mn, 0.5 V, 0.35 Si, bal Fe.
For drills, taps, water hardened.

EXTRA BEST.
M-364; 1 C, bal Fe.
For tools, taps, reamers; water hardening.

EXTRA BEST MH.
M-1340; 1.1 C, 0.25 max Si, 0.25 max Mn, bal Fe.
Annealed: 110,000 TS; 57,000 YS; 18 El; 40 RA; 210 Brin.
For drills, taps, springs, hobs, reamers; Type W1; water hardened.

EXTRA BEST VZH.
M-1340; 1.0 C, 0.25 Mn, 0.1 V, bal Fe.
For springs, taps, reamers, broaches; Type W2; water hardened.

EXTRA BEST Z.
M-1340; 0.85 C, 0.25 max Si, 0.25 max Mn, bal Fe.
Heat treated: 190,000 TS; 145,000 YS; 10 El; 30 RA; 400 Brin.
For springs, taps, drills, reamers, punches; Type W1; water hardened.

EXTRA BEST ZH.
M-1340; 1.0 C, 0.25 max Si, 0.25 max Mn, bal Fe.
Annealed: 100,000 TS; 53,000 YS; 21 El; 42 RA; 200 Brin.
For springs, taps, drills, hobs, reamers; Type W1; water hardened.

EXTRA BEST ZW.
M-1340; 0.70 C, 0.25 max Si, 0.25 max Mn, bal Fe.
Heat treated: 175,000 TS; 128,000 YS; 12 El; 37 RA; 355 Brin.
For springs, rails, punches, axes, crimpers; Type W1; water hardened.

EXTRA CARBON.
M-114; 0.7-1.4 C, bal Fe.
For tools, dies, punches; water hardened.

EXTRA CARBON.
M-73; 1.0 C, 0.25 Mn, 0.25 Si, bal Fe.
For engraving tools, shear blades, taps, drills; water hardened.
AISI W1.

EXTRA CHROME.
M-275; 0.5 C, 4 Cr, bal Fe.
For hot work dies and tools; hot work steel.

EXTRA DRACO.
M-114; 0.7-1.2 C, 0.2 V, bal Fe.
For tools, cold header dies; water hardened.

EXTRA DRACO DV.
M-114; 1.0 C, 0.50 V, bal Fe.
For tools, dies, taps, drills, cutters; Type W2; water hardened.

EXTRA G.
M-782; 0.7-1.1 C, bal Fe.
For tools, drills, taps; water hardened.

EXTRA H.
M-1182; 1.0 C, 0.50 V, bal Fe.
For taps, cutters, drills, tools; Type W2; water hardened.

EXTRA HIGH LEADED BRASS 356.
M-8; 62.5 Cu, 35 Zn, 2.5 Pb.
Hard Sheet: 73,000 TS; 60,000 YS; 7 El; B 80 Rock.
Soft Sheet: 45,000 TS; 17,000 YS; 50 El; B 15 Rock.
For clocks, instruments, screw machine parts, fasteners.
Elect. cond. 26. Hard and strong, good machinability.

EXTRA LION.
M-1415; 0.70 C, 18 W, 4 Cr, 1 V, bal Fe.
For drills, cutters, hobs, broaches, reamers; high speed steel.

EXTRA M.G.
M-363; 1.05 C, 1.1 Mn, 0.5 Cr, 0.25 Si, bal Fe.
For tools and dies, long taps, reamers, screw dies, calipers; non-deforming.

EXTRA MH.
M-1340; 1.1 C, 0.25 max Si, 0.25 max Mn, bal Fe.
Annealed: 110,000 TS; 58,000 YS; 18 El; 40 RA; 210 Brin.
For springs, taps, reamers, cutters; Type W1; water hardened.

EXTRA QUALITY.
M-363; 0.7-1.2 C, bal Fe.
For punches, dies, general tools; water hardening.

EXTRARD.
M-85; 0.35 C, 5 Cr, 1.4 Mo, 1.25 W, 0.2 V, 1 Si, bal Fe.
For heat resistant parts, hot work tools; Type H12; air or oil hardened.

EXTRA S.
M-1182; 0.60-1.4 C, 0.25 V, bal Fe.
For tools, taps, cutters, reamers; Type W2; water hardened.

EXTRA SO.
M-616; 1.05 C, bal Fe.
For tools, drills, taps; water hardened.

EXTRA SPECIAL.
M-364; 1.2 C, bal Fe.
For tools, finishing tools, drills; water hardening.

EXTRA SUPERDURALUMIN.
M-Japan; Al-Cu-Zn-Mn-Mg.
For light alloy parts.

EXTRA SUPER INCOMPARABLE.
M-1409; 0.8 C, 22 W, 4.5 Cr, 1.8 V, 10 Co, 0.7 Mo, bal Fe.
For lathe and planer tools, cutters; high speed steel.

EXTRA TOOL.
M-85; 0.7-1.2 C, bal Fe.
For cutters, dies, general tools; water hardened.

EXTRA TOOL.
M-355; 1.1 C, bal Fe.
For tools, fixtures; water hardened.

EXTRA TOUGH NO. 4.
M-69; 0.7 C, 1.5 Ni, 0.5 Cr, 0.2 Mo, bal Fe.
For punches, shear blades; tough.
AISI L-6.

EXTRA TOUGH NO. 6.
M-69; 0.50 C, 0.8 Mn, 0.95 Cr, 0.2 V, bal Fe.
For gears, bolts, springs, crankshafts; oil hardened, tough.
AISI L-2.

EXTRA TRIPLE CONQUEROR.
M-1406; 1.6 C, 0.6 Cr, 6 W, 0.15 Si, 0.4 Mn, bal Fe.
For turning tools, drawing dies; finishing steel, wear resistant.

EXTRA TRIPLE GRIFFIN.
M-1112; 1.5 C, 0.35 Cr, 6 W, 0.35 Mn, bal Fe.
For drawing dies, cutters; water hardened, hard case and tough core.

EXTRA V.
M-57; 0.90-1.0 C, 0.2 V, bal Fe.
For tools, drills, taps, cutters, reamers; Type W2; water hardened.

EXTRA ZH.
M-1340; 1.0 C, 0.25 max Si, 0.25 max Mn, bal Fe.
Annealed: 100,000 TS; 53,000 YS; 21 El; 42 Ra; 200 Brin.
For springs, taps, drills, hobs; Type W1; water hardened.

EXTRA ZW.
M-1340; 1.0 C, 0.25 max Si, 0.25 max Mn, bal Fe.
Annealed: 100,000 TS; 53,000 YS; 21 El; 42 RA; 200 Brin.
For springs, taps, drills, hobs, reamers; Type W1; water hardened.

EXTRUDAL.
M-249; 0.4-0.8 Si, 0.4-0.8 Mg, bal Al.
Forged: 36,000-43,000 TS; 29,000-36,000 YS; 18-10 El; 75-90 Brin.
For light alloy parts; good formability.

EXTRUDAL 12.
M-541; 0.8-1.3 Cu, 11-13 Si, 0.8-1.2 Mg, 0.8 max Fe, 0.6-1.2 Ni, bal Al.
TA Temper: 55,000-60,000 TS; 42,000-48,000 YS; 5-12 El; 110-135 Brin.
For forged pistons. High corrosion resistance.

EYA-1 CAST IRON.
M-USSR; 1.3-3.2 C, 17.8-19.2 Cr, 8.5-9.2 Ni, 1.7-2.6 Si, bal Fe.
For furnace parts and equipment; corrosion resistant, cast iron.

EYELET BRASS.
M-U.S.; 65-68 Cu, 32-35 Zn.
Hard: 85,000 TS; 60,000 YS; 4 El; 156 Brin.
Soft: 46,000 TS; 20,000 YS; 58 El; 53 Brin.
For eyelets, drawn shells; high ductility.

EZ 9W.
M-336; 0.7 C, 9.5 W, 3.5 Cr, 0.2 V, bal Fe.
For gripper and hot forming dies, punches, piercers; hot work steel.

EZ 14W.
M-336; 0.7 C, 14 W, 3 Cr, 0.2 V, bal Fe.
240 Brin.
For hot dies, tools, extrusion and hot forging dies; high speed steel.

EZ 33A.
M-Various foundries; 2.7 Zn, 3.0 Re (Ce, etc.), 0-7 Zr, bal Mg.
T5-temper: 20,000-23,000 TS; 14,000-15,000 YS; 2-3 El.
At 600°F: 12,000 TS; 8000 YS; 50 El.
Magnesium sand and permanent mold casting with good pressure tightness and weldability. For parts operating at 350-500°F.
ASTM B80-69, B199-68; AMS 4442; QQ-M-55, QQ-M-56; SAE 506.

EZ 33A.
M-43; 2.5-4.0 Re, 2.0-3.1 Zn, 0.45-1.0 Zr, bal Mg.
Melting point: 1189°F.
For welding rod and electrode.

EZ CARB.
M-373; 0.35 C, 0.85 Cr, 0.75 Mn, 0.50 Mo, 0.50 W, bal Fe.
Shock resistant cold work tool steel.

EZ CUT 20.
M-240; 0.18 C, 1.2 Mn, 0.08 P, 0.29 S, 0.23 Si, bal Fe.
67 ksi TS; 38 ksi YS; 31 El; 33 RA.
Free cutting steel for die bases, jigs, molds; case hardenable.

EZ CUT 45.
M-240; 0.46 C, 1.2 Mn, 0.03 P, 0.26 S, 0.18 Si, bal Fe.
92 ksi TS; 45 ksi YS; 20 El; 42 RA.
Direct hardenable, free-cutting steel for dies, jigs, molds.

EZ-DIE.
M-35; 1 C, 5.2 Cr, 1.15 Mo, 0.25 V, bal Fe.
For blanking, coining and forming dies; air hardened; Type A2; non-deforming.

E-Z-DIE SMOOTHCUT.
M-35; 1.0 C, 0.8 Mn, 5 Cr, 1 Mo, 0.25 V, S, bal Fe.
For forming and blanking dies, extrusion equipment, air hardenable; tough and wear resistant.
AISI A2.

E-Z DIE V.
M-35; 2.15-230 C, 1.0-1.3 W, 0.95-1.2 Mo, 5.0-5.5 Cr, 4.6-5.0 V, bal Fe.
Hardened: C 60-66 Rock.
For lamination and forming dies, burnishing rolls, blanking and drawing dies, powder metal dies.
High wear and abrasion resistance. Cold work tool steel, Type A7. Air harden.

E-Z-TEM.
M-1684; 0.35 C, 0.8 Mn, 0.45 Si, 1.0 Cr, 0.4 Mo, 0.5 Cu, bal Fe.
Shock resisting tool steel.

F

F 1 K.
M-912; 0.57 C, 0.30 Si, 0.85 Mn, 0.80 Cr, bal Fe.
For flat and coil springs.
W.-Nr. 7176.

F 1 KB.
M-912; 0.58 C, 0.30 Si, 1.0 Mn, 0.90 Cr, B, bal Fe.
For flat or coil springs.
W.-Nr. 7163.

F 2 K.
M-912; 0.50 C, 0.25 Si, 0.90 Mn, 1.10 Cr, 0.10 V, bal Fe.
For highly stressed flat or coiled springs.
W.-Nr. 8159; similar to AISI 6150.

F 2 KH.
M-912; 0.58 C, 0.25 Si, 0.95 Mn, 1.10 Cr, 0.10 V, bal Fe.
For highly stressed flat or coil springs.
W-Nr. 8161.

F 4 K.
M-912; 0.50 C, 0.25 Si, 0.90 Mn, 1.10 Cr, 0.25 Mo, 0.10 V, bal Fe.
For highly stressed flat or coil springs.
W.-Nr. 7701.

F-17 TRIMETAL see CLEVITE F-17.

F-23 see CLEVITE F-23.

F58.
M-1739; Iron, (some Cu - sintered).
Density: 5.7-5.9 gms/cc; Hardness: VPN5-70.
1-3 Elon: 28-25% porosity.
Hardenable to VPN5 250.
B.S.S. A200; SAE 850.

F58/1.
M-1739; Iron, (some Cu + C, sintered).
Density: 5.7-5.9 gms/cc; Hardness: VPN5 -120.
Elon: 1%; porosity: 28-25%.
Hardenable to VPN5 250.
SAE 855; MPIE F-0010-N.

F68.
M-1739; Iron, sintered.
Density: 6.6-6.9 gms/cc; porosity: 16-12%.
Can be carbo-nitrided to VPN (micro) 650 and Rock 25C core.
B.S.S. A-202; MPIE F-0000-R.

F80/0.
M-1739.
Sintered iron.
Density: 6.2 gms/cc; WTS: 12.5 kg/mm^2;
Hardness: HV5-70.
For motor industry bearings; timing chain adjusters.

F80/1.
M-1739.
Sintered iron with 1% carbon.
Density: 6.2 gms/cc; WTS: 23.5 kg/mm^2;
Hardness: HV5-140.
For brackets, levers, cams, business machine parts.

F-80-S.
M-1766; 0.10 max C, 0.70 max Mn, 0.06 max Si, 0.03 max P, 0.12 max S, bal Fe.
For cold stamping, cold heading and free machining.

F-95.
M-339; 3.2 T.C., 1.6 Si, 0.7 Mn, 1 Ni, 0.25 Cr, bal Fe.
Cast: 108,800 TS; 330 Brin.
For piston rings; wear and fatigue resistant.

FABCO 1CM.
M-1000; weld metal: 0.10 C, 0.70 Mn, 0.25 Si, 1.25 Cr, 0.50 Mo, bal Fe.
As welded: 88,500 psi TS; 78,500 psi YS; 21 El.
Tubular flux cored electrode for CO_2 welding.

FABCO 1MN.
M-1000; weld metal: 0.07 C, 1.33 Mn, 0.26 Si, 0.47 Mo, bal Fe.
As welded: 113,000 psi TS; 103,000 psi YS; 16 El.
Tubular flux cored electrode for CO_2 welding.

FABCO 80.
M-1000; weld metal: 0.10 C, 1.9 Mn, 0.94 Si, bal Fe.
As welded: 101,000 TS; 86,000 YS; 24 El.
Tubular flux cored electrode for CO_2 welding.
AWS-E70T-2.

FABCO 81.
M-1000; weld metal: 0.077 C, 1.45 Mn, 0.72 Si, other metals 0.10, bal Fe.
As welded: 83,000 psi TS; 70,000 psi YS; 28 El.
Tubular flux cored electrode for CO_2 welding.
AWS-E70T-1.

FABCO 82.
M-1000; weld metal: 0.08 C, 1.40 Mn, 0.50 Si, other metals 0.12, bal Fe.
As welded: 88,000 TS; 77,000 YS; 25 El.
Tubular flux cored electrode for CO_2 welding.
AWS-E70T-1.

FABCO 85.
M-1000; weld metal: 0.07 C, 1.40 Mn, 0.50 Si, other metals 0.10, bal Fe.
As welded: 77,000 psi TS; 63,500 psi YS; 28 El.
Tubular flux cored electrode for CO_2 welding.
AWD-E70T-5.

FABCO 115.
M-1000; weld metal: 0.05 C, 1.70 Mn, 0.50 Si, 0.35 Cr, 2.4 Ni, 0.45 Mo, bal Fe.
As welded: 110,500 psi TS; 98,500 psi YS; 22 El.
Tubular flux cored electrode for CO_2 welding.

FABCO 801.
M-1000; weld metal: 0.06 C, 0.70 Mn, 0.22 Si, 2.47 Ni, bal Fe.
As welded: 73,500 psi TS; 62,000 psi YS; 26 El.
Tubular flux cored electrode for CO_2 welding.
AWS E-70T-G; ABS H-3.

FABCO 802.
M-1000; weld metal: 0.05 C, 1.6 Mn, 0.50 Si, bal Fe.
As welded: 84,000-89,000 psi TS; 74,000-77,000 psi YS; 25-27 El.
Tubular welding wire for CO_2 and C_{25} welding.
AWS A5.20-69, Class E70T-1.

FABIS.
M-510; C, bal Fe.
Water hardening tool steel.

FABLOY 1CM.
M-1000; weld metal: 0.07 C, 0.80 Mn, 0.65 Si, 1.2 Cr, 0.50 Mo, bal Fe.
Welded, S.R. 1275°F; 81,000 TS; 67,000 YS; 22 El.
Tubular flux cored electrode for CO_2 welding.
AWS E-8018-B2; ASME A-3.

FABLOY 2CM.
M-1000; weld metal: 0.06 C, 0.82 Mn, 0.67 Si, 2.25 Cr, 1.0 Mo, bal Fe.
Welded, S.R. 1275°F: 91,000 TS; 78,000 YS; 20 El.
Tubular flux cored electrode for CO_2 welding.
AWS E-9018-B3; ASME A-4.

FABLOY 2N.
M-1000; weld metal: 0.05 C, 1.03 Mn, 0.68 Si, 2.39 Ni, bal Fe.
Welded, S.R. 1150°F: 82,000 TS; 70,000 YS; 28 El.
Tubular flux cored electrode for CO_2 welding on 2 1/2 Ni steels requiring impact strength.
AWS E-8018-C1.

FABLOY 3N.
M-1000; weld metal: 0.04 C, 0.97 Mn, 0.56 Si, 3.62 Ni, bal Fe.
Welded, S.R. 1150°F; 81,000 TS; 69,000 YS; 25 El.
Tubular flux cored electrode for CO_2 welding on 3 1/2 Ni steels used at very low temperatures.
AWS E-8018-C2.

FABLOY 5CM.
M-1000; weld metal: 0.05 C, 0.92 Mn, 0.70 Si, 5.30 Cr, 0.57 Mo, bal Fe.
Welded, Ann: 71,000 TS; 40,000 YS; 33 El.
Tubular flux cored electrode for CO_2 welding on 4-6 Cr steels used at high temperature.
AWS E502; ASME A-4.

FABLOY 9CM.
M-1000; weld metal: 0.05 C, 0.91 Mn, 0.79 Si, 10.15 Cr, 0.93 Mo, bal Fe.
As welded: 72,000 TS; 37,500 YS; 29 El.
Tubular flux cored electrode for CO_2 welding on 9-11 Cr Steels.
AWS E505.

FABLOY 410.
M-1000; weld metal: 0.057 C, 0.86 Mn, 0.64 Si, 12.71 Cr, 0.42 Ni, bal Fe.
As welded; 87,000 TS; 65,000 YS; 25.5 El.
Tubular flux cored electrode for CO_2 welding of corrosion resistant steels.
AWS E410.

FABLOY 4130.
M-1000; weld metal: 0.30 C, 0.80 Mn, 0.51 Si, 0.98 Cr, 0.34 Mo, bal Fe.
Welded, Q.T. 750°F: 181,000 psi TS.
Tubular flux cored electrode for CO_2 welding on SAE 4130 and similar steels.

FABLOY T.
M-1000; weld metal: 0.05 C, 2.0 Mn, 0.40 Si, 2.50 Ni, 0.40 Mo, bal Fe.
As welded: 123,000 TS; 108,000 YS; 14 El.
Tubular flux cored electrode for CO_2 welding.
AWS E-12018G.

FABRIALLOY.
M-1171; 0.5-0.6 C, bal Fe.
For welding electrodes; shielded.

FABSHIELD 4.
M-1000; weld metal: 0.15 C, 1.25 Mn, 0.28 Si, 1.15 Al, other metals 0.10, bal Fe.
As welded: 84,000 TS; 64,000 YS; 24 El.
Self shielded, flux cored welding electrode wire.
AWS E70T-4.

FABSHIELD 8.
M-1000; weld metal: 0.08 C, 1.38 Mn, 0.21 Si, 0.73 Al, bal Fe.
As welded: 88,000 TS; 75,000 YS; 23 El.
Self shielded, flux cored welding electrode wire.
AWS E70T-G.

FABSHIELD 8 NI.
M-1000; weld metal: 0.07 C, 1.31 Mn, 0.21 Si, 2.0 Ni, 0.59 Al, bal Fe.
As welded: 90,000 TS; 76,000 YS; 23 El.
Self shielded, flux cored welding electrode wire.
AWS E70T-G.

FABSHIELD 31.
M-1000; weld metal: 0.16 C, 0.63 Mn, 0.10 Si, 0.06 Al, bal Fe.
As welded: 82,000 TS; 64,000 YS; 22.5 El.
Self shielded, flux cored welding electrode wire.
AWS E60T-7.

FABSHIELD 55.
M-1000; weld metal: 0.18 C, 1.35 Mn, 0.50 Si, bal Fe.
As welded: 101,000 psi TS (transverse).
For high speed single pass welding.
AWS E70T-G.

FABTUF 250.
M-1000; weld metal: 0.26 C, 1.73 Mn, 1.85 Cr, 0.32 Si, bal Fe.
For build-up, magnetic, low abrasion resistance, high impact property. Rc 24.

FABTUF 960.
M-1000; weld metal: 0.70 C, 2.0 Mn, 8.0 Cr, 1.0 Si, bal Fe.
For build-up, magnetic, high abrasion resistance. Rc 55-60.

FACE-COR™ 2200.
M-663; weld deposit: 0.80 C, 15.0 Mn, 4.0 Cr, 4.0 Ni, bal Fe.
Wire for joining manganese steel and dissimilar steels and as build-up under base for hard coatings; for crusher jaws, railroad rails and frogs, shovel buckets. Tough, non-magnetic.

FACE-COR™ 2500 RB.
M-663; weld deposit: 0.90 C, 17.5 Mn, 0.50 V, bal Fe.
DC reverse polarity wire for build-up on crusher hammers, dredge pump parts, etc.
Tough, non-magnetic.

FACE-COR™ 2800.
M-663; weld deposit: 0.90 C, 17.5 Mn, 0.50 V, bal Fe.
DC reverse polarity wire for build-up on crusher rolls and jaws, dipper lips and teeth.
Tough, non-magnetic.

FACE-COR™ 5200.
M-663; weld deposit: 3.0 C, 1.5 Mn, 1.5 Si, 16.0 Cr, 1.0 Mo, bal Fe.
DC reverse polarity wire for hard, non-machinable alloy deposits. Magnetic; Rc 50-54 (two passes).
For crusher rolls, impactor breaker bars.

FACE-COR™ 5500 R.F.
M-663; weld deposit: 3.0 C, 1.5 Mn, 1.5 Si, 16.0 Cr, 1.0 Mo, bal Fe.
DC reverse polarity wire for hard, non-machinable alloy deposits. Magnetic; Rc 50-54 (two passes).

FACE-COR™ 6500.
M-663; weld deposit: 5.0 C, 1.5 Mn, 1.5 Si, 30.0 Cr, 1.0 Mo, bal Fe.
Electrode wire for hard build-up. Rc 58-61 (2 passes on mild steel). Non-machinable and cannot be flame cut.
For breaker bars, sand pump parts, mixing plows and scrapers.

FACEWELD 1.
M-578; 4.2 C, 5.4 Mn, 0.71 Si, 0.25 Mo, 22.5 Cr, 0.12 V, bal Fe.
Hard surfacing arc welding electrodes for resistance to severe abrasion.

FACEWELD 12.
M-578; 4.5 C, 1.0 Mn, 1.0 Si, 6.0 Mo, 18.5 Cr, 0.7 V, bal Fe.
Hard surfacing arc welding electrodes for resistance to severe abrasion.

FAGERSTA AD 95.
M-785; 0.70 C, 0.25 Cr, 0.15 V, 1.0 W, bal Fe.
For coining dies; oil hardened.

FAGERSTA ALLOY "30" DRILL.
M-608; 1.0 C, 1.1 Cr, 0.30 Mn, 0.35 Mo, bal Fe.
For tools, drills; water hardened.

FAGERSTA ALLOY CHISEL.
M-385; Cr, W, Mo, C, bal Fe.
For punches, chisels, rivet sets and busters; shock resistant.

FAGERSTA ALLOY SHOE DIE.
M-385; 0.55, C, 0.65 Cr, 0.35 Mo, bal Fe.
For shoe dies, cutting dies for leather and paper and rubber; oil hardened.

FAGERSTA ALLOY SHOE DIE.
M-608; 0.50 C, 0.60 Cr, 0.60 Mn, 0.40 Mo, bal Fe.
For tools, dies; oil hardened.

FAGERSTA BRILLIANT H.H.
M-608; 0.70 C, 3.5 Cr, 14 W, bal Fe.
For tools, cutters, drills; high speed steel.

FAGERSTA BRILLIANT WKE EXTRA.
M-608; 0.70 C, 4.5 Cr, 9 Co, 1.5 V, 19 W, bal Fe.
For tools, cutters, drills; high speed steel.

FAGERSTA BRILLIANT WW.
M-608; 0.70 C, 4.5 Cr, 1.5 V, 18 W, bal Fe.
For tools, cutters, reamers; high speed steel.

FAGERSTA BROACH.
M-608; 1.15 C, 0.40 Cr, 1.25 Mn, bal Fe.
For tools, broaches, reamers; tough.

FAGERSTA BROACH.
M-385; 0.7 C, 18 W, 4 Cr, 1 V, bal Fe.
For broaches, cutters, drills, taps; high speed steel.

FAGERSTA CARBON.
M-385; C, bal Fe.
For general tools; water hardening.

FAGERSTA D61 now FAGERSTA D-161.

FAGERSTA D 161.
M-52; 0.90 C, 1.2 Mn, 0.5 Cr, 0.5 W, 0.2 V, bal Fe.
For drawing and forming dies, punches; oil hardened, non-deforming.

FAGERSTA D-921.
M-52; 0.72 C, 4.0 Cr, 17.8 W, 1.1 V, bal Fe.
High speed steel for cutting tools.
AISI T1.

FAGERSTA D-927.
M-52; 1.25 C, 4.1 Cr, 3.2 Mo, 9.0 W, 3.0 V, 9.0 Co, bal Fe.
Cobalt high speed steel; heat treated: Rock C 67-68.
Good red hardness for tough cutting tool operations. (WKE 4).

FAGERSTA D-930.
M-52; 1.4 C, 4.2 Cr, 3.5 Mo, 8.5 W, 3.4 V, 11.0 Co, bal Fe.
Cobalt high speed steel; heat treated: Rock C 67-68.
Good red hardness and wear resistance for cutting tools on special alloys (WKE 45).

FAGERSTA D-933.
M-52; 0.88 C, 3.7 Cr, 9.5 Mo, 1.7 W, 1.2 V, 8.3 Co, bal Fe.
Co-Mo high speed steel for cutting tools; good red hardness for tough machining jobs.
AISI M33.

FAGERSTA D-941.
M-52; 0.86 C, 4.0 Cr, 5.0 Mo, 6.5 W, 1.9 V, bal Fe.
Mo-W high speed steel for cutting tools; good for lathe rough machining.
AISI M2.

FAGERSTA D-942.
M-52; 0.89 C, 4.2 Cr, 3.0 Mo, 6.4 W, 1.9 V, bal Fe.
High speed steel.

FAGERSTA D-943.
M-52; 0.83 C, 3.9 Cr, 8.7 Mo, 1.8 W, 1.2 V, bal Fe.
Mo high speed steel for drills, cutting tools.
AISI M1.

FAGERSTA D-946.
M-52; 0.89 C, 4.1 Cr, 5.1 Mo, 6.2 W, 1.9 V, 5.0 Co, bal Fe.
Co-Mo-W high speed steel for cutting tools; good red hardness. (WKE-46).

FAGERSTA D-948.
M-52; 1.80 C, 3.7 Cr, 9.4 Mo, 1.5 W, 1.1 V, 8.0 Co, bal Fe.
Special high speed steel; good abrasion resistance and red hardness. For special applications.
Similar to AISI M-42.

FAGERSTA D-950.
M-52; 0.95 C, 4.0 Cr, 5.0 Mo, 1.8 W, 1.2 V, bal Fe.
High speed steel. For twist drills, end mills.

FAGERSTA D-952.
M-52; 0.89 C, 4.1 Cr, 4.4 Mo, 1.2 W, 1.8 V, bal Fe.
High speed steel.

FAGERSTA D-954.
M-52; 0.99 C, 3.9 Cr, 8.8 Mo, 1.7 W, 2.0 V, bal Fe.
Mo-high carbon high speed steel; good abrasion resistance.
AISI M-7.

FAGERSTA D-960.
M-52; 0.97 C, 3.9 Cr, 7.9 Mo, 0.60 W, 1.9 V, bal Fe.
Mo-high carbon, high speed steel; good abrasion resistance.
Similar to AISI M10.

FAGERSTA DIE CASTING.
M-608; 0.50 C, 3 Cr, 0.35 Mn, 0.25 V, bal Fe.
For die casting dies, tools; oil hardened.

FAGERSTA DRAWING DIE STEEL.
M-385; C, Mn, Cr, W, bal Fe.
For wire drawing dies; oil hardening.

FAGERSTA ENGRAVER PLATES.
M-608; 0.35 C, 0.30 Mn, 0.10 Si, bal Fe.
For engraver plates; water hardened.

FAGERSTA ENGRAVERS PLATE.
M-385; 0.35 C, bal Fe.
For engravers plates.

FAGERSTA FB-01 now **FAGERSTA D-161.**

FAGERSTA FB-52.
M-52; 0.89 C, 4.1 Cr, 4.4 Mo, 1.2 W, 1.8 V, bal Fe.
Low cost high speed steel for cutting tools, drills, wood working tools.

FAGERSTA FB-M1 now **FAGERSTA D-943.**

FAGERSTA FB-M2 now **FAGERSTA D-941.**

FAGERSTA FB-M7 now **FAGERSTA D-954.**

FAGERSTA FB-M10 now **FAGERSTA D-960.**

FAGERSTA FB-T1 now **FAGERSTA D-921.**

FAGERSTA FAST FINISHING.
M-385; 1.2 C, 2 W, bal Fe.
For finishing tools; water hardening.

FAGERSTA FB-W1.
M-52; 1.05 C, 0.25 Mn, 0.25 Si, bal Fe.
Ht. Tr.: 208,000 TS; 146,000 YS; 12 El; 460 Brin.
For cold working tools, dies, blanking and coining dies, knurling and threading dies, punches, stamps, taps, reamers.
AISI Type W1, water hardening.

FAGERSTA FB-W1 SPECIAL.
M-52; 1.0-1.2 C, 0.25-0.35 Mn, 0.2-0.3 Si, bal Fe.
AISI Type W1 water hardening tool steel.

FAGERSTA FB-W1 SPECIAL CARBON.
M-52; 1.0-1.2 C, 0.25-0.35 Mn, 0.2-0.3 Si, bal Fe.
AISI Type W1 water hardening tool steel.

FAGERSTA HACK SAW.
M-385; C, W, bal Fe.
For hack saws, water hardening.

FAGERSTA HOLLOW DRILL.
M-608; 0.8 C, 0.3 Mn, bal Fe.
For hollow drills; water hardened.

FAGERSTA HOT DIE.
M-385; C, Cr, W, bal Fe.
For hot work dies; hot work steel.

FAGERSTA HX.
M-52.
Sintered carbide for turning difficult metals.

FAGERSTA NON-DEFORMING.
M-608; 0.95 C, 0.45 Cr, 1.05 Mn, 0.45 W, bal Fe.
For tools, dies; non-deforming.

FAGERSTA OIL HARDENING.
M-385; C, Cr, W, bal Fe.
For blanking, stamping and forming dies and tools; non-deforming.

FAGERSTA OVERCOAT AXE.
M-385, M-52; 1.0 C, 0.25 Mn, bal Fe.
For axes; water hardening.

FAGERSTA PAVEMENT BREAKER.
M-608; 0.65 C, 0.30 Mn, bal Fe.
For pavement breaking tools; water hardened.

FAGERSTA POLHEM WIRE DRAWING.
M-608; 1.8 C, 2.0 Cr, 2.0 Mn, 13 W, bal Fe.
For wire drawing dies; oil hardened.

FAGERSTA R-250.
M-52; 0.06 max C, 17.5 Cr, 0.50 max Ni, bal Fe.
Corrosion resistant steel, ferritic type.
Non-hardenable; magnetic.
AISI 430.

FAGERSTA R-300.
M-52; 0.11 max C, 18.0 Cr, 8.2 Ni, bal Fe.
Austenitic type stainless steel; non-magnetic; non-hardenable.
AISI 302.

FAGERSTA R320.
M-52; 0.08 max C, 17.2-18.7 Cr, 8.5-10.0 Ni, bal Fe.
Annealed: 78,000 min TS; 28,000 min YS; 40 min El; 55 min RA; B 90 Rock.
For equipment in the food and chemical industries, washing machines, fittings, kitchen sinks.
Type 304 Stainless, austenitic.

FAGERSTA R-326.
M-52; 0.06 max C, 20.0 Cr, 10.0 Ni, bal Fe.
Austenitic type stainless steel; non-magnetic; non-hardenable.
AISI 308.

FAGERSTA R-358.
M-52; 0.08 max C, 18.0 Cr, 9.5 Ni, Cb, bal Fe.
Stabilized austenitic stainless steel for welded assemblies.
Similar to AISI 347.

FAGERSTA R-360.
M-52; 0.03 max C, 18.5 Cr, 9.5 Ni, bal Fe.
For welded, austenitic stainless equipment as sinks, chemical tanks, food and beverage processing equipment.
AISI 304L.

FAGERSTA R-366.
M-52; 0.02 max C, 19.5 Cr, 10.0 Ni, bal Fe.
Low carbon grade of AISI 308 for welded equipment, or for welding 18-8 type stainless. Good corrosion resistance and usable in food processing equipment.

FAGERSTA R-380.
M-52; 0.05 max C, 18.0 Cr, 9.0 Ni, 0.30 Mo, S or Se, bal Fe.
Free machining grade of austenitic stainless 304.
For screw machine parts.
Similar to AISI 303.

FAGERSTA R-390.
M-52; 0.05 max C, 18.0 Cr, 11.5 Ni, bal Fe.
Austenitic stainless steel; low work hardening rate; for cold heading, severe drawing, spinning and forming.
AISI 305.

FAGERSTA R-440.
M-52; 0.05 max C, 17.0 Cr, 13.0 Ni, 2.8 Mo, bal Fe.
Better corrosion resistance than 304 alloy; for chemical equipment, food processing.
AISI 316.

FAGERSTA R-460.
M-52; 0.25 max C, 17.0 Cr, 13.0 Ni, 2.8 Mo, bal Fe.
Low carbon type of AISI 316; particularly for welded assemblies.
AISI 316L.

FAGERSTA R-470.
M-52; 0.02 max c, 18.5 Cr, 14.0 Cr, 3.7 Mo, bal Fe.
Low carbon grade of AISI 317; particularly for welded assemblies.
AISI 317L.

FAGERSTA R-575.
M-52; 0.08 max C, 18.0 Cr, 9.0 Ni, 3.5 Cu, bal Fe.
Austenitic stainless steel.

FAGERSTA R-806.
M-52; 0.08 max C, 22.5 Cr, 13.5 Ni, bal Fe.
For high temperature parts and equipment.
AISI 309.

FAGERSTA R-820.
M-52; 0.12 C, 26.0 Cr, 21.0 Ni, bal Fe.
For high temperature parts and equipment.
AISI 310.

FAGERSTA R-823.
M-52; 0.10 C, 23.5 Cr, 20.0 Ni, bal Fe.
For elevated temperature operation; annealing and carburizing boxes; heat treat fixtures.
AISI 314.

FAGERSTA R-860.
M-52; 0.08 max C, 19.0 Cr, 35.0 Ni, bal Fe.
For furnace parts, heat treat fixtures, carburizing boxes.
AISI 330.

FAGERSTA REGULAR.
M-608; 1.0 C, 0.35 Mn, bal Fe.
For tools, drills, taps; water hardened.

FAGERSTA ROLLED AUGER.
M-608; 0.85 C, 0.30 Mn, bal Fe.
For augers, drills, punches; water hardened.

FAGERSTA RRM20 now **FAGERSTA R-250.**

FAGERSTA RRNJ30 now **FAGERSTA R-320.**

FAGERSTA RRNJ31 now **FAGERSTA R-320.**

FAGERSTA RRNJ32 now **FAGERSTA R-320.**

FAGERSTA RRNJ33 now **FAGERSTA R-300.**

FAGERSTA RRNJ36 now **FAGERSTA R-360.**

FAGERSTA RRNJ39 now **FAGERSTA R-300.**

FAGERSTA RRT83 now **FAGERSTA R-820.**

FAGERSTA S25M.
M-52.
Sintered carbide cutting tools for milling cast iron and non-ferrous materials.

FAGERSTA SHOE DIE.
M-385; 0.55 C, 0.65 Cr, 0.35 Mo, bal Fe.
For cutting dies for leather, paper and rubber; water hardened.

FAGERSTA SOLID DRILL.
M-385; 1.0 C, 0.3 Si, 0.3 Mn, bal Fe.
For drills, chisels, taps; water hardened.

FAGERSTA SPECIAL.
M-608; 1.0 C, 0.35 Mn, bal Fe.
For tools, drills, taps; water hardened.

FAGERSTA SSL.
M-52; 0.8 C, 18.5 W, 1.2 Mo, 2.5 Co, 4.5 Cr, 1.6 V, bal Fe.
For cutting and planing tools, milling cutters; high speed steel.

FAGERSTA TWISTED AUGER.
M-385; 0.7 C, 0.3 Si, 0.35 Mn, bal Fe.
For augers for coal and ore mining; water hardened.

FAGERSTA WKE-4.
M-52; 1.25 C, 4.1 Cr, 3.1 V, 9.0 W, 3.1 Mo, 9.0 Co, bal Fe.
Heat treated: Rock C 67.
For cutters, thread tools, lathe and planer tools.
Super high speed steel.

FAGERSTA WKE-45.
M-52; 1.40 C, 4.2 Cr, 3.5 V, 9.0 W, 3.5 Mo, 11.0 Co, bal Fe.
High speed steel for finish machining steel; high hardness, good wear resistance, good red hardness.
Rc 68.

FAHLUN BRILLIANTS (FALUNER DIAMENTEN).
M-German; 60 Sn, 40 Pb.
For ornaments.

FAHR.
Fe-Ni-Cr.
For heat resistant parts; heat resistant.

FAHRALLOY C.
M-994; 56 Ni, 14 Cr, 17 Mo, 5 W, bal Fe.
For pumps, valves, plastic and chemical plant equipment.
Heat and corrosion resistant.

FAHRALLOY CD.
M-994; 0.25 C, 1.0 Si, 27-30 Cr, 3-6 Ni, bal Fe.
Cast: 85,000 TS; 48,000 YS; 16 El; 190 Brin.
At 1600°F: 23,000 TS; 18 El.
For salt pots, furnaces, sintering bars, cracking equipment, recuperators. High oxidation resistance in S-atm. Corrosion resistant casting alloy.
Type HD.

FAHRALLOY CF-4.
M-994; 0.04 C, 1.5 Si, 17-20 Cr, 9-12 Ni, bal Fe.
Annealed: 79,000 TS; 34,000 YS; 71 El.
For computer parts, engine mountings, filter press plates, hardware, oil burner parts, spray nozzles. Stainless casting alloy, austenitic, non-hardenable.

FAHRALLOY CG-12.
M-994; 0.12 C, 1.5 Si, 20-23 Cr, 10-13 Ni, bal Fe.
Annealed: 77,000 TS; 36,000 YS; 50 El; 163 Brin.
For oil refinery and chemical processing equipment, power plants, pumps, valve bodies.
Heat and corrosion resistant casting alloy, austenitic, non-hardenable.

FAHRALLOY CR-W.
M-994; 2.25 C, 11 Cr, 1 max Si, 1 W, bal Fe.
For heat resistant castings; corrosion and heat resistant.

FAHRALLOY D-1.
M-994; 2.5-3 C, 10-12 Cr, 4-5 Mo, 1 max Si, 1 max Mn, bal Fe.
For heat resistant castings; corrosion and heat resistant.

FAHRALLOY F-1.
M-994; 0.35-0.75 C, 37-41 Ni, 17-21 Cr, bal Fe.
For heat resistant castings; heat resistant, Type HU.

FAHRALLOY F-2.
M-994; 0.5 C, 21-24 Cr, 3-6 Ni, 1.75 max Si, 2 max Mn, bal Fe.
For heat resistant castings; corrosion and heat resistant.

FAHRALLOY F-2-B.
M-994; C, 22 Cr, 9 Ni, bal Fe.
For castings; heat resistant.

FAHRALLOY F-2-B.
M-994; 0.8 min C, 16-23 Cr, 7-11 Ni, Si, Al, bal Fe.
For heat and corrosion resistant parts; heat and corrosion resistant.

FAHRALLOY F-3.
M-994; 0.5 C, 26-30 Cr, 4 Ni, 1.5 Si, 1 Mn, bal Fe.
For castings; heat resistant.

FAHRALLOY F3X.
M-994; 3.1-3.5 C, 26.5-28.5 Cr, 2.5-4 Ni, 1.25 max Si, 0.5 max Mn, ba Fe.
For heat resistant castings; corrosion and heat resistant, cast iron.

FAHRALLOY F-5.
M-994; 0.35-0.75 C, 15-19 Cr, 64-68 Ni, 2.5 max Si, 2 max Mn.
For heat resistant castings; corrosion and heat resistant; Type HX.

FAHRALLOY F-5-B.
M-994; 0.51-0.80 C, 49-57 Ni, 16-23 Cr, Mn, Si, bal Fe.
For heat and corrosion resistant parts; heat and corrosion resistant.

FAHRALLOY F-10.
M-994; 0.2-0.5 C, 23-27 Cr, 11-14 Ni, bal Fe.
For heat resistant castings; heat resistant, Type HH.

FAHRALLOY F-10LC.
M-994; 0.2 C, 22-26 Cr, 12-15 Ni, 2 Si, 1.5 Mn, bal Fe.
Annealed: 88,000 TS; 50,000 YS; 38 El; 190 Brin.
For chemical plant equipment, valves, pumps, digester fittings; Type CH 20; stainless, austenitic.

FAHRALLOY F-11.
M-994; C, 14-16 Cr, 49-51 Ni, 19-21 Mo and W, bal Fe.
For heat resistant castings; corrosion and heat resistant.

FAHRALLOY F-12.
M-994; 1.0 C, 28 Cr, 10 Ni, bal Fe.
For furnace parts; heat resistant.

FAHRALLOY F-35.
M-994; 1 C, 33-36 Cr, bal Fe.
For heat resistant castings; heat resistant.

FAHRALLOY F-35 N.
M-994; 2 C, 33-36 Cr, 9-11 Ni, 2 max Si, 2 max Mn, bal Fe.
For heat resistant castings; corrosion and heat resistant.

FAHRALLOY F-0726.
M-994; 0.5 C, 6-8 Cr, 25-27 Ni, 1 max Si, 1.25 max Mn, bal Fe.
For heat resistant castings; corrosion and heat resistant.

FAHRALLOY F-0821.
M-994; 0.5 C, 7-10 Cr, 21-23 Ni, 1.5 max Si, 1 max Mn, bal Fe.
For heat resistant castings; corrosion and heat resistant.

FAHRALLOY F-1260.
M-994; 0.35-0.75 C, 10-14 Cr, 58-62 Ni, 2.5 max Si, 2 max Mn.
For heat resistant castings; corrosion and heat resistant; Type HW.

FAHRALLOY F-1400.
M-994; 0.15 max C, 12-14 Cr, 1.5 max Ni, bal Fe.
Heat treated: 100,000-216,000 TS; 75,000-173,000 YS; 28-6 El; 62-9 RA; 185-415 Brin.
For chemcial plant equipment, heat treating boxes; corrosion resistant; Type CA-15.

FAHRALLOY F-1535.
M-994; 0.35-0.75 C, 13-17 Cr, 33-37 Ni, 2.5 max Si, bal Fe.
Cast: 60,000 TS; 38,000 YS; 20 El; 21 RA; 156 Brin.
Annealed: 84,000 TS; 50,000 YS; 10 El; 18 RA; 200 Brin.
For furnace parts, muffles, heat treat boxes, pots; Type HT; heat and corrosion resistant.

FAHRALLOY F-1800.
M-994; 0.3 max C, 18-22 Cr, 2 Ni, 1.5 Si, 1 Mn, bal Fe.
For castings; stainless; Type CB-30.

FAHRALLOY F-1808.
M-994; 0.2 max C, 18-21 Cr, 8-11 Ni, 2 Si, 1.5 Mn, bal Fe.
For castings; stainless, austenitic; Type CF-20.

FAHRALLOY F-1808CB.
M-994; 0.08 max C, 18-21 Cr, 9-12 Ni, Cb = 8 x C, bal Fe.
Annealed: 85,000 TS; 45,000 YS; 45 El; 165 Brin.
For welded construction, chemical plant equipment; Type 347; stainless, stabilized.

FAHRALLOY F-1808LC.
M-994; 0.08 max C, 18-21 Cr, 8-11 Ni, 2 Si, 1.5 Mn, bal Fe.
Annealed: 78,000-85,000 TS; 42,000-45,000 YS; 50-45 El; 155-165 Brin.
For valves, pumps, mixers, spray nozzles, evaporators; Type CF-8; stainless, austenitic.

FAHRALLOY F-1808MO.
M-994; 0.08 max C, 18-21 Cr, 9-12 Ni, 2 Si, 1.5 Mn, 2-3 Mo, bal Fe.
Annealed: 80,000-88,000 TS; 45,000-48,000 YS; 48-42 El; 160-170 Brin.
For valves, pumps, mixers, spray nozzles, evaporators; Type CF-8M; stainless, austenitic.

FAHRALLOY F-1824.
M-994; 0.08 C, 16-19 Cr, 21-25 Ni, 2 Si, 1.5 Mn, 2-3 Mo, bal Fe.
For heat resistant castings; corrosion and heat resistant; Type CN-7.

FAHRALLOY F-2210.
M-994; C, 20-23 Cr, 9-11 Ni, Al, bal Fe.
For heat resistant castings; corrosion and heat resistant.

FAHRALLOY F-2520.
M-994; 0.25 C, 23-27 Cr, 19-22 Ni, 2 Si, 2 Mn, bal Fe.
For heat resistant castings; corrosion and heat resistant; Type CK-25.

FAHRALLOY F-2520-HK.
M-994; 0.2-0.6 C, 24-28 Cr, 18-22 Ni, 3 max Si, 2 max Mn, bal Fe.
For heat resistant castings; corrosion and heat resistant; Type HK.

FAHRALLOY F-2802.
M-994; 0.5 C, 26-30 Cr, 4 max Ni, 2 max Si, bal Fe.
Cast: 70,000-110,000 TS; 65,000-75,000 YS; 2-19 El; 190-223 Brin.
For ore roasting furnaces, rabble arms, dampers; Type HC; heat resistant.

FAHRALLOY F-2808.
M-994; 0.3 max C, 26-30 Cr, 8-11 Ni, 2 Si, 1.5 Mn, bal Fe.
Cast: 95,000 TS; 55,000 YS; 16 El; 18 RA; 212 Brin.
Annealed: 90,000 TS; 50,000 YS; 18 El; 20 RA; 197 Brin.
For valves, pumps, fittings, Type CE-30; corrosion resistant.

FAHRALLOY F-2810.
M-994; 0.2-0.5 C, 26-30 Cr, 8-11 Ni, 2 max Si, 2 max Mn, bal Fe.
For heat resistant castings; corrosion and heat resistant; Type HE.

FAHRALLOY F-2817.
M-994; 0.25 C, 27-30 Cr, 16-18 Ni, 1.5 Si, 1.5 Mn, bal Fe.
For heat resistant castings; corrosion and heat resistant.

FAHRALLOY F-2817-HI.
M-994; 0.2-0.5 C, 26-30 Cr, 14-18 Ni, 2 max Si, bal Fe.
Cast: 80,000 TS; 45,000 YS; 12 El; 180 Brin.
Aged: 90,000 TS; 65,000 YS; 6 El; 200 Brin.
For valves, fittings, pumps, furnace parts; Type HI; corrosion and heat resistant.

FAHRALLOY F-3020.
M-994; 0.2-0.6 C, 28-32 Cr, 18-22 Ni, 0.3 max Si, 2.0 max Mn, bal Fe.
For heat resistant castings; corrosion and heat resistant; Type HL.

FAHRALLOY-HB.
M-994; 0.40 C, 1.5 Si, 18-22 Cr, 2 max Ni, bal Fe.
Annealed: 95,000 TS; 60,000 YS; 15 El; 195 Brin.
For furnace brackets and hangers, rabble arms, valve parts and bodies. Heat and corrosion resistant casting alloy. Type CB-30.

FAHRALLOY-HT75.
M-994; 0.75 C, 1.75 Si, 13-17 Cr, 33-37 Ni, bal Fe.
Cast: 70,000 TS; 40,000 YS; 10 El; 180 Brin.
At 1400°F; 35,000 TS; 26,000 YS; 10 El.
For air ducts, brazing trays, cyanide pots, fan blades, glass molds.
Heat and abrasion resistant casting alloy. Type HT. High fatigue strength.

FAHRALLOY-HTCB.
M-994; 0.50 C, 1.75 Si, 13-17 Cr, 33-37 Ni, 1.0 Cb, bal Fe.
Cast: 68,000 TS; 39,000 YS; 12 El; 175 Brin.
For glass molds, gear spacers, heat treating fixtures, muffles. Heat and corrosion resistant casting alloy.
Type HT. High fatigue strength.

FAHRALLOY-HU CB.
M-994; 0.50 C, 1.75 Si, 17-21 Cr, 37-41 Ni, 1.0 Cb, bal Fe.
Cast: 70,000 TS; 40,000 YS; 9 El; 170 Brin.
For carburizing retorts, lead pots, resistor guides, muffles, pouring spouts, conveyor screws. Heat and corrosion resistant casting alloy. Type HU.

FAHRENWALD RESISTING GOLD ALLOY.
M-Eng.; 90-60 Au, 10-40 Pd.
For chemical equipment; white gold; high acid resistance.

FAHRIG ANTIFRICTION METAL.
M-Eng.; 90 Sn, 10 Cu.
For bearings; antifriction.

FAHRITE ALLOYS see **OHIOLOY ALLOYS.**

FAIRLEYS.
M-491; 0.7 C, 18 W, 4 Cr, 1 V, bal Fe.
For high speed tools and cutters; high speed steel.

FAKIR FAS2.
M-1312; 0.53 C, 0.90 Si, 0.9 Mn, bal Fe.
For gears, bolts, punches, crimpers; water hardened.

FAKIR FC1.
M-1312; 1.05 C, 1.0 Cr, 0.3 Mn, bal Fe.
For bearings, bushings, liners; water hardened, wear resistant.

FAKIR FGS.
M-1312; 1.05 C, 1.0 Cr, 1.15 W, 0.9 Mn, bal Fe.
For bearings, cutters, punches; water hardened, wear resistant.

FAKIR FJS.
M-1312; 0.9 C, 1.9 Mn, 0.1 V, bal Fe.
For punches, upsetters, blanking dies; oil hardened, nondeforming.

FAKIR FKL.
M-1312; 0.55 C, 1.05 Cr, 0.18 V, 1.85 W, bal Fe.
For cold header and upsetter dies, crimpers; oil hardened, tough.

FAKIR FKO-10.
M-1312; 0.78 C, 10 Co, 4.2 Cr, 0.8 Mo, 1.8 V, 18 W, bal Fe.
For lathe and planer tools, form cutters, hobs; high speed steel, heavy duty.

FAKIR FKP.
M-1312; 0.50 C, 1.05 Cr, 3.25 Ni, bal Fe.
For gears, bolts, crankshafts, fasteners; oil hardened, shock resistant.

FAKIR FP.
M-1312; 0.85 C, 4.1 Cr, 0.85 Mo, 1.6 V, 8.7 W, bal Fe.
For lathe and planer tools, reamers, broaches, taps; high speed steel.

FAKIR FRS27MO.
M-1312; 0.95 C, 4 Cr, W, Mo, V, bal Fe.
For form and milling cutters, reamers, hobs; high speed steel.

FAKIR FSC.
M-1312; 2.1 C, 11.5 Cr, 0.70 W, bal Fe.
For blanking and forming dies, punches; oil hardened, nondeforming.

FAKIR FSD.
M-1312; 0.35 C, 0.18 V, 1.05 Cr, 1.85 W, bal Fe.
For header dies, upsetters, punches, crimpers; oil hardened, tough.

FAKIR FSM.
M-1312; 0.35 C, 0.18 V, 1.05 Cr, 1.85 W, bal Fe.
For header dies, shears, punches, upsetters; oil hardened, tough.

FAKIR FSP.
M-1312; 0.30 C, 2 Co, 2.4 Cr, 0.25 V, 8.5 W, bal Fe.
For upsetters, punches, rivet sets; hot work steel, oil hardened.

FAKIR FSS.
M-1312; 1.05 C, 1.0 Cr, 1.15 W, 0.90 Mn, bal Fe.
For bearings, cutters, bushings; water hardened, wear resistant.

FAKIR FSZ.
M-1312; 1.4 C, 0.3 Cr, 0.1 V, 0.3 Mn, bal Fe.
For engravers tools, milling cutters, reamers; water hardened.

FAKIR FW2.
M-1312; 1.3 C, 0.25 max Si, 0.25 max Mn, bal Fe.
For cutters, drills; Type W1; water hardened.

FAKIR FW3.
M-1312; 1.15 C, 0.25 max Si, 0.25 max Mn, bal Fe.
For cutters, drills, hobs; Type W1; water hardened.

FAKIR FW4.
M-1312; C, bal Fe.
For tools, cutters; water hardened.

FAKIR FW5.
M-1312; C, bal Fe.
For tools, cutters; water hardened.

FAKIR FW6.
M-1312; C, bal Fe.
For tools, cutters; water hardened.

FAKIR FWM.
M-1312; C, bal Fe.
For tools, cutters; water hardened.

FAKIR FWN.
M-1312; 0.55 C, Ni, Cr, Mo, V, bal Fe.
For punches, crimpers, upsetters; oil hardened, shock resistant.

FALCON 4.
M-435; 0.45 C, 2.25 W, 1.5 Cr, 0.25 V, bal Fe.
For chisels, pneumatic tools; shock resistant.

FALCON 6.
M-435; 0.55 C, 2.25 W, 1.25 Cr, 0.25 V, bal Fe.
For punches, shear blades, dies; shock resistant.

FALCON-14.
M-1766; 0.14 C, 0.70 Mn, 0.25 Si, 1.0 Ni, 1.0 Cr, 0.15 Mo, bal Fe.
Ni-Cr-Mo carburizing steel; deep hardening.
IHA F-159.

FALCON-15.
M-1766; 0.15 C, 0.45 Mn, 0.25 Si, 4.0 Ni, 1.0 Cr, 0.25 Mo, bal Fe.
Ni-Cr-Mo carburizing steel; very deep hardening.
IHA F-156; BS EN39B.

FALCON-16.
M-1766; 0.15 C, 0.50 Mn, 0.25 Si, 3.30 Ni, 1.0 Cr, 0.25 Mo, bal Fe.
Ni-Cr-Mo carburizing steel; very deep hardening.
AFNOR 16NCD13; BS EN36B.

FALCON-18.
M-1766; 0.18 C, 0.90 Mn, 0.25 Si, 1.0 Ni, 1.0 Cr, 0.15 Mo, bal Fe.
Ni-Cr-Mo carburizing steel.
IHA F-158; AFNOR 18NCD6; UNI 19NCD4.

FALCON-19.
M-1766; 0.20 C, 0.80 Mn, 0.25 Si, 0.55 Ni, 0.50 Cr, 0.20 Mo, bal Fe.
Ni-Cr-Mo carburizing steel.
AFNOR 20NCD2; AISI 8620.

FALCON-20.
M-1766; 0.36 C, 0.70 Mn, 0.25 Si, 1.1 Ni, 1.0 Cr, 0.25 Mo, bal Fe.
Ni-Cr-Mo structural steel; hardenable to about 48-56 RC.
IHA F-128; DIN 36 CrNiMo 4; BS EN100.

FALCON-25.
M-1766; 0.32 C, 0.60 Mn, 0.25 Si, 2.5 Ni, 0.70 Cr, 0.40 Mo, bal Fe.
Ni-Cr-Mo structural steel; deep hardening.
IHA F-127; BS En25; UNI 30NCD12.

FALCON-27.
M-1766; 0.40 C, 0.70 Mn, 0.25 Si, 0.90 Ni, 0.75 Cr, 0.25 Mo, bal Fe.
Ni-Cr-Mo structural steel; hardenable to about 49-58 RC; 170-209 kg/mm^2 TS.
CENIM F-1282/40 NiCrMo4; UNI 38NCD4.

FALCON-29.
M-1766; 0.40 C, 0.70 Mn, 0.25 Si, 1.8 Ni, 0.80 Cr, 0.25 Mo, bal Fe.
Ni-Cr-Mo structural steel; hardenable to about 50-59 RC; 171-209 kg/mm^2 TS.
CENIM F-1272/40NiCrMo7; AISI 4340.

FALCON-30.
M-1766; 0.33 C, 0.50 Mn, 0.25 Si, 4.0 Ni, 1.25 Cr, 0.35 Mo, bal Fe.
Ni-Cr-Mo structural steel; deep hardening for highly stressed heavy equipment.
IHA F-126; AFNOR 35 NCD16; BS EN30B.

FALCON-40T.
M-1766; 0.40 C, 0.85 Mn, 0.20 Si, 0.55 Ni, 0.50 Cr, 0.20 Mo, bal Fe.
Ni-Cr-Mo structural steel.
AFNOR 40 NCD2; AISI 8640.

FALCON EXTRA.
M-253; high C, bal Fe.
For chisels, punches; see Seminole.

FALK NO. 1 GEARALLOY.
M-862; 0.27-0.37 C, 0.70-1.0 Mn, 0.60 max Si, 0.60-0.90 Cr, 0.90 max Ni, 0.30-0.40 Mo, bal Fe.
Alloy cast steel; hardenable to 335-375 Brin up to 6 in. sections.
For large heavily loaded gearing, flywheels, construction machinery.

FALK NO. 3 GEARALLOY.
M-862; 0.30-0.37 C, 0.70-1.0 Mn, 0.60 max Si, 0.60-0.90 Cr, 0.90 max Ni, 0.40-0.50 Mo, bal Fe.
Alloy cast steel; hardenable to 350-390 Brin up to 4 in. sections.
For large, heavily loaded construction machinery as flywheels, coupling hubs, gearing.

"F" ALLOY.
M-331; 2.5 Cu, 20 Zn, 0.5 Mg, 0.5 Mn, 0.75 Si, bal Al.
Heat treated: 8100 TS; 57,000 YS; 19 El.
For aircraft parts, structures; die casting.

FALLS 4-11-44.
M-92; 30 Ni, 30 Cr, bal Fe.
For strengthening high Si iron.

FALLS BORON ALUMINUM.
M-92; 1 B, bal Al.
For introducing boron in Al alloys; grain refiner.

FALLS BORON COPPER 2%.
M-92; 2 B, bal Cu.
For master alloy for bronzes; grain refiner.

FALLS COPPER ALUMINUM.
M-92; 50 Cu, bal Al.
For Cu addition to Al alloys; alloying.

FALLS COPPER-NICKEL.
M-92; 50 Ni, bal Cu.
For alloying for nickel bronzes.

FALLS FERRO ZINC 5%.
M-92; 5 Fe, bal Zn.
For adding Fe to brasses and bronzes; imparts hardness and creep resistance.

FALLS LITHIUM ALUMINUM 5%.
M-92; 5 Li, bal Al.
For Li additions to Al alloys; diminishes porosity.

FALLS LITHIUM COPPER 2%.
M-92; 2 Li, bal Cu.
For degasifier and deoxidizer for Cu alloys; for dense, non-porous castings.

FALLS MAGNESIUM ZINC. 4%.
M-92; 4 Mg, bal Zn.
For adding Mg to Zn base alloys; imparts hardness and creep resistance.

FALLS MANGANESE COPPER.
M-92; 30 Mn, bal Cu.
For Mn additions to Cu base alloys.

FALLS MANGANESE ZINC 4%.
M-92; 4 Mn, bal Zn.
For adding Mn to Zn base alloys; imparts hardness and creep resistance.

FALLS NICKEL COPPER.
M-92; 65 Ni, bal Cu.
For ladle additions to iron and steel.

FALLS NO. 11.
M-92; Cu alloy.
For high conductivity Cu castings; deoxidizers, degasifier.

FALLS NO. 12 ALUMINUM.
M-92; 7.0-8.5 Cu, bal Al.
Cast: 19,000-22,000 TS; 14,000 YS; 1.5-2.0 El; 50 Brin.
For crankcases, vacuum cleaners; not heat treatable.

FALLS NO. 14.
M-92; Cu alloy.
For pressure tight castings; for Cu base alloys.

FALLS NO. 21.
M-92; Mn-Cu.
For Mn bronze hardener; hardener.

FALLS NO. 31 AL BRONZE HARDENER.
M-92; Fe-Al-Cu.
For producing Al bronze; 89 Cu, 10 Al, 1 Fe.

FALLS NO. 34 AL BRONZE HARDENER.
M-92; Fe-Al-Cu.
For producing Al bronze; 4 Fe, 10 Al, 86 Cu.

FALLS NO. 43 ALUMINUM.
M-92; 5 Si, bal Fe.
Cast: 17,000-19,000 TS; 9000 YS; 6-3 El.
For castings; corrosion resistant.

FALLS NO. 108 ALUMINUM.
M-92; 3 Si, 4 Cu, bal Al.
Cast: 19,000 TS; 14,000 YS; 1.5 El.
For castings of intricate design; pressure tight.

FALLS NO. 154 ALLOY.
M-92; 45 Ni, 15 Cr, bal Fe.
For ladle additions to iron.

FALLS NO. 195 ALUMINUM.
M-92; 4.5 Cu, bal Al.
Cast: 21,000 TS; 14,000 YS; 2.0 El.
For castings of high strength; heat treatable.

FALLS NO. 214 ALUMINUM.
M-92; 4 Mg, bal Al.
Cast: 22,000-25,000 TS; 12,000 YS; 9-6 El.
For castings; corrosion resistant.

FALLS PHOSPHOR COPPER.
M-92; 15 P, bal Cu.
For deoxidizing Cu base alloys.

FALLS SILICON ALUMINUM.
M-92; 50 Si, bal Al.
For Si addition to Al alloys; alloying.

FALLS SILICON COPPER.
M-92; 10-30 Si, bal Cu.
For Si additions to Cu base alloys.

FALLS SPECIAL PATTERN ALUMINUM.
M-92; Si, Ti, bal Al.
For patterns, match plates; small top shrinkage.

FALLS TITANIUM ALUMINUM.
M-92; 2.5-5.0 Ti, bal Al.
For Ti addition to Al alloys; alloying.

FALLS TITANIUM ALUMINUM.
M-92; 5 Ti, bal Al.
For addition of Ti to aluminum alloys; increases fluidity and ductility.

FALLS V-25.
M-92; 25 Ni, bal Cu.
For Ni additions to bronze; hardener.

FAMA 100.
M-Sweden; 24 Ni, 13 Al, 4 Cu, bal Fe.
For permanent magnets, electrical and magnetic equipment. High permeability.

FAMA 600.
M-Sweden; 24 Ni, 13 Al, 4 Cu, bal Fe.
For permanent magnets, electrical and magnetic equipment. High permeability.

FAMA 700.
M-Sweden; 21 Ni, 12 Co, 10 Al, 6 Cu, bal Fe.
For permanent magnets, electrical and magnetic equipment. High permeability.

FAN BLADES.
M-Eng.; 61 Cu, 37.0 Zn, 1.5 Pb.
For fan blades; free-cutting.

FANSTEEL 60 METAL.
M-53; 10 W, bal Ta.
Recrystallized: 94,000 psi TS; 81,000 psi YS; 35 El; 78 R-30T; 16.9 density.
High strength at elevated temperatures; aerospace and chemical process applications; EB melted.

FANSTEEL 61 METAL.
M-53; 7.5 W, bal Ta.
As cold-worked: 165,000 psi TS; 160,000 psi YS; 5 El; 35 Rc, 16.8 density.
High strength for springs or other elastic parts for operation under chlorinating or severe acid corrosive conditions.
P/M material; also called TANTALOY and FANSTEEL TA W ALLOY.

FANSTEEL 63 METAL.
M-53; 2.5 W, 0.15 Cb, bal Ta.
Recrystallized: 56,000 psi TS; 34,000 psi YS; 38 El; 54 R-30T; 16.7 density.
Excellent fabricability and weldability; outstanding corrosion resistance; EB melted.

FANSTEEL 77 METAL.
M-53; 89 W, 7 Ni, 4 Cu.
Sintered; 100,000 psi TS; 2.5 El; 300-370 DPH; 17.1 density.
For rotors, gyroscopes, governors, radiation shields; machinable; excellen resistance to electrical arc erosion in non-oxidizing atmospheres.

FANSTEEL 80 METAL.
M-53; 1 Zr, bal Cb.
Recrystallized: 42,000 psi TS; 27,000 psi YS; 34 El; 8.6 density.
Low neutron absorption cross section; resistance to liquid metals; fabricated readily at room temperature.

FANSTEEL 85 METAL.
M-53; 11 W, 28 Ta, 1 Zr, bal Cb.
Recrystallized: 86,000 psi TS; 66,000 psi YS; 30 El; 10.3 density.
Exceptional resistance to creep up to 2500°F; fabricable, and can be coated to provide resistance to oxidizing environments.

FANSTEEL 103 METAL.
M-53; 10 Hf, 1 Ti, bal Cb.
Recrystallized: 62,000 psi TS; 47,000 psi YS; 35 El; 8.8 density.
Good fabricability and weldability; moderate strength and excellant stability at elevated temperatures; can be coated to provide resistance to oxidizing atmospheres.
For rocket motor skirts.

FANSTEEL 291 METAL.
M-53; 10 W, 10 Ta, bal Cb.
Recrystallized: 66,000 psi TS; 53,000 psi YS; 32 El; 9.6 density.
Good fabricability: high strength up to 3000°F; for rocket nozzle and other aerospace applications.

FANSTEEL COLUMBIUM.
M-53; Cb, commercially pure, unalloyed.
Recrystallized: 38,000 psi TS; 24,000 psi YS; 37 El; 38 R-30T; 8.6 density.
Good fabricability and weldability; corrosion resistance similar to Ta in many environments; low neutron absorption cross section; EB melted.

FANSTEEL CS FOIL.
M-53.
Ta material, special grade.
For foil type Ta electric capacitors; provides relatively constant capacitance over a wide range of operating temperatures.

FANSTEEL SCB-291 see FANSTEEL 291 METAL.

FANSTEEL TANTALUM.
M-53.
Ta, commercially pure, unalloyed.
Recrystallized: 40,000 psi TS; 26,000 psi YS; 38 El; 38 R-30T; 16.6 density.
Excellent fabricability and weldability; outstanding corrosion resistance similar to glass; stable, high dielectric oxide useful in capacitators; EB melted.

FANSTEEL TAW ALLOY see FANSTEEL 61 METAL.

FANSTEEL TPX.
M-53.
Ta material, special grade.
For embedded lead wire in solid Ta electrolytic capacitors and for vacuum furnace hardware; provides resistance to embrittlement during sintering.

FANTASTEC.
M-717.
Electrode for AC-DC metallic arc super-fast welding of all steels including low, medium and high alloy types; 100,000 psi TS.

FARMFACE.
M-9; 4.5 C, 6.0 Mn, 2.0 Si, 30.0 Cr, bal Fe.
Welded: 600 Brin.
For hard facing rod; austenitic, wear resistant.

FARRELL-CHEEK F85-4135C.
M-220; 0.37 C, 0.80 Mn, 0.45 Si, 1.0 Cr, 0.25 Mo, bal Fe.
Cast, N+T: 105,000 psi TS; 85,000 psi YS; 17 El; 210-260 Brin.
ASTM A-148 105-85; SAE 0105.

FARRELL-CHEEK F85-4135D.
M-220; 0.37 C, 0.80 Mn, 0.45 Si, 1.0 Cr, 0.25 Mo, bal Fe.
Cast, Q+T: 105,000 psi TS; 85,000 psi YS; 17 El; 240-290 Brin.

FARRELL-CHEEK F85-4135E.
M-220; 0.37 C, 0.80 Mn, 0.45 Si, 1.0 Cr, 0.25 Mo, bal Fe.
Cast, Q+T: 105,000 psi TS; 85,000 psi YS; 17 El; 300-350 Brin.

FARRELL-CHEEK F85-4135F.
M-220; 0.37 C, 0.80 Mn, 0.45 Si, 1.0 Cr, 0.25 Mo, bal Fe.
Cast, Q+T: 105,000 psi TS; 85,000 psi YS; 17 El; 350-400 Brin.

FARRELL-CHEEK F85-4330F.
M-220; 0.37 C, 0.70 Mn, 0.45 Si, 1.75 Ni, 0.75 Cr, 0.25 Mo, bal Fe.
Cast, Q+T: 105,000 psi TS; 85,000 psi YS; 17 El; 350-400 Brin.

FARRELL-CHEEK F85-4335C.
M-220; 0.37 C, 0.70 mn, 0.45 Si, 1.75 Ni, 0.75 Cr, 0.25 Mo, bal Fe.
Cast, N+T: 105,000 psi TS; 85,000 psi YS; 17 El; 210-260 Brin.
ASTM A-148 105-85; SAE 0105.

FARRELL-CHEEK F85-4335D.
M-220; 0.37 C, 0.70 Mn, 0.45 Si, 1.75 Ni, 0.75 Cr, 0.25 Mo, bal Fe.
Cast, Q+T: 105,000 psi TS; 85,000 psi YS; 17 El; 240-290 Brin.

FARRELL-CHEEK F85-4335E.
M-220; 0.37 C, 0.70 Mn, 0.45 Si, 1.75 Ni, 0.75 Cr, 0.25 Mo, bal Fe.
Cast, Q+T: 105,000 psi TS; 85,000 psi YS; 17 El; 300-350 Brin.

FARRELL-CHEEK F85-8620A.
M-220; 0.20 C, 0.70 Mn, 0.45 Si, 0.55 Ni, 0.50 Cr, 0.20 Mo, bal Fe.
Cast, N+T: 85,000 psi TS; 50,000 psi YS; 22 El; 170-220 Brin.
AISI 8620.

FARRELL-CHEEK F85-8635A.
M-220; 0.35 C, 0.70 Mn, 0.45 Si, 0.55 Ni, 0.50 Cr, 0.20 Mo, bal Fe.
Cast, N+T; 85,000 psi TS; 50,000 psi YS; 22 El, 170-220 Brin.
AISI 8635.

FARRELL-CHEEK F85-8635B.
M-220; 0.35 C, 0.70 Mn, 0.45 Si, 0.55 Ni, 0.50 Cr, 0.20 Mo, bal Fe.
Cast, N+T; 85,000 psi TS; 50,000 psi YS; 22 El; 180-230 Brin.

FARRELL-CHEEK F85-8635C.
M-220; 0.35 C, 0.70 Mn, 0.45 Si, 0.55 Ni, 0.50 Cr, 0.20 Mo, bal Fe.
Cast, HT; 85,000 psi TS; 50,000 psi YS; 22 El; 210-260 Brin.

FARRELL-CHEEK F85-8635D.
M-220; 0.35 C, 0.70 Mn, 0.45 Si, 0.55 Ni, 0.50 Cr, 0.20 Mo, bal Fe.
Cast, HT; 85,000 psi TS; 50,000 psi YS; 22 El; 240-290 Brin.

FARRELL-CHEEK F85-8635E.
M-220; 0.35 C, 0.70 Mn, 0.45 Si, 0.55 Ni, 0.50 Cr, 0.20 Mo, bal Fe.
Cast, NT; 85,000 psi TS; 50,000 psi YS; 22 El; 300-350 Brin.

FARRELL-CHEEK F85LCA.
M-220; 0.24 C, 1.35 Mn, 0.45 Si, 0.25 Mo, bal Fe.
Cast, N+T: 85,000 psi TS; 50,000 psi YS; 22 El; 170-220 Brin.
ASTM A-148 90-60; SAE 090.

FARRELL-CHEEK F85MMB.
M-220; 0.30 C, 1.35 Mn, 0.45 Si, 0.25 Mo, bal Fe.
Cast, N+T: 90,000 psi TS; 60,000 psi YS; 20 El; 180-230 Brin.

FARRELL-CHEEK F85MMC.
M-220; 0.30 C, 1.35 Mn, 0.45 Si, 0.25 Mo, bal Fe.
Cast, Q+T: 105,000 psi TS; 85,000 psi YS; 17 El; 210-260 Brin.

FARRELL-CHEEK F85MMD.
M-220; 0.30 C, 1.35 Mn, 0.45 Si, 0.25 Mo, bal Fe.
Cast, Q+T: 120,000 psi TS; 100,000 psi YS; 14 El; 240-290 Brin.

FARRELL-CHEEK F85MME.
M-220; 0.30 C, 1.35 Mn, 0.45 Si, 0.25 Mo, bal Fe.
Cast, Q+T: 150,000 psi TS; 125,000 psi YS; 10 El; 300-350 Brin.

FARRELL-CHEEK FC-1022.
M-220; 0.22 C, 0.70 Mn, 0.45 Si, bal Fe.
Cast steel, normalized or normalized and tempered; 135-170 Brin.
AISI 1022; ASTM A-216 WCA; SAE 0025.

FARRELL-CHEEK FC-1025.
M-220; 0.20-0.30 C, 0.65-0.85 Mn, 0.35-0.60 Si, bal Fe.
Annealed: 78,000 TS; 38,000 YS; 24 El; 36 RA; 150 Brin.
Heat treated: 80,000 TS; 55,000 YS; 22 El; 35 RA; 200 Brin.
For machine tool castings, gears, housings, shafts; SAE1025, water hardened.

FARRELL-CHEEK FC-1030.
  M-220; 0.25-0.35 C, 0.65-0.85 Mn, 0.35-0.60 Si, bal Fe.
  Annealed: 75,000 TS; 40,000 YS; 24 El; 35 RA; 170 Brin.
  Heat treated: 90,000 TS; 60,000 YS; 20 El; 40 RA; 210 Brin.
  For machine tools castings, gears, shafts, housings; SAE1030, water hardened.

FARRELL-CHEEK FC1030Q.
  M-220; 0.30 C, 0.70 Mn, 0.45 Si, bal Fe.
  Cast, Q+T: 80,000 psi TS; 50,000 psi YS; 22 El; 160-210 Brin.
  ASTM A-148 80-50; SAE 80-50.

FARRELL-CHEEK FC-1045.
  M-220; 0.40-0.50 C, 0.65-0.85 Mn, 0.35-0.60 Si, bal Fe.
  Annealed: 80,000 TS; 40,000 YS; 18 El; 30 RA; 175 Brin.
  Normalized: 85,000 TS; 50,000 YS; 18 El; 26 RA; 195 Brin.
  For machine tool castings, gears, shafts; SAE 1045, water hardened.

FARRELL-CHEEK FC 1045 Q.
  M-220; 0.45 C, 0.70 Mn, 0.45 Si, bal Fe.
  Cast, Q+T: 90,000 psi TS; 60,000 psi YS; 20 El; 180-230 Brin.
  ASTM A-148 90-60; SAE 090.

FASAL-40.
  M-1766; 0.40 C, 0.60 Mn, 0.25 Si, 1.5 Cr, 0.25 Mo, 1.10 Al, bal Fe.
  Nitriding steel.
  IHA F-174; AFNOR 40CAD6-12; BS EN41.

FASALOY 5.
  M-53; 75 Ag, 24.5 Cu, 0.5 Ni.
  Ann: 78 R-15T; Elect. cond.: 79% I.A.C.S.
  For electrical contacts.

FASALOY 7.
  M-53; 85 Ag, 15 Cd.
  Ann: 56 R-15T; Elect. cond.: 34% I.A.C.S.
  For electrical contacts.

FASALOY 13.
  M-53; 90 Pt, 10 Ru.
  Ann: 90 R-15T; Elect. cond.: 4% I.A.C.S.
  For electrical contacts.

FASALOY 14.
  M-53; 90 Pt, 10 Ir.
  Ann: 67 R-15T; Elect. cond.: 7% I.A.C.S.
  For electrical contacts.

FASALOY 16.
  M-53; 85 Pt, 15 Ir.
  Ann: 88 R-15T; Elect. cond.: 6% I.A.C.S.
  For electrical contacts.

FASALOY 19.
  M-53; 73.4 Pt, 18.4 Pd, 8.2 Ru.
  Ann: 90 R-15T; Elect. cond.: 4% I.A.C.S.
  For electrical contacts.

FASALOY 24.
  M-53; 10 Ag, 10 Au.
  Ann: 48 R-15T; Elect. cond.: 47% I.A.C.S.
  For electrical contacts.

FASALOY 31.
  M-53; 92 Pt, 8 Ru.
  Ann: 91 R-15T; Elect. cond.: 8% I.A.C.S.
  For electrical contacts.

FASALOY 35.
  M-53; 72 Pd, 26.2 Ag, 1.4 Cu, 0.4 Ni.
  Ann: 85 R-15T; Elect. cond.: 4% I.A.C.S.
  For electrical contacts.

FASALOY 37.
  M-53; 72 Pd, 26 Ag, 2 Ni.
  Ann: 82 R-15T; Elect. cond.: 4% I.A.C.S.
  For electrical contacts.

FASALOY 38.
  M-53; 40 Pd, 30 Ag, 30 Cu.
  Ann: 81 R-15T; Elect. cond.: 8% I.A.C.S.
  For electrical contacts.

FASALOY 41.
  M-53; 97 Ag, 3 Pd.
  Ann: 55 R-15T; Elect. cond.: 59% I.A.C.S.
  For electrical contacts.

FASALOY 42.
  M-53; 90 Ag, 10 Pd.
  Ann: 64 R-15T; Elect. cond.: 29% I.A.C.S.
  For electrical contacts.

FASALOY 43.
  M-53; 80 Ag, 20 Pd.
  Ann: 70 R-15T; Elect. cond.: 14% I.A.C.S.
  For electrical contacts.

FASALOY 51.
  M-53; 97 Ag, 3 Pt.
  Ann: 56 R-15T; Elect. cond.: 47% I.A.C.S.
  For electrical contacts.

FASALOY 72.
  M-53; 72 Au, 26.2 Ag, 1.8 Ni.
  Ann: 61 R-15T; Elect. cond.: 14% I.A.C.S.
  For electrical contacts.

FASALOY 73.
  M-53; 68.8 Au, 25.9 Ag, 5.3 Pt.
  Ann: 74 R-15T; Elect. cond.: 14% I.A.C.S.
  For electrical contacts.

FASALOY 99.
  M-53; 99.87 Ag, 0.13 Ge.
  Ann: 55 R-15T; Elect. cond.: 80% I.A.C.S.
  For electrical contacts.

FASALOY 115.
M-53; 77 Ag, 22.6 Cd, 0.4 Ni.
Ann: 68 R-15T; Elect. cond.: 30% I.A.C.S.
For electrical contacts.

FASALOY 130.
M-53; 75 Ag, 19.5 Cu, 5 Cd, 0.5 Ni.
Ann: 78 R-15T; Elect. cond.: 50% I.A.C.S.
For electrical contacts.

FASALOY 136.
M-53; 90 Ag, 10 Ni.
Ann: 56 R-15T; Elect. cond.: 90% I.A.C.S.
For electrical contacts.

FASALOY 137.
M-53; 85 Ag; 15 Ni.
Ann: 60 R-15T; Elect. cond.: 85% I.A.C.S.
For electrical contacts.

FASALOY 138.
M-53; 80 Ag, 20 Ni.
Ann: 61 R-15T; Elect. cond.: 81% I.A.C.S.
For electrical contacts.

FASALOY 139.
M-53; 50 Ag, 40 Ni.
Ann: 75 R-15T; Elect. cond.: 65% I.A.C.S.
For electrical contacts.

FASALOY 142.
M-53; 90 Ag, 10 Fe.
Ann: 55 R-15T; Elect. cond.: 92% I.A.C.S.
For electrical contacts.

FASALOY GAH.
M-53; 99.58 Ag, 0.22 MgO, 0.20 Ni.
Sintered: 60 R-30T; Elect. cond.: 75% I.A.C.S.
For electrical contents.

FASALOY RJA.
M-53; 90 Ag, 10 CdO.
Ann: 50 Rf: Elect. cond.: 79% I.A.C.S.

FASALOY RJC.
M-53; 90 Ag, 10 CdO.
Ann: 55 Rf; Elect. cond.: 79% I.A.C.S.
For electrical contacts.

FASALOY ROA.
M-53; 85 Ag, 15 CdO.
Ann: 50 Rf; Elect. cond.: 72% I.A.C.S.
For electrical contacts.

FASALOY RRA.
M-53; 82 Ag, 18 CdO.
Ann: 50 Rf; Elect. cond.: 67% I.A.C.S.
For electrical contacts.

F.A.S. STEEL.
M-521; 0.4 C, 11.5 Cr, bal Fe.
Oil treated: 112,000-123,000 TS; 78,500-90,000 YS; 22-18 El; 43-35 RA; 210-260 Brin.
For aircraft and automobile valves; heat and corrosion resistant; Firth Brown J-180.

FASTBOR HOLLOW DRILL.
M-1067, M-908; 0.8 C, bal Fe.
For hollow drills; tough.

FASTELL 01103 (UB).
M-53; 90 Ag, 10 W.
Coined: 65 Rf; Elect. cond.: 92% I.A.C.S.
For electrical contacts; (special compositions also available).

FASTELL 01153 (UC).
M-53; 85 Ag, 15 W.
Coined: 70 Rf; Elect. cond.: 88% I.A.C.S.
For electrical contacts.

FASTELL 01253 (UE).
M-53; 75 Ag, 25 W.
Sintered: Elect. cond.: 77% I.A.C.S.
For electrical contacts.

FASTELL 01501 (UJ).
M-53; 50 Ag, 50 W.
Sintered: 55 Rb; Elect. cond.: 65% I.A.C.S.
For electrical contacts.

FASTELL 01651 (UM).
M-53; 65 W, 35 Ag.
Sintered: 88 Rb; Elect. cond.: 54% I.A.C.S.
For electrical contacts.

FASTELL 01655 (UM-5).
M-53; 65 W, 35 Ag.
Sintered: 88 Rf; Elect. cond.: 58% I.A.C.S.
For electrical contacts.

FASTELL 01751 (UO).
M-53; 75 W, 25 Ag.
Sintered: 93 Rb; Elect. cond.: 46% I.A.C.S.
For electrical contacts.

FASTELL 01755 (UO-5).
M-53; 75 W, 25 Ag.
Sintered: 96 Rb; Elect. cond.: 48% I.A.C.S.
For electrical contacts.

FASTELL 01801 (UP).
M-53; 80 W, 20 Ag.
Sintered: 100 Rb; Elect. cond.: 42% I.A.C.S.
For electrical contacts.

FASTELL 01901 (UR).
M-53; 90 W, 10 Ag.
Sintered: 100 Rb; Elect. cond.: 32% I.A.C.S.
For electrical contacts; (special compositions also available).

FASTELL 02253 (NE).
M-53; 25 W, 75 Cu.
Ann: 60 Rf; Elect. cond.: 46% I.A.C.S.
For electrical contacts.

FASTELL 02501 (NJ).
M-53; 50 W, 50 Cu.
Sintered: 70 Rb; Elect. cond.: 38% I.A.C.S.
For electrical contacts.

Section I: Alloy Data / 529

FASTELL 02505 (NJ-5).
M-53; 50 W, 50 Cu.
Sintered: 80 Rb; Elect. cond.: 55% I.A.C.S.
For electrical contacts.

FASTELL 02601 (NL).
M-53; 60 W, 40 Cu.
Sintered: 81 Rb; Elect. cond.: 35% I.A.C.S.
For electrical contacts.

FASTELL 02651 (NM).
M-53; 65 W, 35 Cu.
Sintered: 87 Rb; Elect. cond.: 32% I.A.C.S.
For electrical contacts.

FASTELL 02701 (NN).
70 W, 30 Cu.
Sintered: 91 Rb; Elect. cond.: 30% I.A.C.S.
For electrical contacts.

FASTELL 02751 (N).
M-53; 75 W, 25 Cu.
Sintered: 98 Rb; Elect. cond.: 28% I.A.C.S.
For electrical contacts.

FASTELL 02785 (NP-5).
M-53; 78 W, 22 Cu.
Sintered: 100 Rb; Elect. cond.: 44% I.A.C.S.
For electrical contacts.

FASTELL 06501 (BJ).
M-53; 50 Mo, 50 Ag.
Sintered: 68 Rb; Elect. cond.: 55% I.A.C.S.
For electrical contacts.

FASTELL 06601 (BL).
M-53; 60 Mo, 40 Ag.
Sintered: 85 Rb; Elect. cond.: 45% I.A.C.S.
For electrical contacts.

FASTELL 06651 (E).
M-53; 65 Mo, 35 Ag.
Sintered: 97 Rb; Elect. cond.: 42% I.A.C.S.
For electrical contacts.

FASTELL 07103 (RJ).
M-53; 90 Ag, 10 CdO.
Ann: 45 Rf; Elect. cond.: 71% I.A.C.S.
For electrical contacts.

FASTELL 07153 (RO).
M-53; 85 Ag, 15 CdO.
Ann: 50 Rf; Elect. cond.: 62% I.A.C.S.
For electrical contacts.

FASTELL 07173 (RR).
M-53; 82 Ag, 18 CdO.
Ann: 55 Rf; Elect. cond.: 59% I.A.C.S.

FASTELL E see **FASTELL 06651 (E).**

FASTELL N see **FASTELL 02751 (N).**

FASTELL NL see **FASTELL 02601 (NL).**

FASTELL UJ-5.
M-53; 50 Ag, 50 W.
Sintered: 55 Rb; Elect. cond.: 70% I.A.C.S.
For electrical contacts.

FASTELL UM see **FASTELL 01651 (UM).**

FASTELL UP see **FASTELL 01801 (UP).**

FASTELL UR see **FASTELL 01901 (UR).**

FATIGUE-PROOF.
M-1124, M-669; 0.40-0.48 C, 1.35-1.65 Mn, 0.04 max P, 0.24-0.33 S, 0.15-0.30 Si, bal Fe.
Cold finished: 140,000 min TS; 125,000 min YS; 8 El; 30 min Rc; 80% machinability.
For studs, shafts, bolts, axles, machined parts.

FAULTLESS-A BABBITT.
M-88; Sn, Sb, Pb, Cu.
Cast: 21.8 Brin.
For bearings for saw mills and wood working equipment; Babbitt, resists sudden strain.

FAVORIT now **VEW K720.**
M-1308; 0.9 C, 1.9 Mn, 0.1 V, bal Fe.

FB-M1 (AND OTHER "FB" ALLOYS) see **FAGERSTA FB-M1.**

F.B.D.
M-German; 0.15 C, 1 Mn, 1.2 Si, 17.5 Cr, 16 Ni, 1.75 Mo, 2 Cb, 0.12 V, bal Fe.
For supercharger parts, gas engines; high heat resistant.

FC2/0.
M-1739.
Iron-copper, sintered.
Density: 6.2 gms/cc; UTS: 22 kg/mm^2;
Hardness: HV5-70.
Road transport shock absorber parts.

FC3/0.
M-1739.
Iron-copper, sintered.
Density: 6.2 gms/cc; UTS: 25 kg/mm^2.
Hardness: HV5-80.
Shock absorber parts, valves, pistons, levers, spacers.

FC3/1.
M-1739.
Iron-copper, 1% carbon, sintered.
Density: 6.2 gms/cc; UTS: 37 kg/mm^2.
Hardness: HV5-145.
Medium duty structural parts.

FC 85 see **FARRELL-CHEEK FC85.**

FC2/63/0.
M-1739.
Iron (some Cu+C, sintered).
Density: 6.2-6.4 gms/cc; UTS: 23.5 kg/mm^2.
Hardenable to Rock C 25.
B.S.S. A301; MPIE F-0200-P.

FC2/63/1.
M-1739; iron (some Cu+C, sintered).
Density: 6.2-6.4 gms/cc; UTS: 30 kg/mm^2.
Elon: 1%; Porosity: 21-17%.
Hardenable to Rock C 25.
B.S.S. A350; SAE 864B.

FC2/63/1/2.
M-1739.
Iron (some Cu+C, sintered).
Density: 6.2-6.4 gms/cc; UTS: 27.5 kg/mm^2.
Elon: 1%; Porosity: 21-17%.
MPIE FC-0210P.

FC3/58/1.
M-1739; iron (some Cu+C, sintered).
Density: 5.7-6.0 gms/cc: UTs: 27.5 kg/mm^2.
Elon: 1%; Porosity: 28-24%.
Hardenable to Rock C 22.
B.S.S. A350; MPIE FC-0310N.

FC3/61/1.
M-1739; iron (some Cu+C, sintered).
Density: 6.0-6.2 gms/cc; UTS: 31.5 kg/mm^2.
Elon: 1%; Porosity: 24-21%.
Hardenable to Rock C 25.
B.S.S. A350; MPIE FC-0310N.

FC3/63/0.
M-1739; iron (some Cu+C, sintered).
Density: 6.2-6.4 gms/cc; UTS: 25 kg/mm^2.
Elon: 1%; Porosity: 21-17%.
Hardenable to Rock C 25.
B.S.S. A301; MPIE FC-0300P.

FC3/63/1/2.
M-1739; iron (some Cu+C, sintered).
Density: 6.2-6.4 gms/cc; UTS: 31.5 kg/mm^2.
Elon: 1%; Porosity: 21-17%.
B.M.S.A. 31B; MPIE FC-0305P.

FC3/63/1.
M-1739; iron (some Cu+C, sintered).
Density: 6.2-6.4 gms/cc; UTS: 35 kg/mm^2.
Elon: 1%; Porosity: 21-17%.
Hardenable to Rock C 25.
B.S.S. A350; MPIE FC-0310P.

FC3/66/1.
M-1739; iron (some Cu+C, sintered).
Density: 6.5-6.8 gms/cc; UTS: 35 kg/mm^2.
Elon: 1%; Porosity: 16-13%.
Hardenable to Rock C 30.
MPIE FC-0310R.

FC7/63/0.
M-1739; iron (some Cu+C, sintered).
Density: 6.2-6.4 gms/cc; UTS: 31 kg/mm^2.
Elon: 1%; Porosity: 21-17%.
Hardenable to Rock C 25.
B.S.S. A303; MPIE FC-0700P; SAE 862.

FC7/63/1.
M-1739; iron (some Cu+C, sintered).
Density: 6.2-6.4 gms/cc; UTS: 32.5 kg/mm^2.
Elon: 1%; Porosity: 21-17%.
Hardenable to Rock C 25.
B.S.S. A352; MPIE FC-0710P.

FC7/63/1/2.
M-1739; iron (some Cu+C, sintered).
Density: 6.2-6.4 gms/cc; UTS: 31.5 kg/mm^2.
Elon: 1%; Porosity: 21-17%.
B.M.S.A.33; MPIE FC-0705P.

FCB(T).
M-Eng.; 0.12 C, 17.5 Cr, 12 Ni, 1 Cb, bal Fe.
For gas turbine parts; stainless, austenitic, stabilized.

FC-CMS.
M-439, M-1; 0.55-0.65 C, 0.8-1.2 Cr, 0.7 Mn, 0.4 Mo, 0.5 Ni, 0.15 V bal Fe.
For forming dies, oil hardened; tough.

FCHD.
M-1132; 0.14 max C, 17-19 Mn, 10-13 Cr, 0.5-2.0 Si, 0.3-0.8 Ti or 0.1-0.3 N, bal Fe.
For turbine blades; high heat resistant.

FC-HE.
M-220; 0.25-0.35 C, 1.2 Mn, 0.4 Si, 0.2-0.3 Mo, bal Fe.
Normalized: 110,000 TS; 90,000 YS; 20 El; 45 RA.
For hard edge tools, gears; wear and abrasion resistant.

F.C.I. FREE-CUTTING STAINLESS STEEL.
M-521; 0.12 max C, 0.8 max Si, 1.0 max Mn, 11.5-13.5 Cr, 0.3-0.6 Mo, 0.15-0.30 S, bal Fe.
Free machining martensitic stainless steel.
B.S. 970 (PT4) 416S21; similar to AISI 416.

FCI STAINLESS IRON.
M-521; 0.12 C, 0.6 Si, 1.2 Mn, 13.5 Cr, 0.28 Mo, 0.23 S, bal Fe.
Heat treated: 85,000 TS; 58,000 YS; 28 El; 62 RA; 180 Brin.
For hardware, screw machine products; free-cutting, corrosion resistant.

F.C.S. STAINLESS STEEL, FREE CUTTING.
M-521; 0.20-0.28 C, 0.6 Si, 1.2 Mn, 0.23 S, 0.25 Mo, 13.5 Cr, bal Fe.
Heat treated: 92,000 TS; 65,000 YS; 25 El; 54 RA; 220 Brin.
For hardware, screw machine products, bolts, shafts; free cutting, corrosion resistant.

"FD65" STEEL.
M-English; 0.25 C, 0.5 Mn, 3 Ni, 1.2 Cr, 0.45 Mo, 0.2 V, bal Fe.
Hardened: 150,000 TS; 20 El; 65 RA; 320 Brin.
For aircraft engine parts; case hardened.

FE74.
M-1739; iron, sintered.
Density: 7.2-7.9 gms/cc; Porosity: 8-4%.
High purity iron; excellent magnetic properties.
Hardenable and case hardenable.
B.S.S. A203; MPIE F-0000-T.

FE88.
M-1739.
Sintered iron with 1% carbon.
Density: 6.6 gms/cc; UTS: 20 kg/mm^2 6-12%
Elon; Hardness: HV5-70.

FEAL.
M-France; 25 Al, bal Fe.
For atomic reactors.
Outstanding resistance to oxidation and excellent strength at high temperatures. It is particularly transparent to neutrons.

FECHRAL.
M-Russia; 0.06-0.15 C, 17-25 Cr, 4-7 Al, bal Fe.
For resistance wire; heat resistant.

FECRALOY.
M-61; 0.2 C, 13 Cr, 4 Al, bal Fe.
Rolled: 128,000 TS; 104,000 YS; 5 El.
For rheostats, heating elements; high heat resistance.

FEDERALOY A-200.
M-709; 1.0 Cu, 1.0 Ni, 3.0 Cd, 95.0 Al.
Aluminum base bearing liner strip for extra heavy duty gasoline and diesel engines; 40 Brin.

FEDERALOY A-300.
M-709; 1.0 Cu, 6.25 Sn, 0.5 Ni, 1.5 Si, 91.25 Al.
Aluminum base bearing lining strip for connecting rod and main bearings, thrust washers for heavy duty gasoline and diesel engines; 50 Brin.

FEDERALOY AT 4.
M-709; 91.0 Al, 4.0 Sn, 4.0 Si, 1.0 Cu.
Aluminum base bearing liner strip for medium duty on connecting rod and flange main bearings; Rockwell 15T-70.

FEDERALOY AT-7.
M-709; 5.5-7.0 Sn, 0.7-1.3 Cu, 0.7-1.3 Ni, bal Al.
Cast: 21,000 TS; 10,500 YS; 11 El; 44 Brin.
For journal and crankpin bearings; Alcoa 750 TS; heavy duty.

FEDERALOY AT-20.
M-709; 1.0 Cu, 20.0 Sn, 79.0 Al.
Aluminum base bearing liner for connecting rod and main bearings for passenger cars, electric motors. 38 Brin.

FEDERALOY B-100.
M-709; 3-5 Cu, 7-8 Sb, bal Sn.
At 70°F: 105,000 TS; 11 El; 24 Brin.
At 300°F: 42,000 TS; 22 El; 8 Brin.
For reciprocating engine bearings, pump bearings; Babbitt.

FEDERALOY F-1.
M-709; 87.0 Cu, 10.0 Sn, 2.0 Zn, 1.0 max Ni.
Cast bronze bearing metal.

FEDERALOY F-2.
M-709; 80.0 Cu, 10.0 Sn, 10.0 Pb.
Cast bronze bearing metal; SAE 64.

FEDERALOY F-3.
M-709; 84.0 Cu, 5.0 Sn, 5.0 Pb, 5.0 Zn, 1.0 max Ni.
Cast bronze bearing metal.
SAE 40.

FEDERALOY F-5.
M-709; 84.0 Cu, 5.0 Sn, 9.0 Pb, 2.0 max Zn.
Cast bronze bearing metal.
SAE 66.

FEDERALOY F-8.
M-709; 83.0 Cu, 7.0 Sn, 7.0 Pb, 3.0 Zn.
Cast bronze bearing metal.
SAE 660.

FEDERALOY F-12.
M-709; 72.0 Cu, 7.0 Sn, 19.0 Pb, 1.25 max Zn, 1.0 max Ni.
Cast bronze bearing metal.

FEDERALOY F-15.
M-709; 69.0 Cu, 9.0 Sn, 20.0 Pb, 1.0 max Zn, 1.0 max Ni.
Cast bronze bearing metal.

FEDERALOY F-20.
M-709; 89.0 Cu, 11.0 Sn.
Cast bronze bearing metal.

FEDERALOY H-24.
M-709; 21-27 Pb, 1 min Sn, bal Cu.
Cast.
For journal and crankpin bearings; SAE 49; heavy duty.

FEDERALOY H-35.
M-709; 32-37 Pb, bal Cu.
Cast.
For journal and crankpin bearings; SAE 480; heavy duty.

FEDERALOY H-35-LT.
M-709; 7.0 Sn, 28.0 Pb, 65.0 Cu.
Bronze-lead bearing metal liner.

FEDERALOY H-116.
M-709; 73.0 Cu, 3.25 Sn, 23.75 Pb.
Copper-lead bearing liner strip for connecting rod and main bearings for extra heavy duty gasoline and diesel engines; Rockwell 15T-75.

FEDERALOY HF-2.
M-709; 9-11 Sn, 9-11 Pb, 0.75 max Zn, 0.5 max Ni, bal Cu.
Cast.
For heavy duty bushings, bearings; SAE-797.

FEDERALOY HF-3.
M-709; 4 Sn, 4 Zn, 8 Pb, 0.5 max Sb, 0.5 max P, bal Cu.
Cast.
For thrust washers, bearings, bushings; SAE-798.

FEDERALOY HF-16.
M-709; 3-4 Sn, 3 max Zn, 21-25 Pb, bal Cu.
Cast.
For heavy duty bushings, bearings; SAE 799.

FEDERALOY L-200.
M-709; 14-16 Sb, 0.75-1.25 Sn, 0.6 max Cu, 1.0 As, bal Pb.
At 70°F: 8800 TS; 2 El; 21 Brin.
For bearings for engines, compressors, pumps and turbines; Babbitt.

FEDERALOY L-300.
M-709; 9.25-10.75 Sn, 14-16 Sb, 0.5 max Cu, bal Pb.
For bearings, bushings; Babbitt.

FEDERALOY L-301.
M-709; 2-4 Sn, 0.3-1.0 Hg, 0.05-1.0 Ca, 0.15 max Al, bal Pb.
At 70°F: 10,300 TS; 13 El; 23 Brin.
For bearings; Babbitt.

FEDERALOY RS-11.
M-709; 0.25-0.75 Sn, 0.10 max Fe, 88-92 Cu, bal Zn.
For architectural trim; commercial bronze; SAE 795.

FEDERALOY RS-13.
M-709; 66 Cu, 34 Zn.
For ornaments, fixtures, hardware; SAE 70C; yellow brass.

FEDERALOY RS-21.
M-709; 3.5-4.5 Sn, 3.5-4.5 Zn, 3.5-4.5 Pb, bal Cu.
Cast.
For piston pins, bushings, thrust washers; SAE 791; leaded bronze.

FEDERATED F004.
M-815; 4 Mg, bal Al.
Sand Cast: 22,000-26,000 TS; 12,000-14,000 YS; 6-10% El; 50-60 Brin.
For corrosion resistant castings. Tough. Aluminum 214 alloy. Not heat treatable.

FEDERATED F004ZN.
M-815; 4 Mg, 2 Zn, bal Al.
Cast: 24,000-29,000 TS; 16,000-19,000 YS; 3-7 El; 70 Brin.
For cookware, food handling equipment, fittings, marine hardware.
Pressure tight, similar to Aluminum A214.

FEDERATED F008.
M-815; 8 Mg, bal Al.
Cast: 45,000 TS; 28,000 YS; 5 El; 75 Brin.
For general purpose die castings, brake shoes, propellers, cylinder blocks, motor brackets.
Tough, strong, high corrosion resistance. Similar to Aluminum 218.

FEDERATED F024.
M-815; 2 Si, 4 Mg, bal Al.
Cast: 20,000-24,000 TS; 13,000-16,000 YS; 1-2 El; 60 Brin.
For pressure tight castings, cooking utensils, marine fittings.
Aluminum B 214, excellent corrosion resistance.

FEDERATED F050.
M-815; 5 Si, bal Al.
Cast: 22,000 TS; 10,000 YS; 4 El; 50 Brin.
For pressure tight castings, marine hardware, architectural parts.
Aluminum 43 alloy, tough and corrosion resistant.

FEDERATED F070.3.
M-815; 7 Si, 0.3 Mg, bal Al.
Ht. Tr.: 33,000 TS; 24,000 YS; 3 El; 70 Brin.
Sand Cast: 23,000 TS; 12,000 YS; 3 El; 55 Brin.
For complex pressure tight castings.
Aluminum 356. Age hardenable, corrosion resistant.

FEDERATED F090.5.
M-815; 9.5 Si, 0.5 Mg, bal Al.
Cast: 46,000 TS; 24,000 YS; 4 El; 80 Brin.
For general purpose castings, cover plates, instrument housings, castings. Corrosion resistant die castings. Similar to Aluminum 360.

FEDERATED F0120.
M-815; 12 Si, bal Al.
Sand Cast: 27,000 TS; 12,000 YS; 8 El; 55 Brin.
Permanent Mold: 26,000 TS; 12,000 YS; 3 El.
For meter cases, switch boxes, thin walled castings.
Similar to Aluminum 13, not heat treatable, pressure tight, high fluidity.

FEDERATED F100.4.
M-815; 0.8 Cu, 8 Zn, 0.4 Mn, bal Al.
Cast: 29,000 TS; 16,000 YS; 6-7 El; 60 Brin.
For permanent mold castings.

FEDERATED F150.5.
M-815; 1 Cu, 5 Si, 0.5 Mg, bal Al.
Ht. Tr.: 44,000 TS; 42,000 YS; 1 El; 95 Brin.
Cast: 26,000 TS; 17,000 YS; 2 El; 75 Brin.
For complex pressure tight castings, cylinders, housings, hardware.
Age-hardenable, good castability, pressure tight.

FEDERATED F-250.
M-815; 2 Cu, 5 Si, bal Al.
For light alloy castings; age-hardenable.

FEDERATED F360.
M-815; 3 Cu, 6 Si, bal Al.
Ht. Tr.: 35,000 TS; 20,000 YS; 3 El; 80 Brin.
Sand Cast: 28,000 TS; 15,000 YS; 4 El; 60 Brin.
For housings, oil pans, casings.
Heat treatable, high strength.

FEDERATED F391 NI.
M-815; 3.5 Cu, 9 Si, 1 Mg, 1 Ni, bal Al.
Cast: 32,000 TS; 26,000 YS; 2 El; 85 Brin.
For pistons.
Aluminum D 132 alloy.

FEDERATED F401.5 NI.
M-815; 4 Cu, 1.5 Mg, 2 Ni, bal Al.
Ht. Tr.: 45,000 TS; 42,000 YS; 1 El; 100 Brin.
Cast: 34,000 TS; 32,000 YS; 1 El; 86 Brin.
For cylinder heads, pistons.
Aluminum 142 alloy, wear and heat resistant. Age hardenable.

FEDERATED F410.
M-815; 4 Cu, 1 Si, bal Al.
Ht. Tr.: 47,000 TS; 42,000 YS; 1 El; 110 Brin.
Cast: 25,000 TS; 11,000 YS; 4 El; 65 Brin.
For high strength castings, housings, hardware, gear cases, oil pans.
Age-hardenable. Aluminum 195 alloy.

FEDERATED F 430.
M-815; 3.5-4.5 Cu, 3 Si, 0.05 Mg, 0.5 Mn, 1 Fe, 0.3 Ni, 0.3 Zn, bal Al.
Cast: 25,000 TS; 12,000 YS; 3 El; 57 Brin.
Aged: 28,000 TS; 21,000 YS; 2 El; 74 Brin.
For light alloy castings; general purpose alloy.

FEDERATED F450.
M-815; 4 Cu, 5 Si, bal Al.
Ht. Tr.: 45,000 TS; 40,000 YS; 1 El; 105 Brin.
Cast: 25,000 TS; 15,000 YS; 2 El; 70 Brin.
For general purpose castings.
Aluminum A108, heat treatable. Good castability.

FEDERATED F460.
M-815; 3-4 Cu, 5.5-6.5 Si, bal Al.
Ht. Tr.: 45,000 TS; 40,000 YS; 1 El; 105 Brin.
Cast: 25,000 TS; 15,000 YS; 2 El; 70 Brin.
For general purpose castings, grills, reflectors, housings, castings.
Heat treatable, Aluminum 319 alloy.

FEDERATED F-480.
M-815; 4 Cu, 8 Si, bal Al.
For die castings; high strength.

FEDERATED F-720.
M-815; 7 Cu, 2 Si, bal Al.
For castings; general use.

FEDERATED F750.3.
M-815; 6.5 Cu, 5.5 Si, 0.3 Mg, bal Al.
Ht. Tr.: 50,000 TS; 48,000 YS; 0.5 El; 130 Brin.
For pistons.
Aluminum 152 alloy, age-hardenable, high strength.

FEDERATED F1000.3.
M-815; 10 Cu, 0.3 Mg, bal Al.
Ht. Tr.: 44,000 TS; 40,000 YS; 0.5 El; 120 Brin.
For cylinder heads, pistons.
Aluminum 122 alloy. Heat treatable, good high temperature strength.

FEDERATED F1040.3.
M-815; 10 Cu, 4 Si, 0.4 Mg, bal Al.
Ht. Tr.: 46,000 TS; 44,000 YS; 125 Brin.
For cylinder heads, pistons.
Aluminum 138 alloy, age-hardenable.
Good high temperature strength.

FEDERATED F1121 NI.
M-815; 1 Cu, 12 Si, 1 Mg, 2.5 Ni, bal Al.
Stress-Relieved: 32,000-36,000 TS; 28,000-30,000 YS; 0.5% El; 105 Brin.
For general purpose castings.
Aluminum A 132 Alloy. Not heat treatable.

FEDERATED F4110.
M-815; Si, bal Al.
For die castings; high fluidity.

FEDERATED NO. 2.
M-815; 7.5 Sb, 3.5 Cu, 89 Sn.
Cast: 10,900 TS; 8 El; 24 Brin.
For bearings; M.P. 466°F.

FEDERATED NO. 3.
M-815; 83.3 Sn, 8.3 Sb, 8.3 Cu.
Cast: 12,300 TS; 2 El; 26 Brin.
For bearings; antifriction, Babbitt.

FEDERATED NO. 7.
M-815; 75 Pb, 10 Sn, 15 Sb.
Cast: 10,500 TS; 5 El; 20 Brin.
For bearings; antifriction, Babbitt.

FEDERATED NO. 8.
M-815; 80 Pb, 5 Sn, 15 Sb.
Cast: 10,000 TS; 5 El; 20 Brin.
For bearings; antifriction, Babbitt.

FEDERATED NO. 14.
M-815; 83.75 Pb, 0.75 Sn, 12.5 Sb, 3 As.
Cast: 9800 TS; 1.5 El; 22 Brin.
For bearings; antifriction, Babbitt.

FEDERATED NO. 15.
M-815; 83 Pb, 1 Sn, 1 As, 15 Sb.
Cast: 10,350 TS; 2 El; 20 Brin.
For bearings; antifriction, Babbitt.

FEDERATED STEEL MILL ALUMINUM GR. 4.
M-815; 85-90 Al, bal Cu, Fe, Mn, etc.
For deoxidizer in steel mills.

FEDERATED XXXX NICKEL.
M-815; Sn alloy.
For bearings; Babbitt.

FELAX-45.
M-1766; 0.47 C, 0.70 Mn, 1.75 Si, bal Fe.
Si-Mn spring steel; hardenable to 56-61 R.
IHA F-145; AFNOR 4658; DIN 46Si7.

FELAX-55.
M-1766; 0.56 C, 0.75 Mn, 1.75 Si, bal Fe.
Si-Mn spring steel; hardenable to 57-62 Rc.
IHA F-144; DIN 55Si7; AISI 9255; BSEN45.

FELAX-60.
M-1766; 0.60 C, 0.80 Mn, 1.75 Si, bal Fe.
Si-Mn spring steel; hardenable to 58-63 Rc.
DIN 65Si7; AISI 9260; BS EN45A.

FENTON'S ALLOY.
M-Eng.; 80 Zn, 14 Sn, 6 Cu.
For bearings; Babbitt.

FENTON'S ALLOY.
M-Eng.; 80 Zn, 8.5 Cu, 11.5 Sb.
For bearings; Babbitt.

FER.
M-365; 0.7-1.2 C, bal Fe.
For tools, dies, punches; water hardened.

FERAL.
M-Germany; 0.06 C, 0.6 Mn, bal Fe, clad with Al alloy.
For aircraft fire walls; aluminum-clad steel.

FERALITE.
M-145; 3 C, Si, Mn, bal Fe.
For housings, fittings, gears, shafts; malleable iron.

FERALSI.
M-745; Fe-Al-Si.
For magnetic applications; high magnetic permeability.

FERALUN.
M-250; Special iron base + abrasive grains imbedded.
For stairs, platforms; castings; wear resistant.

FERAN.
M-Eng.; sheet iron coated with Al and rolled together.
For engineering construction; corrosion resistance of Al and strength of Fe.

FERCHROMIT.
M-651; 22-30 Cr, 0.5-2.5 Si, 0.1-1.5 C, bal Fe.
Resists oxidation at high temperatures.

FEREX.
M-U.S.; 0.2 C, bal Fe.
For welding rod for armor; coated.

FERHOLZER.
M-1488; 0.005 C, Fe (very pure).
Ann, slow cool: 275 N/mm^2 UTS.
Very soft magnetic material; for solenoids, relays, electromagnets.

FERINOX.
M-101; 0.09 C, 1.2 Si, 1.3 Mn, 17 Cr, 8 Ni, 0.7 Mo, bal Fe.
Rolled: 299,000 TS; 250,000 YS; 8 El.
For watch mainsprings.
Stainless, austenitic, high fatigue strength.

FERMET.
M-Eng.; 4 Cr, 18 Ni, 2.2 Mn, 1 W, 0.3 Cu, 0.35 C, bal Fe.
For corrosion resisting parts; stainless and corrosion resistant.

FERNO.
M-688; 0.70 C, 3.75 Cr, 0.55 V, 0.7 Mo, bal Fe.
For hot work dies, punches, shear blades; good red hardeners.

FERNO EXTRA.
M-688; 0.45 C, 3.75 Cr, 0.4 Mn, 0.7 V, 12 W, bal Fe.
For hot work dies, punches, shear blades; tough and abrasion resistant.

FEROBA-III.
M-210; BaO, 6 Fe$_2$O$_3$.
3500 Br, 2500 Hc, 2,800,000 BH max.
For permanent magnets in electrical and magnetic equipment. High permeability.

FEROVAN.
M-1038; 43 V, 0.6 C, 6.6 Cr, 4.0 Mn, 6.3 Si, 0.10 max Al, bal Fe.
Ferro vanadium alloy for additive in steel melting.

FERRAL.
M-Russia; 0.25 max C, 12-15 Cr, 3.5-5.5 Al, bal Fe.
For electrical and heat resistance elements; applications to 1100°C.

FERRALIUM.
M-1169; 0.08 max C, 24.0-27.0 Cr, 4.5-6.5 Ni, 1.3-4.0 Cu, 2.0-4.0 Mo, 0.10 min N, bal Fe.
Cast or wrought ferritic-austenitic stainless.
820 N/mm^2 TS; 540 N/mm^2 YS; 18-20 El.
Improved strength and corrosion resistance.

FERRO BRONZE.
M-U.S.; 8 Fe, 0.7 Cr, bal Cu.
For hardware; corrosion resistant.

FERROCAL.
M-302; Al alloy.
Chill cast: 23,000-28,000 TS; 4-2 El.
For permanent mold castings; general machinery parts, automobile and motorcycle parts.

FERROCHROME see also ELECTROMET FERROCHROME.

FERROCHRONIN 15-60.
M-303; 15 Cr, 60 Ni, bal Fe.
For electrical equipment, heating elements, resistors; high heat and electrical resistance.

FERROCHRONIN 20-30.
M-303; 20 Cr, 30 Ni, bal Fe.
For electrical equipment, heating elements, resistors; high heat and electrical resistance.

FERROCHRONIN 75-15.
M-303; 14-16 Cr, 72 Ni, 6-9 Fe.
For heaters, combustion liners, manifolds, regenerators; high heat and corrosion resistance.

FERROCITE.
M-799; 0.5 C, bal Fe.
For valves, guides; powder metals.

FERRODUR.
M-508; 3 C, 0.5 Mo, 1.2 Ni, 0.9 Mn, 2 Si, 0.2 Cr, bal Fe.
Heat treated: 50,000 TS; 0 El; 500 Brin.
For pump liners, shaft sleeves; centrifugal casting.

FERROLUM.
M-1193; lead clad steel.
For chemical plant equipment; acid resistant.

FERRO-MANGANESE see also ELECTROMET FERRO-MANGANESE.

FERROMET.
M-1246; Fe alloy.
For machine parts, bearings, magnets; sintered alloy.

FERRO MOLYBDENUM (LOW CARBON).
M-640; 0.50 max C, 60-70 Mo.
For metallurgical applications in steel making; Mo-additions.

FERRO MOLYBDENUM (STANDARD).
M-640; 1.60-2.25 C, 58-65 Mo, bal Fe.
For metallurgical applications in steel making; Mo-additions.

FERRON.
M-329; 50 Fe, 35 Ni, 15 Cr.
For heating elements, heat and corrosion resisting parts; heat and corrosion resistant.

FERROPYR-I.
M-England; 86 Fe, 7 Cr, 7 Al, 1 Mn + Si.
For electrical resistances; used in high temperature range of 2100-2460°F.

FERROSILICON see also ELECTROMET FERROSILICON.

FERROSILID.
M-Russia; 0.5-0.7 C, 12-18 Si, bal Fe.
For cast pumps and conduits; acid resistant, cast.

FERROSTEEL.
M-748; 3 C, 1.8 Si, bal Fe.
Cast: 35,000 TS.
For valves, fittings; Tr.S. 3600; Tr. def. 0.14.

FERROSTEEL see CRANE FERROSTEEL.

FERROTHERM FF 6.
M-72; 0.15 C, 6 Cr, bal Fe.
Heat treated: 78,000 TS; 50,000 YS; 18 El; 50 RA.
For furnace parts, crucibles, autoclaves, recuperators; heat and corrosion resistant to 800°C.

FERROTHERM FF 18.
M-72; 0.15 C, 18 Cr, bal Fe.
Annealed: 78,000 TS; 50,000 YS; 12 El; 12 RA.
For furnace parts, crucibles, autoclaves, recuperators; heat and corrosion resistant to 1000°C.

FERROTHERM FF 30.
M-72; 0.15 C, 30 Cr, bal Fe.
Annealed: 78,000 TS; 58,000 YS; 12 El; 8 RA.
For furnace parts, crucibles, autoclaves, recuperators; heat and corrosion resistant to 1200°C.

FERROTHERM FF 112.
M-72; 1.0 C, 12 Cr, bal Fe.
For furnace parts, crucibles, autoclaves, recuperators; heat and corrosion resistant to 850°C.

FERROTHERM FF 118.
M-72; 1.0 C, 18 Cr, bal Fe.
For furnace parts, crucibles, autoclaves, recuperators; heat and corrosion resistant to 1000°C.

FERROTHERM FF 128.
M-72; 1.0 C, 28 Cr, bal Fe.
For furnace parts, crucibles, autoclaves, recuperators; heat and corrosion resistant to 1100°C.

FERROTHERM FF 228.
M-72; 2.0 C, 28 Cr, bal Fe.
For furnace parts, crucibles, autoclaves, recuperators; heat and corrosion resistant to 1100°C.

FERRO-TIC GR. A.
M-1469; 26 Ti, 7 C, 2 Cr, 2 Mo, bal Fe.
Annealed: 400 Brin.
Heat treated: 700 Brin.
For drawing and forming dies, rolls, cutters; sintered carbide, heat treatable.

FERRO-TIC GR. C.
M-1469; 26 Ti, 7 C, 2 Cr, 2 Mo, bal Fe. (45% carbide by volume).
Ann: 40 Rock C.
Ht. Tr.: 300,000 Tr.S.; 70 Rock C.
For tools, dies, rolling and blanking dies, valves, wear parts.
Sintered carbides with alloy steel matrix, wear resistant; max operating temp. 400°F.

FERRO-TIC GR. CM.
M-1469; 56.5 Fe, 27.6 Ti, 6.6 Cr, 2.0 Mo, 7.4 C. (45% carbide by volume).
Ann: 45 Rock C.
Ht. Tr.: 310,000 Tr.S; 66-70 Rock C.
For tools, dies, valves, forming dies.
Steel bonded carbide; max. operating temp: 1000°F.

FERRO-TIC GR. CN-5.
M-1469; 45% WC by volume, age-hardenable copper-nickel alloy matrix.
For tools and dies, and for wear resistant parts exposed to sea water; corrosion resistant, excellent resistance to sea water.

FERRO-TIC GR. CS-40.
M-1469; 45% TiC by volume, Martensitic stainless steel matrix.
Ann: 50 Rock C.
Hardened: 225,000 Tr.S; 68-70 Rock C.
Good wear and corrosion resistance; for can closing tools, chemical plant valve seats and wear parts.

FERRO-TIC GR. DN-1.
M-1469; 45% TiC by volume, age-hardenable nickel-aluminum alloy matrix.
Ann: 40-41 Rock C.
Age-hardened: 52-53 Rock C.
Good corrosion resistance toward strong chlorides.

FERRO-TIC GR. HT-2.
M-1469; 45% TiC by volume, Ni-Fe-Cr age hardenable alloy matrix.
Ann: 44 Rock C.
Aged: 54 Rock C; 250,000 Tr.S.

FERRO-TIC GR. HT-6.
M-1469; 45% TiC by volume, Ni-Cr age hardenable steel matrix.
Solutionized: 46 Rock C.
Aged (1st step): 51 Rock C.
Aged (fully): 54 Rock C; 310,000 Tr.S.
Excellent oxidation and corrosion resistance and high temperature strength.

FERRO-TIC GR. J.
M-1469; 40% W TiC_2 by volume, high speed steel matrix.
Annealed: 48 Rock C.
Hardened and tempered: 70 Rock C; 250,000 TrS.
For tools for hot working applications as piercing, forming, machining of non-ferrous materials.

FERRO-TIC GR. M-6.
M-1469; 45% TiC by volume; maraging steel matrix.
Solutionized: 51 Rock C.
Aged: 65 Rock C; 280,000 TrS.
Can be nitrided for 75 Rock C surface.

FERRO-TIC GR. M-6A.
M-1469; 50% TiC by volume, maraging steel matrix.
Solutionized: 54 Rock C.
Aged: 67 Rock C; 375,000 TrS.
Very good abrasion resistance; can be nitrided to surface hardness of 74 Rock C.

FERRO-TIC GR. M-6B.
M-1469; 55% TiC by volume; maraging steel matrix.
Ann: 58 Rock C.
Hardened: 68 Rock C; 400,000 TrS.
Excellent wear resistance.

FERRO-TIC GR. MS-5.
M-1469; 45% TiC by volume; age-hardenable martensitic stainless steel matrix.
Solutionized: 51 Rock C.
Aged: 63-64 Rock C; 280,000 TrS.
Good corrosion resistance; can be nitrided for high surface hardness; for abrasion resistance parts in the food, chemical and aerospace industries.

FERRO-TIC GR. S-45.
M-1469; 39 TiC, 11 Cr, 7.3 Ni, 42.7 Fe.
450 Brin.
For valve parts, seal rings, knives, molds; oxidation and heat resistant.

FERRO-TIC GR. S-55.
M-1469; 52 TiC, 8.6 Cr, 5.7 Ni, 33.7 Fe.
550 Brin.
For valve parts, seal rings, knives, molds; oxidation and heat resistant.

FERRO-TIC GR SK.
M-1469; 40% TiC (volume), Matrix = 5 Cr, 4 Mo, 0.5 Ni, 0.5 C, bal Fe.
Ann: 37 Rock C.
Hardened: 64 Rock C.
For applications involving mechanical and thermal shock.

FERROTITE.
M-192; 0.2 C, bal Fe.
For welding rod.

FERROTRODE 2-B.
M-717.
Electrode for AC-DC metallic arc for machinable build-up on steel; resists severe impact; hardness 28-31 Rc.

FERRO TUNGSTEN. (A.S.T.M.).
M-640; 0.60-0.70 C, 70-80 W, 1.0 max Si, bal Fe.
For metallurgical applications in steel making; sum H_2S metals less than 0.25%.

FERRO TUNGSTEN (LOW MELTING).
M-640; 2.5 C, 65-75 W, 1.5 max Si, bal Fe.
For metallurgical applications in steel making; sum H_2S metals less than 0.25%.

FERRO TUNGSTEN (STANDARD).
M-640; 0.60-0.70 C, 76-80 W, 0.75 max Si, bal Fe.
For metallurgical applications in steel making; sum H_2S metals less than 0.25%.

FERROVAC 1020.
M-1214; 0.20 C, 0.01 Ni, 0.003 Mn, 0.005 P, 0.01 Si, bal Fe.
For poles pieces for magnetrons, klystron components; vacuum melted.

FERROVAC 4340.
M-1214, M-38; 0.4 C, 0.8 Cr, 1.8 Ni, 0.25 Mo, bal Fe.
Heat treated: 185,000-285,000 TS; 166,000-215,000 YS; 16-12 El; 60-41 RA; 410-540 Brin.
For bolts, gears, shafts, crankshafts; vacuum melted, shock resistant.

FERROVAC 52100.
M-1214, M-38; 0.95-1.1 C, 1.3-1.6 Cr, bal Fe.
Annealed: 95,000 TS; 65,000 YS; 27 El; 62 RA; 180 Brin.
Heat treated: 200,000 TS; 185,000 YS; 3 El; 30 RA; 400 Brin.
For plug and ring gages, jet engine bearings; vacuum melted, wear resistant.

FERROVAC E.
M-38; 99.9% Fe.
Gas free, high purity iron.
High maximum permeability.
For relays, solenoid plungers, armatures, pole pieces, magnetic core devices.

FERROVAC HALMO.
M-38; 0.58 C, 4.75 Cr, 1.15 Si, 5.25 Mo, 0.55 V, bal Fe.
For high temperature bearings, shears, punches; vacuum melted hot work steel.

FERROWELD.
M-192; 0.1 C, bal Fe.
For welding rods for cast iron; coated.
AWS ESt.

FERROWELD.
M-578; 0.15 C, 0.03 Si, 0.30 Mn, 0.04 P, 0.04 S, bal Fe.
Low carbon steel arc welding electrode.
AWS Class ESt.

FERROXDURE.
M-Dutch; BaO + 6 Fe_2O_3.
For permanent magnets.

FERROXDURE I.
M-1401; $BaFe_{12}O_{19}$.
For magnets for motors and oil filters; magnetic ceramic.

FERROXDURE II.
M-1401; $BaFe_{12}O_{19}$.
For magnets for motors and oil filters; magnetic ceramic.

FERRUL.
M-Eng.; 54.6 Cu, 40 Zn, 5 Pb, 0.4 Al.
For bearings, intricate castings; free-cutting.

FERRY.
M-Eng.; 2 Ba, 1 Ca, 0.25 Hg, bal Pb.
For bearings, solder.

FERRY.
M-86, M-121; 55-56 Cu, 44-45 Ni.
Annealed: 73,000 TS; 34,000 YS; 45 El; 77 RA; 160 Brin.
For electrical resistances, thermocouples.
Max. operating temperature 440°C. Corrosion resistant.

FERRYDUR.
M-1717; 13 Cr, C, others, bal Fe.
Up to 48-65 Rock C hardness. Excellent wear resistance.
For castings to resist wear.
AFNOR: Z 200 C 13 (approx.).

FERRYNOX 8.
M-1717; 0.18-0.25 C, 17-19 Cr, 7-9 Ni, bal Fe.
Corrosion resistant, good resistance to scaling; austenitic.
For ventilator fans, heat exchangers.
AFNOR: Z 20 CN 18.08.

FERRYNOX 8M.
M-1717; 0.18-0.25 C, 16-20 Cr, 8-10 Ni, 3 Mn, bal Fe.
Good corrosion and heat resistance; Austenitic.
For pumps for chemical plants.
AFNOR: Z 20 CNM 18.10.

FERRYNOX 8S.
M-1717; 0.18-0.25 C, 16-20 Cr, 8-10 Ni, 2-3 Mo, bal Fe.
Excellent corrosion resistance, good resistance to scaling at elevated temperatures; austenitic.
For pump rotors, compressors.
AFNOR: Z 20 CND 18.10.

FERRYNOX 10.
  M-1717; 0.05-0.10 C, 16-20 Cr, 8-10 Ni, Ti, bal Fe.
  Corrosion resistant; temperature resistant. Austenitic.
  For food equipment, sugar refineries.
  AFNOR: Z 6 CN 18.10.

FERRYNOX 12.
  M-1717; 0.18-0.25 C, 22-27 Cr, 11-13 Ni, bal Fe.
  Good high temperature properties. Austenitic, corrosion resistant.
  For furnace and heat treat equipment.
  AFNOR: Z 20 CN 25.13.

FERRYNOX 15.
  M-1717; 0.18-0.25 C, 23-27 Cr, 13-16 Ni, bal Fe.
  Good high temperature properties.
  Furnace equipment, jet engine parts.
  AFNOR: Z 20 CN 25.15.

FERRYNOX 20.
  M-1717; 0.18-0.25 C, 24-26 Cr, 19-22 Ni, bal Fe.
  Good high temperature properties.
  For jet engine, gas turbines, heat exchangers.
  AFNOR: Z 20 CN 25.20.
  US-AISI 310 Stainless.

FERRYNOX 35.
  M-1717; 0.18-0.25 C, 14-16 Cr, 34-36 Ni, bal Fe.
  Good high temperature properties in carburizing atmospheres.
  For heat treat equipment, petroleum industry.
  AFNOR: Z 20 NC 35.15.

FERRYNOX 65.
  M-1717; 0.18-0.25 C, 14-16 Cr, 62-66 Ni, bal Fe.
  Good high temperature properties, good corrosion resistance. For chemical plant equipment.
  AFNOR: NC 15 Fe.

FERRYNOX CO 50.
  M-1717; 50 Co, 28 Cr, bal Fe.
  Room temp: 60-70 Kg/mm^2; 2-4 El.
  1000°C: 11-13 Kg/mm^2; 15-17 El.
  Good strength at high temperatures.
  Resistant to oxidation and many chemicals.
  For furnace equipment.

F & G 1.
  M-1548; 3.0-5.5 Cu, 0.3-0.9 Si, 0.3-1.0 Mn, 0.5-1.5 Mg, bal Al.
  Soft: 23,000-31,000 TS; 15-20 El; 50-60 Brin.
  Heat treated: 68,000 TS; 48,000 YS; 10 El; 140 Brin.
  For rivets, hydraulic fittings, hardware, bridges, heavy duty structures.
  Age-hardenable, high strength.

F & G 3.
  M-1548; 5.0-6.0 Cu, 1.0 Si, 0.5 Mn, bal Al.
  Soft: 28,000 TS; 15-25 El; 50-60 Brin.
  Heat Treated: 71,000 TS; 57,000 YS; 2 El; 140 Brin.
  For hydraulic fittings, hardware, aircraft and engine components.
  Age-hardenable, high strength and endurance limit.

F & G 4.
  M-1548; 0.5-1.0 Si, 0.8-1.5 Mn, 0.5-1.0 Mg, bal Al.
  Heat treated: 50,000 TS; 36,000 YS; 12 El; 95 Brin.
  For hardware, structures, tanks, aircraft components.
  Age-hardenable, corrosion resistant.

F & G 5.
  M-1548; 1.5 Mn, 5.0-9.0 Mg, bal Al.
  Soft: 32,000 TS; 12,800 YS; 25 El; 55 Brin.
  Hard: 68,000 TS; 48,000 YS; 10 El; 120 Brin.
  For heat exchangers, fixtures, tanks.
  Non-hardenable, corrosion resistant.

F & G 8.
  M-1548; 1.0-2.0 Mn, bal Al.
  Half hard: 26,000 TS; 21,000 YS; 5 El; 45 Brin.
  Hard: 35,000 TS; 28,000 YS; 2 El; 60 Brin.
  For commercial roofing, heat exchangers, ducts, fixtures, tanks.
  Non-heat treatable, good weldability.

FH "62".
  M-240; 0.58 C, 0.80 Mn, 0.25 Si, bal Fe.
  Induction or flame hardenable to 60/66 Rc for machine ways, gibs, rails, slides, wear strips.

F.I. 20 STEEL.
  M-521; 0.06 C, 0.3 Si, 0.7 Mn, 21 Cr, bal Fe.
  Heat treated: 72,000 TS; 55,000 YS; 34 El; 60 RA; 175 Brin.
  For corrosion resistant parts; corrosion resistant.

FIELD.
  M-Eng.; 10 Si, 3 Cu, 1.5-2 Al, bal Ni.
  For corrosion resisting parts; stainless and corrosion resistant.

FIELD ROLLED.
  M-Eng.; 60 Ni, 29 Fe, 20 Mo.
  Stainless and corrosion resistant.

FIFTEEN PER CENT.
  M-Eng.; 56.66 Cu, 28.33 Zn, 15 Ni.
  For ornamental parts; corrosion resistant.

FIFTY-FIVE.
  M-800; 2.9-3.1 C, 1.2-1.5 Ni, 0.4-0.6 Mo, 1.4-1.6 Si, bal Fe.
  Cast: 55,000-62,000 TS.
  For hydraulic press and diesel engine parts; cast iron.

FIFTY N.
M-604; 0.18 C, 1.30 Mn, 0.35 Si, 0.05 Cb, bal Fe.
Normalized: 70 ksi TS; 50 ksi YS; 23 El.
Readily formed and welded.
For stressed structures at low temperature, arctic and marine structures.
ASTM A633 Grade C.

FILBRITE.
M-Eng.; C, 17 Cr, 35 Ni, bal Fe.
For heat and corrosion resisting parts; stainless and corrosion resistant.

FILE BRONZE.
M-U.S.; 18-31 Sn, 7-8.5 Pb, 0-10 Zn, bal Cu.
For plumbing, hardware; free-cutting.

FILE METAL (GENFER).
64.6-62.0 Cu, 18-20 Sn, 10 Zn, 7.6-8.0 Pb.
For nail file; free-cutting.

FILE METAL (VOGEL).
51-73 Cu, 19-28.5 Sn, 0-7 Zn, 7.0-8.5 Pb.
For nail file; free-cutting.

FILIPOFF LEAD-CALCIUM.
1.37 Cu, 1.9 Ca, 1.0 Sr, 1.1 Ba, 0.1 Na, bal Pb.
For bearings; anti-friction.

FIL-SODER.
M-66; Sn-Pb.
For solder for Al; no flux required.

FINE SILVER.
M-63; 99.9 Ag.
Hard: 41,000 TS; 30,000 YS; 3 El.
Annealed: 23,000 TS; 10,000 YS; 35 El; 38 RA.
For jewelry, alloys; M.P. 1761°F.

FINE SILVER.
M-53; 99.9 Ag.
Ann: 30 R-15T; Elect. cond.: 106% I.A.C.S.
For electrical contacts.

FINIS.
M-485; 0.7 C, 18 W, 4 Cr, 1 V, bal Fe.
For high speed tools and cutters; high speed steel.

FINISHING SPECIAL.
M-85; C, Cr, W, bal Fe.
For tools.

FINKL CUPRODIE.
M-55; 0.55 C, 0.8 Mn, 1.55 Ni, 0.95 Cr, 0.3 Mo, 0.75 Cu, bal Fe.
Prehardened (as required) from 37-50 Rc.
Mech. prop. depend on hardness.
For hammer dies, forging press dies, drop hammer dies. Resists heat checking.

FINKL DC.
M-55; 0.37 C, 0.4 Mn, 1.0 Si, 5.2 Cr, 1.4 Mo, 0.95 V, bal Fe.
Vacuum electric furnace degassed.
As Ann: 229 Brin.
For die casting dies for Al, Mg, Zn, lead and tin base alloys; for plastic mold dies.
AISI Type H 13 Hotwork steel.

FINKL F.
M-55; 0.55 C, Cr-Ni-Mo steel.
Prehardened to ordered hardness.
Economy hot work steel for short runs as hammer dies, sow blocks and other hot work tools.

FINKL FS.
M-55; 0.55 C, 0.75 Mn, 0.27 Si, 1.0 Ni, 1.0 Cr, 0.3 Mo, bal Fe.
Vacuum degassed.
FS is the annealed condition, see Finkl FX for prehardened condition.
For forging dies and other hotworking tools.

FINKL FX see FX-XTRA.

FINKL HB.
M-55; 0.52 C, 1.2 Mn, 0.27 Si, 0.65 Cr, 0.09 S, 0.20 Mo, bal Fe.
Prehardened: C 28-34 Rock (or as required).
For holder blocks, forging and extrusion dies, die casting and plastic molding dies.
Prehardened. Good machinability.

FINKL MD.
M-55; 0.32 C, 0.80 Mn, 0.52 Si, 1.7 Cr, 0.4 Mo, bal Fe.
Prehardened: C 28-34 Rock. Vacuum degassed.
For zinc die casting dies, plastic molds.
Prehardened; Type P20 Tool steel.

FINKL W4X.
M-55; 0.37 C, 0.45 Mn, 1.0 Si, 5.0 Cr, 1.45 Mo, 0.35 V, 1.25 W. Vacuum degassed.
Annealed or hardened as required, 37 to 46 Rc.
For forging press dies, inserts, extrusion press rams, liners, forging machine headers.
AISI Type H-12 Hot work steel.

FINKL WF.
M-55; 0.35 C, 0.55 Mn, 0.5 Si, 0.65 Ni, 2.40 Cr, 0.72 Mo, bal Fe.
Prehardened, as required, from 28 to 50 Rc.
Mech. prop. depend on hardness.
For hammer dies, press dies, punches, headers, gripper dies, insert dies, sow blocks.

FIREARMOR "A".
M-184; 65 Ni, 15 Cr, 1.75 Mn, 0.5 C, 1.0 Si, bal Fe.
Cast: 60,000 TS; 46,000 YS; 2.9 El; 1.2 RA; 190 Brin.
For carburizing boxes; heat and abrasion resisting.

FIREARMOR B.
M-184; 15 Cr, 60 Ni, 0.5 C, 1.7 Mn, 1.2 Si, bal Fe.
Cast: 62,000 TS; 33,000 YS; 1.5 El; 3.0 RA; 200 Brin.
For furnace parts, grates; heat resistant.

FIREDIE.
M-35; 0.38 C, 5 Cr, 0.45 Mn, 0.50 V, 1.4 Mo, 1.0 Si, bal Fe.
For hot work dies, die casting dies; hot work die steel.
AISI H11.

FIREDIE 9.
M-35; 0.38 C, 0.30 Mn, 0.90 Si, 3.6 Cr, 3.0 Mo, 0.65 V, 2.0 Co, bal Fe.
Hot work tool steel; for forging and extrusion dies.

FIREDIE 13.
M-35; 0.38 C, 1.4 Mo, 5.2 Cr, 1.05 V, 1 Si, bal Fe.
Ht. Tr.: 285,000 TS; 225,000 YS; 4 El; C 54 Rock.
For hot extrusion tools and die casting dies for Al, Mg, and Zn alloys.
Hot work tool steel type H13, tough, shock resistant.

FIREX.
M-40; 0.40 C, 4.5 Ni, 1.10 Cr, 0.3 Si, 0.6 Mn, bal Fe.
For cutting tools, hot shears, shock resisting tools, chisels; air or oil hardening.

FIREX SPECIAL.
M-40; 0.40-0.55 C, 0.4-1.0 Cr, 2.75-4.0 Ni, 0.4-0.9 Mn, 0.15-0.20 V, 0.4-0.7 Mo, bal Fe.
For tools, dies, punches; shock resistant, tough.

FIRINIT.
M-381.
For welding and soldering rods.

FIRST QUALITY.
M-363; 0.8-1.2 C, bal Fe.
For tools, punches, dies; water hardening.

FIRTH 16-25-6.
M-57; 0.08 C, 17 Cr, 25 Ni, 6 Mo, 0.15 N, bal Fe.
Rolled: 115,000 TS; 60,000 YS; 35 El; 50 RA; 201 Brin.
Cold drawn: 150,000 TS; 120,000 YS; 19 El; 40 RA; 300 Brin.
For jet engine components, fasteners, bolts; ductile, high temperature uses to 1300°F.

FIRTH 19-9DL.
M-57; 0.28-0.35 C, 18-21 Cr, 8-11 Ni, 1.5 Mo, 1.4 W, 0.50 Cb, 0.4 Ti, 1.2 Mn, bal Fe.
Hot rolled: 118,000 TS; 69,000 YS; 58 El; 55 RA; 216 Brin.
At 1000°F: 89,000 TS; 42,000 YS; 43 El; 52 RA.
For turbine wheels, bolts, turbo superchargers; high temperature uses to 1200°F.

FIRTH 19-9DX.
M-57; 0.3 C, 19 Cr, 9 Ni, 1.5 Mo, 1.2 W, 0.55 Ti, bal Fe.
Hot rolled: 118,000 TS; 69,000 YS; 58 El; 55 RA; 216 Brin.
At 1000°F: 89,000 TS; 42,000 YS; 43 El; 52 RA.
For jet engine fasteners; high temperature uses to 1200°F.

FIRTH A-286.
M-57; 0.05 C, 1.35 Mn, 15 Cr, 26 Ni, 1.2 Mo, 2 Ti, 0.3 V, bal Fe.
Rolled: 135,000-160,000 TS; 85,000-115,000 YS; 20-30 El; 30-55 RA.
For gas turbine discs, jet engine components; austenitic, heat treatable.

FIRTHAG.
M-56; 0.45 C, 0.90 Mn, 0.95 Cr, 0.25 Si, bal Fe.
For punches, gears, bolts; Brit. EN18.

FIRTHALOY.
M-54; W.C.
For wire drawing and extrusion dies; cemented tungsten carbide.

FIRTH "AW".
M-56; 1.20 C, 0.25 Mn, bal Fe.

FIRTH-BROWN 21/4/N.
M-56; 0.48-0.58 C, 0.25 max Si, 8.0-10.0 Mn, 20.0-22.0 Cr, 3.5-4.5 Ni 0.38-0.50 N, C+N = 0.90 min, bal Fe.
Valve steel.

FIRTH-BROWN ANC1.
M-56; 0.3 C, 0.45 Mn, 4.25 Ni, 1.25 Cr, bal Fe.
For shafts, gears; formerly "Brown ATG."

FIRTH-BROWN ANCM.
M-56; 0.3 C, 0.45 Mn, 4.25 Ni, 1.25 Cr, 0.25 Mo, bal Fe.
For shafts, gears; formerly "Atlas AHNC."

FIRTH-BROWN B45CH.
M-56; 0.15 C, 0.60 Ni, 0.65 Cr, 0.12 Mo, bal Fe.
For gears, cams, camshafts, fasteners; case hardened, EN361.

FIRTH-BROWN B55CH.
M-56; 0.20 C, 0.65 Cr, 0.60 Ni, 0.12 Mo, bal Fe.
For gears, cams, camshafts, machinery parts; case hardened, EN362.

FIRTH-BROWN B65CH.
M-56; 0.24 C, 0.60 Ni, 0.65 Cr, 0.12 Mo, bal Fe.
For gears, cams, camshafts; case hardened, EN363.

FIRTH-BROWN CMN1.
M-56; 0.25 C, 1.5 Mn, bal Fe.
For gears, shafts; water hardening.

FIRTH-BROWN CMN 2.
M-56; 0.35 C, 1.5 Mn, bal Fe.
For gears, shafts; water hardening.

FIRTH-BROWN CRM1.
M-56; 0.4 C, 0.65 Mn, 1.2 Cr, 0.25 Mo, bal Fe.
For gears, shafts, bolts; oil hardening.

FIRTH-BROWN CRM2.
M-56; 0.4 C, 0.6 Mn, 1.3 Cr, 0.85 Mo, bal Fe.
For gears, shafts, bolts; tough, oil hardening.

FIRTH-BROWN CRM3.
M-56; 0.27 C, 0.6 Mn, 3.25 Cr, 0.6 Mo, bal Fe.
For gears, shafts, machinery parts; oil hardened, tough.

FIRTH-BROWN CRV1.
M-56; 0.47 C, 0.25 Si, 0.65 Mn, 0.85 Cr, 0.2 V, bal Fe.
For laminated springs; oil hardening.

FIRTH-BROWN CRV2.
M-56; 0.45 C, 0.25 Si, 0.65 Mn, 1.25 Cr, 0.2 V, bal Fe.
For laminated springs; oil hardening.

FIRTH-BROWN DDQ.
M-56; 0.16 C, 0.20 min Si, 2.0 Mn, 11.0-14.0 Ni, 11.0-14.0 Cr, bal Fe.
Acid, rust and heat resisting steel.

FIRTH-BROWN DICROME.
M-56; 0.60 C, 0.60 Cr, 0.60 Mn, 0.25 Si, bal Fe.
For machine tool parts; oil hardened.

FIRTH-BROWN EMS.
M-56; 0.12 max C, 0.20-1.0 Si, 0.50-2.0 Mn, 17.0-19.0 Cr, 8.0-11.0 Ni Ti = 5 X C min - 0.90 max, bal Fe.
Stabilized austenitic stainless.

FIRTH-BROWN F10C.
M-56; 0.10 C, 0.25 Si, 0.65 Mn, bal Fe.
Annealed: 64,000 TS; 40,000 YS; 28 El; 65 RA; 130 Brin.
For nails, rivets, fan blades, case hardened parts; case hardened.

FIRTH-BROWN F15C.
M-56; 0.15 C, 0.25 Si, 0.65 Mn, bal Fe.
Annealed: 70,000 TS; 45,000 YS; 25 El; 60 RA; 145 Brin.
For screws, bolts, case hardened parts; case hardened.

FIRTH-BROWN F20C.
M-56; 0.20 C, 0.25 Si, 0.65 Mn, bal Fe.
Annealed: 73,000 TS; 45,000 YS; 22 El; 58 RA; 150 Brin.
For screws, bolts, gears, case hardened parts; case hardened.

FIRTH-BROWN F25C.
M-56; 0.25 C, 0.25 Si, 0.65 Mn, bal Fe.
Hot rolled: 70,000 TS; 45,000 YS; 31 El; 58 RA; 145 Brin.
For crankshafts, gears, bolts, fasteners; water hardened.

FIRTH-BROWN F27C.
M-56; 0.25-0.35 C, 0.05-0.35 Si, 0.70-0.90 Mn, bal Fe.
Carbon steel.

FIRTH-BROWN F30C.
M-56; 0.32 C, 0.25 Si, 0.65 Mn, bal Fe.
For axles.

FIRTH-BROWN F32C.
M-56; 0.30-0.35 C, 0.05-0.35 Si, 0.70-0.90 Mn, bal Fe.
Carbon steel.

FIRTH-BROWN F35C.
M-56; 0.35 C, 0.25 Si, 0.65 Mn, bal Fe.
Hot rolled: 85,000 TS; 54,000 YS; 30 El; 53 RA; 183 Brin.
For gears, shafts, axles, bolts; water hardened.

FIRTH-BROWN F40C.
M-56; 0.40 C, 0.25 Si, 0.65 Mn, bal Fe.
Hot rolled: 91,000 TS; 58,000 YS; 27 El; 50 RA; 200 Brin.
For gears, shafts, axles, bolts, crankshafts; water hardened.

FIRTH-BROWN F45C.
M-56; 0.45 C, 0.25 Si, 0.65 Mn, bal Fe.
Hot rolled: 98,000 TS; 59,000 YS; 24 El; 45 RA; 212 Brin.
For gears, shafts, bolts, axles, crankshafts; water hardened.

FIRTH-BROWN F45CH.
M-56; 0.15 C, 0.25 Si, 0.70 Mn, 0.80 Ni, 0.60 Cr, bal Fe.
For gears, pinions, shafts, cams, camshafts; case hardened; Brit EN351; shock resistant.

FIRTH-BROWN F47C.
M-56; 0.45-0.50 C, 0.05-0.35 Si, 0.70-1.0 Mn, bal Fe.
Carbon steel.

FIRTH-BROWN F50C.
M-56; 0.50 C, 0.25 Si, 0.65 Mn, bal Fe.
Annealed: 96,000 TS; 52,000 YS; 16 El; 23 RA; 170 Brin.
For gears, shafts, bolts, axles, crankshafts; water hardened.

FIRTH-BROWN F52C.
M-56; 0.45-0.60 C, 0.10-0.40 Si, 0.60-0.80 Mn, bal Fe.
Carbon steel.

FIRTH-BROWN F55C.
M-56; 0.55 C, 0.7 Mo, bal Fe.
For springs; water hardening.

FIRTH-BROWN F55CH.
M-56; 0.18 C, 0.25 Si, 0.70 Mn, 1.0 Ni, 0.80 Cr, bal Fe.
For gears, pinions, cams, camshafts, machine tool parts; case hardened; Brit. EN352; shock resistant.

FIRTH-BROWN F62C.
0.60-0.65 C, 0.05-0.35 Si, 0.40-0.60 Mn, bal Fe.
Carbon steel; for springs and hand tools.

FIRTH-BROWN F65C.
M-56; 0.65 C, 0.25 Si, 0.65 Mn, bal Fe.
Heat treated: 170,000 TS; 116,000 YS; 10 El; 38 RA; 350 Brin.
For die blocks, girders, clutch discs; water hardened.

FIRTH-BROWN F67C.
M-56; 0.65-0.70 C, 0.05-0.35 Si, 0.70-0.90 Mn, bal Fe.
Carbon steel; for springs and hand tools.

FIRTH-BROWN F70C.
M-56; 0.70 C, 0.25 Si, 0.65 Mn, bal Fe.
Heat treated: 174,000 TS; 128,000 YS; 12 El; 37 RA; 352 Brin.
For springs, rails, die blocks, girders: water hardened.

FIRTH-BROWN F75C.
M-56; 0.75 C, 0.65 Mn, bal Fe.
For springs; water hardening.

FIRTH-BROWN F75CH.
M-56; 0.18 C, 0.25 Si, 0.70 Mn, 1.75 Ni, 1.0 Cr, 0.15 Mo, bal Fe.
For gears, cams, camshafts, machine tool parts; case hardened; Brit. EN354; shock resistant.

FIRTH-BROWN F80C.
M-56; 0.70-0.85 C, 0.10-0.40 Si, 0.55-0.75 Mn, bal Fe.
Carbon tool and spring steel.

FIRTH-BROWN F85C.
M-56; 0.85 C, 0.25 Si, 0.65 Mn, bal Fe.
Heat treated: 190,000 TS; 145,000 YS; 12 El; 35 RA; 390 Brin.
For drills, punches, springs, reamers, hobs; Type W1; water hardened.

FIRTH-BROWN F85CH.
M-56; 0.18 C, 0.25 Si, 0.70 Mn, 2 Ni , 1.6 Cr, 0.2 Mo, bal Fe.
For gears, cams, machine tool parts; case hardened; Brit. EN355; shock resistant.

FIRTH-BROWN F117.
M-56; 0.12 C, 1.0 Si, 1.0 Mn, 0.50 Ni, 16-18 Cr, bal Fe.
Ferritic rust resisting steel.

FIRTH-BROWN F120.
M-56; 0.12 C, 1.0 Si, 1.0 Mn, 0.50 Ni, 20-22 Cr, bal Fe.
Ferritic rust resisting steel.

FIRTH-BROWN FCB.
M-56; 0.08 max C, 0.20-1.0 Si, 0.50-2.0 Mn, 17.0-19.0 Cr, 9.0-12.0 Ni Nb = 10 X C min - 1.0 max, bal Fe.
Stabilized austenitic stainless steel.
AISI 347.

FIRTH-BROWN FCCH.
M-56; 0.14 C, 0.6 Mn, bal Fe.
For shafts, gears; formerly "Atlas CCH;" case hardening.

FIRTH-BROWN FCCR.
M-56; 0.45 C, 0.25 Si, 0.90 Mn, 0.95 Cr, bal Fe.
For gears, bolts, crankshafts; water hardened.

FIRTH-BROWN FCI.
M-56; 0.08-0.15 C, 1.0 max Si, 1.5 max Mn, 1.0 max Ni, 11.5-13.5 Cr, 0.15-0.30 S, 0.040 max P, bal Fe.
Free-machining martensitic stainless steel.
AISI 416.

FIRTH-BROWN FCS.
M-56; 0.20-0.28 C, 1.0 max Si, 1.5 max Mn, 1.0 max Ni, 12.0-14.0 Cr, 0.60 max Mo, bal Fe.
Martensitic stainless steel.
Similar to AISI 420.

FIRTH-BROWN FDP.
M-56; 0.12 max C, 0.20-1.0 Si, 0.50-2.0 Mn, 17.0-19.0 Cr, 8.0-11.0 Ni Ti 5 X C min - 0.90 max, bal Fe.
Stabilized austenitic stainless steel.
AISI 321.

FIRTH-BROWN FDP(L).
M-56; 0.08 max C, 0.20-1.0 Si, 0.50-2.0 Mn, 17.0-19.0 Cr, 9.0-12.0 Ni Ti = 5 X C min - 0.70 max, bal Fe.
Stabilized austenitic stainless steel.
AISI 321.

FIRTH-BROWN FG.
M-56; 0.20-0.28 C, 0.80 max Si, 1.0 max Mn, 1.0 max Ni, 12.0-14.0 Cr, bal Fe.
Martensitic stainless steel.
AISI 420.

FIRTH-BROWN FG(L).
M-56; 0.14-0.20 C, 0.80 max Si, 1.0 max Mn, 1.0 max Ni, 11.5-13.5 Cr, bal Fe.
Martensitic stainless steel.

FIRTH-BROWN FH.
M-56; 0.28-0.36 C, 0.80 max Si, 1.0 max Mn, 1.0 max Ni, 12.0-14.0 Cr, bal Fe.
Martensitic stainless steel.
Similar to AISI 420.

FIRTH-BROWN FI.
M-56; 0.09-0.15 C, 0.80 max Si, 0.80 max Mn, 0.50 max Ni, 11.5-13.5 Cr, bal Fe.
Martensitic stainless steel.

FIRTH-BROWN FI.
M-56; 0.08 max C, 0.80 max Si, 1.0 max Mn, 0.50 max Ni, 12.0-14.0 Cr, bal Fe.
Ferritic type stainless steel.
AISI 403.

FIRTH-BROWN FI17.
M-56; 0.10 max C, 0.80 max Si, 1.0 max Mn, 0.50 max Ni, 16.0-18.0 Cr, bal Fe.
Ferritic type stainless steel.
AISI 430.

FIRTH-BROWN FMB.
M-56; 0.07 max C, 0.20-1.0 Si, 0.50-2.0 Mn, 16.5-18.5 Cr, 10.0-13.0 Ni, 2.25-3.0 Mo, bal Fe.
Austenitic stainless steel.
AISI 316.

FIRTH-BROWN FMB (LC).
M-56; 0.03 max C, 0.20-1.0 Si, 0.50-2.0 Mn, 16.5-18.5 Cr, 11.0-14.0 Ni, 2.25-3.0 Mo, bal Fe.
Austenitic stainless steel.
AISI 316L.

FIRTH-BROWN FMBTI.
M-56; 0.08 max C, 0.20-1.0 Si, 0.50-2.0 Mn, 16.5-18.5 Cr, 11.0-14.0 Ni, 2.25-3.0 Mo, Ti = 4 X C min - 0.60 max, bal Fe.
Stabilized austenitic stainless steel.

FIRTH-BROWN FML.
M-56; 0.07 max C, 0.20-1.0 Si, 0.50-2.0 Mn, 16.5-18.5 Cr, 9.0-11.0 Ni 1.25-1.75 Mo, bal Fe.

FIRTH-BROWN FN10.
M-56; 0.34-0.46 C, 0.10-0.35 Si, 0.65-1.05 Mn, 0.70-1.0 Ni, bal Fe.
Structural steel.

FIRTH-BROWN FN30.
M-56; 0.3 C, 0.65 Mn, 3 Ni, bal Fe.
For gears, shafts; tough.

FIRTH-BROWN FN35.
M-56; 0.3 C, 0.6 Mn, 3.6 Ni, bal Fe.
For gears, shafts; oil hardening, tough.

FIRTH-BROWN FNCR.
M-56; 0.35 C, 1.25 Ni, 0.60 Cr, bal Fe.
For gears, pinions, bolts, machinery parts; Brit. EN111; oil hardened, tough.

FIRTH-BROWN FNMC.
M-56; 0.65 C, 0.3 Si, 5.2 Mn, 12 Ni, 3.7 Cr, bal Fe.
For heat and corrosion resistant parts; non-magnetic, austenitic.

FIRTH-BROWN FSL.
M-56; 0.08 C, 0.20-1.0 Si, 2.0 Mn, 17.5-20.0 Cr, 8-11 Ni, bal Fe.
Austenitic stainless steel for wire.

FIRTH-BROWN FSL (LC).
M-56; 0.03 max C, 0.20-1.0 Si, 0.50-2.0 Mn, 17-19 Cr, 9-12 Ni, bal Fe.
Low carbon austenitic stainless steel.
AISI 304L.

FIRTH-BROWN FST.
M-56; 0.12 max C, 17-19 Cr, 8-11 Ni, bal Fe.
Austenitic stainless steel.
AISI 302.

FIRTH-BROWN FST (FC).
M-56; 0.12 max C, 1.0-2.0 Mn, 17-19 Cr, 8-11 Ni, 0.15-0.30 S, 0.045 max P, bal Fe.
Free machining austenitic stainless steel.
AISI 303.

FIRTH-BROWN FST (L).
M-56; 0.06 max C, 0.20-1.0 Si, 0.50-2.0 Mn, 17.5-19.0 Cr, 8-11 Ni, bal Fe.
Low carbon austenitic stainless steel.
AISI 304.

FIRTH-BROWN FV317.
M-56; 0.06 max C, 0.20-1.0 Si, 0.50-2.0 Mn, 17.5-19.5 Cr, 12.0-15.0 Ni, 3.0-4.0 Mo, bal Fe.
Austenitic stainless steel.
AISI 317.

FIRTH-BROWN FV317 (LC).
M-56; 0.03 max C, 0.20-1.0 Si, 0.50-2.0 Mn, 17.5-19.5 Cr, 14.0-17.0 Ni, 3.0-4.0 Mo, bal Fe.
Austenitic stainless steel.

FIRTH-BROWN FV448.
M-56; 0.08-0.16 C, 0.15-0.60 Si, 0.3-1.2 Mn, 0.6-1.2 Ni, 9.8-11.2 Cr, 0.4-0.8 Mo, 0.15-0.45 Nb, 0.10-0.25 V, 0.030-0.075 N, bal Fe.
Heat resisting steel.

FIRTH-BROWN FV 520B.
M-56; 0.07 max C, 0.6 max Si, 1.0 max Mn, 13.2-14.7 Cr, 5.0-5.8 Ni, 1.2-2.0 Mo, 1.2-2.0 Cu, 0.2-0.50 Nb, bal Fe.
Precipitation hardening stainless.

FIRTH-BROWN FV 535 VR.
M-56; 0.06-0.11 C, 0.10-0.70 Si, 0.6-1.1 Mn, 0.20-0.80 Ni, 9.8-11.2 Cr, 0.5-1.0 Mo, 0.004-0.012 B, 5.0-7.0 Co, 0.20-0.45 Nb, 0.10-0.35 V, 0.010-0.035 N, bal Fe.
Heat resisting steel.

FIRTH-BROWN FVS.
M-56; 0.35-0.50 C, 1.0-2.0 Si, 0.50-1.0 Mn, 12.0-15.0 Cr, 12.0-15.0 Ni, 2.0-3.0 W, bal Fe.
Valve steel.

FIRTH-BRONW FVS.
M-56; 0.37-0.47 C, 1.0-2.0 Si, 0.5-1.0 Mn, 13.0-15.0 Cr, 13.0-15.0 Ni 2.2-3.0 W, 0.40-0.70 Mo, bal Fe.
Valve steel.

FIRTH-BROWN G110.
M-56; 0.015 max C, 0.10 max Si, 0.10 max Mn, 17.0-19.0 Ni, 0.25 max Cr, 4.6-5.2 Mo, 0.05-0.15 Al, 7.0-8.5 Co, 0.30-0.60 Ti, bal Fe.
Maraging steel.

FIRTH-BROWN HCM3.
M-56; 0.40 C, 3 Cr, 1.0 Mo, 0.25 V, bal Fe.
For oil refinery equipment; Brit. EN40C; creep resistant.

FIRTH-BROWN HCM5.
M-56; 0.30 C, 3 Cr, 0.4 Mo, bal Fe.
For oil refinery equipment; Brit. EN40B; creep resistant.

FIRTH-BROWN HCM7.
M-56; 0.20 C, 3 Cr, 0.4 Mo, bal Fe.
For oil refinery equipment; Brit. EN40A; creep resistant.

FIRTH-BROWN HRC MAX.
M-56; 0.18-0.45; 1.0-2.0 Si, 1.0 Mn, 6.0-12.0 Ni, 17.0 min Cr, 2.0-4.0 W, bal Fe.
High Cr-Ni-W valve steel.

FIRTH-BROWN LK3.
M-56; 0.35-0.43 C, 0.10-0.45 Si, 0.40-0.65 Mn, 1.4-1.8 Cr, 0.15-0.25 Mo, 0.90-1.30 Al, bal Fe.
Steel to be nitrided.

FIRTH-BROWN LK5.
M-56; 0.27-0.35 C, 0.10-0.45 Si, 0.40-0.65 Mn, 1.4-1.6 Cr, 0.15-0.25 Mo, 0.90-1.30 Al, bal Fe.
Steel to be nitrided.

FIRTH-BROWN MM01.
M-56; 0.32 C, 1.6 Mn, 0.27 Mo, bal Fe.
For gears, shafts; oil hardening.

FIRTH-BROWN MM02.
M-56; 0.33 C, 1.6 Mn, 0.4 Mo, bal Fe.
For gears, shafts; oil hardening.

FIRTH-BROWN N2MCH.
M-56; 0.17 C, 0.5 Mn, 1.8 Ni, 0.25 Mo, bal Fe.
For gears, shafts; case-hardening.

FIRTH-BROWN N3CCH.
M-56; 0.12 C, 0.4 Mn, 3.3 Ni, 0.8 Cr, bal Fe.
For gears, shafts; case-hardening, tough.

FIRTH-BROWN N3CH.
M-56; 0.14 C, 0.25 Si, 0.4 Mn, 3.1 Ni, bal Fe.
For gears, shafts; case-hardening.

FIRTH-BROWN N3CMCH.
M-56; 0.12 C, 3.3 Ni, 0.80 Cr, 0.20 Mo, bal Fe.
For gears, shafts, machinery parts, cams; case-hardened, shock resistant.

FIRTH-BROWN N4CCH.
M-56; 0.14 C, 0.4 Mn, 4.25 Ni, 1.25 Cr, bal Fe.
For gears, shafts; case-hardening, tough.

FIRTH-BROWN N4CMCH.
M-56; 0.14 C, 4.25 Ni, 1.25 Cr, 0.25 Mo, bal Fe.
For gears, shafts, camshafts, cams, bolts; case-hardened, shock resistant.

FIRTH-BROWN N5CH.
M-56; 0.10 C, 0.3 Mn, 5 Ni, bal Fe.
For gears, shafts; case-hardening, tough.

FIRTH-BROWN N5MCH.
M-56; 0.12 C, 0.4 Mn, 5.1 Ni, 0.2 Mo, bal Fe.
For gears, shafts; case-hardening, shock resistant.

FIRTH-BROWN NCM1.
M-56; 0.4 C, 0.6 Mn, 1.6 Ni, 1.0 Cr, 0.5 Mo, bal Fe.
For gears, shafts, bolts; shock resistant, heavy duty.

FIRTH-BROWN NCM3.
M-56; 0.3 C, 0.6 Mn, 2.4 Ni, 0.6 Cr, 0.45 Mo, bal Fe.
For gears, shafts; shock resistant.

FIRTH-BROWN NCM4.
M-56; 0.4 C, 0.6 Mn, 0.65 Cr, 0.6 Mo, 0.3 Ni, bal Fe.
For gears, shafts, bolts, machinery parts; oil hardened, shock resistant.

FIRTH-BROWN NCM5.
M-56; 0.3 C, 0.6 Mn, 3.25 Ni, 0.6 Cr, 0.6 Mo, bal Fe.
For gears, shafts, bolts; shock resistant.

FIRTH-BROWN NCM6.
M-56; 0.3 C, 0.6 Mn, 3.25 Ni, 0.8 Cr, 0.3 Mo, bal Fe.
For bolts, gears, shafts; tough, shock resistant.

FIRTH-BROWN NCMO.
M-56; 0.38 C, 0.25 Si, 0.65 Mn, 1.4 Ni, 1.2 Cr, 0.15 Mo, bal Fe.
For gears, bolts, machine tool parts; Brit. EN110; oil hardened, tough.

FIRTH-BROWN NCR2.
M-56; 0.3 C, 0.6 Mn, 3.25 Ni, 0.75 Cr, bal Fe.
For gears, shafts; tough.

FIRTH-BROWN NMCM.
M-56; 0.40 C, 0.75 Ni, 0.45 Cr, 0.20 Mo, 1.35 Mn, 0.25 Si, bal Fe.
For gears, bolts, machine tool parts, fasteners; Brit. EN100; oil hardened, tough.

FIRTH-BROWN NMM1.
M-56; 0.18 C, 1.4 Mn, 0.5 Ni, 0.3 Mo, bal Fe.
For axles; tough.

FIRTH-BROWN S80.
M-56; 0.12-0.20 C, 0.80 max Si, 1.0 max Mn, 2.0-3.0 Ni, 15.0-18.0 Cr, bal Fe.
Martensitic stainless steel.
AISI 431.

FIRTH-BROWN SLV.
M-56; 0.40-0.50 C, 3.0-3.75 Si, 0.30-0.75 Mn, 7.5-9.5 Cr, 0.50 max Ni, bal Fe.
Valve steel.

FIRTH-BROWN SLV see also S.L.V. STEEL.

FIRTH-BROWN T-374 (FOUNDRY MN 12).
M-56; C, 12 Mn, bal Fe.
200-240 Brin.

FIRTH-BROWN XB.
M-56; 0.74-0.84 C, 1.75-2.25 Si, 1.15-1.65 Ni, 19.0-20.5 Cr, bal Fe.
Hardened and tempered to Rc 47.
For valves.

FIRTH DISCALOY.
M-57; 0.04 C, 13 Cr, 25 Ni, 3 Mo, 1.8 Ti, 0.2 Al, bal Fe.
At 70°F: 145,000 TS; 116,000 YS; 19 El; 23 RA.
At 1000°F: 125,000 TS; 98,000 YS; 16 El; 34 RA.
For gas turbine blades, rotors, jet engine parts; age hardenable, high temperature applications.

FIRTH FS-2.
M-57; 63 TiC, 29.6 Ni, 7.4 Cr.
Sintered: Rockwell A87.
For jet engine components; super refractory, sintered carbides.

FIRTH FS-5.
M-57; 63 TiC, 25.9 Co, 11.1 Cr.
Sintered: Rockwell A90.
For jet engine parts; sintered carbides.

FIRTH FS-8.
M-57; 63 TiC, 22.2 Ni, 7.4 Co, 7.4 Cr.
At 75°F: 26,200 TS; 0 El.
At 800°F: 43,000 TS; 1.3 El.
At 1800°F: 37,800 TS; 2.4 El.
For jet engine components, cutting tools; oxidation, corrosion and heat resistant.

FIRTH FS-9.
M-57; 50 TiC, 30 Ni, 10 Co, 10 Cr.
At 75°F: 30,500 TS; 0 El.
For jet engine components, nozzles, dies; oxidation, corrosion and heat resistant.

FIRTH FS-26.
M-57; 54.3 TiC, 40 Ni, 5.7 Cr_3C_2.
At 80°F: 33,000 TS; 0 El.
At 1200°F: 48,000 TS; 0 El.
At 1800°F: 41,000 TS; 0.8 El.
For jet engine components, nozzles, dies; oxidation, corrosion and heat resistant.

FIRTH FS-27.
M-57; 42.9 TiC, 50 Ni, 7.1 Cr_3C_2.
At 80°F: 75,000 TS; 0 El.
At 800°F: 76,500 TS; 0 El.
At 1800°F: 35,000 TS; 1.7 El.
For jet engine components, cutting tools; oxidation, corrosion and heat resistant.

FIRTH F.S.M-10.
M-57; 0.7 C, 8 Mo, 2 V, bal Fe.
For cutting tools, drills, hobs, reamers; high speed steel.

FIRTH F.S.M-10 MOD.
M-57; 0.65 C, 4 Cr, 8.2 Mo, 2 V, bal Fe.
For hot work tools and dies, punches, shears; hot work steel, oil hardened.

FIRTH GREEK ASCOLOY.
M-57; 0.15 C, 0.3 Si, 0.3 Mn, 13 Cr, 2 Ni, 3 W, 0.12 Al, bal Fe.
Heat treated: 137,000-195,000 TS; 108,000-160,000 YS; 23-14 El; 63-51 RA; 290-400 Brin.
For compressor wheels and blades, bolts, fasteners; heat resistant to 1000°F.

FIRTH HEAVY METAL.
M-57; 95 W, bal Ni + Cu.
Sintered: 120,000 TS; 2 El; 300 Brin.
For counterweights, x-ray tube screens; density 17.5-18.2.

FIRTH H.W.D. MOD.
M-57; 0.55 C, 1 Si, 1.4 W, 5.25 Cr, bal Fe.
For dies; air hardened, non-deforming.

FIRTHITE.
M-57; 97-87 Tungsten carbide, 5-13 Co.
For tipped cutting tools; cemented tungsten carbide.

FIRTHITE GR. TXL.
M-57; 81 WC, 12.5 TiC, 6.5 Co.
Sintered: 790 Brin.
For cutting tools; carbide cement.

FIRTHITE H.
M-57; 6.0 Co, 94 WC.
78 Rockwell-C.
For cutters on cast iron, nonferrous metals and nonmetallics; cemented WC.

FIRTHITE H-23.
M-57; WC, Co.
Sintered: 240,000 TrS; A 91.8 Rock.
Density: 14.8.
Cutting tools for roughing cuts; for machining superalloys.

FIRTHITE HA.
M-57; 6 Co, 3 TaC, 91 Wc.
79 Rockwell-C.
For tool bits, cutters; for semi-finishing hard irons.

FIRTHITE HB.
M-57; 8 Co, 92 WC.
77 Rockwell-C.
For tool bits, cutters; for roughing cast iron.

FIRTHITE HE.
M-57; 95.5 WC, 4.5 Co.
Sintered: Rockwell A92.
For cutting tools; sintered carbide.

FIRTHITE HF.
M-57; WC, 3 Co.
81 Rockwell-C.
For tools, cutters; finishing cuts.

FIRTHITE NHA.
M-57; WC.
For cutters for machining cast iron; cemented carbide.

FIRTHITE NTA.
M-57; WC, Co.
Sintered: 250,000 TrS; A 91.0 Rock.
Density: 12.9.
Cutting tools for rough machining steel.

FIRTHITE T-04.
M-57; 85 WC, 10 Co, 1 TaC, 4 TiC.
For tipped tools for rough cuts; cemented.

FIRTHITE T-22.
M-57; WC, Co.
Sintered: 240,000 TrS; A 92.2 Rock.
Density: 12.8.
Cutting tool for general purpose machining steel.

FIRTHITE T-25.
M-57; WC, Co.
Sintered: 190,000 TrS; A 92.8 Rock.
Density: 12.1.
For semi-finish cuts on ferrous and non-ferrous metals; crater-resistant grade.

FIRTHITE T-31.
M-57; 68.5 WC, 6.5 Co, 25 TiC.
For bits for cutting tools; hardest and most wear resistant, sintered.

FIRTHITE T-41.
M-57; WC, 8 Co.
79 Rockwell-C.
For tools, dies, cutters; cemented WC.

FIRTHITE T-66.
M-57; 60 WC, 12 Co, 28 Ta.
For heavy duty cutters and tools; sintered.

FIRTHITE T-89.
M-57; 12 Co, 59 WC, 22 TaC, 7 TiC.
77 Rockwell-C.
For rough and semi-finish cuts, tipped tools; cemented.

FIRTHITE TXH.
M-57; 83 WC, 8 Co, 9 TiC.
Sintered: Rockwell A91.5.
For cutting tools; sintered carbide.

FIRTHITE TXL.
M-57; 81 WC, 6.5 Co, 12.5 TiC.
Sintered: Rockwell A92.
For cutting tools; sintered carbide.

FIRTHITE WF.
M-57; 70 TiC, 18 MoC, 12 Ni.
Sintered: A 92 Rock; 125,000 TrS.
Sintered carbide tool inserts for cutting steel; high speed fine finishing.

FIRTHITE WL.
M-57; 73 WC, 6 Co, 21 TaC.
Sintered: Rockwell A91.
For cutting tools; sintered carbide.

FIRTH VC.
M-57; 1.1 C, 4 Cr, 2.6 Mo, 2.5 W, 4 V, bal Fe.
For twist drills, reamers, broaches; abrasion resistant.

FIRTH-VICKERS 21/4N STEEL.
M-521; 0.48-0.58 C, 0.25 max Si, 8.0-10.0 Mn, 20.0-22.0 Cr, 3.25-4.5 Ni, 0.38-0.5 N, bal Fe.
Non-magnetic, precipitation hardenable stainless steel.
Ht. Tr. tested at 800°C: 53,000 TS; 35,800 YS.
Good oxidation resistance up to 850°C.
For valves in internal combustion engines.

FIRTH-VICKERS 309.
M-521; 0.15 max C, 1.0 max Si, 2.0 max Mn, 22.0-24.0 Cr, 13.0-15.0 Ni bal Fe.
At room temp: 90,000 TS; 35,000 YS; 50 El.
At 800°C: 26,800 TS; 13,800 YS (0.2% proof).
Good resistance to corrosion and to scaling at elevated temperature; for superheater supports, furnace parts, heat treat boxes and annealing racks. Weldable. AISI 309.

FIRTH-VICKERS 448 STEEL.
M-521; 0.10-0.13 C, 0.50 Si, 1.0 Mn, 11.0 Cr, 0.75 Ni, 0.65 Mo, 0.15-0.30 V, 0.40 Nb, bal Fe.
Air or oil harden; temper 650-700°C.
At Room Temp: 150,000 TS; 127,000 YS; 20 El.
At 700°C: 51,500 TS; 44,000 YS; 28 El.
For aircraft gas turbine discs and similar stressed parts at elevated temperatures.

FIRTH-VICKERS 467.
M-521; 0.2 C, 9.5 Ni, 14 Cr, 2 Mo, 2.5 Cu, 0.7 Ti, bal Fe.
Annealed: 98,000 TS; 43,500 YS; 52 El; 56 RA.
At 500°F: 80,000 TS; 33,600 YS; 31 El; 36 RA.
At 750°F: 61,000 TS; 33,000 YS; 24 El; 27 RA.
For jet engine components; heat and creep resistant, austenitic.

FIRTH-VICKERS 535 STEEL.
M-521; 0.07 C, 0.40 Si, 0.85 Mn, 10.5 Cr, 0.30 Ni, 6.0 Co, 0.75 Mo, 0.20 V, 0.45 Nb, bal Fe.
Air or oil harden; temper 600-650°C: At Room Temp: 160,000 TS; 142,000 YS; 22 El. At 550°C: 118,000 TS; 103,000 YS; 22 El.
For aircraft turbine and compressor disc forgings and wrought rings.

FIRTH-VICKERS 548 STEEL.
M-521; 0.08 C, 0.40 Si, 1.0 Mn, 16.5 Cr, 11.5 Ni, 1.5 Mo, 1.0 Nb, bal Fe.
Austenitic stainless steel; weldable.
Room Temp: 80,000 TS; 33,000 YS; 55 El. At 700°C: 53,000 TS; 18,000 YS; 41 El.
For structural parts and welded assemblies operating at elevated temperatures.

FIRTH-VICKERS 555 STEEL.
M-521; 0.05 C, 0.40 Si, 1.5 Mn, 16.5 Cr, 10.5 Ni, 2.4 Mo, bal Fe.
Austenitic stainless; weldable.
At 700°C: 45,600 TS; 13,800 YS.
For steam pipe, chemical plant equipment.

FIRTH-VICKERS 566.
M-521; 0.12 C, 0.4 Si, 0.8 Mn, 11.5 Cr, 2.3 Ni, 1.4 Mo, 0.15 V, 0.3 Nb, bal Fe.
Air hardened, tempered 630-650°C. Room Temp: 147,000 TS; 124,000 YS; 22 El. At 600°C: 80,000 TS; 66,000 YS; 28 El.
For turbine blading for use up to 450°C.

FIRTH-VICKERS 607 STEEL.
M-521; 0.15 C, 0.30 Si, 0.80 Mn, 11.0 Cr, 0.60 Ni, 0.80 Mo, 0.25 V, bal Fe.
Air or oil hardenable, corrosion resistant steel, primarily for sheet production.
For stamped or formed parts stressed at temperatures up to 600°C.

FIRTH-VICKERS F.A.L.
M-521; 0.12 max C, 0.8 max Si, 10. max Mn, 12.0-14.0 Cr, 3.8-4.8 Al, bal Fe.
Heat treated: 90,000 TS; 67,000 YS; 25 El.
Corrosion resistant steel with high electrical resistivity; for resistance and rheostats.

FIRTH-VICKERS F.A.S.
M-521; 0.43 C, 0.3 Si, 0.5 Mn, 11.5 Cr, bal Fe.
Annealed: 110,000 TS; 79,000 YP; 20 El; 240 Brin.
For stainless ball bearings.
Corrosion resistant.

FIRTH-VICKERS F.C.I.
M-521; 0.12 max C, 0.8 max Si, 1.0-2.0 Mn, 0.15-0.30 S, 11.5-13.5 Cr, 0.3-0.6 Mn, bal Fe.
Martensitic, free-machining, hardenable corrosion resistant steel.
Air or oil hardenable to about 28-38 Rock C.
For shafts, hardware, fasteners, cap screws.
BS: EN.56AM; AISI 416.

FIRTH-VICKERS F.C.S.
M-521; 0.20-0.28 C, 0.8 max Si, 1.0-2.0 Mn, 0.15-0.30 S, 12.0-14.0 Cr 0.3-0.6 Mo, bal Fe.
Martensitic, free-machining, hardenable corrosion resistant steel.
Air or oil hardenable to about 44-52 Rock C.
For spline shafts, gears, studs, hardware.
BS: EN.56 CM, 2S.124; AISI 420 F.

FIRTH-VICKERS F.G.
M-521; 0.20-0.28 C, 0.8 max Si, 1.0 max Mn, 12.0-14.0 Cr, bal Fe.
Martensitic corrosion resistant steel.
Ht. tr. & temp. 400°C: 218,000 TS; 196,000 YS; 18 El; 480 VDH hardness.
For cutlery, surgical instruments, dental tools, turbine equipment, valves.
B.S.: 3S.62, EN56C; AISI 420.

FIRTH-VICKERS F.G. (L).
M-521; 0.12-0.20 C, 0.8 max Si, 1.0 max Mn, 11.5-13.5 Cr, bal Fe.
Martensitic corrosion resistant steel; hardenable to about 42-48 Rock C.
For intricate cutlery and surgical instruments where hot forging may be difficult.
(AISI 420-low carbon range).

FIRTH-VICKERS F.H.
M-521; 0.28-0.36 C, 0.8 max Si, 1.0 max Mn, 12.0-14.0 Cr, bal Fe.
Martensitic corrosion resistant steel.
Air or oil hardenable to 240,000 psi TS and hardness of 598 VDH.
For knives, cutlery, surgical and dental instruments.

FIRTH-VICKERS F.H.M.
M-521; 0.70-0.90 C, 0.8 max Si, 1.0 max Mn, 15.5-17.5 Cr, 0.3-0.7 Mo, bal Fe.
Martensitic corrosion resistant steel.
Air or oil hardenable to about 50-56 Rock C.
For cutlery, dental and surgical instruments, valves, stainless ball and roller bearings.
AISI 440B.

FIRTH-VICKERS FHM NO. 2.
M-521; 0.95-1.20 C, 16.0-18.0 Cr, 0.75 Mo, bal Fe.
Martensitic, air-hardening, stainless steel, AISI 440C.

FIRTH-VICKERS F.I. 17.
M-521; 0.07 C, 17 Cr, 0.80 Mn, bal Fe.
Annealed: 78,400 TS; 49,300 YS; 32 El; 50 RA. 160 Brin.
For auto trim, window frames, grills; corrosion resistant; Type 430.

FIRTH-VICKERS F.I. 17 MO STEEL.
M-521; 0.07 C, 0.40 Si, 0.60 Mn, 16.3 Cr, 0.40 Ni, 1.0 Mo, bal Fe.
Ferritic stainless steel; for motor car hub caps and trim, domestic flatware and holloware.

FIRTH-VICKERS F.I. (AL).
M-521; 0.08 max C, 11.5-14.5 Cr, 0.1-0.3 Al, bal Fe.
Annealed: 65,000 TS; 38,000 YS; 36 El; 145 Brin.
Cold drawn: 85,000 TS; 70,000 YS; 20 El; 185 Brin.
For heat treating boxes, quenching racks, oil refining and chemical plant equipment. Heat and corrosion resistant. AISI Type 405.

FIRTH-VICKERS F.I. MO see **FIRTH-VICKERS MOLYBDENUM STAINLESS STEEL.**

FIRTH-VICKERS F.I. (P) STEEL.
M-521; 0.08 max C, 12.0-14.0 Cr, bal Fe.
Ferritic-martensitic type stainless steel.

FIRTH-VICKERS F.I. (TI) STEEL.
M-521; 0.06 max C, 0.8 max Si, 1.5 max Mn, 10.5-12.0 Cr, 1.0 Ni, 0.2-0.8 Ti, bal Fe.
Soft: 63,000 TS; 34,000 YS; 32 El.
Hard: 72,000 TS; 52,000 YS; 25 El.
For lightly loaded parts to resist scaling up to 700-750°C.
Similar to AISI 405.

FIRTH-VICKERS F. L.
M-521; 0.08 max C, 11.5-14.5 Cr, 0.1-0.3 Al, bal Fe.
Annealed: 70,000 TS; 40,000 YS; 30 El; 60 RA; 160 Brin.
For heat treating boxes, quenching racks, high temperature units.
Type 405 stainless steel.

FIRTH-VICKERS F.M.B.
M-521; 0.08 max C, 16-18 Cr, 10-14 Ni, 2-3 Mo, bal Fe.
Annealed: 85,000-95,000 TS; 35,000-45,000 YS; 60-50 El; 75-60 RA; 150-190 Brin.
For chemical plant equipment, mixers, agitators, filters; Type 316; stainless, austenitic.

FIRTH-VICKERS F.V. 520 (B) STEEL.
M-521; 0.07 max C, 0.7 max Si, 1.0 max Mn, 13.2-14.7 Cr, 5.0-6.0 Ni, 1.2-2.0 Cu, 1.2-2.0 Mo, 0.2-0.7 Nb, bal Fe.
Corrosion resistant maraging type steel.
Aged: 122,000-212,000 TS; 78,000-150,000 YS; 10-20 El; 270-450 VDH (varying treatment).
Weldable; for blading, bolts, fasteners, valves, pumps, water impellers, assemblies.

FIRTH-VICKERS F.V. 520 (S) STEEL.
M-521; 0.04-0.07 C, 0.6 max Si, 0.8-1.8 Mn, 15.3-16.0 Cr, 5.0-5.8 Ni, 1.4-2.1 Cu, 1.2-2.0 Mo, 0.5-0.15 Ti, bal Fe.
Corrosion resistant maraging type steel.
Aged: 142,000-196,000 TS; 116,000-142,000 YS; 6 min El; 320-400 VDH. (varying treatment).
Weldable; for bolts, fasteners, valves, pump parts, water impellers, assemblies.

FIRTH-VICKERS FVS.
M-521; 0.4 C, 13.6 Cr, 14.0 Ni, 2.6 W, bal Fe.
Valve steel.

FIRTH-VICKERS MOLYBDENUM STAINLESS STEEL.
M-521; 0.10 C, 0.30 Si, 0.30 Mn, 12.5 Cr, 0.75 Mo, bal Fe.
Air or oil hardenable; good resistance to scaling and creep at elevated temperatures; for turbine blades and similar applications.

FIRTH-VICKERS MOLYBDENUM VANADIUM STAINLESS STEEL.
M-521; 0.12 C, 0.30 Si, 0.60 Mn, 12.25 Cr, 0.80 Ni, 0.60 Mo, 0.18 V bal Fe.
Air or oil hardenable; tempered 675/700°C; At Room temp: 115,000 TS; 100,000 YS; 29 El.
At 650°C: 48,000 TS; 40,000 YS (0.2% proof).
For stressed parts operating up to 650°C.

FIRTH-VICKERS S. 80.
M-521; 0.16 C, 16.5 Cr, 2.5 Ni, bal Fe.
Oil treated: 100,000-135,000 TS; 78,000-112,000 YS; 25-15 El; 60-45 RA; 240-280 Brin.
For pump spindles, propeller shafts, seaplane and aircraft construction; magnetic, stainless and rust resistant. AISI 431.

FIRTH-VICKERS S.L.V.
M-521; 0.40-0.50 C, 3.0-3.75 Si, 0.3-0.6 Mn, 7.5-9.5 Cr, bal Fe.
Ht. Tr. Temp.: 650°C: 134,000-154,000 TS; 100,000-112,000 YS; 20-10 El.
Good mechanical properties and scale resistance at elevated temperatures.
For valves in internal combustion engines.
BS.EN.52.

FIRTH-VICKERS STAINLESS STEEL TYPE F. I.
M-521; 0.08 C, 0.25 Si, 0.25 Mn, 13.3 Cr, bal Fe.
Stainless steel, for elevated temperature operation; as steam turbine blades.

FIRTH-VICKERS "VIKRO".
M-50, M-521; 60-65 Ni, 15-25 Cr, 1.0-0.10 C, 0.5-1.0 Si, 1.0 max Mn, bal Fe.
Normalized: 102,000-134,000 TS; 65,000-78,000 YS; 25-15 El; 35-25 RA; 180-240 Brin.
At 1000°C: 17,600 TS.
For case hardening boxes, gas or oil burners, furnace muffles, stay bolts; heat resisting; stainless.

FIRTH-VICKERS XB.
M-521; 0.78 C, 2.0 Si, 0.45 Mn, 19.5 Cr, 1.35 Ni, bal Fe.
Hardened: 140,000 TS; 127,000 YS; 16 El; 21 RA.
At 400°C: 110,000 TS; 78,000 YS; 15 El; 28 RA.
For valves. Stainless, heat and corrosion resistant.

FIRTH W.C.R.
M-57; 0.5 C, 7.5 W, 7.5 Cr, bal Fe.
For hot work tools and dies; hot work steel, oil hardened.

FIRTH XDL.
M-57; 0.4 C, 15 W, 3.5 Cr, 0.5 V, bal Fe.
For hot work dies, extrusion rams and liners; hot work steel, oil hardened.

FISCHER VIS-11.
M-1659; 0.25 C, 10.8 Cr, 1.1 Mo, 0.6 Ni, 0.4 V, 0.4 W, bal Fe.
Annealed: 95,000 TS; 45,000 YS; 25 El; B 92 Rock.
Hardened: 245,000 TS; 210,000 YS; 8 El; C 50 Rock.
For valves, gears, cutlery, shafts, bearings, surgical instruments.
Corrosion resistant, hardenable. Cast only.

FISCO 66.
M-627; 0.8 C, 4 Cr, 5.5 W, 1.5 V, 5 Mo, bal Fe.
For tools, cutters; high speed steel.

FISCO AIRDIE.
M-627; 0.6 C, 5 Cr, bal Fe.
For tools, cutters, oil hardening.

FISCO AIRQUENCH.
M-627; 0.95 C, 2.04 Mn, 1.9 Cr, 1 Mo, bal Fe.
For dies, rolls, gauges; air hardening.

FISCO CARBON.
M-627; 1.0 C, bal Fe.
For tools, dies, cutters; water hardening.

FISCO CHROMDIE.
M-627; 1.6 C, 12 Cr, 0.75 Mo, 0.25 V, bal Fe.
For dies, gauges, mandrels, shear blades; air hardening, abrasion resistant.

FISCO COBALT.
M-627; 0.74 C, 4.5 Cr, 1.4 V, 18.8 W, 0.6 Mo, 5 Co, bal Fe.
For tools, cutters, dies; high speed steel.

FISCO DUPLEX.
M-627; 0.8 C, bal Fe.
For tools; water hardening.

FISCO EXCELL.
M-627; 0.8 C, 18 W, 4 Cr, 2 V, 0.7 Mo, bal Fe.
For tools, cutters; high speed steel.

FISCO HIGH SPEED.
M-627; 0.7 C, 18 W, 4 Cr, 1 V, bal Fe.
For tools, cutters; high speed steel.

FISCO HOT WORK.
M-627; 0.6 C, 8.5 Mo, 1.7 V, 3.6 Cr, bal Fe.
For hot work tools and dies; hot work steel.

FISCO MOLY.
M-627; 0.7 C, 1.5 W, 9.5 Mo, 4 Cr, 1 V, bal Fe.
For tools, cutters; high speed steel.

FISCO-MO NO. 1.
M-627; 0.1-0.2 C, 0.4-0.7 Mn, 1.65-2.0 Ni, 0.2-0.3 Mo, bal Fe.
Oil quenched: 120,000 TS; 90,000 YS; 22 El; 60 RA; 248 Brin.
For carburized parts, arbors, gears, pinions, cams, clutches; minimum distortion.

FISCO-MO NO. 3.
M-627; 0.45-0.55 C, 0.6-0.9 Mn, 0.9-1.1 Cr, 0.15-0.25 Mo, bal Fe.
Annealed: 100,000 TS; 68,000 YS; 25 El; 55 RA; 183 Brin.
Heat treated: 240,000 TS; 225,000 YS; 12 El; 30 RA; 460 Brin.
For heavy duty shafts, gears, clutches; good machining qualities.

FISCO NO. 1.
M-627; 0.15 C, 0.5 Mn, 1.8 Ni, 0.25 Mo, bal Fe.
For arbors, gears, pinions; carburizing steel.

FISCO NO. 2.
M-627; 0.3-0.4 C, 0.9-1.2 Mn, bal Fe.
Rolled: 90,000 TS; 65,000 YS; 18 El; 45 RA; 212 Brin.
For gears, shafts, bolts, gears, lead screws, spindles; excellent machining qualities.

FISCO NO. 3.
M-627; C, Mo, Cr, bal Fe.
For axles, bolts, gears, keys, pinions; oil hardening.

FISCO NO. 4.
M-627; 0.45-0.55 C, 0.6-0.9 Mn, 0.8-1.1 Cr, 0.15-0.25 Mo, bal Fe.
Oil quenched: 150,000-130,000 TS; 130,000-100,000 YS; 17-18 El; 55-50 RA; 300-230 Brin.
For shafts, gears, axles; high impact and creep strength, pre-heat treated.

FISCO OILDIE.
M-627; C, bal Fe.
For tools, dies; oil hardening.

FISCO OILHARD.
M-627; 0.9 C, 1.1 Mn, 0.5 Cr, 0.25 V, 0.4 W, bal Fe.
For tools, cutters, dies, gauges; non-deforming.

FISCO OMEGA.
M-627; 0.55 C, 0.8 Mn, 0.5 Mo, 0.25 V, 2.3 Si, bal Fe.
For impact tools; shock resistant, water hardened.

FISCO PNEUMATIC.
M-627; C, bal Fe.
For tools; oil hardening.

FISCO PRECISION.
M-627; C, bal Fe.
For tools; water hardening.

FISCO SEVEN.
    M-627; C, bal Fe.

FISCO SPECIAL.
    M-627; 1.0 C, bal Fe.
    For tools, dies, cutters; water hardening.

FISCO STAR TUNG.
    M-627; C, W, bal Fe.
    For tools, cutters; oil hardening.

FISCO SUPERIOR.
    M-627; 0.8 C, 12 Co, 20 W, 4 Cr, 1 V, 0.6 Mo, bal Fe.
    For tools, cutters; high speed steel.

FISCO TIGER.
    M-627; 1.0 C, 18 W, 4 Cr, 2 V, bal Fe.
    For tools, cutters; high speed steel.

FISCO VANADIUM.
    M-627; 0.9 C, 0.16 V, bal Fe.
    For tools, dies, cutters; water hardening.

FIVE POINT DEEP HARD.
    M-804; 0.4 C, 1.5 Ni, 0.8 Cr, 0.2 Mo, bal Fe.
    For gears, axles, wrist pins, wheels, valves; oil hardening.

F.J.A.B. EXTRA.
    M-365; 0.9-1.4 C, bal Fe.
    For general tools and dies; water hardening.

F.J.A.B. HOLLOW DRILL.
    M-365; 1-1.2 C, bal Fe.
    For mining, quarrying and construction tools; water hardening.

F.J.A.B. REGULAR.
    M-365; 0.9-1.2 C, bal Fe.
    For general tools and dies; water hardening.

F.J.A.B. ROOLED AUGER.
    M-365; 0.7-0.9 C, bal Fe.
    For tools for mining soft ores, coal, etc.; water hardening.

F.J.A.B. SOLID DRILL.
    M-365; 1.1-1.2 C, bal Fe.
    For mining, quarrying, construction work tools; water hardened.

F.J.A.B. SPECIAL.
    M-365; 0.9-1.2 C, bal Fe.
    For special die work tools.

FK(D)M-10.
    M-Germ.; 0.25 C, 3 Cr, 0.4 W, 0.4 Mo, 0.2 V, bal Fe.
    For turbine blades for jet engines; oil hardened.

FLAMALOY.
    M-157; C, alloy, bal Fe.
    For castings; castings, wear resistant.

FLAME HARD see **ISC FLAME HARD.**

FLAME RESISTING ALLOY.
    M-Eng.; 14 Cr, 9.7 Ni, 0.8 Mn, 0.2 Si, 0.2 C, bal Fe.
    For heat resisting alloy parts; corrosion and heat resistant.

FLANGE METAL, FRENCH.
    M-French; 94 Cu, 5.6 Sn, 0.05 Pb.
    For bushings, fittings, flanges; tough.

FLANGE METAL, GERMAN.
    M-German; 92 Cu, 5 Zn, 2.5 Sn.
    For pipes, fittings, flanges; corrosion resistant.

FLASH ALLOY.
    M-915; 0.17-0.22 C, 1.65-2.0 Ni, 0.2-0.3 Mo, bal Fe.
    For sling chains.

FLEETWELD 5.
    M-578; C, bal Fe.
    Steel arc welding electrode.
    AWS Class E6010.

FLEETWELD 5P.
    M-578; C, bal Fe.
    Steel arc welding electrode.
    AWS Class E6010.

FLEETWELD 7.
    M-578; C, bal Fe.
    Steel arc welding electrode.
    AWS Class E6012.

FLEETWELD 35.
    M-578; C, bal Fe.
    Steel arc welding electrodes.
    AWS Class E6011.

FLEETWELD 35LS.
    M-578; C, bal Fe.
    Steel arc welding electrode.
    AWS Class E6011.

FLEETWELD 37.
    M-578; C, bal Fe.
    Steel arc welding electrode.
    AWS Class E6013.

FLEETWELD 47.
    M-578; C, bal Fe.
    Steel arc welding electrode.
    AWS Class E7014.

FLEETWELD 57.
    M-578; C, bal Fe.
    Steel arc welding electrode.
    AWS Class E6013.

FLEETWELD 180.
    M-578; C, bal Fe.
    Steel arc welding electrode.
    AWS Class E6011.

FLETCHER & EMPERER BEARING.
    M-Eng.; 92 Al, 7.5 Cu, 0.3 Sn.
    For bearings; non-hardenable.

FLETCHER'S ALLOY.
M-Eng.; 96 Al, 3 Cu, 1 Sn, 0.5 Sb, P.
For aircraft and light weight parts; non-hardenable.

FLETCHERS BEARING METAL.
90 Al, 7 Cu, 1 Zn.
For bearings.

FLEX now VEW F180.

FLEXALOY.
M-1128; C, Cr, Mn, Ni, bal Fe.
550 Brin.
For hard surfacing electrodes; wear resistant against earth.

FLEXARC ACP-MO.
M-118; C, bal Fe.
For welding electrodes for steel.

FLEXARC AP.
M-118; 0.1 C, bal Fe.
Welded: 62,000 TS; 52,000 YS; 22 El.
For welding electrodes; for low carbon steel, E6012.

FLEXARC AP-MO.
M-118; C, alloy, bal Fe.
For welding electrodes; E-7010.

FLEXARC DH.
M-118; 0.2 C, bal Fe.
Welded: 62,000 TS; 55,000 YS; 25 El.
For welding electrodes; coated, for medium carbon steel; E6020.

FLEXARC DH-MO.
M-118; C, alloy, bal Fe.
Welded: 70,000 TS; 57,000 YS; 25 El.
For welding electrodes; high tensile, E-7020.

FLEXARC FP.
M-118; 0.3 C, bal Fe.
Welded: 68,000 TS; 55,000 YS; 17 El.
For welding electrodes; coated; for medium carbon steel, E6012.

FLEXARC FP-2.
M-118; C, bal Fe.
Welded: 68,000 TS; 55,000 YS; 17 El.
For arc welding electrodes; AWS-E-6012.

FLEXARC GRADE 18.
M-118; C, bal Fe.
For welding rods; Sulcoat E-4510.

FLEXARC LOH-2.
M-118; C, alloy, bal Fe.
Welded: 70,000 TS; 57,000 YS; 22 El.
H_2 coated electrode for alloy steel; E-7016.

FLEXARC SW-2.
M-118; Fe alloy.
Welded: 68,000 TS; 55,000 YS; 17 El.
For welding electrodes; E-6013.

FLEXITE.
M-1684; 0.45 C, 1.1 Mn, 1.1 Cr, 1.75 Ni, 0.2 V, 0.35 Mo, bal Fe.
Oil hardening, low alloy tool steel.

FLEXOR.
M-1433; 0.34 C, 0.85 Cr, 0.50 Mo, 0.45 W, bal Fe.
Heat treated: 160,000-250,000 TS; 125,000-200,000 YS; 14-9 El; 44-37 RA; 300-495 Brin.
For chuck jaws, arbors, motor shafts, gears, racks, arbors, pre-treated machinery steel.

FLINSO.
M-805; Pb, bal Sn.
For solder for Al.

FLINT ALLOY.
M-Eng.; 83 Fe, 12.5 Cr, 0.5 Si, 0.3 C.
For cutlery, stainless parts, corrosion resisting parts; corrosion resistant.

FLINTCAST.
M-192; 3.3 T. C, 2.4 Si, 1 Cr, bal Fe.
For abrasion resisting iron castings; abrasion resisting.

FLINTYPE.
M-507; 0.90 C, 1.0 Si, 0.3 Mn, 9.5 Cr, 0.6 Mo, 3.2 Ni, bal Fe.
Hardfacing electrode.
As arc welded: 55,000 psi, 560 BHN.
For overlay or buildup on plow points, heavy earth moving machinery; abrasion resistant, not designed for edge impact.

FLOCKTON 4% TUNGSTEN.
M-1409; 1.3 C, 4-5 W, 1.0-1.2 Cr, 0.25 V, bal Fe.
For reamers; oil hardened.

FLOCKTON C. T.
M-1409; 0.3 C, 9 W, 2.5 Cr, 0.4 V, bal Fe.
For heading and piercing dies, punches; hot work steel, oil hardened.

FLOCKTON H.C.C.
M-1409; C, Co, bal Fe.
For shear blades, nibblers; air hardened.

FLOCKTON H.D.M.
M-1409; 0.35 C, 5 Cr, 1.5 Mo, 1 Si, 0.4 Mn, 0.4 V, bal Fe.
For extrusion dies, die casting dies; hot work steel, oil hardened.

FLOCKTON I.E.-P.
M-1409; 1.0 C, bal Fe.
Annealed: 100,000 TS; 53,000 YS; 21 El; 42 RA; 200 Brin.
For tools, dies, drills, taps; water hardened; Type W1.

FLOCKTON N.T.C.
M-1409; C, alloy, bal Fe.
For chisels; non-tempering steel, air hardened.

FLOCTON P.B.7.
M-1409; C, alloy, bal Fe.
For taps, dies, drills, reamers; resists severe compression and abrasion.

FLOCTON P-NI.
M-1409; 0.8 C, 1.8 Ni, 1 Cr, 0.35 Mn, bal Fe.
For machinery parts, cold heading dies; oil hardened, shock resistant.

FLOCKTON R.H. D.
M-1409; 0.38 C, 5 Cr, 1.5 Mo, 1.5 W, 1 Si, 0.3 Mn, bal Fe.
For extrusion and heading dies, rams, mandrels; oil or air hardened, hot work steel.

FLOCKTON T. C.
M-1409; 1.5 C, 3.2 W, 0.15 Cr, bal Fe.

FLO-KOTE.
M-374; 0.2 C, 11-13 Mn, Ni, bal Fe.
For welding electrode; tough, wear resistant.

FLOTECTIC SILVER 2.
M-717.
Silver solder type alloy for torch, furnace or induction brazing, ferrous and non-ferrous alloys.
Melts 1295°F; 58,000 TS.

FLUGINOX 51.
M-1120; 0.06 C, 0.4 Ni, 12.8 Cr, 0.5 Mo, 0.25 Al, bal Fe.
Stainless steel, resistant to corrosion and oxidation at elevated temperature; resistant to creep.

FLUGINOX 61.
M-1120; 0.20 C, 0.5 Ni, 12.0 Cr, 1.0 Mo, 0.3 V, bal Fe.
Martensitic stainless steel; resistant to creep up to 600°C, for turbines and hydrocarbon industries.
Modified AISI 420.

FLUGINOX 65.
M-1120; 0.22 C, 0.7 Ni, 11.0 Cr, 0.9 Mo, 0.3 Nb, 0.3 V, 0.015 max S, bal Fe.
Martensitic stainless steel, resistant to oxidation and corrosion and creep up to 500-650°C, steam and hot oil.

FLUGINOX 71.
M-1120; 0.10 C, 2.3 Ni, 12.0 Cr, 1.7 Mo, 0.3 V, bal Fe.
Martensitic stainless steel; resistant to creep up to 500°C.

FLUKS.
M-1260; 55 Sn, 10 Sb, 2.5 Cu, 32.5 Pb.
Cast: 11,400 TS; 22 Brin.
For refrigerator and electric motor bearings; M.P. 355-620°F; good castability.

FLUSH PLATE.
M-Eng.; 65.75 Cu, 32.75 Zn, 1.50 Pb.
For hardware; free-cutting.

FLUXCOR 1/2 MO.
M-663; 0.08 C, 0.8 Mn, 0.5 Si, 0.5 Mo, bal Fe.
Wire for welding and build-up of low alloy steel using CO_2 shielded gas.
Welded, stress ann: 92,000 TS; 81,000 YS; 26 El.

FLUXCOR 1.
M-663; 0.11 C, 1.6 Mn, 0.84 Si, bal Fe.
Wire for welding or build-up on mild steel.
As welded: 101,700 TS; 89,200 YS; 24 El.
AWS E 70T-2.

FLUXCOR 2.
M-663.
For welding steel having residual oil from machining and forming.
AWS E 70T-1.

FLUXCOR CR-MO 1.
M-663; 0.07 C, 0.84 Mn, 0.46 Si, 1.23 Cr, 0.51 Mo, bal Fe.
Wire to join or repair Cr-Mo steel castings and similar Cr-Mo steels.
Welded, stress ann: 103,000 TS; 93,000 YS; 25 El.

FLUXCOR CR-MO 2.
M-663; 0.04 C, 0.83 Mn, 0.48 Si, 2.23 Cr, 1.08 Mo, bal Fe.
Wire for welding or build-up of Cr-Mo castings and wrought steel of similar composition.
Welded, stress ann: 110,000 TS; 100,000 YS; 20 El.
Usually used with CO_2 shielded gas.

FLUXCOR NO. 1.
M-663; weld deposit: 0.06 C, 1.46 Mn, 0.74 Si, 0.06 Ni, 0.07 Cr, 0.07 Mo, bal Fe.
Low alloy cored wire for welding low alloy steel.
As welded: 87,000 psi TS; 74,700 psi YS; 26.5 El.
AWS A5.20 Class E 70T-2.

FLUX-COR NO. 5.
M-663; weld deposit: 0.072 C, 1.27 Mn, 0.39 Si, 0.06 Ni, 0.06 Cu, 0.0 Mo, bal Fe.
As welded: 81,400 psi TS; 64,000 YS; 30 El.
Good impact strength.
MIL-E-24403/1-A CLASS MIL-70T-5.

FLUXRITE.
M-88; Sn, bal Pb.
For solder; soft.

FM-20.
M-1766; 0.19 C, 1.25 Mo, 0.20 Si, bal Fe.
Manganese structural or carburizing steel.
AFNOR 20M5; SAE 1522; BS EN7A.

FM-30T.
M-1766; 0.34 C, 1.25 Mn, 0.20 Si, bal Fe.
Manganese structural steel.
IHA F-411; DIN 30 Mn5; AISI 1536.

FM-100.
 M-1531; 1.25 T.C., 0.92 C.C, 3 Cu, 0.2 Si, 0.5 Mn, 0.9 Mo, bal Fe.
 Sintered: 100,000 TS; Rockwell B95.
 For piston rings; wear resistant.

FMP 035.
 M-1738; 1.05 C, 0.25 Si, 0.35 Mn, 0.35 V, bal Fe.
 Water hardening tool steel.
 AISI W2.

FMP 200.
 M-1738; 0.95 C, 0.30 Si, 1.25 Mn, 0.50 W, 0.50 Cr, 0.17 V, bal Fe.
 Oil hardening cold work tool steel.
 For blanking and forming tools, punches.
 AISI O1.

FMP 328.
 M-1738; 0.40 C, 1.0 Si, 0.30 Mn, 5.25 Cr, 0.40 V, 1.35 Mo, bal Fe.
 Chromium type hot work tool steel.
 For hot forming and forging dies.
 AISI H11.

FMP 329.
 M-1738; 0.40 C, 1.0 Si, 0.35 Mn, 5.25 Cr, 1.0 V, 1.35 Mo, bal Fe.
 Chromium type hot work tool steel.
 Hot forming dies-light metal casting dies.
 AISI H13.

FMP 336.
 M-1738; 1.50 C, 0.40 Si, 0.40 Mn, 12.0 Cr, 0.90 V, 0.80 Mo, bal Fe.
 High carbon-high chrome type cold work tool steel.
 For forming and blanking dies, thread rolling dies, plastic molds.
 AISI D2.

FMP 338.
 M-1738; 2.05 C, 0.40 Si, 0.40 Mn, 13.0 Cr, bal Fe.
 High carbon-high chrome type cold work tool steel; air or oil hardening.
 For blanking and forming dies, gauges, brick mould liners.
 AISI D3.

FMP 348.
 M-1738; 0.43 C, 0.30 Si, 0.40 Mn, 1.55 Cr, 0.30 Mo, 4.0 Ni, bal Fe.
 Shock resisting tool steel.
 For heavy duty shear blades, trimming dies.

FMP 379.
 M-1738; 1.00 C, 0.30 Si, 0.50 Mn, 5.0 Cr, 0.30 V, 1.10 Mo, bal Fe.
 Air hardening cold work tool steel.
 For punching, drawing and coining dies.
 AISI A2.

FMP 399.
 M-1738; 0.50 C, 0.90 Si, 0.35 Mn, 2.25 W, 1.45 Cr, 0.20 V, bal Fe.
 Oil hardening shock resisting tool steel.
 For chisels, shear blades, pneumatic chisels.
 AISI S1.

FMP 455.
 M-1738; 0.80 C, 0.30 Si, 0.30 Mn, 18.5 W, 4.25 Cr, 1.3 V, 5.0 Co, 0.75 max Mo, bal Fe.
 Tungsten-cobalt type high speed tool steel.
 AISI T4.

FMP 470.
 M-1738; 0.80 C, 0.30 Si, 0.30 Mn, 22.0 W, 4.5 Cr, 1.5 V, 1.0 max Mo, bal Fe.
 Tungsten type high speed steel.

FMP 501.
 M-1738; 0.80 C, 0.30 Si, 0.30 Mn, 2.0 W, 3.9 Cr, 1.25 V, 9.0 Mo, bal Fe.
 Molybdenum type high speed tool steel.
 For drills, rough cutting lathe tools.
 AISI M1.

FMP 504.
 M-1738; 1.30 C, 0.30 Si, 0.30 Mn, 5.75 W, 4.4 Cr, 4.0 V, 4.60 Mo, bal Fe.
 Molybdenum type high speed steel.
 AISI M4.

FMP 505.
 M-1738; 0.32 C, 0.30 Si, 0.30 Mn, 9.25 W, 3.25 Cr, 0.50 V, 0.50 Mo, bal Fe.
 Tungsten type hot work tool steel.
 For hot forming and swaging dies.
 AISI H21.

FMP 507.
 M-1738; 0.30 C, 0.30 Si, 0.30 Mn, 9.0 W, 3.0 Cr, 0.30 V, 0.50 Mo, 2.50 Ni, bal Fe.
 Tungsten type hot work tool steel.
 For hot punches and hot heading dies.

FMP 513.
 M-1738; 0.35 C, 1.0 Si, 0.30 Mn, 1.25 W, 5.25 Cr, 0.35 V, 1.5 Mo, bal Fe.
 Chromium type hot work tool steel.
 For hot forming, blanking and extrusion dies.
 AISI H12.

FMP 526.
 M-1738; 1.10 C, 0.30 Si, 0.30 Mn, 6.75 W, 4.25 Cr, 2.0 V, 5.0 Co, 3.75 Mo, bal Fe.
 Molybdenum-Cobalt type high speed steel.
 AISI M41.

FMP 530.
 M-1738; 0.80 C, 0.30 Si, 0.30 Mn, 2.0 W, 4.0 Cr, 1.2 V, 5.0 Co, 8.25 Mo, bal Fe.
 Molybdenum-Cobalt type high speed steel.
 AISI M30.

FMP 536.
M-1738; 1.50 C, 0.30 Si, 0.30 Mn, 6.5 W, 4.0 Cr, 5.0 V, 5.0 Co, 3.5 Mo, bal Fe.
Molybdenum type high speed steel.
AISI M15.

FMP 542.
M-1738; 1.05 C, 0.30 Si, 0.30 Mn, 1.5 W, 3.75 Cr, 1.15 V, 8.0 Co, 9.5 Mo, bal Fe.
Molybdenum-Cobalt type high speed steel.
AISI M42.

FMP 555.
M-1738; 1.50 C, 0.30 Si, 0.30 Mn, 12.75 W, 4.75 Cr, 5.0 V, 5.0 Co, 1.0 max Mo, bal Fe.
Tungsten-Vanadium-Cobalt high speed steel.
AISI T15.

FMP 562.
M-1738; 0.82 C, 0.30 Si, 0.30 Mn, 6.25 W, 4.1 Cr, 2.0 V, 5.0 Mo, bal Fe.
Molybdenum type high speed steel.
AISI M2.

FMP 563.
M-1738; 1.05 C, 0.30 Si, 0.30 Mn, 6.0 W, 4.0 Cr, 2.35 V, 5.0 Mo, bal Fe.
Molybdenum type high speed steel.
AISI M3 Class 1.

FMP 599.
M-1738; 0.70 C, 0.30 Si, 0.30 Mn, 14.0 W, 4.25 Cr, 0.80 V, bal Fe.
Tungsten type high speed tool steel.
For hacksaws, slitting saws, cold punches.

FMP 622.
M-1738; 0.75 C, 0.30 Si, 0.30 Mn, 18.2 W, 4.1 Cr, 1.1 V, 1.0 max Mo, bal Fe.
Tungsten type high speed tool steel.
For cutting tools.
AISI T1.

FMP 682.
M-1738; 0.60 C, 0.30 Si, 0.30 Mn, 4.0 Cr, 2.0 V, 8.25 Mo, bal Fe.
Molybdenum type hot work tool steel.
For hot working tools and dies.
AISI H43.

FMP 808.
M-1738; 0.80 C, 0.30 Si, 0.30 Mn, 18.5 W, 4.25 Cr, 1.6 V, 10.0 Co, 1.0 Mo, bal Fe.
Tungsten-cobalt type high speed tool steel.
Similar to AISI T5 or T6.

FMP 828.
M-1738; 0.80 C, 0.30 Si, 0.30 Mn, 18.5 W, 4.5 Cr, 2.0 V, 8.0 Co, 1.0 max Mo, bal Fe.
Tungsten-cobalt type high speed tool steel.
AISI T5.

FMP 842.
M-1738; 0.84 C, 0.30 Si, 0.30 Mn, 18.5 W, 4.25 Cr, 2.25 V, 1.0 max Mo, bal Fe.
Tungsten type high speed tool steel.
For cutting tools.
AISI T2.

FMP 922.
M-1738; 1.0 C, 0.30 Si, 0.30 Mn, 1.75 W, 4.0 Cr, 2.0 V, 8.75 Mo, bal Fe.
Molybdenum type high speed steel.
AISI M7.

FMP 928.
M-1738; 0.90 C, 0.30 Si, 0.30 Mn, 2.0 W, 4.0 Cr, 2.0 V, 8.0 Co, 8.5 Mo, bal Fe.
Molybdenum-cobalt type high speed steel.
AISI M34.

FMP 929.
M-1738; 1.25 C, 0.30 Si, 0.30 Mn, 1.8 W, 3.75 Cr, 2.0 V, 8.25 Co, 8.75 Mo, bal Fe.
Molybdenum-Cobalt type high speed steel.
AISI M43.

FMP 933.
M-1738; 1.30 C, 0.30 Si, 0.30 Mn, 9.25 W, 4.25 Cr, 3.5 V, 3.75 Mo, 10.0 Co, bal Fe.
Tungsten type high speed tool steel with cobalt.

FMP 948.
M-1738; 0.88 C, 0.30 Si, 0.30 Mn, 4.0 Cr, 2.0 V, 8.25 Mo, bal Fe.
Molybdenum type high speed steel.
AISI M10.

FMP 1850.
M-1738; 0.55 C, 0.30 Si, 0.30 Mn, 18.0 W, 4.10 Cr, 0.70 V, 1.0 max Mo, bal Fe.
Tungsten type hot work tool steel.
For hot forming dies.
AISI H26.

FMS-35.
M-1766; 0.36 C, 1.30 Mn, 1.30 Si, bal Fe.
Mn-Si structural steel; for gears, axles.
Hardenable, water quench, to 48-56 Rc.
DIN 37 Mn Si 5; AFNOR 38MS5.

FO 18. 9.
M-1487; 0.20-0.28 C, 17.0-18.0 Cr, 8.5-9.5 Ni, 1.5 max Mn, 0.50 max Si, bal Fe.
Austenitic stainless steel; good corrosion resistance.
AFNOR Z 25 CN 18.9.

FO 25-12.
M-1487.
Stainless steel; similar to AISI 309.

FO 25-20.
M-1487.
Stainless steel; similar to AISI 310.

FO 25.20 U.
M-1487; 0.03 max C, 19.0-21 Cr, 24.0-28.0 Ni, 4.5 Mo, 2.0 max Mn, 1.0 max Si, 1.5 Cu, bal Fe.
Stainless steel for petroleum refineries, paper making, ammonium sulphate.
AFNOR Z 2 NCDU 25.20.

FO 36-18.
M-1487.
Stainless steel; similar to AISI 330.

FO 90 SUPERVITAC.
M-1488; 0.42 C, 1.0 Cr, 0.20 Mo, bal Fe.
Bars, treated: 880 N/mm^2 UTS; 685 N/mm^2 min YS; 15 El; 262 Brin.
For structural purposes.
Similar to AISI 4142.

FOBALLOY.
M-1801; 95 Sn, 3.5 Ag, 1.5 Sb.
For hermetic sealing of integrated ceramic packages.

FOBES METAL.
M-Eng.; 54 Zn, 46 Cu.
For ornamental castings: low strength.

FOB METAL.
M-Eng.; 87.50 Cu, 12.0 Zn, 0.50 Sn.
For fobs, ornaments; red brass.

FOC O POINTS.
M-1546; 1.25 C, bal Fe.
Hardened: Rock C 65-66.
For files, cutting tools, drills.
Water hardening, wear resistant.

FOC 3 POINTS.
M-1546; 0.90 C, bal Fe.
Annealed: 100,000 TS; 53,000 YP; 2 El; 197 Brin.
Hardened: 216,000 TS; 152,000 YP; 11 El; 600 Brin.
For chisels, bolts, cutting tools, springs, punches, drills. Water hardening, wear resistant.

FOC 4 POINTS.
M-1546; 0.80 C, bal Fe.
Annealed: 120,000 TS; 65,000 YS; 15 El; 223 Brin.
Hardened: 190,000 TS; 142,000 YS; 12 El; 390 Brin.
For chisels, bolts, springs, bearings, liners. Water hardening, wear resistant.

FOCARBO 10.
M-1766; 0.10 C, 0.50 Mn, 0.25 Si, 0.03 max P, 0.03 max S, bal Fe.
Low carbon steel.
DIN CK 10; AFNOR XC 10; AISI 1010.

FOCARBO 15.
M-1766; 0.15 C, 0.50 Mn, 0.25 Si, bal Fe.
Low carbon steel, sometimes for case hardening.
DIN CK-15; AFNOR XC 18.
AISI 1015.

FOCARBO 20.
M-1766; 0.20 C, 0.55 Mn, 0.25 Si, bal Fe.
Low carbon structural steel.
DIN CK-22; AFNOR XC 18.
AISI 1020.

FOCARBO 25.
M-1766; 0.25 C, 0.60 Mn, 0.25 Si, bal Fe.
Structural steel.
IHA F-112; AISI 1025; BS EN 4.

FOCARBO 32.
M-1766; 0.32 C, 0.65 Mn, 0.25 Si, bal Fe.
Structural steel.
AFNOR XC-32; AISI 1030; BS EN 5C.

FOCARBO 35.
M-1766; 0.35 C, 0.65 Mn, 0.25 Si, bal Fe.
Structural steel.
DIN-CK 35; IHA F-113; AISI 1035; UNI C-35.

FOCARBO-35T.
M-1766; 0.36 C, 0.80 Mn, 0.25 Si, bal Fe.
For cold forming, threading and subsequent heat treatment of small screws.
IHA F-113; DIN Cq-35; AISI 1037.

FOCARBO 37.
M-1766; 0.37 C, 0.65 Mn, 0.25 Si, bal Fe.
Medium carbon structural steel.
AFNOR XC 38; AISI 1038; BS EN 6A.

FOCARBO 42.
M-1766; 0.42 C, 0.65 Mn, 0.20 Si, bal Fe.
Medium carbon structural steel.
AFNOR XC 42; AISI 1042; BS EN 8D, UNI C-40.

FOCARBO 45.
M-1766; 0.45 C, 0.65 C, 0.25 Si, bal Fe.
Medium carbon structural steel.
IHA F-114; DIN CK 45; AISI 1045; UNI C-45.

FOCARBO 47.
M-1766; 0.47 C, 0.65 Mn, 0.25 Si, bal Fe.
Medium carbon structural steel.
CENIM F-1142/C47K; AFNOR XC48; AISI 1049.

FOCARBO 55.
M-1766; 0.55 C, 0.65 Mn, 0.25 Si, bal Fe.
Medium-high carbon structural steel.
IHA F-115; DIN CK 55; AFNOR XC 55; AISI 1055.

FOCEM-12.
M-1766; 0.10 C, 0.50 Mn, 0.25 Si, bal Fe.
Low carbon steel for carburizing.
AFNOR XC-10; DIN CK-10; AISI 1010.

FOCEM-17.
M-1766; 0.17 C, 0.50 Mn, 0.25 Si, bal Fe.
Carburizing steel.
IHA F-111; AFNOR XC-18; DIN CK-15; AISI 1017.

FOMO-12.
M-1766; 0.13 C, 0.70 Mn, 0.25 Si, 1.0 Cr, 0.20 Mo, bal Fe.
Cr-Mo carburizing steel.
IHA F-155; DIN 12CrMo4; AFNOR 12Cd4.

FOMO-18.
M-1766; 0.18 C, 0.75 Mn, 0.25 Si, 1.0 Cr, 0.20 Mo, bal Fe.
Cr-Mo carburizing steel.
AFNOR 18CD4; DIN 16CrMo4.

FOMO-20.
M-1766; 0.20 C, 0.75 Mn, 0.25 Si, 0.40 Cr, 0.45 Mo, bal Fe.
Mo-Cr carburizing steel.
CENIM F-1523/20 CrMo2; DIN 20 MoCr4.

FOMO-25.
M-1766; 0.27 C, 0.60 Mn, 0.25 Si, 1.0 Cr, 0.20 Mo, bal Fe.
Cr-Mo structural steel.
IHA F-222; AFNOR 25 CD4; UNI 25CD4.

FOMO-30.
M-1766; 0.30 C, 0.70 Mn, 0.25 Si, 1.0 Cr, 0.20 Mo, bal Fe.
Cr-Mo structural Steel.
UNI 30CD4; AISI 4130.

FOMO-35.
M-1766; 0.35 C, 0.70 Mn, 0.25 Si, 1.0 Cr, 0.20 Mo, bal Fe.
Cr-Mo structural steel.
IHA F-125; AFNOR 35CD4; AISI 4135.

FOMO-35T.
M-1766; 0.37 C, 0.70 Mn, 0.25 Si, 1.0 Cr, 0.20 Mo, bal Fe.
Cr-Mo structural steel.
IHA F-125; AFNOR 38CD4; AISI 4137.

FOMO-40.
M-1766; 0.40 C, 0.75 Mn, 0.25 Si, 1.0 Cr, 0.20 Mo, bal Fe.
Cr-Mo structural steel, for shafts, axles.
DIN 42 CrMo4; AISI 4140; BS En19.

FONI-30.
M-1766; 0.32 C, 0.55 Mn, 0.25 Si, 3.0 Ni, 0.70 Cr, bal Fe.
Ni-Cr structural steel; deep hardening.
IHA F-123; AFNOR 30NC11; BS EN23.

FONIX-10.
M-1766; 0.12 max C, 0.70 max Mn, 0.70 max Si, 13.0 Cr, bal Fe.
Ferritic or martensitic stainless steel.
IHA F-311; DIN X10 Cr13; similar to AISI 410.

FONIX-16.
M-1766; 0.10 max C, 1.0 max Mn, 1.0 max Si, 17.0 Cr, bal Fe.
Ferritic type stainless steel; not hardenable by heat teatment.
DIN X8Cr17; AISI 430; BS EN60.

FONIX-17-2.
M-1766; 0.17 C, 0.70 max M,n, 0.70 max Si, 17.0 Cr, 2.0 Ni, bal Fe.
Martensitic stainless steel; hardenable to 100-135 Kg/mm^2 TS.
For aircraft fittings, marine equipment.

FONIX-18-8.
M-1766; 0.08 max C, 2.0 max Mn, 1.0 max Si, 18.0 Cr, 8.0 Ni, bal Fe.
Austenitic stainless steel; for food, dairy, textile and photographic equipment.
IHA F-314; AISI 304; BS EN58E.

FONIX-18-8-2.
M-1766; 0.08 max C, 2.0 max Mn, 1.0 max Si, 18.0 Cr, 12.0 Ni, 2.25 M Fe.
Austenitic stainless steel; improved corrosion resistance for food, photographic, and beverage equipment.
DIN X5CrNiMo 18-10; AISI 316; BS EN58H.

FONIX-18-8-S.
M-1766; 0.12 max C, 2.0 max Mn, 1.0 max Si, 0.20 S, 18.0 Cr, 9.0 Ni, Fe.
Free-machining, austenitic stainless steel.
For threaded parts.
AISI 303; AFNOR Z10CNF 1809.

FONIX-18-8-TI.
M-1766; 0.08 max C, 2.0 max Mn, 1.0 max Si, 18.0 Cr, 11.0 Ni, Ti = 5 min, bal Fe.
Weldable austenitic stainless steel.
IHA F-332; AISI 321; BS EN58C.

FONIX-18-12-2-TI.
M-1766; 0.08 max C, 2.0 max Mn, 1.0 max Si, 18.0 Cr, 12.0 Ni, 2.25 Mo, Ti = 5 x C min, bal Fe.
Weldable grade of AISI 316 Stainless.
AFNOR Z 8CNDT 18-12; BS En 58 J.

FONIX-20.
M-1766; 0.20 C, 0.70 max Mn, 0.70 max Si, 13.0 Cr, bal Fe.
Martensitic stainless steel; hardenable to 44-50 Rc.
For valves, arbors, plastic molds.
DIN X20Cr13; AFNOR Z20C13; AISI 420.

FONIX-30.
M-1766; 0.30 C, 0.70 max Mn, 0.70 max Si, 13.5 Cr, bal Fe.
Martensitic stainless steel; hardenable to 47-54 Rc.
For valves, parts for chemical equipment.
AFNOR Z30C13; BS EN56D; AISI 420.

FONIX-35.
M-1766; 0.38 C, 0.70 max Mn, 0.70 max Si, 13.5 Cr, bal Fe.
Martensitic stainless steel; hardenable to 48-54 Rc.
For pump shafts, cutlery, shears, valves.
IHA F-312; DIN X40Cr13; AISI 420.

FONTAINMOREAU BRONZE.
M-England; 0-8 Cu, 0-1 Fe, 0-1 Pb, bal Zn.
For ornaments, fittings, die castings.

FONTANOR 11.
M-1645; 3.4 C, 1.7 Cr, 3.6 Ni, bal Fe.
Alloy iron casting.

FONTANOR 15.3.
M-1645; 3.20 C, 15.0 Cr, 3.0 Mo, bal Fe.
Alloy iron casting.

FONTANOR 30.
M-1645; 3.20 C, 30.0 Cr, bal Fe.
Alloy iron casting.

FONTANOR 300.
M-1645; 3.50 C, 16.0 Cr, 1.7 W, bal Fe.
Alloy iron casting.

FONTANOR NI-RESIST.
M-1645; 2.90 C, 1.50 Cr, 15.0 Ni, 5.0 Cu, bal Fe.
Ni-Resist type casting.

FONTE R. A.
M-700; 0.2 C, bal Fe.
For structural parts; carburizing steel.

FOOL-PROOF.
M-80; 0.7 C, 4 W, 2 Cr, bal Fe.
For hand and pneumatic chisels; oil hardening.

FOOTE ELECTROMANGANESE, HYDROGEN REMOVED.
M-1038; 99.5 Mn, 0.028 S, 0.008 C, 0.001 Si, 0.0005 H, 0.013 N, 0.48 O.
For alloying in stainless and special steels.

FOOTE ELECTROMANGANESE, LOW OXYGEN.
99.8 Mn, 0.028 S, 0.008 C, 0.001 Si, 0.024 H, 0.002 N, 0.10 O.
For alloying in stainless and special steels.

FOOTE ELECTROMANGANESE NO. 1.
M-1038; 99.5 Mn, 0.028 S, 0.008 C, 0.001 Si, 0.0005 H, 0.013 N, 0.48 O.
For alloying in stainless and special steels.

FOOTE NITREMANG, GRADE "A".
M-1038; 93.7 Mn, 6.0 N, 0.03 C, 0.032 S, 0.005 Fe, 0.001 P, 0.001 S 0.0005 H, 0.35 O.
Alloying agent for introducing manganese and nitrogen into special steels.

FOOTE NITREMANG, GRADE "B".
M-1038; 94.4 Mn, 5.0 N, 0.03 C, 0.032 S, 0.005 Fe, 0.001 P, 0.001 Si, 0.005 H, 0.35 O.
Alloying agent for introducing manganese and nitrogen into special steels.

FOOTE NODULOY 3 see **NODULOY 3.**

FORAL.
M-1766; 0.10 C, 1.10 Mn, 0.06 max Si, 0.05 max P, 0.30 S, 0.20 Pb, 0.10 max Se, bal Fe.
Free machining, low carbon steel.
IHA F-212; AISI 12L14.

FORD 406.
M-1491; 6.0 Cr, 10.0 Co, 1.0 Mo, 8.5 W, 2.0 Cb, 2.0 Ti, 4.5 Al, 6.0 Ta, bal Ni.
For integrally cast turbine wheels.

FOREZ 2AS.
M-1487; 0.85 C, 6 W, 2 V, 6 Mo, 4.5 Cr, bal Fe.
For lathe and planer tools, drills, reamers, hobs; high speed steel (Z85WD06-06).

FOREZ 2 AS BC.
M-1487; 0.60-0.70 C, 4.0-4.3 Cr, 6.0-6.75 W, 5.25-5.75 Mo, 1.85-2.15 V, bal Fe.
For punches and cold chisels, extrusion mandrils.
AFNOR Z 65 WDV 06.05.02; AISI H42.

FOREZ 2 AS CO.
M-1487; 0.80-0.85 C, 4.0-4.3 Cr, 5.25-5.75 Mo, 6.0-6.75 W, 1.85-2.15 V, 4.75-5.25 Co, bal Fe.
Cobalt-moly-tungsten high speed steel.
Cutting tool for higher speed operation.
AFNOR Z 85 WDKV 06.06.05.02; AISI M 35.

FOREZ 3AS.
M-1487; 0.80 C, 18 W, 5 Cr, 1 Mo, 1.5 V, bal Fe.
For lathe and planer tools, drills, taps, hobs; high speed steel (Z80W18).

FOREZ 4AS.
M-1487; 0.85 C, 18 W, 5 Cr, 1.5 V, 1 Mn, 5 Co, bal Fe.
For lathe and planer tools, milling cutters, hobs; high speed steel (Z85WK-18-05).

FOREZ 333.
M-1487; 1.5-1.6 C, 4.5-5.0 Cr, 12.0-13.0 W, 4.75-5.25 V, 4.75-5.25 Co, bal Fe.
Tungsten-Vanadium-Cobalt high speed steel.
Good wear resistant cutting tool.
AFNOR Z 150 WKV 12.05.05; AISI T 15.

FOREZ 444.
M-1487; 1.25 C, 20 W, 4 Cr, V, Co, bal Fe.
For lathe and planer tools, reams, taps, drills; high speed steel, oil hardened (Z125WK20-15).

FOREZ CH4.
M-1487; 1.2 C, 0.3 Mn, 0.3 Si, bal Fe.
For nail dies, hot snaps, punches, shears; hard and tough (120C2).

FOREZ CH5.
M-1487; 0.80 C, 1.2-1.5 Mn, 0.40 Si, bal Fe.
For reamers, taps, gauges, special dies; oil hardened, non-deforming (80M5).

FOREZ DT.
  M-1487; 2 C, 12 Cr, bal Fe.
  For shear blades, wire drawing and piercing dies; non-deforming, oil hardened (Z200 C12).

FOREZ DT.
  M-1487; 2.0-2.2 C, 12.0-14.0 Cr, bal Fe.
  High carbon-high chrome cold work tool steel for dies for punching and embossing.
  AFNOR Z 200 C 12; AISI D3.

FOREZ DT VA.
  M-1487; 1.50-1.60 C, 11.5-12.5 Cr, 0.75-0.85 Mo, 0.70-0.90 V, bal Fe.
  High carbon-high chrome cold work tool steel for punching and embossing dies.
  AFNOR Z 150 CDV 12; AISI D2.

FOREZ SUPRA EXTRA 4AS.
  M-1487; 0.85 C, 18 W, 5 Cr, 1.5 V, 1 Mo, 10 Co, bal Fe.
  For lathe and planer tools, broaches, reamers, hobs; high speed steel (Z85WK-18-10).

FOREZ TE.
  M-1487; 0.3 C, 3 Cr, 9 W, bal Fe.
  For punches, dies, bolt and nail dies; hot work steel (Z30WC09).

FOREZ TM.
  M-1487; 0.40 C, alloy, bal Fe.
  For hot and cold stamping tools, pneumatic chisels; shock resistant (40WNCD).

FORGE-DIE.
  M-115; 0.25 C, 13.5-14.5 W, 3.5 Cr, 0.5 V, bal Fe.
  For upsetter heads and dies, piercing punches, extrusion dies; resists heat checking. Hot work steel.

FORGE WEAR NO. 1.
  M-275; 1.10 C, 1.0 Si, 5.25 Cr, 1.1 Mo, 4.0 V, bal Fe.
  Hot work tool steel.

FORGE WEAR NO. 2.
  M-275; 0.75 C, 1.0 Si, 5.25 Cr, 1.1 Mo, 2.5 V, bal Fe.
  Hot work tool steel.

FORGE-WELL.
  M-1684; 0.27 C, 1.0 Mn, 1.2 Cr, 0.25 V, 0.3 Mo, bal Fe.
  Oil hardening, low alloy tool steel, for molds,

FORGING BRASS.
  M-Eng.; 60-57 Cu, 40-43 Zn.
  For corrosion resisting forgings; high strength.

FORGING BRASS-377.
  M-8; 60 Cu, 38 Zn, 2 Pb.
  Soft Rod: 54,000 TS; 20,000 YS; 45 El; B 45 Rock.
  For hardware, gears, fasteners, bolts.
  Free-machining leaded brass.

FORGING BRASS NO. 377.
  M-33; 60.0 Cu, 38.0 Zn, 2.0 Pb.
  As extruded: 52,000 TS; 20,000 YS; 45 El; 32,000 shear, 78 Rock F.
  For forgings, tire valve stems.
  ASTM B 124, Alloy 2, B 283.

FORGING RUSSIAN.
  M-USSR; 53.5 Cu, 42 Zn, 4.5 Mn.
  For corrosion resisting forgings; corrosion resistant.

FORMA-1.
  M-111; 0.05 C, 0.05 Si, 0.10 Mn, bal Fe.
  For bakelite molding dies; deep hobbing, case hardened.

FORMA 2.
  M-111; 0.08 C, 0.1 Si, 0.15 Mn, bal Fe.
  For deep drawn parts; deep drawing steel.

FORMALOY.
  M-336; C, Mn, Ni, Cr, Mo, bal Fe.
  For dies, shearing tools; oil hardening.

FORMALOY.
  M-815; 3 Cu, 0.4 Mg, bal Zn.
  Aged: 40,300 TS; 33,100 YS; 2.8 El; 103 Brin.
  Cast: 32,300 TS; 4.7 El; 90 Brin.
  Chilled: 36,000 TS; 2.7 El; 93 Brin.
  For dies for forming aluminum sheets, drop hammer punches and dies; die casting alloy. high strength.

FORMBRITE.
  M-8; Zn, bal Cu.
  For deep drawn parts; good formability.

FORMDIE.
  M-35; 0.51 C, 0.45 Mn, 1.0 Si, 5.2 Cr, 1.5 Ni, 1.05 V, 1.4 Mo, bal Fe.
  For shear blades, extrusion dies, inserts, heavy punches.
  Type A9 air hardening tool steel.

FORMITE NO. 21.
  M-35; 0.33 C, 9.25 W, 3.3 Cr, 0.5 V, bal Fe.
  For hot temperature springs, forging dies, forming dies, gripping dies.
  Type H21 hot work tool steel.

FORMITE NO. 24.
  M-35; 0.51 C, 15.0 W, 3.0 Cr, 0.5 V, bal Fe.
  For die casting dies, punches, piercing tools, blanking dies.
  Type H24 hot work tool steel.

FORMOLD.
  M-38; 0.07 max C, 0.55 Ni, 1.35 Cr, 0.2 Mo, bal Fe.
  For plastic mold dies, cold hubbed; oil hardened, high core strength.

FORTAL.
  M-807; 94.3 Al, 4.0 Cu, 0.5 Mg, 0.5 Mn, 0.7 Si.
  For light alloy parts; age-hardenable.

FORTE-50 M.
M-1766; 0.50 C, 0.70 Mn, 0.40 max Si, bal Fe.
For hand tools as hammers, hatchets, etc.
DIN C45W3.

FORTE-55M.
M-1766; 0.54 C, 0.70 Mn, 0.40 max Si, bal Fe.
For hand tools, axes, hammers.

FORTE-65.
M-1766; 0.65 C, 0.25 Mn, 0.25 max Si, bal Fe.
Water hardening tool steel.
IHA F-512; AFNOR XC-60; AISI W1.

FORTE-65 M.
M-1766; 0.65 C, 0.65 Mn, 0.40 max Si, bal Fe.
For hand tools, saws, knives, wood-working tools.
DIN C67W3.

FORTE-75.
M-1766; 0.75 C, 0.25 Mn, 0.25 max Si, bal Fe.
Water hardening tool steel, as carpenter tools, hand tools.
IHA-513; AFNOR XC-75.

FORTE-75M.
M-1766; 0.75 C, 0.80 Mn 0.40 max Si, bal Fe.
For hand tools, farm tools, chisels, DIN C75W3.

FORTE-90.
M-1766; 0.88 C, 0.25 Mn, 0.25 max Si, bal Fe.
Water hardening tool steel.
IHA F-514; AFNOR XC-85; UNI UC-85.

FORTE-100.
M-1766; 0.97 C, 0.25 Mn, 0.25 max Si, bal Fe.
Water hardening tool steel, for hand tools.
IHA F-515; AFNOR XC100; AISI W1.

FORTE-115.
M-1766; 1.12 C, 0.25 Mn, 0.25 max Si, bal Fe.
Water hardening tool steel.
IHA F-516; DIN C110W1; AISI W1.

FORTE-130.
M-1766; 1.30 C, 0.25 Mn, 0.25 Max Si, bal Fe.
Water hardening tool steel.
IHA F-517; AFNOR XC-120.

FORTEX-4W.
M-1766; 1.35 C, 0.25 Mn, 0.25 Si, 0.70 Cr, 3.75 W, 0.20 V, bal Fe.
For broaches, finish turning tools.
IHA F-531; DIN 130W19.

FORTEX-C.
M-1766; 0.90 C, 0.30 Mn, 0.40 max Si, 0.75 Cr, 0.12 V, bal Fe.
Cold work tool steel; for stamping and coining dies.

FORTEX-W.
M-1766; 1.15 C, 0.25 Mn, 0.20 Si, 1.0 W, bal Fe.
For files, saws, threading tools.
IHA F-532; AFNOR 100 WC 15-04.

FOR TEN 50.
M-747.
HSLA steel, semi-killed or killed.
Meets SAE 950.

FOR TEN 60.
M-747.
HSLA steel; meets SAE 960.

FORTINOX.
M-1724; Ni-Cr-Mn.
Semi-austenitic stainless steel.

FORTISSIMUS.
M-615; 0.7 C, 18 W, 4 Cr, 1 V, bal Fe.
For cutters, tools; high speed steel.

FORTIWELD.
M-1724; 0.10-0.16 C, 0.4-0.6 Mo, 0.6 max Mn, 0.4 max Si, 0.005 max B, bal Fe.
Normalized: 97,500-88,000 TS; 75,000-69,000 YS; 26-17 El; 64 RA.
For bridge cranes, railroad cars, material handling equipment; good formability and weldability.

FORTIWELD PRESSURE VESSEL STEEL.
M-1724; 0.10-0.17 C, 0.10-0.40 Si, 0.40-0.80 Mn, 0.40-0.60 Mo, 0.25 max Cr, 0.30 max Ni, 0.30 max Cu, 0.001-0.005 B, bal Fe.
Weldable; for high strength, elevated temperature pressure vessels.

FORTIWELD STRUCTURAL STEEL.
M-1724; 0.17 max C, 0.40 max Si, 0.80 max Mn, 0.60 max Mo, 0.005 max B, bal Fe.
High yield strength structural steel; weldable.

FORT PITT VANADIUM.
M-275; 1.0 C, 0.2 V, bal Fe.
For tools, drills, taps; water hardened.

FORTUNA 12M.
M-1307; 1.2 C, 12 Mn, bal Fe.
For crusher jaws, impact tools, liners, grab-teeth; wear resistant, work hardened.

FORTUNA A12.
M-1307; 0.10 max C, 12.5 Cr, 12 Ni, bal Fe.
Annealed: 72,000-92,000 TS; 30,000 YS; 60 El; 115-150 Brin.
For clock cases, chemical plant equipment; EN58D Special; stainless, good cold workability.

FORTUNA A18.
M-1307; 0.15 max C, 18 Cr, 9 Ni, bal Fe.
Annealed: 78,000-108,000 TS; 36,000 YS; 55 El; 130-180 Brin.
For chemical plant equipment, vessels, tanks, agitators; Type 302; stainless, austenitic.

FORTUNA A18Z.
M-1307; 0.10 max C, 18 Cr, 10 Ni, Ti, bal Fe.
Annealed: 78,000-108,000 TS; 38,000 YS; 50 El; 140-190 Brin.
For welded chemical plant equipment, tanks, vessels; Type 321; stainless, austenitic.

FORTUNA A18ZN.
M-1307; 0.10 max C, 18 Cr, 10 Ni, Cb + Ta, bal Fe.
Annealed: 78,000-108,000 TS; 38,000 YS; 45-50 El; 140-190 Brin.
For acid resistant welded chemical plant equipment; Type 347; stainless, austentic.

FORTUNA A182Z.
M-1307; 0.10 max C, 18 Cr, 11 Ni, 2.3 Mo, Ti, bal Fe.
Annealed: 78,000-108,000 TS; 38,000 YS; 50 El; 140-190 Brin.
For acid resistant chemical plant equipment; mixers; Type 316 + Ti; stainless, austenitic.

FORTUNA A182ZN.
M-1307; 0.10 max C, 18 Cr, 11 Ni, 2.3 Ti, Cb + Ta, bal Fe.
Annealed: 78,000-108,000 TS; 38,000 YS; 45 El; 140-190 Brin.
For acid resistant chemical plant equipment; Type 316 + Cb; stainless, austenitic.

FORTUNA A183Z.
M-1307; 0.10 max C, 18 Cr, 12 Ni, 2.8 Mo, Ti, bal Fe.
Annealed: 78,000-108,000 TS; 38,000 YS; 50 El; 140-190 Brin.
For chemical and pharmaceutical equipment; EN58J Special; stainless, austenitic.

FORTUNA A183ZN.
M-1307; 0.10 max C, 18 Cr, 12 Ni, 2.8 Mo, Ta + Cb, bal Fe.
Annealed: 78,000-108,000 TS; 38,000 YS; 45 El; 140-190 Brin.
For pharmaceutical and chemical equipment, resists sulfide lye; Type 318; stainless, austenitic.

FORTUNA A2182ZN.
M-1307; 0.06 max C, 18 Cr, 18 Ni, 2.3 Mo, Cu, Ta, Cb, bal Fe.
Annealed: 78,000-108,000 TS; 34,000 YS; 45 El; 140-190 Brin.
For equipment for pickling and sulfuric acid plants; acid resistant, austenitic.

FORTUNA AS18.
M-1307; 0.06 max C, 18 Cr, 10 Ni, bal Fe.
Annealed: 72,000-101,000 TS; 32,000 YS; 55 El; 130-180 Brin.
For chemical plant equipment, tanks, breweries; Type 304; stainless, austenitic.

FORTUNA AS175.
M-1307; 0.06 max C, 17 Cr, 14 Ni, 4.5 Mo, bal Fe.
Annealed: 78,000-108,000 TS; 34,000 YS; 50 El; 140-190 Brin.
For equipment for salt works and soda plants, cooling coils; Type 317; stainless, austenitic.

FORTUNA AS182.
M-1307; 0.06 max C, 18 Cr, 11 Ni, 2.3 Mo, bal Fe.
Annealed: 78,000-108,000 TS; 32,000 YS; 45 El; 140-180 Brin.
For chemical plant equipment, agitators, mixers; tanks; Type 316; stainless, austenitic.

FORTUNA AS183.
M-1307; 0.06 max C, 18 Cr, 12 Ni, 2.8 Mo, bal Fe.
Annealed: 78,000-108,000 TS; 32,000 YS; 50 El; 140-180 Brin.
For acid resistant chemical and textile plant equipment; Type 316; stainless, austenitic.

FORTUNA BS 45.
M-1307; 0.4 C, 1.4 Cr, 4 Ni, 0.5 W, bal Fe.
Annealed: 225 Brin.
For plastic mold dies; oil or air hardened.

FORTUNA BS50.
M-1307; 0.50 C, 1.1 Cr, 3.5 Ni, bal Fe.
For cutlery, blanking dies, hobbing dies, punches; air or oil hardened, shock resistant.

FORTUNA C 1215.
M-1307; 1.65 C, 12 Cr, 0.1 V, bal Fe.
For blanking and drawing dies, piercing punches; air hardened, non-deforming.

FORTUNA C 1215 SUPRA.
M-1307; 1.65 C, 12.5 Cr, 1.4 Co, 1.2 Mo, bal Fe.
For blanking and forming dies, die casting dies; air hardened, non-deforming.

FORTUNA C 1220.
M-1307; 2.1 C, 12 Cr, bal Fe.
For drawing blanking dies, thread rolling dies; oil hardened, non-deforming.

FORTUNA CA 1215.
M-1307; 1.65 C, 12 Cr, 0.8 Mo, 0.5 W, bal Fe.
For blanking and forming dies, die casting dies; air hardened, non-deforming.

FORTUNA CA 1220.
M-1307; 2.1 C, 12 Cr, 0.1 W, bal Fe.
For blanking and forming dies, punches, shears; oil hardened, non-deforming.

FORTUNA CA 1220 SUPRA.
M-1307; 2.1 C, 12 Cr, 0.1 W, Co, Mo, bal Fe.
For blanking and forming dies, shearing knives; oil hardened, non-deforming.

FORTUNA CO300.
M-1307; 0.8 C, 3 Co, 12 W, 4.2 Cr, 0.5 Mo, 1.8 V, bal Fe.
For lathe and planer tools, reamers, broaches; high speed steel.

FORTUNA CO500.
M-1307; 0.8 C, 5 Co, 18 W, 4.2 Cr, 0.7 Mo, 1.5 V, bal Fe.
For lathe and planer tools, drills, taps, reamers; high speed steel.

FORTUNA CO1000.
M-1307; 0.75 C, 10 Co, 18 W, 4.2 Cr, 0.7 Mo, 1.5 V, bal Fe.
For cutting tools, broaches, reamers; high speed steel.

FORTUNA CSV 4.
M-1307; 0.30 C, 1.5 Si, 1.5 Cr, 0.1 V, bal Fe.
For shear blades, dies, stamps, piercers; hot work steel, water hardened.

FORTUNA CSV5.
M-1307; 0.45 C, 1.5 Si, 1.5 Cr, 0.1 V, bal Fe.
Anneald: 225 Brin.
For chisels, shear blades, punches, pneumatic tools; oil hardened, shock resistant.

FORTUNA CSV6.
M-1307; 0.60 C, 1.0 Si, 1.2 Cr, 0.10 V, bal Fe.
Annealed: 235 Brin.
For punches, blanking dies, cold shear blades; oil hardened, shock resistant.

FORTUNA DMO10.
M-1307; 0.24 C, 0.25 Si, 0.60 Mn, 1.2 Cr, 0.25 Mo, bal Fe.
Heat treated: 85,000-115,000 TS; 65,000-72,000 YS; 19-21 El.
For bolts and nuts, oil refinery and chemical plant equipment; creep resistant to 500°C.

FORTUNA DMO11.
M-1307; 0.15 C, 0.25 Si, 0.60 Mn, 0.30 Mo, bal Fe.
Heat treated: 62,000-78,000 TS; 38,000 YS; 24 El.
For flanges, welded collars, oil refinery equipment; creep resistant to 500°C.

FORTUNA DMO14.
M-1307; 0.13 C, 0.25 Si, 0.60 Mn, 1.0 Cr, 0.45 Mo, bal Fe.
Heat treated: 62,000-78,000 TS; 41,000 YS; 24 El.
For flanges, welded collars, oil refinery equipment; creep resistant to 500°C.

FORTUNA DMO20.
M-1307; 0.24 C, 0.25 Si, 0.45 Mn, 1.4 Cr, 0.55 Mo, 0.2 V, bal Fe.
Heat treated: 100,000-135,000 TS; 78,000-85,000 YS; 15-19 El.
For bolts, nut, oil refinery and chemical plant equipment; creep resistant to 500°C.

FORTUNA DMO22.
M-1307; 0.21 C, 0.47 Si, 0.40 Mn, 1.35 Cr, 1.1 Mo, 0.3 V, bal Fe.
Heat treated: 101,000-121,000 TS; 78,500 YS; 19 El.
For bolts and nuts, oil refinery and chemical plant equipment; creep resistant to 500°C.

FORTUNA EC3.
M-1307; 0.15 C, 0.25 Si, 0.50 Mn, 0.60 Cr, bal Fe.
Heat treated: 85,000-120,000 TS; 58,000 YS; 16 El; 45 RA.
For gears, pinions, camshafts; case hardened.

FORTUNA ECMO2H.
M-1307; 0.20 C, 0.35 max Si, 0.70 Mn, 0.60 Cr, 0.35 Mo, bal Fe.
Heat treated: 128,000-170,000 TS; 92,000 YS; 12 El; 45 RA.
For gears, pinions, camshafts, cams; case hardened.

FORTUNA ECMO4.
M-1307; 0.15 C, 0.35 max Si, 0.95 Mn, 1.15 Cr, 0.25 Mo, bal Fe.
Heat treated: 121,000-157,000 TS; 92,000 YS; 12 El; 40 RA.
For gears, pinions, camshafts; case hardened.

FORTUNA ECMO5.
M-1307; 0.20 C, 0.35 max Si, 1.0 Mn, 1.2 Cr, 0.25 Mo, bal Fe.
Heat treated: 156,000-192,000 TS; 107,000 YS; 9 El; 30 RA.
For gears, cams, camshafts, bolts, fasteners; case hardened, tough.

FORTUNA EMC5.
M-1307; 0.16 C, 0.25 Si, 1.2 Mn, 1.0 Cr, bal Fe.
Heat treated: 114,000-157,000 TS; 85,000 YS; 12 El; 40 RA.
For gears, cams, camshafts, crankshafts; case hardened, tough.

FORTUNA EMC5H.
M-1307; 0.20 C, 0.25 Si, 1.3 Mn, 1.2 Cr, bal Fe.
Heat treated: 144,000-186,000 TS; 101,000 YS; 10 El; 35 RA.
For gears, bolts, crankshafts; case hardened, tough.

FORTUNA ENC6.
M-1307; 0.13 C, 0.35 max Si, 0.50 max Mn, 0.20 max Cr, 1.5 Ni, bal Fe.
Heat treated: 85,000-115,000 TS; 58,000 YS; 12 El; 50 RA.
For gears, pinions, camshafts, cams, bolts; case hardened, shock resistant.

FORTUNA ENC10.
M-1307; 0.13 C, 0.35 max Si, 0.50 max Mn, 0.75 Cr, 2.5 Ni, bal Fe.
Heat treated: 114,000-144,000 TS; 78,000 YS; 17 El; 50 RA.
For gears, pinions, camshafts, cams, fasteners; case hardened, shock resistant.

FORTUNA ENC14.
M-1307; 0.13 C, 0.35 max Si, 0.50 max Mn, 0.75 Cr, 3.5 Ni, bal Fe.
Heat treated: 128,000-170,000 TS; 92,000 YS; 11 El; 45 RA.
For gears, pinions, camshafts, cams; case hardened, tough.

FORTUNA ENC18.
M-1307; 0.13 C, 0.35 max Si, 0.50 max Mn, 1.1 Cr, 4.5 Ni, bal Fe.
Heat treated: 177,000-200,000 TS; 128,000 YS; 9 El; 40 RA.
For gears, shafts, camshafts, cams; case hardened, tough.

FORTUNA EW 5H.
M-1307; 0.20 C, 1.3 Mn, 1.2 Cr, bal Fe.
Annealed: 217 Brin.
For plastic mold dies; case hardened.

FORTUNA EW 15 SPECIAL.
M-1307; 0.19 C, 1.3 Cr, 0.20 Mo, 4 Ni, bal Fe.
Annealed: 245 Brin.
For plastic mold dies; case hardened.

FORTUNA EW 52H.
M-1307; 0.20 C, 1 Mn, 1.2 Cr, 0.25 Mo, bal Fe.
Annealed: 217 Brin.
For plastic mold dies; case hardened.

FORTUNA EWX 50.
M-1307; 0.07 max C, 5 Cr, 1 Mo, 0.2 V, bal Fe.
Annealed: 140 Brin.
For plastic mold dies; case hardened.

FORTUNA EX8H.
M-1307; 0.20 C, 0.25 Si, 0.80 Mn, 0.5 Cr, 0.6 Ni, 0.2 Mo, bal Fe.
Heat treated: 135,000-177,000 TS; 92,000 YS; 11 El; 40 RA.
For gears, fasteners, camshafts, cams; case hardened.

FORTUNA EX15.
M-1307; 0.15 C, 0.25 Si, 0.50 Mn, 1.5 Cr, 1.5 Ni, bal Fe.
Heat treated: 128,000-170,000 TS; 92,000 YS; 11 El; 40 RA.
For gears, fasteners, camshafts, cams, bolts; case hardened.

FORTUNA EX17.
M-1307; 0.15 C, 0.25 Si, 0.50 Mn, 1.7 Cr, 1.5 Ni, 0.3 Mo, bal Fe.
Heat treated: 156,000-192,000 TS; 107,000 YS; 10 El; 40 RA.
For gears, axles, shafts, cams, camshafts; case hardened, tough.

FORTUNA EX20.
M-1307; 0.18 C, 0.25 Si, 0.50 Mn, 2.0 Cr, 2.0 Ni, bal Fe.
Heat treated: 170,000-206,000 TS; 114,000 YS; 9 El; 35 RA.
For machine tool parts, gears, crankshafts, camshafts; case hardened, tough.

FORTUNA F13.
M-1307; 0.10 max C, 13 Cr, bal Fe.
Heat treated: 85,000-107,000 TS; 63,000 YS; 26 RA; 170-210 Brin.
For cutlery, fittings, appliances, pump and valve parts; corrosion resistant; Type 410.

FORTUNA F17.
M-1307; 0.08 C, 17 Cr, bal Fe.
Annealed: 62,000-85,000 TS; 43,000 YS; 24 El; 140-180 Brin.
For cutlery, fittings, kitchen appliances and sinks; Type 430; corrosion resistant.

FORTUNA F17A.
M-1307; 0.15 max C, 17 Cr, 0.25 Mo, S, bal Fe.
Heat treated: 100,000-123,000 TS; 63,000 YS; 16 El; 190-240 Brin.
For screw machining products, screws, bolts, gears; free-cutting, corrosion resistant.

FORTUNA F17Z.
M-1307; 0.08 C, 17 Cr, Ti, bal Fe.
Annealed: 68,000-85,000 TS; 43,000 YS; 24 El; 130-170 Brin.
For welded acid resistance and chemical plant equipment; stabilized, welding grade, corrosion resistant.

FORTUNA FU.
M-1307; 1.3 C, 0.25 Mn, bal Fe.
For files, rasps; water hardened.

FORTUNA FU33.
M-1307; 0.37 C, 0.50 Mn, bal Fe.
Hot rolled: 85,000 TS; 54,000 YS; 30 El; 53 RA; 185 Brin.
For files, rasps; case hardened tools.

FORTUNA FU43.
M-1307; 0.47 C, 0.67 Mn, bal Fe.
Hot rolled: 98,000 TS; 60,000 YS; 24 El; 54 RA; 212 Brin.
For files, rasps; water hardened.

Section I: Alloy Data

FORTUNA GB0.
M-1307; 0.10 C, 0.25 Si, 0.40 Mn, 0.035 max S and P, bal Fe.
Heat treated: 60,000-74,000 TS; 36,000 YS; 22 El; 55 RA.
For machine and construction parts, fasteners; case hardened.

FORTUNA GB1.
M-1307; 0.15 C, 0.25 Si, 0.40 Mn, 0.035 max S and P, bal Fe.
Heat treated: 72,000-94,000 TS; 42,000 YS; 19 El; 50 RA.
For bolts, levers, journals, gears; case hardened.

FORTUNA GB2.
M-1307; 0.22 C, 0.25 Si, 0.45 Mn, 0.035 max S and P, bal Fe.
Heat treated: 72,000-92,000 TS; 42,000-52,000 YS; 24-25 El; 45-50 RA.
For machine tool parts, gears, shafts, bolts; water hardened.

FORTUNA GB3.
M-1307; 0.35 C, 0.25 Si, 0.55 Mn, 0.035 max S and P, bal Fe.
Heat treated: 78,000-115,000 TS; 47,000-61,000 YS; 19-24 El; 40-50 RA.
For machine tool parts, gears, fasteners; water hardened.

FORTUNA GB4.
M-1307; 0.45 C, 0.25 Si, 0.65 Mn, 0.035 max S and P, bal Fe.
Heat treated: 85,000-128,000 TS; 52,000-67,000 YS; 17-21 El; 35-45 RA.
For pinions, bolts, transmission gears, shafts; for surface hardening.

FORTUNA GB5H.
M-1307; 0.56 C, 0.30 Si, 0.55 Mn, bal Fe.
Heat treated: 92,000-135,000 TS; 58,000-76,000 YS; 16-19 El; 35-40 RA.
For pistons, machine tool parts; for surface hardening.

FORTUNA GB6.
M-1307; 0.60 C, 0.25 Si, 0.65 Mn, 0.035 max S and P, bal Fe.
Heat treated: 100,000-150,000 TS; 60,000-80,000 YS; 15-18 El; 30-40 RA.
For clutch levers, gears, axles, shafts; water hardened.

FORTUNA GB6H.
M-1307; 0.68 C, 0.32 Si, 0.7 Mn, 0.040 max S and P, bal Fe.
Annealed: 98,000 TS.
For shafts, piston rods, valve stems; wear resistant, oil or water hardened.

FORTUNA GB9.
M-1307; 0.90 C, 0.32 Si, 1.1 Mn, 0.040 max S and P, bal Fe.
Annealed: 145,000 TS; 85,000 YS; 12 El.
Heat treated: 242,000 TS; 199,000 YS; 7 El.
For molds for briquette presses; wear resistant, oil hardened.

FORTUNA GB10.
M-1307; 0.10 C, 0.25 Si, 0.40 Mn, 0.045 max S and P, bal Fe.
Heat treated: 60,000-74,000 TS; 36,000 YS; 22 El; 55 RA.
For bushings, bolts, levers, gears; case hardened.

FORTUNA GB11.
M-1307; 0.15 C, 0.25 Si, 0.40 Mn, 0.045 max S and P, bal Fe.
Heat treated: 72,000-94,000 TS; 42,000 YS; 19 El; 50 RA.
For machine and construction parts, gears, bolts; case hardened.

FORTUNA GB12.
M-1307; 0.22 C, 0.25 Si, 0.45 Mn, 0.045 max S and P, bal Fe.
Heat treated: 60,000-92,000 TS; 33,000-56,000 YS; 24-30 El; 40-45 RA.
For machine tools parts, gears, bolts, shafts; water hardened.

FORTUNA GB13.
M-1307; 0.35 C, 0.25 Si, 0.55 Mn, 0.045 max S and P, bal Fe.
Heat treated: 72,000-114,000 TS; 40,000-62,000 YS; 19-25 El; 35-45 RA.
For shafts, gears, axles, machine tool parts; oil or water hardened.

FORTUNA GB14.
M-1307; 0.45 C, 0.25 Si, 0.65 Mn, 0.045 max S and P, bal Fe.
Heat treated: 85,000-128,000 TS; 49,000-70,000 YS; 17-21 El; 30-40 RA.
For shafts, axles, gears, bolts; oil or water hardened.

FORTUNA GB16.
M-1307; 0.60 C, 0.25 Si, 0.65 Mn, 0.045 max S and P, bal Fe.
Heat treated: 100,000-150,000 TS; 56,000-80,000 YS; 15-18 El; 25-35 RA.
For axles, shafts, spindles, bolts, gears; oil or water hardened.

FORTUNA GFD2.
M-1307; 0.38 C, 1.5 Si, 0.65 Mn, bal Fe.
Heat treated: 170,000-200,000 TS; 150,000 YS; 7-10 El; 30 RA; 350-400 Brin.
For railway and auto springs; water hardened.

FORTUNA GFD3.
  M-1307; 0.48 C, 0.25 Si, 1.75 Mn, 0.050 max S and P, bal Fe.
  Heat treated: 186,000-213,000 TS; 157,000 YS; 7-10 El; 30 RA; 380-435 Brin.
  For railway and auto springs; oil hardened.

FORTUNA GFD4.
  M-1307; 0.46 C, 1.65 Si, 0.65 Mn, 0.050 max S and P, bal Fe.
  Heat treated: 186,000-213,000 TS; 157,000 YS; 7-10 El; 30 RA; 380-435 Brin.
  For railway and auto springs; water hardened.

FORTUNA GFD5.
  M-1307; 0.66 C, 1.65 Si, 0.70 Mn, 0.050 max S and P, bal Fe.
  Heat treated: 186,000-213,000 TS; 157,000 YS; 7-10 El; 30 RA; 380-435 Brin.
  For railway and auto springs; oil hardened.

FORTUNA GFD5W.
  M-1307; 0.56 C, 1.65 Si, 0.70 Mn, 0.050 max S and P, bal Fe.
  Heat treated: 186,000-213,000 TS; 157,000 YS; 7-10 El; 30 RA; 380-435 Brin.
  For railway and auto springs; oil hardened.

FORTUNA GFD7.
  M-1307; 0.65 C, 1.10 Si, 1.0 Mn, 0.050 max S and P, bal Fe.
  Heat treated: 190,000-225,000 TS; 157,000 YS; 7-10 El; 30 RA; 390-450 Brin.
  For railway and auto springs; oil hardened.

FORTUNA GFD8.
  M-1307; 0.67 C, 1.3 Si, 0.50 Mn, 0.50 Cr, bal Fe.
  Annealed: 112,000 TS; 65,000 YS; 16 El.
  Heat treated: 232,000 TS; 190,000 YS; 8 El.
  For highly stressed springs; wear resistant, heavy duty.

FORTUNA GFD8R.
  M-1307; 0.60 C, 1.1 Si, 1.0 Mn, 0.50 Cr, bal Fe.
  Heat treated: 200,000-238,000 TS; 177,000 YS; 7-10 El; 30 RA; 410-435 Brin.
  For highly stressed springs; oil hardened.

FORTUNA GFD530.
  M-1307; 0.30 C, 0.22 Si, 0.30 Mn, 2.35 Cr, 0.6 V, 4.35 W, bal Fe.
  Heat treated: 202,000-248,000 TS; 177,000 YS; 6-9 El; 30 RA.
  For high temperature springs; oil hardened, heat resistant.

FORTUNA GK.
  M-1307; 1.05 C, 0.25 Si, 0.32 Mn, 0.50 Cr, bal Fe.
  Annealed: 215 Brin.
  For ball bearings, roller and needle bearings; oil or water hardened.

FORTUNA GKL.
  M-1307; 1.0 C, 0.25 Si, 0.32 Mn, 1.5 Cr, bal Fe.
  Annealed: 215 Brin.
  For ball bearings, ball races and discs; oil hardened.

FORTUNA GKL0.
  M-1307; 1.0 C, 0.25 Si, 0.32 Mn, 1.5 Cr, bal Fe.
  Annealed: 207 Brin.
  Heat treated: 600 Brin.
  For ball and roller bearings; wear resistant, heavy duty.

FORTUNA GMKL.
  M-1307; 1.0 C, 0.60 Si, 1.1 Mn, 1.5 Cr, bal Fe.
  Annealed: 226 Brin.
  For ball bearing races; oil or water hardened.

FORTUNA GRL.
  M-1307; 1.05 C, 0.25 Si, 0.32 Mn, 1.0 Cr, bal Fe.
  Annealed: 215 Brin.
  For ball bearings, balls and rollers, races; oil or water hardened.

FORTUNA GSB7.
  M-1307; 0.70 C, 0.30 Mn, bal Fe.
  Heat treated: 174,000 TS; 128,000 YS; 12 El; 37 RA; 335 Brin.
  For stone working tools; water hardened.

FORTUNA GSB8.
  M-1307; 0.85 C, 0.30 Mn, bal Fe.
  Heat treated: 185,000 TS; 142,000 YS; 12 El; 32 RA; 390 Brin.
  For stone working tools; water hardened.

FORTUNA GSB10.
  M-1307; 1.0 C, 0.25 Mn, bal Fe.
  Annealed: 100,000 TS; 53,000 YS; 21 El; 42 RA; 200 Brin.
  For stone working tools; water hardened.

FORTUNA GSB11.
  M-1307; 1.1 C, 0.25 Mn, bal Fe.
  Annealed: 110,000 TS; 56,000 YS; 19 El; 40 RA; 210 Brin.
  For stone working tools; water hardened.

FORTUNA GSB63.
  M-1307; 0.62 C, 0.70 Mn, bal Fe.
  Heat treated: 160,000 TS; 113,000 YS; 12 El; 40 RA; 325 Brin.
  For stone working tools; water hardened.

FORTUNA GSP5.
  M-1307; 0.53 C, 1 Si, 1 Mn, bal Fe.
  Rolled: 235-290 Brin.
  For hammer and press saddles, trimming tools; hot work steel, oil hardened.

FORTUNA GSP6.
  M-1307; 0.65 C, 1.0 Si, 1.0 Mn, bal Fe.
  For spring collets, screw drivers, clamping jaws; oil hardened.

FORTUNA GSP8.
M-1307; 0.67 C, 1.31 Si, 0.50 Cr, bal Fe.
For spring collets, screw drivers, clamping jaws; oil hardened.

FORTUNA HSB1.
M-1307; 0.85 C, bal Fe.
For mechanical and hard wood saws; water hardened; Type W1.

FORTUNA HSB2.
M-1307; 0.75 C, 0.70 Mn, bal Fe.
Heat treated: 180,000 TS; 140,000 YS; 14 El; 38 RA; 375 Brin.
For agricultural equipment, saws, cement scrapers; oil or water hardened.

FORTUNA HSB4.
M-1307; 0.80 C, 0.50 Cr, 0.4 V, bal Fe.
For wood working tools, circular saws; oil or water hardened.

FORTUNA M13.
M-1307; 0.40 C, 13 Cr, bal Fe.
Heat treated: 540-560 Brin.
For cutlery, knives, springs, dies, instruments; corrosion resistant, hardenable.

FORTUNA M131.
M-1307; 0.50 C, 14 Cr, 0.5 Mo, bal Fe.
Heat treated: 550-570 Brin.
For cutlery, knives, surgical instruments; corrosion resistant, hardenable.

FORTUNA M171.
M-1307; 0.90 C, 18 Cr, 1.2 Mo, V, bal Fe.
Heat treated: 560-580 Brin.
For gears, bearings, cutters; Type 440B; hardenable, corrosion resistant.

FORTUNA M171H.
M-1307; 1.1 C, 17 Cr, 1 Mo, V, bal Fe.
Heat treated: 590-620 Brin.
For bearing rolls and rings, cutters; Type 440C; hardenable, corrosion resistant.

FORTUNA M-2171.
M-1307; 0.90 C, 17 Cr, 0.5 Mo, V, Co, bal Fe.
Heat treated: 580-600 Brin.
For bearing rolls and races, cutters; hardenable, corrosion resistant.

FORTUNA MO 500.
M-1307; 0.85 C, 5 Mo, 1.8 V, 6.5 W, 4 Cr, bal Fe.
For planing and turning tools, milling cutters, saws; high speed steel.

FORTUNA MO 503.
M-1307; 1.2 C, 5 Mo, 3.3 V, 6.5 W, 4.5 Cr, bal Fe.
For drills, reamers, broaches, saw teeth; high speed steel.

FORTUNA MO 550.
M-1307; 0.8 C, 5 Co, 5 Mo, 1.8 V, 6.5 W, 4.25 Cr, bal Fe.
For heavy milling and planing cutters; high speed steel.

FORTUNA MOG 110.
M-1307; 0.45 C, 1.5 Cr, 0.7 Mo, 0.3 V, bal Fe.
For die casting dies, forging tools; hot work steel, oil hardened.

FORTUNA MOG 111.
M-1307; 0.45 C, 1.5 Cr, 0.5 Mo, 0.8 V, 0.5 W, bal Fe.
For shear blades, forging tools, gripper dies; hot work steel, oil hardened.

FORTUNA MOG 330.
M-1307; 0.30 C, 3 Cr, 3 Mo, 0.6 V, bal Fe.
For piercing mandrels, die inserts, heading dies; hot work steel, oil hardened.

FORTUNA MOG 510.
M-1307; 0.4 C, 1 Si, 5 Cr, 1.5 Mo, 0.6 V, bal Fe.
For forging dies, die inserts, gripping dies; hot work steel, oil or air hardened.

FORTUNA MOG 511.
M-1307; 0.35 C, 1 Si, 5 Cr, 1.5 Mo, 0.3 V, 1.5 W, bal Fe.
For forging dies, die inserts, shear blades; hot work steel, air or oil hardened.

FORTUNA NC6.
M-1307; 0.27 C, 1.5 Cr, 1.0 Al, bal Fe.
Heat treated: 92,000-114,000 TS; 65,000 YS; 19 El; 55 RA.
For gears, shafts, camshafts, cams; nitriding steel.

FORTUNA NC6H.
M-1307; 0.34 C, 1.5 Cr, 1.0 Al, bal Fe.
Heat treated: 114,000-144,000 TS; 85,000 YS; 15 El; 50 RA.
For gears, shafts, camshafts, cams; nitriding steel.

FORTUNA NC7 EXTRA.
M-1307; 0.33 C, 1.7 Cr, 1.0 Ni, 1.0 Al, bal Fe.
Heat treated: 114,000-144,000 TS; 85,000 YS; 17 El; 50 RA.
For gears, shafts, camshafts, cams, piston rods; nitriding steel.

FORTUNA NCM04.
M-1307; 0.32 C, 1.0 Cr, 0.2 Mo, 1.0 Al, bal Fe.
Heat treated: 114,000-135,000 TS; 85,000 YS; 15 El; 50 RA.
For gears, shafts, camshafts, cams; nitriding steel.

FORTUNA NCS5.
M-1307; 0.30 C, 1.2 Cr, 0.9 Al, 0.1 S, bal Fe.
Heat treated: 85,000-115,000 TS; 65,000 YS; 15 El; 50 RA.
For gears, shafts, camshafts; nitriding steel.

FORTUNA NCV9 SPEC.
M-1307; 0.30 C, 2.5 Cr, 0.12 Mo, 0.15 V, bal Fe.
Heat treated: 128,000-164,000 TS; 100,000-114,000 YS; 13-16 El; 45-50 RA.
For gears, shafts, cams, valve spindles, camshafts; nitriding steel.

FORTUNA NG.
M-1307; 0.55 C, 0.7 Cr, 0.2 Mo, 1.7 Ni, 0.1 V, bal Fe.
For forging dies, crankshafts, camshafts; hot work steel, oil hardened.

FORTUNA NG2 SUPRA.
M-1307; 0.55 C, 1 Cr, 0.5 Mo, 1.7 Ni, 0.1 V, bal Fe.
For forging and gripping dies, die inserts; hot work steel, oil or air hardened.

FORTUNA NG3 SUPRA.
M-1307; 0.55 C, 1 Cr, 0.7 Mo, 2.2 Ni, bal Fe.
For drop forging dies, hot mandrels; hot work steel, oil or air hardened.

FORTUNA OB7.
M-1307; 0.71 C, 0.20 Si, 0.22 Mn, bal Fe.
Heat treated: 107,000-144,000 TS; 58,000-80,000 YS; 11-15 El; 30-35 RA.
For machine tool parts, springs, hammers, axes; for surface hardening.

FORTUNA PH5B.
M-1307; 0.10 C, 2.85 Cr, bal Fe.
Heat treated: 65,000-76,000 TS; 32,000 YS; 25 El.
For pipes for oil refineries and high pressure washers; creep and heat resistant.

FORTUNA PH5C.
M-1307; 0.15 C, 2.35 Cr, 0.40 Mn, 0.25 Si, bal Fe.
Heat treated: 72,000-85,000 TS; 43,000 YS; 22 El.
For parts for core pipes for high pressure windings; creep and heat resistant.

FORTUNA PH8N.
M-1307; 0.17 C, 2.65 Cr, 0.25 Mo, 0.15 V, bal Fe.
Heat treated: 92,000-115,000 TS; 65,000 YS; 19 El.
For parts resistant to 400°C; creep and heat resistant.

FORTUNA PH9.
M-1307; 0.20 C, 3.15 Cr, 0.55 Mo, 0.5 V, bal Fe.
Heat treated: 114,000-135,000 TS; 78,000 YS; 17 El.
For parts resistant to 480°C; creep and heat resistant.

FORTUNA PH10.
M-1307; 0.21 C, 2.35 Cr, 0.40 Mo, 0.8 V, 0.37 W, bal Fe.
Heat treated: 114,000-135,000 TS; 78,000 YS; 17 El.
For parts resistant to 520°C; creep and heat resistant.

FORTUNA PH20.
M-1307; 0.15 max C, 5.0 Cr, 0.55 Mo, bal Fe.
Heat treated: 65,000-107,000 TS; 36,000-58,000 YS; 20-23 El.
For pipes for oil refinery plants; creep and heat resistant.

FORTUNA R3.
M-1307; 1.42 C, 0.3 Cr, 0.25 V, 3.0 W, bal Fe.
For thread cutters, scrapers, chisels; water hardened.

FORTUNA R5.
M-1307; 1.3 C, 0.2 Cr, 5 W, bal Fe.
For thread cutters, scrapers, chisels, engraving needles; water hardened.

FORTUNA SC 150.
M-1307; 1.45 C, 1.4 Cr, bal Fe.
For taps, milling cutters, reamers, broaches; oil hardened, wear resistant

FORTUNA SICV.
M-1307; 1.15 C, 0.70 Cr, 0.1 V, bal Fe.
For drills, taps, cutters, files, punches; oil or water hardened.

FORTUNA SIW.
M-1307; 1.2 C, 0.1 V, 1.0 W, bal Fe.
For drills, taps, cutters, files, punches; oil or water hardened.

FORTUNA SMV 200.
M-1307; 0.90 C, 2 Mn, 0.1 V, bal Fe.
For blanking tools, punches, tap drills, chasers; oil hardened, cold work steel.

FORTUNA SW 55.
M-1307; 0.95 C, 1.3 Mn, 0.5 Cr, 0.6 W, bal Fe.
For milling cutters, blanking dies; oil hardened, cold work steel.

FORTUNA SW 100.
M-1307; 1.2 C, 1.0 W, bal Fe.
For drills, taps, cutters, files, punches; oil or water hardened.

FORTUNA SW 111.
M-1307; 1.05 C, 1 Mn, 1 Cr, 1.2 W, bal Fe.
For milling cutters, swages, blanking tools, reamers; oil hardened, cold work steel.

FORTUNA T13.
M-1307; 0.20 C, 13 Cr, bal Fe.
Heat treated: 107,000-130,000 TS; 78,000 YS; 18 El; 210-250 Brin.
For pumps, valves, piston rods, cutlery; Type 420; corrosion resistant, hardenable.

FORTUNA T17.
M-1307; 0.20 C, 17 Cr, 2 Ni, bal Fe.
Heat treated: 114,000-137,000 TS; 85,000 YS; 18 El; 225-280 Brin.
For shafts, axles, valves, pump parts; Type 431; corrosion resistant to sea water.

FORTUNA T131.
M-1307; 0.20 C, 13 Cr, 1.2 Mo, bal Fe.
Heat treated: 107,000-130,000 TS; 78,000 YS; 18 El; 220-260 Brin.
For steam turbine blades, molds; corrosion resistant, hardenable.

FORTUNA T171.
M-1307; 0.35 C, 17 Cr, 1.2 Mo, bal Fe.
Heat treated: 114,000-137,000 TS; 85,000 YS; 18 El; 225-265 Brin.
For axles, spindles, valve seats and cones; corrosion resistant.

FORTUNA TC4.
M-1307; 0.34 C, 0.25 Si, 0.70 Mn, 1.0 Cr, bal Fe.
Heat treated: 114,000-170,000 TS; 78,000-114,000 YS; 13-17 El; 40-50 RA.
For aircraft and auto engine parts; oil hardened, tough.

FORTUNA TC4B.
M-1307; 0.37 C, 0.25 Si, 0.65 Mn, 1.0 Cr, bal Fe.
Heat treated: 114,000-170,000 TS; 78,000-114,000 YS; 13-17 El; 40-50 RA.
For gears, shafts, machinery parts; for surface hardening.

FORTUNA TC4 SPEC.
M-1307; 0.41 C, 0.25 Si, 0.70 Mn, 1.0 Cr, bal Fe.
Heat treated: 219,000-250,000 TS; 185,000 YS; 9 El; 30 RA.
For gears, shafts, crankshafts; oil hardened, shock resistant.

FORTUNA TC6.
M-1307; 0.36 C, 0.25 Si, 0.50 Mn, 1.5 Cr, bal Fe.
Heat treated: 100,000-135,000 TS; 65,000-78,000 YS; 17-18 El; 55-60 RA.
For levers, axles, steering parts, gears, shafts; oil hardened, tough.

FORTUNA TCM04.
M-1307; 0.34 C, 0.25 Si, 0.65 Mn, 1.0 Cr, 0.2 Mo, bal Fe.
Heat treated: 100,000-176,000 TS; 65,000-114,000 YS; 13-18 El; 45-60 RA.
For crankshafts, axles, gears, connecting rods; for surface hardening.

FORTUNA TCM04H.
M-1307; 0.42 C, 0.25 Si, 0.65 Mn, 1.0 Cr, 0.2 Mo, bal Fe.
Heat treated: 107,000-186,000 TS; 78,000 YS; 17 El; 55 RA.
For axle swivels, connecting rods, gears, shafts; oil hardened, shock resistant.

FORTUNA TCM04W.
M-1307; 0.25 C, 0.25 Si, 0.65 Mn, 1.0 Cr, 0.2 Mo, bal Fe.
Heat treated: 92,000-150,000 TS; 60,000-92,000 YS; 15-19 El; 50-65 RA.
For axles, shafts, steering parts; oil hardened, tough.

FORTUNA TCM05.
M-1307; 0.50 C, 1.0 Cr, 0.2 Mo, bal Fe.
Heat treated: 114,000-185,000 TS; 85,000-128,000 YS; 12-16 El; 40-50 RA.
For gears, shafts, crankshafts; for surface hardening.

FORTUNA TCV4.
M-1307; 0.50 C, 0.25 Si, 0.90 Mn, 1.0 Cr, 0.1 V, bal Fe.
Heat treated: 212,000-250,000 TS; 185,000 YS; 10 El; 35 RA.
For axles, levers, gears, shafts; oil hardened, shock resistant.

FORTUNA TCV5.
M-1307; 0.59 C, 0.25 Si, 1.0 Mn, 1.1 Cr, 0.10 V, bal Fe.
Heat treated: 212,000-242,000 TS; 190,000 YS; 7-10 El; 30 RA; 435-495 Brin.
For highly stressed springs; oil hardened.

FORTUNA TCV6.
M-1307; 0.42 C, 0.25 Si, 0.60 Mn, 1.5 Cr, 0.1 V, bal Fe.
Heat treated: 107,000-128,000 TS; 78,000 YS; 17 El; 55 RA.
For gears, shafts, crankshafts, axles; oil hardened, shock resistant.

FORTUNA TCV9 SPEC.
M-1307; 0.30 C, 0.25 Si, 0.55 Mn, 2.5 Cr, 0.2 Mo, 0.15 V, bal Fe.
Heat treated: 175,000-208,000 TS; 145,000-157,000 YS; 11 El; 35-40 RA.
For crankshafts, bolts, gears, screws; oil hardened, shock resistant.

FORTUNA TM4.
M-1307; 0.40 C, 0.35 Si, 1.0 Mn, bal Fe.
Heat treated: 100,000-150,000 TS; 65,000-92,000 YS; 15-18 El; 40-50 RA.
For bolts, spindles, shafts; oil or water hardened.

FORTUNA TM5.
M-1307; 0.30 C, 0.25 Si, 1.4 Mn, bal Fe.
Heat treated: 92,000-135,000 TS; 60,000-75,000 YS; 17-19 El; 45-55 RA.
For forgings, axles, shafts, gears; water hardened.

FORTUNA TMCV4.
M-1307; 0.27 C, 0.25 Si, 1.1 Mn, 0.7 Cr, 0.1 V, bal Fe.
Heat treated: 92,000-150,000 TS; 60,000-92,000 YS; 15-19 El; 50-65 RA.
For crankshafts, axles, levers, gears; oil hardened, tough.

FORTUNA TMS4.
M-1307; 0.53 C, 0.90 Si, 0.90 Mn, bal Fe.
Heat treated: 100,000-170,000 TS; 65,000-114,000 YS; 13-18 El; 35-50 RA.
For gears, pinions, camshafts, connecting rods; for surface hardening.

FORTUNA TMS5.
M-1307; 0.37 C, 1.25 Si, 1.25 Mn, bal Fe.
Heat treated: 100,000-170,000 TS; 65,000-114,000 YS; 13-18 El; 35-50 RA.
For shafts, gears; tough and abrasion resistant.

FORTUNA TMV7.
M-1307; 0.42 C, 0.25 Si, 1.8 Mn, 0.1 V, bal Fe.
Heat treated: 127,000-186,000 TS; 100,000-128,000 YS; 12-15 El; 30-40 RA.
For axles, shafts, gears, bolts, connecting rods; oil hardened, tough.

FORTUNA TNC6H.
M-1307; 0.35 C, 0.30 Si, 0.50 Mn, 0.50 Cr, 1.5 Ni, bal Fe.
Heat treated: 108,000-121,000 TS; 74,000 YS; 18 El; 50 RA.
For axle swivels, axles, shafts, gears; oil hardened, shock resistant.

FORTUNA TNC6W.
M-1307; 0.30 C, 0.30 Si, 0.50 Mn, 0.50 Cr, 1.5 Ni, bal Fe.
Heat treated: 92,000-108,000 TS; 61,000 YS; 21 El; 50 RA.
For construction and machine parts, shafts, gears; oil hardened, shock resistant.

FORTUNA TNC10H.
M-1307; 0.35 C, 0.30 Si, 0.50 Mn, 0.75 Cr, 2.5 Ni, bal Fe.
Heat treated: 114,000-135,000 TS; 78,000 YS; 12 El; 50 RA.
For wheels, hubs, axles, swivels, gears; oil hardened, tough.

FORTUNA TNC10W.
M-1307; 0.30 C, 0.30 Si, 0.50 Mn, 0.75 Cr, 2.5 Ni, bal Fe.
Heat treated: 101,000-121,000 TS; 72,000 YS; 17 El; 50 RA.
For wheel hubs, axle swivels, gears, shafts; oil hardened, shock resistant.

FORTUNA TNC14H.
M-1307; 0.30 C, 0.30 Si, 0.50 Mn, 0.75 Cr, 3.5 Ni, bal Fe.
Heat treated: 128,000-150,000 TS; 101,000 YS; 12 El; 50 RA.
For axles, shafts, levers, gears, connecting rods; oil hardened, shock resistant.

FORTUNA TNC14W.
M-1307; 0.25 C, 0.30 Si, 0.50 Mn, 0.75 Cr, 3.5 Ni, bal Fe.
Heat treated: 108,000-128,000 TS; 78,000 YS; 17 El; 50 RA.
For axles, shafts, levers, gears, connecting rods; oil hardened, shock resistant.

FORTUNA TNC18.
M-1307; 0.35 C, 0.30 Si, 0.50 Mn, 1.3 Cr, 4.5 Ni, bal Fe.
Heat treated: 144,000-166,000 TS; 114,000 YS; 11 El; 50 RA.
For shafts, axles, axle tubes; oil hardened, shock resistant.

FORTUNA TT13.
M-1307; 0.15 C, 13 Cr, bal Fe.
Heat treated: 100,000-123,000 TS; 72,000 YS; 20 El; 190-420 Brin.
For turbine blades, cutlery; Type 420; corrosion resistant, hardenable.

FORTUNA TX10.
M-1307; 0.36 C, 0.25 Si, 0.65 Mn, 1.0 Cr, 1.0 Ni, 0.2 Mo, bal Fe.
For airplane and auto parts, axles, shafts, gears, countershafts; oil hardened, tough.

FORTUNA TX15.
M-1307; 0.34 C, 0.25 Si, 0.55 Mn, 1.5 Cr, 1.5 Ni, 0.2 Mo, bal Fe.
Heat treated: 152,000-186,000 TS; 123,000-135,000 YS; 12 El; 40-45 RA.
For gears, shafts, aircraft and auto parts, machine tool components; oil hardened, tough.

FORTUNA TX20.
M-1307; 0.30 C, 0.25 Si, 0.45 Mn, 2.0 Cr, 0.3 Mo, 2.0 Ni, bal Fe.
Heat treated: 175,000-208,000 TS; 145,000-157,000 YS; 11 El; 35-40 RA.
For aircraft and auto parts, machine tool parts, gears, shafts; subjected to high stresses, oil hardened.

FORTUNA V 300.
M-1307; 0.85 C, 12 W, 2.5 V, 4 Cr, 0.8 Mo, bal Fe.
For milling cutters, lathe and planer tools, saws; high speed steel.

FORTUNA V400.
M-1307; 1.25 C, 12 W, 4 V, 4.2 Cr, 0.8 Mo, bal Fe.
For milling cutters, lathe and planer tools, drills; high speed steel.

FORTUNA V450.
M-1307; 1.35 C, 5 Co, 12 W, 4 V, 4.2 Cr, 0.8 Mo, bal Fe.
For cutters for abrasive material; high speed.

Section I: Alloy Data / 569

FORTUNA VC12.
M-1307; 2.1 C, 0.3 Si, 0.3 Mn, 11.5 Cr, bal Fe.
Heat treated: 114,000-135,000 TS; 72,000 YS; 9 El; 10 RA.
For valve seats; heat and corrosion resistant.

FORTUNA VCN 18-8.
M-1307; 0.42 C, 2.25 Si, 18 Cr, 9 Ni, 1 W, bal Fe.
Heat treated: 114,000-144,000 TS; 58,000 YS; 27 El; 35 RA.
For exhaust valves; high oxidation and heat resistance.

FORTUNA VCN 235.
M-1307; 0.45 C, 1.15 Si, 1.05 Mn, 23 Cr, 2.7 Mo, 5 Ni, bal Fe.
Heat treated: 107,000-135,000 TS; 85,000 YS; 16 El; 25 RA.
For exhaust valves; for service up to 700°C.

FORTUNA VCS2.
M-1307; 0.45 C, 4.0 Si, 0.45 Mn, 2.65 Cr, bal Fe.
Heat treated: 128,000-151,000 TS; 101,000 YS; 16 El; 40 RA.
For exhaust and inlet valves; oil hardened.

FORTUNA VCS9.
M-1307; 0.45 C, 3.05 Si, 0.45 Mn, 9.0 Cr, bal Fe.
Heat treated: 128,000-151,000 TS; 101,000 YS; 16 El; 40 RA.
For exhaust valves; high stressed, oil hardened.

FORTUNA VCS 20.
M-1307; 0.80 C, 2.15 Si, 0.4 Mn, 20 Cr, 1.37 Ni, bal Fe.
Heat treated: 127,000-150,000 TS; 101,000 YS; 7 El.
For exhaust valves; good strength, oxidation resistant.

FORTUNA W3.
M-1307; 0.45 C, 0.70 Mn, bal Fe.
Hot rolled: 98,000 TS; 59,000 YS; 24 El; 45 RA; 212 Brin.
For hammers, axes, pliers, shears, anvils, stamps; water hardened.

FORTUNA W7 EXTRA.
M-1307; 0.70 C, bal Fe.
Heat treated: 174,000 TS; 128,000 YS; 12 El; 37 RA; 352 Brin.
For cold shears, rivet snaps, blanking tools; water hardened.

FORTUNA W7 PRIMA.
M-1307; 0.70 C, bal Fe.
Heat treated: 174,000 TS; 128,000 YS; 12 El; 37 RA; 350 Brin.
For hot and cold work tools, knives, hammers; water hardened.

FORTUNA W8 EXTRA.
M-1307; 0.85 C, bal Fe.
Heat treated: 190,000 TS; 145,000 YS; 10 El; 30 RA; 400 Brin.
For cold punches, leather stampers, gauge tools; water hardened.

FORTUNA W8N.
M-1307; 0.85 C, 0.60 Ni, 0.10 V, bal Fe.
For coining, gripping and heading dies; water hardened.

FORTUNA W8 PRIMA.
M-1307; 0.85 C, bal Fe.
Annealed: 190 Brin.
For shear blades, knives, hammer tools; Type W1; water hardened.

FORTUNA W10 EXTRA.
M-1307; 1.0 C, bal Fe.
Annealed: 95,000 TS; 50,000 YS; 23 El; 45 RA; 200 Brin.
For jaws, shear blades, snaps, cold headers; water hardened.

FORTUNA W10 PRIMA.
M-1307; 1.0 C, bal Fe.
Annealed: 100,000 TS; 53,000 YS; 21 El; 42 RA; 200 Brin.
For cold forging dies, knives, milling cutters, reamers; water hardened.

FORTUNA W10V.
M-1307; 1.0 C, 0.1 V, bal Fe.
For cold forging dies, shear blades, punches; water hardened; Type W2.

FORTUNA W11 EXTRA.
M-1307; 1.1 C, bal Fe.
Annealed: 100,000 TS; 53,000 YS; 21 El; 42 RA; 210 Brin.
For jaws, shear blades, cold heading dies; water hardened, wear resistant.

FORTUNA W11 PRIMA.
M-1307; 1.15 C, bal Fe.
Annealed: 110,000 TS; 56,000 YS; 19 El; 39 RA; 210 Brin.
For wood and leather tools, files, saws, drills; water hardened.

FORTUNA W13 PRIMA.
M-1307; 1.3 C, bal Fe.
Annealed: 210 Brin.
For lathe and planer tools, taps, draw punches; water hardened.

FORTUNA W18.
M-1307; 0.75 C, 18 W, 4 Cr, 1 V, bal Fe.
For lathe and planer tools, reamers, hobs, drills; high speed steel.

FORTUNA W23.
M-1307; 0.15 C, 0.40 Mn, bal Fe.
Annealed: 70,000 TS; 40,000 YS; 25 El; 60 RA; 130 Brin.
For plastic mold dies, rollers; water hardened.

FORTUNA W33.
M-1307; 0.35 C, 0.50 Mn, bal Fe.
Hot rolled: 85,000 TS; 54,000 YS; 30 El; 53 RA; 185 Brin.
For pliers, screw drivers, augers, forks; water hardened.

FORTUNA W63.
M-1307; 0.60 C, 0.70 Mn, bal Fe.
Heat treated: 160,000 TS; 113,000 YS; 12 El; 40 RA; 325 Brin.
For hammers, hatches, knives, vise-jaws, tool holders; water hardened.

FORTUNA W63K.
M-1307; 0.65 C, 1.0 Mn, bal Fe.
For shackles, bolts, liners, hammers; oil hardened, wear resistant.

FORTUNA W73.
M-1307; 0.70 C, 0.70 Mn, bal Fe.
Heat treated: 180,000 TS; 140,000 YS; 14 El; 38 RA; 375 Brin.
For axes, wood working tools, knives; water hardened.

FORTUNA W83K.
M-1307; 0.80 C, 1.0 Mn, bal Fe.
For shackles, bolts, liners, hammers; oil hardened, wear resistant.

FORTUNA W93.
M-1307; 0.90 C, 0.5 Mn, bal Fe.
Heat treated: 190,000 TS; 145,000 YS; 10 El; 30 RA; 400 Brin.
For cement crushers, knives, pressure plates; oil or water hardened.

FORTUNA W93K.
M-1307; 0.90 C, 1.0 Mn, bal Fe.
For shackles, bolts, liners, drag rails, hammers; oil hardened, wear resistant.

FORTUNA WA 235.
M-1307; 0.35 C, 1 Si, 1 Cr, 0.2 V, 2 W, bal Fe.
For hand chisels, cold shears, rivet snaps; hot work steel, water hardened.

FORTUNA WA 245.
M-1307; 0.45 C, 1 Si, 0.3 Mn, 1 Cr, 0.2 V, 2 W, bal Fe.
For hot piercing and trimming tools, cutters; hot work steel, oil hardened.

FORTUNA WA255.
M-1307; 0.55 C, 1.0 Si, 1.0 Cr, 0.2 V, 2.0 W, bal Fe.
For blanking dies, shear blades, punches, knives; oil hardened, tough.

FORTUNA WA 430.
M-1307; 0.30 C, 1 Si, 1 Cr, 0.2 V, 4 W, bal Fe.
For shear blades, gripping dies, mandrels; hot work steel, oil hardened.

FORTUNA WA 530.
M-1307; 0.30 C, 2.5 Cr, 0.6 V, 4.5 W, bal Fe.
For extrusion and tube press tools, liners; hot work steel, oil hardened.

FORTUNA WA 930.
M-1307; 0.30 C, 2.5 Cr, 0.4 V, 9 W, bal Fe.
For extrusion dies, liners and rams, die casting dies; hot work steel, oil or air hardened.

FORTUNA WA 2930.
M-1307; 0.30 C, 2 Co, 2.5 Cr, 0.3 V, 9 W, bal Fe.
For extrusion dies, liners, rams; hot work steel, oil hardened.

FORTUNA WC 6H.
M-1307; 0.34 C, 1 Al, 1.5 Cr, bal Fe.
Annealed: 110 Brin.
For plastic mold dies; nitrided.

FORTUNA WF8.
M-1307; 0.67 C, 1.3 Si, 0.5 Cr, bal Fe.
For crusher parts, hammer mills, brick molds; wear resistant, tough.

FORTUNA WGKL.
M-1307; 1.0 C, 1.5 Cr, bal Fe.
For blanking tools, punches, thread cutting tools; oil hardened, wear resistant.

FORTUNA WM 13.
M-1307; 0.40 C, 13 Cr, bal Fe.
Annealed: 225 Brin.
For plastic mold dies; corrosion resistant.

FORTUNA WO 3.
M-1307; 0.10 C, 0.40 Mn, bal Fe.
Annealed: 140 Brin.
For plastic mold dies; case hardened.

FORTUNA WRL.
M-1307; 0.90 C, 0.80 Cr, bal Fe.
For stamping and coining dies, cold rolls; water hardened.

FORTUNA WSB EXTRA.
M-1307; 1.15 C, 2 W, bal Fe.
For saws; oil hardened.

FORTUNA WT 131.
M-1307; 0.30 C, 13 Cr, 0.4 Mo, bal Fe.
For dies, casting dies, piercing mandrels; hot work steel, corrosion resistant.

FORTUNA ZW.
M-1307; 1.4 C, 0.3 Cr, 0.1 V, bal Fe.
For drawing dies; water hardened, abrasion and wear resistant.

FORTY-TWO N.
  M-604; 0.16 C, 1.20 Mn, 0.35 Si, 0.05 Cb, bal Fe.
  Normalized: 63 ksi TS; 42 ksi YS; 23 El.
  Readily formed and welded.
  For stressed structures at low temperatures, arctic and marine structures.
  ASTM A633 Grade A.

FORVA-50.
  M-1766; 0.50 C, 0.85 Mn, 0.25 Si, 1.0 Cr, 0.15 V, bal Fe.
  Cr-V spring steel; hardenable to 58-63 Rc.
  IHA F-143; DIN 50CrV4; AISI 6150; BS EN47.

FOS-FLO 7.
  M-63; 92.75 Cu, 7.25 P.
  M.P. 1310°F (710°C); Flow. P. 1350°F (730°C).
  For joining copper and copper alloys where joint does not involve critical impact or vibration.
  AWS BCuP-2.

FOURDINIER WIRE.
  M-Eng.; 85-80 Cu, 15-20 Zn, 0-0.4 Sn, 0.01 Pb.
  For wire for Fourdinier screens used in paper manufacture.

FOUR STAR see CARPENTER FOUR STAR.

FOURTEEN PER CENT.
  M-Eng.; 60-58 Cu, 26-28 Zn, 14 Ni.
  For ornamental flatware; corrosion resistant.

FOX NO. 671.
  M-430; 0.35-0.45 C, 1.25-1.75 Ni, 0.8-1.4 Cr, bal Fe.

FPC NONTEMPERING.
  M-365; 0.3-0.4 C, 0.7-0.9 Cr, 0.7 Mn, 0.3-0.6 Mo, bal Fe.
  For chisels, punches, caulking tools; water hardened.

FRANCO.
  M-492; 0.7 C, 18 W, 4 Cr, 1 V, bal Fe.
  For high speed tools and cutters; high speed steel.

FRANKITE.
  M-571; 3.5 C, 2.5 Si, alloy, bal Fe.
  For castings; alloyed gray iron.

FRANKITE E-212.
  M-571; C, bal Fe.
  For hydraulic bodies, compressor cylinders.

FRANKITE E-450.
  M-571; 1.1 min C, 12-15 Ni, Cr, 5-7 Cu, bal Fe.
  25,000-35,000 TS; 145-170 Brin.
  For heat and corrosion resistant parts; heat and corrosion resistant.

FRANKITE E-604.
  M-571; C, 4.5 Ni, 1.5 Cr, bal Fe.
  For mixer blades, ash chutes, scrapers; "Ni-Hard"; abrasion resistant.

FRANKITE E-821.
  M-571; 0.70-1.10 C, 22-25 Cr, bal Fe.
  42,000-50,000 TS; 200-230 Brin.
  For heat and corrosion resistant parts; heat and corrosion resistant.

FRANKITE E-822.
  M-571; 1.75-2.0 C, 22-25 Cr, bal Fe.
  80,000-90,000 TS; 320-350 Brin.
  For heat and corrosion resistant parts; heat and corrosion resistant.

FRANKITE E-830N.
  M-571; C, 30 Cr, 3 Ni, bal Fe.
  For furnace supports, kilns; corrosion and heat resistant.

FRANKITE E-831.
  M-571; 0.80-1.10 C, 30-34 Cr, bal Fe.
  45,000-55,000 TS; 280-300 Brin.
  For heat and corrosion resistant parts; heat and corrosion resistant.

FRANKITE E-832.
  M-571; 1.10 min C, 31-39 Cr, bal Fe.
  For heat and corrosion resistant parts; heat and corrosion resistant.

FRANXA 9.
  M-678; 0.58-0.63 C, 1.5-1.9 Si, 0.5-0.9 Mn, bal Fe.
  Annealed: 120,000 TS; 78,000 YS; 12 El.
  Hardened: 225,000 TS; 141,000 YS; 6 El.
  For tools, dies, springs; shock resistant.

FRAPIMPHY 1.
  M-1488; 0.03 max C, 17.5 Cr, 9.5 Ni, 3.25 Cu, bal Fe.
  Wrought austenitic stainless steel designed for cold heading hexagonal socket heat screws (HSH screw).
  AFNOR Z 6 CNU 18.10 DF.

FRAPIMPHY 304 BC.
  M-1488; 0.03 max C, 18.5 Cr, 11.5 Ni, bal Fe.
  Wrought austenitic stainless steel designed for cold heading bolts and nut for chemical and shipbuilding industries.
  AFNOR Z2CN 18.10 DF; AISI 304L.

FRAPIMPHY 316 BC.
  M-1488; 0.03 max C, 17.0 Cr, 13.0 Ni, 2.3 Mo, bal Fe.
  Wrought austenitic stainless steel designed for cold heading bolts and nuts for chemical and shipbuilding industries.
  AFNOR Z 2 CND 17.13 DF; AISI 316L.

FRAPIMPHY A3.
  M-1488; 0.20 max C, 12.5 Cr, bal Fe.
  Wrought, annealed, martensitic stainless steel for cold-heading balls, pins, hinges.
  AFNOR Z 12C13DF; AISI 410-420.

FRAPIMPHY A5.
M-1488; 0.32 max C, 13.5 Cr, bal Fe.
Wrought, annealed, martensitic stainless steel for cold heading balls, pins, hinges.
AFNOR Z30 C13DF; AISI 420.

FRAPIMPHY B4.
M-1488; 0.08 max C, 17.0 Cr, bal Fe.
Wrought, annealed, ferritic stainless steel for cold heading wood screws and automobile bolts and nuts.
AFNOR Z8C17 DF; AISI 430.

FRAPPANT.
M-1340; 0.9 C, 1.9 Mn, 0.1 V, bal Fe.
For blanking and forming dies, punches; non-deforming, oil hardened.

FREECUT 15.
M-341; 0.20 C, 1.25 Mn, 0.05 Si, 0.25 S, 0.02 P, bal Fe.
Free machining steel plate stock.

FREECUT 45.
M-341; 0.45 C, 1.25 Mn, 0.05 Si, 0.25 S, 0.02 P, bal Fe.
Free machining, flame hardenable plate stock.

FREE CUTTING BRASS-360.
M-8; 61.5 Cu, 35.25 Zn, 3.25 Pb.
Hard: 58,000 TS; 42,000 YS; 20 El; B 70 Rock.
Soft: 47,000 TS; 18,000 YS; 55 El; B 20 Rock.
For hardware, gears, pinions, screw machine products, fasteners.
Free-machining leaded brass.

FREE CUTTING BRASS, COPPER ALLOY NO. 360.
M-33; 62 Cu, 34.75 Zn, 3.25 Pb.
Half hard 1.0 inch; 58,000 TS; 45,000 YS; 25 El; 34,000 shear; 78 Rock B.
Excellent machinability; for automatic machine parts as studs, bolts, nuts shafts.
ASTM B16; CDA 360.

FREE CUTTING BRONZE.
M-U.S.; 89 Cu, 10 Zn, 1.5 Pb.
For screw stock, bolts, automobile radiators; free-cutting.

FREE CUTTING MUNTZ METAL-293.
M-8; 60 Cu, 39 Zn, 1 Pb.
Soft: 54,000 TS; 20,000 YS; 40 El.
Hard: 80,000 TS; 60,000 YS; 6 El.
For screw machine parts; free cutting.

FREE CUTTING MUNTZ METAL-3711.
M-8; 60.0 Cu, 36.75 Zn, 3.25 Pb.
Hard: 60,000 TS; 45,000 YS; 18 El; B 72 Rock.
Soft: 48,000 TS; 20,000 YS; 54 El; B 20 Rock.
For butt hinges, lock bodies, mechanical devices, screw machine products, forging rods; Free-cutting, extrudable.

FREE CUTTING PHOSPHOR BRONZE 610.
M-8; 4 Sn, 4 Zn, 4 Pb, bal Cu.
Hard: 60,000 TS; 50,000 YS; 20 El.
For bearings, bushings, valve and pump parts, gears, pinions; free cutting, wear resistant.

FREE CUTTING STAYBITE F.S.T. STEEL see **STAYBRITE F.S.T. (FC).**

FREE-CUTTING TUBE BRASS-282.
M-8; 66.5 Cu, 31.9 Zn, 1.6 Pb.
Soft: 45,000 TS; 17,000 YS.
Hard: 73,000 TS; 60,000 YS.
For screw machine products; free-cutting.

FREE CUTTING TUBE BRASS-332.
M-8; 66.5 Cu, 31.9 Zn, 1.6 Pb.
Hard Tube: 73,000 TS; 60,000 YS; 10 El; B 80 Rock.
Soft Tube: 45,000 TS; 17,000 YS; 50 El; B 15 Rock.
For screw machine products, ball point pens, plumbing fixtures, musical instruments. Free machining tubes.

FREE-CUTTING YELLOW BRASS 271.
M-8; 3.25 Pb, 35.25 Zn, bal Cu.
Soft: 47,000 TS; 32,000 YS; 60 El; 60 Brin.
Hard: 62,000 TS; 50,000 YS; 20 El; 140 Brin.
For machined parts, hardware; free-cutting.

FREEMACHINEWELD.
M-118; Ni.
For welding electrodes for cast iron; machinable welds.

FREMAX 15.
M-604; 0.13 C, 1.10 Mn, 0.25 S, bal Fe.
Free machining steel; 143 Brin.
For molds, dies, seals, gears.

FREMAX 45.
M-604; 0.45 C, 1.10 Mn, 0.25 S, bal Fe.
Free machining steel; 187 Brin.
Hardenable; for molds, dies, seals, gears.

FRENCH.
M-French; 50 Cu, 31 Zn, 18 Ni.
For ornamental parts, fittings; corrosion resistant.

FRENCH ALLOY.
M-France; 58-50 Cu, 25-30 Zn, 17-20 Ni.
For electrical resistors, ornamental parts; nickel silver.

FRENCH AUTO.
M-Eng.; 75 Pb, 10 Sn, 15 Sb.
For automobile bearings; Babbitt.

FRENCH NAVY ANTIFRICTION METAL.
M-French; 7 Cu, 7.5 Sn, 78.5 Zn.
For bearings, bushings; Babbitt.

FRENCH SILVER SOLDER.
M-France; 66 Ag, 24 Cu, 10 Zn.
For silver solder.

FRENCH TYPE METAL "A".
M-France; 55 Pb, 22 Sn, 23 Sb.
For type metal.

FRENCH TYPE METAL "B".
M-France; 55 Pb, 15 Sn, 30 Sb.
For type metal.

FREUND STEEL.
M-German; 0.7-1.3 Si, 0.3-0.6 Mn, 0.1-0.15 C, Cr, bal Fe.
For highly stressed structural members; high elastic steel.

FRICKE'S HARDER.
69 Cu, 30 Zn, 10 Ni.
For strong corrosion resistant parts; corrosion resistant.

FRICKE'S SILVERY.
50 Cu, 18.8 Zn, 31.2 Ni.
For ornamental and decorative parts; corrosion resistant.

FRICKS ALLOY.
M-Eng.; 69-50 Cu, 18-39 Zn, 5.5-31 Ni.
For ornamental, base for plated ware; nickel silver.

FRICKS ALLOY.
M-Eng.; 55-50 Cu, 31-30 Zn, 19-17 Ni.
For resistances, decorative parts; corrosion resistant.

FRICKS BLUISH YELLOW, HARD.
55.5 Cu, 39 Zn, 5.5 Ni.
For hardware, decorative parts; corrosion resistant.

FRICKS PALE YELLOW, DUCTILE.
62.5 Cu, 31.2 Zn, 6.3 Ni.
For hardware, decorative parts; corrosion resistant.

FRICTION ALLOY (STANDARD).
M-Eng.; 50 Pb, 40 Sn, 10 Sb.
For antifriction metal; Babbitt.

FRICTIONLESS.
M-815; Pb alloy.
For bearings, linings; Babbitt.

FRIDUCTIL 5622 ETC.
M-1759; See Werkstoff Nr. 1.5622 etc.
Steels for toughness at sub-zero temp; 9 grades.

FRIGIDAL.
M-303; 35 Ni, low Cr, bal Fe.
86,000 TS; 32 El.
For electrical resistances; coefficient of expansion (0-40°C.) 0.0000014.

FRILOLIT CRMO.
M-1311; 0.9 C, 18 Cr, 0.1 V, 1.1 Mo, bal Fe.
For bearings, liners, valves; corrosion resistant.

FRILOLIT RF.
M-1311; 0.4 C, 0.4 Si, 13 Cr, bal Fe.
Annealed: 95,000 TS; 50,000 YS; 25 El; 55 RA; 195 Brin.
Cold drawn: 105,000 TS; 85,000 YS; 17 El; 50 RA; 215 Brin.
For valves, cutlery, surgical instruments; Type 420; corrosion resistant.

FRILOLIT RFOO.
M-1311; 0.12 max C, 13 Cr, 0.4 Si, bal Fe.
Annealed: 75,000 TS; 40,000 YS; 35 El; 70 RA; 155 Brin.
Cold drawn: 100,000 TS; 85,000 YS; 17 El; 60 RA; 205 Brin.
For turbine blades, surgical instruments; Type 410; corrosion resistant.

FRILOLIT RF SPEZIAL.
M-1311; 0.85 C, Cr, V, bal Fe.
For cutters, tools; oil or water hardened.

FRILOLIT RFW.
M-1311; 0.2 C, 0.4 Si, 13 Cr, bal Fe.
Annealed: 95,000 TS; 50,000 YS; 25 El; 55 RA; 195 Brin.
Cold drawn: 105,000 TS; 85,000 YS; 17 El; 50 RA; 215 Brin.
For turbine blades, surgical instruments; Type 420; corrosion resistant.

FRISMUTH ALUMINUM SOLDER.
M-Eng.; 67 Sn, 27 Pb, 3 Al.
For aluminum solder.

FRIXTEC 19850.
M-717.
Aluminum bronze powder for metal spraying. Machinable; smooth, bright finish.

FROGALLOY see **MCKAY FROGALLOY.**

FROGALLOY C.
M-849; 0.4-0.6 C, 4 Mn, 18-21 Cr, 9.0-10.5 Ni, 1.2 Mo, bal Fe.
Welded: 117,000 TS; 91,000 YS; 14.5 El; 550 Brin.
For hard surfacing electrodes; work hardenable; wear resistant.

FROGALLOY M.
M-849; 0.5-0.6 C, 4 min Mn, 18-21 Cr, 9.0-10.5 Ni, bal Fe.
Welded: 117,000 TS; 90,000 YS; 16 El; 550 Brin.
For welding electrodes, hard surfacing; stainless, work hardens, shielded arc.

FRONTIER 40E.
M-58; 5.5 Zn, 0.55 Mg, 0.5 Cr, 0.2 Ti, bal Al.
Aged: 32,000-38,000 TS; 22,000-26,000 YS; 3-10 El; 75-90 Brin.
For aircraft parts, pressure tight castings; age-hardenable at room temperature.

FRONTIER BRONZE.
M-106; 0.5-0.75 Mg, 0.1 Ni, 4.8-5.7 Zn, 0.1 Pb, 0.05 Sn, 0.15-0.25 T 0.4-0.6 Cr, bal Al.
Sand cast: 31,200-35,800 TS; 22,400-24,700 YS; 4-6 El; 60-70 Brin, 3 Izod.
Medium strength, reasonable shock resistant sand or gravity die cast aluminum alloy. Ages 21-30 days at room temperature. Somewhat hot short; for simple castings only.

FRONTIER NO. 3.
M-58; 88.5 Cu, 11 Sn, 0.25 Pb, 0.25 P.
35,000-40,000 TS; 20,000-25,000 YS; 6-10 El; 67-85 Brin.
For worm gears, bearings; Iz-2-10.

FRONTIER NO. 5.
M-58; 10 Al, 89 Cu, 1 Fe, traces Ti.
Untreated: 60,000-80,000 TS; 22,000-28,000 YS; 30-15 El; 120-130 Brin.
Heat treated: 80,000-95,000 TS; 50,000-65,000 YS; 10-4 El; 180-200 Brin.
For housing nuts, worm gears, spur gears, bearings, spacer boxes; resists wear and shock; heat resisting.

FRONTIER NO. 6.
M-58; 86 Cu, 9.5 Sn, 2.5 Pb, 2 Zn.
30,000-45,000 TS; 18,000-20,000 YS; 15-20 El; 55-70 Brin.
For general utility bearing bronze; Iz-8-10.

FRONTIER NO. 7.
M-58; 79.5 Cu, 11 Al, 6 Fe, 3.5 Mn.
70,000-80,000 TS; 35,000-45,000 YS; 3-5 El; 150-170 Brin.
For paper mill machinery and valve seats; Iz-3-5.

FRONTIER NO. 8.
M-58; 83.5 Cu, 10 Sn, 2.5 Pb, 3.5 Ni, 0.5 P.
35,000-45,000 TS; 25,000-28,000 YS; 10-15 El; 80-93 Brin.
For heavy duty bearings, worm gears, lead screws; high strength bearing; Iz-3-5.

FRONTIER NO. 9.
M-58; 90 Cu, 10 Sn.
30,000-35,000 TS; 15,000-20,000 YS; 10-15 El; 63-70 Brin.
For acid resisting work; Iz-3-10.

FRONTIER NO. 10.
M-58; 8 Sn, 4 Zn, bal Cu.
Cast: 30,000-45,000 TS; 18,000-23,000 YS; 15-30 El; 65-74 Brin.
For bearings, screw down nuts, worm gears, hydraulic castings; gun metal bronze.

FRONTIER NO. 11.
M-58; 88 Cu, 5 Ni, 5 Sn, 2 Zn.
Cast: 70,000 TS; 50,000 YS; 15 El; 160 Brin.
For bearings, gears, screws, bolts; age-hardenable, tough.

FRONTIER NO. 14.
M-58; 10 Sn, 5 Pb, 0.25 P, bal Cu.
Cast: 30,000-40,000 TS; 19,000-22,000 YS; 15-5 El; 50-61 Brin.
For bearings against soft steel; for moderate speeds.

FRONTIER NO. 15Y.
M-58; 79.75 Cu, 10 Sn, 10 Pb, 0.25 P.
28,000-35,000 TS; 19,000-22,000 TS; 5-10 El; 52-70 Brin.
For small machine tool bearings; Iz-3-8.

FRONTIER NO. 18.
M-58; 85 Cu, 5 Sn, 5 Pb, 5 Zn.
27,000-33,000 TS; 15,000-20,000 YS; 15-20 El; 50-70 Brin.
For general utility bronze; not recommended for bearings: Iz-13-18.

FRONTIER NO. 20.
M-58; 77.75 Cu, 7 Sn, 15 Pb, 0.25 P.
27,000-31,000 TS; 17,000-18,000 YS; 12-18 El; 43-63 Brin.
For bearings, acid resisting alloy in mining machinery; for high speeds; Iz-6-8.

FRONTIER NO. 28.
M-58; 99.6-99.9 Cu.
17,000-20,000 TS; 6000-9000 YS; 40-50 El; 40-50 Brin.
For high electrical conductivity bearings.

FRONTIER NO. 29.
M-58; 0.5 Sn, 41 Zn, 1 Fe, 1 Al, 0.5 Mn, bal Cu.
Cast: 65,000-85,000 TS; 25,000-35,000 YS; 35-20 El; 104-119 Brin.
For propeller blades and hubs, valves; manganese bronze.

FRONTIER NO. 30.
M-58; 20 Zn, 2.5 Fe, 6.5 Al, 2.5 Mn, bal Cu.
Cast: 90,000-110,000 TS; 50,000-65,000 YS; 15-10 El; 196-220 Brin.
For propeller blades, gears; manganese bronze, high strength.

FRONTIER NO. 38.
M-58; 67 Cu, 8 Sn, 24 Pb, 1 Zn.
Cast: 22,500-27,500 TS; 14,500-19,000 YS; 10-15 El; 45-55 Brin.
For castings, bearings; heavy duty.

FS see **FINKL FS.**

FS 85 see **FANSTEEL FS 85, (NOW DISCONTINUED).**

FS M-10.
M-57; 0.85 C, 4.25 Cr, 8.25 Mo, 2 V, bal Fe.
For drills, taps, dies, chasers, broaches; high speed steel.

FSX-414.
  M-1491; 0.25 C, 1.0 max Mn, 1.0 max Si, 29.5 Cr, 10.5 Ni, 7.0 W, 0.012 B, 2.0 max Fe, bal Co.
  Cast alloy for gas turbine vanes.

FSX-418.
  M-1491; 0.25 C, 1.0 max Mn, 1.0 max Si, 29.5 Cr, 10.5 Ni, 7.0 W, 0.012 B, 2.0 max Fe, 0.15 Y, bal Co.
  Cast alloy for gas turbine vanes; improved oxidation resistance.

FSX-430.
  M-1491; 0.40 C, 29.5 Cr, 10.0 Ni, 7.5 W, 0.027 B, 0.9 Zr, 0.5 Y, bal Co.
  Cast alloy for gas turbine vanes.

FUCHS 2000.
  M-1361; 99.90 Cu (contains O_2).
  Good electrical conductivity and formability. For electrical parts as sockets.
  Type E-Cu.

FUCHS 2020.
  M-1361; 0.025-0.25 Ag, bal Cu.
  For armature windings and commutator segments.

FUCHS 2050.
  M-1361; 0.4-1.1 Te, bal Cu.
  High electrical conductivity and good machinability.

FUCHS 2061.
  M-1361; 0.8-1.2 Pb, bal Cu.
  High corrosion resistance and good machinability.

FUCHS 2155.
  M-1361; 53.5-56.0 Cu, 1.0-2.5 Pb, bal Zn.
  For extrusion, hot forming; thin-walled sections.
  Type Cu Zn 44 Pb 2; similar to CDA 380.

FUCHS 2156.
  M-1361; 53.5-56.0 Cu, 1.0-2.5 Pb, 0.5 Al, bal Zn.
  For hot-formability.
  Type Cu Zn 44 Pb 2; similar to CDA 380.

FUCHS 2157.
  M-1361; 57.5-59.0 Cu, bal Zn.
  For extrusions or hot forming; or cold bending.
  Type Cu Zn 42.

FUCHS 2158.
  M-1361; 57.5-59.0 Cu, 2.5-3.3 Pb, bal Zn.
  For hot pressing and forming, good machinability.
  Type CuZn 39 Pb 3; similar to CDA 385, Architectural Bronze.

FUCHS 2159.
  M-1361; 57.5-59.0 Cu, 2.5-3.0 Pb, bal Zn.
  Processed to give improved toughness.
  Type CuZn 39 Pb 3; similar to CDA 385.

FUCHS 2160.
  M-1361; 59.0-61.5 Cu, bal Zn.
  Muntz metal; for hot or cold forming, riveting.
  Type CuZn 40; similar to CDA 280.

FUCHS 2161.
  M-1361; 59.5-61.5 Cu, 0.5-2.0 Pb, bal Zn.
  For hot or cold forming; lock brass.
  Type CuZn 38 Pb1; similar to CDA 377.

FUCHS 2163.
  M-1361; 62.0-64.0 Cu, bal Zn.
  For cold forming, deep drawing; costume jewelry, hollow ware, zip fasteners.
  Type CuZn 37; CDA 274, Yellow Brass 63%.

FUCHS 2167.
  M-1361; 66.0-68.5 Cu, bal Zn.
  Very good cold formability; for radiator strips, tube rivets, wire netting.
  Type CuZn 33; CDA 268, Yellow Brass 66%.

FUCHS 2170.
  M-1361; 69.0-71.0 Cu, bal Zn.
  Excellent cold formability, deep drawing and solderability; for instruments, tubes, bushings, cartridge cases.
  Type CuZn 30; CDA 260 Cartridge Brass.

FUCHS 2171.
  M-1361; 70.0-72.5 Cu, 0.9-1.3 Sn, bal Zn.
  Corrosion resistant, particularly as regards dezincification; for tubes, condenser and heat exchanger plates.
  Type CuZn 28 Sn; CDA 442, Admiralty, Uninhibited.

FUCHS 2180.
  M-1361; 79.0-81.0 Cu, bal Zn.
  Excellent cold formability; for electrical parts and installations, flexible hose.
  Type CuZn 20; CDA Low Brass 80%.

FUCHS 2181.
  M-1361; 79.0-83.0 Cu, 0.6-1.2 Sn, bal Zn.
  Good corrosion resistance, especially as regards dezincification; for tubes, musical instruments as horns.
  Type CuZn 20 Sn; CDA 435, Trumpet Brass.

FUCHS 2185.
  M-1361; 84.0-86.0 Cu, bal Zn.
  Excellent cold formability as spinning, embossing; for jewelry, bushings for spring assembly.
  Type CuZn 15; CDA 230, Red Brass 85%.

FUCHS 2190.
  M-1361; 89.0-91.0 Cu, bal Zn.
  Excellent cold formability; for electrical components, tubes.
  Type CuZn 10; CDA 230, Commercial Bronze 10%.

FUCHS 2195.
M-1391; 95.0-96.0 Cu, bal Zn.
For handicraft articles, tubes.
Type CuZn 5.

FUCHS 2201.
M-1361; 57.0-59.0 Cu, 1.0-2.0 Pb, 0.4-1.8 Mn, bal Zn.
Good hot formability and machinability.
For hot pressing, valves.
Type CuZn 40 Mn Pb.

FUCHS 2202.
M-1361; 58.0-61.0 Cu, 1.5-2.5 Mn, 0.3-1.5 Al, 2.0-3.0 Ni, bal Zn.
Good strength for structural applications.
For appliances, marine engineering.
Type CuZn 35 Ni.

FUCHS 2203.
M-1361; 56.5-59.5 Cu, 0.4-1.8 Mn, 0.4-1.6 Al, bal Zn.
Weather resistant; for architecture, appliances.
Type CuZn 40 Al 1.

FUCHS 2204.
M-1361; 57.0-59.0 Cu, 1.0-2.5 Mn, bal Zn.
Weather resistant, can be soldered and brazed; for appliances, architectural sections, handrails.
Type CuZn 40 Mn.

FUCHS 2205.
M-1361; 55.5-59.0 Cu, 0-0.8 Pb, 0-1.0 Fe, 0.05 Sn, 1.0-2.4 Mn, 1.3-2. Al, 0-0.8 Si, 0-2.0 Ni, bal Zn.
High strength bronze; good weather resistance.
For structural components.
Type CuZn 40 Al 2.

FUCHS 2206.
M-1361; 55.5-59.0 Cu, 0-0.8 Pb, 0-1.0 Fe, 0-0.5 Sn, 1.0-2.4 Mn, 1.3-2.3 Al, 0-0.8 Si, 0-2.0 Ni, bal Zn.
Good strength and bearing properties; for worm gears, sprockets.
Type CuZn 40 Al 2.

FUCHS 2210 THROUGH 2214 similar to **FUCHS 2206.**

FUCHS 2264.
M-1361; 61.0-66.0 Cu, 0.5-3.5 Fe, 2.0-5.0 Mn, 2.5-7.5 Al, 0-0.5 Si, 0 Ni, bal Zn.
Strong, wear-resistant alloy.
Similar to CDA 670, Manganese Bronze B.

FUCHS 2356.
M-1361; 56.0-58.0 Cu, 1.5-2.5 Pb, bal Zn.
For extruded sections.
Type CuZn 41 Pb 2; similar to CDA 380.

FUCHS 2357.
M-1361; 57.5-59.0 Cu, 1.5-2.5 Pb, bal Zn.
Good hot formability and punchability.
Type CuZn 40 Pb 2; similar to CDA 380.

FUCHS 2358.
M-1361; 57.5-59.0 Cu, 2.5-3.3 pb, bal Zn.
For free cutting applications.
Type CuZn 39 Pb 3; similar to CDA 385 Architectural bronze.

FUCHS 2360.
M-1361; 59.5-61.5, 1.5-2.5 Pb, bal Zn.
Forging brass, for hot pressed, punched and machined parts.
Type CuZn 38 Pb 2; similar to CDA 377.

FUCHS 2361.
M-1361; 59.5-61.5 Cu, 1.5-2.5 Pb, bal Zn.
For hot or cold forming, threaded parts, screws.
Type CuZn 37 Pb 2; similar to CDA 377.

FUCHS 2362.
M-1361; 60.0-62.0 Cu, 2.5-3.5 Pb, bal Zn.
Good free-machining properties; for screws, machined parts. Can be hot or cold formed.
Type CuZn 36 Pb 3; similar to CDA 360, Free Cutting Brass.

FUCHS 2363.
M-1361; 62.0-63.5 Cu, 2.0-3.0 Pb, bal Zn.
Good cold forming and machinability; for machined parts, tubes.
Type CuZn 35 Pb 2; CDA 256, Extra High Leaded Brass.

FUCHS 2458.
M-1361; 57.5-59.0 Cu, 2.5-3.3 Pb, bal Zn.
For pressure tight fittings, valve sections.
Type CuZn 39 Pb 3; similar to CDA 385.

FUCHS 2560.
M-1361; 58.0-62.0 Cu, 0-0.5 Sn, 0.1-0.5 Si, bal Zn.
Good formability; for drawn wire, welding wire.

FUCHS A1.
M-1361.
Unalloyed aluminum.
60-130 N/mm^2 TS; 20-110 N/mm^2 YS; 5-27 El.
Type Al 99.8; similar to AA 1080.

FUCHS A2.
M-1361.
Unalloyed aluminum.
60-130 N/mm^2 TS; 20-110 N/mm^2 YS; 5-27 El.
Type Al 99.7; similar to AA 1070.

FUCHS ACID RESISTING ALLOY.
M-Eng.; 73-80 Ag, 13-15 Ni, 13.5-15 Au.
For chemical equipment, corrosion resisting parts; acid resistant.

FUCHS ACID RESISTING GOLD ALLOY.
M-Eng.; 75 Au, 10-15 Ni, 10-15 W.
For chemical equipment; acid resistant.

FUCHS AM, AG, AK, AS, AZ see **AM, AG, AK, AS, AZ.**

FUCHS E.
M-1361; 99.9% Al.
Unalloyed aluminium.
Type Al 99.9; similar to AA 1090.

FUCHS E 05.
M-1361; 0.5 Mg, bal Al.
Good formability and corrosion resistance.
Type Al 99.9 Mg 0.5.

FUCHS E1.
M-1361; 1.0 Mg, bal Al.
Good formability and corrosion resistance.
Type Al 99.9 Mg 1; similar to AA 5657.

FUCHS E 2.
M-1361; 1.8 Mg, bal Al.
Good formability and corrosion resistance.
Type Al 99.9 Mg 2.

FUCHS ES 70.
M-1361; 0.20 Cu, 0.65 Mg, 0.65 Si, bal Al.
130-235 N/mm^2 TS; 80-155 N/mm^2 YS; 14-17 El.
Good formability, surface finish.
Type Al 99.7 Mg Si Cu.

FUCHS ES 90.
M-1361; 0.60 Mg, 0.50 Si, bal Al.
130-235 N/mm^2 TS; 80-155 N/mm^2 YS; 14-17 El.
Good formability, surface finish.
Type Al 99.9 Mg Si.

FUCHS M1.
M-1361.
Unalloyed magnesium.
Good hot formability and weldability.
Type H Mg 99.8; similar to ASTM B92-45.

FUCHS MA3.
M-1363; 3.0 Al, 1.0 Zn, 0.15 Mn, bal Mg.
25 kp/mm^2 TS; 16 kp/mm^2 YS; 10 El.
Medium strength, good hot formability, ductility, weldability and machinability.
Type Mg Al 3 Zn; Bs 3371 MAG-T-111 M. (AZ 31 B).

FUCHS MA6.
M-1361; 6.3 Al, 1.0 Zn, 0.15 Mn, bal Mg.
26-28 kp/mm^2 TS; 18-20 kp/mm^2 YS; 8-10 El.
High strength alloy, medium weldability and excellent machinability.
Type Mg Al 6 Zn; ASTM B107. (AZ 61A).

FUCHS MA8.
M-1361; 8.0 Al, 0.5 Zn, 0.20 Mn, bal Mg.
28-32 kp/mm^2 TS; 20-23 kp/mm^2 YS; 6-10 El.
Heat treatable, high strength, limited weldability.
Type Mg Al 8 Zn. (AZ 80A).

FUCHS MA39.
M-1361; 3.0 Al, 1.0 Zn, 0.35 Mn, bal Mg.
25 kp/mm^2 TS; 16 kp/mm^2 YS; 10 El.
Similar to MA 3 but with lower Fe content.
Type Mg Al 3 Zn.
ASTM B107. (AZ 31).

FUCHS MA74.
M-1361; 7.2 Al, 1.2 Zn, 0.30 Mn, bal Mg.
28-32 kp/mm^2 TS; 20-23 kp/mm^2 YS; 8-10 El.
Heat treatable, high strength alloy, excellent machinability; limited weldability.
Type Mg Al 7 Zn.

FUCHS MG2.
M-1361; 1.2-2.0 Mn, bal Mg.
20-23 kp/mm^2 TS; 15-17 kp/mm^2 YS; 2 El.
Good hot formability and corrosion resistance.
Type Mg Mn 2; ASTM B107.

FUCHS MZ64.
M-1361; 5.5 Zn, 0.6 Zr, bal Mg.
28-32 kp/mm^2 TS; 18-25 kp/mm^2 YS; 4-7 El.
Heat treatable, high strength, limited weldability.
Type Mg Zn 6 Zr. (ZK 60A).

FUCHS R.
M-1361; 99.99% Al.
High purity aluminum.
Type Al 99.99 R; AFNOR A9.

FUCHS R 05.
M-1361; 0.5 Mg, bal Al.
Good formability, corrosion resistance.
Type Al R Mg 0.5.

FUCHS R 1.
M-1361; 1.0 Mg, bal Al.
Good formability and corrosion resistance.
Type Al R Mg 1; AFNOR A 9-G 1.

FUCHS T 2.
M-1361; 0.10 O$_2$, 0.2 max Fe, 0.0125 max H$_2$, 0.08 C, 0.05 N$_2$, 99.5 Ti.
30-42 kp/mm^2 TS; 20 kp/mm^2 min YS; 25 min El.
Good weldability, high corrosion resistance, good formability.
Grade Ti 99.5.

FUCHS T 3.
M-1361; 0.20 O$_2$, 0.25 Max Fe, 0.0125 max H$_2$, 0.08 C, 0.06 N$_2$, 99.4 Ti.
40-55 kp/mm^2 TS; 28 kp/mm^2 min YS; 20 min El.
Similar to FUCHS T2 but slightly stronger.
Gradt Ti 99.4; AMS 4902; DTD 5003 B.

FUCHS T 6.
M-1361; 0.30 O$_2$, 0.35 max Fe, 0.0125 max H$_2$, 0.1 C, 0.07 N$_2$, 99.2 Ti.
55-75 kp/mm^2 TS; 45 kp/mm^2 min YS; 15 min El.
Moderate weldability; highest strength of all unalloyed Ti grades.
Grade Ti 99.2; AMS 4921 A.

FUCHS TA 44.
M-1361; 3.0-5.0 Al, 3.0-5.0 Mo, 1.5-2.5 Sn, 0.3-0.7 Si, 0.2 max Fe, 0.0150 max H$_2$, bal Ti.
Ht Tr; 107-130 kp/mm^2 TS; 92 kp/mm^2 min YS; 9 min El.
Used in British aircraft projects.
Grade Ti Al4 Mo4 Sn2; DTD 5153.

FUCHS TA 52.
M-1361; 4.0-6.0 Al, 2.0-3.0 Sn, 0.20 O$_2$, 0.25 max Fe, 0.020 max H$_2$ 0.08 C, 0.07 N$_2$, bal Ti.
80 kp/mm^2 min TS; 77 kp/mm^2 min YS; 10 min El.
Weldable, good elevated temperature strength.
Grade Ti Al5 Sn 2.5; AMS 4966 B; DTD 5083.

FUCHS TA 64.
M-1361; 5.75-6.75 Al, 3.5-4.5 V, 0.20 O$_2$, 0.25 max Fe, 0.125 max H$_2$, 0.08 C, 0.07 N$_2$, bal Ti.
90 kp/mm^2 min TS; 84 kp/mm^2 min YS; 10 min El.
Most widely used Ti alloy, forgeable, heat treatable, limited weldability. Hardenable to about 110 kp/mm^2 TS.
Grade Ti Al6 V4; AMS 4928 E; DTD 5173.

FUCHS TA 66.
M-1361; 5.0-6.0 Al, 5.0-6.0 V, 1.5-2.5 Sn, 0.35-1.0 Cu, 0.20 O$_2$, 0.35-1.0 Fe, 0.015 max H$_2$, 0.05 C, 0.04 N$_2$, bal Ti.
Ht.Tr.; 112-126 kp/mm^2 TS; 98-119 kp/mm^2 YS; 5 min El.
Premium strength alloy; good forgeability.
Grade Ti Al6 V6 Sn 2; AMS 4971.

FUCHS TA 74.
M-1361; 6.5-7.3 Al, 3.5-4.5 Mo, 0.20 O$_2$, 0.25 max Fe, 0.0125 max H$_2$, 0.08 C, 0.07 N$_2$, bal Ti.
Ht.Tr.:105-119 kp/mm^2 TS; 98-117 kp/mm^2 YS; 6 min El.
For heavy sections requiring heat treat.
Grade Ti Al7 Mo 4; AMS 4970 A.

FUCHS TC 2.
M-1361; 2.0-3.0 Cu, 0.20 O$_2$, 0.20 max Fe, 0.010 max H$_2$, 0.10 C, 0.05 N$_2$, bal Ti.
55 kp/mm^2 min TS; 39 kp/mm^2 min YS; 16 min El.
Weldable, heat treatable to about 80 kp/mm^2 TS.
Grade Ti Cu2; DTD 5123.

FUCHS TP 02.
M-1361; 0.2 Pd, 0.20 O$_2$, 0.25 max Fe, 0.0125 max H$_2$, 0.08 0.06 N$_2$, 99.4 Ti.
45 kp/mm^2 min TS; 35 kp/mm^2 min YS; 16 min El.
Improved resistance to HCl and other reducing acids.
Grade Ti Pd 0.15.

FUEGO.
M-344; 0.5 C, 0.9 Si, 0.5 Mn, Cr, V, bal Fe.
For punches, shears, chisels; water hardened.

F.U.G. 1.
M-1360; 2.5-5.0 Cu, 0.2-1.8 Mg, 0.3-1.5 Mn, bal Al.
Annealed: 27,000 TS; 11,000 YS; 22 El; 47 Brin.
Heat treated: 72,000 TS; 57,000 YS; 130 Brin.
For aircraft structures and fittings, fasteners; age-hardenable.

F.U.G. 4.
M-1360; 0.6-1.4 Mg, 0.6-1.6 Si, 0.6-1.0 Mn, 0.3 Cr, bal Al.
Annealed: 21,000 TS; 8000 YS; 24 El.
For window frames, fan blades, gutters, boats; good forming and welding properties.

F.U.G. 6.
M-1360; 2-4 Mg, 0.4 max Mn, 0.3 max Cr, bal Al.
Soft: 28,000 TS; 13,000 YS; 30 El; 47 Brin.
Hard: 40,000 TS; 35,000 YS; 10 El; 73 Brin.
For aircraft tanks and fittings, fuel lines, marine parts; resists sea water corrosion.

F.U.G. 8.
M-1360; 1.0-1.5 Mn, 0.3 max Cr, bal Al.
Soft: 16,000 TS; 6000 YS; 40 El.
Hard: 29,000 TS; 27,000 YS; 10 El.
For cooking utensils, heat exchangers, tanks, furniture; good forming and welding properties.

FUGI-HIZ.
M-1538; 0.10-0.18 C, 0.6-1.0 Mn, 0.15-0.50 Cu, 0.7-1.0 Ni, 0.4-0.8 Cr 0.4-0.6 Mo, 0.03-0.10 V, 0.002-0.006 B, bal Fe.
Heat Treated: 114,000-135,000 TS; 100,000 min YP; 20% min El.
For car and railroad bodies, agricultural equipment.
High strength-low alloy constructional steel.

FUJI-FTW 42.
M-1538; 0.18 C, 1.5 max Mn, 0.55 max Si, bal Fe.
Rolled: 74,000-88,000 TS; 50,000 min YP; 20 min El.
For building structures, derricks, booms, bridges.
Structural low carbon steel.

FUJI-FTW 58.
  M-1538; 0.18 C, 1.5 max Mn, 0.55 max Si, bal Fe.
  Ht. Tr.: 82,000-97,000 TS; 65,000 min YP; 16 min El.
  For mine cars, bus bodies, booms, derricks, bridges.
  Structural low carbon steel.

FULLER ALLOY.
  M-808; 6 Zn, 1.2 Mg, 2 Fe, bal Al.
  Heat treated: 43,000-48,000 TS; 0-2 El; 110-115 Brin.
  For light alloy castings; heat treatable.

FULTON EGR.
  M-1101; 56 Ag, 22 Cu, 18 Zn, 4 Sn.
  For silver solder, brazing; M.P. 1145-1205°F.

FULTON NO. 110-A.
  M-1101; 5 Ag, 88.5 Cu, 6.5 P.
  For silver solder; self-fluxing, M.P. 1185-1300°F.

FULTON NO. 111.
  M-1101; 9 Ag, 51 Cu, 40 Zn.
  For silver solder; M.P. 1510°F.

FULTON NO. 111A.
  M-1101; 15 Ag, 80 Cu, 5 P.
  For silver solder; M.P. 1185°F.

FULTON NO. 112.
  M-1101; 20 Ag, 45 Cu, 30 Zn, 5 Cd.
  For silver solder, brazing; M.P. 1140-1410°F.

FULTON NO. 112-A.
  M-1101; 25 Ag, 52.5 Cu, 22.5 Zn.
  For silver solder; M.P. 1500-1575°F.

FULTON NO. 113.
  M-1101; 30 Ag, 38 Cu, 32 Zn.
  For silver solder; M.P. 1370°F.

FULTON NO. 113-A.
  M-1101; 35 Ag, 26 Cu, 21 Zn, 18 Cd.
  For silver solder, brazing; M.P. 1125-1295°F.

FULTON NO. 114.
  M-1101; 40 Ag, 30 Cu, 28 Zn, 2 Ni.
  For silver solder, brazing; M.P. 1240°F.

FULTON NO. 114-A.
  M-1101; 45 Ag, 30 Cu, 25 Zn.
  For silver solder; M.P. 1250°F.

FULTON NO. 114-AN.
  M-1101; 45 Ag, 15 Cu, 16 Zn, 24 Cd.
  For silver solder; M.P. 1120°F.

FULTON NO. 114-B.
  M-1101; 40 Ag, 35 Cu, 25 Zn, 5 Ni.
  For silver solder; M.P. 1240-1560°F.

FULTON NO. 114-N.
  M-1101; 40 Ag, 18 Cu, 15 Zn, 27 Cd.
  For silver solder; M.P. 1076°F.

FULTON NO. 115.
  M-1101; 50 Ag, 15.5 Cu, 16.5 Zn, 18 Cd.
  For silver solder; M.P. 1160°F.

FULTON NO. 115-4.
  M-1101; 54 Ag, 40 Cu, 5 Zn, 1 Ni.
  For silver solder; M.P. 1325-1575°F.

FULTON NO. 117-2.
  M-1101; 72 Ag, 28 Cu.
  For silver solder; M.P. 1435-1435°F.

FULTON NO. 118-A.
  M-1101; 85 Ag, 15 Mn.
  For silver solder; M.P. 1760-1778°F.

FULTON NO. 216.
  M-1101; 60 Ag, 25 Cu, 15 Zn.
  For silver solder; M.P. 1260°F.

FULTON NO. 216-A.
  M-1101; 65 Ag, 20 Cu, 15 Zn.
  For silver solder; M.P. 1280-1325°F.

FULTON NO. 217.
  M-1101; 70 Ag, 20 Cu, 10 Zn.
  For silver solder; M.P. 1335°F.

FULTON NO. 218.
  M-1101; 80 Ag, 16 Cu, 4 Zn.
  For silver solder; M.P. 1360°F.

FULTON NO. A-114.
  M-1101; 40 Ag, 36 Cu, 24 Zn.
  For silver solder; M.P. 1330°F.

FULTON NO. A-115.
  M-1101; 50 Ag, 34 Cu, 16 Zn.
  For silver solder; M.P. 1275°F.

FULTON NO. G-112.
  M-1101; 20 Ag, 45 Cu, 35 Zn.
  For silver solder; M.P. 1430-1500°F.

FULTON NO. G4-115.
  M-1101; 50 Ag, 15.5 Cu, 16.5 Zn, 18 Cd.
  For silver solder; M.P. 1160-1175°F.

FULTON NO. G5-115.
  M-1101; 50 Ag, 15.5 Cu, 15.5 Zn, 16 Cd, 3 Ni.
  For silver solder; M.P. 1195-1270°F.

FULTON NO. LA-115.
  M-1101; 50 Ag, 28 Cu, 22 Zn.
  For silver solder; M.P. 1250-1340°F.

FURBALOI.
  M-Eng.; 0.45 Cu, 0.05 Mn, 0.13 C, 13.35 Cr, 0.08 Ni, bal Fe.
  77,000 TS; 38 El; 75 RA.
  For corrosion resisting parts; corrosion resistant.

FUSE-WELL NO. 28.
  M-810; 3.2 C, 2 Si, bal Fe.
  For welding rod for cast iron; flux coated, gas welding.

FUSIBLE.
M-Eng.; 45 Bi, 17 Sn, 30 Pb, 10 HgS.
For solders, binding plugs; M.P. 85°C.

FUSIBLE ALLOY-1.
M-U.S.; 50 Bi, 34 Sn, 17 Cd.
For fusible alloy; M.P. 65°C.

FUSIBLE ALLOY-2.
M-U.S.; 15 Bi, 42 Sn, 43 Pb.
For fusible alloy; low melting point.

FUSIBLE ALLOY-3.
M-U.S.; 33.3 Bi, 33.3 Sn, 33.3 Pb.
For fusible alloy; low melting point.

FUSIBLE METAL "D".
33.3 Bi, 66.7 Sn.
For fusible alloy; M.P. 166°C.

FUSIBLE METAL "E".
13 Bi, 40 Sn, bal Pb.
For fuses, fire extinguishers; M.P. 172°C.

FUSIBLE METAL "F".
12.5 Bi, 39.5 Sn, bal Pb.
For fuses, fire extinguishers; M.P. 178°C.

FUSIBLE METAL "G".
20 Bi, 80 Sn.
For fuses, fire extinguishers; M.P. 200°C.

FUSION NO. 300.
M-1243; 43 Sn, 43 Pb, 14 Bi, plus flux.
Solidus 290°F; Liquidus 310°F.
Paste type solder for very low temperature soldering.

FUSION NO. 360.
M-1243; 60 Sn, 40 Pb, plus flux.
Solidus 361°F; Liquidius 374°F.
Paste type solder.
ASTM B32-70 At 60 B.

FUSION NO. 361.
M-1243; 62 1/2 Sn, 36 Pb, 1 1/2 Ag, plus flux.
Solidus 350°F; Liquidus 372°F.
Paste type solder.

FUSION NO. 365.
M-1243; 63 Sn, 37 Pb, plus flux.
Solidus 361°F; Liquidus 361°F.
Paste type solder, eutectic alloy.
ASTM B32-70AT 63B.

FUSION NO. 430.
M-1243; 96 1/2 Sn, 3 1/2 Ag, plus flux.
Solidus 430°F; Liquidus 430°F.
Paste type solder; eutectic alloy.
ASTM B32-70AT 96.5 TS.

FUSION NO. 440.
M-1243; 45 Sn, 55 Pb, plus flux.
Solidus 361°F; Liquidus 441°F.
Paste type solder.
ASTM B 32-70AT 45B.

FUSION NO. 450.
M-1243; 50 Sn, 50 Pb, plus flux.
Solidus 361°F; Liquidus 421°F.
Paste type solder.
ASTM B32-70AT 50B.

FUSION NO. 455.
M-1243; 40 Sn, 60 Pb, plus flux.
Solidus 360°F; Liquidus 460°F.
Paste type solder.
ASTM B32-70AT 40B.

FUSION NO. 460.
M-1243; 95 Sn, 5 Sb, plus flux.
Solidus 452°F; Liquidus 464°F.
Paste type solder.
ASTM B32-70AT 95 TA.

FUSION NO. 470.
M-1243; 30 Sn, 70 Pb, plus flux.
Solidus 361°F; Liquidus 491°F.
Paste type solder.
ASTM B32-70AT 30B.

FUSION NO. 490.
M-1243; 25 Sn, 75 Pb, plus flux.
Solidus 361°F; Liquidus 511°F.
Paste type solder.
ASTM B32-70AT 25B.

FUSION NO. 500.
M-1243; 100 Sn, plus flux.
Solidus 450°F; Liquidus 450°F.
Paste type solder; pure tin.

FUSION NO. 505.
M-1243; 95 Sn, 5 Ag, plus flux.
Solidus 430°F; Liquidus 473°F.
Paste type solder.

FUSION NO. 560.
M-1243; 5 Sn, 93 Pb, 2 Ag, plus flux.
Solidus 530°F; Liquidus 568°F.
Paste type solider.

FUSION NO. 570.
M-1243; 10 Sn, 88 Pb, 2 Ag, plus flux.
Solidus 514°F; Liquidus 554°F.
Paste type solder.

FUSION NO. 575.
M-1243; 10 Sn, 90 Pb, plus fulx.
Solidus 527°F; Liquidus 572°F.
Paste type solder.
ASTM B32-70AT 10A.

FUSION NO. 1000.
M-1243; 45 Ag, 15 Cu, 16 Zn, 24 Cd, plus flux.
Solidus 1125°F; Liquidus 1145°F.
Paste type braze alloy.
AWS A5.8-69 BAg1.

FUSION NO. 1050.
M-1243; 50 Ag, 15.5 Cu, 16.5 Zn, 18 Cd, plus flux.
Solidus 1160°F; Liquidus 1175°F.
Paste type braze alloy.
AWS A5.8-69 BAg1a; AMS-4770.

FUSION NO. 1100.
M-1243; 35 Ag, 26 Cu, 21 Zn, 18 Cd, plus flux.
Solidus 1125°F; Liquidus 1295°F.
Paste type braze alloy.
AWS A5.8-69 BAg2.

FUSION NO. 1115.
M-1243; 60 Ag, 30 Cu, 10 Sn, plus flux.
Solidus 1095°F; Liquidus 1325°F.
Paste type braze alloy.
AWS A5.8-69 BAg 18.

FUSION NO. 1120.
M-1243; 30 Ag, 27 Cu, 23 Zn, 20 Cd, plus flux.
Solidus 1125°F; Liquidus 1310°F.
Paste type braze alloy.
AWS A5.8-69 BAg Za.

FUSION NO. 1190.
M-1243; 75 Cu, 7.25 P, 17.75 Ag, plus flux.
Solidus 1190°F; Liquidus 1190°F.
Paste type braze alloy.

FUSION NO. 1200.
M-1243; 50 Ag, 15.5 Cu, 15.5 Zn, 16 Cd, 3 Ni, plus flux.
Solidus 1195°F; Liquidus 1270°F.
Paste type braze alloy.
AWS A5.8-69 BAg3.

FUSION NO. 1205.
M-1243; 56 Ag, 22 Cu, 17 Zn, 5 Sn, plus flux.
Solidus 1152°F; Liquidus 1203°F.
Paste type braze alloy.
AWS A5.8-69 BAg 7.

FUSION NO. 1230.
M-1243; 60 Ag, 25 Cu, 15 Zn, plus flux.
Solidus 1260°F; Liquidus 1325°F.
Paste type braze alloy.

FUSION NO. 1235.
M-1243; 65 Ag, 20 Cu, 15 Zn, plus flux.
Solidus 1280°F; Liquidus 1325°F.
Paste type braze alloy.

FUSION NO. 1240.
M-1243; 40 Ag, 30 Cu, 28 Zn, 2 Ni, plus flux.
Solidus 1220°F; Liquidus 1435°F.
Paste type braze alloy.
AWS A5.8-69 BAg 4.

FUSION NO. 1245.
M-1243; 40 Ag, 30 Cu, 25 Zn, 5 Ni, plus flux.
Solidus 1260°F; Liquidus 1550°F.
Paste type braze alloy.

FUSION NO. 1250.
M-1243; 45 Ag, 30 Cu, 25 Zn, plus flux.
Solidus 1250°F; Liquidus 1370°F.
Paste type braze alloy.
AWS A5.8-69 BAg 5.

FUSION NO. 1260.
M-1243; 50 Ag, 20 Cu, 28 Zn, 2 Ni, plus flux.
Solidus 1220°F; Liquidus 1305°F.
Paste type braze alloy.

FUSION NO. 1300.
M-1243; 92.75 Cu, 7.25 P, plus flux.
Solidus 1310°F; Liquidus 1456°F.
Paste type braze alloy.
AWS A5.8-69 BCuP2.

FUSION NO. 1306.
M-1243; 86.75 Cu, 7.25 P, 6 Ag, plus flux.
Solidus 1190°F; Liquidus 1320°F.
Paste type braze alloy.
AWS A5.8-69 BCuP4.

FUSION NO. 1400.
M-1243; 72 Ag, 28 Cu, plus flux.
Solidus 1435°F; Liquidus 1435°F.
Paste type braze alloy; eutectic alloy.
AWS A5.8-69 BAg8.

FUSION NO. 1440.
M-1243; 27.25 Cu, 64.75 Zn, 7.5 Sn, 0.5 Pb, plus flux.
Solidus 1385°F; Liquidus 1440°F.
Paste type braze alloy.

FUSION NO. 1450.
M-1243; 50 Ag, 34 Cu, 16 Zn, plus flux.
Solidus 1260°F; Liquidus 1410°F.
Paste type braze alloy.
AWS A5.8-69 BAg6.

FUSION NO. 1565.
M-1243; 53 Cu, 9 Ag, 38 Zn, plus flux.
Solidus 1450°F; Liquidus 1565°F.
Paste type braze alloy.

FUSION NO. 1600.
M-1243; 51.5 Cu, 4.5 Ag, 44 Zn, plus flux.
Solidus 1410°F; Liquidus 1635°F.
Paste type braze alloy.

FUSION NO. 1610.
M-1243; 89 Ni, 11 P, plus flux.
Solidus 1610°F; Liquidus 1610°F.
Paste type braze alloy.
AWS A5.8-69 BNi6.

FUSION NO. 1630.
M-1243; 77 Ni, 12 Cr, 10 P, plus flux.
Solidus 1630°F; Liquidus 1630°F.
Paste type braze alloy.
AWS A5.8-69 BNi7.

FUSION NO. 1650.
M-1243; 55 Cu, 44.75 Zn, 0.25 Mn, plus flux.
Solidus 1610°F; Liquidus 1635°F.
Paste type braze alloy.

FUSION NO. 1740.
M-1243; 54 Ag, 21 Cu, 25 Pd, plus flux.
Paste type braze alloy.

FUSION NO. 1742.
M-1243; 82 Au, 18 Ni, plus flux.
Solidus 1740°F; Liquidus 1740°F.
Paste type braze alloy.
AWS A5.8-69 BAu4.

FUSION NO. 1761.
M-1243; 100 Ag, plus flux.
Solidus 1761°F; Liquidus 1761°F.
Paste type braze alloy; pure silver.

FUSION NO. 1800.
M-1243; 80 Cu, 20 Sn, plus flux.
Solidus 1470°F; Liquidus 1635°F.
Paste type braze alloy.

FUSION NO. 1830.
M-1243; 90 Cu, 10 Sn, plus flux.
Solidus 1750°F; Liquidus 1830°F.
Paste type braze alloy.

FUSION NO. 1850.
M-1243; 100 Cu_2O, plus flux.
Solidus 2040°F; Liquidus 2100°F.
Paste type braze alloy.

FUSION NO. 1900.
M-1243; 100 Cu, plus flux.
Solidus 1980°F; Liquidus 1980°F.
Paste type braze alloy; pure copper.
AWS A5.8-69 BCu 1a.

FUSION NO. 1900-C.
M-1243; 90 Cu, 10 Cu_2O, plus flux.
Liquidus 1980°F.
Paste type braze alloy.

FUSION NO. 1900-F.
M-1243; 95 Cu, 5 Fe_2O_3, plus flux.
Paste type braze alloy.

FUSION NO. 1900-FC.
M-1243; 90 Cu, 7 Cu_2O, 3 Fe_2O_3, plus flux.
Liquidus 1980°F.
Paste type braze alloy; improved filleting.

FUSION NO. 2412.
M-1243; 48 Au, 29 Ag, 17 Cu, 6 Zn, plus flux.
Solidus 1350°F; Liquidus 1450°F.
Paste type braze alloy; for jewelry.

FUSION NO. 2460.
M-1243; 42 Au, 32 Ag, 16 Cu, 10 Zn, plus flux.
Solidus 1335°F; Liquidus 1380°F.
Paste type braze alloy; for jewelry.

FUSION NO. 2466.
M-1243; 23 Au, 32 Ag, 25 Cu, 1 Zn, 19 Cd, plus flux.
Solidus 1200°F; Liquidus 1285°F.
Paste type braze alloy; for jewelry.

FUSION NO. 2468.
M-1243; 29 Au, 31 Ag, 20 Cu, 1 Zn, 19 Cd, plus flux.
Solidus 1280°F; Liquidus 1400°F.
Paste type braze alloy; for jewelry.

FUSION NO. 4765.
M-1243; 56 Ag, 42 Cu, 2 Ni, plus flux.
Solidus 1420°F; Liquidus 1640°F.
Paste type braze alloy.

FUSION NO. 4772.
M-1243; 54 Ag, 40 Cu, 5 Zn, 1 Ni, plus flux.
Solidus 1325°F; Liquidus 1575°F.
Paste type braze alloy.
AWS A5.8-69 BAg 13; AMS 4772.

FUSION NO. 4774.
M-1243; 63 Ag, 28.5 Cu, 2.5 Ni, 6 Sn, plus flux.
Solidus 1275°F; Liquidus 1475°F.
Paste type braze alloy.
AMS 4774.

FUSION NO. 4775.
M-1243; 74 Ni, 14 Cr, 4.5 Fe, 4.5 Si, 3 B, plus flux.
Solidus 1780°F; Liquidus 1900°F.
Paste type braze alloy.
AWS A5.8-69 BNi1.

FUSION NO. 4776.
M-1243; 74.5 Ni, 15 Cr, 3 Fe, 4.5 Si, 3 B, plus flux.
Solidus 1780°F; Liquidus 1970°F.
Paste type braze alloy.
AMS 4776.
For stainless steels.

FUSION NO. 4777.
M-1243; 82.6 Ni, 7 Cr, 3 Fe, 4.5 Si, 2.9 B, plus flux.
Solidus 1780°F; Liquidus 1830°F.
Paste type braze alloy.
AWS A5.8-69 BNi2.

FUSION NO. 4778.
M-1243; 92.5 Ni, 4.5 Si, 3 B, plus flux.
Solidus 1800°F; Liquidus 1900°F.
Paste type braze alloy.
AWS A5.8-69 BNi 3.

FUSION NO. 4779.
M-1243; 94.5 Ni, 3.5 Si, 2 B, plus flux.
Solidus 1810°F; Liquidus 1935°F.
Paste type braze alloy.
AWS A5.8-69 BNi4.

FUSION NO. 8100.
M-1243; 70.8 Ni, 19 Cr, 10.2 Si, plus flux.
Solidus 1975°F; Liquidus 2075°F.
Paste type braze alloy.
AWS A5.8-69 BNi5.

FUSION NO. 8300.
M-1243; 61 Ni, 19 Cr, 10 Si, 10 Mn, plus flux.
Solidus 1975°F; Liquidus 2075°F.
Paste type braze alloy; for stainless steel.

FUSION NO. 24695.
M-1243; 38 Au, 26 Ag, 19 Cu, 1 Zn, 16 Cd, plus flux.
Solidus 1175°F; Liquidus 1300°F.
Paste type braze alloy; for jewelry.

F.V.S. STEEL.
M-521; 0.42 C, 1.5 Si, 0.7 Mn, 14 Cr, 14 Ni, 2.5 W, bal Fe.
Heat treated: 130,000 TS; 74,000 YS; 23 El; 24 RA.
For stainless parts; austenitic, stainless.

FW 45.
M-1290; 0.42-0.50 C, 0.15-0.40 Si, 0.60-0.90 Mn, bal Fe.
Carbon tool steel.
W.-Nr. 1.1730.

FW 63.
M-1290; 0.58-0.64 C, 0.15-0.40 Si, 0.60-0.90 Mn, bal Fe.
Carbon tool steel.
W.-Nr. 1.1740.

FW 75 C.
M-1290; 0.70-0.80 C, 0.25-0.50 Si, 0.50-0.70 Mn, 0.25-0.40 Cr, bal Fe.
Cold work tool steel.
W.-Nr. 1.2003.

FW 90.
M-1290; 0.85-0.95 C, 0.15-0.40 Si, 0.40-0.60 Mn, bal Fe.
Carbon tool steel.
W.-Nr. 1.1760.

FX XTRA.
M-55; 0.55 C, 0.75 Mn, 1.0 Ni, 1.0 Cr, 0.3 Mo, bal Fe.
Vacuum arc degassed; hot work die steel.
Good high temperature properties and excellent toughness.

G

G 6 AG CU ZN.
M-200; 67% Ag, plus Cu, Zn.
Brazing alloy for silver; melt range: 705-723°C.

G7.
M-1522; 35 Ag, W.
Ann hardness: 165 HV.
Elec. res.: 3.5 $\mu \Omega$ cm.
Elec cond: 49% IACS.
For electrical contacts; good erosion resistance.

G97.
M-249, M-1548; 12.0 Cu, 1.4 Mn, 0.25 Mg, 0.8 Fe, bal Al.
Cast: 24,000-32,000 TS; 21,000-26,000 YS; 0.5-1.0 El.
For pistons. Cast alloy.

G-192 ALLOY.
0.60 C, 22 Cr, 8.5 Mn, 0.35 N, bal Fe.
For high temperature valves, gas turbine parts, nozzles, afterburners; super strength, high temperature alloy.

GADOLINIUM.
M-1755; Gd.
Purities: 99.9% (Special distilled grade) 99.5 + %.
Forms: Ingot, lump, sponge, wire, sheet, foil, turnings, powder, single crystals.

GALAVAN.
M-604; low C, Mn, bal Fe.
82 ksi TS; 80 ksi YS.
Pre-painted galvanized steel sheet for truck and trailer bodies.

GALICAR.
M-405; Pb, Sb, bal Sn.
For antifriction alloy; Babbitt.

GALLIMORE METAL.
M-409; 45 Ni, 28 Cu, 25 Zn, 2 Fe, Si, Mn.
Hard rolled: 156,000 TS; 2.5 El.
Soft rolled: 98,000 TS; 42 El.
For airplane parts, non-corrosive stampings; non-corrosive in sea water.

GALLIUM.
M-1755; Ga.
Purities: 99.99999%, 99.9999%, 99.999%, 99.99%, 99.9%.
Packed in: Sealed poly packets, poly or glass bottles, quartz boats or sealed ampules.

G-ALLOY.
M-311; 18 Zn, 2.5 Cu, 0.35 Mg, 0.35 Mn, 0.20 Fe, 0.75 Si, bal Al.
Hot rolled: 72,000-78,000 TS; 58,000-69,500 YS; 19-17 El.
For cast parts for trucks, airplanes and boats; light weight.

G-ALLOY.
M-314; Pb alloy.
10,000 TS; 22 Brin.
For bearings.

GALVALLOY.
M-813; Zn-Pb-Sn.
For Al solder; no flux required.

GALVANNEALED.
M-604; low C, Mn, bal Fe.
Steel sheet with remelted zinc coating.

GALVA-ONE.
M-604; low C, Mn, bal Fe.
Electro-galvanized steel sheet, zinc coated on one side only.

GALVOBRITE.
M-815; Zn alloy.
For galvanizing industry; improves fluidity of Zn.

GALVOMAG.
M-43; 0.010 max Al, 0.50-1.3 Mn, 0.02 max Cu, 0.001 max Ni, 0.03 max Fe, bal Mg.
Extruded anode.

GALVOROD.
M-43; 2.5-3.5 Al, 0.20 min Mn, 0.7-1.3 Zn, 0.05 max Si, 0.01 max Cu, 0.001 max Ni, 0.002 max Fe, bal Mg.
Extruded anode.

GALV-WELD.
M-1104; Pb, Zn, Sn, Bi.
For solder for regalvanizing; Zn-base alloy.

GAMA.
M-English; 12 Cu, 2 Ni, 0.5 Fe, 0.5 Si, bal Al.
For pistons; low friction.

GAMAN H.
M-38; 0.51 C, 12.0 Mn, 2.7 Si, 21.25 Cr, 0.45 N, bal Fe.
For diesel engine valves.

GAMMA NICKEL STEEL.
M-85; 36 Ni, bal Fe.
Annealed: 70,000 TS; 24,000 YS; 36 El; 68 RA; 143 Brin.
Cold drawn: 90,000 TS; 70,000 YS; 20 El; 60 RA; 185 Brin.
For instruments, chronometers; low thermal expansion.

GANNALOY.
M-1602, M-85; 0.03 C, 1.4 Mn, 0.40 Si, 5.5 Cr, 24.5 Ni, 2.25 Ti, 0.68 Al, 0.003 B, bal Fe.
For corrosion and heat resisting parts.

GAPASIL 9.
M-1493; 9 Pd, 9 Ga, 82 Ag.
Brazing range: 1616-1688°F. (880-920°C).
Uses include brazing Titanium in partial vacuum (500-1000 microns of He or Argon).

G.A. PERCIT EXTRA.
M-72; 1.2-1.4 C, 2-3 Si, 26-28 Cr, 57-63 Fe.
For hard surfacing electrodes; wear and corrosion resistant.

GARANT DOMO.
M-1318; 0.85 C, 4 Cr, W, Mo, V, bal Fe.
For lathe and planer tools, drills, reamers taps; high speed steel.

GARANT DOMO CO.
M-1318; 0.85 C, 4 Cr, Co, Mo, W, V, bal Fe.
For lathe and planer tools, hobs, reamers, taps, drills; high speed steel.

GARANT DOMO V.
M-1318; 0.85 C, 4 Cr, W, Mo, V, bal Fe.
For lathe and planer tools, reamers, broaches, taps; high speed steel.

GARANT EXTRA.
M-1318; 0.74 C, 4 Cr, 1.1 V, 18.5 W, bal Fe.
For lathe and planer tools, reamers, hobs, drills, taps; high speed steel.

GARANT EXTRA 333.
M-1318; 0.95 C, 4 Cr, W, Mo, bal Fe.
For lathe and planer tools, reamers, hobs, drills; high speed steel.

GARANT PRIMA.
M-1318; 0.82 C, 4 Cr, 0.85 Mo, 1.6 V, 8.7 W, bal Fe.
For lathe and planer tools, reamers, hobs, taps; high speed steel.

GARANT REKORD 3.
M-1318; 0.86 C, 2.8 Co, 4.3 Cr, 0.85 Mo, 2.1 V, 12 W, bal Fe.
For lathe and planer tools, hobs, taps, broaches; high speed steel.

GARANT REKORD 5.
M-1318; 0.79 C, 4.7 Co, 4.2 Cr, 0.8 Mo, 1.6 V, 18 W, bal Fe.
For lathe and planer tools, drills, reamers, taps; high speed steel.

GARANT REKORD 10.
M-1318; 0.76 C, 10 Co, 4 Cr, 0.8 Mo, 1.8 V, 18 W, bal Fe.
For lathe and planer tools, hobs, broaches; high speed steel.

GARANT SONDERKLASSE 500.
M-1318; 1.3 C, 4.3 Cr, 0.85 Mo, 3.8 V, 12 W, bal Fe.
For engravers' tools, form cutters, reamers; high speed steel.

GARANT SONDERKLASSE CO.
M-1318; 1.35 C, Co, 4.2 Cr, Mo, W, V, bal Fe.
For engravers' tools, form cutters, reamers; high speed steel.

GARANT SPEZIAL 275.
M-1318; 0.86 C, 4.Cr, 0.85 Mo, 2.5 V, 12 W, bal Fe.
For lathe and planer tools, broaches, reamers, taps; high speed steel.

GARBA.
M-359; 1.0 C, bal Fe.
For drills, taps, drill rod, water hardening.

GARFIELD.
M-1705; 1.0 C, 0.3 Mn, 0.25 Si, 0.5 Cr, 0.2 V, bal Fe.
Water hardening tool steel, AISI W7.

GASID NO. 12.
M-481; 9 Sn, 3 Zn, bal Cu.
30,000 TS; 22 El; 20 RA; 125 Brin.
For high pressure acid pumps in oil refineries; corrosion resistant to hot HCl.

GASID NO. 15.
M-481; 7 Al, 2.5 Fe, bal Cu.
42,000 TS; 20-30 El; 32 RA; 193 Brin.
For chemical apparatus and equipment; corrosion and acid resistant.

GASITE.
M-1251; 3.0-3.6 C, 4.0-4.75 Ni, 1.4-3.5 Cr, 0.4-0.7 Si, bal Fe.
Sand cast: 45,000 TS; 600 Brin.
Permanent mold: 55,000 TS; 675 Brin.
For heavy cams, dies, roller bearing races; white cast iron, corrosion resistant.

GAUSSIT 180.
M-Ger.; 50 Co, 14 V, bal Fe.
8000 Br, 1.4 BH max.
For electromechanical devices, hysteresis motors, digital computers.
Precipitation hardening permanent magnet.

G. BABBITT.
M-815, M 314; 12.5 Sb, 3 As, 0.75 Sn, bal Pb.
At 70°F: 9800 TS; 1.5 El; 22 Brin.
At 392°F: 1900 TS; 70 El.
For Babbitts, bearings; M.P. 486-549°F; heavy duty.

GBN.
M-1655; 0.38 C, 0.40 Mn, 1.50 Si, 0.10 V, bal Fe.
Cold work tool steel for chisels, punches, shearing tools.
W.-Nr. 1.2248.

G-B NO. 35.
M-1108; 40 Ag, 18 Cu, 15 Zn, 27 Cd.
For silver solder, brazing; M.P. 1120-1205°F.

G-B NO. 40.
M-1108; 40 Ag, 20 Cu, 16 Zn, 24 Cd.
For silver solder, brazing; MP 1125-1235°F.

G-B NO. 45.
M-1108; 45 Ag, 15 Cu, 16 Zn, 24 Cd.
For silver solder, brazing; M.P. 1125-1145°F.

G-B NO. 50.
M-1108; 50 Ag, 17.5 Cu, 11.5 Zn, 21 Cd.
For silver solder, brazing; M.P. 1170-1220°F.

G-B NO. 1300.
M-1108; 30 Ag, 29 Cu, 25 Zn, 16 Cd.
For silver solder, brazing; M.P. 1125-1320°F.

G-B NO. 1445.
M-1108; 40 Ag, 18 Cu, 15 Zn, 27 Cd.
For silver solder, brazing; M.P. 1120-1205°F.

G-B SH 7.
M-926; Ag, bal Cu.
For silver solder.

GBV.
M-1655; 0.45 C, 0.10 Mn, 1.5 Si, 1.5 Cr, 0.10 V, bal Fe.
Hot work tool steel.
Pressure casting molds for lead, tin, zinc alloys. Also cold work punches.
W.-Nr. 1.2249.

GBV 6.
M-1655; 0.60 C, 0.80 Mn, 0.90 Si, 1.20 Cr, 0.10 V, bal Fe.
Cold work tool steel for punches and stamping tools.
W.-Nr. 1.2243.

GBZ 10 BRONZE.
M-Ger.; 90 Cu, 10 Sn.
Cast: 45,500 TS; 20,000 YS; 45 El; 62 Brin.
For castings.

GBZ 14 BRONZE.
M-Ger.; 86 Cu, 14 Sn.
Cast: 32,000 TS; 27,000 YS; 7 El; 79 Brin.
For castings.

GBZ 20.
M-Ger.; 80 Cu, 20 Sn.
Cast: 27,000 TS; 0 El; 120 Brin.
For castings.

GCX.
M-836; 0.08 C, 0.45 Mn, 0.05 Si, 0.04 Cb, bal Fe.
For parts to be carbonitrided, as cams, ratchets, transmission levers.

GD15.
M-1522; 85 Ag, CdO.
Ann hardness: 60 HV.
Elec. res.: 2.1 $\mu \Omega$ cm.
Elec. cond.: 82% IACS.
For electrical contacts; good anti-weld properties.

GDH-O1.
M-816; 0.90 C, 1.2 Mn, 0.5 Cr, 0.2 V, 0.5 W, bal Fe.
Oil hardening tool steel; AISI O1.

GDH-O2.
M-816; 0.90 C, 1.5 Mn, 0.25 Si, 0.3 Mo, bal Fe.
Oil hardening tool steel; AISI O2.

GDH-07.
M-816; 1.25 C, 0.3 Mn, 0.35 Si, 0.4 Cr, 0.2 V, 1.4 W bal Fe.
Oil hardening tool steel, high hardness and water resistant; AISI O7.

GDH-3.
M-816; 2.15 C, 0.25 Mn, 0.25 Si, 12.0 Cr, 0.8 V, bal Fe.
Air or oil hardening tool steel for cold working dies, punches; AISI D3.

GDH-3SP.
M-816; 2.10 C, 12.5 Cr, 0.5 Ni, bal Fe.
Air or oil hardening cold work tool steel for dies, punches, gages; AISI D3.

GDH-10.
M-816; 0.40 C, 0.55 Mn, 1.0 Si, 3.25 Cr, 0.35 V, 2.5 Mo, bal Fe.
Air or oil hardening hot work tool and die steel for forging dies, hot forming tools; AISI H10.

GDH-11.
M-816; 0.40 C, 0.4 Mn, 1.0 Si, 5.0 Cr, 0.5 V, 1.35 Mo, bal Fe.
Air or oil hardening hot work tool and die steel for forging dies, hot forming tools; AISI H11.

GDH-12.
M-816; 0.35 C, 0.35 Mn, 1.05 Si, 5.0 Cr, 0.35 V, 1.25 W, 1.35 Mo, bal Fe.
Air or oil hardening hot work tool and die steel for forging dies, hot forming tools; AISI H12.

GDH-13.
M-816; 0.40 C, 0.4 Mn, 1.1 Si, 5.0 Cr, 1.1 V, 1.35 Mo, bal Fe.
Air or oil hardening hot work tool steel, chromium type; AISI H13.

GDH-A2.
M-816; 1.0 C, 0.7 Mn, 0.3 Si, 5.25 Cr, 0.3 V, 1.15 Mo, bal Fe.
Air hardening, medium alloy, cold work tool steel; AISI A2.

GDH-A4.
M-816; 0.95 C, 2.0 Mn, 2.2 Cr, 1.1 Mo, (lead added) bal Fe.
Air hardening, medium alloy, cold work tool steel; AISI A4.

G.D.H. BLUE LABEL.
M-816; 1.55-1.70 C, 11.5-12.5 Cr, 0.15-0.25 V, 0.7-0.9 Mo, bal Fe.
For tools, gauges, punches, dies; oil hardened.

GDH BRAKE DIE.
M-816; 0.50 C, 0.9 Mn 1.0 Cr, 0.20 Mo, bal Fe.
Oil hardening tool steel for shafts, arbors, lathe centers, forming tools.

G.D.H. BRONZE LABEL.
M-816; 0.90-0.95 C, 1.0-1.1 Mn, 0.4-0.5 Cr, 0.4-0.5 W, bal Fe.
Heat treated: 640 Brin.
For tools, gauges, broaches, taps; oil hardened.

GDH-CHW.
M-816; 0.23 C, 0.6 Mn, 1.25 Si, 10.0 Cr, 0.75 Ni, 1.0 V, 0.45 W, 1. Mo, 0.10 N, bal Fe.
Air or oil hardening hot work tool steel, chromium type.

GDH-D2.
M-816; 1.55 C, 0.35 Mn, 0.45 Si, 11.5 Cr, 0.9 V, 0.8 Mo, bal Fe.
Air or oil hardening cold work tool steel, chromium type; for lamination dies, punches, thread forming tools; AISI D2.

GDH-D2 FM.
M-816; 1.55 C, 0.35 Mn, 0.45 Si, 11.5 Cr, 0.9 V, 0.8 Mo, S, bal Fe.
Air or oil hardening cold work tool steel, chromium type; free machining grade. AISI D2.

GDH-D4.
M-816; 2.25 C, 0.35 Mn, 0.5 Si, 11.5 Cr, 0.2 V, 0.8 Mo, bal Fe.
Oil or air hardening cold work tool steel, chromium type; AISI D4.

GDH-D5.
M-816; 1.50 C, 0.4 Mn, 0.4 Si, 12.0 Cr, 0.35 Ni, 0.5 V, 0.9 Mo, 3.2 Co, bal Fe.
Air or oil hardening cold work tool steel, high carbon, high chromium type; AISI D5.

GDH GRAPH-AIR.
M-816; 1.35 C, 1.8 Mn, 1.2 Si, 1.85 Ni, 1.50 Mo, bal Fe.
Air or oil hardening, medium alloy, cold work tool steel; AISI A10.

GDH GRAPH-TUNG.
M-816; 1.50 C, 1.0 Mn, 0.9 Si, 0.5 Ni, 2.75 W, bal Fe.
Water or oil hardening tool steel; good wear resistance and high hardness.

GDH-H24.
M-816; 0.45 C, 0.3 Mn, 0.3 Si, 3.5 Cr, 0.7 V, 14.0 W, bal Fe.
Air or oil hardening hot work tool steel, tungsten type; AISI H24.

GDH-L2.
M-816; 0.50 C, 0.8 Mn, 0.3 Si, 1.0 Cr, 0.2 V, bal Fe.
Oil hardening tool steel for shafts, arbors, lathe centers; AISI L2; similar to AISI 4150.

GDH-L6.
M-816; 0.75 C, 0.75 Mn, 0.9 Cr, 1.75 Ni, 0.35 Mo, bal Fe.
Oil hardening tool steel for shafts, arbors, lathe centers, drill bushings; AISI L6.

GDH LUSTRE DIE.
M-816; 0.50 C, 0.9 Mn, 1.0 Cr, bal Fe.
Oil hardening tool steel for shafts, arbors.

GDH-M2.
M-816; 0.85 C, 4.15 Cr, 1.95 V, 6.4 W, 5.0 Mo, bal Fe.
High speed tool steel, molybdenum-tungsten type; for lathe tools, milling cutters, drills, broaches; AISI M2.

GDH-M41.
M-816; 1.1 C, 0.45 Mn, 0.3 Si, 4.25 Cr, 2.0 V, 6.75 W, 3.75 Mo, 5.0 Co, bal Fe.
High speed steel, molybdenum-tungsten-cobalt type; AISI M41.

GDH NO. 33.
M-816; 0.35 C, 0.8-1.0 Si, 4.5-5.0 Cr, 1.0-1.2 W, 1.3-1.5 Mo, bal Fe.
For forging dies, hot piercing and punching dies; hot work steel.

GDH NO. 212.
M-816; 0.45-0.50 C, 0.85-1.05 Cr, 0.9-1.2 W, 0.15-0.25 Mo, bal Fe.
For chisels, shear blades, die blocks; tough.

GDH NO. 280.
M-816; 0.5 C, 1.3-1.6 Si, 0.3-0.4 Mo, bal Fe.
For tools, punches, shear blades; tough.

GDH NO. 350.
M-816; 1.25-1.35 C, 3.5-3.7 W, bal Fe.
For drills, dies, gear cutters, taps; keen cutting edge.

GDH NON TEMPERING.
M-816; 0.35 C, 0.7 Mn, 0.45 Si, 0.8 Cr, 0.3 Mo, 0.3 Cu, bal Fe.
Oil hardening tool, shock resisting type.

G.D.H. RED LABEL.
M-816; 0.65-0.75 C, 3-4 Cr, 1.0-1.5 V, 18-20 W, bal Fe.
For tools, cutters; high speed steel.

GDH-S1.
M-816; 0.45-0.50 C, 0.2-0.5 Mn, 0.75-1.0 Si, 1.15-1.25 Cr, 0.15-0.25 V, 2.25-2.6 W, bal Fe.
Oil hardening tool, shock resisting type; AISI S1.

GDH-S5.
M-816; 0.6 C, 0.8 Mn, 2.0 Si, 0.25 Cr, 0.2 V, 0.25 Mo, bal Fe.
Water or oil hardening tool steel, shock resisting type; AISI S5.

GDH-S7.
M-816; 0.5 C, 0.7 Mn, 3.25 Cr, 1.4 Mo, bal Fe.
Oil hardening tool steel; shock resisting type; AISI S7.

G.D.H. SILVER LABEL.
M-816; 1.0 C, bal Fe.
Heat treated: 660 Brin.
For tools, punches, shears, dies; water hardened.

GDH-T1.
M-816; 0.75 C, 0.3 Mn, 0.3 Si, 4.0 Cr, 1.15 V, 18.25 W, bal Fe.
High speed steel, tungsten type, general purpose cutting tool; AISI T1.

GDH-W1.
M-816; 1.1 C, 0.3 Mn, 0.25 Si, bal Fe.
Water hardening tool steel; AISI W1.

GDH-W2.
M-816; 1.05 C, 0.25 Mn, 0.25 Si, 0.2 V, bal Fe.
Water hardening tool steel; AISI W2.

GE-17 PS.
M-31; 1.8-2.2 ThO_2, bal W.
For lamp filaments, heaterwire, rocket and missile high temperature parts.

GE 218.
M-31, M-60.
Tungsten base wire.
For operation to 3000°F.

GE 473.
M-31, M-60; 7 W, 3 Re, bal Ta.

GE 1570.
M-31, M-60; 0.20 C, 20.0 Cr, 29.0 Ni, 37.5 Co, 7.0 Mo, 4.2 Ti, 1.5 Fe.
High temperature alloy.

GEANT 3.
M-1118; 0.7 C, 18 W, 5 Co, 4 Cr, 1.3 V, bal Fe.
For tools, dies, cutters, hobs, drills; high speed steel.

GEANT 22.
M-1421; 0.8 C, 10 W, 4 Cr, 1.7 V, 0.9 Mo, bal Fe.
For lathe and planer tools, hobs, reamers; high speed steel.

GEANT 50.
M-1421; 0.75 C, 18 W, 4 Cr, 1 V, 0.7 Mo, bal Fe.
For lathe and planer tools, reamers, broaches, taps; high speed steel.

GEANT 55.
M-1421; 0.85 C, 12 W, 4 Cr, 2 V, 0.9 Mo, 2.8 Co, bal Fe.
For lathe and planer tools, drills, reamers; high speed steel.

GEANT 60.
M-1421; 0.80 C, 18 W, 4 Cr, 1.5 V, 0.75 Mo, bal Fe.
For lathe and planer tools, drills, reamers, hobs; high speed steel.

GEANT 66.
M-1421; 1.3 C, 4 Cr, 12 W, 4.5 V, 0.90 Mo, 4.5 Co, bal Fe.
For blanking and forming dies, engravers tools; high speed steel.

GEANT 77.
M-1421; 0.85 C, 18 W, 4 Cr, 1.5 V, 0.70 Mo, 5 Co, bal Fe.
For lathe and planer tools, milling cutters, drills; high speed steel.

GEANT 88.
M-1421; 0.85 C, 18 W, 4 Cr, 1.5 V, 0.7 Mo, 10 Co, bal Fe.
For lathe and planer tools, cutters, broaches; high speed steel.

GEANT M5.
M-1421; 0.85 C, 6 W, 4 Cr, 2.4 V, 5 Mo, bal Fe.
For lathe and planer tools, reamers; high speed steel.

GEAR BRONZE-1.
M-US; 78-91 Cu, 10-12 Sn, 0-3 Zn, 0.2 Pb, 0.1-0.3 P.
Cast: 38,000 TS; 20,000 YS; 10 El; 80 Brin.
For gears, worm wheels; tough.

GEAR BRONZE-2.
M-US; 88 Cu, 10 Sn, 2 Pb.
For gears; free-cutting.

GEAR BRONZE-3.
M-US; 85 Cu, 13 Sn, 2 Zn.
For gears; tough.

GEARING BRONZE.
M-Eng; 91.3 Cu, 8.7 Sn.
For gears; tough.

GEARS BRONZE.
M-Eng; 85 Cu, 10 Sn, 3 Zn, 2 Pb.
For gears; free-cutting.

GEAR STEEL.
M-Eng; 3.5 Ni, 1.5 Cr, 0.45-0.50 C, bal Fe.
For gears and pinions; oil-hardening.

GEAR STEEL HIGH DUTY.
M-Eng; C, alloy, bal Fe.
For gears; oil-hardening.

GEAR-WHEEL BRONZE.
M-452; 11-12 Sn, 0-1.5 Zn, 0.1-0.3 P, bal Cu.
Cast: 30,000-37,000 TS; 20,000-23,000 YS; 4-1 El; 90-125 Brin.
For high duty worm wheels; working with hardened steel worms.

G.E.C. HEAVY ALLOY.
M-978; 90 W, 7.5 Ni, 2.5 Cu.
Sintered: 90,000 TS; 83,000 YS; 4 El; 290 Brin.
For screens in x-ray tubes; sintered.

GEDGE'S METAL.
M-Eng.; 60 Cu, 38.2 Zn, 1.8 Fe.
For ship sheathing, cylinders for hydraulic presses; malleable at red heat.

GEMCO.
M-1493; 87.75 Cu, 12 Ge, 0.25 Ni.
M.P. 1508-1769°F.
For high temperature brazing.

GEMINOL N.
M-44; 3 Si, bal Ni.
Negative nuclear grade thermocouple.

GEMINOL P.
M-44; 80 Ni, 20 Cr, 1 Si.
Positive nuclear grade thermocouple.

GEMPCO.
M-1063; Sn-Pb-graphite, bal Cu.
For friction material, clutch discs, motor, brushes; sintered, self lubricating.

GENALLOY.
M-1005; 37-40 Ni, 17-21 Cr, C, Si, bal Fe.
Cast: 67,000 TS; 38,000 YS; 6 El; 207 Brin.
For furnace parts; high resistance to heat and corrosion.

GENALLOY B.
M-1005; 0.36-0.50 C, 31-39 Ni, 12-15 Cr, bal Fe.
For heat and corrosion resistant parts; heat and corrosion resistant.

GENCALLOY.
M-US; Pb alloy.
For cable sheathing; corrosion resistant.

GENCALOY.
M-US; Pb alloy.
For cable sheath.

GENELITE.
M-306; 70-73 Cu, 12-14 Sn, 9-10 Pb, 4.5-5.5 graphite.
For bearings for aero engines; synthetic bronze, self-lubricating.

GENERAL ALLOYS CN-1-H.
M-1005; 0.20-0.50 C, 26.0-30.0 Cr, 8.0-11.0 Ni, bal Fe.
Cast: 95,000 TS; 45,000 YS; 20 El; 200 Brin.
For oil burner parts, rabble arms, tube supports. Heat resistant, stainless.

GENERAL ALLOYS Q-10.
M-1005; 0.20 max C, 18-21 Cr, 8-11 Ni, bal Fe.
Annealed: 77,000 TS; 36,000 YS; 50 El; 163 Brin.
For pumps, rolls, valve bodies in chemical plant equipment; stainless, austenitic, ACI CF-20.

GENERAL ALLOYS Q-11.
M-1005; 0.08 max C, 18-21 Cr, 8-11 Ni, bal Fe.
Annealed: 77,000 TS; 37,000 YS; 55 El; 140 Brin.
For spray nozzles, sanitary fittings and chemical plant equipment; stainless, austenitic; ACI CF-8.

GENERAL ALLOYS Q-12.
M-1005; 0.08 max C, 18-21 Cr, 9-12 Ni, 2-3 Mo, bal Fe.
Annealed: 80,000 TS; 42,000 YS; 50 El; 168 Brin.
For mixing propellers; fittings in chemical plant equipment; stainless, austenitic ACI CF-8C.

GENERAL ALLOYS Q-13.
M-1005; 0.08 max C, 19 Cr, 11 Ni, 3-4 Mo, bal Fe.
Annealed: 80,000 TS; 42,000 YS; 50 El; 168 Brin.
For acid resistant casting; type 317; corrosion resistant. Fittings, mixing propellers.

GENERAL ALLOYS Q-14.
M-1005; 0.08 max C, 18-21 Cr, 9-12 Ni, Cb, bal Fe.
Annealed: 77,000 TS; 38,000 YS; 39 El; 149 Brin.
For pump parts, valve bodies in chemical plant equipment; stainless, austenitic; ACI CF-8C.

GENERAL ALLOYS Q-15.
M-1005; 0.16 max C, 18-21 Cr, 9-12 Ni, 0.2-0.35 Se, bal Fe.
Annealed: 77,000 TS; 40,000 YS; 52 El; 150 Brin.
For pumps, castings, valves in chemical plant equipment; free-machining austenitic stainless; ACI CF-16F.

GENERAL ALLOYS Q-16.
M-1005; 0.20 max C, 19 Cr, 9 Ni, 3 Mo, bal Fe.
Annealed: 70,000 TS; 30,000 YS; 35 El; 40 RA; 160 Brin.
For chemical plant equipment; ACI-CH20.

GENERAL ALLOYS Q-21.
M-1005; 0.20 max C, 22-26 Cr, 12-15 Ni, bal Fe.
Annealed: 83,000 TS; 50,000 YS; 38 El; 190 Brin.
For pumps, roasting equipment, furnace parts, heat treating boxes; heat resistant. ACI CH-20.

GENERAL ALLOYS Q-22.
M-1005; 0.10 max C, 22-26 Cr, 12-15 Ni, bal Fe.
Annealed: 86,000 TS; 40,000 YS; 40 El; 170 Brin.
For furnace parts, heat treating boxes, chemical plant equipment; heat resistant; ACI CH-20.

GENERAL ALLOYS Q-23.
M-1005; 0.10 max C, 24 Cr, 12 Ni, 3 Mo, bal Fe.
Annealed: 70,000 TS; 30,000 YS; 45 El; 55 RA; 150 Brin.
For chemical plant equipment; ACI-CH10M, heat and corrosion resistant.

GENERAL ALLOYS Q-30.
M-1005; 0.20 max C, 23-27 Cr, 19-22 Ni, bal Fe.
Annealed: 75,000 TS; 45,000 YS; 20 El; 180 Brin.
For valves, digesters, jet engine parts, furnace parts, heat treating boxes; heat resistant; ACI CK-20.

GENERAL ALLOYS Q-35.
M-1005; 0.20 C, 23-27 Cr, 19.22 Ni, bal Fe.
Annealed: 65,000 TS; 30,000 YS; 30 El; 165 Brin.
For furnace parts, heat treating boxes, chemical plant parts. ACI CK-20.

GENERAL ALLOYS Q-41.
M-1005; 0.07 max C, 16 Cr, 35 Ni, Mo, Cu, bal Fe.
Annealed: 70,000 TS; 35,000 YS; El; 45 RA; 150 Brin.
For chemical plant equipment, furnace fixtures; corrosion and heat resistant.

GENERAL ALLOYS Q-50.
M-1005; 0.08-0.15 C, 12 Cr, 1 max Ni, bal Fe.
Normalized: 100,000-200,000 TS; 75,000-150,000 YS; 30-7 El; 40 RA; 185-390 Brin.
For chemical plant equipment; ACI-CA15, corrosion resistant.

GENERAL ALLOYS Q-51.
M-1005; 0.2-0.4 C, 12 Cr, 1 max Ni, bal Fe.
Heat treated: 110,000-220,000 TS; 67,000-165,000 YS; 16-1 El; 7 RA; 212-470 Brin.
For chemical plant equipment; ACI-CA40, corrosion resistant.

GENERAL ALLOYS Q-56.
M-1005; 0.60-0.75 C, 17 Cr, 1 max Ni, bal Fe.
Heat treated: 400 Brin.
AISI 440A, corrosion resistant.

GENERAL ALLOYS Q-57.
M-1005; 0.95-1.2 C, 17 Cr, 1 max Ni, bal Fe.
Heat treated: 550 Brin.
AISI 440C, corrosion resistant.

GENERAL ALLOYS Q-58.
M-1005; 0.12 max C, 16 Cr, 1 max Ni, bal Fe.
Annealed: 80,000 TS; 45,000 YS; 10 El; 10 RA; 200 Brin.
For chemical plant equipment; AISI 430, corrosion resistant.

GENERAL ALLOYS Q-60.
M-1005; 0.30 max C, 20 Cr, 2 max Ni, bal Fe.
Annealed: 95,000 TS; 60,000 YS; 15 El; 10 RA; 200 Brin.
For chemical plant equipment; ACI-CB20, corrosion resistant.

GENERAL ALLOYS Q-65.
M-1005; 0.30-0.50 C, 28 Cr, 4 max Ni, bal Fe.
Annealed: 97,000 TS; 65,000 YS; 18 El; 212 Brin.
For furnace parts, heat treating boxes; ACI-CC50, corrosion and heat resistant.

GENERAL ALLOYS Q-66.
M-1005; 0.20 max C, 28 Cr, 3 max Ni, bal Fe.
Annealed: 55,000 TS; 45,000 YS.
For furnace parts, heat treating boxes, chemical plant equipment; ACI-CC20, heat and corrosion resistant.

GENERAL ALLOYS Q-67.
M-1005; 0.35 max C, 25 Cr, 1 max Ni, bal Fe.
Annealed: 55,000 TS; 40,000 YS.
For chemical plant equipment; AISI 446, corrosion resistant.

GENERAL ALLOYS Q-80.
M-1005; 0.10 max C, 26-30 Cr, 3-6 Ni, Mo, bal Fe.
Annealed: 60,000 TS; 40,000 YS.
For chemical plant equipment; resists high temperature and high temperature corrosion; AISI 329.

GENERAL ALLOYS Q-81.
M-1005; 0.07 max C, 20 Cr, 25 Ni, 3 Mo, 1 Cu, bal Fe.
Cast.
For chemical plant equipment; corrosion resistant.

GENERAL ALLOYS Q-82.
M-1005; 0.30 max C, 29 Cr, 9 Ni, bal Fe.
Cast: 80,000 TS; 40,000 YS; 20 El; 20 RA.
For furnace parts, chemical plant equipment; ACI-CE30, corrosion and heat resistant.

GENERAL ALLOYS Q-85.
M-1005; 0.07 C, 19-22 Cr, 27-31 Ni, 1.75-2.5 Mo, 3 min Cu, bal Fe.
Annealed: 65,000-75,000 TS; 28,000-38,000 YS; 50-35 El; 120-150 Brin.
For towers, pickling hooks in chemical plant equipment; resists sulphuric acid.

GENERAL ALLOYS Q-86.
M-1005; 0.12 max C, 1 max Cr, 64 Ni, 28 Mo, bal Fe.
Annealed: 80,000 TS; 57,000 YS; 8 El; 179-235 Brin.
For casting; corrosion and heat resistant; Hastelloy B. Chemical plant parts.

GENERAL ALLOYS Q-87.
M-1005; 0.12 max C, 1 max Cr, 55 Ni, 21 Mo, bal Fe.
Annealed: 71,000 TS; 47,000 YS; 12 El; 149-187 Brin.
For castings, corrosion and heat resistant; Hastelloy A.

GENERAL ALLOYS Q-88.
M-1005; 0.15 max C, 16 Cr, 56 Ni, 4 W, 17 Mo, bal Fe.
Annealed: 78,000 TS; 57,000 YS; 10 El; 187-248 Brin.
For castings; corrosion and heat resistant; Hastelloy C. resistant to chlorine.

GENERAL ALLOYS Q-89.
M-1005; 0.12 max C, 1 max Cr, 84 Ni, 9 Si, 4 Cu, bal Fe.
Annealed: 118,000 TS; 118,000 YS; 0-2 El.
Acid resistant casting; maximim resistance to sulphuric acid.
Reference: Hastelloy D.

GENERAL ALLOYS Q-90.
M-1005; 0.20 max C, 22 Cr, 45 Ni, Mo, Co, W, bal Fe.
Anneaded: 70,000 TS; 48,000 YS; 11 El; 170 Brin.
For casting; corrosion and heat resistant; Hastelloy X. Oxidation resistant to 2200°F.

GENERAL ALLOYS-Q91.
M-1005; 0.04 max C, 26 Cr, 5 Ni, 2 Mo, 3 Cu, bal Fe.
Aged: 140,000 TS; 115,000 YS; 18 El; 310 Brin.
Cast: 120,000 TS; 83,000 YS; 27 El; 260 Brin.
For acid resistant and chemical plant equipment, pump, valves, fittings.
ACI-CD4MCu. Corrosion resistant.

GENERAL ALLOYS X-3.
M-1005; 0.20-0.60 C, 28.0-32.0 Cr, 18.0-22.0 Ni, bal Fe.
Cast: 82,000 TS; 52,000 YS; 19 El; 192 Brin.
For carrier fingers, enameling fixtures, stack dampers; Heat resistant, stainless.

GENERAL ALLOYS X-4.
M-1005; 0.20-0.60 C, 24.0-28.0 Cr, 18.0-22.0 Ni, bal Fe.
Cast: 75,000 TS; 50,000 YS; 19 El; 192 Brin.
For cement kiln segments, pier caps, skid rails. Heat resistant, stainless.

GENERAL ALLOYS X-5.
M-1005; 0.50 max C, 26.0-30.0 Cr, 4.0-7.0 Ni, bal Fe.
Cast: 85,000 TS; 48,000 YS; 16 El; 190 Brin.
For brazing furnace parts, cracking equipment, rabble shoes. Heat resistant.

GENERAL ALLOYS X-6.
M-1005; 0.20-0.50 C, 19.0-23.0 Cr, 23.0-27.0 Ni, bal Fe.
Cast: 68,000 TS; 38,000 YS; 1m El; 160 Brin.
For chains, furnace beams, tube supports. Heat resistant. Stainless.

GENERAL ALLOYS X-7.
M-1005; 0.20-0.50 C, 24.0-28.0 Cr, 11.0-14.0 Ni, bal Fe.
Cast: 80,000 TS; 50,000 YS; 25 El; 185 Brin.
For exhaust manifolds, rabble arms, tube hangers.
Heat resistant, stainless.

GENERAL ALLOYS X-8.
M-1005; 0.20.0.40 C, 19.0-23.0 Cr, 9.0-12.0 Ni, bal Fe.
Cast: 85,000 TS; 45,000 YS; 35 El; 165 Brin.
For burner tips, fan housings, wear plates.
Heat resistant, stainless.

GENERAL ALLOYS X-9.
M-1005; 0.20-0.30 C, 26.0-30.0 Cr, 14.0-18.0 Ni, bal Fe.
Cast: 80,000 TS; 45,000 YS; 12 El; 180 Brin.
For furnace rails, lead pots, tube spacers.
Heat resistant, stainless.

GENERAL ALLOYS X-B.
M-1005; 33.37 Ni, 13-17 Cr, 0.35-0.75 C, bal Fe.
Cast: 70,000 TS; 40,000 YS; 10 El; 180 Brin.
For furnace parts, brazing trays, heat treating trays and baskets.
Heat and corrosion resistant.

GENERAL ALLOYS X-B CB.
M-1005; 0.35-0.75 C, 35 Ni, 15 Cr, Cb, bal Fe.
Cast: 70,000 TS; 40,000 YS; 10 El; 180 Brin.
For furnace parts, heat treating boxes, salt pots.
Heat and corrosion resistant.

GENERAL PLATE 154.
M-926; 45 Ag, 17 Cu, 16.5 Zn, 20.5 Cd, 0.5 Ni.
For silver brazing; M.P. 1135-1150°F.

GENERAL PLATE 715.
M-926; 15 Mn, 15 Ni, bal Cu.
Hardened: 100,000-133,000 TS; 40,000-77,000 YS; 12-20 El; 100-320 Brin.
For springs, diaphragms, watch cases; age-hardening, non-magnetic, corrosion resistant.

GENERAL PLATE 720.
M-926; 20 Mn, 20 Ni, bal Cu.
Wrought: 88,000-130,000 TS; 41,000-113,000 YS; 37-0.5 El; 150-300 Brin.
For springs, diaphragms, watch cases; age-hardening, non-magnetic, corrosion resistant.

GENERAL PLATE BH-1.
M-926; 10 Ag, 50 Cu, 40 Zn.
For silver brazing; M.P. 1495-1590°F.

GENERAL PLATE BH-2.
M-926; 10 Ag, 52 Cu, 38 Zn.
For silver brazing.

GENERAL PLATE CH-1.
M-926; 40 Ag, 36 Cu, 24 Zn.
For silver brazing; M.P. 1340-1405°F.

GENERAL PLATE CK-4.
M-926; 45 Ag, 30 Cu, 25 Zn.
For silver brazing.

GENERAL PLATE KA-1.
M-916; 52.5 Ag, 20 Cu, 14 Zn, 11.5 Cd, 2 Ni.
For brazing carbide tips; M.P. 1115-1195°F.

GENERAL PLATE KC-4.
M-926; 54 Ag, 40 Cu, 5 Zn, 1 Ni.
For silver brazing; M.P. 1430-1470°F.

GENERAL PLATE KH-2.
M-926; 50 Ag, 34 Cu, 16 Zn.
For silver brazing.

GENERAL PLATE KH-4.
M-926; 50 Ag, 15.5 Cu, 15.5 Zn, 16 Cd, 3 Ni.
For brazing carbide tips; M.P. 1195-1270°F.

GENERAL PLATE KH-7.
M-926; 50 Ag, 15.5 Cu, 16.5 Zn, 18 Cd.
For brazing ; M.P. 1160-1175°F.

GENERAL PLATE KH-105.
M-926; 50 Ag, 15.5 Cu, 25 Zn, 10 Cd.
For silver brazing.

GENERAL PLATE KK-5.
M-926; 55 Ag, 31.5 Cu, 11.7 Zn, 1.8 Ni.
For furnace brazing of steel parts; M.P. 130-1355°F.

GENERAL PLATE LH-1.
M-926; 20 Ag, 45 Cu, 30 Zn, 5 Cd.
For silver brazing; M.P. 1430-1500°F.

GENERAL PLATE LH-3.
M-926; 19.45 Ag, 47.75 Cu, 32,8 Zn.
For silver brazing; M.P. 1440-1500°F.

GENERAL PLATE LH-4.
M-926; 20 Ag, 45 Cu, 35 Zn.
For silver brazing.

GENERAL PLATE LM-1.
M-926; 27 Ag, 40.15 Cu, 38.25 Zn.
For silver brazing; M.P. 1350-1430°F.

GENERAL PLATE MA-1.
M-926; 72.15 Ag, 22.8 Cu, 5.05 Zn.
For silver brazing; M.P. 1345-1400°F.

GENERAL PLATE MH-4.
M-926; 70 Ag, 20 Cu, 10 Zn.
For silver brazing.

GENERAL PLATE ML.
M-926; 72 Ag, 28 Cu.
For silver brazing; M.P. 1435°F; eutectic alloy.

GENERAL PLATE SB-2.
M-926; 60.5 Ag, 22.5 Cu, 7 Zn, 10 Cd.
For silver brazing; M.P. 1285-1335°F.

GENERAL PLATE SH-7.
M-926; 60 Ag, 20 Cu, 7 Zn, 3 Sn.
For silver brazing; M.P. 1270-1300°F.

GENERAL PLATE SI-1.
M-926; 68 Ag, 27 Cu, 5 Sn.
For silver brazing; M.P. 1370-1400°F.

GENERAL PLATE SK-4.
M-926; 65 Ag, 20 Cu, 15 Zn.
for silver brazing; M.P. 1285-1325°F.
For silver brazing; M.P. 1285-1325°F.

GENERAL PLATE SM-1.
M-926; 66.7 Ag, 28.25 Cu, 5.05 Zn.
For silver brazing; M.P. 1360-1395°F.

GENESEE 100.
M-500; 0.4 C, Cr, Mo, bal Fe.
Heat treated: 105,000-135,000 TS; 85,000-115,000 YS; 15-11 El; 30-25 RA.
For gears, shafts; oil hardened.

GENESEE 180.
M-500; 0.3 C, Mn, bal Fe.
Normalized: 80,000 TS; 45,000 YS; 24 El; 45 RA.
For structures; water hardened.

GENESEE 185.
M-500; 0.27 C, Mn, Mo, bal Fe.
Normalized: 85,000 TS; 53,000 YS; 22 El; 35 RA.
For structures; water hardened.

GENESEE 191.
M-500; 0.3 C, Mn, Cr, Mo, bal Fe.
Rolled: 90,000 TS; 60,000 YS; 22 El; 45 RA.
For abrasion resisting parts; shock resisting.

GENESEE 194.
M-500; 0.20 C, Mo, bal Fe.
Normalized: 65,000 TS; 35,000 YS; 24 El; 35 RA.
For structures; case hardened.

GENESEE 194B.
M-500; 0.25 C, Mo, bal Fe.
Normalized: 70,000 TS; 45,000 YS; 22 El; 35 RA.
For structures; water hardened.

GENESEE 195.
M-500; 0.27 C, Mn, Mo, Cr, bal Fe.
Normalized: 100,000 TS; 65,000 YS; 17 El; 30 RA.
For structures; water hardened.

GENESEE 212.
M-500; 0.2 max C, 12-16 Cr, bal Fe.
For chemical engineering equipment; corrosion resistant.

GENESEE 255.
M-500; C, 4-6 Cr, 0.5 Mo, bal Fe.
For chemical engineering equipment; corrosion resistant.

GENESEE 280.
M-500; C, 28 Cr, 10 Ni, bal Fe.
For corrosion and heat resistant parts; heat and corrosion resistant.

GENESEE 303.
M-500; 0.2 C, 25 Cr, 20 Ni, bal Fe.
75,000 TS; 150 Brin.
For corrosion and heat resistant parts; heat and corrosion resistant to S fuels.

GENESEE 304.
M-500; 35 C, 17 Cr, 35 Ni, 0.65 Mn, bal Fe.
For hearth supports, heat treating baskets, valves, salt pots; heat and corrosion resistant.

GENESEE 305.
M-500; 0.15 C, 17 Cr, 65 Ni, 0.60 Mn, bal Fe.
For hearth supports, heat treating baskets, valves, salt pots; heat and corrosion resistant.

GENESEE 315.
M-500; 2.8 C, Mo, Si, bal Fe.
Cast: 40,000 TS; 250 Brin.
For chemical engineering equipment, stoves, grates; alloy gray iron.

GENESEE 405.
M-500; 0.45 C, 1 Cr, 0.75 Ni, 1.5 Mn 0.3 Mo, bal Fe.
For chemical engineering equipment; oil hardening.

GENESEE 412.
M-500; C, 11-14 Mn, Cr, bal Fe.
200 Brin.
For chemical engineering equipment, crusher plates; wear and abrasion resistant.

GENESEE 460.
M-500; 0.6 C, 1 Cr, 3 Ni, 0.3 Mo, bal Fe.
For chemical engineering equipment; oil hardening.

GENESEE 500.
M-500; 3.1 C, 2 Cr, 4.5 Ni, bal Fe.
550 Brin.
For castings, pulverizers; corrosion and abrasion resisting.

GENESEE KA-2.
M-500; 0.2 C, 18 Cr, 8 Ni, bal Fe.
For chemical engineering equipment, pump parts; stainless.

GENESEE KA4.
M-500; 0.2 max C, Cr, Ni, Mo, bal Fe.
Annealed: 75,000 TS; 160 Brin.
For paper mill equipment; resists sulfites; austenitic, corrosion resistant.

GENESEE NICKEL N-RESIST.
M-500; C, 2 Cr, 16 Ni, bal Fe.
For castings; corrosion resisting.

Section I: Alloy Data / 593

GENESEE NI-HARD.
M-500; 3.2 C, 2 Cr, 4.5 Ni, bal Fe.
550 Brin.
For chemical engineering equipment; wear resistant.

GENESEE NI-RESIST.
M-500; C, 2 Cr, 13 Ni, 6 Cu, bal Fe.
For chemical engineering equipment; heat resistant.

GENSTEEL.
M-1478; C, alloy, bal Fe.
For castings.

GENUINE SILVERINE.
M-586; Pb, Sb, bal Sn.
For bearings; Babbitt.

GENUINE SOVEREIGN BABBITT.
M-586; 86-90 Sn, bal Cu, Sb.
Cast: 28.4 Brin.
For internal combustion engine bearings; upper freezing point 306°C.

GEORO.
M-1493; 88 Au, 12 Ge.
M.P. 673°F. For brazing, Eutectic alloy.

GERMAN ALUMINUM ALLOY.
M-Ger.; 93 Al, 7 Mg.
For pistons; heat treatable.

GERMANIA B. BRONZE.
M-Eng.; 80.4 Zn, 9.6 Sn, 4.4 Cu, 4.7 Pb, 0.8 Fe.
For bearings; will not resist heat or live stream.

GERMANIUM.
M-1755; Ge.
Grades: Intrinsic-30, 40, 50 ohm/cm, First reduction-5 ohm/cm, Epitaxial.
Forms: Semicircular, trapezoidal, circular rods, lumps, powder, single crystals, slices.

GERMAN NAVY ANTIFRICTION METAL.
M-Ger.; 7.5 Cu, 85 Sn, 7.5 Sb.
For bearings, bushings; Babbitt.

GERMAN SILVER, AUSTRIAN.
M-Aus.; 60-50 Cu, 20-25 Zn, 20-25 Ni.
For ornamental purposes, electrical resistances; corrosion resistant.

GERMAN SILVER, AUSTRIAN.
M-Aus; 50 Zn, 4 Cu, 3.3 Sn, 1.2 Pb.
For solder, brazing; nearly white.

GERMAN SILVER BERLIN.
M-Ger.; 54 Cu, 28 Zn, 18 Ni.
For ornamental purposes, electrical resistances; corrosion resistant.

GERMAN SILVER, BEST.
M-Ger.; 46-50 Cu, 20-31 Ni, 29-34 Zn.
For springs and contact points in electrical work; German Silver.

GERMAN SILVER, BIRMINGHAM.
M-Eng.; 62-50 Cu, 32-20 Zn, 12-30 Ni.
For ornamental purposes, electrical resistances; corrosion resistant.

GERMAN SILVER, COMMON FORMULA.
M-Ger.; 55 Cu, 25 Zn, 20 Ni.
For ornamental purposes, electrical resistances; corrosion resistant.

GERMAN SILVER, FRENCH (ARCET).
M-Fr.; 50 Cu, 30-31.3 Zn, 20-18.7 Ni.
For ornamental purposes, electrical resistances; corrosion resistant.

GERMAN SILVER, FRENCH (CHAVAL).
M-Fr.; 58.3 Cu, 25 Zn, 16.7 Ni.
For ornamental purposes, electrical resistances; corrosion resistant.

GERMAN SILVER RUSSIAN.
M-USSR; 45-56 Cu, 18-23.5 Zn, 20-36 Ni.
For ornamental purposes, electrical resistances; corrosion resistant.

GERMAN SILVER, SHEFFIELD.
M-Eng.; 59.3 Cu, 25.9 Zn, 14,8 Ni.
For ornamental purposes, electrical resistances; Common Yellow.

GERMAN SILVER, SHEFFIELD.
M-Eng.; 55.2 Cu, 24.1 Zn, 20.7 Ni.
For ornamental purposes, electrical resistances; Silver White.

GERMAN SILVER, SHEFFIELD.
M-Eng.; 51.6 Cu, 22.6 Zn, 25.8 Ni.
For ornamental purposes, electrical resistances; Electrum, Bluish.

GERMAN SILVER, SHEFFIELD.
M-Eng.; 45.7 Cu, 20 Zn, 31.3 Ni.
For ornamental purposes, electrical resistances; Hard Alloy.

GERMAN SILVER SOLDER-1.
M-Ger.; 20 Ag, 30 Cu, 30 Zn, 20 Cd.
For silver solder.

GERMAN SILVER SOLDER-2.
M-Ger.; 10 Ag, 40 Cu, 30 Zn, 20 Cd.
For silver solder.

GERMAN SILVER SOLDER-3.
M-Ger.; 10 Ag, 3 Cu, 2 Zn, 75 Sn.
For silver solder; corrosion resistant.

GERMAN TYPE METAL "A".
M-Ger.; 75 Pb, 2 Sn, 23 Sb.
For type metal.

GERMAN TYPE METAL "B".
M-Ger.; 60 Pb, 35 Sn, 5 Sb.
For type metal.

GERMAN TYPE METAL "C".
M-Ger.; 60 Pb, 34.6 Sn, 5.4 Sb.
For type metal.

GERMAN TYPE METAL "D".
M-Ger.; 60 Pb, 15 Sn, 25 Sb.
For type metal.

GEROWAL I.
M-1374; 5.5-7.0 Si, 2-4 Cu, 0.4-0.6 Mn, 2 max Zn, 1 max Fe, bal Al.
Heat treated: 36,000 TS; 24,000 YS; 2 El; 80 Brin.
For light alloy parts; age hardenable.

GEROWAL 11.
M-1374; 4-7 Cu, 2-4 Si, 2.5 max Zn, 1.1 max Fe, bal Al.
For light alloy parts; age hardenable.

GEROWI SILBERIT.
M-1374; 1.5-3.5 Mg, 1.3 max Si, 0.6 max Mn, 0.2 Cr, 0.2 Ti, bal Al.
Soft: 28,000 TS; 13,000 YS; 30 El; 47 Brin.
For aircraft tanks and fittings, fuel lines, marine parts; resists sea water corrosion.

GEROWI SILBERIT-K.
M-1374; 2-4 Mg, 0.4 max Mn, 0.3 max Cr, bal Al.
Soft: 28,000 TS; 13,000 YS; 30 El; 47 Brin.
Hard: 40,000 TS; 35,000 YS; 10 El; 73 Brin.
For aircraft tanks and fittings, fuel lines, marine parts; resists sea water corrosion.

GEWA 235.
M-1307; 0.35 C, 0.9 Si, 0.3 Mn, 1 Cr, 0.2 V, 1.8 W, bal Fe.
For cold work tools, upsetters, crimpers; tough, shock resistant.

GEWA 250.
M-1307; 0.45 C, 0.9 Si, 0.3 Mn, 2 W, 1 Cr, 0.4 V, bal Fe.
For punches, chisels, riveters, crimpers, upsetters; tough, shock resistant.

GEWA 255.
M-1307; 0.55 C, 0.9 Si, 0.3 Mn, 1 Cr, 0.2 V, 1.8 W, bal Fe.
For punches, chisels, riveters, crimpers, upsetters; tough, shock resistant.

GEWA 430.
M-1307; 0.3 C, 1 Si, 0.4 Mn, 1.1 Cr, 0.18 V, 3.75 W, bal Fe.
For punches, chisels, riveters, crimpers, upsetters; tough, shock resistant.

GEWA 530.
M-1307; 0.3 C, 0.3 Mn, 2.3 Cr, 0.6 V, 4.2 W, bal Fe.
For punches, upsetters, riveters, shears, crimpers; hot work steel, oil hardened.

GEWA 930.
M-1307; 0.30 C, 0.2 Si, 0.3 Mn, 2.6 Cr, 0.35 V, 8.5 W, bal Fe.
For extrusion rams and liners, punches, upsetters; hot work steel, oil hardened.

GEWA 960.
M-1307; 0.65 C, 3.7 Cr, 0.85 Mo, 0.7 V, 8.5 W, bal Fe.
For drills, reamers, taps, hobs, broaches; high speed steel, oil hardened.

GEWA 2930.
M-1307; 0.30 C, 2 Co, 2.4 Cr, 0.25 V, 8.5 W, bal Fe.
For extrusion rams and liners, punches, shears; hot work steel, oil hardened.

G.F.A. NICKEL.
M-England; 99.0 min Ni + Co.
For anode support wires, valve components; corrosion and heat resistant.

GIANT SPECIAL.
M-1683; 0.35 C, 0.9 Cr, 0.3 Mo, bal Fe.
Oil or water hardening steel designed for plastic molds.

GIBSILOY A-1.
M-650; 95 Ag, 5 Ni.
For contact rivets, discs, screws; low contact resistance, high ductility.

GIBSILOY A 3.
M-650; 50.95 Ag, 5-50 Ni.
For electric contacts; sintered powders, ductile.

GIBSILOY A-4.
M-650; Ag, Ni.
For electrical contacts; circuit breakers.

GIBSILOY A-6.
M-650; Ag, Ni.
For electrical contacts; circuit breakers.

GIBSILOY A8.
M-650; Ag-Ni.
Sintered: 36,000-58,000 TS.
For electrical contacts; disconnecting switch contacts.

GIBSILOY C-1.
M-650; 99 Ag, 1 graphite.
For contact discs; non-sticking.

GIBSILOY C-2.
M-650; Ag, C.
For electrical contacts, disconnecting switches.

GIBSILOY C 4.
M-650; 93-99 Ag, 1-7 graphite.
For electric contacts, sliding; mechanically processed.

GIBSILOY C5.
M-650; Ag-C.
Sintered.
For electrical contacts, circuit breakers; non-welding.

GIBSILOY C-7.
M-650; 93 Ag, graphite.
For contact discs; non-sticking.

GIBSILOY CW-42.
M-650; Ag, W. C.
For electrical contacts, circuit breakers.

GIBSILOY CW-52.
M-650; Ag, W, C.
For electrical contacts, air circuit breakers.

GIBSILOY M 10.
M-650; 30-90 Ag, 10-70 Mo.
For electrical contacts; arcing.

GIBSILOY M-12.
M-650; Ag-Mo.
Sintered: 55,000 TS; 165 Brin.
For electrical contacts for air circuit breakers; arcing.

GIBSILOY NC 22.
M-650; Ag, Ni, graphite.
For electrical contacts; rheostat contacts.

GIBSILOY NC-43.
M-650; Ag, Ni, C.
For electrical contacts, rheostats, contactors.

GIBSILOY ND.
M-650; 5-25 Ni, 1-10 Cd, bal Ag.
For electrical contacts.

GIBSILOY NM.
M-650; 1-35 Ni, 1-30 Mo, bal Ag.
For electrical contacts; for high current densities.

GIBSILOY NW.
M-650; 1-35 Ni, 1-30 W, bal Ag.
For electrical contacts; for high current densities.

GIBSILOY NW-54.
M-650; Ag, Ni, W.
For electrical contacts; D.C. service.

GIBSILOY NW-55.
M-650; Ag, Ni, W.
For electrical contacts; circuit breakers.

GIBSILOY UC-5.
M-650; Ag, Cu, C.
For electrical contacts, rheostats, contactors.

GIBSILOY UT-6.
M-650; Cu-WC.
Rockwell B100.
For electrical contacts, switches; density 12.3; sintered.

GIBSILOY UT8.
M-650; Cu-WC.
Sintered: 230 Brin.
For electrical contacts and facings for resistance welding electrodes; sintered.

GIBSILOY UT-10.
M-650; Cu-WC.
Rockwell B95.
For electrical contacts, switches; density 11.1; sintered.

GIBSILOY UW4.
M-650; Cu-W.
For electrical contacts and facings for resistance welding electrodes; sintered.

GIBSILOY UW-5.
M-650; Cu-W.
Rockwell B100.
For electrical contacts, switches; density 14.3; sintered.

GIBSILOY UW6.
M-650; Cu-W.
B90 Rockwell.
For electric contacts, switches; sintered 50% elec. cond.

GIBSILOY UW-8.
M-650; Cu-W.
Rockwell B85.
For electrical contacts, switches; density 12.6; sintered.

GIBSILOY W 2.
M-650; 30-90 Ag, 10-70 W.
For electrical contacts.

GIBSILOY W-4.
M-650; Ag-W.
For electrical contacts; contactor contacts.

GIBSILOY W 10.
M-650; Ag-W.
For electrical contacts; sintered powders.

GIBSILOY W12.
M-650; Ag, W.
For electrical contacts; air circuit breakers.

GIBSILOY W-13.
M-650; Ag-W.
Sintered: 43,000 TS; 150 Brin.
For electrical contacts for air circuit breakers; arcing.

GIBSILOY W-15.
M-650; Ag-W.
Sintered: 63,000 TS; 200 Brin.
For electrical contacts for air circuit breakers; arcing.

GIBSILOY WC-10.
M-650; Ag, WC.
For electrical contacts for air circuit breakers.

GIBSILOY WID.
M-650; Ag, W.
For electrical contacts; air circuit breakers.

GIBSON S-5.
  M-650; Ag-Zn.
  For thermostats, contactors, switches, relays; electrical contacts.

GIBSON S-10.
  M-650; Ag-Cd.
  For thermostats, contactors, switches, relays; electrical contacts.

GIBSON S-15.
  M-650; Ag-Pd.
  For thermostats, contactors, switches, relays; electrical contacts.

GIBSON S-16.
  M-650; Ag, Pd.
  For electrical contacts, thermostats, switches, relays.

GIBSON S-20.
  M-650; Ag-Pt.
  For thermostats, contactors, switches, relays; electrical contacts.

GIBSON S-25.
  M-650; Ag-Cu.
  For electrical contacts.

GIBSON S-26.
  M-650; Ag, Cu, Ni.
  For electrical contacts, relays, switches.

GIESCHE Z13.
  M-1134; 4 Al, 0.5-1.0 Cu, 0.03 Mg, bal Zn.
  Rolled: 44,000-71,000 TS; 75-140 Brin.
  For die castings, instrument cases, housings; rolled or cast.

GIESCHE Z14.
  M-1134; 0.8-1.0 Al, 0.3-0.4 Cu, bal Zn.
  Rolled: 31,000-42,500 TS; 40-70 Brin.
  For die castings, housings, gears; rolled or cast.

GIESCHE ZL 2.
  M-Ger.; 4 Al, 1.0-1.5 Cu, bal Zn.
  For die or sand castings.

GIESCHE ZL 3.
  M-Ger.; 4 Al, 0.5 Cu, bal Zn.
  For pressed or drawn parts.

GIESCHE ZL7.
  M-1134; 4 Cu, 0.2 Al, bal Zn.
  Rolled: 28,000-59,600 TS; 45-100 Brin.
  For die castings, housings, gears; rolled or cast.

GIESCHE ZO4.
  M-1134; 4 Cu, bal Zn.
  For die castings; rolled or cast.

GIGANT.
  M-617; 0.7 C, 18 W, 4 Cr, 1 V, bal Fe.
  For cutters, tools; high speed steel.

GIGANT 44.
  M-912; 1.3 C, 4.25 Cr, 0.85 Mo, 3.75 V, 12.0 W, bal Fe.
  High speed steel for broaches, lathe tools for finishing cuts.
  W.-Nr. 3302; DIN S 12-1-4.

GIGANT 50.
  M-912; 0.75 C, 4.25 Cr, 1.10 V, 18.0 W, bal Fe.
  High speed steel for lathe tools, milling cutters, taps.
  W.-Nr. 3355; AISI T1.

GIGANT 66.
  M-912; 1.4 C, 4.25 Cr, 0.85 Mo, 3.75 V, 12.0 W, 4.8 Co, bal Fe.
  High speed steel for cutting tools; resistant to wear; for special operations.
  W.-Nr. 3202; similar to AISI T15.

GIGANT 77.
  M-912; 0.80 C, 4.25 Cr, 0.65 Mo, 1.55 V, 18.0 W, 4.80 Co, bal Fe.
  High speed steel for lathe tools and other cutting tools for special operations; resistant to elevated temperature in difficult cutting.
  W.-Nr. 3255; similar to AISI T4.

GIGANT 88.
  M-912; 0.75 C, 4.25 cr, 0.65 Mo, 1.55 V, 18.0 W, 9.5 Co, bal Fe.
  High speed steel; best red-hardness; for lathe tools.
  W.-Nr. 3265; similar to AISI T5.

GIGANT 100.
  M-912; 1.25 C, 4.25 Cr, 3.75 Mo, 3.25 V, 10.0 ws, 12.5 Co, bal Fe.
  High speed steel, particularly for automatic lathe tools.
  W.-Nr. 3207; DIN S 10-4-3-10.

GIGANT 201.
  M-912; 1.10 C, 4.25 Cr, 3.80 Mo, 1.85 V, 6.85 W, 5.0 Co, bal Fe.
  High speed steel; for drills, milling cutters, reamers, taps, counter sinks, lathe tools.
  W.-Nr. 3246; DIN S 7-4-2-5.

GIGANT 301.
  M-912; 1.45 C, 4.25 Cr, 5.0 Mo, 2.4 V, 6.35 W, 8.0 Co, bal Fe.
  High speed steel; for finishing tools and for machining abrasive materials.
  W.-Nr. 3222; DIN S 6-5-2-8.

GIGANT M5.
  M-912; 0.88 C, 4.25 Cr, 5.0 Mo, 1.95 V, 6.35 W, bal Fe.
  High speed steel; for drills, lathe tools, milling cutters. For rough machining.
  W.-Nr. 3343; AISI M2.

GIGANT M 5 CO.
M-912; 0.92 C, 4.25 Cr, 5.0 Mo, 1.95 V, 6.35 W, 4.8 Co, bal Fe.
High speed steel; for special twist drills, milling cutters, sectional tools.
W.-Nr. 3243; Similar to AISI M35.

GIGANT M 5 H.
M-912; 0.96 C, 4.25 Cr, 5.0 Mo, 1.85 V, 6.35 W, bal Fe.
High speed steel for finishing tools and heavy duty drills, milling cutters.
W.-Nr. 3342; AISI M2 (high carbon type).

GIGANT M 5 S.
M-912; 0.88 C, 0.12 S, 4.25 Cr, 5.0 Mo, 1.85 V, 6.35 W, bal Fe.
Free machining Mo-W high speed steel; for special tools or intricate shapes.

GIGANT M 5 V.
M-912; 1.22 C, 4.25 Cr, 5.0 Mo, 3.0 V, 6.35 W, bal Fe.
High speed steel for heavy duty broaches, reamers, milling cutters.
W.-Nr. 3344; AISI M3 Class 2.

GIGANT M 9.
M-912; 0.82 C, 4.0 Cr, 8.60 Mo, 1.15 V, 1.80 W, bal Fe.
High speed steel; for twist drills, lathe tools, milling cutters; for rough machining.
W.-Nr. 3346, AISI M1.

GIGANT M 9 V.
M-912; 1.0 C, 4.0 Cr, 8.60 Mo, 2.0 V, 1.80 W, bal Fe.
High speed steel; milling cutters, lathe tools, for cutting hard materials.
W.-Nr. 3348; AISI M7.

GILDING 4.
M-8; 95 Cu, 5 Zn.
Soft: 35,000 TS; 11,000 YS; 38 El.
Hard: 55,000 TS; 44,000 YS; 5 El.
For jewelry and bullet jackets; fuse caps, primers; gilding metal.

GILDING 95%.
M-33; 95 Cu, 5 Zn.
Annealed: 35,000 TS; 11,000 YS; 45 El.
Hard rolled: 56,000 TS; 50,000 YS; 5 El.
For angles, channels; red brass.

GILDING 210.
M-8; 95 Cu, 5 Zn.
Hard Sheet: 56,000 TS; 46,000 YS; 5 El; B 64 Rock.
Soft Sheet: 36,000 TS; 11,000 YS; 42 El; F 50 Rock.
For coins, costume jewelry, medallions, ornamental trim. 56% elect. cond., malleable.

GILDING FOIL.
M-U.S.; 98 Sn, 2.2 Cu, 0.1 Fe.
For cheap jewelry; corrosion resistant.

GILDING METAL.
M-Eng.; 72-64 Cu, 23-34 Zn, 0.3-2.5 Sn, 0.3 P.
For cheap jewelry, bullet jackets; corrosion resistant.

GILDING METAL 26.
M-141; 95 Cu, 5 Zn.
Hard: 56,000 TS; 50,000 YS; 5 El; 114 Brin.
Soft: 35,000 TS; 11,000 YS; 45 El; 50 Brin.
For jewelry; gilding metal.

GILDING METAL, 95%.
M-33, M-1789; 95 Cu, 5 Zn.
Ann: 35,000 TS; 11,000 YS; 45 El.
Rolled hard: 56,000 TS; 50,000 YS; 5 El.
For angles, coinage, medallions.
Good corrosion resistance.

GILDING METAL 210.
M-279; 95 Cu, 5 Zn.
Ann: 34,000-38,000 TS; 5000-11,000 YS; 45-48 El.
Cold rolled: 37,000-69,000 TS; 14,000-59,000 YS; 2-45 El.
For coins, medallions, bullet jackets, terminals; red color, easily formed good corrosion resistance.

GILGRID 10.
M-897; 45 Ni, 45 Fe, 10 Mo.
Annealed: 90,000 TS; 72,000 YS; 33 El.
For grid wires, electronic tubes; corrosion and heat resistant.

GILGRID 30.
M-897; 67 Ni, 30 Mo, 1 Cr, 1 Si, 1 Mn.
Annealed: 90,000 TS; 63,000 YS; 15 El; 18 RA.
For grid wires, electronic tubes; corrosion and heat resistant.

GILMORE NICKEL ALLOY.
M-817; Cu, bal Ni.
For diamond setting tools, holders.

GILSON BM12C20.
M-1298; 0.2 max C, 22-24 Cr, 12-15 Ni, 2 max Mn, 3.5 max Si, bal Fe.
Annealed: 85,000-95,000 TS; 40,000-50,000 YS; 45-55 El; 150-185 Brin.
For heat treat boxes, furnace parts and equipment; Type 309; austenitic, heat resistant.

GILSON UC-BC14.
M-1298; 0.15 max C, 11.5-13.5 Cr, bal Fe.
Heat treated: 120,000-135,000 TS; 110,000-117,000 YS; 16-15 El; 63-58 RA; 220-240 Brin.
For cutlery, valves, turbine blades; Type 410; corrosion resistant.

GILSON UG.
M-1298; 0.2 max C, 22-24 Cr, 12-15 Ni, 2 max Mn, 3.5 max Si, bal Fe.
Annealed: 85,000-95,000 TS; 40,000-50,000 YS; 45-55 El; 150-185 Brin
For heat treat boxes, furnace parts, refinery equipment; Type 309; austenitic, heat resistant.

GILSON-UGAC14.
M-1298; 0.08 C, 11.5-13 Cr, 0.1-0.3 Al, bal Fe.
Normalized: 71,000 TS; 42,600 YS; 22 El; 150 Brin.
Ht. Tr.: 175,000 TS; 145,000 YS; 21 El; 352 Brin.
For injectors, table flatware, valves, pumps, oil refinery equipment.
Type 405, stainless.

GILSON UG-AM.
M-1298; 0.10 max C, 16-18 Cr, 10-14 Ni, 1.7-2.7 Mo, Cb = 10 x C, bal Fe.
Annealed: 85,000-95,000 TS; 35,000-45,000 YS; 60-50 El; 75-60 RA; 150-190 Brin.
For acid resistant chemical plant equipment; Type 316 Cb; stainless, austenitic.

GILSON UG-AM 8C, 18MO.
M-1298; 0.08 max C, 17-19 Cr, 8-11 Ni, Ti = 5 x C, bal Fe.
Normalized: 93,000 TS; 36,000 YS; 45 El; 60 RA; 165 Brin.
Annealed: 87,000 TS; 33,000 YS; 57 El; 73 RA; 155 Brin.
For chemical plant equipment, muffles, cowls; Type 321; stainless, austenitic.

GILSON UG-CC14.
M-1298; 0.15 min C, 12-14 Cr, bal Fe.
Annealed: 88,000 TS; 40,000 YS; 32 El; 68 RA; 170 Brin.
Heat treated: 256,000 TS; 190,000 YS; 6 El; 10 RA; 540 Brin.
For cutlery, valve trim, springs, turbine blades; Type 420; stainless, hardenable.

GIMAP.
M-1786.
Powdered manganese alloys for coating of welding rods and wire for welding.

GIMEL.
M-1786; 96-98 Mn, 1-3 Fe.
For alloying in steel, copper alloys and light alloys; electrothermic manganese metal.

G-IRON.
M-1001; 4.2 C, 2.4 Si, 0.6 Mn, bal Fe.
For making gray iron castings; graphitized pig iron.

GITTERMETALL.
M-German; 0.2 graphite, 10 Sn, 15 Sb, 1.75 Cu, bal Pb.
For bearings; graphitic.

GKN 58-L.
M-632; 97.0 Fe, 3.0 max Graphite.
Sintered, oil impregnated (26%); 10,400 psi compressive strength.
Bearings for light duty motors.

GKN 59-FM.
M-632; 96.25 Fe, 0.4 S, 3.35 max others.
Sintered, oil impregnated (25%); 17,000 TS; 11,500 compressive strength.
Excellent machining, ductile enough for staking or spinning. For low and medium loaded bearings.
ASTM B439-67, Gr. 1; ASTM B310-67, Type I, Class A2, SAE 850; PMPMA F-0000-N.

GKN 59-I.
M-632; 97.0 Fe, 3.0 max others.
Sintered, oil impregnated (25%); 12,500 psi compressive strength.
For low and medium load bearings.
ASTM B439-67 Gr. 1; ASTM B310-67, Type I, Class A; SAE 850; PMPMA F-0000-N.

GKN 59-PC.
M-632; 0.7 C, 4.0 Cu, 95.3 Fe.
Sintered and oil impregnated (20% oil); 31,000 compressive strength.
Good strength, for products subject to frequent impact or shock loading.
ASTM B426-65 Grade 2, Type 1; SAE 865 A.

GKN 61-A.
M-632; 87.0 Cu, 9.5 Sn, 3.5 Graphite.
Sintered, oil impregnated (20%); 10,000 psi compressive strength.
Bearings for light, non-shock loads.

GKN 61-P.
M-632; 10.0 Cu, 87.0 min Fe, 3.0 max others.
Sintered, oil impregnated (25%), 30,000 compressive strength.
Good load bearing qualities for wheel bearings, automotive, farm equipment
ASTM B439-67, Gr. 3; ASTM B222-61; SAE 862; PMPMA; F-1000-N.

GKN 62-E.
M-632; 38.0 Cu, 4.0 N, 1.0 Grphite, 55.75 min Fe.
Sintered, oil impregnated with approx 23% oil.
Compressive strength 12,500 psi.
For self-aligning bearings in motors, fan motors, agricultural equipment.

GKN 63-H.
M-632; 89 Cu, 9.75 Sn, 1.25 Graphite.
Sintered, oil impregnated (27%); 10,500 psi compressive strength.
For self-aligning bearings in fractional horse power motors.

GKN 63-PC.
M-632; 0.7 C, 4.0 Cu, 95.3 Fe.
Sintered, oil impregnated (8%); 48,000 TS.
High strength grade for units subject to frequent shock or impact loading.

GKN 63-PZ.
M-632; 0.7 C, 20.0 Cu, 76,8 min Fe, 2.5 max others.
As sintered: 47,000 TS; 33,000 compressive stress; density 6.3.
For bearings having intermittant or oscillating loading; needs supplementary lubrication.
ASTM B426 Grade 4, Type II; SAE 867B.

GKN 66-H.
M-632; 89 Cu, 9.75 Sn, 1.25 Graphite.
Sintered, oil impregnated (23%); 12,000 psi compressive strength.
Good load bearing qualities, consumer products and industrial equipment.
ASTM B438-70 Grade 1, Type II; SAE 841; PMPMA BT-0010-R

GKN 66-Q.
M-632; 90 Cu, 10 Sn.
Sintered, oil impregnated (25%); 12,000 psi compressive strength.
Good bearing qualities; can be machined and staked.
ASTM B438-70 Gr. 1, Type II; SAE 841; PMPMA BT-0010-R.

GKN 66-R.
M-632; 87.0 Cu, 9.5 Sn, 3.5 Graphite.
Sintered, oil impregnated (18%); 12,000 psi compressive strength.
For oscillating and reciprocating loads.

GKN 70-H.
M-632; 89 Cu, 9.75 Sn, 1.25 Graphite.
Sintered, oil impregnated (15%); 15,000 psi compressive strength; density 7.0.
Bearings for heavy duty service as construction equipment, agriculture equipment.
ASTM B255-70, Type II; SAE 842; PMPMA BT-0010-S.

GKN 70-Q.
M-632; 90 Cu, 10 Sn.
Sintered, oil impregnated (20%); 15,000 psi compressive strength.
Bearings for shock and high loading.

GKN 70-R.
M-632; 87.0 Cu, 9.5 Sn, 3.5 Graphite.
Sintered, oil impregnated (5%); 18,000 psi compressive strength.
For oscillating and reciprocating loads.

GKN C-000.
M-632; 99.0 min Cu.
As sintered: 20,000-34,000 TS; 10-17 El; 60-90 Rock H; density 7.7-8.5.
Excellent electrical and thermal conductivity.

GKN CT-100.
M-632; 87.5-90.5 Cu, 9.5-10.5 Sn.
As sintered: 13,500-20,000 TS; 1.0-3.0 El; 50-90 Rock H; density 6.4-7.2.
Bearings and structural parts; may be coined for dimensional control.
ASTM B255, Type I or II; CT-0010-R; CT-0010-S.

GKN CZ-103.
M-632; 87.0-90.0 Cu, 1.0-2.0 Pb, 7.35 min Zn.
As sintered: 16,500-19,400 TS; 10-12 El; 45-70 Rock H; density: 7.2-8.0.
Commercial bronze for ornamental parts and lock hardware; good machinability.
CZP-0210-T; CZP-0210-U.

GKN CZ-203.
M-632; 77.0-80.0 Cu, 1.0-2.0 Pb, 17.35 min Zn.
As sintered: 20,000-30,000 TS; 10-19 El; 50-85 Rock H; density 7.2-8.0.
Lock hardware, plates, latches, electrical equipment; good machinability.
ASTM B282, Type I or II; CZP-0200-T; CZP-0220-U.

GKN D-1030.
M-632; 91.0-94.0 Cu, 3.5-4.5 Sn, 0.4-1.4 Zn.
As sintered: 20,000-35,000 TS; 12-22 El; 50-90 Rock H; density 7.4-8.2.
High ductility; good wear and corrosion properties.
May be coined to 8.2 density for increased strength and wear.

GKN I-000.
M-632; 0.25 max C, 97.75 min Fe.
Iron type sintered metal.
21,000-27,000 UTS; 1.5-3.0 El; Rock 40 F; density 6.5-6.9.
For ductile, light duty mechanical components.
ASTM B310 Class A, Type III; F-0000-R.

GKN NS-180.
M-632; 62.5-65.5 Cu, 16.5-19.5 Ni, 14.0 min Zn.
As sintered: 27,000-33,000 TS; 13-16 El; 75-85 Rock H; density 7.3-7.7.
For ornamental parts, lock hardware, structural components. Nickel silver type.
ASTM B458, Grade I, Type I; CZN-1818-U.

GKN S-005.
M-632; 0.25-0.60 C, 97.4 min Fe.
As sintered: 17,000-22,000 psi TS; 1.0-1.5 El; Rock 40-65 F; density 5.7-6.1.
Carbon steel grade; for bearings and low cost structural parts.
ASTM B310, Class B Type 1; F-0005-N.

GKN-S-010.
M-632; 97.0 min Fe, C.
As sintered: 22,000-28,000 TS; 0.5 El; Rock 20-40 B; density 5.6-6.1.
Carbon steel type, hardenable to Rock 85 B min.
For low cost, wear resistant machine parts.
ASTM B310 Class C, Type 1; F-0008-N.

GKN SA-4200.
M-632; 0.5-0.9 C, 0.35 Mn, 0.45 Ni, 0.55 Mo, 97.6 min Fe.
As sintered: 55,000-78,000 TS; 1.0-2.0 El; 70-90 Rb.
Hardened: 90,000-145,000 TS; 0.5-1.0 El; 25 Rc min.
Good hardenability; for structural parts.

GKN SA-4600.
M-632; 0.5-0.9 C, 0.20 Mn, 1.9 Ni, 0.50 Mo, 96.25 min Fe.
As sintered: 50,000-90,000 TS; 10-2.0 El; 75-95 Rb.
Hardened: 85,000-145,000 TS; 0.5-1.0 El; 25 Rc min.
Excellent hardenability and wear resistance.
For heavy duty cams, ratchets, gears, pawls.

GKN SC-210.
M-632; 0.6-1.0 C, 1.5-2.5 Cu, 94.5 min Fe.
As sintered: 60,000-100,000 TS; 0.5-1.5 El; Rock 65-85 B; density 6.4-7.2.
Copper steel grade; hardenable to 15 N 72 min.
For heavy duty structural parts, gears, ratchets.
ASTM B426, Grade 1, Type III or IV; FC-0208-R; FC-0208-S.

GKN SC-310.
M-632; 0.6-1.0 C, 2.5-3.5 Cu, 92.5 min Fe.
As sintered: 37,000-42,000 TS; 0.5 El; Rock 40-60 B; density 5.9-6.2.
Copper steel grade, hardenable to Rock 90 B min.
For structural components as gears, cams, pawls.
ASTM B426, Grade 2, Type I or II.

GKN SC-705.
M-632; 0.25-0.60 C, 6.0-8.0 Cu, 88.4 min Fe.
As sintered: 29,000-34,000 TS; 0.5-1.0 El; Rock 50-80 F; density 5.9-6.2.
Copper steel grade, hardenable.
For structural components.
ASTM B222.

GKN SC-710.
M-632; 0.6-1.0 C, 6.0-8.0 Cu, 88.0 min Fe.
As sintered: 40,000-46,000 TS; 0.5 El; Rock 40-60 B; density 5.9-6.2.
Copper steel grade, hardenable to Rock 90 B min.
For structural components.
ASTM B426, Grade 3, Type I or II.

GKN SN-200.
M-632; 1.25 max Cu, 1.75-2.75 Ni, 93.75 min Fe.
As sintered: 30,000-47,000 TS; 4.0-10.0 El; Rock 60-85 F; density 6.4-7.2.
Nickel steel; good ductility and impact resistance; for firearm parts. Higher density parts can be case carburized for wear resistance.
ASTM B484 Grade 1, Class A, Type I or II; FN-0200-R; FN-0200-S.

GKN SN-208.
M-632; 0.6-0.9 C, 1.75-2.75 Ni, 94.35 min Fe.
As sintered: 46,000-75,000 TS; 1.5-3.5 El; Rock 60-85 B; density 6.4-7.2.
Nickel steel; hardenable to 15 N 72 min and 94,000 to 155,000 psi TS.
High strength; usually heat treated.
ASTM B484, Grade 1, Class C, Type I or II; FN-0208-R; FN-0208-S.

GKN SN-405.
M-632; 0.4-0.7 C, 0.75-1.25 Cu, 3.5-4.5 Ni, 91.55 min Fe.
As sintered: 45,000-74,000 TS; 2.5-5.5 El; Rock 60-75 B; density 6.4-7.2.
Nickel steel; hardenable to 15 N 75 min and 100,000-170,000 psi TS.
Good combination of strength and impact resistance; superior heat treated proprties.
ASTM B484, Grade 2, Class B, Type I or II; FN-0405-R; FN-0405-S.

GKN SS-304.
M-632; 18.0-20.0 Cr, 8.0-12.0 Ni, 62.8 min Fe.
As sintered: 36,000-50,000 TS; 2.0-4.0 El; 40-65 Rb.
AISI 304 composition. Non-magnetic; good corrosion resistance.
ASTM B525, Grade 1, Type I.

GKN SS-316.
M-632; 16.0-18.0 Cr, 10.0-14.0 Ni, 2.0-3.0 Mo, 59.8 min Fe.
As sintered: 40,000-63,500; 3.0-7.0 El; 50-75 Rb.
AISI 316 composition; non-magnetic, ductile, very good corrosion resistance.
ASTM B525, Grade 2, Type 1; SS-316-R.

GKN SS-410.
M-632; 11.5-13.5 Cr, 83.8 min Fe.
Sintered + HT; 65,000-85,000 TS; 55-70 Rock A.
Martensitic, heat treatable stainless; corrosion resistant.
SS-410-P.

GKN SX-000.
M-632; 0.25 max C, 15.0-25.0 Cu, 3.0 max Zn, 69.75-85.0 Fe.
As sintered: 60,000 psi TS; 5.0 El; Rock 40 B min; density 7.1 min.
Infiltrated steel; hardenable to Rock 15 N 70 min.
Good impact resistance, excellent machinability.
ASTM B303, Class A; FX-2000-T.

GKN SX-010.
M-632; 0.6-1.0 C, 15.0-25.0 Cu, 3.0 max Zn, 69.0-85.0 Fe.
As sintered: 100,000 TS; 0.5 El; Rock 75 B min; density 7.1 min.
Infiltrated steel: hardenable to 15 N 75 min.
High strength steel; uniform density in complex designs.
ASTM B303, Class C; FX-2008-T.

GLACIER PRECISION FINISH LEAD BRONZE.
M-818; Sn, Pb, bal Cu.
For machinery parts; free cuttings.

GLASSEAL see also **TECHALLOY GLASSEAL.**

GLASS MOLD ALLOY.
M-Eng.; 65-55 Cu, 12-18 Ni, 11-17 Zn, 8-12 Fe, 0-5-1.0 Si.
For blass molds; U. S. Pat. 1360773.

GLASS SEALING see also **CARPENTER GLASS SEALING.**

GLASS SEALING "42".
M-32; 0.15 max C, 1.0 Mn, 41.5 Ni, bal Fe.
For sealing into glass; low expansion.

GLASS SEALING "49".
M-32; 0.10 C, 49 Ni, bal Fe.
For glass sealing; same coefficient of expansion as glass.

GLIDCOP AL-10.
M-1728; 99.8 Cu, 0.2 Al_2O_3.
Pre-alloyed metal powder for sintering.
Good conductivity; recommended for wire and strip. Meets RWMA spec. for class I material.

GLIDCOP AL-20.
M-1728; 99.6 Cu, 0.4 Al_2O_3.
Pre-alloyed metal powder for sintering.
General purpose grade; good electrical and thermal conductivity.

GLIDCOP AL-35.
M-1728; 99.3 Cu, 0.7 Al_2O_3.
Pre-alloyed metal powder for sintering.
For good conductivity; good strength and wear for long term elevated operation. Meets RWMA spec. for class II materials.

GLIDCOP AL-60.
M-1728; 98.9 Cu, 1.1 Al_2O_3.
Pre-alloyed metal powder for sintering.
Good strength, hardness and wear resistance.
Retains strength at elevated temperatures.
Conductivity: 78% IACS at 68°F.

GLIDDEN 303-L.
M-1728; 17.5 Cr, 12.5 Ni, 0.2 S, 0.7 Si, 0.2 Mn, 0.02 C, bal Fe.
Pre-alloyed metal powders; free-machining austenitic stainless.
Typical sintered properties: 50,000-60,000 psi TS; 3-12 El; 50-64 Rock B.

GLIDDEN 304L.
M-1728; 0.02 C, 18 Cr, 10 Ni, bal Fe.
Pre-alloyed metal powders; stainless.
Typical sintered properties: 50,000-62,000 TS; 3.5-12.4 El; 52-62 Rock B.

GLIDDEN 316L.
M-1728; 0.03 C, 17 Cr, 13 Ni, 2.5 Mo, bal Fe.
Pre-alloyed metal powders; stainless.
Typical sintered properties: 49,000-69,000 TS; 3-14 El; 50-60 Rock B.

GLIDDEN 410L (ANNEALED).
M-1728; 0.02 C, 12 Cr, bal Fe.
Pre-alloyed metal powders; stainless.
Typical sintered properties: 58,000-95,000 TS; 2-3 El; 90-96 Rock B.

GLIDDEN 434-L.
M-1728; 17.0 Cr, 1.0 Mo, 0.8 Si, 0.2 Mo, 0.02 C, bal Fe.
Pre-alloyed metal powders; ferritic type stainless.
Typical sintered properties: 450-550 N/mm^2 TS; 350-400 N/mm^2 YS; 3-5 El; 70-80 Rock B.

GLIDDEN 830.
M-1728; 20.2 Cr, 30.0 Ni, 2.5 Mo, 3.5 Cu, 1.0 Si, 0.2 Mn, 0.02 C, bal Fe.
Pre-alloyed metal powders; corrosion resistant austenitic grade.
Meets chemistry of ACI CN-7M.

GLIDDEN 4600.
M-1728; 0.04 C, 0.6 Mn, 1.9 Ni, 0.25 Mo, bal Fe.
Pre-alloyed metal powders; low alloy steel.
Sintered properties depend on carbon pick-up from graphite and zinc stearate and on heat treatment.

GLIDDEN A-210; B-214.
M-1728.
Iron powder for sintering into pole pieces and other soft P/M parts.

GLIDDEN A-220.
M-1728.
Iron base powder for sintering into permanent magnets.

GLIDDEN CN-1.
M-1728; 70 Ni, 30 Cu.
Pre-alloyed metal powders; Ni-Cu alloy.
Typical sintered properties; 39,000-44,000 TS; 12.5-17.1 El; 34-38 Rock B.
For parts requiring corrosion resistance.

GLIDDEN CN-1; CN-4.
M-1728.
Cu-Ni alloy powders to be sintered for corrosion resistant filters.

GLIDDEN H-120.
M-1728; 50 Fe, 50 Al alloy powder for sintering into permanent magnets.

GLIDDEN HN-1; HN-4.
M-1728.
Ni-base alloy powders to be sintered for corrosion resistant and high temperature filters; and for hard facing.

GLIDDEN LA-100.
M-1728.
Pb-Sn-Sb eutectic alloy powder for use in solder paste for low temperature type metal soldering.

GLIDDEN NF-1.
M-1728; 50 Ni, 50 Fe.
Pre-alloyed metal powders.
Typical sintered properties; 32,000-38,000 TS; 8.4-12.1 El; 26-29 Rock B.
For soft magnetic and electrical applications.

GLIDDEN SF-9.
M-1728; 9 Si, bal Fe.
Pre-alloyed metal powders.
Master alloy for soft magnetic parts.

GLIDDEN SF-9; SF-17.
M-1728.
Pre-alloyed Si-Fe powder for blending into silicon-iron alloys for soft magnetic applications.

GLIDDEN TL-15.
M-1728.
One of several Pb-Sn alloys in powder form; for use as solders.

GLIEVOR BEARING.
M-Eng.; 74 Zn, 9 Sb, 6.7 Sn, 5 Pb, 4.4 Cu, 1.4 Cd.
For bearings; anti-friction.

GLIEVOR BEARING.
M-Eng.; 14 Sb, 8 Sn, 77 Pb, 1.5 Fe.
For bearings; anti-friction.

GLIXEY NO. 1.
M-819; Pb, Sb, bal Sn.
For transmission bearings, pump bearings; anti-friction metal.

GLIXEY NO. 2.
M-819; Pb, Sb, bal Sn.
For bearings; antifriction metal.

GLIXEY NO. 2A.
M-819; Pb, Sb, bal Sn.
For bearings; antifriction metal.

GLIXEY NO. 3.
M-819; Pb, Sb, bal Sn.
For bearings; antifriction metal.

GLM.
M-1290; 1.05-1.15 C, 0.15-0.30 Si, 0.20-0.40 Mn, 1.10-1.30 Cr, 1.20-1.40 W, 0.15-0.25 V, bal Fe.
Cold work tool steel.
W.-Nr. 1.2519.

GLOBLE DRILL STEEL.
M-57; 1.25 C, 0.10 Cr, 0.30 Mn, bal Fe.
For drills, reamers, cutters; water hardened.

GLUHFEST ALLOYS see JUNKER.

GLX-42W.
M-836; 0.21 C, 1.35 Mn, Cb, bal Fe.
Aluminum killed.
Hot rolled plate: 60,000 psi TS; 42,000 psi YS min; 24 El.
For structural parts.

GLX-45W.
M-836; 0.22 C, 1.20 Mn, 0.10 max Si, 0.01-0.04 Cb, bal Fe.
Rolled (min): 60,000 TS; 45,000 YS; 22 El.
For structural parts; weldable.

GLX-50W.
M-836; 0.22 C, 1.2 Mn, 0.10 max Si, 0.01-0.04 Cb, bal Fe.
Rolled (min): 65,000 TS; 50,000 YS; 22 El.
For structural parts; weldable.

GLX-55W.
M-836; 0.22 C, 1.20 Mn, 0.10 max Si, 0.01-0.04 Cb, bal Fe.
Rolled (min): 70,000 TS; 55,000 YS; 20 El.
For structural parts; weldable.

GLX-60W.
M-836; 0.22 C, 1.20 Mn, 0.10 max Si, 0.01-0.04 Cb, bal Fe.
Rolled (min): 75,000 TS; 60,000 YS; 18 El.
For structural parts; weldable.

GLX-65W.
M-366; 0.22 C, 1.20 Mn, 0.10 max Si, 0.01-0.04 Cb, bal Fe.
Rolled (min): 80,000 TS; 65,000 YS; 16 El.
For structural parts; weldable.

GLYCO.
M-240; 15 Sb, 4.5 Sn, 0.5 As, bal Pb.
For bearings, bushings; anti-friction, Babbitt.

GLYCO METAL.
85.5 Zn, 5 Sn, 4.7 Pb, 0.4 Cu, 2 Al.
For bearings, bushings; Babbitt.

GLYCO TURBO.
M-Eng.; 70 Pb, 22 Sb, 8 Sn.
For turbine bearings; Babbitt.

GM 14M.
M-690; 3.2-3.5 Total C, 2.15-2.45 Si, 0.5-0.8 Mn, 0.1-0.4 Cr, 0.15 max P, 0.15 max S, bal Fe.
Ferritic-Pearlitic type gray cast iron for easily machinable automotive castings.
20,000 psi TS; 131-187 Brin.
Similar to ASTM/SAE Gr. G-1800.

GM-4199M.
M-690; 7.5-9.5 Si, 1.3 max Fe, 3.0-4.0 Cu, 0.50 max Mn, 0.10 max Mg, 0.20 max Pb, 0.50 max Ni, 1.0 max Zn, 0.35 max Sn, bal Al.
Cast aluminum alloy.
UNS 13800 (380).

GM-4227 M.
M-690; 5.5-7.0 Si, 1.0 max Fe, 3.0-4.0 Cu, 0.50 max Mn, 0.50 max Mg, 0.35 max Ni, 1.0 max Zn, 0.25 max Ti, bal Al.
Cast aluminum alloy.
UNS A03190 (319).

GM-4334M.
M-690; 8.5-10.5 Si, 1.2 max Fe, 2.0-4.0 Cu, 0.50 max Mn, 0.5-1.5 Mg, 0.20 max Pb, 0.50 max Ni, 1.0 max Zn, 0.25 max Ti, 0.20 max Sn, bal Al.
Cast aluminum alloy.
UNS A63320 (F-132).

GM 6129 M, GR 3815.
M-690; 3.6-3.9 C, 2.2-2.7 Si, 0.25-0.90 Mn, 0.03-07 Mg, bal Fe.
Nodular cast iron, annealed (Ferritic).
38,000 psi YS; 15 El; 126-179 Brin.
90% min ferrite, 10% max pearlite, 3 max carbide.

GM 6129M GR 4010.
M-690; 3.60-3.90 C, 2.2-2.7 Si, 0.25-0.90 Mn, 0.03-0.07 Mg, bal Fe.
Nodular cast iron, as cast Ferritic.
40,000 psi YS; 10 El; 140-217 Brin.
40% min ferrite, 60 max pearlite, 10 max carbide.

GM 6129 GR 5203.
M-690; 3.6-3.9 C, 2.2-2.7 Si, 0.50-1.0 Mn, 0.03-0.07 Mg, bal Fe.
Nodular cast iron, as cast Pearlitic.
52,000 psi YS; 3 El, 187-269 Brin.
50% max ferrite, 50% min pearlite, 10 max carbide.

GM-6209 M.
M-690; 3.2-3.5 total C, 2.15-2.45 Si; 0.05-0.08 Mn, 0.1-0.4 Cr, 0.15 max S, 0.15 max P, bal Fe.
Pearlite gray cast iron for small or medium size automotive castings.
30,000 psi TS; 170-241 Brin.
Similar to ASTM/SAE Gr.G: 3000.

GM-6213 M.
M-690; 3.10-3.4 total C, 2.1-2.4 Si, 0.05-0.07 Mn, 0.2-0.4 Cr, 0.15 max S, 0.15 max P, bal Fe.
Pearlite gray cast iron for cylinder heads, cylinder blocks, engine bearing caps.
35,000 psi TS; 179-255 Brin.
Similar to ASTM/SAE Gr.G-3500.

GM6214 M.
M-690; 3.2-3.5 total C, 2.15-2.45 Si, 0.50-0.80 Mn, 0.1-0.4 Cr, 0.15 max P, 0.15 max S, bal Fe.
Ferritic-Pearlite gray cast iron for lightly loaded automotive castings.
18,000 psi TS; 187 max Brin.
Similar to ASTM/SAE Gr.G-1800.

GMOODIE.
M-165; 4 Al, 3.25 Cu, 0.80 Ni, 0.20 Ti, 0.15 Mg, bal Zn.
For dies for sheet forming.

GMR-235.
M-1286, M-1214, M-1491; 0.15 C, 8-12 Fe, 14-17 Cr, 4.5-6.0 Mo, 2 Ti 2.5-3.5 Al, 0.1 B, bal Ni.
At 70°F: 147,000 TS; 43 El.
At 1500°F: 90,000 TS; 5 El.
At 1800°F: 21,800 TS; 38 El.
For jet engine components, gas turbine parts; high heat resistant to 1500°F.

GMR 235 D.
M-165, M-1491; 0.15 C, 15.5 Cr, 5.0 Mo, 2.5 Ti, 3.5 Al, 0.05 B, bal Ni.
For jet engine parts.

GMR 236.
M-1286; 0.15 C, 15.5 Cr, 5 Mo, 2.2 Ti, 3.25 Al, 25 Fe, 0.06 B, bal Ni.
For heat resistant parts; corrosion and heat resistant.

GMS 58.
M-Ger.; Zn, bal Cu.
For castings; brass.

GMS 63.
M-Ger.; Zn, bal Cu.
For castings; brass.

GMS 67.
M-Ger.; Zn, bal Cu.
For castings; brass.

GN3.
M-1522; 80 Ag, Ni.
Ann: hardness 70 HV.
Elec. res.: 2.2 μ Ω cm.
Elec. cond.: 78% IACS.
For electrical contacts; low contact resistance.

GN5.
M-1522; 90 Ag, Ni.
Ann hardness: 60 HV.
Elec. res.; 2.0 μ Ω cm.
Elec. cond.: 86% IACS.
For electrical contacts; low contact resistance.

GOHI IRON.
M-191; 0.02 C, 0.25 Cu, 0.025 Mn, 0.003 Si, bal Fe.
45,000-50,000 TS; 40-50 El.
For sheet metal construction work, roofing, culverts, pipe, ventilators, skylights; rust resisting.

GOLD.
M-1755; Au.
Purities: Zone refined 99.9999%, 99.999%, 99.99%.
Forms: Sponge, powder, wire, foil, sheet, ingot, single crystals.

GOLD-1.
M-1344; 0.86 C, 2.8 Co, 4.3 Cr, 0.85 Mo, 2.1 V, 12 W, bal Fe.
For lathe and planer tools, reamers, drills, taps; high speed steel.

GOLD-2.
M-1344; 1.3 C, 4.3 Cr, 0.85 Mo, 3.8 V, 12 W, bal Fe.
For engravers' tools, milling cutters, taps, reamers; high speed steel.

GOLD, 8 CARAT.
47 Cu, 33 Au, 20 Ag.
For jewelry; corrosion resistant.

GOLD, 10 CARAT.
38-46 Cu, 42 Au, 12-20 Ag.
For jewelry; corrosion resistant.

GOLD, 14 CARAT.
12 Cu, 58 Au, 30 Ag.
For dental fillings and crowns; corrosion resistant.

GOLD, 14 CARAT.
14-28 Cu, 58 Au, 4-28 Ag.
For jewelry; corrosion resistant.

GOLD, 15 CARAT.
13 Cu, 62 Au, 11 Ag.
For jewelry; corrosion resistant.

GOLD, 16 CARAT.
8-27 Cu, 67 Au, 6.6-26 Ag.
For jewelry.

GOLD, 18 CARAT.
15-5 Cu, 75 Au, 10-20 Ag.
For jewelry.

GOLD, 20 CARAT.
6-8.3 Cu, 84 Au, 8.3-11 Ag.
For jewelry.

GOLD, 22 CARAT.
4.2 Cu, 92 Au, 4.2 Ag.
For jewelry.

GOLD, 22 CARAT.
3.4 Cu, 92 Au, 4.9 Ag.
For dental crown and fillings; corrosion resistant.

GOLDAL.
M-Eng; Au, alloy.
For gold solder, dental alloy; corrosion resistant.

GOLD ANCHOR.
M-342, M 115; 0.68-0.73 C, 17.5 18.5 W, 3.75-4.25 Cr, 0.95-1.15 V, bal Fe.
For chasers, dies, punches, special tools; high speed steel.

GOLD ANCHOR DRILL ROD.
M-342; 0.68-0.73 C, 17.5-18.5 W, 3.75-4.25 Cr, 0.95-1.15 V, bal Fe.
For twist drills, reamers, punches, small tools; high speed steel.

GOLD BRONZE.
M-US; 6.5 Sn, 3 Zn, bal Cu.
For jewelry, ornaments; corrosion resistant.

GOLD BUTTONS.
M-Eng.; 33 Zn, 5 Sn, 3 Pb, bal Cu.
For buttons, jewelry; free-cutting.

GOLD COLOR ELASTIC.
M-349.
Gold color clasp and orthodontic wire.
Fusing temp: 1675°F.

GOLD IMITATION.
M-Eng; 97.8 Cu, 2.0 Al, 0.2 Au.
For gold solder; corrosion resistant.

GOLD LABEL.
M-344, M 343; 0.60-0.7 C, 3-4 Cr, 18-20 W, 2-3 Co, 1.25-1.5 V, bal Fe.
For cutting tools; high speed steel, "Hellers Peerless."

GOLD LABEL (H. & R.).
M-363; 1.2 C, 2.5 W, bal Fe.
For finishing tools, cutters; water or oil hardened.

GOLD LEAF.
M-Eng.; 66-80 Cu, 34-20 Zn.
For gold leaf substitute, signs.

GOLD LEAF, AIX.
M-Eng.; 64,8 Cu, 32.8 Zn, 2.0 Sn, 0.4 Pb.
For gold leaf substitute, signs.

GOLD LEAF, JEMMAPES.
M-Eng.; 64.6 Cu, 33.7 Zn, 1.4 Sn, 0.2 Pb.
For gold leaf substitute, signs.

GOLD LEAF METAL.
M-Eng.; 84 Cu, 16 Zn.
For gold leaf substitute, signs.

GOLDMASTER GM15.
M-1806.
Sintered carbide tool material.
Coated throwaway tips for light cutting on steel and all cutting on cast iron and non-ferrous metals.

GOLDMASTER GM35.
M-1806.
Sintered carbide tool material.
Coated throwaway tips for heavier cutting on steel.

GOLD METAL LEAF.
M-Eng.; 84-66 Cu, 16-34 Zn, 0.-04 Pb.
For gold solder; corrosion resistant.

GOLDPUNKT.
M-1331; 0.86 C, 4.1 Cr, 0.85 Mo, 2.5 V, 12 W, bal Fe.
For lathe and planer tools, drills, taps, hobs; high speed steel.

GOLDSMITH GB-02.
M-1108; 2 Ag, 91 Cu, 7 P.
For silver solder; M.P. 1180-1270°F; self-fluxing.

GOLDSMITH GB-05.
M-1108; 5 Ag, 58 Cu, 37 Zn.
For silver solder; M.P. 1575-1600°F.

GOLDSMITH GB-06.
M-1108; 6 Ag, 86.5 Cu, 7.5 P.
For silver solder; M.P. 1175-1350°F, self-fluxing.

GOLDSMITH GB-07.
M-1108; 7 Ag, 85 Cu, 8 Sn.
For silver solder; M.P. 1450-1565°F.

GOLDSMITH GB09.
M-1108; 9 Ag, 53 Cu, 38 Zn.
Melting range: 1450-1565°F.
Brazing alloy.

GOLDSMITH GB15.
M-1180; 15 Ag, 80 Cu, 5 P.
For brazing; M.P. 1185-1300°F, silver solder.

GOLDSMITH GB20.
M-1108; 20 Ag, 45 Cu, 30 Zn, 5 Cd.
For brazing; M.P. 1140-1500°F, silver solder.

GOLDSMITH GB25.
M-1108; 25 Ag, 32 1/2 Cu, 22 1/2 Zn.
Melting range: 1300-1575°F.
Brazing alloy.

GOLDSMITH GB30.
M-1108; 30 Ag, 38 Cu, 32 Zn.
For silver solder; M.P. 1370-1410°F.

GOLDSMITH GB31.
M-1108; 31 1/2 Ag, 34 Cu, 15 1/2 Zn, 19 Cd.
Melting range: 1165-1390°F.
Brazing alloy.

GOLDSMITH GB35.
M-1108; 35 Ag, 26 Cu, 21 Zn, 18 Cd.
For brazing, M.P. 1125-1295°F, silver solder.

GOLDSMITH GB41.
M-1108; 41 Ag, 17 Cu, 18 Zn, 24 Cd.
For brazing; M.P. 1125-1160F, silver solder.

GOLDSMITH GB45.
M-1108; 45 Ag, 15 Cu, 16 Zn, 24 Cd.
For brazing; M.P. 1125-1145F, silver solder.

GOLDSMITH GB50.
M-1108; 50 Ag, 15.5 Cu, 16.5 Zn, 18 Cd.
For brazing; M.P. 1160-1175°F, silver solder.

GOLDSMITH GB54.
M-1108; 54 Ag, 40 Cu, 5 Zn, 1 Ni.
Melting range: 1375-1575°F.
Brazing alloy; AMS 4772.

GOLDSMITH GB56.
M-1108; 56 Ag, 22 Cu, 17 Zn, 5 Sn.
For silver solder; M.P. 1152-1203F.

GOLDSMITH GB57.
M-1108; 57 Ag, 33 Cu, 3 Mn, 7 Sn.
Melting range: 1120-1345°F.
Brazing alloy.

GOLDSMITH GB60.
M-1108; 60 Ag, 25 Cu, 15 Zn.
For brazing; M.P. 1260-1325F, silver solder.

GOLDSMITH GB65.
M-1108; 65 Ag, 20 Cu, 15 Zn.
For silver solder; M.P. 1280-1325°F.

GOLDSMITH GB70.
M-1108; 70 Ag, 20 Cn, 10 Zn.
For silver solder; M.P. 1335-1390°F.

GOLDSMITH GB72.
M-1108; 72 Ag, 28 Cu.
For silver solder; M.P. 1335-1435F.

GOLDSMITH GB75.
M-1108; 75 Ag, 0.22 Cu, 3 Zn.
For silver solder; M.P. 1365-1490°F.

GOLDSMITH GB80.
M-1108; 80 Ag, 16 Cu, 4 Zn.
For brazing; M.P. 1360-1490F, silver solder.

GOLDSMITH GB85.
M-1108; 85 Ag, 15 Mn.
For silver solder; M.P. 1760-1778F.

GOLDSMITH GB110.
M-1108; 10 Ag, 52 Cu, 38 Zn.
For silver solder; M.P. 1450-1565°F.

GOLDSMITH GB120.
M-1108; 20 Ag, 45 Cu, 35 Zn.
For silver solder; M.P. 1430-1500°F.

GOLDSMITH GB131.
M-1108; 31.5 Ag, 34 Cu, 15.5 Zn, 19 Cd.
For silver solder; M.P. 1165-1390°F.

GOLDSMITH GB140.
M-1108; 40 Ag, 18 Cu, 15 Zn, 27 Cd.
For silver solder; M.P. 1135-1205°F.

GOLDSMITH GB145.
M-1108; 45 Ag, 30 Cu, 25 Zn.
For silver solder; M.P. 1250-1370°F.

GOLDSMITH GB150.
M-1108; 50 Ag, 15 Cu, 25 Zn, 10 Cd.
For silver solder; M.P. 1166-1190°F.

GOLDSMITH GB160.
M-1108; Ag, 30 Cu, 10 Sn.
Melting range: 1095-1325°F.
Brazing alloy; AMS 4773.

GOLDSMITH GB165.
M-1108; 65 Ag, 28 Cu, 5 Mn, 2 Ni.
Melting range: 1385-1445°F.
Brazing alloy.

GOLDSMITH GB175.
M-1108; 75 Ag, 20 Cu, 5 Zn.
For silver solder; M.P. 1350-1425°F.

GOLDSMITH GB240.
M-1108; 40 Ag, 30 Cu, 28 Zn, 2 Ni.
For silver solder; M.P. 1240-1435°F.

GOLDSMITH GB245.
M-1108; 45 Ag, 18 Cu, 21 Zn, 16 Cd.
For silver solder; M.P. 1140-1185°F.

GOLDSMITH GB250.
M-1108; 50 Ag, 34 Cu, 16 Zn.
For brazing; M.P. 1260-1410°F, silver solder.

GOLDSMITH GB260.
M-1108; 60 Ag, 27 Cu, 13 Zn.
Melting range: 1125-1310°F.
Brazing alloy.

GOLDSMITH GB275.
M-1108; 75 Ag, 25 Zn.
For silver solder; M.P. 1300-1325°F.

GOLDSMITH GB340.
M-1108; 40 Ag, 36 Cu, 24 Zn.
For silver solder; M.P. 1330-1445°F.

GOLDSMITH GB350.
M-1108; 50 Ag, 15.5 Cu, 15.5 Zn, 16 Cd, 3 Ni.
For brazing; M.P. 1195-1275°F, silver solder.

GOLDSMITH GB440.
M-1108; 40 Ag, 30 1/2 Cu, 29 1/2 Zn.
Melting range: 1150-1350°F.
Brazing alloy.

GOLDSMITH GB540.
M-1108; 40 Ag, 30 Cu, 25 Zn, 5 Ni.
For brazing; M.P. 1260-1550°F, silver solder.

GOLDSMITH GB650.
M-1108; 50 Ag, 28 Cu, 22 Zn.
For silver solder; M.P. 1250-1340°F.

GOLDSMITH GB95-5.
M-1108; 5 Ag, 95 Cd.
Melting range: 620-750°F.
Solder type alloy.

GOLDSMITH GB1270.
M-1108; 91 Ag, 7 P.
Melting range: 1185-1450°F.
For brazing copper and copper alloys.

GOLD SOLDER.
M-Eng.; Ag, Au, Cu, Zn.
For gold solder; corrosion resistant.

GOLD SOLDER, 8 CARAT.
40 Au, 37 Ag, 23 Cu.
For gold solder.

GOLD SOLDER 8 KT CADMIUM NO. 1.
M-926; Au alloy.
For gold soldering; easy running, yellow.

GOLD SOLDER 8 KT. NO. 10.
M-926; Au alloy.
For gold soldering; hard running, yellow.

GOLD SOLDER 8 KT. WHITE NO. 5.
M-926; Au alloy.
For gold soldering; medium running, white.

GOLD SOLDER 9 1/2 KT. CADMIUM NO. 1.
M-926; Au alloy.
For gold soldering; easy running, yellow.

GOLD SOLDER 9 1/2 KT. NO. 10.
M-926; Au alloy.
For gold soldering; hard running, yellow.

GOLD SOLDER 9 1/2 KT. WHITE NO. 5.
M-926; Au alloy.
For gold soldering; medium running, white.

GOLD SOLDER, 10 CARAT.
41 Au, 37 Ag, 21 Cu, 0.6 brass.
For gold solder.

GOLD SOLDER 10 KT. CADMIUM NO. 1.
M-926; Au alloy.
For gold soldering; easy flowing, yellow.

GOLD SOLDER 10 KT NO. 10.
M-926; Au alloy.
For gold soldering; hard flowing, yellow.

GOLD SOLDER 10 KT. WHITE NO 5.
M-926; Au alloy.
For gold soldering; medium flowing, white.

GOLD SOLDER, 12 CARAT.
50 Au, 35 Cu, 15 Ag.
For gold solder; corrosion resistant.

GOLD SOLDER, 14 CARAT.
50 Au, 33 Ag, 17 Cu.
For gold solder.

GOLD SOLDER, 16 CARAT.
75 Au, 17 Ag, 8.3 Cu.
For gold solder.

GOLD SOLDER, 18 CARAT.
75-63 Au, 13-31 Ag, 6.3-12 Cu.
For gold solder.

GOLD SOLDER, BEST.
63 Au, 23 Ag, 15 Cu.
For gold solder.

GOLD SOLDER, EASY MELT.
55 Ag, 32 Ag, 14 Cu.
For gold solder.

GOLD SOLDER, VERY EASY MELT.
55 Ag, 29 Cu, 12 Au, 5.5 Zn.
For gold solder.

GOLD SPEZIAL.
M-1344; 1.35 C, 2 Co, 4.3 Cr, V, W, bal Fe.
For engravers' tools, milling and form cutters, taps; high speed steel.

GOLD STAR.
M-32; 0.77 C, 13.7 W, 5.0 Co, 3.75 Cr, 2.0 V, bal Fe.
For high speed cutting tools; high speed steel for heavy duty service.

GOLD SUPER.
M-1344; 0.79 C, 4.3 Cr, 0.75 Mo, 1.5 V, 18 W, bal Fe.
For lathe and planer tools, reamers, broaches, taps; high speed steel.

GOLD SUPER KOBALT 10.
M-1344; 0.76 C, 10 Co, 4.2 Cr, 0.8 Mo, 1.8 V, 18 W, bal Fe.
For lathe and planer tools, drills, taps, hobs; high speed steel.

GOLFALLOY.
M-1709, M-1748; 0.08-0.20 C, 15.0-16.5 Cr, 1.5-2.2 Ni, 1.0 max Mn, 1.0 max Si, 0.50 max Cu, 0.50 max Mo, bal Fe.
Cast, HT; 118,000 psi TS; 87,000 psi YS; 21 El; 18-25 Rc hardness.
For golf club heads, investment cast; corrosion resistant.

GOMAK 2.
M-France; 3.5-5.0 Al, 2.5-3.5 Cu, 0.02-0.10 Mg, bal Zn.
Die cast; 46,000-53,000 TS; 5-14 El; 80-100 Brin.
For motor frames, gear housings, hardware; similar to Zamak 2.

GOMAK 3.
M-France; 3.5-4.3 Al, 0.03-0.08 Mg, bal Zn.
Die cast; 35,000-43,000 TS; 4-11 El; 60-85 Brin.
For gear housings, motor frames, fuel pumps; similar to Zamak 3.

GOMAK 5.
M-France; 3.5-4.3 Al, 0.75-1.25 Cu, 0.03-0.06 Mg, bal Zn.
Die cast: 42,000-50,000 TS; 13-3 El; 70-95 Brin.
For motor frames, gear housings, hardware; similar to Zamak 5.

GONG METAL.
M-Eng.; 78 Cu, 22 Zn. For gongs.

GORHAM M-40-U12.
M-820; 0.7 C, 4 Cr, Mo, B, Co, bal Fe.
For tools, dies, cutters; high speed steel, abrasion resistant.

GOST.
Prefix for many USSR alloys.

GOST-0C18N9.
M-USSR; 0.06 C, 18 Cr, 9 Ni, bal Fe.
Annealed: 80,000 TS; 35,000 YS; B 80 Rock.
For tanks, evaporators, agitators, mixers, chemical plant equipment.
Type 304 stainless steel, austenitic.

GOST-0X18H9.
M-USSR; 0.06 C, 19 Cr, 10 Ni, bal Fe.
Annealed: 85,000 TS; 35,000 YS; 150 Brin.
For chemical plant equipment, evaporators, tanks, mixers.
Type 304 stainless steel austenitic.

GOST-1C18N.
M-USSR; 0.05 C, 18 Cr, 12 Ni, 2 Mo, bal Fe.
Annealed: 85,000 TS; 35,000 YS; B 80 Rock.
For chemical plant equipment, acid tanks, evaporators, digesters.
Type 316 stainlesss steel, austenitic, acid resistant.

GOST-1C18N9.
M-USSR; 0.10 C, 18 Cr, 9 Ni, bal Fe.
Annealed: 80,000 TS; 35,000 YS; B 80 Rock.
For chemical and pharmaceutical plant equipment, evaporators, trim, molding, digesters.
Type 302 stainless steel, austenitic.

GOST-1C18N9T.
M-USSR; 0906 C, 18 Cr, 10 Ni, 0.3 Ti, bal Fe.
Annealed: 85,000 TS; 35,000 YS; 185 Brin.
For welded structures, and tanks, mixers, evaporators, agitators.
Type 321 stainless steel, austenitic, stabilized, welding grade.

GOST-1X18H.
M-USSR; 0.06 C, 17 Cr, 12 Ni, 2 Mo, bal Fe.
Annealed: 80,000 TS; 30,000 YS; B 80 Rock.
For chemical plant equipment, valve trim, digesters, evaporators.
Type 316 stainless steel, austenitic, corrosion resistant.

GOST-1X18H9.
M-USSR; 0.12 C, 18 Cr, 9 Ni, bal Fe.
Annealed: 90,000 TS; 40,000 YS; B 85 Rock.
For tanks, chemical and pharmaceutical plant equipment, evaporators.
Type 302 stainless steel, austenic.

GOST-1X18H9T.
M-USSR; 0.06 C, 18 Cr, 11 Ni, 0.3 Cb, bal Fe.
Annealed: 85,000 TS; 35,000 YS; 55 El; 150 Brin.
For welded structures and tanks, mixers, agitators, digesters.
Type 321 stainless steel, stabilized, austenitic, welding grade.

GOST-2C13.
M-USSR; 0.2 C, 13 Cr, bal Fe.
Annealed: 95,000 TS; 50,000 YS; B 92 Rock.
For cutlery, surgical instruments, gears, shafts, ball bearings.
Type 420 stainless steel, heat treatable.

GOST-2X13.
M-USSR; 0.20 C, 12-14 Cr, bal Fe.
Annealed: 95,000 TS; 50,000 YS; B 92 Rock.
For gears, shafts, cutlery, surgical instruments.
Type 420 stainless steel, hardenable.

GOST-C17.
M-USSR; 0.10 C, 17 Cr, bal Fe.
Annealed: 70,000 TS; 40,000 YS; 140 Brin.
For dairy and chemical plant equipment, oil refinery accessories, automobile trim, oil burners, fasteners.
Type 430 stainless steel. Non-hardenable.

GOST-C18N11B.
M-USSR; 0.06 C, 18 Cr, 12 Ni, 0.6 Cb, bal Fe.
Annealed: 90,000 TS; 35,000 YS; B 85 Rock.
For welded structures and tanks, chemical plant equipment, agitators, mixers, evaporators.
Type 348 stainless steel, austenitic, heat and corrosion resistant.

GOST-C18N12M2T.
M-USSR; 0.08 C, 18 Cr, 10 Ni, 2 Mo, bal Fe.
For chemical plant equipment, evaporators, agitators, tanks.
Type 18-10 Mo stainless steel. Acid resistant, austenitic.

GOST-C20N14S2.
M-USSR; 0.16 C, 24 Cr, 14 Ni, bal Fe.
Annealed: 90,000 TS; 40,000 YS; B 84 Rock.
For salt pots, furnace parts, heat treating fixtures.
Type 309B stainless steel, austenitic, heat and corrosion resistant.

GOST-C23N13.
M-USSR; 0.15 C, 23 Cr, 14 Ni, bal Fe.
Annealed: 90,000 TS; 40,000 YS; B 83 Rock.
For heat treating boxes, combustion chambers, salt pots, kilns.
Type 309S stainless steel, heat and corrosion resistant.

GOST-C23N18.
M-USSR; 0.20 C, 25 Cr, 21 Ni, bal Fe.
Annealed: 95,000 TS; 45,000 YS; B 90 Rock.
For salt pots, furnace parts, heat treating equipment.
Type 310S stainless steel, heat and corrosion resistant.

GOST-C25.
M-USSR; 0.20 C, 25 Cr, bal Fe.
Annealed: 85,000 TS; 50,000 YS; 180 Brin.
For salt pots, furnace parts, heat treating equipment and fixtures.
Type 446 stainless steel, heat and corrosion resistant.

GOST-C25N20S2.
M-USSR; 0.20 C, 25 Cr, 20 Ni, bal Fe.
Annealed: 100,000 TS; 50,000 YS; B 90 Rock.
For furnace parts, salt pots, heat treating fixtures.
Type 314 stainless steel, austenitic, heat and corrosion resistant.

GOST-X17.
M-USSR; 0.10 C, 14-18 Cr, bal Fe.
Cold Rolled: 100,000 TS; 70,000 YS; 190 Brin.
For automotive trim, hardware, oil burners, fasteners, soot blowers.
Type 430 stainless steel, corrosion resistant.

GOST-X18H115.
M-USSR; 0.06 C, 18 Cr, 12 Ni, 0.8 Cb, bal Fe.
Annealed: 90,000 TS; 40,000 YS; B 82 Rock.
For exhaust manifolds, storage containers, welded structures, chemical plant equipment.
Type 347 stainless steel, heat and corrosion resistant, austenitic.

GOST-X18H12M2T.
M-USSR; 0.06 C, 17 Cr, 12 Ni, 2 Mo, bal Fe.
Annealed: 85,000 TS; 35,000 YS; B 82 Rock.
For chemical plant equipment, fixtures, mixers, agitators.
Type 18-10 Mo stainless steel, acid resistant, austenitic.

GOST-X20H14C2.
M-USSR; 0.15 C, 18 Cr, 10 Ni, 3 Si, bal Fe.
Annealed: 90,000 TS; 40,000 YS; B 85 Rock.
Heat treating fixtures and equipment, furnace parts, tube supports.
Type 302B stainless steel, heat and corrosion resistant.

GOST-X23H13.
M-USSR; 0.18 C, 23 Cr, 14 Ni, bal Fe.
Annealed: 90,000 TS; 40,000 YS; B 84 Rock.
For heat treating fixtures, furnace parts, salt pots, oil burners.
Type 309S stainless steel, heat and corrosion resistant.

GOST-X23H18.
M-USSR; 0.20 C, 25 Cr, 21 Ni, bal Fe.
Annealed: 95,000 TS; 45,000 YS; B 89 Rock.
For furnace parts and equipment, heat treating boxes and fixtures.
Type 310S stainless steel, heat and corrosion resistant.

GOST-X25.
M-USSR; 0.25 C, 24-30 Cr, bal Fe.
Annealed: 80,000 TS; 48,000 YS; B 85 Rock.
For heat treating boxes, furnace parts, fittings, hardware, valves.
Type 446 stainless steel, heat and corrosion resistant.

GOST-X25H20C2.
M-USSR; 0.20 C, 24 Cr, 20 Ni, bal Fe.
Annealed: 100,000 TS; 50,000 YS; B 90 Rock.
For furnace parts, radiant tubes, heat treating boxes, fixtures.
Type 314 Stainless steel, heat and corrosion resistant.

GOTTINGEN AGG2.
M-1363; 99.5 Al.
Soft: 13,000 TS; 5000 YS; 45 El; 23 Brin.
Hard: 24,000 TS; 22,000 YS; 15 El; 44 Brin.
For roofing, culverts, architectural trim; corrosion resistant.

GOTTINGEN AGG3.
M-1363; 1.0-1.5 Mn, 0.3 max Cr, bal Al.
Soft: 16,000 TS; 6000 YS;40 El.
Hard: 29,000 TS; 27,000 YS; 10 El.
For cooking utensils, heat exchangers, tanks, furniture; good forming and welding properties.

GOTTINGEN AGG4.
M-1363; 1.5-3.0 Mg, 0.5-1.5 Mn, 0.3 max Cr, bal Al.
Soft: 26,000 TS; 10,000 YS; 20 El; 45 Brin.
Hard: 41,000 TS; 36,000 YS; 5 El; 77 Brin.
For roofing, hydraulic tubing, architectural trim; good forming and welding properties.

GOTTINGEN AGG51.
M-1363; 0.6-1.4 Mg, 0.6-1.6 Si, 0.3 max Cr, 0.6-1.0 Mn, bal Al.
Annealed: 21,000 TS; 8000 YS; 24 El.
For window frames, fan blades, gutters, boats; good formability and weldability.

GOTTINGEN AGG57.
M-1363; 2-4 Mg, 0.4 max Mn, 0.3 max Cr, bal Al.
Soft: 28,000 TS; 13,000 YS; 30 El; 47 Brin.
Hard: 40,000 TS; 35,000 YS; 10 El; 73 Brin.
For aircraft tanks and fittings, fuel lines, marine parts; resists sea water corrosion.

GOVERNMENT BRONZE-H.
M-US; 82-84 Cu, 12.5-14.5 Sn, 4.5-2.5 Zn, 0.8-1.0 Pb, 0.1 Fe.
For journals, bushings, bearings; high strength.

GOVERNMENT BRONZE SUBSTITUTE.
90 Cu, 6.5 Sn, 3 Zn, 0.5 Pb.
For structural and marine fittings; corrosion resistant.

GOVERNMENT GENUINE BABBITT.
M-565; Pb, Sb, Sn.
Cast: 7100 psi TS; 0.75 El; 2.8 RA.
For bearings; good shock resistance; Babbitt metal.

GRAC.
M-821; 10 Sn, Sb, As, Cd, Ni, graphite, bal Pb. 27 Brin.
For bearings; M.P. 453-576°F.

GRACITE.
M-1009; 0.9 C, 1.25 Mn, 0.5 Cr, 0.5 W, bal Fe.
For punches, forming and beading dies, upsetters; oil hardened, non-deforming.

GRADE "A" (GRADE 1).
M-822; 0.95 C, 0.3 Mn, 0.2 Si, bal Fe.
Water hardening tool steel; AISI W1.

GRADE S.
M-788; C, Ni, bal Fe.
For sucker rods in oil wells; resists H_2S in water.

GRAINAL NO. 1 see VANADIUM GRAINAL NO. 1.

GRAINAL NO. 79.
M-1038; 20 Ti, 13 Al, 4 Zr, 8 Mn, 5 max Si, 0.5 B, bal Fe.
Used in steel manufacturing; increases hardenability.

GRAINAL 100.
M-1038; 20 Ti, 13 Al, 4 Zr, 8 Mn, 5 max Si, 1.0 B, bal Fe.
For addition in steel melting; increases hardenability.

GRALUR.
M-358; C, bal Fe.
For drills.

GRAMIX.
M-245; Cu-Sn-graphite.
For light machine bearings, bushings; finely powdered mixture; self-lubricating.

GRAMIX GRADE NO. 61.
M-245; Sn, bal Cu. (Bronze type).
Sintered: 29,000 psi (200 MPa) crushing strength; 15,000 psi (103 MPa) TS; 12,000 psi (82 MPa) YS; 2 El; 6.4-6.8 density (treated).
Standard grade for bronze parts.
Note: On all Gramix grades the yield strength (YS) is yield in compression, 0.1% offset.

GRAMIX GRADE NO. 138.
M-245; Cu. Sintered: 50,000 psi (345 MPa) crushing strength; 23,000 psi (158 MPa) TS; 11,000 psi (75 MPa) YS; 12 El; 7.6-8.0 density (dry).
For commutator rings.

GRAMIX GRADE NO. 183.
M-245; Sn, Pb, Ni, bal Cu. (Bronze type).
Sintered: 30,000 psi (207 MPa) crushing strength; 15,000 psi (103 MPa) TS; 13,000 psi (189 MPa) YS; 2 El; 6.4-6.8 density (dry).
Bearings for shaded pole motors.

GRAMIX GRADE NO. 266.
M-245; Cu, Pb, C, bal Fe. (Iron alloy).
Sintered: 46,000 psi (317 MPa) crushing strength; 24,000 psi (165 MPa) TS; 20,000 psi (138 MPa) YS; 0.5 El; 5.7-6.1 density (dry); 88-90 Rf hardness.
For bearings, structural parts, rotors.

GRAMIX GRADE NO. 272.
M-245; Zn, bal Cu. (Brass type).
Sintered: 42,000 psi (289 MPa) crushing strength; 20,000 psi (138 MPa) TS; 10,000 psi YS; 11 El; 7.0-7.4 density (dry).
For brass parts as packing glands.

GRAMIX GRADE NO. 273.
M-245; Ni, Zn, bal Cu. (Nickel-Silver).
Sintered: 49,000 psi (338 MPa) crushing strength; 27,000 psi (186 MPa) TS; 22,000 psi (151 MPa) YS; 10 El; 6.7-7.1 density (dry).
Light duty, corrosion resistant seal face applications.

GRAMIX GRADE NO. 278.
M-245; Sn, C, bal Cu. (Bronze type).
Sintered: 25,000 psi (172 MPa) crushing strength; 13,000 ps (89 MPa) TS; 10,000 psi (69 MPa) YS; 2 El; 6.1-6.5 density (dry).
Bearings for small appliances.

GRAMIX GRADE NO. 353.
M-245; Ni, Sn, bal Cu. (Nickel-Bronze type).
Sintered: 44,000 psi (303 MPa) crushing strength; 24,000 psi (165 MPa) TS; 20,000 psi (138 MPa) YS; 1.5 El; 6.6-7.0 density (dry).
High load bearing and high strength applications.

GRAMIX GRADE NO. 361.
M-245; Fe. (Machinable iron).
Sintered: 37,000 psi (255 MPa) crushing strength; 18,000 psi (124 MPa) TS; 12,500 psi (86 MPa) YS; 1.5 El; 5.8-6.2 density (dry); 43-49 Rf hardness.
For machinable structural parts and pole shoes.

GRAMIX GRADE NO. 363.
M-245; Sn, C, bal Cu. (Bronze type).
Sintered: 23,000 psi (158 MPa) crushing strength; 11,000 psi (75 MPa) TS; 9,000 psi (62 MPa) YS; 0.5 El; 6.3-6.7 density (dry).
Used on bearing applications where quiet operation is major consideration.

GRAMIX GRADE NO. 400.
M-245; Sn, C, bal Cu. (Bronze type).
Sintered: 28,000 psi (193 MPa) crushing strength; 14,000 psi (96 MPa) TS;11,000 ps (75 MPa) YS; 2 El; 6.4-6.8 density (treated).
ASTM B-438-73 Grade 1, Type 2.

GRAMIX GRADE NO. 401.
M-245; Sn, Pb, C, bal Cu. (Bronze type).
Sintered: 28,000 psi (193 MPa) crushing strength; 14,000 psi (96 MPa) TS; 11,000 psi (75 MPa) YS; 2 El; 6.5-6.9 density (treated).
ASTM B-438-73 Grade 2, Type 1.

GRAMIX GRADE NO. 402.
M-245; 0.25 max C, bal Fe. (Iron type).
Sintered: 29,000 psi (200 MPa) crushing strength; 14,000 psi (96 MPa) TS; 10,000 psi (69 MPa) YS; 1.5 El; 5.7-6.1 density (treated); 44-52 Rf hardness.
ASTM B-439-70 Grade 1.

GRAMIX GRADE NO. 403.
M-245; 0.25-0.60 C, bal Fe. (Iron-carbon type).
Sintered: 35,000 psi (241 MPa) crushing strength; 17,000 psi (117 MPa) TS; 15,000 psi (103 MPa) YS; 1 El; 5.7-6.1 density (treated); 5.8-6. Rf hardness.
ASTM B-439-70 Grade 2.

GRAMIX GRADE NO. 404.
M-245; 0.60-1.0 C, bal fe. (Iron-carbon type).
Sintered: 45,000 psi (310 MPa) crushing strength; 22,000 psi (151 MPa) TS; 20,000 psi (138 MPa) YS; 0.5 El; 5.7-6.1 density (treated); 67-80 Rf hardness.
ASTM B-310-70 Class C, Type 1.

GRAMIX GRADE NO. 405.
M-245; Cu, C, bal Fe. (Iron alloy).
Sintered: 53,000 psi (365 MPa) crushing strength; 29,000 psi (200 MPa) TS; 25,000 ps (172 MPa) YS; 0.5 El; 5.8-6.2 density (treated); 80-90 Rf hardness.
ASTM B-439-70 Grade 3.

GRAMIX GRADE NO. 406.
M-245; Zn, Pb, bal Cu. (Brass type).
Sintered: 45,000 psi (310 MPa) crushing strength; 21,000 psi (144 MPa) TS; 11,000 psi (75 MPa) YS; 9 El; 7.2-7.7 density (dry). Lock parts.
ASTM B-282-70 Type 1.

GRAMIX GRADE NO. 407.
M-245; Zn, Pb, bal Cu. (Brass type).
Sintered: 52,000 psi (358 MPa) crushing strength; 24,000 psi (165 MPa) TS; 13,000 psi (89 MPa) YS; 11 El; 7.7 min density (dry). Lock parts.
ASTM B-282-70 Type 2.

GRAMIX GRADE NO. 408.
M-245; Cu, C, bal Fe. (Iron alloy).
Sintered: 68,000 psi (469 MPa) crushing strength; 34,000 psi (234 MPa) TS; 29,000 psi (200 MPa) YS; 1 El; 5.8-6.2 density (dry); 77-87 Rf hardness.
ASTM B-222-70.

GRAMIX GRADE NO. 410.
M-245; Sn, C, bal Cu. (Bronze type).
Sintered: 36,000 psi (248 MPa) crushing strength; 18,000 psi (124 MPa) TS; 17,000 psi (117 MPa) YS; 3 El; 6.8-7.2 density (dry).
ASTM B-255-70 Type 2.

GRAMIX GRADE NO. 411.
M-245; 0.25 max C, bal Fe. (Iron).
Sintered: 31,000 psi (213 MPa) crushing strength; 15,000 psi (106 MPa) TS; 12,500 psi (86 MPa) YS; 2 El; 5.7-6.1 density (dry); 44-52 Rf hardness.
ASTM B-310-70 Type 1, Class A.

GRAMIX GRADE NO. 412.
M-245; 0.25-0.60 C, bal Fe. (Iron-carbon).
Sintered: 39,000 psi (269 MPa) crushing strength; 19,000 psi (131 MPa) TS; 17,000 psi (117 MPa) YS; 1 El; 5.7-6.1 density (dry); 58-68 Rf hardness.
ASTM B-310-70 Type 1, Class B.

GRAMIX GRADE NO. 413.
M-245; 0.60-1.0 C, bal Fe. (Iron-carbon).
Sintered: 48,000 psi (331 MPa) crushing strength; 24,000 psi (166 MPa) TS; 22,500 psi (155 MPa) YS; 0.5 El; 6.1-6.5 density (dry).
For valve plates.

GRAMIX GRADE NO. 414.
M-245; 0.25 max C, bal Fe. (Iron type).
Sintered: 41,000 psi (282 MPa) crushing strength; 20,500 psi (141 MPa) TS; 17,500 psi (120 MPa) YS; 3.5 El; 6.1-6.5 density (dry); 60-70 Rf hardness.
ASTM B-310-70 Type 2, Class A.

GRAMIX GRADE NO. 416.
M-245; 0.60-1.0 C, bal Fe. (Iron-carbon type).
Sintered: 61,000 psi (420 MPa) crushing strength; 31,000 psi (213 MPa) TS; 26,000 psi (179 MPa) YS; 0.5 El; 6.1-6.5 density (dry); 80-90 Rf hardness.
ASTM B-310-70 Type 2, Class C.

GRAMIX GRADE NO. 417.
M-245; 0.25 max C, Cu, bal Fe. (Infiltered iron).
Sintered: 122,000 psi (841 MPa) crushing strength; 65,000 psi (448 MPa TS; 70,000 psi (483 MPa) YS; 1 El; 7.1-7.6 density (dry); 89-93 Rf hardness.
ASTM B-303-70 Class A.

GRAMIX GRADE NO. 419.
M-245; 0.60-1.0 C, Cu, bal Fe. (Infiltrated iron).
Sintered: 165,000 psi(1138 MPa) crushing strength; 85,000 psi (586 MPa TS; 90,000 psi (621 MPa) YS; 0.5 El; 7.1-7.6 density (dry); 98-101 Rf hardness.
ASTM B-303-70 Class C.
Note: On all Gramix grades YS is yield in compression.

GRAMIX GRADE NO. 510.
M-245; Pb, Sn, bal Cu (High lead bronze).
Sintered: 16,000 psi (110 MPa) crushing strength; 8,000 psi (55 MPa) TS; 5,000 psi (34 MPa) YS; 1 El; 6.6-7.0 density (dry).
For bearings requiring a leaded bronze material.

GRAMIX GRADE NO. 514.
M-245; 0.60-1.0 C, Cu, bal Fe. (Iron alloy).
Sintered: 80,000 psi (552 MPa) crushing strength; 45,000 psi (310 MPa) TS; 42,000 psi (289 MPa) YS; 0.5 El; 5.8-6.2 density (dry); 83-84 Rf hardness.
For high strength structural parts.

GRAMIX GRADE NO. 548.
M-245; Fe-C. (Iron-graphite).
Sintered: 9,000 psi (62MPa) crushing strength; 4,500 psi (31 MPa) TS 3000 psi (20 MPa) YS; 1 El; 5.4-5.8 density (dry).
Specially developed iron bearing material for high speed, light duty applications.

GRAMIX GRADE NO. 560.
M-245; Cu, Fe, Mn, Si, bal Ni. (Monel type).
Sintered: 55,000 psi (379 MPa) crushing strength; 30,000 psi (207 MPa) TS; 22,000 psi (151 MPa) YS; 5 El; 7.0-7.4 density (dry).
For corrosion resistant parts.

GRAMIX GRADE NO. 562.
M-245; Ni, C, bal Fe. (Nickel steel).
Sintered: 120,000 psi (828 MPa) crushing strength; 60,000 psi (414 MPa TS; 50,000 psi (345 MPa) YS; 2 El; 6.6-7.0 density (dry); 81-83 Rf hardness.
For high strength structural parts.

GRAMIX GRADE NO. 562-P3.
M-245; Ni, C, bal Fe. (Hardened nickel steel).
Sintered, Ht; 175,000 psi (1207 MPa) crushing strength; 100,000 psi (690 MPa) TS; 100,000 (690 MPa) YS; 0.5 El; 6.6-7.0 density (dry); 25-30 Rc hardness.
For high strength structural parts.
Note: On all Gramix grades YS is yield in compression.

GRAMIX GRADE NO. 563.
M-245; 0.25 max C, bal Fe. (Machinable iron).
Sintered: 36,000 psi (248 MPa) crushing strength; 18,000 psi (124 MPa) TS; 11,000 psi (75 MPa) YS; 3 El; 5.6-6.0 density (dry); 34-40 Rf hardness.
For pole pieces, complex parts.

GRAMIX GRADE NO. 564.
M-245; 0.25-0.60 C, bal Fe. (Machinable iron).
Sintered: 48,000 psi (331 MPa) crushing strength; 24,000 psi (165 MPa) TS; 18,000 psi (124 MPa) YS; 1 El; 5.7-6.1 density (dry); 57-58 Rf hardness.
For shock absorber pistons.

GRAMIX GRADE NO. 565.
M-245; 0.60-1.0 C, bal Fe. (Machinable iron-carbon).
Sintered: 75,000 psi (517 MPa) crushing strength; 38,000 psi (262 MPa) TS; 24,000 psi (165 MPa) YS; 0.5 El; 5.8-6.2 density (dry); 79-83 Rf hardness.
For structural parts.

GRAMIX GRADE NO. 567.
M-245; Ni, sn, bal Cu. (Nickel bronze).
Sintered: 75,000 psi (517 MPa) crushing strength; 40,000 psi (276 MPa) TS; 32,000 psi (220 MPa) YS; 3 El; 6.8-7.2 density (dry).
For structural parts.

GRAMIX GRADE NO. 568.
M-245; 0.25 max C, bal Fe. (Iron).
Sintered: 53,000 psi (365 MPa) crushing strength; 20,000 psi (138 MPa) TS; 21,000 psi (144 MPa) YS; 4 El; 6.5-6.9 density (dry); 66-72 Rf hardness.
Magnetic applications.

GRAMIX GRADE NO. 569.
M-245; 0.60-1.0 C, bal Fe. (Iron-carbon alloy).
Sintered: 100,000 psi (690 MPa) crushing strength; 52,000 psi (358 MPa TS; 41,000 psi (282 MPa) YS; 1 El; 6.4-6.8 density (dry); 88-95 Rf hardness.
For gears and structural parts.

GRAMIX GRADE NO. 571.
M-245; Zn, bal Cu. (Brass type).
Sintered: 26,700 psi (184 MPa) crushing strength; 12,700 psi (87 MPa) TS; 6,000 psi (41 MPa) YS; 7 El; 6.8-7.2 density (dry).

GRAMIX GRADE NO. 572.
M-245; 0.25 max C, bal Fe. (Machinable iron).
Sintered: 48,000 psi (331 MPa) crushing strength; 19,000 psi (131 MPa) TS; 17,000 psi (117 MPa) YS; 3 El; 6.6-7.0 density (dry); 64-72 Rf hardness.
For machinable and magnetic parts.

GRAMIX GRADE NO. 574.
M-245; 0.60-1.0 C, Cu, bal Fe. (Infiltrated iron-carbon).
Sintered: 175,000 psi (1207 MPa) crushing strength; 90,000 psi (621 MPa) TS; 75,000 psi (517 MPa) YS; 1 El; 7.3-7.7 density (dry) 78-84 Rf hardness.
For high strength structural parts.

GRAMIX GRADE NO. 575.
M-245; Cu, bal Fe. (Iron-copper)
Sintered: 50,000 psi (345 MPa) crushing strength; 16,500 psi (113 MPa) TS; 21,000 psi (144 MPa) YS; 0.5 El; 6.0-6.4 density (dry); 48-60 Rf hardness.
For structural parts.
Note: On all Gramix grades YS is yield in compression.

GRAMIX GRADE NO. 576.
M-245; 0.25 max C, bal Fe (Iron).
Sintered: 96,000 psi (662 MPa) crushing strength; 27,600 psi (190 MPa) TS; 24,000 psi (167 MPa) YS; 4 El; 7.0-7.4 density (dry); 89-93 Rf hardness.
High density, magnetic applications.

GRAMIX GRADE NO. 577.
M-245; 0.25-0.60 C, cu, bal Fe. (Iron alloy).
Sintered: 73,200 psi(505 MPa) crushing strength; 38,100 psi (262 MPa) TS; 33,000 psi (227 MPa) YS; 1 El; 5.8-6.2 density (dry); 75-80 Rf hardness.

GRAMIX GRADE NO. 578.
M-245; 0.25 max C, Ni, bal Fe. (Iron-nickel).
Sintered: 69,000 psi (476 MPa) crushing strength; 30,000 psi (207 MPa) TS; 20,000 psi (138 MPa) YS; 2 El; 6.6-7.0 density (dry); 71-75 Rf hardness.

GRAMIX GRADE NO. 580.
M-245; 0.20-0.40 C, bal Fe. (Machinable iron-carbon).
Sintered: 54,000 psi (372 MPa) crushing strength; 16,000 psi (110 MPa) TS; 14,000 psi (96 MPa) YS; 1 El; 6.2-6.6 density (dry); 79-85 Rf hardness.
For structural parts; good machinability.

GRAMIX GRADE NO. 581.
M-245; 0.25-0.60 C, bal Fe. (Machinable iron-carbon).
Sintered: 58,000 psi (400 MPa) crushing strength; 23,300 psi (160 MPa) TS; 19,000 psi (131 MPa) YS; 2 El; 5.7-6.1 density (dry); 64-73 Rf hardness.
Structural parts; good machinability.

GRAMIX GRADE NO. 582.
M-245; 0.60-1.0 C, bal Fe. (Iron-carbon).
Sintered: 70,000 psi (483 MPa) crushing strength; 36,000 psi (248 MPa) TS; 21,000 psi (144 MPa) YS; 2 El; 6.4-6.8 density (dry); 33-50 Rf hardness.
Structural parts.

GRANADA VANADIUM.
M-38; 1 C, 0.3 Mn, 0.2 V, bal Fe.
For blacksmith tools, shears, punches, forming dies; water hardening.

GRANALEC.
M-633; Si, bal Fe.
For laminations; high magnetic permeability.

GRANATOR.
M-633; Si, bal Fe.
For laminations; high magnetic permeability.

GRANATURE.
M-633; Si, bal Fe.
For laminations; high magnetic permeability.

GRANDIOS now VEW S305.

GRANDIOS 5V now VEW S308.

GRANDIOS EXTRA.
M-1314; 0.76 C, 10 Co, 4.2 Cr, 0.8 Mo, 1.8 V, 18 W, bal Fe.
For lathe and planer tools, milling cutters; high speed steel.

GRANE 1.
M-111; 0.55 C, 1 Cr, 3 Ni, 0.3 Mo, bal Fe.
For drop forging dies; tough.

GRANE 2.
M-111; 0.55 C, 1.5 Cr, 3 Ni, bal Fe.
For drop forging dies; tough.

GRANFIN CAST IRON.
M-Eng.; 2.2-2.6 C, 1.7-2.1 graphite; 2.3-2.5 Si, 0.7-1.0 Mn, bal Fe.
For frames, housings, general castings; cast iron.

GRANFORMER.
M-633; Si, bal Fe.
For laminations; high magnetic permeability.

GRANFORMER 52.
M-633; 4.7 Si, bal Fe.
For transformers; high permeability.

GRANFORMER 58.
M-633; 4.0-4.4 Si, bal Fe.
For transformers; high permeability.

GRANFORMER 65.
M-633; 4.0-4.5 Si, bal Fe.
For transformers; low core loss.

GRANFORMER 72.
M-633; 3.0-3.4 Si, bal Fe.
For transformers for radios; high permeability.

GRANIMO.
M-633; Si, bal Fe.
For laminations; high magnetic permeability.

GRANISIL.
M-633; Si, bal Fe.
For laminations; high magnetic permeability.

GRANIT.
M-1318; 2.1 C, 11.5 Cr, bal Fe.
For blanking and forming dies; oil hardening, nondeforming.

GRANIT 2.
M-1290; 0.18-0.24 C, 0.15-0.35 Si, 1.1-1.4 Mn, 1.0-1.3 Cr, bal Fe.
Cold work tool steel.
W.-Nr. 1.2162.

GRANIT CO.
M-1318; 0.65 C, Cr, Co, bal Fe.
For tools, dies; oil hardening.

GRANITE CITY HIGH-YIELD NO 1.
M-633; C, 0.7-0.9 Mn, 0.01 max P, 0.035 S, 0.15-0.20 Si, 0.20 min Cu, 0.15 min Cr, bal Fe.
Annealed: 70,000 TS; 55,000 YS; 25 El.
For transportation equipment, cars, buses, trucks; corrosion resistant.

GRANITE CITY HIGH-YIELD NO. 2.
M-633; C, 1.2-1.6 Mn, 0.01 max P, 0.035 S, 0.15-0.20 Si, 0.2 min Cu, 0.15 min Cr, bal Fe.
Annealed: 85,000 TS; 70,000 YS; 18 El.
For bridges, transportation equipment, cars, buses, trucks; corrosion resistant.

GRANIT W.
M-1318; 2.1 C, 11.5 Cr, 0.7 W, bal Fe.
For thread rolling dies, blanking and forming dies; oil hardened, non-deforming.

GRANT.
M-1705; 0.45 C, 0.55 Mn, 0.2 Si, 0.95 Cr, 0.20 V, bal Fe.
Tool steel for miscellaneous applications as hammers, hatchets. AISI L2.

GRAPH-AIR.
M-376, M-341; 1.35 C, 1.85 Mn, 1.20 Si, 1.85 Ni, 1.50 Mo, bal Fe.
Air hardening tool steel, minimal distortion.

GRAPHALLOY BABBITT.
M-247; graphite impregnated with Pb-base Babbitt.
Sintered: 5000 TS.
For self-lubricating bearings, seal rings, contact shoes; C.S. 19000.

GRAPHALLOY BRONZE.
M-247; graphite impregnated with bronze.
Sintered: 5500 TS.
For self-lubricating bearings, seal rings, contact shoes; C.S. 24000.

GRAPHALLOY CADMIUM.
M-247; graphite impregnated with Cd.
Sintered: 5000 TS.
For bearings, contacts for controllers, switches, seal rings; maximum temprature 250°F; self lubricating.

GRAPHALLOY COPPER.
M-247; graphite impregnated with Cu.
Sintered: 6000 TS.
For slip ring brushes, contacts on controllers, relays, bearings; self lubricating, maximum temperature 700°F; C.S. 25000.

GRAPHALLOY GRADE A.
M-247; 55 graphite, bal Pb Babbitt.
For bearings of light service; Babbitt soft grade.

GRAPHALLOY GRADE N.
M-247; 55 graphite, bal Pb babbitt.
For bearings of medium service; Babbitt medium grade.

GRAPHALLOY GRADE O.
M-247; 55 graphite, bal Pb Babbitt.
For bearings; Babbitt hard grade

GRAPHALLOY GRADE S.
M-247; 55 graphite, bal Pb Babbitt
For bearings; medium hardness.

GRAPHALLOY IRON.
M-247; graphite impregnated with Fe.
Sintered: 6000 TS.
For self-lubricating bearings, submerged bearings; C.S. 25000.

GRAPHALLOY SILVER 8.
M-247; graphite impregnated with Ag.
Sintered: 5500 TS.
For contacts on relays, circuit breakers, bearings, brushes; maximum temperature 700°F; self lubricating; C.S. 24000.

GRAPHEX.
M-824; 87-92 Cu, 9-11 Sn, 4 graphite.
13,000 TS; 11,000 YS; 1 El; 30 Brin.
For bearings, bushings; self lubricating.

GRAPHIDOX.
M-1038; 5-7 Ca, 50-55 Si, 9-11 Ti, 1.0-1.3 Al, 0.15-0.25 C, bal Fe.
Ferroalloy for inoculant for cast iron; also for supplementary deoxidation of steel.

GRAPHITE METAL-1.
M-Eng.; 80 Pb, 20 Sb.
For crucibles, lubricants, lead pencils.

GRAPHITE METAL-2.
M-Eng.; 68 Pb, 17 Sb, 15 Sn.
For foundry facings, electric brush carbons.

GRAPH-MO.
M-376, M-341; 1.5 C, 0.25 Mo, 0.8 Si, 0.4 Mn, bal Fe.
Annealed: 85,000 TS; 50,000 YS; 25 El; 40 RA, 197 Brin.
Heat treated: 218,000 TS; 170,000 YS; 9 El; 14 RA: 388 Brin.
For plug and ring gauge dies, taps, spindles, rolls; wear and abrasion resistant graphitic steel, non-seizing.

GRAPHO BABBITT METAL NO. 1.
M-300; graphite and Babbitt.
19 Brin.
For Babbitt, graphite bearings, transmission and railroad axle bearings, engine bearings; low bearing temperature, reduced friction losses.

GRAPHO BABBITT METAL NO. 2.
M-300; 0.3 graphite, S, Pb, Sb.
26 Brin.
For bearings for electrical motors, blowers, high speed transmissions; high velocity and high pressures.

GRAPHO BABBITT METAL NO. 3.
M-300; 0.3 graphite, Sn, Pb, Sb.
29 Brin.
For bearings for crankshafts in internal combustion engines, mining machinery, machine tools, reciprocating pumps; high pressure and high velocity with alternating load.

GRAPHO BABBITT METAL NO. 4.
M-300; 0.4 graphite, Sn, Pb, Sg.
34 Brin.
For bearings for diesel engines, heavy duty compressors; heavy duty; all kinds of shocks and velocities.

GRAYDAC.
M-712; 0.08 C, 0.31 Mn, 0.29 Si, bal Fe.
Welded: 64,000-72,000 TS; 55,000-62,000 YS; 26-19 El; 45-30 RA.
For welding electrodes; low spatter; E-6013.

GRAY DEVIL.
M-712; 0.09 C, bal Fe.
Welded: 77,000 TS; 65,000 YS; 19 El; 30 RA.
For welding rods for steel; E-6012.

GRAY GOLD.
M-Eng.; 86 Au, 5.7-17.0 Fe, 0-8.6 Ag.
For jewelry; corrosion resistant.

GRAY LABEL.
M-341; 0.95 C, bal Fe.
For cold beading dies; water hardening.

GRAY LABEL STYRIAN.
 M-651; 0.88 C, 0.19 Mn, bal Fe.
 For tools, drills, taps; water hardening.

GREAT WESTERN GW.
 M-1067; 0.85 C, 1 Mn, 0.35 Si, 0.5 Cr, 0.5 W, bal Fe.
 For milling cutters, drills, dies, punches, mandrels; oil hardened, non-deforming.

GREEK ASCOLOY.
 M-57; 0.12 max C, 12-14 Cr, 1.8-2.2 Ni, 2.5-3.5 W, 0.5 max Mo, 0.5 max Cu, bal Fe.
 Heat treated: 137,000-195,000 TS; 108,000-160,000 YS; 23-14 El; 63-51 RA; 290-400 Brin.
 For compressor blades and vanes, fasteners, jet engine components; heat resistant to 1000°F.

GREEN GOLD.
 M-Eng.; 75 Au, 11-25 Ag, 13-0 Cd.
 For jewelry; corrosion resistant.

GREEN LABEL.
 M-343; 0.8-1.1 C, bal Fe.
 For tools depending on temper; "Hellers Extra Tool."

GREEN LABEL.
 M-343; 0.8-0.9 C, 1.2-1.4 Mn, 0.4-0.6 W, 0.4-0.6 Cr, bal Fe.
 For precision tools and files, taps, reamers, chasters, gages; non-deforming; "Stayput" Oil Hardening.

GREEN LABEL.
 M-373; 0.7-1.4 C, bal Fe.
 For tools, drills, cutters; water hardening.

GREEN LABEL.
 M-341; 0.5 C, 1.0 Cr, 0.75 Mn, 0.18 V, bal Fe.
 For gears, crankshafts, springs; oil hardening; shock resisting.

GREEN LABEL.
 M-365; 0.7-0.8 C, bal Fe.
 For tools.

GREEN LABEL.
 M-694; 0.9 C, 1.15 Mn, 0.5 Cr, 0.5 W, 0.2 V, bal Fe.
 For tools, dies, punches; nondeforming, oil hardening.

GREEN LABEL EXTRA.
 M-343; 0.8-1.1 C, bal Fe.
 For tools, dies; water hardening.

GREENLEAF S-6.
 M-1194; WC + Co.
 For tipped cutting tools; cemented carbides.

GRIDUR C25.
 M-380; 2.6 C, 27.0 Cr, bal Fe.
 Weld: Rock C 50-52.
 For worm conveyors, guide bars, slides, drawing dies.
 Hard facing electrode. Wear and corrosion resistant.

GRIDUR C35.
 M-380; 3.5 C, 1.0 Si, 35.0 Cr, bal Fe.
 Weld: Rock C 60-62.
 For machine parts and tools subject to abrasion by sand, gravel, coal, cement. Corrosion and wear resistant. Hard facing electrode.

GRIDUR G5.
 M-380; Granular WC in steel tubes.
 For tools and machine parts, rotary borers, rock drills, dredger teeth.
 Hard facing electrode. Wear resistant.

GRIDUR G10.
 M-380; Granular WC in steel tubes.
 For tools and machine parts, rotary borers, rock drills, dredger teeth.
 Hard facing electrode. Wear resistant.

GRIDUR G15.
 M-380; Granular WC in steel tubes.
 For tool and machine parts, rotary borers, rock drills, dredger teeth, crushing jaws.
 Hard facing electrode. Wear resistant.

GRIDUR K40.
 M-380; 1.2 C, 1.5 Si, 65.0 Co, 27.0 Cr, 5.5 W, bal Fe.
 Weld: Rock C 40-45.
 For steam valves, exhaust valves, bearings, trimming dies.
 Hard facing electrode. Corrosion, heat and wear resistant.

GRIDUR K50.
 M-380; 2.2 C, 55.0 Co, 27.0 Cr, 14.0 W, bal Fe.
 Weld: Rock C 50-54.
 For steam valves, exhaust valves, bearings, trimming dies.
 Corrosion, heat and wear resistant. Hard facing electrode.

GRIDUR K60.
 M-380; 2.7 C, 53.0 Co, 25.0 Cr, 19.0 W, bal Fe.
 Weld: Rock C 56-60.
 For needle valves, impeller parts.
 Hard facing electrode. Wear and corrosion resistant.

GRIDUR SA1.
 M-380; 1.0 C, 4.2 Cr, 2.8 Mo, 2.5 V, 3.0 W, bal Fe.
 Weld: Rock C 48.
 For hot working tools, air cooling systems.
 Hard facing electrode. High speed steel.

GRIDUR SA2.
M-380; 0.85 C, 4.2 Cr, 1.2 V, 18 W, bal Fe.
Weld: Rock C 58-62.
For tough and impact resistant cutting tools.
Hard facing electrode. High speed steel.

GRIDUR SA3.
M-380; 1.0 C, 4.2 Cr, 9.0 Mo, 1.2 V, 1.8 W, bal Fe.
Weld: Rock C 58-62.
For turning tools, milling cutters, reamers, drills, hot shear blades.
Hard facing electrode. High speed steel.

GRIDUR SA4.
M-380; 1.5 C, 8.0 Co, 4.5 Cr, 6.0 Mo, 3.2 V, 8.0 W, bal Fe.
Weld: Rock C 62-65.
For cutting tools with highest wear resistance and hardness at elevated temperatures.
Hard facing electrode. High speed steel.

GRIDUR-SCS.
M-380; 2.0 C, 0.3 Si, 0.3 Mn, 12.0 Cr, 0.8 W, bal Fe.
Weld: 450 Brin.
For welding cutting and blanking dies, punches, shears, broaches, drawing tools.
Hard facing electrode.

GRIDUR SCS-E.
M-380; 2.0 C, 12 Cr, bal Fe.
Weld: Rock C 63-65.
For hard facing cutting and blanking dies, punches, shears, broaches.
Hard facing electrode.

GRIDUR WA1.
M-380; 0.5 C, 2.3 Cr, 0.8 Mo, 1.7 Ni, 0.1 V, bal Fe.
Weld: 300-400 Brin.
For welding on tools requiring high toughness and impact resistance, die blocks, upsetters, piercers to 450°C.
Hard facing electrode. Hot work steel.

GRIDUR WA2.
M-380; 0.3 C, 2.3 Cr, 0.5 Mo, 0.6 V, 4.5 W, bal Fe.
Weld: Rock C 45.
For welding on hot working tools, forging dies, mandrels, punches, bushings.
Hard facing electrode. Hot work steel.

GRIDUR WA3.
M-380; 0.3 C, 2.3 Co, 2.7 Cr, 0.4 V, 9.0 W, bal Fe.
Weld: Rock C 48.
For hot working tools, air cooling systems.
Hard facing electrode. Hot work steel.

GRIESHEIM-EA2.
M-380; 0.5 C, 0.4 Si, 5.0 Mn, bal Fe.
Weld: 500 Brin.
For welding skid rails, crusher jaws, baffle plates, dredger parts.
Hard facing electrode.

GRIESHEIM-EA3.
M-380; 0.45 C, 0.45 Si, 0.4 Mn, 5.8 Cr, bal Fe.
Weld: 600 Brin.
For welding crushing jaws, dies, cams, worms, breakers, chain wheels.
Hard facing electrode.

GRIESHEIM-EA600.
M-380; 0.5 C, 2.4 Si, 0.4 Mn, 9.0 Cr, bal Fe.
Welded: 560-600 Brin.
For welding crushing jaws, dies, cams, worms, breakers, chain wheels.
Hard facing electrode.

GRIESHEIM-EA600W.
M-380; 0.5 C, 0.5 Mn, 2.5 Si, 8.0 Cr, bal Fe.
Weld: 560 Brin.
For welding crusher jaws, dies, breakers, cams, worms, chain wheels.
Hard facing electrode.

GRIESHEIM EA-B.
M-380; 0.22 C, 1.2 Si, 0.8 Mn, bal Fe.
Weld: 250 Brin.
For welding for rails, switch tongues, wheel tires, worms, axes, cams.
Welding electrode, hard surfacing.

GRIESHEIM-EA-V.
M-380; 0.3 C, 1.2 Si, 1.2 Mn, 1.0 Cr, bal Fe.
Weld: 350 Brin.
For welding rails, switch tongues, wheel tires, worms, axes, shafts, cams.
Welding electrode, hard facing.

GRIESHEIM N.
M-380; 0.7 C, 0.6 Si, 13 Mn, 3.5 Ni, bal Fe.
Weld: 200 Brin.
For welding crusher jaws, mill beaters, dredger teeth, rolling mill parts.
Hard facing electrode. Austenitic. Work hardens to 450 Brinell.

GRIESHEIM "RAUCHLOS".
M-380; 0.2 C, bal Fe.
For welding rods.

GRIESHEIM S.
M-380; 1.0 C, 0.4 Si, 13 Mn, bal Fe.
Weld: 200 Brin.
For welding mill beaters, crushing jaws, dredger teeth, rolling mill parts.
Austenitic hard facing electrode. Work hardens to 450 Brinell.

GRIESHEIM "SINI".
M-380; Alloy, bal Cu.
For welding rods for copper welding.

GRIFFIN.
M-350; C, Cr, Mn, bal Fe.
For dies, punches, thread rolling tools; non-deforming.

GRILLO 2105.
M-1134; 5 Al, 1 Cu, 0.02 Ga, bal Zn.
For die castings.

GRILLO 31010.
M-1134; 10 Al, 1 Cu, 0.02 Ga, bal Zn.
For castings.

GRIMM 4S4-ZH.
M-1316; 0.85 C, 0.25 max Si, 0.25 max Mn, bal Fe.
Heat treated: 190,000 TS; 150,000 YS; 10 El; 30 RA; 390 Brin.
For springs, taps, drills, hobs, reamers; Type W1; water hardened.

GRIMM BRF.
M-1316; 0.20 C, 13 Cr, 0.3 Mn, bal Fe.
Annealed: 95,000 TS; 50,000 YS; 25 El; 55 RA; 196 Brin.
For turbine blades, valves, cutlery, surgical instruments; Type 420; stainless.

GRIMM CWL.
M-1316; 2.1 C, 11.5 Cr, 0.70 W, 0.30 Mn, bal Fe.
For blanking and forming dies, punches; oil hardened, non-deforming.

GRIMM DSA.
M-1316; 0.38 C, Cr, V, Si, bal Fe.
For gears, bolts, machine tool parts; oil hardened, tough.

GRIMM DSW.
M-1316; 0.45 C, W, Cr, V, bal Fe.
For header dies, upsetters, crimpers; oil hardened.

GRIMM G4.
M-1316; 0.75 C, 0.25-0.50 Si, 0.3-0.8 Mn, bal Fe.
Heat treated: 185,000 TS; 140,000 YS; 15 El; 40 RA; 375 Brin.
For rails, springs, tools, axes, hammers; water hardened.

GRIMM GC15.
M-1316; 1.05 C, 1.0 Cr, bal Fe.
For bearings, liners, bushings; water hardened, wear and abrasion resistant.

GRIMM GC120.
M-1316; 2.1 C, 11.5 Cr, 0.3 Mn, bal Fe.
For blanking and forming dies; oil hardened, non-deforming.

GRIMM GCK.
M-1316; 1.45 C, 1.4 Cr, 0.6 Mn, bal Fe.
For bearings, bushings, liners; water hardened, wear resistant.

GRIMM GCM.
M-1316; 0.90 C, 1.9 Mn, 0.1 V, bal Fe.
For punches, dies, shears, crimpers; oil hardened, non-deforming.

GRIMM GCS.
M-1316; 0.90 C, Cr, Si, bal Fe.
For bearings, bushings, liners; water hardened, wear resistant.

GRIMM GCZ.
M-1316; 0.90 C, 0.80 Cr, 0.3 Mn, 0.25 Si, bal Fe.
For bearings, cutters, liners, bushings; water hardened, wear resistant.

GRIMM GEE.
M-1316; 0.15 C, 0.25-0.50 Si, 0.30-0.80 Mn, bal Fe.
Annealed: 70,000 TS; 40,000 YS; 25 El; 60 RA; 145 Brin.
For gears, bolts, machine tool parts; case hardening steel.

GRIMM GH4.
M-1316; 0.75 C, 0.25-0.50 Si, 0.30-0.80 Mn, bal Fe.
Heat treated: 180,000 TS; 132,000 YS; 10 El; 34 RA; 375 Brin.
For springs, rails, hammers, tools, cutters; Type W1; water hardened.

GRIMM GH5.
M-1316; 0.60 C, 0.25-0.50 Si, 0.30-0.80 Mn, bal Fe.
Heat treated: 160,000 TS; 113,000 YS; 12 El; 40 RA; 325 Brin.
For gears, springs, fasteners; water hardened.

GRIMM GHM.
M-1316; 0.56 C, Ni, Cr, Mo, V, bal Fe.
For gears, bolts, crankshafts; oil hardened, shock resistant.

GRIMM GK3.
M-1316; 0.86 C, 2.8 Co, 4.3 Cr, 0.85 Mo, 2.1 V, 12 W, bal Fe.
For lathe and planer tools, reamers, broaches; high speed steel.

GRIMM GK5.
M-1316; 0.79 C, 4.75 Co, 4.3 Cr, 0.75 Mo, 1.55 V, 18 W, bal Fe.
For lathe and planer tools, reamers; high speed steel.

GRIMM GK10.
M-1316; 0.76 C, 10 Co, 4.2 Cr, 0.8 Mo, 1.8 V, 18 W, bal Fe.
For lathe and planer tools, reamers, milling cutters; high speed steel.

GRIMM GKS.
M-1316; 1.0 C, 0.1 V, 0.25 Si, 0.25 Mn, bal Fe.
For drills, taps, reamers, broaches; Type W2; water hardened.

GRIMM GLS.
M-1316; 0.55 C, W, Cr, V, bal Fe.
For upsetters, shears, punches; oil hardened, tough.

GRIMM GM.
M-1316; 0.55 C, 0.70 Cr, 0.18 Mo, 1.65 Ni, 0.1 V, bal Fe.
For gears, bolts, crank shafts; oil hardened, shock resistant.

GRIMM GME.
M-1316; 0.40 C, Cr, Mo, Mn, bal Fe.
For gears, bolts, machine tool parts; oil hardened, tough.

GRIMM GR.
M-1316; 0.50 C, 1.05 Cr, 0.95 Mn, 0.1 V, bal Fe.
For springs, bolts, gears, studs; oil hardened, shock resistant.

GRIMM GSM.
M-1316; 0.53 C, 0.90 Si, 0.90 Mn, bal Fe.
For springs, bolts, crankshafts; water or oil hardened.

GRIMM GWP.
M-1316; 0.50 C, 1.05 Cr, 3.25 Ni, bal Fe.
For gears, pinions, countershafts, axles; oil hardened, shock resistant.

GRIMM GWV112B.
M-1316; 0.74 C, 4.1 Cr, 1.1 V, 18.5 W, bal Fe.
For lathe and planer tools, reamers, drills, taps; high speed steel.

GRIMM GWV122 SPEZIAL.
M-1316; 0.82 C, 4 Cr, 0.8 Mo, 1.6 V, 8.7 W, bal Fe.
For lathe and planer tools, reamers, broaches; high speed steel.

GRIMM GWV132 RAPID.
M-1316; 0.86 C, 0.4 Cr, 0.8 Mo, 2.5 V, 12 W, bal Fe.
For lathe and planer tools, reamers, broaches; high speed steel.

GRIMM GWV145 RECORD.
M-1316; 1.3 C, 4.3 Cr, 0.8 Mo, 3.8 V, 12 W, bal Fe.
For form cutters, broaches, reamers; high speed steel.

GRIMM HPM.
M-1316; 0.30 C, 2 Co, 2.4 Cr, 0.25 V, 8.5 W, bal Fe.
For extrusion press dies, rams and liners; oil hardened, tough.

GRIMM HWP.
M-1316; 0.30 C, 2.65 Cr, 0.35 V, 8.5 W, bal Fe.
For extrusion press liners and rams, punches, shears; oil hardened, tough.

GRIMM KSA.
M-1316; 0.45 C, Si, Cr, V, bal Fe.
For dies, bolts, crankshafts, gears, shears; oil hardened, tough.

GRIMM NS12.
M-1316; 0.34 C, 1.1 Al, 1.4 Cr, bal Fe.
For oil refinery equipment; creep resistant.

GRIMM NS13.
M-1316; 0.32 C, 1.1 Al, 1.1 Cr, 0.18 Mo, bal Fe.
For oil refinery equipment; creep resistant.

GRIMM NS15.
M-1316; 0.27 C, 1.1 Al, 1.4 Cr, bal Fe.
For oil refinery equipment; creep resistant.

GRIMM NS23.
M-1316; 0.33 C, 1.1 Al, 1.7 Cr, 1.0 Ni, bal Fe.
For oil refinery equipment; creep resistant.

GRIMM NS48.
M-1316; 0.30 C, 1.1 Cr, 0.2 V, bal Fe.
For gears, pinions, bolts, fasteners; oil hardened, shock resistant.

GRIMM NS54.
M-1316; 0.31 C, 2.5 Cr, 0.18 Mo, 0.13 V, bal Fe.
For gears, pinions, bolts, fasteners, shafts; oil hardened, shock resistant.

GRIMM RSA.
M-1316; 1.42 C, W, V, bal Fe.
For forming and heading dies; oil or water hardened.

GRIMM'S ALUMINUM SOLDER-1.
M-Eng.; 69.1 Sn, 28.8 Pb, 1.44 Zn, 0.72 Ag.
For aluminum solder.

GRIMM'S ALUMINUM SOLDER-2.
M-Eng.; 50 Sn, 25 Pb, 25 Zn.
For aluminum solder.

GRIMM SB.
M-1316; 1.05 C, 1.0 Cr, 1.15 W, 0.90 Mn, bal Fe.
For forming and blanking dies, punches; oil hardened, tough.

GRIMM SE.
M-1316; 0.20 C, 1.25 Mn, 1.15 Cr, bal Fe.
For gears, bolts, camshafts, cams; case hardened.

GRIMM SH4-ZH.
M-1316; 0.75 C, 0.4 Si, 0.6 Mn, bal Fe.
Heat treated: 180,000 TS; 130,000 YS; 10 El; 35 RA; 360 Brin.
For springs, rails, clutch discs; Type W1; water hardened.

GRIMM SH5-ZAH.
M-1316; 0.60 C, 0.4 Si, 0.6 Mn, bal Fe.
Heat treated: 160,000 TS; 114,000 YS; 12 El; 40 RA; 325 Brin.
For wheels, die blocks, rails, axles, bolts; water hardened.

GRIMM SH6-ZW.
M-1316; 0.45 C, 0.4 Si, 0.6 Mn, bal Fe.
Hot rolled: 98,000 TS; 58,000 YS; 24 El; 45 RA; 212 Brin.
For axles, gears, bolts, bushings; water hardened.

GRIMM SSA.
M-1316; 0.95 C, W, Mo, bal Fe.
For tools, dies, cutters; oil hardened.

GRIMM SSCV.
M-1316; 1.35, C, W, Co, bal Fe.
For engravers' tools, blanking dies; oil hardened, wear resistant.

GRIMM WA5.
M-1316; 0.55 C, 0.10-0.40 Si, 0.50-0.70 Mn, bal Fe.
Heat treated: 155,000 TS; 110,000 YS; 14 El; 42 RA; 315 Brin.
For axles, tie rods, bushings, bolts, shafts; water hardened.

GRIMM WE2 HARD.
M-1316; 1.3 C, 0.25 max Si, 0.25 max Mn, bal Fe.
For engravers' tools, reamers, blanking dies; Type W1; water hardened.

GRIMM WE3-MH.
M-1316; 1.0 C, 0.25 max Si, 0.25 max Mn, bal Fe.
Heat treated: 200,000 TS; 125,000 YS; 8 El; 27 RA; 400 Brin.
For springs, taps, reamers, drills, cutters; Type W1; water hardened.

GRIMM WE4-ZH.
M-1316; 0.85 C, 0.25 max Si, 0.25 max Mn, bal Fe.
Heat treated: 190,000 TS; 140,000 YS; 10 El; 30 RA; 390 Brin.
For springs, taps, drills, hobs, reamers; Type W1; water hardened.

GRIMM WE5-ZAH.
M-1316; 0.70 C, 0.25 max Si, 0.25 max Mn, bal Fe.
Heat treated: 174,000 TS; 128,000 YS; 12 El; 37 RA; 350 Brin.
For springs, clutch disks, rails; Type W1; water hardened.

GRIMM WPN.
M-1316; 0.30 C, 1.0 Si, 1.1 Cr, 0.18 V, 3.75 W, bal Fe.
For extrusion rams and liners, punches, upsetters; hot work steel, oil hardened.

GRIMM WPN EXTRA.
M-1316; 0.30 C, 2.35 Cr, 0.6 V, 4.25 W, bal Fe.
For shears, punches, upsetters, dies; hot work steel, oil hardened.

GRIMM WS2.
M-1316; 1.1 C, 0.25 max Si, 0.25 max Mn, bal Fe.
Annealed: 110,000 TS; 58,000 YS; 19 El; 40 RA; 210 Brin.
For springs, taps, drills, hobs, reamers; Type W1; water hardened.

GRIMM WS3-MH.
M-1316; 1.0 C, 0.25 max Si, 0.25 max Mn, bal Fe.
Heat treated: 200,000 TS; 125,000 YS; 10 El; 30 RA; 400 Brin.
For springs, taps, drills, hobs, reamers; Type W1; water hardened.

GRIMM WS5-ZAH.
M-1316; 0.70 C, 0.25 max Si, 0.25 max Mn, bal Fe.
Heat treated: 175,000 TS; 128,000 YS; 12 El; 37 RA; 350 Brin.
For wheels, springs, die blocks, girders, rails; Type W1; water hardened.

GRIMM ZR SPEZIAL.
M-1316; 1.4 C, 0.30 Cr, 0.10 V, bal Fe.
For bearings, cutters, liners; oil hardened, wear resistant.

GRITALLOY 10011.
M-717.
Tungsten carbide and nickel base alloy powder, for hard build up.

GRM 235D.
M-114, M-1491; 0.15 C, 15.5 Cr, 5.2 Mo, 3.6 Al, 2.5 Ti, 4.2 Fe, 0.075 B, bal Ni.
For high temperature cast components in jet engines; high heat and stress-rupture properties.

GROMMET BRASS.
M-US; 70 Cu, 30 Zn.
For grommets; high ductility.

GROSSMAN.
Al alloy.
For light alloy parts.

GS-38.
M-1290; 0.15 C, bal Fe.
Annealed: 54,000 min TS; 28,000 min YS; 25 min El.
Low carbon, unalloyed steel casting.
Werkstoff Nr. 1.0416.

GS-45.
M-1290; 0.22 C, bal Fe.
Annealed: 64,000 min TS; 35,000 min YS; 22 min El.
Low carbon steel casting.
Werkstoff Nr. 1.0443.

GS-52.
M-1290; 0.30 C, bal Fe.
Annealed: 74,000 min TS; 40,000 min YS; 18 min El.
Carbon steel casting.
Werkstoff Nr. 1.0551.

GS-60.
M-1290; 0.37 C, bal Fe.
Annealed: 75,000 min TS; 42,000 min YS; 15 min El.
Carbon steel casting.

GS 62.
M-1290; 0.40 C, bal Fe.
Carbon steel casting.
GS-62.3.

GS-70.
M-1290; 0.45 C, bal Fe.
Annealed: 88,000 min TS; 50,000 min YS; 12 min El.
Carbon steel casting.
Werkstoff Nr. 1.0553.

G.S.N.
M-73; 2.2 C, 13 Cr, bal Fe.
For stamping and forming dies; non-deforming.

G.S.N. + MO.
M-73; 1.5 C, 12 Cr, 0.8 Mo, 1 V, bal Fe.
For stamping and forming dies; non-deforming.

GT2A METAL.
M-France; C, Cr, Ni, bal Fe.
For chemical equipment; corrosion resistant.

GT4A METAL.
M-France; C, cr, Ni, bal Fe.
For chemical equipment; corrosion resistant.

GT-45.
M-100; 0.08 C, 16.7 Cr, 14 Ni, 2.7 Mo, 3 Cu, 0.25 Ti, 0.35 Cb, 0.5 Si, 1.25 Mn, bal Fe.
For high temperature applications, jet engines and turbo superchargers; heat and corrosion resistant.

GUETTIERES BUTTON.
M-Eng.; 56 Cu, 44 Zn.
For buttons.

GUETTIERES BUTTON.
M-Eng.; 61.5 Cu, 29-32 Zn, 6.5-9.7 Sn.
For buttons.

GUILLAUME'S METAL.
M-Eng.; 64.3 Cu, 35.7 Ni.
For tape, chemical equipment; name also applied to "Invar" tape.

GUINEA GOLD.
M-Eng.; 88 Cu, 12 Zn.
For ornamental parts, cheap jewelry; red brass.

GUISHIBUICHI.
M-Jap.; 67-51 Cu, 32-49 Ag, Au, Fe.
For jewelry, ornaments; v. Shibu-ichi.

GULF AIR.
M-1431; 1.0 C, 5 Cr, 1.0 Mo, bal Fe.
For tools and dies; Type A2; air hardened, non-deforming.

GULF GH9.
M-1431; C, alloy, bal Fe.
For machine tool parts; oil hardened.

GULF GH14.
M-1431; C, alloy, bal Fe.
For machine tool parts; oil hardened.

GULF GH-VAN.
M-1431; C, alloy, bal Fe.
For machine tool parts; oil hardened.

GULF GHW.
M-1431; C, alloy, bal Fe.
For machine tool parts; oil hardened.

GULF H.S.
M-1431; 0.70 C, 4 Cr, 1 V, 18 W, bal Fe.
For lathe and planer tools, drills, hobs, reamers; Type T-1; high speed steel.

GULF H.S.5.
M-1431; 0.80 C, 4 Cr, 2 V, 18 W, 8 Co, bal Fe.
For lathe and planer tools, reamers, broaches; Type T-5; high speed steel.

GULF M-2.
M-1431; 0.80 C, 4 Cr, 2 V, 6 W, 5 Mo, bal Fe.
For thread rolling dies, broaches, reamers, cutters; Type M2; high speed steel.

GULF O.H.
M-1431; 0.90 C, 1 Mn, 0.5 Cr, 0.5 W, bal Fe.
For dies, punches, rollers, mandreis; Type O1; oil hardening, non-deforming.

GUN METAL.
M-212; 88 Cu, 8 Sn, 4 Zn.
Cast: 40,000-48,000 TS; 18,000 YS; 50-20 El; 70 Brin.
For steam and structural bronze, gears, bolts; resists salt water corrosion.

GUN METAL NO. 1.
M-1518; 88 Cu, 10 Sn, 2 Zn.
Cast: 36,000 TS; 12 El; 50-65 Brin.
For bearings, gears, marine parts; Gun Metal.

GUN METAL NO. 2.
M-1518; 85 Cu, 5 Sn, 5 Zn, 5 Pb.
Cast: 27,000 TS; 12 El; 45-55 Brin.
For steam valves, bearings, gears; pressure tight, creep resistant.

GUN METAL NO. 3.
M-1518; 86 Cu, 7 Sn, 5 Zn, 2 Pb.
Cast: 31,500 TS; 12 El; 45-60 Brin.
For backing for white metal lined bearings; pressure tight castings, creep resistant.

GUN METAL NO. 4.
M-1518; Sn, Zn, bal Cu.
Cast: 38,100 TS; 25 El; 45-60 Brin.
For bearings, pressure tight castings; pressure tight, not good above 450°F.

GUN METAL-A.
M-Eng.; 90 Cu, 10 Sn.
For electric contacts, valve disks; corrosion resistant.

GUN METAL-B.
M-Eng.; 90 Cu, 6.5 Sn, 2 Zn, 1.5 Pb.
For steam fittings, valves, cocks; free-cutting.

GUN METAL, CHINESE.
M-China; 71.16 Cu, 27.36 Zn, 1.40 Fe.
For early cannons and guns.

GUN METAL, CHINESE.
M-China; 93.2 Cu, 5.05 Sn, 1.72 Fe.
For early cannons and guns.

GUN METAL COCHIN CHINA.
M-China; 77.18 Cu, 3.42 Sn, 5.02 Zn, 13.22 Pb, 1.16 Fe.
For early cannons and guns.

GUN METAL COCHIN CHINA.
M-China; 93.19 Cu, 5.43 Sn, 1.38 Fe.
For early cannons and guns.

GUN METAL, ENGLISH.
M-Eng.; 89-92 Cu, 8-11 Sn.
For early cannons and guns.

GUN METAL, FRENCH.
M-Fr.; 90.1 Cu, 9.9 Sn.
For early cannnons and guns; modern.

GUN METAL, FRENCH.
M-Fr.; 89.44 Cu, 8.91 Sn, 1.31 Zn, 0.16 Pb.
For early cannons and guns; old.

GUN METAL MODIFIED.
M-212; 86 Cu, 9.5 Sn, 2.5 Pb, 2 Zn.
Cast: 32,000-40,000 TS; 25-15 El; 23-12 RA; 63-72 Brin.
For gears, bearings; "Barr Alloy No. 2."

GUN METAL PRUSSIAN.
M-Pruss.; 90.9 Cu, 9.1 Sn.
For cannons and guns; modern.

GUN METAL, RUSSIAN.
M-Russ.; 88-91 Cu, 9-12 Sn, 0.07 Fe.
For early cannons and guns.

GUN METAL, SWISS.
M-Swiss.; 88.93 Cu, 10.37 Sn, 0.42 Zn.
For early cannons and guns; "lucern."

GUN METAL, TURKISH.
M-Turk.; 91-95 Cu, 4.7-8.8 Sn, 0.02 Fe.
For early cannons and guns.

GUN MOUNT BRONZE.
M-US; 17 Zn, 3 Sn, bal Cu.
For gun mounts; high strength.

GURLEY'S METAL.
M-Eng.; 86.5 Cu, 5.4 Zn, 5.4 Sn, 2.7 Pb.
For steam fittings, valves, cocks; free-cutting.

GURNEYS BRONZE.
M-US; 76 Cu, 9 Sn, 15 Pb.
For machine parts; also called "U.S. Government Bronze."

GURONIT.
M-1329; C, alloy, bal Fe.
For machine tool products; oil hardened.

GURONIT 1.1 H.
M-1329; 192 C, 1.3 Si, 29 Cr, bal Fe.
For wear plates, crushers, heat and abrasion resistant.

GURONIT 2.2 H.
M-1329; 1.2 C, 1.3 Si, 29 Cr, 2 Mo, bal Fe.
For wear plates, crushers; corrosion, heat and wear resistant.

GURONIT 14.
M-1329; 0.25 C, 14.5 Cr, 1 max Ni, bal Fe.
Annealed: 85,000 TS; 40,000 YS; 50 El; 65 RA; 185 Brin.
For surgical instruments, gages, valves; Type 420; stainless, austenitic.

GURONIT 18.
M-1329; 0.25 C, 17 Cr, 1.8 max Ni, bal Fe.
For chemical plant equipment; corrosion resistant.

GURONIT 18-8.
M-1329; 0.15 C, 18 Cr, 8.5 Ni, bal Fe.
Annealed: 75,000 TS; 35,000 YS; 55 El; 75 RA; 140 Brin.
For chemical plant equipment, tanks, vessels; Type 302; stainless, austenitic.

GURONIT 18-8E.
M-1329; 0.12 max C, 18 Cr, 9.5 Ni, Ti = 4 x C, bal Fe.
Annealed: 85,000 TS; 35,000 YS; 55 El; 65 RA; 170 Brin.
Cold drawn: 95,000 TS; 60,000 YS; 40 El; 60 RA; 185 Brin.
For welded structures, chemical plant equipment; Type 321; stainles, austenitic.

GURONIT 18-8MO.
M-1329; 0.15 C, 18 Cr, 9.5 Ni, 2 Mo, bal Fe.
Annealed: 85,000 TS; 35,000 YS; 50 El; 75 RA; 150 Brin.
Cold drawn: 150,000 TS; 135,000 YS; 6 El; 300 Brin.
For acid resistant equipment; Type 316; stainless, austenitic.

GURONIT 18-8MOE.
M-1329; 0.12 max C, 2 Mo, 18 Cr, 10.5 Ni, Cb = 8 x C, bal Fe.
Annealed: 85,000 TS; 35,000 YS; 50 El; 75 RA; 150 Brin.
Cold drawn: 150,000 TS; 135,000 YS: 6 El; 300 Brin.
For welded acid resistant equipment; Type 316Cb; stainless, austenitic.

GURONIT 20-25MO.
M-1329; 0.10 C, 24 Ni, 20 Cr, Mo, Cu, bal Fe.
Annealed: 100,000 TS; 45,000 YS; 50 El; 65 RA; 185 Brin.
For acid resistant equipment; corrosion resistant, austenitic.

GURONIT 28.4.
M-1329; 0.4 C, 1.3 Si, 27 Cr, 4 Ni, bal Fe.
Cast: 115,000 TS; 80,000 YS: 15 El; 190 Brin.
For furnace equipment, heat treat boxes; heat resistant.

GURONITE.
M-1329; 0.7-2.3 C, 30-35 Cr, 0.7 Si, Mo, bal Fe.
For heat treating furnaces, pump fittings, valves, grates; heat and corrosion resistant.

GURONIT GS22-10.
M-1329; 0.4 C, 22 Cr, 9.5 Ni, bal Fe.
Cast: 85,000 TS; 45,000 YS; 35 El; 165 Brin.
For heat treating boxes, furnace parts, conveyor belts; Type HF; corrosion and heat resistant.

GURONIT GS25-15.
M-1329; 0.4 C, 26 Cr, 14 Ni, bal Fe.
Cast: 75,000 TS; 47,000 YS; 17 El; 25 RA; 200 Brin.
For heat treating boxes, salt pots, retorts, chains; Type HH; corrosion and heat resistant.

GURONIT GS25-35.
M-1329; 0.5 C, 25 Cr, 30 Ni, bal Fe.
For furnace parts, retorts, pots, heat treating boxes; corrosion and heat resistant.

GURONIT GS28-4.
M-1329; 0.4 C, 1.3 Si, 27 Cr, 4 Ni, bal Fe.
Cast: 90,000 TS; 65,000 YS; 2 El; 212 Brin.
For cylinder liners, bushings, valve seats and bodies; Type CC-50; heat resistant.

GURONIT GS28-10.
M-1329; 0.4 C, Cr, 10 Ni, bal Fe.
For furnace parts, heat treating boxes; Type HE; heat resistant.

GURONIT GS28-20.
M-1329; 0.4 C, 25 Cr, 19 Ni, bal Fe.
Cast: 75,000 TS; 50,000 YS; 17 El; 170 Brin.
For furnace parts, retorts, stack dampers; Type HK; heat resistant.

GURONIT H7.
M-1329; 0.3 C, 6 Cr, 2.2 Si, bal Fe.
For oil refinery equipment; creep and heat resistant.

GURONIT H8.
M-1329; 0.5 C, 1.5 Si, 17 Cr, bal Fe.
For furnace parts, heat treating boxes; heat resistant.

GURONIT H9.
M-1329; 0.3 C, 2.2 Si, 17 Cr, bal Fe.
For furnace parts, heat treating boxes, oil refinery equipment; heat and creep resistant.

GURONIT H10.
M-1329; 0.6 C, 1.5 Si, 22 Cr, bal Fe.
For oil refinery equipment, rollers, crushers; heat and wear resistant.

GURONIT H11.
M-1329; 1.3 C, 1.5 Si, 29 Cr, bal Fe.
For crushers, rollers, grate bars; heat and abrasion resistant.

GUSSBRONZE 6.
M-1420; Sn, bal Cu.
Cast: 21,000-29,000 TS; 18 El; 60 Brin.
For hardware; bronze.

GUSSBRONZE 10.
M-1420; Sn, bal Cu.
Cast: 32,000-36,000 TS; 15 El; 70 Brin.
For hardware; bronze.

GUSSBRONZE 12.
M-1420; Sn, bal Cu.
Cast: 32,000-36,000 TS; 12 El; 85 Brin.
For hardware; bronze.

GUSSBRONZE 14.
M-1420; Sn, bal Fe.
Cast: 29,000-32,000 TS; 3 El; 110 Brin.
For hardware; bronze.

GUSSBRONZE 20.
M-1420; Sn, bal Cu.
Cast: 15,000-21,000 TS; 0 El; 180 Brin.
For hardware; bronze.

GUSS-KORROFESTAL.
M-1376; 9-13 Si, 0.25-0.40 Mg, 0.3-0.5 Mn, bal Al.
For light alloy parts; high corrosion resistance.

GUSSMESSING 63.
M-1420; Zn, bal Cu.
Cast: 22,000-25,000 TS; 10,6 El; 45 Brin.
For air conditioners, housings, machinery parts; brass.

GUSSMESSING 67.
M-1420; Zn, bal Cu.
Cast: 25,000-29,000 TS; 15-12 El; 40 Brin.
For air conditioners, housings, machinery parts; brass.

GUSSMESSING 90.
M-1420; Zn, bal Cu.
Cast: 23,000-25,000 TS; 20-18 El; 60 Brin.
For current carrying equipment, conductors; brass.

GUSS PANTAL.
M-Ger.; 2-4 Mg, 1.2-1.5 Mn, 0-0.2 Sb or Ti, bal Al.
For light alloy castings; resists sea water corrosion.

GUSSTAHL 4H now VEW K970.

GUSSTAHL 4W now VEW K960.

GUSSTAHL 5H now VEW K945.

GUSSTAHL 5W now VEW K935.

GUSSTAHL-D8514.
M-1293; 0.17 C, 0.2 Cb, 11 Cr, 0.6 Mo, 0.55 Ni, 0.6 V, bal Fe.
Annealed: 75,000 TS; 40,000 YS; 30 El; 70 RA; B 82 Rock.
Cold drawn: 95,000 TS; 80,000 YS; 15 El; 60 RA; B 92 Rock.
For springs, table flatware, knives, oil refinery and chemical plant equipment.
Corrosion resistant, hardenable.

GUSSTAHL D8518.
M-1293; 0.22 C, 0.6 Mn, 11.7 Cr, 1.05 Mo, 0.6 Ni, 0.3 V, bal Fe.
Annealed: 95,000 TS; 40,000 YS; 25 El; B 94 Rock.
Hardened: 245,000 TS; 205,000 YS; 9 El; C 50 Rock.
For cutlery, surgical instruments, hardware, gears, shafts.
Corrosion resistant, hardenable.

GUSSTAHL-D8518W.
M-1293; 0.21 C, 12 Cr, 1 Mo, 0.6 Ni, 0.3 V, 0.5 W, bal Fe.
Annealed: 90,000 TS; 38,000 YS; 25 El; B 92 Rock.
Hardened: 240,000 TS; 205,000 YS; 9 El; C 50 Rock.
For surgical instruments, valves, cutlery, gears, shafts.
Corrosion resistant, hardenable.

GUTHRIES ALLOY.
M-Eng.; 47 Bi, 20 Sn, 19 Pb, 13 Cd.
For fire extinguisher plugs; fusible alloy.

GUYS ALLOY.
M-England; 82.5 Ni, 17.5 Al.
At 70°F: 83,800 TS; 0.6 El.
At 1500°F: 52,900 TS; 2.9 El.
For turbine blades, jet engine components; high stress-rupture strength.

G.W.32.
M-1404; 7.6 Zn, 1.7 Mg, bal Al.
Heat treated: 43,000-50,000 TS; 6-4 El; 130-140 Brin.
For light alloy castings, housings; age-hardenable.

GW 280 TUF KUT.
M-1067; 0.60 C, 0.85 Mn, 2.0 Si, 0.25 Cr, 0.25 Mo, 0.2 V, bal Fe.
Annealed: 107,000 TS; 64,000 YS; 27 El; 212 Brin.
Oil Hardened: 145,000-340,000 TS; 127,000-283,000 YS; 5-24 El; 20-44 RA; 293-611 Brin.
For shear blades, pneumatic tools, punches, chisels, caulking tools.
Type S5 shock resisting tool steel.
Oil or water hardening.

G.W. 422 MIRYCAL.
M-1067; 0.5 C, 0.95 Cr, 1 W, 0.2 Mo, bal Fe.
For dies, punches, shear blades; Type S1; oil hardened.

G.W. 6-6-2 HIGH SPEED.
M-1067; 0.8 C, 4 Cr, 6 W, 4.5 Mo, 1.5 V, bal Fe.
For tools, cutters, chasers; high speed steel.

G.W. NO. 99 HOT WORK.
M-1067; 0.35 C, 0.35 Mn, 0.9 Si, 4.7 Cr, 1.1 W, 1.5 Mo, 0.25 V, bal Fe.
For hot work tools and dies, aluminum die casting dies; hot work steel.

G.W. NO. 265 (HIGH PRODUCTION).
M-1067; 1.6 C, 0.3 Mn, 12 Cr, 0.8 Mo, 0.2 V, bal Fe.
For tools, dies, blanking, forming and drawing dies; abrasion resistant, air or oil hardening.

G.W. NO. 310.
M-1067; 0.30 C, 0.25 Mn, 3.25 Cr, 10 W, 0.35 V, bal Fe.
For hot work dies, brass forming dies; hot work steel.

G.W. NO. 313.
M-1067; 0.35 C, 0.30 Mn, 3 Cr, 13.5 W, bal Fe.
Hot work tools and dies; hot work steel, high abrasion resistant.

G.W. NO. 350.
M-1067; 1.3 C, 0.3 Mn, 3.5 W, bal Fe.
For cutting tools; fast finishing tool steel.

G.W. NO. 515.
M-1067; 0.35 C, 1.0 Si, 5 Cr, 5 W, 0.2 Mo, bal Fe.
For hot work tools and dies; hot work steel, air hardening, tough.

G.W. COLD HEADER.
M-1067; 0.95 C, 0.30 Mn, 0.30 Si, bal Fe.
For cold heading tools; water hardening.

G.W. "C.V.M".
M-1067; 1.0 C, 5 Cr, 1 Mo, bal Fe.
For tools, dies, punches; Type A2; air hardened, non-deforming.

G.W. C.W. "OIL".
M-1067; 0.90 C, 1 Mn, 0.50 Cr, 0.50 W, bal Fe.
For dies, punches, rollers, mandrels; Type O1; oil hardened, non-deforming.

G.W. EXTRA.
M-1067; 1.0 C, bal Fe.
Annealed: 100,000 TS; 53,000 YS; 21 El; 42 RA; 200 Brin.
For drills, taps, reamers, lathe and planer tools; Type W1; water hardened.

G.W. OIL HARDENING (G.W.O.H.).
M-1067; 0.9 C, 1.1 Mn, 0.5 Cr, 0.5 W, bal Fe.
For tools, dies; oil hardening.

G.W. PAVEMENT BREAKER.
M-1067; 0.65 C, 0.30 Mn, bal Fe.
For tools, hammers, crushers; water hardening.

G.W. REGULAR.
M-1067; 0.7-1.2 C, 0.3 Mn, bal Fe.
For tools, punches; water hardening.

G.W. SOLID DRILL.
M-1067; 0.85 C, 0.30 Mn, bal Fe.
For tools, pivots; water hardening.

G.W. SPECIAL.
M-1067; 1-1.1 C, 0.30 Mn, bal Fe.
For tools, drills, taps; water hardening.

GYROCAST.
M-977; 0.4 C, 1 Cr, bal Fe.
For straightening rolls, bar mill guides; centrifugally cast castings.

GYROMET.
M-265; 3 Ni, 3 Mo, 3 Cu, 1.5 Fe, bal W.
140,000 psi UTS; 120,000 psi YS; 2 El; 68 Ra.
Density: 16.95 gm/cc.
For rotating inertia members, weights, requiring high strength.

GYROMET 1100.
M-265; 4 Ni, 4 Mo, 2 Fe, bal W.
165,000 psi UTS; 160,000 psi YS; 1.0 El; 38 Rc.
Density: 17.25 gm/cc.
For rotating inertia members, weights, requiring high strength.

H

H1.
M-1522; 61 Ag, Cu, Zn.
Melt range: 690-735°C.
Max stress: 39.4 kgf/mm^2; 23 E.
For silver brazing, high quality.

H2.
M-1769.
Sintered carbide.
400,000 TrS; Density: 12.8-13.0.
Hardness: 82.0-83.5 RA.
Industry code: C-14.

H-3.
M-1769.
Sintered carbide.
400,000 TrS; Density: 12.9-13.1.
Hardness: 83.5-84.5 RA.
Industry code: C-13.

H-4.
M-1769.
Sintered carbide.
400,000 TrS; Density: 13.4-13.6
Hardness: 85.5-86.5 RA.
Industry code: C-12.

H-4-X.
M-750; 0.35 C, 0.8 Cr, 0.5 Mo, 0.7 Mn, 0.2 Si, 0.15 max Ni, bal Fe.
Annealed: 100,000 TS; 80,000 YS; 2 El; 300 Brin.
For gears, shafts, pinions; water hardening.

H5DM.
0.15 C, 1.5 Mn, 0.5 Cr, 0.4 Mo, 4.0 Ni, 0.10 V, bal Fe.
Plate: 107,000 min TS; 92,000 min YS; 13 min El.
For pressure vessels, bridges, mine cars, power shovels, cranes, trucks, trailers.
Shock and wear resistant.

H6.
M-1522.
66.5 Ag, Cu, Zn.
Melt range: 700-720°C.
For silver brazing, Sterling silver.

H-9 DOUBLE HEADER.
M-32; 0.90 C, 0.35 Mn, 0.35 Si, bal Fe.
For header dies, punches, coining and striking dies; water hardened, wear resistant.

H14.
M-1522; 74 Ag, Cu, Zn.
Melt range: 745-778°C.
For silver brazing, Sterling silver.

H15.
M-1522; 75 Ag, Cu, Zn.
Melt range: 740-788°C.
For silver brazing, Sterling silver.

H20.
M-1522; 80 Ag, Cu, Zn.
Melt range: 740-780°C.
For silver brazing, Sterling silver.

H44 see **HEPPENSTALL H44.**

H-46 see **CARPENTER H-46.**

HA-0.
M-Belguim; 0.9 C, 3 Cr, 0.3 Mn, bal Fe.
H max 300, B max 13,500, Br 9900.
Coercive force 66. (BdHd) max 320,000.
For permanent magnets in electrical and magnetic equipment.
Hardened, high permeability.

HA-1.
M-Belgium; 1.0 C, 9 Cr, 3 Co, 1.5 Mo, bal Fe.
For permanent magnets in electrical and magnetic equipment.
Hardened, high permeability.

HA-2.
M-Belgium; 1.0 C, 9 Cr, 6 Co, 1.5 Mo, bal Fe.
For permanent magnets in electrical and magnetic equipment.
Hardened. High permeability.

HA-3.
M-Belgium; 1.0 C, 15 Co, 9 Cr, 1.5 Mo, bal Fe.
For permanent magnets in electrical and magnetic equipment.
Hardened. High permeability.

HA-25 ALLOY.
M-167; 0.15 max C, 1.5 Mn, 0.5 Si, 20.0 Cr, 10.0 Ni, 15.0 W, bal Co.
Heat Treated: 130,000 TS; 60,000 YS; 35 El; C 20 Rock.
At 1500°F: 50,000 TS; 35,000 YS; 17 El.
For jet engine afterburners, liners, nozzles, gas turbine rotors and buckets.
Good oxidation and corrosion resistance.

HA-25 ALLOY see **HAYNES ALLOY NO. 25.**

HACKER CNBVO.
M-1318; 0.36 C, 1.0 Cr, 0.2 Mn, 1.0 Ni, bal Fe.
For gears, bolts, machine tool parts; oil hardened, shock resistant.

HACKER CNBVO/H.
M-1318; 0.34 C, 0.55 Mn, 1.55 Cr, 1.55 Ni, 0.2 Mo, bal Fe.
For gears, bolts, machine tool parts; oil hardened, shock resistant.

HACKER CNBVO/2H.
M-1318; 0.30 C, 2.0 Cr 0.3 Mo, 2.0 Ni, bal Fe.
For gears, bolts, machine tool parts; oil hardened, shock resistant.

HACKER HWG.
M-1318; 0.55 C, 0.70 Cr, 0.18 Mo, 0.1 V, 1.65 Ni bal Fe.
For heading and forging dies, upsetters, punches; oil hardened, tough.

HACKER HWG EXTRA.
M-1318; 0.56 C, 0.85 Cr, 0.2 Mo, 1.8 Ni, 0.1 V, bal Fe.
For forging dies, upsetting dies, punches; oil hardened, tough.

HACKER LD EXTRA EXTRA ZH.
M-1318; 1.0 C, 0.25 max Si, 0.25 max Mn, bal Fe.
Heat treated: 200,000-130,000 TS;150,000-90,000 YS; 10-20 El; 30-50 RA; 420-270 Brin.
For drills, reamers, taps, hobs; Type W1; water hardened.

HACKER LD EXTRA H.
M-1318; 1.3 C, 0.25 max, Si, 0.25 max Mn, bal Fe.
For engravers' tools, blanking and forming dies; Type W1; water hardened.

HACKER LD EXTRA MH.
M-1318; 1.15 C, 0.25 max Si, 0.25 max Mn, bal Fe.
For springs, taps, drills, reamers, broaches; Type W1; water hardened.

HACKER LD EXTRA SEHR ZAH.
M-1318; 0.70 C, 0.25 max Si, 0.25 max Mn, bal Fe.
Heat treated: 175,000-122,000 TS; 130,000-82,000 YS; 12-22 El; 37-52 RA; 360-240 Brin.
For springs, rails, punches, clutch discs; Type W1; water hardened.

HACKER LD EXTRA ZAH.
M-1318; 0.85 C, 0.25 max Si, 0.25 max Mn, bal Fe.
Heat treated: 200,000-130,000 TS; 150,000-90,000 YS; 10-20 El; 30-50 RA; 400-260 Brin.
For drills, reamers, taps, hobs, cutters; Type W1; water hardened.

HACKER LD PRIMA EXTRA ZH.
M-1318; 0.67 C, 0.40 Si, 0.7 Mn, bal Fe.
Heat treated: 170,000-120,000 TS; 125,000-80,000 YS; 15-25 El; 40-55 RA; 350-240 Brin.
For springs, clutch discs, girders, rails, die blocks; water hardened.

HACKER LD PRIMA H.
M-1318; 0.90 C, 0.40 Si. 0.60 Mn, bal Fe.
Heat treated: 200,000-130,000 TS; 150,000-90,000 YS; 10-20 El; 30-50 RA; 400-260 Brin.
For drills, taps, reamers, cutters; Type W1; water hardened.

HACKER LD PRIMA MH.
M-1318; 0.75 C, 0.40 Si, 0.60 Mn, bal Fe.
Heat treated: 175,000-125,000 TS; 130,000-85,000 YS; 10-20 El; 35-50 RA; 360-250 Brin.
For wheels, die blocks, rails, springs, clutches; Type W1; water hardened.

HACKER LD PRIMA SEHR ZAH.
M-1318; 0.35 C, 0.60 Si, 0.60 Mn, bal Fe.
Hot rolled: 85,000 TS; 54,000 YS; 30 El; 53 RA; 185 Brin.
For gears, shafts, axles, bolts, fishplates; water hardened.

HACKER LD PRIMA ZAH.
M-1318; 0.60 C, 0.35 Si, 0.6 Mn, bal Fe.
Heat treated: 160,000-115,000 TS; 113,000-77,000 YS; 12-23 El; 40-54 RA; 320-230 Brin.
For wheels, die blocks, girders, rails; water hardened.

HACKER LD PRIMA ZW.
M-1318; 0.15 C, 0.4 Si, 0.6 Mn, bal Fe.
Annealed: 70,000 TS; 55,000 YS; 25 El; 60 RA; 145 Brin.
For screws, bolts, nuts, camshafts, rivets; case hardening steel.

HACKER LDS.
M-1318; 0.90 C, 1.9 Mn, 0.1 V, bal Fe.
For punches, crimpers, forming dies; oil hardened, non-deforming.

HACKER LD SPEZIAL.
M-1318; 0.75 C, 0.25 Si, 0.25 Mn, bal Fe.
Heat treated: 180,000-125,000 TS; 135,000-85,000 YS; 12-22 El; 36-50 RA; 370-250 Brin.
For springs, rails, punches, girders, crimpers; Type W1; water hardened.

HACKER LD SPEZIAL EXTRA ZH.
M-1318; 1.0 C, 0.25 Si, 0.25 Mn, bal Fe.
Heat treated: 190,000 TS; 120,000 YS; 10 El; 30 RA; 400 Brin.
For springs, tools, reamers, drills; Type W1; Water hardened.

HACKER LD SPEZIAL MH.
M-1318; 1.1 C, 0.25 Si, 0.25 Mn, bal Fe.
Annealed: 110,000 TS; 58,000 YS; 20 El; 40 RA; 210 Brin.
For springs, drills, taps, reamers, broaches; Type W1; water hardened.

HACKER LD SPECIAL ZAH.
M-1318; 0.85 C 0.20 Si, 0.20 Mn, bal Fe.
Heat treated: u80,000-128,000 YS; 135,000-85,000 YS; 12-22 El; 35-50 RA; 400-250 Brin.
For springs, tools, cutters, taps, drills, reamers; Type W1; water hardened.

HACKER LDX.
M-1318; 1.65 C, 11.5 Cr, 0.1 V, bal Fe,
For blanking and forming dies, punches; air hardened, non-deforming.

HACKER MN180.
M-1318; 0.14 C, 1.1 Cr, 4.5 Ni, bal Fe.
For gears, bolts, crankshafts, cams; case hardened, shock resistant.

HACKER ON130.
M-1318; 0.15 C, 0.25 Si, 0.5 Mn, 0.65 Cr, bal Fe.
For gears, bolts, camshafts, cams; case hardened, tough.

HACKER ON160.
M-1318; 0.16 C, 0.25 Si, 1.15 Mn, 0.95 Cr, bal Fe.
For gears, bolts, camshafts, cams; case hardened, tough.

HACKER ON200.
M-1318; 0.20 C, 1.15 Cr, 1.25 Mn, bal Fe.
For gears, bolts, camshafts, cams; case hardened, tough.

HACKER PS.
M-1318; 0.45 C, 0.2 V, 1.85 W, 1.05 Cr, bal Fe.
For punches, rivet sets, upsetters, forging dies; oil hardened, tough.

HACKER PS505.
M-1318; 0.55 C, 0.9 Si, 0.3 Mn, 1.05 Cr, 0.18 V, 1.85 W, bal Fe.
For forging and header dies, upsetters; oil hardened, tough.

HACKER PSN.
M-1318; 0.45 C, Si, Cr, V, bal Fe.
For springs, gears, bolts, crankshafts; oil hardened, shock resistant.

HACKER PS SPEZIAL.
M-1318; 0.45 C, W, Cr, V, bal Fe.
For forging and header dies, upsetters; oil hardened, tough.

HACKER TE.
M-1318; 1.05 C, 0.9 Mn, 1.0 Cr, 1.15 W, bal Fe.
For bearings, cutters, bushings; oil or water hardened, tough.

HACKER WGNH.
M-1318; 0.60 C, 0.40 Si, 0.60 Mn, bal Fe.
Heat treated: 160,000 TS; 113,000 YS; 12 El; 40 RA; 320 Brin.
For wheels, die blocks, clutch discs, rails; water hardened.

HACKER WGNH EXTRA.
M-1318; 0.90 C, 0.40 Si, 0.60 Mn, bal Fe.
Heat treated: 190,000 TS; 145,000 YS; 10 El; 30 RA; 400 Brin.
For springs, taps, drills, hobs, reamers; Type W1; water hardened.

HACKER WGNH SPEZIAL.
M-1318; 0.50 C, 0.90 Si, 0.90 Mn, bal Fe.
For pneumatic tools, punches, dies, shear blades; oil hardened, tough.

HACKER WGW.
  M-1318; 0.85 C, 0.25 max Si, 0.25 Max Mn, bal Fe.
  Heat treated: 188,000 TS; 145,000 YS; 12 El; 35 RA; 400 Brin.
  For springs, drills, reamers, taps, tools, hammers; Type W1; water hardened.

HACKER WHH.
  M-1318; 1.3 C, W, bal Fe.
  For cutters, engravers' tools, cold headers; oil or water hardened.

HACKER WHH1.
  M-1318; 1.1 C, 0.40 Cr, 0.30 Mn, 0.20 Si, bal Fe.
  For bearings, cutters, liners, drills; Type W5; water hardened.

HACKER WHHS.
  M-1318; 1.2 C, 0.20 Cr, 0.1 V, 1.0 W, bal Fe.
  For heading and forming dies, punches; oil hardened, wear resistant.

HACKETT K-COPPER.
  M-825; Cu alloy.
  Cast: 70,000 TS; 20 El; 50 RA; 150 Brin.
  For welding tips and holders; 75-80% conductivity of Cu.

HACKSAW.
  M-350; 0.7 C, 2 W, bal Fe.
  For tools, hacksaws; oil hardened.

HACKSAW "A".
  M-261; bal Fe.
  For hack saws; water hardening.

HACKSAW "B".
  M-261; C, W, bal Fe.
  For hack saws; water hardening.

HACKSAW REG C.
  M-350; 0.7-1.0 C, bal Fe.
  For tools, hacksaws; water hardened.

HADFIELD MANGANESE STEEL see MANGANESE GR. B.

HADFIELDS MANGANESE STEEL.
  M-Eng; 1.0-1.4 C, 10-14 Mn, bal Fe.
  Heat treated: 80,000-120,000 TS; 30,000-50,000 YS; 25-35 El; 20-35 RA; 170-201 Brin.
  For rails, wear plates, castings; non-magnetic, wear resisting.

HAECKER 135H.
  M-1318; 0.40 C, Cr, bal Fe.
  For gears, machine tools parts; water hardened.

HAECKER 135M.
  M-1318; 0.20 C, Cr, bal Fe.
  For gears, cams, machine tool parts; case hardening.

HAECKER 135W.
  M-1318; 0.10 C, Cr, bal Fe.
  For gears, cams, shafts; case hardening.

HAECKER ADS.
  M-1318; 0.38 C, Cr, Si, V, bal Fe.
  For gears, bolts, crankshafts; oil hardened, shock resistant.

HAECKER AE13C.
  M-1318; 2.1 C, 11.5 Cr, bal Fe.
  For blanking and forming dies; oil or air hardened, non-deforming.

HAECKER BSL-SPEZIAL.
  M-1318; 0.40 C, Ni, Cr, Mo, bal Fe.
  For gears, bolts, shafts, studs; oil hardened, shock resistant.

HAECKER F480E.
  M-1318; 0.12 max C, 18 Cr, 10,5 Ni, 2 Mo, Ti = 4 x C, bal Fe.
  Annealed: 90,000 TS; 45,000 YS; 50 El; 60 RA; 180 Brin.
  For chemical plant equipment, mixers, tanks, vessels; Type 316 Ti; stainless, austenitic.

HAECKER KF17.
  M-1318; 0.08 C, Cr, bal Fe.
  For gears, cams, camshafts, case hardened.

HAECKER KF17NI.
  M-1318; 0.22 C, Cr, Ni, bal Fe.
  For gears, cams, camsahfts; case hardened.

HAECKER KF17S.
  M-1318; 0.12 C, 16.5 Cr, 0.25 Mo, 0.2 S, bal Fe.
  Annealed: 80,000 TS; 50,000 YS; 25 El; 50 RA; 150 Brin.
  For chemical plant equipment; corrosion and heat resistant.

HAECKER KF300.
  M-1318; 0.15 max C, 18 Cr, 8.5 Ni, bal Fe.
  Annealed: 80,000 TS; 35,000 YS; 50 El; 65 RA; 160 Brin.
  For chemical plant equipment; Type 302; stainless, austenitic.

HAECKER KF300E.
  M-1318; 0.12 max C, 18 Cr, 9.5 Ni, Ti = 4 x C, bal Fe.
  Annealed: 90,000 TS; 45,000 YS; 45 El; 60 RA; 180 Brin.
  For welded structures, chemical plant equipment; Type 321; stainless, austenitic.

HAECKER KF300S.
  M-1318; 0.07 max C, 18 Cr, 9.5 Ni, bal Fe.
  Annealed: 80,000 TS; 35,000 YS; 55 El; 70 RA; 150 Brin.
  For welded structures, chemical plant equipment; Type 304; stainless, austenitic.

HAECKER KF480S.
M-1318; 0.07 max C, 18 Cr, 2 Mo, 10.5 Ni, bal Fe.
Annealed: 90,000 TS; 45,000 YS; 50 El; 60 RA; 180 Brin.
For acid resistant equipment, tanks, mixers, fitters; Type 316; stainless austenitic.

HAECKER LR.
M-1318; 0.30 C, 2.65 Cr, 0.35 V, 8.5 W, bal Fe.
For extrusion press dies, mandrels, liners, punches; oil hardened, hot work steel.

HAECKER MN130.
M-1318; 0.14 C, 0.7 Cr, 3.5 Ni, bal Fe.
For gears, bolts, camshafts, cams; case hardened, shock resistant.

HAECKER PSBG.
M-1318; 1.0 C, 0.1 V, 0.30 Mn, 0.25 Si, bal Fe.
Heat treated: 185,000 TS; 120,000 YS; 10 El; 30 RA; 400 Brin.
For knives, drills, drawing and stamping dies; oil or water hardened, wear resistant.

HAECKER PSX.
M-1318; 1.4 C, 0.30 Cr, 0.1 V, bal Fe.
For bearings, liners, blanking and forming dies; oil hardened, wear resistant.

HAFNIA.
M-1260; 24.5 Sn, 13 Sb, 0.5 Cu, 62 Pb.
Cast: 12,500 TS; 28 Brin.
DFor refrigerator and electric motor bearings; M.P. 355-535°F; good castability.

HAFNIUM.
M-1755; Hf.
Purities: Spectrographic grade 99.9%, reactor grade, commercial grade.
Forms: Sponge, crystal bar, ingot, powders, wire, sheet, foil, rod, arc melted buttons, zone refined rod, single crystals.

HAGESTA 0060 M.
M-1309; 0.10 C, 28.0 Mo, 65 Ni, bal Fe.
Cast, quenched: 64,000-92,000 TS; 50,000 YS; 10 El; 190 Brin., approx.
Highly corrosion resistant cast alloy.
DIN G-x10NiMo6528.

HAGESTA 1855 MW.
M-1309; 0.10 C, 16 Cr, 17 Mo, 60 Ni, bal Fe.
Cast, quenched: 64,000-92,000 TS; 50,000 YS; 10 El; 190 Brin., approx.
Highly corrosion resistant cast alloy.
DIN G-X10NiMoCr601716.

HAIRSPRINGS.
M-U.S.; 4.5-6.5 Sn, 0.2 P, bal Cu.
For hair springs, phosphor bronze.

HALBERLAND ALLOY.
M-Germ.; 87 Cu, 13 Zn.
For hardware fittings; Red Brass.

HALCOMB.
M-259; 0.9 C, 1 Cr, bal Fe.
For tools, dies; hardened.

HALCOMB 218.
M-38; 0.4 C, 1 Si, 5 Cr, 0.3 V, 1.4 Mo, bal Fe.
Air hardened: 400 Brin.
For brass extrusion mandrels, hot heading and forging dies, Al die casting dies, hot punches; tough, resists heat checking, hot work steel. AISI H11.

HALCOMB 236.
M-38; 0.3 C, 13 Cr, 13 W, 0.9 V bal Fe.
For permanent molds, dies, extension dies; corrosion and heat resistant, hot work steel.

HALCOMB 425.
M-38; 0.40 C, 0.30 Mn, 4.25 Cr, 2.10 V, 4.25 W, 0.4 Mo, 4.25 Co, bal Fe.
At 1200°F: 110,000 TS; 16 El; 36 RA; C 44 Rock..
At 800°F: 180,000 TS; 9 El; 36 RA; C 45 Rock.
For hot work tools, brass extrusion dies and dummy blocks, forging dies and inserts.
Shock and abrasion resistant.
Hot work tool steel Type H19.

HALCOMB 999.
M-362; 0.75 C, 4 Cr, 2 V, 14 W, 5 Co, bal Fe.
For lathe and planer tools, reamers, broaches; Type T8; high speed steel.

HALGRAPH.
M-38; 1.5 C 0.25 Mo, 1 Si, 0.75 Mn, bal Fe.
For machine tool parts; oil hardened; Type 06.

H.A.L.H. STEEL.
M-Eng.; C, alloy, bal Fe.
For tools, cutters; same as "Stellite."

HALLAMITE.
M-1202; 0.7 C, 14 W, 4 Cr, 1 V, bal Fe.
For turning and planing tools, shear blades, saws; high speed steel, oil hardened.

HALLAMITIER.
M-1202; 0.7 C, 18 W, 4 Cr, 1 V, bal Fe.
For saws, saw blades, taps, broaches, reamers; high speed steel, oil hardened.

HALLAMITIEST.
M-1202; 0.7 C, 22 W, 4 Cr, 1 V, bal Fe.
For hacksaw blades, saws, lathe and planer tools; high speed steel, oil hardened.

HALLAMSTEEL NO. 1.
M-1202; 1.0-1.2 C, bal Fe.
For circular saws, turning and shaping tools; Type W1; water hardened.

HALLAMSTEEL NO. 2.
M-1202; 0.9-1.0 C, bal Fe.
For saws, drills, pneumatic chisels, shear blades; Type W1; water hardened.

HALLAMSTEEL NO. 3.
M-1202; 0.8-0.9 C, bal Fe.
For band saws, caulking tools, shear blades; Type W2; water hardened.

HALLAMSTEEL NO. 4.
M-1202; 0.7-0.8 C, bal Fe.
For cutlery, drills, wood saws, stamping dies; Type W1; water hardened.

HALLAMSTEEL NO. 5.
M-1202; 0.6-0.7 C, bal Fe.
For awl blades blacksmith and boilermaker tools; water hardened.

HALLAMSTEEL NO. 6.
M-1202; 0.5-0.6 C, bal Fe.
For scissors, hammers, chisels; water hardened.

HALLIMAX I.
M-1202; 65 Ni, 15 Cr, bal Fe.
Annealed: 100,000 TS; 25-35 El; 55 RA; 147-157 Brin.
For resistances, heating elements; high heat resistant.

HALLIMAX II.
M-1202; 80 Ni, 20 Cr.
Annealed: 95,000 TS; 25-35 El; 55 RA; 142-157 Brin.
For resistances, heating elements; high heat resistance.

HALVAN.
M-38; 0.50 C, 1.0 Cr, 0.7 Mn, 0.2 V, bal Fe.
Hardened: 540 Brin.
For chisels, die holders, fixtures, jigs, punches; shock resistant, tough, AISI L2.

HAMILOY.
M-1514; Si, bal Al.
Die cast: 37,000 TS; 24,000 YS; 5-7 El.
For die castings; corrosion resistant.

HAMILTON ALLOYED IRON.
M-850; Alloy cast iron.
As cast: 25,000-60,000 TS; 20,000-48,000 YS; 145-350 Brin. (as specified).
Higher strength castings.

HAMILTON DUCTILE IRON.
M-850; Several grades.
Cast and heat treated: 60,000-175,000 TS; 40,000-150,000 YS; 143-338 Brin. (as specified)
High strengh and ductility.

HAMILTON DUCTILE NI-RESIST.
M-850; 8 grades, 18-25% Ni castings.
Cast; 55,000-60,000 TS; 28,000-32,000 YS; 130-240 Brin.
Austenitic, ductile, corrosion resistant, with good strength.

HAMILTON GRAY IRON.
M-850; Gray iron castings.
As cast: 20,000-25,000 TS; 16,000-20,000 YS; 130-220 Brin.
For general purpose cast iron.

HAMILTON MEEHANITE.
M-850; Quality cast iron; 24 Types.
Cast: 30,000-70,000 TS; 24,000-56,000 YS; 174-600 Brin. (as specified).
For high quality gray iron castings.

HAMILTON METAL-A.
M-Eng.; 93 Zn, 3.5 Cu, 3.1 Pb, 1.5 Sb, 0.5 P + Sn.
For ornamental castings; similar to "Chrysorin."

HAMILTON METAL-B.
M-Eng.; 33.3 Zn, 66.7 Cu.
For drawn, spun and stamped articles; high ductility.

HAMILTON NI-HARD.
M-850; Ni-Cr cast iron.
Cast: 40,000-75,000 TS; 525-600 Brin.
White cast iron, very good wear properties, not machinable; for pump parts, metal working rolls, coffee grinding burrs.

HAMILTON NI-RESIST.
M-850; 8 grades.
Cast: 20,000-30,000 TS; 100-250 Brin.
Ni or Ni-Cu cast iron; good resistance to corrosion and erosion.

HAMMOND ALLOY NO. 1.
M-992; 80-82 Cu, 2.75-3.50 Sn, 6.5-7.5 Pb, 8-10 Zn.
Cast: 29,000-39,000 TS; 13,700-17,000 YS; 16-30 El; 12-27 RA; 50-60 Brin.
For valve and plumbing castings.

HAMMOND ALLOY NO. 2.
M-992; 83 Cu, 4 Sn, 6 Pb, 7 Zn.
Cast: 30,000 TS; 14,000 YS; 16 El; 55 Brin.
For hydraulic pumps; free-cutting.

HAMMOND ALLOY NO. 3.
M-992; 84-86 Cu, 4-6 Sn, 4-6 Pb, 4-6 Zn.
Cast: 33,000-48,000 TS; 17,000-24,000 YS; 18-35 El; 12-32 RA; 55-65 Brin.
For steam valves, bearings; free cutting.

HAMMOND ALLOY NO. 4.
M-992; 78.5-81.5 Cu, 9-11 Sn, 9-11 Pb, 0.75 max Zn.
Cast: 32,000-38,000 TS; 17,000-22,000 YS; 15-20 El; 6-17 RA; 60-70 Brin.
For motor bearings, castings; tough.

HAMMOND ALLOY NO. 5.
M-992; 86-89 Cu, 9-11 Sn, 1.0 Pb, 1.0 -2.5 Zn.
Cast: 36,000-46,000 TS; 20,000-26,000 YS; 17-25 El; 12-26 RA; 70-80 Brin.
For bearings, bushings, gears; tough.

HAMMOND ALLOY NO. 6.
M-992; 86-89 Cu, 9-11 Sn, 0.2 max Pb, 1-3 Zn.
Cast: 38,000-45,000 TS; 21,000-26,000 YS; 22-28 El; 20-28 RA; 70-80 Brin.
For gears, worms, hydraulic castings; tough.

HAMMOND ALLOY NO. 7.
M-992; 87.0-88.5 Cu, 7.75-8.50 N, 0.3 max Pb, 3-5 Zn, 0.4-0.6 Ni.
Cast: 40,000-50,000 TS; 18,000-23,000 YS; 25-40 El; 32-37 RA; 65-75 Brin.
For gears, bolts, pipe fittings, valves; steam metal.

HAMMOND ALLOY NO. 8.
M-992; 88 Cu, 10 Sn, 2 Pb.
Cast: 36,000 TS; 19,000 YS; 22 El; 65 Brin.
For gears, bearings; free-cutting.

HAMMOND ALLOY NO. 9.
M-992; 75 Cu, 8 Sn, 15 Pb, 2 Zn.
Cast: 26,00 TS; 15,000 YS; 10 El; 50 Brin.
For bearings, bushings; tough.

HAMMOND ALLOY NO. 10.
M-992; 86 Cu, 7 Sn, 7 Pb.
Cast: 31,000 TS; 19,000 YS; 15 El; 55 Brin.
For bearings; acid resisting.

HAMMOND ALLOY NO. 11.
M-992; 84.5-86.5 Cu, 8.5-9.5 Sn, 3.5-4.5 Pb, 1-2 Zn.
Cast: 36,000-38,000 TS; 20,000-26,000 YS; 24-30 El; 65-75 Brin.
For bearings; shock resisting.

HAMMOND ALLOY NO. 12.
M-992; 82 Cu, 17 Sn, 1 max P.
Cast.
For journals, trunions.

HAMMOND ALLOY NO. 13.
M-992; 80-82 Cu, 10.5-11.5 Sn, 5.5-6.5 Pb, 1.5-2.5 Ni, 0.4-0.5 P.
Cast: 38,000-42,000 TS; 20,000-24,000 YS; 15-22 El; 75-85 Brin.
For bearings; shock resisting.

HAMMOND ALLOY NO. 14.
M-992; 80 Cu, 8 Sn, 10 Pb, 0.5 P, 1.5 Ni.
Cast: 32,000 TS; 18,000 YS; 15 El; 70 Brin.
For bearings; high speed.

HAMMOND ALLOY NO. 15.
M-992; Si, bal Cu.
Cast: 45,000 TS; 22,000 YS; 18 El; 90 Brin.
For worms, gears, pinions; acid resisting.

HAMMOND ALLOY NO. A-21.
M-992; 74.5 Cu, 8 Fe, 13 Al, 0.5 Mn, 4 Ni.
Cast: 60,000 TS; 60,000 YS; 0 El; 295 Brin.
For guides, dies, pistons; Al-bronze.

HAMMOND ALLOY NO. A-22.
M-992; 85 Cu, 3 Fe, 12 Al, 0.25 Mn.
Cast: 68,000 TS; 45,000 YS; 1 El; 225 Brin.
For castings; Al-bronze.

HAMMOND ALLOY NO. A-23.
M-992; 87 Cu, 4 Fe, 9 Al, 0.1 Mn, 0.1 Ni.
Cast: 78,000 TS; 27,000 YS; 32 El; 100 Brin.
For castings; Al-bronze.

HAMMOND ALLOY NO. A-24.
M-992; 89 Cu, 1 Fe, 10 Al, 0.1 Mn, 0.25 Ni.
Cast: 80,000 TS; 28,000 YS; 25 El; 115 Brin.
For castings; Al-bronze.

HAMMOND ALLOY NO. A-25.
M-992; 86 Cu, 4 Fe, 10 Al, 0.1 Mn, 0.25 Ni.
Cast: 87,000 TS; 28,000 YS; 18 El; 125 Brin.
For gears, bushings; Al-bronze.

HAMMOND ALLOY NO. A-26.
M-992; 83.5 Cu, 4.5 Fe, 10,75 Al, 1.25 Mn.
Cast: 90,000 TS; 35,000 YS; 10 El; 165 Brin.
For gears; Al-bronze.

HAMMOND ALLOY NO. A-27.
M-992; 79 Cu, 5 Fe, 11 Al, 0.1 Mn, 5 Ni.
Cast: 105,000 TS; 50,000 YS; 5 El; 175 Brin.
For castings; Al-bronze.

HAMMOND ALLOY NO, M-16.
M-992; 55-60 Cu, 38-42 Zn, 1.5 max Sn, 3.5 max Mn.
Cast: 60,000 TS; 28,000 YS; 30 El; 90 Brin.
For valve stems, hubs; free machining.

HAMMOND ALLOY NO. M-17.
M-992; Zn, Fe, Mn, bal Cu.
Cast: 70,000 TS; 30,000 YS; 2 El; 125 Brin.
For castings; corrosion resistant.

HAMMOND ALLOY NO. M-18.
M-992; Zn, Fe, Mn, bal Cu.
Cast: 80,000 TS; 40,000 YS; 20 El; 150 Brin.
For castings; corrosion resistant.

HAMMOND ALLOY NO. M-19.
M-992; Zn, Fe, Mn, bal Cu.
Cast: 90,000 TS; 45,000 YS; 15 El, 170 Brin.
For castings; corrosion resistant.

HAMMOND ALLOY NO. M-20.
M-992; Zn, Fe, Mn, bal Cu.
Cast: 100,000 TS; 60,000 YS; 10 El; 235 Brin.
For castings; high strength.

HAMMONIA METAL.
M-Germ.; 65 Sn, 32 Zn, 3.3 Cu.
For ornamental parts.

HAMPDEN see **CARPENTER HAMPDEN.**

HANCODUR.
M-827; C, alloy, bal Fe.
For valves.

HANDLER BHS.
M-1328; 0.60 C, 0.25-0.50 Si, 0.3-0.8 Mn, bal Fe.
Heat treated: 160,000 TS; 113,000 YS; 12 El; 40 RA; 325 Brin.
For rails, punches, gears, bolts; water hardened.

HANDLER BSD.
M-1328; 0.90 C, 1.9 Cr, 0.1 V, bal Fe.
For bearings, cutters, liners, sleeves; water hardened, wear resistant.

HANDLER BSD SPEZIAL.
M-1328; 1.45 C, 1.4 Cr, 0.6 Mn, bal Fe.
For bearings, cutters, liners, sleeves; water hardened, wear resistant.

HANDLER H18.
M-1309; 0.95 C, 4 Cr, W, V, Mo, bal Fe.
For lathe and planer tools, reamers, broaches; high speed steel.

HANDLER H20.
M-1328; 0.82 C, 4.1 Cr, 0.85 Mo, 1.6 V, 8.7 W, bal Fe.
For lathe and planer tools, reamers, broaches; high speed steel.

HANDLER HC EXTRA AH.
M-1328; 0.85 C, 0.25 max Si, 0.25 max Mn, bal Fe.
Heat treated: 190,000 TS; 145,000 YS; 10 El; 30 RA; 400 Brin.
For springs, cutters, taps, hobs, reamers; Type W1; water hardened.

HANDLER HC EXTRA HART.
M-1328; 1.3 C, 0.25 max Mn, 0.25 max Si, bal Fe.
For engraving tools, cutters, broaches; Type W1; water hardened.

HANDLER HC EXTRA MH.
M-1328; 1.1 C, 0.25 max Si, 0.25 max Mn, bal Fe.
Annealed: 110,000 TS; 55,000 YS; 20 El; 40 RA; 210 Brin.
For springs, cutters, taps, broaches; Type W1; water hardened.

HANDLER HC EXTRA ZH.
M-1328; 1.0 C, 0.25 max Si, 0.25 max Mn, bal Fe.
Annealed: 100,000 TS; 53,000 YS; 21 El; 42 RA; 200 Brin.
For springs, taps, hobs, drills, reamers; Type W1; water hardened.

HANDLER HC PRIMA H.
M-1328; 1.3 C, 0.25 max Mn, 0.25 max Si, bal Fe.
For engraving tools, cutters, broaches; Type W1; water hardened.

HANDLER HC PRIMA MH.
M-1328; 1.15 C, 0.25 max Mn, 0.25 max Si, bal Fe.
Annealed: 110,000 TS; 55,000 YS; 18 El; 40 RA; 210 Brin.
For springs, taps, reamers, broaches; Type W1; water hardened.

HANDLER HC PRIMA WEICH.
M-1328; 0.70 C, 0.25 max Si, 0.25 max Mn, bal Fe.
Heat treated: 175,000 TS; 130,000 YS; 12 El; 37 RA: 355 Brin.
For rails, axes, punches, crimpers; Type W1; water hardened.

HANDLER HC PRIMA ZAH.
M-1328; 0.85 C, 0.25 max Si, 0.25 max Mn, bal Fe.
Heat treated: 190,000 TS; 145,000 YS; 10 El; 30 RA; 400 Brin.
For springs, tools, cutters, taps, drills, hobs; Type W1; water hardened.

HANDLER HC PRIMA ZAHHART.
M-1328; 1.0 C, 0.25 max Mn, 0.25 max Si, bal Fe.
Annealed: 100,00 TS; 53,000 YS; 21 El; 42 RA; 200 Brin.
For springs, tools, cutters, taps, drills, hobs; Type W1; water hardened.

HANDLER IIA GUSSTAHL.
M-1328; 0.35 C, 0.4 Si, 0.6 Mn, bal Fe.
Hot rolled: 85,000 TS; 54,000 YS; 30 El; 53 RA; 185 Brin.
For gears, bolts, machine tool parts; water hardened; SAE 1035.

HANDLER MV2.
M-1328; 0.42 C, 0.1 V, 1.75 Mn, 0.25 Si, bal Fe.
For punches, gears, upsetters, header dies; oil hardened, tough.

HANDLER NI 3.
M-u328; 0.27 C, 1.1 Al, 1.4 Cr, 0.6 Mn, bal Fe.
For oil refinery equipment; heat and creep resistant.

HANDLER NI 4.
M-1328; 0.34 C, 1.1 Al, 1.4 Cr, 0.6 Mn, bal Fe.
For oil refinery equipment; heat and creep resistant.

HANDLER NI 7.
M-1328; 0.32 C, 1.1 Al, 1.1 Cr, 0.18 Mo, bal Fe.
For oil refinery equipment; heat and creep resistant.

HANDLER NI 10.
M-1328; 0.33 C, 1.1 Al, 1.7 Cr, 1.0 Ni, 0.5 Mn, bal Fe.
For oil refinery equipment; heat and creep resistant.

HANDLER NI 19.
M-1328; 0.31 C, 2.35 Cr, 0.18 Mo, 0.13 V, 0.6 Mn, bal Fe.
For oil refinery equipment; heat and creep resistant.

HANDLER SF2.
M-1328; 0.70 C, 0.25 Max Si, 0.25 max Mn, bal Fe.
Heat treated: 174,000 TS; 128,000 YS; 12 El; 37 RA; 350 Brin.
For springs, rails, punches; Type W1 water hardening tool steel.

HANDLER SF3.
M-1328; 0.85 C, 0.25 max Si, 0.25 max Mn, bal Fe.
Heat treated: 190,000 TS; 145,000 YS; 10 El; 30 RA; 400 Brin.
For springs, taps, reamers, drills, cutters; Type W1 water hardening tool steel.

HANDLER SF4.
M-1328; 1.0 C, 0.25 max Si, 0.25 max Mn, bal Fe.
Annealed: 100,000 TS; 53,000 YS; 21 El; 42 RA; 200 Brin.
For springs, taps, reamers; Type W1 water hardening tool steel.

HANDLER SF5.
M-1328; 1.1 C, 0.25 max Si, 0.25 max Mn, bal Fe.
Annealed: 100,000 TS; 53,000 YS; 21 El; 42 RA; 200 Brin.
For springs, taps, reamers, cutters; Type W1; water hardened.

HANDLER SF10.
M-1328; 1.0 C, 1.1 Cr, 0.07 Mn, 0.25 Si, bal Fe.
For bearings, cutters, liners, sleeves; water hardened, wear resistant.

HANDLER SF11.
M-1328; 0.85 C, 0.20 Si, 0.30 Mn, 0.30 Cr, bal Fe.
Heat treated: 190,000 TS; 145,000 yS; 10 El; 30 RA; 400 Brin.
For cutters, bearings, liners, sleeves; water hardened, wear resistant.

HANDLER SF12.
M-1328; 1.0 C, 0.10 V, 0.25 Mn, 0.2 Si, bal Fe.
For bearings, dies, cutters, taps; water hardened, wear resistant.

HANDLER SF15.
M-1328; 0.90 C, Cr, Si, bal Fe.
For bearings, dies, cutters, drills, taps; water hardened, wear resistant.

HANDLER SF16.
M-1328; 1.05 C, 1.0 Cr, 1.15 W, 0.90 Mn, bal Fe.
For drawing and forming dies, cutters; oil hardened, tough.

HANDLER SF18.
M-1328; 0.35 C, 1.05 Cr, 0.18 V, 1.85 W, 0.9 Si, bal Fe.
For heading and forming dies, punches; oil hardened, tough.

HANDLER SF21.
M-1328; 1.42 C, W, V, bal Fe.
For bearings, cutters, liners, sleeves; oil or water hardened.

HANDLER SF22.
M-1328; 0.30 C, 2.65 Cr, 0.35 V, 8.5 W, bal Fe.
For extrusion rams, dies, liners, punches; hot work steel, oil hardened.

HANDLER SF25.
M-1328; 0.74 C, 4.1 Cr, 1.1 V, 18.5 W, bal Fe.
For lathe and planer tools, drills, taps, reamers; high speed steel.

HANDLER SF27.
M-1328; 0.79 C, 4.75 Co, 4.3 Cr, 0.75 Mo, 1.5 V, 18 W, bal Fe.
For lathe and planer tools, drills, taps, reamers; high speed steel.

HANDLER SF28.
M-1328; 0.76 C, 10 Co, 4.2 Cr, 0.8 Mo, 1.8 V, 18 W, bal Fe.
For lathe and planer tools, drills; high speed steel.

HANDLER SF 29.
M-1328; 0.85 C, W, Mo, bal Fe.
For upsetters, punches, dies; hot work steel.

HANDLER SF41.
M-1328; 0.60 C, 0.25-0.50 Si, 0.30 -0.80 Mn, bal Fe.
Heat treated: 160,000 TS; 113,000 YS; 12 El; 40 RA; 320 Brin.
For wheels die blocks, rails, girders, springs; water hardened.

HANDLER SF43.
M-1328; 0.70 C, 1.7 Si, 0.70 Mn, bal Fe.
For springs, punches, pneumatic tools; oil hardened, tough.

HANDLER SF46.
M-1328; 0.50 C, 0.95 Mn, 1.05 Cr, 0.1 V, bal Fe.
For gears, bolts, crankshafts, springs; oil hardened, shock resistant.

HANDLER SF62.
M-1328; 2.1 C, 11.5 Cr, 0.70 W, bal Fe.
For blanking and forming dies, punches; oil hardened, non-deforming.

HANDLER SF85.
M-1328; 0.45 C, Ni, Cr, W, bal Fe.
For forging and heading dies; oil hardened, tough.

HANDLER SF502.
M-1328; 0.12 max C, 0.4 Si, 13 Cr, bal Fe.
Annealed: 75,000 TS; 40,000 YS; 35 El; 70 RA; 155 Brin.
For turbine blades, gages, valves, cutlery; Type 410; stainless.

HANDLER SF511.
 M-1328; 0.2 C, 0.4 Si, 13 Cr, bal Fe.
 Annealed: 95,000 TS; 50,000 YS; 25 El; 55 RA; 196 Brin.
 For cutlery, valves, surgical instruments; Type 420; stainless.

HANDLER SF512.
 M-1328; 0.4 C, 0.4 Si, 13 Cr, bal Fe.
 Annealed: 100,000 TS; 55,000 YS; 22 El; 52 RA; 200 Brin.
 For cutlery, valves, surgical and dental instruments; Type 420; stainless.

HANDLER SF522.
 M-1328; 0.2 C, 1.2 Si, 25 Cr, 4 Ni, bal Fe.
 Aged: 115,000 TS; 80,000 YS; 15 El.
 Cast: 70,000 TS; 65,000 YS; 12 El; 190 Brin.
 For furnace parts, grate bars, salt pots; heat and corrosion resistant.

HANDLER SF527.
 M-1328; 0.12 C, 16.5 Cr, 0.25 Mo, 0.2 S, bal Fe.
 Annealed: 80,000 TS; 50,000 YS; 25 El; 50 RA; 150 Brin.
 For screw machine products, soot blowers; Type 430 F; corrosion resistant.

HANDLER SF553.
 M-1328; 0.07 max C, 18 Cr, 9.5 Ni, bal Fe.
 Annealed: 85,000 TS; 35,000 YS; 60 El; 70 RA; 150 Brin.
 For chemical plant equipment, tanks, agitators; Type 304; stainless, austenitic.

HANDLER SF555.
 M-1328; 0.15 max C, 18 Cr, 8.5 Ni, bal Fe.
 Annealed: 80,000 TS; 35,000 YS; 55 El; 75 RA; 150 Brin.
 For chemical plant equipment, tanks, mixers, agitators; Type 302; stainless, austenitic.

HANDLER SF559.
 M-1328; 0.10 max C, 12.5 Cr, 12.0 Ni, bal Fe.
 For valves, furnace parts; corrosion resistant.

HANDLER SF564.
 M-1328; 0.07 max C, 18 Cr, 2.0 Mo, 10.5 Ni, bal Fe.
 Annealed: 85,000 TS; 35,000 YS; 50 El; 65 RA; 160 Brin.
 For acid resistant chemical plant equipment, tanks; Type 316; stainless, austenitic.

HANDLER SF 566.
 M-1328; 0.12 max C, 18 Cr, 2 Mo, 10.5 Ni, Cb = 8 x C, bal Fe.
 For welded acid resistant chemical plant equipment; Type 316 Cb.
 Stainless, austentic.

HANDLER SF1951.
 M-1328; 0.85 C, 2.5 V, 0.85 Mo, 12 W, bal Fe.
 For cutters, forming and drawing dies; high speed steel.

HANDLER SS212.
 M-1328; 2.1 C, 11.5 Cr, bal Fe.
 For blanking and forming dies, punches; oil hardened, non-deforming.

HANDLER SS308.
 M-1328; 1.25 C, 1.15 Si, 0.70 Mn, 1.2 Cr, bal Fe.
 For blanking and forming dies, headers; oil hardened, tough.

HANDLER SS511.
 M-1328; 0.56 C, Ni, Cr, Mo, bal Fe.
 For forging and heading dies, shears, punches; oil hardened, shock resistant.

HANDLER SS513.
 M-1328; 0.45 C, Si, Cr, V, Bal Fe.
 For springs, gears, bolts, fasteners, crankshafts; oil hardened, shock resistant.

HANDLER SS514.
 M-1328; 0.30 C, 2.35 Cr, 0.6 V, 4.25 W, bal Fe.
 For shear blades, pneumatic tools, chisels; oil hardened, shock resistant.

HANDLER ST MN 2.
 M-1328; 0.90 C, 1.9 Mn, 0.10 V, bal Fe.
 For punches, shears, blanking and forming dies; oil hardened, non-deforming.

HANDY 85 AG-15 MN see PREMABRAZE 130.

HANDY AT SPECIAL see BRAZE 202.

HANDY ATT see BRAZE 200.

HANDY BT see BRAZE 720.

HANDY COIN SILVER.
 M-63; 90 Ag, 10 Cu.
 For coins; corrosion resistant.

HANDY DE see BRAZE 450.

HANDY DT see BRAZE 400.

HANDY EXT see BRAZE 501.

HANDY IT see BRAZE 800.

HANDY NE see BRAZE 250.

HANDY NT see BRAZE 300.

HANDY RE-MN see BRAZE 655.

HANDY RE-MN6 see BRAZE 655.

HANDY RT see BRAZE 600.

HANDY RT-SN see BRAZE 603.

HANDY SILVER SOLDER see BRAZE 650.

HANDY SILVER SOLDER HARD see BRAZE 750.

HANDY SILVER SOLDER MEDIUM see BRAZE 700.

HANDY SN NO 7 see BRAZE 071.

HANDY SS see **BRAZE 403.**

HANDY SS-5 see **BRAZE 404.**

HANDY TE SPECIAL see **BRAZE 051.**

HANDY TL see **BRAZE 090.**

HANFORD 20.
M-1269; 0.07 max C, 29 Ni, 20 Cr, 2-3 Mo, 4 Cu, bal Fe.
Cast: 65,000-75,000 TS; 28,000-38,000 YS; 50-35 El; 50-40 RA; 120-150 Brin.
For chemical plant equipment, mixers; resists mixed acids, austenitic.

HANOVER WHITE METAL.
M-U.S.; 86.8 Sn, 7.6 Sb, 5.6 Cu.
For bearings; anti-friction.

HARBRONZE.
M-1156; Sn, bal Cu.
For bearings.

HARD.
M-388; 1.0 C. bal Fe.
For tools, dies, fixtures; water hardened.

HARD.
M-686; 75 Ag, 22 Cu, 3 Zn.
For silversmithing solder for sterling; MP 1365-1450°F.

HARDALLOY.
M-849; 0.43 C, 0.45 Mn, 0.53 Mo, 5.4 Cr, 0.2 Si, bal Fe.
Cast: 575-600 Brin.
For hard surfacing electrode; resists impact, abrasive wear and corrosion.

HARDALLOY 48.
M-849; Cr, bal Fe.
Welded: Rockwell C55.
For hard facing electrode.

HARDALLOY 48 see **MCKAY HARDALLOY 48.**

HARD BABBITT.
M-Eng.; 83 Sn, 8.4 Cu, 8.3 Sb.
For bearings; anti-friction.

HARD BEARING.
M-Eng.; 98.7 Pb, 1.3 Mg.
For bearings.

HARD BRONZE-1.
M-165; 88 Cu, 2 Pb, 7 Sn, 3 Zn.
30,000 TS; 12 El.
For gears, bushings; tough.

HARD BRONZE-2.
M-165; 88 Cu, 10, Sn, 2 Zn.
For gears, bushings; G.M. NO. 4048 M.

HARD DEVIL.
M-712; 0.85-1.00 C, bal Fe.
250-400 Brin.
For hard facing welding electrodes.

HARDENITE.
M-502; 0.9 C, 1.2 Mn, bal Fe.
For tools, dies; oil hardened.

HARDENTOUGH 1.
M-118; 0.85 C, bal Fe.
300-425 Brin.
For welding electrodes; hard facing wear resistant.

HARDENTOUGH 2.
M-118; C, Mn, Ni, bal Fe.
347 Brin.
For welding electrodes; hard facing wear resistant.

HARDENTOUGH 4.
M-118; C, alloy, bal Fe.
500-600 Brin.
182 hard facing welding electrodes; wear resistant.

HARDENTOUGH 5.
M-118; C alloy, bal Fe.
600-675 Brin.
For hard facing welding electrodes; wear resistant.

HARDENTOUGH 6.
M-118; C, alloy, bal Fe.
700-800 Brin.
For hard facing welding electrodes; wear resistant.

HARDENTOUGH 250.
M-118; C, alloy, bal Fe.
Weld: 230-300 Brin.
For arc welding electrode; hard facing.

HARDENTOUGH 350.
M-118; C, alloy, bal Fe.
Weld: 500 Brin.
For arc welding electrode; hard facing.

HARDENTOUGH 450.
M-118; C, alloy, bal Fe.
Welded: 600 Brin.
Welding electrodes; hard surfacing.

HARDENTOUGH 550.
M-118; C, alloy, bal Fe.
Welded: 400-600 Brin.
For arc welding electrodes; hard surfacing.

HARDENTOUGH FORMWEAR.
M-118; C, alloy, bal Fe.
Welded: 500-650 Brin.
For arc-welding electrodes; hard surfacing.

HARDENTOUGH HA.
M-118; C, alloy, bal Fe.
Welded: 575-650 Brin.
For arc welding eletrodes; hard surfacing.

Section I: Alloy Data / 635

HARDENTOUGH MOLYMANG.
M-118; C, 13 Mn, bal Fe.
Weld hardened: 475 Brin.
For arc welding electrodes; abrasion resistant, work hardens.

HARDENTOUGH WH.
M-118; C, alloy bal Fe.
Cast: 270-380 Brin.
For welding electrodes; hard surfacing.

HARD FACING R 459.
M-9; C, Cr, Mn, Mo, bal Fe.
500-600 Brin.
For hard facing welding rods; wear and heat resistant.

HARDFLEX-11.
M-101; 0.50-0.60 C, 0.6-0.9 Mn, bal Fe.
Heat treated: 128,000-190,000 TS; 107,000-142,000 YS; 10-14 El; 350 DPH; C 36 Rock.
For office and business machine components, electric razors, sewing machines, appliances.
AISI-C1055, Prehardened strip steel, austempered.

HARDFLEX-13M.
M-101; 0.7-0.8 Mn, bal Fe.
Hardened: 190,000-228,000 TS; 149,000-200,000 YS; 6-9 El.
For office and business machine components, electric razors, sewing machines, appliances.
Prehardened strip steel, austempered.
AISI-C1074.

HARD HEAD.
M-Eng.; 90 Sn, 8 Sb, 2 Cu.
For bearings; anti-friction.

HARDITE A.
M-166; 55 Ni, 13 Cr, 1.5-2.0 Si, 2.0 Mn, bal Fe.
At 70°F: 103,000 TS.
At 1800°F: 43,000 TS.
For carburizing boxes, heating elements, lead pots, retorts; heat resisting.

HARDITE B.
M-166; 33-35 Ni, 12-14 Cr, 1.5-2.0 Si, 2.0 Mn, bal Fe.
For carburizing boxes, heating elements, lead pots, retorts; heat resisting.

HARDITE C.
M-166; 93 Ni, 5 Si, 2 Mn.
For heating elements, retorts; high heat resistant.

HARDITE S.
M-166; 88 Ni, 4.5 Si, bal Fe.
For carburizing boxes, heating elements, lead pots, retorts; heat resisting.

HARDITE X.
M-166; 82-86 Ni, 10.13 Cr, 2-4 Si, 2 Mn.
For heating elements, retorts; heat resisting.

HARDITE Z.
M-166; 6-12 Ni, 15-25 Cr, C, 0.4 Si, 1-3 Mn, bal Fe.
For heat treating boxes, furnace parts; stainless.

HARDNAIR.
M-353; 1.0 C, 5 Cr, 1 Mo, 0.4 Mn, 0.4 V, bal Fe.
Annealed: 103,000 TS; 52,000 YS; 26 El; C 18 Rock.
Hardened: 178,000-253,000 TS; 146,000-200,000 YS; 3-12 El; 7-32 RA; C 41-53 Rock.
For blanking and trimming dies, cams, cutters, gauges, lamination dies, shears.
Type A2 air hardening tool steel, nondeforming.

HARD PHOSPHOR BRONZE.
M-U.S.; 7 Sn, 0.2 P, bal Cu.
For springs, bearings, electrical parts, gears; strong, corrosion resistant.

HARDRITE.
M-114, M-275; 1.1 C, 1.7 W, 0.6 Cr, 0.25 V, bal Fe.
For light and heavy press dies, punches, taps, reamers, thread rolling dies, knives, cams, forming dies; non-shrinkable; AISI 07.
Was Vulcan Hardrite.

HARDROCK 33.
M-829; 7 Co, 18 W, 4 Cr, 1 V, C, bal Fe.
For tool bits; high speed steel.

HARD-ROD NO 100.
M-1096; 0.9 C, 0.6 Mn, 0.3 Mo, 1 Cr, bal Fe.
For hard facing rod; for carbon steels.

HARD-ROD NO 450.
M-1096; 0.50-0.60 C, 6-8 Cr, 0.6 Si, bal Fe.
Welded: 500 Brin.
For hard facing electrode; for high impact and abrasion.

HARD-ROD NO 550.
M-1096; 0.6 C, 8 Cr, 0.4 Mn, 0.6 Si, bal Fe.
Welded: 550 Brin.
For hard facing electrode; for high abrasion and impact.

HARD SILVER SOLDER.
M-Eng.; 80 Ag, 13 Cu, 6.8 Zn.
For solder.

HARD SOLDER.
M-Eng.; 57-50 Cu, 43-50 Zn.
For solder.

HARDSTEEL.
M-861; 40 Co, 32 Cr, 18 W, 2 Ni, 2 Fe, trace B, trace Mn.
For drills, tool bits and wear parts.

HARDTEM.
M-64; 0.60 C, 0.70 Mn, 0.61 Cr, Ni, 0.17 V, 0.28 Mo, bal Fe.
Oil treated: 123,000 TS; 101,500 YS; 53 El; 23 RA; 262 Brin.
For die blocks; drop forged and pressed.

HARDTEM.
M-1488; 0.55 C, 0.75 Mn, 0.55 Ni, 1.0 Cr, 0.45 Mo, 0.05 V, bal Fe.
Oil hardening hot work tool steel; for hot forming dies.
AFNOR 55CNDV 4.

HARD-TRAK.
M-240; 0.31 C, 1.4 Mn, 0.24 Si, 0.50 Cr, 0.11 Mo, 0.0004 B, bal Fe.
Heat treated: 360-400 Brin.
For wear bars and rails.

HARDWARE BRONZE 267.
M-8; 85 Cu, 13.25 Zn, 1.75 Pb.
Soft: 40,000 TS; 15,000 YS; 45 El; 72 RA; 62 Brin.
Hard: 50,000 TS; 43,000 YS; 20 El; 49 RA; 218 Brin.
For bolts, screws, nuts, tie rods.

HARDWARE BRONZE-320.
M-8; 85 Cu, 13.25 Zn, 1.75 Pb.
Hard: 52,000 TS; 43,000 YS; 20 El; B 55 Rock.
Soft: 40,000 TS; 15,000 YS; 45 El; B 5 Rock.
For screw machine products, hardware, architectural applications.
Elect. cond. 36, corrosion resistant.

HARD YELLOW SOLDER.
M-Eng.; 58 Cu, 43 Zn, 1.3 Sn, 0.3 Pb.
For solder.

HARDYNE.
M-799; oxides bound with thermoplastic resin.
For permanent magnet; pressed compound oxides.

HARDY NICKEL IRON.
M-435; 0.08 C, 0.20 Mn, 2 Ni, bal Fe.
Annealed: 50,000 TS; 35,000 YS; 35 El; 75 RA; 112 Brin.
For tinning rolls; sub-zero applications.

HARD ZINC (HARTZINK).
M-Germ.; 92 Zn, 5.3 Fe, 2.4 Pb, 0.1 Cu.
For wash boards.

HARGUS.
M-783; 0.9 C, 1.2 Mn, bal Fe.
For punching and blanking dies; non-deforming.

HARGUS.
M-289; 0.9 C, 1.2 Mn, 0.5 Cr, 0.5 W, bal Fe.
For taps, dies, punches, gauges, reamers; non-deforming, oil hardened.

HARLINGTON BRONZE.
M-Eng.; 56 Cu, 43 Zn, 0.9 Sn, 0.6 Fe.
For sheating, condenser tubes, bolts, nuts; high strength.

HARMONIA BRONZE.
M-Eng.; 57-55.7 Cu, 40-41.2 Zn, 0.4-0.46 Pb, 1.8-1.29 Fe. 0-0.5 Sn, 0-0.86 Al.
For fittings, hardware.

HARTALUMIN.
Al alloy.
For light alloy parts.

HARTMETALLE.
M-German; WC.
For tools, dies; sintered WC.

HARVEY 66S.
M-1235; 0.70-1.2 Cu, 0.9-1.8 Si, 0.8-1.4 Mg, 0.6-1.1 Mn, bal Al.
O-temper: 22,000 TS; 12,000 YS; 18 El; 43 Brin.
T6-temper: 62,000 TS; 55,000 YS; 8 El; 120 Brin.
T4-temper: 52,000 TS; 30,000 YS; 18 El; 90 Brin.
For structures, bus and truck bodies, boom scaffolds; age hardenable, good weldability.

HARVEY HA-1900.
M-1235; 0.0125 max H, 0.07 max N, 0.25 max 0, 0.10 max C, 0.30 max Fe, bal Ti.
Bar: 80,000 min TS; 70,000 min YS; 15 El; 30 RA; 10-11 Charpy impact.
For non-structural and moderately stressed aircraft parts, corrosion applications.
Alpha titanium, corrosion resistant.

HARVEY HA-1930.
M-1235; 99.5 Ti.
Annealed: 48,000 TS; 25,000 YS; 35 El.
At 600°F: 20,000 TS; 10,000 YS; 50 El.
For chemical plant equipment, high temperature applications.
Highest formability. Corrosion resistant.

HARVEY HA-1940.
M-1235; 99.2 Ti.
Annealed: 65,000 TS; 43,000 YS; 30 El.
At 600°F: 28,000 TS; 13,000 YS; 45 El.
For aerospace equipment, chemical and marine applications.
Pipe and tubing. Corrosion resistant.

HARVEY HA-1940 PD.
M-1235; 0.15-0.20 Pd, bal Ti.
Annealed: 62,000 TS; 46,000 YS; 27 El.
At 600°F: 28,000 TS; 13,000 YS; 30 El.
At 1000°F: 15,400 TS; 90,000 YS; 31 El.
For chemical industry equipment.
Corrosion resistant grade.

HARVEY HA-1950.
M-1235; 99.0 Ti.
Annealed: 75,000 TS; 63,000 YS; 25 El.
At 600°F: 33,000 TS; 19,000 YS; 33 El.
For marine applications, high temperature applications. Corrosion resistant.

HARVEY HA-1970.
M-1235; 99.0 Ti.
Annealed: 95,000 TS; 80,000 YS; 22 El.
At 600°F: 43,000 TS; 27,000 YS; 28 El.
For marine applications, high temperature components.
Commercially pure titanium. Corrosion resistant.

HARVEY HA-4145.
M-1235; 4 Al, 4 Mn, bal Ti.
Annealed: 148,000 TS; 135,000 YS; 15 El.
At 600°F: 110,000 TS; 90,000 YS; 17 El.
Aged: 162,000 TS; 143,000 YS; 10 El.
For heavy section aircraft components, fasteners, jet engine and guided missile parts.
Alpha-beta titanium alloy.

HARVEY HA-5137.
M-1235; 4.0-6.0 Al, 2.0-3.0 Sn, bal Ti.
Bar: 125,000 TS; 120,000 YS; 18 El; 25 RA; C 35 Rock.
At 600°F: 82,000 TS; 65,000 YS; 19 El.
For structural aircraft parts, missiles.
Weldable, high strength, alpha titanium, corrosion resistant.

HARVEY HA-5137 ELI.
M-1235; 5 Al, 2.5 Sn, low 0, bal Ti.
Annealed: 110,000 TS; 95,000 YS; 20 El.
At 600°F: 78,000 TS; 60,000 YS; 20 El.
For cryogenic applications, missile and aircraft components.
Alpha-titanium, corrosion resistant.

HARVEY HA-5158.
M-1235; 6 Al, 6 V, 2 Sn, 1 (Fe, Cu), bal Ti.
Annealed: 160,000 TS; 155,000 YS; 15 El.
Aged: 190,000 TS; 180,000 YS; 10 El.
For ordnance applications and aircraft components, pressure vessels.
Alpha-beta titanium alloy. Hardenable.

HARVEY HA-6148.
M-1235; 5.5-6.5 Al, 3.5-4.5 V, bal Ti.
Bar: 130,000 min TS; 120,000 min YS; 10 El; 25 RA; C 38 Rock.
For airframe and jet engine compressor parts, ordnance equipment.
Alpha-beta titanium, corrosion resistant.

HARVEY HA-6510.
M-1235; 6 Al, 4 V, bal Ti.
Annealed: 140,000 TS; 130,000 YS; 18 El.
At 600°F: 105,000 TS; 95,000 YS; 11 El.
Aged: 165,000 TS; 155,000 YS; 13 El.
For airframe and jet engine components, ordnance equipment.
Heat treatable to high strength. Alpha-beta type.

HARVEY HA-6510 ELI.
M-1235; 6 Al, 4 V, low 0, bal Ti.
Annealed: 135,000 TS; 125,000 YS; 20 El.
At 600°F: 105,000 TS; 95,000 YS; 12 El.
For cryogenic applications. Aircraft and missile components.
Heat treatable. Alpha-beta alloy.

HARVEY HA-7146.
M-1235; 7 Al, 4 Mo, bal Ti.
Annealed: 160,000 TS; 150,000 YS; 16 El.
At 600°F: 127,000 TS; 108,000 YS; 18 El.
Aged: 185,000 TS; 175,000 YS; 15 El.
For engine and airframe applications, jet engine and missile components.
Heat treatable; High creep strength.

HARVEY HA-8116.
M-1235; 8 Al, 1 Mo, 1 V, bal Ti.
Annealed: 140,000 TS; 130,000 YS; 15 El.
At 1000°F: 95,000 TS; 73,000 YS; 14 El; 30 RA.
For aircraft and jet components, discs, spacers and blades.
High strength, long time creep resistant, high temperature stability.

HARZ REFINED LEAD.
M-382; 99.9 Pb.
For lead tubes and sheets for chemical industry; corrosion resistant.

HAS-ALL.
M-497; 0.7 C, 18 W, 4 Cr, 1 V, bal Fe.
For high speed tools; high speed steel.

HASCROME ALLOY.
M-167; 1.25 C, 12.0 Cr, 3.0 Mn, bal Fe.
Covered electrode for AC-DC welding and build-up.
As welded: 26 Rc; work-hardens to 43 Rc.

HASS.
M-Eng.; 4.5 Cu, 0.75 Mn, 1.0 Si, bal Al.
For aircraft structures; age-hardening alloy.

HASTELLOY ALLOY B.
M-167, M-1169, M-1663; 26-30 Mo, 4-6 Fe, 0.12 max C, 62 Ni.
Cast: 82,000 TS; 57,000 YS; 9 El; 13 RA; 230 Brin.
Rolled: 140,000 TS; 65,000 YS; 45 El; 45 RA; 235 Brin.
For agitators, heating and cooling coils, pump parts, valves; resists HCl and boiling H_3PO_4.

HASTELLOY ALLOY B-2.
M-167; 0.02 max C, 26.0-30.0 Mo, 2.0 max Fe, 1.0 max Cr, 1.0 max Mn, 1.0 max Co, bal Ni.
Plate, Sol. Tr.: 129,000 psi (894 MPa) TS; 59,800 psi (412 MPa) YS; 61 El.
Good strength to 800°F; corrosion resistant (except for ferric or cupric salts).

HASTELLOY ALLOY C.
M-167, M-1663; 16-18 Mo, 13.0-17.5 Cr, 3.7-5.3 W, 4.5-7 Fe, bal Ni.
Cast: 72,000-80,000 TS; 45,000-48,000 YS; 15-10 El; 16-11 RA; 175-215 Brin.
Rolled: 130,000 TS; 65,000 YS; 25 El; 210 Brin.
For pumps and valves for H_2SO_4 at high temperature; resists SO_3, P_2O_5 and Cl_2.

HASTELLOY ALLOY C-4.
M-167; 0.015 max C, 2.0 max Co, 14.0-18.0 Cr, 14.0-17.0 Mo, 0.7 max Ti, 3.0 max Fe, 1.0 max Mn, 0.08 max Si, bal Ni.

1/2 in. plate, welded: 112,700 psi TS; 68,300 psi YS; 40 El.

For corrosion resistance and high-temperature stability.

HASTELLOY ALLOY C-276.
M-167, M-1663; 14.5-16.5 Cr, 15.0-17.0 Mo, 2.5 max Co, 3.0-4.5 W, 4.0-7.0 Fe, 0.02 max C, 1.0 max Mn, bal Ni.

Hardenable by cold work plus aging from 90 Rock B to max of 49 Rock C.

Excellent corrosion resistance even to such chemicals as wet chlorine gas and hypochlorites.

HASTELLOY ALLOY C (POWDER).
M-167; 0.12 max C, 16.5 Cr, 17.0 Mo, 5.5 Fe, 2.5 Co, 4.5 W, bal Ni.

For plasma spray build-up.

HASTELLOY ALLOY F.
M-167; 0.05 max C, 1.0-2.0 Mn, 1.0 max Si, 21.0-23.0 Cr, 44.0-47.0 Ni 2.5 max Co, 5.5-7.5 Mo, 1.0 max W, 1.75-2.5 Cb-Ta, bal Fe.

HASTELLOY ALLOY G.
M-167; 21-23.5 Cr, 5.5-7.5 Mo, 1.0 max W, 1.0-2.0 Mn, 2.5 max Co, 1.0 max Si, 18-21 Fe, 0.05 max C, 1.5-2.5 Cu, 1.75-2.50 Cb + Ta, bal Ni.

Alloy with excellent resistance to hot sulphuric and phosphoric acids. For chemical plant equipment.

HASTELLOY ALLOY N.
M-167; 6-8 Cr, 15-18 Mo, 5 Fe, 0.04-0.08 C, 0.5 Al, 0.01 B, bal Ni.

Annealed: 86,400 TS; 37,300 YS; 17 El.

For containers for molten fluorides; oxidation resistant to 1800°F.

HASTELLOY ALLOY S.
M-167; M-1491; 0.02 max C, 15.5 Cr, 14.5 Mo, 1.0 Fe, 0.20 Al, 0.4 Si, 0.50 Mn, 0.009 B, 0.02 La, bal Ni.

High temperature wrought alloy for applications involving severe cyclic heating conditions. Weldable, can be hot or cold worked.

HASTELLOY ALLOY W.
M-167; 0.12 C, 5 Cr, 24.5 Mo, 2.5 Co, 5.5 Fe, 0.6 V, 1 Mn, 1 Si, bal Ni.

For high temperature applications; corrosion and heat resistant.

HASTELLOY ALLOY X.
M-167, M-1491; 0.05-0.20 C, 20.5-23.0 Cr, 17-20 Fe, 8-10 Mo, 0.5-2.5 Co, 0.2-1.0 W, bal Ni.

At 70°F: 113,000 TS; 55,800 YS; 44 El.
At 1200°F: 83,000 TS; 40,700 YS; 37 El.
At 1800°F: 21,000 TS; 17,000 YS; 43 El.

For aircraft and jet engine components; good creep and stress rupture properties.

HASTELLOY B, C see also JUNKER HASTELLOY B & C.

HASTELLOY C.
M-44; 15 Cr, 5 Fe, 4 W, 15 Mo, bal Ni.

Wire, rod, ribbon for welding wire and fastener stock.

HASTELLOY C WIRE.
M-1507; 0.08 max C, 14.5-16.5 Cr, 15.0-17.0 Mo, 3.0-4.5 W, 2.5 max Co, 0.35 max V, 4.0-7.0 Fe, bal Ni.

For springs requiring resistance to chlorine and sulphur attack.

HASTELLOY DEVELOPMENT ALLOY C-455 see HASTELLOY ALLOY C-4.

HASTELLOY R-235.
M-1491; 0.15 C, 15.5 Cr, 2.5 max Co, 5.5 Mo, 2.5 Ti, 2.0 Al, 10.0 Fe, bal Ni.

High temperature alloy; for gas turbine and engine parts, sheets.

HASTELLOY T.
M-167; M-1491; 0.02 C, 0.20 Mn, 12.0 Cr, 9.0 Mo, 14.0 W, 0.20 Al, 0.02 La, bal Ni.

Corrosion resistant, low expansion alloy.

HASTELLOY W.
M-44; 5 Cr, 5 Fe, 24 Mo, bal Ni.

Wire, rod, ribbon for welding wire and fastener stock.

HASTELLOY X.
M-44; 21 Cr, 1 Co, 18 Fe, 1 W, 9 Mo, bal Ni.

Wire, rod, ribbon for welding wire and fastener stock.

HATHAL-A.
M-1548; 3.0-5.0 Cu, 0.4 Si, 0.4-1.0 Mn, 0.3-1.0 Mg, bal Al.

Heat treated: 23,000-65,000 TS; 10-25 El; 40-120 Brin.

For rivets, hydraulic fittings, hardware.
Age-hardenable. Aluminum 2014.
Good machinability and weldability.

HATHAL-B.
M-1548; 0.2-1.0 Si, 0.2-0.6 Mn, 2.0-8.0 Mg, bal Al.

Heat treated: 31,000-54,000 TS; 5-20 El; 50-85 Brin.

For structural members, marine hardware, fasteners.
Age hardenable. Corrosion resistant.

HATHAL-C.
M-1548; 0.5-1.0 Si, 0.3-1.5 Mn, 0.8-1.5 Mg, bal Al.
Heat treated: 14,000-43,000 TS; 6-22 El; 30-80 Brin.
For structures, marine parts, screw machine products.
Age-hardenable. Corrosion resistant.

HAVAR.
M-1084; 42.5 Co, 13 Ni, 20 Cr, 2 Mo, 0.2 C, 0.04 Be, 1.6 Mn, 2.8 W bal Fe.
Rolled: 260,000-290,000 TS; 200,000-220,000 YS; 48-50 El.
Aged: 330,000-360,000 TS; 260,000-280,000 YS; 56-60 El.
For watch and power springs, valves, flapper valves, electronic parts; age-hardenable, high strength and fatigue resistance.

HAWK 3312.
M-362; 0.08-0.13 C, 0.45-0.60 Mn, 1.4-1.75 Cr, 3.25-3.75 Ni, bal Fe.
Deep hardening carburizing grade alloy steel.

HAWK A-2.
M-362; 1.0 C, 0.7 Mn, 0.3 Si, 5.25 Cr, 0.3 V, 1.15 Mo, bal Fe.
AISI Type A2 air hardening tool steel.

HAWK A-2-S.
M-362; 1.0 C, 0.7 Mn, 0.3 Si, 5.25 Cr, 0.3 V, 1.15 Mo, 0.15 S, bal Fe.
Free machine grade.
AISI Type A2 air hardening tool steel.

HAWK BRAND.
M-362; 0.60-1.40 C, 0.15-0.30 Mn, 0.15-0.30 Si, bal Fe.
For general tools, dies; shallow hardening, fine grained.

HAWK D2.
M-362; 1.55 C, 0.35 Mn, 0.45 Si, 11.5 Cr, 0.9 V, 0.8 Mo, bal Fe.
Air or oil hardening cold work tool and die steel, chromium type; AISI D2.

HAWK D2S.
M-362; 1.55 C, 0.35 Mn, 0.45 Si, 11.5 Cr, 0.9 V, 0.8 Mo, 0.15 S, bal Fe.
Air or oil hardening cold work tool and die steel, chromium type, free machining grade; AISI D2.

HAWK O1.
M-362; 0.90 C, 1.25 Mn, 0.5 Cr, 0.5 W, bal Fe.
AISI Type O1 oil hardening tool steel.

HAWKS BRAND WHITE METAL.
M-410; 53.2 Sn, 7.87 Sb, 0.48 Cu, 38.48 Pb.
For bearings; castings.

HAWK SS EXTRA.
M-362; 1.05 C, 1.2 Cr, 0.3 Mo, bal Fe.
For tools, dies; Type L7.

HAWK STANDARD.
M-362; 0.7-1.2 C, bal Fe.
For tools, dies; water hardening.

HAWK VANADIUM.
M-362; 0.70-1.40 C, 0.15-0.25 V, 0.15-0.30 Mn, 0.15-0.30 Si, bal Fe.
For general tools, dies, cutters, punches; tough.

HAWK W1.
M-362; 0.95-1.05 C, Mn, Si, bal Fe.
AISI Type W1 water hardening tool steel.

HAYNES 5.
M-167; 0.20 C, 2.3 Cr, 0.80 Si, 2.0 Mn, 0.3 Mo, bal Fe.
Covered electrode for AC-DC welding or build-up.
Good impact resistance; 45 Rc.

HAYNES 5-O.
M-167; 0.10 C, 3.5 Cr, 0.30 Si, 2.0 Mn, 0.2 Mo, bal Fe.
Tube wire, open-arc welding or build-up.
Good impact resistance; 44 Rc.

HAYNES 5-S.
M-167; 0.15 C, 3.0 Cr, 0.60 Si, 2.0 Mn, 0.4 Mo, bal Fe.
Tube wire, sub-arc welding.
Good impact resistance, 40 Rc.

HAYNES 6-S.
M-167; 3.10 C, 0.90 Si, 2.0 Mn, 4.5 Mo, 5.0 Ni, bal Fe.
Tube wire, sub-arc welding.
Good impact resistance; 40 Rc.

HAYNES 7-S.
M-167; 0.10 C, 1.7 Cr, 0.50 Si, 1.8 Mn, 0.5 Mo, bal Fe.
Tube wire, sub-arc welding.
Good impact resistance; 38 Rc.

HAYNES 11.
M-167; 3.9 C, 15.0 Cr, 1.4 Si, 1.0 Mn, 0.5 Mo, bal Fe.
Covered electrode for AC-DC welding.
For general purpose and maintenance; 55 Rc.

HAYNES 11-O.
M-167; 3.5 C, 16.0 Cr, 1.0 Si, 1.3 Mn, 0.5 Mo, bal Fe.
Tube wire, open-arc welding.
For general purpose and maintenance; 48 Rc.

HAYNES 52.
M-167; 0.80 C, 6.8 Cr, 0.30 Si, 1.4 Mn, bal Fe.
Covered electrode for AC-DC welding.
For general purpose and maintenance; 54 Rc.

HAYNES 52-S.
M-167; 0.40 C, 5.4 Cr, 0.90 Si, 3.0 Mnm 0.8 Mo, 1.4 W, bal Fe.
Tube wire, sub-arc welding.
For general purpose and maintenance; 52 Rc.

HAYNES 90.
M-167; 2.5 C, 26.0 Cr, 0.25 Si, 0.8 Mn, bal Fe.
Covered electrode for AC-DC welding.
For general purpose and maintence; 52 Rc.

HAYNES 92.
M-167; 3.75 C, 10.0 Mo, bal Fe.
Bare cast rod for AC-CD welding.
For severe wear, but light impact; 64 Rc.

HAYNES 93.
M-167; 3.0 C, 17.0 Cr, 19.0 Mo, 6.3 Co, bal Fe.
Bare cast rod for DC welding.
For severe wear, but light impact: 62 Rc.
Note: HAYNES 93 covered electrode, same but 57 Rc.

HAYNES 94.
M-167; 3.5 C, 31.0 Cr, 0.50 Si, 1.2 Mn, 1.5 Mo, bal Fe.
Covered electrode for AC-DC welding.
For severe wear, but light impact; 57 Rc.

HAYNES 94-G.
M-167; 2.3 C, 29.0 Cr, 1.4 Si, 1.0 Mn, bal Fe.
Bare tube rod for welding.
For severe wear, but light impact; 60 Rc.

HAYNES 94-0.
M-167; 3.25 C, 27.0 Cr, 0.95 Si, 0.75 Mn, 3.0 Co, bal Fe.
Tube wire, open-arc welding.
For severe wear, but light impact; 56 Rc.

HAYNES 420-S.
M-167; 0.30 C, 12.0 Cr, 1.0 Si, 2.0 Mn, bal Fe.
Tube wire, sub-arc welding.
For general purpose and maintenance; 54 Rc.

HAYNES ALLOY NO. 20.
M-167; 21-23 Cr, 4-6 Mo, 25-27 Ni, bal Fe.
Very good corrosion resistance.

HAYNES ALLOY NO. 20 MOD.
M-167; 0.05 max C, 25.0-27.0 Ni, 21.0-23.0 Cr, 4.0-6.0 Mo, 1.0 max Si, 2.5 max Mn, Ti = 4 x C min, bal Fe.
Plate, Sol. Tr.: 94,900 psi TS; 42,000 psi YS; 52 El.
Weldable; resistant to corrosion and to chloride pitting; good low temperature toughness; good high temperature strength.
For boilers and pressure vessels.

HAYNES ALLOY NO. 25.
M-167; 19-21 Cr, 14-16 W, 9-11 Ni, 0.15 max C, 2 max Fe, bal Co.
Annealed: 155,000 TS; 70,000 YS; 50 El; 130 Brin.
For turbine blades and discs, combustion chambers, jet stack; corrosion and heat resistant.

HAYNES ALLOY NO. 31.
M-167; 0.45-0.55 C, 1.0 max Si, 9.5-11.5 Ni, 2.0 max Fe, 24.5-26.5 Cr 7.0-8.0 W, 1.0 max Mn, bal Co.
Sintered: 8.45 density; 120,000 psi TS; 72,000 psi YS; 4 El.
For high temperature operation.

HAYNES ALLOY NO. 40 (POWDER).
M-167; 0.75 C, 14.0 Cr, 4.0 Si, 4.0 Fe, 3.4 B, bal Ni, 2.0 max others.
Powder for flame spray or plasma arc surfacing; 52-56 Rc.
Good erosion and corrosion resistance.

HAYNES ALLOY NO. 40 (ROD).
M-167; 0.75 C, 15.0 Cr, 4.0 Si, 4.0 Fe, bal Ni, 3.5 max others.
Bare cast rod for hard-facing; 57 Rc.
Good erosion and corrosion resistance.

HAYNES ALLOY NO. 41 (POWDER).
M-167; 0.45 C, 12.0 Cr, 3.5 Si, 3.0 Fe, 2.5 B, bal Ni.
Powder for flame spray or plasma arc surfacing: 46-52 Rc.
Good erosion and corrosion resistance.

HAYNES ALLOY NO. 41 (ROD).
M-167; 0.35 C, 12.0 Cr, 3.5 Si, 3.0 Fe, bal Ni, 2.5 max others.
Bare cast rod for hard-facing; 51 Rc.
Good erosion and corrosion resistance.

HAYNES ALLOY NO. 43.
M-167; 0.85 C, 17.0 Cr, 3.9 Si, 2.0 Fe 3.3 B, bal Ni.
Powder for flame spray or plasma arc hard surfacing; 54-60 Rc.
For abrasion resistance.

HAYNES ALLOY NO. 44.
M-167; 0.45 C, 9.0 Cr, 3.0 Si, 3.8 Fe, 2.0 B, bal Ni.
Powder for flame spray or plasma arc surfacing; 40-44 Rc.

HAYNES ALLOY NO. 45.
M-167; 0.10 max C, 0.50 max Cr, 2.5 Si, 1.0 max Fe, 1.5 B, bal Ni.
Powder for manual torch build-up; 20 Rc.

HAYNES ALLOY NO. 46.
M-167; 0.10 max C, 0.50 max Cr, 3.0 Si, 1.0 max Fe, 1.8 B, bal Ni.
Powder for manual torch build-up; 30 Rc.

HAYNES ALLOY NO. 48.
M-167; 0.40 C, 16.0 Cr, 4.0 Si, 2.5 Mo, 3.0 Fe, 4.0 B, 2.5 Cu, bal Ni.
Powder for flame spray surfacing; 55 Rc.
For abrasion resistance.

HAYNES ALLOY NO. 90.
M-167; 2.75 C, 27.0 Cr, bal Fe.
Powder for plasma arc surfacing; 52 Rc.
Good cold abrasion resistance.

HAYNES ALLOY NO. 92.
M-167; 3.75 C, 1.5 max Cr, 10.0 Mo, bal Fe.
Powder for plasma arc surfacing; 60 Rc.
Good erosion and cold abrasion resistance.

HAYNES ALLOY NO. 93.
M-167; 3 C, 15-19 Cr, 13-17 Mo, 4-7 Co, 0.5-3.0 V, bal Fe.
Heat treated: 90,000 TS; 90,000 YS; 0 El; 0 RA; 650-745 Brin.
For hard surfacing electrode; abrasion resistant, high cold hardness.

HAYNES ALLOY NO. 94.
M-167; 32 Cr, 3.5 C, 1.6 Mn, 1 Mo, 0.3 max Si, bal Fe.
Welded: 580 Brin.
For hard facing electrode; resists corrosion and oxidation.

HAYNES ALLOY NO. 150.
M-167; 0.08 C, 0.65 Mn, 0.75 Si, 28.0 Cr, 2.0 Fe, bal Co.
For parts subject to intermittant heating and SO_2 gas.

HAYNES ALLOY NO. 188.
M-167, M-1491, M-1663; 20-24 Cr, 20-24 Ni, 13-16 W, 3 max Fe, 0.05- 0.15 C, 0.20-0.50 Si, 1.25 max Mn, 0.03-0.15 La, bal Co.
Sheet, H.T., Room T.: 139,400 TS; 69,500 YS; 56 El.
Sheet, H.T., 1800°F: 36,800 TS; 23,500 YS; 72 El.
For high temperature applications in aircraft, space, gas turbine, nuclear industries.
Weldable, oxidation and corrosion resistant.

HAYNES ALLOY NO 190.
M-167; 3.0 C, 26.0 Cr, 14.0 W, 1.0 B, 5.0 max Fe, bal Co.
Cobalt base superalloy; for hard facing and powder metallurgy.

HAYNES ALLOY NO. 200.
M-167; 99.2 min Ni, 0.25 max Cu, 0.40 max Fe, 0.35 max Mn, 0.10 max C 0.15 max Si.
Wrought nickel for chemical equipment.
UNS NO2200; ASTM B160 and B162.

HAYNES ALLOY NO. 201.
M-167; 99.0 min Ni, 0.25 max Cu, 0.40 max Fe, 0.35 max Mn, 0.02 max C 0.15 max Si, 0.15 max Mg.
Wrought nickel for chemical equipment.

HAYNES ALLOY NO. 208 PM.
M-167; 2.1-2.7 C, 1.0 max Si, 9.0-11.0 Co, 25.0-27.0 Cr, 9.0-11.0 Mo, 9.0-11.0 W, 11.5-13.5 Fe, 0.75 max Mn, 1.0 max B, bal Ni.
Sintered: 8.4 density; 117,000 psi TrS; 100,000 psi TS; hardness 44 Rc.
For high temperature operation.

HAYNES ALLOY NO. 263.
M-167; 0.04-0.08 C, 19.0-21.0 Cr, 5.6-6.1 Mo, 19.0-21.0 Co, 0.30-0.60 Al, 1.9-2.4 Ti, bal Ni.
Sheet: 144,000 psi (993 MPa) TS; 87,000 psi (600 MPa) YS; 37 El.
Good strength at elevated temperatures.

HAYNES ALLOY NO. 400.
M-167; 63.0-70.0 Ni, 28.0-34.0 Cu, 1.0-2.5 Fe, 1.25 max Mn, 0.15 max C, 0.50 max Al.
Cu-Ni alloy for marine equipment and chemical equipment.
UNS NO4400; ASTM B127 and B164.

HAYNES ALLOY NO. 520.
M-167; 1.0 C, 5.0 Cr, 6.3 Mo, 2.7 V, 7.5 W, bal Fe.
Powder for plasma arc surfacing; 64 Rc.
Good abrasion resistance.

HAYNES ALLOY NO. 525.
M-167; 7.0 C, 41.0 Cr, 1.0 Si, 1.0 Mn, 0.08 B, bal Fe.
Powder for plasma arc surfacing; 62 Rc.
Abrasion resistant.

HAYNES ALLOY 556.
M-167; 0.10 C, 22 Cr, 20 Ni, 20 Co, 3 Mo, 2.5 W, 0.02 La, 0.2 N, 0.1 Cb, 0.9 Ta, 0.3 Al, 0.4 Si, 1.5 Mn, 0.02 Zr, bal Fe.
High temperature alloy; for stressed components operating to 2000°F. (1095°C).

HAYNES ALLOY NO. 589.
M-167; 2.9-3.4 C, 0.5-1.5 Si, 0.5 max Mn, 15.5 -18.5 Cr, 14.5-17.5 Mo, 1.65-2.10 V, bal Fe, 3.0 max others.
Sintered: 7.55 density; 170,000 psi TrS; 125,000 psi TS; hardness 58 Rc.

HAYNES ALLOY NO. 600.
M-167; 72.0 min Ni, 6.0-10.0 Fe, 14.0-17.0 Cr, 0.08 max C, 0.50 max Ti.
Ni-Cr-Fe wrought alloy; good oxidation resistance to 2150°F.
For heat treated, furnace, and tank equipment.
UNS NO6600; ASTM B166 and B168.

HAYNES ALLOY NO. 718.
M-167; 0.08 max C, 1.0 max Co, 17.0-21.0 Cr, 2.8-3.3 Mo, 4.75-5.50 Cb + Ta, 0.20-0.80 Al, 0.65-1.15 Ti, 50.0-55.0 Ni, 0.006 max B, bal Fe.
Sol. Tr. + Aged: 180,000 psi TS; 150,000 psi YS; 15 min El.
Excellent high temperature properties; weldable; used also as filler metal.

HAYNES ALLOY NO. 800.
M-167; 30.0-35.0 Ni, 19.0-23.0 Cr, 0.15-0.60 Al, 0.15-0.60 Ti, 1.0 max Si, 1.5 max Mn, bal Fe.
Fe-Ni-Cr wrought alloy with good resistance to oxidation and carburization at elevated temperatures. Petroleum industry.
UNS NO8800; ASTM B408 and B409.

HAYNES ALLOY NO. 800 H.
M-167; 30.0-35.0 Ni, 19.0-23 Cr, 0.15-0.60 Al, 0.15-0.60 Ti, 1.0 max Si, 1.5 max Mn, bal Fe.
Solution heat-treated version of No. 800 for improved elevated temperature properties.
UNS NO8800; ASTM B408 and B409.

HAYNES ALLOY NO. R-41.
M-167; 0.12 max C, 18-20 Cr, 5 max Fe, 10-12 Co, 3.0-3.3 Ti, 9.0-10.5 Mo, 1.4-1.8 Al, B, bal Ni.
Heat treated: 140,000-206,000 TS; 97,000-154,000 YS; 9-18 El; 400 Brin.
For high temperature applications, jet engine components; age-hardenable, high oxidation resistance.

HAYNES ALLOY NO. X750.
M-167; 0.08 C, 15.5 Cr, 7.0 Fe, 0.5 Cu, 0.7 Al, 2.5 Ti, bal Ni.
Nickel base high temperature alloy; high creep strength to 1500°F.
ASTM A461.

HAYNES CRUSHER.
M-167; 4.1 C, 2.5 Cr, 0.75 Si, 0.25 Mn, 3.0 Mo, bal Fe.
Covered electrode for AC-DC welding.
For severe wear, but light impact; 57 Rc.

HAYNES CRUSHER-O.
M-167; 3.2 C, 23.0 Cr, 0.20 Si, 0.10 Mn, 2.0 Mo, bal Fe.
Tube wire, open-arc welding.
For severe wear, but light impact; 52 Rc.

HAYNES DEVELOPMENT ALLOY NO. 556.
M-167; 20.0 Ni, 20.0 Co, 22.0 Cr, 3.0 Mo, 2.5 W, 0.02 La, 0.10 C, 0.20 N, 1.0 Cb + Ta, 0.3 Al, 0.4 Si, 1.5 Mn, 0.02 Zr, bal Fe.
Good oxidation resistance to 2000°F.
Weldable, hot or cold formed.

HAYNES DEVELOPMENT ALLOY NO. 8117.
M-167; 2.5-3.8 C, 1.0 max Si, 2.5 max Ni, 2.0-5.0 Fe, 27.5-31.5 Cr, 15.0-18.5 W, 1.0-3.0 Mn, 1.0 max B, 3.0-6.5 Cb + Ta, bal Co.
Sintered: 8.8 density; 160,000 psi TrS; 95,000 psi TS; hardness 61 Rc.
For high temperature operations.

HAYNES L-605 Replaced by HAYNES ALLOY NO. 25.

HAYNES MULTIPASS.
M-167; 0.10 C, 2.0 Cr, 0.4 Si, 0.90 Mn, bal Fe.
Covered electrode for AC-DC welding or build-up.
Good impact resistance; 32 Rc.

HAYNES MULTIPASS-O.
M-167; 0.2 C, 0.80 Cr, 1.4 Si, 2.0 Mn, 0.4 Mo, bal Fe.
Tube wire, open arc welding or build-up.
Good impact resistance; 34 Rc.

HAYNES MULTIPASS-S.
M-167; 0.10 C, 0.60 Si, 1.9 Mn, 0.4 Mo, bal Fe.
Tube wire, sub-arc welding.
Good impact resistance; 32 Rc.

HAYNES-NI-MANG.
M-167; 0.60 C, 2.5 Cr, 0.20 Si, 13.5 Mn, 3.5 Ni, bal Fe.
Covered electrode for AC-DC welding and build-up.
As welded: 88 Rb; work-hardens to 46 Rc.
Severe impact resistance; moderate to slight abrasion.

HAYNES NI-MANG-O.
M-167; 0.80 C, 3.8 Cr, 0.40 Si, 14.5 Mn, 3.4 Ni, bal Fe.
Tube wire, open-arc welding and build-up.
As welded: 86 Rb; work-hardens to 44 Rc.
Severe impact resistance; moderate to slight abrasion.

HAYNES PATENT.
M-U.S.; 8-60 Cr, 1.0 max C, bal Fe.
For tools, dies, corrosion and heat resisting parts; stainless and corrosion resistant.

HAYNES STA-MANG.
M-167; 0.60 C, 15.0 Cr, 0.30 Si, 15.0 Mn, 0.3 Mo, 1.5 Ni, bal Fe.
Covered electrode for AC-DC welding and build-up.
As welded: 16 Rc; work-hardens to 47 Rc.
Severe impact resistance; moderate to slight abrasion.

HAYNES STA-MANG-O.
M-167; 0.10 C, 12.0 Cr, 0.20 Si, 14.0 Mn, 0.1 Mo, 1.4 Ni, bal Fe.
Tube wire, open-arc welding and build-up.
As welded: 19 Rc; work-hardens to 48 Rc.
Severe impact resistance; moderate to slight abrasion.

HAYNES STELLITE 98M2 ALLOY.
M-167; 1.7-2.2 C, 1.0 max Si, 2.0-5.0 Ni, 2.5 max Fe, 28.0-32.0 Cr, 0.08 max Mo, 1.0 max Mn, 0.7-1.5 B, 3.7-4.7 V, 17.0-20.0 W, bal Co.
Sintered: 8.45 density; 150,000 psi TrS; 115,000 psi TS; hardness 58 Rc.
For high temperature operations.
(Note: this alloy no longer a cast alloy).

HAYNES STELLITE ALLOY NO. 1 (POWDER).
M-167; 2.5 C, 30 Cr, 1.0 Si, 3.0 max Fe, 3.0 max Ni, 12 W, bal Co.
For plasma arc weld surfacing; 48 Rc.
Excellent metal-to-metal wear.

HAYNES STELLITE ALLOY NO. 1 (SOLID).
M-167; 2.2-2.5 C, 30.0 Cr, 12.0 W, bal Co.
Bare or covered electrode for hard facing.
As welded: 46-54 Rc.
Excellent metal-to-metal wear.

HAYNES STELLITE ALLOY NO. 3 IS NO LONGER A CAST ALLOY see HAYNES STELLITE ALLOY NO. 3 PM.

HAYNES STELLITE ALLOY NO. 3 PM.
M-167; 2.0-2.7 C, 1.0 max Si, 3.0 max Ni, 3.0 max Fe, 29.0-33.0 Cr, 11.0-14.0 W, 1.0 max Mn, 1.0 max B, bal Co.
Sintered: 8.40 density; 140,000 psi TrS; 125,000 psi TS; hardness 54 Rc.
Good strength to above 1400°F.

HAYNES STELLITE ALLOY NO. 6B.
M-167; 1.1 C, 30 Cr, 4.5 W, 3 max Fe, 3 max Ni, bal Co.
Plate: 148,000 TS; 88,000 YS; 7 El; 9 RA; 380 Brin.
Sheet: 165,000 TS; 110,000 YS; 5 El; 430 Brin.
For high temperature applications, valve parts, liners; heat and corrosion resistant.

HAYNES STELLITE ALLOY NO. 6K.
M-167; 1.6 C, 31 Cr, 4.5 W, 3 max Fe, 3 max Ni, bal Co.
Sheet: 176,500 TS; 3.5 El; 460 Brin.
At 1500°F: 70,200 TS; 17.0 El.
For high temperature applications, knives, scrapers; heat and corrosion resistant.

HAYNES STELLITE ALLOY NO. 6 PM.
M-167; 0.90-1.40 C, 1.5 max Si, 3.0 max Ni, 3.0 max Fe, 27.0-31.0 Cr, 3.5-5.5 W, 1.5 max Mo, 1.0 max Mn, 1.0 max B, bal Co.
Sintered: 8.20 density; 250,000 psi TrS; 130,000 psi TS; hardness 48 Rc.
Good strength to above 1400°F.

HAYNES STELLITE ALLOY NO. 6 (POWDER).
M-167; 1.1 C, 28 Cr, 1.0 si, 3.0 max Fe, 3.0 max Ni, 4.0 W, bal Co.
For plasma arc weld surfacing; 37 Rc.
Excellent erosion, corrosion resistance and good metal-to-metal wear.

HAYNES STELLITE ALLOY NO. 6 (SOLID).
M-167; 1.0-1.1 C, 28.0 Cr, 4.0 W, bal Co.
Bare or covered electrode, or tube wire for hard facing.
As welded: 37-41 Rc; work hardens to 49 Rc.
Excellent metal-to-metal wear.

HAYNES STELLITE ALLOY NO. 12 (POWDER).
M-167; 1.4 C, 29 Cr, 1.4 Si, 3.0 max Fe, 3.0 max Ni, 8.0 W, bal Co.
For plasma arc surfacing; 41 Rc.
Excellent erosion, corrosion resistance and good metal-to-metal wear.

HAYNES STELLITE ALLOY NO. 12 (SOLID).
M-167; 1.25-1.35 C, 29.0 Cr, 8.0 W, bal Co.
Bare or covered electrode, or tube wire, for hard facing.
As welded: 40-47 Rc.
Excellent metal-to-metal wear.

HAYNES STELLITE ALLOY NO. 19.
M-167; 1.5-2.1 C, 1.0 max Si, 3.0 max Ni, 3.0 max Fe, 29.5-32.5 Cr, 9.5-11.5 W, 1.0 max Mn, 1.0 max B, bal Co.
Sintered: 8.35 density; 275,000 psi TrS; 150,000 psi TS; hardness 51 Rc.
For high temperature operation.
(Note: this alloy no longer a cast alloy).

HAYNES STELLITE NO. 21 (POWDER).
M-167; 0.25 C, 27 Cr, 5.5 Mo, 2.0 max Fe, 2.8 Ni, bal Co.
For plasma arc surfacing; 28 Rc.
Good erosion and impact resistance and metal-to-metal wear.

HAYNES STELLITE ALLOY NO. 21 (SOLID).
M-167; 0.25 C, 27.0 Cr, 5.0 Mo, 2.8 Ni, bal Co.
Cast rod, tube wire and covered electrode for hard-facing purposes. Soft as cast.
Work hardens to 38-45 Rc.

HAYNES STELLITE ALLOY NO. 31.
M-167; 0.50 C, 25 Cr, 2.0 max Fe, 10.5 Ni, 7.5 W, bal Co.
Powder for plasma spray.

HAYNES STELLITE ALLOY NO. 156.
M-167; 1.6 C, 28 Cr, 1.1 Si, 3.0 max Ni, 4.0 W, bal Co.
Powder for plasma arc surfacing; 43 Rc.

HAYNES STELLITE ALLOY NO. 157.
M-167; 0.07 C, 21 Cr, 1.6 Si, 2.4 B, 4.5 W, bal Co.
Powder for flame spray, plasma arc surfacing, or manual torch; 52-56 Rc.

HAYNES STELLITE ALLOY NO. 158.
M-167; 0.75 C, 26 Cr, 1.2 Si, 0.75 Fe, 3.0 max Ni, 0.70 B, 5.5 W, bal Co.
Powder for plasma arc surfacing; 43 Rc.

HAYNES STELLITE ALLOY NO. 190 PM.
M-167; 3.0-3.5 C, 1.0 max Si, 3.0 max Ni, 5.0 max Fe, 24.0-28.0 Cr, 13.0-15.0 W, 1.0 max Mn, 1.0 max B, bal Co.
Sintered: 8.50 density; 135,000 psi TrS; 90,000 psi TS; hardness 58 Rc.
For high temperature operation.

HAYNES STELLITE ALLOY NO. 1016 (POWDER).
M-167; 2.5 C, 32 Cr, 3.0 max Fe, 2.5 max Ni, 17 W, bal Co.
For plasma arc surfacing; 61 Rc.
Excellent erosion and corrosion resistance and metal-to-metal wear.

HAYNES STELLITE ALLOY NO. 1016 (SOLID).
M-167; 2.5 C, 32.0 Cr, 17.0 W, bal Co.
Bare cast electrode, for hard facing.
As welded: 58 Rc.
Excellent metal-to-metal wear.

HAYNES STELLITE STAR J-METAL PM.
M-167; 2.2-2.7 C, 2.5 max Ni, 1.0 max Si, 3.0 max Fe, 31.0-34,0 Cr, 16.0-19.0 W, 1.0 max Mn, 1.0 max B, bal Co.
Sintered: 8.58 density; 125,000 psi TrS; 75,800 psi TS; hardness 60 Rc.
For high temperature operations.
(Note: this alloy no longer a cast alloy).

HAYSTELLITE 954.
M-167; 90% tungsten carbide, 10% cobalt.
Powder; may be blended with other metal powders.

HAYSTELLITE 956.
M-167; 100% tungsten carbide.
Powder; for blending with other metal powders.

HAYSTELLITE 967.
M-167; 3.9 C, 2.0 max Fe, 12.0 Co, bal W.
For plasma spray; hardness 91 RA.

HAYSTELLITE COMPOSITE NO. 1.
M-167; 50% tungsten carbide, 50% nickel-alloy matrix.
Powder for flame spray hard surfacing; hardness: 91 RA.
To resist hot and cold abrasion.

HAYSTELLITE COMPOSITE NO. 3.
M-167; 15% tungsten carbide, 85% nickel-alloy matrix.
Powder for flame spray hard surfacing; hardness: 91 RA.
To resist hot and cold abrasion.

HAYSTELLITE COMPOSITE NO. 4.
M-167; 60% tungsten carbide, 40% nickel-alloy matrix.
Powder for manual torch hard surfacing; hardness: 91 RA.
To resist hot and cold abrasion.

HAYSTELLITE COMPOSITE NO. 6.
M-167; 35% tungsten carbide, 65% nickel-alloy matrix.
Powder for flame spray hard surfacing; hardness: 91 RA.
To resist hot and cold abrasion.

HAYWOOD'S HIGH TENSILE BRONZE GRADE 8.
M-411; 35 Zn, 7 Al, bal Cu.
Forged: 86,000 TS; 61,000 YS; 36 El; 115 Brin.
For corrosion resisting, high strength cast and forged bronze parts; non-corrosive.

HB see **FINKL HB.**

HB 5.
M-459; 10.5 Al, 4 Fe, 4 Mn, 1 Zn, 0.5 Cr, bal Cu.
For gears; corrosion resistant Al bronze.

HB-18.
M-1000; weld metal: 0.10 C, 1.3 Mn, 0.30 Si, 0.41 Mo, bal Fe.
As welded: 94,000 psi TS; 78,000 psi YS; 20 El.
Welding wire electrode, for welding pipe.
AWS E 70S-1B.

HB-25.
M-1000; weld metal: 0.09 C, 0.64 Mn, 0.28 Si, bal Fe.
As welded: 75,000 psi TS; 60,000 psi YS; 24 El.
Welding wire for welding killed or semi-killed steels.
AWS E70S-3.

HB-28.
M-1000; weld metal: 0.08 C, 1.18 Mn, 0.62 Si, bal Fe.
As welded: 89,000 psi TS; 70,000 psi YS; 22 El.
Good weldability with CO_2 gas.
AWS E70S-6.

"H" BRAND.
M-40; 0.95 C, 0.35 Cr, 0.2 Si, 1.0 Mn, bal Fe.
For dies, punches; non-deforming.

HC250.
M-1099; C, Cr, bal Fe.
Annealed: 85,000 TS; 70,000 YS; 3 El; 350 Brin.
Heat treated: 100,000 TS; 80,000 YS; 750 Brin.
For cylinder liners, fan blades, molds, nozzles; abrasion resistant, hot gas resistant.

HC-250.
M-47; 2.50 C, 0.30 max Ni, 26.0 Cr, 0.75 Mn, 1.0 Si, bal Fe.
Cast heat resisting alloy; good high temperature strength and creep strength; abrasion resistant.

H.C.A.
M-140; 0.30 C, 12 Cr, 1.2 W, 0.9 V, bal Fe.
For master hobs, brass extrusion dies, die casting dies; hot work steel, oil hardened.

HC-HC.
M-432; 1.85 C, 12.25 Cr, bal Fe.
For punches and dies; non-deforming.

HCM9M.
M-1296; 0.08 max C, 0.50 max Si, 0.30-0.70 Mn, 8.0-10.0 Cr, 1.8-2.2 Mo, bal Fe.
High strength boiler tube alloy; intermediate high temperature strength between austenitic stainless steels and commercial low alloy steels.

HCR ALLOY.
M-108; 50 Ni, 50 Fe.
Soft magnetic alloy, rectangular hysteresis loop, for magnetic amplifers.

HCS16.
M-1541; 0.32 C, 0.30 Mn, 0.035 max P, 0.009 max S, 0.80 Si, 15.6 Cr, bal Fe.
97,000 psi TS; 27 El.
Ferritic stainless steel for automotive and motorcycle disc brakes.

HCS27.
M-1541; 0.05 C, 0.50 Mn, 0.03 max P, 0.01 max S, 0.40 Si, 26.5 Cr, bal Fe.
76,000 psi TS; 28 El.
Glass-to-metal sealing alloy.

HD10.
M-1744; 0.32 C, 3.0 Cr, 9.5 W, 0.3 V, 0.3 Mo, bal Fe.
Tungsten hot work steel for gripper and heading dies.

HDA-188.
M-48; 0.08 C, 22.5 Cr, 22 Ni, 15 W, 0.08 La, bal Co.
Cobalt base alloy for high temperature operation.
Same as HAYNES ALLOY No 188.

HDNC-400.
M-1698; Ni-Cu alloy sintered.
Medium strength, high ductility with outstanding corrosion resistance.
Non-magnetic; can be used for filter applications.
Similar to Monel 400.

HDNC-500.
M-1698; Ni-Cu-Si alloy sintered.
High strength; 110,000 psi TS; Rc 20.
Anti-galling properties up to 1100°F; good corrosion resistance; non-magnetic.
Similar to Monel 505.

HDW 2.
M-57; 0.4 C, 5 Cr, 0.25 V, 1.5 Mo, 0.8 Si, bal Fe.
For dies, mandrels, shear blades; hot work steel; tough.

HDW 3.
M-57; 0.4 C, 5 Cr, 1.5 Mo, 1.4 W, 0.8 Si, bal Fe.
For dies, mandrels, shear blades; hot work steel, tough.

HE-30.
M-833; 3.2-3.3 C, 2.2 Si, 0.8 Mn, bal Fe.
Cast: 30,000-40,000 TS; 180-200 Brin.
For motor blocks, machine tool beds, gibs; cast iron.

HE-40.
M-833; 3.1-3.25 C, 1.6-1.8 Si, 0.8 Mn, bal Fe.
Cast: 40,000-50,000 TS; 200-240 Brin.
For radical drill, arms and bases; cast iron.

HE-50.
M-833; 2.9-3.1 C, 1.3-1.5 Si, 0.8 Mn, bal Fe.
Cast: 50,000-60,000 TS; 230-265 Brin.
For machine tool runways, saddles, gears; cast iron.

HE-60.
M-833; 2.9-3.1 C, 1.3-1.5 Si, 0.8 Mn, 1.25-1.5 Ni, 0.5 Mo, bal Fe.
Cast: 60,000-70,000 TS; 240-280 Brin.
For crankshafts, hydraulic rams; cast iron, tough.

HE 2048.
M-U.S.; 0.4 C, 15 Co, 30 Ni, 21 Fe, 26 Cr, 4 Mo, 2 W, 0.15 B.
Cast.
For high temperature applications; high strength, heat resistant.

HEADER DIE DX.
M-32; 1.35 C, 0.30 Mn, 1.0 Si, 6.25 Cr, 1.0 Mo, 6.0 V, bal Fe.
Header dies, brick mold liners.

HEAT RESISTANT STEEL.
M-Eng.; 70 Fe, 15 Cr, 14 Co, 0.5 Mn, 0.5 C, 0.5 Si.
For corrosion and heat resisting parts; corrosion and heat resistant; U. S. Pat. 1357549.

HEAT RESISTING ACID METAL.
M-126; 55 Cu, 6 Pb, 9 Zn, 30 Ni.
For chemical apparatus; acid resistant.

HEAVY AXLE BEARING.
M-Eng.; 47 Zn, 38 Sn, 6 Sb, 4 Pb, 1 Cu.
For axle bearings; anti-friction.

HEAVY BEARING.
M-Eng.; 85 Sn, 7.5 Cu, 7.5 Sb.
For bearings, bushings; anti-friction.

HEAVY DUTY SW-16.
M-605; 0.2 C, bal Fe.
For welding electrodes; for light gage steel.

HEAVY METAL TM-17.
M-1189; Cu, Ni, bal W.
Sintered: 100,000 TS; 90,000 YS; 4 El; 250-300 Brin.
For radioactive shields, pendulums, counterbalance weights; density 16.7.

HEAVY METAL TM-18.
M-1189; Cu, Ni, bal W.
Sintered: 117,000 TS; 108,000 YS; 2.5 El; 250-300 Brin.
For radioactive shields, pendulums, counterbalance weights; density 18.0.

HEAVY METAL TM-170.
M-1189; Cu, Ni, bal W.
Sintered: 110,000 TS; 95,000 YS; 4 El; 250-300 Brin.
For radioactive shields, pendulums, counterbalance weights; density 17.0.

HEAVY METAL TM-175.
M-1189; Cu, Ni, bal W.
Sintered: 117,000 TS; 102,000 YS; 3.5 El; 250-300 Brin.
For radioactive shields, pendulums, counterbalance weights; density 17.5.

HEAVY PRESSURE.
M-88; Sb, Cu, bal Sn.
Cast: 30 Brin.
For bearings for crushers; heavy duty Babbitt.

HEAVY PRESSURE MILL GLYCO.
M-240; Sn, Sb, bal Pb.
For bearings; Babbitt.

HEAVY TUNGSTEN ALLOY see **KENNERTIUM W-2.**

HEBONITE.
M-Eng.; C, alloy, bal Fe.
For heat resistant parts; heat resistant.

HEC.
M-1342; 2.0 C, 2.0 Cr, bal Fe.
As tempered: 60 Rock C.
For wear resistance against abrasion.

HECLA.
M-122; 60 Cu, 1 Pb, 0.7 Sn, bal Zn.
Drawn: 74,000 TS; 43,000 YS; 21 El; 25 RA; 137 Brin.
Normalized: 63,000 TS; 51,000 YS; 43 El; 46 RA; 100 Brin.
For screw machine parts, hot pressings; corrosion resistant, free-cutting.

HECLA.
M-275; 0.5 C, 0.2 V, 0.2 W, 1.0 Cr, bal Fe.
For arbors, spindles, gears, drop hammer clutches, leaf springs, piston rods; resists vibration and shock. AISI L2. Was Vulcan Hecla.

HECLA 135.
M-62; 0.60 C, 0.30 Si, 0.30 Mn, 2.0 Cr, 2.0 Ni, 0.45 Mo, bal Fe.
For hot work extrusion components.

HECLA 138H.
M-62; 0.40 C, 0.30 Si, 0.60 Mn, 0.60 Cr, 2.6 Ni, 0.40 Mo, bal Fe.
For hot work extrusion components.

HECLA 149C.
M-62; 0.30 C, 0.30 Si, 0.35 Mn, 3.5 Cr, 0.45 Mo, 9.0 W, 0.30 V, bal Fe.
For hot work extrusion components.

HECLA 160C.
M-62; 0.35 C, 1.0 Si, 0.35 Mn, 1.5 Cr, 0.45 Mo, 3.5 W, 0.20 V, bal Fe.
For hot work extrusion components.

HECLA 167G.
M-62; 0.40 C, 0.30 Si, 0.60 Mn, 1.5 Cr, 0.70 Ni, 0.40 Mo, bal Fe.
For hot work extrusion components.

HECLA 172.
M-62; 0.25 C, 0.40 Si, 0.3 Mn, 3.0 Cr, 0.60 Mo, 0.60 W, 0.90 Mo.
For hot work extrusion components.

HECLA 174.
M-62; 0.35 C, 1.0 Si, 0.50 Mn, 5.0 Cr, 1.3 Mo, 0.85 V, bal Fe.
For hot work extrusion components.

HECLA 177.
M-62; 0.35 C, 1.0 Si, 0.35 Mn, 5.0 Cr, 1.3 Mo, 1.2 W, 0.20 V, bal Fe.
For hot work extrusion components.

HECLA 179.
M-62; 0.30 C, 0.30 Si, 0.35 Mn, 3.0 Cr, 3.0 Mo, 0.40 V, bal Fe.
For hot work extrusion components.

HECLA 190.
M-62; 0.35 C, 1.0 Si, 0.35 Mn, 1.5 Cr, 1.5 Mo, 1.5 W, 0.20 V, bal Fe.
For hot work extrusion components.

HECLA NO. 7.
M-122; 60 Cu, 1 Pb, 0.7 Sn, bal Zn.
For screw machine products; free cutting, corrosion resistant.

HECLA NO. 13.
M-122; 60 Cu, 1.8 Pb, 0.7 Sn, bal Zn.
Soft: 58,000 TS; 28,000 YS; 35 El; 83 Brin.
1/2 H-temper: 65,000 TS; 38,000 YS; 25 El; 125 Brin.
H-temper: 75,000 TS; 52,000 YS; 20 El; 160 Brin.
For screw machine products, marine hardware; free cutting.

HECLA NO. 69.
M-122; 60 Cu, 0.7 Pb, 0.75 Sn, bal Zn.
Soft: 58,000 TS; 28,000 YS; 40 El; 83 Brin
1/2 H-temper: 65,000 TS; 38,000 YS; 30 El; 125 Brin.
H-temper: 75,000 TS; 52,000 YS; 22 El; 160 Brin.
For screw machine products, marine hardware; free cutting.

HECLA SPECIAL.
M-275; 0.5 C, 1.5 Cr, 0.2 V, 0.2 W, bal Fe.
Heat treated: 130,000-165,000 TS; 125,000-160,000 YS; 19.5-18.0 El; 5 -51 RA; 241-269 Brin.
For gears, forging dies, clutches; shock resistant, oil hardened.

HECNUM.
M-1554; 55 Cu, 45 Ni.
Annealed: 70,000 TS; 30,000 YS; 50 El; 110 Brin.
For electrical resistances, thermocouples.
Negligible temperature coefficient.

HEDDAL.
M-297; 0.2 C, 0.3 Mn, 0.25 Si, bal Fe.
For machinery parts; water hardened.

HEDDAL.
M-1548; 1.5 Mn, bal Al.
Soft: 14,000 TS; 5690 YS; 30-40 El; 28-32 Brin.
Hard: 35,600 TS; 28,500 YS; 3 El; 60 Brin.
For heat exchangers, pressure and storage tanks, chemical equipment.
Similar to Aluminum 3003, not heat-treatable.

HEDDENAL 2.
M-297; 1.5-3.0 Mg, 0.5-1.5 Mn, 0.3 max Cr, bal Al.
Soft: 26,000 TS; 10,000 YS; 20 El; 45 Brin.
Hard: 41,000 TS; 36,000 YS; 5 El; 77 Brin.
For roofing, hydraulic tubing, architectural trim; good forming and welding properties.

HEDDENAL 3.
M-297; 2-4 Mg, 0.4 max Mn, 0.3 max Cr, bal Al.
Soft: 28,000 TS; 13,000 YS; 30 El; 47 Brin.
Hard: 40,000 TS; 35,000 YS; 10 El; 73 Brin.
For aircraft tanks and fittings, fuel lines, marine parts; resists sea water corrosion.

HEDDENAL 5.
M-297; 4.0-5.5 Mg, 0.8 max Mn, 0.3 max Cr, bal Al.
Soft: 42,000 TS; 22,000 YS; 35 El; 65 Brin.
Hard: 60,000 TS; 50,000 YS; 10 El; 105 Brin.
For aircraft and marine parts; good corrosion resistance.

HEDDENAL 7.
M-297; 5.5-7.5 Mg, 0.8 max Mn, 0.3 max Cr, bal Al.
For aircraft and marine parts; good corrosion resistance.

HEDDRONAL 3.5.
M-1548; 0.5 Mn, 3.5 Mg, bal Al.
Rolled: 35,000 TS; 20-25 El; 70-100 Brin.
For structural towers, unfired pressure vessels, rocket motor parts, missile containers.
Similar to Aluminum 5086, non-heat treatable.
Corrosion resistant.

HEDDRONAL 5.
M-1548; 0.5 Mn, 5.0 Mg, bal Al.
Rolled: 32,000-38,000 TS; 15-22 El; 70-100 Brin.
For deck housing, overhead cranes, ship unloaders, heavy duty structures.
Similar to Aluminum 5456. Not heat treatable; corrosion resistant.

HEDDRONAL 7.
M-1548; 0.5 Mn, 7.0 Mg, bal Al.
Rolled: 45,000-52,000 TS; 20-25 El; 80-100 Brin.
For heavy duty structures, overhead cranes, ship unloaders.
Corrosion resistant.

HEDDROXAL.
M-297; 2-4 Mg, 0.4 Mn, 0.3 max Cr, bal Al.
Soft: 28,000 TS; 13,000 YS; 30 El; 47 Brin.
Hard: 40,000 TS; 35,000 YS; 10 El; 73 Brin.
For aircraft tanks and fittings, fuel lines, marine parts; resists sea water corrosion.

HEDDUR.
M-297; 2.5-5.0 Cu, 0.2-1.8 Mg, 0.3-1.5 Mn, bal Al.
Annealed: 27,000 TS; 11,000 YS; 22 El; 47 Brin.
Heat treated: 72,000 TS; 57,000 YS; 130 Brin.
For aircraft structures and fittings, fasteners; age-hardenable.

HEDDUR.
M-983; 0.2-2.0 Mg, 3.5-5.5 Cu, 0.2-1.0 Si, 0.1-1.2 Mn, bal Al.
20,000-31,000 TS; 10,000-14,000 YS; 2-4 El; 60-80 Brin.
For light alloy parts; Duralumin type.

HEDDUR.
M-1545; 3.5-5.0 Cu, 0.3-0.8 Si, 0.2-1.2 Mn, 0.4-1.0 Mg, bal Al.
Heat treated: 68,000-78,000 TS; 50,000-57,000 YS; 12-16 El; 115-130 Brin.
For structural applications, screw machine products, fasteners.
Heat treatable. Similar to Aluminum 2017.

HEDERVAN.
M-73; 0.90 C, V, bal Fe.
For cold heading and striking dies, punches, draw dies; water hardened; Type W2.

HEERDT T7P.
M-1324; 0.7 C, 0.25 max Si, 0.25 max P, bal Fe.
Heat treated: 120,000-174,000 TS; 82,000-128,000 YS; 22-12 El; 52-37 RA; 241-352 Brin.
For crimpers, punches, springs, rails, dies; Type W1; water hardened.

HEERDT T8E.
M-1324; 0.7 C, 0.2 max Si, 0.2 max P, bal Fe.
Heat treated: 120,000-174,000 TS; 82,000-128,000 YS; 22-12 El; 52-37 RA; 241-352 Brin.
For crimpers, punches, dies, springs, rails; Type W1; water hardened.

HEERDT T8V.
M-1324; 0.8 C, 0.2 Si, 0.2 Mn, bal Fe.
Heat treated: 129,000-188,000 TS; 87,000-143,000 YS; 21-12 El; 50-35 RA; 255-388 Brin.
For tools, dies, springs, cutters; Type W1; water hardened.

HEERDT T9E.
M-1324; 0.85 C, 0.25 max Si, 0.25 max Mn, bal Fe.
Heat treated: 130,000-190,000 TS; 88,000-145,000 YS; 20-11 El; 48-32 RA; 260-400 Brin.
For tools, dies, springs, cutters, drills; Type W1; water hardened.

HEERDT T9P.
M-1324; 0.85 C, 0.25 max Si, 0.25 max Mn, bal Fe.
Heat treated: 130,000-190,000 TS; 88,000-145,000 YS; 20-11 El; 48-32 RA; 260-400 Brin.
For tools, dies, springs, cutters, drills; Type W1; water hardened.

HEERDT T10E.
M-1324; 1.0 C, 0.25 max Si, 0.25 max Mn, bal Fe.
Heat treated: 121,000-213,000 TS; 84,000-150,000 YS; 20-11 El; 47-33 RA; 235-535 Brin.
For drills, hobs, cutting tools; Type W1; water hardened.

HEERDT T10P.
M-1324; 1.0 C, 0.25 max Si, 0.25 max Mn, bal Fe.
Heat treated: 121,000-213,000 TS; 84,000-150,000 YS; 20-11 El; 47-33 RA; 235-535 Brin.
For drills, hobs, cutting tools, reamers; Type W1; water hardened.

HEERDT T11E.
M-1324; 1.1 C, 0.25 max Mn, 0.25 max Si, bal Fe.
Heat treated: 122,000-215,000 TS; 85,000-155,000 YS; 18-10 El; 45-30 RA; 240-555 Brin.
For reamers, drills, taps, hobs, lathe tools; Type W1; water hardened.

HEERDT T11P.
M-1324; 1.15 C, 0.25 max Mn, 0.25 max Si, bal Fe.
Heat treated: 122,000-215,000 TS; 85,000-155,000 YS; 18-10 El; 45-30 RA; 240-555 Brin.
For reamers, drills, taps, hobs, lathe tools; Type W1; water hardened.

HEERDT T12E.
M-1324; 1.3 C, 0.25 max Mn, 0.25 max Si, bal Fe.
For reamers, drills, taps, hobs, lathe tools; Type W1; water hardened.

HEERDT T12P.
M-1324; 1.3 C, 0.25 max Si, 0.25 max Mn, bal Fe.
For reamers, drills, hobs, lathe tools; Type W1; water hardened.

H-E IRON.
M-833; C, Si, Mn, bal Fe.
Cast: 30,000-60,000 TS; 180-260 Brin.
For machine tool castings, gears, housings; gray cast iron.

HELBIMPHY 6.
M-772; 0.7-0.8 C, bal Fe.
For tools, springs, punches; water hardened.

HELBIMPHY 7.
M-722; 0.7-0.8 C, bal Fe.
For tools, springs, drills; water hardened.

HELBIMPHY 8.
M-722; 0.9-1.0 C, bal Fe.
For tools, drills, punches, shears; water hardening.

HELBIMPHY 9.
M-772; 1.05-1.15 C, bal Fe.
For tools, reamers, punches; water hardened.

HELBIMPHY 10.
M-772; 1.05-1.15 C, bal Fe.
For tools, taps, drills; water hardened.

HELBIMPHY 12.
M-772; 1.2 C, bal Fe.
For tools, drills, taps; water hardened.

HELBIMPHY 14.
M-772; 1.4 C, bal Fe.
For cutting tools, drills; water hardened.

HELLEFORS 123H.
M-217; 0.4 C, 0.3 Si, 1.25 Mn, bal Fe.
For gears, machinery parts; tough.

HELLEFORS 134 HF.
M-217; 0.55 C, 1.7 Si, 0.7 Mn, bal Fe.
For leaf and coil springs; shock resistant.

HELLEFORS 206A.
M-217; 0.9 C, 0.5 Cr, bal Fe.
For punches, forming dies, headers; water hardened.

HELLEFORS 209A.
M-217; 1.3 C, 0.5 Cr, bal Fe.
For blanking and forming dies, engravers' tools; water or oil hardened.

HELLEFORS 209H.
M-217; 1.5 C, 2.0 Cr, bal Fe.
For bearings, liners, sleeves; oil or water hardened.

HELLEFORS 216 HM.
M-217; 0.95 C, 1.5 Si, 1.0 Cr, bal Fe.
For metal and wood working cutting tools; non-deforming, oil hardening.

HELLEFORS 249A.
M-217; 1.35 C, 0.25 Si, 0.25 Mn, 0.30 Cr, bal Fe.
For blanking and forming dies, engravers' tools; water or oil hardened.

HELLEFORS 264HF.
M-217; 0.50 C, 1.05 Cr, 0.85 Mn, 0.15 V, bal Fe.
For springs, gears, shafts; shock resistant.

HELLEFORS 267B.
M-217; 1.0 C, 0.3 Mn, 0.5 Cr, 0.1 V, bal Fe.
For bearings, punches, liners, forming dies; water hardened.

HELLEFORS 288A.
M-217; 1.2 C, 0.2 Si, 0.3 Mn, 0.55 W, 0.1 V, bal Fe.
For cold work tools, punches, heading dies; oil or water hardened.

HELLEFORS 288 J.
M-217; 1.1 C, 0.3 Cr, 1.1 W, 0.1 V, bal Fe.
For structural parts, cutting tools; drill rod.

HELLEFORS 352H.
M-217; 0.25 C, 0.65 Mn, 1.05 Cr, 0.2 Mo, bal Fe.
For gears, shafts; tough.

HELLEFORS 371H.
M-217; 0.15 C, 0.55 Mn, 0.65 Cr, 1.25 Ni, bal Fe.
For gears, camshafts, crankshafts; case hardened.

HELLEFORS 371 L.
M-217; 0.15 C, 0.6 Cr, 3 Ni, bal Fe.
Hardened: 128,000 TS; 107,000 YS; 16 El; 55 RA; 270 Brin.
For gears, crankshafts, chain balls; case hardened.

HELLEFORS 371L2.
M-217; 0.18 C, 0.55 Mn, 0.75 Cr, 3.0 Ni, bal Fe.
For gears, camshafts, crankshafts; case hardened.

HELLEFORS 371M.
M-217; 0.2 C, 0.55 Mn, 1.8 Cr, 0.25 Mo, bal Fe.
For gears, shafts; case hardened.

HELLEFORS 372H.
M-217; 0.4 C, 0.8 Mn, 0.8 Cr, 1.25 Ni, bal Fe.
For gears, shafts; shock resistant.

HELLEFORS 372J.
M-217; 0.35 C, 1.2 Cr, 2.6 Ni, bal Fe.
Oil hardened: 143,000 TS; 130,000 YS; 19-22 El; 55-60 RA; 300 Brin
For shafts, axles, bolts, gears; tough.

HELLEFORS 372S.
M-217; 0.3 C, 0.55 Mn, 1.25 Cr, 4.25 Ni, bal Fe.
For gears, shafts, machinery parts; tough, shock resistant.

HELLEFORS 372Y.
M-217; 0.30 C, 0.55 Mn, 1.05 Cr, 3.25 Ni, 0.25 Mo, bal Fe.
For gears, shafts, machinery parts; tough, shock resistant.

HELLEFORS 374 H.
M-217; 0.5 C, 0.6 Cr, 1.7 Ni, 0.3 Mo, bal Fe.
For hot and cold working tools; oil hardening.

HELLEFORS 374R.
M-217; 0.55 C, 1.5 Cr, 3.0 Ni, 0.4 Mn, bal Fe.
For cold work dies and tools; oil hardened, shock resistant.

HELLEFORS 374RK.
M-217; 0.55 C, 1 Cr, 3 Ni, 0.3 Mo, 0.4 Mn, bal Fe.
For drop forge dies, cold working tools; oil hardened, shock resistant.

HELLEFORS 394T.
M-217; 0.55 C, 0.7 Cr, 1.75 Ni, 0.7 Mo, 0.7 Mn, bal Fe.
For forging dies, headers, upsetters; oil hardened, shock resistant.

HELLEFORS 408C.
M-217; 1.2 C, 0.2 Si, 0.3 Mn, 1.0 W, bal Fe.
For cold work tools, punches, heading dies; oil or water hardened.

HELLEFORS 605 F.
M-217; 0.8 C, 0.1 V, bal Fe.
For chisels, punches, dies; hard and tough.

HELLEFORS 606F.
M-217; 0.90 C, 0.20 Si, 0.30 Mn, 0.1 V, bal Fe.
Heat treated; 185,000 TS; 118,000 YS; 10 El; 30 RA; 375 Brin.
For springs, taps, reamers, hobs, broaches; Type W2; water hardened.

HELLEFORS 607C.
M-217; 1.0 C, 0.20 Si, 0.30 Mn, 0.1 V, bal Fe.
Annealed: 100,000 TS; 55,000 YS; 20 El; 42 RA; 200 Brin.
For cold work dies and tools, drills, taps; Type W2; water hardened.

HELLEFORS 666.
M-217; 1.2 C, bal Fe.
For structural parts, tools; water hardening.

HELLEFORS 677R.
M-217; 1 C, 0.2 Si, 0.6 Mn, 5.2 Cr, 1.1 Mo, 0.2 V, bal Fe.
For dies, punches, crimpers; air hardened, wear resistant.

HELLEFORS 782R.
M-217; 0.35 C, 1.0 Cr, 2.5 W, bal Fe.
For pneumatic chisels, shears, hot work tools; oil hardened, shock resistant.

HELLEFORS 783 R.
M-217; 0.45 C, 0.9 Si, 1.1 Cr, 2.1 W, bal Fe.
For chisels, punches; oil hardening, tough.

HELLEFORS 786 AM.
M-217; 0.9 C, 1.2 Mn, 0.5 Cr, 0.5 W, bal Fe.
For punches, dies, shears; non-deforming.

HELLEFORS 920A.
M-217; 0.08 max C, 11.5-13 Cr, 0.1-0.3 Al, bal Fe.
Annealed: 71,000 TS; 42,600 YS; 22 El; 70 RA; 150 Brin.
Heat treated: 175,000 TS; 145,000 YS; 21 El; 64 RA; 352 Brin.
For oil refinery and chemical plant equipment; Type 405; corrosion resistant.

HELLEFORS 920-AL.
M-217; 0.1 C, 14 Cr, 10 Ni, bal Fe.
For fasteners, chemical plant equipment. Stainless steel.

HELLEFORS 920B.
M-217; 0.12 max C, 14-18 Cr, bal Fe.
Annealed: 70,000 TS; 40,000 YS; 30 El; 55 RA; 150 Brin.
Cold Drawn: 130,000 TS; 120,000 YS; 2 El; 185 Brin.
For oil refinery equipment, bolts, kitchen sinks; Type 430; stainless, ferritic.

HELLEFORS 920-H.
M-217; 0.08 C, 17 Cr, 10 Ni, bal Fe.
Annealed: 80,000 TS; 38,000 YS; B 80 Rock.
For chemical and pharmaceutical plant equipment, mixers, agitators, tanks, digesters.
Type 17-10 stainless steel, austenitic.

HELLEFORS 920-M.
M-217; 0.15 max C, 11.5-13.5 Cr, bal Fe.
Annealed: 75,000 TS; 40,000 YS; 35 El 155 Brin.
Cold drawn: 100,000 TS; 85,000 YS; 17 El; 205 Brin.
For flat springs, knives, table flatware.
Type 410 stainless steel; hardenable.

HELLEFORS 921A.
M-217; 0.15 max C, 11.5-13.5 Cr, 1 max Mn, bal Fe.
Annealed: 80,000 TS; 40,000 YS; 35 El; 70 RA; 155 Brin.
Cold drawn: 100,000 TS; 85,000 YS; 17 El; 60 RA; 205 Brin.
For valve parts, turbine blades, cutlery, knives; Type 410; corrosion resistant.

HELLEFORS 921C.
M-217; 0.15 min C, 12-14 Cr, bal Fe.
Annealed: 88,000 TS; 40,000 YS; 32 El; 68 RA; 170 Brin.
For cutlery, valve trim, turbine blades; Type 420; stainless, hardenable.

HELLEFORS 922-A.
M-217; 0.15 min C, 12-14 Cr, bal Fe.
Annealed: 95,000 TS; 50,000 YS; 25 El; B 92 Rock.
Oil hardened: 250,000 TS; 192,000 YS; 10 El; C 49 Rock.
For cutlery, surgical instruments, scissors, ball bearings, valves, gears.
Type 420 stainless steel, hardenable.

HELLEFORS 923A.
M-217; 0.38-0.45 C, 13 Cr, bal Fe.
Annealed: 95,000 TS; 50,000 YS; 25 El; 55 RA; 196 Brin.
For valves, cutlery, surgical and dental instruments; Type 420; corrosion resistant.

HELLEFORS 930A.
M-217; 0.08-0.20 C, 2 max Mn, 17-19 Cr, 8-10 Ni, bal Fe.
Annealed: 85,000 TS; 35,000 YS; 60 El; 70 RA; 150 Brin.
Cold drawn: 125,000 YS; 95,000 YS; 22 El; 55 RA; 277 Brin.
For oil refinery and chemical plant equipment; Type 302; stainless, austenitic.

HELLEFORS 930-B.
M-217; 0.15 max C, 17-19 Cr, 8-10 Ni, bal Fe.
Cold rolled: 140,000 TS; 100,000 YS; 30 El; B 100 Rock.
Annealed: 90,000 TS; 40,000 YS; 50 El; B 85 Rock.
For chemical and pharmaceutical plant equipment, valve trim, fasteners.
Type 302 stainless steel, austenitic. Hardenable only by cold work.

HELLEFORS S-7.
M-217; 0.15 C, bal Fe.
Hardened: 100,000 TS; 78,000 YS; 19 El; 40 RA; 200 Brin.
For gears, bolts, pins; case hardened.

HELLEFORS V60.
M-217; 0.60 C, 0.20 Si, 0.30 Mn, bal Fe.
Heat treated: 115,000-160,000 TS; 77,000-113,000 YS; 23-12 El; 54-40 RA; 229-320 Brin.
For wheels, die blocks, rails, girders; water hardened.

HELLEFORS V70.
M-217; 0.70 C, 0.20 Si, 0.30 Mn, bal Fe.
Heat treated: 122,000-174,000 TS; 82,000-128,000 YS; 22-12 El; 52-37 RA; 240-350 Brin.
For springs, rails, girders, clutch discs; water hardened.

HELLEFORS V80.
M-217; 0.80 C, 0.20 Si, 0.30 Mn, bal Fe.
Heat treated: 130,000-188,000 TS; 87,000-143,000 YS; 21-12 El; 50-35 RA; 285-355 Brin.
For drills, reamers, hobs, taps, cutters; water hardened; Type W1.

HELLEFORS V100.
M-217; 1.0 C, 0.20 Si, 0.30 Mn, bal Fe.
Annealed: 100,000 TS; 53,000 YS; 21 El; 42 RA; 200 Brin.
For drills, reamers, taps, hobs, broaches; water hardened; Type W1.

HELLEFORS V110.
M-217; 1.1 C, 0.20 Si, 0.30 Mn, bal Fe.
Annealed: 105,000 TS; 55,000 YS; 20 El; 40 RA; 210 Brin.
For drills, springs, reamers, taps, broaches; water hardened; Type W1.

HELLEFORS V120.
M-217; 1.2 C, 0.20 Si, 0.30 Mn, bal Fe.
Annealed: 110,000 TS; 55,000 YS; 18 El; 38 RA; 220 Brin.
For cutters, taps, reamers, broaches; Type W1; water hardened.

HELLEFORS V140.
M-217; 1.4 C, 0.30 Si, 0.30 Mn, bal Fe.
For engravers' tools, blanking dies, cutters; Type W1; water hardened.

HELLEFORS W406.
M-217; 1.4 C, 0.4 Cr, 5.5 W, 0.5 V, bal Fe.
For finishing tools, cutters; oil hardened.

HELLER 70-20-M.
M-343; 0.6-0.7 C, 0.15-0.25 Mo, bal Fe.
For pneumatic tools; tough.

HELLER 70-20 V.
M-343; 0.6-0.7 C, 0.15-0.25 V, bal Fe.
For pneumatic tools; tough.

HELLER AIR HARDENING DIE STEEL.
M-1221; 0.95-1.05 C, 0.5-0.7 Mn, 0.3-0.5 Si, 5.0-5.5 Cr, 0.9-1.1 Mo, 0.2-0.3 V, bal Fe.
Flat ground die steel.
For punches, dies, gauges, wear resisting tools and parts. AISI type A2 tool steel.

HELLER OIL HARDENING DIE STEEL.
M-1221; 0.85-0.95 C, 1.0-1.25 Mn, 0.2-0.04 Si, 0.4-0.6 Cr, 0.4-0.6 W, 0.1-0.2 V, bal Fe.
Flat ground die steel.
For dies, punches, jigs, gauges, templates, stamps, shims, machine parts. AISI Type 01.

HELLER'S CHROME DIE.
M-343; 0.85-1.0 C, 3-4 Cr, bal Fe.
For tools, dies, shears; now "Brown Label."

HELMENT BRONZE.
M-Eng.; 70 Cu, 30 Zn.
For water pipes, cartridges, shell cases, deep drawn parts; max ductility.

HELMET METAL.
M-Eng.; 72-70 Cu, 28-30 Zn.
For cartridges, shell cases, helmets; deep drawn.

HELUMIN.
M-834; 1.8 Cu, 1.5 Fe, bal Al.
Heat treated: 50,000 TS; 15 El; 50 Brin.
For light alloy parts; non-hardenable.

HELVE.
M-502; 0.7-1.26 C, bal Fe.
For tools, dies; water hardening.

HENCKELS AFH.
M-1287; 0.60 C, 0.4 Si, 0.6 Mn, bal Fe.
Heat treated: 160,000 TS; 113,000 YS; 12 El; 40 RA; 325 Brin.
For crimpers, axes, rails; water hardened.

HENCKELS G5.
M-1287; 0.60 C, 0.35 max Si, 0.30-0.80 Mn, bal Fe.
Heat treated: 165,000 TS; 118,000 YS; 12 El; 40 RA; 330 Brin.
For gears, springs, shafts, axes, hammers; water hardened.

HENCKELS NI 35.
M-1287; 0.50 C, 1.05 Cr, 3.25 Ni, 0.5 Mn, bal Fe.
For gears, bolts, crankshafts, forging dies; oil hardened, tough.

HENCKELS NI 44.
M-1287; 0.40 C, Ni, Cr, Mo, bal Fe.
For gears, bolts, crankshafts; oil hardened, shock resistant.

HENCKELS NI 187.
M-1287; 0.56 C, Ni, Cr, Mo, V, bal Fe.
For forging and heading dies, punches; oil hardened, shock resistant.

HENCKELS SPBA.
M-1287; C, alloy, bal Fe.
For machine tool parts; oil hardened, tough.

HENCKELS SPDM-H.
M-1287; 0.55 C, 0.90 Si, 1.05 Cr, 0.18 V, 1.85 V, 1.85 W, bal Fe.
For pneumatic tools, chisels, punches; oil hardened, shock resistant.

HENCKELS SPDN.
M-1287; 0.45 C, W, Cr, V, bal Fe.
For forging and heading dies, pneumatic tools; oil hardened, tough.

HENCKELS SPDN-A.
M-1287; 0.45 C, Si, Cr, V, bal Fe.
For gears, springs, crankshafts, bolts; oil hardened, shock resistant.

HENCKELS SPDN-W.
M-1287; 0.35 C, 0.90 Si, 1.05 Cr, 0.18 V, 1.85 W, bal Fe.
For pneumatic tools, chisels, rivet sets; oil hardened, shock resistant.

HENCKELS SPEZIAL ZH.
M-1287; 1.45 C, 1.4 Cr, 0.6 Mn, bal Fe.
For blanking and forming dies; oil or water hardened.

HENCKELS SPG/A.
M-1287; 1.05 C, Cr, bal Fe.
For cutters, bearings, liners; oil or water hardened, wear resistant.

HENCKELS SPHF.
M-1287; 1.3 C, 4.75 W, 0.20 max Cr, bal Fe.
For blanking and forming dies, cutters; water hardened.

HENCKELS SPHF-A.
M-1287; 1.0 C, Cr, bal Fe.
For bearings, cutters, liners, sleeves; water hardened, wear resistant.

HENCKELS SPIII VA.
M-1287; C, alloy, bal Fe.
For machine tool parts; oil hardened, tough.

HENCKELS SPJ.
M-1287; 1.4 C, 0.30 Cr, 0.10 V, 0.30 Mn, bal Fe.
For blanking and forming dies, engravers' tools; oil or water hardened, wear resistant.

HENCKELS SPM.
M-1287; 0.90 C, 0.10 V, 1.9 Mn, bal Fe.
For blanking and forming dies, punches, cutters; oil hardened, non-deforming.

HENCKELS SPS.
M-1287; 2.1 C, 11.5 Cr, 0.30 Mn, bal Fe.
For blanking and forming dies, punches; oil hardened, non-deforming.

HENCKELS SPSB.
M-1287; 1.65 C, Cr, Mo, V, bal Fe.
For blanking and forming dies, punches; oil hardened, non-deforming.

HENCKELS SPSE.
M-1287; 1.05 C, 1.15 W, 0.90 Mn, 1.0 Cr, bal Fe.
For blanking and heading dies; oil hardened, tough.

HENCKELS SPSE/A.
M-1287; 1.25 C, 1.15 Si, 0.70 Mn, 1.2 Cr, bal Fe.
For cold work tools; oil hardened, tough.

HENCKELS SPSWO.
M-1287; 2.1 C, 11.5 Cr, 0.70 W, bal Fe.
For blanking and forming dies, shears, punches; oil hardened, non-deforming.

HENCKELS SPWO4.
M-1287; 0.30 C, 1.1 Cr, 0.18 V, 3.75 W, bal Fe.
For pneumatic tools, shears, punches, chisels; oil hardened, tough.

HENCKELS SPWO5VA.
M-1287; 0.30 C, 2.35 Cr, 0.6 V, 4.25 W, bal Fe.
For pneumatic tools, punches, chisels, shears; oil hardened, tough.

HENCKELS SPWO9.
M-1287; 0.30 C, 8.5 W, 0.35 V, 2.65 Cr, bal Fe.
For extrusion dies, rams, liners, punches; oil hardened, hot work steel.

HENCKELS SPWON.
M-1287; 0.30 C, 2.0 Co, 8.5 W, 2.4 Cr, 0.25 V, bal Fe.
For extrusion dies, rams, liners, punches; oil hardened, hot work steel.

HENCKELS SPZII.
M-1287; 1.0 C, 0.10 V, 0.25 Mn, 0.25 Si, bal Fe.
For blanking and heading dies, cutters; water hardened; Type W2.

HENNEQUIN METAL.
C, bal Fe.

HENRICOT HIB.
M-1188; 0.06 max C, 1.0 max Mn, 1.0 max Si, 18.0-21.0 Cr, 14.0-16.0 Mo, 2.0 max Fe, bal Ni.
Cast: 45-60 kg/mm^2 TS; 32 kg/mm^2 min YS; 8 min El.
Good corrosion resistance.

HENRICOT HIF2.
M-1188; 0.25-0.35 C, 0.75 max Mn, 0.60 Max Si, 3.0-4.0 Ni, 1.0-1.5 Cr, 0.16-0.35 Mo, bal Fe.
Cast alloy; hardenable to 420-500 Brin.
For highly stressed parts.

HENRICOT HIF4.
M-1188; 0.30-0.40 C, 0.75 Mn, 0.60 Si, 3.0-4.0 Ni, 1.2-1.7 Cr, 0.15-0.35 Mo, bal Fe.
Cast alloy: hardenable to 500-550 Brin.
For highly stressed parts.

HENRICOT HIG.
M-1188; 0.06 max C, 2.0 max Mn, 1.0 max Si, 46.0-50.0 Ni, 23.0-27.0 Cr, 5.0-7.0 Mo, 2.0-4.0 Cu, Nb, bal Fe.
Cast: 45-60 kg/mm^2 TS; 20 kg/mm^2 min YS; 25 min El.
Good corrosion resistance.
Similar to Hastelloy G.

HENRICOT HIL.
M-1188; 0.06 max C, 1.0 max Mn, 1.0 max Si, 23.0-27.0 Cr, 8.0-10.0 Mo, 2.0 max Fe, 7.0-9.0 Co, bal Ni.
Corrosion resistant casting; for chemical equipment.

HENRICOT HIM2.
M-1188; 0.06 max C, 1.0 max Mn, 1.0 max Si, 26.0-30.0 Mo, 7.0 max Fe, bal Ni.
Corrosion resistant casting; for chemical equipment.
Similar to Hastelloy B.

HENRICOT HIT.
M-1188; 0.06 max C, 21.0-24.0 Cr, 6.0-8.0 Mo, 12.0-14.0 Fe, 2.0-3.5 Co, bal Ni.
Corrosion resistant casting; for chemical equipment.
Similar to Hastelloy F.

HENRICOT HIW.
M-1188; 0.06 max C, 1.0 max Mn, 1.0 max Si, 15.5-17.5 Cr, 16.0-18.0 Mo, 7.0 max Fe, 3.75-5.25 W, bal Ni.
Cast: 45-60 kg/mm² TS; 32 kg/mm² min YS; 5 min El.
Corrosion resistant alloy; for chemical equipment.
Similar to Hastelloy C.

HENRICOT HL1B.
M-1188; 0.25-0.36 C, 0.60-1.0 Mn, 0.20-0.50 Si, 0.50-1.0 Ni, 0.50-1.0 Cr, 0.15-0.25 Mo, bal Fe.
Cast: 286-321 Brin.
Similar to AISI 8630.

HENRICOT HL1C.
M-1188; 0.13-0.18 C, 0.70-1.10 Mn, 0.15-0.40 Si, 0.80-1.20 Ni, 0.60-1. Cr, 0.10 max Mo, bal Fe.
As cast: 220 max Brin.
Low alloy carburizing steel casting.

HENRICOT HL1E.
M-1188; 0.14-0.20 C, 0.50-1.0 Mn, 0.60 max Si, 0.50-1.0 Ni, 0.40-0.80 Cr, 0.40-0.60 Mo, 0.10 max V, bal Fe.
Low alloy carburizing steel casting.

HENRICOT HL10.
M-1188; 0.30-0.40 C, 0.75 max Mn, 0.60 max Si, 1.25-1.75 Ni, 1.25-1.75 Cr, 0.15-0.25 Mo, bal Fe.
Medium carbon alloy steel casting.

HENRICOT HL11.
M-1188; 0.20-0.30 C, 0.75 max Mn, 0.60 max Si, 1.25-1.75 Ni, 1.25-1.75 Cr, 0.15-0.25 Mo, bal Fe.
Alloy steel casting.

HENRICOT HL13.
M-1188; 0.14-0.20 C, 0.75-0.95 Mn, 0.40-0.60 Si, 1.50-2.0 Ni, 0.55-0.7 Cr, 0.35-0.45 Mo, 0.03-0.06 V, bal Fe.
As cast: 286-321 Brin.
Alloy carburizing steel casting.
Similar to AISI 4320.

HENRICOT HM0.
M-1188; 0.35 max C, 1.0-2.0 Mn, 0.60 max Si, 0.50-1.0 Cr, 0.40-0.60 Mo, bal Fe.
Alloy steel casting; hardenable to 425 Brin min.

HENRICOT HM1.
M-1188; 0.35 max C, 1.0-2.0 Mn, 0.60 max Si, 0.50-1.0 Cr, 0.40-0.60 Mo, bal Fe.
Alloy steel casting; hardenable to 450-500 Brin.

HENRICOT HM3.
M-1188; 0.40-0.50 C, 1.0 max Mn, 0.60 max Si, 2.5-3.5 Cr, 0.35-0.45 Mo, bal Fe.
Alloy steel casting; deep hardening.

HENRICOT HM4.
M-1188; 0.60-0.70 C, 0.75-1.25 Mn, 0.60 max Si, 0.50-1.0 Ni, 1.25-2.0 Cr, 0.35-0.45 Mo, bal Fe.
Alloy steel casting; good hardenability.

HENRICOT HM7.
M-1188; 0.90-1.10 C, 0.40-0.60 Mn, 0.60 max Si, 4.5-5.5 Cr, 0.90-1.10 Mo, 0.50 V, bal Fe.
Alloy steel casting.
Similar to AISI A2 tool steel.

HENRICOT HM9.
M-1188; 0.90-1.10 C, 0.40-0.60 Mn, 0.60 max Si, 4.5-5.5 Cr, 0.90-1.10 Mo, V, bal Fe.
Alloy steel casting.
Similar to AISI A2 or A3.

HENRICOT HM51.
M-1188; 0.90-1.0 C, 0.75-1.25 Mn, 0.60 max Si, 2.0-3.0 Cr, 0.25-0.40 Mo, bal Fe.
HT: 450-500 Brin in heavy sections.

HENRICOT HMA.
M-1188; 0.25 max C, 0.50-0.80 Mn, 0.60 max Si, 0.50 max Ni, 0.35 max Cr, 0.45-0.65 Mo, 0.50 max Cu, 0.10 max W, bal Fe.
As cast: 48 kg/mm² TS; 25 kg/mm² YS; 22-24 El.
Creep resistant to 450°C; for oil refineries.
BS. 1398; GS 22 Mo 4.

HENRICOT HMA1.
M-1188; 0.13-0.18 C, 0.50-0.80 Mn, 0.50 max Si, 0.20 max Ni, 0.20 max Cr, 0.30-0.50 Mo, bal Fe.
Low carbon, low alloy steel casting.

HENRICOT HMA2.
M-1188; 0.20-0.25 C, 1.0-1.5 Mn, 0.60 max Si, 0.50 max Ni, 0.30 max Cr, 0.25 max Mo, 0.50 max Cu, bal Fe.
Low alloy casting.

HENRICOT HMA3.
M-1188; 0.16-0.20 C, 1.3-1.7 Mn, 0.30-0.60 Si, 0.25-0.35 Mo, bal Fe.
Manganese alloy steel casting.

HENRICOT HMA4.
M-1188; 0.32-0.37 C, 0.50-0.80 Mn, 0.60 max Si, 0.35 Mo, bal Fe.
Low alloy steel casting.

HENRICOT HMA5.
M-1188; 0.10 C, 1.5 Mn, 0.60 max Si, 0.20 Mo, 0.05 V, bal Fe.
Steel casting.

HENRICOT HMA6.
M-1188; 0.21-0.28 C, 1.15-1.35 Mn, 0.15-0.25 Mo, 0.003-0.006 B, 0.05-0.15 Al, 0.05-0.15 Ti, bal Fe.
Manganese-Boron steel casting.

HENRICOT HMA7.
M-1188; 0.18-0.25 C, 0.50-0.80 Mn, 0.30-0.60 Si, 0.30 max Cr, 0.35-0.4 Mo, bal Fe.
Casting for elevated structural parts.
Similar to W.-Nr. 1.5419.

HENRICOT HME.
M-1188; 0.20-0.30 C, 0.40-0.80 Mn, 0.60 max Si, 0.90-1.2 Cr, 0.15-0.25 Mo, bal Fe.
Alloy steel casting.
Similar to AISI 4130.

HENRICOT HMEV.
M-1188; 0.25-0.35 C, 1.30 max Mn, 0.60 max Si, 0.60 max Cr, 0.25-0.35 Mo, 0.03-0.07 V, bal Fe.
Alloy casting; hardenable to 425-515 Brin.

HENRICOT HMH1.
M-1188; 0.30-0.40 C, 0.50-0.80 Mn, 0.60 Si, 0.90-1.20 Cr, 0.20-0.30 Mo, bal Fe.
Alloy steel casting.
Similar to AISI 4135.

HENRICOT HMH2.
M-1188; 0.35-0.45 C, 0.50-0.80 Mn, 0.60 max Si, 0.90-1.20 Cr, 0.20-0.3 Mo, bal Fe.
Alloy steel casting.
Similar to AISI 4140.

HENRICOT HMH3.
M-1188; 0.48-0.53 C, 0.80-0.95 Mn, 0.60 max Si, 0.80-1.10 Cr, 0.15-0.2 Mo, bal Fe.
Alloy steel casting.
Similar to AISI 4150.

HENRICOT HMV1.
M-1188; 0.20 max C, 0.50-0.80 Mn, 0.50 max Si, 1.0-1.5 Cr, 0.45-0.65 Mo, 0.15-0.25 V, bal Fe.
Alloy steel casting; for oil refinery equipment.

HENRICOT HMV3.
M-1188; 0.15-0.20 C, 0.50-0.80 Mn, 0.60 max Si, 1.2-1.5 Cr, 0.90-1.10 Mo, 0.20-0.30 V, bal Fe.
Alloy steel casting; for oil refinery equipment.

HENRICOT HMX1.
M-1188; 0.20 max C, 0.50-0.80 Mn, 0.50 max Si, 0.50 max res Ni, 1.01-1.20 Cr, 0.45-0.70 Mo, 0.50 max V, bal Fe.
Alloy steel casting; for oil refinery equipment.

HENRICOT HMX3.
M-1188; 0.18 max C, 0.40-0.70 Mn, 0.50 max Si, 2.0-2.5 Cr, 0.90-1.20 Mo, bal Fe.
Alloy steel casting; for oil refinery equipment.

HENRICOT HMX4.
M-1188; 0.16-0.20 C, 0.60-0.90 Mn, 0.30-0.50 Si, 0.30 max Ni, 0.60-0.9 Cr, 0.30-0.40 Mo, bal Fe.
Alloy steel casting.

HENRICOT HN11.
M-1188; 0.13-0.15 C, 0.45-0.65 Mn, 0.60 max Si, 1.65-2.0 Ni, 0.20-0.30 Mo, bal Fe.
Alloy steel casting.
Similar to AISI 4615.

HENRICOT HN12.
M-1188; 0.25 max C, 0.50-0.80 Mn, 0.60 max Si, 2.0-3.0 Ni, bal Fe.
As cast: 49 kg/mm^2 TS; 35 kg/mm^2 YS; 20 El.

HENRICOT HN13.
M-1188; 0.15 max C, 0.50-0.80 Mn, 0.60 max Si, 3.0-4.0 Ni, bal Fe.
Alloy steel casting.
Similar to old SAE 2317 steel.

HENRICOT HN14.
M-1188; 0.12 max C, 0.50-0.80 Mn, 0.60 max Si, 4.0-5.0 Ni, bal Fe.
As cast; 52 kg/mm^2 TS; 35 kg/mm^2 YS; 20 El.
Similar to old SAE 2512.

HENRICOT HO10.
M-1188; 0.12 max C, 0.60 max Mn, 0.60 max Si, bal Fe.
Cast, normalized: 40-45 kg/mm^2 TS; 20 kg/mm^2 YS; 24 El; 45
GS. 38/38-3.

HENRICOT HO20.
M-1188; 0.15 max C, 0.90 max Mn, 0.60 max Si, bal Fe.
Cast, normalizeed: 40-45 kg/mm^2 TS; 25 kg/mm^2 YS; 24 El.

HENRICOT HO30.
M-1188; 0.18 max C, 0.85 max Mn, 0.60 max Si, 0.50 Ni, 0.40 Cr, 0.25 Mo, 0.50 Cu, bal Fe.
Cast, normalized: 45-50 kg/mm^2 TS; 28 kg/mm^2 YS; 22 El.
AAR Gr. A.

HENRICOT HO34.
M-1188; 0.18 max C, 0.85 max Mn, 0.60 max Si, 0.35-0.60 Cu, bal Fe.
Cast, normalized: 45-50 kg/mm^2 TS; 28 kg/mm^2 YS; 22 El.

HENRICOT HO40.
M-1188; 0.25 max C, 1.0 max Mn, 0.60 max Si, 0.50 Ni, 0.40 Cr, 0.25 Mo, 0.50 Cu, bal Fe.
Cast, normalized: 45-55 kg/mm^2 TS; 30 kg/mm^2 YS; 20 El.
ASTM A27-654 Class 70-40.

HENRICOT HO41.
M-1188; 0.28 max C, 0.85 max Mn, 0.60 max Si, 0.50 Ni, 0.40 Cr, 0.25 0.50 Cu, bal Fe.
Cast, normalized; 49 kg/mm^2 TS; 27 kg/mm^2 YS; 20 El.
ASTM A27-65 Class 70-35.

HENRICOT HO42.
M-1188; 0.25 C, 1.1-1.3 Mn, 0.60 max Si, 0.50 Ni, 0.40 Cr, 0.25 Mo, 0.50 Cu, bal Fe.
Cast, normalized; 49 kg/mm^2 TS; 27 kg/mm^2 YS; 20 El.
ASTM A27-65 Class 70-40.

HENRICOT HO43.
 M-1188; 0.20-0.28 C, 0.50-1.20 Mn, 0.60 Si, 0.50 Ni, 0.40 Cr, 0.25 Mo, 0.50 Cu, bal Fe.
 Cast, normalized: 50-60 kg/mm² TS; 30 kg/mm² YS; 15 El.

HENRICOT HO44.
 M-1188; 0.25 max C, 1.0 max Mn, 0.60 max Si, 0.35-0.60 Cu, bal Fe.
 Cast, normalized: 48-55 kg/mm² TS; 30 kg/mm² YS; 20 El.

HENRICOT HO45.
 M-1188; 0.13 C, 0.20-0.55 Mn, 0.25-0.75 Si, 0.07-0.09 P, 0.60 Ni, 0.30-1.25 Cr, 0.20-0.55 Cu, bal Fe.
 Cast, normalized: 45-55 kg/mm² TS; 30 kg/mm² YS; 18 El.

HENRICOT HO46.
 M-1188; 0.21-0.29 C, 0.50-0.80 Mn, 0.30-0.60 Si, bal Fe.
 Cast, normalized: 45.5 kg/mm² TS; 24.5 kg/mm² YS; 20 El.

HENRICOT HO50.
 M-1188; 0.30 max C, 1.50 max Mn, 0.30-0.60 Si, 0.50 Ni, 0.40 Cr, 0.25 Mo, 0.50 Cu, bal Fe.
 Cast, normalized; 55-60 kg/mm² TS; 35 kg/mm² YS; 15 El.
 Similar to ASTM A148-65 Class 80-50.

HENRICOT HO51.
 M-1188; 0.30 max C, 1.0 max Mn, 0.30-0.60 Si, 0.50 Ni, 0.40 Cr, 0.25 Mo, 0.50 Cu, bal Fe.
 Cast, normalized: 55-60 kg/mm² TS; 28 kg/mm² YS.
 Similar to Bs 592 Gr. C.

HENRICOT HO52.
 M-1188; 0.28 max C, 1.20 max Mn, 0.30-0.60 Si, 0.50 Ni, 0.40 Cr, 0.25 Mo, 0.50 Cu, bal Fe.
 Cast, normalized: 52-60 kg/mm² TS; 30 kg/mm² YS; 18 El.

HENRICOT HO53.
 M-1188; 0.25-0.35 C, 0.50-0.78 Mn, 0.25-0.50 Si, bal Fe.
 Carbon steel casting.

HENRICOT HO54.
 M-1188; 0.30 max C, 1.50 max Mn, 0.60 max Si, 0.35-0.60 Cu, bal Fe.
 Cast, normalized: 55-60 kg/mm² TS; 35 kg/mm² YS; 15 El.

HENRICOT HO60.
 M-1188; 0.25 max C, 2.0 max Mn, 0.60 max Si, bal Fe.
 Cast: 60-70 kg/mm² TS; 35-45 kg/mm² YS; 18 El.

HENRICOT HO61.
 M-1188; 0.45 max C, 1.20 max Mn, 0.60 max Si, bal Fe.
 Cast: 60-70 kg/mm² TS; 35-45 kg/mm² YS: 16 El.

HENRICOT HO62.
 M-1188; 0.20-0.28 C, 1.50 max Mn, 0.60 max Si, 0.20-0.50 Cr, bal Fe.
 Cast: 60-65 kg/mm² TS; 40 kg/mm² YS; 18 El.
 Similar to ASTM A148 Class 60-90.

HENRICOT HO63.
 M-1188; 0.30 max C, 1.60 max Mn, 0.60 max Si, 0.20-0.50 Cr, bal Fe.
 Cast: 68-78 kg/mm² TS; 50 kg/mm² YS; 18 El.

HENRICOT HO64.
 M-1188; 0.30 max C, 1.50 max Mn, 0.60 max Si, 0.20-0.50 Ni, 0.40-0.80 Cr, bal Fe.
 Cast, tempered: 270-320 Brin.

HENRICOT HO65.
 M-1188; 0.35 max C, 1.50 max Mn, 0.60 max Si, bal Fe.
 Cast: 65-75 kg/mm² TS; 35-45 kg/mm² YS; 15 El.

HENRICOT HO66.
 M-1188; 0.30 max C, 1.60 max Mn, 0.60 max Si, bal Fe.
 Cast: 63 kg/mm² TS; 42 kg/mm² YS; 22 El.

HENRICOT HO70.
 M-1188; 0.32-0.40 C, 1.25-1.65 Mn, 0.30-0.60 Si, bal Fe.
 Cast: 70-75 kg/mm² TS; 40-45 kg/mm² YS; 16 El.

HENRICOT HO71.
 M-1188; 0.40-0.50 C, 0.60-0.90 Mn, 0.30-0.60 Si, bal Fe.
 Cast carbon steel.

HENRICOT HO72.
 M-1188; 0.35-0.45 C, 1.0-1.25 Mn, 0.60 max Si, bal Fe.
 Cast carbon steel.

HENRICOT HO80.
 M-1188; 0.35-0.45 C, 1.75 max Mn, 0.60 max Si, bal Fe.
 Norm. & Temp: 80-85 kg/mm² TS; 48 kg/mm² YS.
 Carbon-manganese steel casting.

HENRICOT HO81.
 M-1188; 0.50 max C, 1.0 max Mn, 0.60 max S; bal Fe.
 Norm. & Temp: 75-85 Kg/mm² TS; 45 Kg/mm² YS.
 Carbon-Manganese steel casting.

HENRICOT HO82.
M-1188; 0.50 max C, 1.50 max Mn, 0.60 max Si, bal Fe.
Norm. & Temp: 75-85 Kg/mm² TS; 45 kg/mm² YS; 10 El.
Carbon-Manganese steel casting.

HENRICOT HO83.
M-1188; 0.35-0.45 C, 1.40 max Mn, 0.60 max Si, 0.10-0.20 Mo, bal Fe.
Norm. & Temp: 75-85 Kg/mm² TS; 45 kg/mm² YS.
Carbon-Manganese steel casting.

HENRICOT HO85.
M-1188; 0.60 max C, 0.85 max Mn, 0.60 max Si, bal Fe.
Norm & Temp: 75-85 kg/mm² TS; 45 kg/mm² YS; 10 El.
Carbon steel casting.

HENRICOT HO90.
M-1188; 0.70 max C, 0.85 max Mn, 0.60 max Si, bal Fe.
Norm. & Temp: 80-90 kg/mm² TS; 48 Kg/mm² YS; 8 El.
Carbon steel casting.

HENRICOT HOT.
M-1188; 0.30 min C, 0.70 max Mn, 0.60 max Si, bal Fe.
Norm. & Temp: 50-60 kg/mm² TS; 30 kg/mm² YS; 16 El.
Carbon steel casting.
SIS 1505.

HENRICOT HOT1.
M-1188; 0.35 min C 0.65 max Mn, 0.60 max Si, bal Fe.
Norm. & Temp: 60-70 kg/mm² TS; 35 kg/mm² YS; 15 El.
Carbon steel casting.

HENRICOT HOT2.
M-1188; 0.45 min C, 0.60 max Mn, 0.60 max Si, bal Fe.
Norm. & Temp: 65-75 kg/mm² TS; 38 kg/mm² YS; 14 El.
Carbon steel casting.

HENRICOT HOT3.
M-1188; 0.50 min C, 0.55 max Mn, 0.60 max Si, bal Fe.
Norm. & Temp: 75-85 kg/mm² TS; 45 kg/mm² YS.
Carbon steel casting.

HENRICOT HPL.
M-1188; 0.50-0.60 C, 1.0 max Mn, 0.60 max Si, 2.0-2.5 Ni, 1.4-2.0 Cr, 0.15-0.25 Mo, bal Fe.
Alloy cast steel; wear resistant.

HENRICOT HR2F4.
M-1188; 1.35-1.40 C, 12.0-13.0 Cr, 1.2-1.4 Mo, bal Fe.
Corrosion resistant steel casting.

HENRICOT HR2R.
M-1188; 0.25-0.35 C, 0.85 max Mn, 1.0 max Si, 1.0 Ni (res), 12.0-14.0 Cr, 0.50 max Mo, bal Fe.
Q&T. 650°C: 85-95 kg/mm² TS; 68 kg/mm² YS; 10 El.
ASTM A296 Type CA-40.

HENRICOT HR2R2.
M-1188; 0.40-0.50 C, 1.0 max Mn, 1.0 max Si, 1.0 Ni(res), 12.0-14.0 Cr, bal Fe.
Corrosion resistant steel casting.

HENRICOT HR.3M3.
M-1188; 0.95-1.20 C, 1.0 max Mn, 1.0 max Si, 16.0-18.0 Cr, 0.75 max Mo, bal Fe.
Corrosion resistant steel casting.
AISI 440C.

HENRICOT HR.3M5.
M-1188; 3.0-3.5 C, 15.0-18.0 Cr, 3.0-3.2 Mo, bal Fe.
Corrosion resistant alloy casting.

HENRICOT HR.4M.
M-1188; 0.10 max C, 2.0 max Mn, 1.0 max Si, 4.5-6.0 Ni, 25.0-27.0 Cr, 1.3-2.0 Mo, bal Fe.
Corrosion resistant alloy steel casting.
Similar to AISI 329; SIS 2324.

HENRICOT HR.5A.
M-1188; 2.5-3.50 C, 1.50 max Mn, 1.0 max Si, 25.0-29.0 Cr, bal Fe.
Alloy steel casting for high temperature operation.

HENRICOT HR.5M.
M-1188; 1.0 max Mn, 1.0 max Si, 27.0-31.0 Cr, 4.0-5.0 Mo, bal Fe.
Alloy steel casting for chemical equipment and high temperature operation.

HENRICOT HR5M.
M-1188; 2 C, 30 Cr, Mo, bal Fe.
For chemical plant equipment; cast iron; corrosion and wear resistant.

HENRICOT HR24.
M-1188; 0.15 max C, 1.0 max Mn, 1.5 max Si, 1.0 max Ni, 11.5-14.5 Cr, 0.50 max Mo, bal Fe.
Q&T. 720°C: 70-85 kg/mm² TS; 50 kg/mm² min YS; 15 min El.
ASTM A296 Type CA-15.

HENRICOT HR25.
M-1188; 0.08-0.12 C, 1.0 max Mn, 1.5 max Si, 0.80-1.2 Ni, 12.2-13.2 Cr, bal Fe.
Q+T. 720°C: 65-75 kg/mm² TS; 50 kg/mm² min YS; 15 min El.
Corrosion resistant steel casting.

HENRICOT HR26.
 M-1188; 0.18 max C, 1.0 max Mn, 0.60 max Si, 12.0-14.0 Cr, bal Fe.
 Q+T. 720°C: 70-80 kg/mm² TS; 45 kg/mm² min YS; 15 min El; 50 min RA.
 Corrosion resistant steel casting.

HENRICOT HR27.
 M-1188; 0.12 C, 1.0 Ni, 13.0 Cr, 1.0 Mo, bal Fe.
 Corrosion resistant steel casting.

HENRICOT HR28.
 M-1188; 0.07 max C, 0.75 max Mn, 0.60 max Si, 3.25-4.25 Ni, 11.5-14.0 Cr, 1.0 max Mo, bal Fe.
 Q+T. 630°C: 80-95 kg/mm² TS; 60 kg/mm² min YS; 14 min El.
 Corrosion resistant steel casting.

HENRICOT HR30.
 M-1188; 0.15 max C, 1.0 max Mn, 1.0 max Si, 1.0 max Ni(res), 17.0-19.0 Cr, bal Fe.
 Corrosion resistant steel casting.

HENRICOT HR.33.
 M-1188; 0.15 max C, 1.0 max Mn, 1.0 max Si, 2.0-3.0 Ni, 17.0-19.0 Cr, bal Fe.
 Similar to ASTM A296 CB-30.

HENRICOT HR.34.
 M-1188; 0.08 max C, 1.0 max Mn, 1.0 max Si, 3.0-5.0 Ni, 15.5-17.5 Cr, 3.0-5.0 Cu, 0.10-0.50 Nb, bal Fe.
 Precipitation hardening stainless steel casting; 17-4 PH Type.

HENRICOT HR.43.
 M-1188; 0.50 max C, 1.0 max Mn, 3.0 max Si, 2.0-4.0 Ni, 26.0-30.0 Cr, 0.50 max Mo(res), bal Fe.
 Heat resistant alloy casting.
 Similar to ASTM A297 HC.

HENRICOT HR.44.
 M-1188; 0.30-0.50 C, 1.0 max Mn, 1.5 max Si, 3.5-4.5 Ni, 26.0-28.0 Cr, bal Fe.
 Heat resistant alloy casting.
 Similar to ASTM A297 HC; W.-Nr. 1.4340.

HENRICOT HR.46.
 M-1188; 0.50 C, 1.0 max Mn, 3.0 max Si, 5.0-7.0 Ni, 26.0-30.0 Cr, 0.50 max Mo(res), bal Fe.
 Heat resistant alloy casting.
 Similar to ASTM A297 HD.

HENRICOT HR.51.
 M-1188; 2.50 max C, 0.75 max Mn, 0.75 max Si, 27.0-30.0 Cr, bal Fe.
 Alloy steel casting for high temperature application, as furnace parts.

HENRICOT HR.56.
 M-1188; 1.0-1.5 C, 1.0 max Mn, 1.0-2.5 Si, 27.0-30.0 Cr, bal Fe.
 Alloy steel casting for high temperature operation. W.-Nr. 1.4777.

HENRICOT HR60.
 M-1188; 0.20 C, 1.0 Mn, 1.0 Si, 27.0-30.0 Cr, bal Fe.
 Corrosion resistant alloy steel casting.

HENRICOT HR80.
 M-1188; 0.20 max C, 0.40-0.70 Mn, 0.75 Si, 0.50 Ni, 4.0-6.5 Cr, 0.45-0.65 Mo, 0.50 Cu, 1.0 W, bal Fe.
 Q&T. 720°C; 65-80 kg/mm² TS; 45 kg/mm² YS; 15 El.
 Alloy steel casting for pressure vessels for high temperature service.
 ASTM A217 C5.

HENRICOT HS0.
 M-1188; 0.95-1.15 C, 11.5-14.0 Mn, 0.65 max Si, bal Fe.
 Austenitic manganese steel castings; ASTM A128.

HENRICOT HS1.
 M-1188; 0.95-1.15 C, 11.5-14.0 Mn, 0.65 max Si, 1.0-1.25 Cr, bal Fe.
 Ann: 160 Brin; Cold worked: 375 Brin.
 Austenitic manganese steel castings. ASTM A128.

HENRICOT HS2.
 M-1188; 1.10-1.30 C, 11.5-14.0 Mn, 1.0 Si, 1.1-1.3 Cr, bal Fe.
 Ann: 190 Brin; Cold worked 425 Brin.
 Austenitic manganese steel castings. ASTM A128.

HENRICOT HS3.
 M-1188; 1.2-1.4 C, 12.0-14.0 Mn, 1.0 max Si, 1.5-2.0 Cr, bal Fe.
 Ann: 200 Brin; Cold worked 450 Brin.
 Austenitic manganese steel castings. ASTM A128.

HENRICOT HS10.
 M-1188; 1.0-1.15 C, 11.5-14.0 Mn, 0.65 Si, 0.070 P, bal Fe.
 Ann: 160 Brin; Cold worked 375 Brin.
 Austenitic manganese steel castings. ASTM A128.

HENRICOT HS20.
 M-1188; 1.1-1.3 C, 11.5-14.0 Mn, 1.0 max Si, 0.070 max P, bal Fe.
 Ann: 180 Brin; Cold worked: 400 Brin.
 Austenitic manganese steel castings. ASTM A128.

HENRICOT HS25.
 M-1188; 1.30-1.50 C, 18.0-20.0 Mn, 1.0 max Si, 1.7-2.3 Cr, bal Fe.
 Stainless steel casting.

HENRICOT HS26.
 M-1188; 1.30-1.50 C, 18.0-20.0 Mn, 1.0 max Si, 2.3-2.9 Cr, bal Fe.
 Stainless steel casting.

HENRICOT HS30.
M-1188; 1.25-1.40 C, 12.0-14.0 Mn, 1.0 Si, 0.070 max P, bal Fe.
Ann: 200 Brin; Cold worked: 450 Brin.
Austenitic manganese steel castings. ASTM A128.

HENRICOT HT10.
M-1188; 0.20-0.40 C, 2.0 max Mn, 2.0 max Si, 15.0-17.0 Ni, 18.0-20.0 Cr, bal Fe.
Cast: 48-65 kg/mm^2 TS; 25 kg/mm^2 YS; 15 El.
Stainless steel casting for high temperature operation.
Similar to W.-Nr. 1.4832.

HENRICOT HT12.
M-1188; 0.20-0.40 C, 2.0 max Mn, 2.0 max Si, 8.0-12.0 Ni, 18.0-23.0 Cr, 0.50 max Mo, bal Fe.
Cast: 49 kg/mm^2 min TS; 24 kg/mm^2 YS; 20 El.
Austenitic stainless steel casting for high temperature operation.
Similar to: W.-Nr.1.4826; ASTM A297-HF.

HENRICOT HT14.
M-1188; 0.20-0.40 C, 1.0 max Mn, 1.0 max Si, 7.5-9.5 Ni, 17.0-19.0 Cr, bal Fe.
Cast: 50-65 kg/mm^2 TS; 25 kg/mm^2 YS; 20 El.
Stainless steel casting; B.S. 1648.D.

HENRICOT HT20.
M-1188; 0.20-0.40 C, 2.0 max Mn, 2.0 max Si, 18.0-22.0 Ni, 24.0-28.0 Cr, 0.50 max Mo(res), bal Fe.
Cast: 48 kg/mm^2 min TS; 26 kg/mm^2 YS; 10 El.
Heat resistant cast alloy.
W.-Nr. 1.4848; ASTM A297 HK.

HENRICOT HT21.
M-1188; 0.30 max C, 2.0 max Mn, 2.5 max Si, 24.0-28.0 Ni, 24.0-28.0 Cr, bal Fe.
Cast: 43 kg/mm^2 min TS; 28 kg/mm^2 min YS; 12 El.
Heat resistant alloy casting.

HENRICOT HT23.
M-1188; 0.35-0.45 C, 2.0 max Mn, 1.5 max Si, 19.0-22.0 Ni, 24.0-27.0 Cr, 0.50 max Mo(res), bal Fe.
Cast: 45.5 kg/mm^2 min TS; 25 kg/mm^2 min YS; 10 min El.
Heat resistant cast alloy.
Similar to: W.-Nr. 1.4848; B.S. 1648-F; ASTM A297-HK.

HENRICOT HT24.
M-1188; 0.35-0.45 C, 2.0 max Mn, 1.5 max Si, 19.0-22.0 Ni, 24.0-27.0 Cr, 0.50 max Mo(res), bal Fe.
Cast: 45 kg/mm^2 min TS; 25 kg/mm^2 min YS; 8 min El.
Heat resistant cast alloy.
Similar to: W.-Nr. 1.4848; B.S. 1648-F; ASTM A297-HK.

HENRICOT HT25.
M-1188; 0.25-0.35 C, 2.0 max Mn, 1.5 max Si, 23.0-25.0 Ni, 23.0-25.0 Cr, 1.0-2.0 Cu, bal Fe.
Cast: 45-65 kg/mm^2 TS; 25 kg/mm^2 YS; 10 El.
Heat resistant cast alloy.

HENRICOT HT30.
M-1188; 0.35-0.45 C, 2.0 max Mn, 2.5 max Si, 33.0-37.0 Ni, 13.0-17.0 Cr, 0.50 max Mo(res), bal Fe.
Cast: 45-65 kg/mm^2 TS; 25 kg/mm^2 YS; 4 El.
Heat resistant alloy; for furnace parts.
Similar to: W.-Nr. 1.4865; ASTM A297-HT.

HENRICOT HT31.
M-1188; 0.20 max C, 2.0 max Mn, 0.60 max Si, 33.0-37.0 Ni, 19.0-23.0 Cr, 0.50 max Mo(res), bal Fe.
Cast, Ann: 50 kg/mm^2 TS; 21 kg/mm^2 YS; 25 El.
Heat and oxidation resistant alloy.
Similar to: W.-Nr. 1.4861; Incoloy 800.

HENRICOT HT32.
M-1188; 0.35-0.45 C, 2.0 max Mn, 2.5 max Si, 37.0-41.0 Ni, 17.0-21.0 Cr, 0.50 max Mo(res), bal Fe.
Cast, Ann: 45-65 kg/mm^2 TS; 25 kg/mm^2 YS; 4 El.
Heat and oxidation resistant alloy; similar to: W.-Nr. 1.4865; ASTM A297-HU.

HENRICOT HT33.
M-1188; 0.45-0.55 C, 1.0 max Mn, 2.5 max Si, 45.0-55.0 Ni, 26.0-30.0 Cr, 0.50 max Mo, 4.0-6.0 W, bal Fe.
42-60 kg/mm^2 TS.
Heat and oxidation resistant casting.
Similar to: W.-Nr. 2.4879.

HENRICOT HT35.
M-1188; 0.15 max C, 2.0 max Mn, 1.5 max Si, 31.0-33.0 Ni, 19.0-22.0 Cr, 0.5-1.5 Co, bal Fe.
Heat and oxidation resistant casting.

HENRICOT HT36.
M-1188; 0.12 max C, 1.0 max Mn, 1.0 max Si, 26.0-30.0 Cr, 48.0-52.0 Co, bal Fe.
50/65 kg/mm^2 TS; 28 kg/mm^2 YS; 10 El.
Heat and oxidation resistant casting.

HENRICOT HT37.
M-1188; 0.25-0.40 C, 0.50-1.0 Mn, 0.50-1.0 Si, 27.0-29.0 Cr, 48.0-52.0 Co, 1.8-2.5 Nb, bal Fe.
Heat and oxidation resistant casting.

HENRICOT HT42.
M-1188; 0.20-0.50 C, 1.50 max Mn, 2.0 max Si, 11.0-14.0 Ni, 24.0-28.0 Cr, 0.50 max Mo, bal Fe.
Cast: 52 kg/mm^2 TS; 24 kg/mm^2 YS; 25 El.
Heat and oxidation resistant castings. W.-Nr. 1.4837; ASTM A297-HH.

HENRICOT HT43.
M-1188; 0.20-0.45 C, 2.0 max Mn, 1.75 max Si, 11.0-14.0 Ni, 23.0-28.0 Cr, 0.50 max Mo, 0.20 N, bal Fe.
Cast: 56 kg/mm^2 TS; 4 El.
Heat and oxidation resistant alloy casting.
Similar to ASTM A447 Type 2.

HENRICOT HT46.
M-1188; 0.30 max C, 1.50 max Mn, 2.0 max Si, 8.0-11.0 Ni, 26.0-30.0 Cr, 0.50 max Mo, bal Fe.
Cast, tempered: 50 kg/mm^2 TS; 28 kg/mm^2 YS; 10 El.
Heat and oxidation resistant alloy casting.
Similar to ASTM A296-CE30.

HENRICOT HT50.
M-1188; 0.05 max C, 0.60 max Mn, 0.45 max Si, 48.0-52.0 Cr, 2.0 max Fe, bal Ni.
Cast: 55-70 kg/mm^2 TS; 35 kg/mm^2 YS; 15 El.
Heat and oxidation resistant alloy casting.

HENRICOT HT51.
M-1188; 0.05 max C, 0.60 max Mn, 0.45 max Si, 48.0-52.0 Cr, 2.0 max Fe, 2.0 max Nb, bal Ni.
Cast: 50-75 kg/mm^2 TS; 40 kg/mm^2 YS; 25 El.
Heat and oxidation resistant alloy casting.

HENRICOT HT60.
M-1188; 0.25-0.35 C, 2.0 max Mn, 2.5 max Si, 58.0-62.0 Ni, 16.0-20.0 Cr, 0.50 max Mo, bal Fe.
Cast: 45 kg/mm^2 TS.
Heat and oxidation resistant alloy casting.
Similar to ASTM A297-HW.

HENRICOT HT70.
M-1188; 0.10 max C, 1.0 max Mn, 1.5 max Si, 14.0-17.0 Cr, 6.0-10.0 Fe 1.0-2.0 Nb, 0.10 max Co, 0.50 max Cu, bal Ni.
Cast: 48 kg/mm^2 TS; 18 kg/mm^2 YS; 35 El.
Heat and oxidation resistant alloy casting.

HENRICOT HT75.
M-1188; 0.20 max C, 2.0 max Mn, 2.0 max Si, 75.0-79.0 Ni, 12.5-14.5 Cr, bal Fe.
Heat and oxidation resistant casting.
Similar to W.-Nr. 2.4640.

HENRICOT HV9.
M-1188; 0.04 max C, 2.0 max Mn, 1.0 max Si, 24.0-26.0 Ni, 20.0-23.0 C 4.0 Mo, Nb, bal Fe.
Heat and corrosion resistant wrought steel.

HENRICOT HV9A.
M-1188; 0.04 max C, 2.0 max Mn, 1.0 max Si, 24.0-26.0 Ni, 20.0-23.0 C 4.0-5.0 Mo, Cu, Nb, bal Fe.
Heat and corrosion resistant wrought steel.

HENRICOT HV12.
M-1188; 0.08 max C, 1.5 max Mn, 0.75 max Si, 8.0-11.0 Ni, 18.0-20.0 Cr, bal Fe.
Cast: 48 kg/mm^2 TS; 20 kg/mm^2 YS; 35 El.
Corrosion resistant steel casting.
W.-Nr. 1.4308; ASTM A296-CF8.

HENRICOT HV13.
M-1188; 0.03 max C, 1.5 max Mn, 1.0 max Si, 8.0-12.0 Ni, 18.0-20.0 Cr, bal Fe.
Cast: 49 kg/mm^2 TS; 21 kg/mm^2 YS; 35 El.
Corrosion resistant steel casting.
Similar to W.-Nr. 1.4306; ASTM A296-CF3.

HENRICOT HV30.
M-1188; 0.08 max C, 1.5 max Mn, 1.5 max Si, 10.5-13.5 Ni, 16.5-18.5 C 2.0-3.0 Mo, bal Fe.
Cast, Ann: 1100°C: 49 kg/mm^2 TS; 21 kg/mm^2 YS; 30 El.
Corrosion resistant cast steel.
Similar to W.-Nr. 1.4580.

HENRICOT HV30B.
M-1188; 0.06 max C, 2.0 max Mn, 1.0 max Si, 11.0-14.0 Ni, 16.0-19.0 Cr, 1.5-3.0 Mo, bal Fe.
Corrosion resistant steel casting.

HENRICOT HV31.
M-1188; 0.08 max C, 1.5 max Mn, 2.0 max Si, 9.0-12.0 Ni, 18.0-21.0 Cr, 2.0-3.0 Mo, bal Fe.
Cast, Ann: 1100°C: 49 kg/mm^2 TS; 21 kg/mm^2 YS; 30 El.
Corrosion resistant steel casting.
Similar to: W.-Nr. 1.4408; ASTM A296-CF8M; AISI 316.

HENRICOT HV32.
M-1188; 0.06 max C, 2.0 max Mn, 1.0 max Si, 10.0-13.5 Ni, 16.5-19.0 Cr, 2.5-3.0 Mo, bal Fe.
Corrosion resistant steel casting.
Similar to W.-Nr. 1.4437.

HENRICOT HV36.
M-1188; 0.06 max C, 2.0 max Mn, 1.0 max Si, 13.0-14.0 Ni, 18.0-20.0 Cr, 3.0-4.0 Mo, bal Fe.
Cast, Ann: 1100°C: 47 kg/mm^2 TS; 20 kg/mm^2 YS; 30 El.
Corrosion resistant steel casting.
Similar to: W.-Nr. 1.4448; B.S. 1632-A; AISI 317.

HENRICOT HV37.
M-1188; 0.03 max C, 2.0 max Mn, 1.0 max Si, 10.0-14.0 Ni, 16.0-18.0 Cr, 2.0-3.0 Mo, bal Fe.
Corrosion resistant steel casting; Similar to: ASTM A296-CF3M; AISI 316L.

HENRICOT HV60.
M-1188; 0.08 max C, 1.5 max Mn, 0.75 max Si, 9.0-12.0 Ni, 18.0-21.0 Cr, Nb = 8 x C, bal Fe.
Cast, Ann: 49 kg/mm^2 TS; 21 kg/mm^2 YS; 35 El.
Corrosion resistant steel casting.
Similar to: W.-Nr. 1.4552; ASTM A296-CF8C.

HENRICOT HV70.
M-1188; 0.07 max C, 2.0 max Mn, 1.0 max Si, 19.0-21.0 Ni, 15.5-17.5 Cr, 4.0 Mo, Nb, bal Fe.
Cast; Ann: 48 kg/mm^2 TS; 23 kg/mm^2 YS; 30 El.
Corrosion resistant steel casting.

HENRICOT HV90.
M-1188; 0.06 max C, 2.0 max Mn, 1.5 max Si, 27.0-29.0 Ni, 20.0-23.0 Cr, 4.0-5.0 Mo, Nb, 2.0-4.0 Cu, bal Fe.
Corrosion resistant steel casting.

HENRICOT HV91.
M-1188; 0.07 max C, 1.5 max Mn, 1.3 max Si, 27.0-32.0 Ni, 19.0-22.0 Cr, 2.0-3.0 Mo, 3.0-4.0 Cu, bal Fe.
Cast, Ann: 45 kg/mm^2 TS; 21 kg/mm^2 YS; 35 El.
Corrosion resistant steel casting.
Similar to ASTM A296-CN7M.

HENRICOT HV98.
M-1188; 0.04 max C, 2.0 max Mn, 1.0 max Si, 24.0-26.0 Ni, 20.0-23.0 Cr, 4.0 Mo, 0.30 Nb, bal Fe.
Corrosion resistant steel casting.

HENRICOT HV181.
M-1188; 0.06 max C, 2.0 max Mn, 1.5 max Si, 8.0-11.0 Ni, 18.0-21.0 Cr, 0.20 Co, 1.0 Cu, 0.15 Nb, 0.020 N, bal Fe.
Cast: 50 kg/mm^2 TS; 21 kg/mm^2 YS; 30 El.
Corrosion resistant steel casting.
Similar to ASTM A296-CF8. For nuclear energy.

HENRICOT HV182.
M-1188; 0.08, 1.50 max Mn, 2.0 max Si, 8.0-11.0 Ni, 18.0-21.0 Cr, 0.20 Co, bal, Fe.
Cast: 54 kg/mm^2 TS; 24.5 kg/mm^2 YS; 35 El.
Corrosion resistant steel casting.
Similar to ASTM A296-CF8. For nuclear energy.

HENRICOT HV382.
M-1188; 0.06 max C, 1.5 max Mn, 1.5 max Si, 9.0-12.0 Ni, 18.0-21.0 Cr, 2.3-2.8 Mo, 0.20 Co, 1.0 C, 0.15 Nb + Ta, bal Fe.
Corrosion resistant steel casting.

HENRICOT HVD1.
M-1188; 0.08 max C, 2.0 max Mn, 1.0 max Si, 7.0-9.0 Ni, 24.0-26.0 Cr, 2.0-3.0 Mo, 1.0-2.0 Cu, bal Fe.
Heat and corrosion resistant wrought steel.

HENRICOT HVD10.
M-1188; 0.08 max C, 2.0 max Mn, 1.0 max Si, 7.0-9.0 Ni, 24.0-26.0 Cr, 2.0-3.0 Mo, 1.0-2.0 Cu, bal Fe.
Cast: 70-85 kg/mm^2 TS; 50 kg/mm^2 YS; 20 El.
Heat and corrosion resistant steel casting.

HENRICOT HVD13.
M-1188; 0.10 max C, 2.0 max Mn, 1.0 max Si, 7.0-9.0 Ni, 19.5-22.0 Cr, 2.0-3.0 Mo, 1.0-2.0 Cu, bal Fe.
Cast: 65 kg/mm^2 TS; 35 kg/mm^2 YS; 30 El.
Corrosion resistant steel casting.

HENRICOT HVD20.
M-1188; 0.08 max C, 2.0 max Mn, 1.5 max Si, 7.0-9.0 Ni, 25.0-27.0 Cr, 2.0-3.0 Mo, 1.0-2.0 Cu, bal Fe.
Heat and corrosion resistant steel casting.

HENRICOT MONEL.
M-1188; 0.10-0.30 C, 0.50-1.50 Mn, 1.25-2.0 Si, 28.0-34.0 Cu, 3.0 Fe, bal Ni.
Cast: 150 Brin.

HEPOSIL.
M-Germany; 21-22 Si, 1.5 Cu, 1.5 Ni, 0.7 Mn, 1.2 Co, 0.5 Mg, bal Al.
For pistons in engines, liners.
Low expansivity, high corrosion resistance.

HEPPENSTALL 2 C 30.
M-64; 0.35 C, 0.7 Mn, 0.8 Cr, 0.45 Mo, 1.75 Ni, bal Fe.
Hardened: 135,000 TS; 115,000 YS; 20 El; 50 RA; 293 Brin.
For piston rods, crankshafts, rams; upsetter.

HEPPENSTALL 2V72.
M-64; 0.75 C, 0.4 Mn, 0.1 V, 0.6 Si, bal Fe.
For coining and crimping dies, striking dies, shear knives, trimmers; water hardening tool steel; AISI W2.

HEPPENSTALL 3C40.
M-64; 0.43 C, 2.6 Ni, 1.35 Cr, 0.55 Mo, 0.18 V, bal Fe.
Annealed: 250 Brin.
Heat treated: 470 Brin.
For die blocks; oil hardened, shock resistant.

HEPPENSTALL 5H50.
M-64; 0.55 C, 0.90 Cr, 0.43 Mo, 0.06 V, bal Fe.
Annealed: 210 Brin.
Heat treated: 450 Brin.
For die blocks, inserts; oil hardened, tough.

HEPPENSTALL 6H55.
M-64; 0.55 C, 1.05 Cr, 0.47 Mo, 0.13 V, bal Fe.
Annealed: 210 Brin.
Heat treated: 500 Brin.
For die blocks, inserts; oil hardened, tough.

HEPPENSTALL 9C60.
M-64; 0.60 C, 1.25 Ni, 0.65 Cr, 0.45 Mo, bal Fe.
Annealed: 250 Brin.
Heat treated: 540 Brin.
For miscellaneous forgings, shear knives; oil hardened, shock resistant.

HEPPENSTALL B76.
M-64; 0.97 C, 1.15 Mn, 0.53 Cr, 0.20 V, 0.50 W, bal Fe.
Annealed: 250 Brin.
Heat treated: 620 Brin.
For stamping, forming and blanking dies; shear knives; oil hardened, tough AISI W1.

HEPPENSTALL C50.
M-64; 0.5 C, 0.55 Mn, 1 Cr, 1.5 Ni, bal Fe.
Oil hardened: 115,000-125,000 TS; 85,000-100,000 YS; 20-18 El; 50-45 RA; 248-293 Brin. For large mill pinions, couplings; high tensile forgings.

HEPPENSTALL C55.
M-64; 0.55 C, 0.90 Cr, 0.30 Mo, 1.5 Ni, bal Fe.
For hot work dies, punches; oil hardened, tough.

HEPPENSTALL C93.
M-64; 0.57 C, 1.2 Ni, 2.65 Cr, 0.45 Mo, bal Fe.
Annealed: 250 Brin.
Heat treated: 580 Brin.
For cold shear knives; oil hardened, shock resistant.

HEPPENSTALL CR T77.
M-64; 0.30-0.35 C, 0.20-0.25 V, 0.2-0.4 Mn, 3.75-4.15 Cr, 11.0-12.0 W bal Fe.
Heat treated: Rock C, 40-53; 220,000-240,000 TS; 180,000-215,000 YS; 10-12 El; 37-42 RA.
For die casting dies, forging and press dies, extrusion and gripper dies, piercers, punches, extrusion mandrels.
Hot work steel, Type H21. High resistance to heat. Wear and abrasion resistant.

HEPPENSTALL GR 14A60.
M-64; 0.55-0.65 C, 0.7-0.9 Mn, 0.15-0.35 Si, 0.2-0.3 Cr, bal Fe.
Heat Treated: 285-461 Brin.
For trimmers, die and sow blocks, axles, crankpins, rollers, shafting. Shallow hardening, wear resistant. Type W4 water hardening tool steel.

HEPPENSTALL GR 2V90.
M-64; 0.90-0.95 C, 0.3-0.5 Mn, 0.5-0.7 Si, 0.08-0.10 V, bal Fe.
Water hardened: 216,000 TS; 152,000 YS; 11 El; 600 Brin.
For coining and crimping dies, embossing dies, and rolls, heading and forming dies, swaging and trimming dies.
Type W2 water hardening tool steel, wear and abrasion resistant.

HEPPENSTALL GR 5M21.
M-64; 0.18-0.23 C, 3.0-3.25 Ni, 3.25-3.50 Mo, 0.15 max C, 0.7 Mn, 0.3 Si, bal Fe.
For press and upsetter dies, piercers, punches, die casting dies; supplied in two grades, hardened and annealed.

HEPPENSTALL GR 6E14GV.
M-64; 0.15 C, 3.5 Ni, 1.6 Cr, 0.05 V, bal Fe.
Ht. Tr.: 320 Brin; Annealed: 230 Brin.
For piston rods, cams, camshafts.
Case hardening and shock resisting.

HEPPENSTALL GR 9C68.
M-64; 0.65-0.70 C, 0.4-0.5 Mo, 1.0-1.5 Ni, 0.5-0.8 Cr, 0.5-0.8 Mn, 0.2-0.35 Si, bal Fe.
Oil hardened: 208,000 TS; 180,000 PL; C 45 Rock.
For spindles, punches, brake dies, collets, plastic mold dies, idler rolls, rocker plates, stamping dies, shear blades.
Type L6 tool steel. Deep oil hardening.
Tough and wear resistant.

HEPPENSTALL GR 9R40.
M-64; 0.40 C, 0.95 Cr, 0.45 Mo, bal Fe.
Annealed: 220 Brin; Ht. Tr.: 370 Brin.
For piston rods, shafts, counter-shafts, gears.
Oil hardening, shock resistant.

HEPPENSTALL GR A110.
M-64; 1.05-1.15 C, 0.25-0.35 Mn, 0.25-0.35 Si, bal Fe.
Annealed: 100,000 TS; 54,000 YS; 21 El; 197 Brin.
Water Hardened: 216,000 TS; 152,000 YS; 11 El; 600 Brin.
For cold header and striking dies, arbors, coining and broaching tools, mandrels, pipe cutters, tube drawing dies.
Water hardening Type W1 tool steel, wear and abrasion resistant.

HEPPENSTALL GRADE C.
M-64; 0.55 C, Ni, Cr, Mo, C, bal Fe.
Oil treated: 305,000-206,000 TS; 288,000-195,000 YS; 6-18 El; 17-52 RA; 555 Brin.
For die blocks, inserts; drop forged.

HEPPENSTALL GR C57.
M-64; 0.52-0.57 C, 2.0-2.2 Ni, 0.53-0.68 Mn, 0.80-0.95 Cr, 0.5-0.7 Si 0.70-0.75 Mo, bal Fe.
For hammer dies, press and upsetter dies, gripper and header dies, piercing mandrels, shear blades.
Hot work steel. High compressive strength.

HEPPENSTALL GR C58.
M-64; 0.42-0.47 C, 4.1-4.5 Ni, 1.25-1.65 Cr, 0.7-0.8 Mo, 0.12-0.16 V, bal Fe.
Heat treated: 218,000 TS; 190,000 YS; 12 El; 33 RA; C 45 Rock.
At 1000°F: 144,000 TS; 133,000 YS; 9 El; 39 RA.
For press and hammer dies, insert and upsetter dies, deep drawing dies.
Resists softening at elevated temperatures.
High compressive strength.

HEPPENSTALL GR H44.
M-64; 0.95-1.05 C, 0.2-0.4 Si, 0.5-0.7 Mn, 0.95-1.05 Mo, 4.9-5.3 Cr, 0.2-0.3 V, bal Fe.
Annealed: 104,000 TS; 51,000 YS; 26 El; C 18 Rock.
Air Hardened: 253,000 TS; 200,000 YS; 3 El; C 53 Rock.
For cold work trimming and blanking dies, forming and drawing dies, gauges, rolls, punches, thread rolling dies.
Cold work Type A2 tool steel. Deep air hardening, tough, wear resistant.

HEPPENSTALL GR H230.
M-64; 0.32 C, 3.3 Cr, 2.4 Mo, 0.37 V, bal Fe.
Air Hardened: 180,000-296,000 TS; 160,000-247,000 YS; 10-14 El; 16-4 RA; C 40-59 Rock.
For hot extrusion dies, punches, shear blades, gauges.
Hot work tool and die steel, Type H10, air hardening.

HEPPENSTALL GR H720.
M-64; 0.37-0.42 C, 5.0-5.5 Cr, 0.23-0.38 Mn, 1.0-1.3 Mo, 0.85-1.10 Si 0.9-1.0 V, bal Fe.
At 80°F: 217,000 TS; 184,000 YS; 13 El; 40 RA.
At 1100°F: 110,000 TS; 88,000 YS; 21 El; 69 RA.
For extrusion rams and liners, mandrels, punches, piercers, upsetting and forging dies, die casting dies. Type H13 hot work steel.

HEPPENSTALL GR H722.
M-64; 0.37-0.42 C, 5.0-5.5 Cr, 0.23-0.38 Mn, 1.0-1.2 Mo, 0.85-1.1 Si, 0.45-0.55 V, bal Fe.
At 80°F: 217,000 TS; 184,000 YS; 13 El; 40 RA.
At 1100°F: 110,000 TS; 88,000 YS; 21 El; 69 RA.
For extrusion rams and liners, mandrels, punches, piercers, upsetting dies, die casting dies.
Air hardening and non-deforming. Type H11. High resistance to wash and erosion and heat checking.

HEPPENSTALL GR T72.
M-64; 0.45-0.50 C, 1.2-1.5 Cr, 1.8-2.05 W, 0.20-0.25 V, bal Fe.
For chiesels, punches, compression and trimming dies, hot swaging and heading dies, insert dies.
Hot and cold work tool steel.
Oil hardened. Fatigue resistant.

HEPPENSTALL GR T74.
M-64; 0.47-0.53 C, 1.3-1.5 Cr, 2.25-2.50 W, 0.18-0.23 V, 0.40 max Mo, bal Fe.
For chisels, punches, compression and trimming dies, hot swaging and heading dies, insert dies.
Hot and cold work tool steel. Oil hardened. Fatigue resistant.

HEPPENSTALL GR T716.
M-64; 0.44-0.50 C, 0.70-0.90 Cr, 0.4-0.6 Mn, 0.20-0.35 Si, bal Fe.
Heat treated: 415-601 Brin.
For punches, chisels, pneumatic tools, rivet busters, hubbing dies, hot heading and swaging dies, hot forming dies.
Type S3 tool steel for cold and hot working.

HEPPENSTALL GR T719.
M-64; 0.62-0.67 C, 0.90-1.10 Si, 1.4-1.6 Mo, 0.2-0.4 Mn, 4.5-5.0 Cr, 1.0-1.2 W, bal Fe.
For cold work plastic dies and forming dies.
For hot work gripper and compression dies, trimming dies, aluminum extrusion dies. Air or oil hardened. Tough and wear resistant.

HEPPENSTALL GR T721.
M-64; 0.34-0.40 C, 0.25-0.45 Mn, 0.85-1.15 Si, 4.75-5.25 Cr, 1.25-1.65 Mo, 1.05-1.45 W, 0.25-0.45 V, bal Fe.
At 80°F: 217,000 TS; 184,000 YS; 13 El; 40 RA.
At 1100°F: 110,000 TS; 88,000 YS; 21 El; 69 RA.
For extrusion rams and liners, mandrels, punches, piercers, dummy blocks, shear blades, aluminum die casting dies.
Type H12 hot work steel. Resists heat checking.

HEPPENSTALL H41.
M-64; 0.95 C, 3.75 Cr, 0.23 Mo, 0.16 V, bal Fe.
Annealed: 220 Brin.
Heat treated: 510 Brin.
For hot and cold trimmers; oil hardened.

HEPPENSTALL H340.
M-64; 0.41 C, 3.3 Cr, 2.25 Mo, 0.38 V, bal Fe.
Annealed: 250 Brin.
For extrusion parts; oil hardened, tough.

HEPPENSTALL R43.
M-64; 1.55 C, 11.5 Cr, 0.55 Mo, 0.55 W, bal Fe.
Annealed: 250 Brin.
Heat treated: 610 Brin.
For shear knives, cold work dies; air hardened, non-deforming.

HEPPENSTALL R45.
M-64; 0.83 C, 11.5 Cr, 0.43 Mo, bal Fe.
Annealed: 250 Brin.
Heat treated: 560 Brin.
For shear knives, cold work dies; air hardened, non-deforming.

HEPPENSTALL R97.
M-64; 0.55 C, 1.05 Mn, 2.1 Si, 0.3 Cr, 0.38 Mo, bal Fe.
Annealed: 250 Brin.
Heat treated: 570 Brin.
For shear knives, cold work dies; oil hardened, shock resistant. AISI S5.

HEPPENSTALL R718.
M-64; 0.33 C, 4.75 Cr, 1.87 Mo, bal Fe.
Annealed: 250 Brin.
Heat treated: 510 Brin.
For extrusion parts, shear knives; oil hardened.

HEPPENSTALL SPECIAL.
M-64; 0.45 C, 4.3 Ni, 1.45 Cr, 0.75 Mo, 0.14 V, bal Fe.
For hot work dies, trimmers, punches.
Tough, oil hardening.

HEPPENSTALL T51.
M-64; 0.63 C, 1.35 Ni, 0.70 Cr, 2.12 W, bal Fe.
Annealed: 250 Brin.
Heat treated: 580 Brin.
For hot and cold trimmers; oil hardened, tough.

HEPPENSTALL T71.
M-64; 0.50 C, 1.12 Cr, 2.35 W, bal Fe.
Annealed: 250 Brin.
Heat treated: 560 Brin.
For cold shear knives, hot heading dies; oil hardened, tough.

HEPPENSTALL T73.
M-64; 0.28 C, 3.4 Cr, 0.23 V, 8.75 W, bal Fe.
Annealed: 250 Brin.
Heat treated: 500 Brin.
For shear knives, inserts, hot work dies; oil hardened, tough.

HEPPENSTALL T75.
M-64; 0.68 C, 4 Cr, 1 V, 18 W, bal Fe.
Heat treated: 650 Brin.
For shear knives; oil hardened, high speed steel.

HEPPENSTALL T78.
M-64; 0.85 C, 4.25 Cr, 5 Mo, 1.85 V, 6.25 W, bal Fe.
Annealed: 250 Brin.
Heat treated: 630 Brin.
For shear knives; high speed steel, oil hardened.

HEPPENSTALL T79.
M-64; 0.43 C, 5.25 Cr, 0.22 V, 4 W, bal Fe.
Annealed: 250 Brin.
Heat treated: 600 Brin.
For ramsk, die casting dies, shear knives; oil hardened.

HEPPENSTALL T717.
M-64; 0.33 C, 4.75 Cr, 1.3 Mo, 1.1 W, bal Fe.
Annealed: 250 Brin.
Heat treated: 530 Brin.
For shear knives, dies, punches, mandrels; oil hardened.

HEPPENSTALL T745.
M-64; 0.45 C, 8.5 Cr, 1.1 Mo, 1.1 W, bal Fe.
Annealed: 250 Brin.
Heat treated: 580 Brin.
For chipper knives; oil hardened.

HEPPENSTALL T746.
M-64; 0.68 C, 9.5 Cr, 1.1 Mo, 1.1 W, bal Fe.
Annealed: 250 Brin.
Heat treated: 620 Brin.
For flaking knives; oil hardened.

HERBOHN BELL METAL.
M-Germany; 60-71.43 Cu, 35-26.4 Sn, 5-2.7 Zn.
For bells; corrosion resistant.

HERC-ALLOY.
M-835; 0.4 C, 1.5 Ni, 0.20 Mo, bal Fe.
Heat treated: 125,000-150,000 TS; 275-290 Brin.
For chains, fittings, joining links, sling chains; wear and impact resistant.

HERCULES.
M-114; 1.0 C, 0.5 Cr, 0.2 V, bal Fe.
For blanking, forming and drawing dies; fatigue resistant. AISI W7.

HERCULES BRONZE-A.
M-U.S.; 86 Cu, 10 Sn, 2.5 Al, 2 Zn.
For hardware, worm gears; high strength.

HERCULES BRONZE-B.
M-U.S.; 54 Cu, 36 Zn, 7.5 Fe, 2.5 Al.
For hardware.

HERCULOY 420 now **HERCULOY 655.**

HERCULOY 421 now **HERCULOY 651.**

HERCULOY 651.
M-238; 1.5-2.0 Si, 0.25 Mn, bal Cu.
Hard: 70,000 TS; 55,000 YS; 15 El; 880 Brin.
For bolts, hardware; corrosion resistant.

HERCULOY 655.
M-238; 3.0-3.25 Si, 1.0 Mn, bal Cu.
Hard: 92,000 TS; 55,000 YS; 22 El; 890 Brin.
For bolts, hardware; corrosion resistant.

HERGERMUHL BRASS.
M-England; 62-72 Cu, 0.2-1.0 Sn, 0.8 max Pb, bal Zn.
For hardware, fittings, sheathing; good workability.

HERKULES BS.
M-1322; 2.1 C, 0.30 Si, 0.35 Mn, 11.5 Cr, bal Fe.
For forming and blanking dies, punches; oil or air hardened, non-deforming.

HERKULES FZ.
M-1322; 2.1 C, 11.5 Cr, 0.3 Mn, 0.3 Si, bal Fe.
For forming and blanking dies, punches; oil or air hardened, non-deforming.

HERKULES ME.
M-1322; 2.1 C, 0.3 Si, 0.35 Mn, 11.5 Cr, bal Fe.
For forming and blanking dies, punches; oil or air hardened, non-deforming.

HERMES 90 now VEW K985.
M-1308; 0.90 C, 0.25-0.50 Si, 0.30-0.80 Mn, bal Fe.
For drills, taps, hobs, reamers, springs, cutters; Type W1; water hardened.

HETZEL L SN 25.
M-1181; 25 Sn, bal Pb.
Solder; soft.

HETZEL L SN 40.
M-1181; 40 Sn, bal Pb.
For solder; soft.

HETZEL L ZN 98.
M-1181; 98 Zn, 2 Cu.
For solder; M.P. 790°F.

HEUSLER ALLOY-1.
M-Germ; 11.1 Al, 66.5 Cu, 22.4 Mn.
For electrical machinery; ferromagnetic.

HEUSLER ALLOY-2.
M-Germ; 10 Al, 4 Pb, 68 Cu, 18 Mn.
For electrical machinery; ferromagnetic.

HEUSLER ALLOY-3.
M-Germ; 4-15 Al, 54-76 Cu, 16-30 Mn.
For electrical machinery; ferromagnetic.

HEUSLER ALLOY-4.
M-Germ; 61 Cu, 13 Al, 26 Mn.
For electrical machinery; ferromagnetic.

HEVA-5B4.
M-1768; 0.40 C, 0.40 Mn, 1.0 Si, 5.3 Cr, 1.5 Mo, 1.0 V, bal Fe.
Ht Tr 48/57 Rc.
For tools, dies, hot working mandrels for extrusion.

HEVA-12A.
M-1768; 1.60 C, 0.35 Mn, 0.30 Si, 11.5 Cr, 0.70 Mo, 0.45 V, 0.50 W, bal Fe.
Air hardenable to 58/63 Rc.
Cold work tool steel.
Similar to AISI D2.

HEVA-17CN.
M-1768; 0.10-0.20 C, 15.0-3.0 Ni, bal Fe.
Hardenable stainless steel for combination of high strength for paper machinery and sea water resistance.
Similar to AISI 431.

HEVA-ACROM 35.
M-1768; 0.32-0.38 C, 0.60-0.90 Mn, 0.15-0.40 Si, 0.85-1.15 Cr, 0.15- 0.25 Mo, bal Fe.
Q+T: 75-120 kg/mm^2 TS; 50 kg/mm^2 YS; 11 El.
For gears, shafts, bolts; oil or water hardening.

HEVA-ALS.
M-1768; 0.09 max C, 1.15 Mn, 0.05 max Si, 0.33 S, bal Fe.
Normalized: 38 kg/mm^2 min TS.
For machining steel; for bolts, screws.

HEVA-ALSP.
M-1768; 0.09 max C, 1.15 Mn, 0.05 max Si, 0.33 S, 0.25 Pb, bal Fe.
Normalized: 38 kg/mm^2 min TS.
Free cutting leaded steel; for screw machine products.

HEVA-BT1.
M-1768; 0.42 C, 0.30 Mn, 1.0 Si, 1.2 Cr, 2.0 W, bal Fe.
Ht Tr 52-56 Rc.
For punches, pneumatic tools; shock resistant.

HEVA-CNE.
M-1768; 0.11-0.16 C, 0.35-0.65 Mn, 0.15-0.40 Si, 0.60-0.90 Cr, 2.5-3.0 Ni, bal Fe.
Q+T: 90-130 kg/mm^2 TS; 70 kg/mm^2 YS; 8 El.
Case hardening-carburizing steel.
For gears, shafts, pinions.

HEVA-CTM.
M-1768; 0.92 C, 1.30 Mn, 0.30 Si, 0.50 Cr, 0.12 V, 0.50 W, bal Fe.
Oil hardenable to 58-62 Rc.
Cold work tool steel.
Similar to AISI 01.

HEVA-DF-35-B.
M-1768; 0.32-0.38 C, 0.45-0.75 Mn, 0.15-0.40 Si, 0.003 min B, bal Fe.
Ann: 60 kg/mm^2 max TS; Ht Tr 80 kg/mm^2 min TS.
For cold headed bolts and parts.

HEVA-DF-35-M.
M-1768; 0.32-0.38 C, 1.1-1.35 Mn, 0.15-0.40 Si, bal Fe.
Ann: 62 kg/mm^2 max TS; Ht Tr 80 kg/mm^2 min TS.
For cold headed bolts and parts.

HEVA-DF-100.
M-1768; 0.35-0.41 C, 0.50-0.80 Mn, 0.15-0.40 Si, 0.85-1.15 Cr, 0.15-0.25 Mo, bal Fe.
Ann: 68 kg/mm^2 max TS; Ht Tr: 100 kg/mm^2 min TS.
For cold headed, oil hardened bolts and parts.

HEVA-DF-120.
M-1768; 0.37-0.43 C, 0.65-0.95 Mn, 0.15-0.40 Si, 0.35-0.65 Cr, 0.40-0.70 Ni, 0.15-0.25 Mo, bal Fe.
Ann: 68 kg/mm^2 max TS; Ht Tr: 120 kg/mm^2 min TS.
For cold headed, oil hardened bolts and parts.

HEVA-DTA.
M-1768; 0.29-0.35 C, 0.45-0.75 Mn, 0.15-0.40 Si, 1.1-1.4 Cr, 4.0-4.5 Ni, bal Fe.
Q+T: 90-120 kg/mm^2 TS; 70 kg/mm^2 min YS; 10 El.
For gears, shafts; air hardening; high tensile.

HEVA-ELASTIC.
M-1768; 0.29-0.35 C, 0.45-0.75 Mn, 0.15-0.40 Si, 0.50-0.80 Cr, 2.25-2.75 Ni, 0.45-0.55 Mo, bal Fe.
Q.+T.: 100-125 kg/mm^2 TS; 70 kg/mm^2 min YS; 10 El.
For gears, shafts; oil hardening, high tensile.

HEVA-EM-3.
M-1768; 0.32 C, 0.30 Mn, 0.30 Si, 3.0 Cr, 2.8 Mo, 0.50 V, bal Fe.
Heat treatment 46/50 Rc.
Containers and mandrels for extrusion of non-ferrous metals.
Similar to AISI H10.

HEVA-ENIMO.
M-1768; 0.20 C, 0.70 Mn, 0.30 Si, 3.15 Ni, 3.4 Mo, bal Fe.
Precipitation hardening steel for tools and dies; heat treatment 36/42 Rc.

HEVA-EST-EXTRA.
M-1768; 0.55 C, 0.70 Mn, 0.22 Si, 1.1 Cr, 1.7 Ni, 0.50 Mo, 0.10 V, bal Fe.
Heat treatment 35/45 Rc.
For tools, containers for rod and tube extrusion, hot work steel for drop forging dies.

HEVA-F-1282.
M-1768; 0.37-0.43 C, 0.50-0.80 Mn, 0.15-0.40 Si, 0.60-0.90 Cr, 0.70-1. Ni, 0.15-0.30 Mo, bal Fe.
Q+T.: 75-105 kg/mm^2 TS; 60 kg/mm^2 min YS; 10 El.
For gears, shafts, bolts; oil hardening.

HEVA-FCA.
M-1768; 2.2 C, 0.35 Mn, 0.50 Si, 11.5 Cr, 0.80 Mo, 0.20 V, bal Fe.
Air hardenable to 60-65 Rc.
Cold work tool steel.
Similar to AISI D4.

HEVA-HI.
M-1768; 0.15 max C. 11.5-14.0 Cr, bal Fe.
Q+T: 70-90 kg/mm^2 min TS; 55 kg/mm^2 min YS.
General purpose heat treatable type for machine parts, pump shafts.
Similar to AISI 403 and 410.

HEVA-IDF.
M-1768; 0.10 max C, 17.0-19.0 Cr, 11.0-13.0 Ni, bal Fe.
50-70 kg/mm^2 TS; 20 kg/mm^2 min YS; 40 min El.
For corrosion parts. Austenitic stainless steel for cold heading.
Similar to AISI 305.

HEVA-IF.
M-1768; 0.10 max C, 16.0-18.0 Cr, bal Fe.
45-65 kg/mm^2 TS; 26 kg/mm^2 min YS; 18 min El.
General purpose non-hardenable stainless steel, chromium type.
AISI 430.

HEVA-IM-3.
M-1768; 0.03 max C, 16.0-18.0 Cr, 10.0-14.0 Ni, 2.0-3.0 Mo, bal Fe.
45-65 kg/mm^2 TS; 20 kg/mm^2 min YS; 40 min El.
For corrosion parts. Low carbon for restriction of carbide precipitation during welding.
AISI 316L.

HEVA-IM-8.
M-1768; 0.08 max C, 16.0-18.0 Cr, 10.0-14.0 Ni, 2.0-3.0 Mo, bal Fe.
50-70 kg/mm^2 TS; 20 kg/mm^2 min YS; 40 min El.
For corrosion parts. Austenitic stainless steel.
AISI 316.

HEVA-INOX 20.
M-1768; 0.16-0.25 C, 12.0-16.0 Cr, bal Fe.
70-90 kg/mm^2 TS; 55 kg/mm^2 min YS; 13 min El.
Higher carbon modification of HEVA-HI, often used for cutlery; hardenable to 40-50 Rc.
Similar to AISI 420.

HEVA-INOX 42.
M-1768; 0.36-0.45, 12.5-14.5 Cr, bal Fe.
Hardenable to 45-55 Rc.
Martensitic stainless, high carbon modification of HEVA-HI and Inox 20, often used for cutlery and surgical instruments.

HEVA-KLT.
M-1768; 0.30 C, 0.30 Mn, 0.30 Si, 2.6 Cr, 8.8 W, 0.35 V, bal Fe.
Heat treatment 48/52 Rc.
For tools, dies, hot work tools.

HEVA-LCH-3.
M-1768; 0.03 max C, 18.0-20.0 Cr, 8.0-12.0 Ni, bal Fe.
45-65 kg/mm^2 TS; 18 kg/mm^2 min YS; 40 min El.
For corrosion parts. Low carbon for restriction of carbide precipitation during welding.
AISI 304L.

HEVA-LCH-8.
M-1768; 0.08 max C, 18.0-20.0 Cr, 8.0-10.5 Ni, bal Fe.
50-70 kg/mm^2 TS; 20 kg/mm^2 min YS; 40 min El.
For corrosion parts. Austenitic stainless.
AISI 304.

HEVA-MCV.
M-1768; 0.45-0.55 C, 0.60-0.90 Mn, 0.10-0.35 Si, 0.85-1.35 Cr, 0.10-0.20 V, bal Fe.
Q & T.: 80-130 kg/mm² TS; 60 kg/mm² min YS; 8 El.
Oil hardening; for laminating springs.

HEVA-MOLICORT.
M-1765; 0.85 C, 0.30 Mn, 0.25 Si, 4.0 Cr, 5.0 Mo, 6.25 W, 1.85 V, bal Fe.
Heat treatment 63/66 Rc.
High speed steel, for tools, drills, cutters, cold forming dies.
AISI M2.

HEVA-MP.
M-1768; 0.30 C, 0.80 Mn, 0.50 Si, 1.65 Cr, 0.40 Mo, bal Fe.
Heat treatment 45/55 Rc.
Plastic mold steel.

HEVA-PERFOR.
M-1768; 1.17 C, 0.30 Mn, 0.30 Si, 1.0 W, bal Fe.
Heat treatment 60/64 Rc.
For drills, water hardening.

HEVA-RAPID.
M-1768; 0.90 C, 0.30 Mn, 0.25 Si, 4.0 Cr, 5.0 Mo, 6.25 W, 1.85 V, 4.75 Co, bal Fe.
Heat treatment 63/66 Rc.
High speed steel, for tools, cutters.

HEVA-SM-75.
M-1768; 0.32-0.38 C, 1.1-1.4 Mn, 1.1-1.4 Si, bal Fe.
Q & T.: 70-100 kg/mm² TS; 50 kg/mm² min YS; 12 El.
For gears, shafts; water hardening.

HEVA-SPT.
M-1768; 0.09 max C, 1.15 mn, 0.05 max Si, 0.33 S, 0.25 Pb, 0.035 Te, bal Fe.
Normalized: 38 kg/mm² min TS.
Free cutting S-Pb-Te steel; for screw machine products.

HEVA-SUPER BONO.
M-1768; 0.78 C, 0.30 Mn, 0.25 Si, 4.25 Cr, 18.0 W, 1.0 V, bal Fe.
Heat treatment 63/65 Rc.
High speed steel, for tools, dies, cutters.
AISI T1.

HEVA-TC1.
M-1768; 0.47 C, 0.50 Mn, 0.22 Si, 1.30 Cr, 4.0 Ni, 0.20 Mo, bal Fe.
Heat treatment 40/50 Rc.
For cold tools, shock resistant; air hardening.

HEVA-TSD.
M-1768; 0.29-0.35 C, 0.45-0.75 Mn, 0.15-0.40 Si, 0.50-0.80 Cr, 2.80-3.25 Ni, bal Fe.
Q. & T.; 95-125 kg/mm² TS; 60 kg/mm² min YS; 10 El.
For gears, shafts; oil hardening.

HEVA-V-214.
M-1768; 0.49-0.57 C, 8.0-10.0 Mn, 20.0-22.0 Cr, 3.3-4.3 Ni, 0.36-0.50 N, bal Fe.
100-120 kg/mm² TS; 60 kg/mm² YS; 8 min El.
For auto exhaust valves; resists leaded fuels.
Precipitation hardening steel.

HEVA-VSC.
M-1768; 0.38-0.48 C, 2.5-3.2 Si, 8.6-9.5 Cr, bal Fe.
90-105 kg/mm² TS; 70 kg/mm² min YS; 15 min El.
For auto inlet and exhaust valves; heat resistant.

HEVA-XKW.
M-1768; 0.76 C, 0.30 Mn, 0.25 Si, 4.25 Cr, 18.5 W, 1.0 Mo, 1.55 V, 9.5 Co, bal Fe.
Heat treatment 63/66 Rc.
High speed steel, for tools, dies, cutters.
Similar to AISI T5.

HEVI-DUTY SW-14.
M-605; 0.3 C, bal Fe.
For welding rod; high tensile.

HEVIMET see CARBOLOY HEVIMET.

HEWITT COPPER-HARD.
M-959; 4.5 Cu, 4.5 Sb, bal Sn.
Cast: 11,050 TS; 24 Brin.
For heavy duty bearings, bushings; Babbitt, M.P. 365-590°F.

HEWITT GENUINE.
M-959; 7.5 Sb, 3.5 Cu, bal Sn.
Cast: 15,150 TS; 29 Brin.
For heavy duty bearings, bushings; Babbitt, M.P. 460-682°F.

HEWMET.
M-959; 5 Sn, 15 Sb, bal Pb.
Cast: 7580 TS; 23.7 Brin.
For bearings, bushings; Babbitt, heavy loads.

HFI.
M-Germany; 73.5 WC, 24 TiC, 2.5 Co.
Sintered: A 92 Rock.
For cutters to machine steel, wear dies.
Sintered carbides; hard and abrasion resistant.

HH1.
M-Germany; 97.5 WC, 2.5 Co.
Sintered: A 92 Rock.
For cutters to machine cast iron, form tools.
Sintered carbide. Hard and abrasion resistant.

HI 1.
M-1655; 0.10 C, 0.80 Si, 0.80 Al, 6.5 Cr, bal Fe.
For elevated temperature operations, as pyrometer sheath tubes; heat resisting steel.
W.-Nr. 1.4713.

HI 2.
M-1655; 0.10 C, 1.0 Si, 1.0 Al, 13.0 Cr, bal Fe.
For elevated temperature operation, as furnace rails and supports.
W-Nr. 1.4724.

HI 3.
M-1655; 0.12 C, 1.0 Si, 0.7 Mn, 1.0 Al, 18.0 Cr, bal Fe.
For elevated temperature equipment as furnace fittings.
W.-Nr. 1.4742.

HI 4.
M-1655; 0.12 C, 1.4 Si, 0.7 Mn, 1.45 Al, 24.0 Cr, bal Fe.
For high temperature operation as steam boiler equipment.
W.-Nr. 1.4762.

HI 10.
M-1655; 0.18 C, 1.2 Si, 0.7 Mn, 25.0 Cr, 4.0 Ni, bal Fe.
For high temperature equipment; furnace parts.
W-Nr. 1.4821.

HI 20.
M-1655; 0.15 C, 0.40 Si, 1.6 Mn, 18.0 Cr, 10.5 Ni, 0.40 Ti, bal Fe.
For high temperature operation, as annealing boxes.
W-Nr. 1.4878.

HI 21.
M-1655; 0.15 C, 2.0 Si, 0.70 Mn, 20.0 Cr, 12.0 Ni, bal Fe.
For high temperature operation, as annealing boxes.
W.-Nr. 1.4828; similar to AISI 309.

HI 22.
M-1655; 0.15 C, 2.0 Si, 0.70 Mn, 25.0 Cr, 20.0 Ni, bal Fe.
For high temperature equipment; annealing pots, thermocouple housings.
W.-Nr. 1.4841; similar to AISI 310.

HI 23.
M-1655; 0.10 C, 1.8 Si, 1.4 Mn, 16.0 Cr, 35.5 Ni, bal Fe.
For high temperature equipment, as furnace parts.
W.-Nr. 1.4864.

HI 24.
M-1655; 0.10 Cr, 2.0 Si, 1.8 Mn, 21.0 Cr, 1.8 Mo, 15.5 Ni, 1.2 Nb, bal Fe.
For high temperation equipment, as support bars for bright annealing.
W.-Nr. 1.4885.

HI 50.
M-1655; 0.10 C, 1.0 Si, 1.0 Mn, 28.0 Cr, 48.0 Co, bal Fe.
For high temperature equipment.
W.-Nr. 2.4778.

HI-120.
M-118; Cb-Ti.
For superconducting wires.
High magnetic fields at low reduced power use.

HIBBO NO. 100.
M-215; Al, bal Cu.
50,000 TS; 100 Brin.
For light bearings and bushings; heavy duty.

HIBBO NO. 125.
M-215; Al, bal Cu.
Cast: 81,000 TS; 29,000 YS; 24 El; 22 RA; 120 Brin.
For gears, worm wheels, bearings; for general service.

HIBBO NO. 150.
M-215; Al, bal Cu.
Cast: 83,000 TS; 35,000 YS; 13 El; 19 RA; 140 Brin.
For bearings, gears, bushings, pinions; heavy duty, shock resistant.

HIBBO NO. 175.
M-215; Al, bal Cu.
Cast: 82,000 TS; 36,000 YS; 9 El; 10 RA; 180 Brin.
For gears, pinions, shafts; heavy compressive loading.

HIBBO NO. 200.
M-215; Al, bal Cu.
Cast: 96,4000 TS; 48,000 YS; 6 El; 65 RA; 190-220 Brin.
For bushings, bearings, gears, welder dies and jaws; wear and corrosion resistant, Al-bronze.

HIBBO NO. 225.
M-215; Al, bal Cu.
85,000 TS; 190-240 Brin.
For forming dies, bearing slides, cam rollers, non-sparking tools; resists shock, fatigue and corrosion.

HIBBO NO. 250.
M-215; Al, bal Cu.
Cast: 81,000 TS; 76,000 YS; 0.5 El; 0 RA; 230-260 Brin.
For bearing slides, cam rollers, wear strips; high wear resistance.

HIBBO NO. 275.
M-215; Al, bal Cu.
90,000 TS; 250 Brin.
For dies, rolls, pins, cams, bushings; resists shock, fatigue and corrosion.

HIBBO NO. 300.
M-215; Al, bal Cu.
270-300 Brin.
For forming and drawing dies; very hard, tough to machine.

HI-CHROME.
M-1473; 2.6 C, 27 Cr, 1.0 Mn, 1.0 Si, 0.6 Mo, 0.3 V, bal Fe.
Hardened: 60 Rc.
For pump and cylinder liners; abrasion resistant.

HICKORY NO. 7.
M-1423; 0.50 C, 1.5 Cr, 2.5 W, bal Fe.
For punches, rivet sets; Type S1; shock resistant.

HICRO.
M-822; 1.5 C, 12 Cr, 0.8 Mo, 0.35 V, bal Fe.
For cold forming dies; oil hardened.

HICRO-150.
M-822; 1.5 C, 12 Cr, 0.4 V, 1.0 Mo, bal Fe.
Hardened: 278,000 TS; 214,000 YS; 1 El; 567 Brin.
For blanking and drawing dies, wire drawing and stamping dies, punches, gauges, hobs.
Type D2 air hardening, nondeforming, cold work tool steel, tough.

HICRO 200.
M-822; 2.0 C, 0.7 Mn, 0.3 Si, 13.0 Cr, 1.2 W, bal Fe.
Air or oil hardening cold work tool steel, chromium type; AISI D3.

HICRO T.
M-822; 2.10 C, 0.7 Mn, 0.3 Si, 13.0 Cr, 1.3 W, bal Fe.
Air or oil hardenable cold work tool and die steel, chromium type; AISI D6.

HIDALGO.
M-614; 0.7 C, 14 W, 4 Cr, 2 V, bal Fe.
For cutters, tools; high speed steel.

HIDALGO I.
M-614; 0.55 C, 0.9 Si, 1 Cr, 0.18 V, 1.85 W, bal Fe.
For cold work tools, punches, headers; oil hardened, tough.

HIDALGO II.
M-614; 0.40 C, 0.9 Si, 1 Cr, 0.18 V, 1.85 W, bal Fe.
For cold work tools, punches; oil hardened, tough.

HIDALGO III.
M-614; 0.35 C, 0.9 Si, 1.05 Cr, 0.18 V, 1.85 W, bal Fe.
For cold work tools, headers, punches, upsetters; oil hardened, tough.

HIDALGO NO. 48 ALLOY.
M-344; 0.55 C, 0.20 Cr, 0.75 Mn, 0.20 V, 2.0 Si, bal Fe.
For chisels, shear blades, punches; shock resistant.

HIDALGO RL.
M-344; 1.25 C, 0.4-0.5 Cr, 5.5 W, bal Fe.
For cutting tools; for finishing cuts on hard material.

HI-DI 5.
M-336; 1.0 C, 5.0 Cr, 1.0 Mo, 0.4 Mn, 0.4 V, bal Fe.
Annealed: 103,000 TS; 51,000 YS; 26 El; C 18 Rock.
Hardened: 178,000-255,000 TS; 145,00-200,000 YS; 3-12 El; 7-32 RA; C 41-53 Rock.
For blanking and trimming dies, shear blades, punches gauges, master tools, rolling dies, broaches.
Type A2 air hardening, nondeforming cold work tool steel, tough.

HIDUMINIUM 00.
M-426; 0.75-2.5 Cu, 9.0-11.5 Si, 3 max Ni, bal Al.
Sand cast: 16,000 TS.
Die cast: 29,000 TS; 13,500 YS; 2 El; 85 Brin.
For general purpose castings of thin wall sections; high fluidity.

HIDUMINIUM 01.
M-137, m-426; 4 Cu, 0.6 Mg, 0.5 Mn, bal Al.
Forged: 54,000 TS; 30,500 YS; 15 El.
Extruded: 54,000 TS; 31,500 YS; 12 El.
Tube: 62,000 TS; 47,000 YS; 12 El.
For general engineering components; age-hardenable.

HIDUMINIUM 1A.
M-426; 98.8 Al.
O-temper: 11,200 TS; 35 El.
1/2 H-temper: 13,500 TS; 8 El.
H-temper: 18,000 TS; 5 El.
For light alloy parts.

HIDUMINIUM 1B.
M-426; 99.5 Al.
O-temper: 13,500 TS; 30 El.
1/2 H-temper: 14,500 TS; 8 El.
H-temper: 19,000 TS; 5 El.
For light alloy parts.

HIDUMINIUM 1C.
M-426; 99.0 Al.
-temper: 14,500 TS; 5600 YS; 30 El.
1/2 H-temper: 15,700 TS; 14,500 YS; 7 El.
H-temper: 20,000 TS; 19,000 YS; 3 El.
For light alloy parts.

HIDUMINIUM 02.
M-426; 4 Cu, 1.5 Mg, 2 Ni, bal Al.
Heat treated: 49,000-58,000 TS; 32,000-31,500 YS; 20-8 El; 115 Brin.
For pistons, cylinder heads; age hardenable, high temperature uses.

HIDUMINIUM 03.
M-239, m-426; 1 Cu, 1 Mg, 1 Si, bal Al.
T4-temper: 38,000 TS; 22,400 YS; 15 El.
T6-temper: 56,000 TS; 45,000 YS; 10 El.
For aircraft forgings and extrusions.

HIDUMINIUM 05.
M-239, m-426; 5 Mg, bal Al.
Annealed: 38,000 TS; 18,000 YS; 18 El.
1/2 Hard: 40,500 TS; 31,500 YS; 5 El.
For marine structures; corrosion resistant.

HIDUMINIUM 07.
M-239, m-426; 7 Mg, 0.6 Mn, 0.6 max Fe, bal Al.
Annealed: 45,000 TS; 18,000 YS; 18 El; 20 RA.
1/2 Hard: 56,000 TS; 36,000 YS; 5 El.
For hardware, sporting goods, machine tool parts; high corrosion resistance.

HIDUMINIUM 08.
M-426; 1 Cu, 1 Mg, 11 Si, 1 Ni, bal Al.
Heat treated: 47,000 TS; 33,500 YS; 5 El; 115 Brin.
For pistons, cylinder heads; age hardenable.

HIDUMINIUM 10.
M-426; 10-13 Si, 0.2 max Ti, 0.1 max Cu, 0.6 max Fe, bal Al.
Sand cast: 26,000 TS; 8000 YS; 8 El; 55 Brin.
Die cast: 30,500 TS; 10,000 YS; 10 El; 65 Brin.
For general purpose castings of intricate shape; good fluidity.

HIDUMINIUM 11.
M-426; 0.15 max Cu, 0.6 max Si, 0.75 max Fe, 1-1.5 Mn, bal Al.
Soft temper: 15,000 TS; 12,000 YS; 25 El; 27 Brin.
3/4 H temper: 22,000 TS; 19,000 YS; 5 El; 55 Brin.
For structures; not heat treatable, corrosion resistant.

HIDUMINIUM-12.
M-426; 1.3 Mg, bal Al.
Hard drawn: 36,000 TS; 35,000 YS; 4 El; H 105 Rock.
For deep drawn parts.
For trim molding, washing machine tubs, vacuum cleaner hoods, fan blades.
Corrosion resistant, low strength.
Similar to Aluminum 5050.

HIDUMINIUM-14.
M-426; 1.0 Mg, 0.5 Mn, bal Al.
Hard drawn: 40,000 TS; 36,000 YS; 6 El; 77 Brin.
Annealed: 26,000 TS; 10,000 YS; 25 El; 45 Brin.
For general purpose applications, commercial roofing, vessels, tanks.
Corrosion resistant. Similar to Aluminum 3004.

HIDUMINIUM 20.
M-426; 2-4 Cu, 3-6 Si, 0.3-0.7 Mn, 0.5 max Zn, 0.8 max Fe, bal Al.
Sand cast: 18,000 TS; 8000 YS; 2 El; 55 Brin.
Die cast: 20,000 TS; 9000 YS; 2 El; 80 Brin.
For marine castings; age-hardenable.

HIDUMINIUM-21.
M-426; 3.0 Cu, 5.0 Si, 0.5 Mn, 0.25 Mg, bal Al.
Heat treated: 36,000 TS; 24,000 YS; 2 El; 80 Brin.
Cast: 27,000 TS; 18,000 YS; 2 El; 70 Brin.
For sand and permanent mold castings, crankcases, housings, cylinder heads, oil pans.
Heat treatable, pressure-tight.

HIDUMINIUM 22.
M-426; 0915 max Cu, 1.5-2.5 Mg, 0.6 max Si, 0.75 max Fe, bal Al.
Soft temper: 24,000 TS; 12,000 YS; 18 El; 45 Brin.
1/2 H temper: 35,000 TS; 27,000 YS; 5 El; 85 Brin.
For fuel pipes; high corrosion resistance.

HIDUMINIUM 24.
M-426; 2 Mg, bal Al.
O-temper: 24,500-27,000 TS; 10,500 YS; 18 El.
For deep drawn and formed parts; high ductility.

HIDUMINIUM 29.
M-426; 6.5 Sn, 1.0 Cu, 0.8 Ni, bal Al.
Sand cast: 18,000 TS; 7800 YS; 8 El.
Permanent mold: 20,000 TS; 18,000 YS; 15 El.
For bearings; shock resistant.

HIDMINIUM 33.
M-426; 3-4 Mg, 1.0 max Mn, 0.15 max Cu, 0.7 max Fe, 0.5 max Cr, bal Al.
Soft temper: 28,000 TS; 14,000 YS; 18 El; 55 Brin.
1/2 H temper: 36,000 TS; 30,000 YS; 5 El; 100 Brin.
For marine parts; highest corrosion resistance.

HIDUMINIUM 35.
M-426; 4.25 Mg, bal Al.
O-temper: 32,000 TS; 15,700 YS; 18 El.
M-temper: 38,000 TS; 18,000 YS; 12 El.
For light alloy parts; high corrosion resistance and ductility.

HIDUMINIUM 40.
M-426; 0.2-0.8 Mg, 4.5-6 Si, 0.2 max Ti, bal Al.
Cast: 17,000 TS; 10,000 YS; 2,5 El; 50 Brin.
W-temper: 22,000 TS; 14,000 YS; 2,5 El; 50 Brin.
WP-temper: 30,000 TS; 26,000 YS; 100 Brin.
For marine castings; high fluidity, pressure tight.

HIDUMINIUM 42.
M-426; 0.75 Mg, 1 Si, bal Al.
W-temper: 26,7000 TS; 15,700 YS; 18 El.
WP-temper: 43,600 TS; 40,300 YS; 10 El.
For light alloy parts; good corrosion resistance and formability; age-hardenable.

HIDUMINIUM 44.
M-426; 0.5-1.2 Mg, 0.75-1.3 Si, 1 max Mn, 0.6 max Fe, bal Al.
Annealed and tempered: 18,000 TS; 10,000 YS; 27 El; 30 Brin.
WP-temper: 46,000 TS; 36,00/ YS; 8 El; 110 Brin.
For marine applications; heat treatable, resists sea-water corrosion.

HIDUMINIUM 46.
M-426; 0.7 Mg, 0.5 Si, 0.05 Ti, bal Al.
W-temper: 27,100 TS; 16,200 YS; 25 El.
WP-temper: 35,800 TS; 27,00/ YS; 18 El.
For structural extrusions; heat treatable; corrosion resistant.

HIDUMINIUM 51.
M-426; 0.8-2.0 Cu, 0.05-0.2 Mg, 0.75-2.8 Si, 0.25-1.4 Fe, 0.8-1.75 Ni, bal Al.
Sand cast: 23,000 TS; 11,000 YS; 2 El; 55 Brin.
Die cast: 25,000 TS; 13,000 YS; 2 El; 80 Brin.
For general purpose castings; fair corrosion resistance.

HIDUMINIUM 55.
M-426; 1.75-2.7 Cu, 0.5-1.3 Mg, 0.4-1.5 Si, 1.4 max Fe, 1.4 max Ni, 0 max Ti, bal Al.
Forged: 50,000 TS; 36,000 YS; 8 El; 148 Brin.
WP-temper: 44,000 TS; 36,000 YS; 8 El.
For high strength parts, aircraft structures; heat treatable.

HIDUMINIUM 66.
M-426; 3.5-4.8 Cu, 0.3-0.6 Mg, 0.1-1.5 Si, 1.2 max Mn, bal Al.
WP-temper: 64,000 TS; 52,000 YS; 8 El; 160 Brin.
For high strength parts, aircraft structures; heat treatable.

HIDUMINIUM 72.
M-426; 4.6 Cu, 1.3 Mg, 0.25 Fe, 0.7 Mn, bal Al.
Heat treated: 67,000 TS; 51,000 YS; 8 El.
For aircraft and general engineering parts; age-hardenable.

HIDUMINIUM 80.
M-426; 4-5 Cu, 0.1-0.25 Ti, 0.25 max Si, bal Al.
W-temper: 33,000 TS; 18,000 YS; 10 El; 60 Brin.
WP-temper: 48,000 TS; 27,000 YS; 10 El; 95 Brin.
For brackets, levers, housings, aircraft castings; age-hardenable, hot-short.

HIDUMINIUM 90.
M-426; 9.5-11.0 Mg, 0.15 max Cu, 0.25 max Si, bal Al.
Sand cast: 45,000 TS; 22,000 YS; 15 El; 90 Brin.
Die cast: 47,000 TS; 25,000 YS; 15 El; 90 Brin.
For gasoline flow-meters; age-hardenable, corrosion resistant.

HIDUMINIUM 100.
M-239, M-426, M-86; Al_2O_3 + Al.
Sintered.
For components operating above 250°C; high temperature resistance.

HIDUMINIUM DU BRAND.
M-426; 3.5-4.8 Cu, 0.6 Mg, 0.5 Mn, 0.3 Ti, bal Al.
Rolled: 54,000 TS; 35,000 YS; 18 El; 120 Brin.
For light alloy parts, seamless tubes, airscrews; age-hardened.

HIDUMINIUM R.R. 50.
M-239, M-86, M-426, M-137; 0.8-2 Cu, 0.05-2 Mg, 0.2 Ti, 1.5-2.8 Si, 0.8-1.4 Fe, 0.8-1.7 Ni, bal Al.
Heat treated: 25,000-30,000 TS; 3 El; 5 RA; 72 Brin.
For cylinder blocks, cylinder heads, pistons; crankcases; heat treatable.

HIDUMINIUM R R 53.
M-137; 1.5-2.5 Cu, 0.5-2.0 Ni, 1.4-1.8 Mg, 1.2-1.5 Fe, 0.2-0.12 Ti, 2 max Si, bal Al.
Heat treated: 50,000 TS; 44,000 YS; 2 El; 140 Brin.
For pistons, cylinder heads; die casting.

HIDUMINIUM R.R.53.
M-239, M-426; 2.2 Cu, 1.3 Ni, 1.6 Mg, 1.4 Fe, 0.08 Ti, 1.2 Si, bal Al.
Die cast: 31,300 TS; 28,000 YS; 3 El; 4 RA; 80 Brin.
Heat treated: 56,000 TS; 50,400 YS; 1 El; 1.5 RA; 130 Brin.
For pistons for automobile and aircraft engines, cylinder heads; die cast; Rolls-Royce automobile.

HIDUMINIUM RR 53 B.
M-137; 1.5 Cu, 1-2 Ni, 0.6-1.0 Mg, 0.8-1.5 Fe, 0.7-0.9 Si, bal Al.
Cast: 21,000 TS; 14,000 YS; 3 El; 78 Brin.
Heat treated: 42,000 TS; 37,000 YS; 2 El; 138 Brin.
For levers, brackets, textile and food machinery, pistons; high temperature strength.

HIDUMINIUM RR 53 C.
M-426; 0.8-2.0 Cu, 0.5-1.5 Ni, 0.3-0.8 Mg, 0.8-1.4 Fe, 2-3 Si, 0.3 Ti, bal Al.
Cast: 30,000 TS; 20,000 YS; 2 El; 75 Brin.
Heat treated: 44,000 TS; 40,000 YS; 2 El; 115 Brin.
For stand and die castings; light alloy.

HIDUMINIUM R.R. 56.
M-239, M-426; 2 Cu, 1.25 Ni, 1.2 Fe, 0.8 Mg, 0.08 Ti, 0.6 Si, bal Al.
Heat treated: 60,000-72,000 TS; 54,000-58,000 YS; 10-20 El; 14-25 R 120-160 Brin.
For automobile forgings, connecting rods, supercharger rotors; forgings; Rolls-Royce automobile.

HIDUMINIUM R.R. 57.
M-426; 5.75-6.25 Cu, 0.2 max Si, 0.2-0.3 Mn, 0.1-0.15 Ti, bal Al.
At 70°F: 54,000 TS; 33,000 YS; 8 El; 121 Brin.
At 650°F; 14,000 TS; 9000 YS; 27 El.
For high temperature applications, aircraft structures; heat treatable.

HIDUMINIUM R.R. 58.
M-426; 1.5-3 Cu, 1.2-1.8 Mg, 1-1.5 Fe, 0.5-1.5 Ni, 0.2 max Ti, bal Al.
At 70°F: 60,000 TS; 44,000 YS; 10 El; 148 Brin.
At 650°F: 12,500 TS; 10,000 YS; 27.5 El.
For high temperature applications, aircraft structures; heat treatable.

HIDUMINIUM R.R. 59.
M-239, M-426; 2.2 Cu, 1.35 Ni, 1.25 Mg, 1.35 Fe, 0.08 Si, bal Al.
Forged: 52,000-65,000 TS; 50,000-56,000 YS; 6-10 El; 10-20 RA; 120-150 Brin.
Heat treated: 55,000 TS; 47,000 YS; 8 El; 17 RA 127 Brin.
For pistons, cylinder heads; forgings; Rolls Royce automobile; aero-engines.

HIDUMINIUM R.R. 60.
M-426; 1.5-3.0 Cu, 0.3-1.5 Si, 0.5-1.5 Mg, 0.5-1.5 Fe, 2 max Ni, 0.3 Ce, bal Al.
For light alloy parts; antifriction properties.

HIDUMINIUM RR-66.
M-426; 0.5 Cu, 0.33 Ni, 4.8 Mg, 0.33 Fe, bal Al.
Annealed: 45,000 TS; 22,000 YS; 24 El; 80 Brin.
Cold worked: 60,500 TS; 49,000 YS; 10 El; 121 Brin.
For automobile parts, airplanes, boats; corrosion resistant.

HIDUMINIUM RR72.
M-426; 94 Al, 4 Cu, 1.2 Mg, 0.8 Mn.
Extruded: 60,000 TS; 15 El; 148 Brin.
For extruded sections, solid drawn tubes; heat treatable.

HIDUMINIUM RR 75.
M-426; 97.6 Al, 2 Cu, 0.4 Mg.
For wire, rivets.

HIDUMINIUM RR 77.
M-426; 1.5 Cu, 4-6 Zn, 2-4 Mg, 0.6 Fe, 0.6 Si, 1.0 max Ni, 0.3 Ti, bal Al.
Annealed: 32,000 TS; 19,000 YS; 20 El; 65 Brin.
Heat treated: 85,000 TS; 74,000 YS; 16 El; 180 Brin.
For extruded and drawn tubing, rolled sheet; light alloy, high strength.

HIDUMINIUM RR 82.
M-426; Cu, Si, bal Al.
For fuel pipes.

HIDUMINIUM RR88.
M-426; 3 max Cu, 4 max .Mg, 4.0-8.5 Zn, 1 max Cr, bal Al.
Aged: 89,000 TS; 78,000 YS; 7 El; 180 Brin.
For aircraft construction, high strength applications; age hardenable.

HIDUMINIUM RR250.
M-239, M-426; 5 Cu, 0.2 Ti, 0.25 Mn, 0.25 Ce, 1 Ni, 0.25 Sb, bal Al.
Heat treated: 36,000 TS; 22,500 YS; 2 El.
For engine components; age hardenable; sand castings.

HIDUMINIUM RR257.
M-426; 6 Cu, 0.25 Co, 1 Ni, 0.25 Sb, 0.25 Mn, 0.2 Ti, bal Al.
WP-temper: 54,000 TS; 33,500 YS; 8 El.
For engine castings; age hardenable.

HIDUMINIUM RRAC9A.
M-239, M-426; Al alloy.
For light alloy parts.

HIDUMINIUM S 12.
M-426; 12 Si, 1.0 Cu, 1.0 Ni, 1.2 Mg, bal Al.
Forged: 38,000 TS; 7 El; 105 Brin.
For aircraft pistons; forging alloy.

HIDUMINIUM SR.
M-426; 2.5 Cu, 0.1 Mg, 1.0 Fe, 0.75 Ni, 0.1 Ti, bal Al.
60-85 Brin.
For gasket sealing rings; corrosion resistant.

HIDUMINIUM Y.
M-426; 4 Cu, 1.5 Mg, 2 Ni, 0.6 Fe, 0.5 Mn, bal Al.
Forged: 55,000 TS; 35,000 YS; 2 El; 120 Brin.
Cast: 40,000 TS; 34,000 YS; 1 El; 105 Brin.

HIDURAX 1 CAST.
M-1169; 8.5-10.5 Al, 4.0-5.5 Fe, 4.0-5.5 Fe, 4.0-5.5 Ni, bal Cu.
Cast: 90,00-100,000 TS; 40,000-45,000 YS; 20-12 El; 170-180 Brin.
For shafts, gears, spindles, valve seats; resists corrosion and cavitation erosion.

HIDURAX 1 WROUGHT.
M-1169; 8.5-10.5 Al, 4-6 Fe, 4-6 Ni, 0.5 Mn, bal Cu.
Forged: 100,000-116,000 TS; 56,000-76,000 YS; 25-15 El; 180-240 Brin.
For shafts, gears, spindles, valve seats; resists corrosion and cavitation.

HIDURAX SPECIAL.
M-1169; 2-4 Al, 1-3 Fe, 13-16 Ni, bal Cu.
Brin.
Resistant.

HIDURAX 2-CAST.
M-1169; 8.5-10.5 Al, 1.5-3.5 Fe, 0.5 Mn, bal Cu.
Cast: 72,000-85,000 TS; 25,000-30,000 YS; 30-20 El; 110-140 Brin.
For pumps, valves, gears, bushings; resists wear, corrosion.

HIDURAX 2-WROUGHT.
M-1169; 8.5-10.5 Al, 0.5-2.5 Fe, 1-3 Ni, bal Cu.
Rolled: 85,000-108,000 TS; 45,000-56,000 YS; 25-18 El; 149-212 Brin.
For shafts, spindles, gears, liners, bushings; resists wear, corrosion and erosion.

HIDURAX 4.
M-1169; 11 Al, 4-5 Fe, 4 Ni, bal Cu.
Heat treated: 110,000-135,000 TS; 67,000-90,000 YS; 12-5 El; 218-270 Brin.
For valve inserts, plastic molding dies, nozzles; tough, hardenable, corrosion resistant.

HIDUREL-5.
M-1169; .0-3.5 Ni, 0.4-0.8 Si, bal Cu.
Bar: 85,000-117,000 TS; 61,000-105,000 YS; 15-25 El; 160-210 Brin.
Strip: 85,000-108,000 TS; 61,000-94,000 YS; 15-25 El; 160-210 Brin.
For switch gears, contacts, gears, current carrying parts; high conductivity and high strength.

HIDUREL 5-CAST.
M-1169; 1.8-2.6 Ni, 0.3-5 Si, bal Cu.
Heat treated: 63,000-72,000 TS; 40,000-54,000 YS; 15-18 El; 140-170 Brin.
For switchgears, bushings, electrical conductors; high strength and conductivity.

HIDUREL 6.
M-1169; 0.4-1.2 Cr, 0.2 others, bal Cu.
Heat treated: 40,000-80,000 TS; 27,000-78,000 YS; 30-15 El; 140-170 Brin.
For switchgears, resistance welding electrodes; high strength at high temperature.

HIDUREL 640.
M-1164; Cr, Zr, bal Cu.
High conductivity bronze bar.
Ann: 21-24 tons/sq. in TS; 100-120 Brin.
Cold worked; 28-34 tons/sq.in TS; 125-170 Brin.
For bus bars.

HIDUREX 7.
M-1169; 6.0-6.4 Al, 2.0-2.4 Si, 0.5-0.7 Fe, bal Cu.
Wrought: 34 tons/sq.in TS; 130-180 Brin.
Cast: 30 tons/sq.in TS; 110-150 Brin.
For pump shafts, bolts, studs, impellers.

HIDURIT 15.
M-1169; Zn, bal Cu.
Rolled: 67,000-85,000 TS; 34,000-45,000 YS; 30-15 El; 120-180 Brin.
For general castings; BS-250; high strength brass.

HIDURON 102.
M-1169; 10 Ni, 90 Cu.
Wrought: 18 tons/sq.in; 30 min El.
BS. 2875-CN 102.

HIDURON 107.
M-1169; 30 Ni, 70 Cu.
Wrought: 20 tons/sq.in; 30 min El.
BS. 2875-CN 107.

HIDURON 191.
M-1169; wrought cupro-nickel, + Al, Fe, Mn.
46 tons/sq.in UTS; 27 tons/sq.in proof strength.
For improved strength over normal cupro-nickel.

HIDURON 501.
M-1169; 12 Ni, Al, Fe, Mn, Cr, bal Cu.
Cast cupro-nickel with addions.
30 tons/sq.in UTS; 17 tons/sq.in proof stress.
For improved strength; marine use.

HIFLEX.
M-1431; C, alloy, bal Fe.
For tools, dies, machinery parts; fatigue resistant.

HI-FORM 40.
M-67; 0.10 max C, 0.60 Max Mn, 0.15 max P, 0.035 max S, 0.05 max Cb, Fe.
Sheet: 50 ksi TS; 40 ksi YS; 28 El.
For transportation and mobile equipment parts requiring good formability and weldability.

HI-FORM 50.
M-67; 0.15 max C, 1.40 max Mn, 0.005-0.15 Cb, bal Fe.
Sheet: 60 ksi TS; 50 ksi YS; 25 El.
For transportation and mobile equipment parts requiring good formability and weldability.

HI-FORM 60.
M-67; 0.15 max C, 1.40 max Mn, 0.005-0.15 Cb, bal Fe.
Sheet: 70 ksi TS; 60 ksi YS; 21 El.
For transportation and mobile equipment parts requiring good formability and weldability.

HI-FORM 70.
M-67; 0.15 max C, 1.40 max Mn, 0.005-0.15 Cb, bal Fe.
Sheet: 80 ksi TS; 70 ksi YS; 18 El.
For transportation and mobile equipment parts requiring good formability and weldability.

HI-FORM 80.
M-67; 0.15 max C, 1.40 max Mn, 0.005-0.15 Cb, bal Fe.
Sheet: 90 ksi TS; 80 ksi YS; 16 El.
For transportation and mobile equipment parts requiring good formability and weldability.

HIGH BRASS.
M-8; 65 Cu, 35 Zn.
Rolled: 47,000-75,000 TS; 20,000-60,000 YS; 50-60 El; 50-75 RA; 45-180 Brin.
For spun parts, tanks, vessels; high strength, good ductility.

HIGH BRASS 262.
M-279; 68,5 Cu, 31.5 Zn.
Ann: 44,000-61,000 TS; 13,000-34,000 YS; 35-65 El.
Cold rolled: 49,000-110,000 TS; 20,000-100,000 YS; 1-52 El.
A general purpose yellow brass.
For formed parts, hardware, electrical assemblies.

HIGH CARBON ALLOY.
M-389; 0.9 C, 0.2 V, bal Fe.
For tools, drills, taps; water hardening.

HIGH CHROME-NICKEL.
M-200; 0.33-0.45 C, 0.40-0.75 Cr, 1.0-1.35 Mn, 0.90-1.20 Ni, 0.40-0.75 Si, bal Fe.
Annealed: 85,000-100,000 TS; 60,000 YS; 18 El; 23 RA.
For gears, shafts; tough.

HIGH CONDUCTIVITY BRONZE 405.
M-279; 95 Cu, 4 Zn, 1 Sn.
Ann: 38,000-45,000 TS; 9000-17,000 YS; 40-50 El.
Cold rolled: 41,000-82,000 TS; 15,000-74,000 YS; 20-45 El.
Moderate strength high conductivity bronze for fuse clips, meter clips, electrical assemblies.

HIGH DUTY.
M-1409; 2.2 C, 0.6 W, 12.8 Cr, 0.4 Mn, bal Fe.
For blanking and thread rolling dies, molds; oil or air hardened, abrasion resistant.

HIGH DUTY GREY IRON.
M-106; 3.2-3.4 Total C, 2.1-2.3 Si, 0.60-0.90 Mn, 0.55-0.75 Comb. C, 0.20-0.60 Cr, 0.15 max P, 0.12 max S, bal Fe.
230-280 N/mm^2 TS; 385-485 N/mm^2 TrS; (1.2 in. dia test bar)
Mainly for cylinder blocks.

HIGH EXPANSION.
M-32; 0.10 C, 0.50 Mn, 0.25 Si, 3.1 Cr, 2.20 Ni, bal Fe.
Annealed: 70,000 TS; 40,000 YS; 35 El; B 74 Rock.
For applications requiring high thermal expansion, bimetals.
High thermal expansion, austenitic, non-magnetic.

HIGH LEADED BRASS 62.
M-141; 62.25 Cu, 35.75 Zn, 2 Pb.
Hard: 74,000 TS; 60,000 YS; 7 El; 150 Brin.
Soft: 49,000 TS; 17,000 YS; 52 El; 61 Brin.
For clock gears, key stock; high leaded brass.

HIGH LEADED BRASS 62%.
M-33; 62 Cu, 36 zn, 2 Pb.
Ann: 49,000 TS; 18,00 YS; 50 El.
Rolled hard: 74,000 TS; 63,000 YS; 9 El.
For hardware, clocks, gears, locks, meters.
Very good machinability.

HIGH LEADED BRASS-342.
M-8; 64.0 Cu, 34.0 Zn, 2.0 Pb.
Annealed: 49,000 TS; 17,000 YS; 52 El; F 68 Rock.
Half hard: 61,000 TS; 50,000 YS; 20 El; B 70 Rock.
For plaques, hinges, gears, pinions, wheels, valve stems, rivets.

HIGH LEADED BRASS 353.
M-8; 61,5 Cu, 36.7 Zn, 1.8 Pb.
Annealed: 49,000 TS; 17,000 YS; 52 El; F 68 Rock.
Half Hard: 61,000 TS; 50,000 YS; 20 El; B 70 Rock.
For plaques, hinges, gears, pinions, wheels, valve stems, rivets.
Corrosion resistant.
Electrical conductivity 26, free-cutting.

HIGH LEADED BRASS 353.
M-279; 61 Cu, 37 Zn, 2 Pb.
Ann: 45,000-54,000 TS; 12,000-24,000 YS; 45-60 El.
Cold rolled: 49,000-99,000 TS; 20,000-80,000 YS; 1-50 El.
High strength yellow brass, good machinability.
For clock and watch plates, and gears, brass keys.

HIGH LEADED BRASS 3531.
M-8; 37 Zn, 2 Pb, bal Cu.
Hard: 58,000 TS; 42,000 YS; 22 El; 125 Brin.
Soft: 47,000 TS; 18,000 YS; 58 El; 62 Brin.
For hardware, clock parts, screw macine parts; free-cutting.

HIGH LEADED BRASS 3532.
M-8; 38 Zn, 1.5 Pb, bal Cu.
Hard: 73,000 TS; 60,000 YS; 7 El; 150 Brin.
Soft: 45,000 TS; 17,000 YS; 50 El; 60 Brin.
For hardware, clock parts; free cutting.

HIGH LEADED TUBE BRASS.
M-33; 66 Cu, 32.25 Zn, 1.75 Pb.
Annealed: 52,000 TS; 20,000 YS; 50 El.
Drawn: 75,000 TS; 60,000 YS; 7 El.
For screw machine products; free cutting.

HIGH MANGANESE NICKEL.
M-Eng; 98-94 Ni, 2-6 Mn.
For spark plug wire; heat resistant.

HIGH PERMEABILITY 49.
M-32; 0.1 max C, 49 Ni, bal Fe.
Rolled: 25 El; 62 RA; 200 Brin.
For transformer cores; high permeability.

HIGH PHOSPHORUS COPPER see **COPPER DEOXIDIZED D H P.**

HIGH SILICON BRONZE A.
M-33; 97 Cu, 3 Si.
Ann: 58,000 TS; 22,000 YS; 60 El.
Drawn 36%; 92,000 TS; 55,000 YS; 22 El.
Marine and pole line hardware, bolts, shafts.

HIGH SPEED N M HOT ROD.
M-507.
Mild steel welding rod with alloy coating.
ASW 6010.

HIGH SPEED TOOL-A.
M-Eng.; 5-12 W, 6-12 Ti, 3-6 Si, 3-5 Al, up to 1.0 B, bal Ni.
For cutting tools, dies; cast nonferrous.

HIGH STRENGTH.
18.5-21.5 Zn, 2-3 Fe, 6-7 Al, 1.5-2.25 Ni, bal Cu.
For high strength corrosion resistant parts; corrosion resistant.

HIGH STRENGTH COMMERCIAL BRONZE.
M-8; Cu, 7.65 Zn, 1.75 Pb, 0.1 P, 1.0 Ni.
Hard: 68,000 TS; 58,000 YS; 10 El; B 75 Rock
For cable clamps, pole-line hardware, screw machine products.
High strength, corrosion resistance. Elect, cond. 32.

HIGH TEMPERATURE BRONZE.
M-US; 6.3 Zn, 2.7 Sn, bal Cu.
For fittings, hardware; high strength.

HIGH TENSILE BRASS NO. 1.
M-1518; Zn, Sn, bal Cu.
Cast: 67,200 TS; 20 El; 120-165 Brin.
For high strength castings; corrosion resistant.

HIGH TENSILE BRASS NO. 2.
M-1518; Zn, Sn, bal Cu.
Cast: 85,200 TS; 15 El; 170-220 Brin.
For heavy castings; high strength.

HIGH TENSILE BRASS NO. 3.
M-1518; Zn, Sn, bal Cu.
Cast: 107,500 TS; 12 El; 180-229 Brin.
For machine tool parts; high strength.

HIGH TENSILE BRONZE.
M-US; 68.5 Cu, 6.5 Al, 2.2 Mn, 2.5 Fe.
Forged: 101,000-110,000 TS; 50,000-57,000 YS; 10-5 El; 200-220 Brin.
For propellers, nuts, bolts; corrosion resistant.

HIGHTENSILE BRONZE NO. 1.
M-508; 60-70 Cu, 21-25 Zn, 3-7 Al, 3-7 Mn, 2-5 Fe.
Cast: 120,000 TS; 55,000 YS; 15 El; 220 Brin.
For pump liners, sleeves, bushings, gears, pinions; pressure tight castings.

HIGHTENSILE BRONZE NO. 2.
M-508; 60-70 Cu, 21-25 Zn, 3-7 Al, 3-7 Mn, 2-5 Fe.
Cast: 100,000 TS; 50,000 YS; 20 El; 200 Brin.
For pump liners, sleeves, bushings; pressure tight castings.

HIGHTENSILE BRONZE NO. 3.
M-508; 60-70 Cu, 21-25 Zn, 3-7 Mn, 2-5 Fe.
Cast: 90,000 TS; 45,000 YS; 25 El; 180 Brin.
For pump liners, sleeves, bushings; centrifugal castings.

HIGH TEST.
M-68; 2.75-3.15 T.C., 1.0-1.25 Ni, 0.6-1.0 Mn, 0.9-1.1 Si, bal Fe.
For brake drums, valve bodies; cast iron.

HIGH TIN COMMERCIAL BRONZE see **TIN BRASS 2% TIN.**

HIGHWAY.
M-131; C, Cu, bal Fe.
For building construction.

HIGH YIELD.
M-633; C, bal Fe.
For transportation equipment, buses, bridges.

HIGH YIELD see **RODNEY 270 & 290.**

HIGH YIELD 70/30.
M-141; 31.0 Ni, 0.5 Fe, 0.75 Mn, bal Cu.
Cold drawn: 72,000 TS; 50,000 YS; 15 El; 125 Brin.
For condenser and heat exchanger tubes, feedwater heaters; high strength, corrosion resistant.

HILANIC.
M-897; 79 Ni, 18 Co, 1 Si, 2 Fe.
Annealed: 85,000 TS; 60,000 YS; 35 El.
For cathodes and filaments in electronic tubes; heat resistant.

HILLS-MCCANNA NO. 1.
M-168; 89 Cu, 11 Sn.
35,000-40,000 TS; 22,000-25,000 YS; 6-10 El; 7-9 RA; 75-85 Brin.
For gears, worm wheels; known as "Stones English Worm Gear Bronze."

HILLS-MCCANNA NO. 2.
M-168; 88 Cu, 10 Sn, 2 Zn.
32,000-45,000 TS; 19,000-23,000 YS; 15-30 El; 12-25 RA; 65-75 Brin.
For gears, superheated steam and hydraulic castings, bearings; known as "Gun Metal" or "G-Bronze."

HILLS-MCCANNA NO. 10.
M-168; 80 Cu, 10 Sn, 10 Pb.
28,000-33,000 TS; 19,000-22,000 YS; 5-10 El; 6-11 RA; 52-60 Brin.
For bearings for heavy duty; resists shock and vibration.

HILLS-MCCANNA NO. 11.
M-168; 88 Cu, 10 Sn, 2 Pb.
30,000-40,000 TS; 19,000-23,000 YS; 15-25 El; 12-20 RA; 65-70 Brin.
For centrifugal pump parts, mine pump bodies, rods, gears; C.E.L. 15000

HILLS-MCCANNA NO. 20.
M-168; 85 Cu, 5 Sn, 5 Pb, 5 Zn.
27,000-33,000 TS; 15,000-19,000 YS; 16-20 El; 15-20 RA; 50-60 Brin.
For pump bodies, valves, bearings; known as "Ounce Metal" or "Red Composition."

HILLS-MCCANNA NO. 20 P.
M-168; 2 Si, 0.5 Fe, bal Cu.
Cast: 40,000 TS; 19,000 YS; 35 El; 30 RA; 70 Brin.
For bronze castings; corrosion resistant.

HILLS-MCCANNA NO. 21.
M-168; 82 Cu, 4 Sn, 6 Pb, 8 Zn.
28,000-33,000 TS; 14,000-16,000 YS; 15-20 El; 20-26 RA; 55-60 Brin.
For general use, bearings; C.E.L. 10000.

HILLS-MCCANNA NO. 22.
M-168; 65 Cu, 2 Pb, 30 Zn, 1 special element.
Cast: 30,000-35,000 TS; 25-35 El; 20-30 RA; 40-50 Brin.
For plumbing fixtures; SAE-41.

HILLS-MCCANNA NO. 25.
M-168; 99.6-99.9 Cu.
17,000-20,000 TS; 6000-9000 YS; 50-40 El; 70-60 RA; 30-40 Brin.
For electrical installations, parts of electric welding machines; C.E.L 4000.

HILLS-MCCANNA NO. 30.
M-168; 56 Cu, 41 Zn, 1 Fe, 1 Al, 0.5 Mn, 0.5 Sn.
65,000-85,000 TS; 26,000-33,000 YS; 20-35 El; 18-30 RA; 104-119 Brin.
For propeller blades, hubs, valve stems, engine frames, bearings, gears; C.E.L. 28000.

HILLS-MCCANNA NO. 30 A.
M-168; 39 Zn, 0.5 Mn, 1.5 Al, bal Cu.
Cast: 57,000-62,000 TS; 32,000-35,000 YS; 24-35 El; 27-30 RA; 96-99 Brin.
For bronze castings; Mn bronze.

HILLS-MCCANNA NO. 30 B.
M-168; 40 Zn, 1 Al, 1.5 Fe, bal Cu.
Cast: 71,000-80,000 TS; 30,000-35,000 YS; 25-30 El; 25-30 RA; 130-135 Brin.
For bronze castings; SAE-43.

HILLS-MCCANNA NO. 30 C.
M-168; 3, Zn, 1 Al, 0.5 Mn, bal Cu.
cast: 115,000-119,000 TS; 99,000-102,000 YS; 9-9.5 El; 8.4-8.6 RA; 220-235 Brin.
For Mn bronze castings; corrosion resistant.

HILLS-MCCANNA NO. 35.
M-168; 92 Al, 8 Cu.
16,000-22,000 TS; 11,000-13,000 YS; 1-2 El; 1-2 RA; 50-55 Brin.
For crankcases, housings, automobile castings; C.E.L. 10000.

HILLS-MCCANNA NO. 39.
M-168; special hardener, 3 Si, 1 Fe, bal Cu.
Cast: 57,000 TS; 26,000 YS; 27 El; 27 RA; 100 Brin.
For castings; same as "Everdur D."

HILLS-MCCANNA NO. 40.
M-168; 90 Cu, 10 Al.
65,000-80,000 TS; 23,000-28,000 YS; 20-30 El; 21-29 RA.
For structural parts; non-magnetic; wear resistant.

HILLS-MCCANNA NO. 41.
M-168; 77 Cu, 11 Al, 5 Fe, 7 Mn.
80,000-97,800 TS; 28,000-26,200 YS; 30-10 El; 29-12 RA; 190 Brin.
For structural parts; wear resistant; C.E.L. 28300.

HILLS-MCCANNA NO. 42.
M-168; 88 Cu, 8 Al, 4 Fe.
65,000-80,000 TS; 23,000-28,000 YS; 20-30 El; 21-29 RA; 92-100 Brin.
For gears, bushings, bolts; wear resistant; C.E.L. 19000.

HILLS-MCCANNA NO. 43.
M-168; special hardener, 10 Al, 1 Fe, bal Cu.
Cast: 84,000 TS; 33,000 YS; 22 El; 25 RA; 134 Brin.
For bearings, bushings; Al bronze.

HILLS-MCCANNA NO. 45.
M-168; 89 Cu, 10 Al, 1 Fe.
65,000-80,000 TS; 23,000-28,000 YS; 20-30 El; 21-29 RA; 92-100 Brin.
For equipment in oil refineries; wear and corrosion resistant; C.E.L. 19000.

HILLS-MCCANNA NO. 46.
M-168; 81 Cu, 11 Al, 5 Fe, 3 Mn.
80,000-94,300 TS; 28,000-26,400 YS; 30-10 El; 29-11 RA; 176 Brin.
For packing tools in the explosive industry; wear resistant; C.E.L. 26100.

HILLS-MCCANNA NO. 47.
M-168; 80 Cu, 12 Al, 8 Fe.
65,000-80,000 TS; 23,000-28,000 YS; 20-30 El; 21-29 RA.
For bearings, gears to resist heavy loads and shocks; wear resistant; C.E.L. 22100.

HILLS-MCCANNA NO. 48.
M-168; special hardener, 9 Al, bal Cu.
Cast: 72,000 TS; 55,000 YS; 3 El; 3 RA; 229 Brin.
For construction parts; Al bronze.

HILLS-MCCANNA NO. 49.
M-168; special hardener, 10 Al 4 Fe, bal Cu.
Cast: 55,000 TS; 55,000 YS; 0 El; 0 RA; 300 Brin.
For dies for forming tools; Al bronze.

HILLS-MCCANNA NO. 50.
M-168; 52 Cu, 31 Ni, 12 Fe, 5 Cr.
92,300-75,000 TS; 26,400-42,000 YS; 8.5-4.5 El; 8.5-7.8 RA; 20l-190 Brin.
For hydraulic valve seats; corrosion resistant; C.E.L. 33040.

HILLS-MCCANNA NO. 50A.
M-168; 60.5 Cu, 23.5 Ni, 2 Sn, 3 Pb, 11 Zn.
38,000-45,000 TS; 19,000-22,000 YS; 13-21 El; 15-20 RA; 85-90 Brin.
For vessels to resist H_2SO_4 corrosion resistant.

HILLS-MCCANNA NO. 52.
M-168; 74.5 Cu, 16.5 Ni, 9 Al.
85,000 TS; 36,000 YS; 4 El; 3.5 RA; 213 Brin.
For vessels to resist hot oils; corrosion resistant; C.E.L. 29610.

HILLS-MCCANNA NO. 53.
M-168; 80 Cu, 10 Al, 5 Ni, 5 Fe.
Untreated: 95,000 TS; 44,000 YS; 9 El; 18 RA; 170 Brin.
Heat treated: 110,000 TS; 77,000 YS; 3 El; 9 RA; 235 Brin.
For spark proof tools; corrosion resistant.

HILLS-MCCANNA NO. 54.
M-168; 60-65 Ni, 23-28 Cu, 3.5 Fe, 2 Mn, 0.75 Si, 0.25 C.
65,000-70,000 TS; 35,000-40,000 YS; 30-25 El; 100 Brin.
For chemical apparatus; corrosion resistant; C.E.L. 29610.

HILLS-MCCANNA NO. 55.
M-168; 65-67 Ni, 30-32 Cu, 2.5-3.5 Si, 2.5-3.5 Fe.
75,000-80,000 TS; 44,000-48,000 YS; 10.5-10 El; 16-13 RA; 170 Brin.
For still plugs in oil industries; corrosion resistant.

HILLS-MCCANNA NO. 102.
M-168; 88 Cu, 10 Sn, 2 Zn.
Cast: 48,000 TS; 24,000 YS; 42 El; 35 RA; 85 Brin.
For castings; SAE-62.

HILLS-MCCANNA NO. 102 P.
M-168; special hardener, 2 Si, 1 Fe, bal Cu.
Cast: 52,000 TS; 26,000 YS; 17 El; 19 RA; 119 Brin.
For castings; Si bronze.

HILLS-MCCANNA NO. 105.
M-168; 85.75 Cu, 10 Sn, 2.5 Pb, 1.75 Ni.
Cast: 44,500 TS; 33,500 YS; 12 El; 15 RA; 96 Brin.
For gears.

HILLS-MCCANNA NO. 110.
M-168; 80 Cu, 10 Sn, 10 Pb.
Cast: 45,000 TS; 25,000 YS; 32 El; 24 RA; 64 Brin.
For bearings; SAE-64.

HILLS-MCCANNA NO. 111.
M-168; 88 Cu, 10 Sn, 2 Pb.
Cast: 49,000 TS; 24,000 YS; 49 El; 36 RA; 81 Brin.
For bearings; SAE-63.

HILO.
M-61; 75 Ni, 18 Co, 2 Ti, bal Fe.
Annealed: 106,000-90,000 TS; 30-20 El.
For radio tube filaments; magnetic.

HILO.
M-897; 75 Ni, 18 Co, 2 Ti, 5 Fe.
Annealed: 105,000 TS; 75,000 YS; 30 El.
For cathodes and filaments in electronic tubes; heat resistant.

HI MAG PERM.
M-1425; 0.03-0.05 C, 0.005-0.009 P, 0.01-0.02 Si, 0.03-0.07 Cr, 0.006 -0.01 Al, 0.04-0.07 Mo, bal Fe.
Ann: 40,000 TS; 20,000 YS; 40 El; 78 RA; 69 Brin.
For electrical applications, magnetic control devices, magnetic clutches and chucks, pole pieces, armatures.
Higgh magnetic permeability. Vacuum degassed.

HI-MAN.
M-67; 0.25 C, 1.35 Mn, 0.30 Si, 0.20 Cu, bal Fe.
Rolled: 75,000 TS; 50,000 YS; 20 El.
For railroad and agriculture equipment, mine cars, auto bodies; high strength, low alloy construction steel.

HI-MAN 440.
M-67; 0.28 C, 1.35 Mn, 0.30 Si, 0.20 Cu, bal Fe.
Rolled: 70,000 TS; 50,000 YS.
For railroad and agriculture equipment, mine cars; high strength, low alloy construction steel.

HI-MO.
M-57; 0.83 C, 1.5 W, 8.75 Mo, 4 Cr, 1.25 V, bal Fe.
For cutting tools, reamers; high speed steel.

HINGE.
M-Eng.; 62.0-63.5 Cu, 1-2 Pb, bal Zn.
Soft: 50,000 TS; 50 El; 65 Brin.
Hard: 85,000 TS; 0 El; 180 Brin.
Engraving brass.

HIPERCO 50-FM see CARPENTER HIPERCO 50-FM.

HIPERSIL.
M-118; 3-4 Si, bal Fe.
Cast: 50,000 TS; 45,000 YS; 6 El.
For transformer cores; high permeability.

HI-PHY KIRKSITE.
 M-88, M-794; Cu, Al, Mg, bal Zn.
 For forming dies; refined grain.

HI-PRO.
 M-275; 2.2 C, 12 Cr, 0.25 V, bal Fe.
 For blanking and deep drawing dies, punches, trimming dies; shear blades, ring gages, liners, lamination dies; high resistance to wear and abrasion; AISI D3.
 Was Vulcan Hi-Pro.

HI-PROOF 304.
 M-1724; 0.06 max C, 17.5-19.0 Cr, 8.0-11.0 Ni, 0.25 max N, bal Fe.
 High proof stress version of 304.
 Cryogenic, storage, and pressure vessels.
 Similar to AISI 304.

HI-PROOF 304L.
 M-1724; 0.03 max C, 17.5-19.0 Cr, 9.0-12.0 Ni, 0.25 max N, bal Fe.
 High proof stress version of 304l.
 For cryogenic, storage, and pressure vessels.
 Similar to AISI 304l.

HI-PROOF 316.
 M-1724; 0.07 max C, 16.5-18.5 Cr, 10.0-13.0 Ni, 2.25-3.0 Mo, 0.25 max N, bal Fe.
 High proof stress version of 316.
 For cryogenic storage and pressure vessels.

HI-PROOF 316L.
 M-1724; 0.03 max C, 16.5-18.5 Cr, 11.0-14.0 Ni, 2.25-3.0 Mo, 0.25 max N, bal Fe.
 High proof stress version of 316l.
 Cryogenic storage and pressure vessels.

HI-PROOF 347.
 M-1724; 0.08 max C, 17.0-19.0 Cr, 9.0-12.0 Ni, Nb = 10 x C/1.00, 0.15-0.25 N, bal Fe.
 High proof strength version of 347.

HI-QUA-LED.
 M-1378; leaded alloy steels.
 For screw machine products; free-cutting.

HI-QUA-LED 10L45.
 M-1378; 0.43-0.50 C, 0.6-0.9 Mn, Pb, bal Fe.
 Heat treated: 110,000 TS; 68,500 YS; 22 El; 44 RA; 230 Brin.
 For gears, bolts, machine tool parts; free-cutting, leaded steel.

HI-QUA-LED 10L50.
 M-1378; 0.48-0.50 C, 0.6-0.9 Mn, Pb, bal Fe.
 Heat treated: 117,000 TS; 72,500 YS; 22 El; 43 RA; 235 Brin.
 For crankshafts, gears, tie rods, bushings; free-cutting, leaded steel.

HI-QUA-LED 10L60.
 M-1378; 0.55-0.65 C, 0.7-1.0 Mn, Pb, bal Fe.
 Heated treated: 142,000 TS; 81,000 YS; 16 El; 32 RA; 270 Brin.
 For crankshafts, die blocks, girders; free-cutting, leaded steel.

HI-QUA-LED 10L70.
 M-1378; 0.65-0.75 C, 0.9 Mn, Pb, bal Fe.
 Heat treated: 145,000-157,000 TS; 87,000-104,000 YS; 14.0-15.5 El; 26.5-37.0 RA; 321-285 Brin.
 For punches, hammers, tools; free-cutting leaded steel.

HI-QUA-LED 41L30.
 M-1378; 0.28-0.35 C, 0.4-0.6 Mn, 0.8-1.1 Cr, 0.15-0.25 Mo, Pb, bal Fe.
 Heat treated: 126,000 TS; 105,000 YS; 18 El; 52 RA; 262 Brin.
 For gears, bolts, crankshafts; free-cutting, leaded steel.

HI-QUA-LED 41L37.
 M-1378; 0.35-0.40 C, 0.8-1.1 Cr, 0.15-0.25 Mo, Pb, bal Fe.
 Heat treated: 122,000 TS; 100,000 YS; 23 El; 62 RA; 248 Brin.
 For gears, bolts, crankshafts; free-cutting, leaded steel.

HI-QUA-LED 41L40.
 M-1378; 0.38-0.43 C, 0.75-1.0 Mn, 0.8-1.1 Cr, 0.20 Mo, Pb, bal Fe.
 Heat treated: 118,000-132,000 TS; 92,000-106,000 YS; 20-17 El; 57-46 RA; 235-277 Brin.
 For gears, bolts, machine tools parts; free-cutting, leaded steel.

HI-QUA-LED 43L40.
 M-1378; 0.38-0.43 C, 1.6-2.0 Ni, 0.7-0.9 Cr, 0.3 Mo, Pb, bal Fe.
 Heat treated: 145,000 TS; 125,000 YS; 16 El; 48 RA; 310 Brin.
 For machine tool parts; free-cutting leaded steel.

HIROX.
 M-118; 6-10 Al, 3-9 Cr, 0-4 Mn, B, Zr, bal Fe.
 At 70°F: 118,850 TS; 111,650 YS; 15 El; 42 RA.
 At 1300°F: 15,000 TS; 14,950 YS; 94 El; 98 RA.
 For resistances, heating elements; high electrical resistivity and high oxidation resistance to 2300°F.

HI-RUN.
 M-335; 1.55 C, 11.5 Cr, 0.8 Mo, 0.9 V, bal Fe.
 Air hardening.
 For tools, dies.
 AISI D2 Hi Chrome-Hi Carbon tool steel.

HI SHOCK 60 see **CARPENTER HI SHOCK 60.**

HI-TEM-IRON NO. 7.
 M-23; 3 C, 1.5 Si, Ni, Cr, bal Fe.
 For retorts, dye and paint equipment; corrosion and heat resisting, cast iron.

HI-TENSILE.
M-801; 7-8 Zn, 0.2-0.5 Mg, 0.4-1.0 Cu, 0.2 Ti, bal Al.
T5-temper: 35,000 TS; 26,000 YS; 6 El; 75 Brin.
For aircraft and machine tool casting; room temperature aging properties.

HITENSILOY NO. 11.
M-122; 57 Cu, 1 Pb, 1.75 Ni, bal Zn.
Rolled: 80,000 TS; 45,000 YS; 165 Brin.
For pump rods, valve stems, bolts, nuts, die castings; nickel brass.

HITENSO 162.
M-8; 99.0 Cu, 1.0 Cd.
(formerly Hitenso 961)
Soft (rod) 37,000 TS; 12,000 YS; 50 El; 47 Rock F.
Hard (rod) 58,000 TS; 50,000 YS; 15 El; 65 Rock B.
For electrical equipment, trolley wire, contact shoes.
High electrical conductivity, wear resistant.
Copper Alloy No. 162.

HITENSO 165.
M-8; 0.8 Cd, 0.02 Si, 0.60 Sn, bal Cu.
Hard: 65,000 TS; 55,000 YS; 15 El; B 75 Rock.
Soft: 40,000 TS; 14,000 YS; 55 El; B 6 Rock.
Wire: 95,000 TS; 1.5 El.
For electrical equipment, trolley wire, contact shoes, electrical welding tips and wheels. 58% electrical conductivity, high strength. Copper Alloy No. 165.

HI-TENSO 961 see HITENSO 162.

HI-TENSO 965 see HI-TENSO 165.

HI-TENSO 1622.
M-8; 1 Cd, bal Cu.
Hard wire; 90,000 TS; 1.5 El.
For low load transmission lines, trolley wire, spring contracts.
Corrosion and wear resistant.
Copper Alloy No. 1622.

HITEN-SPEED 45.
M-478; 0.33-0.41 C, 1.3-1.6 Mn, bal Fe.
For gears, bolts, crankshafts, axles; oil hardened, tough.

HITEN-SPEED 55.
M-478; 0.30-0.41 C, 1.3-1.9 Mn, 0.25-0.34 Mo, bal Fe.
For gears, bolts, crankshafts, axles; oil hardened, tough.

HI-VAN.
M-341; 1.1 Si, 5.25 Cr, 1.2 Mo, 0.9 V, bal Fe.
For punches, upsetters, forming dies; hot work steel, oil hardened.

HI-VAN NO. 28.
M-0.85; C, 4 Cr, 19 W, 2 V, 0.75 Mo, bal Fe.
For tools, cutters, reamers, taps, gauges; high speed steel.

HI-WEAR 64 see CARPENTER HI-WEAR 64.

HI-YAW-TEN.
M-1540; 0.12 max C, 0.25-0.75 Si, 0.2-0.5 Mn, 0.06-0.12 P, 0.25-0.5 Cu, 0.65 max Ni, 0.4-1.0 Cr, 0.15 max Ti, bal Fe.
Ann: 67,000 min TS; 50,000 min YS.
Rolled: 71,000 min TS; 57,000 min YS.
For buildings, rolling stock, buses, mine cards, bridges.
Resists atmospheric corrosion.

HI-YIELD 42.
M-633; 0.21 C, 0.9 Mn, 0.01 min Cb, bal Fe.
Plate: 63,000 TS; 42,000 YS; 24 El.
For trucks, mine cars, derricks, bridges, plow frames, pressure vessels.
Good fabricability and weldability.

HI-YIELD 45.
M-633; 0.22 C, 1.25 Mn, 0.01 min Cb, bal Fe.
Plate: 60,000 TS; 45,000 YS; 22 El.
For trucks, mine cars, derricks, bridges, plow frames, pressure vessels.
Good fabricability and weldability.

HI-YIELD 50.
M-633; 0.22 C, 1.25 Mn, 0.01 min Cb, bal Fe.
Plate: 65,000 TS; 50,000 YS; 20 El.
For trucks, mine cars, derricks, bridges, plow frames, pressure vessels.
Good fabricability and weldability.

HI-YIELD 55.
M-633; 0.25 C, 1.35 Mn, 0.01 min Cb, bal Fe.
Plate: 70,000 TS; 55,000 YS; 18 El.
For trucks, mine cars, derricks, bridges, plow frames, pressure vessels.
Good fabricability and weldability.

HI-Z.
M-1538; 0.10-0.18 C, 0.60-1.0 Mn, 0.15-0.35 Si, 0.002-0.006 B, 0.4-0.8 Cr, 0.15-0.50 Cu, 0.4-0-6 Mo, 0.7-1.0 Ni, 0.03-0.10 V, bal Fe.
Plate: 114,000-135,000 TS; 100,000 min YS; 20 min El.
For pressure vessels, bridges, mine cars, power shovels, cranes, trucks, trailers.
Shock and wear resistant. Heat treated by mill.

HIZUTIT 700.
M-1760; 0.10 max C, 1.5-1.8 Si, 1.0 max Mn, 1.5-2.0 Cr, bal Fe.
Heat resisting steel.
W.-Nr. 1.4700.

HIZUTIT 701.
M-1760; 0.12 max C, 2.0-2.5 Si, 1.0 max Mn, 5.5-6.5 Cr, bal Fe.
Heat resisting steel.
W.-Nr. 1.4712.

HIZUTIT 702.
M-1760; 0.12 max C, 0.5-1.0 Si, 1.0 max Mn, 6.0-7.0 Cr, 1.0 Al, bal Fe.
Heat resisting steel.
W.-Nr. 1.4713.

HIZUTIT 703.
M-1760; 0.12 max C, 1.9-2.4 Si, 1.0 max Mn, 12.0-14.0 Cr, bal Fe.
Heat resisting steel.
W.-Nr. 1.4722.

HIZUTIT 704.
M-1760; 0.12 max C, 0.90-1.40 Si, 1.0 max Mn, 12.0-14.0 Cr, 1.2 Al, bal Fe.
Heat resisting steel.
W.-Nr. 1.4724.

HIZUTIT 705.
M-1760; 0.12 max C, 1.9-2.4 Si, 1.0 max Mn, 17.0-19.0 Cr, bal Fe.
Heat resisting steel.
W.-Nr. 1.4741.

HIZUTIT 706.
M-1760; 0.12 max C, 0.70-1.20 Si, 1.0 max Mn, 17.0-19.0 Cr, 1.2 Al, bal Fe.
Heat resisting steel.
W.-Nr. 1.4742.

HIZUTIT 707.
M-1760; 0.12 max C, 1.2-1.5 Si, 1.0 max Mn, 23.0-25.0 Cr, 1.7 Al, bal Fe.
Heat resisting steel.
W.-Nr. 1.4762.

HIZUTIT 709.
M-1760; 0.15-0.25 C, 0.80-1.30 Si, 2.0 max Mn, 24.0-26.0 Cr, 3.3-5.5 Ni, bal Fe.
Heat resisting steel.
W.-Nr. 1.4821.

HIZUTIT 710.
M-1760; 0.20 Max C, 1.8-2.3 Si, 2.0 max Mn, 19.0-21.0 Cr, 11.0-13.0 Ni, bal Fe.
Heat resisting steel.
W.-Nr. 1.4828.

HIZUTIT 711.
M-1760; 0.20 max C, 1.8-2.3 Si, 2.0 max Mn, 24.0-26.0 Cr, 19.0-21.0 Ni, bal Fe.
Heat resisting steel.
W.-Nr. 1.4841.

HIZUTIT 712.
M-1760; 0.15 max C, 0.75 max Si, 2.0 max Mn, 24.0-26.0 Cr, 19.0-22.0 Ni, bal Fe.
Heat resisting steel.
W.Nr. 1.4845.

HIZUTIT 713.
M-1760; 0.15 max C, 1.5-2.0 Si, 2.0 max Mn, 15.0-17.0 Cr, 34.0-37.0 Ni, bal Fe.
Heat resisting steel.
W.-Nr. 1.4864.

HIZUTIT 714.
M-1760; 0.15 C, max C, 1.0 max Si, 2.0 max Mn, 17.0-19.0 Cr, 9.0-11.0 Ni, 0.70 Ti, bal Fe.
Heat resisting steel.
W.-Nr. 1.4878.

HIZUTIT G 1.
M-1760; 0.20-0.40 C, 1.0-2.5 Si, 1.0 max Mn, 6.0-8.0 Cr, bal Fe.
Heatproof steel castings.
W.-Nr. 1.4710.

HIZUTIT G 2.
M-1760; 0.30-0.60 C, 1.0-2.5 Si, 1.0 max Mn, 12.0-14.0 Cr, bal Fe.
Heat proof steel castings.
W.-Nr. 1.4729.

HUZUTIT G 3.
M-1760; 0.40-0.60 C, 1.0-2.5 Si, 1.0 max Mn, 16.0-18.0 Cr, bal Fe.
Heat proof steel castings. W.-Nr. 1.4740.

HIZUTIT G 4.
M-1760; 0.30-0.60 C, 1.0-2.5 Si, 1.0 max Mn, 22.0-24.0 Cr, bal Fe.
Heatproof steel castings.
W.-Nr. 1.4745.

HIZUTIT G 5.
M-1760; 0.30-0.60 C, 1.0-2.5 Si, 1.0 max Mn, 27.0-30.0 Cr, bal Fe.
Heatproof steel castings. W. Nr. 1.4776.

HIZUTIT G 6.
M-1760; 1.2-1.4 C, 1.0-2.5 Si, 1.0 max Mn, 27.0-30.0 Cr, bal Fe.
Heatproof steel castings.
W.-Nr. 1.4777.

HIZUTIT G 7.
M-1760; 0.30-0.50 C, 1.0-2.0 Si, 1.0 max Mn, 26.0-28.0 Cr, 3.5-5.5 Ni, bal Fe.
Heatproof steel castings.
W.-Nr. 1.4823.

HIZUTIT G 8.
M-1760; 0.30-0.50 C, 1.0-2.5 Si, 1.5 max Mn, 21.0-23.0 Cr, 9.0-11.0 Ni, bal Fe.
Heatproof steel castings.
W.-Nr. 1.4826.

HIZUTIT G 9.
M-1760; 0.15-0.35 C, 1.0-2.5 Si, 1.5 max Mn, 19.0-21.0 Cr, 13.0-15.0 Ni, bal Fe.
Heatproof steel castings.
W.-Nr. 1.4832.

HIZUTIT G 10.
M-1760; 0.20-0.50 C, 1.0-2.5 Si, 1.5 max Mn, 25.0-28.0 Cr, 13.0-16.0 Ni, bal Fe.
Heatproof steel castings.
W.-Nr. 1.4846.

HIZUTIT G 11.
M-1760; 0.20-0.50 C, 1.0-2.5 Si, 1.5 max Mn, 24.0-27.0 Cr, 19.0-21.0 Ni, bal Fe.
Heatproof steel castings.
W.-Nr. 1.4848.

HIZUTIT G 12.
M-1760; 0.20-0.50 C, 1.0-2.5 Si, 1.5 max Mn, 16.0-19.0 Cr, 38.0-39.0 Ni, bal Fe.
Heatproof steel castings.
W-Nr. 1.4865.

HIZUTIT G 13.
M-1760; 1.40-1.80 C, 1.0-2.5 Si, 1.0 max Mn, 17.0-19.0 Cr, bal Fe.
Heatproof steel castings.
W.-Nr. 1.4743.

HIZUTIT G 14.
M-1760; 0.20-0.50 C, 1.0-2.5 Si, 1.5 max Mn, 24.0-26.0 Cr, 11.0-14.0 Ni, bal Fe.
Heatproof steel castings.
W.-Nr. 1.4837.

HIZUTIT G 15.
M-1760; 0.15-0.35 C, 1.0-2.5 Si, 1.5 max Mn, 17.0-19.0 Cr, 8.0-10.0 Ni, bal Fe.
Heatproof steel castings.
W.-Nr. 1.4825.

HIZUTIT G 16.
M-1760; 0.10-0.20 C, 1.0-2.0 Si, 1.5 max Mn, 24.0-27.0 Cr, 19.0-21.0 Ni, bal Fe.
Heatproof steel castings.
W.-Nr. 1.4849.

HK 4 M.
M-1296; 0.2-0.3 C, 0.75 max Si, 1.5 max Mn, 24-26 Cr, 24-26 Ni, 0.2-0.6 Ti, 0.2-0.6 Al, bal Fe.
75,400 min TS; 34,100 min YS; 25 min El.
For high temperature service on petrochemical plants, as reformer tubes and cracking tubes.

HK31A.
M-43; 3.0 Th, 0.7 Zr, bal Mg.
O-temper: 30,000-33,000 TS; 18,000-20,000 YS; 12-23 El.
H 24-temper: 34,000-39,000 TS; 24,000-31,000 YS; 4-14 El.
Sheet and plate have good formability and weldability; used at 400°F to 600°F in aircraft and missiles.
ASTM B90-69; AMS 4384, 4385.

HK31A.
M-Various foundries; 0.7 Zr, 3.0 Th, bal Mg.
T6-temper: 27,000-32,000 TS; 13,000-15,000 YS; 4-8 El.
At 700°F: 13,000 TS; 8000 YS; 26 El.
Magnesium sand castings with good pressure tightness, weldability and corrosion resistance. For parts operating at 400-700°F.
ASTM B80-69; AMS 4445; SAE 507; QQ-M-56.

HK-40CB.
M-1748; 0.40 C, 23 Cr, 21 Ni, 1.5 Cb, bal Fe.
As cast: 86,000 psi TS; 44,000 psi YS; 16 El; 19 RA.
Cast high temperature furnace fittings.
Improved creep strength and thermal fatigue resistance.

HM21A.
M-43; 0.6 Mn, 2.0 Th, bal Mg.
T8-temper: 33,000-35,000 TS; 18,000-23,000 YS; 6-11 El; 56 Brin. Sheet and plate have good properties 400-750°F; good formability and weldability.

HM31A (EXTRUSIONS).
M-43; 1.2 Mn, 3.0 Th, bal Mg.
T5-temper: 37,000-44,000 TS; 26,000-38,000 YS; 4-13 El; 63 Brin.
Extrusions: good elevated temperature properties, 400-800°F; used in aircraft and missiles.
AMS 4388,4389.

HMS-CHROME.
M-1038; 22-24 Cr, 1.4-1.55 C, 29-31 Si, bal Fe.
Ferro alloy for source of chromium and silicon, particularly in stainless steels.

HNS 40.
M-1488; 0.18 C, 1.1 Mn, 0.30 max Si, bal Fe.
23 mm max, normalized: 510-610 N/mm^2 TS; 355 N/mm^2 min YS; El.
For chains.
AFNOR 20 Mn 5.

HNS 50.
M-1488; 0.18 C, 1.1 Mn, 0.30 max Si, bal Fe.
23 mm max Q + T.: 785-880 N/mm^2 TS; 685 N/mm^2 min YS; 15
For chains.
AFNOR 20 Mn 5.

HNS 60.
M-1488; 0.20 C, 1.10 Mn, 0.30 max Si, 1.0 Cr, bal Fe.
23 mm max Q + T.: 930-1030 N/mm^2 TS; 835 N/mm^2 YS; 16 El.
For chains.
AFNOR 20 MC 4.

HNS 80.
M-1488; 0.22 C, 1.10 Mn, 0.30 max Si, 1.0 Cr, 0.20 Mo, bal Fe.
23 mm max, Q + T.: 1280-1380 N/mm^2 TS; 1175 N/mm^2 min YS; El.
For chains.
AFNOR 22 MCD 4.

HO 124.
M-1364; Al Si 12 Cu Ni Mg T6.
Chill cast: 180 N/mm^2 TS; 130 Brin.
Sand cast: 170 N/mm^2 TS; 85 Brin.
Aluminum castings with good elevated temperature strength and wear resistance, as engine pistons.

HOB-A-DIE.
M-289; 0.06 C, 0.3 Mn, 0.2 Si, 0.25 Mo, 1.0 Cr, bal Fe.
For Zn, casting dies; resists heat checking.

HOB-A-FORM.
0.06 max C, 0.15 Mn, 0.10 Si, bal Fe.
For plastic mold dies; water hardening, hobbed cavity.

HOBALITE.
M-40; C, alloy, bal Fe.
Air hardened.
For mold dies, hobs; air hardening.

HOBALITE.
M-289; 0.05 C, 0.15 Mn, bal Fe.
For plastic mold dies; case hardening steel.

HOBALITE.
M-783; 0.05 C, 0.16 Mn, bal Fe.
For mold dies; water hardened.

HOBALITE 500A.
M-217; 0.05 C, 0.05 Si, 0.10 Mn, bal Fe.
For deep hobbing steel for bakelite dies; case hardened.

HOBART 10.
M-1000; weld metal: 0.07 C, 0.50 Mn, 0.25 Si, 0.016 P, 0.020 S, bal Fe.
As welded: 62,000-72,000 psi TS; 51,000-67,000 psi YS.
AWS E-6010 all position coated electrode; DCEP.

HOBART 12.
M-1000; weld metal: 0.08 C, 0.38 Mn, 0.20 Si, 0.023 P, 0.026 S, bal Fe.
As welded: 67,000 psi TS; 55,000 psi YS; 17 El; 25-50 RA.
AWS E-6012 general purpose electrode; AC or DCEN.

HOBART 12A.
M-1000; weld metal: 0.07 C, 0.50 Mn, 0.13 Si, 0.016 P, 0.025 S, bal Fe.
As welded: 67,000 psi TS; 55,000 psi YS; 17 El.
AWS E-6012 electrode; DCEN.

HOBART 13A.
M-1000; weld metal: 0.09 C, 0.50 Mn, 0.41 Si, 0.015 P, 0.024 S, bal Fe.
As welded: 67,000 psi TS; 55,000 psi YS; 17 El.
AWS E-6013; AC, DCEN or DCEP.

HOBART 14A.
M-1000; weld metal: 0.09 C, 0.46 Mn, 0.32 Si, 0.018 P, 0.022 S, 0.039 Ni, 0.054 Cr, 0.020 Mo, 0.023 V, bal Fe.
As welded: 72,000 psi TS; 60,000 psi YS; 17 El.
AWS E-7014; DCEN or AC.

HOBART 16.
M-1000; weld metal: 0.08 C, 0.80 Mn, 0.35 Si, 0.015 P, 0.020 S, bal Fe.
As welded: 72,000 psi TS; 60,000 psi YS; 22 El.
AWS E-7016; DCEP or AC.

HOBART 24.
M-1000; weld metal: 0.04 C, 0.60 Mn, 0.50 Si, 0.035 P, 0.035 S, 0.052 Ni, 0.055 Cr, 0.022 Mo, 0.018 V, bal Fe.
As welded: 72,000 psi TS; 60,000 psi YS; 17 El.
AWS E-7024; AC or DCEN.

HOBART 24H.
M-1000; weld metal: 0.084 C, 0.76 Mn, 0.61 Si, 0.018 P, 0.010 S, bal Fe.
As welded: 83,500 psi TS; 71,000 psi YS; 25 El.
AWS E-7024; electrode for shielded metal arc welding.

HOBART 25 P.
M-1000; 0.11 C, 1.12 Mn, 0.50 Si, bal Fe.
Solid wire consumable electrode for electroslag welding of mild steel.

HOBART 27.
M-1000; weld metal: 0.08 C, 0.62 Mn, 0.17 Si, 0.010 P, 0.015 S, bal Fe.
As welded: 62,000 psi TS; 50,000 psi YS; 25 El.
AWS E-6027; AC or DCEN.

HOBART 27H.
M-1000; weld metal: 0.08 C, 1.0 Mn, 0.42 Si, 0.017 P, 0.016 S, bal Fe.
As welded: 71,000 psi TS; 61,000 psi YS; 28 El.
AWS E-6027; electrode for shielded metal arc welding.

HOBART 70AP.
M-1000; weld metal: 0.10 C, 0.30 Mn, 0.15 Si, 0.015 P, 0.010 S, 1.5 Ni, 0.17 Mo, bal Fe.
As welded: 70,000 psi TS; 57,000 psi YS; 22 El.
AWS E-7010-G; DCEP.
For welding pipe steels, storage tanks, drill platforms, ships.

HOBART 101P.
M-1000; weld metal: 0.11 C, 0.50 Mn, 0.20 Si, 0.10 P, 0.022 S, bal Fe.
As welded: 64,000-74,500 psi TS; 53,000-64,000 psi YS; 22 El; 35-60 RA.
AWS E-6010 all position electrode; DCEP.

HOBART 212A.
M-1000; weld metal: 0.06 C, 0.25 Mn, 0.15 Si, 0.018 P, 0.019 S, bal Fe.
As welded: 67,000 psi TS; 55,000 psi YS; 17 El.
AWS E-6012; DCEN or AC.

HOBART 308-15 & 308-16.
M-1000; weld metal: 0.07 max C, 19.0 Cr, 9.5 Ni, 1.6 Mn, 0.50 Si, bal Fe.
As welded: 85,000-95,000 psi TS; 40-50 El.
For welding AISI 301, 302, 304, 308 stainless.

HOBART 308L.
M-1000; 0.25 max C, 1.8 Mn, 0.40 Si, 20.5 Cr, 10.0 Ni, bal Fe.
80,000 psi TS; 40 El. Solid stainless wire.
For welding types 304, 308, 321 and 347 steels.
AWS A5.9-69 ER 308L.

HOBART 308L-15 & 308L-16.
M-1000; weld metal: 0.04 max C, 19.0 Cr, 9.5 Ni, 1.0 Mn, 0.30 Si, bal Fe.
As welded: 80,000-90,000 psi TS; 40-50 El.
For welding AISI 304 ELC stainless steel.

HOBART 308L HISIL.
M-1000; 0.25 max C, 1.8 Mn, 0.85 Si, 20.5 Cr, 10.0 Ni, bal Fe.
80,000 psi TS; 40 El.
Solid stainless steel welding electrode wire.
ER-308L Si.

HOBART 309.
M-1000; 0.25 max C, 1.8 Mn, 0.40 Si, 24.0 Cr, 13.5 Ni, bal Fe.
80,000 psi TS; 38 El.
Solid stainless steel welding electrode wire.
For welding 309 steels; ER-309.

HOBART 309-15 & 309-16.
M-1000; weld metal: 0.10 max C, 23.0 Cr, 13.0 Ni, 1.6 Mn, 0.50 Si, bal Fe.
As welded: 85,000-95,000 psi TS; 34-45 El.
For welding AISI 309 stainless steel.

HOBART 309CB-15 & 309CB-16.
M-1000; weld metal: 0.10 max C, 23.0 Cr, 13.0 Ni, 1.6 Mn, 0.80 Cb, 0.60 Si, bal Fe.
As welded: 85,000-95,000 psi TS; 30-40 El.
For welding AISI 347, 321 stainless clad steels.

HOBART 309MO-15.
M-1000; weld metal: 0.10 max C, 23.0 Cr, 13.0 Ni, 1.7 Mn, 2.2 Mo, 0.50 Si, bal Fe.
As welded: 85,000-95,000 psi TS; 35-45 El.
For welding AISI 316 stainless clad steel.

HOBART 310.
M-1000; 0.12 C, 1.8 Mn, 0.45 Si, 26.0 Cr, 21.0 Ni, bal Fe.
80,000 psi TS; 30 El.
Solid stainless steel welding electrode wire.
For welding AISI 310 stainless; ER-310.

HOBART 310-15 & 310-16.
M-1000; weld metal: 0.20 max C, 26.0 Cr, 21.0 Ni, 1.8 Mn, 0.40 Si, bal Fe.
As welded: 85,000-95,000 psi TS; 35-45 El.
For welding AISI 410, 430, 502 stainless steels.

HOBART 310CB-15 & 310CB-16.
M-1000; weld metal: 0.12 max C, 26.0 Cr, 21.0 Ni, 1.8 Mn, 0.80 Cb, 0.40 Si, bal Fe.
As welded: 85,000-95,000 psi TS; 30-40 El.
For welding AISI 310, 321, 347 stainless steels.

HOBART 310MO-15 & 310MO-16.
M-1000; weld metal: 0.12 max C, 26.0 Cr, 21.0 Ni, 1.8 Mn, 2.0 Mo, 0.40 Si, bal Fe.
As welded: 85,000-95,000 psi TS; 35-45 El.
For welding AISI 310 and 316 stainless steels.

HOBART 312-15 & 312-16.
M-1000; weld metal: 0.15 max C, 29.0 Cr, 9.5 Ni, 1.9 Mn, 0.50 Si, bal Fe.
As welded: 110,000-120,000 psi TS; 80,000-90,000 psi YS; 22-75 El.
Highest "as-welded" strength of any austenitic stainless weld.

HOBART 316-15 & 316-16.
M-1000; weld metal: 0.07 max C, 18.0 Cr, 13.0 Ni, 2.25 Mo, 1.7 Mn, 0.40 Si, bal Fe.
As welded: 85,000-95,000 psi TS; 35-45 El.
For welding AISI 316 stainless steel.

HOBART 316L.
M-1000; 0.025 max C, 1.8 Mn, 0.35 Si, 19.5 Cr, 13.0 Ni, 2.2 Mo, bal Fe.
78,000 psi TS; 30 El.
Solid stainless steel welding electrode wire.
For welding AISI 316 and 316L stainless; ER-316L.

HOBART 316L-15 & 316L-16.
M-1000; weld metal: 0.04 max C, 18.0 Cr, 13.0 Ni, 2.25 Mo, 1.0 Mn, 0.30 Si, bal Fe.
As welded: 80,000-90,000 psi TS; 35-45 El.
For welding AISI 316L stainless steel.

HOBART 316L HISIL.
M-1000; 0.025 max C, 1.8 Mn, 0.85 Si, 1..5 Cr, 13.0 Ni, 2.7 Mo, bal Fe.
78,000 psi TS; 30 El.
Solid stainless steel welding electrode wire.
For improved welding 316 and 316L steels; ER-316L-Si.

HOBART 317-15 & 317-16.
M-1000; weld metal: 0.07 max C, 19.0 Cr, 13.0 Ni, 3.5 Mo, 1.7 Mn, 0.50 Si, bal Fe.
As welded: 85,000-95,000 psi TS; 35-45 El.
For welding AISI 317 stainless steel.

HOBART 318-15 & 318-16.
M-1000; weld metal: 0.07 max C, 18.0 Cr, 12.0 Ni, 2.25 Mo, 1.6 Mn, 0.80 Cb, 0.60 Si, bal Fe.
As welded: 85,000-95,000 psi TS; 30-40 El.
For welding Type 318 stainless steel.

HOBART 320CB-15.
M-1000; weld metal: 0.07 max C, 20.0 Cr, 29.0 Ni, 3.0 Cu, 2.0 Mo, 0.5 Cb, 1.5 Mn, 0.40 Si, bal Fe.
As welded: 75,000-85,000 psi TS; 35-40 El.
For welding Carpenter No 20 or Durimet 20 and similar stainless steels.

HOBART 330-15 & 330-16.
M-1000; weld metal: 0.25 max C, 15.0 Cr, 35.0 Ni, 1.6 Mn, 0.30 Si, bal Fe.
As welded: 75,000-85,000 psi TS; 25-35 El.
For welding Type 330 stainless steel.

HOBART 330HC-15.
M-1000; weld metal: 0.30 C, 15.0 Cr, 35.0 Ni, 2.0 Mo, 0.30 Si, bal Fe.
As welded: 75,000-85,000 psi TS; 25-35 El.
For welding high carbon 330HC stainless.

HOBART 335A.
M-1000; weld metal: 0.10 C, 0.45 Mn, 0.20 Si, 0.015 P, 0.035 S, bal Fe.
As welded: 62,000 psi TS; 50,000 YS; 22 El; 22-63 RA.
AWS E-6011 all position electrode; AC, DCEP or DCEN.

HOBART 347.
M-1000; 0.06 C, 1.3 Mn, 0.40 Si, 19.5 Cr, 9.5 Ni, 0.80 Cb, bal Fe.
82,000 psi TS; 30 El.
Solid stainless steel welding electrode wire.
For welding AISI 347 and 321 stainless steels.
ER-347.

HOBART 347-15 & 347-16.
M-1000; weld metal: 0.07 max C, 19.0 Cr, 9.5 Ni, 0.80 Cb, 1.6 Mn, 0.60 Si, bal Fe.
As welded: 85,000-95,000 psi TS; 35-45 El.
For welding AISI 347 stainless steels.

HOBART 349-15 & 349-16.
M-1000; weld metal: 0.13 max C, 19.0 Cr, 9.0 Ni, 1.4 W, 0.50 Mo, 1.0 Cb, 1.5 Mn, 0.70 Si, bal Fe.
As welded: 105,000-110,000 psi TS; 80,000-85,000 psi YS; 27-37 El.
For welding material of similar analysis on turbo-jet engines.

HOBART 410-15 & 410-16.
M-1000; weld metal: 0.10 max C, 12.5 Cr, 0.60 Mn, 0.40 Si, bal Fe.
Welded and Ann: 80,000-90,000 psi TS; 55,000-60,000 psi YS; 30-35 El
For welding AISI 410 stainless steel.

HOBART 413.
M-1000; weld metal: 0.08 C, 0.50 Mn, 0.56 Si, 0.019 P, 0.020 S, bal Fe.
As welded: 67,000 psi TS; 55,000 psi YS; 17 El.
AWS E-6013; AC, DCEN or DCEP.

HOBART 430-15 & 430-16.
M-1000; weld metal: 0.10 max C, 16.0 Cr, 0.60 Mn, 0.60 Si, bal Fe.
Welded and Ann: 75,000-80,000 psi TS; 40,000-45,000 psi YS; 30-35 El.
For welding AISI 430 Stainless steels.

HOBART 447A.
M-1000; weld metal: 0.12 C, 0.41 Mn, 0.30 Si, 0.019 P, 0.022 S, bal Fe.
As welded: 67,000 psi TS; 55,000 YS; 17 El.
AWS E-6013; AC, DCEN or DCEP.

HOBART 718.
M-1000; weld metal: 0.08 C, 0.65 Mn, 0.37 Si, 0.011 P, 0.013 S, 0.010 Ni, 0.088 Cr, 0.018 Mo, 0.014 V, 0.011 Cu, bal Fe.
As welded: 72,000 pai TS; 60,000 psi YS; 22 El.
AWS E-7018; DCEP or AC.

HOBART 718 SR.
M-1000; weld metal: 0.04 C, 0.78 Mn, 0.32 Si, 0.21 P, 0.013 S, 0.015 Ni, 0.038 C, 0.160 Mo, 0.011 V, 0.013 Cu, bal Fe.
As welded: 72,000 psi TS; 60,000 psi YS; 22 El.
AWS E-7018; DCEP.

HOBART 1100.
M-1000; 0.05-0.20 Cu, 0.05 Mn, 1.0 Sit Fe; 0.10 Zn, 0.0008 B, 99.0 min Al.
Solid aluminum welding electrode wire; 25,000 psi min TS.
AWS A5.10-69 ER-1100.

HOBART 1139.
M-1000; weld metal: 0.06 C, 0.33 Mn, 0.31 Si, 0.020 P, 0.022 S, bal Fe.
As welded: 60,000 psi TS min.
AWS E-6022; DCEN or AC.

HOBART 4043.
M-1000; 0.30 Cu, 0.05 Mg, 0.05 Mn, 4.5-6.0 Si, 0.8 Fe, 0.10 Zn, 0.20 Ti, 0.0008 B, bal Al.
Solid aluminum welding electrode wire; 27,000 psi min TS; ER-4043.

HOBART 4047.
M-1000; 0.30 Cu, 0.10 Mg, 0.15 Mn, 11.0-13.0 Si, 0.08 Fe, 0.20 Zn, 0.0008 B, bal Al.
Solid aluminum welding electrode wire; 30,000 psi min TS.
ER-4047.

HOBART 5183.
M-1000; 0.10 Cu, 4.3-5.2 Mg, 0.5-1.0 Mn, 0.05-0.25 Cr, 0.40 Si, 0.40 Fe, 0.25 Zn, 0.15 Ti, 0.0008 B, bal Al.
Solid aluminum welding electrode wire; 42,000 psi min TS.
ER-5183.

HOBART 5356.
M-1000; 0.10 Cu, 4.5-5.5 Mg, 0.05-0.20 Mn, 0.05-0.20 Cr, 0.50 Si, Fe 0.10 Zn, 0.06-0.20 Ti, 0.0008 B, bal Al.
Solid aluminum welding electrode wire; 42,000 psi min TS.
ER-5356.

HOBART 5554.
M-1000; 0.10 Cu, 2.4-3.0 Mg, 0.5-1.0 Mn, 0.05-0.20 Cr, 0.40 Si, Fe, 0.25 Zn, 0.05-0.20 Ti, 0.0008 B, bal Al.
Solid aluminum welding electrode wire; 30,000 psi min TS.
ER-5554.

HOBART 5654.
M-1000; 0.05 Cu, 3.1-3.9 Mg, 0.01 Mn, 0.15-0.35 Cr, 0.45 Si, Fe, 0.20 Zn, 0.05-0.15 Ti, 0.0008 B, bal Al.
Solid aluminum welding electrode wire; 30,000 psi min TS.
ER-5654.

HOBART LH-718-MO.
M-1000; weld metal: 0.05 C, 0.75 Mn, 0.56 Si, 0.53 Mo, bal Fe (low hydrogen).
As welded: 79,000 psi TS; 68,000 YS; 31 El.
E7018-A1; for welding low alloy, high tensile steels of 50,000 psi min YS.

HOBART LH-818-CM.
M-1000; weld metal: 0.05 C, 0.68 Mn, 0.60 Si, 1.24 Cr, 0.49 Mo, bal Fe. (low hydrogen).
Welded, stress relieved 1275°F: 92,000 psi TS; 83,000 psi YS; 27 El; 62 RA.
E8018-B2; for welding low alloy Mo and Cr-Mo steels.

HOBART LH-818-N1.
M-1000; weld metal: 0.04 C, 1.04 Mn, 0.31 Si, 2.37 Ni, bal Fe (low hydrogen).
As welded: 88,500 psi TS; 73,800 psi YS; 28 El.
E8018-C1; for welding nickel steels that operate at low temperatures.

HOBART LH-818-N2.
M-1000; weld metal: 0.05 C, 0.84 Mn, 0.37 Si, 3.3 Ni, bal Fe. (low hydrogen).
As welded: 94,000 psi TS; 83,000 psi YS; 25 El.
E8018-C2; for welding nickel steels that operate at low temperatures.

HOBART LH-818-N3.
M-1000; weld metal: 0.05 C, 1.06 Mn, 0.38 Si, 1.04 Ni, bal Fe (low hydrogen).
As welded: 84,000 psi TS; 73,500 psi YS; 30 El.
E8018-C3; for welding 70,-80,000 psi steels for operation down to -60°F.

HOBART LH-918-CM.
M-1000; weld metal: 0.05 C, 0.78 Mn, 0.60 Si, 2.2 Cr, 1.05 Mo, bal Fe. (low hydrogen).
Welded, stress relieved 1275°F: 96,000 psi TS; 83,000 psi YS; 25 El; 67 RA.
E9018-B3; for welding steels up to 2% Cr and 1% Mo.

HOBART LH-918M.
M-1000; weld metal: 0.05 C, 1.11 Mn, 0.32 Si, 1.7i Ni, 0.28 Mo, bal Fe (low hydrogen).
As welded: 94,400 psi TS; 84,600 psi YS; 27 El.
E9018-M; for welding Hy-80 and Hy-90 and similar high tensile and heat treated steels.

HOBART LH-1018.
M-1000; weld metal: 0.06 C, 1.77 Mn, 0.68 Si, 0.44 Mo, bal Fe (low hydrogen).
As welded: 106,000 psi TS; 101,00 psi YS; 22 El.
E10018-D2; for welding high strength steels.

HOBART LH-1018M.
M-1000; weld metal: 0.05 C, 1.2 Mn, 0.54 Si, 1.73 Ni, 0.33 Mo, bal Fe (low hydrogen).
As welded: 103,000 psi TS; 96,000 psi YS; 24 El.
E10018-M; for strong welds in highly stressed military equipment.

HOBART LH-1118.
M-1000; weld metal: 0.06 C, 1.53 Mn, 0.27 Si, 0.31 Cr, 0.42 Mo, 1.88 Ni, bal Fe (low hydrogen).
As welded: 115,000 psi TS; 103,000 psi YS; 22 El.
E11018-M; for welding high strength steel.

HOBART LH-1218.
M-1000; weld metal: 0.05 C, 1.90 Mn, 0.25 Si, 0.85 Cr, 0.50 Mo, 2.0 Ni, bal Fe (low hydrogen).
As welded: 132,000 psi TS; 120,000 psi YS; 20 El.
E12018-M; for welding high strength steels. DIN high stressed assemblies.

HOBART LH-4130.
M-1000; weld metal: 0.25 C, 0.60 Mn, 0.22 Si, 0.80 Cr, 0.25 Mo, bal Fe (low hydrogen).
As welded: 121,000 psi TS; 93,500 psi YS; 16 El.
For welding SAE 4130 and 8630 and similar steels that are heat treated after welding.

HOBART LH-4340.
M-1000; weld metal: 0.37 C, 0.83 Mmn, 0.33 Si, 0.67 Cr, 0.24 Mo, 1.68 Ni, bal Fe. (low hydrogen).
As welded: 137,000 psi TS; 95,000 psi YS; 6 El.
For welding SAE 4140, 4330 and 4340 and similar steels that are to be heat treated after welding; weld metal will harden to same value as base metal.

HOBART PS-588.
M-1000; weld metal: 0.07 C, 1.1 Mn, 0.50 Si, 0.50 Cr, 0.80 Ni, 0.45 Cu, bal Fe.
Tubular welding wire for welding weathering steels.

HOBART QUIKFIL 525.
M-1000; 0.02 C, 0.20 Mn, 0.01 Si, 1.8 Ni, 0.25 Mo, bal Fe.
Granular material to enrich weld deposit in one pass submerged arc welding.

HOBB DIE STEEL.
M-627; 0.07 max C, 0.15 Mn, bal Fe.
For plastic molds, hobs; carburizing steel.

Section I: Alloy Data / 685

HOBRITE.
M-333; 0.05 max C, 0.11 Mn, bal Fe.
For hobbing molds; plastic mold dies; water hardening.

HOBSON NON-SHRINK.
M-340; 0.9 C, 1.2 Mn, 0.2 V, bal Fe.
For dies; and tools; non-deforming.

HOBSON'S CHOICE.
M-340; 1.48 C, 0.34 Mn, bal Fe.
For tools, cutters; water hardened.

HOBSON'S CHOICE DRILL RODS.
M-340; 1.1 C, bal Fe.
For tools, drills, taps; water hardened.

HOBSON'S CHOICE XX.
M-340; 0.7-1.2 C, 0.3 Mn, 0.3 Si, bal Fe.
For drills, punches, taps, reamers, hobs, broaches; water hardened; Type W1.

HOBSON'S FAST FINISHING.
M-340; 1.2 C, 0.2 V, bal Fe.
For fast finishing tools, cutters; water or oil hardened.

HOBSON'S OIL HARDENING.
M-340; 0.9 C, 1.2 Mn, bal Fe.
For tools, dies, punches; non-deforming.

HOBSON'S SPECIAL TAP RODS.
M-340; 1 C, bal Fe.
For tools, taps; water hardened.

HOBSON'S WARRANTED BEST.
M-340; 0.7-1.2 C, bal Fe.
For general tools; water hardening.

HOCKLEISTEUNGSTAHL.
M-Germ.; 10 Cr, 2 W, 0.5 V, 1.5 C, 1.2 Si, 0.4 Mn, bal Fe.
For shear blades, dies.

HOCO EXTRA EXTRA MH.
M-1338; 1.1 C, 0.25 max Si, 0.25 max Mn, bal Fe.
Annealed: 110,000 TS; 58,000 YS; 18 El; 40 RA; 210 Brin.
For springs, taps, cutters, drills, hobs; Type W1; water hardened.

HOCO EXTRA EXTRA SEHR ZAH.
M-1338; 0.70 C, 0.25 max Mn, bal Fe.
Heat treated: 175,000 TS; 128,000 YS; 12 El; 37 RA; 355 Brin.
For springs, axes, punches, crimpers; type w1; water hardened.

HOCO EXTRA EXTRA ZAH.
M-1388; 0.85 C, 0.25 max Si, 0.25 max Mn, bal Fe.
Heat treated: 190,000 TS; 145,000 YS; 10 El; 30 RA; 400 Brin.
For springs, taps, drills, hobs, reamers; type W1; water hardened.

HOCO EXTRA EXTRA ZH.
M-1338; 1.0 C, 0.25 max Si, 0.25 max Mn, bal Fe.
Annealed: 100,000 TS; 53,000 YS; 21 El; 42 RA; 200 Brin.
For springs, drills, reamers, broaches; type W1; water hardened.

HOCO EXTRA HART.
M-1338; 1.3 C, 0.25 max Si, 0.25 max Mn, bal Fe.
For engravers tools, forming dies, reamers; type W1; water hardened.

HOCO EXTRA MH.
M-1338; 1.15 C, 0.25 max Si, 0.25 max Mn, bal Fe.
Annealed: 110,000 TS; 58,000 YS; 20 El; 40 RA; 210 Brin.
For springs, tools, reamers, hobs, drills, taps; type W1; water hardened.

HOCO EXTRA SEHR HART.
M-1338; 0.70 C, 0.25 max Si, 0.25 max Mn, bal Fe.
Heat treated: 175,000 TS; 128,000 YS; 12 El; 37 RA; 355 Brin.
For springs, rails, punches, axes, hammers; type W1; water hardened.

HOCO EXTRA WEICH SCHWEISSBAR.
M-1338; 0.55 C, 0.10-0.40 Si, 0.50-0.70 Mn, bal Fe.
Heat treated: 160,000 TS; 115,000 YS; 12 El; 40 RA; 325 Brin.
For gears, bolts, shafts, machine tool parts; water hardened.

HOCO EXTRA ZAH.
M-1338; 0.85 C, 0.25 max Si, 0.25 max Mn, bal Fe.
Heat treated: 190,000 TS; 145,000 YS; 10 El; 30 RA; 400 Brin.
For springs, taps, reamers, drills, broaches; type W1; water hardened.

HOCO EXTRA ZH.
M-1338; 1.0 C, 0.25 max Si, 0.25 max Mn, bal Fe.
Annealed: 100,000 TS; 53,000 YS; 21 El; 42 RA; 200 Brin.
For springs, tools, cutters, drills; type W1; water hardened.

HOCO HSSP.
M-1338; 0.70 C, 1.7 Si, 0.70 Mn, bal Fe.
For springs; oil hardened, tough.

HOCO HWA.
M-1338; 0.30 C, 0.25 Si, 0.25 Mn, 2.5 Cr, 0.15 Mo, 0.1 V, bal Fe.
For gears, bolts, crankshafts; oil hardened, tough.

HOCO HWA50.
M-1338; 0.30 C, 2.35 Cr, 0.6 V, 4.25 W, bal Fe.
For extrusion dies, rams and liners; oil hardened, tough.

HOCO HWA90.
M-1338; 0.30 C, 2.65 Cr, 0.35 V, 8.5 W, bal Fe.
For extrusion dies, rams and liners, hot work steel, oil hardened.

HOCO HWA95K.
M-1338; 0.30 C, 2.4 Cr, 2.0 Co, 0.25 V, 8.5 W, bal Fe.
For extrusion dies and liners; hot work steel, oil hardened.

HOCO HWA100M.
M-1338; 0.65 C, 3.75 Cr, 0.85 Mo, 0.7 V, 8.5 W, bal Fe.
For lathe planer tools, taps, reamers; high speed steel.

HOCO HWAA40.
M-1338; 0.30 C, 1.0 Si, 1.1 Cr, 0.1, V, 3.75 W, bal Fe.
For punches, upsetters, riveters; oil hardened, tough.

HOCO HWAMV.
M-1338; 0.45 C, 1.4 Cr, 0.7 Mo, 0.3 V, bal Fe.
For header and forging dies; oil hardened, tough.

HOCO HWR4.
M-1338; 1.42 C, W, v, bal Fe.
For bearings, bushing, cutters; water or oil hardened, wear resistant.

HOCO HZR.
M-1338; 1.4 C, 0.30 Cr, 0.1 V, bal Fe.
For bearings, bushings, cutters; water or oil hardened, wear resistant.

HOCO KG.
M-1338; 1.0 C, 1.55 Cr, 0.35 Mn, bal Fe.
For bearings, cutters, liners; water or oil hardened, shock resistant.

HOCO PRIMA-MH.
M-1338; 0.90 C, 0.25-0.50 Si, 0.30-0.80 Mn, bal Fe.
Heat treated: 185,000 TS; 118,000 YS; 10 El; 30 RA: 375 Brin.
For springs, taps, drills, reamers; Type W1; water hardened.

HOCO PRIMA SEHR WEICH.
M-1338; 0.35 C, 0.25-0.50 Si, 0.30-0.80 Mn, bal Fe.
Hot rolled: 84,000 TS; 54,000 YS; 30 El; 53 RA; 185 Brin.
For gears, shafts, axles, bolts, screws; water hardened.

HOCXO PRIMA WEICH.
M-1338; 0.45 C, 0.25-0.50 Si, 0.30-0.80 Mn, bal Fe.
Hot rolled: 98,000 TS; 60,000 YS; E1; 54 RA; 212 Brin.
For axles, gears, bolts, shafts; water hardened.

HOCO PRIMA ZAH.
M-1338; 0.60 C, 0.25-0.50 Si, 0.30-0.80 Mn, bal Fe.
Heat treated: 160,000-115,000 TS; 113,000-77,000 YS; 12-23 El; 40-54 RA; 320-230 Brin.
For wheels, die blocks, girders, rail; water hardened.

HODI.
M-435; 0.28 C, 9 W, 4 Cr, 0.4 V, bal Fe.
For hot gripping and swaging dies.

HOESCH ARK 35.
M-1767; 0.12 C, 0.35 Si, 0.70 Mn, 3.8 Ni, 0.025 P, 0.025 S, (all max), Al, bal Fe.
Normalized: 440-610 N/mm^2 TS; 345 N/mm^2 min YS; 24 Min El.
Welded construction for low temperatures.

HOESCH ARK 50.
M-1767; 0.10 C, 0.35 Si, 0.70 Mn, 5.25 Ni, 0.025 P, 0.025 S, (all max), Al, bal Fe.
Normalized: 490 N/mm^2 TS; 390 N/mm^2 min YS; 25 min El.
Welded construction for low temperatures.

HOESCH ARK 90.
M-1767; 0.10 C, 0.30 Si, 0.70 Mn, 10.0 Ni, 0.025 P, 0.025 S, (all max), Al, bal Fe.
Normalized: 640-830 N/mm^2 TS; 490 N/mm^2 min YS; 18 min El.
Welded construction for low temperatures.

HOESCH ARK 6-29.
M-1767; 0.13 C, 0.35 Si, 1.50 Mn, 0.80 Ni, 0.025 P, 0.025 S, (all max), Al, bal Fe.
Normalized: 410-530 N/mm^2 TS; 285 N/mm^2 min YS; 24 min El.
Containers for liquified gas; for low temperatures.

HOESCH ARK 6-32.
M-1767; 0.14 C, 0.35 Si, 1.60 Mn, 0.80 Ni, 0.025 P, 0.025 S, (all max), Al, bal Fe.
Normalized: 440-560 N/mm^2 TS; 315 N/mm^2 min YS; 23 min El.
Tanks and ships; good at low temperatures.

HOESCH ARK 6-36.
M-1767; 0.15 C, 0.35 Si, 1.65 Mn, 0.80 Ni, 0.025 P, 0.025 S, (all max), Al, bal Fe.
Normalized: 490-610 N/mm^2 TS; 355 N/mm^2 min YS; 22 min El.
Containers for use at low temperatures.

HOESCH NOVAR K 260 PR.
M-1767; 0.10 C, 0.50 Si, 0.80 Mn, (all max), Nb, V, Ti, Al, bal Fe.
Cold rolled and process annealed: 350-480 N/mm^2 TS; 260 N/mm^2 YS; 26 min El.
For cold forming.

HOESCH NOVAR K 300 PR.
M-1767; 0.10 C, 0.50 Si, 1.0 Mn, (all max), Nb, V, Ti, Al, bal Fe.
Cold rolled and process annealed: 380-510 N/mm^2 TS; 300 N/mm^2 min YS; 24 min El.
For cold forming.

HOESCH NOVAR K 340 PR.
M-1767; 0.10 C, 0.50 Si, 1.20 Mn, (all max), Nb, V, Ti, Al, bal Fe.
Cold rolled and process annealed: 410-540 N/mm^2 TS; 340 M/mm^2 min YS; 22 min El.
For cold forming.

HOESCH NOVAR K 380 PR.
M-1767; 0.10 C, 0.50 Si, 1.30 Mn, (all max), Nb, V, Ti, Al, bal Fe.
Cold rolled and process annealed: 440-580 N/mm^2 TS; 380 N/mm^2 min YS; 20 min El.
For cold forming.

HOESCH NOVAR K 420 PR.
M-1767; 0.10 C, 0.50 Si, 1.40 Mn, (all max), Nb, V, Ti, Al, bal Fe.
Cold rolled and process annealed: 470-620 N/mm^2 TS; 420 N/mm^2 min YS; 18 min El.
For cold forming.

HOESCH NOVAR X 52 PR.
M-1767; 0.12 C, 0.25 Si, 1.40 Mn, 0.08 V, 0.05 Nb, (all max), Al, bal Fe.
Treated: 510-630 N/mm^2 TS; 365 min YS; 20 min El.
For welded pipes.

HOESCH NOVAR X 56 PR.
M-1767; 0.12 C, 0.25 Si, 1.40 Mn, 0.08 V, 0.05 Nb, (all max), Al, bal Fe.
Treated: 530-680 N/mm^2 TS; 385 N/mm^2 min YS; 19 min El.
For welded pipes.

HOESCH NOVAR X 60 PR.
M-1767; 0.12 C, 0.30 Si, 1.45 Mn, 0.08 V, 0.05 Nb, (all max), Al, bal Fe.
Treated: 550-710 N/mm^2 TS; 410 N/mm^2 min YS; 18 min El.
For welded pipes.

HOESCH NOVAR X 65 PR.
M-1767; 0.12 C, 0.35 Si, 1.50 Mn, 0.08 V, 0.05 Nb, (all max), Al, bal Fe.
Treated: 560-710 N/mm^2 TS; 445 N/mm^2 min YS; 18 min El.
For welded pipe.

HOESCH NOVAR X 70 PR.
M-1767; 0.12 C, 0.35 Si, 1.50 Mn, 0.08 V, 0.05 Nb, (all max), Al, bal Fe.
Treated: 600-750 N/mm^2 TS; 480 N/mm^2 min YS; 18 min El.
For welded pipe.

HOESCH NOVAR W 340 PR.
M-1767; 0.10 C, 0.50 Si, 0.80 Mn, 0.08 V, (all max), 0.02 min Nb, Al, bal Fe.
Treated: 420-540 N/mm^2 TS; 340 N/mm^2 min YS; 19 min El.
For cold forming.

HOESCH NOVAR W 38D PR.
M-1767; 0.10 C, 0.50 Si, 0.90 Mn, 0.08 V, (all max), 0.02 min Nb, Al, bal Fe.
Treated: 450-590 N/mm^2 TS; 380 N/mm^2 min YS; 18 min El.
For cold forming.

HOESCH NOVAR W 420 PR.
M-1767; 0.10 C, 0.50 Si, 1.0 Mn, 0.08 V, (all max), 0.02 min Nb, Al, bal Fe.
Treated: 480-620 N/mm^2 TS; 420 N/mm^2 min YS; 16 min El.
For cold forming.

HOESCH NOVAR W 460 PR.
M-1767; 0.10 C, 0.50 Si, 1.20 Mn, 0.08 V, (all max), 0.02 min Nb, Al, bal Fe.
Treated: 520-670 N/mm^2 TS; 460 N/mm^2 min YS; 14 min El.
For cold forming.

HOESCH NOVAR W 500 PR.
M-1767; 0.10 C, 0.50 Si, 1.40 Mn, 0.08 V, (all max), 0.02 min Nb, Al, bal Fe.
Treated: 550-700 N/mm^2 TS; 500 N/mm^2 min YS; 12 min El.
For cold forming.

HOESCH RESISTASTAHL 37.
M-1767; 0.13 C, 0.40 Si, 0.50 Mn, 0.50 Cu, 0.80 Cr, 0.40 Ni, (all max) Al, bal Fe.
360-440 N/mm^2 TS; 235 N/mm^2 YS; 25 min El.
For bridges, contraction, vehicles; resistant to atmospheric corrosion.

HOESCH RESISTASTAHL 37 EXTRA.
M-1767; 0.13 C, 0.40 Si, 0.50 Mn, 0.045 P, 0.50 Cu, 0.80 Cr, 0.40 N (all max), Al, bal Fe.
360-440 N/mm^2 TS; 235 N/mm^2 min YS; 25 min El.
For steel construction, penstock pipes; resistant to atmospheric corrosion.

HOESCH RESISTASTAHL 52.
M-1767; 0.12 C, 0.75 Si, 0.50 Mn, 0.55 Cu, 1.25 Cr, 0.65 Ni, (all max) Al, bal Fe.
510-610 N/mm^2 TS; 355 N/mm^2 min YS; 22 min El.
For bridge construction, industrial plants; resistant to atmospheric corrosion.

HOESCH RESISTASTAHL 52 EXTRA.
M-1767; 0.15 C, 0.40 Si, 1.30 Mn, 0.45 P, 0.50 Cu, 0.80 Cr, 0.40 Ni 0.10 V, (all max), Al, bal Fe.
510-610 N/mm^2 TS; 355 N/mm^2 min YS; 22 min El.
For construction machinery, vehicles; resistant to atmospheric corrosion.

HOESCH UNION 26 AK.
M-1767; 0.16 C, 0.40 Si, 1.30 Mn, (all max), Al, bal Fe.
Normalized: 360-480 N/mm^2 TS; 255 N/mm^2 min YS; 25 min El.
Special non-ageing quality with low temperature toughness.

HOESCH UNION 26 G.
M-1767; 0.18 C, 0.40 Si, 1.30 Mn, (all max), Al, bal Fe.
Normalized: 360-480 N/mm^2 TS; 255 N/mm^2 min YS; 25 min El (26 kg/mm^2 min YS).
Standard quality weldable structural steel.

HOESCH UNION 26 W.
M-1767; 0.18 C, 0.40 Si, 1.30 Mn, (all max), Al, bal Fe.
Normalized: 360-480 N/mm^2 TS; 255 N/mm^2 min YS; 25 min El.
Weldable structural steel; for resisting high temperature.

HOESCH UNION 29 AK.
M-1767; 0.16 C, 0.40 Si, 1.4 Mn, (all max), Al, bal Fe.
Normalized: 390-510 N/mm^2 TS; 285 N/mm^2 min YS; 24 min El.
Special non-ageing quality with low temperature toughness.

HOESCH UNION 29 G.
M-1767; 0.18 C, 0.40 Si, 1.40 Mn, (all max), Al, bal Fe.
Normalized: 390-510 N/mm^2 TS; 285 N/mm^2 min YS; 24 min El.
Standard quality, weldable, fine-grained structural steel.

HOESCH UNION 29 W.
M-1767; 0.18 C, 0.40 Si, 1.40 Mn, (all max), Al, bal Fe.
Normalized: 390-510 N/mm^2 TS; 285 N/mm^2 min YS; 24 min El.
Weldable structural steel; for resisting high temperature.

HOESCH UNION 32 AK.
M-1767; 0.16 C, 0.45 Si, 1.50 Mn, (all max), Al, bal Fe.
Normalized: 440-560 N/mm^2 TS; 315 N/mm^2 min YS; 23 min El.
Non-ageing quality with low temperature toughness for liquified gas containers, etc.

HOESCH UNION 32 G.
M-1767; 0.18 C, 0.45 Si, 1.50 Mn, (all max), Al, bal Fe.
Normalized: 440-560 N/mm^2 TS; 315 N/mm^2 min YS; 23 min El.
For bridges, steel structures, vehicles.

HOESCH UNION 32 W.
M-1767; 0.18 C, 0.45 Si, 1.50 Mn, (all max), Al, bal Fe.
Normalized: 440-560 N/mm^2 TS; 315 N/mm^2 min YS; 23 min El.
For boiler construction, engine parts, etc., for resisting high temperature.

HOESCH UNION 36 AK.
M-1767; 0.18 C, 0.50 Si, 1.60 Mn, (all max), Al, bal Fe.
Normalized: 490-630 N/mm^2 TS; 355 N/mm^2 min YS; 22 min El.
Non-ageing quality for low temperature operations.

HOESCH UNION 36 G.
M-1767; 0.20 C, 0.50 Si, 1.60 Mn, (all max), Al, bal Fe.
Normalized: 490-630 N/mm^2 TS; 355 N/mm^2 min YS; 22 min El.
For vehicles, steel structures, spheres.

HOESCH UNION 36 NB.
M-1767; 0.18 C, 0.50 Si, 1.60 Mn, (all max), 0.02 min Nb, Al, bal Fe.
Normalized: 490-630 N/mm^2 TS; 355 N/mm^2 min YS; 22 min El.
For tanks and containers.

HOESCH UNION 36 W.
M-1767; 0.20 C, 0.50 Si, 1.60 Mn, (all max), Al, bal Fe.
Normalized: 490-630 N/mm^2 TS; 355 N/mm^2 min YS; 22 min El.
For equipment to operate at elevated temperatures.

HOESCH UNION 39 AK.
M-1767; 0.18 C, 0.60 Si, 1.60 Mn, 0.80 Ni, 0.20 V or Ti, (all max), bal Fe.
Normalized: 500-650 N/mm^2 TS; 380 N/mm^2 min YS; 20 min El.
Storage and transport of liquified gas.

HOESCH UNION 39 G.
M-1767; 0.18 C, 0.60 Si, 1.60 Mn, 0.80 Ni, 0.20 V or Ti, (all max), bal Fe.
Normalized: 500-650 N/mm^2 TS; 380 N/mm^2 min YS; 20 min El.
Steel structures, vehicles, tanks.

HOESCH UNION 39 W.
M-1767; 0.18 C, 0.60 Si, 1.6 Mn, 0.80 Ni, 0.20 V or Ti, (all max), bal Fe.
Normalized: 500-650 N/mm^2 TS; 380 N/mm^2 min YS; 20 min El.
Equipment for elevated temperatures.

HOESCH UNION 43 AK.
M-1767; 0.18 C, 0.60 Si, 1.70 Mn, 0.80 Ni, 0.20 V or Ti, (all max), bal Fe.
Normalized: 530-680 N/mm² TS; 420 N/mm² min YS; 19 min El.
Non-ageing quality with low temperature toughness.

HOESCH UNION 43 G.
M-1767; 0.18 C, 0.60 Si, 1.70 Mn, 0.80 Ni, 0.20 V or Ti, (all max), bal Fe.
Normalized: 530-680 N/mm² TS; 420 N/mm² min YS; 19 min El.
For structures, transport equipment.

HOESCH UNION 43 W.
M-1767; 0.18 C, 0.60 Si, 1.70 Mn, 0.80 Ni, 0.20 V or Ti, (all max), bal Fe.
Normalized: 530-680 N/mm² TS; 420 N/mm² min YS; 19 min El.
Equipment to resist high temperature.

HOESCH UNION 47 AK.
M-1767; 0.20 C, 0.60 Si, 1.70 Mn, 0.80 Ni, 0.20 V or Ti, (all max), bal Fe.
Normalized: 560-730 N/mm² TS; 460 N/mm² min YS; 17 min El.
Non-ageing quality with low temperature toughness.

HOESCH UNION 47 G.
M-1767; 0.20 C, 0.60 Si, 1.70 Mn, 0.80 Ni, 0.20 V or Ti, (all max), bal Fe.
Normalized: 560-730 N/mm² TS; 460 N/mm² min YS; 17 min El.
For penstock construction, bridges, vehicles.

HOESCH UNION 47 W.
M-1767; 0.20 C, 0.60 Si, 1.70 Mn, 0.80 Ni, 0.20 V or Ti, (all max), bal Fe.
Normalized: 560-730 N/mm² TS; 460 N/mm² min YS; 17 min El.
Equipment for resisting high temperature.

HOESCH UNION H 60 L.
M-1767; 0.24 C, 0.60 Si, 1.6 Mn, 1.2 Cr, Al, bal Fe.
Normalized: 590 N/mm² min TS; 175 min Brin.
Wear resistant, hardenable alloy for construction, agricultural equipment and mixers.

HOESCH UNION H 90 L.
M-1767; 0.22 C, 0.40 Si, 1.3 Mn, 0.70 Cr, 0.2 Ti, Al, bal Fe.
Normalized: 880 N/mm² min TS; 265 min Brin.
For parts not subject to abrasion either normalized or hardened.

HOESCH UNION H 90 U.
M-1767; 0.50 C, 0.60 Si, 2.0 Mn, Al, bal Fe.
Normalized: 880 N/mm² min TS; 265 min Brin.
Wear resistant and hardenable alloy for agricultural equipment, chutes.

HOESCH UNION Q 260.
M-1767; 0.16 C, 0.50 Si, 1.20 Mn, 0.2 Ti, (all max), Al, bal Fe.
Normalized: 370-490 N/mm² TS; 260 N/mm² min YS; 30 min El.
For cold forming and bending; ageing resistant.

HOESCH UNION Q 300.
M-1767; 0.16 C, 0.50 Si, 1.40 Mn, 0.2 Ti, (all max), Al, bal Fe.
Normalized: 400-520 N/mm² TS; 300 N/mm² min YS; 27 min El.
For cold forming and bending; ageing resistant.

HOESCH UNION Q 340.
M-1767; 0.16 C, 0.50 Si, 1.50 Mn, 0.2 Ti, (all max), Al, bal Fe.
Normalized: 460-580 N/mm² TS; 340 N/mm² min YS; 25 min El.
For cold forming and bending; ageing resistant.

HOESCH UNION Q 340 TM.
M-1767; 0.12 C, 0.50 Si, 1.30 Mn, (all max), Al, V, Nb, Ti, bal Fe.
Treated: 420-540 N/mm² TS; 340 N/mm² min YS; 25 min El.
For cold forming and bending; ageing resistant.

HOESCH UNION Q 360.
M-1767; 0.18 C, 0.50 Si, 1.60 Mn, 0.2 Ti, (all max), Al, bal Fe.
Normalized: 480-610 N/mm² TS; 360 N/mm² min YS; 25 min El.
For cold forming and bending; ageing resistant.

HOESCH UNION Q 380.
M-1767; 0.18 C, 0.50 Si, 1.60 Mn, 0.2 Ti, (all max), Al, bal Fe.
Normalized: 500-640 N/mm² TS; 380 N/mm² min YS; 25 min El.
For cold forming and bending; ageing resistant.

HOESCH UNION Q 380 TM.
M-1767; 0.12 C, 0.50 Si, 1.40 Mn, (all max), Al, V, Nb, Ti, bal Fe.
Treated: 450-580 N/mm² TS; 380 N/mm² min YS; 23 min El.
For cold forming and bending; ageing resistant.

HOESCH UNION Q 420.
M-1767; 0.20 C, 0.50 Si, 1.60 Mn, 0.2 Ti, (all max), Al, bal Fe.
Normalized: 530-670 N/mm² TS; 420 N/mm² min YS; 23 min El.
For cold forming and bending; ageing resistant.

HOESCH UNION Q 420 TM.
M-1767; 0.12 C, 0.50 Si, 1.50 Mn, (all max), Al, V, Nb,Ti, bal Fe.
Treated: 480-620 N/mm² TS; 420 N/mm² min YS; 21 min El.
For cold forming and bending; ageing resistant.

HOESCH UNION Q 460.
M-1767; 0.21 C, 0.50 Si, 1.70 Mn, 0.2 Ti, (all max), Al, bal Fe.
Normalized: 550-700 N/mm^2 TS; 460 N/mm^2 min YS; 21 min El.
For cold forming and bending; ageing resistant.

HOESCH UNION Q 460 TM.
M-1767; 0.12 C, 0.50 Si, 1.60 Mn, (all max), Al, V, Nb, Ti, bal Fe.
Treated: 520-670 N/mm^2 TS; 460 N/mm^2 min YS; 19 min El.
For cold forming and bending; ageing resistant.

HOESCH UNION Q 500.
M-1767; 0.22 C, 0.50 Si, 1.70 Mn, 0.2 Ti, (all max), Al, bal Fe.
Normalized: 580-730 N/mm^2 TS; 500 N/mm^2 min YS; 19 min El.
For cold forming and bending; ageing resistant.

HOESCH UNION Q 500 TM.
M-1767; 0.12 C, 0.50 Si, 1.70 Mn, (all max), Al, V, Nb, Ti, bal Fe.
Treated: 550-700 N/mm^2 TS; 500 N/mm^2 min YS; 17 min El.
For cold forming and bending, ageing resistant.

HOESCH UNION Q 550.
M-1767; 0.24 C, 0.50 Si, 1.70 Mn, 0.2 Ti, (all max), Al, bal Fe.
Normalized: 590-750 N/mm^2 TS; 550 N/mm^2 min YS; 18 min El.
For cold forming and bending; ageing resistant.

HOFORS-1.
M-224; 1.0 C, 1.1 Mn, 0.6 Si, 1.0 Cr, bal Fe.
For large bell and roller bearing rings; oil hardened; non-deforming.

HOFORS-3.
M-224; 1.0 C, 1.5 Cr, bal Fe.
For ball bearings, master gauges, taps.

HOFORS-7.
M-224; 1.1 C, 0.6 Cr, bal Fe.
For taps, twist drills, dies, punches, reamers.

HOFORS-9.
M-224; 1.1 C, 0.5 Cr, bal Fe.
For bearings, rollers, reamers, taps.

HOFORS-13.
M-224; 1.05 C, 1.0 Cr, bal Fe.
For ball bearing rings, rollers, punches, taps.

HOFORS-22.
M-224; 1.0 C, 1.1 Cr, 0.35 Mo, bal Fe.
For cutting tools, chisels, drill, rollers, taps; wear resistant.

HOFORS-46.
M-224; 0.90 C, 1.15 Mn, 0.5 Cr, 0.5 W, 0.1 V, bal Fe.
For blanking dies, broaches, form tools, punches, taps; oil hardened; tough.

HOFORS-100.
M-224; 1.0 C, bal Fe.
For twist drills, punches, shear blades; water hardened.

HOFORS-711.
M-224; 0.50 C, 0.70 Si, 1.15 Cr, 2.5 W, 0.15 V, bal Fe.
For broaches, milling cutters, blanking and heading dies; oil hardened.

HOFORS KN-13.
M-224; 0.18 C, 0.3 Si, 0.55 Mn, 0.75 Cr, 3 Ni, bal Fe.
For gears, camshafts, crankshafts; case hardened.

HOFORS KN-14.
M-224; 0.4 C, 0.3 Si, 0.8 Mn, 0.8 Cr, 1.25 Ni, bal Fe.
For gears, shafts; shock resistant.

HOFORS KN-16.
M-224; 0.15 C, 0.3 Si, 0.55 Mn, 0.65 Cr, 1.25 Ni, bal Fe.
For gears, camshafts, crankshafts; case hardened.

HOFORS KN-17.
M-224; 0.12 C, 0.75 Cr, 3.0 Ni, bal Fe.
For gears, camshafts, crankshafts; case hardened.

HOFORS KN-18.
M-224; 0.35 C, 0.3 Si, 0.55 Mn, 1.15 Cr, 2.6 Ni, bal Fe.
For gears, machinery parts; shock resistant.

HOFORS KN-19.
M-224; 0.3 C, 0.3 Si, 0.55 Mn, 1.25 Cr, 4.25 Ni, bal Fe.
For gears, shafts; tough, oil hardened.

HOFORS KNY-1.
M-224; 0.3 C, 0.55 Mn, 1.05 Cr, 3.25 Ni, 0.25 Mo, bal Fe.
For gears, machinery parts, shafts; tough.

HOFORS KY-2.
M-224; 0.25 C, 0.65 Mn, 1.05 Cr, 0.2 Mo, bal Fe.
For gears, machinery parts; tough.

HOFORS M-11.
M-224; 0.40 C, 0.30 Si, 1.25 Mn, bal Fe.
For machinery parts; tough.

HOFORS NY-1.
M-224; 0.2 C, 0.55 Mn, 1.8 Ni, 0.25 Mo, bal Fe.
For gears, shafts; case hardened.

HOFORS SM-5.
M-224; 0.55 C, 1.75 Si, 0.75 Mn, bal Fe.
For springs; oil hardened.

HOHENZOLLERN BRASS.
M-Germ.; 30-40 Zn, bal Cu.
For gears, worm wheels; a high tensile brass.

HOLDER BLOCK.
M-694; 0.30 C, Cr, Ni, Mo, bal Fe.
Prehardened steel, tempered to 250-310 Brin.
Used primarily in die casting dies and plastic molds.

HOLDERTEM.
M-64; 0.40 C, 0.85 Mn, 0.95 Cr, 0.2 Mo, bal Fe.
Prehardened low alloy steel for shafts, arbors, structural parts, and special purpose tools.

HOLFOS AB1 AS CAST.
M-169; 878 Cu, 9.5 Al, 2.5 Fe.
Cast: 72,000-78,000 TS; 25,000-31,000 YS; 20 El; 120-140 Brin.
For high strenght components in food handling, acid resistant fittings, flanges.

HOLFOS AB2 AS CAST.
M-169; 80 Cu, 10 Al, 4.5 Fe, 5.5 Ni.
Cast: 90,000-95,000 TS; 36,000 YS; 15 El; 165 Brin.
For gears, shafts, components in marine equipment, food and oil industry.
BS 1400 AB2-C; BS 1073; ASTM B148-65T-9D.

HOLFOS B O.
M-169; 87.8 Cu, 11.5 Sn, 0.7 P.
Cast: 38,000-54,000 TS; 27,000-47,000YS; 5-7 El; 90-130 Brin.
(Properties vary with method of casting).
Good wear and abrasion characteristics.
For gears, shafts, wear parts.
BS 1400 PB1-C and PB2-C; BS 1059.

HOLFOS B O H T.
M-169; 87.8 Cu, 11.5 Sn, 0.7 P.
Centrifugally-spun and continously cast.
As cast: 45,000-54,000 TS; 22,400-23,500 YS; 20-40 El; 85-90 Brin.
Good wear and abrasion resistance, resistant to sea water.
For elevating nuts, bearings, small worm gears.
DTD 900/4454/A.

HOLFOS CT1.
M-169; 89.85 Cu, 10.0 Sn, 0.15 max P.
Cast tin bronze; 70-130 Brin. depending on casting method.
For bearings, pump bodies, piston shafts, marine equipment.
BS 1400 CT1-C.

HOLFOS G1.
M-169; 88 Cu, 10 Sn, 2 Zn.
Cast; 38,000-50,000 TS; 18,000-27,000 YS; 3-15 El; 85-100 Brin. (Properties vary with method of casting).
(Known as Admiralty Gunmetal).
For pumps, shaft liners, hydraulic parts.
BS 1400 G1-C; BS 383: SAE 62; AMS 4845D.

HOLFOS G3 (AS CAST).
M-169; 85.2 Cu, 7 Sn, 2.3 Zn, 5.5 Ni.
Cast tin-nickel bronze (gunmetal); 70-130 Brin depending on casting methods and conditions.
Good wear resistance.
For gear wheels, actuating nuts, spindles.
BS 1400 G3-C.

HOLFOS G 3-TF (FULLY HEAT TREATED).
M-169; 85.2 Cu, 7 Sn, 2.3 Zn, 5.5 Ni.
Heat treated cast tin-nickel bronze; improved strength and wear properties.
160-200 Brin.
BS 1400 G3-TF.

HOLFOS HTB 1.
M-169; 60 Cu, 36 Zn, 1 Al, 2 Mn, 1 Fe.
Cast: 67,000-70,000 TS; 25,000-31,000 YS; 20 El; 110-120 Brin.
For screw down nuts in rolling mills and screw presses, neck bushings and stuffing boxes.
BS 1400 HTB1-C; SAE 43; AMS 4860A.

HOLFOS HTB 3.
M-169; 60 Cu, 30 Zn, 5 Al, 3 Mn, 2 Fe.
Cast: 107,000 TS; 60,000 YS; 12-15 El; 180-185 Brin.
For highly stressed hardware and components.
BS 1400 HTB3-C.

HOLFOS J H 17.
M-169; 84.7 Cu, 14.0 Sn, 1.3 P.
Centrifugally spun and continuously cast.
As cast: 49,000-63,000 TS; 43,000-58,000 YS; 1-1.5 El; 130-160 Brin.
Good wear resistance and resistant to sea water.
For aircraft auxiliary driving gears, timing gears.
(Ger) DIN 1705 G-SnBz14.

HOLFOS LB1.
M-169; 76 Cu, 9 Sn, 15 Pb.
Cast: 25,000-34,000 TS; 11,200-22,400 YS; 3-8 El; 65-90 Brin. (Properties vary with method of casting).
For bearings and bushings, acid resistant fttings.
BS 1400 LB1-C; ASTM B144-52-3D.

HOLFOS LB2.
M-169; 80 Cu, 10 Sn, 10 Pb.
Cast; 27,000-40,000 TS; 11,200-25,000 YS; 3-10 El; 70-95 Brin. (Properties vary with method of casting).
Good sliding properties, corrosion resistant.
For bearings for heavy duty.
BS 1400 LB2-C; BS 963: SAE 64.

HOLFOS LB4.
M-169; 85 Cu, 5 Sn, 10 Pb.
Cast: 22,400-38,000 TS; 9,000-22,400 YS; 5-25 El; 65-75 Brin. (Properties vary with method of casting).
For shafts, bushes and bearings.
BS 1400 LB4-C; SAE 66.

HOLFOS LB5.
M-169; 75 Cu, 5 Sn, 20 Pb.
Cast: 22,400-27,000 TS; 9,000-22,400 YS; 6-8 El; 60-70 Brin. (Properties vary with method of casting).
For bearings for agricultural and railroad equipment, flour mill equipment.
BS 1400 LB5-C; ASTM B144-52-3E.

HOLFOS LG1.
M-169; 83 Cu, 3 Sn, 9 Zn, 5 Pb.
Cast: 25,000-38,000 TS; 13,500-18,000 YS; 12-40 El; 55-70 Brin. (Properties vary with method of casting).
For ligtly loaded marine hardware.
BS 1400 LGI-C; BS 1159; ASTM B145-63-5A.

HOLFOS LG2.
M-169; 85 Cu, 5 Sn, 5 Zn, 5 Pb.
Cast: 25,000-38,000 TS; 13,500-22,400 YS; 7-25 El; 65-80 Brin. (Properties may vary with method of casting).
For pump castings, water pump impellers, hose couplings.
BS 1400 LG2-C; SAE 40; AMS 4855 B.

HOLFOS LG4.
M-169; 87 Cu, 7 Sn, 3 Zn, 3 Pb.
Cast: 36,000-43,000 TS; 18,000-22,400 YS; 5-20 El; 65-75 Brin. (Properties vary with method of casting).
For hydraulic pressure control equipment, petrol pump meters, valve bodies, compressor parts.
BS 1400 LS4-C.

HOLFOS LG773.
M-169; 83 Cu, 7 Sn, 3 Zn, 7 Pb.
Cast: 31,000-54,000 TS; 14,5000-22,400 YS; 12-18 El; 75-90 Brin. (Properties vary with methopd of casting).
For shafts, bearings, bushings, sliding surfaces, marine equipment.
SAE 660; ASTM B144-52-3B..

HOLFOS LPB1.
M-169; 88.2 Cu, 7.5 Sn, 4 Pb, 0.03 min P.
Cast: 27,000-46,000 TS; 11,200-27,000 YS; 2-30% El; 65-90 Brin. (Properties vary with method of casting).
Free machining bronze for bushings, valves, steam pressure fittings.
BS 1400 LPB1-C; BS 1061.

HOLFOS PB1.
M-169; 89.5 Cu, 10 Sn, 0.5 Min P.
Cast: 34,000-66,000 TS; 18,000-45,000 YS; 2-20 El; 70-120 Brin. (Properties vary with method of casting).
Good bearing properties.
For shafts, guide wheels, blade wheels for pumps and water turbines.
BS 1400 PB1-C, BS 2B8.

HOLFOS PB2.
M-169; 88.35 Cu, 11.5 Sn, 0.15 min P.
Cast: 36,000-54,000 TS; 20,000-47,000 YS; 6-10 El; 85-130 Brin. (Properties vary with method of casting).
Good wear and abrasion resistance.
For bearings, feed nuts operating under load, coupling blocks, worm wheels.
BS 1400 PB2-C; BS 421; SAE 65.

HOLFOS PB4.
M-169; 89.5 Cu, 10.0 Sn, 0.5 P.
Cast tin bronze; 70-130 Brin. depending on casting method.
For bearings with hard shafts, medium duty gears, marine equipment.
BS 1400 PB4-C.

HOLFOS W W.
M-169; 88.3 Cu, 11.5 Sn, 0.2 P.
Sand cast: 36,000 TS; 20,000 YS; 10 El; 895 Brin.
Good wear properties.
For gears, shafts, splined couplings.
BS 1400 PB2-C; BS 421 SAE 65.

HOLLO.
M-344; C, bal Fe.
For rock drills; water hardening.

HOLLOBAR.
M-341; 1.0 C, 1.2 Cr, 0.3 Mo, bal Fe.
For tools; water hardened.

HOLLOBAR.
M-822; 1.05 C, 0.38 Mn, 1.46 Cr, 0.28 Si, bal Fe.
For cold work dies; oil hardened.

HOLLOW BAR.
M-80; C, Cr, bal Fe.
For dies, punches; oil hardening.

HOLLOW BLUE BAND.
M-844; 0.10 max C, 25 Cr, 12 Ni, bal Fe.
For welding electrodes for stainless steels; stainless; coated.

HOLLOW DRILL.
M-289; 1.2 C, bal Fe.
For hollow drills; water hardened.

HOLLOW DRILL.
M-783; 1.0-1.1 Cr, 0.35 Mn, C, 0.30 Si, bal Fe.
For rock drills; water hardened..

HOLLUP GR. 30-S.
M-844; 0.10-0.15 C, bal Fe.
Welded: 50,000 TS; 8 El.
For welding electrodes; coated.

HOLLUP GR. 30XL.
M-884; C, bal Fe.
Welded: 60,000 TS; 12 El.
For welding rods; coated.

HOLLUP GRADE 1.
M-844; 0.06 max C, bal Fe.
Cast: 45,000 TS; 13 El.
For welding electrodes for low carbon steel; coated.

HOLLUP GRADE 3.
M-844; 0.10-0.15 C, bal Fe.
Cast: 60,000 TS; 12 El.
For welding electrodes for boiler flues and fire boxes; coated.

HOLLUP GRADE 3R.
M-844; 0.10-0.15 C, bal Fe.
For welding electrodes for boiler tubes; special coating.

HOLLUP GRADE 5.
M-844; 0.18-0.25 C, bal Fe.
Cast: 70,000 TS.
For welding electrodes for building up bearing surfaces; coated, wear resistant.

HOLLUP GRADE 6.
M-844; C, Ni, bal Fe.
For welding electrodes; coated, shock resistant.

HOLLUP GRADE 7.
M-844; 0.85-1.10 C, bal Fe.
For welding electrodes for high carbon steel parts; coated.

HOLLUP GRADE 9.
M-844; 0.75-0.95 C, 12-14 Mn, 3-4 Ni, bal Fe.
190-500 Brin.
For welding electrodes for high manganese steels; coated.

HOLLUP GRADE 10.
M-844; 0.06 max C, bal Fe.
Cast: 45,000 TS; 13 El.
For welding electrodes; coated.

HOLLUP GRADE 10X.
M-844; 0.06 max C, bal Fe.
For welding electrodes; coated, ductile.

HOLLUP GRADE 11.
M-844; 0.06 max C, bal Fe.
For welding rod for wrought iron and steel; Cu coated.

HOLLUP GRADE 15.
M-844; 0.08-0.10 C, bal Fe.
Cast: 45,000 TS 13 El.
For welding electrodes; coated.

HOLLUP GRADE 20.
M-844; Cu alloy.
Cast: 40,000 TS.
For coated bronze welding rod for copper alloys.

HOLLUP GRADE 20B.
M-844; Cu alloy.
Cast: 40,000 TS.
For coated bronze welding rod, for light gauge sheets.

HOLLUP GRADE CI-3.
M-844; C, bal Fe.
For welding electrodes for cast iron; coated.

HOLLUP GRADE CI-8.
M-844; C, bal Fe.
For welding electrodes for cast iron; coated.

HOLLUP GRADE CI-9.
M-844; C, Ni, Cu, bal Fe.
For welding electrodes for cast iron; coated.

HOLLUP GRADE CI-10.
M-844; C, bal Fe.
435 Brin.
For welding electrodes for hard surfacing cast iron: coated.

HOLLUP GRADE NO 22.
M-844; C, bal Fe.
For welding electrodes.

HOLLUP GRADE NO 30.
M-844; 0.10-0.15 C, bal Fe.
Cast: 60,000 TS; 12 El.
For welding electrodes for flues, boilers and fire boxes; coated.

HOLLUP GR. NO 30SB.
M-844; 0.10-0.15 C, bal Fe.
Welded: 55,000 TS; 10 RA.
For welding electrodes; coated.

HOLLUP GR. NO 30-X.
M-844; 0.10-0.15 C, bal Fe.
For welding electrodes; coated.

HOLLUP GR. NO.30XB.
M--844; 0.10-0.15 C, bal Fe.
For welding electrodes; coated.

HOLLUP GR. NO. 33.
M-884; 0.08-0.10 C, bal Fe.
For welding electrodes; copper coated.

HOLLUP GR. NO. 50.
M-844; 0.18-0.25 C, bal Fe.
Welded: 70,000 TS.
For welding electrodes; coated.

HOLLUP GR. NO. 50A.
M-844; 0.18-0.25 C, bal Fe.
For welding electrodes; coated.

HOLLUP GR. NO. 55.
M-844; 0.2 C, bal Fe.
For welding rod; for pipe and pressure weld.

HOLLUP GR. NO. 70A.
M-844; 0.85-1.1 C, bal Fe.
For welding electrodes; coated.

HOLLUP GR. NO. 120S.
M-844; 0.1 C, 18 Cr, 8 Ni, bal Fe.
For welding rod; stainless.

HOLLUP GR. NO. 150.
M-844; 0.08-0.10 C, bal Fe.
Welded: 45,000 TS; 13 El.
For welding electrodes; coated.

HOLLUP GR. YOLOY.
M-844; 0.15 C, 0.5 Cu, bal Fe.
For welding electrodes; coated.

HOLLUP H.S.V.
M-844; 0.2 C, 1 Ni, 0.6 Cr, 0.15 V, bal Fe.
For coated welding electrodes.

HOLLUP I.H.S.
M-844; C, alloy, bal Fe.
For coated welding electrodes.

HOLLUP MCM.
M-844; 0.2 C, 1.2 Cr, 0.2 Mo, bal Fe.
Cast: 85,000 TS; 27 El.
For coated welding electrodes; for aircraft tubing.

HOLLUP N.C.
M-844; C, bal Fe.
For coated welding electrodes.

HOLLUP NO. 1217.
M-844; Al alloy.
For coated Al welding rod.

HOLLUP NO. 2320.
M-844; 0.10 C, 3.5 Ni, bal Fe.
For coated welding electrodes for Ni steels; shock resistant.

HOLLUP NO. 2512.
M-844; 0.2 C, 5 Ni, bal Fe.
Cast: 100,000 TS; 25 El.
For coated welding electrodes for Ni steels; tough.

HOLLUP NO. 6723.
M-844; 23 min Cu, 3.5 max Fe, bal Ni.
50,000 TS.
For monel metal welding rod; corrosion resistant.

HOLLUP PHOS-COPPER.
M-844; 7-9 P, bal Cu.
For P-Cu brazing rod for copper alloys; self-fluxing.

HOLLUP R.D.S.
M-844; 0.2 C, bal Fe.
For coated welding electrodes.

HOLLUP RED BAND.
M-844; 0.1 C, 18 Cr, 8 Ni, bal Fe.
Cast: 90,000 TS; 30 El.
For stainless steel welding electrodes; stainless.

HOLLUP YELLOW BAND.
M-844; 0.10 C, 4-6 Cr, 0.5 Mo, bal Fe.
Cast: 75,000 TS; 35 El.
For welding electrodes for 5 Cr steel.

HOLMIUM.
M-1755; Ho.
Purities: 99,9% (special distilled grade), 99.5+%.
Forms: Ingot, lump, sheet, foil, wire, fillings, sponge, powder, single crystals.

HOLTO 1948.
M-1546; 0.28 C, 1.5 Cr, 0.6 Mo, 5.0 , 0.15 V, 5.0 Co, bal Fe.
Heat treated: 128,000-270,000 Ts.
For die casting dies, brass forging dies. Hot work tool steel, deep hardening.

HOLTO A.C.M.
M-1546; 0.60 C, 1.6 Ni, 0.8 Cr, 0.30 Mo, 0.10 V, bal Fe.
Heat treated: 150,000-220,000 TS; 130,000-200,000 YS; 12-20 El; 40-58 RA; 150-400 Brin.
For hot work tools, punches, hot shears, upsetters, caulking tools. Hot work steel, oil hardening, shock resistant.

HOLTO A.F.L.M.
M-1546; 0.60 C, 2.5 Ni, 0.9 Cr, 0.35 Mo, 0.15 V, bal Fe.
For hot work tolls, punches, hot shears, upsetters. Hot work steel, oil hardening, tough.

HOLTO C.R.V.
M-1546; 1.5 C, 1.8 Cr, 0.25 V, bal Fe.
For gauges, taps, threading dies, cold cutting and stamping tools, plastic molds.
Cold working tool steel, oil hardening.

HOLTO D.R.B.
M-1546; 1.0 C, 1.25 Cr, bal Fe.
Oil hardened: 288,000 TS: 278,000 YS; 540 Brin.
For rollers, bearings, taps punches, cutters, pivot pins, stamping dies, plug gages.
Cold working steel, oil hardening, deep hardening, medium toughness.

HOLTO N.B.2.
M-1546; 0.90 C, 1.3 Mn, 0.5 Cr, bal Fe.
Annealed: 85,000 TS; 60,000 YS; 26 El; 185 Brin. Oil hardened: 280,000 TS; 272,000 El; 535 Brin.
For dies, punches, gauges, cutters, cold headers, stamping dies, broaches. Non-deforming tool steel, oil hardening, shock resistant, tough.

HOLTO R.B.L.
M-M-1546; 1.4 C, 0.70 Cr, bal Fe.
For cutters, bearings, cold heading dies, hand taps, bearings, reamers, boring tools. Cold working steel. Oil hardening.

HOLTO R.B.O.
M-1546; 0.95 C, 0.75 Cr, bal Fe.
Annealed: 85,000 TS; 46,000 YS; 175 Brin.
Hardened: 260,000 TS; 240,000 YS; 525 Brin.
For bearings, rollers, bushings, gauges, mandrels, punches, cold heading tools.
Cold work tool steel, oil or water harden, wear resistant.

HOLTO SPECIAL B.M.
M-1546; 0.70 C, 1.40 Cr, 0.70 Mo, 0.15 V, bal Fe.
For punches, shears, die casting dies.
Hot and cold work tool steel, oil hardening, shock resisting.

HOLTO SPECIAL D.
M-1546; 0.28 C, 9.5 W, 3.1 Cr, 0.3 V, bal Fe.
Heat treated: 142,000-350,000 TS; 120,000-180,000 YS; 6-13 El.
For drawing dies, extrusion rams and dies, valve heads, stamping dies. Hot work tool steel. High hot hardness above 650°C..

HOLTO SPECIAL D.A.
M-1546; 0.40 C, 3.25 Cr, 2.5 W, 0.4 V, bal Fe.
Hardened: 250,000 TS; 200,000 YS; 8 El; 12 RA.
For press dies, upsetters, rivet sets, forging and punching dies, dummy blocks. Hot work steel up to 600°C applications.

HOLTO SPECIAL D.C.
M-1546; 0.21 C, 8.0 W, 2.6 Cr, 2.0 Co, 1.0 Mo, 0.4 V, bal Fe.
Heat treated: 135,000-340,000 TS; 110,000-214,000 YS; 10-14 El.
For wire drawing dies, hot upsetters, stamping dies, forging dies. Hot work tool steel. High hot hardness.

HOLTO SPECIAL E.X.
M-1546; 0.45 C, 4.60 Ni, 1.75 W, 0.40 Cr, 0.30 Mo, bal Fe.
For forging and riveting dies, shears, punches. Hot and cold work steel. Shock resistant. Oil hardening.

HOLTO SPECIAL F.A.41.
M-1546; 0.45 C, 4.75 Cr, 3.75 W, 0.45 V, bal Fe.
Air hardened: 270,000 TS:; 208,000 YS; 5 El; 9 RA.
For forging dies and inserts, extrusion dies and rams, dummy blocks. Hot work steel, oil or air hardening, non-deforming, deep hardening.

HOLTO SPECIAL F.A.48.
M-1546; 0.42 C, 4.75 W, 2.0 Co, 4.75 Cr, 0.45 V, bal Fe.
Heat treated: 170,000-300,000 TS; 156,000-256,000 YS; 4-8 El.
For thread rolling dies, spinning tools, dies. Hot work steel, oil hardening. Resists cyclic heating.

HOLTO SPECIAL F.A.82.
M-1546; 0.38 C, 5.0 Cr, 1.3 Mo, 1.0 Si, 0.5 V, bal Fe.
Annealed: 90,000 TS; 46,000 YS; 18 El. Air hardened: 217,000 TS; 184,000 YS; 13 El; 40 RA; C 53-55 Rock.
For die casting dies, forging dies and inserts, stamping dies, extrusion dies, hot punches. Hot work steel, oil hardening, non-deforming, shock resistant.

HOLTO SPECIAL M.
M-1546; 0.42 C, 1.6 Si, 0.8 Cr, 0.8 Mo, bal Fe.
Water hardened: 190,000-300,000 Ts; 6-16 El; 15-50 RA; C 40-55 Rock.
For cold headers, punches, chisels, rivet sets, upsetters, pneumatic tools. Cold working steel. Shock resistant. Oil hardening.

HOLTO SPECIAL M.D.
M-1546; 0.55 C, 1.90 , 1.10 Cr, 0.9 Se, bal Fe.
For chisels, rivet sets, upsetters, pneumatic tools. Cold work steel, shock resistant.

HOLTO SPECIAL M.O.V.
M-1546; 0.45 C, 1.9 Cr, 0.6 Mo, 0.35 V, bal Fe.
Annealed : 100,000 TS; 48,000 YS; 22 El; 197 Brin.
Hardened: 290,000 TS; 245,000 YS; 10 El; 580 Brin.
For hot work tools, hot upsetters, rivet sets, punches, forging dies, inserts, molds for plastics. Hot work steel, oil hardening, shock resistant.

HOLTO SPECIAL M.O.V.2.
M-1546; 0.37 C, 3.0 W, 2.5 Cr, 0.70 Mo, 0.25 V, bal Fe.
For wire cutters, forging die inserts, die casting dies, hot heading bolts.
Hot work steel, oil hardening, wear resistant.

HOLTO SPECIAL P.
M-1546; 0.42 C, 4.60 Ni, 1.15 Mo. 0.30 Cr, bal Fe.
For hot work tools, punches, upsetters.
Hot work steel, oil hardening.

HOLTO SPECIAL R.
M-1546; 1.0 C, 1.65 Cr, 0.15 V, bal Fe.
Hardened: 280,000 TS; 2558000 YS; 555 Brin.
Annealed: 90,000 TS; 50,000 YS; 28 El; 185 Brin.
For rolling mill cylinders, cold headers, gauges, bearings, bushings, drawing dies, punching, drills. Cold work tools, oil hardening, deep hardening.

HOLTO SPECIAL R.K.
M-1546; 1.4 C, 1.45 Cr, 0.30 Mo, 0.15 V, bal Fe.
For gauges, headers, drawing dies, punches. Cold work tool steel, oil hardening.

HOLTO SPECIAL R.U.
M-1546; 0.95 C, 0.90 Cr, 0.45 Mo, bal Fe.
For bearings, bushings, gauges, rollers, cold header dies. Cold work steel, oil hardening.

HOLTO SPECIAL S.C.1.
M-1546; 1.2 C, 1.25 W, bal Fe.
Heat treated: 280,000 TS; 270,000 YS; 540 Brin.
Annealed: 85,000 TS; 22 El; 45 RA; 185 Brin.
For fast finishing cutters, drills, taps, cold cutting tools, augers, paper knives. Cold work steel, oil or water hardening, wear resistant.

HOLTO S.P.2.
M-1546; 0.28 C, 5.0 Mo, 0.25 V, 6.0 Cr, 5.0 W, 0.6 Co, bal Fe.
Hot rolled: 114,000 TS.
Heat treated: Rockwell C 47-56.
For die casting dies, extrusion rams and liners, brass forging dies.
Hot work steel. High hot hardness and resistance to thermal shock.

HOLTO SPECIAL-U.
M-1546; 0.9 C, 2.0 Mn, bal Fe.
Hardened: C 64-65 Rock.
For stamping dies, gauges, punches, taps, cold headers, broaches, trimming dies.
Non-deforming, oil hardening, shock resistant, Type 02 cold work tool steel.

HOLTO V.D.L.D.M.
M-1546; 0.35 C, 4.0 Ni, 1.7 Cr, 0.5 Mo, bal Fe.
For hot work tools, punches, hot shears, upsetters.
Hot work steel, oil hardening, tough, shock resistant.

HOMBERGS ALLOY.
33.3 Pb, 33.3 Sn, 33.3 Bi.
For sprinkler plugs, fusible metal; M.P. 251°F.

HONALIUM 5.
M-1364; AlMg5.
Chill cast: 150 N/mm² TS; 55 Brin.
Sand cast: 140 N/mm² TS; 50 Brin.
For corrosion resistant aluminum castings; resists seawater. Good decorative anodized finish.

HONALIUM 31.
M-1364; AlMg3Si T6.
Chill cast: 220 N/mm² TS. 90 Brin.
Sand cast: 180 N/mm² TS. 60 Brin.
For corrosion resistant aluminum castings; resist seawater. Good decorative anodized finish.

HONALIUM 51.
M-1364; AlMg5Si.
Chill cast: 150 N/mm² TS; 60 Brin.
Sand cast: 140 N/mm² TS; 55 Brin.
Aluminum castings; resistant to seawater; better castability than HONALIUM 5. Good decorative anodized finish.

HONALIUM 411.
M-1364; AlMg4Si1Mn.
Chill cast, F: 160 N/mm² TS; 50 Brin.
Chill cast, T6: 320 N/mm² TS; 100 Brin.
Aluminum casting for elevated temperature, as cylinder heads.

HONALIUM-ELEKTRAL.
M-1364; AlSi5Mg T6.
Casting: 220 N/mm² TS; 60-85 Brin.
Aluminum casting with high electrical conductivity. (>27 m/ohm mm²).

HONALIUM S.
M-1364; AlMg10Si3 T2.
Chill cast: 190 N/mm² TS; 105 Brin.
Sand cast: 170 N/mm² TS; 80 Brin.
For aluminum castings with good corrosion resistance.

HONDA NEW.
M-Japan; 6.7 Ti, 3.7 Al, 18 Ni, 27 Co, 45 Fe.
For permanent magnets.

HONSEL HO 3.
M-1364; AlMn.
Soft: 100-120 N/mm² TS; 30-35 Brin.
Hard: 160-230 N/mm² TS; 40-55 Brin.
Sheet and plate for cooking utensils, chemical equipment, storage tanks, builders' hardware.

HONSEL HO E 10.
M-1364; AlMg1.
Soft: 100-130 N/mm² TS; 30-35 Brin.
Hard: 160-240 N/mm² TS; 50-65 Brin.
Sheet and plate for architectural trim, builders' hardware, appliances.

HONSEL HO E 20.
M-1364; AlMg2.
Soft: 150-180 N/mm² TS; 40-55 Brin.
Hard: 210-280 N/mm² TS; 60-75 Brin.
Aluminum alloy sheet and plate for architectural and furniture trim, traffic signs.

HONSEL HO E 25.
M-1364; AlMg2.5.
Soft: 170-215 N/mm²; 50 Brin.
Hard: 250-290 N/mm²; 80 Brin.
Aluminum sheet and plate for metal work, auto and appliance trim.

HONSEL HO S.
M-1364; AlMgMn.
Soft: 180-220 N/mm²; 45-60 Brin.
Hard: 260-350 N/mm²; 75-90 Brin.
Aluminum sheet and plate for moderate strength structures, storage tanks, chemical equipment.

HONTAL.
M-1364; AlCu4Ti T6.
Chill cast: 280 N/mm²; 90 Brin.
Sand cast: 250 N/mm²; 85 Brin.
Aluminum casting with good strength and high pressure tightness; good castability.

HONTAL S.
M-1364; AlCu4TiMg T6.
Chill cast: 300 N/mm² TS; 95 Brin.
Sand cast: 280 N/mm² TS; 90 Brin.
Aluminum casting with good strength and high pressure tightness.

HONTRON A6.
M-1364; MgA16.
Sand cast: 150 N/mm² TS; 50-65 Brin.
Magnesium casting alloy with high tear resistance, as for motor car wheels.

HOOKER BRASS.
M-Eng; 61 Cu, 3m Zn, 2 Pb.
For hot forgings, brass parts; free-cutting.

HOPKINSON ALLOY.
75 Fe, 24.5 Ni, traces Si.
160,000 TS; 88,000 YS; 45 El; 68 RA.
For corrosion and heat resistant parts; Sc.-21.

HORBACH HH80.
M-1377; 0.12 C, 13 Cr, 2.2 Si, bal Fe.
Annealed: 75,000 TS; 40,000 YS; 35 El; 70 RA; 155 Brin.
For valves, cutlery, pump bodies, bolts; Type 410; corrosion resistant.

HORBACH HH100.
M-1377; 0.12 max C, 1 Al, 18 Cr, bal Fe.
For oil refinery equipment; heat and creep resistant.

HORBACH HH110A.
M-1377; 0.15 C, 19.5 Cr, 9.5 Ni, bal Fe.
Annealed: 80,000 TS; 35,000 YS; 55 El; 75 RA; 160 Brin.
For chemical plant equipment, tanks, mixers; Type 302; stainless, austenitic.

HORBACH HH120.
M-1377; 0.12 max C, 1.5 Al, 24 Cr, bal Fe.
For oil refinery equipment; heat and creep resistant.

HORBACH HH120A.
M-1377; 0.15 C, 24 Cr, 19 Ni, bal Fe.
Annealed: 100,000 TS; 45,000 YS; 50 El; 65 RA; 185 Brin.
For valves, pumps, furnace parts, turbine and jet parts; Type 310; stainless, austenitic.

HORBACH HN1.
M-1377; 0.15 max C, 18 Cr, 8 Ni, bal Fe.
Annealed: 80,000 TS; 35,000 YS; 55 El; 75 RA; 150 Brin.
For chemical plant equipment tanks, mixers, agitators; Type 302; stainless, austenitic.

HORBACH HN1 EXTRA.
M-1377; 0.12 max C, 1, Cr, 9.5 Ni, Ti = 4 x C, bal Fe.
Annealed: 85,000 TS; 35,000 YS; 55 El; 65 RA; 150 Brin.
For welded chemical plant equipment, tanks, mixers; Type 321; stainless, austenitic.

HORBACH HN1 SUPRA.
M-1377; 0.07 max C, 18 Cr, 9.5 Ni, bal Fe.
Annealed: 85,000 TS; 35,000 YS; 60 El; 70 RA; 150 Brin.
For chemical plant equipment, tanks, vessels; Type 304; stainless, austenitic.

HORBACH HN2 EXTRA.
M-1377; 0.12 max C, 18 Cr, 2 Mo, 10.5 Ni, Ti = 4 x C, bal Fe.
Annealed: 85,000 TS; 35,000 YS; 50 El; 65 RA; 160 Brin.
For chemical plant equipment, welded structures; Type 316 Ti; stainless, austenitic.

HORBACH HN2 SUPRA.
M-1377; 0.07m max C, 18 Cr, 10.5 Ni, 2 Mo, bal Fe.
Annealed: 85,000 TS; 35,000 YS; 50 El; 65 RA; 160 Brin.
For acid resistant chemical plant equipment; Type 316; stainless, austenitic.

HORBACH HNH.
M-1377; 0.4 C, 13 Cr, 0.4 Si, bal Fe.
Annealed: 95,000 TS; 50,000 YS; 25 El; 55 RA; 196 Brin.
For valves, cutlery, surgical and dental instruments; Type 420; stainless.

HORBACH HNM.
M-1377; 0.2 C, 0.4 Si, 13 Cr, bal Fe.
Annealed: 95,000 TS; 50,000 YS; 25 El; 55 RA; 196 Brin.
For valves, cutlery, surgical instruments; Type 420; stainless.

HORBACH HNW.
M-1377; 0.12 max C, 0.4 Si, 13 Cr, bal Fe.
Annealed: 75,000 TS:; 40,000 YS; 35 El; 70 RA; 155 Brin.
For turbine blades, cutlery, valves; Type 410; stainless.

HORSEHEAD POWDERS.
M-91.
Series of alloy powders, mostly copper base, for brazing, infiltrating, powder metals.

HORSEHEAD SPECIAL ZINC ANODES.
M-91; 99.99 pure Zn.
Plates, bars, balls.
For electroplating; electro galvanizing.

HORSEHEAD ZAMAK NO 3 see ZAMAK NO 3.

HOSKINS 717 see CHROMEL 1A.

HOSKINS ALLOY 400.
M-65; 67 Ni, 31 Cu, 1 Fe, 1 Mn.
Drawn wire: 60,000 psi TS.
For corrosion resistance.

HOSKINS ALLOY 600.
M-65; 76 Ni, 15.5 Cr, 8 Fe, 0.5 Mn, 0.2 Si.
For corrosion resistance; heat and oxidation resistant.

HOSKINS ALLOY 750.
M-65; 15.0 Cr, 4.0 Al, 0.5 Si, 0.1 C, bal Fe.
Elec. resistance: 750 ohms/circular mil ft.
Resistance wire for furnaces, kilns, other heating devices; requires adequate support at high temperatures. Useful to 2050°F.

HOSKINS ALLOY 800.
M-65; 35 Ni, 21 Cr, 1.6 Si, bal Fe.
Drawn wire: 100,000 psi TS.
For furnace fixtures; heat resistance applications; heat and oxidation resistant.

HOSKINS ALLOY 815.
M-65; 0.1 C, 0.5 Si, 4.6 Al, 22.5 Cr, bal Fe.
Hot rolled: 143,000 TS; 12 El; 234 Brin.
Annealed: 108,000 TS; 27 El; 190 Brin.
For resistances, heating elements, rheostats; heat resistant to 2150°F.

HOSKINS ALLOY 831.
M-65; 15 Cr, 7.5 Fe, bal Ni.
Drawn: 100,000 TS.
For spark plugs. Electrode alloy, oxidation resistant.

HOSKINS ALLOY 875.
M-65; 22.5 Cr, 5.5 Al, 0.5 Si, 0.1 C, bal Fe.
Hot rolled: 143,000 TS; 12 El; 185 Brin.
Annealed: 108,000 TS; 27 El; 190 Brin.
For electrical resistances, heating elements; high heat resistance to 2350°F.

HOT.
M-1522; 60 Ag, Cu, Zn.
Melt range: 600-720°C.
Max stress: 34.6 kgf/mm^2; 5 El.
For silver brazing in protective atmospheres.

HOT DIE.
M-350; C, W, bal Fe.
For tools; oil hardened.

HOT DIE.
M-435; 0.3 C, 11 W, 3 Cr, 0.5 V, bal Fe.
For hot die tools, punches; hot die steel.

HOT DIE 593.
M-1; 0.38 C, 0.25 Mo, 5 W, 5 Cr, 0.2 V, 0.25 Mn, 0.45 Co, 0.9 Si, bal Fe.
Heat treated: 270,000 TS; 208,000 YS; 5 El; 9 RA. At 900°F: 217,000 TS; 172,000 YS; 10 El; 31 RA.
Uses: extrusion dies, hot upset punches, insert forging dies, dummy blocks. Type h14 Hot work steel.

HOTFORM NO 1.
M-115, M-342; 0.32 C, 1.5 Mo, 0.9 Si, 1.0 W, 4.75 Cr, 0.25 Mn, bal Fe.
For tools and dies for Al die castings; hot work steel.

HOTFORM NO 2.
M-115; 0.33 C, 0.9 Si, 4.7 Cr, 1.3 Mo, 0.5 V, bal Fe.
For punches, upsetters, hot work tools and dies, extrusion press; Type h11; hot work steel.

HOTFORM NO 3.
M-115; 0.52 C, 0.9 Si, 1.4 Mo8 4.75 Cr, bal Fe.
For punches, upsetters, hot dies and tools; Type h12; air hardened.

HOT FORM DRILL ROD.
M-342; 0.35 C, 1 Si, 1.25 W, 5 Cr, 0.5 V, 1.45 Mo, bal Fe.
For die casting die tools, high temperature fasteners; hot work steel, oil hardened.

HOTFORM V.
M-115; 0.37-0.43 C, 1.0 Si, 5.0-5.5 Cr, 1.2 Mo, 1.1 V, bal Fe.
For Al die casting dies, shear blades, forging dies; hot work steel oil hardened.

HOTPRESS.
M-115; 0.35 C, 2 Cr, 9.5 W, 0.5 V, bal Fe.
For hot work dies, upsetter dies; tough and heat resistant; hot work steel.

HOT STAMPING ALLOY DIE.
M-261; C, alloy, bal Fe.
For hot stamping dies, tools; oil hardened.

HOT STAMPING ALLOY DIE.
M-389; 0.5 C, 3 Cr, bal Fe.
For hot stamping dies;: hot work steel.

HOT WORK 8.
M-24; 0.6 C, 8.5 Mo, 1.5 V, 3.2 Cr, bal Fe.
For hot work tools and dies, punches, blanking dies; hot work steel, wear resistant.

HOT WORK NO 2.
M-9; 0.3 C, 27 Cr, 5.5 Mo, 2.8 Ni, 62 Co.
Weld hardness 32 Rc; work hardens.
Hard facing rod and electrode; resistant to heat, corrosion and impact.
For build-up and repair valves and dies.

HOT WORK NO 22.
0.4 C, 3 Cr, 9 W, 0.2 V, bal Fe.
For tools, dies; hot work steel.

HOT WORK NO 23.
0.45 C, 3 Cr, 14 W, 0.3 V, bal Fe.
For tools, cutters, taps, gauges, reamers; high speed steel.

HOUGHTON'S SHAVING PACKING.
M-Eng; 5.97 Sn, trace Cu, 94.02 Pb.
For piston rod packing for steam engines.

HOVER 91C.
M-1306; 0.55 C, 1 Cr, 0.18 V, 1.85 W, bal Fe.
For cold work tools, hammers, upsetters; oil or water hardened.

HOVER 91KA.
M-1306; 0.45 C, 1 Cr, 0.2 V, bal Fe.
For gears, pinions, shafts, bolts, crankshafts; oil hardened, shock resistant.

HOVER 123.
M-1306; 1.45 C, 1.4 Cr, bal Fe.
For bearings, liners, races, sleeves; water hardened.

HOVER 151.
M-1306; 0.61 C, 1.18 Cr, 0.1 V, bal Fe.
For springs, shafts, crankshafts, punches, crimpers; oil or water hardened.

HOVER 210.
M-1306; 1.25 C, 1.15 Si, 0.7 Mn, 1.2 Cr, bal Fe.
For dies, punches, shock resistant cutters; oil hardened, tough.

HOVER 350.
M-1306; 1.0 C, 0.1 V, 0.25 Mn, bal Fe.
For cutters, tools, dies, taps, drills; Type W2; water hardened.

HOVER 401.
M-1306; 1.05 C, 0.9 Mn, 1.0 Cr, 1.15 W, bal Fe.
For fast finishing cutters, reamers; water hardened.

HOVER 550.
M-1306; 0.7 C, 1.7 Si, 0.7 Mn, bal Fe.
For punches, upsetters, riveters, springs; oil hardened, tough.

HOVER 702.
M-1306; 0.28 C, Ni, Cr, Mo, V, bal Fe.
For gears, bolts machine tool parts; oil hardened, shock resistant.

HOVER A18A.
M-1306; 0.12 C, 18 Cr, 8 Ni, bal Fe.
Annealed: 75,000 TS; 35,000 YS; 50 El; 65 RA; 150 Brin.
For chemical plant equipment; Type 302; stainless, austenitic.

HOVER A18N.
M-1306; 0.12 max C, 18 Cr, 9.5 Ni, Cb = 8 x C, bal Fe.
Annealed: 90,000 TS; 35,000 YS; 50 El; 65 RA; 160 Brin. Cold drawn: 100,00 TS; 65,000 YS; 40 El; 60 RA; 212 Brin.
For welded structures, chemical plant equipment; Type 347; stainless, austenitic.

HOVER A18T.
M-1306; 0.12 max C, 18 Cr, 9.5 Ni, Ti = 4 x C, bal Fe.
Annealed: 85,000 TS; 35,000 YS; 55 El; 65 RA; 150 Brin. Cold drawn: 95,000 TS; 60,000 YS; 40 El; 60 RA; 185 Brin.
For welded structures, chemical plant equipment; Type 321; stainless, austenitic.

HOVER A18Z.
M-1306; 0.07 max C, 18 Cr, 9.5 Cr, Ni, bal Fe.
Annealed: 85,000 TS; 35,000 YS; 60 El; 70 RA; 150 Brin. Cold drawn: 1808 000 TS; 125,000 YS; 10 El: 330 Brin.
For welded structures, chemical plant equipment; Type 304; stainless, austenitic.

HOVER A18ZS.
M-1306; 0.1 max C, 18 Cr, 8.5 Ni, bal Fe.
Annealed: 75,000 TS; 35,000 YS; 55 El; 75 RA; 155 Brin.
For chemical plant equipment, welded tanks and vessels; Type 304; stainless, austenitic.

HOVER A20T.
M-1306; 0.12 max C, 18 Cr, 10.5 Ni, 2 Mo, Ti = 4 x C, bal Fe.
Annealed: 85,000 TS; 35,000 YS; 50 El; 75 RA; 150 Brin.
Cold drawn: 150,000 TS; 130,000 YS; 6 El; 300 Brin.
For acid resistant equipment, welded structures; Type 316 Ti; stainless, austenitic.

HOVER A20Z.
M-1306; 0.07 C, 18 Cr, 10.5 Ni, 2 Mo, bal Fe.
Annealed: 85,000 TS; 35,000 YS; 50 El; 75 RA; 150 Brin. Cold drawn: 150,000 TS; 130,000 YS; 6 El; 300 Brin.
For acid resistant equipment; Type 316; stainless; austenitic.

HOVER A20ZS.
M-1306; 0.1 max C, 18 Cr, 9.5 Ni, 2 Mo, 2.2 Si, bal Fe.
Annealed: 85,000 TS; 35,000 YS; 50 El; 75 RA; 150 Brin. Cold drawn: 150,000 TS; 135,000 YS; 6 El; 300 Brin.
For acid resistant equipment, chemical plant apparatus; Type 316; stainless, austenitic.

HOVER A21Z13.
M-1306; 0.07 max C, 17 Cr, 4.75 Mo, 13 Ni, bal Fe.
Annealed: 90,000 TS; 40,000 YS; 45 El; 70 RA; 160 Brin.
For acid resistant equipment, chemical plant apparatus; Type 317; stainless, austenitic.

HOVER GHK243.
M-1306; 0.40 C, Cr, Mn, Mo, bal Fe.
For gears, bolts, machine tool parts; oil hardened, tough.

HOVER H15.
M-1306; 0.15 C, 1.55 Cr, 1.55 Ni, bal Fe.
For gears, cams, camshafts; case hardening.

HOVER H20.
M-1306; 0.18 C, 2.0 Ni, 2.0 Cr, 0.50 Mn, bal Fe.
For gears, bolts, camshafts, cams; case hardened, tough.

HOVER H10.
M-1306; 1.1 C, 0.25 max Si, 0.25 max Mn, bal Fe.
Annealed: 110,000 TS; 56,000 YS; 20 El; 40 RA; 210 Brin.
For springs, cutters, drills, taps, teamers; Type W1; water hardened.

HOVER H20.
M-1306; 1.0 C, 0.25 max Si, 0.25 max Mn, bal Fe.
For springs, tools, drills, taps, reamers; Type W1; water hardened.

HOVER H127.
M-1306; 1 C, 0.25 max Mn, 0.25 max Si, bal Fe.
Annealed: 100,00 TS; 53,000 YS; 21 El; 42 RA; 200 Brin.
For drills, taps, reamers, hobs; Type W1; water hardened.

HOVER H391.
M-1306; 0.90 C, 0.80 Cr, 0.30 Mn, bal Fe.
For drills, punches, springs, taps, reamers; water hardened.

HOVER H680.
M-1306; 0.50 C, 0.85 Mn, 1 Cr, 3.1 V, bal Fe.
For gears, shafts, crankshafts; oil hardened, shock resistant.

HOVER H1000.
M-1306; 0.15 C, 2 Si, 19.5 Cr, 9.5 Ni, bal Fe.
Annealed: 85,000 TS; 40,000 YS; 50 El; 65 RA; 160 Brin.
For chemical and oil refinery equipment; Type 302; stainless, austenitic.

HOVER H1050-ON.
M-1306; 0.12 max C, 18 Cr, 0.95 V, bal Fe.
Annealed: 80,000 TS; 35,000 YS; 50 El; 65 RA; 160 Brin.
For chemical plant and oil refinery equipment; corrosion and heat resistant.

HOVER H1200.
M-1306; 0.15 C, 24 Cr, 19 Ni, bal Fe.
Annealed: 90,000 TS; 40,000 YS; 55 El; 70 RA; 165 Brin.
For heat treating boxes, furnace parts and equipment; heat resistant, austenitic.

HOVER HED16.
M-1306; 0.15 C, 0.25 Si, 0.37 Mn, bal Fe.
Annealed: 70,000 TS; 40,000 YS; 25 El; 60 RA; 145 Brin.
For gears, bolts, machine tool parts; case hardened.

HOVER HK904.
M-1306; 0.30 C, 2.35 Cr,0.6 V, 4.25 W, bal Fe.
For extrusion press dies, rams, liners; hot work steel, oil hardened.

HOVER HNRO7.
M-1306; 0.15 max C, 8.5 Ni, 1, Cr, bal Fe.
Annealed: 80,000 Ts; 35,000 Ys; 55 El; 75 Ra; 150 Brin.
For chemical plant equipment, tanks; Type 302; stainless, austenitic.

HOVER HNRO 8.
M-1306; 0.07 max C, 18 Cr, 9.5 Ni, bal Fe.
Annealed: 85,000 TS; 35,000 YS; 60 El; 70 RA; 150 Brin.
For chemical plant equipment, tanks, vessels; Type 304; stainless, austenitic.

HOVER HNRO 9.
M-1306; 0.07 max C, 18 Cr, 2 Mo, 10.5 Ni, bal Fe.
Annealed: 85,000 TS; 35,000 YS; 50 El; 65 RA; 160 Brin.
For acid resistant chemical plant equipment, tank; Type 316; stainless, austenitic.

HOVER HNRO 11.
M-1306; 0.20 C, Cr, Mo, bal Fe.
For chemical plant equipment; stainless.

HOVER HNRO 12.
M-1306; 0.1 max C, 1, Cr, 9.5 Ni, 2 Mo, bal Fe.
Annealed: 85,000 TS; 50 El; 65 RA; 160 Brin.
For acid resistant chemical plant equipment, tanks; Type 316; stainless, austenitic.

HOVER HNRO 14.
M-1306; 0.12 max C, 18 Cr, 9.5 Ni, Ti = 4 x C, bal Fe.
Annealed: 85,000 TS; 35,000 YS; 55 El; 65 RA; 150 Brin.
For welded chemical plant equipment, tanks; Type 321; stainless, austenitic.

HOVER HNRO 16.
M-1306; 0.22 C, 17 Cr, 1.5 Ni, bal Fe.
Annealed: 125,000 TS; 90,000 YS; 20 El; 55 RA; 260 Brin.
For pumps marine hardware, valves; Type 321; stainless.

HOVER HNRO 17.
M-1306; 0.12 max C, 18 Cr, 2 Mo, 10.5 Ni, Ti = 4 x C, bal Fe.
Annealed: 85,000 TS; 35,000 YS; 50 El; 65 RA; 160 Brin.
For welded acid resistant chemical plant equipment; Type 316 Ti; stainless, austenitic.

HOVER HNRO 19.
M-1306; 0.35 C, 1.15 Mo, 16.5 Cr, bal Fe.
For oil refinery equipment; heat resistant.

HOVER HNRO 20.
M-1306; 0.8 C, 17 Cr, bal Fe.
Heat treated: 280,000 TS; 270,000 YS; 3 El; 15 RA; 555 Brin.
For bearings, surgical instruments, cutlery; corrosion resistant, wear resistant; Type 440B.

HOVER HNRO 24.
M-1306; 0.1 max C, 17 Cr, 1.8 Mo, Ti = 7 x C, bal Fe.
Annealed: 125,000 TS; 95,000 YS; 20 El; 55 RA; 260 Brin.
For welded structures, pump parts; corrosion resistant; Type 431 Ti.

HOVER HNRO 27.
M-1306; 0.12 max C, 18 Cr, 2 Mo, 10.5 Ni, Cb = 8 x C, bal Fe.
Annealed: 90,000 TS; 45,000 YS; 55 El; 65 RA; 160 Brin.
For welded chemical plant equipment, tanks, mixers; Type 347; stainless, austenitic.

HOVER HSE.
M-1306; 0.20 C, 1.25 Mn, 1.15 Cr, bal Fe.
For camshafts, cams, gears, bolts; case hardened.

HOVER HVU 2 1/2.
M-1306; 0.22 C, 0.25 Si, 0.45 Mn, bal Fe.
Annealed: 75,000 TS; 42,000 YS; 20 El; 55 RA; 150 Brin.
For gears, bolts, machine tool parts; case hardened.

HOVER HVU 3 1/2.
M-1306; 0.35 C, 0.25 Si, 0.55 Mn, bal Fe.
Hot rolled: 85,000 TS; 54,000 YS; 30 El; 53 RA; 185 Brin.
For gears, bolts, machine tool parts; water hardened.

HOVER HVU 4 1/2.
M-1306; 0.45 C, Cr, Mo, bal Fe.
For chemical plant equipment; stainless.

HOVER HVU 6.
M-1306; 0.45 C, 0.25 Si, 0.65 Mn, bal Fe.
Hot rolled: 98,000 TS; 60,000 YS; 24 El; 45 RA; 212 Brin.
For gears, bolts, machine tool parts; water hardened.

HOVER HW.
M-1306; 0.53 C, 0.90 Si, 0.90 Mn, bal Fe.
For machine tool parts, punches; oil hardened.

HOVER I-EBN.
M-1306; 0.15 C, 0.65 Cr, 0.5 Mn, bal Fe.
For gears, cams, camshafts, fasteners; case hardening steel.

HOVER I-EC.
M-1306; 0.15 C, 1.15 Cr, 0.25 Mo, bal Fe.
For gears, cams, camshafts, fasteners, case hardening steel.

HOVER I-ECN.
M-1306; 0.16 C, 1.15 Mn, 0.95 Cr, bal Fe.
For gears, cams, camshafts, fasteners; case hardening steel.

HOVER I-ED.
M-1306; 0.20 C, 1.15 Cr, 0.25 Mo, bal Fe.
For gears, cams, camshafts fasteners; case hardening steel.

HOVER I-EDN.
M-1306; 0.20 C, 1.25 Mn, 1.15 Cr, bal Fe.
For gears, cams, camshafts; case hardening steel.

HOVER II-VA.
M-1306; 0.25 C, 0.65 Mn, 1 Cr, bal Fe.
For gears, pinions, bolts, fasteners; water hardened.

HOVER II-VB.
M-1306; 0.33 C, 0.65 Mn, 1 Cr, bal Fe.
For gears, shafts, crankshafts; water hardened.

HOVER II-VC.
M-1306; 0.33 C, 0.65 Mn, 1 Cr, bal Fe.
For gears, shafts, crankshafts; water hardened.

HOVER II-VD.
M-1306; 0.42 C, 1 Cr, 0.2 Mo, bal Fe.
For gears, shafts, crankdhafts; oil hardened, shock resistant.

HOVER II-VE.
M-1306; 0.50 C, Cr, bal Fe.
For gears, shafts, crankshafts, bolts; oil or water hardened.

HOVER ILO-35H.
M-1306; 0.86 C, 2.8 Co, 4.3 Cr, 0.85 Mo, 2.1 V, 12 W, bal Fe.
For lathe and planer tools, reamers, drills, taps; high speed steel.

HOVER ILO-45H.
M-1306; 0.80 C, 4 Cr, Co, V, W, Mo, bal Fe.
For lathe and planer tools, reamers, broaches, taps; high speed steel.

HOVER ILO-55H.
M-1306; 1.35 C, W, Cr, Co, bal Fe.
For blanking and forming dies, engravers tools; high speed steel.

HOVER ILO-105H.
M-1306; 0.76 C, 11 Co, 4.2 Cr, 0.8 Mo, 1.8 V, 18 W, bal Fe.
For lathe and planer tools, drills, taps, hobs, reamers; high speed steel.

HOVER ILO-ABC II.
M-1306; 0.82 C, 4.1 Cr, 0.85 Mo, 1.6 V, 8.7 W, bal Fe.
For lathe and planer tools, reamers, hobs, taps, drills; high speed steel.

HOVER ILO-BCO.
M-1306; 0.79 C, 4.7 Co, 4.3 Cr, 0.75 Mo, 1.5 V, 18 W, bal Fe.
For drills, taps, hobs, reamers, lathe tools; high speed steel.

HOVER ILO-EXTRA.
M-1306; 0.86 C, 4.1 Cr, 0.85 Mo, 2.5 V, 18 W, bal Fe.
For lathe and planer tools, drills, taps, reamers; high speed steel.

HOVER ILO-PRIMA.
M-1306; 0.95 C, 4.1 Cr, 0.85 Mo, V, W, bal Fe.
For lathe and planer tools, drills, reamers; high speed steel.

HOVER ILO-SUPER B.
M-1306; 1.3 C, 4.3 Cr, 0.85 Mo, 3.8 V, 12 W, bal Fe.
For blanking and forming dies, form cutters; high speed steel.

HOVER ILO-ULTRA W.
M-1306; 0.74 C, 4.1 Cr, 1.1 V, 18.5 W, bal Fe.
For lathe and planer tools, drills, taps, hobs; high speed steel.

HOVER IV-SF EXTRA QUALITY.
M-1306; 0.55 C, 0.3 Si, 0.6 Mn, bal Fe.
Heat treated: 110,000-160,000 TS; 75,000-112,000 YS; 25-13 El; 55-42 RA; 230-325 Brin.
For gears, axles, crankshafts; water hardened.

HOVER M48.
M-1306; 0.38 C, Si, Cr, V, bal Fe.
For gears, bolts, machine tool parts; oil hardened, tough.

HOVER P12.
M-1306; 0.56 C, Ni, Cr, Mo, V, bal Fe.
For forging and heading dies; oil hardened, tough.

HOVER P14.
M-1306; 0.55 C, 0.70 Cr, 0.18 Mo, 1.65 Ni, 0.1 V, bal Fe.
For forging and heading dies; oil hardened, tough.

HOVER P15.
M-1306; 0.45 C, Cr, Ni, bal Fe.
For gears, bolts, crankshafts; oil hardened, tough.

HOVER P16.
M-1306; 1.1 C, 0.2 Si, 0.4 Cr, 0.3 Mn, bal Fe.
For bearings, cutters, liners, sleeves; water hardened, wear resistant.

HOVER P35.
M-1306; 0.50 C, 1.05 Cr, 3.25 Ni, bal Fe.
For gears, bolts, crankshafts; oil hardened, shock resistant.

HOVER PRIMA SEHR ZAH.
M-1306; 0.60 C, 0.25-0.50 Si, 0.30-0.80 Mn, bal Fe.
Heat treated: 160,000-115,000 TS; 113,000-77,000 YS; 12-23 El; 40-54 RA; 320-230 Brin.
For wheels, die blocks, rails, girders; water hardened.

HOVER PRIMA WEICH.
M-1306; 0.75 C, 0.25-0.50 Si, 0.30-0.80 Mn, bal Fe.
Heat treated: 180,000-125,000 TS; 140,000-85,000 YS; 12-20 El; 38-52 RA; 360-240 Brin.
For springs, clutch discs, rails, hammers; Type W1; water hardened.

HOVER PRIMA ZH.
M-1306; 0.90 C, 0.25-0.50 Si, 0.30-0.80 Mn, bal Fe.
Heat treated: 180,000 TS; 118,000 YS; 10 El; 30 RA; 375 Brin.
For springs, cutters, hobs, drills, taps; Type W1; water hardened.

HOVER R5.
M-1306; 0.30 C, 0.25 Si, 0.55 Mn, 2.5 Cr, 0.2 Mo, 0.15 V, bal Fe.
For die casting dies, upsetters, rivet sets; oil hardened, tough.

HOVER R10.
M-1306; 0.45 C, 1.4 Cr, 0.70 Mo, 0.30 V, 0.7 Mn, bal Fe.
For die casting dies, forging and heading dies; oil hardened, tough.

HOVER R11.
M-1306; 1.4 C, 0.1 V, 0.3 Cr, 0.3 Mn, bal Fe.
For engravers' tools, textile needles, cutters; oil or water hardened.

HOVER SBL1.
M-1306; 0.12 max C, 0.4 Si, 13 Cr, bal Fe.
Annealed: 75,000 TS; 40,000 YS; 35 El; 70 RA; 155 Brin.
For turbine blades, valves, cutlery; Type 410; stainless.

HOVER SBL2.
M-1306; 0.2 C, 0.4 Si, 13 Cr, bal Fe.
Annealed: 95,000 TS; 50,000 YS; 25 El; 55 RA; 195 Brin.
For turbine blades, valves, cutlery, knives; Type 420; stainless.

HOVER SBL4.
M-1306; 0.4 C, 0.4 Si, 13 Cr, bal Fe.
Annealed: 100,000 TS; 55,000 YS; 20 El; 50 RA; 210 Brin.
For valves, cutlery, surgical and dental instruments; Type 420; stainless.

HOVER SBL9.
M-1306; 0.90 C, 0.4 Si, 18 Cr, 1.15 Mo, 1.0 V, bal Fe.
Annealed: 110,000 TS; 65,000 YS; 16 El; 30 RA; 230 Brin.
For cutlery, ball bearings; Type 440BMo; stainless.

HOVER SP50.
M-1306; 1.15 C, 0.65 Cr, 0.10 V, 0.30 Mn, bal Fe.
For heading and forming dies, cutters; oil hardened, wear resistant.

HOVER SPEZIAL MH.
M-1306; 1.1 C, 0.25 max Si, 0.25 max Mn, bal Fe.
Annealed: 100,000 TS; 53,000 YS; 21 El; 42 RA; 200 Brin.
For springs, cutters, reamers, drills, broaches; Type W1; water hardened.

HOVER SPEZIAL SEHR HART.
M-1306; 0.70 C, 0.25 max Si, 0.25 max Mn, bal Fe.
Heat treated: 174,000 TS; 128,000 YS; 12 El; 37 RA; 352 Brin.
For springs, rails, girders, dies; Type W1; water hardened.

HOVER SPEZIAL ZAH.
M-1306; 0.85 C, 0.25 max Si, 0.25 max Mn, bal Fe.
Heat treated: 188,000 TS; 145,000 YS; 10 El; 30 RA; 400 Brin.
For drills, taps, reamers, broaches, hobs; Type W1; water hardened.

HOVER SPEZIAL ZH.
M-1306; 1.0 Cx, 0.25 max Si, 0.25 max Mn, bal Fe.
Annealed: 100,000 TS; 53,000 YS; 21 El; 42 RA; 200 Brin.
For drills, springs, taps, reamers, broaches; Type W1; water hardened.

HOVER WA18.
M-1306; 0.55 C, 0.90 Si, 0.30 Mn, 1.05 Cr, 0.18 V, 1.85 W, bal Fe.
For heading and forming dies, punches; cold work steel, tough.

HOWEGE EXTRA.
M-1348; 0.55 C, 0.7 Cr, 0.18 Mo, 1.85 W, 0.1 V, bal Fe.
For cold work tools, headers, upsetters; oil hardened, tough.

HOWEGE EXTRA I.
M-1348; 0.56 C, 0.7 Cr, 0.18 Mo, 1.85 W, 0.1 V, bal Fe.
For cold work tools, headers, upsetters; oil hardened, tough.

HOWEGE SPEZIAL.
M-1348; 0.40 C, Cr, Mo, bal Fe.
For gears, shafts, bolts, studs, crankshafts; oil hardened, tough.

HOWES SHEAR BLADE.
M-38; 0.90 C, 0.30 Mo, bal Fe.
For shear blades; water hardening.

HOWMET-25.
M-1512; 0.12 C, 1.5 Mn, 1.0 Si, 20 Cr, 10 Ni, 51 Co, 15 W, 1 Fe.
Cast: 106,800 TS; 24.1 El; C 18-25 Rock.
Heat treated: 145,000-165,000 TS; 65,000-80,000 YS; 55-70 El; B 95-C 25 Rock.
For gas turbine rotors and buckets, nozzles, afterburners, valves.
Similar to L605 Alloy. Investment cast.

HOWMET-50.
M-1512; 0.05-0.12 C, 0.18 Ti, 0.6 Cb, 26-30 Cr, 47-52 Co, bal Fe.
Cast: 135,000 TS; 48,000 YS; 10 El; 10 RA; 250 Brin.
For furnace baffles, burner tips, sintering grates, quench baskets.
Corrosion, heat and thermal shock resistant.

HOWMET FA see FA.

HOWMET NO 3.
M-1512; 2.45 C, 31.0 Cr, 12.5 W, 3 max Fe, 3 max Ni, bal Co.
Cast: 64,000 TS; 0 El; 0 RA; C 53 Rock.
For cutting tools, ball bearing rolls, sleeves, bushings, wear strips, valve seat inserts, scraper blades, valves.
High heat and abrasion resistance, non-galling; Corrosion and oxidation resistant.

HOWMET STANDARD.
M-1512; 2.45 C, 31 Cr, 12.5 W, 3 max Fe, 3 max Ni, bal Co.
Cast: 64,000 TS; 0 El; 0 RA; 530 Brin.
For cutting tools, high temperature parts.
Oxidation and wear resistant.

HOWMET STANDARD NO 6.
M-1512; 0.9-1.4 C, 3 max Fe, 3 max Ni, 1.5 max Mo, 27-31 Cr, 3.5-5.5 W, bal Co.
Cast: 115,000 TS; 96,000 YS; 3 El; 410 Brin.
At 1500°F: 70,000 TS; 5 El; 8 RA.
For turbine blades, valve parts, hot work punches.
Heat and thermal shock resistant.

HOWMET SUPER-3.
M-1512; 2.45 C, 31 Cr, 12.5 W, 6 max Fe, 6 max Ni, bal Co.
Cast: 62,000 TS; 0 El; 0 RA; 560 Brin.
For cutting tools, high temperature parts, Oxidation and wear resistant.

HOWMET SUPER-6.
M-1512; 0.9-1.4 C, 6 max Fe, 6 max Ni, 1.5 max Mo, 27-31 Cr, 3.5-5.5 W, bal Co.
Cast: 114,000 TS; 101,000 YS; 1.5 El; 430 Brin.
At 1500°F: 72,000 TS; 7 El; 7 RA; 230 Brin.
For turbine blades, valve parts, hot work punches.
Heat and thermal shock resistant.

HOWORD A.
M-373; 0.35 C, 5.0 Cr, 0.4 V, 1.5 Mo, bal Fe.
Air or oil hardening hot work tool and die steel for forging dies, hot forming tools.
AISI H11.

HOWORD B.
M-373; 0.35 C, 5.0 Cr, 0.4 V, 1.5 W, 1.5 W, 1.5 Mo, bal Fe.
Air or oil hardening hot work tool and die steel for forging dies, hot work forming tools. AISI H12.

HOWORD C.
M-373; 0.35 C, 5.0 Cr, 1.0 V, 1.5 Mo, bal Fe.
Air or oil hardening hot work tool and die steel; AISI H13.

HOYLES METAL.
M-U.S.; 46 Sn, 12 Sb, 42 Pb.
For bearings; Babbitt.

HOYT.
M-88; Sn, Sb, bal Fe.
For bearings; Babbitt.

HOYT 11.
M-88; Pb, Cu, Sb, bal Sn.
For bearings; Babbitt.

HOYT ARROW.
M-317; Cu, Pb, bal Sn.
For bearings for heavy loads and high speeds; antifriction metal; tough.

HOYT I C E.
M-317; Pb, Sb, bal Sn.
For bearings for internal combustion engines; antifriction metal.

HOYT METAL.
M-88; 10 Cu, 78 Sn, 12 Pb, 26 Brin.
For bearings for internal combustion engines and pumps; bronze Babbitt; M.P. 360°C.

HOYT NO 1.
M-317; Sn, Sb, bal Pb.
For bearings for heavy loads and medium speeds; antifriction metal, tough.

HOYT NO 3M.
M-317; Sn, Sb, bal Pb.
For bearings; antifriction metal.

HOYT NO 14.
M-317; Sn, Sb, bal Pb.
For stern tubes; zinc-free Babbitt.

HOYT NO 40.
M-317; Sn, Sb, bal Pb.
For stern tubes and outer bearings; Babbitt.

HOYT NO 71.
M-317; Sn, Sb, bal Pb.
For engine bearings; Babbitt.

HOYT NO 133C.
M-317; Sn, Sb, bal Pb.
For bearings; Babbitt.

HOYT NO 175.
M-317; Pb, Sb, bal Sn.
Cast: 12,000 TS; 10,500 YS; 1.6 El; 35 Brin.
For bearings for locomotives, crushers and tube mills; Babbitt.

HOYT NO 400.
M-317; Sn, Sb, bal Pb.
For bearings operating in sea water; Babbitt.

HOYT NUMBER ELEVEN D.
M-317.
Tin base, fine grain White metal.
Compressive proof strength 56.37 N/mm^2.
For bearings for marine steam engines and heavy duty compressors.

HOYT NO ELEVEN R.
M-317; 4 Sb, 4 Cu, 0.5 Ni, 0.2 Pb, bal Sn.
Cast: 12,200 TS; 9000 YS; 11.7 El; 32 Brin.
For bearings for compressors and engines; Babbitt.

HOYT NO ELEVEN Z3.
M-317; 4 Sb, 4 Cu, 0.5 Ni, 0.2 Pb, bal Sn.
Cast: 13,500 TS; 9300 YS; 13.9 El; 30 Brin.
For bearings for compressors and gasoline engines; Babbitt.

HOYT STAR BRAND.
M-317; Sn, Sb, bal Pb.
For bearings for fans, stone crushers, mining machinery; antifriction metal.

HOYT STAR M.
M-317; Sb, Cu, Pb, bal Sn.
Cast: 11,500 TS; 11,200 YS; 1.5 El; 30 Brin.
For machine tool bearings; Babbitt.

HP 9-4-20 ETC see REPUBLIC HP 9-4-20 ETC.

H P D.
M-289; 0.35 C, 1.05 Si, 5.15 Cr, 0.3 V, 1.25 W, 1.55 Mo, bal Fe.
Air or oil hardening hot work tool and die steel; AISI H12.

HP NICKEL see JELLIFF HP NICKEL.

HRA 376.
M-England; 0.03-0.15 C, 4-8 Cr, 3.5-10.5 W, 5-7 Al, 3-9.5 Ta, 0.005-0.04 B, 10-14 Co, 4 max Mo, 3.5 max Cb, 0.15 max Zr, bal Ni.
For cast turbine blades.
Vacuum cast. High creep fatigue resistance.

H & R A4.
M-363; 0.95 C, 2.0 Mn, 0.35 Si, 2.2 Cr, 1.1 Mo, (lead added) bal Fe.
Free machining grade of air hardening cold work tool steel; AISI A4.

H RAPID 199.
M-1328; 0.86 C, 2.8 Co, 4.3 Cr, 0.85 Mo, 2.1 V, 12 W, bal Fe.
For lathe and planer tools, reamers, broaches; high speed steel.

H RAPID 200.
M-1328; 1.35 C, 4 Cr, W, Co, bal Fe.
For blanking and forming dies, engravers tools; high speed steel.

H RAPID 201.
M-1328; 0.79 C, 4.3 Cr, 4.75 Co, 0.75 Mo, 1.5 V, 18 W, bal Fe.
For lathe and planer tools, cutters, reamers, taps; high speed steel.

H & R BRAKE DIE.
M-363; 0.50 C, 1 Mn, 0.95 Cr, 0.20 Mo, bal Fe.
For brake dies; pre-heat treated.

H & R CARBON.
M-363; 0.9-1.05 C, bal Fe.
For punches, dies, tools; water hardening.

H R C MAX.
M-521; 0.25 max C, 18-20 Cr, 24-26 Ni, bal Fe.
Annealed: 85,000-95,000 TS; 35,000-45,000 YS; 60-50 El; 75-60 RA 150-190 Brin.
For furnace parts, heat treating boxes, valves, pumps; Type 311; austenitic, heat resistant.

H & R COBALT HIGH SPEED STEEL.
M-363; 0.75-0.80 C, 4 Cr, 18 W, 5 Co, 1 V, bal Fe.
For cutting tools; high speed steel.

H & R COBALT MOLY.
M-363; 0.88 C, 4.1 Cr, 6 W, 6 Mo, 1.9 V, 9 Co, bal Fe.
For cutters, tools; high speed steel.

H REKORD EMINENT.
M-1328; 0.76 C, 10 Co, 4.3 Cr, 0.8 Mo, 1.8 V, 18 W, bal Fe.
For lathe and planer tools, reamers, broaches, taps; high speed steel.

H REKORD SUPERIOR.
M-1328; 0.74 C, 4.1 Cr, 1.1 18.5 W, bal Fe.
For lathe and planer tools, reamers; high speed steel.

H & R GOLD LABEL.
M-363; 1.33 C, 0.35 Mo, 4.25 W, bal Fe.
For drawing dies, finishing tools; keen cutting edge.

H & R GRAY LABEL.
M-363; 0.90-1.05 C, bal Fe.
For punches, dies, stamp, headers.

H & R HEADING DIE STEEL.
M-363; 0.9-1.0 C, bal Fe.
For embossing dies, cold heading dies; water hardening.

H & R HOT WORK.
M-363; 0.30-0.35 C, 3.0-3.5 Cr, 0.3-0.5 V, 10.0-11.5 W, bal Fe.
For extrusion and swedging dies, shear blades; hot work steel.

H & R HOT WORK 7.
M-363; 0.55 C, 1.2 W, 1.2 Mo, 0.95 Si, 0.3 Mn, bal Fe.
For dies, hot work tools, punches; air or oil hardened, shock resistant.

H & R HOT WORK NO 4.
M-363; 0.97 C, 3.9 Cr, bal Fe.
For upsetting dies, gripper dies; oil or air hardening.

H & R HOT WORK NO 5.
M-363; 0.35 C, 1.0 Si, 5 Cr, 0.4 V, 1.0 Mo, bal Fe.
For die casting dies, tools, hot punches; hot work steel.

H & R WORK 5V.
M-363; 0.35 C, 5 Cr, 1 V, 1.5 Mo, bal Fe.
For hot work steel, extrusion rams and liners; Type H13; hot work steel.

H & R HOT WORK NO 6.
M-363; 0.35 C, 1.0 Si, 5 Cr, 1.35 W, 1.75 Mo, bal Fe.
For extrusion dies, piercing mandrels; hot work steel.

H & R HOT WORK NO 10.
M-363; 0.23 C, 0.6 Mn, 1.25 Si, 10.0 Cr, 0.75 Ni, 1.0 V, 0.45 W, 1. Mo, 0.10 N, bal Fe.
Air or oil hardening hot work tool and die steel, chromium type.

H & R HOT WORK NO 12.
M-363; 0.30 C, 0.35 Mn, 0.5 Si, 12.0 Cr, 0.9 V, 12.0 W, bal Fe.
Air or oil hardening hot work tool and die steel; tungsten type; AISI H23.

H & R HOT WORK NO 15A.
M-363; 0.25 C, 14-16 W, 3.75-4.25 Cr, 0.4-0.6 V, bal Fe.
For hot heading dies, gripper dies, high temperature springs; hot work steel, oil hardened.

H & R-K.
M-363; 2.25-2.40 C, 0.25-0.40 Mn, 12.75-13.25 Cr, 0.15-0.25 V, bal Fe.
For plug gauges, punches, forming dies; non-deforming.

H & R K2.
M-363; 1.5 C, 11.5 Cr, 0.75 Mo, 0.25 V, bal Fe.
For blanking and forming dies, lamination dies; non-deforming.

H & R K2L.
M-363; 0.85 C, 1 Ni, 0.45 Mo, 11.5 Cr, 0.3 V, bal Fe.
For shear blades, punches, dies; shock and abrasion resistant, non-deforming.

H & R K3.
M-363; 2.4 C, 12.75 Cr, 4 V, 1.1 Mo, bal Fe.
For mold liners, lamination dies; oil hardened, non-deforming.

H & R K4.
M-363; 1.4 C, 4.1 Cr, 4.1 V, 4.25 Mo, bal Fe.
For cutting tools; machining scaly material.

H & R M4.
M-363; 1.28 C, 4.5 Cr, 4.0 V, 5.5 W, 4.5 Mo, bal Fe.
Molybdenum-tungsten-vanadium grade of high speed steel for cutting tools; AISI M4.

H & R M7.
M-363; 1.0 C, 3.75 Cr, 2.05 V, 1.75 W, 8.75 Mo, bal Fe.
High speed tool steel, molybdenum type; AISI M7.

H & R M42.
M-363; 1.07 C, 3.75 Cr, 1.15 V, 1.5 W, 9.5 Mo, 8.0 Co, bal Fe.
High speed tool steel, molybdenum-cobalt type, AISI M42.

H & R M43.
M-363; 1.20 C, 3.75 Cr, 1.6 V, 2.7 W, 8.0 Mo, 8.2 Co, bal Fe.
High speed tool steel, molybdenum-cobalt type; AISI M-43.

H & R MOLYHI.
M-363; 0.8 C, 4 Cr, 1.5 W, 8.5 Mo, 1.15 V, bal Fe.
For cutters, tools; high speed steel.

H & R MOLY VAN.
M-363; 0.8 C, 4 Cr, 9 Mo, 2.2 V, bal Fe.
For cutters, tools; high speed steel.

H & R MULTIMOLD.
M-363; 0.35 C, 0.80 Cr, 0.30 Mo, 0.70 Mn, bal Fe.
For plastic mold dies; for high mold pressures.

H & R MY.
M-363, M-27; 0.42 C, 1.4 Cr, 1.5 Si, 0.25 V, bal Fe.
For pneumatic tools, chisels, snaps, hot dies, hot shears; shock resistant.

H & R MY-A.
M-363; 0.40 C, 1.5 Cr, 1.5 Si, 0.25 V, bal Fe.
For rivet sets, hand chisels, punches, shears; tough, shock resistant.

H & R-MY EXTRA.
M-363; 0.4 C, 0.9 Si, 1 Cr, 2 W, bal Fe.
For pneumatic tools, chisels, snaps, hot dies, hot shears; water hardening.

H & R N175.
M-363; 0.75 C, 0.75 Mn, 0.3 Si, 0.9 Cr, 1.75 Ni, 0.42 Mo, bal Fe.
Oil hardening low alloy tool steel for shafts, arbors, lathe centers; AISI L6.

H & R NO 1.
M-363; 0.55-0.75 C, 3.8-4.3 Cr, 0.90-1.25 V, 17.75-18.85 W, bal Fe.
For lathe and planer tools, cutters; high speed steel.

H & R NO 2.
M-363; 0.80-0.85 C, 4.0-4.5 Cr, 2.0-2.25 V, 18-19 W, 0.6-0.8 Mo, bal Fe.
For lathe tools, drills, taps, reamers; high speed steel.

H & R NO 3.
M-363; 1.0 C, 0.25 Mn, 4 Cr, 18 W, 0.8 Mo, 3.4 V, bal Fe.
For cutters, tools; high speed steel.

H & R NO 4.
M-363; 0.8 C, 18.5 W, 4.5 Cr, 1.75 V, 7.5 Co, 0.8 Mo, bal Fe.
For cutters, reamers; high speed steel.

H & R NO 7.
M-363; 1.3 C, 5.5 W, 4.5 Cr, 4.5 Mo, 4 V, bal Fe.
For broaches, reamers, chasers; high speed steel.

H & R NO 8.
M-363; 0.55 C, 0.85 Mn, 0.25 Cr, 0.25 V, 2.1 Si, bal Fe.
For carbide tool shanks; water or oil hardened.

H & R NO 8M.
M-363; 0.55 C, 0.75 Mn, 0.2 Mo, 2.0 Si, bal Fe.
For carbide tool shanks; water or oil hardened.

H & R NO 14.
M-363; 0.75 C, 4 Cr, 2 V, 14 W, 5 Co, bal Fe.
For lathe and planer tools, reamers, broaches; Type T8; high speed steel.

H & R NO 15.
M-363; 0.37 C, 3 Mo, 15.5 W, 4 Cr, 2 Ni, bal Fe.
For extrusion dies, hot piercing punches; hot work steel.

H & R NO 19.
M-363; 0.9 C, 1.5 Mn, 0.3 Mo, bal Fe.
For precision tools, dies; oil hardening.

H & R NO 44.
M-363; 0.80 C, 4 Cr, 4.2 Mo, 1.1 V, 6 W, bal Fe.
For cutting tools, drills, hobs, reamers; high speed steel.

H & R NO 45.
M-363; 0.65 C, 4 Cr, 2 V, 6 W, 5 Mo, bal Fe.
For drills, reamers, lathe and planer tools; Type M2; high speed steel.

H & R NO 50.
M-363; 0.57 C, 18 W, 4 Cr, 1.15 V, bal Fe.
For drills, reamers, taps, milling cutters; high speed steel.

H & R NO 55.
M-363; 0.35 C, 5.2 Cr, 5.2 W, 0.2 Mo, 0.2 V, 0.5 Co, 0.9 Si, bal Fe.
For die casting dies; oil hardening.

H & R NO 57.
M-363; 0.8 C, 3.75 Cr, 5.75 W, 4.5 Mo, 1.5 V, bal Fe.
For taps, chasers, broaches, reamers; high speed steel.

H & R NO 59.
M-363; 0.88 C, 4.1 Cr, 1.8 V, 4.25 Mo, 6 W, bal Fe.
For thread chasers, drills, pipe taps, tools; high speed steel.

H & R NO 60.
M-363; 1.2 C, 0.7 Cr, 0.25 Mo, 1.6-0.2 V, bal Fe.
For taps, broaches, punches, gauges; non distorting.

H & R NO 61.
M-363; 1.3 C, 12 Cr, 0.6 Mo, 3 Co, 0.5 Si, bal Fe.
For blanking and forming dies, cold heading dies; air hardened.

H & R NO 80.
M-363; 1.0 C, 0.5 Cr, 0.7 Mn, 0.25 V, 1.25 Mo, bal Fe.
For lamination dies, gages, reamers; oil hardened.

H & R NO 85.
M-363; 0.50 C, 0.95 Cr, 0.20 V, bal Fe.
For die casting dies, shear blades; shock resistant.

H & R NO 150.
M-363; 0.7 C, 0.85 Cr, 0.42 Mo, 1.4 Ni, bal Fe.
For carbide tool shanks; oil hardened.

H & R NO 225.
M-363; 0.5 C, 1.5 Cr, 2.5 W, 0.25 V, bal Fe.
For dies, shear blades, cutters; oil hardening.

H & R NO 434.
M-363; 1.18 C, 4.25 Mo, 3.1 V, 4.1 Cr, bal Fe.
For cutting tools; for cutting abrasive materials.

H & R NO 444.
M-363; 1.4 C, 4.1 Cr, 4.25 Mo, 4.15 V, bal Fe.
For cutting tools; abrasion resistant.

H & R NO 445.
M-363; 1.5 C, 13.5 W, 4.5 Cr, 4.75 V, 5 Co, 0.5 Mo, bal Fe.
For cutting tools; high speed steel.

H & R NON-TEMPERING.
M-363; 0.35 C, 0.7 Mn, 0.45 Si, 0.8 Cr, 0.3 Mo, 0.3 Cu, bal Fe.
Oil hardening tool steel, shock resisting type.

H & R 06 GRAPHITIC.
M-363; 1.45 C, 1.0 Mn, 1.25 Si, 0.25 Mo, bal Fe.
AISI Type 06 oil hardening tool steel.

H & R OIL HARDENING.
M-363; 0.90-1.00 C, 0.90-1.00 Mn, 0.5-0.6 Cr, 0.2 V, bal Fe.
For cold forming tools and dies, punches, broaches, gauges.

H & R PISTON.
M-363; 1.14 C, 0.58 Cr, 0.19 V, 0.2 Si, 0.3 Mn, bal Fe.
For pistons; water hardening.

H & R PLASTIC MOLD-A.
M-363; 0.07 C, 4.5 Cr, 0.45 Mo, 0.4 Mn, 0.25 Si, bal Fe.
For plastic mold dies; hobbing die steel, air hardened.

H & R PLASTIC MOLD-B.
M-363; 0.06 C, 1 Cr, 0.25 Mo, B, bal Fe.
For plastic mold dies; for cold hobbing, oil hardened.

H & R PLASTIC MOLD-C.
M-363; 0.1 C, bal Fe.
For plastic mold dies; case hardened.

H & R PLASTIC MOLD L.
M-363; 0.50 C, 1.0 Mn, 0.3 Si, 1.1 Cr, 0.25 Mo, bal Fe.
Oil hardening tool and die steel designed for plastic molds.

H & R SILICO 1.
M-363; 0.5 C, 0.5 Mo, 0.2 V, 1.1 Si, bal Fe.
For impact tools, punches, chisels; shock resistant: oil or water hardened.

H & R SILICO 2.
M-363; 0.55 C, 0.7 Mn, 0.45 Mo, 0.2 V, 2.15 Si, bal Fe.
For impact tools, punches, chisels; shock resistant, oil or water hardened.

H & R SPECIAL CARBON.
M-363; 1.0-1.2 C, 0.3 Mn, 0.2 Si, bal Fe.
For drills, taps, hobs, reamers; Type W1; water hardened.

H & R SPECIAL HARDENING DIE.
M-363; 0.9-1.0 C, 0.2 Mn, 0.3 Si, bal Fe.
For hardening dies; water hardened.

H & R SPECIAL HEADING DIE.
M-363; 0.9-1.0 C, 0.2 Mn, 0.3 Si, bal Fe.
AISI Type W1 water hardening tool steel.

H & R SUPER COBALT.
M-363; 0.8 C, 20-21 W, 4-4.5 Cr, 1.3 V, 12 Co, 0.6 Mo, bal Fe.
For cutters, hobs, millers; high speed steel.

H & R SUPER MOLYHI.
M-363; 0.8 C, 4 Cr, 1.5 W, 8.5 Mo, 1.2 V, 5 Co, bal Fe.
For cutters, tools; high speed steel.

H & R TUNGSTEN OIL HARDENING.
M-363; 0.9 C, 1.1 Mn, 0.5 W, 0.2 V, 0.5 Cr, bal Fe.
For hobs, reamers, broaches, gauges; fast finishing tool steel.

H & R VANADIUM.
M-363; 1.05-1.15 C, 0.18 V, bal Fe.
For striking and heading dies, swedging tools; water hardened.

H & R VH.
M-363; 1.4 C, 0.4 Mn, 0.35 Si, 0.15 max Cr, 3.5 V, 0.1 max Mo, bal Fe.
Water hardening tool steel.

HS-1.
M-1542; 0.18 max C, 0.14 max Mn, 0.55 max Si, bal Fe.
Rolled: 71,000-85,000 TS; 47,000 min YP; 23 min El.
For bridges, booms, buildings, bus and truck bodies. Constructional steel.

HS-21.
M-1491; 0.25 C, 0.60 Mn, 0.60 Si, 27.0 Cr, 3.0 Ni, 5.0 Mo, 1.0 Fe, bal Co.
Cast alloy for gas turbine parts.

HS-25.
M-1491; 0.10 C, 1.5 Mn, 0.50 Si, 20.0 Cr, 10.0 Ni, 15 W, bal Co.
For jet engine parts, sheets.

HS 25.
M-73; Composition similar to HSS Type M-1 except higher carbon content.
Hardened: C 63-67 Rock.
For lathe tools and other cutting tools and dies requiring extra high hardness and wear resistance.
AISI Type M-1 (mod) High speed tool steel.

HS 31 see HAYNES STELLITE ALLOY NO 31.

HS-31.
M-1491; 0.50 C, 0.50 Mn, 0.50 Si, 25.0 Cr, 10.0 Ni, 7.5 W, 1.5 Fe, bal Co.
Cast alloy for gas turbine parts, nozzle vans.

HS-35.
M-67; 0.25 max C, 0.90 max Mn, bal Fe.
Sheet, plate: 50 ksi TS; 35 ksi YS; 22 El.
Structural equipment where little or no forming is required.

HS-40.
M-67; 0.25 max C, 0.90 max Mn, bal Fe.
Sheet, plate: 55 ksi TS; 40 ksi YS; 21 El.
Structural equipment where little or no forming is required.

HS-45.
M-67; 0.25 max C, 0.90 max Mn, 0.010-0.013 N, bal Fe.
Sheet, plate: 60 ksi TS; 45 ksi YS; 20 El.
Structural equipment where little or no forming is required.

HS-50.
M-67; 0.25 max C, 0.90 max Mn, 0.10-0.13 N, bal Fe.
Sheet, plate: 65 ksi TS; 50 ksi YS; 18 El.
Structural equipment where little or no forming is required.

HS 100.
M-73; 1.08 C, 0.30 Si, 1.50 W, 3.75 Cr, 1.10 V, 9.50 Mo, 80.0 Co, bal Fe.
Super high-speed steel, hardenable to 68 Rc, for long production runs or heavy duty machining, particularly for lathe tools.
AISI M 42.

HS-151.
M-1491; 0.50 C, 1.0 max Mn, 1.0 max Si, 20.0 Cr, 12.7 W, 0.05 B, bal Co.
Cast alloy for gas turbine blades, vanes.

HS-151 see also HAYNES ALLOY 151.

HS 188 see HAYNES ALLOY 188.

HS-220.
M-376; 0.30 C, 2 Ni, 1.2 Cr, 0.45 Mo, bal Fe.
Heat treated: 237,000-258,000 TS; 201,000-212,000 YS; 11-13 El; 36-51 RA; 500-540 Brin.
For aircraft structures, landing gears; tough, shock resistant, oil hardened.

HS-260.
M-376; 0.40 C, 1.25 Cr, 2.2 Ni, 0.50 Mo, bal Fe.
Heat treated: 271,000-280,000 TS; 221,000-235,000 YS; 7-10.5 El; 24.5-38.5 RA; 510-530 Brin.
For aircraft structures landing gears; tough, shock resistant, oil hardened.

H S C 350.
M-908; 1.3 C, 3.5 W, 0.3 Mn, 0.45 Si, bal Fe.
For cutters, tools, cold extrusion dies; fast finishing steel.

H S C 6-6-2.
M-9008; 0.8 C, 4 Cr, 5.75 W, 5 Mo, 1.5 V, bal Fe.
For tools, drills, cutters, chasers, broaches; high speed steel.

H S C 18-4-1.
M-908; 0.70 C, 4 Cr, 18 W, 1 V, bal Fe.
For tools, taps, cutters, chasers, punches; high speed steel.

H S C ALLOY SHOE DIE.
M-908; 0.5 C, 0.6 Cr, 0.6 Mn, 0.4 Mo, bal Fe.
For shoe dies; wear resistant; oil hardening.

H S C COBALT 5.
M-908; 0.7 C, 4 Cr, 18 W, 0.5 Mo, 5 Co, 1 V, bal Fe.
For tools, cutters, broaches, chasers, reamers; high speed steel.

H S C COLD HEADER.
M-908; 0.95 C, 0.30 Mn, bal Fe.
For cold headed dies, tools.

H S C CUTLERY.
M-908; 1.1 C, 0.2 Mn, bal Fe.
For tools, cutlery, pocket knives; water hardening.

HSC-CVM.
M-908; 1.0 C, 0.7 Mn, 5.25 Cr, 1.1 Mo, 0.25 V, bal Fe.
For cold work dies, shears, punches; air hardening.

HSC NO 33.
M-908; 0.30-0.35 C, 0.9 Si, 5 Cr, 1.2 W, 1.5 Mo, bal Fe.
For hot work tools, Al-die casting dies; hot work steel.

HSC NO 265.
M-908; 1.55-1.70 C, 11.5-12.5 Cr, 0.7-0.9 Mo, 0.15-0.25 V, bal Fe.
For dies, tools, blanking and forming dies; non-deforming.

HSC NO 280.
M-908; 0.55 C, 1.5 Si, 0.35 Mo, bal Fe.
For tools, punches, chisels; shock resistant.

H S C NO 310.
M-908; 0.30 C, 3.25 Cr, 10 W, 0.35 V, bal Fe.
For hot work dies, brass forging dies; oil and air hardening.

H S C NO 313.
M-908; 0.35 C, 3 Cr, 13.5 W, bal Fe.
For hot work dies, hot punching and extruding dies; hot work steel.

HSC NO 422.
M-908; 0.45-0.50 C, 0.85-1.05 Cr, 0.9-1.2 W, 0.15-0.25 Mo, bal Fe.
For tools, punches, shear blades; shock resistant.

H S C NO 515.
M-908; 0.35 C, 1.0 Si, 5 Cr, 5 W, 0.2 Mo, bal Fe.
For hot work dies; hot work steel, air hardening.

H S C OVERCOAT AXE.
M-908; 1.0 C, 0.25 Mn, 0.15 Si, bal Fe.
For axes, tools; water hardening.

H S C PAVEMENT BREAKER.
M-908; 0.65 C, 0.3 Mn, 0.1 Si, bal Fe.
For tools, sledges, pavement breakers; tough.

H S C REGULAR.
M-908; 0.7-1.1 C, 0.3 Mn, bal Fe.
For tools, cold work tools; water hardened.

H S C SHOE DIE.
M-908; 0.9 C, 0.35 Mn, 0.15 Si, bal Fe.
For shoe dies; water hardening.

H S C SOLID DRILL.
M-908; 0.85 C, bal Fe.
For drills; tough.

H S C SPECIAL.
M-908; 1.1 c, 0.3 Mn, 0.3 Si, bal Fe.
For tools, cold work tools, taps, blanking dies; water hardened.

HSC-SS EXTRA.
M-908; 1.0-1.1 C, 0.3 Mn, 0.3 Si, bal Fe.
For drills, taps, hobs, reamers, broaches; Type 1; water hardened.

HSM COPPER 194.
M-279; 97.5 Cu, 2.35 Fe, 0.12 Zn, 0.03 P.
Ann: 40,000-60,000 TS; 24,000-45,000 YS; 20-35 El.
Cold rolled: 53,000-85,000 TS; 48,000-78,000 YS; 1-13 El.
For electrical connectors and terminals, springs, tubular products; high conductivity, strength, corrosion resistance.

HSM/S.
M-1083; 0.25-0.45 S, bal Cu.
Bar: 36,000-47,000 TS; 16-21 El; 75-90 Brin.
Forged: 25,000-38,000 TS; 20-30 El; 45-55 Brin.
Free machining copper, 95-98% electrical conductivity.

H SUPER REKORD.
M-1328; 1.3 C, 4.3 Cr, 0.85 Mo, 3.8 V, 12 W, bal Fe.
For engravers tools, blanking and forming tools; high speed steel.

H SUPER REKORD CO.
M-1328; 1.35 C, Cr, Co, V, bal Fe.
For engravers tools, blanking and forming dies; high speed steel.

HTB-1.
M-1779; 1.02 C, 0.35 Mn, 0.40 Si, 1.45 Cr, 1.0 Al, bal Fe.
For jet engine bearings; for high temperature applications.

HTB-2.
M-1779; 0.58 C, 0.30 Mn, 1.15 Si, 4.75 Cr, 5.25 Mo, 0.55 V, bal Fe.
For aircraft bearings and other high strength components operating up to 900°F.

HT IRON-40.
M-637; 3.1 T.C., 0.2 Si, 0.7 Mn, 1.5 Ni, bal Fe.
Cast: 40,000 TS; 220 Brin.
For roller bearing cages; cast iron; Tr.S. 3800.

HTM.
M-89; 3 C, Mn, Si, bal Fe.
For tractor parts; high strength malleable iron.

HT MOLYBDENUM.
M-1246; 100 Mo.
For electronic, nuclear, chemical, aerospace, metallurgical and electrical industries, vacuum equipment.
Ductile after exposure to 2700-3600°F.

HTP 50.
M-1541; 0.12 C, 1.10 Mn, 0.011 P, 0.007 S, 0.025 Ti, bal Fe.
74,700 psi YS; 33 El.
High strength low alloy steel for structural purposes.

HTP 60.
M-1541; 0.16 C 1.40 Mn, 0.010 P, 0.006 S, 0.045 Ti, bal Fe.
92,000 psi TS; 74,700 psi YS; 28 El.
High strength low alloy steel for structural purposes.

HTP 60 W.
M-1541; 0.16 C, 1.48 Mn, 0.020 P, 0.009 S, 0.47 Si, 0.21 Cr, 0.043 V, 0.025 Cb, bal Fe.
96,000 psi TS; 76,700 psi YS; 40 El.
High strength low alloy normalized steel for weldable structures and construction equipment.

HTP 80 E.
M-1541; 0.12 C, 1.98 Mn, 0.010 P, 0.006 S, 0.22 Si, 0.138 Ti, 0.044 Cb, bal Fe.
119,200 psi TS; 108,200 psi YS; 23 El.
High strength low alloy steel for structural purposes.

HTP100E.
M-1541; 0.09 C, 2.23 Mn, 0.006 P, 0.005 S, 0.20 Si, 0.22 Ti, 0.039 Cb, 0.65 Mo, bal Fe.
147,000 psi TS; 120,000 psi YS; 18 El.
High strength low alloy steel for structural purposes.

H T S 33.
M-845; Al alloy.
For castings; not hardenable.

HTW.
M-24; 0.9 C, 1.7 Cr, 0.2 V, bal Fe.
For bearing races; wear resistant.

HUBARD SPECIAL.
M-225, M-451; C, Ni, Cr, bal Fe.
For rolls, guides, castings; high wear resistance.

HUB BUSHING-HB-1.
M-264; 83 Cu, 0.9 Pb, 11.0 Ni, 4.5 Fe, 0.8 Si.
Cast: 63,000 psi TS; 90-145 Brin.
For valve components where iron is cast around the bronze bushing in core.
Good corrosion resistance; Wears well.

HUB METAL-H-1.
M-264; 80.5 Cu, 5 Sn, 4.5 Zn, 7 Pb.
Cast: 30,000 TS; 45-50 Brin.
For pressure containing parts at temperatures to 400°F; wears well with silicon brass stems.

HUCKINGERHUTTE HH428U.
M-1381; 0.15 C, 0.60 Mn, 0.30 Mn, 0.30 Mo, bal Fe.
For gears, bolts, shafts, cams, fasteners; case hardened.

HUCKINGERHUTTE HH432.
M-1381; 0.19 C, 0.50 Si, 1.15 Mn, bal Fe.
For gears, pinions, camshafts, cams, bolts; case hardened, tough.

HUCKINGER KI.
M-1381; 0.17 max C, 0.35 max Si, 0.35 Mn, bal Fe.
Annealed: 70,000 TS; 55,000 YS; 25 El; 60 RA; 143 Brin.
Cold drawn: 72,000 TS; 60,000 YS; 22 El; 58 RA; 146 Brin.
For screws, bolts, gears, shafts, nails, bushings; case hardened.

HUCKINGER KIW.
M-1381; 0.16 max C, 0.35 max Si, 0.40 max Mn, bal Fe.
Annealed: 70,000 TS; 55,000 YS; 25 El; 60 RA; 143 Brin.
Cold drawn: 72,000 TS; 60,000 YS; 22 El; 58 RA; 146 Brin.
For screws, bolts, gears, shafts, nails, bushings; case hardened.

HUCKINGER KII.
M-1381; 0.23 max C, 0.25 max Si, 0.35 max Mn, bal Fe.
Hot rolled: 70,000 TS; 45,000 YS; 31 El; 58 RA; 143 Brin.
Cold drawn: 80,000 TS; 69,000 YS; 18 El; 48 RA; 162 Brin.
For crankshafts, gears, bolts, armature shafts; water hardened.

HUCKINGER KIIW.
M-1381; 0.20 max C, 0.35 max Si, 0.50 max Mn, bal Fe.
Cold drawn: 78,000 TS; 68,000 YS; 20 El; 55 RA; 149 Brin.
For fan blades, bushings, gears, bolts; water hardened.

HUCKINGER KIIIW.
M-1381; 0.22 max C, 0.35 max Si, 0.55 max Mn, bal Fe.
Cold drawn: 78,000 TS; 68,000 YS; 20 El; 55 RA; 159 Brin.
Annealed: 73,000 TS; 61,000 YS; 22 El; 58 RA; 149 Brin.
For gears, bolts, crankshafts, fan blades; water hardened.

HUCKINGER KIVW.
M-1381; 0.26 max C, 0.35 max Si, 0.60 max Mn, bal Fe.
Hot rolled: 70,000 TS; 45,000 YS; 31 El; 58 RA; 143 Brin.
Cold drawn: 80,000 TS; 69,000 YS; 18 El; 48 RA; 162 Brin.
For crankshafts, gears, bolts, armature shafts; water hardened.

HUGHES ALLOY.
M-161; 17 Sn, 70 Pb, 13 Sb, 30 Brin.
For metallic packing.

HUNDRED METAL.
M-England; Ni, Cr.
For high temperature applications; heat and corrosion resistant.

HUNTER-DOUGLAS ZG73.
M-1389; Al alloy.
For extrusion; heat treatable, resists fuming nitric acid.

HURBENIUM B.
M-892; 91 Pb, 8 Sn, 1 Bi.
For hot dip plating; corrosion resistant.

HURON.
M-1779; 2.1 C, 12.5 Cr, 1.0 V, 0.32 Mn, 0.45 Si, bal Fe.
Heat treated: 744-418 Brin.
For high performance blanking and cold forming dies, punches, rolls and gages. High resistance to abrasion.
AISI D-3.

HURON ALUMINUM ALLOY.
M-Eng.; 3-7 Cu, 0-0.6 Mn, 0.5 Mg, 0-1.25 Ni, Cr, Co, Cd, bal Al.
For light alloy parts; heat treatable.

HURON ALUMINUM ALLOY A ROLLED.
3.5 Cu, 0.6 Mn, 0.5 Mg, 0.1 Cr, bal Al.
For light alloy parts; heat treatable.

HURON ALUMINUM ALLOY B ROLLED.
4 Cu, 0.6 Mn, 0.5 Mg, 0.1 Cr, bal Al.
For light alloy parts; heat treatable.

HURON ALUMINUM ALLOY D ROLLED.
4 Cu, 0.5 Mg, 0.6 Cr, bal Al.
For light alloy parts; heat treatable.

HURON ALUMINUM CASTINGS A-5 ALLOY.
6.6 Cu, 0.5 Mg, 1.25 Ni, 0.25 Co, 0.5 Mn + Sn + Cd, bal Al.
For light alloy parts; heat treatable.

HURON-EZ.
M-1; 2.1 C, 12.5 Cr, 1 V, 0.3 Mn, 0.4 Si, bal Fe.
For blanking and forming dies; rolls, gages; Type D3; oil hardened, abrasion resistant.

HURON METAL CAST.
6.6 Cu, 1.3 Ni, 0.5 Mg, 0.3 Co, bal Al.
For light alloy parts; heat treatable.

HURON METAL ROLLED.
3.5-4.0 Cu, 0.5 Mg, 0-0.6 Mn, 0.1-0.6 Cr, bal Al.
For light alloy parts; heat treatable.

HUSMANN METAL.
M-Eng.; 74 Sn, 11 Sb, 11 Pb, 4 Cu, 0.2 Fe, 0.2 Zn.
For bearings, bushings; anti-friction.

HUSQUARNA HVA.
M-1271; 0.05 max C, 0.02 max Si, 0.06 max Mn, 0.01 max S, bal Fe.
Sintered.
For sintered parts; electrolytic powdered iron.

HV 5.
M-1744; 1.5 C, 12.5 W, 4.75 Cr, 0.5 Mo, 5.0 V, 5.0 Co, bal Fe.
Cobalt-vanadium high speed steel; extra high abrasion resistance.

H V BLUE CHIP.
M-57; 0.82 C, 4 Cr, 2 V, 18 W, bal Fe.
For finishing tools and cutters; high speed steel.

HW-8.
M-24; 0.55 C, 4.0 Cr, 8.0 Mo, 2.0 V, bal Fe.
Hot work tool steel. (same as HOT WORK 8).

HW10C.
M-1522; 49.75 Ag, WC, Co.
Ann hardness: 130 HV.
Elec. res.: 2.8 μ % cm.
Elec. cond.: 62% IACS.
For electrical contacts; good erosion resistance.

H W A.
M-40; 0.6 C, 5 Cr, 1 Mo, 1 V, bal Fe.
Heat treated: 440-550 Brin.
For die casting dies, hot working tools and dies; hot work steel, air or oil hardening.

H W D 2.
M-57; 0.38 C, 5.25 Cr, 0.5 V, 1.35 Mo, 0.4 Mn, bal Fe.
For hot punches and dies; hot work steel, oil hardened.

H W D 3.
M-57; 0.4 C, 1.05 V, 5.25 Cr, 1.25 Mo, bal Fe.
At 950°F: 153,000 TS; 138,000 YS; 17 El; 63 RA; 260 Brin.
At 1150°F: 86,000 TS; 78,000 YS; 27 El; 78 RA; 155 Brin.
For hot punches, shear blades, die casting dies; Type H13; hot work steel, oil hardened.

HWD MOD.
M-57; 0.55 C, 1.4 W, 5 Cr, 1.5 Mo, 0.3 V, bal Fe.
For shear blades, forging dies, punches; air hardened, non-deforming.

H W D NO 1.
M-57; 0.4 C, 0.5 Cr, 0.15 V, 1.1 W, 1.5 Mo, 0.8 Si, bal Fe.
At 950°F: 141,000 TS; 130,000 YS; 19 El; 61 RA; 275 Brin.
At 1150°F: 80,000 TS; 68,000 YS; 29 El; 80 RA; 165 Brin.
For hot punches, mandrels, shear blades; Type H12; hot work steel.

HWF 22.
M-1797; 0.20 C, 12.0 Cr, 1.2 max Mo, 0.80 max Ni, 0.35 max V, bal Fe.
For high temperature parts as steam power plant equipment.

HWF 81.
M-1797; 0.10 max C, 1.5 max Mn, 0.60 max Si, 16.5 Cr, 2.0 max Mo, 16. Ni, Nb = 10 x c, bal Fe.
High temperature stainless; for steam and gas turbines.

HWS see DARWIN HWS.

HY 43.
M-116; 4.5 Zn, 3.5 Mg, 0.3 Mn, 0.1 max Cr, 0.4 Cu, 0.2-0.6 V, bal Al.
Aged: 71,000-77,000 TS; 64,000-68,000 YS; 8 El.
For light alloy parts; high strength, age-hardenable.

HY-65.
M-U.S.; 0.12 C, 0.48 Mn, 2.16 Ni, 0.39Mo, 0.70 Cu, bal Fe.
For pressure vessels; tough, corrosion resistant.

HY-80.
M-England; 0.22 max C, 1.9-3.3 Ni, 0.8-1.9 Cr, 0.13-0.63 Mo, bal Fe.
Heat treated: 105,000 TS; 80,000 YS; 20 El; 72 RA.
For structural and pressure vessels; good weldability.

HY-80.
M-604, M-176, M-1780, M-1602; 0.18 C, 0.1-0.4 Mn, 2.0-3.25 Ni, 1.0-1.8 Cr, 0.2-0.6 Mo, 0.002 Ti, 0.02 V, 0.25 Cu, bal Fe.
Plate: 80,000 TS; 84,800 YS; 20 El; 55 RA; 50 Charpy at-120°F.
For pressure vessels, submarine hulls.
Tough, good weldability.

HY-80 see also BIRDSBORO HY80.

HY-100.
M-604, M-176, M-1780; 0.2 C, 0.1-0.4 Mn, 2.25-3.5 Ni, 1.0-1.8 Cr, Cr, 0.2-0.6 Mo, 0.02 Ti, 0.02 V, 0.25 Cu, bal Fe.
Plate: 100,000-120,000 YS; 18 min El; 50 min RA; 105,700 CYS.
For warship hulls, structures, deck railings, heat exchangers, submarine hulls, pressure vessels.
Supplied in quenched and tempered condition.

HY-130.
M-604, M-176; 0.12 C, 0.6-0.9 Mn, 4.75-5.25 Ni, 0.4-0.7 Cr, 0.3-0.65 Mo, 0.02 Ti, 0.05-0.10 V, 0.15 Cu, bal Fe.
Plate: 130,000 TYS; 141,000 CYS; 15 El; 50 RA.
For submarine hulls, pressure vessels.
Tough, high impact resistance.

HY-130/140 MOD.
M-1780; 0.10-0.17 C, 0.60-0.90 Mn, 0.15-0.35 Si, 0.45-0.65 Cr, 0.45-0.65 Mo, 4.75-5.25 Ni, 0.05-0.10 V, bal Fe.
Plate, Quenched & tempered: 145,000 psi YS; 15 El.
For large, highly stressed structural equipment.

HY-180.
M-604; 0.1 C, 10 Ni, 8 Co, 2 Cr, 1 Mo, bal Fe.
Plate: 180,000 YS; 95 Charpy v-notch.

HY-511.
M-Germany; 5 Mg, 1 Si, Mn, Fe, 0.2 Ti, bal Al.
Cast.
For cylinder heads; heat resistant.

HY-BB.
M-1635; 0.25-0.33 C, 0.60-0.90 Mn, 0.15-0.30 Si, 2.5-0.0 Ni, 0.90-1.10 Cr, 0.40-0.60 Mo, 0.10-0.15 V, bal Fe.
For bleach rings and breech blocks.
MIL-S-10185A.
High yield and impact strength.

HYBNICKEL A.
M-194; 12-25 Ni, 20-24 Cr, bal Fe.
Cast: 50,000 TS; 40,000 YS; 0-10 El.
Forged: 100,000 TS; 70,000 YS; 20 El.
For carburizing boxes, hearth plated, furnace conveyor parts; heat resisting; maximum operating temperature 2100°F.

HYBNICKEL A-1.
M-194; 31-39 Ni, 16-23 Cr, 0.51-0.65 C, bal Fe.
For furnace parts; heat resistant.

HYBNICKEL B.
M-194; 5-7 Ni, 18-22 Cr, bal Fe.
Cast: 50,000 TS; 40,000 YS; 0-15 El.
Forged: 100,000 TS; 70,000 YS; 25 El.
For carburizing boxes, hearth plates, furnace conveyor parts; heat resisting; maximum operating temperature 1800°F.

HYBNICKEL C.
M-194; 15-35 Ni, 25-35 Cr, bal Fe.
Cast: 50,000 TS; 40,000 YS; 0.15 El.
Forged: 100,000 TS; 70,000 YS; 25 El.
For carburizing boxes, hearth plates, furnace conveyor parts; acid resisting; maximum operating temperature 2200°F.

HYBNICKEL D.
M-194; 0.25-0.50 C, 20-30 Cr, 5-10 Ni, bal Fe.
For heat and corrosion resistant parts; heat and corrosion resistant.

HYBNICKEL R.
M-194; 0.2 C, 31-39 Ni, 7-11 Cr, bal Fe.
For heat and corrosion resistant parts; heat resistant.

HYBNICKEL S.
M-194; 25 Ni, 20 Cr, bal Fe.
Cast: 50,000 TS; 40,000 YS; 0-15 El.
Forged: 100,000 TS; 20,000 YS; 25 El.
For furnace parts; acid and alkali resistant.

H Y C C.
M-38; 2.2 C, 12 Cr, 0.25 V, 0.8 Mo, bal Fe.
For forming and crimping rolls, perforating dies, plug gages; wear resistant; air or oil hardened.

HYCO.
M-365; 0.7 C, 18 W, 4 Cr, 1-2 V, 3.0-5.5 Co, bal Fe.
For tools, cutters; high speed steel.

HYCO.
M-688; 2 C, 12.5 Cr, 1.0 V, 0.75 Mo, bal Fe.
For dies, punches, rivet sets, rolls; wear and abrasion resistant.

HYCO 1.
M-688; 1.50 C, 0.35 Mn, 0.3 Si, 12.0 Cr, 0.9 V, 0.8 Mo, bal Fe.
Air or oil hardening cold work tool steel, chromium type; AISI D2.

HYCO 2.
M-688; 2.25 C, 0.3 Si, 11.5 Cr, 0.2 V, 0.8 Mo, bal Fe.
Air or oil hardening cold work tool and die steel chromium type; AISI D3.

HYCOMAX-I.
M-650; 21 Ni, 20 Co, 9.5 Al, 2 Cu, bal Fe.
Annealed: 850 Hc, 9000 Br.
For electrical and magnetic equipment. Permanent magnet. High permeability.

HYCOMAX-II.
M-210; 21 Ni, 20 Co, 9.5 Al, 2 Cu, bal Fe.
For permanent magnets in electrical and magnetic equipment.
High permeability. Permanent magnet.

HYCOMAX-II.
M-650; 14 Ni, 29 Co, 3 Cu, 7 Al, 4 Ti, 2 Cb, bal Fe.
8500 Br, 1200 Hc, 4,000,000 (BH) max.
For electrical and magnetic equipment. Permanent magnet.

HYCOMAX-III.
M-650; 14 Ni, 34 Co, 4 Cu, 7 Al, 5 Ti, bal Fe.
8800 Br, 1500 Hc, 5,300,000 (BH) max.
For electrical and magnetic equipment.
Permanent magnet. Similar to Alnico VIII.
High permeability.

HYCOMAX-IV.
M-650; 14 Ni, 40 Co, 3 Cu, 7.5 Al, 8 Ti, bal Fe.
7400 Br, 2100 Hc, 6,000,000 (BH) max.
For electrical and magnetic equipment. Permanent magnet.

HYCRO V80.
M-England; 0.2 C, 3 Cr, 0.6 Mo, 0.5 W, 0.8 V, bal Fe.
For gas turbine components; creep resistant.

HYCV.
M-38; 2.45 C, 3.75 V, 1 Mo, 12.25 Cr, bal Fe.
Heat treated: 660 Brin.
For brick mold liners, porcelain molds, punches; air or oil hardened, non-deforming.

HY-DI 5.
M-336, M-387; 1 C, 0.3 Si, 0.7 Mn, 5.2 Cr, 1-1 Mo, 0.2 V, bal Fe.
For dies, cutters, cold working tools; air hardened, nondeforming.

HYDRAULIC 56.
M-32; 0.85 C, 0.3 Mn, 0.3 Si, 1.75 Cr, 0.2 V, bal Fe.
Oil hardening tool steel for shafts, arbors, drill bushings, lathe centers; AISI L3.

HYDRAULIC BRONZE.
M-U.S.; 72.5 Cu, 19.5 Zn, 1.75 Sn, 6.5 Pb.
For pumps, cocks, valves; pressure tight.

HYDRAULIC BRONZE.
M-U.S.; 83 Cu, 11 Sn, 6 Zn, 0.1 Pb.
For pumps, cocks, valves; pressure tight.

HYDRONALIUM.
M-1372, M-1375; 6-10 Mg, 0.2-0.7 Mn, 1.5 max Fe, bal Al.
Heat treated: 48,000 TS; 26,000 YS; 16 El; 75 Brin.
For light alloy parts; age hardenable.

HYDRONALIUM.
M-1420; 1.5-3.5 Mg, bal Al.
Sand cast: 19,900-27,000 TS; 11,400-14,000 YS; 8-3 El; 50-60 Brin.
For marine castings; corrosion resistant.

HYDRONALIUM 2.
M-1375; 1.3 Mg, 0.4 max Mn, 0.3 max Cr, bal Al.
Soft: 28,000 TS; 13,000 YS; 30 El; 47 Brin.
Hard: 42,000 TS; 37,000 YS; 8 El; 77 Brin.
For marine products, aircraft structures; resists sea water corrosion.

HYDRONALIUM 3.
M-1375; 2-4 Mg, 0.4 max Mn, 0.3 max Cr, bal Al.
Soft: 30,000 TS; 15,000 YS; 28 El; 50 Brin.
For marine and aircraft parts; resists sea water corrosion.

HYDRONALIUM 5.
M-78; 95 Al, 5 Mg.
Extruded: 33,000-36,000 TS; 13,000-14,000 YS; 16-22 El; 60-70 Brin.
For seaplane pistons, propellers; resists sea water.

HYDRONALIUM 5.
M-1375; 4.0-5.5 Mg, 0.8 max Mn, 0.3 max Cr, bal Al.
Soft: 42,000 TS; 22,000 YS; 35 El; 65 Brin.
Hard: 60,000 TS; 50,000 YS; 15 El; 100 Brin.
For marine hardware, aircraft parts; corrosion resistant.

HYDRONALIUM 7.
M-78; 7 Mg, 0.45 Mn, bal Al.
Rolled: 52,000 TS; 32,000 YS; 12 El; 115 Brin.
For light alloy parts, seaplane pontoons and propellers; resists sea water corrosion.

HYDRONALIUM 7.
M-1375; 5.5-7.5 Mg, 0.8 max Mn, 0.3 max Cr, bal Al.
For light alloy parts; corrosion resistant.

HYDRONALIUM 9.
M-78; 9 Mg, bal Al.
Rolled: 60,000 TS; 40,000 YS; 10 El.
Extruded: 54,000 TS; 25,000 YS; 20 El; 45 RA; 82 Brin.
For seaplane pontoons and propellers; resists sea water corrosion.

HYDRONALIUM 10.
M-78; 1.0 Mg, 0.5 Mn, 0.2-1.5 Si, bal Al.
Forged: 54,000 TS; 32,000 YS; 15 El; 93 Brin.
For seaplane pontoons and propellers; resists sea water corrosion.

HYDRONALIUM 21.
M-1375; 0.6-1.4 Mg, 0.6-1.6 Si, 0.6-1.0 Mn, 0.3 max Cr, bal Al.
Soft: 28,000 TS; 13,000 YS; 30 El; 47 Brin.
For marine products, aircraft structures; corrosion resistant.

HYDRONALIUM 43.
M-78; 4.5 Zn, 3.5 Mg, bal Al.
Wire: 70,000 TS; 45,000 YS; 16 El.
Sheet: 63,000 TS; 40,500 YS; 18.5 El.
For light alloy parts; heat treatable.

HYDRONALIUM 43.
M-1375; Mg, Zn, bal Al.
For aircraft and light alloy parts.

HYDRONALIUM 51.
M-78; 0.2-1.5 Si, 5-12 Mg, 0.2-0.5 Mn, bal Al.
Cast: 23,000-27,000 TS; 13,000-18,000 YS; 2-5 El; 60-70 Brin.
For light alloy parts, aircraft components; corrosion resistant.

HYDRONALIUM 51.
M-1375; 5-12 Mg, 0.2-1.5 Si, 0.2-0.5 Mn, Cr, Ti, bal Al.
Sand cast: 23,000-27,000 TS; 13,000-18,000 YS; 2-5 El; 60-70 Brin.
For gasoline flow meters, engine components; corrosion resistant.

HYDRONALIUM 71.
M-78; 7 Mg, bal Al.
Cast: 32,000-36,000 TS; 17,000-20,000 YS; 5-8 El; 70-80 Brin.
For gasoline flow meters and aircraft parts; corrosion resistant.

HYDRONALIUM 71.
M-1375; 7 Mg, Si, Mn, Cr, Ti, bal Al.
Permanent mold: 32,000-36,000 TS; 17,000-20,000 YS; 5-8 El; 70-80 Brin.
For marine and aircraft parts; corrosion resistant.

HYDRONALIUM 91.
M-1375; Mg, Si, Mn, Cr, Ti, bal Al.
For light alloy parts.

HYDRONALIUM 511.
M-78; 0.3 Fe, 0.2 Mn, 5 Mg, 1 Si, 0.1 Ti, bal Al.
For castings; sand.

HYDRONALIUM 5112.
M-78; 1 Cu, 0.3 Fe, 0.1 Mn, 5 Mg, 1 Si, 0.1 Ti, 0.18 Be, bal Al.
For sand castings; heat treatable.

HYDRONALIUM D B A.
M-78; Al alloy.
For light alloy parts, aircraft parts, corrosion resistant.

HYDRONALIUM HY25.
 M-1548; 0.2-1.0 Si, 0.2-0.5 Mn, 3.0-12.0 Mg, bal Al.
 Soft: 27,000 TS; 12,800 YS; 24 El; 50 Brin.
 Half hard: 36,000 TS; 26,000 YS; 10 El; 60 Brin.
 For hardware, aircraft parts.
 Not heat treatable.

HYDRONALIUM HY51.
 M-1420; 5-12 Mg, 0.2-1.5 Si, 0.2-0.5 Mn, bal Al.
 Sand cast: 23,000-28,000 TS; 11,400-14,400 YS; 5.0-2.5 El; 55-70 Brin.
 For aircraft and marine equipment; corrosion resistant to sea water.

HYDRONALIUM HY71.
 M-1420; 7 Mg, Si, Mn, bal Al.
 Permanent mold: 32,000-36,000 TS; 17,000-20,000 YS; 8-5 El; 70-80 Brin.
 For aircraft and marine equipment; corrosion resistant to sea water.

HYDRONE.
 M-Eng.; 67 Pb, 33 Sb.
 For bearings; light shocks only.

HYDRONE.
 M-308; 31 Na, bal Pb.
 Used as a deoxidizer; for non-ferrous metals.

HYDRO-T-METAL.
 M-1505; 0.4-0.7 Cu, 0.08-0.16 Ti, 0.002-0.010 Mn, 0.003-0.02 Cr, bal Zn.
 Rolled: 26,500 TS; 20,000 YS; 36 El.
 For auto trim, furniture, rain gutters, hardware, electrical fixtures; stable, corrosion resistant.

HYDRO-T-METAL 200.
 M-1505, M 226; 0.3-1.25 Cu, 0.1-0.4 Ti, bal Zn.
 Sheet: 24,000 TS; 10 El; 55 Brin.
 For roofing, gutters, trim, fuses, housings, curtain walls. High creep resistance.

HYFLUX ALNICO K-7.
 M-1076; 8 Al, 14 Ni, 24 Co, bal Fe.
 Energy product (B_dH_c) max x 10^6 7.5; Residual induction Br (gauss) 13,400; Coercive force H_c (oersted) 3000.
 Permanent magnet.

HYFLUX ALNICO-9.
 M-1076; 7 Al, 15 Ni, 35 Co, 34 Fe, 4 Cu, 5 Ti.
 Energy product 10,000,000 gauss-oersteds.
 Coercive force 1600 oersteds.
 For straight field focusing devices, repulsion and torque transmitters, motors, space age applications.
 Permanent magnet. Resists demagnetization.

HYKRO V-80.
 M-England; 0.2 C, 3 Cr, 0.6 Mo, 0.5 W, 0.8 V, bal Fe.
 For oil refinery equipment; corrosion and heat resistant.

HYLASTIC.
 M-473; 0.30 C, 1.6 Mn, 0.10 V, 0.4 Si, bal F.
 Normalized: 100,000-90,000 TS; 76,000-60,000 YS; 28-25 El; 58-50 RA; 180 Brin.
 For railroad and structural work where high physicals are required; Iz-40.

HYLITE-1.
 M-261; 100 Ti.
 Rolled: 78,000 TS; 38,000 YS; 30 El.
 For the chemical industries.
 Commercially pure Ti.

HYLITE 10.
 M-261; Ti.
 Bar: 60,500 TS; 36,000 YS; 30 El; 150 Brin.
 Sheet: 61,600 TS; 44,800 YS; 20 El; 160 Brin.
 For aircraft parts; good form- and weldability.

HYLITE 15.
 M-261; Ti.
 Bar: 71,700 TS; 44,800 YS; 25 El.
 Sheet: 76,200 TS; 56,000 YS; 18 El.
 For aircraft parts; alpha type.

HYLITE-15H.
 M-261; 100 Ti.
 Rolled: 98,000 TS; 78,000 YS; 27 El.
 For the chemical industry.
 Commercially pure Ti with addition of oxygen to raise strength.

HYLITE 20.
 M-261; 5 Al, 2.5 Sn, bal Ti.
 Bar: 125,500 TS; 103,000 YS; 18 El.
 For aircraft and missile components; alpha type.

HYLITE-25.
 M-261; 2.5 Cu, bal Ti.
 Rolled: 85,000 TS; 64,000 YS; 27 El.
 For the chemical industry.
 Alpha-eutectoid alloy. Good weldability, and ductility.

HYLITE 30.
 M-261; 2 Mn, 2 Al, bal Ti.
 Bar: 98,600 TS; 80,600 YS; 20 El.
 For compressor discs and blades, fasteners, fuel systems; heat treatable.

HYLITE 40.
 M-261; 4 Mn, 4 Al, bal Ti.
 Bar: 138,900 TS; 130,000 YS; 18 El.
 For compressor discs and blades, fasteners, fuel systems; heat treatable.

HYLITE 45.
 M-261; 6 Al, 4 V, bal Ti.
 Bar: 141,200 TS; 130,000 YS; 22 El.
 For compressor discs and blades, fasteners, fuel systems; heat treatable.

HYLITE-50.
M-261; 4 Al, 4 Mo, 2 Sn, 0.5 Si, bal Ti.
Bar: 161,300 TS; 143,400 YS; 15 El.
For compressor discs and blades, fuel systems, fasteners.
Heat treatable.

HYLITE-51.
M-261; 4 Al, 4 Mo, 4 Sn, 0.5 Si, bal Ti.
Heat treated: 200,000 TS; 179,000 YS; 15 El.
For heavy duty service aircraft structural members.
Alpha-beta alloy. Good creep properties.

HYLITE-55.
M-261; 6 Sn, 5 Zr, 3 Al, 0.5 Si, bal Ti.
Heat teated: 130,000 TS; 116,000 YS; 16 El.
For tubine compressor blades and discs.
Alpha titanium alloy. Good properties, to over 500°C. High temperature strength.

HYLITE-60.
M-261; 3 Al, 6 Sn, 5 Zr, 2 Mo, 0.4 Si, bal Ti.
At 70°F: 157,000 TS; 135,000 YS; 14 El.
At 1000°F: 90,000 TS; 81,000 YS.
For turbine compressor blades and discs up to 500°C.
Good creep and tensile properties. Alpha-beta titanium alloy.

HYLITE-65.
M-261; 6 Sn, 5 Zr, 3 Al, 0.5 Mo, 0.5 Si, bal Ti.
Heat treatd: 155,000 TS; 136,000 YS; 13 El; 27 RA.
At 600°C: 91,400 TS; 69,500 YS; 16.5 El; 49 RA.
For jet engine compressor components, discs, spacers, blades.
High creep resistance, weldable.
Near alpha alloy, heat treatable.

HY-LO.
M-336; 0.85 C, 2.24 Mn, 0.1 V, bal Fe.
For dies; non-shrinking.

HY-LO.
M-712; 0.12 C, 0.77 Mn, 0.72 Si, 0.8 Mo, bal Fe.
Welded: 79,000-85,000 TS; 67,000-72,000 YS; 26-20 El; 50-35 RA.
For welding; A.W.S.-E7015, low H_2 content.

HYMAN.
M-Eng.; 58.3 Ag, 16.7 Cu, 16.7 Zn, 8.3 Ni.
For jewelry, ornaments; Nickel silver.

HYMAN ALLOY.
0.8 Si, 3 Cu, 0.5 Mg, 0.5 Ni, bal Al.
Heat treated: 36,000-43,000 TS; 6-9 El.
For light alloy parts; age-hardenable.

HYMAX 3% COBALT MAGNET.
M-210; 3 Co, 1.15 C, 10 Cr, 1.5 Mo, bal Fe.
Cast.
For magnet steel; Br-7000-6750; Hc-140-130; B.H. max 410,000.

HYMAX 6% COBALT MAGNET.
M-210; 1.15 C, 6 Co, 10 Cr, 1.5 Mo, bal Fe.
Cast.
For magnet steel; Br-7250-7000; Hc-160-145; B.H. max 460,000.

HYMAX 9% COBALT MAGNET.
M-210; 9 Co, 1.15 C, 10 Cr, 1.5 Mo, bal Fe.
Cast.
For magnet steel; Br-7550-7250; Hc-175-165; B.H. max 527,000.

HYMAX 15% COBALT MAGNET.
M-210; 15 Co, 1.15 C, 10 Cr, 1.5 Mo, bal Fe.
Cast.
For magnet steel; Br-7850-7600; Hc-200-220; B.H. max 648,000.

HYMAX 35% COBALT MAGNET.
M-210; 35 Co, 0.85 C, 4.5 W, 6 Cr, bal Fe.
Cast.
For magnet steel; Br-8800-8350; Hc-265-240.

HYMU 80.
M-32; 0.1 max C, 80 Ni, 4 Mo, bal Fe.
Cold Drawn: 97,000 TS; 69,000 YS; 37 El; 71 RA; 220 Brin.
For transformer cores, magnetic shields; hydrogen annealed.

HYMU-800.
M-32; 4 Mo, 79 Ni, bal Fe.
For laminations for motor cores, toroids.
High magnetic permeability, soft magnet.

HYNICAL.
M-England; 31 Ni, 12.5 Al, 0.4 Ti, bal Fe.
For permanent magnets, electrical and magnetic equipment. High permeability.

HYPERM 4.
M-72; C, 2.5-4.5 Si, bal Fe.
For precision transformers; magnetic alloy.

HYPERM 6.
M-72; C, Si, bal Fe. For magnets, electrical apparatus; magnetic alloy, high induction.

HYPERM 36.
M-72; C, alloy, bal Fe.
For magnets, electrical apparatus; magnetic alloy, high permeability.

HYPERM 50 Y.
M-72; C, high Ni, bal Fe.
For magnets, electrical apparatus; magnetic alloy, high permeability.

HYPERM-CO 27.
  M-72; 0.006 C, 27.5 Co, 0.002 Oxy., bal Fe.
  Annealed: 80,000 TS; 45,000 YS; 15 El; B 85 Rock.
  Cold Rolled: 100,000 TS; 90,000 YS; 1 El; C 30 Rock.
  For torque motors, generators, pole pieces, relays, alternators, transformer rectifiers.
  Soft magnetic alloy, high magnetic saturation.
  Operates at high flux levels.

HYPERM-CO 50.
  M-72; 49 Co, 2 V, bal Fe.
  Annealed: 80,000 TS; 48,000 YS; 0.5 El; B 97 Rock.
  Cold rolled: 195,000 TS; 185,000 YS; 1 El; C 35 Rock.
  For electric motors, computors, transformers, magnetic amplifiers, magnetostriction reducers.
  Magnetically soft for high flux levels.
  High permeability at high inductions.

HYPERM-O.
  M-72; Ni, bal Fe.
  For electrical equipment.

HYPERNOM.
  M-118; 79 Ni, 4 Mo, bal Fe.
  Hot rolled: 125,000 TS; 112,000 YS; 17.5 El; 165 Brin; 85 Rock B.
  For telephone loading coils, instrument transformers, electronic shields, magnetometers; high permeability, low fields.

HYPERNOM see **CARPENTER HYPERNOM.**

HYPERSILICIE.
  M-English; 3 Cu, 0.5 Fe, 18 Si, bal Al.
  For pistons; low density, cast.

HYPERSILID.
  M-846; C, 14-16 Si, bal Fe.
  For castings; cast iron, acid resistant.

HYPERTHERMOS 150.
  M-1495; 0.25-0.35 C, 1.0 Max Si, 1.0 max Mn, 27-29 Cr, 5.0-6.0 Mo, 1.5-3.0 Ni, bal Co.
  For high temperature equipment.

HYPERTHERMOS 190.
  M-1495; 0.07-0.12 C, 1.0 max Si, 1.0 max Mn, 16.0-18.0 Cr, 12.0-14.0 Ni, 2.5-4.0 W, Ti = 4 x C min, bal Fe.
  Austenitic stainless steel; for elevated temperature service.

HYPERTHERMOS 304.
  M-1495; 0.07 max C, 1.0 max Si, 2.0 max Mn, 17.0-19.0 Cr, 8.0-10.0 Ni, bal Fe.
  Austenitic stainless steel.
  AFNOR Z 6 CN 18 09; AISI 304.

HYPERTHERMOS 316.
  M-1495; 0.07 max C, 1.0 max Si, 2.0 max Mn, 16.0-18.0 Cr, 10.0-12.0 Ni, 2.0-2.5 Mo, bal Fe.
  Austenitic stainless steel; for chemical equipment.
  AFNOR Z 6 CND 17-11; AISI 316.

HYPERTHERMOS 316 ST.
  M-1495; 0.10 max C, 1.0 max Si, 2.0 max Mn, 16.0-18.0 Cr, 11.0-13.0 Ni, 2.0-2.5 Mo, Ti = 5 x C to 0.80, bal Fe.
  Stabilized austenitic stainless steel; for welded chemical equipment.
  AFNOR Z 8 CNDT 17-12.

HYPERTHERMOS 321.
  M-1495; 0.08 max C, 1.0 max Si, 2.0 max Mn, 17.0-19.0 Cr, 10.0-12.0 Ni, Ti = 5 x C to 0.80, bal Fe.
  Stabilized austenitic stainless steel.
  AFNOR Z 6 CNT 18-11; AISI 321.

HYPERTHERMOS 347.
  M-1495; 0.08 max C, 1.0 max Si, 2.0 max Mn, 17.0-19.0 Cr, 10.0-12.0 Ni, Nb + Tb = 10 x C to 1.0, bal Fe.
  Stabilized austenitic stainless steel.
  AFNOR Z 8 CNNb 18-11; AISI 347.

HYPLUS 23.
  M-1724; 0.20 max C, 0.50 max Si, 1.5 max Mn, 0.10 max Nb, bal Fe.
  High strength structural steel.
  For bridges, ships, cranes, pipelines.

HYPLUS 29.
  M-1724; 0.22 max C, 0.15-0.50 Si, 1.60 max Mn, 0.20 max V, bal Fe.
  High strength structural steel.

HYPOFLEX.
  M-945.
  Needle drawn stainless tubing to AISI 304 composition.

HYPROCRODE.
  M-494; C, 12.5-14.5 Ni, 6.5-7.5 Cu, 4.75-5.75 Cr, bal Fe.
  25,000-30,000 TS; 150 Brin.
  For hydraulic parts, pump bodies, impellers, plungers, liners, furnace parts, burners, fire bars; austenitic cast iron; corrosion, heat and erosion resistant.

HY-PRESS 20.
  M-1485, M-1724; 0.06 C, 0.58 Mn, 0.16 Cu, 0.02 Nb, bal Fe.
  As rolled: 63,000-72,000 TS; 47,000-54,000 YS; 30-34 El.
  Sheet and strip for stamped and formed parts; weldable.

HY-PRESS 23.
M-1724; 0.08 C, 0.56 Mn, 0.16 Cu, 0.025 Nb, bal Fe.
As rolled: 67,000-76,000 TS; 51,000-58,000 YS; 28-34 El.
Sheet and strip for stamped and formed parts; weldable.

HY-PRESS 26.
M-1724; 0.10 C, 0.90 Mn, 0.17 Cu, 0.024 Nb, bal Fe.
As rolled: 74,000-85,000 TS; 56,000-65,000 YS; 25-31 El.
Sheet and strip for stamped and formed parts; weldable.

HY-PRESS 29.
M-1724; 0.12 C, 1.2 Mn, 0.17 Cu, 0.03 Nb, bal Fe.
As rolled: 78,000-87,000 TS; 65,000-72,000 YS; 24-29 El.
Sheet and strip for stamped and formed parts; weldable.

HYPRO 61 AIR HARDENING.
M-336; 1.5 C, 11.5 Cr, 0.8 Mo, 2 V, bal Fe.
For gages, drawing dies, trimmers; air hardening.

HYPRO-62-O.H.
M-336; 2.2 C, 12 Cr, 1 V, bal Fe.
Hardened: C 64-67 Rock.
For blanking and forming dies, burnishing tools and rolls, plug gauges, slitting cutters, drawing dies.
Type D3 oil hardening cold work steel, deep hardening, abrasion resistant.

HYPRO 63.
M-336; 1.7 C, 18 Cr, bal Fe.
For dies, punches, rotary shears, gauges; non-shrink, deep hardening.

HYPRO NO. 62.
M-336; 2.0 C, 12 Cr, 0.6 Mn, bal Fe.
122,000 TS; 54,000 YS; 7.6 El; 257 Brin.
For general tools, high production tools, stamping and blanking dies; non-deforming.

HYREM RADIOMETAL.
M-108.
Soft magnetic alloy, square hysteresis loop, with high remanance for transformers.

HYRHO RADIOMETAL.
M-108; Soft magnetic alloy, high resistivity for A.C. uses, low eddy current losses for transformers.

HY-SPEED.
M-226; 88 Cu, 10 Sn, 2 Pb.
For bushings, bearings; heavy duty.

HYSTAL 77.
M-1724; 0.08-0.17 C, 0.10-0.35 Si, 1.1-1.4 Mn, 0.25-0.45 Mo, 0.002-0.006 B, 0.10 max Ti, bal Fe.
Q+T.: 790-930 N/mm^2 TS; 695 N/mm^2 min YS.
For structural hollow sections.

HYTEMPCO.
M-44; 70 Ni, bal Fe.
Wrought: 150,000-70,000 TS.
For immersion heaters, heater pads, electrical resistances; max operating temperature 500°C.

HY-TEN AH-70-FM.
M-119; 0.70 C, 1.15 Mn, 3.0 Ni, 1.5 Cr, 0.25 Mo, 0.05 Nb, bal Fe.
Air hardening tool steel; non-deforming, tough, good machinability.

HY-TEN "A" NO. 15.
M-119; 0.15 C, Ni, Mo, bal Fe.
Heat treated: 118,000 TS; 85,000 YS; 25 El; 65 RA; 235 Brin.
For carburized arbors, camshafts, piston pins, ratchets, worms, valve tappets, pinions; similar to SAE 4615.

HY-TEN "A" TEMPER NO. 20.
M-119; 0.20 C, 0.60 Ni, 0.55 Cr, 0.20 Mo, bal Fe.
Heat treated: 167,000 TS; 138,000 Ys; 11 El; 52 RA; 352 Brin.
For gears, worms, clutches, arbors, spindles, rolls, toggle pins.
Carburizing alloy steel, shock resistant.

HYTEN B-43.
M-119; 0.40-0.45 C, 0.65-0.85 Cr, 1.5-2.0 Ni, 0.3 Mo, bal Fe.
Heat treated: 326,000-132,000 TS; 294,000-117,000 YS; 10-61 El; 32-69 RA; 627-285 Brin.
For heavy duty gears, shafts, pinions, bolts, oil hardened.

HY-TEN "B" NO. 2.
M-119; 0.40 C, Mn, Cr, bal Fe.
Hot rolled: 100,000 TS; 62,000 YS; 25 El; 47 RA; 201 Brin.
For shafts, spindles, tap shanks, sleeves, worms, arbors, gears; free machining; tough.

HY-TEN "B" NO. 3 X.
M-119; 0.5 C, Cr, Mn, Mo, bal Fe.
Rolled: 135,000 TS; 70,000 YS; 18 El; 45 RA; 293 Brin.
Heat treated: 240,000 TS; 220,000 YS; 10 El; 40 RA; 477 Brin.
For clutches, gears, drive shafts, pinions, connectiong rods, mandrels; readily machined at high hardness.

HYTENAL BRONZE.
M-Eng.; 60 Cu, 10 Al, 6 Fe, 1. Zn, 5 Mn.
105,000 TS; 65,000 YS; 20 El.
For staybolts, pump rods, valves, valve spindles; corrosion resistant.

HY-TEN D-2.
M-119; 1.5 C, 11.5 Cr, 0.9 V, 0.4 Mn, 0.8 Mo, bal Fe.
For forming and drawing dies, punches; air hardened, non-deforming.

HY-TEN M.
M-119; 0.7 C, 0.6 Mn, 1.3 Ni, 0.65 Cr, 0.25 Mo, bal Fe.
Annealed: 95,000 TS; 55,000 YS; 25 El; 55 RA; 180 Brin.
Heat treated: 230,000 TS; 200,000 YS; 10 El; 35 RA; 444 Brin.
For cams, arbors, gauges, springs, clutches, dies; Type 16; shock shock and wear resistant.

HY-TEN "MC" TEMPER.
M-119; 0.70 C, 0.80 Cr, 0.50 Ni, 0.25 Mo, 0.08 V, bal Fe.
Annealed: 96,000 TS; 71,000 YS; 24 El; 53 RA; 200 Brin.
Heat treated: 302,000 TS; 279,000 YS; 653 Brin.
For gears, arbors, punches, clutches; oil hardened, shock resistant.

HY-TEN MOLD.
M-119; 0.35 C, Mn, Cr, Mo, bal Fe.
Heat treated: 160,000-240,000 TS; 132,000-205,000 YS; 19-10 El; 58-37 RA; 311-475 Brin.
For plastic molds, Zn die cast molds, tie rods; shock and impact resistant.

HY-TEN PNEUMATIC CHISEL.
M-119; 0.60 C, 0.85 Mn, 2 Si, 0.30 Cr, bal Fe.
For chisels, punches, cold cutters, pick points; oil hardened, shock resistant.

HYTENSIL ALUMINUM NO. 375.
M-847; 98 Al.
Heat treated: 48,000 TS; 3 El; 100 Brin.
For light alloy parts; corrosion resistant.

HYTENSILITE.
M-847; 98 Al, bal other elements.
Cast: 32,000 TS; 28,000 YS; 15 El; 40 Brin.
Heat treated: 45,000 TS; 41,000 YS; 5 El; 100 Brin.
For light alloy parts; corrosion resistant.

HY-TEN-SL BRONZE GRADE 1.
M-129; 60-68 Cu, 20-24 Zn, 3-7 Al, 2.5-5.0 Mn, 2-4 Fe.
Cast: 108,000 TS; 65,000 YS; 14 El; 14 RA; 220 Brin.
For pump and fan impellers, gears, valves; wear and corrosion resistant.

HY-TEN-SL BRONZE GR. 2.
M-129; 60-68 Cu, 20-24 Zn, 3-7 Al, 2.5-5.0 Mn, 2-4 Fe.
Cast: 100,000 TS; 55,000 YS; 16 El; 16 RA; 200 Brin.
For pump and fan impellers, gears, valves; wear and corrosion resistant.

HY-TEN-SL BRONZE GR. 3.
M-129; 60-68 Cu, 20-24 Zn, 3-7 Al, 2.5-5.0 Mn, 2-4 Fe.
Cast: 90,000 TS; 45,000 YS; 20 El; 20 RA; 175 Brin.
For impellers, gears, valves; wear and corrosion resistant.

HY-TEN-SL BRONZE GR. 4.
M-129; 60-68 Cu, 20-24 Zn, 3-7 Al, 2.5-5.0 Mn, 2-4 Fe.
Rolled: 85,000 TS; 40,000 YS; 25 El; 25 Ra; 150 Brin.
For impellers, gears, valves; wear and corrosion resistant.

HY-TEN-SL GR. 1 A.
M-129; 60-68 Cu, 20-24 Zn, 3-7 Al, 2-5 Mn, 2-4 Fe.
Sand cast: 115,000 TS; 70,000 YS; 12 El; 12 RA; 210 Brin.
Forged: 120,000 TS; 73,000 YS; 8 El; 8 RA; 240 Brin.
For gears, worm wheels, bearings; wear and corrosion resistant.

HY-TENSO M20.
M-775; 2.4 C, 1.2 Si, alloy, bal Fe.
Cast: 75,000 TS; 50,000 YS; 5 El; 200 Brin.
For stoker parts, stove and grate parts, dampers; cast iron, pearlitic.

HY-TENSO M-51.
M-775; 2.3 C, 1.2 Si, bal Fe.
Cast: 65,000 TS; 40,000 YS; 4-8 El; 185 Brin.
For chain links, connecting rods, gears; refined pearlitic malleable iron.

HY-TEST-ROD NO. 85.
M-1096; 0.10-0.15 C, 0.95-1.05 Mn, 0.25-0.35 Si, 0.4-0.6 Mo, bal Fe.
Welded: 85,000-95,000 TS; 65,000-75,000 YS; 18-22 El; 25-35 RA.
For welding rod; AWS-E8011.

HY-TEST-ROD NO. 110.
M-1096; 0.12 C, 0.6 Mn, 0.3 Mo, 0.6-1.0 Cr, bal Fe.
Welded: 110,000 TS; 100,000 YS; 15 El; 20 RA; 200 Brin.
For hard facing electrode; for moderate shock resistance.

HY-TUF.
M-38; 0.25 C, 1.3 Mn, 1.5 Si, 1.8 Ni, 0.4 Mo, bal Fe.
Heat treated: 234,000 TS; 193,000 YS; 13 El; 50 RA; 470 Brin.
For engineering parts, gears, shafts; tough.

HZ10.
M-1797; 0.12 max C, 1.0-1.5 Si, 18.0 Cr, 1.0 Al, bal Fe.
Heat resisting steel; for high temperature equipment.

HZ 11.
M-1797; 0.10-0.20 C, 1.0 Si, 25.0 Cr, 4.5 Ni, bal Fe.
Heat resisting steel; for high temperature equipment.

HZ 16.
M-1797; 0.20 max C, 2.0 Si, 25.0 Cr, 20.0 Ni, bal Fe.
Heat resisting stainless steel; for high temperature equipment.
Similar to AISI 310.

HZ32A.
M-Various foundries; 2.1 Zn, 0.7 Zr, 3.0 Th, bal Mg.
T5-temper: 27,000-30,000 TS; 13,000-14,000 YS; 4-7 El; 57 Brin; 68 Rock E.
At 700°F: 10,000 TS; 7000 YS; 29 El.
Magnesium sand castings with good pressure tightness and weldability; some permanent mold.
For parts operating at 350-700°F.
ASTM B80-69; AMS 4447; QQ-M-56.

I

145X.
M-1751; 0.22 C, 1.35 Mn, 0.04 P, 0.05 S, 0.30 Si, 0.005 min Cb, 0.01 min V, bal Fe (Cu min Opt).
Plate: 60,000 psi TS; 45,000 psi YS; 25 El.
For applications requiring higher strengths and higher strength with good formability and weldability.
ASTM A607, A572; SAE J410c.

150F.
M-1751; 0.15 C, 1.65 Mn, 0.025 P, 0.035 S, 0.60 Si, 0.005 min Cb, 0.01 min V, bal Fe.
Plate: 60,000 psi TS; 50,000 psi YS; 24 El.
For applications requiring higher strengths and higher strength with good formability and weldability.
ASTM A715.

150X.
M-1751; 0.23 C, 1.35 Mn, 0.04 P, 0.05 S, 0.30 Si, 0.005 min Cb, 0.01 min V, bal Fe (Cu min Opt).
Plate: 65,000 psi TS; 50,000 psi YS; 22 El.
For applications requiring higher strengths and higher strength with good formability and weldability.
ASTM A607, A572; SAE J410c.

155X.
M-1751; 0.25 C, 1.35 Mn, 0.04 P, 0.05 S, 0.30 Si, 0.005 min Cb, 0.01 min V, bal Fe (Cu min Opt).
Plate: 70,000 psi TS; 55,000 psi YS; 20 El.
For applications requiring higher strengths and higher strength with good formability and weldability.
ASTM A607, A572; SAE J410c.

160F.
M-1751; 0.15 C, 1.65 Mn, 0.025 P, 0.035 S, 0.60 Si, 0.005 min Cb, 0.01 min V, bal Fe.
Plate: 70,000 psi TS; 60,000 psi YS; 22 El.
For applications requiring higher strengths and higher strength with good formability and weldability.
ASTM A715.

160X.
M-1751; 0.26 C, 1.35 Mn, 0.04 P, 0.05 S, 0.30 Si, 0.005 min Cb, 0.01 min V, bal Fe (Cu min Opt).
Plate: 75,000 psi TS; 60,000 psi YS; 18 El.
For applications requiring higher strengths and higher strength with good formability and weldability.
ASTM A607, A572; SAE J410c.

170F.
M-1751; 0.15 C, 1.65 Mn, 0.025 P, 0.035 S, 0.60 Si, 0.005 min Cb, 0.01 min V, bal Fe.
Plate: 80,000 psi TS; 70,000 psi YS; 20 El.
For applications requiring higher strengths and higher strength with good formability and weldability.

180F.
M-1751; 0.15 C, 1.65 Mn, 0.025 P, 0.035 S, 0.60 Si, 0.005 min Cb, 0.01 min V, bal Fe.
Plate: 90,000 psi TS; 80,000 psi YS; 18 El.
For applications requiring higher strengths and higher strength with good formability and weldability.
ASTM A715.

I.A.S. INVINCIBLE.
M-1406; 1.2 C, 1.6 W, 0.75 Cr, 0.25 Mo, 0.25 V, bal Fe.
For finishing tools, paper and wood working knives; tough, cold work steel.

I.B.D.
M-1749; 0.32 C, 0.40-0.60 Mn, 1.0-1.5 Cr, 4.0-4.5 Ni, 0.3-0.4 Mo, 0.20 Si, bal Fe.
Nickel chrome mould steel; hardenable to 50 Rock C, can also be cyanide hardened or nitrided.

IBIS.
M-1742; 45% Sn content White Metal.
Heavy pressure moderate severity.

IBIS WHITE NAVY.
M-1742; 75% Sn content White Metal.
Medium/high speed-stone crushing plant, cold rolling mill bearings.

ICI 13518.
M-282; 0.25 C, 27.5 Cr, 5.5 Mo, 2.75 Ni, 3 max Fe, bal Co.
Cast: 52,000 TS; 10 El; 340 Brin.
For high temperature applications; heat resistant.

ICI 13530.
M-282; 0.5 C, 25.5 Cr, 2 max Fe, 10.5 Ni, 7.5 W, bal Co.
Cast: 50,000 TS; 10 El; 340 Brin.
For high temperature applications; heat resistant.

I.C.I. TITANIUM 115.
M-282; 0.010 max H_2, 0.20 max Fe, bal Ti.
Bar: 58,300 max TS; 30 min El.
For high temperature applications, corrosion resistant parts.
Commercially pure titanium.

I.C.I. TITANIUM 120.
M-282; 0.010 max H_2, 0.20 max Fe, bal Ti.
Sheet: 67,000 max TS; 25 min El.
For high temperature components, corrosion resistant parts.
Commercially pure titanium.

I.C.I. TITANIUM 125.
M-282; 0.010 max H_2, 0.20 max Fe, bal Ti.
Sheet: 56,000-78,500 TS; 22 min El.
For high temperature components, corrosion resistant parts.
Commercially pure titanium.

I.C.I. TITANIUM 130.
M-282; 0.010 max H_2, 0.20 max Fe, bal Ti.
Sheet: 56,000-79,000 TS; 22 min El.
For high temperature components, corrosion resistant parts.
Commercially pure titanium.

I.C.I. TITANIUM 150.
M-282; 0.010 max H_2, 0.20 max Fe, bal Ti.
Bar: 79,000-100,000 TS; 18 El.
For high temperature components, corrosion resistant parts.
Commercially pure titanium.

I.C.I. TITANIUM 160.
M-282; 0.010 max H_2, 0.20 max Fe, bal Ti.
Bar: 89,000-112,000 TS; 18 El; 40 RA.
For high temperature applications, corrosion resistant parts, non-structural aircraft parts.
Commercially pure titanium. DTD 5063 spec.

I.C.I. TITANIUM 230.
M-282; 1.8-2.8 Cu, bal Ti.
Bar: 71,000-100,000 TS; 20 min El.
For high temperature applications, corrosion resistant parts.
Good strength retained up to 300°C.

I.C.I. TITANIUM 314A.
M-282; 3.0-5.0 Al, 3.0-5.0 Mn, bal Ti.
Bar: 139,000 min TS; 15 min El.
For jet engine and guided missile components.
High strength and good creep resistance, to 300°C.

I.C.I. TITANIUM 314C.
M-282; 2 Al, 2 Mn, bal Ti.
Bar: 94,000-117,000 TS; 20 min El.
For high temperature applications, missile and jet engine components.
Heat and corrosion resistant.

I.C.I. TITANIUM 317.
M-282; 4.5-5.5 Al, 2.0-3.0 Sn, bal Ti.
Bar: 116,000 min TS; 10 min El.
For aircraft and guided missile components.
Good retention of strength and creep resistance up to 400°C. Good weldability.

I.C.I. TITANIUM 318A.
M-282; 5.5-6.5 Al, 3.5-4.5 V, bal Ti.
Bar: 139,000 min TS; 10 min El.
For jet engine components, airframe forgings and fasteners.
Good retention of strength and creep resistance to 300°C.

I.C.I. TITANIUM 679.
M-282; 2.0-2.5 Al, 0.8-1.2 Mo, 0.1-0.5 Si, 10.5-11.5 Sn, 4.0-6.0 Zr, bal Ti.
Bar: 150,000 min TS; 12 min El.
For compressor blades.
High creep resistance.

ICL 001.
M-1488; 0.12 max C, 18 Cr, 9 Ni, bal Fe.
Wrought, annealed: 490-690 N/mm² UTS.
Austenitic stainless steel for kitchen equipment, furniture springs, cycle spokes.
AFNOR Z10CN 18.09; AISI 302.

ICL 004 BC.
M-1488; 0.045 C, 18 Cr, 7 Ni, 7.5 Mn, 0.2 N, bal Fe.
Wrought, annealed: 650 N/mm² min UTS.
Austenitic stainless steel for containers of gases and liquified gases.
AFNOR Z3CMN 18.8.7 (non-standard).

ICL 164.
M-1488; 0.08 max C, 17.5 Cr, 12 Ni, 2.2 Mo, bal Fe.
Wrought, annealed: 520 N/mm² min UTS.
Austenitic stainless steel, for equipment for food industry, chemical industry, petroleum.
AFNOR Z6 CND 17.11; AISI 316.

ICL 164 BC.
M-1488; 0.03 max C, 17.5 Cr, 12.0 Ni, 2.2 Mo, bal Fe.
Wrought, annealed: 490 N/mm² min UTS.
Austenitic stainless steel, weldable, for best corrosion resistance in food, petroleum, chemical industries.
AFNOR Z2CND 17.12; AISI 316L.

ICL 164 FLUAGE.
M-1488; 0.06 C, 17.0 Cr, 13.0 Ni, 2.2 Mo, B, bal Fe.
Water quenched: 245 N/mm² min YS; 590 N/mm² min UTS; 45 El.
AFNOR Z6CND 18-12B.

ICL 164 HE.
M-1488; 0.03 max C, 17.5 Cr, 12.5 Ni, 2.8 Mo, 0.17 N, bal Fe.
Wrought, annealed: 600 N/mm² min UTS.
Austenitic stainless steel, improved strength, for automotive, rail and ship equipment that require welding and good corrosion resistance.
AFNOR Z2CND 17.13.3 + N; AISI 316N.

ICL 164 NB.
M-1488; 0.08 max C, 17.5 Cr, 12.0 Ni, 2.2 Mo, Nb = 8 x C, bal Fe.
Wrought, annealed: 520 N/mm² min UTS.
Stabilized austenitic stainless steel for welded equipment for food, petroleum and chemical industries.
AFNOR Z6CNDNb 17.12.

ICL 164 T.
  M-1488; 0.08 max C, 17.5 Cr, 12.0 Ni, 2.2 Mo, Ti = 5 x C, bal Fe.
  Wrought, annealed: 520 N/mm² min UTS.
  Stabilized austenitic stainless steel for welded equipment for food, petroleum and chemical industries.
  AFNOR Z6CNDT 17.12; AISI 316T.

ICL 166 BC.
  M-1488; 0.03 max C, 17.0 Cr, 13.0 Ni, 2.7 Mo, bal Fe.
  Wrought, annealed: 490 N/mm² min UTS.
  Austenitic stainless steel; good resistance to dilute sulphuric acid, bisulphites, alkaline earth chlorides, organic acids; weldable.
  AFNOR Z2CND 17.13.

ICL 167 CN.
  M-1488; 0.04 max C, 17.5 Cr, 12.0 Ni, 2.5 Mo, 0.50 max Cu, 0.20 max Co, 0.15 max Ta, 0.08 max N, bal Fe.
  Wrought, annealed: 540 N/mm² min UTS.
  Austenitic stainless for parts in light-water nuclear power plants.
  AFNOR Z3CND 17-12.

ICL 168 BC.
  M-1488; 0.03 max C, 18.5 Cr, 15.0 Ni, 3.3 Mo, bal Fe.
  Wrought, annealed: 490 N/mm² min UTS.
  Austenitic stainless steel; good resistance to dilute sulphuric acid, bisulphites, alkaline earth chlorides, organic acids.
  AFNOR Z3CND 19.15; AISI 317L.

ICL 472.
  M-1488; 0.07 max C, 18.5 Cr, 10 Ni, bal Fe.
  Wrought, annealed: 520 N/mm² min UTS.
  Austenitic stainless steel for petroleum industry, food and beverage equipment.
  AFNoR Z6CN 18.09; AISI 304.

ICL 472 BC.
  M-1488; 0.03 max C, 18.5 Cr, 10.5 Ni, bal Fe.
  Wrought, annealed: 480 N/mm² min UTS.
  Low carbon austenitic stainless steel, weldable; for petroleum, food industries.
  AFNOR Z2CN 18.10; AISI 304L.

ICL 472 HE.
  M-1488; 0.03 max C, 18.5 Cr, 10.0 Ni, 0.15 N, bal Fe.
  Wrought, annealed: 550 N/mm² min UTS.
  Low carbon austenitic stainless steel for household appliances, transportation equipment.
  AFNOR Z2CN 18.10 plus N; AISI 304 N.

ICL 472 NB.
  M-1488; 0.08 max C, 18.5 Cr, 10.5 Ni, Nb = 8 x C, bal Fe.
  Wrought, annealed: 520 N/mm² min UTS.
  Stabilized austenitic stainless steel, weldable, for chemical industry, food and beverage, petroleum industry.
  AFNOR Z6CN Nb 18.11; AISI 347.

ICL 472 T.
  M-1488; 0.08 max C, 18.5 Cr, 10.5 Ni, Ti = 5 x C, bal Fe.
  Wrought, annealed: 520 N/mm² min UTS.
  Stabilized austenitic stainless steel, weldable, for chemical industry, food and beverage, petroleum industry.
  AFNOR Z6CNT 18.11; AISI 321.

ICL 472 U.
  M-1488; 0.12 max C, 18 Cr, 9 Ni, 0.60 max Mn, 0.15 min S, bal Fe.
  Wrought, annealed: 510 N/mm² min UTS.
  Free machining austenitic stainless steel.
  AFNOR Z10CNF 18.09; AISI 303.

ICL 473 BC.
  M-1488; 0.04 max C, 19.9 Cr, 9.5 Ni, 0.50 max Cu, 0.20 max Co, 0.15 max Ta, 0.08 max N, bal Fe.
  Wrought, annealed: 540 N/mm² min UTS.
  Austenitic stainless for parts in light-water nuclear power plants.
  AFNOR Z3CN19-10.

I.C.N.
  M-1749; 0.18-0.25 C, 1.0 Si, 1.0 Mn, 14.0 Cr, 1.0 Ni, bal Fe.
  Plastic extrusion die steel; air or oil hardenable, corrosion resistant.

ICN 001.
  M-1302; 0.08-0.20 C, max Mn, 17-19 Cr, 8-10 Ni, bal Fe.
  Annealed: 85,000 TS; 35,000 YS; 60 El; 70 RA; 150 Brin.
  Cold drawn; 125,000 TS; 95,000 YS; 25 El; 55 RA; 277 Brin.
  For oil refinery and chemical plant equipment; Type 302; stainless, austenitic.

ICN 164.
  M-1302; 0.10 max C, 16-18 Cr, 10-14 Ni, 2-3 Mo, bal Fe.
  Annealed: 85,000-95,000 TS; 35,000-45,000 YS; 60-50 El; 75-60 RA; 150-190 Brin.
  For acid-resistant chemical plant equipment; Type 316; stainless, austenitic.

ICN 164 T (TI).
  M-1302; 0.10 C, 16-18 Cr, 10-14 Ni, 1.7-2.7 Mo, Ti, bal Fe.
  Annealed: 85,000-95,000 TS; 35,000-45,000 YS; 60-50 El; 75-60 RA; 150-190 Brin.
  For acid resistant chemical plant equipment; Type 316 Ti; stainless, austenitic.

ICN 472.
M-1302; 0.08 max C, 18-20 Cr, 8-11 Ni, 2 max Mn, bal Fe.
Annealed: 90,000 TS; 45,000 YS; 60 El; 135 Brin.
Cold drawn: 180,000 TS; 150,000 YS; 10 El; 330 Brin.
For chemical plant equipment, welded structures; Type 304; stainless, austenitic.

ICN 472T.
M-1302; 0.08 max C, 17-19 Cr, 8-11 Ni, Ti = 5 x C, bal Fe.
Normalized: 93,000 TS; 36,000 YS; 45 El; 60 RA; 165 Brin.
Annealed: 87,000 TS; 33,000 YS; 57 El; 73 RA; 155 Brin.
For chemical plant equipment; Type 321; stainless, austenitic.

I.C.S.
M-1749; cast steel for tools: 0.60-0.70 C, bal Fe, for mason tools, Smith's tool, hand punches; 0.80-0.90 C, bal Fe, for drills, taps, punches; 0.90-1.0 C, bal Fe, for lathe centers, mandrels; 1.1-1.2 C, bal Fe, keen edge for brass and woodworking tools.

I.C.W.
M-1749; 2.0 C, 0.50 Mn, 13.0 Cr, 0.60 Si, bal Fe.
Heavy duty die steel; air hardenable.
For lamination dies, blanking dies, master hobs.

IDEAL BC12V.
M-1346; 2.1 C, 11.5 Cr, bal Fe.
For blanking and forming dies, punches; oil or air hardened, nondeforming.

IDEAL BC13V.
M-1346; 1.65 C, 11.5 Cr, 0.1 V, 0.35 Si, 0.3 Mn, bal Fe.
For blanking and forming dies, punches; nondeforming, air hardened.

IDEAL BCK1.
M-1346; 1.45 C, 1.4 Cr, bal Fe.
For fast finishing cutters, bearings, liners; water hardened, wear resistant.

IDEAL BCK2.
M-1346; 1.0 C, 1.55 Cr, bal Fe.
For bearings, bushings, liners; water hardened, wear resistant.

IDEAL BCK3.
M-1346; 1.05 C, 1.0 Cr, bal Fe.

IDEAL BCK4.
M-1346; 1.0 C, 1.1 Cr, bal Fe.
For bearings, liners, bushings, sleeves, cutters; water hardened.

IDEAL BCR.
M-1346; 1.05 C, 1.0 Cr, 0.3 Mn, bal Fe.
For bearings, liners, pivots, sleeves; water hardened, wear resistant.

IDEAL BCS.
M-1346; 0.90 C, Cr, Si, bal Fe.
For bearings, liners, sleeves; water hardened, wear resistant.

IDEAL BCZ1.
M-1346; 1.4 C, 0.3 Cr, 0.1 V, bal Fe.
For bearings, liners, bushings, sleeves, cutters; water hardened.

IDEAL BCZ2.
M-1346; 0.90 C, 0.80 Cr, bal Fe.
For bearings, liners, bushings, sleeves, cutters; water hardened.

IDEAL BDS.
M-1346; 0.5 C, 0.9 Mn, 1.05 Cr, 0.1 V, bal Fe.
For gears, shafts, springs, crankshafts; oil hardened, shock resistant.

IDEAL BEE.
M-1346; 0.15 C, 0.2 Si, 0.4 Mn, bal Fe.
Annealed: 70,000 TS; 40,000 YS; 25 El; 60 RA; 145 Brin.
For gears, pinions, cams, camshafts; case hardening steel.

IDEAL BFM.
M-1346; 0.70 C, 1.7 Si, 0.70 Mn, bal Fe.
For springs, punches; oil hardened, tough.

IDEAL BG1.
M-1346; 1.05 C, 1 Cr, 1.15 W, bal Fe.
For bearings, cold work tools, drawing dies; water hardened.

IDEAL BG2.
M-1346; 0.45 C, Ni, Cr, W, bal Fe.
For cold work tools, upsetters; oil hardened, tough.

IDEAL BGW.
M-1346; 0.5 C, 1.05 Cr, 3.25 Ni, bal Fe.
For gears, shafts, crankshafts, bolts, fasteners; oil hardened, shock resistant.

IDEAL BHM.
M-1346; 0.56 C, Ni, Cr, Mo, V, bal Fe.
For gears, shafts, crankshafts, bolts, studs; oil hardened, shock resistant.

IDEAL BKS.
M-1346; 0.45 C, Si, Cr, V, bal Fe.
For gears, springs, crankshafts, axles; oil hardened, shock resistant.

IDEAL BKV.
M-1346; 1.0 C, 0.1 V, 0.25 Mn, bal Fe.
For drills, taps, reamers, broaches; Type W2; water hardened.

IDEAL BME.
M-1346; 0.40 C, Cr, Mn, Mo, bal Fe.
For gears, pinions, crankshafts; oil hardened, shock resistant.

IDEAL BMK.
M-1346; 0.85 C, Cr, bal Fe.
For tools, dies, cutters, taps, bearings; water hardened.

IDEAL BMS.
M-1346; 0.53 C, 0.8 Si, 1.05 Mn, bal Fe.
For gears, axles, crankshafts; water hardened.

IDEAL BMV.
M-1346; 0.90 C, 1.9 Mn, 0.1 V, bal Fe.
For dies, punches, upsetters, cutters; oil hardened, nondeforming.

IDEAL BPK2.
M-1346; 0.2 C, 1.15 Cr, 1.25 Mn, bal Fe.
For cams, fasteners, camshafts; case hardened.

IDEAL BRF1.
M-1346; 0.2 C, 13 Cr, bal Fe.
Annealed: 95,000 TS; 50,000 YS; 25 El; 55 RA; 195 Brin.
Cold drawn: 105,000 TS; 85,000 YS; 17 El; 50 RA; 215 Brin.
For turbine blades, valves, cutlery, surgical instruments; Type 420; stainless.

IDEAL BRF2.
M-1346; 0.4 C, 13 Cr, 0.4 Si, 0.3 Mn, bal Fe.
Annealed: 95,000 TS; 50,000 YS; 25 El; 55 RA; 195 Brin.
For valves, cutlery, surgical and dental instruments; Type 420; stainless.

IDEAL BSA.
M-1346; 0.38 C, Si, Cr, V, bal Fe.
For gears, punches, bolts, machine tool parts; oil hardened, tough.

IDEAL BSE.
M-1346; 0.15 C, 0.65 Cr, 0.25 Si, 0.5 Mn, bal Fe.
For gears, cams, camshafts, fasteners; case hardening steel, tough.

IDEAL BW6.
M-1346; 1.2 C, 0.2 Cr, 0.1 V, 1 W, bal Fe.
For bearings, cutters, sleeves; water hardened, wear resistant.

IDEAL BWP EXTRA.
M-1346; 0.30 C, 2.35 Cr, 0.6 V, 4.25 W, bal Fe.
For extrusion dies, upsetters, header dies; oil hardened, tough.

IDEAL BWPKO.
M-1346; 0.30 C, 2.4 Cr, 2 Co, 0.25 V, 8.5 W, bal Fe.
For shears, punches, upsetters, header dies; hot work steel, oil hardened.

IDEAL BWP-PRIMA.
M-1346; 0.30 C, 1 Si, 1.1 Cr, 0.18 V, 3.75 W, bal Fe.
For header dies, crimpers, punches; oil hardened, tough.

IDEAL BWP SPEZIAL.
M-1346; 0.30 C, 2.65 Cr, 0.35 V, 8.5 W, bal Fe.
For header dies, extrusion dies and rams, shears; hot work steel, oil hardened.

IDEAL BWPU.
M-1346; 0.45 C, 1.4 Cr, 0.7 Mo, 0.3 V, 0.7 Mn, bal Fe.
For gears, bolts, crankshafts, fasteners; oil hardened, tough.

IDEAL EXTRA HART 1.
M-1346; 1.3 C, 0.25 max Si, 0.25 max Mn, bal Fe.
For engravers' tools, forming and blanking dies; Type W1; water hardened.

IDEAL EXTRA HART 2.
M-1346; 1.15 C, 0.25 max Si, 0.25 max Mn, bal Fe.
Annealed: 110,000 TS; 58,000 YS; 18 El; 38 RA; 220 Brin.
For springs, drills, reamers, taps, tools; Type W1; water hardened.

IDEAL EXTRA HART 3.
M-1346; 1.0 C, 0.25 max Si, 0.25 max Mn, bal Fe.
Annealed: 100,000 TS; 53,000 YS; 21 El; 42 RA; 200 Brin.
For springs, tools, drills, taps, cutters; Type W1; water hardened.

IDEAL EXTRA HART 4.
M-1346; 0.85 C, 0.25 max Si, 0.25 max Mn, bal Fe.
Heat treated: 190,000 TS; 145,000 YS; 10 El; 30 RA; 400 Brin.
For springs, drills, cutters, tools, punches; Type W1; water hardened.

IDEAL EXTRA HART 5.
M-1346; 0.70 C, 0.25 max Si, 0.25 max Mn, bal Fe.
Heat treated: 175,000 TS; 128,000 YS; 12 El; 37 RA; 355 Brin.
For springs, rails, punches, tools, axes; Type W1; water hardened.

IDEAL NIT 1.
M-1346; 0.26 C, 1.4 Cr, 1.1 Al, 0.6 Mn, bal Fe.
For oil refinery equipment; creep and heat resistant.

IDEAL NIT 3.
M-1346; 0.34 C, 1.4 Cr, 1.1 Al, 0.6 Mn, bal Fe.
For oil refinery equipment; creep and heat resistant.

IDEAL NIT 7.
M-1346; 0.32 C, 1.1 Cr, 1.1 Al, 0.18 V, 0.6 Mn, bal Fe.
For oil refinery equipment; creep and heat resistant.

IDEAL NIT 9.
M-1346; 0.33 C, 1.7 Cr, 1.0 Ni, 0.5 Mn, bal Fe.
For dies, upsetters, punches, shafts; oil hardened, shock resistant.

IDEAL NIT 11.
M-1346; 0.31 C, 0.6 Mn, 2.35 Cr, 0.18 Mo, bal Fe.
For gears, dies, crankshafts; oil hardened, tough.

IDEOR.
M-40, M-487; 0.4 C, 2 W, 1 Cr, 1 Si, 0.3 mn, bal Fe.
For chisels, punches, rivet sets, shears, picks; Type S1; tough.

IDEOR SPECIAL.
M-487; 0.5 C, 1 Cr, 2 W, 0.15 V, bal Fe.
For blanking and piercing dies; shock resistant, oil hardened.

I.D.I.
M-1749; 1.30 C, 1.0-1.2 Cr, 1.25-1.5 Mn, 0.50 W, bal Fe.
Oil hardening, non-distorting press tool steel.

IDIART.
3 C, 2 Si, bal Fc.
For compressor pistons; cast iron.

IDRONAL.
M-Ital.; 7 Mg, bal Al.
For marine parts; resists sea water corrosion.

IDRONAL.
M-Ital.; 5 Mg, 1 Si, bal Al.
For marine parts; resists sea water corrosion.

IGEDUR.
M-78; 0.2-2.0 Mg, 3.5-5.5 Cu, 0.2-1.0 Si, 0.1-1.2 Mn, bal Al.
For light alloy parts; Duralumin type.

IGETALLOY.
M-1515; WC.
For cutting tools; sintered carbide.

IGNIDUR PC8.
M-1325; 0.3 C, 2.2 Si, 6 Cr, bal Fe.
For oil refinery equipment; creep and heat resistant.

IGNIDUR PC9.
M-1325; 0.5 C, 1.5 Si, 17 Cr, bal Fe.
Annealed: 100,000 TS; 60,000 YS; 20 El; 38 RA; 210 Brin.
For valve parts, cylinder liners, furnace parts; Type 440; stainless, hardenable.

IGNIDUR PC10.
M-1325; 0.6 C, 1.5 Si, 22 Cr, bal Fe.
For cylinder liners, rollers, bushings, furnace parts; heat resistant.

IGNIDUR PC11.
M-1325; 1.3 C, 1.5 Si, 29 Cr, bal Fe.
For bushings, valve parts, rollers; heat and water resistant.

IGNIDUR PC12.
M-1325; 0.6 C, 1.5 Si, 29 Cr, bal Fe.
Cast: 90,000 TS; 65,000 YS; 2 El; 212 Brin.
For cylinder liners, bushings, valve seats, furnace parts; heat resistant.

IGNIDUR PK4.
M-1325; 0.4 C, 1.3 Si, 27 Cr, 4 Ni, bal Fe.
Cast: 70,000 TS; 65,000 YS; 2 El; 190 Brin.
Heat treated: 115,000 TS: 80,000 YS; 15 El.
For furnace parts, grate bars, salt pots, baffles; heat resistant.

IGNIDUR PK9.
M-1325; 0.4 C, 2 Si, 22 Cr, 9.5 Ni, bal Fe.
Cast: 85,000 TS; 45,000 YS; 35 El; 165 Brin.
For heat treating boxes, baskets, chains, conveyors, belts; corrosion and heat resistant; Type HF.

IGNIDUR PK14.
M-1325; 0.4 C, 2 Si, 26 Cr, 14 Ni, bal Fe.
Cast: 75,000 TS; 47,000 YS; 17 El; 25 RA; 200 Brin.
For furnace parts, heat treating boxes, salt pots, belts; Type HH; corrosion and heat resistant.

IGNIDUR PK19.
M-1325; 0.4 C, 2 Si, 25 Cr, 19 Ni, bal Fe.
Cast: 75,000 TS; 50,000 YS; 17 El; 170 Brin.
Aged: 85,000 TS; 50,000 YS; 10 El; 190 Brin.
For furnace parts, retorts, skids, rabble arms; Type HK; corrosion and heat resistant.

IGNIDUR PK30.
M-1325; 0.5 C, 1.8 Si, 25 Cr, 30 Ni, bal Fe.
For furnace parts, heat treating boxes; heat and corrosion resistant.

IGNITION PIN ALLOY-1.
M-Eng.; 61 Ce, 39 Fe.
For ignition pins; pyrophoric.

IGNITION PIN ALLOY-2.
M-Eng.; 73-70 Ce, 1.6-6.0 Fe, 17-24 Zn, 0-2.4 Al.
For ignition pins; pyrophoric.

IH-50.
M-1613; 0.22 max C, 1.5 max Mn, 0.7 max Si, 0.20 min Cu, bal Fe.
Plate: 75,000 min TS; 50,000 min YP; 20 El.
For earth moving and agricultural equipment, tractors, trucks, trailers, bridges, bus bodies.
Good fabricability and weldability.
Good fatigue strength and ductility.

IH-60.
M-1613; 0.22 C, 1.65 Mn, 0.70 Si, 0.005 min Cb or 0.01 min V, bal Fe.
Wrought: 80,000 psi TS; 60,000 psi YS; 18 El.
HSLA steel.

IH-65.
M-1613; 0.22 max C, 1.5 max Mn, 0.7 max Si, 0.20 min Cu, 0.01 min Cb, bal Fe.
Plate: 90,000 min TS; 65,000 min YP; 18 El.
For earth moving and agricultural equipment, tractors, trucks, trailers.
Good fabricability and weldability.
Good fatigue strength and ductility.

IHCP.
M-669.
Cold finished, induction-hardened, chrome-plated carbon steel bars; as per order.
For piston rods.

IH-WX STEEL.
M-1613; 0.09 max C, 0.75-1.05 Mn, 0.05-0.10 P; 0.24-0.33 S, bal Fe.
For screw machine products, fasteners, hardware, cams.
Free machining, case-hardening.

IHX-42.
M-1613; 0.21 C, 1.35 Mn, 0.01 min Cb or V, bal Fe.
60,000 min TS; 42,000 min YS; 24 min El.
For structural parts.

IHX-45.
M-1613; 0.20 max C, 1.0 max Mn, 0.10 max Si, 0.01 min Cb or V, bal Fe.
Plate: 60,000 min TS; 45,000 min YS; 22 El.
For earth moving and agricultural equipment, trucks, trailers, tractors, mine cars.
High strength, low alloy steel.

IHX-50.
M-1613; 0.22 max C, 1.1 max Mn, 0.1 max Si, 0.01 min Cb or V, bal Fe.
Plate: 65,000 min TS; 50,000 min YP; 20 El.
For earth moving and agricultural equipment, mine cars, trucks, trailers, derricks, booms.
Shock resistant.

IHX-55.
M-1613; 0.24 max C, 1.4 max Mn, 0.3 max Si, 0.01 min Cb or V, bal Fe.
Plate: 70,000 min TS; 55,000 min YP; 18 El.
For earth moving and agricultural equipment, trucks, trailers, tractors, booms, mine cars.
Shock resistant.

IHX-60.
M-1613; 0.26 max C, 1.55 max Mn, 0.3 max Si, 0.01 min Cb or V, bal Fe.
Plate: 75,000 min TS; 60,000 YP; 18 El.
For earth moving and agricultural equipment, trucks, trailers, tractors, booms, mine cars.
Shock resistant.

IHX-65.
M-1613; 0.26 max C, 1.6 max Mn, 0.3 max Si, 0.01 min Cb or V, bal Fe.
Plate: 80,000 min TS; 65,000 min YP; 17 El.
For earth moving and agricultural equipment, tractors, trailers, booms, mine cars.
Shock resistant.

IHX-70.
M-1613; 0.26 max C, 1.65 max Mn, 0.30 max Si, 0.01 min Cb or V, bal Fe.
Plate: 85,000 min TS; 70,000 min YP; 16 El.
For earth moving and agricultural equipment, tractors, trailers, mine cars, buildings.
Shock resistant.

IK.
M-France; 4.5 Cu, bal Al.
64,000 TS; 5 El; 110 Brin.
For light alloy parts.

IK STEELS.
M-Ger.; C, Cr, bal Fe.
For heat and corrosion resistant parts; chromized steel, surface layer.

ILLIUM 98.
M-1262; 55 Ni, 28 Cr, 8.0 Mo, 5.0 Cu, 0.05 C.
Cast: 54,000 TS; 41,000 YS; 18 El; 22 RA; 143 Brin.
For acid pumps, hardware, impellers, valves, nozzles; resists sulphuric acid.

ILLIUM 98 HF.
M-1262; 50.25-62.50 Ni, 26.0-30.0 Cr, 7.5-9.0 Mo, 4.0-6.5 Cu, 1.5 max Mn, 0.50 max Si, 1.5 max Fe, 0.07 max C.
Cast: 54,000 TS; 41,000 YS; 18 El; 22 RA; 160 Brin.
Solution Heat treated: 78,000 TS; 43,000 YS; 43 El; 39 RA; 150 Brin.
For fluorine service. Corrosion resistant casting alloy.

ILLIUM 98 LSI now **ILLIUM 98HF.**

ILLIUM-B.
M-1262; 28 Cr, 0.05 C, 8 Mo, 5 Cu, 5 Fe, bal Ni.
Cast: 60,000 min TS; 41,000 min YS; 0.5 El; 221-420 Brin.
Hardenable; max wear and galling resistance; resists sulphuric acid.
For pumps, valves, chemical plant equipment.

ILLIUM CD-4MCU.
M-1262; 0.03 C, 26 Cr, 5.5 Ni, 2.0 Mo, 3.0 Cu, bal Fe.
Cast: 108,000 TS; 80,000 YS; 25 El; 250 BHN.
Corrosion resistance greater than CF-8 or CF-8M, with double the yield strength of these alloys.

ILLIUM-G.
M-1262; 0.2 C, 22 Cr, 5 Fe, 6 Mo, 5 Cu, bal Ni.
Cast: 60,000 min TS; 35,000 min YS; 5 El; 159 Brin.
Excellent resistance to corrosion from hot H_2SO_4, H_3PO_4, acid-salt mixtures, sea water.
For pumps, valves, chemical and marine equipment.

ILLIUM-H.
M-1262; 0.1 C, 0.18 Ti, 0.6 Cb, 26-30 Cr, 47-52 Co, bal Fe.
Cast: 135,000 TS; 48,000 YS; 10 El; 10 RA; 257 Brin.
Wrought; 132,000 TS; 61,000 YS; 7 El; 6 RA; 350 Brin.
For furnace baffles, burner tips, sintering grates, quench baskets.
Corrosion, heat and thermal shock resistant.

ILLIUM-M.
M-1262; 0.08 C, 28 Mo, bal Ni.
Cast: 72,000 min TS; 46,000 min YS; 61 El; 190 Brin.
Resists halogen chemicals and HCl.
For chemical plant equipment.

ILLIUM-P.
M-1262; 28 Cr, 8 Ni, 3 Cu, 2 Mo, bal Fe.
Cast: 99,000-110,000 TS; 65,000-81,000 YS; 10-18 El; 217-255 Brin.
Resists erosion and corrosion by mixed acids; resists phosphoric acid.
For chemical plant equipment.

ILLIUM-PD.
M-1262; 0.08 C, 26 Cr, 5 Ni, 2 Mo, bal Fe.
Cast: 95,000 min TS; 65,000 min YS; 30 El; 212 Brin.
High erosion-corrosion resistance; for equipment in pulping liquors, dairy and chemical plants.

ILLIUM S.
M-1262; 0.10 C, 0.9 Mn, 9.0 Si, 85.0 Cr, 2.0 Al, 3.0 Cu.
Cast corrosion resistant alloy.

ILLIUM-W.
M-1262; 0.15 C, 6 Fe, 4 W, 17 Mo, 16 Cr, bal Ni.
Cast: 72,000 TS; 46,000 YS; 4 El.
For chemical equipment; resists Cl_2, H_2SO_4, and HNO_3.

ILLIUM-X.
M-1262; 0.75 C, 29 Cr, 14 W, 55 Co.
Cast: 100,000 min TS; 90,000 min YS; 2 El; 293 Brin.
Corrosion and abrasion resistant, particularly for equipment for dry battery manufacture.

ILSSA AT-15.
M-1452; 0.25 max C, 25 Cr, 20 Ni, 1.5 max Si, bal Fe.
Annealed: 100,000 TS; 45,000 YS; 50 El; 65 RA; 185 Brin.
For furnace parts, valves, pumps, jet engine parts; Type 310; corrosion and heat resistant.

ILSSA IC.
M-1452; 0.09-0.12 C, 13 Cr, 0.5 max Si, bal Fe.
Annealed: 75,000 TS; 40,000 YS; 35 El; 70 RA; 155 Brin.
For turbine blades, valves, cutlery, surgical instruments; Type 403 and 410; corrosion resistant.

ILSSA ICC.
M-1452; 0.13-0.25 C, 17 Cr, bal Fe.
Annealed: 90,000 TS; 55,000 YS; 20 El; 45 RA; 170 Brin.
For oil refinery equipment, oil burners and heaters; Type 430; corrosion resistant.

ILSSA ICMS.
M-1452; 0.10 max C, 17 Cr, 1.5 Mo, bal Fe.
Annealed: 90,000 TS; 55,000 YS; 20 El; 45 RA; 170 Brin.
For oil refinery equipment, oil burners and heaters; corrosion resistant.

ILSSA ICS.
M-1452; 0.07-0.12 C, 17 Cr, bal Fe.
Annealed: 80,000 TS; 50,000 YS; 25 El; 50 RA; 150 Brin.
For oil refinery equipment, oil burners and heaters; Type 430; corrosion resistant.

ILSSA IN.
M-1452; 0.11-0.16 C, 18 Cr, 8 Ni, bal Fe.
Annealed: 80,000 TS; 35,000 YS; 55 El; 75 RA; 150 Brin.
For chemical plant equipment, tanks, mixers, agitators; Type 302; stainless, austenitic.

ILSSA INF 20.
M-1452; 0.35 max C, 25 Cr, bal Fe.
Annealed: 85,000 TS; 50,000 YS; 30 El; 55 RA; 180 Brin.
For furnace parts and equipment, heat treating boxes; Type 446; heat resistant.

ILSSA INF 25/5.
M-1452; 0.12 max C, 26 Cr, 4 Ni, bal Fe.
Cast: 90,000 TS; 65,000 YS; 2 El; 212 Brin.
For cylinder liners, bushings, valve seats and bodies; Type CC50; corrosion and heat resistant.

ILSSA INI.
M-1452; 0.08 max C, 18 Cr, 8 Ni, bal Fe.
Annealed: 85,000 TS; 35,000 YS; 60 El; 70 RA; 150 Brin.
For chemical plant equipment, tanks, mixers, filters; Type 304; stainless, austenitic.

ILSSA INI/BC.
M-1452; 0.05 max C, 18 Cr, 8 Ni, bal Fe.
Annealed: 85,000 TS; 35,000 YS; 60 El; 70 RA; 150 Brin.
For chemical plant equipment, tanks, mixers; Type 304; stainless, austenitic.

ILSSA INL.
M-1452; 0.08-0.20 C, 18 Cr, 9 Ni, 2.5 Si, bal Fe.
Annealed: 85,000 TS; 40,000 YS; 50 El; 70 RA; 160 Brin.
For chemical plant and oil refinery equipment; Type 302B; stainless, austenitic.

ILSSA INMI.
M-1452; 0.12 max C, 18 Cr, 10 Ni, 2.5 Mo, bal Fe.
Annealed: 85,000 TS; 35,000 YS; 50 El; 65 RA; 160 Brin.
For acid resistant chemical plant equipment; Type 316; stainless.

ILSSA INS.
M-1452; 0.08-0.15 C, 18 Cr, 9 Ni, P, S, Se, Zr, bal Fe.
Annealed: 80,000 TS; 35,000 YS; 40 El; 60 RA; 160 Brin.
For screw machine products, shafts; Type 303; stainless, free-cutting.

ILSSA MIC.
M-1452; 0.26-0.37 C, 13 Cr, bal Fe.
Annealed: 95,000 TS; 50,000 YS; 25 El; 55 RA; 195 Brin.
For valves, cutlery, surgical and dental instruments; Type 420; corrosion resistant.

ILZRO-12.
M-91.
Zinc base alloy for gravity casting for prototype work and production runs.

ILZRO-12.
M-91, M-130, M-1661; 11-13 Al, 0.5-1.25 Cu, 0.01-0.03 Mg, bal Zn.
Chill cast: 50,000-53,000 TS; 30,000-31,000 YS; 5-7 El.
Sand cast: 41,000-45,000 TS; 30,000-31,000 YS; 2-3.5 El.
For general purpose castings, housings, instrument cases, fuel pumps, automotive parts. Insensitive to cooling rate. Low casting temperature.

ILZRO-14.
M-91.
Zinc base alloy for die casting in cold chamber machines.

ILZRO-14.
M-91, M-130; 0.01-0.03 Al, 1.0-1.5 Cu, 0.25-0.30 Ti, bal Zn.
Die cast: 33,000 TS; 20,000 YS; 5.0-6.0 El; 71-77 Brin.
Good creep strength at ambient and elevated temperatures.

ILZRO-14.
M-1661; 1.0-1.5 Cu, 0.25-0.30 Ti, 0.01-0.03 Al, bal Zn.
Die cast: 33,000 TS; 5-6 El; 20,000 YS; 79-86 Vickers.
For die castings, carburetors, fuel pumps, valve bodies. Creep resistant.

ILZRO-16.
M-1661; 1.0-1.5 Cu, 0.15-0.25 Ti, 0.10-0.20 Cr, 0.1-0.04 Al, bal Zn.
Die cast: 33,000 TS; 5 El; 84 Vickers.
For automobile die castings. Good creep resistance at high temperatures.

IM 001.
M-1488; 0.12 max C, 1.5 max Mn, 2.0 max Si, 8.0-10.0 Ni, 17.0-19.5 Cr bal Fe.
Cast, annealed; 490 N/mm^2 min UTS.
For faucets, pump parts; corrosion resistant.
AFNOR Z10CN 18.9 M; similar to ACI CF 20.

IM 004 BC.
M-1488; 0.04 max C, 7.0-9.0 Mn, 1.5 max Si, 6.0-8.0 Ni, 17.0-19.0 Cr, 0.15-0.25 N, bal Fe.
Cast, annealed: 540 N/mm^2 min UTS.
Austenitic stainless for food industry, cryogenic operations.
AFNOR Z3CMN 18.8.7 AzM.

IM 164.
M-1488; 0.08 max C, 1.5 max Mn, 1.5 max Si, 9.0-12.0 Ni, 18.0-21.0 Cr 2.0-3.0 Mo, bal Fe.
Cast, annealed: 490 N/mm^2 min UTS.
Austenitic stainless for pumps, valves, faucets; improved resistance to corrosion.
AFNOR Z5CND 20.10 M; similar to ACI CF-8M.

IM 164 BC.
M-1488; 0.03 max C, 1.5 max Mn, 1.5 max Si, 9.0-13.0 Ni, 17.0-21.0 Cr, 2.0-3.0 Mo, bal Fe.
Cast, annealed: 490 N/mm^2 min UTS.
Austenitic stainless, weldable, for pumps, valves, faucets; improved resistance to corrosion.
AFNOR Z3CND 20.10 M; ACI CF-3M.

IM 164 NB.
M-1488; 0.08 max C, 1.5 max Mn, 1.5 max Si, 10.5-12.5 Ni, 17.0-19.5 Cr, 2.0-2.5 Mo, Nb = 8 x C, bal Fe.
Cast, annealed: 490 N/mm^2 min UTS.
Austenitic stainless, weldable, for pumps, valves, faucets; improved resistance to corrosion.
AFNOR Z4CND Nb 18.12 M; W. Nr. 1.4581.

IM 167 CN.
M-1488; 0.045 max C, 1.5 max Mn, 1.5 max Si, 10.5-11.5 Ni, 17.0-21.0 Cr, 2.3-2.8 Mo, 0.20 max Co, 0.08 max N, 0.5 max Cu, bal Fe.
Cast, annealed: 520 N/mm^2 min UTS.
Austenitic stainless for special applications in nuclear processes.
AFNOR Z3CND 19.10 M.

IM 168.
M-1488; 0.05 max C, 1.5 max Mn, 1.5 max Si, 12.0-15.0 Ni, 17.5-20.5 Cr, 3.0-3.5 Mo, bal Fe.
Cast, annealed: 490 N/mm² min UTS.
Austenitic stainless for pumps, valves, faucets; improved resistance to corrosion.
AFNOR Z4CND 19.13 M; similar to AISI 317.

IM 472.
M-1488; 0.08 max C, 1.5 max Mn, 2.0 max Si, 8.0-11.0 Ni, 18.0-21.0 Cr, bal Fe.
Cast, annealed: 490 N/mm² min UTS.
Austenitic stainless for pumps, valves, parts requiring corrosion resistance.
AFNOR Z6CN 19.9 M; ACI CF-8.

IM 472 BC.
M-1488; 0.03 max C, 1.5 max Mn, 2.0 max Si, 8.0-12.0 Ni, 18.0-21.0 Cr bal Fe.
Cast, annealed: 490 N/mm² min UTS.
Austenitic stainless steel, weldable, for faucets, valves, pumps.
AFNOR Z3CN 19.9 M; ACI CF-3.

IM 472 NB.
M-1488; 0.08 max C, 1.5 max Mn, 2.0 max Si, 9.0-12.0 Ni, 18.0-21.0 Cr Nb = 8 x C, bal Fe.
Cast, annealed: 490 N/mm² min UTS.
Stabilized austenitic stainless, preferred for welded pumps, valves, faucets.
AFNOR Z4CN Nb 19.10 M; ACI CF-8C.

IM 473 BC.
M-1488; 0.04 max C, 1.5 max Mn, 1.5 max Si, 9.0-10.0 Ni, 18.5-21.0 Cr, 0.20 max Co, 0.50 max Cu, 0.15 max Ta, 0.08 max N, bal Fe.
Cast, annealed: 510 N/mm² min UTS.
Austenitic stainless steel, weldable, for special nuclear applications.
AFNOR Z3CN 19.10 M; similar to ACI CF-3.

IMI 100.
M-1578, M-1805; 99.95 Cu (including Silver), oxygen free. Bar, sheet, strip, wire.
For thermionic valves, transistors, welding and brazing wire, glass-metal seals; high conductivity and ductility; not subject to hydrogen embrittlement during annealing and brazing.
ASTM B170; BS 1861 2873.C103 (wire).

IMI 103.
M-1578, M-1805; 99.9 min Cu (including silver).
Electrolytic tough pitch copper wire, sheet, strip. Standard copper for electrical purposes and stampings, spinning, etc. Ductile, high conductivity. BS 2870 C101; 2873.C101(Wire).

IMI 111.
M-1578, M-1805; 99.9 min Cu, (including silver).
Fire refined tough pitch copper.
For electrical applications and cold heading wire, and rods for bus bars.
BS1037/2870.C102 etc; ASTM B5, B124 etc.

IMI 125.
M-1578, M-1805; Sulphur addition to copper.
Sulphur copper bar and rod.
Improved machinability; good conductivity.

IMI 131.
M-1578, M-1805; Cu, phosphorus deoxidized, non-arsenical.
Ductile and corrosion resistant; for severe heading and forming.
Drawn shapes, rod, bar, sheet and strip. ASTM B152 (DHP); BS 2870.C106.

IMI 145.
M-1578; 99.8 Cu, 0.2 Cd.
High conductivity copper for radiator fin manufacture.

IMI 146.
M-1578; 99.85 min Cu, 0.3-0.5 As, 0.013-0.05 P.
Phosphorized deoxidize arsenical copper.
Used in some chemical plants where resistance to scaling is desirable; rods, strip and sheet.
BS 2870-C.107.

IMI 149.
M-1578; 0.1 Sn, 99.9 Cu.
O Temper 55 VHN; H Temper 130 VHN.
For fin material in car radiators.

IMI 151.
M-1578; 99.85 Cu, 0.05 Ag.
Higher softening temperature to resist softening during tinning and soldering; for electrical commutators, for printers' engraving plates.
ASTM B152-STP.

IMI 153 (KUFIL).
M-1578; 99 Cu, 1 Ag.
Rods, spooled wire or slittings; for oxyacetylene welding of copper.
BS 1453.C1.

IMI 156.
M-1578, M-1805; 99.4 Cu, 0.3 Al, 0.3 Ti.
For nitrogen arc welding of copper.
BS 2901.C8.

IMI 161 (BOROFIL).
M-1578, M-1805; 99.92 Cu, 0.08 B.
For argon arc welding of copper, retaining high conductivity. BS 2901.C21.

IMI 166.
M-1578, M-1805; Cd addition to copper.
Increased strength and hardness while still retaining high conductivity.
For line wire, trolley wires, overhead conductors, welding tips.
BS 23,175,672,2755,2870.C108,2873.C108.

IMI 171 (KUMIUM).
M-1578; Chromium addition to copper.
Heat treatable alloy retaining strength up to 350°C with high electric conductivity.
For electrode tips for resistance welding; for certain electronic applications. ASTM F268 (Chromium copper).

IMI 176 (ARGOFIL).
M-1578, M-1805; 99.5 Cu, 0.25 Si, 0.25 Mn.
For argon arc welding of copper.
BS 2901.C7; ASTM B225,B259.

IMI 180.
M-1805; 99.5 Cu, 0.5 Te.
1/2 hard: 310 N/mm^2 TS; 12 El.
Good machinability; for automatic screw machines.
BS 2874 C109; UNS C14500.

IMI 195.
M-1578; Copper containing small amounts Pb.
Improved machinability while still retaining good electrical and thermal conductivity.
Called leaded copper.

IMI 215.
M-1805; 85 Cu, 15 Zn.
Gilding metal; 6 grades of hardness; 65-150 VHN.
For jewelry and other fancy goods.
BS 2873 CZ102; UNSC23000.

IMI 220.
M-1805; 80 Cu, 20 Zn.
Low brass; 6 grades of hardness; 70-160 VHN.
BS 2873 CZ103; UNS C24000.

IMI 233.
M-1805; 67 Cu, 33 Zn.
Yellow brass; 6 grades of hardness; 70-175 VHN.
For stamping and forming.
BS 2873 CZ107; UNS C27000.

IMI 235.
M-1578, M-1805; 65 Cu, 35 Zn.
As sheet and strip for general commercial presswork and general coppersmith work.
BS 2870 CZ.107; ASTM B36-268.

IMI 237.
M-1578, M-1805; 63 Cu, 37 Zn.
Wire and drawn shapes for cold forming of screws, rivets and cold headed products.
BS 265;2870.CZ108;2873.CZ108;ASTM B19.

IMI 239.
M-1805; 61 Cu, 39 Zn.
Often called "pin wire" because of use in making brass pins.

IMI 240.
M-1805; 60 Cu, 40 Zn.
Wire and rod for brazing ferrous and nonferrous materials; yellow brass.
BS 1845.C23.

IMI 260.
M-1805; 70 Cu, 30 Zn.
Known as "Cartridge Brass"; very ductile when soft.
For drawn parts.
BS 2874 CZ106; UNS C26000.

IMI 276.
M-1805; 70 Cu, 30 Zn, 0.03 As.
Inhibited deep drawing brass; 6 grades of hardness; 70-170 VHN.
BS 2871 CZ105.

IMI 307 (ALUMBRO).
M-1578; 76 Cu, 22 Zn, 2 Al, 0.4 As.
Arsenic inhibited bronze strip.
BS 2870 CZ 110; ASTM B111-687; CDA and UNS C68700.

IMI 307 ("ALUMBRO").
M-1578; 76 Cu, 22 Zn, 2 Al, 0.04 As.
Arsenic inhibited aluminum brass to meet stringent operating conditions common in marine condensers; also evaporators, boat sheathing and instrument bellows.
BS 2870. CZ.110; ASTM B111.

IMI 329.
M-1805; 57.5 Cu, 0.5 Fe, 2.25 Pb, 2.0 Mn, 1.0 Sn, 0.5 Al, bal Zn.
Rod and bar: 520-540 N/mm^2 TS; 20-25 El; 170 VHN; 85% machinability.
High tensile brass for valve spindles.

IMI 333.
M-1805; 57.5 Cu, 0.5 Fe, 1.25 Pb, 1.5 Mn, 0.75 Sn, bal Zn.
Rod and bar: 530-580 N/mm^2 TS; 20-30 El; 175-190 VHN; 65% machinability.
High tensile brass; may be soldered.
BS 2874 CZ114.

IMI 334.
M-1805; 57.0 Cu, 1.0 Fe, 1.25 Pb, 1.5 Mn, 0.4 Sn, 1.25 Al, bal Zn.
Rod and bar: 600-660 N/mm^2 TS; 20-25 El; 195-210 VHN; 60% machinability.
High strength bar.
BS 2874 CZ114.

IMI 341 (COIN BRONZE).
M-1578; 97 Cu, 2.5 Zn, 0.5 Sn.
Strip-used for stamping coins in Great Britain, Australia, New Zealand and Finland.

IMI 345.
M-1578; Zn, Sn, bal Cu.
Wire for safety pins, decorative pins, spectacle frames.

IMI 360 (NAVAL BRASS).
M-1578, M-1805; 62 Cu, 1.25 Sn, bal Zn.
Sheet, strip, bar for underwater fittings for ships and other marine hardware.
BS 250, 409 2872.CZ112, 2874.CZ112.
ASTM B171.

IMI 365.
M-1578; 60 Cu, 0.75 Sn, bal Zn.
Sheet and strip for underwater fittings or ships and general marine hardware.
BS 2870.CZ.113; ASTM B21A.

IMI 365.
M-1805; 60.5 Cu, 0.8 Sn, bal Zn.
Rod and bar: 410-450 N/mm^2 TS; 20-30 El; 115-135 VHN; 50% machinability.
American Naval brass; UNS C46400.

IMI 388.
M-1578, M-1805; 59.4 Cu, 0.3 Si, 40.3 Zn.
Wire and rod for brazing ferrous and nonferrous materials; called "silicon brass."
BS 1845.C26.

IMI 410.
M-1578; 64 Cu, 1 Pb, bal Zn.
Leaded brass strip, particularly for stamped and machined watch, clock and instrument parts; called "clock and matrix brass."
BS 2870.CZ.116.

IMI 432.
M-1805; Leaded brass wire.
Called "nipple wire."
For cold heading, riveting and cold forming.
BS 2873.CZ119; ASTM B146.

IMI 433.
M-1805; 62 Cu, 2.0 Pb, bal Zn.
Rod and bar: 360-380 N/mm^2 TS; 30-40 El. 75% machinability; for turning, riveting, spinning and cold swaging.
BS 2873; UNS 35300.

IMI 441.
M-1805; 61.5 Cu, 3.25 Pb, bal Zn.
Rod and bar: 370-430 N/mm^2 TS; 22-35 El. 100% machinability; for free-turning brass.
BS 2874; ASTM B16; UNS C36000.

IMI 442.
M-1805; 61.5 Cu, 2.5 Pb, bal Zn.
Rod and bar: 370-400 N/mm^2 TS; 25-45 El. 90% machinability; for high speed machining and riveting.

IMI 443.
M-1578; 61 Cu, 0.25 Pb, bal Zn.
Low leaded wire called "shoe rivet wire," main use is upsetting to shoe rivets, and heading operations where some machining is required.

IMI 452 (MUNTZ METAL).
M-1805; 60 Cu, 0.5 max Pb, bal Zn.
Rod and extruded or drawn sections for cold work, bending, riveting, machining.
BS 1949 2872.CZ123; 2874.CZ123.

IMI 453.
M-1578, M-1805; 60 Cu, 0.7 max Pb, bal Zn.
Rod and extruded or drawn sections, called "special bending brass," for cold forming and machining.
BS 1949 2872.CZ123; 2874.CZ123.

IMI 455.
M-1805; 59.25 Cu, 2.25 Pb, bal Zn.
Rod and bar: 420-460 N/mm^2 TS; 20-30 El. 85% machinability; free-machining brass; may be hot forged.
ASTM B124; UNS C37700.

IMI 456.
M-1578; 59 Cu, 0.75 Pb, bal Zn.
Rod and extruded or drawn sections called "riveting quality brass." For upsetting, stamping and limited turning.
BS CZ9B/12872.CZ119; 2874.CZ119.

IMI 458.
M-1578; 59 Cu, 2 Pb; bal Zn.
Strip, called "clock brass."
For stamped parts requiring machining to make clock and instrument parts.
BS 2870.CZ120.

IMI 463.
M-1805; 58.0 Cu, 2.25 Pb, bal Zn.
Extruded: 330-400 N/mm^2 TS; 30-38 El. Hot stamping brass; 90% machinability.
UNS 38000.

IMI 466.
M-1805; 57.5 Cu, 4.25 Pb, bal Zn.
Rod and bar: 450-515 N/mm^2 TS; 19-25 El. 100 plus machinability; for free-turning brass.

IMI 467.
M-1805; 57.75 Cu, 2.25 Pb, bal Zn.
Rod and extruded or drawn sections for hot stamping and forging; free machining.
BS 218 2872.CZ122; 2874.CZ122.

IMI 469.
M-1805; 57.5 Cu, 3.25 Pb, bal Zn.
Rod and extruded or drawn sections for easy machining to bolts, nuts, screws and fittings.
BS 249 2872.CZ121; 2874.CZ121.

IMI 470.
M-1805; 58.25 Cu, 2.75 Pb, bal Zn.
Rod and bar: 435-495 N/mm^2 TS; 21-28 El. 100% machinability; for free-turning brass.
BS 2874.

IMI 471.
M-1805; 57.0 Cu, 1.5 Pb, bal Zn.
Rod and shapes: 140 VHN; 85% machinability.
General purpose section brass.

IMI 473.
M-1805; 58.0 Cu, 3.0 Pb, bal Zn.
Rod and bar: 435-495 N/mm² TS; 20-28 El.
100% machinability; for free-turning brass.
BS 2874; DIN 17660 CuZn 39 Pb3.

IMI 475.
M-1805; 57.0 Cu, 3.0 Pb, 0.3 Al, bal Zn.
Sections: 160 VHN; 90% machinability.
General purpose section; not for brazing.

IMI 476.
M-1805; 56.0 Cu, 1.5 Pb, 0.5 Al, bal Zn.
Sections; 150 VHN; 75% machinability.
For thin sections.

IMI 500 (NICKEL BRASS COIN).
M-1578; 79 Cu, 20 Zn, 1 Ni.
Sometimes used for coins.

IMI 506 (NICKEL SILVER).
M-1578, M-1805; 65 Cu, 18 Ni, 17 Zn.
Wire and drawn shapes for decorative purposes, springs, fishing tackle; ductile, malleable, machinable, corrosion resistant.
BS 790, 1824, 2870, 2873; ASTM B122, B206.

IMI 511 (NICKEL SILVER).
M-1578, M-1805; 62 Cu, 10 Ni, 28 Zn.
Wire, drawn shapes, sheet for decorative purposes, jewelry, tableware, fishing tackle; ductile, malleable, machinable, corrosion resistant.
BS 2870 NS.103; ASTM B122, B206; BS 790, 1824, 2873.

IMI 512 (NICKEL SILVER).
M-1578, M-1805; 62 Cu, 12 Ni, 26 Zn.
Wire, drawn shapes, sheet for decorative purposes, jewelry, slide fastener nameplates; ductile, malleable, machinable, corrosion resistant.
BS 790, 1824, 2870, 2873; ASTM B122, B206.

IMI 513 (NICKEL SILVER).
M-1578, M-1805; 62 Cu, 15 Ni, 23 Zn.
Wire and drawn shapes for decorative purposes, tableware, jewelry; ductile, malleable, machinable, corrosion resistant.
BS 790, 1824, 2870, 2873; ASTM B122, B206.

IMI 514 (NICKEL SILVER).
M-1578, M-1805; 62 Cu, 18 Ni, 20 Zn.
Wire, drawn shapes and sheet for decorative purposes, tableware, name plates, engraving plates, springs; ductile, malleable, machinable, corrosion resistant.
BS 790, 1824, 2870, 2873; ASTM B122, B206.

IMI 515 (NICKEL SILVER).
M-1578; 62 Cu, 20 Ni, 18 Zn.
Wire and drawn shapes for decorative purposes, tableware, springs for electronics and telecommunications; ductile, malleable, machinable, corrosion resistant.
BS 790, 1824, 2870, 2873; ASTM B122, B206.

IMI 525.
M-1578; Brass plus 10% Ni.
Nickel brass rod, stamping grade; good white color for sanitary fittings.
ASTM B124-55 alloy 14.

IMI 551.
M-1578; Free machining brass plus 10% Ni.
Nickel brass rod, free turning grade; good white color for sanitary fittings.
BS (near) 2872.NS101; 2874.NS101.

IMI 575 ("KUTHERM" 3).
M-1578, M-1805; Resistance alloy wire, maximum continuous operating temperature 150°C.
For electric blankets; under-road, under-ramp heating.

IMI 576 ("KUTHERM" 10).
M-1578, M-1805; Resistance alloy wire, maximum continuous operating temperature 250°C.
For Cold "tails" electric blankets; under-road, under-ramp heating.

IMI 577 (KUTHERM X).
M-1805; Cu-Ni alloy used in electric blankets, controlled resisters for black heat applications.
Resistivity: 10 micro-ohm cm.

IMI 579 ("KUTHERM" 26).
M-1578, M-1805; Resistance alloy wire, maximum continuous operating temperature 250°C.
For electric blanket heating cables, fixed resistance, space heating systems.

IMI 581 ("KUTHERM" 14).
M-1578, M-1805; Resistance alloy, intermediate electrical resistivity, outstanding stability up to 350°C.
For heating cables, thermocouples, space heating systems.

IMI 583 (KUTHERM 49).
M-1805.
Cupro-nickel alloy with low temperature coefficient. Used for instrument shunts, field regulators and all types of black heat resistances.
Resistivity: 49 micro omh cm.

IMI 591 (KUTHERM 15).
M-1805.
Cupro-nickel alloy with low temperature coefficient.
Resistivity: 15 micro omh cm.

IMI 651.
M-1578; 99 Cu, 1 Sn.
Conductivity bronze wire; for high strength conductors, rivets and chains.

IMI 657.
M-1578, M-1805; 95 Cu, 5 Sn, P.
Phosphor bronze wire, various tempers.
For springs, electrical contacts, switch parts, chains, welding wire.
BS 369, 384, 407 etc; ASTM B103, B159, B225.

IMI 658.
M-1578; 93 Cu, 7 Sn, P.
Phosphor bronze, various tempers; for springs and stampings.
BS 2870 PB 103.

IMI 659.
M-1578, M-1805; 93 Cu, 7 Sn, P.
Phosphor bronze wire, various tempers; for flat and coiled springs, weldin wire, Fourdrinier cloth warp.
BS 2873.PB103 etc; ASTM B103, B159, B225.

IMI 660 (CAROBRONZE).
M-1578; 92 Cu, 8 Sn.
Tin bronze wire and drawn shapes for bushes and piston rings.

IMI 661.
M-1578; 62 Cu, 8 Sn, P.
Phosphor bronze wire; for Fourdrinier cloth warp.

IMI 670.
M-1805; 99 Cu, 0.75 Sn, 0.25 Cd.
Sometimes called "Post office bronze".

IMI 675.
M-1805; 93 Cu, 7.0 Sn.
Wire; for manufacture of hair springs.
BS 2873 PB103; UNS C52100.

IMI 705 ("EVERDUR" A).
M-1578, M-1805; 96 Cu, 3 Si, 1 Mn.
Good strength bars and rods, weldable, and corrosion resistant to weak acids.
For screws, bolts, nuts, particularly marine welding wire.
BS 1948 2872.CS101; 2874.CS101; ASTM B98.

IMI 757.
M-1578, M-1805; 93 Cu, 7 Al.
Aluminum bronze welding wire; for welding aluminum bronzes. BS 2901.C12

IMI 758 ("SILVABRONZE").
M-1578; 92.6 Cu, 7 Al, 0.3 Sn, 0.1 Ag.
Bronze strip, good resistance to flue gas condensates and to stress corrosion, particularly in chloride-contaminated waters.

IMI 830 (CUPRONIC).
M-1805.
Cupro-nickel alloy with controlled thermo emf used as compensating leads for thermocouples.

IMI 837 (CUPRO-NICKEL).
M-1578; 94 Cu, 6 Ni.
Wire and drawn shapes; good strength, ductility and resistance to corrosion.

IMI 838 (CUPRO-NICKEL).
M-1578; 90 Cu, 10 Ni.

IMI 842 (CUPRO-NICKEL).
M-1578; 80 Cu, 20 Ni.
Wire and drawn shapes; good strength, ductility and resistance to corrosion.
BS 374; ASTM B122.

IMI 849 (CUPRO-NICKEL).
M-1578; 70 Cu, 30 Ni.
Wire and drawn shapes; good strength, ductility and resistance to corrosion.
BS 374; ASTM B122.

IMI 850.
M-1805.
Cupro-nickel with controlled emf used in thermocouples.

IMI 885 ("KUNIFER" 10).
M-1560; 87 Cu, 10 Ni, 2 Fe, 1 Mn.
Sheet and strip for marine parts; good strength and corrosion resistance, especially against sea water; also for heat exchangers. BS 2870 CN.102.

IMI 891 ("KUNIFER 5T").
M-1578; 5 Ni, 1.3 Fe. 0.2-0.5 Ti, bal Cu.
Welding wire, for Argon arc welding of "Kunifer" 5 and 95/5 cupro-nickel.
BS 2901.C19.

IMI 892 ("KUNIFER 10T").
M-1578, M-1805; 10 Ni, 1.5 Fe, 0.8 Mn, 0.2-0.5 Ti, bal Cu.
Welding wire, for Argon arc welding of "Kunifer" 10 and 90/10 cupro-nickel.

IMI 893 ("KUNIFER 30 T").
M-1560; 30 Ni, 0.5 Fe, 0.8 Mn, 0.2-0.5 Ti, bal Cu.
Welding wire for Argon arc welding of "Kunifer" 30 and 70/30 cupro-nickel.
BS 2901.C18.

IMI 895 ("KUNIFER" 30).
M-1578; 68.25 Cu, 30 Ni, 0.75 Fe, 1.0 Mn.
Sheet and strip for marine parts; good strength and corrosion resistance, especially against sea water. Also for heat exchangers.
BS 2870 CN.107.

IMI 981.
M-1578; Tin coated tough pitch or oxygen free HC copper.
For electrical windings and conductors that have to be soldered.
BS 128; ASTM B33.

IMI-RM 236.
M-1578; 64 Cu, 36 Zn.
Ductile alloy; as sheet or strip is used in automobile radiator tanks.
BS 2870 CZ108; ASTM B36-272.

IMITA GOLD.
M-Eng.; 89 Cu, 10.3 Al, 0.33 Fe, 0.25 Ni, 0.23 Sn, 0.2 Mn.
For cheap jewelry, ornaments; corrosion resistant.

IMI TITANIUM 115.
M-1578; Commercially pure titanium.
Sheet, strip, bar, wire, extrusions.
42,500-60,000 TS; 29,000 YS; 25 El.
Ductile, formable, corrosion resistant.
BS TA 1; DTD 501 3B.

IMI TITANIUM 125.
M-1578; Commercially pure titanium.
Sheet, bar, tubes. 54,000-78,000 TS; 42,500 YS; 22 El.
Ductile, formable, corrosion resistant.
BS TA 2,3,4,5; AMS 4902 etc.

IMI TITANIUM 130.
M-1578; Commercially pure titanium.
Sheet, strip, bar, wire, extrusions, tubes (welded). 67,000-90,000 TS; 49,000 YS; 20 El.
Corrosion resistant, formable.
DTD 5003B, 5023B; AMS 4900B.

IMI TITANIUM 155.
M-1578; Commercially pure titanium.
Sheet: 83,000-105,000 TS; 67,000 YS; 15 El.
Corrosion resistant; slight forming only.
BS TA6; AMS 4901, 4921.

IMI TITANIUM 160.
M-1578; Commercially pure titanium.
Bar, wire, extrusions. 78,000-108,000 TS; 63,000 YS; 15 El.
Corrosion resistant.
BS TA 7,8,9; AMS 4921.

IMI TITANIUM 230.
M-1578; 2.5 Cu, bal Ti.
Sheet, bar, wire, extrusions.
Annealed: 78,000-112,000 TS; 67,000 YS; 15 El.
Sol. tr. & age: 110,000-134,000 TS; 80,000 YS; 10 El. Fatigue limit: 60-65% of TS.
Formable, corrosion resistant.
DTD 5233,5243,5253,5263; BS TA21,22,23,24.

IMI TITANIUM 260.
M-1578; Pd (slight), bal Ti.
Sheet: 42,500-60,000 TS; 29,000 YS; 25 El.
Improved resistance to non-oxidizing acids.

IMI TITANIUM 262.
M-1578.
Commercially pure titanium with 0.15 min Pd.
Sheet, strip: 390-540 MPa TS; 290 MPa min YS; 22 min El.

IMI TITANIUM 315.
M-1578; 2 Al, 2 Mn, bal Ti.
Bar: 94,000-116,000 TS; 67,000 YS; 20 El.
Fatigue limit: 60-65% of TS.
DTD 5043B.

IMI TITANIUM 317.
M-1578; 5 Al, 2.5 Sn, bal Ti.
Sheet, bar, extrusions. 118,000-157,000 TS; 112,000 YS; 10 El.
Weldable, high strength.
BS TA14,15,16,17; AMS 4910, 4926 etc.

IMI TITANIUM 318.
M-1578; 6 Al, 4 V, bal Ti.
Sheet, bar, wire, extrusions, rod.
Rod: 130,000-168,000 TS; 120,000 YS; 8 El.(variable properties with varying heat treatment).
Fatigue limit: 50-60% of TS.
Most popular titanium alloy; weldable, machinable, corrosion resistant.
BS TA10,11,12,13,28; AMS 4911, 4928 etc.

IMI TITANIUM 550.
M-1578; 4 Al, 4 Mo, 2 Sn, 0.5 Si, bal Ti.
Bar: 164,000 min TS; 145,000 YS; 9 El.
High strength alloy, creep resistant up to 400°C.
DTD 5103, 5153. (Hylite 50).

IMI TITANIUM 551.
M-1578; 4 Al, 4 Mo, 4 Sn, 0.5 Si, bal Ti.
Bar: 179,000 min TS; 159,000 YS; 8 El.
Very high strength titanium alloy.
DTD 5203, 5223 (Hylite 51).

IMI TITANIUM 679.
M-1578; 11 Sn, 5 Zr, 2.25 Al, 1.0 Mo, 0.2 Si, bal Ti.
Bar, quenched and aged: 160,000-195,000 TS; 141,000 YS; 8 El.
High strength alloy; creep resistant up to 450°C.
BS TA18,19,20,25,26,27; AMS 4974.

IMI TITANIUM 680.
M-1578; 11 Sn, 4 Mo, 2.25 Al, 0.2 Si, bal Ti.
Bar, quenched and aged: 179,000 min TS; 156,000 YS; 10 El.
High strength alloy.
DTD 5213, M160.

IMI TITANIUM 684.
M-1578; 6 Al, 5 Zr, 1 W, 0.3 Si, bal Ti.
Bar: 143,000 TS; 127,000 YS; 6 El.
Good strength alloy, weldable; creep resistant up to 520°C.
DTD M200.

IMI TITANIUM 685.
M-1578; 6 Al, 5 Zr, 0.5 Mo, 0.3 Si, bal Ti.
Forged; 20°C: 990 MPa TS; 85 MPa YS; 6 El.
520°F: 62 MPa TS; 48 MPa YS; 9 El.
Compressor blades for turbofan engine.

IMI TITANIUM 700.
  M-1578; 6 Al, 5 Zr, 4 Mo, 1 Cu, 0.2 Sn, bal Ti.
  Air-cooled and aged, bar: 183,000 min TS; 166,000 YS; 8 El.
  Quenched and aged, bar: 195,000 min TS; 179,000 YS; 6 El.
  Fatigue limit: 45-55% of TS.
  Ultra high strength alloy, creep resistant up to 400°C.
  DTD M201.

IMI TITANIUM 829.
  M-1578; 5.5 Al, 3.5 Sn, 3 Zr, 1 Nb, 0.30 Mo, 0.3 Si, bal Ti.
  Room Temp: 960 MPa min TS; 820 MPa min YS; 10 El.
  540°C: 600 MPa min TS; 460 MPa min YS; 12 El.
  Good high temperature creep strength.

IMMACULATE 2W.
  M-521; 0.30 C, 1.3 Si, 0.8 Mn, 20 Cr, 7.5 Ni, 2 W, bal Fe.
  Annealed: 119,000 TS; 54,500 YS; 49 El; 56 RA; 220 Brin.
  For valves, furnace equipment; austenitic, stainless.

IMMACULATE 5.
  M-56; 0.15 max C, 0.20-1.0 Si, 0.50-2.0 Mn, 23.0-26.0 Cr, 19.0-22.0 Ni, bal Fe.
  Austenitic stainless; AISI 310.

IMMACULATE 5.
  M-521; 0.1 C, 23.0 Cr, 21.0 Ni, bal Fe.
  Stainless steel for elevated temperature operation.

IMMACULATE 5T.
  M-56; 0.15 C, 0.20-1.0 Si, 0.5-2.0 Mn, 22.0-23.0 Cr, 13.0-16.0 Ni, Ti = 4 x C min, bal Fe.
  Stabilized heat resisting steel.

IMMACULATE 5T.
  M-521; 0.15 max C, 1.0 max Si, 2.0 max Mn, 21.5-23.5 Cr, 16.0-17.5 Ni, 1.0 Ti, bal Fe.
  Ann, Room Temp: 90,000 TS; 40,000 YS; 45 El. At 800°C: 42,000 TS; 52 El.
  Good strength and resistance to scaling at elevated temperatures. For radiant tube supports, furnace brackets.

IMMUNIT 2E.
  M-1331; 0.12 max C, 18 Cr, 9.5 Ni, Ti = 4 x C, bal Fe.
  Annealed: 85,000 TS; 35,000 YS; 55 El; 65 RA; 150 Brin.
  For welded chemical plant equipment; Type 321; stainless, austenitic.

IMMUNIT 2G.
  M-1331; 0.15 C, 18 Cr, 8.5 Ni, bal Fe.
  Annealed: 80,000 TS; 35,000 YS; 55 El; 75 RA; 150 Brin.
  Cold drawn: 180,000 TS; 150,000 YS; 10 El; 250 Brin.
  For chemical plant equipment, mixers, filters, tanks; Type 302; stainless austenitic.

IMMUNIT 2N.
  M-1331; 0.15 max C, 18 Cr, 8.5 Ni, bal Fe.
  Annealed: 80,000 TS; 35,000 YS; 55 El; 75 RA; 150 Brin.
  Cold drawn: 180,000 TS; 150,000 YS; 10 El; 250 Brin.
  For chemical plant equipment, mixers, filters, tanks; Type 302; stainless austenitic.

IMMUNIT 2NA.
  M-1331; 0.15 max C, 18 Cr, 8.5 Ni, 0.2 S, bal Fe.
  Annealed: 80,000 TS; 35,000 YS; 45 El; 65 RA; 150 Brin.
  For screw machine products, bolts, fasteners, shafts; Type 303; stainless free-cutting.

IMMUNIT 2S.
  M-1331; 0.07 max C, 18 Cr, 9.5 Ni, bal Fe.
  Annealed: 85,000 TS; 35,000 YS; 60 El; 70 RA; 150 Brin.
  Cold drawn: 180,000 TS; 135,000 YS; 10 El; 330 Brin.
  For welded chemical plant equipment, tanks, dryers; Type 304; stainless, austenitic.

IMMUNIT 2X(G).
  M-1331; 0.12 max C, 18 Cr, 9.5 Ni, Cb = 8 x C, bal Fe.
  Annealed: 90,000 TS; 45,000 YS; 56 El; 65 RA; 160 Brin.
  Cold drawn: 100,000 TS; 65,000 YS; 40 El; 60 RA; 205 Brin.
  For welded structures, chemical plant equipment, tanks; Type 347; stainless, austenitic.

IMMUNIT 4E.
  M-1331; 0.12 max C, 18 Cr, 2 Mo, 10.5 Ni, Ti = 4 x C, bal Fe.
  Annealed: 85,000 TS; 35,000 YS; 50 El; 65 RA; 160 Brin.
  For welded chemical plant equipment; Type 316 Ti; stainless, austenitic.

IMMUNIT 4G.
  M-1331; 0.15 C, 18 Cr, 2 Mo, 9.5 Ni, bal Fe.
  Annealed: 85,000 TS; 35,000 YS; 50 El; 65 RA; 160 Brin.
  For chemical plant equipment, mixers, agitators, filters; Type 316; stainless, austenitic.

IMMUNIT 4N.
M-1331; 0.1 C, Si, 18 Cr, 8 Ni, bal Fe.
Annealed: 85,000 TS; 35,000 YS; 50 El; 70 RA; 150 Brin.
For chemical plant equipment; corrosion resistant, austenitic.

IMMUNIT 4S.
M-1331; 0.07 max C, 18 Cr, 10.5 Ni, 2 Mo, bal Fe.
Annealed: 85,000 TS; 35,000 YS; 50 El; 65 RA; 160 Brin.
Cold drawn: 150,000 TS; 135,000 YS; 6 El; 300 Brin.
For acid resistant chemical plant equipment; Type 316; stainless, austenitic.

IMMUNIT 4SI.
M-1331; 0.1 C, 18 Cr, 9.5 Ni, 2 Mo, bal Fe.
Annealed: 85,000 TS; 35,000 YS; 50 El; 65 RA; 160 Brin.
Cold drawn: 150,000 TS; 135,000 YS; 6 El; 300 Brin.
For acid resistant chemical plant equipment, tanks; Type 316; stainless, austenitic.

IMMUNIT 4X(G).
M-1331; 0.12 max C, 18 Cr, 10.5 Ni, 2 Mo, Cb = 8 x C, bal Fe.
Annealed: 85,000 TS; 35,000 YS; 45 El; 60 RA; 170 Brin.
For welded chemical plant equipment, mixers, tanks; Type 316 Cb; stainless, austenitic.

IMMUNIT 8E.
M-1331; 0.12 max C, 18 Cr, 2 Mo, 10.5 Ni, Ti = 4 x C, bal Fe.
Annealed: 85,000 TS; 35,000 YS; 45 El; 60 RA; 170 Brin.
For welded chemical plant equipment, tanks, mixers; Type 316 Ti; stainless, austenitic.

IMMUNIT 14S.
M-1331; 0.07 max C, 17 Cr, 4.75 Mo, 13 Ni, bal Fe.
Annealed: 90,000 TS; 40,000 YS; 40 El; 55 RA; 180 Brin.
For acid resistant chemical plant equipment, tanks; Type 317; stainless, austenitic.

IMMUNIT 16S.
M-1331; 0.07 max C, 17.5 Cr, 17.5 Ni, 2 Cu, 2 Mo, Cb = 8 x C, bal Fe.
Annealed: 100,000 TS; 50,000 YS; 40 El; 50 RA; 190 Brin.
For acid resistant chemical plant equipment, valves; stainless, austenitic.

IMMUNIT 68G.
M-1331; 0.25 C, 1.18 Cr, 0.1 V, 0.7 Mn, 0.8 Si, bal Fe.
For gears, shafts, cams, bolts, camshafts; case hardening steel, tough.

IMMUNIT C58G.
M-1331; 0.25 C, 14.5 Cr, 1 max Ni, bal Fe.
Annealed: 95,000 TS; 50,000 YS; 25 El; 55 RA; 200 Brin.
For turbine blades, cutlery, surgical and dental instruments; Type 420; corrosion resistant.

IMMUNIT-R1013 MONI.
M-1331; 0.10 C, 12.5 Cr, 0.4 Mo, 0.45 Ni, bal Fe.
Annealed: 70,000 TS; 35,000 YS; 40 El; 80 RA; 145 Brin.
Cold drawn: 90,000 TS; 80,000 YS; 20 El; 65 RA; 205 Brin.
For oil refinery and chemical plant equipment, table hardware, flatware, knives.
Corrosion resistant.

IMMUNIT-R1213 MOA.
M-1331; 0.15 max C, 13 Cr, 0.3 Mo, bal Fe.
Annealed: 80,000 TS; 40,000 YS; 25 El; B 93 Rock.
For springs, shafts, table flatware, oil refinery equipment.
Corrosion resistant, hardenable.

IMMUNIT-R1513 MO.
M-1331; 0.15 C, 13 Cr, 1 Mo, bal Fe.
Annealed: 80,000 TS; 40,000 YS; 25 El; B 93 Rock.
Cold drawn: 100,000 TS; 80,000 YS; 15 El; B 96 Rock.
For springs, table flatware, oil and chemical plant equipment.
Corrosion resistant, hardenable.

IMMUNIT-R2013 MO.
M-1331; 0.20 C, 13 Cr, 1 Mo, bal Fe.
Annealed: 95,000 TS; 40,000 YS; 25 El; B 92 Rock.
Heat treated: 240,000 TS; 200,000 YS; 10 El; C 50 Rock.
For cutlery, bearings, knives, shafts, surgical instruments, scissors.
Corrosion resistant, hardenable.

IMMUNIT-R2213.
M-1331; 0.20 C, 12 Cr, 1 Mo, 0.3 V, 0.5 W, bal Fe.
Annealed: 90,000 TS; 40,000 YS; 25 El; B 92 Rock.
Hardened: 240,000 TS; 210,000 YS; 9 El; C 50 Rock.
For valves, bearings, cutlery, surgical instruments.
Corrosion resistant.

IMPACTO.
M-1636; 0.20 C, 0.8 Mn, 0.55 Ni, 0.50 Cr, 0.20 Mo, bal Fe.
Heat treated: 121,000 TS; 84,500 YS; 19 El; 248 Brin.
For carburized splined shafts, piston pins, cam shafts, guide pins, ratchets.
Tough, case hardening steel.

IMPACTO.
M-435; 0.15 C, 0.5 Mn, 0.25 Mo, 1.75 Ni, bal Fe.
Normalized: 80,000 TS; 45,000 YS; 30 El; 65 RA; 163 Brin.
For case hardened parts, gears, pinions, cams; carburizing steel.

IMPALCO 740.
M-1547; 0.3-0.7 Cu, 2.2-3.2 Mg, 0.5 max Si, 0.5 max Fe, 0.3-0.7 Mn, 5.2-6.2 Zn, bal Al.
Heat treated: 72,000 min TS; 7 min El.
For machining applications of high strength parts, structural members.
Poor weldability. Heat treatable.

IMPALCO 750.
M-1547; 0.3-0.7 Cu, 2.2-3.2 Mg, 5.2-6.2 Zn, 0.08-0.25 Cr, bal Al.
Heat treated: 72,000 min TS; 7 min El.
For machining applications of high strength parts, structural members. Poor weldability. Heat treatable.

IMPALCO 760.
M-1547; 0.8-1.4 Cu, 2.2-3.2 Mg, 0.3-0.7 Mn, 5.0-6.0 Zn, bal Al.
Heat treated: 78,000 min TS; 5 min El.
For machining applications of high strength parts, structural members. Heat treatable. Poor weldability.

IMPALCO C66.
M-1547; 3.5-4.8 Cu, 0.85 max Mg, 0.9 max Si, 1.0 max Fe, 1.2 max Mn, bal Al.
Heat treated: 69,000 min TS; 8 min El.
At 200°C: 55,000 TS; 7.5 El.
For highly stressed structural members.
Heat treatable. Not corrosion resistant. Fair ductility and weldability.

IMPALCO C66A.
M-1547; 3.5-4.8 Cu, 0.85 max Mg, 0.90 max Si, 1.0 max Fe, 1.2 max Mn, bal Al.
T-4 Temper: 54,000 min TS; 15 min El.
T-6 Temper: 58,000 min TS; 8 min El.
For stressed skin structures where corrosion resistance is important.
Clad with pure aluminum.
High strength. Heat treatable.

IMPALCO C69.
M-1547; 1.0-2.0 Cu, 0.5-1.2 Mg, 0.8-1.3 Si, 0.7 max Fe, 1.0 max Mn, bal Al.
Heat treated: 58,000 min TS; 8 min El.
For medium and high strength applications.
Heat treatable. Poor resistance to atmospheric attack.

IMPALCO C80.
M-1547; 5.0-6.0 Cu, 0.4 max Si, 0.7 max Fe, 0.2-0.6 Pb, 0.2-0.6 Bi, 0.3 max Zn, bal Al.
Heat treated: 42,000 min TS; 12 min El.
For screw machine products, fasteners, shafts, screws.
Free-machining. Poor resistance to atmospheric attack. Age-hardenable.

IMPALCO M31.
M-1547; 0.15 max Cu, 0.4-0.8 Mg, 0.15 max Si, 0.2 max Fe, 0.25 Mn, bal Al.
O-Temper: 18,000 max TS.
H-Temper: 24,600 min TS.
For bright trim, automobile panels, architectural applications.
Good formability and weldability. Non-heat treatable.

IMPALCO M32.
M-1547; 0.15 max Cu, 0.8-1.2 Mg, 0.15 max Si, 0.2 max Fe, 0.25 Mn, bal Al.
O-Temper: 22,400 max TS.
H-Temper: 29,200 min TS.
For bright trim, architectural applications, automobile panels.
Good ductility, formability and weldability. Non-heat treatable.

IMPALCO M32X.
M-1547; 0.15 max Cu, 0.8-1.2 Mg, 0.15 max Si, 0.2 max Fe, 0.25 Mn, bal Al.
O-Temper: 22,400 max TS.
H-Temper: 29,200 min TS.
For automobile trim, architectural applications, paneling.
Good ductility, formability and weldability. Non-heat treatable.

IMPALCO M34.
M-1547; 0.1 max Cu, 0.8-1.2 Mg, 0.35 max Si, 0.5 max Fe, 0.15-0.30 Mn, bal Al.
O-Temper: 18,000 min TS; 20 min El.
H-Temper: 33,600 min TS; 29,200 min YS; 4 min El.
For tubular furniture, television antennae.
Non-heat treatable. Good formability and weldability.

IMPALCO M35/1.
M-1547; 0.10 max Cu, 1.8-2.7 Mg, 0.6 max Si, 0.7 max Fe, 0.5 max Mn, 0.5 max Cr, bal Al.
At 25°C: 25,700 TS; 10,000 YS; 29 El.
At 200°C: 20,700 TS; 10,700 YS; 50 El.
For architectural and marine applications, panelling, containers, welded structures.
Non-heat treatable. Good formability and weldability.

IMPALCO M35/2.
M-1547; 0.10 max Cu, 3.0-4.0 Mg, 0.6 max Si, 0.7 max Fe, 1.0 max Mn, 0.5 max Cr, bal Al.
At 25°C: 32,000 TS; 10,000 YS; 29 El.
At 200°C: 23,500 TS; 11,200 YS; 53 El.
For welded structures, casings.
Good formability and weldability. Work hardens rapidly. Non-heat treatable.

IMPALCO M35/3.
M-1547; 0.10 max Cu, 3.5-5.5 Mg, 0.7 max Si, 0.7 max Fe, 1.0 max Mn, 0.5 max Cr, bal Al.
M-Temper: 38,100 min TS; 12 min El.
O-Temper: 38,100 min TS; 18 min El.
For riveted and welded vessels or containers subjected to relatively high stresses.
Non-heat treatable. Good formability and weldability. Work hardens rapidly.

IMPALCO M36.
M-1547; 0.10 max Cu, 4.5-5.5 Mg, 0.6 max Si, 0.7 max Fe, 1.0 max Mn, 0.5 max Cr, bal Al.
M-Temper: 38,100 min TS; 18 min El.
1/4 H-Temper: 42,000 min TS; 8 min El.
For marine, chemical and food industries, aircraft, welded structures.
Non-heat treatable. Good formability and weldability. Work hardens rapidly.

IMPALCO M38.
M-1547; 0.04 max Cu, 0.4-0.9 Mg, 0.3 -0.7 Si, 0.5 max Fe, bal Al.
At 25°C: 33,000 TS; 17 El.
At 150°C: 21,000 TS; 20 El.
For bus bars.
Heat treatable. Good ductility and weldabilty.
High electrical conductivity of 55% min.

IMPALCO M39/1.
M-1547; 0.10 max Cu, 0.4-0.9 Mg, 0.3-0.7 Si, 0.6 max Fe, 0.3 max Mn, bal Al.
O-Temper: 15,700 min TS; 15 El.
Heat treated: 27,000 min TS; 12 min El.
For glazing bars, window frames, curtain walls.
Heat treatable. Good formability and weldability.

IMPALCO M39/2.
M-1547; 0.10 max Cu, 0.4-1.5 Mg, 0.6-1.3 Si, 0.6 max Fe, 0.4-1.0 Mn, 0.3 max Cr, bal Al.
Heat treated: 42,600 min TS; 10 min El.
O-Temper: 15,700 min TS; 15 min El.
For automobile and bus bodies, coke skips, crossbearers.
Heat treatable. Good weldability and ductility. Resists stress and shock.

IMPALCO M40.
M-1547; 0.15 -0.40 Cu, 0.8-1.2 Mg, 0.4-0.8 Si, 0.7 max Fe, 0.2-0.8 Mn, 0.15-0.35 Cr, bal Al.
Heat treated: 40,400 min TS; 10 min El.
O-Temper: 15,700 min TS; 15 min El.
For tubular furniture.
Good formability and weldability. Heat treatable.

IMPALCO M42.
M-1547; 0.10 max Cu, 0.4-1.5 Mg, 0.6-1.3 Si, 0.6 max Fe, 0.2 max Mn, 0.1 max Cr, bal Al.
Heat treated: 40,400 min TS; 10 min El.
O-Temper: 15,700 min TS; 15 min El.
For general purpose applications requiring good surface finish and medium strength.
Heat treatable. Good ductility and weldability. Not shock resistant.

IMPALCO P3.
M-1547; 0.02 max Cu, 0.15 max Si, 0.15 max Fe, 99.8 min Al.
O-Temper: 11,200 max TS; 35 min El.
1/2H-Temper: 13,500-16,800 TS; 8 min El.
For chemically brightened and anodized components, costume jewelry.
High ductility, poor machinability, low strength, high corrosion resistance.

IMPALCO P5.
M-1547; 0.05 max Cu, 0.3 max Si, 0.4 max Fe, 0.05 max Mn, 99.5 min Al.
M-Temper: 9000 min TS; 25 min El.
O-Temper: 13,500 max TS; 30 min El.
For food, pharmaceutical and chemical processing equipment, containers.
High corrosion resistance, high ductility, and workability.

IMPALCO P5E.
M-1547; 0.05 max Cu, 0.5 max Cu + Si + Fe, 99.5 min Al.
O-Temper: 10,000-14,500 TS; 25 min El.
For electrical applications, bus bars.
Low strength, high conductivity.

IMPALCO P10.
M-1547; 0.1 max Cu, 0.5 max Si, 0.7 max Fe, 0.1 max Mn, 99.0 min Al.
At 25°C: 12,800 TS; 5200 YS; 45 El.
At 200°C: 6000 TS; 2500 YS; 70 El.
H-Temper: 20,200 min TS; 3 El.
For processing equipment for chemical, pharmaceutical, petroleum and food industries.
High corrosion resistance, good ductility, low strength.

IMPALCO PA15.
M-1547; 10.0-13.0 Si, 0.10 max Cu, 0.6 max Fe, 0.5 max Mn, bal Al.
For brazing applications. Filler alloy.
Non-heat treatable. Good fluidity. Good corrosion resistance.

IMPALCO PA16.
M-1547; 0.10 max Cu, 4.5-6.0 Si, 0.6 max Fe, 0.5 max Mn, bal Al.
For brazing applications. Filler alloy.
Non-heat treatable. Good fluidity and corrosion resistance.

IMPALCO PA17.
M-1547; 0.10 max Cu, 7.0-8.0 Si, 0.6 max Fe, 0.5 max Mn, bal Al.
For brazing applications. Filler alloy.
Non-heat treatable. Good fluidity and corrosion resistance.

IMPALCO PA19.
M-1547; 0.15 max Cu, 0.6 max Si, 0.7 max Fe, 1.0-1.5 Mn, bal Al.
At 25°C: 15,200 TS; 5000 YS; 40 El.
At 200°C: 11,200 TS; 4480 YS; 43 El.
For panelling of buildings, or vehicles, containers, holloware, roofing.
Not hardenable by heat treatment. Good ductility, fair machinability.

IMPAX.
M-111, M-111a; 0.36 C, 1.4 Cr, 1.4 Ni, 0.2 Mo, bal Fe.
Heat treated: 300-330 Brin.
For molds, dies, tools, die casting tools for zinc, plastic molds.
Prehardened Type P20 tool steel.

IMPERATOR 15.
M-1345; 0.3 C, 1 Si, 0.4 Mn, 1.1 Cr, 0.18 V, 3.7 W, bal Fe.
For upsetters, cold headers, crimpers; oil hardened, tough.

IMPERATOR 200.
M-1345; 0.30 C, 2.35 Cr, 0.6 V, 4.25 W, bal Fe.
For upsetters, header dies, crimpers; oil hardened, tough.

IMPERIAL.
M-49; 80 Cu, 20 Ni.
For condenser tubes; corrosion resistant.

IMPERIAL.
M-210; 1.4 C, 0.75 Cr, 4.5 W, bal Fe.
For cutting tools, drills, reamers, textile needles.
Water hardening. Wear resistant.

IMPERIAL ALL-METAL SOLDER.
M-786; Pb, Sn.
For Al solder; high strength.

IMPERIAL C.R. 17.
M-210; 0.12 C, 17 Cr, 0.3 Mn, bal Fe.
Annealed: 78,000 TS; 49,000 YS; 32 El; 50 RA; 160 Brin.
For chemical plant equipment, oil burner parts; corrosion resistant, ferritic.

IMPERIAL C.T.
M-210; 0.32 C, 13 Cr, 0.3 Mn, bal Fe.
Heat treated: 202,000 TS; 190,000 YS; 15 El; 35 RA; 450 Brin.
For surgical instruments, valve bodies, pump parts; impellers; corrosion and heat resistant, hardenable.

IMPERIAL E.Q.
M-210; 0.20 C, 12-14 Cr, bal Fe.
Heat treated: 100,000 TS; 80,000 YS; 30 El; 65 RA: 200 Brin.
For steam valves, pump bodies; stainless, hardenable.

IMPERIAL MAJOR.
M-365; 21-22 W, 4-5 Cr, 13 Co, 1.5 V, 0.5 Mo, bal Fe.
For tools, cutters; high speed steel.

IMPERIAL MANGANESE.
M-210; 12 Mn, 1.2 C, bal Fe.
123,000 TS; 200 Brin.
For tracks, dredger buckets, dipper teeth, gears, sheave wheels, sprockets, tube mill liners; wear resistant; difficult to machine.

IMPERIAL R. 1.
M-210; 0.12 C, 13 Cr, 0.3 Mn, bal Fe.
Heat treated: 85,000 TS; 56,000 YS; 35 El; 75 RA; 175 Brin.
Annealed: 67,000 TS; 38,000 YS; 38 El; 77 RA; 140 Brin.
For shafts, housings, valve bodies, pump casings; corrosion resistant.

IMPERIAL R10.
M-210; C, Ni, Cr, bal Fe.
For plastic mold dies; stainless, hobbed dies.

IMPERIAL S.61.
M-210; 0.3 C, 13 Cr, bal Fe.
For turbine blades, valves, cutlery; stainless, hardenable.

IMPERIAL S.80.
M-210; 0.20 C, 18 Cr, 1.25 Ni, bal Fe.
Heat treated: 130,000 TS; 107,000 YS; 20 El; 45 RA; 270 Brin.
Annealed: 112,000 TS; 80,000 YS; 14 El; 245 Brin.
For steam valves, pump bodies; stainless, hardenable.

IMPERIAL STAINLESS IRON.
M-210; 0.10 C, 13 Cr, bal Fe.
Oil treated: 161,000-116,500 TS; 12-20 El; 37-67 RA; 360-220 Brin.
For dental mirrors, nitric acid equipment, dye vats, heat exchangers, stove fittings; acid and heat resistant, malleable.

IMPERIAL STEEL.
M-Eng.; 6.4 W, 2.1 Mn, 1.6 C, bal Fe.
For tools, cutters; oil hardening.

IMPERIAL TOOL.
M-210; 1.4 C, 0.75 Cr, 4.5 W, bal Fe.
For fast finishing cutters; water hardened.

IMPHY 304 see ICL 472.

IMPHY 304L see ICL 472 BC.

IMPHY 316 see ICL 164.

IMPHY 316L see **ICL 164 BC.**

IMPHY 317L see **ICL 168 BC.**

IMPHY 321 see **ICL 472 T.**

IMPHY 410 see **SOLEIL A2.**

IMPHY 430 see **SOLEIL B4.**

IMPHY A8.
M-771; 0.06 C, 12-14 Cr, 0.2 Al, bal Fe.
Annealed: 70,000 TS; 40,000 YS; 30 El; 160 Brin.
Cold drawn: 85,000 TS; 70,000 YS; 20 El; 185 Brin.
For heat treating boxes, oil refinery equipment, quenching racks.
Type 405 stainless steel, ferritic.

IMPHY AFK.
M-771; Si, bal Fe.
For magnetic equipment; soft magnet for strong or moderate field.

IMPHY ARC-2266 see **ARCM 2266 ETC.**

IMPHY ATG-M.
M-771; C, alloy, bal Fe.
For heat resistant parts, oil refinery equipment; heat resistant.

IMPHY ATG-S.
M-771; C, alloy, bal Fe.
For oil refinery equipment; heat resistant.

IMPHY ATG-S4.
M-771; C, alloy, bal Fe.
For oil refinery equipment; heat resistant.

IMPHY ATG-S7.
M-771; C, alloy, bal Fe.
For oil refinery equipment; heat resistant.

IMPHY ATG-T.
M-771; C, alloy, bal Fe.
For oil refinery equipment; heat resistant.

IMPHY ATG-Z.
M-771; C, alloy, bal Fe.
For oil refinery equipment; heat resistant.

IMPHY ATV-1.
M-771; C, alloy, bal Fe.
For oil refinery equipment; oxidation resistant.

IMPHY ATV-3.
M-771; C, alloy, bal Fe.
For oil refinery equipment; oxidation resistant.

IMPHY ATV-R.
M-771; C, alloy, bal Fe.
For oil refinery equipment, oxidation resistant.

IMPHY ATV-S.
M-771; C, alloy, bal Fe.
For oil refinery equipment; oxidation resistant.

IMPHY CCA 1007 see **SOLEIL 33.**

IMPRESS.
M-336; 0.45 C, 1.0 Si, 1.4 Cr, 2.2 W, 0.2 V, 0.3 Mo, bal Fe.
For impression dies; non-deforming.

IMPROVED KS.
M-Japan; 20 Ni, 15 Ti, 20 Co, bal Fe.
For magnetic and electrical equipment; permanent magnet.

IN-40.
M-1570, M-1540; 0.10 C, 0.25 Si, 1.0 Mn, 0.04-0.075 N, bal Fe.
Rolled: 69,100 TS; 48,600 YS; 32 El.
Heat treated: 60,700 TS; 45,800 YS; 44 El; 79 RA.
For bridges, structures, booms, bus bodies.
Tough and readily welded.

IN-50.
M-1540, M-1570; 0.14 C, 0.30 Si, 1.4 Mn, 0.04-0.075 N, bal Fe.
Rolled: 77,500 TS; 54,400 YS; 29 El.
Heat treated: 72,100 TS; 56,000 YS; 38 El; 75.6 RA.
For bridges, structures, booms, mine and railroad cars.
Good fabricability and weldability.

IN-60.
M-1540, M-1570; 0.17 C, 0.35 Si, 1.3 Mn, 0.3 Cr, 0.04-0.075 N, bal
Rolled: 87,300 TS; 59,700 YS; 35 El; 69 RA.
Heat treated: 90,700 TS; 72,800 YS; 32 El; 68.5 RA.
For bridges, structures, bus and truck bodies, booms, mine cars.
High strength-low alloy steel. Good weldability.

IN-80.
M-1540, M-1570; 0.16 C, 0.25 Si, 0.85 Mn, 1.0 Cr, 0.4 Mo, 0.25 Cu, 0.04-0.075 N, bal Fe.
Heat treated: 107,000 TS; 102,000 YS; 23 El.
For structures, buildings, bridges, mine cars, truck and bus bodies.
High strength-low alloy steel, good weldability.

IN-80N.
M-1540, M-1570; 0.15 C, 0.25 Si, 0.40 Mn, 2.25 Ni, 1.2 Cr, 0.35 Mo, 0.04-0.075 N, bal Fe.
Heat treated: 120,600 TS; 105,200 YS; 27 El; 73 RA.
For mine cars, crushers, bridges, truck and bus bodies, booms.
High strength-low alloy steel, good weldability.

IN-80V.
  M-1540, M-1570; 0.14 C, 0.25 Si, 0.85 Mn, 0.85 Ni, 0.50 Cr, 0.45 Mo, 0.07 V, 0.3 Cu, 0.06 N, bal Fe.
  Heat Treated: 113,000 TS; 103,000 YS; 24 El; 67 RA.
  For bridges, booms, structures, truck and bus bodies, mine cars.
  High strength-low alloy steel, tough and shock resistant.

IN-100.
  M-1491; 8-11 Cr, 13-17 Co, 2-4 Mo, 0.7-1.2 V, 5-6 Al, 4.5-5.5 Ti, 0.18 C, 0.01 B, 0.06 Zr, bal Ni.
  Cast: 142,000 TS; 122,000 YS; 8 El.
  At 1800°F: 75,000 TS; 60,000 YS; 5 El.
  For jet engine parts, turbine blades; for operating temperatures up to 1900°F.

IN-100.
  M-1540, M-1570; 0.16 C, 0.25 Si, 0.85 Mn, 1.2 Ni, 0.6 Cr, 0.55 Mo, 0.07 V, 0.25 Cu, 0.06 N, bal Fe.
  Heat treated: 144,200 TS; 138,000 YS; 20 El.
  For mine cars, bus and truck bodies, booms, derricks.
  High strength-low alloy steel, tough and shock resistant.

IN-102.
  M-1491; 7 Fe, 15 Cr, 3 Cb, 3 Mo, 3 W, 0.5 Al, 0.5 Ti, 0.06 C, B, Zr, bal Ni.
  Annealed: 130,000 TS; 60,000 YS; 45 El.
  At 1200°F: 103,000 TS; 50,000 YS; 52 El; 42 RA.
  For high-temperature, high-pressure steam turbines; for long service up to 1200°F, high strength and ductility.

IN-162.
  M-1491; 0.12 C, 10.0 Cr, 4.0 Mo, 2.0 W, 1.0 Ti, 1.0 Cb, 6.5 Al, 2.0 Ta, 0.02 B, 0.1 Zr, bal Ni.
  Cast superalloy; for gas turbine blades and vanes.

IN-643.
  M-68; 0.50 C, 0.30 Si, 3.0 Fe, 25 Cr, 12 Co, 9 W, 0.5 Mo, 2 Nb, Ti, Zr, Mg, bal Ni.
  Cast: 56-60 h bar TS; 28-30 h bar YS; 8-15 El (1 h bar = 10 MPa).
  Heat resisting alloy for high temperature operation including carburization equipment.

IN-731.
  M-68, M-1499, M-1491; 0.18 C, 0.2 max Mn, 0.2 max Si, 9.5 Cr, 10.0 Co, 2.5 Mo, 4.65 Ti, 5.5 Al, 0.015 B, 0.06 Zr, 0.5 max Fe, 0.95 V, bal Ni.
  Modified IN-100.
  Jet engine blades and wheels.

IN 738.
  M-68, M-1499, M-1491; 0.17 C, 16 Cr, 8.5 Co, 1.75 Mo, 2.6 W, 0.9 Cb, 3.4 Ti, 3.4 Al, 0.01 B, 0.10 Zr, 1.75 Ta, bal Ni.
  Cast good hot corrosion resistant alloy.

IN-744.
  M-68; 0.06 max C, 26 Cr, 6.5 Ni, 0.40 Mn, 0.40 Si, Ti = 5 x C, bal Fe.
  At 1000°F: 53,000 psi TS UTS; 45,000 YS; 27 El.
  Good corrosion resistance; high temperature properties; weldable.

IN-787.
  M-68; 0.04 C, 0.5 Mn, 0.3 Si, 0.90 Ni, 0.60 Cr, 0.20 Mo, 1.2 Cu, 0.05 Cb, bal Fe.
  Precipitation hardened and aged: 90,000 TS; 85,000 YS; 25 El.
  Plate or welded pipe for gas or liquid transmission; good low temperature properties.

IN-787.
  M-1602; 0.05 C, 0.55 Mn, 0.26 Si, 0.91 Ni, 0.75 Cr, 0.20 Mo, 1.23 Cu, 0.06 Cb, bal Fe.
  Age-hardened: 78,000-88,000 psi TS; 65,000-75,000 psi YS; 28 El; 78 RA; 180-200 Charpy (-50°F).
  Weld neck flanges for low temperature service.

IN-792.
  M-68, M-1491; 12.7 Cr, 9.0 Co, 2.0 Mo, 3.9 W, 3.9 Ta, 3.2 Al, 4.2 Ti, 0.02 B, 0.10 Zr, 0.21 C, bal Ni.
  1600°F: 115,000 psi TS; 85,000 YS; 15 El.
  Good high stress rupture properties.
  (See also PM IN-792).

IN-833.
  M-68; 0.03 max C, 1.0 max Si, 0.5 max Mn, 7.0 Ni, 11.5 Cr, 0.1 Al, 0.1 Ti, bal Fe.
  Cast, aged: 120,000-150,000 psi TS; 100,000-137,000 psi YS; 14-19 El.
  Age hardening, weldable, corrosion resistant; good elevated temperature properties.

IN-853 (MA-753).
  M-1491; 0.06 C, 19.7 Cr, 2.3 Ti, 0.88 Al, 0.007 B, 0.07 Zr, 1.18 Y_2O_3, 1.12 Al_2O_3, bal Ni.
  Mechanically alloyed powder, for turbine blades and vanes.

INAFOND C4S.
  M-1445; 4.5 Cu, 2-3 Si, 0.45 max Fe, 0.1 max Ti, bal Al.
  Heat treated: 36,000-40,000 TS; 18,000-23,000 YS; 6-3 El; 75-85 Brin.
  For light alloy parts; age-hardenable.

INAFOND C5.
  M-1445; 4.2-5.0 Cu, 0.3 Mg, 0.2 Ti, bal Al.
  Heat treated: 47,000-54,000 TS; 28,000-37,000 YS; 14-7 El; 100-115 Brin.
  For light alloy parts; age-hardenable.

INAFOND C8.
M-1445; 7.0-8.5 Cu, 0.6 max Fe, 0.5 max Si, 0.2 Ti, bal Al.
Cast: 20,000-26,000 TS; 11,000-13,500 YS;; 5-2 El; 55-70 Brin.
For light alloy parts.

INAFOND C41.
M-1445; 4.5 Cu, 0.8 max Fe, 1 max Si, 0.2 max Ti, bal Al.
Heat treated: 36,000-43,000 TS; 28,000-33,000 YS; 4-2 El; 75-95 Brin
For light alloy parts; age-hardenable.

INAFOND S7.
M-1445; 7 Si, 0.3 Mg, 0.5 Mn, 0.15 Ti, bal Al.
Cast: 37,000-43,000 TS; 25,000-30,000 YS; 10-6 El; 90-110 Brin.
For light alloy parts; corrosion resistant.

INAFOND S9.
M-1445; 9 Si, 0.35 Mg, 0.5 Mn, bal Al.
Cast: 35,000-43,000 TS; 28,000-37,000 YS; 5.5-3.5 El; 80-95 Brin.
For light alloy parts; corrosion resistant.

INAFOND S12.
M-1445; 11.5-13.0 Si, 0.3 Mg, 0.5 Mn, bal Al.
Cast: 34,000-46,000 TS; 28,000-40,000 YS; 4-1 El; 80-110 Brin.
For light alloy parts; corrosion resistant.

INAFOND S13.
M-1445; 12.75-13.25 Si, bal Al.
Cast: 31,000-34,000 TS; 14,000-17,000 YS; 3.0-1.5 El; 60-80 Brin.
For light alloy parts; corrosion resistant.

INAFOND S52.
M-1445; 1.3 Cu, 5 Si, 0.5 Mg, bal Al.
Heat treated: 50,000-57,000 TS; 40,000-46,000 YS; 5-2 El; 110-140 Brin.
For light alloy parts; age-hardenable.

INAFOND S71.
M-1445; 7 Si, 0.3 Mg, bal Al.
Cast: 37,000-43,000 TS; 25,000-30,000 YS; 10-6 El; 90-110 Brin.
For light alloy parts; corrosion resistant.

INAFOND S131.
M-1445; 0.8 Cu, 12.0-13.3 Si, 0.3 Mn, bal Al.
Cast: 31,000-37,000 TS; 17,000-20,000 YS; 3.0-1.5 El; 60-80 Brin.
For light alloy parts; corrosion resistant.

INAFOND S132.
M-1445; 1.75-2.25 Cu, 12.0 -13.3 Si, 0.3 Mn, bal Al.
Cast: 33,000-23,000 TS; 18,000-23,000 YS; 2.5-1.0 El; 65-85 Brin.
For light alloy parts; corrosion resistant.

INAFOND Z5F.
M-1445; 1 Fe, 0.6 Mg, 5 Zn, 0.2 Ti, bal Al.
Heat treated: 42,000-50,000 TS; 24,000-30,000 YS; 14-9 El; 90-105 Brin.
For light alloy parts; heat treatable.

INALIUM.
M-623; 0.5 Si, 1.2 Mg, 1.7 Cd, bal Al.
26,000-29,000 TS; 10,000-18,500 YS; 22-18 El; 50-60 Brin.
For light alloy parts.

INALIUM.
M-1394; 0.8 Mg, 0.25 Fe, 0.45 Si, 0.18 Zn, 0.08 W, bal Al.
For light alloy parts.

INAMEL.
M-67; 0.008 C, 0.60 max Mn, 0.040 max S, bal Fe.
Decarburized steel intended for 1 coat enameling application.

INCO 13.
M-68; 5 Al, bal Ni.
For jet engine components; corrosion and heat resistant.

INCO 220 NICKEL now **NICKEL 220.**

INCO 425.
M-68; 0.04 C, 1.3 Mn, 0.75 Si, 5.5 Cr, 25.5 Ni, 2.4 Ti, 0.65 Al, 63.8 Fe.

INCO 546 now **INCONEL ALLOY 721.**

INCO 550 now **INCONEL ALLOY 751.**

INCO 700.
M-68; 0.14 C, 15.0 Cr, 44.0 Ni, 30.0 Co, 3.0 Mo, 2.35 Ti, 3.10 Al, 1.5 Fe, 0.005 B, 0.05 Zr.

INCO A-2.
M-68; 3.4 T.C., 2.3 Si, 0.5 Mn, 1.0 Ni, 1.0 Cr, bal Fe.
For bottom plates in chain drag conveyors; heat resistant.

INCO A-15.
M-68; 3.3 C, 1.5 Ni, 2.0 Si, bal Fe.
Cast: 40,000 TS.
For machine tool castings; cast iron.

INCO "B" MONEL now **MONEL ALLOY 400.**

INCO F-NICKEL.
M-68; 5-6 Si, 1-2 Fe, 0.2 Cu, 0.025 S, 90-92 Ni.
For alloying cast iron; Ni additions.

INCO IN-100.
M-68; 8-11 Cr, 13-17 Co, 2-4 Mo, 0.7-1.2 V, 5-6 Al, 4.5-5.5 Ti, 0.18 C, 0.01 B, 0.06 Zr, bal Ni.
Cast: 142,000 TS; 122,000 YS; 8 El.
At 1800°F: 75,000 TS; 60,000 YS; 5 El.
For jet engine parts, turbine blades; for operating temperatures up to 1900°F.

INCO IN-102.
M-68, M-1491; 7 Fe, 15 Cr, 3 Cb, 3 Mo, 3 W, 0.5 Al, 0.5 Ti, 0.06 C, B, Zr, bal Ni.
Ann.: 130,000 TS; 60,000 YS; 45 El.
At 1200°F: 103,000 TS; 50,000 YS; 52 El.
For high-temperature, high-pressure steam turbines; for long service up to 1200°F; high strength and ductility.

INCO IN-732.
M-68; 30 Ni, 2.8 Cr, bal Cu.
Air Cooled: 55,000 YS; 150 Charpy Impact; 90,000 TS; 30 El; 60 RA.
For marine hardware.
High strength, tough, corrosion resistant, good weldability.

INCO IN-738.
M-68; 0.17 C, 8.5 Co, 16 Cr, 1.75 Mo, 2.6 W, 1.75 Ta, 0.9 Cb, 3.4 Al, 3.4 Ti, 0.01 B, 0.1 Zr, bal Ni.
Heat treated: 160,000 TS; 140,000 YS; 4 El; 4 RA.
At 1200°F: 155,000 TS; 130,000 YS; 3 El; 5 RA.
For jet engine and gas turbine components, blades, vanes, integral-wheel configuration.
Precipitation hardenable. Sulfidation resistant. High rupture strength at elevated temperatures.

INCOLOY now INCOLOY ALLOY 800.

INCOLOY 65 (FILLER METAL).
M-121; 0.03 C, 42.0 Ni, 0.7 Mn, 30.0 Fe, 0.3 Si, 0.007 S, 1.7 Cu, 21.0 Cr, 1.0 Ti, 3.0 Mo.
Filler metal for gas-shielded arc welding INCOLOY alloy 825.
MIL-E-21562.

INCOLOY 135 (WELDING ELECTRODE).
M-121; 0.05 C, 36.0 Ni, 2.0 Mn, 26.0 Fe, 0.008 S, 0.4 Si, 1.8 Cu, 29.0 Cr, 3.75 Mo.
Welding electrode for shielded metal-arc welding INCOLOY alloy 825.

INCOLOY ALLOY 800.
M-1499, M-121, M-1491; 32.5 Ni, 0.05 C, 0.75 Mn, 46.0 Fe, 0.008 S, Si, 0.38 Cu, 21.0 Cr, 0.38 Al, 0.38 Ti.
Cold drawn: 100,000-150,000 TS; 75,000-125,000 YS; 30-10 El; 180-300
For heat exchangers and process piping; carburizing fixtures and retorts, electric range element sheathing; resistant to elevated temperature oxidation and carburization.

INCOLOY ALLOY 800H.
M-1499; 32.5 Ni, 0.08 C, 0.8 Mn, 46.0 Fe, 0.008 S, 0.5 Si, 0.4 Cu, 21.0 Cr, 0.4 Al, 0.4 Ti.
Ann.: 80,000 TS; 30,000 YS; 50 El.
For petrochemical process piping with steam-generator components.
Variation of INCOLOY 800 with controlled carbon content and grain size.

INCOLOY ALLOY 800 H.
M-121; 0.8 C, 32.0 Ni, 21.0 Cr, 45.0 Fe, 0.4 Ti, 0.3 Al.
Sol. Treat 1150°C: 460 N/mm^2 TS; 190 N/mm^2 YS; 16 El.
For high temperature applications.

INCOLOY ALLOY 801.
M-1499, M-121, M-1491; 32.0 Ni, 0.05 C, 0.75 Mn, 44.5 Fe, 0.008 S; 0.50 Si, 0.25 Cu, 20.5 Cr, 1.13 Ti.
For jet engine combustion liners and transition pieces. High temperature tensile and rupture strength. Formerly INCOLOY T.

INCOLOY ALLOY 802.
M-1499, M-121, M-1491; 32.5 Ni, 0.35 C, 0.75 Mn, 46.0 Fe, 0.008 S, 0.38 Si, 21.0 Cr, 0.58 Al, 0.75 Ti.
Hot finished and annealed: 80,000-105,000 TS; 35,000-50,000 YS; 47-18 El; 140-188 Brin.
For hot sizing dies for aerospace industry, connecting pins for cast link heat-treating furnace belts; good high temperature strength and corrosion resistance.

INCOLOY ALLOY 804.
M-1499, M-1491; 41.0 Ni, 0.25 C, 0.75 Mn, 25.4 Fe, 0.008 S, 0.38 Si Cu, 29.5 Cr, 0.30 Al, 0.60 Ti.
Annealed: 95,000 TS; 45,000 YS; 40 El.
High temperature strength; resistant to carburization and sulfidation.

INCOLOY ALLOY 805.
M-1499; 36.0 Ni, 0.12 C, 0.6 Mn, 54.5 Fe, 0.008 S, 0.50 Si, 0.10 Cu, 7.5 Cr, 0.5 Mo.

INCOLOY ALLOY 825.
MIL-E-21562.

INCOLOY ALLOY 825.
M-1499, M-121, M-1491; 42.0 Ni, 0.03 C, 0.50 Mn, 30.0 Fe, 0.015 S, 2.25 Cu, 21.5 Cr, 0.10 Al, 0.90 Ti, 3.0 Mo.
Annealed: 85,000-105,000 TS; 35,000-65,000 YS; 50-30 El; 120-180 Bri
For phosphoric acid evaporators, pickling tank heaters, pickling hooks and equipment, propeller shafts, tank trucks; corrosion resistant.
Formerly Ni-O-NEL alloy 825.

INCOLOY ALLOY 825 CP.
M-121; 0.04 C, 42.0 Ni, 2.2 Cu, 21.5 Cr, 30.0 Fe, 3.0 Mo, 0.9 Nb.
Very good corrosion resistance; equipment for phosphoric and sulphuric acids, sea water, nuclear fuel element recovery plant.

INCOLOY ALLOY 840.
M-1499; 21.0 Ni, 58.0 Fe, 19.0 Cr, 0.8 Si, 0.03 C.
Ann.: 90,000 TS; 50,000 YS; 35 El.
For electric heater element sheathing.

INCOLOY ALLOY 901.
 M-1499, M-1491; 0.05 C, 12.8 Cr, 43 Ni, 5.7 Mo, 2.4 Ti, 35 Fe.
 Heat treated: 168,000 TS; 110,000 YS; 23 El.
 At 1200°F: 125,000 TS; 95,000 YS; 5 El.
 For aircraft gas turbine blades, turbine discs; age hardenable, up to 1400°F service.

INCOLOY ALLOY 903.
 M-121, M-1280; 0.02 C, 38.0 Ni, 41.0 Fe, 15.0 Co, 1.4 Ti, 0.7 Al, 3.0 Nb.
 Low coefficient of expansion, high strength, constant modulus of elasticity and resistance to thermal fatigue and shock from -240 to + 650°C.
 For instrumentation, rocket engines.

INCOLOY ALLOY 903.
 M-1499; 38.0 Ni, 15.0 Co, 0.7 Al, 1.4 Ti, 3.0 Cb, 41.0 Fe.
 Heat treated: 190,000 TS; 160,000 YS; 14 El.
 At 1200°F: 145,000 TS; 130,000 YS; 18 El.
 For rocket-engine thrust chambers, steam-turbine bolts, and ordnance hardware.

INCOLOY ALLOY 904.
 M-121; 0.02 C, 33.0 Ni, 50 Fe, 14.0 Co, 1.7 Ti.
 Low coefficient of expansion; good strength.
 For compensating members in gas turbine engines.

INCOLOY ALLOY MA 956.
 M-121; 20.0 Cr, 74.4 Fe, 0.5 Ti, 4.5 Al, 0.5 Y_2O_3.
 At 700°C; 330 N/mm² TS; 15 El.
 Dispersion strengthened alloy as sheet; resistant to oxidation, carburization and hot corrosion.
 For gas turbine combustion chambers.

INCOLOY ALLOY MA 956.
 M-1499; 75.0 Fe, 20.0 Cr, 4.5 Al, 0.35 Ti, 0.5 Y_2O_3.
 For combustion chambers in gas turbines and components in fossil-fuel burners.
 High-temperature oxidation and sulfidation resistance. Melting point approximately 2700°F.

INCOLOY DS.
 M-121; 0.15 max C, 2.0-2.5 Si, 36-39 Ni, 17-19 Cr, bal Fe.
 At 68°F: 105,300 TS; 38 El; 50 RA.
 At 1112°F: 66,100 TS; 45 El; 38 RA.
 At 1832°F: 12,300 TS; 124 El; 83 RA.
 For furnace parts, high temperature equipment; useful to 950°C in oxidizing atmosphere.

INCOLOY FILLER METAL 65.
 M-1499; 42.0 Ni, 0.03 C, 0.70 Mn, 30.0 Fe, 0.007 S, 0.30 Si, 1.7 Cu, 21.0 Cr, 1.0 Ti, 3.0 Mo.
 For gas-shielded arc welding.

INCOLOY T now **INCOLOY ALLOY 801.**

INCOLOY WELDING ELECTRODE 135.
 M-1499; 36.0 Ni, 0.05 C, 2.0 Mn, 26.0 Fe, 0.008 S, 0.40 Si, 1.8 Cu, 29.0 Cr, 3.75 Mo.
 Electrode for shielded metal-arc welding of Incoloy alloy 825 to itself and to steels.

INCOMET NICKEL.
 M-68; 95.0 Ni, 1.3 Co, 0.4 Cu, 0.4 Fe, 1.1 O_2.
 Intermediate quality granular nickel for most commercial alloying uses.

INCOMPARABLE.
 M-1409; 0.8 C, 22 W, 4.5 Cr, 1.5 V, bal Fe.
 For form tools, cutters, lathe and planer tools; high speed steel.

INCONEL now **INCONEL ALLOY 600.**

INCONEL 62 see **INCONEL FILLER METAL 62.**

INCONEL 62 (FILLER METAL).
 M-121; 0.02 C, 74.0 Ni, 0.1 Mn, 7.5 Fe, 1.0 Si, 0.005 S, 0.03 Cu, 16.0 Cr, 2.25 Nb.
 Filler metal for gas-shielded arc welding INCONEL alloy 600 and overlaying on steel.
 AWS A5.14 (Class ERNiCrFe-5); AMS 5679.

INCONEL 69 (FILLER METAL).
 M-121; 0.04 C, 73.0 Ni, 0.55 Mn, 6.5 Fe, 15.2 Cr, 0.007 S, 0.3 Si, 0.05 Cu, 0.7 Al, 2.5 Ti, 0.85 Nb.
 Filler metal for gas tungsten-arc welding INCONEL alloy X-750.
 AWS A5.14 (Class ERNiCrFe-7); AMS 5778.

INCONEL 82 (FILLER METAL).
 M-121; 0.02 C, 72.0 Ni, 3.0 Mn, 1.0 Fe, 0.2 Si, 0.04 Cu, 20.0 Cr, 0.55 Ti, 2.5 Nb, 0.007 S.
 Filler metal for gas-shielded arc welding INCONEL alloy 600 and INCOLOY alloy 800 to themselves and to steel.
 AWS A5.14 (Class ERNiCr-3); BS 2901-NA-35.

INCONEL 92 (FILLER METAL).
 M-121; 0.03 C, 71.0 Ni, 2.3 Mn, 6.6 Fe, 0.1 Si, 0.007 S, 0.04 Cu, 16.4 Cr, 3.2 Ti.
 Filler metal for gas-shielded arc welding dissimilar alloys, such as austenitic and ferritic steels, to each other and to high nickel alloys, and overlaying on steel.
 AWS A5.14 (Class ERNiCrFe-6); BS 2901-NA-39.

INCONEL 112.
 M-121; 0.05 C, 0.3 Mn, 4.0 Fe, 0.01 S, 0.4 Si, 21.5 Cr, 9.0 Mo, 3.65 Nb + Ta, Ni 61.0.
 Welding electrode for shielded-arc welding INCONEL alloy 625 and dissimilar nickel and iron-base alloys.
 MIL-E-22200/3.

INCONEL 132.
 M-897; 0.2 max C, 6-10 Fe, 14-17 Cr, bal Fe.
 For heat resistant parts; high heat resistance.

INCONEL 132 (WELDING ELECTRODE).
M-121; 0.04 C, 0.75 Mn, 8.5 Fe, 73.0 Ni, 0.006 S, 0.2 Si, 0.04 Cu, 15.0 Cr, 2.1 Nb.
Welding electrode for shielded metal-arc welding INCONEL alloy 600.
AWS A5.11 (Class ENiCrFe-1); AMS 5684.

INCONEL 182 (WELDING ELECTRODE).
M-121; 0.05 C, 67.0 Ni, 7.75 Mn, 7.5 Fe, 0.008 S, 0.5 Si, 0.1 Cu, 14.0 Cr, 0.4 Ti, 1.75 Nb.
Welding electrode for shielded metal-arc welding INCONEL alloy 600, dissimilar nickel and iron-base alloys, and overlaying.
AWS A5.11 (Clas ENiCrFe-3); ASME SFB 5.11.

INCONEL 601 (FILLER METAL).
M-121; 0.05 C, 60.5 Ni, 0.5 Mn, 14.1 Fe, 0.25 Si, 0.007 S, 0.5 Cu, 23.0 Cr, 1.35 Al.
Filler metal for gas tungsten-arc welding INCONEL alloy 601.

INCONEL 604.
M-1280; 0.04 C, 74.0 Ni, 0.20 Mn, 7.2 Fe, 0.20 Si, 0.10 Cu, 15.8 Cr Cb.
For woven belts for furnaces, steam turbine nozzle parts.

INCONEL 625 (FILLER METAL).
M-121; 0.05 C, 61.0 Ni, 0.25 Mn, 2.5 Fe, 0.25 Si, 0.008 S, 21.5 Cr, 0.2 Al, 0.2 Ti, 9.0 Mo, 3.65 Nb + Ta.
Filler metal for gas-shielded arc welding INOCONEL alloy 625, dissimilar alloys, and for overlaying on steel.
AMS 5837.

INCONEL 718 (FILLER METAL).
M-121; 0.04 C, 52.5 Ni, 0.2 Mn, 18.5 Fe, 0.3 Si, 0.007 S, 0.07 Cu, 18.6 Cr, 0.4 Al, 0.9 Ti, 3.1 Mo, 5.0 Nb.
Filler metal for gas tungsten-arc welding INCONEL alloys 718 and X-750.
AMS 5832.

INCONEL ALLOY 600.
M-1499, M-121, M-1491; 76.0 Ni, 0.08 C, 0.5 Mn, 8.0 Fe, 0.008 S, 0. 0.25 Cu, 15.5 Cr.
Annealed: 80,000-100,000 TS; 30,000-50,00 YS; 50-35 El; 120-170 Brin.
Cold drawn: 105,000-150,000 TS; 80,000-125,000 YS; 30-10 El; 180-300 Brin.
For furnace muffles, electronic components, heat exchanger tubing, nuclear reactors, springs, jet engine parts; good high temperature properties.

INCONEL ALLOY 601.
M-1499, M-121, M-1491; 60.5 Ni, 0.05 C, 0.5 Mn, 14.1 Fe, 0.007 S, 0.25 Si, 0.50 Cu, 23.0 Cr, 1.35 Al.
HR, Annealed: 86,000-107,000 TS; 29,000-48,000 YS; 46-60 El.
For heat treating baskets and fixtures, radiant furnace tubes, thermocouple protection tubes, furnace muffles and retorts; good high temperature strength and corrosion resistance.

INCONEL ALLOY 617.
M-121, M-1280; 0.07 C, 54.0 Ni, 22.0 Cr, 12.5 Co, 9.0 Mo, 1.0 Al.
At 1000°C: 150 N/mm^2 TS; 120 N/mm^2 YS; 120 El.
Good high temperature strength and oxidation resistance; for equipment such as gas turbines.

INCONEL ALLOY 617.
M-1499; 54.0 Ni, 22.0 Cr, 12.5 Co, 9.0 Mo, 1.0 Al, 0.07 C.
Ann.: 110,000 TS; 45,000 YS; 60 El.
For gas turbine engines and nuclear reactor components.

INCONEL ALLOY 625.
M-1499, M-121; 61.0 Ni, 0.05 C, 0.25 Mn, 2.5 Fe, 0.008 S, 0.25 Si, 21.5 Cr, 0.2 Al, 0.2 Ti, 9.0 Mo, 3.65 Cb + Ta.
Sol ann: 105,000-130,000 TS; 42,000-60,000 YS; 65-40 El; 116-194 Brin.
As rolled: 120,000-160,000 TS; 60,000-110,000 YS; 60-30 El; 175-240 Brin.
For ducting systems, combustion systems, fuel nozzles, after burners; high strength and corrosion resistance.

INCONEL ALLOY 671.
M-121; 0.05 C, 51.0 Ni, 48.0 Cr, 0.35 Ti.
At 700°C: 480 N/mm^2 TS; 290 N/mm^2 YS.
Resists high temperature corrosion.
For parts subject to fuel ash; pulp and paper industries.

INCONEL ALLOY 671.
M-1499; 52.5 Ni, 46.5 Cr, 0.45 Ti, 0.07 C.
Ann.: 125,000 TS; 70,000 YS; 25 El.
For boiler superheater tube shields, soot-blower tubes, and hangers.
High-temperature corrosion resistance, particularly fuel-ash corrosion in atmospheres containing sulphur and vanadium.

INCONEL ALLOY 690.
M-1499; 60.0 Ni, 30.0 Cr, 9.5 Fe, 0.03 C.
Ann.: 105,000 TS; 50,000 YS; 40 El.
For equipment for steel pickling and reprocessing of spent nuclear fuels.
Resistance to oxidizing chemicals, high-temperature sulphur-containing gases, and nitric or nitric/hydrofluoric acid solutions.

INCONEL ALLOY 702.
M-1499, M-1491; 79.5 Ni, 0.05 C, 0.50 Mn, 1.0 Fe, 0.005 S, 0.35 Si, Cu, 15.5 Cr, 3.25 Al, 0.63 Ti.

For afterburner liners, furnace components and fixtures; oxidation resistant to 2400°F.

INCONEL ALLOY 706.
M-1499, M-121, M-1491; 41.5 Ni, 0.3 C, 0.18 Mn, 40.0 Fe, 0.008 S, 0.18 Si, 0.15 Cu, 16.0 Cr, 0.20 Al, 1.75 Ti, 2.9 Cb + Ta.

Sol. treated and aged: 185,000-193,000 TS; 143,000-161,000 YS; 20-19 El.

For gas turbine components; high strength, machinable and weldable.

INCONEL ALLOY 718.
M-1499, M-121, M-1491, M-1602; 52.5 Ni, 0.04 C, 0.18 Mn, 18.5 Fe, 0.008 S, 0.18 Si, 0.15 Cu, 19.0 Cr, 0.50 Al, 0.90 Ti, 3.05 Mo, 5.13 Cb + Ta.

HR & Aged: 203,500 TS; 181,000 YS; 16 El; 393 Brin.

For jet engines, pump bodies and parts, rocket motors and thrust reversers, space craft.

High strength, weldable in aged condition.

INCONEL ALLOY 721.
M-1499; 71.0 Ni, 0.04 C, 2.25 Mn, 4.0 Fe, 0.005 S, 0.08 Si, 0.10 Cu, 16.0 Cr, 3.05 Ti.

For internal combustion engine valves; age-hardenable; high creep strength.

Formerly INCONEL M.

INCONEL ALLOY 722.
M-1499; 75.0 Ni, 0.04 C, 0.50 Mn, 7.0 Fe, 0.005 S, 0.35 Si, 0.25 Cu 15.5 Cr, 0.7 Al, 2.38 Ti.

For jet engine components; age hardenable; high temperature strength and oxidation and corrosion resistant.

Formerly INCONEL W.

INCONEL ALLOY 751.
M-1499; 72.5 Ni, 0.05 Cr, 0.5 Mn, 7.0 Fe, 0.005 S, 0.25 Si, 0.25 Cu, 15.5 Cr, 1.2 Al, 2.3 Ti, 0.95 Cb + Ta.

For diesel exhaust valves; high rupture strength at 1600°F.

INCONEL ALLOY MA 754.
M-121; 0.05 C, 20.0 Cr, 0.5 Ti, 0.3 Al, 0.6 Y_2O_3, bal Fe.

At 700°C: 440 N/mm² TS; 45 El.

Dispersion hardened alloy; for gas turbine engine guide vanes.

INCONEL ALLOY MA 754.
M-1499; 78.0 Ni, 20.0 Cr, 0.5 Ti, 0.3 Al, 0.05 C, 0.6 Y_2O_3.

For vanes in aircraft gas-turbine engines.

High stress rupture strength to 2100°F.

INCONEL ALLOY X-750.
M-1499, M-121, M-1491; 73.0 Ni, 0.04 C, 0.50 Mn, 7.0 Fe, 0.005 S, 0.25 Si, 0.25 Cu, 15.5 Cr, 0.7 Al, 2.5 Ti, 0.95 Cb + Ta.

Hot finished and aged: 170,000-206,000 TS; 100,000-163,000 YS; 30-15 El; 302-400 Brin.

For aviation and industrial gas turbine parts, springs for steam service and nuclear reactors, bolts, heat treating fixtures; corrosion and oxidation resistant; age hardenable; high creep-rupture strength.

INCONEL CY 40.
M-775; 0.40 max C, 1.5 max Mn, 3.0 max Si, 14.0-17.0 Cr, 11.0 max Fe, bal Ni.

Alloy casting; corrosion and temperature resistant.

ACI CY-40; ASTM A-296 CY-40.

INCONEL FILLER METAL 62.
M-1499; 74.0 Ni, 0.02 C, 0.10 Mn, 7.5 Fe, 0.005 S, 0.10 Si, 0.03 Cu 16.0 Cr, 2.25 Cb.

For gas-shielded arc welding of Inconel alloy 600 and Incoloy alloy 800; overlaying on steel.

AWS A 5.14 Class ERNiCrFe-5; AMS 5679.

INCONEL FILLER METAL 69.
M-1499; 73.0 Ni, 0.04 C, 0.55 Mn, 6.5 Fe, 0.007 S, 0.30 Si, 0.05 Cu 15.2 Cr, 0.7 Al, 2.5 Ti, 0.85 Cb.

For gas tungsten-arc welding of Inconel alloys 722 and X750.

AWS A 5.14 Class ERNiCrFe-7; AMS-5778.

INCONEL FILLER METAL 82.
M-1499; 72.0 Ni, 0.02 C, 3.0 Mn, 1.0 Fe, 0.007 S, 0.20 Si, 0.04 Cu, 20.0 Cr, 0.55 Ti, 2.5 Cb.

For gas-shielded arc welding of Inconel alloy 600, Incoloy alloy 800 Inconel alloy 600 to steel, other dissimilar combinations of nickel-base and iron-base alloys; overlaying on steel.

AWS A 5.14 Class ERNiCr-3; ASME SFB 5.14.

INCONEL FILLER METAL 92.
M-1499; 71.0 Ni, 0.03 C, 2.3 Mn, 6.6 Fe, 0.007 S, 0.10 Si, 0.04 Cu, 16.4 Cr, 3.2 Ti.

For gas shielded arc welding of dissimilar alloys such as austenitic and ferritic steels, to each other and to high-nickel alloys; overlaying on steel.

AWS A 5.14 Class ERNiCrFe-6; ASME SFB 5.14.

INCONEL FILLER METAL 601.
M-1499; 60.5 Ni, 0.05 C, 0.50 Mn, 14.1 Fe, 0.007 S, 0.25 Si, 0.50 C 23.0 Cr, 1.35 Al.

For gas tungsten-arc welding of Inconel alloy 601.

INCONEL FILLER METAL 617.
M-1499; 54.0 Ni, 22.0 Cr, 12.5 Co, 9.0 Mo, 1.0 Al, 0.07 C.

For gas-tungsten-arc welding of INCONEL Alloy 617.

INCONEL FILLER METAL 625.
M-1499; 61.0 Ni, 0.05 C, 0.25 Mn, 2.5 Fe, 0.008 S, 0.25 Si, 21.5 Cr 0.2 Al, 0.2 Ti, 9.0 Mo, 3.65 Cb + Ta.
For gas-shielded arc welding of Inconel alloy 625.
AMS 5837.

INCONEL FILLER METAL 718.
M-1499; 52.5 Ni, 0.04 C, 0.20 Mn, 18.5 Fe, 0.007 S, 0.30 Si, 0.07 C 18.6 Cr, 0.40 Al, 0.90 Ti, 5.0 Cb, 3.1 Mo.
For gas tungsten-arc welding of Inconel alloys 718 and X-750.
AMS 5832.

INCONEL M now INCONEL ALLOY 721.

INCONEL W now INCONEL ALLOY 722.

INCONEL X now INCONEL ALLOY X750.

INCONEL X550 now INCONEL 751.

INCONEL WELDING ELECTRODE 112.
M-1499; 61.0 Ni, 0.05 C, 0.3 Mn, 4.0 Fe, 0.010 S, 0.40 Si, 21.5 Cr, 9.0 Mo, 3.65 Cb + Ta.
Electrode for shielded metal-arc welding of Inconel alloy 625 and dissimilar nickel-base and iron-base alloys.
MIL-E-22200/3.

INCONEL WELDING ELECTRODE 113.
M-1499; 61.0 Ni, 0.05 C, 1.0 Mn, 4.0 Fe, 0.4 Si, 21.5 Cr, 9.0 Mo, 2.75 Cb + Ta.
For shielded metal-arc welding of nickel steels for cryogenic applications.

INCONEL WELDING ELECTRODE 132.
M-1499; 73.0 Ni, 0.04 C, 0.75 Mn, 8.5 Fe, 0.006 S, 0.20 Si, 0.04 Cu 15.0 Cr, 2.1 Cb, 0.05 Co, 0.05 Ta.
Electrode for shielded metal-arc welding of Inconel alloy 600.
AWS A 5.11 Class ENiCrFe-1; AMS 5684.

INCONEL WELDING ELECTRODE 182.
M-1499; 67.0 Ni, 0.05 C, 7.75 Mn, 7.50 Fe, 0.008 S, 0.50 Si, 0.10 C 14.0 Cr, 0.40 Ti, 1.75 Cb, 0.08 Co, 0.03 Ta.
Electrode for shielded metal-arc welding of Inconel alloy 600; welding dissimilar combinations of nickel-base and iron-base alloys; overlaying on steel.
AWS A 5.11 Class ENiCrFe-3; ASME SFB 5.11.

INCOR 16.
M-1076; 33 Re, 67 Co.
Energy product (B_dH_d) max x 10^6, 16.0; Residual induction Br (gauss), 8100; Coercive force Hc (oersted), 7900; Intrinsic coercive force Hci (oersted), 18,000. Permanent magnet.

INCO-WELD A ELECTRODE.
M-121, M-1499; 70.0 Ni, 0.03 C, 2.0 Mn, 9.0 Fe, 0.008 S, 0.30 Si, 0.06 Cu, 15.0 Cr, 2.0 Cb, 1.5 Mo, 0.05 Co, 0.03 Ta.
Electrode for shielded metal-arc welding of dissimilar alloys such as austenitic and ferritic steels to each other and to high-nickel alloys; welding Incoloy alloy 800 to itself.
AWS A5.11 Class ENiCrFe-2; ASME SFB 5.11.

INCO-WELD "B".
M-121; 0.2 C, 2.25 Mn, 9.0 Fe, 0.02 S, 0.7 Si, 0.5 Cu, 15.0 Cr, 2.2 Mo, 2.25 Nb, bal Ni.
An AC electrode for manual metal-arc welding 9% nickel steel and dissimilar alloys.

INCO-WELD B ELECTRODE.
M-1499; 70.0 Ni, 0.1 C, 2.0 Mn, 9.0 Fe, 0.3 Si, 15.0 Cr, 2.5 Cb, 2.0 Mo.
For shielded metal-arc welding of 9% nickel steel, using AC.

INCURO 20.
M-1493; In, Cu, Au.
For brazing alloy; M.P. 970-1015°C.

INCURO-60.
M-1493; 60 Au, 32 Cu, 3 In.
For brazing Kovar.
M.P. 860-900°C. Low vapor pressure.

INCRA 4040-6.
M-1037; 13.5-16.5 Ni, 9.5-10.5 Al, 0.4-1.0 Fe, 1.0-2.0 Co, bal Cu.
At R.T.: 91,000 TS; 60,000 YS; 5 El; 195 Brin.
For molds and accessory components in the glass industry.
Thermal shock and fatigue resistant. Good oxidation and growth resistance.

INCRAMET-800.
M-Mfg.: Licensees of International Copper Res. Assoc. 13.5-16.5 Ni, 10.75-11.50 Al, 0.4-1.0 Fe, 1.0-2.0 Co, bal Cu.
Cast: 91,000 TS; 60,000 YS; 5 El; 195 Brin.
At 1100°F: 25,000 TS; A 30 Rock.
For glass molds, sheet glass rolls.
Heat and corrosion resistant.

INCRAMUTE I AND II.
M-1256; Cu-Mn.
Alloy with high damping characteristics.

INCRAMUTE 699.
M-279; 44 Mn, 1.9 Al, bal Cu.
HT: 81,000 TS; 44,000 YS; 27 El.
Vibration damping; noise damping; good strength.

INCROLOY.
M-1076; 15 Co, 28 Cr, 1 Si, bal Fe.
Energy product (B_dH_d) max x 10^6, 4.0; Residual induction Br (gauss), 13,000; Coercive force Hc (oersted), 550.
Permanent magnet.

INCUS.
  M-111; 0.55 C, 0.7 Mn, 0.7 Cr, 1.75 Ni, 0.7 Mo, bal Fe.
  For drop forging dies; tough.

INCUSIL 10.
  M-1493; 63 Ag, 27 Cu, 10 In.
  For brazing glass headers.
  M.P. 685-730°C.

INCUSIL 15.
  M-1493; 61.5 Ag, 24 Cu, 14 In.
  For brazing glass headers.
  M.P. 630-705°C.

INDALLOY-1.
  M-1041; 50 In, 50 Sn.
  Cast: 1720 TS; 83 El.
  For solder; M.P. 125°C.

INDALLOY 1E.
  M-1041; 52 In, 48 Sn.
  Melting point 118°C.

INDALLOY 2.
  M-1041; 80 In, 5 Ag, 15 Pb.
  Cast: 2550 psi TS; M.P. 149°C.

INDALLOY-3.
  M-1041; 90 In, 10 Ag.
  Cast: 1650 TS; 61 El.
  For solder; M.P. 237°C.

INDALLOY-4.
  M-1041; 100 In.
  Cast: 575 TS; 41 El.
  For solder; M.P. 157°C.

INDALLOY 5.
  M-1041; 25 In, 37.5 Sn, 37.5 Pb.
  M.P. 181°C.

INDALLOY 6.
  M-1041; 4.76 In, 2.38 Ag, 92.86 Pb.
  M.P. 300°C.

INDALLOY-7.
  M-1041; 50 In, 50 Pb.
  Cast: 4670 TS; 55 El.
  For solder; M.P. 209°C.

INDALLOY 8.
  M-1041; 44 In, 42 Sn, 14 Cd.
  M.P. 93°C.

INDALLOY 9.
  M-1041; 12 In, 70 Sn, 18 Pb.
  Cast: 5320 psi TS; M.P. 162°C.

INDALLOY-10.
  M-1041; 25 In, 75 Pb.
  Cast: 5450 TS; 47.5 El.
  For solder; M.P. 264°C.

INDALLOY-11.
  M-1041; 5 In, Pb.
  Cast: 4330 TS; 52 El.
  For solder; M.P. 314°C.

INDALLOY 12.
  M-1041; 5 In, 5 Ag, 90 Pb.
  M.P. 310°C.

INDALLOY 13.
  M-1041; 70 In, 15 Sn, 9.6 Pb, 5.4 Cd.
  M.P. 125°C.

INDALLOY 15.
  M-1041; 18.33 In, 42.9 Bi, 21.7 Pb, 7.97 Sn, 5.09 Cd, 4 Hg.
  M.P. 43°C.

INDALLOY 16.
  M-1041; 16.1 In, 44.7 Bi, 22.6 Pb, 11.3 Sn, 5.3 Cd.
  M.P. 52°C.

INDALLOY 17.
  M-1041; 20.89 In, 49.14 Bi, 17.92 Pb, 11.55 Sn, 0.5 Cd.
  M.P. 56°C.

INDALLOY 18.
  M-1041; 61.72 In, 30.78 Bi, 7.5 Cd.
  M.P. 61.5°C.

INDALLOY 25.
  M-1041; 41.5 In, 48.5 Bi, 10 Cd.
  M.P. 77.5°C.

INDALLOY 27.
  M-1041; 29.68 In, 54.02 Bi, 16.3 Cd.
  M.P. 81°C.

INDALLOY 51.
  M-1041; 21.5 In, 62.5 Ga, 16 Sn.
  M.P. 10.7°C.

INDALLOY 53.
  M-1041; 33 In, 67 Bi.
  M.P. 109°C.

INDALLOY 60.
  M-1041; 24.5 In, 75.5 Ga.
  M.P. 15.7°C.

INDALLOY 70.
  M-1041; 40 In, 40 Sn, 20 Pb.
  M.P. 130°C.

INDALLOY 71.
  M-1041; 48 In, 52 Sn.
  M.P. 131°C.

INDALLOY 87.
  M-1041; 42 In, 58 Sn.
  M.P. 145°C.

INDALLOY 88.
  M-1041; 99.3 In, 0.7 Ga.
  M.P. 150°C.

INDALLOY 90.
  M-1041; 99.4 In, 0.6 Ga.
  M.P. 152°C.

INDALLOY 91.
M-1041; 99.5 In, 0.5 Ga.
M.P. 153°C.

INDALLOY 92.
M-1041; 99.6 In, 0.4 Ga.
M.P. 154°C.

INDALLOY 117.
M-1041; 19.1 In, 44.7 Bi, 22.6 Pb, 8.3 Sn, 5.3 Cd.
M.P. 47°C.

INDALLOY 136.
M-1041; 21 In, 49 Bi, 18 Pb, 12 Sn.
M.P. 58°C.

INDALLOY 146.
M-1041; 20 In, 79 Pb, 1 Sb.
M.P. 270°C.

INDALLOY 147.
M-1041; 5 In, 47.5 Bi, 25.4 Pb, 12.6 Sn, 9.5 Cd.
M.P. 65°C.

INDALLOY 150.
M-1041; 19 In, 81 Pb.
M.P. 280°C.

INDALLOY 158.
M-1041; 50 Bi, 26.7 Pb, 13.3 Sn, 10 Cd.
M.P. 70°C.

INDALLOY 162.
M-1041; 66.3 In, 33.7 Bi.
M.P. 72°C.

INDALLOY 164.
M-1041; 5 In, 92.5 Pb, 2.5 Ag.
M.P. 300°C.

INDALLOY 174.
M-1041; 26 In, 57 Bi, 17 Sn.
M.P. 79°C.

INDALLOY 178.
M-1041; 20 In, 80 Au.
M.P. 485°C.

INDALLOY 179.
M-1041; 15 In, 61 Ag, 24 Cu.
M.P. 705°C.

INDALLOY 203.
M-1041; 95 In, 5 Bi.
M.P. 150°C.

INDALLOY 204.
M-1041; 70 In, 30 Pb.
Cast: 3450 psi TS; M.P. 174°C.

INDALLOY 205.
M-1041; 60 In, 40 Pb.
Cast: 4150 psi TS; M.P. 185°C.

INDALLOY 206.
M-1041; 40 In, 60 Pb.
Cast: 5000 psi TS; M.P. 225°C.

INDALLOY 253.
M-1041; 74 In, 26 Cd.
M.P. 123°C.

INDALLOY 290.
M-1041; 97 In, 3 Ag.
M.P. 143°C.

INDEFATIGABLE.
M-French; 0.26 C, 4 Ni, 0.6 Cr, 0.9 Mo, bal Fe.
Heat treated: 230,000 TS; 190,000 YS; 6 El.
Annealed: 120,000 TS; 100,000 YS; 16 El.
For rail and switch parts, street car tracks; oil hardening.

INDEX.
M-1322; 1.0 C, 0.2 Si, 0.25 Mn, 0.1 V, bal Fe.
For springs, tools, cutters, reamers; Type W2; water hardened.

INDIE ZP.
M-64; 0.35 C, 0.8 Mn, 0.45 Si, 1.65 Cr, 0.43 Mo, bal Fe.
Oil hardening tool and die steel designed for molds; AISI P20.

INDIUM.
M-1755; In.
Purities: zone refined 99,9999%, 99.999%, 99.99%, 99.9%.
Forms: rod, ingot, shot, powders, wires, sheets, foils, single crystals.

INDOX 1.
M-1076; Sr 0.6 Fe_2O_3 (ceramic permanent magnet).
Normal peak energy product (B_dH_d) max x 10^6, 1.0; Residual induction B_r (gauss), 2200; Coercive force H_c (oerst 1825; Intrinsic coercive force H_{ci} (oersted), 3750.

INDOX 3.
M-1076; Sr 0.6 Fe_2O_3 (ceramic permanent magnet).
Peak normal energy product (B_dH_d) max x 10^6, 2.6; Residual induction B_r (gauss), 3350; Coercive force H_c (oerst 2350; Intrinsic coercive force H_{ci} (oersted), 2550.

INDOX 5.
M-1076; Sr 0.6 Fe_2O_3 (ceramic permanent magnet).
Peak normal energy product (B_dH_d) max x 10^6, 3.4; Residual induction B_r (gauss), 3800; Coercive force H_c (oerst 2400; Intrinsic coercive force H_{ci} (oersted), 2450.

INDOX 6.
M-1076; Sr 0.6 Fe_2O_3 (ceramic permanent magnet).
Peak normal energy product (B_dH_d) max x 10^6, 2.45; Residual induction B_r (gauss), 3300; Coercive force Hc (oersted), 2800; Intrinsic coercive force Hci (oersted), 3300.

INDOX 7.
M-1076; Sr 0.6 Fe$_2$O$_3$ (ceramic permanent magnet).
Normal peak energy product (B$_d$H$_d$) max x 10^6, 2.8; Residual induction B$_r$ (gauss), 3450; Coercive force H$_c$ (oerst 3250; Intrinsic coercive force H$_{ci}$ (oersted), 4000.

INDOX 8.
M-1076; Sr 0.6 Fe$_2$O$_3$ (ceramic permanent magnet).
Normal peak energy product (B$_d$H$_d$), max x 10^6, 3.5; Residual induction B$_r$ (gauss), 3850; Coercive force H$_c$ (oerst 3050; Intrinsic coercive force H$_{ci}$ (oersted), 3150.

INDUSTRIAL CRO-TUNG.
M-440; 0.30 max C, 4.5-6.5 Cr, 0.75-1.0 W, bal Fe.
193,000-100,000 TS; 167,000-85,000 YS; 5-24 El; 6-62 RA; 490-210 Brin.
For corrosion resistant parts; corrosion resistant.

INDUSTRIAL NO. 12.
M-440; 0.15 max C, 11.5-13.5 Cr, bal Fe.
Annealed: 85,000 TS; 60,000 YS; 35 El; 75 RA; 135 Brin.
For valves, cutlery, turbine blades; Type 410; corrosion resistant.

INDUSTRIAL NO. 12T.
M-440; 0.15 max C, 11.5-13.0 Cr, 1 max Mn, bal Fe.
Annealed: 85,000-100,000 TS; 60,000-75,000 YS; 35-25 El; 75-60 RA; 135-165 Brin.
Heat treated: 100,000-200,000 TS; 60,000-175,000 YS; 30-10 El; 75-50 RA; 200-390 Brin.
For chemical plant equipment; Type 403; corrosion resistant.

INDUSTRIAL NO. 17.
M-440; 0.12 max C, 14-18 Cr, 1 max Mn, 1 max Si, bal Fe.
Cold drawn: 130,000 TS; 120,000 YS; 3 El; 10 RA; 270 Brin.
Annealed: 75,000 TS; 35,000 YS; 40 El; 60 RA; 145 Brin.
For auto trim, kitchen sinks, oil refinery equipment; Type 430; corrosion resistant.

INDUSTRIAL NO. 17-7.
M-440; 0.09-0.20 C, 16-18 Cr, 7-9 Ni, 1.2 max Mn, bal Fe.
Annealed: 100,000 TS; 40,000 YS; 60 El; 160 Brin.
1/2 H-temper: 150,000 TS; 110,000 YS; 15 El; 320 Brin.
For chemical plant equipment, structural members; austenitic, stainless.

INDUSTRIAL NO. 18.
M-440; 0.12 max C, 18 Cr, bal Fe.
For corrosion resistant parts; corrosion resistant.

INDUSTRIAL NO. 18-8S.
M-440; 0.08 max C, 18-20 Cr, 8-12 Ni, 2 max Mn, bal Fe.
Annealed: 80,000 TS; 35,000 YS; 60 El; 70 RA; 135 Brin.
Cold drawn: 180,000 TS; 150,000 YS; 10 El; 330 Brin.
For kitchen equipment, welded structures, process equipment; stainless, austenitic; Type 304.

INDUSTRIAL NO. 18-10CB.
M-440; 0.08 max C, 17-19 Cr, 9-13 Ni, Cb = 10 x C, bal Fe.
Annealed: 80,000-90,000 TS; 35,000-45,000 YS; 60-55 El; 70-60 RA; 135-185 Brin.
Cold drawn: 100,000-180,000 TS; 50,000-120,000 YS; 50-10 El; 180-300 Brin.
For chemical plant equipment, welded structures; Type 347; austenitic, stainless.

INDUSTRIAL NO. 18-10 TI.
M-440; 0.08 max C, 17-19 Cr, 9-12 Ni, Ti = 5 x C, bal Fe.
Annealed: 80,000 TS; 35,000 YS; 60 El; 70 RA; 135 Brin.
Cold rolled: 180,000 TS; 120,000 YS; 10 El; 300 Brin.
For welded chemical plant equipment; Type 321; austenitic, stabilized.

INDUSTRIAL NO. 18-12MO.
M-440; 0.08 max C, 16-18 Cr, 10-14 Ni, 2-3 Mo, bal Fe.
Cold drawn: 100,000-150,000 TS; 50,000-125,000 YS; 55-15 El; 180-200 Brin.
Annealed: 80,000-90,000 TS; 35,000-50,000 YS; 70-60 El; 70-60 RA; 135-185 Brin.
For chemical plant equipment, mixers, agitators; Type 316; austenitic, acid resistant.

INDUSTRIAL NO. 19-12-3MO.
M-440; 0.10 max C, 18-20 Cr, 11-15 Ni, 3-4 Mo, bal Fe.
Cold drawn: 100,000-150,000 TS; 50,000-125,000 YS; 50-15 El; 180-300 Brin.
Annealed: 80,000-90,000 TS; 35,000-50,000 YS; 70-60 El; 70-60 RA; 135-185 Brin.
For chemical plant equipment, mixers, agitators; Type 317; austenitic, acid resistant.

INDUSTRIAL NO. 20-10S.
M-440; 0.08 max C, 19-22 Cr, 10-12 Ni, 2 max Mn, bal Fe.
Annealed: 90,000 TS; 40,000 YS; 50 El; 65 RA; 150 Brin.
For chemical plant equipment; stainless, austenitic; Type 308.

INDUSTRIAL NO. 20-25.
M-440; 0.25 max C, 19-21 Cr, 24-26 Ni, bal Fe.
Annealed: 100,000 TS; 50,000 YS; 45 El; 60 RA; 180 Brin.
For furnace parts, heat treat equipment; Type 311; austenitic, heat resistant.

INDUSTRIAL NO. 21.
M-440; 0.35 max C, 18-23 Cr, bal Fe.
Cold drawn: 70,000-80,000 TS; 45,000-55,000 YS; 40-55 El; 60-55 RA; 150-180 Brin.
For furnace parts, chemical plant equipment; Type 442; corrosion resistant.

INDUSTRIAL NO. 25-12.
M-440; 0.20 max C, 22-24 Cr, 12-15 Ni, bal Fe.
Cold drawn: 100,000-250,000 TS; 50,000-200,000 YS; 45-5 El; 60-50 R 160-400 Brin.
Annealed: 90,000-110,000 TS; 35,000-45,000 YS; 55-40 El; 65-50 RA; 170-200 Brin.
For chemical plant equipment, furnace parts; Type 309; stainless, austenitic.

INDUSTRIAL NO. 25-12S.
M-440; 0.08 max C, 22-26 Cr, 12-14 Ni, bal Fe.
Cold drawn: 100,000-250,000 TS; 50,000-200,000 YS; 45-5 El; 60-50 R 160-400 Brin.
For heat treat boxes, furnace parts, kiln linings; Type 309S; austenitic heat resistant.

INDUSTRIAL NO. 25-20.
M-440; 0.25 max C, 24-26 Cr, 19-22 Ni, 2 max Mn, bal Fe.
Annealed: 95,000 TS; 45,000 YS; 50 El; 65 RA; 170 Brin.
For furnace parts, heat treat equipment; Type 310; austenitic, heat resistant.

INDUSTRIAL NO. 27.
M-440; 0.35 max C, 23-30 Cr, bal Fe.
Cold drawn: 75,000-85,000 TS; 50,000-60,000 YS; 35-50 El; 55-50 RA; 150-200 Brin.
For furnace parts, chemical plant equipment; Type 443; corrosion resistant.

INDUSTRIAL NO. 35.
M-440; 0.30-0.45 C, 12-14 Cr, 0.5 max Ni, 0.5 max Mo, bal Fe.
Annealed: 95,000 TS; 60,000 YS; 30 El; 50 RA; 190 Brin.
Heat treated: 300,000 TS; 250,000 YS; 1 El; 3 RA; 620 Brin.
For surgical instruments, valves, cutlery; Type 420; corrosion resistant.

INDUSTRIAL NO. 65.
M-440; 0.60-0.70 C, 16-17 Cr, bal Fe.
Heat treated: 250,000-143,000 TS; 223,000-115,000 YS; 4-13 El; 8-42 RA; 512-302 Brin.
For corrosion resistant parts, cutlery; corrosion resistant.

INDUSTRIAL NO. 100.
M-440; 0.95-1.2 C, 16-18 Cr, 0.75 max Mo, bal Fe.
Annealed: 95,000 TS; 60,000 YS; 30 El; 50 RA; 190 Brin.
Heat treated: 300,000 TS; 250,000 YS; 1 El; 3 RA; 620 Brin.
For bearings, valves, cutlery; corrosion and wear resistant; Type 440C.

INDUSTRIAL NO. 100FM.
M-440; 0.95-1.2 C, 16-18 Cr, 0.75 max Ni, 0.6 max Mo, bal Fe.
Annealed: 95,000 TS; 60,000 YS; 30 El; 50 RA; 190 Brin.
Heat treated: 300,000 TS; 250,000 YS; 1 El; 3 RA; 620 Brin.
For bearings, cutlery, valves; Type 440F; wear resistant.

INDUSTRIAL NO. 188.
M-440; 0.15 max C, 17-19 Cr, 8-10 Ni, bal Fe.
Annealed: 80,000 TS; 35,000 YS; 60 El; 70 RA; 135 Brin.
Cold rolled: 180,000 TS; 150,000 YS; 10 El; 330 Brin.
For chemical plant equipment, tanks, mixers; Type 302; austenitic, stainless.

INDUSTRIAL NO. 188-M.
M-440; 0.2 C, 18 Cr, 8 Ni, 2-4 Mo, bal Fe.
For stainless and corrosion resistant parts and equipment; stainless and corrosion resistant.

INDUSTRIAL NO. 188-U.
M-440; 0.07 max C, 17.5-18.5 Cr, 8.5-9.5 Ni, bal Fe.
113,000-90,000 TS; 84,000-35,000 YS; 40-61 El; 60-68 RA; 223-135 Brin.
For stainless parts, chemical plant equipment; stainless, austenitic.

INDUSTRIAL NO. 512.
M-440; 0.15 max C, 12-14 Cr, 0.15 min S, 0.06 max P, 1.2 Mn, bal Fe.
Heat treated: 100,000-200,000 TS; 60,000-125,000 YS; 30-10 El; 75-50 RA; 200-390 Brin.
Annealed: 85,000-100,000 TS; 60,000-75,000 YS; 35-25 El; 75-60 RA; 135-165 Brin.
For screw machine products, valve trim, pump shafts, Type 416; corrosion resistant.

INDUSTRIAL NO. 512 FREE MACHINING.
M-440; 0.12 max C, 11.5-13.0 Cr, 0.3-0.4 S, bal Fe.
175,000-40,000 TS; 160,000-25,000 YS; 15-35 El; 60-75 RA; 375-150 Brin.
For corrosion resistant parts, new machine products; corrosion resistant; free-cutting.

INDUSTRIAL NO. 517.
M-440; 0.12 max C, 14-18 Cr, 0.07 min S or Se, 0.6 max Mo, bal Fe.
Cold drawn: 75,000-100,000 TS; 45,000-90,000 YS; 34-8 El; 62-58 RA; 140-250 Brin.
For screw machine products, shafts, pumps; Type 430F; corrosion resistant.

INDUSTRIAL NO. 530.
M-440; 0.21-0.35 C, 12-15 Cr, S, bal Fe.
For stainless and corrosion resistant parts; corrosion resistant.

INDUSTRIAL NO. 5188.
M-440; 0.15 max C, 17-19 Cr, 8-10 Ni, 0.15 min S, bal Fe.
Cold drawn: 150,000 TS; 120,000 YS; 10 El; 330 Brin.
Annealed: 80,000 TS; 35,000 YS; 60 El; 75 RA; 130 Brin.
For screw machine products; free-cutting, stainless; Type 303.

INGERSOLL I-50.
M-328; 0.50 C, 0.75 Mn, bal Fe.
For knives, saws; water hardened.

INGERSOLL I-60.
M-328; 0.60 C, 0.75 Mn, bal Fe.
For knives, saws; water hardened.

INGERSOLL I-75.
M-328; 0.75 C, 0.75 Mn, bal Fe.
For knives, saws; water hardened.

INGERSOLL I-85.
M-328; 0.85 C, 0.85 Mn, bal Fe.
For knives, saws; water hardened.

INGERSOLL IC-3C.
M-328; 0.80 C, 0.65 Mn, 0.35 Cr, bal Fe.
For knives, saws; water hardened.

INGERSOLL IC-66.
M-328; 0.70 C, 0.65 Mn, 0.45 Cr, bal Fe.
For knives, saws; water hardened.

INGERSOLL IC-77.
M-328; 0.75 C, 0.75 Mn, 0.45 Cr, bal Fe.
For knives, saws; water hardened.

INGERSOLL IC-88.
M-328; 0.85 C, 0.45 Mn, 0.45 Cr, bal Fe.
For knives, saws; water hardened.

INGERSOLL IC-M.
M-328; 0.90 C, 0.90 Mn, 0.80 Cr, bal Fe.
For knives and saws; water hardened.

INGERSOLL ICN-70.
M-328; 0.70 C, 0.25 Mn, 0.35 Cr, 1.35 Ni, bal Fe.
For knives, saws; oil hardened.

INGERSOLL ICN-75.
M-328; 0.75 C, 0.35 Mn, 0.35 Cr, 1.25 Ni, bal Fe.
For knives, saws, valve discs; oil hardened.

INGERSOLL ICN-80.
M-328; 0.80 C, 0.40 Mn, 0.25 Cr, 0.60 Ni, bal Fe.
For circular saws, hog knives, cordwood saws; water or oil hardened.

INGERSOLL ICN-100.
M-328; 1.0 C, 0.40 Mn, 0.75 Cr, 1.35 Ni, bal Fe.
For circular and hack saws, hog and meat knives; oil hardened.

INGERSOLL ICNM-55.
M-328; 0.70 C, 0.50 Mn, 0.60 Cr, 0.80 Ni, 0.20 Mo, bal Fe.
For circular and hack saws, knives, valve discs; oil hardened.

INGERSOLL ICNM-CS.
M-328; 0.75 C, 0.30 Mn, 0.20 Cr, 0.70 Ni, 0.15 Mo, bal Fe.
For circular and hack saws, knives; oil hardened.

INGERSOLL ICNV-LA.
M-328; 1.25 C, 0.30 Mn, 1.0 Cr, 0.55 Ni, 0.25 V, bal Fe.
For hack and circular saws, knives; oil hardened.

INGERSOLL ICV-50.
M-328; 0.60 C, 0.45 Mn, 1.0 Cr, 0.20 V, bal Fe.
For circular and hack saws, knives; water or oil hardened.

INGERSOLL ICV-98.
M-328; 0.90 C, 0.35 Mn, 0.40 Cr, 0.20 V, bal Fe.
For circular and hack saws, knives; oil or water hardened.

INGERSOLL ICW-90.
M-328; 0.90 C, 1.1 Mn, 0.50 Cr, 0.50 W, bal Fe.
For hack and circular saws, hog knives; oil hardened, tough.

INGERSOLL IN-8.
M-328; 0.75 C, 0.40 Mn, 2.05 Ni, bal Fe.
For knives, saws; oil hardened.

INGERSOLL IN-8B.
M-328; 0.75 C, 0.85 Mn, 2.05 Ni, bal Fe.
For knives, saws; oil hardened.

INGERSOLL IN-9.
M-328; 0.75 C, 0.40 Mn, 2.75 Ni, bal Fe.
For knives, saws; oil hardened.

INGERSOLL IN-9B.
M-328; 0.75 C, 0.85 Mn, 2.75 Ni, bal Fe.
For knives, saws; oil hardened.

INGERSOLL IN-40.
M-328; 0.40 C, 0.80 Mn, 3.5 Ni, bal Fe.
For knives, saws; oil hardened.

INGERSOLL INM-45.
M-328; 0.45 C, 0.50 Mn, 1.4 Ni, 0.35 Mo, bal Fe.
For saws, knives, friction saws; oil hardened.

INGERSOLL INV-75.
M-328; 0.75 C, 0.35 Mn, 2.0 Ni, 0.10 V, bal Fe.
For circular saws, meat knives, drag saws; oil hardened.

INGERSOLL IV-85.
M-328; 0.85 C, 0.25 Mn, 0.25 V, bal Fe.
For friction and hack saws, knives; water hardened.

INGERSOLL M-2.
M-328; 0.8 C, 6 W, 5 Mo, 2 V, 4 Cr, bal Fe.
For hack saw blades; high speed steel.

INHIBITED ADMIRALTY see also **ADMIRALTY BRASS, INHIBITED.**

INHIBITED ALUMINUM BRASS see **ALUMINUM BRASS.**

INJECTOR BRONZE.
M-U.S.; 8.5 Sn, 2.5 Pb, bal Cu.
For injectors, fittings, hardware; free-cutting.

INKUS.
M-111; 0.56 C, 0.85 Cr, 0.2 Mo, 1.8 Ni, 0.10 V, bal Fe.
For forging dies, punches; oil hardened, tough.

INLAND AA.
M-67; 0.10 C, 0.65 Mn, 0.010 P, 0.20 S, 0.10 Si, 0.005 Al, 0.010 N_2, bal Fe.
Cold Rolled: 45,000 YS; 64,000 TS; 26.1 El.
Aged: 70,000 YS; 15 El.
For formed sheet products, sills, containers, vessels, trunk hinges. Age-hardens, readily fabricated.

INLAND COPPER BEARING STEEL.
M-67; 0.05-0.15 C, 0.2-0.3 Cu, bal Fe.
For corrosion resistance applications, roofing; corrosion resistant.

INLAND LEDLOY.
M-67; C, 0.15-0.35 Pb, bal Fe.
For machinery parts; free cutting.

INMANITE.
M-1749; 0.80 C, 4.5 Cr, 20.0 W, 1.0 Mo, 1.5 V, 5.0 Co, bal Fe.
5% cobalt-tungsten high speed tool steel.
For cutting tools for very high speed and for hard materials.

INMET.
M-Poland; Cu, Mn, Ni, Fe, bal Al.
For light alloy parts; age hardened.

INNERSHIELD NR-1.
M-578; C, bal Fe.
Self-shielded flux-cored arc welding electrode.

INNERSHIELD NR-5.
M-578; C, bal Fe.
Self-shielded flux-cored arc welding electrode.

INNERSHIELD NR-131.
M-578; C, bal Fe.
Self-shielded, flux-cored arc welding electrode.
As welded (single pass): 72,000-75,000 TS.
AWS 70T-G.

INNERSHIELD NR-202.
M-578; C, bal Fe.
Self-shielded, flux-cored arc welding electrode.
AWS Class E60T-7.

INNERSHIELD NR-203.
M-578; C, bal Fe.
Self-shielded flux-cored arc welding electrode.
AWS Class E60T-7.

INNERSHIELD NR-203M.
M-578; C, bal Fe.
Self-shielded, flux-cored arc welding electrode.
As welded, (aged): 72,000-78,000 psi TS; 22-23 El; 70-120 Charpy V-Notch.
AWS E70T-G.

INNER SHIELD NR-203-NICKEL.
M-578; C, 2 Ni, bal Fe.
Self-shielded, flux-cored arc welding electrode.
For welding low-temperature alloys.
AWS E70T-G.

INNERSHIELD NR-211.
M-578; C, bal Fe.
Self-shielded, flux-cored arc welding electrode.
As welded (aged): 72,000-78,000 TS; 60,000-67,000 YS; 22-26 El.
AWS E70T-G.

INNERSHIELD NR-301.
M-578; C, bal Fe.
Self-shielded, flux-cored arc welding electrode.
As welded (aged): 72,000-81,000 TS; 60,000-69,000 YS; 22-29 El; 20-40 Charpy V-notch.
AWS E70T-G.

INNERSHIELD NR-311.
M-578; C, bal Fe.
Self-shielded flux-cored arc welding electrode.
AWS Class E70T-G.

INNERSHIELD NR 431.
M-578; C, bal Fe.
Self-shielded, flux-cored arc welding electrode.
As welded: 78,000-82,000 TS approx.

INNERSHIELD NS-3M.
M-578; C, bal Fe.
Self-shielded flux-cored arc welding electrode.
AWS Class E70T-4.

INNOSEIN.
M-France; 5-6 Cu, bal Al.
For light alloy parts; age-hardenable.

INOCULOY 63.
M-1038; 60-65 Si, 9-12 Mn, 1.5-3.0 Ca, 4-6 Ba, 1.0-1.5 Al, bal Fe.
For reducing chill in gray cast iron.
Cast iron inoculant; Ductile Iron inoculant.

INOFO 1.
M-1487.
Stainless steel; similar to AISI 410.

INOFO 1S.
M-1487.
Stainless steel; similar to AISI 416.

INOFO 2.
M-1487.
Stainless steel; similar to AISI 420.

INOFO 4.
M-1487.
Stainless steel; similar to AISI 430.

INOFO 4S.
M-1487.
Stainless steel; similar to AISI 430F.

INOFO 5.
M-1487.
Stainless steel; similar to AISI 431.

INOFO 10.
M-1487.
Stainless steel; similar to AISI 302.

INOFO 10B.
M-1487.
Stainless steel; similar to AISI 304.

INOFO 10TI.
M-1487.
Stainless steel; similar to AISI 321.

INOFO 11.
M-1487.
Stainless steel; similar to AISI 304L.

INOFO 12.
M-1487.
Stainless steel; similar to AISI 316.

INOFO 12B.
M-1487.
Stainless steel; similar to AISI 316 L.

INOFO 12TI.
M-1487.
Stainless steel; similar to AISI 316 Ti.

INOFO 13S.
M-1487.
Stainless steel; similar to AISI 303.

INOR-8.
M-118; 0.06 C, 7 Cr, 17 Mo, 5 max Fe, 0.8 max Mn, 0.5 max Al + Ti, bal Ni.
Heat treated: 115,100 TS; 45,500 YS; 50.7 El.
At 1600°F: 35,000 TS; 25,000 YS; 30 El.
For high temperature applications, resists hot fluoride salts; high oxidation resistance, resists aging and embrittlement.

INOSSIDALFA.
M-658; Al alloy.
For light alloy parts.

INOX 1.
M-884; 0.15 max C, 13 Cr, 0.5 max Ni, bal Fe.
Annealed: 75,000 TS; 40,000 YS; 35 El; 70 RA; 155 Brin.
For turbine blades, cutlery, valves; Type 410; stainless.

INOX 1 F.
M-884; 0.15 max C, 13.0 Cr, 0.20 S, + Mo, bal Fe.
Free machining martensitic stainless.
AISI 416.

INOX 2.
M-884; 0.20 C, 13.5 Cr, 0.5 max Ni, bal Fe.
Annealed: 95,000 TS; 50,000 YS; 25 El; 55 RA; 195 Brin.
For turbine blades, cutlery, valves, surgical instruments; Type 420; stainless, hardenable.

INOX 3.
M-884; 0.30 C, 13.5 Cr, 0.5 max Ni, bal Fe.
Annealed: 100,000 TS; 55,000 YS; 22 El; 52 RA; 210 Brin.
For turbine blades, valves, cutlery, surgical instruments; Type 420; stainless, martensitic.

INOX 16.
M-884; 0.08 C, 16 Cr, 0.5 max Ni, bal Fe.
Annealed: 80,000 TS; 50,000 YS; 25 El; 50 RA; 150 Brin.
For bolts, shafts, oil refinery equipment, heaters; Type 430; stainless, ferritic.

INOX 430 F.
M-884; 0.10 C, 16.5 Cr, 0.20 S, + Mo, bal Fe.
Free machining ferritic type stainless steel.
AISI 430 F.

INOXALIUM.
M-787; 0.2 C, bal Fe.
For machinery parts; water hardenable.

INOXARGENT.
M-1120; 0.12 max C, 12-14 Cr, 12-14 Ni, Cu, bal Fe.
Annealed: 70,000-78,000 TS; 28,000-44,000 YS; 45-50 El.
For clocks, decorative trim, furniture; corrosion resistant.

INOXESCO 16.
M-1495; 0.12 max C, 1.0 max Mn, 14-18 Cr, 0.5 max Ni, bal Fe.
Annealed: 60,000 TS; 35,000 YS; 20 El; 190 Brin.
For oil refinery tubes and equipment; stainless, ferritic.

INOXIUM 18.10 N.
M-1117, M-1117a; 0.06 C, 18.5 Cr, 9.0 Ni, bal Fe.
Austenitic stainless steel; AISI 302.

INOXIUM 18.10 R.
M-1117, M-1117a; 0.06 C, 18.0 Cr, 9.5 Ni, bal Fe.
Austenitic type stainless; AISI 304.

INOXIUM 18.10 S.
M-1117, M-1117a; 0.02 C, 18.5 Cr, 10.5 Ni, bal Fe.
Austenitic type stainless; AISI 304 LC.

INOXIUM 18.10 T.
M-1117, M-1117a; 0.05 C, 17.5 Cr, 9 Ni, Ti, bal Fe.
Stabilized austenitic type stainless; AISI 321.

INOXIUM 18.13 D 25 R.
M-1117, M-1117a; 0.05 C, 17.5 Cr, 11.0 Ni, 2.5 Mo, bal Fe.
Austenitic type stainless steel; improved corrosion resistance.
AISI 316.

INOXIUM 18.13 D 25 T.
M-1117, M-1117a; 0.05 C, 17.5 Cr, 11 Ni, 2.5 Mo, Ti, bal Fe.
Austenitic type stainless steel; improved corrosion resistance.
AISI 316 Ti.

INOXIUM 18.8 C.
M-1117, M-1117a; 0.10 C, 18.0 Cr, 7.5 Ni, bal Fe.
Austenitic type stainless; cold rolled for increased strength.

INOXIUM 18.8 E.
M-1117, M-1117a; 0.10 C, 17.5 Cr, 8.0 Ni, bal Fe.
Austenitic type stainles; cold rolled for increased strength.

INOXIUM 130 AB.
M-1117, M-1117a; 0.06 C, 13 Cr, bal Fe.
Ferritic type stainless steel.

INOXIUM 130 TI.
M-1117, M-1117a; 0.06 C, 13 Cr, Ti, bal Fe.
Ferritic type stainless steel for mufflers.

INOXIUM 170.
M-1117, M-1117a; 0.06 C, 16.5 C, bal Fe.
Ferritic stainless steel; for decoration and trim.
AISI 430.

INOXIUM 170 MO.
M-1117, M-1117a; 0.06 C, 16.5 Cr, Mo, bal Fe.
Ferritic stainless steel; for automotive bumpers, wheel covers.
AISI 434.

INOXYDA BRONZE.
1-9 Al, 82-89 Cu, 0-1 Zn.
For steam fittings; corrosion resistant.

IN-STEEL.
M-1570; 0.18 max C, 1.5 max Ni, 0.4-0.8 Cr, 0.6 max Mo, 0.10 max V, 0.15-0.35 Si, 0.6-1.2 Mn, bal Fe.
Plate: 138,000-163,500 TS; 128,000 YS; 15 El.
For heavy duty welded structures and equipment, mine cars, bus bodies. Ultra-high strength, excellent weldability.

INSTRUMENT BRONZE.
M-U.S.; 13 Sn, 5 Zn, bal Cu.
For instruments, utensils; corrosion resistant.

INTAL N 54.
M-5; 2 Mg, 0.8 Mn, bal Al.
For cast door fittings.

INTAL N 89.
M-5; 50 Zn, 48 Al, 2 Cu.
Cast: 49,000 psi TS; 38,000 psi YS.
For bearings, sand and permanent mould castings. Was MAIN METAL.

INTENSIV.
M-1339; 2.0 C, 11.5 Cr, 0.1 V, bal Fe.
For blanking and forming dies, punches; air hardened, non-deforming.

INTENSIVE 25.
M-1339; 2.5 C, 12 Cr, bal Fe.
Cold work tool steel; AISI D3.

INTENSIVE PLATT.
M-1339; 1.65 C, 12.0 Cr, 0.5 Mo, 1.3 Co, bal Fe.
Cold work tool steel.

INTENSIVE SUPRA.
M-1339; 1.5 C, 12.0 Cr, 0.7 Mo, 1.0 V, bal Fe.
Cold work tool steel; similar to AISI D2.

INTENSIVE SUPRA V.
M-1339; 2.2 C, 12.5 Cr, 0.9 Mo, 2.2 V, bal Fe.
Cold work tool steel.

INTENSIV L SUPRA.
M-1339; 1.0 C, 12.0 Cr, 0.9 Mo, 0.9 V, bal Fe.
Cold work tool steel; similar to AISI D2.

INTENSIV SPEZIAL.
M-1339; 1.65 C, 11.5 Cr, 0.6 Mo, 0.5 W, 0.3 V, bal Fe.
For blanking and forming dies, punches; air hardening tool steel.

INTENSIV W CO.
M-1339; 2.1 C, 12.0 Cr, 0.4 Mo, 0.7 V, 1.0 Co, bal Fe.
Cold work tool steel.

INTENSIV WO.
M-1339; 2.1 C, 11.5 Cr, 0.7 W, bal Fe.
For blanking and forming dies, punches; oil hardened.

INTENSIV Z.
M-1339; 1.65 C, 11.5 Cr, 0.1 V, bal Fe.
For blanking and forming dies, punches; air hardened, non-deforming.

INTENSIV ZAHDUR.
M-1339; 1.0 C, 5.0 Cr, 1.0 Mo, 0.2 V, bal Fe.
Hot work tool steel.

INVAR.
M-73; 0.15 C, 38 Ni, 0.8 Mn, bal Fe.
Annealed: 70,000 TS; 24,000 YS; 36 El; 68 RA; 143 Brin.
Cold drawn: 90,000 TS; 70,000 YS; 20 El; 60 RA; 185 Brin.
For bimetal thermostats, geodetic instruments; low coefficient of expansion.

INVAR.
M-1733; 0.08 max C, 36.0 Ni, bal Fe.
Low expansion steel.
DIN 1.3912.

INVAR see also **CARPENTER INVAR.**

INVAR 36.
M-108; 36 Ni, bal Fe.
Low expansion alloy.

INVAR 36 FREE-CUT.
M-32, M-114; 0.12 C, 0.90 Mn, 0.35 Si, 0.20 Se or S, 36 Ni, bal Fe.
Annealed: 65,000 TS; 40,000 YS; 35 El; B 70 Rock.
For geodetic instruments, thermostats, temperature control and indicating devices, component parts requiring minimum expansivity.
Free-cutting, low coefficient expansion.

INVAR 42.
M-108; 42 Ni, bal Fe.
Low expansion alloy.

INVARIANT.
M-Eng.; 47 Ni, 53 Fe.
For surveyors' tape, piston struts; corrosion resistant.

INVAR M63.
M-1488; 36 Ni, bal Fe.
Low coefficient of thermal expansion.
Good at low temperatures.

INVARO-B.
M-57; 0.94 C, 1.2 Mn, 0.53 W, 0.22 V, 0.5 Cr, bal Fe.
For punches, forming dies, shears; tough, non-deforming, oil hardened.

INVARO NO. 1.
M-57; 0.88 C, 1.0-1.25 Mn, 0.5 Cr, 0.2 V, 0.5 W, bal Fe.
For tools, hobs, milling cutters, taps; Type O1; oil hardened.

INVINCIBLE.
M-1406; 0.70 C, 18 W, 4 Cr, 1 V, bal Fe.
For turning tools, reamers, drills, chasers, taps; high speed steel.

INVINCIBLE 22%.
M-1406; 0.75 C, 22 W, 4 Cr, 1.2 V, bal Fe.
For lathe tools, reamers, drills, hobs; high speed steel.

INVINCIBLE DRILL ROD.
M-339; 1-1.2 C, bal Fe.
For drills, general tools; water hardening.

INX-42.
M-67; 0.2 C, 0.9-1.35 Mn, 0.01 Cb, bal Fe.
Rolled (min): 63,000 TS; 42,000 YS; 20 El.
For trucks, plow frames, pressure vessels, building components, automobiles, trailers, railroad accessories.
Good weldability, tough.

INX-45.
M-67; 0.2 C, 1.2 Mn, 0.02 Cb, bal Fe.
Rolled: 45,000 min YS; 65,000 min TS; 19 min El.
For trucks, plow frames, pressure vessels, automobiles, trailers, buildings. Good weldability, tough.

INX-50.
M-67; 0.2 C, 1.3 Mn, 0.01 Cb, bal Fe.
Rolled (min): 70,000 TS; 50,000 YP; 18 El.
For trucks, plow frames, pressure vessels, building components, water heaters, railroad accessories. Good weldability, tough.

INX-55.
M-67; 0.22 C, 1.35 Mn, 0.01 Cb, bal Fe.
Rolled (min): 75,000 TS; 55,000 YP; 17 El.
For trucks, plow frames, pressure vessels, construction equipment, water heaters, railroad accessories. Good weldability, tough.

INX-60.
M-67; 0.25 C, 1.35 Mn, 0.01 Cb, bal Fe.
Rolled (min): 80,000 TS; 60,000 YS; 16 El.
For trucks, plow frames, pressure vessels, building components, pipe, railroad accessories, automobiles. Good weldability, tough.

INX-65.
M-67; 0.25 C, 1.3 Mn, 2 Si, 0.03 Cb, bal Fe.
Rolled: 85,000 min TS; 65,000 min YS; 15 min El.
For trucks, plow frames, pressure vessels, automobiles, railroad accessories.
Good weldability.

INX-70.
M-67; 0.26 C, 1.3 Mn, 0.2 Si, 0.03 Cb, 0.01 V, bal Fe.
Plate: 90,000 min TS; 70,000 min YP; 14 min El.
For structures, plow shares, pressure vessels, railroad accessories, building components, water heaters.
Good weldability and forming properties.

IOCHROME.
M-1492; 99.997 Cr.
Rolled: 44 El; 78 RA.
Alloying ingredient for high Cr alloys; for high temperature service.

IPM 338-IH.
M-1764.
High strength, low alloy steel, sintered.
Density: 7.4 g/cc.
Sintered, heat treated: 148,000 TS; 2 El; Rc 43.
Unnotched Charpy impact: 13 ft. lbs.
For high strength and impact resistance.

IPM FG-982.
M-1764.
Iron graphite, sintered.
As sintered: 5000 TS; K Factor 11,500; density: 5.8 g/cc.
For P/M bearings to also include lubricating properties.

IPM FP-1.
M-1764.
High phosphorus, soft magnetic material, sintered.
Density: 6.8 g/cc.
At 10,000 gauss Hc = .9, Br = 9200, Hm = 2.0.
Electrical resistivity: 27 microhm-cm.
Max permeability: 5500.

IPM SS-100.
M-1764.
Austenitic stainless steel, sintered.
As sintered: 47,000 TS; 16 El; Rf 67; density: 6.4 g/cc.
For corrosion resistance comparable to AISI 316.

IPM SS-101.
M-1764.
Austenitic stainless steel, sintered.
As sintered: 38,000 TS; 9 El; Rf 70; density 6.4.
For corrosion resistance comparable to AISI 304; excellent machinability.

IRALITE.
M-229; 3.1 C, 1.2 Cr, 0.3 Mo, bal Fe.
Cast: 36,000-52,000 TS; 150-180 Brin.
For sheaves, sprockets, plungers, glass rolls, brake drums, conveyor parts.
Pearlitic cast iron, wear resistant.

IRCAMET.
M-1224; 0.07 C, 20 Cr, 29 Ni, 2.75 Mo, 1 Si, 4 Cu, 0.8 Mn, bal Fe.
Cast: 65,000 TS; 30,000 YS; 30-45 El; 40-50 RA; 140 Brin.
Rolled: 85,000 TS; 35,000 YS; 35-50 El; 50-70 RA; 170 Brin.
For furnace parts, chemical plant equipment; stainless, austenitic, nonmagnetic.

IRIDAL.
M-1397; 0.3 B, 5 Si, 0.7 Mg, bal Al.
For architectural uses; heat treatable.

IRIDIUM.
M-Eng.; 77-83 Zn, 1.21-1.25 Cu, 21.6-15.7 Sn, trace Sb.
For bearings; Babbitt.

IRIDIUM.
M-1755; Ir.
Purities: 99.999%, 99.99%, 99.9%.
Forms: sponge, powder, rod, sheet, foil, single crystals, crucibles.

IRIDIUM II.
M-Eng.; 77.25 Zn, 21.63 Sn, 1.12 Cu.
For bearings; will not resist heat or live steam.

IRIDIUM B.M.
M-Eng.; 83 Zn, 15.75 Sn, 1.25 Cu, trace Sb.
For bearings; will not resist heat or live steam.

IRIDIUM EXTRA SPECIAL.
M-613; 0.7 C, 18 W, 4 Cr, 2 V, 5 Co, bal Fe.
For cutters, tools; high speed steel.

IRIDOSMINE.
M-Eng.; 57 Os, 34 Rh + Ir, 8 Ru.
For pen points; "Osmiridium."

I.R. METAL.
M-788; 2.5-3.25 C, 0.7-1.1 B, 0.5-1.5 Si, 0.5-1.25 Mn, 3.5-4.5 Ni, bal Fe.
For oil field accessories; abrasion resistant cast iron.

IRON.
M-1755; Fe.
Purities: zone and chemical refined 99.999%, 99.99%, 99.9+%.
Forms: sponge, rod, chips, powder, wire, foil, single crystals.

IRONAC.
M-Eng.; 13.2 Si, 0.77 Mn, 1.1 C, 0.8 P, bal Fe.
For pumps, valves, ejectors, nozzles, fans; acid resistant cast iron.

IRON BRONZE.
M-2, M-2a; 8.6 Sn, 4.4 Zn, 4 Fe, bal Cu.
Cast: 20-35 El; 70-85 Brin.
Pressed: 20-35 El; 80-90 Brin.
For shafts, piston rods, screws, marine parts; resists sea water corrosion.

IRONIERS BRONZE.
M-Eng.; Sn, 1 Hg, bal Cu.
For special applications; corrosion resistant.

IRON OILITE.
M-211; 97.25 min Fe, 0.25 max C.
Sintered: 10,000-15,000 TS; 5.7-6.1 density.
Oilite bearing material.
ASTM B439-67 Gr 1; SAE 850; PMPMA F-0000-N.

IRON OILITE-1.
M-211; 97 min Fe, 0.6-1.0 C.
Sintered: 24,000 TS; 5.7-6.1 density.
ASTM B310-67 Type I Cl.C; SAE 852.

IRON OILITE-1 IM.
M-211; 15-25 Cu, 69.0-84.4 Fe, 0.6-1.0 C.
Sintered: 85,000 TS; 7.1 min density. 40,000 shear; 17,000 fatigue.
ASTM B303-67 Cl.C; SAE 872. PMPMA FX-2010-T.

IRON OILITE-M.
M-211; 0.5 C, 0.5 S, 99 Fe.
Sintered: 30,000-37,000 TS; 25,000-32,000 YS; 1.0-1.5 El; 30 Rb; density 6.1-6.8.
Machinable grade oilite bearing material.

IROQUOIS.
M-435; 1.25 C, 1.4 W, bal Fe.
For taps, cartridge dies, punches; water hardening.

IRRUBIGO, ETC see MARKER IRR, ETC.

ISA-13.
M-789; 96 Cu, 2 Mn.
Hard: 100,000-114,000 TS.
Annealed: 42,000-50,000 TS.
Temp. coef. resistance per °C: + 0.0004.
For electrical equipment and instruments.
Resistance alloy. Max working temperature 300°C.

ISA-50.
M-789; 81 Cu, 5 Ni, 12 Mn, 1.5 Al.
Hard: 115,000-128,000 TS.
Soft: 58,000-65,000 TS.
Temp. coef. resist. per °C: ± 0.00002.
For electrical equipment and instruments.
Resistance alloy. Max working temperature 500°C.

ISABELLIN.
M-789; Cu, Mn, Al.
For electrical resistance parts.

ISABELLIN-A.
M-789; 84 Cu, 13 Mn, 3 Al.
Hard: 128,000-144,000 TS.
Soft: 72,000-78,000 TS.
For electrical equipment and instruments.
Resistance alloy. Max operating temperature to 400°C.

ISA-NICKEL.
M-789; 32 Cu, 67 Ni.
Hard: 120,000-135,000 TS.
Soft: 64,000-78,000 TS.
Thermal coef. resist. per °C: + 0.0005.
For electrical equipment and instruments.
Resistance alloy. Max operating temperature to 600°C.

ISA-OHM.
M-789; 3 Cu, 71 Ni, 21 Cr.
Annealed: 157,000-170,000 TS.
Temp. coef. resistance per °C: ± 0.00001.
For electrical equipment and instruments.
Resistance alloy. Max working temperature 250°C.

ISA-SPRAY.
M-789; 88 Cu, 10 Mn, 2 Al.
Hard: 128,000-144,000 TS.
Soft: 64,000-72,000 TS.
For electrical equipment and instruments.
Resistance alloy. Max working temperature to 400°C.

ISA-ZIN.
M-789; 77 Cu, 21 Ni, 1.5 Mn.
Hard: 78,000-85,000 TS.
Soft: 50,000-60,000 TS.
Temp. coef. resist. per °C: + 0.00025.
For electrical equipment and instruments.
Resistance alloy. Max operating temperature to 400°C.

ISC 15-3.
M-15-3; 3.0 C, 0.70 Mn, 0.50 Si, 15.0 Cr, 3.0 Mo, bal Fe.
550-650 Brin; extreme wear resistance, poor impact.

ISC 1020.
M-1482; 0.20 C, 0.60 Mn, 0.50 Si, bal Fe.
Cast: 60,000 psi TS; 35,000 psi YS; 24 El; 35 RA.
General purpose cast steel.

ISC 1045.
M-1482; 0.45 C, 0.60 Mn, 0.50 Si, bal Fe.
Cast: 85,000 psi TS; 40,000 psi YS; 16 El; 24 RA.
Steel casting; can be hardened by heat treat.

ISC 4335.
M-1482; 0.35 C, 0.70 Mn, 0.50 Si, 1.75 Ni, 0.70 Cr, 0.35 Mo, bal Fe.
Alloy steel casting, hardenable to 200-400 Brin in comparatively heavy sections.
For components requiring high strength.

ISC 6145.
M-1482; 0.45 C, 0.60 Mn, 0.50 Si, 0.95 Cr, 0.15 V, bal Fe.
For gears, cams, shafts; flame or induction hardened.

ISC 8630.
M-1482; 0.30 C, 0.70 Mn, 0.50 Si, 0.50 Ni, 0.50 Cr, 0.20 Mo, bal Fe.
Steel casting, hardenable to 180-350 Brin.

ISC AIR HARD.
M-1482; 1.5 C, 0.8 Mo, 0.5 V, 0.7 Mn, 12 Cr, bal Fe.
For dies, cams, punches, blanking and forming tools; air hardened, non-deforming.

ISCAR IC 2.
M-1745; Sintered carbide.
330,000 psi TrS; A 91.9 Rock.
For machining soft cast iron, non-ferrous alloys, wood, plastic and other non-metallics.
Code C2.

ISCAR IC 4.
 M-1745; Sintered carbide.
 250,000 psi TrS; A 93.0 Rock.
 For semi-finishing and finishing abrasive alloys and non-metallics.
 Code C4.

ISCAR IC 20.
 M-1745; Sintered carbide.
 300,000 psi TrS; A 92.5 Rock.
 For machining cast irons, silicon aluminum alloys and non-metallics.
 Code C3.

ISCAR IC 28.
 M-1745; Sintered carbide.
 450,000 psi TrS; A 90.5 Rock.
 For roughing cuts on grey cast iron, austenitic and high temperature alloys.
 Code C1.

ISCAR IC 50.
 M-1745; Sintered carbide.
 315,000 psi TrS; A 91.5 Rock.
 For rough and semi-finish machining of steel, steel castings and malleable iron.
 Code C5.

ISCAR IC 50M.
 M-1745; Sintered carbide.
 340,000 psi TrS; A 91.7 Rock.
 For rough and semi-finish milling of steel, steel castings and malleable cast iron.
 Code C6.

ISCAR IC 54.
 M-1745; Sintered carbide.
 380,000 psi TrS; A 90.7 Rock.
 For planing, turning, heavy roughing, and difficult cutting of steel and iron and steel castings.

ISCAR IC 70.
 M-1754; Sintered carbide.
 275,000 psi TrS; A 92.6 Rock.
 For roughing and semi-finishing steel, steel castings and malleable cast iron.
 Code C7, C6.

ISCAR IC 78.
 M-1745; Sintered carbide.
 240,000 psi TrS; A 92.4 Rock.
 For general machining carbon and alloy steels.
 Code C7.

ISCAR IC 80 T.
 M-1745; Sintered titanium carbide.
 210,000 psi TrS; A 92.9 Rock.
 For semi-finish machining carbon and alloy steels.
 Code C7, C8.

ISCAR IC 424 (ISCOTIC).
 M-1745; Sintered TiC coated carbide.
 For machining cast irons, manganese steels, and high temperature resistant alloys.
 Code C4, C2.

ISCAR IC 636 (ISOTIC).
 M-1745; Sintered TiC/N coated carbide.
 Tough grade for roughing of tool steels, iron and steel castings.
 Code C5, C6.

ISCAR IC 757 (ISCOTIC).
 M-1745; Sintered TiC coated carbide.
 For rough and finish turning of steel casting, malleable and nodular iron.
 Code C7, C5.

ISC FLAME HARD.
 M-1482; 0.55 C, 1.2 Mn, 1.25 Cr, 0.4 Mo, 0.1 V, bal Fe.
 For trim dies; oil or flame hardened.

ISC H-13.
 M-1482; 0.35 C, 0.50 Mn, 0.50 Si, 5.0 Cr, 1.5 Mo, 0.90 V, bal Fe.
 Cast hot work tool steel; air hardening.

ISC OIL HARD.
 M-1482; 0.90 C, 1.25 Mn, 0.5 Cr, 0.3 Mo, 0.25 V, bal Fe.
 For forming and blanking dies, punches, headers; oil hardened, non-deforming.

ISC WEAR HARD.
 M-1482; 0.30 C, 1.40 Mn, 0.50 Si, 1.0 Cr, 0.50 Mo, B, bal Fe.
 Cast (hardened): 450-500 Brin; high wear resistance; good impact properties.

ISERLOHN.
 M-Eng.; 70.1 Cu, 29.9 Zn.
 For cartridges, shell cases, condenser tubes; maximum ductility.

ISERLOHN.
 64 Cu, 34 Zn, 2.4 Sn, 0.3 Pb.
 For fittings, hardware; corrosion resistant.

ISERLOHN BRASS.
 M-Eng.; 70 Cu, 30 Zn.
 1/4 H-temper: 54,000 TS; 40,000 YS; 43 El; 100 Brin.
 1/2 H-temper: 62,000 TS; 52,000 YS; 23 El; 116 Brin.
 For cartridge cases, condenser tubes; high ductility and workability.

ISO-450.
 M-162; 0.05 max C, 14.5-16.5 Cr, 5.5-7.0 Ni, 0.5-1.0 Mo, 1.25-1.75 Cu, Cb = 8 x C min, bal Fe.
 Sol. ann: 128 ksi TS; 118 ksi YS; 15 El; 53 RA.
 Martensitic age-hardenable stainless.

ISO-CAST 1.
M-162; 0.2-0.3 C, 0.6-0.8 Mn, 0.35-0.50 Si, bal Fe.
Annealed: 70,000 TS; 40,000 YS; 25 El; 40 RA; 130 Brin.
For castings; general purpose.

ISO-CAST 1A.
M-162; 0.3-0.4 C, 0.5-0.7 Mn, 0.35-0.50 Si, bal Fe.
Annealed: 75,000 TS; 45,000 YS; 25 El; 40 RA; 165 Brin.
For castings; general purpose.

ISO-CAST 1B.
M-162; 0.4-0.5 C, 0.6-0.8 Mn, 0.35-0.50 Si, bal Fe.
Annealed: 85,000 TS; 55,000 YS; 20 El; 30 RA; 174 Brin.
For castings; general purpose.

ISO-CAST 2.
M-162; 0.5-0.6 C, 0.6-0.8 Mn, 0.35-0.50 Si, bal Fe.
Annealed: 95,000 TS; 60,000 YS; 15 El; 20 RA; 202 Brin.
For castings; high strength.

ISO-CAST 2A.
M-162; 0.6-0.7 C, 0.6-0.8 Mn, bal Fe.
Annealed: 105,000 TS; 65,000 YS; 12 El; 15 RA; 222 Brin.
For castings; high strength.

ISO-CAST 2B.
M-162; 0.70-0.80 C, 0.6-0.8 Mn, 0.35-0.50 Si, bal Fe.
Annealed: 112,000 TS; 75,000 YS; 10 El; 12 RA; 235 Brin.
For castings; abrasion resistant.

ISOCAST 3.
M-162; 0.25-0.35 C, 1.25-1.5 Mn, bal Fe.
Heat treated: 90,000 TS; 55,000 YS; 25 El; 40 RA; 200 Brin.
For machine and railroad parts; castings.

ISO-CAST 3 S.
M-162; 0.33-0.47 C, 1.25-1.5 Mn, bal Fe.
Cast steel.

ISOCAST 4.
M-162; 0.15-0.25 C, 0.5-0.7 Mn, 0.4-0.6 Mo, bal Fe.
Annealed: 75,000 TS; 45,000 YS; 25 El; 40 RA.
For turbines, locomotives, valves, pumps; castings.

ISOCAST 5.
M-162; 0.25-0.35 C, 1.0-1.2 Mn, 0.35-0.45 Mo, bal Fe.
Heat treated: 80,000 TS; 55,000 YS; 25 El; 40 RA.
For cars, excavating machinery sheaves; castings.

ISO-CAST 6A.
M-162; 0.15-0.25 C, 0.6-0.8 Mn, 2.75-3.50 Ni, bal Fe.
Tempered: 85,000 TS; 55,000 YS; 20 El; 40 RA; 159 Brin.
For castings, machinery parts.

ISOCAST 7.
M-162; 0.85 C, 0.5 Mn, 0.5 Si, 2.5 Cr, 0.2 V, bal Fe.
Heat treated: 95,000-165,000 TS; 60,000-135,000 YS; 15 El; 30 RA; 179-302 Brin.
For machine tools, sheaves, brick making equipment; wear resistant castings.

ISO-CAST 7A.
M-162; 0.53-0.67 C, 0.5 Mn, 0.5 Si, 2.5 Cr, 0.2 V, bal Fe.
Cast low alloy steel; for mixing blades, crushing equipment, ball mills.
Hard and abrasion resistant.

ISOCAST 8.
M-162; 0.25-0.35 C, 0.6-0.9 Mn, 1.0-1.25 Cr, 0.2-0.3 Mo, bal Fe.
Heat treated: 90,000-105,000 TS; 55,000-65,000 YS; 15-25 El; 30-45 RA; 210 Brin.
For railroads, connecting rods, valves; castings.

ISO-CAST 8A.
M-162; 0.35-0.45 C, 1 Cr, 0.3 Mo, 0.7 Mn, 0.5 Si, bal Fe.
Normalized: 90,000-105,000 TS; 60,000-80,000 YS; 18-24 El; 35-50 RA 207-220 Brin.
For gears, shafts, machine tool housings; SAE 4142; tough and wear resistant.

ISO-CAST 8B.
M-162; 0.2 max C, 1.0-1.5 Cr, 0.45-0.65 Mo, bal Fe.
Normalized: 75,000-90,000 TS; 45,000-65,000 YS; 30-25 El; 50-40 RA; 159-170 Brin.
For steam service castings; for strength at elevated temp.

ISO-CAST 8C.
M-162; 0.18 max C, 0.40-0.70 Mn, 0.60 max Si, 2.0-2.75 Cr, 0.90-1.20 Mo, bal Fe.
Cast: 75,000-95,000 TS; 45,000-65,000 YS; 25-30 El; 40-50 RA.
For strength at elevated temperatures, steam service.

ISOCAST 9.
M-162; 0.2-0.3 C, 0.5-0.7 Ni, 0.5-0.7 Cr, 0.1-0.2 Mo, bal Fe.
Heat treated: 110,000-120,000 TS; 70,000-100,000 YS; 18-12 El; 35-30 RA; 230-270 Brin.
For valve bodies, turbines, gears, shafts, housings; SAE 8630; castings.

ISOCAST 9A.
M-162; 0.25-0.35 C, 1.0-1.2 Ni, 1.0-1.5 Cr, 0.1-0.2 V, bal Fe.
Heat treated: 115,000 TS; 65,000 YS; 20 El; 40 RA; 235 Brin.
For impellers, pump casings, structural castings; ASTM 4C4; abrasion resistant.

ISOCAST 20.
M-162; 0.07 max C, 19-22 Cr, 27.5-30.0 Ni, 2 min Mo, 3 min Cu, bal Fe.
Cast: 65,000 TS; 30,000 YS; 35 El; 140 Brin.
For pumps, valves, chemical plant equipment; austenitic, stainless, resist H_2SO_4.

ISOCAST 40.
M-162; 0.04 max C, 25-57 Cr, 4.7-6.0 Ni, 1.7-2.2 Mo, 2.7-3.2 Cu, bal Fe.
Annealed: 100,000 TS; 80,000 YS; 30 El; 250 Brin.
Heat treated: 145,000 TS; 120,000 YS; 10 El; 350 Brin.
For high temperature applications; age-hardenable, corrosion resistant.

ISO-CAST 85.
M-162; 0.2 max C, 4.0-6.5 Cr, 0.45-0.65 Mo, bal Fe.
Normalized: 90,000-105,000 TS; 60,000-70,000 YS; 24-18 El; 50-35 RA; 192-200 Brin.
For pump valves; for strength at elevated temp.

ISO-CAST-89.
M-162; 0.20 max C, 0.35-0.65 Mn, 8.0-10.0 Cr, 0.90-1.2 Mo, 1.0 max Si bal Fe.
Cast: 90,000-105,000 TS; 60,000-70,000 YS; 18-24 El; 35-50 RA.
For strength at elevated temperatures, steam service.

ISO-CAST B.
M-162; 0.12 max C, 1.0 max Cr, 63 Ni, 28 Mo, 6 Fe.
Cast: 80,000 TS; 57,000 YS; 8 El; 11 RA; 200 Brin.
For equipment handling HCl in all concentrations.
Corrosion resistant to H_2SO_4 and H_3PO_4.

ISO-CAST C.
M-162; 0.12 max C, 17 Cr, 55 Ni, 17 Mo, 6 Fe, 4.25 W, 0.8 Co, 0.3 V.
Cast: 78,000 TS; 57,000 YS; 7 El; 10 RA; 200 Brin.
For equipment handling nitric acid, solutions containing Cl_2. Corrosion resistant.

ISO-CAST CA6NM.
M-162; 0.06 C, 11.5 Cr, 3.5 Ni, 0.40 Mo, bal Fe.
Corrosion resisting alloy cast iron.
ASTM A 296 CA-6NM.

ISO-CAST CA-15.
M-162; 0.15 C, 1 Ni, 11-14 Cr, bal Fe.
Quenched and treated: 95,000 TS; 65,000 YS; 20 El; 45 RA; 200 Brin.
For valves, pumps, castings; corrosion resistant, hardenable.

ISO-CAST CA-40.
M-162; 0.2-0.4 C, 1 Ni, 11.5-14.0 Cr, bal Fe.
Quenched and treated: 110,000 TS; 75,000 YS; 18 El; 35 RA; 220 Brin.
For valves, pumps, castings; corrosion resistant, hardenable.

ISO-CAST CB7CU.
M-162; 0.07 max C, 1.0 max Mn, 1.0 max Si, 15.5-17.0 Cr, 3.6-4.6 Ni, 2.3-3.3 Cu, bal Fe.
Cast grade of 17-4 PH; age hardenable to about 200,000 psi TS.

ISO-CAST CB-30.
M-162; 0.3 C, 2 Ni, 18-22 Cr, bal Fe.
Cast: 90,000 TS; 40,000 YS; 200 Brin.
For valves and heat resisting castings; corrosion and heat resistant.

ISOCAST CC-50.
M-162; 0.5 max C, 26-30 Cr, 4 max Ni, bal Fe.
Cast: 60,000 TS; 190 Brin.
For ore roasting equipment; stainless.

ISO-CAST CE-30.
M-162; 0.3 C, 8-11 Ni, 26-30 Cr, bal Fe.
Cast: 90,000 TS; 55,000 YS; 25 El; 25 RA; 200 Brin.
For corrosion and heat resisting castings; corrosion and heat resistant.

ISOCAST CF3.
M-162; 0.03 max C, 18-21 Cr, 9-12 Ni, bal Fe.
Annealed: 75,000 TS; 36,000 YS; 45 El; 140 Brin.
For chemical plant equipment, food processing equipment; Type 304L; stainless, austenitic.

ISOCAST CF3M.
M-162; 0.03 max C, 18-21 Cr, 10-13 Ni, 2-3 Mo, bal Fe.
Annealed: 78,000 TS; 40,000 YS; 44 El; 150 Brin.
For acid resistant chemical plant castings; Type 316L; stainless, austenitic.

ISO-CAST CF-8.
M-162; 0.08 C, 8-11 Ni, 18-21 Cr, bal Fe.
Annealed: 75,000 TS; 40,000 YS; 45 RA; 140 Brin.
For stainless castings, pumps, valves, processing equipment; stainless, austenitic.

ISOCAST CF-8C.
M-162; 0.08 max C, 18-21 Cr, 9-12 Ni, 1.0 max Cb, bal Fe.
Cast: 80,000 TS; 44,000 YS; 40 El; 160 Brin.
For dairy and chemical plant equipment, valves, tanks; Type 347; stainless, austenitic.

ISO-CAST CF-8M.
M-162; 0.08 C, 9-12 Ni, 18-21 Cr, 2-3 Mo, bal Fe.
Annealed: 80,000 TS; 40,000 YS; 50 El; 156 Brin.
For stainless castings, valves, pumps; stainless, austenitic.

ISOCAST CF-16F.
M-162; 0.08 max C, 18-21 Cr, 9-12 Ni, Cb = 8 x C, bal Fe.
Cast: 80,000 TS; 44,000 YS; 40 El; 156 Brin.
For stainless and heat resistant parts; stainless and heat resistant; welding grade.

ISOCAST CF16FA.
M-162; 0.16 max C, 18-21 Cr, 9-12 Ni, 0.20-0.35 Se or 0.2-0.4 S, bal Fe.
Annealed: 75,000 TS; 35,000 YS; 40 El; 160 Brin.
For chemical plant equipment; free-cutting, stainless, austenitic.

ISOCAST CF-20.
M-162; 0.20 max C, 20 Cr, 8-11 Ni, bal Fe.
Cast: 80,000 TS; 40,000 YS; 45 El; 160 Brin.
For valves, pumps; austenitic, corrosion resistant.

ISOCAST CF-30.
M-162; 0.30 max C, 17-20 Cr, 8-11 Ni, bal Fe.
Annealed: 80,000 TS; 40,000 YS; 40 El; 160 Brin.
For chemical plant equipment; stainless, austenitic.

ISO-CAST CG-8M.
M-162; 0.08 max C, 10-14 Ni, 18-21 Cr, 3-4 Mo, bal Fe.
Cast: 80,000 TS; 44,000 YS; 33 El; 165 Brin.
For paper and textile equipment, bleaching equipment; Type 317; stainless, austenitic.

ISOCAST CG-12.
M-162; 0.12 max C, 20-23 Cr, 10-13 Ni, bal Fe.
Cast: 80,000 TS; 40,000 YS; 40 El; 156 Brin.
For paper and textile processing equipment; stainless.

ISOCAST CH-20.
M-162; 0.20 max C, 24-26 Cr, 12-15 Ni, bal Fe.
Cast: 85,000 TS; 50,000 YS; 30 El; 170 Brin.
For paper and pulp and chemical plant equipment; Type 309; stainless, austenitic.

ISOCAST CK-20.
M-162; 0.20 max C, 23-27 Cr, 19-22 Ni, bal Fe.
Cast: 75,000 TS; 45,000 YS; 30 El; 190 Brin.
For paper and pulp and chemical plant equipment; Type 310; stainless, austenitic.

ISO-CAST HA.
M-162; 0.20 max C, 0.50 Mn, 1.0 max Si, 8.0-10.0 Cr, 0.90-1.20 Mo, bal Fe.
Cast corrosion resistant steel; wear and abrasion resistant.
ACI HA.

ISOCAST HC.
M-162; 0.5 max C, 26-30 Cr, 4 max Ni, bal Fe.
Cast: 60,000 TS; 200 Brin.
For ore roasting equipment; austenitic, heat resistant.

ISO-CAST HD.
M-162; 0.50 max C, 1.5 Mn, 2.0 max Si, 26.0-30.0 Cr, 4.0-7.0 Ni, 0.5 max Mo, bal Fe.
Cast corrosion and heat resistant alloy.
ACI HD.

ISOCAST HE.
M-162; 0.20-0.50 C, 26-30 Cr, 8-11 Ni, bal Fe.
Cast: 90,000 TS; 55,000 YS; 20 El; 200 Brin.
For oil still parts; austenitic, corrosion and heat resistant.

ISOCAST HF.
M-162; 0.2-0.4 C, 18-23 Cr, 8-12 Ni, bal Fe.
Cast: 85,000 TS; 40,000 YS; 45 El; 160 Brin.
At 1400°F: 37,200 TS; 19,750 YS; 20 El.
For valves, pumps, chemical plant equipment; corrosion resistant, austenitic.

ISO-CAST HH.
M-162; 0.25-0.50 C, 24-28 Cr, 11-14 Ni, bal Fe.
Cast: 85,000 TS; 50,000 YS; 35 El; 170 Brin.
For corrosion and heat resisting castings, furnace parts; corrosion and heat resistant.

ISO-CAST HI.
M-162; 0.20-0.50 C, 2.0 Mn, 2.0 Si, 26-30 Cr, 14-18 Ni, 0.5 max Mo, bal Fe.
Cast corrosion and heat resistant alloy.
ACI HI.

ISO-CAST HK.
M-162; 0.20-0.60 C, 18-22 Ni, 24-28 Cr, bal Fe.
Cast: 80,000 TS; 45,000 YS; 25 El; 190 Brin.
For corrosion and heat resisting castings, furnace parts; corrosion and heat resistant, austenitic.

ISO-CAST HL.
M-162; 0.20-0.60 C, 2.0 max Mn, 2.0 max Si, 28-32 Cr, 18-22 Ni, 0.5 max Mo, bal Fe.
Cast corrosion and heat resistant alloy.
ACI HL.

ISO-CAST HN.
M-162; 0.20-0.50 C, 2.0 max Mn, 2.0 max Si, 19-23 Cr, 23-27 Ni, 0.5 max Mo, bal Fe.
Cast corrosion and heat resistant alloy.
ACI HN.

ISO-CAST HT.
M-162; 0.35-0.75 C, 33-37 Ni, 13-17 Cr, bal Fe.
Cast: 75,000 TS; 7 El; 180 Brin.
For corrosion and heat resisting castings, furnace parts; corrosion and heat resistant.

ISO-CAST HU.
M-162; 0.35-0.75 C, 2.0 max Mn, 2.5 max Si, 17-21 Cr, 37-41 Ni, 0.5 max Mo, bal Fe.
Cast corrosion and heat resistant alloy.
ACI HU.

ISO-CAST HW.
M-162; 0.35-0.75 C, 58-62 Ni, 10-14 Cr, bal Fe.
Cast: 65,000 TS; 180 Brin.
At 1800°F: 10,000 TS; 8000 YS; 40 El.
For furnace equipment, retorts, pots; heat and corrosion resistant.

ISO-CAST HX.
M-162; 0.35-0.75 C, 2.0 max Mn, 2.5 max Si, 15-19 Cr, 64-68 Ni, 0.5 max Mo, bal Fe.
Cast corrosion and heat resistant alloy.
ACI HX.

ISO-CAST "MONEL-E".
M-162; 0.15 C, 63 Ni, 30 Cu, 1.5 Si, 1.0 Cb.
As cast: 65,000 TS; 32,000 YS; 30 El.
Used in contact with reducing compounds as boric acid, Ca, K, Na hydroxides.

ISO-CAST NICKEL.
M-162; 1.0 max C, 97 Ni, 2.0 max Si, 1.5 max Mn, 3.0 max Fe.
As cast: 53 ksi TS; 25 ksi YS; 25 El.
Used for resistance to strong caustic solutions.

ISOPERM-1.
M-621; 40-55 Ni, bal Fe.
For magnets; high permeability.

ISOPERM 2.
M-621; 35-50 Ni, 9-15 Cu, bal Fe.
For magnets; high permeability.

ISOPERM 3.
M-621; 40-60 Ni, 3-4 Al, bal Fe.
For magnets; high permeability.

ISOROD.
M-507; 0.2 C, 0.5 Si, 0.2 Mn, 15.0 Cr, 8.0 Mo, 1.0 Ti, bal Fe.
Coated, hardfacing electrode. As arc welded: 185,000 psi, 520-580 BHN.
For overlay or build up on tractor sprockets, shovel teeth, dredge pump impellers, heavy hammers; strong and tough.

ISOTAN.
M-789; 55 Cu, 44 Ni.
Annealed: 65,000-72,000 TS.
Hard: 107,000-121,000 TS.
Temp. coef. resistance per °C (-0.00008 to + 0.00004).
For electrical equipment and instruments. Resistance alloy. Max working temperature 600°C.

ISTEG STEEL.
M-471; C, bal Fe.
For reinforcing steel; mild steel, pre-stressed.

ITALSIL.
M-Italy; 5 Si, bal Al.
For light alloy parts; die and sand castings.

I.W. 1.
M-772; 0.5 C, 1.5 Cr, 2.2 W, 0.2 V, bal Fe.
For tools, dies; hot work steel.

I.W. 2.
M-772; 0.6 C, 9.5 W, 2.5 Cr, 0.1 V, bal Fe.
For hot work tools and dies; hot work steel.

IWI see **DARWIN IWI.**

IZETT STEEL NO. 1.
M-72; 0.125 C, 0.56 Mn, 0.018 P, 0.15 S, 0.10 Si, bal Fe.
At 20°C: 62,600 TS; 39,000 YS; 43 El; 68 RA.
At 400°C: 50,300 TS; 22,000 YS; 39 El; 70 RA.
For boilers; tough, Al deoxidized; non-aging.

IZETT STEEL NO. 2.
M-72; 0.15 C, 0.54 Mn, 0.025 P, 0.039 S, 0.06 Si, 10.06 Al, bal Fe.
At 20°C: 91,800 TS; 44,500 YS.
At 400°C: 68,000 TS; 23,000 YS.
For boilers; tough, Al deoxidized; non-aging.

IZETT STEEL NO. 3.
M-72; 0.355 C, 0.49 Mn, 0.024 P, 0.024 S, 0.10 Si, bal Fe.
At 20°C: 98,500 TS; 51,000 YS.
At 400°C: 77,000 TS; 29,000 YS.
For boilers; tough, Al deoxidized; non-aging.

IZETT STEEL NO. 4.
M-72; 0.34 C, 0.48 Mn, 0.025 P, 0.025 S, 0.11 Si, bal Fe.
At 20°C: 105,000 TS; 54,000 YS.
At 400°C: 81,000 TS; 30,000 YS.
For boilers; tough, Al deoxidizing; non-aging.

J

J-4 CHISEL.
M-261, M-486; 0.44 C, 1.3 Cr, 2.3 W, 0.15 V, bal Fe.
For chisels, shear blades, punches; shock resistant.

J-100 see **JORGENSEN 100.**

J 1300.
  M-31, M-60, M-1491; 0.08 C, 14.0 Cr, 33.0 Ni, 4.0 Mo, 6.5 W, 2.0 Ti, 0.25 Al, 0.25 Zr, bal Fe.
  For gas turbine blades and parts.

J-1500.
  M-31, M-40, M-1491; 0.15 C, 20.0 Cr, 10.0 Co, 10.0 Mo, 3.0 Ti, 1.0 Al, bal Ni.
  For gas turbine blades and parts.

J-1530.
  M-31, M-60, M-1491; 0.08 C, 19.5 Cr, 57.0 Ni, 13.5 Co, 4.3 Mo, 3.1 Ti, 1.3 Al.
  For gas turbine parts.

J-1570.
  M-31, M-60, M-1491; 0.20 C, 20.0 Cr, 28.0 Ni, 6.0 W, 4.0 Ti, bal Co.
  For jet engine parts.

J-1600.
  M-31, M-60, M-1491; 0.10 C, 19.0 Cr, 19.0 Co, 4.0 Mo, 3.0 Ti, 3.0 Al, bal Ni.
  For high temperature equipment.

J-1650.
  M-31, M-60, M-1491; 0.20 C, 19.0 Cr, 27.0 Ni, 12.0 W, 3.8 Ti, 0.02 B, 2.0 Ta, bal Co.
  For jet engine parts.

J.A.C.-ACM.
  M-1179; C, 28 Cr, Mo, bal Fe.
  Normalized: 100,000 TS.
  For glass, chemical and dying industries; acid resistant.

J.A.C.-ACN 1.
  M-1179; C, 18 Cr, 8 Ni, bal Fe.
  Normalized: 100,000-120,000 TS; 50,000-60,000 YS; 25 El.
  For food, chemical, petroleum industries; stainless.

J.A.C.-ACN 2.
  M-1179; C, 26 Cr, 10 Ni, bal Fe.
  Heat treated: 110,000-120,000 TS; 50,000-60,000 YS; 18 El.
  For furnace parts, grates; heat resistant.

J.A.C.-ACN 3.
  M-1179; C, Cr, Ni, Mo, bal Fe.
  Heat treated: 110,000-130,000 TS; 50,000-70,000 YS; 18 El.
  For furnace parts, grates; heat resistant, scale resistant.

J.A.C.-ACR.
  M-1179; C, 30 Cr, bal Fe.
  Heat treated: 84,000-100,000 TS.
  For furnace parts; scale and heat resistant.

J.A.C.-AMH.
  M-1179; C, 12-14 Mn, bal Fe.
  Heat treated: 180,000 TS; 80,000 YS; 35 El; 40 RA.
  For crusher plates, hammers, harrowers, mixers; wear resistant.

J.A.C.-AMH 2.
  M-1179; C, 12-14 Mn, Cr, bal Fe.
  Heat treated: 180,000-200,000 TS.
  For crusher plates, hammers, harrowers, mixers; wear and abrasion resistant.

J.A.C.-ARF.
  M-1179; C, Ni, Cr, Mo, Bal Fe.
  Heat treated:; 1608000-180,000 TS; 120,000-140,000 YS; 12 El; 25 RA.
  For crusher plates, hammers, harrowers, mixers; easily machined and welded.

J.A.C.-B 1.
  M-1179; C, bal Fe.
  Annealed: 80,000 TS; 50,000 YS; 18 El; 35 RA.
  For general usage; Bessemer steel.

J.A.C.-B 1 S.
  M-1179; C, bal Fe.
  Annealed: 80,000 TS; 50,000 YS; 22 El; 40 RA.
  For general usage; Bessemer steel.

J.A.C.-B 2.
  M-1179; C, bal Fe.
  Annealed: 90,000 TS; 60,000 YS; 16 El; 30 RA.
  For general usage; Bessemer steel.

J.A.C.-B 2 S.
  M-1179; C, bal Fe.
  Annealed: 90,000 TS; 60,000 YS; 20 El; 35 RA.
  For general usage; Bessemer steel.

J.A.C.-B 3.
  M-1179; C, bal Fe.
  Annealed: 110,000 TS; 66,000 YS; 12 El; 25 RA.
  For general usage; Bessemer steel.

J.A.C.-E 1.
  M-1179; C, 0.03 P., 0.3 S, bal Fe.
  Normalized: 71,000 TS; 48,000 YS; 25 El; 50 RA.
  For motor frames, rotors, dynamo parts; high magnetic permeability.

J.A.C.-E 2.
  M-1179; C, bal Fe.
  Normalized 80,000 TS; 52,000 YS; 22 El; 45 RA.
  For wheels, gears, axles; high yield strength.

J.A.C.-E 3.
  M-1179; C, bal Fe.
  Normalized: 90,000 TS; 60,000 YS; 20 El; 35 RA.
  For bridge bearings, rudder frames, gears, naval construction.

J.A.C.-E 4.
  M-1179; C, bal Fe.
  Normalized: 100,000 TS; 64,000 YS; 20 El; 35 RA.
  For machine tool parts, bridge gear boxes.

J.A.C.-E 5.
M-1179; C, bal Fe.
Normalized: 110,000 TS; 72,000 YS; 17 El; 32 RA.
For gears, axles, couplings; water hardening.

J.A.C.-E 6.
M-1179; C, bal Fe.
Normalized: 120,000 TS; 80,000 YS; 15 El; 25 RA.
For pinions, block dies, gears; wear resistant.

J.A.C.-EA 1.
M-1179; C, Cr, Mn, bal Fe.
Normalized: 120,000-160,000 TS; 80,000-110,000 YS; 15-7 El.
For gears, excavator parts, tractors, cranes, dies; tough, wear resistant.

J.A.C.-EA 2.
M-1179; C, Ni, bal Fe.
Normalized: 120,000 TS; 96,000 YS; 14 El.
For drills, pinions, roller mill cylinders, brake drums; wear and shock resistant.

J.A.C.-EA 3.
M-1179; C, Cr, Ni, bal Fe.
Normalized: 150,000 TS; 120,000 YS; 12 El.
For drills, pinions, roller mill cylinders; wear and shock resistant.

J.A.C.-EA 4.
M-1179; C, Cr, Ni, Mo, bal Fe.
Normalized: 160,000-180,000 TS; 110,000-150,000 YS; 10 El.
For rollers, gears, valves; heat resistant to 500°F.

J.A.C.-EA 5.
M-1179; C, Cr, Mo, bal Fe.
Normalized: 130,000-150,000 TS; 90,000-100,000 YS; 10 El.
For rollers, gears, valves; shock and wear resistant.

J.A.C.-EA 6.
M-1179; C, Cr, Mo, bal Fe.
Normalized: 100,000-120,000 TS; 80,000-90,000 YS; 15 El.
For valves, collectors, piping; maximum operating temp. 550°F.

JACKMANIZED STEEL.
M-477; 0.5 C, bal Fe.
For dredger bucket pins, bushes, tin rolls.

JACKSBERG.
M-447; 0.7 C, 18 W, 4 Cr, 1 V, bal Fe.
For high speed tools and cutters; high speed steel.

JACKSONS BUTTON ALLOY.
M-England; 30-35 Zn, 2-5 Sb, bal Cu.
For ornamental and architectural parts; high strength.

JACOBY METAL.
M-Eng.; 85 Sn, 10 Sb, 5 Cu.
For bearings; liners; Babbitt.

JACOL 9.
M-1488; 0.02 max C, 7.0 Mn, 0.6 max Si, 16.5 Cr, 12.5 Ni, 3.8 W, bal Fe.
Austenitic stainless steel; weldable.
AFNOR Z 2 CNMW 17-13-7.

JACOL 1886.
M-1546; 0.08 C, 1.4 Cr, 1.1 Mo, 0.6 V, bal Fe.
Heat treated: 107,000 TS; 93,000 YS; 15 El.
For gas turbine and steam turbine rotors and discs.
Holds properties up to 600°C. Resists oxidation up to 550°C.

JACOL-CVM6.
M-1546; 0.17 C, 5.5 Cr, 0.55 Mo, 0.25 V, bal Fe.
Annealed: 64,000 TS; 40,000 YS; 32% El.
Heat treated: 107,000 TS; 88,000 YS; 17% El.
For piping equipment in oil refineries.
Resistant to oxidation to 750°C.

JACONA METAL.
M-Eng.; 70 Pb, 20 Sb, 10 Sn.
For bearings, liners; Babbitt.

JADE.
M-1120; 1.0 C, bal Fe.
For reamers, taps, punches, cold work tools; Type W1, water hardened.

JADOT J1.
M-1297; 0.08 max C, 13 Cr, bal Fe.
Annealed: 75,000 TS; 40,000 YS; 35 El; 70 RA; 155 Brin.
For valves, cutlery, surgical instruments; Type 403; corrosion resistant.

JADOT J1.
M-1297; 0.15 max C, 11.5-13.5 Cr, bal Fe.
Heat treated: 120,000-135,000 TS; 110,000-117,000 YS; 16-15 El; 63-58 RA; 220-240 Brin.
For cutlery, valves, turbine blades, fasteners; Type 410; corrosion resistant.

JADOT J2.
M-1297; 0.26-0.45 C, 13 Cr, bal Fe.
Annealed: 95,000 TS; 50,000 YS; 25 El; 55 RA; 196 Brin.
For valves, cutlery, surgical and dental instruments; Type 420; corrosion resistant.

JADOT JA.
M-1297; 0.08-0.20 C, 17-19 Cr, 8-10 Ni, 2 max Mn, bal Fe.
Annealed: 85,000 TS; 35,000 YS; 60 El; 70 RA; 150 Brin.
Cold drawn: 125,000 TS; 958000 YS; 25 El; 55 RA; 277 Brin.
For oil refinery and chemical plant equipment; Type 302; stainless, austenitic.

JADOT JA1.
M-1297; 0.10 max C, 16-18 Cr, 10-14 Ni, 2-3 Mo, bal Fe.
Annealed: 85,000-95,000 TS; 35,000-45,000 YS; 60-50 El; 75-60 RA; 150-190 Brin.
For chemical plant equipment, mixers, agitators, filters; Type 316; austenitic, stainless.

JADOT JA2.
M-1297; 0.10 C, 18 Cr, 10 Ni, 2.5 Mo, bal Fe.
Annealed: 85,000 TS; 35,000 YS; 50 El; 65 RA; 160 Brin.
For acid resistant chemical plant equipment, mixers; Type 316; stainless, austenitic.

JADOT JA2E.
M-1297; 0.10 C, 19 Cr, 12 Ni, 3.5 Mo, bal Fe.
Annealed: 90,000 TS; 40,000 YS; 45 El; 60 RA; 180 Brin.
For acid resistant chemical plant equipment; mixers; Type 317; stainless, austenitic.

JADOT JAA.
M-1297; 0.07 C, 17 Cr, 13 Ni, 4.5 Mo, bal Fe.
Annealed: 90,000 TS; 40,000 YS; 45 El; 60 RA; 180 Brin.
For acid resistant chemical plant equipment, mixers; Type 317; stainless, austenitic.

JADOT JAS.
M-1297; 0.12 C, 18 Cr, 8 Ni, bal Fe.
Annealed: 85,000 TS; 35,000 YS; 60 El; 70 RA; 150 Brin.
For chemical plant equipment, tanks, mixers, filters; Type 304; stainless, austenitic.

JADOT JD.
M-1297; 0.25 max C, 15 Cr, 35 Ni, bal Fe.
For furnace parts and equipment, salt pots; Type 330; heat resistant.

JADOT JH.
M-1297; 0.12 max C, 14-18 Cr, bal Fe.
Annealed: 70,000 TS; 40,000 YS; 30 El; 55 RA; 150 Brin.
Cold drawn: 130,000 TS; 120,000 YS; 2 El; 185 Brin.
For oil refinery equipment, bolts, kitchen sinks; Type 430; stainless, ferritic.

JADOT JH1.
M-1297" 0.2 max C, 24-26 Cr, 1 max Si, bal Fe.
Annealed: 75,000-85,000 TS; 50,000-60,000 YS; 12 El; 180 Brin.
For furnace parts, heat treating boxes; Type 442; corrosion and heat resistant.

JADOT JH1A.
M-1297; 0.35 C, 23-27 Cr, 0.25 max N, bal Fe.
Annealed: 90,000 TS; 60,000 YS; 20 El; 45 RA; 180 Brin.
Cold drawn: 175,000 TS; 155,000 YS; 2 El; 25 RA; 250 Brin.
For furnace parts, preheaters, annealing boxes; Type 446; corrosion resistant.

JADOT JH2.
M-1297; 0.2 max C, 22-24 Cr, 12-15 Ni, 2 max Mn, 3.5 max Si, bal Fe.
Annealed: 85,000-95,000 TS; 40,000-50,000 YS; 45-55 El; 150-185 Brin.
For heat treat boxes, furnace parts and equipment; Type 309; austenitic, heat resistant.

JADOT JH3.
M-1297; 0.25 max C, 24-26 Cr, 19-22 Ni, bal Fe.
Annealed: 958000 TS; 45,000 YS; 50 El; 65 RA; 180 Brin.
At 1200°F: 57,000 TS; 22,000 YS; 32 El; 45 RA.
For furnace parts, valves, pumps, engine parts; Type 310; heat resistant, austenitic.

JADOT JH3AS.
M-1297; 0.12 C, 20 Cr, 20 Ni, bal Fe.
For furnace parts, salt pots; corrosion and heat resistant.

JADOT JH4.
M-1297; 0.25 max C, 24-26 Cr, 19-22 Ni, bal Fe.
Annealed: 95,000 TS; 45,000 YS; 50 El; 65 RA; 180 Brin.
At 1200°F: 57,000 TS; 22,000 YS; 32 El; 45 RA.
For furnace parts, valves, pumps, engine components; Type 310; austenitic heat resistant.

JADOT JH6.
M-1297; 0.10 max C, 18-20 Cr, 11-14 Ni, 3-4 Mo, bal Fe.
Annealed: 85,000-95,000 TS; 35,000-45,000 YS; 60-50 El; 75-60 RA; 150-190 Brin.
For acid resistant chemical plant equipment; Type 317; stainless, austenitic.

JAE METAL.
M-121; 70 Ni, 30 Cu.
Annealed: 63,000 TS; 22,000 YS; 52 El; 81 RA; 110 Brin.
For compensating shunts on magnetic instruments; high reg. temp. coef. of magnetic permeability.

JALCASE 100.
M-173; 0.40-0.48 C, 1.35-1.65 Mn, 0.24-0.33 S, bal Fe.
Cold drawn: 115,000 TS; 100,000 YS; 10-20 El; 30-50 RA; 241 Brin.
For screw machine products, gears, bolts; free-cutting stress stabilized.

J-ALLOY.
0.76 C, 23 Cr, 6 Ni, 6 Mo, 2 Ta, bal Co.
For jet engine and gas turbine parts; high heat resistant.

JALLOY 280.
M-173; 0.25-0.31 C, 1.35-1.65 Mn, 0.1-0.2 Mo, 0.2 min Cu, Al, B, Ti bal Fe.
Heat treated: 115,000 TS; 100,000 YS; 15 El; 260-300 Brin.
For liners, power shovels, mining equipment, fan blades; resists impact and abrasion.

JALLOY 320.
M-173; 0.25-0.31 C, 1.35-1.65 Mn, 0.1-0.2 Mo, 0.2 min Cu, Al, B, Ti, bal Fe.
Heat treated: 300-340 Brin.
For liners, power shovels, mining equipment, fan blades, stone crushers; resists impact and abrasion.

JALLOY AR.
M-173; 0.25-0.31 C, 1.65 max Mn, 0.15-0.30 Si, 0.35 max Mo, 0.20 min Cu, 1.20 max Cr, 0.0005 min B, bal Fe.
90,000 min YS.
Good abrasion resistance, atmosphere corrosion resistance.
For snowmobile skis and cleats.

JALLOY AR-280.
M-173; 0.25-0.31 C, 1.65 max Mn, 0.15-0.30 Si, 0.35 max Mo, 0.20 min Cu, 1.20 max Cr, 0.0005 min B, bal Fe.
130,000 psi YS (typical).
Good abrasion resistance, atmosphere corrosion resistance.
For floor plates, coal chutes, mining equipment.

JALLOY AR-320.
M-173; 0.25-0.31 C, 1.65 max Mn, 0.15-0.30 Si, 0.35 max Mo, 0.20 min Cu, 1.20 max Cr, 0.0005 min B, bal Fe.
140,000 psi YS (typical).
Good abrasion resistance, atmosphere corrosion resistance.
For mining equipment, off-highway equipment.

JALLOY AR-360.
M-173; 0.25-0.31 C, 1.65 max Mn, 0.15-0.30 Si, 0.35 max Mo, 0.20 min Cu, 1.20 max Cr, 0.0005 min B, bal Fe.
157,000 psi YS. (typical).
Excellent abrasion resistance.
For coal chutes, floor plates, mining equipment.

JALLOY AR-400.
M-173; 0.25-0.31 C, 1.65 max Mn, 0.15-0.30 Si, 0.35 max Mo, 0.21 min Cu, 1.20 max Cr, 0.0005 min B, bal Fe.
184,000 psi YS (typical).
Excellent abrasion resistance.
For snowmobile skis and cleats, coal haulers.

JALLOY AR-Q.
M-173; 0.25-0.31 C, 1.65 max Mn, 0.15-0.30 Si, 0.35 max Mo, 0.20 min Cu, 1.20 max Cr, 0.0005 min B, bal Fe.
217,000 psi YS (typical).
Good abrasion resistance, atmosphere corrosion resistance.
For highly stressed highway equipment.

JALLOY S-90.
M-173; 0.10-0.20 C, 1.50 max Mn, 0.50 max Si, 0.30 max Mo, 1.50 max Cr, 0.0005 min B, bal Fe.
90,000 psi min YS.
Good-forming, weldability, impact resistance.
For tractor components, railroad car equipment.
ASTM A-514.

JALLOY S-100.
M-173; 0.10-0.20 C, 1.50 max Mn, 0.50 max Si, 0.30 max Mo, 1.50 max Cr, 0.0005 min B, bal Fe.
100,000 psi min YS.
Good cold-forming, weldability, impact resistance.
For dump trucks, agricultural equipment.
ASTM A-514, A-517.

JALLOY S-110.
M-173; 0.10-0.20 C, 1.50 max Mn, 0.50 max Si, 0.30 max Mo, 1.50 max Cr, 0.0005 min B, bal Fe.
110,000 psi min YS.
Good cold-forming, weldability, impact resistance.
For mobile cranes, booms, tractor components.
ASTM A-514, A-517.

JALLOY S-340.
M-173; 0.10-0.20 C, 1.50 max Mn, 0.50 max Si, 0.30 max Mo, 1.50 max Cr, 0.000 B, bal Fe.
150,000 psi YS (typical).
Good Weldability, abrasion resistance, impact resistance.
For dump trucks, coal chutes, tractor parts.

JALTEN NO. 1.
M-173; 0.15 max C, 1.3 max Mn, 0.20 min Cu, 0.035-0.065 V, bal Fe.
Rolled: 65,000 TS; 50,000 YS; 22 El.
For structural parts, truck bodies, mine cars; good formability and weldability.
ASTM A441.

JALTEN NO. 2.
M-173; 0.15 max C, 1.4 max Mn, 0.14 max P, 0.30 min Cu, bal Fe.
Hot rolled: 70,000 TS; 50,000 YS; 22 El.
For structural parts, truck bodies, railroad cars; good formability and weldability.

Section I: Alloy Data / 769

JALTEN NO. 3.
M-173; 0.25 max C, 1.6 max Mn, 0.20 min Cu, bal Fe.
Hot rolled: 70,000 TS; 50,000 YS; 22 El.
For construction equipment, truck bodies; moderate formability.
ASTM A440.

JALWELD.
M-173; 0.15 C, bal Fe.
For welding rods.

JAMAG.
M-1640; 2.2 C, 1.6 Si, 0.3 Mn, 0.1 Mg, bal Fe.
Heat treated: 50,000-110,000 YS.
For electrical parts, where high rotation speeds are involved.
High magnetic permeability, malleable iron.

JAMALEX.
M-1640; 2.2 C, 1.6 Si, 0.3 Mn, 0.05 Mg, bal Fe.
Heat treated: 65,000-140,000 TS; 50,000-110,000 YS; 2-9 El.
For gears, sprockets, hubs, cams, bearings, wear resistant parts.
Malleable iron, high magnetic permeability.

JAMAPPES BRASS.
M-Eng.; 0.2 Sn, 1.4 Pb, 33.4 Zn, bal Cu.
For hardware, fittings, forgings; good workability, free-cutting.

JAMISON HOBBING IRON.
M-1423; 0.05 C, bal Fe.
For plastic molding hobs.

JAMISON K46.
M-1423; 0.85 C, 1.2 Mn, 1.5 Cr, 0.5 W, 0.1 V, bal Fe.
For blanking and forming dies; reamers, punches; oil hardened, non-deforming.

JAMISON K-46 O.H.
M-1423; 0.90 C, 1 Mn, 0.5 Cr, 0.5 W, bal Fe.
For reamers, broaches, punches, cutters; Type 01; oil hardened, non-deforming.

JAMISON SPECIAL.
M-1423; 0.7-1.4 C, 0.3 Mn, 0.3 Si, bal Fe.
For drills, reamers, hobs, planer tools, springs; Type W1; water hardened.

JANO.
M-344; C, 4 Cr, 2 V, 1. W, 0.6 Mo, bal Fe.
For tools, cutters; high speed steel.

JANUS.
M-344; C, bal Fe.
For cold dies; very tough.

JANUS EXTRA ZH.
M-614; 1.0-1.1 C, 0.25 max Mn, 0.25 max Si, bal Fe.
For drills, taps, reamers, broaches, hobs; Type W1; water hardened.

JANUS GHC17M.
M-614; 0.35 C, 17 Cr, 2 Mo, bal Fe.
For acid-resistant chemical plant equipment; corrosion resistant.

JANUS GR15.
M-614; 0.15 max C, 0.75 Si, 13 Cr, 1.8 max Ni, bal Fe.
Annealed: 75,000 TS; 40,000 YS; 35 El; 70 RA; 150 Brin.
Cold drawn: 100,000 TS; 85,000 YS; 1m El; 60 RA; 205 Brin.
Type 410; corrosion resistant.

JANUS GR18.
M-614; 0.25 max C, 0.7 Si, 17 Cr, 1.8 max Ni, bal Fe.
For furnace parts, chemical plant equipment; corrosion resistant.

JANUS GR25.
M-614; 0.25 C, 0.7 Si, 14.5 Cr, 1.8 max Ni, bal Fe.
Annealed: 95,000 TS; 50,000 YS; 25 El; 55 RA; 195 Brin.
Cold drawn: 105,000 TS; 85,000 YS; 17 El; 50 RA; 215 Brin.
For cutlery turbine blades, surgical instruments; Type 420; corrosion resistant.

JANUS GSA.
M-614; 0.15 C, 1.5 Si, 18 Cr, 8.5 Ni, bal Fe.
Annealed: 85,000 TS; 40,000 YS; 55 El; 65 RA; 160 Brin.
For chemical plant equipment, tanks, mixers, vessels; Type 302; stainless austenitic.

JANUS GSAT.
M-614; 0.12 max C, 18 Cr, 9.5 Ni, Cb = 8 x C, bal Fe.
Annealed: 90,000 TS; 45,000 YS; El; 60 RA; 170 Brin.
For chemical plant equipment welded structures; Type 346; stainless, austenitic.

JANUS GSAW.
M-614; 0.07 max C, 18, Cr, 9.5 Ni, bal Fe.
Annealed: 85,000 TS; 40,000 YS; 55 El; 65 RA; 160 Brin.
For chemical plant equipment, welded structures; Type 304; stainless, austenitic.

JANUS GSB.
M-614; 0.15 C, 2 Si, 18 Cr, 9.5 Ni, 2 Mo, bal Fe.
Annealed: 90,000 TS; 45,000 YS; 50 El; 60 RA; 180 Brin.
For acid-resistant chemical plant equipment; Type 316; stainless, austenitic.

JANUS GSBT.
M-614; 0.12 max C, 18 Cr, 10.5 Ni, 2 Mo, Cb = 10 x C, bal Fe.
Annealed: 85,000 TS; 40,000 YS; 50 El; 60 RA; 180 Brin.
For acid resistant, welded structures, mixers, filters; Type 316 Cb; stainless, austenitic.

JANUS GSBW.
M-614; 0.07 max C, 18 Cr, 10.5 Ni, 2 Mo, bal Fe.
Annealed: 80,000 TS; 35,000 YS; 55 El; 65 RA; 160 Brin.
For chemical plant equipment, mixers, agitators, filters; Type 316L; stainless, austenitic.

JANUS HC17M.
M-614; 0.35 C, 17 Cr, 2 Mo, bal Fe.
For acid-resistant chemical plant equipment; corrosion resistant.

JANUSIT-A.
M-614; 0.4 C, 1.3 Si, 27 Cr, 4 Ni, bal Fe.
For furnace parts and equipment, heat treating boxes; heat resistant.

JANUSIT-B.
M-614; 0.4 C, 2 Si, 22 Cr, 9.5 Ni, bal Fe.
For furnace parts, heat treating boxes; corrosion and heat resistant.

JANUSIT-C.
M-614; 0.4 C, 2 Si, 26 Cr, 14 Ni, bal Fe.
For furnace parts and equipment, heat treating boxes; corrosion and heat resistant, austenitic.

JANUSIT-D.
M-614; 0.4 C, 2 Si, 25 Cr, 19 Ni, bal Fe.
For furnace parts and equipment, heat treating boxes; corrosion and heat resistant, austenitic.

JANUSIT-E.
M-614; 0.5 C, 1.8 Si, 25 Cr, 30 Ni, bal Fe.
For furnace parts and equipment, heat treating boxes; corrosion and heat resistant, austenitic.

JANUS MH.
M-614; 1.0 C, 0.25 max Si, 0.25 max Mn, bal Fe.
For drills hobs, reamers, taps, punches; Type W1; water hardened.

JANUS R3V.
M-614; 0.30 C, 14 Cr, Ni, V, bal Fe.
Annealed: 100,000 TS; 60,000 YS; 40 El; 50 RA; 200 Brin.
For oil refinery equipment; corrosion and creep resistant.

JANUS R15.
M-614; 0.15 max C, 0.4 Si, 13 Cr, bal Fe.
Annealed: 75,000 TS; 40,000 YS; 35 El; 70 RA; 155 Brin.
Cold drawn: 100,000 TS; 85,000 YS; 1m El; 60 RA; 205 Brin.
For valves, turbine blades, tableware, hardware; Type 410; corrosion resistant.

JANUS R15W.
M-614; 0.12 max C, 0.4 Si, 13 Cr, bal Fe.
Annealed: 75,000 TS; 40,000 YS; El; 70 RA; 155 Brin.
Cold drawn: 100,000 TS; 85,000 YS; 17 El; 60 RA; 205 Brin.
For valves, turbine blades, tableware, hardware; Type 410; corrosion resistant.

JANUS R17.
M-614; 1.0 C, 18 Cr, W, Mo, V, bal Fe.
Annealed: 110,000 TS; 65,000 YS; 17 El; 32 RA; 22 Brin.
Heat treated: 280,000 TS; 270,000 YS; 3 El; 15 RA; 550 Brin.
For bearings, sleeves, liners, corrosion resistant, wear resistant.

JANUS R18.
M-614; 0.22 C, 0.4 Si, 17 Cr, 1.5 Ni, bal Fe.
Annealed: 120,000 TS; 90,000 YS; 15 El; 20 RA; 260 Brin.
Heat treated: 180,000 TS; 130,000 YS; 10 El; 15 RA; 420 Brin.
For pump shafts, surgical trusses, valve trim; Type 431; corrosion resistant.

JANUS R25.
M-614; 0.2 C, 0.4 Si, 13 Cr, bal Fe.
Annealed: 95,000 TS; 50,000 YS; 25 El; 55 RA; 195 Brin.
Cold drawn: 105,000 TS; 85,000 YS; 17 El; 50 RA; 215 Brin.
For cutlery, valves, surgical instruments; Type 420; stainless, hardenable.

JANUS R45.
M-614; 0.40 C, 14 Cr, 0.4 Si, bal Fe.
Annealed: 90,000 TS; 45,000 YS; 50 El; 65 RA; 180 Brin.
For oil refinery equipment, cutlery, valves; Type 420; corrosion resistant.

JANUS R50.
M-614; 0.50 C, 16 Cr, 0.4 Si, bal Fe.
For oil refinery equipment; corrosion resistant.

JANUS SA.
M-614; 0.15 max C, 0.4 Si, 18 Cr, 8.5 Ni, bal Fe.
Annealed: 85,000 TS; 40,000 YS; 50 El; 65 RA; 160 Brin.
For chemical plant and oil refinery equipment; Type 302; austenitic, stainless.

JANUS SAT.
M-614; 0.12 max C, 18 Cr, 9.5 Ni, Cb = 8 x C, bal Fe.
Annealed: 90,000 TS; 45,000 YS; 45 El; 60 RA; 170 Brin.
For chemical plant equipment, welded structures; Type 346; stainless, austenitic.

JANUS SAW.
M-614; 0.07 max C, 18 Cr, 9.5 Ni, bal Fe.
Annealed: 85,000 TS; 40,000 YS; 55 El; 65 RA; 160 Brin.
For chemical plant equipment, welded structures; Type 304; stainless, austenitic.

JANUS SB.
M-614; 0.15 max C, 18 Cr, 8.5 Ni, bal Fe.
Annealed: 80,000 TS; 35,000 YS; 55 El; 65 RA; 160 Brin.
For chemical plant equipment tanks; Type 302; stainless, austenitic.

JANUS SBT.
M-614; 0.12 max C, 18 Cr, 10.5 Ni, 2 Mo, Cb = 8 x C, Fe.
Annealed: 85,000 TS 40,000 YS; 50 El; 60 RA; 180 Brin.
For acid resistant welded structures; Type 316 Cb; stainless, austenitic.

JANUS SBW.
M-614; 0.07 max C, 18 Cr, 10.5 Ni, 2 Mo, bal Fe.
Annealed: 80,000 TS; 35,000 YS; 55 El; 65 RA; 160 Brin.
For chemical plant equipment, mixers, agitators, tanks; Type 316L; stainless, austenitic.

JANUS SPEZIAL.
M-614; 1.4 C, 13 W, V, 0.30 Mn, bal Fe.
For cutters, tools; fast finishing.

JANUS SPEZIAL EXTRA.
M-614; 1.3 C, 0.2 max Cr, 4.7 W, bal Fe.
For cutters, tools; fast finishing.

JANUS V.
M-614; 0.9 C, 1.9 Mn, 0.1 V, bal Fe.
For dies; punches, cutters, crimpers; oil hardened, nondeforming.

JANUS ZAH.
M-614; 0.70 C, 0.2k max Mn, Si, bal Fe.
For drills, punches, crimpers, springs; Type W1; water hardened.

JANUS ZH.
M-614; 0.85 C, 0.25 max Si, 0.25 max Mn, bal Fe.
For drills, taps, reamers, hobs; Type W1; water hardened.

JAPAN 2H-SUPER.
M-1539; 0.08-0.16 C, 0.6-1.2 Mn, 1.0 max Ni, 0.5 max Cu, 0.4 max Mo, bal Fe.
Heat treated: 110,000-114,000 TS; 90,000 min YP; 22 min El.
For mine cars, agricultural equipment, railroad car and automobile bodies.
High-strength low-alloy constructional steel.

JAPAN 2H-ULTRA.
M-1539; 0.08-0.16 C, 0.6-1.2 Mn, 0.15-0.50 Cu, 1.5 max 0.8 max Cr, 0.7 max Mo, 0.1 max V, 0.006 max B, bal Fe.
Heat treated: 114,000-135,000 TS; 100,000 min YP; 20 min El.
For mine and railroad cars, booms, agricultural equipment, structures. High-strength low-alloy constructional steel.

JAPAN BRASS.
M-Japan; 66.6 Cu, 33.4 Zn.
For hardware, fixtures, radiators; high strength.

JBR ALLOY.
M-790; C, 6 Co, W, Cr, V, bal Fe.
For tools, cutters; tipped tools.

J.C./C.
M-359, M-584.
For general tools; water hardening.

JEFFALOY 55M.
M-170; 2.85-3-05 T.C., 0.5-0.7 Mn, Si, 0.7-0.9 Mo, bal Fe.
Cast: 50,000-60,000 TS; 248-269 Brin.
For castings; wear resisting.

JEFFALOY A.
M-170; 2.85-3.5 T.C., 0.3-0.7 Mn, Si, bal Fe.
Cast: 35,000-50,000 TS; 192-248 Brin.
For castings, pump parts, sprockets, housings; wear resistant cast iron, processed.

JELENKO NO. 7.
M-791; Au.
For dental work; cast.

JELLIFF MANGANIN.
M-231; 86 Cu, 12 Mn, 2 Ni.
Resistivity 290 ohms/c.m.f. at 25°C.
Wire for resistance boxes, potentiometers.

JELLIFF ALLOY 30.
M-231; 2 Ni, bal Cu.
Annealed: 30,000 TS; 30 El.

JELLIFF ALLOY 45.
M-231; 45 Ni, bal Cu.
Annealed: 60,000 TS; 30 El.
Hard: 135,000 TS; 3 El.
For thermocouples, resistance coils, low temperature heaters; temperature coefficient of res.; resists oxidation and corrosion at lower temperature; max operating temperature 930°F.

JELLIFF ALLOY 60.
M-231; 6 Ni, max Cu.
Annealed: 35,000 TS; 30 El.
Hard: 70,000 TS; 3 El.
For relays, low resistance resistors, rheostats; non-magnetic, easily soldered.

JELLIFF ALLOY 90.
M-231; 12 Ni, bal Cu.
Hard: 75,000 TS; 3 El.
Annealed: 35,000 TS; 30 El.
For resistance wire, rheostats, potentiometers; maximun operating temperature 850°F.

JELLIFF ALLOY 180.
M-231; 22 Ni, bal Cu.
Hard: 100,000 TS; 3 El.
Annealed: 50,000 TS; 30 El.
For resistance wire, rheostats, potentiometer; max operating temperature 750°F; low temp coef. of resistance.

JELLIFF ALLOY 800.
M-231; 20 Cr, 1m.5 Mn, 1.5 Mo, bal Ni.
Annealed: 150,000 TS; 30 El.
Hard: 250,000 TS; 3 El.
For precision resistors, potentiometers; electrical resistance, 800 ohms/c.m.f.

JELLIFF ALLOY A.
M-231; 80 Ni, 20 Cr.
Annealed: 95,000 TS; 30 El.
Hard: 105,000 TS; 3 El.
For industrial furnace elements, voltmeter windings, electrical instruments; max operating temperature 2100°F; high resistance to oxidation and corrosion.

JELLIFF ALLOY C.
M-231; 60 Ni, 16 Cr, bal Fe.
Annealed: 95,000 TS; 30 El.
Hard 200,000 TS; 3 El.
For heating elements, rheostats, resistors, potentiometers; max operating temperature 1700°F; heat and corrosion resistant.

JELLIFF ALLOY C.O.J.
M-231; Ni, Cr, Mn, Mo.
Fine wire with very high elec. resistivity.
Elec. resistivity: 1040 ohms/cmf.
Temp. coef. of resistance, -65 to + 150°C, ±5 ppm/°C.
Thermal EMF vs Copper, max.: 0.010 mV/°C.
Magnetic attraction: none.

JELLIFF ALLOY K.
M-231; 15.5 Cr, 5.5 Al, bal Fe.
Annealed: 100,000 TS; 20 El.
Hard: 200,000 TS; 3 El.
For high temperature heating elements, resistors; maximum operating temperature 2200°F, magnetic.

JELLIFF ALLOY K20.
M-231; 15.5 Cr, 5.5 Al, bal Fe.
Annealed: 100,000 TS; 20 El.
Hard: 200,000 TS; 3 El.
For precision resistors, power resistors; zero temperature co-efficient of resistance.

JELLIFF 800 LN (LOW NOISE).
M-231; Ni, Cr, Mn, Mo, (Similar to Jelliff 800).
Resistance alloy; 800 ohms per C.M.F.
For potentiometers and power resistors.

JELLIFF HP (HIGH PURITY) NICKEL.
M-231; 99.9% Ni, no minor consitituent over 0.05%.
Resistivity 44 ohms/c.m.f. at 25°C.
High temperature co-efficient of 0.006 per degree C.
For ballast applications and resistance thermometers.

JESSAIR.
M-69; 0.70 C, 2.0 Mn, 0.3 Si, 1.0 Cr, 1.35 Mo, bal Fe.
Heat Treated: 293,000 TS; 265,000 YS; 1 El; 2 RA; C 55 Rock.
For blanking and forming dies; trimming and notching dies; master hubs, shear blades, bending tools, mandrels, heavy duty punches.
Type A6 air hardening tool steel.

JESSOP 2B (HC).
M-69; 0.48 C, 2.75 Cr, 0.3 V, 11.25 W, bal Fe.
For piecing punches, mandrels, grippers, heading dies; hot work steel.

JESSOP 2B (LC).
M-69; 0.30 C, 3.15 Cr, 0.3 V, 10 W, bal Fe.
For forming dies; hot work steel.

JESSOP 2 B (MC).
M-69; 3 Cr, 0.25 Mn, 0.3 V, 13.5 W, bal Fe.
For tools, bolts and rivet dies; mandrels; tough, hot work steel.

JESSOP 3 C (AISI D2 CAST-TO-SHAPE).
M-69; 2.25 C, 12 Cr, 0.2 V, 0.8 Mo, bal Fe.
For coining, crimping and cutting dies, hobs, punches;; wear resistant, oil hardening.

JESSOP 3 C SPECIAL (AISI D5 CAST-TO-SHAPE).
M-69; 1.45 C, 12.5 Cr, 0.8 Mo, 3.2 Co, 0.4 Si, bal Fe.
For die casting dies; air hardening.

JESSOP 6-STAR VANADIUM.
M-261, M 486; 1 C, 0.3 Mn, 0.2 V, bal Fe.
For coining and cutlery dies, cold-heading dies; tough.

JESSOP 302.
M-69; 0.08-0.20 C, 17-19 Cr, 8-10 Ni, bal Fe.
Annealed: 85,000 TS; 40,000 YS; 50 El; 65 RA; 150 Brin.
For springs, screens, chemical plant equipment; Type 302; stainless, austenitic.

JESSOP 302-B.
M-69; 0.15 C, 2.0 Mn, 17.0-19.0 Cr, 8.0-10.0 Ni, 2.0-3.0 Si, bal Fe.
Improved scale resistance over AISI 302; for furnace parts.

JESSOP 302-S.
M-69; 0.08 max C, 17.0-18.0 Cr, 8.0-10.0 Ni, bal Fe.
Modified AISI 304.

JESSOP 303.
M-69; 0.15 max C, 17-19 Cr, 8-10 Ni, 0.07 min P, S, or Se, bal Fe.
Annealed: 85,000 TS; 40,000 YS; 40 El; 55 RA; 150 Brin.
For screw machine products, bolts, fasteners; Type 303; stainless, free-cutting.

JESSOP 304.
M-69; 0.08 max C, 18-20 Cr, 8-11 Ni, bal Fe.
Annealed: 85,000 TS; 35,000 YS; 60 El; 70 RA; 150 Brin.
Cold drawn: 180,000 TS: 125,000 YS; 10 El; 330 Brin.
For architectural molding and trim, kitchen equipment; Type 304; stainless, austenitic.

JESSOP 304L.
M-69; 0.03 max C, 2.0 max Mn, 1.0 max Si, 18.0-20.0 Cr, 8.0-12.0 Ni, bal Fe.
For boilers and pressure vessels; weldable.

JESSOP 304 N.
M-69; 0.08 max C, 2.0 Mn, 18.0-20.0 Cr, 8.0-10.5 Ni, 1.0 Si, 0.10-0.1 N, bal Fe.
Increased strength over AISI 304.

JESSOP 309.
M-69; 0.2 max C, 22-24 Cr, 12-15 Ni, bal Fe.
Annealed: 90,000 TS; 40,000 YS; 50 El; 65 RA; 180 Brin.
For furnace parts, tube supports, heat treat boxes; Type 309; stainless, austenitic.

JESSOP 309-CB.
M-69; 0.20 C, 22.0-24.0 Cr, 12.0-15.0 Ni, 2.0 Mn, 1.0 Si, Cb, bal Fe.
Columbium stabilized AISI 309 to prevent carbide precipitation at elevated temperatures.

JESSOP 309-MOD.
M-69.
Similar to AISI 309 but modified to give very hot strength in high temperature applications.

JESSOP 310.
M-69; 0.25 max C, 24-26 Cr, 19-22 Ni, bal Fe.
Annealed: 100,000 TS; 45,000 YS; 50 El; 65 RA; 180 Brin.
For furnace parts and equipment, valves, pumps, baffles; Type 310; austenitic, heat resistant.

JESSOP 316.
M-69; 0.1 max C, 16-18 Cr, 10-14 Ni, 2-3 Mo, bal Fe.
Annealed: 80,000 TS; 30,000 YS; 60 El; 80 RA; 135 Brin.
Cold drawn: 150,000 TS; 135,000 YS; 6 El; 300 Brin.
For chemical plant equipment, agitators, digesters, valve trim; Type 316; austenitic, stainless.

JESSOP 316-L.
M-69; 0.03 max C, 2.0 max Mn, 1.0 max Si, 160-18.0 Cr, 10.0-14.0 Ni, 2.0-3.0 Mo, bal Fe.
Increased corrosion and temperature resistance over AISI 316.

JESSOP 317.
M-69; 0.1 max C, 18-20 Cr, 11-14 Ni, 3-4 Mo, bal Fe.
Annealed: 85,000 TS; 40,000 YS; 50 El; 70 RA; 150 Brin.
For chemical plant equipment agitators, digesters, valve trim; Type 317; stainless, austenitic.

JESSOP 317-L.
M-39; 0.03 max C, 2.0 Mn, 1.0 Si, 18.0-20.0 Cr, 11.0-15.0 Ni, 3.0-4.0 Mo, bal Fe.
Low carbon grade of AISI 317 for welded assemblies.

JESSOP 319.
M-69; 0.07 max C, 18.0 Cr, 12 Ni, 2.5 Mo, bal Fe.
Modified AISI 316 with improved corrosion resistance.

JESSOP 319-L.
M-69; 0.03 max C, 18.0 C, 12.0 Ni, 2.5 Mo, bal Fe.
Low carbon modification of Jessop 319.

JESSOP 321.
M-69; 0.08 max C, 17-19 Cr, 8-18 Ni, Ti = 5 x C, bal Fe.
Annealed: 85,000 TS; 33,000 YS; 58 El; 75 RA; 165 Brin.
For jet aircraft, chemical plant equipment, refinery tubes; Type 321; stabilized, stainless.

JESSOP 347.
M-69; 0.08 max C, 17-19 Cr, 9-12 Ni, Cb = 10 x C, bal Fe.
Annealed: 91,000 TS; 39,50 YS; 50 El; 71 RA; 200 Brin.
For welded structures, chemical plant equipment, vessels, tanks; Type 347 stabilized, stainless.

JESSOP 348.
M-69; 0.08 max C, 2.0 max Mn, 1.0 max Si, 17.0-19.0 Cr, 9.0-13.0 Ni, Cb+Ta = 10 x C, bal Fe.
Stabilized austenitic stainless steel.

JESSOP 409.
M-69; 0.08 C, 1.0 Mn, 10.5-11.75 Cr, 1.0 Si, Ti = 6 x C, bal Fe.
Chromium stainless steel.

JESSOP 410.
M-69; 0.15 max C, 11.5-13.5 Cr, bal Fe.
Annealed: 75,000 TS; 40,000 YS; 35 El; 70 RA; 155 Brin.
Cold drawn: 100,000 TS; 85,000 YS; 17 El; 60 RA; 205 Brin.
For tableware, hardware, flat springs; Type 410; corrosion resistant.

JESSOP 416.
M-69; 0.15 max C, 12-14 Cr, 0.07 min P, S or Se, bal Fe.
Annealed: 75,000 TS; 40,000 YS; 30 El; 60 RA; 155 Brin.
Heat treated: 110,000 TS; 85,000 YS; 18 El; 55 RA; 230 Brin.
For screw machine products, shafts, fasteners, gears: type 416; stainless, free-cutting.

JESSOP 440A.
M-69; 0.60 0.75 C, 16-18 Cr, 0.75 max Mo, bal Fe.
Annealed: 95,000 TS; 55,000 YS; 20 El; 220 Brin.
Heat treated: 225,000 TS; 240,000 YS; 2 El; 550 Brin.
For needle valves, ball bearings, pivots, shafts, surgical instruments; Type 440A; stainless, hardenable.

JESSOP 440B.
M-69; 0.75-0.95 C, 16-18 Cr, 0.75 max Mo, bal Fe.
Annealed: 107,000 TS; 62,000 YS; 18 El; 35 RA; 220 Brin.
Heat treated: 280,000 TS; 270,000 YS; 3 El; 15 RA; 555 Brin.
For dental and surgical instruments, pivots, ball bearings, valves; Type 440B; stainless, hardenable.

JESSOP 440C.
M-69; 0.95-1.2 C, 16-18 Cr, 0.75 max Mo, bal Fe.
Annealed; 110,000 TS; 70,000 YS; 15 El; 30 RA; 225 Brin.
For ball bearings, valve parts, pivots, type 440C; stainless, hardenable.

JESSOP 446.
M-69; 0.35 max C, 23-27 Cr, 0.25 max N, bal Fe.
Annealed: 75,000 TS; 45,000 YS; 35 El; 65 RA; 160 Brin.
Cold drawn: 175,000 TS; 155,000 YS; 2 El; 25 RA; 250 Brin.
For annealing boxes, oil burner and furnace parts; Type 446; stainless and heat resistant.

JESSOP 494E.
M-261, M-486; 1.1 C, 0.3 Mn, 0.5 W, bal Fe.
For taps, drills, reamers; tough core.

JESSOP A-1.
M-261, M-486; 0.15 C, 0.6 Mn, 0.25 S, bal Fe.
60,000 TS; 15 El.
For bolts, screws, shafts, bushes; free-cutting, case-hardening.

JESSOP A-2 A.
M-261, M-486; 0.15 C, 0.6 Mn, bal Fe.
Rolled: 65,000 TS; 20 El.
For camshafts, tappets, rollers, spindles; case-hardening.

JESSOP A-3.
M-261, M-486; 0.22 C, 0.7 Mn, bal Fe.
For turbine casings, engine frames; good moldability.

JESSOP A-4.
M-486; 0.3 C, 0.5 Mn, bal Fe.
Rolled: 76,200-81,000 TS; 45,000-51,000 YS; 35-30 El; 60-55 RA; 150-160 Brin.
For bolts, studs, brake drums; water hardening.

JESSOP A-5.
M-486; 0.4 C, 0.8 Mn, bal Fe.
Rolled: 85,200-89.600 TS; 49,000-65,000 YS; 26 El; 55 RA; 170 Brin.
For axles, axle tubes, bolts; low stress applications.

JESSOP ALLOY C.
M-261; 2.3 C, 0.35 Mn, 13 Cr, bal Fe.
For shear blades, liners, thread rolling dies; cold work steel, nondeforming.

JESSOP A-M-2.
M-261, M-486; 0.15 C, 1.5 Mn, bal Fe.
Rolled: 80,000 TS; 20 El.
For rifle parts, general engineering parts; case-hardening.

JESSOP A-M-2 S.
M-261, M-486; 0.15 C, 1.5 Mn, 0.10 S, bal Fe.
Rolled: 80,000 TS; 18 El.
For general engineering parts; case-hardening, free-cutting.

JESSOP AM-3.
M-261, M-486; 0.25 C, 1.6 Mn, bal Fe.
For welded structures; tough.

JESSOP AM-4.
M-261, M-486; 0.35 C, 1.4 Mn, bal Fe.
For axles, shafts, gas cylinders; water hardened.

JESSOP AM-4 NICKEL.
M-261, M-486; 0.35 C, 1.3 Mn, 0.5 Ni, bal Fe.
For axles, shafts, gas cylinders; tough.

JESSOP AOO.
M-261, M-486; 0.08 C, bal Fe.
For plastic molding dies; hobbing steel.

JESSOP B-1.
M-261, M-486; 0.55 C, bal Fe.
For die blocks; oil or air hardening.

JESSOP B-1 NICKEL.
M-261, M-486; 0.55 C, 0.6 Mn, 0.5 Ni, bal Fe.
For rifle barrels, gear wheels, housings; water hardening.

JESSOP BLACK LABEL.
M-261; 0.7-1.4 C, 0.3 Mn, 0.3 Si, bal Fe.
For tools, punches, shear blades, taps, drills; water hardened; Type W1.

JESSOP BLUE LABEL.
M-261; 0.7-1.2 C, 0.3 Mn, bal Fe.
For taps, mandrels, punches, drills; water hardening.

JESSOP BX-3 (AISI A7).
M-69; 2.45 C, 5 Cr, 4.5 V, 1.1 Mo, bal Fe.
For blanking and forming dies; headers; cold work steel. Wear resistant, AISI A7.

JESSOP C.
M-486; 1.65 C, 0.45 Mn, 13 Cr, 0.7 Mo, 0.3 V, bal Fe.
For burnishing rolls, shear blades, gauges; tough, nondeforming.

JESSOP CNS NO. 1 (AISE D2).
M-69; 1.55 C, 12.5 Cr, 0.8 V, 0.8 Mo, bal Fe.
For blanking and forming dies; punches; air hardened, non-deforming. AISI D2.

JESSOP CNS-3 (AISE D4).
M-69; 2.25 C, 12.0 Cr, 0.2 V, 0.8 Mo, bal Fe.
Hardened: C 65-67 Rock.
For coining and crimping dies, forming and extruding dies, punches, hobs, taps, gauges, lamination dies. Air hardening, nondeforming, high wear resistance. Type D4 tool steel.

JESSOP CORROSION RESISTANT R-1.
M-261; 0.15 max C, 12 Cr, 1 max Ni, bal Fe.
For corrosion resistant parts; corrosion resistant.

JESSOP CORROSION RESISTANT R-2.
M-261; 0.30 C, 14 Cr, 1 max Ni, bal Fe.
For corrosion resistant parts; corrosion resistant.

JESSOP CORROSION RESISTANT R-3.
M-261; 0.12 C, 18 Cr, 11 max Ni, bal Fe.
For stainless parts; stainless.

JESSOP CORROSION RESISTANT R-4.
M-261; 0.25 C, 18 Cr, 1 Ni, bal Fe.
For corrosion resistant parts; corrosion resistant.

JESSOP E-1 A.
M-261, M-486; 0.3 C, 3.25 Ni, bal Fe.
At 70°F: 102,000 TS; 25 El.
At 1100°F: 32,000 TS; 50 El.
At 1300°F: 12,000 TS; 70 El.
For inlet valves, exhaust valves; low to medium temperature applications.

JESSOP E-4.
M-261, M-486; 0.12 C, 0.5 Mn, 5 Ni, bal Fe.
Rolled: 80,000 TS; 20 El.
For gears, pinions, shafts, collets; shock resistant, case-hardening.

JESSOP E-4 MOLYBDENUM.
M-261, M-486; 0.12 C, 0.35 Mn, 5 Ni, 0.25 Mo, bal Fe.
Rolled: 130,000 TS; 13 El.
For rocker arms, crankshafts, gears, pinions; shock resistant, case-hardening.

JESSOP E-5.
M-261, M-486; 0.6 C, 1.25 Ni, bal Fe.
For dies; high core toughness.

JESSOP E-6.
M-261, M-486; 0.15 C, 0.5 Mn, 2 Ni, 0.25 Mo, bal Fe.
Rolled: 90,000 TS: 18 El.
For pinions, shafts, rifle parts; shock resistant, case-hardening.

JESSOP E-8.
M-261, M-486; 0.55 C, 0.9 Ni, 0.25 Cr, 0.95 Mn, bal Fe.
For mandrel bars, resists heat checking.

JESSOP E-9.
M-261, M-486; 0.25 C, 0.5 Mn, 2 Ni, 0.25 Mo, bal Fe.
Rolled: 110,000 TS; 15 El.
For rocker arms, camshafts, gears, pinions; shock resistant.

JESSOP F 2.
M-261; 0.57 C, 0.7 Cr, 0.2 V, bal Fe.
For die casting dies; hot die steel.

JESSOP F-5.
M-261, M-486; 0.44 C, 1.3 Cr, 2.3 W, 0.15 V, bal Fe.
For mandrel bars; resists heat checking.

JESSOP F 7.
M-261, M-486; 0.48 C, 0.7 Ni, 1.5 Cr, 0.2 Mo, 0.12 V, 1.2 W, bal Fe.
For punches, chisels, shear blades, dies; shock resistant, durable cutting edge.

JESSOP G-2.
M-261; 0.4 C, 13 Ni, 13 Cr, 2.5 W, bal Fe.
At 70°F: 106,000 TS; 31 El.
At 1100°F: 66,000 TS; 35 El.
For valves, dies for extruding steel tubes, exhaust valves; high heat resistant, shock resistant.

JESSOP G-3.
M-261, M-486; 0.55 C, 1.5 Ni, 0.65 Cr, bal Fe.
For dies; tough.

JESSOP G-5 SPECIAL.
M-261, M-486; 0.32 C, 1.55 Ni, 1.1 Cr, 0.2 Mo, bal Fe.
For mandrel bars, gears, connecting rods; resists heat checking.

JESSOP G-7.
M-261, M-486; 0.55 C, 1.75 Ni, 0.6 Cr, 0.12 Mo, bal Fe.
For stamping dies; deep hardening, tough.

JESSOP G-8.
M-261, M-486; 0.25 C, 3 Ni, 1.2 Cr, 0.4 Mo, bal Fe.
For mandrel bars, crankshafts; resists heat checking.

JESSOP G-11.
M-261, M-486; 0.3 C, 0.6 Mn, 2.5 Ni, 0.8 Cr, 0.6 Mo, bal Fe.
Hardened: 110,000-200,000 TS; 88,000-170,000 YS; 18-10 El; 248-444 Brin.
For crankshafts, connecting rods, engine parts; shock resistant.

JESSOP G-12.
M-261, M-486; 0.4 C, 0.6 Mn, 2.5 Ni, 0.6 Cr, 0.6 Mo, bal Fe.
For connecting rods, gears, turbine parts; shock resistant.

JESSOP G 14.
M-261, M-486; 0.39 C, 0.5 Mn, 4.1 Ni, 1.3 Cr, 0.3 Mo, bal Fe.
For cold punches, shear blades, chisels; air hardening, shock resistant.

JESSOP G-15.
M-261, M-486; 0.12 C, 0.45 Mn, 3 Ni, 1 Cr, bal Fe.
Heat treated: 110,000-160,000 TS; 15-13 El. For gears, cams, pinions, camshafts; case-hardening, shock resistant.

JESSOP G-16.
M-261, M-486; 0.15 C, 0.5 Mn, 4.25 Ni, 1.3 Cr, 0.3 Mo, bal Fe.
Heat treated: 170,000 TS; 12 El.
For gears, pinions, tappets, valve rockers; shock resistant, case-hardening.

JESSOP G-18B.
M-486; 0.4 C, 0.8 Mn, 1 Si, 13o Ni, 13 Cr, 10 Co, 2.5 W, 2 Mo, 3 Cb, bal Fe.
At 70°F: 92,000 TS; 40 El.
At 1500°F: 44,000 TS; 9 El.
For turbine discs, rotor blades, engine exhaust valves; high heat resistant.

JESSOP G19.
M-261; 0.4 C, 0.8 Mn, 1 Si, 13 ni, 1. Cr, 2 Mo, 2.5 W, 3 Cb, 10 Co, bal Fe.
Heat treated: 102,500 TS; 30,000 YS; 42 El; 50 RA
For nozzle guide vanes in gas turbines; high heat and oxidation resistant.

JESSOP G-20.
M-261, M-486; 0.40 C, 1.35 Mn, 0.65 Ni, 0.45 Cr, 0.2 Mo, bal Fe.
For gears, shafts, drop forgings; shock resistant.

JESSOP G-21.
M-261, M-486; 0.42 C, 1.3 Si, 13 Cr, 13 Ni, 2.5 W, 1.0 Nb, bal Fe.
106,000 TS; 50,000 YS; 31 El.
For furnace parts, bolts, stays, mandrels; high hot strength.

JESSOP G-24.
M-261, M-486; 3 C, 1.5 Si, 13 Ni, 5 Cr, 5 Cu, 1.2 Mn, bal Fe.
Annealed.
For impellers, pump parts; austenitic, castings, stainless.

JESSOP G-25.
M-261, M-486; 0.17 C, 0.5 Mn, 2 Ni, 2 Cr, 0.2 Mo, bal Fe.
Heat treated: 170,000 TS; 12 El.
For camshafts, gears, bearings; shock resistant, case-hardening.

JESSOP G-26.
M-261, M-486; 0.4 C, 0.55 Mn, 1.55 Ni, 1.15 Cr, 0.15 Mo, bal Fe.
Heat treated: 92,000-150,000 TS; 70,000-120,000 YS; 22-15 El; 240-360 Brin.
For forging bars, stamping; shock resistant.

JESSOP G-27.
M-261, M-486; 0.30 C, 0.55 Mn, 3.5 Ni, 0.95 Cr, 0.4 Mo, bal Fe.
Heat treated: 120,000-160,000 TS; 100,000-140,000 YS; 17-14 El; 300-400 Brin.
For crankshafts, connecting rods, gears; shock resistant.

JESSOP G29.
M-261; 0.35-0.45 C, 11-13.5 Ni, 19-20.5 Cr, 7-10 Co, bal Fe.
Normalized: 110,000 TS; 43,000 YS; 40 El; 210 Brin.
For gas turbine components, furnace parts; high creep resistance, good to 1740°F.

JESSOP G30.
M-261; C, alloy, bal Fe.
For machine tool parts; oil hardened.

JESSOP G32.
M-261; 0.3 C, 12 Ni, 19 Cr, 16 Fe, 2.8 V, 2 Mo, 1.2 Cb, bal Co.
144,000 TS; 85,000 YS; 10 El.
For gas turbine rotors; high creep and fatigue strength.

JESSOP G-32.
M-261; 0.4 C, 1.0 Si, 13 Cr, 13 Ni, 15 Co, Cb, Mo, W, bal Fe.
For jet engine parts; high heat resistant.

JESSOP G-32.
M-486; 0.27 C, 19 Cr, 10.5 Ni, 2.2 Mo, 1.4 Cb, 0.8 Mn, 3 C, 16 Fe, bal Co.
Rolled: 126,000 TS; 78,000 YS; 10 El.
For gas turbine disc shafts, rotor blades; heat and oxidation resistant.

JESSOP G-33.
M-261, M-486; 0.4 C, 1.4 Ni, 0.6 Cr, 0.25 Mo, bal Fe.
For mandrel bars; resists heat checking.

JESSOP G34.
M-261; 1 max C, 0.8 Mn, 0.3 Si, 13 Ni, 19 Cr, 2 Mo, 46 Co, 1.2 Cb, 2.8 V, bal Fe.
Cast: 89,600 TS; 58,200 YS; 2.5 El; 2.5 RA; 300 Brin.
For turbine blades, nozzle guide vanes; high creep and heat resistance, shock resistant.

JESSOP G39.
M-261; 0.5 C, 20 Cr, 3 W, 3 Mo, 1.5 Cb, 1.5 Ta, bal Ni.
Cast: 73,200 TS; 5 El; 5 RA ; 190 Brin.
For nozzle guide vanes in gas turbines; high heat and oxidation resistant.

JESSOP G.42.
M-261; C, 19 Cr, 15 Ni, 25 Co, bal Fe.
For high temperature applications; corrosion and heat resistant.

JESSOP G42B.
M-261; 0.26 C, 25 Co, 20 Cr, 15 Ni, plus carbide formers.
Heat treated: 120,000 TS; 91,000 YS; 10 El; 12 RA; 260 Brin.
For turbine blades; high heat and creep resistance, used to 1650°F.

JESSOP G56.
M-261; 0.2 C, 13 Cr, 23 Ni, bal Fe.
Heat treated: 151,000 TS; 94,700 YS; 26 El; 27 RA.
For turbine discs; good creep and heat resistance, stainless.

JESSOP GB-18.
M-261; 0.4 C, 1.0 Si, 13 Cr, 13 Ni, 10 Co, 3 Cb 2 Mo, 2.5 W, bal Fe.
Age-hardened: 107,500 TS; 62,700 YS; 12 El; 15 RA.
For high temperature parts, gas turbine discs; heat resistant.

JESSOP G.I. SPECIAL.
M-261; C 4.1 Ni, 1.3 Cr, 0.2 Mo, bal Fe.
Hardened: 215,000 TS; 190,000 YS; 12 El; 477 Brin.
For plastic molding dies, mandrels; shock resist, air hardening.

JESSOP GO.
M-261, M-486; 0.33 C, 0.6 Mn, 3.5 Ni, 0.75 Cr, bal Fe.
For collets, cold punches, dies; oil hardening.

JESSOP H-3.
M-486; 0.6 C, 1.55 Si, 5.8 Cr, bal Fe.
At 70°F: 110,000 TS; 24 El.
At 1100°F: 40,000 TS; 50 El.
At 1500°F: 8000 TS; 140 El.
For inlet and exhaust valves; heat resistant; Silchrom type steel.

JESSOP H3A.
M-261; 0.6 C, 1.2 Si, 6 Cr, 0.5 Mo, bal Fe.
Heat treated: 135,000 TS; 100,000 YS; 20 El; 40 RA.
For turbine disks; heat resistant.

JESSOP H 4.
M-261, M-480; 0.85 C, 1.9 Mn, bal Fe.
For press tools, gauges taps, dies; non-distorting.

JESSOP-H4 SPECIAL.
M-261; 0.90 C, 1.6 Mn, bal Fe.
For dies, punches, upsetters, crimpers; Type 02; oil hardened, non- deforming.

JESSOP H-18.
M-261, M-486; 3.3 Si, 8.5 Cr, 1.2 Mo, bal Fe.
At 70°F: 114,000 TS; 24 El.
At 1100°F: 36,000 TS; 60 El.
For auto inlet and exhaust valves; resists scaling.

JESSOP H-22.
M-216, M-486; 0.2 C, 1.55 Cr, 0.5 Mo, bal Fe.
For small stamping dies; shock resistant.

JESSOP H-23.
M-261, M-486; 0.47 C, 0.6 Mn, 1.0 Si, 1.4 Cr, bal Fe.
For shear blades, punches, chipping chisels; oil hardening.

JESSOP H-24.
M-261, M-486; 0.4 C, 0.65 Mn, 1.2 Cr, 0.3 Mo, bal Fe.
Heat treated: 90,000-140,000 TS; 70,000-120,000 YS; 22-15 El; 220-360 Brin.
For gun barrels, superheater tubes; creep and shock resistant.

JESSOP H27.
M-261; 0.4 C, 0.6 Mn, 0.3 Si, 3 Cr, 0.8 Mo, bal Fe.
Oil hardened: 135,000 TS; 112,000 YS; 18 El; 40 RA.
For turbine disks; heat resistant.

JESSOP H-28.
M-261, M-486; 0.4 C, 0.8 Mn, 1 Cr, bal Fe.
Heat treated: 90,000-110,000 TS; 64,000-88000 YS; 22-18 El 240-302 Brin.
For general engineering components; water hardening.

JESSOP H-29.
M-261, M-486; 0.8 C, 2 Si, 1.5 Ni, 1..5 Cr, bal Fe.
At 70°F: 120,000 TS; 15 El.
At 1100°F: 40,000 TS; 40 El.
At 1500°F: 10,000 TS; 90 El.
For inlet and exhaust valves; resists leaded fuels and scaling.

JESSOP H30.
M-261, M-486; 0.6 C, 0.65 Mn, 0.6 Cr, bal Fe.
For chisels, rivet snaps, expander mandrels; wear and shock resistant.

JESSOP H31.
M-261; 0.4 C, 0.4 Mn, 0.3 Si, 1.1 Cr, 0.7 Mo, bal Fe.
Oil hardened: 130,000 TS; 96,000 YS; 20 El; 40 RA.
For turbine disks; heat resistant.

JESSOP H-32.
M-261; M-486; 0.3 C, 0.6 Mn, 0.55 Mo, 3.22 Cr, bal Fe.
Heat treated: 90,000-200,000 TS; 34,000-170,000 YS; 22-10 El; 248-444 Brin.
For crankshafts, cylinder liners; shock resistant, nitriding steel.

JESSOP H33.
M-261, M-486; 1 C, 0.35 Mn, 6.25 Cr, 1 Mo, bal Fe.
For shear blades, press tools, taps, dies; abrasion resistant.

JESSOP H-34.
M-261, M-486; 0.35 C, 0.6 Mn, 1 Cr, 0.7 Mo, bal Fe.
Heat treated: 90,000-14,000 TS; 70,000-120,000 YS; 22-15 El; 230-360 Brin.
For superheater tubes, gears; creep and shock resistant.

JESSOP H-35.
M-261, M-486; 0.3 C, 0.5 Mn, 1 Cr, 1.2 Mo, bal Fe.
Heat treated: 110,000 TS; 90,000 YS; 18 El; 293 Brin.
For crankshafts; shock resistant, nitriding steel.

JESSOP H-36.
M-261, M-486; 0.35 C, 0.5 Mn, 1.5 Cr, 0.2 Mo, 1 Al, bal Fe.
Heat treated: 70,000-90,000 TS; 48,000-84,000 YS; 24-17 El; 207-302 Brin.
For crankshafts, cylinder liners; shock resistant, nitriding steel.

JESSOP H-38.
M-261, M-486; 0.5 C, 3. Si, 7.5 Cr, 0.5 Mo, bal Fe.
At 70°F: 135,000 TS; 25 El.
At 1100°F: 44,000 TS; 55 El.
At 1500°F: 7000 TS; 110 El.
For auto inlet and exhaust valves; resists scaling.

JESSOP H39.
M-261; 1 C, 1.4 Cr, 0.4 Mo, bal Fe.
For bearings, heading dies; Type L7.

JESSOP H40.
M-261; 0.25 C, 0.4 Mn, 0.4 Si, 3 Cr, 0.5 Mo, 0.5 W, 0.75 bal Fe.
Oil hardened: 135,000 TS; 112,000 YS; 18 El; 45 RA.
For turbine disks; heat resistant.

JESSOP H 42.
M-261; M-486; 1.65 C, 0.45 Mn, 13 Cr, 0.7 Mo, 0.3 V, bal Fe.
For drawing and extrusion dies, gauges, shear blades; shock resistant.

JESSOP H 44.
M-261, M-486; 1 C, 0.7 Mn, 1.45 Cr, bal Fe.
For shear blades, punches, press tools; deep hardening, abrasion resistant.

JESSOP H-46.
M-261, M-486; 0.2 C, 0.6 Mn, 12 Cr, 0.6 Mo, 0.15 Cb, 0.75 V, bal Fe.
Annealed: 120,000 TS; 90,000 YS; 20 El; 241 Brin.
For aircraft gas turbine discs and blades, furnace parts; corrosion and heat resistant, hardenable.

JESSOP H49.
M-261; 0.07 max C, 5 Cr, Mo, V, bal Fe.
For hobbed molds; soft, deep hardening, hobbing steel.

JESSOP H50.
M-261; 0.37 C, 1 Si, 5 Cr, 1.1 V, 1.35 Mo, bal Fe.
For extrusion and die casting dies; hot work steel; Type H13.
Hot punches, shear blades.

JESSOP HP.
M-261, M-486; 0.9 C, 0.9 Mn, 1.15 C, 1.5 W, bal Fe.
For press tools, cold punches, gauges, taps; abrasion resistant.

JESSOP J 13.
M-261, M-486; 1.2 C, 4.35 Cr, 0.25 Mo, 4.5 V, 14 W, bal Fe.
For broaches, reamers, drills, taps; high speed steel.

JESSOP J 18.
M-261, M-486; 0.84 C, 4.5 Cr, 5.5 Mo, 1.4 V, 6 W, bal Fe.
For milling cutters, drills, taps, reamers, broaches; high speed steel.

JESSOP J 23.
M-261; 0.36 C, 0.4 Si, 2.6 Cr, 1 V, 1.8 W, 4.3 Mo, bal Fe.
For die casting dies, extrusion dies; hot die steel.

JESSOP J 27.
  M-261, M-486; 0.42 C, 3.25 Cr, 0.25 V, 14 W, bal Fe.
  For die casting dies; resists heat checking; hot die steel.

JESSOP J 32.
  M-261, M-486; 0.35 C, 4 Ni, 1.5 Cr, 5.8 W, 0.35 V, bal Fe.
  For mandrels, punches, hot stamping dies; hot work steel.

JESSOP J34.
  M-261; 0.80 C, 4 Cr, 2 V, 6 W, 5 Mo, bal Fe.
  For drills, taps, reamers, broaches, cutters; high speed steel; Type M4.

JESSOP J35.
  M-261; 0,80 C, 4 Cr, 1 V, 1.5 W, 8 Mo, bal Fe.
  For drills, taps, reamers, hobs, lathe cutters; high speed steel; Type M1.

JESSOP JS700.
  M-69; 0.03 C, 25.0 Ni, 21.0 Cr, 4.5 Mo, 1.7 Mn, 0.50 Si, 0.30 Cb bal Fe.
  Bar: 85,000 TS; 39,000 YS; 45 El; 50 RA; 170 Brin.
  For agitators, centrifuges, mixers, filters, separators, seals, valves, tanks; austenitic, stainless, resists phosphoric acid in presence of halogens.

JESSOP JS-777.
  M-69; 0.03 max C, 21.0 Cr, 25.0 Ni, 4.5 Mo, 0.30 Cu, bal Fe.
  Improved corrosion resistance over JS-700.

JESSOP K4 SPECIAL.
  M-261; 0.90 C, 1 Mn, 0.5 Cr, 0.5 W, bal Fe.
  For dies, cutters, punches, hobs, broaches; Type 01; oil hardened, non-deforming.

JESSOP K15.
  M-261, M-486; 1.25 C, 1.1 Cr, 0.2 V, 4.5 W, bal Fe.
  For broaches, reamers, taps, punches; abrasion resistant.

JESSOP K 18.
  M-261; 0.32 C, 0.95 Si, 5 Cr, 1.4 W, 1.7 Mo, bal Fe.
  For die casting dies, swaging dies; hot die steel.

JESSOP M-1.
  M-261, M-486; 0.37 C, 1.5 Mn, 0.25 Mo, bal Fe.
  Heat treated: 90,000-130,000 TS; 70,000-105,000 YS; 22-16 El; 241-341 Brin.
  For general engineering parts; shock resistant.

JESSOP M-2.
  M-261, M-486; 0.35 C, 1.5 Mn, 0.5 Mo, bal Fe.
  Heat treated: 90,000-130,000 TS; 70,000-105,000 YS; 22-16 El; 241-341 Brin.
  For general engineering components; shock resistant.

JESSOP M-3 TYPE 1.
  M-69; 1.15 C, 4.15 Cr, 3.25 V, 6.00 W, 6.00 Mo, bal Fe.
  Hardened: C 64-66 Rock.
  For broaches, chasers, drills, hobs, lathe and milling tools, reamers, tap and dies.
  Good red-hardness.
  High speed steel. High abrasion resistance.

JESSOP M-3 TYPE 2.
  M-69; 1.15 C, 4.15 Cr, 3.25 V, 6 W, 6 Mo, bal Fe.
  Hardened: C 64-66 Rock.
  For broaches, chasers, drills, hobs, lathe and planer tools, reamers, taps and dies.
  Good red-hardness.
  High speed steel. High abrasion resistance.

JESSOP M-4.
  M-69; 1.3 C, 5.5 W, 4.5 Mo, 4.0 Cr, 4.0 V, bal Fe.
  High speed steel for cutting tools, abrasion and wear resisting.
  AISI M-4.

JESSOP M-7.
  M-69; 1.0 C, 1.75 W, 8.75 Mo, 4.0 Cr, 2.0 V, bal Fe.
  Molybdenum high speed steel for cutting tools; good wear resistance.
  AISI m-7.

JESSOP M-10.
  M-69; 0.85-1.10 C, 8.0 Mo, 4.0 Cr, 2.0 V, bal Fe.
  Molybdenum high speed steel for cutting tools.
  AISI M-10.

JESSOP NO. 9 NON-MAGNETIC.
  M-69; 0.35 C, 12.4 Mn, 3.25 Ni, 4.15 Cr, 0.5 Mo, bal Fe.
  Rolled: 115,000-160,000 TS; 55,000-100,000 YS; 45-35 El; 50-40 RA.
  For armor plate, coal chutes, mine cars; austenitic, tough and wear resistant.

JESSOP NO. 91.
  M-69; 0.7 C, 0.5 Mn, 0.7 Ni, 0.5 Cr, 0.15 Mo, bal Fe.
  For circular wood saws, segment plates; oil hardened.

JESSOP NO. 139B.
  M-69; 0.82 C, 0.35 Mn, 2.6 Ni, 0.25 Cr, bal Fe.
  Heated treated: 320-620 Brin.
  For woodworking knives, saws; oil hardened, tough.

JESSOP NO. 200.
  M-69; 0.35 C, 11.5 Mn, 7.5 Ni, 0.5 max Cr, bal Fe.
  Annealed: 80,000-95,000 TS; 30,000-60,000 YS; 50-30 El; 60-30 RA.
  For circuit breakers, transformers, motor, motor shafts; nonmagnetic, austenitic.

JESSOP NO. 259 (AISI S5).
M-69; 0.54 C, 0.75 Mn, 2.0 Si, 0.15 Cr, 0.2 V, 0.5 Mo, bal Fe.
Shock resisting tool steel, AISI S5.

JESSOP NO. 4824.
M-261; 0.40 C, 1.5 Cr, 0.65 Ni, 0.5 Mo, 0.27 V, 1 Mn, 1.6 Si, bal Fe.
Heat treated: 275,000 TS; 220,000 YS; 12 El; 37 RA.
For aircraft parts, gears, shafts; oil hardened, high tensile.

JESSOP NO. 4828.
M-261; 0.26 C, 0.6 Mn, 1.8 Si, 2.9 Ni, 0.47 Mo, 1.2 Cr, bal Fe.
Heat treated: 246,000 TS; 189,000 YS; 12.5 El; 40 RA.
For aircraft parts, gears, shafts; oil hardened, high tensile.

JESSOP 0-6.
M-69; 1.45 C, 1.0 Si, 0.25 Mo, bal Fe.
Oil hardening tool steel for coining dies, trim dies, cams.
AISI 0-6.

JESSOP OK.
M-261; 0.44 C, 0.35 Mn, 1.3 Cr, 0.15 V, 2.3 W, bal Fe.
For punches, chipping chisels, pneumatic tools; tough, shock resistant.

JESSOP R-1.
M-261, M-486; 0.10 C, 13 Cr, bal Fe.
70,000-160,000 TS; 25-15 El; 153-360 Brin.
For fittings, golf club heads, hardware; hardenable, corrosion resistant.

JESSOP R 1A.
M-261, M-486; 0.08 C, 13 Cr, bal Fe.
Hardened: 125,000 TS; 110,000 YS; 16 El; 277 Brin.
For plastic molding dies; corrosion resistant.

JESSOP R-2.
M-261, M-486; 0.20 C, 13 Cr, bal Fe.
90,000-200,000 TS; 20-10 El; 207-444 Brin.
For steam valves, knives, piston rods, bolts; hardenable, corrosion resistant.

JESSOP R-3.
M-261, M-486; 0.12 C, 10.5 Ni, 17 Cr, bal Fe.
Annealed: 70,000 TS; 30 El.
For fittings, chemical plant equipment; austenitic, stainless.

JESSOP R-4.
M-261, M-486; 0.16 C, 1.5 Ni, 16.5 Cr, bal Fe.
110,000-180,000 TS; 15-10 El; 241-400 Brin.
For seaplane parts, propeller shafts, steam valves; corrosion resistant, hardenable.

JESSOP R-9.
M-261, M-486; 0.12 C, 1.2 Si, 10 Ni, 18 Cr, 2 Mo, 0.5 Ti, bal Fe.
Annealed: 70,000 TS; 30 El.
For chemical plant equipment, textile equipment; austenitic, stainless.

JESSOP R-10.
M-261, M-486; 0.12 C, 12 Ni, 12 Cr, bal Fe.
Annealed: 66,000 TS; 35 El.
For cooking utensils, holloware; austenitic, stainless.

JESSOP R-11.
M-261, M-486; 0.15 C, 1.2 Si, 14 Ni, 23 Cr, 0.8 Mo, bal Fe.
For furnace parts, annealing boxes, dampers; heat resistant, austenitic.

JESSOP R-12.
M-261, M-486; 0.12 C, 1.25 Si, 1.25 Ni, 27 Cr, 0.7 Mo, 0.4 Ti, bal Fe.
For furnace linings, flame tubes, tubes, cowling; high scale resistant.

JESSOP R-13.
M-261, M-486; 0.30 C, 0.5 Mn, 6.5 Ni, 18 Cr, 2 W, 1.6 Si, bal Fe.
For salt pots, lead baths, recuperators; heat and corrosion resistant.

JESSOP R-15.
M-261, M-486; 0.17 C, 14 Cr, S and Zr, bal Fe.
90,000 TS; 20 El; 192-269 Brin.
For food processing equipment, nuts, bolts, rivets; corrosion resistant.

JESSOP R-16.
M-261, M-486; 0.10 C, 1.2 Si, 9.5 Ni, 18.5 Cr, 0.7 Ti, bal Fe.
Annealed: 70,000 TS; 30 El.
For exhaust manifolds and cowling; austenitic, stainless.

JESSOP R-17.
M-261, M-486; 0.35 C, 0.6 Mn, 1.4 Si, 25 Ni, 18 Cr, 2 W, bal Fe.
For salt pots, carburizing boxes; austenitic, heat resistant.

JESSOP R-18.
M-261, M-486; 0.35 C, 1.8 Si, 18 Cr, 8 Ni, 3.5 W, bal Fe.
For furnace parts, impellers; scale resistant, austenitic.

JESSOP R-19.
M-261, M-486; 0.30 C, 14 Cr, S and Zr, bal Fe.
100,000 TS; 20 El; 227-277 Brin.
For shafts, steam valves, chemical industries; corrosion resistant.

JESSOP R-20.
M-261, M-486; 0.10 C, 14 Ni, 19 Cr, 1.7 Nb, bal Fe.
For exhaust manifolds, furnace parts, cowlings; austenitic, heat resistant.

JESSOP R-22.
M-486; 0.30 C, 0.6 Mn, 25 Cr, 14 Ni, 1.35 Si, 3 W, bal Fe.
For furnace parts, valves, impellers; austenitic, heat resistant.

JESSOP R-23.
M-261, M-486; 0.12 C, 22 Ni, 25 Cr, 1.7 Si, bal Fe.
For gas burners, pyrometer tubes; resists heat, austenitic.

JESSOP R-24.
M-261, M-486; 0.10 C, 10.5 Ni, 17.5 Cr, Zr, S or Se, bal Fe.
Annealed: 70,000 TS; 30 El.
For valve and pump parts, chemical plant; austenitic, stainless, free-cutting.

JESSOP R-25.
M-261, M-486; 0.12 C, 9.5 Ni, 18.5 Cr, 2.75 Mo, bal Fe.
Annealed: 70,000 TS; 30 El.
For chemical plant and textile equipment; austenitic, stainless.

JESSOP R-28.
M-261, M-486; 0.16 C, 1.9 Ni, 17 Cr, S and Zr, bal Fe.
110,000 TS; 13 El; 241-302 Brin.
For steam valves, seaplane parts, bolts; corrosion resistant.

JESSOP R-29.
M-261, M-486; 0.12 C, 0.9 Si, 21 Cr, 0.45 Mo, 0.35 Ti, bal Fe.
For nitriding boxes, cowlings for gas turbines; scale resistant.

JESSOP RTS.
M-69; 0.55 C, 0.8 Si, 0.5 Mn, 0.45 Mo, bal Fe.
Heat treated: 290,000 TS; 270,000 YS; 5 El; 15 RA; 60 Rockwell C.
For chisels, punches, dies, sledges; water hardening. Tough, shock resistant.

JESSOP S-7.
M-69; 0.50 C, 1.40 Mo, 3.25 Cr, bal Fe.
Shock resisting tool steel for punches, chisels, rivet sets, shear blades.
AISI S-7.

JESSOP-SAVILLE 6 STAR VANADIUM.
M-261; 1.0 C, 0.3 Mn, 0.25 Si, 0.12 V, bal Fe.
For coining and medal dies, striking dies, parts for cold heading operations.
Water hardening, wear resistant.

JESSOP-SAVILLE G 30.
M-261; 0.05-0.12 C, 27-29 Cr, 48-52 Co, 0.5-1.0 Si, 0.3-1.0 Mn, bal Fe.
Forged; 134,000 TS; 88,000 YS; 10 El; 350 DPH.
At 1290°F: 47,000 TS; 33,000 YS; 21 El; 18 RA.
For furnace parts, burner tips, quenching baskets, sintering machines.
Good oxidation resistance at 2200°F.

JESSOP-SAVILLE G 31.
M-261; 0.25-0.30 C, 2.0-2.2 Cb, 27-29 Cr, 48-52 Co, 0.5-1.0 Si, 0.5-1.0 Mn, bal Fe.
Cast: 89,500 TS; 64,000 YS; 4 El; 280 DPH.
For furnace baffles, grates, clinker coolers, rolls. High stress rupture properties. High resistance to thermal shock and oxidation.

JESSOP-SAVILLE G 35.
M-261; 0.15 C, 0.50 Si, 12.0 Ni, 19.5 Cr, bal Fe.
Annealed: 90,000 TS; 40,000 YS; 60 El; B 82 Rock.
At 1200°F: 45,000 TS; 30,000 YS; 32 El.
For gas turbine rotor blades and nozzle guide vanes.
Austenitic, heat and corrosion resisting, non-hardening.

JESSOP-SAVILLE G 38.
M-261; 0.15 C, 0.50 Si, 12.0 Ni, 16.0 Cr, bal Fe.
Annealed: 85,000 TS; 35,000 YS; 60 El; B 80 Rock.
Cold Drawn: 100,000 TS; 65,000 YS; 40 El; B 95 Rock.
For gas turbine and steam turbine rotors and blades, superchargers.
Austenitic, high creep and heat resistance.

JESSOP-SAVILLE G 40.
M-261; 0.20 C, 0.30 Si, 25.0 Ni, 20.0 Cr, bal Fe.
Annealed: 95,000 TS; 45,000 YS; 50 El; B 89 Rock.
Normalized: 108,000 max TS; 30 El.
For gas turbine rotor blades, jet engine components, heat exchangers. Austenitic. Heat and corrosion resisting.

JESSOP-SAVILLE G 41.
M-261; 0.17 C, 12.0 Ni, 16.0 Cr, bal Fe.
Normalized: 98,500 max TS; 36 El.
For gas turbine blades, high temperature bolts, jet engine components.
Austenitic, heat and corrosion resistant.

JESSOP-SAVILLE G 46.
M-261; 0.16 C, 0.30 Si, 11.5 Cr, Mo, V, Cb, bal Fe.
Heat treated: 146,000 max TS; 20 El.
For aircraft and gas turbine discs.
High creep strength. Good scale resistance up to 1380°F.

JESSOP-SAVILLE G 47.
M-261; 0.17 C, 0.80 Mn, 1.0 Ni, 0.80 Cr, bal Fe.
Heat treated: 124,000 max TS; 15 El.
For gudgeon pins, pinions, shafts, rifle components.
Case hardening steel. Shock resisting.

JESSOP-SAVILLE G 48.
M-261; 0.18 C, 0.80 Mn, 1.25 Ni, 0.90 Cr, 0.11 Mo, bal Fe.
Heat treated: 145,000 max TS; 12 El.
For rocker arms and shafts, camshafts, gears, clutch plates, levers, bushings.
Case hardening steel, shock resisting.

JESSOP-SAVILLE G 49.
M-261; 0.18 C, 0.80 Mn, 1.75 Ni, 0.90 Cr, 0.15 Mo, bal Fe.
Heat treated: 168,000 max TS; 12 El.
For gears, pinions, cams, camshafts, fasteners.
Case hardening steel, shock resisting.

JESSOP-SAVILLE G 50.
M-261; 0.18 C, 0.60 Mn, 2.0 Ni, 1.6 Cr, 0.2 Mo, bal Fe.
Heat treated: 190,000 max TS; 12 El.
For gears, pinions, shafts, camshafts, levers, clutch plates, magneto drive wheels.
Case hardening steel. Shock resisting.

JESSOP-SAVILLE G 51.
M-261; 0.15 C, 0.80 Mn, 0.60 Ni, 0.65 Cr, 0.11 Mo, bal Fe.
Heat treated: 101,000 max TS; 18 El.
For rocker arms and shafts, gears, camshafts, rifle components, levers, bushings, pinions.
Case hardening steel. Shock resisting.

JESSOP-SAVILLE G 52.
M-261; 0.20 C, 0.80 Mn, 0.60 Ni, 0.65 Cr, 0.11 Mo, bal Fe.
Heat treated: 124,000 max TS; 15 El.
For high duty gears and pinions, bearings, camshafts, clutch plates, tappets.
Case hardening steel. Shock resisting.

JESSOP-SAVILLE G 53.
M-261; 0.25 C, 0.80 Mn, 0.60 Ni, 0.65 Cr, 0.11 Mo, bal Fe.
Heat treated: 145,000 max TS.
For rocker arms and shafts, gears, camshafts, bushings, levers.
Case hardening. Shock resisting.

JESSOP-SAVILLE G 64.
M-261; 0.1 C, 11 Cr, 3 Mo, 3.5 W, 2 Cb, 6 Al, B, bal Fe.
For high temperature castings, turbine blades, gas turbine rotors, turbo-blower impellers.
Heat and corrosion resistant.

JESSOP-SAVILLE G 68.
M-261; 0.05 C, 1.5 Mn, 0.5 Si, 25.4 Ni, 14.0 Cr, 1.4 Mo, 2.35 Ti, 0.3 V, 0.35 Al, 0.005 B, bal Fe.
Heat treated: 16,000 TS; 115,000 YS; 20 El; 30 RA; 250-325 Brinell.
For jet engines and superchargers, turbine wheels and blades, afterburners. Precipitation hardening.
Austenitic. Similar to A286.

JESSOP-SAVILLE G 84.
M-261; 0.14 C, 10 Cr, 15 Co, 2.5 Mo, 1.25 W, 5 Al, 5.2 Ti, B, bal Ni.
For high temperature castings, turbine blades, gas turbine rotors, turbo-blower impellers.
Heat and corrosion resistant.

JESSOP-SAVILLE G 94.
M-261; 0.06 C, 6 Al, 9 Cr, 4 Cb, 10 Co, 4 W, 4 Mo, B, Zr, bal Ni.
For high temperature castings, turbine blades, gas turbine rotors, turbo-blower impellers.
Heat and corrosion resistant.

JESSOP-SAVILLE G 110.
M-261; 0.02 max C, 0.25 max Cr, 17-19 Ni, 7.0-8.5 Co, 4.6-5.2 Mo, 0.3-0.6 Ti, 0.05-0.15 Al, 0.02 Zr, 0.003 B, 0.05 Ca, bal Fe.
Heat treated: 258,000 TS; 247,000 YS; 10 El; 40 RA; 520 Vickers.
For rocket motor cases and accessories, bolts, fasteners, pressure vessels.
Maraging steel. Tough and ductile at 250,000 YS.

JESSOP-SAVILLE G 125 D V.
M-261; 0.01 C, 18.5 Ni, 9 Co, 4.85 Mo, 0.75 Ti, 0.10 Al, 0.05 Si, 0.05 Mn, bal Fe.
Aged: 318,000 TS; 294,000 YS; 8 El; 50 RA.
For rotor motor cases and accessories, fasteners, aircraft structures, pressure vessels.
Maraging stel, weldable, hardened by aging. Good notch toughness.

JESSOP-SAVILLE H 48.
M-261; 0.45 C, 0.6 Si, 4.7 Ni, 24.0 Cr, 2.9 Mo, bal Fe.
At 70°F: 150,000 max TS; 11 El.
At 1470°F: 34,000 max TS; 50 El.
For exhaust valves for aircraft and automobile engines.
Corrosion and heat resisting. Air hardening.

JESSOP-SAVILLE H 51.
M-261; 0.19 C, 0.25 Si, 0.55 Mo, 0.25 V, bal Fe.
Heat treated: 130,000 max TS; 24 El.
For gas turbine rotors and discs, high temperature stresses components.
High creep strength up to 1020°F.; shock resistant.

JESSOP-SAVILLE H 52.
M-261; 0.12 C, 0.35 Si, 16.0 Cr, bal Fe.
Bar: 90,000 max TS; 15 El.
For steam turbine and gas turbine components.
Non-hardenable. Good creep resistance. Suitable for welding.

JESSOP-SAVILLE H 53.
M-261; 0.06 C, 11 Cr, 0.8 Mo, 0.45 Cb, 0.5 V, 6.5 Co, 0.8 W, B, bal Fe.
At 75°F: 148,000 TS; 130,000 YS; 18 El; 62 RA.
At 1200°F: 56,000 TS; 50,200 YS; 28 El; 76 RA.
For gas turbine discs, compressor discs.
High heat resistance. Martensitic stainless steel.

JESSOP-SAVILLE H 55.
M-261; 0.55 C, 0.8 Mn, 1.0 Cr, 0.45 Mo, 0.5 Ni, 0.2 V, bal Fe.
Ht. Tr.: 265,000 TS; 250,000 YS; 10 El; 514 Brin.
For die blocks, forging dies, punches, turbine rotors, rollers.
High strength, tough, shock resistant.

JESSOP-SAVILLE H 59.
M-261; 0.07 C, 0.15 Cb, 11 Cr, 1.5 Mo, 3.0 Ni, 0.3 V, bal Fe.
Annealed: 80,000 TS; 38,000 YS; 24 El; B 95 Rock.
For table flatware, oil refinery and chemical plant equipment.
Corrosion resistant, non-hardenable.

JESSOP-SAVILLE-H.D.S.
M-261; 0.28 C, 3.2 Cr, 9.5 W, 0.3 V, bal Fe.
Hardened: 243,000 TS; 215,000 YP; 12 El; 37 RA; Rock C 50.
For extrusion dies and liners, forging dies, trimming dies, punches, rotar shear blades.
Hot work steel, AISI Type H21.

JESSOP-SAVILLE J 12.
M-261; 0.26 C, 3.0 Ni, 2.75 Cr, 10.0 W, 0.50 Mo, 0.30 V, bal Fe.
For forging dies for brass, die inserts.
Tough, oil hardening, hot work steel.

JESSOP-SAVILLE J-36.
M-261; 1.5 C, 4.75 Cr, 5 V, 12.5 W, 5 Co, bal Fe.
Hardened: C 64-68 Rock.
For form and planing tools, roll turning tools, broaches, milling cutters, end mills, gear cutters.
High speed steel. Highest wear resistant. High red-hardness.

JESSOP-SAVILLE J O SPECIAL.
M-261; 0.28 C, 3.2 Cr, 9.5 W, 0.3 V, bal Fe.
Heat treated: 224,000 TS; 191,000 YS; 12 El; C 51 Rock.
At 600°C: 134,000 TS; 103,000 YS; 10 El; 28 RA.
For extrusion dies and liners, forging and trimming dies, hot punches, rotary shear blades.
Hot work steel, Type H21.

JESSOP-SAVILLE K.S.A.
M-261; 0.90 C, 1.85 Mn, 0.25 Si, 0.45 Cr, 0.45 W, bal Fe.
Hardened: C 60-65 Rock.
For gages, taps, dies, chasers, cams, blanking and forming dies.
Oil hardening, tough, shock resisting, non-deforming.

JESSOP-SAVILLE R 27.
M-261; 0.42 C, 3.25 Cr, 14 W, 0.25 V, bal Fe.
For forging dies, extrusion dies, punches, shear blades.
Hot work steel. Oil hardening, good red-hardness.

JESSOP-SAVILLE R 34.
M-261; 0.30 C, 13.0 Cr, bal Fe.
Heat treated: 246,000 max TS; 10 El; C 52 Rock.
Annealed: 95,000 TS; 50,000 YS; 25 El; 196 Brin.
For knife blades, cutlery, surgical instruments, needle valves, gauges.
Hardenable, corrosion resisting steel.

JESSOP-SAVILLE R 40.
M-261; 0.21 C, 1.0 Si, 11.5 Ni, 22.5 Cr, 1.0 W, bal Fe.
For unstresses furnace parts, valves, impellers, dampers, roller hearths, burner nozzles.
Heat and corrosion resistant. Not hardenable, austenitic.

JESSOP-SAVILLE R 41.
M-261; 0.10 max C, 12.0 Ni, 17.0 Cr, 2.35 Mo, bal Fe.
Annealed: 78,000 TS; 30 El; B 78 Rock.
Cold Rolled: 125,000 TS; 100,000 YS; 25 El; 250 Brin.
Chemical plant equipment, paper making and textile plant equipment.
Stainless, acid resisting, austenitic, non-hardenable.

JESSOP-SAVILLE R 42.
M-261; 0.08 max C, 8.5 Ni, 18.0 Cr, bal Fe.
Annealed: 78,000 max TS; 30 El; B 80 Rock.
Cold Rolled; 150,000 TS; 100,000 YS 25 El; B-100 Rock.
For chemical plant equipment, aiscrews, fish knives, fittings, cooking utensils.
Austenitic, stainless steel, Type 304, non-heat treatable.

JESSOP-SAVILLE SPECIAL-BB.
M-261; 0.28 C, 3.20 Cr, 9.5 W, 0.3 V, bal Fe.
Heat treated: 224,000 TS; 191,000 YS; 12 El; C 51 Rock.
At 600°C: 134,000 TS; 103,000 YS; 10 El; 28 RA.
For extrusion dies and liners, forging dies, hot trimming dies, and punches, rotary shear blades.
Hot work steel, Type H21.

JESSOP-SAVILLE S.V.L.
M-261; 0.32 C, 1.3 Cr, 0.3 Mo, 0.5 Mn, 0.25 Si, bal Fe.
For cold punches, trimming dies, shear blades, chisels.
Oil or air hardening. Shock resistant.

JESSOP-SAVILLE W.P.S.
M-261; 2.3 C, 0.35 Mn, 0.30 Si, 13 Cr, bal Fe.
Hardened: C 65-67 Rock.
For blanking and forming dies, gauges, shear blades, thread rollers, mandrels.
Oil or air hardening. Nondeforming, wear and abrasion resistant.

JESSOP-SAVILLE ZIRCONIUM.
M-261; 1.5 Sn, 0.12 Fe, 0.10 Cr, 0.05 Ni, bal Zr.
At 70°F: 74,000 TS; 42,700 YS; 28 El; 38 RA.
At 930°F: 24,200 TS; 12,000 YS; 40 El; 60 RA.
For canning and structural material for use in pressurized water reactors.
Corrosion resistant and low neutron absorption.

JESSOPS E-0.
M-261, M-486; 0.12 C, 0.4 Mn, 3 Ni, bal Fe.
Rolled: 90,000 TS; 18 El.
For gears, pinions, shafts, collets; tough, case-hardening.

JESSOP SHEET SPRING.
M-261; 0.95 C, 0.30 Mn, bal Fe.
For tools, springs; water hardened.

JESSOPS "J-4" STEEL.
M-261; 0.5 C, 1.5 Cr, 2.2 W, 0.25 V, bal Fe.
For hand and pneumatic chisels, boiler cups, punches, shear blades, tools oil hardened.

JESSOP SPECIAL ALLOY B.
M-261, M-486; 1.6 C, 0.6 Cr, 5.5 W, bal Fe.
For drawing dies, working tools, stone cutting tools; wear resistant.

JESSOP SPECIAL H 4.
M-261; 0.9 C, 1.85 Mn, 0.45 Cr, 0.45 W, bal Fe.
For gauges, taps, dies, chasers, cams, pawls; non-deforming, oil hardening

JESSOP SPECIAL K 4.
M-261, M-486; 0.92 C, 1.25 Mn, 0.45 Cr, 0.45 W, bal Fe.
For gauges, taps, dies, chasers, cams; non-deforming, oil hardening.

JESSOP STEEL A-2.
M-261; 0.10-0.18 C, 0.6-0.09 Mn, bal Fe.
Case hardened core: 85,000 TS; 56,000 YS; 25 El; 60 RA; 180 Brin.
For case hardened parts; case-hardening steel.

JESSOP STEEL A-4 (S M-30).
M-261; 0.25-0.35 C, 0.6-0.8 Mn, bal Fe.
76,200-80,100 TS; 45,000-51,500 YS; 30-35 El; 55-60 RA; 150-160 Brin.
For shafts, axles, machinery parts; water hardened.

JESSOP STEEL A-5 (S M-40).
M-261; 0.35-0.45 C, 0.6-0.8 Mn, bal Fe.
85,200-89,600 TS; 49,000-65,000 YS; 26 El; 55 RA; 170 Brin.
For gears, pinions, shafts, machinery parts; water hardening.

JESSOP STEEL B (S M-50).
M-261; 0.5-0.6 C, 0.5-0.7 Mn, bal Fe.
89,600-112,000 TS; 51,500-56,000 YS; 20-25 El; 45-50 RA; 187-217 Brin.
For axles, shafts, wheels, shock resistant tools; water hardening.

JESSOP STEEL C-1 (S M-60).
M-261; 0.6-0.7 C, 0.4-0.6 Mn, bal Fe.
98,600 TS; 51,500 YS; 21 El; 40 RA; 200 Brin.
For blanking, drawing and trimming dies, punches, hobs, gages; abrasive resistance, non-deforming.

JESSOP STEEL C-2 (S M-80).
M-261; 0.75-0.85 C, 0.4-0.6 Mn, bal Fe.
Heat treated: 134,000-156,800 TS; 100,800-123,200 YS; 12-18 El; 40 RA; 320 Brin.
For blanking, drawing and trimming dies, punches, hobs, gages; resists scaling, corrosion and abrasion.

JESSOP STEEL E.
M-261; 0.10-0.15 C, 0.2-0.6 Mn, 3-3.5 Ni, bal Fe.
Case hardened core: 100,800-136,700 TS; 71,700-85,200 YS; 15-20 El; 4 RA; 280-300 Brin.
For case-hardened parts, gears; case-hardening steel.

JESSOP STEEL E-1.
M-261; 0.4-0.45 C, 0.4-0.85 Mn, 3-3.5 Ni, bal Fe.
Heat treated: 112,000-125,000 TS; 71,700-85,200 YS; 22-27 El; 47 RA 225 Brin.
For gears, pinions, crankshafts, structural parts; high impact strength.

JESSOP STEEL F.
M-261; 0.5-0.6 C, 0.7-0.8 Mn, 0.6-0.8 Cr, 0.25 V, bal Fe.
Heat treated: 132,200 TS; 91,800 YS; 25-30 El; 62 RA; 240 Brin.
For tools, wheels, jaws, clutches, springs; oil hardening.

JESSOP STEEL G.
M-261; 0.28-0.35 C, 0.45-0.60 Mn, 2.5-3.5 Ni, 0.75-1.0 Cr, bal Fe.
Heat treated: 112,000 TS; 78,400 YS; 20-25 El; 50 RA; 240 Brin.
For gears, pinions, shafts; oil hardening.

JESSOP STEEL G-1.
M-261; 0.28-0.35 C, 4-5 Ni, 1-1.5 Cr, 0.2 Mo, bal Fe.
Heat treated: 123,000 TS; 100,800 YS; 20 El; 45 RA; 260-270 Brin.
For gears, shafts, pinions, clutches, jaws; air hardening.

JESSOP STEEL H-2.
M-261; 0.95-1.05 C, 0.3 max Si, 0.25-0.40 Mn, 1.3-1.5 Cr, bal Fe.
For balls, shear blades; abrasion resistant.

JESSOP STEEL H-3.
M-261; 0.55-0.65 C, 1.4-1.6 Si, 0.3-0.6 Mn, 6-7 Cr, bal Fe.
Heat treated: 118,700-130,000 TS; 69,000-80,500 YS; 24-28 El; 50-60 RA; 230 Brin.
For valves, inlet and exhaust; air hardening.

JESSOPS SUPERIOR OIL HARDENING.
M-261; 0.95 C, 1.2 Mn, 0.5 Cr, 0.5 W, bal Fe.
For tools and dies, punches; non-deforming steel.

JESSOP T-8.
M-69; 0.75 C, 14.0 W, 4.0 Cr, 2.0 V, 5.0 Co, bal Fe.
Tungsten-Cobalt type high speed steel for cutting tools.
AISI T-8.

JESSOP T-15.
M-69; 1.50 C, 12.0 W, 4.0 Cr, 5.0 V, 5.0 Co, bal Fe.
High carbon, tungsten-vanadium-cobalt high speed for cutting tools. Good wear resistance.
AISI T-15.

JESSOP TOOL.
M-261; 0.7-1.2 C, bal Fe.
For tools, drills, taps; water hardened.

JESSOP TTQ.
M-261; 0.8 C, 5 Cr, 20 W, 1.6 V, 11 Co, bal Fe.
For broaches, reamers, planer and lathe cutters; high speed steel, oil hardened.

JESSOP WEX 491.
M-261; 1.0 C, 2 Cr, 2 Mn, 1 Mo, bal Fe.
For tool shanks, punches; air hardened, nondeforming.

JETALLOY 209.
M-1530; 0.02 C, 20 Cr, 10 Ni, 15 W, 2 Ti, 1 Fe, bal Co.
For high temperature applications; high creep resistance, oxidation and wear resistant.

JETALLOY 249.
M-1530; cobalt alloy.
For corrosion resistant applications; resists corrosion and abrasion.

JET FORGE.
M-115; 0.45-0.50 C, 1.3 Mo, 1.4 V, 0.9 Si, 0.3 Mn, 0.25-0.80 Cr, bal Fe.
For forging dies; air hardened, hot work steel.

JETHETE M-151.
M-1724; 0.10 C, 1.5 Mn, 1.25 Ni, 12.0 Cr, 0.6 Mg, 0.3 V, bal Fe.
Heat treated: 134,400 TS; 112,000 YS; 15 El.
For gas turbine discs and blades; creep resistant.

JETHETE M-152.
M-1724; 0.12 C, 0.7 Mn, 2.5 Ni, 12.0 Cr, 1.7 Mo, 0.3 V, bal Fe.
Heat treated: 134,400 TS; 112,000 YS; 20 El.
For gas turbine discs and blades; creep resistant.

JETHETE M 153.
M-430, M-1724; 0.10 C, 1.5 Mn, 1.5 Ni, 12.0 Cr, 1.25 Mo, bal Fe.
Heat treated: 134,000-190,000 TS; 103,000-146,000 YS; 25-12 El; 64-25 RA.
For high temperature applications; heat resistant.

JETHETE M 154.
M-430, M-1724; 0.08-0.15 C, 0.5-1.0 Mn, 2-3 Ni, 11-13 Cr, 1.5-2.0 Mo 0.25-0.40 V, 0.05 N, bal Fe.
Hardened: 147,000 TS; 129,000 YS; 10 El.
For turbine blading, bolting, gas turbine components.
Good oxidation resistance and high strength at elevated temperatures.

JETHETE M 160.
M-1724; 0.20 max C, 11-13 Cr, 1.0 max Mo, 1.0 max V, 0.7 max Cb, 1.25 max Ni, bal Fe.
At 70°F: 143,000 TS; 121,000 YS; 22 El; 56 RA.
At 570°F: 124,000 TS; 106,000 YS; 18 El; 51 RA.
At 750°F: 116,000 TS; 97,000 YS; 22 El; 65 RA.
For gas turbine compressor blades, corrosion and heat resistant.

JET LH 72.
M-578; C, bal Fe.
Steel arc welding electrode.
AWS E-7018.

JET-LH 8018-C1.
M-578; 0.12 C, 1.20 Mn, 0.03 P, 0.04 S, 0.60 Si, 2.50 Ni, bal Fe.
Steel arc welding electrode. AWS Class E8018-C1.

JET-LH 8018-C3.
M-578; 0.12 C, 0.4 Mn, 0.03 P, 0.03 S, 0.80 Si, 1.0 Ni, 0.15 Cr, 0.35 Mo, 0.05 V, bal Fe.
Steel arc welding electrode AWS Class E8018-C3.

JET-LH BU-90.
M-578; 0.14 C, 1.15 Mn, 0.60 Si, 1.40 Cr, bal Fe.
Hard surfacing arc welding electrode, for resistance to metal to metal wear.

JETWELD 1.
M-578; C, bal Fe.
Steel arc welding electrode.
AWS Class E7024.

JETWELD 2.
M-578; C, bal Fe.
Steel arc welding electrode.
AWS Class E6027.

JETWELD 2HT.
M-578; C, bal Fe.
Steel arc welding electrode.
AWS Class E7027-Al.

JETWELD 3.
M-578; C, bal Fe.
Steel arc welding electrode.
AWS Class E7024.

JETWELD LH70.
M-578; C, bal Fe.
Steel arc welding electrode.
AWS Class E7018.

JETWELD LH-90.
M-578; 0.12 C, 0.90 Mn, 0.03 P, 0.04 S, 0.06 Si, 1.25 Cr, 0.5 Mo, bal Fe.
Steel arc welding electrode.
AWS Class E8018-B2.

JETWELD LH-110M.
M-578; 0.10 C, 1.3 Mn, 0.03 P, 0.03 S, 0.60 Si, 2.0 Ni, 0.40 Cr, 0.35 Mo, 0.05 V, bal Fe.
Steel arc welding electrode.
AWS Class E 11018-M.

JETWELD LH3800.
M-578; C, bal Fe.
Steel arc welding electrode.
AWS Class E7028.

JEWEL CAST IRON.
M-Eng.; 0.88 combined C, 1.24 graphite, 0.14 Mn, 0.81 Si, bal Fe.
Cast.
For castings, gears, housings; cast iron.

JEWELER'S METAL.
M-Eng.; 91.5-83 Cu, 6.5-17 Zn, 0-2 Sn.
For jewelry; corrosion resistant.

JEWELRY BRONZE 87.5.
M-33, M-1789; 87.5 Cu, 12.5 Zn.
Annealed: 40,000 TS; 13,000 YS; 46 El.
Hard rolled: 66,000 TS; 56,000 YS; 5 El.
For costume jewelry, slide fasteners; containers; red brass.

JEWELRY BRONZE 226.
M-279; 87 Cu, 13 Zn.
Ann: 37,000-45,000 TS; 10,000-21,000 YS; 35-46 El.
Cold rolled: 42,000-86,000 TS; 16,000-75,000 YS; 2-40 El.
For jewelry, buckles, slide fasteners.
Gold color, easily buffed and finished.

JISCON.
M-1032; 2.3-2.5 C, 6.0-6.25 Si, 1.25 Cu, 0.5 Cr, bal Fe.
For high temperature castings; resists growth and scaling to 1700°F.

JISCO SILVERY.
M-1032; 5-17 Si, 1-2 Mn, 0.8-3.0 C, bal Fe.
For foundry steel and iron castings; pig iron.

JJ HOT WORK see JESSOP JJ.

J L.
M-620; 4.5 Cu, bal Al.
64,000 TS; 5 El; 120 Brin.
For light alloy parts; age-hardenable.

J & L-1211.
M-173; 0.08-0.13 C, 0.07-0.12 P, 0.08-0.15 S, 0.6-0.9 Mn, bal Fe.
Cold drawn: 82,000 TS; 80,000 YS; 14 El; 47 RA; 174 Brin.
For screw machine products, bolts, shafts; AISI-C1211, free-cutting.

J & L-1212.
M-173; 0.13 max C, 0.16-0.23 S, 0.07-0.12 P, 0.7-1.0 Mn, bal Fe.
Cold drawn: 82,000 TS; 80,000 YS; 14 El; 47 RA; 174 Brin.
For screw machine products, bolts, shafts, screws; AISI-c1212, free cutting.

J & L-1213.
M-173; 0.09 max C, 0.24-0.33 S, 0.07-0.12 P, 0.7-1.0 Mn, bal Fe.
Cold drawn: 82,000 TS; 80,000 YS; 14 El; 47 RA; 174 Brin.
For screw machine products, bolts, screws, shafts; AISI-C1213, free cutting.

J & L-12L13.
M-173; 0.09 max C, 0.70-1.00 Mn, 0.07-0.12 P, 0.25-0.35 S, 0.15-0.35 Pb, bal Fe.
For screw machine products, fasteners, hardware cams, camshafts.
Free machining, case hardening.

J & L 1215.
M-173; 0.09 max C, 0.75-1.05 Mn, 0.04-0.09 P, 0.25-0.35 S, bal Fe.
For screw machining products, fasteners, cams, gears, camshafts.
Free-machining, case hardening.

J & L COR-TEN.
M-173; 0.12 max C, 0.20-0.50 Mn, 0.70-0.15 P, 0.050 max S, 0.25-0.55 Cu, 0.30-1.25 Cr, 0.65 max Ni, 0.25-0.75 Si, bal Fe.
50,000 psi min YS.
Good weldability, atmosphere corrosion resistance. For guard rails, building materials.
ASTM A-242.

JLX-36.
M-173; 0.15 max C, 0.90 max Mn, bal Fe.
Sheet: 50,000 TS; 36,000 YS; 28 El (minimum); B 63-75 Rock.
For construction applications, bridges, tanks.
Good formability, weldability and notch toughness.

JLX-42.
M-173; 0.20 max C, 1.0 max Mn, 0.30 max Si, 0.01 min Cb or V, bal Fe.
42,000 psi min YS.
Good weldability and impact toughness.
For bumpers, agricultural equipment.

JLX-45.
M-173; 0.20 max C, 1.10 max Mn, 0.30 max Si, 0.01 min Cb or V, bal Fe.
45,000 psi min YS.
Good weldability and impact toughness.
For bumpers, railroad car components.

JLX-50.
M-173; 0.22 max C, 1.20 max Mn, 0.30 max Si, 0.1 min Cb or V, bal Fe.
50,000 psi min YS.
Good weldability. For agricultural equipment.
ASTM A-572.

JLX-50 CC.
M-173; 0.12 max C, 0.90 max Mn, 0.01 max P, 0.25 max S, 0.02 max Si, 0.01 min Cb, bal Fe.
50,000 psi min YS.
Good weldability and impact toughness.
For automobile bumpers, couplings, agricultural equipment.
ASTM A-572.

JLX-55.
M-173; 0.24 max C, 1.20 max Mn, 0.30 max Si, 0.01 min Cb or V, bal Fe.
55,00 psi min YS.
Good weldability. For agricultural equipment.

JLX-60.
M-173; 0.25 max C, 1.35 max Mn, 0.30 max Si, 0.01 min Cb or V, bal Fe.
60,000 psi min YS.
Good weldability. For railroad car components.

JLX-65.
M-173; 0.26 max C, 1.50 max Mn, 0.30 max Si, 0.01 min Cb or V, bal Fe.
65,000 psi min YS.
Good weldability.
For automobile bumpers, agriculture equipment.

JLX-70.
M-173; 0.26 max C, 1.65 max Mn, 0.30 max Si, 0.01 min Cb or V, bal Fe.
70,000 psi min YS.
Good weldability.
For truck frames, dump trucks, farm equipment.

JLX-W.
M-173; 0.20 max C, 0.50-1.0 Mn, 0.10 max Si, 0.01 min Cb, bal Fe.
Gr. 45: 60,000 TS; 45,000 YP; 24 El; B 79 Rock.
Gr. 50: 65,000 TS; 50,000 YP; 22 El; B 82 Rock.
Gr. 55: 70,000 TS; 55,000 YP; 20 El; B 83 Rock.
Gr. 60: 75,000 TS; 60,000 YP; 18 El; 8 85 Rock.
For water wheel generators, freight cars, trucks, storage tanks.
Produced to minimum yield points indicated.

JMC 77.
M-200; 40 Cu, plus Pd.
For wiping contacts.

JMC 625.
M-200; Cu, Ag, Au.
For wiping contacts.

JMC 1715 MG NI.
M-200; Ag, Mg, and others.
For electrical spring contact; conductivity; 55% IACS.

JMM 77.
M-200; Pt, Pd, Au, Ag.
Precious metal electrical contacts.

JOBBINS ALMAG 35 now USCO-JOBBINS ALMAG 35 (535.2).

JOHNSON.
9-36 Ni, 10-30 Cr, 1-10 Si, bal Fe.
For heat and corrosion resisting parts; heat and corrosion resistant.

JOHNSON BRONZE NO. 29 now SAE NO. 67.

JOHNSON BRONZE NO. 53 now SAE NO. 63.

JOHNSON LOCOMOTIVE BRONZE.
M-Eng.; 7.8 Sn, 5 Zn, bal Cu.
For railroad hardware, fittings, injectors; high strength.

JOHNSONS (A) STAINLESS STEEL SOLDER.
M-393.
For solder for stainless steel; M.P. 450°F.

JONAS & COLVER.
M-1195; 1.36 C, 0.41 Mn, bal Fe.
For tools, cutters; water hardened.

JORGENSEN 100.
M-1780; 0.16-0.20 C, 0.60-0.90 Mn, 0.20-0.30 Si, 0.50-0.75 Cr, 0.40-0.60 Mo, 1.1-1.40 Ni, 0.06-0.12 V, bal Fe.
Plate; Quenched & tempered: 115,000 psi TS; 100,000 psi YS; 16 El.
For large, highly stressed structural equipment.

JORGENSEN J-20.
M-1780; 0.20 C, 1.25 Mn, 0.04 max P, 0.25 S, 0.20 Si, bal Fe.
Free machining steel, may be carburized and hardened for gears, cams, sprockets, molds.
Similar to AISI 1119.

JORGENSEN J-45.
M-1780; 0.45 C, 1.25 Mn, 0.04 max P, 0.25 S, 0.15 Si, bal Fe.
Medium carbon free machining steel; may be hardened for use as cams, gears, trimming dies, molds.
Similar to AISI 1144.

JORGENSEN NI-MO.
M-1780; 0.28 C, 0.15-0.45 Mn, 0.15-0.35 Si, 0.25-0.60 Mo, 2.75-3.50 Ni, 0.08 V, bal Fe.
Plate; Quenched & tempered: 110,000 psi TS; 75,000 psi YS; 20 El.
For stressed structural equipment.

JOSLYN STAINLESS AISI TYPE 304.
M-560; 0.08 max C, 2.0 max Mn, 1.0 max Si, 18-20 Cr, 8-11 Ni, bal Fe.
Ann: 85,000 TS; 35,000 YS; 60 El; 149 Brin.
Austenitic, non-magnetic, corrosion resistant, hardenable by cold work only.
For shafts, dairy fittings, meat hangers and hooks, food processing equipment.
AISI 304; AMS 5639; ASTM A193.

JOSLYN STAINLESS AISI TYPE 316.
M-560 0.08 max C, 2.0 max Mn, 1.0 max Si, 0.03 max S, 16-18 Cr, 10-14 Ni, 2.0-3.0 Mo, bal Fe.
Ann: 80,000 TS; 30,000 YS; 60 El; 149 Brin.
Austenitic, non-magnetic, weldable, excellent corrosion resistance, hardenable only by cold work.
For food equipment, chemical equipment, paper, textile and photographic equipment, weld rod.
AISI 316; AMS 5648; SAE 30316.

JOSLYN STAINLESS TYPE 15-5 PH.
M-560; 0.07 max C, 1.0 max Mn, 1.0 max Si, 14.0-15.5 Cr, 3.5-5.2 Ni, 2.5-4.5 Cu, 0.15-0.45 Cb+Ta, bal Fe.
Precipitation hardening stainless steel.
Cond. H 900: 200,000 psi TS; 185,000 psi YS; 14 El; 450 Brin.
For valves, shafts, gears, bolts, studs.

JOSLYN STAINLESS TYPE 17-4 PH.
M-560; 0.07 max C, 1.0 max Mn, 1.0 max Si, 15.5-17.5 Cr, 3.0-5.0 Ni, 3.0-5.0 Cu, 0.15-0.45 Cb+Ta, bal Fe.
Precipitation hardening stainless steel.
Cond: H 900: 200,000 psi TS; 185,000 psi YS; 14 El; 420 Brin.
For valves, shafts, gears, fasteners, studs.

JOSLYN STAINLESS TYPE 303.
M-560; 0.15 max C, 2.0 max Mn, 1.0 max Si, 0.15 min S, 17-19 Cr, 8.0-10.0 Ni, 0.60 max Mo, bal Fe.
Ann: 90,000 TS; 35,000 YS; 50 El; 160 Brin.
Austenitic, free machining, non-magnetic, hardenable only by cold work, corrosion resistant.
For studs, bolts, nuts, gears, hardware fittings, screw machine parts.
AISI 303; AMS 5640 Type 1; SAE 30303.

JOSLYN STAINLESS TYPE 303 FORGING QUALITY.
M-560; 0.15 max C, 2.0 max Mn, 0.15 min S, 1.0 max Si, 17-19 Cr, 8.0-10.0 Ni, 0.60 max Mo, bal Fe.
Ann: 90,000 TS; 35,000 YS; 50 El; 160 Brin.
Forging grade, austenitic, free machining, hardenable only by cold work, non-magnetic.
For hot headed bolts, aircraft fittings, forged and machined fittings, and fasteners.
AISI 303; AMS 5640 Type 1; SAE 30303.

JOSLYN STAINLESS TYPE 303 PB.
M-560; 0.15 max C, 2.0 max Mn, 0.12-0.30 S, 1.0 max Si, 17-19 Cr, 8.0-10.0 Ni, 0.60 max Mo, 0.12-0.30 Pb, bal Fe.
Ann: 90,000 TS; 35,000 YS; 50 El; 160 Brin.
Austenitic, free machining, non-magnetic, hardenable only by cold work, corrosion resistant.
For gears, shafts, studs, bolts, nuts, screw machine parts, Swiss automatic parts.
QQ-S-764; AMS 5635; ASTM A581.

JOSLYN STAINLESS TYPE 303 SE.
M-560; 0.15 max C, 2.0 max Mn, 0.06 max S, 1.0 max Si, 17-19 Cr, 8.0-10.0 Ni, 0.50 max Mo, 0.15 min Se, bal Fe.
Ann: 90,000 TS; 35,000 YS; 50 El; 160 Brin.
Austenitic, free machining, non-magnetic, hardenable only by cold work, corrosion resistant.
For gears, shafts, rivets, screw machine parts, bolts, nuts, studs, ordnance parts.
AISI 303 Se; AMS 5640 Type 2; SAE 30303 Se.

JOSLYN STAINLESS TYPE 304L.
M-560; 0.030 max C, 2.0 max Mn, 1.0 max Si, 18-20 Cr, 8-11 Ni, bal Fe.
Ann: 80,000 TS; 30,000 YS; 60 El; 140 Brin.
Austenitic, weldable, non-magnetic, hardenable only by cold work, corrosion resistant.
For weld rod, welded assemblies, food processing equipment.
AISI 304L; AMS 5647; SAE 30304L.

JOSLYN STAINLESS TYPE 304 MAX MACHINABILITY.
M-560; 0.08 max C, 2.0 max Mn, 0.03 max S, 1.0 max Si, 18-20 Cr, 8-12 Ni, bal Fe.
Ann: 85,000 TS; 35,000 YS; 60 El; 149 Brin.
Austenitic, non-magnetic, corrosion resistant, hardenable by cold work only.
For shafts, dairy fittings, valve parts, food and laundry equipment.
AISI 304; AMS 5639; SAE 30304.

JOSLYN STAINLESS TYPE 304 N.
M-560; 0.08 max C, 2.0 max Mn, 1.0 max Si, 18.0-20.0 Cr, 8.0-10.0 Ni, 0.18-0.30 N, bal Fe.
Ann: 115,000 TS; 70,000 YS; 45 El; 65 RA; 229 Brin.
Higher strength than AISI 304.

JOSLYN STAINLESS TYPE 308.
M-560; 0.08 max C, 2.0 max Mn, 1.0 max Si, 19-21 Cr, 10-12 Ni, bal Fe.
Ann: 85,000 YS; 30,000 YS; 55 El; 149 Brin.
Austenitic, non-magnetic, corrosion resistant. Less subject to work hardening, weldable.
For weld rod, welded assemblies, heat treat equipment, furnace parts.
AISI 308; ASTM A276; SAE 30308.

JOSLYN STAINLESS TYPE 308L.
M-560; 0.030 max C, 2.0 max Mn, 1.0 max Si, 19-21 Cr, 9-12 Ni, bal Fe.
Ann: 85,000 TS; 35,000 YS; 55 El; 149 Brin.
Austenitic, weldable, corrosion resistant.
Main use is welding rod and parts to be welded.

JOSLYN STAINLESS TYPE 309.
M-560; 0.20 max C, 2.0 max Mn, 1.0 max Si, 0.030 max S, 22-24 Cr, 12-15 Ni, bal Fe.
Ann: 95,000 TS; 40,000 YS; 45 El; 160 Brin.
Austenitic, corrosion resistant, good resistance to attack at high temperatures to 2000°F.
For furnace parts, heat treating baskets and equipment, weld rod, pump parts.
AISI 309; ASTM A276; SAE 30309.

JOSLYN STAINLESS TYPE 309S.
M-560; 0.08 max C, 2.0 max Mn, 1.0 max Si, 0.030 max S, 22-24 Cr, 12-15 Ni, bal Fe.
Ann: 95,000 TS; 40,000 YS; 45 El; 160 Brin.
Austenitic, weldable, good high temperature resistance to corrosion and oxidation.
For furnace parts, heat treating equipment, cement kiln chain, weld rod.

JOSLYN STAINLESS TYPE 310.
M-560; 0.25 max S, 2.0 max Mn, 1.50 max Si, 0.030 max S, 24-26 Cr, 19-22 Ni, bal Fe.
Ann: 95,000 TS; 45,000 YS; 50 El; 185 Brin.
Austenitic, corrosion resistant, weldable, non-magnetic, good heat resistance.
For furnace parts, gas turbine parts, pump parts, welded assemblies, weld rod.
AISI 310; AMS 5694; SAE 30310.

JOSLYN STAINLESS TYPE 310S.
M-560; 0.08 max C, 2.0 max Mn, 1.5 max Si, 0.030 max S, 24-26 Cr, 19-22 Ni, bal Fe.
Ann: 95,000 TS; 45,000 YS; 50 El; 185 Brin.
Austenitic, corrosion resistant, weldable, non-magnetic, good heat resistance.
For furnace parts, oil refinery equipment, weld rod, jet engine rings.
AISI 310 S; AMS 5651; SAE 30310 S.

JOSLYN STAINLESS TYPE 316F.
M-560; 0.08 max C, 1.0-2.0 Mn, 0.11-0.17 P, 0.10-0.20 S, 1.0 max Si, 17-19 Cr, 12-14 Ni, 1.75-2.50 Mo, bal Fe.
Ann: 80,000 TS; 30,000 YS; 55 El; 149 Brin.
Austenitic, non-magnetic, free machining, excellent corrosion resistance.
For automatic screw machine parts requiring good corrosion resistance.

JOSLYN STAINLESS TYPE 316L.
M-560; 0.030 max C, 2.0 max Mn, 1.0 max Si, 0.030 max S, 16-18 Cr, 10-14 Ni, 2.0-3.0 Mo, bal Fe.
Ann: 80,000 TS; 30,000 YS; 60 El; 140 Brin.
Austenitic, non-magnetic, excellent corrosion resistance, weldable.
Parts for special application in food, paper, photographic and chemical processing equipment.
AISI 316 L; AMS 5653; SAE 30316 L.

JOSLYN STAINLESS TYPE 316 MAX MACHINABILITY.
M-560; 0.080 max C, 2.0 max Mn, 1.0 max Si, 0.030 max S, 16-18 Cr, 10-14 Ni, 2.0-3.0 Mo, bal Fe.
Ann: 80,000 TS; 30,000 YS 60 El; 149 Brin.
Austenitic, non-magnetic, improved machinability, excellent corrosion resistance, weldable.
For automatic screw machine parts for food, chemical, paper and textile equipment.
AMS 5648; ASTM A276; SAE 30316.

JOSLYN STAINLESS TYPE 317.
M-560; 0.08 max C, 2.0 max Mn, 1.0 max Si, 0.030 max S, 18-20 Cr, 11-15 Ni, 3-4 Mo, bal Fe.
Ann: 85,000 TS; 40,000 YS; 50 El; 160 Brin.
Austenitic, excellent corrosion resistance, useful to 1700°F, weldable.
For parts for paper, food, chemical, refining, photographic and textile industries.
AISI 317; ASTM A479; SAE 30317.

JOSLYN STAINLESS TYPE 317L.
M-560; 0.030 max C, 2.0 max Mn, 1.0 max Si, 0.030 max S, 18-21 Cr, 11-15 Ni, 3-4 Mo, bal Fe.
Ann: 85,000 TS; 40,000 YS; 50 El; 160 Brin.
Austenitic, excellent corrosion resistance, useful to 1700°F; weldable.
For parts for food, paper, chemical, refining, photographic and textile industries.

JOSLYN STAINLESS TYPE 321.
M-560; 0.08 max C, 2.0 max Mn, 1.0 max Si, 0.030 max S, 17-19 Cr, 9-12 Ni, Ti = 5 x C min, bal Fe.
Ann: 85,000 TS; 35,000 YS; 55 El; 149 Brin.
Austenitic, non-magnetic, weldable.
For welded assemblies, for parts operated at temperatures up to 1600°F
AISI 321; AMS 5645; SAE 30321.

JOSLYN STAINLESS TYPE 330.
M-560; 0.08 max C, 2.0 max Mn, 0.75-1.50 Si, 17-20 Cr, 34-37 Ni, bal Fe.
Ann: 85,000 TS; 45,000 YS; 35 El; 179 Brin.
Austenitic, non-magnetic, weldable, good at elevated temperatures.
For furnace parts, heat treating boxes and baskets, jet engine parts, petroleum plants.
AMS 5716; SAE 30330.

JOSLYN STAINLESS TYPE 347.
M-560; 0.08 max C, 2.0 max Mn, 1.0 max Si, 17-19 Cr, 9-13 Ni, Ta + Cb = 10 x C min, bal Fe.
Ann: 90,000 TS; 35,000 YS; 50 El; 160 Brin.
Austenitic, non-magnetic, weldable.
For parts subject to temperatures to 1600°F, welded assemblies, weld rods.
AISI 347; AMS 5646, 5680; SAE 30347.

JOSLYN STAINLESS TYPE 347 FM (SE).
M-560; 0.08 max C, 2.0 max Mn, 0.11-0.17 P, 0.030 max S, 1.5 max Si, 17-19 Cr, 9-12 Ni, Cb + TA = 10 x C min, 0.15-0.35 Se, 0.50 max Mo, 0.50 max Cu, bal Fe.
Ann: 90,000 TS; 35,000 YS; 50 El; 160 Brin.
Austenitic, non-magnetic; welding not recommended.
For machined parts for assemblies operating at 800-1500°F.
AMS 5642 Type 2.

JOSLYN STAINLESS TYPE 348.
M-560; 0.08 max C, 2.0 max Mn, 1.0 max Si, 17-19 Cr, 9-13 Ni, Cb + Ta = 10 x C min, 0.10 max Ta, 0.20 max Co, bal Fe.
Ann: 90,000 TS; 35,000 YS; 50 El; 160 Brin.
Austenitic, non-magnetic, heat resistant, weldable.
For welded construction for radio active systems, and operating temperatures 800-1500°F.
AISI 348; SAE 30348; MIL-S-23195.

JOSLYN STAINLESS TYPE 403.
M-560; 0.15 max C, 1.0 max Mn, 0.5 max Si, 11.5-13.0 Cr, bal Fe.
Ann: 75,000 TS; 40,000 YS; 35 El; 155 Brin.
Q&T, 1200°F; 100,000 TSf 85,000 YS; 23 El; 225 Brin.
Magnetic, heat treatable, weldable, corrosion resistant.
For jet engine parts, corrosion resistant hardware, steam and gas turbine blading.
AISI 403; AMS 5614; SAE 51403.

JOSLYN STAINLESS TYPE 405.
M-560; 0.08 max C, 1.0 max Mn, 1.0 max Si, 11.50-14.50 Cr, 0.10-0.30 Al, bal Fe.
Ann: 70,000 TS; 40,000 YS; 30 El; 149 Brin.
Magnetic, not normally hardenable, oxidation resistant up to 1500°F.
For annealing boxes, forged fittings, quench racks.
AISI 405; SAE 51405.

JOSLYN STAINLESS TYPE 410.
M-560; 0.15 max C, 1.0 max Mn, 1.0 max Si, 11.50-13.50 Cr, bal Fe.
Ann: 75,000 TS; 40,000 YS; 35 El; 155 Brin.
Q&T, 1200°F: 100,000 TS; 85,000 YS; 35 El; 225 Brin.
Magnetic, heat treatable, weldable, corrosion resistant.
For cutlery, corrosion resistant hardware, steam and gas turbine parts, pump parts.
AISI 410; AMS 5613; SAE 51410.

JOSLYN STAINLESS TYPE 410 CB.
M-560; 0.15 max C, 0.60 max 0.50 max Si, 11.5-13.5 Cr, 0.20 max Cb, bal Fe.
Ann: 75,000 TS; 40,000 YS; 35 El; 155 Brin.
Q&T, 1200°F: 124,000 TS; 100,000 YS; 20 El.
Magnetic, hardenable, weldable, corrosion resistant.
For steam turbine blades, valve parts, aircraft and missile components.
AMS 5609.

JOSLYN STAINLESS TYPE 410 LOW CARBON.
M-560; 0.08 max C, 1.0 max Mn, 1.0 max Si, 12.5-14.0 Cr, bal Fe.
Ann: 75,000 TS; 40,000 YS; 35 El; 155 Brin.
Magnetic weldable corrosion resistant, only partially hardenable.
For welded assemblies.

JOSLYN STAINLESS TYPE 416 FORGING QUALITY.
M-560; 0.15 max C, 1.25 max Mn, 0.06 max P, 0.15 min S, 1.0 max Si, 12-14 Cr, 0.5 max Ni, 0.6 max Mo, bal Fe.
Processed to produce good quality forgings.
Hardenable to Rc 35 min.
Magnetic corrosion resistant, free machining.
For hot headed bolts, forged and machined parts.
AISI 416; AMS 5610; SAE 51416.

JOSLYN STAINLESS TYPE 416 HIGH HARDENABILITY.
M-560; 0.15 max C, 1.25 max Mn, 0.15 min S, 1.0 max Si, 12-14 Cr, 0.5 max Ni, 0.60 max Mo, bal Fe.
Ann, CF: 90,000 TS; 80,000 YS; 15 El; 197 Brin.
Guaranteed hardenable to 35 Rc min.
(ave:41 Rc).
Magnetic, corrosion resistant, free machining.
For fasteners, shafts, impellers gears.
AISI 416; AMS 5610; SAE 51416.

JOSLYN STAINLESS TYPE 416 MACHINE SPECIAL.
M-560; 0.15 max C, 1.25 max Mn, 0.15 min S, 1.0 max Si, 12-14 Cr, 0.50 max Ni, 0.60 max Mo, bal Fe.
Ann, CF: 90,000 TS; 80,000 YS; 15 El; 197 Brin.
Q & T, 1200°F: 110,000 TS; 85,000 YS; 18 El; 225 Brin.
Magnetic, hardenable, corrosion resistant, free machine.
For pump shafts, motor shafts, valve trim, valve stems, fasteners, fittings.
AISI 416; AMS 5610; SAE 51416.

JOSLYN STAINLESS TYPE 416 SE.
M-560; 0.15 max C, 1.25 max Mn, 0.06 max P, 0.06 max S, 1.0 max Si, 12-14 Cr, 0.15 min Se, bal Fe.

JOSLYN STAINLESS TYPE 416 XS.
M-560; 0.10 max C, 1.0 max Mn, 0.06 max P, 0.35 min S, 1.0 max Si, 13-15 Cr, 0.50 max Ni, 0.30-0.60 Mo, bal Fe.
Ann, CF: 90,000 TS; 80,000 YS; 15 El; 197 Brin.
Magnetic, corrosion resistant, free machining, not recommended for welding or quench hardening.
For oil burner nozzles, motor shafts, screw machine parts.

JOSLYN STAINLESS TYPE 420.
M-560; over 0.15 C, 1.0 max Mn, 1.0 max Si, 12-14 Cr, bal Fe.
Ann: 95,000 TS; 50,000 YS; 25 El; 195 Brin.
Q&T, 600°F: 230,000 TS; 195,000 YS; 8 El; 500 Brin.
Magnetic, corrosion resistant.
For cutlery, surgical and dental instruments, valves, shafts, gears, knife blades, springs.
AISI 420; AMS 5621; SAE 51420.

JOSLYN STAINLESS TYPE 420 F.
M-560; over 0.15 C, 1.25 max Mn, 0.06 max P, 0.15 min S, 1.0 max Si, 12-14 Cr, 0.60 max Mo, bal Fe.
Ann: 95,000 TS; 50,000 YS; 25 El; 195 Brin.
Q&T, 600°F: 230,000 TS; 195,000 YS; 8 El; 500 Brin.
Magnetic, corrosion resistant, free machining.
For gears, shafts, surgical and dental instruments, knife blades.
AISI 420; AMS 5620; SAE 51420.

JOSLYN STAINLESS TYPE 422.
M-560; 0.20-0.25 C, 1.0 max Mn, 1.0 max Si, 12.0-14.0 Cr, 0.5-1.0 Ni, 0.75-1.25 Mo, 0.75-1.25 W, 0.20-0.50 V, bal Fe.
Martensitic type corrosion resisting steel.
Ht, Temp.1000°F; 220,000 psi TS; 158,000 psi YS; 460 Brin.
For steam turbine blades & buckets, valves, high temperature bolting.
AISI 422.

JOSLYN STAINLESS TYPE 430.
M-560; 0.12 max C, 1.0 max Mn, 1.0 max Si, 14-18 Cr, bal Fe.
Ann: 75,000 TS; 45,000 YS; 30 El; 156 Brin.
Ferritic type magnetic, corrosion resistant, not hardenable.
For automotive fasteners, kitchen utensils, cold headed parts, food handling equipment.
AISI 430; AMS 5627; SAE 51430.

JOSLYN STAINLESS TYPE 430 F.
M-560; 0.12 max C, 1.25 max Mn, 0.06 max P, 0.15 min S, 1.0 max Si, 14-18 Cr, 0.6 max Mo, bal Fe.
Ann: 80,000 TS; 45,000 YS; 25 El; 163 Brin.
Ferritic, magnetic, free machining, corrosion resistant, not hardenable, weldability-poor.
For machined parts, bolts and nuts, fasteners, shafts, burner parts.
AISI 430 F; SAE 51430 F.

JOSLYN STAINLESS TYPE 430 F SOLENOID QUALITY.
M-560; 0.06 max C, 0.50 max C, 0.50 max Mn, 0.25-0.35 S, 0.50 max Si, 17.25-18.25 Cr, 0.30 max Ni, 0.50 max Mo, 0.25 max Cu, bal Fe.
Ann: 80,000 TS; 45,000 YS; 25 El; 163 Brin.
Magnetic, ferritic, free machining.
For solenoid valve plungers.

JOSLYN STAINLESS TYPE 615 (GREEK ASCOLOY).
M-560; 0.15-0.20 C, 0.50 max Mn, 0.50 max Si, 12-14 Cr, 1.8-2.2 Ni, 0.50 max Mo, 2.5-3.5 W, bal Fe.
Ann: 140,000 TS; 130,000 YS; 15 El; 277 Brin.
Q&T, 1200°F; 135,000 TS; 108,000 YS; 20 El; 277 Brin.
Martensitic, magnetic, hardenable to Rc 45 min.
For steam turbine parts, gas turbine parts, jet engine parts, compressor blades and wheels.

J.P.S. STEEL.
M-261; 0.90-1.0 C, 1.1-1.3 W, 0.40-0.60 Cr, bal Fe.
For shock tools, threading, taps, reamers, broaches; oil hardening.

JS700.
M-69; 0.03 C, 1.7 Mn, 0.5 Si, 21.0 Cr, 25.0 Ni, 4.5 Mo, Cb = 8 x C, bal Fe.
Ann: prop. 85,000 TS; 39,000 YS; 45 El; 50 RA; 170 Brin.
Special engineering alloy for superior corrosion resistance; for agitators piping, mixers, vessels, vaporizers, tanks and stills.

JS 700; JS 777 see **JESSOP JS 700; JESSOP JS-777.**

J.S. PUNCH.
M-57; 0.55 C, 1.25 Cr, 2.75 W, 0.2 V, bal Fe.
Heat treated: 231,000-315,000 TS;
209,000-275,000 YS; 9-11 El; 27-42 RA; C 48-56 Rock.
For header and swaging dies, chipping chisels, rivet busters, track tools.
Type S1 shock resisting tool steel, good fatigue resistance.

"J" TEMPER RED (CUT SUPERIOR).
M-115; 0.50 C, 4 Cr, 1 V, 18 W, bal Fe.
For hot work tools and dies; hot work steel, oil hardened.

JUNKER A 8 M NB.
M-1347; 0.04 C, 18.0 Cr, 11.0 Ni, 2.2 Mo, Cb, bal Fe.
Corrosion resistant casting.
65,000 min TS; 30,000 min YS; 20 min El.

JUNKER A 8 M S.
M-1347; 0.06 C, 18.0 Cr, 11.0 Ni, 2.2 Mo, bal Fe.
Corrosion resistant casting.
65,000 min TS; 30,000 min YS; 20 min El.
Type CF-8M.

JUNKER A 8 M SS.
M-1347; 0.02, 18.0 Cr, 11.0 Ni, 2.2 Mo, bal Fe.
Corrosion resistant casting.
57,000 min TS; 27,000 min YS; 30 min El.
Type CF-3M.

JUNKER A 8 NB.
M-1347; 0.04 C, 18.0 Cr, 10.0 Ni, Cb, bal Fe.
Cast: 64,000 TS; 25,000 YS; 20 El; 130-200 Brin.
Corrosion resistant steel casting; Type CF-8C.

JUNKER A 8 NB TT.
M-1347; 0.04 max C, 18.0 Cr, 10.0 Ni, Cb, bal Fe.
65,000 min TS; 28,500 min YS; 20 min El.
For castings for low temperature, applications to -195°C.

JUNKER A 8 S.
M-1347; 0.06 C, 18.0 Cr, 10.0 Ni, bal Fe.
Corrosion resistant casting.
65,000 min TS; 28,500 min YS; 20 min El.
Type CF-8.

JUNKER A 8 SS.
M-1347; 0.02 C, 18.0 Cr, 10.5 Ni, bal Fe.
Corrosion resistant casting.
57,000 min TS; 25,500 min YS; 30 min El.
Type CF-3.

JUNKER A 8 S TT.
M-1347; 0.6 C, 18.0 Cr, 10.0 Ni, bal Fe.
65,000 min TS; 28,500 min YS; 20 min El.
For castings for low temperature applications, to -195°C.

JUNKER A 12 MS.
M-1347; 0.06 C, 17.0 Cr, 13.5 Ni, 4.5 Mo, bal Fe.
Corrosion resistant casting.
57,000 min TS; 28,500 min YS; 15 min El.

JUNKER A 12 MSSN.
M-1347; 0.02 C, 17.0 Cr, 14.0 Ni, 4.4 Mo, bal Fe.
Cast: 71,000 TS; 30,000 YS; 20 El; 150-200 Brin.
Corrosion resistant steel casting.

JUNKER A 13 MS.
M-1347; 0.06 C, 17.5 Cr, 12.5 Ni, 2.8 Mo, bal Fe.
Corrosion resistant casting.
65,000 min TS; 30,000 min YS; 20 min El.

JUNKER A 17 MKN.
M-1347; 0.02 C, 18.0 Cr, 20.0 Ni, 2.2 Mo, 2.0 Cu, bal Fe.
Cast: 64,000 TS; 27,000 YS; 20 El; 130-200 Brin.
Corrosion resisting steel casting.

JUNKER A 25 MKN.
M-1347; 0.02 C, 20.0 Cr, 25.0 Ni, 3.2 Mo, 1.8 Cu, bal Fe.
Cast: 64,000 TS; 27,000 YS; 20 El; 130-200 Brin.
Corrosion resistant steel casting.

JUNKER A 26 MKN.
M-1347; 0.02 C, 20.0 Cr, 25.0 Ni, 4.1 Mo, 1.8 Cu, bal Fe.
Cast: 64,000 TS; 27,000 YS; 20 El; 130-200 Brin.
Corrosion resistant steel casting.

JUNKER A 28 MKN.
M-1347; 0.02 C, 21.0 Cr, 27.5 Ni, 4.1 Mo, 3.0 Cu, bal Fe.
Cast: 64,000 TS; 27,000 YS; 20 El; 130-200 Brin.
Corrosion resistant steel casting.

JUNKER A 29 MKN.
M-1347; 0.02 C, 20.0 Cr, 29.0 Ni, 2.2 Mo, 3.2 Cu, bal Fe.
Cast: 64,000 TS; 27,000 YS; 20 El; 130-200 Brin.
Corrosion resistant steel casting.

JUNKER A 35 MKN.
M-1347; 0.02 C, 20.0 Cr, 35.0 Ni, 2.2 Mo, 3.2 Cu, bal Fe.
Cast: 64,000 TS; 27,000 YS; 20 El; 130-200 Brin.
Corrosion resistant steel casting.

JUNKER A 42 MKN.
M-1347; 0.01 C, 21.0 Cr, 42.0 Ni, 3.0 Mo, 2.0 Cu, bal Fe.
Cast: 64,000 TS; 27,000 YS; 20 El; 130-200 Brin.
Corrosion resistant steel casting.

JUNKER A 57 M.
M-1347; 0.03 max C, 16.5 Cr, 16.5 Mo, bal Ni.
Cast: 72,000 TS; 33,000 YS; 15 El; 140-200 Brin.
Corrosion resistant steel casting.

JUNKER A 65 M.
M-1347; 0.02 C, 27.0 Mo, bal Ni.
Cast: 72,000 TS; 40,000 YS; 10 El; 160-220 Brin.
Corrosion resistant steel casting.

JUNKER A 75.
M-1347; 0.05 C, 15.5 Cr, 8.0 Fe, bal Ni.
Cast: 58,000 psi TS; 19,000 psi YS; 20 El.
Corrosion resistant steel casting.

JUNKER A 1050.
M-1347; 0.25 C, 18.0 Cr, 9.0 Ni, bal Fe.
Cast: 64,000 min TS; 15 min El; 130-200 Brin.
Heat resisting casting.

JUNKER A 1101.
M-1347; 0.35 C, 22.0 Cr, 10.0 Ni, bal Fe.
Cast: 64,000 min TS; 12 min El; 150-220 Brin.
Heat resisting casting; Type HF.

JUNKER A 1102.
M-1347; 0.25 C, 20.0 Cr, 14.0 Ni, bal Fe.
Cast: 64,000 min TS; 12 min El; 150-220 Brin.
Heat resisting casting.

JUNKER A 1151.
M-1347; 0.30 C, 26.0 Cr, 14.0 Ni, bal Fe.
Cast: 64,000 min TS; 8 min El; 150-220 Brin.
Heat resisting casting; Type HI.

JUNKER A 1152.
M-1347; 0.40 C, 17.5 Cr, 37.5 Ni, bal Fe.
Cast: 58,000 min TS; 6 min El; 150-220 Brin.
Heat resistant casting; Type HU.

JUNKER A 1153.
M-1347; 0.40 C, 25.0 Cr, 12.0 Ni, bal Fe.
Cast: 64,000 min TS; 8 min El; 130-200 Brin.
Heat resisting casting; similar to Type HH.

JUNKER A 1201.
M-1347; 0.40 C, 25.0 Cr, 20.0 Ni, bal Fe.
Cast: 64,000 min TS; 8 min El; 150-220 Brin.
Heat resisting casting; Type HK.

JUNKER A 1205.
M-1347; 0.50 C, 29.0 Cr, 49.0 Ni, 4.5 W, bal Fe.
Cast: 58,000 psi min TS; 32,000 psi min YS; 3 min El; 150-220 Brin.
Heat resisting casting.

JUNKER A 1234 NB.
M-1347; 0.08 C, 20.0 Cr, 32.0 Ni, Cb, bal Fe.
Cast: 64,000 psi min TS; 25,000 psi min YS; 20 min El.
Heat resisting casting.

JUNKER A 1237.
M-1347; 0.35 C, 25.0 Cr,35.0 Ni, bal Fe.
Cast; 64,000 psi min TS; 32,000 psi min YS; 8 min El.
Heat resisting casting.

JUNKER AF 22 MN.
M-1347; 0.03 C, 22.3 Cr, 6.0 Ni, 1.5 Mo, bal Fe.
Corrosion resistant steel casting.

JUNKER AF 22 NMNB.
M-1347; 0.06 C, 23.3 Cr, 9.1 Ni, 2.1 Mo, Cb, bal Fe.
Cast: 85,000 TS; 50,000 YS; 20 El; 180-230 Brin.
Corrosion resistant steel casting.

JUNKER AF 25 N.
M-1347; 0.05 max C, 25.0 Cr, 8.0 Ni, bal Fe.
Corrosion resistant casting.
85,000 min TS; 57,000 min YS; 20 min El.

JUNKER AF 25 N M K.
M-1347; 0.03 C,25.0 Cr, 5.5 Ni, 2.5 Mo, 3.0 Cu, bal Fe.
Cast: 108,000 TS; 79,000 YS; 15 El; 240 min Brin.
Corrosion resistant steel casting.

JUNKER AF 27 NC.
M-1347; 0.40 C, 27.0 Cr, 4.5 Ni, bal Fe.
Corrosion resistant casting.
71,000 min TS.
Type CC-50.

JUNKER AF 27 NMN.
M-1347; 0.05 C, 24.0 Cr, 5.7 Ni, 1.4 Mo, bal Fe.
Cast: 94,000 TS; 60,000 YS; 20 El; 200-250 Brin.
Corrosion resistant steel casting.

JUNKER AF 1101.
M-1347; 0.40 C, 27.0 Cr, 4.5 Ni, bal Fe.
Cast: 71,000 min TS; 4 min El; 200-300 Brin.
Heat resisting casting; Type HD.

JUNKER AF 1152.
M-1347; 0.28 C, 28.0 Cr, 10.0 Ni, bal Fe.
Cast: 71,000 min TS; 150-250 Brin.
Heat resisting casting; Type HE.

JUNKER AUN 9.
M-1347; 0.04 C, 16.5 Cr, 9.0 Ni, N_2, bal Fe.
For non-magnetizable castings.
65,000 min TS; 28,500 min YS; 20 min El.

JUNKER AUN 11.
M-1347; 0.04 C, 16.5 Cr, 11.0 Ni, bal Fe.
Cast: 64,000 min TS; 25,000 min YS; 20 min El; 150-190 Brin.
Non-magnetic casting.

JUNKER F 901 S.
M-1347; 0.20 C, 7.0 Cr, bal Fe.
Cast: 71,000 psi min TS; 4 min El; 200-280 Brin.
Heat resisting casting.

JUNKER F 1001.
M-1347; 0.35 C, 17.0 Cr, bal Fe.
Cast: 71,000 min TS; 2 min El; 200-300 Brin.
Heat resisting casting.

JUNKER F 1002 S.
M-1347; 1.6 C, 18.0 Cr, bal Fe.
Cast: 250-350 Brin.
Heat resisting casting.

JUNKER F 1051.
M-1347; 0.40 C, 23.0 Cr, bal Fe.
Cast: 200-300 Brin.
Heat resisting casting.

JUNKER F 1101.
M-1347; 0.40 C, 29.0 Cr, bal Fe.
Cast: 200-300 Brin.
Heat resisting casting.

JUNKER F 1102.
M-1347; 1.30 C, 29.0 Cr, bal Fe.
Cast: 250-350 Brin.
Heat resisting casting.

JUNKER G-NI 95.
M-1347; 0.45 C, bal Ni.
Cast: 45,000 psi TS; 17,000 psi YS; 12 El; 80 min Brin.
Nickel casting.

JUNKER G-NI CU 30 NB.
M-1347; 0.14 C, 30.0 Cu, 2.2 Fe, Cb, bal Ni.
Cast: 57,000 TS; 25,000 YS; 25 El; 120 min Brin.
Nickel-copper casting.

JUNKER HASTELLOY B.
M-1347; 0.05 C, 27.0 Mo, 5.0 Fe, 0.3 V, bal Ni.
Cast: 72,000 TS; 43,000 YS; 6 El; 180-240 Brin.
Corrosion resistant steel casting.

JUNKER HASTELLOY C.
M-1347; 0.05 C, 16.5 Cr, 16.5 Mo, 0.3 V, 5.0 Fe, 4.5 W, bal Ni.
Cast: 72,000 TS; 43,000 YS; 5 El; 180-240 Brin.
Corrosion resistant steel casting.

JUNKER IN 657.
M-1347; 0.04 C, 50.0 Cr, Cb, bal Ni.
Cast: 78,000 psi min TS; 39,000 psi min YS; 8 min El.
Heat resisting casting.

JUNKER M 13 N (GR. I).
M-1347; 0.05 C, 12.5 Cr, 4.5 Ni, 0.4 Mo, bal Fe.
Cast: 113,000 TS; 92,000 YS; 15 El; 230-285 Brin.
Corrosion resistant cast steel.

JUNKER M 13 N (GR. II).
M-1347; 0.05 C, 12.5 Cr, 4.5 Ni, 0.4 Mo, bal Fe.
Cast: 128,000 TS; 121,000 YS; 12 El.
Corrosion resistant steel casting.

JUNKER M 14.
M-1347; 0.19 C, 13.0 Cr, bal Fe.
Corrosion resistant casting.
85,000 min TS; 64,000 min YS; 12 min El.
Type CA-40.

JUNKER M 14S.
M-1347; 0.12 C, 13.0 Cr, 0.7 Ni, bal Fe.
Corrosion resistant casting.
85,000 min TS; 57,000 min YS; 15 min El.
Type CA-15.

JUNKER M 18.
M-1347; 0.22 C, 17.0 Cr, 1.2 Ni, bal Fe.
Corrosion resistant casting.
113,000 min TS; 85,000 min YS; 4 min El.

JUNKER NI CR 50.
M-1347; 0.07 C, 50.0 Cr, bal Ni.
Cast: 92,000 psi min TS; 57,000 psi min YS; 6 min El.
Heat resisting casting.

JUNKER NICRO 50 now **JUNKER NICR 50.**

JUNKER NI MO 59.
M-1347; 0.02 max C, 20.0 Cr, 60.0 Ni, 17.0 Mo.
Corrosion resistant casting.
71,000 min TS; 40,000 min YS; 30 min El.

JUNKER NI MO 69.
M-1347; 0.02 max C, 70.0 Ni, 28.0 Mo.
Corrosion resistant casting.
71,000 min TS; 38,000 min YS; 20 min El, 140-200 Brin.

JUNKER THERMO 50.
M-1347; 0.10 C, 28.0 Cr, 50.0 Co, bal Fe.
Cast: 71,000 psi min TS; 35,000 psi min YS; 6 min El.
Heat resisting casting; resistant to thermal shock and some high temperature corrosion.

JUNKER THERMO 51.
M-1347; 0.30 C, 28.0 Cr, 50.0 Co, Cb, bal Fe.
Cast: 78,000 psi min TS; 42,000 psi min YS; 3 min El.
Heat resisting casting; resistant to thermal shock and some high temperature corrosion.

JUNKER X.
M-Eng.; 6 Fe, 0.5 Mg, bal Al.
For bearings; for camshafts.

JUNKER Z.
M-Eng.; 6.5 Ni, 0.5 Ti, bal Al.
For bearings; for camshafts.

K

K-4.
M-Ger.; Zn, Mg, bal Al.
For bearings.

K-4ONKHM.
M-USSR; 0.05-0.12 C, 40 Co, 20 Cr, 15 Ni, 6 Mo + W, 2 Mn, bal Fe.
Heat treated: 358,000 TS; 248,000 YS; C 58 Rock.
For spiral springs in corrosive media and operating up to 400°C.
High elastic properties imparted by cold work and subsequent tempering at 500 C.

K-9 SPECIAL OIL HARDENING.
M-365; 0.90 C, 0.5 W, 0.7 Cr, 1.0 Mn, 0.2 Si, bal Fe.
For tools, cutters; oil hardening.

K-9 STEEL.
  M-210; 1.0 C, 0.85 Mn, 0.75 Cr, 0.40 W, bal Fe.
  Annealed: 100,000 TS; 54,000 YP; 21 El; 197 Brin.
  Ht.Tr.: 220,000 TS; 155,000 YP; 10 El; 600 Brin.
  For dies, milling cutters, plugs, gauges, circular cutters, master tools.
  Non-deforming, oil hardening, wear resistant.

K-42-B.
  M-118; 42 Ni, 22 Co, 14 Fe, 18 Cr, 2 Ti, 0.6 Al.
  At 70°F: 165,000 TS; 19 El; 37 Brin.
  At 1110°F: 127,100 TS; 21 El; 22 Brin.
  Hardened: 185,000 TS; 90,000 YS; 30 El.
  For dies, valves, steam fitting and turbine blades, springs, bolts; retain strength at high temperature.

K 059.
  M-1290; 1.10-1.25 C, 0.20-0.40 Si, 0.20-0.40 Mn, 1.2-1.5 Cr, bal Fe.
  Cold work tool steel.
  W.-Nr. 1.2059.

K 833.
  M-1290; 0.95-1.05 C, 0.15-0.25 Si, 0.15-0.30 Mn, 0.05-0.15 V, bal Fe.
  Cold work tool steel.
  W.-Nr. 1.2833.

K-390.
  M-114; 0.25 C, 2.75 Cr, 10 W, 2 Ni, bal Fe.
  For brass forging and die casting dies, hot punches; hot work steel.

KABEL OWS.
  M-1310; 0.38 C, Si, Cr, V, bal Fe.
  For bolts, springs, crankshafts; oil hardened, shock resistant.

KABEL SBN.
  M-1310; 1.1 C, 0.20 Si, 0.30 Mn, 0.4 Cr, bal Fe.
  For springs, bearings, drills, cutters; water or oil hardened, wear resistant.

KA HO LOY.
  M-517, M-131; 0.07-0.10 C, 0.2-0.3 Cu, 0.35-0.50 Mn, bal Fe.
  32,000-37,000 TS; 18,000-23,000 YS.
  For roofs; resists atmosphere corrosion.

KAISALOY 42 CV.
  M-1148; 0.20 C, 0.90 Mn, 0.30 Si, 0.01 min Cb or V, bal Fe.
  Plate: 60,000 TS; 42,000 min YS; 24 El; 140 Brin.
  For bridges, booms, mine and railroad cars, derricks.
  High strength-low alloy steels.
  SAE J410C, Gr. 942X.

KAISALOY 45-CV.
  M-1148; 0.22 C, 1.0 Mn, 0.3 Si, 0.01 min Cb, bal Fe.
  Plate: 60,000 TS; 22 El; B 80 Rock.
  For bridges, booms, truck and bus bodies, derricks, structures.
  High strength-low alloy steel.
  SAE J410, Gr. 945X.

KAISALOY 45-FG.
  M-1148; 0.12 C, 0.60 Mn, 0.50 Si, 0.30 Cu, 0.10 Mo, 0.25 Cr, 0.60 Ni, 0.02 min V, 0.005 min Ti, bal Fe.
  60,000 psi min TS; 45,000 min YS; 25 min El.
  High ductility and strength; for drawn frame members, formed channels. Easily welded.
  SAE J410C, Grade 945A.

KAISALOY 50-CR.
  M-1148; 0.20 C, 1.25 Mn, 0.25-0.75 Si, 0.20-0.35 Cu, 0.15 Mo, 0.10-0.25 Cr, 0.30-0.60 Ni, 0.02-0.10 V, 0.005-0.030 Ti, bal Fe.
  70,000 min TS; 50,000 psi min YS; 21 min El.
  Good weldability, abrasion and corrosion resistance.
  For material handling equipment, truck and trailer equipment.

KAISALOY 50-CV.
  M-1148; 0.22 C, 1.1 Mn, 0.3 Si, 0.01 min Cb, bal Fe.
  Plate: 65,000 TS; 50,000 YS; 22 El; B 80 Rock.
  For booms, bridges, bus and truck bodies, derricks.
  High strength-low alloy steel.
  SAE J410C, Gr. 950X.

KAISALOY 50 MM.
  M-1148; 0.27 C, 1.10-1.60 Mn, 0.30 Si, 0.20 min Cu, bal Fe.
  63,000-70,000 psi min TS; 42,000-50,000 min YS; (depending on thickness), 21 min El.
  For booms, bridges (riveted), mixers, earth moving equipment.
  SAE J410C, Grade 950C.

KAISALOY 50 MV.
  M-1148; 0.22 C, 0.85-1.25 mn, 0.30 Si, 0.20 min Cu, 0.02 min V, bal Fe.
  60,000-70,000 min TS; 40,000-50,000 psi min YS; (depending on thickness), 21 min El.
  Good weldability; for welded bridges and buildings.
  SAE J410C, Grade 950B.

KAISALOY 55-CV.
  M-1148; 0.25 C, 1.35 Mn, 0.3 Si, 0.01 min Cb, bal Fe.
  Plate: 70,000 TS; 55,000 YS; 20 El; 160 Brin.
  For bridges, structures, mine cars, booms, bus and truck bodies.
  High-strength, low-alloy steel.
  SAE J410C, Gr. 955X.

KAISALOY 60-CV.
M-1148; 0.26 C, 1.35 Mn, 0.3 Si, 0.01 min Cb, bal Fe.
Plate: 75,000 TS; 60,000 YS; 18 El; B 85 Rock.
For mine cars, bridges, booms, truck and bus bodies, derricks.
High strength, low alloy steel.
SAE J410 C, Gr, 960x.

KAISALOY 60 SG.
M-1148; 0.20 C, 1.25 Mn, 0.35 Si, 0.80 Cu, 0.25 Mo, 0.90 Ni, 0.005 min V, bal Fe.
80,000 psi min TS; 60,000 psi min YS; 20 min El.
Weldable; good wear and impact resistance.
For dump trucks, freight cars, coal chutes, cranes, barges.

KAISALOY AR.
M-1148; 0.35-0.50 C, 1.5-2.0 Mn, 0.15-0.35 Si, bal Fe.
200-280 Brin (normal).
Good abrasion resistance; for parts requiring good surface wear as chutes, road machinery.

KAISALOY SPECIAL FORMING see KAISALOY 45 FG.

KAISER K2.
M-1147; Al-Mg-S. Alloy aluminum casting.
Develops integral color anodic coatings When properly anodized. Kalcolor trade name.

KAISER K3.
M-1147; Al-Mg-Si-Cr. Alloy aluminum casting.
Develops clear anodic coatings when properly anodized. Kalcolor trade name.

KALCHOIDS.
Cu-Sn-Zn.
For bearings.

KALKOS.
M-73; 0.3 C, 0.35 Mn, 12 Cr, 12 W, 0.9 V, 0.5 Si, bal Fe.
For hot forming and working tools; hot work steel.

"K" ALLOY.
M-Eng; 11.3 Cu, 1.1 Mn, 0.5 Mg, 0.79 Si, bal Al.
For light alloy parts; leak-proof castings.

KALOR.
M-111; 0.3 C, 1.25 Cr, 4.25 Ni, 5.5 W, 0.15 V, bal Fe.
For extrusions, dies, rams and liners; hot work steel, oil hardened.

KALOR 2.
M-111; 0.30 C, 1.0 Si, 1.1 Cr, 0.18 V, 3.75 W, bal Fe.
For extrusion press dies, liners and rams, punches; oil hardened, tough.

KAMARSCH BEARING.
85 Sn, 5 Sb, 3.6 Cu, 1.6 Bi, 1.4 Zn.
For bearings; anti-friction.

KAMARSCH BEARING.
71 Sn, 7.2 Sb, 21 Cu.
For bearings; ati-friction.

KAMARSCH BEARING.
71 Sn, 20 Sb, 9.5 Cu.
For bearings; anti-friction.

KANTHAL "A".
M-70; 23.4 Cr, 6.2 Al, 1.9 Co, 0.06 C, bal Fe.
Annealed: 106,000-114,000 TS; 80,000 YS; 20 El; 65 RA; 230 Brin.
For heating elements, furnaces; resists oxidation to 1300°C.

KANTHAL A.
M-70, M-1150; 0.05 C, 22.0 Cr, 5.0 Al, 1.0 max Co, bal Fe.
Ann: 106,000-114,000 psi TS; 80,000 psi YS; 20 El; 65 RA; 230 Brin.
For heating elements, furnaces; resists oxidation to 1330°C.

KANTHAL A-1.
M-70, M-1150; 0.05 C, 22.0 Cr, 5.5 Al, 1.0 max Co, bal Fe.
Ann: 114,000 psi TS; 80,000 psi YS; 20 El; 65 RA; 230 Brin.
For heating elements, furnaces; resists oxidation to 1375°C.

KANTHAL "A-1".
M-70; 0.06 C, 23 Cr, 2 Co, 5.7 Al, 69 Fe.
Annealed: 114,000 TS; 80,000 YS; 20 El; 65 RA; 230 Brin.
For heating elements, furnaces; heat resistant, non-scaling.

KANTHAL "D".
M-70; 0.09 C, 22.6 Cr, 2 Co, 4.5 Al, bal Fe.
Annealed; 104,000 TS; 80,000 YS; 20 El; 65 RA; 200 Brin.
For heating elements; heat resistant, tough; 1350°C max.

KANTHAL D.
M-1150; 4.5 Al, 22 Cr, 0.5 Co, bal Fe.
Wire, ribbon, strip, foil for heating elements; max operating temp. 2335°F. (1280°C).

KANTHAL DR.
M-70; 20 Cr, 4.5 Al, 0.5 Co, bal Fe.
Annealed: 100,000-140,000 TS; 3-12 El.
For resistors, potentiometers; high electrical resistance.

KANTHAL DS.
M-1150, M-70; 4.5 Al, 22 Cr, 0.5 Co, bal Fe.
For resistace and heating elements in appliances and furnaces, toasters.
Max. operating temperature 2100°F.

KANTHAL DSD.
M-70, M-1150; 0.05 C, 22.0 Cr, 4.5 Al, 1.0 max Co, bal Fe.
Ann: 114,000 psi TS; 80,000 psi YS; 20 El; 65 RA; 230 Brin.
For heating elements, furnaces; resists oxidation to 1280°C.

KANTHAL SUPER.
M-1150; Mo Si_2 + SiO_2.
Cermet, 1200 Vickers, for heating and resistance elements.
At 1500°C. TS = 145.

KANTHAL SUPER.
M-70, M-1150; Mo Si_2 + Si O_2.
For heating and resistance elements for service up to 1800°C.

KARABUK.
M-Turkey; 0.12-0.65 C, 0.35-0.85 Mn, 0.15-1.0 Si, bal Fe.
For structures, rails, gears, shafts, machinery parts. Constructional steel.

KARMA.
M-44; 20 Cr, 3 Al, 3 Fe, bal Ni.
Heat treated: 150,000 TS; 35 El.
For resistaces, shunts, potentiometers; high specific resistance; vacuum melted.

KARONI 8-18.
M-1310; 0.15 max C, 18 Cr, 8.5 Ni, 0.4 Si, bal Fe.
Annealed: 80,000 TS; 35,000 YS; 55 El; 75 RA; 150 Brin.
Cold drawn: 180,000 TS; 150,000 YS; 10 El; 250 Brin.
For chemical plant equipment, tanks, mixers, filters; stainless, austenitic; Type 302.

KARONI 8-18 EXTRA.
M-1310; 0.12 max C, 18 Cr, 19.5 Ni, Ti = 4 x C, bal Fe.
For welded chemical plant equipment; corrosion and heat resistant.

KARONI 8-18 SU.
M-1310; 0.07 max C, 0.4 Si, 18 Cr, 9.5 Ni, bal Fe.
Annealed: 85,000 TS; 35,000 YS; 60 El; 70 RA; 150 Brin.
Cold drawn: 180,000 TS; 125,000 YS; 10 El; 330 Brin.
For welded chemical plant equipment; Type 304; stainless, austenitic.

KARONI 10-18 MO.
M-1310; 0.10 max C, 18 Cr, 9.5 Ni, 2 Mo, 2.2 Si, bal Fe.
Annealed: 85,000 TS; 35,000 YS; 50 El; 65 RA; 160 Brin.
Cold drawn: 150,000 TS; 135,000 YS; 6 El; 300 Brin.
For acid resistant chemical plant equipment; Type 316; stainless, austenitic.

KARONI 10-18 MO EXTRA.
M-1310; 0.12 max C, 18 Cr, 2 Mo, 10.5 Ni, Ti = 4 x C, bal Fe.
Annealed: 90,000 TS; 40,000 YS; 45 El; 60 RA; 180 Brin.
For welded chemical plant equipment; Type 316 Ti; corrosion resistant.

KARONI 10-18 MOCU EXTRA.
M-1310; 0.07 max C, 0.4 Si, 17.5 Ni, 17.5 Cr, 2 Cu, Ti = 7 x C, 2 Mo, bal Fe.
For valves, pumps, chemical plant equipment; corrosion and heat resistant.

KARONI 10-18 MO SU.
M-1310; 0.07 max C, 18 Cr, 10.5 Ni, 2 Mo, bal Fe.
Annealed: 85,000 TS; 35,000 YS; 50 El; 65 RA; 160 Brin.
For acid resistant chemical plant equipment; Type 316; stainless, austenitic.

KARONI 12-12 SU.
M-1310; 0.10 max C, 12.5 Cr, 12 Ni, bal Fe.
For valves, pumps, turbine parts; corrosion and heat resistant.

KARONI 13-17 MO SU.
M-1310; 0.07 max C, 17 Cr, 13 Ni, 4.75 Mo, bal Fe.
Annealed: 85,000 TS; 35,000 YS; 50 El; 65 RA; 160 Brin.
For acid resistant chemical plant equipment; Type 317; stainless, austenitic.

KARONI 17-17 MOCU EXTRA.
M-1310; 0.07 max C, 17.5 Cr, 17.5 Ni, 2 Mo, 2 Cu, Ti = 7 x C, bal Fe.
For valves, pumps, turbine parts; corrosion resistant.

KARONI 20-13.
M-1310; 0.12 max C, 13 Cr, bal Fe.
Annealed: 75,000 TS; 40,000 YS; 35 El; 70 RA; 155 Brin.
Cold drawn: 100,000 TS; 85,000 YS; 17 El; 60 RA; 205 Brin.
For pumps, valves, turbine and jet parts; Type 310; corrosion resistant.

KARONI 20-13S.
M-1310; 0.12 max C, 13 Cr, 0.2 S, bal Fe.
Annealed: 75,000 TS; 40,000 YS; 35 El; 70 RA; 155 Brin.
For screw machine products, shafts, cutlery; Type 410F; corrosion resistant.

KARONI 20-16.
M-1310; 0.08 C, 17 Cr, bal Fe.
Annealed: 95,000 TS; 50,000 YS; 25 El; 55 RA; 196 Brin.
Cold drawn: 105,000 TS; 85,000 YS; 17 El; 50 RA; 215 Brin.
For furnace parts, heat treating boxes, baffles; Type 430; corrosion resistant.

KARONI 20-16 MOS.
M-1310; 0.12 C, 16.5 Cr, 0.25 Mo, 0.2 S, bal Fe.
Annealed: 95,000 TS; 50,000 YS; 25 El; 55 RA; 195 Brin.
For screw machine products, shafts; Type 430F; corrosion resistant.

KARONI 40-13.
M-1310; 0.2 C, 0.4 Si, 13 Cr, bal Fe.
Annealed: 95,000 TS; 50,000 YS; 25 El; 55 RA; 195 Brin.
Cold drawn: 105,000 TS; 85,000 YS; 17 El; 50 RA; 215 Brin.
For turbine blades, cutlery, valves, knives; Type 420; corrosion resistant.

KARONI 40.13 MO.
M-1310; 0.20 C, 13 Cr, 1.2 Mo, bal Fe.
Annealed: 95,000 TS; 40,000 YS; 25 El; 55 RA; B 92 Rock.
Heat treated: 240,000 TS; 200,000 YS; 9 El; 26 RA; C 50 Rock.
For cutlery, knives, surgical instruments, oil refinery and chemical plant equipment.
Corrosion resistant, hardenable.

KARONI 40-17.
M-1310; 0.22 C, 17 Cr, 1.5 Ni, bal Fe.
Annealed: 125,000 TS; 95,000 YS; 20 El; 55 RA; 260 Brin.
Cold drawn; 130,000 TS; 110,000 YS; 15 El; 35 RA; 270 Brin.
For marine hardware, valves, pumps; Type 431; corrosion resistant.

KARONI 70-16 MO.
M-1310; 0.35 C, 0.4 Si, 16.5 Cr, 1.15 Mo, bal Fe.
For chemical plant equipment, valves, pumps; corrosion resistant.

KARONI 80-13.
M-1310; 0.4 C, 0.4 Si, 13 Cr, bal Fe.
Annealed: 95,000 TS; 50,000 YS; 25 El; 55 RA; 195 Brin.
For turbine blades, cutlery, valves; Type 420; stainless.

KARONI 80-13 EXTRA.
M-1310; 0.85 C, Cr, V, bal Fe.
For bearings, liners; water hardened.

KARONI 210-15 MO.
M-1310; 1.1 C, Cr, Mo, S, bal Fe.
For bearings, liners, sleeves; water hardened.

KATADYN SILVER.
M-Ger.; Ag, Cu.
For jewelry; corrosion resistant.

KAYEM-1.
M-French; Al, Cu, bal Zn.
For die castings.

KAYEM-2.
M-French; Al, Cu, bal Zn.
For die castings.

KB 90.
M-1739; Infiltrated, sintered iron.
Typical UTS: 43 kg/mm^2; Rock B 30 min.
Not recommended for hardening.
SAE 870; ASTM B303-58T Class-A.

KB90/ 3/4.
M-1739; Infiltrated, sintered iron.
Typical UTS: 45.5 kg/mm^2; Hardness HV5-240.
Hardenable to Rock C 35 approx.
For small hand tools, high duty gears.
SAE 872; ASTM B303-58T Class C.

KB KELLEY BRONZE.
M-900.
Nickel aluminum bronze for glass molds.

KBV1.
M-1739; Copper infiltrated sintered Fe.
Density: 7.2 gms/cc; UTS: 55 kg/mm^2.
Hardness: HV5 250 min; Good conductivity.
For exhaust valve seats of internal combustion engines.

KBV2.
M-1739; Copper infiltrated sintered Fe. Higher copper than KBV1.
Density: 7.3 gms/cc; UTS: 61.5 kg/mm^2.
Hardness: HV5 250 min; good conductivity.
For exhaust valve seats.

K.C.M. NICKEL BRONZE.
M-1518; Ni, Sn, bal Cu.
Cast: 33,600 TS; 10-15 El; 100-120 Brin.
For slip rings for electrical motors; high conductivity.

K.D. NO. 6.
M-344; C. Co, W, bal Fe.
For high speed lathe tools, twist drills, reamers; high speed steel.

K.D. NO. 10.
M-344; C, W, Co, bal Fe.
For high speed lathe tools, twist drills, reamers; high speed steel.

K.D. NO. 16.
M-344; 0.7 C, 4.5 Cr, 10 Co, 2.2 V, 16 W, 1.5 Mo, bal Fe.
For reamers, drills, broaches, taps, hobs; high speed steel, oil hardened.

K.E. 4L.
M-510; high C, high Cr, bal Fe.
For springs; oil hardened.

K.E. 15.
M-510; 0.07-0.10 C, 13 Cr, bal Fe.
Annealed: 78,400 TS; 54,000 YS; 35 El; 70 RA; 152 Brin.
Heat treated: 123,000 TS; 105,000 YS; 25 El; 65 RA; 255 Brin.
For cutlery, stainless parts; stainless, martensitic.

Section I: Alloy Data / 799

K.E. 35.
M-510; 0.35 C, 13 Cr, bal Fe.
Heat treated: 99,000-274,000 TS; 68,000-242,000 YS; 30-9 El; 61-30 RA; 197-532 Brin.
For cutlery, valves, surgical instruments; stainless, martensitic.

K.E. 40 A.
M-510; 0.10 C, 12.5 Cr, Se, bal Fe.
Free cutting martensitic corrosion resistant steel, hardenable; for pump shafts, hardware, fasteners, gears, axles.

K.E. 169.
M-510; 0.14 C, 0.85 Cr, 3.4 Ni, 0.15 Mo, bal Fe.
Nickel-chrome-moly case hardening steel.
Core hardenable to 135,000-157,000 TS.
For gears, spline shafts, axles, plastic mold dies.

K.E. 355.
M-510; 0.32 C, 1.2 Cr, 4.1 Ni, 0.2 Mo, bal Fe.
For plastic mold dies.
Oil or air hardening, tough and strong.

K.E. 339 see SABEX.

K.E. 396.
M-510; 0.63 C, 1.0 Cr, 1.45 Ni, 0.20 Mo, bal Fe.
Nickel-chrome-moly oil hardening tool steel. AISI L6.
For blanking and forming dies, shear blades.

K.E. 595.
M-510; 1.25 C, 0.85 Mn, 1.2 Cr, 1.3 W, bal Fe.
Oil hardening alloy tool steel; for cold drawing and swaging dies; oil or water hardening.

K.E. 672.
M-510; 0.95 C, 1.2 Mn, 0.55 Cr, 0.65 W, 0.15 V, bal Fe.
Oil hardening alloy tool steel. AISI O1.
For press and blanking tools.

K.E. 805.
M-510; 0.38 C, 1 Cr, 1.5 Ni, 0.2 Mo, bal Fe.
Ht. Tr.: 121,000-280,000 YS; 90,000-246,000 YS; 10-26 El; 35-68 RA; 250-515 Brin.
For gears, machinery parts, shafts, machinery tools, fasteners.
Oil hardening, wear resistant.

K.E. 839.
M-510; 1.05 C, 0.50 Mn, 1.4 Cr, bal Fe.
Oil hardening alloy tool steel; for ball bearings, cutters, gages, taps, swaging dies, shear blades. AISI L1.

K.E. 896.
M-510; 0.5 C, 0.65 Mn, 1.1 Cr, 0.2 V, bal Fe.
For arbors, axles, boring bars, connecting rods.
Type L2 tool steel, oil hardening, shock resistant.

K.E. 897.
M-510; 0.3 C, 1.25 Cr, 4.25 Ni, 0.2 Mo, bal Fe.
Ht. Tr.: 224,000-270,000 TS; 10-14 El; 30-45 RA; 444-512 Brin.
For gears, shafts, bolts, fasteners.
Oil hardening, shock resistant.

K.E. 960.
M-510; 0.50 C, 0.65 Si, 1.5 Cr, 2.25 W, 0.20 V, bal Fe.
Extra tough alloy oil hardening tool steel.
For shear blades, pneumatic tools, stamping and riveting tools. AISI S1.

K.E. 970.
M-510; 2.1 C, 13.5 Cr, 0.5 W, bal Fe.
Chrome high duty tool steel; for plug and ring gages, punches, lamination and blanking dies. AISI D3.

K.E. 1006.
M-510; 1.05 C, 0.20 Si, 0.20 Mn, 0.15 V, bal Fe.
Water hardening tool steel, soft center steel for cold heading dies. AISI W2.

K.E.A. 23.
M-510; 0.07 C, 18.0 Cr, 9.0 Ni, Ti, bal Fe.
Weldable austenitic stainless; for welded structures and for operation at elevated temperatures.

K.E.A. 28.
M-510; 0.45 C, 13.25 Cr, 1.0 Ni, bal Fe.
Martensitic corrosion resistant steel, hardenable.
For plastic molds.

K.E.A. 108.
M-510; 1.0 C, 0.55 Mn, bal Fe, plus free cutting agent.
Free cutting water hardening carbon tool steel.

K.E.A. 138.
M-510; 0.35 C, 12.0 Cr, 12.0 W, 1.0 V, ba Fe.
Chrome-Tungsten hot work tool steel.
For pressure die casting dies, extrusion dies, punches. AISI H23.

K.E.A. 145.
M-510; 0.38 C, 1.05 Si, 5.25 Cr, 1.35 Mo, 1.0 V, bal Fe.
5% Chrome-Vanadium-Moly die casting steel.
For die casting dies, extrusion dies, hot piercing and gripping dies. AISI H13.

K.E.A. 180.
M-510; 1.55 C, 12.0 Cr, 0.85 Mo, 0.30 V, bal Fe.
High duty tool steel, air hardening. AISI D2.

K.E.A. 205.
M-510; 1.4 C, 0.3 Si, 0.4 Mn, 3.45 V, bal Fe.
Cold heading punch and die steel.
For dies, punches, headers.

K.E.A. 218.
M-510; 0.32 C, 0.50 Si, 0.35 Mn, 3.0 Cr, 2.8 Mo, 0.35 V, 3.0 Co, bal Fe.
Hot work tool steel.
B.S. 4659 Type BH10A.

K.E.A. 221.
M-510; 0.09 C, 18.5 Cr, 10.5 Ni, 2.7 Mo, bal Fe.
Annealed: 85,000 TS; 35,000 YS; 50 El; 165 Brin.
Cold rolled: 150,000 TS; 135,000 YS; 6 El; C 32 Rock.
For chemical and pharmaceutical equipment, digesters, valve trim tanks.
AISI 316, stainless, austenitic, acid resistant.

K.E.A. 227.
M-510; 0.50 C, 1.2 Mn, 0.65 Cr, 0.20 Mo, bal Fe.
Easy machining bolster steel.
For molds, press components.

K.E.A. 505.
M-510; 0.17 C, 17 Cr, 1.75 Ni, Se, bal Fe.
Annealed: 125,000 TS; 95,000 YS; 20 El; 260 Brin.
Cold Drawn: 130,000 TS; 110,000 YS; 15 El; 270 Brin.
For marine hardware, pump shafts, valve trim, aircraft structural components.
AISI 431 Se, corrosion resistant, free cutting.

K.E.A. 507.
M-510; 0.09 C, 18.0 Cr, 8.5 Ni, Se, bal Fe.
Free cutting austenitic stainless steel.
For stainless parts requiring appreciable machining. AISI 303 Se.

K.E.A. 521.
M-510; 0.07 C, 18.0 Cr, 11.5 Ni, 2.65 Mo, Se, bal Fe.
Free machining austenitic stainless steel with very good corrosion resistance.
For machined parts in food, beverage, laundry plants. AISI 316 Se.

K.E. CARLISLE.
M-510; 1.0 C, bal Fe.
For taps, drills, reamers, general tools; water hardened.

K.E. DIAMOND.
M-510; 1.1 C, bal Fe.
For taps, drills, reamers, general tools; water hardened.

K.E. DIAMOND NO. 10.
M-510; 1.5 C, 0.40 Mn, 0.60 Cr, 5.75 W, bal Fe.
High carbon tungsten alloy tool steel.
For reamer blades, roll turning tools, form cutters; water hardenable.

KEENE'S ALLOY.
75 Cu, 16 Ni, 2.8 Sn, 2.3 Zn, 2 Co, 1.5 Fe, 0.5 Al.
For corrosion resistant and ornamental parts; corrosion resistant.

KEEN-KUT.
M-15; C, alloy, bal Fe.
For hollow drills; non-tempering.

KEEWATIN.
M-435; 0.90 C, 1 Mn, 0.5 Cr, 0.5 W, bal Fe.
For dies, cutters, punches, reamers, broaches; Type O1; oil hardened, non-deforming.

KEINMAYER'S AMALGAM.
50 Hg, 25 Sn, 25 Zn.
For amalgam.

KELCALOY.
M-792; C, alloy, bal Fe.
For oil refinery equipment; corrosion resistant.

KEL-CAST.
M-1163, M-900; 3.6 C, 1.25-1.5 Si, 0.7 Mn, 0.3 Cr, bal Fe.
Cast: 45,000 TS; 200-220 Brin.
For bearings, gears, valve seats; cast iron.

KEL-CAST MM.
M-1163; 3.25 C, 2.5 Si, 0.7 Mn, 0.3 Cr, 0.4 Mo, bal Fe.
Cast: 45,000 TS; 200-220 Brin.
For bushings, cams, gears, valves; cast iron.

KELCAST 5MM.
M-900.
Alloy gray iron castings, with Cr, Mo, V.
For I.S. plungers, neck rings, bottom plates, baffles.

KELCAST NO. 4.
M-900.
Gray iron casting; Class 25.
25,000 psi min TS; 160-190 Brin.
General purpose containers mold metal; for molds, baffles, blanks, bottom plates.
ASTM A48-56.

KELCAST NO. 5.
M-900.
Gray iron casting; Class 35.
35,000 psi min TS; 180-210 Brin.
Good strength general purpose casting.
ASTM A48-56.

KELCAST NO. 5 MMP.
M-900.
Gray iron casting; Class 45.
45,000 psi min TS; 190-235 Brin.
Specially alloyed and processed for Hartford I.S. 28 paste mold blanks, plungers, half bars and ring stock.
ASTM A48-56.

KELCAST NO. 6XXX.
M-900.
Alloy gray iron castings.
Specially alloyed and processed for packers ware tumbler plungers.

KELCAST NO. 7 FIRE X.
 M-900.
 Alloy cast iron.
 Heat resistant iron used principally for paste-mold plungers, glazier burners.

KELCAST NO. 10 DUCTILE IRON.
 M-900.
 Ductile iron casting.
 Used as press mold plunger for forming boro-silicate ware and for container molds.

KELCAST TYPE 3.
 M-900.
 Alloy gray iron castings with Mo, V.
 Long life contaianer and press mold iron.
 For boro-silicate glass operation.

KELNOD NO. 6.
 M-900.
 Ductile iron, Type 60-45-10.
 60 ksi TS; 45 ksi YS; 10 El; 160-180 Brin.
 ASTM A339-55.

KELNOD NO. 8.
 M-900.
 Ductile iron, Type 80-60-03.
 80 ksi TS; 60 ksi YS; 3 El; 200-250 Brin.
 ASTM A339-55.

KELNOD NO. 10.
 M-900.
 Ductile iron, Type 100-70-03.
 100,000 psi TS; 70,000 psi YS; 3 El; 230-270 Brin.

KELMET.
 70.5-68.0 Cu, 22.5-25.5 Pb, 6.4-6.6 Sn, 0.03 S, 0.03 Zn, 0.04 Ni.
 For bearings; heavy duty.

KELOCK 237.
 M-510; 0.7 C, 18 W, 4 Cr, 1 V, bal Fe.
 Hardened: C 64-67 Rock.
 For die cores, lathe and planer tools, drills, chasers, reamers, broaches.
 Type T1 high-speed steel, high red-hardness.

KELOCK 795.
 M-510; 0.7 C, 14 W, 4 Cr, 0.6 V, bal Fe.
 Hardened: C 64-67 Rock.
 For lathe and planer tools, cutters, hot and cold shear blades, drills, reamers, punches.
 High-speed steel, high red-hardness.

KELOCK 873.
 M-510; 0.7 C, 4.25 Cr, 18.25 W, 0.75 Mo, 1 V, 5 Co, bal Fe.
 Hardened: C 64-66 Rock.
 For lathe and planer tools, cutters, reamers,broaches, drills, form tools.
 Type T4 high-speed steel, high red-hardness.

KELOCK 1021.
 M-510; 0.7 C, 18.2 W, 4 Cr, 0.75 Mo, 1.45 V, 10 Co, bal Fe.
 Hardened: C 64-67 Rock.
 For lathe and planer tools, cutters, reamers, broaches, form cutters.
 High-speed steel, Type T5, high red-hardness.

KELOCK A157.
 M-510; 0.8 C, 6 W, 4 Cr, 2 V, 5 Mo, bal Fe.
 Hardened: C 64-67 Rock.
 For lathe and planer tools, drills, hobs, reamers, taps, milling cutters.
 High-speed steel, high red-hardness. AISI M2.

KELOCK A182.
 M-510; 0.80 C, 3.9 Cr, 8.5 Mo, 1.5 W, 1.15 V, bal Fe.
 High speed tool steel.
 B.S. 4659 Type BM 1; AISI M1.

KELOCK A229.
 M-510; 1.05 C, 3.75 Cr, 1.5 W, 9.5 Mo, 1.15 V, 8.25 Co, bal Fe.
 High speed steel for cutting tools that require extra red hardness. AISI M42.

KELVAN see ELECTRITE KELVAN.

KEMLER.
 M-Eng; 76 Zn, 9 Cu, 15 Al.
 For ornamental parts.

KEMLET.
 67 Zn, 15 Al, 9 Cu.
 For bearings.

KEN AIR A-6.
 M-1700; 0.70 C, 2.0 Mn, 1.0 Cr, 1.3 Mo, bal Fe.
 AISI A6 air hardening tool steel.

KEN CHROME D-2.
 M-1700; 1.5 C, 12.0 Cr, 0.8 V, 0.75 Mo, bal Fe.
 Air or oil hardening cold work tool and die steel; AISI D2. High carbon, high chromium type.

KEN DIE A-2.
 M-1700; 1.0 C, 0.5 Mn, 0.25 Si, 5.0 Cr, 0.3 V, 1.25 Mo, bal Fe.
 Air hardening and cold work tool steel; AISI A2.

KEN GRAPH OIL 06.
 M-1700; 1.45 C, 0.8 Mn, 1.15 Si, 0.25 Mo, bal Fe.
 Graphitic type oil hardening cold work tool steel; AISI O6.

KEN H 13.
 M-1700; 0.40 C, 0.4 Mn, 1.0 Si, 5.0 Cr, 1.0 V, 1.5 Mo, bal Fe.
 Air or oil hardening hot work tool and die steel, chromium type; AISI H13.

KEN LUSTRE MOLD.
  M-1700; 0.55 C, 1.0 Mn, 0.3 Si, 1.1 Cr, 0.25 Mo, bal Fe.
  Oil hardenable tool steel, mainly for molds.

KEN M-2.
  M-1700; 0.84 C, 4.2 Cr, 1.95 V, 6.35 W, 5.0 Mo, bal Fe.
  High speed tool steel, Mo-W type; AISI M2.

KEN P-20.
  M-1700; 0.30 C, 0.8 Mn, 0.5 Si, 1.7 Cr, 0.4 Mo, bal Fe.
  Oil hardenable mold steel; AISI P20.

KEN T-1.
  M-1700; 0.72 C, 4.3 Cr, 1.1 V, 18.0 W, bal Fe.
  High speed tool steel, tungsten type; AISI T1.

KEN W-1.
  M-1700; 1.0-1.1 C, 0.3 Mn, 0.25 Si, bal Fe.
  Water hardening tool steel; AISI W1.

KENNAMETAL.
  M-793.
  A series of sintered hard carbide alloys with WC base and various binder metals; for metal cutting, metal forming, rock cutting, wear and abrasion resistant parts and structural parts.

KENNAMETAL GR. 420.
  M-793; WC, TiC, TaC, Co.
  Sintered: Rock A, 91.3.
  For metal cutting of steels.

KENNAMETAL GR. K1.
  M-793; WC, 11.5 Co.
  Sintered: Rock A 89.7.
  For metal cutting, wear parts.

KENNAMETAL GR. K2S.
  M-793; WC, TiC, TaC, Co.
  Sintered: Rock A 91.5.
  For metal cutting of steels.

KENNAMETAL GR. K4H.
  M-793; WC, TiC, TaC, Co.
  Sintered: Rock A 92.0.
  For metal cutting of steels.

KENNAMETAL GR. K4H.
  M-793; WC, TiC, TaC, Co.
  Sintered: Rock A 92.0.
  For metal cutting of steels.

KENNAMETAL GR. K5H.
  M-793; WC, TiC, TaC, Co.
  Sintered: Rock A 93.0.
  For metal cutting of steels.

KENNAMETAL GR. K6.
  M-793; 5.5 Co.
  Sintered: Rock A 92.0.
  For metal cutting, wear parts.

KENNAMETAL GR. K7H.
  M-793; WC, TiC, TaC, Co.
  Sintered: Rock A 93.5.
  For metal cutting of steels.

KENNAMETAL GR. K7H.
  M-793; WC, TiC, TaC, Co.
  Sintered: Rock A 93.5.
  For metal cutting of steels.

KENNAMETAL GR. K8.
  M-793; WC, 3.8 Co.
  Sintered: Rock A 92.9.
  For metal cutting.

KENNAMETAL GR. K9.
  M-793; WC, 8.8 Co.
  Sintered: Rock A 89.0.
  For structural parts, high modulus and strength.

KENNAMETAL GR. K11.
  M-793; WC, 2.8 Co.
  Sintered: Rock A 93.0.
  For metal cutting.

KENNAMETAL GR. K21.
  M-793; WC, TiC, TaC, Co.
  Sintered: Rock A 91.0.
  For metal cutting of steels.

KENNAMETAL GR. K45.
  M-793; WC, TiC, TaC, Co.
  Sintered: Rock A 92.5.
  For metal cutting of steels.

KENNAMETAL GR. K68.
  M-793; WC, 5.8 Co.
  Sintered: Rock A 92.6.
  For metal cutting.

KENNAMETAL GR. K82.
  M-793; WC, TiC, TaC, Co.
  Sintered: Rock A 90.0.
  For metal forming dies and punches.

KENNAMETAL GR. K84.
  M-793; WC, TiC, TaC, Co.
  Sintered: Rock A 91.0.
  For metal forming dies and punches.

KENNAMETAL GR. K86.
  M-793; WC, TiC, TaC, Co.
  Sintered: Rock A 91.7.
  For metal forming dies and punches.

KENNAMETAL GR. K90.
  M-793; WC, 25 Co.
  Sintered: Rock A 85.0.
  For impact forming of metals.

KENNAMETAL GR. K91.
  M-793; WC, 19.5 Co.
  Sintered: Rock A 86.8.
  For impact forming of metals.

KENNAMETAL GR. K92.
M-793; WC, 15.5 Co.
Sintered: Rock A 88.5.
For punches and dies blanking metals.

KENNAMETAL GR. K94.
M-793; WC, 11.5 Co.
Sintered: Rock A 89.7.
For punches and dies blanking metals.

KENNAMETAL GR. K95.
M-793; WC, 9 Co.
Sintered: Rock A 90.2.
For wear resistant parts.

KENNAMETAL GR. K96.
M-793; WC, 5.5 Co.
Sintered: Rock A 92.0.
For wear resistant parts.

KENNAMETAL GR. K602.
M-793; WC, TaC, under 1.5% binder.
Sintered: Rock A 94.3.
For wear and corrosion resistant parts.

KENNAMETAL GR. K701.
M-793; WC, Cr, Co.
Sintered: Rock A 92.0.
For severe abrasion-corrosion resistant parts.

KENNAMETAL GR. K703.
M-793; WC, Cr, Co.
Sintered; Rock A 91.5.
For abrasion-corrosion resistant parts.

KENNAMETAL GR. K714.
M-793; WC, TiC, Cr, Co.
Sintered: Rock A 92.5.
For abrasion-corrosion resistant parts.

KENNAMETAL GR. K801.
M-793; WC, Ni.
Sintered: Rock A 90.0.
For wear parts exposed to radiation or mineralized water.

KENNAMETAL GR. K803.
M-793; WC, Ni.
Sintered: Rock A 91.0.
For wear parts which must be non-magnetic.

KENNAMETAL GR. K2884.
M-793; WC, TiC, TaC, Co.
Sintered: Rock A 92.0.
For metal cutting (milling) of steels.

KENNAMETAL GR. K3047.
M-793; WC, 10.5 Co.
Sintered: Rock A 88.5.
For wear-shock resistant parts.

KENNAMETAL GR. K3109.
M-793; WC, 12.2 Co.
Sintered: Rock A 88.0.
For wear-shock resistant parts.

KENNAMETAL GR. K3404.
M-793; WC, 5.5 Co.
Sintered: Rock A 90.5.
For wear-shock resistant parts.

KENNAMETAL GR. K3406.
M-793; WC, 7.8 Co.
Sintered: Rock A 89.5.
For wear-shock resistant parts.

KENNAMETAL GR. K3411.
M-793; WC, 9.5 Co.
Sintered: Rock A 88.7.
For wear-shock resistant parts.

KENNAMETAL GR. K8735.
M-793; WC, TiC, Co.
Sintered: Rock A 92.0.
For metal cutting of alloy irons.

KENNAMETAL GR. KC210.
M-793; WC, Co plus multi-coating.
Sintered and coated.
For metal cutting of alloy irons.

KENNAMETAL GR. KC810.
M-793; WC, TiC, TaC, Co, plus multi-coating.
Sintered and coated.
For metal cutting of steels.

KENNERTIUM.
M-793.
A series of sintered heavy tungsten alloys for use as radiation shielding, intertial device components; machinable.

KENNERTIUM W-2.
M-793; 97.5 W, 2.5 Ni plus Cu.
Sintered: density 18.5.
For parts of inertial devices, balancing slugs, weights and counterbalance.

KENNERTIUM W-10.
M-793; 90 W, 7.5 Ni plus Cu.
Sintered: density 17.0.
For parts of intertial devices, radiation shielding, weights and counterbalances.

KEN OIL O-1.
M-1700; 0.09 C, 1.2 Mn, 0.3 Si, 0.5 Cr, 0.2 V, 0.5 W, bal Fe.
Oil hardening tool steel; AISI O1.

KENT.
M-1705; 2.15 C, 0.25 Mn, 0.25 Si, 12.0 Cr, 0.8 V, bal Fe.
High carbon, high chromium tool and die steel; air or oil hardening AISI D3.

KENTANIUM.
M-793.
A series of sintered hard carbide alloys with TiC base and Ni-Mo binders: for metal cutting, metal forming, wear and abrasion resistant parts.

KENTANIUM K151A.
M-793; TiC, Ni, Mo.
Sintered: Rock A 90.0.
For high temperature abrasion and wear resisting parts.

KENTANIUM K162B.
M-793; TiC, Ni, Mo.
Sintered: Rock A 89.5.
For wear and abrasion resistant parts combined with light weight.

KENTANIUM K165.
M-793; TiC, Ni, Mo.
Sintered: Rock A 93.5.
For metal cutting and wear resistant parts.

KEN TOUGH S-7.
M-1700; 0.50 C, 0.7 Mn, 3.25 Cr, 1.4 Mo, bal Fe.
Air or oil hardening, shock resistng tool steel; AISI S7.

KEN TOUGH S-7 MOLD QUALITY.
M-1700; 0.50 C, 0.7 Mn, 3.25 Cr, 1.4 Mo, bal Fe.
Air or oil hardening tool and die steel, mainly for molds.

KENTUCKY.
M-191; C, Cu, bal Fe.
For construction steel.

KEOKUK ELECTRO-SILVERY.
M-1039; 15 Si, 6 C, bal Fe.
For iron and steel making; addition agent.

KERCHSTEEL.
0.12-0.18 As, C, bal Fe.
For construction steel.

KERN'S HYDRAULIC BRONZE.
78 Cu, 12 Sn, 10 Zn.
For vessels, pressure castings, valves, fittings; high strength.

KERROFESTAL.
M-Ger.; 1.0 Si, 0.3 Fe, 0.7 Mn, 0.65 Mg, bal Al.
For architecture, window frames, chemical industry; light alloy.

KETOS.
M-38; 0.90 C, 0.25 Si, 1.3 Mn, 0.5 Cr, 0.5 W, bal Fe.
For tools, dies, taps, reamers, broaches, gauges, hobs; non-deforming, oil hardened.

KEY ALLOY-B.
65.2 Cu, 20.15 Zn, 12.70 Ni, 0.40 Fe, 1.10 Pb, 0.15 Sn.
For hardware keys; free-cutting.

KEY STOCK-A.
60-65 Cu, 26-22 Zn, 12 Ni, 2-1 Pb.
For hardware, keys; free-cutting.

KEYSTONE.
M-339; 0.90 C, 1 Mn, 0.5 Cr, 0.5 W, bal Fe.
For blanking and forming dies, punches, gauges; Type O1; oil hardened, non-deforming.

KEYSTONE C62.
M-989.
High graphite bronze.
Sintered: 11,000 psi TS; 16% porosity; 6.0-6.4 density; Rock H 25.
For bearings requiring quiet motor operation.

KEYSTONE C64 5.8-6.2.
M-989; Bronze 90 Cu, 10 Sn.
Sintered: 11,000 psi TS; 25% porosity; Roch H 25; density: 5.8-6.2 for light loaded bearings.
ASTM B438-66T Gr. 1 Type I; SAE 840.

KEYSTONE C64 6.4-6.8.
M-989; Bronze 90 Cu, 10 Sn.
Sintered: 13,000 psi TS; 18% porosity, Rock H 40; density; 6.4-6.8 for bearings.
ASTM B438-66T Gr. 1, Type II; SAE 841.

KEYSTONE C64, 6.81-7.2 DENSITY.
M-989; Sintered Cu-Sn bronze.
6.81-7.2 density; 20,000 psi TS; 3.0 El; Rock H 60.
Self-lube for moving parts.
ASTM B255-61 Type II.

KEYSTONE C65.
M-989; Staking grade Phos bronze.
Sintered: 25,000 psi TS; 10.0 El; 14% porosity; 7.0-7.4 density; Rock H 60.
High density bearing; good ductility, can be staked.

KEYSTONE C 71.
M-989; Leading bronze.
Sintered: 12,000 psi TS; 18% porosity; 1.5% El; density 6.5-6.9; Rock H 30.
For bearings difficult to keep oiled.
ASTM B438-66T Gr. 2, Type I; SAE 843.

KEYSTONE D10.
M-989; Graphite bronze.
Sintered: 10,000 psi TS; 1.0 El; Rock H 30; density 6.5-6.9.
Oilless bearing for light loads, slow speeds.

KEYSTONE D69.
M-989; Diluted bronze 75-25.
Sintered: 17,000 psi TS; 18% porosity; density 6.2-6.6; Rock H 50.
For oscillatory-type bearings.

KEYSTONE DRILL ROD.
M-1703; 0.88-0.93 C, 1.1-1.3 Mn, 0.2-0.35 Si, 0.15-0.20 Cr, 0.15-0.20 V, 0.4-0.6 W, 0.02 max P, 0.02 max S, bal Fe.
Oil hardening drill rod; AISI 01.

KEYSTONE F 11.
M-989; Iron Copper 99-1.
Sintered: 20,000 psi TS; 18% porosity; 5.7-6.1 density; Rock H 70.
Low cost bearing for moderate loading.

KEYSTONE F-20, BEARING GRADE.
M-989; Iron copper 90-10.
Sintered: 30,000 psi TS; 18% porosity; 5.8-6.2 density; Rock H 80.
Low cost iron bearing for moderate loading.

KEYSTONE F-20, STRUCTURAL GRADE.
M-989. Sintered Fe-Cu.
5.8-6.2 density; 31,000 psi TS; Rock H 85.
Standard grade for medium stressed structural parts.
ASTM B-222-61.

KEYSTONE P-7.
M-989; 50 Fe, 50 Ni, sintered.
6.8 density; coercive force 1.2-2.5 oersteds.
Remanence: 5160-8500 gauss.
Soft magnetic material for corrosion resistant cores and solenoids.

KEYSTONE TO4.
M-989. Sintered stainless steel to AISI 316 composition.
6.40 min density; 55,000 psi TS; 7.0 El; Rock B 55.
For corrosion resistant structural parts.

KEYSTONE T-1, FILTER GRADE.
M-989. Sintered stainless steel to AISI 316 composition.
5.5 nominal density; 10,000 psi TS min.
Pore size: 30 micron (0.0012 in).
Permeability: 5.0 CFM air/sq. in. in a 10 psi press. diff. for 0.100 in. filter thickness.

KEYSTONE T-2, FILTER GRADE.
M-989. Sintered stainless steel to AISI 316 composition.
5.7 nominal density; 15,000 psi TS min.
Pore size: 10 micron (0.0004 in.).
Permeability: 1.8 CFM sq/in. in a 10 psi press. diff. for 0.100 filter thickness.

KEYSTONE T-3, FILTER GRADE.
M-989. Sintered stainless steel to AISI 316 composition.
6.0 nominal density; 15,000 psi TS min.
Pore size: 5 micron (0.0002 in.).
Permeability: 0.7 CFM sq/in. in a 10 psi press. diff. for 0.100 in filter thickness.

KEYSTONE T06.
M-989.
Sintered stainless steel to AISI 304 composition.
6.40 min desnity; 55,000 psi TS; 7.0 El; Rock B 55.
For corrosion resistant structural parts.

KEYSTONE T-9, FILTER GRADE.
M-989.
Sintered stainless steel to AISI 316 composition.
4.5 nominal density; 5,000 psi TS min.
Pore size 60 micron (0.0024 in).
Permeability: 14.0 C.F.M. air/sq. in. in a 10 psi press. diff for 0.100 in. filter thickness.

KEYSTONE V19.
M-989.
White bronze.
Sintered: 11,000 psi TS; 5 El; 18% porosity; density 6.5-6.9; Rock H 40.
For bearings; low coefficient of friction; improved elevated temperature operation.

KEYSTONE V03, 7.5-7.99 DENSITY.
M-989.
Sintered nickel silver.
7.5-7.99 density; 32,000 psi TS; Rock H 80.
Pleasant appearing structural parts; corrosion resistant.

KEYSTONE V03, 8.0 MIN DENSITY.
M-989.
Sintered nickel silver.
8.0 min density; 36,000 psi TS; 10.0 El; Rock H 75.
Pleasant appearing structural parts; corrosion resistant.

KEYSTONE Y11.
M-989.
Sintered high conductivity copper.
8.3 density; 25,000 psi TS; 4.0 El; Rock H 80.
For parts requiring high electrical conductivity.

KEYSTONE Z02 6 85-7 15 DENSITY.
M-989.
Sintered steel to AISI 4630 composition.
As sintered: Rock B 65.
Ht Tr: Rc 25, 110,000 psi TS.
For structural parts.

KEYSTONE Z02 716-7 50 DENSITY.
M-989.
Sintered steel to AISI 4630 composition.
As sintered: Rock B 75.
Ht Tr: Rc 35; 160,000 psi TS.
For highly stressed structural parts, as hydraulic pump gears.

KEYSTONE Z04 LOW DENSITY.
M-989.
Sintered iron.
5.7-6.1 density; 16,000 psi TS; Rock H 50.
For lightly loaded structural parts.
ASTM B310-64T Type I Class A; SAE 850.

KEYSTONE Z04 HIGH DENSITY.
M-989.
Sintered iron.
6.91-7.29 density; 30,000 psi TS; Rock B 35; hardenable to 51,000 psi TS; RA 65.
For heavier loaded structural parts.
ASTM B310-64T Type IV, Class A.

KEYSTONE Z04 MEDIUM DENSITY.
M-989.
Sintered iron.
6.11-6.50 density; 20,000 psi TS; Rock H 60.
For lightly loaded structural parts.
ASTM B310-64T Type II, Class A; SAE 853.

KEYSTONE Z04 MEDIUM HIGH DENSITY.
M-989.
Sintered iron.
6.51-6.90 density; 26,000 psi TS; Rock B 30; hardenable to RA 60.
For medium loaded structural parts.
ASTM B310-64T Type III, Class A.

KEYSTONE Z-4M MAGNETIC GRADE.
M-989; 100 Fe, sintered.
6.0-6.5 density; coersive force 2.8-2.9 oersteds.
Remanence 6,500-8,100 gauss.
Soft magnetic material for pole pieces, rotor and stator cores for small motors.

KEYSTONE Z-23.
M-989.
Sintered Fe-Cu-P alloy.
6.1-6.5 density; 60,000 psi TS; Rock B 45.
Standard grade for structural parts.

KEYSTONE Z-24.
M-989.
Sintered electrolytic iron, high density.
7.3 min density; 40,000 psi TS; Rock B 40.
Can be hardened or case hardened, and plated.
For many types of structural parts.
ASTM B310-64T Type V, Class A.

KEYSTONE Z-28.
M-989.
Diluted bronze 40-60. (40 bronze, 60 Fe)
Sintered: 14,000 psi TS; 18% porosity; 5.8-6.2 density; Rock H 40.
Lower priced bearing.

KEYSTONE Z-29M.
M-989; 99 + Fe, sintered.
6.0-6.9 density; coercive force 2.5-2.7 oersteds.
Remanence: 6,900-8,500 gauss.
Soft magnetic material for cores for AC relays, rotor and stator cores for small AC motors and generators.

KEYSTONE Z33, 5.7-6.1 DENSITY.
M-989.
Sintered carbon steel, class C.
5.7-6.1 density; 26,000 psi TS; Rock B 35; hardenable to 40,000 psi TS; RA 55.
For structural parts.
ASTM B310-64T Type I; SAE 852.

KEYSTONE Z33, 6.11-6.50 DENSITY.
M-989.
Sintered carbon steel, class C.
6.11-6.50 density; 34,000 psi TS; Rock B 55; hardenable to 50,000 psi TS; Rock A 60.
For structural parts.
ASTM B310-64T Type II; SAE 855.

KEYSTONE Z33, 6.51-6.90 DENSITY.
M-989.
Sintered carbon steel, class C.
6.51-6.90 density; 44,000 psi TS; Rock B 70; hardenable to 64,000 psi TA; Rock A 64.
For structural parts.
ASTM B310-64T Type III.

KEYSTONE Z33, 6.91-7.30 DENSITY.
M-989.
Sintered carbon steel, class C.
6.91-7.30 density; 60,000 psi TS; Rock B 85; hardenable to 80,000 psi TS; Rock A 68.
For structural parts.
ASTM B310-64T Type IV.

KEYSTONE Z34, 5.70-6.10 DENSITY.
M-989.
Sintered carbon copper steel, class C.
5.7-6.1 density; 30,000 psi TS; Rock B 35; hardenable to 40,000 psi TS; Rock A 55.
For structural parts; recommended for parts requiring induction hardening.
ASTM B426-65 Type I, Gr. 1; SAE 846A.

KEYSTONE Z34, 6.11-6.50 DENSITY.
M-989.
Sintered carbon copper steel, class C.
6.11-6.50 density; 41,000 psi TS; Rock B 50; hardenable to 50,000 psi TS; Rock A 60.
For structural parts; recommended for parts requiring induction hardening.
ASTM B426-65 Type II, Gr. 1; SAE 864B.

KEYSTONE Z46, 5.7-6.1 DENSITY.
M-989.
Sintered carbon copper steel, class C; 3-6 Cu.
5.70-6.10 density; 35,000 psi TS; Rock B 45; hardenable to 50,000 psi TS; Rock A 55.
All purpose grade for structural parts.
ASTM B426-65 Type I, Gr.2; SAE 865A.

KEYSTONE Z46, 6.11-6.5 DENSITY.
M-989.
Sintered carbon copper steel, class C, 3-6 Cu.
6.11-6.5 density; 50,000 psi TS; Rock B 60; hardenable to 75,000 psi TS; Rock A 60.
All purpose grade for structural parts.
ASTM B426-65 Type II, Gr.2; SAE 865B.

KEYSTONE Z46, 6.51-6.90 DENSITY.
M-989.
Sintered carbon copper steel, class C, 3-6 Cu.
6.51-6.90 density; 60,000 psi TS; Rock B 65; hardenable to 85,000 psi TS; Rock A 64.
All purpose grade for structural parts.
ASTM B426-65 Gr.2 except density.

KEYSTONE Z63, 6.3-6.7 DENSITY.
M-989.
Sintered nickel steel.
6.3-6.7 density; 40,000 psi TS; Rock B 45.
Hardened and tempered: 65,000 psi TS; Rock A 60.
For structural parts; may be oil impregnated.

KEYSTONE Z63, 6.75-7.20 DENSITY.
M-989.
Sintered nickel steel.
6.75-7.20 density; 55,000 psi TS; Rock B 60.
Hardened and tempered: 110,000 psi TS; Rock A 65.
For structural parts.

KEYSTONE Z63, 7.21-7.60 DENSITY.
M-989.
Sintered nickel steel.
7.21-7.60 density; special process 100,000 psi TS; Rock B 80.
Hardened and tempered: 180,000 psi TS; Rock A 70.
For structural parts.

K F N 1.
M-1541; 0.06 C, 0.34 Mn, 0.008 P, 0.0125 S, 0.007 max B, bal Fe.
46,900 psi TS; 28,400 psi YS; 48 El.
Low alloy steel for deep drawing purposes.

K F N 2.
M-1541; 0.005 c, 0.10 Mn, 0.010 P, 0.004 S, 0.007 max B, bal Fe.
39,800 psi TS; 25,600 psi YS; 54 El.
Low alloy steel for extra deep drawing purposes.

KH 15N25.
M-USSR; 0.2 C, 15 Cr, 25 Ni, bal Fe.
For furnace parts, heat treating boxes; heat and corrosion resistant.

KH 28.
M-USSR; 1.7-2.5 C, 1.3-1.7 Si, 28-32 Cr, bal Fe.
For chemical plant equipment; corrosion and heat resistant, cast.

KH 31 (STANDARD ALLOY).
M-Eng.; 2.5-4.0 Th, 0.5-0.7 Zr, bal Mg.
H24-temper: 38,900 TS; 29,800 YS; 6 El.
T6-temper: 37,100 TS; 19,900 YS; 17 El.
For airframes, rockets, missile components; high creep resistance to 700°F.

KH34.
M-USSR; 1.5-2.2 C, 1.3-1.7 Si, 32-36 Cr, bal Fe.
For cast rabbles for furnaces, ore roasters; corrosion and heat resistant, cast.

KHG.
M-USSR; 1.44 C, 0.62 Mn, 1.29 Mn, 0.22 Si, bal Fe.
For cutters, tools, punches; oil hardened.

KHN 80.
M-USSR; 0.25 max C, 1.5 max Mn, 1.2 max Si, 18-22 Cr, bal Ni.
For electrical resistance and heating elements; application to 1000°C.

KICK PLATE BRASS.
M-Eng; 84 Cu, 15 Zn, 1 Pb.
For hardware, water pipe, fittings; free-cutting.

KIND EXTRA HART.
M-1352; 1.3 C, 0.25 max Si, 0.25 max Mn, bal Fe.
For forming and blanking dies, engravers tools; Type W1; water hardened.

KIND EXTRA MH.
M-1352; 1.15 C, 0.25 max Si, 0.25 max Mn, bal Fe.
Annealed: 110,000 TS; 58,000 YS; 20 El; 40 RA; 210 Brin.
For springs, tools, drills, taps, reamers; Type W1; water hardened.

KIND EXTRA SCHWEISSBAR.
M-1352; 0.55 C, 0.10-0.40 Si, 0.50-0.70 Mn, bal Fe.
Heat treated: 150,000 TS; 110,000 YS; 15 El; 45 RA; 320 Brin.
For gears, bolts, machine tool parts; water hardened.

KIND EXTRA SEHR ZAH.
M-1352; 0.70 C, 0.25 max Si, 0.25 max Mn, bal Fe.
Heat treated: 175,000 TS; 128,000 YS; 12 El; 37 RA; 355 Brin.
For springs, rails, punches, crimpers; Type W1; water hardened.

KIND EXTRA ZAH.
M-1352; 0.85 C, 0.25 max Si, 0.25 max Mn, bal Fe.
Heat treated: 190,000 TS; 145,000 YS; 10 El; 30 RA; 400 Brin.
For springs, tools, cutters, taps, drills, hobs; Type W1; water hardened.

KIND EXTRA ZH.
M-1352; 1.0 C, 0.25 max Si, 0.25 max Mn, bal Fe.
Annealed: 100,000 TS; 53,000 YS; 21 El; 42 RA; 210 Brin.
For springs, tools, drills, taps; Type W1; water hardened.

KIND KW.
M-1352; 0.85 C, 1.75 Cr, 0.35 Mn, 0.30 Si, bal Fe.
For bearings, cutters, dies, tools; oil or water hardened, wear resistant.

KIND KWS.
M-1352; 0.85 C, 1.75 Cr, 0.35 Mn, bal Fe.
For bearings, piercing dies, cutters; oil or water hardened, wear resistant.

KIND PRIMA MH.
M-1352; 0.90 C, 0.25-0.50 Si, 0.30-0.80 Mn, bal Fe.
Heat treated: 180,000 TS; 115,000 YS; 10 El; 30 RA; 370 Brin.
For springs, taps, reamers, broaches, drills; Type W1; water hardened.

KIND PRIMA SEHR ZAH.
M-1352; 0.45 C, 0.25-0.50 Si, 0.30-0.80 Mn, bal Fe.
Hot rolled: 98,000 TS; 60,000 YS; 24 El; 54 RA; 212 Brin.
For axles, gears, bolts, shafts, crankpins; water hardened.

KIND PRIMA WEICH.
M-1352; 0.35 C, 0.25-0.50 Si, 0.30-0.80 Mn, bal Fe.
Hot rolled: 85,000 TS; 54,000 YS; 30 El; 53 RA; 183 Brin.
For gears, plates, shafts, bolts, axles; water hardened.

KIND PRIMA ZAH.
M-1352; 0.60 C, 0.25-0.50 Si, 0.30-0.80 Mn, bal Fe.
Heat treated: 160,000-115,000 TS; 113,000-77,000 YS; 12-23 El; 40-54 RA; 320-230 Brin.
For wheels, die blocks, girders, rails; water hardened.

KIND PRIMA ZH.
M-1352; 0.75 C, 0.25-0.50 Si, 0.30-0.80 Mn, bal Fe.
Heat treated: 180,000 TS; 140,000 YS; 12 El; 35 RA; 375 Brin.
For springs, punches, clutch discs; Type W1; water hardened.

KIND SPEZIAL MH.
M-1352; 1.1 C, 0.25 max Si, 0.25 max Mn, bal Fe.
Annealed: 100,000 TS; 53,000 YS; 21 El; 42 RA; 200 Brin.
For springs, taps, reamers, hobs; Type W1; water hardened.

KIND SPEZIAL SEHR ZAH.
M-1352; 0.70 C, 0.25 max Si, 0.25 max Mn, bal Fe.
Heat treated: 175,000 TS; 130,000 YS; 12 El; 37 RA; 355 Brin.
For wheels, die blocks, rails, girders, springs; Type W1; water hardened.

KIND SPEZIAL ZAH.
M-1352; 0.85 C, 0.25 max Si, 0.25 max Mn, bal Fe.
Heat treated: 188,000 TS; 143,000 YS; 12 El; 35 RA; 390 Brin.
For springs, taps, drills, hobs, cutters; Type W1; water hardened.

KIND SPEZIAL ZH.
M-1352; 1.0 C, 0.25 max Si, 0.25 max Mn, bal Fe.
Annealed: 100,000 TS; 53,000 YS; 21 El; 42 RA; 200 Brin.
For springs, taps, reamers, drills, hobs; Type W1; water hardened.

KINGHORN METAL.
M-Eng.; 58.5 Cu, 39.3 Zn, 0.96 Sn, 1.14 Fe.
For screws, nuts, hardware; high strength.

KINGSTON BRONZE.
M-282; 83 Cu, 4 Sn, 0.5 Fe, bal Zn.
Drawn: 94,500 TS; 35,000 YS; 10.5 El; 183 Brin.
For condenser tubes; high strength.

KINITE.
M-174; 0.55 Si, 1.5 C, 12.5-14.5 Cr, 1.1 Mo, 0.7 Co, 0.5 Mn, 0.4 Ni, bal Fe, trace to 0.5 B.
Cast: 100,000 TS.
Rolled: 200,000 TS.
For dies, anvils, cutters, mandrels, press tools, blanking, forming and drawing dies, shear blades; abrasion and compression resistant.

KINNALLOY.
M-546; 3.3 C, 2.5 Si, 1.5 Ni, 0.8 Cr, bal Fe.
40,000 TS; 500 Brin.
For castings; cast iron.

KINNITE.
M-546; 2.5 C, 1.5 Cr, bal Fe.
Cast: 40,000 TS; 500 Brin.
For abrasion resisting castings; wear resistant.

KINSALLOY.
M-68; 21-24 Mo, 7-8 Al, 0.1 max C, bal Ni.
Cast: 100,000-130,000 TS.
For high temperature applications; precision cast, heat resistant.

KINSALOY B.
M-68; 70 Ni, 22 Mo, 8 Al.
For jet engine components, turbine rotor blades; heat resistant.

KIN-SHIBU-ICHI.
M-Jap.; one part "Shaku-do", two parts "Shibu-ichi."
For jewelry, ornaments; corrosion resistant.

KIRKALLOY.
M-794; 3 Al, 8 Cu, 0.04 Be, bal Zn. (max .007 Pb, .003 Sn, .003 Cd, .10 Fe)
Sand cast: 42,500 TS; 2.0 El; 7.6 impact; 124 Brin.

KIRKSITE see also **HI-PHY KIRKSITE.**

KIRKSITE A.
M-88; 3.5 Cu, 4 Al, 0.04 Mg, bal Zn.
Cast: 38,000 TS; 5 El; 107 Brin.
Rolled: 62,000 TS; 3 El.
For stamping dies, forming dies; M.P. 717°F.

KISKI.
M-140; 0.9-1.0 C, 1.0-1.2 Mn, 0.43-0.57 Cr, 0.15-0.25 V, 0.45-0.75 W, bal Fe.
For tools, dies, blanking and forming dies, thread gauges; non-deforming.

K K K.
M-1744; 2.10 C, 0.65 Mn, 0.75 Si, 0.85 Ni, 13.0 Cr, 0.65 Mo, bal Fe.
13% chromium cold work tool steel, for press tools, moulding dies, profilin rolls.

KL 33 A.
M-1541; 0.06 C, 1.32 Mn, 0.011 P, 0.013 S, 0.21 Si, 0.25 Ni, bal Fe.
64,000 psi TS; 49,800 psi YS; 43 El.
Low alloy normalized steel for structural and pressure vessels for low temperature service.

KL 33 B.
M-1541; 0.06 C, 1.30 Mn, 0.011 P, 0.014 S, 0.21 Si, 0.25 Ni, bal Fe.
65,400 psi TS; 54,000 psi YS; 41 El.
Low alloy quenched and tempered steel for low temperature pressure vessels and structures.

KL 36 B.
M-1541; 0.05 C, 1.44 Mn, 0.008 P, 0.009 S, 0.40 Si, 0.21 Ni, 0.10 Cr, bal Fe.
71,300 psi TS; 57,200 psi YS; 51 El.
Low alloy quenched and tempered steel for low temperature pressure vessels and structures.

KL 85.
M-1290; 0.75-0.85 C, 0.25-0.40 Si, 0.30-0.50 Mn, 0.40-0.70 Cr, 0.15-0.25 V, bal Fe.
Cold work tool steel.
W.-Nr. 1.2235.

KLOSTER V-76.
M-335; 0.9 C, 1.2 Mn, 2.0 Si, bal Fe.

KLN 3A.
M-1541; 0.09 C, 0.66 Mn, 0.008 P, 0.006 S, 0.21 Si, 3.64 Ni, bal Fe.
78,700 psi TS; 64,900 psi YS; 38 El.
Normalized 3.5% Ni steel for low temperature use.

KLN 3B.
M-1541; 0.08 C, 0.68 Mn, 0.006 P, 0.009 S, 0.21 Si, 3.65 Ni, bal Fe.
82,600 psi TS; 73,700 psi YS; 35 El.
Quenched and tempered 3.5% Ni steel for low temperature use.

KLN 9.
M-1541; 0.05 C, 0.49 Mn, 0.007 P, 0.006 S, 0.30 Si, 9.16 Ni, bal Fe.
110,260 psi TS; 107,200 YS; 32 El.
Quenched and tempered 9% Ni steel for low temperature use.

KLS.
M-1290; 0.95-1.05 C, 0.15-0.30 Si, 0.25-0.40 Mn, 1.4-1.7 Cr, bal Fe.
Cold work tool steel.
W.-Nr. 1.2067.

KLS.
M-1655; 1.25 C, 0.70 Mn, 1.20 Si, 1.20 Cr, bal Fe.
For planing tools, threading tools, counter bores, grooving tools.
W.-Nr. 1.2109.

KM.
M-351; Al, Cu, bal Zn.
For die castings.

KM 80.
M-Ger.; 0.2 C, 14 Cr, 2 Mo, 1 V, bal Fe.
For turbine rotors; heat resistant.

KM 609.
M-1290; 1.55-1.75 C, 0.25-0.40 Si, 0.20-0.40 Mn, 11.0-12.0 Cr, 0.50-0.70 Mo, 0.40-0.60 W, 1.1-1.3 V, bal Fe.
Cold work tool steel.
W.-Nr. 1.2609.

K-MONEL see MONEL ALLOY K-500.

KMV.
M-1405; 1.5 C, 12.0 Cr, 0.85 Mo, 0.50 V, bal Fe.
Cold work tool steel.
For thread rolling, blanking and forming dies.
BS 4659 BD2; AISI D2.

KN.
M-1583; 0.32-0.40 C, 0.60 Mn, 2.25-2.75 Ni, 0.55-0.95 Cr, bal Fe.
Low alloy steel castings.
W.-Nr. 5736.

KN 6 G.
M-1583; 0.6-0.12 C, 0.40 Si, 0.75 Mn, 1.3-1.8 Ni, bal Fe.
Low alloy steel castings.
W.-Nr. 5621.

KNA.
M-1583; 0.30-0.38 C, 0.35 Mn, 3.5-4.0 Ni, 1.5-1.9 Cr, bal Fe.
Alloy steel castings.
W.-Nr. 5952.

KNC 4 E.
M-1583; 0.10-0.17 C, 0.40 Mn, 3.30-3.60 Ni, 0.65-0.85 Cr, bal Fe.
Austenitic steel castings, for high temperature operations.
W.-Nr. 2735.

KNC 5 E.
M-1583; 0.10-0.17 C, 0.40 Mn, 4.2-4.7 Ni, 0.90-1.20 Cr, bal Fe.
Austenitic steel castings, for high temperature operations.
W.-Nr. 2745.

KNCO 4 E.
M-1583; 0.16-0.22 C, 0.4 Mn, 3.8-4.3 Ni, 1.1-1.4 Cr, 0.15-0.25 Mo, bal Fe.
Low alloy steel castings.
W.-Nr. 2764.

KNEISS METAL.
50-40 Zn, 25-42 Pb, 25-15 Sn, 0-3 Cu.
For bearings; anti-friction.

KN GS 10NI6.
M-1583; 0.06-0.12 C, 0.40 Si, 0.50-0.80 Mn, 1.30-1.80 Ni, bal Fe.
Low carbon, nickel alloy steel castings.
W.-Nr. 5621.

KN GS 38.
M-1583; 0.25 max C, 0.20-0.60 Si, 0.20-0.50 Mn, bal Fe.
Carbon steel castings.
W.-Nr. 0416.

KN GS 45.
M-1583; 0.25 max C, 0.60 max Si, 0.35 Mn, 0.007 max N, bal Fe.
Low carbon steel castings.
W.-Nr. 0443.

KN GS 52.
M-1583; 0.30 C, 0.45 Si, 0.40 Mn, bal Fe.
Carbon steel castings.
W.-Nr. 0551.

KN GS 60.
M-1583; 0.40 C, 0.45 Si, 0.40 Mn, bal Fe.
Carbon steel castings.
W.-Nr. 0553.

KN GS 70.
M-1583; 0.50 C, 0.45 Si, 0.40 Mn, bal Fe.
Carbon steel castings.
W.-Nr. 0554.

KNIFE BOLSTERS.
M-Eng.; 68 Cu, 16.5 Zn, 15 Ni, 0.5 P.
For knife bolsters; corrosion resistant.

KNIFE BOLSTERS.
M-Eng.; 56 Cu, 28 Zn, 16 Ni.
For ornamental parts; nickel silver.

K-O-1.
M-1642; 4.8 Cu, 0.23 Mg, 0.27 Ti, 0.64 Ag, bal Al.
Sand Cast: 62,000-72,000 TS; 52,000-65,000 YS; 3-9 El.
Permanent Mold: 64,000-70,000 TS; 50,000-60,000 YS; 6-14 El.
For landing gear struts, gear box housings, aircraft and auto/truck components, aerospace equipment.
High strength casting alloy.

KO-1 ALLOY.
M-1719; 4.8 Cu, 0.5 Ag, 0.25 Mg, bal Al.
Heat treatable cast alloy.
T6: 58,000-68,000 TS; 47,000-58,000 YS; 4-10 El.
For high strength castings, aerospace housings, transmission line fittings, cylinder heads and pistons, turbine impellers.
AA X201.0; AMS 4228 (T6) Tentative.

KOBALT-1.
M-1496; 0.85 C, 18 W, 4 Cr, 2 V, 12 Co, 0.8 Mo, bal Fe.
Hardened: C 64-66 Rock.
For lathe and planer tools, hobs, broaches, reamers, milling cutters.
Type T-6 tool steel, high red-hardness, abrasion resistant.

KOBALT-II.
M-1496; 0.70-0.74 C, 18-19 W, 4.0-4.25 Cr, 1.0-1.2 V, 4.5-5.5 Co, 0.5 Mo, bal Fe.
Hardened: C 64-66 Rock.
For boring and forming tools, gear cutters, lathe and planer cutters, taps drills, reamers.
Type T4 high speed steel, high red-hardness.

KOBALT III N.
M-1331; 0.80 C, 4.3 Cr, Co, W, V, Mo, bal Fe.
For lathe and planer tools, reamers, broaches, taps; high speed steel.

KOBALT 3 ETC see **CORONA KOBALT 3 ETC.**

KOBALT 5.
M-1655; 1.30 C, 4.5 Cr, 1.0 Mo, 12.0 W, 3.8 V, 5.0 Co, bal Fe.
High speed steel.
Finishing tools; cutting tools for higher speed operation and greater wear resistance.
W.-Nr. 1.3202.

KOBALT 5 W.
M-1655; 0.80 C, 4.5 Cr, 1.0 Mo, 18.0 W, 1.7 V, 5.0 Co, bal Fe.
High speed steel; lathe and milling cutters for tough work.
W.-Nr. 1.3255; similar to AISI T4.

KOBALT 10.
M-1655; 0.80 C, 4.5 Cr, 1.0 Mo, 18.0 W, 1.6 V, 10.0 Co, bal Fe.
High speed steel; for lathe and planing tools for tough jobs.
W.-Nr. 1.3265.

KOBALT 10 SPEZIAL.
M-1655; 1.25 C, 4.0 Cr, 3.8 Mo, 10.5 W, 3.2 V, 10.5 Co, bal Fe.
High speed steel; automatic lathe tools.
W.-Nr. 1.3207.

KOBALT 18.
M-1655; 0.70 C, 4.5 Cr, 1.0 Mo, 18.0 W, 1.5 V, 18.0 Co, bal Fe.
High cobalt high speed steel.
For special difficult machining work.

KOBALT 30.
M-1331; 0.86 C, 2.8 Co, 4.3 Cr, 0.85 Mo, 2.1 V, 12 W, bal Fe.
For lathe and planer tools, hobs, reamers; high speed steel.

KOBALT 50.
M-1331; 0.80 C, Co, 4.3 Cr, V, W, bal Fe.
For lathe and planer tools, reamers, drills, taps, hobs; high speed steel.

KOBALT 100.
M-1315; 0.76 C, 10 Co, 4.2 Cr, 0.8 Mo, 1.8 V, 18 W, bal Fe.
For lathe and planer tools, reamers, hobs, taps: high speed steel.

KOBALT 150.
M-1315; 0.80 C, W, Co, Cr, V, Mo, bal Fe.
For lathe and planer tools, reamers, broaches, taps; high speed steel.

KOBALT V33.
M-1315; 0.86 C, 2.8 Co, 4.3 Cr, 0.85 Mo, 2.1 V, 12 W, bal Fe.
For lathe and planer tools, reamers, taps, drills, hobs; high speed steel.

KOBALT V65.
M-1315; 1.35 C, W, Cr, Co, V, bal Fe.
For engravers' tools, forming and blanking dies; high speed steel.

KOBITALIUM.
M-Eng.; 1-5 Cu, 0.2-2.0 Ni, 0.25-2.0 Mn, 1-2 Fe, 0.5-2.0 Si, 0.4-2.0 Mg, 0.08-0.12 Ti, bal Al.
For light alloy castings; also rolled or forged.

KOCHLINS BEARING.
90 Cu, 10 Sn.
For bearings; Bronze.

KOCH WHITE GOLD.
M-Eng.; 75 min Au, 3.3-24.75 Ni, 0.25-5.0 Mn.
For jewelry, ornaments; corrosion resistant.

KOERFLEX-20.
M-72; 30 Co, 15 Cr, bal Fe.
For electromechanical devices, motors, generators.
Permanent magnet, high permeability.

KOERFLEX-30.
M-72; 30 Co, 15 Cr, bal Fe.
For electrical and magnetic equipment, generators, motors.
Permanent magnet, high permeability.

KOERFLEX-200.
M-72; 52 Co, 10 V, bal Fe.
For hysteresis motors, automation devices, electromechanical equipment.
Permanent magnet, similar to Vicalloy I. High permeability.

KOERFLEX-300.
M-72; 52 Co, 8 Cr, 4 V, bal Fe.
1000 Br, 2.8 (BH) max.
For electromechanical devices, speedometers, recorders, hysteresis motors, computers.
Precipitation hardening, permanent magnet, similar to Vicalloy II.

KOERVER-UMCO50.
M-1583; 0.05-0.12 C, 0.18 Ti, 0.6 Cb, 26-30 Cr, 47-52 Co, bal Fe.
Cast: 135,000 TS; 48,000 YS; 10 El; 10 RA; 250 Brin.
Wrought: 132,000 TS; 61,000 YS; 7 El; 6 RA; 350 Brin.
For furnace baffles, burner tips, sintering grates, quench baskets.
Corrosion, heat, thermal shock resistant.

KOERZIT 250S.
M-72; 8 Al, 19 Ni, 24 Co, 5 Ti, bal Fe.
For electrical and magnetic equipment.
Permanent magnet. High permeability.

KOERZIT 350.
M-72; 15 Ni, 30 Co, 4 Cu, 7 Al, 5 Ti, bal Fe.
8500 Br, 1150 H_c, 3,6000,000 (BH) max.
For permanent magnets in electrical and magnetic equipment.
High permeability.

KOERZIT 400K.
M-72; 25 Co, 2.5 Cb, bal Fe.
For permanent magnets in electrical and magnetic equipment.
Alnico type, high permeability.

KOERZIT 450.
M-72; 34 Co, 5 Ti, bal Fe.
For permanent magnets in electrical and magnetic equipment.
Alnico type magnet. High permeability.

KOERZIT-HI.
M-72; 52 Co, 7 Cr, 3 V, bal Fe.
For electric and magnetic equipment, hysteresis motors.
Magnetically soft; high permeability.

KOERZIT-H2.
M-72; 52 Co, 38 Fe, 10 V.
For hysteresis motors, automation devices, electro and magnetic devices. Requires cold rolling and tempering to develop the magnetic properties.
Permanent magnet. Similar to Vicalloy I.

KOERZIT-T.
M-72; 52 Co, 8 Cr, 4 V, bal Fe.
10,000 Br, 2.8 (BH) max.
For electromechanical devices, speedometers, recorders, hysteresis motors, digital computers.
Precipitation hardening. Permanent magnet. Similar to Vicalloy II.

KOERZIT-VS55.
M-72; 38 Co, 7 Ti, bal Fe, Al, Ni, Cu.
For permanent magnets in electrical and magnetic equipment.
Alnico type magnet, high permeability.

K O H.
M-174; C, alloy, bal Fe.
For tools; oil hardening.

KOLASSAL.
M-1338; 2.1 C, 11.5 Cr, 0.3 Mn, bal Fe.
For forming and blanking dies, punches; oil hardening, non-deforming.

KOLASSAL EXTRA.
M-1338; 2.1 C, 11.5 Cr, 0.7 W, bal Fe.
For forming and blanking dies, punches; oil hardened, non-deforming.

KOLASSAL SUPRA.
M-1338; 1.65 C, 11.5 Cr, 0.1 V, bal Fe.
For blanking and forming dies, punches; air hardened, non-deforming.

KOLTCHOUGALUMIN.
M-Eng.; 91.7-93.9 Al, 6-4 Cu, 0.6 Mn, 0.4 Mg, 0.8 Fe, 0.2 Si, 0.11 Cr.
For light alloy parts, aircraft structures; similar to "Dural."

KOLTSCHUG.
M-USSR; 4 Cu, 0.3 Ni, bal Al.
For light alloy parts; age-hardenable.

KOMALP 3 HERZ now **VEW S200.**

KOMALP 3 HERZ EXTRA M now **VEW S205.**

KOMALP 300 now **VEW S610.**

KOMALP WM now **VEW S600.**

KOMET FRS27.
M-1312; 0.86 C, 4.1 Cr, 0.85 Mo, 2.5 V, 12 W, bal Fe.
For lathe and planer tools, reamers, hobs, taps, drills; high speed steel.

KOMO 205.
M-1496; 0.80 C, 4 Cr, 5 Mo, 6 W, 2 V, 5 Co, bal Fe.
Hardened: 63-66 Rock C.
For reamers, hobs, lathe and planer tools, drills, broaches.
Type M35, high speed steel, good red-hardness. Wear resistant.

KOMO 310.
M-1496, M-459; 1.25 C, 9.5 W, 4.2 Cr, 3.2 V, 3.15 Mo, 10 Co, bal Fe
Hardened: 64-68 Rock C.
For reamers, broaches, chasers, hobs, lathe and planer tools, milling cutters.
High-speed steel, high red-hardness.

KONALLOY C-COCR20NI20W.
M-1583; 0.35-0.45 C, 1.0 max Si, 1.5 max Mn, 19.0-21.0 Cr, 19.0-21.0 Ni, 3.5-4.5 Mo, 3.5-4.5 W, 3.5-4.5 Nb, 5.0 max Fe, bal Co.
Components for gas turbines, combustion chambers, steel castings.
W.-Nr. 4989.

KONALLOY C-COCR20W15NI.
M-1583; 0.05-0.15 C, 1.0 max Si, 1.0-2.0 Mn, 19.0-21.0 Cr, 9.0-11.0 Ni, 14.0-16.0 W, 3.0 max Fe, bal Co.
High strength and corrosion resistance at elevated temperature; steel castings.
W.-Nr. 4967.

KONALLOY COCR20W15N.
M-1583; 0.5-0.15 C, 1.0 max Si, 1.0-2.0 Mn, 19.0-21.0 Cr, 9.0-11.0 Ni, 14.0-16.0 W, 3.0 max Fe, bal Co.
High temperature alloy; for valves, gas turbine parts; steel castings.
W.-Nr. 4964.

KONALLOY COCR25NIW.
M-1583; 0.45-0.60 C, 1.2 max Si, 1.0 max Mn, 22.0-28.0 Cr, 9.0-12.0 Ni, 9.0-12.0 W, 2.0 max Fe, bal Co.
High strength and corrosion resistance at elevated temperature; steel castings.
W.-Nr. 4966.

KONALLOY COCR28FE.
M-1583; 0.10 max C, 1.0 max Si, 1.0 max Mn, 45.0-50.0 Co, 26.0-30.0 Cr, bal Fe.
High strength and corrosion resistance at elevated temperature; steel castings.
W.-Nr. 4778.

KONALLOY COCR28MO.
M-1583; 0.25-0.35 C, 1.0 max Si, 1.0 max Mn, 27.0-29.0 Cr, 5.0-6.0 Mo 1.5-3.0 Ni, bal Co.
High strength and corrosion resistance at elevated temperatures; steel castings.
W.-Nr. 4979.

KONALLOY CRNI25 20.
M-1583; 0.20 max C, 1.5-2.5 Si, 2.0 max Mn, 22.0-25.0 Cr, 19.0-22.0 Ni, bal Fe.
Austenitic stainless steel for high temperature operation; steel castings.
Similar to AISI 310; W.-Nr. 4843.

KONALLOY NI36.
M-1583; 0.10 max C, 0.50 max Mn, 35.0-37.0 Ni, bal Fe.
High temperature steel castings.
W.-Nr. 3912.

KONALLOY NI38.
M-1583; 0.10 max C, 1.0 max Mn, 36.0-40.0 Ni, bal Fe.
High temperature steel castings.
W.-Nr. 3913.

KONALLOY NI42.
M-1583; 0.05 max C, 1.0 max Mn, 41.0-43.0 Ni, bal Fe.
High temperature steel castings.
W.-Nr. 3917.

KONALLOY NI43.
M-1583; 0.10 max C, 1.0 max Mn, 1.0 max Cr, 42.0-44.0 Ni, bal Fe.
High temperature steel castings.
W.-Nr. 3918.

KONALLOY NI46.
M-1583; 0.10 max C, 1.0 max Mn, 1.0 max Cr, 45.0-47.0 Ni, bal Fe.
A high temperature steel castings.
W.-Nr 3920.

KONALLOY NI48.
M-1583; 0.05 max C, 0.50 max Mn, 46.0-50.0 Ni, bal Fe.
High temperature steel castings.
W.-Nr. 3926.

KONALLOY NI49.
M-1583; 0.05 max C, 1.0 max Mn, 0.70-1.0 Cr, 48.0-50.0 Ni, bal Fe.
High temperature steel castings.
W.-Nr. 3921.

KONCOR.
M-73; 1.0 Si, 5.25 Cr, 4.0 V, 1.1 Mo, bal Fe.
Hardened: 64-67 Rock C.
For upsetter and forging dies, extrusion and pointing dies, chipping chisels, punches, die casting inserts.
Air hardening, non-deforming, abrasion resistant, tough, heat resistant.

KONDUR 6.
M-1583; 0.30-0.35 C, 0.60 max Si, 0.60 max Mn, 26.0-28.0 Cr, 4.5-5.0 W, bal Co.
Wear resistant steel castings.
W.-Nr. 4519.

KONDUR 21.
M-1583; 0.30-0.50 C, 1.0 max Si, 1.0 max Mn, 26.0-30.0 Cr, 4.0-5.0 Mo, bal Co.
Wear resistant steel castings.
W.-Nr. 4520.

KONEL.
M-118; 73.07 Ni, 17.16 Co, 8.8 Ti, 0.55 Si, 0.26 Al, 0.16 Mn.
100,000 TS; 40,000 YS; 35 El; 55 RA; 140 Brin.
For lamp filaments, valves, valve stems; substitute for platinum; heat and corrosion resisting; hot short above 1250°C.

KONIK.
M-631; 0.15-0.25 C, 0.2 Cu, 0.4 Ni, 0.3 Cr, bal Fe.
Cold drawn: 81,000 TS; 79,000 YS; 3 El; 5 RA.
For cross chains, culvert sheets; case hardening steel.

KONIT B.
M-1583; 0.10 max C, 1.0 max Si, 1.0 max Mn, 1.0 max Cr, 26.0-30.0 Mo, 4.0-7.0 Fe, 62.0 min Ni.
Stainless and heat resisting steel castings.
W.-Nr. 4600.

KONIT C.
M-1583; 0.10 max C, 1.0 max Si, 1.0 max Mn, 14.0-18.0 Cr, 3.0-5.0 W, 7.0 max Fe, 15.0-18.0 Mo, 52.0 min Ni.
Stainless and heat resisting steel castings.
W.-Nr. 4537.

KONIT D.
M-1583; 0.04-0.08 C, 10.0 Si, 1.0 Mn, 3.5 max Cu, 2.0 max Fe, bal Ni,
Stainless and heat resisting steel castings.
W.-Nr. 4566.

KONIT F.
M-1583; 0.10 max C, 1.0 max Mn, 1.0 max Si, 20.0-23.0 Cr, 8.0-10.0 Mo 0.2-10 W, 17.0-20.0 Fe, bal Ni.
Stainless and heat resisting steel castings.
W.-Nr. 4613.

KONIT X.
M-1583; 0.08 max C, 1.0 max Si, 1.0-2.0 Mn, 21.0-23.0 Cr, 5.5-7.5 Mo, 1.75-2.50 Nb, 44.0-47.0 Ni, bal Fe.
Stainless and heat resisting steel castings.
W.-Nr. 4557.

KONOLOY 800.
M-1583; 0.20 max C, 3.0-4.0 Si, 1.5 max Mn, 20.0-22.0 Cr, 28.0-31.0 Ni, bal Fe.
Heat resisting steel casting.
W.-Nr. 4860.

KONOLOY 825.
M-1583; 0.15 C, 1.0 max Si, 1.0 max Mn, 20.5-23.0 Cr, 0.50-2.50 Co, 6.0-10.0 Mo, 0.20-1.0 W, 17.0-20.0 Fe, bal Ni.
Austenitic stainless steel castings for high temperature operations.
W.-Nr. 4972.

KONOMAG KN 3944.
M-1583; 0.07 max C, 1.0 max Si, 2.0 max Mn, 16.0-18.0 Cr, 10.0-12.0 Ni, bal Fe.
Austenitic stainless steel castings.
Similar to AISI 302.

KONONEL 600.
M-1583; 0.15 C, 0.50 max Si, 1.0 max Mn, 14.0-17.0 Cr, 6.0-10.0 Fe, 72.0 min Ni.
Austenitic stainless steel castings for high temperature operation.
W.-Nr. 4816.

KONOX 4008.
M-1583; 0.08-0.15 C, 1.0 max Si, 1.0 max Mn, 12.0-14.0 Cr, 0.5-1.5 Mo, bal Fe.
Martensitic stainless steel castings.
Similar to ACI CA-15.

KONSTANT.
M-1331; 0.90 C, 1.9 Mn, 0.1 V, 0.25 Si, bal Fe.
For dies, punches, upsetters, crimpers; oil hardened, non-deforming.

KONSTANT 15.
M-1583; 1.40-1.60 C, 0.60 Mn, 1.30-1.60 Cr, bal Fe.
Wear resisting steel casting.
W.-Nr. 2063.

KONSTANTAN.
M-297; 54 Cu, 45 Ni, 1 Mn.
Annealed: 83,000 TS; 30 El.
Hard: 120,000 TS; 1 El.
For precision resistances, reducing rheostats, shunts, electrical instruments; maximum operating temperature 600°C., M.P. 1276°C.

KONTHERM KN 4710.
M-1583; 0.20-0.40 C, 1.0-2.5 Si, 1.0 max Mn, 6.0-8.0 Cr, bal Fe.
Heat resisting steel castings.

KONTRAB 2062.
M-1583; 0.50-0.60 C, 0.45 Si, 0.45 Mn, 1.30-1.60 Cr, bal Fe.
Wear resisting steel casting.

KOPPERS B-18.
M-399; 78-82 Cu, 18-19.5 Sn, 0.5 max Pb, 0.5 max Fe, 0.5 max Zn.
Cast: 45,000-55,000 TS; 165-195 Brin.
For oil sealing rings, piston and pump rings; corrosion and acid resisting.

KOPPERS B-19.
M-399; 78-82 Cu, 12-14 Sn, 4-6 Pb, 0.75-1.25 Ni.
Cast: 30,000 TS; 100 Brin.
For locomotive main cylinder packing and segmental rings; pressure tight.

KOPPERS B-23.
M-399; 78-82 Cu, 15-17 Sn, 4-6 Pb, 0.5 max Fe, 0.5 max Zn.
Cast: 30,000 TS; 130-156 Brin.
For hydraulic sealing rings, piston rings; pressure tight.

KOPPERS B-48.
M-399; 64 Cu, 6 Al, 4 Mn, 3 Fe, bal Zn.
Cast: 110,000 TS; 60,000 YS; 12 El; 225 Brin.
For propellers; strong and corrosion resistant.

KOPPERS F-17.
M-399; 3.5 TC., 2.5 Si, 2.25 Ni, 1 Mo, bal Fe.
Cast: 46,000 TS; 200 Brin.
For piston rings; diesel rings; castings.

KOPPERS F-88.
M-399; 3.2 C, 0.5 Mo, 1.4 Si, 0.5 Mo, 0.5 Cu, bal Fe.
Heat treated: 88,000 TS; 265 Brin.
For piston rings; high strength cast iron.

KOPPERS K-6B.
M-399; 2.8 C, 2 Si, 1.3 Mn, 15 Ni, 6.5 Cu, 1.9 Cr, bal Fe.
Annealed: 25,000 TS; 145 Brin.
For cylindrical shapes; austenitic, centrifugally cast iron.

KOPPERS K-6E.
M-399; 3.8 C, 2.9 Si, 0.6 Mn, 0.3 Cr, 0.45 Mg, bal Fe.
Cast: 30,000 TS; 193 Brin.
For piston rings; centrifugally cast gray iron.

KOPPERS K-8.
M-399; 3.3 C, 2.3 Si, 0.65 Mn, 1.5 max Ni, 0.4 max Cr, bal Fe.
Cast: 27,000 TS; 180 Brin.
For pistons ring inserts; centrifugally cast gray iron.

KOPPERS K-10.
M-399; 2.6 C, 2 Si, 1.2 Mn, 15 Ni, 6.5 Cu, 2.1 Cr, bal Fe.
Cast: 32,000 TS; 137 Brin.
For corrosion resistant castings; Type 1, Ni-Resist, ASM 5392.

KOPPERS K-13.
M-399; 1.75 C, 1.75 Si, 0.9 Ni, 14 Cr, 0.4 Mo, bal Fe.
Annealed: 90,000 TS; 270 Brin.
For heat resistant castings; heat and abrasion resistant.

KOPPERS K-14.
M-399; 3.6 C, 2.3 Si, 0.6 Mn, 1.1 Ni, 1.1 Mo, 0.3 Cr, bal Fe.
Heat treated: 45,000 TS; 270 Brin.
For heavy duty piston rings; cast iron.

KOPPERS K-15.
M-399; 3.4 C, 2.75 Si, 0.5 Mn, bal Fe.
Annealed: 60,000 TS; 45,000 YS; 10 El; 140 Brin.
For general castings, gears, housings; ductile iron, ferritic.

KOPPERS K-16.
M-399; 3.4 C, 2.2 Si, 0.5 Mn, bal Fe.
Heat treated: 90,000 TS; 65,000 YS; 2 El; 240 Brin.
For gears, shafts, housings; pearlitic ductile iron.

KOPPERS K-25.
M-399; 3.25 C, 2.25 Si, 0.5 Mn, 0.75 Ni, bal Fe.
Annealed: 60,000 TS; 40,000 YS; 15 El; 185 Brin.
For gears, shafts, housings, ferritic ductile iron.

KOPPERS K-26.
M-399; 3.1 C, 1.3 Si, 0.75 Mn, 0.5 Mo, 0.5 Cu, bal Fe.
Heat treated: 95,000 TS; 400 Brin.
For aircraft piston rings; cast iron.

KOPPERS K-27.
M-399; 3.3 C, 2.2 Si, 0.5 Mn, 0.5 Mo, 0.5 Cu, bal Fe.
Heat treated: 110,000 TS; 80,000 YS; 1 El; 265 Brin.
For cylindrical shapes; alloy ductile iron.

KOPPERS K-28.
M-399; 3.3 C, 2.2 Si, 0.5 Mn, 0.5 Mo, 0.5 Cu, bal Fe.
Heat treated: 120,000 TS; 100,000 YS; 1 El; 400 Brin.
For cylindrical shapes; alloy ductile iron.

KOPPERS K-35.
M-399; 2.3 C, 1.2 Si, 0.5 Mn, 30 Ni, 3 Cr, bal Fe.
Cast: 55,000 TS; 33,000 YS; 7 El; 140 Brin.
For corrosion and abrasion resistant castings; Ni-Resist Type 3.

KOPPERS K-37.
M-399; 2.7 C, 2.6 Si, 1.2 Mn, 20.0 Ni, 2.1 Cr, bal Fe.
55,000 psi TS; 170 Brin.
Centrifugally or static cast austenitic ductile iron used at temperatures up to 1400°F on applications such as engine exhaust seal rings, valve guides, turbocharger and gas turbine parts.

KOPPERS K-46.
M-399; 3.0 C, 3.15 Si, 7.75 Mn, 5.0 Ni, bal Fe.
35,000 psi TS; 155 Brin.
Coefficient of expansion: 10.5×10^{-6} in/in/°F.
Centrifugally cast austenitic Mn-Ni alloy cast iron used for aluminum piston inserts where a high coefficient of expansion is required.

KOPPERS K-53.
M-399; 3 C, 2.1 Si, 0.7 Mn, 0.6 Ni, 0.5 Mo, 0.5 Cr, 0.75 Cu, bal Fe.
Cast: 40,000 TS; 229 Brin.
For cylinder liners; centrifugally cast gray iron.

KOPPERS K-60.
M-399; 0.07 max C, 16.5 Cr, 4 Ni, 0.45 Cb, 4 Cu, 0.05 N, bal Fe.
Heat treated: 150,000 TS; 140,000 YS; 6 El; 350 Brin.
For corrosion resistant hard castings; Armco 17-4PH, stainless, age hardened.

KOPPERS K-61.
M-399; 0.15 max C, 12.5 Cr, 0.75 Ni, 0.5 Mo, 0.5 Cu, bal Fe.
Cast: 125,000 TS; 110,000 YS; 10 El; 310 Brin.
For corrrosion resistant castings; Type 410; stainless.

KOPPERS K-66.
M-399; 0.5 C, 25.5 Cr, 10.5 Ni, max Fe, 54.5 Co.
Cast: 110,000 TS; 95,000 YS; 2 El; 370 Brin.
For chemical plant equipment, high temperature castings; Haynes Stellite No. 31, heat and corrosion resistant.

KOPPERS K-76.
M-399; 3.15 C, 4.75 Si, bal Fe.
85,000 psi TS; 250 Brin.
Centrifugally cast ductile iron used for exhaust manifold seal rings and other applications where resistance to oxidation and growth are necessary.

KOPPERS K-IRON.
M-399; 3.7 C, 2.85 Si, 0.6 Mn, bal Fe.
Cast: 25,000 TS; 180 Brin.
For piston rings; cast iron.

KOPPERS XLS.
M-399; 3.5 C, 1.75 Si, 0.7 Mn, 0.35 Cr, bal Fe.
Cast: 20,000 TS; 150 Brin.
For large engine piston rings; gray cast iron.

KORA.
M-618; 0.15 -0.18 C, 0.10-0.15 Si, 0.7-0.9 Mn, bal Fe.
Hot rolled: 83,000 TS; 67,000 YS; 37 El; 55 RA; 163 Brin.
Heat treated: 95,000 TS; 67,000 YS; 28 El; 60 RA; 180 Brin.
For gears, camshafts, machinery parts; case hardening steel.

KORA NO. 2.
M-510; 0.16 C, 0.12 Cr, 0.06 max S, 0.05 max P, bal Fe.
Case hardening steel.

KORROFESTAL.
M-1376; 0.6-1.4 Mg, 0.6-1.6 Si, 0.6-1.0 Mn, 0.3 max Cr, bal Al.
Annealed: 21,000 TS; 8000 YS; 24 El.
For window frames, gutters, fan blades, boats; good forming and welding properties.

KORRONIT 17.
M-1340; 0.08 C, 17 Cr, bal Fe.
Annealed: 80,000 TS; 50,000 YS; 25 El; 50 RA; 150 Brin.
For oil refinery and food processsing equipment, sinks; Type 430; corrosion resistant, ferritic.

KORRONIT 17E.
M-1340; 0.1 max C, 17.5 Cr, Ti = 7 x C, bal Fe.
Annealed: 85,000 TS; 55,000 YS; 20 El; 4 RA; 180 Brin.
For oil refinery and food processing equipment, weldments; Type 430 Ti; corrosion resistant.

KORRONIT 17S.
M-1340; 0.12 C, 16.5 Cr, 0.25 Mo, 0.2 S, bal Fe.
Annealed: 80,000 TS; 50,000 YS; 25 El; 50 RA; 160 Brin.
For screw machine products, fasteners; Type 430F; stainless, free-cutting.

KORRONIT 18M.
M-1340; 0.15 C, 18 Cr, 0.4 Si, bal Fe.
Annealed: 80,000 TS; 50,000 YS; 25 El; 50 RA; 160 Brin.
For oil refinery equipment, food processing equipment; Type 430; stainless, ferritic.

KORRONIT 188.
M-1340; 0.15 max C, 18 Cr, 8.5 Ni, bal Fe.
Annealed: 80,000 TS; 35,000 YS; 55 El; 75 RA; 150 Brin.
Cold drawn: 180,000 TS; 150,000 YS; 10 El; 250 Brin.
For chemical plant equipment, tanks; Type 302; stainless, austenitic.

KORRONIT 188E.
M-1340; 0.12 max C, 18 Cr, 8.5 Ni, Ti = 4 x C, bal Fe.
Annealed: 85,000 TS; 35,000 YS; 55 El; 65 RA; 150 Brin.
Cold drawn: 95,000 TS; 60,000 YS; 40 El; 60 RA; 185 Brin.
For welded chemical plant equipment, tanks, mixers; Type 321; stainless, austenitic.

KORRONIT 188P.
M-1340; 0.07 max C, 18 Cr, 9.5 Ni, bal Fe.
Annealed: 85,000 YS; 35,000 YS; 60 El; 70 RA; 150 Brin.
Cold drawn: 180,000 TS; 125,000 YS; 10 El; 330 Brin.
For chemical plant equipment, tanks, mixers; Type 304; stainless, austenitic.

KORRONIT 189.
M-1340; 0.1 max C, 2.2 Si, 18 Cr, 2 Mo, 9.5 Ni, bal Fe.
Annealed: 85,000 TS; 35,000 YS; 50 El; 65 RA; 160 Brin.
Cold drawn: 150,000 TS; 135,000 YS; 6 El; 300 Brin.
For acid resistant chemical plant equipment; Type 316; stainless austenitic.

KORRONIT 189E.
M-1340; 0.12 max C, 18 Cr, 2 Mo, 10.5 Ni, Ti = 4 x C, bal Fe.
Annealed: 85,000 TS; 40,000 YS; 50 El; 65 RA; 160 Brin.
For welded chemical plant equipment, tanks; Type 316 Ti; stainless, austenitic.

KORRONIT 189P.
M-1340; 0.07 max C, 18 Cr, 2 Mo, 10.5 Ni, bal Fe.
Annealed: 85,000 TS; 35,000 YS; 50 El; 65 RA; 160 Brin.
For acid resistant chemical plant equipment, tanks; Type 316; stainless, austenitic.

KORRONIT 1212 T.
M-1340; 0.10 max C, 12.5 Cr, 12 Ni, bal Fe.
For valves, pumps, oil refinery equipment; corrosion and heat resistant.

KORRONIT 1717.
M-1340; 0.07 max C, 17.5 Cr, 2 Mo, 17.5 Ni, 2 Cu,Ti = 7 x C, bal Fe.
For valves, pumps, chemical plant equipment; corrosion resistant.

KORRONIT-C.
M-1340; 0.90 C, 18 Cr, 1.15 Mo, 1.0 V, bal Fe.
Annealed: 105,000 TS; 60,000 YS; 18 El; 35 RA; 220 Brin.
Heat treated: 280,000 TS; 270,000 YS: 3 El; 15 RA; 555 Brin.
For cutlery, valves, ball bearings, surgical instruments; Type 440 B; corrosion resistant.

KORRONIT-H.
M-1340; 0.4 C, 0.4 Si, 13 Cr, bal Fe.
Annealed: 95,000 TS; 50,000 YS; 25 El; 55 RA; 195 Brin.
For cutlery, valves, surgical instruments; Type 420; stainless, hardenable.

KORRONIT-M.
M-1340; 0.2 C, 0.4 Si, 13 Cr, bal Fe.
Annealed: 95,000 TS; 50,000 YS; 25 El; 55 RA; 195 Brin.
For valves, cutlery, oil refinery equipment; Type 420; corrosion resistant.

KORRONIT-S.
M-1340; 0.22 C, 0.4 Si, 17 Cr, 1.5 Ni, bal Fe.
Annealed: 125,000 TS; 95,000 YS; 20 El; 55 RA; 260 Brin.
For marine hardware, pumps, valves; Type 431; corrosion and heat resistant.

KORRONIT-W.
M-1340; 0.12 max C, 0.4 Si, 13 Cr, bal Fe.
For turbine blades, valves, cutlery, surgical instruments; Type 410; stainless.

KORROSIL 1.
M-1323; 0.12 max C, 0.4 Si, 13 Cr, bal Fe.
Annealed: 75,000 TS; 40,000 YS; 35 El; 70 RA; 155 Brin.
Cold drawn: 100,000 TS; 85,000 YS; 17 El; 60 RA; 205 Brin.
For turbine blades, surgical instruments; Type 410; stainless.

KORROSIL 2.
M-1323; 0.2 C, 0.4 Si, 13 Cr, bal Fe.
Annealed: 95,000 TS; 50,000 YS; 25 El; 55 RA; 195 Brin.
Cold drawn: 100,000 TS; 85,000 YS; 17 El; 50 RA; 215 Brin.
For cutlery, valves, turbine blades, knives; Type 420; stainless.

KORROSIL-2M.
M-1650; 0.20 C, 13 Cr, 1.2 Mo, bal Fe.
Annealed: 95,000 TS; 40,000 YS; 25 El; 55 RA; B 92 Rock.
Heat Treated: 240,000 TS; 210,000 YS; 8 El; 25 RA; C 50 Rock.
For cutlery, knives, surgical instruments,, shafts, gears.
Corrosion resistant, hardenable.

KORROSIL 4.
M-1323; 0.4 C, 0.4 Si, 13 Cr, bal Fe.
Annealed: 100,000 TS; 50,000 YS; 23 El; 52 RA; 200 Brin.
For cutlery, valves, surgical and dental instruments; Type 420; stainless.

KORROSIL 6A.
M-1323; 0.12 C, 16.5 Cr, 0.25 Mo, 0.2 S, bal Fe.
Annealed: 80,000 TS; 50,000 YS; 25 El; 50 RA; 150 Brin.
For screw machine products, shafts, fasteners; Type 430F; stainless, free cutting.

KORROSIL 6U.
M-1323; 0.1 max C, 17.5 Cr, Ti = 7 x C, bal Fe.
Annealed: 80,000 TS; 50,000 YS; 25 El; 50 RA; 150 Brin.
For welded furnace parts and heat treating boxes; Type 430 Ti; corrosion resistant.

KORROSIL 60.
M-1323; 0.22 C, 17 Cr, 1.5 Ni, bal Fe.
Annealed: 125,000 TS; 95,000 YS; 20 El; 55 RA; 260 Brin.
For marine hardware, pumps, valves; Type 431.

KOULTCHOOG-ALUMINUM (KAULTCHAAG).
4-6 Cu, Mn, Mg, Si, bal Al.
For light alloy parts; similar to Duralumin, age-hardenable.

KOVAL.
M-928; 72 Ni, 17 Co, 2.2 Ti, 6.25 Fe.
For internal combustion valves, molds and machine parts; high temperature resistant.

KOVAR.
M-928; 29 Ni, 17 Co, 0.2 Mn, bal Fe.
Annealed: 77,500 TS; 59,500 YS; 25 El.
For electronic tubes, radio and X-ray tubes; same coefficient of expansion as glass; M.P. 100°C.

KOVAR see also **CARPENTER VACUMET KOVAR.**

KP.
M-1655; 1.0 C, 0.20 Mn, 0.20 Si, 0.10 V, bal Fe.
Water hardening tool steel.
W.-Nr. 1.2833; similar to AISI W1.

KP-1.
M-USSR; 0.13 C, 0.7 Mn, 0.2 Si, 12.2 Cr, 0.7 Mo, bal Fe.
Annealed: 75,000 TS; 35,000 YS; 25 El; B 92 Rock.
For springs, table flatware, oil refinery equipment.
Corrosion resistant.

KP EXTRA.
M-1655; 0.95 C, 0.40 Mn, 0.30 Si, 0.40 V, bal Fe.
Water hardening tool steel.
W.-Nr. 1.2835; similar to AISI W2.

"K.P." STYRIAN.
M-Eng.; 1.15 C, 0.24 Mn, 0.04 Cr, 4.67 W, bal Fe.
For tools, dies; oil hardened.

KRAFFTS ALLOY.
M-Swedish; 8 Sn, 26 Pb, 66 Bi.
M.P. 100°C.

KRAMER 7 1/2% PHOSPHOR COPPER.
M-346; 7.5 P, bal Cu.
Phosphor additive to copper alloys and brazing alloys.

KRAMER 12% NICKEL SILVER.
M-346; 12 Ni, 22 Zn, 10 Pb, 3 Sn, 53 Cr.
Cast: 38,000 TS; 20,000 YS; 18 El; 65 Brin.
Corrosion resistant; for hardware, plumbing, fixtures, valves, ornamental castings.

KRAMER 15% NICKEL DIE CAST ALLOY.
M-346; 15 Ni, Zn, bal Cu.
Sand Cast: 80,000 TS; 45,00 YS; 8 El; 175 Brin.
For ornamental castings, railroad car fittings.

KRAMER 15% PHOSPHOR COPPER.
M-346; 15 P, bal Cu.
Phosphor additive to copper alloys and brazing alloys.

KRAMER 68-1-3-28 ALLOY.
M-346; 68 Cu, 1 Sn, 3 Pb, 28 Zn.
Cast: 34,000 TS; 12,500 YS; 40 El; 55 Brin.
General purpose yellow brass, good machinability; for furniture, hardware, ornamental castings.

KRAMER 70/30 HARDENER.
M-349; Cu-Mn.

KRAMER 76-2 1/2-6 1/2-15 ALLOY.
M-346; 76 Cu, 2.5 Sn, 6.5 Pb,15 zn.
Cast: 36,000 TS; 14,500 YS; 36 El; 58 Brin.
For plumbing fixtures, faucets; good machinability.

KRAMER 78-7-15 ALLOY.
M-346; 78 Cu, 7 Sn, 15 Pb.
Cast: 34,000 TS; 17,500 YS; 30 El; 61 Brin.
Anti-acid metal corrosion resistant for chemical pumps, mining equipment, bearings, and bushings.

KRAMER 80-10-10 ALLOY.
M-346; 80 Cu, 10 Sn, 1 Pb.
Cast: 41,000 YS; 18,500 YS; 35 El; 65 Brin.
For bushings, bearings, pump bodies, and impellers, heavy duty; corrosion resistant.

KRAMER 81-3-7-9 ALLOY.
M-346; 81 Cu, 3 Sn, 7 Pb, 9 Zn
Cast: 34,000 TS; 17,000 YS; 27 El; 55 Brin.
For low pressure valves and fittings; good machinability.

KRAMER 83-7-7-3 ALLOY.
M-346; 83 Cu, 7 Sn, 7 Pb, 3 Zn.
Cast: 40,000 TS; 18,000 YS; 39 El; 67 Brin.
General purpose bearing alloy, corrosion resistant, for bushings, bearing, heavy duty auto fittings.

KRAMER 85-5-5-5 ALLOY.
M-346; 85 Cu, 5 Sn, 5 Pb 5 Zn.
Cast: 37,000 TS; 16,500 YS; 34 El; 65 Brin.
Red brass, good machinability; for low pressure valves and fittings.

KRAMER 85/15 HARDENER.
M-346; Cu-Si.

KRAMER 86-6-1 1/2-4 1/2 ALLOY.
M-346; 88 Cu, 6 Sn, 1.5 Pb, 4.5 Zn.
Cast: 42,000 TS; 18,000 YS; 43 El; 70 Brin.
Corrosion resistant; for valves, fittings, flanges, gears; for high temperatures and pressures.

KRAMER 87-8-1-4 ALLOY.
M-346; 8 Sn, 1 Pb, 4 Zn, bal Cu.
Cast: 45,000 TS; 19,000 YS; 40 El,; 70 Brin.
Corrosion resistant; for valves, steam pressure castings; resists high pressure.

KRAMER 88-6-1 1/2-4 1/2 ALLOY.
M-346; 88 Cu, 6 Sn, 1.5 Pb, 4.5 Zn.
Cast: 42,000 TS; 18,000 YS; 43 El; 70 Brin.
Corrosion resistant; for valves, fittings, flanges, gears; for high temperatures and pressures.

KRAMER 88-8-0-4 ALLOY.
M-346; 88 Cu, 8 Sn, 4 Zn.
Cast: 47,000 TS; 20,000 YS; 39 El; 70 Brin.
Corrosion resistant; for valves, pumps.

KRAMER 88-12 ALUMINUM BRONZE.
M-346; Al, bal Cu.
Cast: 55,000 TS; 40,000 YS; 1 El; 230 Brin.
For welding fixtures; 20% IACS.

KRAMER ALLOY NO. 66.
M-346; Mg, Mn, Ca, P, bal Cu.
For use as a deoxidizer and degassifier for Cu-Ni alloys.

KRAMER ALLOY NO. 77.
M-346; Mg, Mn, Ca, Si, Ti, Al, bal Cu.
For use as an aluminum bronze degassifier and to improve humidity.

KRAMER A MANGANESE BRONZE.
M-346; 58 Cu, 1 Fe, 0.5 Mn, 1 Al, bal Zn.
Cast: 70,000-75,000 TS; 28,000-35,000 YS; 22-35 El; 120-140 Brin.
For piston rods, valve stems, worm gears, propellers, structural castings, marine fittings.
Corrosion resistant; substitute for steel and malleable iron where corrosion resistance is required.

KRAMER A MANGANESE BRONZE.
M-346; 58 Cu, 1 Fe, 0.5 Mn, 1 Al, bal Zn.
Cast: 70,000-75,000 TS; 28,000-35,000 YS; 22-35 El; 120-140 Brin.
For piston rods, valve stems, worm gears, propellers, structural castings, marine fittings.
Corrosion resistant; substitute for steel and malleable iron where corrosion resistance is required.

KRAMER ANTI-ACID METAL.
M-346; 75 Cu, 10 Sn, 15 Pb.
For chemical apparatus; acid resistant.

KRAMER AX MANGANESE BRONZE.
M-346; 58 Cu, 1 Mn, 1.5 Fe, 1.5 Al, bal Zn.
Cast: 80,000-88,000 TS; 38,000-40,000 YS; 18-24 El; 160-165 Br.
Corrosion resistant, for structural castings, valve stems, propellers, marine castings.

KRAMER C ALUMINUM BRONZE.
M-346; 10 Al, 4 Fe, 2 Ni, bal Cu.
Cast: 88,000-96,000 TS; 33,000-36,000 YS; 25-10 El; 30-7 RA; 150-175 Brin.
Heat treated: 105,000-125,000 TS; 45,000-75,000 YS; 22-2 El; 20-3 RA; 180-300 Brin.
For gears, worm wheels, bushings; corrosion resistant.

KRAMER CONTACT METAL.
M-346; Sn, Pb, Zn, bal Cu.
Cast: 32,000 TS; 9500 YS; 30 El; 50 Brin.
For electrical fittings; 32% IACS.

KRAMER DIE CAST YELLOW BRASS.
M-346; 1 Sn, 1 Pb, 35 Zn, Al, bal Cu.
Cast: 55,000 TS; 30,000 YS; 15-20 El; 120-130 Brin.
For plumbing fixtures, ornamental castings; permanent mold castings.

KRAMER DIE MOLD BRONZE.
M-346; Ni, Al, Zn, bal Cu.
Sand cast: 90,000 TS; 60,000 YS; 2.5 El; 200 Brin.
For glass molds and high temperature applications; sand cast.

KRAMER EVERDUR.
M-346; Cu, Si, Mn.
Corrosion resistant silicon bronze.

KRAMER G ALUMINUM BRONZE.
M-346; Al, Fe, bal Cu.
Cast: 69,000 TS; 49,000 YS; 1 El; 265 Brin.
For wear plates and guides.

KRAMER GLASS MOLD ALLOY.
M-346; Cu, Zn, Ni, Al.
Used for making glass objects.

KRAMER I ALUMINUM BRONZE.
M-346; 1.0 Fe, 10.0 Al, bal Cu.
As cast: 67,000-77,000 TS; 26,000-34,000 YS; 12-26 El; 10-24 RA; 130-150 Brin.
Heat treated: 90,000-95,000 TS; 40,000-55,000 YS; 10-25 El; 160-220 Brin.
Heat treatable aluminum bronze, corrosion resistant; for structural castings, gears, machine parts, steel mill nuts.

KRAMER L ALUMINUM BRONZE.
M-346; 4 max Fe, 10 min Al, 0.5 max Mn, bal Cu.
As cast: 82,000-93,000 TS; 29,000-38,000 YS; 12-32 El; 140-165 Brin.
Heat treated: 100,000-110,000 TS; 35,000-55,000 YS; 10-30 El; 160-210 Brin.
Heat treatable aluminum bronze, corrosion resistant; for structural castings, gears, worm wheels.

KRAMER M ALUMINUM BRONZE.
M-346; 3-4 Fe, 9.0-9.5 Al, bal Cu.
Cast: 75,000-85,000 TS; 23,000-33,000 YS; 25-45 El; 26-47 RA; 125-140 Brin.
General utility aluminum bronze, corrosion resistant; for structural castings, valves, gears, nuts, pumps.

KRAMER N ALUMINUM BRONZE.
M-346; 4-5 Fe, 10-11 Al, 4-5 Ni, 1 max Mn, bal Cu.
As cast: 100,000-108,000 TS; 40,000-50,000 YS; 5-18 El; 6-20 RA; 180-210 Brin.
Heat treated: 110,000-125,000 TS; 60,000-80,000 YS; 5-12 El; 220-260 Brin.
Heat treatable aluminum bronze, corrosion resistant; for structural castings, pump and valve parts, gears, aircraft parts, marine casting.

KRAMER NICKEL ALUMINUM BRONZE.
M-346; Fe, Al, Ni, Mn, bal Cu.
Cast: 90,000 TS; 40,000 YS; 20 El; 160 Brin.
Corrosion resistant; for marine propellers.

KRAMER R ALUMINUM BRONZE.
M-346; 4 Fe, 10 Al, bal Cu.
As cast: 85,000 TS; 35,000 YS; 15 El; 160 Brin.
Heat treated: 105,000 TS; 50,000 YS; 10 El; 200 Brin.
Heat treatable aluminum bronze; corrosion resistant; for structural castings, valves, gears, bushings.

KRAMER SILICON-ALUMINUM BRONZE.
M-346; Al, Si, bal Cu.
Sand cast: 70,000 TS; 30,000 YS; 25 El; 130 Brin.
For valve stems, structural castings; corrosion resistant.

KRAMER SILICON BRONZE.
M-346; Si, Zn or Mn, bal Cu.
Cast: 60,000 TS; 25,000 YS; 40 El; 88 Brin.
Corrosion resistant; for valves, pumps.

KRAMER S MANGANESE BRONZE.
M-346; 60 Cu, 1 Fe, 0.5 Mn, 0.75 Al, 1 Sn, 1 Pb, bal Zn.
Cast: 60,000-65,000 TS; 25,000-30,000 YS; 15-20 El; 100-120 Brin.
Corrosion resistant, free machining.
For piston rods, valve stems, worm gears, propellers, valve bodies, fittings, structural castings, marine fittings.

KRAMER SPECIAL 20% NICKEL SILVER.
M-346; 20 Ni, Zn, bal Cu.
Sand cast: 50,000 TS; 25,000 YS; 20 El; 80 Brin.
For valves, dairy and laundry castings, ornamental parts; corrosion resistant.

KRAMER SPECIAL 25% NICKEL SILVER.
M-346; 25 Ni, Zn, bal Cu.
Sand cast: 60,000 TS; 36,000 YS; 14 El; 135 Brin.
For valve seats; corrosion resistant.

KRAMER SX MANGANESE BRONZE.
M-346; Zn, Mn, Al, Fe, bal Cu.
Cast: 95,000-104,000 TS; 45,000-55,000 YS; 22-18 El; 22-18 RA; 180-195 Brin.
For marine and aircraft castings; corrosion resistant.

KRAMER X MANGANESE BRONZE.
M-346; 60 Cu, 1.5 Fe, 1.5 Mn, 3 Al, bal Zn.
Cast: 85,000-95,000 TS; 45,000 YS; 18-20 El; 175 Brin.
Corrosion resistant; for hubcaps, piston rods, valve stems, propellers, marine fittings, structural castings.

KRAMER XX MANGANESE BRONZE.
M-346; 63 Cu, 2 Fe, 2.5 Mn, 5 Al, bal Zn.
Cast: 110,000-115,000 TS; 60,000-72,000 YS; 13-18 El; 225 Brin.
Heavy duty high strength alloy.
For gears, cams, bridge bearings, screw down nuts, structural castings.

KREIDLER MS 57/2.
M-324; 58 Cu, 2 Pb, bal Zn.
Good machinability and hot formability.
Cu Zn 40 Pb 2.

KREIDLER MS 58.
M-324; 58 Cu, 3 Pb, bal Zn.
Good machinability and hot formability.
Cu Zn 39 Pb 3.

KREIDLER MS 58A.
M-324; 58 Cu, 2 Pb, bal Zn.
For automatic machining, good machinability; good hot forming.
Cu Zn 40 Pb 2.

KREIDLER MS 59/2.
M-324; 59 Cu, 2 Pb, bal Zn.
Free machining brass.
Cu Zn 39 Pb 2.

KREIDLER MS 60 R.
M-324; 60.5 Cu, 1.5 Pb, 38 Zn.
Free machining brass; good hot and cold forming.
Cu Zn 38 Pb 1.5.

KREIDLER MO 60 R-NIET.
M-324; 60.5 Cu, 1.5 Pb, 38 Zn.
Ductile, free-machining brass; for rivets.

KREIDLER MS 63 D.
M-324; 63 Cu, 37 Zn.
For cold forming operations.
Cu Zn 37.

KREIDLER SU 4.
M-324; 57-59 Cu, 1.0-2.5 Mn, bal Zn.
45-50 kp/mm^2 TS; 18-28 kp/mm^2 YS; 18-20 El.
Brass for appliance parts; good solderability.
Cu Zn 40 Mn.

KREIDLER SU 10.
M-324; 58 Cu, 2 Al, 40 Zn.
55-65 kp/mm^2 TS; 24-32 kp/mm^2 YS; 10-18 El.
High strength aluminium brass.
Cu Zn 40 Al 2.

KRIEDLER SU 81.
M-324; 58-61 Cu, 2-3 Ni, bal Zn.
45-55 kp/mm^2 TS; 20-40 kp/mm^2 YS; 12-20 El.
Medium strength brass for marine applications.
Cu Zn 35 Ni.

KROKOLOY.
M-157; C-Cr-Co, bal Fe.
For dies for forming, blanking, drawing and coining, shear blades; cast to shape, air hardening.

KROMAIR see REPUBLIC A2.

KROMAL NO. 1.
M-750; 0.9 C, 1.2 Mn, 0.5 Cr, bal Fe.
For punches, taps, dies, crimpers; non-deforming, oil hardened.

KROMAL NO. 2.
M-750; 0.9 C, 1.5 Mn, 0.4 Cr, 0.3 V, bal Fe.
For tools, punches; non-deforming.

KROMAL NO. 3.
M-750; 0.5 C, 0.9 Mn, 0.9 Cr, 0.2 Mo, bal Fe.
For maintenance and repair; oil hardening.

KROMAL NO. 4.
M-750; 0.48 C, 1.85 Cr, 0.4 Mo, 1.25 Ni, bal Fe.
For maintenance and repair; oil hardening.

KROM-ALOY.
M-147; Cr, Sn, Pb, Al, Mn, Ni, Zn, bal Cu.
77,210 TS; 54,600 YS; 10 El; 9 RA; 153 Brin.
For marine hardware, propellers, pump valves and seats, trimmings; corrosion resistant.

KROMAX.
80 Ni, 20 Cr.
For resistance alloy; heat resistant.

KROMITE 14% MANGANESE PLATE.
M-15; 1.3 C, 13.5 Mn, 0.20 Si, 0.20 Mo, bal Fe.
Austenitic manganese steel.
As shipped: 190 Brin; will work harden to 580 Brin. Weldable with 14% Mn electrodes.

KROMITE BRAKE DIE.
M-15; C, alloy, bal Fe.
For brake dies, bead-forming dies; preheat treated, tough and wear resistant.

KROMITE NO. 2.
M-15; C, alloy, bal Fe.
For chain links, mandrels springs, tongs; shock resistant.

KROMITE NO. 3.
M-15; C, alloy, bal Fe.
Heat treated: 156,000 TS; 130,000 YS; 17 El; 56 RA; 270 Brin.
For shafts, gears, bolts, tie rods; preheat treated.

KROMITE NO. 4.
M-15; C, alloy, bal Fe.
Heat treated: 211,000-304,000 TS; 450-560 Brin.
For punches, picks, swages, vice jaws, bolts; tough, shock resistant.

KROMITE WEAR PLATE.
M-15; 0.39 C, 2.3 Mn, 0.28 Si, 0.90 Cr, 0.34 Mo, 0.50 Cu, bal Fe.
As delivered: 164,000 psi TS; 129,000 psi YS; 380-400 Brin.
For chute liners, stamping dies, mixing blades, scraper blades.

KROMOX.
M-1771; 2.6-2.9 C, 15.5 Cr, 1.9 Mo, 1.1 Cu, bal Fe.
Heat treated iron 650 min Brin.
For high abrasion resistant cement and ore grinding.

KRONA.
M-1705; 0.90 C, 1.2 Mn, 0.5 Cr, 0.2 V, 0.5 W, bal Fe.
Oil hardening tool steel.
AISI O1.

KROPP 23.
M-1135; 0.30 C, 0.70 Mn, 0.35 Si, 0.96 Cr, 0.2 Mo, bal Fe.
For plastic mold dies; water hardened.

KROPP 23C.
M-1135; 0.32 C, 0.75 Mn, 0.35 Si, 0.9 Cr, 0.2 Mo, bal Fe.
For zinc die casting dies; water hardened.

KROPP 56.
M-1135; 0.60 C, 0.65 Mn, 1.8 Ni, 1.6 Cr, 0.35 Mo, bal Fe.
For gorging dies, shear blades, high pressure cylinders; oil hardening.

KROPP 61.
M-1135; 0.10 C, 0.5 Mn, 3.5 Ni, 1.5 Cr, bal Fe.
For gears, pinions, spindels, cams, bearing rolls; tough.

KROPP 93.
M-1135; 0.35 C, 1.0 Si, 5 Cr, 1 Mo, 0.15 V, 1 W, bal Fe.
For shear blades, extrusion and upsetting dies, mandrels; hot work steel.

KROPUNCH.
M-750; 0.4 C, 2.5 W, Cr, 0.3 V, 2.0 Mo, 1.1 Si, bal Fe.
For shear blades, punches, trim dies, mandrels; hot and cold work steel, air hardening.

KROTUNG.
M-750; 0.3 C, 5 Cr, 1.5 Mo, 1.3 W, 0.3 V, bal Fe.
For hot forging dies and tools, nut crowners; high red hardness, tough.

KRUPP 15 M.
M-72; 1.05 C, 14 Mn, bal Fe.
Austenitic manganese steel, for heavy wear resistance.

KRUPP 150.
M-72; 0.2 C, 0.35 Si, 1.0 Mn, 0.2 Ti, bal Fe.
For hard facing electrodes; tough.

KRUPP 204.
M-72; 1.05 C, 1.0 Cr, 0.30 Mn, bal Fe.
For bearings, liners, sleeves, bushings; water hardened, wear resistant.

KRUPP 206.
M-72; 1.05 C, 0.35 Mn, 1.2 cr, bal Fe.
For bearings, liners, sleeves, bushings; water hardened, wear resistant.

KRUPP 250.
M-72; 0.25 C, 0.35 Si, 1.6 Mn, 0.2 Ti, bal Fe.
For hard facing electrodes; tough.

KRUPP 303.
M-72; 1.4 C, 0.1 V, 0.3 Cr, bal Fe.
For engravers' tools, cutters, bearings, form dies; water hardened, wear resistant.

KRUPP A3P.
M-72; 0.15 C, 0.40 Mn, 0.25 Si, bal Fe.
Annealed: 70,000 TS; 40,000 YS; 25 El; 60 RA; 145 Brin.
For gears, cams, camshafts, bolts, fasteners; case hardening steel.

KRUPP A3PO.
M-72; 0.15 C, 0.25 Si, 0.35 Mn, bal Fe.
Annealed: 70,000 TS; 40,000 YS; 25 El; 60 RA; 145 Brin.
For gears, cams, camshafts, machine tool parts; case hardening steel.

KRUPP A11K.
M-72; 0.52-0.58 C, 0.5 Mn, bal Fe.
For forging die blocks; water hardened.

KRUPP A12.0.
M-72; 0.6 C, 0.4 Si, 0.6 Mn, bal Fe.
Heat treated: 160,000 TS; 113,000 YS; 12 El; 40 RA; 321 Brin.
For punches, rails, crimpers; water hardened.

KRUPP A12-14P.
M-72; 0.7 C, 0.25 Si, 0.25 Mn, bal Fe.
Heat treated: 174,000 TS; 128,000 YS; 12 El; 37 RA; 355 Brin.
For rails, crimpers, punches, springs; water hardened.

KRUPP A14K.
M-72; 0.60 C, 0.40 Si, 0.60 Mn, bal Fe.
Heat treated: 160,000-115,000 TS; 113,000-77,000 YS; 12-13 El; 40-54 RA; 321-230 Brin.
For punches, rails, crimpers; water hardened.

KRUPP A15-17P.
M-72; 0.85 C, 0.25 Si, 0.25 Mn, bal Fe.
Heat treated: 190,000 TS; 145,000 YS; 10 El; 30 RA; 400 Brin.
For lathe and planer tools, drills, taps, hobs; Type W1; water hardened.

KRUPP A16K.
M-72; 0.75-0.83 C, 0.5 Mn, bal Fe.
For forging die blocks; water hardened.

KRUPP A17 G.
M-72; 0.8-0.9 C, 0.25 -0.4 Si, 0.45-0.6 Mn, bal Fe.
For tools, drills, springs; water hardening.

KRUPP A 18 AZ.
M-72; 0.10 C, 18 Cr, 10 Ni, Ti, S, bal Fe.
Austenitic stainless steel.

KRUPP A18-20P.
M-72; 1.0 C, 0.25 Mn, 0.25 Si, bal Fe.
Annealed: 100,000 TS; 53,000 YS; 21 El; 42 RA; 200 Brin.
For drills, taps, reamers, tools, dies; Type W1; water hardened.

KRUPP A22P/A23P.
  M-72; 1.15 C, 0.25 max Si, 0.25 max Mn, bal Fe.
  Annealed: 110,000 TS; 60,000 YS; 18 El; 38 RA; 210 Brin.
  For reamers, drills, taps, broaches, hobs, tools; Type W1; water hardened.

KRUPP A24P.
  M-72; 1.3 C, 0.25 max Si, 0.25 max Mn, bal Fe.
  For engravers tools, form cutters, milling cutters; Type W1; water hardened.

KRUPP A26P.
  M-72; 1.3 C, 0.25 max Si, 0.25 max Mn, bal Fe.
  For engravers' tools, form and milling cutters; Type W1; water hardened.

KRUPP A30.
  M-72; 0.15 C, 0.25 Si, 0.40 Mn, bal Fe.
  Cold drawn: 72,000 TS; 60,000 YS; 22 El; 58 RA; 145 Brin.
  For gears, cams, camshafts, fasteners; carburizing steel.

KRUPP A50.
  M-72; 0.22 C, 0.25 Si, 0.45 Mn, bal Fe.
  Annealed: 73,000 TS; 41,000 YS; 22 El; 58 RA; 150 Brin.
  For gears, cams, bolts, fasteners, camshafts; case hardening steel.

KRUPP A70.
  M-72; 0.35 C, 0.25 Si, 0.45 Mn, bal Fe.
  Hot rolled; 85,000 TS; 54,000 YS; 30 El; 53 RA; 185 Brin.
  For gears, pinions, shafts, bolts, fasteners; water hardened.

KRUPP A90.
  M-72; 0.45 C, 0.25 Si, 0.45 Mn, bal Fe.
  Hot rolled: 98,000 TS; 59,000 YS; 24 El; 45 RA; 212 Brin.
  For gears, pinions, shafts, bolts, fasteners; water hardened.

KRUPP A 163.
  M-72; 0.13 C, 16 Cr, 5 Ni, 3 Mo, bal Fe.
  Stainless steel.

KRUPP ALC 184 H.
  M-72; 0.02 C, 19 Cr, 17 Ni, Mo, N, bal Fe.
  Austenitic stainless steel.

KRUPP ALC 187 H.
  M-72; 0.02 C, 20 Ni, 18 Cr, Mo, N, bal Fe.
  Austenitic stainless steel.

KRUPP ALC 250.
  M-72; 0.01 C, 25 Ni, 21 Ni, bal Fe.
  Stainless steel for high temperature operation.

KRUPP AM 17.
  M-72; 0.10 C, 17 Cr, 7 Mn, Ni, bal Fe.
  Cr-Mn stainless steel; austenitic.

KRUPP AM 18.
  M-72; 0.08 C, 18 Cr, 9 Mn, Ni, bal Fe.
  Cr-Mn stainless steel.

KRUPP AM 18 H.
  M-72; 0.05 C, 18 Cr, 9 Mn, Ni, N, bal Fe.
  Cr-Mn stainless steel.

KRUPP AM 213.
  M-72; 0.03 C, 21 Cr, 15 Ni, 7 Mn, 3 Mo, bal Fe.
  Stainless steel.

KRUPP ANALYSIS.
  M-72, M-68; 0.08-0.16 C, 1.2-1.7 Cr, 3.8-4.3 Ni, bal Fe.
  For armor plate, gears; tough, case-hardening alloy steel.

KRUPP AP 17 7.
  M-72; 0.06 C, 17 Cr, 7 Ni, Al, Ti, bal Fe.
  Austenitic stainless steel, for cold drawn springs.

KRUPP AP 173.
  M-72; 0.09 C, 17 Cr, 5 Ni, 3 Mo, N, bal Fe.
  Stainless steel.

KRUPP AS 183 A.
  M-72; 0.03 C, 17 Cr, 11 Ni, Mo, S bal Fe.
  Stainless steel.

KRUPP B126 P.
  M-72; 0.62-0.68 C, 1.7 Si, 0.6 Mn, bal Fe.
  For tools, springs, punches; shock resistant.

KRUPP B136M.
  M-72; 0.7 C, 1.7 Si, 0.7 Mn, bal Fe.
  For punches, crimpers, upsetters, springs; oil hardened, shock resistant.

KRUPP BC40.
  M-72; 0.37 C, 1.25 Si, 1.25 Mn, bal Fe.
  For punches, crimpers, upsetters; oil hardened.

KRUPP BC 1044.
  M-72; 0.53 C, 0.8 Si, 1.05 Mn, bal Fe.
  For springs, punches, upsetters; oil hardened.

KRUPP BC1244.
  M-72; 0.57-0.63 C, 0.9-1.1 Si, 0.9-1.1 Mn, bal Fe.
  For tools, springs, punches; shock resistant.

KRUPP BDF30.
  M-72; 0.30-0.36 C, 0.8-1.0 Si, 0.3 Mn, 0.9-1.1 Cr, 0.20 V, 1.9-1.1 C 0.20 V, 1.7-2.0 W, bal Fe.
  For tools, dies, hot work tools; hot work steel.

KRUPP BDF50.
  M-72; 0.40-0.46 C, 0.8-1.0 Si, 0.9-1.1 Cr, 1.7-2.0 W, bal Fe.
  For tools, hot work dies; hot work steel.

KRUPP BDFD.
  M-72; 0.5 C, 1.5 Cr, 2.25 W, 0.2 V, bal Fe.
  For tools, dies, punches; oil hardened.

KRUPP BDF SPEZIAL.
  M-72; 0.54-0.60 C, 0.8-1.1 Si, 1.7-2.0 W, bal Fe.
  For tools, hot work dies; hot work steel.

KRUPP BEARING.
M-72; 87 Al, 8 Cu, 5 Sn.
For bearings.

KRUPP BF1552.
M-72; 0.67 C, 1.3 Si, 0.5 Cr, 0.5 Mn, bal Fe.
For springs; oil hardened, tough.

KRUPP BF2555.
M-72; 1.2-1.3 C, 1.05-1.2 Si, 0.7 Mn, 1.1-1.3 Cr, bal Fe.
For tools, dies, bearings; oil hardening.

KRUPP BFD-50G.
M-72; C, alloy, bal Fe.
For tools.

KRUPP BFG.
M-72; 0.95 C, 1.2 Mn, 0.5 Cr, 0.5 W, bal Fe.
For tools, dies; non-deforming.

KRUPP BFM755.
M-72; 0.30-0.38 C, 1.4-1.6 Si, 1.2-1.5 Cr, 0.1-0.2 V, bal Fe.
For tools, punches, dies; shock resistant.

KRUPP BFM SPEZIAL.
M-72; 0.45 C, 1.4 Si,1.2 Cr, 0.1 V, bal Fe.
For tools, punches, chisels; shock resistant.

KRUPP BFV.
M-72; 0.65 -0.70 C, 0.15 -0.25 Si, 0.5-0.6 Mn, bal Fe.
For tools, springs, punches; water hardening.

KRUPP BMH.
M-72; 0.42-0.48 C, 1.5 -1.7 Si, 0.6 Mn, bal Fe.
For tools, springs, punches; shock resistant.

KRUPP BS1.
M-72; 0.95 C, bal Fe.
For tools, drills, taps; water hardened.

KRUPP BS2.
M-72; 0.95 C, bal Fe.
For tools, drills, taps; water hardened.

KRUPP BS3.
M-72; 1.1 C, 0.25 Si, 0.25 Mn, bal Fe.
For drills, taps, reamers, cutters; Type W1; water hardened.

KRUPP BS4.
M-72; 1.0 C, 0.25 Si, 0.25 Mn, bal Fe.
For springs, drills, taps, punches, reamers; Type W1; water hardened.

KRUPP BS5.
M-72; 0.85 C, 0.25 max Si, 0.25 max Mn, bal Fe.
Heat treated: 190,000 TS; 145,000 YS; 10 El; 30 RA; 400 Brin.
For drills, taps, reamers, punches, cutters; Type W1; water hardened.

KRUPP BS6.
M-72; 0.70 C, 0.25 max Si, 0.25 max Mn, bal Fe.
Heat treated: 175,000 TS; 128,000 YS; 12 El; 37 RA; 355 Brin.
For rails, punches, hammers, axes; Type W1; water hardened.

KRUPP BS 45 W.
M-72; 0.34 C, 4 Ni, Cr, W, bal Fe.
Hot work tool or mold steel.

KRUPP BVR 1.
M-72; 0.28 C, 2.0 Ni, 1.3 Cr, 0.45 Mo, bal Fe.
For deep hardening shafts and gears.
W. Nr. 1.6932.

KRUPP BVR 1S.
M-72; 0.26 C, 2.75 Ni, 1.5 Cr, 0.45 Mo, bal Fe.
For deep hardening structural parts.
W.Nr. 1.6948.

KRUPP BVR 1 SS.
M-72; 0.26 C, 3.5 Ni, 1.5 Cr, 0.45 Mo, bal Fe.
For very deep hardening structural parts.

KRUPP C7E.
M-75; 0.15 C, 0.65 Cr, 0.5 Mn, 0.25 Si, bal Fe.
For gears, pinions, machine tool parts; cast hardening steel.

KRUPP C66.
M-72; 0.30 C, 1.35 Mn, 0.25 Si, bal Fe.
For gears, shafts, machine tool parts; water hardened, tough.

KRUPP CA 1220 SPEZIAL.
M-72; 2.0 c, 12 Cr, 1.0 V, Mo, bal Fe.
High chromium cold work tool steel.

KRUPP CARBON.
M-371; 0.7-1.0 C, bal Fe.
For tools, drills, taps; water hardened.

KRUPP CMV SPECIAL.
M-72; C 18 Mn, 12.5 Cr, 1 V, 0.2 Ni, bal Fe.
For blading for gas turbine on jet engine; corrosion and heat resistant.

KRUPP DF35.
M-72; 0.9 C 0.45 Cr, bal Fe.
For tools, drills, taps; water hardening.

KRUPP DF51.
M-72; 1.05 C, 0.9 Mn, 1.0 Cr, 1.15 W, bal Fe.
For cutters, bearings, dies, tools; water hardened.

KRUPP DF108C.
M-72; 1.5 C, 11.5 Cr, 0.75 Mo, bal Fe.
For tools, dies; non-deforming.

KRUPP DF109CN.
M-72; 0.8 C, 9.5 W, 2.5 Cr, 0.1 V, bal Fe.
For hot work tools and dies; hot work steel.

KRUPP DF109CW.
 M-72; 0.3 C, 2.65 Cr, 0.35 V, 8.5 W, 0.3 Mn, bal Fe.
 For extrusion rams and liners, punches; hot work steel, oil hardened.

KRUPP DF168.
 M-72; 2.1 C, 11.5 Cr, 0.7 W, bal Fe.
 For blanking and forming dies; oil hardened, non-deforming.

KRUPP DFM.
 M-72; 0.7 C, 18 W, 5 Co, 4 Cr, 1.3 V, bal Fe.
 For tools, dies, cutters; high speed steel.

KRUPP DFM EXTRA SPECIAL.
 M-72; 0.7 C, 18 W, 9 Co, 1 Mo, 4 Cr, 2 V, bal Fe.
 For tools, dies, cutters; high speed steel.

KRUPP DFM SUPRA.
 M-72; 0.79 C, 4.75 Co, 4.3 Cr, 0.75 Mo, 1.5 V, 18 W, bal Fe.
 For lathe and planer tools, reamers, hobs, taps, drills; high speed steel, heavy duty.

KRUPP DFM SUPRA SPEZIAL.
 M-72; 0.76 C, 10 Co, 4.2 Cr, 0.8 Mo, 1.8 V, 18 W, bal Fe.
 For lathe and planer tools, reamers, broaches; high speed steel.

KRUPP DFM ZWEIKARRO.
 M-72; 0.74 C, 4.1 Cr, 1.1 V, 18.5 W, bal Fe.
 For lathe and planer tools, drills, taps, hobs; high speed steel.

KRUPP DFMC.
 M-72; 0.7 C, 18 W, 4 Cr, 1 V, bal Fe.
 For tools, dies, cutters; high speed steel.

KRUPP DFMN E-SPEZIAL.
 M-72; 0.86 C, 4.1 Cr, 0.85 Mo, 2.5 V, 12 W, bal Fe.
 For lathe and planer tools, reamers, broaches, taps; high speed steel.

KRUPP DFMN SUPRA.
 M-72; 0.86 C, 2.8 Co 4.3 Cr, 0.85 Mo, 2.1 V, 12 W, bal Fe.
 For lathe and planer tools, reamers, broaches; high speed steel.

KRUPP DFMN ZWEIKARRO.
 M-72; 0.82 C, 4.1 Cr, 0.85 Mo, 1.6 V, 8.7 W, bal Fe.
 For lathe and planer tools, broaches; high speed steel.

KRUPP DFMV3.
 M-72; 0.86 C, 4.1 Cr, Mo, W, V, bal Fe.
 For lathe and planer tools, reamers, drills, hobs; high speed steel.

KRUPP DFMV 5.
 M-72; 1.35 Co, 0.3 Si, 0.3 Mn, 4.25 Cr, 10.5 W, 4.35 V, bal Fe.
 For hard facing electrodes; wear resistant.

KRUPP E33E.
 M-72; 0.2 C, 1.5 Ni, bla Fe.
 For gears, pinions, machine tool parts; case hardening steel.

KRUPP E67.
 M-72; 0.18 C, 2 Cr, 2 Ni, bal Fe.
 For gears, cams, camshafts; case hardened.

KRUPP EB7.
 M-72; 0.55 C, 0.50-0.70 Mn, 0.10-0.40 Si, bal Fe.
 Heat treated: 160,000 TS; 113,000 YS; 12 El; 40 RA; 320 Brin.
 For gears, bolts, machine tool parts; water hardened.

KRUPP EF23V.
 M-72; 0.28 C, 1.5 Ni, 0.5 Cr, bal Fe.
 For bolts, gears, machine tool parts; oil hardened, shock resistant.

KRUPP EF24V.
 M-72; 0.35 C, 0.5 Cr, 1.5 Ni, bal Fe.
 For bolts, gears, machine tool parts; oil hardened, shock resistant.

KRUPP EF29V.
 M-72; 0.28 C, 0.7 Cr, 2.5 Ni, bal Fe.
 For bolts, gears, machine tool parts; oil hardened, shock resistant.

KRUPP EF30V.
 M-72; 0.35 C, 0.7 Cr, 2.5 Ni, bal Fe.
 For gears, bolts, crankshafts; oil hardened, shock resistant.

KRUPP EF35E.
 M-72; 0.13 C, 0.7 Cr, 2.5 Ni, bal Fe.
 For gears, cams, camshafts, fasteners; case hardening steel, shock resistant.

KRUPP EF48V.
 M-72; 0.22 C, 0.7 Cr, 3.5 Ni, 0.6 Mn, bal Fe.
 For gears, cams, camshafts, fasteners; case hardening steel, shock resistant.

KRUPP EF49V.
 M-72; 0.30 C, 0.7 Cr, 3.5 Ni, bal Fe.
 For gears, bolts, crankshafts, fasteners; oil hardened, shock resistant.

KRUPP EF58E.
 M-72; 0.13 C, 0.7 Cr, 3.5 Ni, bal Fe.
 For gears, cams, camshafts, machine tool parts; case hardening steel, shock resistant.

KRUPP EF59E.
 M-72; 0.13 C, 1.1 Cr, 4.5 Ni, bal Fe.
 For gears, pinions, camshafts, fasteners; case hardened, shock resistant.

KRUPP EF62V.
M-72; 0.35 C, 1.3 Cr, 4.5 Ni, bal Fe.
For gears, bolts, crankshafts, machine tool parts; oil hardened, shock resistant.

KRUPP EF1514.
M-72; 0.8 C, 0.7 Mn, 1-1.2 Cr, 0.5 Ni, bal Fe.
For tools, dies; tough.

KRUPP EFS.
M-72; 0.45 C, 0.7 Cr, 2.5 Ni, bal Fe.
For gears, bolts, crankshafts, fasteners; oil hardened, shock resistant.

KRUPP EFS SPECIAL S.
M-72; 0.45-0.50 C, 0.5 -0.8 Mn, 1.2 -1.5 Cr, 1.1-1.4 Ni, bal Fe.
For tools, gears, shafts; oil hardening.

KRUPP EFWP.
M-72; 0.50-0.55 C, 0.9-1.2 Cr, 3-3.5 Ni, bal Fe.
For tools, gears, shafts; oil hardening.

KRUPP EV420.
M-72; 0.08 C, 1-1.25 Mn, bal Fe.
Welded: 55,000 TS; 45,000 YS; 33-40 El; 60-80 RA.
For welding rod.

KRUPP EXTRA.
M-72; 0.35 C, 2.5 Cr, 0.65 V, 4.5 W, bal Fe.
For tools, hot work dies; hot work steel.

KRUPP EXTRA SPEZIAL P.
M-72; 0.5 C, 0.85 Mn, 1.5 Cr, 0.65 Mo, 0.3 V, bal Fe.
For tools, gears, shafts; oil hardening.

KRUPP F13.
M-72; 0.12 max C, 2.2 Si, 13 Cr, bal Fe.
Annealed: 75,000 TS; 40,000 YS; 35 El; 70 RA; 155 Brin.
For valves, cutlery, surgical instruments, knives; Type 410; stainless.

KRUPP F 16.
M-72; 0.07 C, 16 Cr, bal Fe.
Ferritic type stainless steel.

KRUPP F 17 A ZN.
M-72; 0.04 C, 17 Cr, Nb, Mo, Si, bal Fe.
For high temperature operation.

KRUPP F 17 B.
M-72; 0.05 C, 17 Cr, bal Fe.
Ferritic type stainless steel.

KRUPP F20S.
M-72; 1.6 C, 0.9 Si, 0.2 Mn, 20 Cr, bal Fe.
For hard facing electrodes; corrosion resistant.

KRUPP F20SH.
M-72; 2.5 C, 0.9 Si, 0.2 Mn, 20 Cr, bal Fe.
For hard facing electrodes; corrosion resistant.

KRUPP F22.
M-72; 0.13 C, 0.5 Cr, 0.3 Mn, 0.2 Si, bal Fe.
For gears, bolts, machine tool parts; case hardened.

KRUPP F28S.
M-72; 1.6 C, 0.9 Si, 0.55 Mn, 28 Cr, bal Fe.
For hard facing electrodes; corrosion resistant.

KRUPP F28SH.
M-72; 2.1 C, 0.9 Si, 0.55 Mn, 28 Cr, bal Fe.
For hard facing electrodes; corrosion resistant.

KRUPP F33.
M-72; 0.15 C, 0.65 Cr, 0.25 Si, 0.5 Mn, bal Fe.
For gears, bolts, machine tool parts; case hardened.

KRUPP F74.
M-72; 0.41 C, 0.25 Si, 0.65 Mn, 1.0 Cr, bal Fe.
For gears, bolts machine tool parts; water hardened.

KRUPP F153.
M-72; 1.4 C, 0.30 Cr, 0.30 Mn, 0.25 Si, 0.1 V, bal Fe.
For bearings, cutters, tools, dies; water hardened, wear resistant.

KRUPP F 161.
M-72; 0.06 C, 16 Cr, Mo, bal Fe.
Ferritic type stainless steel.

KRUPP F168.
M-72; 0.82-0,87 C, 1.7 -1.9 Cr, bal Fe.
For tools, bearings; water hardening.

KRUPP F182.
M-72; 0.9 C, 0.45 Cr, bal Fe.
For tools, drills, taps; water hardening.

KRUPP F193.
M-72; 0.9-1.0 C, 0.7-0.9 Cr, bal Fe.
For tools, drills, taps; water hardening.

KRUPP F202.
M-72; 1.1 C, 0.30 Mn, 0.4 Cr, 0.20 Si, bal Fe.
For cutters, tools, bearings, drills; water hardened, wear resistant.

KRUPP F 206 G.
M-72; 0.95-1.05 C, 1.25-1.5 Cr, bal Fe.
For tools, bearings; wear resistant.

KRUPP F261.
M-72; 1.26-1.32 C, 0.2-0.3 Cr, 0.03 S, bal Fe.
For tools, cutters, drills; water hardening.

KRUPP F261 P.
M-72; 1.26-1.32 C, 0.025 S, 0.025 P, 0.2-0.3 Cr, bal Fe.
For tools, cutters, drills; water hardening.

KRUPP F283.
M-72; 1.35-1.45 C, 0.6-0.8 Cr, bal Fe.
For tools, cutters, hobs; water hardening.

KRUPP F306.
 M-72; 1.4-1.5 C, 1.3-1.5 Cr, bal Fe.
 For tools, bearings; wear resistant.

KRUPP F306 G.
 M-72; 1.4-1.5 C, 1.3-1.5 Cr, bal Fe.
 For tools, bearings; wear resistant.

KRUPP F1248 C.
 M-72; 1.65 C, 11.5 Cr, 0.1 V, bal Fe.
 For blanking and forming dies, punches; air hardened, non-deforming.

KRUPP F1448C.
 M-72; 1.65 C, 11.5 Cr, 0.1 V, bal Fe.
 For blanking and forming dies, punches; air hardened, non-deforming.

KRUPP F2048.
 M-72; 2.25 C, 13 Cr, bal Fe.
 For tools, dies; non-deforming.

KRUPP F3014-LCS.
 M-72; 1.4-1.5 C, 1.4-1.6 Si, 3.5 Cr, bal Fe.
 For tools, dies; tough.

KRUPP F4048.
 M-72; 2.0-2.2 C, 11-12 Cr, bal Fe.
 For tools, dies; non-deforming.

KRUPP FALC 223.
 M-72; 0.03 max C, 22 Cr, 5.5 Ni, 3.0 Mo, bal Fe.
 Stainless steel for high temperature operation.

KRUPP FC29.
 M-72; 0.20 C, 1.15 Cr, 1.25 Mn, bal Fe.
 For gears, cams, camshafts; case hardening steel.

KRUPP FEDERSTAHLDRAHT A.
 M-72; 0.30-0.70 Mn, 0.20-0.30 Si, C, bal Fe.
 Spring steel wire.
 W. Nr. 1.0500.

KRUPP FEDERSTAHLDRAHT B.
 M-72; 0.30-0.70 Mn, 0.10-0.30 Si, C, bal Fe.
 Spring steel wire.
 W. Nr. 1.0600.

KRUPP FF6.
 M-72; 0.12 max C, 2.3 Si, 6 Cr, bal Fe.
 For oil refinery equipment; creep and heat resistant.

KRUPP FF6N.
 M-72; 0.3 C, 2.2 Si, 6 Cr, bal Fe.
 For oil refinery equipment; creep and heat resistant.

KRUPP FF13.
 M-72; 0.12 max C, 2.2 Si, 13 Cr, bal Fe.
 For turbine blades, cutlery, valves, surgical instruments; corrosion and heat resistant.

KRUPP FF18.
 M-72; 0.12 max C, 18 Cr, 2 Si, bal Fe.
 Annealed: 80,000 TS; 50,000 YS; 25 El; 50 RA; 150 Brin.
 For oil refinery parts, soot blowers, bolts; Type 430; corrosion resistant.

KRUPP FF25.
 M-72; 0.35 C, 23-27 Cr, bal Fe.
 Annealed: 90,000 TS; 60,000 YS; 20 El; 45 RA; 180 Brin.
 For furnace parts, preheaters, heat treating boxes; Type 446; corrosion resistant.

KRUPP FF30.
 M-72; 0.35 C, 23-27 Cr, bal Fe.
 Annealed: 90,000 TS; 60,000 YS; 20 El; 45 RA; 180 Brin.
 For furnace parts, preheaters, heat treating boxes; Type 446; corrosion resistant.

KRUPP FF112.
 M-72; 0.3 C, 2.2 Si, 6 Cr, bal Fe.
 For oil refinery equipment; heat and creep resistant.

KRUPP FF120.
 M-72; 0.6 C, 1.5 Si, 22 Cr, bal Fe.
 For furnace grates, wear plates, rabble arms; heat and creep resistant.

KRUPP FF128.
 M-72; 0.6 C, 1.5 Si, 29 Cr, bal Fe.
 For frunace grates, wear plates, rabble arms; heat resistant.

KRUPP FF228.
 M-72; 1.3 C, 1.5 Si, 29 Cr, bal Fe.
 For wear blades, scrapers, crushers, rollers; heat and abrasion resistant.

KRUPP FFE25.
 M-72; 0.2 C, 1.2 Si, 25 Cr, 4 Ni, bal Fe.
 Cast: 90,000 TS; 65,000 YS; 2 El; 212 Brin.
 For cylinder liners, bushings, valves, corroson and heat resistant.

KRUPP FFE28.
 M-72; 0.2 C, 1.2 Si, 25 Cr, 4 Ni, bal Fe.
 Cast: 90,000 TS; 65,000 YS; 2 El; 212 Brin.
 For cylinder liners, bushings, valve seats and bodies; corrosion and heat resistant.

KRUPP FFP6.
 M-72; 0.12 C, 0.8 Si, 0.5 Cr, 0.8 Al, bal Fe.
 For oil refinery equipment; heat and creep resistant.

KRUPP FFP13.
 M-72; 0.12 max C, 1.2 Si, 1 Al, 13 Cr, bal Fe.
 For oil refinery equipment; heat and creep resistant.

KRUPP FFP18.
M-72; 0.12 max C, 1 Al, 1, Cr, bal Fe.
Annealed: 120,000 TS; 90,000 YS; 20 El; 55 RA; 250 Brin.
For oil refinery equipment; heat and creep resistant.

KRUPP FFP24.
M-72; 0.12 max C, 1.5 Si, 1.5 Al, 24 Cr, bal Fe.
For oil refinery equipment; heat and creep resistant.

KRUPP FK15.
M-72; 0.15 C, Cr, Mo, bal Fe.
For gears, shafts, machine tool parts; case hardening steel, tough.

KRUPP FK24.
M-72; 0.20 C, Cr, Mo, bal Fe.
For machine tool parts, gears, shafts; case hardening steel, tough.

KRUPP FK30.
M-72; 0.25 C, 0.65 Mn, 1 Cr, 0.2 Mo, bal Fe.
For machine tool parts, gears; oil hardened, tough.

KRUPP FK34.
M-72; 0.33 C, 1 Cr, 0.65 Mn, 0.2 Mo, bal Fe.
For machine tool parts, gears, fasteners; oil hardened, tough.

KRUPP FK44.
M-72; 0.42 C, 1 Cr, 0.2 Mo, 0.65 Mn, bal Fe.
For gears, bolts, fasteners, shafts, studs; oil hardened, tough.

KRUPP FKM45.
M-72; 0.42 C, Cr, Mo, bal Fe.
For gears, bolts, fasteners, shafts; oil hardened, tough.

KRUPP FKM54.
M-72; 0.30 C, Cr, Mo, V, bal Fe.
For gears, bolts, machine tool parts; oil hardened, tough.

KRUPP FM21.
M-72; 0.50 C, 1.0 Cr, 0.09 V, bal Fe.
For springs, gears, bolts, shafts; oil hardened, shock resistant.

KRUPP FM881.
M-72; 0.34-0.39 C, 1.6-1.9 Cr, 0.15 V, bal Fe.
For gears, springs; tough.

KRUPP FM1041.
M-72; 0.50 C, 1.0 Cr, 0.09 V, bal Fe.
For gears, bolts, springs, shafts; oil hardened, shock resistant.

KRUPP FM1251.
M-72; 0.6-0.68 C, 0.7-1.0 Si, 0.7-1.0 Mn, 1.1-1.3 Cr, 0.1-0.18 V bal Fe.
For tools, gears, springs; tough.

KRUPP FM2432.
M-72; 1.15-1.25 C, 0.6-0.8 Cr, 0.15 V, bal Fe.
For tools, bearings; water hardening.

KRUPP FMC 24.
M-72; 0.27 C, Cr, V, Mn, bal Fe.
For gears, bolts, machine tool parts; oil hardened, tough.

KRUPP FP12.
M-72; 0.34 C, 1.1 Al, 1.4 Cr, bal Fe.
For oil refinery equipment; creep and heat resistant.

KRUPP FP15.
M-72; 0.27 C, 1.1 Al, 1.4 Cr, bal Fe.
For oil refinery equipment; creep and heat resistant.

KRUPP FPE23.
M-72; 0.33 C, 1.1 Al, 1.7 Cr, 1.0 Ni, bal Fe.
For oil refinery equipment; creep and heat resistant.

KRUPP FPK13.
M-72; 0.32 C, 1.1 Al, 1.1 Cr, 0.18 Mo, bal Fe.
For oil refinery equipment; heat and creep resistant.

KRUPP GA 350.
M-72; 0.45 C, 0.35 Si, 2.1 Mn, 1.15 Cr, 0.2 Ti, bal Fe.
For hard facing electrodes; wear resistant.

KRUPP GA 500.
M-72; 0.9 C, 0.35 Si, 2.1 Mn, 1.3 Cr, 0.2 Ti, bal Fe.
For hard facing electrodes; wear resistant.

KRUPP GB 7 C.
M-72; 0.75 C, 0.6 Mn, 0.5 Cr, bal Fe.
For hand tools.

KRUPP HCT1.
M-72; 0.2 max C, 22-24 Cr, 12-15 Ni, 2 max Mn, 3.5 max Si, bal Fe.
Annealed: 85,000-95,000 TS; 40,000-50,000 YS; 45-55 El; 150-185 Brin.
For heat treating boxes, oil refinery and chemical plant equipment; Type 309; austenitic, heat resistant.

KRUPP HCT3.
M-72; 0.25 max C, 24-26 Cr, 19-22 Ni, bal Fe.
Annealed: 95,000 TS; 45,000 YS; 50 El; 65 RA; 180 Brin.
At 1200°F: 57,000 TS; 22,000 YS; 32 El; 45 RA.
For furnace parts and equipment, heat treating boxes; Type 310; austenitic, heat resistant.

KRUPP HGS.
M-72; 0.55-0.60 C, 0.6-0.8 Cr, 1.5-1.8 Ni, 0.15-0.20 Mo, bal Fe.
For forging die blocks; oil hardened, tough.

KRUPP HGS SPEZIAL.
M-72; 0.34 C, 0.68 Cr, 0.15 Mo, bal Fe.
For tools, gears, shafts; oil hardening.

KRUPP HGSE.
M-72; 0.55 C, 0.7 Mn, 0.8 Cr, 1.8 Ni, 0.5 Mo, bal Fe.
For forging die blocks; oil hardening.

KRUPP HIGH SPEED.
M-371; C, W, alloy, bal Fe.
For high speed tools and cutters; high speed steel.

KRUPP HM1.
M-72; C, bal Fe.
For tools.

KRUPP HM2.
M-72; C, bal Fe.
For tools.

KRUPP HM3.
M-72; 1.1 C, bal Fe.
For tools, drills, taps; water hardened.

KRUPP HM4.
M-72; 1.1 C, bl Fe.
For tools, drills, taps; water hardened.

KRUPP HM4 G.
M-72; 0.9-1.0 C, 0.5 Cr, bal Fe.
For tools, bearings, drills; water hardening.

KRUPP HSB 8-SO.
M-72; 0.72 C, 2.5 Ni, bal Fe.
Tool or mold steel.

KRUPP HSK.
M-72; 0.90 C, 0.25 -0.50 Si, 0.3-0.8 Mn, bal Fe.
Heat treated: 190,000 TS; 145,000 YS; 10 El; 30 RA; 400 Brin.
For drills, taps, springs, tools, reamers; Type W1; water hardened.

KRUPPIN.
M-72; 28 Ni, C, bal Fe.
For resistantce alloy; heat resistant.

KRUPP K.A.-2.
M-72; 7-10 Ni, 16-20 Cr, 0.75 Si, 0.15 C, bal Fe.
For heat and corrosion resisting parts; for high temperature use; heat and corrosion resistant.

KRUPP KFM.
M-72; 0.95 C, 0.3 Si, 0.3 Mn, 3.75 Cr, 1.3 W, 0.12 Ti, 2.8 V, 2.35 Mo, bal Fe.
For hard facing electrodes; wear resistant.

KRUPP KFM EXTRA.
M-72; 0.85 C, W, Mo, bal Fe.
For cutters, dies; oil or water hardened.

KRUPP KFM ZWEISTERN.
M-72; 0.95 C, W, Mo, bal Fe.
For cutters, dies; oil or water hardened.

KRUPP M 13 H.
M-72; 0.90 C, 13 Cr, bal Fe.
Martensitic stainless steel; for tools, stainless ball bearings; punches.

KRUPP M 131 CO.
M-72; 0.65 C, 14 Cr, 1 Co, Mo, V, bal Fe.
For surgical scissors and similar tools.

KRUPP M 131 V.
M-72; 0.79 C, 13 Cr, 2 V, Mo, bal Fe.
Cold work tool steel; for stamping and forming dies.

KRUPP M 171 W.
M-72; 0.68 C, 17 Cr, bal Fe.
Martensitic type stainless steel; hardenable to above 52 Rc.

KRUPP M202.
M-72; 0.9-1.0 C, 0.3 V, bal Fe.
For tools, drills, taps; water hardening.

KRUPP MINIMUM R.
M-72; 0.04 max C, 0.20 max Mn, 0.05-0.15 Al, bal Fe.
For magnet cores.
W. Nr. 1.1004.

KRUPP MINIMUM RA.
M-72; 0.04 max C, 0.12-0.20 Mn, bal Fe.
For magnet cores.
W. Nr. 1.1005.

KRUPP MOG 310.
M-72; 0.51 C, 3.5 Cr, 3.5 Mo, V, bal Fe.
Hot or cold work tool steel.

KRUPP N48.
M-72; 0.30 C, Cr, V, bal Fe.
For gears, bolts, machine tool parts; oil hardened, tough.

KRUPP N54.
M-72; 0.31 C, 2.35 Cr, 0.18 Mo, 0.13 V, bal Fe.
For gears, bolts, machine tool parts; oil hardened, tough.

KRUPP NCT1A.
M-72; 0.15 C, 2 Si, 19.5 Cr, 9.5 Ni, bal Fe.
Annealed: 80,000 TS; 35,000 YS; 55 El; 75 RA; 150 Brin.
For chemical plant equipment, tanks, mixers; Type 302; stainless, austenitic.

KRUPP NCT1A GUSS.
M-72; 0.4 C, 22 Cr, 9.5 Ni, bal Fe.
Cast: 85,000 TS; 45,000 YS; 35 El; 165 Brin.
For heat treat boxes, furnace parts, conveyors; Type HF; corrosion and heat resistant.

KRUPP NCT2 GUSS.
M-72; 0.4 C, Cr, Ni, Si, bal Fe.
For furnace parts, heat treat boxes; corrosion and heat resistant.

KRUPP NCT3.
M-72; 0.15 C, 24 Cr, 19 Ni, bal Fe.
Annealed: 100,000 TS; 45,000 YS; 50 El; 65 RA; 185 Brin.
For valves, pumps, turbine and jet parts; Type 310; stainless, austenitic.

KRUPP NCT3/133.
M-72; 20 Cr, 25 Ni, bal Fe.
For welding electrodes; stainless, austenitic.

KRUPP NCT3 GUSS.
M-72; 0.4 C, 25 Cr, 19 Ni, bal Fe.
Cast: 75,000 TS; 50,000 YS; 17 El; 170 Brin.
Aged: 85,000 TS; 50,000 YS; 10 El; 190 Brin.
For heat treat boxes, burner tips, hearth plates; Type HK; austenitic, stainless.

KRUPP NCT3A GUSS.
M-72; 0.4 C, 26 Cr, 14 Ni, bal Fe.
Cast: 75,000 TS; 50,000 YS; 15 El; 170 Brin.
For heat treat boxes, furnace parts, hearth plates; Type Hi; austenitic, heat resistant.

KRUPP NCT6.
M-72; 0.15 C, Ni, Cr, bal Fe.
For chemical plant equipment; stainless.

KRUPP NCT8.
M-72; 0.15 C, Ni, Cr, bal Fe.
For chemical plant equipment; stainless.

KRUPP NCT30.
M-72; 0.15 C, Ni, Cr, bal Fe.
For chemical plant equipment; stainless.

KRUPP NFKC.
M-72; 1.4 C, 0.2 Si, 0.3 Mn, 0.45 Cr, 3.25 W, 0.2 V, bal Fe.
For hard facing electrodes; wear resistant.

KRUPP NI 36.
M-72; 0.10 max C, 36.0 Ni, bal Fe.
Low co-efficient of thermal expansion.

KRUPP NI 48.
M-72; 0.05 max C, 48 Ni, bal Fe.
For glass to metal seals, or soft magnetic parts as magnet cores.

KRUPP NIROSTAGUSS 1.
M-72; 1-2 C, 24 min Cr, bal Fe.
Cast: 58,000 TS; 250 Brin.
For chemical equipment, HNO_3 and pulp industries; corrosion resistant.

KRUPP NIROSTAGUSS 2.
M-72; 1-2 C, 24 min Cr, bal Fe.
Cast: 65,000 TS; 300 Brin.
For chemical equipment, HNO_3 and pulp industries; corrosion resistant.

KRUPP NIROSTAGUSS 3.
M-72; 1-2 C, 24 min Cr, bal Fe.
Cast: 72,000-114,000 TS; 400-500 Brin.
For paper drying cylinders, chemical equipment; corrosion resistant.

KRUPP NITROPLAT.
M-72; composite sheet Krupp V2A on C steel.
For corrosion resistant parts; corrosion resistant.

KRUPP NON-DEFORMING.
M-371; C, alloy, bal Fe.
For tools, dies; non-deforming.

KRUPP P77.
M-72; 0.90 C, 1.9 Mn, 0.1 V, bal Fe.
For punches, dies, shears, upsetters; oil hardened, non-deforming.

KRUPP P141.
M-72; 0.85 C, W, Mo, Cr, V, bal Fe.
For tools, dies, cutters; oil hardened.

KRUPP P193.
M-72; C, Ni, bal Fe.
For jet engine components, turbine rotor blades; austenitic, heat resistant.

KRUPP P427.
M-72; 1.65 C, Cr, Co, bal Fe.
For blanking and forming dies; oil hardened, wear resistant.

KRUPP PFM.
M-72; 0.95-1.05 C, 0.3 Cr, 0.15 V, bal Fe.
For tools, drills, taps; water hardening.

KRUPP PRESSGUETE 260.
M-72; 0.16 max C, 0.50 max Si, 1.20 max Mn, Ti, bal Fe.
For beams, frame construction.
W.Nr. 1.8941; QSTE 260 N.

KRUPP PRESSGUETE 340.
M-72; 0.16 max C, 0.50 max Si, 1.50 max Mn, Ti or Cb or V, bal Fe.
For beams, frame construction.
W. Nr. 1.8945; QSTE 340 N.

KRUPP PRESSGUETE 380.
M-72; 0.18 max C, 0.50 max Si, 1.60 max Mn, 0.12-0.20 Ti, bal Fe.
For beams, frame construction.
W. Nr. 1.8950; QSTE 380 N.

KRUPP PRESSGUETE 420.
M-72; 0.20 max C, 0.50 max Si, 1.60 max Mn, 0.12-0.20 Ti, bal Fe.
For beams, frame construction.
W.Nr. 1.8952; QSTE 420 N.

KRUPP PRESSGUETE 460.
M-72; 0.21 max C, 0.50 max Si, 1.70 max Mn, 0.12-0.20 Ti, bal Fe.
For beams, frame construction.
W.Nr. 1.8955; QSTE 460 N.

KRUPP PRESSGUETE 500.
M-72; 0.22 max C, 0.50 max Si, 1.70 max Mn, 0.12-0.20 Ti, bal Fe.
For beams, frame construction.
W.Nr. 1.8957; QSTE 500 N.

KRUPP REZ EXTRA.
M-72; 0.35 C, 0.8 Cr, 0.55 Mo, bal Fe.
For tools, gears, shafts; oil hardening.

KRUPP RNI 24.
M-72; 0.15 max C, 1.0 max Mn, 1.0 max Si, 38.0 Ni, bal Fe.
For high temperature operation, and for glass to metal seals.
W.Nr. 1.3911.

KRUPP SC2448.
M-72; 1.3 C, 12 Mn, bal Fe.
For tools, wear resistant parts; non-deforming, austenitic.

KRUPP SK 135.
M-72; 1.35 C, 1.2 W, Cr, Mo, V, bal Fe.
Cold work tool steel.

KRUPP SKV 73.
M-72; 0.77 C, 3.0 W, Cr, V, bal Fe.
Tool steel.

KRUPP SKV 81.
M-72; 0.80 C, 0.7 W, V, bal Fe.
Cold work tool steel.

KRUPP SONDERSTAHL.
M-72; 0.12 max C, 0.40-1.0 Mn, 0.10-0.16 Ti, bal Fe.
For special applications.
W.Nr. 1.8882.

KRUPP SPECIAL.
M-371; C, alloy, bal Fe.
For tools; oil hardened.

KRUPP SPEZIAL S.
M-72; 0.45 C, Cr, Ni, bal Fe.
For machine tool parts; oil hardened.

KRUPP STE 47.
M-72; 0.20 max C, 1.40 Mn, 0.50 Ni, 0.35 Cu, 0.15 V, bal Fe.
Fine grain structural steel.
W.Nr. 1.8905.

KRUPP STE 690 V.
M-72; 0.15 C, 0.80 Mn, 0.85 Ni, 0.55 Cr, 0.50 Mo, V, Cu, B, bal Fe.
Steel for case hardening.
W.Nr. 1.8920.

KRUPP T 13 A.
M-72; 0.22 C, 14 Cr, S, bal Fe.
Martensitic type stainless steel; hardenable to above 150,000 psi TS.

KRUPP TNC 18 + MO.
M-72; 0.35 C, 4.5 Ni, 1.2 Cr, Mo, bal Fe.
For axles, shafts, truck and tractor parts.

KRUPP TT 17.
M-72; 0.16 C, 17 Cr, Ni, bal Fe.
Chromium stainless steel.

KRUPP TT 171 N.
M-72; 0.03 C, 16 Cr, 6 Ni, Mo, bal Fe.
Austenitic stainless steel; for cold drawn springs.

KRUPP TTSTE 47.
M-72; 0.20 C, 1.40 Mn, 0.50 Ni, 0.40 Cu, 0.15 V, bal Fe.
Fine grain structural steel.
W. Nr. 1.8915.

KRUPP TYPE METAL.
M-Ger.; 60 Pb, 18 Sb, 12 Sn, 4.7 Cu, 4.7 Ni, 1 Bi.
For type metal.

KRUPP UK17F EXTRA.
M-72; 0.1 max C, 17 Cr, 1.8 Mo, Ti = 7 x C, bal Fe.
For welded oil refinery and chemical plant equipment; corrosion resistant.

KRUPP V1M.
M-72; 0.15 C, 2 Ni, 15 Cr, 0.4 Mn, 0.5 Si, bal Fe.
Annealed: 85,000 TS; 55,000 YS; 30 El; 150 Brin.
Heat treated: 213,000 TS; 160,000 YS; 450 Brin.
For furnace equipment, cutlery, turbine parts; Type 431; corrosion resistant.

KRUPP V2A.
M-72; 23 Cr, 9.5 Ni, 0.4 C, bal Fe.
85,000-105,000 TS; 35,000 YS; 50 El; 150-200 Brin.
For heat and corrosion resisting parts, chemical equipment; heat and corrosion resistant; austenitic.

KRUPP V2AB SUPRA.
M-72; 0.1 max C, 18 Cr, 8.5 Ni, bal Fe.
Annealed: 80,000 TS; 35,000 YS; 55 El; 75 RA; 150 Brin.
For chemical plant equipment, tanks, mixers, filters; Type 302; stainless austenitic.

KRUPP V2AE.
M-72; 0.08 max C, 17-19 Cr, 8-11 Ni, Ti = 5 x C, bal Fe.
Normalized: 93,000 TS; 36,000 YS; 45 El; 60 RA; 165 Brin.
Annealed: 87,000 TS; 33,000 YS; 57 El; 73 RA; 155 Brin.
For chemical plant equipment, muffles, welded parts; Type 321; stainless austenitic.

KRUPP V2A EXTRA.
M-72; 0.1 C, 18 Cr, 8-9.5 Ni, Ti = 4 x C, bal Fe.
Annealed: 83,000 TS; 29,000 YS; 45 El; 55 RA.
For crucibles, autoclaves, chemical plant equipment; Type 321; stainless, Austenitic.

KRUPP-V2AFH.
 M-72; 0.10 C, 16-18 Cr, 6-8 Ni, bal Fe.
 Annealed: 110,000 TS; 40,000 YS; 60 El; B 85 Rock.
 For aircraft structural members, railroad cars, household utensils, diaphragms, springs.
 Type 301 stainless steel, austenitic, non-hardenable.

KRUPP V2A NIROSTA.
 M-72; 0.2 C, 20 Cr, 7 Ni, bal Fe.
 Annealed: 85,000 TS; 35,000 YS; 60 El; 70 RA; 150 Brin.
 Rolled: 125,000 TS; 95,000 YS; 25 El; 55 RA; 277 Brin.
 For oil refinery and chemical plant equipment; Type 302; stainless, austenitic.

KRUPP V2A NORMAL.
 M-72; 0.11-0.16 C, 18 Cr, 8 Ni, bal Fe.
 For chemical plant equipment, mixers, Type 302; stainless, austenitic.

KRUPP V2AS.
 M-72; 0.08 max C, 18-20 Cr, 8-11 Ni, 2 max Mn, bal Fe.
 Annealed: 80,000 TS; 45,000 YS; 60 El; 135 Brin.
 Cold drawn: 180,000 TS; 150,000 YS; 10 El; 330 Brin.
 For chemical plant equipment, welded structures; Type 304; stainless, austenitic.

KRUPP V2A SPECIAL.
 M-72; 0.08 C, 18 Cr, 8 Ni, bal Fe.
 Annealed: 90,000 TS; 40,000 YS; 40 El; 60 RA; 160 Brin.
 For chemical plant equipment, architectural trim; Type 304; stainless, austenitic.

KRUPP V2A SUPRA.
 M-72; 0.07 max C, 18 Cr, 9.5 Ni, bal Fe.
 Annealed: 78,000 TS; 31,000 YS; 45 El; 60 RA; 145 Brin.
 For chemical and nitric acid plant equipment; Type 304; stainless, austenitic.

KRUPP V2AX EXTRA.
 M-72; 0.12 max C, 18 Cr, 9.5 Ni, Cb = 8 x C, bal Fe.
 Annealed: 90,000 TS; 45,000 YS; 56 El; 65 RA; 160 Brin.
 For welded chemical plant equipment; Type 347; stainless, austenitic.

KRUPP V3M.
 M-72; 0.4 C, 14 Cr, 0.5 Ni, bal Fe.
 Heat treated: 256,000 TS; 190,000 YS; 6 El; 10 RA; 540 Brin.
 For cutlery, razors, knives; Type 420; corrosion resistant.

KRUPP V3ME.
 M-72; 0.75-0.95 C, 16-18 Cr, 0.75 max Mo, bal Fe.
 Annealed: 107,000 TS; 18 El; 35 RA; 220 Brin.
 Heat treated: 280,000 TS; 270,000 YS; 3 El; 15 RA; 555 Brin.
 For cutlery, valves, bearings, instruments; Type 440B; corrosion resistant.

KRUPP V3M EXTRA.
 M-72; 1.0 C, 12-18 Cr, bal Fe.
 Annealed: 210 Brin.
 Heat treated: 620 Brin.
 For ball and roller bearings, disks for meat grinders; corrosion resistant.

KRUPP V3MS.
 M-72; 0.15 max C, 12-14 Cr, 1.25-2.5 Ni, 0.07 min P, S, or Se, bal Fe.
 Annealed: 75,000 TS; 40,000 YS; 30 El; 60 RA; 155 Brin.
 Cold drawn: 100,000 TS; 85,000 YS; 13 El; 50 RA; 205 Brin.
 For screw machine products, valve trim, pump shafts; Type 416; free-cutting, stainless.

KRUPP V4A.
 M-72; 0.07 max C, 10.5 Ni, 18 Cr, 2.5 Mo, bal Fe.
 Rolled: 85,000-108,000 TS; 36,000-70,000 YS; 50-45 El; 150-185 Brin.
 For chemical and textile plant equipment; Type 316; stainless, austenitic.

KRUPP V4AB SUPRA.
 M-72; 0.1 C, 18 Cr, 9.5 Ni, 2 Mo, bal Fe.
 Annealed: 85,000 TS; 35,000 YS; 50 El; 65 RA; 160 Brin.
 For acid resistant chemical plant equipment; Type 316; stainless, austenitic.

KRUPP V4 AE-TI.
 M-72; 0.10 max C, 16-18 Cr, 10-14 Ni, 1.7-2.7 Mo, Cb = 10 x C, bal Fe.
 Annealed: 85,000-95,000 TS; 35,000-45,000 YS; 60-50 El; 75-60 RA; 150-190 Brin.
 For acid resistant chemical plant equipment; Type 316 Cb; stainless, austenitic.

KRUPP V4A EXTRA.
 M-72; 0.12 max C, 18 Cr, 10 Ni, 2 Mo, Ti = 4 x C, bal Fe.
 Annealed: 85,000 TS; 39,000 YS; 50 El; 60 RA; 160 Brin.
 For sulphide, paper and pulp and textile industries; stainless, austenitic; Type 316 Ti.

KRUPP V4AS.
 M-72; 0.10 max C, 16-18 Cr, 10-14 Ni, 2-3 Mo, bal Fe.
 Annealed: 85,000-95,000 TS; 35,000-45,000 YS; 60-50 El; 75-60 RA; 150-190 Brin.
 For chemical plant equipment, mixers, agitators, filters; Type 316; austenitic, stainless.

KRUPP V4A SUPRA.
M-72; 0.12 max C, 18 Cr, 8 Ni, 2.5 Mo, bal Fe.
Annealed: 78,000 TS; 32,000 YS; 50 El; 60 RA; 140 Brin.
For sulphide, paper and pulp industries, textile equipment; Type 316; stainless, austenitic.

KRUPP V4AX EXTRA.
M-72; 0.12 max C, 18 Cr, 2 Mo, 10.5 Ni, Cb = 8 x C, bal Fe.
Annealed: 85,000 TS; 35,000 YS; 45 El; 60 RA; 165 Brin.
For welded acid resistant chemical plant equipment; Type 318; stainless, austenitic.

KRUPP V5M.
M-72; 0.2 C, 13 Cr, bal Fe.
Annealed: 95,000 TS; 50,000 YS; 25 El; 55 RA; 195 Brin.
For turbine blades, valves, cutlery, surgical instruments; Type 420; corrosion resistant.

KRUPP V6A STEEL.
M-72; 20 Cr, 7 Ni, 0.2 C, 2.5 Cu, bal Fe.
For chemical equipment; corrosion resistant; resists NH_4Cl.

KRUPP V8A EXTRA.
M-72; 0.12 max C, 18 Cr, 10.5 Ni, 2 Mo, Ti = 4 x C, bal Fe.
Annealed: 85,000 TS; 35,000 YS; 45 El; 60 RA; 165 Brin.
For welded acid resistant chemical plant equipment; Type 316 Ti; stainless, austenitic.

KRUPP V10A.
M-72; 0.12 C, 8 Mn, 0.8 Si, 19.5 Cr, 9 Ni, bal Fe.
Welded: 85,000 TS; 45,000 YS; 45 El; 45 RA.
For welding rod; stainless, austenitic.

KRUPP V12A NORMAL.
M-72; 0.10 max C, 12.5 Cr, 12 Ni, bal Fe.
For valves, cutlery; corrosion resistant.

KRUPP V12A SUPRA.
M-72; 0.08 C, 18 Cr, 8 Ni, bal Fe.
Annealed: 78,000 TS; 31,000 YS; 50 El; 60 RA; 140 Brin.
For architectural trim, tanks, vessels, stainless, austenitic.

KRUPP V13F.
M-72; 0.12 C, 13 Cr, 0.2 Ni, bal Fe.
Heat treated: 150,000 TS; 20 El; 250 Brin.
For tableware, instruments; Type 403; corrosion resistant.

KRUPP V13FA1.
M-72; 0.05 C, 11.5-14.5 Cr, 0.1-0.3 Al, bal Fe.
Annealed: 70,000 TS; 40,000 YS; 30 El; 160 Brin.
For annealing boxes, quenching racks, furnace parts, oil refinery equipment.
Type 405 stainless steel, heat and corrosion resistant.

KRUPP V14AS.
M-72; 0.10 max C, 18-20 Cr, 11-14 Ni, 3-4 Mo, bal Fe.
Annealed: 85,000-95,000 TS; 35,000-45,000 YS; 60-50 El; 75-60 RA; 150-190 Brin.
For acid resistant chemical plant equipment; Type 317; stainless, austenitic.

KRUPP V14A SUPRA.
M-72; 0.10 max C, 18-20 Cr, 10-14 Ni, 2-3 Mo, bal Fe.
Annealed: 85,000 TS; 31,000 YS; 50 El; 60 RA; 140 Brin.
For hypochlorite plant equipment; stainless, austenitic; Type 316.

KRUPP V15F.
M-72; 0.08 C, Cr, bal Fe.
For corrosion resistant parts; corrosion resistant.

KRUPP V16A EXTRA.
M-72; 0.07 max C, 17.5 Cr, 17.5 Ni, 2 Mo, Ti = 7 x C, bal Fe.
For valves, acid resistant equipment, pumps; stainless, austenitic.

KRUPP V16 SUPRA.
M-72; 0.15 C, 18 Cr, 8 Ni, bal Fe.
Rolled: 78,000-101,000 TS; 55-65 El; 140-175 Brin.
For stainless parts, chemical plant equipment; austenitic, stainless.

KRUPP V16A SUPRA.
M-72; 0.10 max C, 18-20 Cr, 10-14 Ni, 3-4 Mo, bal Fe.
Annealed: 90,000 TS; 35,000 YS; 50 El; 60 RA; 160 Brin.
For chemical plant equipment; Type 317; stainless, austenitic.

KRUPP V17F.
M-72; 0.1 C, 17 Cr, 0.2 Ni, 0.2 Ti, bal Fe.
Annealed: 78,000-93,000 TS; 50,000-55,000 YS; 20-18 El; 137-150 Brin.
For chemical plant equipment; Type 430; corrosion resistant.

KRUPP V17F EXTRA.
M-72; 0.1 C, 17.5 Cr, Ti = 7 x C, bal Fe.
Annealed: 95,000 TS; 50,000 YS; 25 El; 55 RA; 200 Brin.
For welded corrosion resistant parts; Type 430 Ti, corrosion resistant.

KRUPP V17FS.
M-72; 0.12 max C, 14-18 Cr, 0.07 min S or Se, bal Fe.
Annealed: 75,000-85,000 TS; 40,000-55,000 YS; 30-25 El; 137-150 Brin.
For chemical plant equipment, screw machine products; Type 430F; stainless, free-cutting.

KRUPP V17FX EXTRA.
M-72; 0.08 C, 17 Cr, 8 Ni, 0.6 Cb, bal Fe.
Annealed: 85,000 TS; 40,000 YS; 50 El; B 82 Brin.
For welded structures, agitators, evaporators, tanks, digesters, chemical plant equipment.
Type 17-8 Cb stainless steel, austenitic, stabilized.

KRUPP V44A SUPRA.
M-72; 0.06 C, 17 Cr, 12 Ni, 2-3 Mo, bal Fe.
Annealed: 85,000 TS; 35,000 YP; 50 El; B 80 Rock.
For chemical plant equipment, tanks, evaporators, digesters.
Type 316 stainless steel, austenitic.

KRUPP VCM 21-10.
M-72; 0.60 C, 21 Cr, 10 Mn, Nb, Mo, V, bal Fe.
For exhaust valves.

KRUPP VCMO 18 3.
M-72; 0.85 C, 17.5 Cr, 2.25 Mo, 0.50 V, bal Fe.
For valves.
W. Nr. 1.4748.

KRUPP VCN 21-2.
M-72; 0.55 C, 8 Mn, 21 Cr, 2 Ni, 0.30 N, bal Fe.
For automotive exhaust valves.
W.Nr. 1.4785.

KRUPP VCN 21-4.
M-72; 0.53 C, 21 Cr, 4 Ni, 9 Mn, 0.50 N, bal Fe.
For exhaust valves.
W.Nr. 1.4871.

KRUPP VCN 21-7.
M-72; 0.70 C, 6.5 Mn, 21 Cr, 1.7 Ni, 0.23 N bal Fe.
For exhaust valves.
W.Nr. 1.4881.

KRUPP VCN 188.
M-72; 0.45 C, 2.5 Si, 18 Cr, 9 Ni, 1 W, bal Fe.
For automotive and aviation exhaust valves.
W.Nr. 1.4873.

KRUPP VCS 2.
M-72; 0.45 C, 4.0 Si, 3.0 Cr, bal Fe.
For valves.
W.Nr. 1.4704.

KRUPP VCS 9.
M-72; 0.45 C, 3.25 Si, 9.0 Cr, bal Fe.
For exhaust valves.
W.Nr. 1.4718

KRUPP VCS 10.
M-72; 0.40 C, 2.5 Si, 10.0 Cr, 1.0 Mo, bal Fe.
For exhaust valves.
W.Nr. 1.4731.

KRUPP VCS 15.
M-72; 0.80 C, 15 Cr, 1.0 Mo, 0.75 Ni, 1.0 W, bal Fe.
For automotive valves.
W.Nr. 1.4732.

KRUPP VCS 20.
M-72; 0.80 C, 2.25 Si, 20 Cr, 1.5 Ni, bal Fe.
For automotive exhaust valves.
W.Nr. 1.4747.

KRUPP VK3M EXTRA.
M-72; 0.90 C, Mo, Cr, V, W, bal Fe.
For lathe and planer tools, reamers; high speed steel.

KRUPP VK5M.
M-72; 0.20 C, 13 Cr, 1.15 Mo, bal Fe.
For valves, cutlery, surgical and dental instruments; Type 420 Mo; stainless.

KRUPP V M STEEL.
M-72; 11-15 Cr, low Ni, bal Fe.
For chemical equipment; corrosion resistant.

KRUPP W 6.
M-72; 1.0 C, 1.75 C, 1.25 Mo, bal Fe.
For shears, chisels, knives.

KRUPP W 53.
M-72; 0.53 C, 0.30 Si, 0.70 Mn, bal Fe.
Tool steel for hand tools as hammers, shears, screwdrivers.

KRUPP W 93.
M-72; 0.90 C, 0.70 Mn, bal Fe.
For knives, hand saws, shears.

KRUPP WA.
M-72; 0.4 C, 1.2 Cr, 0.4 Mo, 0.75 V, 0.4 W, bal Fe.
For tools, gears, shafts; water hardening.

KRUPP WA5 M.
M-72; 0.17-0.22 C, 12.5-13.5 Cr, bal Fe.
For corrosion resistant parts; corrosion resistant.

KRUPP WA100.
M-72; 0.43 C, 2.3 Si, 1.0 Mn, 17.5 Cr, 8.5 Ni, 0.9 Ti, bal Fe.
For stainless parts; corrosion resistant, austenitic.

KRUPP WA 342.
M-72; 0.45 C, 1.0 Mn, 1.7-2.0 Cr, 0.2 Mo, bal Fe.
For forging die blocks; oil hardened.

KRUPP WA402.
M-72; 0.45-0.50 C, 0.9-1.1 Mn, 1.1-1.3 Cr, 0.2 Mo, bal Fe.
For forging die blocks; oil hardened.

KRUPP WA594.
M-72; 0.36-0.32 C, 0.8-1.0 Si, 2.2-2.5 Cr, 0.4 V, bal Fe.
For dies, punches; shock resistant.

KRUPP WA650.
M-72; 0.3-038 C, 1.4-1.6 Si, 1.2-1.5 Cr, 0.1-0.2 V, bal Fe.
For tools, punches, dies; shock resistant.

KRUPP WA904.
M-72; 0.25-0.30 C, 1.0-1.2 Cr, 0.2 V, 3.5-4.0 W, bal Fe.
For tools, hot work dies; hot work steel.

KRUPP WA904 EXTRA.
M-72; 0.30 C, 2.35 Cr, 0.6 V, 4.25 W, bal Fe.
For pneumatic tools, upsetters, shear blades; oil hardened, tough.

KRUPP WA930 N.
M-72; 0.26 C, 9 W, 3 Cr, Ni, V, bal Fe.
For die casting molds.

KRUPP WAGS.
M-72; 0.36 C, 0.95 Si, 0.45 Mn, 0.025 P, 0.019 S, 5 Cr, 1.37 Mo, 0.33 V, bal Fe.
For extrusion press dies; tough, heat resistant.

KRUPP WAM03.
M-72; 0.29 C, 0.33 Si, 0.36 Mn, 2.91 Cr, 3.05 Mo, 0.53 V, bal Fe.
For extrusion press dies; tough, heat resistant.

KRUPP WB97 M.
M-72; 0.42-0.48 C, 1.0-1.3 Si, 0.5-0.7 Mn, bal Fe.
For tools, springs, punches; shock resistant.

KRUPP WB136 M.
M-72; 0.62-0.68 C, 1.6 Si, 0.6 Mn, bal Fe.
For tools, springs, punches; shock resistant.

KRUPP WBC1255.
M-72; 0.6-0.7 C, 1.0-1.3 Si, 0.9-1.2 Mn, bal Fe.
For tools, springs, punches; shock resistant.

KRUPP WBF1552.
M-72; 0.67-0.73 C, 1.2-1.4 Si, 0.4-0.55 Cr, bal Fe.
For tools, punches, dies; oil hardening.

KRUPP WC 128.
M-72; 0.7 C, 1.9 Si, 0.07 S, bal Fe.
For tools, chisels, dies; tough.

KRUPP WCV 3.
M-72; 0.48 C, 0.90 Cr, V, bal Fe.
Cold work tool steel. Similar to AISI 6150.

KRUPP WCV 4 SPEZIAL.
M-72; 0.56 C, 1.0 Cr, Mo, V, bal Fe.
For axes, chisels, hand tools.

KRUPP WF 3.
M-72; 0.53 C, 1.75 Mn, bal Fe.
For hand tools.

KRUPP WF33.
M-72; 0.12-0.18 C, 0.4-0.6 Mn, 0.6-0.8 Cr, bal Fe.
For gears, shafts; case hardening.

KRUPP WF50.
M-72; 0.55 C, Ni, Cr, W, bal Fe.
For forging and heading dies; shear blades; oil hardened, tough.

KRUPP WF100.
M-72; 0.45 C, Cr, Ni, W, bal Fe.
For forging and heading dies, punches; oil hardened, tough.

KRUPP WF100 D.
M-72; 0.38 C, 0.52 Mn, 1.84 Si, 14.8 Cr, 12.9 Ni, 0.23 Mo, 2.5 W, bal Fe.
For valves; high heat and corrosion resistance.

KRUPP WF202.
M-72; 1.0-1.15 C, 0.3-0.5 Cr, bal Fe.
For tools, drills, taps; water hardening.

KRUPP WF204.
M-72; 1.0-1.1 C, 0.8-1.0 Cr, bal Fe.
For tools, drills, taps; water hardening.

KRUPP WF206.
M-72; 0.95-1.05 C, 1.25-1.5 Cr, bal Fe.
For tools, bearings; wear resistant.

KRUPP WFC27.
M-72; 0.14-0.19 C, 1-13 Mn, 0.8-1.1 Cr, bal Fe.
For tools, gears, shafts; case hardening.

KRUPP WFC29.
M-72; 0.17-0.22 C, 1.1-1.4 Mn, 1-1.3 Cr, bal Fe.
For tools, dies, bearings; case hardening.

KRUPP WFC31.
M-72; 0.20-0.25 C, 1.3-1.6 Mn, 1.2-1.5 Cr, bal Fe.
For tools, bearings; case hardening.

KRUPP WFF24.
M-72; 0.15 C, 1.0-1.3 Si, 23-25 Cr, 2 Al, bal Fe.
For corrosion resistant parts; corrosion resistant.

KRUPP WFM48.
M-72; 0.28-0.32 C, 2.2-2.5 Cr, 0.2 V, bal Fe.
For tools, dies.

KRUPP WFM961.
M-72; 0.43-0.48 C, 1.3-1.5 Cr, 0.2 V, bal Fe.
For tools, gears, springs; tough.

KRUPP WFM1041.
M-72; 0.47-0.55 C, 0.8-1.0 Mn, 1.2 Cr, 0.18 V, bal Fe.
For tools, springs, gears; shock resistant.

KRUPP WFM1141.
M-72; 0.55-0.62 C, 0.9-1.1 Mn, 1.2 Cr, 0.18 V, bal Fe.
For tools, gears, springs; shock resistant.

KRUPP WFP12.
M-72; 0.30-0.35 C, 1.3-1.5 Cr, 1-1.2 Al, bal Fe.
For tools, gears, shafts; oil hardening.

KRUPP WK 143 C.
M-72; 0.75 C, 1.35 W, Ni, Mo, bal Fe.
Tool or mold steel.

KRUPP WM 131 V.
M-72; 0.79 C, 13 Cr, 2 V, Mo, bal Fe.
Cold work tool steel, or mold steel.

KRUPP WN48.
M-72; 0.28-0.33 C, 2.2-2.5 Cr, 0.2 V, bal Fe.
For tools, dies.

KRUPP WSTE 47.
M-72; 0.20 max C, 1.50 Mn, 0.80 max Ni, 0.70 max Cu, Ti, V, Nb, bal Fe.
Fine grain structural steel.
W.Nr. 1.8935.

KRUPP WT10.
M-72; C, Si, Mn, bal Fe.

KRUPP WW.
M-72; 0.4 C, 0.01 Si, 0.14 Mn, 0.05 Cr, 0.15 Ni, bal Fe.
For tools, hubbing dies; case hardened.

KRUPP WX 25.
M-72; 0.25 C, 4.0 Ni, 1.25 Cr, Mo, bal Fe.
Hot work tool or mold steel.

KRUPP WX 70.
M-72; 0.72 C, 1.0 Si, Ni, Mo, V, bal Fe.
Tool steel, shock resisting.

KRUPP ZF.
M-72; 0.85 C, 0.25 max Si, 025 max Mn, bal Fe.
Heat treated: 190,000 TS; 145,000 YS; 10 El; 30 RA; 400 Brin.
For springs, taps, hobs, reamers, broaches; Type W1; water hardened.

KS 1.
M-912; 0.08 C, 0.12 Si, 0.5 Mn, bal Fe.
For welded chains.
W.-Nr. 0208.

KS 1 R.
M-912; 0.08 C, 0.12 Si, 0.5 Mn, bal Fe.
For welded chains.
W.-Nr. 0209.

KS 2.
M-912; 0.16 C, 0.15 Si, 0.80 Mn, bal Fe.
For welded chains.
W.-Nr. 0847.

KS 3.
M-912; 0.21 C, 0.15 Si, 1.0 Mn, bal Fe.
For welded chains.
W. Nr. 0848.

KS 4.
M-912; 0.21 C, 0.40 Si, 1.4 Mn, bal Fe.
For welded chains.
W.-Nr. 0849.

KS 5.
M-912; 0.27 C, 0.40 Si, 1.4 Mn, bal Fe.
For welded chains.
W.-Nr. 0840.

KS 9.
M-912; 0.23 C, 0.20 Si, 1.5 Mn, 0.35 Cr, 0.27 Mo, 0.62 Ni, bal Fe.
Ht Tr: 1080 min N/mm^2 TS.
For strong welded chains.
W.-Nr. 6542.

KS 10.
M-912; 0.24 C, 0.20 Si, 1.5 Mn, 0.27 Cr, 0.52 Mo, 1.05 Ni, bal Fe.
Ht Tr: 1080 min N/mm^2 TS.
For strong, heavy, welded chains.
W.-Nr. 6753.

KS 13.
M-English; 6-8 Sb, bal Al.
For bearings.

KS 1275 KOLBENLEG.
M-English; 1 Mg, bal Al.
For light alloy parts; corrosion resistant.

KS ALUSIL.
M-196; 20 Si, 1.5-2.0 Cu, 0.5 Ni, bal Al.
23,000 TS; 21,000 YS; 0.25 El; 100 Brin.
For pistons for general use; low temperature coefficient; coefficient of expansion 0.000018.

KS MAGNET STEEL.
M-Japan; 30-40 Co, 1.5-3.0 Cr, 5-9 W, 0.35 Mn, 0.12 Si, 0.4-0.8 C, bal Fe.
Hardened: 444-652 Brin.
For tools, magnets; corrosion and heat resistant; Br 10,000-10,500; Hc-200-240.

KS MAGNET STEEL.
M-Japan; 0.7-1.0 C, 6-8 W, 1-2 Co, bal Fe.
For magnets; magnetos; high permeability.

KS NO. 83A.
M-196; Al alloy.
For light alloy parts.

KS NO. 245.
M-196; 14 Si, 4.5 Cu, 1.5 Ni, 0.7 Mg, 1.0 Mn, bal Al.
29,000 TS; 25,000 YS; 0.2 El; 130 Brin.
For pistons for small diesel motors and motorcycles; high wear resistance; coefficient of expansion 0.00002.

KS NO. 280.
M-196; 21 Si, 1.5 Cu, 0.7 Mn, 1.5 Ni, 0.5 Mg, 1.2 Co, bal Al.
27,000 TS; 26,000 YS; 0.25 El; 120 Brin.
For pistons for air-cooled motors; coefficient of expressions 0.000018.

KS NO. 280B.
M-196; Si, bal Al.
For light alloy parts; corrosion resistant.

KS NO. 282.
M-196; 23-25 Si, 0.81.3 Cu, 0.8-1.3 Ni, 0.2 Ti, 0.2 Mn, 0.3-0.5 Co, 0.7 Fe, 0.2 Zn, 0.8-1.3 Mg, bal Al.
For pistons in internal combustion engines. Medium thermal stressing, corrosion resistant.

KS NO. 283.
M-196; 17-19 Si, 4.7-5.3 Cu, 3.8-4.2 Ni, 0.1 Ti, 0.6-0.8 Mn, 0.5-0.7 Co, 0.7 Fe, 0.2 Zn, 0.4-0.6 Mg, bal Al.
For pistons in internal combustion engines. High thermal stressing, corrosion resistant.

KS NO. 411B.
M-196; Si, bal Al.
For light alloy parts; corrosion resistant.

KS NO. 837.
M-196; Si, bal Al.
For light alloy parts; corrosion resistant.

KS NO. 1275.
M-196; Si, Cu, Ni, bal Al.
For light alloy parts; corrosion resistant.

KS RED.
M-English; 16.5 Cu, 0.8 Ni, 0.5 Fe, 0.5 Si, bal Al.
For pistons; cast.

KS SEEWASSER.
M-196, M-116; 1.3 Mn, 2.2 Mg, 0.2 Sb, 0.7 Si, bal Al.
Sand cast: 26,000 TS; 12,800 YS; 2.5 El; 5 RA; 60 Brin.
Die cast: 32,720 TS; 12,800 YS; 1.6 El; 81 Brin.
Rolled: 45,500 TS; 36,000 YS; 2 El; 35 RA; 83 Brin.
For furniture, interior light fixtures, wire castings, ship parts, chemical industry; light alloy; resists sea water corrosion.

KS SEEWASSER JUB.
M-Japan; Al alloy.
For light alloy parts.

KS SPECIAL PISTON ALLOY.
M-196; 15 Cu, 0.6 Si, 0.6 Fe, 0.6 Ni, 0.3 Mg, bal Al.
26,000 TS; 0.2 El; 120 Brin.
For pistons for Zeppelin motors and diesel engines; coefficient of expansion 0.000024.

KS "Y" ALLOY.
M-196; 4 Cu, 0.4 Si, 2 Ni, 0.4 Fe, 1.5 Mg, bal Al.
35,000 TS; 28,500 YS; 1.0 El; 90 Brin.
For pistons for steam engines; coefficient of expansion 0.0000245.

K-SPUN.
M-399; 2.85-3.50 T.C., 0.95-1.45 Si, 0.4-0.8 Mn, bal Fe.
Heat treated: 70,000 TS; 285 Brin.
For piston rings, cylinder liners; wear resistant cast iron.

K-TEN 60.
M-1604; 0.14 C, 0.38 Si, 1.26 Mn, 0.015 P, 0.010 S, 0.20 Ni, 0.05 Mo, 0.04 bal Fe.
32 mm thick plate: 94,000 TS; 78,000 YS; 45 El.
High strength low alloy steel for structural purposes.

K-TEN 70.
M-1604; 0.12 C, 0.31 Si, 1.03 Mn, 0.013 P, 0.010 S, 0.21 Cu, 0.62 Ni, 0.31 Cr, 0.33 Mo, 0.24 V, bal Fe.
50 mm thick plate: 110,000 TS; 99,000 YS; 27 El.
High strength low alloy steel for structural purposes.

K-TEN 80.
M-1604; 0.13 C, 0.32 Si, 0.86 Mn, 0.009 P, 0.0010 S, 0.24 Cu, 0.92 Ni, 0.45 Cr, 0.40 Mo, 0.03 V, 0.002 B, bal Fe.
50 mm thick plate: 118,000 TS; 110,000 YS; 44 El.
High strength low alloy steel for structural purposes.

KU-112.
M-1517; 90 W, bal Ni-Cu.
Density 17; 110,000 TS; 70,000 YS; 5 El; C 25 Rock.
High density alloy for counterweights, balancing and shielding applications.

KU-112-18.
M-1517; 95 W, bal Ni-Cu.
Density 18; 105,000 TS; 90,000 YS; 3 El; C 28 Rock.
High density alloy for counterweights, balancing and shielding applications.

KU A-5.
M-1517; 95 W, bal Ni-Fe.
Density 18; 120,000 TS; 92,000 YS; 12 El; C 30 Rock.
High density alloy for counterweights, balancing and shielding applications.

KU A-10.
M-1517; 90 W, bal Ni-Fe.
Density 17; 125,000 TS; 90,000 YS; 15 El; C 27 Rock.
High density alloy for counterweights, balancing and shielding applications.

KUFIL.
M-282; 99 Cu, 1 Ag.
Welded: 29,000 TS.
For welding rod for Cu; free flowing.

KUHBIER A12.
M-1327; 0.10 max C, 12.5 Cr, 12 Ni, bal Fe.
For chemical plant equipment, valves; heat and corrosion resistant.

KUHBIER A18.
M-1327; 0.15 max C, 18 Cr, 9 Ni, bal Fe.
Annealed: 80,000 TS; 35,000 YS; 55 El; 75 RA; 150 Brin.
For chemical plant equipment, tanks; Type 302; stainless, austenitic.

KUHBIER F13.
M-1327; 0.12 max C, 0.4 Si, 13 Cr, bal Fe.
Annealed: 75,000 TS; 40,000 YS; 35 El; 70 RA; 155 Brin.
For valves, cutlery, surgical instruments; Type 410; stainless.

KUHBIER F17.
M-1327; 0.8 C, 17 Cr, bal Fe.
Annealed: 107,000 TS; 62,000 YS; 18 El; 35 RA; 220 Brin.
For bearings, instrument pivots, liners; corrosion resistant, hardenable.

KUHBIER F17A.
M-1327; 0.12 C, 16.5 Cr, 0.25 Mo, 0.2 S, bal Fe.
Annealed: 80,000 TS; 50,000 YS; 20 El; 40 RA; 150 Brin.
For screw machine products, shafts; free-cutting; Type 430 Mo; corrosion resistant.

KUHBIER F17T.
M-1327; 0.1 max C, 17.5 Cr, Ti = 7 X C, bal Fe.
Annealed: 80,000 TS; 50,000 YS; 25 El; 50 RA; 150 Brin.
For oil refinery equipment, welded structures; corrosion and creep resistant.

KUHBIER F19T.
M-1327; 0.1 max C, 17 Cr, 1.8 Mo, Ti = 7 X C, bal Fe.
Annealed: 125,000 TS; 95,000 YS; 20 El; 55 RA; 260 Brin.
For oil refinery equipment, welded structures; corrosion and creep resistant.

KUHBIER I.
M-1327; 0.15 C, 24 Cr, 19 Ni, bal Fe.
Annealed: 100,000 TS; 45,000 YS; 50 El; 65 RA; 185 Brin.
For furnace parts, heat treat boxes; heat resistant, austenitic.

KUHBIER IA.
M-1327; 0.15 C, 19.5 Cr, 9.5 Ni, bal Fe.
Annealed: 80,000 TS; 35,000 YS; 55 El; 75 RA; 150 Brin.
For chemical plant equipment, tanks, vessels; Type 302; stainless, austenitic.

KUHBIER M13.
M-1327; 0.2 C, 0.4 Si, 13 Cr, bal Fe.
Annealed: 95,000 TS; 50,000 YS; 25 El; 55 RA; 195 Brin.
For turbine blades, valves, cutlery, surgical instruments; Type 420; stainless.

KUHBIER M13.4.
M-1327; 0.4 C, 0.4 Si, 13 Cr, bal Fe.
Annealed: 100,000 TS; 55,000 YS; 20 El; 50 RA; 210 Brin.
For surgical instruments, valves, cutlery, knives; Type 420; stainless.

KUHBIER M15.
M-1327; 0.20 C, 0.4 Si, 13 Cr, 1.15 Mo, bal Fe.
Annealed: 100,000 TS; 55,000 YS; 20 El; 50 RA; 210 Brin.
For turbine blades, valves, cutlery, knives; Type 420 Mo; stainless.

KUHBIER M18.
M-1327; 0.22 C, 0.4 Si, 17 Cr, 1.5 Ni, bal Fe.
Annealed: 125,000 TS; 95,000 YS; 20 El; 55 RA; 260 Brin.
For pumps, marine hardware, valves, gauges; Type 431; corrosion and heat resistant.

KUHBIER M18.9.
M-1327; 0.90 C, 18 Cr, 1.15 Mo, 1.0 V, bal Fe.
Annealed: 110,000 TS; 65,000 YS; 17 El; 32 RA; 230 Brin.
For cutlery, valves, bearings, surgical instruments; corrosion and wear resistant.

KUHBIER M19.
M-1327; 0.35 C, 16.5 Cr, 1.15 Mo, bal Fe.
For chemical plant and oil refinery equipment; corrosion and heat resistant.

KUHNE PHOSPHOR BRONZE.
M-Eng.; 78 Cu, 11 Sn, 10 Pb, 0.3 Ni, 0.6 P.
For hard bearings; heavy duty.

KUKI EXTRA MH.
M-1320; 1.1 C, 0.25 max Si, 0.25 max Mn, bal Fe.
Annealed: 100,000 YS; 53,000 YS; 21 El; 42 RA; 200 Brin.
For drills, hobs, taps, springs; Type W1; water hardened.

KUKI EXTRA ZAH.
M-1320; 0.85 C, 0.25 max Mn, 0.25 max Si, bal Fe.
Heat treated: 190,000 TS; 145,000 YS; 12 El; 35 RA; 320 Brin.
For drills, punches, cutters, springs; Type W1; water hardened.

KUKI EXTRA ZH.
M-1320; 1.0 C, 0.25 max Si, 0.25 max Mn, bal Fe.
Annealed: 100,000 TS; 53,000 YS; 21 El; 42 RA; 200 Brin.
For drills, taps, reamers, cutting tools; Type W1; water hardened.

KUKI PRIMA H.
M-1320; 1.3 C, 0.25 max Si, 0.25 max Mn, bal Fe.
For cutters, reamers, broaches; Type W1; water hardened.

KUKI PRIMA MH.
M-1320; 1.15 C, 0.25 max Mn, 0.25 max Si, bal Fe.
Annealed: 110,000 TS; 55,000 YS; 20 El; 40 RA; 210 Brin.
For cutters, hobs, reamers, broaches; Type W1; water hardened.

KUKI PRIMA ZAH.
M-1320; 1.0 C, 0.25 max Si, 0.25 max Mn, bal Fe.
Annealed: 100,000 TS; 53,000 YS; 21 El; 42 RA; 200 Brin.
For cutters, hobs, reamers, broaches; Type W1; water hardened.

KUKI PRIMA ZH.
M-1320; 0.85 C, 0.25 max Si, 0.25 max Mn, bal Fe.
Heat treated: 190,000 TS; 145,000 YS; 12 El; 45 RA; 330 Brin.
For cutters, tools, springs, punches, drills; Type W1; water hardened.

KUKI PRIMA ZW.
M-1320; 0.70 C, 0.25 max Si, 0.25 max Mn, bal Fe.
Heat treated: 175,000 TS; 128,000 YS; 12 El; 37 RA; 355 Brin.
For punches, crimpers, springs; Type W1; water hardened.

KUKI SPEZIAL H.
M-1320; 0.90 C, 0.25-0.50 Si, 0.3-0.8 Mn, bal Fe.
Heat treated: 190,000 TS; 145,00 YS; 10 El; 30 RA; 400 Brin.
For punches, crimpers, drills, springs; Type W1; water hardened.

KUKI SPEZIAL MH.
M-1320; 0.75 C, 0.4 Si, 0.6 Mn, bal Fe.
Heat treated: 175,000 TS; 130,000 YS; 12 El; 36 RA; 355 Brin.
For punches, drills, springs, crimpers; Type W1; water hardened.

KUKI SPEZIAL ZAH.
M-1320; 0.67 C, 0.4 Si, 0.6 Mn, bal Fe.
Heat treated: 170,000 TS; 125,000 YS; 15 El; 38 RA; 350 Brin.
For punches, springs, crimpers; Type W1; water hardened.

KUKI SPEZIAL ZH.
M-1320; 0.6 C, 0.3-0.5 Si, 0.5-0.8 Mn, bal Fe.
Heat treated: 160,000 TS; 113,000 YS; 12 El; 40 RA; 325 Brin.
For punches, crimpers; water hardened.

KUKI SPEZIAL ZW.
M-1320; 0.45 C, 0.4 Si, 0.6 Mn, bal Fe.
Hot rolled: 98,000 TS; 59,000 YS; 24 El; 45 RA; 212 Brin.
For gears, shafts, crankshafts; water hardened.

KULGRID 28.
M-963.
For high temperature wire; heat resistant to 1100°F.

KULGRID "C".
M-856, M-796; Ni, clad Cu.
For lead-in wires for vacuum tubes; resists high temperatures.

KULITE KLT-115.
M-1517; Pb, W.
Sintered: 18,000-20,000 TS; 200-280 Brin.
For shielding for nuclear materials; high density.

KUMANAL.
M-282; 88 Cu, 10 Mn, 2 Al.
Annealed: 50,000 TS; 30 El; 73 Brin.
For resistance wire, instrument shunts; low temperature coefficient.

KUMANIC.
M-282; 60 Cu, 20 Mn, 20 Ni.
Annealed: 70,000 TS; 21,000 YS; 35 El; 97 Brin.
Heat treated: 155,000 TS; 130,000 YS; 2 El; 320 Brin.
For springs, contacts; corrosion resistant.

KUMIUM.
M-282; 99.5 Cu, 0.5 Cr.
Annealed: 33.600 TS; 16,700 YS; 60 El; 59 Brin.
Heat treated: 70,000 TS; 60,000 YS; 14 El; 127 Brin.
For welding tips, electrical contacts; hardenable, high conductivity.

KUNHEIM METAL.
M-Ger.; Hydrides of Misch metal, 36 Ce, 49 La, 10 Di, 1 Mg.
For cigarette lighters; pyrophoric.

KUNHEIM METAL.
M-Ger.; 86 rare earths, 12 Mg, 2 Al.
For cigarette lighters.

KUNIAL ALLOYS.
M-282; Cu-Zn alloys.
For engineering applications; can be heat treated.

KUNIAL BRASS.
M-282; 72.5 Cu, 6 Ni, 1.5 Al, bal Zn.
Heat treated: 108,000 TS; 81,000 YS; 11 El; 240 Brin.
For nuts, bolts, valves, primers, fuse bodies, keys, springs; hardened and strengthened by heat treatment.

KUNIAL BRONZE.
M-282; Sn, bal Cu.
For hardware, plumbing, pipes; tough.

KUNIAL COPPER.
M-282; 90 Cu, Ni, Al.
Fo general use, tubes, wire, plate, sheets; temper hardening.

KUNIAL NICKEL SILVER.
M-282; Ni, Zn, bal Cu.
For hardware, utensils; corrosion resistant.

KUNIFER 5.
M-1805; 5 Ni, 1 Fe, bal Cu.
Annealed: 40,000-45,000 TS; 20,000-25,000 YS; 50-40 El; 68-80 Brin.
For marine equipment, hardware, fasteners; corrosion resistant.

KUNIFER 10.
M-1805; 10 Ni, 90 Cu.
Annealed: 45,000 TS; 22,400 YS; 55 El; 90 Brin.
1/2 Hard: 62,000 TS; 45,000 YS; 25 El; 110 Brin.
Hard: 83,000 TS; 81,000 YS; 15 El; 190 Brin.
For condenser tubes; corrosion resistant, no season cracking.

KUNIFER 10 (COPPER-NICKEL).
M-286; 10.5 Ni, 1.7 Fe, 0.8 Mn, bal Cu.
Ann: 52,200 psi TS; 23,200 psi YS; 48 El; 95 DPN (88 Brin).
Tubes for condensers, hydraulic and sea water pipelines. Corrosion resistant.

KUNIFER 30.
M-1805; 30 Ni, 1 Mn, 1 Fe, bal Cu.
Annealed: 63,000 TS; 25,000 YS; 55 El; 100 Brin.
1/2 Hard: 75,000 TS; 42,000 YS; 25 El; 150 Brin.
Hard: 100,000 TS; 94,000 YS; 10 El; 210 Brin.
For condenser tubes; corrosion resistant, no season cracking.

KUNIFER 30A.
M-1805; 66 Cu, 30 Ni, 2 Mn, 2 Fe.
For condenser tubes; corrosion resistant.

KUNIFER 30 (COPPER-NICKEL).
M-286; 31 Ni, 0.6 Fe, 0.8 Mn, bal Cu.
Ann: 56,500 psi TS; 26,100 psi YS; 48 El; 100 DPN (97 Brin).
Tubes for condensers and hydraulic lines. Erosion-corrosion resistant.

KUPFER-NICKEL 54/45.
M-297; 54 Cu, 45 Ni.
For electrical resistances; constantan.

KUPFER-NICKEL 67/30/3.
M-297; 67 Cu, 30 Ni, 3 Mn.
For electrical resistances.

KUPFERNICKEL 75/25.
M-297; 25 Ni, 75 Cu.
Annealed: 45,000 TS; 18,500 YS; 42 El; 62 Brin.
For coinage; corrosion resistant.

KUPFERNICKEL 80/20.
M-297; 80 Cu, 20 Ni.
Rolled: 62,500 TS; 59,700 YS; 12 El; 120 Brin.
Annealed: 45,500 TS; 17,100 YS; 41 El; 64 Brin.
For condenser tubes; corrosion resistant.

KUPFER-SILUMIN.
M-299; 12 Si, 0.8 Cu, 0.3 Mn, bal Al.
Sand cast: 25,000 TS; 14,000 YS; 2-4 El; 60 Brin.
For wheels, rolls, engine blocks, motor housing; high fluidity.

KUPRODUR.
M-Ger.; 0.5 Si, 0.7 Ni, bal Cu.
For staybolts, propeller parts; age-hardenable, corrosion resistant.

KUROMI.
M-Japan; Cu + Sn + Co. Japanese alloy.

KUTERN.
M-282; 99.5 Cu, 0.5 Te.
Annealed: 33.600 TS; 5500 YS; 50 El; 48 Brin.
For machined parts; free-cutting.

KUTHERM see IMI 575 KUTHERM.

KUT KOST.
M-797; 0.7 C, 1.5 B, 18 W, 4 Cr, 1 V, bal Fe.
For tools, dies, tipped tools; high speed steel.

KUT KOST.
M-797; C, W, Cr, V, bal Fe.
For tipped tools and cutters; high speed steel.

KUT KOST GRADE V.
M-797; C, W-Co,1.5 B, bal Fe.
For tools, cutters; for heavy cuts.

KUT KOST GRADE X.
M-797; C, W-Co-1.5 B, bal Fe.
For tools, cutters; centrifugally cast.

KUT KOST GRADE XV.
M-797; C, W, Co, 1.5 B, bal Fe.
For tools, cutters; centrifugally cast.

KUTKOST GRADE XX.
M-797; W, Co, 1.5 B bal Fe.
For tools, cutters; light cuts.

K-V7.
 M-USSR; 0.10-0.15 C, 10.5-12.5 Cr, 0.6-0.8 Mo, 0.8 max Ni, 0.2-0.3 V, 3.7-4.3 W, bal Fe.
 Annealed: 85,000 TS; 45,000 YS; 20 El; B 95 Rock.
 For cutlery, surgical instruments, chemical plant equipment. Corrosion resistant.

KWIK-KUT.
 M-207; 1.1 C, 0.5 Cr, bal Fe.
 For hollow drills; fatigue resistant.

KXA STEEL.
 M-1525; C, Cr, Mo, bal Fe.
 For die casting machine parts, plungers; oil hardened.

KYNAL C65.
 M-282; 4 Cu, 0.6 Mn, 0.7 Fe, 0.5 Si, bal Al.
 Heat treated: 56,000-60,500 TS; 40,500-49,000 YS; 10-6 El; 100 Brin.
 For aircraft construction; age-hardened, high strength.

KYNAL C66.
 M-282; 4 Cu, 0.85 Si, 0.85 Mg, 1.2 Mn, bal Al.
 Heat treated: 56,000-62,700 TS; 33,600-51,500 YS; 15-8 El; 120-150 Brin.
 For aircraft construction; age-hardened, high strength.

KYNAL C 67.
 M-282; 4 Cu, Mn, Si, Mg, bal Al.
 H.T, Temper: 63,840 TS; 39,200 YS; 15 El; 138 Brin.
 For high strength applications; hardenable.

KYNAL C69.
 M-282; 1-2 Cu, 0.5-1.25 Mg, 1 Si, 1 Mn, bal Al.
 Heat treated: 38,100-52,000 TS; 29.,100-41,000 YS; 15-6 El; 105 Brin.
 For aircraft construction; age-hardened, high strength.

KYNAL C70.
 M-282; 2.5-4.0 Cu, 0.25-0.75 Mg, 0.3-1.0 Sb, bal Al.
 Rolled: 36,000 TS; 15,700 YS; 10 El.
 For fuel caps, structures; free-cutting.

KYNAL C71.
 M-282; 1.3-3.0 Cu, 0.2-0.5 Mg, 0.1-0.5 Sn, bal Al.
 Heat treated: 38,000 TS.
 For rivets, structural members; heat treatable.

KYNAL M33.
 M-282; 0.9-1.1 Mg, 0.05 max Cu, 0.05 max Mn, bal Al.
 Soft: 17,000 TS; 7500 YS; 22 El.
 1/2 H-temper: 25,000 TS; 20,000 YS; 4 El.
 H-temper: 29,200 TS; 29,200 YS; 2 El.
 For medium strength structures; good ductility and drawability.

KYNAL M 35/1.
 M-282; 2 Mg, 0.5 Mn, bal Al.
 Annealed: 24,650 TS; 13,400 YS; 18 El; 45 Brin.
 Hard: 33,600 TS; 26,880 YS; 5 El; 67 Brin.
 For medium strength applications; good ductility and corrosion resistance.

KYNAL M 35/2.
 M-282; 3 Mg, 1 Mn, bal Al.
 Annealed: 31,360 TS; 15,680 YS; 18 El; 50 Brin.
 Hard: 40,320 TS; 33,600 YS; 5 El; 75 Brin.
 For medium strength applications; good ductility and corrosion resistance.

KYNAL M 35/3.
 M-282; 3.75-4.25 Mg, 0.6 max Si, 0.7 max Fe, 0.2 max Ti, bal Al.
 Soft: 38,000 TS; 15,700 YS.
 For marine applications; corrosion resistant.

KYNAL M36.
 M-282; 5 Mg, 1 Mn, bal Al.
 Annealed: 38,000 TS; 17,920 YS; 18 El; 55 Brin.
 Hard: 44,800 TS; 38,080 YS; 5 El; 80 Brin.
 For marine parts; good corrosion resistance.

KYNAL M37.
 M-282; 7 Mg, 1 Mn, bal Al.
 Annealed: 49,300 TS 20,200 YS; 20 El; 60 Brin.
 For marine parts; corrosion resistant.

KYNAL M39.
 M-282; 0.7-1.0 Si, 0.5-1.0 Mg, bal Al.
 Soft: 16,000 TS; 25 El.
 Heat treated: 38,000 TS; 31,000 YS; 10 El.
 For brazing alloy; heat treatable.

KYNAL M39/1.
 M-282; 0.4-1.5 Mg, 0.3-0.7 Si, bal Al.
 Heat treated: 33,600-38,100 TS; 20,00-31,600 YS; 22-14 El; 65-80 Brin.
 For aircraft construction; corrosion resistant, age-hardened.

KYNAL M39/2.
 M-282; 0.5-1.0 Mg, 0.7-1.3 Si, 1.0 Mn, 0.6 Fe, 0.2 Ti, bal Al.
 Heat treated: 33,600-44,800 TS; 20,200-41,500 YS; 22-10 El; 64-95 Brin.
 For aircraft construction; corrosion resistant, age-hardened.

KYNAL M40.
 M-282; 0.15-0.40 Cu, 0.2-0, Mn, 0.8-1.2 Mg, 0.4-0.8 Si, 0.7 max Fe, bal Al.
 Heat treated: 36,000-41,000 TS; 30,000-34,000 YS; 14-7 El.
 For structural members; heat treatable, corrosion resistant.

KYNAL M41.
M-282; 0.4-1.5 Mg, 0.6-1.3 Si, 0.2 max Ti, 0.5 max Cr, bal Al.
Heat treated: 36,000-41,000 TS; 30,000-34,000 YS; 14-7 El.
For light alloy parts; heat treatable, corrosion resistant.

KYNAL P1.
M-282; 99.99 Al
Soft: 8950 TS; 45 El.
1/2 H-temper: 13,000 TS; 12 El.
H-temper: 14,500 TS; 6 El.
For heat and light reflectors; corrosion resistant and ductile.

KYNAL P3.
M-282; 99.8 Al, 0.02 max Cu, 0.15 max Si, 0.15 max Fe.
Soft: 10,000 TS.
1/2 H-temper: 16,000 TS
H-temper: 18,000 TS.
For heat and light reflectors; corrosion resistant and ductile.

KYNAL P5.
M-282; 99.6 min Al.
Annealed: 11,200 TS; 40 El; 21 Brin.
Hard: 19,000 TS; 7 El; 38 Brin.
For chemical plant equipment; high resistance to corrosion.

KYNAL P10.
M-282; 0.1 Cu, 0.5 Si, 0.7 Fe, 0.1 Mn, bal Al.
Soft: 12,300 TS; 7500 YS; 35 El; 21 Brin.
Hard: 20,100 TS; 16,000 YS; 5 El; 38 Brin.
For trim, architectural applications; good ductility.

KYNAL PA 19.
M-282; 1.25 Mn, bal Al.
Annealed: 15,680 TS; 25 El; 25 Brin.
Hard: 24,700 TS; 5 El; 47 Brin.
For cold formed and welded parts; corrosion resistant parts.

KYNAL PA20.
M-282; 0.9-1.1 Zn, bal Al.
For cladding sheet for Z93A alloy.

KYNAL S57.
M-282; 0.7-1.3 Cu, 0.8-1.5 Mn, 10.5-13.5 Si, 0.7-1.3 Ni, bal Al.
Heat treated: 40,000 TS; 3 El.
For medium strength applications; heat treatable, corrosion resistant.

KYNAL Y88.
M-282; 1.5-4.0 Cu, 0.3-1.5 Mg, 0.5-1.3 Si, 0.6-1.5 Fe, 2 Ni, 0.3 Ce, 0.3 Ti, bal Al.
Heat treated: 60,500 TS; 47,100 YS; 10 El; 130 Brin.
For aircraft construction, pistons, cylinder heads; age-hardenable.

KYNAL Y89.
M-282; 1.5-3.0 Cu, 1.2-1.8 Mg, 0.55-1.25 Si, 0.5-1.5 Ni, 0.2 Ti, bal Al.
WP-temper: 58,000-60,500 TS; 44,500-47,500 YS; 10-8 El.
Forgings for high temp. applications; age-hardened, high strength.

KYNALCORE C65A.
M-282; 3.5-5.0 Cu, 0.4-1.2 Mg, 0.4-1.2 Mn, 0.3 Ti, 0.7 Fe, bal Al.
Heat treated: 54,000 TS; 31,500 YS; 15 El.
For structural members; heat treatable, Al-clad.

KYNALCORE C66A.
M-282; 3.5-4.8 Cu, 0.8 Mg, 0.9 Si, 1 Fe, 1.2 Mn, 0.3 Ti, bal Al.
W-temper: 53,800 TS; 31,400 YS; 15 El.
WP-temper: 58,200 TS; 44,800 YS; 8 El.
For structural members; age-hardened, Al-clad.

KYNALCORE C68A.
M-282; 4.4 Cu, 0.6 Mg, 0.7 Si, 0.6 Mn, bal Al.
T4-temper: 56,000 TS; 33,000 YS; 15 El.
T6-temper: 60,100 TS; 47,000 YS; 8 El.
For aircraft structures; Al-coated.

KYNALCORE Z93A.
M-282; 1.5 max Cu, 2.0-3.5 Mg, 4.5-6.5 Zn, 0.8 Mn, bal Al.
Heat treated: 74,000 TS; 62,700 YS; 9 El; 170 Brin.
For structural members; heat treatable, Al-clad.

KYNALCORE Z93C.
M-282; 1.5 Cu, 2.0-3.5 Mg, 0.5 Si, 0.5 Fe, 4.5-6.5 Zn, 0.2 Ti, bal Al.
WP-temper: 72,000 TS; 60,000 YS; 8 El.
For structural members; age-hardened, high strength.

L

L2 ALLOY.
M-USSR; 89.2 WC, 8 Co, 2 TaC, 0.8 Cr_2O_3.
Sintered: Tr.S. 165,000; Rock A 90.
For cutting tools to machine cast iron, high alloy steels, hard plastics.
Sintered carbide, wear resistant.

L 3 B.
M-1653; 1.05-1.20 C, 0.35 max Si, 0.35 max Mn, bal Fe.
Water hardening carbon tool Steel.
UC 112 KU; AISI W1 (1.1 C).

L 4 B.
M-1653; 1.20-1.35 C, 0.35 Mn, 0.20-0.35 Cr, bal Fe.
Water hardening carbon tool steel.
W.-Nr. 1.2002.

L 5.
M-137; 12.5-14.5 Zn, 2.5-3.0 Cu, 0.8 max Fe, 0.7 max Si, 0.2 max Ti, bal Al.
Cast: 20,000 TS; 11,000 YS; 4 El; 65 Brin.
For crankcases, gear boxes, fans, brackets; sand or permanent mold castings.

L 8.
M-137; 11-13 Cu, 0.8 max Fe, 0.7 max Si, 0.2 max Ti, bal Al.
Cast: 22,000 TS; 16,000 YS; 1.5 El; 80 Brin.
For carburetors, automobile pistons; high pressure castings.

L 11.
M-137; 6-8 Cu, 1.0 max Sn, 0.8 max Fe, 0.2 max Ti, bal Al.
Cast: 22,000 TS; 16,000 YS; 4 El; 60 Brin.
For gear boxes, cylinder heads; sand or permanent mold castings.

L13.
M-1522; 20 Ag, Cu, Zn.
Melt range: 770-810°C.
Max stress: 50.4 kgf/mm^2; 4 El.
For silver brazing; general use.

L13S.
M-1522; 20 Ag, Cu, Zn, Si.
Melt range: 690-810°C.
Max stress: 52.3 kgf/mm^2; 30 El.
For silver brazing; general use.

L 14.
M-1290; 0.95-1.05 C, 0.10-0.25 Si, 0.10-0.25 Mn, bal Fe.
Carbon tool steel.
W.-Nr. 1.1540.

L17.
M-1522; 24 Ag, Cu, Zn.
Melt range: 740-780°C.
Max stress: 35.3 kgf/mm^2; 13 El.
For silver brazing; general use.

L 23.
M-1290; 1.05-1.15 C, 0.10-0.30 Si, 0.10-0.35 Mn, bal Fe.
Carbon tool steel.
W.-Nr. 1.1650.

L 24.
M-1290; 0.95-1.05 C, 0.10-0.30 Si, 0.10-0.35 Mn, bal Fe.
Carbon tool steel.
W.-Nr. 1.1640.

L 25.
M-1290; 0.80-0.90 C, 0.10-0.30 Si, 0.10-0.35 Mn, bal Fe.
Carbon tool steel.
W.-Nr. 1.1630.

L-35.
M-40; 0.9 C, 0.4 Mn, 0.35 V, bal Fe.
For cold heading dies; water hardened.

L-605.
M-44; 10 Ni, 20 Cr, 15 W, bal Co.
Wire and rod for welding wire and fastener stock.

L-605 see UDIMET L-605; ALLOY L-605; AND UNITEMP L-605.

LA-1 ALLOY.
M-USSR; 0.16 C, 15 Cr, 15 Ni, 3 Co, 2 Mo, 1 W, bal Fe.
Heat resisting alloy.

LA-4 ALLOY.
M-USSR; 0.12 C, 15 Cr, 15 Ni, 3 Co, 2 Mo, 1 W, bal Fe.
Heat resisting alloy.

LA-5 ALLOY.
M-USSR; 0.16 C, 15 Cr, 16 Ni, 3 Co, 2 Mo, 1 W, 1 Cb, bal Fe.
Heat resisting alloy.

LA 21.
M-Swiss; 6 Cu + Ni, Mn, Mg, Si, bal Al.
Cast: 60,000-89,000 TS; 56,000-80,000 YS; 3-1 El; 100-120 Brin.
Wrought: 63,000-76,000 TS; 44,000-60,000 YS; 4-2 El; 80-100 Brin.
For bearings.

LA 31.
M-Swiss; 8 Sn, Cu, Ni, Mg, bal Al.
Cast: 44,000-60,000 TS; 22,000-31,000 YS; 8-3 El; 40-55 Brin.
Wrought: 50,000-63,000 TS; 28,000-41,000 YS; 12-6 El; 45-60 Brin.
For bearings.

LA BELLE 2-70.
M-38; 0.6 C, 2 Si, 0.8 Mn, bal Fe.
For punches, chisels, stamps; tough.

LA BELLE COLD HEADER DIE.
M-38; 0.95 C, bal Fe.
For cold header dies; water hardening.

LA BELLE COLD STRIKING DIE.
M-38; 0.95 C, bal Fe.
For cold forging dies; water hardening.

LA BELLE EXTRA.
M-38; 0.95 C, bal Fe.
For cold work dies, cold forming dies; water hardened.

LABELLE HT.
M-38; 0.43 C, 1.3 Mn, 2.25 Si, 1.3 Cr, 0.3 V, 0.4 Mo, bal Fe.
Heat treated: 314,000 TS; 255,000 YS; 9 El; 29 RA; 560 Brin.
For shear blades, chisels, punches; oil hardened, shock resistant.

LA BELLE SILICON 2.
M-38; 0.6 C, 0.8 Mn, 1.9 Si, 0.3 Mo, 0.25 Cr, bal Fe.
For punches, chisels, rivet sets; tough.

LA BOUR R-55.
M-263; 4 Si, 23 Cr, 6 Cu, 4 Mo, 2 W, 0.2-0.3 C, 52 Ni, 8 Fe.
For pump parts; modification of "R-50."

LAC 10.
M-137; 9.0-10.5 Cu, 0.15-0.35 Mg, 0.3-1.0 Fe, 0.6 max Si, bal Al.
30,000-44,000 TS; 28,000-40,000 YS; 110-130 Brin.
For light alloy parts for high temperature service; hardenable.

LAC 10.
M-British; 0.15-0.35 Mg, 0.6 Si, 9-10.5 Cu, 0.6 Mn, 0.3-1.0 Fe, bal Al.
For pistons; non-hardenable.

LAC 112.
M-British; 0.3 Mg, 7-13 Si, 2-3 Cu, 0.5 Mn, 1.5 Ni, 1.2 Zn, bal Al.
For die castings; high fluidity.

LAC 112A.
M-137; 0.75-2.5 Cu, 9-11.5 Si, bal Al.
For die castings.

LAC 113A.
M-British; 1.3 Si, 2.5-4.5 Cu, 1.0 Fe, 9-13 Zn, bal Al.
For sand castings; non-hardenable.

LACOLITE.
M-1480; 3 C, Si, Mn, alloy, bal Fe.
For gears, shafts, machine tool parts; cast iron.

LACTOVAC.
M-679; 0.2-0.5 C, bal Fe.
For machinery parts; water hardening.

LADISH D6A.
M-1527.
Same composition as LADISH D6AC, but normally not vacuum melted.
For aircraft structural components, stone crushers, gas turbine compressor discs, thrust bearings.

LADISH D6AC.
M-1527; 0.42-0.50 C, 0.60-0.90 Mn, 0.15-0.30 Si, 0.40-0.70 Ni, 0.90-1. Cr, 0.90-1.10 Mo, 0.05-0.15 V, bal Fe.
Vacuum melted; hardenable to 300,000 psi TS.
Used in aerospace industry for airframe components and solid rocket motor boosters.

LADISH D-11.
M-1527; 0.42-0.50 C, 0.6-0.9 Mn, 0.25-0.45 Si, 0.4-0.7 Ni, 0.9-1.2 Cr 1.9-2.15 Mo, 0.45-0.6 V, bal Fe.
Air or oil hardening; good high temp. strength.
For hot work tool and die applications.

LAFOND'S AXLE BEARING.
80 Cu, 18 Sn, 2 Zn.
For axle bearings; heavy duty.

LAFOND'S HEAVY BEARING.
83 Cu, 15 Sn, 1.5 Zn, 0.5 Pb.
For heavy duty bearings; high strength.

LAFONDS MALLEABLE BRONZE.
M-France; 2 Sn, bal Cu.
For electrical equipment; corrosion resistant.

LAFONDS PUMP BRONZE.
M-France; 10 Sn, 2 Zn, bal Cu.
For pump parts, gears, worm wheels, shafts; tough, corrosion resistant.

LAGAL.
M-Germany; Al alloy.
For bearings; Zn-free.

LAGERBRONZE.
M-557; 4-5 Sn, 3.5-5.0 Zn, 3.5-4.5 Pb, bal Cu.
Cast: 50,000-58,000 TS; 50 El; 70-85 Brin.
For machine and engine parts; corrosion resistant bronze.

LAITON YELLOW BRASS.
M-Eng.; 70-60 Cu, 27-40 Zn, 5.3 Pb, 0-1 Sn.
For plumbing, hardware, bolts, nuts; free-cutting.

LAKE'S METAL.
87 Pb, 6 Sn, 7 Sb.
For bearings for heavy loads; anti-friction.

LA-LED.
M-669; 0.08-0.13 C, 0.8-1.1 Mn, 0.2-0.3 Si, 0.15-0.35 Pb, bal Fe.
Drawn: 70,000 TS; 60,000 YS; 15 El; 45 RA; 140 Brin.
For screw machine products; free-cutting, carburizing steel.

LA-LED X.
M-669; 0.15 C, 0.8-1.2 Mn, 0.04-0.09 P, 0.25 S, 0.15-0.35 Pb, Te, bal Fe.
Bar: 70,000 TS; 60,000 YS; 15 El; 45 RA; 140 Brin.
For screw machine products, fasteners, hardware.
Free machining.

LALL.
M-Swiss; Mg, Zn, bal Al.
Cast: 44,000-56,000 TS; 19,000-34,000 YS; 12-4 El; 35-45 Brin.
Wrought: 44,000-60,000 TS; 19,000-28,000 YS; 22-14 El; 35 RA; 50 Brin.
For bearings.

LANARK.
M-73; 0.55 C, 0.2 Cr, 0.90 Mn, 0.3 V, 1.2 Mo, 1.9 Si, bal Fe.
For chisels, punches, shear blades, stamps; shock resistant.

LAN-CER-AMP.
M-1213; 45-50 Ce, 30 min La, 20-24 Nd + Pr.
For alloying ferrous and nonferrous alloys.

LANDERIG'S SPECULUM.
70 Cu, 30 Sn.
For mirrors, reflectors; bronze.

LANGALLOY 1R.
M-1169; 0.5 max C, 14 Cr, 75 Ni, 8 Fe, 2.5 max Si, 0.7 max Mn.
For furnace equipment, salt pots, heat treating boxes; high heat resistance, good strength.

LANGALLOY 1V.
M-1169; 0.2 C, 17.5-19.5 Cr, 10-12 Ni, Cb = 10 x C, bal Fe.
Sand cast: 70,000 TS; 30,000 YS; 20 El; 200 Brin.
For valves, pump parts, welded structures; stainless, austenitic, stabilized.

LANGALLOY 3V.
M-1169; 17.5-19.0 Cr, 10-12 Ni, 2.5-3.0 Mo, 0.2 C, bal Fe.
Sand cast: 70,000 TS; 30,000 YS; 15 El; 200 Brin.
For valves, pump parts, chemical plant equipment; Type 317, stainless, austenitic.

LANGALLOY 4V.
M-1169; 0.2-0.3 C, 10.5-12.5 Ni, 21.0-24.0 Cr, 2.0-3.0 W, bal Fe.
Cast: 78,500 TS; 33,000 YS; 20 El; 165-185 Brin.
For oil refinery cracking plant, gas turbine parts; stainless steel casting, heat and creep resistant.

LANGALLOY 5V.
M-1169; 0.12 C, 18 Cr, 15 Ni, 3 Mo, bal Fe.
Cast.
For specialized instrumentation applications; stainless, non-magnetic.

LANGALLOY 7R.
M-1169; 23 Cr, 6 Mo, 6 Cu, 5 Fe, 2 W, bal Ni.
Sand cast: 56,000 TS; 40,000 YS; 6 El; 180 Brin.
For valves, valve seats, pump parts; resists H_2SO_4 acids.

LANGALLOY 7V.
M-1169; 0.15 max C, 8 Ni, 18 Cr, 3 Mo, bal Fe.
For chemical plant equipment; acid resistant.

LANGALLOY 8R.
M-1169; 0.75 max C, 60 Ni, 15 Cr, 2 max Si.
For heat treating equipment, salt pots; high thermal fatigue and hot gas resistance.

LANGALLOY 8V.
M-1169; 0.15 max C, 8 Ni, 18 Cr, 3 Mo, S, 2 max Mn, bal Fe.
For chemical plant equipment; acid resistant, free-cutting.

LANGALLOY 9R.
M-1169; 0.5 max C, 37 Ni, 18 Cr, 3 max Si, 2 max Mn, bal Fe.
For heat treating equipment, furnace parts; good for cyclic heating up to 1100°C.

LANGALLOY 12V.
M-1169; 0.08 max C, 12 Ni, 18 Cr, 3.5 Mo, S, bal Fe.
For chemical plant equipment; acid resistant, free-cutting.

LANGALLOY 13V.
M-1169; 0.08 max C, 12 Ni, 18 Cr, 3.5 Mo, bal Fe.
For chemical plant equipment; acid resistant.

LANGALLOY 14V.
M-1169; 18 Cr, 10 Ni, 3 Mo, 1 Nb, bal Fe.
Cast: 460 N/mm^2 TS; 240 N/mm^2 YS; 12 El.
Niobium stabilized austenitic for chemical equipment. Free machining casting.

LANGALLOY 15V.
M-1169; 18 Cr, 8 Ni, bal Fe.
Austenitic stainless steel casting.
Similar to ASTM A296 CF-8.

LANGALLOY 16V.
M-1169; 0.08 max C, 8 Ni, 18 Cr, 3 Mo, bal Fe.
For chemical plant equipment; acid resistant.

LANGALLOY 18V.
M-1169; 0.08 max C, 8 Ni, 18 Cr, 3 Mo, S, bal Fe.
For chemical plant equipment; acid resistant, free-cutting.

LANGALLOY 20V.
M-1169; 0.07 max C, 20 Cr, 29 Ni, 3 Mo, 1 Cb, bal Fe.
For sulphuric acid equipment; stainless casting for H_2SO_4.

LANGALLOY 22V.
M-1169; 18 Cr, 10 Ni, 3 Mo, 1 Ni, bal Fe.
Stabilized austenitic stainless steel casting.
For weldable chemical equipment.
BS 3100 318C17.

LANGALLOY 25V.
M-1169; 0.5 max C, 12 Ni, 25 Cr, 2 max Si, 2 max Mn, bal Fe.
For superheaters in oil refineries, furnace parts, heat treating equipment oxidation resistant to 1100°C, good hot strength.

LANGALLOY 33V.
M-1169; 18 Cr, 10 Ni, 3 Mo, bal Fe.
Austenitic stainless steel casting. For chemical equipment.
Similar to ASTM A296 CF-8M.

LANTHANUM.
M-1755; La.
Purities: 99.9%, 99.6%, 99+%.
Forms: Ingot, lump, rod, sheet, foils, wire, powder.

LANTHANUM METAL.
M-1497.
Comparatively pure La, for alloying.

LANZ CAST IRON.
M-527; 3.4 C, 2.8 Si, 0.7 Mn, bal Fe.
56,500 TS; 26 El.
For housings, frames, gears, castings, high strength castings.

LAPCO.
M-1481; Al alloy.
For light alloy parts.

LAPELLOY.
M-114, M-32; 0.25-0.35 C, 11-12 Cr, 2.5-3.0 Mo, 0.3 V, 0.5 max Ni, bal Fe.
At 20°F: 155,000 TS; 140,000 YS; 17 El; 35 RA.
At 1000°F: 60,000 TS; 50,000 YS; 35 El; 85 RA.
For high temperature bolts, valve stems, turbine buckets and blades; scale and oxidation resistant to 1400°F.

LAPELLOY C.
M-32; 0.20-0.25 C, 11-12 Cr, 2.5-3.0 Mo, 1.75-2.25 Cu, 0.1 N, bal Fe.
Heat treated: 135,000-203,000 TS; 105,000-170,000 YS; 18-17 El; 55-47 RA.
For turbine shafts, compression wheels and buckets; heat and oxidation resistant to 1400°F.

LA SALLE 1541 A,B see COLFORM E.T.D. 1541 A,B.

LA SALLE ETD 150.
M-669; 0.4 C, Si, Mn, 1 Cr, 0.2 Mo, bal Fe.
Heat treated: 150,000 TS; 130,000 YS; 10-20 El; 35-45 RA; 302 Brin.
For gears, pinions, shafts, fasteners; preheat treated alloy steel.

LA-SALLE E.T.D. 180 see "E.T.D." 180.

LATROBE BR-3.
M-73; 2.8 C, 5.2 Cr, 4.5 V, bal Fe.
For cold working dies; for cold working applications.

LATROBE BR-4.
M-73; 2.4 C, 12.75 Cr, 0.4 Mn, 4 V, 1.1 Mo, bal Fe.
For shaving, stamping and deep draw dies; abrasion resistant.

LATROBE CLW NO. 1.
M-73; 0.3 C, 9 W, 3.3 Cr, 0.5 V, bal Fe.
For hot headers, punches, extrusion rams and dies; oil hardened, hot work steel.

LATTEN (LAITON)-1.
M-Eng.; 70-60 Cu, 27-40 Zn, 5.3-0 Pb, 0-1 Sn.
For hardware, fixtures; yellow brass.

LATTEN (LAITON)-2.
M-Eng.; 71 Cu, 28.53 Zn, 0.25 Pb, 0.02 Sn, 0.20 Fe.
Annealed: 56,000 TS; 45,000 YS; 35 El; 65 RA.
For hardware, fixtures; yellow brass.

LAUTAL-1.
M-116; 4.5-5.5 Cu, 0.2-0.5 Si, bal Al.
Wrought: 35,000-63,000 TS; 18,000-40,000 YS; 25-20 El; 65-110 Brin.
For aircraft construction; age-hardening.

LAUTAL-2.
M-116; 4.5 Cu, 0.75 Si, 0.75 Mn, bal Al.
Annealed: 23,000-35,000 TS; 7,000-12,000 YS; 20-12 El; 45-55 Brin.
For electric cables; light alloy.

LAVELSSIERE BRONZE.
61 Cu, 38 Zn, 1 Sn.
For condenser tubing, marine parts; corrosion resistant.

LAVIN 7A MANGANESE BRONZE.
M-801; 59 Cu, 0.75 Sn, 0.75 Pb, 1 Fe, 0.3 Mn, 1 Al, bal Zn.
Cast: 60,000 TS; 20,000 YS; 15 El; 80-95 Brin.
For valve stems, propellers, worm gears; corrosion resistant.

LAVIN 8A MANGANESE BRONZE.
M-801; 57 Cu, 1 Fe, 1 Al, 0.25 Mn, bal Zn.
Cast: 65,000 TS; 25,000 YS; 20 El; 90-120 Brin.
For valves, propellers, worm gears, valve stems; corrosion resistant.

LAVIN 8C MANGANESE BRONZE.
M-801; 64 Cu, 3 Fe, 5 Al, 4 Mn, bal Zn.
Cast: 110,000 TS; 60,000 YS; 12 El; 190-235 Brin.
For screwdown nuts, gears, bridge parts; corrosion resistant.

LAVIN 9A ALUMINUM BRONZE.
M-801; 87.5 Cu, 3.5 Fe, 9 Al.
Cast: 65,000 TS; 25,000 YS; 20 El; 110-140 Brin.
For worm wheels, bearings, bushings; corrosion resistant.

LAVIN 9B ALUMINUM BRONZE.
M-801; 89 Cu, 1 Fe, 10 Al.
Cast: 65,000 TS; 25,000 YS; 20 El; 110 Brin.
Heat treated: 80,000 TS; 40,000 YS; 12 El; 160 Brin.
For valve seats, stripper nuts, gears; corrosion resistant, heat treatable.

LAVIN 9C ALUMINUM BRONZE.
M-801; 85 Cu, 4 Fe, 11 Al.
Cast: 75,000 TS; 30,000 YS; 12 El; 150 Brin.
Heat treated: 90,000 TS; 45,000 YS; 6 El; 190 Brin.
For worm wheels, pump parts, bushings; corrosion resistant, heat treatable.

LAVIN 9D ALUMINUM BRONZE.
M-801; 81 Cu, 4 Ni, 4 Fe, 11 Al.
Cast: 90,000 TS; 40,000 YS; 6 El; 190 Brin.
Heat treated: 110,000 TS; 60,000 YS; 5 El; 200 Brin.
For worm wheels, pump parts, bushings; corrosion resistant, heat treatable.

LAVIN 10A NICKEL SILVER.
M-801; 12 Ni, 20 Zn, 9 Pb, 2 Sn, 57 Cu.
Cast: 30,000 TS; 15,000 YS; 8 El; 50-60 Brin.
For hardware fittings, valves, plumbing; corrosion resistant.

LAVIN 10B NICKEL SILVER.
M-801; 16 Ni, 16 Zn, 5 Pb, 3 Sn, 60 Cu.
Cast: 35,000 TS; 17,000 YS; 15 El; 65-80 Brin.
For plumbers fittings, statuary, bolts; corrosion resistant.

LAVIN 11A NICKEL SILVER.
M-801; 20 Ni, 8 Zn, 4 Pb, 4 Sn, 64 Cu.
Cast: 30,000 TS; 17,000 YS; 8 El; 76-120 Brin.
For hardware, fittings, plumbing, valves; corrosion resistant.

LAVIN 11B NICKEL SILVER.
M-801; 25 Ni, 2 Zn, 1.5 Pb, 5 Sn, 66.5 Cu.
Cast: 45,000 TS; 22,000 YS; 15 El; 120-150 Brin.
For hardware and plumbing fixtures; corrosion resistant.

LAVIN CF NO. 4.
M-801; Cu alloy.
Deoxidizer for copper alloys; densifier.

LAVIN-NDZ BRONZE.
M-801; 5 max Zn, 5.5 max Ni, 5.5 max Fe, 2 max Al, 2 max Si, 0.25 max Pb, bal Cu.
Cast: 65,000-69,000 TS; 30,000-35,000 YS; 20-34 El; 17-31 RA; 123-134 Brin.
Heat treated: 78,900 TS; 63,400 YS; 5 El; 179 Brin.
For valve stems and bodies, valve bonnets, propeller wheels, marine outboard gears.
High corrosion resistance with good strength.

LAVIN NDZ-S BRONZE.
M-801; 5 max Zn, 5.5 max Ni, 5.5 max Fe, 2 max Al, 2 max Si, 0.25 max Pb, bal Cu.
Cast: 69,000-75,000 TS; 40,000-45,000 YS; 13-23 El; 10-20 RA; 143-149 Brin.
Heat treated: 86,400 TS; 62,400 YS; 8 El; 196 Brin.
For valve stems and bodies, valve bonnets, propeller wheels, marine outboard gears.
High corrosion and dezincification resistance.
High yield strength.

LAVIN NSF NO. 5.
M-801; alloy.
Degasifier for nickel-silver castings; densifier.

LAVIN SPECIAL.
M-801; Cu, Ni, Al, bal Zn.
Cast: 70,000-102,000 TS; 40,000-68,000 YS; 6-1 El; 165-227 Brin.
For neck rings and plungers for glass molding machines; high thermal conductivity.

LAWS PHOSPHOR BRONZE.
M-England; 9.5-11 Sn, 0.7-1.0 P, bal Cu.
For bearings, gears, bushings; heavy duty.

L C H S CHISEL.
M-261; C, alloy, bal Fe.
For tools, chisels, punches; tough.

LC LOW CARBON.
M-370; 0.20-0.25 C, Mn, bal Fe.
AISI C-1020 precision ground flats and squares.

LCN-1.
M-843; 55 Cu, 41 Ni, 4 In.
Metallic spray coating for D-Gun.
High strength and anti-galling; 300-DPH.

L C N-155.
M-114; 0.08-0.16 C, 1-2 Mn, 20.0-22.5 Cr, 19-21 Ni, 18.5-21.0 Co, 2.5 3.5 Mo, 2-3 W, 0.75-1.25 Cb + Ta, 0.10-0.20 N_2, bal Fe.
Heat treated: 80,000-150,000 TS; 40,000-110,000 YS; 50-15 El; 55-20 RA.
For rotors, blades, bolts, buckets for gas engine and jet engines; high oxidation and heat resistance.

LC-NICKEL 99.2.
M-297; 99.2 Ni, 0.02 max C.
For general engineering, especially production and processing of caustic soda.

LC-NICKEL 99.6.
M-297; 99.6 min Ni.
For manufacturing and processing of mineral products, especially caustic alkalis.

LCU-2.
M-843; 90 Cu, 10 Al.
Metallic spray coating for Plasma Torch.
Machinable; 175 DPH.
For gear bushings.

LEAD.
M-1755; Pb.
Purities: Zone refined 99.9999%, 99.9995%, 99.999%, 99.99%.
Forms: Rod, bar, shot, powders, wires, sheets, foils, single crystals.

LEAD ALLOY.
75 Sn, 20 Sb, 5 Pb.
For bearings; anti-friction.

LEADBEATER BR.
M-1408; high C, med Cr, bal Fe.
For dies; oil hardened.

LEADBEATER L S 11.
M-1408; 2.0 C, 12 Cr, bal Fe.
For blanking and wire drawing dies, press tools; oil hardened, non-deforming.

LEADBEATER L S 25.
M-1408; C, W, bal Fe.
For hot heading, swaging and piercing dies; hot die steel, shock and abrasion resistant.

LEADBEATER L S 54.
M-1408; high C, Cr, Mn, bal Fe.
For taps, reamers, chasers, dies; oil hardened.

LEADBEATER L S 3000.
M-1408; C, 7 W, bal Fe.
For drawing dies; oil hardened.

LEADBEATER N S O H.
M-1408; high C, Mn, bal Fe.
For punches, dies; oil hardened, non-deforming.

LEADBEATER P B.
M-1408; 0.6 C, Cr, bal Fe.
For taps, reamers, threading dies, punches, chasers; oil hardened.

LEADBEATER S P C.
M-1408; high C, alloy, bal Fe.
For chisels, punches, marking dies; shock resistant, oil hardened.

LEAD BRONZE NO. 1.
M-1518; 9 Sn, 15 Pb, bal Cu.
Cast: 22,400 TS; 4 El; 50-70 Brin.
For soft shaft bearings; for poor lubrication.

LEAD BRONZE NO. 2.
M-1518; 10 Sn, 10 Pb, bal Cu.
Cast: 24,600 TS; 4 El; 60-75 Brin.
For bearings for medium hard shafts and heavy loads; good anti-friction properties.

LEAD BRONZE NO. 3.
M-1518; 10 Sn, 5 Pb, bal Cu.
Cast: 26,900 TS; 5 El; 65-75 Brin.
For bearings; good corrosion resistance, high strength.

LEAD-CALCIUM.
97 Pb, 0.79 Ca, 0.66 Ba.
For bearings; anti-friction.

LEADED BEARING BRONZE 544.
M-279; 89 Cu, 4 Sn, 4 Pb, 3 Zn.
Ann: 42,000-52,000 TS; 14,000-27,000 YS; 40-50 El.
Cold rolled: 47,000-103,000 TS; 20,000-92,000 YS; 2-40 El.
For bearings, bushing, thrust washers.
Excellent bearing properties.

LEADED BRASS 244.
M-8; 37 Zn, 1 Pb, bal Cu.
Hard: 73,000 TS; 60,000 YS; 8 El; 150 Brin.
Soft: 45,000 TS; 17,000 YS; 50 El; 60 Brin.
For hardware, screw machine parts; free-cutting.

LEADED BRONZE.
M-U.S.; 10-35 Zn, 1.5-2.5 Pb, bal Cu.
For hardware, screws, fittings, bolts; free-cutting.

LEADED COMMERCIAL BRONZE.
M-33; 89 Cu, 9 Zn, 2 Pb.
Drawn: 52,000 TS; 45,000 YS; 18 El; 60 Brin.
For hardware, screw machine parts; free cutting.

LEADED COMMERCIAL BRONZE-201.
M-8; 9.5 Zn, 0.5 Pb, bal Cu.
Soft: 37,000 TS; 12,000 YS; 40 El.
Hard: 62,000 TS; 47,000 YS; 6 El.
For percussion caps, gilding; good workability.

LEADED COMMERCIAL BRONZE 202.
M-8; 88.5 Cu, 9.25 Zn, 2.25 Pb.
Hard: 54,000 TS; 45,000 YS; 15 El; B58 Rock.
Soft: 37,000 TS; 12,000 YS; 40 El; B1 Rock.
For screw machine parts, hardware; free cutting.

LEADED COPPER-187.
M-8; 99.00 Cu, 1.0 Pb.
Hard: 48,000 TS; 40,000 YS; 12 El; B 50 Rock.
Soft: 32,000 TS; 10,000 YS; 45 El; B 45 Rock.
For current carrying studs, nuts, bolts, fasteners. Free-cutting, 98% electrical conductivity.

LEADED FLANGING BRASS.
M-33; 62.5 Cu, 35.6 Zn, 1.9 Pb.
For screw machine products; flange grade.

LEADED GUN METAL.
M-Eng.; gun metal + 1.0 Pb.
For corrosion resistant parts; improved machinability.

LEADED HIGH BRASS.
M-8; 65-78 Cu, 0.3-0.8 Pb, bal Zn.
For tanks, vessels, containers; good ductility and strength.

LEADED MONEL METAL.
M-Eng.; 60 Ni, 32 Cu, 2.2 Pb, 2.2 Fe, 2.0 Mn, 0.9 Si, 0.2 C.
For corrosion resistant parts; improved machinability.

LEADED MUNTZ METAL-365.
M-8; 60 Cu, 39.35 Zn, 0.65 Pb.
Soft: 54,000 TS; 20,000 YS; 45 El; B 45 Rock.
For industrial condensers.
Strong, stiff and elastic, free-cutting.

LEADED NAVAL BRASS see also **NAVAL BRASS, HIGH LEADED.**

LEADED NAVAL BRASS 29.
M-141; 60 Cu, 1.75 Pb, 0.75 Sn, bal Zn.
Hard: 75,000 TS; 53,000 YS; 15 El.
Soft: 57,000 TS; 25,000 YS; 40 El.
For hardware; free-cutting, corrosion resistant.

LEADED NAVAL BRASS 482.
M-8; 38.55 Zn, 0.75 Sn, 0.70 Pb, bal Cu.
Hard: 63,000 TS; 35,000 YS; 28 El; 116 Brin.
Soft: 56,000 TS; 25,000 YS; 38 El; 83 Brin.
For marine hardware, screw machine products; free-cutting, forgeable.

LEADED NAVAL BRASS-485.
M-8; 60 Cu, 37.5 Zn, 1.75 Pb, 0.75 Sn.
Hard: 63,000 TS; 35,000 YS; 25 El; B 65 Rock.
Soft: 56,000 TS; 25,000 YS; 35 El; B 50 Rock.
For marine hardware, valve stems, screw machine products.
Free-machining, high strength.

LEADED NICKEL COPPER-831.
M-8; 1 Pb, 1 Ni, 0.2 P, bal Cu.
Heat treated: 80,000 TS; 70,000 YS; 7 El.
For screw machine products, fasteners; free-cutting, corrosion resistant.

LEADED NICKEL COPPER 7021.
M-8; 97.8 Cu, 1.0 Pb, 1.0 Ni, 0.20 P.
Rod: 85,000 TS; 75,000 YS; 5 El; 55% elec. cond.; 80% machinability.
For electrical contacts, connectors, control elements for power tubes.

LEADED NICKEL SILVER 10%-796.
M-8; 42 Zn, 1 Pb, 10 Ni, 2 Mn, bal Cu.
Hard: 70,000 TS; 40,000 YS; 15 El; 137 Brin.
Soft: 60,000 TS; 20 El.
For screw machine products, architectural parts; free-cutting, corrosion resistant.

LEADED NICKEL SILVER 10%-823.
M-8; 2.75 Pb, 10 Ni, 40.6 Zn, 0.15 Mn, bal Cu.
For architectural trim, forgings; extruded.

LEADED PHOSPHOR BRONZE.
M-58; 80 Cu, 10 Sn, 10 Pb.
Cast: 30,000-37,000 TS; 16,000-23,000 YS; 18-5 El; 16-6 RA; 55-60 Brin.
Chilled: 30,000-33,000 TS; 19,000-21,000 YS; 7-4 El; 10-7 RA.
For bearings for high speeds and heavy pressures; resists shock and vibrations.

LEADED PHOSPHOR BRONZE (B) 379.
M-8; 5 Sn, 1 Pb, 0.10 P, bal Cu.
Hard: 65,000 TS; 55,000 YS; 25 El.
For tubes, bushings, perforated sheets, clutch plate; resists fatigue and corrosion.

LEADED PHOSPHOR BRONZE GR B see also PHOSPHOR BRONZE B-1.

LEADED TUBE BRASS 331.
M-8; 33 Zn, 1 Pb, bal Cu.
Hard: 73,000 TS; 60,000 YS; 10 El; 150 Brin.
Soft: 45,000 TS; 17,000 YS; 55 El; 60 Brin.
For screw machine parts; free-cutting tubes.

LEADED TUBE BRASS 3301.
M-8; 33.25 Zn, 0.25 Pb, bal Cu.
Hard: 73,000 TS; 60,000 YS; 10 El; 150 Brin.
Soft: 45,000 TS; 17,000 YS; 45 El; 60 Brin.
For plumbing goods, tubes; yellow brass.

LEAD FOIL.
86 Pb, 6.9 Fe, 5.5 Al, 1.9 Sn.
For lead foil.

LEAD FOIL (CALIN).
86 Pb, 13 Sn, 1 Cu.
For lead foil; corrosion resistant.

LEAD NO. 1 HARD.
M-88; 90 Pb, 10 Sb.
8,220 TS; 17 El; 17 Brin.
For type metal; M.P. 486°F.

LEAD NO. 1 HARD.
M-558; 99.25 Pb, 0.25 Cd, 0.5 Sb.
For valves, cocks, cable sheathing.

LEAD NO. 2 HARD.
M-88; 85 Pb, 15 Sb.
9,000 TS; 11.7 El; 17 Brin.
For type metal; M.P. 476°F.

LEAD NO. 2 HARD.
M-558; 98.25 Pb, 0.25 Cd, 1.5 Sn.
For valves, cocks, tank lining.

LEAD SHOT.
99.8 Pb, 0.2 As.
For lead shot.

LEAD TAPE.
95 Pb, 4.5 Sb, 0.5 Sn.
For lead tape; hard.

LEAKPRUF.
M-1073; 40-60 Pb, bal Sn.
For soft solder for Ni, Monel and stainless steels; acid flux filled.

LEBANON 15.5.
M-74; 0.06 max C, 14-15.5 Cr, 4.1-5.1 Ni, 0.15-0.25 Cb, 2.5-3.5 Cu, bal Fe.
Precipitation hardenable stainless steel casting.
As heat treated: 180,000 TS; 150,000 YS; 6 El.
For pumps, impellers, fuel controls. (Armco 15-5 PH).

LEBANON 17-4.
M-74; 0.06 max C, 15.5-16.7 Cr, 3.6-4.6 Ni, 0.10-0.35 Cb, 2.5-3.2 Cu, bal Fe.
Precipitation hardenable stainless steel casting.
As heat treated: 180,000 TS; 150,000 YS; 6 El.
For pumps, impellers, fuel controls.
ACI CB-7 Cu. (Armco 17-4 PH).

LEBANON 17-22.
M-74; 0.25-0.35 C, 0.3-0.6 Si, 0.45-0.70 Mn, 1.0-1.5 Cr, 0.4-0.6 Ni, bal Fe.
Low carbon, low alloy steel casting.
Nor. + T.: 125,000 TS; 100,000 YS; 12 El; 262-306 Brin.
For aircraft brake components. Reference: Timken 17-22 A.

LEBANON 33.
M-74; 0.07 C, 18-21 Cr, 22-25 Ni, 2.5-3.0 Mo, 1.5-2.0 Cu, bal Fe.
Non-magnetic stainless steel casting.
Not hardenable by heat treatment.
For paper pulp machinery, food machinery. ACI CN-7M.

LEBANON 431.
M-74; 0.16-0.22 C, 15.0-16.5 Cr, 1.5-2.5 Ni, bal Fe.
Corrosion resistant steel casting; oil or air hardenable.
For special purpose hardenable components as impellers, pumps, fuel controls. AISI 431; SAE 51431.

LEBANON 440 C.
M-74; 0.95-1.20 C; 16-18 Cr, 0.75 max Mo, bal Fe.
Corrosion resistant steel casting.
Oil or air hardenable to above 55 Rock C.
For stainless ball bearings, races.
AISI 440 C; SAE 51440 C.

LEBANON 1005 replaced by **LEBANON 1010.**

LEBANON 1010.
M-74; 0.13 max C, 0.7 max Si, 0.5 max Mn, bal Fe.
For electro magnet components.

LEBANON 1040.
M-74; 0.35-0.45 C, 0.6 max Si, 0.6-0.8 Mn, bal Fe.
Medium carbon steel casting.
No. + T.: 80,000 TS; 40,000 YS; 18 El. ASTM A148-73 Gr 80-40.

LEBANON 4140.
M-74; 0.35-0.45 C, 0.60 max Si, 0.6-0.8 Mn, 0.8-1.2 Cr, 0.2-0.3 Mo, bal Fe.
Medium carbon low alloy steel casting.
Nor. + T: 100,000 TS; 80,000 YS; 15 El.
Can be oil hardened to about 200,000 psi TS.
Structural parts; AISI 4140.

LEBANON 4330.
M-74; 0.25-0.35 C, 0.60 max Si, 0.6-0.8 Mn, 0.6-1.25 Cr, 1.25-2.50 Ni 0.25-0.50 Mo, bal Fe.
Medium carbon low alloy steel casting.
Oil hardenable to 140,000-180,000 TS.
For structural parts, gears, dies, wheels, housings.
AISI 4330.

LEBANON 4335.
M-74; 0.30-0.35 C, 0.60 max Si, 0.6-0.8 Mn, 0.60-1.25 Cr, 1.25-2.50 Ni, 0.25-0.50 Mo, bal Fe.
Medium carbon low alloy steel casting.
Oil hardenable to 180,000-240,000 psi TS.
For structural parts requiring good strength and toughness. AISI 4335.

LEBANON 4340.
M-74; 0.35-0.45 C, 0.60 max Si, 0.6-0.8 Mn, 0.60-1.25 Cr, 1.25-2.50 Ni, 0.25-0.50 Mo, bal Fe.
Medium carbon low alloy steel casting.
Oil hardenable in fairly heavy sections to 200,000-260,000 psi TS.
For heavily loaded structural parts. AISI 4340.

LEBANON 8615.
M-74; 0.10-0.20 C, 0.60 max Si, 0.6-0.9 Mn, 0.4-0.9 Cr, 0.4-1.1 Ni, 0.15-0.25 Mo, bal Fe.
Low carbon low alloy steel casting.
For structural and carburized parts. AISI 8615.

LEBANON 8630.
M-74; 0.25-0.35 C, 0.60 max Si, 0.60-0.95 Mn, 0.4-0.9 Cr, 0.4-1.1 Ni, 0.15-0.25 Mo, bal Fe.
Medium carbon low alloy steel casting.
Nor. + Tem: 90,000 TS; 60,000 YS; 20 El.
Weldable; water or oil quench. Structural parts.
ASTM A148-73 Gr 90-60.

LEBANON 8630/1.
M-74; 0.25-0.35 C, 0.60 max Si, 0.60-0.95 Mn, 0.4-0.9 Cr, 0.4-1.1 Ni, 0.15-0.25 Mo, bal Fe.
Medium carbon low alloy steel casting.
Quench and Temper condition: 105,000 TS; 85,000 YS; 17 El; 220-227 Brin.
Structural parts; can be re-heat treated; water or oil quench. ASTM A148-73 Gr 105-85.

LEBANON 8630/2.
M-74; 0.25-0.35 C, 0.60 max Si, 0.60-0.95 Mn, 0.4-0.9 Cr, 0.4-1.1 Ni, 0.15-0.25 Mo, bal Fe.
Medium carbon low alloy steel castings.
Quench and Temper Condition: 120,000 TS; 95,000 YS; 14 El; 250-311 Brin.
Structural parts; can be re-heat treated; water or oil quench.
ASTM A148-73 Gr 120-95.

LEBANON 8630/3.
M-74; 0.25-0.35 C, 0.60 max Si, 0.60-0.95 Mn, 0.4-0.9 Cr, 0.4-1.1 Ni, 0.15-0.25 Mo, bal Fe.
Medium carbon low alloy steel castings.
Quench and Temper Condition: 150,000 TS; 125,000 YS; 9 El; 310-375 Brin.
For structural parts, gears, rollers, sprockets, rollers, housings.
ASTM A148-73 Gr 150-125.

LEBANON C5.
M-74; 0.20 max C, 0.75 max Si, 0.4-0.7 Mn, 4.0-6.5 Cr, 0.45-0.65 Mo, bal Fe.
Low carbon alloy steel casting.
Nor. + T; 90,000 TS; 60,000 YS; 18 El.
Air hardenable.
For valves, bonnets, fittings, yokes, flanges, to 1200°F.
ASTM A217-75 Gr C5; ASME SA 217 Gr. C5.

LEBANON C12.
M-74; 0.20 max C, 1.0 max Si, 0.35-0.65 Mn, 8.0-10.0 Cr, 0.9-1.2 Mo, bal Fe.
Low carbon, chromium steel castings.
Nor. + T; 90,000 TS; 60,000 YS; 18 El.
Air hardenable.
For valves, fittings for moderate corrosion.
ASTM A217-75 Gr C12; ASME SA 217 Gr. C12.

LEBANON CA6NM.
M-74; 0.06 max C, 1.0 max Si, 12.0 Cr, 4.0 Ni, 0.5 Mo, bal Fe.
For compressors, pump impellers, valves, and turbine castings.
ASTM A296-76 Gr. CA6NM; ASME SA487 Gr. CA6NM.

LEBANON CA15.
M-74; 0.15 max C, 11.5-14.0 Cr, 1.0 max Ni, bal Fe.
Corrosion resistant steel casting.
Oil or air hardenable, weldable.
For valves, pump parts, oil refinery equipment.
ACI CA-15; AISI 410; ASME SA 217 Gr. Ca15.

LEBANON CA15M.
M-74; 0.15 max C, 11.5-14.0 Cr, 1.0 max Ni, 0.40-0.80 Mo, 0.20-0.35 Se, bal Fe.
Free machining, corrosion resistant steel casting.
Oil or air hardenable.
For valve and pump parts requiring considerable machining.
Similar to AISI 416 Se.

LEBANON CA40.
M-74; 0.20-0.30 C, 11.5-14.0 Cr, 1.0 max Ni, bal Fe.
Corrosion resistant steel castings.
Oil or air hardenable to 180,000-240,000 psi.
For cylinder liners, valves, cutter blades.
ACI CA-40; AISI 420.

LEBANON CA40A replaced by **LEBANON CA40B.**

LEBANON CA40B.
M-74; 0.30-0.40 C, 12.0-14.0 Cr, 1.0 Mo, bal Fe.
Corrosion resistant steel castings.
Oil or air hardenable to above 48 Rc.
For glass molds and plungers of mirror quality finish.

LEBANON CC50.
M-74; 0.50 max C, 26-30 Cr, 4.0 max Ni, bal Fe.
Corrosion resistant steel casting.
Excellent resistance to dilute sulphuric acid, in mine waters, and other dilute acids.
For mine pumps, chemical plants.
AISI CC-50; AISI 446.

LEBANON CD.
M-74; 0.04 max C, 26.0 Cr, 5.5 Ni, 2.0 Mo, 3.0 Cu, bal Fe.
Corrosion resistant steel casting.
Precipitation hardening alloy; very good corrosion resistance.
ACI CD-4MCu.

LEBANON CE30.
M-74; 0.30 max C, 26-30 Cr, 8-11 Ni, bal Fe.
Stainless steel casting; resistant to sulphurous acid.
For pulp and paper mill equipment, mine equipment, pumps and valves.
ACI CE-30.

LEBANON CF3.
M-74; 0.03 max C, 18-21 Cr, 8-11 Ni, bal Fe.
Non-magnetic stainless steel casting.
Weldable, not hardenable by heat treatment.
For valves, pumps, headers for chemical plants.
ACI CF3; AISI 304L; ASME SA 351 Gr. CF3.

LEBANON CF3A.
M-74; 0.03 max C, 17-21 Cr, 8-12 Ni, bal Fe.
Non-magnetic stainless steel casting; solution treated. Weldable, not hardenable by heat treatment.
Good resistance to stress corrosion cracking.
For valves, pumps in chemical plants.
ASTM A351-76 Gr CF-3A; ASME SA 351 Gr. CF3A.

LEBANON CF3M.
M-74; 0.03 max C, 18-21 Cr, 9-12 Ni, 2.0-3.0 Mo, bal Fe.
Non-magnetic stainless steel casting.
Weldable, not hardenable by heat treating.
Exceptional corrosion resistance; pumps, valves, impellers, fittings.
ACI CF-3M; AISI 316L; ASME SA 351 Gr. CF3M.

LEBANON CF8.
M-74; 0.08 max C, 18-21 Cr, 8.0-11.0 Ni, bal Fe.
Non-magnetic stainless steel casting; quenched.
Not hardenable by heat treatment.
General purpose stainless casting.
ACI CF-8; AISI 304; ASME SA351 Gr. CF8.

LEBANON CF8A.
M-74; 0.08 C, 18-21 Cr, 8-11 Ni, bal Fe.
Non-magnetic stainless steel casting; solution treated.
Not hardenable by heat treatment.
Sol. treat increases ferrite content to improve resistance to stress corrosion cracking.
ASTM A351-64 Gr CF-8A.

LEBANON CF8C.
M-74; 0.08 max C, 18-21 Cr, 9-12 Ni, Cb/Ta = 8 x %C min, bal Fe.
Non-magnetic stainless steel casting; stabilized; Weldable without reheat treatment.
For welded stainless assemblies. AISI 347; ACI CF-8C; ASME SA 351 Gr. CF8C.

LEBANON CF8F.
M-74; 0.08 max C, 18-21 Cr, 9-12 Ni, 1.5 max Mo, 0.20-0.35 Se, 0.15 max P, bal Fe.
Non-magnetic stainless steel casting, free machining type; not hardenable heat treatment; not recommended for welding.
For castings requiring much machining; AISI 303 Se.

LEBANON CF8M.
M-74; 0.08 max C, 18-21 Cr, 9-12 Ni, 2.0-3.0 Mo, bal Fe.
Non-magnetic stainless steel casting. Not hardenable by heat treatment. Exceptional resistance to corrosion.
For chemical plant equipment. ACI CF-8M; AISI 316; ASME SA 351 Gr. CF8M.

LEBANON CF 20.
M-74; 0.20 max C, 18-21 Cr, 8-11 Ni, bal Fe.
Non-magnetic stainless steel casting. Not hardenable by heat treatment.
General purpose stainless cast components.
ACI CF-20; AISI 302.

LEBANON CG8M.
M-74; 0.08 max C, 18-21 Cr, 9-12 Ni, 3.0-3.5 Mo, bal Fe.
Non-magnetic stainless steel casting. Not hardenable by heat treatment. Excellent corrosion resistance, especially in reducing environment. AISI 317.

LEBANON CH20.
M-74; 0.20 max C, 22-26 Cr, 12-15 Ni, bal Fe.
Non-magnetic stainless steel casting. Not hardenable by heat treatment. Improved corrosion resistance in hot dilute sulphuric acid.
For digester parts, pumps, strainers.
ACI CH-20; AISI 309; ASME SA 351 Gr. CH20.

LEBANON CHW.
M-74; 0.15-0.30 C, 22-25 Cr, 11-14 Ni, 2.5-3.5 W, bal Fe.
Non-magnetic stainless steel casting. Not hardenable by heat treatment.
For flanged rings and tail cone rings requiring both corrosion and heat resistance.

LEBANON CK20.
M-74; 0.25 max C, 23-27 Cr, 19-22 Ni, bal Fe.
Non-magnetic stainless steel casting. Not hardenable by heat treatment. Improved resistance to corrosion and high temperatures; weldable.
For jet engine parts, pumps, valves.
ACI CK-20 AISI 310; ASME SA 351 Gr. CK20.

LEBANON CK45.
M-74; 0.35-0.45 C, 23-27 Cr, 19-22 Ni, bal Fe.
Non-magnetic stainless steel casting. Not hardenable by heat treatment.
Heat resistant alloy for use in valves and fittings for chemical and refinery parts.

LEBANON CN3M.
M-74; 0.03 max C, 19-22 Cr, 27.5-30.5 Ni, 2.0-3.0 Mo, 3.0-4.0 Cu, bal Fe.
Non-magnetic stainless steel casting. Weldable; not hardenable by heat treatment. Good resistance to hot sulphuric acid.
For paper pulp machinery, food machinery.

LEBANON CN7M.
M-74; 0.07 max C, 19-22 Cr, 27.5-30.5 Ni, 2.0-3.0 Mo, 3.0-4.0 Cu, bal Fe.
Non-magnetic stainless steel casting. Not hardenable by heat treatment. Good resistance to hot sulphuric acid.
For paper pulp machinery, food machinery.
ACI CN-7M; ASTM A296-76 Gr CN-7M; ASME SA 351 Gr. CN7M.

LEBANON CU-NI.
M-74; 0.15 max C, 28-32 Ni, 0.25-1.0 Fe, 1.5 max Cb, bal Cu.
Corrosion resistant alloy casting for marine pumps, valves, fittings ASTM B369-72.

LEBANON H810 (INCOLOY).
M-74; 20.0 Cr, 32.0 Ni, bal Fe.
Heat resistant steel casting. Maximum operating temperature 2000°F (1093°C).

LEBANON HAB.
M-74; 0.12 max C, 26-30 Mo, 4-7 Fe, 0.6 max V, 2.5 max Co, 1.0 max Cr, bal Ni.
Corrosion resistant alloy casting; for valves, pump parts, fittings in hydrochloric acid environment.
Similar to Hastelloy B; ASTM A494-76.

LEBANON HAC.
M-74; 0.12 max C, 15.5-17.5 Cr, 16-18 Mo, 4.5-7.0 Fe, 3.75-5.25 W, 0.2-0.4 V, 2.5 max Co, bal Ni.
Corrosion resistant alloy casting for use with wet chlorine gas, hypochloric acid and other acids as valves and fittings.
ASTM A494-76; Hastelloy C.

LEBANON HE.
M-74; 20.0 Cr, 9.0 Ni, bal Fe.
Heat resistant steel casting. Maximum operating temperature 2000°F (1093°C).
ACI HE; SAE 70312; ASTM 297-76 Gr He.

LEBANON HH.
M-74; 25.0 Cr, 12.5 Ni, bal Fe.
Heat resistant steel casting. Maximum operating temperature 2000°F (1093°C).
ACI HH: SAE 70309; ASTM A297-76 Gr HH.

LEBANON HK.
M-74; 26.0 Cr, 20.0 Ni, bal Fe.
Heat resistant steel casting. Maximum operating temperature 2100°F.
ACI HK; SAE 70310; ASTM A297-63 Gr HK.

LEBANON HP.
M-74; 0.50 C, 26.0 Cr, 35.0 Ni, bal Fe.
Heat resisting steel casting. Service temperature 2000°F (1093°C)

LEBANON HR.
M-74; 23.0 Cr, 11.5 Ni, bal Fe.
Heat resistant steel casting.

LEBANON HT.
M-74; 15.0 Cr, 35.0 Ni, bal Fe.
Heat resistant steel casting. Maximum operating temperature 2000°F (1093°C).
ACI HT; SAE 70330; ASTM A297-76 Gr HT.

LEBANON HX.
M-74; 0.40 C, 18.0 Cr, 65.0 Ni, bal Fe.
Heat resistant steel casting for use in furnace parts and environments of highly corrosive nature.
ASTM A296-76.

LEBANON HY-80.
M-74; 0.20 max C, 0.5 max Si, 0.55-0.75 Mn, 1.35-1.65 Cr, 2.50-3.25 Ni, 0.3-0.6 Mo, bal Fe.
Low carbon alloy steel casting. Oil hardenable to 100,000-160,000 TS.
For valves, fittings; good strength with high impact energy. MIL-S-23008 B HY-80.

LEBANON HY-100.
M-74; 0.22 max C, 0.5 max Si, 0.55-0.75 Mn, 1.35-1.85 Cr, 2.75-3.50 Ni, 0.3-0.6 Mo, bal Fe.
Low carbon alloy steel casting; 100,000-120,000 yield. Oil hardenable.
For valves, fittings; good strength with good impact values; low notch sensitivity.
MIL-S-23008 B HY-100.

LEBANON INC.
M-74; 0.10 max C, 14-17 Cr, 6-10 Fe, 1.0-2.0 Cb/Ta, bal Ni.
Corrosion resistant alloy casting, resistant to alkalies, oxidizing and organic acids.
For food processing and textile processing equipment, high temperature service. ACI CY-40; (Inconel).

LEBANON LC2.
M-74; 0.25 max C, 0.6 max Si, 0.5-0.8 Mn, 2.0-3.0 Ni, bal Fe.
Low carbon, nickel alloy steel casting.
Nor., Q & T.: 65,000 TS; 40,000 YS; 24 El.
For valves for cryogenic service to -100°F (-73°C).
ASTM A352-76 Gr LC2; ASME SA352 Gr. LC2.

LEBANON LC3.
M-74; 0.15 max C, 0.6 max Si, 0.5-0.8 Mn, 3.0-4.0 Ni, bal Fe.
Low carbon, nickel alloy steel casting.
Nor.; Q+T: 65,000 TS; 40,000 YS; 24 El.
Valves for cryogenic service to -150°F (-101°C).
ASTM A352-76 Gr. LC3; ASME SA352 Gr. LC3.

LEBANON LCB.
M-74; 0.30 max C, 0.6 max Si, 1.0 max Mn, bal Fe.
Low carbon steel casting.
Nor. + T.: 65,000 TS; 35,000 YS; 24 El.
General purpose cast housings.
ASTM A352-76 Gr. LCB; ASME SA352 Gr. LCB.

LEBANON ME.
M-74; 0.30 max C, 60 min Ni, 26-33 Cu, 1.5-3.5 Fe, 1.0-3.0 Cb/Ta.
Corrosion resistant alloy casting, resistant to mineral and organic acids, sea water, caustic solutions.
For valves and fittings for marine service.
Fed: QQ-N-288-Comp E; (Monel).

LEBANON NI.
M-74; 1.0 max C, 1.25 max Cu, 1.25 max Fe, bal Ni.
Nickel casting. As cast: 45,000 TS; 20,000 YS; 15 El.
Resistant to many industrial chemicals.
For equipment for food processing, pharmaceuticals.
ACI CZ-100.

LEBANON WC1.
M-74; 0.25 max C, 0.6 max Si, 0.5-0.8 Mn, 0.45-0.65 Mo, bal Fe.
Low carbon, low alloy steel casting.
For valves, fittings, turbines and other pressure castings to 1100°F.
ASTM A217-75 Gr WC1: A352-76 Gr LC1; ASME SA352 Gr. LC

LEBANON WC6.
M-74; 0.20 max C, 0.6 max Si, 0.5-0.8 Mn, 1.0-1.5 Cr, 0.45-0.65 Mo, bal Fe.
Low carbon, low alloy steel casting.
Nor. & T.: 70,000 TS; 40,000 YS; 20 El.
For valves, fittings, turbines and other pressure castings for service to 1200°F (649°C).
ASTM A217-75 Gr WC6; ASME SA217 Gr. WC6.

LEBANON WC6A.
M-74; 0.25 max C, 0.6 max Si, 0.7 max Mn, 0.4-0.7 Cr, 0.4-0.6 Mo, bal Fe.
Low carbon, low alloy steel casting.
Nor. & T.: 70,000 TS; 40,000 YS; 22 El.
For valves, fittings, turbines and other pressure castings for service below 1200°F (649°C).
ASTM A356-75 Gr 5.

LEBANON WC6B.
M-74; 0.30 max C, 0.75 max Si, 1.0 max Mn, 1.5-2.25 Cr, 0.45-0.65 Mo, bal Fe.
Medium carbon low alloy steel casting.
Nor. Q & T.: 105,000 TS; 85,000 YS; 15 El.
For valves, fittings, turbines and other pressure castings for service below 1200°F (649°C).

LEBANON WC9.
M-74; 0.18 max C, 0.6 max Si, 0.4-0.7 Mn, 2.0-2.75 Cr, 0.9-1.2 Mo, bal Fe.
Low carbon alloy steel casting. Nor. & T.: 70,000 TS; 40,000 YS; 20 El.
For valves, fittings, turbines, and other pressure castings for service below 1200°F (649°C).
ASTM A217-75 Gr WC9; ASME SA217 Gr. WC9.

LEBANON WCA.
M-74; 0.25 max C, 0.6 max Si, 0.7 max Mn, bal Fe.
Low carbon steel casting. Nor. & T.: 60,000 TS; 30,000 YS; 24 El.
ASTM A216-75 Gr WCA; AISI 1020; ASME SA216 Gr. WCA.

LEBANON WCB.
M-74; 0.30 max C, 0.60 max Si, 1.0 max Mn, bal Fe.
Low carbon steel casting.
70,000-95,000 psi (485-655 MPa) TS; 36,000 psi (250 MPa) YS.
ASTM A216-75 Gr. WCB; ASME SA216 Gr. WCB.

LEBANON WCB/2 replaced by LEBANON WCB.

LEBANON WCC.
M-74; 0.25 max C, 1.2 max Mn, bal Fe.
Low carbon steel casting.
70,000-95,000 psi (485-655 MPa) TS; 40,000 psi (275 MPa) YS.
For valves, fittings, turbines for service to 1100°F (593°C).
ASTM A216-75 Gr. WCC; ASME SA216 Gr. WCC.

LECHESNE.
90-60 Cu, 10-40 Ni, 0.2-0.05 Al.
For chemical engineering equipment; corrosion resistant.

LECO.
M-688; 0.33 C, 0.75 Cr, 0.30 Ni, 0.75 Mo, 0.75 Cu, 0.15 Ti, bal Fe.
For chisels, dies, punches, pneumatic tools; oil hardened, shock resistant.

LECO EXTRA.
M-688; 0.44 C, 0.75 Cr, 0.35 Ni, 0.4 Mn, 0.75 Mo, 0.65 Si, 0.75 Cu, 0.15 Ti, bal Fe.
For dies, rivet sets, chisels, punches, cutters; fatigue and abrasion resistant.

LECO NON-TEMPERING.
M-57; C, alloy, bal Fe.
For dies, tools; water hardened, non-tempering.

LEDDEL ALLOY.
90 Zn, 5 Cu, 5 Al.
For die castings.

LEDDEL BEARING.
87.5 Zn, 6.26 Cu, 6.25 Al.
For bearings; anti-friction.

LEDEBUR'S BEARING-1.
85 Zn, 10 Sb, 5 Cu.
For bearings; anti-friction.

LEDEBUR'S BEARING-2.
77 Zn, 5.5 Cu, 18 Sn.
For bearings; anti-friction.

LEDLOY 300.
M-240; 0.12 C, 1.0 Mn, 0.07 P, 0.30 S, 0.15-0.35 Pb, bal Fe.
79 ksi TS; 71 ksi YS; 16 El; 52 RA.
Free cutting; for screw machine products.
AISI 12L14.

LEDLOY 375.
M-240; 0.09 C, 1.0 Mn, 0.04-0.09 P, 0.26-0.35 S, 0.15-0.35 Pb, 0.03 Te, bal Fe.
79 ksi TS; 71 ksi YS; 16 El; 52 RA.
Free machining steel, particularly for threaded parts, gears.

LEDLOY 1018.
M-67; 0.15-0.20 C, 0.60-0.90 Mn, 0.15-0.35 Pb, bal Fe.
Free-machining steel.

LEDLOY 1117.
M-67; 0.14-0.20 C, 1.0-1.3 Mn, 0.040 max P, 0.08-0.13 S, 0.15-0.35 Pb, bal Fe.
Free machining steel.

LEDLOY 1137.
M-67; 0.32-0.39 C, 1.35-1.65 Mn, 0.040 max P, 0.08-0.13 S, 0.15-0.35 Pb, bal Fe.
Free machining medium carbon steel.

LEDLOY 1144.
M-67; 0.40-0.48 C, 1.35-1.65 Mn, 0.040 max P, 0.24-0.33 S, 0.15-0.35 Pb, bal Fe.
Free machining medium carbon steel.

LEDLOY 1215.
M-67; 0.09 max C, 0.75-1.05 Mn, 0.04-0.09 P, 0.26-0.35 S, 0.15-0.35 Pb, bal Fe.
Free machining low carbon steel.

LEDLOY 4140.
M-67; 0.39 C, 0.80 Mn, 0.03 S, 0.24 Si, 0.84 Cr, 0.09 Ni, 0.18 Mo, 0.15-0.35 Pb, bal Fe.
Heat treated: 158,000 TS; 144,000 YS; 16 El; 56 RA; 321 Brin.
For fasteners, gears, camshafts, housings, machine tools, crankshafts.
Free-cutting. Oil hardening.

LEDLOY A.
M-67; 0.09 max C, 0.85-1.15 Mn, 0.04-0.09 P, 0.26-0.35 S, 0.15-0.35 Pb, bal Fe.
Free machining low carbon steel.

LEDLOY-AX.
M-516, M 67; 0.08 C, 1.07 Mn, 0.075 P, 0.32 S, 0.02 Si, 0.003 Ni, 0.21 Pb, 0.048 Te, bal Fe.
Bar: 75,600 TS; 71,900 YS; 13 El; 37 RA; B 91 Rock.
For screw machine products, fasteners, screws, bolts, spark wheels. Free machining.

LEDLOY C-1045.
M-67; 0.40-0.50 C, 0.15-0.35 Pb, bal Fe.
Rolled: 98,000 TS; 65,000 YS; 28 El; 52 RA.
For machinery parts; water hardening; free-cutting.

LEDRITE 2.
M-141; 61 Cu, 1.8 Pb, bal Zn.
Hard: 55,000 TS; 42,000 YS; 30 El.
For screw machine parts, hardware; free-cutting.

LEDRITE 6.
M-141; 61 Cu, 3.4 Pb, bal Zn.
Hard: 58,000 TS; 45,000 YS; 25 El.
For screw machine parts, hardware; free-cutting.

LEDRITE BRASS.
M-141; 61 Cu, 3.25 Pb, bal Zn.
Soft: 45,000 TS; 35 El.
Hard: 80,000 TS; 10 El.
For screw machine parts; free cutting.

"LEDUCT" 370-17.
M-1762; 3.5-3.8 C, 2.0-2.6 Si, 0.10-0.60 Mn, 0.005-0.01 S, 0.06 max P 0.03-0.055 Mg, bal Fe.
Ann: 370 N/mm^2 (53,660 psi) TS; 230 N/mm^2 (33,360 psi) YS; El; 115-179 Brin.
Spheroidal graphitic iron for automotive, truck and tractor industries.

"LEDUCT" 420/12.
M-1762; 3.5-3.8 C, 2.0-2.6 Si, 0.10-0.60 Mn, 0.005-0.01 S, 0.06 max P, 0.03-0.055 Mg, bal Fe.
As cast: 420 N/mm^2 (60,920 psi) TS; 250 N/mm^2 (36,260 psi) 12 El; 149-201 Brin.
Spheroidal graphitic iron for automotive, truck and tractor industries.

"LEDUCT" 500-7.
M-1762; 3.5-3.8 C, 2.0-2.6 Si, 0.10-0.60 Mn, 0.005-0.01 Si, 0.06 max P, 0.03-0.055 Mg, bal Fe.
As cast: 500 N/mm^2 (72,520 psi) TS; 310 N/mm^2 (44,960 psi) 7 El; 170-241 Brin.
Spheroidal graphitic iron for automotive, truck and tractor industries.

"LEDUCT" 600-3.
M-1762; 3.5-3.8 C, 2.0-2.6 Si, 0.10-0.60 Mn, 0.005-0.01 S, 0.06 max P 0.03-0.055 Mg, bal Fe.
As cast: 600 N/mm^2 (87,020 psi) TS; 350 N/mm^2 (50,760 psi) 3 El; 192-269 Brin.
Spheroidal graphitic iron for automotive, truck and tractor industries.

"LEDUCT" 700-2.
M-1762; 3.5-3.8 C, 2.0-2.6 Si, 0.10-0.60 Mn, 0.005-0.01 S, 0.06 max P 0.03-0.055 Mg, bal Fe.
As cast: 700 N/mm^2 (101,530 psi) TS; 400 N/mm^2 (58,020 psi) 2 El; 229-302 Brin.
Spheroidal graphitic iron for automotive, truck and tractor industries.

"LEDUCT" 800-2.
M-1762; 3.5-3.8 C, 2.0-2.6 Si, 0.10-0.60 Mn, 0.005-0.01 S, 0.06 max P 0.03-0.055 Mg, bal Fe.
As cast: 800 N/mm^2 (116,030 psi) TS; 460 N/mm^2 (66,720 psi) 2 El; 248-352 Brin.
Spheroidal graphitic iron for automotive, truck and tractor industries.

LEGAL.
M-199; 0.5-2.0 Mg, 0.3-1.5 Si, 0-15 Mn, bal Al.
For light alloy parts; corrosion resistant, hardenable.

LEGA-Y.
M-1176; 3.8-4.2 Cu, 1.3-1.7 Mg, 1.8-2.3 Ni, bal Al.
Heat treated: 36,000-47,000 TS; 32,000-43,000 YS; 0.5-0.3 El; 95-115 Brin.
For light alloy parts; age-hardened.

LEHIGH DD1.
M-688; C, alloy, bal Fe.
For machine parts, gears, shafts, mandrels; oil hardened.

LEHIGH H.
M-24; 1.6 C, 11.5 Cr, 0.8 Mo, 0.4 V, bal Fe.
For gages, punches, master hobs; Type D2; oil hardened.

LEHIGH L.
M-24; 0.85 C, 11.5 Cr, 1.05 Ni, 0.3 V, 0.45 Mo, bal Fe.
For dies, shear blades, punches; abrasion resistant, non-deforming.

LEHIGH N.C.
M-688; 0.70 C, 1 Cr, 1.6 Ni, 0.35 Cu, 0.4 Mn, 0.2 Si, bal Fe.
For shear blades, rivet sets, punches, dies; oil hardened, shock resistant.

LEHIGH SPECIAL NO. 3.
M-688; 0.7-1.2 C, bal Fe.
For tools and dies; general use.

LEHIGH SPECIAL NO. 5.
M-688; 0.7-1.4 C, bal Fe.
For tools and dies; abrasion resistant.

LEHIGH SS.
M-688; 0.75 C, 4.5 Cr, 12 Co, 2 V, 20 W, bal Fe.
For lathe and planer tools, reamers, taps, broaches, hobs; high speed steel, oil hardened.

LEHIGH XXX.
M-688; 0.7 C, 4 Cr, 1 V, 18 W, bal Fe.
For lathe and planer tools, drills, taps, reamers; high speed steel; Type T2.

LEHIGH ZINC.
M-91; 99.99 Zn.
For hot dip and continuous line galvanizing; brass mill products.
ASTM B6; Fed: QQ-Z-351b.

LEICHTMETALL MN 20.
M-457; 1-2 Mn, bal Al.
Soft: 13,000 TS; 6000 YS; 35 El; 35 Brin.
Hard: 43,000 TS; 36,000 YS; 4 El; 70 Brin.
For light alloy parts; corrosion resistant.

LEICHTMETALLWERKE M115.
M-116; 1.0-1.5 Mn, 0.3 Cr, bal Al.
Soft: 16,000 TS; 6000 YS; 40 El.
Hard: 29,000 TS; 27,000 YS; 10 El.
For cooking utensils, tanks, furniture; good forming and welding properties.

LEICHTMETALLWERKE MZB.
M-116; 2.5-5.0 Cu, 0.2-1.8 Mg, 0.3-1.5 Mn, 0.5-2.5 Pb, Sn, Cd, Bi, bal Al.
For screw machine products; free-cutting.

LEKTROCAST.
M-157; 3.0-3.4 C, 2 Si, 1.5-2.5 Ni, bal Fe.
Cast: 50,000 TS; 225-280 Brin.
For castings, draw dies for metal stamping; wear resistant.

"LEMAG" see "LEDUCT".

LEMARQUANDS ALLOY.
39 Cu, 7 Ni, 37 Zn, 9 Sn, 8 Co.
For cheap jewelry; corrosion resistant.

LE MATS METAL.
80 Cu, 5 Zn, 16 Ni, 5 Fe, 2 Sn, 1 Co.
For cheap jewelry; v. "Lutecin."

"LEMAX" HEAT TREATED PEARLITIC 38/30/4.
M-1762; 2.3-2.6 C, 1.3-1.55 Si, 0.4-0.57 Mn, 0.15-0.25 S, 0.08 max P, bal Fe.
Ht Tr pearlitic malleable iron castings.
85,100 TS; 67,200 YS; 4 El; 197-255 Brin.
For automotive, truck and tractor industries.

"LEMAX" HEAT TREATED PEARLITIC 45/40/2.
M-1762; 2.3-2.6 C, 1.3-1.55 Si, 0.4-0.57 Mn, 0.15-0.25 S, 0.08 max P, bal Fe.
Ht Tr pearlitic malleable iron castings.
100,800 TS; 89,600 YS; 2 El; 241-269 Brin.
For automotive, truck and tractor industries.

LE MOYNE ALCOCAN.
M-1196; 0.7 C, Ni, Cr, bal Fe.
For hand tools, chisels; oil or air hardened.

LE MOYNE BEST.
M-1196; 1.2 C, 0.3 Cr, 1.2 W, bal Fe.
For broaches, reamers, dies, cutters; water hardened.

LE MOYNE EXTRA.
M-1196; 0.7-1.0 C, bal Fe.
For tools, drills, punches; water hardened.

LE MOYNE FINISHING.
M-1196; 1.3 C, 4 W, 4 Mo, bal Fe.
For finishing tools and cutters; water or oil hardened.

LE MOYNE HAK.
M-1196; 0.90-1.05 C, 3.7-4.1 Cr, bal Fe.
Annealed: 90,000 TS; 78,000 YS; 28 El; 57 RA; 195 Brin.
Heat treated: 205,000 TS; 168,000 YS; 2 El; 12 RA; 412 Brin.
For riveters, gripper dies, compression dies; wear and heat resistant.

LE MOYNE HS.
M-1196; 0.56-0.75 C, 18 W, 4 Cr, 1 V, bal Fe.
Annealed: 105,000 TS; 75,000 YS; 14 El; 17 RA; 255 Brin.
For chasers, taps, reamers, drills, broaches; high speed steel, oil hardened.

LE MOYNE K760.
M-1196; 0.72 C, 17.5 W, 4 Cr, 4.5 Co, 1 V, 0.5 Mo, bal Fe.
For reamers, taps, hobs, drills, lathe tools; high speed steel, oil hardened.

LE MOYNE NON-SHRINK.
M-1196; 0.9 C, 1.2 Mn, 0.5 Cr, 0.5 W, bal Fe.
For punches, tools, crimpers, cutters; oil hardened, nondeforming.

LE MOYNE PYRO.
M-1196; 0.33 C, 10.2 W, 3.5 Cr, 0.45 V, bal Fe.
For punches, shears, hot dies; hot work steel, oil hardened.

LE MOYNE SPECIAL.
M-1196; 0.8-1.1 C, bal Fe.
For tools, drills, punches, cutters; water hardened.

LE MOYNE SUPREME.
M-1196; 0.7 C, W, Co, Mo, bal Fe.
For tools, dies, cutters; high speed steel.

LENIN-T58.
M-651; 0.16-0.20 C, 11.5-12.5 Cr, 0.5 Mo, 0.5-1.0 Ni, 0.15-0.25 V, 2.0-2.5 W, bal Fe.
Annealed: 80,000 TS; 40,000 YS; 24 El; B 95 Rock.
Hardened: 230,000 TS; 195,000 YS; 12 El; C 48 Rock.
For valves, bearings, cutlery, surgical instruments, gears; corrosion resistant, hardenable.

LENNALSIL.
M-1375; 0.6-1.4 Mg, 0.6-1.6 Si, 0.6-1.0 Mn, 0.3 max Cr, bal Al.
Annealed: 21,000 TS; 8000 YS; 24 El.
For window frames, fan blades, gutters, boats; good forming and welding properties.

LENNEDUR.
M-1375; 2.5-5.0 Cu, 0.2-1.8 Mg, 0.3-1.5 Mn, bal Al.
Annealed: 27,000 TS; 11,000 YS; 22 El; 47 Brin.
Heat treated: 72,000 TS; 57,000 YS; 130 Brin.
For aircraft structures and fittings, fasteners; age hardenable.

LENOX.
0.9 C, 1.5 Mn, 0.3 Mo, 0.3 Si, bal Fe.
Annealed: 200 Brin.
For tools, dies, punches, knives; water hardened.

LEONARD.
M-261; 0.9-1.0 C, bal Fe.
For tools, drills, punches; water hardened.

"LEPAZ" PEARLITIC 30/22/8.
M-1762; 2.3-2.6 C, 1.3-1.55 Si, 0.4-0.57 Mn, 0.15-0.25 S, 0.08 max P, bal Fe.
Pearlitic malleable iron castings.
67,200 psi TS; 49,280 psi YS; 8 El; 163-207 Brin.
For automotive, truck and tractor industries.

"LEPAZ" PEARLITIC 36/24/7.
M-1762; 2.3-2.6 C, 1.3-1.55 Si, 0.4-0.57 Mn, 0.15-0.25 S, 0.08 max P, bal Fe.
Pearlitic malleable iron castings.
80,600 TS; 53,800 YS; 7 El; 179-217 Brin.
For automotive, truck and tractor industries.

LEPPE HWF 48.
M-1797; 0.06 C, 18 Cr, 11 Ni, bal Fe.
Stainless steel for high temperature pipelines, pressure vessels.

LEPPESTAHL.
M-1306; 2.1 C, 11.5 Cr, bal Fe.
For blanking and forming dies, punches; oil hardened, non-deforming.

LEPPESTAHL EXTRA 6.
M-1306; 2.1 C, 11.5 Cr, 0.7 W, bal Fe.
For blanking and forming dies, punches; non-deforming, oil hardened.

LEPPESTAHL X.
M-1306; 1.65 C, 11.5 Cr, 0.1 V, bal Fe.
For blanking and forming dies, punches; air hardened, non-deforming.

LE PROVINOX.
M-French; C, alloy, bal Fe.
50,000-57,000 TS; 35,000-45,000 YS; 36 El.
For general engineering applications; semi-non-oxidizing steel.
Oil Q & Temp. 900°F: C 45 Rock.
For gas turbine compressor wheels and structural members.

LESCALLOY 300 M.
M-73; 0.42 S, 0.75 Mn, 1.65 Si, 1.8 Ni, 0.8 Cr, 0.4 Mo, 0.07 V, bal Fe.
Heat treated: 285,000-290,000 TS; 238,000-243,000 YS; 10-13 El; 32-47 RA.
For landing gears, air frames, missile cases, pressure vessels.
High strength structural steel, tough, shock resistant.

LESCALLOY 4330 VAC ARC.
M-73; 0.30 C, 0.85 Mn, 0.3 Si, 0.35 Cr, 1.8 Ni, 0.4 Mo, 0.07 V, bal Fe.
Vacuum arc structural steel.
Hardened: Up to 235,000 TS; 195,000 YS; 11 El; 47-50 Rock C.
For highly stressed structural parts.

LESCALLOY 4335 VAC ARC.
M-73; 0.35 C, 0.75 Mn, 0.5 Si, 0.8 Cr, 1.85 Ni, 0.35 Mo, 0.2 V, bal Fe.
Vacuum arc melted structural steel.
Hardened: Up to 275,000 TS; 240,000 YS; 10 El.
Deep hardening; for highly stressed structural parts.

LESCALLOY 4340 VAC ARC.
M-73; 0.40 C, 0.75 Mn, 0.3 Si, 0.8 Cr, 1.8 Ni, 0.2 V, 0.25 Mo, bal
Vacuum arc melted structural steel.
Hardened: Up to 276,000 TS; 222,000 YS; 11 El in 3 1/2 inch round sections.
Very deep hardening; for highly stressed structural parts.

LESCALLOY 6304 VAC ARC.
M-73; 0.45 C, 0.25 Si, 0.60 Mn, 1.0 Cr, 0.25 V, 0.50 Mo, bal Fe.
Quenched and tempered to 100,000-150,000 psi TS. It is used for high temperature bolts, jet engine shafts, fittings and high temperature lances.

LESCALLOY 8620, VAC ARC.
M-73; 0.18-0.23 C, 0.40-0.70 Ni, 0.40-0.60 Cr, 0.15-0.25 Mo, bal Fe.
Low alloy steel for case hardening for gears, cam shafts. AISI 8620.

LESCALLOY 9310 (AGT).
M-73; 0.10 C, 0.5 Mn, 0.25 Si, 3.25 Ni, 1.2 Cr, 0.12 Mo, bal Fe.
Vacuum arc melted, wrought.
Ht. Tr.: 331-363 Brin.
To be carburized for aircraft engine gears and pinions; oil quench.
ASM 6265 A.

LESCALLOY 52100 VAC ARC.
M-73; 1.0 C, 0.35 Mn, 0.3 Si, 1.5 Cr, bal Fe.
Vacuum arc melted, wrought.
Hardenable to 60-66 Rc, oil or water quench.
For aircraft engine bearings and instrument bearings for use below 400-450°F.
AMS 6444.

LESCALLOY BG 42 VAC ARC.
M-73; 1.15 C, 14.5 Cr, 4.0 Mo, 1.0 V, bal Fe.
For bearings operating in corrosive environment and at high temperature; stainless, retains hot hardness.

LESCALLOY D6AC.
M-73; 0.46 C, 0.75 Mn, 0.25 Si, 1.10 Cr, 0.60 Ni, 1.0 Mo, 0.12 V, bal Fe.
Heat treated: 228,000-280,000 TS; 195,000-250,000 YS; 7 El; 23-25 RA; C 46-53 Rock.
For solid fuel rocket motor cases, gears, crankshafts.
High strength structural steel, tough, shock resistant.

LESCALLOY EXPANDAL.
M-73; 0.60 C, 5.75 Mn, 0.2 Si, 10.0 Ni, bal Fe.
High thermal expansion alloy; co-efficient of thermal expansion 72°-600°F: 11.5-12.5x10^{-6}in/in/°F.
Austenitic, cold drawn: 140,000 TS.

LESCALLOY HP 9-4-20 VAC ARC.
M-73; 0.20 C, 0.10 max Si, 0.30 Mn, 0.75 Cr, 0.10 V, 9.0 Ni, 1.0 Mo 4.50 Co, bal Fe.
180,000 psi YS (typical); good weldability.
For pressure vessels, rocket motor cases, hulls for deep submersible vessels, aircraft components.

LESCALLOY HP 9-4-30 VAC ARC.
M-73; 0.30 C, 0.30 Mn, 0.05 Si, 1.0 Cr, 0.10 V, 1.0 Mo, 7.50 Ni, 4.50 Co, bal Fe.
Low alloy, tough, high strength steel, hardenable to 50-52 Rc; resists softening at elevated temperature; weldable.
For armor plate, underwater pressure vessels, rocket motor cases, aircraft structural parts.

LESCALLOY JETHETE M-152.
M-73; 0.12 C, 0.7 Mn, 0.12 Si, 11.25 Cr, 2.9 Ni, 1.6 Mo, 0.3 V, 0.035 N, bal Fe.
Nitrogen bearing, vacuum melted, hardenable stainless.
Oil Q & Temp. 900°F: C 45 Rock.
For gas turbine compressor wheels and structural members.

LESCALLOY M50 VAC ARC.
M-73; 0.85 C, 0.30 Mn, 0.20 Si, 4.10 Cr, 4.52 Mo, 1.0 V, bal Fe.
Hardenable to 63-64 Rc; used widely for aircraft engine bearings. Stands elevated temperatures.
AISI M50.

LESCALLOY MARVAC 250.
M-73; 0.01 C, 18.25 Ni, 8.0 Co, 5.0 Mo, 0.40 Ti, 0.10 Al, bal Fe.
Vacuum induction melt plus vacuum consumable electrode, maraging steel.
Ann: 290 Brin; Sol. tr & age: 250,000-265,000 TS; 240,000-255,000 YS; 9-13 El.
For high strength missile and aircraft components; high fatigue properties; weldable.

LESCALLOY MARVAC 300.
M-73; 0.01 C, 18.25 Ni, 9.0 Co, 5.0 Mo, 0.65 Ti, 0.10 Al, bal Fe.
Double vacuum melted maraging steel.
Ann: 300 Brin; Sol. tr & age: 290,000-300,000 TS; 280,000-290,000 YS; 6-9 El.
For high strength missile and aircraft components; high fatigue properties weldable.

LESCALLOY MP35N.
M-73; 35 Ni, 35 Co, 20 Cr, 10 Mo.
Annealed: 132,000 TS.
Work strengthened and aged: 200,000-300,000 TS; 160,000-290,000 YS; 9-18 El; 17-95 Ft. lbs. Charpy.
Good high temp. strength and corrosion resistance.

LESCALLOY NITRALLOY 135 M, VAC ARC.
M-73; 0.40 C, 0.6 Mn, 0.3 Si, 1.6 Cr, 1.2 Al, 0.2 Mo, bal Fe.
Vacuum arc melted nitriding steel.
Oil Q & Temp. 1200°F: 135,000 TS; 100,000 YS; 16 El; 280-340 Brin.
High case hardness develops when nitrided.
For aircraft gears and shafts.

LESCALLOY NITRALLOY N, VAC ARC.
M-73; 0.24 C, 0.3 Si, 0.6 Mn, 1.15 Cr, 1.25 Al, 3.5 Ni, 0.25 Mo, bal Fe.
Vacuum arc melted, age hardenable, nitriding steel.
Oil Q & Temp. 1200°F: C 29 Rc.
Nitride (and age) 975°F; Case: C 60 Rock min.
Core: C 38-43 Rock.
For aircraft gears and shafts.

LESCALLOY-UT18, VAC ARC.
M-73; 0.40 C, 0.60 Mn, 0.25 Si, 0.95 Mo, 0.2 Ni, 0.20 V, 3.25 Cr, bal Fe.
Ht. Tr.: 209,000 TS; 173,000 YS; 16 El; 60 Ra; 388-429 Brin.
For turbine and compressor rotor shafts.
High strength and good impact resistance up to 1000°F. Heat resistant.

LESCALLOY-UT19, VAC ARC.
M-73; 0.16 C, 0.45 Mn, 0.23 Si, 1.25 Cr, 4.25 Ni, 0.25 Mo, bal Fe.
Ht. Tr.: 212,000 TS; 145,000 YS; 18 El; 67 RA; 388-444 Brin. (core).
For gears, pinions, shafts, cams, aircraft components.
Carburizing steel, tough and shock resistance.

LESCO HW 108.
M-73; 0.38 C, 1.0 Si, 0.30 Mn, 3.75 Cr, 3.60 Mo, 0.50 V, 2.0 Co, bal Fe.
Hot work tool steel; for forging dies, hot working tools.

LESCO NINETEEN.
M-73; 0.40 C, 0.3 Si, 0.3 Mn, 4.25 Cr, 4.25 W, 2.0 V, 4.25 Co, bal Fe.
Hot work die steel; retains 50 rock C, hardness to 1100°F.
For forging dies, brass extrusion tooling, hot punches, brass die casting dies.
AISI Type H19. Hot work steel.

LESJOFORS GLA.
M-1459; 0.45 C, 1 Si, 1 Mn, bal Fe.
For springs; oil hardened.

LESJOFORS LB8MN.
M-1459; 0.40 C, 1.5 Mn, bal Fe.
For springs; water hardened, tough.

LESJOFORS LB11.
M-1459; 0.55 C, 0.5 Mn, bal Fe.
For wire ropes; water or oil hardened.

LESJOFORS LB12.
M-1459; 0.60 C, 0.5 Mn, bal Fe.
For wire ropes; water or oil hardened.

LESJOFORS LB13 "LEWI".
M-1459; 0.65 C, 0.6 Mn, bal Fe.
Heat treated: 160,000 TS; 114,000 YS; 12 El; 40 RA; 325 Brin.
For springs; water or oil hardened.

LESJOFORS LB13MN.
M-1459; 0.65 C, 1.5 Mn, bal Fe.
For springs; oil hardened, tough.

LESJOFORS LB14.
M-1459; 0.70 C, 0.6 Mn, bal Fe.
Heat treated: 175,000 TS; 128,000 YS; 12 El; 37 RA; 355 Brin.
For springs; oil hardened.

LESJOFORS LB15.
M-1459; 0.75 C, 0.5 Mn, bal Fe.
Heat treated: 180,000 TS; 130,000 YS; 10 El; 35 RA; 375 Brin.
For springs; oil hardened.

LESJOFORS LB16.
M-1459; 0.80 C, 0.4 Mn, bal Fe.
Heat treated: 188,000 TS; 145,000 YS; 12 El; 35 RA; 390 Brin.
For tools, drills, taps, hobs, reamers; Type W1; water hardened.

LESJOFORS LB50.
M-1459; 0.50 C, 0.8 Mn, 0.1 V, Cr, bal Fe.
For springs; oil hardened, tough.

LESJOFORS LB85.
M-1459; 0.40 C, 0.8 Mn, 0.8 Cr, 1.25 Ni, bal Fe.
For gears, shafts, machine tool parts; oil hardened.

LESJOFORS LB145.
M-1459; 0.65 C, 1.3 Si, 0.5 Mn, 0.5 Cr, bal Fe.
For springs, punches; oil hardened, tough.

LESJOFORS LERO.
M-1459; 0.1 C, 18 Cr, 8 Ni, bal Fe.
For springs; Type 302; stainless.

LESJOFORS LSMN.
M-1459; 0.55 C, 1.75 Si, 1.5 Ni, 0.85 Mn, bal Fe.
For springs; oil hardened, tough.

LESJOFORS LSMO.
M-1459; 0.55 C, 1.75 Si, 0.8 Mn, 0.2 Cr, bal Fe.
For springs; oil hardened, tough.

LESTEM.
M-688; C, alloy, bal Fe.
For dies, tools; water hardened, non-tempering.

L.F.M. 4-6 CR-MO.
M-573; 0.13-0.20 C, 4-7 Cr, Mo, bal Fe.
For corrosion resistant parts; corrosion resistant.

LFM GR.A.
M-573; 0.15-0.22 C, 0.65-0.85 Mn, 0.35-0.50 Si, bal Fe.
Annealed: 73,000 TS; 41,000 YS; 22 El; 58 RA; 140 Brin.
For nails, rivets, gears, fasteners, fan blades; Type 1020; case hardened.

LFM GR.B.
M-573; 0.20-0.26 C, 0.65-0.85 Mn, 0.35-0.50 Si, bal Fe.
Annealed: 75,000 TS; 45,000 YS; 20 El; 55 RA; 160 Brin.
For rivets, gears, fan blades, fasteners; Type 1022; water hardened.

LFM GR.C.
M-573; 0.25-0.35 C, 0.65-0.80 Mn, 0.35-0.50 Si, bal Fe.
Hot rolled: 80,000 TS; 50,000 YS; 30 El; 56 RA; 165 Brin.
For gears, bolts, crankshafts, brackets; Type 1030; water hardened.

LFM GR.D.
M-573; 0.30-0.40 C, 0.65-0.80 Mn, 0.35-0.50 Si, bal Fe.
Hot rolled: 85,000 TS; 72,000 YS; 26 El; 51 RA; 180 Brin.
For armature shafts, gears, axles, bolts; screws; Type 1035; water hardened.

LFM GR.E.
M-573; 0.40-0.50 C, 0.65-0.80 Mn, 0.35-0.50 Si, bal Fe.
Hot rolled: 98,000 TS; 59,000 YS; 24 El; 45 RA; 212 Brin.
For axles, gears, bolts, tie rods, rails; Type 1045; water hardened.

LFM GR.G.
M-573; 0.26-0.32 C, 1.1-1.3 Mn, 0.35-0.50 Si, 0.2-0.3 Mo, bal Fe.
For gears, bolts, crankshafts; oil hardened, tough.

LFM GR.H.
M-573; 0.4-0.5 C, 0.9-1.2 Mn, 0.3-0.4 Mo, bal Fe.
For gears, bolts, crankshafts; oil hardened, tough.

LFM GR.I.
M-573; 0.2-0.3 C, 0.65-0.80 Mn, 2.0-2.25 Ni, bal Fe.
For gears, bolts, shafts, machine tool parts; oil hardened, tough.

LFM GR.J.
M-573; 0.22-0.33 C, 0.15-0.20 Mo, 0.65-0.80 Mn, bal Fe.
For gears, bolts, machine tool parts; Type 4027; water hardened.

LFM GR.K.
M-573; 0.13-0.20 C, 0.4-0.7 Mn, 4.0-6.5 Cr, 0.45-0.65 Mo, bal Fe.
For oil refinery equipment; Type 502; creep resistant.

LFM GR.L.
M-573; 0.28-0.33 C, 1.0-1.5 Cr, 0.15-0.25 Mo, bal Fe.
For gears, bolts, machine tool parts; Type 4130; oil hardened.

LFM GR.M.
M-573; 0.35-0.45 C, 1.0-1.5 Cr, 0.15-0.25 Mo, bal Fe.
For gears, bolts, crankshafts, machinery parts; Type 4140; oil hardened.

LFM GR.N.
M-573; 0.13-0.20 C, 1.0-1.5 Cr, 0.45-0.65 Mo, bal Fe.
For gears, bolts, camshafts, cams; Type 4115; case hardened.

LFM GR.O.
M-573; 0.15-0.20 C, 1.25-1.75 Cr, 2.25-2.75 Ni, 0.2-0.4 Mo, bal Fe.
For gears, bolts, camshafts, cams; case hardened, tough.

LFM GR.P.
M-573; 0.25-0.35 C, 1.2-1.7 Cr, 2.2-2.7 Ni, 0.2-0.4 Mo, bal Fe.
For gears, bolts, crankshafts; oil hardened, tough.

LFM GR.R.
M-573; 0.55-0.65 C, 2.0-2.5 Cr, 0.35-0.40 Mo, bal Fe.
For dies, crimpers, upsetters; oil hardened, tough.

LFM GR.S.
M-573; 0.28-0.35 C, 2.75-3.25 Cr, 0.20-0.30 Mo, bal Fe.
For oil refinery equipment; creep resistant.

LFM GR.T.
M-573; 0.16-0.25 C, 0.45-0.65 Mo, 0.65-0.80 Mn, bal Fe.
For housings, machine tool parts; Type WC1; water hardened.

LIBERTY PISTONS.
77 Al, 21 Zn, 1.1 Cu, 0.5 Fe.
For pistons; non-heat treatable.

LICHTENBERG FUSIBLE ALLOY.
M-Germany; 50 Bi, 20 Sn, 30 Pb.
For boiler safety plugs, fire extinguishers; M.P. 75°C.

LICO.
M-802; 0.3 C, bal Fe.
For castings; water hardening.

LIDDELS ALLOY.
90-88 Zn, 6.5-5.0 Cu, 6.5 Al.
For die castings.

LIDEOX.
M-1521; Hg, Be, S, Li, bal Cu.
For condensers; anti-biofouling.

LIEBKNECHT WHITE GOLD.
80 Au, 13.9 Ni, 5 Zn, 1 Cu, 0.1 Pd.
For jewelry; corrosion resistant.

LIGHT DUTY BEARINGS-1.
90-80 Pb, 10-17 Sb, 10-0 Sn, 0-1 Cu.
For bearings; light duty.

LIGHT DUTY BEARINGS-2.
83.3 Pb, 9.8 Sb, 6.9 Sn.
For light duty bearings; anti-friction.

LIGMALLOY.
 M-1159.
 For molded bearings.

LILY BRAND NO. 1.
 M-405; 10 Sn, 2 Zn, 0.5 max Pb, 1.0 max Ni, bal Cu.
 Cast: 44,800 TS; 20,200 YS; 40 El; 75 Brin.
 For bearings, hardware, machine tool parts; Gun Metal BS1400G1C.

LILY BRAND NO. 2.
 M-405; 10 min Sn, 0.05 min Zn, 0.5 min P, bal Cu.
 Cast; 40,400 TS; 22,400 YS; 14 El; 95 Brin.
 For bearings, hardware; Bronze BS1400PB1C.

LILY BRAND NO. 3.
 M-405; 8 Sn, 4 Zn, 0.5 max Pb, 1.0 max Ni, bal Cu.
 Cast: 44,800 TS; 20,200 YS; 38 El; 70 Brin.
 For bearings, hardware, machine tool parts; Gun Metal BS1400G2C.

LILY BRAND NO. 4.
 M-405; 5 Sn, 5 Zn, 5 Pb, 1 max Ni, bal Cu.
 Cast: 36,000 TS; 15,700 YS; 25 El; 60 Brin.
 For hardware, machine tool parts; Gun Metal BS1400LG2C.

LILY BRAND NO. 5.
 M-405; 7 Sn, 5 Zn, 2 Pb, 1.0 max Ni, bal Cu.
 Cast: 40,400 TS; 18,000 YS; 35 El; 60 Brin.
 For hardware, plumbing; Gun Metal BS1400LG3C.

LILY BRAND NO. 7.
 M-405; 10 min Sn, 0.05 min Zn, 0.5 min P, bal Cu.
 Cast: 49,300 TS; 26,900 YS; 3 El; 110 Brin.
 For bearings, hardware; Bronze BS1400PB1C.

LILY BRAND NO. 8.
 M-405; 12 Sn, 0.3 max Zn, 0.5 max Pb, 0.5 max Ni, 0.15 min P, bal Cu.
 Cast: 40,400 TS; 20,200 YS; 10 El; 100 Brin.
 For bearings, hardware; Bronze BS1400PB2C.

LILY BRAND NO. 9.
 M-405; 12 Sn, 0.3 max Zn, 0.5 max Pb, 0.15 min P, bal Cu.
 Cast: 44,800 TS; 22,400 YS; 4 El; 120 Brin.
 For bearings, hardware; Bronze BS1400PB2C.

LILY BRAND NO. 10.
 M-405; Sn, P, Ni, bal Cu.
 Cast: 42,600 TS; 25,700 YS; 8 El; 115 Brin.
 For bearings, hardware; P-Bronze.

LILY BRAND NO. 12.
 M-405; 10 Sn, 10 Pb, 1.5 max Ni, 0.3 max P, bal Cu.
 Cast: 35,900 TS; 13,500 YS; 8 El; 85 Brin.
 For bearings, hardware; free-cutting, Leaded Bronze BS1400LB2C.

LILY BRAND NO. 13.
 M-405; 10 Sn, 10 Pb, 1.5 max Ni, 0.3 max P, bal Cu.
 Cast: 40,400 TS; 4 El; 100 Brin.
 For bearings, hardware; Leaded Bronze BS1400LB1C.

LILY BRAND NO. 15.
 M-405; 10 Al; 5 Ni, 4 Fe, 1.5 max Mn, bal Cu.
 Cast: 94,100 TS; 38,100 YS; 15 El; 170 Brin.
 For bearings, propellers, hardware, machine tool parts; Al-Bronze BS1400AB2C.

LILY BRAND NO. 16.
 M-405; Al, bal Cu.
 Cast: 92,000 TS; 74,000 YS; 6 El; 300 Brin.
 For non-sparking tools; Al-Bronze, corrosion resistant.

LILY BRAND NO. 17.
 M-405; 1.5 max Sn, 0.5 max Pb, 2.5 max Al, 1.2 Fe, 3 max Mn, 55 min Cu, bal Zn.
 Cast: 76,200 TS; 36,000 YS; 24 El; 130 Brin.
 For propellers, hardware, marine parts; Mn-Bronze BS1400HTB1C.

LILY BRAND NO. 18.
 M-405; 1 max Sn, 1-5 Pb, 66-73 Cu, bal Zn.
 Cast: 33,600 TS; 35 El.
 For hardware, Commercial Brass BS1400B2.

LILY BRAND NO. 19.
 M-405; Zn, bal Cu.
 Cast: 33,600 TS; 35 El.
 For hardware; Commercial Brass BS1400B3.

LILY PHOSPHOR BRONZE.
 M-576; 5 Sn, 0.05 P, bal Cu.
 For bushings, bearings; chill cast.

LINCO see **LESCALLOY LINCO.**

LINCOLNWELD L-50.
 M-578; C, bal Fe.
 Submerged arc welding electrode. AWS Class EM13K.

LINCOLNWELD L-60.
 M-578; C, bal Fe.
 Submerged arc welding electrode. AWS Class EL 12.

LINCOLNWELD L-61.
 M-578; C, bal Fe.
 Submerged arc welding electrode. AWS Class EM 12K.

LINCOLNWELD L-70.
 M-578; C, bal Fe.
 Submerged arc welding electrode.

LINEAR METAL.
 M-494, M-1485; 3.3 C, 1 Si, 0.2 P, bal Fe.
 Cast: 180-220 Brin.
 For castings, cylinder liners; cast iron.

LINOTYPE.
85 Pb, 11 Sb, 3.5 Sn.
11,700 TS; 9 El; 21 Brin.
For type metal; cheap alloy; M.P. 476°F.

LION.
M-389; 0.7-0.9 C, bal Fe.
For tools, drills, taps; water hardening.

LION.
M-1415; 0.70 C, 14 W, 3.5 Cr, 0.6 V, bal Fe.
For tools, reamers, hobs, drills, broaches; high speed steel.

LION 22%.
M-1415; 0.75 C, 22 W, 4.5 Cr, 1.2 V, bal Fe.
For lathe and planer tools, hobs, reamers, drills; high speed steel.

LION (AISI W-1 GR3).
M-69; 0.6-1.4 C, bal Fe.
For tools, general tools, broaches, chisels, dies, knives, files; water hardened.

LION BRAND ANTIFRICTION METAL.
M-413; 5-15 Sn, 10-15 Sb, bal Pb.
For bearings for rolling mills and stone crushers; for heavy pressures and high speeds.

LION BRAND PLASTIC WHITE METAL.
M-413; 3-10 Sb, 1-3 Cu, 10-17 Pb, bal Sn.
For main bearings for marine engines; Babbitt.

LION EXTRA (AISI W1 GR 2).
M-69; 0.6-1.3 C, 0.25 Si, 0.3 Mn, bal Fe.
For drills, hobs, reamers, taps, broaches; Type W1; water hardened.

LION G.19.F.
M-1415; 0.90 C, 2 Mn, bal Fe.
For taps, dies, reamers, forming dies; oil hardened, non-deforming.

LION G.19.H.
M-1415; 0.90 C, 0.45 Si, 1.25 Mn, 0.5 W, 0.5 Cr, bal Fe.
For taps, dies, reamers, forming dies; oil hardened, non-deforming.

LION H.18.V.
M-1415; 1.0 C, 1.5 Cr, 0.2 V, bal Fe.
For taps, dies, forming rolls, broaches, reamers; oil hardened, wear resistant.

LION K.15.
M-1415; 0.30 C, 13 Cr, S, bal Fe.
Annealed: 95,000 TS; 50,000 YS; 25 El; 55 RA; 196 Brin.
For stainless screw machine products; free-cutting, corrosion resistant.

LION P.16.
M-1415; 1.0 C, 1.0 Mn, 0.75 W, 0.75 Cr, bal Fe.
For blanking and forming dies, reamers, broaches; oil hardened, non-deforming.

LION PEG STEEL 5.5.0.
M-1415; 0.5 C, 14 W, 3 Cr, 0.5 V, bal Fe.
For hot punches, core and die inserts; hot work steel, oil hardened.

LION R.19.
M-1415; 0.25 C, 13 Cr, bal Fe.
Annealed: 95,000 TS; 50,000 YS; 25 El; 55 RA; 196 Brin.
For dies, cutlery, valves; Type 420; corrosion resistant.

LION REGENT.
M-1415; 0.8 C, 20 W, 5 Cr, 1.25 V, 18 Co, 0.75 Mo, bal Fe.
For cutting tools, broaches, milling cutters; high speed steel.

LION SPECIAL.
M-1415; 0.75 C, 19 W, 4.25 Cr, 1.25 V, 8 Co, 0.75 Mo, bal Fe.
For lathe and planer tools, hobs, drills; high speed steel.

LION VANADIUM.
M-69; 0.60-1.0 C, 0.25 V, bal Fe.
AISI W2 Gr. 1.

LISCO.
M-802; 0.35 C, bal Fe.
For castings; water hardening.

LITHALOYS.
M-1050; Li-Cu.
For master alloy for Cu castings; for sound castings.

LITHIUM.
M-1755; Li.
Purities: 99.99%, Reactor grade 99.9+%, 99.8%.
Forms: Ingot, rod, sheet, foil, wire, shot, powders, billets, ribbon.

LITHIUM-CALCIUM.
M-315; 50 Li, 50 Ca.
For deoxidizer, metallurgical applications.

LITHIUM-CONDUCTIVITY BRONZE.
M-315; 98 min Cu, 2 max Cd, Sn, Si.
For high strength conductors; bronze treated with Li with increased conductivity.

LITHIUM-COPPER.
M-315; Li, bal Cu.
Annealed: 31,500-36,500 TS; 72-60 El.
For electrical conductors; high conductivity, oxygen free.

LITHOBRAZE 720 (BT).
M-63; 71.3-72.3 Ag, 0.15-0.30 Li, bal Cu.
For brazing alloy for stainless steels; M.P. 1435°F; eutectic alloy. AWS BAg-8a.

LITHOBRAZE 925.
M-63; 92.5 Ag, 7.3 Cu, 0.2 Li.
For high temperature brazing; M.P. 1435-1635°F; for short time operation to 900°F, continuous to 500°F. AWS BAg-19.

LITINUM BRONZE.
M-Eng; 90 Cu, 4.7 Sn, 3.86 Zn, 1.6 Pb.
For nuts, bushings, bearings, screw machine parts; free-cutting.

LITTITE.
M-1479; 3 C, Si, Mn, alloy, bal Fe.
For gears, shafts, machine tool housings; cast iron.

LITTLES SPECULUM.
65 Al, 31 Sn, 2.3 Zn, 1.9 As.
For mirrors, reflectors; high polish.

L-IV.
M-116; 4 Cu, 2 Si, bal Al.
Heat treated: 43,000 TS; 36,000 YS; 4 El.
For light alloy parts; same as "Lautal."

LK4.
M-USSR; 0.20 C, 26.5 Cr, 5 Mo, 3.3 Ni, bal Co.
Cast cobalt-base superalloy.
For nozzle guide vanes.

LK4YA.
M-USSR; 0.26 C, 26.5 Cr, 5 W, 0.02 B, 3 max Ni, bal Co.
Cast cobalt base superalloy.

LK66YA.
M-USSR; 0.30 C, 22.5 Cr, 9.5 W, 0.02 B, 1.75 Cb, 2 max Ni, bal Co.
Cast cobalt base superalloy.

L.M.
M-620; 0.75 Si, 4.75 Cu, 0.75 Mn, bal Al.
58,000-65,000 TS; 20-18 El.
For light alloy parts; age-hardenable.

LM 0.
M-Eng.; 99.5 Al, 0.40 max Fe.
Aluminium casting.
AFNOR A5; BS 1490 LM0.

L.M.-0.
M-302; 4 Si, 1.5 Cu, bal Al.
23,000-28,000 TS; 4-3 El.
For light alloys, food handling machinery; age-hardenable.

LM 1.
M-28; 7 Cu, 3 Si, 3 Zn, bal Al.
Sand cast: 19,000 TS; 12,300 YS; 73 Brin.
Permanent mold: 24,600 TS; 15,700 YS; 1 El; 80 Brin.
For general purpose castings; Brit. DTD428.

LM 2 TYPE 1.
M-28; 1.1 Cu, 10.3 Si, bal Al.
Sand cast: 21,300 TS; 9,800 YS; 2 El; 70 Brin.
For pressure tight castings; corrosion resistant.

LM 2 TYPE 2.
M-28; 2.1 Cu, 10.3 Si, 0.9 Zn, bal Al.
Permanent mold: 25,500 TS; 12,000 YS; 2 El; 78 Brin.
For pressure tight castings; corrosion resistant.

LM 3.
M-Eng.; 2.5-4.5 Cu, 1.3 max Si, 9.0-13.0 Zn, 1.0 max Fe, 0.5 max Mn, 0.5 max Ni, 0.3 max Pb, bal Al.
Cast aluminium alloy.
BS 1490 LM3.

LM 4.
M-28; 3 Cu, 5 Si, 0.5 Mn, bal Al.
Sand cast: 22,400 TS; 12,300 YS; 2.5 El; 65 Brin.
Permanent mold: 23,500 TS; 12,300 YS; 2.5 El; 70 Brin.
For pressure tight castings; Brit. DTD424, age hardenable.

LM 5.
M-106, M-28; 3.0-6.0 Mg, 0.3 Si, 0.6 Fe, 0.3-0.7 Mn, bal Al.
Sand cast: 140 MPa TS; 90 MPa YS; 3 El; 55 Brin.
Special purpose cast aluminum alloy; high corrosion resistance-for marine applications.
BS 1490 LM5; AFNOR A-G 3T.

LM 6.
M-28; 11.3 Si, bal Al.
Sand cast: 25,700 TS; 9000 YS; 7 El; 55 Brin.
Die cast: 34,000 TS; 10,200 YS; 10 El; 60 Brin.
For pressure tight thin wall castings; Brit. BS3L33, corrosion resistant.

LM6M ETC see MILLS LM6M ETC.

LM 7.
M-Eng.; 1.0-2.5 Cu, 1.5-3.5 Si, 0.3-1.4 Fe, 0.5-1.7 Ni, 0.05-0.20 Mg 0.05-0.30 Ti, bal Al.
Cast aluminium alloy.
BS 1490 LM7.

LM 8.
M-Eng.; 3.5-6.0 Si, 0.6 max Fe, 0.3-0.8 Mg, 0.5 max Mn, bal Al.
Cast aluminium alloy.
BS 1490 LM8; AFNOR A-S 4G.

LM 9.
M-106, M-28; 0.2-0.6 Mg, 10.0-13.0 Si, 0.6 Fe, 0.3-0.7 Mn, bal Al.
Sand cast, precipitation treated: 170 MPa TS; 120 MPa YS; 1.5 El; 60 Brin.
Special purpose cast aluminum alloy; high strength, rigid, good corrosion resistance.
BS 1490 LM9; AFNOR A-S 9G.

LM 10.
M-106, M-28; 9.5-11.0 Mg, 0.25 Si, 0.35 Fe, bal Al.
Sand cast, sol. treated: 280 MPa TS; 170 MPa YS; 8 El; 70 Brin.
Special purpose cast aluminum alloy; high proof stress, resistant to shock and corrosion.
BS 1490 LM10; AFNOR A-G 10.

LM 10A.
M-200; Ag, Cu, Sn.
Soft solder; melt range: 214-275°C.

LM 11.
M-Eng.; 4.0-5.0 Cu, 0.25 max Si, 0.25 max Fe, 0.05-0.30 Ti, bal Al.
Cast aluminium alloy.
BS 1490 LM11.

LM 12.
M-28; 9.0-11.0 Cu, 2.5 max Si, 0.8 max Zn, 1.0 max Fe, 0.6 max Mn, 0.5 max Ni, 0.2-0.4 Mg, bal Al.
Cast aluminium alloy.
BS 1490 LM12; AFNOR A-S 12.

LM 13.
M-28; 0.7-1.5 Cu, 10.0-12.0 Si, 0.5 max Zn, 1.0 max Fe, 0.5 max Mn, 1.5 max Ni, 0.8-1.5 Mg, bal Al.
Cast aluminium alloy.
BS 1490 LM13; AFNOR A-S 12UN.

LM 14.
M-Eng.; 3.5-4.5 Cu, 0.6 max Si, 0.6 max Fe, 0.6 max Mn, 1.8-2.3 Ni, 1.2-1.7 Mg, bal Al.
Cast aluminium alloy.
BS 1490 LM14; AFNOR A-U 4 NT.

LM 15.
M-200; Zn, plus Ag, Cd.
Zinc base soft solder; melt range: 280-320°C.

LM 15.
M-Eng.; 1.3-3.0 Cu, 0.6-2.0 Si, 0.8-1.4 Fe, 0.5-2.0 Ni, 0.5-1.7 Mg, 0.05-0.30 Ti, bal Al.
Cast aluminium alloy.
BS 1490 LM15.

LM 16.
M-106, M-28; 1.0-1.5 Cu, 0.4-0.6 Mg, 1.5-5.5 Si, 0.25 Ni, 0.6 Fe, 0.5 Mn, bal Al.
Sand cast, sol. tr. & aged: 230 MPa TS; 220 MPa YS; 90 Brin.
General purpose alloy; good pressure tightness and castability.

LM 18.
M-28; 4.5-6.0 Si, 0.6 max Fe, 0.5 max Mn, bal Al.
Cast aluminium alloy.
BS 1490 LM18.

LM 20.
M-28; 11.5 Si, bal Al.
Sand cast: 25,200 TS; 8900 YS; 5.0 El; 55 Brin.
Die cast: 31,400 TS; 9500 YS; 7.5 El; 60 Brin.
For pressure tight castings; Brit. LM20M, corrosion resistant.

LM 21.
M-28; 3 Cu, 5 Si, 0.5 Mn, bal Al.
Sand cast: 20,200 TS; 1.5 El.
Permanent mold: 22,400 TS; 1.5 El.
For general engineering castings; Brit. LM21M, corrosion resistant.

LM 22.
M-106, M-28; 2.8-3.8 Cu, 4.0-6.0 Si, 0.6 Fe, 0.2-0.6 Mn, bal Al.
Chill cast, sol. treated: 245 MPa TS; 110 MPa YS; 8 El; 70 Brin.
Permanent mould casting only; for heavy duty applications.

LM 24.
M-28; 3.5 Cu, 8.5 Si, bal Al.
Permanent mold: 25,700 TS; 1.5 El.
For general engineering castings; Brit. LM24M, corrosion resistant.

LM 24 TYPE A.
M-Eng.; 3.0-4.0 Cu, 7.5-9.5 Si, 1.5 max Zn, 1.3 max Fe, 0.5 max Mn, 0.5 max Ni, 0.3 max Mg, 0.3 max Pb, 0.2 max Sn, 0.2 max Ti, bal Al.
Cast aluminium alloy.
BS 1490 LM24-A; AFNOR A-S 9U 3/A.

LM 24 TYPE B.
M-Eng.; 3.0-4.0 Cu, 7.5-9.5 Si, 3.0 max Zn, 1.3 max Fe, 0.5 max Mn, 0.5 max Ni, 0.3 max Mg, 0.3 max Pb, 0.2 max Sn, 0.2 max Ti, bal Al.
Cast aluminium alloy.
BS 1490 LM24-A; AFNOR A-S 9U 3/B.

LM 25.
M-106, M-28; 0.2-0.45 Mg, 6.5-7.5 Si, bal Al.
Sand cast, sol. treated and aged: 230 MPa TS; 215 MPa YS; 80 Brin.
General purpose alloy with good casting qualities.

LM 26.
M-28; 2.0-4.0 Cu, 8.5-10.5 Si, 1.0 max Zn, 1.2 max Fe, 0.5 max Mn, 1.0 max Ni, 0.5-1.5 Mg, 0.2 max Pb, 0.2 max Ti, bal Al.
Cast aluminium alloy.
BS 1490 LM26.

LM 27.
M-106, M-28; 1.5-2.5 Cu, 0.3 Mg, 6.0-8.0 Si, 0.8 Fe, 0.2-0.6 Mn, 0.3 Ni, 1.0 Zn, bal Al.
Sand cast: 140 MPa TS; 90 MPa YS; 1 El; 75 Brin.
Good castability; pressure tight.

LM 28.
M-28; 1.3-1.8 Cu, 17-20 Si, 0.2 max Zn, 0.7 max Fe, 0.6 max Mn, 0.8-1.5 Ni, 0.8-1.5 Mg, 0.6 max Cr, 0.5 max Co, bal Al.
Cast aluminium alloy.
BS 1490 LM28; AFNOR A-S 20U.

LM 29.
M-28; 0.8-1.3 Cu, 22-25 Si, 0.2 max Zn, 0.7 max Fe, 0.6 max Mn, 0.8-1.3 Ni, 0.8-1.3 Mg, 0.6 max Cr, 0.5 max Co, bal Al.
Cast aluminium alloy.
BS 1490 LM29; AFNOR A-S 22 UNK.

LM 30.
M-28; 4.0-5.0 Cu, 16-18 Si, 0.2 max Zn, 1.1 max Fe, 0.3 max Mn, 0.4-0.7 Mg, 0.2 max Ti, bal Al.
Cast aluminium alloy.
BS 1490 LM30.

LM 630.
M-1031; 6 Al, 3 Zn, 0.2 Mn, bal Mg.
Cast: 27,000 TS; 5 El; 50 Brin.
Heat treated: 38,000 TS; 5 El; 73 Brin.
For light alloy castings; heat treatable.

LM 920.
M-1031; 9 Al, 2 Zn, 0.1 Mn, bal Mg.
Cast: 24,000 TS; 2 El; 65 Brin.
Heat treated: 40,000 TS; 2 El; 87 Brin.
For light alloy castings; heat treatable.

L.M.S.
M-Eng; Pb-Sn-Sb.
For bearings; anti-friction.

LMW.
M-1779; 0.83 C, 4.0 Cr, 1.5 W, 8.5 Mo, 1.1 V, bal Fe.
Mo-type high speed steel for taps, drills, reamers, planer tools, milling cutters, and wood working tools.
AISI M-1.

LMW-V.
M-1779; 1.0 C, 3.75 Cr, 2.0 V, 1.75 W, 8.75 Mo, bal Fe.
Mo-type high speed steel for drills, end mills, reamers, form tools, milling cutters, planer and boring tools.
AISI M-7.

LN-2.
M-843; 99 min Ni.
Metallic spray coating for Plasma Torch.
For build-up of worn parts; 200 DPH.

LOADED CENTRICAST.
M-494, M-1485; 3.2 C, 2.8 Si, 0.8 Cr, bal Fe.
Cast: 36,000 Ts: 240-300 Brin.
For cylinder liners; centrifugal.

LOADED IRON.
M-494, M-1485; 3.3 C, 2.2 Si, 1 Cr, 0.4 P, bal Fe.
320 Brin.
For cylinder liners; cast iron.

LO-AIR.
M-114; 0.7 C, 2.25 Mn, 1.0 Cr, 1.35 Mo, bal Fe.
Ht. Tr.: 293,000 TS; 264,000 YS; 560 Brin.
For blanking, piercing and embossing dies.
Type A6 Air hardening, wear resistant.

LOCKALLOY.
M-646; 38 Al, 62 Be.
Sheet: 56,000 TS; 44,000 YS; 8 El.
At 400°F: 44,000 TS; 40,000 YS; 11 El.
For nuclear fuel canning, gyro cages, aircraft brakes, computer memory drums and discs, spacecraft parts, missiles, satellite launch vehicles.
High modulus, low density alloy.

LOCKPORT SPECIAL.
M-373; 0.7 C, 18 W, 4 Cr, 2 V, bal Fe.
For tools, cutters; high speed steel.

LOCOMOTIVE TUBE.
M-Eng; 97 Cu, 3 Ni.
For tubes for locomotives; corrosion resistant.

LO-EX.
M-299; 0.5-1.3 Cu, 0.8-1.5 Mg. 11-13 Si, 2-3 Ni, bal Al.
Sand cast: 37,000 TS; 35,900 YS; 0 El; 125 Brin.
Permanent mold: 44,800 TS; 42,600 YS; 0 El; 130 Brin.
For pistons; age-hardenable.

LO-EX.
M-1220, M-541; 0.5-1.0 Cu, 12.8 Si, 0.8 Mg, 2.2 Ni, bal Al.
Cast: 36,000-47,000 TS; 36,000-46,000 YS; 0.2-0.5 El; 120-135 Brin.
For pistons, sleeves, bushings; low thermal expansion.

LO-EX ALLOY.
M-137; 11-13 Si, 1.0-2.5 Ni, 1.0 Mg, 0.7 Cu, bal Al.
Cast: 26,000 TS; 23,000 YS; 0.5 El; 75 Brin.
Heat treated: 38,000 TS; 33,000 YS; 0.5 El; 140 Brin.
For pistons; low thermal expansion.

LOFLEX.
M-237; bimetal.
For thermostatic bimetal; maximum temperature 800°F.

LOGAN.
M-802; 0.4 C, bal Fe.
For castings; water hardening.

LOGAN FORGING BRASS.
M-122; 58.5 Cu, 2 Pb, bal Zn.
Extruding: 61,000 TS; 32,000 YS; 40 El; 104 Brin.
For brass forgings, hot pressings. CA 377.

LOGAN (W2 GRADE 4).
M-1705; 1.0 C, 0.25 Mn, 0.25 Si, bal Fe.
Water hardening tool steel; AISI W2.

LOHM.
M-44; 7.0-7.5 Ni, bal Cu.
Wrought: 50,000-100,000 TS.
For electrical equipment, radio rheostats. Low electrical resistance, maximum operating temperature 200°C.

LOHMANIT.
W + Mo carbides (chiefly W_2C.)
For hard cutting tools and dies; cast alloy.

LOHMANN 120.
M-1311; 1.2 C, 0.25 max Si, 0.25 max Mn, bal Fe.
Annealed: 115,000 TS; 60,000 YS; 18 El; 38 RA; 225 Brin.
For springs, taps, drills, cutters, broaches; Type W1; water hardened.

LOHMANN A37.
M-1311; 0.85 C, Cr, bal Fe.
For tools, springs, bearings, dies; water hardened.

LOHMANN BCNI.
M-1311; 0.55 C, 1.05 Cr, 3.25 Ni, bal Fe.
For gears, bolts, fasteners, crankshafts; oil hardened, shock resistant.

LOHMANN C112/50.
M-1311; 1.05 C, Cr, bal Fe.
For bearings, liners, cutters; water hardened, wear resistant.

LOHMANN CO 3.
M-1311; 0.86 C, 2.8 Co, 4.3 Cr, 0.85 Mo, 2.1 V, 12 W, bal Fe.
For lathe and planer tools, reamers, broaches; high speed steel.

LOHMANN CO 5.
M-1311; 0.79 C, 4.7 Co, 4.3 Cr, 0.75 Mo, 1.5 V, 18 W, bal Fe.
For lathe and planer tools, reamers, broaches, drills, taps; high speed steel.

LOHMANN CO 10.
M-1311; 0.76 C, 10 Co, 4.2 Cr, 0.8 Mo, 1.8 V, 18 W, bal Fe.
For lathe and planer tools, reamers, drills, cutters; high speed steel.

LOHMANN E75.
M-1311; 0.75 C, 0.25 max Si, 0.25 max Mn, bal Fe.
Heat treated: 175,000 TS; 130,000 YS; 12 El; 35 RA; 355 Brin.
For springs, rails, punches, tools; water hardened.

LOHMANN E85.
M-1311; 0.85 C, 0.25 max Si, 0.25 max Mn, bal Fe.
Heat treated: 190,000 TS; 145,000 YS; 10 El; 30 RA; 400 Brin.
For springs, tools, punches, drills, taps; Type W1; water hardened.

LOHMANN E95.
M-1311; 0.95 C, 0.25 max Si, 0.25 max Mn, bal Fe.
Heat treated: 195,000 TS; 150,000 YS; 8 El; 28 RA; 420 Brin.
For drills, taps, reamers, broaches; Type W1; water hardened.

LOHMANN E120.
M-1311 1.1 C, 0.25 max Si, 0.25 max Mn, bal Fe.
Annealed: 110,000 TS; 55,000 YS; 20 El; 40 RA; 210 Brin.
For springs, taps, drills, hobs, reamers; Type W1; water hardened.

LOHMANN GP35.
M-1311; 1.5 C, Cr, Si, bal Fe.
For engravers' tools, forming and blanking dies; water or oil hardened.

LOHMANN GS125.
M-1311; 1.2 C, 0.20 Cr, 0.10 V, 1.0 W, bal Fe.
For fast finishing cutters, bushings; water or oil hardened.

LOHMANN GSNI.
M-1311; 0.56 C, Ni, Cr, Mo, V, bal Fe.
For gears, bolts, crankshafts, machine tool parts; oil hardened, shock resistant.

LOHMANN HCS.
M-1311; 1.25 C, 1.15 Si, 0.7 Mn, 1.2 Cr, bal Fe.
For header dies, punches, bushings; water or oil hardened.

LOHMANN HKL.
M-1311; 0.58 C, 0.95 Mn, 1.0 Cr, 0.09 V, bal Fe.
For gears, springs, crankshafts; oil hardened, shock resistant.

LOHMANN HKM.
M-1311; 1.0 C, 0.1 V, 0.25 Mn, bal Fe.
For header dies, punches, bearings; Type W2; water or oil hardened.

LOHMANN HPW.
M-1311; 0.38 C, Si, Cr, V, bal Fe.
For machine tool parts; oil hardened, tough.

LOHMANN HPW EXTRA.
M-1311; 0.45 C, Si, Cr, V, bal Fe.
For springs, gears, crankshafts; oil hardened, tough.

LOHMANN HPWH.
M-1311; 0.55 C, 0.9 Si, 1.05 Cr, 0.18 V, 1.85 W, bal Fe.
For header dies, shears, grippers; oil hardened, tough.

LOHMANN HPWW.
M-1311; 0.35 C, 0.9 Si, 1.05 Cr, 0.18 V, 1.85 W, bal Fe.
For header dies, shears, grippers; oil hardened, tough.

LOHMANN HSL.
M-1311; 2.1 C, 11.5 Cr, 0.30 Mn, bal Fe.
For blanking and forming dies, punches; oil hardened, non-deforming.

LOHMANN HSL EXTRA.
M-1311; 1.65 C, 11.5 Cr, 0.1 V, 0.3 Mn, bal Fe.
For blanking and forming dies, punches; air hardened, non-deforming.

LOHMANN HSL SPEZIAL.
M-1311; 2.1 C, 11.5 Cr, 0.3 Mn, bal Fe.
For blanking and forming dies, punches; oil hardened, non-deforming.

LOHMANN HSO-111.
M-1311; 1.45 C, 1.4 Cr, 0.6 Mn, bal Fe.
For bearings, punches, bushings; water hardened, wear resistant.

LOHMANN HSO-121.
M-1311; 1.05 C, 1.0 Cr, 1.15 W, 0.9 Mn, bal Fe.
For header dies, cutters, upsetters; oil or water hardened, tough.

LOHMANN HSO EXTRA.
M-1311; 1.15 C, 0.1 V, 0.65 Cr, 0.3 Mn, bal Fe.
For dies, cutters, punches; oil hardened, tough.

LOHMANN HSO SPEZIAL.
M-1311; 0.90 C, 1.9 Mn, 0.1 V, bal Fe.
For punches, dies, shears, crimpers; oil hardened, non-deforming.

LOHMANN HWR.
M-1311; 1.42 C, W, V, bal Fe.
For header dies, cutters; oil or water hardened.

LOHMANN HWS.
M-1311; 0.30 C, 1.1 Cr, 0.18 V, 3.75 W, bal Fe.
For extrusion rams and liners, punches, riveters; oil hardened, tough.

LOHMANN HWS EXTRA.
M-1311; 0.30 C, 2.35 Cr, 0.6 V, 4.25 W, bal Fe.
For extrusion rams and liners, punches; oil hardened, tough.

LOHMANN HWS SPEZIAL.
M-1311; 0.30 C, 2.65 Cr, 0.35 V, 8.5 W, bal Fe.
For extrusion rams and liners, punches; oil hardened, tough.

LOHMANN K12.
M-1311; 2.1 C, 11.5 Cr, 0.30 Mn, bal Fe.
For blanking and forming dies, punches; oil hardened, non-deforming.

LOHMANN KL1.
M-1311; 1.0 C, 1.55 Cr, 0.35 Mn, 0.35 Si, bal Fe.
For bearings, cutters, forming dies; oil or water hardened, wear resistant.

LOHMANN KM12.
M-1311; 1.65 C, 11.5 Cr, 0.1 V, bal Fe.
For blanking and forming dies, punches; air hardened, non-deforming.

LOHMANN KS15V.
M-1311; 1.45 C, 1.4 Cr, 0.6 Mn, bal Fe.
For bearings, liners, bushings; water hardened, wear resistant.

LOHMANN KS85.
M-1311; 1.15 C, 0.65 Cr, 0.10 V, bal Fe.
For cutters, header dies, blanking dies; oil hardened, tough.

LOHMANN KST.
M-1311; 0.58 C, 1.0 Cr, 0.1 V, bal Fe.
For springs, gears, forging dies, crankshafts; oil hardened, shock resistant.

LOHMANN KWS45.
M-1311; 0.45 C, 1.05 Cr, 0.18 V, 1.85 W, bal Fe.
For header dies, upsetters, shears, punches; oil hardened, tough.

LOHMANN KWS55.
M-1311; 0.55 C, 0.9 Si, 1.05 Cr, 0.18 V, 1.85 W, bal Fe.
For header dies, punches, tools, shears; oil hardened, tough.

LOHMANN KWS75.
M-1311; 0.35 C, 1.05 Cr, 1.85 W, 0.18 V, bal Fe.
For heading dies, upsetters, piercers, tools; oil hardened, tough.

LOHMANN KZV30.
M-1311; 0.80 C, W, Cr, V, bal Fe.
For tools, cutters; oil hardened.

LOHMANN L161.
M-1311; 1.65 C, 11.5 Cr, 0.3 V, bal Fe.
For blanking and forming dies, punches; air hardened, non-deforming.

LOHMANN LC SPEZIAL.
M-1311; 2.1 C, 11.5 Cr, 0.7 W, 0.3 Mn, bal Fe.
For blanking and forming dies; oil hardened, non-deforming.

LOHMANN LMS125.
M-1311; 1.2 C, W, bal Fe.
For cutters, forming rolls, header dies; oil hardened.

LOHMANN LMS200.
M-1311; 1.15 C, W, bal Fe.
For header dies, forming rolls, cutters; oil hardened.

LOHMANN MC 100.
M-1311; 1.05 C, Mn, Cr, bal Fe.
For bearings, liners, bushings; water hardened, wear resistant.

LOHMANN MSV4.
M-1311; 0.7 C, 0.25 max Si, 0.25 max Mn, bal Fe. Heat treated: 175,000 TS; 128,000 YS; 12 El; 37 RA; 355 Brin.
For springs, rails, tools, hammers, axes; Type W1; water hardened.

LOHMANN MSVE.
M-1311; 1.0 C, 0.25 max Si, 0.25 max Mn, bal Fe.
Annealed: 100,000 TS; 53,000 YS; 21 El; 42 RA; 200 Brin.
For springs, tools, cutters, drills, taps, reamers; Type W1; water hardened.

LOHMANN N2.
M-1311; 0.82 C, 4.1 Cr, 0.85 Mo, 1.6 V, 8.7 W, bal Fe.
For lathe and planer tools, drills, hobs, broaches; high speed steel.

LOHMANN N3.
M-1311; 0.95 C, 4.1 Cr, Mo, W, V, bal Fe.
For lathe and planer tools, reamers, broaches, hobs; high speed steel.

LOHMANN N4.
M-1311; 0.85 C, 4.1 Cr, Mo, W, V, bal Fe.
For lathe and planer tools, reamers; high speed steel.

LOHMANN NE.
M-1311; 0.74 C, 4.1 Cr, 1.1 V, 18.5 W, bal Fe.
For lathe and planer tools, reamers, drills, taps, hobs, broaches; high speed steel.

LOHMANN NR.
M-1311; 1.3 C, 4.3 Cr, 0.85 Mo, 3.8 V, 12 W, bal Fe.
For blanking and forming dies, cutters, reamers; high speed steel.

LOHMANN NR5.
M-1311; 1.35 C, W, Co, Mo, V, Cr, bal Fe.
For engravers' tools, blanking and forming dies; high speed steel.

LOHMANN NSP.
M-1311; 0.86 C, 4.1 Cr, 0.85 Mo, 2.5 V, 12 W, bal Fe.
For lathe and planer tools, reamers, broaches, taps; high speed steel.

LOHMANN P75.
M-1311; 0.70 Cr, 0.25 max Si, 0.25 Mn, bal Fe.
Heat treated: 174,000 TS; 128,000 YS; 12 El; 37 RA; 355 Brin.
For springs, rails, clutch discs; Type W1; water hardened.

LOHMANN P85.
M-1311; 0.85 C, 0.25 max Mn, bal Fe.
Heat treated: 190,000 TS; 145,000 YS; 10 El; 30 RA; 400 Brin.
For springs, tools, cutters, drills, taps:; Type W1; water hardened.

LOHMANN P95.
M-1311; 1.0 C, 0.25 max Si, 0.25 max Mn, bal Fe.
Annealed: 100,000 TS; 53,000 YS; 21 El; 42 RA; 200 Brin.
For springs, taps, drills, hobs, cutters; Type W1; water hardened.

LOHMANN PM90.
M-1311; 0.90 C, 1.9 Mn, 0.1 V, 0.25 Si, bal Fe.
For punches, dies, crimpers, upsetters, cutters; oil hardened, nondeforming.

LOHMANN PM150.
M-1311; 1.05 C, 1.0 Cr, 1.15 W, 0.90 Mn, bal Fe.
For cutters; water hardened.

LOHMANN PV15.
M-1311; 0.61 C, 1.18 Cr, 0.1 V, 0.75 Mn, 0.85 Si, bal Fe.
For springs, gears, bolts, crimpers, punches; oil hardened, shock resistant.

LOHMANN PZ105.
M-1311; 1.05 C, 1.0 Cr, 0.30 Mn, bal Fe.
For bearings, sleeves, liners, bushings; water hardened, wear resistant.

LOHMANN RF.
M-1311; 0.40 C, 13 Cr, 0.30 Mn, bal Fe.
Annealed: 100,000 TS; 55,000 YS; 20 El; 50 RA; 200 Brin.
For valves, cutlery, surgical and dental instruments; Type 420; stainless.

LOHMANN-RFK.
M-1311; 0.20 C, 13 Cr, 1 Mo, bal Fe.
Annealed: 95,000 TS; 50,000 YS; 25 El; B 92 Rock. Heat treated: 250,000 TS; 215,000 YS; 8 El; C 52 Rock.
For cutlery, surgical instruments, gears, shafts, needle valves, gauges. Corrosion resistant, hardenable.

LOHMANN RFW.
M-1311; 0.20 C, 13 Cr, 0.3 Mn, bal Fe.
Annealed: 95,000 TS; 50,000 YS; 25 El; 55 RA; 195 Brin.
For valves, cutlery, turbine blades; Type 420; stainless.

LOHMANN S39.
M-1311; 0.85 C, 0.10-0.40 Si, 0.5-0.7 Mn, bal Fe.
Heat treated: 190,000 TS; 145,000 YS; 10 El; 30 RA; 400 Brin.
For springs, tools, cutters, drills, taps; Type W1; water hardened.

LOHMANN SCV14.
M-1311; 0.38 C, Cr, V, Si, bal Fe.
For gears, bolts, machine tool parts; oil hardened, tough.

LOHMANN SM4.
M-1311; 0.35 C, 0.25-0.50 Si, 0.3-0.8 Mn, bal Fe.
Hot rolled: 85,000 TS; 54,000 YS; 30 El; 53 RA; 183 Brin.
For gears, bolts, machine tool parts; water hardened.

LOHMANN SM4F.
M-1311; 0.45 C, 0.25-0.50 Si, 0.3-0.8 Mn, bal Fe.
Hot rolled: 98,000 TS; 59,000 YS; 24 El; 45 RA; 212 Brin.
For gears, bolts, machine tool parts; water hardened.

LOHMANN SM5.
M-1311; 0.60 C, 0.25-0.50 Si, 0.3-0.8 Mn, bal Fe.
Heat treated: 160,000 TS; 113,000 YS; 12 El; 40 RA; 325 Brin.
For machine tool parts, punches, springs; water hardened.

LOHMANN SM7.
M-1311; 0.75 C, 0.25-0.50 Si, 0.3-0.8 Mn, bal Fe.
Heat treated: 180,000 TS; 130,000 YS; 10 El; 35 RA; 360 Brin.
For springs, rails, punches, hammers; Type W1; water hardened.

LOHMANN SM9.
M-1311; 0.90 C, 0.25-0.50 Si, 0.3-0.8 Mn, bal Fe.
Annealed: 100,000 TS; 53,000 YS; 20 El; 40 RA; 200 Brin.
For springs, dies, tools, cutters, punches; Type W1; water hardened.

LOHMANN SNI.
M-1311; 0.45 C, Cr, Ni, bal Fe.
For gears, bolts, machine tool parts; oil hardened, tough.

LOHMANN SV.
M-1311; 0.80 C, Cr, V, bal Fe.
For bearings, cutters, form dies, punches; oil or water hardened, wear resistant.

LOHMANN T130.
M-1311; 1.3 C, 0.25 max Si, 0.25 max Mn, bal Fe.
For engravers' tools, cutters, taps, reamers; Type W1; water hardened.

LOHMANN WM4.
M-1311; 0.30 C, 1.0 Si, 1.1 Cr, 0.18 V, 3.75 W, bal Fe.
For header dies, upsetters, crimpers; oil hardened, tough.

LOHMANN WM5.
M-1311; 0.30 C, 2.35 Cr, 0.6 V, 4.25 W, bal Fe.
For extrusion press rams and liners, upsetters, punches; oil hardened, tough.

LOHMANN WM10.
M-1311; 0.30 C, 2.65 Cr, 0.35 V, 8.5 W, bal Fe.
For extrusion press rams and liners, punches, shears; hot work steel, oil hardened.

LOHMANN WMCO3.
M-1311; 0.30 C, 2 Co, 2.4 Cr, 0.25 V, 8.5 W, bal Fe.
For extrusion press rams and liners, punches, shears; hot work steel, oil hardened.

LOHMANN WORI EXTRA.
M-1311; 1.42 C, W, V, bal Fe.
For cutters, engravers' tools; water hardened.

LOHMAN WSCR60.
M-1311; 0.45 C, 1.15 Si, 1.2 Cr, 0.1 V, bal Fe.
For springs, gears, bolts; oil hardened.

LOHMANN WSP.
M-1311; 1.25 C, 1.15 Si, 1.2 Cr, 0.7 Mn, bal Fe.
For liners, bushings; oil hardened.

LOHMANN WWMV.
M-1311; 0.45 C, 1.35 Cr, 0.45 Mo, 0.80 V, 0.45 W, bal Fe.
For punches, crimpers; oil hardened.

LOHMANN Z90.
M-1311; 1.0 C, 0.25 Mn, 0.1 V, bal Fe.
For springs, cutters, drills, reamers, taps; Type W2; water hardened.

LOHMANN ZCR.
M-1311; 1.4 C, Cr, V, bal Fe.
For blanking and forming dies, cutters, bearers; oil or water hardened, abrasion resistant.

LOIRE-ALC6.
M-1546; 0.08 max C, 23.0 Cr, 5.2 Al, bal Fe.
For heating elements, industrial furnaces, parabolic radiators. High heat resistant to 1100°C.

LOIRE-AMCR.
M-1546; 0.30 C, 10.0 Cr, 18.0 Mn, 1.0 Ni, bal Fe.
Forged: 106,700 TS; 49,800 YS; 45 El. Cast: 92,500 TS; 42,700 YS; 28 El.
For marine hardware, diesel engine motor frames. Nonmagnetic, corrosion resistant steel.

LOIRE-AMCRN.
M-1546; 0.65 C, 8 Ni, 8 Mn, 4 Cr, bal Fe.
Rolled; 154,000 TS; 114,000 YS; 20 El.
For electrical construction. Non-magnetic, corrosion resistant steel.

LOIRE A.S.
M-1546; 0.50 C, bal Fe.
Ht. Tr: 120,000 TS; 77,200 YS; 22 El; 245 Brin.
For cutting tools, chisels, hammers, gears, punches, shafts, axles. Water hardening.

LOIRE CMY6.
M-1546; 0.15 C, 0.5 Cr, 0.5 Mo, bal Fe.
Heat treated: 71,000 TS; 42,700 YS; 20 El.
For furnace tubes, oil refineries, boilers. Resists heat to 600°C.

LOIRE-CMY16.
M-1546; 0.12 C, 5.5 Cr, 0.6 Mo, bal Fe.
Heat treated: 78,000-107,000 TS; 35,600-92,500 YS; 14-23 El.
For oil refining and hot hydrogenization equipment. Corrosion resistant to sour crude oils.

LOIRE-CMY17.
M-1546; 0.13 C, 2.3 Cr, 1.0 Mo, bal Fe.
Heat treated: 71,000-135,000 TS; 34,000-121,000 YS; 15-25 El.
For oil refining equipment, hydrogenization equipment, gas turbine housings.
Max. operating temperature 600°C.

LOIRE-CMY18.
M-1546; 0.13 C, 1.0 Cr, 0.5 Mo, bal Fe.
Heat treated: 85,000-128,000 TS; 57,000-92,500 YS; 12-20 El;
For rotors, blades and arbors in steam and gas turbines.
Resists oxidation to 550°C.

LOIRE CMY22.
M-1546; 0.30 C, 3.0 Cr, 0.5 Mo, bal Fe.
Heat treated: 135,000 TS; 121,000 YS; 15 El.
For alternator and turbine parts, rotors.
Resists oxidation to 600°C.

LOIRE-CMYV.
M-1546; 0.15 C, 1.0 Cr, 1.0 Mo, 0.2 V, bal Fe.
Heat treated: 88,000 TS; 58,000 YS; 16 El.
For steam and gas turbine rotors and blades, superheated steam units.
Resists oxidation to 600°C.

LOIRE-COQ2.
M-1546; 1.0 C, bal Fe.
Annealed: 100,000 TS; 54,000 YP; 20 El; 190 Brin.
Ht. Tr.: 216,000 TS; 152,000 YP; 11 El; 601 Brin.
For chisels, rivet sets, fixtures, bearings. Water hardening, wear resistant.

LOIRE-HKMV.
M-1546; 0.14 C, 0.55 Cr, 0.55 Mo, 0.12 V, bal Fe.
Annealed: 78,200 TS; 57,000 YS; 22 El.
For flanges for condensers, steam turbine discs and arbors, superheater tubes.
Free from temper embrittlement.

LOIRE-HR200.
M-1546; 0.38 C, 5.0 Cr, 1.3 Mo, 1.0 Si, 0.5 V, bal Fe.
Ht. Tr.: 300,000 TS; 250,000 YS; 6 El; C 55 Rock.
Annealed: 102,000 TS; 66,000 YS; 28 El; B 94 Rock.
For aircraft landing gears, rocket motors, bolts, missile structures, cylinder liners, dies.
Martensitic steel. Hardenable.

LOIRE-ICN 001.
M-1546; 0.10 C, 18.0 Cr, 9.0 Ni, bal Fe.
Annealed: 92,500 TS; 31,300 YS; 50 El.
Rolled: 170,000 TS; 149,000 YS; 20 El.
For decorations, trim, cooking utensils, chemical plant and food processing equipment.
Corrosion resistant. Austenitic, nonhardenable by heat treatments.

LOIRE-ICN 162T.
M-1546; 0.08 max C, 17.0 Cr, 11.5 Ni, 2.2 Mo, 0.4 Ti, bal Fe.
Annealed: 85,300 TS; 35,600 YS; 40 El.
For chemical plant equipment, acid mixing tanks, vessels, agitators.
Austenitic, stainless, welding grade.

LOIRE-ICN 164.
M-1546; 0.07 max C, 17.0 Cr, 11.5 Ni, 2.2 Mo, bal Fe.
Annealed: 192,500 TS; 35,600 YS; 45 El.
For chemical plant equipment, acid mixing tanks, agitators, vessels. Austenitic stainless, Type 316.

LOIRE-ICN 164BC.
M-1546; 0.03 max C, 16.5 Cr, 13.5 Ni, 2.2 Mo, bal Fe.
Annealed: 85,300 TS; 28,400 YS; 50 El.
For chemical plant equipment, acid mixing tanks, agitators, vessels.
Austenitic, stainless, non heat-treatable.

LOIRE-ICN 164BCN.
M-1546; 0.03 max C, 17.5 Cr, 13.5 Ni, 2.8 Mo, bal Fe.
Annealed: 78,200 TS; 31,300 YS; 40 El.
For chemical plant equipment, acid mixing tanks, agitators, vessels.
Austenitic, stainless, not heat treatable.

LOIRE-ICN 164BCS.
  M-1546; 0.03 max C, 18.5 Cr, 14.0 Ni, 3.3 Mo, bal Fe.
  Annealed: 78,200 TS; 32,700 YS; 38 El.
  For chemical plant equipment, acid mixing tanks, agitators, vessels.
  Austenitic, stainless, Type 317.

LOIRE-ICN 164K.
  M-1546; 0.04 max C, 17.0 Cr, 11.5 Ni, 2.2 Mo, bal Fe.
  Annealed: 85,300 TS; 31,300 YS; 45 El.
  For chemical plant equipment, acid mixing tanks, agitators.
  Austenitic, stainless, Type 316.

LOIRE ICN 164T.
  M-1546; 0.07 max C, 17.0 Cr, 11.5 Ni, 2.8 Mo, 0.4 Ti, bal Fe.
  Annealed: 85,300 TS; 35,600 YS; 40 El.
  For chemical plant equipment, acid mixing tanks, vessels, agitators.
  Austenitic, stainless, welding grade.

LOIRE-ICN 212.
  M-1546; 0.10 max C, 12.5 Cr, 12.0 Ni, bal Fe.
  Rolled: 36%: 135,100 TS; 128,000 YS; 15 El.
  Annealed: 78,200 TS; 35,600 YS; 50 El.
  For trim, watch cases, instrument housings, jewelry, decorations.
  Corrosion resistant.

LOIRE-ICN 472.
  M-1546; 0.07 max C, 18.0 Cr, 9.5 Ni, bal Fe.
  Annealed: 82,500 TS; 35,600 YS; 48 El; B 80 Rock.
  Cold drawn: 150,000 TS; 100,000 YS; 25 El; C 25 Rock.
  For chemical plant equipment, tanks, vessels, mixers, agitators.
  Stainless Type 304. Austenitic.

LOIRE-ICN 472BC.
  M-1546; 0.03 max C, 18.5 Cr, 11.5 Ni, bal Fe.
  Annealed: 78,200 TS; 29,900 YS; 48 El; B 72 Rock.
  Cold drawn: 125,000 TS; 80,000 YS; 30 El; C 20 Rock.
  For chemical plant equipment, tanks, agitators, mixers, welded structures.
  Stainless Type 304L, austenitic.

LOIRE-ICN 472K.
  M-1546; 0.04 max C, 18.0 Cr, 10.5 Ni, bal Fe.
  Annealed: 78,200 TS; 29,900 YS; 48 El.
  Cold drawn: 130,000 TS; 85,000 YS; 28 El; C 20 Rock.
  For chemical plant equipment, tanks, vessels, welded structures.
  Stainless, Type 304L. Austenitic.

LOIRE-ICN 472NB.
  M-1546; 0.06 max C, 17.5 Cr, 12.5 Ni, 1.0 Cb, bal Fe.
  Annealed: 82,500 TS; 35,600 YS; 48 El; B 80 Rock.
  At 1000°F: 55,000 TS; 35,000 YS; 36 El; 68 RA.
  For chemical plant equipment, tanks, agitators, mixers, vessels, welded structures.
  Type 347 stainless steel. Austenitic.
  Welding grade. Stabilized.

LOIRE-ICN 472T.
  M-1546; 0.07 max C, 18.0 Cr, 10.5 Ni, 0.4 Ti, bal Fe.
  Annealed: 78,200 TS; 29,900 YS; 48 El; B 80 Rock.
  Cold drawn: 95,000 TS; 60,000 YS; 40 El; 185 Brin.
  For chemical plant equipment, cold headed parts, tanks, vessels, agitators.
  Corrosion resistant. Welding grade.
  Type 321. Austenitic.

LOIRE-IMC 201.
  M-1546; 0.10 C, 17.0 Cr, 8.0 Mn, 4.0 Ni, 0.2 N_2, bal Fe.
  Annealed: 102,000 TS; 54,000 YS; 45 El.
  For food and chemical processing equipment, trim, decorations.
  Stainless, austenitic. Low work hardenability.

LOIRE-IMC 202.
  M-1546; 0.05 C, 18.0 Cr, 9.0 Mn, 5.0 Ni, 0.2 N_2, bal Fe.
  Annealed: 78,200 TS; 45,500 YS; 45 El.
  For food and chemical processing equipment, trim, decorations.
  Austenitic, stainless. Weldable grade.

LOIRE-NYS.
  M-1546; 0.10 max C, 78.0 Ni, 20.0 Cr, 1.0 max Fe.
  Annealed: 106,000 TS; 50,000 YS.
  For heating elements, furnace parts, combustion chambers of gas and turbine reactors.
  High heat resistance to 1200°C.

LOIRE-NYSR.
  M-1546; 0.1 max C, 45.0 Ni, 23.0 Cr, 1.2 Si, 1.2 Mn, bal Fe.
  For radiators, industrial furnaces, heating elements.
  High heat resistant to 1050°C.

LOIRE-RBL.
  M-1546; 1.4 C, 0.7 Cr, bal Fe.
  Harden: C 62-64 Rock.
  For files.
  Water hardening, wear and abrasion resistant.

LOIRE-RM3.
M-1546; 0.20 C, 27.0 Cr, 3.0 Ni, bal Fe.
Annealed: 85,300 TS; 64,000 YS; 12 El.
For furnace parts, heat treat boxes, retorts, salt pots, lead pots, burners.
Cast heat resistant steel.

LOIRE-RM6.
M-1546; 0.32 C, 26.0 Cr, 5.5 Ni, 1.8 Si, bal Fe.
Annealed: 99,600 TS; 78,200 YS; 18 El.
For furnace parts, salt and lead pots, oil burners.
Austenitic-ferritic cast steel. Corrosion and heat resistant to 1100°C.

LOIRE-RM12.
M-1546; 0.30 C, 25.0 Cr, 12.0 Ni, 1.5 Si, 1.2 Mn, bal Fe.
Ht. Tr.: 110,000 TS; 60,000 YS; 50 El; B 90 Rock.
For furnaces, heat treat boxes, rails, furnace hearths, salt and lead pots.
Austenitic, heat and corrosion resistant.

LOIRE-RM20.
M-1546; 0.30 C, 25.0 Cr, 20.0 Ni, 1.8 Si, 1.3 Mn, bal Fe.
Annealed: 95,000 TS; 45,000 YS; 50 El; B 90 Rock.
For furnace parts, heat treat boxes, salt and lead pots, glass industry molds.
Austenitic, heat and corrosion resistant.

LOIRE-RM35.
M-1546; 0.45 C, 35.0 Ni, 15.0 Cr, 2.2 Si, 1.2 Mn, bal Fe.
Cast: 70,000 TS; 40,000 YS; 10 El; 180 Brin.
At 1600°F: 18,800 TS; 15,000 YS; 26 El.
For furnace parts, salt and lead pots, heat treat boxes, pyrometer tubes, grids, glass making equipment.
Austenitic, corrosion and heat resistant to 1150°C. ACI Type HT.

LOIRE-RM60.
M-1546; 0.45 C, 60.0 Ni, 12.0 Cr, 2.2 Si, 1.2 Mn, bal Fe.
Annealed: 80,000 TS; 35,600 YS; 4 El.
For cyanide and salt baths, furnace parts, heat treat boxes, glass furnace parts.
Corrosion and heat resistant.

LOIRE-SCH4.
M-1546; 0.40 C, 9.5 Cr, 0.8 Mo, 2.4 Si, bal Fe.
Heat treated: 114,000-135,000 TS; 71,000-99,600 YS; 16 El.
For inlet and exhaust valves for internal combustion engines.
Martensitic, heat resistant to 900°C.

LOIRE S I C.
M-1546; 0.98 C, 0.5 Cr, bal Fe.
Ht. Tr.:. 200,000 TS; 180,000 YS; 3 El; 390 Brin.
For bearings, rollers, bushings, gauges, mandrels.
Cold heading tool steel, oil or water hardening.

LOIRE-SP20.
M-1546; 0.77 C, 19.5 Cr, 1.5 Ni, 2.0 Si, bal Fe.
Heat treated: 121,000-143,000 TS; 71,100-114,000 YS; 12-14 El.
For exhaust valves, jet engine components.
Martensitic, corrosion and heat resistant to 1000°C.

LO-LUMINIUM.
M-Eng.; 55 Sn, 33 Zn, 11 Al, 1 Cu.
For aluminum solder.

LOMINIUM STEEL.
M-210; 0.47 C, 0.6 Mn, 1.75 Cr, 0.2 V, bal Fe.
For die casting die steel; oil hardened.

LONG TERNE.
M-604.
Steel sheet, coated with lead-tin alloy for roofing, siding.

LOPHOS.
M-289; C, V, bal Fe.
For cold heading dies.

LOSTA.
M-1311; 0.90 C, 1.9 Mn, 0.1 V, bal Fe.
For punches, dies, upsetters, cutters; oil hardened, non-deforming.

LO-TEMP.
M-1044; 40 Zn, bal Cu.
For repairing castings; brass.

LOTMESSING 60.
M-297; 38 Zn, 2 Si, bal Cu.
Soft: 52,000 TS; 48 El; 75 Brin.
1/2 H-temper: 65,000 TS; 30 El; 105 Brin.
For hardware, bolts, fasteners; corrosion resistant.

LOTUNG.
M-289; 0.30 C, 3 Cr, 9 W, 0.45 V, bal Fe.
For Al die casting dies, brass extrusion dies; hot work steel, oil hardened.

LOTUS.
M-77; 10 Sn, 15 Sb, 75 Pb.
For bearings; Babbitt; medium bearing loads.

LOW BRASS.
M-33, M-8, M-1789; 80 Cu, 20 Zn.
Annealed: 44,000 TS; 10,000 YS; 60 El.
Rolled: 65,000 TS; 45,000 YS; 12 El.
For formed and drawn parts requiring high finish; high ductility.

LOW BRASS 5.
M-141; 80 Cu, 20 Zn.
Hard: 74,000 TS; 59,000 YS; 7 El; 156 Brin.
Soft: 44,000 TS; 14,000 YS; 50 El; 56 Brin.
For hardware, diaphragms; corrosion resistant.

LOW BRASS-32.
M-8; 80 Cu, 20 Zn.
Soft: 43,000 TS; 16,000 YS; 50 El.
Hard: 73,000 TS; 60,000 YS; 8 El.
For formed and drawn parts; high ductility.

LOW BRASS 240.
M-279; 80 Cu, 20 Zn.
Ann: 44,000-54,000 TS; 12,000-29,000 YS; 42-58 El.
Cold rolled: 48,000-97,000 TS; 18,000-84,000 YS; 1-35 El.
Low zinc brass with good corrosion resistance; for bellows, flexible hose, water meters.

LOW CARBON NICKEL replaced by **NICKEL 201.**

LOW CARBON TATMO.
M-73; 0.75 C, 1.50 W, 3.75 Cr, 1.10 V, 8.50 Mo, bal Fe.
High speed steel for punches, thread rolling dies, pipe taps, hot and cold heading tools. Improved toughness.

LOW CARBON TNW.
M-73; 0.65 C, 4.0 Cr, 8.0 Mo, 1.9 V, bal Fe.
Low carbon high speed steel for cold heading inserts, punches, pipe taps, forging die inserts, and trimming dies. Good toughness.

LOW CHROME-NICKEL.
M-200; 0.35-0.45 C, 0.25-0.50 Cr, 0.80-1.00 Mn, 0.50-0.75 Ni, 0.30-0.4 Si, bal Fe.
Annealed: 100,000 TS; 60,000 YS; 20 El; 18 RA; 200 Brin.
Heat treated: 120,000-200,000 TS; 100,000-175,000 YS; 35-10 El; 20-5 RA.
For gears, shafts, pinions; tough.

LOW LEADED BRASS.
M-33; 66.5 Cu, 33 Zn, 0.5 Pb.
For drain tubes, plumbing goods, pump liners; free-cutting.

LOW LEADED BRASS-226.
M-8; 35 Zn, 0.5 Pb, bal Cu.
Soft: 45,000 TS; 17,000 YS; 55 El.
Hard: 73,000 TS; 60,000 YS; 9 El.
For hardware; good formability.

LOW LEADED TUBE BRASS-330.
M-33, M-8; 66.5 Cu, 33.0 Zn, 0.5 Pb.
Hard: 73,000 TS; 60,000 YS; 10 El; B 80 Rock.
Soft: 45,000 TS; 17,000 YS; 55 El; B 15 Rock.
For pump and power cylinders and liners, plumbing parts, munitions.
Free-cutting yellow brass.

LOWMOOR IRON.
M-803; 3.2 C, 2.6 Si, bal Fe.
For castings; cast iron.

LOW PHOSPHORUS COPPER.
M-33; 99.9 + Cu, 0.007 P.
For electrical conductors, waveguides; high conductivity, deoxidized copper.

LOW PHOSPHORUS COPPER see **COPPER, DEOXIDIZED, DLP.**

LOWROFF PHOSPHOR BRONZE.
M-England; 5-16 Pb, 4-13 Sn, 0.5-1.0 P, bal Cu.
For bearings, bushings; heavy duty.

LOW SILICON BRONZE.
M-33; 98.5 Cu. 1.5 Si.
Ann: 40,000 TS; 15,000 YS; 50 El.
Drawn 50%: 90,000 TS; 67,000 YS; 12 El.
Bolts, marine and pole line hardware.

LOW TEMPERATURE AIR HARDENING.
M-1424; 0.7 C, 2.1 Mn, 0.3 Si, 1.0 Cr, 1.3 Mo, 0.12 S, bal Fe.
Free machining grade. AISI Type A6 air hardening tool steel.

LOYCON N.
M-1724; 0.15 max C, 0.25 max Si, 1.2 max Mn, 0.35 max Mo, 0.40 max Cr, 1.6 max Ni, 0.08-0.12 V, bal Fe.
Used mainly for large pressure vessel or boiler construction.

LOYCON QT.
M-1724; 0.15 max C, 0.25 max Si, 1.2 max Mn, 0.35 max Mo, 0.80 max Cr, 1.60 max Ni, bal Fe.
Good strength, good impact strength at low temperatures; weldable.
For crane parts, storage tanks, mining equipment.

L P D.
M-73; 0.35 C, 5 Cr, 1.5 W, 1.5 Mo, 1 Si, bal Fe.
For hot forging dies, punches, mandrels; hot work steel.

LR 80.
M-1295; 80 Au, bal Pt, Cu, Ag.
Resistance wire for precision potentiometer uses, bridgewire for electroexplosive devices.

LS-31.
M-843; 25 Cr, 10 Ni, 7 W, bal Co.
Metallic spray coating for Plasma Torch.
Good wear resistance, 350 DPH.
For jet engine ducts.

LS 33 ETC see **MILLS LS 33 ETC.**

LT-75 see **LUKENS LT-75.**

LTA-60.
M-531; 0.07 C, 0.40 Mn, 0.01 P, 0.025 S, bal Fe.
60,000 psi YS.
For high strength cold rolled steel sheet applications.

LTA-70.
M-531; 0.07 C, 0.40 Mn, 0.01 P, 0.025 S, bal Fe.
70,000 psi YS.
For high strength cold rolled steel sheet applications.

L T A H.
M-510; 0.70 C, 0.30 Si, 2.0 Mn, 1.0 Cr, 1.45 Mo, bal Fe.
Cold work tool steel.
B.S. 4659 Type BA6; AISI A6.

L T C.
M-1295; 65 Au, bal Ni-Cr.
Precision potentiometer resistance wire, bridgewire for electro explosive devices.

LUBALLOY X ALLOY 80.
M-279; 1.75 Sn, 8.25 Zn, bal Cu.
Hard: 70,000 TS; 4 El.
Soft: 45,000 TS; 50 El.
For contact springs, radio tube sockets, fuse clips; corrosion resistant.

LUBALOY 411.
M-279; 90 Cu, 9.5 Zn, 0.5 Sn.
Ann: 38,000-44,000 TS; 9000-18,000 YS; 40-44 El.
Cold rolled: 42,000-87,000 TS; 17,000-78,000 YS; 1-40 El.
For bushings, washers, clutch plates, fuse clips; wear resistant and good corrosion resistance.

LUBALOY X425.
M-279; 88 Cu, 10 Zn, 2 Sn.
Ann: 40,000-50,000 TS; 12,000-24,000 YS; 44-55 El.
Cold rolled: 45,000-106,000 TS; 20,000-88,000 YS; 1-40 El.
For electrical springs, connectors, terminals.
High strength, fatigue-corrosion resistant.

LUBRAL.
M-1397; 6.5 Sn, 1.07 Cu, 1 Ni, 0.14 Si, 0.19 Fe, bal Al.
Cast: 14,000 TS; 6.8 El; 37 Brin.
For bearings.

LUBRAL SN6.
M-1445; 1 Cu, 1.2 Si, 0.12 Ti, 1 Ni, 6 Sn, bal Al.
Rolled: 21,000-27,000 TS; 9,000-12,000 YS; 12-6 El; 45-55 Brin.
For light alloy parts.

LUBRICO NO. 1.
M-256; 75-70 Cu, 20-22 Pb, 5-10 Sn.
For bearings and bushings; heavy duty.

LUBRICO NO. 2.
M-256; 70-75 Cu, 20-22 Pb, 5-10 Sn, Ni.
For bearings and bushings; heavy duty.

LUBRICO NO. 3.
M-256; 70-75 Cu, 20-22 Pb, 5-10 Sn, Ni.
For bearings and bushings; heavy duty.

LUBRI-DIE.
M-289; 1.45 C, 0.8 Mn, 1.15 Si, 0.25 Mo, bal Fe.
Annealed: 84,500 TS; 49,500 YS; 25 El; 197 Brin.
Heat treated: 164,000 TS; 136,000 YS; 13 El; 302 Brin.
Heat treated: 218,000 TS; 177,000 YS; 8 El; 388 Brin.
For blanking and forming dies, punches, gauges, hobs, cams, taps, pneumatic hammers, piercing dies.
Type 06, oil hardening, cold work tool steel.

LUBRONZE 422.
M-279; 87 Fu, 12 Zn, 1 Sn.
Ann: 40,000-50,000 TS; 10,000-20,000 YS; 44-50 El.
Cold rolled: 47,000-94,000 TS; 20,000-83,000 YS; 1-40 El.
For electrical terminals, connectors, clips, chains; strength, stress-corrosion resistance.

LUBROTEC 19985.
M-717.
Alloy powder for metal spraying final coat; machinable but tough and wear resistant.

LUCALOX.
M-U.S.; Al_2O_3.
For cutting tools; sintered.

LUCAS NIFAL.
25 Ni, 12 Al, 63 Fe.
For permanent magnets.

LUCERNO.
68-65 Ni, 27-30 Cu, 0-2.4 Fe, 2.2-5.0 Mn.
For resistance alloy; heat resistant.

LUCKY 7.
M-1215; C, V, bal Fe.
For tools; water hardened.

LUDENSCHEIDT BUTTON METAL.
M-Germany; 20 Cu, bal Zn.
For ornamental parts; die castings.

LUDLOY.
M-1422; C, alloy bal Fe.
Heat treated: 265,000 TS; 13 El; 53 Rock C.
For chisels, sledges, arbors, dies, punches; shock resistant, water hardening.

LUDLUM.
13-17 Cr, 1 Si, 1 Mo, 0.4 C, bal Fe.
For corrosion resisting parts, cutlery; corrosion resistant.

LUDLUM 602.
M-1779; 0.50 C, 0.40 Mo, 1.5 Si, 0.12 V, bal Fe.
Hardened: 330,000 TS; 270,000 YS; 9 El; 28 RA; 578 Brin.
For chisels, punches; shock resistant.

LUDLUM 609.
M-1779; 0.6 C, 0.85 Mn, 2.0 Si, 0.25 Cr, 0.2 Mo, 0.2 V, bal Fe.
For shear blades, punches, pneumatic tools; shock resistant.

LUDLUM DBL2-EZ.
M-1779; 0.8 C, 4 Cr, 2 V, 6 W, 5 Mo, bal Fe.
For lathe and planer tools, reamers, drills, taps; high speed steel, oil hardened.

LUDLUM DBL-3.
M-1779; C, Cr, V, W, bal Fe.
For lathe and planer tools, reamers, broaches; high speed steel.

LUDLUM LMW-EZ.
M-1779; 0.7 C, 3.25-4.25 Cr, 1.2-2.0 W, 0.75-1.0 V, 7.5-9.5 Mo, bal Fe.
For reamers, lathe tools, drills, broaches; high speed steel, oil hardened.

LUDLUM VLM-EZ.
M-1779; 0.85 C, 8 Mo, 4 Cr, 2 V, bal Fe.
For reamers, drills, hobs, taps, lathe cutters; high speed steel, oil hardened.

LUDLUM XCM.
M-1779; 1.2 C, 1.4 Cr, 0.45 Mo, bal Fe.
For tools, taps, gauges, lathe centers; oil hardened.

LUETECIN.
M-France; 80 Cu, 16-6 Ni, 5 Zn, 5 Fe, 2 Sn, 1 Co.
For cheap jewelry; corrosion resistant.

LUKENS 45.
M-176; 0.2-0.25 C, 1.2-1.35 Mn, 0.15-0.3 Si, bal Fe.
Plate: 65,000 TS; 45,000 min YS; 24 min El.
For bridges, barges, field-erected tanks, heavy industrial equipment.
High strength carbon steel plate.

LUKENS 50.
M-176; 0.2-0.25 C, 1.35 Mn, 0.2 Si, bal Fe.
Plate: 70,000 min TS; 50,000 min YS; 24 min El.
For bridges, barges, field-erected tanks, heavy industrial equipment.
High strength carbon steel plate.

LUKENS 55.
M-176; 0.23 C, 0.25 Si, 1.3-1.6 Mn, bal Fe.
Plate: 75,000 min TS; 55,000 min YS; 23 min El.
For bridges, barges, field-erected tanks, heavy industrial equipment.
High strength carbon steel plate.

LUKENS 60.
M-176; 0.23 C, 1.6 Mn, 0.25 Si, bal Fe.
Plate: 80,000 min TS; 60,000 min YS; 23 min El.
For bridges, barges, field erected tanks, heavy industrial equipment.
High strength carbon steel plate.

LUKENS 80.
M-176; 0.20-0.25 C, 1.1-1.6 Mn, 0.2-0.6 Si, bal Fe.
Heat treated: 95,000-120,000 TS; 75,000-80,000 YS; 19 El (min).
For welded structures, heavy industrial equipment, bridges.
Good low temperature notch toughness.

LUKENS A440.
M-176; 0.28 C, 1.1-1.6 Mn, 0.20 min Cu, bal Fe.
Plate: 70,000 TS; 50,000 YS; 18 El.
For derricks, booms, bridges, mine cars, bus and truck bodies.
High strength low alloy steel.

LUKENS A441.
M-176; 0.22 C, 1.25 Mn, 0.20 min Cu, 0.02 min V, bal Fe.
Plate: 70,000 TS; 50,000 YS; 22 El.
For bridges, booms, derricks, mine cars, bus and truck bodies.
High strength low alloy steel.

LUKENS AR-300.
M-176; 0.28 C, 1.4 Mn, 0.2-0.5 Si, 0.2 Cu, bal Fe.
Hardened: 285-321 Brin; 137,000-158,000 TS; 123,000-148,000 YS; 10-11 El.
For materials handling equipment, reactor vessels, floor wear plates.
Quenched and tempered carbon steel plate.
Wear and abrasion resistant.

LUKENS AR-350.
M-176; 0.3 C, 1.0-1.6 Mn, 0.4 Si, 0.20 Cu, bal Fe.
Ht. Tr.: 171,000 TS; 165,000 YS; 13 El; 320-380 Brin.
For materials handling equipment, reactor vessels, floor wear plates.
Quenched and tempered carbon steel plate.
Wear and abrasion resistant.

LUKENS CROMANSIL (GRADE A).
M-176; 0.17 max C, 1.05-1.40 Mn, 0.3-0.6 Cr, 0.6-0.9 Si, bal Fe.
Rolled: 75,000-90,000 TS; 45,000-54,000 YS.
For high tensile plate work; tough.

LUKENS CROMANSIL STEEL (GRADE B).
M-176; 0.25 max C, 1.05-1.40 Mn, 0.3-0.6 Cr, 0.6-0.9 Si, bal Fe.
Rolled: 85,000-100,000 TS; 47,000-55,000 YS.
For high tensile plate work; tough.

LUKENS FROSTLINE (CLASS 1).
M-176; 0.22 max C, 1.0-1.6 Mn, 0.15-0.30 Si, 0.01-0.05 Cb, bal Fe.
70,000-90,000 psi TS; 50,000 psi min YS; 21 El.
For LNG ship components, coal handling equipment.

LUKENS FROSTLINE (CLASS 2).
M-176; 0.22 max C, 1.0-1.6 Mn, 0.15-0.30 Si, 0.01-0.05 Cb, bal Fe.
80,000-115,000 psi TS; 65,000 psi min YS; 21 El.
For pipeline fittings, off road construction equipment.

LUKENS HP.
M-176; 0.06 C, 0.35 Mn, 0.05 Si, 0.025 S, 0.025 P, bal Fe.
Plate: 35,000 TS; 25,000 YS; 28 El.
For magnet cores in linear accelerators, cyclotrons, synchrotrons, bubble chambers.
Soft magnet steel, high permeability, high flux density, low hysteresis loss.

LUKENS LT-75.
M-176; 0.24 max C, 0.70-1.35 Mn, 0.15-0.30 Si, bal Fe.
LT-75 N: 65,000-90,000 TS; 46,000-50,000 YS; 23-24 El.
For low temperature applications, cold storage tanks, rail tank cars.
Low temperature steel, good notch toughness.

LUKENS LT-75HS.
M-176; 0.22 max C, 1.1-1.6 Mn, 0.2-0.6 Si, bal Fe.
Plate: 75,000 min YS; 95,000-115,000 TS.
For heavy construction equipment, off-shore drilling platforms, rail tank-cars, cold storage tanks, bridges. Good notch toughness at -75°F, quenched and tempered steel plate.

LUKENS LT-75QT.
M-176; 0.24 max C, 0.70-1.35 Mn, 0.15-0.30 Si, bal Fe.
Plate: 75,000-100,000 TS; 56,000-60,000 YS; 23-24 El.
For low temperature applications, cold storage tanks, rail tank cars.
Good low temperature notch toughness.

LUKENS MANGANESE-VANADIUM STEEL.
M-176; 0.18 max C, 1.0-1.45 Mn, 0.08-0.18 V, bal Fe.
Rolled: 80,000-95,000 TS; 50,000 YS.
For structural purposes; welds satisfactorily.

LUKENS NINE NICKEL.
M-176; 0.13 max C, 9 Ni, 0.80 max Mn, bal Fe.
Normalized: 90,000 TS; 60,000 YS; 22 El.
For storage and process vessels for handling liquid N_2 and O_2 for low temperature service.
ASTM A203.

LUKENS SP-40.
M-176; 0.24-0.31 C, 1.0-1.40 Mn, 0.15-0.30 Si, bal Fe.
Normalized: 40,000 psi YS.
For press bases.

LUKENS SPA-90.
M-176; 0.14-0.21 C, 0.95-1.30 Mn, 0.15-0.35 Si, 1.0-1.5 Cr, 1.2-1.5 Ni, 0.40-0.55 Mo, 0.03-0.08 V, bal Fe.
Plate, Q+T.: 105,000-135,000 psi TS; 90,000 psi min YS, 14-16 El.
For earthmoving and transport equipment, booms, bridges, tower and building members.

LUKENS T-1.
M-176; 0.10-0.20 C, 0.60-1.0 Mn, 0.15-0.35 Si, 0.40-0.65 Cr, 0.70-1.0 Ni, 0.40-0.60 Mo, 0.15-0.50 Cu, 0.03-0.08 V, 0.002-0.006 B, bal Fe.
Plate, Q+T.: 115,000-135,000 psi TS; 100,000 psi min YS.
For bridge, tower members, pressure vessels.
ASTM A514 Type F; A517 Grade F.

LUKENS T-1A.
M-176; 0.15-0.21 C, 0.70-1.0 Mn, 0.20-0.35 Si, 0.40-0.65 Cr, 0.15-0.25 Mo, 0.03-0.08 V, 0.01-0.03 Ti, 0.0005-0.005 B, bal Fe.
Plate, Q+T: 115,000-135,000 psi TS; 100,000 psi min YS.
For earth moving equipment, truck frames and bodies, storage tanks, oil field rigs.
ASTM A514 Type B; A517 Grade B.

LUKENS T-1 TYPE B.
M-176; 0.12-0.21 C, 0.95-1.30 Mn, 0.20-0.35 Si, 0.40-0.65 Cr, 0.30-0.70 Ni, 0.20-0.30 Mo, 0.20-0.40 Cu, 0.03-0.08 V, 0.0005 min B, bal Fe.
Plate, Q+T.: 115,000-135,000 psi TS; 100,000 psi min YS.
For bridges, buildings, mining equipment.
ASTM A514 Type H; A517 Grade H.

LUKENS TRANSLINE BCV 42 ETC see BCV-42 ETC.

LUKENS TRANSLINE UCV-60 see UCV-60.

LUMDIE.
M-73; 0.4 C, 1.0 Si, 0.2-0.4 Mn, 4.5 W, 5 Cr, 0.5 Co, bal Fe.
For dies for Al die casting, header gripper dies on bolt machines; hot work steel.

LUMEN 00-A.
M-77; 80 Cu, 20 Sn.
Cast: 33,000 TS; 21,000 YS; 0.5 El; 0.4 RA; 143 Brin.
For bells and bearings; alloy deoxidized with P.

LUMEN 00-C.
M-77; 84 Cu, 16 Sn.
Cast: 32,000 TS; 27,000 YS; 1-1.5 El; 0.8-1.2 RA; 70-80 Brin.
For bearing metal, washers, dies, trunnions; alloy deoxidized with P.

LUMEN 1.
M-77; 88 Cu, 8-10 Sn, 2-4 Zn.
Cast: 40,000 TS; 20,000 YS; 15-25 El; 14-23 RA; 57-74 Brin.
For machine parts, spur and bevel gears, air valves, pumps; alloy deoxidized with Zn.

LUMEN-1A.
M-77; 88 Cu, 8 Sn, 4 Zn.
Cast: 40,000 TS; 18,000 YS; 20 El; 18 RA; 57 Brin.
For bearings, bushings, liners.
Corrosion resistant.

LUMEN 2.
M-77; 86 Cu, 9.5 Sn, 2.5 Pb, 2.0 Zn.
Cast: 40,000 TS; 19,000 YS; 15-25 El; 12-23 RA; 52-65 Brin.
For bearing bronze, worm gears; tough.

LUMEN 4.
M-77; 80 Cu, 10 Sn, 10 Pb.
Cast: 36,000 TS; 20,000 YS; 6-8 El; 5-7.5 RA; 46-70 Brin.
For bearings, bushings, feed nuts; for high speeds.

LUMEN 5.
M-77; 85 Cu, 5 Sn, 5 Pb, 5 Zn.
Cast: 37,000 TS; 18,600 YS; 15-20 El; 15-20 RA; 45-50 Brin.
For medium and low pressure valve bodies, carburetors, general brass castings; Iz-6-7.

LUMEN 6.
M-77; 78 Cu, 8 Sn, 14 Pb.
Chill cast: 32,000 TS; 18,000 YS; 10-15 El; 8-15 RA; 48-65 Brin.
For bushings, pumps, mine and acid machines; high speed heavy duty bushings; Iz-4-5.

LUMEN 9.
M-77; 57.5 Cu, 0.77 Sn, 40 Zn, 1.0 Fe, 1.0 Al, 0.25 Mn.
Cast: 65,000 TS; 25,000 YS; 10-35 El; 9-30 RA; 109-120 Brin.
For propeller blades and hubs, valves, pump bodies; known as Manganese Bronze; Iz-20-40.

LUMEN 11 C.
M-77; 89 Cu, 1 Fe, 10 Al.
Cast: 75,000 TS; 30,000 YS; 15-25 El; 15-25 RA; 140-114 Brin.
For gears, stripper nuts; shock resisting Al Bronze; Iz-30-36.

LUMEN 14.
M-77; 90 Cu, 6.5 Sn, 2.0 Zn, 1.5 Pb.
Cast: 40,000 TS; 18,000 YS; 15-33 El; 14-33 RA; 44-48 Brin.
For valve bodies, carburetors; oil pump; known as Valve or Steam Bronze Iz-11-15.

LUMEN 15.
M-77; 88.75 Cu, 11 Sn, 0.25 Pb.
Sand cast: 40,000 TS; 23,000 YS; 5-10 El; 5-10 RA; 63-70 Brin.
For worm wheels, heavy duty bearings, spur gears; deoxidized with P.

LUMEN 15A.
M-77; 11 Sn, 1.5 Pb, bal Cu.
Sand cast: 36,000 TS; 20,400 YS; 12 El; 10 RA; 54 Brin.
Permanent mold: 45,000 TS; 23,000 YS; 10 El; 8 RA; 72 Brin.
For gears, worms, bearings; P-deoxidizer.

LUMEN 19.
M-77; 62.5 Cu, 28 Zn, 2.5 Fe, 3 Mn, 4 Al.
Cast: 90,000 TS; 45,000 YS; 18 El; 17 RA; 179 Brin.
For propellers, shafts, gears; Mn-bronze.

LUMEN 20.
M-77; 63.75 Cu, 23 Zn, 2.75 Fe, 3.75 Mn, 6.75 Al.
Cast: 120,000 TS; 75,000 YS; 5-10 El; 4-8 RA; 240-270 Brin.
For spur gears, gibs, cams; known as Super Mn Bronze; Iz-7-11.

LUMEN 31.
M-77; 70 Cu, 9 Sn, 21 Pb.
Cast: 27,000-29,000 TS; 17,000-19,000 YS; 17-14 El; 59-63 Brin.
For bearings; heavy duty.

LUMEN 33.
M-77; 70 Cu, 4 Sn, 26 Pb.
Cast: 18,000-23,000 TS; 12,000-16,000 YS; 20-12 El; 41-45 Brin.
For bearings; heavy duty.

LUMEN 43.
M-77; 10.5 Sn, 1.5 Ni, bal Cu.
Cast: 50,000 TS; 27,000 YS; 15 El; 13 RA; 74 Brin.
For bearings, worm gears, nuts; corrosion resistant.

LUMEN 48.
M-77; 84 Cu, 10 Sn, 2.5 Pb, 3.5 Ni.
Chill cast: 45,000 TS; 26,000 YS; 7 El; 8 RA; 83 Brin.
For bearings, worms, gears, nuts; deoxidized with P.

LUMEN 54.
M-77; 85 Cu, 10 Sn, 5 Pb.
Sand cast: 34,000 TS; 22,000 YS; 12 El; 10 RA; 50 Brin.
Permanent mold: 40,000 TS; 20,000 YS; 8 El; 7 RA; 70 Brin.
For elevator worm gears, feed nuts, tool bearings; IZ 4-7.

LUMEN 90.
M-77; 81 Cu, 14 Al, 5 Fe.
Cast: 302 Brin.
For die blocks; Al-bronze, wear resistant.

LUMEN 91.
M-77; 80 Cu, 15 Al, 5 Fe.
Cast: 340 Brin.
For die blocks; Al-bronze, wear resistant.

LUMEN 96.
M-77; 8.5 Al, 3.5 Fe, bal Cu.
Cast: 73,000 TS; 25,000 YS; 120 Brin.
For gears, bolts, hardware, bearings; Al-bronze, corrosion resistant.

LUMEN 97.
M-77; 85 Cu, 11 Al, 4 Fe.
Cast: 80,000 TS; 35,000 YS; 8 El; 6 RA; 163 Brin.
Heat treated: 90,000 TS; 45,000 YS; 3 El; 3 RA; 200 Brin.
For gears, feed nuts, shifter forks; Al-bronze, hardenable.

LUMEN 98.
M-77; 84 Cu, 12 Al, 4 Fe.
Cast: 80,000 TS; 37,000 YS; 2 El; 1 RA; 212 Brin.
For feed shoes, gibs, cams, gears, forming dies; Al bronze. Wear resistant.

LUMEN BRONZE.
M-77; 10 Cu, 86 Zn, 4 Al.
Cast: 36,000 TS; 36,000 YS; 0 El; 0 RA; 119 Brin.
For bearings for electric motors and lathes; will not resist excessive heat or live steam.

LUMINARC 2-S.
M-811; 99 min Al.
For Al welding rod.

LUNKENHEIMER H-1.
M-264; Sn, alloy, bal Fe.
For bonnets, bushings, valves; valve bronze.

LUNKENHEIMER HB-1.
M-264; Sn, alloy, bal Fe.
For key cocks, valves; valve bronze.

LUNKENHEIMER ML-3.
M-264; 60 min Ni, 23 min Cu, 3.5 max Fe, 2 max C and Si.
Cast: 65,000-90,000 TS; 32,000-40,000 YS; 45-25 El; 125-150 Brin.
For valve parts, discs, stems, seat rings; corrosion resistant.

LUNKENHEIMER ML-5.
M-264; 60 min Ni, 23 min Cu, 3.5 max Fe, 0.5 max Al, 2.0 max C + Si.
Cast: 65,000-90,000 TS; 32,000-40,000 YS; 45-25 El.
For valve parts, seat rings, discs, stems; corrosion resistant, cast Monel.

LUNKENHEIMER N3.
M-264; 67 Cu, 30 Ni.
Cast: 45,000 psi TS; 70 Brin (500).
For pressure containing valve components; corrosion resistant, wear resistant, ductile.

LUNKENHEIMER N-31.
M-264; 63 Cu, 28 Ni.
Cast: 40,000 psi TS; 70 Brin (500).
For valve components; corrosion resistant, ductile, wear resistant, withstands high temperature for short exposures.

LUNKENHEIMER NO. 25 IRON.
M-264; 2.6 C, 1.0-1.5 Ni, bal Fe.
Cast: 29,000 TS.
For valves, flanges, pipe fittings; cast iron.

LUNKENHEIMER NO. 28 IRON.
M-264; 2.6 C, 1.0-1.5 Ni, bal Fe.
Cast: 39,000 TS.
For valve, flanges, pipe fittings; cast iron.

LUNKENHEIMER NO. 30 IRON.
M-264; 2.6 C, 1.0-1.5 Ni, bal Fe.
Cast: 43,000 TS.
For valves, flanges, pipe fittings; cast iron.

LUNKENHEIMER NO. 50 IRON.
M-264; 2.6 C, 1.5 Ni, 0.3 Cr, 0.7 Mo, bal Fe.
Cast: 56,000 TS.
For valves, flanges, pipe fittings; alloy cast iron.

LUNKENHEIMER NS-5.
M-264; 47 Ni, 44 Cu.
Cast: 110,000 TS; 301 Brin.
For valve components such as seat rings and discs; good hardness and corrosion resistance.

LUNKENHEIMER NT-4.
M-264; 66 Cu, 27 Ni.
Cast: 65,000 psi TS; 97 Rock B.
For valve seats and discs; excellent resistance to wear and corrosion.

LUNKENHEIMER NT-7.
 M-264; 50 Ni, 37 Cu.
 Cast: 60,000 psi TS; 201 Brin.
 For valve components such as seat rings; good wearing and non-galling in high temperature steam service.

LUNKENHEIMER S-1.
 M-264; 6 Sn, 1.7 Pb, 3.7 Zn, 0.47 Ni, bal Cu.
 Cast: 42,500 TS; 18,900 YS; 50 El.
 For valves; corrosion resistant, free-cutting.

LUNKENHEIMER T-1.
 M-264; 5 Sn, 4.6 Pb, 4.6 Zn, 0.75 Ni, bal Cu.
 Cast: 38,200 TS; 18,300 YS; 35 El.
 For valves; corrosion resistant, free-cutting.

LUNKENHEIMER WC1.
 M-264; 0.19 C, 0.47 Si, 0.66 Mn, 0.56 Mo, bal Fe.
 At 70°F: 77,800 TS; 52,200 YS; 29.9 El; 48.5 RA.
 At 950°F: 63,000 TS; 37,000 YS; 30 El; 65 RA.
 For castings, body bonnets.

LUNKENHEIMER WC4.
 M-264; 0.15 C, 0.5 Mo, 0.7 Cr, 0.9 Ni, bal Fe.
 Cast: 86,000 TS; 60,000 YS; 29 El; 54 RA.
 For valves, castings; resists graphitization.

LUNKENHEIMER WC 5.
 M-264; 0.14 C, 0.93 Mo, 0.79 Cr, 0.88 Ni, bal Fe.
 Cast: 88,000 TS; 61,000 YS; 27 El; 54 RA.
 For castings, valves; resists graphitization.

LUNKENHEIMER WCB.
 M-264; 0.27 C, 0.68 Mn, 0.43 Si, bal Fe.
 At 70°F: 75,600 TS; 49,100 YS; 30 El; 47.2 RA.
 At 750°F: 64,000 TS; 27,000 YS; 35 El; 54 RA.
 For castings; good weldability.

LUNORIUM.
 14.9 Cr, 0.9 Co, 55.6 Ni, 18.5 Mo, 4 W, 5 Fe, 0.2 Mn, 0.4 Si, 0.2 C.
 Cast: 69,000 TS; 32,000 YS; 1.4 El.

LUNZ IRON.
 M-534; 1.5 C, 13 Cr, bal Fe.
 For glass molds; non-warping.

LURGIMETALL.
 M-299; 96.5 Pb, 2.8 Ba, 0.4 Ca, 0.3 Na.
 For bearing metals; anti-friction.

LURIUM 107B.
 M-1285; 0.5-1.0 Mg, 0.2-1.0 Si, bal Al.
 Heat treated: 28,500-42,700 TS; 21,400-35,600 YS; 15-5 El; 80-100 Brin.
 For jewelry, reflectors, auto trim, lighting fixtures; takes good polishing and bright anodizing.

LURIUM L.
 M-1285; 99.99 Al.
 O: 6000 TS; 2100 YS; 60 El; 13 Brin.
 1/2 H: 13,000 TS; 10,000 YS; 8 El; 25 Brin.
 H: 20,000 TS; 18,500 YS; 4 El; 35 Brin.
 For reflectors, domestic appliances; high reflectivity.

LURIUM L5.
 M-1285; 0.4-0.6 Mg, bal Al.
 Soft: 10,000 TS; 2800 YS; 40 El; 22 Brin.
 1/2 H-temper: 21,400 TS; 20,000 YS; 6 El; 45 Brin.
 H-temper: 27,000 TS; 25,800 YS; 2 El; 50 Brin.
 For reflectors, ordnance and domestic appliances; high reflectivity.

LURIUM L10.
 M-1285; 0.08-1.2 Mg, bal Al.
 Soft: 14,200 TS; 4200 YS; 20 El; 30 Brin.
 1/2 H-temper: 24,200 TS; 21,400 YS; 8 El; 50 Brin.
 H-temper: 35,500 TS; 30,000 YS; 2 El; 60 Brin.
 For reflectors, ornamental trim, domestic appliances; high reflectivity.

LURIUM L20.
 M-1285; 1.5-2.5 Mg, bal Al.
 Soft: 21,400 TS; 7000 YS; 35 El; 40 Brin.
 1/2 H-temper: 35,000 TS; 32,600 YS; 6 El; 75 Brin.
 H-temper: 50,000 TS; 45,500 YS; 2 El; 80 Brin.
 For reflectors, ornamental trim, domestic appliances; high reflectivity.

LUSCO.
 M-1422; C, alloy, bal Fe.
 For pneumatic tools, wear and shock resisting tools and dies; oil hardened.

LUSTER.
 M-812; C, Cr, Ni, bal Fe.
 For stainless parts; corrosion resistant.

LUSTRE-DIE.
 M-24, M-341; 0.5 C, 1.0 Mn, 0.3 Si, 0.25 Mo, 1.1 Cr, bal Fe.
 302-352 Brin; for plastic mold dies; good weldability.

LUTETIUM.
 M-1755; Lu.
 Purities: 99.9% (Special distilled grade), 99.5 + %, low tantalum contained.
 Forms: Ingot, lumps, turnings, sheet, wire, sponge.

LUXAL 63/25.
 M-116; 2-4 Mg, 0.4 max Mn, 0.3 max Cr, bal Al.
 Soft: 28,000 TS; 13,000 YS; 30 El; 47 Brin.
 Hard: 40,000 TS; 35,000 YS; 10 El; 73 Brin.
 For aircraft tanks and fittings, fuel lines, marine parts; resists sea water corrosion.

L-VICTRIX.
M-96; 0.26 C, 2.3 Cr, 1.3 Ni, 0.15 Mo, 0.1 V, bal Fe.
For gears, shafts; case hardening.

L-VICTRIX EXTRA.
M-96; 0.26 C, 2.45 Cr, 1.75 Ni, 0.23 Mo, bal Fe.
For gears, shafts; case hardening.

L-VICTRIX SPECIAL.
M-96; 0.15 C, 1 Cr, 3.65 Ni, 0.35 Mo, bal Fe.
For gears, shafts; case hardening.

L-VICTRIX SPECIAL 32.
M-96; 0.26 C, 0.75 Cr, 0.65 W, 3.4 Ni, 0.3 Mo, bal Fe.
For gears, shafts; case hardening.

LW-1.
M-843; 91 WC, 9 Co.
Cemented carbide spray coating for D-Gun.
Extreme wear resistance; 1300 DPH.

LW-1N40.
M-843; 85 WC 15 Co.
Cemented carbide spray coating for D-Gun.
Wear resistance-better impact resistance than LW-1; 1050 DPH.

LW-11 B.
M-843; 88 WC, 12 Co.
Cemented carbide spray coating for Plasma Torch.
Good wear resistance.
750 DPH.

LW-15.
M-843; 86 WC, 10 Co, 4 Cr.
Cemented carbide spray coating for D-Gun.
Wear and corrosion resistant.
1100 DPH.

LW 2325.
M-Germany; 0.10 C, 0.42 Si, 0.26 Mn, 15.68 Cr, 5.25 W, 1.38 Mo, 2.1 Al, 1.58 Ti, 1.68 Fe, bal Ni.
For hot extrusion dies.
Good high temperature properties and wear resistance.

LW 2326.
M-Germany; 0.18 C, 0.34 Si, 0.39 Mn, 14.91 Cr, 1.39 W, 4.35 Al, 1.6 Fe, bal Ni.
For hot extrusion dies.
Good high temperature properties and wear resistance.

LX 8.
M-1522; 15 Ag, Cu, Zn, Cd.
Melt range: 700-780°C.
Max stress: 39.4 kgf/mm^2; 10 El.
For silver brazing; general use.

LXX.
M-1779; 0.7 C, 18 W, 4 Cr, 1 V, bal Fe.
Original 18-4-1 general purpose high speed steel for wide variety of cutting requirements.
AISI T-1.

LYNITE.
M-92; 9-13.5 Cu, bal Al.
For castings, pistons, connecting rods; piston alloy.

LYNITE.
M-92; 2 Cu, 1.5 Mg, 0.8 Fe, 0.02 Si, bal Al.
For pistons; age-hardenable.

LYNITE 43.
M-92; 95 Al, 5 Si.
Sand cast: 19,000 TS; 9,000 YS; 4 El; 50 Brin.
For light alloy automotive and airplane bodies; see ALCOA 43.

LYNITE 109.
M-92; 11.5-13.5 Cu, 1.7 Fe + Mn + Mg + Si, bal Al.
19,000-28,500 TS; 15,000 YS; 0.5-1.0 El; 70 Brin.
For pressure castings, pistons, chill castings; pressure castings; registered name.

LYNITE 112.
M-92; 90 Al, 7-8 Cu, 1.5 Zn, 1.3 Fe.
Sand cast: 22,000 TS; 14,000 YS; 2 El; 70 Brin.
For crankcase; see ALCOA 112.

LYNITE 122.
M-92; 9.25-10.75 Cu, 2 Fe + Mn + Mg + Si, bal Al.
28,000-40,000 TS; 20,000 YS; 1.5-0.5 El; 115 Brin.
For castings to be used at high temperatures, pistons; registered name.

LYNITE NO. 145.
M-92; 87.7 Al, 3 Cu, 8 Zn, 1.25 Fe.
27,500 TS; 4.5 El; 65 Brin.
For light alloy castings; casting, shock resistant.

LYNITE NO. 146.
M-92; 91-93 Al, 7-8.5 Cu, 1.3 Fe, 0.1-1.7 other elements.
26,000 TS; 1.4 El.
For light alloy castings; casting.

LYNUX BRONZE.
M-Eng.; 89 Cu, 7.2 Fe, 3.8 Al.
For corrosion resisting hardware and fittings.

LYON'S GOLD.
M-Eng.; 27 Zn, 72 Cu.
For cartridge cases, condenser tubes, brazing; "Tombac."

M

M1.
M-1488; 0.15 C, 0.50 Mn, 0.35 max Si, bal Fe.
Carburizing steel; for cams, small machine parts.
Italie: UNI 16.

M1-CARBIDE.
M-Ger.; 84 WC, 10 TiC + TaC, 6 Co.
For hard tools, cutters, dies.
Sintered carbides, wear and abrasion resistant.

M2A.
M-118; Mo alloy.
For spot welding tips.

M-2A STEEL.
M-1191; 0.25-0.30 C, 1.15-1.25 Ni, bal Fe.
Cast: 80,000-95,000 TS; 53,000-58,000 YS; 30-24 El; 50-35 RA; 175-200 Brin.
For sheave wheels, gears; oil hardened, shock resistant.

M2-CARBIDE.
M-Ger.; 82 WC, 10.5 TiC + TaC, 7.5 Co.
For hard tools, cutters, dies.
Sintered carbides, wear and abrasion resistant.

M-2 DREADNAUGHT.
M-362; 0.80 C, 4 Cr, 2 V, 6 W, 5 Mo, bal Fe.
For lathe and planer tools, reamers, broaches; high speed steel; Type M2

M-2 STEEL.
M-1191; 0.30 C, 1.15 Mn, 1.25-1.5 Ni, bal Fe.
Cast: 90,000-110,000 TS; 55,000-65,000 YS; 30-20 El; 50-40 RA; 200-250 Brin.
For lifting forks for trucks; oil hardening, shock resistant.

M5.
M-1769.
Sintered Carbide.
400,000 TrS; Density: 14.4-14.6; Hardness: 87.5-88.5 RA.
Industry code: C-16.

M5T.
M-1522; 35 Ag, Cu, Zn, Sn.
Melt range: 645-735°C.
Max stress: 47.5 kgf/mm^2; 10 El.
For silver brazing - general engineering purposes.

M6.
M-1769.
Sintered carbide.
400,000 TrS; Density: 14.4-14.6; Hardness: 88.3-89.3 RA.
Industry code: C-16.

M7.
M-1769.
Sintered carbide.
360,000 TrS; Density: 14.5-14.7; Hardness: 89.0-90.0 RA.
Industry code: C-16.

M9T.
M-1522; 39 Ag, Cu, Zn, Sn.
Melt range: 635-710°C. Max stress: 49.0 kgf/mm^2, 10 El.
For silver brazing,-general engineering purposes.

M-11.
M-1105; 0.7 C, 18 W, 4 Cr, 1 V, 5 Co, bal Fe.
For taps, threading tools; high speed steel.

M13.
M-1653; 1.20 C, 13 Mn, bal Fe.
Austenitic manganese steel.
X 120 Mn 13.

M21.
M-1769.
Sintered carbide.
325,000 TrS; Density: 14.75-15.0; Hardness: 90.5-91.5 RA. Industry code C-1-9; ISO K30.

M22.
M-68, M-1499; 0.13 C, 5.7 Cr, 2.0 Mo, 11.0 W, 6.3 Al, 0.6 Zr, 3.0 Ta, bal Ni.
High temperature alloy.

M22VC.
M-86; 0.08-0.16 C, 5.0-6.5 Cr, 5.9-6.6 Al, 1.5-2.5 Mo, 10.5-11.5 W, 2.6-3.4 Ta, 0.4-0.8 Zr, 1.0 max Fe, bal Ni.
Cast: 106,700 TS; 99,000 YS; 5.6 El; 14 RA.
At 600°C: 117,000 TS; 109,000 YS; 4.5 El; 13.8 RA.
For aircraft gas turbines, spinners for glass fibre production.
High temperature vacuum casting alloy; Creep and oxidation resistant.

M25T.
M-1522; 55 Ag, Cu, Zn, Sn.
Melt range: 630-660°C.
Max stress: 44.7 kgf/mm^2; 24 El.
For silver brazing; excellent fluidity.

M31.
M-1769.
Sintered carbide.
350,000 TrS; Density: 14.15-14.45; Hardness: 88.3-89.3 RA.
Industry code: C-11; ISO K40.

M40.
M-1769.
Sintered carbide.
400,000 TrS; Density: 14.15-14.35; Hardness: 85.5-86.5 RA.
Industry code: C-16.

M41.
M-1769.
Sintered carbide.
400,000 TrS; Density: 14.15-14.45; Hardness: 86.8-87.8 RA.
Industry code: C-16.

M45.
M-1488; 0.46 C, 0.70 Mn, 0.40 max Si, bal Fe.
Carbon structural steel.
AISI-SAE 1049.

M50 see **VASCO M50 CVM; CRUCIBLE M50 VAR; AND CARPENTER CONSUMET M-50.**

M56.
M-1769.
Sintered carbide.
400,000 TrS; Density: 14.4-14.6; Hardness: 87.7-88.7 RA.
Industry code: C-16.

M115.
M-1548; 1.5 Mn, bal Al.
Half Hard: 22,000 TS; 16,000 YS; 7-14 El; 40-45 Brin.
Hard: 35,000 TS; 31,000 YS; 3-6 El; 60 Brin.
For roofing, structures, commercial vehicles.
Corrosion resistant, non-hardenable.

M200.
M-1290; 0.85-0.95 C, 0.15-0.30 Si, 1.9-2.1 Mn, 0.2-0.5 Cr (op), 0.05- 0.15 V, bal Fe.
Cold work tool steel.
W.-Nr. 1.2842.

M-203.
M-31, M-60; 0.07 C, 1.95 Cr, 24.5 Ni, 36.5 Co, 12.0 W, 1.5 Cb, 2.15 Ti, 0.75 Al, 1.6 Fe.
High temperature alloys, heat and corrosion resistant.

M-204.
M-31, M-60; 0.07 C, 18.5 Cr, 24.5 Ni, 40.5 Co, 12.0 W, 1.2 Cb, 1.6 Fe, 0.22 B.
High temperature alloy; heat and corrosion resistant.

M-205.
M-31, M-60; 0.07 C, 18.5 Cr, 24.5 min Ni, 37.5 Co, 12.0 W, 1.2 Cb, 2.75 Al, 1.6 Fe, 0.22 B.
High temperature alloy; heat and corrosion resistant.

M-250.
M-24; 0.01 C, 0.07 Mn, 0.07 Si, 18.0 Ni, 8.0 Co, 4.75 Mo, 0.4 Ti, 0.1 Al, bal Fe.
High temperature alloy; heat and corrosion resistant.

M-252.
M-44; 19 Cr, 10 Co, 10 Mo, 2 Ti, 1 Al, bal Ni.
Wire and rod for welding wire and fastener stock.

M-252.
M-114, M-1491; 0.15 C, 0.5 Mn, 0.5 Si, 20 Cr, 10 Co, 10 Mo, 2.6 Ti 1.0 Al, 0.005 B, bal Ni.
Bar: 180,000 TS; 122,000 YS; 16 El.
At 1400°F: 137,000 TS; 104,000 YS; 10 El.
For high temperature service, bolts, jet engine and gas turbine parts.
Heat and corrosion resistant.

M-252 see also **UDIMET M-252; ALLOY M-252; ALLVAC M-252; CARPENTER PYROMET M-252; CRUCIBLE M-252; AND UNITEMP M-252.**

M308.
M-318, M-60; 0.08 C, 14.0 Cr, 33.0 Ni, 4.0 Mo, 6.5 W, 2.0 Ti, 0.25 Al, 0.25 Zr, bal Fe.
High temperature alloy; heat and corrosion resistant.

M331.
M-365; 0.32 C, 3.4 Cr, 0.4 V, 2.5 Mo, bal Fe.
Air or oil hardening hot work tool steel.

M333.
M-365; 0.30 C, 3.0 Cr, 0.6 V, 3.0 Mo, 2.25 Co, bal Fe.
Air or oil hardening hot work tool steel.

M-600.
M-31, M-60; 0.08 C, 19.0 Cr, 7.0 Mo, 2.3 Ti, 1.1 Al, 13.0 Fe, 55.5 min Ni.
For high temperature structural parts.

M-813.
M-31, M-60; 0.08 C, 18 Cr, 35 Ni, 4 Mo, 2.25 Ti, 1.4 Al, bal Fe.
For aircraft gas turbine parts, high temperature bolting.

M-841.
M-1405; 1.05 C, 4.0 Cr, 9.5 Mo, 1.2 V, 1.5 W, 8.0 Co, bal Fe.
Molybdenum-cobalt high speed steel.
For drills, reamers, lathe tools.
AISI M-42; BS 4659 BM 42.

M 921.
M-1405; 0.80 C, 4.0 Cr, 9.0 Mo, 1.10 V, 1.50 W, bal Fe.
Molybdenum high speed steel.
For twist drills, taps, lathe tools.
BS 4659 BT1; AISI M1.

MA-18NICOMO.
M-68; C, 18 Ni, Co, Mo, bal Fe.
Heat treated: 220,000-305,000 TS; 215,000-300,000 YS; 64-60 RA.
Ductile and weldable.

MA-754.
M-1491; 0.04 C, 20.0 Cr, 0.5 Ti, 0.3 Al, 0.6 Y_2O_3 bal Ni.
Mechanically alloyed powder for blades and vanes.

MACADAMITE.
M-92; 74 Al, 3 Cu, 23 Zn.
34,500 TS; 1.0 El.
For pistons; sand castings.

MACALLOY.
M-1724; 0.60 C, 0.75 Cr, bal Fe.
1000 N/mm^2 normal UTS.
Steel bar for prestressed concrete.

MACCO 33.
M-80; 0.4 C, 5.5 Cr, 1.4 Mo, 1 V, 1 Si, bal Fe.
For die casting dies; air hardening.

MACCO 35.
M-80; C, alloy, bal Fe.
For tools, air-hardened.

MACCO 35 AIR HARD.
M-80; 1d C, 0.6 Mn, 5.25 Cr, 1.15 Mo, 0.25 V, bal Fe.
For punch and cold work dies; air hardening.

MACCO 99.
M-80; 0.35 C, 0.8 Mn, 0.85 Cr, 0.35 Mo, 0.6 Si, bal Fe.
For die casting dies; water or oil hardening.

MACCO ALLOY RAZOR BLADE.
M-80; 0.7 C, 1.2 Cr, bal Fe.
For razor blades; water hardened.

MACCO B-29.
M-80; 0.98 C, 0.26 Mn, 0.2 V, bal Fe.
For cold heading and forming dies; water hardening.

MACCO BELT KNIFE.
M-80; C, Cr, Ni, V, Si, Mn, bal Fe.
For knives, leather splitting belt knives; oil hardened.

MACCO BRAKE DIE.
M-80; 0.50 C, 0.90 Mn, 1.02 Cr, 0.22 Mo, bal Fe.
For brake dies; oil hardened.

MACCO BRAND SWEDISH.
M-80; C, bal Fe.
For tools; special grade of tool steel.

MACCO CARBON.
M-80; 1.0 C, 0.20 Mn, 0.16 Si, bal Fe.
Annealed: 100,000 TS; 53,000 YS; 21 El; 42 RA; 200 Brin.
For drills, taps, chasers, lathe and planer tools; Type W1; water hardened.

MACCO CERTIFIED.
M-80; C, bal Fe.
For general purpose tools; water hardening.

MACCO ENORMOUS.
M-80; 0.8 C, 5 Cr, 17 W, 1.5 V, 5 Co, bal Fe.
For taps, dies, twist drills, cutters; high speed steel.

MACCO FOOLPROOF O.H.
M-80; 0.55 C, 1.25 Cr, 2.75 W, 0.2 V, bal Fe.
Heat treated: 231,000-315,000 TS; 209,000-275,000 YS; 9-11 El; 27-42 RA; C 48-56 Rock.
For header and swaging dies, chipping chisels, rivet busters, track tools.
Type S1 shock resisting tool steel, good fatigue resistance.

MACCO HARDTUF.
M-80; 0.6 C, 0.45 Mo, 0.20 V, 1.85 Si, bal Fe.
For plastic mold dies; oil hardened.

MACCO HOBOMOLD A.
M-80; 0.07 C, 4.5 Cr, 0.45 Mo, 0.4 Mn, bal Fe.
For plastic mold dies; Type P4; air hardened, hobbing steel.

MACCO HOBOMOLD B.
M-80; 0.06 C, 0.03 Mn, 1 Cr, 0.20 Si, bal Fe.
For plastic mold dies; Type P5; oil hardened, hobbing steel.

MACCO HOBOMOLD C.
M-80; 0.04 C, 0.2 Mn, 0.16 Si, bal Fe.
For plastic mold dies; case-hardening.

MACCO HOLLOW-DRILL.
M-80; C, bal Fe.
For hollow drills, tools, rock drills; water hardened.

MACCO KROMAX.
M-80; 2.5 C, 0.6 Si, 0.5 Ni, 12-14 Cr, bal Fe.
For dies, tools; does not possess red hardness.

MACCO KROMAX 1.
M-80; 1.5 C, 11.9 Cr, 0.7 V, bal Fe.
For blanking, forming and punching dies; air hardened, non-deforming.

MACCO KROMAX 2.
M-80; 1.4 C, 12.5 Cr, 0.85 Mo, 3.25 Co, 0.4 Si, bal Fe.
For blanking, punching, forming dies; air hardening, cold forming.

MACCO KROMAX SPECIAL.
M-80; 1.6 C, 12.4 Cr, 0.72 Mo, bal Fe.
For tools, dies; non-deforming.

MACCO LENS MOLD.
M-80; 0.4 C, 5.25 Cr, 4.65 W, 1 Si, bal Fe.
For die casting lens mold; oil hardening.

MACCOMAX.
M-80; 0.7 C, 18 W, 4 Cr, 1 V, bal Fe.
For tools, cutters; high speed steel.

MACCO ML.
M-80; 0.35 C, 5 Cr, 1.5 W, 1.65 Mo, 1 Si, bal Fe.
For hot work dies; hot work steels.

MACCO M.L.V.
M-80; 0.35 C, 1.05 Si, 1.55 W, 1.65 Mo, 5.15 Cr, bal Fe.
Hardened: 216,000 TS; 185,000 YS; 14 El; 53 RA.
At 1000°F: 144,000 TS; 110,000 YS; 19 El; 64 RA.
For die casting dies, hot shear blades, forging and heading dies, punches.
Type H12 hot work tool steel, tough and shock resistant.

MACCO NLS.
M-80; 0.9 C, 0.6 Mn, 5.2 Cr, 1.2 Mo, 0.3 V, bal Fe.
For plastic molding dies; oil hardening.

MACCO NON-TEMP.
M-80; 0.35 C, 0.30 Cu, 0.45 Si, 0.70 Mo, 0.80 Cr, 0.30 Mo, bal Fe.
For general purpose tools, dies, punches; oil or water hardened, shock resistant.

MACCO P-125.
M-80; 0.25 C, 4.2 Cr, 14.5 W, 0.5 V, bal Fe.
For hot work dies; hot work steels.

MACCO P-150.
M-80; 0.5 C, 2.9 Cr, 15.3 W, 0.6 V, bal Fe.
For hot work dies; high speed steel.

MACCO P-175.
M-80; 0.3 C, 3.3 Cr, 9.5 W, 0.5 V, 0.4 Si, bal Fe.
For hot work dies; hot work steel.

MACCO RADIO.
M-80; 0.8 C, 4 Cr, 6 W, 5.5 Mo, 1.8 V, bal Fe.
For tools, cutters; high speed steel.

MACCO ROYAL CROWN.
M-80; 1.0 C, 1.0 Cr, 0.5 W, 0.5 Mn, bal Fe.
For dies, punches, taps, drills, reamers, drawing dies; non-deforming steel.

MACCO SILICON MANGANESE.
M-80; 0.63 C, 0.78 Mn, 2 Si, bal Fe.
For punches, chisels, crimpers; Type S4; oil hardened, shock resistant.

MACCO SILVER DIE.
M-80; 0.52 C, 1 Mn, 1.1 Cr, 0.25 Mo, bal Fe.
For dies for plastic molds; oil hardened.

MACCO SPECIAL.
M-80; 1.0-1.1 C, bal Fe.
For tools, drills, taps; water hardening.

MACCO SUPERIOR HIGH SPEED.
M-80; 0.72 C, 4 Cr, 1 V, 18.5 W, bal Fe.
For tools, cutters, drills; high speed steel.

MACCO SUPERIOR SWEDISH.
M-80; C, bal Fe.
For tools, special punches, dies; oil hardened.

MACCO SUPER MOLY.
M-80; 0.8 C, 3.9 Cr, 1.7 W, 8.8 Mo, 1.1 V, bal Fe.
For cutting tools, reamers; high speed steel.

MACCO SWEDISH CHROME TUNGSTEN.
M-80; C, Cr, W, bal Fe.
For tools, special taps, dies; water hardened.

MACCO SWEDISH MAGNETIC IRON.
M-80; 0.2 C, 0.02 Si, bal Fe.
For magnetic parts for all kinds of electrical instruments; high permeability.

MACCO WJF.
M-80; 1.4 C, 4 W, 0.35 V, 0.6 Cr, bal Fe.
For cutters, taps, reamers; water hardened, fast finishing.

MACH 5.
M-604; 0.09 max C, 0.85-1.15 Mn, 0.04-0.09 P, 0.26-0.35 S, 0.15-0.35 Pb, 0.05-0.10 Bi, bal Fe.
Cold-drawn: 73 ksi TS; 68 ksi YS; 15.8 El.
Free machining steel.

MAC HEMPITE GRADE A.
M-229; 1.5-3.5 Ni, 0.7-4 Mn, 0-1.25 Cr, 0-0.75 Mo, 0.4-3.0 C, bal Fe.
Soft: 100,000 TS; 65,000 YS; 25 El; 45 RA.
Hard: 600 Brin.
For jaws, rollers, tires, balls, liners, impellers; heavy crusher grade; resists wear and spalling.

MAC HEMPITE GRADE B.
M-229; 1.5-3.5 Ni, 0.7-4 Mn, 0-0.75 Mo, 0.4-3.0 C, bal Fe.
Soft: 90,000 TS; 65,000 YS; 25 El; 45 RA; 200 Brin.
For heavy duty gears, pinions; medium hardened grade; shock resistant.

MAC HEMPITE GRADE C.
M-229; 1.5-3.5 Ni, 0.7-4 Mn, 0-1.25 Cr, 0-0.75 Mo, 0.4-3.0 C, bal Fe.
For gears, pinions, hammers, beaters, conveyor parts, brake-drums; hardened grade.

MACHINE BRONZE.
M-77; 50 Cu, 25 Ni, 25 Sn.
For bearings; wear resisting.

MACHINERY 30.
M-435; 0.3 C, 0.8 Mn, 0.2 Si, bal Fe.
For machinery parts; construction steel.

MACHINERY 40.
M-435; 0.4 C, 0.8 Mn, 0.2 Si, bal Fe.
For machinery parts; construction steel.

MACHINERY BRASS.
M-Eng.; 16 Zn, 1 Sn, bal Cu.
For machinery parts; corrosion resistant.

MACHS ALLOY.
98-90 Al, 2-10 Mg.
For light alloy parts.

MACHS SPECULUM.
69 Al, 31 Mg.
For light alloy parts.

MACHTS METAL.
57 Cu, 43 Zn.
For brazing metal, hardware.

MACHTS YELLOW METAL.
57 Cu, 43 Zn.
For hardware, bolts, nuts; high strength.

MAC-IT.
M-1081; 0.4 C, 0.8 Cr, 1.5 Ni, bal Fe.
For screws, threaded fasteners; oil hardening.

MACKENITE METAL.
M-177; 0.2 C, 20 Cr, 12 Ni, bal Fe.
For annealing pots, furnace pots; heat resistant.

MACKENZIE METAL.
70-68 Pb, 17-16 Sb, 13-16 Sn.
For bearings; anti-friction.

MACLOY B.
M-521; 0.27 C, 0.3 Si, 1.25 Mn, 11 Cr, 36 Ni, bal Fe.
Heat treated: 90,000 TS; 64,000 YS; 30 El; 55 RA; 200 Brin.
For corrosion and heat resistant parts; austenitic, stainless.

MACLOY G.
M-521; 0.5 C, 1.8 Si, 0.6 Mn, 17 Cr, 37 Ni, bal Fe.
94,000 TS; 54,000 YS; 30 El; 40 RA; 200 Brin.
For heat and corrosion resistant parts; austenitic, heat and corrosion resistant.

MACOLOY TYPE E.
M-1107; P, bal Cu.
For brazing; M.P. 1150°F.

MACOLOY TYPE F.
M-1107; P, bal Cu.
For brazing; M.P. 1180°F.

MACOLOY TYPE G.
M-1107; P, bal Cu.
For brazing; M.P. 1600-1900°F.

MACROMAL.
M-625; 0.2 C, 12-13 Mn, bal Fe.
Rolled: 100,000-135,000 TS; 35,600 YS; 30 El.
For compass housings, deck superstructures, transformer shields; austenitic, nonmagnetic.

MACROMAL S.
M-625; 0.2 C, 17.18 Mn, bal Fe.
Rolled: 100,000-135,000 TS; 35,600 YS; 30 El.
For compass housings, deck superstructures, transformer shields; austenitic, monmagnetic.

MACROSIL.
M-297; 0.15 max C, 11-13 Cr, 1.5-2.5 Ni, 0.3-0.6 Mo, 17-19 Mn, bal Fe.
For chemical plant equipment; corrosion resistant.

MADISON-KIPP NO 400.
M-1051; 8 Si, bal Al.
For die castings; corrosion resistant.

MAFERITE 6.
M-1509; 0.1 C, 6 Cr, bal Fe.
Rolled: 200 Brin.
For oil refinery equipment; scale resistant to 750°C.

MAFERITE 14.
M-1509; C, 14 Cr, bal Fe.
200 Brin.
For oil refinery equipment; scale resistant to 850°C.

MAFERITE 20.
M-1509; C, 20 Cr, bal Fe.
230 Brin.
For oil refinery equipment, furnace parts; scale resistant to 1000°C.

MAFERITE 25.
M-1509; C, 25 Cr, bal Fe.
230 Brin.
For furnace parts; scale resistant to 1050°C.

MAFERITE 28N.
M-1509; C, 28 Cr, 4 Ni, bal Fe.
For furnace parts; scale resistant to 1150°C.

MAFERITE 30.
M-1509; C, 30 Cr, bal Fe.
250 Brin.
For furnace parts; scale resistant to 1150°C.

MAFERITE 30P.
M-1509; C, 30 Cr, Mo, bal Fe.
260 Brin.
For furnace parts; scale resistant to 1150°C.

MAG 1.
M-106; 7.5-9.0 Al, 0.3-1.0 Zn, 0.15-0.4 Mn, bal Mg.
Sand cast: 140 MPa TS; 85 MPa YS; 2 El; 55 Brin.
General purpose cast magnesium alloy.

MAG 2.
M-106; 7.5-9.0 Al, 0.3-1.0 Zn, 0.15-0.7 Mn, bal Mg.
Sand cast: 140 MPa TS; 85 MPa YS; 2 El; 55 Brin.
Special purpose cast magnesium alloy; has high intrinsic corrosion resistance.

MAG 3.
M-106; 9.0-10.5 Al, 0.3-1.0 Zn, 0.15-0.4 Mn, bal Mg.
Sand cast: 125 MPa TS; 95 MPa YS; 55 Brin.
General purpose cast magnesium alloy; for intricate castings.

MAG 4.
M-106; 3.5-5.5 Zn, 0.15 Mn, 0.4-1.0 Zr, bal Mg.
Sand cast, stress relieved: 230 MPa TS; 145 MPa YS; 7 El; 65 Brin.
General purpose cast magnesium alloy.
High strength for use up to 150°C.

MAG 5.
M-106; 3.5-5.0 Zn, 0.15 Mn, 0.4-1.0 Zr, 0.75-1.75 rare metals, bal Mg.
Sand cast, prec. tr,: 200 MPa TS; 135 MPa YS; 3 El, 65 Brin.
Special purpose cast magnesium alloy.
For thin, narrow sectioned castings.

MAG 6.
M-106; 0.8-3.0 Zn, 0.15 Mn, 0.4-1.0 Zr, 2.5-4.0 rare metals, bal Mg.
Sand cast, prec. tr.: 155 MPa TS; 110 MPa YS; 3 El; 50 Brin.
Special purpose cast magnesium alloy.
Good creep resistance up to 250°C.

MAG 7.
M-106; 7.5-9.5 Al, 0.3-1.5 Zn, 0.15-0.8 Mn, 0.35 Cu, 0.4 Si, bal Mg.
Sand cast, sol. tr, aged: 185 MPa TS; 110 MPa YS; 70 Brin.
General purpose cast magnesium alloy.

MAG 8.
M-106; 1.7-2.5 Zn, 0.15 Mn, 0.4-1.0 Zr, 0.10 rare metals, 2.5-4.0 Th, bal Mg.
Sand cast, prec. tr.: 185 MPa TS; 85 MPa YS; 5 El; 50 Brin.
Special purpose cast magnesium alloy.
Good creep resistance up to 350°C.

MAG 9.
M-106; 5.0-6.0 Zn, 0.15 Mn, 0.4-1.0 Zr, 0.20 rare metals, 1.5-2.3 Th, bal Mg.
Sand cast, prec. tr.: 255 MPa TS; 155 MPa YS; 5 El; 65 Brin.
Special purpose cast magnesium alloy; for heavy duty structural use.

MAGAL.
M-275; 0.38 C, 5 Cr, 1.35 Mo, 0.3 V, bal Fe.
Heat treated: 215,000 TS; 184,000 YS; 13 El; 40 RA.
For punch inserts, gripper and header dies, punches; hot work steel; Type H11.

MAGALOY.
M-1241; 3.2 C, 2 Si, 0.5 Mg, 0.7 Mn, bal Fe.
For rolls; ductile cast iron.

MAGALUMA.
M-622; 3.4 Mg, 0.15 Mn, bal Al.
Soft: 29,000-36,000 TS; 10,000-14,500 YS; 20-15 El; 50-60 Brin.
Hard: 42,500-50,000 TS; 36,000-42,500 YS; 6-2 El; 85-100 Brin.
For light alloy parts; heat treatable, corrosion resistant.

MAGDAL.
M-871; 1 Mg, 1.25 Mn, bal Al.
Soft: 21,000 TS; 16 El.
Hard rolled: 34,000 TS; 8 El.
For light alloy parts; wrought.

MAGEX A.
M-1432; C, alloy, bal Fe.
Heat treated: 155,000 TS.
For pump shafts, gears, drive shafts; preheat treated.

MAGEX B.
M-1432; C, alloy, bal Fe.
For cams, chucks, reamers; oil hardened.

MAG METAL.
M-870; Cu, Ni.
Rolled: 55 Brin.
For jewelry; Pt substitute.

MAGNADURE.
M-1232; BaO + Fe_2O_3.
For TV ring magnets; permanent magnets.

MAGNADUR-3.
M-Eng.; $BaFe_{12}O_{19}$.
For permanent magnets in metering devices, magnetic and electrical equipment.
Permanent magnet. High permeability.
Barium and iron oxides.

MAGNALITE.
M-227; 94.2 Al, 2.5 Cu, 1.5 Ni, 0.5 Zn, 1.3 Mg.
26,000 TS; 2.5 El.
For airplane construction and engine pistons; casting alloy.

MAGNALIUM.
M-Eng.; 0-2.5 Cu, 1-5.5 Mg, 0-1.2 Ni, 85-95 Al, 0-3 Sn, 0.2-0.6 Si, 0-0.9 Fe, 0-0.3 Mn.
41,800 TS; 34 El; 29 RA.
14,900 TS; 31 El; 69 RA.
For light castings, ornamental, commercial shapes; very brittle; M.P. 1110-1280°F.

MAGNALIUM ALLOY NO 1.
M-Eng.; 95.5 Al, 1.75 Mg, 1.75 Cu, 1.0 Ni.
For light alloy parts; cast alloy.

MAGNALIUM ALLOY NO 2.
M-Eng.; 95.5 Al, 1.75 Mg, 1.75 Cu, 1.0 Pb.
For light alloy parts; cast alloy.

MAGNALIUM (CAST).
M-Eng; 85 Al; 15 Mg.
For light alloy castings.

MAGNALIUM (CAST X).
M-Eng.; 95 Al, 1.8 Cu, 1.6 Mg, 1.2 Ni.
For light alloy castings.

MAGNALIUM (CAST Y).
M-Eng.; 97 Al, 1.8 Cu, 1.5 Mg, Sn + Pb.
For light alloy castings.

MAGNALIUM (CAST Z).
M-Eng.; 95 Al, 3.2 Sn, 1.6 Mg, 0.2 Cu, 0.7 Pb.
For light alloy castings.

MAGNALIUM (ORIGINAL).
M-Eng.; 95-70 Al, 5-30 Mg.
For light alloy castings.

MAGNALIUM (SHEET).
M-Eng.; 95 Al, 5 Mg.
For pistons.

MAGNALLOY 1151.
M-1721; 85 Cu, 5 Sn, 5 Pb, 5 Zn.
Cast: 45,000 TS; 21,400 YS; 28 El; 72 Brin.
Leaded red brass.
ASTM B145 (4A); B271 (4A); SAE 40; CDA 836; AMS 4855.

MAGNALLOY 1152.
M-1721; 85 Cu, 5 Sn, 5 Pb, 5 Zn.
Cast: 55,000 TS; 24,200 YS; 13.2 El.

MAGNALLOY 1232.
M-1721; 80 Cu, 3 Sn, 7 Pb, 9 Zn, 1 Ni.
Cast: 34,000 TS; 18,000 YS; 22 El; 62 Brin.
Leaded semi-red brass.
ASTM B145 (5A), B271 (5A); CDA 844.

MAGNALLOY 2051.
M-1721; 89 Cu, 11 Sn.
Cast: 51,500 TS; 29,000 YS; 18 El; 95 Brin.
Tin bronze; for bearings, bushings.
SAE 65; CDA 907.

MAGNALLOY 2101.
M-1721; 88 Cu, 10 Sn, 2 Zn.
Cast: 51,500 TS; 29,000 YS; 18 El; 92 Brin.
Tin bronze, formerly called gun metal.
SAE 62; ASTM B22(D), B143(1A).
AMS 4845; CDA 905.

MAGNALLOY 2251.
M-1721; 88 Cu, 8 Sn, 4 Zn.
Cast: 49,000 TS; 23,000 YS; 18 El; 77 Brin.
ASTM B143(1B); SAE 620; CDA 903.

MAGNALLOY 2451.
M-1721; 88 Cu, 6 Sn, 1 Pb, 4 Zn, 1 Ni.
Cast: 45,500 TS; 23,000 YS; 35 El; 76 Brin.
ASTM B143(2A), B271(2A); SAE 622, CDA 922.

MAGNALLOY 3051.
M-1721; 80 Cu, 10 Sn, 10 Pb.
Cast: 41,000 TS; 24,000 YS; 10 El; 80 Brin.
High leaded tin bronze.
ASTM B144(3A), B271(3A): SAE 64.
AMS 4842; CDA 937.

MAGNALLOY 3111.
M-1721; 83 Cu, 8 Sn, 8 Pb, 1 Ni.
Cast: 37,400 TS; 22,000 YS; 19.8 El.
High leaded tin bronze.
Fed. Std. 00153:E8; QQ-B-1005(8); CDA 934.

MAGNALLOY 3151.
M-1721; 83 Cu, 7 Sn, 7 Pb, 3 Zn.
Cast: 44,000 TS; 27,000 YS; 16 El; 72 Brin.
ASTM B144(3B); SAE 660; CDA 932.

MAGNALLOY 3191.
M-1721; 78 Cu, 6 Sn, 16 Pb.
Cast: 34,200 TS; 23,000 YS; 12 El; 62 Brin.
High leaded tin bronze.
SAE 67; CDA 941(Similar specs).

MAGNALLOY 3221.
M-1721; 70 Cu, 5 Sn, 25 Pb.
Cast: 23,100 TS; 13,200 YS; 7.7 El.
High leaded tin bronze.
ASTM B144(3E); CDA 943.

MAGNALLOY 3242.
M-1721; 68 Cu, 2 Sn, 30 Pb.
Cast: 19,800 TS; 11,000 YS; 5.5 El.

MAGNALLOY 3261.
M-1721; 84 Cu, 5 Sn, 9 Pb, 9 Zn.
Cast: 38,000 TS; 21,000 YS; 20 El; 66 Brin.

MAGNALLOY 3271.
M-1721; 75 Cu, 5 Sn, 19 Pb.
Cast: 28,700 TS; 22,800 YS; 8.0 El; 57 Brin.
High leaded tin bronze.
SAE 67; CDA 941.

MAGNALLOY 4151.
M-1721; 88 Cu, 3 Fe, 9 Al.
Cast: 71,500 TS; 27,500 YS; 22 El.
Aluminum bronze.
ASTM B148(9A); SAE 68A; CDA 952.

MAGNALLOY 4152.
M-1721; 88 Cu, 1 Fe, 11 Al.
Cast: 71,500 TS; 27,500 YS; 22 El.
Aluminum bronze.
ASTM B148(9B); SAE 68B; CDA 953.

MAGNALLOY 4153.
M-1721; 86 Cu, 4 Fe, 10 Al.
Cast: 102,700 TS; 45,800 YS; 18.8 El; 190 Brin.
Aluminum bronze.
ASTM B148(9C); CDA 954 (Similar).

MAGNALLOY 4211.
M-1721; 58 Cu, 1 Sn, 38 Zn, 1 Fe, 1 Al, 1 Mn.
Cast: 71,500 TS; 27,500 YS; 22 El.
ASTM B147(8C); AMS 4860; CDA 865.

MAGNALLOY 4231.
M-1721; 64 Cu, 26 Zn, 2 Fe, 4 Al, 4 Mn.
Cast: 99,000 TS; 49,500 YS; 22 El.
Manganese bronze.
ASTM B147(8B); CDA 862.

MAGNALLOY 4241.
M-1721; 64 Cu, 24 Zn, 2 Fe, 6 Al, 4 Mn.
Cast: 121,000 TS; 66,000 YS; 13.0 El.
Manganese-aluminum bronze.
ASTM B147 8 B or 8 C; CDA 862; 863.
AMS 4862.

MAGNALLOY 4401.
M-1721; 88 Cu, 5 Sn, 2 Zn, 5 Ni.
Cast: 80,000 TS; 55,000 YS; 10 El.
Nickel-tin bronze.
ASTM B292-A; CDA 947.

MAGNA METAL.
Mg alloy.
For light alloy parts; same as "Electron."

MAGNE 1.
M-111; 0.9 C, 3.25 Cr, bal Fe.
For permanent magnets.

MAGNE 2.
M-111; 0.9 C, 6.25 Cr, bal Fe.
For permanent magnets.

MAGNEL.
M-89; 3 C, 1.5 Si, bal Fe.
Annealed: 52,000 TS; 33,000 YS; 13 El; 120 Brin.
For electrical and electromagnetic instruments, magnet cores; high permeability, low hysteresis loss.

MAGNESIL.
M-1626; 3.25 Si, bal Fe.
Alloy strip, soft magnetically. Often used for magnetic shielding.

MAGNESIL N.
M-1626; 3.25 Si, bal Fe.
Insulated soft magnetic iron strip; non-oriented.

MAGNESIL-O.
M-1626; 3.25 Si, bal Fe.
Insulated soft magnetic strip; with pronounced directional magnetic properties in the direction of rolling.

MAGNESIUM.
M-1755; Mg.
Purities: Special distilled 99.99%, 99.9+% (Nuclear grade), 99.5% (commercial).
Forms; ingot, rod, powders, wire, sheet, foil, granules, single crystals.

MAGNESIUM A10.
M-Eng.; 10 Al, 0.10 Mn, ba Mg.
Cast: 12,000 TS; 12,000 YS; 2 El; 53 Brin.
T4-temper: 40,000 TS; 13,000 YS; 10 El; 52 Brin.
T6-temper: 40,000 TS; 19,000 YS; 1 El; 69 Brin.
For high-strength castings; age-hardened.

MAGNESIUM ALLOY NO 23 (6% AL).
M-243; 6 Al, Mn, bal Mg.
For oxy-acetylene or TIG welding magnesium alloy extrusions, forgings and castings.
BS 2901 (D3).

MAGNESIUM AZ31X.
M-1239; 2.5-3.5 Al, 0.6-1.4 Zn, 0.20 min Mn, bal Mg.
Rolled: 32,000-40,000 TS; 16,000-29,000 YS; 12-4 El.
For light alloy parts.

MAGNESIUM AZ51.
M-Eng.; 4.8 Al, 0.8 Zn, 0.25 Mn, bal Mg.
Soft: 40,000 TS; 21,000 YS; 19 El; 57 Brin.
Hard: 45,000 TS; 34,000 YS; 10 El; 71 Brin.
For light alloy structures; sheet and plate.

MAGNESIUM-AZ61A.
M-1130; 6.5 Al, 1.0 Zn, 0.20 Mn, bal Mg.
Tubes: 40,000 TS; 22,000 YS; 15 El; 55 Brin.
Bars: 45,000 TS; 32,000 YS; 15 El; 56 Brin.
For light alloy structures, extrusion and forgings.

MAGNESIUM-AZ63A.
M-1130; 6 Al, 3 Zn, 0.2 Mn, bal Mg.
Cast: 29,000 TS; 14,000 YS; 6 El; 50 Brin.
T6-Temper: 40,000 TS; 19,000 YS; 5 El; 73 Brin.
For airplane wheels and brakes, oil pumps, crankcases.
Age hardenable casting alloy.

MAGNESIUM AZ80.
M-England; 8.5 Al, 0.5 Zn, 0.15 Mn, bal Mg.
Forged: 46,000 TS; 31,000 YS; 8 El; 69 Brin.
Aged: 50,000 TS; 34,000 YS; 6 El; 72 Brin.
For high-strength forgings, aircraft parts; age-hardened.

MAGNESIUM AZ81A.
M-43; 7.0-8.1 Al, 0.4-1.0 Zn, 0.13 min Mn, bal Mg.
F-temper: 28,000 TS; 13,000 YS; 6 El.
T4-temper: 40,000 TS; 30,000 YS; 12 El; 55 Brin.
For aircraft parts, structural members; heat treatable.

MAGNESIUM AZ91A.
M-43; 9 Al, 0.7 Zn, 0.2 Mn, bal Mg.
Die cast: 33,000 TS; 22,000 YS; 3 El; 60 Brin.
For instrument housings, portable tools; general die castings.

MAGNESIUM AZ92.
M-Eng.; 9 Al, 2 Zn, 0.1 Mn, bal Mg.
Cast: 25,000 TS; 14,000 YS; 2 El; 65 Brin.
T5-temper: 25,000 TS; 17,000 YS; 1 El; 69 Brin.
T4-temper: 40,000 TS; 14,000 YS; 10 El; 63 Brin.
T6-temper: 40,000 TS; 22,000 YS; 3 El; 81 Brin.
For high-strength casting; age-hardened.

MAGNESIUM AZ916.
M-Eng.; Mg alloy
T6-temper: 19,000 TS.
For aircraft and general purpose castings; age-hardenable.

MAGNESIUM EM22.
M-Ger.; 2-6 rare earths, 1.5-2.0 Mn, bal Mg.
For aircraft engine components; high temp. use.

MAGNESIUM EM42.
M-Ger.; 2-6 rare earths, 1.5-2.0 Mn, bal Mg.
For aircraft engine components; high temp. use.

MAGNESIUM-IA6.
9 Li, 3 Th, 2 Zn, 4 Al, 4 Ag, 1 Mn, bal Mg.
Heat treated: 45,000 min TS; 35,000 min YS; 15 min El.
For cryogenic applications.
Good low temperature properties to -423°F.

MAGNESIUM-I14.
7 Li, 1 Zr, 3 Th, 6 Zn, 5 Cd, 6 Ag, bal Mg.
Rolled: 45,000 min TS; 35,000 min YS; 15 min El.
At -452°F: 8 min El.
For cryogenic applications.
Good low temperature properties to -423°F.

MAGNESIUM INGOT (PRIMARY), GRADE MG-1.
M-43; 0.10 max Mn, 0.05 max Fe, 0.02 max Cu, 0.001 max Ni, 0.01 max Pb, 0.01 max Sn, 99.8 min Mg.
For remelting.
ASTM B92-71a. Gr. 9980 A.

MAGNESIUM INGOT (PRIMARY), GRADE MG-2.
M-43; 0.01 max Mn, 0.05 max Fe, 0.02 max Cu, 0.001 max Ni, 0.01 max Pb, 0.01 max Sn, 99.9 min Mg.
For remelting.

MAGNESIUM INGOT (PRIMARY), GRADE MG-4.
M-43; 0.004 max Mn, 0.04 max Fe, 0.003 max Al, 0.005 max Si, 0.004 max Ca, 0.001 max Ni, 0.005 max Pb, 0.005 max Sn, 0.00007 max B, 0.003 max Cu, 99.90 min Mg.
For remelting.

MAGNESIUM K1A.
M-43; 0.7 Zr, bal Mg.
Cast: 26,000 TS; 80,000 YS; 19 El.
Cast: 26,000 TS; 80,000 YS; 19 El.
For light alloy parts, missiles; fine grained.

MAGNESIUM M-1.
M-Eng.; 1.5 Mn, bal Mg.
Soft: 33,000 TS; 18,000 YS; 17 El.
Hard: 35,000 TS; 26,000 YS; 7 El.
Extruded: 38,000 TS; 28,000 YS; 10 El.
For structural members, aircraft parts; corrosion resistant.

MAGNESIUM MSR.
2 rare earths, 2.5 Ag, 0.7 Zr, bal Mg.
Heat treated: 40,000 TS; 30,000 YS; 4 El; 70-85 Brin.
For high temperature applications, missile components; good creep strength, heat treatable.

MAGNESIUM MSR-A.
M-1130; 2.5 Ag, 0.7 Zr, 1.7 rare earths, bal Mg.
Heat treated: 78,400 TS; 50,400 YS; 4 El; 70-90 Brin.
For high strength castings for high temperature; age-hardenable, cast alloy.

MAGNESIUM MSR-B.
M-1130; 2.5 Ag, 0.7 Zr, 2.5 rare earths, bal Mg.
Heat treated: 78,400 TS; 54,900 YS; 2 El; 70-90 Brin.
For high strength castings for high temperature; age-hardenable, cast alloy.

MAGNESIUM MTZ.
M-Eng.; 3 Th, 0.7 Zr, bal Mg.
Heat treated: 65,000 TS; 24,600 YS; 5 El; 50-60 Brin.
For aircraft parts; high creep resistance.

MAGNESIUM -ZE63A.
M-1130; 5.5-6.0 Zn, 2.0-3.0 R.E., 0.4 min Zr, bal Mg.
86 Temper: 42,000-45,000 TS; 29,000-32,000 YS; 5-10 El.
At 257°F: 30,000 TS; 19,000 YS; 34 El.
At 347°F: 21,000 TS; 16,000 YS; 32 El.
For helicopter structural and transmission castings, gearboxes, aircraft landing wheels.
Castings, operating up to 300°F. High fatigue strength.

MAGNESIUM-ZE63B.
M-1130; 5.5-6.0 Zn, 2.0-3.0 R.E., 0.75-1.25 Ag, 0.4 min Zr, bal Mg.
T6 Temper: 43,000-46,000 TS; 30,000-34,000 YS; 4-8 El.
For helicopter structural and transmission castings, gearboxes, aircraft landing wheels.
Cast alloy. High fatigue strength.

MAGNESIUM ZK60A.
M-4; 4.8-6.2 Zn, 0.45 min Zr, bal Mg.
F-temper: 49,000 TS; 38,000 YS; 12 El.
TS-temper: 51,000 TS; 42,000 YS; 10 El.
T6-temper: 53,000 TS; 48,000 YS.
For brake housings, fuel meter bodies, bulkheads; heat treatable, extrusions, high strength and tough.

MAGNESIUM ZK61.
M-1239; 6 Zn, 0.8 Zr, bal Mg.
Cast: 21,700 TS; 10,900 YS; 12 El.
Heat treated: 228,000 TS; 14,100 YS; 8 El.
Aged: 255,000 TS; 16,400 YS; 10 El.
For engine components; age hardenable.

MAGNESIUM-ZK61A.
M-1130; 6 Zn, 0.6 Zr, bal Mg.
T6 Temper: 45,000 TS; 28,000 YS; 8 El; 75 Brin.
T5 Temper: 40,000 TS; 26,000 YS; 8 El; 65 Brin.
For ordnance vehicles, missiles, aircraft components. Heat treatable castings.

MAGNESIUM ZK62.
M-Eng.; 6 Zn, 2 Misch Metal, 0.5 Zr, bal Mg.
Extruded: 50,000 TS; 45,000 YS; 12 El.
For aircraft components.

MAGNESIUM-ZLH972.
7 Li, 9 Zn, 2 Th, bal Mg.
For cryogenic applications.
Good low temperature properties to -423°F.

MAGNESIUM-ZTY.
M-1130; 0.5 Zn, 0.75 Th, 0.6 Zr, bal Mg.
Forging: 43,000 TS; 26,000 YS; 18 El.
Sheet: 38,000 TS; 24,000 YS; 12 El.
For compressor cases, valve covers, pistons.
Good creep and thermal properties.
Hot formable, weldable.

MAGNET C.
M-96; 0.95 C, 1.35 Si, 4 Cr, bal Fe.
For permanent magnets; high coercive force.

MAGNETHERM.
M-1786; 99.8-99.85 Mg.
(And other magnesium alloys) for sand and shell casting.

MAGNETICS ROUND PERMALLOY 80.
M-1626; 80.0 Ni, 4.5 Mo, 0.25 Si, 0.35 Mn, bal Fe.
High initial and maximum permeability with low coercive force, low hysteresis loss, low eddy-current loss, and low magneto striction.
For filters, relays, recording heads, etc.

MAGNETOFLEX.
M-1631; Fe, Co, V.
Deformable permanent magnet alloy, e.g. for rings of hysteresis motors.

MAGNETOFLEX-12.
M-Ger.; 68 Cu, 20 Ni, bal Co.
For permanent magnets in electrical and magnetic equipment.
Age hardenable, high permeability.

MAGNETOFLEX 20.
M-1630; 20 Fe, 20 Ni, 60 Cu.
Wire: 100,000-120,000 TS; 200 Brin.
For magnets in electrical and electronic equipment; age-hardenable, permanent magnet, same as Cunife.

MAGNEWIN.
M-869; Al, bal Mg.
For light alloy parts.

MAGNEWIN 3515.
M-Ger.; 7.3 Al, 0.75 Zn, 0.12 Mn, bal Mg.
For light alloy parts; heat treatable.

MAGNICO.
M-U.S.; 12 Co, 6 Cu, 10 Al, 18 Ni, bal Fe.
For permanent magnets.

MAGNICO.
M-USSR; 50 Fe, 24 Co, 14 Ni, 9 Al, 3 Cu.
For electrical and magnetic equipment; magnet.

MAGNIKO.
M-USSR; 25.3 Co, 14.6 Ni, 2.8 Cu, 8.5 Al, bal Fe.
For electrical and magnetic equipment.
Permanent magnet, high magnetic permeability.

MAGNIL.
M-1595; 0.10 C, 15.5 Mn, 0.75 max Ni, 18.0 Cr, bal Fe.
Cold rolled: 219,400 TS; 187,400 YS; 6 El.
Heat treated: 296,000 TS; 268,000 YS.
For electronics, computer, instrument and control industries, high modulus diaphragms, bellows, springs, wear plates. Austenitic, nonmagnetic, stainless.

MAGNO.
M-89; 0.05 C, bal Fe.
Annealed: 60,000 TS; 22 El.
For magnet cores; high permeability, low hysteresis loss.

MAGNO-90.
M-1766; 0.90 C, 1.90 Mn, 0.40 max Si, 0.40 Cr, 0.15 V, bal Fe.
Cold work tool steel; for small tools as reamers, punches, hand stamps.
DIN 90Mn8; AFNOR 80M8; UNI U85MV8.

MAGNO-522.
M-1766; 0.95 C, 1.10 Mn, 0.40 max Si, 0.50 Cr, 0.50 W, 0.15 V, bal Fe.
Oil hardening tool steel for dies, gages.
IHA F-522; DIN 105 MnCr4; AISI O1.

MAGNO (ELALCO).
M-Eng.; 95 Ni, 5 Mn.
Hard drawn: 140,000 TS.
Soft: 65,000 TS; 22 El.
For electrical resistance wire, electromagnetic uses; resistance alloy.

MAGNOLIA 120 BRONZE.
M-178; 75 Cu, 20 Pb, 5 Sn.
Cast: 25,500-28,000 TS; 12,000-18,000 YS; 10-9 El; 55-67 Brin.
For bearings for light loads; resists many acid solutions.

MAGNOLIA AA BRONZE.
M-178; 88 Cu, 10 Sn, 2 Zn.
Cast: 34,000-40,000 TS; 26,000-30,000 YS; 8 El; 85 Brin.
For bearings for underate loads; resists brine solutions.

MAGNOLIA ANTIFRICTION METAL.
M-178; 10-15 Sb, 2-7 Sn, bal Pb.
Cast: 18,000 TS; 8400 YS; 23 Brin.
For high speed and heavy pressure bearings; antifriction metal; 825-900°F.

MAGNOLIA CADMIUM NICKEL.
M-436; Cd, bal Ni.
For bearings, liners; high loads.

MAGNOLIA CONTINUOUS CAST BRONZE.
M-178; 80 Cu, 6-8 Sn, 11-13 Pb, Zn less than 1.
As cast: 44,000 TS; 26,000 YS; 17 El; 80 Brin.
For bearings, bushings.

MAGNOLIA ISOTROPIC BRONZE.
M-178; 10 Sn, 10 Pb, 80 Cu.
Cast: 31,250 TS; 26,000 YS; 8.5 El; 70 Brin.
For die cast bronze bearings; die cast.

MAGNOLIA METAL.
M-178; 84-78 Pb, 15-16 Sb, 0.03 Fe, 0-7 Sn.
Cast: 15,000 TS; 3400 YS; 21 Brin.
For antifriction bearing metal; for bearings subjected to moderate loads.

MAGNOLIA MODIFIED SAE 64 BRONZE.
M-178; 78-82 Cu, 6-9 Sn, 11-13 Pb, 1.0 max Zn.
223250 psi TS; 20,000 psi YS; 8.5 El; 70 Brin.
Good machinability bearing bronze with improved bearing characteristics.

MAGNOWIN.
M-869; Mg alloy.
For light alloy parts.

MAGNOX-C.
M-1130; 0.7-0.9 Al, 0.005 Be, bal Mg.
At 212°F: 12,800 TS; 8 El.
At 575°F: 3000 TS; 53 El.
At 750°F: 1000 TS; 88 El.
For light alloy parts; structural members.

MAGNUMINIUM 127.
M-426; 9 Al, 1 Zn, 0.2 Mn, bal Mg.
Sand cast: 20,200 TS; 9000 YS; 2 El.
Permanent mold: 26,900 TS; 9000 YS; 4 El.
For aircraft parts, crankcases, pressure tight castings and housings; heat treatable.

MAGNUMINIUM 133.
M-556; 0.2 Al, 0.2 Zn, 2.5 Mn, 0.4 Si, 0.2 Cu, bal Mg.
30,000 TS; 16,000 YS; 10 El; 50 Brin.
For light alloy parts, fuel tanks, structures; weldable.

MAGNUMINIUM 155.
M-556; 9 max Al, 1.5 max Zn, 1 max Mn, 0.3 max Cu, bal Mg.
40,000 TS; 20,000 YS; 10 El; 55 Brin.
For light alloy parts; non-hardenable.

MAGNUMINIUM 166.
M-556; 5-11 Al, 1.5 max Zn, 1.0 max Mn, bal Mg.
40,000-45,000 TS; 16,000-28,000 YS; 14-10 El; 50-60 Brin.
For light alloy forgings and stampings; heat treatable.

MAGNUMINIUM 177.
M-556; 8.5 max Al, 3.5 Zn, bal Mg.
Sand cast: 20,000 TS; 9000 YS; 2-8 El; 50 Brin.
For die castings; M.P. 610°C.

MAGNUMINIUM 181.
M-556; 8.5 max Al, 3.5 max Zn, 0.5 max Mn, bal Mg.
Cast: 23,000 TS; 10,000 YS; 8 El; 45 Brin.
Heat treated: 32,000 TS; 10,000 YS; 12 El; 55 Brin.
For light castings; heat treatable.

MAGNUMINIUM 199.
M-556; 8.5 Al, 3.5 Zn, 0.4 Si, 0.4 Mn, bal Mg.
Cast: 25,000-30,000 TS; 8-4 El; 45-55 Brin.
For sand and die light alloy castings; heat treatable.

MAGNUMINIUM 220.
M-556; 9-11 Al, 3.5 max Zn, 0.5 max Mn, 1.5 max impurities, bal Mg.
Sand cast: 24,000 TS; 12,000 YS; 2-4 El; 50-60 Brin.
Die cast: 26,000 TS; 2-5 El; 50-60 Brin.
For sand and die light alloy castings; casting alloy, heat treatable.

MAGNUMINIUM 266.
M-556; 11 max Al, 1.5 max Zn, 1 max Mn, bal Mg.
45,000-50,000 TS; 20,000-30,000 YS; 10-14 El; 50-60 Brin.
For light alloy extruded bars and shapes; for structures.

MAGNUMINIUM 288 A.
M-556; 11 max Al, 2 max Zn, 1 max Mn, bal Mg.
35,000-48,000 TS; 20,000-33,000 YS; 12-5 El; 80-90 Brin.
For aircraft parts; heat treatable.

MAGNUMINIUM 299.
M-137, M-556; 2.5-6.0 Al, 4.5-7.5 Sn, 0.2-2 Ag, 0.1-0.4 Mn, bal Mg.
Cast: 22,000 TS; 10,000 YS; 5-9 El; 45 Brin.
Aged: 31,400 TS; 10,000 YS; 4-14 El; 50 Brin.
For airplane wheels, motor and instrument housings; pressure tight castings.

MAGNUMINIUM 299.
M-556; 2.0-6.0 Al, 3-10 Sn, 0.25-4.00 Ag, bal Mg.
Cast: 18,000-26,000 TS; 10,000-12,000 YS; 4-5 El; 40-50 Brin.
For light castings, meter parts; heat treatable.

MAGOTTEAUX-UM CO50.
M-1581; 0.05-0.12 C, 0.18 Ti, 0.6 Cb, 26-30 Cr, 47-52 Co, bal Fe.
Cast: 135,000 TS; 48,000 YS; 10 El; 10 RA; 250 Brin.
Wrought: 132,000 TS; 61,000 YS; 7 El; 6 RA; 350 Brin.
For furnace baffles, burner tips, sintering grates, quench baskets.
Corrosion, heat, thermal shock resistant.

MAHLE 124.
 M-1365; 11-13 Si, 0.8-1.5 Cu, 1.3 Ni, 0.8-1.3 Mg, 0.7 Fe, bal Al.
 Cast: 28,000-35,000 psi TS; 90-125 Brin.
 Forged: 41,000-51,000 psi TS; 90-125 Brin.
 For pistons, cylinders.

MAHLE 138.
 M-1365; 17-19 Si, 0.8-1.5 Cu, 1.3 Ni, 0.8-1.3 Mg, 0.7 Fe, bal Fe.
 Cast: 25,000-31,000 psi TS; 90-125 Brin.
 Forged: 31,000-41,000 psi TS; 90-125 Brin.
 For pistons, cylinders.

MAHLE 244.
 M-1365; 23-26 Si, 0.8-1.5 Cu, 1.3 Ni, 0.8-1.3 Mg, 0.7 Fe, 0.6 Cr, bal Al.
 Cast: 24,000-29,000 psi TS; 90-125 Brin.
 For pistons, cylinders.

MAHLE Y.
 M-1365; 3.5-4.5 Cu, 1.75-2.25 Ni, 1.25-1.75 Mg, 0.6 Fe, 0.5 Si, bal Al.
 Cast: 31,000-38,000 psi TS; 95-125 Brin.
 Forged: 48,000-58,000 psi TS; 95-125 Brin.
 For pistons, cylinders.

MAHLE Y.
 M-1365; 3.8-4.2 Cu, 1.7-2.2 Ni, 1.5 Mg, 0.3 Si, 0.7 Fe, bal Al.
 For pistons, cylinder heads; high temperature use.

MAILLECHORT.
 67-65 Cu, 13.5 Zn, 13-19 Ni, 0.5-3.2 Fe, Sn + Pb.
 For ornamental white metal parts, nickel silver.

MAILLECHORT, GERMAN.
 M-Ger.; 65.4 Cu, 13.4 Zn, 16.8 Ni, 3.4 Fe.
 For hardware, fittings; corrosion resistant.

MAILLECHORT, PARIS.
 M-France; 66.24 Cu, 13.42 Zn, 16.42 Ni, 3.2 Fe.
 For hardware, fittings; corrosion resistant.

MAILLECHORT, VIENNA.
 M-Austria; 66.6 Cu, 13.6 Zn, 19.3 Ni, 0.48 Fe.
 For hardware, fittings; corrosion resistant.

MAIN METAL see **INTAL N 89.**

MAINTENAL.
 M-1432; C, alloy, bal Fe.
 Heat treated: 155,000 TS; 132,000 YS; 17 El; 57 RA; 280 Brin.
 For wrenches, drills, gears, pump shafts, bolts; preheat treated, fatigue resistant.

MAJOR.
 M-365; 0.7 C, 21 W, 4 Cr, 1.5 V, 13 Co, 0.5 Mo, bal Fe.
 For tools, cutters; high speed steel.

MAJOR METAL.
 3 Cu, 2 Fe, 0.4 Zn, 0.4 Ni, bal Al.
 For light alloy parts; hardenable.

MAL-ARC.
 M-265; Cr-Mo-Co.
 For hard facing welding rod; abrasion resistant.

MALAX A.
 M-750; 0.83 C, 5 Mo, 4 Cr, 1.9 V, 6.4 W, bal Fe.
 For drills, taps, lathe and planer tools; oil hardening.

MALAX AA.
 M-750; 0.75 C, 4.25 Cr, 19 W, 1 V, bal Fe.
 For tools, cutters, reamers; high speed steel.

MALAX AAA.
 M-750; 0.75 C, 4.25 Cr, 19 W, 2 V, 5 Co, bal Fe.
 For tools, cutters, lathe and planer tools; high speed steel.

MALAX AAAA.
 M-750; 0.8 C, 4.2 Cr, 19 W, 2 V, 1 Mo, 7.5 Co, bal Fe.
 For cutters, forming tools; high speed steel.

MAL COLLOY.
 M-Japan; 15 Al, 19.8 Ni, bal Co.
 Coercive force 1500. Residual induction 3200. Max energy product 1,450,000.
 For electrical and magnet equipment and instruments.
 Permanent magnet after heat treatment.

MALCROME.
 M-750; C, Cr, Mo, W, bal Fe.
 For tools, dies; oil hardened.

MAL-DIE.
 M-750; 1 C, 5 Cr, 1.1 Mo, bal Fe.
 For blanking and forming dies; air-hardened.

MALGA.
 M-750; C, Cr, Mo, W, bal Fe.
 For tools, dies; oil hardened.

MALGA ELEKTRO SPECIAL.
 M-750; C, Cr, W, bal Fe.
 For tools, dies; water or oil hardened.

MALGALOY.
 M-750; Si, Cr, Mo, V, W, C, bal Fe.
 For shear blades; shock resistant.

MALGA MRO SPECIAL.
 M-750; C, Mo, W, Cr, bal Fe.
 For tools, dies; wear and shock resistant.

MALGA SPECIAL NON-TEMPERING.
 M-750; C, W, Mo, Cr, bal Fe.
 For tools, dies; no tempering after quenching.

MALLEABLE CAST IRON "BLACK HEART".
 M-Eng.; 1.5-2.5 Carbon, 0.1-0.35 Mn, 0.1 S, 0.2 P, 0.5-1.2 Si, bal Fe.
 35,000-45,000 TS; 10-5 El.
 For railway and automotive castings; machines easily.

MALLEABLE CAST IRON "WHITE HEART".
M-Eng.; <0.4 combined C, <0.8 total C, 0.1-0.4 Mn, 0.25 S, 0.1 P, 0.5-1.1 Si, bal Fe.
40,000-55,000 TS; 5-3 El.
For railway and automotive castings; machines easily.

MALLEABLE IRON 35018.
M-775; 2.2 T.C., 1.0 Si, 0.4 Mn, bal Fe.
Cast: 53,000 TS; 35,000 YS; 18 El; 115-135 Brin.
For hardware, agriculture and auto parts; tough, shock resistant.

MALLEABLE IRON 32510.
M-775; 2.5 T.C., 1.0 Si, 0.4 Mn, bal Fe.
Cast: 50,000 TS; 32,500 YS; 10 El; 115-135 Brin.
For hardware, agriculture and auto parts; tough, shock resistant.

MALLEABLE IRON GM11M.
M-609; 2.55 C, 1.4 Si, 0.45 Mn, 0.12 S, 0.05 P, bal Fe.
Malleable Iron.
50,000 TS; 32,500 YS; 10 El; 156 max Brin.
For lightly stressed parts; e.g. steering gear housings.
SAE & ASTM Grade 32510.

MALLEABLE IRON GM 11M.
M-690.
50,000 psi TS; 32,500 psi YS; 10 El; 150 max Brin.
For less highly stressed parts; e.g. steering gear housing.

MALLET ALLOY.
75 Zn, 25 Cu.
For bearings.

MALLIX.
M-89; 1.8 C, 1.0 Si, 0.5 Mn, bal Fe.
Cast: 78,000 TS; 55,000 YS; 5 El; 180 Brin.
For grate bars, elevator buckets; pearlitic malleable iron.

MALLORY 1000 ETC see CMW 1000 ETC.

MALLORY ALLOYS see CMW ALLOYS.

MALLORY D-54.
M-265; Ag-CdO.
Sintered: 18,000 TS.
For electrical contacts; 80% cond.

MALLORY L-2748.
5 Al, 5 Cr, bal Ti.
Rolled: 165,000 TS; 153,000 YS; 10 El; Rockwell A71.
For high temperature applications; corrosion and heat resistant.

MALLORY L-2749.
Al, Cr, bal Ti.
Rolled: 90,000 TS; 75,000 YS; 18 El; Rockwell A62.
For high temperature applications; corrosion and heat resistant.

M-ALLOY.
M-1258; C, alloy, bal Fe.
For foundry sling liners; wear resistant.

MALLOY.
M-1405; 0.60 C, 0.45 Mn, 1.1 Si, 1.1 Cr, 0.25 Mo, bal Fe.
Shock resisting tool steel; for punches and dies, shear blades, collets.

MALLOYDIUM.
60 Cu, 23 Ni, 13 Zn, 0.9 Fe.
For tableware; acid resisting.

MALUMINUM.
German; 87 Al, 6.4 Cu, 4.8 Zn, 1.4 Fe, 0.2 Si, 0.1 Mn, 0.2 Pb.
21,000 TS; 1.5 El.
For light alloy parts.

MAMMUT SPECIAL, now VEW S200.

MAMMUT SPECIAL KN, now VEW S205.

MAN.
M-1291; 0.3 C, 0.25 Si, 1.35 Mn, bal Fe.
For gears, shafts, machine tool parts; water hardened.

MANAURITE 8S.
M-1509; 0.2 C, 8 Ni, 18 Cr, bal Fe.
Annealed: 72,000 TS; 39,000 YS; 20 El.
For chemical plant equipment; resists scaling to 900°C, stainless austenitic.

MANAURITE 10.
M-1509; 0.2 C, 10 Ni, 22 Cr, bal Fe.
Annealed: 72,000 TS; 37,000 YS; 18 El.
For chemical plant and oil refinery equipment; resists scaling to 1050°C, stainless.

MANAURITE 12.
M-1509; C, 12 Ni, 25 Cr, bal Fe.
Annealed: 77,000 TS; 40,000 YS; 20 El.
For furnace parts, heat treating equipment; resists scaling to 1080°C, heat resistant.

MANAURITE 20.
M-1509; C, 20 Ni, 25 Cr, bal Fe.
Annealed: 80,000 TS; 36,000 YS; 22 El.
For furnace parts, heat treating equipment; resists scaling to 1150°C, heat resistant.

MANAURITE 35.
M-1509; C, 35 Ni, 15 Cr, bal Fe.
Annealed: 72,000 TS; 39,000 YS; 22 El.
For heat treating boxes, salt pots; resists scaling to 1200°C, heat resistant.

MANAURITE 60.
M-1509; C, 60 Ni, 15 Cr, bal Fe.
Annealed: 64,000 TS; 39,000 YS; 22 El.
For salt pots, heat treating equipment; resists scaling to 1200°C, heat resistant.

MANCRO.
M-111a; 0.15 C, 0.25 Si, 0.50 Mn, 0.75 Cr, bal Fe.
For gears, shafts; case hardening.

MANCRO 4.
M-111a; 0.20 C, 1.35 Cr, 1.35 Mn, bal Fe.
For gears, shafts; case hardening.

MANCRO 8.
M-111; 0.38 C, 0.65 Si, 1.15 Mn, 1.15 Cr, bal Fe.
For gears, shafts; tough.

MANCRO 32.
M-111a; 0.15 C, 1.2 Mn, 1.0 Cr, bal Fe.
For gears, shafts; case hardening.

MANCRO 71.
M-111; 0.34 C, 1.05 Mn, 1.05 Cr, bal Fe.
For gears, shafts; oil hardening.

MANCRO 72.
M-111; 0.40 C, Cr, Mn, Mo, bal Fe.
For gears, shafts, machine tool parts; water hardened.

MANDUR.
M-1548; 3.5-5.5 Cu, 0.3-1.0 Si, 0.5-1.0 Mn, 0.5-1.2 Mg, bal Al.
Heat treated: 78,000-85,000 TS; 64,000-71,000 YS; 10-15 El; 130-150 Brin.
For oil pans, crankcases, housings, engine cylinder heads.
Age-hardenable. High strength.

MANELEC.
M-691; 1 Si, bal Fe.
For electrical equipment, motors; high permeability.

MANGABRAZE.
M-1510; 0.35 C, 1.92 Mn, 0.32 Mo, 0.28 Si, 0.42 Cu, bal Fe.
Heat treated: 157,000 TS; 146,000 YS; 16 El; 58 RA; 360-385 Brin.
For chutes, hoppers, conveyors, screens, liners, scrapers, buckets; wear and impact resistant, mill heat treated.

MANGA-KOTE.
M-507; 0.80 C, 0.8 Si, 14.0 Mn, 4.7 Cr, 3.24 Mo, 3.5 Ni, 0.10 B, bal Fe.
Coated hardfacing electrode.
For overlay or buildup of bucket teeth, grading buckets, tractor rollers, crusher parts.

MANGAL.
M-116; 1.5 Mn, bal Al.
Annealed: 17,000 TS; 20 El.
For light alloy parts; corrosion resistant.

MANGAL.
M-1548; 1.5 Mn, bal Al.
Soft: 14,000 TS; 20-30 El; 20-25 Brin.
Hard: 36,000 TS; 28,000 YS; 2 El; 60 Brin.
For commercial vehicles, roofing, structures.
Non-hardening, corrosion resistant.

MANGALAL replaced by ALCAN GB-3S.

MANGALOY.
M-Eng.; Ni, Mn, bal Fe.
For electrical resistor; heat resistant.

MANGALOY.
M-1128; C, Mn, Ni, Cr, bal Fe.
For hard surfacing electrodes; for manganese steel.

MANGAN.
M-72; 1.05-1.25 C, 13.5-14.5 Mn, bal Fe.
For hard surfacing electrodes; wear resistant.

MANGANAL.
M-374; 0.60-0.90 C, 11-14 Mn, 2.5-3.5 Ni, bal Fe.
155,000-140,000 TS; 60,000-55,000 YS; 55-72 El; 35-54 RA; 180-500 Brin.
For welding rod, and for resurfacing broken and worn high Mn steel parts; abrasion resistant.

MANGANEND 1 M.
M-677; C, Mo, bal Fe.
Welded: 90,000-100,000 TS; 70,000-85,000 YS; 30-20 El; 70-40 RA.
For welding electrodes; for steel, low H_2.

MANGANEND 2M.
M-677; 0.10 C, 1.6 Mn, 0.35 Mo, bal Fe.
Welded: 125,000 TS; 105,000 YS; 10 El; 15 RA.
For welding electrodes for steel; low H_2.

MANGANEND 13A.
M-677; 0.85 C, 0.8 Si, 13.5 Mn, 1.0 Mo, bal Fe.
Welded: 16 Rc; work hardened: 48 Rc.
For hardfacing electrodes; austenitic 13.5% manganese steel.

MANGANESE.
M-1755; Mn.
Purities: 99.99+%, 99.9%, dehydrogenated.
Forms: flake, powder, foil, vacuum melted lump.

MANGANESE ANTIFRICTION.
M-565; Mn, Sb, Sn, Pb.
5,670 TS.
For machinery bearings; Babbitt metal; C.S. 19450; self-lubricating.

MANGANESE BRASS 73.
M-141; 70 Cu, 1.25 Mn, bal Zn.
Annealed: 50,000 TS; 55 El.
Drawn: 90,000 TS; 6 El.
For resistance seam and spot welding.

MANGANESE BRASS 510.
M-8; 29 Zn, 1 Mn, bal Cu.
Hard: 76,000 TS; 62,000 YS; 10 El; 160 Brin.
Soft: 47,000 TS; 16,000 YS; 65 El; 60 Brin.
For strip for resistance spot and seam welded products.

MANGANESE BRASS 667.
M-279; 70 Cu, 28.8 Zn, 1.2 Mn.
Ann: 47,000-67,000 TS; 13,000-37,000 YS; 30-60 El.
Cold rolled: 51,000-105,000 TS; 2-40 El.
Seam and spot welding brass; for communication equipment, welded assemblies.

MANGANESE BRONZE.
86-82 Cu, 6-17 Sn, 0.2-2.7 Mn, 0-5 Zn, 0.3 Pb.
For gears, general castings; tough.

MANGANESE BRONZE.
M-129; 56-60 Cu, 0.4-1.5 Fe, 0.5-1.0 Al, 0.4 max Pb, bal Zn.
Cast: 65,000 TS; 30,000 YS; 28 El; 25 RA; 100 Brin.
For propellers, hubs, valves, pump bodies; corrosion resistant.

MANGANESE BRONZE 19.
M-141; 58.5 Cu, 0.3 Mn, 0.7 Sn, 1 Fe, bal Zn.
Hard: 83,000 TS; 55,000 YS; 25 El.
Soft: 72,000 TS; 30,000 YS; 45 El.
For bolts, valve parts, tie rods; corrosion resistant.

MANGANESE BRONZE 937.
M-8; 57-62 Cu, 36-40 Zn, 0.5-1.5 Sn, 0.5-1.0 Fe, 0.5 Mn.
Soft: 60,000 TS; 30,000 YS; 30 El; 30 RA.
Hard: 80,000 TS; 45,000 YS; 20 El; 15 RA.
For valve stem forgings, slotted and perforated screens; high strength and toughness; resists action of salt water.

MANGANESE BRONZE 984.
M-8; 38 Zn, 1 Sn, 0.5 Fe, 0.5 Mn, bal Cu.
For welding rod; for steel, cast iron, Cu and Ni alloys; M.P. 1598°F.

MANGANESE BRONZE (A) COPPER ALLOY NO. 675.
M-33; 58.5 Cu, 39.25 Zn, 1.0 Sn, 1.0 Fe, 0.25 Mn.
Half hard: 72,000 TS; 40,000 YS; 35 El; 70 Rock B.
For balls, forging, valve stems, welding rod.
ASTM B124 Alloy 4; CDA 675.

MANGANESE BRONZE A 675.
M-8; 58.5 Cu, 39.25 Zn, 1.0 Sn, 0.25 Mn, 1.0 Fe.
Hard rod: 75,000 TS; 45,000 YS; 20 El; B 85 Rock.
Soft rod: 65,000 TS; 30,000 YS; 30 El; B 65 Rock.
For clutch discs, pump rods, shafts, balls, valve stems and bodies.
Readily machined and welded.

MANGANESE BRONZE CAST.
M-Eng.; 58 Cu, 2 Mn, 40 Zn.
For marine propellers; corrosion resistant.

MANGANESE BRONZE E-77.
M-1191; 57-60 Cu, 0.5-1.5 Sn, 0.25 max Al, 0.5 max Mn, bal Zn.
Forged: 75,000-85,000 TS; 40,000-45,000 YS; 20-15 El; 120-142 Brin.
For marine parts, hardware, propellers; corrosion resistant, high strength

MANGANESE BRONZE ROLLED.
M-Eng.; 59 Cu, 1 Sn, 0.3 Mn, 31 Zn.
For marine parts, nuts, bolts; corrosion resistant.

MANGANESE CASTING BRASS.
M-165; 58.5 Cu, 40 Zn, 0.85 Sn, 0.5 Al, 1.5 max Pb, 0.15 Mn.
70,000 TS; 20 El.
A substitute for malleable iron; high strength.

MANGANESE COPPER "A".
M-Eng.; 29.2 Cu, 51.65 Mn, 9.68 Fe, 6.25 Al 3.23 C.
For resistances, heat and corrosion resistant parts; heat and corrosion resistant.

MANGANESE COPPER "B".
M-Eng.; 56.3 Cu, 40.9 Mn, 1.5 Fe, 1.1 Si.
For resistances, heat and corrosion resistant parts; heat and corrosion resistant.

MANGANESE COPPER "C".
M-Eng.; 75 Cu, 25 Mn.
For resistances, heat and corrosion resistant parts; heat and corrosion resistant.

MANGANESE COPPER "D".
M-Eng.; 75.3 Cu, 22.4 Mn, 2.15 Fe.
For resistances, heat and corrosion resistant parts; heat and corrosion resistant.

MANGANESE COPPER "E".
M-Eng.; 85 Cu, 10.92 Mn, 1.83 Fe, 2.0 Zn.
For resistances, heat and corrosion resistant parts; heat and corrosion resistant.

MANGANESE COPPER "F".
M-Eng.; 89.7 Cu, 8.72 Mn, 1.54 Fe.
For resistances, heat resisting parts; heat resistant.

MANGANESE COPPER "G".
M-Eng.; 85.55 Cu, 10.66 Mn, 2.66 Fe, 0.39 Sn, 0.45 Pb.
For resistances; heat resistant.

MANGANESE COPPER "H".
M-Eng.; 84.33 Cu, 10.61 Mn, 2.31 Fe, 2.1 Zn, 0.4 Sn, 0.3 Pb.
For resistances; heat resistant.

MANGANESE GRADE A.
M-604; 0.80 C, 13.0 Mn, 0.50-0.80 Si, 3.0-3.50 Ni, bal Fe.
Work hardened: 450 Brin.
Weldable; for severe impact, crushers, wear plates; non-magnetic.

MANGANESE GRADE B.
M-604; 0.80 C, 12.0 Mn, 0.20 Si, 0.50 max Cr, bal Fe.
Work hardened: 500 Brin.
For severe impact, crushers, wear plates.

MANGANESE GRADE C.
M-604; 0.80 C, 13.0 Mn, 0.65 Si, 2.0 Ni, 0.50 Mo, 0.50 max Cr, bal Fe.
Work hardened: 450 Brin.
Weldable; for severe impact, hammers, crushers.

MANGANESE M1.
M-1017; C, Mn, Ni, bal Fe.
Welded: 350 Brin.
For arc welding electrodes, hard facing; forgeable, wear resistant.

MANGANESE NICKEL.
M-Eng.; 85-52 Cu, 14-31 Mn, 3-16 Ni.
For heat resistant parts; heat resistant.

MANGANESE NICKEL-2%.
M-61; 3 Mn, bal Ni.
Annealed: 75,000 TS; 35,000 YS; 40 El; 140 Brin.
For lead wires for electrical appliances; heat resistant.

MANGANESE NICKEL 5%.
M-61; 5 Mn, 95 Ni.
Annealed; 75,000 TS; 35,000 YS; 40 El; 140 Brin.
For lead wires for electrical appliances; heat resistant.

MANGANESE NICKEL SILVER.
M-Ger.; 73-60 Cu, 2.4-20 Mn, 10-17 Ni, 0-10 Sn, 0-8.8 Zn.
For white metal parts; corrosion resistant.

MANGANESE RED BRASS 507.
M-8; 14 Zn, 1 Mn, bal Cu.
Hard: 69,000 TS; 55,000 YS; 7 El; 140 Brin.
Soft: 40,000 TS; 15,000 YS; 45 El; 55 Brin.
For strip for resistance spot and seam welded products.

MANGANESE S3.
M-1017; C, Mn, Ni, bal Fe.
Welded: 180 Brin.
Work hardened: 450 Brin.
For hard facing electrodes; wear resistant.

MANGANESE STEEL PEARLITIC.
M-Eng.; 1.4-3.5 Mn, 0.2-0.3 Si, 0.2-0.6 C, bal Fe.
For gears, shafts; British Patent 131980.

MANGANIN.
M-44; 10.5-13.5 Mn, 1-2 Ni, bal Cu.
Annealed: 70,000 TS.
For instrument resistors and shunts; low thermal EMF.

MANGANIN.
M-61, M-897, M-1523; 4 Ni, 12-14 Mn, bal Cu.
Annealed: 80,000 TS; 45,000 YS; 50 El.
For resistors, thermocouples; low temperature coefficient of resistance.

MANGANIN.
M-231; 86 Cu, 12 Mn, 2 Ni.
Resistance alloy; 290 ohms per C.M.F. for resistance boxes, Wheat stone bridges.

MANGANINGOT see **SPECIALLOY 5025 MANGANINGOT TM.**

MANGAN-NEUSILBER.
M-Ger.; 73-59 Cu, 10-18 Ni, 2.4-20 Mn, 5-20 Zn.
For white metal parts; corrosion resistant.

MANGANNICKEL 1.
M-297; 1.5 Mn, 0.25 C, bal Ni.
Annealed; 74,000 TS; 36 El; 95 Brin.
Drawn: 132,000 TS; 1 El; 250 Brin.
For German silver, candlesticks, water meter parts; corrosion resistant.

MANGANNICKEL 2.
M-297; 1.5 Mn, bal Ni.
Rolled: 135,000 TS; 1 El; 270 Brin.
Annealed: 76,800 TS; 35 El; 100 Brin.
For thermocouples; corrosion resistant.

MANGANNICKEL 4.
M-297; 4 Mn, bal Ni.
Rolled: 142,200 TS; 1 El; 280 Brin.
Annealed: 81,000 TS; 34 El; 110 Brin.
For thermocouples, spark plugs; corrosion resistant.

MANGANNICKEL 5.
M-297; 4.5-5.5 Mn, bal Ni.
For spark plug electrodes; heat resistant.

MANGA-TONE N.M.
M-507; 2.0 C, 0.8 Si, 26.2 Mn, 8.3 Ni, bal Fe.
Filler rod for repair or buildup of austenitic manganese steel parts as shovel teeth, railroad frogs, crossovers; pump impellers, tractor bottom rollers, crusher plates.

MANGCRAFT A.
M-1201; 0.5-0.8 C, 11-14 Mn bal Fe.
For welding rod for 14% Mn steel; austenitic, wear resistant.

MANGCRAFT B.
M-1201; 0.5-0.8 C, 11-14 Mn, 4.5-5.5 Ni, bal Fe.
For welding rod for 14% Mn-Ni steel; austenitic, abrasion resistant.

MANGJET.
M-578; 0.65 C, 14.5 Mn, 0.14 Si, 1.15 Mo, bal Fe.
Hardsurfacing arc welding electrodes for resistance to severe impact.

MANGO-PLATE.
M-1432; C, alloy, bal Fe.
Heat treated: 153,000 TS; 141,000 YS; 14 El; 54 RA; 390 Brin.
For coal chutes, conveyor lines, scraper blades; work hardens, abrasion resistant.

MANG-ROD NO. 250.
M-1096; 0.9-1.05 C, 0.9-1.2 Cr, 0.4 Si, 14 Mn, bal Fe.
Welded: 250 Brin.
For hard facing electrode; for Mn steels.

MANHARDT'S ALLOY.
M-Eng.; 83 Al, 10 Sn, 6.2 Cu, 0.1 Mg, P.
For light alloy parts; non-hardenable.

MANIFLEX see CARPENTER MANIFLEX.

MANNESMANNSTAHL-F12.
M-1381; 0.20 C, 12 Cr, 1 Mo, 0.4 Ni, 0.3 V, bal Fe.
Annealed: 90,000 TS; 38,000 YS; 26 El; B 92 Rock.
Hardened: 230,000 TS; 190,000 YS; 11 El; C 46 Rock.
For valves, bearings, cutlery, surgical instruments.
Corrosion resistant, hardenable.

MANNHEIM GOLD.
M-314; 89-80 Cu, 7-20 Zn, 0.9 Sn.
25,200 TS; 25 El; 69 Brin.
For cheap jewelry; moderately corrosion resistant.

MANOFORT 160.
M-1509; 0.30 C, 2.0 Ni, 1.0 Cr, 0.40 Mo, bal Fe.
Heat treated: 212,000-242,000 TS; 163,000-200,000 YS; 9-7 El.
For gears, shafts, crankshafts; oil hardened, shock resistant.

MANOFORT 180.
M-1509; 0.40 C, 3.5 Ni, 1.4 Cr, 0.4 Mo, bal Fe.
Heat treated: 228,000-275,000 TS; 185,000-214,000 YS; 6-4 El.
For gears, shafts, countershafts; oil hardened, shock resistant.

MANOIR ABRADUR 220.
M-1509; 0.70 C, 1.0 Mn, bal Fe.
Heat treated: 100,000-115,000 TS; 65,000-78,000 YS; 8-6 El; 220-260 Brin.
For rails, tools, hammers; water or oil hardened.

MANOIR ABRADUR 240.
M-1509; 0.50 C, 1.0 Mn, 1.0 Cr, bal Fe.
Heat treated: 100,000-114,000 TS; 65,000-78,000 YS; 8-6 El; 220-260
For camshafts, countershafts, gears, shafts; wear resistant, oil hardened.

MANOIR ABRADUR 400.
M-1509; 0.30 C, 1.8 Ni, 1.0 Cr, 0.35 Mo, bal Fe.
Heat treated: 186,000-228,000 TS; 156,000-186,000 YS; 6-4 El; 400-480 Brin.
For gears, shafts, countershafts; oil hardened, tough, shock resistant.

MANOIR ABRADUR 500.
M-1509; 0.40 C, 3.5 Ni, 1.4 Cr, 0.40 Mo, bal Fe.
Heat treated: 144,000-156,000 TS; 92,000-114,000 YS; 12-10 El; 280-32 Brin.
For gears, shafts, countershafts, mining equipment; oil hardened, tough, shock resistant.

MANOIR ABRADUR 600.
M-1509; C, Cr, W, V, bal Fe.
600 Brin.
For abrasion resistant parts, sand blasting nozzles; wear and abrasion resistant.

MANOIR ABRADUR M14.
M-1509; 1.2 C, 13 Mn, bal Fe.
Rolled: 120,000-157,000 TS; 50,000-65,000 YS; 35-30 El.
For shovels, dippers, frogs, cross tracks; wear resistant.

MANOIR ABRADUR M14K.
M-1509; 1.2 C, 13 Mn, Cr, bal Fe.
For dippers, shovels, cross tracks; wear and abrasion resistant.

MANIOR APS10.
M-1509; 0.12 C, 2 Cr, 0.4-0.9 Al, Mo, bal Fe.
Normalized: 69,000-92,000 TS; 42,000-54,000 YS; 16-12 El.
For oil refinery equipment; corrosion resistant.

MANOIR APS20.
M-1509; 0.12 C, 4 Cr, 0.4-0.9 Al, Mo, bal Fe.
Normalized: 69,000-92,000 TS; 42,000-54,000 YS; 16-12 El.
For oil refinery equipment: corrosion resistant.

MANOIR APS25.
M-1509; 0.12 C, 4.Cr, 0.9 Al, 0.9 Ni, bal Fe.
Normalized: 114,000-135,000 TS; 85,000-107,000 YS; 11-7 El.
For oil refinery equipment; corrosion resistant.

MANOIR EL38M.
M-1509; low C, Mn, bal Fe.
Annealed: 55,000 TS; 26,000 YS; 22 El; 120 Brin.
For motors, electrical equipment; high magnetic permeability.

MANOIR EL40.
M-1509; low C, Mn, bal Fe.
Annealed: 58,000 TS; 30,000 YS; 22 El; 120 Brin.

MANOIR EL45.
 M-1509; C, Mn, bal Fe.
 Annealed: 65,000 TS; 32,000 YS; 20 El; 140 Brin.

MANOIR EL50.
 M-1509; C, Mn, bal Fe.
 Annealed: 72,000 TS; 36,000 YS; 18 El; 155 Brin.

MANOIR EL55.
 M-1509; C, Mn, bal Fe.
 Annealed: 79,000 TS; 40,000 YS; 15 El; 170 Brin.
 For bed plates.

MANOIR EL65.
 M-1509; C, Mn, bal Fe.
 Annealed: 92,000 TS; 50,000 YS; 10 El; 210 Brin.

MANOIR P17.
 M-1509; 0.12 C, Mn, Mo, V, bal Fe.
 Rolled: 76,000-92,000 TS; 58,000-65,000 YS; 22-18 El.
 For gears, cams, camshafts; case hardened, tough.

MANOIR PF-0.
 M-1509; 0.15 max C, 0.50 Mo, bal Fe.
 Annealed: 65,000 TS; 37,000 YS; 24 El.
 For oil refinery equipment.

MANOIR PF-1.
 M-1509; 0.15 max C, 0.50 Cr, 0.50 Mo, bal Fe.
 Annealed: 78,000 TS; 40,000 YS; 18 El.
 For oil refinery equipment.

MANOIR PF-2.
 M-1509; 0.20 max C, 1.0 Cr, 0.50 Mo, 0.15 V, bal Fe.
 Annealed: 78,000 TS; 46,000 YS; 16 El.
 For oil refinery equipment.

MANOIR PF-5.
 M-1509; 0.15 max C, 5.0 Cr, 0.50 Mo, bal Fe.
 Annealed: 78,000 TS; 36,000 YS; 18 El.
 For oil refinery equipment, stills; corrosion resistant.

MANOIR PF-6.
 M-1509; 0.15 max C, 2.2 Cr, 1.0 Mo, bal Fe.
 Annealed: 85,000 TS; 36,000 YS; 15 El.
 For oil refinery equipment; corrosion resistant.

MANOIR PF-15.
 M-1509; 0.20 max C, 0.80 Cr, 0.50 Mo, bal Fe.
 Annealed: 92,000 TS; 58,000 YS; 16 El.
 For oil refinery equipment.

MANOIR PFV-55.
 M-1509; 0.15 max C, 2.0 Cr, 0.35 Mo, V, Al, bal Fe.
 Annealed: 78,000 TS; 36,000 YS; 15 El.
 For oil refinery equipment; corrosion resistant.

MANOIR PM35.
 M-1509; 0.17 C, Mn, Cr, Mo, V, bal Fe.
 Normalized: 143,000-170,000 TS; 107,000-121,000 YS; 10-7 El.
 For gears, cams, camshafts; case hardened, tough.

MANOIR RS1.
 M-1509; 0.45 C, 0.70 Mn, 1.8 Si, bal Fe.
 Heat treated: 177,000-200,000 TS; 156,000-177,000 YS; 4 El.
 For punches, air hammers, chisels; shock resistant, oil hardened.

MANSIL.
 M-275; 0.90 C, 1 Mn, 0.5 Cr, 0.5 W, bal Fe.
 For punches, cutters, dies, mandrels; type O1; oil hardened, non-deforming.

MANSILOY.
 M-48; 60-63 Mn, 28-31 Si, 0.07 max C, 0.05 max P.
 For production of stainless steels; reduces metal oxides from the slag.

MAN-VAN.
 M-US; 0.95 C, 0.15 Mo, 0.10 V, bal Fe.
 For cutting tools, dies; water hardening.

MAPLE LEAF.
 M-435; 0.8-1.2 C, 0.25 Si, 0.25 Mn, bal Fe.
 Water hardened: 165,000-215,000 TS; 110,000-150,000 YS; 11-15 El; 32-37 RA; 330-600 Brin.
 For stamps, knurls, drills, taps, mandrels, reamers, cutters.
 Type W1, water hardening.

MAPLE LEAF 8.
 M-435; 0.80 C, bal Fe.
 For blacksmith tools; water hardened.

MARADAMIT.
 Al alloy.
 For light alloy parts.

MARATHON-000 EXTRA.
 M-1496; 0.80 C, 4 Cr, 18 W, 2 V, bal Fe.
 Hardened: C 64-66 Rock.
 For lathe and planer tools, drills, hobs, reamers, chasers, form cutters.
 Type T-2 tool steel, high red-hardness.

MARATHON CRM SPEZIAL.
 M-1496; 1.0 C, 1.0 Mn, 1.0 Cr, 0.25 Mo, bal Fe.
 Ht. Tr.: 288,000 TS; 278,000 YS; 540 Brin.
 For precision and plug gauges, bearings, liners, arbors.
 High toughness and wear resistance.
 Type L-5 tool steel.

MARATHON-CRS.
M-1496; 1.05 C, 1.4 Cr, bal Fe.
Ht. Tr.: 200,000 TS; 185,000 YS; 3 El; 390 Brin.
For gauges, bushings, knurls, taps, dies, arbors, rolls, bearings.
Oil hardening, tough and wear resistant.
Type L-1 tool steel.

MARATHON-E38 MO.
M-1496; 0.35 C, 1 Si, 5 Cr, 1 V, 1.5 Mo, bal Fe.
Ht. Tr.: 290,000 TS; 227,000 YS; 3 El; C 55 Rock.
For forging and drawing dies, die casting dies, hot piercing and forming dies, swaging and gripping dies.
Type H13 tool steel, air hardening.
High resistance to heat checking.

MARATHON-E38V.
M-1496; 0.35 C, 5 Cr, 1 V, 1.2 Mo, bal Fe.
Ht. Tr.: 290,00 TS; 225,000 YS; 3 El; C 55 Rock.
For die casting dies, swaging and gripping dies, forging and extrusion dies.
Type H-13 tool steel, air hardening.
High resistance to heat checking.

MARATHON-E38W.
M-1496; 0.36 C, 5 Cr, 1.4 Mo, 1.4 W, 0.5 V, bal Fe.
Ht. Tr.: 290,000 TS; 235,000 YS; 8 El; C 54 Rock.
For die casting dies, hot punches and heading dies, shear blades.
Type H12 tool steel, hot work steel, high toughness and wear resistance.

MARATHON E612.
M-1496; 0.40 C, 1.15 Si, 5.25 Cr, 4.25 W, bal Fe.
Hardened: 269,000 TS; 207,000 YS; 5 El; 9 RA.
At 900°F: 217,000 TS; 172,000 YS; 10 El; 31 RA.
For die casting dies, forging and extrusion tools, upsetters, dummy blocks.
Type H14 hot work steel, deep hardening.

MARATHON MO10.
M-1496; 0.80 C, 4 Cr, 8 Mo, 1.5 W, 1 V, bal Fe.
Hardened: C 64-66 Rock.
For lathe and planer tools, hobs, drills, reamer, broaches, form cutters.
Type M1 high speed steel, high red-hardness, wear resistant.

MARATHON MO19.
M-1496; 1.0 C, 4 Cr, 8.75 Mo, 1.75 W, 2 V, bal Fe.
Hardened: C 64-67 Rock.
For reamers, hobs, chasers, lathe and planer tools.
Type M7 high-speed steel, high red-hardness.
Abrasion resistant.

MARATHON MO20.
M-1496; 0.85 C, 5 Mo, 6.5 W, 4 Cr, 2 V, bal Fe.
Hardened: C 64-66 Rock.
For lathe and planer tools, drills, taps, chasers, drawing dies, punches, hobs, form cutters.
Type M2 high speed steel, high red-hardness, tough.

MARATHON MO20S.
M-1496; 0.62-0.68 C, 3.8-4.4 Cr, 1.8-2.1 V, 6.5-6.7 W, 4.7-5.2 Mo, bal Fe.
Hardened: C 58-62 Rock.
For hot extrusion dies, punches, shear blades, forging mandrels.
Type H42 high speed steel for hot working.

MARATHON PW16.
M-1496; 0.50 C, 4 Cr, 18 W, 1 V, bal Fe.
Hardened: C 60-65 Rock.
Ht. Tr.: 130,000 TS; 95,000 YS; 4 El; C 52 Rock.
For punches, hot work tools and dies, extrusion and die casting dies.
Type H26 hot work steel, tough, good high temperature strength.

MARATHON-SA.
M-1496; 0.55 C, 0.9 Mn, 2.0 Si, 0.3 Cr, 0.2 V, bal Fe.
Hardened: 338,000 TS; 281,000 YS; 5 El; 600 Brin.
For punches, shear blades, pneumatic tools, rivet busters, knurling tools.
Type S-4 tool steel, deep hardening, shock and wear resistant.

MARATHON SA200.
M-1496; 0.75 C, 4 Cr, 14 W, 2 V, bal Fe.
Hardened: C 64-66 Rock.
For drills, reamers, taps, punches, lathe and planer tools.
Type T7 high speed steel, high red-hardness.

MARATHON SA900.
M-1496; 1.5 C, 13 W, 4.5 Cr, 5 V, 5 Co, bal Fe.
Hardened: C 64-66 Rock.
For cutting tools, broaches, drills, hobs, lathe and planer cutters, reamers.
Type T15 high speed steel, high-red-hardness, abrasion resistant.

MARATHON SS4.
M-501; 1.05-1.15 C, bal Fe.
For tools, taps, drills; water hardened.

MAR-CON 660.
M-1671; 7 Sn, 7 Pb, 3 Zn, bal Cu.
Cast: 44,000 TS; 27,000 YS; 16 El; 14 RA; 73 Brin.
For bearings, bushings, liners.
Continuous cast bronze.

MARINA 50.
M-1612; 1.5 Mn, 1 Cb or V, 0.8 Cr, 0.55 Al, 0.35 Cu, 0.4 max Ni, C, bal Fe.
High tensile, low alloy steel for sea-water equipment.

MARINE ALLOY.
M-US; 2 Cu, 40 Sn, 48 Pb, 10 Sb.
For bearings: submerged bearings.

MARINE BABBITT.
72 Pb, 21 Sn, 7 Sb.
For marine bearings; anti-friction.

MARINE BRONZE.
M-US; 57.5 Cu, 0.8 Ni, 0.15 Sn, 0.50 Al, bal Zn, 83,500 TS; 15 El.
For marine parts.

MARINE BRONZE NO. 8.
M-481; Mn, Zn Sn, Al, bal Cu.
Cast: 65,000 TS; 20-25 El; 29 RA; 150 Brin.
For marine propellers, ship parts, etc., subjected to salt water corrosions; resists corrosion.

MARINE GLYCO.
M-240; Sn, Sb, bal Pb.
For marine bearings; Babbitt.

MARINE NICKEL.
M-540; Pb, Ni, bal Sn.
For bearings; Babbitt metal.

MARINER.
M-604; 0.22 C, 0.80 Mn, 0.11 P, bal Fe.
Sheet, as rolled: 70 ksi TS; 50 ksi YS; 18 El.
Sheet piling for use in marine environment, ASTM A690.

MARK 2 ETC see CENTRICAST MARK 2 ETC.

MARK 12KH14A.
M-USSR; 0.12 C, 14 Cr, bal Fe.
Annealed: 75,000 TS; 40,000 YS; 35 El; 70 RA; 155 Brin.
For valves, cutlery, valve turbines; type 410; stainless.

MARK 18KH14A.
M-USSR; 0.18 C, 14 Cr, bal Fe.
Annealed: 95,000 TS; 50,000 YS; 25 El; 55 RA; 195 Brin.
For valves, cutlery, surgical and dental instruments; Type 420; stainless.

MARK 50.
M-USSR; 0.50 C, bal Fe.
Annealed: 96,000 TS; 52,000 YS; 16 El; 23 RA; 170 Brin.
For gears, bolts, fasteners; water hardened.

MARK E169.
M-USSR; 14 Cr, 14 Ni, 2 W, bal Fe.
For valves, cutlery; corrosion resistant.

MARKER 215 M.
M-1305; 0.45 C, 0.9 Si, 0.3 Mn, 1.05 Cr, 0.2 V, 1.85 W, bal Fe.
Cold work tool and die steel; oil hardenable to 54-58 Rock C.
For pneumatic hammers, chisels, riveting tools.
Werkstoff Nr. 1.2542.

MARKER 220 M.
M-1305; 0.60 C, 0.9 Si, 0.3 Mn, 1.05 Cr, 0.2 Vi, 1.85 W, bal Fe.
Cold work tool and die steel; oil or water hardenable to 57-61 Rock C.
Punching and trimming dies, shear knives to 9 mm thick, wood working tools.
Werkstoff Nr. 1.2550.

MARKER 430 M.
M-1305; 1.0 C, 0.60 Cr, 0.60 W, 0.10 V, Mn, bal Fe.
Tool steel for hand punches, chisels.
W-Nr 1.2510.

MARKER 465 M.
M-1305; 0.90 C, 1.90 Mn, 0.1 V, bal Fe.
Cold work tool and die steel; oil hardenable to 63-64 Rock C.
For cutting dies of complicated shapes for thin metal to 3 mm thick, threading dies.
Werkstoff Nr. 1.2842.

MARKER 476 MEL.
M-1305; 0.96 C, 12.0 Cr, 0.90 Mo, 2.2 V, bal Fe.
High chromium cold work tool steel.
W.-Nr. 1.2376.

MARKER 476 M EXTRA.
M-1305; 1.65 C, 12.0 Cr, 0.6 Mo, 0.1 V, 0.5 W, bal Fe.
Cold work tool and die steel; air or oil hardenable to 63-65 Rock C.
For cold forming and stamping laminations, saw blades for precision parts as gages; some corrosion resistance.
Werkstoff Nr. 1.2601.

MARKER 477 M.
M-1305; 1.05 C, 0.9 Mn, 1.0 Cr, 1.15 W, bal Fe.
Cold work tool and die steel; oil hardenable to 63-66 Rock C.
For punching and trimming dies for thin sheet and strip up to 4 mm thick, shears.
Werkstoff Nr. 1.2419.

MARKER 480 M.
M-1305; 1.05 C, 0.90 Cr, Mn, bal Fe.
Tool steel for punches, cutting knives.
W.Nr. 1.2127.

MARKER 4712 ETC see WERKSTOFF 1.4712 ETC FOR COMPOSITION, PROPERTIES AND USES.

MARKER 4922.
M-1305; 0.20 C, 11.5 Cr, 1.0 Mo, 0.60 Ni, 0.30 V, bal Fe.
Temperature resisting and corrosion resisting steel.
Martensitic type.
W.-Nr. 1.4922.

MARKER 4948.
M-1305; 0.06 C, 18.0 Cr, 11.0 Ni, 0.30 Mo, bal Fe.
Austenitic stainless steel.
W.-Nr. 1.4948.

MARKER 4961.
M-1305; 0.08 C, 16.0 Cr, 13.0 Ni, 1.0 Nb, bal Fe.
Stabilized austenitic stainless steel.
W.-Nr. 1.4961.

MARKER 4980.
M-1305; 0.08 C, 15.0 Cr, 26.0 Ni, 1.25 Mo, 0.30 V, 2.1 Ti, Al, B, bal Fe.
Special alloy for high temperature operation.
W.-Nr. 1.4980.

MARKER 4981.
M-1305; 0.08 C, 16.5 Cr, 16.5 Ni, 1.8 Mo, 1.0 Nb, bal Fe.
W.-Nr. 1.4981.

MARKER 4988.
M-1305; 0.08 C, 16.5 Cr, 13.5 Ni, 1.3 Mo, 0.75 V, 1.0 Nb, bal Fe.
W.-Nr. 1.4988.

MARKER AW50.
M-1305; 0.45 C, 1.65 Cr, 0.50 Mo, 0.90 V, 0.60 W, bal Fe.
Hot work tool and die steel; oil or water hardenable to 250,000-270,000 TS.
Pressing dies for hot forming of screws, hot shears.
Werkstoff No. 1.2603.

MARKER B 12.
M-1305; 0.20 C, 13.0 Cr, bal Fe.
Cold work tool steel.
W.-Nr. 1.2082.

MARKER C 3.
M-1305; 1.40 C, 0.25 Si, 0.30 Mn, 0.30 Cr, 0.10 V, bal Fe.
Water hardening tool steel, high surface hardness possible; for drawing dies and tools.
Werkstoff Nr. 1.2206.

MARKER C 73.
M-1305; 1.45 C, 0.25 Si, 0.60 Mn, 1.40 Cr, bal Fe.
Oil hardening tool steel; for woodworking tools, files, rubber cutting knives.
Werkstoff Nr. 1.2063.

MARKER CDD.
M-1305; 0.90 C, 0.25 Si, 0.30 Mn, 0.80 Cr, bal Fe.
Water or oil hardening tool steel; for drawing dies, embossing tools, cold forming press and bend rollers.
Werkstoff Nr. 1.2056.

MARKER D690.
M-1305; 0.06 C, 15.0 Cr, 1.25 Mo, 26.0 Ni, 0.30 V, Ti, B, Al, bal Fe.
Steel for parts operated at high temperature.
W.-Nr. 1.2779.

MARKER DCM.
M-1305; 0.35 C, 1.0 Si, 5.15 Cr, 1.45 Mo, 0.50 V, bal Fe.
Hot work tool and die steel; oil, water or air hardenable to 256,000-274,000 TS.
For forging dies, pressure casting molds.
Werkstoff Nr. 1.2343; AISI H 11.

MARKER DCMX.
M-1305; 0.27 C, 1.0 Si, 5.15 Cr, 1.45 Mo, 0.50 V, bal Fe.
Hot work tool and die steel; air or oil hardenable to 240,000-270,000 TS.
For pressure casting molds, forging dies; better toughness than Marker DCM.

MARKER DCS.
M-1305; 0.40 C, 1.0 Si, 0.4 Mn, 5.15 Cr, 1.2 Mo, 1.05 V, 0.15 S, bal Fe.
Hot work tool and die steel.
Free machining grade similar to Marker DCV.

MARKER DCV.
M-1305; 0.40 C, 1.0 Si, 5.15 Cr, 1.35 Mo, 1.0 V, bal Fe.
Hot work tool and die steel; oil or water hardenable to 256,000-276,000 TS.
For hot pressing dies for light metals, pressure casting molds, forging dies.
Werkstoff Nr. 1.2344; AISI H 13.

MARKER DCW.
M-1305; 0.37 C, 1.0 Si, 5.15 Cr, 1.45 Mo, 0.30 V, 1.50 W, bal Fe.
Hot work tool and die steel; oil or water hardenable to 280,000-300,000 TS.
For hot shears, punches, forming tools for screws, rivets; hot tools for welding machines, forging dies.
Werkstoff Nr. 1.2606; AISI H 12.

MARKER-ED12.
M-1335; 0.20 C, 11.5 Cr, 1 Mo, 0.3 V, bal Fe.
Annealed: 90,000 TS; 40,000 YS; 25 El; B 92 Rock.
Hardened: 240,000 TS; 200,000 YS; 10 El; C 50 Rock.
For valves, bearings, cutlery, surgical instruments.
Corrosion resistant, hardenable.

MARKER-ED12G.
M-1335; 0.18 max C, 11 Cr, 1.05 Mo, 0.75 Ni, 0.3 V, bal Fe.
Annealed: 90,000 TS; 38,000 YS; 26 El; B 92 Rock.
Hardened: 235,000 TS; 200,000 YS; 10 El; C 48 Rock.
For valves, bearings, cutlery, surgical instruments.
Corrosion resistant, hardenable.

MARKER-ED12W.
M-1335; 0.20 C, 12 Cr, 1 Mo, 0.3 V, 0.5 W, bal Fe.
Annealed: 90,000 TS; 40,000 YS; 25 El; B 92 Rock.
Hardened: 240,000 TS; 205,000 YS; 9 El; C 50 Rock.
For cutlery, valves, bearings, surgical instruments.
Corrosion resistant, hardenable.

MARKER EUZONIT 60.S.
M-1305; 0.02 C, 20.0 Cr, 15.5 Mo, Fe, bal Ni.
Nickel-chrome-moly stainless alloy.
Werkstoff-Nr. 2.4811.

MARKER EUZONIT 70.5.
M-1305; 0.02 C, 28.0 Mo, Fe, bal Ni.
Nickel-moly stainless alloy.
Werkstoff-Nr. 2.4810.

MARKER EUZONIT G-60.
M-1305; 0.05 C, 17.0 Cr, 18.0 Mo, Fe, bal Ni.
Stainless nickel base steel casting.
Werstoff-Nr. 2.4537.

MARKER EUZONIT G-4810.
M-1305; 0.05 C, 28.0 Mo, Fe, bal Ni.
Stainless nickel-moly casting.
Werkstoff-Nr. 2.4810.

MARKER FS.
M-1305; 0.70 C, 1.65 Si, 0.70 Mn, bal Fe.
Water or oil hardening steel; for wrenches, pliers, screw drivers.
Werkstoff Nr. 1.2823.

MARKER G-4008.
M-1305; 0.10 C, 13.5 Cr, bal Fe.
Rustproof chrome steel casting.
ACI CA 15; Werkstoff Nr. 1.4008.

MARKER G-4027.
M-1305; 0.20 C, 14.0 Cr, bal Fe.
Rustproof chrome steel casting.
ACI CA 15; Werkstoff. Nr. 1.4027.

MARKER G-4034.
M-1305; 0.45 C, 15.0 Cr, bal Fe.
Rustproof chrome steel casting.
ACI CA 40.

MARKER G-4059.
M-1305; 0.25 C, 17.0 Cr, 1.5 Ni, bal Fe.
Rustproof chrome steel casting.
ACI CB 30.

MARKER G-4085.
M-1305; 0.70 C, 29.0 Cr, bal Fe.
Rustproof and acid resisting steel casting.
Werkstoff 1.4085.

MARKER G-4086.
M-1305; 1.10 C, 29.0 Cr, bal Fe.
Rustproof and acid resisting steel casting.
Werkstoff.-Nr. 1.4086.

MARKER G-4088.
M-1305; 1.70 C, 18.0 Cr, bal Fe.
Rust resistant and wear resistant steel casting.

MARKER G-4122.
M-1305; 0.40 C, 17.0 Cr, 1.0 Ni, 1.0 Mo, bal Fe.
Rustproof chrome steel casting.
Werkstoff-Nr. 1.4122.

MARKER G-4136.
M-1305; 0.70 C, 29.0 Cr, 2.2 Mo, bal Fe.
Rustproof and acid resisting steel casting.
Werkstoff-Nr. 1.4136.

MARKER G-4138.
M-1305; 1.10 C, 29.0 Cr, 2.2 Mo, bal Fe.
Rustproof and acid resisting steel casting.
ACI CC 50; Werkstoff 1.4138.

MARKER G-4306.
M-1305; 0.03 C, 19.0 Cr, 10.0 Ni, bal Fe.
Stainless chrome-nickel steel casting.
ACI CF 3; Werkstoff-Nr. 1.4306.

MARKER G-4308.
M-1305; 0.07 C, 20.0 Cr, 10.0 Ni, bal Fe.
Stainless chrome-nickel steel casting.
ACI CF 8; Werkstoff-Nr. 1.4308.

MARKER G-4312.
M-1305; 0.10 C, 18.0 Cr, 9.0 Ni, bal Fe.
Stainless chrome-nickel steel casting.
ACI CF 12; Werkstoff-Nr. 1.4312.

MARKER G-4313.
M-1305; 0.08 C, 13.0 Cr, 3.5 Ni, 0.50 Mo, bal Fe.
Rustproof chrome steel casting.
Werkstoff-Nr. 1.4313.

MARKER G-4340.
M-1305; 0.40 C, 27.0 Cr, 4.0 Ni, bal Fe.
Rustproof and acid resisting steel casting.
ACI CC 50; Werkstoff-Nr. 1.4340.

MARKER G-4347.
M-1305; 0.08 C, 25.0 Cr, 7.0 Ni, bal Fe.
Stainless chrome-nickel steel casting.
Werkstoff-Nr. 1.4347.

MARKER G-4404.
  M-1305; 0.03 C, 18.0 Cr, 11.0 Ni, 2.2 Mo, bal Fe.
  Stainless chrome-nickel steel casting.
  ACI CF 3M; AISI 316L; Werkstoff-Nr. 1.4404.

MARKER G-4405.
  M-1305; 0.07 C, 16.0 Cr, 4.5 Ni, 1.0 Mo, bal Fe.
  Rustproof chrome steel casting.
  Werkstoff-Nr 1.4405.

MARKER G-4408.
  M-1305; 0.70 C, 20.0 Cr, 10.0 Ni, 2.2 Mo, bal Fe.
  Stainless chrome-nickel steel casting.
  ACI CF 8 M; AISI 316; Werkstoff-Nr. 1.4408.

MARKER G-4410.
  M-1305; 0.10 C, 18.0 Cr, 10.0 Ni, 2.2 Mo, bal Fe.
  Stainless chrome-nickel steel casting.
  Werkstoff-Nr. 1.4410.

MARKER G-4457.
  M-1305; 0.25 C, 25.0 Cr, 9.0 Ni, 2.2 Mo, bal Fe.
  Stainless chrome-nickel steel casting.

MARKER G-4460.
  M-1305; 0.10 C, 25.0 Cr, 7.0 Ni, 1.5 Mo, bal Fe.
  Stainless chrome-nickel steel casting.

MARKER G-4464.
  M-1305; 0.35 C, 26.0 Cr, 5.0 Ni, 2.5 Mo, bal Fe.
  Rustproof and acid resisting steel casting.
  Werkstoff-Nr. 1.4464.

MARKER G-4500.
  M-1305; 0.08 C, 20.0 Cr, 25.0 Ni, 3.0 Mo, Nb, Cu, bal Fe.
  Stainless chrome-nickel steel casting.
  ACI CN 7 M; Werkstoff-Nr. 1.4500.

MARKER G-4552.
  M-1305; 0.08 C, 18.0 Cr, 10.0 Ni, Nb, bal Fe.
  Stainless chrome-nickel steel casting.
  ACI CF 8 C; Werkstoff-Nr. 1.4552.

MARKER G-4559.
  M-1305; 0.08 C, 20.0 Cr, 42.0 Ni, 5.0 Mo, Nb, Cu, bal Fe.
  Stainless chrome-nickel steel casting.
  Werkstoff-Nr. 2.4557.

MARKER G-4579.
  M-1305; 0.08 C, 17.5 Cr, 13.5 Ni, 4.5 Mo, Nb, bal Fe.
  Stainless chrome-nickel steel casting.
  ACI CG 8 M; Werkstoff-Nr. 1.4579.

MARKER G-4581.
  M-1305; 0.08 C, 18.0 Cr, 11.0 Ni, 2.2 Mo, Nb, bal Fe.
  Stainless chrome-nickel steel casting.
  Werkstoff-Nr. 1.4581.

MARKER G-4585.
  M-1305; 0.08 C, 18.0 Cr, 20.0 Ni, 2.2 Mo, Nb, Cu, bal Fe.
  Stainless chrome-nickel steel casting.
  Werkstoff-Nr. 1.4585.

MARKER IRR 4006 ETC see WERKSTOFF NR. 1.4006 ETC. FOR COMPOSITION, PROPERTIES AND USES.

MARKERITE.
  M-Ger.; carbide.
  For tool tips, cutters; hard metal.

MARKER K 6 R.
  M-1305; 0.60 C, 0.85 Si, 0.75 Mn, 1.20 Cr, 0.10 V, bal Fe.
  Cold work tool steel; oil hardenable to 59-62 Rock C.
  For hole punching tools in heavy metal up to 4 mm thick, wood working tools as planes, saws.
  Werkstoff Nr. 1.2243.

MARKER K 8 R.
  M-1305; 0.85 C, 1.15 Si, 0.70 Mn, 1.20 Cr, bal Fe.
  Cold work tool steel; oil hardenable to 63-65 Rock C.
  For punches, chisels, woodworking tools, milling cutters.
  Werkstoff Nr. 1.2108.

MARKER K 97.
  M-1305; 0.06 C, 0.20 Si, 0.25 Mn, 3.75 Cr, 0.50 Mo, 0.10 V, bal Fe.
  Case hardening steel; case hardness expected: C 62-65 Rock.
  Core strength: 112,000-142,000 psi TS.

MARKER KEW.
  M-1305; 0.60 C, 3.75 Cr, 0.85 Mo, 0.70 V, 9.0 W, bal Fe.
  Hot work tool steel, oil or air hardenable to 60-62 Rock C.
  For hot punching, stamping and cutting, tube pressing mandrels.
  Werkstoff Nr. 1.2622.

MARKER KL 9.
  M-1305; 1.00 C, 0.25 Si, 0.40 Mn, 1.55 Cr, bal Fe.
  Oil hardening tool steel; used for cutters for meat grinders, for ball and roller bearings and races.
  Werkstoff Nr. 1.2067.

MARKER KO 11.
  M-1305; 0.78 C, 9.50 Co, 4.15 Cr, 0.65 Mo, 1.55 V; 18.0 W, bal Fe.
  High speed steel for turning, planing, and drilling steel and cast iron.
  (AISI T5); W. Nr. 1.3265.

MARKER KO 12.
  M-1305; 1.32 C, 10.50 Co, 4.50 Cr, 4.00 Mo, 3.25 V, 10.0 W, bal Fe.
  High speed steel for finishing tools, for inserted cutters, for cold flow press tools.
  W. Nr. 1.3207.

MARKER KO 13.
M-1305; 1.10 C, 8.0 Co, 3.8 Cr, 9.5 Mo, 1.2 V, 1.5 W, bal Fe.
High speed steel for lathe tools, form cutters, planing tools, drills, reamers.
Good high temperature hardness.
AISI M42; W. Nr. 1.3247.

MARKER KO 55.
M-1305; 1.32 C, 4.75 Co, 4.15 Cr, 0.85 Mo, 3.75 V, 12.0 W, bal Fe.
High speed steel for finishing tools for turning, reaming, form cutting, for inserted cutters.
W. Nr. 1.3202.

MARKER KO 109.
M-1305; 0.80 C, 4.75 Co, 4.15 Cr, 0.65 Mo, 1.55 V, 18.0 W, bal Fe.
High speed steel for turning and punching tools for steel and cast iron; for deep hole drills.
AISI T4; W. Nr. 1.3255.

MARKER KSP.
M-1305; 0.50 C, 0.30 Si, 0.65 Mn, 0.80 Cr, 0.30 Mo, 1.80 Ni, 0.10 V, bal Fe.
Oil hardening steel, to 56-58 Rock C.
For forming tools, can also be nitrided for 600-650 Vickers hardness; smooth finish possible.

MARKER KSPE.
M-1305; 0.50 C, 1.1 Cr, 0.50 Mo, 1.7 Ni, 0.10 V, bal Fe.

MARKER LW 10.
M-1305; 1.20 C, 0.25 Si, 0.30 Mn, 0.20 Cr, 0.10 V, 1.00 W, bal Fe.
Water hardening tool steel, high surface hardness possible; for drills, drawing tools, knives for pipe cutting.
Werkstoff Nr. 1.2516.

MARKER MAT.
M-1305; 0.03 C, 18.0 Ni, 9.0 Co, 4.9 Mo, Ti, bal Fe.
W.-Nr. 1.2709.

MARKER MEWE.
M-1305; 0.30 C, 2.35 Cr, 0.60 V, 4.25 W, bal Fe.
Hot work tool and die steel; air or oil hardenable to 240,000-265,000 TS.
For hot forming and thread rolling of screws, nuts, rivets; for hot shears for pressure casting molds.
Werkstoff Nr. 1.2567.

MARKER MH.
M-1305; 0.35 C, 1.40 Cr, 0.30 Mo, 4.1 Ni, bal Fe.
Hot work tool steel as hot pressing dies.
W.-Nr. 1.2766.

MARKER MNSK.
M-1305; 0.62 C, 0.60 Cr, Mn, Si, bal Fe.
Tool steel for punches, shear knives.
W-Nr. 1.2101.

MARKER MO 5.
M-1305; 0.85 C, 4.15 Cr, 5.10 Mo, 1.85 V, 6.35 W, bal Fe.
High speed steel for tools for general purpose cutting, lathe turning, milling, drilling of steel and non-ferrous metals; tough.
W Nr. 1.3343; AISI M2.

MARKER MO 5 CO.
M-1305; 0.85 C, 4.75 Co, 4.15 Cr, 5.10 Mo, 1.85 V, 6.35 W, bal Fe.
High speed steel for turning tools, threading tools, deep hole drills, cutting tough iron and steel.
W Nr. 1.3243; AISI M35.

MARKER MO 5 CO H.
M-1305; 1.10 C, 5.0 Co, 4.25 Cr, 3.75 Mo, 2.0 V, 6.75 W, bal Fe.
High speed steel for fast finishing tools for turning, planing, deep hole drilling.
W. Nr. 1.3246; AISI M41.

MARKER MO 5 H.
M-1305; 1.00 C, 4.15 Cr, 5.10 Mo, 1.85 V, 6.35 W, bal Fe.
High speed steel for finish cutting lathe, milling and drilling tools.
W. Nr. 1.3342

MARKER MO 9 CO 8.
M-1305; 0.94 C, 8.0 Co, 4.0 Cr, 8.5 Mo, 2.0 V, 2.0 W, bal Fe.
High speed steel for turning and planing steel and cast iron under difficult conditions.
AISI M34; W. Nr. 1.3249.

MARKER MO 9 V.
M-1305; 1.05 C, 3.80 Cr, 9.0 Mo, 2.0 V, 1.75 W, bal Fe.
High speed steel for turning, milling, drilling steel and non-ferrous metals.
W Nr. 1.3348; AISI M7.

MARKER MOV 4.
M-1305; 1.23 C, 4.15 Cr, 5.10 Mo, 2.90 V, 6.35 W, bal Fe.
High speed steel for turning, reaming, gear cutting of steel and non-ferrous metals; high hardness.
W. Nr. 1.3344; AISI M3(2).

MARKER MSV.
M-1305; 1.00 C, 0.20 Si, 0.25 Mn, 0.15 V, bal Fe.
Water hardening tool steel; for small cold forming dies for making screws, rivets, coins; for drawing dies.

MARKER P 42 W.
M-1305; 0.85 C, 26.0 Cr, 15.0 W, Fe, Nb, bal Co.
Special alloy for high temperature operation.

MARKER P 63.
M-1305; 0.30 C, 28.0 Cr, 5.5 Mo, 2.5 Ni, Fe, bal Co.
Special alloy for high temperature operation.
W.-Nr. 2.4979

MARKER PHM.
M-1305; 0.45 C, 1.4 Cr, 0.7 Mo, 0.3 V, bal Fe.
Hot work tool steel; oil or water hardenable to 300,000 TS.
For centrifugal casting molds for zinc alloys, hot forming tools.
Werkstoff Nr. 1.2323.

MARKER PSH.
M-1305; 0.50 C, 0.25 Si, 0.50 Mn, 1.05 Cr, 3.50 Ni, bal Fe.
Oil hardening tool steel, to 55-58 Rock C.
For dies for forming emblems, plaques, buttons, commemorative coins, cutlery.
Werkstoff Nr. 1.2767.

MARKER PSHB.
M-1305; 0.45 C, 0.25 Si, 0.50 Mn, 1.25 Cr, 0.20 Mo or 0.50 W, 3.75 Ni, bal Fe.
Oil hardening tool steel, to 53-56 Rock C.
For pressing dies requiring deep impressions, thread rolling dies, embossing tools.
Werkstoff Nr. 1.2767.

MARKER PT.
M-1305; 1.0 C, 5.2 Cr, 1.10 Mo, 0.20 V, bal Fe.
Air hardening tool steel for thread rolling dies, shear punches, trimming tools.
W.-Nr. 1.2363.

MARKER PT 8.
M-1305; 0.50 C, 9.0 Cr, 1.2 Mo, 1.2 W, bal Fe.
Cold work tool steel for shear knives and cutting dies.
W.-Nr. 1.2631.

MARKER PT 13.
M-1305; 0.80 C, 13.5 Cr, 1.1 Co, bal Fe.
W.-Nr. 1.2883.

MARKER PTE.
M-1305; 0.63 C, 5.3 Cr, 1.2 Mo, 0.30 V, bal Fe.
Air hardening tool steel.
W.-Nr. 1.2362.

MARKER PTM.
M-1305; 0.81 C, 4.0 Cr, 43 Mo, 1.0 V, bal Fe.
Air hardening cold work tool steel.
W.-Nr. 1.2369.

MARKER PWC.
M-1305; 0.45 C, 4.25 Cr, 0.50 Mo, 2.25 V, 4.25 W, 4.25 Co, bal Fe.
Hot work tool steel, air or oil hardenable to 240,000-265,000 TS.
For hot punching and stamping operations.
Werkstoff Nr. 1.2678; AISI H 19.

MARKER PWC 2.
M-1305; 0.20 C, 9.5 Cr, 2.0 Mo, 5.5 W, 10.0 Co, bal Fe.
Hot work tool steel.
W.-Nr. 1.2888.

MARKER R 17.
M-1305; 0.38 C, 0.40 Si, 0.80 Mn, 17.0 Cr, 1.25 Mo, 1.0 Ni, bal Fe.
Air or oil hardenable to 50-55 Rc.
For surgical and dental tools, cutlery, kitchen tools; corrosion resistant.

MARKER REM.
M-1305; 0.45 C, 13.5 Cr, 13.0 Ni, 0.50 V, 2.5 W, bal Fe.
Austenitic, hot work tool steel.
For dies and forming tools for working copper-nickel and nickel alloy that are difficult to work.
Werkstoff Nr. 1.2731.

MARKER REM SPEZIAL.
M-1305; 0.50 C, 4.0 Cr, 0.70 Mo, 11.5 Ni, 1.1 V, 12.5 W, 1.7 Co, bal Fe.
Hot work tool steel.
W.-Nr. 1.2758.

MARKER RH.
M-1305; 0.40 C, 0.40 Si, 0.30 Mn, 13.0 Cr, bal Fe.
Air or oil hardening high chromium steel; for synthetic resin molds, cold stamping and punching dies, surgical tools; corrosion resistant.
Werkstoff Nr. 1.2083.

MARKER S 7 A.
M-1305; 0.20 C, 0.25 Si, 1.25 Mn, 1.15 Cr, bal Fe.
Case hardening steel.
Case hardness expected: C 62-64 Rock.
Core strength: 142,000-182,000 psi TS.
Werkstoff Nr. 1.2162.

MARKER SL 8.
M-1305; 0.05 C, 20.0 Cr, 2.0 Fe, 1.2 Al, 2.2 Ti, bal Ni.
Special alloy for high temperature operation.
W.-Nr. 2.4952; Similar to Nimonic 80 A.

MARKER SL 10.
M-1305; 0.06 C, 20.0 Cr, 18.0 Co, Fe, Ti, Al, bal Ni.
Special alloy for high temperature operation.
W.-Nr. 2.4969; Similar to Nimonic 90.

MARKER SL 15.
M-1305; 0.08 C, 19.0 Cr, 10.0 Mo, 11.0 Co, Fe, Ti, Al, bal Ni.
Special alloy for high temperature operation.
W.-Nr. 2.4973.

MARKER S.N.18.
M-1335; 0.08 C, 18 Cr, 18 Ni, 2 Mo, 2 Cu, bal Fe.
For chemical plant equipment, pump and valve parts; corrosion resistant, austenitic.

MARKER S.N.25.
M-1335; 0.10 C, 18 Cr, 25 Ni, 4 Mo, 2 Cu, bal Fe.
For chemical plant equipment, pump and valve parts; corrosion resistant, austenitic.

MARKER S.N.42.
M-1335; 0.1 C, 18 Cr, 42 Ni, 5 Mo, 2 Cu, bal Fe.
For chemical plant equipment; corrosion and heat resistant.

MARKER SPM.
M-1305; 0.30 C, 2.35 Cr, 0.25 V, 9.0 W, 2.05 Co, bal Fe.
Hot work tool steel, oil or air hardenable to 240,000-265,000 TS max.
For dies for hot forming bronze, brass and other copper alloys.
Werkstoff Nr. 1.2662.

MARKER SRS.
M-1305; 0.55 C, 0.70 Cr, 0.30 Mo, 1.65 Ni, 0.10 V, bal Fe.
Hot work tool and die steel; oil hardenable to 300,000 TS.
For drop forge dies, jaws for welding machines, shears.
Werkstoff Nr. 1.2713.

MARKER SRSE.
M-1305; 0.55 C, 1.1 Cr, 0.5 Mo, 1.65 Ni, 0.10 V, bal Fe.
Hot work tool and die steel; air, oil or water hardenable to 270,000-300,000 TS.
For casting molds for zinc, tin and lead; hot forming tools.
Werkstoff Nr. 1.2714.

MARKER SRS SPEZIAL.
M-1305; 0.57 C, 1.10 Cr, 0.80 Mo, 1.7 Ni, 0.10 V, bal Fe.
Hot work tool steel.
W.-Nr. 1.2744.

MARKER SS 19.
M-1305; 0.74 C, 4.15 Cr, 1.10 V, 18.0 W, bal Fe.
High speed steel for general purpose tools for turning, planing, milling and drilling steel and non-ferrous metals.
W. Nr. 1.3355; AISI T1.

MARKER VK-25.
M-1305; 1.70 C, 25.0 Cr, bal Fe.
Rust resistant and wear resistant steel casting.

MARKER W5/0.
M-1305; 0.30 C, 0.85 Si, 1.05 Cr, 0.20 V, 3.75 W, bal Fe.
Hot work tool and die steel; oil or water hardenable to 210,000-250,000 TS.
For hot forming rivets; thread rolling screws; mandrels, jaws for electric welders.
Werkstoff Nr. 1.2564.

MARKER W 11.
M-1305; 0.30 C, 2.65 Cr, 0.35 V, 9.0 W, bal Fe.
Hot work tool steel, air or oil hardenable to 240,000-265,000 TS.
For hot stamping dies, punches, shears, hot extrusion dies.
Werkstoff Nr. 1.2581.

MARKER WAGT.
M-1305; 0.40 C, 1.25 mn, 1.85 Cr, 0.20 Mo, bal Fe.
Hot work tool and die steel; oil hardenable to 270,000 TS.
For drop hammer forging dies, hot punches, frames for casting molds.
Werkstoff Nr. 1.2311.

MARKER WAGT S.
M-1305; 0.40 C, 0.3 Si, 1.25 Mn 1.85 Cr, 0.20 Mo, 0.10 S, bal Fe.
Hot work tool steel; free machining grade of MARKER WAGT.

MARKER WM 25.
M-1305; 0.22 C, 1.4 Cr, 1.1 Mo, 0.30 V, bal Fe.
Hot work tool steel.

MARKER WM 28.
M-1305; 0.32 C, 3.0 Cr, 2.80 Mo, 0.60 V, bal Fe.
Hot work tool and die steel; oil or water hardenable to 235,000-250,000 TS.
For hot forming heavy and light metals, hot shears, thread rolling, forming railroad spikes, pressure casting molds.
Werkstoff Nr. 1.2365; AISI H 10.

MARKER WM 28 S.
M-1305; 0.25 C, 2.85 Cr, 2.55 Mo, 0.45 V, bal Fe.
Hot work tool steel.

MARKER WM 30.
M-1305; 0.40 C, 5.0 Cr, 3.0 Mo, 0.90 V, bal Fe.
Hot work tool steel.
W.-Nr. 1.2367.

MARKER WP 0.
M-1305; 0.14 C, 0.25 Si, 0.40 Mn, 0.75 Cr, 3.50 Ni, bal Fe.
Case hardening steel; for heavy sections.
Case hardness expected: C 60-64 Rock.
Core strength: 112,000-170,000 psi TS.
For large gears and shafts requiring good strength and high surface hardness.
Werkstoff Nr. 1.2735.

MARKER WP 1.
M-1305; 0.19 C, 0.25 Si, 0.40 Mn, 1.25 Cr, 0.20 Mo or W, 3.75 Ni, bal Fe.
Case hardening steel; for heavy sections.
Case hardness expected: C 60-64 Rock.
Core strength: 155,000-200,000 psi TS.
For case hardening heavy drive gears, tractor transmission gears, aircraft gears and shafts.
Werkstoff Nr. 1.2764.

MARKER ZES.
M-1305; 2.10 C, 12.0 Cr, bal Fe.
Cold work tool and die steel; air or oil hardenable to 63-65 Rock C.
For stamping and forming tools on thin metal sheet and strip; for rolls and presses for processing rubber, graphite, artificial resins and iron powder.
Werkstoff Nr. 1.2080.

MARKER ZESE.
M-1305; 1.65 C, 12.0 Cr, 0.1 V, bal Fe.
Cold work tool and die steel; air or oil hardenable to 63-65 Rock C.
For dies and shears for cutting thin sheet and strip up to 5 mm thick, for broaches.
Werkstoff Nr. 1.2201.

MARKER ZESEK.
M-1305; 1.55 C, 12.0 Cr, 0.70 Mo, 1.0 V, bal Fe.
High carbon-high chrome cold work tool steel for cutting and punching dies.
W.-Nr. 1.2379.

MARKER ZESV.
M-1305; 2.2 C, 12.5 Cr, 0.90 Mo, 2.2 V, bal.
High carbon-high chrome cold work tool steel; for punching and stamping dies.
W.-Nr. 1.2378.

MARKET BRASS.
M-U.S.; 65 Cu, 35 Zn.
For condenser tubes, hardware; corrosion resistant.

MARK EYA-1T.
M-USSR; 0.12 C, 18 Cr, 8 Ni, Ti = 7 x C, bal Fe.
Annealed: 85,000 TS; 35,000 YS; 55 El; 65 RA; 150 Brin.
For welded chemical plant equipment; Type 321; stainless, austenitic.

MARKEY BRONZE ALLOY NO M-1 ETC see **MARKEY M-1 ETC.**

MARKET M-1.
M-1671; 83 Cu, 7 Sn, 7 Pb, 3 Zn.
Cast leaded bronze; SAE 660; Copper alloy No. 932.
ASTM B144-3B.

MARKEY M-2 replaced by **MARKEY M-55.**

MARKEY M-3.
M-1671; 85 Cu, 5 Sn, 5 Zn, 5 Pb.
Cast leaded red brass; SAE 40; Copper alloy No. 836.
ASTM B145-4A.

MARKEY M-4.
M-1671; 85 Cu, 5 Sn, 9 Pb, 1 Zn.
Cast leaded bronze; SAE 66; Copper alloy No. 935.
ASTM B144-3C.

MARKEY M-6.
M-1671; 88 Cu, 10 Sn, 2 Zn.
Cast bronze; SAE 62; Copper alloy No. 905.
ASTM B143-1A.

MARKEY M-7.
M-1671; 88 Cu, 10 Sn, 2 Pb.
Cast low lead bronze; SAE 63; Copper alloy No. 927.

MARKEY M-15.
M-1671; 85 Cu, 14 Sn, 1 Zn.
Cast tin bronze; Copper alloy No. 910.

MARKEY M-16.
M-1671; 78 Cu, 6 Sn, 16 Pb.
Cast high lead tin bronze; Copper alloy No. 939.

MARKEY M-17.
M-1671; 87 Cu, 11 Sn, 1 Pb, 1 Ni.
Cast leaded tin bronze; SAE 640; Copper alloy No. 925.

MARKEY M-29.
M-1671; 81 Cu, 3 Sn, 7 Pb, 9 Zn.
Cast leaded semi-red brass; copper alloy No. 884; ASTM B145-5A.

MARKEY M-31.
M-1671; 73-77 Cu, 4.75-6.0 Sn, 18-20 Pb, 3 max Zn.
Cast high leaded tin bronze.

MARKEY M-32.
M-1671; 82.5-86.0 Cu, 9.3-11.0 Sn, 2.25-3.25 Pb, 3.0-4.0 Ni, 0.75 max P.
Leaded tin bronze casting.

MARKEY M-33.
M-1671; 86.25-89.0 Cu, 10.3-12.0 Sn, 0.5 max Pb, 1.25-2.0 Ni, 0.10-0.30 P.
Tin bronze casting.
Similar to UNS C90500 (former ASTM B-143-1A).

MARKEY M-35.
M-1671; 89 Cu, 11 Sn, 0.1-0.3 P.
Cast phosphor gear bronze; SAE 65; Copper alloy No. 907.

MARKEY M-36.
M-1671; 70 Cu, 3.5-5 Sn, 19-23 Pb. 0.50 max P, others 1.0 max. Cast high leaded tin bronze.

MARKEY M-37.
M-1671; 1.5-2.5 Sn, 28-29 Pb, 0.50 max Zn, 0.25-0.75 Ni, bal Cu. Cast copper lead alloy.

MARKEY M-38.
M-1671; 82.3-85.0 Cu, 7.25-9.0 Sn, 7.25-9.0 Pb, 0.75 max Zn, 1.0 max Ni, 0.50 max P.
Leaded tin bronze casting.
Similar to UNS C 93200 (former ASTM B-144-3B).

MARKEY M-40.
M-1671; 73.0-76.0 Cu, 5.0-6.0 Sn, 18.0-20.0 Pb, 0.50 max Zn, 0.50 max Ni, 0.05 max P.
Leaded tin bronze casting.

MARKEY M-41.
M-1671; 68-5.73.5 Cu, 4.5-6.0 Sn, 22-26 Pb.
Cast high leaded bronze; Copper alloy No. 943. ASTM B144-3E.

MARKEY M-42.
M-1671; 76-78 Cu, 2-4 Sn, 9-11 Pb, 2.0-2.5 Ni, 6-9 Zn. Cast high leaded bronze.

MARKEY M-43.
M-1671; 86-89 Cu, 9-11 Sn, 0.3 max Pb, 1-3 Zn.
Cast tin bronze; SAE 62; Copper alloy No. 905.

MARKEY M-44.
M-1671; 84-89 Cu, 9-13.5 Sn, 1.0 max Pb, 1-3 Zn.
Cast tin bronze.

MARKEY M-45.
M-1671; 86-89 Cu, 7.5-9 Sn, 0.30 Pb, 3-5 Zn, 1.0 Ni, max.
Cast in bronze; Copper alloy No. 903; ASTM B143-1B.

MARKEY M-46.
M-1671; 86-89 Cu, 5.5-6.5 Sn, 1-2 Pb, 3-5 Zn, 1.0 Ni, max.
Cast leaded tin bronze; copper alloy No. 922; ASTM B143-2A.

MARKEY M-47.
M-1671; 78-82 Cu, 2-3 Sn, 9-11 Pb, 6.5-8.5 Sn, 0.04 max P.
Cast leaded brass.

MARKEY M-50.
M-1671; 66.5-73.5 Cu, 8.5-11.5 Sn, 18-20 Pb.
Cast high leaded tin bronze; 48-78 BHN.

MARKEY M-53.
M-1671; 78.0-82.0 Cu, 6.0-8.0 Sn, 11.0-13.0 Pb, 0.75 max Zn, 0.25 max P.
Leaded tin bronze casting.

MARKEY M-55.
M-1671; 78.0-82.0 Cu, 9.5-11.0 Sn, 9.0-11.0 Pb, 1.0 max Zn, 0.50 max Ni, 0.25 max P.
Leaded tin bronze casting.
Similar to UNS C93700 (former ASTM B-144-3A).

MARKEY M-56.
M-1671; 77.25-81.0 Cu, 10.25-11.0 Sn, 8.25-9.5 Pb, 0.75 max Zn, 0.50 max Ni, 0.05 max P.
Leaded tin bronze casting.
Similar to UNS C93700 (former ASTM B-144-3A).

MARKEY M-57.
M-1671; 78-82 Cu, 4.0-6.0 Sn, 12.0-14.0 Pb, 3.0 max Zn, 1.0 max Ni.
Leaded tin bronze casting; high lead content.

MARKEY M-58.
M-1671; 75.0-79.0 Cu, 6.3-6.7 Sn, 13.0-16.0 Pb, 0.7 max Zn, 0.7 max Ni, 0.05 max P.
Leaded tin bronze casting; high lead content.
Similar to UNS C93800 (former ASTM B-144-3D).

MARKEY M-59.
M-1671; 73.0-76.5 Cu, 4.75-6.0 Sn, 18.0-20.0 Pb, 0.50-1.0 Ni, 0.05 max P.
Leaded tin bronze casting; high lead content.

MARKEY M-61.
M-1671; 66.0-74.0 Cu, 6.5-9.5 Sn, 19.5-24.5 Pb, 0.50 max Zn, 0.25 max Ni, 0.05 max P.
Leaded tin bronze casting; high lead content.

MARK I.
M-494, M-1485; 3.3 C, 1.4 Si, 0.2 P, bal Fe.
Cast: 180-250 Brin.
For medium section castings; cast iron.

MARK III.
M-494, M-1485; 3.2 C, 1.9 Si, 1.4 Ni, 0.4 Cr, bal Fe.
Cast: 220-280 Brin.
For heat treated castings, gears; cast iron.

MARK VII.
M-494, M-1485; 3.4 C, 1.4 Si, 0.2 P, bal Fe.
Cast: 170-220 Brin.
For piston rings; cast iron.

MARK U7.
M-USSR; 0.70 C, 0.25 max Mn, 0.25 max Si, bal Fe.
Heat treated: 174,000 TS; 128,000 YS; 12 El; 37 RA; 352 Brin.
For springs, rails, clutch discs; water hardened.

MARK U8.
M-USSR; 0.80 C, 0.25 max Si, 0.25 max Mn, bal Fe.
Heat treated: 188,000 TS; 143,000 YS; 12 El; 35 RA; 388 Brin.
For springs, taps, reamers, drills, hobs; water hardened.

MARKUS ALLOY.
M-Eng.; Cu, Ni, Zn.
For decorative parts; corrosion resistant.

MAR-M-247.

M-1536; 0.15 C, 8.25 Cr, 10.0 Co, 10.0 W, 0.70 Mo, 3.0 Ta, 5.5 Al, 1.0 Ti, 1.5 Hf, 0.015 B, 0.05 Zr, bal Ni.

Cast high temperature alloy; for turbine wheels and blades. (Same as M-M-0011 ALLOY).

MAR-M 905.

M-1536; 0.05 C, 20 Cr, 20 Ni, 0.5 Ti, 0.10 Zr, 7.5 Ta, bal Co.

Sheet alloy for use to 1400°F.

MAR-M ALLOY 200.

M-1536, M-1491; 0.12-0.17 C, 8-10 Cr, 11.5-13.5 W, 9-11 Co, 0.05 Zr, 0.75-1.25 Cb, 4.75-5.25 Al, 1.75-2.25 Ti, 0.015 B, 1.5 max Fe, bal Ni.

As Cast: 135,000 TS; 120,000 YS; 7 El.

At 1400°F: 135,000 TS; 122,500 YS; 3.5 El.

For turbine blades and vanes in aircraft gas turbine engines.

High strength and oxidation resistant to 1900°F.

MAR-M ALLOY 211.

M-1536, M-1491; 0.15 C, 9 Cr, 10 Co, 2.5 Mo, 5.5 W, 2.7 Cb, 2.0 Ti 5 Al, 0.015 B, 0.05 Zn, bal Ni.

For integrally cast turbine wheels and blades. Cast alloy. High stress-rupture strength. Corrosion and heat resistant.

MAR-M ALLOY 246.

M-1536, M-1491; 0.15 C, 9 Cr, 10 Co, 10 W, 2.5 Mo, 1.5 Ta, 1.5 Ti, 5.5 Al, 0.015 B, 0.05 Zr, bal Ni.

Cast: 139,000 TS; 122,000 YS; 4.5 El.

At 1600°F: 132,000 TS; 100,000 YS; 4.5 El.

For turbine vanes, nozzles, jet engine components.

Precipitation hardening. For service up to 1900°F. High oxidation resistant.

MAR-M ALLOY 302.

M-1536, M-1491; 0.78-0.93 C, 1.5 max Fe, 0.01 max B, 20-23 Cr, 9-11 W, 8-10 Ta, 0.1-0.3 Zr, bal Co.

Cast: 140,000 TS; 100,000 YS; 2 El; C 40 Rock.

At 1600°F: 56,000-78,000 TS; 40,000-50,000 YS; 7-14 El.

For turbine vanes, nozzle guide vanes and buckets in gas turbines.

High oxidation and thermal shock resistance. For service to 2100°F.

MAR-M ALLOY 322.

M-1536, M-1491; 0.9-1.1 C, 20-23 Cr, 8-10 W, 4-5 Ta, 2.0-2.5 Zr, 0.65-0.85 Ti, 1.5 max Fe, bal Co.

Cast: 121,000 TS; 91,000 YS; 3 El; 4 RA; C 36 Rock.

At 1500°F: 95,000 TS; 55,000 YS; 10 El; 10 RA.

For turbine vanes and blades, jet engine components.

High temperature strength and ductility. Oxidation resistant to 2000°F.

MAR-M ALLOY 421.

M-1536, M-1491; 0.15 C, 15.5 Cr, 1.75 Mo, 10 Co, 3.5 W, 1.75 Cb, 1.75 Ti, 4.25 Al, 0.05 Zr, 0.015 B, 1 max Fe, bal Ni.

As Cast: 132,000 TS; 115,000 YS; 6 El; 10 RA.

Heat treated: 150,000 TS; 132,000 YS; 4.4 El; 6.5 RA.

Wrought: 198,000 TS; 136,000 YS; 20 El; 26 RA.

For turbine rotors and discs, jet engine components.

Precipitation hardening, high strength and sulfidation resistant.

MAR-M ALLOY 432.

M-1536, M-1491; 20 Co, 15.5 Cr, 4.3 Ti, 2.8 Al, 3 W, 2 Ta, 2 Cb, 0.015 B, 0.05 Zr, bal Ni.

Cast: 180,000 TS; 150,000 YS; 5 El; 6.5 RA.

For integrally cast turbine wheels, jet engine components, turbine blades. Sulphidation resistant.

Resists creep rupture and hot corrosion.

MAR-M ALLOY 509.

M-1536, M-1491; 0.6 C, 23.5 Cr, 10 Ni, 7 W, 3.5 Ta, 0.2 Ti, 0.5 Zr bal Co.

At 70°F: 113,000 TS; 85,000 YS; 3.5 El; 5.8 RA.

At 1600°F: 68,000 TS; 45,000 YS; 7 El; 13 RA.

For aircraft and industrial turbines.

Cast superalloy, high strength, low creep rate. Resists shock and oxidation.

MAR-M ALLOY 918.

M-1536, M-1491; 20 Ni, 20 Cr, 7.5 Ta, 0.1 Zr, 0.05 C, bal Co.

Bar: 130,000 TS; 56,000 YS; 48 El.

For burner cans and afterburner liners in gas turbine engines.

High heat, oxidation, and corrosion resistance.

MARQUES AGPV see AGPV 11.

MARREL 5NS.

M-1180; 0.12-0.16 C, 0.2 Cr, 5-5.5 Ni, bal Fe.

Hardened: 170,000-200,000 TS; 108,000 YS; 6 El; 40 RA; 350-420 Brin.

For motor valves, boilers, gears; case hardening steel, shock resistant.

MARREL 50SS.

M-1180; 0.18-0.25 C, 0.8-1.0 Mn, bal Fe.

Normalized: 74,000-83,000 TS; 53,000 YS; 24 El; 58 RA; 149-166 Brin.

For machinery and welded parts; parts workable at low temperature.

MARREL AMMO.

M-1180; 0.20 max C, 0.9-1.2 Mn, 0.2 Cr, 0.4-0.6 Mo, bal Fe.

Normalized: 84,000-90,000 TS; 50,000 YS; 18 El; 50 RA; 150-180 Brin.

For sheet iron parts for superchargers; for use at temperatures up to 500°C.

MARREL ARM.
M-1180; 1.0-1.15 C, 1.4-1.6 Cr, bal Fe.
Hardened: 630-740 Brin.
For cutlery, stamping dies, cams, tools; oil hardening.

MARREL ASK1.
M-1180; 0.15-0.21 C, 0.6-0.9 Mn, 0.6-1.0 Cr, bal Fe.
Hardened: 135,000-165,000 TS; 86,000 YS; 8 El; 33 RA; 280-340 Brin.
For gears, shafts; case hardening steel.

MARREL ASK2.
M-1180; 0.20-0.25 C, 0.6-0.9 Mn, 0.85-1.00 Cr, bal Fe.
Hardened: 162,000-191,000 TS; 103,000 YS; 5 El; 28 RA.
For gears, shafts; case hardening steel.

MARREL ASK3.
M-1180; 0.30-0.40 C, 0.6-0.9 Mn, 0.85-1.00 Cr, bal Fe.
Hardened: 114,000-128,000 TS; 102,000 YS; 15 El; 62 RA; 230-272 Brin.
For mechanical parts, gears, shafting; oil hardening.

MARREL ASK4.
M-1180; 0.40-0.48 C, 0.6-0.9 Mn, 0.85-1.00 Cr, bal Fe.
Hardened: 128,000-142,000 TS; 115,000 YS; 14 El; 58 RA; 260-310 Brin.
For forged parts, gears; oil hardening.

MARREL ASK5.
M-1180; 0.48-0.58 C, 0.6-0.9 Mn, 0.85-1.15 Cr, bal Fe.
Hardened: 140,000-155,000 TS; 128,000 YS; 13 El; 50 RA; 290-330 Brin.
For wear rings, grinding parts; wear resistant.

MARREL ASM3.
M-1180; 0.29-0.36 C, 0.45-0.55 Mn, 0.4-0.6 Cr, 0.4-0.6 Ni, bal Fe.
Hardened: 114,000-128,000 TS; 100,000 YS; 13 El; 60 RA; 235-265 Brin.
For connecting rods, center bits, cylinders; oil hardening.

MARREL ASM4.
M-1180; 0.36-0.44 C, 0.5-0.7 Mn, 0.4-0.6 Cr, 0.4-0.6 Ni, bal Fe.
Hardened: 128,000-141,000 TS; 114,000 YS; 12 El; 56 RA; 260-295 Brin.
For mechanical parts, gears, shafts; oil hardening.

MARREL ASM5.
M-1180; 0.44-0.54 C, 0.5-0.7 Mn, 0.4-0.6 Ni, 0.4-0.6 Cr, bal Fe.
Hardened: 142,000 TS; 125,000 YS; 11 El; 50 RA; 290-330 Brin.
For gears, shafts; oil hardening.

MARREL ASMY.
M-1180; 0.35-0.40 C, 0.4-0.6 Mn, 0.4-0.6 Cr, 0.4-0.6 Ni, 0.15-0.20 Mo bal Fe.
Hardened: 121,000-135,000 TS; 114,000 YS; 15 El; 61 RA; 260-298 Brin.
For compressed gas bottles and cylinders; oil hardening.

MARREL AST.
M-1180; 0.32-0.38 C, 0.5-0.7 Mn, 1.6-2.0 Cr, 3.6-4.0 Ni, bal Fe.
Hardened: 240,000-320,000 TS; 158,000 YS; 6 El; 10 RA; 470-560 Brin.
For rollers, gears, connecting rods; oil hardening, tough.

MARREL C2N.
M-1180; 0.045-0.095 C, 0.3-0.5 Mn, 1.7-2.0 Ni, bal Fe.
Rolled: 78,000-95,000 TS; 57,000 YS; 15 El; 62 RA; 150-210 Brin.
For axles, connecting rods; case hardening steel.

MARREL C14A.
M-1180; 0.10-0.15 C, 13 min Cr, bal Fe.
Heat treated: 114,000-128,000 TS; 100,000 YS; 14 El; 50 RA; 230-270 Brin.
For pump shafts, steam turbine parts; corrosion resistant.

MARREL C14B.
M-1180; 0.15-0.23 C, 13 min Cr, bal Fe.
Hardened: 136,000-150,000 TS; 121,000 YS; 12 El; 48 RA; 270-320 Brin.
For motor valves, steam turbine parts; corrosion resistant.

MARREL C14C.
M-1180; 0.23-0.32 C, 13 min Cr, bal Fe.
Hardened: 140,000-170,000 TS; 128,000 YS; 10 El; 42 RA; 300-360 Brin.
For motor valves, pump parts; corrosion resistant.

MARREL C14M.
M-1180; 0.32-0.40 C, 13 min Cr, bal Fe.
Hardened: 165,000-200,000 TS; 142,000 YS; 8 El; 36 RA; 360-420 Brin.
For cutlery, surgical instruments; corrosion resistant, hardenable.

MARREL CH1.
M-1180; 0.08-0.11 C, 0.6-0.8 Cr, 3.3-3.7 Ni, bal Fe.
Hardened: 135,000-156,000 TS; 108,000 YS; 12 El; 48 RA; 270-340 Brin.
For gears, shafts, pinions; case hardening steel.

MARREL CH2.
M-1180; 0.11-0.15 C, 0.6-0.8 Cr, 3.3-3.7 Ni, bal Fe.
Hardened: 156,000-185,000 TS; 111,000 YS; 9 El; 33 RA; 340-430 Brin.
For gears, shafts, pinions; case hardening steel.

MARREL CH2D.
 M-1180; 0.15-0.20 C, 0.6-0.8 Cr, 3.3-3.7 Ni, bal Fe.
 Hardened: 185,000-210,000 TS; 128,000 YS; 8 El; 30 RA; 390-460 Brin.
 For gears, pinions, shafts; case hardening steel.

MARREL CH3.
 M-1180; 0.21-0.29 C, 0.7-0.9 Cr, 3.3-3.7 Ni, bal Fe.
 Hardened: 114,000-129,000 TS; 110,000 YS; 16 El; 60 RA; 230-270 Brin.
 For spindles, shafts, gears; oil hardening.

MARREL CH4.
 M-1180; 0.29-0.38 C, 0.7-0.9 Cr, 3.3-3.7 Ni, bal Fe.
 Hardened: 114,000-141,000 TS; 121,000 YS; 15 El; 58 RA; 260-300 Brin.
 For spindles, shafts, gears; oil hardening.

MARREL CNCO.
 M-1180; 0.06-0.10 C, 0.5-0.6 Cr, 2.6-3.0 Ni, bal Fe.
 Hardened: 107,000-135,000 TS; 83,000 YS; 13 El; 50 RA; 210-280 Brin.
 For gears, axles, camshafts, connecting rods; case hardening steel.

MARREL CNC1.
 M-1180; 0.10-0.13 C, 0.6-0.8 Cr, 2.6-3.0 Ni, bal Fe.
 Hardened: 135,000-165,000 TS; 90,000 YS; 10 El; 45 RA; 270-350 Brin.
 For automotive and aircraft parts, gears, shafts; case hardening steel.

MARREL CNC2.
 M-1180; 0.13-0.17 C, 0.6-0.8 Cr, 2.6-3.0 Ni, bal Fe.
 Hardened: 162,000-185,000 TS; 107,000 YS; 9 El; 33 RA; 340-430 Brin.
 For automotive and aircraft parts, gears, shafts; case hardening steel.

MARREL CNC2D.
 M-1180; 0.17-0.21 C, 0.6-0.8 Cr, 2.6-3.0 Ni, bal Fe.
 Hardened: 185,000-210,000 TS; 114,000 YS; 7 El; 30 RA; 390-460 Brin.
 For gears, shafts; case hardening steel.

MARREL CNC3.
 M-1180; 0.22-0.30 C, 0.6-0.8 Cr, 2.6-3.0 Ni, bal Fe.
 Hardened: 114,000-128,000 TS; 109,000 YS; 17 El; 66 RA; 230-270 Brin.
 For gears, shafts, axles; oil hardening.

MARREL CNC4.
 M-1180; 0.30-0.38 C, 0.5-0.7 Cr, 0.5-0.7 Mn, 2.6-3.0 Ni, bal Fe.
 Hardened: 128,000-141,000 TS; 121,000 YS; 15 El; 60 RA; 260-300 Brin.
 For gears, shafts, axles; oil hardening.

MARREL CNC5.
 M-1180; 0.38-0.46 C, 0.5-0.7 Mn, 0.5-0.7 Cr, 2.6-3.0 Ni, bal Fe.
 Hardened: 142,000-157,000 TS; 128,000 YS; 13 El; 48 RA; 300-350 Brin.
 For gears, shafts, axles; oil hardening.

MARREL CNPO.
 M-1180; 0.30-0.35 C, 0.5-0.7 Mn, 0.6-0.8 Cr, 2.5-2.8 Ni, 0.25-0.35 Mo bal Fe.
 Hardened: 135,000-150,000 TS; 128,000 YS; 14 El; 60 RA; 270-320 Brin.
 For shafts, cranks, gears; tough, oil hardening.

MARREL CNY1.
 M-1180; 0.13-0.17 C, 0.6-0.9 Mn, 0.5-0.7 Cr, 1.1-1.3 Ni, 0.16-0.20 Cr bal Fe.
 Hardened: 135,000-165,000 TS; 92,000 YS; 9 El; 38 RA; 270-340 Brin
 For gears, shafts; case hardening steel.

MARREL CNY2.
 M-1180; 0.17-0.20 C, 0.6-0.9 Mn, 1.1-1.3 Ni, 0.5-0.7 Cr, 0.16-0.20 Mo bal Fe.
 Hardened: 155,000-185,000 TS; 107,000 YS; 7 El; 30 RA; 340-390 Brin.
 For gears, crankshafts; case hardening steel.

MARREL CNY2D.
 M-1180; 0.20-0.24 C, 0.6-0.9 Mn, 0.5-0.7 Cr, 1.1-1.3 Ni, 0.16-0.20 Mo bal Fe.
 Hardened: 185,000-225,000 TS; 110,000 YS; 6 El; 28 RA; 390-460 Brin.
 For gears, crankshafts; case hardening steel.

MARREL CNY3.
 M-1180; 0.22-0.28 C, 0.6-0.9 Mn, 0.5-0.7 Cr, 1.1-1.3 Ni, 0.16-0.20 Mo bal Fe.
 Hardened: 113,000-128,000 TS; 110,000 YS; 16 El; 60 RA; 230-270 Brin.
 For center bits, shafts, gears, propellers; oil hardening.

MARREL CNY4.
 M-1180; 0.28-0.36 C, 0.6-0.9 Mn, 0.5-0.7 Cr, 1.1-1.3 Ni, 0.16-0.20 Mo bal Fe.
 Hardened: 128,000-142,000 TS; 120,000 YS; 14 El; 63 RA; 260-300 Brin.
 For shafts, gears, propellers; oil hardening.

MARREL CNY5.
 M-1180; 0.36-0.46 C, 0.6-0.9 Mn, 0.5-0.7 Cr, 1.1-1.3 Ni, 0.16-0.20 Mo bal Fe.
 Hardened: 141,000-155,000 TS; 124,000 YS; 13 El; 58 RA; 300-350 Brin.
 For shafts, axles, gears; oil hardening.

MARREL CTD.
 M-1180; 0.05-0.10 C, 0.4 Mn, bal Fe.
 Normalized: 72,000-93,000 TS; 57,000 YS; 28 El; 70 RA; 140-187 Brin.
 For gears, cams, crankshafts; case hardening steel.

MARREL M1.
M-1180; 0.65-0.75 C, 0.45-0.65 Mn, bal Fe.
Normalized: 104,000 TS; 64,000 YS; 14 El; 29 RA; 207-229 Brin.
For springs, hammers, files, chisels; oil or water hardening.

MARREL M2.
M-1180; 0.55-0.65 C, 0.5-0.7 Mn, bal Fe.
Normalized: 107,000-121,000 TS; 66,000 YS; 14 El; 36 RA; 207-258 Brin.
For hammers, springs, punches; water hardening.

MARREL M3.
M-1180; 0.45-0.55 C, 0.5-0.7 Mn, bal Fe.
Normalized: 93,000-107,000 TS; 60,000 YS; 17 El; 45 RA; 180-224 Brin.
For forgings, axes, shafts; water hardening.

MARREL M3W.
M-1180; 0.48-0.55 C, 0.45 Mn, bal Fe.
Normalized: 93,000-107,000 TS; 60,000 YS; 17 El; 45 RA; 180-224 Brin.
For cutlery, table flatware; water hardening.

MARREL M4.
M-1180; 0.32-0.38 C, 0.5-0.7 Mn, bal Fe.
Normalized: 78,000-93,000 TS; 55,000 YS; 20 El; 51 RA; 156-190 Brin.
For gears, pinions, shafts; water hardening.

MARREL M5.
M-1180; 0.18-0.23 C, 0.5-0.7 Mn, bal Fe.
Normalized: 72,000 TS; 48,500 YS; 23 El; 57 RA; 140-160 Brin.
For axles, cotter pins, bolts, forged parts; water hardening.

MARREL M6.
M-1180; 0.10-0.175 C, 0.5-0.7 Mn, bal Fe.
Normalized: 64,000-72,000 TS; 43,000 YS; 25 El; 65 RA; 120-146 Brin.
For gears, pinions, shafts; case hardening steel.

MARREL M7.
M-1180; 0.09-0.125 C, 0.5-0.7 Mn, bal Fe.
Normalized: 59,000-64,000 TS; 40,000 YS; 27 El; 72 RA; 112-128 Brin.
For bolts, rivets, stampings; water hardening.

MARREL MO.
M-1180; 0.95-1.05 C, 0.3 Mn, bal Fe.
Hardened: 160,000-114,000 TS; 100,000-72,000 YS; 10 El; 25 RA; 290-265 Brin.
For tools, dies, files, punches, hammers; water hardening.

MARREL MSG.
M-1180; 0.45-0.50 C, 0.4-0.8 Mn, 1.8-2.0 Si, bal Fe.
Hardened: 200,000-220,000 TS; 185,000 YS; 6 El; 28 RA; 420-500 Brin.
For springs; oil hardening.

MARREL NK1.
M-1180; 0.07-0.11 C, 0.6-0.9 Mn, 0.9-1.25 Cr, 1.2-1.6 Ni, bal Fe.
Hardened: 142,000-180,000 TS; 120,000 YS; 8 El; 42 RA; 290-390 Brin.
For gears, pinions, shafts, camshafts; case hardening steel.

MARREL NK2.
M-1180; 0.12-0.18 C, 0.6-0.9 Mn, 0.9-1.25 Cr, 1.2-1.6 Ni, bal Fe.
Hardened: 180,000-210,000 TS; 142,000 YS; 7 El; 40 RA; 370-450 Brin.
For aircraft gears, pinion, shafts; case hardening steel.

MARREL NK2D.
M-1180; 0.16-0.22 C, 0.6-0.9 Mn, 0.9-1.25 Cr, 1.2-1.6 Ni, bal Fe.
Hardened: 185,000-230,000 TS; 155,000 YS; 13 El; 36 RA; 390-500 Brin.
For aircraft gears, pinions, shafts; case hardening steel.

MARREL NK3.
M-1180; 0.22-0.29 C, 0.6-0.9 Mn, 0.75-1.10 Cr, 1.2-1.6 Ni, bal Fe.
Hardened: 195,000-240,000 TS; 170,000 YS; 6 El; 24 RA; 420-540 Brin.
For gears, shafts, machinery parts; oil hardening.

MARREL NK4.
M-1180; 0.3-0.38 C, 0.6-0.9 Mn, 0.75-1.1 Cr, 1.2-1.6 Ni, bal Fe.
Hardened: 225,000-280,000 TS; 200,000 YS; 5 El; 22 RA; 480-630 Brin.
For gears, pinions, axles; oil hardening.

MARREL SKM1.
M-1180; 0.11-0.15 C, 0.65-0.80 Mn, 0.85-1.15 Cr, 0.2-0.3 Mo, bal Fe.
Hardened: 135,000-170,000 TS; 86,000 YS; 10 El; 35 RA; 220-350 Brin.
For gears, pinions, cams, shafts; case hardening steel.

MARREL SKM 2.
M-1180; 0.15-0.19 C, 0.65-0.80 Mn, 0.85-1.15 Cr, 0.2-0.3 Mo, bal Fe.
Hardened: 160,000-185,000 TS; 107,000 YS; 7 El; 31 RA; 340-400 Brin.
For bolts, cylinder liners, gears; case hardening steel.

MARREL SKM 2 D.
M-1180; 0.19-0.23 C, 0.65-0.80 Mn, 0.85-1.15 Cr, 0.2-0.3 Mo, bal Fe.
Hardened: 182,000-210,000 TS; 111,000 YS; 6 El; 28 RA; 390-460 Brin.
For gears, shafts, camshafts, bolts; case hardening steel.

MARREL SKM 3.
  M-1180; 0.23-0.29 C, 0.65-0.80 Mn, 0.85-1.15 Cr, 0.2-0.3 Mo, bal Fe.
  Hardened: 114,000-128,000 TS; 100,000 YS; 15 El; 63 RA; 230-270 Brin.
  For shafts, axles, gears; oil hardening.

MARREL SKM 3 S.
  M-1180; 0.21-0.28 C, 0.65-0.80 Mn, 0.85-1.15 Cr, 0.2-0.3 Mo, bal Fe.
  Hardened: 128,000-156,000 TS; 99,500 YS; 10 El; 50 RA; 260-340 Brin.
  For gears, aircraft parts; good weldability.

MARREL SKM 4.
  M-1180; 0.32-0.38 C, 0.65-0.80 Mn, 0.85-1.15 Cr, 0.2-0.3 Mo, bal Fe.
  Hardened: 128,000-141,000 TS; 118,000 YS; 14 El; 60 RA; 260-300 Brin.
  For shafts, pneumatic hammer pistons, torsion bars; oil hardening.

MARREL SKM 5.
  M-1180; 0.38-0.48 C, 0.65-0.80 Mn, 0.85-1.15 Cr, 0.2-0.3 Mo, bal Fe.
  Hardened: 141,000-156,000 TS; 131,000 YS; 13 El; 53 RA; 300-340 Brin.
  For shafts, torsion bars, pneumatic hammer pistons; oil hardening.

MARREL SKM 5 B.
  M-1180; 0.40-0.47 C, 0.7-0.8 Mn, 1-1.3 Cr, 0.4-0.5 Mo, bal Fe.
  Hardened: 141,000-156,000 TS; 132,000 YS; 13 El; 53 RA; 300-340 Brin.
  For cutting pliers, tools, dies; oil hardening, tough.

MARREL SKMC.
  M-1180; 0.10-0.15 C, 0.65-0.90 Mn, 0.6-0.9 Cr, 0.45-0.60 Mo, bal Fe.
  Rolled: 68,000-83,000 TS; 43,000 YS; 20 El; 70 RA; 130-170 Brin.
  For tubes for high temperature and steam pipes; good weldability.

MARREL SKMO.
  M-1180; 0.08-0.11 C, 0.65-0.80 Mn, 0.85-1.15 Cr, 0.2-0.3 Mo, bal Fe.
  Hardened: 114,000-135,000 TS; 83,000 YS; 10 El; 40 RA; 235-285 Brin.
  For gears, pinions, cams, shafts; case hardening steel.

MARSHALL CRAT.
  M-1149; 0.18 C, 0.50 Mn, 0.20 Si, bal Fe.
  Annealed: 72,000 TS; 40,000 YS; 22 El; 58 RA; 140 Brin.
  For jigs, fixtures, patterns, machinery parts; Type 1018 steel, killed.

MARSH CH5.
  M-753; 0.34-0.40 C, 0.9 Si, 5 Cr, 1.25-1.5 Mo, 0.5-0.7 V, bal Fe.
  For hot forging dies, punches; hot work steel, oil hardened.

MARSH CMN.
  M-753; 0.50-0.55 C, 0.15-0.20 Si, 1.5 Cr, bal Fe.
  For drop stamping dies; oil hardened.

MARSH CTC 1.
  M-753; 0.45 C, 1.25 Cr, 2 W, 0.75 Si, bal Fe.
  For pneumatic chisels, boiler making tools; oil hardened, shock resistant.

MARSH CTH.
  M-753; 0.43 C, 0.25 Cr, 0.50 W, bal Fe.
  For hot forging dies and punches; hot work steel, oil hardened.

MARSH CTHV.
  M-753; 0.30 C, 3.4 Cr, 0.34 V, 8.4 W, bal Fe.
  For hot forging dies, punches; hot work steel, oil hardened.

MARSH CYC.
  M-753; 0.7 C, 18 W, 4 Cr, 1 V, bal Fe.
  For tools, cutters, reamers, hobs, drills; high speed steel.

MARSH CYC 5% COBALT.
  M-753; 0.70 C, 4.5 Cr, 1 Mo, 1.5 V, 18 W, 5 Co, bal Fe.
  For lathe and boring tools, drills, shapers, cutters; high speed steel.

MARSH CYC275.
  M-753; 0.8 C, 20 W, 4.5 Cr, 1.25 Mo, 1.2 V, 10 Co, bal Fe.
  For lathe and planer tools, hobs, reamers, milling cutters.

MARSH CYC EXTRA.
  M-753; 0.7 C, 18 W, 4 Cr, 1 V, 5 Co, bal Fe.
  For drills, hobs, reamers, broaches, lathe and planer cutters; high speed steel; Type T4.

MARSH CYC SPECIAL 14%.
  M-753; 0.7 C, 14 W, 4 Cr, 2 V, bal Fe.
  For lathe and planer tools, drills, reamers, taps; high speed steel.

MARSH H.F.C.
  M-753; 2.15 C, 12.5 Cr, 0.6 Si, 0.3 Mn, bal Fe.
  For forming and drawing dies, punches, shears; oil hardened, non-deforming.

MARSH M.N.
  M-753; 0.9 C, 1.7 Mn, 0.2 Si, 0.25 Cr, bal Fe.
  For dies, gauges, precision tools, taps, drills; oil hardened, non-deforming.

MARSH N3.
  M-753; 0.4 C, 3.0-3.5 Ni, 0.4 Si, 0.6 Mn, bal Fe.
  For chisels, gears, punches, crimpers; oil hardened, shock resistant.

MARSH NC.
  M-753; 0.45 C, 3.75 Ni, 1.1 Cr, 0.5 Mn, 0.35 Si, bal Fe.
  For chisels, snaps, gears, crimpers, punches; oil hardened, shock resistant.

MARSH NC1.
 M-753; 0.40 C, 1.5 Ni, 1 Cr, bal Fe.
 For spring collets; oil hardened, tough.

MARSH PATENT.
 75 Ni, 25 Cr.
 For electrical resistances, heating elements; heat resistant.

MARSH SPECIAL.
 M-753; 1.2 C, 0.3 Mn, bal Fe.
 For wood working tools, drills; water hardened; Type W1.

MARSH WTS (0.70C).
 M-753; 0.70-0.80 C, 0.10-0.20 Si, 0.30-0.40 Mn, bal Fe.
 Heat treated: 185,000 TS; 140,000 YS; 13 El; 37 RA; 375 Brin.
 For lathe centers, vice jaws, press tools; Type W1; water hardened.

MARSH WTS (0.95C).
 M-753; 0.95-1.05 C, 0.30-0.40 Mn, 0.1-0.2 Si, bal Fe.
 Annealed: 100,000 TS; 53,000 YS; 21 El; 42 RA; 200 Brin.
 For blanking tools, punches, shears, press tools; Type W1; water hardened.

MARS SUPERIEUR.
 M-1120; 0.7 C, 18 W, 4 Cr, 1 V, bal Fe.
 For tools, dies, cutters; high speed steel.

MARTIES ALLOY.
 18 Zn, 35 Ni, 10 Fe, 10 Sn, bal Cu.
 For resistances, heat and corrosion resistant parts; heat and corrosion resistant.

MARTIES' NON-OXIDIZABLE ALLOY.
 35 Ni, 18 Zn, 17 Cu, 10 Fe, 10 Sn.
 For heat and corrosion resistant parts; also called "Marlies."

MARTIES' NON-OXIDIZABLE ALLOY.
 35 Ni, 10 Fe, 10 Zn, 20 Sn, 25 brass.
 For heat and corrosion resistant parts; heat and corrosion resistant.

MARTIN.
 M-157; 1.4-1.6 C, 12-14 Cr, 0.6-0.8 Co, 0.8-0.9 Mo, 0.35-0.40 V, bal Fe.
 For dies, rotary shears, gages, punches, shear blades, liners for dust mills; castings; cast to shape; air hardening.

MARTINEL.
 M-348; C, 0.6-0.9 Mn, 2.75 min, Ni, bal Fe.
 85,000 TS; 50,000 YS; 20 El; 40 RA.
 For shipbuilding purposes in strength members, flat plate keel, bilge, deck and side plates; special method of rolling.

MARTINSITE M130.
 M-67; 0.04-0.22 C, 0.20-0.60 Mn, bal Fe.
 130,000 psi TS; (900 MPa).
 For fasteners, tubing, and miscellaneous parts.

MARTINSITE M160.
 M-67; 0.04-0.22 C, 0.20-0.60 Mn, bal Fe.
 160,000 psi TS (1100 MPa).
 For fasteners, various automotive and appliance parts.

MARTINSITE M190.
 M-67; 0.04-0.22 C, 0.20-0.60 Mn, bal Fe.
 190,000 psi TS (1300 MPa).
 For miscellaneous parts.

MARTINSITE M220.
 M-67; 0.04-0.22, 0.20-0.60 Mn, bal Fe.
 220,000 psi TS (1500 MPa).
 Fasteners and miscellaneous parts for appliances and automotive.

MARTIN STEEL.
 M-Eng.; 0.73 C, 0.4 Si, bal Fe.
 For blacksmith tools, dies, hammers; water hardening.

MARWEDUR-F11.
 M-1381; 0.16-0.23 C, 12 Cr, 0.5 Ni, 1.0 Mo, 0.5 W, 0.3 V, 0.6 Mn, 0.3 Si, bal Fe.
 For furnace parts, oil refinery equipment, cutlery.
 Corrosion and heat resistant, hardenable.

MARWEDUR-F-12.
 M-1381; 0.20 C, 12 Cr, 1 Mo, 0.4 Ni, 0.3 V, bal Fe.
 Annealed: 95,000 TS; 40,000 YS; 25 El; B 92 Rock.
 Hardened: 240,000 TS; 195,000 YS; 10 El; C 50 Rock.
 For valves, cutlery, bearings, surgical instruments.
 Corrosion resistant, hardenable.

MARVAC 250 ETC see **LESCALLOY MARVAC 250 ETC.**

MARVAL see **AUBERT & DUVAL MARVAL.**

MARVEL.
 M-115, M-342; 0.33 C, 10 W, 3.5 Cr, 0.45 V, bal Fe.
 For cutting tools, forming dies, punches, shears; hot die steel.

MARWIN.
 M-848; 0.25 C, 0.3 Si, bal Fe.
 For machinery parts; free machining.

MARWIN SUPASTUFF.
 M-848; 0.3 C, bal Fe.
 For machinery parts, bolts; water hardened.

MASSALLOY.
 M-467; C, Cr, Ni, Mo, bal Fe.
 Heat treated: 135,000 TS; 110,000 YS; 17 El; 30 RA; 286 Brin.
 For steel castings; oil hardened.

MASSILLON.
  M-467; C, Mn, bal Fe.
  Normalized: 100,000 TS; 60,000 YS; 20 El; 35 RA; 207 Brin.
  For stokerworms; cast steel.

MASTALLOY 1151 ETC see MAGNALLOY 1151 ETC.

MASTER.
  M-358; C, Mn, bal Fe.
  For tools and dies; water hardening.

MASTER ALLOY 123.
  M-91; 12.45 Al, 0.12 Mg, 0.30 max Cu, 0.10 max Fe, bal Zn.
  For remelting; one part MASTER ALLOY 123 to two parts pure Zn to produce ZAMAK 3.

MASTER METAL.
  M-565; Sn, Cu, Pb.
  For machinery bearings; Babbitt.

MATADOR 91.
  M-1348; 0.82 C, 4.1 Cr, 0.85 Mo, 1.6 C, 8.7 W, bal Fe.
  For lathe and planer tools, reamers, hobs, drills; high speed steel.

MATADOR 122.
  M-1348; 0.86 C, 4.1 Cr, 0.85 Mo, 2.5 V, 12 W, bal Fe.
  For lathe and planer tools, drills, taps, hobs; high speed steel.

MATADOR 181.
  M-1348; 0.74 C, 4.1 Cr, 1.1 V, 18.5 W, bal Fe.
  For lathe and planer tools, reamers, drills, broaches, taps; high speed steel.

MATADOR 185.
  M-1348; 0.79 C, 4.7 Co, 4.3 Cr, 0.75 Mo, 1.5 V, 18 W, bal Fe.
  For drills, taps, reamers, hobs, broaches; high speed steel.

MATADOR 333.
  M-1348; 0.95 C, 4 Cr, V, W, Mo, bal Fe.
  For cutters, taps, broaches, reamers, drills; high speed steel.

MATHESIUS METAL.
  Pb, 3 Ca, 1-2 alkali earth metals.
  For bearings; anti-friction.

MATOBAR.
  M-1724.
  Hard drawn steel wire or bar.

MATRIX-35.
  M-1766; 0.33 C, 0.35 Mn, 1.0 Si, 1.2 Cr, 4.0 W, 0.25 Mo, bal Fe.
  Hot work tool steel; dies, molds.
  IHA F-527; DIN 30 WCrV15.

MATRIX-54.
  M-1766; 0.55 C, 0.40 Si, 0.60 Mn, 1.10 Cr, 0.50 Mo, 0.15 V, bal Fe.
  Special purpose steel for hot work operation; drop hammer dies for deep engraving.
  IHA F-528; DIN 56 NiCrMoV7.

MATRIX BRASS.
  62 Cu, 37 Zn, 1.5 Pb.
  For engraving, hardware; free-cutting.

MATRIX-II.
  M-115; 0.50-0.55 C, 12 W-Mo-Cr-V, bal Fe.
  Annealed: 95,000 TS; 48,000 YS; 25 El; 55 RA; B 90 Rock.
  Heat treated: 350,000 TS; 290,000 YS; 8 El; 33 RA; C 60 Rock.
  Heat treated: 404,000 TS; 363,000 YS; 18 RA; 6 El.
  For rocket motor cases, high speed rotors, gears, pressure vessels, engine mounts, nuclear applications, molds.
  High temperature strength and heat resistance.

MATRIX-W.
  M-1766; 0.30 C, 0.40 Mn, 0.5 max Si, 2.5 Cr, 9.0 W, 0.25 V, bal Fe.
  Hot work tool steel; for hot blanking dies, punches, mandrels, forming dies.
  IHA F-526; similar to AISI H21.

MATTHEY 3.
  M-200; Cu, plus Cr.
  Cast, for resistance welding electrodes for mild steel; conductivity: 80% IACS.

MATTHEY 20S.
  M-200; 27.5 Ag, 72.5 W.
  Sintered; for spot welding electrodes.
  Conductivity: 43% IACS.

MATTHEY 35S.
  M-200; 35 Ag, 64 W.
  Sintered; for electrical contacts.
  Conductivity: 52% IACS.

MATTHEY 50S.
  M-200; 49 Ag, 51 W.
  Sintered; for electrical contacts.
  Conductivity: 61% IACS.

MATTHEY 53.
  M-200; Cu, plus Si, Ni.
  For welding electrodes.

MATTHEY 73.
  M-200; 1.8 Be, 0.1 Co, bal Cu.
  Beryllium copper castings; conductivity: 20% IACS.

MATTHEY 100.
  M-200; Cu, plus Be, Co.
  Beryllium copper casting; conductivity: 45% IACS.

MATTHEY 328.
M-200; Cu, plus Cr, Zr.
For resistance welding electrodes; conductivity: 80% IACS.

MATTHEY 1000.
M-200; W, Cu.
Sintered alloy for spot welding electrodes; conductivity: 14% IACS.

MATTHEY A 5.
M-200; Pb, plus Ag.
Lead base soft solder; melt range 304-370°C.

MATTHEY A 25.
M-200; Pb, plus Ag.
Lead base soft solder; melt point 304°C.

MATTHEY D54.
M-200; Ag, Cd.
For electrical contacts; conductivity: 82% IACS.

MATTHEY D54L.
M-200; Ag, Cd.
For electrical contacts; conductivity: 82% IACS.

MATTHEY D54X.
M-200; Ag, Cd.
For electrical contacts; conductivity: 82% IACS.

MATTHEY D55X.
M-200; Ag, Cd.
For electrical contacts; conductivity: 75% IACS.

MATTHEY D56.
M-200; Ag, Ni,
For electrical contacts; conductivity: 72% IACS.

MATTHEY D58/1.
M-200; 1 Graphite, bal Ag.
For sliding contacts; conductivity: 96% IACS.

MATTHEY D58/2.
M-200; 2 Graphite, bal Ag.
For sliding contacts; conductivity: 87% IACS.

MATTHEY D510.
M-200; Ag, Ni.
For electrical contacts; conductivity: 87% IACS.

MATTHEY D520.
M-200; Ag, Ni.
For electrical contacts; conductivity: 57% IACS.

MATTIBEL "9".
M-200; 78.3 Au, plus Pt.
For dental purposes.

MATTIBRAZE 34.
M-200; Ag, Cu, Zn.
Brazing alloy; melting range: 612-668°C.

MATTIDENT "E".
M-200; 63 Au, plus Pt.
For dental purposes.

MATTIDENT "R".
M-200; 76 Au, plus Pt.
For dental purposes.

MATTINAX "9A".
M-200; 92 min Au.
For dental purposes.

MAUSTINOX A.
M-1509; 0.12 max C, 18 Cr, 8 Ni, bal Fe.
Annealed: 70,000-85,000 TS; 30,000-37,000 YS; 30 El; 140-190 Brin.
For chemical and textile plant equipment; stainless, austenitic.

MAUSTINOX B.
M-1509; 0.12 max C, 17 Cr, 9 Ni, 2.5 Mo, bal Fe.
Annealed: 71,000-92,000 TS; 30,000-37,000 YS; 30 El; 150-200 Brin.
For acid resistant and chemical plant equipment; stainless, austenitic.

MAUSTINOX C.
M-1509; 0.12 max C, 20 Cr, 8 Ni, 2.5 Mo, bal Fe.
Annealed: 71,000-85,000 TS; 40,000-50,000 YS; 25 El; 150-200 Brin.
For acid resistant chemical plant equipment; stainless, austenitic.

MAUSTINOX D.
M-1509; 0.20 max C, 25 Cr, 5 Ni, bal Fe.
Annealed: 92,000 TS; 65,000 YS; 16 El; 190-200 Brin.
For chemical plant equipment, furnace parts; corrosion and heat resistant.

MAUSTINOX F.
M-1509; 0.20 max C, 24 Cr, 13 Ni, bal Fe.
Annealed: 78,000-100,000 TS; 30,000-37,000 YS; 30 El; 150-200 Brin.
For furnace parts, salt pots, heat treat boxes; heat and corrosion resistant, austenitic.

MAUSTINOX H.
M-1509; 0.20 max C, 25 Cr, 20 Ni, bal Fe.
Annealed: 65,000-92,000 TS; 30,000-37,000 YS; 30 El; 140-180 Brin.
For oil refinery equipment, furnaces, salt pots, heat treat boxes; austenitic, heat and corrosion resistant.

MAUSTINOX SA.
M-1509; 0.12 max C, 18 Cr, 8 Ni, Ti, bal Fe.
Annealed: 75,000-85,000 TS; 30,000-37,000 YS; 30 El; 140-190 Brin.
For welded chemical and textile plant equipment, tanks, agitators; stainless, austenitic, stabilized.

MAUSTINOX SB.
M-1509; 0.12 max C, 17 Cr, 9 Ni, 2.5 Mo, Ti, bal Fe.
Annealed: 71,000-92,000 TS; 30,000-37,000 YS; 30 El; 150-200 Brin.
For welded acid resistant and chemical plant equipment; stainless, austenitic, stabilized.

MAUSTINOX Y.
M-1509; 0.12 max C, 20 Cr, 20 Ni, 4.5 Mo, bal Fe.
Annealed: 65,000-78,000 TS; 30,000-40,000 YS; 32 El.
For heat and corrosion resistant parts; austenitic.

MAUSTINOX X.
M-1509; 0.12 max C, 15 Cr, 30 Ni, Mo, bal Fe.
Annealed: 65,000-74,000 TS; 29,000-36,000 YS; 35 El.
For heat and corrosion resistant parts; austenitic.

M-A, VM.
M-115; 0.53 C, 12 W, Mo, Cr, V, bal Fe.
Hardened: 361,000 TS; 292,000 YS; 20 RA; 6 El; 130,000 fatigue strength.
For airframes, fasteners, pressure vessels, gears, engine mounts.
Tough and fatigue resistant.

MAX. 4.
M-210; 0.15 C, 20 Ni, 25 Cr, 0.6 Mn, 1 Si, bal Fe.
Annealed: 95,000 TS; 45,000 YS; 50 El; B 90 Rock.
For furnace and heat treating equipment, furnace linings, heat exchangers, retorts.
Type 310 stainless; austenitic, corrosion and heat resistant.

MAXAL CO.
M-Italy; 24 Co, 14 Ni, 8 Al, 3 Cu.
For permanent magnet, magnetic and electrical equipment.
High permeability.

MAXCHIP NO. 1.
M-210; C, Ni, Cr, bal Fe.
For hand tools, chisels, punches; air hardened, tough, shock resistant.

MAXCHIP NO. 2.
M-210; C, Cr, bal Fe.
For chisels, punches; oil hardened, tough, shock resistant.

MAX-EL 1-B.
M-38; 0.2 C, 1 Mn, 0.2 Mo, bal Fe.
Rolled: 70,000 TS; 50,000 YS; 30 El; 55 RA; 156 Brin.
Heat treated: 135,300 TS; 114,000 YS; 18 El; 54 RA; 332 Brin.
For shafts, gears, pinions, bolts, studs; case hardening.

MAX-EL 2-B.
M-38; 0.4 C, 1 Mn, 0.2 Mo, bal Fe.
Rolled: 90,000 TS; 55,000 YS; 20 El; 45 RA.
Heat treated: 119,500 TS; 85,000 YS; 20 El; 53 RA; 251 Brin.
For shafts, spindles, worms, racks, gears, pinions; oil hardening.

MAX-EL 3 1/2.
M-38; 0.5 C, 1.25 Mn, 0.6 Cr, 0.15 Mo, bal Fe.
Heat treated: 230,000 TS; 225,000 YS; 9.5 El; 36 RA; 514 Brin.
For screws, racks, shafts, gears, pinions, spindles; oil hardening.

MAX-EL BRAKE DIE.
M-38; 0.5 C, 1.5 Mn, 0.18 Mo, 0.65 Cr, 0.08 S, bal Fe.
For press brake dies, vee dies, bending and flanging dies; free-cutting, wear and impact resistant.

MAXHETE-3.
M-617; M-210; C, high Cr, bal Fe.
For rabble arms and blades in roasting furnaces; heat resistant to 1150°C.

MAXHETE-6.
M-617, M-210; C, Ni, Cr, bal Fe.
For furnace skids, superheater supports; heat resistant, austenitic.

MAXHETE NO. 1.
M-210; 0.35 C, 18.5 Cr, 8.5 Ni, 2.5 W, bal Fe.
Cast: 137,000 TS; 35 El.
At 900°C: 38,000 TS; 47 El.
For burner tubes, pipe unions; austenitic, heat resistant to 1000°C.

MAXHETE NO. 1A.
M-210; 0.4 max C, 13.5 Cr, 13.5 Ni, 2.2 W, 1.5 Si, 0.6 Mn, bal Fe.
At 20°C: 106,000 TS; 27 El; 44 RA.
At 90°C: 61,000 TS; 45 El; 54 RA.
For furnace parts, heat treating boxes, nozzles; heat resistant, austenitic.

MAXHETE NO. 2.
M-210; 0.25 max C, 25.0 Cr, 12.0 Ni, 0.6 Mn, 0.85 Si, bal Fe.
For furnace parts, heat treat equipment, conveyors; resists heat and scaling to 1100°C.

MAXHETE NO. 4.
M-210; 0.10 max C, 25.0 Cr, 33.5 Ni, 0.4 Mn, 0.75 Si, bal Fe.
For furnace parts, hearth plates, grids; austenitic, high temperature strength.

MAXHETE NO. 5.
M-210; 0.2 max C, 65 Ni, 15 Cr, 0.85 Mn, bal Fe.
For carburizing boxes, retorts, oil burner parts; high corrosion and cyclic heat resistance.

MAXHETE NO. 7.
M-210; 0.12 C, 20 Cr, 0.5 Si, bal Fe.
For furnace parts and equipment, heat treating boxes.
Heat resistant. Maximum operating temperature 1150°C.

MAXHETE NO. 8.
M-210; 0.10 max C, 79 Ni, 19.5 Cr, 0.85 Mn, bal Fe.
For electrical resistances; high heat resistance.

MAXI-FORM 50.
M-97; 0.12 C, 0.90 Mn, 0.01 min Cb, 0.02 min Al, bal Fe.
Plate: 90,000 psi TS; 50,000 psi min YS; 18 El.
High strength, low alloy steel.

MAXI-FORM 60.
M-97; 0.12 C, 1.20 Mn, Cb, V, Al, bal Fe.
As rolled: 70,000 psi min TS; 60,000 psi min YS; 23 El.
High strength low alloy steel.

MAXI-FORM 70.
M-97; 0.12 C, 1.40 Mn, Cb, V, Al, bal Fe.
As rolled: 80,000 psi min TS; 70,000 psi min YS; 20 El.
High strength low alloy steel.

MAXI-FORM 80.
M-97; 0.09 C, 1.60 Mn, 0.60 Si, 0.06-0.15 Cb, 0.08 max V, 0.02 min Al, bal Fe.
As rolled: 90,000 psi min TS; 80,000 psi min YS; 18 El.
High strength low alloy structural steel.

MAXILVRY.
M-210; 0.12 C, 18.5 Cr, 8.5 Ni, bal Fe.
Annealed: 83,000 TS; 38,000 YS; 48 El; 60 RA; 143 Brin.
Cold drawn: 140,000 TS; 100,000 YS; 21 El; 45 RA; 190 Brin.
For fittings, cutlery, valve plates, chemical retorts; Type 302; stainless, austenitic.

MAXILVRY ADS.
M-210; 0.1 C, 12 Cr, 12 Ni, 0.5 Cu, bal Fe.
Annealed: 90,000 TS; 38,000 YS; 40 El; 45 RA; 160 Brin.
For deep drawn parts, chemical plant equipment; stainless, austenitic.

MAXILVRY AM.
M-210; 0.07 C, 18 Cr, 13 Ni, 0.2 Cu, 3 Mo, bal Fe.
Annealed: 100,000 TS; 40,000 YS; 40 El; 45 RA; 190 Brin.
For chemical equipment to resist H_2SO_4, mixers, tanks; Type 316; stainless, austenitic.

MAXILVRY AT.
M-210; 0.2 C, 18 Cr, 8 Ni, bal Fe.
Annealed: 100,000 TS; 40,000 YS; 40 El; 45 RA; 190 Brin.
For chemical plant equipment; Type 302; stainless, austenitic.

MAXILVRY AWP.
M-210; 0.10 max C, 18 Cr, 8.5 Ni, 0.5 Ti, bal Fe.
Annealed: 100,000 TS; 40,000 YS; 40 El; 45 RA; 190 Brin.
For chemical plant equipment; Type 321; stainless, austenitic.

MAXILVRY C.B.
M-210; 0.10 C, 18.5 Cr, 8.5 Ni, 1 Cb, bal Fe.
Annealed: 78,000-100,000 TS; 33,000-40,000 YS; 40-60 El; 45-65 RA; 160-195 Brin.
For impellers, tanks, chemical plant equipment; Type 347; stainless, austenitic.

MAXILVRY M.B.T.
M-210; 0.04 C, 17.5 Cr, 12 Ni, 2.75 Mo, 0.75 Cb, bal Fe.
Annealed: 78,000-100,000 TS; 45,000-67,000 YS; 60-40 El; 75-60 RA; 170-200 Brin.
For chemical plant equipment, welded structures; stainless, austenitic.

MAXILVRY ML.
M-210; M-617; C, Ni, Cr, bal Fe.
For cutlery, valves; stainless.

MAXILVRY SPECIAL.
M-210; 0.06 C, 18.5 Cr, 8.5 Ni, bal Fe.
Annealed: 78,000-100,000 TS; 33,000-40,000 YS; 60-40 El; 65-45 RA; 160-190 Brin.
For chemical plant equipment, mixers, agitators, tanks; austenitic, stainless; Type 304.

MAXIMOLD.
M-289; 0.4 C, 5.2 Cr, 1.0 Si, 1.0 V, 1.2 Mo, bal Fe.
For Al and Zn die casting dies; resists heat checks, air hardened.

MAXIMUM.
M-1322; 0.74 C, 4.1 Cr, 1.1 V, 18.5 W, bal Fe.
For lathe and planer tools, drills, reamers, taps; high speed steel.

MAXIMUM SPEZIAL 30.
M-1322; 0.86 C, 2.8 Co, 4.3 Cr, 0.85 Mo, 2.1 V, 12 W, bal Fe.
For lathe and planer tools, drills, reamers, taps; high speed steel.

MAXIMUM SPEZIAL 55.
M-1322; 0.76 C, 10 Co, 4.2 Cr, 0.8 Mo, 1.8 V, 18 W, bal Fe.
For lathe and planer tools, taps, broaches, drills; high speed steel.

MAXIMUM SPEZIAL 55G.
M-1332; 0.80 C, Co, Cr, W, Mo, V, bal Fe.
For lathe and planer tools, reamers, broaches; high speed steel.

MAXIMUM SPEZIAL G.
M-1322; 0.82 C, 4.1 Cr, 0.85 Mo, 1.6 V, 8.7 W, bal Fe.
For lathe and planer tools, reamers, broaches, taps; high speed steel.

MAXIMUM SPEZIAL G EXTRA.
M-1322; 0.86 C, 4.1 Cr, 0.85 Mo, 2.5 V, 12 W, bal Fe.
For lathe and planer tools, drills, taps, hobs; high speed steel.

MAXIMUM SPEZIAL MO.
M-1322; 0.95 C, Cr, W, Mo, V, bal Fe.
For lathe and planer tools, drills; high speed steel.

MAXINIUM STEEL.
M-210; 0.32 C, 1.2 Si, 5.25 Cr, 5.0 W, 0.5 Mo, bal Fe.
For die casting dies, molds. Oil hardening.

MAXITE.
M-35; 0.8 C, 14 W, 4 Cr, 2 V, 5.2 Co, 0.6 Mo, bal Fe.
Annealed: 228 Brin.
For cutting tools, drills, reamers, boring and shaping tools.
Type T8 high-speed steel. High red hardness.

MAXITE 15.
M-35; 1.57 C, 12.65 W, 0.65 Mo, 4.75 Cr, 5.0 V, 5.0 Co, bal Fe.
For milling cutters, lathe tools, shaper tools, broaches.
Type T15 high speed tool steel.

MAXMITH.
M-210; 0.40 C, 0.60 max Mn, 3.25 Ni, bal Fe.
For chisels, riveting tools, caulking and beading tools; requires no tempering, tough.

MAXNAP.
M-210; 0.30 C, 0.60 Mn, 1.05 Cr, 0.2 V, bal Fe.
For pneumatic hammers, riveters, nut piercers; tough and fatigue resistant.

MAXTACK.
M-365; 2.25 C, 10 W, 2 Cr, 2.5 Mn, 1.0 Si, bal Fe.
For tools, cutters; oil or air hardened.

MAXTACK.
M-210; C, Cr, W, Mn, bal Fe.
For nail and tack dies, wood working tools; air hardened.

MAXTENSILE.
M-163; 3.3 C, 2.4 Si, 1.5 Ni, 0.8 Cr, bal Fe.
For hydraulic castings, sliding parts, spindles, couplings, sprockets, mill rolls; discontinued; alloy cast iron.

MAXTUFF.
M-289; 0.50 C, 0.3 Mn, 0.75 Si, 1.15 Cr, 0.2 V, 2.50 W, bal Fe.
Type S1 shock resisting tool steel.

MAYARI R.
M-24; 0.12 max C, 0.5-1.0 Mn, 0.5-0.7 Cu, 0.25-0.75 Ni, 0.2-1.0 Cr, 0.08-0.12 P, bal Fe.
Rolled: 70,000 TS; 50,000 YS; 22 El; 150 Brin.
For transportation equipment, structures; good weldability.

MAYARI R-50.
M-24; 0.10-0.20 C, 0.75-1.25 Mn, 0.2-0.4 Cu, 0.4-0.7 Cr, 0.25-0.50 Ni 0.01-0.10 V, bal Fe.
Plate: 50,000 min YS; 70,000 min TS; 21 min El.
For structural railroad and general manufacturing applications.
Low alloy-high tensile structural steel.

MAYARI R-60.
M-24; 0.1-0.2 C, 0.75-1.35 Mn, 0.2-0.4 Cu, 0.4-0.7 Cr, 0.25-0.50 Ni, 0.01-0.10 V, bal Fe.
Plate: 60,000 min YS; 80,000 min TS; 16 min El.
For structural, architectural, railroad and general manufacturing applications.
Low alloy-high strength structural steel.

MAZAK.
M-351; Al, Cu, bal Zn.
For die castings.

MAZAK 2.
M-414, M-87; 4.1 Al, 2.7 Cu, 0.03 Mg, bal Zn.
Die cast: 47,300 TS; 83 Brin.
For die castings free from hot shortness; IZ-15, C.U.S. 93100.

MAZAK 3.
M-137; 3.9-4.3 Al, 0.03-0.06 Mg, bal Zn.
Die cast: 42,000 TS; 15 El; 80 Brin.
For instrument cases, housings, ornamental grills; impact strength and dimensional stability.

MAZAK 3.
M-414; 4 Al, 0.04 Mg, bal Zn.
For gears, frames, hardware, die castings.

MAZAK 5.
M-414; 4 Al, 1 cu, 0.3 Mg, bal Zn,
Die cast: 42,00 TS; 3 El; 65 Brin.
For gears, die castings.

MAZAK 5.
M-137; 3.9-4.3 Al, 0.75-1.25 Cu, 0.03-0.06 Mg, bal Zn.
Die cast: 42,000 TS; 3 El; 65 Brin
For motor frames, gears, instrument cases; corrosion resistant to atmosphere.

MAZAK 6.
M-414; 4 Al, 1.25 Cu, bal Zn.
Die cast: 42,000 TS; 5 El; 65 Brin.
For die castings; maximum fluidity.

MC 102.
M-86; 0.02-0.06 C, .1-0.4 Si, 0.1-0.5 Mn, 4.0 max Fe, 19.0-20.5 Cr, 5.0 max Co, 5.5-6.5 Mo, 6.2-6.7 Cb, 2.0-3.0 W, bal Ni.
Heat treated: 103,000 TS; 85,000 YS; 10 El; 340 DPN.
For gas turbine stator blades, diesel hot plugs, turbine rotors.
Good oxidation resistance and strength to 900°C.
High temperature casting alloy. Age-hardenable.

MCADAMITE.
12-18 Zn, 3.1 Cu, 0.2 Mg, bal Al.
For strong light alloy parts; non-hardenable.

MCADAMS ALLOY "A".
60 Al, 11-55 Cu, 10-43 Cr, 20 Zn.
For light alloy parts; non-hardenable.

MCADAMS ALLOY "B".
69 Al, 7.7 Cu, 0.6 Ni, 23 Zn.
For light alloy parts; non-hardenable.

MCADAMS ALLOY "C".
70 Al, 3 Cu, 22 Zn, 5 Sb.
For light alloy parts; non-hardenable.

MCADAMS ALLOY "D".
80 Al, 8 Cd, 4 Ag, 8 Sn.
For light alloy parts; non-hardenable.

MCADAMS ALLOY "E".
82 Al, 12 Cu, 5 Cd, 1 Ag.
For light alloy parts; non-hardenable.

MCADAMS ALLOYS.
82-60 Al, 0-55 Cu, 0-8 Cd, 0-4 Ag, 0-8 Sn, 0-43 Cr, 0-23 Zn, 0-5 Sb
For light alloy parts.

M.C.C.
M-Eng.; 10 Cu, 1.3 Mg, 2.0 Ni, 0.5 Fe, 0.5 Si, 2.5 Cr, bal Al.
For pistons; self-aging.

MCFARLAND & HARDER ALLOY "A".
10 Cr, 48 Ni, 43 Cu.
For heat and corrosion resistant parts; stainless and corrosion resistant.

MCFARLAND & HARDER ALLOY "B".
16 Cr, 29 Ni, 55 Cu.
For heat and corrosion resistant parts; stainless and corrosion resistant.

MCFARLAND & HARDER ALLOY "C".
30 Cr, 59 Ni, 11 Cu.
For heat and corrosion resistant parts; stainless and corrosion resistant.

MCFARLAND & HARDER ALLOY "D".
43 Cr, 46 Ni, 11 Cu.
For heat and corrosion resistant parts; stainless and corrosion resistant.

MCFARLAND & HARDER ALLOYS.
59-29 Ni, 11-55 Cu, 10-43 Cr.
For corrosion and heat resistant parts; stainless and corrosion resistant.

M-CHROME.
M-73; 1.0 C, 1.5 Cr, bal Fe.
For tools, dies, gauges; deep hardening.

MCKAY 1.
M-849; 2.0 C, 30.0 Cr, 50.0 Co, 12.0 W.
Coated electrode. For hard-surfacing. Hardness about 52 Rc; corrosion and wear resistant; retains hardness above 1500°F. For rocker arms, cams, wire drawing blocks, slag ladles.
Meets AWS A5.13; Class E CoCr-C.

MC KAY 6.
M-849; 1.0 C, 30.0 Cr, 61.0 Co, 4.5 W.
Coated electrode.
For hard surfacing. Hardness about 42 Rc; corrosion and metal-to-metal wear resistant, retains hardness above 1200°F. For exhaust valves, pistons, steam valves, hot punches.

MCKAY 16-8-2.
M-849; 0.06 C, 1.80 Mn, 0.25 Si, 15.8 Cr, 8.2 Ni, 1.6 Mo, bal Fe.
Bare wire for welding.

MCKAY 16-8-2.
M-849; Coated stainless steel weld rod.
Typical deposit analysis: 0.08 C, 2.25 Mn, 0.35 Si, 15.3 Cr, 8.2 Ni, .5 Mo.
Welded: 94,000 TS; 70,500 YS; 40 El.
For welding Types 316, 317 & 347 stainless steels for high temperature operation.
AWS A5-4; ASTM A298.

MC KAY 17-4 PH see MC KAY 630.

MCKAY 17-4 PH.
M-849; 0.04 C, 0.60 Mn, 0.50 Si, 16.7 Cr, 4.5 Ni, 0.40 Cb, 3.5 Cu, bal Fe.
Bare wire for welding.

MCKAY 17-4 PH.
M-849; Coated stainless steel weld rod.
Typical deposit analysis: 0.035 C, 0.48 Mn, 0.40 Si, 16.35 Cr, 4.9 Ni, 0.24 Cb, 3.6 Cu.
Weld metal strength depends on heat treatment.
AMS 5827.

MCKAY 21.
M-849; 0.25 C, 28.0 Cr, 56.0 Co, 5.75 Mo.
Coated electrode.
For build-up requiring corrosion resistance and good strength at elevated temperatures up to 1500°F. For hot shears, hot trim dies, hot extrusion dies, pressure valves.

MCKAY 25.
M-849; 0.05 C, 20.5 Cr, 45.0 Co, 15.5 W, 10.0 Ni.
Coated electrode.
For build-up requiring good wear and corrosion resistance and strength at temperatures to and above 1200°F. Machinable with carbide tools.

MCKAY 308.
M-849; 0.040 C, 1.85 Mn, 0.45 Si, 20.5 Cr, 9.6 Ni, bal Fe.
Bare wire for welding.
AWS ER308.

MCKAY 308.
M-849; Coated stainless steel weld rod.
Typical deposit analysis: 0.06 C, 1.8 Mn, 0.50 Si, 19.8 Cr, 10.0 Ni.
Welded: 86,000 TS; 62,000 YS; 45 El.
For welding or build-up on austenitic 18-8 stainless steel. AWS E308.

MCKAY 308 HC.
M-849; Coated stainless steel weld rod.
Typical deposit analysis: 0.09 C, 1.8 Mn, 0.50 Si, 19.5 Cr, 10.2 Ni.
Welded: 93,500 TS; 68,000 YS; 40 El.
For welding or build-up on austenitic 18-8 stainless steel.

MCKAY 308 HI SIL.
M-849; 0.04 C, 1.75 Mn, 0.80 Si, 20.8 Cr, 10.0 Ni, bal Fe.
Bare wire for welding.

MCKAY 308 L.
M-849; 0.02 C, 1.75 Mn, 0.45 Si, 20.8 Cr, 10.0 Ni, bal Fe.
Bare wire for welding.

MCKAY 308 L.
M-849; Coated stainless steel weld rod.
Typical deposit analysis: 0.030 C, 1.2 Mn, 0.50 Si, 20.0 Cr, 10.0 Ni.
Welded: 79,000 TS; 59,000 YS; 45 El.
For welding Type 308L. Stainless. AWS E308L.

MCKAY 308 L HI SIL.
M-849; 0.02 C, 1.75 Mn, 0.80 Si, 20.8 Cr, 10.0 Ni, bal Fe.
Bare wire for welding.

MCKAY 309.
M-849; 0.04 C, 1.90 Mn, 0.46 Si, 23.5 Cr, 13.0 Ni, bal Fe.
Bare wire for welding.
AWS ER309.

MCKAY 309.
M-849; Coated stainless steel weld rod.
Typical deposit analysis: 0.09 C, 1.8 Mn, 0.45 Si, 23.8 Cr, 13.0 Ni.
Welding: 81,000 TS; 65,000 YS; 40 El.
For welding and build-up on type 309 stainless.
AWS E309.

MCKAY 309 CB.
M-849; 0.06 C, 1.60 Mn, 0.40 Si, 24.5 Cr, 13.0 Ni, 0.60 Cb, bal Fe.
Bare wire for welding.

MCKAY 309 CB.
M-849; Coated stainless steel weld rod.
Typical deposit analysis; 0.08 C, 1.20 Mn, 0.40 Si, 23.5 Cr, 13.0 Ni, 0.85 Cb.
Welded: 100,000 TS; 80,000 YS; 35 El.
For welding and build-up on Type 309 stainless steel. AWS E309 Cb.

MCKAY 309 L.
M-849; 0.021 C, 1.95 Mn, 0.40 Si, 24.0 Cr, 13.0 Ni, bal Fe.
Bare wire for welding.

MCKAY 309 L.
M-849.
Coated stainless steel weld rod.
Typical deposit analysis: 0.035 C, 1.2 Mn, 0.50 Si, 23.5 Cr, 13.6 Ni.
Welded: 79,000 TS; 64,000 YS; 41 El.
For welding Type 309 stainless.

MCKAY 309 MO.
M-849; coated stainless steel weld rod.
Typical deposit analysis: 0.08 C, 1.8 Mn, 0.45 Si, 23.5 Cr, 13.0 Ni, .25 Mo.
Welded: 93,000 TS; 70,000 YS; 36 El.
For improved weld strength when welding Type 309 stainless steel.

MCKAY 310.
M-849; 0.09 C, 1.90 Mn, 0.43 Si, 27.0 Cr, 21.1 Ni, bal.
Bare wire for welding.
AWS ER 310.

MCKAY 310.
M-849; Coated stainless steel weld rod.
Typical deposit analysis: 0.11 C, 2.25 Mn, 0.45 Si, 26.0 Cr, 21.0 Ni.
Welded: 85,000 TS; 60,000 YS; 40 El.
For welding and build-up on Type 310 and similar metals. AWS E310.

MCKAY 310 CB.
M-849; Coated stainless steel weld rod.
Typical deposit analysis: 0.11 C, 2.25 Mn, 0.45 Si, 26.0 Cr, 21.0 Ni, 0.85 Cb.
Welded: 90,000 TS; 65,000 YS; 35 El.
Improved quality welds in Type 310 stainless steel. AWS 310 Cb.

MCKAY 310 H.
M-849.
Coated stainless steel weld rod.
Typical deposit analysis: 0.40 C, 2.25 Mn, 0.40 Si, 25.6 Cr, 20.9 Ni.
Welded: 118,000 TS; 90,000 YS; 21 El.
For welding Type 310 casting for repair.

MCKAY 312.
M-849; 0.10 C, 1.60 Mn, 0.40 Si, 30.0 Cr, 9.0 Ni, bal Fe.
Bare wire for welding. AMS 5784.

MCKAY 312.
M-849; Coated stinless steel weld rod.
Typical deposit analysis: 0.12 C, 1.30 Mn, 0.60 Si, 29.0 Cr, 10.0 Ni.
Welded 112,000 TS; 90,000 YS; 32 El.
For welding Type 312 materials and some dissimilar metal joining.
AWS E312.

MCKAY 312 MO.
M-849; Coated stainless steel weld rod.
Typical deposit analysis: 0.12 C, 1.30 Mn, 0.60 Si, 29.0 Cr, 9.5 Ni,
Welded: 120,000 TS; 95,000 YS; 29 El.
Improved weld quality on Type 312 stainless steel.

MCKAY 316.
M-849; Coated stainless steel weld rod.
Typical deposit analysis: 0.06 C, 1.85 Mn, 0.38 Si, 18.0 Cr, 13.0 Ni, Mo.
Welded: 80,000 TS; 68,000 YS; 45 El.
For welding Type 316 stainless steel. AWS 316.

MC KAY 316.
M-849; 0.030 C, 1.70 Mn, 0.48 Si, 18.7 Cr, 12.7 Ni, 2.1 Mo, bal Fe.
Bare wire for welding.
AWS ER316.

MC KAY 316 HF.
M-849.
Coated stainless steel weld rod.
Typical deposit analysis: 0.04 C, 1.20 Mn, 0.35 Si, 19.3 Cr, 12.0 Ni, 2.35 Mo.
Welded: 88,000 TS; 70,000 YS; 40 El.

MC KAY 316 L.
M-849; 0.017 C, 1.80 Mn, 0.50 Si, 18.9 Cr, 12.8 Ni, 2.2 Mo, bal Fe.
Bare wire for welding.
AWS ER316L.

MCKAY 316 L.
M-849; Coated stainless steel weld rod.
Typical deposit analysis: 0.030 C, 1.65 Mn, 0.38 Si, 18.0 Cr, 13.0 Ni.
Welded: 78,000 TS; 60,000 YS; 40 El.
For welding Type 316L stainless steel. AWS E316L.

MC KAY 316 L-HF (BARE).
M-849; 0.020 C, 1.60 Mn, 0.45 Si, 19.3 Cr, 11.6 Ni, 2.2 Mo, bal Fe.
Bare wire for welding.
AWS ER316L.

MC KAY 316 L-HF (COATED).
M-849.
Coated stainless steel weld rod.
Typical deposit analysis: 0.030 C, 1.25 Mn, 0.35 Si, 19.5 Cr, 11.6 Ni 2.25 Mo.
Welded: 85,000 TS; 66,000 YS; 44 El.

MCKAY 316 L HI SIL.
M-849; 0.020 C, 1.75 Mn, 0.85 Si, 19.0 Cr, 13.0 Ni, 2.3 Mo, bal Fe.
Bare wire for welding.

MCKAY 317.
M-849; Coated stainless steel weld rod.
Typical deposit analysis: 0.06 C, 2.0 Mn, 0.40 Si, 18.5 Cr, 13.3 Ni, .65 Mo.
Welded: 95,000 TS; 70,000 YS; 35 El.
Higher Mo improves weld quality over Type 316. AWS E317.

MCKAY 317 L.
M-849; 0.020 C, 1.75 Mn, 0.45 Si, 19.3 Cr, 14.0 Ni, 3.6 Mo, bal Fe.
Bare wire for welding.

MCKAY 317 L.
M-849; Coated stainless steel weld rod.
Typical deposit analysis: 0.035 C, 1.5 Mn, 0.40 Si, 18.5 Cr, 13.3 Ni, Mo.
Welded: 95,000 TS; 70,000 YS; 33 El.
For welding Types 316, 316L, 317, 317L stainless steel.

MCKAY 318.
M-849; Coated stainless steel weld rod.
Typical deposit analysis: 0.05 C, 1.75 Mn, 0.40 Si, 19.5 Cr, 12.5 Ni, .3 Mo, 0.55 Cb.
Welded: 95,000 TS; 75,000 YS; 35 El.
Use similar to 316 but Cb addition improves weld quality. AWS E318.

MC KAY 320 (BARE).
M-849; 0.033 C, 0.50 Mn, 0.35 Si, 19.9 Cr, 33.8 Ni, 2.1 Mo, 0.70 Cb, 3.3 Cu, bal Fe.
Bare wire for welding.
AWS ER320. (Was McKay Stainless 20).

MC KAY 320 (COATED).
M-849.
Coated stainless steel weld rod.
Typical deposit analysis; 0.05 C, 2.25 Mn, 0.25 Si, 19.7 Cr, 32.9 Ni, 2.15 Mo, 0.50 Cb, 3.1 Cu.
Welded: 84,000 TS; 54,000 YS; 39 El.
For Welding Carpenter 20 and 20 Cb-3 stainless.

MCKAY 330.
M-849; 0.20 C, 1.75 Mn, 0.35 Si, 16.0 Cr, 35.0 Ni, bal Fe.
Bare wire for welding.

MCKAY 330.
M-849; Coated stainless steel weld rod.
Typical deposit analysis: 0.20 C, 2.25 Mn, 0.50 Si, 14.5 Cr, 34.0 Ni.
Welded: 86,000 TS; 58,000 YS; 40 El.
For welding Type 330 stainless steel for heat and creep resisting applications. AWS E330.

MCKAY 347.
M-849; Coated stainless steel weld rod.
Typical deposit analysis: 0.05 C, 1.5 Mn, 0.50 Si, 19.5 Cr, 10.0 Ni, 0.65 Cb.
Welded: 96,000 TS; 64,000 YS; 36 El.
For welding Types 321 and 347 stainless. AWS E347.

MCKAY 347.
M-849; 0.046 C, 1.76 Mn, 0.50 Si, 19.6 Cr, 9.5 Ni, 0.85 Cb, bal Fe.
Bare wire for welding.
AWS ER347.

MCKAY 347 HC.
M-849; Coated stainless steel weld rod.
Typical deposit analysis: 0.09 C, 1.85 Mn, 0.45 Si, 19.5 Cr, 10.2 Ni, 0.95 Cb.
Welded: 90,500 TS; 72,500 YS; 36 El.
For welding types 321 and 347 for elevated temperature operation. MIL-E-22200/2.

MCKAY 349.
M-849; Coated stainless steel weld rod.
Typical deposit analysis: 0.10 C, 1.20 Mn, 0.60 Si, 19.5 Cr, 9.0 Ni, .50 Mo, 1.0 Cb, 1.5 W, 0.10 Ti.
Welded: 110,000 TS; 84,000 YS; 30 El.
For welds for high temperature applications such as the jet-aircraft industry. AWS E349.

MCKAY 363.
M-849; 0.030 C, 0.30 Mn, 0.24 Si, 11.5 Cr, 4.2 Ni, 0.40 Ti, bal Fe.
Bare wire for welding.

MCKAY 410.
M-849; 0.079 C, 0.50 Mn, 0.40 Si, 12.9 Cr, 0.35 Ni, 0.15 Mo, bal Fe.
Bare wire for welding.
AWS ER410.

MCKAY 410.
M-849; Coated stainless steel weld rod.
Typical deposit analysis: 0.08 C, 0.50 Mn, 0.45 Si, 12.0 Cr.
Welded, stress rel. at 1350°F: 77,000 TS; 58,000 YS; 34 El.
For welding Type 410 stainless steel; weld and base metal hardenable. A S E410.

MCKAY 410 NI MO.
M-849.
Coated stainless steel weld rod.
Typical deposit analysis: 0.05 C, 0.60 Mn, 0.70 Si, 11.5 Cr, 4.5 Ni, 0.50 Mo.
Welded: 155,000 TS; 138,000 YS; 14 El.
For repair Type 410 castings.

MCKAY 502.
M-849; 0.035 C, 0.46 Mn, 0.42 Si, 5.6 Cr, 0.18 Ni, 0.55 Mo, bal Fe.
Bare wire for welding.
AWS ER502.

MCKAY 502.
M-849; Coated stainless steel weld rod.
Typical deposit analysis: 0.07 C, 0.55 Mn, 0.38 Si, 5.25 Cr, 0.50 Mo.
Welded, stress rel. 1350°F: 75,000 TS; 50,000 YS; 30 El.
For welded assemblies subject to high temperature conditions; AWS E502.

MCKAY 502-18.
M-849; Low carbon chrome alloy, coated weld rod.
Typical deposit analysis: 0.06 C, 0.80 Mn, 0.60 Si, 5.0 Cr, 0.50 Mo.
Welded and stress relieved at 1350°F: 100,000 TS; 67,000 YS; 23 El.
For welding Cr-Mo steels for elevated service conditions. AWS E502.

MCKAY 505-18.
M-849; Low carbon chrome alloy, coated weld rod.
Typical deposit analysis: 0.06 C, 0.55 Mn, 0.50 Si, 9.25 Cr, 1.05 Mo.
Welded and stress relieved at 1575°F; 71,000 TS; 47,500 YS; 40 El.
For welding CR-MO steels for high temperature service conditions; AWS 505-18.

MCKAY 630 (BARE).
M-849; 0.040 C, 0.60 Mn, 0.50 Si, 16.6 Cr, 4.5 Ni, 0.45 Cb, 3.5 Cu, bal Fe.
Bare wire for welding.
AWS ER630. (Was McKay 17-4 PH).

MCKAY 630 (COATED).
M-849.
Coated stainless steel weld rod.
Typical deposit analysis; 0.035 C, 0.48 Mn, 0.40 Si, 16.35 Cr, 4.75 N 0.20 Cb, 3.30 Cu.
Precipitation hardenable coating.
Similar to 17-4 PH.

MCKAY 6010.
M-849; Low C, bal Fe.
Weld rod, Dc. Welded: 68,000 TS; 58,000 YS; 26 El.
For mild steel welding. AWS E6010.

MCKAY 6011.
M-849; Low C, bal Fe.
Weld rod, AC. Welded: 67,000 TS; 58,000 YS; 29 El.
For welding mild steel. AWS E6011.

MCKAY 6012.
M-849; Low C, bal Fe.
Weld rod. Welded: 72,000 Ts; 60,000 Ys; 20 El.
For general purpose low carbon steel welding, particularly for DC with poor fit-up. AWS E6012.

MCKAY 6013.
M-849; Low C, bal Fe.
Weld rod. Welded: 82,000 TS; 73,000 YS; 20 El.
General purpose welding low carbon steel; AC or DC; AWS E6013.

MCKAY 7014.
M-849; Low C, bal Fe.
Weld rod. Welded: 71,000 TS; 64,000 YS; 26 El.
General purpose iron powder electrode. AWS E7014.

MCKAY 7016.
M-849.
Mild steel electrode.
Typical weld analysis: 0.08 C, 0.80 Mn, 0.30 Si, bal Fe.
Welded: 78,000 TS; 71,000 YS; 32 El.
For welding mild and free-machining steel with AC or DC (reverse polarity).
AWS E-7016.

MCKAY 7018.
M-849; mild steel weld rod.
Typical deposit analysis: 0.06 C, 1.05 Mn, 0.65 Si.
Welded: 78,000 TS; 70,000 YS; 32 El.
For mild steel welding, AC or DC. AWS E7018.

MCKAY 7018-AL.
M-849; Mild steel weld rod.
Typical deposit analysis: 0.07 C, 0.75 Mn, 0.60 Si, 0.50 Mo.
Welded: 84,000 TS; 69,000 YS; 30 El.
For welding carbon-molybdenum high temperature piping. AWS E7018-Al.

MCKAY 7018-HC.
M-849; Higher carbon mild steel electrode.
Typical deposit analysis: 0.21 C, 1.0 Mn, 0.60 Si.

MCKAY 7024.
M-849; Mild steel weld rod.
Welded: 87,000 TS; 76,500 YS; 18 El.
For horizontal or flat fillet welds. AWS E7024.

MCKAY 8016-C3.
M-849; Low carbon, low alloy weld rod.
Typical deposit analysis: 0.09 c, 0.85 Mn, 0.40 Si, 0.95 Ni.
Welded: 91,000 TS; 77,000 YS; 27 El.
For welding 1% nickel alloy steels for low temperature service. AWS E8 16-C3.

MCKAY 8018 B2L.
M-849; Low C, alloy steel weld rod.
Typical deposit analysis: 0.045 C, 0.70 Mn, 0.60 Si, 1.25 Cr, 0.50 Mo
Welded: 99,500 TS; 83,500 YS; 26 El.
Primarily for welding chromium molybdenum piping. ASW E8018 B2L.

MCKAY 8018-C1.
M-849; Low C, nickel alloy steel weld rod.
Typical deposit analysis: 0.05 C, 0.80 Mn, 0.45 Si, 2.50 Ni.
Welded: 84,000 TS; 71,000 YS; 32 El.
For welding 2 1/2% nickel steels. AWS E8018-C1.

MCKAY 8018-C2.
M-849; Low C, nickel alloy steel weld rod.
Typical deposit analysis: 0.055 C, 0.80 Mn, 0.45 Si, 3.20 Ni.
Welded: 91,000 TS; 75,000 YS; 29 El.
For welding 3 1/2 % nickel steels. AWS E8018-C2.

MCKAY 8018-C3.
M-849; Low C, low alloy steel weld rod.
Typical deposit analysis: 0.05 C, 0.80 Mn, 0.45 Si, 1.0 Ni.
Welded: 83,000 TS; 70,000 YS; 32 El.
For welding low alloy, high strength steels requiring tough welds for sub-zero operation. AWS E8018-C3.

MCKAY 8018-G.
M-849; Low C, low alloy steel weld rod.
Typical deposit analysis: 0.085 C, 1.3 Mn, 0.60 Si.
Welded: 86,200 TS; 73,250 YS; 32 El.
For welding low alloy steels in the 80,000-85,000 tensile range. AWS 8018-G.

MCKAY 8018-W.
M-849.
Low carbon, alloy steel weld rod.
Typical deposit analysis: 0.07 C, 0.80 Mn, 0.45 Si, 0.60 Cr, 0.50 Ni, 0.50 Cu, bal Fe.
Welded: 86,000 TS; 73,600 YS; 26 El.
For welding weathering steels as ASTM A588, A242 (and COR-TEN, MAYARI R).
AWS E8018-G.

MCKAY 9018.
M-849; Low carbon alloy steel weld rod.
Typical deposit analysis: 0.07 C, 1.0 Mn, 0.40 Si, 1.6 Ni, 0.15 Mo.
Welded 94,000 TS; 84,000 YS; 29 El.
For welding low alloy high strength steels in the 90,000 tensile strength range.
For MIL 9018 military requirements. AWS E9018-M.

MCKAY 9018-B3.
M-849; Low carbon alloy steel weld rod.
Typical deposit analysis: 0.045 C, 0.70 Mn, 0.35 Si, 2.25 Cr, 1.05 Mo.
Welded 124,500 TS; 110,000 YS; 22 El.
Use for welding Cr-Mo piping of similar composition. AWS E9018-B3.

MCKAY 9018-B3L.
M-849.
Low carbon alloy steel weld rod.
Typical deposit analysis: 0.04 C, 0.80 Mn, 0.35 Si, 2.25 Cr, 1.05 Mo, bal Fe.
Welded: 112,000 TS; 96,000 YS; 21 El.
For welding Cr-Mo piping of similar composition. Lower carbon reduces crack sensitivity.
AWS E9018-B3L.

MCKAY 10016 G.
M-849; Low carbon alloy steel weld rod.
Typical deposit analysis: 0.09 C, 0.90 Mn, 0.45 Si, 1.75 Ni, 0.30 Mo, 0.08 V.
Welded: 108,000 TS; 99,000 YS; 23 El.
For welding low alloy, high strength steels in the 100,000 psi tensile strength range. AWS E10016-G.

MCKAY 10018.
M-849; low carbon alloy weld rod.
Typical deposit analysis: 0.06 C, 1.25 Mn, 0.45 Si, 0.12 Cr, 1.55 Ni, Mo.
Welded: 104,000 TS; 95,000 YS; 24 El.
For welding low alloy, high strength steels in the 100,000 psi tensile strength range. For MIL 10018 military requirements. AWS E10018-M.

MCKAY 10018-D2.
M-849; Low carbon alloy steel weld rod.
Typical deposit analysis: 0.11 C, 1.85 Mn, 0.45 Si, 0.35 Mo, 0.75 Ni, Fe.
Welded: 110,000 TS; 98,000 YS; 26 El.
For welding low alloy, high strength steels in the 100,000 psi tensile strength range. AWS E10018-D2.

MCKAY 11016.
M-849; Low carbon alloy steel weld rod.
Typical deposit analysis: 0.08 C, 1.25 Mn, 0.45 Si, 1.20 Cr, 2.0 Ni, Mo.
Welded: 117,000 TS; 106,000 YS; 23 El.
For welding low alloy, high strength steels in the 110,000 psi tensile strength range. AWS E11016-G.

MCKAY 11018-M.
M-849; Low carbon alloy steel weld rod.
Typical deposit analysis: 0.05 C, 1.6 Mn, 0.40 Si, 1.6 Ni, 0.4 Mo.
Welded: 115,000 TS; 104,000 YS; 24 El.
For welding low alloy, high strength steels in the 110,000 psi tensile strength range.
AWS E 11018-M.

MCKAY 12018,-M.
M-849; Low carbon alloy steel weld rod.
Typical deposit analysis: 0.05 C, 1.70 Mn, 0.45 Si, 0.45 Cr, 2.0 Ni, Mo.
Welded: 130,000 TS; 117,000 YS; 22 El.
For welding low alloy high strength steel in the 120,000 psi tensile strength range.
For MIL 12018 military requirements. AWS E 12018-M.

MCKAY 14018.
M-849; Low carbon alloy steel weld rod.
Typical deposit analysis: 0.08 C, 1.0 Mn, 0.40 Si, 0.50 Cr, 3.50 Ni, Mo.
Welded: 147,000 TS; 141,000 YS; 18 El.
For welding high yield strength steels.

MCKAY 14018 HT.
M-849. Low carbon alloy steel weld rod.
Typical deposit analysis: 0.085 c, 0.60 Mn, 0.45 Si, 0.40 Cr, 8.20 Ni, 0.50 Mo, 0.08 V, bal Fe.
Welded: 165,000 TS; 155,000 YS; 16 El.
For welding HY-130 steel prior to heat treatment.

MCKAY ARMORLOY A-8.
M-849; Stainless steel weld rod.
Typical deposit analysis: 0.10 C, 4.3 Mn, 0.45 Si, 19.3 Cr, 9.9 Ni, 1 bal Fe.
Welded: 90,000 TS; 67,000 YS; 44 El.
For welding armor plate, dissimilar metals, and "difficult-to-weld" steel; weld metal has excellent toughness and crack resistance.
MIL-E-13080; U.S.Army 57-203-3 Grade V.

MCKAY ARMORLOY A-9.
M-849; Stainless steel weld rod.
Typical deposit analysis: 0.11 C, 1.9 Mn, 0.60 Si, 18.9 Cr, 9.4 Ni, 2
Welded: 100,000 TS; 73,000 YS; 37 El.
For welding armor plate, dissimilar metals and "difficult-to-weld" steels; weld metal has excellent toughness and crack resistance. U.S. Ar 57-203-3 Grade VI.

MCKAY BARE STAINLESS STEEL WELD WIRE.
M-849; 12 grades conforming to AISI Specs: 308, 309, 310, 316, 316L, 32 347-348, 410, 430, 442, 446 and 502.
(Non-AISI grades have individual listing).

MCKAY CAST ALLOY 60 see MCKAY NICKALLOY 60.

MCKAY CHROME MANG.
M-849; Hard facing electrode, AC or DC.
For joining or build-up; deposit has strength of about 141,000 psi TS; is work hardenable, and crack resistant. For shovel tracks, jaw crusher plates.

MCKAY FROGALLOY.

M-849; Hard facing electrode; DC reverse.
For build-up and hard facing; deposit is high carbon, high strength austenitic stainless; for surfacing bucket teeth, railroad frogs and switches.

MCKAY HARDALLOY 32.

M-849; Hard surfacing electrode, AC or DC.
For build up carbon steel and cast iron; hardness 17-30 Rock C; machinable.

MCKAY HARDALLOY 40 TIC.

M-849; Hard surfacing electrode, AC or DC.
Titanium bearing, for building up hard surface on carbon steels and also austenitic manganese steels for hardness 45-50 Rock C. Not machinable.
For dipper teeth, screw conveyors, hammer mill hammers.

MCKAY HARDALLOY 44.

M-849; Hard surfacing electrode, coated rod.
For overlaying carbon steel and austenite manganese steel for wear resistant surface.
Work hardens to 40-55 Rock C; for crusher screens, dredge pump impellers.

MCKAY HARDALLOY 48.

M-849; Hard surfacing electrode.
For overlaying carbon steel with wear resistant surface of 38-40 Rock C; for shafts, dies, friction clutches, sand chutes, conveyor screws.

MCKAY HARDALLOY 52.

M-849; Hard surfacing electrode, AC or DC.
For overlaying carbon steel with hard, impact resisting surface up to 57-6 Rock C.
For gear teeth, roll ends, mine car wheels, mine rails, cams.

MCKAY HARDALLOY 55.

M-849; Hard surfacing electrode, AC or DC.
For overlaying carbon steel or austenitic manganese steel with hard surface up to 55 Rock C, resistant to gouging type of wear.
For bucket teeth, pusher shoes, muller tires, mower shoes, plow shares.

MCKAY HARDALLOY 55 TIC.

M-849; Hard surfacing electrode.
For overlaying carbon steel and austenitic manganese steel with hard surface with extreme resistance to abrasion.
For crusher rolls, brick dies, coke pusher shoes, and scraper blades.

MCKAY HARDALLOY 58.

M-849; Hard surfacing electrode, AC or DC.
For overlaying carbon steel with hard (55-60 Rock C) surface to resist abrasion and impact.
For drag line bucket tips, dipper lips, crane hooks, coupling boxes and spindles.

MCKAY HARDALLOY 58 TIC.

M-849; Hard surfacing electrode.
For overlaying carbon steel with hard (59-61 Rock C) surface for resistance to severe abrasion and impact.
For cement conveyor screws, augers, scraper blades.

MCKAY HARDALLOY 61.

M-849; Hard surfacing electrode.
For depositing hardenable (up to 61 Rock C) alloy on steel for wear and abrasion resistance up to 1000°F.
For mandrels, shear blades, scraper knives, shaper and planer tools.
AWS-E Fe5-B.

MCKAY HARDALLOY 118.

M-849; Hard surfacing electrode.
For depositing austenitic manganese steel on carbon and manganese steel.
For crusher rolls, dipper teeth, railroad frogs and crossovers; work hardens rapidly.

MCKAY HARDALLOY 119.

M-849.
Coated alloy (26%) austenitic manganese electrode.
For overlaying wear resistant surface on construction, crushing or railroad industries.
Work hardens to 50-55 Rc.

MCKAY HARDALLOY 120.

M-849; Welding electrode, AC & DC.
For joining dissimilar metals and hard-to-weld steels; and for build-up.
Deposits have good corrosion resistance and strength of about 125,000 psi TS.
Work hardens to about 40-45 Rc.

MCKAY IN-FLUX STEEL TUBULAR WELDING WIRES.

M-849; 7 grades for inert gas welding; MIG: 308 L-G, 309 L-G, 316 347-G, 309-G, 310-G, 430-G.
See similar AISI numbers for compositions.
Note: Suffix "S" in place of "G" are designed for submerged arc.

MCKAY MON-ALLOY.

M-849; Monel base weld rod.
Typical deposit analysis: 0.12 C, 1.8 Mn, 0.8 Si, 65.0 Ni, 1.8 Fe, bal Cu.
Welded: 96,000 TS; 49,000 YS; 37 El; 150 Brin.
For welding Monel and Ni-Cu alloys to steel, or for surfacing steel. A NiCu-1.

MCKAY N-155.

M-849; 0.1-0.2 C, 20-22 Cr, 19-21 Ni, 0.75-1.25 Cb, 2.75-3.75 Mo, 2-3 0.1-0.2 N, 18.5-21.0 Co.
Welded: 106,000 TS; 75,000 YS; 24 El.
For welding electrodes for high temperature service; shielded arc, high heat resistant.

MCKAY NICKALLOY.
M-849; Pure nickel weld rod.
Welded: 45,000 TS; 2 El.
For machinable welds on cast iron. AWS E Ni-C1.

MCKAY NICKALLOY 60.
M-849; Nickel-Iron weld rod.
Typical deposit analysis: 59 Ni, 41 Fe.
Welded: 77,000 TS; 61,000 YS; 8 El.
For machinable welds on cast iron. AWS E NiFe-C1.

MCKAY NICKEL MANGANESE.
M-849; Hard facing electrode: DC reverse.
For build-up on carbon steels or manganese austenitic steels where severe impact is expected. For railroad car castings, crusher screens, railroad frogs and switches.

MCKAY PLURALLOY 70 AC.
M-849; Mild steel weld rod.
Typical deposit analysis: 0.08 C, 0.80 Mn, 0.30 Si.
Welded: 78,000 TS; 71,000 YS; 32 El.
For welding mild steel and free machining steel. AWS E7016.

MCKAY STAINLESS 20 see MCKAY 320.

MCKAY TOOL-AGE 400.
M-849; 0.045 C, 0.50 Mn, 0.30 Si, 5.0 Cr, 12.5 Mo, 19.0 Co, bal Fe.
Coated AC-DC electrode, all position.
Deposit: 44-51 Rc; Aged: 56-64 Rc.
Age hardenable deposit; good impact, erosion, thermal shock resistant.
For build-up of hot punches, forging dies, extrusion dies.

MCKAY TOOL-AGE 400-S.
M-849.
Similar to TOOL-AGE 400 but is 1/8 in. wire for submerged arc.

MCKAY TOOL-ALLOY "822".
M-849; Coated Ni-C alloy steel electrode, AC or DC.
For build-up or repair of forming and drawing dies, large drive gears.

MCKAY TOOL-ALLOY "A".
M-849; Air hardening weld rod and wire.
Typical weld metal comp. (rod); 0.85 C, 1.0 Mn, 0.5 Si, 5.2 Cr, 1.5 M 0.35 V. (deposit from wire is slightly different).
For repair and build-up air hardening tools and dies.

MCKAY TOOL-ALLOY C.
M-849; 0.03 C, 0.60 Mn, 0.60 Si, 15.5 Cr, 16.0 Mo, 3.75 W, 6.0 Fe, bal Ni.
Low hydrogen, AC-DC coated electrode.
As welded: 200-225 Brin; work hardens to 36-40 Rc.
For repair and build-up forging equipment.

MCKAY TOOL-ALLOY C-O.
M-849; 0.03 C, 0.65 Mn, 0.60 Si, 15.0 Cr, 16.0 Mo, 3.5 W, 5.0 Fe, bal Ni.
Open arc weld wire, as deposited: 200-225. Brin; work hardens to 36-40 Rc.
Excellent corrosion resistance.

MCKAY TOOL-ALLOY C-S.
M-849; 0.03 C, 0.60 Mn, 0.60 Si, 14.5 Cr, 15.0 Mo, 3.3 W, 4.0 Fe, bal Ni.
Wire for submerged arc welding.
Deposit: 200-225 Brin.; Work hardens to 36-40 Rc.
Excellent corrosion resistance.

MCKAY TOOL-ALLOY "FH-30".
M-849; Coated steel electrodes, AC or DC.
As welded hardness: 34 Rc; flame hardenable to 53 Rc.
For repair of plastic molds, SAE 4130, 6145 steel casting and for build-up of soft material.

MCKAY TOOL-ALLOY "FH-45".
M-849; Coated steel electrodes, AC or DC.
As welded hardness; 44 Rc; flame hardenable to 60 Rc.
For repair of plastic molds, build-up and surface harden of softer material

MCKAY TOOL-ALLOY "GP".
M-849; Coated stainless steel electrode.
As welded: 120,000 TS: 90,000 YS; 27 El.
For welding, build-up or repair of many types of dies or tools.

MCKAY TOOL-ALLOY "HS".
M-849; High speed steel rod and wire.
Typical weld metal comp. (rod): 0.65 C, 0.8 Mn, 0.45 Si, 3.95 Cr, 8.75 0.9 V, 1.8 W. (deposit from wire is slightly different).
For repair and build-up of parts made of AISI M1 tool steel or to deposit such material on carbon steel parts.

MCKAY TOOL-ALLOY "HW".
M-849; Alloy steel welding rod and wire.
Typical weld metal comp. (rod); 0.30 C, 0.5 Mn, 0.5 Si, 4.75 Cr, 1.75 0.25 V, 1.3 W (deposit from wire is slightly different).
For repair and build-up of AISI H-12 and similar hot work tool and die materials.

MCKAY TOOL-ALLOY "HW-2".
M-849; Alloy steel coated electrode for DC reverse; and bare wire for TiG and MiG welding.
For repair or build-up of various hot-work tool steel dies or tools.

MCKAY TOOL-ALLOY "HW-2FC".
M-849; Flux cored open arc wire.
For repair or build-up of various hot-work tool steel dies or tools.

MCKAY TOOL-ALLOY "O".
M-849; Oil hardening weld rod and wire.
Typical weld metal comp. (rod): 0.65 C, 1.12 Mn, 0.4 Si, 1.35 Cr, 1.12 0.5 W, 0.35 V. (wire is slightly different).
For repair and build-up of oil-hardening tools and dies.

MCKAY TOOL-ALLOY "W".
M-849; Water hardening weld rod and wire.
Typical weld metal comp. (rod): 0.8 C, 0.6 Mn, 0.45 Si, 0.4 Cr, 0.4 M
For repair and build-up of water-hardening tools and dies.

MCKAY TOOL-FAB.
M-849; Low alloy coated electrode.
As welded: 96,000 TS; 85,500 YS; 30 El.
For repairing broken or cracked tool steel dies and for build-up.

MCKAY TOOL-FORGE.
M-849; Cr-Mo-V steel flux cored wire.
For repairing sow blocks, rams etc; as welded hardness 35-45 Rock C.

MCKAY TOOL-FORGE 29.
M-849; 0.08 C, 1.20 Mn, 0.60 Si, 0.85 Cr, 2.0 Ni, 0.60 Mo, bal Fe.
Deposit: 129,000 TS; 118,000 YS; 19 El; 23-27 Rc.
Weld wire for GMA welding (CO_2 shielding).
For build-up and repair forge dies, sow blocks, shafts; used also for undercoat.

MCKAY TOOL FORGE 36.
M-849; 0.095 C, 1.10 Mn, 0.60 Si, 5.40 Cr, 1.15 Ni, 1.48 Mo, 0.40 W 0.40 V, bal Fe.
Deposit: 192,000 TS; 163,000 YS; 14 El; 38-42 Rc.
Weld wire for GMA Welding, (CO_2 shielding).
For build-up and repair gears, shafts, forging die blocks.

MCKAY TOOL-MAR "300".
M-849; Ni-Co-Mo-Fe bare wire for TIG welding or build-up of maraging steel, or of H11, H12, H13 hot work tool steels.

MCKAY TOOL-S7-T.
M-849; 0.50 C, 0.80 Mn, 0.50 Si, 3.25 Cr, 1.45 Mo, bal Fe.
Welding wire for build-up of shock-resisting tools. Heat treatable.

MCKAY TUBE-ALLOY 204-0.
M-849.
Low carbon, 2.5% alloy steel wire.
Deposit: 90,000 TS; 70,000 YS; 23 El.
Crack resistant, machinable build-up on steel shafts, gears, crane wheels, back-up collets.

MCKAY TUBE-ALLOY 218-0.
M-849; Low phos. austenitic manganese 19.5% alloy steel wire.
For work-hardenable build-up and surfacing of shovel teeth, crusher jaws, breaker bars, railroad frogs.

MCKAY TUBE-ALLOY 218-S.
M-849; Low phos. austenitic manganese 19.5% alloy steel wire.
For work-hardenable build-up and surfacing of shovel buckets, crusher rolls, hammer mill hammers, railroad frogs and switches.

MCKAY TUBE-ALLOY 219-0.
M-849; C-Cr austenitic manganese steel; total alloy content over 26%.
For welding or hard facing as on drag line teeth or dredge pump shells.

MCKAY TUBE-ALLOY 230-0.
M-849.
High carbon, Cr-Mn alloy (21%) weld wire.
Deposit: 28-37 Rc; work hardens to 50 Rc.
For build-up on bucket teeth, impactor bars.

MCKAY TUBE-ALLOY 236-S.
M-849; 5 Ni, 5 Mo, hard facing wire.
For build-up and hard facing carbon steels.

MCKAY TUBE-ALLOY 240-0.
M-849; High chromium cast iron alloy wire.
For weld or build-up of hammer mill hammers, shovel teeth, grader blades.

MCKAY TUBE-ALLOY 240 TIC-0.
M-849; High chromium cast iron alloy wire.
Contains TiC; for hard facing hammer mill hammers, crusher rolls.

MCKAY TUBE-ALLOY 242-0.
M-849; Low carbon. 5% alloy wire.
For hard surfacing tractor rollers and idlers, mine car wheels, crane wheels.

MCKAY TUBE-ALLOY 242-S.
M-849; Low carbon 5% alloy wire.
For hard surfacing crane wheels, mine car wheels, steel shafts.

MCKAY TUBE-ALLOY 244-0.
M-849; High carbon, chromium, 10.5% alloy wire; for hard facing and reclamation of dredge pump shell. Will stand moderate to heavy impact.

MCKAY TUBE-ALLOY 250-S.
M-849; 13.5% Cr-C alloy steel wire.
For heat treatable surfacing of crane wheels, coiler rolls, plungers, pinch rolls.

MCKAY TUBE-ALLOY 252-0.
M-849; 6.5% alloy steel hard-facing wire,
For surfacing tractor track rollers and idlers, shovel rollers, brake drum ditcher rolls, mine car wheels.

MCKAY TUBE-ALLOY 252-S.
M-849; 6.5% alloy steel hard-facing wire.
For surfacing skip and crane wheels, charging car wheels, brake drums, tractor rollers and idlers.

MCKAY TUBE-ALLOY 255-O.
M-849; High carbon chrome cast iron 34% alloy hard facing wire.
For surfacing dredge pump impellers, dredge cutters, bucket teeth and lips, non-machinable.

MCKAY TUBE-ALLOY 258-O.
M-849; Cr-Mo-W heat treatable 11.5% alloy steel wire.
For hard surfacing pinch rolls, scrap press plungers, table rolls, cable sheaves.

MCKAY TUBE-ALLOY 258-S.
M-849; Cr-Mo-W heat treatable 11.5% alloy steel wire.
For hard surfacing dredge ladder rolls, blocker rolls, cable sheaves.

MCKAY TUBE-ALLOY 258 TIC-O.
M-849; 17% Cr-C alloy with 11% TiC (by volume) wire.
For hard surfacing augers, dredge pump parts, bucket teeth and lips, fan blades.

MCKAY TUBE-ALLOY 263-O.
M-849.
High carbon, high Cr alloy (31%) cast iron rod. Deposit: 57-63 Rc; holds 50 Rc to 1500°F. Non-machinable, non-forgeable.
For sintering plant parts; pusher shoes, screw conveyers.

MCKAY TUBE-ALLOY 821-S.
M-849.
Cr-Mo-W alloy (10.5%) heat treatable wire.
Deposit hardness: 38-45 Rc.
For crack-free, machinable (carbide tools) build-up. Good metal to metal wear.

MCKAY TUBE-ALLOY 828-S.
M-849; Medium carbon 3.5% alloy steel wire.
For medium hardness deposits on gears, sheaves, steel shafting, conveyor rolls.

MCKAY TUBE-ALLOY 829-O.
M-849; 18% Hadfield manganese alloy with 10% TiC (by volume) wire.
For non-machinable, work-hardening deposits of high wear resistance, for bucket lips and teeth, breaker bars, shovel pads.

MCKAY TUBE-ALLOY AP-O.
M-849; High strength Cr-Mn work-hardenable austenitic stainless steel wire.
For surfacing rail frogs and crossovers, shovel buckets, shovel teeth, impeller bars.

MCKAY TUBE-ALLOY BU-A2.
M-849.
High Cr-Mn alloy (36%) stainless steel wire.
For high speed automatic surfacing of large areas of crusher cones, car shredders, dredge pump shells, crusher rolls.
Work hardens to about 50-55 Rc.

MCKAY TUBE-ALLOY BU-C1.
M-849.
Low carbon, 1 1/2% alloy steel wire.
For machinable build-up of worn parts, often used as under-coat.

MCKAY TUBE-ALLOY BU-O.
M-849; Low carbon, 3% alloy steel wire.
For build-up on weldable steels as on tractor rails, rollers and idlers, cable drums, gears, mixer blades, rail ends.

MCKAY TUBE-ALLOY BU-S.
M-849; Low carbon, 3% alloy steel wire.
For build-up on weldable steels, as on cable drums, steel shafts, trunnion gears, mixer blades, rail ends.

MCKECHNIE 4.
M-914; 58 Cu, 3.0 Pb, bal Zn.
Drawn: 410 N/mm² TS; 200 N/mm² YS; 20 El; 130-150 V.P.N.
Rod for high speed machining.
BS 2874 CZ121.

MCKECHNIE 4 STAR.
M-914; 58 Cu, 4.25 Pb, bal Zn.
Drawn: 430 N/mm² TS; 210 N/mm² YS; 15 El; 140-170 V.P.N.
Excellent machinability.
BS 2874 CZ121.

MCKECHNIE 20.
M-914; 61.5 Cu, 2.0 Pb, bal Zn.
Soft: 320 N/mm² TS; 140 N/mm² YS; 30 El; 80-110 V.P.N.
Drawn: 380 N/mm² TS; 210 N/mm² YS; 25 El; 110-130 V.P.N.
Moderate hot working, moderate machinability.
BS 2874 CZ 119.

MCKECHNIE 21.
M-914; 65 Cu, bal Zn.
Soft: 320 N/mm² TS; 100 N/mm² YS; 40 El; 80-100 V.P.N.
Drawn: 350 N/mm² TS; 140 N/mm² YS; 35 El; 100-120 V.P.N.
Fair hot working, excellent cold working.
BS 2874 CZ107.

MCKECHNIE 65.
M-914; 66.5 Cu, 0.75 Fe, 4.75 Al, 0.75 Mn, bal Zn.
Extruded: 610 N/mm² TS; 300 N/mm² YS; 15 El; 130 V.P.N.
Good hot working, fair machinability.
Super high-strength bronze.
BS 2874 CZ114.

MCKECHNIE 75.
M-914; 77.5 Cu, 5.0 Fe, 11.5 Al, 1.0 Mn, 5.0 Ni.
Extruded: 900 N/mm^2 TS; 460 N/mm^2 YS; 5 El; 250 V.P.N.
Good hot working, fair machinability.
Super high-strength aluminum bronze.
Very good wear resistance.

MCKECHNIE 115.
M-914; 56 Cu, 3 Pb, bal Zn.
Extruded: 380 N/mm^2 TS; 180 N/mm^2 YS; 20 El 90-120 V.P.N.
For high speed machining.

MCKECHNIE 160.
M-914; 90.5 Cu, 9.5 Al.
Drawn: 540 N/mm^2 TS; 250 N/mm^2 YS; 20 El; 150 V.P.N.
Moderate hot working, fair machinability.
Corrosion resistant aluminum bronze.

MCKECHNIE 164.
M-914; 87.7 Cu, 1.5 Fe, 9.3 Al, 1.5 Ni.
Drawn: 590 N/mm^2 TS; 310 N/mm^2 YS; 15 El; 170 V.P.N.
Moderate hot working, fair machinability.
Medium strength aluminum bronze.

MCKECHNIE 197.
M-914; 81 Cu, 4.5 Fe, 10.0 Al, 4.5 Ni.
Drawn: 770 N/mm^2 TS; 430 N/mm^2 YS; 15 El; 220 V.P.N.
Good hot working, fair machinability.
High strength, corrosion resistant aluminum bronze.

MCKECHNIE 210.
M-914; 57 Cu, 2.5 Pb, bal Zn.
Extruded: 380 N/mm^2 TS; 180 N/mm^2 YS; 25 El; 90-120 V.P.N.
Drawn: 410 N/mm^2 TS; 200 N/mm^2 YS; 20 El; 120-150 V.P.N.
Excellent hot working and good machinability.
Similar to BS 2874 CZ122.

MCKECHNIE 226.
M-914; 59.5 Cu, 2.0 Pb, bal Zn.
Extruded: 360 N/mm^2 TS; 140 N/mm^2 YS; 30 El; 90-120 V.P.N.
Drawn: 380 N/mm^2 TS; 200 N/mm^2 YS; 25 El; 110-140 V.P.N.
Good hot working, good machinability.
ASTM B124 Alloy 2 (UNS C 37700).

MCKECHNIE 310.
M-914; 80 Cu, 0.2 Pb, bal Zn.
Soft: 290 N/mm^2 TS; 100 N/mm^2 YS; 40 El; 80-100 V.P.N.
Drawn: 320 N/mm^2 TS; 140 N/mm^2 YS; 30 El; 90-120 V.P.N.
Fair hot working, excellent cold working. Gilding brass.
BS 2874 CZ104.

MCKECHNIE 311.
M-914; 70 Cu, bal Zn.
Soft: 290 N/mm^2 TS; 100 N/mm^2 YS; 50 El; 80-100 V.P.N.
Drawn: 350 N/mm^2 TS; 140 N/mm^2 YS; 40 El; 100-120 V.P.N.
Fair hot working, excellent cold working.
For severe bending and cold forming.
BS 2874 CZ106.

MCKECHNIE 312.
M-914; 60 Cu, bal Zn.
Soft: 320 N/mm^2 TS; 140 N/mm^2 YS; 35 El; 90-110 V.P.N.
Drawn: 380 N/mm^2 TS; 180 N/mm^2 YS; 30 El; 110-140 V.P.N.
Good hot working, moderate cold working.
"Muntz" metal.
BS 2874 CZ109.

MCKECHNIE 540.
M-914; 57.5 Cu, 0.75 Sn, 1.25 Pb, 0.75 Fe, 1.25 Mn, bal Zn.
Extruded: 460 N/mm^2 TS; 185 N/mm^2 YS; 20 El; 100 V.P.N.
Drawn: 500 N/mm^2 TS; 215 N/mm^2 YS; 15 El; 130 V.P.N.
Good hot working, good machinability.
Manganese bronze.
BS 2874 CZ115.

MCKECHNIE 651.
M-914; 89.2 Cu, 2.0 Fe, 8.8 Al.
Drawn: 540 N/mm^2 TS; 240 N/mm^2 YS; 35 El; 150 V.P.N.
Moderate hot working, fair machinability.
Tough, corrosion resistant aluminum bronze.

MCKECHNIE 830.
M-914; 60 Cu, 0.4 Sn, 0.4 Si, bal Zn.
Melt range: 890-900°C.
Low fuming, free flowing brazing rod.
BS 1453/C2.

MCKECHNIE 850.
M-914; 59 Cu, 0.2 Si, bal Zn.
Melt range 890-900°C.
Tin free, low fuming welding or brazing rod.
Similar to BS 1845/10.

MCKECHNIE AB.
M-914; 88.7 Cu, 1.5 Pb, 9.8 Al.
Drawn: 560 N/mm^2 TS; 280 N/mm^2 YS; 15 El; 200 V.P.N.
Moderate hot working, good machinability.
Free machining aluminum bronze.

MCKECHNIE CON.
M-914; 58 Cu, 0.5 Sn, 0.5 Pb, 1.0 Fe, 1.5 Al, 2.75 Mn, 1.75 Ni, bal Zn.
Extruded: 520 N/mm^2 TS; 230 N/mm^2 YS; 20 El; 120 V.P.N.
Drawn; 560 N/mm^2 TS; 280 N/mm^2 YS; 15 El; 150 V.P.N.
Corrosion resistant pump and valve stock.

MCKECHNIE CSC STAR.
M-914; 62.5 Cu, bal Zn.
Soft: 320 N/mm² TS; 100 N/mm² YS; 40 El; 80-100 V.P.N.
Drawn: 350 N/mm² TS; 140 N/mm² YS; 35 El; 100-130 V.P.N.
Moderate hot working, excellent cold working. For rivets and pin wire.
BS 2874 CZ108.

MCKECHNIE DO.
M-914; 99.95 Cu, 0.03 P.
Elec. cond: 85% IACS.
Deoxidized, non-arsenical copper; suitable for welding.
BS 2874 C106.

MCKECHNIE E78.
M-914; 90 Cu, bal Zn.
Soft: 250 N/mm² TS; 90 N/mm² YS; 40 El; 80-100 V.P.N.
Drawn 320 N/mm² TS; 140 N/mm² YS; 30 El; 90-120 V.P.N.
Fair hot working, good cold working. Gilding brass.
BS 2874 CZ101.

MCKECHNIE E86.
M-914; 91 Cu, 7 Al, 2 Si.
Drawn: 590 N/mm² TS; 310 N/mm² YS; 30 El; 150 V.P.N.
Moderate hot working, moderate machinability. Corrosion resistant, tough, aluminum bronse.

MCKECHNIE E102.
M-914; 59 Cu, 0.7 Pb, 1.0 Al, 2.0 Mn, 0.5 Si, bal Zn.
Extruded: 530 N/mm² TS; 250 N/mm² YS; 15 El; 130 V.P.N.
Good hot working, moderate machinability.
Hard, wear resistant, forgeable bearing bronze.

MCKECHNIE EC.
M-914; 99.99 Cu.
Half hard: 250 N/mm² TS; 150 N/mm² YS; 30 El; 90 V.P.N.
Hard: 300 N/mm² TS; 240 N/mm² YS; 15 El; 105 V.P.N.
Electrolytic tough pitch copper.
BS 2874 C101.

MCKECHNIE ETS.
M-914; 60 Cu, 0.75 Sn, 0.1 Pb, bal Zn.
Drawn: 380 N/mm² TS; 200 N/mm² YS; 20 El; 100-130 V.P.N.
Good hot working, fair machinability.
American type naval brass.
ASTM B21 Alloy A (CDA 464).

MCKECHNIE F7S.
M-914; 61.5 Cu, 1.2 Sn, 0.4 Pb, bal Zn.
Drawn: 380 N/mm² TS; 140 N/mm² YS; 25 El; 100-130 V.P.N.
Good hot working, moderate machinability.
Standard admiralty naval brass.
BS 2874 CZ112.

MCKECHNIE F58.
M-914; 58.5 Cu, 0.9 Sn, 0.4 Pb, bal Zn.
Drawn: 380 N/mm² TS; 210 N/mm² YS; 25 El; 100-130 V.P.N.
Good hot working, moderate machinability.
Naval brass, for stamping.
BS 2874 CZ113.

MCKECHNIE FX.
M-914; 60 Cu, 0.75 Si, 1.75 Pb, bal Zn.
Drawn: 380 N/mm² TS; 200 N/mm² YS; 20 El; 100-130 V.P.N.
Good hot working, good machinability. Naval brass.
ASTM B21 Alloy C (CDA 485).

MCKECHNIE GB.
M-914; 58 Cu, 3.75 Pb, bal Zn.
Extruded: 380 N/mm² TS; 180 N/mm² YS; 20 El; 90-120 V.P.N.
Excellent machinability; "Gill" brass for deep drilling operations.
BS 2874 CZ121.

MCKECHNIE HC.
M-914; 99.95 Cu, 0.04 O.
Elec. cond: 98% IACS.
Fire refined, tough pitch, high conductivity copper.
BS 2874 C102.

MCKECHNIE K.
M-914; 57.5 Cu, 0.75 Sn, 1.25 Pb, 0.75 Fe, 0.25 Al, 1.25 Mn, bal Zn.
Extruded: 460 N/mm² TS; 215 N/mm² YS; 20 El; 110 V.P.N.
Drawn: 500 N/mm² TS; 250 N/mm² YS; 15 El; 140 V.P.N.
Standard free-machining manganese bronze.
BS 2874 CZ114.

MCKECHNIE KP.
M-914; 59 Cu, 0.75 Sn, 0.75 Pb, 0.25 Fe, 0.25 Al, 1.25 Mn, bal Zn.
Drawn: 460 N/mm² TS; 215 N/mm² YS; 20 El; 130 V.P.N.
Good hot working, moderate machinability.
For propeller shafting.

MCKECHNIE KS.
M-914; 57.5 Cu, 0.75 Pb, 0.25 Fe, 1.75 Mn, bal Zn.
Extruded: 430 N/mm² TS; 185 N/mm² YS; 25 El; 100 V.P.N.
Good hot working, moderate machinability.
Architectural (manganese) bronze.

MCKECHNIE KW.
M-914; 58 Cu, 2.0 Pb, 1.0 Mn, bal Zn.
Extruded: 430 N/mm^2 TS; 185 N/mm^2 YS; 20 El; 100 V.P.N.
Good hot working, good machinability.
Free machining manganese bronze.

MCKECHNIE KZ.
M-914; 58 Cu, 0.5 Sn, 0.75 Pb, 0.75 Fe, 0.75 Al; 0.75 Mn, bal Zn.
Extruded: 500 N/mm^2 TS; 220 N/mm^2 YS; 20 El; 110 V.P.N.
Drawn: 530 N/mm^2 TS; 280 N/mm^2 YS; 15 El; 140 V.P.N.
High strength forging and machining stock.
BS 2874 CZ114.

MCKECHNIE NS.
M-914; 46 Cu, 0.1 Mn, 10.0 Ni, 0.2 Si, bal Zn.
Melt range: 905-915°C.
Low fuming nickel bronze welding rod.
BS 1453/C5.

MCKECHNIE PW.
M-914; 58.5 Cu, 0.75 Sn, 0.3 Fe, 0.04 Mn, 0.2 Ni, 0.1 Si, bal Zn.
Melt range: 885-895°C.
Low fuming manganese bronze filler rod.
Similar to BS 1453/C4.

MCKECHNIE S.
M-914; 57 Cu, 3 Pb, 0.3 Al, bal Zn.
Extruded: 410 N/mm^2 TS; 200 N/mm^2 YS; 20 El; 100-130 V.P.N.
For high speed machining.
Similar to BS 2874 CZ121.

MCKECHNIES BRONZE.
M-Eng.; 57 Cu, 41 Zn, 1 Sn, 0.5 Pb, 1 Fe.
For rods, nuts, bolts; high strength.

MCKECHNIE SC.
M-914; 99.7 Cu, 0.3 S.
Elec cond: 95% IACS.
Good machinability.

MCKECHNIE SIB.
M-914; 59 Cu, 1.75 Al, 3.0 Mn, 1.0 Si, bal Zn.
Extruded: 580 N/mm^2 TS; 280 N/mm^2 YS; 15 El; 130 V.P.N.
Good hot working, fair machinability.
Hard, wear resistant, forgeable bearing bronze.
CDA 674.

MCKECHNIE SS.
M-914; 59 Cu, 1.2 Pb, bal Zn.
Extruded: 360 N/mm^2 TS; 140 N/mm^2 YS; 30 El; 100-120 V.P.N.
Drawn: 380 N/mm^2 TS; 200 N/mm^2 YS; 25 El; 120-150 V.P.N.
Good hot working, moderate machinability.
Similar to BS 2874 CZ122.

MCKECHNIE SS1.
M-914; 59 Cu, 0.5 Pb, bal Zn.
Soft: 320 N/mm^2 TS; 140 N/mm^2 YS; 35 El; 90-110 V.P.N.
Drawn: 380 N/mm^2 TS; 180 N/mm^2 YS; 30 El; 110-140 V.P.N.
Good hot working, fair machinability.
BS 2874 CZ123.

MCKECHNIE SS SPECIAL.
M-914; 60.5 Cu, 0.5 Pb, bal Zn.
Soft: 320 N/mm^2 TS; 140 N/mm^2 YS; 35 El; 90-110 V.P.N.
Drawn: 380 N/mm^2 TS; 180 N/mm^2 YS; 30 El; 110-140 V.P.N.
Good hot working, fair machinability.
BS 2874 CZ123.

MCKECHNIE SWM.
M-914; 45 Cu, 1.5 Pb, 1.0 Mn, 8.5 Ni, bal Zn.
Extruded: 530 N/mm^2 TS; 250 N/mm^2 YS; 15 El; 130 V.P.N.
Good hot working, good machinability.
For architectural purposes.

MCKECHNIE TC.
M-914; 99.5 Cu, 0.5 Te, trace P.
Drawn: 300 N/mm^2 TS; 240 N/mm^2 YS; 15 El; 100 V.P.N.
Elec. cond: 90% IACS.
Excellent machinability.
BS: 2874 C109.

MCKECHNIE W.
M-914; 58 Cu, 2.0 Pb, bal Zn.
Extruded: 380 N/mm^2 TS; 200 N/mm^2 YS; 25 El; 90-120 V.P.N.
Drawn: 410 N/mm^2 TS; 210 N/mm^2 YS; 20 El; 120-150 V.P.N.
Excellent hot working and good machinability.
BS 2874 CZ122.

MCKECHNIE WM.
M-914; 46 Cu, 2.0 Pb, 0.3 Mn, 9.5 Ni, bal Zn.
Drawn: 530 N/mm^2 TS; 280 N/mm^2 YS; 15 El; 150 V.P.N.
Good hot working, good machinability.
Standard free machining nickel brass.
BS 2874 NS101.

MCKECHNIE WMW.
M-914; 47 Cu, 1.5 Pb, 0.5 Mn, 8.0 Ni, bal Zn.
Drawn: 530 N/mm^2 TS; 280 N/mm^2 YS; 15 El; 150 V.P.N.
Moderate hot working, good machinability.
Nickel brass for decorative hardware.

MCKECHNIE XX.
M-914; 50 Cu, bal Zn.
Melt range: 875-885°C.
Hard brazing rod.
BS 1845/8.

MCKECHNIE YR.
M-914; 61.5 Cu, 1.5 Pb, bal Zn.
Soft: 350 N/mm² TS; 140 N/mm² YS; 30 El; 80-110 V.P.N.
Drawn: 380 N/mm² TS; 180 N/mm² YS; 25 El; 110-140 V.P.N.
Moderate hot working, moderate machinability.
BS 2874 CZ119.

MCKECHNIE YS.
M-914; 61 Cu, 3.0 Pb, bal Zn.
Soft: 350 N/mm² TS; 150 N/mm² YS; 25 El; 80-110 V.P.N.
Drawn: 400 N/mm² TS; 180 N/mm² YS; 20 El; 110-140 V.P.N.
Standard grade free-machining brass.
BS 2874 CZ124; ASTM B16.

MCKINNEY ALLOYS.
M-Eng.; 95-97 Al, 2-3 Cu, 1-2 Mn.
For light alloy parts exposed to sea water; resists sea water corrosion.

MCLOUTH ML-50.
M-1592; 0.13 C, 0.48 Mn, 0.01 P, 0.015 S, 0.17 Si, 0.01 Cb, 0.22 Cu Mo, 0.40 Cr, 0.61 Ni, bal Fe.
Bars: 73,000 min TS; 50,000 min YS; 22 El.
For chutes, crane booms, derricks, dump bodies, marine parts, pressure tanks.
Good formability and weldability. Low alloy high-strength steel.

MCLOUTH ML-60.
M-1592; 0.13 C, 0.90 Mn, 0.01 P, 0.016 S, 0.20 Si, 0.29 Cu, 0.01 Mo Cr, 0.68 Ni, 0.014 Cb, bal Fe.
Bars: 75,000 min TS; 60,000 min YS; 22 min El.
For chutes, crane booms, derricks, dump bodies, marine parts, pressure tanks.
Good formability and weldability. Low alloy high-strength steel.

MCLOUTH ML-70.
M-1592; 0.16 C, 0.95 Mn, 0.01 P, 0.021 S, 0.17 Si, 0.30 Cu, 0.01 Mo, 0 0.029 Cb, bal Fe.
Bars: 85,000 min TS; 70,000 min YS; 20 min El.
For chutes, pressure tanks, crane booms, derricks, dump bodies, marine parts.
Good formability and weldability. Low-alloy high-strength steel.

MCLOUTH ML-F.
M-1592; 0.22 max C, 1.25 max Mn, 0.2 min Cu, 0.02 min V, 0.012 Cb, bal Fe.
Plate: 70,000 TS; 50,000 YS; 22 El.
For automobile parts, buckets, chutes, dump bodies, truck frames, wheels, transmission towers, derricks. Good formability and weldability.
Low alloy-high strength steel.

MCLOUTH-MLX.
M-1592; 0.26 max C, 1.5 max Mn, 0.005 min Cb, 0.02 min V, bal Fe.
Gr. 45: 60,000 TS; 45,000 YS; 22 El.
Gr. 50: 65,000 TS; 50,000 YS; 22 El.
Gr. 55: 70,000 TS; 55,000 YS; 22 El.
Gr. 60: 75,000 TS; 60,000 YS; 20 El.
For derricks, transmission towers, chutes, crane booms, buckets, dump bodies.
Good formability and weldability. High strength-low alloy steel.

MCLURE ALLOY.
M-Eng.; 85 Al, 8.2 Cu, 5-6 Sn, 0.9 Fe, 0.3 Si, 0.2 Mn.
For light alloy parts; non-hardenable.

M C V.
M-510; 0.46 C, 0.60 Cr, 0.55 Ni, 0.25 Mo, 1.2 Mn, B, Ti, Zr, V, bal Fe.
Oil hardening structural steel.

MD see **FINKL MD.**

MD-22.
M-1775; 96.7 Al, 2.0 Cu, 1.0 Mg, 0.3 Si.
Powder to be compacted and sintered to make powder metal parts.

MD-24.
M-1775; 93.8 Al, 4.4 Cu, 0.5 Mg, 0.9 Si, 0.4 Mn.
Powder to be compacted and sintered to make powder metal parts.
Equivalent to AA 2014.

MD-69.
M-1775; 98.05 Al, 0.25 Cu, 1.0 Mg, 0.6 Si, 0.1 Cr.
Powder to be compacted and sintered to make powder metal parts.
Equivalent to AA 6061.

MD-76.
M-1775; 90.1 Al, 1.6 Cu, 2.5 Mg, 0.2 Cr, 5.6 Zn.
Powder to be compacted and sintered to make powder metal parts.
Equivalent to AA 7075.

MD-9824.
M-1775; 45 Ag, 30 Cu, 25 Zn.
Powder (100 mesh) for making special solders.

MECO.
50 Cu, 25 Ni, 20 Zn.
For chemical equipment; corrosion resistant.

MEDAL BRONZE-1.
92 Cu, 8 Sn.
For medals, ornaments; corrosion resistant.

MEDAL BRONZE-2.
95 Cu, 4 Sn, 1 Zn.
For medals, ornaments; corrosion resistant.

MEDAL BRONZE-3.
97 Cu, 1 Sn, 2 Zn.
For medals, ornaments; corrosion resistant.

MEDAL METAL.
84 Cu, 16 Zn.
For medals, ornaments; corrosion resistant.

MEDIUM.
M-686; 70 Ag, 20 Cu, 10 Zn.
For silversmithing solder for silver; M.P. 1335-1390°F.

MEDIUM HARD.
M-388; 0.90 C, bal Fe.
For tools, dies, fixtures; water hardened.

MEDIUM LEADED BRASS 229.
M-8; 64 Cu, 35 Zn, 1 Pb.
Soft; 45,000 TS; 17,000 YS; 57 El.
Hard: 73,000 TS; 60,000 YS; 8 El.
For hardware, bolts; moderate cold working, free-cutting.

MEDIUM LEADED BRASS-340.
M-8; 64.5 Cu, 34.5 Zn, 1.0 Pb.
Hard Rod; 65,000 TS; 50,000 YS; 15 El; B 75 Rock.
Soft Rod: 46,000 TS; 17,000 YS; 60 El; B 15 Rock.
For plaques, hinges, gears, wheels, ratchets, pinions, valve stems, rivets, channel plates. Corrosion resistant. Elect. cond 26.

MEDIUM LEADED BRASS 350.
M-279; 61 Cu, 38 Zn, 1 Pb.
Ann: 45,000-55,000 TS; 12,000-26,000 YS; 45-60 El.
Cold rolled: 49,000-99,000 TS; 20,000-79,000 YS; 1-44 El.
Good machinability and formability.
For hose coupling nuts, sink strainers, bearing cages.

MEDIUM SILVER SOLDER.
75-70 Ag, 20-23 Cu, 5-7.5 Zn.
For silver solder; corrosion resistant.

MEDUSA.
M-1431; 0.50 C, 1.5 Cr, 2.5 W, bal Fe.
For rivet sets, punches, pneumatic tools; Type S1; shock resistant.

MEECHITE.
M-1167; 0.2 C, 29 Cr, bal Fe.
For melting pots; nonferrous metals, corrosion resistant.

MEEHANITE ALMANITE TYPE W.
M-82.
Austenitic-martensitic white cast iron.
W_1 500-600 Brin; pearlitic matrix; W_2 500-600 Brin; martensitic matrix; W_4 400-700 Brin; austenitic as cast, but can be converted to martensitic.
For parts requiring severe abrasive wear.

MEEHANITE ALMANITE TYPE W5.
M-82.
Martensitic cast iron with nodular graphite.
415-552 N/mm^2 (60,000-80,000 psi) TS; 2-4 El; good impact strength.
For crusher jaws, pulverizers, hammers.

MEEHANITE ALMANITE TYPE WSH.
M-82.
Austenitic cast iron with nodular graphite.
690 N/mm^2 (100,000 psi) TS; 350-500 Brin, 4-10 El.
Work hardens readily; for crusher liners, dredge buckets, dipper teeth.

MEEHANITE-HE.
M-82, M-850; 3.55 max C, 1.9 min Si, Cu,Cr, Mn, bal Fe.
Cast: 30,000 TS; 220 min Brin.
For slag pots, furnace castings, ingot molds, sinter grates.
Heat resistant cast iron.

MEEHANITE HR.
M-82, M-850; 3.15 min C, 1.0 max Si, Cu, Cr, bal Fe.
Cast: 40,000 TS; 300 Brin.
For glass molds, furnace and burner parts, rolls; heat resistant to 1550°F, alloy iron.

MEEHANITE HS.
M-82; 2.8 max C, 5.0 min Si, Mn, others, bal Fe.
Cast: 60,000 TS; 196 Brin.
For oil refinery supports, blast furnace parts, trays, dampers: heat resisting cast iron up to 1700°F.

MEEHANITE TYPE AQ.
M-82.
Gray iron casting type; hardenable.
As cast: N/mm^2 345 (50,000 psi) TS; 280 Brin.
Heat treated: N/mm^2 448 (65,000 psi) TS; up to 500 Brin.
For cams, dies, punches, rollers.

MEEHANITE TYPE AQS.
M-82.
Hardenable cast iron, as cast.
As cast: N/mm^2 550 TS; 225 Brin.
Hardened: N/mm^2 480 TS; up to 500 Brin.
For wear and abrasion resisting parts.

MEEHANITE TYPE CC.
M-82.
Moderately corrosion resistant cast iron.
276 N/mm^2 (40,000 psi) TS; 200 Brin.
For pumps, valves, evaporators, filter presses.

MEEHANITE TYPE CR.
M-82.
Austenitic cast iron with flake graphite.
Good corrosion resistance; for handling acid and alkali solutions to 700°C.
25,000 psi TS; 131-183 Brin.
ASTM A436-72a.

MEEHANITE TYPE CRS.
M-82.
Austenitic cast iron with nodular graphite.
58,000 psi min TS; 8.0 min El.
Good corrosion resistance, for handling acids and alkali solutions to 700°C.

MEEHANITE TYPE GA-350.
M-82.
Gray iron casting type.
N/mm² 350 (50,000 psi) TS; Brin 220.
General purpose cast iron, including diesel engine cylinders and liners.
ASTM A48-74; QQ-I-652C.

MEEHANITE TYPE GC-275.
M-82.
Gray iron casting type.
N/mm² 275 (40,000 psi) TS; 190 Brin.
For small and medium size castings.
ASTM A48-74; Federal QQ-I-652C.

MEEHANITE TYPE GE-200.
M-82.
Gray iron casting type.
N/mm² 200(30,000 psi) TS; 180 Brin.
Replaces ordinary cast iron.
ASTM A48-74; Federal QQ-I-652c.

MEEHANITE TYPE GF-150.
M-82.
Gray iron casting type.
N/mm² 150 (120,000 psi) TS; 160 Brin.
For lightly loaded castings requiring much machining.

MEEHANITE TYPE GM-400.
M-82.
Gray iron casting type.
N/mm² 400 (55,000 psi) TS; Brin 230.
General purpose cast iron, including pressing, blanking and header dies.
ASTM A48-74; Federal QQ-I-652c.

MEEHANITE TYPE H5V.
M-82.
Heat resisting type cast iron.
670/828 N/mm² (100,000-120,000 psi) TS; 2-10 El; 200 Brin.
For hot forming dies, turbo and supercharger castings, furnace parts.

MEEHANITE TYPE SF-400.
M-82.
Cast duct iron, essentially Ferritic.
N/mm² 400 min TS (60,000 psi TS); N/mm² 310 min YS (45,000 psi YS); 15-20 El; Brin. Up to 160.
For steel weldments; replace malleable iron.
ASTM A-395-74; AMS 5315.

MEEHANITE TYPE SH-700 (SH 100).
M-82.
Cast ductliron.
100,000 psi TS; 65,000 YS; 1-5 El; 240 Brin.
Hardenable, often surface hardened for cams, dies, brake drums.
ASTM A-536-72; Mil-I-11466 B (MR).

MEEHANITE TYPE SH-800 (SH-100).
M-82.
Cast ductliron, heat treated.
Properties from 100,000-170,000 psi TS (N/mm² 700-1190 TS); Brin 263-600.

MEEHANITE TYPE SP-600 (SP 80).
M-82.
Cast ductliron.
80,000-100,000 psi TS; 60,000-75,000 psi YS; 3-10 El; 200 Brin (approx). Pearlitic.
For automotive connecting rods, crankshafts, gears, cams, car journal boxes.
ASTM A536-72; ASM 5316.

MEIGH METAL.
M-1556; 9 Al, 4 Fe, 5 Ni, bal Cu.
Cast: 101,000 TS; 43,000 YS; 14 El.
Non-magnetic, non-sparking corrosion resistant bronze.
BS ABCD; DTD 412.

MELCHOIR WIRE-1.
57.3 Cu, 41.6 Ni, 1.1 Mn.
For white metal wire; corrosion resistant.

MELCHOIR WIRE-2.
62.7 Cu, 25.9 Zn, 10.8 Ni, 0.6 Mn.
For white metal wire; corrosion resistant.

MELCHOR WIRE.
57.3 Cu, 41.6 Ni, 1.1 Mn.
For wire; corrosion resistant.

MELCLIF A5.
M-1130; 5 Al, 0.5 Zn, 0.4 Mn, bal Mg.
Cast: 29,700 TS; 12,000 YS; 11 El; 50 Brin.
For light alloy parts; wrought.

MELKHIOR.
M-USSR; Ni, bal Cu.

MELMAG 75 BATTERY PLATE.
M-1130; 6.6-7.6 Tl, 4.6-5.6 Al, 0.25 max Mn, bal Mg.
For electrochemical applications.

MELMAG AP 65 BATTERY PLATE.
M-1130; 6.0-7.0 Al, 4.5-5.0 Pb, 0.4-1.5 Zn, 0.15-0.3 Mn, bal Mg.
High voltage alloy for electrochemical applications.

MELMAG AZ61 BATTERY PLATE.
M-1130; 5.8-7.2 Al, 0.4-1.5 Zn, 0.15-0.25 Mn, bal Mg.
General purpose alloy for electrochemical applications.

MELOTTE FUSIBLE ALLOY.
M-England; 50 Bi, 31 Sn, 19 Pb.
For fire extinguishers; M.P. 99.5°C.

MELTRITE.
M-1072; 4 C, bal Fe.
For making steel and cast iron; pig iron.

MEL-TROL see also **CARPENTER MEL-TROL.**

MELTROL HAMDEN see **CARPENTER HAMPDEN.**

MEL-TROL HAMDEN.
M-32; 2.1 C, 0.35 Mn, 0.25 Si, 12.5 Cr, 0.5 Ni, bal Fe.
Hardened: C 63-65 Rock.
For spindles, hubs, cold rolls, slitting cutters, master tools, blanking and forming dies, lamination dies.
Type D3. Oil hardening, nondeforming.

MELTRON A8.
M-1130; 7.5-9.0 Al, 0.3-1.0 Zn, 0.15-0.40 Mn, bal Mg.
Cast: 25,000 TS; 13,500 YS; 2 El; 60 Brin.
Heat treated: 38,000 TS; 14,000 YS; 6 El; 70 Brin.

MELTRON AZ31.
M-1130; 3 Al, 1 Zn, 0.3 Mn, bal Mg.
O-temper: 37,000 TS; 22,000 YS; 21 El; 56 Brin.
H24-temper: 42,000 TS; 32,000 YS; 16 El; 73 Brin.
F-temper: 37,000 TS; 22,000 YS; 21 El.
For truck bodies, aircraft cowling and frames; good formability.

MELTRON AZ855.
M-1130; 5.5-8.5 Al, 1.5 max Zn, 0.15-0.40 Mn, bal Mg.
Forged: 44,800 TS; 29,200 YS; 10 El; 70 Brin.
For bearing housings, cylinder heads, control levers; age hardenable, forgings and extrusions.

MELTRON ZTX.
M-487; 2.5 Th, 1 Zn, 0.6 Zr, bal Mg.
Sheet: 16,200 TS; 8000 YS; 18 El.
At 300°C: 5200 TS; 4200 YS; 43 El.
For aircraft and missile components; creep resistant and good properties to 600°F.

MERAL.
M-Swiss; 3.2 Cu, 0.8 Mg, 0.3 Mn, 1 Ni, bal Al.
56,000 TS; 36,000 YS; 18 El; 110 Brin.
For light alloy parts; age-hardenable.

MERCOLOY.
M-519; Cu alloy.
For valves; corrosion resistant.

MERCURY.
M-1755; Hg.
Purities: U.H.P. (99.99999%), 99.9999%, 99.999%, triple distilled (99.99%).
Packed in: Vacuum sealed containers, flasks, poly bottles.

MERICO-1.
M-1436; 0.35 C, 0.77 Cr, 0.5 Co, 0.75 Mo, 0.73 Cu, bal Fe.
For blacksmith tools, chisels, punches; water hardened, non-tempering.

MERICO-2.
M-1436; 0.40 C, 0.77 Cr, 0.75 Mo, 0.75 Cu, 0.6 Si, 0.4 Mn, bal Fe.
For shear blades, rivet sets, pneumatic tools; oil hardened, non-tempering.

MERICO TOOL STEEL.
M-1436; 0.35 C, 0.40 Mn, 0.64 Si, 0.77 Cr, 0.50 Ni, 0.76 Mo, 0.73 Cu, bal Fe.
159,000 psi TS; 98,000 psi YS; 22 El; 34 Rc.
Hot work tool steel.

MERICROME ALLOY STEEL.
M-1436; 0.40 C, 1.7 Mn, 0.33 Mo, 0.25 max Cr + Ni, bal Fe.
Ht: 155,000 psi TS; 135,000 psi YS; 21 El; 315 Brin.

MERIDIAN ABRASION RESISTANT STEEL.
M-1436; 0.45-0.50 C, 0.90-1.0 Mn, 0.60-0.70 Cr, 0.25-0.35 Mo, 0.30-0.4 Si, 0.70-0.80 Ni, 0.05 Ti, 0.10 Zr, bal Fe.
200,000-210,000 psi TS; 180,000-186,000 YS; 16 El; 45 RA; 42 Rc.
For parts to resist wear and abrasion.

MERIDIAN AR STEEL.
M-1436; 0.60-1.4 C, 0.25 V, bal Fe.
For drills, reamers, lathe and planer tools, hobs; Type W2; water hardened.

MERIDIAN CARBIDE HIGH SPEED.
M-1436; 0.65 C, 4.5 Cr, 18 W, 2 V, 0.75 Mo, 12 Co, bal Fe.
For cutting tools, drills, reamers, hobs; high speed steel.

MERIDIAN DIE STEEL.
M-1436; 1 C, 5 Cr, 1 Mo, bal Fe.
For dies, rolls, spindles, punches; Type A2; oil hardened, non-deforming

MERIDIAN DIE "O" STEEL.
M-1436; 0.90 C, 1 Mn, 0.5 W, bal Fe.
For cold work tools, header and blanking dies; Type O1; oil hardened, Non-deforming.

MERIDIAN HS STEEL.
M-1436; 0.80 C, 4 Cr, 2 V, 18 W, bal Fe.
For drills, taps, hobs, reamers, lathe cutters; Type T2; high speed steel.

MERIDIAN SUPER C HIGH SPEED TOOL BITS.
M-1436; 1.25 C, 4.1 Cr, 3.1 Mo, 3.1 V, 9.0 W, 12.0 Co, bal Fe.
High speed steel; Rc 67.

MERIT METAL.
M-815; Sb, Sn, bal Pb.
For bearings; Babbitt.

MERMAID.
M-1737; 0.75 C, 4.25 Cr, 18.0 W, 1.1 V, bal Fe.
High speed steel for cutting tools.
AISI T1.

MERTEN.
M-1436; C, alloy, bal Fe.
For machine tool parts; fatigue and wear resistant..

MERTEN ALLOY STEEL.
M-1436; 0.40 C, 0.85 Mn, 0.35 Si, 1.0 Cr, 1.0 Ni, 0.45 Mo, bal Fe.
165,000 psi TS; 142,000 psi YS; 18 El; 325 Brin.
Hot work tool steel.

MESSING 63.
M-297; 37 Zn, bal Cu.
Soft: 48,000 TS; 55 El; 70 Brin.
1/2H-temper: 58,000 TS; 32 El; 100 Brin.
Hard: 69,000 TS; 20 El; 135 Brin.
For hardware; corrosion resistant, brass.

MESSING 63PB.
M-297; 37 Zn, 2 Pb, bal Cu.
Soft: 48,000 TS; 55 El; 70 Brin.
1/2Hard: 58,000 TS; 32 El; 100 Brin.
Hard: 69,000 TS; 20 El; 136 Brin.
For hardware, screw machine products; free-cutting, leaded brass.

MESSING 72.
M-297; 28 Zn, bal Cu.
Soft: 47,000 TS; 54 El; 70 Brin.
1/2H-temper: 58,000 TS; 30 El; 100 Brin.
Hard: 65,000 TS; 16 El; 130 Brin.
For hardware; corrosion resistant, brass.

MESSING 75.
M-297; 25 Zn, bal Cu.
Soft: 47,000 TS; 54 El; 70 Brin.
1/2H-temper: 58,000 TS; 29 El; 100 Brin.
Hard: 66,000 TS; 15 El; 125 Brin.
For hardware; corrosion resistant, brass.

MESSING 80.
M-297; 20 Zn, bal Cu.
Soft: 44,000 TS; 63 El; 65 Brin.
1/2H-temper: 52,000 TS; 28 El; 95 Brin.
Hard: 62,000 TS; 15 El; 120 Brin.
For hardware, ornamental parts; corrosion resistant, brass.

MESSING 85.
M-297; 15 Zn, bal Cu.
Soft: 42,000 TS; 50 El; 65 Brin.
1/2H-temper: 52,000 TS; 26 El; 95 Brin.
Hard: 61,000 TS; 12 El; 115 Brin.
For ornamental and decorative parts; corrosion resistant, brass.

MESSING 90.
M-297; 10 Zn, bal Cu.
Soft: 40,000 TS; 48 El; 60 Brin.
1/2H-temper: 48,000 TS; 24 El; 90 Brin.
Hard: 56,000 TS; 10 El; 110 Brin.
For ornamental and decorative parts; corrosion resistant, brass.

MESTA SPECIAL.
M-852; 0.4 C, 1.5 Ni, 0.8 Cr, bal Fe.
For rolls; cast steel.

METALINE.
M-990; 10 Sn, bal Cu.
For oil-less bronze bearings; sintered.

METALJOINER.
M-744; Pb-Sn-Cd.
For solder; M.P. 650°F.

METALLIC PACKING.
M-U.S.; 82.25 Pb, 4.75 Sn, 13 Sb.
For metallic packing.

METALLINE.
M-U.S.; 35 Co, 10 Fe, 30 Cu, 25 Al.
For tools; heat and corrosion resistant.

METAMIC 247.
M-1519; 75 Ni, 25 mullite.
Sintered: 200-300 Brin.
For bearings for 500-900°C operating temperature; cermet.

METAMIC LT-1.
M-1519; 70 Cr, 30 Al_2O_3.
Sintered: 350 Brin.
For gas turbine blades; sintered, refractory.

METARSAL.
M-1783; 76-79 Cu, 2 Al, 0.2-1.0 As, bal Zn.
Heat exchanger tubing.
Cu Zn 22 Al 2 As.

METARSIC.
M-1783; 7.0 Cu, 0.5 As, bal Zn.
Tube for heat exchanger operations.
Cu Zn 30 As.

METARSTAN.
M-1783; 70 Cu, 1 Sn, 0.2-0.10 As, bal Zn.
Heat exchanger tubing.
Cu Zn 29 Sn 1 As; ASTM B 111, Alloy 443.

METCO 12C.
M-853; 2.5 Fe, 10 Cr, 0.15 C, 2.5 Si, 2.5 B, bal Ni.
Corrosion resistant metallic powders for spraying hard coating.

METCO 14E-14F.
M-853; 4 Fe, 14 Cr, 0.6 C, 3.5 Si, 2.75 B, bal Ni.
Corrosion resistant metallic powders for spraying hard coating.

METCO 15E-15F.
 M-853; 4 Fe, 17 Cr, 1.0 C, 4 Si, 3.5 B, bal Ni.
 Corrosion resistant metallic powder for spraying hard coating.

METCO-15F.
 M-853; 17 Cr, 4 Fe, 4 Si, 3.5 B, 1 C, bal Ni.
 Cast: Rockwell C 60.
 For hard facing application.
 Self fluxing metal spray.
 M.P. 1875°F., wear resistant.

METCO 16C.
 M-853; 2.5 Fe, 16 Cr, 0.5 C, 4 Si, 4 B, 3 Cu, 3 Mo, bal Ni.
 Corrosion resistant metallic powders for spraying hard coating.

METCO 18C.
 M-853; 2.5 Fe, 18 Cr, 0.2 C, 3.5 Si, 3 B, 6 Mo, 40 Co, bal Ni.
 Corrosion resistant metallic powders for spraying hard coating.

METCO 19E.
 M-853.
 Self-fluxing, hard facing alloy powder for flame spraying. Hardness 55-60 Rc.
 Excellent wear resistance; for plug gages, fuel rod mandrels, cam followers.

METCO 31C.
 M-853; 2.5 Fe, 46 Ni, 11 Cr, 0.50 C, 2.5 Si, 2.5 B, 35 WC-Co aggregate.
 Metallic powder for spraying; gives extreme wear-resistant coating.

METCO 32C.
 M-853; 0.8 Fe, 14 Ni, 3.5 Cr, 0.1 C, 0.8 Si, 0.8 B, 80 WC-Co aggreate.
 Metal powder for spraying; gives extreme wear resistant coating.

METCO 34F-34FP.
 M-853; 3.5 Fe, 33 Ni, 9 Cr, 0.5 C, 2 Si, 2 B, 50 WC-Co aggregate.
 Metal powder for spraying; gives extreme wear resistant coating.

METCO 41C.
 M-853; 12 Ni, 17 Cr, 0.1 C, 1 Si, 2.5 Mo, bal Fe.
 Corrosion resistant alloy for metal spraying.

METCO 42C.
 M-853; 2 Ni, 16 Cr, 0.2 C, bal Fe.
 Corrosion resistant alloy for metal spraying.

METCO 43C, 443CNS, 43F, 43F-NS.
 M-853; 80 Ni, 80 Cr.
 Corrosion resistant alloy for metal spraying.

METCO 54.
 M-853; 99.0 + Al.
 For metal spraying.

METCO 55.
 M-853; 99.0 + Cu.
 For metal spraying.

METCO 56F-NS.
 M-853; 99.3 Ni.
 For metal spraying.

METCO 63 63 NS.
 M-853; 99.0 + Mo.
 For metal spraying.

METCO 70C-NS.
 M-853; 99.0 + Chrome carbide.
 For metallic spraying.

METCO 71-NS.
 M-853; 1 Fe, 4 C, 12 Co, bal WC-Co aggregate.
 For metal spraying.

METCO 72F-NS.
 M-853.
 Tungsten carbide/cobalt powder.
 For spraying guillotine knives, can-making seaming chucks and rails, jet engine parts.

METCO 80-NS.
 M-853; 12 Ni, 3 Cr, 85 Chrome carbide.
 For metal spraying.

METCO 81-NS.
 M-853; 20 Ni, 5 Cr, 75 Chrome carbide.
 For metal spraying.

METCO 404.
 M-853; 80 Ni, 20 Al.
 Metal powder for spraying; self-bonding.

METCO 439.
 M-853; 1.5 Fe, 6 Cr, 0.5 C, 1.5 Si, 1.0 B, 3 Al, 50 WC-Co aggregate, bal Ni.
 Metal powder for spraying; self-bonding and extreme wear resistance.

METCO 450.
 M-853; 4.5 Al, bal Ni.
 Metal powder for spraying; self-bonding.

METCO 451.
 M-853; 9.5 Cr, 2.5 Si, 1.5 B, 0.5 Al, bal Ni.
 Metal powder for spraying; self-bonding.

METCOLOY NO. 1.
 M-853; C, Cr, Ni, bal Fe.
 For metal spraying; stainless steel wire.

METCOLOY NO. 2.
 M-853; C, Cr, Ni, bal Fe.
 For metal spraying; stainless steel wire.

METCO MONEL METAL.
 M-863; 67 Ni, 23 Cu.
 For metal spraying; Monel metal wire.

METEOR.
 M-57; 1.25 C, 0.25 Cr, 0.15 V, 1.5 W, bal Fe.
 For taps, punches, dental burrs; oil hardening.

METILLURE.
 M-France; 14-15 Si, 0.7 Mn, bal Fe.
 For drains, anodes, evaporators; acid resistant, brittle.

METONAL 10P.
 M-1783; 9 Al, 2 Ni, 1 Fe, bal Cu.
 Cu-Al condenser tube plates.

METONAL 11.
 M-1783; 6 Al, bal Cu.
 Cu-Al tubing for heat exchanger applications.
 Cu Al 6; ASTM B111, Alloy 608.

METONAL 15.
 M-1783; 9 Al, 5 Ni, 3 Fe, bal Cu.
 Cu-Al-Ni alloy for condenser tube plates.
 Cu Al 9 Ni 5 Fe 3; BSS 2875, CA 105.

METONAL 20.
 M-1783; 7 Al, 2 Fe, bal Cu.
 Cu-Al tubing for heat exchanger applications.
 Cu Al 7 Fe 2; BSS 2871, Alloy CA 102.

METONIC 10.
 M-1783; 9.0-11.0 Ni, 1.0-1.8 Fe, 1.0 max Mn, bal Cu.
 Tubing for heat-exchanger applications.
 Cu Ni 10 Fe 1 Mn; ASTM B111, Alloy 706.

METONIC 30.
 M-1783; 30 Ni, 1 Mn, 1 max Fe, bal Cu.
 Tubing for heat exchanger applications.
 Cu Ni 3 Mn 1 Fe; ASTM B111, Alloy 715.

METORITE.
 98-94 Al, 1-4 P, 1-2 Zn.
 For light alloy parts; non-hardenable.

METROL NO 610-FM see **CARPENTER NO 610-FM.**

METSPEC 117.
 M-1753.
 Bismuth base low melting alloy.
 M.P. 117°F (47.5°C); 5400 psi TS; 12 Brin.

METSPEC 136.
 M-1753.
 Bismuth base low melting alloy.
 M.P. 136°F (58°C); 6300 psi TS; 14 Brin.

METSPEC 158.
 M-1753.
 Bismuth base low melting alloy.
 M.P. 158°F (70°C); 5990 psi TS; 9 Brin.

METSPEC 158/190.
 M-1753.
 Bismuth base low melting alloy.
 Melt range 159-190°F (70-88°C); 5400 psi TS; 9 Brin.

METSPEC 255.
 M-1753.
 Bismuth base low melting alloy.
 M.P. 255°F (124°C); 6400 psi TS; 10 Brin.

METSPEC 281.
 M-1753.
 Bismuth base low melting alloy.
 M.P. 281°F (138.5°C); 8000 psi TS; 22 Brin.

METSPEC 281/338.
 M-1753.
 Bismuth base low melting alloy.
 Melt range: 281-338°F, (138.5-170°C); 8000 psi TS; 22 Brin.

MFA (LEAD) see **SJMFA.**

MFRS.
 M-912; 0.40 C, 1.50 Mn, 1.95 Cr, 0.20 Mo, 0.10 V, bal Fe.
 Hardenable steel for plastic moulds.
 W.-Nr. 2312.

M.G.R.
 M-73; 0.55 C, 1 Si, 0.3 Mn, 1.25 W, 5 Cr, 1.25 Mo, bal Fe.
 Air hardening tool steel, tough and wear resistant, for punches, rivet sets, tools and dies.
 AISI A8

M.H. ALLOY NO. 6.
 M-212; 77.5 Cu, 7 Sn, 14 Pb, 1.5 Ni.
 30,000-36,000 TS; 17,000-19,000 YS; 20-28 El; 21-25 RA; 64 Brin.
 For machinery bearings for high speed and medium pressures; "Bar Alloy No. 6;" acid resistant.

M H ALLOYS see **BARR ALLOYS.**

M-H COMPOSITE.
 M-299; C, alloy, bal Fe.
 For heavy duty rollers; in sheet mills.

M-H SPECIAL ALLOY STEEL.
 M-229; C, alloy, bal Fe.
 For rolls for slabbing mill; water hardened.

MIAMI FAST CUT.
 M-U.S.; 1-2-1.3 C, bal Fe.
 For cutters, broaches; fast finishing.

MIAMI NO CHARGE.
 M-U.S.; 0.9 C, 1.2 Mn, bal Fe.
 For tools, dies, broaches, punches, shears; non-deforming.

MICHALLOY.
 M-U.S.; C, Ni, Cr, bal Fe.
 For grinding balls, mill liners, pumps; high abrasion resistance; Ni-Hard.

MICHIANA 49.
 M-184; 18 Cr, 8 Ni, 0.2 C, bal Fe.
 For chemical engineering equipment; stainless.

MICHIANA 55.
M-184; 0.5 C, 30 Cr, 2 Ni, bal Fe.
Cast: 45,000 TS; 30,000 YS; 1 El; 2 RA; 250 Brin.
For parts in contact with brass and copper at rolling temperature; heat and corrosion resistant.

MICHIANA 111.
M-184; 25 Cr, 12 Ni, 3.5 Mo, 0.2 C, bal Fe.
For chemical engineering equipment; stainless, heat resistant.

MICHIANA NO. 48.
M-184; 28 Cr, 8 Ni, 0.50 Max C, bal Fe.
Cast: 75,000 TS; 55,000 YS; 1 El; 1 RA; 220 Brin.
For resistance to sulfurous gases at elevated temperatures; corrosion and heat resisting.

MICHIANA NO. 49 A.
M-184; 0.2 C, 1.5 max Mn, 2.0 max Si, 18-21 Cr, 8-11 Ni, bal Fe.
Cast: 70,000 TS; 30,000 YS; 30 El.
For castings; stainless.

MICHIANA NO. 49 CB.
M-184; 0.08 C, 1.5 max Mn, 2.0 max Si, 18-21 Cr, 9-12 Ni, C = 8 x Cb, bal Fe.
Cast: 70,000 TS; 30,000 YS; 35 El.
For welded castings; stainless, stabilized.

MICHIANA NO. 49 MO.
M-184; 0.08 C, 1.5 max Mn, 1.5 max Si, 18-21 Cr, 9-12 Ni, 2-3 Mo, bal Fe.
Cast: 70,000 TS; 30,000 YS; 30 El.
For castings; stainless.

MICHIANA NO. 63.
M-184; C, 28 Cr, 15 Ni, bal Fe.
Cast.
For resistance to sulfurous gases at high temperatures and also molten salts; heat and corrosion resistant.

MICHIANA NO. 100.
M-184; 25 Cr, 12 Ni, 0.50 max C, bal Fe.
Cast: 63,000-65,000 TS; 40,000-45,000 YS; 2.5-3.5 El; 3-5 RA; 210 Brin.
For grids, furnace parts; corrosion and heat resisting.

MICHIANA NO 100S.
M-184; 0.21-0.35 C, 24-30 Cr, 0.25 S, 12-15 Ni, bal Fe.
For heat anc corrosion resistant parts; heat and corrosion resistant.

MICHIANA NO. 119.
M-184; 25 Cr, 20 Ni, 1.25 Si, 1.5 Mn, 0.2 C, bal Fe.
Cast.
For heat and corrosion resisting parts; heat and corrosion resisting.

MICHIANA NO. 122.
M-184; 0.2 C, 18 Cr, 8 Ni, 3.5 Mo, bal Fe.
For use in sulfite paper industry; corrosion and heat resistant.

MICHIANA NO. 147.
M-184; 0.15 max C, 1 max Mn, 1.5 max Si, 11-14 Cr, 1 max Ni, bal Fe.
Cast: 90,000 TS; 65,000 YS; 18 El.
For castings; corrosion resistant.

MICHIANA NO. 233.
M-184; 20 Ni, 3 Cr, 1.5 Si, 1.5 Mn, 2.2-2.4 C, bal Fe.
For chemical handling equipment; corrosion resisting.

MICHIANA NO. 241.
M-184; 2.6-2.8 C, 28 Cr, 1.25 Si, 1.0 Mn, bal Fe.
For coal-coke handling equipment; abrasion resisting.

MICHIGAN APEX.
M-545; 86.5 Cu, 3.5 Fe, 10 Al.
Cast: 85,000 TS; 35,000 YS; 20 El; 160 Brin.
For castings, gears; Al bronze.

MICHIGAN GRADE A.
M-545; 56 Cu, 3 hardener, bal Zn.
Cast: 70,000-80,000 TS; 30,000-35,000 YS; 25-40 El; 25-34 RA; 114-140 Brin.
For castings; Mn bronze.

MICHIGAN GRADE AX.
M-545; 57 Cu, 4 hardener, bal Zn.
Cast: 80,000-90,000 TS; 40,000-46,000 YS; 25-40 El; 20-30 RA; 120-130 Brin.
For castings; Mn bronze.

MICHIGAN GRADE C.
M-545; 56 Cu, 2.5 hardener, bal Zn.
Cast: 60,000-70,000 TS; 25,000-40,000 YS; 15-25 El; 30-40 RA; 80-95 Brin.
For castings; Mn bronze.

MICHIGAN GRADE X.
M-545; 56 Cu, 6 hardener, bal Zn.
Cast: 85,000-95,000 TS; 40,000-45,000 YS; 20-30 El; 20-25 RA; 135-155 Brin.
For castings; Mn bronze.

MICHIGAN GRADE XX.
M-545; 57 Cu, 9 hardener, bal Zn.
Cast: 95,000-110,000 TS; 45,000-55,000 YS; 20-30 El; 18-25 RA; 150-175 Brin.
For castings; Mn bronze.

MICHIGAN GRADE XXX.
M-545; 62 Cu, 13 hardener, bal Zn.
Cast: 110,000-125,000 TS; 75,000-90,000 YS; 12-18 El; 12-17 RA; 200-240 Brin.
For castings; Mn bronze.

MICHIGAN NO. 90.
M-545; 90 Cu, 10 Al.
Cast: 77,000 TS; 30,000 YS; 30 El; 120 Brin.
Heat treated: 90,000 TS; 55,000 YS; 5 El; 160 Brin.
For castings, gears; Al bronze.

MICHIGAN NO. 90H.
M-545; 84.5 Cu, 3.5 Fe, 12 Al.
Cast: 85,000 TS; 40,000 YS; 10 El; 160 Brin.
Heat treated: 95,000 TS; 50,000 YS; 5 El; 200 Brin.
For castings; Al bronze.

MICHIGAN NO. 90M.
M-545; 85.5 Cu, 3.5 Fe, 11 Al.
Cast: 80,000 TS; 40,000 YS; 10 El; 166 Brin.
Heat treated: 90,000 TS; 50,000 YS; 7 El; 200 Brin.
For castings; Al bronze.

MICHIGAN NO. 90S.
M-545; 88 Cu, 3.5 Fe, 8.5 Al.
Cast: 65,000-85,000 TS; 22,000-30,000 YS; 20-30 El; 115-135 Brin.
For castings, gears; Al bronze.

MICHIGAN NO. 90V.
M-545; 79 Cu, 5 Fe, 5 Ni, 11 Al.
Cast: 90,000-105,000 TS; 45,000-60,000 YS; 3-7 El; 190-215 Brin.
For castings; Al bronze.

MICRO.
M-282; 98 Cu, 2 Ni.
Cold worked: 37,500 TS; 23,500 YS; 39 El; 80 Brin.
Extruded: 33,600 TS; 6000 YS; 55 El; 59 Brin.
For locomotive boiler tubes and plates; corrosion resistant.

MICROFLAT.
M-1484.
Cold rolled, stretcher leveled sheet and coils.

MICROFLEX.
M-1484.
Soft tempered austenitic stainless sheet for roofing, flashing and architectural work.

MICRO MACH.
0.08-0.12 C, 17.3 Cr, 6.2 Ni, bal Fe.
Long: 200,000 TS; 180,000 YS; 12 El.
Trans.: 209,000 TS; 175,000 YS; 13 El.
For aircraft and missile wing and skin surfaces; austenitic, stainless.

MICROROLD.
M-1484.
Sendzimir mill precision rolled sheet and strip steel products; all products.

MICROSIL.
M-1089; 97 Fe, 3 Si; grain oriented.
Generally used for high power, relatively low frequency applications in high performance power transformers, saturable reactors, inverter transformers, magnetic amplifiers.

MIDFLEX.
M-237; bimetal.
For thermo-metal; bimetal element.

MIDLING HARD 115.
M-1182; 1.1 C, 0.25 Si, bal Fe.
For tools; drill rod.

MIDOHM.
M-44; 22-23 Ni, bal Cu.
Wrought: 100,000-50,000 TS.
For resistances, rheostats; load banks; heat resistant; max operating temperature 200°C.

MIDVAC-422.
M-85; 0.25 C, 12 Cr, 0.9 Mo, 0.9 Ni, 0.2 V, 0.9 W, bal Fe.
Annealed: 85,000 TS; 42,000 YS; 22 El; B 95 Rock.
Hardened: 240,000 TS; 205,000 YS; 9 El; C 50 Rock.
For cutlery, surgical instruments, hardware, bearings, valves.
Corrosion resistant, hardenable.

MIDVALOY 20.
M-85; 0.07 max C, 29 Ni, 20 Cr, 2-3 Mo, 4 Cu, bal Fe.
Cast: 65,000-75,000 TS; 28,000-38,000 YS; 50-35 El; 50-40 RA; 120-150 Brin.
For chemical plant equipment, tanks; resists mixed acids, austenitic.

MIDVALOY 1300.
M-85; 0.07-0.40 C, 13 Cr, bal Fe.
Wrought: 70,000-200,00 TS; 40,000-165,000 YS; 28-13 El; 70-53 RA; 130-400 Brin.
For corrosion resisting parts; engineering construction; corrosion and abrasion resisting.

MIDVALOY 1808.
M-85; 0.07-0.20 C, 18 Cr, 8 Ni, bal Fe.
For stainless parts, chemical plant equipment; stainless, austenitic.

MIDVALOY 1808 M.
M-85; 0.07 C, 18 Cr, 8 Ni, 2.1-3.5 Mo, bal Fe.
Annealed: 70,000 TS; 35,000 YS; 40 El; 60 RA; 160 Brin.
For corrosion resistant parts, chemical plant equipment; stainless, austenitic.

MIDVALOY 2512-7.
M-85; 0.07 C, 23-26 Cr, 11-13 Ni, bal Fe.
Rolled: 70,000-115,000 TS; 40,000-60,000 YS; 9-45 El; 240-230 Brin.
For heat and corrosion resisting parts; heat and corrosion resistant.

MIDVALOY 2512-10.
M-85; 0.10 C, 12 Ni, 25 Cr, bal Fe.
Forged: 75,000-115,000 TS; 45,000-65,000 YS; 25-45 El; 30-50 RA; 150-240 Brin.
Cast: 65,000-85,000 TS; 40,000-50,000 YS; 30-35 El; 30-40 RA; 130-195 Brin.
Annealed: 113,000 TS; 62,700 YS; 43 El; 49 RA; 196 Brin.
For apparatus to resist nitric and acetic acids, sulfite liquors and sulfurous acids; corrosion and heat resisting.

MIDVALOY 2512-16.
M-85; 0.16 C, 25 Cr, 12 Ni, bal Fe.
For automatic furnaces, skid rails, walking beams; heat resistant to 1850°F.

MIDVALOY 2520.
M-85; 25 Cr, 0.25 max C, 20 Ni, 0.18 Mo, bal Fe.
Cast: 70,000-80,000 TS; 30,000-36,000 YS; 25-50 El; 20-50 RA; 145 Brin.
Wrought: 75,000-165,000 TS; 30,000-120,000 YS; 48-5 El; 55-30 RA; 160 Brin.
For corrosion resisting apparatus; corrosion resisting, austenitic.

MIDVALOY ATV-3.
M-85; 26 Ni, 14.5 Cr, 1.2 Mn, 0.45 C, 4 W, bal Fe.
Cast: 80,000 TS; 43,000 YS; 20 El; 20 RA;
Wrought: 100,000 TS; 50,000 YS; 33 El; 45 RA; 185 Brin.
For exhaust valves, gas turbine rotors; austenitic; heat resisting.

MIDVALOY NO. 11.
M-85; 0.35-0.5 C, 0.6-0.9 Cr, 3.0 Ni, bal Fe.
Heat treated: 275,000-135,000 TS; 245,000-110,000 YS; 10-25 El; 38-58 RA; 500-275 Brin.
For airplane crankshafts, gears, splines, transmission units; shock resistant; (0.45-0.50 C alloy).

MIDVALOY NO. 11-MO.
M-85; 0.4 C, 0.6-0.9 Cr, 2.6-3.2 Ni, 0.2-0.3 Mo, bal Fe.
Heat treated: 125,000-160,000 TS; 100,000-140,000 YS; 22-15 El; 55-45 RA; 250-325 Brin.
For shafts, gears; tough.

MIGRA IRON NO. 1.
M-403; 1-3 Si, 0.4-0.9 Mn, 3.8-4.1 C, bal Fe.
Cast.
For pistons and cylinders; cast iron.

MIGRA IRON NO. 2.
M-403; 1-3 Si, 0.4-0.9 Mn, 3.8-4.1 C, bal Fe.
For pistons and cylinders; cast iron.

MIGRA IRON NO. 3.
1-3 Si, 3.8-4.1 C, 1-2 Ni, 0.5-1.0 Cr, 1.0-1.5 Mn, bal Fe.
For pistons and cylinders; cast iron.

MIKADO.
M-1306; 1.45 C, 1.4 Cr, 0.6 Mn, 0.25 Si, bal Fe.
For bearings, bushings, liners, sleeves; water hardened, wear resistant.

MIL 48.
M-365; 0.55 C, 2.0 Si, 0.25 Cr, 0.25 V, bal Fe.
Water hardening tool steel, shock resisting type; AISI S4.

MILBRITE.
M-365; 0.50 C, 1.0 Mn, 0.3 Si, 1.1 Cr, 0.25 Mo, bal Fe.
Oil hardening steel designed for molds; similar to AISI 4150.

MILCO 9.
M-365; 0.73 C, 0.28 Mn, 4 Cr, 1.75 V, 18.25 W, 8.5 Co, bal Fe.
For tools, cutters, reamers; high speed steel.

MILLALOY.
M-855; 0.4 C, 4 Ni, 1.5 Cr, bal Fe.
Heat treated: 312,000 TS; 272,000 YS; 11 El; 35 RA; 532 Brin.
For shear blades; oil hardened.

MILL BRASS MIX.
M-741; Sn, Pb, bal Cu.
For bearings, bushings; tough.

MILLENITE.
M-857; 0.3 C, 1-5 Ni, 0.8 Cr, bal Fe.
For high duty castings; oil hardened.

MILLER 200-PLUS.
M-1144; 4.8-9.8 Sn, 0.15 P, bal Cu.
Rolled: 96,000-118,00 TS; 70,000-85,00 YS; 24-20 El; 210-240 Brin.
For clips, springs, bellows, diaphragms; phosphor bronze.

MILLER 200-PLUS GR.A.
M-1144; 4.82 Sn, 0.18 P, 95 Cu.
Spring: 96,000 TS; 70,000 TS; 22 El; 210 Brin.
For diaphragms, bellows, springs, clips, fasteners; high ductility, corrosion resistant.

MILLER 200 PLUS GR. C.
M-1144; 7.82 Sn, 0.18 P, 92 Cu.
Spring: 110,000 TS; 80,000 YS; 24 El; 235 Brin.
For diaphragms, bellows, springs, clips, fasteners; high ductility, corrosion resistant.

MILLE 200-PLUS GR.D.
M-1144; 9.85 Sn, 0.15 P, 90 Cu.
Spring: 118,000 TS; 85,000 YS; 20 El; 240 Brin.
For diaphragms, bellows, springs, clips, fasteners; high ductility, corrosion resistant.

MILLING.
M-Eng.; 54-56 Cu, 27.5-31 Zn, 15-18 Ni, 0.5-1.0 Pb.
For white metal parts; easy to machine.

MILLING SILVER-1.
 M-Eng.; 56 Cu, 31 Zn, 12 Ni, 1 Pb.
 For white metal parts; easy to machine.

MILLING SILVER-2.
 M-Eng.; 56 Cu, 27.5 Zn, 16 Ni, 0.50 Pb.
 For white metal parts; easy to machine.

MILLS DTD 361B.
 M-187; 4.5 Cu, bal Al.
 Aluminium alloy casting; chill cast, solution treated and aged.
 BS Aerospace DTD 361B.

MILLS DTD 716A.
 M-187; 0.5 Mg, 5.0 Si, bal Al.
 Aluminium alloy casting; chill cast.
 BS Aerospace DTD 716A.

MILLS DTD 722A.
 M-187; 0.5 Mg, 5.0 Si, bal Al.
 Aluminium alloy casting; chill cast and aged.
 BS Aerospace DTD 722A.

MILLS DTD 727A.
 M-187; 0.5 Mg, 5.0 Si, bal Al.
 Aluminium alloy casting; chill cast and solution treated.
 BS Aerospace DTD 727A.

MILLS DTD 735A.
 M-187; 0.5 Mg, 5.0 Si, bal Al.
 Aluminium alloy casting; chill cast, solution treated and aged.
 BS Aerospace DTD 735A.

MILLS DTD 741A.
 M-187; 4.2 Cu, 2.0 Mg, 0.7 Co, 0.2 Nb, bal Al.
 Aluminium alloy casting; chill cast, solution treated and aged.
 BS Aerospace DTD 741A.

MILLS DTD 5008A.
 M-187; 0.6 Mn, 5.0 Zn, 0.2 Ti, 0.5 Cr, bal Al.
 Aluminium alloy casting; chill cast and aged.
 BS Aerospace DTD 5008A; AA D712.

MILLS DTD 5018.
 M-187; 7.6 Si, 0.2 Mn, 1.2 Zn, bal Al.
 Aluminium alloy sand casting.
 BS Aerospace DTD 5018.

MILLS DTD 5028.
 M-187; 7.0 Si, 0.3 Mg, bal Al.
 Aluminium alloy sand casting.
 BS Aerospace DTD 5028; AA A356.

MILLS L33.
 M-187; 11.0 Si, bal Al.
 Aluminium alloy, chill cast.
 BS 1490 LM-6M.

MILLS L35.
 M-187; 4.0 Cu, 1.5 Mg, 2.0 Ni, 0.2 Ti, bal Al.
 Aluminium alloy casting (Y Alloy).
 BS Aerospace L35; AA 242.

MILLS L51.
 M-187; 1.0 Cu, 2.5 Si, 0.9 Ni, 1.0 Fe, 0.2 Ti, bal Al.
 Aluminium alloy casting; chill cast and aged.
 BS Aerospace L51.

MILLS L52.
 M-187; 2.5 Cu, 1.0 Mg, Ni, Si, Fe, bal Al.
 Aluminium alloy casting; chill cast.

MILLS L53.
 M-187; 10.0 Mg, bal Al.
 Aluminium alloy casting; chill cast, solution treated.
 BS 1490 LM 10TB; AA 520.

MILLS L78.
 M-187; 1.25 Cu, 0.5 Mg, 5.0 Si, bal Al.
 Aluminium alloy casting; chill cast.
 BS 1490 LM-16TF; AA 355.

MILLS L91.
 M-187; 4.5 Cu, bal Al.
 Aluminium alloy sand casting.
 BS Aerospace L91; AA 295.

MILLS L92.
 M-187; 4.5 Cu, bal Al.
 Aluminium alloy sand casting.
 BS Aerospace L92; AA 295.

MILLS L99.
 M-187; 0.3 Mg, 7.0 Si, bal Al.
 Aluminium alloy sand casting.
 BS Aerospace L99; AA A356.

MILLS LMOM.
 M-187; 99.5 Al.
 Aluminium casting.
 BS 1490 LM 0M.

MILLS LM2M.
 M-187; 1.5 Cu, 10.0 Si, bal Al.
 Silicon-aluminium die casting, as cast.
 BS 1490 LM2M.

MILLS LM4M.
 M-187; 3.0 Cu, 5.0 Si, 0.5 Mn, bal Al.
 Aluminium die casting, as cast.
 BS 1490. LM4M; AA 319.

MILLS LM4TF.
 M-187; 3.0 Cu, 5.0 Si, 0.5 Mn, bal Al.
 Aluminium die casting alloy; solution treated and precipitation treated.
 BS 1490 LM4WP.

MILLS LM5M.
 M-187; 5.0 Mg, 0.5 Mn, bal Al.
 Sand or die cast aluminium alloy, as cast.
 BS 1490 LM5M.

MILLS LM6M.
 M-187; 11 Si, bal Al.
 Aluminium sand or die casting alloy.
 As chill cast.
 BS 1490 LM 6M.

MILLS LM9M.
M-187; 0.4 Mg, 12.0 Si, 0.5 Mn, bal Al.
Aluminium sand or die casting; as chill cast.
BS 1490 LM9.

MILLS LM9TE.
M-187; 0.4 Mg, 12.0 Si, 0.5 Mn, bal Al.
Aluminium alloy casting; precipitation treated.

MILLS LM9TF.
M-187; 0.4 Mg, 12.0 Si, 0.5 Mn, bal Al.
Aluminium alloy casting; solution treated and precipitation treated.
BS 1490 LM9WP.

MILLS LM10TB.
M-187; 10 Mg, bal Al.
Magnesium-Aluminium sand or die casting. Solution treated.
BS 1490 LM10W.

MILLS LM12M.
M-187; 10.0 Cu, 0.3 Mg, bal Al.
Aluminium alloy casting; as chill cast.
AA 222.

MILLS LM13TE.
M-187; 1.0 Cu, 1.0 Mg, 12.0 Si, 2.0 Ni, bal Al.
Aluminium alloy sand or die casting; precipitation treated.
AA A332.

MILLS LM13TF.
M-187; 1.0 Cu, 1.0 Mg, 12.0 Si, 2.0 Ni, bal Al.
Aluminium alloy sand or die casting; chill cast, solution treated, and aged.
BS 1490 LM13WP.

MILLS LM13TF7.
M-187; 1.0 Cu, 1.0 Mg, 12.0 Si, 2.0 Ni, bal Al.
Aluminium alloy sand or die casting; full heat treatment plus stabilization.

MILLS LM16TB.
M-187; 1.0 Cu, 0.5 Mg, 5.0 Si, bal Al.
Aluminium alloy sand or die casting; chill cast and solution treated.
AA 355.

MILLS LM16TF.
M-187; 1.0 C, 0.5 Mg, 5.0 Si, bal Al.
Aluminium alloy sand or die casting; chill cast, solution treated and aged.
BS 1490 LM16WP.

MILLS LM18M.
M-187; 5 Si, bal Al.
Aluminium sand or die casting alloy; as chill cast.
BS 1490 LM18m.

MILLS LM20M.
M-187; 12 Si, bal Al.
Aluminium die casting alloy; as chill cast.
BS 1490 LM20M.

MILLS LM21M.
M-187; 4.0 Cu, 6.0 Si, 0.5 Mn, bal Al.
Aluminium alloy sand or die casting; as chill cast.
BS 1490 21M; AA319.

MILLS LM22TB.
M-187; 3.0 Cu, 5.0 Si, 0.5 Mn, bal Al.
Aluminium alloy casting; solution treated.
BS 1490 LM22W.

MILLS LM24M.
M-187; 3.5 Cu, 8.0 Si, bal Al.
Aluminium alloy die casting; as cast.
BS 1490 LM24M.

MILLS LM25M.
M-187; 0.3 Mg, 7.0 Si, bal Al.
Aluminium alloy sand casting; as cast.
BS 1490 LM25; AA 356.

MILLS LM25TB7.
M-187; 0.3 Mg, 7.0 Si, bal Al.
Aluminium alloy sand casting; full heat treatment plus stabilization.

MILLS LM25TE.
M-187; 0.3 Mg, 7.0 Si, bal Fe.
Aluminium alloy sand casting; precipitation treated.

MILLS LM25TF.
M-187; 0.3 Mg, 7.0 Si, bal Al.
Aluminium alloy sand casting; solution treated and precipitation treated.

MILLS LM26TE.
M-187; 3.0 Cu, 1.0 Mg, 9.5 Si, bal Al.
Aluminium alloy casting; precipitation treated.
AA332.

MILLS LM27M.
M-187; 2.0 Cu, 7.0 Si, 0.4 Mn, bal Al.
Aluminium alloy casting; as chill cast.

MILLS LM28TE.
M-187; 1.5 Cu, 1.2 Mg, 19 Si, 1.2 Ni, bal Al.
Aluminium alloy casting; precipitation treated.

MILLS LM28TF.
M-187; 1.5 Cu, 1.2 Mg, 19 Si, 1.2 Ni, bal Al.
Aluminium alloy casting; solution treated and precipitation treated.

MILLS LM29TE.
M-187; 1.1 Cu, 1.1 Mg, 23 Si, 1.1 Ni, bal Al.
Aluminium alloy casting; precipitation treated.

MILLS LM29TF.
M-187; 1.1 Cu, 1.1 Mg, 23 Si, 1.1 Ni, bal Al.
Aluminium alloy casting; solution treated and precipitation treated.

MILLS LM30M.
M-187; 4.5 Cu, 0.5 Mg, 17 Si, bal Al.
Aluminium alloy casting; as chill cast.

MILLS LM30TS.
M-187; 4.5 Cu, 0.5 Mg, 17 Si, bal Al.
Aluminium alloy casting; chill cast and stress relieved.

MILMOLD.
M-365; 0.30 C, 0.8 Mn, 0.5 Si, 1.7 Cr; 0.4 Mo, bal Fe.
Oil hardening tool steel designed for molds; AISI P 20.

MILNAIR.
M-365; 0.95 C, 2 Mn, 2 Cr, 1 Mo, bal Fe.
For tools; abrasion resistant.

MILNAIR 4.
M-365; 0.95 C, 2.0 Mn, 0.35 Si, 2.2 Cr, 1.1 Mo, bal Fe.
Air or oil hardening cold work tool steel; AISI A4.

MILNAIR 5.
M-365; 1.0 C, 5.25 Cr, 0.25 V, 1.1 Mo, bal Fe.
For tools, dies; air hardening, nondeforming.

MILNE 3074.
M-365; 0.35 C, 4 Cr. 9.25 W, 0.5 V, bal Fe.
For punches, shears, extrusion dies; hot work steel, oil hardened.

MILNE CMV.
M-365; 0.38 C, 5.25 Cr, 1.25 Mo, 1.05 V, bal Fe.
For hot work tools, punches: hot work steel, oil hardened.

MILNE CMW.
M-365; 0.37 C, 5 Cr, 1.5 Mo, 0.3 V, 1.3 W, bal Fe.
For hot work tools, punches; hot work steel, oil hardened.

MILNE DOUBLE SIX.
M-365; 2.25 C, 12 Cr, bal Fe.
For blanking and forming dies; Type D3; oil hardened.

MILNE HOLLOW DIE STEEL.
M-365; 0.8-1.0 C, bal Fe.
For hollow dies; water hardened.

MILNE M-330.
M-365; 0.30 C, 3 Cr, 3 Mo, 0.6 V, bal Fe.
For punches, crimpers, upsetters; hot work steel, resists heat checking.

MILNE M-331.
M-365; 0.40 C, 3.3 Cr, 2.25 Mo, 0.4 V, bal Fe.
For punches, upsetters, dies, shears; hot work steel, resists softening.

MILNE M-333.
M-365; 0.30 C, 3 Cr, 3 Mo, 0.6 V, 2.25 Co, bal Fe.
For heavy duty hot work tools, punches, dies; hot work steel, oil hardened.

MILNE MM6+6.
M-365; 0.85 C, 4.15 Cr, 6.4 W, 5 Mo, 1.9 V, bal Fe.
For tools, cutters, dies; high speed steel.

MILNE MMCO.
M-365; 0.7 C, 5.8 W, 5.2 Mo, 4 Cr, 2 V, 9 Co, bal Fe.
For lathe and planer tools, reamers, hobs, cutters; high speed steel.

MILNE MT-9.
M-365; 1.5 C, 12.5 Cr, bal Fe.
For blanking and drawing dies; oil hardening; non-deforming.

MILNE MX-15.
M-365; 0.50 C, 12 W, 4 Cr, 12 Ni, 1 V, bal Fe.
For extrusion dies and mandrels, punches; hot work steel, oil hardened.

MILNE ORANGE LABEL.
M-365; 1.01-1.05 C, bal Fe.
For tools, drills, taps; water hardening.

MILNE RED LABEL.
M-365; 0.95-1.1 C, 0.3 Mn, 0.25 Si, bal Fe.
For tools, drills, hobs, taps, punches; Type W1; water hardened.

MILNE WHITE LABEL.
M-365; 0.9-1.0 C, bal Fe.
For tools, drills, springs; water hardening.

MILO.
M-344; 0.4 C, 1.1 Mn, bal Fe.
For flogging tools, scarifer and fire tools; water hardened.

MILO 35.
M-344; 0.7-1.1 C, bal Fe.
For pneumatic tools; water hardening.

MILO 38.
M-344; 0.7-0.9 C, bal Fe.
For pneumatic tools, chisels, punches; water hardened; Type W1.

MILRITE.
M-88; Sn, Sb, Cu, bal Pb.
For bearings; Babbitt.

MILTUFF.
M-365; 0.50 C, 0.7 Mn, 0.25 Si, 3.25 Cr, 1.4 Mo, bal Fe.
Air or oil hardening tool steel, shock resisting type; AISI S7.

MILVAN.
M-365; 0.7 C, 19 W, 4.25 Cr, 2.25 V, bal Fe.
For tools, cutters; high speed steel.

MILVAN NO. 1.
M-365; C, 19 W, 4 Cr, 2.25 V, bal Fe.
For tools, cutters; high speed steel.

MILWALOY 1 1/4 MN.
M-475; 0.3-0.4 C, 1.1-1.4 Mn, bal Fe.
Cast: 75,000-95,000 TS; 45,000-60,000 YS; 23-30 El; 30-55 RA; 160-196 Brin.
For tractor parts, road machinery, sprockets; water hardened.

MILWALOY 7.
M-475; 0.30-0.40 C, 1.50-1.75 Cr, 0.60-0.70 V, bal Fe.
For nitriding; nitralloy steel.

MILWALOY 13.
M-473; 0.08-0.12 C, 12-14 Cr, bal Fe.
For stainless parts; stainless.

MILWALOY 18.
M-475; 0.30-0.4 C, 13-15 Ni, 3.9-4.5 Mn, bal Fe.
For circuit breakers, switches; austenitic, non-magnetic.

MILWALOY 18-8.
M-475; 0.08-0.15 C, 8-10 Ni, 17-20 Cr, bal Fe.
Cast: 70,000 TS; 25,000 YS; 40 El; 35 RA; 160 Brin.
Heat treated: 80,000 TS; 40,000 YS; 70 El; 60 RA; 187 Brin.
For food machinery; stainless.

MILWALOY 26.
M-475; 0.08-0.12 C, 18-20 Cr, 8-10 Ni, bal Fe.
For corrosion resistant parts; A Krupp-Nirosta Steel; 18-8.

MILWALOY 29-9.
M-475; 0.10-0.20 C, 8-10 Ni, 27-31 Cr, bal Fe.
Cast: 80,000 TS; 40,000 YS; 35 El; 35 RA; 187 Brin.
Heat treated: 90,000 TS; 50,000 YS; 50 El; 60 RA; 207 Brin.
For paper mill machinery; heat resistant.

MILWALOY 35-15.
M-475; 0.2-0.4 C, 32-38 Ni, 13-17 Cr, bal Fe.
Cast.
For heat resisting castings; heat resistant.

MILWALOY 38.
M-475; 0.12-0.15 C, 28-30 Cr, 8-10 Ni, bal Fe.
For corrosion resistant parts; A Krupp-Nirosta 29-9 steel.

MILWALOY 50.
M-475; 0.30-0.40 C, 15-18 Cr, 32-36 Ni, bal Fe.
For furnace parts to resist high temperatures; heat and corrosion resistant.

MILWALOY CC.
M-475; 0.4-0.5 C, 0.5-0.8 Mn, 0.9-1.1 Cr, bal Fe.
Cast: 85,000-100,000 TS; 50,000-70,000 YS; 17-22 El; 25-45 RA; 179-210 Brin.
For crusher machinery, screen plates, wear segments; wear resistant.

MILWALOY COMMERCIAL.
M-475; 0.25-0.35 C, 0.5-0.8 Mn, bal Fe.
Cast: 65,000-80,000 TS; 38,000-55,000 YS; 25-35 El; 40-55 RA; 140-165 Brin.
For structural and machine castings; water hardened.

MILWALOY CR NI.
M-475; 0.35-0.45 C, 0.6-0.9 Mn, 1.25-1.75 Ni, 0.6-0.9 Cr, bal Fe.
Cast: 90,000-110,000 TS; 55,000-75,000 YS; 18-25 El; 30-50 RA; 180-220 Brin.
For gears, cams, rollers; wear resistant.

MILWALOY CR NI MO.
M-475; 0.35-0.45 C, 0.5-0.8 Mn, 1.25-1.75 Ni, 0.6-0.9 Cr, 0.3-0.4 Mo, bal Fe.
Cast: 95,000-155,000 TS; 65,000-135,000 YS; 12-24 El; 25-50 RA; 189-255 Brin.
For castings; wear resistant.

MILWALOY DYNAMO.
M-475; 0.05-0.15 C, 0.20 max Mn, bal Fe.
Cast: 50,000-65,000 TS; 25,000-37,000 YS; 30-37 El; 50-70 RA; 130-150 Brin.
For magnet bodies, armature frames, electrical machinery; high magnetic permeability.

MILWALOY HI-CARBON.
M-475; 0.4-0.5 C, 0.5-0.8 Mn, bal Fe.
Cast: 80,000-95,000 TS; 47,000-65,000 YS; 20-30 El; 30-50 RA; 170-196 Brin.
For gears, racks, sprockets; water hardened.

MILWALOY KA2SMO.
M-475; 0.4-0.6 C, 8-10 Ni, 18-20 Cr, 2-4 Mo, bal Fe.
Cast: 70,000 TS; 30,000 YS; 35 El; 45 RA; 178 Brin.
Heat treated: 80,000 TS; 40,000 YS; 50 El; 70 RA; 196 Brin.
For corrosion resisting castings; corrosion resistant.

MILWALOY MN BO.
M-475; 0.3-4.0 C, 1.1-1.4 Mn, 0.003 B, bal Fe.
Cast: 80,000-95,000 TS; 50,000-65,000 YS; 20-30 El; 35-55 RA; 174-196 Brin.
For tractors, road machinery; wear resisting.

MILWALOY MN MO BO.
M-475; 0.3-0.4 C, 1.1-1.4 Mn, 0.15-0.20 Mi, 0.003 B, bal Fe.
Cast: 90,000-110,000 TS; 70,000-80,000 YS; 20-30 El; 35-55 RA; 179-217 Brin.
For tractor parts, road machinery; wear resisting.

MILWALOY NIT.
M-475; 0.20-0.30 C, 0.5-0.9 Mn, 2.5-3.0 Cr, 0.35-0.45 Mo, 0.2 V, bal Fe.
Cast: 80,000-95,000 TS; 70,000-80,000 YS; 20-30 El; 30-55 RA; 170-2 Brin.
For pistons, cylinders, valve parts; nitriding alloy.

MINARGENT.
57-46 Cu, 40-32 Ni, 28-0 W, 0.2-0.5 Al.
For silver solder, substitute for silver in silverware.

MINEOR.
M-40; 1.0 C, 5 Cr, 1 Mo, bal Fe.
For punches, mandrels, rolls, dies; Type A2; air hardened, non-deforming.

MINEOR FM.
M-40; 1.0 C, 5 Cr, 1 Mo, 0.2 S, bal Fe.
For punches, mandrels, rolls; air hardening, free-cutting.

MINERVA CHISEL.
M-365; 0.5 C, 1.75 Cr, 1.9-2.0 W, 0.2 V, bal Fe.
For tools, dies; oil hardened.

MINERVA H.C.
M-210; 0.53 C, 1.8 Cr, 1.9 W, 0.2 V, bal Fe.
For chisels, shear blades, extrusion dies; hot work steel, oil hardened.

MINERVA L.C.
M-210; 0.43 C, 1.8 Cr, 1.9 W, 0.2 V, bal Fe.
For chisels, bits, screwdrivers, scarfing tools; oil hardened, shock resistant.

MINERVA SPECIAL.
M-210; 0.5 C, 1.8 Cr, 2.2 W, 0.25 V, bal Fe.
For pneumatic chisels, cold chisels, rivet busters; shock resistant.

MINE TALABOT SPECIAL.
M-1119; 0.75 C, bal Fe.
For tools, springs, punches; water hardened.

MINIMAX NO. 178.
M-858; Ag-Hg.
For dental amalgams; corrosion resistant.

MINOFOR.
M-Eng.; 69-66 Sn, 18-20 Sb, 9-10 Zn, 3-4 Cu, 0.1 bal Fe.
For bearings; anti-friction.

MINOVAR.
M-68; 2.4 max T.C., 1-2 Si, 34-36 Ni, 0.10 max Cr, 0.5 max Mn, bal Fe.
Cast: 20,000-25,000 TS; 100-125 Brin.
For electrical equipment; formerly Invar Cast Iron, low coefficient of expansion.

MINOX 10.
M-1509; 0.10 C, 13 Cr, bal Fe.
Annealed: 69,000-85,000 TS; 40,000-48,000 YS; 18 El; 140-160 Brin.
For turbine blades, surgical instruments, cutlery; corrosion resistant, hardenable.

MINOX-15.
M-884; 0.20 C, 13 Cr, 1 Mo, bal Fe.
Cast: 90,000 TS; 45,000 YS; 15 El; B 95 Rock.
Hardened: 200,000 TS; 180,000 YS; 10 El; C 45 Rock.
For valves, cutlery, oil and chemical plant equipment. Corrosion resistant castings, hardenable.

MINOX 15.
M-1509; 0.15 C, 13 Cr, bal Fe.
Annealed: 74,000-92,000 TS; 42,000-52,000 YS; 14 El; 150-180 Brin.
For turbine blades, surgical instruments, cutlery; corrosion resistant, hardenable.

MINOX 20.
M-1509; 0.20 C, 14 Cr, bal Fe.
Annealed: 78,000-107,000 TS; 46,000-54,000 YS; 12 El; 160-200 Brin.
For turbine blades, surgical instruments, cutlery corrosion resistant, hardenable.

MINOX 30.
M-1509; 0.30 C, 14 Cr, bal Fe.
Annealed: 85,000-114,000 TS; 52,000-68,000 YS; 10 El; 160-200 Brin.
For surgical instruments, cutlery; corrosion resistant, hardenable.

MINOX 1820.
M-1509; 0.20 C, 18 Cr, 2 Ni, bal Fe.
Annealed: 85,000-106,000 TS; 58,000-92,000 YS; 12 El; 200-240 Brin.
For chemical plant equipment; resists nitric acid.

MIN-OX-GRADE 51 C.
M-25; 2.5 C, 3 Si, 3 Al, 0.2 V, 2.5 Cr, 0.15 Mn, bal Fe.
Natural: 45,000 TS; 320 Brin.
Heat treated: 600 Brin.
For glass molds, dies for tile pressing, oven enameling racks; heat and wear resistant.

MIN-OX-GRADE D-V.
M-25; 62 Cu, 7 Al, 7 Zn, 22 Ni.
80,000 TS; 0.5-2.0 El; 0 RA.
For glass molds, cams, machine parts; heat and wear resistant.

MINT DIE STEEL.
M-261, M-486; C, bal Fe.
For coining, cold-heading and embossing dies; water hardening.

MIRACULOY.
M-288; 0.35 C, 1.25 Mn, 0.40 Si, 0.65 Cr, 1.5 Ni, 0.30 Mo, bal Fe.
Normalized: 115,000 TS; 85,000 YS; 18 El; 40 RA; 275 Brin.
Heat treated: 125,000-245,000 TS; 90,000-190,000 YS; 20-8 El; 50-10 RA; 250-650 Brin.
For heavy duty castings; high strength.

MIRA METAL.
M-U.S.; 75 Cu, 16.3 Pb, 6.8 Sb, 0.24 Ni, 0.43 Fe, 0.62 Zn, 0.91 Sn.
For valves, pipes; acid and corrosion resisting.

MIRAMINT.
W, C+Co.
For hard cutting tools and dies; sintered.

MIR-O-COL.
M-988; Cr, Mo, W, Co.
For hard surfacing welding electrode; wear resistant.

MIR-O-COL BR.
M-988; C, Cr, Ni, Mo, bal Co.
For hard facing rod; abrasion and impact resistant.

MIR-O-COL NO. 1.
M-988; 3-4 C, 11-13 Cr, 1.0-1.5 Mn, 2-3 Si, 4-6 Ni, 8-10 Mo, bal Fe
Welded: 600-650 Brin.
For welding rod for plowshares, shovels, root cutters; austenitic, abrasion resistant.

MIR-O-COL NO. 3.
M-988; 1.75-2.2 C, 0.5-1.0 Mn, 0.6-0.9 Si, 8-9 Cr, bal Fe.
Cast: 45,000 TS; 0 El; 480 Brin.
For hard facing welding rod; impact and abrasion resistant, tough, hard.

MIR-O-COL NO. 4.
M-988; 1.0-1.5 C, 12-14 Mn, 0.8 Si, 0.5 1.0 Mo, bal Fe.
Welded: 60,000 TS; 22 El; 250 Brin.
For hard facing rod; martensitic impact and abrasion resistant.

MIR-O-COL NO. 4T-S.
M-988; Cr, Mo, W, Co.
For hard facing electrode; ductile.

MIR-O-COL NO. 5.
M-988; 0.7-1.0 C, 0.75-1.25 Mn, 3.0-3.5 Ni, 6.5-8.0 Mo, 0.4 Co, 0.8 B, 1.2 Si, 3.5-4.5 Cr, bal Fe.
Welded: 500 Brin.
For welding rod for hot cutting dies, shears and tools; corrosion, heat and abrasion resistant.

MIR-O-COL NO. 6.
M-988; 0.9-1.2 C, 26-30 Cr, 0.5 max Ni, 0.5 max Mo, 3 max Fe, 3.5-5.0 W, bal Co.
Welded: 400-440 Brin.
For welding rod for diesel engine valves; corrosion, abrasion and heat resistant.

MIR-O-COL NO. 11.
M-988; C, Co, Cr, W.
Cast: 500-520 Brin.
For hard facing electrode; corrosion and abrasion resistant.

MIRRALOY.
M-15; C, bal Fe.
For journals, shafting; turned, ground and polished.

MIRROMOLD see CARPENTER MIRROMOLD.

MIRYCAL.
M-1067; 0.5 C, 0.25 Mn, 0.95 Cr, 1.0 W, 0.2 Mo, bal Fe.
For tools, dies, punches, shear blades, boiler makers' tools, shock resistant.

MIRYCAL CHISEL.
M-434; 0.45 C, 1.0-1.25 W, 0.85-0.95 Cr, 0.15-0.20 Mo, 0.2-0.3 Mn, bal Fe.
For battering tools, chipping chisels, beading tools, rivet busters, cold sets, swages, track chisels, hot dies; tough, shock resistant.

MISCHMETAL.
M-111; 50 Ce, 45 Lathanum + didymium.
For pyrophoric alloy in cigarette lighters; removes gas from radio tubes.

MISCHMETAL see also CERALLOY MISCHMETAL.

MISCO 16.
M-84; 0.2 max C, 16 Cr, bal Fe.
For heat and corrosion resistant parts; heat and corrosion resistant.

MISCO 16-8.16C.
M-84; 0.16 max C, 8-10 Ni, 18-20 Cr, 1 max Mn, 2 max Si, bal Fe.
Annealed: 80,000 TS; 40,000 YS; 50 El; 60 RA; 170 Brin.
For stainless castings; corrosion resistant.

MISCO 18-8.
M-84; 0.16 C, 18 Cr, 8 Ni, bal Fe.
85,000 TS; 45,500 YS; 40 El; 50 RA.
For stainless parts, chemical engineering equipment; resists sulfurous acid.

MISCO 18-8.07C.
M-84; 0.07 max C, 8-10 Ni, 18-20 Cr, 1 max Mn, 2 max Si, bal Fe.
Annealed: 75,000 TS; 42,000 YS; 55 El; 65 RA; 156 Brin.
For stainless castings; corrosion resistant.

MISCO 18-8CB.
M-84; 0.10 max C, 8-10 Ni, 18-20 Cr, Cb = 8 + C, bal Fe.
Cast: 70,000 TS; 35,000 YS; 40 El; 40 RA; 190 Brin.
Annealed: 82,000 TS; 42,000 YS; 40 El; 40 RA; 179 Brin.
For chemical and plastic plant equipment; Type 347; stainless, austenitic.

MISCO 18-8 MO.
M-84; 0.10 C, 18 Cr, 8 Ni, 3 Mo, bal Fe.
Annealed: 85,000 TS; 45,000 YS; 50 El; 60 RA; 179 Brin.
For stainless parts; stainless.

MISCO 18-8 SE.
M-84; 0.16 max C, 8-10 Ni, 18-20 Cr, 0.2-0.3 Se, bal Fe.
Annealed: 80,000 TS; 40,000 YS; 50 El; 50 RA; 183 Brin.
For corrosion resistant parts, castings; corrosion resistant.

MISCO 18-8 TI.
M-84; 0.10 max C, 8-10 Ni, 18-20 Cr, Ti = 4 x C, bal Fe.
Annealed: 70,000 TS; 35,000 YS; 40 El; 40 RA.
For stainless castings; stainless.

MISCO 18-8.10C.
M-84; 0.10 max C, 8-10 Ni Cr, 1 max Mn, 2 max Si, bal Fe.
Annealed: 75,000 TS; 42,000 YS; 55 El; 65 RA; 156 Brin.
For atainless castings; corrosion resistant.

MISCO 20.
M-84; 0.07 max C, 27.5-30.5 Ni, 19-22 Cr, 2-3 Mo, 3.5-4.5 Cu, 0.7 Mn, 1.5 max Si, bal Fe.
Annealed: 70,000 TS; 34,000 YS; 38 El; 45 RA; 143 Brin.
For corrosion resistant castings; resists H_2SO_4.

MISCO 22.
M-84; 0.18-0.25 C, 0.6-0.7 Mn, 0.4-0.5 Si, bal Fe.
Cast: 60,000 TS; 30,000 YS; 26 El; 38 RA; 120-149 Brin.
For gears, pinions, shafts, housings; water hardened.

MISCO 25.
M-84; 0.22-0.28 C, 0.7 Mn, 0.5 Si, bal Fe.
Cast: 70,000 TS; 38,000 YS; 24 El; 36 RA; 143-179 Brin.
For gears, shafts, housings, machinery parts; water hardened.

MISCO 25-20.
M-84; 0.3-0.4 C, 19-21 Ni, 24-26 Cr, bal Fe.
Cast: 229 Brin.
For heat resistant castings; corrosion and heat resistant.

MISCO 27.
M-84; 0.24-0.30 C, 0.8 Mn, 0.5 Si, bal Fe.
Cast: 75,000 TS; 43,000 YS; 22 El; 30 RA; 149-187 Brin.
For machine tool castings, gears, housings, shafts; water hardened.

MISCO 28.
M-84; 0.2 max C, 28 Cr, bal Fe.
For heat and corrosion resistant parts; heat and corrosion resistant.

MISCO 28-15.
M-84; 0.4 max C, 14.5-16.5 Ni, 27-30 Cr, bal Fe.
Cast: 70,000 TS; 40,000 YS; 12 El; 15 RA.
For heat resistant castings; corrosion and heat resistant.

MISCO 28-20.
M-84; 0.4 max C, 19-21 Ni, 27-30 Cr, bal Fe. Cast.
For heat resistant castings; corrosion ahd heat resistant.

MISCO 30-30.
M-84; C, 30 Cr, 30 Ni, bal Fe.
For chemical engineering; heat and corrosion resistant.

MISCO 430.
M-84; 0.2 C, 17 Cr, bal Fe.
For corrosion resistant castings; corrosion resistant; Type 430.

MISCO B.
M-84; 0.25 C, 25 Cr, 13 Ni, bal Fe.
For corrosion and heat resisting parts, chemical engineering equipment, corrosion and heat resistant.

MISCO B-1.
M-84; 0.5 C, 25 Cr, 13 Ni, bal Fe.
For heat and corrosion resisting parts; heat and corrosion resistant.

MISCO "C".
M-84; 0.2-0.3 C, 28-30 Cr, 8-10 Ni, 0.6-0.8 Si, bal Fe.
Cast: 90,000-95,000 TS; 60,000-65,000 YS; 28-23 El; 36-30 RA; 197-212 Brin.
For use in sulfite industry, castings, valves; corrosion resistant.

MISCO C-1.
M-84; 0.5 C, 29 Cr, 9 Ni, bal Fe.
For heat and corrosion resisting parts; heat and corrosion resistant.

MISCO C-MO.
M-84; 0.18-0.25 C, 0.7 Mn, 0.5 Si, 0.5 Mo, bal Fe.
Cast: 70,000 TS; 45,000 YS; 24 El; 35 RA; 143-179 Brin.
For machine tool castings, gears, housings, shafts; water hardened.

MISCO CROMO.
M-84; 0.27-0.33 C, 1.4 Mn, 0.5 Si, 0.9 Cr, 0.3 Mo, bal Fe.
Cast: 100,000 TS; 70,000 YS; 18 El; 30 RA; 202-235 Brin.
For machinery castings, gears, shafts, housings; oil hardened.

MISCO GRADE A.
M-84; 0.5-0.7 C, 35-37 Ni, 15-17 Cr, bal Fe.
For furnace parts, retorts, carburizing boxes; resists heat to 1950°F.

MISCO "H.N".
M-84; 0.6-0.7 C, 15-18 Cr, 60-65 Ni, bal Fe.
For heat treating boxes, furnace parts, valves, stills; resists H_2SO_4; heat, wear and corrosion resistant.

MISCO HN-1.
  M-84; 0.5-0.8 C, 16-20 Cr, 66-70 Ni, bal Fe.
  Cast: 60,000 TS; 36,000 YS; 3 El; 217 Brin.
  For stainless steel castings, furnace equipment; acid resistant.

MISCO HN-2.
  M-84; 0.7 C, 18 Cr, 65 Ni, bal Fe.
  For chemical engineering equipment; resists H_2SO_4.

MISCO K.
  M-84; 0.2 C, 25 Cr, 20 Ni, bal Fe.
  For stainless and heat resistant castings; Type 310; corrosion and heat resistant.

MISCO METAL.
  M-84; 0.5 C, 38 Ni, 18 Cr, 1.5 Si, bal Fe.
  Cast: 68,000 TS; 40,000 YS; 10 El; 10 RA; 187 Brin.
  For furnace parts, annealing boxes, retorts; heat and corrosion resistant.

MISCO MS.
  M-84; 0.45 C, 18 Cr, 38 Ni, bal Fe.
  Cast: 70,000 TS; 45,000 YS; 5 El; 6 RA; 190 Brin.
  For castings subject to cyclic heating; resists thermal fatigue.

MISCO N.
  M-84; 0.4 C, 9 Cr, 21 Ni, bal Fe.
  75,000 TS; 45,000 YS; 30 El; 40 RA.
  For chemical engineering equipment; resists sulfuric acid.

MISCO N-5.
  M-84; 0.4 C, 30 Ni, 4 Si, bal Fe.
  For chemical engineering equipment; resists H_2SO_4.

MISCROME 1.
  M-84; 0.25-0.35 C, 13-15 Cr, bal Fe.
  Cast: 10,000 TS; 72,000 YS; 17 El; 25 RA; 200 Brin.
  Heat treated: 123,000 TS; 98,000 YS; 20 El; 44 RA.
  For chemical engineering equipment; resists HNO_3.

MISCROME 2.
  M-84; 0.25 C, 21 Cr, bal Fe.
  For chemical engineering equipment; resists HNO_3.

MISCROME 3.
  M-84; 0.25 C, 28 Cr, bal Fe.
  For chemical engineering equipment; resists HNO_3.

MISCROME 5.
  M-84; 1.4-1.6 C, 12-14 Cr, 0.9-1.1 Mo, 0.5 max Ni, 0.5-0.7 Mn, bal Fe.
  Annealed: 269 Brin.
  Heat treated: 131,000 TS; 352 Brin.
  For wear and corrosion resistant castings; castings, hard, wear resistant.

MISCROME CR.
  M-84; 2.5 C, 16 Cr, bal Fe.
  For heat and corrosion resisting parts, grids; heat and corrosion resistant.

MISCROME KR.
  M-84; 1.6 min C, 24-30 Cr, bal Fe.
  For heat and corrosion resistant parts; heat and corrosion resistant.

MISCROME NO. 4.
  M-84; 0.15 C, 13 Cr, bal Fe.
  Heat treated: 185,000 TS; 138,000 YS; 1 El; 4 RA; 328 Brin.
  For corrosion resisting parts; corrosion resistant.

MISHIMA STEEL.
  M-Japan; 10 Al, 25 Ni, bal Fe.
  For permanent magnets; high coercive force, sintered.

MITCHALLOY A.
  M-536; 2.9-3.1 T.C., 2-2.5 Ni, 0.3-0.6 Cr, bal Fe.
  45,000-50,000 TS; 240-280 Brin.
  For brake drums; wear resistant.

MITCHALLOY B.
  M-536; 2.8-3.5 C, 4-5 Ni, 1.5-2.0 Cr, bal Fe.
  Sand cast: 40,000-50,000 TS; 550-650 Brin.
  For pulverizers, grinders, mixers; wear and abrasion resistant.

MITCHALLOY C.
  M-536; 2.7-3.0 T.C., 12-15 Ni, 5-7 Cu, 1.5-4.0 Cr, bal Fe.
  Cast: 20,000-35,000 TS; 140-190 Brin.
  For heat and corrosion resistant parts; austenitic.

MITCHALLOY D.
  M-536; 3.3 C, 2.6 Si, 1.5 Ni, 0.8 Cr, bal Fe.
  For gratebars; alloy cast iron.

MITIS IRON.
  M-Eng.; 0.06-0.27 Al, C, bal Fe.
  For pipes, fittings; wrought iron deoxidized with Al.

MITSUBISHI SH-IS-54.
  M-1543; 0.12-0.17 C, 0.8-1.1 Mn, 0.35-0.60 Si, 0.40 max Cu, bal Fe.
  Rolled: 77,000-88,000 TS; 50,000 min YP; 20 min El.
  For railroad and mine cars, bridges, agricultural equipment.
  High-tensile, low alloy constructional steel.

MITSUBISHI SH-IS-60.
M-1543; 0.15-0.21 C, 0.95-1.25 Mn, 0.50-0.80 Si, bal Fe.
Rolled: 85,000-100,000 TS; 54,000 min YP; 17 min El.
For railroad and mine cars, bridges, agricultural equipment.
Constructional steel, tough.

MIXEND 60.
M-677; 60 Ni, 40 Fe.
Welded: 262 Brin.
For welding rod; machinable welds on high P-cast iron.

MIXEND 99.
M-677; 99 Ni.
Welded: 150 Brin.
For welding rod; machinable welds on cast iron.

M-K-9.
M-859; 4 Cu, bal Al.
Rolled: 60,000 TS; 30 El; 70 Brin.
For light alloy parts; age-hardenable.

MK9-AS.
M-859; 4 Si, 5 Cu, bal Al.
Cast: 35,000 TS.
Wrought: 47,500 TS.
For light alloy parts; age-hardenable.

M.K. STEEL.
M-Japan; 10-40 Ni, 1-20 Al, bal Fe.
For magnets; not forged readily.

M.K. STEEL.
M-Japan; 25 Ni, 10 Al, bal Fe.
For magnets; must be cast to shape.

ML.
M-1779; 0.8 C, 18 W, 4 Cr, 2 V, 0.75 Mo, bal Fe.
Hardened: C 64-66 Rock.
18-4-2 high speed steel for cutting applications where added abrasion resistance is needed.

ML.
M-Italy; 0.8 C, 18 W, 4 Cr, 2 V, 0.75 Mo, bal Fe.
Hardened: C 64-66 Rock.
For lathe tools, reamers, hobs, broaches, milling cutters.
Type T2 high speed steel.

ML25.
M-1653; 0.22-0.28 C, 0.50-0.80 Mn, 0.80-1.10 Cr, 0.15-0.25 Mo, bal Fe.
Cr-Mo structural steel. 25 Cr Mo 4.

ML30.
M-1653; 0.27-0.33 C, 0.40-0.70 Mn, 0.80-1.10 Cr, 0.15-0.25 Mo, bal Fe.
Cr-Mo structural steel. 30 Cr Mo 4.
AISI 4130.

ML35.
M-1653; 0.32-0.38 C, 0.60-0.90 Mn, 0.80-1.10 Cr, 0.15-0.25 Mo, bal Fe.
Cr-Mo structural steel. 35 Cr Mo 4.
AISI 4135.

ML40.
M-1653; 0.37-0.44 C, 0.7-1.0 Mn, 0.90-1.20 Cr, 0.15-0.25 Mo, bal Fe.
Cr-Mo structural steel. 40 Cr Mo 4.
Similar to AISI 4140.

ML50.
M-1653; 0.47-0.55 C, 0.60-0.90 Mn, 0.8-1.1 Cr, 0.15-0.25 Mo, bal Fe.
Cr-Mo structural steel.
50 Cr Mo 4; AISI 4150.

ML1700.
M-31, M-60; 0.20 C, 25.0 Cr, 15.0 W, 0.4 B, bal Co.
Corrosion resistant, high temperature alloy.

ML-ALLOY.
M-U.S.; 4 Cu, 2 Ni, 2 Mg, 0.3 Mn, 0.05 V, 0.1 Ti, bal Al.
Cast: 33,000 TS.
At 600°F: 17,000 TS.
For aircraft castings, cylinder heads, pistons; age-hardenable, high temperature use.

ML-ALUMINUM.
M-U.S.; 4 Cu, 2 Ni, 2 Mg, 0.1 Ti, 0.3 Cr, 0.1 V, 0.3 Mn, bal Al.
Cast: 36,000 TS; 31,000 YS; 1.2 El.
At 600°F: 17,000 TS; 14,500 YS; 9.0 El.
For aircraft parts, cylinder heads, pistons; good high temperature properties.

MLV.
M-1653; 0.45 C, 1.0 Cr, 0.35 Mo, 0.25 V, bal Fe.
Low alloy steel, good for high temperature bolts.

MM001.
M-1536; 0.12 C, 10 Cr, 10 Co, 3 Mo, 0.5 V, 3.8 Ti, 5.5 Al, 1.4 Hf, 0.015 max B, 0.05 max Zr, bal Ni.
High temperature alloy.

MM002.
M-1536, M-1491; 0.15 C, 9.0 Cr, 10.0 Co, 10.0 W, 2.5 Ta, 1.5 Ti, 5.5 Al, 1.5 Hf, 0.05 Zr, 0.015 B, bal Ni.
Cast Aged (1600°F); 145,000 psi TS; 125,000 psi YS; 8 El.
High temperature alloy; for turbine wheels and blades.

MM004.
M-1536, m-1491; 0.05 C, 12.0 Cr, 4.5 Mo, 2.0 Cb, 5.9 Al, 0.60 Ti, 0.10 Zr, 1.3 Hf, 0.01 B, bal Ni.
High temperature alloy; for turbine wheels and blades.

MM-007.
M-1491; 0.10 C, 8.0 Cr, 10.0 Co, 6.0 Mo, 0.10 max W, 0.10 Max Cb, 1.0 Ti, 6.0 Al, 0.015 B, 0.075 Zr, 4.25 Ta, 1.3 Hf, bal Ni.
Hafnium modified B-1900; improved ductility for turbine parts.

MM-008.
M-1491; 0.07 C, 14.6 Cr, 15.2 Co, 4.4 Mo, 3.35 Ti, 4.3 Al, 0.015 B, 0.03 Zr, 1.3 Hf, bal Ni.
Hafnium containing alloy; improved ductility for turbine parts.

M-M-0011 ALLOY.
M-1536; 0.15 C, 8.25 Cr, 10.0 Co, 10.0 W, 0.70 Mo, 3.0 Ta, 5.5 Al, 1.0 Ti, 1.5 Hf, 0.015 B, 0.5 Zr, bal Ni.
Cast high temperature alloy; for turbine wheels and blades.
Same as MAR-M-247.

M-M-1.
M-365; 0.65-0.85 C, 7.5-9.5 Mo, 1.25-2.25 W, 3.5-4.5 Cr, 0.9-1.5 V, bal Fe.
For tools, cutters; high speed steel.

MM 6 & 6.
M-365; 0.83 C, 4.15 Cr, 1.9 V, 6.35 W, 5.0 Mo, bal Fe.
High speed steel for cutting tools; Mo-W type; AISI M2.

M.M.M.
M-230, M-636; 57.5 min, Ni, 26.0 min Cu, 9.25 max Sn, bal Fe, Si, Mn.
Cast: 70,000 TS; 45,000 YS; 17.5 El.
For valves for superheated steam; modified "Monel" Metal.

MMV.
M-365; 1.15 C, 4.0 Cr, 2.8 V, 6.0 W, 5.25 Mo, bal Fe.
High speed steel for cutting tools, molybdenum-tungsten type; AISI M3 Class 2.

MNC.
M-1655; 1.05 C, 1.10 Mn, 0.30 Si, 0.90 Cr, bal Fe.
For punching and stamping dies for light work.
W.-Nr. 1.2127.

MNMO 18.
M-1740; 0.16-0.20 C, 0.30-0.60 Si, 1.3-1.7 Mn, 0.40 max Ni, 0.25 max Cr, 0.25-0.35 Mo, 0.30 max Cu, bal Fe.

MNMO 38.
M-1740; 0.36-0.40 C, 0.30-0.60 Si, 1.35-1.55 Mn, 0.40 max Ni, 0.25 max Cr, 0.30-0.35 Mo, 0.30 max Cu, bal Fe.
BS 1458 Gr A&B.

MNV3.
M-US; 0.45 C, 9 Cr, 3 Si, bal Fe.
Valve steel.

MO.
M-1522; 30 Ag, Cu, Zn.
Melt range: 680-770°C.
Max stress: 49 kgf/mm^2, 5 El.
For silver brazing; general use.

MO-0.5 TI.
M-151; 0.03 C, 0.5 Ti, bal Mo.
High Young's modulus-for boring bar shanks.

MO-1.
M-Japan; 23 Mo, 2.5 Cr, bal Fe.
For permanent magnets in magnetic and electrical equipment. High permeability.
Precipitation hardening.

MO 18.
M-912; 0.18 C, 0.25 Si, 0.85 Mn, 1.0 Cr, 0.25 Mo, bal Fe.
Alloy carburizing steel; for gears, pinions, shafts.
W.-Nr. 7242.

MO18.
M-1740; 0.16-0.20 C, 0.20-0.50 Si, 0.50-1.0 Mn, 0.40 max Ni, 0.25 max Cr, 0.40-0.70 Mo, 0.30 max Cu, bal Fe.
Carbon-molybdenum steel for operations up to 450°C.
BS 1398 Gr A.

MO26.
M-912; 0.25 C, 0.25 Si, 0.75 Mn, 0.50 Cr, 0.45 Mo, 0.020-0.035 S, ba Fe.
Alloy carburizing steel; for gears, pinions, spline couplings.
W.-Nr. 7325, 7326.

M.O. 81 STEEL.
M-1116; 0.87 C, 4 Cr, trace W; 8.5 Mo, 2 V, bal Fe.
Air or oil harden to 61-65 Rc.
For twist drills, "hand" taps, dies, extra long drills.
AFNOR: Z 90 DV 09.02; US-AISI M10.

M.O. 82 STEEL.
M-1116; 0.84 C, 4 Cr, 1.4 W, 8 Mo, 1.5 V, bal Fe.
Air or oil harden to 62-66 Rc.
For twist drills, end mills, reamers, taps, counter bores, hobbing cutters.
AFNOR: Z 85 DVW 08.02.01; US-AISI M1; Germany: S 2.9.1 (B Mo 9); W.Nr 1.3346.

M.O. 83 STEEL.
M-1116; 1.0 C, 4 Cr, 1.4 W, 8.5 Mo, 1.9 V, bal Fe.
Air or oil harden to 62-66 Rc.
For drills (deep hole), taps (blind holes), counter bores, end mills.
AFNOR: Z 100 DVW 08.02.01;
US-AISI M7;
Germany: S 2.9.2 (B Mo 9V); W.Nr 1.3348.

M.O. 88 STEEL.
M-1116; 1.10 C, 3.8 Cr, 1.5 W, 9.5 Mo, 1.2 V, 8 Co, bal Fe.
Air or oil harden to 63-70 Rc.
For turning, drilling, milling, planing extra hard or tough materials.
AFNOR: Z 110 DKWV 10.08.02.01;
US-AISI M 42.

MO330.
M-912; 0.32 C, 0.25 Si, 0.40 Mn, 3.10 Cr, 1.0 Mo, 0.30 V, bal Fe.
Alloy steel, hardenable to 1520-1620 N/mm^2 UTS; 1300 N/mm^2 YS.
For highly stressed parts; deep hardening.
W.-Nr. 7765.

MOCAR.
M-959; 10 Sn, 15 Sb, bal Pb.
For bearings for connecting rods and camshafts; Babbitt.

MOCASCO.
M-601; 3.2 C, 2.5 Si, 1 Cr, bal Fe.
For castings; cast iron.

MOCASCO 30.
M-601; 3.25-3.35 T.C., 2.0-2.2 Si, 1.0-1.35 Ni, 0.25-0.30 Cr, bal Fe
Cast: 30,000 TS; 200 Brin.
For air compressor cylinders, motor blocks and heads; wear and heat resistant.

MOCASCO 40.
M-601; C, 0.35 Mo, bal Fe.
40,000 TS; 200 Brin.
For cylinder heads and liners, gears, valves.

MOCASCO 50.
M-601; C, 1.5 Ni, 0.6 Mo, bal Fe.
Cast: 50,000 TS; 241 Brin.
For castings, pistons, gears, valves; cast iron.

MOCASCO 60.
M-601; C, 1.5 Ni, Cr, 0.90 Mo, bal Fe.
Cast: 60,000 TS; 228-260 Brin.
For gears, brake drums, pump and hydraulic castings; cast iron.

MOCASCO CG GRADE 60.
M-601.
Special cast iron.
Cast: 60,000 TS; 40,000 YS; 3 El; 180 Brin.

MOCASCO CG GRADE 68.
M-601.
Special cast iron.
Cast; 68,000 TS; 45,000 YS; 2.5 El; 220 Brin.

MOCASCO CG GRADE 70.
M-601.
Special cast iron.
Cast: 70,000 TS; 50,000 YS; 2 El; 260 Brin.

MO-CHIP.
M-57; 0.7 C, 8 Mo, 2.5 Co, 0.5 Cr, 1.0 V, bal Fe.
For tools, cutters, lathe and planer tools; high speed steel, oil hardened.

MOCK GOLD-1.
M-Eng.; 80-67 Cu, 20-29 Pt, 0-4 Zn.
For ornaments; corrosion resistant.

MOCK GOLD-2.
M-Eng.; 6 Ni, 1 Pt, 1 Ag, 1 Brass.
For ornaments; corrosion resistant.

MOCK SILVER.
M-Eng.; 84 Al, 10 Sn, 5.5 Cu, P.
For instruments, fittings, ornamental; corrosion resistant.

MOCUT.
M-140; 0.60-0.85 C, 3.5-4.0 Cr, 0.9-1.3 V, 1.3-1.8 W, 7.75-9.0 Mo, bal Fe.
For twist drills, lathe and planer cutters, reamers, hobs; Type M 1; high speed steel.

MOD-1.
M-604; 0.90-1.05 C, 0.95-1.25 Mn, 0.90-1.15 Cr, 0.50-0.70 Si, bal Fe.
High hardness after heat treating; for tools or wear resistant parts.

MOD-2.
M-604; 0.85-1.0 C, 1.4-1.7 Mn, 1.4-1.7 Cr, 0.60-0.80 Si, bal Fe.
High hardness after heat treatment; for tools or wear resistant parts.

MODERN.
M-613; 0.7 C, 18 W, 4 Cr, 1 V, bal Fe.
For cutters, tools; high speed steel.

MODIFIED MONEL.
M-230; 60-65 Ni, 24-27 Cu, 9-11 Sn, 1-3 Fe.
Cast: 70,000 TS; 45,000 YS; 17 El.
For valves for super heated steam; corrosion resistant.

MODIFIED MONEL METAL.
M-230; 60-65 Ni, 24-27 Cu, 9-11 Sn, 1-3 Fe, Mn, Si.
For valves for superheated steam; v.M.M.M.

MODIFIED "OUNCE" METAL.
M-126; 84 Cu, 5 Sn, 5 Pb, 5 Zn, 1 Ni.
For fittings, valves, hardware castings; pressure tight.

MOGUL, (AISI M2).
M-69; 0.78 C, 4 Cr, 1.5 W, 1.15 V, 8.7 Mo, bal Fe.
Annealed: 228-241 Brin.
For reamers, drills, lathe and planer tools, hobs; Type M1; high speed steel.

MOHAWK HOT DIE.
M-1779; 0.45 C, 14 W, 3.5 Cr, 0.60 V, bal Fe.
Oil hardened: 310,000-268,000 TS; 1-23 El; 5-79 RA; 600-321 Brin.
Heavy duty tungsten hot work tool steel suitable for hot shear blades, hot punches, long run hot forging and extrusion dies.
AISI H-24.

MOHICAN.
M-435; 0.8 C, 1.5 W, 4 Cr, 1 V, 8.5 Mo, bal Fe.
For tools, cutters, broaches, milling cutters; high speed steel.

MOHICAN 6.
M-435; 0.62 C, 1.7 W, 3.75 Cr, 8.7 Mo, 1 V, bal Fe.
At 400°F: 243,000 TS; 3.6 El; 8.1 RA; 477 Brin.
For hot punches, shear blades, forming dies; shock and wear resistant, high red hardness.

MOHICAN 8.
M-435; 0.80 C, 1.5 W, 4 Cr, 9 Mo, 1.2 V, bal Fe.
For twist drills, hobs, mills, hot punches; tough, high speed steel.

MOLDALOY.
M-1056; Bi, Pb, Sn, Sb.
Cast: 11,500 TS; 22 Brin.
For molds, forming dies, forging dies, chuck jaws; M.P. 430°F; molding alloy.

MOLDIE.
M-336; 0.9 C, 2.25 Mn, 0.1 V, bal Fe.
For Bakelite and rubber molds; steam resistant.

MOLDTEM.
M-64; C, Cr, Mo, V, bal Fe.
Heat treated: 340 Brin.
For zinc die casting dies, plastic mold dies; prehardened, hot work steel.

MOLECULOY.
M-1523; 20 Cr, 3 Al, 2 Co, bal Ni.
Resistance wire: 800 ohms per circular mil ft.
Ann: 130,000 psi ave; hard drawn: 250,000 psi.
For winding precision resistors and potentiometers; max. operating temp.: 250°C.

MOLECULOY III.
M-1523; 20 Cr, 4 Al, 5 Mn, 1 Si, bal Ni.
Resistance wire: 835 ohms per cir.mil.ft.
Ann: 130,000 psi; hard drawn: 260,000 psi.
Max. oper. temp.: 250°C.

MOLEGRAIN.
M-1455; 3.3 C, alloy, Si, Mn, bal Fe.
For gears, shafts, housings; cast iron.

MOLEL.
M-432; C, alloy, bal Fe.
For tools, cutters, punches; oil hardened.

MOLEX-A.
M-15; C, alloy, bal Fe.
For dies, gauges, broaches, shears, reamers; wear resistant, oil hardened.

MOLEX NO. 5.
M-15; C, alloy, bal Fe.
For crushers, punches, wrenches, wedges; shock and abrasion resistant.

MOLEX NO. 6.
M-15; C, alloy, bal Fe.
For gages, chuck jaws, knives, reamers; oil hardened, shock and wear resistant.

MOLEX NO. 7.
M-15; C, alloy, bal Fe.
For bushings, cam rollers, shear blades, forming dies; oil hardened, wear resistant.

MOLEX NO. 8.
M-15; 1.1 C, 0.75 Mn, 0.5 Si, 5.5 Cr, 0.3 V, 1.3 Mo, 0.015 P, 0.015 S, bal Fe.
Air hardening tool steel for cold working tools.

MOLEX NO. 8 MODIFIED.
M-15; 0.55 C, 0.9 Mn, 0.25 Si, 3.5 Cr, 0.25 Ni, 0.25 V, 1.4 Mo, bal Fe.
Air hardening tool steel.

MOLEX-O.
M-15; C, alloy, bal Fe.
Annealed: 202 Brin.
For blanking and drawing dies, shear blades; oil hardened, non-deforming.

MOLFOR-35.
M-1766; 0.35 C, 0.40 Mn, 1.0 Si, 5.0 Cr, 1.3 Mo, 0.40 V, bal Fe.
Hot work tool steel; die casting dies, forging dies and inserts.
IHA F-537; similar to AISI H13.

MOLFOR-40.
M-1766; 0.30 C, 0.50 Mn, 1.0 max Si, 3.0 Cr, 2.8 Mo, 0.50 V, bal Fe
Hot work tool steel; pressure casting molds, punches, extrusion cylinders.
DIN XC 32 CrMo 33.

MOLIBLOC.
M-1488; 0.35 C, 0.80 Mn, 0.50 Si, 1.6 Cr, 0.4 Mo, bal Fe.
Oil hardening tool steel for molds.
AFNOR 34 CD 7; similar to AISI P20.

MOLIN METAL.
M-1399; Al, bal Cu.
Cast.
For gears, shafts, bearings; aluminum bronze.

MOLIN METAL.
M-1456; Al, bal Cu.
For molds, dies; Al-bronze.

MOLITE 2.
M-35; 0.86 C, 6.25 W, 5.0 Mo, 4.15 Cr, 1.9 V, bal Fe.
For milling cutters, twist drills, broaches, reamers; AISI M2 high speed steel.

MOLITE 2 SMOOTHCUT.
M-35; 0.86 C, 6.25 W, 5.0 Mo, 4.15 Cr, 1.9 V, bal Fe, plus free machining additions.
For gear cutters, twist drills, reamers, broaches; AISI M2 high speed tool steel.

MOLITE 3.
M-35; 1.03 C, 6.25 Mo, 6.25 W, 4.0 Cr, 2.5 V, bal Fe.
Mo-W high speed tool steel; for broaches, taps, form tools, twist drills, blanking dies.
AISI M3 Class 1.

MOLITE 3 SMOOTHCUT TYPE 1.
M-35; 1.03 C, 6.25 W, 6.25 Mo, 4.0 Cr, 2.5 V, sulphides, bal Fe.
Hardened: C 64-66 Rock.
For broaches, taps, dies, reamers; from tools, hobs, slitting saws.
High speed steel; high red-hardness.
Good machinability rating. AISI M3 (class 1).

MOLITE 3 TYPE 2.
M-35; 1.2 C, 6.25 W, 6.25 Mo, 4.0 Cr, 3 V, bal Fe.
Hardened: C 64-66 Rock.
For broaches, form tools, taps, dies, milling cutters, cut-off tools.
High speed steel; high red-hardness.
AISI Type M3 (Class 2).

MOLITE 4.
M-35; 1.28 C, 5.50 W, 4.5 Mo, 4.5 Cr, 4.0 V, bal Fe.
Hardened: C 64-66 Rock.
For broaches, counterbores, drills, reamers, milling cutters.
High speed steel. High red-hardness.
AISI Type M4.

MOLITE 42.
M-35; 1.06 C, 1.6 W, 9.5 Mo, 3.75 Cr, 1.15 V, 8.0 Co, bal Fe.
Mo-Co high speed tool steel.
For broaches, shaving tools, lathe tools.
AISI M42.

MOLMANG.
M-374; C, 11-13.5 Mn, bal Fe.
Welded: 500-600 Brin.
For welding electrodes for build-up work; austenitic, work hardens.

MO-LO.
M-1456; 3 C, Mn, Si, bal Fe.
For piston rings, cylinder liners, bushings; cast iron.

MOLTROP.
0.3-0.5 C, bal Fe.
For gears, shafts; water hardened.

MOLVA C.
M-373; 1.02 C, 3.75 Cr, 1.75 W, 8.50 Mo, 1.9 V, bal Fe.
Molybdenum high speed steel for drills, reamers, lathe tools.
AISI M7.

MOLVA-T.
M-373; 0.8 C, 4 Cr, 2 V, 5 Mo, 6 W, bal Fe.
For lathe and planer tools, drills, form cutters, hot punches; high speed steel.

MOLY 8.
M-1433; 0.60 C, 3.6 Cr, 1.75 V, 8.5 Mo, bal Fe.
For hot work tools, shears, punches; Type H12; hot work steel.

MOLY ARK.
M-261; 0.7 C, 5.5 W, 4 Mo, 4 Cr, 1.5 V, bal Fe.
For tools, cutters; high speed steel.

MOLY ASCOLOY.
M-1509; 0.08 C, 13 Cr, 2 Mo, bal Fe.
For jet engine parts, afterburners, nozzles, bolts; high strength and high heat resistance.

MOLY ASTROLOY see **CARPENTER MOLY ASTROLOY.**

MOLY B 100.
M-1246; 50 Mo, 50 W.
For high vacuum tubes; sintered alloy.

MOLYBDENITE.
M-451; C, Cr, Mo, bal Fe.
For mill pinions, guides, rolls.

MOLYBDENUM.
M-1755; Mo.
Purities: Zone and chemical refined 99.995% 99.95%.
Forms: Powders, rods, wires, sheets, tubing, sintered, wrought bars, foils, arc castings, single crystals.

MOLYBDENUM-50W.
M-1246; 0.008 C, 49.3 W, bal Mo.
Stress Relieved: 144,000 TS; 133,800 YS; 14 El.
Recrystallized: 97,700 TS; 0 El.
For high temperature applications, heat engines. High heat and corrosion resistant.

MOLYBDENUM CHISEL.
M-365; 0.6 C, 0.5 Mo, bal Fe.
For chisels, tools; oil hardened.

MOLYBDENUM PERMALLOY.
M-1089; 80 Ni, 16 Fe, 4 Mo.
Soft magnetic alloy.

MOLYBDENUM PERMALLOY 2-81.
M-22, M-205; 2 Mo, 81 Ni, bal Fe.
For inductance and loading coils; low magnetic core loss.

MOLYBDENUM PERMALLOY 2-81.
M-269; 2 Mo, 81 Ni, 17 Fe.
For inductance and loading coils, low magnetic core loss, high permeability.

MOLYBDENUM PERMALLOY, 4-79.
M-22, M-205; 4 Mo, 79 Ni, 17 Fe.
For high frequency and electrical apparatus; high permeability and resistivity.

MOLYBDENUM SILICON.
M-347; C, Mo, Si, bal Fe.
Heat treated: 100,000-110,000 TS; 197-228 Brin.
For pistons, brake drums; wear resistant.

MOLYBDENUM STEEL.
M-Eng.; 0.15-0.40 Mo, 0.1-0.45 C, 0.6-1.3 Mn, 0.3-0.5 Si, bal Fe.
For machinery parts, gears, shafts; tough.

MOLYBESCO 1.
M-1495; 0.16 C, 0.30 Cr, 0.3 Mo, bal Fe.
Good tensile strength, good creep properties up to 550°C. For boilers, super heaters.

MOLYBESCO 2.
M-1495; 0.10 C, 0.15 Cr, 0.55 Mo, bal Fe.
High tensile strength; for heat exchangers, condensers, furnace tubes.

MOLYBESCO 4.
M-1495; 0.10 C, 0.15 Cr, 0.55 Mo, bal Fe.
High tensile strength; for heat exchangers, condensers, furnace tubes.

MOLYBESCO 5.
M-1495; 0.12 C, 0.15 Cr, 0.55 Mo, bal Fe.
High tensile strength; for heat exchangers, condensers, furnace tubes.

MOLY-IRON.
M-278; C, 1 Mo, 1 Cr, bal Fe.
Cast: 55,000 TS; 240 Brin.
For heat and abrasion resistant castings; heat and abrasion resistant.

MOLY-MANG.
M-118; C, 11-13 Mn, bal Fe.
For welding rod for high Mn steel; austenitic, wear and abrasion resistant.

MOLYNEAUX FUSIBLE ALLOY.
M-France; 41.5 Bi, 16.7 Sn, 25 Pb, 16.7 Cd.
For fuses, fire extinguishers; M.P. 60°C.

MOLY-NICKEL.
M-810; C, Mo, Ni, bal Fe.
For welding rod.

MOLY PERMINVAR-3.
M-22; 33.7 Fe, 34 Ni, 29 Co, 3 Mo, 0.3 Mn.
For magnetic amplifiers.

MOLY TELASTIC.
M-862; 0.30-0.40 C, 0.70-1.0 Mn, 0.60 max Si, 0.40-0.60 Cr, 0.15-0.25 Mo, bal Fe.
Low alloy cast steel; hardenable to 300-340 Brin in 1 in. dia. size.
Weldable, tough, for gears, construction machinery, pressure vessels.

MO-MANG.
M-9; 0.7-0.9 C, 12-14 Mn, bal Fe.
For welding rod; for Mn steel.

MO-MANG.
M-642; 0.8 C, 14 Mn, 0.2 Mo, bal Fe.
Cast: 200 Brin.
For welding electrodes; work hardens to 550 Brinell.

MOMARC.
M-115; 1.0 C, 1.4 Cr, 1.0 Mo, bal Fe.
Normalized: 185,000 TS; 139,000 YS; 13 El; 360 Brin.
For ball bearings, special rolls, precision gauges, arbors, dies.
Wear and abrasion resistant, oil or water hardening.

MONACA.
M-755; 0.5 C, bal Fe.
For gears, shafts; water hardening.

MONAR 816.
M-677; 0.09 C, 0.8 Mn, 0.08 Si, 65.5 Ni, 2.5 Ti, 1.0 Al, bal Cu.
For bare Monel welding wires: class ER-NiCu-7.

MONARCH.
M-502; C, alloy, bal Fe.
For tools, dies; oil hardening.

MONARCH NO. 10.
M-863; Pb, bal Cu.
Cast: 30,000 TS; 22,000 YS; 60 Brin.
For bearings for pumps, electric motors and machine tools; for high speed and moderate load.

MONARCH NO. 12.
M-863; Pb, bal Cu.
Cast: 30,900 TS; 25,100 YS; 70 Brin.
For bearings for large presses, nail and wire machines; high speeds and moderate, heavy loads.

MONARCH NO. 16.
M-863; Pb, bal Cu.
Cast: 37,000 TS; 29,800 YS; 75 Brin.
For bearings for large presses, cranes; extra heavy duty at low speeds.

MONARCH METAL.
M-863; 20 Pb, 5 Sn, bal Cu.
For bearings; leaded.

MONARK.
M-435; 0.6 C, 2 Si, 0.3 Cr, 0.2 Mo, bal Fe.
For tools, chisels; shock resisting.

MONARK-1.
M-435; 0.5 C, 0.4 Mn, 1.2 Si, 0.5 Mo, bal Fe.
For chisels, pneumatic tools; water hardened.

MONARK-2.
M-435; 0.6 C, 0.75 Mn, 2 Si, 0.3 Cr, 0.2 Mo, bal Fe.
For chisels, pneumatic tools; water hardened.

MOND.
M-86; 26 Cu, 70 Ni, 4 Mn.
For turbine blades, pump rods, valve stems; corrosion resistant.

MONEL 40 now MONEL FILLER METAL 40.
M-1499; 0.3 max C, 2.0 max Mn, 63-70 Ni, bal Cu.
For acetylene welding rod.

MONEL 60 now MONEL FILLER METAL 60.
M-1499; 23 min Cu, 60 min Ni, 3.5 max Fe, 3.5 max Mn.
For inert gas metal arc welding Monel.

MONEL 67 (FILLER METAL).
M-121; 0.02 C, 0.75 Mn, 31.0 Ni, 0.5 Fe, 67.5 Cu, 0.3 Ti 0.1 Si, 0.005 S.
For gas shielded arc welding 70/30, 80/20 and 90/10 Cu-Ni alloys, etc.
Filler metal.
AWS A 5.6 (Class E Cu Ni); ASME SFB 5.6 and 5.7.

"MONEL" 130.
M-897; 28-34 Cu, bal Ni.
For corrosion resistant parts; corrosion resistant.

MONEL 187.
M-121; 0.02 C, 32.0 Ni, 2.0 Mn, 0.15 Si, 65.0 Cu.
Welding electrode for shielded metal-arc welding 70/30, 80/20 and 90/10 Cu-Ni alloys and the clad side of Cu-Ni-clad steels.
AWS A5.6 (Class E Cu Ni); ASME SFB 5.6.

MONEL 190.
M-121; 0.01 C, 65.0 Ni, 3.1 Mn, 0.3 Fe, 0.75 Si, 0.007 S, 30.5 Cu, 0.15 Al, 0.55 Ti.
Welding electrode for shielded metal-arc welding. Monel alloy 400 to itself and to steel; for over laying on steel.
AWS A5.11 (Class E Cu Ni-2); ASME SFB 5.11.

MONEL ALLOY 400.
M-121; 0.12 C, 65.0 Ni, 32.0 Cu, 1.5 Fe, 1.0 Mn.
As cast: 550/mm² TS; 200 N/mm² YS; 40 El; 90-12 VHN.
Good corrosion resistance for valves, pumps, propeller shafts and marine hardware.
BS 3072-76; NA13; ASTM B127; ASME SB 127; ASM 4544 Etc.

MONEL ALLOY 400.
M-1499, M-1663, M-1602; 66.5 Ni, 0.15 C, 1.0 Mn, 1.25 Fe, 0.012 S, 0.25 Si, 31.5 Cu.
Annealed: 70,000-90,000 TS; 25,000-50,000 YS; 60-35 El; 110-149 Brin
Cold drawn, St.Rel.: 84,000-120,000 TS; 55,000-100,000 YS; 40-22 El 160-225 Brin.
For valves and pumps, marine fixtures, tanks, piping, heat exchangers; corrosion resistant.

MONEL ALLOY 401.
M-1499, M-121; 42.5 Ni, 0.05 C, 1.13 Mn, 0.38 Fe, 0.008 S, 0.13 Si, bal Cu.
For specialized electrical and electronic applications, wire wound resistors; bimetal contacts.

MONEL ALLOY 404.
M-1499; 54.5 Ni, 0.08 C, 0.05 Mn, 0.25 Fe, 0.012 S, 0.05 Si, 44.0 Cu, 0.03 Al.
For wave guides, metal to ceramic seals, transistor capsules, power tubes. Low magnetic permeability; excellent brazing characteristics.

MONEL ALLOY 502.
M-1499; 66.5 Ni, 0.05 C, 0.75 Mn, 1.0 Fe, 0.005 S, 0.25 Si, 28.0 Cu, 3.0 Al, 0.25 Ti.
Cold drawn: 87,000 TS; 55,000 YS; 42 El; 158 Brin.
Cold drawn and aged: 140,000 TS; 94,000 YS; 25 El; 255 Brin.
For fasteners, pump and propeller shafts, valve stems; machinable; age hardenable.

MONEL ALLOY K-500.
M-1499, M-121; 66.5 Ni, 0.13 Cu, 0.75 Mn, 1.0 Fe, 0.005 S, 29.5 Cu, 3.0 Al, 0.63 Ti.
Annealed: 90,000-110,000 TS; 40,000-60,000 YS; 45-25 El; 140-185 Brin.
Annealed and aged: 130,000-165,000 TS; 85,000-120,000 YS; 35-20 El; 250-315 Brin.
For pump shafts and impellers, oil well drill collars and instruments, electronic components; corrosion resistant.

MONEL ALLOY R-405.
M-1499, M-121; 66.5 Ni, 0.15 C, 1.0 Mn, 1.25 Fe, 0.043 S, 0.25 Si, 31.5 Cu.
Annealed: 70,000-85,000 TS; 25,000-40,000 YS; 50-35 El; 110-140 Brin
Cold drawn: 85,000-115,000 TS; 50,000-105,000 YS; 35-15 El; 160-245 Brin.
For water meter parts, screw machine products, valve seat inserts, fasteners for nuclear applications. Corrosion resistant, free machining.

MONEL FILLER METAL 40.
M-1499; 66.0 Ni, 0.10 C, 0.90 Mn, 1.35 Fe, 0.005 S, 0.15 Si, 31.5 Cu.
For oxyacetylene welding of Monel alloys 400 and 404.
AWS A 5.14 Class RNiCu-5; ASME SFB 5.14.

MONEL FILLER METAL 60.
M-1499; 65.0 Ni, 0.03 C, 3.5 Mn, 0.20 Fe, 0.005 S, 1.0 Si, 27.0 Cu, 2.2 Ti.
For gas-shielded arc welding of Monel alloys 400 and 404.
AWS A 5.14 Class ERNiCu-7; ASME SFB 5.14.

MONEL FILLER METAL 67.
M-1499; 31.0 Ni, 0.02 C, 0.75 Mn, 0.50 Fe, 0.005 S, 0.10 Si, 67.5 Cu, 0.30 Ti.
For oxyacetylene and gas-shielded arc welding of 70/30, 80/20, 90/10, copper-nickel alloys and clad side of copper-nickel clad steel.
AWS A 5.6 Class ECuNi; ASME SFB 5.6.

"MONEL" K.
M-897; 30 Cu, 2 Al, bal Ni.
For heat resistant parts; corrosion and heat resistant.

"MONEL" K44.
M-897; 30 Cu, 1 Mn, 2 Al, 2 Fe, bal Ni.
For corrosion and heat resistant parts; corrosion and heat resistant.

"MONEL" K64.
M-897; 30 Cu, 1 Mn, 2 Al, 2 Fe, bal Ni.
For corrosion and heat resistant parts; corrosion and heat resistant.

MONEL K-502.
M-1280; 0.10 C, 65.0 Ni, 1.0 Mn, 1.0 Fe, 0.50 Si, 29.5 Cu, 2.8 Al, 0.50 Ti.
Age hardenable, corrosion resistant alloy for gyroscope components, small machined parts.
QQ-N-286A.

MONEL R405.
M-44; 30 Cu, bal Ni.
Machineable grade wire and rod.

MONEL WELDING ELECTRODE 187.
M-1499; 32.0 Ni, 0.02 C, 2.0 Mn, 0.60 Fe, 0.01 S, 0.15 Si, 65.0 Cu.
Electrode for shielded metal-arc welding of 70/30, 80/20, 90/10 copper-nickel alloys and clad side copper-nickel clad steel.
AWS A 5.6 Class E CuNi; MIL-E-22200/4.

MONEL WELDING ELECTRODE 190.
M-1499; 65.0 Ni, 0.01 C, 3.1 Mn, 0.30 Fe, 0.007 S, 0.75 Si, 30.5 Cu 0.15 Al, 0.55 Ti.
Electrode for shielded metal-arc welding of Monel alloys 400 and 404; overlaying on steel.
AWS A 5.11 Class E NiCu-2; ASME SFB 5.11.

MONEND see MONEND 806.

MONEND 806.
M-677; 0.03 C, 3.7 Mn, 0.8 Si, 64 Ni, 28 Cu, 0.3 Fe, 0.8 Ti, 0.4 Al.
For welding electrodes; monel, meeting classes ENiCu-1 (3N10), ENiCu-2 (4N10), ENiCu-4 (8N10), also 9N10.

MONEVAL 806.
M-677; 64-68 Ni, 27-30 Cu, 1.0 max Fe.
Welded: 75,500 TS; 42,500 YS; 42 El; 60 RA.
For welding electrodes for Monel; Monel type.

MONIK-30.
M-1766; 0.32 C, 0.40 Mn, 0.50 max Si, 4.2 Ni, 0.50 Cr, 1.10 Mo, 0.15 V, bal Fe.
Hot work tool steel; mandrels for hot fabrication of tubes.
DIN 28 Ni Mo 179

MONIKROM.
M-185; 3.1-3.4 C, 2.0-2.4 Si, 0.7 Mn, 0.8-1.0 Cr, 0.15-0.25 Ni, 0.15-0.25 Mo, bal Fe.
Cast: 43,000 TS; 240-280 Brin.
For camshafts, cam sleeves; alloy cast iron; Tr.S. 80,000.

MONIX.
M-912; 0.16 C, 0.25 Si, 0.75 Mn, 0.60 Cr, 0.15 Mo, 1.15 Ni, bal Fe.
Alloy carburizing steel; for highly stressed case hardened parts.

MONIX 2.
M-912; 0.30 C, 0.25 Si, 0.50 Mn, 1.95 Cr, 0.45 Mo, 1.95 Ni, bal Fe.
Deep hardening steel for highly stressed parts in automotive and heavy equipment.
W.-Nr. 6580.

MONIX 3.
M-912; 0.35 C, 0.25 Si, 0.40 Mn, 1.35 Cr, 0.25 Mo, 3.55 Ni, bal Fe.
Deep hardening alloy steel; hardenable to 1300-1500 N/mm^2 UTS.
For highly stressed heavy parts.

MONIX 3 K.
M-912; 0.22 C, 0.25 Si, 0.40 Mn, 1.35 Cr, 0.30 Mo, 3.55 Ni, bal Fe.
Heat treated: 800-1050 N/mm^2.
For parts operating up to 600°C.

MONIX 4.
M-912; 0.42 C, 0.25 Si, 0.40 Mn, 1.35 Cr, 0.50 Mo, 4.55 Ni, bal Fe.
Deep hardening alloy steel; hardenable to 1300-1500 N/mm^2 UTS in heavy section.

MONIX 10.
M-912; 0.37 C, 0.25 Si, 0.65 Mn, 1.05 Cr, 0.20 Mo, 1.05 Ni, bal Fe.
Alloy steel, hardenable to 900-1050 N/mm^2 UTS.
For automotive parts.
W.-Nr. 6511.

MONIX 15.
M-912; 0.35 C, 0.25 Si, 0.50 Mn, 1.55 Cr, 0.25 Mo, 1.55 Ni, bal Fe.
Alloy steel, hardenable to 1000-1200 N/mm^2 UTS for automotive equipment.
W.-Nr. 6582.

MONIX 30.
 M-912; 0.30 C, 0.30 Si, 0.80 Mn, 0.50 Cr, 0.20 Mo, 0.55 Ni, bal Fe.
 Alloy steel, hardenable to 850-1000 N/mm² for automotive equipment.
 W.-Nr. 6545; AISI 8630.

MONIX 40.
 M-912; 0.40 C, 0.30 Si, 0.90 Mn, 0.50 Cr, 0.20 Mo, 0.55 Ni, bal Fe.
 Alloy steel, hardenable to 950-1150 N/mm² for automotive parts and equipment.
 W.-Nr. 6546; AISI 8640.

MONIX E.
 M-912; 0.21 C, 0.25 Si, 0.80 Mn, 0.50 Cr, 0.20 Mo, 0.50 Ni, bal Fe.
 Alloy carburizing steel; for stressed case hardened parts as automotive parts.
 W.-Nr. 6523; AISI 8620.

MONIX F.
 M-912; 0.17 C, 0.25 Si, 0.50 Mn, 1.65 Cr, 0.30 Mo, 1.55 Ni, bal Fe.
 Alloy carburizing steel for highly stressed case hardened parts.
 W.-Nr. 6587.

MONIX H.
 M-912; 0.17 C, 0.25 Si, 0.80 Mn, 0.90 Cr, 0.20 Mo, 1.45 Ni, bal Fe.
 Alloy carburizing steel for highly stressed case hardened parts.
 W.-Nr. 6566; similar to AISI 4320.

MONO-LOY.
 M-507; 1.4 C, 0.9 Si, 31.0 Mn, 0.5 Cr, 0.5 Mo, 5.5 Ni, bal Fe.
 Hard facing electrode.
 For overlaying or buildup on crusher rolls, dredge buckets, tractor sprockets, hammers. Good shock resistance.

MONOTYPE STANDARD.
 M-Eng.; 74 Pb, 18 Sb, 8 Sn.
 For type metal; slightly expansive on solidifying.

MONOWELD see ALL-STATE MONOWELD.

MONTAN.
 M-1331; 0.70 C, 0.25 max Si, 0.35 max Mn, bal Fe.
 For rails, tools, springs, axes; water hardened.

MONTAN 70.
 M-1331; 0.70 C, 0.25 max Si, 0.25 max Mn, bal Fe.
 Heat treated: 175,000 TS; 130,000 YS; 12 El; 37 RA; 355 Brin.
 For rails, punches, springs, axles, crimpers; water hardened.

MONTANIUM.
 M-Eng.; 2.5-3.5 Cu, 0.5 Mg, bal Al.
 51,000 TS; 16 El; 50 RA.
 For light alloy parts; similar to "Duralumin."

MONTEGAL.
 M-299; 0.95 Mg, 0.8 Si, 0.2 Ca, bal Al.
 50,000 TS; 6 El; 100 Brin.
 For light alloy parts; non-hardenable.

MOP 16.
 M-1488; 0.38 C, 4.0 Ni, 1.7 Cr, 0.50 Mo, bal Fe.
 Air or oil hardening mold steel; shock resistant.
 AFNOR Y 35NCD 16.

MOP 82.
 M-1488; 0.38 C, 1.0 Si, 5.25 Cr, 1.35 Mo, 0.5 V, bal Fe.
 Air hardening hot work tool steel for die casting or pressure casting molds.
 AFNOR Z38 CDV 5; AISI H11.

MOP 115 VITAC.
 M-1488; 0.34 C, 1.0 Mn, 1.8 Cr, 0.4 Mo, bal Fe.
 Oil hardening for dies for plastic molds.
 AFNOR 34 CD 7; similar to AISI P20.

MOPERMALLOY 4-79.
 M-61; 79 Ni, 17 Fe.
 Saturation inductance 8700.
 For laminated cores for communication inductors, transformers, electro-magnetic shields.
 Soft magnet, high permeability.

MO-PERMALLOY 4-79 ALLOY.
 M-61; 79 Ni, 4.3 Mo, bal Fe.
 For laminated cores for communication inductors, transformers, electro-magnetic shields.
 Soft magnet, high permeability.

MORAINE.
 M-839; Sn, bal Cu.
 For bearings; sintered.

MO RAPID EXTRA 3 now VEW S610.

MO-RE 1 ETC see DBK MO-RE 1 ETC.

MOREX.
 M-985; 0.7 C, 0.6 W, Cr, 4-6 Mo, bal Fe.
 For tools, cutters; high speed steel.

MOREX.
 M-487; C, Mo, W, bal Fe.
 For tools, cutters; high speed steel.

MORFLEX.
 M-237; high expanding side 22 Mn, 18 Cu, 10 Ni and low expanding side 36 Ni, bal Fe.
 For thermostatic bimetal; maximum temperature 500°F.

MORINS CHINESE BRONZE.
 M-Eng.; 83 Cu, 10 Pb, 5 Sn, 2 Zn.
 For ornaments.

MOSAIC GOLD.
 M-Eng.; 63 Cu, 37 Zn.
 46,000 TS.
 For brass ornaments; yellow brass.

MOSIL.
M-115; 0.57 C, 1.9 Si, 0.8 Mn, 0.25 Cr, 0.33 Mo, bal Fe.
For punches, chisels; shock resistant.

MOTAL.
M-302; 5 Si, bal Al.
21,500-27,000 TS; 6-3 El.
For light alloys, automobile parts, motor housings; sand cast alloy.

MO-TECHNICKILL.
M-229; C, Mo, bal Fe.
For heavy duty rollers; wear resistant.

MOTELEC.
M-691; 2.5 Si, bal Fe.

MOTEMP.
M-140; 0.88 C, 4 Cr, 8 Mo, 2 V, bal Fe.
Hardened: C 64-66 Rock.
For circular saws, taps, drills, broaches, counterbores, milling cutters, hobs, chasers, form tools.
Type M-10 high speed steel, high red-hardness.

MOTEMP RSP.
M-140; 0.68 C, 8.0 Mo, 4.0 Cr, 2.0 V, bal Fe.
Molybdenum hot work tool steel; for dies, hot forming tools.
AISI H43.

MOTOR MAGNUS.
M-1410; 0.80 C, 4 Cr, 2.0 V, 6 W, 5.5 Mo, bal Fe.
For drills, reamers, hobs, broaches, shear blades; high speed steel.
AISI M2.

MOTOR MAXIMUM.
M-1410; 0.75 C, 4.5 Cr, 1.2 V, 18 W, bal Fe.
For die casting dies, bushings, ejectors; high speed steel.
AISI T1.

MOTOR O6S ETC see CARR'S QUALITY O6S ETC.

MOTOR SPECIAL.
M-1410; 0.70 C, 4 Cr, 14 W, bal Fe.
For drills, slitters, shear blades, chisels; high speed steel.

MOTUF.
M-140; 1.0 C, 3.75 Cr, 2.1 V, 1.75 W, 8.75 Mo, bal Fe.
Hardened: C 65-66 Rock.
For broaches, burnishing tools, reamers, milling cutters, counterbores, taps, end mills, roll turning tools.
Type M-7 high-speed steel. High wear resistance and high red-hardness.

MO-TUNG.
M-114; 0.65-0.85 C, 0.2-0.35 Mn, 7.5-8.5 Mo, 1.5 W, 3.75-4.5 Cr, 1.0-1.25 V, bal Fe.
For high speed steel cutting tools, reamers, broaches, drills, lathe tools high speed steel.

MO-TUNG 652.
M-114; 0.8 C, 4 Cr, 2 V, 5 Mo, 6.5 W, bal Fe.
For lathe and planer tools, drills, form cutters, hot punches; high speed steel.

MOTUNG-CV.
M-114; 1.0 C, 8.75 Mo, 1.75 W, 4.0 Cr, 2.0 V, bal Fe.
Hardened: C 64-66 Rock.
For cutters, drills, reamers, end mills, slotting saws, routers, wood working tools, milling cutters. Good wear resistance.
AISI-M7 High speed steel.

MOTUNG P & D.
M-114; 0.74 C, 8.7 Mo, 1.65 W, 1.15 V, 4 Cr, bal Fe.
Annealed: 228 Brin.
For trim dies; cavity punches; high speed steel, oil hardened.

MOULTREX.
M-912; 0.55 C, 0.65 Mn, 0.80 Cr, 0.30 Mo, 0.165 Ni, 0.10 V, bal Fe.
Oil hardenable steel for plastic moulds.

MOUREY WHITE GOLD.
M-Eng.; 50 Au, 35 Ag, 15 Pd.
For jewelry; corrosion resistant.

MOUSSETS SILVER.
M-Eng.; 27.5 Ag, 59.5 Cu, 9.5 Zn, 3.5 Ni.
For instruments and cheap jewelry; corrosion resistant.

MP35N.
M-522, M-1439, M-73, M-1491; 35 Ni, 35 Co, 20 Cr, 10 Mo.
Ht. Tr.: 286,300 TS; 225,000 YS; 11 El; 39 RA.
Annealed: 146,000 TS; 61,000 YS; 70 El; B 100 Rock.
For high strength fasteners, bolts, aircraft and spacecraft components, steam turbine parts.
Corrosion resistant, high fatigue strength.

MP35N see MULTIPHASE MP35N.

MP35N-MULTIPHASE MP35N.
M-1439; 35 Ni, 35 Co, 20 Cr, 10 Mo.
Ultra high strength, 260 ksi UTS.
Tough, ductile, corrosion resistant, high fatigue strength.
For springs, fasteners, tubing; to 700°F.
AMS 5758, 5844, 5845.

MP159-MULTIPHASE MP159.
M-1439; 25 Ni, 36 Co, 19 Cr, 7 Mo, 2.9 Ti, 9 Fe, 0.5 Cb, 0.2 Al.
At 1000°F: 200,000 psi TS (typical).
Good high temperature properties, including creep rupture. For high temperature fasteners.
AMS 5841, 5842, 5843.

MRL-A8.
M-1719; 6-8 Mg, 0.5 max Si, 0.5 max Fe, 0.35 max Mn, 0.25 max Cu, 0.05-0.30 Cr, 0.20 max Zn, 0.001-0.05 B, 0.02 max Be, 0.015 max Ti, bal Al.
O cond: 50,000 TS; 22,000 YS; 25 El.
H38: 69,000 TS; 54,000 YS; 10 El.
Good forming prop.; especially good corrosion resistance, including marine atmosphere.
AA-X5090.

MRL-A11.
M-1719; 0.6-0.9 Si, 0.7 max Fe, 0.4-0.6 Mg, 0.25 max Zn, 0.2 max Cu, 0.15 max Ti, 0.05-0.15 Mn, 0.05-0.15 Cr, 0.05-0.15 Zr, bal Al.
T5: 38,000 TS; 35,000 YS; 8 El.
Good impact strength: 20-60 V-notch Charpy, generally above 30 ft. lb.
For highway guard rails, semi-hollow type.
Hardenable.

MS4V.
M-1653; 0.45-0.52 C, 1.30-1.60 Si, 0.40-0.80 Mn, bal Fe.
Silico-manganese steel.
48 Si 5; Similar to AISI 9254.

MS5VE.
M-1653; 0.55-0.65 C, 1.8-2.2 Si, 0.70-1.0 Mn, bal Fe.
Silico-manganese spring steel.
60 Si 8; Similar to AISI 9260.

MS5Y.
M-1653; 0.50-0.60 C, 1.8-2.2 Si, 0.70-1.0 Mn, (0.15-0.45 Cr), bal Fe.
Silico-manganese spring steel.
55 Si 8; AISI 9255.

MS-6.
M-Switzerland; 1.0 C, 9 Cr, 6 Co, 1.5 Mo, bal Fe.
Hardened: C 60-64 Rock.
For permanent magnets, in magnetic and electrical equipment. Hardened. High permeability.

MS7V.
M-1653; 0.55-0.65 C, 1.8-2.2 Si, 0.70-1.0 Mn, 0.25-0.40 Cr, bal Fe.
Silico-manganese spring steel.
60 Si Cr 8.

MS8Y.
M-1653; 0.42-0.55 C, 1.5-2.0 Si, 0.40-0.80 Mn, bal Fe.
Silicon shock resisting steel.
50 Si 7.

MS-15.
M-Switzerland; 1.0 C, 15 Co, 9 Cr, 1.5 Mo, bal Fe.
Hardened: C 60-64 Rock.
For permanent magnets in magnetic and electrical equipment.
Hardened. High permeability.

MS-35.
M-Switzerland; 0.9 C, 35 Co, 6 Cr, 5 W, bal Fe.
For permanent magnets in magnetic and electrical equipment.
Hardened: High permeability.

MS-250.
M-1619; 0.03 max C, 16.0-17.5 Ni, 9.5-11.0 Co, 4.4-4.9 Mo, 0.05-0.15 Al, 0.15-0.45 Ti, 0.03 max Zr, 0.005 max B, bal Fe.
Heat Treated: 275,000 TS; 263,000 YS; 6 El; 16 RA.
For compressor rotors, high impact pressure devices.
High strength and ductility. Maraging cast alloy steel.

MS ALLOY.
M-Japan; 30-60 Ni, 1-18 Cr, bal Fe.
For magnetic shunts; magnetically soft.

M.S. BRONZE.
M-1068; 96 Cu, 3.5 Sn, 0.25 P.
For springs; for watches.

M S M ALLOY.
M-365; 0.5 C, 0.7 Mn, 1.8-2.0 Si, 0.25 Mo, bal Fe.
For shear blades, chisels, punches; tough.

MST ALLOYS see RMI ALLOYS.

M/S VENUS.
M-1260; 83.5 Sn, 8 Sb, 6.5 Cu, 2 Pb.
Cast: 17,100 TS; 31 Brin.
For diesel engine bearings; M.P. 355-710°F; wear resistant.

MT-6 see DARWIN MT6.

MT-17.
M-Eng.; 0.07 C, 1.5 Mn, 21 Cr, 30 Ni, 21 Co, 3 Mo, 2.2 W, 1.6 Ti, bal Fe.
For high temperature applications; heat resistant.

MTR.
M-1180; 0.13 C, 0.65-0.95 Mn, 0.4-0.7 Cr, 0.25 Cu, 0.4-0.6 Mo, 1.1-1. Ni, 0.04-0.10 V, bal Fe.
Plate 107,000-135,000 TS; 78,000 min YS; 13 min El.
For pressure vessels, bridges, mine cars, power shovels, cranes, trucks, trailers.
Shock and wear resistant.

MT-STEEL.
M-1566; 8 Al, 2 C, 0.20 max Si, 0.6 Co, bal Fe.
Heat treated: Br = 5000 gauss, H_c = 200 Oersted.
For electrical and magnetic equipment.
Permanent magnet. Quenched and tempered for max. magnetic properties.

MUELLER 14S see **MUELLER 2014**.

MUELLER 17S see **MUELLER 2017**.

MUELLER 200 see **MUELLER 3770**.

MUELLER 201 see **MUELLER 3600**.

MUELLER 203 see **MUELLER 4640**.

MUELLER 204 see **MUELLER 4851**.

MUELLER 207 see **MUELLER 4850**.

MUELLER 210 see **MUELLER 4643**.

MUELLER 211 see **MUELLER 3771**.

MUELLER 212 see **MUELLER 3450**.

MUELLER 213 see **MUELLER 3801**.

MUELLER 214 see **MUELLER 1100**.

MUELLER 214A see **MUELLER 1220**.

MUELLER 224C see **MUELLER 6230**.

MUELLER 224E see **MUELLER 6180**.

MUELLER 224H see **MUELLER 6240**.

MUELLER 224K see **MUELLER 6300**.

MUELLER 241A see **MUELLER 6750**.

MUELLER 241W see **MUELLER 6780**.

MUELLER 245 see **MUELLER 4820**.

MUELLER 246 see **MUELLER 3602**.

MUELLER 248 see **MUELLER 3530**.

MUELLER 600 see **MUELLER 6741**.

MUELLER 601 see **MUELLER 6680**.

MUELLER 602 see **MUELLER 6730**.

MUELLER 603 see **MUELLER 6731**.

MUELLER 604 see **MUELLER 6732**.

MUELLER 605 see **MUELLER 6733**.

MUELLER 721 see **MUELLER 6700**.

MUELLER 799 see **MUELLER 1450**.

MUELLER 802 see **MUELLER 6420**.

MUELLER 902 see **MUELLER 1920**.

MUELLER 1020.
M-266; 99.95 Cu, 0.04 O_2.
Half Hard: 53,000 TS; 46,000 YS; 5 El; F 90 Rock.
Soft: 32,000 TS; 10,000 YS; 40 El; F 35 Rock.
For high conductivity tubing and forgings.
CDA 102, Formerly Mueller 2140F.
Oxygen Free-High Conductivity Copper.

MUELLER 1100.
M-266; 99.9 Cu, 0.04 Oxy.
Hard Rod: 48,000 TS; 44,000 YS; 16 El; F 87 Rock.
For trolley wire, cables, conductors; connectors, terminals, fittings, pipes.
Tough pitch copper. CDA 110, Formerly Mueller 214 ETP.

MUELLER 1220.
M-266; 99.90 Cu, 0.015-0.040 P.
Annealed: 32,000 TS; 10,000 YS; 45 El; F 40 Rock.
Hard Drawn: 55,000 TS; 50,000 YS; 8 El; B 60 Rock.
For air conditioners, refrigerators, kettles, oil coolers, rotating bands.
Phosphorus deoxidized copper.
CDA 122. Formerly Mueller 214 DHP.

MUELLER 1450.
M-266; 0.5 Te, 0.01 P, 99.49 Cu.
Tellurium copper rod and screw machine products; free machining.
CDA 145. (Was Mueller 799W).

MUELLER 1470.
M-266; 0.3 S, bal Cu.
Sulphurized copper.
Half hard: 43,000 TS; 40,000 YS; 18 El; 42 Rb.
Hard: 48,000 TS; 46,000 YS; 13 El; 50 Rb.
96% conductivity; 85% machinability.
Free machining copper; for screw machined parts requiring properties similar to copper.
CDA No. 147. Was Mueller 771W.

MUELLER 1476.
M-266; 0.20-0.40 S, 99.9 min Cu + Ag + S.
Sulphur bearing copper, free cutting.
CDA 14720.

MUELLER 1477.
M-266; 0.20 max S, 99.90 min Cu + Ag + S.
Sulphur bearing copper, free machining.
CDA 14710.

MUELLER 1820.
M-266; 1 Cr, 0.02 Si, bal Cu.
Heat treated: 78,000 TS; 70,000 YS; 13 El; B 80 Rock.
For resistance welding tips, holders and wheels.
CDA 182 and 184. Formerly Mueller 902.
Heat treatable. High strength.

MUELLER 2014.
M-266; 4.4 Cu, 0.8 Mn, 0.4 Mg, 0.8 Si, bal Al.
Heat treated: 70,000 TS; 60,000 YS; 14 El; 140 Brin.
For hardware; forging alloy.

MUELLER 2017.
M-266; 4 Cu, 0.5 Mn, 0.5 Mg, 0.75 Si, bal Al.
Heat treated: 60,000 TS; 37,500 YS; 22 El; 105 Brin.
For hardware; forging alloy.

MUELLER 2300.
M-266; 84-86 Cu, 0.05 max Pb, bal Zn.
Red brass tubing and pipe.
CDA 230.

MUELLER 2600.
M-266; 63-68.5 Cu, 0.15 max Pb, bal Zn.
Yellow brass tubing.
CDA 270.

MUELLER 2740.
M-266; 61-64 Cu, 0.10 max Pb, bal Zn.
Yellow brass tubing.
CDA 274.

MUELLER 3300.
M-266; 65-68 Cu, 0.20-0.80 Pb, bal Zn.
Low leaded brass tube.
CDA 330.

MUELLER 3320.
M-266; 65-68 Cu, 1.3-2.0 Pb, bal Zn.
High leaded brass tube.
CDA 332.

MUELLER 3400.
M-266; 1 Pb, 65 Cu, bal Zn.
HH.Temper: 58,000 TS; 45,000 YS; 30 El; B 70 Rock.
For cold headed fasteners, hardware, marine hardware.
CDA 340. Formerly Mueller 314.
Leaded Brass, good machinability.

MUELLER 3406 now MUELLER 3400.

MUELLER 3450.
M-266; 2 Pb, 62.5 Cu, bal Zn.
1/4 Hard: 53,000 TS; 35,000 YS; 35 El; B 68 Rock.
For thread rolled fasteners, marine hardware.
CDA 345. Formerly Mueller 212.
Leaded Brass, good cold workability.

MUELLER 3470.
M-266; 63 Cu, 1.5 Pb, bal Zn.
Medium leaded brass.
3/8 H: 55,000 TS; 35,000 YS; 30 El; 68-70 Rb.
Thread rolling brass; good machining and cold working characteristics. 85% machinability.
CDA No. 347. Formerly Mueller 312.

MUELLER 3500.
M-266; 61.7 Cu, 1.0 Pb, bal Zn.
Thread rolling brass.
1/4 Hard: 53,000 TS; 35,000 YS; 35 El; 70 Rb.
For roll threaded brass parts.
CDA No. 350. Formerly Mueller 251.

MUELLER 3530.
M-266; 35.25 Zn, 2.75 Pb, bal Cu.
Cast.
For screw machine products; commercial brass.

MUELLER 3600.
M-266; 3 Pb, 61.5 Cu, bal Zn.
HH Temper: 60,000 TS; 45,000 YS; 25 El; B 72 Rock.
Hard Temper: 75,000 TS; 55,000 YS; 10 El; B 78 Rock.
For screw machine products, fasteners, bearings, bushings.
CDA 360. Formerly Mueller 201.
Leaded brass, free-cutting.

MUELLER 3602.
M-266; 2.75 Pb, 61.5 Cu, bal Zn.
3/8 Hard: 55,000 TS; 40,000 YS; 25 El; B 68 Rock.
For fasteners, marine hardware.
CDA 360. Formerly Mueller 246.
Leaded brass, free-cutting.

MUELLER 3770.
M-266; 1.5-2.5 Pb, 58-61 Cu, 0.3 max Fe, bal Zn.
Forgings: 50,000-60,000 TS; 20,000-25,000 YS; 40-50 El; B 40-60 Rock.
Extruded: 58,000 TS; 23,000 YS; 42 El; B 48 Rock.
For hardware, valve bodies, automobile parts, plumbing fittings.
Brass forging alloy. Formerly Mueller 200.

MUELLER 3771.
M-266; 2 Pb, 60 Cu, bal Zn.
Forged: 58,000 TS; 23,000 YS; 42 El; B 48 Rock.
For forged hardware, shafts, fasteners.
CDA 377. Formerly Mueller 211.
Leaded forging brass.

MUELLER 3801.
M-266; 2 Pb, 56 Cu, bal Zn.
Forged: 65,000 TS; 30,000 YS; 15 El; B 60 Rock.
For forged hardware, fasteners.
Formerly Mueller 213.
Leaded forging brass.

MUELLER 4430.
M-266; 70.0-73.0 Cu, 0.07 max Pb, 0.8-1.2 Sn, 0.02-0.10 As, bal Zn.
Admiralty arsenical copper tubing.
CDA 443.

MUELLER 4450.
 M-266; 70.0-73.0 Cu, 0.07 max Pb, 0.8-1.2 Sn, 0.02-0.10 P, bal Zn.
 Admiralty phosphorized copper tubing.
 CDA 445.

MUELLER 4620.
 M-266; 62-65 Cu, 0.5-1.0 Sn, bal Zn.
 HH Temper: 65,000 TS; 45,000 YS; 30 El; B 72 Rock.
 For fasteners, bolts, marine hardware, rivets, clamps, connectors.
 Corrosion resistant. Naval Brass.
 CDA 462. Formerly Mueller 308.

MUELLER 4640.
 M-266; 0.75 Sn, 60.5 Cu, bal Zn.
 Forged: 64,000 TS; 26,000 YS; 40 El; B 55 Rock.
 Hard: 75,000 TS; 60,000 YS; 20 El; B 80 Rock.
 For marine products, gears, valve stems, light bearings, screw machine products.
 Naval Brass.
 CDA 464. Formerly Mueller 203.

MUELLER 4643.
 M-266; 60 Cu, 0.3 max Pb, 0.75 Sn, bal Zn.
 1/2 H-temper: 68,000 TS; 45,000 YS; 22 El; 150 Brin.
 Soft: 54,000 TS; 22,000 YS; 40 El; 83 Brin.
 For propeller shafts, gears, bolts, screw machine products; Tobin Bronze.

MUELLER 4820.
 M-266; 0.75 Pb, 0.75 Sn, 60.5 Cu, bal Zn.
 Forged: 60,000 TS; 30,000 YS; 40 El; B 55 Rock.
 Hard: 70,000 TS; 48,000 YS; 30 El; B 78 Rock.
 For marine parts, gears, screw machine products.
 Leaded Naval Brass, free-turning.
 CDA 482. Formerly Mueller 245.

MUELLER 4850.
 M-266; 2 Pb, 0.75 Sn, 60.5 Cu, bal Zn.
 HH Temper: 65,000 TS; 48,000 YS; 15 El; B 72 Rock.
 For marine products, gears, valve stems, light bearings, screw machine products.
 Leaded Naval Brass.
 CDA 485. Formerly Mueller 207.

MUELLER 4851.
 M-266; 2 Pb, 0.75 Sn, 59.5 Cu, bal Zn.
 Forged: 62,000 TS; 24,000 YS; 40 El; B 55 Rock.
 For light bearings, forged hardware.
 CDA 485. Formerly Mueller 204.
 Leaded Naval Brass, free-cutting.

MUELLER 6162 now **MUELLER 6181.**

MUELLER 6180.
 M-266; 10 Al, 1 Fe, 89 Cu.
 HH Temper; 100,000 TS; 65,000 YS; 16 El; 208 Brin.
 For gears, valve parts, worm wheels, marine hardware.
 Aluminum Bronze, corrosion resistant.
 CDA 617. Formerly Mueller 224 E-75.

MUELLER 6181 TUF-STUF ALUMINUM BRONZE.
 M-266; 90 Cu, 1 Fe, 9 Al.
 1/2 Hard: 82,000 TS; 50,000 YS; 30 El; 175 Brin (1000 kg).
 Forged: 75,000 TS; 36,000 YS; 20 El; 140 Brin (1000 kg).
 High strength, heat treatable bronze for shafts and other sliding parts.
 CDA No. 617. Formerly Mueller 224 E-30.

MUELLER 6230.
 M-266; 88 Cu, 3 Fe, 9 Al.
 HH Temper: 95,000 TS; 62,000 YS; 9 El; 185 Brin.
 Forged: 84,000 TS; 40,000 YS; 40 El; 135 Brin.
 For gears, worm wheels, marine hardware, shafts.
 Aluminum bronze. Not heat treatable.
 CDA 616. Formerly 224C.

MUELLER 6236.
 M-266; 8.0-10.0 Al, 2.0-4.0 Fe, 0.25 max Si, 0.20 max Sn, bal Cu.
 Aluminum bronze rod.
 CDA 623.

MUELLER 6246.
 M-266; 85.5 Cu, 3.6 Fe, 10.6 Al.
 Aluminum bronze, heat treatable.
 Ext: 100,000 TS; 50,000 YS; 7 El; 195 Brin (3000 Kg).
 Moderately high strength aluminum bronze, possessing excellent corrosion and high temperature properties.
 CDA No. 616. Formerly Mueller 224-11.

MUELLER 6250.
 M-266; 82.8 Cu, 4.3 Fe, 12.9 Al.
 Aluminum bronze.
 Ext: 105,000 TS; 65,000 YS; 1.5 El; 285 Brin; (3000 kg).
 Heat treatable, good strength, very good wear resistance; for gears, worm wheels, wear plates.
 Formerly Mueller 224-13.

MUELLER 6300.
 M-266; 5 Ni, 1 Mn, 10 Al, 3 Fe, 81 Cu.
 Annealed: 120,000 TS; 75,000 YS; 15 El; 241 Brin.
 Heat treated: 114,000 TS; 66,000 YS; 15 El; 223 Brin.
 For gears, aviation landing gears, shafts, marine hardware.
 Aluminum Bronze, heat treatable.
 CDA 628. Formerly Mueller 224K.

MUELLER 6390 now **MUELLER 6420.**

MUELLER 6391 now **MUELLER 6421.**

MUELLER 6420.
M-266; 7 Al, 2 Si, 91 Cu.
Hard: 98,000 TS; 65,000 YS; 25 El; B 95 Rock.
Forged: 75,000 TS: 35,000 YS; 45 El; B 85 Rock.
For nuts, bolts, pole line hardware.
Aluminum Bronze, corrosion resistant.
CDA 639. Formerly Mueller 802.

MUELLER 6421.
M-266; 91 Cu, 6.7 Al, 1.9 Si.
Alumunum silicon bronze.
Hard: 85,000 TS; 45,000 YS; 40 El; 84 Rb.
As forged: 75,000 TS; 35,000 YS; 45 El; 80 Rb.
Machinability 50%.
Free turning, good strength, non-magnetic, corrosion resistant; for shafts, nuts, bolts, pole line hardware.
CDA No. 639. Formerly Mueller 808.

MUELLER 6680.
M-266; 60.5 Cu, 1 Si, 2.5 Mn, bal Zn.
HH Temper: 70,000-85,000 TS; 40,000-65,000 YS; 15-25 El; 70-87 Roc
For bearings, gears, connecting rods, marine hardware.
Bearing Bronze. Ductile.
Formerly Mueller 601.

MUELLER 6700.
M-266; 3 Fe, 5 Al, 4 Mn, 64 Cu, bal Zn.
Soft: 90,000 TS; 50,000 YS; 20 El; B 84 Rock.
Hard: 124,000 TS; 72,000 YS; 18 El; B 95 Rock.
For rollers, rotors, valve stems.
Manganese Bronze, shock resistant.
CDA 670. Formerly Mueller 721B.

MUELLER 6701.
M-266; 3 Fe, 5 Al, 4 Mn, 64 Cu, bal Zn.
HH Temper: 112,000 TS; 67,000 YS; 13 El; B 94 Rock.
For rollers, rotors, valve stems.
Manganese Bronze, shock resistant.
CDA 670. Formerly Mueller 721E.

MUELLER 6702.
M-266; 64 Cu, 3 Fe, 5 Al, 4 Mn, bal Zn.
Forged: 115,000 TS; 67,000 YS; 18 El; B 95 Rock.
For rotors, rollers, valve stems.
Manganese Bronze, shock resistant.
CDA 670. Formerly Mueller 721x.

MUELLER 6730.
M-266; 1.0 Pb, 2.5 Mn, 1.0 Si, 60.5 Cu, bal Zn.
HH-Temper: 70,000-85,000 TS; 40,000-65,000 YS; 15-25 El; B 70-87 Rock.
For gears, sleeve and thrust bearings, bushings, cams, marine hardware.
Good bearing characteristics. Formerly Mueller 602.

MUELLER 6731.
M-266; 59.75 Cu, 1 Pb, 1 Si, 2.5 Mn, bal Zn.
Heat treated: 75,000 TS; 45,000 YS; 20 El; 82 Rock.
For gears, hydraulic pump cylinder liners, bushings, bearings.
Bearing Bronze, free-cutting.
CDA 673. Formerly Mueller 603.

MUELLER 6732.
M-266; 61.5 Cu, 2.5 Pb, 1 Si, 2.5 Mn, bal Zn.
HH Temper: 65,000-80,000 TS; 40,000-60,000 YS; 10-20 El; 75-86 Rock B.
For bearings used against soft or hard mating surfaces.
Bearing Bronze. Can be soldered.
CDA 673. Formerly Mueller 604

MUELLER 6733.
M-266; 62 Cu, 0.6 Pb, 1 Si, 2.5 Mn, bal Zn.
HH Temper: 70,000-75,000 TS; 45,000-55,000 YS; 18-25 El; 75-82 Roc
For bearing mating with soft or hard members.
CDA 673. Formerly Mueller 605.

MUELLER 6741.
M-266; 58 Cu, 1 Si, 1.5 Al, 2.5 Mn, bal Zn.
HH Temper: 68,000-100,000 TS; 34,000-65,000 YS; 12-18 El; B 78-88 Rock.
Forged: 68,000 TS; 34,000 YS; 18 El; B 78 Rock.
For bearings against hardened surfaces, bushings, connecting rods, hardware.
Bearing Bronze, corrosion resistant.
CDA 674. Formerly Mueller 600.

MUELLER 6750.
M-266; 59 Cu, 1 Sn, 1 Fe, 0.25 Mn, bal Zn.
Hard: 83,000 TS; 56,000 YS; 20 El; B 80 Rock.
Forged: 68,000 TS; 35,000 YS; 27 El; B 65 Rock.
For screw machine products, valve stems, airplane parts, balls.
Manganese Bronze, corrosion resistant.
CDA 675. Formerly Mueller 241A.

MUELLER 6780.
M-266; 57 Cu, 1 Fe, 1 Al, 0.25 Mn, bal Zn.
Forged: 80,000 TS; 29,000 YS; 35 El; 130 Brin.
For counterweights for airplane engine crankshafts.
Manganese Bronze. Formerly Mueller 241W.

MUELLER 6940.
M-266; 80-83 Cu, 0.3 max Pb, 3.5-4.5 Si, bal Zn.
Rod: 90,000 TS; 57,000 YS; 38 El.
For valve stems.
Silicon Red Brass, corrosion resistant.
CDA 694.

MUELLER 6970.
M-266; 77 Cu, 1 Pb, 3 Si, bal Zn.
Silicon brass.
1/4 hard: 80,000 TS; 46,000 YS; 42 El; 82 Rock B.
For shafts, bolts, nuts, cap screws, screw machine parts; good machinability and corrosion resistance; valve stem alloy.
CDA No 697. Formerly Mueller 804.

MUELLER TUF-STUF 6240.
M-266; 86 Cu, 11 Al, 3 Fe.
Forged: 80,000 TS; 45,000 YS; 10 El; 170 Brin.
Heat treated: 95,000 TS; 70,000 YS; 3 El; 200 Brin.
For valve seat inserts, gears, worm wheels; forging alloy, heat treatable; "Tuf-Stuf H."

MUFLEX.
M-237; iron-invar.
For thermostatic bimetal; high-iron, low-invar.

MU-GUARD 48.
M-1626; 48 Ni, bal Fe.
Soft magnetic alloy strip for magnetic shielding; medium effectiveness.

MU-GUARD 80.
M-1626; 80 Ni, 4.5 Mo, bal Fe.
Soft magnetic alloy strip. Very high magnetic shielding value.

MULTI-ALLOY.
M-Eng.; 0.25 C, 46.5 Ni, 20.5 Cr, 2.7 Mo, 3.3 Co, 1.2 Ti, 2.9 Cb, 3.5 W, bal Fe.
For gas turbine parts; heat resistant.

MULTICUT LEADED.
M-97; 0.09 max C, 0.85-1.15 Mn, 0.04-0.09 P, 0.26-0.35 S, 0.15-0.35 Pb, bal Fe.
Free machining carbon steel bars.

MULTICUT LEADED WITH SELENIUM.
M-97; 0.09 max C, 0.85-1.15 Mn, 0.04-0.09 P, 0.26-0.35 S, 0.15-0.35 Pb, Se, bal Fe.
Free machining carbon steel bars.

MULTICUT REGULAR.
M-97; 0.09 max C, 0.75-1.05 Mn, 0.04-0.09 P, 0.26-0.35 S, bal Fe.
Free machining carbon steel bars.

MULTIMET ALLOY.
M-167; 0.2 max C, 18-22.5 Cr, 18-22 Ni, 2.75-3.75 Mo, 2-3 W, 18-22 Co, 0.75-1.5 Cb, 0.1-0.2 N, bal Fe.
Cast: 83,000 TS; 55,000 YS; 12 El; 14 RA.
Forged: 110,000-140,000 TS; 30-15 El; 48-45 RA.
For turbine blading, jet and combustion chambers, welding rod; high heat resistant.

MULTIPASS see HAYNES MULTIPASS.

MULTIPHASE - MP20N.
M-1439; 20 Ni, 50 Co, 20 Cr, 10 Mo.
Heat treated: 290,000 TS; 260,000 YS; 10 El.
For high strength bolts and fasteners.
High fatigue strength; corrosion resistant.

MULTIPHASE MP-159.
M-73; 25.5 Ni, 35.5 Co, 19 Cr, 7 Mo, 9 Fe, 3 Ti, 0.6 Cb, 0.2 Al.
Room temp; 260,000 psi TS.
1100°F: 200,000 psi TS.
High temperature super alloy.

MULTOLE.
M-336; 0.6 C, 0.7 Mn, 1.9 Si, bal Fe.
For punches, shear blades, knives; keen edge tools.

MUMETAL.
M-108; 77 Ni, 14 Fe, 5 Cu, 4 Mo.
Soft magnetic alloy, high permeability at low field strengths for transformers, chokes, etc.

MUMETAL 40.
M-108; 5 Cu, 4 Mo, 14 Fe, bal Ni.
Soft magnetic alloy; high permeability.

MUMETAL PLUS.
M-108.
Soft magnetic alloy, high permeability, low losses for current transformer etc.

MUNGOOSE.
M-135; Cu, 12-15 Ni, Zn.
For domestic utensils, ornaments; nickel silver.

MUNTZ METAL.
M-33; 60 Cu, 40 Zn.
Annealed: 54,000 TS; 21,000 YS; 45 El.
Rolled: 80,000 TS; 60,000 YS; 5 El.
For trim, brazing rod, valve stems, condenser tubes; high strength.

MUNTZ METAL.
M-188; 60 Cu, 40 Zn.
Cast: 48,200 TS; 54 El; 52 RA; 93 Brin.
Hard sheet: 80,000 TS; 9.5 El.
For bolts, pins, spindles, wire, condenser tubes; high corrosion resistance.

MUNTZ METAL see also IMI 452.

MUNTZ METAL NO 3.
M-122; 61 Cu, bal Zn.
For forgings; 58,000 TS; 25,000 YS; 45 El; Rockwell B55.
For forgings, hardware; yellow brass, high strength.
CA 280.

MUNTZ METAL-280.
M-8; 60 Cu, 40 Zn.
Soft: 54,000 TS; 20,000 YS; 45 El.
For architecture, trim, perforated sheets; high strength.

MUNTZ PATENTS.
M-Eng.; 63-56 Cu, 37-42 Zn, 0-4 Pb.
For tubes, hardware; free-cutting.

MUNZBRONZE.
M-297; 4 Sn, 1 Zn, bal Cu.
Soft: 48,000 TS; 44 El; 68 Brin.
Hard: 82,000 TS; 5 El; 156 Brin.
For chemical plant equipment; corrosion resistant.

MUREX 4-6 CHROME.
M-529; C, 3.21 Cr, 0.5 Mo, bal Fe.
83,000 TS; 62,000 YS; 26 El; 45 RA.
For welding electrodes; corrosion resistant.

MUREX ALTERNEX-A.
M-529; 0.1 C, bal Fe.
Welded: 76,400 TS; 66,200 YS; 24 El.
For welding electrodes; for mild steel; Type E-6013.

MUREX FILLEX.
M-529; 0.2 C, bal Fe.
Welded: 66,000 TS; 53,000 YS; 28 El.
For welding electrodes; for mild steel; Type E-6020.

MUREX GENEX-M.
M-529; 0.2 C, bal Fe.
Welded: 68,000 TS; 54,000 YS; 28 El.
For welding electrodes; Type E-6012.

MUREX HARDEX 25.
M-529; 0.2 C, 1.2 Mn, 0.7 Cr, 0.9 Si, bal Fe.
For hard facing electrodes; for moderate abrasion and high impact.

MUREX HARDEX 45.
M-529; 0.6 C, 1.4 Mn, 2.7 Cr, 0.8 Si, bal Fe.
For hard facing electrodes; for high abrasion and moderate impact.

MUREX MOLEX.
M-529; 0.15 C, bal Fe.
Welded: 75,000 TS; 62,000 YS; 25 El.
For welding electrodes; for C-Mo steel; Type E-7010 Al.

MUREX TYPE FHP.
M-529; 0.08-0.13 C, 0.2-0.4 Mn, bal Fe.
Welded: 67,000 TS; 56,000 YS; 28 El.
For welding, electrodes; shielded arc; Type E-6020.

MURMANS ALLOY-1.
M-Eng.; 72 Al, 15 Zn, 14 Mg.
For light alloy parts; non-hardenable.

MURMANS ALLOY-2.
M-Eng.; 92 Al, 4.4 Zn, 3.6 Mg.
For light alloy parts; non-hardenable.

MUSHET STEEL.
M-Japan; 1.5-2.0 C, 4-6 W, 0.25-0.30 Cr, 0.3-0.5 Mn, bal Fe.
For tools, dies; cannot be cold worked.

MUSIC SPRING.
M-120; 0.80 C, 0.25 Mn, 0.25 Si, bal Fe.
For springs; Swedish steel.

MUSTANG-LC (AISI H-42).
M-69; 0.65 C, 4.0 Cr, 2.0 V, 6.5 W, 5.0 Mo, bal Fe.
As quenched: C 58-59 Rock.
Tempered 1000°F: C 60-61 Rock.
For hot forming and swaging dies, nut piercers, hot punches, extrusion dies, forging mandrels, chipper dies.
Good resistance to high temperature softening.
High speed steel. Type H-42.

MUSTANG M-2 (AISI M2).
M-69; 0.84 C, 4.2 Cr, 6.5 W, 2 V, 5 Mo, bal Fe.
For cutting tools, broaches, chasers, drills, hobs, reamers; high speed steel.

MV 350.
M-1290; 1.4-1.5 C, 0.20-0.35 Si, 0.30-0.50 Mn, 3.0-3.5 V, bal Fe.
Cold work tool steel.
W.-Nr 1.2838.

MVD.
M-1653; 0.45 C, Cr, Mo, V, bal Fe.
Steel for high temperature bolting 48 Cr Mo V 4.

M. V. "C" ALLOY.
M-83; 7.5-13 Si, bal Al.
Sand cast: 18,000-27,000 TS; 10,000-11,000 YS; 15-5 El; 50 Brin.
Chill cast: 22,000-30,000 TS; 10 El.
For airplane and automobile parts, marine parts, tubes; great resistance to atmospheric and sea water corrosion; modified Al-Si alloy.

M. W. METAL.
M-Eng.; 90-95 Mg, 3-7 Al, 2-5 Zn, 0.5 Mn.
36,000-39,000 TS; 33,000 TS; 16-13 El.
For light alloy parts; age-hardenable.

MX-2.
M-US; 0.39 C, 1.1 Cr, 1.0 Co, 1.0 Si, 0.7 Mn, 0.15 V, 0.25 Mo, bal Fe.
Modified AISI 4037 (4037 + Co).

MX8N.
M-1522; 38 Ag, Cu, Zn, Cd, Ni.
Melt range: 640-670°C.
Max stress; 45.7 kgf/mm^2; 30 El.
For silver brazing; WC tipped tools.

MX10.
M-1522; 40 Ag, Cu, Zn, Cd.
Melt range: 605-635°C.
Max stress: 44.9 kgf/mm^2; 15 El.
For silver brazing; good fluidity.

MX18N.
M-1522; 48 Ag, Cu, Zn, Cd, No.
Melt range; 640-660°C.
Max stress: 47.2 kgf/mm^2; 35 El.
For silver brazing; WC tipped tools.

MX20.
M-1522; 50 Ag, Cu, Zn, Cd.
Melt range: 620-640°C.
Max stress: 45.7 kgf/mm^2; 35 El.
For silver brazing; optimum fluidity.

MX20N.
M-1522; 50 Ag, Cu, Zn, Cd, Ni.
Melt range: 47.2 kgf/mm^2; 35 El.
For silver brazing; WC tipped tools.

MYA.
M-363; 0.48 C, 0.3 Mn, 1.45 Si, 1.45 Cr, 0.25 V, bal Fe.
Oil hardening tool steel, shock resisting type.

MY-A-CHROME.
M-363; 0.7 C, 1.0 Cr, bal Fe.
For tools, drills, punches; oil or water hardening.

MYSTIC METAL.
Eng.; 89 Pb, 11 Sn, 0.1 Bi.
For solder.

N

N 1 ETC see **EASTERN N 1 ETC.**

N-2 see **ALOYCO N-2.**

N-3 see **ALOYCO N-3.**

N-3 ALLOY.
M-U.S.; 0.4-0.7 C, 2.0 max Si, 1.5 max Mn, 20-26 Cr, bal Fe.
Cast: 60,000 TS; 40,000 YS; 0 EL; 200 Brin.
For furnace parts, grids, combustion chambers.
High heat and corrosion resistance.

N-3 METAL.
M-264; 32 Ni, 2.75 Pb, 9.5 Si, 0.75 Mn, bal Cu.
For hardware; free-cutting, corrosion resistant.

N10 (NEW BIDE).
M-1203; Sintered carbide tool material.
For heavy roughing cuts on cast iron, non-ferrous metals and non-metallics.

N 12.
M-1740; 0.10-0.14 C, 0.30-0.60 Si, 0.60-0.80 Mn, 3.4-3.6 Ni, 0.25 max Cr, 0.15 max Mo, 0.30 max Cu, bal Fe.
Low temperature, impact resistant.
B.S. 1504-503; ASTM A352-66 Gr. LC3.

N18K8M3T.
M-USSR; 0.05 C, 18.3 Ni, 8.5 Co, 3.5 Mo, 0.34 Ti, 0.14 Al, bal Fe.
Ht. Tr.; 300,000 TS; 250,000 YP.
For jet engine and spacecraft components, landing gears, missiles, crimping tools.
Maraging steel, tough, shock resistant.

N20 (NEW BIDE).
M-1203; Sintered carbide tool material.
For general purpose machining and milling of cast iron and non-ferrous metals.

N22 (NEW BIDE).
M-1203; Sintered carbide tool material.
For machining of high temperature alloys, non-ferrous and non-metallics.

N30 (NEW BIDE).
M-1203; Sintered carbide tool material.
For semi-finishing and finishing of cast iron non-ferrous and non-metallic materials, and high temperature alloys.

N35K6.
M-USSR; 0.01 C, 35 Ni, 5 Co, 2.3 Ti, 0.2 Cu, 0.4 Si, 0.6 Mn, bal Fe.
Hardened: 142,000 TS.
For electrical equipment, instruments, pumps.
Precipitation hardenable.
Low expansion alloy.

N40 (NEW BIDE).
M-1203; Sintered carbide tool material.
For high speed finishing and precision boring of cast iron, non-ferrous metals and non-metallics.

N50 (NEW BIDE).
M-1203; Sintered carbide tool material.
For heavy duty rough machining and milling of carbon and alloy steels.

N 52 (NEW BIDE).
M-1203; 72 WC, 11 Co, 8TiC, 10 TaC.
Sintered: 375,000 TrS, RA 91.0.
For tools.

N60 (NEW BIDE).
M-1203; Sintered carbide tool material.
For general purpose turning, planning and milling of carbon and alloy steel.

N70 (NEW BIDE).
M-1203; Sintered carbide tool material.
For light roughing and finishing cuts on carbon and alloy steel.

N72 (NEW BIDE).
M-1203; Sintered carbide tool material.
For moderate roughing and finishing of carbon and alloy steels.

N80 (NEW BIDE).
M-1203; Sintered carbide tool material.
For fast finishing and precision boring of carbon and alloy steels.

N80.
M-884; 0.12 C, 1.5 Mn, 0.2 Mo, bal Fe.
Tube: 100,000 min TS; 80,000 min YS.
For gas well piping and tubing.
Resists sulphide stress corrosion.

N93 (NEW MET).
M-1203; Sintered carbide tool material.
For normal to high velocity machining of carbon and alloy steel where extra toughness is required.

N95 (NEW MET).
M-1203; Sintered carbide tool material.
For normal to high velocity machining of carbon and alloy steel to close tolerance finishes.

N-153 ALLOY.
M-U.S.; 0.35-0.45 C, 1.6 Mn, 15-16 Ni, 15-16 Cr, 1.25-1.75 W, 4 Mo Co, 0.8-1.2 Cb, bal Fe.
Rolled: 176,000 TS; 146,000 YS; 15 El; 30 RA.
For jet engine parts; heat resistant.

N-155.
M-44; 10 Ni, 23 Cr, 20 Co, 2 W, 3 Mo, 1 Cb, bal Fe.
Wire, rod, ribbon for welding wire and fastener stock.

N-155 see also: **UDIMET N-155; ALLOY N-155; CARPENTER PYR N-155.**

N-155.
M-167, M-373, M-1491; 0.1-0.2 C, 18-22 Ni, 18-22 Cr, 2.5-3.5 Mo, 2.3 0.7-1.2 Cb, 0.1-0.2 N_2, 18-22 Co.
Solution treated: 30,000-50,000 TS.
For jet engine parts, gas turbine blades, ship propulsion blades; heat resistant, similar to "Multimet."

N-238.
M-Eng.; Ni, Cr.
For high temperature applications; heat and corrosion resistant.

NA-1.
M-189; 0.10-0.50 C, 28 Cr, 10 Ni, bal Fe.
For heat and corrosion resistant parts, furnace parts; heat and corrosion resistant to S atw.

NA-2.
M-189; 0.35-0.70 C, 15 Cr, 35 Ni, bal Fe.
Cast: 70,000 TS; 35,000 YS; 10 El; 10 RA; 175 Brin.
For furnace parts; retorts, heat treating boxes; Type HT; corrosion and resistant.

NA-2T.
M-189; 0.4-0.6 C, 36-39 Ni, 16-19 Cr, 1.5 Si, 1.2 Mn, bal Fe.
Cast: 68,000 TS; 40,000 YS; 10 El; 10 RA; 187 Brin.
For high temperature chains and conveyors, heat treating boxes; Type HU good thermal fatigue.

NA 3 ETC see **DBK NA-3 ETC.**

NA6LC ALLOY.
M-189; C, Ni, Cr, bal Fe.
For pickling hooks, hangers, frames; corrosion resistant.

NA-7.
M-189; 0.3-0.6 C, 24-30 Cr, 18-22 Ni, bal Fe.
Cast: 75,000 TS; 50,000 YS; 15 RA; 175 Brin.
For furnace parts; ACI-HK; nonmagnetic, high creep resistant to 2000°F.

NA-12.
M-189; 0.15-0.40 C, 11.5-14 Cr, 1 max Ni, bal Fe.
Annealed: 95,000 TS; 65,000 YS; 25 El; 190 Brin.
For glass mold dies, steam turbine parts, valves, pumps; Type CA40; corrosion resistant.

NA-18.
M-189; 17 Cr, 0.30 max C, bal Fe.
For corrosion resisting parts; corrosion resistant.

NA-19.
M-189; 26-19 Cr, 0.50-0.70 C, 3 Ni, bal Fe.
Cast: 65,000 TS; 50,000 YS; 2 El; 2 Ra; 175 Brin.
For stainless parts, heat resisting parts, Mg pots, pipe; corrosion resistant; heat resistant; ACI-CCHC.

NA-20.
M-189; 0.5 max C, 26-30 Cr, 4-7 Ni, 0.5 max Mo, bal Fe.
Cast: 85,000 TS; 48,000 YS; 16 El; 190 Brin.
For furnace blowers, copper melting equipment; Type HD; resists flue gases.

NA22H.
M-1222; 0.44 C, 46 Ni, 26.3 Cr, 5.3 W, 1.4 Mn, 1 Si, bal Fe.
At 70°F: 64,500 TS; 3.5 El; 2.7 Ra.
At 1800°F: 18,000 TS; 32.0 El; 48.0 RA.
For radiant tubes for furnaces, gas generator retorts; high heat resistance to 2200°F.

NA 26.
M-45.
Special alloy limited to chutes (or snouts) in continuous galvanizing lines.

NA-34.
M-189; 0.10 max C, 18-20 Cr, 8-10 Ni, 2-4 Mo, bal Fe.
Heat treated: 75,000 TS; 35,000 YS; 50 El; 50 RA; 200 Brin.
For castings, pipe, paper mill equipment; corrosion resistant.

NA-60.
M-189; 0.20 C, 10-14 Cr, 58-62 Ni, bal Fe.
Cast: 68,000 TS; 38,000 YS; 4 El; 179 Brin.
Aged: 75,000 TS; 38,000 YS; 4 El; 205 Brin.
For retorts, heat treating fixture, salt pots; Type HW; heat resistant.

NA-65.
M-189; 18-20 Cr, 65-70 Ni, 0.30 max C, bal Fe.
Cast: 75,000 TS; 45,000 YS; 15 El; 15 RA; 200 Brin.
For high temperature service castings, pipe, retorts; heat resistant to 2100°F; ACI-HX.

NA226.
M-137; 4-5 Cu, 0-0.2 Ti, bal Al.
32,000-42,000 TS; 14,000-28,000 YS; 12-5 El; 65-100 Brin.
For light alloy parts; hardenable.

NA350.
M-137; 9.5-10.5 Cu, 0.15 max Cu, 0.25 max Si, bal Al.
40,000 TS; 21,000 YS; 10 El; 75 Brin.
For light alloy parts; hardenable.

NACO 1S.
M-1237, M-1238; 99.5 min Al.
O-temper: 12,000 TS; 4500 YS; 37 El.
1/2 Hard: 16,000 TS; 14,000 YS; 10 El.
Hard: 20,500 TS; 19,500 YS; 9 El.
For light weight structure; corrosion resistant.

NACO 2S.
M-1237, M-1238; 1 Fe + Si max, 99.0 min Al.
O-temper: 13,500 TS; 5500 YS; 35 El; 23 Brin.
Hard temper: 24,000 TS; 23,000 YS; 5 El; 44 Brin.
For light weight structures, tanks, containers; corrosion resistant.

NACO 3S.
M-1237, M-1238; 1.2 Mn, bal Al.
O-temper: 17,000 TS; 7500 YS; 30 El; 28 Brin.
Hard temper: 29,000 TS; 27,000 YS; 4 El; 55 Brin.
For structures, tank cars, containers; resists reducing and oxidizing atmospheres.

NACO 50S.
M-1237, M-1238; 0.7 Mg, 0.4 Si, bal Al.
O-temper : 16,000 TS; 9000 YS; 26 El.
T-temper: 33,000 TS; 29,000 YS; 18 El.
For architectural shapes, trim, molding; heat treatable, corrosion resistant.

NACO 54S.
M-1237, M-1238; Al alloy.
For light weight parts.

NACO 57S.
M-1237, M-1238; 2.5 Mg, 0.3 Cr, bal Al.
O-temper: 29,000 TS; 14,000 YS; 25 El; 45 Brin.
Hard Temper: 43,000 TS; 38,000 YS; 7 El; 85 Brin.
For marine construction and hardware; resists sea water corrosion.

NACO 123.
M-1237, M-1238; 5 Si, 0.2 max Ti, bal Al.
Cast: 19,000-24,000 TS; 9000-19,000 YS; 10-6 El; 40-50 Brin.
For instrument housings, cases; corrosion resistant, good castability.

NACO 160.
M-1237, M-1238; 12 Si, bal Al.
Die cast: 37,000 TS; 18,000 YS; 2 El.
For instrument housings, cases; corrosion resistant; for thin wall sections.

NACO 162.
M-1237, M-1238; 1 Cu, 1 Mg, 12 Si, bal Al.
Cast: 36,000 TS; 28,000 YS; 0.5 El; 105 Brin.
For pistons: low coefficient of expansion.

NACO 320.
M-1237, M-1238; 4 Mg, 0.5 Si, 0.2 max Ti, bal Al.
Cast: 25,000 TS; 12,000 YS; 9 El; 50 Brin.
For marine parts, architectural trim; corrosion resistant.

NACO 350.
M-1237, M-1238; 10 Mg, bal Al.
Heat treated: 45,000 TS; 25,000 YS; 14 El; 75 Brin.
For gasoline fuel meters, aircraft and marine castings; heat treatable, corrosion resisting.

NACO STEEL.
M-89; 0.3-0.4 C, 1.5 Mn, bal Fe.
110,000 TS; 70,000 YS; 8 El; 15 RA; 250 Brin.
For car coupler knuckles, car wheels, anchor chains; water hardening.

NALOY.
M-864; 0.7 C, 18 W, 4 Cr, 1 V, bal Fe.
For broaches, form tools; high speed steel.

NAPAC.
M-836; 0.07 C, 0.32 Mn, 0.003 Cb, 0.07 Al, bal Fe.
Rolled: 52,000-60,000 TS; 38,000-45,000 YS; 35 El; 62 Rock B. (as ordered).
For structural parts such as automobile bumpers.

NAPAC-35.
M-836; 0.10 C, 0.35 Mn, Cb, bal Fe.
Aluminum-killed.
Hot rolled sheet: 35,000 psi min YS.
Weldable; for automotive equipment.

NAPAC-40.
M-836; 0.10 C, 0.55 Mn, Cb, bal Fe.
Aluminum killed.
Hot rolled sheet: 40,000 psi min YS.
Weldable; for automotive equipment.

NAPAC-45.
M-836; 0.10 C, 0.55 Mn, Cb, bal Fe.
Aluminum killed.
Hot rolled sheet: 45,000 psi min YS.
Weldable; for automotive equipment.

NAPAC-50.
M-836; 0.10 C, 0.55 Mn, Cb, bal Fe.
Aluminum killed.
Hot rolled sheet: 50,000 psi min YS.
Weldable; for automotive equipment.

NAPAC-F-45.
M-836; 0.12 C, 0.75 Mn, Cb, Zr, bal Fe.
Aluminum killed.
Hot rolled sheet: 55,000 min TS; 45,000 min YS.
Weldable; for automotive equipment.

NAPAC-F-50.
M-836; 0.12 C, 0.75 Mn, Cb, Zr, bal Fe.
Aluminum killed.
Hot rolled sheet: 60,000 min TS; 50,000 min YS.
Weldable; for automotive equipment.

NAPAC-F-55.
M-836; 0.12 C, 0.75 Mn, Cb, Zr, bal Fe.
Aluminum killed.
Hot rolled sheet: 65,000 min TS; 55,000 min YS.
Weldable; for automotive equipment.

NAPAC-F-60.
M-836; 0.12 C, 0.75 Mn, Cb, Zr, bal Fe.
Aluminum killed.
Hot rolled sheet: 70,000 min TS; 60,000 min YS.
Weldable; for automotive equipment.

NAPAC-S-45.
M-836; 0.15 C, 0.75 Mn, Cb, bal Fe.
Zirconium killed.
Hot rolled sheet: 60,000 min TS; 45,000 min YS.
Weldable; for automotive equipment.

NAPAC-S-50.
M-836; 0.15 C, 0.75 Mn, Cb, bal Fe.
Aluminum killed.
Hot rolled sheet: 65,000 min TS; 50,000 min YS.
Weldable; for automotive equipment.

NAPRALOY.
M-952; 4.5 C, 24 Cr, bal Fe.
Cast: 600 Brin.
For welding rod, hard facing electrodes; wear resistant.

NARCOLOY.
M-1256; 7 Al, Sn, Co, bal Cu.
Cold forging aluminum bronze rod.

NARITE.
M-1256; 14.0 Al, 1.0 Ni, 4.5 Fe, bal Cu.
Casting for deep draw dies; superior in compression.

NARITE HT.
M-1256; 11.25 Al, 5.0 Ni, 5.0 Fe, 1.0 Mn, bal Cu.
Heat treatable to 400 Brin.
For non-magnetic, non-sparking tools.

NARLOY.
M-U.S.; 3 Ag, bal Cu.
Brazing alloy.

NARRMAC II.
M-1256; 9.5 Al, bal Cu.
35 Tons/sq.in. UTS; 25 El.
Wrought aluminum bronze.
DTD 160; BS 2032 CA 103.

NARRMAC III.
M-1256; 9.75 Al, 2.5 Fe, bal Cu.
32 Tons/sq.in. UTS; 20 El.
Wrought aluminum bronze.
DTD 197.

NARRMAC V.
M-1256; 9.5 Al, 4.5 Ni, 4.0 Fe, 0.5 Mn, bal Cu.
45 Tons/sq.in. UTS; 15 El.
Wrought aluminum bronze.
DTD 197; BS 2033 CA 104.

NARRMAC HNA.
M-1256; 10.5 Al, 4.8 Ni, 4.0 Fe, 0.5 Mn, bal Cu.
50 Tons/sq.in. UTS; 12 El.
Wrought aluminum bronze.

NARTRODE E.
M-1256; 9.25 Al, 4.5 Ni, 3.25 Fe, bal Cu.
Weld metal: 40 Tons/sq.in. UTS; 15 El.
Welding wire for automatic processes.

NARTRODE S.
M-1256; 9.5 Al, 1.0 Fe, bal Cu.
Weld metal: 38 Tons/sq.in. UTS; 25 El.
Welding wire for automatic processes.

NARVE.
M-111; 0.65 C, 1 Si, 0.5 Mn, 1.1 Cr, 0.6 Mo, bal Fe.
For dies, rams, shafts, tools; cold work steel.

NASA CO-W-RE.
M-1491; 0.40 C, 3.0 Co, 25.0 W, 1.0 Ti, 1.0 Zr, 2.0 Re, bal Co.
Cast alloy for high temperature space applications.

NASA-TRW VI A see **TRW VI A.**

NASCO.
M-207; 0.5 C, 0.8 Mn, 1 Cr, 0.4 Mo, bal Fe.
Hardened: 263,000-274,000 TS; 209,000-220,000 YS; 550-600 Brin.
For tools, chisels, punches, wrenches; non-tempering, shock resistant, tough.

NASCOLOY O.
M-207; 0.5 C, 0.8 Mn, 1.8 Si, bal Fe.
For pneumatic tools, chisels, rivet sets; wear resistant, tough, oil hardening, requires tempering.

NASCOLOY W.
M-207; 0.4 C, 0.7 Mn, 0.2 Si, 0.8 Cr, 0.3 Mo, 0.3 Cu, bal Fe.
For pneumatic tools, chisels, rivet sets; wear resistant, tough, water hardening, no tempering.

NATALLOY NO. 2.
M-1446; 0.45 C, 0.80 Mn, 0.95 Cr, 0.06 Mo, 0.15 V, bal Fe.
Heat treated: 108,000 TS; 88,000 YS; 16 El; 34 RA; 225-300 Brin.
For plastic and die casting dies; oil hardened, AISI 6145.

NATIONAL.
M-432; C, 18 W, Cr, V, bal Fe.
For high speed steel cutters and tools; high speed steel.

NATIONALLOY 1 see AISI 303.

NATIONALLOY 2.
M-483; inactive; for reference only.

NATIONALLOY 3 see AISI 410.

NATIONALLOY 7.
M-483; 0.30-0.40 C, 0.60-0.90 Mn, 0.15 max P, 0.015 max S, 0.40 max Si, 2.25-3.0 Ni, 0.80-1.20 Cr, 0.30-0.50 Mo, 0.20 max V, bal Fe.
Medium hardening alloy steel.
Modified AISI 4335.

NATIONALLOY 14.
M-483; 0.28-0.38 C, 0.40-0.70 Mn, 0.015 max P, 0.015 max S, 0.40 max Mn, 3.0-4.0 Ni, 0.8-1.20 Cr, 0.40-0.80 Mo, 0.05-0.20 V, bal Fe.
Deep hardening low alloy, low carbon steel.

NATIONALLOY 22.
M-483; 0.28-0.36 C; 0.60-0.90 Mn, 0.015 max P, 0.015 max S, 0.40 max Si, 1.50-2.0 Ni, 0.80-1.20 Cr, 0.20-0.30 Mo, add 0.05 V, bal Fe.
Modified AISI 4330 steel.

NATIONAL 217.
M-844; C, alloy, bal Fe.
For hard facing electrodes; wear resistant.

NATIONAL 459.
M-844; C, alloy, bal Fe.
For hard facing electrodes; wear resistant.

NATIONAL BEARING SILVER BABBITT NO. 397.
M-539; 1.5-2.6 Ag, 2.5-4.0 Sn, 9-11 Sb, bal Pb.
Cast: 9500-10,000 TS; 5 El; 15.3 Brin.
For bearings; C.S. 8000; Babbitt.

NATIONAL ECONOMY.
M-844; C, 10 alloy, bal Fe.
For hard facing electrodes; wear resistant.

NATIONAL GRAPHITIC STEEL.
M-89; 0.3 C, 0.5 graphite, bal Fe.
Annealed: 75,000 TS; 200 Brin.
For steel castings; abrasion resistant.

NATIONAL HTM.
M-89; 3 C, Mn, Si, bal Fe.
Cast: 70,000-110,000 TS; 48,000-85,000 YS; 12-7 El; 163-302 Brin.
For gears, shafts, housings, staybolts; pearlitic malleable iron.

NATIONAL NO. 1.
M-844; 0.06 C, bal Fe.
For welding electrodes for low carbon steel; copper coated.

NATIONAL NO. 6.
M-844; 0.1 C, alloy, bal Fe.
For welding electrodes; for high pressure piping.

NATIONAL TOOLFACE.
M-844; C, alloy, bal Fe.
For hard facing electrodes; wear resistant.

NATIONAL TUBE DIAMOND METAL.
M-844; WC.
For hard facing electrodes; wear resistant.

NAUBUC.
M-Eng.; 58 Cu, 16.25 Zn, 25 Ni, 0.75 Fe.
For knives; corrosion resistant.

NAUTAL.
M-Hungary; 4.5 Mn, 0.4 Mg, bal Al.
For light alloy parts; corrosion resistant.

NAVAL.
61 Cu, 1.0-1.5 Sn, bal Zn.
Drawn: 52,000 TS; 30,000 YS; 35 El; 100 Brin.
For extrusions; corrosion resistant.

NAVAL ALUMINUM.
M-U.S.; 1.5 Cu, 0.9 Mn, 0.4 Ni, Fe + Si, bal Al.
For instruments and fittings; resists corrosion.

NAVAL ALUMINUM BRONZE.
87-85 Cu, 7-9 Al, 2.5-4.5 Fe.
For marine parts, propellers, pumps; tough.

NAVAL BRASS 2.
M-141; 60 Cu, 0.75 Sn, bal Zn.
Hard: 80,000 TS; 52,000 YS; 15 El.
Soft: 55,000 TS; 25,000 YS; 50 El.
For hardware, marine parts, valve parts; resists sea water corrosion.

NAVAL BRASS 28.
M-141; 60 Cu, 0.6 Pb, 0.75 Sn, bal Zn.
Hard: 75,000 TS; 53,000 YS; 20 El.
Soft: 57,000 TS; 25,000 YS; 47 El.
For hardware; strong.

NAVAL BRASS, 64%.
M-33; 64 Cu, 0.75 Sn, bal Zn.

NAVAL BRASS-462.
M-8; 60 Cu, 39.25 Zn, 0.75 Sn.
Soft: 56,000 TS; 22,000 YS.
For condenser plates, marine hardware; corrosion resistant.

NAVAL BRASS-464.
M-8; 60 Cu, 39.25 Zn, 0.75 Sn.
Hard: 63,000 TS; 35,000 YS; 30 El; B 65 Rock.
Soft; 56,000 TS; 25,000 YS; 40 El; B 50 Rock.
For nuts, bolts, fasteners, rivets, valve stems, pump shafts, marine hardware.
Corrosion resistant to sea water.

NAVAL BRASS E-24.
M-1191; 58.5-60.0 Cu, 0.5-1.0 Sn, bal Zn.
Forged: 60,000-65,000 TS; 25,000-30,000 YS; 45-3 El; 83-94 Brin.
For hardware, bolts, machinery parts; corrosion resistant.

NAVAL BRASS, HIGH LEADED.
M-33; 60 Cu, 37.5 Zn, 1.8 Pb, 0.7 Sn.
Ann: 57,000 TS; 25,000 YS; 40 El.
Drawn 20%: 75,000 TS; 53,000 YS; 15 El.
Marine hardware, valve stems, screw machine products.

NAVAL BRASS NO. 63.
M-1191; 57.5-60.5 Cu, 1.25-1.75 Sn, 0.25-0.75 Pb, bal Zn.
Forged: 60,000-65,000 TS; 26,000-30,000 YS; 25-20 El; 95-110 Brin.
For hardware, bolts, machinery parts, fasteners; free-cutting, leaded bronze.

NAVAL BRASS, UNINHIBITED.
M-33; 60 Cu, 0.75 Sn, bal Zn.
Hot rolled: 55,000 TS; 25,000 YS; 50 El; 30 RA.
Cold rolled: 70,000 TS; 58,000 YS; 17 El; 40 RA.
For condenser tubes, bolts, spindles; resists sea water corrosion.

NAVAL BRONZE.
M-8; 88 Cu, 8 Sn, bal Zn.
For steam and structural parts, expansion joints, gears, valves; tough bronze.

NAVAL BRONZE NO. 4.
44 Pb, 36 Sn, 16 Sb, 4 Cu.
For bearings; anti-friction.

NAVAL GUN METAL "G".
M-U.S.; 88 Cu, 10 Sn, 2 Zn.
For gears, pistons, bearings, bushings; high strength.

NAVALIUM.
M-Eng.; 96.6 Al, 0.1 Cu, 0.7 Sn, 0.6 Fe, 2.3 Mn, 0.24 Si.
For light alloy parts.

NAVAL JOURNAL BEARING, SPEC. "H".
M-U.S.; 83 Cu, 14 Sn, 3.5 Pb.
50,000 TS; 9.5 El.
Bearings; heavy duty.

NAVAL JOURNAL BEARING, SPEC "HX".
M-U.S.; 83 Cu, 14 Sn, 3.5 Pb.
For bearings; heavy duty.

NAVAL NO. 6.
M-122; 60 Cu, 0.75 Sn, bal Zn.
Soft: 58,000 TS; 28,000 YS; 42 El.
1/2H-temper: 65,000 TS; 38,000 YS; 37 El.
H-temper: 75,000 TS; 52,000 YS; 25 El.
For marine hardware, bolts, rivets, propeller shafts; corrosion resistant. CA 464.

NAVAL NO. 95.
M-122; 63 Cu, 0.7 Sn, bal Zn.
For marine hardware, bolts, propeller shafts; corrosion resistant.

NAVALOY.
M-Eng.; Sn, Sb, bal Pb.
For bearings; Babbitt metal.

NAVAL PHOSPHOR BRONZE (P-C) CAST.
M-U.S.; 88 Cu, 8 Sn, 4.0 Zn, 0.5 Pb.
57,000-35,000 TS; 31,000 YS; 18 El.
For bearings, gears, marine parts; resists sea water corrosion; U.S.N.-46 B5f.

NAVAL PHOSPHOR BRONZE (P-R) ROLLED.
M-U.S.; 94 Cu, 3.5 Sn, 0.5 P.
80,000 TS; 60,000 YS; 12 El.
For pump parts, valve stems, bolts; resists sea water corrosion; U.S.N.-46 B14d.

NAVAL VALVE BRONZE.
M-U.S.; 88 Cu, 6.5 Sn, 4 Zn, 1.5 Pb.
58,000-32,000 TS; 29,000 YS; 9-17 El; 80 Brin.
For valve stems, valve seats, valve bodies; "Composition Mn"; U.S.N.-46 B8d.

NAVAN.
M-1405; C, bal Fe.
For gears, shafts, wear resistant parts; case hardened; B.S.I. 5005/101.

NAVIBRONZE.
M-1687; 8 Sn, 0.25 max P, bal Cu.
Bronze, wrought.

NAVY.
M-1260; 74 Sn, 9 Sb, 4 Cu, 13 Pb.
Cast: 14,200 TS; 28 Brin.
For steam turbine bearings; M.P. 355-665°F; wear resistant.

NAVY ALUMINUM ALLOY.
M-U.S.; 1.5 Cu, 0.9 Mn, 0.4 Ni, 0.4 Fe, 0.3 Si, bal Al.
For instruments, fittings, light alloy parts; non-hardenable.

NAVY ANTIFRICTION METAL GRADE 1.
M-561; 90-93 Sn, 3.5-5 Sb, 0.5 Pb, 3.5-5 Cu.
10,770 TS; 18-30 Brin.
For antifriction metal, bearings; Babbitt metal, aircraft engine bearings Babbitt.

NAVY ANTIFRICTION METAL GRADE 2.
M-561; 87.5-89.5 Sn, 7-8 Sb, 0.35 Pb, 3.5-4.5 Cu, 0.10 As.
For antifriction metal, genuine Babbitt, automotive engine bearings; for moderately severe service.

NAVY ANTIFRICTION METAL GRADE 3.
M-561; 83-85 Sn, 7.5-8.5 Sb, 0.35 Pb, 7.5-8.5 Cu, 0.10 As.
For hard Babbitt, bearings; for moderately heavy pressures.

NAVY ANTIFRICTION METAL GRADE 4.
M-561; 80.5-82.5 Sn, 12-14 Sb, 0.25 Pb, 5-6 Cu, 0.10 As.
For hard bearings; for heavy pressure and high speeds.

NAVY ANTIFRICTION METAL GRADE 5.
M-561; 61-63 Sn, 9.5-10.5 Sb, 24-26 Pb, 2.5-3.5 Cu, 0.15 As.
For electric motor bearings; for low pressures and high speeds.

NAVY ANTIFRICTION METAL GRADE 6.
M-561; 4.5-5.5 Sn, 14-16 Sb, 79-81 Pb, 0.50 Cu, 0.20 As.
For cheap Babbitt bearings; for light service.

NAVY ANTIFRICTION METAL GRADE 7.
M-561; 9-10 Sn, 14-16 Sb, 74-76 Pb, 0.50 Cu, 0.20 As.
For cheap Babbitt bearing; for light service.

NAVY BEARING.
M-U.S.; 91-80 Sn, 4.5-15 Sb, 3.7-5.0 Cu.
For bearings, bushings; anti-friction.

NAVY BEARING, HARD.
M-U.S.; 80 Sn, 15 Sb, 5 Cu.
For bearings; anti-friction.

NAVY COMPOSITION "W".
M-U.S; 89 Sn, 7.3 Sb, 3.7 Cu.
For bearings; anti-friction.

NAVY GEAR BRONZE.
M-U.S.; 84-86 Cu, 13-15 Sn, 1.5 Zn, 0.5 P.
30,000 TS; 1-8 El.
For gears and worm wheels; tough.

NAVY "N" ALLOY.
M-U.S.; 3 Mn, 6 Cu, bal Al.
20,000 TS; 8 El.
For general castings; corrosion resisting.

NAVY NO. 4 ALLOY.
M-U.S.; 96 Al, 4 Cu.
15,000 TS; 5 El.
For light alloy parts, boxes, covers, face plates.

NAVY TOMBASIL.
M-346; 5 Si, 6 Zn, bal Cu.
Cast: 60,000 TS; 35,000 YS; 15 El; 110 Brin.
Corrosion resistant; for valve stems.

N-A-X.
M-1071; 0.12 C, 0.80 Mn, 0.70 Si, 0.55 Cr, 0.08 Zr, bal Fe.
HSLA hot rolled strip or sheet.
70,000 psi min TS; 50,000 psi min YS; 22 min El.
For automotive and structural applications.
SAE 950 A.

N-A-X 80.
M-836; 0.12 C, 1.0 Mn, 0.80 Cu, Cb, bal Fe, Zirconium treated.
Sheet: 90,000 psi min TS; 80,000 psi min YS.
For automotive, truck, agricultural and railroad applications.
ASTM A-715; SAE J410c, Grade 980 XK.

NAX 9112.
M-836; 0.10-0.15 C, 1.10 max Mn, 0.50-0.90 Si, 0.50-0.80 Cr, 0.05-0.15 Zr, bal Fe.
Water or oil hardenable; for structural parts.

NAX 9115.
M-836; 0.13-0.18 C, 1.10 max Mn, 0.50-0.90 Si, 0.50-0.80 Cr, 0.05-0.15 Zr, bal Fe.
Water or oil hardenable; for structural parts.

NAX 9120.
M-836; 0.18-0.23 C, 1.10 max Mn, 0.50-0.90 Si, 0.50-0.80 Cr, 0.05-0.15 Zr, bal Fe.
Water or oil hardenable; for structural parts.

NAX 9120 MOD.
M-836; 0.15-0.20 C, 1.10 max Mn, 0.50-0.90 Si, 0.50-0.80 Cr, 0.05-0.15 Zr, bal Fe.
Water or oil hardenable; for structural parts.

NAX 9130.
M-836; 0.28-0.33 C, 1.10 max Mn, 0.50-0.90 Si, 0.50-0.80 Cr, 0.05-0.15 Zr, bal Fe.
Water or oil hardenable; for structural parts.

NAX FINE GRAIN.
M-836; 0.22 max C, 1.10 Mn, 0.40-0.90 Si, 0.03-0.15 Zr, bal Fe.
Rolled (min): 70,000 TS; 50,000 YS; 22 El; 140-160 Brin.
For structural parts; weldable.

NAX HIGH TENSILE.
M-836; 0.18 max C, 0.50-0.90 Mn, 0.60-0.90 Si, 0.40-0.80 Cr, 0.03-0.12 Zr, bal Fe.
Rolled (min): 70,000 TS; 50,000 YS; 22 El; 140-160 Brin.
For structural parts; improved corrosion resistance over most carbon steels; weldable.

N-A-XTRA 80.
M-836; 0.21 max C, 0.60-1.10 Mn, 0.40-0.90 Si, 0.50-0.80 Cr, 0.30 max Mo, 0.05-0.15 Zr, 0.0025 max B, bal Fe.
Hardened, min: 95,000 TS; 80,000 YS; 20 El.
For structural parts.

N-A-XTRA 90.
M-836; 0.21 max C, 0.60-1.10 Mn, 0.40-0.90 Si, 0.50-0.80 Cr, 0.30 max Mo, 0.05-0.15 Zr, 0.0025 max B, bal Fe.
Hardened, min: 105,000 TS; 90,000 YS; 18 El.
For structural parts.

N-A-XTRA 100.
M-836; 0.21 max C, 0.60-1.10 Mn, 0.40-0.90 Si, 0.50-0.80 Cr, 0.30 max Mo, 0.05-0.15 Zr, 0.0025 max B, bal Fe.
Hardened, min: 115,000 TS; 100,000 YS; 18 El.
For structural parts.

N-A-XTRA 110.
M-836; 0.21 max C, 0.60-1.10 Mn, 0.40-0.90 Si, 0.50-0.80 Cr, 0.30 max Mo, 0.05-0.15 Zr, 0.0025 max B, bal Fe.
Hardened, min: 125,000 TS; 110,000 YS; 18 El.
For structural parts.

NAX X9115.
M-836; 0.13-0.18 C, 1.10 max Mn, 0.50-0.90 Si, 0.50-0.80 Cr, 0.10-0.20 Mo, 0.05-0.15 Zr, bal Fe.
Water or oil hardenable; for structural parts.

NAX X9120.
M-836; 0.18-0.23 C, 1.10 max Mn, 0.50-0.90 Si, 0.50-0.80 Cr, 0.10-0.20 Mo, 0.05-0.15 Zr, bal Fe.
Water or oil hardenable; for structural parts.

NAX X9120 MOD.
M-836; 0.15-0.20 C, 1.10 max Mn, 0.50-0.90 Si, 0.50-0.80 Cr, 0.10-0.20 Mo, 0.05-0.15 Zr, bal Fe.
Water or oil hardenable; for structural parts.

NAX X9130.
M-836; 0.28-0.33 C, 1.10 max Mn, 0.50-0.90 Si, 0.50-0.80 Cr, 0.10-0.20 Mo, 0.50-0.15 Zr, bal Fe.
Water or oil hardenable; for structural parts.

NB.
M-1577; 0.04 C, 65.32 Ni, 28.20 Mo, 5.18 Fe.
For corrosion and heat resistant parts; chemical plant equipment, pump and valve parts.
Corrosion and heat resistant.

NBD CUPRO NICKEL.
M-539; 65-67 Cu, 30-31.5 Ni, 0.6-0.8 Fe, 0.7-1.0 Cb, 0.25-0.50 Si, 1.0-1.25 Mn, 0.01 max Pb.
60,000 psi TS; 32,000 psi YS; 20 El.

NBD NO. 1 STANDARD BABBITT.
M-539; Sn, Sb, bal Pb.
For bearings; Babbitt metal.

NBD NO. 2.
M-539; Sn, Pb, Ni, bal Cu.
Cast: 33,000 TS; 21,000 YS; 7 El; 7 RA; 75 Brin.
For bearings, engine and machinery castings; C.Y.P. 14000.

NBD NO. 2 STANDARD BABBITT.
M-539; Pb, Sn, Sb.
For large, slow running bearings; Babbitt metal.

NBD NO. 3.
M-539; Sn, Pb, bal Cu.
Cast: 32,000 TS; 21,000 YS; 10 El; 10 RA; 74 Brin.
For bushings, gearings; C.Y.P. 14000.

NBD NO. 3 A.
M-539; 80 Cu, 10 Sn, 10 Pb.
Cast: 25,000 TS; 12,000 YS; 8 El; 60 Brin.
For bearings; SAE 64.

NBD NO. 3 B.
M-539; 83 Cu, 7 Sn, 7 Pb, 3 Zn.
Cast: 30,000 TS; 14,000 YS; 12 El; 60 Brin.
For bearings; SAE 660.

NBD NO. 3 D.
M-539; 79.75 Cu, 9 Sn, 10 Pb, 0.75 P, 0.25 Ni.
Cast: 25,000 TS; 12,000 YS; 8 El; 60 Brin.
For bearings; heavy duty.

NBD NO. 3 STANDARD BABBITT.
M-539; Pb, Sn, Sb.
For bearings; Babbitt metal.

NBD NO. 4.
M-539; Sn, Zn, bal Cu.
Cast: 42,000 TS; 23,000 YS; 21 El; 22 RA; 88 Brin.
For gears, pinions, worm wheels, nuts; C.Y.P. 16000.

NBD NO. 4 F.
M-539; 83 Cu, 11 Sn, 3 Pb, 3 Zn.
Cast: 35,000 TS; 16,000 YS; 16 El; 60 Brin.
For bearings; heavy duty.

NBD NO. 4 H.
M-539; 87 Cu, 7 Sn, 3 Pb, 3 Zn.
Cast: 30,000 TS; 15,000 YS; 15 El; 60 Brin.
For bearings; heavy duty.

NBD NO. 4 I.
M-539; 85 Cu, 5 Sn, 5 Pb.
Cast: 30,000 TS; 14,000 YS; 20 El; 60 Brin.
For bearings; SAE 40.

NBD NO. 4 J.
M-539; 87 Cu, 9 Sn, 2 Pb, 2 Zn.
Cast: 30,000 TS; 15,000 YS; 15 El; 60 Brin.
For bearings; modified Gun Bronze.

NBD NO. 4 K.
M-539; 88 Cu, 10 Sn, 2 Zn.
Cast: 40,000 TS; 18,000 YS; 20 El; 70 Brin.
For bearings; SAE 62; Gun Bronze.

NBD NO. 4 L.
M-539; 88 Cu, 8 Sn, 4 Zn.
Cast: 40,000 TS; 18,000 YS; 20 El; 70 Brin.
For bearings; SAE 620; Navy "G".

NBD NO. 4 STANDARD BABBITT.
M-539; Pb, Sn, Sb.
For severe service bearings; Babbitt metal.

NBD NO. 5.
M-539; Sn, bal Cu.
Cast: 37,000 TS; 21,000 YS; 10 El.
For gears; acid resistant.

NBD NO. 6.
M-539; Sn, Pb, bal Cu.
Cast: 27,500 TS; 17,000 YS; 16 El; 15 RA; 47 Brin.
For gearings with poor lubrication; C.Y.P. 9700.

NBD NO. 6 A.
M-539; 77 Cu, 7 Sn, 15 Pb.
Cast: 25,000 TS; 14,000 YS; 10 El; 50 Brin.
For bearings; hard bronze.

NBD NO. 6 G.
M-539; 78 Cu, 6 Sn, 16 Pb.
Cast: 30,000 TS; 16,000 YS; 12 El; 55 Brin.
For bearings; heavy duty.

NBD NO. 6-H.
M-539; 74 Cu, 6 Sn, 20 Pb.
Cast: 28,000 TS; 14,000 YS; 10 El; 50 Brin.
For bearings, bushings; leaded bronze.

NBD NO. 6 I.
M-539; 78 Cu, 6 Sn, 16 Pb.
Cast: 25,000 TS; 14,000 YS; 10 El; 50 Brin.
For bearings; heavy duty.

NBD NO. 6M.
M-539; 75 Cu, 7 Sn, 18 Pb.
For bearings, bushings, sleeves; heavy duty.

NBD NO. 6S.
M-539; 5 Sn, 24 Pb, bal Cu.
Cast: 21,000 TS; 7 El.
For bearings, bushings, liners; heavy duty.

NBD NO. 7.
M-539; Sn, Pb, Ni, bal Cu.
Cast: 40,000 TS; 27,000 YS; 12 El; 15 RA; 90 Brin.
For bearings; for heavy loads.

NBD NO. 7A.
M-539; 10 Sn, 2 Pb, 3 Ni, bal Cu.
Cast: 42,000 TS; 18,000 YS; 15 El; 80 Brin.
For railway bearings; leaded nickel bronze.

NBD NO. 9A.
M-539; 10 Al, 1 Fe, bal Cu.
Cast heat treated: 65,000 TS; 25,000 YS; 20 El; 11 RA; 120 Brin.
For bearings, wear plates; C.Y.P. 15000; SAE-68B.

NBD NO. 9-AF.
M-539; 86 Cu, 10.5 Al, 3.5 Fe.
Cast: 75,000 TS; 30,000 YS; 12 El; 150 Brin.
Heat treated: 90,000 TS; 45,000 YS; 6 El; 190 Brin.
For bearing segments, wearing plates, screwdown nuts; Al-bronze, corrosion and wear resistant.

NBD NO. 9B.
M-539; Al, Fe, bal Cu.
Cast: 64,000 TS; 28,000 YS; 11 El; 12 RA; 144 Brin.
For acid resisting parts, crates, pickling racks; C.Y.P. 24000.

NBD NO. 9C.
M-539; Al, Fe, bal Cu.
Cast: 84,000 TS; 33,000 YS; 25 El; 24 RA; 137 Brin.
For bushings, bearings; C.Y.P. 24000.

NBD NO. 9D.
M-539; Al, Fe, bal Cu.
Cast: 88,000 TS; 43,000 YS; 13 El; 13 RA; 149 Brin.
For bushings, bearings; C.Y.P. 30000.

NBD NO. 9E.
M-539; Al, Fe, bal Cu.
Cast: 90,000 TS; 43,000 YS; 12 El; 12 RA; 170 Brin.
For gears, pinions, worms, worm wheels; C.Y.P. 35000.

NBD NO. 9F.
M-539; 9 Al, 3 Fe, bal Cu.
Cast heat treated: 65,000 TS: 25,000 YS; 20 El; 130 Brin.
For slides, gibs, gears; C.Y.P. 40000; SAE-68A.

NBD NO. 9G.
M-539; Al, Fe, bal Cu.
Cast: 55,500 TS; 55,500 YS; 0 El; 0 RA; 300 Brin.
For forming or drawing dies; C.Y.P. 61000.

NBD NO. 9H.
M-539; Al, Fe, bal Cu.
Cast: 120,000 TS; 101,000 YS; 3 El; 5.5 RA; 235 Brin.
For propellers, high strength castings; heat treatable.

NBD NO. 9K.
M-539; 10 Al, 4 Fe, 2 Ni, bal Cu.
Heat treated: 75,000-90,000 TS; 30,000-45,000 YS; 12-6 El; 150-190 Brin.
For propellers, bearings, gears; Al-bronze, heat treatable.

NBD NO. 9L.
M-539; 10 Al, 5 Fe, 5 Ni, bal Cu.
Heat treated: 90,000-110,000 TS; 40,000-60,000 YS; 6-5 El; 109-200 Brin.
For bearings, propellers, gears; Al-bronze, heat treatable.

NBD NO. 10A.
M-539; 60 Cu, 1 Al, 3 Mn, 1 Fe, bal Zn.
Cast: 60,000 TS; 20,000 YS; 15 El; 110 Brin.
For high strength castings.

NBD NO. 10B.
M-539; 64 Cu, 6 Al, 4 Mn, 3 Fe, bal Zn.
Cast: 90,000 TS; 45,000 YS; 20 El; 180 Brin.
For high strength castings; C.Y.P. 64000.

NBD NO. 10C.
M-539; 1.2 Al, 1 Mn, 1.2 Fe, 58 Cu, bal Zn.
Cast: 65,000 TS; 25,000 YS; 20 El; 100 Brin.
For bearings, bushings, sleeves; SAE 43; manganese bronze.

NBD NO. 10D.
M-539; 64 Cu, 6 Al, 3 Mn, 3 Fe, bal Zn.
Cast: 110,000 TS; 60,000 YS; 12 El; 200 Brin.
For bearings, worm wheels, gears, valves; SAE 430B; manganese bronze.

NBD NO. 11.
M-539; Sn, Pb, bal Cu.
Cast: 33,000 TS; 16,000 YS; 12 El; 58 Brin.
For bearings, bushings.

NBD NO. 11A.
M-539; 88 Cu, 8 Sn, 4 Pb.
Cast: 28,000 TS; 16,000 YS; 12 El; 60 Brin.
For bearings; heavy duty.

NBD NO. 11 C.
M-539; 86 Cu, 8 Sn, 2 Pb, 4 Zn.
Cast: 34,000 TS; 16,000 YS; 20 El; 65 Brin.
For bearings; heavy duty.

NBD NO. 11D.
M-539; 5 Sn, 4 Zn, 3 Ni, bal Cu.
Cast: 40,000 TS; 17,000 YS; 25 El; 65 Brin.
For bearings, bushings, sleeves; heavy duty nickel bronze.

NBD NO. 12.
M-539; Sn, Pb, bal Cu.
Cast: 40,000 TS; 21,000 YS; 19 El; 18 RA; 81 Brin.
For bearings, bushings; C.Y.P. 16600.

NBD NO. 13.
M-539; Sn, Pb, Zn, bal Cu.
Cast: 30,000 TS; 19,000 YS; 13 El; 18 RA; 69 Brin.
For bearings, bushings; for steam and water pressure.

NBD NO. 13 A.
M-539; 80 Cu, 8 Sn, 12 Pb.
Cast: 25,000 TS; 12,000 YS; 8 El; 60 Brin.
For bearings; heavy duty.

NBD NO. 15A.
M-539; 5 Sn, 20 Pb, 3 Zn, bal Cu.
Cast: 21,000 TS; 10,000 YS; 7 El; 45 Brin.
For bearings, bushings; sleeves; heavy duty leaded bronze.

NBD NO. 20 A.
M-539; 97 Cu, 2 Sn, 1 Zn.
For bearings, copper, electrodes.

NBD NO. 20B.
M-539; 99.7 Cu.
For bearings; blast furnace copper.

NBD NO. 20-C.
M-539; 99.9 Cu.
Cast: 25,000 TS; 7000 YS; 45 El; 35 Brin.
For electrode holders, water cooled linings; 80% electrical conductivity.

NBD NO. 20-H.
M-539; 99.3 Cu, 0.7 Cr.
Heat treated: 55,000 TS; 35,000 YS; 25 El; 100 Brin.
For castings, electrode holders; heat treatable, corrosion resistant.

NBD NO. 20-K.
M-539; 98.2 Cu, 1.5 Ni, 0.3 Be.
Heat treated: 75,000 TS; 50,000 YS; 3 El; 200 Brin.
For castings; age-hardenable, corrosion resistant.

NBD NO. 20L.
M-539; 99.9 Cu.
Cast: 25,000 TS; 8500 YS; 45 El; 35 Brin.
For conductors; 90% electrical conductivity.

NBD. NO. 21 A.
M-539; 68 Cu, 2 Pb, 30 Zn.
Cast: 30,000 TS; 11,000 YS; 20 El; 40 Brin.
For bearings; heavy duty.

NBD NO. 21 B.
M-539; 63 Cu, 1 Sn, 2 Pb, 34 Zn.
Cast: 40,000 TS; 14,000 YS; 15 El; 40 Brin.
For bearings; heavy duty.

NBD NO. 21 C.
M-539; 72 Cu, 1 Sn, 2 Pb, 25 Zn.
Cast: 35,000 TS; 12,000 YS; 25 El; 40 Brin.
For bearings; heavy duty.

NBD NO. 22.
M-539; 88 Cu, 2 Sn.
Cast: 25,00 TS; 7000 YS; 40 El; 45 Brin.
For bells; welding bell copper.

NBD NO. 22A.
M-539; 90 Cu, 10 Sn.
Cast: 35,000 TS; 21,000 YS; 10 El; 90 Brin.
For bearings, gears; SAE 65.

NBD NO. 22 B.
M-539; 93 Cu, 7 Sn.
Cast: 35,000 TS; 20,000 YS; 10 El; 85 Brin.
For bearings; heavy duty.

NBD NO. 22-C.
M-539; 91 Cu, 9 Sn.
Cast.
For plumbing, hardware; acid bronze.

NBD NO. 22 D.
M-539; 84 Cu, 16 Sn.
Cast: 18,000 TS.
For bearings; bridge bronze.

NBD NO. 22 E.
M-539; 80 Cu, 20 Sn.
Cast: 24,000 TS.
For bearings; bridge bronze.

NBD NO. 22 F.
M-539; 91 Cu, 5.5 Sn, 3.5 Zn.
Cast: 37,000 TS; 14,000 YS; 30 El; 40 Brin.
For bearings, trolley wheels; heavy duty.

NBD NO. 46.
M-539; 84 Cu, 8 Sn, 8 Pb.
Cast: 25,000 TS; 12,000 YS; 8 El.
For bearings.

NBD NO. 63.
M-539; 88 Cu, 10 Sn, 2 Pb.
Cast: 35,000 TS; 10 El.
For plumbing, hardware, bearings; SAE63; leaded tin bronze.

NBD NO. 64.
M-539; 80 Cu, 10 Sn, 10 Pb.
Cast: 25,000 TS; 12,000 YS; 8 El; 50 Brin.
For bearings, liners, sleeves; leaded bronze; SAE64.

NBD NO. 65.
M-539; 88 Cu, 12 Sn.
Cast: 35,000 TS; 21,000 YS; 10 El; 90 Brin.
For gears, worm wheels, bearings; SAE65; gear bronze.

NBD NO. 65-N.
M-539; 87.5 Cu, 11 Sn, 1.5 Ni.
Cast: 35,000 TS; 21,000 YS; 10 El; 90 Brin.
For gears; nickel gear bronze.

NBD NO. 66.
M-539; 85 Cu, 5 Sn, 9 Pb, 1 Zn.
Cast: 25,000 TS; 12,000 YS; 8 El.
For bearings; SAE 66.

NBD NO. 197.
M-539; 95 Cu, 4 Si, 1 Mn.
Cast: 45,000 TS; 18,000 YS; 20 El; 80 Brin.
For gears, bolts, shafts, fasteners; silicon bronze.

NBD NO. 198.
M-539; 91 Cu, 1 Sn, 5 Zn, 3 Si.
Cast: 45,000 TS; 18,000 YS; 20 El.
For hardware, bolts, shafts; silicon bronze.

NBD NO. 199.
M-539; 82 Cu, 14 Zn, 4 Si.
Cast: 60,000 TS; 24,000 YS; 16 El.
For hardware, valves; silicon bronze.

NBD NO. 295.
M-539; 91 Cu, 9 Zn.
Cast.
For ornamental accessories; gilding bronze.

NBD NO. 622.
M-539; 88 Cu, 6 Sn, 1.5 Pb, 4 Zn.
Cast: 34,000 TS; 16,000 YS; 22 El; 50 Brin.
For valves, plumbing; valve bronze; SAE622.

NBD NO. 660.
M-539; 83 Cu, 7 Sn, 7 Pb, 3 Zn.
Cast: 30,000 TS; 14,000 YS; 12 El; 50 Brin.
For plumbing, hardware; SAE660; modified red brass.

NBD "A" GRADE.
M-539; Pb, Sb, bal Sn.
For low grade bearings; Babbitt metal.

NBD ALUMINUM BABBITT.
M-539; Al, bal Sn.
Cast: 12,100 TS; 8.8 El; 6.6 RA; 28.6 Brin.
For heavy pressure bearings; Babbitt metal; C.S. 23200.

NBD ARCTIC BRONZE.
M-539; Sb, Sn, bal Cu.
For locomotive bearings.

NBD ARMATURE BABBITT.
M-539; Sn, Cu, Ni, Sb.
Cast: 13,600 TS; 6.1 El; 7.0 RA; 33.8 Brin.
For armature bearings, marine turbine bearings; Babbitt metal; C.S. 26700.

NBD BETA CRUSHER BABBITT.
M-539; Sb, Sn, Pb, Cu.
Cast: 11,300 TS; 4.9 El; 5.3 RA; 27.5 Brin.
For bearings for cement mills and mines; Babbitt metal; resists heavy loads.

NBD CRESCENT BABBITT.
M-539; Cu, Sn, Sb.
Cast: 8,500 TS; 1.4 El; 2.1 RA; 25.1 Brin.
For heavy duty bearings for rolling mills, paper mills; Babbitt metal; C.S. 18300.

NBD ENGINE BABBITT.
M-539; 9-11 Sb, 4-6 Cu, bal Sn.
Cast: 10,000-15,000 TS; 5 El.
For bearings, bushings, liners; Babbitt; C.S. 10600.

NBD EXTRA COPPER-HARDENED BABBITT.
M-539; Cu, Sn, Sb.
Cast: 8,500 TS; 5.0 El; 5.2 RA; 27.1 Brin.
For bearings; Babbitt metal; C.S. 7400.

NBD GENUINE BABBITT (ORIGINAL).
M-539; 88.9 Sn, 3.7 Cu, 7.4 Sb.
Cast: 13,000 TS; 10 El; 14 RA; 28.7 Brin.
For high temperature resistant bearings; Babbitt metal; C.S. 21200.

NBD GENUINE BABBITT (SPECIAL).
M-539; Cu, Sb, bal Sn.
Cast: 10,000 TS; 5.0 El; 12.9 RA; 28.9 Brin.
For heavy duty bearings; Babbitt metal; C.S. 22200.

NBD HOO-HOO.
M-539; Sn, Cu, Ni, Sb.
Cast: 12,800 TS; 11,1 El; 16 RA; 28 Brin.
For bearings; Babbitt metal; C.S. 25600.

NBD IMPROVED BABBITT.
M-539; Cu, Sn, Sb.
Cast: 9,600 TS; 2.2 El; 1.5 RA; 24.1 Brin.
For general utility bearings; Babbitt metal; C.S. 19500.

NBD NICKEL BABBITT.
M-539; Sn, Cu, Ni.
Cast: 11,900 TS; 5.0 El; 15.8 RA; 28.9 Brin.
For bearings; high speed and high pressures; C.S. 10600.

NBD PHOSPHOR BRONZE BABBITT.
M-539; P, Cu, bal Sn.
Cast: 11,150 TS; 5.6 El; 7.1 RA; 28.1 Brin.
For marine and stationary engine bearings; Babbitt metal; C.S. 18100.

NBD REGENT BABBITT.
M-539; 14 Sb, 5 Sn, bal Pb.
Cast: 9,000 TS; 2.0 El; 20 Brin.
For relining railroad bearings; Babbitt metal.

NBD REX BABBITT.
M-539; 10 Sn, 15 Sb, bal Cu.
Cast: 10,000 TS; 2.5 El; 1.8 RA; 24.8 Brin.
For medium high speed and heavy pressure bearings; Babbitt metal; C.S. 10100.

NBD SPECIAL MOTOR BABBITT.
M-539; Sb, Sn.
Cast: 9,100 TS; 2.2 El; 3.6 RA; 26.5 Brin.
For motor bearings; Babbitt metal; C.S. 18000.

NBD SPECIAL NO. 1 BABBITT.
M-539; Sn alloy.
Cast: 7,500 TS; 1.0 El; 1.7 RA; 24.2 Brin.
For bearings for rolling and paper mills; Babbitt metal; C.S. 14200.

NBD TIGER BRONZE.
M-539; Pb, Sn, bal Cu.
Cast: 33,000 TS; 18,000 YS; 19 El; 15 RA; 53 Brin.
For engine castings, bearings; shock resistant.

NC.
M-1577; 0.05 C, 15.32 Cr, 58.07 Ni, 13.73 Mo, 5.68 Fe, 4.24 W.
For chemical plant equipment, pump and valve parts.
Corrosion and heat resistant.

NC-4.
M-French; 24 Ni, 3 Cr, bal Fe.
For thermostatic bimetal.

NC 80/20.
M-121; 0.26 C, 1.2 Mn, 0.5 Fe, 0.5 Si, 0.2 Cu, 19.6 Cr, bal Ni.
Filler metal for gas-shielded arc welding BRIGHTRAY alloys, INCONEL alloy 600, INCOLOY ALLOYS DS and NIMONIC alloy 75 for high temperature applications.
BS 2901-NA-34; DIN S-NiCr 20.

N.C. ALLOY.
M-688; C, alloy, bal Fe.
For tools, dies; oil hardening.

N.C.A. (NON-CORRODIBLE ALUMINUM).
M-411; 2.2 Ni, 3 Cu, 0.2 Mg, 0.2 N, bal Al.
Rolled: 45,000 TS.
Sand cast: 24,000-31,000 TS; 12,000-15,000 YS; 7-5 El; 63 Brin.
For light alloy parts; resists sea-water corrosion.

N.C.C. PIG.
M-68; 56 min Ni, 23 min Cu, 7.5 min Cr, 1.5 max C, 1.75 max Si.
For foundry alloy for Ni-Resist Iron; M.P. 2300°F.

NCM.
M-1405; 0.40 C, 0.50 Mn, 0.30 Si, 1.25 Cr, 0.20 Mo, 1.50 Ni, bal Fe.
Oil hardening tool steel; for shear blades, collets, plastic moulds.

NCM.
M-1740; 0.25-0.35 C, 0.35-0.60 Si, 0.70-1.0 Mn, 0.60-1.1 Ni, 0.60-0.90 Cr, 0.30-0.50 Mo, bal Fe.
Low alloy structural steel.

NCM 30.
M-1740; 0.25-0.35 C, 0.30-0.60 Si, 0.60-0.80 Mn, 1.5-1.8 Ni, 0.90-1.20 Cr, 0.30-0.40 Mo, 0.30 max Cu, bal Fe.
High tensile steel with high abrasion and impact resistance.
B.S. 1458 Gr A&B; ASTM A148-65 Cr, 105-85; SAE 0105; Similar to AISI 4330.

NCM 35.
M-1740; 0.33-0.37 C, 0.25-0.60 Si, 0.60-0.80 Mn, 1.65-2.0 Ni, 0.70-0.8 Cr, 0.20-0.30 Mo, bal Fe.
BHN: 250-320.
Similar to AISI 4335.

NCM 70.
M-1740; 0.65-0.75 C, 0.30-0.70 Si, 0.025 max S, 0.025 max P, 0.60-0.80 Mn, 0.65-0.85 Ni, 1.3-1.7 Cr, 0.25-0.35 Mo, bal Fe.
Oil hardenable, low alloy steel.

NCM 75.
M-1740; 0.70-0.80 C, 0.30-0.60 Si, 0.70-0.90 Mn, 0.80-1.20 Ni, 1.50-2. Cr, 0.25-0.35 Mo, bal Fe.

NCMV.
M-1740; 0.25-0.30 C, 0.20-0.50 Si, 0.60-0.90 Mn, 1.45-1.80 Ni, 1.3-1.7 Cr, 0.30-0.40 Mo, 0.15-0.25 V, bal Fe.
Yield strength: 40 tons/sq.in. min; U.T.S.; 65 tons/sq.in. min; BHN 300 min.

NDHTC-OILITE.
M-211; Fe alloy.
Sintered: 80,000 TS.
For bearings; ferrous, porous.

N.D.S.
M-73; 0.75 C, 1.0 Cr, 1.75 Ni, bal Fe.
For tools, shear blades; shock resisting.

NDZ BRONZE.
M-801; 0.25 max Pb, 5.0 max Zn, 5.5 max Ni, 5.5 max Fe, 2.0 max Al, 2.0 max Si, bal Cu.
As sand cast: 65,000-69,000 TS; 30,000-35,000 YS; 20-34 El; 123-134 Brin.
For valve stems, propeller wheels, gears for marine and outboard industry, marine parts requiring resistance to dezincification and for dealuminization, water works equipment.
Heat treatable to slightly higher properties.

NDZ-S BRONZE.
M-801; 0.25 max Pb, 2.0 max Zn, 5.5 max Ni, 5.5 max Fe, 2.0 max Al, 2.0 max Si, bal Cu.
As sand cast: 69,000-75,000 TS; 40,000-45,000 YS; 13-23 El; 143-149 Brin.
For high strength valve stems, propeller wheels, gears, marine hardware requiring resistance to dezincification and/or dealuminization.
Heat treatable to slightly higher properties.

NEALLOY.
M-727; 0.6 Cr, 1.0 max Ni, 0.12 max C, 0.6 Si, 0.5 max Cu, bal Fe.
For woven wire conveyor belts, wire cloth and slings.

NEASCO.
M-1265; Si, Fe, bal Cu.
For master alloy; for silicon bronze.

NEATRO.
M-115; 1.27 C, 55 W, 4.5 Cr, 4 V, 4.5 Mo, bal Fe.
For cutting tools, broaches, chasers, hobs; high speed steel.

NEBALOY.
M-1720; 62.0-65.0 Cu, 0.07 max Pb, 0.05 max Fe, bal Zn.
Elec. cond.: 26.5% IACS at 68°F.
Good cold working properties.
For electrical terminals and connectors, hardware, jewelry, washers, shells, stampings.
Copper alloy No. 272.

NEB-BRONZE see **1% NEB-BRONZE.**

NECOMICLE.
M-Japan; C, Ni, Cr, Mo, bal Fe.
For turbine blades, chemical apparatus; stainless.

NEEDLE METAL.
M-U.S.; 85 Cu, 8 Sn, 5.3 Zn, 1.7 Pb.
For needles, valves, fittings; free-cutting.

NEELIUM.
M-1513; Bi, Te, Se, Sb.
For thermoelectric cooling; semi-conductor.

NELOY.
M-232; 0.3-0.4 C, 1.0-1.25 Mn, bal Fe.
Heat treated: 190,000-220,000 TS; 175,000-195,000 YS; 14-8 El; 45-25 RA; 364-477 Brin.
For gears, crankshafts; tough; hard after heat treatment.

NELOY NO. 1.
M-232; C, alloy, bal Fe.
Cast: 85,000-95,000 TS; 23-30 El; 50-60 RA; 163-170 Brin.
For steel castings, gears; tough.

NELOY NO. 2.
M-232; 0.4 C, 0.4 Si, 0.75 Mn, bal Fe.
Drawn: 90,000-93,000 TS; 65,000-75,000 YS; 23-28 El; 55-65 RA; 174-198 Brin.
For steel castings, gears; tough.

NELOY NO. 3.
M-232; 0.4 C, 0.4 Si, 0.75 Mn, bal Fe.
Heat treated: 100,000-110,000 TS; 80,000-90,000 YS; 20-26 El; 50-60 RA; 202-240 Brin.
For steel castings, gears; tough.

NELOY NO. 4.
M-232; 0.4 C, 0.4 Si, 0.75 Mn, bal Fe.
Heat treated: 115,000-130,000 TS; 95,000-110,000 YS; 15-20 El; 40-45 RA; 240-268 Brin.
For steel castings, gears; tough.

NELOY NO. 5.
M-232; C, alloy, bal Fe.
Heat treated: 125,000-135,000 TS; 110,000-125,000 YS; 12-20 El; 35-45 RA; 268-288 Brin.
For steel castings, gears; tough.

NELOY NO. 6.
M-232; C, alloy, bal Fe.
Heat treated: 135,000-150,000 TS; 120,000-135,000 YS; 9-18 El; 30-40 RA; 286-302 Brin.
For steel castings, gears; tough.

NELOY MOLYBDENUM.
M-232; 0.3 C, 0.6 Ni, 0.6 Cr, 0.2 Mo, bal Fe.
Annealed: 85,000-95,000 TS; 55,000-65,000 YS; 20-25 El; 30-40 RA; 170-197 Brin.
For steel castings, gears; tough.

NELOY MOLYBDENUM NO. 2A.
M-232; 0.3 C, 0.6 Ni, 0.6 Cr, 0.2 Mo, bal Fe.
100,000-115,000 TS; 70,000-90,000 YS; 15-20 El; 35-45 RA; 190-220 Brin.
For steel castings, gears; tough.

NELOY MOLYBDENUM NO. 3A.
M-232; 0.3 C, 0.6 Ni, 0.6 Cr, bal Fe.
Heat treated: 125,000-140,000 TS; 115,000-130,000 YS; 12-15 El; 30-40 RA; 250-280 Brin.
For steel castings, gears; tough.

NELOY MOLYBDENUM NO. 5A.
M-232; 0.3 C, 0.6 Ni, 0.6 Cr, bal Fe.
Heat treated: 140,000-160,000 TS; 130,000-145,000 YS; 10-13 El; 20-30 RA; 302-331 Brin.
For steel castings, gears; tough.

NELOY MOLYBDENUM NO. 6A.
M-232; 0.3 C, 0.6 Ni, 0.6 Cr, bal Fe.
Heat treated: 160,000-180,000 TS; 145,000-160,000 YS; 7.5-10 El; 15-25 RA; 341-375 Brin.
For steel castings, gears; tough.

NELSON-BOHNALITE.
M-196; 10 Cu, 0.2 Si, 0.3 Mg, 0.3 Ni, bal Al.
25,500 TS; 23,400 YS; 0.2 El; 110 Brin.
For pistons for motor vehicles; coefficient of expansion 0.000024.

NELSON-BOHNALITE.
M-1548; 9.0-11.0 Cu, 0.2 Si, 0.3 Mg, bal Al.
For general castings, fittings, hardware.
Good machinability, good strength.

NEMICLE.
M-Japan; C, Ni, Cr, Mo, bal Fe.
For turbine blades, chemical apparatus; stainless.

NEMICLE C.
M-Japan; C, Ni, Cr, Mo, bal Fe.
For turbine blades, chemical apparatus; stainless.

NEMICLE F.
M-Japan; C, Ni, Cr, Mo, bal Fe.
For turbine blades, chemical apparatus; stainless.

NEOCHRAN.
M-651; 22-30 Cr, 0.5-2.5 Si, 1.0-1.5 C, bal Fe.
For stainless castings; corrosion resistant.

NEODYMIUM.
M-1755; Nd.
Purities: 99.9%, 99.5+%
Forms: Ingot, lump, sheet, wire, turnings, foil, rod.

NEOGEN.
M-U.S.; 58 Cu, 27 Zn, 12 Ni, 2 Sn, 0.5 Al, 0.5 Bi.
For ornamental and structural parts; corrosion resistant.

NEOMAGNAL A.
M-German; Mg, Zn, bal Al.
For bearings.

NEONALIUM.
M-283; 86-94 Al, 6-14 Cu, 0.4-1.0 other elements.
Heat treated: 22,800-34,000 TS; 11,000-16,000 YS; 1.0 RA; 80-120 Brin.
For light alloy parts; good heat resistance.

NEONALIUM.
M-1548; 6.0-14.0 Cu, bal Al.
Sand Cast: 26,000 TS; 18,000 YS; 0.8 El; 90 Brin.
Chill cast: 32,000 TS; 26,000 YS; 0.3 El; 110 Brin.
For housings, casings, general castings.
Good machinability.

NEOR.
M-40, M-289; 2.3 C, 13 Cr, 0.6 Si, 0.6 Ni, 0.4 Mn, bal Fe.
For press tools, punches, dies, reamers, broaches, gages; remarkable resistance to abrasion.

NEOREX.
M-1488, A.C.L. Italy; 2.0 C, 13.0 Cr, bal Fe.
Cold work tool steel; air or oil hardening.
For blanking and coining dies.
Italy: UNI X210 Cr 13 KU.

NEPTUNE.
M-1432; 0.30 C, 0.75 Mn, 0.50 Si, 1.7 Cr, 0.4 Mo, bal Fe.
Oil hardening steel designed for molds. AISI P20.

NERGANDIN.
M-188, M-282; 70 Cu, 28 Zn, 2 Pb.
89,000 TS; 10 El; 135 Brin.
For condenser tubes; resistant to sea water corrosion.

NERO 3.
M-1138; 1.05 C, 1.0 Cr, bal Fe.
For bearings, liners, sleeves, punches, cutters; water hardened, wear resistant.

NERO EXTRA.
M-1338; 1.0 C, 1.1 Cr, 0.07 Mn, 0.25 Si, bal Fe.
For bearings, liners, sleeves, punches, cutters; water hardened, wear resistant.

NERO EXTRA SPEZIAL.
M-1338; 1.45 C, 1.4 Cr, 0.6 Mn, 0.25 Si, bal Fe.
For bearings, liners, dies; water hardened, wear resistant.

NERO HBK.
M-1338; 1.05 C, 1 Cr, 1.15 W, 0.9 Mn, 0.25 Si, bal Fe.
For bearings, forming and blanking dies; oil hardened, abrasion resistant.

NERO HFG.
M-1338; 1.25 Cu, 1.15 Si, 0.7 Mn, 1.2 Cr, bal Fe.
For bearings, liners, punches, dies; water or oil hardened, wear resistant.

NERO HWF1.
M-1338; 1.2 C, 0.1 V, 1.0 W, 0.28 Mn, 0.25 Si, bal Fe.
For wear plates, punches, tools, dies; oil hardened.

NERO KST.
M-1338; 0.9 C, 1.9 Mn, 0.25 Si, 0.1 V, bal Fe.
For punches, shears, crimpers, dies, upsetters; oil hardened, nondeforming.

NERO SPEZIAL.
M-1338; 1.05 C, Cr, bal Fe.
For bearings, liners, tools, dies; water hardened, wear resistant.

NES-3.
M-1651; 15 Cr, 17 Mo, 5 Fe, 4 W, bal Ni.
For valves, pumps, pharmaceutical and chemical plant equipment. Corrosion and heat resistant.

N.E.S. 70.
M-97; 0.18 Cu, 1.5 Ni, 1.0 Cu, 0.2 Mo, 0.8 Mn, bal Fe.
Rolled: 110,000 TS; 70,000 YS.
For mine cars, cages, skips, coal conveyors; abrasion and corrosion resistant.

NESALOY.
M-865; Li alloy.
To disperse Pb in Cu; alloying.

NETIC-S3.
M-1567; Ni, bal Fe.
Sheet: 67,100 TS; 29,500 Yp; 7.5 El; B 68-71 Rock.
For magnetic shields.
Low magnetic retentivity.

NETIC S3-5.
M-1567; Ni, bal Fe.
Sheet: 46,100 TS; 28,200 YS; 25 El; B 45 Rock.
For magnetic shielding.
Low magnetic retentivity.

NETIC S3-6.
M-1567; Ni, bal Fe.
Sheet: 40,000-45,000 TS; 28,000 YP; B 45 Rock.
For magnetic shields.
Low magnetic retentivity.

NEUMAL BD.
M-1366; 2.5-5.0 Cu, 0.2-1.8 Mg, 0.3-1.5 Mn, 0.5-2.5 Pb, Sn, Cd, Bi, bal Al.
For screw machine products; free-cutting.

NEUMAL D3.
M-1366; 2.5-5.0 Cu, 0.2-1.8 Mg, 0.3-1.5 Mn, bal Al.
Annealed: 27,000 TS; 11,000 YS; 22 El; 47 Brin.
Heat treated: 72,000 TS; 57,000 YS; 130 Brin.
For aircraft structures and fittings, fasteners; age-hardenable.

NEUMAL-S.
M-1366; 0.6-1.4 Mg, 0.6-1.6 Si, 0.6-1.0 Mn, 0.3 max Cr, bal Al.
Annealed: 21,000 TS; 8000 YS; 24 El.
For window frames, gutters, fan blades, boats. Good forming and welding properties.

NEUSILBER 47-11 PB.
M-297; 47 Cu, 11 Ni, 1.5 Pb, bal Zn.
Soft: 58,000 TS; 25 El; 100 Brin.
Hard: 82,000 TS; 0.5 El; 185 Brin.
For optical and camera parts, hardware, tableware; leaded nickel silver, corrosion resistant.

NEUSILBER 57-12 PB.
M-297; 57 Cu, 12 Ni, 1.7 Pb, bal Zn.
Soft: 58,000 TS; 35 El; 80 Brin.
Hard: 82,000 TS; 8 El; 160 Brin.
For optical and camera parts, tableware; leaded nickel silver, corrosion resistant.

NEUSILBER 60-25.
M-297; 60 Cu, 25 Ni, bal Zn.
Soft: 65,000 TS; 30 El; 90 Brin.
Hard: 90,000 TS; 7 El; 170 Brin.
Spring: 114,000 TS; 0.5 El; 200 Brin.
For optical and camera parts, tableware, jewelry; nickel silver, corrosion resistant.

NEUSILBER 62-18.
M-297; 62 Cu, 18 Ni, bal Zn.
Soft: 61,000 TS; 40 El; 85 Brin.
Hard: 85,000 TS; 7 El; 165 Brin.
Spring: 108,000 TS; 1 El; 195 Brin.
For optical and camera parts, tableware, jewelry; nickel silver, corrosion resistant.

NEUSILBER 62-18 PB.
M-297; 62 Cu, 18 Ni, 2 Pb, bal Zn.
Soft: 58,000 TS; 35 El; 85 Brin.
Hard: 82,000 TS; 8 El; 160 Brin.
For optical and camera parts, hardware, tableware; leaded nickel silver, corrosion resistant.

NEUSILBER 65-12.
M-297; 65 Cu, 12 Ni, bal Zn.
Soft: 58,000 TS; 40 El; 85 Brin.
Hard: 80,000 TS; 10 El; 160 Brin.
Spring: 103,000 TS; 1.5 El; 190 Brin.
For tableware, jewelry, camera parts; nickel silver, corrosion resistant.

NEUSILBER 71-7.
M-297; 71 Cu, 8 Ni, bal Zn.
Soft: 58,000 TS; 35 El; 80 Brin.
Hard: 80,000 TS; 8 El; 150 Brin.
Spring: 103,000 TS; 3 El; 180 Brin.
For tableware, costume jewelry, hardware; nickel silver, corrosion resistant.

NEUSILBER B. UND D. I.
M-297; 57 Cu, 12 Ni, 2 Pb, bal Zn.
Rolled: 65,400 TS; 8 El; 120 Brin.
Annealed: 46,800 TS; 35 El; 70 Brin.
For instruments, optical frames; corrosion resistant, free-cutting.

NEUSILBER B. UND D. II.
M-297; 54 Cu, 8 Ni, 1 Pb, bal Zn.
Rolled: 79,500 TS; 4 El; 180 Brin.
Annealed: 56,800 TS; 48 El; 85 Brin.
For instruments, optical frames; free-cutting, corrosion resistant.

NEUSILBER B. UND D. III.
M-297; 62 Cu, 15 Ni, 1 Pb, bal Zn.
Rolled: 81,000 TS; 5 El; 170 Brin.
Annealed: 58,300 TS; 37 El; 95 Brin.
For instruments, optical frames; free cutting, corrosion resistant.

NMEUSILBER B. UND D. IV.
M-297; 47 Cu, 11 Ni, 1 Pb, bal Zn.
Rolled: 120,800 TS; 2 El; 210 Brin.
Annealed: 85,200 TS; 30 El; 140 Brin.
For instruments, optical frames; free-cutting, corrosion resistant.

NEUSILBER, ENAMEL QUALITY.
M-297; 20 Ni, 5 Zn, bal Cu.
For bullet jackets, condenser tubes; German silver.

NEUSILBER EXCELSIOR.
M-297; 25 Ni, 15 Zn, 60 Cu.
Annealed: 57,000 TS; 32 El; 85 Brin.
For food handling equipment; German silver.

NEUSILBER, NICKELIN.
M-297; 22 Ni, 20 Zn, 56 Cu.
For hardware, electrical resistances; German silver.

NEUSILBER, PRIMA.
M-297; 12 Ni, 23 Zn, 65 Cu.
Annealed: 50,000 TS; 35 El; 75 Brin.
For ornamental parts; German silver.

NEUSILBER, PRIMA-PRIMA.
M-297; 18 Ni, 20 Zn, 62 Cu.
Annealed: 50,000 TS; 35 El; 75 Brin.
For white metal parts; German silver.

NEUSILBER, QUARTA.
M-297; 7 Ni, 22 Zn, 71 Cu.
Annealed: 51,200 TS; 45 El; 70 Brin.
For ornaments; German silver.

NEUSILBER, SEKUNDA.
M-297; 10 Ni, 24 Zn, 66 Cu.
For ornamental parts; German silver.

NEUSILBER, TERTIA.
M-297; 9 Ni, 25 Zn, 66 Cu.
For ornaments; German silver.

NEUSILBER, TUBES.
M-297; 16 Ni, 19.5 Zn.
For German silver tubes; German silver.

NEUSTADT.
M-Ger.; 71.5 Cu, 28.5 Zn.
For condenser tubing; deep drawn.

NEUTRALEISEN.
M-Ger.; 14-15 Si, 0.7 Mn, bal Fe.
For insoluble anodes, crucibles, condensers, evaporators; acid resistant, brittle.

NEUTRALLOY.
M-866; 75 Ni, 15 Cr, 10 Fe, Si, Ti, Mn.
For castings; corrosion and acid resistant.

NEUTROLOY.
M-1523; 55 Cu, 45 Ni.
Resistance wire: 300 Ohms per cir.mil.ft.
Low temp. coefficient of resistance.
Ann: 50,000 psi; hard drawn: 140,000 psi.
Max. oper. temp.: 500°C.

NEUTRON FLUX TI-CU ALLOY.
M-1187; 1% Cu, bal Ti.
For neutron flux density measurement.

NEVADA.
M-1704; 1.1 C, 0.75 Mn, 0.5 Si, 5.5 Cr, 0.3 V, 1.3 Mo, 0.015 P, 0.015 S, bal Fe.
Air hardening tool steel for cold work tools.

NEVADA MODIFIED.
M-1704; 0.55 C, 0.9 Mn, 0.25 Si, 3.5 Cr, 0.25 Ni, 0.25 V, 1.4 Mo, bal Fe.
Air hardening tool steel.

NEVADA SILVER.
M-U.S.; Cu-Ni.
For ornaments, electrical resistances; nickel silver.

NEVASTAIN C. "A".
M-15; 12.5-13.5 Cr, 0.30-0.35 C, bal Fe.
Annealed: 90,000-105,000 TS; 20-25 El; 40-50 RA: 195-210 Brin.
Heat treated: 225,000-145,000 TS; 9-14 El; 25-47 RA; 420-270 Brin.
For cutlery, turbine blades; corrosion resistant; see Silcrome L-12.

NEVASTAIN C. "B".
M-15; 16.5-17.5 Cr, 0.65-0.70 C, bal Fe.
For cutlery, knives; corrosion resistant; see Silcrome M-17.

NEVASTAIN "D".
M-15; 0.15 C, 1.0 Si, 20-21 Cu, bal Fe.
At 70°F: 65,000-75,000 TS; 40,000-45,000 YS; 32 El; 63 RA.
At 1500°F: 10,000 TS; 8,000 YS; 90 El; 98 RA.
For furnace linings, conveyors, heat treating boxes; heat resistant; see Silcrome 21.

NEVASTAIN E. "Z".
M-15; 13.5-15 Cr, 0.12 max C, bal Fe.
For corrosion resisting parts; corrosion resisting; see Silcrome 12-EZ.

NEVASTAIN, GRADE A.
M-15; 0.3-0.35 C, 12.5-13.5 Cr, bal Fe.
Annealed: 90,000-105,000 TS; 40,000-50,000 YS; 20-25 El; 40-50 RA; 195-210 Brin.
Hardened: 225,000 TS; 185,000 YS; 9 El; 25 RA; 420 Brin.
For stainless cutlery, tanks; corrosion resisting; see Silcrome L-12.

NEVASTAIN, GRADE B.
M-15; 0.65-0.70 C, 16.5-17.5 Cr, bal Fe.
Annealed: 90,000-95,000 TS; 40,000-45,000 YS; 26 El; 50-45 RA; 180-210 Brin.
For stainless cutlery; corrosion resisting; see Silcrome M-17.

NEVASTAIN, GRADE RA.
M-15; 0.10 max C, 16 Cr, 1.0 Cu, 1.0 Si, 0.40 Mn, bal Fe.
Annealed: 75,000 TS; 40,000 YS; 40 El; 75 RA; 150 Brin.
Oil treated: 90,000-103,000 TS; 50,000-98,000 YS; 30-25 El; 60-63 RA; 170-217 Brin.
For general fabricated stainless products; stainless; see Silcrome R.A.

NEVASTAIN "H".
M-15; 1.0-1.1 C, 16-18 Cr, 0.60 Si, bal Fe.
Annealed: 115,000 TS; 40,000-50,000 YS; 12 El; 17-18 RA; 195-240 Brin.
Air cooled: 140,000-240,000 TS; 140,000-220,000 YS; 0-2.8 El; 0-13.5 RA.
For furnace parts, heat treating equipment; see Silcrome H-17.

NEVASTAIN K.A. "2".
M-15; 17-18.5 Cr, 8.25-10 Ni, 0.08-0.16 C, bal Fe.
Heat treated: 112,000-86,000 TS; 90,000-37,000 YS; 41-75 El; 64-82 RA.
For stainless parts, chemical plant equipment; corrosion resisting; see Silcrome KA2.

NEVASTAIN K.A. 2 "S".
M-15; 17-18.5 Cr, 8.25-10 Ni, 0.07 max C, bal Fe.
Heat treated: 132,000 TS; 42,000 YS; 50 El; 76 RA.
At 1600°F: 20,000 TS; 7,000 YS; 28 El; 26 RA.
For stainless parts, chemical plant equipment; corrosion resisting; see Silcrome KA2S.

NEVASTAIN "K.N.C.-3".
M-15; 24-26 Cr, 19-21 Ni, 0.15 max C, bal Fe.
For furnace parts; heat and corrosion resisting; see Silcrome 25-20.

NEVASTAIN RA (CINCINNATI STEEL CASTING CO).
M-437; 0.10 C, 16 Cr, 1 Cu, 1 Si, bal Fe.
For rustless and stainless parts; stainless.

NEVASTAIN "S".
M-15; 11.5-13 Cr, 0.12 max C, bal Fe.
Rolled: 125,000-150,000 TS; 8-10 El; 25-35 RA; 280-320 Brin.
Heat treated: 130,000 TS; 21 El; 68 RA; 237 Brin.
Annealed: 50,000 TS; 35 El; 60 RA; 140 Brin.
For turbine blading, pump rods, machine parts, valves, spoons, and forks; high resistance to shock and impact; see Silcrome 12.

NEVEROIL-21.
M-U.S.; 32.5 Ni, 64.5 Cu.
For corrosion resistant parts; corrosion resistant.

NEVYANSKITE.
M-USSR; 58-44 Ir, 27-49 Os, 0-10 Pt, 0-6 Ru, 1.5-3 Rh, Pd + Fe + Cu.
For fountain pen points; mined by U.S.S.R.

NEW-BIDE see N20 ETC.

NEW CAPITAL STEEL.
M-Eng.: 14 W, 3.7 Cr, 0.1 V, 0.6 C, bal Fe.
For high speed tools, reamers, cutters, punches, gages; high speed steel.

NEWHALL.
M-510; 0.95 C, 1.20 Mn, 0.55 Cr, 0.55 W, 0.20 V, bal Fe.
Cold work tool steel; oil hardening.
B.S. 4659 Type B01; AISI 01.

NEWLOY.
M-310; 64 Cu, 35 Ni, 1 Sn.
For base metal for tableware, resistance wire; good corrosion and acid resistance.

NEWMAX.
M-289; 0.45-0.53 C, 0.75-1.00 Mn, 0.8-1.1 Cr, 0.15-0.25 Mo, bal Fe.
Oil hardening steel, designed for molds; Similar to AISI 4150.

NEW MET.
M-1203; Mo, Ti, carbides.
For cutters; sintered carbides.

NEW MET N93.
M-1203; 61 TiC, 8 Ni, 31 MoC.
Sintered: 175,000 Tr.S.; Rock. A 93.5.
For tools, bearings and seals.
Resists heat, oxidation and wear.

NEW MET N95.
M-1203; 64 TiC, 2 Ni, 34 MoC.
Sintered: 150,000 Tr.S.; Rock A 95.
For tools, bearings and seals.
Resists heat, oxidation and wear.

NEWPORT ARMATURE.
M-191; Si, bal Fe.
For electrical generators; high permeability.

NEWPORT ELECTRICAL "A".
M-191; Si, bal Fe.
For electric generators; high permeability.

NEWPORT ELECTRICAL "B".
M-191; Si, bal Fe.
For motors, armatures; high permeability.

NEWPORT ELECTRICAL "C".
M-191; Si, bal Fe.
For motors, armatures; high permeability.

NEWPORT FIELD.
M-191; Si, bal Fe.
For fields, armatures, electrical equipment; high permeability.

NEWPORT TRANSFORMER.
M-191; Si, bal Fe.
For transformers; high permeability.

NEWPORT TRANSFORMER EXTRA SPECIAL.
M-191; Si, bal Fe.
For transformers; high permeability.

NEWPORT TRANSFORMER SPECIAL.
M-191; Si, bal Fe.
For transformers; high permeability.

NEW PROCESS COLD HEADER.
M-69; 1.0 C, bal Fe.
Annealed: 100,000 TS; 53,000 YS; 21 El; 42 RA; 200 Brin.
For cutters, drills, taps, reamers, broaches; Type W1; water hardened.

NEW PROCESS COLD HEADER DIE STEEL.
M-69; 1.0 C, 0.25 Mn, 0.18 Si, bal Fe.
For cold header junches and dies; water hardened.

NEW RAPID.
M-363; C, W, bal Fe.
For punches, dies, cutting tools; high speed steel.

NEW RYCUT 50 see RYCUT 50.

NEWTON FUSIBLE ALLOY.
M-England; 50 Bi, 18.75 Sn, 31.25 Pb.
For fire and signal alarms, fire extinguisher plugs; M.P. 95°C.

NEW TOOL STEEL CAST.
M-Eng.; 58 Ni, 20 Zn, 12 Al, 10 Si.
For cuttings tools and dies; corrosion resistant.

NEY 76.
M-349.
Dental casting alloy.
Melt range: 1650-1750°F; Casting temp: 1800°F.
200 Brin, can be reduced to 140 Brin by heat treatment.
Economical; for white crown and bridge.

NEY 90.
M-349; 10 Cu, 90 Ag.
130 Brin.
For sliding contacts; low resistivity and high corrosion resistance.

NEYCAST.
M-349.
Dental solder; gold color.
Flows at 1425°F; color matched to NEYCAST III casting gold.

NEYCAST III.
M-349.
Dental casting alloy; gold color.
Melt range: 1600-1790°F; casting temp: 1950°F.
175 Brin.

NEYDIUM GOLD CERAMIC.
M-349.
Dental alloy; white gold color.
Melt range: 2100-2225°F; 230 Brin.

NEYDIUM NO. 90.
M-349; 10 Cu, bal Ag.
For slip rings, commutator segments, rivet head contacts.
Corrosion resistant. High electrical conductivity.

NEYDIUM NON PRECIOUS.
M-349.
Dental Alloy.
Melt range: 2220-2430°F; 195 Brin.
For porcelain fused to metal restorations.

NEYDIUM NP.
M-349.
Dental solder (Non-precious).
Melt range: 2050-2100°F; for use with NEYDIUM non-precious alloy.

NEYLASTIC H. F.
M-349.
Gold color wire for orthodontic use.
Fusing temp: 1830°F.

NEY-ORO 5.
M-349.
Dental alloy.
Melt range 1595-1705°F; 220 Brin.
For partial dentures.

NEY-ORO 6.
M-349.
Dental alloy.
Melt range: 1550-1635°F; 235 Brin.
For partial dentures.

NEY-ORO 28.
M-349; 25 Ag, 75 Au.
Cold worked: 60,000 TS; 35,000 P.L.; 100 Brin.
For contact brushes, used against coin silver slip rings.
Low contact resistance.

NEY-ORO 28A.
M-349; 75 Au, 23.5 Ag, 1.5 Ni.
Rolled: 90,000 TS; 50,000 YP; 140 Brin.
Annealed: 51,000 TS; 84 Brin.
For sliding contacts, electrical brush contacts.
72-75 ohms/cmf elect. resistivity.

NEY-ORO 28B.
M-349; 75 Au, 22 Ag, 3 Ni.
Rolled: 75,000 TS; 40,000 YP; 130 Brin.
For sliding contacts on Constantan, make and break contacts.
71-75 ohms/cmf (circular mil feet) elect. resistivity.

NEY-ORO 69.
M-349; 69 Au, 25 Ag, 6 Pt.
Annealed: 40,000 TS; 18,000 YP; 85 Brin.
Work Hardened: 70,000 TS; 30,000 YP; 120 Brin.
For make and break contacts, telephone relays, slip rings. Corrosion resistant.

NEY-ORO "A-1".
M-349; 80 Au-Pt, bal 20 Cu, Ag, Zn.
Cast: 55,000 TS; 27,000 YS; 25 El; 95 Brin.
For dental inlays, soft material.

NEY-ORO A-A.
M-349.
Dental alloy; casting gold.
Melt range: 1840-1960°F; 65 Brin.

NEY-ORO "B-2".
M-349; 78 Au-Pt, bal Cu, Ag, Zn.
Cast: 65,000 TS; 38,000 YS; 19 El; 120 Brin.
For dental inlays and dental bridges; hard.

NEY-ORO B-20.
M-349; Au, Ag dental alloy.
Melt range: 1600-1720°F; casting temp. 1850°F.
Hardened to 170 Brin by slow cooling after casting.

NEY-ORO CB.
M-349.
Dental casting alloy; gold color.
Melt range: 1550-1640°F; casting temp: 1750°F.
200 Brin, can be reduced to 145 Brin by heat treatment.

NEY-ORO "G".
M-349; 8.5 Pt, 4.5 Ag, 14.5 Cu, 1 Zn, 71.5 Au.
Ht. Tr.: 185,000 TS; 165,000 YS; 1 El; 280 Brin.
For pivots in instruments bearings, slip rings, commutator bars, make and break contacts.
Age-hardenable, corrosion and wear resistant.

NEY-ORO "G-3".
M-349; 75 Au-Pt, bal Cu, Ag, Zn.
Cast: 73,500 TS; 39,000 YS; 20 El; 220 Brin.
Age hardenable; for dental inlays, partial dentures; extra hard.

N.G.F. ALLOY.
M-U.S.; 1-1.5 Cu, 0.75-2.0 Mn, bal Al.
25,000 TS; 18,000 YS; 4 El.
For light alloy castings.

NGSA.
M-912; 0.43 C, 1.85 Cr, 0.50 Mo, 0.80 V, 0.40 W, bal Fe.
Hot work tool steel; metal upsetting tools, shear knives, pressing punches.
W.-Nr. 2603; AFNOR 45 CVD8.

NH 11.
M-912; 0.15 C, 2.0 Si, 2.0 max Mn, 20.0 Cr, 12.0 Ni, bal Fe.
Heat resisting steel; for annealing trays, hardening boxes.
W.-Nr. 4828.

NH 22.
M-912; 0.15 C, 2.0 Si, 2.0 max Mn, 25 Cr, 20 Ni, bal Fe.
Heat resisting steel; for annealing pots, enameling grates.
W.-Nr. 4841; Similar to AISI 310.

NH 40.
M-912; 0.15 max C, 16.0 Cr, 35 Ni, bal Fe.
Heat resisting steel; for furnace parts.
W.-Nr. 4864.

NH PIG.
M-68; 44-47 Ni, 15-17 Cr, bal Fe.
For foundry alloys; M.P. 2250°F.

NI 14.
M-1697; 88.0 WC, 12.0 Ni.
300,000 TrS; A 87.5 Rock; Density 14.40.
Sintered carbide tool material.

NI-20 CR-2THO$_2$ see TD NICR.

NIAG.
M-U.S.; 46.7 Cu, 40.7 Zn, 9.1 Ni, 2.8 Pb, 0.3 Mn.
For white metal parts; corrosion resistant.

NIAGRA BRAND FERRO-CHROME.
M-626; 66-70 Cr, bal Fe.
For metallurgical applications in steel; Cr-additions.

NIAGRA BRAND FERRO-SILICON.
M-626; 15-90 Si, bal Fe.
For metallurgical applications in steel; Si-additions.

NIAGRA BRAND SILICO MANGANESE.
M-626; 12-20 Si, 65-70 Mn, bal Fe.
For steel metallurgical applications; Mn-additions.

NIAL.
M-210; 24 Ni, 13 Al, 4 Cu, bal Fe.
For loud speakers, lighting and ignition equipment.
Permanent magnet. High permeability.

NIAL I.
M-61; 2.5 Mn, 2.0 Al, 1.0 Si, bal Ni.
For negative thermoelement of standard Type K thermocouple.

NIALCO.
M-1120; 19 Ni, 12 Co, 10 Al, 6 Cu, bal Fe.
For permanent magnets, electrical and magnetic equipment.
High magnetic permeability.

NIALCO I.
M-1120; 12 Ni, 10 Al, 4-20 Co, 2 Cu, bal Fe.
6500 Br, 530 Hc, 1.4 Bh max., C 45 Rock.
For magnets in wattmeters.
Permanent magnet. High permeability.

NIALCO II.
M-1120; 12 Ni, 10 Al, 4-20 Co, 2 Cu, bal Fe.
6300 Br, 650 Hc, 1.5 BH max., C 45 Rock.
For magnets in electrical relays.
Permanent magnet. High permeability.

NIALCO III.
M-1120; 12 Ni, 10 Al, 4-20 Co, 2 Cu, bal Fe.
7000 Br, 690 Hc, 1.7 BH max, C 45 Rock.
For magnets in magnetos.
Permanent magnet, high permeability.

NIALCO IV.
M-1120; 12 Ni, 10 Al, 4-20 Co, 2 Cu, bal Fe.
5700 Br, 1000 Hc, 1.9 BH max, C 58 Rock.
For magnets in electrical equipment.
Permanent magnet, high permeability.

NIALCO 200.
M-Austria; 10 Al, 20 Ni, 15 Co, 3 Cu, bal Fe.
For permanent magnets.

NIALCO 400.
M-Austria; 8 Al, 14 Ni, 24 Co, 3 Cu, bal Fe.
For permanent magnets.

NIALITE.
M-309; 10 Al, 5 Ni, 5 Fe, 1.5 Mn, bal Cu.
Cast: 85,700 TS; 39,000 YS; 24 El; 220 Brin.
For pumps, valves, propellers; corrosion and cavitation resistant.

NI-BAR IRON.
M-637; 3.3 T.C., 0.6 Mn, 1.5 Si, 0.6 Cr, 1.5 Ni, bal Fe.
36,000 TS; 210 Brin.
For grate bars, stoker links.

NIBORIUM B.
M-1693.
Electrical contact alloy.
Ann: 65,000 psi TS; 40,000 psi YS; 40 El.
Spring temper: 135,000 psi TS; 133,000 psi YS; 1 El.
Solderable, weldable, machinable, corrosion resistant electrical contact material.

NIBSI.
M-1493; 3.5 Si, 1.8 B, bal Ni.
Brazing powder, melt range: 1800-1950°F.
For brazing stainless and high temperature alloys.
AMS 4779.

NI C 1.5.
M-303; 1.5 Mn, C, bal Ni.
For chemical equipment.

NI C 4.
M-303; 4.0 Mn, C, bal Ni.
For spark plug electrodes.

NI C 5.
M-303; 5.0 Mn, C, bal Ni.
For spark plug electrodes.

NICA-0.
M-1449; 0.08 max C, 18 Cr, 8 Ni, bal Fe.
Annealed: 85,000 TS; 35,000 YS; 60 El; 70 RA; 150 Brin.
For chemical plant equipment, tanks, mixers; Type 304; stainless, austenitic.

NICA-00.
M-1449; 0.05 max C, 18 Cr, 8 Ni, bal Fe.
Annealed: 85,000 TS; 35,000 YS; 60 El; 70 RA; 150 Brin.
For chemical plant equipment, tanks, mixers; Type 304; stainless, austenitic.

NICA-1.
M-1449; 0.11-0.16 C, 18 Cr, 8 Ni, bal Fe.
Annealed: 80,000 TS; 35,000 YS; 55 El; 75 RA; 150 Brin.
For chemical plant equipment, tanks, mixers, filters; Type 302; stainless austenitic.

NICA-2.
M-1449; 0.17-0.25 C, 18 Cr, 8 Ni, bal Fe.
Annealed: 85,000 TS; 40,000 YS; 50 El; 70 RA; 160 Brin.
For chemical plant equipment, tanks, mixers, filters; Type 301 and 302; stainless, austenitic.

NICALLOY.
M-Eng.; 47 Ni, 53 Fe.
For electrical equipment; magnetically soft, high permeability.

NICALUN.
M-250; Ni, Cu, alloy base + abrasive grains imbedded.
For elevator door sills, stair treads; castings; wear resistant.

NICAR.
M-677; 3.5-4 C, 1.0 Si, 12-18 Cr, 65-75 Ni, 0.2 Co, 2.5-4.5 B, 4 Fe.
For hardfacing bare rod; class RNiCr-C.

NI CHILLITE.
M-229; Ni, Cr, Mo, C, bal Fe.
For rolls; chilled iron.

NI CHILLITE NO. 2.
M-229; C, alloy, bal Fe.
For rod mill rods; heavy duty.

NICHROFRY 152.
M-1302; 0.25 max C, 24-26 Cr, 19-22 Ni, bal Fe.
Annealed: 95,000 TS; 45,000 YS; 50 El; 65 RA; 180 Brin.
At 1200°F: 57,000 TS; 22,000 YS; 32 El; 45 RA.
For furnace parts, valves, pumps, heat treating boxes; Type 310; austenitic, heat resistant.

NICHROLLOY I.
M-1027; 0.3 C, 23 Ni, 20 Cr, 1 Mn, 1 V, 0.5 Al, bal Fe.
Annealed: 60,000 TS.
For heat treating boxes, resistance wire; high heat resistance.

NICHROLOY II.
M-1027; 40 Ni, 7 Cr, 3 Mn, bal Fe.
For resistance wire, heat treating boxes; heat and corrosion resistant.

NICHROLOY III.
M-1027; 75 Ni, 16 Cr, 3 Mn, bal Fe.
105,000-112,000 TS; 60,000-80,000 YS; 45-25 El; 65-59 RA; 180-210 Brin.
For resistance wire, heat treating boxes; heat and corrosion resistant.

NICHROLOY 37.
M-1027; 0.2 C, 25 Cr, 12 Ni, bal Fe.
Cast: 70,000 TS; 30,000 YS; 35 El; 60 RA; 150 Brin.
For furnace parts, cast heat treating boxes; heat resistant to 2000°F.

NICHROLOY 45.
M-1027; 0.4 C, 25 Cr, 20 Ni, bal Fe.
Cast: 75,000 TS; 50,000 YS; 17 El; 170 Brin.
For furnace parts, retorts, stills; heat resistant to 2100°F.

NICHROLOY 50.
M-1027; 0.4 C, 15 Cr, 35 Ni, bal Fe.
Cast: 70,000 TS; 40,000 YS; 10 El; 12 RA; 170 Brin.
For salt pots, furnace parts, heat treating boxes; heat resistant to 2100°F.

NICHROLOY 72.
M-1027; 0.4 C, 12 Cr, 60 Ni, bal Fe.
Cast 70,000 TS; 40,000 YS; 6 El.
For lead and cyanide pots, furnace equipment; heat resistant.

NICHROLOY A.
M-1510; 0.51 C, 1.05 Cr, 0.25 Mo, 0.53 Ni, 0.21 V, 0.97 Mn, bal Fe.
Heat treated: 165,000 TS; 150,000 YS; 20 El; 59 RA; 280-310 Brin.
For arbors, axles, bolts, cams, gears, hubs, die liners; shock and fatigue resistant, preheat treated at mill.

NICHROLOY L.
M-1510; less than 0.5 C, 1.05 Cr, 0.25 Mo, 0.53 Ni, 0.21 V, 0.97 Mn, bal Fe.
For arbors, axles, bolts, cams, gears, hubs, die liners; preheat treated at mill, shock and fatigue resistant.

NICHROME.
M-44; 60 Ni, 15 Cr, 25 Fe, 1 Si.
Wire and ribbon for heating elements up to 1900°F.

NICHROME 62-16 see **WIRESPRAY NICHROME 62-16.**

NICHROME V.
M-44; 80 Ni, 20 Cr.
95,000-175,000 TS; 50,000 YS; 35 El.
For heating elements in electric furnaces, electric ranges; max operating temperature 1150°C; high heat and oxidation resistant.

NICHROME V-245 see **D-H NO 245.**

NICHROME V-242 see **D-H NO 242.**

NICHROTHERM NCT-1.
M-72; 0.15 C, 15 Ni, 20 Cr, bal Fe.
Untreated: 100,000-114,000 TS; 78,000-93,000 YS; 35-25 El; 55-45 Brin.
For furnace parts, crucibles, autoclaves, recuperators; heat and corrosion resistant to 1050°C.

NICHROTHERM-NCT-1A.
M-72; 0.15 C, 2.5 Si, 1.5 Mn, 18 Cr, 10 Ni, bal Fe.
Bar: 90,000 TS; 40,000 YS; 50 El; B 85 Rock.
For heat treating fixtures, annealing boxes, tube supports, furnace parts.
Type 302B stainless steel, austenitic.

NICHROTHERM NCT-3.
M-72; 0.15 C, 20 Ni, 25 Cr, bal Fe.
Untreated: 100,000-114,000 TS; 65,000-78,000 YS; 35-25 El; 55-45 RA.
For furnace parts, crucibles, autoclaves, recuperators; heat and corrosion resistant to 1200°C.

NICHROTHERM NCT-6.
M-72; 0.15 C, 60 Ni, 15 Cr, bal Fe.
Untreated: 85,000-100,000 TS; 35,000-50,000 YS; 35-25 El; 55-45 RA.
For furnace parts, crucibles, autoclaves, recuperators; heat and corrosion resistant to 1150°C.

NICHROTHERM NCT-6A.
M-72; Fe, Cr, Ni.
For furnace parts, crucibles, autoclaves, recuperators; heat and corrosion resistant to 1150°C.

NICHROTHERM NCT-8.
M-72; 0.15 C, 80 Ni, 17 Cr, bal Fe.
Untreated: 85,000-100,000 TS; 35,000-50,000 YS; 25-35 El; 45-55 RA.
For furnace parts, crucibles, autoclaves, recuperators; heat and corrosion resistant; austenitic, max temperature 1300°C.

NICHROTHERMSTEEL.
M-118; Ni, bal Fe.
Low coefficient of expansion.

NI-CHRO-ZINK.
M-Ger.; Ni, Cr, Zn.
For die casting alloy.

NICKAHL IRON.
M-249; 1.5 C, 0.8 Cr, 2.2 Si, bal Fe.
For die to stamp fenders, door panels and seat sides for automobiles; tough and wear resisting cast iron.

NICKEL.
M-1755; Ni.
Purities: Zone and chemical refined 99.999%, 99.99%, 99.95%.
Forms: Powders, rod, pellets, sheets, wires, foils, platelletts, cathodes, single crystals.

NICKEL 61.
M-121; 0.06 C, 96.0 Ni, 0.3 Mn, 0.1 Fe, 0.4 Si, 0.02 Cu, 3.0 Ti, 0.005 S.
Filler metal for gas-shielded arc welding Nickel 200 and 201, and overlaying on steel.
AWS A5.14 (class ERNi-3); BS 2901-NA-32.

NICKEL 99.2.
M-297; 99.2 min Ni.
For manufacturing and processing of mineral products, especially caustic alkalis.

NICKEL 99.6/99.6 K.
M-297; 99.6 min Ni.
For manufacturing and processing of mineral products, especially caustic alkalis.

NICKEL-131.
M-897; 0.15 max C, bal Ni.
For welding rod.

NICKEL 141.
M-121, M-897; 0.03 C, 96.0 Ni, 0.3 Mn, 0.05 Fe, 0.6 Si, 2.5 Ti, 0.0 Cu, 0.25 Al, 0.005 S.
Welding electrode for shielded metal-arc Welding Nickel 200 and 201, and overlaying on steel, and joining nickel to steel.
AWS A5.11 (class ENi-1); ASME SFB 5.11.

NICKEL 200.
M-1499; 99.5 Ni, 0.08 C, 0.18 Mn, 0.2 Fe, 0.005 S, 0.18 Si, 0.13 Cu.
Annealed: 55,000-80,000 TS; 15,000-30,000 YS; 55-40 El; 90-120 Brin.
Cold drawn: 65,000-110,000 TS; 40,000-100,000 YS; 35-10 El; 140-230 Brin.
Hot finished: 60,000-85,000 TS; 15,000-45,000 YS; 55-35 El; 90-150 Brin.
Commercially pure nickel; for chemical handling, food processing, electronic equipment; corrosion resistant.

NICKEL 201.
M-1499, M-121; 99.5 Ni, 0.01 C, 0.18 Mn, 0.2 Fe, 0.005 S, 0.18 Si, 0.13 Cu.
Annealed: 50,000-60,000 TS; 10,000-25,000 YS; 60-40 El; 75-100 Brin.
Cold drawn: 60,000-100,000 TS; 39,000-90,000 YS; 35-10 El; 125-200 Brin.
Lower carbon than Nickel 200; for caustic evaporators, plater bars, combustion boats; preferred for application above 600°F.

NICKEL 205.
M-1499, M-121; 99.5 Ni, 0.08 C, 0.18 Mn, 0.10 Fe, 0.004 S, 0.08 Si, 0.08 Cu, 0.03 Ti, 0.05 Mg.
Annealed: 50,000 TS; 13,000 YS; 45-40 El; 77 Brin.
Cold rolled: 95,000 TS; 90,000 YS; 3 El; 210 Brin.
For electrical and electronic applications.

NICKEL 211.
M-1499; 96.85 Ni, 0.10 C, 4.75 Mn, 0.38 Fe, 0.008 S, 0.08 Si, 0.13 Cu.
For electronic applications, Formerly "D" Nickel.

NICKEL 212.
M-121; 98.0 Ni, 0.10 C, 2.0 Mn, 0.05 Fe, 0.005 S, 0.05 Si, 0.03 Cu.
Electron tube supports; Formerly "E" Nickel.

NICKEL 220.
M-1499; 99.5 Ni, 0.04 C, 0.10 Mn, 0.05 Fe, 0.004 S, 0.03 Si, 0.05 Cu, 0.03 Ti, 0.05 Mg.
Annealed: 70,000 TS; 20,000 YS; 40 El; 100 Brin.
For electronic receiving tube cathodes.

NICKEL 222.
M-121; 0.01 C, 99.8 Ni, 0.05 Mg.
Cathode nickel; also used for sleeves of indirectly heated, oxide coated cathodes in radio valves.
ASTM F239; BS 3504.

NICKEL 230.
M-1499; 99.5 Ni, 0.05 C, 0.08 Mn, 0.05 Fe, 0.004 S, 0.02 Si, 0.05 Cu, 0.003 Ti, 0.06 Mg.
Annealed: 70,000 TS; 20,000 YS; 40 El; 100 Brin.
Electron tube applications.

NICKEL 233.
  M-1499; 99.5 Ni, 0.09 C, 0.18 Mn, 0.05 Fe, 0.005 Si, 0.03 Si, 0.03 Cu, 0.003 Ti, 0.07 Mg,

NICKEL 240.
  M-121; 95.0 Ni, 1.7 Cr, 0.3 Ti, 2.0 Mn, 0.45 Si, 0.15 Zr.
  Special nickel grades for spark plug centre and earth electrode.

NICKEL 241.
  M-121; 90.0 Ni, 5.0 Cr, 3.0 Mn, 1.7 Si.
  Special nickel grade for spark plug centre and earth electrode.

NICKEL 270.
  M-1499, M-121; 99.98 Ni, 0.01 C, 0.003 Mn.
  High purity nickel; for electronic applications; heat exchangers.

NICKEL 400.
  M-1; 0.3 max C, 25-32 Cu, 63 Ni.
  Ni-Cu alloy with good high strength, excellent corrosion resistance and good weldabiltiy.
  UNS N04400.

NICKELALLOY.
  M-849; 100 Ni.
  Welded: 45,000 TS; 3 El.
  For welding electrodes; shielded arc.

NICKEL ALUMINUM BRONZE.
  M-U.S.; 88 Cu, 10 Ni, 2 Al, Sn.
  For heat and corrosion resisting parts; high strength.

NICKEL ALUMINUM BRONZE.
  M-U.S.; 10 Cu, 40 Ni, 30 Al, 20 Sn.
  For ornaments; corrosion resistant.

NICKEL ALUMINUM BRONZE.
  M-129; 78 min Cu, 9-11.5 Al, 3.0-5.5 Ni, 3.0-5.0 Fe, 1.5 max Mn.
  Cast: 80,000 TS; 35,000 YS; 15 El; 15 RA; 140 Brin.
  For ship propellers, pump and turbine parts; resists cavitation and corrosion, tough.

NICKEL-ARC.
  M-1713; high C, Ni, bal Fe.
  For welding electrodes for cast iron; machinable cast iron.

NICKEL-ARC 55.
  M-1713; 55 Ni-45 Fe.
  As welded: 49,000 psi TS (approx.)
  For welding cast iron; machinable weld metal.
  AWS class ENiFe-C1.

NICKEL-ARC 99.
  M-1713; 99 + Nickel.
  As welded: 32,800 psi (approx.)
  For welding and overlay on cast iron.
  AWS class ENi-C1.

NICKEL-AT.
  M-897; 0.15 max C, bal Ni.
  For welding rod.

NICKEL BEARING.
  M-Eng.; 50 Cu, 25 Ni, 25 Sn.
  For bearings; heavy duty.

NICKEL BORON STEEL.
  2.8-3.6 Ni, 0.1-0.5 B, 0.15-0.7 C, bal Fe.
  For dynamically stressed parts; water or oil hardened.

NICKEL BRASS.
  M-U.S.; 50-54 Cu, 35-44 Zn, 1.5 Ni, 0.5 Fe, Al.
  For condenser tubes, hardware; corrosion resistant.

NICKEL BRASS.
  M-538; Ni coated brass.
  For fabricated parts; easily stamped, formed, drawn.

NICKEL BRONZE.
  M-126; 82 Cu, 8 Sn, 2 Zn, 8 Ni.
  For superheated steam parts; corrosion resistant.

NICKEL-BRONZE.
  M-309; 85-90 Cu, 3-6 Sn, 3-7 Ni, 1-3 Zn.
  Heat treated: 50,000-80,000 TS; 25,000-60,000 YS; 15-25 El; 15-25 RA; 95-170 Brin.
  For gears, screw down nuts; castings.

NICKEL "C".
  M-303; 1.5 Mn, bal Ni.
  69,000 TS; 45 El.
  For chemical apparatus, thermocouple element; heat and corrosion resistant to salt solutions and caustics.

NICKELCAST.
  M-1000; weld metal: 2.0 C, 4.0 Si, 1.0 Mn, 8.0 max Fe, 85.0 max Ni, 2.5 max Cu, 1.0 max other.
  As welded: 40,100 psi TS; 38,200 psi YS; 3.6 El.
  For welding ductile iron and cast iron.
  AWS A5.15-69 (E Ni-CI).

NICKELCAST 55.
  M-1000; weld metal: 2.0 C, 4.0 Si, 1.0 Mn, 45.0-60.0 Ni, 2.5 Cu, (other 1.0), bal Fe.
  As welded: 57,000-84,000 psi TS; 43,000-63,000 psi YS; 6-13 El.
  For welding ductile iron and cast iron.
  AWS A5.15-69 (E NiFeCI).

NICKEL CAST IRON NO. 1.
  M-68; 3.0 Ni, 1.5 Si, 0.8-1.0 Cr, C, bal Fe.
  50,000 TS; 385 Brin.
  For cast gears; wear resisting casting.

NICKEL CERIUM STEEL.
  M-Eng.; 2.2-3.0 Ni, 0.1-0.9 Ce, 0.4-0.75 C, bal Fe.
  For machinery parts; oil hardening.

NICKEL CHROME ALUMINUM.
M-Eng.; 88 Ni, 12 Al, 8 Cr.
For corrosion and heat resistant parts; corrosion and heat resistant.

NICKEL CHROME CAST IRON.
1.5-1.75 Ni, 0.6-0.8 Cr, 3-3.4 C, 0.9-1.75 Si, 0.5-0.7 Mn, bal Fe.
Cast: 35,000-48,000 TS; 0 El; 0 RA; 170-230 Brin.
For grids; corrosion and abrasion resisting.

NICKEL-CHROME-COPPER.
M-Eng.; 80-85 Ni, 20-25 Cr, 15-20 Cu.
For electric irons, percolators.

NICKELCHROMEIGHT.
M-118; C, Cr, Ni, bal Fe.
For welding electrodes; for 18/8 stainless steel.

NICKEL CHROME PEERLESS.
M-Eng.; 16.5 Cr, 3.0 Fe, 2.0 Mn, 0.1 C, bal Ni.
For heat resisting parts; heat resistant.

NICKEL CHROME PREMIER.
M-Eng.; 25 Fe, 11 Cr, 3.0 Mn, bal Ni.
For heat resisting parts; heat resistant.

NICKEL CHROME SUPERIOR.
M-Eng.; 19.5 Cr, 0.5 Fe, 2.0 Mn, 0.2 C, bal Ni.
For heat resisting parts; heat resistant.

NICKELCHROMTWELVE.
M-118; C, Ni, Cr, bal Fe.
For welding electrodes; for 25/12 stainless steel.

NICKEL COBALT 9-H.
M-580; Ni-Co.
For anodes; electroplating.

NICKEL COINAGE U.S.A.
M-U.S.; 75 Cu, 25 Ni.
For coinage; corrosion resistant.

NICKEL COPPER.
M-538; Ni coated Cu.
For fabricated parts; easily stamped, formed, drawn.

NICKEL COPPER STEEL.
M-U.S.; 1-25 Ni, 0.4-10 Cu, 0.15-0.8 C, bal Fe.
For structures, machinery parts, fences, gates; resists soil corrosion.

NICKEL-COPPER-TITANIUM.
M-173; 0.15 max C, 1.0 max Mn, 0.50 max Si, 0.70 max Ni, 0.30 min Cu, 0.05 max Ti, bal Fe.
50,000 psi min YS.
Good cold-forming, weldability, atmosphere corrosion resistance.
For guard rails, automobile bumpers.

NICKEL CZ100.
M-775; 1.0 max C, 1.5 max Mn, 2.0 max Si, 1.25 max Mo, 3.0 max Fe, bal Ni.
Alloy castings; corrosion and temperature resistant.
ACI CZ 100; ASTM A-296 CZ-100.

NICKELDUR.
M-508; 80 Cu, 10 Pb, 5 Ni, 5 Sn.
Cast: 19,000 TS; 85 Brin.
For centrifugal cast liner; nickel bronze, corrosion resistant.

NICKELEISEN 36K.
M-297; 36-38 Ni, bal Fe.
For magnetic and electrical equipment, motors; soft magnet, high permeability.

NICKELEISEN 36W.
M-297; 36-38 Ni, bal Fe.
For magnetic and electrical equipment, motors; soft magnet, high permeability.

NICKELENE-1.
M-Eng.; 5-30 Ni, 52-80 Cu, 10-35 Zn.
For electrical instruments, resistance wires, thermocouples; German silver.

NICKELENE-2.
M-Eng.; 13 Ni, 55 Cu, 21 Zn, 2 Sn, 10 Pb.
For resistance wires; heat resistant.

NICKEL FILLER METAL 61.
M-1499; 96.0 Ni, 0.06 C, 0.30 Mn, 0.10 Fe, 0.005 S, 0.40 Si, 0.02 Cu, 3.0 Ti.
For gas shielded arc welding of Nickel 200 and Nickel 201; overlaying on steel.
AWS A 5.14 Class ERNi-3; ASME SFB 5.14.

NICKELIN.
M-297; 58 Cu, 22 Ni, 20 Fe.
For regulating and checking electrical resistances; German silver.

NICKELIN-40.
M-789; 67 Cu, 30 Ni, 3 Mn.
Hard: 115,000-128,000 TS.
Soft: 50,000-60,000 TS.
Temp. coef. resist. per °C: + 0.00014.
For electrical equipment and instruments.
Resistance alloy. Max operating temperature 400°C.

NICKELIN "A".
M-Eng.; 18 Ni, 62 Cu, 20 Zn.
For resistance alloy; heat resistant.

NICKELIN "B".
M-Eng.; 32 Ni, 68 Cu.
For resistance alloy; heat resistant.

NICKELIN "C".
M-Eng.; 31.5 Ni, 55.3 Cu, 13.1 Zn.
For resistance alloy; heat resistant.

NICKELINE.
55-75 Cu, 18-32 Ni, 0-20 Zn, 0.20-0.45 Fe.
For electrical resistance; heat resistant.

NICKELINE.
M-88; Pb, Sb, bal Sn.
24 Brin.
For Babbitt, bearings; Babbitt metal.

NICKELIN I.
M-303; 33.3 Ni, bal Cu.
For chemical equipment, fruit presses, boilers, kettles; corrosion resistant.

NICKELIN I.
M-297; 54 Cu, 26 Ni, 20 Zn.
Hard: 121,000 TS; 1.5 El.
Annealed: 85,000 TS; 30 El.
For starting rheostats for motors, field regulators, loading resistances; max. operating temperature 500°C.

NICKELIN II.
M-297; 67 Cu, 30 Ni, 3 Mn.
Hard: 100,000 TS; 2 El; 170 Brin.
Annealed: 70,000 TS; 44 El; 92 Brin.
For starting rheostats, field regulators, loading resistances; max. operating temperature 500°C.

NICKELIN II.
M-303; 25 Ni, bal Cu.
For fruit presses, coinage; corrosion resistant.

NICKELIN III.
M-303; 20 Ni, bal Cu.
For medals, ornaments; corrosion resistant.

NICKELIN III.
M-297; 58 Cu, 22 Ni, 20 Zn.
Hard: 118,700 TS; 1 El.
Annealed: 72,000 TS; 34 El.
For regulating and control resistances; max. operating temperature 500°C.

NICKELIN IV.
M-303; 85 Cu, 15 Ni.
45,000 TS; 48 El.
For projectiles or bullets to be nickel-jacketed; tough.

NICKELIN V.
M-303; 10 Ni, 90 Cu.
For armature wires, electrical equipment; corrosion resistant.

NICKELIN RESISTANCE.
M-297; 54 Cu, 20 Zn, 26 Ni.
For motor starters, electrical equipment; nickel silver.

NICKEL MALLEABLE.
M-68; 99.4 Ni, 0.15 Mn, 0.1 Cu, 0.05 C, 0.15 Fe.
For coinage, constituent of alloys; corrosion resistant.

NICKELMANG.
M-118; C, alloy, bal Fe.
For welding electrodes for Mn steel; bare and coated.

NICKEL-MANGANESE.
M-897; 2-5 Mn, bal Ni.
Annealed: 78,000 TS; 50,000 YS; 30 El.
For grid wires, electronic tubes; corrosion and heat resistant.

NICKEL MANGANESE BRONZE.
M-U.S.; 53 Cu, 39 Zn, 2.6 Sn, 2.5 In, 1.7 Mn, Al, Pb.
For corrosion resistant strong castings; corrosion resistant.

NICKEL MOLYBDENUM STEEL.
M-U.S.; 1.5-3.0 Ni, 0.1-0.7 Mo, 0.4 C, Mn, bal Fe.
For gears, shafts; oil hardened.

NICKEL N. 93.
M-England; 2 W, 1 Al, 0.2 C, bal Ni.
For tube cathodes and valve components; corrosion and heat resistant.

NICKEL N. 100.
M-England; 2 W, 1 Al, 0.2 C, 0.2 Mg, bal Ni.
For tube cathodes and valve components; corrosion and heat resistant.

NICKEL-NI-C.
M-897; 2.0 W, C, bal Ni.
For electron tube elements; heat resistant.

NICKELOID.
M-135; 45-40 Ni, bal Cu.
For non-rusting and corrosion resistant parts; corrosion resistant.

NICKELOID.
M-553; Pb, Sb, Ni, bal Sn.
For bearings; Babbitt metal.

NICKELOID.
M-538; Ni coated Zn.
For construction; corrosion resistant.

NICKELOY.
M-Eng.; 93.8 Al, 4.15 Cu, 1.41 Ni.
20,000 TS; 5 RA.
For automotive engine parts; age-hardenable.

NICKELOY-ROD NO. 2512.
M-1096; 0.10-0.15 C, 0.35-0.55 Mn, 0.25-0.35 Si, 4.75-5.25 Ni, bal Fe
Welded: 80,000-90,000 TS; 65,000-75,000 YS; 25-30 El; 50-55 RA.
For all-position welding rod; low temperature toughness.

NICKEL-SILC-ON.
M-461; Ni, Si.
Cold rolled: 70,000-85,000 TS; 60,000-70,000 YS; 25-15 El; 55-45 RA; 149-170 Brin.
For pistons.

NICKEL SILICON STEEL.
M-U.S.; 2.8-3.3 Ni, 0.5-2.2 Si, 0.35-0.50 Cu, bal Fe.
For machinery parts, gears, shafts, axles; oil hardening.

NICKEL SILVER.
M-322; 18 Ni, 65 Cu, 18 Zn.
59,900 TS; 31,800 YS; 34 El; 65 RA.
For silver plated table ware; Rockwell "B"-50.

NICKEL SILVER 8%.
M-1426; 7.0-9.0 Ni, 63-66 Cu, bal Zn.
CDA 741,743.

NICKEL SILVER 8% 65-8 743.
M-279; 65 Cu, 8 Ni, 27 Zn.
Ann: 53,000 TS; 23,000 YS; 40 El.
Hard: 85,000 TS; 73,000 YS; 4 El.
For hollow ware, flat wave, optical goods.

NICKEL SILVER 10%.
56-65 Cu, 25-34 Zn, 10 Ni.
For ornaments, plated ware; corrosion resistant.

NICKEL SILVER 10% 740.
M-279; 71 Cu, 10 Ni, 19 Zn.
Ann: 50,000-60,000 TS; 15,000-33,000 YS; 33-43 El.
Cold rolled: 56,000-95,000 TS; 30,000-86,000 YS; 1-34 El.
A high copper nickel silver, good corrosion resistance; for musical instruments, electronic parts.

NICKEL SILVER 10% 745.
M-8; 65 Cu, 24.75 Zn, 10 Ni, 0.25 Mn.
Hard sheet: 87,000 TS; 70,000 YS; 6 El; B 87 Rock.
Soft sheet: 55,000 TS; 20,000 YS; 42 El; B 30 Rock.
For optical goods, costume jewelry, holloware, radio dials, nameplates, camera parts.
Corrosion resistant, high strength.

NICKEL SILVER 12% 738.
M-279; 70 Cu, 12 Ni, 18 Zn.
Ann: 48,000-58,000 TS; 12,000-30,000 YS; 35-45 El.
Silver color; for coinage.

NICKEL SILVER 12% 762.
M-279; 59 Cu, 12 Ni, 29 Zn.
Ann: 57,000-75,000 TS; 21,000-51,000 YS; 35-50 El.
Cold rolled: 65,000-125,000 TS; 36,000-115,000 YS; 1-40 El.
High strength corrosion fatigue resistant, for relay springs, contacts, connectors, clips.

NICKEL SILVER 12%-766.
M-8; 31.25 Zn, 12 Ni, 0.25 Mn, bal Cu.
Hard: 97,000 TS; 65,000 YS; 5 El; 176 Brin.
Soft: 60,000 TS; 24,000 YS; 40 El; 80 Brin.
For springs; corrosion resistant.

NICKEL SILVER 13%-776.
M-8; 43.25 Cu, 43.6 Zn, 13 Ni, 0.15 Mn.
Extruded: 60,000 TS; 20,000 YS; 20 El.
For architectural decoration; warm silver color.

NICKEL SILVER 14%.
60-56 Cu, 26-28 Zn, 14 Ni.
For ornaments, plated ware; corrosion resistant.

NICKEL SILVER 15%.
64-57 Cu, 21-28 Zn, 15 Ni.
For ornaments, plated ware; corrosion resistant.

NICKEL SILVER 15% 767.
M-8; 56.5 Cu, 28.25 Zn, 15 Ni, 0.25 Mn.
Hard Sheet: 100,000 TS; 75,000 YS; 4 El; B 93 Rock.
Soft Sheet: 60,000 TS; 24,000 YS; 40 El; B 50 Rock.
For architectural panel and trim, nameplates, dials, musical instruments, watch cases.
Corrosion resistant, high strength.

NICKEL SILVER 18%.
65-55 Cu, 17-27 Zn, 18 Ni.
For ornaments, plated ware; corrosion resistant.

NICKEL SILVER 18%-719.
M-8; 17 Zn, 18 Ni, 0.25 Mn, bal Cu.
Soft: 58,000 TS; 40 El.
Hard: 70,000 TS.
For hardware, marine trim; corrosion resistant.

NICKEL SILVER 18% 735.
M-8, M-279; 72 Cu, 18 Ni, 10 Zn.
Ann: 48,000-56,000 TS; 14,000-24,000 YS; 34-44 El.
Cold rolled: 57,000-90,000 TS; 28,000-80,000 YS; 1-25 El.
Good deep drawing white brass; for ferrules, ink cartridges, condenser cans, instruments.

NICKEL SILVER 18% 752.
M-8, M-279; 65 Cu, 18 Ni, 17 Zn.
Ann: 51,000-61,000 TS; 16,000-29,000 YS; 35-42 El.
Cold rolled: 59,000-104,00 TS; 22,000-96,000 YS; 1-35 El.
Silver color, good formability.
For flatware, holloware, musical instruments, coins.

NICKEL SILVER 18% 770.
M-279; 55 Cu, 18 Ni, 27 Zn.
Ann: 61,000-75,000 TS; 22,000-40,000 YS; 35-48 El.
Cold rolled: 72,000-126,000 TS; 40,000-116,000 YS; 2-35 El.
High strength spring nickel silver.
For relay springs, contact springs, diaphragms.

NICKEL SILVER 18%-770.
M-8; 27 Zn, 18 Ni, 0.25 Mn, bal Cu.
Soft: 60,000 TS; 22,000 YS; 45 El.
Hard: 80,000 TS; 60,000 YS; 20 El.
For springs, hardware; corrosion resistant.

NICKEL SILVER 18%-7641.
M-8; 20.25 Zn, 18 Ni, 0.25 Mn, bal Cu.
Hard: 80,000 TS; 6,000 YS; 8 El; 137 Brin.
Soft: 60,000 TS; 25,000 YS; 35 El; 83 Brin.
For tubes; corrosion resistant.

NICKEL SILVER 18% (A) COPPER ALLOY NO. 752.
M-33; 65 Cu, 18 Ni, 17 Zn.
Strip, Ann: 58,000 TS; 25,000 YS; 40 El; 45 Rock B.
Strip, Hard: 85,000 TS; 74,000 YS; 3 El; 87 Rock B.
For tableware, hollow ware, base for silver plated parts, welding rod.
ASTM: B122 Alloy 2, B151 Alloy A, B206 Alloy A; CDA 752.

NICKEL SILVER 20%.
64-53 Cu, 16-27 Zn, 20 Ni.
For ornaments, plated ware; corrosion resistant.

NICKEL SILVER 25%.
55 Cu, 25 Ni, 20 Zn.
For ornaments, plated ware; corrosion resistant.

NICKEL SILVER 30%.
65-47 Cu, 30 Ni, 5-23 Zn.
For ornaments, plated ware; corrosion resistant.

NICKEL SILVER 55-18.
M-8, M-33, M-103; 53.5-56.5 Cu, 27 Zn, 16.5-19.5 Ni.
Soft: 60,000 TS; 27,000 YS; 40 El.
Hard: 100,000 TS; 85,000 YS; 3 El.
For optical goods, springs, resistance wire; corrosion resistant, good workability.

NICKEL SILVER 65-10.
M-8, M-33, M-103; 63.5-68.5 Cu, 25 Zn, 9-11 Ni.
1/4-H temper: 65,000 TS; 45,000 YS; 25 El.
1/2-H temper: 73,000 TS; 60,000 YS; 12 El.
H temper: 86,000 TS; 75,000 YS; 4 El.
For hardware, optical parts, holloware; corrosion resistant, good formability.

NICKEL SILVER 65-10 now CDA C78800.

NICKEL SILVER 65-12 now CDA C75700.

NICKEL SILVER 65-12.
M-8, M-33, M-103; 63.5-66.5 Cu, 23 Zn, 11-13 Ni.
1/4 H-temper: 65,000 TS; 45,000 YS; 23 El.
1/2 H-temper: 73,000 TS; 60,000 YS; 11 El.
H-temper: 85,000 TS; 75,000 YS; 4 El.
For fasteners, hardware, optical parts; corrosion resistant, good formability.

NICKEL SILVER 65-15.
M-8, M-33, M-103; 63.5-66.5 Cu, 14-16 Ni, 20 Zn.
1/4 H-temper: 65,000 TS; 49,000 YS; 21 El.
1/2 H-temper: 74,000 TS; 62,000 YS; 10 El.
H-temper: 85,000 TS; 75,000 YS; 3 El.
For optical goods, jewelry, hardware; corrosion resistant, good workability.

NICKEL SILVER 65-15 now CDA C75400.

NICKEL SILVER 65-18 now CDA C75200.

NICKEL SILVER 65-18.
M-8, M-33, M-103; 63.0-66.5 Cu, 17 Zn, 16.5-19.5 Ni.
Soft: 58,000 TS; 25,000 YS; 40 El.
1/2 H-temper: 74,000 TS; 62,000 YS; 8 El.
Hard: 85,000 TS; 74,000 YS; 3 El.
For hardware, optical goods, hollowware; corrosion resistant, good workability.

NICKEL SILVER 548.
M-141; 48 Cu, 9.5 Ni, bal Zn.
For braze welding rods; corrosion resistant.

NICKEL SILVER 565-18% (A).
M-141; 65 Cu, 18 Ni, bal Zn.
Annealed: 60,000 TS; 32 El.
Drawn: 85,000 TS; 3 El.
For hardware, hollowware, jewelry; corrosion resistant.

NICKEL SILVER 567-10%.
M-141; 65 Cu, 10 Ni, bal Zn.
Annealed: 50,000 TS; 45 El.
Drawn: 90,000 TS; 3 El.
For hardware, hollowware, jewelry; corrosion resistant.

NICKEL SILVER 828.
M-8; 40 Zn, 10.2 Ni, 0.15 Si, 0.02 P, bal Cu.
For welding rod; low fuming, nickel silver, M.P. 1690°F.

NICKEL SILVER CASTING.
70-56 Cu, 13-20 Ni, 5.6-24 Zn, 0-4 Sn, 0-3.5 Pb.
For plumbing fixtures, fittings; corrosion resistant.

NICKEL SILVER, ROLLING.
49 Cu, 39 Zn, 12 Ni.
For ornamental sheets and rolled parts; corrosion resistant.

NICKEL SILVER, TURNING.
66-59 Cu, 22-29 Zn, 12 Ni, 0-5 Pb.
For screws, bolts; machining grade.

NICKEL SILVER WELDING ROD.
M-1770; 48 Cu, 10 Ni, 42 Zn.
Rod for welding nickel silver alloys.
CDA 773.

NICKEL SPECIAL.
M-789; 99.4 Ni, 0.2-0.6 Fe.
For electrical equipment and instruments.
Resistance alloy. Max working temperature to 150°C.

NICKEL STAYBOLT.
M-435; 0.08 C, 0.25 Mn, 2.25 Ni, bal Fe.
For boiler staybolts; tough.

NICKEL STEEL.
M-538; Ni bonded to steel.
For floor plates, reflectors, hardware, stampings; resists heat to 1050°F; Ni bonded to steel.

NICKEL-TUNGSTEN.
75-50 W, 25-50 Ni.
For chemical apparatus; resistant to acids.

NICKEL URANIUM STEEL.
M-U.S.; 0.3-0.4 Ni, 0.2-0.4 U, 0.2-0.8 C, bal Fe.
183,000-209,000 TS; 159,000-175,000 YS; 8.5-9.0 El; 34-31 RA.
For general engineering applications; corrosion resistant.

NICKELVAC 90 now NICKELVAC N90.

NICKELVAC 600.
M-1490; 0.07 C, 15.5 Cr, 0.3 Ti, 0.2 Al, 9.0 Fe, 0.5 Mn, bal Ni.
For high temperature equipment.

NICKELVAC 625.
M-1490; 0.05 C, 21.5 Cr, 9.0 Mo, 0.1 Ti, 0.1 Al, 3.8 Cb + Ta, bal Ni.

NICKELVAC-700 now ALLVAC 700.

NICKELVAC-751 now NICKELVAC X751.

NICKELVAC-901.
M-1490; 0.06 C, 12.5 Cr, 6.1 Mo, 43 Ni, 3 Ti, 0.015 B, bal Fe.
At R.T. 175,000 TS; 130,000 YS; 14 El.
At 1400°F: 105,000 TS; 92,000 YS; 19 El.
For aircraft and gas turbine components, rotors and compressor discs, fasteners, bolts.
High heat and corrosion resistant. High creep and rupture strength.

NICKELVAC A-286.
M-1490; 0.06 C, 14.75 Cr, 1.25 Mo, 25.5 Ni, 2.1 Ti, 0.2 Al, 1.5 Mn, bal Fe.
At R.T. 146,000 TS; 105,00 YS; 25 El.
At 1400°F: 64,000 TS; 62,000 YS; 19 El.
For jet engine and supercharger parts, turbine wheels and blades, after-burners, bolting.
Austenitic, high heat and corosion resistant.

NICKELVAC B replaced by NICKELVAC H-B.

NICKELVAC-C now NICKELVAC H-C.

NICKELVAC C-263.
M-1490; 0.07 C, 20.0 Cr, 20.0 Co, 5.8 Mo, 2.1 Ti, 0.5 Al, 0.005 B, 0.3 Si, 0.4 Mn, 0.06 Zr, bal Ni.
For gas turbine and heat engine components.

NICKELVAC H-B.
M-1490; 0.10 C, 0.8 Mn, 0.7 Si, 0.6 Cr, 2.5 Co, 28 Mo, 5 Fe, 0.3 V, bal Ni.
Annealed: 127,000 TS; 56,000 YS; 52 El.
At 1200°F: 75,000 TS; 42,000 YS; 13 El.
For chemical and oil refinery equipment, valves, pumps, bolts, gas turbine components.
Acid resistant, austenitic, tough.
Good creep and high temperature properties.

NICKELVAC H-B2.
M-1490; 0.01 C, 26.5 Mo, 0.1 Ti, 0.1 Al, 0.75 Mn, 1.0 Fe, bal Ni.
Corrosion resistant alloy.

NICKELVAC H-C.
M-1490; 0.04 C, 15.5 Cr, 15.5 Mo, 0.1 Ti, 0.1 Al, 6.5 Fe, 3.5 W, 0. Mn, 0.8 Si, 0.2 V, bal Ni.
Corrosion and heat resistant; for chemical and oil refinery equipment.

NICKELVAC H-C276.
M-1490; 0.01 C, 15.5 Cr, 15.5 Mo, 0.1 Ti, 0.1 Al, 6.5 Fe, 3.8 W, 0. Mn, 0.2 V, bal Ni.
For chemical and oil refinery equipment.

NICKELVAC HN.
M-1490; 0.06 C, 7.0 Cr, 16.5 Mo, 0.1 Ti, 0.1 Al, 2.5 Fe, bal Ni.
For chemical plant equipment.

NICKELVAC H-W.
M-1490; 0.02 C, 5.0 Cr, 24.2 Mo, 0.1 Ti, 0.1 Al, 0.003 B, 5.8 Fe, 0.3 V, bal Ni.
For gas turbine and heat engine components.

NICKELVAC H-X.
M-1490; 0.10 C, 21.5 Cr, 1.5 Co, 9.0 Mo, 0.1 Ti, 0.1 Al, 19.0 Fe, bal Ni.
For gas turbines and heat engine components.

NICKELVAC N replaced by NICKELVAC HN.

NICKELVAC-N now NICKELVAC HN.

NICKELVAC N-80A.
M-1490; 0.05 C, 20.0 Cr, 2.4 Ti, 1.3 Al, bal Ni.
High temperature alloy.

NICKELVAC N-90.
M-1490; 0.07 C, 20.0 Cr, 16.0 Co, 2.5 Ti, 1.5 Al, bal Ni.
High temperature alloy.

NICKELVAC-W now NICKELVAC HW.

NICKELVAC W-722.
M-1490; 0.06 C, 15.5 Cr, 2.4 Ti, 0.7 Al, 8.0 Fe, bal Ni.

NICKELVAC-X now NICKELVAC HX.

NICKELVAC X-750.
M-1490; 0.06 C, 15.5 Cr, 2.55 Ti, 0.75 Al, 8.0 Fe, 1 Cb, bal Ni.
At R.T.: 162,000 TS; 92,000 YS; 24 El.
At 1400°F: 70,000 TS; 62,000 YS; 9 El.
For gas turbine blades and wheels, springs, turbochargers, afterburners, fasteners.
Age hardenable. High heat and corrosion resistance.

NICKELVAC X-751.
M-1490; 0.06 C, 16.0 Cr, 2.5 Ti, 1.2 Al, 8.5 Fe, 1.0 Cb, 0.007 B, bal Ni.
For exhaust valves.

NICKEL VANADIUM STEEL.
3-3.0 Ni, 0.1-0.45 V, 0.36 C, 0.2-0.4 Si, bal Fe.
For machinery parts, gears, pinions, shafts, crankshafts; tough, shock resistant.

NICKEL WELD FILLER.
M-108; 70 Cu, 30 Ni.
For brazing.

NICKEL WELDING ELECTRODE 141.
M-1499; 96.0 Ni, 0.03 C, 0.30 Mn, 0.05 Fe, 0.005 S, 0.60 Si, 0.03 Cu, 0.25 Al, 2.5 Ti.
Electrode for shielded metal-arc welding of Nickel 200 and Nickel 201, welding clad side of nickel-clad steel, joining nickel to steel, overlaying on steel.
AWS A 5.11 Class ENi-1; ASME SFB 5.11.

NICKEL-ZIRCONIUM.
86 Ni, 6 Si, 1.5 Zr, 0.1 C.
For chemical apparatus; corrosion resistant.

NICKEL ZIRCONIUM STEEL.
3 Ni, 0.24 Zr, 0.4 C, 2.4 Si, bal Fe.
For crankshafts, axles, machinery parts; oil hardening.

NICKEND 2.
M-677; C, 2.25 Ni, bal Fe.
Welded: 80,000-90,000 TS; 65,000-75,000 YS; 30-20 El; 60-45 RA.
For welding electrodes; for steel, low H^2.

NICKEND 3.
M-677; 0.06 C, 0.8 Mn, 0.3 Si, 3.4 Ni, bal Fe.
For welding electrodes; class E8016-C2, for welding 3 1/4 nickel steel.

NICKOLITE.
60 Cu, 3 Sn, 6 Pb, 20 Ni, 11 Zn.
For typewriter parts, door knobs; nickel silver.

NICKRALEX K5.
M-1687; 3 Al, 14 Ni, bal Cu.
Nickel-Aluminium-Copper alloy; wrought or cast. hardness 205-275 Brin.
AFNOR UN 14 A2.

NICKRALEX KC1.
M-1687; 3 Al, 16 Ni, 2 Cr, bal Cu.
Copper-Nickel alloy; Wrought; hardness 250-310 Brin.

NICKREL.
M-1687; 67 Ni, 1.5 Fe, 1.5 Mn, bal Cu.
Nickel-Copper alloy; wrought; work-hardenable; hardness 100-200 Brin.
ASTM B164-70.

NICKRELK.
M-1687; 3 Al, 66 Ni, 1.5 Fe, 1.5 Mn, bal Cu.
Nickel-Copper alloy; wrought; hardness: 240-290 Brin.

NICKROFRY 345.
M-1302; 0.2 max C, 22-24 Cr, 12-15 Ni, 2 max Mn, 1 max Si, bal Fe.
Annealed: 85,000-95,000 TS; 40,000-50,000 YS; 45-55 El; 150-185 Brin.
For heat treating boxes, oil refinery and chemical plant equipment; Type 309; austenitic, heat resistant.

NICKROTHERM.
M-72; C, Cr, Ni, Si, bal Fe.
For chemical plant equipment; corrosion and heat resistant.

NICKROTHERM "F.F".
M-72; 0.2 C, 1.5 Si, 18 Cr, bal Fe.
For heating muffles, autoclaves, roasting furnaces, rabbles, boilers, grates, recuperators, dampers; heat and corrosion resistant.

NICKROTHERM "N.C.T".
M-72; 0.2 C, 19 Cr, 9 Ni, bal Fe.
For heating muffles, autoclaves, roasting furnaces, rabbles, boilers, grates, recuperators, dampers; heat and corrosion resistant.

NICLOY-36.
M-447; 36 Ni, bal Fe.
Annealed: 65,000 TS; 40,000 YS; 35 El; B 70 Rock.
Cold Worked: 105,000 TS; 95,000 YS; 8 El; 217 Brin.
For hairsprings, time pieces, precision instruments, bourdon tubes, bimetals.
Low coefficient of expansion. Constant modulus. Corrosion resistant.

NIC-MOTAL.
M-554; Pb, Sb, Ni, bal Sn.
For bearings; Babbitt metal.

NICO.
M-542; 0.2 C, 0.5 Cu, 0.8 Ni, bal Fe.
For corrugated culverts; rust resistant.

NICO.
M-685; 0.1 C, bal Fe.
For cast iron welding electrodes; ductile.

NICO NO. 1.
M-Eng; up to 23 Sb, 4-5 Sn, 2-3 Ni, bal Pb.
36 Brin.
For bearings; White Metal.

NICO NO. 2.
M-Eng; 10 Sb, 10 Sn, 1 Ni, bal Pb.
29 Brin.
For bearings; White Metal.

NICOL.
M-1392; 10 Al, 5 Ni, 5 Fe, 1.5 Mn, bal Cu.
Cast: 93,000 TS; 40,000 YS; 20 El; 137 Brin.
For ship propellers; heat treatable, aluminum bronze, tough.

NICOLOY 3 1/2 ETC see **B&W NICOLOY 3 1/2 ETC.**

NICOL-ROD NO. 44.
M-1096; Cu, Ni.
Welded: 221 Brin.
For welding rod; for malleable iron.

NICOL-ROD NO. 99.
M-1096; Ni.
For welding rod for cast iron; special iron coating.

NICORO.
M-1493; 35 Au, 62 Cu, 3 Ni.
For brazing Kovar, copper, nickel and steel.
M.P. 1832-1886°F. Corrosion resistant.

NICORROS.
M-297; 63 min Ni, 0.5-2.5 Fe, 2 max Mn, bal Cu.
For nuclear technology, manufacture and processing of mineral products.

NICORROS-AL.
M-297; 63 min Ni, 27-34 Cu, 0.5-2.0 Fe, 1.5 max Mn, 2.0-4.0 Al.
For corrosion resistant components for scrapers, machine shafts and electrical elements.

NICORROS LC.
M-297; 63 min Ni, 0.5-2.5 Fe, 2 max Mn, 0.04 max C, bal Cu.
For nuclear technology, manufacture and processing of mineral products.

NICOSEAL see **KOVAR.**

NICOSEAL see **CARPENTER KOVAR.**

NICRAL "A".
M-93; 0.1 Ni, 0.5 Cr, 0.5 Cu, 0.5 Mg, bal Al.
Annealed: 20,000 TS; 10,000 YS; 24 El; 40 Brin.
Heat treated: 46,000 TS; 41,000 YS; 8 El; 120 Brin.
For light Al alloy parts; high strength and workability.

NICRAL AD.
M-93; Al alloy.
For light alloy parts.

NICRAL AD018.
M-93; Al alloy.
For rivets.

NICRAL AD2012.
M-93; Al alloy.
For airplane wing ribs, bus body panels; sheets.

NICRAL AD-8412.
M-93; Al alloy.
For light alloy parts.

NICRAL AO.
M-93; Al alloy.
For rolled trim, shells.

NICRAL "B".
M-93; 0.5 Ni, 0.25 Cr, 0.25 Cu, 0.25 Mg, bal Al.
Annealed: 17,000 TS; 6,000 YS; 27 El.
Heat treated: 38,000 TS; 32,000 YS; 12 El.
For light Al alloy parts; for stamping and forming.

NICRAL BO.
M-93; Al alloy.
For bezels, reflectors, switch covers; deep drawing temper.

NICRAL C.
M-1488; 0.07 C, 33.5 Ni, 21 Cr, 0.35 Ti, 0.30 Al, bal Fe.
High temperature corrosion resistant alloy.
For heat exchangers, furnace parts.
AFNOR 5 NC 35 20; W.Nr. 2.4856; Similar to INCOLOY 800.

NICRAL "C".
M-93; 0.5 Ni, 0.5 Cr, 0.7 Cu, 0.2 Mg, bal Al.
For light alloy parts.

NICRAL C (SIRIUS 35).
M-1488; 0.20 max C, 18 Cr, 37 Ni, bal Fe.
Wrought, annealed: 590 N/mm² min UTS.
Austenitic stainless for high temperature operation, as furnace parts.
AFNOR Z12NC37-18; AISI 330.

NICRAL D (SIRIUS 3).
M-1488; 0.25 max C, 25 Cr, 20 Ni, 2 Si, bal Fe.
Wrought, annealed: 540 N/mm² min UTS.
Austenitic stainless for high temperature operation.
AFNOR Z12CNS 25-20; AISI 310.

NICRAL E.
M-771; 0.35 C, 30 Cr, bal Fe.
Rolled: 85,000 TS; 58,000 YS; 20 El.
For carburizing boxes, furnace parts; Type 416; resists S.

NICRAL "FM".
M-93; Mg, Ni, Cr, bal Al.
Extruded: 22,000 TS; 12,000 YS; 22 El.
Heat treated: 41,000 TS; 36,000 YS; 11 El.
For structural and extruded architectural sections; corrosion resistant.

NICRAL FM09.
M-93; Al alloy.
For sash assemblies; extruded.

NICRAL H (SIRIUS 345).
M-1488; 0.20 max C, 23 Cr, 13.5 Ni, bal Fe.
Wrought, annealed: 540 N/mm² min UTS.
Austenitic stainless for high temperature operation.
AFNOR Z15CN24.13; AISI 309.

NICRALIUM D2018.
M-93; 0.45 Cu, 0.48 Mg, 0.19 Mn, 0.42 Fe, 0.3 Si, 1 Ni, 0.2 Cr, 0.17 Mo, bal Al.
For light alloy parts.

NICRAL K 25.
M-1488; 0.03 C, 41 Ni, 21 Cr, 3 Mo, 2 Cu, 0.90 Ti, bal Fe.
Corrosion resistant alloy; for use with phosphoric acid and other chemicals.
AFNOR NC 21 Fe DU; INCOLOY 825.

NICRAL T.
M-1488; 0.12 C, 17.0 Cr, 13.0 Ni, 3.0 W, Ti, bal Fe.
Water quenched: 540 N/mm² min UTS; 225 N/mm² min YS; 40 El.
AFNOR Z 12 CNWT 17-13 B.

NICRAL "X".
M-93; 1.0 Ni, 0.5 Cr, 1.0 Cu, 0.5 Mg, bal Al.
Extruded: 35,000 TS; 24,000 YS; 16 El.
Heat treated: 53,000 TS; 43,000 YS; 15.5 El.
For light Al alloy parts, extruded structural shapes; formerly known as Super-Hyblum.

NICRAL XD018.
M-93; Al alloy.
For structural shapes.

NICRAL Z.
M-1488; 0.04 C, 15.5 Cr, 10 max Fe, bal Ni.
High temperature alloy.
For furnace muffles, heat exchanger tubing, jet engine parts.
AFNOR NC 15 Fe; W.Nr. 2.4640; similar to INCONEL 600.

NICRFE.
M-U.S.; 0.08 C, 0.65 Mn, 0.015 max P, 0.65 Si, 17 Cr, 2 max Cb, 0. max Cu, 8 Fe, bal Ni.
(Inactive - for reference only).
Similar to Inconel 604.

NI-CR-MO ABRASION RESISTANT STEEL.
M-604; 0.30 C, 0.30 Mn, 0.25 Si, 3.0 Ni, 1.0 Cr, 0.30 Mo, bal Fe.
W.Q.+T.: 400 Brin.
Will take mild forming and welding.
For heavy impact and abrasion; chutes, wear plates, dump trucks.

NICRO 31.
M-111a; 0.15 C, 0.55 Mn, 0.65 Cr, 1.25 Ni, bal Fe.
For gears, crankshafts, camshafts; case hardened.

NICRO 33.
M-111a; 0.12 C, 0.55 Mn, 0.75 Cr, 3.0 Ni, bal Fe.
For gears, crankshafts, camshafts; case hardened.

NICRO 34.
M-111a; 0.12 C, 0.55 Mn, 1.25 Cr, 4.5 Ni, bal Fe.
For gears, crankshafts, camshafts; case hardened.

NICRO 63.
M-111; 0.3 C, 0.6 Mn 0.75 Cr, 3.5 Ni, bal Fe.
For gears, shafts, bolts; shock resistant.

NICRO 64.
M-111; 0.3 C, 0.65 N, 1.25 Cr, 4.25 Ni, bal Fe.
For gears, shafts, bolts; shock resistant.

NICRO 71.
M-111; 0.37 C, 0.6 Mn, 0.7 Cr, 1.5 Ni, bal Fe.
For gears, shafts; tough.

NICRO 81.
M-111; 0.38 C, 0.8 Mn, 0.8 Cr, 1.25 Ni, bal Fe.
For gears, shafts; tough.

NICRO 82.
M-111; 0.38 C, 0.55 Mn, 1.15 Cr, 2.6 Ni, bal Fe.
For gears, shafts; oil hardening.

NICRO 632.
M-111; 0.3 C, 0.65 Mn, 1 Cr, 3.25 Ni, 0.25 Mo, bal Fe.
For gears, shafts, bolts; shock resistant.

NICRO 642.
M-111; 0.3 C, 0.5 Mn, 1.25 Cr, 4.25 Ni, 0.25 Mo, bal Fe.
For gears, shafts, bolts; shock resistant.

NICROBRAZ 10.
M-963; 0.10 max C, 11 P, bal Ni.
For heat resistant brazing alloy; M.P. 1610°F.
BNi-6.

NICROBRAZ 30.
M-963; 0.10 max C, 10 Si, 19 Cr, bal Ni.
For heat resistant brazing alloy; M.P. 1975-2075°F, oxidation resistant to 1800°F.
BNi-5.

NICROBRAZ 35.
M-963; 19.5 Cr, 9.5 Mn, 9.8 Si, bal Ni.
Brazing alloy for heat resistant joints; M.P. 1975-2025°F.
B50T50.

NICROBRAZ 50.
M-963; 14.0 Cr, 10.0 P, bal Ni.
Melt: 1630°F: Braze: 1800°-1950°F.
Low melting, free flowing for atmosphere brazing honeycomb structures; used for joining 400 series stainless parts.
For solubility with base metals; usable with stainless steel cutlery. AW B Ni-7

NICROBRAZ 51.
M-963; 25 Cr, 10 P, bal Ni.
Braze temp.: 1800-2000°F; 980-1095°C.
For build-up or brazing joints on stainless and other alloys for improved corrosion and oxidation resistance.

NICROBRAZ 65.
M-963; 23.0 Mn, 4.5 Cu, 7.0 Si, bal Ni.
Melt: 1800-1850°F; Braze: 1850°-2000°F.
Low melting, free flowing for atmosphere brazing of ferrous, nickel and cobalt alloys. BNi-8.

NICOBRAZ-120.
M-963; 13.5 Cr, 3.5 B, 4.5 Si, 4.5 Fe, 0.8 C, bal Ni.
For heat resistant brazing alloy; M.P. 1760-1875°F.

NICROBRAZ 125.
M-963; 0.7 C, 14.0 Cr, 4.5 Si, 4.5 Fe, 3.0 B, bal Ni.
Melt range: 1780°F-1900°F.
Suggested brazing temp.: 2150°F.
For atmosphere brazing stainless steels, nickel and cobalt base high temp. alloys.
Good strength and corrosion resistance.
AMS 4775; AWS B Ni-1.

NICROBRAZ 130.
M-963; 0.06 max C, 4.5 Si, 1.8-3.5 B, bal Ni.
For heat resistant brazing alloy; M.P. 1800-1900°F, free flowing. A 4778. BNi-3.

NICROBRAZ 135.
M-963; 0.06 max C, 2.0 B, 3.5 Si, bal Ni.
Melt: 1810°-1935°F; Brazing: 1950°-2150°F.
For atmosphere brazing stainless and high temperature alloys; wide brazing range; machinable joints, Corrosion resistant.
AMS 4779; AWS B Ni-4.

NICROBRAZ 150.
M-963; 0.15 max C, 15 Cr, 3.5 B, bal Ni.
For heat resistant brazing alloy; M.P. 1930°F, corrosion resistant.

NICROBRAZ 160.
M-963; 11.0 Cr, 2.25 B, 3.5 Si, 3.5 Fe, 0.5 C, bal Ni.
M.P. 1780-2120°F.
For brazing stainless steels and nickel alloys; bridges, wide joints, corrosion and heat resistant.

NICROBRAZ 170.
M-963; 0.55 C, 12 Cr, 2.5 B, 3.5 Si, 3.5 Fe, 16.0 W, bal Ni.
High temperature brazing alloy.
Braze temp.: 2100-2200°F.
For brazing base metals containing cobalt, tungsten and molybdenum for high strength joints at high temprature.

NICROBRAZ 171.
M-963; 0.40 C, 10.0 Cr, 3.5 Si, 12.0 W, 3.5 Fe, 2.5 B, bal Ni.
Melt: 1780°-2000°F; Brazing: 2100°2200°F.
For atmosphere brazing of high temperature alloys containing cobalt, tungsten and molybdenum.
Corrosion resistant; good high temp. strength.

NICROBRAZ 200.
M-963; 7.0 Cr, 4.5 Si, 6.0 W, 3.0 Fe, 3.2 B, bal Ni.
Melt: 1790°-1900°F; Braze: 1950-2150°F.
For atomsphere brazing on hardenable base metals; high creep and stress rupture strength.

NICROBRAZ 210.
M-963; 0.40 C, 19.0 Cr, 17.0 Ni, 8.0 Si, 4.0 W, 0.8 B, bal Co.
Melt: 2025°-2100°F; Braze: 2100°2250°F.
For atmosphere brazing of cobalt base alloys and similar metals; high temperature strength and low base metal penetration. BCo-1; AMS 4783.

NICROBRAZ 300.
M-963; 0.80 C, 21.0 Cr, 3.0 Si, 17.0 Ni, 10.0 W, 3.25 B, bal Co.
Melt: 1900°-2050°F; Braze: 2150°-2250°F.
For atmosphere brazing of T joints or wide gap joints where maximum strength is needed.

NICROBRAZ 1351.
M-963; 3.0 Si, 1.5 B, bal Ni.
Braze temp: 2000-2150°F; 1095-1175°C.
Modified NICROBRAZE 135.

NICROBRAZ 3001.
M-963; 11.5 Cr, 6.0 Si, bal Ni.
Braze temp: 2225-2250°F; 1220-1230°C.
Heat and corrosion resistant brazing alloy; J-8101.
Meets brazing process spec. P50T9D.

NICROBRAZ 3002.
M-963; 15.0 Cr, 8.0 Si, bal Ni.
Braze temp: 2150-2200°F; 1175-1205°C.
For brazing thin gauge honeycomb.
B50TF143 (J-8102).

NICROBRAZ 3003.
M-963; 17.0 Cr, 9.5 Si, 0.10 B, bal Ni.
Braze temp: 2100-2150°F; 1150-1175°C.
Similar to NICROBRAZ 30 but with improved flow. B50TF142 (J-8103).

NICROBRAZ 3004.
M-963; 11.5 Cr, 7.0 Si, 0.40 B, bal Ni.
Braze temp: 2125-2150°F; 1165-1175°C.
Modified NICROBRAZ 30; J-8104.
Meets brazing process spec. P50T9K.

NICROBRAZ 3005.
M-963; 13.0 Cr, 8.0 Si, 0.35 B, bal Ni.
Braze temp: 2125-2150°F; 1165-1175°C.
Modified NICROBRAZE 30; J-8105.
Meets brazing process spec. P50T9L.

NICROBRAZE 5040.
M-96; 5.0 Cr, 2.1 Si, 0.6 Fe, 4.0 P, 1.2 B, bal Ni.
Braze temp: 1850-2100°F; 1010-1150°C.
For brazing thin walled and delicate structures where heavy and ductile fillets are desired.

NICROBRAZE 5060.
M-963; 8.0 Cr, 1.4 Si, 0.4 Fe, 6.0 P, 0.8 B, bal Ni.
Braze temp: 1850-2050°F; 1010-1120°C.
For brazing thin walled and delicate structures where heavy and ductile fillets are desired.

NICROBRAZE 5075.
M-963; 10.0 Cr, 0.9 Si, 0.25 Fe, 7.5 P, 0.5 B, bal Fe.
Braze temp: 1800-2000°F; 980-1095°C.
For brazing thin walled and delicate structures where heavy and ductile fillets are desired.

NICROBRAZE STANDARD.
M-963; 13.5 Cr, 3.5 B, 4.5 Si, 4.5 Fe, 0.8 C, bal Ni.
For brazing stainless steel and nickel base alloys; melt range 1790-1900°F.

NICROBRAZ LC.
M-963; 0.06 max C, 15.0 Cr, 3.0 Fe, 4.5 Si, 3.0 B, bal Ni.
Melt: 1780-1975°F; Braze Temp: 1950-2200°F.
For high strength, heat resistant joints on stainless and super alloys. AMS 4776; BNi-1a.

NICROBRAZ L. M.
M-963; 0.06 max C, 7.0 Cr, 3.1 B, 4.5 Si, 3.0 Fe, bal Ni.
Heat resistant brazing alloy; brazing range 1850-2150°F; corrosion and heat resistant. AMS 4777; BNi-2.

NICROBRAZ LM01.
M-963; 3.0 Cr, 3.0 Si, 1.3 Fe, 1.9 B, bal Ni.
Braze temp: 2000-2110°F; 1095-1155°C.
Modified NICROBRAZE LM; J-8201.
Meets brazing process spec. P50T9A.

NICROBRAZ LM02.
M-963; 2.25 Cr, 2.75 Si, 1.0 Fe, 1.7 B, bal Ni.
Braze temp: 2000-2110°F; 1095-1155°C.
Modified NICROBRAZE LM; J-8202.
Meets brazing process spec. P50T9B.

NICROBRAZ LM03.
M-963; 2.5 Cr, 3.0 Si, 1.1 Fe, 1.8 B, bal Ni.
Braze temp: 2000-2110°F; 1095-1155°C.
Modified NICROBRAZE LM; J-8203.
Meets brazing process spec. P50T9C.

NICROBRAZE LM04.
M-963; 2.3 Cr, 3.0 Si, 1.2 Fe, 1.8 B, bal Ni.
Braze temp: 2000-2120°F; 1095-1160°C.
Modified NICROBRAZE LM; J-8204.
Meets brazing process spec. P50T9E.

NICROBRAZE LM05.
M-963; 2.7 Cr, 3.0 Si, 1.2 Fe, 1.8 B, bal Ni.
Braze temp: 2000-2120°F; 1095-1160°C.
Modified NICROBRAZE LM; J-8205.
Meets brazing process spec. P50T9F.

NICROBRAZE LM06.
M-963; 2.3 Cr, 2.4 Si, 1.0 Fe, 1.5 B, bal Ni.
Braze temp: 2000-2125°F; 1095-1165°C.
Modified NICROBRAZE LM; J-8206.
Meets brazing process spec. P5oT9G.

NICROBRAZE LM07.
M-963; 5.6 Cr, 3.6 Si, 2.4 Fe, 2.5 B, bal Ni.
Braze temp: 2050-2120°F; 1120-1160°C.
Modified NICROBRAZE LM; J-8207.
Meets brazing process spec. P50T9N.

NICROBRAZ LM08.
M-963; 3.8 Cr, 3.5 Si, 1.7 Fe, 2.2 B, bal Ni.
Braze temp: 2000-2110°F; 1095-1155°C.
Modified NICROBRAZE LM; J-8208.

NICROCOAT 1.
M-963; 13.5 Cr, 73.2 Ni, 13.3 other.
Fusing temp: 2000-2150°F.
For build-up on Inconel 600, Hastelloy X, 300 series stainless steel.

NICROCOAT 2.
M-963; 4.5 Si, 92.35 Ni, 3.15 other.
Fusing temp: 1900-1950°F.
For build-up of thin-walled structures of 300 series stainless.

NICROCOAT 3.
M-963; 13.0 Cr, 76.85 Ni, 10.15 other.
Fusing temp: 1800-1900°F.
For build-up on mild steel and low carbon steel as mufflers, heat exchangers.

NICROCOAT 4.
M-963; 15.0 Cr, 81.5 Ni, 3.5 other.
Fusing temp: 1950-2050°F.
For build-up on jet engine combustion chambers, etc., made of Hastelloy X, Inconel 600, 304 stainless.

NICROCOAT 5.
M-963; 7.0 Cr, 3.0 Fe, 82.5 Ni, 7.5 other.
Fusing temp: 1900-2000°F.
For build-up on 300 series stainless for operation up to 1500°F.

NICROCOAT 6.
M-963; 7.0 Cr, 6.0 W, 77.0 Ni, 10.0 other.
Fusing temp: up to 1950°F.
For build-up on jet engine components containing W and Mo.

NICROCOAT 7.
M-963; 3.5 Si, 94.5 Ni, 2.0 other.
Fusing temp: 1950 and up.
For build-up on 300 series stainless and low alloy steel for petro-chemical equipment, glass molds.

NICROCOAT 8.
M-963; 10 Cr, 83.0 Ni, 7.0 other.
Fusing temp: 2000°F and up.
For build-up on 300 series stainless for operation up to 1500°F.

NICROCOAT 9.
M-963; 19.0 Cr, 10.0 Si, 66.0 Ni, 5.0 other.
Fusing temp: 2050°F and up.
For build-up on turbine blades, vanes and other jet engine parts.

NICROCOAT 610.
M-963; 13.5 Cr, 76.4 Ni, 10.1 TiB_2.
Fusing treat: 15-30 min. 2000-2050°F, vacuum.
For build-up on high temperature alloys to resist heat and erosion.

NICROCOAT 620.
M-963; 17.9 Cr, 4.9 Si, 68.6 Ni, 8.6 $TiSi_2$.
Fusing treat: 15-30 min. 2080-2100°F, vacuum.
For build-up on high temperature alloys to resist heat and erosion.

NICROCOAT 630.
M-963; 16.9 Cr, 5.3 Si, 64.4 Ni, 6.7 $TiSi_2$, 6.7 TiN.
Fusing treat: 15-30 min. 2080-2100°F, vacuum.
For build-up on high temperature alloys for maximum resistance to oxidation and erosion.

NICROCOAT 700.
M-963; 13.0 Cr, 72.0 Ni, 10.0 TiB_2, 5.0 Al_2O_3.
Fusing treat: 15-30 min. 2100-2150°F, vacuum.
For build-up on high temperature alloys for maximum resistance to sulfidation.

NICRODIE.
M-35; 0.72 C, 0.60 Mn, 0.35 Si, 0.25 Mo, 0.90 Cr, 1.50 Ni, bal Fe.
Oil hardening cold work tool steel.
For forming dies, shear blades, brake dies.
AISI L6.

NICROFER 3220.
M-297; 30 Ni, 20 Cr, Ti, bal Fe.
For nuclear technology, heat-resisting components, furnace and steam boiler constructions.

NICROFER 3220 H.
M-297; 30 Ni, 20 Cr, 0.06-0.1 C, Ti, bal Fe.
For nuclear technology and heat-resistng components, for furnaces & boilers.

NICROFER 3220 LC.
M-297; 30 Ni, 20 Cr, 0.025 max C, Ti, bal Fe.
For nuclear technology and heat-resistant components, for furnaces and boilers.

NICROFER 3620 NB.
M-297; 37 Ni, 20 Cr, 3.5 Cu, 2.5 Mo, bal Fe.
For sulphuric acid plants.

NICROFER 3718.
M-297; 35 Ni, 16 Cr, Si, bal Fe.
Structural components for production and utilization of exothermic gas for bright annealing.

NICROFER 4221.
M-297; 42 Ni, 21 Cr, 2 Cu, 3 Mo, Ti, bal Fe.
For chemical engineering, heat exchangers, pumps and heating coils, cellulose industry, sulphuric acid pickling baths.

NICROFER 6023.
M-297; 60 Ni, 23 Cr, 1.5 Al, bal Fe.
For industrial furnaces, petrochemical engineering, detoxication of exhaust gases.

NICROFER 7216.
M-297; 75 Ni, 16 Cr, 8 Fe.
For nuclear technology and chemical engineering.

NICROFER 7216 LC.
M-297; 75 Ni, 16 Cr, 8 Fe, 0.025 max C.
For nuclear technology and chemical engineering.

NICROFER 7520.
M-297; 72 Ni, 20 Cr, 5 Fe.
For industrial furnaces and heat resisting components in the aircraft industry.

NICROGAP 106.
M-963; 0.35 Si, 0.2 B, bal Ni.
For use with copper or silver alloys or NICROBRAZ 50 to fill wide joints in mild and low alloy steels.

NICROGAP 108.
M-963; 7.0 Fe, 15.0 Cr, 0.75 Si, 0.2 B, bal Ni.
For use with NICROBRAZ 150 to fill wide joints in stainless and heat resistant alloys.

NICROGAP 112.
M-963; 8.0 Ni, 18.0 Cr, 1.25 Si, 0.2 B, bal Fe.
For use with NICROBRAZ 150 to fill wide joints in 300 series stainless.

NICROGAP 114.
M-963; 12.0 Cr, 1.25 Si, 0.2 B, bal Fe.
For use with NICROBRAZ 150 to fill wide joints in 400 series stainless.

NICROMA 17.
M-1315; 0.22 C, 17 Cr, 1.5 Ni, bal Fe.
Annealed: 85,000 TS; 50,000 YS; 30 El; 55 RA; 180 Brin.
For furnace parts, grids, conveyors; corrosion and heat resistant.

NICROMAN.
M-275; 1.0 C, 1.25 Cr, bal Fe.
For blanking and forming dies; shock resistant, oil hardened; Type L-6.

NICROMANG see AMSCO NICROMANG.

NICROMAZ C1.
M-1488; 0.04 max C, 2.0 max Mn, 1.0 max Si, 24.0-27.0 Ni, 19.0-22.0 Cr, 4.0-4.8 Mo, 2.0-3.0 Cu, bal Fe.
Cast, ann: 450 N/mm² min UTS.
For chemical industries; phosphoric and sulphuric acids, petrochemicals.
AFNOR Z3 NCDU 25.20 M; similar to ACI CN-7M.

NICROMAZ CM.
M-1488; 0.08 max C, 1.5 max Mn, 1.5 max Si, 24.0-27.0 Ni, 19.0-22.0 Cr, 4.0-4.8 Mo, 1.5-2.5 Cu, 0.5-0.8 Nb, bal Fe.
Cast, ann: 460 N/mm² min UTS.
For chemical industry; phosphoric and sulphuric acids, petro-chemicals.
AFNOR Z6 NCDU Nb 25.20 M; similar to ACI CN-7M.

NICROMAZ SP B.
M-1488; 0.20 max C, 2.0 max Mn, 4.5-5.5 Si, 39.0-42.0 Ni, 13.0-15.0 Cr, 4.0-6.0 Mo, bal Fe.
Cast, ann: 180-228 Brin.
Resists corrosion by concentrated sulphuric acid up to 150°C.
AFNOR Z8 NCDS 40.14.5.5 M.

NICROMINA-2.
M-1763; 0.12 max C, 2.0 max Mn, 1.0 max Si, 17.0-19.0 Cr, 10.5-13.0 Ni, bal Fe.
Austenitic stainless steel.
AISI 305.

NICROMINA-2 + MO.
M-1763; 0.08 max C, 2.0 max Mn, 1.0 max Si, 16.0-18.0 Cr, 10.0-14.0 Ni, 2.0-3.0 Mo, bal Fe.
Austenitic stainless steel, for chemical equipment.
AISI 316.

NICROMINA-2 + NB.
M-1763; 0.08 max C, 2.0 max Mn, 1.0 max Si, 17.0-19.0 Cr, 9.0-13.0 Ni, Nb-Ta = 10 X C min, bal Fe.
Stabilized austenitic stainless steel.
AISI 347.

NICROMINA-2 + TI.
M-1763; 0.08 max C, 2.0 max Mn, 1.0 max Si, 17.0-19.0 Cr, 9.0-12.0 Ni, Ti = 5 X C min, bal Fe.
Stabilized austenitic stainless steel.
AISI 321.

NICROSIL.
M-USSR; 1.7-2.0 C, 1.8-3.0 Cr, 16-20 Ni, 0.8-1.3 Mn, bal Fe.
For furnace equipment; corrosion resistant, cast iron.

NICROSILAL.
M-86; 1.8 C, 6.0 Si, 18 Ni, 2 Cr, 1 N, bal Fe.
36,000 TS; 3 El; 110-250 Brin.
For furnace grids, generator boxes, annealing pots; heat resistant; does not glow at high temperature; austenitic.

NICROSILAL.
M-494, M-1485; 1.8 C, 5 Si, 18 Ni, 2 Cr, bal Fe.
Cast: 36,000 TS; 140-200 Brin.
For furnace grids, heat treating boxes, corrosion resistant, non-magnetic.

NICROSILAL 5 CR.
M-England; 2.1 C, 4.9 Si, 0.9 Mn, 22.8 Ni, 4.6 Cr, 0.05 P, bal Fe.
For grids; heat resistant, cast iron.

NICROTUNG.
M-118, M-1591; 0.08-0.13 C, 0.02-0.08 B, 0.02-0.08 Zr, 11-13 Cr, 9-11 Co, 7.0-8.5 W, 3.75-4.75 Al, 3.75-4.75 Ti, bal Ni.
Cast: 130,000 TS; 120,000 YS; 5 El; 9 RA; 380 Brin.
At 1800°F: 67,000 TS; 52,000 YS; 6 El.
For missile and rocket engine components, gas turbines; heat resistant to 2000°F.

NICUAR 813.
M-677; 0.03 C, 0.8 Mn, 0.1 Si, 30.5 Ni, 0.6 Fe, 0.4 Ti, 0.2 Al, bal Cu.
For bare cupro-nickel welding wire, class RCuNi.

NICUEND.
M-677; 70 Cu, 30 Ni.
For welding electrodes; for Cu-Ni alloys.

NICUITE.
M-214; 10 Sn, 3.5 Ni, 2.5 Zn, bal Cu.
Cast: 50,000 TS; 25,000 YS; 24 El; 70 Brin.
Heat treated: 95,000 TS; 75,000 YS; 6 El; 160 Brin.
For bearings, worm gears, slippers, slides; heat treatable, heavy loads, slow speeds.

NICULOY.
M-1265; Cu, Ni, Fe.
For master alloy; for making alloys.

NICUSIL-3.
M-1493; 71.15 Ag, 28.10 Cu, 0.75 Ni.
M.P. 1436-1463°F.
For brazing.
Good wetting and filleting.

NICU STEEL.
M-Eng.; 2.2 Ni, 0.6 Mn, 0.5 Cu, 0.3 C, bal Fe.
For machinery parts, structural work; oil hardening.

NI-FE 30.
M-210; 30 Ni, bal Fe.
For shunts in electrical equipment.
Temperature compensating alloy.
Low coefficient of expansion, high permeability.

NIFER.
M-926; low C-steel clad on both sides with Ni.
For electron tubes.

NI-HARD.
M-460; 4.5 Ni, 1.5 Cr, 1.5 Mn, 2.75-3.75 C, bal Fe.
Chilled: 600-750 Brin.
For abrasion resisting applications, crushers, rolls, feeder vanes; cast iron, abrasion resisting, formerly "Super Manga Iron."

NI-HARD.
M-494, M-1485; 3 C, 3.5-4.5 Ni, 1.5 Cr, bal Fe.
Cast: 40,000 TS; 500-650 Brin.
For castings; abrasion resisting.

NI-HARD.
M-752; 3.2-3.8 C, 0.3-0.8 Si, 0.3-0.8 Mn, 3.5-5.5 Ni, 1.5-2.5 Cr, bal Fe.
Abrasion resistant casting; 500 min Brin.

NI-HARD, TYPE 1.
M-68; 4.2-4.7 Ni, 3.0-3.60 C, 1.4-2.5 Cr, 0.5 Si, 0.3-0.7 Mn, bal Fe
Cast: 40,000-50,000 TS; 550-650 Brin.
Chilled: 80,000 TS; 700 Brin.
For die casting pots, pump plungers, roller bearing races, chilled rolls and liners; corrosion and wear resistant; tough.

NI-HARD TYPE 2.
M-68; 2.9 max T.C.; 0.5-0.8 Si, 0.4-0.6 Mn, 4.25-4.75 Ni, 1.4-2.5 Cr, bal Fe.
Cast: 45,000-55,000 TS; 525-565 Brin.
For ball mill liners, crushers, pump parts; abrasion and corrosion resistant.

NI-HARD TYPE 3.
M-68; 1.0-1.6 C, 0.4-0.7 Si, 0.4-0.7 Mn, 4.0-4.75 Ni, 1.4-1.6 Cr, bal Fe.
Sand cast: 75,000-125,000 TS; 350-500 Brin.
Chill cast: 90,000-140,000 TS; 300-600 Brin.
For jaw crushers, grinding balls, mill liners, slurry pumps. Wear resistant, tough.

NI-HARD TYPE 4.
M-68; 3.0-3.6 TC, 0.1 max G.C., 1.5-2.0 Si, 0.4-0.7 Mn, 5.5-6.5 Ni, 7.0-10.0 Cr, bal Fe.
Sand cast: 75,000-85,000 TS; 5000-6000 Tr.S., 0.08-0.11 def., 580 min D.P.N.
For jaw crushers, hammer mills, grinding bals, mill liners.
Wear resistant martensitic white iron with improved resistance to fracture under impact.

NI-HARD TYPE 4.
M-1460; 2.5-3.6 C, 1.3 max Mn, 1.0-2.2 Si, 5.0-7.0 Ni, 7.0-11.0 Cr, 1.0 max Mo, 0.10 max P, 0.15 max S, bal Fe.
Sand cast: 550 min Brin.
Chill cast: 600 min Brin.
For large section cast iron with greater abrasion resistance.
ASTM A 532 Cl. I Type D.

NI-HARD TYPE A.
M-1460; 3.0-3.6 C, 1.3 Mn, 0.8 max Si, 3.3-3.5 Ni, 1.4-4.0 Cr, 1.0 max Mo, 0.30 max P, 0.15 max S, bal Fe.
Sand cast: 550 min Brin. Chill cast: 600 min Brin.
Abrasion resistant cast iron - for medium size sections.
ASTM A 532 Cl. I Type A.

NI-HARD TYPE D.
M-1460; 2.5-3.6 C, 1.3 max Mn, 1.0-2.2 Si, 5.0-7.0 Ni, 7.0-11.0 Cr, 1.0 max Mo, 0.10 max P, 0.15 max S, bal Fe.
Sand cast: 550 min Brin.
Chill cast: 600 min Brin.
Abrasion resistant cast iron - for tough large section size castings.
ASTM A 532 Cl. I Type D.

NI-HARD TYPE N.
M-1460; 3.0-3.6 C, 1.3 Mn, 0.8 max Si, 3.3-3.5 Ni, 1.4-4.0 Cr, 1.0 max Mo, 0.30 max P, 0.15 max S, bal Fe.
Sand cast: 550 min Brin.
Chill cast: 600 min Brin.
Abrasion resistant cast iron - for larger size sections.
ASTM A 532 Cl. I Type A.

NI-HTC.
M-897; 0.04 Fe, 99.9 Ni.
For heating elements; useful operating temperature to 100°C.

NIKA.
M-Japan; Pb, Sn, bal Cu.
For bearings; tough and non-corrosive.

NIKE-3.
M-1449; 0.10 max C, 18 Cr, 8 Ni, bal Fe.
Annealed: 80,000 TS; 35,000 YS; 55 El; 75 RA; 150 Brin.
For chemical plant equipment, tanks, mixers, filters; Type 302; stainless, austenitic.

NIKE-B.
M-1449; 0.08 max C, 18 Cr, 8 Ni, Ti, bal Fe.
Annealed: 85,000 TS; 35,000 YS; 55 El; 65 RA; 150 Brin.
For welded chemical plant equipment, tanks, vessels; Type 321; stainless, austenitic.

NIKE-M.
M-1449; 0.12 max C, 18 Cr, 10 Ni, 2.5 Mo, bal Fe.
Annealed: 85,000 TS; 35,000 YS; 50 El; 65 RA; 160 Brin.
For acid resistant chemical plant equipment, mixers, tanks; Type 316; stainless, austenitic.

NIKON.
M-1277; 3.3 C, 0.7 Mn, 2.2 Si, bal Fe.
For machinery castings; cast iron.

NIKRO M.
M-115; 0.55-0.75 C, 1.4 Ni, 0.85 Cr, 0.4 Mo, bal Fe.
Heat treated: 290,000 TS.
For collets, races, arbors, gears.

NIKROME 285.
M-240; 0.41 C, 0.75 Mn, 1.90 Ni, 0.82 Cr, 0.26 Mo, bal Fe.
H.T.: 130 ksi TS; 110 ksi YS; 15 El; 45 RA; 285-341 Brin.
Pre-heat treated steel; for shafts, axles.

NIKROME 302.
M-240; 0.41 C, 0.75 Mn, 1.90 Ni, 0.82 Cr, 0.26 Mo, bal Fe.
H.T.: 140 ksi TS; 120 ksi YS; 14 El; 42 RA; 302-363 Brin.
Pre-heat treated steel; for shafts, axles.

NIKROME M see NIKROME 285.

NIKROTHAL see NIKROTHAL 40.

NIKROTHAL-4.
M-1150; 35 Ni, 20 Cr, bal Fe.
Elect. resistivity 600 ohms/cmf.
For unsupported heating elements, resistors and potentiometers.
Max. operating temperature 2000°F.

NIKROTHAL-6.
M-1150; 16 Cr, 60 Ni, bal Fe.
Annealed: 90,000-200,000 TS; 40,000-50,000 YS; 22-33 El.
For resistors, heating elements; high electrical resistance, max operating temperature 1800°F.

NIKROTHAL-8.
M-70, M-1150; 20 Cr, bal Ni.
Annealed: 95,000-200,000 TS; 40,000-50,000 YS; 22-33 El.
For resistors, heating elements; high electrical resistance, max operating temperature 2100°F.

NIKROTHAL 20.
M-70; 25 Cr, 55 Fe, 20 Ni.
Ann: 95,000-114,000 psi TS; 50,000 psi YS.
For resistors, heating elements, max operating temp. 1050°C.

NIKROTHAL 40.
M-70; 20 Cr, 45 Fe, 35 Ni.
Ann: 95,000-114,000 psi TS; 50,000 psi YS.
For resistors, heating elements, max operating temp. 1100°C.

NIKROTHAL 60.
M-70; 15 Cr, 25 Fe, 60 Ni.
Ann: 95,000-114,000 psi TS; 50,000 psi YS.
For resistors, heating elements, max operating temp. 1125°C.

NIKROTHAL 80.
M-70; 20 Cr, 80 Ni.
Ann: 95,000-114,000 psi TS; 50,000 psi YS.
For resistors, heating elements, max operating temperature 1200°C.

NIKROTHAL LX.
M-70, M-1150; 75 Ni, 20 Cr, plus Al, Si, Mn, + Cu.
Ann: 135,000-200,00 psi TS; 115,000-175,000 psi YS; 3-25 El.
Wire for wire-wound resistors. Temperature coefficient of resistance 0 ± 5 ppm for deg. C.

NIKROTHAL-LX.
M-70, M-1150; 20 Cr, 5 additives, 75 Ni.
Wire: 155,000-200,00 TS; 5-25 El. Elect. resistivity 800 ohms/cmf.
For rheostats, resistors, potentiometers, electric appliances, precision and vitreous enamel resistors.
High electrical resistivity, low temperature coefficient. Max. operating temperature 572°F.

NILGRO 36.
M-487; 36 Ni, bal Fe.
For thermostats; controlled expansion.

NILGRO 42.
M-487; 42 Ni, bal Fe.
For thermostats; controlled expansion.

NILO 36.
M-121, M-1499; 36 Ni, bal Fe.
Annealed: 70,000 TS; 24,000 YS; 36 El; 68 RA; 143 Brin.
For thermostats, glass-to-metal seals; controlled expansion.

NILO 40.
M-86; 40 Ni, 60 Fe.
For glass to metal seals, thermostats; controlled coefficient of expansion.

NILO 42.
M-121, M-1499; 42 Ni, bal Fe.
Annealed: 78,000 TS; 41,000 YS; 45 El; 73 RA; 143 Brin.
For thermostats, bimetals; controlled low expansion.

NILO 48.
M-121; 48 Ni, bal Fe.
For thermostats, glass to metal seals, instrument components. Controlled and low coefficient of expansion.

NILO 50.
M-86; 50 Ni, 50 Fe.
For glass to metal seals, thermostats; conrolled coefficient of expansion.

NILO 51.
M-121; 0.05 C, 51 Ni, 48 Fe.
Controlled expansion alloy; for sealing to soft glasses, e.g. in reed relay switch blades. AFNOR Fe-N 505.

NILO 475.
M-86, M-121; 47 Ni, 5 Cr, bal Fe.
For glass to metal seals, themostats; controlled coefficient of expansion.

NILO ALLOY 45.
M-121; 0.05 C, 45.0 Ni, 54 Fe.
Magnetic alloy with high saturation flux density.
Initial permeability at 20°C: 6000-10,000.
For rocking armature telephone receivers.

NILO K.
M-121; 29 Ni, 17 Co, bal Fe.
Annealed: 77,200 TS; 54,000 YS; 41 El; 72 RA; 170 Brin.
For glass to metal seals; expansion of medium hard glass.

NILOMAG 36.
M-121; 47 Ni, 3 Mo, bal Fe.
For transformers, alternators; soft magnet for low power losses.

NILOMAG 45.
M-121; 64.3 Ni, 34.7 Fe, 1.0 Mo.
For magnetic amplifiers, pulse transformers; soft magnet, domain oriented.

NILOMAG 77.
M-121; 77 Ni, 14 Fe, 5 Cu, 4 Mo.
For cores for telephone transformers, magnetic amplifiers; soft magnet, high permeability.

NILOY 48.
48 Ni, bal Fe.
Annealed: 50,000 TS; 33,000 YS; 46 El; 67 RA; 143 Brin.
For thermostats, bimetals; controlled low expansion.

NILSTAIN 302.
M-61; 18-20 Cr, 8-10 Ni, 0.20 C, 2 Mn, bal Fe.
Rolled: 85,000 TS; 35,000 YS; 5 El; 9 RA; 140 Brin.
For stainless parts, springs; corrosion resistant.

NILSTAIN 304.
M-61; 8-10 Ni, 18-20 Cr, 0.08 max C, 2.0 max Mn, bal Fe.
Rolled: 85,000-250,000 TS; 35,000-200,000 YS; 60-5 El; 70-55 RA; 140-455 Brin.
For springs, screws, bolts; stainless wire.

NILSTAIN 305.
M-61; 0.12 max C, 17-19 Cr, 10-13 Ni, bal Fe.
Annealed: 85,000 TS; 38,000 YS; 50 El; 150 Brin.
For cold headed parts; corrosion resistant, austenitic.

NILSTAIN 330.
M-61; 33-36 Ni, 14-16 Cr, 0.25 max C, bal Fe.
Cold drawn: 75,000 TS; 15-25 El.
For heat resisting parts; heat resistant.

NILSTAIN 430.
M-61; 14-18 Cr, 0.12 max C, bal Fe.
Cold drawn: 75,000-100,000 TS; 50,000-90,000 YS; 35-10 El; 75-60 RA; 140-250 Brin.
For instruments; corrosion resistant.

NILVAR.
M-44; 36 Ni, bal Fe.
Wrought: 150,000-70,000 TS.
For bimetals, measuring tapes, length standards, thermostats; low expansion alloy; maximum operating temperature 200°C.

NIMAR 700.
M-677; 0.02 C, 0.06 Mn, 0.06 Si, 18.2 Ni, 4.8 Mo, 0.6 Ti, 0.1 Al, 9.8 Co, bal Fe.
For bare welding wire; maraging steel grade 250, for gas metal-arc welding.

NIMAR 701.
M-677; 0.02 C, 0.06 Mn, 0.06 Si, 18.2 Ni, 4.8 Mo, 1.0 Ti, 0.1 Al, 9.8 Co, bal Fe.
For bare welding wire; maraging steel grade 250, for submerged arc welding.

NIMAG 60.
M-1626; 76.0 Ni, 0.10 Cu, 7.20 Fe, 0.20 Mn, 0.20 Si, 15.8 Cr.
Resists oxidation at high temperatures; resists corrosion. Non-magnetic.

NIMAG 100.
M-1626; 99.5 Ni, 0.06 C, 0.15 Fe, 0.25 Mn, others 0.105.
Good mechanical properties; corrosion resistant.

NIMAG 101.
M-1626; 99.5 Ni, 0.01 C, 0.15 Fe, 0.20 Mn, others 0.15.
Good mechanical properties. Corrosion resistant.

NIMAG 104.
M-1626; 95.2 Ni, 4.5 Co, others 0.30.
Good magnetostrictive properties.

NIMAG 105.
  M-1626; 99.5 Ni, 0.06 C, 0.20 Mn, 0.10 Fe, 0.04 Mg, 0.02 Ti, others 0.08.
  Easily formed or drawn. Low oxidation at high temperatures. High Curie temperature.

NIMAG 111.
  M-1625; 95.0 Ni, 4.75 Mn, others 0.25.
  Good strength and base hardness. Resistant to sulphur compounds at high temperature.

NIMAG 120.
  M-1626; 99.5 Ni, 0.06 C, 0.12 Mn, 0.04 Mg, 0.02 Ti.

NIMAG 130.
  M-1626; 99.5 Ni, 0.09 C, 0.10 Mn, 0.06 Mg, 0.05 Fe, 0.01 Cu, 0.03 Si, 0.003 Ti.

NIMAG 133.
  M-1626; 99.5 Ni, 0.09 C, 0.18 Mn, 0.07 Mg, 0.003 Ti, 0.05 Fe, 0.03 Cu, 0.03 Si.
  For vacuum tube anodes, oxide coated and cold cathodes, structural tube parts.

NIMARK see **CARPENTER NIMARK.**

NIMBUS DC 02.
  M-1377; 1.35 C, 4.2 Cr, 12.0 W, 3.75 V, 0.8 Mo, 5.0 Co, bal Fe.
  High speed steel; for finishing; good abrasion resistance.
  W. Nr. 1.3202.

NIMBUS DC 07.
  M-1377; 1.25 C, 4.2 Cr, 10.5 W, 3.25 V, 3.75 Mo, 10.5 Co, bal Fe.
  High speed steel; for lathe tools; good red hardness and abrasion resistance.
  W. Nr. 1.3207.

NIMBUS DC 55.
  M-1377; 0.80 C, 4.2 Cr, 18.0 W, 1.6 V, 0.7 Mo, 5.0 Co, bal Fe.
  High speed steel; for lathe and planing tools for severe service.
  W. Nr. 1.3255; similar to AISI T4.

NIMBUS DC 65.
  M-1377; 0.76 C, 4.2 Cr, 18.0 W, 1.6 V, 0.7 Mo, 9.0 Co, bal Fe.
  High speed steel; for heavy duty lathe tools.
  W. Nr. 1.3265; similar to AISI T5.

NIMBUS DD 16.
  M-1377; 0.82 C, 4.2 Cr, 9.0 W, 1.6 V, 0.8 Mo, bal Fe.
  High speed steel; for milling cutters, twist drills, end mills.
  W. Nr. 1.3316.

NIMBUS DD 18.
  M-1377; 0.95 C, 4.2 Cr, 12.0 W, 2.5 V, 0.9 Mo, bal Fe.
  High speed steel; for lathe tools, taps, threading cutters.
  W. Nr. 1.3318.

NIMBUS DD 33.
  M-1377; 1.0 C, 4.2 Cr, 2.7 Mo, 2.8 W, 2.4 V, bal Fe.
  High speed steel; for twist drills, milling cutters, broaches, reamers.
  W. Nr. 1.3333.

NIMBUS DD 43.
  M-1377; 0.88 C, 4.2 Cr, 5.0 Mo, 6.3 W, 1.8 V, bal Fe.
  High speed steel; for milling cutters, rough cutting lathe tools, broaches.
  W. Nr. 1.3343; similar to AISI M2.

NIMBUS DD 55.
  M-1377; 0.75 C, 4.2 Cr, 18.0 W, 1.1 V, bal Fe.
  High steed steel; for lathe and planer tools, taps, broaches, twist drills.
  W. Nr. 1.3355; AISI T1.

NIMEND.
  M-677; 0.1-0.15 C, 14-18 Cr, 4-5 W, 15-18 Mo, 5.3 Fe, bal Ni.
  Welded: 20-25 Rc; Aged 50 hrs at 1475°F: 38-40 Rc.
  For hardfacing electrodes; similar to Hastelloy C.

NIMO.
  M-40; 0.5-0.6 C, 0.3-0.5 Cr, 0.10-0.15 V, 0.4-0.5 Mo, 2.5-2.8 Ni, bal Fe.
  For tools, dies; oil hardening.

NIMOCAST 75.
  M-1485; 0.1 C, 20 Cr, 5 max Fe, 0.4 Ti, 0.2 Al, bal Ni.
  Cast: 75,000 TS; 29,400 YS; 34.8 El; 164 Brin.
  Heat treated: 81,500 TS; 27,500 YS; 40.5 El; 172 Brin.
  For high temperature applications; high oxidation resistance.

NIMOCAST 80.
  M-121; 0.05 C, 20 Cr, 5 max Fe, 2.4 Ti, 1.2 Al, bal Ni.
  Cast: 72,000 TS; 66,700 YS; 3 El; 253 Brin.
  Heat treated: 111,600 TS; 78,000 YS; 14 El; 270 Brin.
  For high temperature applications; high creep resistance to 750°C.

NIMOCAST 90.
  M-1485; 0.1 C, 16 Co, 5 max Fe, 2.4 Ti, 1.2 Al, 20 Cr, bal Ni.
  Cast: 92,800 TS; 72,700 YS; 8.1 El; 280 Brin.
  Heat treated: 106,000 TS; 79,600 YS; 12.6 El; 291 Brin.
  For high temperature applications; high creep resistance to 870°C.

NIMOCAST 235D.
M-86; 14-17 Cr, 4.5-6.0 Mo, 3.5-5.0 Fe, 3.25-4.0 Al, 2.0-3.0 Ti, 5.6-6.5 Al + Ti, 0.3 max Si, 0.10-0.20 C, 0.10 max Mn, 0.05-0.10 B, bal Ni.
At 1200°F: 83,400 Fatigue St.
At 1650°F: 65,000 Fatigue St.
For jet engine and aerospace equipment and parts.
High heat and corrosion resistance.

NIMOCAST 242.
M-121; 0.3 C, 10 Co, 20 Cr, 03 max Ti, 0.2 max Al, 10 Mo, bal Ni.
Cast: 70,000 TS; 43,000 YS; 8 El; 220 Brin.
For high temperature applications; high resistance to thermal shock.

NIMOCAST 257.
M-1485; 0.8 C, 16 Co, 20 Cr, 2 max Fe, 1.6 Ti, 0.9 Al, bal Ni.
Cast.
For gas turbine engine rings; good properties to 650°C.

NIMOCAST 713.
M-86, M-121; 14 Cr, 4.5 Mo, 1 Ti, 6 Al, 2 Cb + Ta, 1.5 max Fe, bal Ni.
For gas turbine stator and rotor blades.
Creep resistant for service to 1000°C.

NIMOCAST ALLOY 263.
M-121; 0.06 C, 51.0 Ni, 20.0 Cr, 20.0 Co, 5.9 Mo, 2.2 Ti, 0.5 Al.
Vacuum melted cast alloy, - similar to wrought alloy NIMONK alloy 263.

NIMOCAST ALLOY 713LC.
M-121; 0.05 C, 74.0 Ni, 12.0 Cr, 4.5 Mo, 0.7 Ti, 6.0 Al.
At 700°C: 1010 N/mm^2 TS; 775 N/mm^2 YS; 13 El.
Low carbon grade of NIMOCAST alloy 713.
For gas turbine rotor blades, turbine and turbocharger rotors.

NIMOCAST ALLOY 738.
M-121; 0.18 C, 61.0 Ni, 16.0 Cr, 8.5 Co, 1.8 Mo, 3.4 Ti, 3.4 Al, 1.6 Ta, 2.5 W.
At 1000°C: 450 N/mm^2 TS; 325 N/mm^2 YS; 16 El.
Good high temperature creep strength and hot corrosion reisitance.
Similar to IN-738.

NIMOCAST ALLOY 738 LC.
M-121; 0.11 C, 61.0 Ni, 16.0 C, 8.5 Co, 1.8 Mo, 3.4 Ti, 3.4 Al, 1.6 Ta, 2.5 W.
At 700°F; 1010 N/mm^2 TS; 500 N/mm^2 YS; 4 El.
Low carbon version of NIMOCAST alloy 738.

NIMOCAST ALLOY 739.
M-121; 0.15 C, 48.0 Ni, 22.4 Cr, 19.0 Co, 3.7 Ti, 1.9 Al, 1.4 Ta, 2.0 W, 1.0 Nb.
At 700°F: 980 N/mm^2 TS; 590 N/mm^2 YS; 7 El.
Vacuum melted, cast alloy; good high temperature strength and corrosion resistance; for blades and vanes in marine and industrial gas turbines.
Similar to IN-939.

NIMOCAST ALLOY PD21.
M-121; 0.10 C, 73.0 Ni, 6.0 Cr, 2.0 Mo, 6.0 Al, 10.5 W.
At 1000°C: 575 N/mm^2 TS; 375 N/mm^2 YS; 4 El.
High stress rupture properties up to 1050°C.
For stator blades.
Similar to IN-M-21.

NIMOCAST-C 242.
M-86; 0.3 C, 0.3 Si, 0.3 N, 20 Cr, 10 Co, 1 max Fe, 0.3 max Ti, 0.2 max Al, 10 Mo, bal Ni.
Cast: 69,700 TS; 43,000 YS; 8 El; 220 DPN.
For high temperature cast components, turbine nozzle guide vanes.
High resistance to thermal shock.
High corrosion and heat resistant castings.

NIMOCAST-MC 57.
M-86; 0.08 C, 0.4 Si, 0.3 Mn, 20 Cr, 16 Co, 2 max Fe, 1.6 Ti, 0.9 Al, bal Ni.
Ht. Tr.: 100,000 TS; 65,000 YS; 16 El.
For high temperature cast components, gas turbine engine rings, jet engine parts.
Cast high temperature alloy.
Good combination of yield strength and ductility at 650°C. Heat treatable.

NIMOCAST-MC 58.
M-86; 0.2 C, 0.4 Mn, 0.4 Si, 10 Cr, 20 Co, 2 max Fe, 3.7 Ti, 4.8 Al, 5 Mo, bal Ni.
Ht. Tr.: 117,500 TS; 110,00 YS; 4.5 El; 383 DPN.
Cast: 122,00 TS; 114,00 YS; 3.4 El; 375 DPN.
For high temperature cast components, turbine and jet engine components.
Good creep resistance up to 1000°C.
Casting alloy, good castability. Heat treatable.

NIMOCAST PE 10.
M-121, M-86; 20 Cr, 6 Mo, 2.5 W, 6.5 Cb + Ta, bal Ni.
For turbo-charger rotors, gas turbine components, diesel engine pre-combustion chambers.
Creep resistant for service to 870°C, oxidation and corrosion resistant.

NIMOCAST PK 24.
M-86, M-121; 0.18 C, 15 Co, 10 Cr, 3 Mo, 5.2 Ti, B, Zr, 5.6 Al, bal Ni.
For gas turbines, engines, turbine-stator blades, turbocharger rotors.
High temperature characteristics for max. service about 1040°C.

NIMOL.
M-86; 75 cast iron, up to 14 Cr, 14 Ni, 6 Cu. 20,000-27,000 TS; 2.5-1.5 El; 120-240 Brin.
For retorts for fuel carbonization, pump and engine liners, pans for fusing caustic alkalis; superseded by Ni-Resist.

NIMOLOY ALLOY PK37.
M-121; 0.12 C, 60.0 Ni, 18.5 Cr, 17.0 Co, 2.2 Ti, 1.2 Al.
At 700°C: 970 N/mm² TS; 680 N/mm² YS; 12 El.
High strength alloy with good abrasion and shock resistance up to 850°C. For forging press anvils, hot shear blades, other hot working tools.

NIMONIC.
M-86; 20 Cr, 2 Ti, 2 Al, bal Ni.
For high temperature applications; heat resistant.

NIMONIC 58.
M-86; 11 Cr, 5 Mo, 20 Co, 5 Al, 2 Ti, bal Ni.
Cast: 123,000 TS; 114,000 YS; 3.4 El; 375 DPN.
For high temperature castings, jet engine and gas turbine parts.
Heat and corrosion resistant.
High creep and rupture strength.

NIMONIC 75.
M-86, M-121; 0.08-0.15 C, 0.2-0.6 Ti, 18-21 Cr, 1.0 max Si, 1.0 max M max Cu, 5-11 Fe, bal Ni + Co.
Solution treated: 103,000-96,000 TS; 60,000 YS; 50 El; 50 RA.
For combustion chambers, turbines; age-hardenable, good resistance to creep and oxidation.

NIMONIC 75 F.
M-86; 0.08-0.15 C, 0.2-0.6 Ti, 18-21 Cr, 1.0 max Si, 1.0 max Mn, 0.5 max Cu, 5-11 Fe, bal Ni + Co.
For turbines, combustion chambers; heat resistant, age-hardenable, creep and oxidation resistant.

NIMONIC 80.
M-86; 0.04 C, 0.47 Si, 75 Ni, 21 Cr, 0.56 Mn, 2.45 Ti, 0.63 Al.
Heat treated: 147,000 TS; 41 El; 35 RA; 350 Brin.
At 20°C: 132,000 TS; 80,000 YS; 45 El; 36 RA.
At 800°C: 63,000 TS; 53,000 YS; 8 El; 10 RA.
For rotor blades, gas turbine engines; high creep resistance, age-hardenable.

NIMONIC 80A.
M-86, M-121; 0.04 C, 21 Cr, 0.6 Mn, 2.5 Ti, 0.7 Al, bal Ni.
At 20°C: 132,000 TS; 80,000 YS; 45 El; 36 RA.
At 800°C: 62,000 TS; 53,000 YS; 8 El; 10 RA.
For gas turbine blades, and valves; high creep resistant.

NIMONIC 90.
M-86, M-121; 0.1 max C, 0.8-1.8 Al, 1.8-3.0 Ti, 5 max Fe, 15-21 Co, Cr, bal Ni.
At 20°C: 166,000 TS; 101,000 YS; 39 El; 20 RA.
At 600°C: 132,000 TS; 90,000 YS; 26 El; 21 RA.
at 800°C: 85,000 TS; 63,000 YS; 7 El; 4 RA.
For turbine blades, combustion chambers, rotor dies, and valves; age hardened, super heat and creep resistance.

NIMONIC 90 (FILLER METAL).
M-121; 0.1 C, 1.0 Mn, 3.0 Fe, 1.5 Si, 20.0 Cr, 2.5 Ti, 1.5 Al, bal Fe.
Filler metal for gas-shielded arc welding, NIMONiC alloys 80A and 90.
BS 2901-NA-36.

NIMONIC 93.
M-86; 0.10 C, 19.5 Cr, 2.75 Ti, 1.5 Al, 18 Co, 0.3 max Mo, 0.08 Zr, 0.008 B, bal Ni.
Heat treated: 185,000 TS; 120,000 YS; 28 El; 41 RA.
At 600°C: 160,000 TS; 105,000 YS; 24 El; 23 RA.
For jet engine and gas turbine components.
Age-hardenable, heat and oxidation resistant.

NIMONIC 100.
M-86; 11 Cr, 5 Mo, 20 Co, 1.5 Ti, 5 Al, bal Ni.
Heat treated: 320-400 Brin.
For gas turbine rotor blades; creep resistant to 1000°C.

NIMONIC 105.
M-121; 15 Cr, 20 Co, 5 Mo, 4.5 Al, 1.4 Ti, 0.15 C, bal Ni.
At 20°C: 144,000 TS; 116,000 YS; 7 El; 7 RA.
At 800°C: 107,000 TS; 78,500 YS; 16 El; 17 RA.
For gas turbine components, jet engines, combustion chambers. Corrosion and heat resistance.

NIMONIC 110.
M-86; 0.15 C, 15 Cr, 20 Co, 1.75 Ti, 5.75 Al, 5.0 Mo, bal Ni.
Hot Rolled: 180,000 TS; 124,000 YS; 18 El; 17 RA.
At 800°F: 113,00 TS; 84,000 YS; 37 El; 35 RA.
For high temperature applications, aircraft gas turbines, high temperature fasteners, gas turbine blades.
Creep and high oxidation resistance.

NIMONIC 115.
M-121; 0.15 C, 15 Cr, 20 Co, 5 Mo, 4 Al, bal Ni.
For gas turbine engine blades; high creep and rupture strength at high temperature.

NIMONIC 118.
M-86; 0.16 C, 15 Cr, 3.85 Ti, 4.8 Al, 14.9 Co, 3.5 Mo, 0.045 Zr, 0.016 B, 0.7 max Fe, 0.5 max Mn, bal Ni.
Heat treated: 170,000 TS; 125,000 YS; 29 El; 45 RA.
At 600°C: 150,000 TS; 110,000 YS; 25 El; 34 RA.
For gas turbine and jet engine components, missiles, aircraft.
Age-hardenable, heat and oxidation resistant.

NIMONIC 263 (FILLER METAL).
M-121; 0.03 C, 0.4 Mn, 0.75 Fe, 0.25 Si, 20.0 Cr, 5.9 Mo, 2.15 Ti, 0.45 Al, 0.2 Cu, 0.02 Zr, 0.001 B, bal Ni.
Filler metal for gas-shielded arc welding NIMONIC alloy 263 and other high temperature materials.
BS 2901-NA-38.

NIMONIC B.
M-86; 0.1 max C, 1.8-2.7 Ti, 0.8-1.8 Al, 5 max Fe, 15-21 Co, 18-21 Cr, bal Ni.
Rolled: 160,000 TS; 90,000 YS; 39 El; 20 RA.
At 1500°F: 76,000 TS; 56,000 YS; 7 El; 4 RA.
For turbine gland springs; heat resistant, heat treatable.

NIMONIC C.
M-86; 0.5-1.8 Al, 1.8-2.7 Ti, 18-21 Cr, bal Ni.
At 20°C: 134,000 TS; 80,000 YS; 45 El; 36 RA.
At 800°C: 64,000 TS; 54,000 YS; 8 El; 10 RA.
For heat exchangers, valves, valve inserts, gas turbine parts; heat and corrosion resistant.

NIMONIC C75.
M-86; 0.08-0.15 C, 5 max Fe, 0.2-0.6 Ti, 18-21 Cr, 0.5-max Cu, 0.2 Al, bal Ni.
Cast: 74,500 TS; 30,000 YS; 35 El; 164 DPN.
Ht. Tr.: 82,000 TS; 28,000 YS; 41 El; 172 DPN.
For furnace trays, gas turbine parts, combustion chambers, jet engine components.
Resists scaling and thermal shock.
Heat and corrosion resistant.

NIMONIC C242.
M-86; 0.3 C, 20 Cr, 10 Co, 10 Mo, 1 max Fe, bal Ni.
Cast: 70,000 TS; 44,000 YP; 8 El; 220 DPN.
For high temperature castings, nozzle guide vanes, jet engine and gas turbine parts.
Heat and corrosion resistant.
High resistance to creep and thermal shock.

NIMONIC ALLOY 81.
M-121; 0.03 C, 66.0 Ni, 30.0 Cr, 1.8 Ti, 1.4 Al.
At 700°C: 790 N/mm² TS; 500 N/mm² YS; 25 El.
Precipitation hardening alloy; for exhaust valves in internal combustion engines.

NIMONIC ALLOY 86.
M-121; 0.05 C, 64.5 Ni, 25.0 Cr, 10.0 Mo, 0.15 Mn, 0.03 Ce.
At 700°C: 500 N/mm² TS; 260 N/mm² YS; 74 El.
Alloy sheet with good ductility, high creep strength, and resistance to cyclic oxidation up to 1050°C.
For gas turbine components and heat treat equipment.

NIMONIC ALLY 91.
M-121; 0.08 C, 47.5 Ni, 28.5 Cr, 20.0 Co, 2.3 Ti, 1.2 Al.
At 700°C: 945 N/mm² TS; 580 N/mm² YS; 28 El.
Precipitation hardening alloy; improved corrosion resistance to salt and sulphur environments.
For gas turbine blades in engines burning impure fuels.

NIMONIC ALLOY 263.
M-121; 0.06 C, 51.0 Ni, 20.0 Cr, 20.0 Co, 5.9 Mo, 2.0 Ti, 0.5 Al.
At 700°C: 750 N/mm² TS; 460 N/mm² YS; 23 El.
Precipitation hardening, creep-resisting alloy for gas turbine rings and sheet components for use to 850°C.
BS HR10; ANOR NCK 20D.

NIMONIC ALLOY 901.
M-121; 0.04 C, 42.5 Ni, 12.5 Cr, 35.0 Fe, 5.7 Mo, 2.9 Ti, 0.3 Al.
At 700°C: 910 N/mm² TS; 810 N/mm² YS; 12 El.
Precipitaion hardening alloy; maximum service temperature about 600°C. For gas turbine discs and shafts.
AMS 5660; Werkstoff Nr. 2.4662.

NIMONIC ALLOY C263.
M-86; 0.03 C, 20 Cr, 2.15 Ti, 0.45 Al, 20 Co, 5.9 Mo, 0.02 max Zr, max B, 0.007 max S, bal Ni.
Heat treated: 146,000 TS; 90,000 YS; 45 El; 42 RA.
At 600°C: 120,000 TS; 70,000 YS; 44 El; 50 RA.
For jet engine and gas turbine components, missiles.
Age-hardenable, heat and oxidation resistant.

NIMONIC-CB.
M-86; 1.8-2.7 Ti, 0.8-1.8 Al, 5 max Fe, 15-21 Co, 18-21 Cr, bal Ni.
Ht. Tr.: 106,000 TS; 80,000 YS; 13 El; 290 DPN.
For gas turbine rotor blades.
High creep and oxidation resistant. Cast alloy.

NIMONIC C.C.
M-86; 0.10 max C, 0.5-1.8 Al, 5 max Fe, 1.8-2.7 Ti, 18-21 Cr, 2 max Co, bal Ni.
Ht. Tr.: 112,000 TS; 80,000 YS; 14 El; 270 DPN.
For gas engine and jet engine and gas turbine components.
High creep resistance to 900°C.
Cast alloy, oxidation resistant.

NIMONIC CF.
M-86; 0.08-0.15 C, 0.2-0.6 Ti, 5-11 Fe, 8-21 Cr, bal Ni.
For jet engine components; cast alloy.

NIMONIC D.
M-86; 37 Ni, 18 Cr, 2 Si, bal Fe.
At 70°F: 94,000 TS; 48,00 YS; 38 El; 50 RA.
At 2000°F: 10,000 TS; 124 El; 83 RA.
For gas turbine parts, impulse blades; heat resistant.

NIMONIC F.
M-86; 0.08-0.15 C, 0.2-0.6 Ti, 5-11 Fe, 18-21 Cr, bal Ni.
For gas turbine flame tube; oxidation and heat resistant.

NIMONIC MC 57.
M-86; 0.08 C, 16 Co, 20 Cr, 0.9 Al, 1.6 Ti, bal Ni.
Ht. Tr.: 180,000 TS; 65,000 YS; 16 El.
For high temperature castings, gas turbine components.
Heat and corrosion resistant.
Good properties to 650°C.

NIMONIC PE 7.
M-86; 0.1 C, 18 Cr, 5 Mo, 37 Ni, 2.0 max Co, 1.2 Ti, 1.2 Al, bal Fe.
For power generating equipment, gas turbines, steam plant, casings and compressors.
Creep and high temperature resistance. For sevice up to 580°C.

NIMONIC PE 11.
M-86; 18 Cr, 5.2 Mo, 2.3 Ti, 0.8 Al, 38 Ni, bal Fe.
Heat treated: 160,000 TS; 100,00 YS; 21 El; 35 RA.
At 600°C: 140,000 TS; 100,00 YS; 23 El; 34 RA.
For gas turbine thrust reversers, noise suppressors, jet pipes.
Creep resistant. Heat and corrosion resistant; age-hardenable.

NIMONIC PE 13.
M-86, M-121; 0.05-0.15 C, 20.5-23.0 Cr, 17.0-20.0 Fe, 0.5-2.5 Co, 8.0-10.0 Mo, 0.2-1.0 W, bal Ni.
Annealed: 116,500 TS; 51,500 YS; 43 El; 200 VPN.
At 1000°C; 20,200 TS; 13,500 YS; 44 El.
For elevated temperature applications, missile and jet engine components, furnace parts.
High temperature, heat resistant.

NIMONIC PE13 (FILLER METAL).
M-121; 0.1 C, 1.0 Mn, 17.2 Fe, 1.0 Si, 0.02 S, 22.0 Cr, 8.1 Mo, 0.6 W, 0.002 Pb, bal Ni.
Filler metal for gas-shielded arc welding NIMONIC alloy PE13.
BS 2901-NA-40.

NIMONIC PE 16.
M-86, M-121; 16.5 Cr, 2.0 max Co, 3.2 Mo, 1.2 Ti, 1.2 Al, 43.5 Ni, bal Fe.
Heat Treated: 121,000 TS; 65,000 YS; 29 El.
At 600°C: 99,000 TS; 54,000 YS; 30 El.
For gas turbine compressor delivery casings, turbine casings.
Creep and oxidation resistant for service up to 750°C., age-hardenable.

NIMONIC PK 25.
M-86, M-121; 19 C, 18 Co, 4.2 Mo, 3 Ti, 2.7 Al, bal Ni.
Heat Treated: 197,000 TS; 18 El; 22 RA.
At 1400°F: 155,000 TS; 20 El; 28 RA.
For gas turbine blades and other components, bolts, valves, jet engine components.
Creep and heat resistant to 940°C.

NIMONIC PK 31.
M-86; 14 Co, 0.06 C, 20 Cr, 4.5 Mo, 2.3 Ti, 0.6 Al, 5.0 Cb, bal Ni.
Heat Treated: 190,000 TS; 140,000 YS; 25 El; 34 RA.
At 600°C: 160,000 TS; 120,00 YS; 25 El; 34 RA.
For rotor discs in gas turbines, jet engine and missile components.
High tensile strength and creep resistant.
High oxidation and corrosion resistance.

NIMONIC PK 33.
M-86, M-121; 14 Co, 0.05 C, 19 Cr, 7.5 Mo, 2.0 Ti, 2.0 Al, bal Ni.
Heat Treated: 165,000 TS; 95,000 YS; 33 El; 41 RA.
At 600°C: 140,000 TS; 88,000 YS; 32 El; 40 RA.
For engine rotor and sheet components, gas turbine parts, jet pipes for service up to 950°C.
High oxidation and corrosion resistance. Age-hardenable.

NIO-O-NEL ALLOY 825 replaced by **INCOLOY ALLOY 825.**

NIORO.
M-1493; 82 Au, 18 Ni.
For brazing W, Mo, Cu, Ni, and Kovar.
M.P. 950°C. Excellent flow and wetting properties.

NIPERMAG.
M-867; 12 Al, 32 Ni, 0.4 Ti, bal Fe.
Cast: 10,500 TS; 23,000 Tr.S.; C 45 Rock. 5600 Br, 660 Hc, 3400 Bo.
For loud speakers, motors, generators.
Permanent magnet, high permeability.

NIPIGON.
M-435; 0.8 C, 19 W, 2 V, 4 Cr, 1 Mo, 9 Co, bal Fe.
For tools, dies, cutting tools; high speed steel.

NIPPERT ALLOY N 4 see ZIRCONIUM COPPER N-4.

NIPPON EVERSHINING STEEL.
M-716; 0.15 C, 18 Cr, 8 Ni, bal Fe.
For stainless parts; stainless.

NIPPON NST-M1.
M-1651; 0.07-0.12 C, 0.3-0.6 Mn, 0.5 max Si, 12.0-13.5 Cr, 0.4-0.6 Mo, 0.5 max Ni, bal Fe.
Annealed: 75,000 TS; 35,000 YS; 25 El; B 92 Rock.
For table flatware, springs, oil refinery equipment. Corrosion resistant.

NIPPON NST-M2.
M-1651; 0.10-0.20 C, 0.2-0.8 Mn, 0.5 max Si, 11.5-13.0 Cr, 0.60-1.5 Mo, 0.5 max Ni, bal Fe.
Annealed: 75,000 TS; 40,000 YS; 25 El; B 92 Rock.
For springs, table flatware, oil refinery equipment, surgical instruments, valves. Corrosion resistant, hardenable.

NIPURE.
M-61; 0.01 max Si, 0.05 max Fe, 0.02 max Mn, 0.01 max Mg, 0.04 max Cu 0.02 max C, 0.01 max Al, 0.01 max Ti, 0.05 max Cr, 0.01 max Pb, bal Ni.
For electronic components.
Vacuum melted high purity nickel.

NIRANIUM.
28.8 Cr, 64.2 Co, 4.3 Ni, 2 W, 0.1 Si, 0.2 C, 0.7 Al.
At 1000°C: 82,000 TS; 43,000 YS; 0.7 El.
At 700°C: 85,000 TS; 47,000 YS; 1.3 El.

NI-RESIST.
M-494, M-1485; 2.8 C, 14 Ni, 7 Cu, 2 Cr, bal Fe.
Cast: 28,000 TS; 140-200 Brin.
For castings; heat and corrosion resisting.

NIRESIST.
M-Russia; 2.7 C, 3.0-3.4 Cr, 12-15 Ni, 5-8 Cu, 1.5 Si, bal Fe.
Cast: 20,000-35,000 TS; 125-200 Brin.
For heat and corrosion resistant parts; corrosion and heat resistant, cast iron.

NI-RESIST I.
M-1188; 2-3 C, 14 Ni, 2 Cr, 6 Cu, bal Fe.
Cast: 25,000-30,000 TS; 120-180 Brin.
For corrosion and heat resistant castings; stainless cast iron, austenitic.

NI-RESIST II.
M-1188; 3 C, 20 Ni, 2 Cr, 2 Si, bal Fe.
Cast: 25,000-30,000 TS; 120-180 Brin.
For corrosion and heat resistant castings; stainless cast iron, austenitic.

NI-RESIST III.
M-1188; 2.7 C, 30 Ni, 2 Si, bal Fe.
Cast: 25,000-35,000 TS; 110-170 Brin.
For corrosion and heat resistant castings; stainless cast iron, austenitic.

NI-RESIST IV.
M-1188; C, 30 Ni, 3 Si, bal Fe.
Cast: 25,000-35,000 TS; 130-200 Brin.
For food industry equipment; stainless cast iron.

NI-RESIST V.
M-1188; 2.4 C, 36 Ni, 1-2 Si, bal Fe.
Cast: 20,000-25,000 TS; 110-160 Brin.
For heat resistant castings; stainless cast iron, low coefficient of expansion.

NI-RESIST G.
M-500; 2.7 C, Ni, Cu, Cr, bal Fe.
Annealed: 30,000 TS; 150 Brin.
For chemical plant equipment; resists acids, oxidation.

NI-RESIST N.
M-500; 2.7 C, Ni, Cr, bal Fe.
Annealed: 30,000 TS; 150 Brin.
For chemical plant equipment; resists caustic, ammonia.

NI-RESIST TYPE 1.
M-763; 2.5-3.5 C, 1.2 Si, 0.75-1.2 Mn, 12-15 Ni, 1-3 Cr, 5-7 Cu, bal Fe.
Cast: 25,000-30,000 TS; 130-160 Brin.
For corrosion and heat resistant castings; resists corrosion, heat and wear.

NI-RESIST TYPE 1.
M-68; 1.7-2.5 Cr, 1.0-2.5 Si, 1-1.5 Mn, 13-17 Ni, 5-7 Cu, 3 max T.C., bal Fe.
Cast: 25,000-30,000 TS; 2 El; 0 RA; 130-160 Brin.
For pipes, valves, pumps, propellers, hydraulic turbines, oil burners, automotive engine pistons and sleeves; corrosion and heat resisting; austenitic cast iron; non-magnetic.

NI-RESIST TYPE 1-B.
M-68; 3 max T.C., 1.0-2.8 Si, 1.0-1.5 Mn, 13.5-17.5 Ni, 2.5-4.0 Cr, bal Fe.
Cast: 25,000 TS; 1 El; 140-190 Brin.
For pumps, valves, filter presses, impellers, nozzles; non-magnetic, erosion and corrosion resistant cast iron.

NI-RESIST, TYPE 2.
M-68; 18-22 Ni, 2-4 Cr, 2.2-3.0 C, 1-1.5 Mn, 1.0-2.5 Si, bal Fe.
Cast: 25,000-30,000 TS; 2 El; 0 RA; 120-170 Brin.
For heat resistant parts, grids, furnace parts, pumps, pipe; corrosion resisting and heat resisting.

NI-RESIST TYPE 2.
 M-763; 3 max C, 1.7-2.5 Cr, 18-22 Ni, 1-2.5 Si, 0.8-1.5 Mn, bal Fe.
 Cast: 25,000-30,000 TS; 130-160 Brin.
 For corrosion and heat resistant castings; resists heat, corrosion and wear.

NI-RESIST TYPE 2.
 M-994; 3 max C, 1.7-2.5 Cr, 18-22 Ni, 1-2.5 Si, 0.8-1.5 Mn, bal Fe.
 Cast: 25,000-30,000 TS; 130-160 Brin.
 For castings; corrosion and heat resistant.

NI-RESIST TYPE 2A.
 M-994; 2.8 max C, 1.7-2.5 Cr, 18-22 Ni, 1.5-2.7 Si, 0.8-1.5 Mn, bal Fe.
 Cast: 30,000-50,000 TS; 145-190 Brin.
 For castings; corrosion and heat resistant.

NI-RESIST TYPE 2-B.
 M-68; 3 max T.C., 1.0-2.8 Si, 0.8-195 Mn, 18-22 Ni, 2.5-4.0 Cr, bal Fe.
 Cast: 25,000 TS; 1 El; 130-190 Brin.
 For turbocharger casings, manifolds, steam turbine nozzles; non-magnetic, heat resistant cast iron.

NI-RESIST TYPE 2B.
 M-994; 3 max C, 3-6 Cr, 18-22 Ni, 1.2 Si, 0.8-1.5 Mn, bal Fe.
 Cast: 25,000-45,000 TS; 170-250 Brin.
 For castings; corrosion and heat resistant.

NI-RESIST TYPE 3.
 M-994, M-68; 2.75 max C, 2.5-3.5 Cr, 28-32 Ni, 1-2 Si, 0.4-0.8 Mn, bal Fe.
 Cast: 25,000-35,000 TS; 120-150 Brin.
 For heat and corrosion resistant castings; corrosion and heat resistant.

NI-RESIST TYPE 3.
 M-763; 2.75 max C, 2.5-3.5 Cr, 28-32 Ni, 1-2 Si, 0.4-0.8 Mn, bal Fe.
 Cast: 25,000-35,000 TS; 120-150 Brin.
 For corrosion and heat resistant castings; resists heat, corrosion and wear.

NI-RESIST TYPE 4.
 M-763; 2.6 max C, 4.5-5.5 Cr, 29-32 Ni, 5-6 Si, 0.4-0.8 Mn, bal Fe.
 Cast: 25,000-35,000 TS; 150-180 Brin.
 For corrosion and heat resistant castings; resists heat, corrosion and wear.

NI-RESIST TYPE 4.
 M-994, M-68; 2.6 max C, 4.5-5.5 Cr, 29-32 Ni, 5-6 Si, 0.4-0.8 Mn, bal Fe.
 Cast: 25,000-35,000 TS; 150-180 Brin.
 For heat resistant castings, stove stops, cookware; corrosion and heat resistant.

NI-RESIST TYPE 5.
 M-68; 2.4 max T.C., 1-2 Si, 0.4-0.8 Mn, 34-36 Ni, 0.5 max Cu, 0.1 max Cr, bal Fe.
 Cast: 20,000-25,000 TS; 100-125 Brin.
 For gages, glass molds, paper dies, chemical equipment; very low thermal expansion.

NI-RESIST TYPE 5.
 M-994; 2.4 max C, 3 max Cr, 34-36 Ni, 1-2 Si, 0.4-0.8 Mn, bal Fe.
 Cast: 20,000-25,000 TS; 100-140 Brin.
 For heat resistant castings; corrosion and heat resistant.

NI-RESIST TYPE D-2.
 M-68; 3 max T.C., 1.5-3.0 Si, 0.7-1.2 Mn, 18-22 Ni, 1.75-2.75 Cr, bal Fe.
 Cast: 60,000 TS; 30,000 YS; 8-20 El; 140-200 Brin.
 For pumps, valves, pipe fittings, paper rolls; corrosion and heat resistant cast iron.

NI-RESIST TYPE D-2B.
 M-68; 3 max T.C., 1.5-3.0 Si, 0.7-1.2 Mn, 18-22 Ni, 2.7-4.0 Cr, bal Fe.
 Cast: 60,000 TS; 30,000 YS; 7-15 El; 150-210 Brin.
 For impellers, pumps, valves, engine parts; corrosion and heat resistant cast iron.

NI-RESIST TYPE D-2C.
 M-68; 2.9 T.C. 1-3 Si, 1.8-2.4 Mn, 28-32 Ni, bal Fe.
 Cast: 60,000 TS; 30,000 YS; 20-40 El; 120-170 Brin.
 For switch gears, pumps, valves, bearings, seals; corrosion and heat resistant cast iron.

NI-RESIST TYPE D-3.
 M-68; 2.6 T.C., 1.0-2.8 Si, 28-32 Ni, 2.5-3.5 Cr, bal Fe.
 Cast: 55,000 TS; 30,000 YS; 6-20 El; 140-200 Brin.
 For steam turbines, engines, liners, valves, kettles; corrosion and heat resistant cast iron.

NI-RESIST TYPE D-3A.
 M-68; 2.6 T.C., 1.0-2.8 Si, 28-32 Ni, 1.5 Cr, bal Fe.
 Cast: 55,000 TS; 30,000 YS; 10-20 El; 130-190 Brin.
 For liners, bearings, valve guides; corrosion and heat resistant cast iron.

NI-RESIST TYPE D-4.
 M-68; 2.6 T.C., 5-6 Si, 28-32 Ni, 4.5-5.5 Cr, bal Fe.
 Cast: 60,000 TS; 200-270 Brin.
 For cookware, range tops; nonmagnetic, corrosion resistant cast iron.

NI-RESIST TYPE D-5.
M-68; 2.4 max T.C., 1.0-2.8 Si, 34-36 Ni, bal Fe.
Cast: 55,000 TS; 30,000 YS; 20-40 El; 130-280 Brin.
For dies, glass molds, ingot molds; corrosion and heat resistant cast iron.

NI-RESIST TYPE D-5B.
M-68; 2.4 max T.C., 1.0-2.8 Si, 34-36 Ni, 2-3 Cr, bal Fe.
Cast: 55,000 TS; 30,000 YS; 6-15 El; 140-190 Brin.
For dies, glass molds, ingot molds; corrosion and heat resistant cast iron.

NI-ROD FC 55 CORED WIRE.
M-1499; 50.0 Ni, 44.0 Fe, 1.0 C, 4.2 Mn, 0.6 Si.
For automatic and semi-automatic welding of gray, malleable and ductile cast irons.

NIROSTA see also **B + W NIROSTA**.

NIRESULT.
M-84; 1.1 min C, 24-30 Cr, 12-15 Ni, 5-7 Cu, bal Fe.
For heat and corrosion resistant parts; heat and corrosion resistant.

NIREX.
M-44; 70 min Ni, 11-15 Cr, 10 max Fe.
For heat resising parts; stainless, heat resistant.

NI-ROD see **NI-ROD WELDING ELECTRODE**.

NI-ROD 55 WELDING ELECTRODE.
M-1499; 53 Ni, 1.5 C, 0.3 Mn, 45.0 Fe, 0.005 S, 0.50 Si, 0.10 Cu.
Electrode for shielded metal-arc welding of cast and ductile irons; cast irons to wrought alloys.
AWS A 5.15 Class E NiFe-CI.

NI-ROD WELDING ELECTRODE.
M-1499; 95.0 Ni, 1.0 C, 0.2 Mn, 3.0 Fe, 0.005 S, 0.7 Si, 0.10 Cu.
Electrode for shielded metal-arc welding of cast iron.
AWS A 5.15 Class E Ni-CI.

NIROMET 36.
M-61; 36 Ni, 64 Fe.
Wire: (Annealed) 70,000 TS. (Cold worked) 150,000 TS.
For bimetals, precision springs, time devices. Low expansion.

NIROMET 42.
M-61; 42 Ni, bal Fe.
Bar: 70,000-150,000 TS; 50,000 YS; 35 El.
For glass-to-metal hermetic sealing, leads and terminals for resistors.
Matches Corning 1075 glass. Controlled expansion.

NIROMET 44.
M-61; 44 Ni, 56 Fe.
Wire: 70,000 TS (Annealed); 150,00 TS (Cold worked).
For special glass sealing and fiber optics. Controlled expansion.

NIROMET 46.
M-61; 46 Ni, bal Fe.
Bar: 70,000-150,000 TS; 50,000 YS; 35 El.
For terminal bands in vitreous enameled resistors, cores and armatures for relays, motors, transformers; controlled expansion alloy.

NIROMET 48.
M-61; 48 Ni, 52 Fe.
Wire: 70,000-150,000 TS.
For glass-to-metal seals. Controlled expansion.

NIROMET 426.
M-61; 42 Ni, 6 Cr, bal Fe.
Wire: 80,000-150,00 TS.
For glass to metal seals (Corning 0120 glass). Controlled expansion.

NIRON 52.
M-61; 51 Ni, 49 Fe.
Annealed: 70,000 TS; 50,000 YS; 35 El.
For glass to metal seals; controlled expansion.

NIRONITE.
M-229; C, alloy, bal Fe.
For rolls for strip mills; heavy duty.

NIRONITE B.
M-229; 3.2 C, 2 Si, 1 Cr, bal Fe.
For rolling mill casting; special iron alloy.

NIRONITE C.
M-229; C, alloy, bal Fe.
For rolls for bar and billet mills; heavy duty.

NIRONITE D.
M-229; C, alloy, bal Fe.
For rolls for strip mills; heavy duty.

NIRONZE 635.
M-141; 1.9 Ni, 0.6 Ci, bal Cu.
Annealed: 40,000 TS; 12,000 YS; 50 El; 90 RA; 56 Brin.
Aged: 88,000 TS; 70,000 YS; 12 El; 20 RA; 170 Brin.
For cold headed bolts, fasteners, switch gear, marine hardware; age-hardenable, corrosion resistant.

NIROSTA 18-8.
M-184; 0.6-1.25 Si, 8-10 Ni, 16-19 Cr, 0.35 max C, bal Fe.
Corrosion resisting articles; now Michiana No. 49.

NIROSTA 19-9-4.
M-184; 0.6-1.25 Si, 8-12 Ni, 18-20 Cr, 0.35 C, 2-4 Mo, bal Fe.
For hot sulfuric acid tanks; corrosion resistant and hot H_2SO_4 resistant, under pressure.

NIROSTA 25-20.
M-184; 1.5 max Si, 18-22 Ni, 24-27 Cr, 0.35 max C, bal Fe.
Applicable to temperature and corrosion work where temperature exceeds 1400°F; heat and corrosion resistant up to 1400°F.

NIROSTA KA2.
M-15, M-72; 0.16 max C, 0.50 max Mn, 0.50 Si, 17-20 Cr, 7-10 Ni, bal Fe.
85,000-95,000 TS; 30,000-40,000 YS; 60-55 El; 75-50 RA; 135-150 Brin.
For cutlery, stainless articles; non-magnetic, austenitic; scaling point 1700°F.

NIROSTA KA4.
M-105; 0.25 C, 0.50 Mn, 0.5-0.75 Si, 17-20 Cr, 8-10 Ni, 2-4 Mo, bal Fe.
At 70°F: 105,000 TS; 49,000 YS; 55 El; 62 RA.
At 1110°F: 70,000 TS; 22,000 YS; 30 El.
At 1830°F: 9,000 TS; 4,500 YS; 47 El.
For pipe fittings, pump castings, impellers, shafts; high corrosion resistance; austenitic; same as "Standard alloy KA4."

NIROSTA VK5M.
M-72; 0.20 C, 13 Cr, 1.2 Mo, bal Fe.
Annealed: 95,000 TS; 40,000 YS; 25 El; 55 RA; B 92 Rock.
Cold Drawn: 105,000 TS; 85,000 YS; 17 El; 50 RA; B 95 Rock.
Heat Treated: 250,000 TS; 215,000 YS; 8 El; 25 RA; C 50 Rock.
For cutlery, surgical instruments, oil refinery equipment, valves.
Corrosion resistant, hardenable.

NIROSTA VK7M.
M-72; 0.15 C, 13 Cr, 1.2 Mo, bal Fe.
Annealed: 80,000 TS; 40,000 YS; 25 El; B 93 Rock.
Heat Treated: 200,000 TS; 160,000 YS; 12 El; C 45 Rock.
For surgical instruments, cutlery, oil refining equipment, table flatware.
Corrosion resistant, hardenable.

NIRUS TYPE 1.
M-1744; 0.25 C, 0.65 Mn, 0.25 Si, 3.25 Cr, 23.5 Ni, bal Fe.
Non-magnetic steel.
UTS: 35 tons/sq. in. min; YS: 18 tons/sq. in. min.
For bus bars, studs, nuts, bolts.

NIRUS TYPE 3.
M-1744; 0.25 C, 1.25 Mn, 0.25 Si, 0.15 Cr, 25.0 Ni, bal Fe.
Non-magnetic steel, similar to NIRIUS TYPE 1 but is readily weldable.

NI RW 4 ZR.
M-Germany; 4.2 W, 0.1 Zr, bal Ni.
For thermionic valves.
Heat and corrosion resistant.

NISILOY.
M-68; 60 Ni, 30 Si, 10 Fe.
For iron inoculant; for gray iron castings; M.P. 1800°F.

NI-SPAN ALLOY C-902.
M-121; 0.03 C, 42.5 Ni, 49.0 Fe, 5.3 Cr, 2.4 Ti, 0.5 Al.
Low thermal expansion alloy.
Thermal expansion at 20-100°C: 6.2×10^{-6}/K.

NI-SPAN C.
M-237; 41-43 Ni, 2.4 Ti, 5.1-5.7 Cr, 0.6 max C, 0.6 Al, 0.8 Si, bal Fe.
Annealed: 90,000 TS; 35,000 YS; 40 El; 145 Brin.
Aged: 200,000 TS; 180,000 YS; 7 El; 395 Brin.
For instruments, springs, diaphragms; age-hardenable, constant modulus.

NI-SPAN-C ALLOY 902.
M-1499, M-1798; 42.25 Ni, 0.03 C, 0.40 Mn, 48.5 Fe, 0.02 S, 0.50 Si 0.05 Cu, 5.33 Cr, 0.55 Al, 2.58 Ti.
Hot rolled and aged: 175,000 TS; 110,000 YS; 25 El.
For tuning forks and other mechanical resonators, electromechanical filter watch and clock hairsprings; age-hardenable; controllable thermo-elastic coefficient.

NI-SPAN HI.
M-68; 28-30 Ni, 2.4 Ti, 8-9 Cr, 0.06 max C, 0.5 Si, 0.4 Mn, 0.6 Al, bal Fe.
Aged: 140,000 TS; 90,000 YS; 20 El.
For thermostats, bimetals; high coefficient of expansion, age-hardenable.

NI-SPAN LO 42.
M-68; 40.5-42.5 Ni, 2.2-2.6 Ti, 0.06 max C, 0.4 Mn, 0.5 Si, 0.6 Al, bal Fe.
Annealed: 90,000 TS; 40,000 YS; 32 El; 330 Brin.
Hardened: 165,000 TS; 120,000 YS; 14 El; 330 Brin.
For thermostats, bimetals; age-hardenable, low coefficient of expansion.

NI-SPAN LO 45.
M-68; 44.4-46.5 Ni, 2.4 Ti, 0.06 max C, 0.4 Mn, 0.5 Si, 0.6 Al, bal Fe.
For thermostats, bimetals; low coefficient of expansion, age-hardenable.

NI-SPAN LO 52.
M-68; 51-53 Ni, 2.4 Ti, 0.06 max C, 0.4 Mn, 0.5 Si, 0.6 Al, bal Fe.
Annealed: 85,000 TS; 35,000 YS; 27 El; 125 Brin.
Hardened: 120,000 TS; 95,000 YS; 17 El; 305 Brin.
For thermostats, bimetals; low coefficient of expansion, age-hardenable.

NITEC 10224.
M-717.
Nickel base alloy powder for cast iron and steel brazing or build-up.

NITECTIC 222.
M-717.
Electrode for DC welding of nickel alloys and dissimilar alloys.
Tensile strength 100,000 psi.

NI TENSILIRON.
M-68; 2.5-3.1 C, 0.5-0.9 Mn, 1.2-2.75 Si, 1-4 Ni, 0-1 Mo, 0-0.5 Cr, bal Fe.
Cast: 40,000-100,000 TS; 220-350 Brin.
For gears, turbines, casings and rotors, valves, bushings; Ni "Tensile;" high strength cast iron.

NI TENSYLE.
M-494, M-1485; 2.8 C, 1.5-2.0 Ni, 1.25-1.75 Si, bal Fe.
Cast: 56,000 TS; 200 Brin.
Heat treated: 66,000 TS; 280 Brin.
For flywheels; cast iron.

NITINOL.
M-US; 40-45 Ti, 55-60 Ni.
54.5% Ni: 110,000-124,000 TS; 40,000-55,000 YS; 15.5 El; 16 RA; Rock A 42-52.
For nonmagnetic tools, sensing devices, chemical plant equipment, space components.
Nonmagnetic, corrosion resistant, hardenable, alloy with a memory.

NITINOL-55.
M-US; 55 Ni, bal Ti.
Bar: 120,000 TS; 50,000 YS; 15 El; 16 RA; A 50 Rock.
For non-magnetic tools, sensing devices, chemical plant equipment.
Alloy with a memory. Non-magnetic, corrosion resistant.

NITRALLOY NO. 135 CVM.
M-386; 0.30-0.40 C, 0.9-1.4 Cr, 0.15-0.25 Mo, 1.0-1.4 Al, bal Fe.
Oil quenched and drawn: 225,000-105,000 TS; 180,000-90,000 YS; 10-27 El; 37-62 RA; 440-175 Brin.
For nitrided parts, gears, shafts; Nitralloy G.

NITRALLOY 230.
M-386; 0.25-0.35 C, 0.4-0.6 Mn, 0.2-0.3 Si, 1-1.5 Al, 0.6-1.0 Mo, bal Fe.
For nitrided parts, gears, shafts, pinions; nitriding steel, Nitralloy N.

NITRALLOY 640.
M-386; 0.35-0.45 C, 0.4-0.6 Mn, 0.2-0.3 Si, 1.4-1.6 Cr, 0.45 min V, bal Fe.
For nitrided parts, gears, shafts, pinions; nitriding steel, hard case.

NITRALLOY EZ.
M-24, M-1, M-240; 0.3-0.4 C, 1.25 Cr, 1.1 Al, 0.2 Mo, 0.15-0.25 Se or S, bal Fe.
Heat treated: 155,000-200,000 TS; 135,000-175,000 YS; 18-16 El; 55-50 RA; 300-400 Brin.
For nitrided parts, gears, shafts; nitriding steel.

NITRALLOY GK5.
M-56; 0.25 C, 0.5 Mn, 2.0 Cr, 0.25 Mo, 0.15 V, bal Fe.
For gears, shafts; nitriding steel.

NITRALLOY GR.
M-386; 1.25-1.5 C, 0.4-0.6 Mn, 1.25-1.5 Si, 0.2-0.4 Cr, 0.2-0.3 Mo, 1.0-1.5 Al, bal Fe.
Rolled: 108,500 TS; 84,000 YS; 17.5 El; 19.4 RA; 363 Brin.
For seal rings, cylinder liners; self lubricating, free-cutting.

NITRALLOY HCM3.
M-56; 0.4 C, 0.5 Mn, 0.3 Ni, 3 Cr, 1 Mo, 0.25 V, bal Fe.
For gears, shafts; nitriding steel.

NITRALLOY HCM5.
M-56; 0.3 C, 0.45 Mn, 0.5 Ni, 3 Cr, 0.4 Mo, bal Fe.
For gears, shafts; nitriding steel.

NITRALLOY HCM7.
M-56; 0.2 C, 0.45 Mn, 0.5 Ni, 3 Cr, 0.4 Mo, bal Fe.
For gears, shafts; nitriding steel.

NITRALLOY LK3.
M-56; 0.4 C, 0.65 Mn, 1.6 Cr, 0.2 Mo, 1.1 Al, bal Fe.
For gears, shafts; nitriding steel.

NITRALLOY LK5.
M-56; 0.3 C, 0.65 Mn, 1.6 Cr, 0.2 Mo, 1.1 Al, bal Fe.
For gears, shafts; nitriding steel.

NITRALLOY N.
M-1779, M-24; 0.20-0.27 C, 0.4-0.7 Mn, 1.0-1.3 Cr, 0.2-0.3 Mo, 3.25-3.75 Ni, 1.1-1.4 Al, bal Fe.
For nitrided parts; nitriding steel.

NITREX see **CARPENTER NITREX.**

NITREX-I.
M-32; 0.38-0.45 C, 0.6 Mn, 0.3 Si, 1.4-1.8 Cr, 0.85-1.2 Al, 0.30-0.45 Mo, bal Fe.
Tempered 1100°F: 181,000 TS; 165,000 YS; 15 El; 54 RA; Rock C 41.
Tempered 1000°F: 206,000 TS; 182,000 YS; 13 El.
For cylinder liners, gears, cams, thread guides, camshafts.
Nitriding steel, wear and abrasion resistant case.

NITREX-II.
M-32; 0.20-0.27 C, 0.6 Mn, 0.3 Si, 1.0-1.5 Cr, 0.8-1.2 Al, 0.2-0.3 M 3.25-3.75 Ni, bal Fe.
Tempered 1000°F: 198,000 TS; 191,000 YP; 14 El; 45 RA; 38 Rc.
Tempered 1100°F: 145,000 TS; 130,000 YP; 21 El; 58 RA; 30 Rc.
After Nitriding: 190,000 TS; 180,000 YP; 6-15 El; 43 RA; core Rc 41; case 94 R 15N.
For cylinder liners, bushings, shafts, gears, cams, camshafts.
Precipitation hardening nitriding steel.

NITRICAST.
M-1748; 2.5 C, 2.5 Si, 1 Cr, 0.2 Mo, 1 Al, bal Fe.
As nitrided: 50,000 TS; 277 Brin (core); file hard case.
For oilwell tooling, sleeves, liners, cams.

NITRICASTIRON.
M-386; 2.75 T.C., 1.89 G.C., 0.86 C.C., 2.58 Si, 1.22 Cr, 0.16 V, 0.24 Mo, 1.01 Al, bal Fe.
Annealed: 56,000-67,000 TS; 302 Brin.
For nitrided cast iron, cams, cylinders, valves; wear resistant.

NITRIDING G (135).
M-97; 0.3-0.4 C, 0.4-0.7 Mn, 0.9-1.4 Cr, 0.15-0.25 Mo, 0.85-1.2 Al, bal Fe.
Heat treated: 105,000-225,000 TS; 80,000-177,000 YS; 28-11 El; 60-35 RA; 200-445 Brin.
For cylinder liners, gears, camshafts, connecting rods, bushings; nitriding steel.

NITRIDING G-MOD.
M-97; 0.38-0.43 C, 0.5-0.7 Mn, 1.4-1.8 Cr, 0.95-1.3 Al, 0.30-0.4 Mo, bal Fe.
Heat treated: 158,000-206,000 TS; 141,000-181,000 YS; 20-13 El; 64-45 RA; 285-415 Brin.
For cylinder liners, gears, bolts, crankshafts; nitriding steel.

NITRIDING H (125).
M-97; 0.2-0.3 C, 0.9-1.4 Cr, 0.15-0.25 Mo, 0.85-1.2 Al, bal Fe.
Heat treated: 102,000-178,000 TS; 85,000-155,000 YS; 26-12 El; 70-46 RA; 225-400 Brin.
For cylinder liners, bushings, shafts, gears, rolls; nitriding steel.

NITRIDING N.
M-97; 0.20-0.27 C, 0.4-0.7 Mn, 1.0-1.3 Cr, 0.85-1.2 Al, 0.2-0.3 Mo, 3.25-3.75 Ni, bal Fe.
Heat treated: 198,000 TS; 191,000 YS; 14 El; 45 RA; 390 Brin.
For gears, cams, crankshafts, camshafts; age-hardenable.

NITRIDING NTR.
M-1488; 0.38 C, 0.60 Mn, 0.40 max Si, 1.0 Al, 1.65 Cr, 0.32 Mo, bal Fe.
For nitrided parts.
Similar to Nitralloy 135.

NITRIDING STEEL 135 MODIFIED.
M-38, M-97; 0.4 C. 0.5 Mn, 1 Al, 1.75 Cr, 0.4 Mo, bal Fe.
Heat treated: 158,700 TS; 141,000 YS; 17 El; 56 RA; 320 Brin.
For cylinder barrels, nitrided parts; nitriding steel.

NITRIX 65.
M-501, M-912; 0.27 C, 1.1 Al, 1.4 Cr, bal Fe.
For oil refinery equipment; creep and heat resistant.

NITRIX 65A.
M-501, M-912; 0.30 C, 0.9 Al, 1.2 Cr, bal Fe.
For oil refinery equipment, bolts; creep and heat resistant.

NITRIX 71.
M-501, M-912; 0.32 C, 1.1 Al, 1.1 Cr, 0.18 Mo, bal Fe.
For oil refinery equipment, bolts, fasteners; creep and heat resistant.

NITRIX 72.
M-501, M-912; 0.34 C, 1.1 Al, 1.4 Cr, bal Fe.
For oil refinery equipment; heat and creep resistant.

NITRIX 73.
M-501, M-912; 0.31 C, 2.35 Cr, 0.18 Mo, 0.13 V, bal Fe.
For gears, shafts, crankshafts, bolts, studs; oil hardened, shock resistant.

NITRIX 80.
M-501, M-912; 0.32 C, 1.1 Al, 0.18 Mo, 1.1 Cr, bal Fe.
For oil refinery equipment; heat and creep resistant.

NITRIX 1470.
M-501, M-912; 0.31 C, 2.35 Cr, 0.18 Mo, 0.13 V, bal Fe.
For gears, shafts, crankshafts, bolts, fasteners; oil hardened, shock resistant.

NITRIX 1471.
M-501, M-912; 0.32 C, 1.1 Al, 1.1 Cr, 0.18 Mo, bal Fe.
For oil refinery equipment; creep and heat resistant.

NITRIX 1472.
  M-501, M-912; 0.34 C, 1.1 Al, 1.1 Cr, bal Fe.
  For oil refinery equipment; creep and heat resistant.

NITRIX 1473.
  M-501, M-912; 0.30 C, Cr, V, 0.6 Mn, bal Fe.
  For gears, shafts, bolts, fasteners; oil hardened, shock resistant.

NITRO.
  M-32; 1.0 C, 0.35 Mn, 0.25 Si, 0.2 V, bal Fe.
  For drills, taps, springs, cutters; Type W2.

NITROFIL.
  M-1805; 0.2 Al, 0.2 Ti, bal Cu.
  For filler rod and wire for nitrogen-arc and inert gas shielded metal arc welding of copper.

NITRONIC see **ARMCO NITRONIC.**

NITRONIC 33 see **ARMCO NITRONIC 33.**

NITROSA STEEL.
  M-57; 18 Cr, 8 Ni, 0.35 Mn, 0.35 C, bal Fe.
  Cold worked: 200,000 TS; 100,000 YS; 13 El.
  For stainless parts, chemical apparatus; resists corrosion of acids and gases; resists high temperature to 1650°F.

NITTANY (FREE CUTTING) BRASS.
  M-122; 61 Cu, 3 Pb, bal Zn.
  Cold drawn: 66,000 TS; 50,000 YS; 29 El; 46 RA; 126 Brin.
  For screw machine parts, bolts, screws; free-turning yellow brass. CA 360.

NITTANY NO. 2.
  M-122; 61.5-62 Cu, 2.2-2.5 Pb, bal Zn.
  51,000-68,000 TS; 23,000-37,000 YS; 29-54 El; 46-50 RA; 58-110 Brin
  For screw machine parts; free cutting.

NITTANY NO. 4.
  M-122; 62-62.5 Cu, 3.0-3.25 Pb, bal Zn.
  50,000-65,000 TS; 27,000-37,000 YS; 29-54 El; 46-51 RA; 63-110 Brin
  For screw machine parts; free cutting.

NITTANY NO. 7.
  M-122; 62-62.5 Cu, 1.25-1.5 Pb, bal Zn.
  48,000-60,000 TS; 19,000-36,000 YS; 26-44 El; 62-75 RA; 63-90 Brin.
  For screw machine parts; free cutting.

NITTANY NO. 30.
  M-122; 63 Cu, 1.2 Pb, bal Zn.
  Rolled: 67-79 Brin.
  For screw machine products, hardware; free cutting. CA 350.

NITTANY NO. 31.
  M-122; 63 Cu, 1.75 Pb, bal Zn.
  Rolled: 67-79 Brin.
  For screw machine products, hardware; free cutting. CA 353.

NITTANY NO. 35.
  M-122; 64 Cu, 0.85 Pb, bal Zn.
  Rolled: 100-107 Brin.
  For hardware, screw machine products; for spinning and swaging parts.

NITTANY NO. 38.
  M-122; 62 Cu, 2.3 Pb, bal Zn.
  Rolled: 107-125 Brin.
  For hardware, screw machine products; free cutting. CA 356.

NITTANY NO. 49.
  M-122; 61 Cu, 3 Pb, bal Zn.
  Soft: 50,000 TS; 22,000 YS; 32 El; 83 Brin.
  1/2 H-temper: 60,000 TS; 42,000 YS; 22 El; 125 Brin.
  H-temper: 85,000 TS; 56,000 YS; 7 El; 160 Brin.
  For screw machine products, screws, bolts; free-cutting.

NITTANY NO. 77.
  M-122; 62.5 Cu, 3 Pb, bal Zn.
  For screw machine products, hardware, bolts; free-cutting.

NITUNG.
  M-1433; 0.30 C, 2.75 Cr, 0.30 Mo, 9.5 W, 1.6 Ni, bal Fe.
  For extrusion rams and liners, hot work tools; hot work steel, oil hardened.

NIVAC.
  M-1214, M-38; 99.92 Ni.
  For equipment handling fluoride at high temperatures of 500-600°C.
  Corrosion and heat resistant.

NIVAC 50 NI.
  M-38; 50 Ni, bal Fe.
  Annealed: 58,000 TS; 19,000 YP; 27 El.
  Coercive force 0.06 for 10,000 gauss.
  For magnetic temperature compensator, speedometers, tachometers, voltage regulators, watt-hour meters.
  High permeability, magnetically soft.

NIVAC-77 NI.
  M-38; 77 Ni, bal Fe.
  For telephone loading coils, magnetic shielding, senitive relays, pulse transformers.
  High permeability, soft magnet.

NIVAC-78 NI.
  M-38; 78 Ni, bal Fe.
  For telephone loading coils, magnetic shielding, sensitive relays, pulse transformers.
  High permeability, soft magnet.

NIVAC 79 NI.
  M-38; 79 Ni, bal Fe.
  For telephone loading coils, magnetic shielding, sensitive relays, pulse transformers.
  High permeability.

NIVAC 80 NI.
M-38; 80 Ni, bal Fe.
For telephone loading coils, magnetic shielding, sensitive relays, pulse transformers.
High permeability.

NIVAC P.
M-1214; 0.007 C, 0.005 O_2, 0.005 Si, 0.002 P, 0.13 Co, 0.005 C 0.010 Fe, bal Ni.
For vacuum tubes, diaphragms, resistance wire; high purity nickel.

NIVAN.
M-357; C, 1.4 Ni, 0.35 Cr, 0.2 V, bal Fe.
For cutlery; tough.

NIVCO.
M-118; 0.02 C, 0.35 Mn, 0.15 Si, 22.5 Ni, 1.1 Zr, 1.8 Ti, 0.22 Al, 1.0 Fe, bal Co.
At R.T.: 165,000 TS; 110,000 YS; 25 El.
At 1200°F: 105,000 TS; 75,000 YS; 20 El.
For steam turbine blading high heat strength to 1200°F.

NIVCO-10.
M-118; 0.02 C, 1.1 Zr, 1.8 Ti, 22.5 Ni, bal Co.
At 70°F: 165,000 TS; 110,000 YS; 25 El.
At 1200°F: 105,000 TS; 75,000 YS; 20 El.
At 1300°F: 85,000 TS; 65,000 YS; 26 El.
For steam turbine blades; used where vibratory stresses are critical.

NI-VEE.
M-1007; 5 Ni, 5 Sn, 2 Zn, bal Cu.
Aged: 80,000 TS; 50,000 YS; 10 El; 150 Brin.
For gears, valves, construction castings; age-hardenable, corrosion resistant.

NI-VEE 1.
M-214; 88 Cu, 5 Ni, 5 Sn, 2 Zn.
Cast: 50,000 TS; 22,000 YS; 40 El; 85 Brin.
Heat treated: 65,000-85,000 TS; 40,000-55,000 YS; 10-8 El; 130-180 Brin.
For gears, cams, valves; fine grain, heat treatable.

NI-VEE 2.
M-214; 80 Cu, 5 Ni, 5 Sn, 5 Pb, 5 Zn.
Cast: 40,000 TS; 20,000 YS; 20 El; 80 Brin.
Heat treated: 50,000 TS; 30,000 YS; 5 El; 130 Brin.
For pressure castings, valves, fittings, plumbing parts; fine grain, pressure tight.

NI-VEE 3.
M-214; 80 Cu, 5 Ni, 5 Sn, 10 Pb.
Cast: 35,000 TS; 20,000 YS; 10 El; 80 Brin.
Heat treated: 40,000 TS; 25,000 YS; 5 El; 110 Brin.
For bearings, bushings, acid resistant castings; pressure tight, fine grain.

NI-VEE 4.
M-214; 70 Cu, 5 Ni, 5 Sn, 20 Pb.
Cast: 25,000 TS; 18,000 YS; 10 El; 70 Brin.
Heat treated: 30,000 TS; 22,000 YS; 5 El; 80 Brin.
For bearings, bushings, acid resistant castings; pressure tight, fine grain.

NI-VEE L2.
M-1007; 5.Ni, 5 Sn, 2.5 Pb, 0.5 max Zn, bal Cu.
Cast: 44,000 TS; 23,000 YS; 30 El; 92 Brin.
Heat treated: 55,000 TS; 40,000 YS; 10 El; 135 Brin.
For gears, bushings, bearings, construction castings; fine grain, heat treatable, good castability.

NI-VEE L2-Z2.
M-1007; 5 Ni, 5 Sn, 2.5 Pb, 2 max Zn, bal Cu.
Cast: 42,000 TS; 24,000 YS; 25 El; 90 Brin.
Heat treated: 49,000 TS; 35,000 YS; 8 El; 120 Brin.
For gears, bearings, bushings, construction castings, fine grain, heat treatable, good castability.

NI-VEE L5.
M-1007; 5 Ni, 5 Sn, 5 Pb, 0.5 max Zn, bal Cu.
Cast: 41,000 TS; 20,000 YS; 22 El; 92 Brin.
Heat treated: 47,000 TS; 35,000 YS; 7 El; 130 Brin.
For bearings, bushings, acid resistant castings; fine grain, heat treatable, good castability.

NI-VEE L5-Z5.
M-1007; 5 Ni, 55 Sn, 5 Pb, 5 Zn, 0.05 P, bal Cu.
Cast: 40,000 TS; 20,000 YS; 20 El; 80 Brin.
Heat treated: 50,000 TS; 30,000 YS; 5 El; 120 Brin.
For pumps, valves, fittings; corrosion resistant.

NI-VEE-L10.
M-1007; 80 Cu, 5 Sn, 10 Pb, 5 Ni.
Cast: 35,000 TS; 10 El; 80 Brin.
Heat treated: 40,000 TS; 5 El; 110 Brin.
For valves, fittings; pessure tight.

NI-VEE L15.
M-1007; 5 Ni, 5 Sn, 15 Pb, 0.5 max Zn, bal Cu.
Cast: 30,000 TS; 21,000 YS; 12 El; 78 Brin.
Heat treated: 35,000 TS; 24,000 YS; 5 El; 88 Brin.
For bearings, bushings, acid resistant castings; fine grain, heat treatable, good castability.

NI-VEE-L20.
M-1007; 70 Cu, 5 Ni, 5 Sn, 20 Pb.
Cast: 25,000 TS; 10 El; 70 Brin.
Heat treated: 30,000 TS; 5 El; 80 Brin.
For pressure castings, bearings; free-cutting.

NI-VEE TYPE A.
M-68; 88 Cu, 5 Ni, 5 Sn, 2 Zn.
Cast: 50,000 TS; 22,000 YS; 40 El; 50 RA; 85 Brin.
Tempered: 65,000 TS; 40,000 YS; 10 El; 130 Brin.
Heat treated: 85,000 TS; 55,000 YS; 10 El; 26 RA; 180 Brin.
For machine tools, cams, rollers, guides, gears; age-hardenable, corrosion resistant.

NI-VEE TYPE B.
M-68; 87 Cu, 5 Ni, 5 Sn, 1 Pb, 2 Zn.
Cast: 45,000 TS; 20,000 YS; 30 El; 80 Brin.
Tempered: 60,000 TS; 30,000 YS; 8 El; 120 Brin.
For machine tools, cylinder, cams, gears, rollers; age-hardenable, corrosion resistant.

NI-VEE TYPE C.
M-68; 80 Cu, 5 Ni, 5 Pb, 5 Zn, 5 Sn.
Cast: 40,000 TS; 20,000 YS; 15 El; 15 RA; 80 Brin.
Tempered: 50,000 TS; 30,000 YS; 5 El; 2.5 RA; 130 Brin.
For pressure castings, pumps, valves, fittings; age-hardenable, corrosion resistant.

NI-VEE TYPE D.
M-68; 80 Cu, 5 Ni, 5 Sn, 10 Pb.
Cast: 35,000 TS; 20,000 YS; 10 El; 10 RA; 80 Brin.
Tempered: 40,000 TS; 25,000 YS; 2 El; 3 RA; 110 Brin.
For bearings, bushings, liners; age-hardenable, corrosion resistant.

NI-VEE TYPE E.
M-68; 70 Cu, 5 Ni, 5 Sn, 20 Pb.
Cast: 25,000 TS; 18,000 YS; 5 El; 5 RA; 70 Brin.
Tempered: 30,000 TS; 22,000 YS; 2 El; 2 RA; 80 Brin.
For bearings, bushings, liners; age-hardenable, corrosion resistant.

NI-WELD.
M-38; 100 Ni.
For welding rod for cast iron; machinable welds.

NJZ BRASCO.
M-91.
Co-brass alloy powder.
For hard and strong brass powder metal parts.

NJZ BRASCO 1170.
M-91; 67-70 Cu, 3.2-3.8 Co, bal Zn.
Cobalt-Brass alloy powder for sintering.

NJZ NO. 1.
M-91; Zn plus Pb, Cd, Fe.
High ductility zinc sheet and strip for deep drawing and spinning.
For cosmetic containers, jewelry, hardware.

NJZ NO. 5.
M-91; Zn plus Pb, Cd.
Medium hardness zinc sheet and strip for impact extrusion.
For battery cans, condenser cans, hardware.

NJZ NO. 10.
M-91; Zn plus Pb, Cd, Fe.
Standard commercial zinc sheet and strip for battery cells, laundry tags, embossing, hardware.

NJZ NO. 15.
M-91; Zn plus Pb, Cu, Cd, Mg.
Zinc sheet and strip of good stiffness for sporting goods, weather-stripping, washers.

NJZ NO. 25.
M-91; Zn plus Cu, Ti.
Low expansion zinc sheet and strip for automotive trim, heat sinks, roofing.

NJZ NO. 30.
M-91; Zn plus Pb, Cd.
Commercial zinc sheet and strip, for battery cans, hardware, medallions.

NJZ NO. 35.
M-91; Zn plus Pb, Cd, Fe.
Zinc alloy sheet and strip, good strength but ductile. For jar caps and hardware.

NJZ NO. 45.
M-91; Zn plus Cu.
Zinc - copper sheet and strip. Strong, bends and forms readily; work hardens.
For hardware, medallions, addressing plates.

NJZ NO. 49.
M-91; Zn plus Pb, Cu, Fe, Cd.
Zinc alloy sheet and strip; for organ pipes.

NJZ NO. 50.
M-91; Zn plus Pb, Cd, Fe.
Zinc alloy plate with good ductility.
For addressing plates, novelties, jewelry.

NJZ NO. 55.
M-91; Cd, bal Zn.
Zinc alloy sheet and strip; good strength and stiffness.
For soldered battery cans, jewelry, hardware.

NJZ NO. 60.
M-91; Zn (very pure).
Soft, low tensile sheet and strip.
Primarily for cathodic protection.

NK-AC 90.
M-1612; 0.18 C, 0.60 Mn, 0.90 Cr, 0.18 Mo, bal Fe.
Ht Tr: 90,000-105,000 psi YS.
Seamless tube and pipe.
Good impact resistance at low temperature; resistant to hydrogen sulfide corrosion cracking.

NKBH 40.
M-1612.
Low carbon, low alloy strip and sheet.
Ann, cold rolled: 40 kg (56,800 psi) TS; 24 kg/mm^2 (34,100 psi) Y 30 min El.
0.6-1.2 mm thick, good formability.

NKBH 45.
M-1612.
Low carbon, low alloy strip and sheet.
Ann, cold rolled: 45 kg/mm^2 (63,900 psi) TS; 28 kg/mm^2 (39,800 psi) YS; 26 min El.
0.6-1.2 mm thick, good formability.

NKBH 50.
M-1612.
Low carbon, low alloy strip and sheet.
Ann, cold rolled: 50 kg/mm^2 (71,000 psi) TS; 32 kg/mm^2 (45,500 psi) YS; 23 min El.
0.6-1.2 mm thick, good formability.

NKBH 55.
M-1612.
Low carbon, low alloy strip and sheet.
Ann, cold rolled: 55 kg/mm^2 (78,100 psi) TS; 36 kg/mm^2 (51,100 psi) YS; 20 min El.
0.6-1.2 mm thick, good strength and formability.

NKBH 60.
M-1612.
Low carbon, low alloy strip and sheet.
Ann, cold rolled: 60 kg/mm^2 (63,900 psi) TS; 40 kg/mm^2 (39,800 psi) YS; 17 min El.
0.6-1.2 mm thick, good strength and formability.

NKCA 40.
M-1612.
Low carbon, low alloy strip and sheet.
Ann, cold rolled: 40 kg/mm^2 (56,800 psi) TS; 24 kg/mm^2 (34,100 psi) YS; 30 min El.
0.6-2.0 mm thick, good formability.

NKCA 45.
M-1612.
Low carbon, low alloy strip and sheet.
Ann, cold rolled: 45 kg/mm^2 (63,900 psi) TS; 28 kg/mm^2 (34,100 psi) YS; 26 min El.
0.6-2.0 mm thick, good formability.

NKCA 50.
M-1612.
Low carbon, low alloy strip and sheet.
Ann, cold rolled: 50 kg/mm^2 (71,000 psi) TS; 32 kg/mm^2 (45,500 psi) YS; 23 min El.
0.6-2.0 mm thick, good formability.

NKCA 55.
M-1612.
Low carbon, low alloy strip and sheet.
Ann, cold rolled: 55 kg/mm^2 (78,100 psi) TS; 36 kg/mm^2 (51,100 psi) YS; 20 min El.
0.6-2.0 mm thick, good strength and formability.

NKCA 60.
M-1612.
Low carbon, low alloy strip and sheet.
Ann, cold rolled: 60 kg/mm^2 (85,200 psi) TS; 40 kg/mm^2 (56,800 psi) YS; 17 min El.
0.6-2.0 mm thick, good strength and formability.

NK-CMV12.
M-1612; 0.20 C, 0.55 Mn, 0.55 Ni, 12.0 Cr, 1.0 Mo, 0.30 V, bal Fe.
Heat and oxidation resisting seamless pipe and tube; for operation at 500-550°C.

NKHA 50.
M-1612.
Hot rolled strip and sheets; comp. by agreement.
As rolled, 1.6-3.2 mm thick: 50 kg/mm^2 (71,000 psi) TS; 35 kg/mm^2 (49,700 psi) YS; 22-24 El.
For structural purposes; good strength and formability.

NKHA 55.
M-1612.
Hot rolled strip and sheets; comp. by agreement.
As rolled, 1.6-3.2 mm thick: 55 kg/mm^2 (78,100 psi) TS; 38 kg/mm^2 (54,000 psi) YS; 21-23 El.
For structural purposes; good strength and formability.

NKHA 60.
M-1612.
Hot rolled strip and sheets; comp. by agreement.
As rolled, 1.6-3.2 mm thick: 60 kg/mm^2 (85,200 psi) TS; 45 kg/mm^2 (63,900 psi) YS; 19-21 El.
For structural purposes; good strength and formability.

NKHF 50.
M-1612; 0.18 max C, 0.55 max Si, 1.50 max Mn, 0.10 max Nb, 0.10 max Ti, bal Fe.
Hot rolled strip and sheets, as rolled, 3.2-8.0 mm thick: 50 kg/mm^2 (71,000 psi) TS; 36 kg/mm^2 (51,000 psi) YS; 24 El.
For structural applications, good formability.

NKHF 55.
M-1612; 0.18 max C, 0.55 max Si, 1.50 max Mn, 0.10 max Nb, bal Fe.
Hot rolled strip and sheets, as rolled, 3.2-8.0 mm thick: 55 kg/mm^2 (78,000 psi) TS; 40 kg/mm^2 (56,800 psi) YS; 24 El.
For structural applications, good strength and formability.

NKHF 60.
M-1612; 0.18 max C, 0.55 max Si, 1.50 max Mn, 0.10 max Nb, 0.20 max Zr, bal Fe.
Hot rolled strip and sheets, as rolled, 3.2-8.0 mm thick: 60 kg/mm^2 (85,200 psi) TS; 46 kg/mm^2 (65,300 psi) YS; 20-24 El.
For structural applications, good strength and formability.

NKHF 70.
  M-1612; 0.15 max C, 0.70 max Si, 1.0-1.5 Mn, 0.50 max Cu, 0.50 max Ni, 0.50 max Cr, 0.15 max V, 0.10 max Nb, 0.20 max Zr, bal Fe.
  Hot rolled strip and sheets, as rolled, 3.2-8.0 mm thick: 70 kg/mm^2 (99,400 psi) TS; 56 kg/mm^2 (79,500 psi) YS; 14-18 El.
  For structural applications; good strength and formability.

NKHF 80.
  M-1612; 0.10 max C, 0.70 max Si, 1.5-2.20 Mn, 0.80 max Cu, 0.80 max Ni, 0.50 max Cr, 0.15 max V, 0.10 max Nb, 0.20 max Zr, 0.15 max Ti, bal Fe.
  Hot rolled strip and sheets, as rolled, 3.2-8.0 mm thick: 80 kg/mm^2 (113.600 psi) TS; 70 kg/mm^2 (99,400 psi) YS; 12-16 El.
  For structural applications; good strength and formability.

NK-HITEN 60.
  M-1612; 0.18 max C, 0.55 max Si, 1.50 max Mn, (others as Cu, Cr, Mo, Ti as required) bal Fe.
  Plates, Q+T; 60-72 kg/mm^2 (83,500-102,400 psi) TS; 46 kg/mm^2 (65,400 psi) YS; 20 min El.
  6-100 mm thick plate for structural purposes.

NK-HITEN 60.
  M 1542, M 1564; 0.16 max C, 1.35 max Mn, 0.60 max Ni, 0.30 max Mo, 0.15 max V, bal Fe.
  Normalized: 85,000 min TS; 65,000 min YP; 16 min El.
  Heat Treated: 114,000 TS; 100,000 YS; 18-20 El.
  For agricultural equipment, mine and railroad cars.
  High strength low-alloy constructional steel.
  Tough, shock resistant.

NK-HITEN 60C.
  M-1612; 0.18 max C, 0.70 max Si, 1.60 max Mn, 0.80 max Cu, 0.80 max Ni, 0.09 V or Nb, bal Fe.
  Plates, as rolled or normalized: 60-72 kg/mm^2 (85,300-102,400 psi) T 46 kg/mm^2 (65,000 psi) YS; 20-28 El.
  6-50 mm thick plate for structural purposes.

NK-HITEN 60U.
  M-1612; 0.05-0.12 C, 0.15-0.40 Si, 0.90-1.40 Mn, 0.30 max Cu, 0.30 max Cr, 0.20 max Mo, 0.08 max V, 0.003 max B, bal Fe.
  Q+T, 6-50 mm thick plate: 60-72 kg/mm^2 (83,300-102,400 psi) TS; 46 kg/mm^2 (65,400 psi) YS; 20-28 El.
  Plates for structural purposes.

NK-HITEN 62.
  M-1612; 0.18 max C, 0.55 max Si, 1.50 max Mn, (others as Cu, Cr, Mo, Ti as required) bal Fe.
  Plates, Q+T; 62-74 kg/mm^2 (88,200-105,300 psi) TS; 50 kg/mm^2 (71,000 psi) YS; 19-29 El.
  6-100 mm thick plate for structural purposes.

NK-HITEN 62C.
  M-1612; 0.18 max C, 0.70 max Si, 1.60 max Mn, 0.80 max Cu, 0.80 max Ni, 0.09 V or Nb, bal Fe.
  Plates, as rolled or normalized: 62-74 kg/mm^2 (88,200-105,300 psi) T 50 kg/mm^2 (71,100 psi) YS; 19-29 El.
  6-40 mm thick plate for structural purposes.

NK-HITEN 62U.
  M-1612; 0.05-0.12 C, 0.15-0.40 Si, 0.90-1.40 Mn, 0.30 max Cu, 0.30 max Cr, 0.20 max Mo, 0.08 max V, 0.003 max B, bal Fe.
  Q+T, 6-50 mm thick plate: 62-74 kg/mm^2 (88,200-105,300 psi) TS; 50 kg/mm^2 (71,100 psi) YS; 19-29 El.
  Plates for structural purposes.

NK-HITEN 68.
  M-1612; 0.16 max C, 0.35 max Si, 1.20 max Mn, 0.30 max Cu, 0.80 max Ni, 0.70 max Cr, 0.40 max Mo, 0.07 max V, 0.003 max B, bal Fe.
  Plates, Q+T: 68-82 kg/mm^2 (96,700-116,600 psi) TS; 56 kg/mm^2 (79,700 psi) YS; 18-26 El.
  6-100 mm thick plate for structural purposes.

NK-HITEN 70B.
  M-1612; 0.14 max C, 0.35 max Si, 1.0 max Mn, 0.30 max Cu, 1.30 max Ni 0.70 max Cr, 0.40 max Mo, 0.07 max V, 0.003 max B, bal Fe.
  Plates, Q+T: 70-85 kg/mm^2 (99,600-120,900 psi) TS; 63 kg/mm^2 (89,600 psi) YS; 17-25 El.
  6-100 mm thick plate for structural purposes.

NK-HITEN 72.
  M-1612; 0.16 max C, 0.35 max Si, 1.20 max Mn, 0.30 max Cu, 0.80 max Ni, 0.70 max Cr, 0.40 max Mo, 0.07 max V, 0.003 max B, bal Fe.
  Plates, Q+T: 72-86 kg/mm^2 (102,400-122,300 psi) TS; 63 kg/mm^2 (89,600 psi) YS; 17-25 El.
  6-100 mm thick plate for structural purposes.

NK-HITEN 80.
  M-1612; 0.18 max C, 0.35 max Si, 1.0 max Mn, 0.15-0.50 Cu, 1.0 max Ni, 0.80 max Cr, 0.60 max Mo, 0.10 max V, 0.006 B, bal Fe.
  Plates, Q+T: 80-95 kg/mm^2 (113,800-135,100 psi) TS; 70 kg/mm^2 (89,600 psi) YS; 16-24 El.
  6-100 mm thick plate for structural purposes.

HK-HITEN 80.
  M-1564; 0.18 C, 1.0 Mn, 0.15-0.35 Si, 0.006 B, 0.8 Cr, 0.15-0.50 Cu, 0.6 Mo, 1.0 Ni, 0.1 V, bal Fe.
  Plate: 114,000 min TS; 100,000 min YS; 18 min El.
  For mine cars, bus bodies, trailers, pressure vessels, cranes, bridges.
  Wear and shock resistant.

NK-HITEN 80A.
M-1612; 0.18 max C, 0.60 max Si, 1.0 max Mn, 0.15-0.50 Cu, 1.20 max Cr, 0.60 max Mo, 0.10 max V, 0.006 max B, bal Fe.
Plates, Q+T: 80-95 kg/mm^2 (113,800-135,100 psi) TS; 70 kg/mm^2 (99,600 psi) YS; 16-24 El.
6-100 mm thick plate for structural purposes.

NK-HITEN 80B.
M-1612; 0.14 max C, 0.35 max Si, 1.20 max Mn, 0.30 max Cu, 1.50 max Ni, 0.70 max Cr, 0.40 max Mo, 0.10 max V, 0.003 max B, bal Fe.
Plate, Q+T: 80-95 kg/mm^2 (113,800-135,100 psi) TS; 70 kg/mm^2 (99,600 psi) YS; 16-24 El.
6-100 mm thick plate for structural purposes.

NK-HITEN 100.
M-1612; 0.18 max C, 0.55 max Si, 1.20 max Mn, 0.15-0.50 Cu, 1.20 max Ni, 0.80 max Cr, 0.70 max Mo, 0.15 max V, 0.15 max Zr, 0.006 max B, bal Fe.
Plate, Q+T: 97-115 kg/mm^2 (138,000-163,000 psi) TS; 90 kg/mm^2 (128,000 psi) YS; 12-19 El.
6-26 mm thick plate for structural purposes.

N.K.L. METAL.
M-417.
For superheated steam.

NK-MARINE-50F.
M-1612; 0.15 max Si, 0.55 max Si, 1.50 max Mn, 0.20-0.50 Cu, 0.40 max 0.50-0.80 Cr, 0.10 max Nb, 0.15-0.55 Al, bal Fe.
Hot rolled strip and sheets, as rolled, 1.4-13.0 mm thick: 50-62 kg/mm^2 TS; 37 kg/mm^2 YS; 25 El.
Sea water resistant steel.

NKS.
M-Japan; 3.7 Al, 17.7 Ni, 27.2 Co, 6.7 Ti, bal Fe.
For permanent magnets, electrical and magnetic equipment. High permeability.

NKS-1.
M-Japan; 8 Al, 15 Ni, 3 Cu, 1.25 Ti, 24 Co, bal Fe.
Cast: 20,000 TS; For electrical and magnetic equipment.
Similar to Alnico VI, permanent magnet.

NKS MAGNET.
M-Japan; 8 Al, 14 Ni, 24 Co, 3 Cu.
For permanent magnets; high coercive force.

NK STS-50.
M-1612; 0.15 max C, 0.20-0.50 Si, 1.0-1.5 Mn, 0.020 max P, 0.020 max S, bal Fe.
High strength plates to meet ship building code.

NK-T 95.
M-1612; 0.24 C, bal Fe.
Quenched, tempered and straightened seamless pipe and tube: 110,000 psi min TS; 95,000-125,000 psi YS.
High collapse value.

NM-100.
M-1629; 17.5 Cr, 10.5 W, 9.5 Co, 1.25 C, 0.75 V, bal Fe.
Powder metal alloy. Retains hardness better than Powder metal alloy.
AISI 44C at elevated temperatures.

NM-100.
M-1665, M-1629; 1.25 C, 17.5 Cr, 10.5 W, 9.5 Co, 0.75 V, bal Fe.
Hardened: Room temp: 290,000 TS; 245,000 YS; 1000°F: 240,000 TS 175,000 YS.
For aircraft bearings and other bearings and parts subject to dry friction at elevated temperatures; resistant to galling, corrosion and high temperature softening and oxidation.

N.M.C.
M-Eng.; 12 Ni, 5 Mn, 3.5 Cr, 0.5 C, 0.5 Si, bal Fe.
For valve inserts; heat resistant.

NMHG.
M-104; C, 29 Ni, 3 Cr, bal Fe.

NO. 000 DOUBLE EXTRA CARBON.
M-350; 1.1-1.3 C, bal Fe.
For intricate tools, broaches, drawing and threading dies; water hardening.

NO. 2 DIE METAL.
M-1391; 3.5-4.5 Al, 2.5-3.5 Cu, 0.02-1.0 Mg, bal Zn.
Sand cast: 37,000 TS; 3 El.
For stamping and drop hammer dies; subject to intergranular attack.

NO. 3 MINE TALABOT SPECIAL D.
M-1119; 0.95 C, bal Fe.
For tools, springs, taps, drills; water hardened.

NO. 6 MINE TULIPE SPECIAL D.
M-1119; 0.75 C, bal Fe.
For tools, springs, punches; water hardened.

NO. 7 ALLOY (MATHER & PLATT).
M-454; 3.0 C, 4-5 Cu, 12-14 Ni, 1.0-1.2 Mn, 4-6 Cr, 1.2-1.5 Si, bal Fe.
25,000-30,000 TS; 200-220 Brin.
For valves, centrifugal pumps; austenitic cast iron; abrasion and corrosion resistant.

NO. 50 SOLDER ALLOY.
M-1136; In-Pb.
For solder; M.P. 600°F.

NO.55 ALLOY.
M-Eng.; 27-30 Cr, bal Fe.
For glass-to-metal seals; same coefficient of expansion as glass.

NO. 155 PREMIUM SILVER BRAZING ROD.
M-1713; 55 Ag, bal Cu.
Cast: 50,000 TS.
For silver brazing; 1155°F working temperature.

NO. 155 PREMIUM SILVER BRAZING ROD
changed to **ALL-STATE NO. 155**.

NO. 444 ALLOY see **CLEVITE S 56**.

NO. 446 ALLOY.
M-Eng.; 27-30 Cr, bal Fe.
For glass-to-metal seals.

NO. 500 ALLOY.
M-U.S; Al alloy.
Cast: 27,000 TS; 20,000 YS; 2 El.
For casting; light weight.

NO. 712 ALLOY.
M-18; 29 Os, 40-50 Ir, bal Pt.
For contact points; corrosion resistant.

NO. 812.
M-357; 1.8 C, 12.25 Cr, bal Fe.
For blanking and coining dies; taps, reamers; non-deforming.

NO.999.
M-340; 0.7 C, 18 W, 4 Cr, 1 V, 5 Co, bal Fe.
For hogging cuts for fast speeds, cutters; high speed steel.

NO 1002.
M-U.S.; 22 Cr, 16 Ni, 7W, 1.5 Fe, 0.3 Al, 0.2 Ti, 0.3 Zr, 0.6 C, 0.5 Si, 0.7 M, 0.05 La, bal Co.
Good stress rupture at elevated temp. and resistant to oxidation.

NO. 1040 ALLOY.
M-Ger.; 15 Cu, 1 Mn, 3 Mo, 71 Ni, 10 Fe.
For electrical equipment; magnetic alloy.

NO. 3074 HOT WORK.
M-365; 0.40-0.45 C, 8-10 W, 2.5 Cr, 0.10-0.15 V, bal Fe.
For hot work tools and dies, forging mandrels: hot work steel.

NO. 9500-1.
M-216; 95 Bi, 5 Sn.
For solder; soft.

NOBELOY.
Au alloy.
For jewelry; corrosion resistant.

NOBILIUM.
M-US; 65 Co, 28 Cr, 0.1 Ni, 5 Mo, 0.4 C, 1 V, 0.5 Fe, 0.1 Mn, 0.05 Si.
Cast: 168,000 TS; 88,000 YS; 4 El.
For dentures; corrosion resistant.

NO-CHAT.
M-265; 90 W, 6 Ni, 4 Cu.
Sintered: 112,000 TS; 6 El; 250 Brin.
Density: 16.96.
Modulus of rigidity 19.2 x 10^6 psi.
For boring bars, grinding quills or arbors, cut-off tools.

NODULITE 60-40-18.
M-850; 3.3 C, 2.5 Si, 0.7 Mn, 0.05 Mg, bal Fe.
Cast: 60,000-80,000 TS; 40,000-60,000 YS; 18-25 El; 137-192 Brin.
For gears, pinions, cams, dies, idlers, track rollers, pumps.
Resists thermal shock; tough, ductile iron.

NODULITE 65-45-12.
M-850; 3.3 C, 2.5 Si, 0.7 Mn, 0.05 Mg, bal Fe.
Cast: 65,000-85,000 TS; 45,000-60,000 YS; 12-20 El; 143-207 Brin.
For pressure castings, pipe fittings, valves, cylinders, pump bodies.
Ductile iron, tough, high strength.

NODULITE 80-55-06.
M-850; 3.3 C, 2.5 Si, 0.7 Mn, 0.05 Mg, bal Fe.
Cast: 80,000-100,000 TS; 55,000-75,000 YS; 6-12 El; 179-269 Brin.
For gears, cams, bearings, pistons, crankshafts, sheaves, sprockets.
Ductile iron, high strength, wear resistant, tough.

NODULITE 100-70-03.
M-850; 3.3 C, 2.5 Si, 0.7 Mn, 0.05 Mg, bal Fe.
Cast: 100,000-120,000 TS; 70,000-90,000 YS; 3-10 El; 241-302 Brin.
For gears, crankshafts, pistons, camshafts, track shoes, brake drums.
Ductile iron, high strength and wear resistant, tough.

NODULITE 120-90-02.
M-850; 3.3 C, 2.5 Si, 0.7 Mn, 0.05 Mg, bal Fe.
Cast: 120,000-175,000 TS; 90,000-150,000 YS; 2-7 El; 269-388 Brin.
For pinions, gears, cams, machine guides, dies; pumps. Wear resistant ductile iron.

NODULOY 3.
M-1038; 2.8-3.2 Mg, 44-48 Si, 0.80-1.30 Ca, 1.20 max Al, bal Fe.
Magnesium-ferrosilicon additive for nodularization of ductile iron. Also known as NODULOR type.

NODULOY 3R.
M-1038; 2.8-3.3 Mg, 44-48 Si, 0.70-1.0 rare earths, 0.80-1.30 Ca, 1.2 max Al, bal Fe.
Magnesium-ferrosilicon with rare earths (NODULOY R group) for nodularization of ductile iron.

NODULOY 5.
M-1038; 5.0-6.0 Mg, 44-48 Si, 0.80-1.30 Ca, 1.2 max Al, bal Fe.
Magnesium-ferrosilicon additive for nodularization of ductile iron. Also known as NODULOY.

NODULOY 5-1C.
M-1038; 5.0-6.0 Mg, 44-48 Si, 0.90-1.1 Ce, 0.80-1.3 Ca, 1.2 max Al, bal Fe.
Magnesium-ferrosilicon with cerium (NODULOY C group) for nodularization of ductile iron.

NODULOY 5C.
M-1038; 5.0-6.0 Mg, 44-48 Si, 0.50-0.75 Ce, 0.80-1.3 Ca, 1.2 max Al, bal Fe.
Magnesium-ferrosilicon with cerium (NODULOY C group) for nodularization of ductile iron.

NODULOY 5LC.
M-1038; 5.0-6.0 Mg, 44-48 Si, 0.30-0.40 Ce, 0.80-1.30 Ca, 1.2 max Al, bal Fe.
Magnesium-ferrosilicon with cerium (NODULOY C group) for nodularization of ductile iron.

NODULOY 5R-1.
M-1038; 5.0-6.0 Mg, 44-48 Si, 0.60-0.80 rare earths, 0.80-1.30 Ca, 12 max Al, bal Fe.
Magnesium-ferrosilicon with rare earths (NODULOY R group) for nodularization of ductile iron.

NODULOY 5R-2.
M-1038; 5.0-6.0 Mg, 44-48 Si, 0.90-1.20 rare earths, 0.80-1.30 Ca, 1.20 max Al, bal Fe.
Magnesium-ferrosilicon with rare earths (NODULOY R group) for nodularization of ductile iron.

NODULOY 5R-3.
M-1038; 5.0-6.0 Mg, 44-48 Si, 1.7-2.0 rare earths, 0.80-1.30 Ca, 1.20 max Al, bal Fe.
Magnesium-ferrosilicon with rare earths, (NODULOY R group) for nodularization of ductile iron.

NODULOY 9.
M-1038; 8.5-10.0 Mg, 44-48 Si, 1.0-1.5 Ca, 1.20 max Al, bal Fe.
Magnesium-ferrosilicon additive for nodularization of ductile iron. Also known as NODULOY type.

NODULOY 9C.
M-1038; 8.5-10.0 Mg, 44-48 Si, 0.50-0.70 Ce, 1.0-1.5 Ca, 1.2 max Al, bal Fe.
Magnesium-ferrosilicon with cerium (NODULOY C group) for nodularization of ductile iron.

NODULOY 9LC.
M-1038; 8.5-10.0 Mg, 44-48 Si, 0.30-0.40 Ce, 1.0-1.5 Ca, 1.2 max Al, bal Fe.
Magnesium-ferrosilicon with cerium (NODULOY C group) for nodularization of ductile iron.

NODULOY 9R.
M-1038; 8.5-10.0 Mg, 44-48 Si, 0.70-1.0 rare earths, 0.80-1.30 Ca, 1.20 max Al, bal Fe.
Magnesium-ferrosilicon with rare earths (NODULOY R group) for nodularization of ductile iron.

NO-DU-MAG.
M-54; 3.3 C, 10-11 Ni, 5.6 Mn, 0.17 Mg, 2.5 Si, bal Fe.
Cast: 53,700-62,700 TS; 40,300-49,300 YS; 8-12 El; 13-18 RA; 260 Brin.
For switch gears, resistant grids, magnetic chucks.
Austenitic, ductile cast iron.

NOGROTH.
M-630; 1.85 T.C., 1.05 Si, 0.16 P, 1.5 Ni, 0.33 Cr, 0.35 Mn, bal Fe.
Annealed: 102,000 TS; 85,000 YS; 9.5 El; 250 Brin.
For pump liner, piston, valves; cast iron.

NOHEET.
M-581; 98 Pb, 1.4 Na, 0.11 Sb, 0.1 Sn.
For bearings, metal or die cast; anti-friction metal.

NOIL.
M-134; 20 Sn, 80 Cu.
158 Brin.
For piston rings.

NOMAG.
M-54; M-86; 3 C, 6 Mn, 2.3 Si, 10 Ni, bal Fe.
Cast: 22,000-26,000 TS; 158-160 Brin.
For switch covers, resistance grids; austenitic, cast iron, non-magnetic.

NO. MR-100 BRONZE.
M-1174; Cu composite.
For bearings; sintered; 25-30% porosity.

NON-CORRODITE NO. 25.
M-186; 22.2 Cr, 0.4 Ni, 1.57 Si, 1 Mn, 0.15 Cu, 0.34 C, bal Fe.
For castings, corrosion resisting parts; non-corrosive; stainless.

NONGRAM.
87 Cu, 11 Sn, 2-3 Zn.
35,000 TS; 16 El; 62 Brin.
For bushings, bearings, valves.

NON-GRAN BRONZE.
M-1153; Sn, bal Cu.
Sintered: 40,000 TS; 22,000 YS; 15 El; 11 RA; 80 Brin.
For bolt headers, pulleys, machinery; nongranular structure.

NON-HARDENING (VALCAN).
M-275; C, alloy, bal Fe.
For tools, drawing dies; tough.

NON-OXIDIZABLE.
62 Fe, 25 Cr, 10 Mn, 1.1 C, 0.95 Si.
For corrosion and heat resisting parts, abrasion resistant parts; U.S.P 1,333,151; corrosion, heat and abrasion resistant.

NON-PAREIL.
M-562; 78 Pb, 17 Sb, 5 Sn.
24 Brin.
For bearings, solders; antifriction metal; M.P. 300°C.

NON-SCALING-1.
30-40 Ni, 15-20 Cr, 3.5 Si, 1.25 Cu, bal Fe.
For furnace parts; heat resisting.

NON-SCALING-2.
24 Ni, 24 Cr, 3 Si, bal Fe.
For furnace parts; heat resisting.

NON-SHRINKABLE.
M-115; 0.9 C, 1.2 Mn, 0.5 Cr, 0.5 W, bal Fe.
For general tools, broaches; non-deforming.

NON-SHRINKABLE.
M-275; 0.9 C, 1.2 Mn, 0.5 Cr, bal Fe.
For general tools, dies; non-deforming.

NON-SHRINKABLE DRILL ROD.
M-342, M-115; 0.95 C, 1.2 Mn, 0.5 Cr, 0.5 W, 0.2 V, bal Fe.
For tools, punches; non-deforming.

NON-SHRINKING PATENT.
87 Pb, 6 Sb, 6 Sn, 1.3 Cd.
For impressions, type.

NONSULITE.
M-184; 0.36-0.50 C, 28 Cr, 8 Ni, bal Fe.
For heat and corrosion resistant parts; now Michiana No. 48.

NON-TARNISHABLE.
63.6 Cu, 31 Zn, 3.25 Sn, 2 Pb.
For corrosion and tarnish resisting parts; corrosion resisting.

NON-TEMPERING.
M-750; 0.35 C, 0.75 Mn, 0.45 Si, 0.8 Cr, 0.25 W, 0.5 Mo, bal Fe.
Water or oil hardening shock resisting tool steel.

NON-TEMPERING.
M-24; 0.35 C, 0.7 Mn, 0.8 Cr, 0.3 Mo, 0.3 Cu, bal Fe.
For chisels, punches, blacksmith tools; shock resisting.

NONVAR.
M-56; 0.92 C, 1.75 Mn, 0.3 Si, bal Fe.
For tools, dies; oil hardening, non-deforming.

NO-OX.
M-1678; C, 6 Al, Si, Mn, bal Fe.
Bar annealed: B78.8 Rock; 68,000 TS; 50,000 YS; 30 El.
For pack carburizing containers, baffles and flame deflectors, burners, furnace parts, retorts, muffles.
High temperature alloy. Aluminized. Oxidation resistant to 2200°F.

NO. PS-10 SINTERED IRON.
M-1174; Fe.
For pole pieces, stators and rotors in motors; for electromagnetic uses, sintered.

NO. R-35 SINTERED IRON.
M-1174; Fe.
For bearings; sintered; 18% porosity.

NORANDA-1 now **NORANDA 2680.**

NORANDA-3 now **NORANDA 3300.**

NORANDA-5 now **NORANDA 2400.**

NORANDA-6 now **NORANDA 3600.**

NORANDA-18 now **NORANDA 3300.**

NORANDA-19 now **NORANDA 6750.**

NORANDA-20 now **NORANDA 4430.**

NORANDA-24 now **NORANDA 4640.**

NORANDA-25 now **NORANDA 2200.**

NORANDA-26 now **NORANDA 2100.**

NORANDA-28 now **NORANDA 4820.**

NORANDA-29 now **NORANDA 4850.**

NORANDA-30 now **NORANDA 4430.**

NORANDA-32 now **NORANDA 5050.**

NORANDA-35 now **NORANDA 5210.**

NORANDA-46 now **NORANDA 4620.**

NORANDA-53 now **NORANDA 6080.**

NORANDA-54 now **NORANDA 6870.**

NORANDA-62 now **NORANDA 3530.**

NORANDA-63 now **NORANDA 3400.**

NORANDA-64 now **NORANDA 3320.**

NORANDA-69 now **NORANDA 2600.**

NORANDA-73 now **NORANDA 6670.**

NORANDA-85 now **NORANDA 2300.**

NORANDA-89 now **NORNADA 3140.**

NORANDA-92 now **NORANDA 4250.**

NORANDA-102 now **NORANDA 1100.**

NORANDA-106 now **NORANDA 1200.**

NORANDA-110 now **NORANDA 1220.**

NORANDA-136 now **NORANDA 3770.**

NORANDA-520 now **NORANDA 7100.**

NORANDA-531 now **NORANDA 7150.**

NORANDA-555 now **NORANDA 7700.**

NORANDA-565 now **NORANDA 7520.**

NORANDA-567 now **NORANDA 7450.**

NORANDA-580 now **NORANDA 7930.**

NORANDA-609 now **NORANDA 6510.**

NORANDA-707 now **NORANDA 6371.**

NORANDA 1100.
 M-1608; 99.9 min Cu.
 Electrolytic tough pitch copper; for bus bars, conductors, wave guides, copper to brass seals.
 CDA and UNS C11000; was NORANDA 102.

NORANDA 1140.
 M-1608; 99.9 Cu, 0.034 min Ag, (10 oz/ton).
 Tough pitch copper with silver.
 CDA & UNS C11400.

NORANDA 1150.
 M-1608; 99.9 min Cu, 0.054 min Ag, (16 oz/ton).
 Tough pitch copper with silver; resists softening by heat.
 CDA and UNS C11500; was NORANDA 3111.

NORANDA 1160.
 M-1608; 99.9 Cu, 0.085 min Ag, (25 oz/ton).
 Tough pitch copper with silver; resists softening by heat; for commutator segments.
 CDA and UNS C11600.

NORANDA 1200.
 M-1608; 99.9 min Cu, 0.012 max P.
 Phosphorus deoxidized copper.
 For radiators, commutators, switches.
 CDA and UNS C12000; was NORANDA 106.

NORANDA 1220.
 M-1608; 99.9 min Cu, 0.02 P.
 Phosphorus deoxidized copper.
 For air conditioners, gas lines, hydraulic and oil lines.
 CDA & UNS C12200; was NORANDA 110.

NORANDA 1450.
 M-1608; 99.9 min Cu, 0.4-0.6 Te, 0.012 max P.
 Tellurium bearing phosphorus deoxidized copper.
 For improved machining.
 CDA & UNS C14500.

NORANDA-1502 now **NORANDA 5190.**

NORANDA-1532 now **NORANDA 5210.**

NORANDA 2100.
 M-1608; 94-96 Cu, 4-6 Zn.
 Gilding brass (5%); for jewelry, tokens.
 CDA and UNS C21000; was NORANDA 26.

NORANDA 2200.
 M-1608; 90 Cu, 10 Zn.
 Commercial bronze; for costume jewelry, ornamental trim, weather stripping.
 CDA & UNS C22000; was NORANDA 25.

NORANDA 2300.
 M-1608; 85 Cu, 15 Zn.
 Red brass (85%); for radiator cores, conduit, pump lines, trim.
 CDA & UNS C23000; was NORANDA 85.

NORANDA 2400.
 M-1608; 80 Cu, 20 Zn.
 Low brass (80%); for ornamental metal work, medallions, pump liners.
 CDA and UNS C24000; was NORANDA 5.

NORANDA 2600.
 M-1608; 30.5 Zn, 69.5 Cu.
 Cartridge brass; good strength and ductility.
 For grillwork, lamp fixtures, cartridge cases.
 CDA and UNS C26000; was NORANDA 69.

NORANDA 2680.
 M-1608; 66 Cu, 34 Zn.
 Yellow brass; good strength, particularly after cold work.
 For hardware, reflectors, plumbing parts.
 CDA and UNS C26800; was NORANDA 1.

NORANDA-3111 now **NORANDA 1150.**

NORANDA 3140.
 M-1608; 2.0 Pb, 8.5 Zn, bal Cu.
 Leaded commercial bronze; free-machining for screw machine parts.
 CDA and UNS C31400; was NORANDA 89.

NORANDA 3200.
 M-1608; 2.0 Pb, 85.0 Cu, 13 Zn.
 Leaded red brass; free-machining, for screw machine operations.
 CDA and UNS C32000.

NORANDA 3300.
 M-1608; 0.5 Pb, 32.5 Zn, 67.0 Cu.
 Low leaded brass; for tubing, plumbing, pumps, power cylinders. Good machining.
 CDA and UNS C33000; was NORANDA 18.

NORANDA 3320.
 M-1608; 1.75 Pb, 67.0 Cu, bal Zn.
 High leaded brass; good machinability.
 For screw machine parts.
 CDA and UNS C33200; was NORANDA 64.

NORANDA 3400.
 M-1608; 1.1 Pb, 64.5 Cu, bal Zn.
 Medium leaded brass; good machinability.
 For hardware, clock plates, fasteners.
 CDA and UNS C34000; was NORANDA 63.

NORANDA 3530.
 M-1608; 2.0 Pb, 62.0 Cu, bal Zn.
 High leaded brass (62%); for hardware, watch cases, clock and watch parts.
 CDA and UNS C35300; was NORANDA 62.

NORANDA 3600.
M-1608; 3.4 Pb, 61.5 Cu, bal Zn.
Free cutting brass; for automatic screw machine parts; clock gears.
CDA and UNS C36000; was NORANDA 6.

NORANDA 3770.
M-1608; 2.0 Pb, 59.5 Cu, bal Zn.
Forging brass; for forged marine hardware, valve stems, fasteners, bolts, plumbing.
CDA and UNS C37700.

NORANDA 4250.
M-1608; 1.9 Sn, 9.1 Zn, bal Cu.
Tin commercial bronze; for heat exchangers.
CDA and UNS C42500; was NORANDA 92.

NORANDA 4430.
M-1608; 1.0 Sn, 0.02-0.10 As, 71.0 Cu, bal Zn.
Arsenical Admiralty grade; for marine hardware.
CDA and UNS C44300; was NORANDA 30.

NORANDA 4620.
M-1608; 0.5-1.0 Sn, 64.0 Cu, bal Zn.
Naval brass; for hardware, fixtures.
CDA and UNS C46200; was NORANDA 46.

NORANDA 4640.
M-1608; 0.65 Sn, 0.20 Pb, 60.5 Cu, bal Zn.
Naval brass, uninhibited; for marine hardware, valve stems, shafts.
CDA and UNS C46400; was NORANDA 24.

NORANDA 4820.
M-1608; 0.75 Sn, 0.70 Pb, 60.5 Cu, bal Zn.
Naval brass, medium leaded; for marine hardware, screw machine products.
CDA and UNS C48200; was NORANDA 28.

NORANDA 4850.
M-1608; 0.75 Sn, 1.75 Pb, 60.5 Cu, bal Zn.
Naval brass, high leaded; free machining, for screw machine products.
CDA and UNS C48500; was NORANDA 29.

NORANDA 5050.
M-1608; 1.3 Sn, 0.3 Zn, 0.35 max P, bal Cu.
Phosphor bronze (1.25% E); for electrical contacts, flexible hose, pole-line hardware.
CDA and UNS C50500; was NORANDA 32.

NORANDA 5100.
M-1608; 5.0 Sn, 0.30 Zn, 0.35 max P, bal Cu.
Phosphor bronze (5% A); for hardware.
CDA and UNS C51000.

NORANDA 5190.
M-1608; 6.0 Sn, 0.35 max P, 0.30 Zn, bal Cu.
Phosphor bronze; for springs, clutch discs, hardware.
CDA and UNS C51900; was NORANDA 1502.

NORANDA 5210.
M-1608; 8.0 Sn, 0.35 max P, 0.20 Zn, bal Cu.
Phosphor bronze (8% C); high strength.
For springs, switch parts, wire brushes.
CDA and UNS C52100; was NORANDA 1532.

NORANDA 5211.
M-1608; 8.0 Sn, 0.1 P, bal Cu.
Phosphor bronze; for Bourdon tubing, springs, clutch discs, sleeve bushings.
Was NORANDA 35.

NORANDA 6080.
M-1608; 5.5 Al, 0.25 As, bal Cu.
Aluminum bronze; for fasteners, structural components, condensers.
CDA and UNS C60800; was NORANDA 53.

NORANDA 6371.
M-1608; 7.15 Al, 2.0 Si, bal Cu.
Aluminium-silicon bronze; good strength.
For bushings, fasteners, marine fittings.
Was NORANDA 707.

NORANDA 6510.
M-1608; 1.9 Si, bal Cu.
Low silicon bronze; for hydraulic pressure lines, bolts, pole line hardware.
CDA and UNS C65100; was NORANDA 609.

NORANDA 6670.
M-1608; 1.25 Mn, 70.0 Cu, bal Zn.
Manganese brass; for shafting, valve stems, pump rods, hardware.
CDA and UNS C66700; was NORANDA 73.

NORANDA 6750.
M-1608; 0.3 Mn, 1.0 Fe, 1.0 Sn, 58.0 Cu, bal Zn.
Manganese bronze A; for automotive clutch discs, shafting, hardware.
CDA and UNS C67500; was NORANDA 19.

NORANDA 6810.
M-1608; 1.0 Sn, 1.0 Fe, 0.5 max Mn, 58.0 Cu, bal Zn.
Bronze, low fuming.
CDA and UNS C68100.

NORANDA 6870.
M-1608; 2.0 Al, 0.10 max As, 78.0 Cu, bal Zn.
Aluminum brass, arsenical; for condenser tubes, heat exchangers.
CDA and UNS C68700; was NORANDA 54.

NORANDA 7060.
M-1608; 0.75 Mn, 1.0 Fe, 10.0 Ni, bal Cu.
Copper-nickel, 10%; for fasteners, decorative trim.
CDA and UNS C70600;

NORANDA 7100.
M-1608; 0.75 Mn, 0.40 Fe, 20.0 Ni, bal Cu.
Copper-nickel 20%; for valves, condenser plates, evaporators, fasteners.
CDA and UNS C71000; was NORANDA 520.

NORANDA 7150.
M-1608; 0.75 Mn, 0.40 Fe, 30.0 Ni, bal Cu.
Copper-nickel, 30%; for valves, pumps, tanks, fasteners.
CDA and UNS C71500; was NORANDA 531.

NORANDA 7450.
M-1608; 10.0 Ni, 65 Cu, 0.15 Mn, bal Zn.
Nickel silver, 10%; for costume jewelry, camera parts, tableware.
CDA and UNS C74500; was NORANDA 567.

NORANDA 7520.
M-1608; 18.0 Ni, 0.15 Mn, 65.0 Cu, bal Zn.
Nickel silver 18%; for diaphragms, springs, slide fasteners, jewelry.
CDA and UNS C75200; was NORANDA 565.

NORANDA 7620.
M-1608; 12.0 Ni, 60.0 Cu, bal Zn.
Nickel silver; for decorative hardware.
CDA and UNS C76200.

NORANDA 7700.
M-1608; 18.0 Ni, 0.15 Mn, 55.0 Cu, bal Zn.
Nickel silver, 55-18; for tableware, springs, fixtures, hardware.
CDA and UNS C77000; was NORANDA 555.

NORANDA 7930.
M-1608; 12.0 Ni, 1.5 Pb, 25.0 Zn, 0.15 Mn, bal Cu.
Leaded nickel silver; good machinability.
For costume jewelry, tableware, hardware.
CDA and UNS C79300; was NORANDA 580.

NORBIDE.
M-532; 78 B, 0.14 Fe, 21 C, (B,C).
For wire drawing dies; drilling W C die nibs, nozzles for abrasive blasting; cemented boron-carbide.

NORELCO CONTACT 15.
M-1139; 0.05-0.10 C, 0.5-1.0 Mn, 0.1-0.3 Si, bal Fe.
Welded: 75,000 TS; 65,000 YS; 35 El.
For welding electrodes; lime coated.

NORELCO CONTACT 18.
M-1139; 0.08-0.15 C, 0.5-0.7 Mn, 0.1-0.2 Si, bal Fe.
For contact welding electrodes; arc welding, organic coating.

NORELCO CONTACT 20.
M-1139; 0.08-0.15 C, 0.5-0.7 Mn, 0.1-0.2 Si, bal Fe.
Welded: 67,000 TS; 53,000 YS; 25 El.
For welding electrodes; iron oxide coating.

NORGRIP.
M-1787.
WC plus alloy.
For anti-skid tire studs and studded straps for snow and ice chains.

NORIS CME.
M-1338; 0.20 C, 1.25 Mn, 1.15 Cr, bal Fe.
For gears, cams, camshafts, mandrels; case hardening steel.

NORMANNA.
M-333; C, alloy, bal Fe.
For tools; oil hardening.

NORMAR.
M-289; 0.9 C, 0.5 Cr, 1.5 Mn, 0.25 Mo, bal Fe.
For dies, punches, crimpers; non-deforming, shock resistant.

NORO 30.
M-1338; 0.95 C, W, Co, Cr, V, bal Fe.
For lathe and planer tools, reamers, broaches; high speed steel.

NORO 40.
M-1338; 0.82 C, 4 Cr, 0.85 Mo, 1.6 V, 8.7 W, bal Fe.
For lathe and planer tools, hobs, reamers, taps; high speed steel.

NORO 50.
M-1338; 0.85 C, Cr, V, W, Mo, bal Fe.
For lathe and planer tools, drills, hobs, taps; high speed steel.

NORO 60.
M-1338; 0.86 C, 4.1 Cr, 2.5 V, 0.85 Mo, 12 W, bal Fe.
For lathe and planer tools, hobs, broaches; high speed steel.

NORO 60 EXTRA.
M-1338; 1.3 C, 4.3 Cr, 0.85 Mo, 3.8 V, 12 W, bal Fe.
For engraving tools, form dies, taps; high speed steel.

NORO 60 SPEZIAL.
M-1338; 0.74 C, 4.1 Cr, 1.1 V, 18.5 W, bal Fe.
For lathe and planer tools, reamers, hobs, taps, drills; high speed steel.

NORO EXTRA D.
M-1338; 0.35 C, 1.05 Cr, 0.18 V, 1.85 W, 0.9 Si, bal Fe.
For cold work tools, upsetters, headers, dies; oil hardened, tough.

NORO EXTRA L.
M-1338; 0.55 C, 1.05 Cr, 0.18 V, 1.8 W, 0.9 Si, bal Fe.
For cold work tools, upsetters, headers, dies; oil hardened, tough.

NORO MKW.
M-1338; 0.38 C, Si, Cr, V, bal Fe.
For gears, pinions, shafts, bolts, crankshafts; oil hardened, shock resistant.

NORO MRO.
M-1338; 0.45 C, Si, Cr, V, bal Fe.
For gears, springs, shafts, arbors, crankshafts; oil hardened, shock resistant.

NORO REKORD 30.
  M-1338; 0.86 C, 2.8 Co, 4.3 Cr, 0.85 Mo, 2.1 V, 12 W, bal Fe.
  For lathe and planer tools, reamers, hobs, taps; high speed steel.

NORO REKORD 50.
  M-1138; 0.80 C, Co, Cr, V, Mo, W, bal Fe.
  For lathe and planer tools, drills, taps, reamers; high speed steel.

NORO REKORD 53.
  M-1138; 0.79 C, 4.75 Co, 4.3 Cr, 0.75 Mo, 1.5 V, 18 W, bal Fe.
  For lathe and planer tools, hobs, reamers; high speed steel.

NORO REKORD 53V.
  M-1338; 1.3 C, W, Co, V, Cr, Mo, bal Fe.
  For engraving tools, forming dies; reamers; high speed steel.

NORO REKORD 110.
  M-1138; 0.76 C, 10 Co, 4.2 Cr, 0.8 Mo, 1.8 V, 18 W, bal Fe.
  For lathe and planer tools, cutters, reamers, hobs; high speed steel.

NORTH STAR.
  M-333; C, W, bal Fe.
  For dies; non-deforming.

NORTON 90-10.
  M-532; 90 TA, 10 W.
  High temp. strength and corrosion resistance.
  For missile components, electronics, high temperature devices.

NORTON BORON MASTER ALLOY.
  M-532; 8.5 B, 2.7-3.3 Si, 81-84 Fe, 1-1.7 C, bal Ti, Al.
  For steel inoculant; increase depth of hardening of steel.

NORTON T-111.
  M-532; 90 Ta, 8 W, 2 Hf.
  Weldable, ductile; very high strength to weight ratio at high temperatures.
  For missile hardware, supersonic air and spacecraft, liquid reactors.

NOVALITE.
  M-Eng.; 12.5 Cu, 0.3 Mg, 1.4 Ni, 0.5 Fe, 0.5 Si, bal Al.
  For pistons; cast.

NOVAR B.
  M-1307; 0.1 max C, 17 Cr, 1.8 Mo, Ti = 7 x C, bal Fe.
  For welded chemical plant and oil refinery equipment; stainless, stabilized.

NOVOKONSTANT.
  M-297; 82.5 Cu, 3 Al, 13.5 Mn, 1 Fe.
  Rolled: 142,200 TS; 0.5 El.
  Annealed: 78,200 TS; 25 El.
  For electrical resistances; heat resistant.

NOVONIT 4773.
  M-72; 0.10 max C, 30 Cr, 2.0 max Mo, bal Fe.
  Stainless for high temperature operation.

NOVONIT FALC 233.
  M-72; 0.02 C, 23 Cr, 6 Ni, Mo, bal Fe.
  Stainless steel for high temperature operation.

NOVONOX-A17.
  M-1307; 0.10 C, 17 Cr, 7 Ni, bal Fe.
  Annealed: 110,000 TS; 40,000 YS; 60 El; B 85 Rock.
  Hard: 185,000 TS; 140,000 YS; 8 El; C 41 Rock.
  For aircraft structural members, diaphragms, household utensils.
  Type 301 stainless steel, good ductility.
  Work hardens.

NOVONOX-A18.
  M-1307; 0.12 C, 18 Cr, 9 Ni, bal Fe.
  Annealed; 90,000 TS; 40,000 YS; 50 El; B 85 Rock.
  For chemical and textile plant equipment, food processing apparatus, tanks, vessels, agitators.
  Type 302 Stainless steel, austenitic.

NOVONOX-A18Z.
  M-1307; 0.06 C, 18 Cr, 11 Ni, 0.4 Cb, bal Fe.
  Annealed: 85,000 TS; 30,000 YS; 55 El; B 80 Rock.
  For welded structures, chemical and pharmaceutical plant equipment.
  Type 321 stainless steel, stabilized, austenitic.

NOVONOX-A18ZN.
  M-1307; 0.06 C, 12 Ni, 18 Cr, 0.5 Cb, bal Fe.
  Annealed: 85,000 TS; 35,000 YS; 60 El; B 80 Rock.
  For welded structures, chemical and pharmaceutical plant equipment.
  Type 347 stainless steel, stabilized, austenitic.

NOVONOX-ALC 18.
  M-1307; 0.03 max C, 19 Cr, 10 Ni, bal Fe.
  Annealed: 77,000 TS; 30,000 YS; 60 El; 140 Brin.
  For architectural trim, chemical plant equipment, food processing equipment.
  Type 304L stainless steel, austenitic, non-hardenable.

NOVONOX-AS18.
  M-1307; 0.08 max C, 19 Cr, 10 Ni, bal Fe.
  Annealed: 85,000 TS; 35,000 YS; 60 El; 150 Brin.
  For architectural trim, chemical plant equipment, tanks, food processing equipment.
  Type 304 stainless steel, austenitic.
  Non-hardenable.

NOVONOX-AS 182.
M-1307; 0.05 C, 17 Cr, 11 Ni, 3 Mo, bal Fe.
Annealed: 85,000 TS; 35,000 YS; 50 El; B 80 Rock.
For chemical plant equipment, kettles, agitators, evaporators, tanks, valve trim.
Type 316 stainless steel, austenitic, acid resistant.

NOVONOX-AS 183.
M-1307; 0.06 C, 17 Cr, 12 Ni, 2.5 Mo, bal Fe.
Annealed: 80,000-90,000 TS; 30,000-40,000 YS; 40-60 El; 70-80 RA; B 78-85 Rock.
For chemical plant equipment, agitators, evaporators, tanks, valve trim.
Type 316-319 stainless steels, austenitic, acid resistant.

NOVONOX-F 13 A1.
M-1307; 0.06 C, 13 Cr, 0.2 Al, bal Fe.
Annealed: 70,000 TS; 40,000 YS; 30 El; 160 Brin.
For annealing boxes, oil refining equipment, quenching racks.
Type 405 stainless steel, ferritic.

NOVONOX-F17.
M-1307; 0.10 C, 16 Cr, bal Fe.
Annealed: 70,000 TS; 40,000 YS; 30 El; 140 Brin.
For automotive trim, hardware, oil burners, fasteners.
Type 430 stainless steel, high heat and corrosion resistance.

NOVONOX-FA26.
M-1307; 0.12 C, 28 Cr, 5 Ni, 1 Mo, bal Fe.
Annealed: 105,000 TS; 80,000 YS; 25 El; 230 Brin.
Ht. Tr.: C 45-50 Rock.
For valves, valve fittings, pumps.
Type 329 stainless steel, precipitation hardening, corrosion and heat resistant.

NOVONOX-T13.
M-1307; 0.20 C, 13 Cr, bal Fe.
Annealed: 95,000 TS; 50,000 YS; 25 El; B 92 Rock.
Cold drawn: 105,000 TS; 85,000 YS; 17 El; B 95 Rock.
For cutlery, surgical and dental equipment, gears, shafts.
Type 420 stainless steel, hardenable.

NOVONOX-T131.
M-1307; 0.20 C, 13 Cr, 1.2 Mo, bal Fe.
Annealed: 95,000 TS; 40,000 YS; 25 El; 55 RA; B 92 Rock.
Heat Treated: 250,000 TS; 215,000 YS; 8 El; 25 RA; C 50 Rock.
For cutlery, surgical instruments, ball bearings, valves, hardware.
Corrosion resistant, hardenable.

NOVONOX-TT131.
M-1307; 0.15 C, 13 Cr, 1.2 Mo, bal Fe.
Annealed: 80,000 TS; 40,000 YS; 25 El; B 93 Rock.
Heat Treated: 200,000 TS; 160,000 YS; 12 El; C 45 Rock.
For surgical instruments, cutlery, table flatware, oil refinery equipment.
Corrosion resistant, hardenable.

NOVOSTON.
M-452; 1.5-5.0 Ni, 7-9 Al, 11-13 Mn, 2-4 Fe, bal Cu.
Cast.
For ship propellers; Al bronze, corrosion resistant.

NOVOTHERM 7A.
M-72; 0.10 C, 18 Mn, 12 Cr, bal Fe.
For special oil refinery equipment.

NOVOTHERM 8A.
M-1332; 0.12 max C, 0.6 Si, 18 Cr, 10 Ni, Ti, bal Fe.
Annealed: 78,000-108,000 TS; 38,000 YS; 40 El; 140-190 Brin.
For welded chemical plant equipment; austenitic, resists scaling to 800°C.

NOVOTHERM 8F.
M-1332; 0.10 max C, 0.8 Si, 0.8 Al, 6.5 Cr, bal Fe.
Annealed: 64,000-85,000 TS; 36,000 YS; 20 El; 140-185 Brin.
For oil refinery equipment; AISI 501 and 502; resists scaling to 800°C.

NOVOTHERM 9F.
M-1332; 0.10 max C, 1.2 Si, 1.0 Al, 13 Cr, bal Fe.
Annealed: 72,000-92,000 TS; 43,000 YS; 15 El; 160-205 Brin.
For oil refinery equipment; resists scaling to 950°C.

NOVOTHERN 10A.
M-1332; 0.15 max C, 2 Si, 20 Cr, 12 Ni, bal Fe.
Annealed: 85,000-108,000 TS; 43,000 YS; 40 El; 145-190 Brin.
For heat treating boxes, furnace parts; AISI 309; austenitic, resists scaling to 1050°C.

NOVOTHERM 10F.
M-1332; 0.10 max C, 1.0 Si, 1.0 Al, 18 Cr, bal Fe.
Annealed: 72,000-92,000 TS; 43,000 YS; 12 El; 165-210 Brin.
For oil refinery equipment; AISI 405; resists scaling to 1050°C.

NOVOTHERM 10 FZ.
M-72; 0.12 C, 21 Cr, Ti, bal Fe.
Stainless for high temperature operation, and oil refinery equipment.

NOVOTHERM 11FA.
M-1332; 0.20 max C, 1.0 Si, 25 Cr, 4.0 Ni, bal Fe.
Annealed: 85,000-107,000 TS; 58,000 YS; 25 El; 175-200 Brin.
For heat treating boxes, furnace parts, oil refinery equipment; AISI 446; resists scaling to 1100°C.

NOVOTHERM 12A.
M-1332; 0.15 max C, 2 Si, 25 Cr, 20 Ni, bal Fe.
Annealed: 85,000-108,000 TS; 43,000 YS; 40 El; 175-190 Brin.
For furnace parts and equipment; AISI 310; austenitic, resists scaling to 1200°C.

NOVOTHERM 12F.
M-1332; 0.10 max C, 1.5 Si, 1.5 Al, 24 Cr, bal Fe.
Annealed: 72,000-92,000 TS; 50,000 YS; 10 El; 170-215 Brin.
For oil refinery equipment; resists scaling to 1200°C.

NOVOTHERM 70.
M-1307; 0.12 max C, 0.8 Al, 6.5 Cr, bal Fe.
For oil refinery equipment; heat and creep resistant.

NOVOTHERM 85.
M-1307; 0.12 max C, 1.2 Si, 1 Al, 13 Cr, bal Fe.
Annealed: 75,000 TS; 40,000 YS; 35 El; 70 RA; 155 Brin.
For oil refinery equipment; heat and creep resistant.

NOVOTHERM 85 F.
M-72; 0.08 C, 17 Cr, bal Fe.
Ferritic stainless steel, for oil refinery equipment.

NOVOTHERM 100.
M-1307; 0.12 max C, 1 Si, 1 Al, 18 Cr, bal Fe.
Annealed: 80,000 TS; 50,000 YS; 25 El; 50 RA; 150 Brin.
For oil refinery equipment; heat and creep resistant.

NOVOTHERM 105.
M-1307; 0.15 C, 2 Si, 19.5 Cr, 9.5 Ni, bal Fe.
Annealed: 80,000 TS; 35,000 YS; 55 El; 75 RA; 150 Brin.
For chemical plant equipment, tanks, mixers, filters; Type 302; stainless. austenitic.

NOVOTHERM 110.
M-1307; 0.12 max C, 1.5 Si, 1.5 Al, 24 Cr, bal Fe.
For furnace parts, oil refinery equipment; heat and creep resistant.

NOVOTHERM 120 A.
M-72; 32 Co, 25 Cr, Ni, bal Fe.
High temperature alloy.

NOVOTHERM 130.
M-1307; 0.15 C, 2 Si, 24 Cr, 19 Ni, bal Fe.
Annealed: 100,000 TS; 45,000 YS; 50 El; 65 RA; 185 Brin.
For furnace parts, valves, pumps, turbine parts; Type 310; stainless, austenitic.

NOVOTHERM NC36.
M-1332; 0.12 max C, 1.8 Si, 16 Cr, 36 Ni, bal Fe.
Annealed: 78,000-108,000 TS; 38,000 YS; 40 El; 140-185 Brin.
For furnace parts and equipment; austenitic, resists scaling to 1100°C.

NO. X-10 COMPOSITE METAL.
M-1174; Fe composite.
For machine parts; sintered.

NO. X-20 COMPOSITE METAL.
M-1174; Fe composite.
Hardened: 500 Brin.
For machine parts; sintered, hardenable.

NOXIDA-C.
M-72; 63 Ni, 16 Cr, 16 Mo, 6 max Fe.
For pharmaceutical and chemical plant equipment, valves, pumps.
Heat and corrosion resistant.

NOXIS.
M-1115; 0.3-0.4 C, 13-15 Ni, 13-15 Cr, 2-3 W, bal Fe.
Heat treated: 107,000 TS; 50,000 YS; 40 El.
For valves; austenitic, corrosion and heat resistant.

NP-464.
M-10; 0.35 C, 2.3 Mn, 0.20 Si, 25.0 Cr, 5.5 Ni, 0.40 N, bal Fe.
For diesel exhaust valves.

N.P.L. "Y" ALLOY.
M-311, M-338; 93 Al, 2 Ni, 1.5 Mg, 4 Cu.
Chill cast: 29,000 TS; 25,000 YS; 1 El; 80 Brin.
Heat treated: 43,000 TS; 34,000 YS; 3 El; 105 Brin.
For airplane castings, pistons, crankcases, cylinder heads; corrosion resistant.

NR-106 see SYLVANIA NR-106.

NR-203 ETC see also INNERSHIELD NR-203 ETC.

NR-NICKEL.
M-303; low C, bal Ni.

NS.
M-1083; 2.0-3.0 Co, 0.37-0.70 Be, 0.08-0.15 Si, bal Cu.
Bar: 90,000-130,000 TS; 9-20 El; 170-230 Brin.
Forged; 90,000-130,000 TS; 9-20 El; 170-230 Brin.
For machine parts and resistance welding equipment. High strength and hardness.
Age-hardenable.
BS 4577 and ISO 5182 alloy A/3/1.

NS-5.
M-264; 50 Ni, 46 Cu, 2.2 Si, 1.9 Mn.
100,000 TS; 82,500 YS; 2 El; 300 Brin.
For valve seats and disks; wear, abrasion and corrosion resistant.

NS-18-2.
M-1507; 0.10 C, 12 Mn, 0.50 Si, 18.0 Cr, 1.6 Ni, 0.34 N, bal Fe.
Low nickel austenitic stainless steel wire.
Drawn (0.018" dia): 320,000 TS; 3.2 El.
Non-magnetic after severe cold work.
Hardens by cold work.
For non-magnetic stainless steel springs.

NS-18-2.
M-1507; 0.10 C, 12.0 Mn, 0.50 Si, 18.0 Cr, 1.60 Ni, 0.34 N, bal Fe.
Wire for springs; non-magnetic after cold-work; excellent corrosion resistance.
Cold drawn (0.062 inch) 260,000-290,000 psi TS.

NS-22.
M-1507; 0.06 max C, 4.0-6.0 Mn, 1.0 max Si, 20.5-23.5 Cr, 11.5-13.5 Ni, 1.5-3.0 Mo, 0.20-0.40 N, 0.10-0.30 Cb, 0.10-0.30 V, bal Fe.
Wire for springs; non-magnetic after cold work; excellent chloride resistance.
Spring temper (0.062 inch): 255,000-285,000 psi TS.

N-S 25.
M-189; 0.05-0.15 C, 19-21 Cr, 19-11 Ni, 14-16 W, 46-53 Co, 3 max Fe, Mn.
Cold drawn wire: 178,500-207,000 TS; 45,000-50,000 YS; 24-7 El.
For springs; high temperature applications.

NS-101.
M-1507; 0.10 C, 1.0 Mn, 0.55 Si, bal Fe.
Welding wire: as welded: 76,000 TS; 60,000 YS; 28 El; 130 Brin.
For inert gas welding mild steel.

NS-102.
M-1507; 0.08 C, 1.95 Mn, 0.60 Si, 0.50 Mo, bal Fe.
Welding wire; as welded: 102,000 TS; 84,000 YS; 21 El; 163 Brin.
For inert gas welding 4130 and some high strength steels.

NS-103.
M-1507; 0.05 C, 1.25 Mn, 0.50 Si, 0.10 Al, 0.09 Zr, 0.10 Ti, bal Fe.
Welding wire; as welded: 83,000 TS; 71,000 YS; 27.5 El; 140 Brin.
For improved quality welds in inert gas welding mild steel.

NS-107.
M-1507; 0.15 C, 1.2 Mn, 0.55 Si, 0.65 Al, bal Fe.
Welding wire; as welded: 80,000 TS; 64,000 YS; 25 El; 130 Brin.
For inert gas welding oily and rusty mild steel without porosity.

NS-115.
M-1507; 0.10 C, 1.70 Mn, 1.0 Si, bal Fe.
Welding wire; as welded: 90,000 TS; 68,000 YS; 28 El; 160 Brin.
For producing neat appearing welds when inert gas welding mild steel.

NS-116.
M-1507; 0.10 C, 1.95 Cr, 0.55 Mn, bal Fe.
Welding wire; as welded: 87,000 TS; 71,000 YS; 26 El; 42 ft. lbs. Charpy V notch.
For inert gas welding of mild steel.

NS 190.
M-French; C, alloy, bal Fe.
For heat, creep and oxidation resistant.

NS-308-HISI.
M-1507; 0.05 C, 1.70 Mn, 0.85 Si, 0.015 P, 0.010 S, 21.0 Cr, 10.25 Ni, bal Fe.
Bare stainless steel wire for welding 308 and similar steels.

NS-308L HISI.
M-1507; 0.022 C, 1.70 Mn, 0.85 Si, 0.015 P, 0.010 S, 21.0 Cr, 10.25 Ni, bal Fe.
Bare stainless steel wire for welding 308L and similar steels.

NS-309 HISI.
M-1507; 0.05 C, 1.70 Mn, 0.85 Si, 0.015 P, 0.010 S, 24.5 Cr, 13.5 Ni, bal Fe.
Bare stainless steel wire for welding 309 and similar steels.

NS-316L HISI.
M-1507; 0.022 C, 1.70 Mn, 0.85 Si, 0.015 P, 0.010 S, 19.0 Cr, 13.0 Ni, 2.25 Mo, bal Fe.
Bare stainless steel wire for welding 316L and similar steels.

NS-750X.
M-1507; 1 max Co, 14-17 Cr, 69 min Ni, 5-9 Fe, 0.08 max C, 0.4-1.0 Al, 2.0-2.5 Ti.
Drawn (15% red.) and heat treated wire has: 190,000-205,000 TS.
Corrosion resistant.
For springs operating at 550-1100°F.

N-S 1000.
M-189; 0.37-0.43 C, 4.75-5.25 Cr, 1.2-1.4 Mo, bal Fe.
Drawn: 200,000 TS; 1-2 El.
For dies; tools; oil or air hardened.

NS-A286.
M-1507; 13.5-16 Cr, 24-27 Ni, 1.0-1.75 Mo, 0.35 max Al, 1.75-2.25 Ti, 0.10-0.50 V, bal Fe.
Drawn (68% red.) and heat treated wire has: 210,000-236,000 TS.
Austenitic, corrosion resistant.
For springs operating at 600-1000°F.

NS-L605.
M-1507; 50 Co, 19-21 Cr, 9-11 Ni, 3 max Fe, 0.05-0.15 C, 14-16 W.
Drawn (28% red.) and heat treated wire has: 205,000-225,000 TS.
Corrosion resistant.
For springs operating at 1100-1500°F range; for wire holding parts together for dip brazing at 1400°F.

NSCD.
M-1495; 0.03 C, 17.5 Cr, 16.0 Ni, 5.0 Mo, 3.0 Co, bal Fe.
Resistant to pitting.
For seawater treatment.

NSK.
M-1655; 1.0 C, 0.30 Mn, 0.25 Si, 1.5 Cr, bal Fe.
Cold work tool steel for punches, stamping dies, trimming dies.
W.-Nr. 1.2067.

N-S STAINLESS STEEL WELDING WIRE.
M-1507.
Welding wire of 22 stainless compositions for arc welding stainless steels, bare wire.

NSZ.
M-1655; 1.45 C, 0.60 Mn, 0.20 Si, 1.5 Cr, bal Fe.
Cold work tool steel for reamers, threading cutters, broaches.
W.-Nr. 1.2063.

NT-2.
M-1203.
Titanium carbide coated N-2 carbide tool material.

NT-5.
M-1203.
Titanium carbide coated N-52 carbide tool material.

NT-6.
M-1203.
Titanium carbide coated N-60 carbide tool material.

NTK-M7.
M-1577; 9-11 Cr, 16-18 Ni, 6-8 Mo, 0.06 max C, 1.0 max Si, 2.0 max Mn, 0.03 max S & P, bal Fe.
Rolled: 106,000 TS; 86,000 YS; 34 El; 43 RA.
Annealed: 79,000 TS; 35,000 YS; 60 El; 62 RA.
For chemical and food processing equipment, pumps, impellers.
Stainless. Resists organic chlorination media. Resists moderate concentrations of hot HCl.

NTK-T1.
M-1577; 1-3 Mo, bal TiC.
Sintered: 120,000-140,000 Tr. Si, Rock A 93-94.
For tools, bearings, seals.
Resists heat and wear.

NTK-T3.
M-1577; 13-16 Mo, bal TiC.
Sintered: 155,000-185,000 TS; Rock A 92-93.
For tools, bearings, seals.
Resists heat and wear.

N-TUF-CR 196.
M-1732; 0.13 max C, 0.9-1.5 Mn, 5.0-6.0 Ni, 0.1-0.3 Mo, 0.60 Cr, bal Fe.
For welded pressure vessels.

NUALL.
M-1409; C, alloy, bal Fe.
For chuck jaws, reamers, drills; oil hardened.

NU-BRAZE III.
M-1002; Ag, P, Cu.
For brazing of non-ferrous metals; B.P. 1300°F.

NU-BRAZE VI.
M-1002; Ag, Cu, Zn, Cd.
For brazing of non-ferrous metals; M.P. 1170 °F.

NU-BRAZE SUPER FLO.
M-1002; Ag, bal Cu.
For silver brazing; M.P. 1076°F.

NU-BRONZE.
M-Eng; 95.4 Cu, 3.25 Ni, 0.25 Mn, 0.73 Si.
For corrosion resistant parts; corrosion resistant.

NUCALLOY 41.
M-202; 12 Cr, 2 B, 4 Si, 0.50 C, bal Ni.
Centrifugal castings; as cast: 44 Rc.
For high temperature and corrosion application and good wear.

NUCALLOY 45.
M-202; 14 Cr, 3 B, 4.5 Si, 0.65 C, bal Ni.
Centrifugal castings; as cast: 53 Rc.
For high temperature and corrosion application and good wear.

NUCAST.
M-1075; 99 + Ni.
For nickel anodes.

NUCUT.
M-1221; C, alloy, bal Fe.
For dies; oil hardened.

NU-DIE.
M-38; 0.4 C, 1.2 Si, 5.2 Cr, 1.5 Mo, bal Fe.
For die casting dies, plastic mold dies; oil hardening.

NU-DIE.
M-1409; 0.4 C, 1.2 Si, 5.2 Cr, 1.5 Mo, bal Fe.
For plastic molds, die casting dies.
Oil or air hardening.

NU-DIE CASTING.
M-362; 0.35 C, 5 Cr, 0.4 V, 1.5 Mo, bal Fe.
For die casting dies, punches; Type H11; hot work steel.

NU-DIE DENSIFIED HOT WORK STEEL.
M-38; 0.40 C, 0.4 Mn, 1.0 Si, 5.0 Cr, 0.55 V, 1.35 Mo, bal Fe.
Air hardened: 40-55 Rock C.
Designed for extrusion dies for aluminum and magnesium.
AISI H11 Hot work tool steel.

NU-DIE V.
M-38; 0.3-0.4 C, 5 Cr, 1.4 Mo, 1.1 V, 1 Si, bal Fe.
For die casting dies for Al-base alloys; air hardening.

NUERAL.
M-555; Si, Cu, Ni, bal Al.
For cylinder heads; heat resistant.

NUERAL 132.
M-555; 0.8-1.0 Cu, 13-14 Si, 1.9-2.4 Ni, bal Al.
Sand cast: 32,000 TS; 0.3 El.
For cylinder heads; heat treatable.

NUERENBERGER GOLD.
M-Ger.; 2-7.5 Al, 0.2-2.5 Au, bal Cu.
For gold substitute; corrosion resistant.

NUGILD.
M-103; 13 Zn, bal Cu.
Soft: 43,000 TS; 56 Brin.
1/2 Hard: 54,000 TS; 116 Brin.

NU-GOLD.
M-Eng.; 87.73 Cu, 12.22 Zn.
For jewelry, ornaments; red brass.

NULOY.
M-288; 0.3 C, 1.5 Ni, 0.8 Cr, bal Fe.
For steel castings; tough.

NUMETAL.
M-1; 77 Ni, 4.5 Cu, 1.5 Cr, bal Fe.
For audio transformers, sensitive relays; high permeability.

NU-MOL.
M-1062; 0.9 C, alloy, bal Fe.
For hacksaws; tough.

NU-PYR-LOY.
M-1432; C, alloy, bal Fe.
Heat treated: 309,000-342,000 TS; 264,000-282,000 YS; 570-600 Brin.
For pneumatic and shock tools, chisels, shear blades; tough and shock resistant, oil hardened.

NURAL.
M-1548; 0.2-1.0 Si, 0.2-0.5 Mn, 3.0-12.0 Mg, bal Al.
Chill Cast: 32,000-37,000 TS; 5-10 El; 78-80 Brin.
For aircraft fittings, car frames, marine parts, lever brackets, hardware, general housings.
Permanent mold and die cast, corrosion resistant.

NURAL 25.
M-555; 1-2 Mg, 5.6 Si, 0.1-0.4 Mn, 0.7 Fe, 0.5 Cu, 0.2 Ti, 0.2 Zn, bal Al.
Sand: 19,000 TS; 14,000 YS; 3 El.
Cast 26,000 TS; 19,000 YS; 1 El.
Hardened: 40,000 TS; 36,000 YS; 1 El.
For sand and permanent mold castings; hardenable.

NURAL 30.
M-555; 6-7.5 Cu, 0.5-2.0 Si, bal Al.
Sand cast: 18,000-28,000 TS; 2.0-0.8 El.
For light alloy parts; non-hardenable.

NURAL 43.
M-555; 4.5-6 Si, bal Al.
Sand cast: 18,000 -24,000 TS; 5-2 El.
For light alloy parts; corrosion resistant.

NURAL 43.
M-1548; 4.5-6.0 Si, 1.2 Fe, bal Al.
Sand cast: 17,000-24,000 TS; 2-5 El; 42-50 Brin.
Die cast: 20,000-28,000 TS; 3-5 El; 50-60 Brin.
For marine castings, manifolds, meter housings, carburetors, food handling equipment.
Sand, permanent mold and die cast.

NURAL 77.
M-555; 4-7 Cr, 4-6 Zn, 1-3 Si, 1.5 max Fe, bal Al.
Sand cast: 18,000-25,000 TS; 2-0.5 El.
For light alloy parts; heat treatable.

NURAL 85.
M-555; 2-4 Cu, 5-7 Si, 0.4-0.6 Mn, bal Al.
Chill cast: 28,000-35,000 TS; 4.5-1.0 El.
For light alloy parts; heat treatable.

NURAL 85 S.
M-555; 2-3 Cu, 4-6 Si, 2.0 max Fe, 0.5 max Mn, bal Al.
Sand cast: 20,000-27,000 TS; 4-1 El.
For light alloy parts; heat treatable.

NURAL 93.
M-555; 4-4.5 Cu, 1.5-2.5 Si, 3-4.5 Ni, bal Al.
Sand cast: 20,000-26,000 TS; 1.2-0.5 El.
For light alloy parts; heat treatable.

NURAL 122.
M-555; 9.5-10.2 Cu, 0.15-0.35 Mg, 0.8-1.5 Fe, bal Al.
Sand cast: 22,000-30,000 TS; 1.5-3.3 El.
For light alloy parts; non-hardenable.

NURAL 122.
M-1548; 9.5-10.2 Cu, 0.15-0.35 Mg, 0.8-1.1.5 Fe, bal Al.
Chill Cast: 26,000-37,000 TS; 1 El; 70-90 Brin.
For sole plates for electric hand irons, general castings.
Permanent mold cast, good machinability.

NURAL 132.
M-1548; 0.8-2.0 Cu, 12.5-14.2 Si, 0.8-1.0 Mg, 0.5 Fe, 0.8-2.4 Ni, bal Al.
Chill Cast: 28,000-37,000 TS; 0.5-1.2 El.
For meter cases, switches boxes, manifolds, fittings, aircraft parts.
Permanent mold cast, good castability and corrosion resistant.

NURAL 132 A.
M-555; 0.8-1.0 Cu, 12.2-12.8 Si, 0.8-1.0 Mg, 0.5 Fe, 1.9-2.4 Ni, bal Al.
Chill cast: 29,000-37,000 TS; 1.2-0.5 El.
For light alloy parts; corrosion resistant.

NURAL 132 B.
M-555; 2 Cu, 12.5 Si, 0.8-1.0 Mg. 0.5 Fe, 0.8-1.0 Ni, bal Al.
Chill cast: 29,000-37,000 TS; 1.2-0.5 El.
For light alloy parts; corrosion resistant.

NURAL 142.
M-555; 4-4.5 Cu, 1.3-1.8 Mg, 1.8-2.2 Ni, bal Al.
Sand cast: 21,000-28,000 TS; 4-1 El.
For light alloy parts; age-hardenable.

NURAL 142.
M-1548; 4.0-4.5 Cu, 1.3-1.8 Mg, 1.8-2.2 Ni, bal Al.
Chill Cast: 23,000-32,000 TS; 1-5 El; 70-85 Brin.
Heat Treated: 26,000-35,000 TS; 1-6 El; 80-110 Brin.
For cylinder heads, pistons, generator housings, fittings.
Permanent mold castings, heat treatable, for high temperature service.

NURAL 195.
M-555; 4-5.5 Cu, 0.5-1.5 Si, 0.4-1.0 Fe, bal Al.
Sand cast; 23,000-36,000 TS; 8-2 El.
For light alloy parts; age-hardenable.

NURAL 200.
M-555; 14-16.8 Cu, 0.5-1.0 Fe, 0.15-0.35 Mg, bal Al.
Sand cast: 20,000-28,000 TS; 0.5-0.3 El.
For light alloy parts; heat treatable.

NURAL 511.
M-555; 4.5-5.5 Mg, 0.1-0.5 Mn, 0.6-1.5 Si, 0.2 Ti, bal Al.
Sand cast: 26,000 TS; 14,000 YS; 3 El; 65 Brin.
For light alloy castings; corrosion resistant.

NURAL 1761.
M-555; 0.8-1.2 Cu, 16.4-17.5 Si, 3.2-3.6 Ni, 0.5 Mn, 0.7-1.2 Mg, 0.5 Cr, bal Al.
Heat treated: 28,000-36,000 TS; 0.3-0.5 El.
For auto engine pistons, permanent mold castings; low density.

NURAL 1761P.
M-555; 0.8-1.2 Cu, 16.4-17.5 Si, 3.2-3.6 Ni, 0.4-0.6 Mn, 0.7-1.2 Mg, 0.5 Cr, bal Al.
Permanent mold: 29,000-33,000 TS; 28,000-30,000 YS; 0.8-0.3 El; 90-14 Brin.
For pistons; low density.

NURAL 2361.
M-555; 0.8-1.2 Cu, 22-25 Si, 0.8-1.0 Ni, 0.7-1.2 Mg, 0.5 Cr, bal Al.
Permanent mold: 29,000-33,000 TS; 28,000-30,000 YS; 0.3-0.1 El; 90-12 Brin.
For pistons; low density.

NURAL 3210.
M-555; 0.8-1.5 Cu, 11-13 Si, 0.8-1.3 Ni, 0.8-1.3 Mg, bal Al.
Permanent mold: 31,000-35,000 TS; 26,000-30,000 YS; 1-0.3 El; 90-125 Brin.
For light alloy parts; corrosion resistant.

NURAL AL-MG-MN.
M-555; 0.2-1.3 Si, 0.6-1.5 Mn, 1.5-3.0 Mg, bal Al.
Chill cast: 32,000-36,000 TS; 10-5 El.
For light alloy parts; age-hardenable.

NURAL DZNAL4CUL.
M-555; 0.6-1.0 Cu, 3.7-4.3 Al, 0.02-0.05 Mg, bal Zn.
Die cast: 36,000-47,000 TS; 5-2 El; 80-100 Brin.
For housings, cases, machinery casting; Zamak 5.

NURAL GALCU6SI3.
M-555; 4-7 Cu, 2-4 Si, 2.5 Zn, 1.1 Fe, bal Al.
Sand cast: 29,000 TS; 23,000 YS; 0.5 El; 100 Brin.
Permanent mold: 32,000 TS; 28,000 YS; 0.2 El; 110 Brin.
For light alloy castings; age hardenable.

NURAL GALSI12.
M-555; 11-13.5 Si, 0.3-0.5 Mn, bal Al.
Sand cast: 31,000 TS; 13,000 YS; 4 El; 60 Brin.
For light alloy castings; corrosion resistant.

NUREMBERG GOLD.
90 Cu, 7.5 Al, 2.5 Au.
For cheap jewelry; gold color.

NUREX.
M-89; C, Cr, Mn, bal Fe.
For mill balls, linings; corrosion and abrasion resisting.

NUROX.
M-89; 3.3 C, Si, Mn, bal Fe.
For gears, shafts, housings; cast iron.

NUSHANK.
M-435; 0.45 C, 0.6 Mn, 0.4 Cr, 3.0 Ni, 0.25 Mo, bal Fe.
For tools, shanks; wear resistant.

NUSITE.
M-1454; 0.9 C, 13 Co, bal Fe.
For permanent magnets; heat treated.

NUTHERM.
M-435; 0.70 C, 2 Mn, 1 Cr, 1.35 Mo, 0.30 Si, bal Fe.
For cold work tools and dies; low temperature air hardening.

N.W.S.
M-487, M-352; 0.9 C, 1.2 Mn, bal Fe.
For tools, dies; non-shrinkable.

NY.
M-French; 0.07-0.20 C, 0.35 Mn, 4-7 Ni, bal Fe.
Heat treated: 78,000-190,000 TS; 58,000-156,000 YS; 25-8 El.
For construction work, mine car wheels.

NX-188.
M-1491; 0.04 C, 18.0 Mo, 8.0 Al, bal N.
Cast alloy for jet engine vanes.
Good high temperature strength.
Directionally solifified.

NYBLADE.
M-510; 0.48 C, 1.0 Si, 0.65 C, 3.0 Ni, bal Fe.
Cold work tool steel for chisels, punches, shear blades.

NYBY 18-8EL.
M-1448; 0.05 max C, 18 Cr, 8 Ni, bal Fe.
Annealed: 85,000 TS; 35,000 YS; 60 El; 70 RA; 150 Brin.
For chemical plant equipment, tanks, mixers; Type 304; stainless, austenitic.

NYBY 18-8L.
M-1448; 0.06-0.08 C, 18 Cr, 8 Ni, bal Fe.
Annealed: 85,000 TS; 35,000 YS; 60 El; 70 RA; 150 Brin.
For chemical plant equipment, tanks, mixers; Type 304; stainless, austenitic.

NYBY 18-8-LNB.
M-1448; 0.08 max C, 18 Cr, 10 Ni, Cb = 10 x C, bal Fe.
Annealed: 78,000 TS; 31,000 YS; 40 El; 50 RA; 190 Brin.
For welded chemical plant equipment, tanks, mixers, agitators, stills.
Austenitic Type 347 stainless steel, stabilized, welding grade.

NYBY 18-8 LT.
M-1448; 0.08 max C, 18 Cr, 10 Ni, Ti = 5 x C, bal Fe.
Annealed: 78,000 TS; 31,000 YS; 40 El; 50 RA; 190 Brin.
For welded chemical plant equipment, tanks, vessels, agitators, mixers.
Austenitic stainless steel Type 321, stabilized, welding grade.

NYBY 18-8 UL.
M-1448; 0.03 max C, 19 Cr, 11 Ni, bal Fe.
Annealed: 78,000 TS; 29,000 YS; 50 El; 55 RA; 180 Brin.
For chemical plant equipment, tanks, vessels, mixers, filters, agitators.
Austenitic stainless steel. Type 304 L., welding grade.

NYBY 18-8 ULN.
M-1448; 0.03 max C, 18 Cr, 9 Ni, 0.15 N, bal Fe.
Ann: 95,000 TS; 44,000 YS; 50 El; 190 Brin.
For LNG tanks and other low temperature equipment.
Austenitic stainless steel, Type 304 LN, with high yield strength and good low temperature properties.

NYBY 18-10 E MO.
M-1448; 0.06 max C, 17 Cr, 11 Ni, 2.3 Mo, bal Fe.
Annealed: 78,000 TS; 29,000 YS; 50 El; 55 RA; 180 Brin.
For chemical plant equipment, vats, filters, agitators, digestors, tanks.
Austenitic stainless steel, acid resistant, not hardenable.

NYBY 18-10L MONB.
M-1448; 0.08 C, 17 Cr, 12 Ni, 2.3 Mo, Cb = 10 x C, bal Fe.
Annealed: 78,000 TS; 34,000 YS; 40 El; 50 RA; 190 Brin.
For welded chemical plant equipment, mixers, tanks, vessels, retorts.
Type 318 stainless steel, austenitic, stabilized, acid resistant.

NYBY 18-10 LMOT.
M-1448; 0.08 C, 17 Cr, 12 Ni, 2.3 Mo, Ti = 5 x C, bal Fe.
Annealed: 78,000 TS; 32,000 YS; 40 El; 50 RA; 190 Brin.
For welded chemical plant equipment, tanks, mixers, agitators, vessels.
Austenitic stainless steel, stabilized, welding grade, not hardenable.

NYBY 18-10 UMO.
M-1448; 0.03 max C, 17 Cr, 11 Ni, 2.2 Mo, bal Fe.
Ann: 78,000 TS; 29,000 YS; 50 El; 160 Brin.
For chemical plant equipment.
Austenitic stainless steel; Type 316L.

NYBY 18-12 EMO.
M-1448; 0.06 C, 17 Cr, 12 Ni, 2.8 Mo, bal Fe.
Annealed: 78,000 TS; 30,000 YS; 50 El; 55 RA; 180 Brin.
For acid resistant chemical plant equipment, tanks, mixers, vats.
Type 316 stainless steel, austenitic, acid resistant.

NYBY 18-12 LMOT.
M-1448; 0.08 max C, 17 Cr, 13 Ni, 2.8 Mo, Ti = 5 x C, bal Fe.
Annealed: 78,000 TS; 30,000 YS; 40 El; 50 RA; 190 Brin.
For welded chemical plant equipment, tanks, vessels, agitators.
Austenitic stainless steel, stabilized, welding grade.

NYBY 18-12 UMO.
M-1448; 0.03 max C, 17 Cr, 12 Ni, 2.8 Mo, bal Fe.
Annealed: 78,000 TS; 30,000 YS; 50 El; 55 RA; 180 Brin.
For acid resistant chemical plant equipment, tanks, mixers, vats.
Type 316 L, stainless steel, austenitic, acid resistant.

NYBY 18-12 UMON.
M-1448; 0.03 max C, 17 Cr, 13 Ni, 2.7 Mo, 0.15 N, bal Fe.
Ann: 100,000 TS; 45,000 YS; 50 El; 190 Brin.
For low temperature purposes in marine atmosphere.
Austenitic, corrosion resistant stainless steel, Type 316 LN, with high yield strength and good low temperature properties.

NYBY 18-14 EMO.
M-1448; 0.06 max C, 19.0 Cr, 14.0 Ni, 3.5 Mo, bal Fe.
Annealed: 78,000 TS; 29,000 YS; 45 El; 45 RA; 190 Brin.
For chemcial plant equipment, vats, tanks.
Austenitic, stainless steel, acid resistant. Type 317 stainless steel, acid resistant.

NYBY 18-14 UMO.
M-1448; 0.03 max C, 18 Cr, 14 Ni, 3.5 Mo, bal Fe.
Ann: 80,000 TS; 30,000 YS; 45 El; 165 Brin.
For chemical plant equipment.
Austenitic stainless steel; Type 317L.

NYBY 18-15 EMO.
M-1448; 0.06 C, 17 Cr, 15 Ni, 4.5 Mo, bal Fe.
Annealed: 78,000 TS; 29,000 YS; 45 El; 45 RA; 190 Brin.
For chemical plant equipment, vats, tanks.
Austenitic, stainless, acid resistant.
Type 317 stainless steel, not hardenable.

NYBY 18-15 UMO.
M-1448; 0.03 max C, 17 Cr, 15 Ni, 4.5 Mo, bal Fe.
Ann: 80,000 TS; 32,000 YS; 45 El; 165 Brin.
For chemical plant equipment.
Austenitic stainless steel with high resistance to general corrosion and pitting.

NYBY 20-12 SI.
M-1448; 0.20 max C, 20 Cr, 12 Ni, 2 Si, bal Fe.
Annealed: 85,000 TS; 36,000 YS; 45 El; 45 RA; 190 Brin.
For heat treating furnaces and boxes, retorts, salt pots.
Austenitic. Heat and corrosion resistant. Type 309B stainless steel.

NYBY 20-25 UMOCU.
M-1448; 0.025 max C, 20 Cr, 25 Ni, 4.5 Mo, 1.5 Cu, bal Fe.
Ann: 85,000 TS; 33,000 YS; 40 El; 165 Brin.
For use in high-corrosive environments such as H_2SO_4.
Austenitic stainless steel with very good resistance to general corrosion and pitting.

NYBY 23-14L.
M-1448; 0.08 max C, 23 Cr, 13 Ni, bal Fe.
Annealed: 78,000 TS; 31,000 YS; 35 El; 50 RA; 190 Brin.
For heat treating boxes, furnace parts and equipment, retorts, valves.
Type 309S stainless steel. Austenitic, heat resistant.

NYBY 24-6 UMO.
M-1448; 0.03 max C, 24 Cr, 6 Ni, 1.5 Mo, 0.10 N, bal Fe.
Ann: 100,000 TS; 68,000 YS; 25 El; 240 Brin.
For chemical plant equipment as heat-exchangers, tanks, mixers.
Ferritic-austenitic stainless steel with high yield strength and good resistance to stress corrosion and pitting.

NYBY 25-21 L.
M-1448; 0.08 max C, 25 Cr, 20 Ni, bal Fe.
Annealed: 78,000 TS; 31,000 YS; 45 El; 45 RA; 190 Brin.
For furnace parts and equipment, valves, pumps, gas turbine components, retorts.
Type 310 S stainless steel, austenitic. Resists corrosion and oxidation.

NYBY 25-21 SI.
M-1448; 0.20 C, 25 Cr, 21 Ni, 2 Si, bal Fe.
Annealed: 85,000 TS; 36,000 YS; 30 El; 35 RA; 200 Brin.
For furnace parts, heat treating boxes, retorts.
Type 314 stainless steel, austenitic. Resists oxidation and carburization.

NYBY 25-22 UMON.
M-1448; 0.02 max C, 25 Cr, 22 Ni, 2 Mo, 0.12 N, bal Fe.
Ann: 78,000 TS; 38,000 YS; 35 El; 180 Brin.
For chemical plant equipment, particularly urea fabrication.
Austenitic stainless steel with high resistance to general corrosion and intercrystalline corrosion.

NYBY 1803 MOT.
M-1448; 0.025 max C, 18 Cr, 2 Mo, 0.40 Ti, 0.025 max N, bal Fe.
Ann: 79,000 TS; 52,000 YS; 30 El; 180 Brin.
For water heaters, hot water pipes, heat exchangers.
Ferritic stainless steel wiith good weldability, immune to stress corrosion cracking and good resistance to pitting and general corrosion.

NYBY 1803 T.
M-1448; 0.025 max C, 18 Cr, 0.40 Ti, 0.025 max N, bal Fe.
Ann: 73,000 TS; 40,000 YS; 35 El; 160 Brin.
For water heaters, hot water pipes, coolers.
Ferritic stainless steel with good weldabiltiy and immune to stress corrosion cracking.

NYBYLOID-1.
M-667; Core wire of Ni and a covering containing 13 Cr, 10 Fe, Mn, C, Mo, bal Ni.
Weld: 96,000-101,000 TS.
For welding 9% nickel steel.
Low hydrogen welding electrode.

O

O-6.
M-24; 1.45 C, 0.8 Mn, 1.05 Si, 0.25 Mo, bal Fe.
For cold working, forming, shaping and drawing dies.
AISI Type O6 oil hardening tool steel.

OAKES TRODALOY 1.
M-1094; 0.4 Be, 2.6 Co, bal Cu.
Cast: 90,000 TS; 50,000 YS; 10-15 El.
Wire: 125,000 TS; 60,000 YS; 8-15 El.
For springs; age hardenable.

O B ALLOY NO 1 A.
M-94; 88 Cu, 8 Sn, 4 Zn.
Cast: 41,400 TS; 20,300 YS; 28 El; 70 Brin.
Used in conjunction with non-ferrous catenary materials; electric motor bearings; Iz-14.2.

O B ALLOY NO 11.
M-94; 94 Ci, 2 Sn, 3 Pb, 4 Zn.
Cast: 31,000 TS; 11,000 YS; 35 El; 49 Brin.
For electric apparatus; free-cutting.

O B ALLOY NO 16.
M-94; 95.75 Cu, 5 Sn, 0.22 Pb.
Cast: 36,300 TS; 16,400 YS; 2l El; 64 Brin.
For trolley wheels; Iz-20.

O B ALLOY NO 25.
M-94; 90 Cu, 7 Al, 3 Si.
Cast: 74,500 TS; 54,600 YS; 68 El; 160 Brin.
Used in conjunction with catenary construction; Iz-9.7.

O B ALLOY NO 26.
M-94; 89 Cu, 6 Sn, 1.5 Pb, 3.5 Zn.
Cast: 38,000 TS; 17,000 YS; 30 El; 65 Brin.
For valves; Navy M.

O B ALLOY NO 27.
M-94; 81.5 Cu, 14.5 Zn, 4 Si.
Cast: 65,000 TS; 34,000 YS; 26 El; 160 Brin.
For trolley shoes, valve parts; Tombasil.

O B ALLOY U-5.
M-94; 85 Cu, 5 Sn, K, Pb, 5 Zn.
34,750 TS; 14,800 YS; 25 El; 59 Brin.
For valves, electrical appliances; Ounce metal.

ODESSA.
33.25 Ag, 42.5 Cu, 15.75 Zn, 8.5 Ni.
For silver solder; corrosion resistant.

OERDERLIN FMB.
M-1420; 0.1 C, 18 Cr, 8 Ni, 2.5-3.0 Mo, bal Fe.
Rolled: 58,000-78,000 TS; 29,000-32,000 YS; 12-18 El; 170-200 Brin.
For chemical plant equipment; stainless, acid resistant, austenitic.

OERDERLIN FST.
M-1420; 0.1 C, 1, Cr, 8 Ni, Si, Mn, W, bal Fe.
Annealed: 65,000-85,000 TS; 3l,000-40,000 YS; 15-20 El; 160-200 Brin.
For chemical palnt equipment; stainless, austenitic.

OERSTIT 35.
M-1496; 1 C, 4.5 Cr, 1 Mn, bal Fe.
For permanent magnets; workable, max service temperature 175°F.

OERSTIT 90G.
M-1496; 0.04 C, 22 Ni, 2 Cu, 11.5 Al, 0.5 Ti, bal Fe.
For permanent magnets; cast.

OERSTIT 160K.
M-1496; 22 Ni, 17 Co, 10 Al, 3 Cu, 1 Ti, bal Fe.
For permanent magnets.

OERSTIT 160R.
M-1496; 22 Ni, 14 Co, 10 Al, 3 Cu, 1 Ti, bal Fe.
For permanent magnets.

OERSTIT 190.
M-1496; 20 Ni, 16 Co, 9 Al, 3 Cu, 1 Ti, bal Fe.
For permanent magnets.

OERSTIT 300.
M-1496; 17 Ni, 2 Co, 7.5 Al, 3 Cu, 2.5 Ti, bal Fe.
For permanent magnets.

OERSTIT 350.
M-1496; 15 Ni, 30 Co, 7.5 Al, 4 Cu, 5 Ti, bal Fe.
For permanent magnets.

OERSTIT 400K.
M-1496; 14 Ni, 24 Co, 8 Al, 3 Cu, 0.5 Ti, bal Fe.
For permanent magnets.

OERSTIT 400R.
M-1496; 14 Ni, 24 Co, 8 Al, 3 Cu, 0.5 Ti, bal Fe.
For permanent magnets.

OERSTIT 500.
M-1496; 14 Ni, 24 Co, 8 Al, 3 Cu, 0.5 Ti, bal Fe.
For permanent magnets; high permeability.

OERSTIT 600.
M-1496; 14 Ni, 24 Co, 8 Al, 3 Cu, 0.5 Ti, bal Fe.
For magnets; maximum service temperature 660°F, cast and sintered alloy.

OFHC COPPER.
M-1778; 99.98 Cu.
For electrical equipment; oxygen free, high conductivity.

OFHC COPPER 101.
M-8; 99.9 + Cu.
Hard: 48,000 TS; 40,000 YS; 15 El; 83 Brin.
Soft: 32,000 TS; 10,000 YS; 50 El; 42 Brin.
For electrical apparatus; high conductivity, oxygen free.

OFHC COPPER 102.
M-8; 99.95 min Cu.
Hard: 55,000 TS; 50,000 YS; 10 El; B 60 Rock.
Soft: 32,000 TS; 10,000 YS; 55 El; F 40 Rock.
For radar and electronic components, waveguide tubes, Dumet wire, connectors. Elect. cond. 101. Corrosion resistant.

OFHC SULFUR COPPER.
M-8; 0.20-0.60 S, 99.90 min Cu + S.
1/2 H-temper: 42,000 TS; 36,000 YS; 25 El; 73 Brin.
Hard: 48,000 TS; 42,000 YS; 15 El; 83 Brin.
For contact pins and inserts, screw machine products; high thermal and electrical conductivity, free-cutting, corrosion resistant.

O.H. 38.
M-841; 15 Zn, 2-3 Cu, 2 Si, bal Al.
Cast: 30,000 TS; 17,500 YS; 2.5 El; 65 Brin.
For light alloy parts, gear guards.

OHIO DIE STEEL.
M-34, M-115; 1.55 C, 12 Cr, 0.85 V, 0.8 Mo, bal Fe.
105,000 TS; 60,000 YS; 13 El; 18 RA; 212 Brin.
For dies, roller threading dies, punches, gages, shear blades; non-warping wear resisting.

OHIOLOY 2300.
M-95; 0.50 C, 45 Ni, 25 Cr, 1.5 Si, 1.5 Mn, W, Co, bal Fe.
High temperautre alloy; non-magnetic.
Cast, room temp: 60,875 TS; 60,875 YS; 2 El; 220 Brin.
At 1800°F: 23,500 TS; 18,000 YS; 27 El.
Weldable (certain conditions); for high temperature applications; resistant to scaling and to creep; for heat treat equipment.

OHIOLOY A-6.
M-95; 1.0-1.2 C, 1.0 max Mn, 1.5 max Si, 25-31 Cr, 1.5 max Mo, 3-6 W max Fe, bal Co.
Abrasion and wear resistant.
For worms, sleeves, spacers.
Reference: Stellite 6; formerly FAHRITE C-6.

OHIOLOY A-28.
M-95; 3.0 max C, 1.25 Mn, 2.0 Si, 26-30 Cr, 4.0 max Ni, bal Fe.
Abrasion resistant and corrosion resistant casting: for catalyst plates, cement mill, refinery service.
Formerly FAHRITE HC 28.

OHIOLOY B.
M-95; 0.12 C, 1.0 Mn, 1.0 max Si, 1.0 max Cr, 26-30 Mo, 4.0-6.0 Fe, V, 2.5 max Co, bal Ni.
Heat and corrosion resistant casting.
Cast: 72,000 TS; 46,000 YS; 6 El.
For turbine blades, bolting material, impellers, shafting: resistant to hot hydrochloric acid.
ASTM B332-58T; reference: HASTELLOY B.
Formerly FAHRITE C-708.

OHIOLOY C.
M-95; 0.12 C, 1.0 Mn, 1.0 max Si, 15.5-17.5 Cr, 16-18 Mo, 2.5 max Co 3.75-5.25 W, 4.5-7.5 Fe, 0.2-0.6 V, bal Ni.
Heat and corrosion resistant casting.
Cast: 72,000 TS; 46,000 YS; 4 El.
High temperature trays, shafts, conveyor chains, equipment for chemical plants.
ASTM B332-58T; Reference: HASTELLOY C.
Formerly FAHR

OHIOLOY CA-15.
M-95; 0.15 max C, 1.0 Mn, 1.5 Si, 11.5 -14.0 Cr, 1.0 max Ni, 0.5 max bal Fe.
Corrosion resistant casting, weldable.
Annealed: 908000 TS; 65,000 YS; 18 El.
Hardenable to 189-400 Brin.
For pump castings, liners, valve trim, fittings; burner tips, skimmer ladles.
ACI CA-15; AISI 410; SAE 60410.
Formerly FAHRITE C-12.

OHIOLOY CA-40.
M-95; 0.20-0.40 C, 1.0 Mn, 1.5 Si, 11.5-14.0 Cr, 1.0 max Ni, 0.5 max bal Fe.
Corrosion resitant casting.
Annealed: 90,000 TS; 65,000 YS; 18 El.
Hardenable up to 500 Brin.
For choppers, cutting blades, cylinder liners, pump parts, molds, dies, valves trim.
AIC CA-40; AISI 420; SAE 60420.
Formerly FAHRITE C-12-40.

Section I: Alloy Data / 1039

OHIOLOY CB-7CU.
M-95; 0.06 max C, 0.70 max Mn, 0.5-1.0 Si, 15.7-16.7 Cr, 3.5-4.75 Ni, 0.10-0.35 Cb, 2.5-3.5 Cu, bal Fe.
Corrosion resistant, age hardenable.
Cast 170,000 TS; 155,000 YS; 10 El.
For valves, shafts, fuel controls, compressor parts, pulp and paper equipment.
ACI CB-7Cu; SAE 60432; Reference 17-4PH.
Formerly FAHRITE C-17-4C.

OHIOLOY CB-30.
M-95; 0.30 C, 1.0 Mn, 1.5 Si, 18-21 Cr, 2.0 max Ni, bal Fe.
Corrosion resistant casting, weldable.
Annealed: 65,000 TS; 30,000 YS.
Hardenable up to 500 Brin.
For furnace brackets, hangers, pump parts, valve parts.
ACI CB-30; AISI 431; SAE 60442.
Formerly FAHRITE C-18.

OHIOLOY CC-50.
M-95; 0.50 C, 1.0 Mn, 1.5 Si, 2l-30 Cr, 4.0 max Ni, bal Fe.
Corrosion resistant casting.
Annealed: 55,000 TS.
For bushings, cylinder liners, pump castings, impellers; good corrosion resistance.
ACI CC-50. AISI 446; SAE 60446.
Formerly FAHRITE C-28.

OHIOLOY CD-4MCU.
M-95; 0.04 max C, 1 max Mn, 1 max Si, 25-27 Cr, 4.75-6.0 Ni, 1.75-2.2 2.75-3.25 Cu, bal Fe.
Corrosion and heat resistant, age hardenable.
Cast: 110,000 TS; 95,000 YS; 10 El.
For valve trim, pumps gears, high pressure parts, pulp and paper equipment.
ACI CD-4mCu; SAE 60328.
Formerly FAHRITE C-4.

OHIOLOY CF-3.
M-95; 0.03 max C, 1.5 Mn, 2.0 Si, 17-21 Cr, 8-12 Ni, bal Fe.
Corrosion resistant casting, weldable.
Cast: 65,000 TS; 28,000 YS; 35 El.
For autoclaves, pump parts, spray nozzles, heating coils, welded assemblies.
ACI CF-3 AISI 304L.
Formerly FAHRITE C-8-3.

OHIOLOY CF-3M.
M-95; 0.03 max C, 1.5 max Mn, 1.5 max Si, 17-21 Cr, 9-13 Ni, 2-3 Mo, Fe.
Austenitic, stainless, heat resistant.
Cast: (min.) 70,000 TS; 30,000 YS; 30 El.
For centrifuges, fittings, pump parts.
ACI CF-3M; AISI 316L.
Formerly FAHRITE C-8-3M.

OHIOLOY CF-8.
M-95; 0.08 C, 1.5 Mn, 2.0 Si, 18-21 Cr, 8-11 Ni, bal Fe.
Austenitic corrosion resistant casting.
Cast: 70,000 TS; 30,000 YS; 35 El.
For pump parts, valve parts, spray nozzles, filter press plates, autoclaves.
ACI CF-8; AISI 304; SAE 60304.
Formerly FAHRITE C-8-7.

OHIOLOY CCF-8C.
M-95; 0.08 C, 1.5 Mn, 2.0 Si, 18-21 Cr, 9-12 Ni, Cb = 8 x C min, bal Fe.
Stabilized austenitic stainless steel casting.
Cast: 70,000 TS; 30,000 YS; 30 El.
For pump parts, valves, stainless welded assemblies, chemical plant equipment.
ACI CF-8C; AISI 347; SAE 60347.
Formerly FAHRITE C-8Cb.

OHIOLOY CF-8M.
M-95; 0.08 C, 1.5 Mn, 1.5 Si, 18-21 Cr, 9-12 Ni, 2-3 Mo, bal Fe.
Austenitic, stainless, heat resistant.
Cast: 70,000 TS; 30,000 YS; 30 El.
For agitators, spray nozzles, valve bodies, for food and chemical plants; elevated temperature equipment.
ACI CF-8M; AISI 316; SAE 60316.
Formelry FAHRITE C-8-7M.

OHIOLOY CF-16F.
M-95; 0.16 max C, 1.5 max Mn, 2 max Si, 0.17 max P, 0.04 max S, 18-21 9-12 Ni, 1.5 max Mo, 0.20-0.35 Se, bal Fe.
Austenitic, stainless, free-machining.
Cast: (min) 70,000 TS; 30,000 YS; 25 El.
For bearings, bushings, hardware, pump and machinery parts, valves.
ACI-CF-16F; AISI 303; SAE 60303.
Formerly FAHRITE C-8-16F.

OHIOLOY CF-16FA.
M-95; 0.15 max C, 1.5 max Mn, 2 max Si, 0.04 max P, 0.2-0.4 S, 18-21 9-12 Ni, 0.4-0.8 Mo, bal Fe.
Free-machining, stainless, austenitic.
Cast: (min) 70,000 TS; 30,000 YS; 25 El.
For bushings, bearings, hardware, pump and machinery parts, valves.
ACI-CF-16FA; AISI 303; SAE 6030A.
Formerly FAHRITE CF-16 FA.

OHIOLOY CF-20.
M-95; 0.20 C, 1.5 Mn, 2.0 Si, 18-21 Cr, 8-11 Ni, bal Fe.
Austenitic, corrosion resistant casting.
Cast: 70,000 TS; 30,000 YS; 30 El.
For cylinder liners, pumps, valve parts, street markers, pulp and paper equipment.
ACI CF-20; AISI 302; SAE 60302.
Formerly FAHRITE C-8-20.

OHIOLOY CG-12.
M-95; 0.12 C, 1.5 Mn, 2.0 Si, 20-23 C, 10-13 Ni, bal Fe.
Austenitic, stainless heat resisting.
Cast: 70,000 TS; 28,000 YS; 35 El.
For pump parts, valve parts, furnace and heat treat equipment.

OHIOLOY CH-20.
M-95; 0.20 C, 1.5 Mn, 2.0 Si, 22-26 Cr, 12-15 Ni, bal Fe.
Austenitic, stainless, heat resisting.
Cast: 70,000 TS; 30,000 YS; 30 El.
For digester fittings, valves, chemical processing equipment, furnace and heat treat parts.
ACI CH-20; AISI 309; SAE 60309.
Formerly FAHRITE C-3.

OHIOLOY CK-20.
M-95; 0.20 C, 2.0 Mn, 2.0 Si, 23-27 Cr, 19-22 Ni, bal Fe.
Austenitic, stainless, heat resisting.
Cast: 65,000 TS; 28,000 YS; 30 El.
For filter press parts, chemical and food equipment, furnace and heat treat parts.
ACI CK-20; AISI 310; SAE 60310.
Formerly FAHRITE C-63.

OHIOLOY CN-7M.
M-95; 0.07 C, 1.5 Mn, 1.5 Si, 19-22 Cr, 28-30 Ni, 2-2.75 Mo, 3.0-3.5 bal Fe.
Austenitic, very good stainless properties.
Cast: 62,500 TS; 25,000 YS; 30 El.
For pickling rolls equipment for food processing, pulp and paper, plastics paint industries.
ACI CN-7M; SAE 60332.
Formerly FAHRITE C-20.

OHIOLOY H-611.
M-95; 0.20-0.5 C, 2.0 max Mn, 2.0 Si, 14-16 Cr, 70-72 Ni, bal Fe.
High temperature corrosion resistant casting.
For muffle tubes, heat exchangers, oxygen nozzles in glass making, furnace parts.
Reference-Inconel.
Formerly FAHRITE N-88.

OHIOLOY H-810.
M-95; 0.45 max C, 2.0 max Mn, 2.0 Si, 19-21 Cr, 31-33 Ni, bal Fe.
High temperature corrosion resistant casting.
For parts used at high temperature, heat treat parts, furnace parts.
Reference-Incoloy.
Formerly FAHRITE N-53.

OHIOLOY HC.
M-95; 0.50 max C, 1.0 Mn, 2.0 Si, 26-30 Cr, 4 max Ni, 0.50 max Mo, bal Fe.
Ferritic, corrosion resistant, heat resistant.
Cast 55,000 min TS.
For salt pots, electrodes, tubes, parts for elevated temperature operation.
ACI HC; AISI 446; SAE 70446.
Formerly FAHRITE C-28.

OHIOLOY HD.
M-95; 0.50 max C, 1.5 Mn, 2.0 Si, 26-30 Cr, 4-7 Ni, 0.50 max Mo, bal Fe.
High temperature, corrosion resistant casting.
Cast: 75,000 TS; 35,000 YS; 8 El.
For furnace parts, salts pots; resists sulphur atmospheres: light loads to 1900°F.
ACI HD; AISI 327; SAE 70327.
Formerly FAHRITE N-83.

OHIOLOY HE.
M-95; 0.20-0.50 C, 2.0 Mn, 2.3 Si, 26-30 Cr, 8-11 Ni, 0.50 max Mo, bal Fe.
High temperature, corrosion resistant casting.
Cast: 85,000 TS; 40,000 YS; 9 El.
For furnace parts, smelting, and elevated temperature operation; resists sulphur atmosphere and scaling; usable to 2000°F.
ACI HE; AISI 312; SAE 70312.
Formerly FAHRITE M-73.

OHIOLOY HF.
M-95; 0.20-0.40 C, 2.0 Mn, 2.0 Si, 18-23 Cr, 8-12 Ni, 0.5 max Mo, bal Fe.
High temperature, corrosion resistant casting.
Cast: 70,000 TS; 35,000 YS; 25 El.
For tube supports, beams, brackets in cement mills furnaces, glass equipment, refineries.
Usable to 1600°F.
ACI HF; AISI 302B; SAE 70308.
Formerly FAHRITE N-8.

OHIOLOY HH.
M-95; 0.20-0.50 C, 2.0 Mn, 2.0 Si, 24-28 Cr, 11-14 Ni, 0.5 max Mo, bal Fe.
High temperature, corrosion resistant casting.
Cast: 75,000 TS; 35,000 YS; 10 El.
Usable to 2000°F.
For tube supports and beams in oil refineries, heat treat plants, ore roasting equipment.
ACI HH; AISI 309; SAE 70309. Formerly FAHRITE N-3.

OHIOLOY HH-45.
M-95; 0.35-0.45 C, 2.0 Mn, 1.75 Si, 23-28 Cr, 10-14 Ni, bal Fe.
High temperature, corrosion resistant casting.
Aged 24 hours, 1400°F: 80,000 TS; 4 El.
Usable to 2000°F.
For equipment operating at high temperature in glass factories, furnaces, petroleum plants.
ACI HH; AISI 309; SAE 70309.
Formerly FAHRITE N-3-45.

OHIOLOY HI.
M-95; 0.20-0.50 C, 2.0 Mn, 2.0 Si, 26-30 Cr, 14-18 Ni, 0.5 max Mo, bal Fe.
High temperature, corrosion resistant casting.
Cast 70,000 TS; 35,000 YS; 10 El.
For cast retorts for cyaniding, heat treating, brazing fixtures, melting magnesium; weldable, usable to 2150°F.
ACI HI.
Formerly FAHRITE N-4.

OHIOLOY HK.
M-95; 0.20-0.60 C, 2.0 Mn, 2.3 Si, 24-28 Cr, 18-22 Ni, 0.5 max Mo, bal Fe.
High temperature, corrosion resistant casting.
Cast: 65,000 TS; 35,000 YS; 10 El.
For tube supports, beams, fittings, tubing in chemical and oil refineries, furnaces; weldable, usable to 2100°F.
ACI HK; AISI 310; SAE 703109.
Formerly FAHRITE N-63.

OHIOLOY HL.
M-95; 0.20-0.60 C, 2.0 Mn, 2.0 Si, 28-32 Cr, 18-22 Ni, 0.5 max Mo, bal Fe.
High temperature corrosion resistant casting.
Cast: 65,000 TS; 35,000 YS; 10 El.
For gas dissociation equipment, furnace equipment, chemical and petroleum plants; weldable, usable to 2100°F.
ACI HL; SAE 70310A.
Formerly FAHRITE N-43.

OHIOLOY HN.
M-95; 0.20-0.50 C, 2.0 Si, 19-23 Cr, 23-27 Ni, 0.5 max Mo, bal Fe.
High temperature corrosion resistant casting.
Cast: 63,000 TS; 8 El.
For brazing fixture, trays, furnace parts, chains, beams, radiant tubes; usable to 2150°F.
ACI HN.
Formerly FAHRITE N-41.

OHIOLOY HT.
M-95; 0.35-0.75 C, 2.0 Mn, 2.5 Si, 13-17 Cr, 33-37 Ni, 0.5 max Mo, bal Fe.
High temperature corrosion resistant casting.
Cast: 65,000 TS; 4 El.
For heat treat and carburizing boxes, fixtures, fan blades, furnace parts.
Usable to 2000°F reducing; 2100°F oxidizing.
ACI HT; AISI 330; SAE 70330.
Formerly FAHRITE N-1.

OHIOLOY HU.
M-95; 0.35-0.75 C, 2.3 Mn, 2.5 Si, 17-21 Cr, 37-41 Ni, 0.5 max Mo, bal Fe.
High temperature corrosion resistant casting.
Cast: 65,000 TS; 4 El.
For parts requiring high strength and corrosion resistance at high temperatures. Usable to 2000°F reducing, 2100°F oxidizing.
ACI HU; ASTM A297-60T HU; SAE 70331.
Formerly FAHRITE N-51.

OHIOLOY HW.
M-95; 0.35-0.75 C, 2.0 Mn, 2.5 Si, 10-14 Cr, 58-62 Ni, 0.5 max Mo, bal Fe.
High temperature corrosion resistant casting.
Cast: 60,000 TS.
For furnace muffles, hearths, retorts, trays, enameling boxes, quenching fixtures. Usable to 1900°F in non-sulphur reducing; 2100°F oxidizi
ACI HW; ASTM A297-60T HW; SAE 70334.
Formerly FAHRITE N-61.

OHIOLOY HX.
M-95; 0.35-0.75 C, 2.0 Mn, 2.5 Si, 15-19 Cr, 64-68 Ni, 0.5 max Mo, bal Fe.
High temperature corrosion resistant casting.
Cast: 60,000 TS.
For high temperature application as cyanide pots, salt pots, muffles, annealing trays.
Usable to 1900°F in non-sulphur reducing; 2100°F oxidizing.
ACI HX; ASTM A297-60T HX; SAE 70335.
Formerly FAHRITE N-71.

OHMAL.
87.5 Cu, 9 Mn, 3.5 Ni.
For electrical resistors; heat resistant.

OHMALLOY.
M-897; 15 Cr, 5 Al, 80 Fe.
Annealed: 100,000 TS; 70,000 YS; 25 El.
For cathodes and filaments in electronic tubes; heat resistant.

OHMALOY now FECRALOY.

OILCRAT.
M-1149; 0.95 C, 1.25 Mn, 0.15 V, 0.5 Cr, 0.5 W, bal Fe.
For taps, punches, spindles, crimpers; oil hardened, non-deforming.

OIL CUPS.
88 Cu, 5 Sn, 7 Zn.
For oil cups, fittings.

OIL DIE SMOOTHCUT.
M-35; 1.05 C, 0.80 Mn, 0.50 W, 1.60 Cr, bal Fe.
Oil hardening cold work tool steel.
For blanking and forming dies, punches, hobs.
Type 03.

OILGRAPH.
M-1779; 1.45 C, 0.8 Mn, 1.15 Si, 0.2 Cr, 0.25 Mo, bal Fe.
Graphitic oil hardening tool steel of outstanding machinability. For blanking and forming dies, punches, gages, fixtures and machine parts.
Usual heat treated hardness: Rc 58/62.
AISI 05.

OIL HARDENING.
M-1424; 0.90 C, 1.25 Mn, 0.3 Si, 0.5 Cr, 0.2 V, 0.5 W, bal Fe.
Oil hardening tool steel; AISI O1.

OIL HARDENING GAUGE PLATE.
M-1749; 0.85-0.90 C, 1.6-1.7 Mn, 0.50 Cr, 0.25 V, bal Fe.
Supplied precision ground; for gauges.

OILITE 148 HARDENED.
M-211; 0.5 C, 2 Ni, 97 Fe.
Sintered, hardened, tempered at 500°F: 110,000 psi TS; 103,000 psi YS; 32 Rc.
MPIF FN-0205-S; ASTM B-484 Gr. 1, Class B, Type II.

OILITE 160 HARDENED.
M-211; 0.5 C, 2 Ni, 0.5 Mo, 97 Fe.
Sintered, hardened, tempered at 500°F; 110,000 psi TS; 103,000 psi YS; 32 Rc.

OILITE 182 HARDENED.
M-211; 0.5 C, 0.5 Ni, 0.5 Mo, 0.4 Mn, 98 Fe.
Sintered, hardened, tempered at 500°F: 110,000 psi TS; 103,000 psi YS; 32 Rc.

OILITE 304.
M-211; 19 Cr, 10 Ni, 71 Fe.
Sintered: 35,000 TS; 32,000 YS; 1.0 El; Density 61-64.
Austenitic stainless powder metal.
MPIF SS-304-P.

OILITE 316.
M-211; 17 Cr, 12 Ni, 2 Mo, 69 Fe.
Sintered: 38,000 TS; 3,000 YS; 1.0 El; Density 6.1-6.4.
Austenitic stainless powder metal.
MPIF SS-316-P.

OILITE 410.
M-211; 12 Cr, 88 Fe.
Sintered: 55,000 TS; 54,000 YS; 1.0 El; Density 6.1-6.4.
Chromium type stainless steel.
MPIF SS-410-P.

OILITE 993.
M-211; 1.0-3.0 Cu, 0.25-0.60 C, 3.0-5.0 Ni, bal Fe.
Sintered and heat treated: 120,000 TS; 6.8 min density; 40,000 fatigue.

OILITE ALUMINUM.
M-211; 4.0-5.0 Cu, 3.0-4.0 Pb, 3.0-4.0 Sn, 85-87 Al, 3.0 max other.
Sintered: 15,000 TS; 2.2-2.4 density.
Oilite bearing material.

OILITE BRASS.
M-211; 77.0-80.0 Cu, 0.25 max Fe, 1.0-2.0 Pb, 0.10 max Sn, bal Zn.
Sinter: 20,000 TS; 9 El; 7.2-7.7 density, 25,000 shear; 8500 fatigue.
ASTM B282-60 Class A; SAE 890; PMPMA BZ-0218-T.

OLITE BRONZE (WAS OILITE).
M-211; 87.5-90.5 Cu, 1.0 max Fe, 1.75 max C, 9.5-10.5 Sn.
Sintered: 14,000 TS; 11,000 YS; 6.4-6.8 density.
Oilite bearing material.
ASTM B438-67 Gr 1 Type 11; SAE 841.
PMPMA BT-0010-R.

OILITE COPPER.
M-211; 98.0 min Cu, 0.10 max Fe.
Sintered: 20,000 TS; 8 El; 8.0 min density.
MIL-B-20296.

OILITE IM-20.
M-211; 0.8 C, 20 Cu, 79 Fe.
Sintered: 75,000 YS; 1.0 El; 80 Rb; Density 7.2-7.6.
Infiltered type powder metal.
MPIF FX-2008-T; SAE 872; ASTM B-303 Class C.

OILITE IN-10.
M-211; 0.8 C, 10 Cu, 89 Fe.
Sintered: 90,000 TS; 75,000 YS; 2.5 El; 80 Rb; Density 7.2-7.6.
Infiltered type powder metal.
MPIF FX-1008-T.

OILITE LEAD BRONZE.
M-211; 76-78.5 Cu, 1.0 max Fe, 14-16 Pb, 1.0 max C, 6.5-8.5 Sn.
Sintered: 8000 TS; 6.8-7.2 density.
Oilite bearing material.

OILITE NICKEL SILVER.
M-211; 63.0-66.5 Cu, 16.5-19.5 Ni, 16.5-19.5 Zn.
Sintered: 30,000-37,500 TS; 10-15 El; 7.5-8.80 density.
ASTM B458-67 Gr 1, Type 1; PMPMA BZN-1818-V.

OILITE R-112 FRICTION MATERIAL.
M-211.
Sintered: 2500 TS; Density 5.8-6.2; Hardness 55 RL; 1.0 El.
Coefficient of friction 0.30-0.35.

OILITE STAINLESS STEEL.
M-211; 0.08-0.20 C, 17-19 Cr, 8-9 Ni, 70.5-73.0 Fe.
Sintered: 35,000 TS; 6.0-6.4 density. 20,000 shear 26,000 fatigue.
PMPMA SS-304-L-P.

OILITE STESSITE BRONZE.
M-211; 93.0-96.0 Cu, 4.0-6.0 Sn, 2.5 max others.
Sintered 20,000 TS; 10 El; 7.2-7.8 density. 10,000 fatigue stength.

OIL PUMP.
M-Eng.; 85 Cu, 3 Sn, 9 Zn, 3 P.
For oil pump parts.

OJIBWAY.
M-435; 0.95 C, 0.3 Mn, 0.3 Si, bal Fe.
For open headed dies; water hardened.

OKADUR 6.
M-462; 2.5-5.0 Cu, 0.2-1.8 Mg, 0.3-1.5 Mn, bal Al.
For light alloy structural parts; age hardenable.

OKADUR 10.
M-462; 2.5-5.0 Cu, 0.2-1.8 Mg, 0.3-1.5 Mn, bal Al.
For light alloy structural parts; age hardenable.

OKADUR 58.
M-462; 2.5-5.0 Cu, 0.2-1.8 Mg, 0.3 -1.5 Mn, 0.5-2.5 Pb, Sn, Cd, Bi,
For screw machine products; free-cutting.

OKADURPLAT.
M-462; 2.5-5.0 Cu, 0.2-1.8 Mg, 0.3-1.5 Mn, bal Al.
For aircraft structures; clad, age hardenable.

OKER-1.
M-Eng.; 69 Cu, 30 Zn, 0.97 Pb.
Rolled: 40,000 TS; 35 El; 37 Brin.
For tubes, sheets.

OKER-2.
M-Eng.; 55 Cu, 45 Zn, 0.5 Sn.
Rolled 50,000 TS.
For sheets, architectural purposes; corrosion resistant.

OKER BRASS.
M-Ger.; 64.25 Cu, 35.25 Zn, 0.39 Sn, 0.12 Bp.
For hardware bolts, fittings; high strength.

OKER BRASS.
M-Eng.; 44.5 Zn, 0.5 Sn, bal Cu.
For architectural structures and trim; high strength, corrosion resistant.

OKER I BRASS.
M-Eng.; 30 Zn, 1 Pb, bal Cu.
For fittings, hardware; good workability, free-cutting.

OKER-CAST.
M-Eng.; 72 Cu, 24 Zn, 1.1 Pb, 2.3 Fe.
Cast: 60,000 TS; 33 El; 30.6 RA; 86 Brin.
For turbine parts; free-cutting.

OLD GENUINE BABBITT.
M-77; 89 Sn, 7.5 Sb, 3.5 Cu.
For bearings; heavy duty.

OLDS BEARING BRONZE.
M-1007; 60-65 Cu, 17-19 Pb, 9-11 Zn, 4.5-6.0 Ni, 4.5-6.5 Sn.
Cast: 22,000 TS; 16,000 YS; 3 El; 56 Brin.
For formed parts, bearings, bushings; corrosion resistant.

OLDS HI TENSIL BRONZE NO. 7025.
M-1007; Cu alloy.
Rolled: 75,000 TS; 32,000 YS; 31.5 El; 140 Brin.
For hardware, gears; tough.

OLDS HI TENSIL BRONZE NO. 9025.
M-1007; Cu alloy.
Rolled: 92,000 TS; 42,000 YS; 28 El; 180 Brin.
For hardware, gears; tough.

OLDS HI TENSIL BRONZE NO. 11515.
M-1007; Cu alloy.
Rolled: 118,000 TS; 80,000 YS; 15 El; 215 Brin.
For hardware gears; tough.

OLDS LEADED BRONZE, OA-10.
M-1007; 70 Cu, 10 Sn, 20 Pb.
As cast: 54-58 Brin (500 kg).
For bearings.
QQ-B691 Comp 7; SAE 794.

OLDSMOLOY.
M-1007; 15 Ni, 35 Zn, Cr, Mn, Sn, 45 Cu.
Rolled: 70,000-76,000 TS; 15-20 El; 21 RA; 150-160 Brin.
For bearings, gears, hardware, food machinery; resists salt water and food acids.

OLIN OTHER NUMBERS refer to **CDA NUMBERS**.

OLIN ALCOLOY 688.
M-279; 22.7 Zn, 3.4 Al, 0.40 Co, bal Cu.
Spring Temper: 120,000 TS; 108,000 YS; 1.5 El.
Annealed: 78,000-82,000 TS; 47,000-55,000 YS; 36-39 El.
1/2 Hard: 90,000 TS; 81,000 YS; 18 El.
For springs, diaphragms, connectors.
Readily deep drawn, good corrosion and stress corrosion resistance.

OLIN ALLOY 263-HIGH BRASS.
M-279; 68.5 Cu, 31.5 Zn.
Ann: 45-61 ksi TS; 10-33 ksi YS; 52-62 El.
Rolled: 49-102 ksi TS; 21-93 ksi YS.
For belt buckles, cabinet handles, electrical connectors, strike plates, table lamps, weatherstrip.

OLIN ALLOY 425-LUBALOY X see **LUBALOY X 425**.

OLIN ALLOY 638, CORONZE see **CORONZE**.

OLIN ALLOY 664 see **COBRON 664**.

OLIN ALLOY 688 ALCALOY see **ALCALOY 688**.

OLIN ALLOY NO. 195-STRESCON.
M-279; 97 Cu, 1.5 Fe, 0.1 P, 0.8 Co, 0.6 Sn.
Ann: 50-60 ksi TS; 22 ksi YS; 20 El.
Rolled: 60-100 ksi TS; 45-97 ksi YS; 23-1 El.
For electric and electronic terminals and connectors, lead frames, switch spring contacts, thermostats.

OLIN ALLOY NO. 411 see **LUBALOY 411**.

OLIN ALLOY NO. 422 see **LUBRONZE 422**.

OLIN COBRAZE 6991 see **COBRAZE 6991**.

OLIN COBRON 664 see **COBRON 664**.

OLIN COPPER, LOW OXYGEN ETP 1102.
M-279; 99.90 min Cu, 0.02 max 0.
Ann: 26,000-36,000 TS; 6,000-12,000 YS; 20-44 El.
Cold rolled: 34,000-60,000 TS; 30,000-58,000 YS; 1-33 El.
For deep drawn parts.

OLIN COPPER, LOW OXYGEN STP 1142.
M-279; 99.90 min Cu, 0.044 min Ag (13 oz/ton), 0.02 max O.
Ann: 26,000-36,000 TS; 6000-12,000 YS; 20-44 El.
Cold rolled: 34,000-60,000 TS; 30,000-58,000 YS; 1-33 El.
For commutator and collector rings.
Resists softening at elevated temperatures.

OLIN COPPER, LOW OXYGEN STP 1162.
M-279; 99.90 min Cu, 0.085 min Ag (25 oz/ton), 0.02 max O.
Ann: 26,000-36,000 TS; 6000-12,000 YS; 20-44 El.
Cold rolled: 34,000-60,000 TS; 30,000-58,000 YS; 1-33 El.
For commutator bars, electronic components.
Resist softening at elevated temperatures.

OLIN COPPER NO. 1092.
M-279; 99.90 Cu, (incl. Ag) min, 0.02 max O.
Ann: (soft): 30,000 TS; 10,000 YS; 30 El.
Rolled: 34-60 ksi TS; 26-58 ksi YS; 33-1 El.
For electrical parts requiring good formability.
Low oxygen electrolytic copper.

OLIN COPPER NO. 1093.
M-279; 99.90 Cu (incl Ag) min, 0.044 Ag min, 0.02 max O.
Ann (soft): 30,000 TS; 10,000 YS; 30 El.
Rolled: 34-60 ksi TS; 26-58 ksi YS; 33-1 El.
For electronics, lead frames, commutator rings.
Low oxygen electrolytic copper with silver.

OLIN COPPER NO. 1094.
M-279; 99.90 Cu (incl Ag) min, 0.085 Ag min, 0.02 max O.
Ann (soft): 30 ksi TS; 10 ksi YS; 30 El.
Rolled: 34-60 ksi TS; 26-58 ksi YS; 33-1 El.
For electronics, lead frames.
Low oxygen electrolytic copper with silver.

OLIN CORONZE 638.
M-279; 2.8 Al, 1.8 Si, 0.4 Co, bal Cu.
Annealed: 82,000 TS; 54,000 YS; 36 El.
Spring Temper: 125,000 TS; 111,000 YS; 4 El.
At 400°C, Spring Temper: 25,000 TS; 18,000 YS; 100 El.
For heat exchangers, glass to metal seals, springs, electrical terminals.
Aluminum bronze, high strength corrosion resistant, oxidation resistant.

OLIN HSM COPPER 194.
M-279; 97.5 Cu, 2.35 Fe, 0.12 Zn, 0.03 P.
Ann: 40,000-60,000 TS; 24,000-45,000 YS; 20-35 El.
Cold rolled: 53,000-85,000 TS; 48,000-78,000 YS; 1-13 El.
For electrical connectors, terminals, springs, tubular products.

OLIN INCRAMUTE 699 see INCRAMUTE 699.

OLIN STRESCON 195.
M-279; 97 Cu, 1.5 Fe, 0.8 Co, 0.6 Sn, 0.1 P.
Ann: 50,000-65,000 TS; 22,000-45,000 YS; 22-35 El.
Cold rolled: 60,000-100,000 TS; 35,000-87,000 YS; 3-24 El.
For electrical terminals, connectors, contact springs.

OLYMPIC.
M-1702, M-73; 1.5 C, 12 Cr, 1 V, 0.7 Mo, bal Fe.
For dies, tools, lamination dies, punches, forming dies; air hardening, abrasion resistant, non-deforming.

OLYMPIC.
M-484; 0.7 C, 18 W, 4 Cr, 1 cv, bal Fe.
For high speed tools and cutters; high speed steels.

OMAN.
M-842; Pb-Cu.
For bearings;: self-lubricating.

OMANITBRONZE.
M-330; 81 Cu, 10 Al, bal Ni, Fe.
10 El.
For valve gauge fittings; corrosion resistant.

OMAN METAL.
M-714; 25 Pb, bal Cu.
For bearings; heavy duty.

OMC 3A1-2.5V.
M-1532; 3 Al, 2.5 V, bal Ti.
Wrought, annealed, min: 125,000 TS; 105,000 YS; 10.0 El.
Weldable, suitable for fabricating into welded titanium tubing or seamless tubing.
AMS 4943.

OMC 5AL-2.5 SN.
M-1532; 5 Al, 2.5 Sn, bal Ti.
Wrought ann, min: 115,000 TS; 110,000 YS; 10.0 El; 25.0 RA.
For aircraft tailcones, compressor caps, housings, stiffeners, cryogenic tankage.
Weldable, forgeable, good creep resistance to 900°F.
AMS 4910, 4926, 4953, 4966.
MIL-T-9046, 9047, 81556.

OMC 6AL-2SN-4ZR-2MO.
M-1532; 6 Al, 2 Sn, 4 Zr, 2 Mo, bal Ti.
Wrought ann, min: 130,000 TS; 120,000 YS; 10.0 El; 25.0 RA.
Compressor blades and wheels, good strength and toughness to 800-900°F heat treatable.
AMS 4975, 4976; MIL-T-9046, 9047.

OMC 6AL-2SN-4ZR-6MO.
M-1532; 6 Al, 2 Sn, 4 Zr, 6 Mo, bal Ti.
Wrought, STA, min: 170,000 TS; 160,000 YS; 12.0 El; 30.0 RA.
Engine discs and compressor blades, deep hardenable.
PWA 1216; AMS 4981.

OMC 6AL-4V.
M-1532; 6 Al, 4 V, bal Ti.
Wrought, annealed, min: 130,000 TS; 120,000 YS; 10.0 El, 25.0 RA.
For aircraft and engine forgings, compressor wheels and blades, spacers, cryogenic and marine equipment; good response to heat treatment in thinner sections.
AMS 4928; MIL-T-9046, 9047, 81556.

OMC 6AL-6V-2SN.
M-1532; 6 Al, 6 V, 2 Sn, bal Ti.
Wrought, annealed, min.: 140,000 TS; 130,000 YS; 8.0 El; 20.0 RA.
For pressure vessels, rocket motor cases, airframes, ordance equipment; heat treatable.
AMS 4971, 4978; MIL-T-9046, 9047, 81556.

OMC 8AL-1MO, 1V.
M-1532; 8 Al, 1 Mo, 1 V, bal Ti.
Wrought, annealed, min: 130,000 TS; 120,000 YS; 10.0 El, 20.0 RA.
High temperature jet engine forging alloy.
For airframes, turbine parts, blades, vanes and discs.
AMS 4915, 4972, 4973.
MIL-T-9046, 9047, 81556.

OMC-55 now OMC GRADE III.

OMC-70 now OMC GRADE IV.

OMC 85W-15MO (CAST).
M-1532; 15 Mo, bal W.
For rocket throat inserts, rocket engine vanes exposed to solid fuel exhausts.

OMC 105 COMP I GRADE 2 (CAST).
M-1532; Ti unalloyed.
Cast, min: 50,000 TS; 40,000 YS; 20.0 El.
For aircraft equipment, marine and chemical processing equipment.

OMC 105 COMP. I GRAD 3 (CAST).
M-1532.
Ti unalloyed.
Cast, min: 65,000 TS; 55,000 YS; 15.0 El.
For heat exchangers, valve trim, anodizing racks, chemical plant equipment.

OMC 105 COMP. I GRADE 4 (CAST).
M-1532.
Ti unalloyed.
Cast, min: 80,000 TS; 70,000 YS; 12.0 El.
For aircraft equipment, marine equipment, brackets, housings, anodizing racks.

OMC 105 COMP. II (CAST).
M-1532; 0.15 Pd, bal Ti.
Cast, min: 65,000 TS; 55,000 YS; 15.0 El.
For chemical process equipment for oxidizing and mildly reducing environments.

OMC 105 COMP. IV (CAST).
M-1532; 5 Al, 2.5 Sn, bal Ti.
Cast, min.: 115,000 TS; 105,000 YS; 10.0 El.
Compressor case housings, stiffeners, cryogenic tankage.

OMC 163 (CAST).
M-1532; 6 Al, 4 V, bal Ti.
Cast, min.: 130,000 TS; 120,000 YS; 6 El.
For compressor wheels, blades, rocket cases, cryogenic equipment, marine equipment.

OMC 164 now OMC GRADE 163 (CAST).

OMC 302 (CAST).
M-1532; Zr, unalloyed.
Cast, min.: 55,000 TS; 40,000 YS; 12 El.
Chemical process equipment, dry chlorine environments.

OMC GRADE I.
M-1532; Ti, unalloyed.
Wrought ann, min.: 35,000 TS; 25,000 YS; 24.0 El; 30.0 RA.
For heat exchangers, aircraft ducting, good corrosion resistance.
ASTM B265, B348.

OMC GRADE II.
M-1532; Ti unalloyed.
Wrought ann, min.: 50,000 TS; 40,000 YS; 20 El; 30.0 RA.
For heat exchangers, aircraft ducting, good corrosion resistance.
ASTM B265, B348.

OMC GRADE III.
M-1532; Ti unalloyed.
Wrought, ann, min.: 65,000 TS; 55,000 YS; 18.0 El; 30.0 RA.
For heat exchangers, valve trim, welding rods, anodizing racks, and chemical plant equipment.
AMS 4900.

OMC GRADE IV.
M-1532; Ti unalloyed.
Wrought ann, min.: 80,000 TS; 70,000 YS; 15.0 El; 25.0 RA.
For aircraft equipment, ammunition boxes, anodizing racks, shroud spacers.
AMS 4901, 4921.

OMC PD now OMC TI-PD.

OMC TI-17.
M-1532; 5 Al, 2 Zr, 2 Sn, 4 Mo, 2 Cr, bal Ti.
Wrought, STA, min.: 160,000 TS; 150,000 YS; 5 El; 10.0 RA.
For engine discs, heat treatable.
GE C50TF44, TF 57, TF 62.

OMC TI-PD.
M-1532; 0.15 Pd, bal Ti.
Wrought, ann, min.: 50,000 TS; 40,000 YS; 20.0 El; 35.0 RA.
For chemical processing equipment, oxidizing and mildly reducing environments.

OMEGA.
M-24; 0.6 C, 0.7 Mn, 1.85 Si, 0.25 V, 0.5 Mo, bal Fe.
Heat treated: 340,000-320,000 TS; 310,000-280,000 YS; 5-6 El; 11-21 RA.
For pneumatic chisels, rivet sets, punches, blacksmith tools; Iz-7-15; resists shock and fatigue.

OMEGA EXTRA.
M-1339; 0.55 C, 0.7 Cr, 0.18 Mo, 1.6 Ni, 0.1 V, bal Fe.
For gears, machine tool parts, crankshafts; oil hardened, shock resistant.

OMEGA EXTRA MO.
M-1339; 0.55 C, 1.0 Cr, 1.65 Ni, 0.80 Mo, bal Fe.
Hot work tool steel.

OMEGA EXTRA V.
M-1339; 0.56 C, 1.1 Cr, 0.18 Mo, 1.65 Ni, 0.1 V, bal Fe.
For gears, machine tool parts, crankshafts; oil hardened, shock resistant.

OMMET A.
M-1246; 50 Ni, 20 Mo, 22 Fe.
For amplifiers and transmitter tubes; corrosion and heat resistant.

OM METAL.
M-815; 63 Cu, 2.0 Pb, 0.5 Ni, 0.5 Mn, 0.3 Al, 1.0 Si, 0.5 Sn, 0.5 Sb, bal Zn.
53,000 psi TS; 22,300 psi YS; 25 El.
Yellow brass; resists dezincification.

OMMET "A" ALLOY.
M-837; Mo-Ni-Fe.
280 Brin.
For machinery parts; sintered alloy, heat resistant.

OMMET-B.
M-1246; Ni-Mo-Fe.
For amplifiers and transmitter tubes, chemical plant equipment; sintered, corrosion resistant.

OMMET-FE.
M-1246; Fe.
For structural parts for amplifers, transmitter tubes, rectifiers, valves; sintered, highest purity.

OMMET IRON.
M-837; 0.001 C, 0.01 Si, 0.01 P, 0.01 Mn, 99.98 Fe.
For vacuum tubes; sintered iron.

OMMET-SIVAR 48.
M-837; Fe-Co-Ni.
For glass to metal seals, vacuum tubes; sintered alloy.

OMMET-SIVAR 60.
M-837; Fe-Co-Ni.
For glass to metal seals, vacuum tubes; sintered alloy.

OMMET-SIVAR 90.
M-837; Fe-Co-Ni.
For glass to metal seals; sintered alloy.

ON.
M-366; 0.7 C, 18 W, 4 Cr, 1 V, bal Fe.
For tools, high speed cutters; high speed steel.

ONE FIVE ONE.
M-1405; 1.0 C, 0.50 Mn, 5.0 Cr, 1.0 Mo, 0.30 V, bal Fe.
Air hardening cold work tool steel.
For punching dies, slitting cutters, knurling tools.
BS 4659 BA2; AISI A2.

ONE TON BRASS.
M-Eng.; 61 Cu, 38 Zn, 1 Sn.
For marine parts, condenser tubes; corrosion resistant.

ONEIDA.
M-15; 0.90 C, 1.2 Mn, 0.5 Cr, 0.2 V, 0.5 W, bal Fe.
Oil hardening tool steel; AISI 01.

ONERAL.
M-Eng.; 0.9 C, 10 Mo, 0.03 Ti, 6 Ni, 0.03 Zr, 4 Fe, 28 Cr, bal Co.
For cast turbine blades, high temperature components.
High heat resistance.

O-NICKEL.
M-Eng.; 99.5 min Ni + Co, controlled Mg.
For valve components, tube cathodes; corrosion and heat resistant.

ONION FUSIBLE ALLOY.
M-Eng.; 50 Bi, 2 Sn, 30 Pb.
For fuses, safety plugs; M.P. 92°C.

ONPLUS.
M-366; 0.7 C, 18 W, 4 Cr, 2 V, bal Fe.
For cutting tools; high speed steel.

ONTARIO.
M-1779; 1.5 C, 12 Cr, 0.9 V, 1.0 Mo, bal Fe.
High performance cold work die steel; high carbon-high chromium type for long run blanking and forming dies, punches, thread rolling dies. Typical heat treated hardness: 58-61 Rc.
AISI D-2.

ONTARIO-EZ.
M-1779; 1.5 C, 12 Cr, 0.9 V, 1.0 Mo, 0.12 S, bal Fe.
Same as ONTARIO except free-machining grade.

ONTOP.
M-366; 0.7 C, 18 W, 4 Cr, 1 V, 5 Co, bal Fe.
For tools, high speed cutters; high speed steel.

ONYX SPRING STEEL.
M-38; 0.7 C, bal Fe.
For springs; water hardening.

O.P.
M-Japan.
For permanent magnets; high coercive force.

OPTAL 3 now WIELAND A13.

OPTAL 5 now WIELAND A15.

OPTAL 7.
M-764; 6.8 Mg, 0.15 Mn, 0.3 Cr, bal Al.
1/2 H-temper: 50,000 TS; 31,000 YS; 9 El; 90 Brin.

OPTICAL WIRE ALLOY.
M-U.S.; 54 Cu, 28 Zn, 18 Ni.
Annealed: 33,000 TS; 20,000 YS; 33 El; 38 RA; 50 Brin.
For optical instruments; German Silver.

OPTIMUM.
M-1340; 2.1 C, 11.5 Cr, bal Fe.
For forming and blanking dies, punches; oil or air hardened, non-deforming.

OPTIMUM CO20.
M-1340; 1.65 C, 12 Cr, Co, bal Fe.
For forming and blanking dies, punches; air hardened, non-deforming.

OPTIMUM W.
M-1340; 1.65 C, 11.5 Cr, 0.1 V, bal Fe.
For forming and blanking dies, punches; air hardened, non-deforming.

OPTIMUM Z.
M-1340; 2.1 C, 11.5 Cr, 0.7 W, bal Fe.
For forming and blanking dies, punches; oil or air hardened, non-deforming.

ORA-FUTURA II.
M-1748; 0.10 C, 1.2 Si, 14 Cr, 1.1 Mo, 0.50 Cb, 1.7 Al, 2.3 Ti, 0.20 Ce + La, bal Ni.
As cast: 101,000 psi TS; 79,000 psi YS; 7 El; 16 RA; 262 BHN.
Dental alloy; for bridges and restorations; polished or porcelainized.

ORALLOY.
M-838; Au alloy.
For dentures.

ORANGE LABEL.
M-343; 0.45-0.55 C, 1.25-1.75 Cr, 2-3 W, 0.2-0.3 V, bal Fe.
For chisels, punches, rivet sets; "Special Alloy Tool."

ORANGE LABEL.
M-365; 1.0-1.05 C, bal Fe.
For tools; precision cast.

ORANGE LABEL SPECIAL ALLOY TOOL STEEL.
M-343; C, alloy, bal Fe.
For chisels, punches, heading tools, rivet sets; fatigue resistant.

ORANIUM BRONZE.
M-141; 90 Cu, 10 Al.
Cast: 65,000 TS; 20 El. Rolled: 80,000 TS.
For castings, gears; Al-Bronze.

ORANIUM BRONZE, "H.H".
89 Cu, 11 Al.
For gears, propellers; Al-bronze, tough.

ORANIUM BRONZE, "H.X".
88.5 Cu, 11.5 Al.
For gears, bearings; Al-bronze, tough.

ORANIUM BRONZE "M.H".
92 Cu, 8 Al.
130,000 TS; 4 El.
For hardware; medium hard-soft.

ORANIUM BRONZE "S".
97 Cu, 3 Al.
30,000 TS; 60 El.
For hardware; soft-soft.

ORBIS.
M-111; 0.95 C, 1.5 Si, 0.75 Mn, 1 Cr, 0.1 V, bal Fe.
For dies; for cold work.

ORBIT.
M-38; 0.7 C, 2 Mn, 0.15 S, 1.0 Cr, 1.35 Mo, bal Fe.
For dies for blanking and forming, rim rolls, master hubs; tough, air hardened, free-machining.

ORDIX.
M-912; 0.58 C, 1.80 Si, 0.75 Mn, 0.40 Cr, bal Fe.
Cold work tool steel; for chuck jaws, rivet sets, springs.
W.-Nr. 2103; AFNOR 60 SC 7.

ORDIX SPECIAL.
M-912; 0.55-0.65 C, 0.8-1.1 Mn, 0.9-1.2 Cr, 0.07-0.12 V, bal Fe.
For tools, dies, crimpers, punches; oil hardening, tough.
W.-Nr. 2242.

OREIDE "A".
M-Eng.; 68 Cu, 32 Zn, 0.5 Sn.
60,000 TS; 30 El.
For hardware; Tobin Bronze.

OREIDE "B".
M-Eng.; 90-81 Cu, 10-15 Zn, 0-4 Sn.
French gold, carriage and harness hardware, ornamental work.

OREIDE, BRUNSWICK.
M-Eng.; 68 Cu, 32 Zn, 0.5 Sn.
For condenser tubing; corrosion resistant.

OREIDE, FRENCH GOLD.
90-81 Cu, 10-15 Zn, 0-9 Sn, 1-2 Pb.
For plumbing, hardware; free-cutting.

ORELLOY-42.
M-1603; 0.21 C, 1.35 Mn, 0.30 Si, 0.005 min Cb, bal Fe.
Wrought: 60 ksi TS; 42 ksi YS; 24 El.
HSLA steel.
ASTM A572.

ORELLOY 45.
M-1603; 0.22 C, 1.35 Mn, 0.30 Si, 0.005 min Cb, bal Fe.
Wrought: 60 ksi TS; 45 ksi YS; 22 El.
HSLA steel.
ASTM A572.

ORELLOY 50.
M-1603; 0.23 C, 1.35 Mn, 0.30 Si, 0.005 min Cb, bal Fe.
Wrought: 65 ksi TS; 50 ksi YS; 21 El.
HSLA steel.
ASTM A572.

ORELLOY 50 LT.
M-1603; 0.20 C, 1.50 Mn, 0.025 P, 0.025 S, 0.30 Si, 0.001 min C, bal Fe.
Wrought: 70 ksi TS; 50 ksi YS; 25 El.
High strength low alloy steel.

ORELLOY 55.
M-1603; 0.25 C, 1.35 Mn, 0.30 Si, 0.005 min Cb, bal Fe.
Wrought: 70 ksi TS; 55 ksi YS; 20 El.
HSLA steel.
ASTM A572.

ORELLOY 60.
M-1603; 0.26 C, 1.35 Mn, 0.30 Si, 0.005 min Cb, bal Fe.
Wrought: 75 ksi TS; 60 ksi YS; 18 El.
HSLA steel.
ASTM A572.

ORELLOY 65.
M-1603; 0.26 C, 1.35 Mn, 0.30 Si, 0.005 min Cb, bal Fe.
Wrought: 80 ksi TS; 65 ksi YS; 17 El.
HSLA steel.
ASTM A572.

ORELLOY 70.
M-1603; 0.26 C, 1.60 Mn, 0.30 Si, 0.005 min Cb, bal Fe.
Wrought: 85 ksi TS; 70 ksi YS; 16 El.
HSLA steel.
ASTM A572.

ORELLOY 70 FG.
M-1603; 0.15 C, 1.60 Mn, 0.025 P, 0.025 S, 0.30 Si, 0.001 min Cb, bal Fe.
Wrought: 80 ksi TS; 70 ksi YS; 20 El.
High strength low alloy steel.

ORELLOY 100.
M-1603; 0.10-0.20 C, 1.1-1.5 Mn, 0.15-0.30 Si, 0.2-0.3 Mo, 0.001-0.005 B, bal Fe.
Q + T.: 321 Brin.
Abrasion resistant steel.

ORELLOY 100A.
M-1603; 0.10-0.20 C, 1.1-1.5 Mn, 0.15-0.30 Si, 0.2-0.3 Mo, 0.001-0.005 B, bal Fe.
Q + T.: 110-130 ksi TS; 100 ksi YS; 18 El.
HSLA steel, extra high strength.
ASTM A514.

ORELLOY 100B.
M-1603; 0.10-0.20 C, 1.1-1.5 Mn, 0.15-0.30 Si, 0.2-0.3 Mo, 0.001-0.005 B, bal Fe.
Q + T.: 115-135 ksi TS; 100 ksi YS; 16 El.
HSLA steel, extra high strength.
ASTM A517.

ORELLOY 242.
M-1603; 0.10 C, 0.50 Mn, 0.09 P, 0.03 S, 0.48 Si, 0,86 Cr, 0.43 Cu,
Rolled: 70,000 TS; 50,000 YS; 22 El.
For structural members, bridges, booms, derricks, tanks.
High strength low-alloy steel.

ORELLOY 242 1.
M-1603; 0.10 C, 0.50 Mn, 0.09 P, 0.03 S, 0.48 Si, 0.43 Cu, 0.43 Ni, Cr, bal Fe.
Wrought: 70 ksi TS; 50 ksi YS; 22 El.
HSLA; improved atmosphere corrosion resistance.
ASTM A242.

ORELLOY 242 2.
M-1603; 0.17 C, 1.05 Mn, 0.20 S, 0.25 Cu, 0.50 Cr, 0.06 Ni, bal Fe.
Wrought: 70 ksi TS; 50 ksi YS; 18 El.
HSLA steel; improved atmosphere corrosion resistance.
ASTM A242.

ORELLOY 440.
M-1603; 0.25 C, 1.1-1.6 Mn, 0.30 Si, 0.20 min Cu, bal Fe.
Wrought: 70 ksi TS; 50 ksi YS; 21 El.
HSLA steel; improved atmosphere corrosion resistance.
ASTM A440.

ORELLOY 441.
M-1603; 0.22 C, 1.25 Mn, 0.30 Si, 0.20 min Cu, 0.02 min V, bal Fe.
Wrought: 70 ksi TS; 50 ksi YS; 21 El.
HSLA steel; improved atmosphere corrosion resistance.
ASTM A441.

ORELLOY 588.
M-1603; 0.19 C, 1.25 Mn, 0.30 Si, 0.25-0.40 Cu, 0.40-0.65 Cr, 0.02 mi bal Fe.
Wrought: 70 ksi TS; 50 ksi YS; 21 El.
HSLA steel; improved atmosphere corrosion resistance.
ASTM A588.

ORELLOY AR.
M-1603; 0.43 C, 1.5 Mn, 0.29 Si, 0.08 Cu, bal Fe.
Hot rolled: 235 Brin.
Abrasion resistant steel.

ORELLOY AR 320.
M-1603; 0.31 C, 1.65 Mn, 0.15-0.30 Si, 0.2-0.3 Mo, 0.001-0.005 B, bal Fe.
Q+T.: 320 Brin.
Abrasion resistant steel.

ORELLOY AR 340.
M-1603; 0.31 C, 1.65 Mn, 0.15-0.30 Si, 0.2-0.3 Mo, 0.001-0.005 B, bal Fe.
Q+T: 340 Brin.
Abrasion resistant steel.

ORELLOY AR 360.
M-1603; 0.31 C, 1.65 Mn, 0.15-0.30 Si, 0.2-0.3 Mo, 0.001-0.005 B, bal Fe.
Q+T: 360 Brin.
Abrasion resistant steel.

ORELLOY AR 400.
M-1603; 0.31 C, 1.65 Mn, 0.15-0.30 Si, 0.2-0.3 Mo, 0.001-0.005 B, bal Fe.
Q+T: 400 Brin.
Abrasion resistant steel.

ORIENTED M-6W.
M-10; Si, bal Fe.
For transformers; formerly Tran-Cor 4W-O; grain oriented.

ORIENTED M-7X.
M-10; Si, bal Fe.
For transformers; formerly Tran-Cor 3X-O; grain oriented.

ORIENTED M-7W.
M-10; Si, bal Fe.
For transformers; formerly Tran-Cor 3W-O; grain oriented.

ORIENTED M-8X.
M-10; Si, bal Fe.
For transformers; formerly Tran-Cor 2X-O; grain oriented.

ORIENTED T.
M-10; Si, bal Fe.
For armatures, electric generators; high permeability; formerly Tran-Cor T-O.

ORIENTED T-S.
M-10; Si, bal Fe.
For wound type transformers and reactors; high permeability; formerly Tran-Cor T-O-S.

ORIGINAL MONEL METAL.
M-557; 23 Cu, 2.5 Si, bal Ni.
Sand cast: 56,000-72,000 TS; 28,000-43,000 YS; 30-20 El; 120-140 Brin.
For pump parts, chemical industries; corrosion resistant.

ORION.
M-414; 0.5 C, 1.0 Cr, 0.2 V, bal Fe.
Heat treated: 240,000-115,000 TS; 195,000-80,000 YS; 5-23 El; 15-65 440-262 Brin.
For machine parts, tools, automobile springs and gears, forging dies, shear blades; resists heavy vibrations.

ORION 26-1.
M-1488; 0.01 max C, 26.0 Cr, 1.0 Mo, 0.015 max N, bal Fe.
Wrought, annealed: 400 N/mm^2 min UTS.
Stainless, ferritic grade; good with organic acids; resistant to stress corrosion.
AFNOR Z-01CD26-1; W.-Nr. 1.4131.

ORION-100 CAST.
M-1546; 0.35 C, 27 Cr, 1.2 Si, bal Fe.
Cast: 95,000 TS; 65,000 YS; 2 El; 212 Brin.
For furnace parts, heat treating boxes, salt pots, oil burners, exhaust manifolds.
Ferritic stainless steel, heat and oxidation resistant.

ORION-100 FORGE.
M-1546; 0.15 C, 27 Cr, bal Fe.
Annealed: 92,500 TS; 57,000 YS; 25 El.
For furnace parts, heat treating boxes, salt pots.
Ferritic stainless steel. Heat and oxidation resistant.

ORION-120.
M-1546; 0.02 max C, 27 Cr, bal Fe.
Annealed: 80,000 TS; 50,000 YS; 25 El; B 82 Rock.
For furnace parts, heat treating boxes, salt pots, oil burners.
Ferritic stainless steel. Heat resistant to 1150°C in oxidizing atmosphere.

ORKAN.
M-1796; 1.3-1.4 C, 4.2 Cr, 1.0 Mo, 4.0 V, 12 W, 5 Co, bal Fe.
Co-W high speed steel, for finishing tools.

ORKAN-S.
M-1796; 1.2 C, 4.0 Cr, 4.0 Mo, 10 W, 3.0 V, 10 Co, bal Fe.
Co-W high speed steel; lathe tools for finish machining.

ORLEANS.
M-373; 0.60 C, 0.80 Mn, 2 Si, 0.25 Cr, 0.2 V, 0.45 Mo, bal Fe.
For pneumatic chisels, punches, concrete breakers, cutters; Type S5; oil hardened, shock resistant.

ORNAL 0.5.
M-324; 0.5 Mg, 0.5 Si, bal Al.
13-27 kp/mm^2 TS; 7-20 kp/mm^2 YS; 10-15 El.
Parts for instruments and optical equipment.
Al Mg Si 0.5.

ORNAL 0.8.
M-324; 0.8 Mg, 0.8 Si, bal Al.
13-28 kp/mm^2 TS; 7-20 kp/mm^2 YS; 12-16 El.
Parts for instruments and optical equipment.
Al Mg Si 0.8.

OROBRAZE.
M-200; 80 Au, Cu, Fe.
Brazing alloy; melt range: 908-910°C.

OROBRAZE 910.
M-200; 80 Au, Cu, Fe.
Brazing alloy; melt range: 908-910°C.

OROBRAZE 940.
M-200; 62.5 Cu, bal Au.
Brazing alloy; melt range: 930-940°C.

OROBRAZE 950.
M-200; 17.5 Ni, bal Au.
Brazing alloy; melt point: 950°C.

OROBRAZE 990.
M-200; 25 Ni, bal Au.
Brazing alloy; melt range: 950-990°C.

OROBRAZE 1040.
M-200; 30 Ag, bal Au.
Brazing alloy; melt range: 1030-1040°C.

ORO-FUTURA I.
M-1748; 0.10 C, 1.2 Si, 15 Cr, 0.50 Mo, 1.5 Al, 2.0 Ti, 0.20 Ce+La, bal Ni.
As cast: 99,000 psi TS; 72,000 psi YS; 11 El; 17 RA; 229 BHN.
For dental use; porcelain coating restorations, crowns.

ORTHOMETAL.
M-108.
Soft magnetic alloy, square hysteresis loop for low level magnetic amplifiers, etc.

ORTHOMUMETAL.
M-108.
Soft magnetic alloy.

ORTHONAL, ROUND; ORTHONAL, SQUARE.
M-1626; 50 Ni, 50 Fe.
Grain oriented thin strip soft magnetic alloy. For magnetic amplifiers, flux counters, flux switching, etc.

ORTHONIK.
M-10; 50 Ni, 53 Fe.
For magnetic and electrical equipment, magnetic tape and foil.
Magnetically soft, high permeability.

ORVAR 1.
M-111, M-11a; 0.3 C, 0.9 Si, 0.5 Mn, 5.25 Cr, 1.4 Mo, 0.1 V, bal Fe.
For punches, crimpers, upsetters, riveters; hot work steel, oil hardened.
AISI H13.

OSBORN O1.
M-233; 0.90 C, 1.25 Mn, 0.50 Cr, 0.50 W, 0.20 V, bal Fe.
Oil hardening cold work tool steel.
B.S. 4659 BO.1; similar to AISI O1.

OSBORN 303.
M-233; 0.08 max C, 18.0 Cr, 10.0 Ni, 0.20 S, bal Fe.
Free cutting 18/8 stainless steel.
BS 970 303SZ1; EN 58 AM; similar to AISI 303.

OSBORN 304.
M-233; 0.06 max C, 18.0 Cr, 10.0 Ni, bal Fe.
Austenitic 18/8 type stainless steel, for food and dairy equipment.
B.S. 970 304S15; En 58 E; similar to AISI 304.

OSBORN 304 L.
M-233; 0.03 max C, 18.0 Cr, 11.0 Ni, bal Fe.
Low carbon 18/8 type austenitic stainless steel; preferred for welded assemblies.
B.S. 970 304S12; similar to AISI 304 L.

OSBORN 309.
M-233; 0.12 max C, 24.0 Cr, 15.0 Ni, bal Fe.
Heat resisting stainless steel.
B.S. 970 309S24; similar to AISI 309.

OSBORN 310.
M-233; 0.12 max C, 25.0 Cr, 21.0 Ni, bal Fe.
Heat resisting stainless steel; for mildly carburizing atmospheres up to 1100°C.
B.S. 970 310S24; similar to AISI 310.

OSBORN 316.
M-233; 0.07 C, 18.0 Cr, 11.0 Ni, 2.5 Mo, bal Fe.
Acid resisting stainless steel for severe conditions.
B.S. 970 PT4 316S16; EN 58 J; similar to AISI 316.

OSBORN 316 L.
M-233; 0.03 max C, 18.0 Cr, 13.0 Ni, 2.5 Mo, bal Fe.
Extra low carbon grade of 316 for welded assemblies.
B.S. 970 PT4 316S12; similar to AISI 316 L.

OSBORN 317.
M-233; 0.06 max C, 19.0 Cr, 14.0 Ni, 3.5 Mo, bal Fe.
Austenitic stainless steel for highly corrosive conditions.
B.S. 970 317S16; similar to AISI 317.

OSBORN 320.
M-233; 0.08 max C, 18.0 Cr, 11.0 Ni, 2.5 Mo, Ti, bal Fe.
Titanium stabilized weldable version of 316.
B.S. 970 PT4 320S17; EN 58 J.

OSBORN 321.
M-233; 0.08 max C, 18.0 Cr, 11.0 Ni, Ti, bal Fe.
Titanium stabilized weldable grade of 18/8 stainless steel.
B.S. 970 PT4 321S12; EN 58 B/C; similar to AISI 321.

OSBORN 325.
M-233; 0.08 C, 18.0 Cr, 10.0 Ni, 0.20 S, Ti, bal Fe.
Free maching version of 321.
B.S. 970 PT4 325S21; EN 58 M.

OSBORN 330.
M-233; 0.10 max C, 19.0 Cr, 38.0 Ni, 2.0 Si, bal Fe.
Heat resisting steel for furnace equipment.

OSBORN 347.
M-233; 0.08 max C, 18.0 Cr, 10.0 Ni, Nb, bal Fe.
Niobium stabilized weldable grade of 18/8 stainless steel.
B.S. 970 PT4 347S17; EN 58 F/G; similar to AISI 347.

OSBORN 403.
M-233; 0.08 max C, 13.0 Cr, bal Fe.
Ferritic-martensitic chromium stainless steel for lightly stressed engineering fittings.
B.S. 970 PT4 403S17; similar to AISI 403.

OSBORN 405.
M-233; 0.08 max C, 13.0 Cr, Al, bal Fe.
Non-hardening weldable ferritic stainless steel.
B.S. 970 PT4 405S17; similar to AISI 405.

OSBORN 410.
M-233; 0.12 max C, 12.5 Cr, bal Fe.
Martensitic stainless steel for mildly corrosive conditions.
B.S. 970 PT4 410S21; EN 56 A; similar to AISI 410.

OSBORN 416A.
M-233; 0.12 C, 12.5 Cr, 0.25 S, bal Fe.
Free machining martensitic stainless steel.
B.S. 970 416S21; EN 56 AM; similar to AISI 416.

OSBORN 416C.
M-233; 0.25 C, 13.0 C, 0.25 S, bal Fe.
High carbon, free machining, martensitic stainless steel.
B.S. 970 416S37; EN 56 CM; similar to AISI 420 F.

OSBORN 420 C.
M-233; 0.23 C, 12.5 Cr, bal Fe.
Martensitic stainless steel.
B.S. 970 420S37; EN 56 C; similar to AISI 420.

OSBORN 420 D.
M-233; 0.32 C, 12.5 Cr, bal Fe.
Martensitic stainless steel; hardened to 550 HV it is used for cutlery and edge tools.
B.S. 970 420S45; EN 56 D; similar to AISI 420.

OSBORN 430.
M-233; 0.10 max C, 17.0 Cr, bal Fe.
Ferritic type stainless, non-hardenable.
B.S. 970 PT4 430S15; EN 60; similar to AISI 430.

OSBORN 431.
M-233; 0.16 max C, 16.0 Cr, 2.3 Ni, bal Fe.
Martensitic stainless; hardenable; improved corrosion resistance.
B.S. 970 PT4 431S29; EN 57; similar to AISI 431.

OSBORN A2.
M-233; 1.0 C, 5.0 Cr, 1.05 Mo, 0.30 V, bal Fe.
Air hardening cold work tool steel.
B.S. 4659 BA 2; similar to AISI A2.

OSBORN A7.
M-233; 2.3 C, 5.3 Cr, 1.05 W, 1.05 Mo, 4.7 V, bal Fe.
Air hardening cold work tool steel.
B.S. 4659 BA 2; similar to AISI A7.

OSBORN D2.
M-233; 1.5 C, 12.0 Cr, 0.9 Mo, 0.9 V, bal Fe.
Air or oil hardening cold work tool steel.
B.S. 4659 BD-2; similar to AISI D2.

OSBORN D3.
M-233; 2.2 C, 12.5 Cr, 0.25 V, bal Fe.
Air or oil hardening cold work tool steel.
B.S. 4659 BD 3; similar to AISI D3.

OSBORN D4.
M-233; 2.1 C, 12.25 Cr, 0.8 Mo, 0.65 V, bal Fe.
Air or oil hardening cold work tool steel.
Similar to AISI D4.

OSBORN H10A.
M-233; 0.37 C, 3.0 Cr, 2.75 Mo, 0.45 V, 3.0 Co, bal Fe.
Hot work tool steel, chromium type.
B.S. 4659 BH10A; similar to AISI H10.

OSBORN H12.
M-233; 0.35 C, 5.0 Cr, 1.4 W, 1.6 Mo, 0.25 V, bal Fe.
Hot work tool steel, chromium type.
B.S. 4659 BH12; similar to AISI H12.

OSBORN H13.
M-233; 0.39 C, 5.0 Cr, 1.4 Mo, 1.0 V, bal Fe.
Hot work tool steel, chromium type.
B.S. 4659 BH13; similar to AISI H13.

OSBORN H21.
M-233; 0.30 C, 3.0 Cr, 8.8 W, 0.30 V, bal Fe.
Hot work tool steel, tungsten type.
B.S. 4659 BH21; similar to AISI H21.

OSBORN H21 N.
M-233; 0.25 C, 3.0 Cr, 8.8 W, 0.30 V, 2.3 Ni, bal Fe.
Hot work tool steel, tungsten type.
B.S. 4659 BH21A.

OSBORN LV10N.
M-233; 0.4 C, 3.0 Si, 8.0 Ni, 19.0 Cr, 0.20 N^2, bal Fe.
Austenitic type valve steel, for exhaust valves for diesel engines.
U.S. Si10N.

OSBORN LV20.
M-233; 0.8 C, 2.0 Si, 1.4 Ni, 19.5 Cr, bal Fe.
Martensitic steel for exhaust valves.
B.S. 443S65; U.S. XB.

OSBORN LV21/2N.
M-233; 0.55 C, 8.5 Mn, 2.0 Ni, 20.0 Cr, 0.30 N^2, bal Fe.
Austenitic type valve steel.
U.S. 21/2N.

OSBORN LV21/4N.
M-233; 0.5 C, 9.0 Mn, 4.0 Ni, 21.0 Cr, 0.45 N^2 bal Fe.
Austenitic type valve steel for exhaust valves in petrol engines.
B.S. 970 349S52; U.S. 21/4N.

OSBORN LV21/4NS.
M-233; 0.05 C, 9.0 Mn, 4.0 Ni, 21.0 Cr, 0.45 N^2, 0.05 S, bal Fe.
Austenitic type valve steel for exhaust valves in petrol engines.
B.S. 970 349S54.

OSBORN LV21/12N.
M-233; 0.2 C, 11.0 Ni, 21.0 Cr, 0.20 N^2, bal Fe.
Austenitic type valve steel, for exhaust valves for diesel engines.
B.S. 970 381S34; U.S. 21/12N.

OSBORN LV21/42.
M-233; 0.05 C, 9.0 N, 4.0 Ni, 21.0 Cr, 0.45 N^2, 2.0 Nb, bal Fe.
Austenitic type valve steel for exhaust valves; improved stress-rupture properties.
B.S. 970 352S529.

OSBORN LV21/43.
M-233; 0.05 C, 9.0 Mn, 4.0 Ni, 21.0 Cr, 0.45 N^2, 2.0 Nb, 1.0 W, bal Fe.
Austenitic type valve steel.

OSBORN LV52.
M-233; 0.45 C, 3.4 Si, 8.2 Cr, bal Fe.
Martensitic type steel for inlet valves in petrol engines.
B.S. 970 401S45; U.S. Si1.

OSBORN LV54A.
M-233; 0.4 C, 1.5 Si, 14.0 Ni, 13.5 Cr, 2.7 W, 0.5 Mo, bal Fe.
Austenitic type steel for making hollow valves; easily hard faced.
B.S. 970 331S42; AMS 5700.

OSBORN LV55.
M-233; 0.45 C, 2.3 Si, 1.2 Mn, 9.0 Ni, 17.2 Cr, 1.0 W, bal Fe.
Austenitic steel for hollow valves, - to be hard faced.
W.-Nr. 1.4873.

OSBORN M1.
M-233; 0.80 C, 4.0 Cr, 1.8 W, 8.75 Mo, 1.1 V, bal Fe.
Molybdenum high speed steel.
B.S. 4659 BM 1; similar to AISI M1.

OSBORN M2.
M-233; 0.85 C, 4.0 Cr, 6.3 W, 5.0 Mo, 1.9 V, bal Fe.
Mo-W high speed steel.
B.S. 4659 BM 2; similar to AISI M2.

OSBORN M2 S.
M-233; 0.85 C, 4.0 Cr, 6.3 W, 5.0 Mo, 1.9 V, 0.1 S, bal Fe.
Mo-W high speed steel, free-machining.

OSBORN M15.
M-233; 1.55 C, 4.7 Cr, 6.5 W, 3.0 Mo, 5.0 V, 5.0 Co, bal Fe.
Mo-W-Co high speed steel.
B.S. 4659 BM 15.

OSBORN M35.
M-233; 0.9 C, 4.0 Cr, 6.3 W, 5.0 Mo, 1.9 V, 5.0 Co, bal Fe.
Mo-W-Co high speed steel.
W.-Nr. 1.3243.

OSBORN M42.
M-233; 1.07 C, 3.8 Cr, 1.5 W, 9.7 Mo, 1.1 V, 8.3 Co, bal Fe.
Mo-Co high speed steel.
B.S. 4659 BM 42; similar to AISI M42.

OSBORN MN.
M-233; 1.2 C, 12 Mn, bal Fe.
12% manganese austenitic steel with good wear resistance for wearing plates, shutes, screens and security applications.

OSBORN QJ.
M-233; 1.1 C, 1.3 Cr, bal Fe.
Cold work tool steel; hardenable to 60-64 Rc.

OSBORN RAB1.
M-233; 0.31 C, 4.1 Ni, 1.3 Cr, 0.3 Mo, bal Fe.
Supplied at 285 max Brin, but can be rehardened to 53-26 Rc. For moulds.
W.-Nr. 1.2766.

OSBORN RAB20.
M-233; 0.40 C, 1.5 N, 2.0 Cr, 0.2 Mo, bal Fe.
Supplied prehardened to 30 Rc, but can be rehardened to 53-26 Rc. For moulds.

OSBORN S1.
M-233; 0.52 C, 1.4 Cr, 2.2 W, 0.2 V, bal Fe.
Cold work tool steel, shock resisting type.
B.S. 4659 BS 1; similar to AISI S1.

OSBORN S510.
M-233; 0.21 C, bal Fe.
Carbon steel with 21 ton/sq. in. tensile strength. Aircraft grade.
B.S. 2S510.

OSBORN S511.
M-233; 0.10 C, bal Fe.
A deep drawing carbon steel; aircraft grade.
B.S. 2S511.

OSBORN S514/5.
M-233; 0.21 C, 1.5 Mn, bal Fe.
Carbon-manganese steel with 30-50 ton/sq. in. tensile strength; aircraft grade.
B.S. 2S514/5.

OSBORN S516/7.
M-233; 0.46 C, 1.5 Mn, bal Fe.
Carbon-manganese steel with 60-75 tons/sq. in. tensile strength; aircraft grade.
B.S. 2S516/7.

OSBORN S524/5.
M-233; 0.08 C, 18.0 Cr, 10.0 Ni, Ti, Nb, bal Fe.
Stabilized corrosion resisting steel, cold rolled to 52 ton/sq. in. TS. Aircraft grade.
B.S. 2S524/5.

OSBORN S526/7.
M-233; 0.08 C, 18.0 Cr, 10.0 Ni, Ti, Nb, bal Fe.
Stabilized corrosion resistant steel, cold rolled to 35 ton/sq. in. tensile strength; aircraft grade.
B.S. 2S526/7.

OSBORN S530/1.
M-233; 0.12 C, 24.5 Cr, 17.5 Ni, Ti, Nb, bal Fe.
Stabilized heat resisting steel; aircraft grade.
B.S. 2S530/1.

OSBORN S532/3.
M-233; 0.05 C, 15.6 Cr, 5.4 Ni, 1.7 Mo, Ti, Cu, bal Fe.
Corrosion resisting steel, precipitation hardening to 64/76 ton/sq. in.
B.S. 2S532/3.

OSBORN S534/5.
M-233; 0.25 C, 1.0 Cr, 0.2 Mo, bal Fe.
Chrome-molybdenum steel with 57-74 ton/sq. in. tensile, suitable for welding.
B.S. 2S534/5.

OSBORN T1.
M-233; 0.75 C, 4.0 Cr, 18.0 W, 1.1 V, bal Fe.
Tungsten high speed steel.
B.S. 4659 BT1; similar to AISI T1.

OSBORN T4.
M-233; 0.78 C, 4.0 Cr, 18.0 W, 0.8 Mo, 1.1 V, 5.0 Co, bal Fe.
Tungsten-cobalt high speed steel.
B.S. 4659 BT4; similar to AISI T4.

OSBORN T6.
M-233; 0.80 C, 4.0 Cr, 20.5 W, 0.8 Mo, 1.5 Co, bal Fe.
Tungsten-cobalt high speed steel.
B.S. 4659 BT6; similar to AISI T6.

OSBORN WJ.
M-233; 0.04 C, 21.0 Cr, 25.0 Ni, 4.5 Mo, Ti, bal Fe.
Austenitic stainless for severe conditions.

OSCILLUMIN.
M-Ger.; 0.8 Cu, 0.4-0.6 Fe, 12.6-13.2 Si, bal Al.
28,000 TS; 4 El; 65 Brin.
For light alloy castings; high fluidity.

OSMAGAL.
M-462; 1.8 Mn, bal Al.
For sheets; rolled.

OSMAGAL.
M-1548; 1.8 Mn, bal Al.
Soft; 15,000 TS; 6400 YS; 30 El; 25 Brin.
Hard; 35,000 TS; 28,000 YS; 2 El; 60 Brin.
For heat exchangers, truck panels, fixtures, duct work.
Non-heat treatable.

OSMAGAL S.
M-462; 1.5-3.0 Mg, 0.5-1.5 Mn, 0.3 max Cr, bal Al.
For aircraft and structural parts; resists sea water corrosion.

OSMIRIDIUM (NATURAL) NEVYANSKITE.
58-44 Ir, 27-49 Os, 0-10 Pt, 0-6 Ru, 1.5-3.0 Rh.
For pen points; corrosion resistant.

OSMIRIDIUM (NATURAL) SISERSKITE.
57 Os, 34 Rh + Ir, 8 Ru.
For pen points; corrosion resistant.

OSMIRIDIUM SISEROKITE.
57 Os, 34 Rh + Ir, 8 Ru.
For pen points.

OSMIUM.
M-1755; Os.
Purities: 99.999%, 99.99%, 99.9%.
Forms: Sponge, powder, arc melted buttons.

OSNALIUM 3.
M-462; 2-4 Mg, 0.4 max Mn, 0.3 max Cr, bal Al.
Soft: 28,000 TS; 13,000 YS; 30 El; 47 Brin.
Hard: 42,000 TS; 37,000 YS; 8 El; 77 Brin.
For light alloy parts, marine hardware; resists sea water corrosion.

OSNALIUM 5.
M-462; 4.0-5.5 Mg, 0.8 max Mn, 0.3 max Cr, bal Al.
Soft: 42,000 TS; 22,000 YS; 35 El; 65 Brin.
Hard: 60,000 TS; 50,000 YS; 15 El; 100 Brin.
For aircraft structures, light alloy parts; corrosion resistant.

OSNALIUM 7.
M-462; 5.5-7.5 Mg, 0.8 max Mn, 0.3 max Cr, bal Al.
For light alloy parts; corrosion resistant.

OSSENBERG ESS, ETC see ESS ETC.

OSTERMANN CAST BRONZE NO. 10.
M-330; 90 Cu, 10 Sn.
29,000 TS; 15 El.
For general construction of machinery fittings, apparatus; corrosion resistant.

OSTERMANN CAST BRONZE NO. 14.
M-330; 86 Cu, 14 Sn.
29,000 TS; 3 El.
For bearings, gears, high pressure hydraulic apparatus; wear resistant.

OSTERMANN CAST BRONZE NO. 20.
M-330; 80 Cu, 20 Sn.
21,000 TS.
For step-bearings, slide valves, wearing plates, bells; for parts with high friction loading.

OSTERMANN LEAD BRONZE NO. 10.
M-330; 86 Cu, 10 Sn, 4 Pb.
28,000 TS; 15 El.
For bearings for hot rolling works and electrical machinery; heavy duty.

OSTERMANN LEAD TIN BRONZE NO. 8.
M-330; 80 Cu, 8 Sn, 12 Pb.
23,000 TS; 8 El.
For bearings with high compressive loading and cold rolling mills; heavy duty.

OSTERMANN RED BRASS NO. 4.
M-330; 93 Cu, 4 Sn, 2 Zn, 1 Pb.
29,100 TS; 25 El.
For pipe flanges; known as "Flange Bronze."

OSTERMANN RED BRASS NO. 5.
M-330; 85 Cu, 5 Sn, 7 Zn, 3 Pb.
21,000 TS; 10 El.
For railroad machinery fittings to be bright finished or polished; free-cutting.

OSTERMANN RED BRASS NO. 8.
M-330; 82 Cu, 8 Sn, 7 Zn, 3 Pb.
21,000 TS; 6 El.
For machine fittings to be bright finished or polished; free-cutting.

OSTERMANN RED BRASS NO. 9.
M-330; 85 Cu, 9 Sn, 6 Zn.
29,000 TS; 12 El.
For railroad bearings, fittings; heavy duty.

OSTERMANN RED BRASS NO. 10.
M-330; 86 Cu, 10 Sn, 4 Zn.
29,000 TS; 10 El.
For pipe lines and fittings, general machinery construction; also called "Machine Bronze."

OSTERMANN WROUGHT BRONZE NO. 6.
M-330; 94 Cu, 6 Sn.
For wire, sheet strip; tin bronze.

OT4.
M-USSR; 3 Al, 1.5 Mn, bal Ti.
Alpha titanium alloy.

OT4-0.
M-USSR; 1 Al, 1.5 Mn, bal Ti.
Alpha titanium alloy.

OT4-1.
M-USSR; 2 Al, 1.5 Mn, bal Ti.
Alpha titanium alloy.

OT4-2.
M-USSR; 6 Al, 1.5 Mn, bal Ti.
Titanium alloy.

OTISCOLOY.
M-840; 0.08-0.12 C, 0.90-1.25 Mn, 0.35 Cu, 0.10 Ni, 0.05 Cr, bal Fe.
Drawn: 70,000 TS; 50,000 YS; 25 El; 55 Brin.
For deep drawn parts.

OTISEL O-12.
M-331; 0.12-0.35 C, 11-14 Cr, bal Fe.
Cast: 75,000-130,000 TS; 39,000-90,000 YS; 10-25 El; 140-150 Brin.
For acid resistant castings; corrosion resistant.

OTISEL O-16.
M-331; 0.12 C, 14-18 Cr, bal Fe.
Cast: 70,000-80,000 TS; 39,000-43,000 YS; 20-15 El; 140-175 Brin.
For acid resistant castings; corrosion resistant.

OTISEL O-18.
M-331; 0.08-0.20 C, 18-20 Cr, 8-10 Ni, bal Fe.
Cast: 78,000-85,000 TS; 39,000-42,000 YS; 60-65 El; 135-160 Brin.
For acid resistant castings; stainless.

OTISEL 0-24.
M-331; 0.20 max C, 22-25 Cr, 10-13 Ni, bal Fe.
Cast: 80,000-85,000 TS; 40,000-43,000 YS; 45-50 El; 140-150 Brin.
For heat resistant castings; heat resistant.

OTISEL 0-30.
M-331; 0.25 max C, 25-30 Cr, 1.3 max Mn, bal Fe.
Cast: 45,000-99,000 TS; 35,000-60,000 YS; 2-3 El; 170-190 Brin.
For heat resistant castings; resists heat to 2150°F.

OTISEL 0-S.
M-331; C, alloy, bal Fe.
For castings.

OTISEL K-4 MOLYBDENUM.
M-331; 0.13-0.35 C, 5-7 Cr, bal Fe, Mo.
For corrosion resistant parts; corrosion resistant.

OTISEL K-5 TUNGSTEN.
M-331; 0.13-0.35 C, 5-7 Cr, W, bal Fe.
For corrosion resistant parts; corrosion resistant.

OTISEL NO. 1.
M-331; 18 Cr, 8 Ni, bal Fe.
For heat and corrosion resisting parts; heat and corrosion resisting alloy.

OTISEL NO. 2.
M-331; 0.10-0.15 C, 12-16 Cr, bal Fe.
For corrosion resistant parts; corrosion resistant.

OTISEL NO. 3.
M-331; 0.10-0.15 C, 15-18 Cr, bal Fe.
For corrosion resistant parts; corrosion resistant.

OTISEL NO. 4.
M-331; 0.20 max C, 20-30 Cr, 12-20 Ni, bal Fe.
For heat and corrosion resistant parts; heat and corrosion resistant.

OTISEL NO. 5.
M-331; 0.25 max C, 26-30 Cr, bal Fe.
For heat and corrosion resistant parts; heat and corrosion resistant.

OTOTANI-ALLOY GRADE 2.
M-1718; 35-50 Si, 20-30 Ca, 5-15 Mn, 10-20 Fe, 1.5 max Al, 0.08 max P, 0.05 max S.
For melting or ladle additions to produce cast steel with nodular graphite to increase the wear resistance, as on steel rolls for rolling mills.

OTOTANI-ALLOY GRADE 3.
M-1718; 35-50 Si, 20-25 Ca, 10-25 Fe, 5-15 Mn, 4-10 Al, 0.3 max C, 0 max P, 0.05 max S.
For melting or ladle additions to produce fine grain steel.

OTTAWA.
M-435; 0.8 C, 0.2 Mn, 0.15 Si, bal Fe.
For mining drills and rods; hollow drill steel.

OTTO'S SPECULUM.
M-Eng.; 69 Cu, 31 Sn.
40,000 TS; 10,000 YS; 38 El.
For telescope reflectors, mirrors.

OUNCE METAL.
M-126; 85 Cu, 5 Sn, 5 Zn, 5 Pb.
27,000-33,000 TS; 20-15 El; 20-15 RA; 50-59 Brin.
For bearing metal, casting valves, carburetor and pump parts; Composition Brass, also called "Std. Red Composition."

OV.
M-1655; 0.50 C, 1.0 Mn, 0.25 Si, 1.0 Cr, 0.10 V, bal Fe.
Cold work tool steel for shear knives, hand tools, shanks for carbide tipped lathe tools.
W.-Nr. 1.2241.

OVH.
M-1655; 0.58 C, 1.0 Mn, 0.25 Si, 1.0 Cr, 0.10 V, bal Fe.
Cold work stamps, punches, split chucks.
W.-Nr. 1.2242.

OXALLOY-28.
M-856; 99.92 min Cu.
For high temperature wire.
Heat resistant to 400°F.

OXWELD NO. 1 HT.
M-843; C, alloy, bal Fe.
Welded 70,000 TS.
For welding rod for alloy steel.

OXWELD NO. 7.
M-843; C, bal Fe.
Welded: 45,000 TS.
For welding rod for steel; copper coated.

OXWELD NO. 9.
M-843; 3.2 C, 2.2 Si, bal Fe.
For cast iron welding rod; for gray cast iron.

OXWELD NO. 19 CUPRO.
M-843; Cu, alloy.
Welded: 35,000 TS.
For bronze welding rod; phosphor bronze.

OXWELD NO. 23.
M-843; 5 Si, bal Al.
For Al welding rod.

OXWELD NO. 25 M.
M-843; Sn, Zn, bal Cu.
Welded: 50,000 TS; 96 Brin.
For bronze welding rod.

OXWELD NO. 26.
M-843; Si-Cu alloy, 2.5 Si, bal Cu.
For bronze welding rod.

OXWELD NO. 28.
M-843; 0.08 C, 18 Cr, 8 Ni, Cb, bal Fe.
For stainless steel welding rod.

OXWELD NO. 31 T.
M-843; Sn, Zn, bal Cu.
For bronze welding rod.

OXWELD NO. 32 CMS.
M-843; C, bal Fe.
Welded; 98,000 TS.
For welding rod.

OXWELD NO. 38.
M-843; Cu alloy.
For Cu welding rod.

OXYGEN FREE COPPER.
M-33; 99.98 Cu.
Annealed: 32,000 TS; 10,000 YS; 45 El.
Rolled 50,000 TS; 45,000 YS; 6 El.
For electrical conductors; high conductivity.

P

P2 ALLOY.
M-137; 3.0-4.5 Cu, 4-5 Si, 2-4 Fe, 1.75-2.5 Ni, bal Al.
Cast; 25,000 TS; 23,000 YS; 2 El; 101 Brin.
For carburetors, brake shoes, cameras, electric meters; pressure die casting.

P6 ALLOY.
M-61; 4.8 V, 6.1 Ni, 45 Co, bal Fe.
For hysteresis motors.
Saturation induction 16,000 gausses.
$H_c = 60$ oe, magnetically semihard.

P-50 NICKEL.
M-68; 0.04 max Cu, 0.02 max Mn, 0.05 max Fe, 0.05 max C, 0.01 max Mg, max Si, bal Ni + Co.
For cathodes in electron tubes; high purity nickel.

P.1000 see **CARR'S QUALITY P.1000.**

PACKING.
82 Pb, 4.8 Sn, 13 Sb.
20 Brin.
For metallic packing; pouring temperature 324°C.

PACKING METAL.
51.8 Cu, 14.3 Zn, 15 Ni, 17 Pb, 1.8 Sn.
For metallic packing.

PACKING PISTON.
M-258; 73-76 Pb, 12-14 Sn, 10-15 Sb.
11,000 TS; 26 Brin.
For metallic packing.

PACKING RINGS, FRENCH.
M-French; 51.8 Cu, 14.3 Zn, 15 Ni, 17 Pb, 1.8 Sn.
For packing rings.

PACKING RUSSIAN.
M-USSR; 99 Zn, 0.9 Sn, 0.3 Pb, 0.2 Fe.
For packing metal.

PACKING (VALVE).
71 Sn, 24 Sb, 5 Cu.
For valve packing: anti-friction.

PAGE 3 1/2 NI.
M-872; C, 3.5 Ni, bal Fe.
For gas welding rod; corrosion resistant.

PAGE-ALLEGHENY 12% CR.
M-872; 12 Cr, 0.12 max C, bal Fe.
Annealed: 70,000 TS; 35,000 YS; 25 El; B 80 Rock.
For welding electrodes; corrosion resistant.

PAGE-ALLEGHENY 16% CR.
M-872; 0.12 max C, 16 min Cr, bal Fe.
For welding electrodes; corrosion resistant.

PAGE-ALLEGHENY 18% CR.
M-872; 0.35 max C, 18 min Cr, bal Fe.
For welding electrodes; corrosion resistant.

PAGE-ALLEGHENY 23% CR.
M-872; 0.16 max C, 23 Cr, bal Fe.
Annealed: 85,000 TS; 25 El; B 84 Rock.
For welding electrodes; corrosion resistant.

PAGE-ALLEGHENY 33 GRADE C1.
M-872; 0.12 max C, 12-14 Cr, bal Fe.
For fans, blowers, condensers, engine parts; heat resistant to 1500°F.

PAGE-ALLEGHENY 33 N.H.
M-872; 0.08 max C, 11.5-13.5 Cr, 0.10-0.20 Al, bal Fe.
For welding electrodes; stainless.

PAGE-ALLEGHENY 33 T.Q.
M-872; 0.12 max C, 11.5-13 Cr, bal Fe.
For welding electrodes; stainless.

PAGE-ALLEGHENY 34.
M-872; 0.12 max C, 14-16 Cr, bal Fe.
For fans, blowers; corrosion resistant.

PAGE-ALLEGHENY 44.
M-872; 0.20 max C, 22-26 Cr, 11-13 Ni, bal Fe.
For furnace parts, boiler baffles, pumps; heat resistant to 2100°F.

PAGE-ALLEGHENY 46.
M-872; 0.10 min C, 4-6 Cr, bal Fe.
For welding electrodes; for 4-6 Cr steel.

PAGE ALLEGHENY 4-6 CHROMIUM MOLY.
M-872; 0.10 max C, 4 Cr, Mo, bal Fe.
Annealed: 70,000 TS; 30,000 YS; 37 El; B 70 Rock.
For welding electrodes; corrosion resistant.

PAGE-ALLEGHENY 55.
M-872; 0.35 max C, 23-30 Cr, bal Fe.
For furnace parts, boiler baffles; heat resistant to 2100°F.

PAGE-ALLEGHENY 66 GRADE B.
M-872; 0.12 min C, 15-18 Cr, bal Fe.
For fans, blowers, condensers, evaporators; heat resistant to 1600°F.

PAGE-ALLEGHENY 66 GRADE C2.
M-872; 0.12 max C, 16-18 Cr, bal Fe.
For corrosion resisting parts; corrosion resistant.

PAGE-ALLEGHENY 66W.
M-872; 0.12 max C, 16-18 Cr, 2.5-3.5 W, bal Fe.
For welding electrodes; corrosion resistant.

PAGE-ALLEGHENY 67.
M-872; 0.35 max C, 18-23 Cr, bal Fe.
For welding electrodes; corrosion resistant.

PAGE-ALLEGHENY 17-7.
M-872; 0.10-0.20 C, 16-18 Cr, 7-8.5 Ni, bal Fe.
Annealed: 110,000 TS; 40,000 YS; 35 El; B88 Rockwell.
Drawn: 270,000 TS; 240,000 YS; 1 El; C47 Rockwell.
For trim, household articles; Type 301X; stainless, austenitic.

PAGE-ALLEGHENY 18-8.
M-872; 18 Cr, 8 Ni, 0.07 C, 1.3 Cb, bal Fe.
Annealed: 90,000 TS; 35,000 YS; 55 El; B 80 Rockwell.
Drawn: 350,000 TS; 175,000 YS; C45 Rockwell.
For welding electrodes, strand springs, rope; stainless steel.

PAGE-ALLEGHENY 18-8 EZ.
M-872; 0.15 max C, 17-19 Cr, 8-10 Ni, 0.7 min P or S or Se, bal Fe.
Annealed: 90,000-120,000 TS; 35,000 YS; 50-35 El; 130-150 Brin.
For screw machine products, hardware; stainless, austenitic; Type 303; free-cutting.

PAGE-ALLEGHENY 18-8 MO.
M-872; 0.07 max C, 18-20 Cr, 12-15 Ni, 2.0-2.5 Mo, bal Fe.
Annealed: 90,000 TS; 30,000 YS; 50 El.
Drawn: 210,000 TS; 190,000 YS; 0.5 El.
For welding electrodes; austenitic, stainless.

PAGE-ALLEGHENY 18-8S.
M-872; 0.03 C, 19-22 Cr, 10-12 Ni, bal Fe.
For welding electrodes; Type 304; stainless.

PAGE-ALLEGHENY 18-8 TYPE 307.
M-872; 0.07-0.15 C, 4 Mn, 0.5 Si, 19.5-22 Cr, 9-10.5 Ni, bal Fe.
Annealed: 80,000 TS; 30,000 YS; 50 El; 60 RA; 180 Brin.
For stainless parts; Type 307; stainless, austenitic.

PAGE-ALLEGHENY 19-9.
M-872; 19 Cr, 9 Ni, 0.07 C, bal Fe.
Annealed: 80,000 TS; 30,000 YS; 50 El; B 80 Rockwell.
For welding electrodes; stainless steel.

PAGE-ALLEGHENY 430F.
M-872; 0.12 max C, 14-18 Cr, 0.07 min P, S, Se, or 0.6 max Mo, Zn, bal Fe.
For welding rod; free-cutting, stainless.

PAGE-ALLEGHENY 8-10 CR-MO.
M-872; 0.08 max C, 8.5-10.5 Cr, 1.25-1.75 Mo, bal Fe.
For welding electrodes; martensitic stainless steel.

PAGE-ALLEGHENY 12-14 CR.
M-872; 0.12 min C, 12-15 Cr, bal Fe.
Annealed: 85,000 TS; 40,000 YS; 30 El; B95 Rock.
For combustion and steam engine parts, fans; heat resistant to 1500°F.

PAGE-ALLEGHENY 15-35.
M-872; 0.25 max C, 15 Cr, 35 Ni, bal Fe.
For welding electrodes for stainless steel; coated, stainless.

PAGE-ALLEGHENY 18-85.
M-872; 0.03 C, 19-22 Cr, 10-12 Ni, bal Fe.
For welding electrodes; stainless steel.

PAGE-ALLEGHENY 25-20.
M-872; 0.25 max C, 27 Cr, 21 Ni, bal Fe.
Annealed: 90,000 TS; 40,000 YS; 50 El; B80 Rockwell.
Drawn: 200,000 TS; 150,000 YS; C36 Brin.
For welding electrodes; stainless steel.

PAGE-ALLEGHENY 25-20 CB.
M-872; 0.06-0.10 C, 27.5 min Cr, 20-22 Ni, 1.2-1.4 Cb, bal Fe.
Annealed: 75,000 TS; 30,000 YS; 40 El; 50 RA; 180 Brin.
For welding electrodes; stainless, austenitic, stabilized.

PAGE-ALLEGHENY 25-20 MO.
M-872; 0.06-0.10 C, 1.5-2.0 Mn, 27.5 min Cr, 20-22-Ni, 2 Mo, bal Fe.
For welding electrodes; heat and corrosion resistant.

PAGE-ALLEGHENY ALLOY 22 CR.
M-872; 0.08-0.20 C, 19-22 Cr, 9-12 Ni, bal Fe.
For digesters, pipes, tanks, agitators, strainers; corrosion resistant.

PAGE-ALLEGHENY AMO.
M-872; 0.10 max C, 16-19 Cr, 14 max Ni, 2-4 Mo, bal Fe.
For welding electrodes; stainless.

PAGE-ALLEGHENY A-TI.
M-872; 0.10 max C, 17-20 Cr, 7-10 Ni, Ti = 6 x C, bal Fe.
For welding electrodes; stainless.

PAGE-ALLEGHENY "B" SP.
M-872; 0.11 max C, 18-20 Cr, 8-10 Ni, bal Fe.
For welding electrodes; stainless.

PAGE-ALLEGHENY METAL A.
M-872; 0.11 max C, 17-19 Cr, 7-9 Ni, bal Fe.
For welding electrodes; stainless.

PAGE-ALLEGHENY METAL B.
M-872; 0.08-0.20 C, 18-20 Cr, 8-10 Ni, bal Fe.
For welding electrodes; stainless.

PAGE-ALLEGHENY METAL C.
M-872; 0.08-0.20 C, 17-19 Cr, 7-9 Ni, bal Fe.
For stainless parts; Type 302.

PAGE-ALLEGHENY METAL CB.
M-872; 0.10 max C, 17-20 Cr, 8-12 Ni, Cb = 10 X C, bal Fe.
For welding electrodes; stainless.

PAGE-ALLEGHENY METAL FM.
M-872; 0.20 max C, 1m-19 Cr, 7-9.5 Ni, S or Se or Mo, bal Fe.
For welding electrodes; stainless.

PAGE-ALLEGHENY OHMALOY.
M-872; 0.12 max C, 12-14 Cr, 4-4.5 Al, bal Fe.
For welding electrodes; stainless.

PAGE-ARMCO.
M-872; 0.03 max C, bal Fe.
Weled: 55,000 TS; 8 El.
For welding electrodes; bare and coated.

PAGE A-S-3 1/2N.
M-872; 0.15-0.20 C, 0.3-0.6 Mn, 0.15-0.75 Si, 3.25-3.75 Ni, bal Fe.
For welding wire for Ni and alloy steels; bare.

PAGE A-S-6.
M-872; C, bal Fe.
For welding wire for submerged arc welding; bare.

PAGE A-S-10.
M-872; C, bal Fe.
For welding wire for submerged arc welding; bare.

PAGE A-S-15.
M-872; C, bal Fe.
For welding wire for submerged arc welding; bare.

PAGE A-S-15-MO.
M-872; C, alloy, bal Fe.
For welding wire for submerged arc welding; bare.

PAGE A-S-20.
M-872; C, bal Fe.
For welding wire; bare.

PAGE A-S-65.
M-872; C, bal Fe.
For welding wire; bare.

PAGE A-S-110.
M-872; C, bal Fe.
For welding wire; bare.

PAGE-AUTO.
M-872; 0.06 C, bal Fe.
For welding electrodes.

PAGE B.
M-872; 0.13-0.18 C, bal Fe.
Welded: 55,000 TS; 7 El.
For welding electrodes; rust and lime coated.

PAGE C.
M-872; 0.06 max C, 0.15 max Mn, bal Fe.
Welded: 52,000-60,000 TS; 27-23 El.
For gas welding rod, radio tube leads; AWS-GA-50, gas welding.

PAGE CE.
M-872; 0.06 C, bal Fe.
Welded: 50,000 TS; 8 El.
For welding electrodes; rust and lime coated.

PAGE DENTAL ALLOY.
93 Al, 2.25 Cu, 4.75 Au.
For dental applications.

PAGE E.
M-872; 0.13-0.18 C, bal Fe.
Welded: 55,000 TS; 5 El.
For welding electrodes; rust and lime coated.

PAGE HARD FACING.
M-872; 0.6 C, 12-14 Mn, 2 Ni, bal Fe.
For hard facing electrodes; high manganese steel.

PAGE HC.
M-872; 0.9 C, 1.4 Mn, bal Fe.
For welding electrodes; abrasion resistant.

PAGE HIGH CARBON.
M-872; 0.9-1.1 C, 0.6 Mn, bal Fe.
For hard facing electrodes; wear resistant.

PAGE HI-TENSILE C.
M-872; 0.07 C, bal Fe.
Welded: 75,000 TS; 20 El.
For welding electrodes; shielded arc.

PAGE HI-TENSILE F.
M-872; 0.06 C, bal Fe.
Welded: 80,000 TS; 17 El.
For welding electrodes; shielded arc.

PAGE HI-TENSILE G.
M-872; 0.11 C, bal Fe.
Welded: 85,000 TS; 17 El.
For welding electrodes; shielded arc.

PAGE HI-TENSILE M.
M-872; 0.06-0.08 C, 0.32-0.42 Mn, bal Fe.
Welded: 70,000-85,000 TS; 55,000 YS; 32-25 El.
For welding electrodes for low alloy steels; E-6015.

PAGE HI-TENSILE SHIELDED ARC.
M-872; 0.07 C, bal Fe.
Welded: 80,000 TS; 25 El.
For welding electrodes; shielded arc.

PAGE HT.
 M-872; 0.2 C, bal Fe.
 For welding electrodes.

PAGE MANGANESE BRONZE.
 M-872; 0.3 Mn, 40 Zn, 0.75-0.85 P, bal Cu.
 Welded: 60,000-65,000 TS; 25-20 El.
 For gas welding wire; for cast iron welding.

PAGE MANGANESE NICKEL.
 M-872; 0.8-1.0 C, 12-14 Mn, 3.0-3.5 Ni, 0.2 Mo, 0.8-1.1 Si, bal Fe.
 Welded: 150 Brin.
 For welding electrode, hard facing electrode; shielded arc, austenitic.

PAGE MC.
 M-872; 0.36 C, 1.Mn, 1.2 Si, bal Fe.
 For welding electrodes; mild, wear resistant.

PAGE MEDIUM CARBON.
 M-872; 0.32-0.45 C, 0.9-1.1 Mn, bal Fe.
 For hard facing electrodes; tough.

PAGE NAVAL BRONZE.
 M-872; 59-61 Cu, 0.5-1.0 P, bal Zn.
 Welded: 49,000-55,000 TS; 35-25 El.
 For gas welding wire; for cast iron welding.

PAINTWELL.
 M-604; Low C, Mn, bal Fe.
 Zinc coated steel, chemically treated for painting.

PALAURAL.
 M-200; 34 Au, Pt, Pd.
 For dental purposes.

PAITUNG WHITE COPPER.
 26-40 Cu, 1-32 Zn, 16-37 Ni, 0-2.6 Fe.
 For ornamental white metal parts; corrosion resistant.

PAKTONG, COOKSON.
 40.9 Cu, 45 Zn, 11.1 Ni, 2.5 Fe, 0.16 Co.
 For domestic utensils; corrosion resistant.

PAKTONG, FYFE.
 40.4 Cu, 25.4 Zn, 31.6 Ni, 2.6 Fe.
 For tableware, domestic utensils; corrosion resistant.

PAKTONG, KEFERSTEIN.
 26.3 Cu, 36.8 Zn, 36.8 Ni.
 For tableware; corrosion resistant.

PAKTONG (PACKFONG).
 M-Eng.; 32-41 Ni, 26-40 Cu, 16-37 Zn, 0-2.6 Fe.
 130,000 TS; 2 El.
 For tableware and ornamental uses; Nickel Silver.

PAKTONG, PEAT.
 57.9 Cu, 32.2 Zn, 7.7 Ni, 2.5 Fe.
 For ornamental ware, domestic utensils; corrosion resistant.

PAKTONG, THURSTON.
 43.8 Cu, 40.6 Zn, 15.6 Ni.
 For tableware; corrosion resistant.

PAKTONG, TUTENAG.
 44-45.7 Cu, 16-39.6 Zn, 17.4-40 Ni.
 For domestic ware, ornamental parts; corrosion resistant.

PALAU-A.
 M-Eng.; 60 Ni, 20 Pt, 10 Pd, 10 V.
 For jewelry, chemical apparatus; Pt substitute, white gold.

PALAU-B.
 M-Eng.; 20 Pd, 80 Au.
 For jewelry, chemical apparatus; Pt substitute, white gold.

PALCO.
 M-1493; 65 Pd, 35 Co.
 For brazing molybdenum and tungsten.
 For cathode structures.
 M.P. 2244-2250°F.

PALCRONIRY T-49.
 M-1493; 41 Au, 27 Pd, 22 Ni, 10 Cr.
 M.P. 2150-2200°F.
 For high temperature brazing.
 Good flowability and wettability.

PALCUSIL-5.
 M-1493; 68 Ag, 27 Cu, 5 Pd.
 M.P. 1485-1490°F.
 For brazing Kovar to ceramic seals.
 Good wettability.

PALCUSIL-10.
 M-1493; 58 Ag, 32 Cu, 10 Pd.
 M.P. 1515-1566°F.
 For high temperature brazing of Kovar to ceramic seals.
 Good wettability.

PALCUSIL-15.
 M-1493; 65 Ag, 20 Cu, 15 Pd.
 M.P. 1562-1652°F.
 For high temprature brazing for Kovar and ceramic seals.
 Good flowability and wettability.

PALCUSIL-25.
 M-1493; 54 Ag, 21 Cu, 25 Pd.
 M.P. 1652-1742°F.
 For high temperature brazing for Kovar.
 High vapor pressure. Does not embrittle Kovar.

PALE YELLOW GOLD.
 92 Au, 0.8-3 Ag, 0-8.3 Fe.
 For jewelry; corrosion resistant.

PALID.
 M-Ger.; 11 Sb, 7 As, bal Pb.
 For bearings; anti-friction.

PALINEY 4.
M-349.
Palladium casting gold dental alloy.
Melt range: 1670-1810°F; 205 Brin.
White gold color.

PALINEY 9.
M-349; 30 Ag: Pd + Pt + Au = 55.
Stress relieved: 155,000 TS; 120,000 YS; 300 Knoop.
For telephone spring wire relays.
Note: "PALINEY" is a registered trade mark.

PALINEY CB.
M-349; Au, Ag + others. Dental alloy.
Melt range: 1600-1740°F; casting temp: 1900°F. 235 Brin. White gold color.

PALINEY M.
M-349; Pt, Au, Ag, bal Pd.
Wire: 60,000 TS; 24,000 YS; 23 El; 95 Brin.
For sliding contacts in potentimeters; corrosion and tarnish resistant, resistance wire.

PALINEY NO. 6.
M-349; 44 Pd, 38 Ag, 1 Pt, 16 Cu, 1 Zn.
Ht. Tr.: 165,000 TS; 110,000 YS; 15 El; 260 Brin.
Ann.; 110,000 TS; 63,000 YS; 24 El; 150 Brin.
For pivots, springs, bearings, electrical contacts. Heat treatable, non-magnetic, corrosion resistant.

PALINEY NO. 7.
M-349; 35 Pd, 10 Pt, 30 Ag, 14 Cu, 10 Au, 1 Zn.
Ht. Tr.: 185,000 TS; 150,000 YS; 10 El; 280 Brin.
Ann.; 120,000 TS; 89,000 YS; 24 El; 180 Brin.
For springs, pivots, bearings, contacts, potentiometers.
Non-magnetic, corrosion resistant.

PALINEY NO. 8.
M-349; 45 Pt, 38 Ag, bal Pd.
Wire, Ann.: 95,000-125,000 TS; 14 min El.
Sheet, Ht. Tr.: 160,000-200,000 TS; 0.5 min El; 110,000 PL; 390 Knoop.
For sliding contact in potentiometer windings of Nichrome. Corrosion and wear resistant.

PALINEY MEDIUM FUSING.
M-349.
Dental solder; white gold solder.
Melt range: 1450-1525°F.

PALIUM A.
M-French; 1 Sn, 4 Pb, 4 Cu, 0.8 Mg, 0.3 Mn, 22 Zn, bal Al.
For bearings; antifriction.

PALIUM Z.
M-French; 2.6 Sn, 4 Pb, 4.5 Cu, 0.6 Mg, 0.3 Mn, 0.3 Zn, bal Al.
For bearings; antifriction.

PALLABRAZE-810.
M-200; 5 Pd, bal Ag + Cu.
For brazing thermionic valves, magnetrons, klystrons.
M.P. 870-810°C.
High temperature brazing alloy.

PALLABRAZE-840.
M-200; 10 Pd, Ag, bal Cu.
For brazing thermionic valves, magnetrons, klystrons.
M.P. 830-840°C.
High temperature brazing alloy.

PALLABRAZE-850.
M-200; 10 Pd, Cu, bal Ag.
For brazing thermionic valves, magnetrons, klystrons.
M.P. 824-850°C.
High temperature brazing alloy.

PALLABRAZE-880.
M-200; 15 Pd, Cu, bal Ag.
For brazing thermionic valves, magnetrons, klystrons.
M.P. 856-880°C.
High temperature brazing alloy.

PALLABRAZE-900.
M-200; 20 Pd, Cu, bal Ag.
For brazing thermionic valves, magnetrons, klystrons.
M.P. 876-900°C.
High temperature brazing alloy.

PALLABRAZE-950.
M-200; 25 Pd, Cu, bal Ag.
For brazing thermionic valves, magnetrons, klystrons.
M.P. 901-950°C.
High temperature brazing alloy.

PALLABRAZE-1010.
M-200; 5 Pd, bal Ag.
For brazing thermionic valves, magnetrons, klystrons.
M.P. 970-1010°C.
High temperature brazing alloy.

PALLABRAZE -1090.
M-200; 18 Pd, bal Cu.
For brazing thermionic valves, magnetrons, klystrons.
M.P. 1080-1090°C.
High temperature brazing alloy.

PALLABRAZE-1225.
M-200; 30 Pd, bal Ag.
For brazing thermionic valves, magnetrons, klystrons.
M.P. 1150-1225°C.
High temperature brazing alloy.

PALLABRAZE-1237.
 M-200; 60 Pd, bal Ni.
 For brazing thermionic valves, magnetrons, klystrons.
 M.P. 1237°C.
 High temperature brazing alloy.

PALLADENT.
 40 Pd, bal Al.

PALLADIUM.
 M-53; 99.9 Pd.
 Ann: 66 R-15T; Elect. cond.: 16% I.A.C.S. For electrical contacts.

PALLADIUM.
 M-1755; Pd.
 Purities: 99.999%, 99.99%, 99.9%.
 Forms: Sponge, rod, wire, ingot, sheet, foil, single crystals.

PALLADIUM ALLOY-1.
 10 Rh, 90 Pd.
 For jewelry; corrosion resistant.

PALLADIUM ALLOY-2.
 67 Pd, 33 Ag.
 For jewelry; corrosion resistant.

PALLADIUM GOLD-1.
 90 Au, 10 Pd.
 For jewelry; white gold, Pt substitute.

PALLADIUM GOLD-2.
 40 Cu, 31 Au, 10 Ag, 10 Pd.
 For jewelry; white gold, Pt subsititue.

PALLAS.
 M-1306; 1.05 C, Cr, bal Fe.
 For bearings, cutters, liners, sleeves; water hardened, wear resistant.

"P" ALLOY.
 M-Eng.; 0.11 Fe, 0.11 Si, bal Al.
 8,000 TS; 34 El.
 For light alloy parts.

PALMANSIL-5.
 M-1493; 75 Ag, 20 Pd, 5 Mn.
 M.P. 1846-1962°F.
 For high temperature brazing.
 Good flowability and weldability.

PALNI.
 M-1493; 60 Pd, 40 Ni.
 M.P. 2260°F.
 For high temperature brazing.
 Good flowability and wettability.

PALNIRO-1.
 M-1493; 50 Au, 25 Pd, 25 Ni.
 M.P. 2016-2050°F.
 For high temperature brazing.
 Wets and flows well on Mo-W and stainless steel.

PALNIRO-4.
 M-1493; 30 Au, 34 Pd, 36 Ni.
 M.P. 2075-2136°F.
 For high temperature brazing.
 Wets and flows well on Mo-W and stainless steel.

PALNIRO-7.
 M-1493; 70 Au, 8 Pd, 22 Ni.
 M.P. 1841-1899°F.
 For high temperature brazing.
 Wets W-Mo and stainless steel.

PALORO.
 M-1493; 92 Au, 8 Pd.
 For brazing on molybdenum and tungsten.
 M.P. 2192-2264°F.

PALSIL-10.
 M-1493; 90 Ag, 10 Pd.
 M.P. 1835-1950°F.
 For high temperature brazing.
 Good wettability and flowability.

PAN-7.
 M-1218, M-1218a; 81.9 Cu, 10 Sn, 1 Ni, 7 P, 0.1 P.
 Cast: 22,500-29,000 TS; 13,500-20,000 YS; 11-5 El; 65-75 Brin.
 For connecting rod bearings, liners, sleeves; leaded bronze, anti-friction.

PAN-444.
 M-1218, M-1218a; 83.75 Cu, 4 Sn, 4 Ni, 4 Pd, 0.25 P.
 Cast: 27,000-36,000 TS; 16,000-20,000 YS; 18-8 El; 65-80 Brin.
 For rocker bracket bushings in aircraft engines; leaded bronze, high ductility.

PAN-B.
 M-1218, M-1218a; 72.45 Cu, 9.5 Sn, 3 Ni, 15 Pd, 0.05 P.
 Cast: 22,500-38,000 TS; 13,500-22,500 YS; 15-6 El; 60-75 Brin.
 For bearings, bushings, liners, sleeves; leaded phosphor bronze, corrosion resistant.

PAN-BS.
 M-1218, M-1218a; 37 Pd, 0.5 Sn, 0.05 P, 0.5 Ni, bal Cu.
 Cast.
 For main bearings, locomotive slippers; leaded bronze.

PANDALOY.
 M-M-1; C, Cr, Ni, bal Fe.
 For paper making equipment, feed screws; stainless.

PAN-H.
M-1218, M-1218a; 66,85 Cu, 10 Sn, 3 Ni, 20 Pd, 0.15 P.
Cast: 20,000-27,000 TS; 11,500-18,000 YS; 13-6 El; 60-70 Brin.
For rotors, sleeves, bearings, bushings; leaded bronze, corrosion resistant.

PANSERI.
M-Eng.; 1 Cu, 0.4 Mg, 4.5 Ni, 0.5 Fe, 11.5 Si, bal Al.
For pistons; high temperature resistance.

PANTAL.
M-116; 0.5-1.0 Mg, 0.4-1.4 Mn, 0.8-2.0 Si, 0.3 Ti, bal Al.
Annealed: 16,000 TS; 14,000 YS; 25 El; 50 Brin.
Aged: 60,000 TS; 50,000 YS; 10 El; 100 Brin.
For sheathing, body construction, architectural use; corrosion resistant, age-hardenable.

PANTAL.
M-935; 0.7 Si, 1.4 Mg, 0.9 Mg, 0-0.2 T, bal Al.
Heat treated: 43,000-50,000 TS; 26,000-36,000 YS; 12-15 El; 70-95 Brin.
For light alloy parts.

PANTAL.
M-1548; 0.5-1.0 Si, 0.4-1.4 Mn, 0.8-2.0 Mg, bal Al.
Half Hard: 22,000 TS; 17,000 YS; 8 El; 45 Brin.
Hard: 36,000 TS; 28,000 YS; 4 El; 60 Brin.
For general structures, scaffolds, booms, transmission towers.
Corrosion resistant.

PANTAL 5.
M-299; 9-13 Si, 0.25-0.40 Mg, 0.3-0.5 Mn, bal Al.
For light alloy parts; high corrosion resistance.

PANTAL-5.
M-1548; 5.0 Si, 0.7 Mn, 0.7 Mg, bal Al.
Hard: 26,000-37,000 TS; 21,000-33,000 YS; 4-8 El; 55-70 Brin.
For general purposes, welding rods.
Corrosion resistant.

PANTANAX 1273 ETC.
M-1759.
See Werkstoff Nr. 1.1273 etc.
Wear resistant steels.

PANTHER 5.
M-1779; 1.5 C, 4.7 Cr, 12.5 W, 5 V, 5 Co, bal Fe.
For heavy duty cutting tools requiring extra red hardness and abrasion resistance. Usual heat treated hardness: Rc 66-68.
W-C-Co high speed steel. AISI T-15.

PANTHER EXTRA 655 now VEW S705.

PATHER EXTRA SPEZIAL now VEW S305.

PANTHER SPECIAL.
M-1779; 0.75 C, 19 W, 4 Cr, 1 V, 5 Co, bal Fe.
18-4-1+5 Co high speed steel for heavy duty single point cutting applications; lathe and planer tools. AISI T-4.

PANTHER ULTRA now VEW S300.

PANZER.
M-615; 0.7 C, 19 W, 4 Cr, 2 V, bal Fe.
For cutters, tools; high speed steel.

PAR 1.
M-153; Cr-W-Co, Fe.
For heat and wear resisting parts; resists wear at high temperature to 2000°F.

PAR 2.
M-153; 0.3 C, 1.5 Cr, 2.5 Ni, 0.3 Mo, bal Fe.
For gears, pinions, shafts; wear and shock resistant.

PAR 2C.
M-153; C, Cr, Ni, bal Fe.
For heat and corrosion resisting castings; heat, wear and corrosion resistant.

PAR 3.
M-153; 0.35 C, 3 Cr, bal Fe.
For machinery parts; wear and abrasion resistant.

PAR 4.
M-153; 0.4 C, 25 Cr, 2 Ni, bal Fe.
For heat and corrosion resisting parts; heat and acid resistant.

PAR 5.
M-153; 0.3 C, 18 Cr, 8 Ni, bal Fe.
For heat and corrosion resisting parts; heat and acid resistant.

PAR 6.
M-153; 0.4 C, 28 Cr, 10 Ni, bal Fe.
For heat and corrosion resisting parts; heat and acid resistant.

PAR 7.
M-153; 0.4 C, 16 Cr, 35 Ni, bal Fe.
For heat and corrosion resisting parts; heat and acid resistant.

PAR 8.
M-153; 0.25 C, 18 Cr, 65 Ni, bal Fe.
For heat and corrosion resisting parts; heat and acid resistant.

PAR 9.
M-153; 0.40 C, 30 Cr, bal Fe.
For heat and corrosion resisting parts; heat and acid resistant.

PAR 9 A.
M-153; C, Ni, Cr, bal Fe.
For heat and corrosion resisting castings; heat, wear and corrosion resistant.

PAR 10.
M-153; 0.8 C, 15 Cr, 0.45 Mo, 0.50 W, 0.04 Al, bal Fe.
For guides for hot work; corrosion and heat resistant.

PAR 10A.
M-163; C, Ni, Cr, bal Fe.
For heat and corrosion resisting castings; heat, wear and corrosion resistant.

PAR 11.
M-153; 0.55 C, 0.75 Cr, 1.5-0.3 Mo, bal Fe.
For gorging dies; oil hardened.

PAR 11A.
M-153; C, Ni, Cr, bal Fe.
For heat and corrosion resisting castings; heat, wear and corrosion resistant.

PARA.
M-114; 1.25 C, 0.4 Cr, 1.6 W, 0.2 V, bal Fe.
For cutters, dies, taps, drills, shear blades, punches; dense deep hardening steel.

PARALLOY.
M-873; 3 C, 1 Mn, 2.5-3 Ni 0.5-0.75 Cr, 2 Si, 0.2-0.25 Mo, bal Fe.
Cast: 50,000 TS; 250 Brin.
For castings, cylinders, dies; cast iron.

PARALOY NO. 2.
M-873; 3 C, 1.5 Si, 2-4 Ni, 0.75-1.0 Cr, bal Fe.
Cast: 40,000-50,000 TS; 250-300 Brin.
For cast dies for sheet metal stamping and drawing; cast iron.

PAR-EXC.
M-115; 0.50 C, 2 W, 1.65 Cr, 0.25 V, bal Fe.
Annealed: 70,00 TS; 30 El; 64 RA; 165 Brin.
Heat treated: 245,000 TS; 5 El; 18 RA; 525 Brin.
For punches, dies, chisels, shear blades, cutters, pneumatic tools, hot working dies; extremely tough.

PARISON ALLOY.
M-Eng.; 69 Cu, 19.5 Ni, 6.5 Zn, 5 Cd.
For cheap jewelry; gilt finish.

PARKALOY.
M-874; 0.3 C, 1.2 Ni, 7 Cr, bal Fe.
For screws, bolts; cold forging alloy.

PARKERS CHROME ALLOY.
60 Cu, 20 Zn, 10 Ni, 10 Cr.
For fountain pen points; corrosion resistant.

PARK GATE.
M-488; 0.7 C, 18 W, 4 Cr, 1 V, bal Fe.
For high speed tools and cutters; high speed steel.

PARK SPECIAL.
M-38; C, alloy, bal Fe.
For tools.

PARR.
M-Eng.; 80 Ni, 15 Cr, 5 Cu.
73,000 TS; 15 RA; 137 Brin.
For chemical machinery parts; stainless.

PARR.
M-Eng.; 66.6 Ni, 18 Cr, 8.5Cu, 3.3 W, 2 Al, 1 Mn, 0.2 Ti, 0.2 B.
For chemical machinery parts; corrosion resistant.

PARSON'S ALLOY.
M-129; 56 Cu, 41.5 Zn, 1.2 Fe, 0.7 Sn, 0.1 Mn, 0.46 Al.
70,000 TS; 25 El; 119 Brin.
For propeller blades, valve stems, engine frames, machine parts; tough.

PARSON'S WHITE BRASS.
M-309; 74 Sn, 5 Cu, 7 Sd, 14 Pb.
Cast: 12,250 TS; 108700 YS; 3 El.
For marine and automobile bearings; Babbitt.

PARSONS WHITE BRASS DA.
M-309; 3.7 Cu, 7.5 Sb, bal Sn.
Cast: 10,500 TS; 9400 YS; 3.5 El; 3.9 RA; 29 Brin.
For bearings for diesel turbine and gasoline engines; Babbitt.

PARSTEEL.
M-367; 0.5 C, 3 Ni, 1 Cr, bal Fe.
For gears, shafts; tough.

PAR-TEN.
M-604; 0.12 C, 0.75 Mn, 0.10 Si, 0.04 V, bal Fe.
Rolled: 65,000 TS; 45,000 YS; 28 El.
For railroad and agricultural equipment, mine cars, auto bodies; high strength, low alloy construction steel.

PARTINIUM-1.
M-Eng.; 89 Al, 7.4 Cr, 1.7 Zn, 1.3 Fe, 1.1 Si.
19,000 TS; 12,000 YS; 2.5 El; 50 Brin.
For light aircraft and automobile parts; similar to Alcoa No. 113

PARTINIUM-2.
M-Eng.; 96 Al, 2.4 Sb, 0.8 W, 0.6 Cu, 0.2 Sn, 65 Brin.
For light aircraft and automobile parts.

PATENT COBALT-CHROME STEEL.
M-M40; 1.27-1.43 C, 11.75-13.75 Cr, 2.7-3.3 Co, 0.60 max Ni, 0.55-0.85 Mo, 0.35 Mn, 0.6 max Si.
For air cooled aeronautic engine valves, blanking, trimming, forming and shearing tools; air hardening, non-deforming; resists wear.

PATRIUS CNM.
M-1388; 0.56 C, 0.8 Cr, 0.2 Mo, 1.75 Ni, 0.1 V, bal Fe.
For forging and heading dies; oil hardened, tough.

PATRIUS V.
M-1388; 0.55 C, 0.70 Cr, 0.18 Mi, 1.65 Ni 0.1 V bal Fe.
For forging and heading dies; oil hardened, tough.

PATTERN ALLOY-1.
M-Eng.; 90 Al, 8 Cu, 2 Sn.
20,000-29,000 TS; 6-4 El.
For pistons; similar to Alcoa No. 12.

PATTERN ALLOY-2.
M-Eng.; 87 Pb, 13 Sb.
18,00 TS; 11 El; 10 RA.
For bearings; hard.

PATTERN METAL.
30-40 Zn, 15-40 Sn, 20-42 Pb, 0-3 Cu.
For bearings; anti-friction.

PAULITE.
M-875; 3.2 C, 2.5 Si, 1.5 Ni, 0.8 Cr, bal Fe.

PAX NO 2.
M-510; 0.50 C, 0.80 Si, 1.5 Cr, 2.25 W, 0.20 V, bal Fe.
Shock resisting tool steel.
B.S. 4659 Type BS1; AISI S1.

PAX NON-BREAK.
M-510; 0.40 C, 1.5 Cr, 2.2 W, 0.20 V, bal Fe.
For chisels and other shock tools.

PAX NON BREAK.
M-618; 0.5 C, 1.5 Cr, 2.25 W, 0.25 V, bal Fe.
For hot work dies; hot work steel.

P.B.
M-368; 0.7-0.9 C, bal Fe.
For tools; water hardening.

PD 135.
M-319; Cr-Cd, bal Cu.
Pyec, hardened plus cold work: 65,000 psi TS; 60,000 psi YS; 12 El; good conductivity.
For springs.

PD-135.
M-1598.
Copper base precipitation hardening alloy.
Sol. tr, cold work, plus aging gives high strength, spring properties, and good flex life.

PD 135 FM.
M-319.
Free machining grade of PD 135.

PD-135 FM.
M-1598.
Free machining grade of PD-135.

PDOF NO 101 AND 102.
M-319; 99.95 min Cu, 10 parts per million max Cu.
High conductivity.

PDOF NO 104.
M-319; 99.95 min Cu, 8 oz/ton min Ag. Oxygen free.
High conductivity; resists softening with heat.

PDOF NO 105.
M-319; 99.95 min Cu, 10 oz/ton min Ag. Oxygen free.
Resists softening with heat; high conductivity.

PDOF NO 107.
M-319; 99.95 min Cu, 25 oz/ton min Ag. Oxygen free.
Resists softening with heat; good conductivity.
Commutator segments.

PDRL 162.
M-1491; 0.12 C, 10 Cr, 4 Mo, 2 W, 1 Cb, 1 Ti, 6.5 Al, 0.02 B, 0.10 Zr, 2.0 Ta, bal Ni.
Cast: 146,000 YS; 118,000 YS; 7 El.
At 1400°F: 146,000 TS; 123,000 YS; 5.5 El.
For high temperature service jet engine and turbine components.
Heat and corrosion resistant. Casting alloy.

PDRL 163.
M-1491; 16.7 Cr, 6.3 Al, 0.1 Ti, 1.6 Mo, 2 W, 1 Cb, 0.05 C, 2 Ta, 0.02 B, 0.10 Zr, 0.30 max Fe, bal Ni.
For jet engine and gas turbine components, high temperature parts.
High heat and oxidation resistance. Cast alloy.

P-D TUNGSTEN.
M-432; C, 9.5 W, alloy, bal Fe.
For hot work and punch dies; high speed steel for maximum red-hardness.

PE 8.
M-Russia; 92 WC, 8 Co.
For cutting tools; cemented.

PE 50.
M-Italy; 5 Mg, bal Al.
Annealed: 42,000 TS; 22,000 YS; 35 El; 65 Brin.
For rivets.
Non-heat treatable.

PECHKO WHITE GOLD.
60 Au, 30 Pd, 10 Pt, 0.1-2.0 Ir.
For jewelry, ornaments; corrosion resistant.

PECKRITE.
M-876; 2.8-3.0 TC., 1.6-1.8 Si, 0.75 Mn, 1.2-1.5 Ni, bal Fe.
Cast: 50,000 TS.
Heat treated: 70,000 TS; 450 Brin.
For bushings, gears, piston rings, cams; cast iron.

PECO now VEW K244.

PEERLESS.
16.5 Cr, 3 Fe, 2 Mn, bal Ni.
For heating elements; resistance alloy.

PEERLESS 750.
  M-38; 0.45 C, 0.25 Mn, 0.95 Si, 7.5 Cr, 1.0 W, 1.0 Mo, bal Fe.
  Hot work tool steel; forging dies.

PEERLESS "A".
  M-38; 0.3 C, 9 W, 3.25 Cr, 0.25 V, bal Fe.
  For hot-heading and forging dies, gripping dies, swedges; hot work steel.

PEERLESS LLCT.
  M-38; 0.25 C, 0.3 Mn, 0.3 Si, 4.0 Cr, 15.0 W, 0.5 V, bal Fe.
  Hot work tool steel.

PEERLESS SPECIAL.
  M-343; C, Co, bal Fe.
  For tools, dies; oil hardened.

PEGASE.
  M-884.
  Free machining medium carbon steel.

PELCOLOY.
  M-1523; 70 Ni, 30 Fe.
  Resistance wire: 120 ohms per cir.mil.ft.
  Ann: 70,000 psi.; Hard drawn: 150,000 psi.
  High temp. coefficient of resistance: 4500 ppm between 0° and 100°C.
  Max. op. temp.: 590°C.

PENAIR 5.
  M-341; 1.0 C, 0.6 Mn, 5.25 Cr, 1.1 Mo, 0.25 V, bal Fe.
  For dies, rollers, punches; air hardened, non-deforming. AISI A2.

PENCO.
  0.95-1.05 C, bal Fe.
  For tools, drills, taps; water hardened.

PENCO ACS.
  M-341; 0.07 max C, 0.4 Mn, 0.25 Si, 4.5 Cr, bal Fe.
  For hubbed cavity plastic mold; oil hardening.

PENCO-OCS.
  M-341; 0.06 C, 0.30 Mn, 1.0 Cr, 0.25 Mo, B, bal Fe.
  For hobbed cavity molds; case hardened.

PEN HOB.
  0.10 max C, 0.60 Cr, 1.25 Ni, bal Fe.
  For hobbed cavity molds; case hardened.

PEN METAL-1.
  M-Eng.; 67 Au, 25 Cu, 8 Ag.
  145,000 TS.
  For pen points; corrosion resistant.

PEN METAL-2.
  M-Eng.; 85 Cu, 13 Zn, 2 Sn.
  For pen points; corrosion resistant.

PENN-AIR.
  M-1433; 1.0 C, 2 Mn, 1 Cr, 1 Mo, bal Fe.
  For punches, dies, blanking and forming tools; Type A2; air hardened.

PENNANT.
  M-432; 0.90 C, 1.6 Mn, 0.20 V, bal Fe.
  For tools, blanking and forming dies, broaches, taps, reamers; non-shrinkable.

PENN-CUT.
  M-1433; 0.70 C, 4 Cr, 1 V, 18 W, bal Fe.
  For taps, hobs, reamers, lathe cutters; Type T1; high speed steel.

PENN-CUT 5.
  M-1433; 0.80 C, 4 Cr, 1 V, 18 W, 5 Co, bal Fe.
  For lathe and planer tools, form cutters, hobs; Type T5; high speed steel.

PENN-CUT MOLY.
  M-1433; 0.80 C, 4 Cr, 2 V, 6.4 W, 5 Mo, bal Fe.
  For lathe and planer tools, form cutters, hobs; Type M2; high speed steel.

PENN-FLEX.
  M-1433; 0.33 C, 0.72 Mn, 0.25 Si, 0.85 Cr, 0.42 W, 0.45 Mo bal Fe.
  Oil or water hardening tool steel, designed for plastic molds.

PENNROLD-10.
  M-1252; 0.40-0.60 Be, 2.35-2.60 Co or Ni, bal Cu.
  Annealed: 40,000 TS; 25,000 YS; 30 El; 40 RA.
  Heat treated: 150,000 TS; 140,000 YS; 1 El.
  For current carrying springs, switch parts, circuit breaker parts; age-hardenable, fatigue and corrosion resistant.

PENNROLD-25.
  M-1252; 1.8-2.05 Be, 0.20 min Co or Ni, 0.60 max Co plus Ni, bal Cu.
  Annealed: 60,000 TS; 28,000 YS; 60 El.
  Hard: 120,000 TS; 112,000 YS; 2 El.
  Heat treated: 215,000 TS; 205,000 YS; 1 El.
  For current carrying springs, diaphragms, switch blades, contacts, bellows; age-hardenable, fatigue and corrisoin resistant.

PENNROLD-165.
  M-1252; 1.6-1.8 Be, 0.20 min Co or Ni, 0.60 max Co plus Ni, bal Cu.
  Annealed: 60,000 TS; 25,000 YS; 60 El.
  Hard: 120,000 TS; 110,000 YS; 2 El.
  For current carrying and mechanical springs; age-hardenable, fatigue and corrosion resistant.

PENNSYLVANIA L.C.D.
  M-1433; 0.40 C, 4.5 Cr, 1 Mo, 0.8 W, 0.4 mn, bal Fe.
  For shears, punches, blanking and trimming dies; oil hardened, non-deforming.

PENNSYLVANIA L.T.A.
M-1433; 1 C, 2 Mn, 0.9 Cr, 0.9 Mo, bal Fe.
For chisels, punches, dies, mandrels, crimpers; air hardened, non-deforming.

PENNSYLVANIA P.B. DRILL ROD.
M-368; 1.0-1.2 C, bal Fe.
Annealed: 100,000 TS; 53,000 YS; 21 El; 42 RA; 200 Brin.
For drills, taps, reamers; Type W1; water hardened.

PENNSYLVANIA P.H. 9.
M-1433; 0.35 C, 2.75 Cr, 0.30 V, 9 W, bal Fe.
For extrusion rams, liners, die casting dies; hot work steel, oil hardened.

PENNSYLVANIA P.H. 14.
M-1433; 0.42 C, 3.5 Cr, 0.3 V, 14 W, bal Fe.
For extrusion dies, liners, and rams, hot punches; hot work steel, oil hardened.

PENNSYLVANIA P.H. VAN.
M-1433; 0.40 C, 1.1 Si, 5.25 Cr, 0.9 V, 1.2 Mo, bal Fe.
For hot work tools and dies; hot work steel, oil hardened.

PENNSYLVANIA P.H.W.
M-1433; 0.35 C, 1.5 Mn, 5 Cr, 1 Si, 1.25 W, 1.25 Mo, 0.2 V, bal Fe.
For hot work tools and dies; hot work steel, oil hardened.

PENNTEMP 165.
M-1252; 1.6-1.8 Be, 0.20-0.35 Co, bal Cu.
AM-temper: 100,000-110,000 TS; 75,000-90,000 YS; 18-22 El.
XHM-temper: 160,000-175,000 YS; 135,000-150,000 YS; 3-7 El.
For contacts, switch members, clips, springs; age hardenable.

PEN-O-FOUR.
M-341; 0.75 C, 0.7 Mn, 0.9 Cr, 1.75 Ni, 0.35 Mo, bal Fe.
For tools, dies; oil hardening.

PENSTOCK A.
M-176; 0.25 max C, 0.90-1.35 C, 0.15-0.30 Si, bal Fe.
As rolled: 63,000-70,000 psi TS; 42,000-50,000 psi YS; 18-22 El.
Used for penstock.

PENSTOCK B.
M-176; 0.25 C, 0.90-1.35 Mn, 0.15-0.30 Si, bal Fe.
Normalized, fine grain: 63,000-70,000 psi TS; 42,000-50,000 psi YS; 18-22 El.
Fine grain grade, for penstocks.

PENSTOCK-B.
M-176; 0.5 max C, 0.90-1.35 Mn, 0.15-0.30 Si, bal Fe.
Normalized: 70,000 min TS; 42,00 min YS; 23 min El.
For welded structures, hydroelectric penstocks. Shock resistant.

PEN-VAN NO. 12.
0.55 C, 1 Cr, 0.15 V, bal Fe.
For tools, chisels, punches; oil hardening.

PER 2.
M-French; 0.25 C, 22 Cr, 2 Ti, 2 Al, bal Ni.
For gas turbine compontents; heat and oxidation resistant.

PERALUMAN 1.
M-249, M-624; 0.5-1.5 Mg, 0.5-1.5 Mn, bal Al.
Soft: 25,000 TS; 10,000 YS; 20 El; 45 Brin.
Hard: 41,000 TS; 38,000 YS; 8 El; 75 Brin.
Extruded: 26,000 TS; 14,000 YS; 20 El; 45 Brin.
For paneling, roofing, containers; corrosion resistant, non-hardenable.

PERALUMAN 2.
M-249, M-624; 2.15 Mg, 1.4 Mn, bal Al.
Soft: 31,000-36,000 TS; 15,000-21,000 YS; 20-16 El.
Hard: 50,000-65,000 YS; 45,000-58,000 YS; 5-2 El.
For paneling in architecture, ship building, automotive parts; corrosion resistant.

PERALUMAN 2.
M-1548; 1.3-1.5 Mn, 2.0-2.3 Mg, bal Al.
Soft: 31,000 TS; 14,000 YS; 20 El; 50 Brin.
Hard: 60,000 Y; 57,000 YS; 2 El; 105 Brin.
For vessels, storage tanks, welded structures. Non-hardenable, corrosion resistant.

PERALUMAN 3.
M-541, M-1220; 3.0-3.3 Mg, 0.30 Mn, bal Al.
Cast: 21,400-28,500 TS; 9000-12,000 YS; 6 -10 El; 45-50 Brin.
For light alloy castings; corrosion resistant.

PERALUMAN 3.
M-1420; 2.3 Mg, 0-0.4 Mn, bal Fe.
Sand cast: 22,800 TS; 11,400 YS; 5 El; 50 Brin.
Permanent mold: 29,000 TS; 11,400 YS; 10 El; 55 Brin.
For aircraft and marine equipment; corrosion resistant to sea water.

PERALUMAN 3.
M-249, M-493; 2-3 Mg, 0-0.7 Mn, bal Al.
Soft: 32,000 TS; 16,000 YS; 25 El; 60 Brin.
Hard: 45,000 TS; 40,000 YS; 7 El; 90 Brin.
For architectural trim, containers, ship building; corrosion resistant.

Section I: Alloy Data / 1067

PERALUMAN 3.
M-1176; 3 Mg, 0.3 Mn, bal Al.
Cast: 21,000-29,000 TS; 8,000-12,000 YS; 10-6 El; 45-55 Brin.
For light alloy parts; corrosion resistant.

PERALUMAN 3G.
M-1605; 0.2-0.4 Si, 0.25-0.35 Mn, 3.0-3.3 Mg, 0.08-0.15 Ti, bal Al.
As cast: 27,000 TS; 14,000 YS; 3-8 El; 45-60 Brin.
Heat treated: 33,000 TS; 20,000 YS; 2-8 El; 50-70 Brin.
For sand and die castings, marine hardware, crankcases, oil pans, housing.
Fair castability, high corrosion resistance. Resists sea water. Heat treatable.

PERALUMAN 5.
M-249, M-1220, M-493; 5 Mg, 0.5 Mn, bal Al.
Soft: 41,000 TS 21,000 YS; 24 El; 70 RA; 160 Brin.
For architectural trim; corrosion resisant.

PERALUMAN 5.
M-1176; 5 Mg, 0.4 Mn, bal Al.
Heat treated: 28,000-33,000 TS; 13,000-17,000 YS; 12-8 El; 60-80 Brin.
For light alloy parts; corrosion resistant.

PERALUMAN 5G.
M-1605; 0.8-1.0 Si, 0.25-0.35 Mn, 5.0-5.5 Mg, 0.08-0.15 Ti, bal Al.
Sand cast: 28,500 TS; 15,500 YS; 2-5 El; 55-70 Brin.
Die cast: 36,000 TS; 16,000 YS; 3-8 El; 60-80 brin.
For marine hardware, crankcases, instrument cases, fuel gages.
Fair castability, high strength and corrosion resistant to sea water.

PERALUMAN 7.
M-624, M-249; 7 Mg, 0.5 Mn, bal Al.
Half hard: 60,000 TS; 43,000 YS; 15-20 El; 90-95 Brin.
Annealed: 43,000 TS; 29,000 YS; 26-30 El; 80-90 Brin.
For light alloy parts, aircraft parts; corrosion resistant.

PERALUMAN 7.
M-1176; 7 Mg, 0.4 Mn, bal Al.
Heat treated: 37,000-43,000 TS; 19,000-22,000 YS; 11-5 El; 70-80 Brin.
For marine hardware; resists sea water corrosion, age-hardenable.

PERALUMAN 7.
M-1548; 0.3-0.5 Mn, 7.0 Mg, bal Al.
Soft: 42,000 TS; 21,000 YS; 24 El; 75 Brin.
Hard: 64,0000 TS; 51,000 YS; 4 El; 130 Brin.
For rivets.
Non-heat treatable.

PERALUMAN 9.
M-541, M-1220; 8-10 Mg, 0.3 Mn, bal Al.
Cast: 35,600-40,000 TS; 21,400-27,000 YS; 6-13 El; 75-80 Brin.
For fuel meters; corrosion resistant.

PERALUMAN 9.
M-1176; 9.3 Mg, 0.3 Mn, 0.02 Be, bal Al.
Heat treated: 28,000-35,000 TS; 14,000-19,000 YS; 2.5-1.5 El; 70-80 Brin.
For marine hardware, resists sea water corrosion, age-hardenable.

PERALUMAN 9G.
M-1605; 0.25 -0.35 Mn, 9.0-9.5 Mg, 0.03 Be, bal Al.
Die cast 38,000 TS; 23,000 YS; 1-3 El; 60-80 brin.
For general die castings, marine hardware.
Fair castability, good strength, high corrosion resistance. Heat treatable.

PERALUMAN 10.
M-1176; 10 Mg, 0.4 Mn, bal Al.
Heat treated: 50,000-56,000 TS; 21,000-26,000 YS; 13-6 El; 75-85 Brin.
For marine and aircraft parts; heat treatable, resists sea water corrosion.

PERALUMAN 15.
M-249, M-541; 0-0.3 Mn, 1.3-1.8 Mg, bal Al.
Soft: 21,000 TS; 8,000 YS; 28 El; 40 Brin.
Hard: 32,000 TS; 30,000 YS; 6 El; 65 Brin.
For panels, containers.
Corrosion resistant.

PERALUMAN 25.
M-624, M-541; 2.2-2.8 Mg, 0.15.0.45 Mn + Cr, 0.4 max Fe, 0.3 max Si, bal Al.
Annealed: 22,000-31,300 TS; 11,400-17,000 YS; 22-30 El; 45-45 Brin.
Hard: 37,000-41,200 TS; 34,000-40,000 YS; 5-9 El; 70-85 Brin.
For boat hulls, guided missile containers.
Corrosion resistant.

PERALUMAN-30.
M-249; 0.4 max Mn, 3-4 Mg, bal Al.
1/2 H-temper: 39,750 TS; 35,500 YS; 8 El; 80 Brin.
H-temper: 49,500 TS.
For chemical and food processing equipment, ship building; resists sea water corrosion.

PERALUMAN 30G.
M-249, M-1634; 2-4 Mg, 0.25-0.35 Mn, bal Al.
Annealed: 21,400-28,500 TS; 9,000-12,000 YS; 20-6 El; 45-50 Brin.
For architecture, shipbuilding; corrosion resistant.

PERALUMAN 35.
M-624, M-541; 3.2-3.8 Mg, 0.2-0.4 Mn, 0.1 max Ti, 0.4 max Fe, 0.05 max Cu, bal Al.
Annealed: 30,000-37,000 TS; 12,800-20,000 YS; 20-30 El; 55-70 Brin.
Hard: 41,200-47,000 TS; 35,600-44,100 YS; 4-9 El; 85-100 Brin.
For boat hulls, guided missile containers. Corrosion resistant.

PERALUMAN 50.
M-249, M-541; 0.2-0.4 Mn, 4.5-5.6 Mg, bal Al.
1/2 Htemper: 45,5000 TS; 25,000 YS; 30 El; 80 Brin.
H-25 temper: 57,000 TS; 48,000 YS; 8-13 El; 110-95 Brin.
For chemical and food processing equipment; resists sea water corrosion.

PERALUMAN 50G.
M-249, M-1634; 4-6 Mg, 0.3 Mn, 0.7-1.0 Si, bal Al.
Annealed: 41,000 TS; 21,000 YS; 24 El; 70 Brin.
For architectural trim, corrosion resistant.

PERALUMAN-100.
M-493; 1.0 Mg, bal Al.
Ann: 105 MPa TS; 27 El.
Hard: 210 MPa TS; 3 El.
Decorative plate.

PERALUMAN-401.
M-493; 4.0 Mg, bal Al.
Ann: 240 MPa TS; 18 El.
Hard: 330 MPa TS; 4 El.

PERAX.
M-753; C, alloy, bal Fe.
For punches, dies, tools; oil hardened, tough.

PERCEIT.
M-Eng.; 55-80 Co, 20-35 Cr, 0.10 W.
40,000-130,000 TS; 570 Brin.
For high speed tools and dies; high speed steel.

PERCIT Cr-Co-W-Ni.
For tools; similar to stellite.

PERCIT EXTRA.
M-72; 1.1 C, 2.5 Si, 0.4 Mn, 27 Cr, 60 Co, 4.3 W, bal Fe.
For hard facing electrodes; heat resistant.

PERCIT SPECIAL.
M-72; 1.5 C, 1.45 Si, 0.4 Mn, 27.5 Cr, 32.5 Co, 4.25 W, bal Fe.
For hard facing electrodes; heat resistant.

PERCY ALUMINUM.
M-Eng; 90-86 Cu, 7.5-13 Al, 0-2 Pb, 0-1.5 Mn.
Rolled: 53,000 TS; 9 El.
Cast: 12,000 TS; 30 El; 30 RA.
For bearings, stripper nuts; heavy duty.

PERDURO.
M-170; 1.6-1.8 T.C., 0.4-0.6 C.C., 0.5-0.7 Mn, 1.5 Si, 0.9-1.1 Cu, bal Fe.
Heat treated: 80,000-90,000 TS; 60,000-70,000 YS; 8-6 El; 179-207 Brin.
For cast chains, sprockets, chain links; resists heat to 1100°F; malleable iron.

PEREKS CC.
M-40; 0.66-0.80 C, 49-57 Ni, 24-30 Cr, Si, Mn, bal Fe.
For heat and corrosion resistant parts; heat and corrosion resistant.

PERFECTION ANTIFRICTION.
M-88; Pb, Sn, Sb.
For bearings; babbitt.

PARFORCE SPECIAL 3 now VEW M310.

PERFORM 300 K.
M-72; 0.12 max C, 0.50 max Si, 1.20 max Mn, Al or Nb or Ti or V, bal Fe.
Structural steel, for plates, beams.

PERFORM 340 K.
M-72; 0.12 max C, 0.50 max Si, 1.50 max Mn, Al, or Nb, or Ti or V, bal Fe.
Structural steel, for plates, girders, W. Nr. 1.0548.

PERFORM 340 W.
M-72; 0.12 max C, 0.50 max Si, 1.50 max Mn, bal Fe.
Structural steel, for plates, girders, W. Nr. 1.0534.

PERFORM 420 K.
M-72; 0.12 max C, 0.50 max Si, 1.60 max Mn, Al or Nb or Ti or V, ba Fe.
Structural steel, for palates, girders. W. Nr. 1.0556.

PERFORM 420 W.
M-72; 0.12 max C, 0.50 max Si, 1.60 max Mn, bal Fe.
Structural steel, for beams, girders. W. Nr. 1.0590.

PERFORM 460 W.
M-72; 0.12 max C, 0.50 max Si, 1.7 max Mn, bal Fe.
Structural steel, for beams, girders. W. Nr. 1.0592.

PERFORM 500 W.
M-72; 0.12 max C, 0.50 max Si, 1.70 max Mn, bal Fe.
Structural steel, for plates, beams, girders. W. Nr. 1.0596.

PERKING BRASS.
M-Eng.; 76.2-80 Cu, 23.9-19.8 Sn, 0-0.14 Zn.
For ornamental parts, reflectors.

PERLIT NICKEL CAST IRON.
  1.7-3.0 C, <1 Si, Ni, Mn bal Fe.
  For casting, frames, housings; U.S.P. 1564284; sufficient Ni to ppt. graphite.

PERLIT (PERLITGUSS).
  M-160, M-527; 2.4-3.5 C, 0.5-1.5 Si, 0.2-0.6 P, 0.6-1.0 Mn, Ni, Cr, bal Fe.
  30,000 TS; 200 Brin.
  For automotive engine blocks, brake drums; nickel cast iron; wear, resistant, tough.

PERMADUR H.
  M-1340; 0.55 C, 1.05 Cr, 0.18 V, 1.85 W, bal Fe.
  For cold work tools, upsetters, crimpers, punches: oil hardened, tough.

PERMADUR W.
  M-1340; 1.05 Cr, 0.18 V, 1.85 W, bal Fe.
  For cold work tools, upsetters, punches, dies; oil hardened tough.

PERMAG see ALLOYMET PERMAG.

PERMAL.
  M-878; 5 Si, bal Al.
  For die casting; good fluidity.

PERMALLOY.
  M-205; 39-81 Ni, 61-19 Fe.
  For loading material for signaling conductors, audio frequency transformers; greater magnetic qualities than iron.

PERMALLOY 45.
  M-22, M 205; 45 Ni, 0.3 Mn, bal Fe.
  Sheet: 65,800 TS; 21,000 YS; 30 El; 10, Brin.
  Initial permeability 2500; max permeability 25,000.
  For audio frequency apparatus and output transformers; high permeability.

PERMALLOY 45.
  M-61; 45 Ni, 55 Fe.
  Saturation induction: 16,000.
  For cores and armatures in relays, motors, inductors, transformers, electromagnetic shields. Soft magnet.

PERMALLOY 49.
  M-61; 49-51 Ni, bal Fe.
  Saturation induction 16,000.
  For leads and armatures for glass shielded switches, laminated cores for magnetic amplifiers, motors transformers, shielding. Soft magnet.

PERMALLOY-68.
  M-22; 68 Ni, bal Fe.
  For magnetic apparatus; high permeability.

PERMALLOY 78.
  M-22, M 205; 78.5 Ni, 21.1 Fe, 0.3 Mn.
  Sheet: 75,000 TS; 15,000 YS; 35 El; 115 Brin.
  Initial permeability 8000; max permeability 100,000; Hc = 0.05.
  For sensitive D.C. appaaratus, relays; high permeability.

PERMALLOY-78.5.
  M-61; 78.5 Ni, 21.5 Fe.
  Saturation induction 10,700.
  For reeds in mercury wetted switches, relay armatures. Soft magnet.

PERMALLOY 80.
  M-1626; 80.0 Ni, 4.5 Mo, bal Fe.
  Soft magnetic alloy strip, very high shielding value.

PERMALLOY 80 see also MAGNETICS ROUND PERMALLOY 80.

PERMALLOY 4-79 MO.
  M-205; 0.3 Mn, 4 Mo, 16.7 Fe, bal Ni.
  85,000 TS; 24,000 YS.
  For laminated cores, inductors, transformers, magnetic field detectors.

PERMALLOY 12.5-80 MO.
  M-205; 12.5 Mo, 7.5 Fe, 80 Ni.
  For magnets; soft.

PERMALLOY 3.8-78.5 CR.
  M-US; 3.8 Cr, 0.6 Mn, 17.1 Fe, bal Ni.
  For soft magnets; high permeability.

PERMALLOY B.
  M-269; 46 Ni, 54 Fe.
  For telecommunications; low coefficient of expansion.

PERMALLOY D.
  M-269; 36 Ni, 64 Fe.
  For magnetic equipment; low coefficient of expansion.

PERMALLOY F.
  M-269; 65 Ni, 35 Fe.
  Domain oriented alloy for electrical equipment; high magnetic permeability.

PERMALLOY G.
  M-269; 46 Ni, 54 Fe.
  Domain oriented alloy for electrical equipment; high magnetic permeability.

PERMALLOY STANDARD.
  M-22; 78.5 Ni, 21.5 Fe, 0.04 C, 0.37 Co, 0.022 Mn.
  For loading submarine cables; high elec. resisting.

PERMALLOY SUPER C.
  M-269; 77.4 Ni, 13.3 Fe, 3.7 Mo, 5 Cu.
  For transformers, chokes, magnetic shielding. μ (.001) >50,000.
  Soft magnet; high permeability.

PERMAN.
M-1331; 1.05 C, 1.0 Cr, 1.15 W, bal Fe.
For bearings, cutters, liners, sheaves; water hardened, wear resistant.

PERMANENT see also **SODING PERMANENT**.

PERMANENT 35.
M-1309; 0.80 C, 5.0 Co, 4.5 Cr, 0.8 Mo, 1.7 V, 18.5 W, bal Fe.
High speed steel, for lathe and planer tools, reamers, drills, hobs.

PERMANENT 70.
M-1309; 1.40 C, 4.5 Cr, 5.0 Co, 4.0 V, 12.5 W, 1.0 Mo, bal Fe.
High speed steel; for lathe and planer tools, reamers, hobs, taps.

PERMANICKEL renamed PERMANICKEL ALLOY 300.

PERMANICKEL 300.
M-44; 98.5 Ni, C, Mn, Fe, Si, Cu, Ti, Mg.
For high temperature age hardenable springs.

PERMANICKEL ALLOY 300.
M-1499, M-1280; 98.5 Ni, 0.20 C, 0.25 Mn, 0.30 Fe, 0.005 S. 0.18 Si, 0.13 Cu, 0.40 Ti, 0.35 Mg.
Hot-finished: 90,000-120,000 TS; 35,000-65,000 YS; 45-20 El; 140-230 Brin.
Hot-finished and aged: 160,000-200,000 TS; 120,000-150,000 YS; 20-10 El; 285-360 Brin.
For grid lateral winding wires; magnetostriction devices; thermostat contact arms, solid-state capacitors; grid slide rods, diaphragms, springs; clips, fuel cells.

PERMANIT 30 now **VEW P702**.

PERMANIT 35 now **VEW P702**.

PERMANIT 40 now **VEW P712**.

PERMANIT 120 now **VEW P752**.

PERMANIT 160 now **VEW P754**.

PERMANIT 400 now **VEW P760**.

PERMANIT 500 now **VEW P758**.

PERMANITE.
M-U.S.; Co, Cr, W, bal Fe.
For permanent magnet steel; high coercive force.

PERMAN N.
M-1331; 1.0 C, Mn, W, Cr, V, bal Fe.
For bearings, cutters, sleeves, liners; water hardened, wear resistant.

PERMAS.
M-378; 0.25 max C, 24-26 Cr, 19-22 Ni, bal Fe.
Rolled: 95,000 TS; 45,000 YS; 50 El; 65 RA; 185 Brin.
For balance weights; Type 311; corrosion resistant.

PERMAT.
45 Cu, 30 Co, 25 Ni.
For magnets for electrical equipment; magnetically soft.

PERMAX.
M-775; C, alloy, bal Fe.
Cast: 60,000 TS; 38,000 YS; 10 El; 150 Brin.
For bell yokes, direct current magnets.

PERMAX M.
M-1631.
Soft magnetic 50% Ni-Fe alloy for transducers, leakage current protection switches, transformers.

PERMENDUR.
M-205; 50 Co, 50 Fe.

PERMENDUR 2-V.
M-22, M-205; 48.8 Co, 48.8 Fe, 0.4 Mn, 1.7 V.
Sheet: 87,000 TS; 60,000 YS.
For receiver diaphragms.

PERMENDUR 24.
M-108; 24 Co, bal Fe.
Soft magnetic alloy, high saturation and good mechanical properties.

PERMENDUR 49.
M-108; 49 Fe, 49 Co, 2 V.
Soft magnetic alloy, high saturation, used for stators, etc.

PERMENDUR-50 KF.
M-USSR; 49.95 Co, 1.4 V, 0.04 Si, 0.22 Mn, 0.02 C, 0.007 S, 0.00 P, bal Fe.
For magnetic and electrical equipment.
Soft magnetic alloy. High permeability. High magnetostriction saturation.

PERMENORM 3601.
M-1631.
Soft magnetic 36% Ni-Fe alloy for transformers, chokes, relay components.

PERMENORM 5000 H2.
M-1631.
Soft magnetic approx 50% Ni-Fe alloy for transducers, leakage current protective switches, magnetic shields, systems for measuring instruments.

PERMENORM 5000 Z.
M-1631.
Soft magnetic 50% Ni-Fe alloy for magnetic amplifiers, counting and storage cores, chokes, pulse transformers.

PERMET PF-2.
M-38; 30 Co, 70 Fe.
For magnets; 1,520,000 BH max. Br 6000; 3830 Bo; Hc 625.

PERMINVAR 7-70.
M-70; 7 Co, 23 Fe, 70 Ni.
Permeability 850-4000, coercive force 0.6, saturation induction 12,500.
For magnets for electrical and magnetic instruments.
Curie temperature 650°C.
Soft magnetic alloy.

PERMINVAR 25-45.
M-22, M-269; 45 Ni, 25 Co, 30 Fe.
1.2 Hc, 400 initial permeability, 2000 max permeability, B = 10,000.
For electrical communication, circuits, magnetic circuits.
High permeability, soft magnet.

PERMINVAR 43-23.
M-205; 0.3 Mn, 43 Ni, 23 Co, 33.7 Fe.
For magnetic coils, transformers; soft.

PERMINVAR 7.5-70.
M-22, M-205; 7.5 Co, 70 Ni, 22.7 Fe, 0.3 Mn.
For communication equipment for modulus cores; high initial permeability.

PERMINVAR 7.5-45-25 MO.
M-205; 0.6 Mn, 7.5 Mo, 45 Ni, 25 Co, 21.9 Fe.
Initial permeability 850;
Max. permeability 4000; Hc = 0.6.
Soft magnetic material.

PERMITE.
M-6; 0.5 Si, 10 Cu, 1 Fe, 0.4 Mg, bal Al.
40,000 TS; 38,000 YS; 0.5 El; 120 Brin.
For internal combustion engine pistons; coef. exp. 0.0000127.

PERMITE 1002 (2002).
M-6; 9-11 Cu, 0.4-0.6 Si, 0.25-0.40 Mg, 1.0-1.5 Fe, bal Al.
Cast: 26,000-45,000 TS; 21,000-35,000 YS; 1.0 El; 60-150 Brin.
For light alloy parts, pistons, permanent mold castings; non-hardenable.

PERMITE 1005.
M-6; 7 Cu, 1.2 Fe, 1.7 Si, bal Al.
For permanent mold castings; similar to Alcoa 12.

PERMITE 1011 (2011).
M-6; 0.75 Fe, 4-5 Si, 0.5 Zn, 0.1 Mg, bal Al.
Chill cast: 21,000 TS; 2.5 El; 50 Brin.
Sand cast: 17,000 TS; 3.0 El; 45 Brin.
For general castings; corrosion resistant.

PERMITE 1018.
M-6; 1.0-1.5 Cu, 0.5 max Si, 0.6 max Fe, 0.7-1.2 Mn, bal Al.
23,000-30,000 TS; 3,500-10,000 YS; 12-8 El; 45-70 Brin.
For light alloy parts; non-hardenable.

PERMITE 1020.
M-6; 7 Si, 0.3 Mg, bal Al.
Heat treated and T-6 temper: 38,000 TS; 5 El; 90 Brin.
For permanent mold castings; similar to Alcoa 356.

PERMITE 1021.
M-6; 1.8 Zn, 3.8 Mg, bal Al.
Cast: 25,000 TS; 8 El; 60 Brin.
For permanent mold castings; similar to Alcoa A214.

PERMITE 1024.
M-6; 7 Cu, 5.5 Si, 0.3 Mg, bal Al.
T-551 Temper: 31,000 TS; 1 El; 110 Brin.
For permanent mold castings; similar to Alcoa 152.

PERMITE 1027.
M-6; 4.5 Cu, 5.5 Si, bal Al.
Cast: 28,000 TS; 4 El; 70 Brin.
For permanent mold castings; similar to Alcoa A 108.

PERMITE 1029.
M-6; 3.5 Cu, 6.0 Si, bal Al.
Temper: 38,000 TS; 4 El; 95 Brin.
For permanent mold castings; similar to Alcoa 319.

PERMITE 1031.
M-6; 1.0 Cu, 1.0 Ni, 6.5 Sn, bal Al.
T-533 Temper: 20,000 TS; 10 El; 100 Brin.
For permanent mold castings; similar to Alcoa 750.

PERMITE 1034.
M-6; 3.0 Cu, 9.5 Si, 1 Mg, 1 Ni, bal Al.
Heat treated: 34,000 TS; 10 El; 105 Brin.
For permanent mold castings; similar to Alcoa D 132.

PERMITE 2003.
M-6; 0.8 Cu, 0.8 Fe, 12 Si, 1 Mg, 2.5 Ni, bal Al.
For sand castings; similar to Alcoa A 132 alloy.

PERMITE 2004.
M-6; 7 Cu, 1.2 Fe, 1.7 Zn, bal Al.
Cast: 23,000 TS; 3 El; 70 Brin.
For sand castings; similar to Alcoa 112.

PERMITE 2010.
M-6; 4-5 Cu, 1.0 Fe, 1.0 Si, 0.25 Zn, 0.20 Mg, bal Al.
Sand cast: 27,000 TS; 2.0 El; 50 Brin.
Heat treated: 28,000-38,000 TS; 5.0-1.0 El; 100-60 Brin.
For general castings; age-hardened.

PERMITE 2025.
M-6; 4.0 Cu, 3.0 Si, bal Al.
Cast: 21,000 TS; 2.5 El; 55 Brin.
For sand castings; similar to Alcoa 108.

PERMITE 3011.
M-6; 5 Si, bal Al.
Cast: 30,000 TS; 14,000 YS; 7 El.
For die castings; similar to Alcoa 43.

PERMITE 3012.
M-6; 12 Si, bal Al.
Cast: 37,000 TS; 18,000 YS; 1.8 El.
For die castings; similar to Alcoa 13.

PERMITE 3027.
M-6; 4 Cu, 5 Si, bal Al.
Cast: 40,000 TS; 22,000 YS; 3.5 El.
For die castings; similar to Alcoa 85.

PERMITE 3032.
M-6; 8 Mg, bal Al.
Cast: 42,000 TS; 23,000 YS; 7 El.
For die castings; similar to Alcoa 218.

PERMITE 3033.
M-6; 9.5 Si, 0.5 Mg, bal Al.
Cast: 42,000 TS; 23,000 YS; 1.8 El.
For die castings; similar to Alcoa 360.

PERMITE 3034.
M-6; 3.5 Cu, 8.5 Si, bal Al.
Cast: 45,000 TS; 25,000 YS; 2.0 El.
For die castings; similar to Alcoa 380.

PERMITE A-2023 SPECIAL.
M-6; 0.20 max Si, 9.5-10.5 Mg, max Fe, 0.4-0.6 Zn, bal Al.
27,000 TS; 18,000 YS; 4 El; 60-70 Brin.
For light alloy parts, pistons, permanent mold castings, pumps; heat treatable.

PERMITE NO 1003 (LOW EXPANSION ALLOY).
M-6; 1.0 Cu, 1.0 Fe, 13.5 Si, 0.25 Zn, 1.0 Mg, 0.10 Mn, 2.5 Ni, bal Al.
Heat treated: 28,000-35,000 TS; 0.5-1.0 El; 100-90 Brin.
Cast: 25,000 TS; 1.0 El; 85 Brin.
For low expansion alloy; coefficient of expansion 0.0000110.

PERMITE NO 1004 (2004).
M-6; 6-8 Cu, 1.5 Fe, 2.0 Si, 2.5 Zn, 0.2 Mg, 0.2 Mn, bal Al.
Chill cast: 23,000 TS; 1.0 El; 80 Brin.
Sand cast: 20,000-16,000 TS; 1.0-2.5 El; 60 Brin.
For general castings; coefficient of expansion 0.000013.

PERMITE NO 1006 (2006).
M-6; 11.0-13.5 Cu, 0.75 Fe, 0.5 Si, 0.25 Zn, bal Al.
Chill cast: 25,000 TS; 1.0 El; 100 Brin.
Sand cast: 23,000 TS; 1.0 El; 90 Brin.
For leak proof castings; high fluidity.

PERMITE NO 1007 (2007).
M-6; 6-8 Cu, 1.0 Fe, 1-2 Si, 0.25 Zn, 0.20 Mg, 0.20 Mn, bal Al.
Chill cast: 26,000 TS; 2.5 El; 75 Brin.
Sand cast: 24,000 TS; 2.0 El; 70 Brin.
For general castings; good machinability.

PERMITE NO 1008 (2008).
M-6; 4-5 Cu, 1.2 Fe, 2-3 Si, 0.25 Zn, bal Al.
Chill cast: 25,000 TS; 2.0 El; 70 Brin.
Heat treated: 35,000-38,000 TS; 4.0-1.0 El; 70-100 Brin.
For general castings; age-hardened.

PERMITE NO 1009 (2009).
M-6; 4-5 Cu, 0.5 Fe, 0.5 Si, bal Al.
Sand cast: 25,000 TS; 2 El.
Heat treated: 28,000-36,000 TS; 6.0-1.0 El; 60-75 Brin.
For general castings; age-hardened.

PERMITE NO 1010.
M-6; 4 Cu, 1 Si, bal Al.
For machine parts; age-hardenable.

PERMITE NO 1014 (2014) (Y-ALLOY).
M-6; 4.0 Cu, 0.75 Fe, 0.75 Si, 0.2 Zn, 1.0-1.5 Mg, 2.0 Ni, bal Al.
Sand cast: 23,000 TS; 0.5 El; 90 Brin.
Heat treated: 30,000 TS; 0 El; 110 Brin.
For general castings; coefficient of expansion 0.0000136.

PERMITE NO 1015 (2015).
M-6; 2.5 Cu, 0.75 Fe, 1.0 Si, 30-31 Zn, bal Al.
Sand cast: 25,000 TS; 3 El; 70 Brin.
Chill cast: 30,000 TS; 3 El; 70 Brin.
For general castings; non-hardenable.

PERMITE NO 1016.
M-6; 10 Cu, 1.0 Fe, 3-4 Si, 0.25 Zn, 0.20 Mg, bal Al.
Chill cast: 22,000 TS; 1.0 El; 85 Brin.
For general castings; non-hardenable.

PERMITE NO 1019 (2019).
M-6; 1.25 Cu, 1.0 Fe, 4.5-5.5 Si, 0.25 Zn, 0.5 Mg, bal Al.
Heat treated: 27,000-35,000 TS; 4.5-1.0 El; 60-90 Brin.
For general castings; age-hardened.

PERMITE NO 2005.
M-6; 6.0-8.5 Cu, 0.75 Fe, 0.50 Si, 0.25 Zn, bal Al.
Sand cast: 20,000 TS; 2.0 El; 65 Brin.
For general castings; non-hardenable.

PERMITE NO 2012.
M-6; 0.75 Fe, 9-11 Si, 0.5 Zn, 0.1 Mg, bal Al.
Sand cast: 23,000 TS; 5.0 El; 50 Brin.
For general castings; corrosion resistant.

PERMITE NO 2013.
M-6; 0.75 Fe, 13 Si, 0.5 Zn, 0.1 Mg, bal Al.
Sand cast: 28,000 TS; 8.0 El; 50 Brin.
For general castings; corrosion resistant.

PERMITE NO 2017 (PATTERN ALLOY).
 M-6; 6-8 Cu, 1.0 Fe, 1.0 Si, 2.5 Zn, 1.0 Sn, bal Al.
 Sand cast: 18,000 TS; 2.0 El; 70 Brin.
 For general castings; good machinability.

PERMITE NO 2018.
 M-6; 2.5 Cu, 0.5-2.0 Mn, bal Al.
 Sand cast: 20,000 TS; 8 El.
 Heat treated: 28,000-30,000 TS; 6-4 El.
 For general castings; high ductility.

PERMITE NO 2020.
 M-6; 1.0 Fe, 6.5-7.5 Si, 0.25 Zn, 0.35 Mg, bal Al.
 Heat treated: 26,000-30,000 TS; 5-3 El; 55-70 Brin.
 For general castings; age-hardened.

PERMITE NO 2021.
 M-6; 0.1 Cu, 0.5 Fe, 0.5 Si, 3-4 Mg, bal Al.
 Sand cast: 22,000 TS; 6 El; 50 Brin.
 For general castings; corrosion resistant.

PERMITE NO 2022.
 M-6; 0.1 Cu, 0.5 Fe, 0.5 Si, 5-6 Mg, bal Al.
 Sand cast: 25,000 TS; 4 El; 60 Brin.
 For general castings; corrosion resistant.

PERMITE NO 2023.
 M-6; 0.1 Cu, 0.5 Fe, 0.5 Si, 9-10 Mg, bal Al.
 Sand cast: 27,000 TS; 3 El; 60 Brin.
 Heat treated: 38,000 TS; 11 El; 75 Brin.
 For general castings; corrosion resistant.

PERMIUM NO 205.
 M-879; Os, Ru, Rh, Co, Ni.
 For jewelry, instrument jewels; sintered.

PERMO.
 M-880; Os, Rh, Ru.
 For instrument bearings, fountain pen tips; corrosion and wear resistant.

PERMOMETAL NO 11.
 M-880; Os, Ru, Ir.
 For instrument pivots; wear and corrosion resistant.

PERMOMETAL NO 81.
 M-880; Os-Ru.
 For instrument pivots, fountain pens, needles; corrosion resistant.

PERMOMETAL NO 115.
 M-880; Os, Ru, Ir.
 For instrument pivots; corrosion resistant.

PERPLEX.
 M-1331; 1.05 C, 1 Cr, 1.15 W, 0.9 Mn, bal Fe.
 For bearings, cutters, liners, sleeves; water hardened, wear resistant.

PERUNAL.
 M-249, M-493, M-1634; 6 Zn, 2 Mg, 1.5 Cu, bal Al.
 74,000-83,000 TS; 64,000-72,000 YS; 10-5 El; 155 Brin.
 For light alloy parts.

PERUNAL 215.
 M-493; 5.5 Zn, 2.5 Mg, 1.5 Cu, bal Al.
 Sol. Ht: 530 MPa TS; 7 El.
 For highly stressed construction of vehicles and machines.

PETERSON 52100.
 M-1053; 0.95-1.1 C, 0.25-0.45 Mn, 1.3-1.6 Cr, bal Fe.
 Annealed: 90,000 TS; 51,000 YS; 31 El; 61 RA; 180 Brin.
 Hardened: 300,000 TS; 650 Brin.
 For tools, dies, bearings, bushings; oil hardening.

PETITE IRON.
 M-153; Cr, Ni, V, bal Fe.
 For cams, forming and spinning rolls, machine slides, forming dies; wear resisting.

PEWTER.
 M-U.S.; 85-90 Sn, bal Pb.
 For household utensils, dishes, ornamental articles; soft, ductile, easily worked.

PF.
 M-1290; 0.50-0.60 C, 0.15-0.35 Si, 0.50-0.80 Mn, 0.60-0.80 Cr, 0.25-0.35 Mo, 1.5-1.8 Ni, 0.07-0.12 V, bal Fe.
 Cold work tool steel. W-Nr. 1.2713.

PF1.
 M-1763; 0.35 C, 5.0 Cr, 1.5 Mo, 0.40 V, bal Fe.
 Hot work tool steel, for dies. AISI H11.

PF1 + V.
 M-1763; 0.35 C, 5.0 Cr, 1.5 Mo, 1.0 V, bal Fe.
 Hot work tool steel, for dies.
 AISI H13.

PF1+W.
 M-1763; 0.35 C, 5.0 Cr, 1.5 Mo, 1.5 W, 0.40 V, bal Fe.
 Hot work tool steel, for dies.
 AISI H12.

PFA.
 M-1713; weld metal: 0.09 C, 0.45 Mn, 0.28 Si, bal Fe.
 As welded: 77,500 psi TS; 65,500 psi YS; 22 El.
 Covered electrode, AC-DC, straight polarity, for form equipment, metal furniture and general welding.

PF-EXTRA.
 M-1290; 0.50-0.60 C, 0.15-0.35 Si, 0.60-0.80 Mn, 1.0-1.2 Cr, 0.45-0.55 Mo, 1.5-1.8 Ni, 0.07-0.12 V, bal Fe.
 Cold work tool steel.
 W-Nr. 1.2714.

PFIZER INC DUCTILE COBALT.
M-1623; 95 Co, 5 Fe.
High cobalt content with excellent ductility 80,000 TS; 50 El.
Sheath for cobalt base composite welding rods.

PFIZER INC HP COBALT.
M-1623; 99.9 Co + Ni.
High purity cobalt in strip form. 110,000 TS; 20 El.
For electroplating anodes, high temperature magnetic applications, cobalt 60 gamma radiation sources, sheath for cobalt base composite welding rods, x-ray tube targets.

PFIZER INC HP NICKEL.
M-1623; 99.9 min Ni. (6 grades of hardness).
Soft: 45,000-50,000 TS.
Half hard: 65,000-75,000 TS.
Full hard: 85,000-100,000 TS.
High purity, ductile, corrosion resistant, gas free.
For electronic tube components, deep drawing and stamping small parts, magnetostrictive transducers, heat exchange fins.

PFIZER INC NICKEL CLAD STEEL.
M-1623; Nickel on one or both sides of carbon steel base. (as per customer requirements),

PFIZER INC PM 36 ALLOY.
M-1623.
Ni-Fe alloy.
Controlled expansion alloy.

PFIZER INC PM 42 ALLOY.
M-1623; 42 Ni, bal Fe.
80,000 TS; 34,000 YS. controlled expansion.
For glass to metal sealing on hard or soft glass.

PFIZER INC PM 46 ALLOY.
M-1623; 46 Ni, bal Fe.
82,000 TS; 34,000 YS. Controlled expansion.
For glass to metal seals, especially terminal caps and bonds on vitreous enameled resistors.

PFIZER INC PM 48 ALLOY.
M-1623.
Ni-Fe alloy.
Controlled expansion alloy.

PFIZER INC PM 52 ALLOY.
M-1623.
Ni-Fe alloy.
Controlled expansion alloy.

PFIZER INC PM NICKEL-COPPER ALLOY.
M-1623; 65 Ni, 1 Fe, bal Cu.
Corrosion resistant, especially in marine environments.
For tubing, banding, small parts.

PFIZER INC SEAL VAR TM.
M-1623; 30 Ni, 15 Co, bal Fe.
Controlled thermal expansivity to match hard glasses.
For hermetic glass to metal seals (hard glass) and ceramic to metal seals.
ASTM F-15.

PFIZER INC STAINLESS STEEL CLAD ALUMINUM.
M-1623; Al or Al alloys clad on one or both sides with a stainless steel (as per customer requirements).

PFIZER INC STAINLESS CLAD CARBON STEEL.
M-1623; Stainless steel on one or both sides of carbon steel base. (as per customer requirements).

PFIZER INC TITANIUM CLAD ALUMINUM.
M-1623.
For chemical processing applications.

PH-2.
M-USSR; 7-10 W, bal Cb.
Resists molten lithium.

PH13-8 MO see CARPENTER PH 13-8 MO; REPUBLIC PH13-8 MO; AND A

PH 15-7 MO see ARMCO PH15-7 MO; AND REPUBLIC PH 15-7 MO.

PH-55 A.
M-36; 0.04 C, 0.5 Mn, 3.0 Si, 20 Cr, 8.8 Ni, 4 Mo, bal Fe.
Corrosion resistant casting.
Same as COOPER PH-55 A.

PHENIX.
M-68; 75 Fe, 25 Ni.
Rolled: 104,000 TS; 56,000 YS; 45 El; 68 RA.
For resistance alloy; Sc-21.

PHI 84.
M-1488; 0.38 C, 1.0 Si, 5.25 Cr, 1.30 Mo, 1.0 V, bal Fe.
Air hardening hot work tool steel for die casting or pressure casting dies.
AFNOR Z 38 CDV5; AISI H 13.

PHILO BRAND FERROMANGANESE.
M-398; .7 C, Si, 0.30 P, 78-82 Mn, bal Fe.
For metallurgical applications; Mn-additions.

PHILO BRAND FERROSILICON 50% GRADE.
M-398; 50 Si, 48.5 Fe, 1.5 others.
For metallurgical applications; Si-additions.

PHILO BRAND FERROSILICON 75% GRADE.
M-398; 75 Si, 22.5 Fe, 2.5 others.
For metallurgical applications; Si-additions.

PHILO BRAND FERROSILICON 85% GRADE.
M-398; 85 Si, 12.5 Fe, 2.5 others.
For metallurgical applications; Si-additions.

PHILO BRAND FERROSILICON 90% GRADE.
M-398; 91 Si, 6.5 Fe, 2.5 others.
For metallurgical applications; Si-additions.

PHOENIX.
M-88; Sn, Sb, bal Vu.
Cast: 24 Brin.
For bearings for crossheads; heavy loads and high speeds.

PHOENIX A.
M-369; C, alloy, bal Fe.
For steel castings.

PHOENIX A SPECIAL.
M-369; C, alloy, bal Fe.
For steel castings.

PHOENIX MANGEAR.
M-1724; 0.12 C, 0.20 Si, 1.50 Mn, bal Fe.
Used in manufacture of chains.

PHOENIX METAL.
M-88; Pb alloy.
For bearings; solidification temperature 234°C.

PHOENIX METAL.
M-369; 3.2 C, 1.5 Ni, 1 Cr, 1.7 Si, bal Fe.
Cast.
For rolls; hard cast iron.

PHOENIX MOLYBDENUM CHILL.
M-369; 3.2 C, 1.8 Si, 0.5 Mo, bal Fe.
For castings; cast iron.

PHOENIX NICKEL CHILL.
M-369; 3.2 C, 1.8 Si, 1.5 Ni, bal Fe.
For castings, wear plates; cast iron.

PHOENIX PIROCO.
M-369; C, alloy, bal Fe.
For steel castings.

PHOENIXITE.
M-Ger.; For cutting tools; hard metal.

PHOENIX K.
M-369; 3.2 C, 1.8 Si, 1 Cr, bal Fe.
For castings; cast iron.

PHOENIXLOY.
M-68, M-369; 5 Ni + Cr, bal Fe.
For mill rolls for heavy duty; super hard alloy cast iron; Sc 80-90.

PHONO BRONZE.
M-141; 98.74 Cu, 1.25 Sn, 0.008 Fe.
Hard: 90,000 TS; 65,000 YS; 4 El; 70 RA.
For trolley wire, hardware; resists air corrosion.

PHONO CADMIUM 955.
M-141; 0.85 Cd, 0.55 Sn, bal Cu.
Annealed: 40,000 TS; 50 El.
Drawn: 85,000 TS; 10 El.
For trolley wire, cable; 55% conductivity.

PHONO ELECTRIC 865.
M-141; 0.4 Sn, bal Cu.
Annealed: 35,000 TS; 50 El.
Drawn: 63,000 TS; 10 El.
For trolley wire, marine hardware; 65% conductivity.

PHONO ELECTRIC BRONZE 840.
M-141; 1.4 Sn, bal Cu.
Annealed: 40,000 TS; 50 El.
Drawn: 85,000 TS; 10 El.
For trolley wire, hardware; 40% min conductivity.

PHONO ELECTRIC WIRE.
98.55 Cu, 1.40 Sn, 0.05 Si.
For trolley wire, electrical conductors; high electrical conductivity.

PHONO HI CONDUCTIVITY.
M-141; 0.9 Cd, bal Cu.
Annealed: 37,000 TS; 50 El.
Drawn: 65,000 TS; 6 El.
For trolley wire, hardware; 80% min conductivity.

PHONO HI-CONDUCTIVITY 985.
M-141; 0.9 Cd, bal Cu.
Annealed:; 37,000 TS; 50 El.
Drawn: 80,000 TS; 6 El.
For cable, trolley wire; 85% conductivity.

PHONO-HI-STRENGTH.
M-141; 97 Cu, 2 Sn, 1 Si.

PHONO HI-STRENGTH 715.
M-141; 2.7 Al, 0.35 Si, bal Cu.
Annealed: 50,000 TS; 50 El.
Drawn: 125,000TS; 5 El.
For cable, hardware, bolts; 15% conductivity.

PHONO TELEPHONE 830.
M-141; 1.7 Sn, bal Cu.
Annealed: 40,000 TS; 50 El.
Drawn: 90,000 TS; 10 El.
For cable, drop wire; 30% conductivity.

PHOS O.
M-1522; Cu, P.
Melt range: 705-800 + °C.
Max stress: 50.4 kgf/mm^2; 2 El.
For brazing copper; fluid, self-fluxing.

PHOS 2.
M-1522; 2 Ag, Cu, P.
Melt range: 645-740 +°C.
Max stress: 44.1 kgf/mm^2; 5 El.
For silver brazing copper; self-fluxing.

PHOS 5.
M-1522; 5 Ag, Cu, P.
Melt range: 645-730+°C.
Max stress: 47.2 kgf/mm^2; 6 El.
For silver brazing copper; self fluxing.

PHOS 15.
M-1522; 14.5 Ag, Cu, P.
Melt range: 645-700+ °C.
dMax stress: 66.9 kgf/mm²; 25 El.
For silver brazing copper; self-fluxing.

PHOSCO.
M-1013; 7.5 P, 92.5 Cu.
For brazing non-ferrous metals; M.P. 1320-1450°F; self-fluxing.

PHOS-COPPER.
M-118; 8 P, bal Cu.
Welded: 85,000 TS; 17 El.
For brazing alloy for Cu and brass parts; M.P. 1305-1460°F.

PHOS-COPPER-5.
M-118; 4.8-5.2 P, bal Cu.
For filler metal for electrical connections, brazing; M.P. 1640-1310°F; AWS C1.BCuP-1.

PHOSON-0.
M-1013; 7.2 P, bal Cu.
Cast: 93,000 TS; 18 El.
For brazing alloy for copper and brass parts for close joint tolerance.
M.P. 1305-1485°F.

PHOSON 2.
M-1013; 2 Ag, 7 P, 91 Cu.
Melt range: 1185-1450°F; Braze: 1300-1500°F.
90,000 psi TS; 18 El; 96 Rock 15 T.
For various silver brazing applications.

PHOSON 5.
M-1013; 5 Ag, 6 P, 89 Cu.
Melt range; 1190-1485°F; Braze: 1300-1500°F.
92,000 psi TS; 24 El; 86 Rock 15 T.
For various silver brazing applications.

PHOSON 6.
M-1013; 7 P, 6 Ag, bal Cu.
Cast: 95,000 TS.
For silver brazing alloy; M.P. 1190-1380°F; self-fluxing.

PHOSON-15.
M-1013; 15 Ag, 5 P, 85 Cu.
Cast: 86,000 TS.
For brazing alloy for copper and Cu alloys; M.P. 1185-1500°F; self fluxing.

PHOSPHATIZED (PAINTBOND).
M-604; low C, Mn, bal Fe.
Zinc coated steel, chemically treated for painting.

PHOSPHOR BRONZE.
80 Cu, 8 Sn, 10 Pb, 2 P-Sn.
For railroad bearings; heavy duty.

PHOSPHOR BRONZE 1.25% E.
M-33; 98.55 Cu, 1.25 Sn, 0.2 P.
Ann: 40,000 TS; 14,000 YS; 48 El.
Rolled 60%: 75,000 TS; 4 El.
Electrical contacts, pole line hardware, flexible hose.
CDA 505.

PHOSPHOR BRONZE 1.25% E, 505.
M-279; 98.7 Cu, 1.25 Sn, 0.04 P.
Ann: 37,000-42,000 TS; 8000-17,000 YS; 40-47 El.
Cold rolled: 41,000-84,000 TS; 16,000-77,000 YS; 1-45 El.
For flexible metal hose, pole line hardware.
Corrosion resistant and corrosion fatigue resistant.

PHOSPHOR BRONZE 1.5%.
M-1426; 1.5 Sn, 0.04 P, bal Cu.
UNS C50200; C50500.

PHOSPHOR BRONZE 1.5%.
M-1426; 1.5 Sn, 0.04 P, bal Cu.
UNS C50200; C50500.

PHOSPHOR BRONZE 4%, 511.
M-279.
40,000-52,000 TS; 12,00-24,000 YS; 45-58 El.
Cold rolled: 45,000-109,000 TS; 18,000-96,000 YS; 1-44 El.
High conductivity spring phosphor bronze.
For electric terminals, clips, connectors.

PHOSPHOR BRONZE 5% A 510.
M-33, M-279; 94.8 Cu, 5 Sn, 0.2 P.
Ann: 46,000-56,000 TS; 19,000-29,000 YS; 48-62 El.
Cold rolled: 49,000-122,000 TS; 21,000-110,000 YS; 1-50 El.
High strength, fatigue and corrosion resistant.
For bellows, electric spring contacts, connectors, terminals.

PHOSPHOR BRONZE 7%.
M-1426; 7 Sn, 0.04 P, bal Cu.

PHOSPHOR BRONZE 8% C 521.
M-33, M-279; 91.8 Cu, 8 Sn, 0.2 P.
Ann: 56,000-65,000 TS; 23,000-34,000 YS; 60-67 El.
Cold rolled: 63,000-134,000 TS; 33,000-122,000 YS; 1-60 El.
Very high strength spring bronze.
For diaphragms, springs, bellows, pen clips.

PHOSPHOR BRONZE 10% D.
M-33; 90 Cu, 9.85 Sn, 0.15 P.
Annealed: 66,000 TS; 28,000 YS; 68 El.
Hard rolled: 100,000 TS; 13 El.
For springs, bridge bearing plates; tough.

PHOSPHOR BRONZE 301.
M-8; 3 Sn, 0.25 P, bal Cu.
Hard: 90,000 TS; 70,000 YS; 4 El.
For Fourdrinier wire.

PHOSPHOR BRONZE-314.
M-8; 4 Sn, 0.25 Mn, 0.8 P, bal Cu.
Soft: 48,000 TS; 20,000 YS; 50 El.
Hard: 65,000 TS; 55,000 YS; 30 El.
For hot work parts; strong.

PHOSPHOR BRONZE 320.
M-8; 6.5 Sn, 0.30 P, bal Cu.
Hard: 120,000 TS; 2 El.
Soft: 57,000 TS; 40 El.
For Fourdrinier wire.

PHOSPHOR BRONZE-356.
M-8; 1.25 Sn, 0.05 P, bal Cu.
Soft: 40,000 TS; 14,000 YS; 48 El.
Hard: 65,000 TS; 50,000 YS; 6 El.
For electrical conductors, metal hose; corrosion resistant.

PHOSPHOR BRONZE-507.
M-8; 1.75 Sn, 0.01 P, bal Cu.
Soft: 45,000 TS.
Hard: 105,000 TS.
For trolley and line wires; strong.

PHOSPHOR BRONZE (A)-305.
M-8; 5 Sn, 0.05 P, bal Cu.
Hard: 65,000 TS; 57,000 YS; 20 El; 150 Brin.
Soft: 48,000 TS; 20,000 YS; 50 El; 67 Brin.
For strips and tubes; corrosion resistant.

PHOSPHOR BRONZE (A) 351.
M-8; 5 Sn, 0.25 P, bal Cu.
Hard drawn: 130,000 TS.
Soft: 45,000 TS; 50 El.
For tubes, sheets, wire, general parts; resists fatigue and corrosion and abrasion.

PHOSPHOR BRONZE A510.
M-8; 94.8 Cu, 5.0 Sn, 0.20 P.
Hard sheet: 80,000 TS; 65,000 YS; 10 El; B 86 Rock.
Soft sheet: 48,000 TS; 20,000 YS; 50 El; B 28 Rock.
For chemical hardware, springs, bridge bearing plates, Bourdon tubes.
Good resilience and fatigue resistance. Corrosion resistance.

PHOSPHOR BRONZE (A) 5090.
M-8; 4.0 Sn, 0.05 P, bal Cu.
Sheet, hard: 80,000 TS; 65,000 YS; 8 El; 86 Rb.
Soft: 48,000 TS; 20,000 YS; 48 El; 28 Rb.
For drawing into containers as kettles, electrical terminals; weldable and solderable.

PHOSPHOR BRONZE B-1.
M-33; 93.8 Cu, 5 Sn, 1 Pb, 0.2 P.
Drawn 20%: 70,000 TS; 58,000 YS; 25 El.
For bearings, bushings, gears, spindles, CDA 534.

PHOSPHOR BRONZE B-2.
M-33; 88d Cu, 4 Pb, 4 Sn, 4 Zn, P.
Ann: 44,000 TS; 19,000 YS; 50 El.
Drawn 35%: 75,000 TS; 63,000 YS; 15 El.
For bearings, gears, pinions, valve parts.
CDA 544.

PHOSPHOR BRONZE BEARINGS-1.
83 Cu, 14 Sn, 1 P, 2 Zn.
For bearings, bushings; heavy duty.

PHOSPHOR BRONZE BEARINGS-2.
88.1 Cu, 8 Sn, 0.15 P, 4.7 Pb.
For bearings, bushings; heavy duty.

PHOSPHOR BRONZE BRIDGE-1.
80 Cu, 20 Sn, 1.0-0.2 P.
For bearings, bushings; high strength.

PHOSPHOR BRONZE BRIDGE-2.
85 Cu, 15 Sn, 1 P.
For bearings, bushings; high strength.

PHOSPHOR BRONZE (C) 353.
M-8; 8 Sn, 0.25 P, bal Cu.
Hard drawn: 130,000 TS.
Hard sheet: 110,000 TS; 3 El.
For tubes, sheets, wire general parts; resists fatigue and corrosion.

PHOSPHOR BRONZE (D) 354.
M-8; 10 Sn, 0.25 P, bal Cu.
Hard: 130,000 TS; 5 El.
Soft: 60,000 TS; 65 El.
For tubes, sheets, wire, general parts; resists fatigue and corrosion.

PHOSPHOR BRONZE-D524.
M-8; 89.75 Cu, 10.0 Sn, 0.25 P.
Hard sheet: 102,000 TS; 70,000 YS; 12 El; B 97 Rock.
Soft sheet: 66,000 TS; 28,000 YS; 65 El; B 55 Rock.
For clips, beater bars, chemical hardware, springs, condenser tubes.
High strength, resilience and resistance to fatigue. Corrosion resistant.

PHOSPHOR BRONZE ENGLISH.
79.2 Cu, 10.2 Sn, 9.6 Pb, 0.97 P.
For tubes, hardware; free-cutting.

PHOSPHOR BRONZE GEAR-1.
88 Cu, 10 Sn, 2 Pb, 0.1 P.
For gears; tough.

PHOSPHOR BRONZE GEAR-2.
85 Cu, 13 Sn, 2 Zn, 0.1 P.
For gears; tough.

PHOSPHOR BRONZE HAIRSPRINGS.
93 Cu, 6.6 Sn, 0.12-0.2 P.
For hairsprings; high strength.

PHOSPHOR BRONZE NO 1.
M-1518; 90 Cu, 10 min Sn, 0.50 min P.
Cast: 27,000 TS; 1.5 El; 65-90 Brin.
For bearings with hardened shafts, heavy loads and high speeds.

PHOSPHOR BRONZE NO 2.
M-1518; 88.0 Cu, 12 Sn, 0.15 P.
Cast: 31,500 TS; 7 El; 69 Brin.
For gears, bushings, bearings with hardened shafts; heavy loads and high speeds.

PHOSPHOR BRONZE NO 3.
M-1518; 9-11 Sn, 0.03-0.25 P, bal Cu.
Cast: 36,000 TS; 10 El.
For pressure tight castings; corrosion resistant.

PHOSPHOR BRONZE NO 4.
M-1518; 7.5 Sn, 0.3 P, 2-5 Pb, bal Cu.
Cast: 27,000 TS; 3 El; 60-80 Brin.
For medium duty castings; good bearing qualities.

PHOSPHOR BRONZE PRR-B.
76.8 Cu, 8 Sn, 15 Pb, 0.2 P.
For railroad bearings; heavy duty.

PHOSPHOR BRONZE PRR-P.
86.5 Cu, 9.85 Sn, 3.77 Zn, 0.05 P.
For railroad bearings; heavy duty.

PHOSPHORBRONZE SN BZ 4.
M-297; 4 Sn, 0.3 P, bal Cu.
Rolled: 82,000 TS; 5 El; 156 Brin.
Annealed: 48,300 TS; 44 El; 70 Brin.
For springs; tough.

PHOSPHORBRONZE SN BZ 6.
M-297; 6.5 Sn, 0.3 P, bal Cu.
Rolled: 113,800 TS; 2 El; 210 Brin.
Annealed: 59,700 TS; 70 El; 75 Brin.
For springs, contacts; tough.

PHOSPHORBRONZE SN BZ 8.
M-297; 8 Sn, 0.3 P, bal Cu.
Rolled: 156,200 TS; 1 El.
Annealed: 71,100 TS; 55 El.
For springs; contacts; tough.

PHOSPHOR BRONZE S PA.
M-U.S.; 79.7 Cu, 10 Sn, 9.5 Pb, 0.8 P.
For heavy duty bearings; high strength.

PHOSPHOR COPPER GRADE A.
M-165; 0.14 min P, 0.15 max Fe, 99.75 min P + Cu.
25,000 TS; 8 El.
For bearings and machine castings; corrosion resistant.

PHOSPHOR COPPER GRADE B.
M-165; 0.10 min P, 0.15 max Fe, 99.75 min P + Cu.
For bearings and machine castings; corrosion resistant.

PHOSPHORIZED ADMIRALTY.
M-1770; 71 Cu, 1.0 Sn, 28 Zn, 0.03 P.
Tubing for naval equipment.
CDA 445.

PHOSPHORIZED ARSENICAL COPPER-142.
M-8; 99.68 Cu, 0.02 P, 0.30 As.
Hard tube: 45,000 TS; 40,000 YS; 10 El; B 50 Rock.
Soft tube: 33,000 TS; 10,000 YS; 45 El; F 45 Rock.
For condensers and heat exchangers.
Elect. cond. 45. Pitting and corrosion resistant.

PHOSPHORIZED COPPER-122.
M-8; 99.9 Cu, 0.02 P.
Hard sheet: 48,000 TS; 40,000 YS; 6 El; B 50 Rock.
Soft sheet: 33,000 TS; 10,000 YS; 45 El; F 45 Rock.
For refrigeration and air conditioning units, plumbing and heating units, oil carriers, hydraulic and gas lines.
Elect. cond. 85. hydrogen embrittlement.

PHOS SIL-0.
M-1483; 92.8 Cu, 7.2 P.
Cast: 93,000 TS.
For brazing alloy for copper to copper.
B CuP-2 Spec; M.P. 1305°F.
Corrosion resistant.

PHOS SIL-2.
M-1483; 91 Cu, 2 Ag, 7 P.
Cast: 93,000 TS.
For brazing alloy for copper to copper.
B CuP Spec.; M.P. 1190°F.
Corrosion resistant.

PHOS SIL-6.
M-1483; 87.75 Cu, 6 Ag, 6.25 P.
Cast: 92,000 TS.
For brazing alloy for joints with poor fit-up.
B CuP-3 Spec.; M.P. 1185°F.
Very ductile.

PHOS SIL-6F.
M-1483; 6 Ag, 86.75 Cu, 7.25 P.
Cast: 90,000 TS.
For brazing alloy for critical joints.
B CuP-4 Spec.; M.P. 1190°F.
Good penetration.

PHOS SIL-15.
M-1483; 15 Ag, 80 Cu, 5 P.
Cast: 86,000 TS.
For brazing alloy for non-ferrous metals.
B CuP-5 Spec.; M.P. 1185°F.
Rapid penetration.

PHOS-SILVER.
M-118; 6 Ag, 7.8 P, bal Cu.
Cast: 85,000 TS.
For brazing for copper alloys; self-fluxing, M.P. 1185-1230°F.

PHOS-SILVER-2.
M-118; 6.9-7.1 P, 1.8-2.2 Ag, bal Cu.
For brazing alloy; M.P. 1190-1145°F.

PHOS-SILVER-6.
M-118; 7.20-7.35 P, 5.8-6.2 Ag, bal Cu.
For brazing alloy; M.P. 1190-1330°F; for close fit-up work; AWS Cl-BCuP-4.

PHOS-SILVER-6M.
M-118; 6.05-6.20 P, 5.8-6.2 Ag, bal Cu.
For brazing alloy; M.P. 1190-1465°F; very ductile; AWS Cl.BCuP-3.

PHOS-SILVER-15.
M-118; 4.9-5.1 P, 14.8-15.2 Ag, bal Cu.
For brazing alloy; M.P. 1190-1485°F; AWS Cl.BCuP-5.

PHOS-SILVER 18.
M-118; 17.5-18.5 Ag, 7.0-7.5 P, bal Cu.
For brazing copper, brass and bronze.
Lowest melting point 1190°F.
Brazing temperature 1200-1250°F.

PHOS-SILVER-65.
M-118; 4.9-5.1 P, 5.8-6.2 Ag, bal Cu.
For brazing alloy for lap joints; M.P. 1190-1595°F.

PHOS-TRODE.
M-13; 7-9 Sn, 0.35 P, 0.5 max others, bal Cu.
Cast: 55,000 TS; 29,000 YS; 35 El; 33 RA; 89 Brin.
For shielded arc welding rod.
Grade C. Phosphor Bronze.

P H S.
M-210; 0.10 C, 0.10 Si, 0.40 Mn, bal Fe.
For plastic mold dies; hobbing steel.

P H VAN.
M-1433; 0.35 C, 1.0 Si, 5 Cr, 1 V, 1.5 Mo, bal Fe.
Annealed: 98,000 TS; 74,000 YS; 28 El; 210 Brin.
Hardened; 135,000-290,000 TS; 100,000-228,000 YS; 3-16 El; 7-48 RA; C 27-55 Rock.
For forging and heading dies, compression tools, casting dies, hot piercing and forming punches, bolt dies, swaging dies.
Type H13 hot work steel, red-tough, shock and impact resistant.

P.H.W.
M-1433; 0.35 C, 1.05 Si, 1.55 W, 1.65 Mo, 5.15 Cr, bal Fe.
Hardened: 216,000 TS; 185,000 YS; 14 El; 53 RA.
At 1000°F: 144,000 TS; 110,000 YS; 19 El; 64 RA.
For die casting dies, hot shear blades, forging and heading dies, punches.
Type H12 hot work tool steel, tough and shock resistant.

PIERROT METAL, BEUGNOT.
M-83 Zn, 8.3 Cu, 7.6 Sn, 3.5 Sb, 3 Pb.
For bearings; anti-friction.

PIERROTS B.M.
M-England; 83.3 Zn, 7.6 Sn, 2.3 Cu, 3.8 Sb, 3.0 Pb.
For bearings: will not resist heat or live steam.

PINCHBECK.
88-94 Cu, 6-12 Zn.
For cheap jewelry; red brass.

PINKUS BRASS.
M-Eng.; 88.1 Cu, 6.9 Zn, 2.5 Sn, 1.8 Pb, 0.3 Ni, 0.32 Sb.
For hardware, fittings; free-cutting.

PINKUS BRONZE.
M-England; 14.7 Sn, 1.5 Zn, 8.8 Pb, 2.5 Sb, bal Cu.
For bearings, bushings; heavy duty.

PINSBAC.
M-882; 0.3-0.5 C, bal Fe.
For machinery parts, gears; water hardening.

PIONEER.
M-1705; 0.55 C, 0.3 Mn, 0.9 Si, 5.1 Cr, 1.25 W, 1.45 Mo, bal Fe.
Air hardening tool steel, AISI A8.

PIONEER 921-T.
M-1403; 2 Si, 3.5 Cu, 0.5 max Fe, 0.08 Ti, bal Al.
Cast: 20,000 TS; 18,000 YS; 1.0 El; 70 Brin.
T6-temper: 45,000 TS; 22,000 YS; 0.5 El; 104 Brin.
For tooling plates; heat treatable, dimensionally stable.

PIONEER HEAT RESISTING METAL.
M-326; C, 20 Cr, 20 Ni, 1 Mo, bal Fe.
For heat treating equipment between 1700-2000°F; resists heat up to 2000°F.

PIONEER METAL.
M-326; 65 Ni, Cr-Mo, bal Fe.
Cast: 74,000 TS; 36,5000 YS; 42 El; 150 Brin.
For castings, fittings, pumps, valve parts; acid resistant.

PIONEER METAL (PIONEER "A" ACID RESISTING.).
M-326; 35 Ni, 25 Cr, <5 Mo, 0.2-0.5 C, bal Fe.
Cast: 74,000 TS; 36,500 YS; 42 El; 150 Brin.
Heat treated: 65,000 TS; 45,000 YS; 20 El.
For valves, fittings, pump parts, castings; corrosion and acid resistant.

PIREKS 12/25.
M-487; C, 12 Ni, 25 Cr, bal Fe.
For furnace parts and equipment; heat resistant to 1100°C.

PIREKS 12/25.
M-1725.
Heat resisting alloy; good against sulphurous atmospheres.

PIREKS 20/25.
M-487; C, 20 Ni, 25 Cr, bal Fe.
For furnace parts and equipment; creep and heat resistant to 950°C.

PIREKS 20/25.
M-1725;
Heat resisting alloy; good creep resistance up to 1050°C.

PIREKS 25/20.
M-487; C, 25 Ni, 20 Cr, bal Fe.
For furnace parts and equipment; heat resistant to 950°C.

PIREKS 25/20.
M-1725.
Heat resisting alloy; for continuous service to 1050°C.

PIREKS 35/15.
M-487; C, 35 Ni, 15 Cr, bal Fe.
For furnace parts and equipment; heat resistant to 950°C.

PIREKS 35/15.
M-1725.
Nickel-chromium heat resisting alloy for cycling use in carburizing atmospheres.

PIREKS 37/18/2.
M-1725.
Heat resisting nickel-chromium-niobium alloy for cycling use in carburizing atmospheres, as carburizing boxes.

PIREKS 60.
M-487; C, alloy, bal Fe.
For quenching jigs, and fixtures; heat resistant to 1050°C.

PIREKS 60.
M-1725.
Heat resisting alloy; for continuous use up to 1050°C.

PIREKS 60/13.
M-487; C, alloy, bal Fe.
For furnace trays, containers, electrical resistors, high heat resistance.

PIREKS 60/13.
M-1725.
Heat resisting alloy; for cycle heating and cooling to about 1050 °C.

PIREKS 228.
M-87; C, alloy, bal Fe.
For damper plates, pyrometer tubes; heat resistant to S atmosphere.

PIREKS 529.
M-1725.
Heat resisting alloy, ferritic type, good resistance to attack from high sulphur.

PIREKS METAL.
M-418; 0.25 C, 18 Cr, 8 Ni, bal Fe.
At 20°C: 80,500 TS; 250 Brin.
At 1000°C: 34,000 TS.
For annealing and carburizing boxes, heat treating appliances; cast alloy; resists heat up to 1000°C.

PIREKS RCC.
M-40; 0.21-0.35 C, 49-57 Ni, 24-30 Cr, Si, Mn, bal Fe.
For heat and corrosion resistant parts; heat and corrosion resistant.

PIREKS-REACTAL.
M-40; 0.6 C, 20 Cr, 65 Ni, 2 Si, bal Fe.
For furnace parts, carburizing and annealing boxes; heat resistant.

PIRO-R.
M-1766; 0.10 C, 1.10 Mn, 0.06 max Si, 0.05 max P, 0.30 S, bal Fe.
Free-machining, low carbon steel.
AISI 1215; BS EN1A; DIN 95 Mn z3.

PIRSCH'S GERMAN SILVER.
80-71 Cr, 16-17 Ni, 1-7.5 Zn, 1-2.5 Zn, 1-2.8 Sb, 1-2 Co, 1-1.5 Fe, 0-0.5 Al.
For ornaments, tableware; corrosion resistant.

PISTONS-1.
M-83 Cu, 16 Zn, 1 Sn.
For pistons; corrosion resistant.

PISTONS-2.
93.5 Al, 2.45-3.4 Cu, 1.39 Mg, 0.4 Si, 0-0.28 Zn, 0-1.47 Ni.
For pistons, cylinder heads; age-hardenable.

PITALOY NO. 90.
M-236; 0.3 C, 0.1 V, 0.9 Mn, 0.35 Si, 1.6 Ni, bal Fe.
Cast: 90,000 TS; 60,000 YS; 25 El; 50 RA.
For general castings, locomotive frames, roll mill machinery, cross-heads; tough.

PITALOY NO. 100.
M-236; 0.35-0.40 C, 1.0 V, 0.9 Mn, 0.35 Mo, 0.35 Si, bal Fe.
Cast: 100,000 TS; 70,000 YS; 22 El; 40 RA.
For general castings, locomotive frames; tough.

PITTSBURG.
M-370; 1.0 C, 5.25 Cr, 0.5 Mn, 1.1 Mo, 0.25 V, bal Fe.
Air hardenable to 64 Rc.
For tools dies, jigs, precison parts.
Type A2 air hardening tool steel.

PITTSBURGH.
M-883; 0.1 C, bal Fe.
For welding rod.

PITT-TEN A441.
M-883; 0.22 C, 1.0 Mn, 0.30 Si, 0.20 Cu, 0.02 Min V, bal Fe.
Wrought: 70 ksi TS; 50 ksi YS; 18 El.
HSLA steel.
ASTM A441.

PITT-TEN NO. 1.
M-883; 0.12 max C, 0.10 max Si, 0.7 Mn, 0.8 Cr, 0.7 Ni, 0.07 max P, 0.05 max S, bal Fe.
For truck and bus bodies, mine cars; high strength structural steel.

PITT-TEN NO. 2.
M-83; 0.15 max C, 0.75 max Mn, 0.10 max Si, 0.07 max P, 0.05 max S, 0.05 Cr, bal Fe.
For trucks and bus bodies, mine cars; high strength structural steel.

PITT-TEN "X".
M-833; 0.10-0.20 C, 0.50-1.0 Mn, 0.01 min Cb, bal Fe.
Rolled: 60,000-75,000 TS; 45,000-60,000 YS; 18-24 El.
For agricultural equipment, bus and truck bodies.
High strength structural steel.

PITT-TEN X45W.
M-883; 0.20C, 1.0 Mn, 0.10 Si, 0.01 min Cb, 0.01 min V, bal Fe.
Wrought: 60 ksi TS; 45 ksi YS; 24 El.
HSLA steel.
ASTM A572.

PITT-TEN X50W.
M-883; 0.20 C, 1.0 Mn, 0.10 Si, 0.01 min Cb, 0.01 min V, bal Fe.
Wrought: 65 ksi TS; 50 ksi YS; 22 El.
HSLA steel.
ASTM A572.

PITT-TEN X55W.
M-883; 0.20 C, 1.0 Mn, 0.10 Si, 0.01 min Cb, 0.01 min V, bal Fe.
Wrought: 70 ksi TS; 55 ksi YS; 20 El.
HSLA steel.
ASTM A572.

PITT-TEN X60W.
M-883; 0.20 C, 1.0 Mn, 0.10 Si, 0.01 Min Cb, 0.01 min V, bal Fe.
Wrought: 75 ksi TS; 60 ksi YS; 18 El.
HSLA steel.
ASTM A572.

PLACET.
M-Eng.; 60 Ni, 20 Fe, 15 Cr, 5 Mn.
Cast: 50,000 TS; 1 El; 179 Brin.
For resistance alloy; heat resistant.

PLANCHER.
M-289; 0.55-0.65 C, 0.5 Mo, 0.8 Mn, 2.0 Si bal Fe.
For chisels, punches, shear blades, plastic master hobs; oil hardened, shock resistant.

PLANET CHOICE.
M-372; 0.7-1.2 C, bal Fe.
For tools, drills, taps; water hardened.

PLANET COLD ROLLED.
M-372; 0.7-0.9 C, bal Fe.
For tools; water hardening.

PLANET DRILL ROD.
M-372; 1.2 C, bal Fe.
For drill rods, drills, tools; cold drawn.

PLANET EXTRA.
M-372; 0.3-1.2 C, bal Fe.
For tools, drills, taps; water hardened.

PLANET HIGH SPEED.
M-372; 0.7 C, 18 W, 4 Cr, 1 V, bal Fe.
For cuttting tools, taps; high speed steel.

PLANET REGULAR.
M-372; 0.8-1.2 C, bal Fe.
For tools, drills, taps: water hardened.

PLANET SHEFFOIL.
M-372; 0.7 C, 18 W, 4 Cr, 1 V, bal Fe.
For high speed tools, cutters; high speed steel.

PLANET SPECIAL.
M-372; 0.7-1.0 C, 0.2 V, bal Fe.
For dies and tools; water hardening.

PLANSEE WZ 12D.
M-1246; 35 TiC, 39 Ni,13 Co,13 Cr.
For jet engine components; high heat resistance.

PLASMEX 1000.
M-207; 0.30 C, 0.75 Mn, 0.50 Si, 1.7 Cr, 0.40 Mo, bal Fe.
Oil hardening tool and steel, for molds.

PLASTIC BRONZE.
M-126; 66 Cu, 5 Sn, 28 Pb, 1 Ni.
For bearings; heavy duty.

PLASTIC HOBBING.
M-210; 0.10 max C; 1.0 Si, 0.4 Mn, bal Fe.
Brinell 110.
For hobbed dies & plastic molds.

PLASTIC METAL.
M-Eng.; 81 Sn, 9.5 Cr, 8.6 Sb, 1.4 Fe.
For bearings, bushings; Babbitt.

PLASTIFORM 1.
M-1752.
Flexible permanent magnet.
Coercive Force (Hc): 1650 oersteds.
Residual Inductance (Br): 2150 gauss.
Flexible and easily cut.
For instruments and electrical systems and controls.

PLASTIFORM 1 H.
M-1752.
Flexible permanent magnet.
Coercive Force (Hc): 1940 oersteds.
Residual Inductance (Br): 2150 gauss.
Flexible and easily cut.
For instruments and electrical systems and controls.

PLASTIFORM 1.4H.
M-1752.
Flexible permanent magnet.
Coercive Force (Hc): 2200 oersteds.
Residual Inductance (Br): 2450 gauss.
Flexible and easily cut.
For instruments and electrical systems and controls.

PLAST-IRON.
M-1090; 0.008 C, 0.001 Mn, 0.005 Si, 99.98 Fe.
For magnets, radio cores; powdered metal.

PLASTO C.
M-1331; 0.15 C, Cr, bal Fe.
For gears, cams, camshaft, fasteners; case hardening steel.

PLASTO CC.
M-1331; C, Cr, bal Fe.
For gears, cams, camshafts, fasteners; case hardening steel.

PLASTO MC.
M-1331; 0.2 C, 1.25 Mn,1.15 Cr, bal Fe.
For gears, cams, camshafts, fasteners; case hardening steel.

PLASTO MCW.
M-1331; 0.15 C, 1.0 Cr,1.25 Mn, 0.25 Si, bal Fe.
For gears, cams, camshafts, fasteners; case hardening steel.

PLASTO NI.
M-1331; 0.19 C, 1.75 Cr, 0.2 Mo, 3.75 Ni, bal Fe.
For gears,cams, camshafts, fasteners; case hardening steel, shock resistant.

PLASTO RR.
M-1331; 0.4 C, 13 Cr, 0.3 Mn , 0.4 Si, bal Fe.
Annealed: 100,000 TS; 55,000 YS; 22 El; 52 RA; 200 Brin.
For cutlery, valves, oil refinery equipment; Type 420; stainless.

PLASTO U.
M-1331; 0.15 C, 0.15-0.35 Si, 0.25-0.50 Mn, bal Fe.
Annealed: 70,000 TS; 40,000 YS25 El; 60 RA; 145 Brin.
For gears, cams, fasteners, bolts; case hardening steel.

PLASTO V.
M-1331; 0.5 C, Ni, Cr, V, bal Fe.
For gears, bolts, crankshafts; oil hardened, shock resistant.

PLATALARGAN.
Pt, Al, Ag.
For pen points; corrosion resistant.

PLATA METAL NO. 5.
M-567; Sb, Sn, Pb.
For bearings; Babbitt metal.

PLATE AJ 30/H.
M-1337; 0.61 C, 0.90 Si, 0.80 Mn, 1.20 Cr, 0.10 V, bal Fe.
Oil hardening tool steel.
For cold punches for heavy sheet, large embossing tools, staking tools.
DIN 61CrSiV5; Werkstoff Nr. 1.2243.

PLATE AR 40/1.
M-1337; 0.45C, 1.0 Si, 0.35 Mn, 1.1 Cr, 0.20 V, 2.0 W, bal Fe.
Oil hardening tool steel.
For pneumatic tools, chisels, staking tools, riveting hammers.
DIN 45 WCrV7; Werkstoff Nr. 1.2542

PLATE BM 260.
M-1337; 0.40 C, 0.40 Si, 0.30 Mn, 13.5 Cr, bal Fe.
Air or oil hardening tool steel.
For plastic molds for corrosive plastics.
DIN X40Cr13; Werkstoff Nr. 1.2083.

PLATE BP 20 E.
M-1337; 0.21 C, 0.30 Si, 1.2 Mn, 1.2 Cr, bal Fe.
For case hardening gears, spline shafts, pinions, crank shafts for small engines.
Din 21 MnCr5.

PLATE BS 30 E.
M-1337; 0.15 C, 0.25 Si,0.50 Mn, 1.55 Cr, 1.55 Ni, bal Fe.
For case hardening cog wheels, chain drives, splined shafts, gears.
Din 15 CrNi6.

PLATE BS 70 E.
M-1337; 0.14 C, 0.30 Si, 0.40 Mn,0.7 Cr, 3.50 Ni, bal Fe.
For case hardening parts, such as gears, cam shafts, splined coupling of middle stress.
DIN 14NiCr14; W9 Nr, 1.5752.

PLATE BS 90 E.
M-1337; 0.14 C, 0.30 Si, 0.40 Mn, 1.10 Cr, 4.50 Ni, bal Fe.
For case hardening parts as crankshafts, highly stressed gears and shafts of large section as for trucks and ordnance.
Din 14Ni, Cr18; W. Nr. 1.5860.

PLATE BS 90 E MO.
M-1337; 0.19 C, 0.20 Si, 0.40 Mn, 1.25 Cr, 0.25 Mo, 4.0 Ni, bal Fe.
For case hardening parts such as truck and tractor gears and splined shaft for case hardened ordnance parts.
Din 19NiCrMo15; W.Nr. 1.6587.

PLATE CLIMAX TZM.
M-1337; 99.3 Mo, 0.5 Ti, 0.1 Zr.
For hot extrusion dies, mandrels, turbine components, casting dies and molds.

PLATE E 5 S.
M-1337; 0.06 max C, 0.20 max Si, 0.20 Mn, 4.50 Cr, 0.50 Mo, bal Fe.
For case hardening heavy gears, axles, shafts, tractor, railroad and ordnance equipment.
DIN X6CrMo5.

PLATE K18.
M-1337; 0.85 C, 0.30 max Si, 0.35 max Mn, bal Fe.
Water hardening tool steel; for stone tools for middle hard stones, hammers, leather and spoon dies.
DIN C85W2; Werkstoff Nr. 1.1630.

PLATE K 20.
M-1337; 1.00 C, 0.30 max Si, 0.35 max Mn, bal Fe.
Water hardening tool steel; for stone tools, for hard stone, embassing tools, scythes.
DIN C100 W2; Werkstoff 1.1640.

PLATE KM 20 V.
M-1337; 1.0 C, 0.20 Si, 0.25 Mn, 0.10 V, bal Fe.
Water hardening tool steel.
For piercing dies, upsetting dies, small shear blades, shovels, hand tools.
DIN 100 V1; Werkstoff Nr. 1.2833.

PLATE KMV SUPRA.
M-1337; 1.45 C, 0.20-0.35 Si, 0.30-0.50 Mn, 3.0-3.5 V, bal Fe.
Cold work tool steel.
Workstoff Nr.1.2838.

PLATE KS 9.
M-1337; 0.85 C, 0.25 max Si, 0.25 max Mn, bal Fe.
Water hardening tool steel; for cold impact tools, cold cutting and punching dies, strainers, snap dies.
DIN C85W1; Werkstoff Nr. 1.1530.

PLATE KS 10.
M-1337; 1.00 C, 0.25 max Si, 0.25 max Mn, bal Fe.
Water hardening tool steel; for cutting and punching dies, shear blades, hollow and massive embossing dies.
DIN C100W1; Werkstoff Nr. 1.1540.

PLATE KS 66.
M-1337; 0.50 C, 0.30 Si, 0.50 Mn, 1.1 Cr, 0.20 Mo, 3.25 Ni, bal Fe.
Oil hardening tool steel.
For cutlery dies, artificial resin molding dies, dies for tableware.
DIN (Similar to) 50 NiCrl3 and Werkstoff 1.2721.

PLATE KW 83.
M-1337; 0.45 C, 0.30 Si, 0.65 Mn, 1.4 Cr, 0.50 Mo, 4.0 Ni, 0.15 V, 0.50 W, bal Fe.
Oil or air hardening tool steel.
For large, tough embossing tools, cold upsetting tools, air hardening dies.
DIN (Similar to) X45NiCrMo4 and Werkstoff Nr. 1.2767.

PLATE KW 83 SPEZIAL.
M-1337; 0.40-0.50 C, 1.2-1.5 Cr, 0.15-0.35 Mo, 3.8-4.3 Ni, 0.50 W.
Cold work tool steel.
Werkstoff Nr. 1.2767.

PLATE-LOY.
Pb.
For hot lead plating.

PLATE MSS.
M-1337; 0.53 C, 0.90 Si, 0.90 Mn, bal Fe.
Oil hardenable to 54-60 Rock C.
For axles, pins, staking tools, cold chisels.
DIN 53MnSi4; Werkstoff Nr. 1.2825.

PLATE NM 30.
M-1337; 1.05 C, 0.30 Si, 0.20 Mn, 1.40 Cr, bal Fe.
Water or oil hardening tool steel for small dies, punches, gages.
DIN 105Cr5; Werkstoff Nr. 1.2060.

PLATE NM 110 MO.
M-1337; 1.00 C, 0.30 Si, 0.50 Mn, 5.20 Cr, 1.15 Mo, 0.30 V, bal Fe.
Air hardening tool steel.
For thread rolling dies, trimming dies, shear punches.
DIN X100CrMoV51; Werkstoff Nr. 1.2363.

PLATE NM 150 V.
M-1337; 1.5-1.6 C, 11.5-12.5 Cr, 0.60-0.80 Mo, 0.90-1.10 V, bal Fe.
Cold work tool steel.
Werkstoff Nr. 1.2379; AISI D2.

PLATE NM 240.
M-1337; 2.10 C, 0.30 Si, 0.30 Mn, 12.0 Cr, bal Fe.
Oil or air hardening tool steel.
For heavy duty stamping dies, broaches, cold shears, wood milling cutters.
DIN X210Cr12; Werkstoff Nr. 1.2080.

PLATE NM 240 CO.
M-1337; 2.10 C, 0.30 Si, 0.40 Mn, 12.0 Cr, 0.40 Mo, 0.15 V, 0.70 W, 1.0 Co, bal Fe.
Air or oil hardening.
For thread rolling dies, punching and forming dies, broaches.
DIN X210CrCoW12; Werkstoff Nr. 1.2884.

PLATE NM 240 V.
M-1337; 2.10 C, 0.30 Si, 0.30 Mn, 12.0 Cr, 0.80 W, bal Fe.
Oil or air hardening tool steel.
For heavy duty stamping dies, punches, broaches, die rings, plastic molds.
DIN X210CrW12; Werkstoff Nr. 1.2436.

PLATE NM 240 W.
M-1337; 1.65 C, 0.30 Si, 0.30 Mn, 12.0 Cr, 0.10 V, bal Fe.
Air or oil hardening tool steel.
For heavy duty and long wearing stamping and punching dies, broaches.
DIN X165CrV12; Werkstoff 1.2201.

PLATE NM 240 WMO.
M-1337; 1.65 C, 0.30 Si, 0.35 Mn, 12.0 Cr, 0.70 Mo, 0.35 V, 0.50 W, bal Fe.
Air or oil hardening.
For heavy duty punching and stamping dies and die blocks, broaches, plastic molds for corrosive plastics.
DIN X165CrMoV12; Werkstoff Nr. 1.2601.

PLATE NN 20.
M-1337; 1.05 C, 0.25 Si, 1.10 Mn, 0.90 Cr, bal Fe.
Oil hardening tool steel.
For threading tools, cutting and punching dies for medium duty.
DIN 105MnCr4; Werkstoff Nr. 1.2127.

PLATE NN 40.
M-1337; 0.90 C, 0.20 Si, 2.0 Mn, 0.10 V, bal Fe.
Oil hardening tool steel.
For difficult cutting dies and punches for sheets up to 3 mm thick.
DIN 90MnV8; Werkstoff Nr. 1.2842.

PLATE NN 90.
M-1337; 0.85-0.95 C, 1.05-1.25 Si, 0.60-0.90 Mn, 1.1-1.3 Cr, bal Fe.
Cold work tool steel.
Werkstoff Nr. 1.2108.

PLATE NR 40/H.
M-1337; 0.60 C, 0.60 Si, 0.30 Mn, 1.10 Cr, 0.20 V, 2.0 W, bal Fe.
Oil hardening tool steel.
For highly stressed perforating dies, trimming dies, lower dies.
DIN 60WCrV7; Werkstoff Nr. 1.2550.

PLATE ON 22.
M-1337; 1.15 C, 0.20 Si, 0.35 Mn, 0.70 Cr, 0.10 V, bal Fe.
Water or oil hardening tool steel.
For twist drills, reamers, punches, taps.
DIN 115CrV3; Werkstoff Nr. 1.2210.

PLATE OR 33 M.
M-1337; 1.05 C, 0.25 Si, 1.0 Mn, 1.0 Cr, 1.2 W, bal Fe.
Oil hardening tool steel.
For cutting and stamping dies for sheet steel up to 5 mm thick.
DIN 105WCr6; Werkstoff Nr. 1.2419.

PLATE OW20V.
M-1337; 1.2 C, 0.10 V, 1.0 W, 0.28 N, bal Fe.
For cutters, bearings, liners; water hardened.

PLATE OW 90.
M-1337; 1.42 C, 0.20 Si, 0.30 Mn, 0.35 Cr, 0.25 V, 3.25 W, bal Fe.
Water or oil hardening tool steel.
For lathe and planing tools, engraving needles, scrapers, serrating steels.
DIN 142WV13; Werkstoff Nr. 1.2562.

PLATE PA 188 TI.
M-1337; 0.10 max C, 18.0 Cr, 11.0 Ni, Ti, bal Fe.
Ann: 71,000-106,000 TS; 35,000 min YS; 40 min El; 130-190 Brin.
Austenitic, stainless, weldable.
For parts and welded assemblies for use in food, beverage, dairy and chemical industries.
DIN X10CrNiTi1810; Werkstoff Nr. 1.4541.

PLATE PA 188 W.
M-1337; 0.07 C, 18.0 Cr, 10.0 Ni, bal Fe.
Ann: 71,000-100,000 TS; 31,000 min YS; 50 min El; 130-180 Brin.
Austenitic, stainless, weldable.
For parts and welded assemblies in food, dairy, beverage, and chemical industries.
DIN X5CrNi189; Werkstoff Nr. 1.4301.

PLATE PA 1810.
M-1337; 0.10 max C, 1.0 max Si, 2.0 max Mn, 17.0-19.0 Cr, 9.0-11.5 Ni, Nb=8x%C, bal Fe.
Stabilized austenitic stainless steel.
Werkstoff Nr. 1.4550; AISI 347.

PLATE PAO 188/2 TI.
M-1337; 0.10 max C, 18.0 Cr, 11.0 Ni, 2.3 Mo, Ti, bal Fe.
Ann: 71,000-106,000 TS; 35,000 min YS; 40 min El; 130-190 Brin.
Austenitic, weldable very good stainless properties.
For parts and welded assemblies in paper, textile and chemical industries.
DIN X10CrNiMoTi810; Werkstoff Nr. 1.4571.

PLATE PAO 188/2 W.
M-1337; 0.07 max C, 18.0 Cr, 11.0 Ni, 2.3 Mo, bal Fe.
Ann: 71,000-100,00 TS; 28,000 min YS; 45 min El: 130-180 Brin.
Austenitic, very good stainless properties.
For parts and assemblies in paper, textile, and chemical industries.
DIN X5CrNiMo1810; Werkstoff Nr. 1.4401.

PLATE PCL.
M-1337; 1.05 C, Cr, bal Fe.
For bearings, cutters, liners; water hardened.

PLATE PFC 141.
M-1337; 0.08 max C, 14.0 Cr, bal Fe.
Ann: 71,000-92,000 TS; 43,000 YS; 20 min El; 140-180 Brin.
Corrosion resistant steel, not hardenable by heat treatment, ferritic.
For structural parts, building fittings.
DIN X7Cr14; Werkstoff Nr. 1.4001.

PLATE PFC 171.
M-1337; 0.10 max C, 17.0 Cr, bal Fe.
Ferritic corrosion resistant steel; not hardenable by heat treating.
Ann: 64,000-85,000 TS; 42,000 min YS; 20 min El; 130-170 Brin.
For building fittings, corrosion resistant hardware, spoons.
DIN X8Cr17; Werkstoff Nr. 1.4016.

PLATE PFC 171 S.
M-1337; 0.12 C, 17.0 Cr, 0.25 Mo, S, bal Fe.
Oil quenched: 100,000-120,000 TS; 64,000 min YS; 12 min El; 190-235 Brin.
Ferritic, corrosion resistant steel, free machining grade.
For machined parts as corrosion resistant screws, bolts, studs, nuts.
DIN X12CrMoS17; Werkstoff Nr. 1.4104.

PLATE PFMC 14.
M-1337; 0.20 C, 13.0 Cr, bal Fe.
Heat treated: 113,000-135,000 TS; 78,000 min YS; 5 min El; 225-275 Brin.
Martensitic corrosion resistant steel, oil hardenable to 40-48 Rock C.
High strength structural parts for pumps, impellers, turbine wheels.
DIN X20Cr13; Werkstoff Nr. 1.4021.

PLATE PGS 4.
M-1337; 0.45 C, 0.35 Si, 0.70 Mn, bal Fe.
Water hardening tool steel; for hammers, forks, axes, knives, shears, screw drivers.
DIN C45W3; Werkstoff Nr. 1.1730.

PLATE PGS 6.
M-1337; 0.60 C, 0.35 Si, 0.70 Mn, bal Fe.
Water or oil hardening tool steel; for shanks for tools, bars, needle beds, stone breakers, hammers.
DIN C60W3; Werkstoff Nr. 1.1740.

PLATE PLATIT 40.
M-1337; 1.0 C, 31.0 Cr, 14 W, 53 Co.
For parts for high temperature operation.

PLATE PLATIT HH.
M-1337; 0.25 C, 27.0 Cr, 6.0 Mo, 55 Co, bal Fe.
For high temperature operations.

PLATE PM 512.
M-1337; 0.37 C, 1.0 Si, 0.5 Mn, 5.3 Cr, 1.5 Mo, 0.2 V, 1.3 W, bal Fe.
Hot work tool steel; air or oil hardening.
For forging and pressing dies, ferrous and nonferrous.
DIN X37CrMoW51; Werkstoff Nr. 1.2606.

PLATE PM 524.
M-1337; 0.38 C, 1.0 Si, 0.4 N, 5.3 Cr, 1.4 Mo, 0.6 V, bal Fe.
Hot work tool steel; air or oil hardening.
For pressure casting molds for light alloys.
DIN X40CrMoV51; Werkstoff Nr. 1.2344.

PLATE PMC 145.
M-1337; 0.40 C, 13.0 Cr, bal Fe.
Martensitic corrosion resistant steel.
Air or oil hardenable to 45-55 Rc.
For shafts, spline shafts, impellers, turbine wheels, pump parts, cutlery.
DIN X40Cr13; Werkstoff Nr. 1.4034.

PLATE PMC NI 162.
M-1337; 0.22 C, 17.0 Cr, 2.0 Ni, bal Fe.
Heat treated: 115,000-135,000 TS; 85,00 min YS; 4 min El; 225-275 Brin.
Corrosion resistant.
For structural parts requiring good strength and non-rusting properties.
DIN X22CrNi17; Werkstoff Nr. 1.4057.

PLATE PMCO 174.
M-1337; 0.38 C, 17.0 Cr, 1.20 Mo, 0.50 Ni, bal Fe.
Ht. Tr. R.T.: 115,000-135,000 TS; 85,000 min YS; 14 min El; 225-275 Brin.
At 400°C: 71,000 min YS.
Corrosion resistant, good temperature resistance.
For arbors, shafts, spindels, bolts operating up to 400°C.
DIN X35CrMo17; Werkstoff Nr. 1.4122.

PLATE PMCO 189.
M-1337; 0.90 C, 18.0 Cr, 1.20 Mo, V, bal Fe.
Martensitic, corrosion resistant, air or oil hardenable to 50-60 Rock C.
For valves, knives, cutlery, ball bearings and races, shafts, couplings, surgical instruments, dental tools.
DIN X90CrMoV18; Werkstoff Nr. 1.4112.

PLATE PN 24.
M-1337; 0.50 C, 0.25 Si, 0.90 Mn, 1.0 Cr, 0.10 V, bal Fe.
Oil hardening tool steel.
For cold punches.
DIN 50CrV4; Werkstoff Nr. 1.2241.

PLATE PP 21.
M-1337; 0.19-0.24 C, 0.35-0.55 Si, 0.30-0.50 Mn, 1.3-1.5 Cr, 1.0-1.2 Mo, 0.25-0.35 V, bal Fe.
Alloy steel, low-carbon; for carburizing.
Werkstoff Nr. 1.2352.

PLATE PP 32.
M-1337; 0.40 C, 0.3 Si, 1.5 Mn, 2.0 Cr, 0.2 Mo, bal Fe.
Hot work tool steel; oil hardening.
For punches, centrifugal casting molds.
DIN 48CrMoV67; Werkstoff Nr. 1.2323.

PLATE PS 300.
M-1337; 0.50 C, 1.3 Si, 0.7 Mn, 14.0 Cr, 13.0 Ni, 1.2 V, 1.3 W, bal Fe.
Hot work tool steel; not heat treatable.
For pressing dies of simple profile; Austenitic.
DIN X50NiCrWV1313; Werkstoff Nr. 1.2731.

PLATE PU 37.
M-1337; 0.55 C, 0.3 Si, 0.7 Mn, 1.0 Cr, 0.5 Mo, 1.7 Ni, 0.1 V, bal Fe.
Hot work tool steel; oil or air hardening.
For drop forge hammer dies.
DIN 56NiCrMoV7; Werkstoff Nr. 1.2714.

PLATE PVX 1212.
M-1337; 0.45 C, 0.6 Si, 0.6 Mn, 4.5 Cr, 0.8 Mo, 12.0 Ni, 1.2 V, 12. W, 0.5 Co, bal Fe.
Hot work tool steel; not heat treatable.
For high temperature pressing using light loads.

PLATE PVX 2615.
M-1337; 0.08 C, 1.0 max Si, 1.0-2.0 Mn, 13.5-16.0 Cr, 1.0-1.5 Mo, 24.0-27.0 Ni, 1.9 Ti, 0.5 max V, 0.35 max Al, B, bal Fe.
For high temperature applications.
Werkstoff Nr: 1.4980.

PLATE PW 78.
M-1337; 0.45 C, 0.6 Si, 0.7 N, 1.7 Cr, 0.6 Mo, 0.8 V, 0.8 W, bal Fe.
Hot work tool steel; oil hardening.
Shear blades, pressing dies and punches.
DIN 45CrVMoW58; Werkstoff Nr. 1.2603.

PLATE PW 100.
M-1337; 0.30 C, 0.2 Si, 0.3 Mn, 2.4 Cr, 0.6 V, 4.3 W, bal Fe.
Hot work tool steel; oil hardening.
For pressure casting molds and dies for non-ferrous processing.
DIN X30WCrV53; Werkstoff Nr. 1.25679.

PLATE PW 190.
M-1337; 0.30 C, 0.2 Si, 0.3 Mn, 2.6 Cr, 0.4 V, 8.5 W, bal Fe.
Hot work tool steel; oil or air hardening.
Hot extrusion dies, pressure casting molds and dies for non-ferrous.
DIN X30WCrV93; Werkstoff Nr. 1.2581.

PLATE PW 1096.
M-1337; 0.17-0.23 C, 9.5-10.5 Co, 9.0-10.0 Cr, 1.8-2.2 Mo, 5.0-6.0 W, bal Fe.
Steel for high temperature applications.
Werkstoff Nr. 1.2888.

PLATE SVX 33.
M-1337; 0.95 C, 4.0 Cr, 2.7 Mo, 2.4 V, 3.0 W, bal Fe.
For twist drills, milling cutters, broaches, band saws, hack saw blades.
High speed steel.
Werkstoff Nr. 1.3333; DIN S3-3-2.

PLATE SVX 34.
M-1337; 0.85 C, 4.0 Cr, 0.9 Mo, 1.6 V, 9.0 W, bal Fe.
Milling cutters, twist drills and taps for working on soft materials.
High speed steel.
Werkstoff Nr. 1.3316; DIN S9-1-2.

PLATE SVX 54.
M-1337; 0.85 C, 4.0 Cr, 0.9 Mo, 2.5 V, 12.0 W, bal Fe.
For lathe tools, milling knives, relieved cutters for working on hard materials.
High speed steel.
Werkstoff Nr. 1.3318; DIN S12-1-2.

PLATE SVX 90.
M-1337; 1.25 C 4.2 Cr, 0.9 Mo, 3.8 V, 12.0 W, bal Fe.
For finishing and milling tools, pinion type cutters, steel shapes, broaches.
High speed steel.
Werkstoff Nr. 1.3302; DIN S12-1-4.

PLATE SVX 360.
M-1337; 0.75 C, 4.0 Cr, 0.5 Mo, 1.1 V, 18.0 W, bal Fe.
For twist drills, threading tools, milling cutters, gear shapers.
High speed steel.
Werkstoff Nr. 1.3355; DIN S18-0-1.

PLATE SVX 526.
M-1337; 0.85 C, 4.2 Cr, 5.0 Mo, 2.0 V, 6.3 W, bal Fe.
For twist drrills, broaches, milling cutters, segments for circular saws, lathe tools for roughing cuts.
High speed steel.
Werkstoff Nr. 1.3343; DIN S6-5-2.

PLATE SVX 536.
M-1337; 1.20 C, 4.2 Cr, 5.0 Mo, 3.3 V, 6.3 W, bal Fe.
For heavy duty milling cutters, highly stressed broaches, reamers.
High speed steel.
Werkstoff Nr. 1.3344; DIN S6-5-3.

PLATE SVX 811.
M-1337; 0.82 C, 3.8 Cr, 8.6 Mo, 1.2 V, 1.8 W, bal Fe.
For twist drills, reamers, milling cutters, threading tools and taps.
High speed steel.
Werkstoff Nr. 1.3346; DIN S2-9-1.

PLATE SVZ 245.
M-1337; 1.30 C, 4.2 Cr, 0.9 Mo, 3.8 V, 12.0 W, 4.8 Co, bal Fe.
For finishing and roughing cutting tools for best wear characteristics. High speed steel.
Werkstoff Nr. 1.3202; DIN S12-1-4-5.

PLATE SVZ 365.
M-1337; 0.80 C, 4.2 Cr, 0.7 Mo, 1.6 V, 18.0 W, 4.8 Co, bal Fe.
Lathe and planing tools with eminent cutting strength and toughness.
High speed steel.
Werkstoff Nr. 1.3255; DIN S18-1-2-5.

PLATE SVZ 410.
M-1337; 0.76 C, 4.2 Cr, 0.7 Mo, 1.6 V, 18.0 W, 9.5 Co, bal Fe.
Lathe and planing tools for heavy work requiring good red hardness, and for machining austenitic steels.
High speed steel.
Werkstoff Nr. 1.3265; DIN S18-1-2-10.

PLATE SVZ 526 CO.
M-1337; 0.82 C, 4.2 Cr, 5.0 Mo, 2.0 V, 6.3 W, 4.8 Co, bal Fe.
For lathe tools, drills, milling cutters for tough austenitic steels.
High speed steel, good red hardness.
Werkstoff Nr. 1.3243; DIN S6-5-2-5.

PLATE US MO.
M-1337; 0.30 C, 0.3 Si, 0.3 N, 3.0 Cr, 2.8 Mo, 0.5 V, bal Fe.
Hot work tool steel; oil hardening.
Pressure casting molds for heavy metal alloy, pressing dies.
DIN X32CrMoV33; Werkstoff Nr. 1.2365.

PLATE US MO K.
M-1337; 0.30 C, 0.2 Si, 0.4 Mn, 2.8 Cr, 2.8 Mo, 0.5 V, bal Fe.
Hot work tool steel; oil or air hardening.
For pressure castings molds, pressing dies.
DIN X3CrMoV33; Werkstoff Nr. 1.2365.

PLATE WV 27.
M-1337; 0.55 C, 0.30 Si, 0.60 Mn, 0.70 Cr, 0.30 Mo, 1.70 Ni, 0.10 V, bal Fe.
Oil hardening tool steel, for hot work.
For coining dies, embossing tools, counterbores, large mandrels.
DIN 55NiCrMoV6; Werkstoff Nr. 1.2713.

PLATE ZM 14.
M-1337; 1.40 C, 0.30 Si, 0.30 Mn, 0.35 Cr, 0.10 V, bal Fe.
Water hardening tool steel.
For punches, cold drawing dies, cutting tools that operate cold, bearings.
DIN 140CrV1; Werkstoff Nr. 1.2206.

PLATE ZW 100.
M-1337; 1.30 C, 0.25 Si, 0.30 Mn, 0.30 Cr, 5.0 W, bal Fe.
Water or oil hardening tool steel.
For drawing dies, highly stressed cold forming dies.
DIN X130W5; Werkstoff Nr. 1.2453.

PLATINA, BIRMINGHAM.
M-Eng.; 46.6 Cu, 53.15 Zn, 0.25 Fe.
For hardware, ornaments; corrosion resistant.

PLATINA, BIRMINGHAM.
M-Eng.; 20.25 Cu, 79.4 Zn, 0.33 Fe.
For hardware, ornaments.

PLATINA, PLATING.
M-Eng.; 20.25 Cu, 79.4 Zn, 0.33 Fe.
For hardware ornaments.

PLATINAX-II.
M-200; 23.3 Co, bal Pt.
6400 Br, 9.2 BH max.
For permanent magnets, metering devices.
Can be machined, rolled and drawn.
Outstanding magnetic properties obtained by a 2 stage heating treatment. Corrosion resistant to H_2SO_4 and HNO_3.

PLATINE.
M-Eng.; 57 Zn, 43 Cu.
For ornaments; weak and brittle.

PLATINE-AU-TITRE (PROPLATINUM).
M-France; 83-65 Ag. 17-35 Pt.
For substitute for platinum; corrosion resistant.

PLATINEL 1503.
M-686; 65 Au, 35 Pd.
For high temperature thermocouples.
Negative leg.

PLATINEL 1786.
M-686; 3 Au, 83 Pd, bal Pt.
For high temperature thermocouples.
Positive leg.

PLATINEL 1813.
M-686; 55 Pd, 31 Pt, 14 Au.
For high temperature thermocouples.

PLATINIRIDIUM.
M-France; 53 Pt, 28 Ir, 7 Rh, 3 Cu, 4 Fe, traces Pd + As.
For pen nibs, ornamental parts; mined in Russia.

PLATINIT.
M-Eng.; 46 Ni, bal Fe.
For chemical apparatus; corrosion resistant.

PLATINOID.
50-90 Cu, 3-40 Ni, 0.10 Al, 0.40 Zn.
For chemical equipment and construction; corrosion resistant.

PLATINOID A.
M-Eng.; 60 Cu, 24 Zn, 14 Ni, 2 W.
57,000 TS; 33 El.
High resistant alloy, heating elements, thermocouples; Tungsten, German Silver.

PLATINOID B.
M-Eng.; 54 Cu, 20 Zn, 25 Ni, W, 0.5 Fe, 0.2 Mn.
72,000 TS; 30 El; 89 Brin.
For high resistance alloys, heating elements, thermocouples; Tungsten, German Silver.

PLATINOR.
45 Cu, 18 Pt, 18 brass, 9 Ag, 9 Ni.
For cheap jewelry, ornaments; corrosion resistant.

PLATINUM.
M-53; 99.9 Pt.
Ann: 60 R-15T; Elect. cond. : 15% I.A.C.S. for electrical contacts.

PLATINUM.
M-1755; Pt.
Purities: 99.999%, 99.99%, 99.9%.
Forms: Sponge, powder, rod, wire, sheet foil, single crystals.

PLATINUM ALLOY.
2-5 parts Ag, 1 part Pt, 0-1 part Cu.
For ornaments, jewelry; corrosion resistant.

PLATINUM ALLOY NO. 479.
M-1295; 92 Pt, 8 W.
Cold drawn: 300,000 TS.
For potentiometers, variable resistors; high electrical and heat resistance.

PLATINUM ALLOY NO. 851.
M-1295; 79 Pt, 15 Rh, 5 Ru.
Cold drawn: 300,00 TS.
For potentiometers, variable resistors; high electrical and heat resistance.

PLATINUM GOLD-1.
70 Au, 30 Pt.
For ornaments, jewelry; white, corrosion resistant.

PLATINUM GOLD-2.
58 Pt, 25 Ag, 17 Au.
For jewelry; corrosion resistant.

PLATINUM GOLD-3.
60 Au, 40 Pt.
For jewelry; corrosion resistant.

PLATINUM IRIDIUM-1.
5.0 Ir, 95 Pt.
170 Brin.
For jewelry; corrosion resistant.

PLATINUM IRIDIUM-2.
30 Ir, 70 Pt.
400 Brin.
For surgical instruments; corrosion resistant.

PLATINUM LEAD (BIRMINGHAM PLATINUM).
79-53 Zn, 20-47 Cu, 0.3 Fe.
For ornamental parts; castings.

PLATINUM RHODIUM.
100-80 Pt, 0-20 Rh.
For thermocouples; heat resistant.

PLATINUM SILVER.
66.7 Ag, 33.3 Pt.
For ornaments, jewelry; corrosion resistant.

PLATINUM SOLDER.
73 Ag. 27 Pt.
For solder for platinum alloys; corrosion resistant.

PLATINUM SUBSTITUTE.
23.6-24.0 Al, 3.7 Bi, 0.7 Au, 72 Ni.
For ornamental white metal parts; corrosion resistant.

PLATINUM SUBSTITUTE COOPERS-1.
70 Ag, 25 Pd, 5 Co.
For jewelry; corrosion resistant.

PLATINUM SUBSTITUTE COOPERS-2.
70 Ag, 25 Pt, 5 Ni.
For jewelry; corrosion resistant.

PLATINUM SUBSTITUTE ELECTRICAL-1.
70 Au, 25 ag, 5 Ni or Pt.
For electrical contacts; heat resistant.

PLATINUM SUBSTITUTE ELECTRICAL-2.
6, A, 25 Ag, 7.5 Pt.
For electrical contacts; heat resistant.

PLATINAM.
M-995; 50 Ni, 35 Cu, 10 Sn, 5.0 Fe, 0.3 Al.
For steam valves up to 480°C, valve disks and seats; heat resisting, abrasion and erosion resistant.

PLATNIK.
Pt-Ni.
For ornaments, jewelry; corrosion resistant.

PLETTENBERG CMV.
M-1344; 0.45 C, Ni, Cr, W, bal Fe.
For forging dies, upsetters, header dies; oil hardened, tough.

PLETTENBERG EXTRA MH.
M-1344; 1.1 C, 0.25 max Si, 0.25 max Mn, bal Fe.
Annealed: 110,000 TS; 58,000 YS;18 El; 38 RA; 210 Brin.
For springs, tools, drlls, taps, reamers; Type W1; water hardened.

PLETTENBERG EXTRA WEICH.
M-1344; 0.70 C, 0.25 max Si, 0.25 max N, bal Fe.
Heat treated: 175,00 TS; 128,000 YS; 12 El; 37 RA; 355 Brin.
For springs, rails, punches, axes, tools; Type W1; water hardened.

PLETTENBERG EXTRA ZAH.
M-1344; 0.85 C, 0.25 max Si, 0.25 max Mn; bal Fe.
Heat treated: 190,000 TS; 145,00 YS; 10 El; 30 RA; 400 Brin.
For springs, tools, cutters, taps, drills; Type W1; water hardened.

PLETTENBERG EXTRA ZH.
M-1344; 1.0 C, 0.25 max Si, 0.25 max Mn, bal Fe.
Annealed: 100,000 TS; 53,000 YS; 21 El; 42 RA; 200 Brin.
For springs, drills, taps, hobs, reamers; Type W1; water hardened.

PLETTENBERG GC.
M-1344; 0.40 C, Cr, Mn, Mo, bal Fe.
For gears, bolts, machine tool parts; oil hardened, shock resistant.

PLETTENBERG GCN1.
M-1344; 0.55 C, 0.7 Cr, 0.18 Mo, 1.65 Ni, 0.1 V, bal Fe.
For springs, gears, crankshafts, bolts, studs; oil hardened, shock resistant.

PLETTENBERG GCN2.
M-1344; 0.56 C, Ni, Cr, Mo, V, bal Fe.
For gears, bolts, crankshafts; oil hardened, shock resistant.

PLETTENBERG GMS.
M-1344; 0.53 C, 0.90 Si, 0.90 Mn, bal Fe.
For upsetters, punches, crimpers; water hardened.

PLETTENBERG HW.
M-1344; 0.67 C, 1.3 Si, 0.5 N, 0.5 Cr, bal Fe.
For upsetters, punches, die blocks; oil hardened, tough.

PLETTENBERG KCR2.
M-1344; 0.20 C, 1.25 Mn, 1.15 Cr, bal Fe.
For cams bolts, camshafts; case hardened.

PLETTENBERG KCR13.
M-1344; 0.40 C, 13 Cr, 0.3 Mn, bal Fe.
Annealed: 100,00 TS; 55,000 YS; 20 El; 50 RA; 200 Brin.
For valves, cutlery, surgical and dental instruments; Type 420; stainless.

PLETTENBERG KL.
M-1344; 0.55 C, 0.9 Si, 1.05 Cr, 0.18 V, 1.85 W, bal Fe.
For header dies, forging dies; oil hardened, tough.

PLETTENBERG KLV.
M-1344; 0.61 C, 0.85 Si, 0.75 Mn, 1.18 Cr, 0.1 V, bal Fe.
For crankshafts, header dies, punches; oil hardened.

PLETTENBERG KS.
M-1344; 0.45 C, W, Cr, V, bal Fe.
For forging and header dies, upsetters; oil hardened, tough.

PLETTENBERG LC13.
M-1344; 2.1 C, 11.5 Cr, 0.30 Mn, bal Fe.
For blanking and forming dies, punches; oil hardened, nondeforming.

PLETTENBERG LC13S.
M-1344; 1.65 C, 11.5 Cr, 0.1 V, 0.30 Mn, bal Fe.
For blanking and forming dies, punches; air hardened, nondeforming.

PLETTENBERG LC13SK.
M-1344; 1.65 C, 11.5 Cr, Co, 0.30 Mn, bal Fe.
For blanking and forming dies, punches; air hardened, nondeforming.

PLETTENBERG LC13W.
M-1344; 2.1 C, 11.5 Cr, 0.70 W, 0.30 Mn, bal Fe.
For blanking and piercing dies, punches; oil hardened, nondeforming.

PLETTENBERG MO6.
M-1344; 0.85 C, W, Mo, bal Fe.
For cutters, tools, dies; oil hardened.

PLETTENBERG MO10.
M-1344; 0.80 C, Cr, Mo, W, V, bal Fe.
For cutters, tools, dies; oil hardened.

PLETTENBERG PCR1.
M-1344; 0.45 C, 0.7 Mo, 0.3 V, 1.4 Cr, bal Fe.
For forging dies, headers, upsetters; oil hardened, tough.

PLETTENBERG PCR2.
M-1344; 0.90 C, 0.8 Cr, 0.30 Mn, bal Fe.
For bearings, cutters, bushings, liners, races; water or oil hardened.

PLETTENBERG PD.
M-1344; 0.35 C, 0.9 Si, 1.05 Cr, 0.18 V, 1.85 W, bal Fe.
For heading and forging dies, punches, upsetters; oil hardened, tough.

PLETTENBERG PF.
M-1344; 0.70 C, 1.7 Si, 0.70 Mn, bal Fe.
For springs, punches; oil hardened, tough.

PLETTENBERG PG.
M-1344; 0.15 C, 0.35 Si, 0.3 Mn, bal Fe.
Annealed: 70,000 TS; 55,000 YS; 25 El; 60 RA; 145 Brin.
For screws, bolts, gears, machine tool parts; case hardened.

PLETTENBERG PG65.
M-1344; 0.45 C. 0.25-0.50 Si, 0.30-0.80 Mn, bal Fe.
Hot rolled: 98,000 TS; 59,000 YS; 24 El; 45 RA; 212 Brin.
For axles, gears, bolts, tie rods, crankshafts; water hardened.

PLETTENBERG PG85.
M-1344; 0.60 C, 0.25-0.50 Si, 0.30-0.80 Mn, bal Fe.
Heat treated: 160,000-115,000 TS; 113,000-77,000 YS; 12-23 El; 40-54 RA; 320-230 Brin.
For wheels, die blocks, griders, tie rods, springs; water hardened.

PLETTENBERG PG90.
M-1344; 0.67 C, 0.25-0.50 Si, 0.30-0.80 Mn, bal Fe.
Heat treated: 174,000-122,000 TS; 128,000-82,000 YS; 12-22 El; 37-52 RA; 352-240 Brin.
For springs, clutch discs, die blocks, girders, rails; water hardened.

PLETTENBERG PG95.
M-1344; 0.75 C, 0.25-0.50 Si, 0.30-0.80 Mn, bal Fe.
Heat treated: 180,000-125,000 TS; 130,000-85,000 YS; 10-20 El; 35-50 RA; 360-245 Brin.
For springs, clutch discs, die blocks, rails; water hardened.

PLETTENBERG PG100.
M-1344; 0.90 C, 0.25-0.50 Si, 0.30-0.80 Mn, bal Fe.
Heat treated: 180,000-130,000 TS; 120,000-80,000 YS; 10-20 El; 30-47 RA; 375-270 Brin.
For drills, hobs, taps, springs, reamers; water hardened; Type W1.

PLETTENBERG PN.
M-1344; 0.50 C, 1.05 Cr, 3.25 Ni, 0.50 N, bal Fe.
For gears, bolts, crankshafts; oil hardened, shock resistant.

PLETTENBERG PRIMA-MH.
M-1344; 1.15 C, 0.25 max Si, 0.25 max Mn, bal Fe.
Annealed: 110,000 TS; 58,000 YS; 18 El; 38 RA; 210 Brin.
For springs, taps, reamers, drills, hobs; Type W1; water hardened.

PLETTENBERG PRIMA WEICH.
M-1344; 0.70 C, 0.25 max Si, 0.25 max Mn, bal Fe.
Heat treated: 174,000-122,000 TS; 128,000-82,000 YS; 12-22 El; 37-52 RA; 350-240 Brin.
For wheels, die blocks, girders, springers; Type W1; water hardened.

PLETTENBERG PRIMA ZAH.
M-1344; 0.85 C, 0.25 max Si, 0.25 max Mn, bal Fe.
Heat treated: 188,000-129,000 TS; 143,000-87,000 YS; 12-21 El; 35-50 RA; 388-235 Brin.
For springs, taps, reamers, drills, cutters; Type W1; water hardened.

PLETTENBERG PRIMA-ZH.
M-1344; 1.0 C, 0.25 max Si, 0.25 max Mn, bal Fe.
Heat treated: 200,000 TS; 150,000 YS; 8 El; 28 RA; 410 Brin.
For springs, taps, reamers, drills, hobs; Type W1; water hardened.

PLETTENBERG RAPID EXTRA III.
M-1344; 0.95 C, W, Mo, bal Fe.
For cutters, dies, tools; oil hardened.

PLETTENBERG S111.
M-1344; 1.05 C, 1.0 Cr, 1.15 W, 0.90 Mn, bal Fe.
For heading dies, drawing and forming dies; oil hardened, tough.

PLETTENBERG S113.
M-1344; 1.45 C, 1.4 Cr, 0.60 Mn, bal Fe.
For blanking and forming dies, bearings; oil hardened, abrasion and wear resistant.

PLETTENBERG SK.
M-1344; 1.0 C, 0.1 V, 0.30 Mn, bal Fe.
For blanking and forming dies, reamers; Type W2; water hardened.

PLETTENBERG SP.
M-1344; 0.45 C, Si, Cr, V, bal Fe.
For gears, bolts, crankshafts; oil hardened, shock resistant.

PLETTENBERG SP35.
M-1344; 0.38 C, Si, Cr, V, bal Fe.
For gears, bolts, crankshafts; oil hardened, shock resistant.

PLETTENBERG SVM2.
M-1344; 0.90 C, 1.9 Mn, 0.10 V, bal Fe.
For punches, shears, blanking and forming dies; oil hardened, non-deforming.

PLETTENBERG WCR13.
M-1344; 0.20 C, 13 Cr, 0.30 Mn, bal Fe.
Annealed: 95,000 TS; 50,000 YS; 25 El; 55 RA; 195 Brin.
For valves, cutlery, surgical and dental instruments; Type 420; stainless.

PLETTENBERG WP5.
M-1344; 0.30 C, 2.35 Cr, 0.6 V, 4.25 W, bal Fe.
For shear blades, upsetters, forging dies; hot work steel, oil hardened.

PLETTENBERG WP9.
M-1344; 0.30 C, 2.6 Cr, 0.35 V, 8.5 W, bal Fe.
For extrusion rams and linrs, shears, punches; hot work steel, oil hardened.

PLETTENBERG WPK.
M-1344; 0.30 C, 2 Co, 2.4 Cr, 0.25 V, 8.5 W, bal Fe.
For extrusion rams and liners, punches; hot work steel, oil hardened.

PLETTENBERG WP SPEZIAL EXTRA.
M-1344; 0.30 C, 1.1 Cr, 0.18 V, 3.75 W, bal Fe.
For forging and heading dies, upsetters, shears; hot work steel, oil hardened.

PLETTENBERG WPV.
M-1344; 0.65 C, 3.75 Cr, 0.85 Mo, 0.70 V, 8.5 W, bal Fe.
For shear blades, cutters, upsetters; oil hardened, hot work steel.

PLOWFACE.
M-712; 4.5 C, 6.5 Mn, 2 Si, 30 Cr, bal Fe.
550 Brin.
For hard surfacing electrodes; abrasion and impact resistant.

PLOW STEEL.
0.6-0.9 C, bal Fe.
For plow shares, blacksmith tools; water hardening.

PLUMBERS SOLDER.
67 Pb, 33 Sn.
For solder; soft solder.

PLUMBERS WHITE NO 1.
54 Cu, 27 Zn, 17 Ni, 2 Pb.
For plumbing fixtures; free-cutting.

PLUMBERS WHITE NO 2.
54 Cu, 25 Zn, 13 Ni, 7 Pb, 1 Sn.
For faucets, cocks; free-cutting.

PLUMBERS WHITE NO 3.
58 Cu, 25 Zn, 15 Ni, 1 Pb, 1 Fe, 0.3 Mn.
For faucets, cocks; corrosion resistant.

PLUMBIC BRONZE.
M-England; 26 Pb, 1.7 Mn, 1.5 Sn, 1.2 Fe, bal Cu.
For bearings, bushings; heavy duty.

PLUMBSOL.
M-200; Ag, Sn.
Cast: 3350 TS; 60 El.
For soft solder for plumbing installations; M.P. 221-225°C.

PLUMRITE NO 67.
M-141; 67 Cu, 32.5 Zn, 0.5 Pb.
For water pipes, plumbing; free-cutting.

PLUMRITE 85.
M-141; 85 Cu, 15 Zn.
Annealed: 45,000 TS; 45 El.
For water pipes, plumbing; red brass.

PLUMRITE COPPER.
M-141; 99.9 Cu.
Annealed: 30,000 TS; 40 El.
Drawn: 48,000 TS; 4 El.
For piping, water tube; corrosion resistant.

PLUMRITE STANDARD 85.
M-141; 85 Cu, 15 Zn.
Hard: 70,000 TS; 57,000 YS; 5 El.
Soft: 40,000 TS; 12,000 YS; 47 El.
For flexible hose, deep drawn parts; corrosion resistant.

PLURALLOY 70.
M-849; 0.06-0.10 C, 0.6-0.8 Mn, bal Fe.
Welded: 75,000 TS; 64,000 YS; 31 El; 137 Brin.
For weldng electrodes; AWS-E6015; shielded arc, low hydrogen.

PLURALLOY 100.
M-849; 0.06-0.10 C, 0.6-0.8 Mn, 0.7-0.9 Mo, bal Fe.
Welded: 95,000 TS; 83,000 YS; 27 El; 197 Brin.
For welding electrodes; AWS-E9015; shielded arc.

PLURALLOY 120.
M-849; 0.06-0.10 C, 0.6-0.8 Mn, 0.2-0.4 Mo, 1.75-2.25 Ni, 0.1-0.2 V, bal Fe.
Welded: 110,000 TS; 106,000 YS; 22 El; 220 Brin.
For welding electrodes; shielded arc.

PLUTEOUS 1000 see CARR'S QUALITY P.1000.

PLUTOCRAT.
M-1410; 0.8 C, 4.5 Cr, 1 V, 22 W, bal Fe.
For milling cutters, drills, reamers, boring tools; high speed steel.

PLUTO G now VEW 5200.

PLUTOIL 704 see CARR'S QUALITY P.704.

PLUTONIC 151 ETC see CARR'S QUALITY P.151 ETC.

PLUTO PARAMOUNT.
M-1410; 0.78 C, 4.5 Cr, 18.75 W, 0.70 Mo, 1.25 V, 10.0 Co, bal Fe.
10% Cobalt-Tungsten high speed steel.
For tableware; corrosion resistant.

PLUTO PEERLESS.
M-1410; 1.40 C, 4.0 Cr, 3.2 Mo, 9.0 W, 3.1 V, 9.5 Co, bal Fe.
Tungsten-Molybdenum-Cobalt high speed steel.
AISI T42.

PLUTO PERFECTUM.
M-1410; 0.85 C, 5 Cr, 18 W, 5-6 Co, bal Fe.
For milling cutters, lathe and planer tools; high speed steel.

PLUTO PLUS.
M-1410; 1.1 C, 3.75 Cr, 9.5 Mo, 1.5 W, 1.2 V, 8.0 Co, bal Fe.
Molybdenum-Cobalt high speed steel.
AISI M42.

PLUTO PREMIUM.
M-1410; 1.55 C, 4.5 Cr, 3.5 Mo, 6.5 W, 5.0 V, 5.0 Co, bal Fe.
High speed steel.

PLYKROME.
M-72, M-318; composite steel plus welded stainless surface high Cr, high Ni Ferrous alloy.
For stainless steel parts, tanks, vessels; stainless steel sheet welded onto steel surface.

PLYMITE.
M-193; W alloy.
For wear resistant parts.

PM 18.
M-1740; 0.16-0.20 C, 0.30-0.50 Si, 1.35-1.55 Mn, 0.40 max Ni, 0.25 max Cr, 0.15 max Mo, 0.30 max Cu, bal Fe.
Pearlitic manganese steel.
B.S. 2772.

PM 23.
M-1740; 0.18-0.23 C, 0.50 max Si, 1.2-1.6 Mn, bal Fe.
Pearlitic manganese steel.
B.S. 1456 Gr. A+B.

PM 18/8/6.
M-1342; 0.20 max C, 18 Cr, 8 Ni, 6 Mn, bal Fe.
Wire or electrodes: 45-65 kp/mm^2 TS; 18 kp/mm^2 min YS; 15 min El.
For welding of comparable base metals.

PM 20/70 NB.
M-1342; 0.05 max C, 20 Cr, 70 Ni, 2.5 Nb, 0.5 Ti, bal Fe.
Wire or electrodes: 45-65 kp/mm^2 TS; 22 kp/mm^2 min YS; 10 min El.
For welding of high Nickel alloys.

PM 24/24 NB.
M-1342; 0.35 C, 24 Cr, 24 Ni, 1.5 Nb, bal Fe.
Wire or electrodes: 50-70 kp/mm^2 TS; 25 kp/mm^2 min YS; 8 min El.
For welding of comparable base metals.

PM 25/20 R.
M-1342; 0.40 C, 25 Cr, 20 Ni, bal Fe.
Wire or electrodes: 50-70 kp/mm^2 TS; 25 kp/mm^2 min YS; 5 min El.
For welding of comparable base metals.

PM 25/35 NB.
M-1342; 0.40 C, 25 Cr, 35 Ni, 1.5 Nb, bal Fe.
Wire or electrodes: 50-70 kp/mm^2 TS; 25 kp/mm^2 min YS; 8 min El.
For welding of comparable base metals.

PM 25/35 R.
M-1342; 0.40 C, 25 Cr, 35 Ni, bal Fe.
Wire or electrodes: 50-70 kp/mm^2 TS; 25 kp/mm^2 min YS; 5 min El.
For welding of comparable base metals.

PM 26/36 MO.
M-1342; 0.45 C, 26 Cr, 36 Ni, 1.5 Ma, bal Fe.
Wire or electrodes: 50-70 kp/mm^2 TS; 25 kp/mm^2 min YS; 5 min El.
For welding of comparable base metals.

PM 28/48/5.
M-1342; 0.10 max C, 28 Cr, 48 Ni, 5 W, bal Fe.
Wire or electrodes: 45-65 kp/mm^2 TS; 18 kp/mm^2 min YS; 3 min El.
For welding of comparable base metals.

PM 30/30.
M-1342; 0.45 C, 30 Cr, 30 Ni, bal Fe.
Wire or electrodes: 50-70 kp/mm^2 TS; 25 kp/mm^2 min YS; 4 min El.
For welding of comparable base metals.

PMB 20 ETC see BOHNALLOY PMB 20 ETC.

PMD-45.
M-435; 0.40 C, 0.45 Mn, 0.3 Si, 5.0 Cr, 2.25 Mo, 1.0 V, bal Fe.
Hot work tool steel.

PM IN-792.
M-68; 0.10 max C, 12.5 Cr, 2.0 Mo, 9.3 Co, 3.2 Al, 4.3 Ti, 4.0 W, 0.015 B, 0.1 Zr, bal Ni.
Ht. Tr. room: 1400 N/mm^2 UTS; 1100 N/mm^2 YS; 8 El.
1400°F: 1000 N/mm^2 UTS; 950 N/mm^2 YS; 20 El.
For special elevated temperature operations.

PMP-Z-70.
M-1098; Cu alloy.
For cams, latches, keys, bushings; powder metals; sintered.

PM UMCO 50.
M-1342; 0.10 max C, 28 Cr, 50 Co, bal Fe.
Wire or electrodes: 45-65 kp/mm^2 TS; 18 kp/mm^2 min YS; 3 min El.
For welding of comparable base metals.

P N BRONZE.
M-1169; 3.0-4.0 Ni, 0.05-0.30 P, 11.5-13.0 Sn, bal Cu.
Cast: 31,000-45,000 TS; 5-20 El; 70-100 Brin.
For pump and valve components.
Good for high pressure work.

PNEUMO.
M-1405; 0.45 C, 1.25 Cr, 0.2 W, 0.35 Mn, 0.75 Si, bal Fe.
For upsetters, rivet sets; oil hardened, tough.

POBEDIT.
M-USSR; 90 WC, 10 Co.
For cutting tools; cemented.

POBEDIT-ALPHA.
M-USSR; TiC, Co.
For cutting tools; cemented.

POFORS SR1855.
M-1184; 1 C, 1.5 Si, 1 Cr, bal Fe.
For cutting and threading tools; oil hardened.

POLARIS 2.
M-1705; 0.06 C, 0.3 Mn, 0.15 Si, 0.95 Cr, 0.25 Mo, (boron added), bal Fe.
Steel for plastic molds. AISI P2.

POLARIS 20.
M-1705; 0.37 C, 1.0 Mn, 0.3 Si, 1.25 Cr, 0.15 V, 0.35 Mo, bal Fe.
Steel for plastic molds. AISI P 20.

POLDI.
M-96; 1.47 C, 0.23 Mn, 4.11 W, bal Fe.
For cutting tools; oil hardening.

POLDI 2.
M-96; 1.2-1.35 C, 0.3 Mn, 0.2 Si, bal Fe.
For tools, files, drills; water hardening.

POLDI 3.
M-96; 1.05-1.2 C, 0.3 Mn, 0.2 Si, bal Fe.
For tools, files, drills; water hardening.

POLDI 3 C.
M-96; 1.15 C, 0.15 Cr, bal Fe.
For tools, drills, taps; water hardening.

POLDI 4.
M-96; 0.90-1.05 C, 0.3 Mn, 0.2 Si, bal Fe.
For tools, drills, files; water hardening.

POLDI 5.
M-96; 0.75-0.90 C, 0.32 Mn, 0.20 Si, bal Fe.
For tools, files, drills; water hardening.

POLDI 6.
M-96; 0.65-0.75 C, 0.35 Mn, 0.20 Si, bal Fe.
For tools, files, drills; water hardening.

POLDI 212 D2.
M-96; 0.3 C, 2.75 Cr, 4.35 W, 0.55 V, bal Fe.
For tools, dies; hot work steel.

POLDI 425.
M-96; 0.3 C, 1.1 Si, 1.45 Cr, 3.75 W, bal Fe.
For tools, dies; hot work steel.

POLDI 702D.
M-96; 0.45 C, 0.4 Mn, 3.1 Si, 0.85 Cr, bal Fe.
For gears, shafts; oil hardening.

POLDI 1555.
M-96; 0.12 C, 0.45 Mn, 5 Cr, 0.43 Mo, bal Fe.
For corrosion resistant parts; corrosion resistant.

POLDI 1888.
M-96; 0.70 C, 0.55 Mn, 0.05 V, bal Fe.
For tools, springs, punches; water hardening.

POLDI 2002 E S.
M-96; 1.5 C, 13 Cr, bal Fe.
For highly stressed tools subject to severe abrasive wear; oil or air hardening.

POLDI 2002 SPECIAL.
M-96; 2.2 C, 13 Cr, bal Fe.
For tools, dies; non-deforming.

POLDI 2514.
M-96; 0.28-0.35 C, 1.2-1.5 Mn, bal Fe.
For shafts, dies; water hardening.

POLDI 2518.
M-96; 0.33-0.40 C, 1.1-1.4 Mn, 1.1-1.4 Si, bal Fe.
For dies, chisels, punches; shock resistant.

POLDI 2526.
M-96; 0.36 C, 1.7 Mn, 0.3 Cr, bal Fe.
For dies, punches; oil hardening, non-deforming.

POLDI 2526N.
M-96; 0.33-0.40 C, 1.6-1.9 Mn, bal Fe.
For dies, punches; non-deforming.

POLDI AK-1.
M-96; 0.10 C, 13 Cr, bal Fe.
For corrosion resistant parts; corrosion resistant.

POLDI AK1B.
M-96; 0.10 C, 16 Cr, 0.3 Ni, 0.2 Mo, 1.8 Cu, bal Fe.
Annealed: 75,000-85,000 TS; 40,000-55,000 YS; 30-25 El; Rockwell B75-80.
For chemical plant equipment, trim; stainless.

POLDI AK1B EXTRA.
M-96; 0.09 C, 17.5 Cr, 0.5 Ti, bal Fe.
Annealed: 75,000-85,000 TS; 40,000-55,000 YS; 30-25 El; Rockwell B75-80.
For chemical plant equipment; Type 430; corrosion resistant.

POLDI AK1V.
M-96; 0.12 max C, 14-18 Cr, 0.07 min S or Se, bal Fe.
Annealed: 75,000-85,000 TS; 40,000-55,000 YS; 30-25 El; Rockwell B75-80.
For chemical plant equipment, screw machine products; Type 430 F; free-cutting, stainless.

POLDI AK1W.
M-96; 0.07 C, 0.5 Mn, 13 Cr, bal Fe.
For corrosion resistant parts, corrosion resistant.

POLDI-AK2.
M-96; 0.20 C, 16 Cr, bal Fe.
Annealed: 75,000 TS; 45,000 YS; 28 El; 145 Brin.
Cold Rolled: 120,000 TS; 90,000 YS; 10 El; 195 Brin.
For automotive trim, oil burner parts, septic tanks, hardware. Corrosion resistant.

POLDI-AK2MV.
 M-96; 0.20 C, 0.9 Mn, 0.1 Si, 11.2 Cr, 2.4 Mo, 0.1 Ni, 0.25 V, bal Fe.
 Annealed: 100,000 TS; 45,000 YS; 20 El; 50 RA; B 95 Rock.
 Heat Treated: 250,000 TS; 215,000 YS; 8 El; 25 RA; C 52 Rock.
 For surgical instruments, cutlery, gears, shafts. Corrosion resistant, hardenable.

POLDI AK2S.
 M-96; 0.22 C, 0.5 Mn, 13 Cr, bal Fe.
 For corrosion resistant parts; corrosion resistant.

POLDI AK 2 SPECIAL.
 M-96; 0.22 C, 0.5 Mn, 15.2 Cr, bal Fe.
 For corrosion resistant parts; corrosion resistant.

POLDI AK3S.
 M-96; 0.33 C, 13 Cr, 0.5 Mn, bal Fe.
 Annealed: 88,000 TS; 40,000 YS; 32 El; 68 RA; 170 Brin.
 Heat treated: 256,000 TS; 190,000 YS; 6 El; 10 RA; 540 Brin.
 For valve trim, surgical instruments, cutlery; Type 420; hardenable, corrosion resistant.

POLDI AK-5.
 M-96; 0.50 C, 15.5 Cr, 0.40 Mo, bal Fe.
 For corrosion resistant parts; corrosion resistant.

POLDI AKC.
 M-96; 0.20 C, 24 cr, 19.5 Ni, bal Fe.
 Annealed: 95,000 TS; 45,000 YS; 50 El; 65 RA; 180 Brin.
 At 1200°F: 57,000 TS; 22,000 YS; 32 El; 45 RA.
 For furnace parts and equipment, heat treating boxes; Type 310; austenitic, heat resistant.

POLDI AKCF.
 M-96; 0.10 C, 21 Cr, 10 Ni, bal Fe.
 Annealed: 85,000 TS; 35,000 YS; 55 El; 65 RA; 150 Brin.
 For valve trim, oil refinery and chemical plant equipment; Type 308; austenitic, corrosion resistant.

POLDI AKCM.
 M-96; 0.45 C, 18 Cr, 19.5 Ni, 3.25 Mo, bal Fe.
 For corrosion and heat resistant parts; corrosion and heat resistant.

POLDI AKCR.
 M-96; 0.15 C, 25 Cr, 14.5 Ni, bal Fe.
 Annealed: 85,000-95,000 TS; 40,000-50,000 YS; 45-55 RA; 150-185 Brin.
 For refinery and chemical plant equipment; Type 309; austenitic, heat resistant.

POLDI AK-H.
 M-96; 1.0 C, 16 Cr, 1 Mo, 1.3 Co, bal Fe.
 For corrosion resistant parts; corrosion resistant.

POLDI AKL1.
 M-96; 0.15 C, 0.6 Mn, 12.3 Cr, 10.5 Ni, bal Fe.
 For stainless parts; corrosion resistant.

POLDI AKM.
 M-96; 0.07 C, 18 Mn, 0.8 Si, 10 Cr, 1 Ni, 1 Mo, bal Fe.
 For wear and corrosion resistant parts; corrosion resistant.

POLDI AKR.
 M-96; 0.45 C, 1.5 Si, 12.5 Cr, 2.2 W, 12.5 Ni, bal Fe.
 For stainless parts; corrosion resistant.

POLDI AKS.
 M-96; 0.35 C, 0.5 Mn, 4.5 Cr, 22 Ni, bal Fe.
 For stainless parts; corrosion resistant.

POLDI AKS 2.
 M-96; 0.10 C, 4.8 Cr, 2 W, 25 Ni, 3.5 Mo, 2.5 Cu, bal Fe.
 For acid resistant equipment; corrosion resistant.

POLDI AKV.
 M-96; 0.15 C, 18.5 Cr, 9 Ni, bal Fe.
 Annealed: 85,000 TS; 35,000 YS; 60 El; 70 RA; 150 Brin.
 Cold drawn: 125,000 TS; 95,000 YS; 25 El; 55 RA; 277 Brin.
 For chemical and plastic plant equipment; Type 302; stainless, austenitic.

POLDI AKV EXTRA.
 M-96; 0.07 C, 18.5 Cr, 9 Ni, 1.65 Mo, bal Fe.
 Annealed: 95,000 TS; 45,000 YS; 50 El; 60 RA; 195 Brin.
 For chemical and plastic plant equipment; Type 316; austenitic, stainless.

POLDI AKV EXTRA S.
 M-96; 0.10 max C, 16-18 Cr, 10-14 Ni, 1.7-2.7 Mo, Cb = 10 x C, bal Fe.
 Annealed: 85,000-95,000 TS; 35,000-45,000 YS. 60-50 El; 75-60 RA; 150-190 Brin. For acid resistant chemical plant equipment; Type 316 Cb; stainless, austenitic.

POLDI AKVH.
 M-96; 0.08 max C, 18-20 Cr, 8-11 Ni, 2 max Mn, bal Fe.
 Annealed: 90,000 TS; 45,000 YS; 60 El; 135 Brin.
 Cold drawn: 180,000 TS; 150,000 YS; 10 El; 330 Brin.
 For chemical plant equipment, welded structures; Type 304; stainless, austenitic.

POLDI AKVM.
 M-96; 0.10 C, 18 Cr, 9 Ni, bal Fe.
 Annealed: 90,000 TS; 35,000 YS; 60 El; 150 Brin.
 For chemical plant equipment, tanks, digesters, agitators, molding and trim.
 Type 302 stainless steel. Austenitic.

POLDI AKVN.
  M-96; 0.10 C, 18.5 Cr, 9 Ni, bal Fe.
  Annealed: 85,000 TS; 35,000 YS; 60 El; 70 RA; 150 Brin.
  Cold drawn: 125,000 TS; 95,000 YS; 25 El; 55 RA; 260 Brin.
  For chemical plant equipment, oil refinery tanks; Type 302; austenitic, stainless.

POLDI AKVS.
  M-96; 0.10 C, 18.5 Cr, 8.5 Ni, 0.6 Ti, bal Fe.
  Annealed: 85,000 TS; 35,000 YS; 60 El; 70 RA; 150 Brin.
  For welded structures, chemical plant equipment; Type 321; austenitic, stainless.

POLDI AKVU.
  M-96; 0.15 C, 18.5 Cr, 9 Ni, bal Fe.
  Annealed: 85,000 TS; 35,000 YS; 50 El; 65 RA; 150 Brin.
  For chemical plant equipment; Type 302; austenitic, stainless.

POLDI AKX.
  M-96; 0.10 C, 0.5 Mn, 25 Cr, bal Fe.
  For heat resistant parts; heat resistant.

POLDI AL 14.
  M-96; 0.44 C, 1.5 Cr, 0.23 Mo, 0.95 Al, bal Fe.
  For gears, cams, shafts; nitriding steel.

POLDI AL 16.
  M-96; 0.35 C, 0.75 Mn, 1.65 Cr, 1.1 Al, bal Fe.
  For gears, cams, shafts; nitriding steel.

POLDI AM.
  M-96; 0.7 C, 8.9 Mn, 3.15 Cr, 7.7 Ni, bal Fe.
  For corrosion resistant parts; corrosion resistant.

POLDI AMK.
  M-96; 0.4 C, 18 Mn, 3.3 Cr, bal Fe.
  For wear and heat resistant parts; wear and heat resistant.

POLDI AMS.
  M-96; 0.55 C, 5.25 Mn, 3.55 Cr, 12.15 Ni, bal Fe.
  For heat and corrosion resistant parts; heat and corrosion resistant.

POLDI AUTOR.
  M-96; 0.40-0.50 C, 1.4-1.7 Cr, 0.6 Mn, 0.4 max Si, bal Fe.
  For bolts, gears, crankshafts; oil hardened, tough.

POLDI B 200.
  M-96; 0.38 C, 1.9 Mn, 3.75 Cr, 0.18 V, bal Fe.
  For dies, punches; oil hardening.

POLDI BO-4.
  M-96; 0.35 C, 0.6 Cr, 1.45 Ni, bal Fe.
  For gears, shafts; shock resistant.

POLDI BOZ.
  M-96; 0.39 C, 0.6 Mn, 1.1 Cr, 2 Ni, 0.15 Mo, bal Fe.
  For gears, shafts; oil hardening.

POLDI BRAND NO 4.
  M-96; 1.0 C, bal Fe.
  For general tools, cold work dies, stamps, punches; water hardening.

POLDI CE.
  M-96; 0.10 C, 0.55 Mn, 0.75 Cr, bal Fe.
  For gears, shafts; case hardening.

POLDI CM 2.
  M-96; 0.20 C, 1.05 Mn, 1.2 Cr, 0.23 Mo, bal Fe.
  For gears, shafts; tough.

POLDI CM 3.
  M-96; 0.22-0.30 Cv, 0.9-1.2 Cr, 0.15-0.25 Mo, bal Fe.
  For gears, shafts; tough.

POLDI CNB.
  M-96; 0.38 C, 1.5 Cr, 0.6 W, 4.25 Ni, 0.3 Mo, bal Fe.
  For tools, gears, shafts; tough, oil hardening.

POLDI CNBD.
  M-96; 0.38 C, 1.5 Cr, 3.5 Ni, bal Fe.
  For tools, gears, shafts; tough, shock resistant.

POLDI CNF.
  M-96; 0.35 C, 0.9 Cr, 4.85 Ni, bal Fe.
  For tools, gears, shafts; tough, oil hardening.

POLDI CNH SPECIAL.
  M-96; 0.55 C, 0.5 Mn, 0.75 Cr, 2.6 Ni, bal Fe.
  For tools, gears, shafts; tough, shock resistant.

POLDI CNL.
  M-96; 0.3 C, 0.85 Cr, 4.85 Ni, bal Fe.
  For gears, shafts; oil hardening.

POLDI CNLW.
  M-96; 0.27 C, 0.4 Mn, 0.85 Cr, 4.5 Ni, bal Fe.
  For gears, shafts; oil hardening.

POLDI CNS.
  M-96; 0.33 C, 0.8 Cr, 3.35 Ni, bal Fe.
  For gears, shafts; oil hardening, tough.

POLDI CNSW.
  M-96; 0.25 C, 0.8 Cr, 3.35 Ni, bal Fe.
  For gears, shafts; tough, shock resistant.

POLDI C R.
  M-96; 0.9 C, 0.75 Cr, 0.1 V, bal Fe.
  For dies, punches, cold rolls; deep hardening, water quenched.

POLDI CR 1.
  M-96; 0.90 C, 0.35 Mn, 1.0 Cr, 1 V, bal Fe.
  For tools, bearings, dies; wear resistant.

POLDI CR 2.
  M-96; 0.8 C, 0.35 Mn, 1.7 Cr, bal Fe.
  For tools, dies, cutter; oil hardening.

POLDI CR2W.
  M-96; 0.7 C, 0.35 Mn, 1.7 Cr, bal Fe.
  For tools, dies; oil hardening.

POLDI CV.
M-96; 0.6 C, 1.0 Cr, 0.2 V, 0.5 Mn, bal Fe.
For tools, punches; oil hardened.

POLDI CX.
M-96; 0.3 C, 0.9 Si, 2.45 Cr, 0.35 V, bal Fe.
For tools, punches, dies; tough, oil hardening.

POLDI D.
M-96; 1.05-1.2 C, 0.3 Mn, bal Fe.
For tools, files, drills; water hardening.

POLDI DIAMANTHART.
M-96; 1.35 C, 0.55 Cr, 4.9 W, 0.35 V, bal Fe.
For tools, dies.

POLDI DS SPECIAL.
M-96; 1.15 C, 1.0 Cr, 0.18 V, bal Fe.
For tools, cutters, bearings; oil hardening.

POLDI DUPLEX EXTRA.
M-96; 1.05 C, 0.25 Cr, 1.15 W, 0.2 V, bal Fe.
For tools, dies.

POLDI E.
M-96; 0.90-1.05 C, 0.3 Mn, bal Fe.
For tools, files, drills; water hardening.

POLDI EPAZ.
M-96; 1.25 C, 0.12 Cr, 0.15 V, bal Fe.
For tools, cutters, bearings; water hardening.

POLDI ESH.
M-96; 0.6 C, 0.9 Mn, 1.9 Si, bal Fe.
For punches, dies, chisels; tough.

POLDI ESH 1.
M-96; 0.65 C, 0.75 Mn, 1.75 Si, bal Fe.
For punches, dies, chisels; tough.

POLDI EXTRA F-GUB.
M-96; 1.7 C, 0.4 Si, 18 Cr, 1.25 W, 1.4 Ni, bal Fe.
For chemical plant equipment; corrosion resistant.

POLDI E Z H.
M-96; 0.65-1.05 C, bal Fe.
Annealed: 91,000 TS; 180 Brin.
For cold dies, punches, tools, short shear blades; tool steel.

POLDI EZH SPECIAL.
M-96; 1.0 C, 0.35 Mn, 0.15 V, bal Fe.
For tools, drills, taps; water hardening.

POLDI GFE.
M-96; 0.6 C, 25 Cr, bal Fe.
For heat and corrosion resistant parts; heat resistant.

POLDI GS3.
M-96; 0.38 C, 1.4 Mn, 2.5 Cr, 0.28 Mo, bal Fe.
For gears, dies, shafts; oil hardening.

POLDI HP51.
M-96; 0.5 C, 2 W, 0.8 Cr, 0.2 V, bal Fe.
For hot work tools and dies; hot work steel.

POLDI H P S.
M-96; 0.5 C, 3 W, bal Fe.
For tools for hot work; hot work steel.

POLDI HPS 2.
M-96; 0.25 C, 2.25 Cr, 9.5 W, 1.5 Ni, 0.1 V, bal Fe.
For tools, dies; hot work steel.

POLDI HS.
M-96; 1.3 C, 13.75 Mn, 0.25 Si, bal Fe.
For wear resistant parts; austenitic, wear resistant.

POLDI K 1.
M-96; 1.2-1.35 C, 0.25 Mn, 0.12 Si, bal Fe.
For tools, drills, files; water hardening.

POLDI K 2.
M-96; 1.0-1.2 C, 0.25 Mn, 0.12 Si, bal Fe.
For tools, files, drills; water hardening.

POLDI KAPTOR.
M-96; 0.35 C, 1.5 Si, 12.5 Cr, 2.6 W, 12.5 Ni, 1.25 V, bal Fe.
For stainless parts; corrosion resistant.

POLDI KO.
M-96; 1.35-1.50 C, 0.2 Mn, 0.12 Si, bal Fe.
For tools, drills, files; water hardening.

POLDI L-AK2.
M-96; 0.20 C, 0.5 Mn, 15.2 Cr, bal Fe.
For corrosion resistant parts; corrosion resistant.

POLDI L-AKMF 3.
M-96; 0.09 C, 17.5 Mn, 1.5 Si, 11.5 Cr, 0.5 Ti, bal Fe.
For corrosion and wear resistant parts; corrosion resistant.

POLDI L-AKR.
M-96; 0.45 C, 1.5 Si, 12.5 Cr, 2.2 W, 12.5 Ni, bal Fe.
For stainless parts; corrosion resistant.

POLDI L-AKRD.
M-96; 0.45 C, 0.75 Mn, 1.5 Si, 15 Cr, 2.3 W, 13 Ni, bal Fe.
For stainless parts; corrosion resistant.

POLDI L-AL 12.
M-96; 0.33 C, 0.45 Mn, 1.8 Cr, 0.18 Mo, 1 Ni, 1.1 Al, bal Fe.
For machinery parts, gears, shafts; nitriding steel.

POLDI L-AL 14.
M-96; 0.44 C, 0.55 Mn, 1.5 Cr, 0.23 Mo, 0.95 Al, bal Fe.
For machinery parts, gears, shafts; nitriding steel.

POLDI L-AL 16.
M-96; 0.35 C, 0.75 Mn, 1.65 Cr, 1.1 Al, bal Fe.
For machinery parts, gears, shafts; nitriding steel.

POLDI L-AL 30.
 M-96; 0.32 C, 0.70 Mn, 1.1 Cr, 0.18 Mo, 1.15 Mo, bal Fe.
 For machinery parts, gears, shafts; nitriding steel.

POLDI L-CNL.
 M-96; 0.3 C, 0.85 Cr, 4.85 Ni, bal Fe.
 For gears, shafts; oil hardening.

POLDI L-CNS.
 M-96; 0.27 C, 0.55 Cr, 3.25 Ni, bal Fe.
 For gears, shafts; tough, shock resistant.

POLDI L-CVMW.
 M-96; 0.27-0.34 C, 1.9-2.7 Cr, 0.25 Mo, 0.1-0.3 V, bal Fe.
 For gears, shafts; oil hardening, tough.

POLDI LDH.
 M-96; 0.30 C, 0.7 Cr, 4.25 Ni, 1.5 Mo, bal Fe.
 For tools, gears, shafts; oil hardening, tough.

POLDI LDH 3 SPECIAL.
 M-96; 0.3 C, 0.7 Cr, 2.4 Ni, 0.75 Mo, 0.3 V, bal Fe.
 For tools, gears, shafts; tough, oil hardening.

POLDI LDH 4 SPECIAL.
 M-96; 0.3 C, 0.7 Cr, 1.4 Ni, 0.45 Mo, 0.22 V, bal Fe.
 For tools, gears, shafts; oil hardening, tough.

POLDI LDS.
 M-96; 0.45 C, 0.5 Mn, 1.45 Si, 0.6 Cr, bal Fe.
 For tools, dies, shears; tough, oil hardening.

POLDI L-MS.
 M-96; 0.29 C, 18 Mn, 0.5 Si, 1.3 Cr, bal Fe.
 For wear resistant parts; corrosion resistant, austenitic.

POLDI L-MV 4.
 M-96; 0.38-0.42 C, 1.5-2 Mn, 0.4 Si, 0.10-0.20 V, bal Fe.
 For dies, punches; non-deforming.

POLDI L-NIT 2.
 M-96; 0.30 C, 0.6 Mn, 2.5 Cr, 0.25 Mo, 0.25 V, bal Fe.
 For machinery parts, gears, shafts; nitriding steel.

POLDI L-NIT 4.
 M-96; 0.27 C, 0.6 Mn, 2.5 Cr, 0.30 V, bal Fe.
 For machinery parts, gears, shafts; nitriding steel.

POLDI MAXIMUM SPECIAL.
 M-96; 0.7 C, 19 W, 4 Cr, 1.2 V, bal Fe.
 For turning, planing tools, twist drills, milling cutters, boring and milling tools, screwing dies; high speed steel.

POLDI MAXIMUM SPECIAL G.
 M-96; 0.8 C, 4 Cr, 38 W, 2 V, bal Fe.
 For tools, cutters; high speed steel.

POLDI MAXIMUM SPECIAL G EXTRA.
 M-96; 0.85 C, 4.2 Cr, 9.75 W, 2.4 V, bal Fe.
 For tools, cutters; high speed steel.

POLDI MAXIMUM SPECIAL H.
 M-96; 0.8 C, 4.2 Cr, 17.5 W, 0.6 Mo, 1.35 V, bal Fe.
 For tools, cutters; high speed steel.

POLDI MAXIMUM SPECIAL MO.
 M-96; 1.0 C, 5.2 Cr, 1.7 W, 2.4 Mo, 2.85 V, bal Fe.
 For tools, cutters; high speed steel.

POLDI MAXIMUM SPECIAL NO 30.
 M-96; 0.7 C, 17.5 W, 4.2 Cr, 0.6 Mo, 1.4 V, 2.7 Co, bal Fe.
 For turning, planing and slotting tools, twist drills, milling cutters, boring and milling tools; high speed steel.

POLDI MAXIMUM SPECIAL NO 55.
 M-96; 0.7 C, 17.5 W, 4 Cr, 0.6 Mo, 1.4 V, 5 Co, bal Fe.
 For turning, planing and slotting tools, flat drills, twist drills, milling cutters, boring and milling tools; high speed steel.

POLDI M K.
 M-96; 0.65 C, 18 W, 5 Co, 4 Cr, 1.3 V, 0.7 Mo, bal Fe.
 For turning, planing, drilling and milling tools, flat drills, twist drills, boring and milling cutters; high speed steel.

POLDI MO.
 M-96; 0.3-0.4 C, 0.5 Mn, 0.2 Si, bal Fe.
 For welding rod.

POLDI NO 25 BEST.
 M-96; 0.85 C, bal Fe.
 For tools, drills, springs; water hardened.

POLDI NO 212.
 M-96; 0.3 C, 9.5 W, 2.5 Cr, 0.1 V, bal Fe.
 For hot work tools; hot work steel.

POLDI NO 2002.
 M-96; 2 C, 12 Cr, bal Fe.
 Annealed: 102,000 TS; 205 Brin.
 For cutting dies, punches, milling cutters, reamers, broaches; for highly stressed parts.

POLDI ORI.
 M-96; 1.25 C, 0.7 Mn, 1.15 Si, 1.2 Cr, bal Fe.
 For tools, chisels; shock resistant.

POLDI RADECO D.
 M-96; 0.85 C, 4.2 Cr, 12.5 W, 0.6 Mo, 2 V, 5 Co, bal Fe.
 For tools, cutters, hobs; abrasion resistant.

POLDI RCR-1.
 M-96; 1.5 C, 0.25 Mn, 0.75 Cr, bal Fe.
 For tools, cutters; wear resistant.

POLDI REDI.
M-96; 0.45 C, 0.5 Mn, 1.45 Si, 0.6 Cr, bal Fe.
For tools, dies, shears; tough, oil hardening.

POLDI S.
M-96; 0.70-0.80 C, 0.3 Mn, bal Fe.
For tools, files, drills; water hardening.

POLDI S 4.
M-96; 0.6-1.05 C, 0.05 Si, bal Fe.
For welding rod.

POLDI S 6.
M-96; 0.65-0.75 C, 0.05 Si, bal Fe.
For welding rod.

POLDI SC.
M-96; 0.6 C, 0.7 Mn, 1.55 Si, 0.8 Cr, bal Fe.
For tools, dies, shears; tough, oil hardening.

POLDI SCH.
M-96; 0.55 C, 0.7 Mn, 1.45 Si, 0.6 Cr, bal Fe.
For punches, dies, chisels; oil hardening.

POLDI SCM.
M-96; 0.4-0.55 C, 1.5-2.0 Mn, bal Fe.
For shafts, dies; oil hardening.

POLDI SP.
M-96; 1.2 C, 0.7 W, bal Fe.
For tools, dies.

POLDI SPS.
M-96; 1.2 C, 0.4 Cr, 1.15 W, bal Fe.
For tools, dies.

POLDI SR 4.
M-96; 0.90-1.05 C, 0.2 Si, bal Fe.
For welding rod.

POLDI S SPECIAL.
M-96; 0.75 C, 0.3 Mn, 0.15 V, bal Fe.
For tools, punches, springs; water hardening.

POLDI S S T.
M-96; 1 C, 1.2 W, bal Fe.
For taps, dies, punches, shop tools; water hardening.

POLDI T3.
M-96, M-1322, M-1323; 1.1 C, 0.25 max Mn, 0.25 max Si, bal Fe.
Heat treated: 125,000-220,000 TS; 88,000-155,000 YS; 18-10 El; 45-30 RA; 240-555 Brin.
For drills, taps, reamers, broaches; Type W1; water hardenend.

POLDI T4.
M-96, M-1322, M-1323; 1.0 C, 0.25 max Si, 0.25 max Mn, bal Fe.
Heat treated: 121,000-213,000 TS; 84,000-150,000 YS; 20-11 El; 47-33 RA; 235-534 Brin.
For drills, taps, reamers, broaches; Type W1; water hardened.

POLDI T5.
M-96, M-1322, M-1323; 0.85 C, 0.25 max Mn, 0.25 max Si, bal Fe.
Heat treated: 188,000-130,000 TS; 143,000-87,000 YS; 12-21 El; 35-50 RA; 388-255 Brin.
For tools, cutters, springs; Type W1; water hardened.

POLDI T5 HART.
M-1322, M-96; 0.61 C, 0.65 Mn, 0.25 Si, bal Fe.
Hot rolled: 118,000 TS; 70,000 YS; 16 El; 35 RA; 229 Brin.
For punches, crimpers, axles; water hardened.

POLDI T5W.
M-96, M-1322; 0.60 C, 0.5 Si, 0.6 Mn, bal Fe.
Hot rolled: 118,000 TS; 70,000 YS; 16 El; 35 RA; 229 Brin.
For punches, crimpers, axles; water hardened.

POLDI T5W EXTRA.
M-96; 0.52-0.60 C, 0.25 Si, bal Fe.
For welding rod.

POLDI T6.
M-96, M-1322; 0.7 C, 0.25 max Si, 0.25 max Mn, bal Fe.
Heat treated: 122,000-174,000 TS; 82,000-128,000 YS; 22-12 El; 52-37 RA; 241-352 Brin.
For punches, crimpers, axles; water hardened.

POLDI T6H.
M-96, M-1322; 0.45 C, 0.25 Si, 0.65 Mn, bal Fe.
Hot rolled: 94,000 TS; 60,000 YS; 26 El; 50 RA; 200 Brin.
For gears, pinions, shafts, crankshafts; water hardened.

POLDI T6HB.
M-96; 0.48-0,54 C, 0.5 Mn, 0.2 Si, bal Fe.
For welding rod.

POLDI T6H EXTRA.
M-96; 0.42-0.50 C, 0.75 Mn, 0.25 Si, bal Fe.
For welding rod.

POLDI T6W.
M-96, M-1322; 0.35 C, 0.25 Si, 0.55 Mn, bal Fe.
Hot rolled: 85,000 TS; 54,000 YS; 30 El; 53 RA; 183 Brin.
For gears, shafts, crankshafts; water hardened.

POLDI T6W EXTRA.
M-96, M-1322; 0.35 C, 0.25-0.60 Si, 0.3-0.8 Mn, bal Fe.
Hot rolled: 85,000 TS; 54,000 YS; 30 El; 53 RA; 183 Brin.
For gears, shafts, crankshafts, bolts; water hardened, SAE 1035.

POLDI T7.
M-96, M-1322; 0.22 C, 0.25 Si, 0.45 Mn, bal Fe.
Heat treated: 71,000-80,000 TS; 36,000-50,000 YS; 30-20 El; 69-60 RA; 143-163 Brin.
For gears, bolts, shafts, camshafts; case hardening steel.

POLDI T8.
 M-96, M-1322; 0.15 C, 0.25 Si, 0.45 Mn, bal Fe.
 Hot rolled: 61,000 TS; 46,000 YS; 39 El; 61 RA; 126 Brin.
 For gears, pinions, camshafts, cams; case hardening steel.

POLDI TBOS.
 M-96; 0.05 C, 0.8 Cr, 1.45 Ni, bal Fe.
 For gears, shafts; oil hardening.

POLDI TO EXTRA SIMPLEX.
 M-96; 0.8 C, 1 Mn, 0.6 Si, 0.4 Cr, bal Fe.
 For tools, dies, cutters; water hardening.

POLDI W8.
 M-96; 0.06-0.13 C, 0.5 max Mn, 0.35 max Si, bal Fe.
 Annealed: 58,000 TS; 45,000 YS; 30 El; 66 RA; 125 Brin.
 For nails, rivets, screws, bolts; water and case hardened.

POLDI WP.
 M-96; 0.4 C, 0.65 Mn, 2.5 Cr, 0.6 Mo, 0.4 V, bal Fe.
 For axes, shafts; oil hardened.

POLISH 2H18N9.
 M-Poland; 0.2 C, 0.7 Mn, 18 Cr, 8.3 Ni, 0.9 Mo, bal Fe.
 Annealed: 90,000 TS; 40,000 YS; 55 El; 155 Brin.
 For chemical plant equipment, agitators, digesters, tanks.
 Stainless, austenitic, nonmagnetic.

POLITAL.
 M-457; 0.5-2.0 Mg, 0.3-1.5 Si, 0-1.5 Mn, bal Al.
 For light alloy parts; corrosion resistant, hardenable.

POLITAL.
 M-1548; 0.5-1.5 Si, 0.4-1.0 Mn, 0.4-1.0 Mg, bal Al.
 Hardened: 40,000-50,000 TS; 26,000-36,000 YS; 10-20 El; 70-100 Brin.
 For general structures, scaffolds, booms, transmission towers.
 Heat treatable, high strength.

POLITAL 38.
 M-457; 2.5-4.0 Mg, bal Al.
 Soft: 26,500 TS; 12,000 YS; 26 El; 60 Brin.
 Hard: 43,000 TS; 37,000 YS; 4 El; 85 Brin.
 For light alloy parts; corrosion resistant.

POLITIT.
 M-1338; 0.40 C, 13 Cr, 0.3 Mn, 0.4 Si, bal Fe.
 Annealed: 95,000 TS; 50,000 YS; 25 El; 55 RA; 196 Brin.
 For turbine blades, cutlery, valves, knives, fasteners; Type 420; stainless, hardenable.

POLITIT 13M.
 M-1338; 0.2 C, 13 Cr, 1.15 Mo, bal Fe.
 Annealed: 95,000 TS; 50,000 YS; 25 El; 55 RA; 200 Brin.
 For acid resistant and oil refinery equipment; corrosion resistant.

POLITIT 17M.
 M-1338; 0.35 C, 16.5 Cr, 1.15 Mo, bal Fe.
 For chemical plant and oil refinery equipment; corrosion resistant.

POLITIT 17N.
 M-1338; 0.22 C, 17 Cr, 1.5 Ni, bal Fe.
 Annealed: 125,000 TS; 95,000 YS; 20 El; 55 RA; 240 Brin.
 For furnace parts and accessories; corrosion resistant, hardenable.

POLITIT 18CM.
 M-1338; 0.90 C, 18 Cr, 1 V, 1.15 Mo, bal Fe.
 Annealed: 108,000 TS; 62,000 YS; 18 El; 35 RA; 220 Brin.
 Heat treated: 280,000 TS; 270,000 YS; 3 El; 15 RA; 555 Brin.
 For bearings, valves, gauges, sleeves; stainless, wear resistant.

POLITIT 13/1.
 M-1338; 0.12 max C, 13 Cr, 0.4 Si, bal Fe.
 Annealed: 75,000 TS; 40,000 YS; 35 El; 70 RA; 155 Brin.
 Cold drawn: 100,000 TS; 85,000 YS; 17 El; 60 RA; 205 Brin.
 For turbine blades, valve trim; Type 410; stainless.

POLITIT 13/2.
 M-1338; 0.2 C, 0.4 Si, 13 Cr, bal Fe.
 Annealed: 95,000 TS; 50,000 YS; 25 El; 55 RA; 195 Brin.
 Cold drawn: 105,000 TS; 85,000 YS; 17 El; 50 RA; 215 Brin.
 For turbine blades, valve trim, cutlery, surgical instruments; Type 420; stainless, hardenable.

POLITIT 13/4.
 M-1338; 0.4 C, 13 Cr, 0.4 Si, bal Fe.
 Annealed: 95,000 TS; 50,000 YS; 25 El; 55 RA; 195 Brin.
 For cutlery, knives, surgical instruments; Type 420; stainless, hardenable.

POLITIT 18/8.
 M-1338; 0.15 max C, 18 Cr, 8.5 Ni, bal Fe.
 Annealed: 80,000 TS; 35,000 YS; 55 El; 75 RA; 150 Brin.
 For chemical plant equipment, tanks, dryers, mixers; Type 302; stainless, austenitic.

POLITIT 18/8T.
M-1338; 0.12 max C, 18 Cr, 9.5 Ni, Ti = 4 x C, bal Fe.
Annealed: 85,000 TS; 35,000 YS; 55 El; 65 RA; 150 Brin.
For welded chemical plant equipment, tanks, mixers; Type 321; stainless, austenitic.

POLITIT 18/9.
M-1338; 0.07 C, 18 Cr, 9.5 Ni, bal Fe.
Annealed: 85,000 TS; 35,000 YS; 60 El; 70 RA; 150 Brin.
Cold drawn: 180,000 TS; 125,000 YS; 10 El; 330 Brin.
For welded chemical plant equipment; Type 304; stainless, austenitic.

POLITIT 18/10M.
M-1338; 0.07 max C, 18 Cr, 2 Mo, 10.5 Ni, bal Fe.
Annealed: 85,000 TS; 35,000 YS; 50 El; 65 RA; 160 Brin.
Cold drawn: 150,000 TS; 135,000 YS; 6 El; 300 Brin.
For acid resistant chemical plant equipment; Type 316; stainless, austenitic.

POLITIT 18/10MT.
M-1338; 0.12 max C, 18 Cr, 2 Mo, 10.5 Ni, Ti = 4 x C, bal Fe.
Annealed: 90,000 TS; 40,000 YS; 45 El; 60 RA; 180 Brin.
For welded chemical plant equipment; Type 316 Ti, stainless, austenitic.

POLIVIT.
M-German; 97.4 Al, 0.6 Cu, 1.8 Mn, 0.2 Ag.
For light alloy parts.

POLYMET A-G.
M-1567; Ag + plastic lubricant.
For instrument bearings, valve seats, rings and brushes.
Low friction. Withstands temperatures to 700°F.

POMET 111.
M-885; 5-10 Sn, bal Cu.
For mechanical and electrical parts; powder metals.

POMET 117B.
M-885; 94-96 Cu, 4-6 Sn.
Sintered: 18 El.
For precision parts; sintered.

POMET 117C.
M-885; 89-92 Cu, 8-11 Sn.
Sintered: 39,000 TS.
For instrument parts; sintered, good wear resistance.

POMET 141.
M-885; 0.2 C, 18 Cr, 8 Ni, bal Fe.
For mechanical parts; sintered, stainless.

POMET 300.
M-885; Fe.
For magnetic cores, pole shoes; powder metallurgy.

POMET 309.
M-885; 0.2-0.3 C, bal Fe.
Sintered: 38,000 TS.
For machine and instrument parts; sintered alloy.

POMET 309H.
M-885; 0.3 C, bal Fe.
Sintered and heat treated: 400 Brin.
For clutch parts, cams, gears; sintered.

POMET 389.
M-885; 0.10 C, bal Fe.
For general parts; powdered metals.

POMET 560.
M-885; 5 Cu, bal Al.
For light alloy parts; powder metals.

POMOLOY C-2.
M-886; 3.2 C, 2.2 Si, bal Fe.
Cast: 40,000 TS; 215 Brin.
For pump casings; gray cast iron.

POMOLOY C-3.
M-886; 3.2 C, 2.0 Si, bal Fe.
Cast: 40,000 TS; 210 Brin.
For pump casings; gray cast iron.

POMPEY 10 CNK1.
M-884; 0.09 C, 0.75 Mn, 0.25 Si, 1.4 Ni, 1.1 Cr, bal Fe.
For gears, bolts, cams, camshafts; AFNOR 10NC6, case hardening steel, tough.

POMPEY 10 M1.
M-884; 0.15 C, 1.0 Mn, 0.25 Si, bal Fe.
Carburizing steel.
AFNOR A 42 FP 2.

POMPEY 12 CNK3.
M-884; 0.14 C, 3 Ni, 0.75 Cr, bal Fe.
For gears, cams, camshafts, machine tool parts; AFNOR 14NC12, case hardening steel, tough.

POMPEY 14 CNK 1.
M-884; 0.14 C, 0.50 Mn, 3.0 Ni, 0.75 Cr, bal Fe.
Alloy carburizing steel, deep hardening.
AFNOR 14 NC 12.

POMPEY 14 N3.
M-884; 0.14 C, 0.45 Mn, 0.20 Si, 3.5 Ni, bal Fe.
Nickel carburizing steel; deep hardening.
AFNOR 12 N 14 et 3.5% Ni.

POMPEY 16 CNK1.
M-884; 0.15 C, 0.75 Mn, 0.25 Si, 1.4 Ni, 1.1 Cr, bal Fe.
For gears, bolts, cams, machine tool parts; AFNOR 16NC6, case hardening steel, tough.

POMPEY 16 NC 4.
  M-884; 0.16 C, 0.85 Mn, 0.25 Si, 1.0 Ni, 1.0 Cr, bal Fe.
  Low alloy carburizing steel.
  AFNOR 16 NC 4.

POMPEY 17 M 1.
  M-884; 0.18 C, 1.10 Mn, 0.25 Si, bal Fe.
  Carburizing steel.
  AFNOR A 48 Fp 2.

POMPEY 18 M 2.
  M-884; 0.18 C, 1.35 Mn, 0.25 Si, bal Fe.
  Manganese carburizing steel.
  AFNOR 52 Fp 2.

POMPEY 20 CD 2.
  M-884; 0.20 C, 1.0 Mn, 0.25 Si, 0.50 Cr, 0.20 Mo, bal Fe.
  Low alloy carburizing steel.
  AFNOR 20 CD 2.

POMPEY 20 MN 5.
  M-884; 0.20 C, 1.25 Mn, 0.25 Si, bal Fe.
  Manganese carburizing steel.
  AFNOR 20 Mn 5.

POMPEY 20 NC 4.
  M-884; 0.20 C, 0.90 Mn, 0.25 Si, 1.0 Ni, 1.0 Cr, bal Fe.
  Low alloy carburizing steel.
  AFNOR 20 Nc 4.

POMPEY 20 NK1.
  M-884; 0.20 C, 1.4 Ni, 1.1 Cr, bal Fe.
  For gears, bolts, cams, machine tool parts; AFNOR 20NC6, case hardening steel, tough.

POMPEY 20NKD.
  M-884; 0.20 C, 0.6 Ni, 0.5 Cr, 0.2 Mo, bal Fe.
  Hot rolled: 93,000 TS; 65,000 YS; 25 El; 63 RA; 162 Brin.
  For gears, bolts, machine tool parts; case hardening, shock resistant, AFNOR 20 NCD2.

POMPEY 21 CDV 5.
  M-884; 0.21 C, 0.40 Mn, 0.45 Si, 1.3 Cr, 0.25 V, bal Fe.
  Cr-V carburizing steel.
  AFNOR 20 CDV 5-08.

POMPEY 22 M 1.
  M-884; 0.22 C, 1.2 Mn, 0.25 Si, bal Fe.
  Manganese carburizing steel.
  AFNOR A.58.F.P.2.

POMPEY 23 D 5 E.
  M-884; 0.23 C, 0.65 Mn, 0.20 Si, 0.55 Mo, bal Fe.
  Molybdenum carburizing steel.
  AFNOR 23 D 5.

POMPEY 28 CD 12.
  M-884; 0.30 C, 0.85 Mn, 0.25 Si, 3.0 Cr, 0.40 Mo, bal Fe.
  Cr-Mo medium carbon steel.
  AFNOR 30 CD 12.

POMPEY 29 NKD 2.
  M-884; 0.28 C, 0.50 Mn, 0.25 Si, 0.020 max P, 0.020 max S, 2.35 Ni, 0.60 Cr, 0.50 Mo, bal Fe.
  Ni-Cr-Mo structural steel.
  AFNOR 25 NCD 9.

POMPEY 30NK3.
  M-884; 0.29 C, 2.8 Ni, 0.75 Cr, bal Fe.
  Heat treated: 143,000-175,000 TS; 128,000 YS.
  For gears, bolts, crankshafts, fasteners; AFNOR 30NC11, oil hardened, shock resistant.

POMPEY 30NKD2.
  M-884; 0.30 C, 2 Ni, 2 Cr, 0.35 Mo, bal Fe.
  Heat treated: 170,000-200,000 TS; 143,000 YS; 8 El.
  For bolts, gears, machine tool parts; oil hardened, shock resistant.

POMPEY 30NKD3.
  M-884; 0.30 C, 3 Ni, 0.6 Cr, 0.35 Mo, bal Fe.
  Heat treated: 135,000-185,000 TS; 128,000 YS; 12 El.
  Fro bolts, gears, crankshafts, fasteners; oil hardened, shock resistant.

POMPEY 35 NK1.
  M-884; 0.34 C, 1.45 Ni, 1.0 Cr, bal Fe.
  For gears, bolts, crankshafts, axles, studs; AFNOR 35NC6, oil hardened, shock resistant.

POMPEY 35 NKD 2.
  M-884; 0.34 C, 0.75 Mn, 0.25 Si, 1.4 Ni, 1.0 Cr, 0.25 Mo, bal Fe.
  Ni-Cr-Mo structural steel.
  AFNOR 35 NCD 6.

POMPEY 38 NKD.
  M-884; 0.38 C, 0.65 Mn, 0.25 Si, 0.85 Ni, 0.80 Cr, 0.25 Mo, bal Fe.
  Ni-Cr-Mo structural steel.
  AFNOR 40 NCD 3.

POMPEY 40 CD 2.
  M-884; 0.40 C, 1.0 Mn, 0.25 Si, 0.55 Cr, 0.20 Mo, bal Fe.
  Cr-Mo structural steel.
  AFNOR 40 CD 2.

POMPEY 48 MN 5.
  M-884; 0.48 C, 1.25 Mn, 0.25 Si, bal Fe.
  Manganese structural steel.
  AFNOR 48 M 5.

POMPEY 1144.
  M-884; 0.45 C, 1.50 Mn, 0.25 Si, 0.040 P, 0.28 S, bal Fe.
  Manganese structural steel.
  AFNOR 45 MF 6; Free machining; AISI 1144.

POMPEY 1541.
  M-884; 0.40 C, 1.50 Mn, 0.25 Si, bal Fe.
  Managanes structural steel.
  AFNOR 40 M 6; AISI 1541.

POMPEY 4142.
  M-884; 0.42 C, 0.85 Mn, 0.25 Si, 1.0 Cr, 0.20 Mo, bal Fe.
  Cr-Mo structural steel.
  AFNOR (42 CD 4); AISI 4142.

POMPEY 4.330.
  M-884; 0.30 C, 0.65 Mn, 0.25 Si, 1.8 Ni, 0.75 Cr, 0.25 Mo, bal Fe.
  Ni-Cr-Mo structural steel. AFNOR 30 NCD 7; (AISI 4330).

POMPEY 4.340.
  M-884; 0.40d C, 0.65 Mn, 0.25 Si, 1.8 Ni, 0.75 Cr, 0.25 Mo, bal Fe.
  Ni-Cr-Mo structural steel. AFNOR 40 NCD 7; AISI 4340.

POMPEY ABRADUR M14.
  M-884; 1.2 C, 13 Mn, bal Fe.
  For pulverizers, crushers, rock breakers; high wear and abrasion resistance.

POMPEY APS10.
  M-884; 0.12 C, 2.0-2.5 Cr, 0.35-0.90 Al, 0.35 max Mo, bal Fe.
  Normalized: 92,000 TS; 52,000 YS; 20 El.
  For evaporators, transportation tanks; corrosion resistant.

POMPEY-APS10M.
  M-884; 0.12 C, 2.0 Cr, 0.9 Al, 0.12 Mo, bal Fe.
  Normalized: 72,000-88,000 TS; 46,000 YS.
  For evaporators, dryers, graters, nitrate plants, chemical equipment.
  Corrosion resistant to stress cracking due to nitrate.

POMPEY-APS20A.
  M-884; 0.12 C, 4.0 Cr, 0.9 Al, bal Fe.
  Normalized: 69,000-85,000 TS; 43,000 YS.
  For oil refinery and coking equipment.
  Corrosion resistant to industrial and marine atmospheres.

POMPEY APS25.
  M-884; 0.12 C, 4 Cr, 0.9 Al, 0.9 Ni, 0.2 Cu, 0.12 Mo, bal Fe.
  Normalized: 121,000-157,000 TS; 85,000 YS; 14 El.
  For automobile bodies, railroad cars, diesel motor parts; resists marine and industrial atmospheres.

POMPEY B 1021.
  M-884; 0.21 C, 0.75 Mn, 0.25 Si, B, bal Fe.
  Boron-Carbon structural steel.
  AFNOR 21.B.3.

POMPEY B1038.
  M-884; 0.37 C 0.75 M, 0.25 Si, B, bal Fe.
  Boron-Carbon structural steel.
  AFNOR 38 B 3.

POMPEY B 1419.
  M-884; 0.18 C, 1.15 Mn, 0.25 Si, B, bal Fe.
  Boron-Mn carburizing steel.
  AFNOR 18 MB 4.

POMPEY B 1422.
  M-884; 0.20 C, 1.25 Mn, 0.25 Si, B, bal Fe.
  Boron-Mn carburizing steel.
  AFNOR 20 MB 5.

POMPEY B 1435.
  M-884; 0.35 C, 1.0 Mn, 0.25 Si, B, bal Fe.
  Boron-Mn structural steel.
  AFNOR 35 MB 4.

POMPEY B 1440.
  M-880; 0.40 C, 1.0 Mn, 0.25 Si, B, bal Fe.
  Boron-Mn structural steel.
  AFNOR 40 MB 4.

POMPEY B 8620.
  M-884; 0.20 C, 0.80 Mn, 0.25 Si, 0.55 Ni, 0.50 Cr, 0.20 Mo, B, bal Fe.
  AFNOR 19 NCDB 2; AISI 8620+B.

POMPEY C.1.MK.
  M-884; 0.16 C, 1.15 Mn, 0.25 Si, bal Fe.
  Managanese carburizing steel.
  AFNOR 16 MC 5.

POMPEY C 2 MK.
  M-884; 0.20 C, 1.15 Mn, 0.25 Si, bal Fe.
  Manganese carburizing steel.
  AFNOR 20 MC 5.

POMPEY CKD2.
  M-884; 0.18 C, 1.0 Cr, 0.25 Mo, bal Fe.
  Heat treated: 157,000-193,000 TS; 130,000 YS.
  For gears, bolts, machine tool parts; AISI 4120, case hardening steel, tough.

POMPEY CNKD2.
  M-884; 0.17 C, 1.4 Ni, 1.0 Cr, 0.20 Mo, bal Fe.
  Heat treated: 170,000-206,000 TS; 143,000 YS; 9 El.
  For gears, cams, camshafts, fasteners; AFNOR 18NCD6, case hardening, shock resisting.

POMPEY CNKD3.
  M-884; 0.17 C, 3.2 Ni, 1.0 Cr, 0.25 Mo, bal Fe.
  Heat treated: 186,000-221,000 TS; 143,000 YS; 8 El.
  For gears, cams, camshafts, fasteners; AFNOR 16NCD13, case hardening, shock resistant.

POMPEY D 1 S.
  M-884; 0.11 C, 0.75 Mn, 0.20 Si, 0.040 P, 0.185, bal Fe.
  Free machining low carbon steel.
  AFNOR. 10.F.2.

POMPEY D3SS.
  M-884; 0.35 C, 1.15 Mn, 0.25 Si, 0.15 S, bal Fe.
  For screw machine products, bolts, fasteners; free cutting.

POMPEY D 15 SS.
M-884; 0.17 C, 1.15 Mn, 0.25 Si, 0.040 P, 0.11 S, bal Fe.
Free machining steel.
AFNOR 17 MF 4; AISI 1117.

POMPEY D 37 SS.
M-884; 0.36 C, 1.50 Mn, 0.25 Si, 0.040 P, 0.11 S, bal Fe.
Free machining steel; hardenable.
AFNOR 35 MF 6; AISI 1137.

POMPEY D 38 K.
M-884; 0.38 C, 0.75 Mn, 0.25 Si, 0.45 Cr, bal Fe.
Low Cr structural steel.
AFdNOR 38 C 2.

POMPEY D 42 K.
M-884; 0.43 C, 0.75 Mn, 0.25 Si, 0.45 Cr, bal Fe.
Low Cr structural steel.
AFNOR 42.C.2.

POMPEY D 46 SS.
M-884; 0.45 C, 1.0 Mn, 0.25 Si, 0.040 P, 0.11 S, bal Fe.
Free machining structural steel.
AFNOR 45 MF 4; AISI 1146.

POMPEY D 52 L.
M-884; 0.18 C, 1.35 Mn, 0.25 Si, bal Fe.
Low C structural steel.
AISI 1518.

POMPEY DA 42 L.
M-884; 0.15 C, 1.0 Mn, 0.25 Si, bal Fe.
Low C structural steel.

POMPEY DA 48 L.
M-884; 0.18 C, 1.10 Mn, 0.25 Si, bal Fe.
Low C structural steel.

POMPEY DK3.
M-884; 0.32 C, 0.75 Mn, 1 Cr, bal Fe.
For gears, bolts, machine tool parts.
AFNOR 32C4, oil hardened.

POMPEY DK4.
M-884; 0.40 C, 0.75 Mn, 1, Cr, bal Fe.
For gears, bolts, machine tool parts.
AFNOR 38C4, oil hardened.

POMPEY DK 35.
M-884; 0.35 C, 0.80 Mn, 0.25 Si, 1.0 Cr, bal Fe.
Chromium structural steel; hardenable.
AFNOR 35 C 4.

POMPEY DMS.
M-884; 0.37 C, 1.25 Mn, 1.25 Si, bal Fe.
Structural steel, hardenable.
AFNOR 37 MS.5.

POMPEY D Z S.
M-884; 0.18 C, 0.75 Mn, 0.20 Si, 0.040P, 0.18 S, bal Fe.
Free machining low carbon steel.
AFNOR 20.F.2.

POMPEY EP 7.
M-884; 0.7 C, 0.6 Mn, 0.030 max P, 0.025 max S, bal Fe.
For radial tyre mesh in tyre industry.

POMPEY E7.
M-884; 0.7 C, 0.04 max S, and P, bal Fe.
Heat treated: 174,000 TS; 128,000 YS; 12 El; 37 RA; 352 Brin.
For cams, forging tools, nippers, lifters; water hardened.

POMPEY E7FX.
M-884; 0.75 C, 0.015 S, 0.025 P, bal Fe.
Heat treated: 180,000 TS; 140,000 YS; 14 El; 38 RA; 370 Brin.
For crimpers, punches, tools; water hardened.

POMPEY E8FX.
M-884; 0.8 C, 0.015 S, 0.025 P, bal Fe.
Heat treated: 188,000 TS; 143,000 YS; 12 El; 35 RA; 390 Brin.
For tools, dies, drills, taps; water hardened.

POMPEY EF10.
M-884; 1.0 C, 0.025 max S and P, bal Fe.
Annealed: 100,000 TS; 53,000 YS; 21 El; 42 RA; 200 Brin.
For cams, knives, cutters, drills, gages; water hardened; Type W1.

POMPEY EGALIT.
M-884; 0.85 C, 2.0 Mn, 0.2 V, bal Fe.
For drills, taps, cutting tools, stamps; water hardened.

POMPEY ETAD 1.
M-884; 0.30 C, 0.40 Mn, 0.25 Si, 3.75 Ni, 1.35 Cr, 0.50 Mo, bal Fe.
Ni-Cr-Mo structural steel, deep hardening.
AFNOR 30 NCD 16.

POMPEY ETAD 3.
M-884; 0.34 C, 0.45 Mn, 0.25 Si, 0.030 max P, 0.025 max S, 4.0 Ni, 1.8 Cr, 0.40 Mo, bal Fe.
Alloy structural steel; deep hardening.
AFNOR 35 NCD 16.

POMPEY ETANO.
M-884; 0.35 C, 3.75 Ni, 1.7 Cr, bal Fe.
Heat treated: 250,000 TS; 192,000 YS; 6 El; 40 RA.
For gears, pinions, crankshafts.
AFNOR 35NC15, oil hardened, shock resistant.

POMPEY FF 1.
M-884; 0.10 C, 0.45 Mn, 0.20 Si, Al, bal Fe.
AFNOR XC 10 FF.

POMPEY FF 2.
M-884; 0.19 C, 0.55 Mn, 0.20 Si, Al, bal Fe.
AFNOR XC 20 FF.

POMPEY FF 2 H.
M-884; 0.20 C, 0.85 Mn, 0.20 max si, 0.25 Cr, Al, bal Fe.
AFNOR 20 C 1 FF.

POMPEY FF 3 H.
  M-884; 0.38 C, 0.70 Mn, 0.20 Si, 0.25 Cr, Al, bal Fe.
  AFNOR 38 C 1 FF.

POMPEY FF 15.
  M-884; 0.15 C, 0.70 Mn, 0.20 Si, Al, bal Fe.
  AFNOR XC 15 FF.

POMPEY FF 32.
  M-884; 0.32 C, 0.65 Mn, 0.20 max Si, Al, bal Fe.
  AFNOR XC 32 FF.

POMPEY FF 35.
  M-884; 0.35 C, 0.65 Mn, 0.20 Si, Al, bal Fe.
  AFNOR X C 35 FF.

POMPEY FF 38.
  M-884; 0.38 C, 0.65 Mn, 0.20 max Si, Al, bal Fe.
  AFNOR XC 38 FF.

POMPEY FF 42 H.
  M-884; 0.42 C, 0.70 Mn, 0.20 Si, 0.25 Cr, Al, bal Fe.
  AFNOR 42 C 1 FF.

POMPEY FFC.
  M-884; 0.06 C, 0.40 Mn, 0.10 max Si, Al, bal Fe.
  AFNOR XC 6 FF.

POMPEY FFC 1.
  M-884; 0.10 C, 0.45 Mn, 0.10 max Si, Al, bal Fe.
  AFNOR XC 10 FF.

POMPEY FFC 2.
  M-884; 0.18 C, 0.55 Mn, 0.10 max Si, Al, bal Fe.
  AFNOR XC 18 F.F.

POMPEY FFC 3.
  M-884; 0.34 C, 0.40 Mn, 0.20 max Si, Al, bal Fe.
  AFNOR XC 35 FF.

POMPEY FFC 15.
  M-884; 0.15 C, 0.70 Mn, 0.10 max Si, Al, bal Fe.
  AFNOR XC 15 FF.

POMPEY FFC 22.
  M-884; 0.22 C, 0.60 Mn, 0.10 max Si, Al, bal Fe.
  AFNOR XC 22 FF.

POMPEY KD2.
  M-884; 0.26 C, 1 Cr, 0.25 Mo, bal Fe.
  For gears, bolts, machine tool parts; Type 4125; oil or water hardened, tough.
  AFNOR 25 Cd4.

POMPEY KD2S.
  M-884; 0.23 C, 1 Cr, 0.25 Mo, bal Fe.
  For gears, bolts, machine tool parts; Type 4125; case hardening steel, tough.
  AFNOR 25 CD 2 S.

POMPEY KD3.
  M-884; 0.35 C, 1.0 Cr, 0.25 Mo, bal Fe.
  For gears, bolts, crankshafts, machine tool parts; Type 4135; oil hardened, tough.
  AFNOR 35 CD 4.

POMPEY KD4.
  M-884; 0.42 C, 1.0 Cr, 0.25 Mo, bal Fe.
  hardened, tough.
  AFNOR 42 CD 4.

POMPEY KD 16.
  M-884; 0.16 C, 0.75 Mn, 0.25 Si, 1.1 Cr, 0.25 Mo, bal Fe.
  Cr-Mo case hardening steel.
  AFNOR 18 CD 4.

POMPEY KD 20.
  M-884; 0.19 C, 0.75 Mn, 0.25 Si, 1.1 Cr, 0.25 Mo, bal Fe.
  Cr-Mo case hardening steel.
  AFNOR 20 CD 4; AISI 4120.

POMPEY KD 27.
  M-884; 0.27 C, 0.75 Mn, 0.25 Si, 1.1 Cr, 0.25 Mo, bal Fe.
  Cr-Mo structural steel.
  AFNOR 27 CD 4.

POMPEY KD 30.
  M-884; 0.30 C, 0.75 Mn, 0.25 Si, 1.1 Cr, 0.25 Mo, bal Fe.
  Cr-Mo structural steel.
  AFNOR 30 CD 4; AISI 4130.

POMPEY KD 33.
  M-884; 0.33 C, 0.75 Mn, 0.25 Si, 1.1 Cr, 0.25 Mo, bal Fe.
  Cr-Mo structural steel.
  AFNOR 33 CD 4; AISI 4130-4135.

POMPEY KD 39.
  M-884; 0.39 C, 0.75 Mn, 0.25 Si, 1.0 Cr, 0.25 Mo, bal Fe.
  Cr-Mo structural steel.
  AFNOR 39 CD 4; AISI 4140.

POMPEY KR6.
  M-884; 1.0 C, 1.4 Cr, bal Fe.
  Heat treated: 120,000-237,000 TS; 95,000-226,000 YS; 14-3 El; 47-30 RA; 227-444 Brin.
  For bearings, sleeves, liners; SAE52100; AFNOR 100C6; wear resist

POMPEY MSV.
  M-884; 0.06-0.11 C, 0.30-0.50 Mn, 0.10 Si, bal Fe.
  Annealed: 65,000 TS; 40,000 YS; 28 El; 65 RA; 130 Brin.
  For nails, rivets, screws, bolts, bushing; Type 1008; case hardened.

POMPEY MSV3.
  M-884; 0.26-0.35 C, 0.40-0.90 Mn, 0.12 Si, bal Fe.
  Hot rolled: 80,000 TS; 50,000 YS; 30 El; 56 RA; 165 Brin.
  For gears, bolts, armature shafts, keys, brackets; Type 1030; water hardened.

POMPEY MSVD.
M-884; 0.15-0.23 C, 0.40-0.60 Mn, 0.12 Si, bal Fe.
Annealed: 73,000 TS; 42,000 YS; 22 El; 58 RA; 140 Brin.
For screws, bolts, gears, cams; Type 1020; case hardened.

POMPEY PF1.
M-884; 0.15 C, 0.55 Cr, 0.55 Mo, bal Fe.
Normalized: 72,000 TS; 25,000 YS; 18 El.
For oil refinery equipment; creep resistant.
AFNOR 13CD2.

POMPEY PF 2.
M-884; 0.12 C, 0.50 Mn, 0.60 Si, 1.1 Cr, 0.55 Mo, bal Fe.
Cr-Mo structural steel, for oil refinery equipment; creep resistant.
AFNOR 12 CD4-05.

POMPEY PF 2 A.
M-884; 0.13 C, 0.60 Mn, 0.25 Si, 1.0 Cr, 0.50 Mo, bal Fe.
Cr-Mo carburizing or structural steel.
AFNOR 15 CD 4-05.

POMPEY PF5.
M-884; 0.13 C, 5 Cr, 0.55 Mo, bal Fe.
Normalized: 85,000 TS; 25,000 YS; 25 El.
For oil refinery equipment; creep and heat resistant. AFNOR z15CD5-05.

POMPEY PF6.
M-884; 0.13 C, 2.2 Cr, 1 Mo, bal Fe.
Normalized: 82,000 TS; 30,000 YS; 21 El.
For oil refinery equipment; creep resistant.
AFNOR 10CD9.

POMPEY PFOA.
M-884; 0.16 C, 0.65 Mn, 0.25 Si, 0.55 Mo, bal Fe.
Molybdenum carburizing steel.
AFNOR 15 D 6.

POMPEY PFOB.
M-884; 0.15 C, 0.65 Mn, 0.20 Si, 0.30 Mo, bal Fe.
Molybdenum carburizing steel.
AFNOR 15 D 3.

POMPEY PFV 55.
M-884; 0.12 C, 0.40 Mn, 0.35 Si, 2.1 Cr, 0.35 Mo, 0.08 V, 0.45 Al, bal Fe.
Cy-Mo structural steel; for oil refinery equipment; creep resistant.
AFNOR 12 CADV 8.

POMPEY RKV.
M-884; 0.50 C, 0.80 Mn, 1 Cr, 0.15 V, bal Fe.
Heat treated: 170,000-206,000 TS; 8 El; 35 RA.
For gears, bolts, springs, crankshafts; oil hardened, shock resistant.
AFNOR 50CV4.

POMPEY RMK.
M-884; 0.48 C, 0.80 Mn, 0.30 Si, 1.0 Cr, bal Fe.
Heat treated: 193,000-221,000 TS; 171,000 YS; 5 El; 30 RA.
For springs, torsion bars; oil hardened.
AFNOR 48C4.

POMPEY RS.
M-884; 0.45 C, 0.65 Mn, 1.8 Si, bal Fe.
Heat treated: 200,000-235,000 TS; 170,000 YS; 6 El; 30 RA.
For springs; water hardened.
AFNOR 45S7.

POMPEY RS 2.
M-884; 0.51 C, 0.65 Mn, 1.80 Si, bal Fe.
Silicon structural steel; for springs.
AFNOR 51 S 7.

POMPEY RS 2 K.
M-884; 0.56 C, 0.75 Mn, 1.80 Si, 0.30 Cr, bal Fe.
Silico-manganese steel for springs.
AFNOR 56 SC 7; AISI 9255.

POMPEY RS3.
M-884; 0.55 C, 0.75 Mn, 1.7 Si, bal Fe.
Heat treated: 200,000-235,000 TS; 177,000 YS; 5 El; 25 RA.
For springs; oil hardened.
AFNOR 55S7.

POMPEY RS3K.
M-884; 0.60 C, 0.85 Mn, 1.8 Si, 0.40 Cr, bal Fe.
Heat treated: 213,000-242,000 TS; 185,000 YS; 5 El; 25 RA.
For springs, coils; oil hardened. AFNOR 61SC7.

POMPEY RS 4.
M-884; 0.60 C, 0.85 Mn, 1.80 Si, bal Fe.
Silico-manganese steel for springs.
AFNOR 60 S 7; AISI 9260.

POMPEY RS 4 K.
M-884; 0.60 C, 0.75 Mn, 1.60 Si, 0.60 Cr, bal Fe.
Silico-manganese steel for springs.
AFNOR 61 SC 7; similar to AISI 9260.

POMPEY RSKD.
M-884; 0.48 C, 0.65 Mn, 1.50 Si, 0.60 Cr, 0.25 Mo, bal Fe.
Special alloy steel for springs, torsion bars.
AFNOR 45 SCD 6.

POMPEY RSOB.
M-884; 0.38 C, 0.65 Mn, 1.50 Si, bal Fe.
Silicon structural steel; for springs; refineries.
AFNOR 38.S.6.

POMPEY S3.
M-884; 0.35 C, bal Fe.
Hot rolled: 85,000 TS; 54,000 YS; 30 El; 53 RA; 185 Brin.
For gears, shafts, housings; water hardened.

POMPEY S3F.
 M-884; 0.35 C, bal Fe.
 Hot rolled: 85,000 TS; 54,000 YS; 30 El; 53 RA; 185 Brin.
 For gears, shafts, axles, mowers; water hardened.

POMPEY S4.
 M-884; 0.45 C, bal Fe.
 Hot rolled: 98,000 TS; 59,000 YS; 24 El; 45 RA; 212 Brin.
 For machine tool parts, gears, shafts; water hardened.

POMPEY S4F.
 M-884; 0.40 C, bal Fe.
 Hot rolled: 91,000 TS; 58,000 YS; 27 El; 50 RA; 200 Brin.
 For gears, shafts, axles, mowers; water hardened.

POMPEY S4M.
 M-884; 0.9 C, bal Fe.
 Heat treated: 190,000 TS; 145,000 YS; 10 El; 30 RA; 400 Brin.
 For cutters, dies; water hardened.

POMPEY S4S.
 M-884; 0.4 C, 0.035 S, 0.035 P, bal Fe.
 For rolled: 91,000 TS; 58,000 YS; 27 El; 50 RA; 200 Brin.
 For agricultural equipment, plows; water hardened.

POMPEY S5M.
 M-884; 1.1 C, bal Fe.
 Annealed: 110,000 TS; 56,000 YS; 20 El; 40 RA; 210 Brin.
 For cutters, dies; water hardened.

POMPEY S6F.
 M-884; 0.62 C, bal Fe.
 Heat treated: 160,000 TS; 113,000 YS; 12 El; 40 RA; 325 Brin.
 For agricultural equipment, mowers; water hardened.

POMPEY S7F.
 M-884; 0.68 C, bal Fe.
 Heat treated: 174,000 TS; 128,000 YS; 12 El; 37 RA; 355 Brin.
 For agricultural equipment, mowers; water hardened.

POMPEY S7 TRIPLEX.
 M-884; 0.7 C, bal Fe.
 Heat treated: 174,000 TS; 128,000 YS; 12 El; 37 RA; 352 Brin.
 For agricultural equipment, plows; water hardened.

POMPEY S65.
 M-884; 0.65 C, bal Fe.
 Heat treated: 170,000 TS; 116,000 YS; 10 El; 38 RA; 340 Brin.
 For shafts, mandrels, dies, tools; water hardened.

POMPEY S75.
 M-884; 0.75 C, bal Fe.
 Heat treated: 180,000 TS; 140,000 YS; 13 El; 36 RA; 370 Brin.
 For crimpers, punches, dies, tools; water hardened.

POMPEY S90.
 M-884; 0.9 C, 0.035 S, 0.035 P, bal Fe.
 Heat treated: 190,000 TS; 145,000 YS; 10 El; 30 RA; 400 Brin.
 For agricultural equipment, mowing machines; water hardened.

POMPEY SH4.
 M-884; 0.48 C, bal Fe.
 Annealed: 96,000 TS; 52,000 YS; 16 El; 23 RA; 170 Brin.
 For machine tool parts, gears, shafts; water hardened.

POMPEY SH5.
 M-884; 0.54 C, bal Fe.
 Annealed: 100,000 TS; 55,000 YS; 15 El; 22 RA; 180 Brin.
 For shafts, axles, tools; water hardened.

POMPEY TR 42.
 M-884; 0.42d C, 0.65 Mn, 0.25 Si, bal Fe.
 Structural steel for shafts, bolts.
 AFNOR XC 42 TS; similar to AISI 1040.

POMPEY TR4M.
 M-884; 0.45 C, 0.15 Cr, 0.9 Mn, 0.25 Si, bal Fe.
 For machine tool parts, shafts, gears; induction or flame hardened parts.
 AFNOR XC 45TS.

POMPEY TR5.
 M-884; 0.52 C, 0.7 Mn, 0.25 Si, 0.15 Cr, bal Fe.
 For machine tool parts, shafts, gears; induction or flame hardened parts.
 AFNOR XC 52 TS.

POMPEY TR5M.
 M-884; 0.52 C, 0.8 Mn, 0.25 Si, 0.15 Cr, bal Fe.
 For machine tool parts, shafts, gears; induction or flame hardened parts.
 AFNOR XC 52TS.

POMPEY TRJ.
 M-884; 0.45 Cr, 0.15 Cr, 0.7 Mn, 0.25 Si, bal Fe.
 For machine tool parts, shafts, gears; induction or flame hardened parts.
 AFNOR XC 45TS.

POMPEY TRKD.
 M-884; 0.42 C, 0.7 Mn, 0.3 Si, 1 Cr, 0.25 Mo, bal Fe.
 For machine tool parts, shafts, gears; induction or flame hardened parts.
 AFNOR 42CD 4TS.

POMPEY TRNKD.
  M-884; 0.40 C, 0.75 Mn, 0.85 Ni, 0.75 Cr, 0.25 Mo, bal Fe.
  Ni-Cr-Mo structural steel; oil hardenable for highly stressed parts.
  AFNOR 40 NCD 3 TS.

POMPEY XC-10, ETC, AND OTHER NUMBERS see **AFNOR TABLES.**

POMPTON.
  M-1779; 0.7-1.4 C, 0.25 Mn, 0.25 Si, bal Fe.
  For threading dies, shear knives, chisels, drills, punches, trimming and blanking dies, cold heading dies, pneumatic tools. Water hardening.
  Typical heat treated hardness: Rc 56-60.
  AISI W-1.

POMPTON EXTRA.
  M-1779; 0.7-1.4 C, 0.25 Mn, 0.25 Si, bal Fe.
  For threading dies, shear knives, chisels, drills, punches, trimming and blanking dies, cold heading dies, pneumatic tools. Water hardening.
  Typical heat treated hardness: Rc 56-60.
  AISI W-1.

POMPTON SPECIAL.
  M-1779; 0.7-1.4 C, 0.25 Mn, 0.25 Si, bal Fe.
  For threading dies, shear knives, chisels, drills, punches, trimming and blanking dies, cold heading dies, pneumatic tools. Water hardening.
  Typical heat treated hardness: Rc 56-60.
  AISI W-1.

PONSARD'S HIGH MANGANESE BRASS.
  M-Eng.; 75-50 Cu, 20-25 Mn, 2-15 Zn, 0-16 Fe.
  For corrosion resisting and marine parts; corrosion resistant.

POPES ISLAND METAL.
  M-France; 70 Cu, 15 Zn, 14 Ni, 1 Sn.
  For tableware, base for plated ware; generic name of a series of French alloys.

POREX.
  M-839; 5-11 Sn, bal Cu.
  For filters, porous membranes, diffusers; powdered metals.

PORO BRONZE.
  M-525; 13 Sb, 7 Cu, bal Sn.
  For bearings, bushings; Babbitt.

POROSINT BRONZE.
  M-1739.
  Bronze, sintered.
  7 grades (A-G) having various porosity for filtering; max. particle size to pass from 0.0001-0.0024 in.

POROSINT CUPRO-NICKEL.
  M-1739; 67-70 Cu, 6.5-7.5 Sn, bal Ni.
  Sintered; for filtering; Grade CN 4 will pass 0.001 in particles max.

POROSINT RIGID MESH.
  M-1739.
  Similar to Porosint Stainless Steel but produced in sheets with wide range of porosity.
  Material: usually AISI 316, but others (Monel, AISI 310 and 95/5 Bronze) available.

POROSINT STAINLESS STEEL.
  M-1739.
  AISI 316L powders, sintered.
  6 grades (P/2 1/2-P/40) having various porosity for filtering; max particle size to pass from 0.0001-0.0016 in.

POSITIVE-ROD NO. 105.
  M-1096; 0.08-0.12 C, 0.45-0.65 Mn, 0.15-0.25 Si, 0.03 S, 0.03 P, bal Fe.
  Welded: 62,000-70,000 TS; 50,000-55,000 YS; 25-35 El; 40-45 RA.
  For all-position general purpose welding rod; AWS-E6010.

POTASSIUM.
  M-1755; K.
  Purities: Special distilled grade 99.99%, 99.9%, (low sodium), commercial grade.
  Packaging: Bottles, ampules, cylinders (under vacuum, inert gas, petroleum distillate).

POTASSIUM AMALGAM.
  $Hg_{24}K$.
  For organic reduction.

POTERIE D'ETAIN.
  90 Sn, 9 Sb, 1 Cu.
  For tableware, bearings; anti-friction.

POTINGRIS.
  Potinjaune plus Pb, + Sn.
  For hardware.

POTINJAUNE.
  72 Cu, 2 Zn, 2 Pb, 1.2 Sn.
  For hardware; French yellow brass.

POT METAL.
  80-67 Cu, 20-33 Pb.
  For bearings; heavy duty.

POTOMAC.
  M-1779; 0.35 C, 0.85 Si, 1.25 W, 5 Cr, 0.2 V, 1.5 Mo, bal Fe.
  Air hardening hot work die steel for hot forging dies, hot nut forming tools, and hot extrusion tooling.
  AISI H-12.

POTOMAC A.
M-1779; 0.40 C, 0.30 Mn, 0.90 Si, 5.0 Cr, 1.3 Mo, 0.5 V, bal Fe.
Air hardenable to 290,000 psi TS; 55 Rc.
Maintains high strength to 1000°F.
For aircraft and missile cases, rocket cases; also for general hot work tooling.
AISI H-12.

POTOMAC M.
M-1779; 0.40 C, 1.0 Si, 5.25 Cr, 1.0 V, 1.0 Mo, bal Fe.
General purpose hot work die steel for aluminum and zinc die casting dies, hot extrusion and forging; also cold heading and plastic molding.
Air hardening.
AISI H-13.

POTOSI SILVER.
Ni-Cu.
For ornamental and corrosion resistant parts; nickel silver.

POWDER BRAZE.
M-799; Zn, bal Cu.
For brazing; powdered metals.

POWDIRON 59 I ETC see GKN 59-I ETC.

POWER 52.
4-5 Si, bal Fe.
For laminations for power units; high permeability.

POWER 58.
4.25-4.75 Si, bal Fe.
For laminations for power units; high permeability.

POWER 65.
4.24-4.75 Si, bal Fe.
For laminations for power units; high permeability.

POWER 72.
3.75-4.25 Si, bal Fe.
For laminations for power units; high permeability.

POWER 82.
3.0-3.5 Si, bal Fe.
For laminations for power units; high permeability.

POWER 101.
2.25-2.75 Si, bal Fe.
For laminations for power units; high permeability.

POWER 117.
1.0-1. Si, bal Fe.
For laminations for power units; high permeability.

POWER 130.
0.5-0.75 Si, bal Fe.
For laminations for power units; high permeability.

POWER 145.
0.25-0.5 Si, bal Fe.
For laminations for power units; high permeability.

POWER 165.
Under 0.25 Si, bal Fe.
For laminations for power units; high permeability.

POWER NICKLE GENUINE BABBITT.
M-178; 83-85 Sn, bal Sb and Cu.
Cast: 17,500 TS; 10,500 YS; 27 Brin.
For bearings; for high pressures and temperatures.

POWERSTEEL.
M-1052; 0.9 C, 4 Co, bal Fe.
For tools, cutters; cast.

POWHATAN.
M-435; 0.83 C, 18 W, 4 Cr, 1.7 V, 10.5 Co, 1 Mo, bal Fe.
For cutting tools, lathe and planer tools; high speed steel.

PRASEODYMIUM.
M-1755; Pr.
Purities: 99.9%, 99.5 + %.
Forms: Ingot, lump, sheet, rod, wire, foil, turnings.

PRECEDENT 71 GRADE A now USCO PRECEDENT 71A (771.2).

PRECEDENT 71 GRADE B now USCO-PRECEDENT 71B (B771.2).

PRECEDENT NO. 356.
M-585; 6.5-7.5 Si, 0.20-0.40 Mg, bal Al.

PRECEDENT NO. 356A.
M-585; 7 Si, 0.3 Mg, bal Al.

PRECIBRONZE.
M-1675; 8 Sn, bal Cu.
Tin bronze; Wrought; Ann: 80-100 Brin; work hardenable to 210 Brin.
AFNOR UE 9P; ASTM B103-77.

PRECISION.
M-740; 0.3 C, bal Fe.
For structural parts.

PRECISION A-12.
M-887; 11-13 Si, 0.6 max Cu, 2 max Cu, 2 max Fe, 0.5 max Ni, bal Al.
Cast: 33,000 TS; 1.5 El; 80 Brin.
For light alloy die casting; corrosion resistant.

PRECISION A-50.
M-887; 4.5-6.0 Si, 0.6 max Cu, 2 max Fe, bal Al.
Cast 29,000 TS; 4 El; 70 Brin.
For die casting; high impact strength.

PRECISION A-54.
M-887; 4.5-5.5 Si, 3.4-4.5 Cu, 2 max Fe, bal Al.
Cast: 32,000 TS; 2 El; 75 Brin.
For die casting; high strength.

PRECISION A-94.
M-887"; 3.5-4.5 Si, 3.0-4.0 Cu, 1.3 max Fe, 0.6 max Zn, bal Al.
Die cast: 38,000 TS; 2 El; 77 Brin.
For die casting; high strength.

PRECISION A-218.
M-887; 8 Mg, 1.8 max Fe, 0.2 max Cu, 0.3 max Mn, 0.1 max Ni, bal Al.
For die casting; corrosion resistant.

PRECISION A-360.
M-887; 8-10 Si, 0.5 Mg, 2 max Fe, bal Al.
Die cast: 42,000 TS; 23,000 YS; 1.8 El.
For die castings; corrosion resistant.

PRECISION BR-1.
M-887; 63-67 Cu, 32-36 Zn, 0.7-1.3 Si.
Die cast: 70,000 TS; 35,000 YS; 25 El; 120 Brin.
For die castings; high strength.

PRECISION BR-2.
M-887; 81.5 Cu, 14.5 Zn, 4 Si.
Die cast: 85,000 TS; 50,000 YS; 8 El; 170 Brin.
For die casting; high strength.

PRECISION BR-3.
M-887; 83 Cu, 10 Zn, 5 Si, 1 Mn, 1 Al.
Die cast: 105,000 TS; 60,000 YS; 5 El; 190 Brin.
For die castings; high strength.

PRECISION M-13.
M-887; 8,3-9. Al 0.4-10 Zn, bal Mg.
Die cast: 29,000-34,000 TS; 20,000 YS; 2-5 El; 66 Brin.
For die castings.

PRECISION TIE BARS.
M-1724; 0.25 max C, 1.30 max Mn, bal Fe.
High strength tie bars; 550 N/mm² TS.

PRECISION ZN-5.
M-887; 3.5-4.3 Al, 0.75-1.25 Cu, bal Zn.
Cast: 42,000 TS; 4 El; 75 Brin.
For die castings.

PRECISION ZN-6.
M-887; 3.5-4.3 Al, 0.10 max Cu, bal Zn.
Cast: 36,000 TS; 5 El; 65 Brin.
For die castings.

PRECISION ZN-7.
M-887; 3.5-4.3 Al, 2.5-3.5 Cu, bal Zn.
Cast: 45,000 TS; 5 El; 80 Brin.
For die castings.

PREMABRAZE 130.
M-63; 82 Au, 18 Ni.
M.P. & Flow P. 1742°F.
For brazing stainless steel, Inconel X, other nickel base alloys; good oxidation resistance to 1500°F. AWS BAu-4.

PREMABRAZE 615.
M-63; 61.5 Ag, 24 Cu, 14.5 In.
M.P. 115-1305°F; for ferrous and non-ferrous and in vacuum tube brazing.

PREMAG.
M-526; 0.8 Ag, bal Cu.
For welding rod; copper welds.

PREMALOY 301.
M-63; Pt, Au, Ag, Cu, bal Pd.
For resistors; high electrical resistance.

PREMIER NO. 1.
0.10-0.20 C, 0.65-0.85 Cr, 0.10 max V, bal Fe.
For case hardened parts; carburizing steel.

PREMIER NO. 6.
M-Eng.; 0.6-0.7 C, 0.7-0.9 Cr, 0.10 max V, bal Fe.
For hammers, dies, blacksmith tools; water hardened.

PRESCOLOY.
M-583; 0.2 C, 0.9 Ni, bal Fe.
90,000 TS; 60,000 YS; 25 El; 40 RA.
For truck side frames, bolsters, freight car castings; high strength.

PRESNEAL.
M-64; 0.3 C, Ni, Mo, bal Fe.
Annealed.
For dies, punches; hot work, press and upsetter.

PRESSANT.
M-1340; 0.15 C, 0.2 Si, 0.4 Mn, bal Fe.
Annealed: 70,000 TS; 40,000 YS; 25 El; 60 RA; 145 Brin.
For gears, pinions, cams, camshafts; case hardening steel.

PRESSANT ECR80.
M-1340; 0.15 C, 0.25 Si, 0.5 Mn, 0.65 Cr, bal Fe.
Annealed: 70,000 TS; 40,000 YS; 25 El; 60 RA; 145 Brin.
For gears, pinions, cams, camshafts; case hardening steel.

PRESSANT EM212.
M-1340; 0.20 C, 125 Mn, 1.15 Cr, bal Fe.
For gears, cams, camshafts; case hardening steel.

PRESSCO NO. 1.
M-1168; 85-88 Cu, 1-1.5 Fe, 10-11 Al.
Die cast: 75,000-85,000 TS; 36,000 YS; 26-17 El; 140-180 Brin.
For marine hardware, propellers; heat treatable; Al-bronze.

PRESSCO NO.2.
M-1168; 80-83 Cu, 13-16 Zn, 3.5-4.5 Si.
Die cast: 75,000-85,000 TS; 40,000 YS; 160-200 Brin.
For gears, shafts, pump parts; silicon bronze.

PRESSCO NO. 3.
M-1168; 55-59 Cu, 1 Mn, 0.2 max Fe, 1 max Sn, 0.5 max Pb, bal Zn.
Cast: 76,000 TS; 33,000 YS; 15 El; 102 Brin.
For fittings, spiders, brackets; SAE 43; manganese bronze.

PRESSCO NO. 4.
M-1168; 58-62 Cu, 36-41 Zn, 1 max Pb, 1 max Sn.
Die cast: 50,000-60,000 TS; 35,000 YS; 12-10 El.
For die castings; corrosion resistant.

PRESSCO NO. 5.
M-1168; 3.5-5.5 Ni, 3-5 Fe, 10-11 Al, 3.5 max Mn, bal Cu.
Cast: 90,000 TS; 40,000 YS; 10 El, 157 Brin.
For pump parts, bushings, bearings; Al bronze, heat treatable.

PRESSCO NO 6.
M-1168; 59 min Cu, 0.5 max Al, bal Zn.
Cast 40,000 TS; 11,00 YS; 25 El; 79 Brin.
For hardware, die castings; yellow brass.

PRESSCO NO. 7.
M-1168; 61.5-63.5 Cu, 35-38 Zn, 0.75-1.25 Pb, 0.9-1.2 Al, 0.5 max Sn, 0.5 Fe.
Cast: 63,000 TS; 35,000 YS; 30 El, 102 Brin.
For hardware, die castings; free-cutting.

PRESSCO NO. 8.
M-1168; 4.5-6.5 Fe, 6.5-7.5 Ni, 11.5-12.5 Al, bal Cu.
Cast: 115,000 TS; 82,00 YS; 3 El; 220 Brin.
For hardware, die castings; Al-bronze.

PRESSCO NO. 9.
M-1168; 58 min Cu, 1.3 max Pb, bal Zn.
Cast: 49,000 TS; 12,000 YS; 5 El; 70 Brin.
For hardware, die castings; SAE 41; leaded brass.

PRESSCO NO. 10.
M-1168; 60 max Cu, 1 min Si, bal Zn.
Cast: 74,000 TS; 38,000 YS; 19 El; 130 Brin.
For hardware, die castings; high strength.

PRESSOFOND S5C.
M-1445; 4 Cu, 5.5 Si, bal Al.
Cast: 28,000-32,000 TS; 20,000-23,000 YS; 4-1.5 El; 55-75 Brin.
For light alloy parts; heat treatable.

PRESSOFOND S9CF.
M-1445; 3.5 Cu 0.9 Fe, 8.5 Si, bal Al.
Cast: 41,000-47,000 TS; 24,000-29,000 YS; 5.0-2.5 El; 80-100 Brin.
For light alloy parts; corrosion resistant.

PRESSURDIE 1.
M-140; 0.3-0.4 C, 0.8-1.0 Si, 4.75-5.25 Cr 2 V, 5 W, 6 Co, 0.25 Mo, bal Fe.
For extrusion dies, dummy blocks, punches; hot work steel, oil hardened.

PRESSURDIE 2.
M-140; 0.35 C, 1.5 Mo, 1.25 W, 5 Cr, 0.15 V, bal Fe.
For extrusion and forging dies, punches; Type H12; hot work steel.

PRESSURDIE 3.
M-140; 0.3-0.4 C, 0.85-1.0 Si, 4.9-5.25 Cr, 0.10-0.15 V, 0.45-0.60 Mo, bal Fe.
For tools, aluminum die casting dies; oil hardening.

PRESSURDIE 3-L.
M-140; 0.37 C, 0.35 Mn, 1.0 Si, 5.5 Cr, 0.45 V, 1.15 Mo, bal Fe.
For dies; oil hardening.

PRESSURDIE 6.
M-140; 0.40 C. 0.50 Mn, 1.0 C, 3.2 Cr, 0.30 V, 2.5 Mo, bal Fe.
For aluminum extrusion dies and mandrels, hot shears; hot work steel, tough.

PRESSURDIE 16.
M-140; 0.55 C, 0.90 Si, 1.25 W, 5 Cr, 1.25 Mo, bal Fe.
For trimming and forming dies, forging die inserts; hot work steel, oil hardened.

PRESSURDIE "C".
M-140; 0.38-0.43 C, 4.0-4.5 Cr, 4.0-4.5 W, 4.0-4.5 Co, 1.9-2.2 V, 0.4-0.7 Mo, bal Fe.
For dies, dummy blocks, valve dies, extrusion mandrels; hot work steel, air hardened.

PRESSURE DIE CASTING.
M-122; 62 Cu, 0.75 Pb, 0.5 Sn, bal Zn.
Die cast: 58,000 TS; 36,000 YS; 6 El; 7 RA; 124 Brin.
For plumbing fixtures, hardware; pressure die casting.

PRESSURE TITE IRON.
M-508; 2.85 T.C., 0.65 C.C., 1.25 Ni, 0.8 Mn, 0.4 Cr, 1.6 Si, bal Fe.
Cast: 55,000 TS; 285 Brin.
For liners, engine sleeves, oil pumps; centrifugal cast.

PRESS-X.
M-55; 0.2 C, 0.70 Mn, 3.0 Ni, 2.0 Cr, 1.1 Mo, 0.5 V, bal Fe.
A precipitation hardened steel prehardened to 37 to 46 Rc.
For forging press dies; flat open die forging dies, upsetter dies, inserts and piercing and punching dies.

PRESTEM.
M-1488; 0.20 C, 3.15 Ni, 3.4 Mo, bal Fe.
Hot work tool steel for forging dies.
AFNOR 20 DN 33-12.

PRESTEM.
M-64; 0.3 C, Ni, Mo, bal Fe.
For dies punches; hot work, press and upsetter, prehardened.

PRESTO 2 KRONEN.
M-1339; 0.74 C, 4.1 Cr, 1.1 V, 18.5 W, bal Fe.
For lathe and planer tools, reamers, broaches, taps; high speed steel. AISI T1.

PRESTO C 3.
M-1339; 1.0 C, 4.0 Cr, 2.6 Mo, 2.3 V, 3.0 W, bal Fe.
High speed steel cutting tool.

PRESTO CHROME BEARING STEEL.
M-32; 1.0 C, 0.3 Mn, 0.25 Si, 1.4 Cr, bal Fe.
Oil or water hardenable to 60-66 Rock C.
For ball and roller bearings and races. G-15216

PRESTO ENORM.
M-1339; 0.82 C, 4.1 Cr, 0.85 Mo, 1.6 V, 8.7 W, bal Fe.
For lathe and planer tools, drills, taps, hobs; high speed steel.

PRESTO MO 8.
M-1339; 1.1 C, 8.5 Mo, 1.5 W, 1.1 V, 8.0 Co, 4.0 Cr, bal Fe.
High speed tool steel.

PRESTO MO 10.
M-1339; 0.90 C, 9.5 Mo, 4.0 Cr, 2.0 Mo, 1.7 W, 8.0 Co, bal Fe.
High speed tool steel.

PRESTO MO 166.
M-1339; 1.2 C, 4.0 Cr, 5.0 Mo, 6.3 W, 3.0 V, bal Fe.
High speed tool steel.
AISI M3 class 2.

PRESTO MO PLUS.
M-1339; 0.85 C, 4.0 Cr, 5.0 Mo, 6.2 W, 2 V, bal Fe.
High speed steel.
For cutting tools.
AISI M2.

PRESTO MO PLUS CO.
M-1339; 0.90 C, 4.2 Cr, 5.0 Mo, 6.3 W, 2.0 V, 5.0 Co, bal Fe.
High speed tool steel.
Similar to AISI M41.

PRESTO MO PLUS K.
M-1339; 1.0 C, 4.2 Cr, 5.0 Mo, 6.5 W, 2.0 V, bal Fe.
High speed tool steel.
Similar to AISI M2.

PRESTO MO SUPERIOR.
M-1339; 0.80 C, 4.0 Cr, 8.6 Cr, 1.7 W, 1.2 V, bal Fe.
Mo high speed steel.
For cutting tools.
AISI M1.

PRESTO UNIKUM.
M-1339; 0.79 C, 4.7 Co, 4.3 Cr, 0.75 Mo, 1.5 V, 18 W, bal Fe.
For lathe and planer tools, reamers, broaches, taps; high speed steel.

PRESTO UNIKUM 3.
M-1339; 0.86 C, 2.8 Co, 4.3 Cr, 0.85 Mo, 2.1 V, 12 W, bal Fe.
For lathe and planer tools, reamers, broaches, hobs; high speed steel.

PRESTO UNIKUM 5.
M-1339; 0.79 C, 4.7 Co, 4.3 Cr, 0.75 Mo, 1.5 V, 18 W, bal Fe.
For lathe and planer tools, reamers, broaches, drills; high speed steel.

PRESTO UNIKUM 125.
M-1339; 0.80 C, 4.3 Cr, Co, W, V, Mo, bal Fe.
For lathe and planer tools, reamers, broaches, hobs; high speed steel.

PRESTO UNIKUM SPEZIAL.
M-1339; 1.25 C, 4.2 Cr, 3.7 Mo, 10.0 W, 3.2 V, 10 Co, bal Fe.
High speed tool steel.

PRESTO UNIKUM SUPERB.
M-1339; 0.76 C, 10 Co, 4.2 Cr, 0.8 Mo, 1.8 V, 18 W, bal Fe.
For lathe and planer tools, form cutters; high speed steel.

PRESTO V22.
M-1339; 0.86 C, 4.1 Cr, 0.8 Mo, 2.5 V, 12 W, bal Fe.
For lathe and planer tools, reamers, broaches, hobs; high speed steel.

PRESTO V205.
M-1339; 1.3 C, 4.3 Cr, 0.85 Mo, 3.8 V, 12 W, bal Fe.
For engraving tools, form cutters, forming dies; high speed steel.

PRESTO V210.
M-1339; 1.35 C, 4.2 Cr, 12.0 W, 0.80 Mo, 3.75 V, 4.8 Co, bal Fe.
High speed tool steel.

PRESTO VA MO SUPERIOR.
M-1339; 1.02 C, 4.2 Cr, 8.5 Mo, 1.8 W, 2.0 V, bal Fe.
High speed tool steel. W-nr. 1.3348.
AISI M7.

PREUSS ALLOY.
30 Co, 0.2-1.5 Si, 0.04-0.06 C, bal Fe.
For magnetic circuits; high permeability.

PREWESTA 1.
 M-1294; 0.55 C, 1 Ni, 0.18 V, 1.8 W, bal Fe.
 For upsetters, crimpers, cold punches; oil hardened, tough.

PREWESTA 2.
 M-1294; 0.45 C, 1 Ni, 0.18 V, 1.85 W, bal Fe.
 For upsetters, crimpers, cold punches; oil hardened, tough.

PREWESTA 3.
 M-1294; 0.35 C, 1.05 Cr, 0.1, V, 1.85 W, bal Fe.
 For upsetters, crimpers, cold punches; oil hardened, tough.

PRG 35.
 M-1340; 0.5 C, 1.05 Cr, 3.25 Ni, bal Fe.
 For gears, bolts, springs, crankshafts; oil hardened, shock resistant.

PRIMA 4ZH.
 M-1322; 1.1 C, 0.25 max Si, 0.25 max Mn, bal Fe.
 For drills, taps, hobs, cutters, lathe tools; Type Wl; water hardened.

PRIMA 5ZAH.
 M-1322; 0.85 C, 0.25 max Si, 0.25 max Mn, bal Fe.
 Heat treated: 190,000 TS; 145,000 YS; 10 El; 30 RA; 400 Brin.
 For springs, tools, punches, drills; Type Wl; water hardened.

PRIMA 45.
 M-1340; 0.45 C, 0.5 Si, 0.6 Mn, bal Fe.
 Hot rolled: 98,000 TS; 59,000 YS; 24 El; 45 RA; 212 Brin.
 For gears, bolts, shafts; water hardened.

PRIMA 55.
 M-1340; 0.55 C, 0.2 Si, 0.6 Mn, bal Fe.
 Heat treated: 150,000 TS; 110,000 YS; 14 El; 45 RA; 320 Brin.
 For gears, springs crimpers; axles, shafts; water hardened.

PRIMA 60.
 M-1340; 0.60 C, 0.5 Si, 0.6 Mn, bal Fe.
 Heat treated: 160,000 TS; 113,000 YS; 12 El; 40 RA; 325 Brin.
 For springs, punches, crimpers, axes, hammers; water hardened.

PRIMA 75.
 M-1340; 0.75 C, 0.4 Si, 0.6 Mn, bal Fe.
 Heat treated: 180,000 TS; 140,000 YS; 14 El; 36 RA; 375 Brin.
 For crimpers, punches, springs, rails; water hardened.

PRIMA 90.
 M-1340; 0.9 C, 0.4 Si, 0.6 Mn, bal Fe.
 Heat treated: 190,000 TS; 145,000 YS; 10 El; 30 RA; 40 Brin.
 For tools, cutters, springs, drills; water hardened; Type Wl.

PRIMA H.
 M-1344; 1.3 C, 0.25 max Si, 0.25 max Mn, bal Fe.
 For engravers tools, drills, reamers, form cutters; Type Wl; water hardened.

PRIMA HARTE 3.
 M-1346; 0.9 C, 0.4 Si, 0.6 Mn, bal Fe.
 Heat treated: 190,000 TS; 145,000 YS; 10 El; 30 RA; 400 Brin.
 For springs, tools, cutters, dies; Type Wl; water hardened.

PRIMA HARTE 4.
 M-1346; 0.60 C, 0.4 Si, 0.6 Mn, bal Fe.
 Heat treated: 160,000 TS; 113,000 YS; 12 El; 40 RA; 325 Brin.
 For crankshafts, axes, axles, crimpers, punches, gears; water hardened.

PRIMA HARTE 5.
 M-1346; 0.45 C, 0.4 Si, 0.6 Mn, bal Fe.
 Hot rolled: 98,000 TS; 59,000 YS; 24 El; 45 RA; 215 Brin.
 For gears, pinions, bolts, studs, crankshafts; water hardened.

PRIMA HARTE 6.
 M-1346; 0.35 C, 0.3-0.5 Si, 0.3-0.8 Mn, bal Fe.
 Hot rolled: 85,000 TS; 54,000 YS; 30 El; 53 RA; 185 Brin.
 For gears, bolts, fasteners, crankshafts; water hardened.

PRIMA HARTE 3/4.
 M-1346; 0.75 C, 0.4 Si, 0.6 Mn, bal Fe.
 Heat treated: 175,000 TS; 126,000 YS; 10 El; 35 RA; 375 Brin.
 For springs, tools, hammers, rails, axes; Type Wl; water hardened.

PRIMA MH.
 M-1344; 1.15 C, 0.25 max Si, 0.25 max Mn, bal Fe.
 Annealed: 100,000 TS; 56,000 YS; 20 El; 40 RA; 210 Brin.
 For drills, taps, reamers, hobs, cutters; Type Wl; water hardened.

PRIMA MH.
 M-1352, M-1338; 0.90 C, 0.25-0.50 Si, 0.3-0.8 Mn, bal Fe.
 Heat treated: 190,000 TS; 145,000 YS; 10 El; 30 RA; 400 Brin.
 For springs, tools, drills, taps, reamers, hobs; Type Wl; water hardened.

PRIMER GILDING.
97 Cu, 3 Zn, traces Pb, Fe.
For primers, base for fire-enameled parts.

PRINCE.
M-365; 1 C, 1 Cr, bal Fe.
For ballbearing; water hardening.

PRINCES METAL NO. 1.
83-61 Cu, 17-39 Zn.
For hardware; brass.

PRINCES METAL NO. 2.
85 Cu, 15 Zn.
For flexible hose; good workability.

PRIZE RIBBON.
M-959; Cu, Sb, bal Sn.
Cast: 15,200 TS; 34.4 Brin.
For bearings; Babbitt.

PRK-33.
M-40, M-157; 3.7 Co, 13.6 Cr, 79.5 Fe, 0.84 Mo, 1.5 C, 0.60 Si, 0.25 V, 0.30 Mn, 0.50 Ni.
Air hardened: 602-654 Brin.
For milling cutters, twist drills, press tools; non-scaling steel, non-deforming.

PRK 33 COBALTCROM.
M-289; 1.4 C, 13 Cr, 3.3 Co, 0.5 Ni, 0.6 Mo, bal Fe.
For blanking and forming dies, trimmers, punches; air hardened, non-deforming, wear resistant.

PRK-SH.
M-40; C, Cr, W, bal Fe.
For self hardening welding electrodes; for hardened tool and die steels.

PROBEDIT.
M-USSR; W.C.
For cutting tools and dies; sintered.

PRODUCTION-ROD NO. 140.
M-1096; 0.06-0.10 C, 0.55 Mn. 0.03 S, 0.01 P, 0.10-0.25 Si, bal Fe.
Welded: 68,000-75,000 TS; 55,000-65,000 YS; 23-27 El; 50-55 RA.
For all-purpose welding rod; AWS-E6013.

PROFERALL.
M-347; 2.2-2.35 Si, 0.5-0.6 Mn, 0.55-1.0 combined C, 3.15 total C, 0.8-1.0 Cr, 0.4-0.5 Mo, 0.4-0.5 Ni, bal Fe.
60,000-75,000 TS; 250-300 Brin.
For diesel engine parts, crankshafts, cylinders, pistons, refrigeration parts; high test cast iron; wear resistant.

PROFERALL A.
M-347; 3.3 C, 1.5 Ni, 0.9 Cr, 0.2 Mo, 2 Si, bal Fe.
Cast: 35,000 TS; 207-241 Brin.
For oil pump gears; cast iron.

PROGEN.
M-969; 0.33-0.40 C, 0.5-0.7 Mo, 0.55-0.75 Cr, 0.45-0.65 Cu, 0.5-0.7 Si, 0.1 Ti, bal Fe.
Annealed: 97,000 TS; 57,000 YS; 22 El; 42 RA; 197 Brin.
Hardened: 285,000 TS; 245,000 YS; 11 El; 40 RA; 555 Brin.
For tools, chisels, picks, dies for die casting; tough, shock resistant, self-tempering tool steel.

PROJECT-70.
M-32; 0.15 max C, 12-14 Cr, bal Fe, 0.15 S.
Annealed: 103,000 TS; 75,000 YS; 22 El; 216 Brin.
For fittings, gears, housings, valve stems, valve trim.
Corrosion resistant, free-cutting.

PROJECTILE STEEL.
M-U.S.; 2.4 Cr, 0.8 C, 0.4 Mn, 0.2 Si, bal Fe.
110,000 TS; 80,000 YS; 18 El.
For projectiles; oil hardening.

PROMAL.
M-89; 1.8 T.C., 0.4 C.C., 1.0 Si, bal Fe.
Cast: 65,000-70,000 TS; 45,000-50,000 YS; 14-10 El; 170-190 Brin.
For sprockets, gears, chains, brake drums; malleable iron.

PROMAL.
M-228; 1.8 total C, 0.4 combined C, 1.0 Si, 0.3 Mn, bal Fe.
Cast: 65,000-70,000 TS; 45,000-50,000 YS; 14-10 El; 170-190 Brin.
For cast chains, sprockets, parts, valve parts, wrench gears, brake drums; malleable cast iron.

PROMET 2S.
M-956; 70-75 Cu, 4.0-5,5 Sn, 19.0-23.0 Pb, 0.2-0.8 Ni.
Sand cast: 27,000 TS; 15,000 YS; 16 Dl; 45-55 Brin.
For bearings, bushings, sleeves.
AMS 4840; ASTM B144-3E.

PROMET 6SK.
M-956; 75-79 Cu, 6.5-8.0 Sn, 12.5-15.5 Pb, 1.25-1.75 Ni.
Sand Cast: 36,000 TS; 19,000 YS; 18 El; 60 Brin.
For bearings, bushings, liners, sleeves, pump bodies; free-cutting.

PROMET 80A.
M-956; 78-82 Cu, 9-11 Sn, 8-11 Pb, 0.75 max Zn, 0.75 max Ni.
Sand cast: 39,000 TS; 18,000 YS; 30 El; 67 Brin.
High leaded tin bronze for bearing applications.

PROMET 83A.
M-956; 81-85 Cu, 6.25-7.5 Sn, 6.0-8.0 Pb, 2.0-4.0 Zn, 0.50 max Ni.
Sand cast: 34,000 TS; 18,000 YS; 15 El; 65 Brin.
Leaded, general utility bearing bronze for bearings, bushings.
SAE 660. ASTM B144-3B.

PROMET 85A.
M-956; 84-86 Cu, 4.0-6.0 Sn, 4.0-6.0 Pb, 4.0-6.0 Zn, 0.8 max Ni.
Sand Cast: 35,000 TS; 15,000 YS; 32 El; 54-67 Brin.
For bearings, bushings.
SAE 40; ASTM B145-4A.

PROMET 89-S.
M-956; 86-89 Cu, 7.0-9.0 Sn, 2.0-4-0 Pb, 0.25-0.75 Ni.
Cent. chill cast: 40,000 TS; 20,000 YS; 30 El.
For heavy duty bearings, gears, worm wheels.

PROMET 91-CK.
M-956; 83 min Cu, 9.0-10.5 Sn, 1.3-3.0 Pb, 0.75-1.5 Ni, 1.5 max Zn.
Cent. chill cast: 50,000 TS; 22,000 YS; 15 El.
For heavy duty bearings, gears, worm wheels.

PROMET 91-SK replaced by PROMET 91-CK.

PROMET 93 AB.
M-956; 8.5-9.5 Al, 2.5-4.0 Fe, 86.0 min Cu.
Sand cast: 79,000 TS; 28,000 YS; 30 El; 130 Brin.
Alpha phase aluminum bronze.
For worm wheels, bearings and bushings, slides, pump parts, utility machine parts.

PROMET 104 AB.
M-956; 10-11 Al, 3.5 Fe, 1.0 max Ni, 0.5 max Mn, 83.0 min Cu.
Sand Cast (min): 80,000 TS; 30,000 YS; 12 El; 145 Brin.
Centrifugal (min): 90,000 TS; 32,000 YS; 12 El; 150 Brin.
For gears, worm wheels, rolling mill bearings, and slippers, bearing races.
Aluminum bronze, wear and fatigue and corrosion resistant.

PROMET 115-N.
M-956; 78 min Cu, 3.0-5.0 Fe, 3.0-5.5 Ni, 10.0-11.5 Al, 3.5 max Mn.
Cast: 100,000 TS; 45,000 YS; 8 El; 195 Brin.
Heat Treated: 118,000 TS; 80,000 YS; 7 El; 240 Brin.
For bearings, gears, cams, valve and pump parts, castings.
Corrosion resistant, heat treatable. ASM 4880.

PROMET 150.
M-956; 60-68 Cu, 0.2 max Sn, 0.2 max Pb, 2.0-4.0 Fe, 3.5-5.0 Al, 2.5-5.0 Mn, bal Zn.
Cast (min): 90,000 TS; 45,000 YS; 18 El.
For gears, cams, hydraulic cylinder parts, bearings.

PROMET 200.
M-956; 60-68 Cu, 3.5-7.0 Al, 2.0-4.0 Fe, 2.5-5.0 Mn, 0.2 max Sn, 0.2 max Pb, bal Zn.
Sand cast: 110,000-119,000 TS; 60,000-80,000 YS; 12-14 El; 200-225 Brin.
For valve stems, gears, cams, bridge parts, heavy load bearings, hydraulic cylinder parts.
SAE 430B ASTM B147-8C.

PROMET 406.
M-956; 62 Cu, 1.0 Sn, 1.0 Pb, 36 Zn (nominal).
Cast: 40,000 TS; 14,000 YS; 15 El.
For bushings, hardware fittings, ornaments.
QQ-B-621 Class A.

PROMET 712.
M-956; 67-72 Cu, 8.5-11.0 Sn, 18-22 Pb, 0.25-1.0 Ni.
Sand cast: 33,000 TS; 17,000 YS; 18 El; 53-63 Brin.
For bearings, especially under boundary or mixed film lubricants.

PROMET 782.
M-956; 65-72 Cu, 7.0-9.0 Sn, 20-24 Pb, 0.25-1.0 Ni.
Sand cast: 30,000 TS; 17,000 YS; 15 El; 45-60 Brin.
High lead tin bronze; for bearings, especially under boundary or mixed film lubricants; Can deform to correct bad fit.

PROMET NO. 3.
M-956; 85-89 Cu, 7.5-9.0 Sn, 0.5-1.5 Pb, 0.5-1.0 Ni, 2.5-5.0 Zn.
Bearing and gear bronze.
Sand cast: 40,000 TS; 20,000 YS; 28 El; 70 Brin.
For high duty bearings, gears, pump impellers, piston rings, steam fittings.

PROMET NO. 6.
M-956; 76-79 Cu, 4.5-6.5 Sn, 12.5-15.5 Pb, 1.0-1.7 Ni.
Cast: 30,000 TS; 20,000 YS; 16 El; 64 Brin.
For bearings, bushings, castings; heavy duty.

PROMET NO. 101-AB.
M-956; 86-90 Cu, 0.75-1.5 Fe, 9-11 Al, 0.50 max Mn, 0.5 max Ni.
Cast: 85,000 TS; 45,000 YS; 25 El.
For gears, propellers, shafts; Al-bronze, corrosion resistant.

PROMETAL.
M-700; C, Cr, Ni, bal Fe.
For stainless parts; heat and corrosion resistant.

PROMETAL CHROME INOXYDABLE R.F.
M-700; C, Cr, Ni, bal Fe.
For stainless parts; heat and corrosion resistant.

PROMETAL INOXYDABLE R.F.-3.
M-700; C, Cr, Ni, bal Fe.
For stainless parts; heat and corrosion resistant.

PROMETHIUM.
67 Cu, 30 Zn, 3 Al.
For condenser tubes; high ductility.

PROMET M-640.
M-956; 84-88 Cu, 10-12 Sn, 0.5-1.25 Pb, 1.25-1.75 Ni, 0.10-0.30 P.
Centrifugal Chill Cast: 55,000 TS; 27,000 YS; 12 El.
For heavy duty bearings, gears, worm wheels.

PROMET P-48N.
M-956; 9.25-10.75 Sn, 2.0-3.5 Pb, 3.0-4.0 Ni, (others 1.25), bal Cu.
Leaded nickel tin bronze.
Sand Cast: 50,000 TS; 28,000 YS; 18 El; 85 Brin.
For heavy duty bearings, nuts, gears, worm wheels, ASTM B427; Alloy D.

PROMET X.
M-956; 75 Sb, 3.5 Cu, bal Sn.
Cast: 10,900 TS; 8 El.
For bearings, bushings; Babbitt; C.S. 13625.

PROMET XX.
M-956; 7.5 Pb, 15 Sb, 10 Sn.
Cast: 10,500 TS; 4 El.
For bearings, bushings; Babbitt; C.S. 17150.

PROMET XXX.
M-956; 17.5 Sb, 1.4 Sn, 1.5 As, 0.5 Cu, bal Pb.
Cast: 10,350 TS; 2 El.
For bearings, bushings; Babbitt; C.S. 14200.

PROOF AG.
M-1522; 99.9 Ag.
Melt range: 960-960°C.
Max stress: 12.6 kgf/mm^2; 45 El.
For vacuum brazing, fuses, electronic applications.

PROPELLER BUSHING.
69 Zn, 19 Sn, 7 Sb, 5 Cu.
For propeller bushings.

PROPLATINUM.
72 Ni, 23.6 Ag, 3.7 Bi, 0.7 Au.
For corrosion resistant parts; platinum substitute.

PROTECTIVE DECK PLATE.
3.5 Ni, 1.5 Cr, 0.2-0.3 C, bal Fe.
For armor plate; case-hardened.

PROTECTIVE NETTING.
27.8 Ni, 0.4 C, bal Fe.
For armor steel for torpedo defense, netting; corrosion resistant.

PROTOCHROME.
M-1201; 2.0-2.5 C, 24-27 Cr, bal Fe.
Welded: 600 Brin.
For hard surfacing electrode; heat and abrasion resistant.

PROTOLOY.
M-1523; 80 Ni, 20 Cr.
Resistance Wire: 650 ohms per cir. mil. ft.
Ann: 100,000 psi; hard drawn: 225,000 psi.
For resistors and heating elements.
Max. oper. temp.: 1100°C.

PROTOLOY B.
M-1523; 80 Ni, 20 Cr.
Resistance wire: 680 ohms per cir. mil. ft.
Ann: 100,000 psi; hard drawn: 225,000 psi.
For resistors.
Max. oper. temp.: 1100°C.

PRV 13.
M-1495; 0.11 C, 13.0 Cr, 0.7 Mn, 2.0 Si, bal Fe.
Resistant to oxidation reactions in sulphurous media up to 850°C.
For heat exchangers, pyrometric sheaths.

PRV 18.
M-1495; 0.10 C, 18.5 Cr, 0.6 Mn, 1.85 Si, bal Fe.
Resistant to sulphurous media to 1000°C.
For heat exchangers, heating cells.

PRV 25.
M-1495; 0.10 C, 24.5 Cr, 1.2 Mn, 2.0 Si, bal Fe.
Resistant to sulphurous media to 1100°C.
For heat exchangers, pyrometric sheaths.

PRV 25-12.
M-1495; 0.15 C, 23 Cr, 14 Ni, bal Fe.
Resistant to creep rupture and oxidation up to 1050°C.
For heat exchangers, reactors.
Similar to AISI 309.

PRV 29.
M-1495; 0.11 C, 29 Cr, 1.0 Si, bal Fe.
Resistant to sulphurous media to 1150°C.
For heat exchangers, pyrometric sheaths.

PRV 18-10.
M-1495; 0.07 C, 18 Cr, 11 Ni, 0.5 Ti, bal Fe.
Resistant to creep rupture and oxidation to 850°C.
For heat exchangers, reactors, heating cells.
Similar to AISI 321.

PRV 25-20.
M-1495; 0.12 C, 25 Cr, 20 Ni, bal Fe.
For high temperature operation to 1150°C.
Similar to AISI 310.

PRV 33-20.
M-1495; 0.10 C, 21.0 Cr, 32.5 Ni, bal Fe.
For high temperature operation to 1150°C.

PRV 75-15.
M-1495; 0.06 C, 15.5 C, 72.0 Ni, bal Fe.
Oxidation resistant to 1200°C.
For furnace equipment, pyrometer sheaths, heat exchangers.

PSI NO. 1.
M-1053; 0.92-1.02 C, 0.95-1.2 Mn, 0.5-0.7 Si, 0.9-1.15 Cr, bal Fe.
For master tools, gauges, ceramic molds; non-deforming, wear resistant.

PSI NO. 2.
M-1053; 0.87-0.97 C, 1.4-1.7 Mn, 0.6-0.8 Si, 1.4-1.7 Cr, bal Fe.
For master tools and gauges, clutch liners, pump sleeves; non-deforming, wear resistant.

P-TIN ALLOY.
M-1557; 55.5 Sn, 41.1 Pb, 3.4 Sb.
For soldering.
Soft solder.

PULSUS.
M-912; 0.35-0.42 C, 4.7-5.0 Cr, 1.5 Ni, 0.5 Mo, 0.2 V, bal Fe.
For tools, dies; hot work steel.

PUMA.
M-1796; 0.65 C, 4.25 Cr, 0.80 Mo, 18.0 W, 1.6 V, 15 Co, bal Fe.
Co-W high speed steel; for rough machining, especially on austenitic steels.

PUMP COCKS.
72 Zn, 7 Cu, 21 Sn.
For pump cocks; castings.

PURALLOY.
M-1466; 3 C, Si, Mn, bal Fe.
For rolls for paper and allied industries; cast iron.

PURCO CHISEL.
M-372; 0.7 C, 2 Si, 0.9 Mn, bal Fe.
For chisels, punches; tough.

PURE ORE AIRCHROME see AIR CHROME.

PURE ORE CLIPPER see CLIPPER.

PURE ORE EXTRA.
M-335; C as specified, bal Fe.
For tools, dies; water hardening; AISI W1.

PURE ORE HI-RUN see HI-RUN.

PURE ORE SPECIAL.
M-335; C as specified, bal Fe.
Water hardening tool steel; for dies and special tools; AISI W1.

PURE ORE STANDARD.
M-335; C as specified; bal Fe.
Water hardening tool steel; for tools, dies, AISI W2.

PURE ORE SUPERALLOY see SUPERALLOY.

PURE SILVER SOLDER.
72 Ag, 28 Cu.
For silver solder; corrosion resistant.

PURETUNG.
M-725; 99.9 W.
For welding electrodes.

PURO 17/7.
M-1339; 0.12 max C, 1.0 max Si, 2.0 max Mn, 16.0-18.0 Cr, 7.0-9.0 Ni, bal Fe.
Austenitic stainless steel; work-hardens easily; for springs.
W.-Nr. 1.4310; similar to AISI 301.

PURO 18/8.
M-1339; 0.15 max C, 18 Cr, 8.5 Ni, bal Fe.
Annealed: 80,000 TS; 35,000 YS; 55 El; 75 RA; 150 Brin.
For chemical plant equipment, tanks; Type 302; stainless, austenitic.

PURO 18/8 E.
M-1339; 0.10 max C, 1.0 max Si, 2.0 max Mn, 17.0-19.0 Cr, 9.0-11.5 Ni, Ti = 5 x C min, bal Fe.
Stabilized austenitic stainless steel; weldable.
W.-Nr. 1.4541; similar to AISI 321.

PURO 18/8 L SUPRA.
M-1339; 0.03 max C, 1.0 max Si, 2.0 max Mn, 17.0-20.0 Cr, 10.0-12.5 Ni, bal Fe.
Austenitic stainless, for food industries.
W.-Nr. 1.4306; AISI 304 L.

PURO 18/8 NB.
M-1339; 0.10 max C, 1.0 max Si, 2.0 max Mn, 17.0-19.0 Cr, 9.0-11.5 Ni Nb = 10xC min, bal Fe.
Stabilized austenitic stainless steel; weldable.
W.-Nr. 1.4550; similar to AISI 347.

PURO 18/8 NB SG.
M-1339; 0.07 max C, 17-20 Cr, 9.0-11.5 Ni, Nb = 10xC min, bal Fe.
Stabilized austenitic stainless steel.
W.-Nr. 1.4543; AISI 347.

PURO 18/8 S.
M-1335; 0.15 max C, 1.0 max Si, 2.0 max Mn, 17.0-19.0 Cr, 8.0-10.0 Ni, 0.15-0.35 S, bal Fe.
Free machining type austenitic stainless.
W.-Nr. 1.4305; AISI 303.

PURO 18/8 SUPRA.
M-1339; 0.07 max C, 1.0 max Si, 2.0 max Mn, 17.0-20.0 Cr, 8.0-10.0 Ni, bal Fe.
Austenitic stainless steel.
W.-Nr. 1.4301; AISI 304.

PURO 18/10 L SUPRA.
M-1339; 0.03 max C, 1.0 max Si, 2.0 max Mn, 16.5-18.5 Cr, 11.0-14.0 Ni, 2.0-2.5 Mo, bal Fe.
Austenitic stainless for equipment in chemical and cellulose industries.
W.Nr. 1.4404; AISI 316L.

PURO 18/10 SUPRA.
M-1339; 0.07 max C, 1.0 max Si, 2.0 max Mn, 16.5-18.5 Cr, 10.5-13.5 Ni, 2.0-2.5 Mo, bal Fe.
Austenitic stainless for equipment in chemical and textile industries.
W.-Nr. 1.4401; AISI 316.

PURO 18/12 LM SUPRA.
M-1339; 0.03 max C, 1.0 max Si, 2.0 max Mn, 16.5-18.5 Cr, 12.5-15.0 Ni, 2.5-3.0 Mo, bal Fe.
Austenitic stainless steel; improved corrosion resistance for chemical equipment.
W.-Nr. 1.4435; similar to low carbon AISI 317.

PURO 18/14 M4.
M-1339; 0.07 max C, 1.0 Max Si, 2.0 max Mn, 16.0-18.0 Cr, 12.5-14.5 Ni, 4.0-5.0 Mo, bal Fe.
Austenitic stainless steel; extra resistance to corrosion and pitting for chemical equipment.
W.-Nr. 1.4449; similar to AISI 317.

PURO 18/18 M NB.
M-1339; 0.07 max C, 17.5 Cr, 20.0 Ni, 2.0 Cu, 2.2 Mo, Cb = 8 x C min, bal Fe.
Stabilized austenitic stainless steel.
W.-Nr. 1.4505.

PURO 19/11.
M-1339; 0.07 max C, 1.0 max Si, 2.0 max Mn, 17.0-20.0 Cr, 10.0-12.0 Ni, bal Fe.
Austenitic stainless steel for chemical industry.
W.-Nr. 1.4303; similar to AISI 305.

PURO 25/25 M.
M-1339; 0.07 max C, 25.0 Cr, 25.0 Ni, 2.2 Mo, Ti = 10 x C min, bal Fe.
Austenitic stainless steel.
For chemical equipment.
W.-Nr.1.4577.

PURO 275 N.
M-1339; 0.10 C, 26.0-28.0 Cr, 4.0-5.0 Ni, 1.3-2.0 Mo, bal Fe.
Stainless steel for high temperature operations.
W.-Nr. 1.4460.

PURO ANTIOXYDUR.
M-1339; 0.15-0.23 C, 1.0 max Si, 1.0 max Mn, 16.0-18.0 Cr, 1.5-2.5 Ni, bal Fe.
Martensitic stainless steel.
W.-Nr. 1.4057; similar to AISI 431.

PURO CR.
M-1339; 0.10 max C, 1.0 max Si, 1.0 max Mn, 15.5-17.5 Cr, bal Fe.
Ferritic stainless steel.
W.-Nr. 1.4016; AISI 430.

PURO CRTI.
M-1339; 0.10 C, 16.0-18.0 Cr, Ti = 7 x C min, bal Fe.
Stabilized stainless steel.
Not hardenable by heat treating; weldable.
W.Nr. 1.4510.

PURO H.
M-1339; 0.4 C, 13 Cr, 0.4 Si, bal Fe.
Annealed: 100,000 TS; 55,000 YS; 25 El; 55 RA; 200 Brin.
For valve trim, gauges, valves, turbine blades; Type 420; corrosion resistant.

PURO H EXTRA.
M-1339; 0.90 C, 17.0-19.0 Cr, 1.0-1.3 Mo, 0.07-0.12 V, bal Fe.
Martensitic stainless steel.
W.Nr. 1.4112; similar to AISI 440B.

PURO H MO2.
M-1339; 0.38 C, 16.5 Cr, 1.15 Mo, bal Fe.
Martensitic stainless steel.
W.-Nr. 1.4122.

PURO H MO EXTRA.
M-1339; 1.05 C, 16.0-18.0 Cr, 0.60 Mo, bal Fe.
Martensitic stainless steel.
For stainless ball bearings.
W-Nr. 1.4125; AISI 440C.

PURO M.
M-1339; 0.2 C, 0.4 Si, 13 Cr, bal Fe.
Annealed: 95,000 TS; 50,000 YS; 28 El; 58 RA; 195 Brin.
For valve trim, gauges, cutlery, turbine blades; Type 420; stainless, hardenable.

PURO M 2.
M-1339; 0.12-0.17 C, 1.0 max Si, 1.0 max Mn, 12.0-14.0 Cr, bal Fe.
Martensitic stainless steel.
W.-Nr. 1.4024.

PURO-MMO.
M-1652, M-1339; 0.20 C, 13 Cr, 1 Mo, bal Fe.
Annealed: 95,000 TS; 50,000 YS; 25 El; B 92 Rock.
Ht. Tr.: 250,000 TS; 215,000 YS; 8 El; C 52 Rock.
For cutlery, surgical instruments, gears, shafts, needle valves, gages.
Corrosion resistant, hardenable.

PURO-MMO 2.
M-1652, M-1339; 0.15 C, 13 Cr, 1 Mo, bal Fe.
Annealed: 70,000 TS; 35,000 YS; 30 El; B 80 Rock.
Cold Drawn: 95,000 TS; 80,000 YS; 15 El; B 92 Rock.
For chemical plant and oil refinery equipment.
Corrosion resistant.

PURON.
M-118; 99.95 Fe.
For spectroscopic and magnetic standards; high purity iron.

PURO W.
M-1339; 0.10 C, 0.4 Si, 13 Cr, bal Fe.
Annealed: 75,000 TS; 40,000 YS; 35 El; 70 RA; 155 Brin.
Cold drawn: 100,000 TS; 85,000 YS; 17 El; 60 RA; 200 Brin.
For valve trim, cutlery, surgical instruments; Type 410; stainless.

PURO W2.
M-1339; 0.08 max C, 1.0 max Mn, 1.0 max Si, 12.0-14.0 Cr, bal Fe.
Chromium stainless steel.
W.-Nr. 1.4000.

PURO W2 AL.
M-1339; 0.08 max C, 1.0 max Si, 1.0 max Mn, 12.0-14.0 Cr, 0.10-0.30 Al, bal Fe.
Ferritic type stainless steel.
AISI 405; W.Nr. 1.4002.

PURO WA.
M-1339; 0.15 max C, 1.0 max Si, 1.0 max Mn, 12.0-13.0 Cr, 0.25 S, bal Fe.
Martensitic stainless steel.
Free machining type.
W.-Nr. 1.4005; AISI 416..

PURO WS.
M-1339; 0.10-0.17 C, 15.5-17.5 Cr, 0.2-0.3 Mo, 0.15-0.35 S, bal Fe.
Free machining ferritic type stainless steel.
W.-Nr. 4104; AISI 430 F.

PURO WW.
M-1339; 0.08 max C, 1.0 max Si, 1. max Mn, 13.0-15.0 Cr, bal Fe.
Chromium stainless steel.
W.-Nr. 1.4001.

PURPLE LABEL (AISI T4).
M-69; 0.74 C, 5 Co, 18.5 W, 4.3 Cr, 1.5 V, bal Fe.
For high speed cutting tools, lathe and planer tools; high speed steel.

PURPLE LABEL.
M-343; 0.5-1.1 C, 0.2-0.9 Mn, bal Fe.
Water hardened.
For tools.

PURPLE LABEL EXTRA.
M-69; 0.78 C, 0.8 Mo, 18.5 W, 8 Co, 2 V, 4 Cr, bal Fe.
Annealed: 228-241 Brin.
For heavy roughing cutters, lathe and planer tools; Type T5; high speed steel.

PW 190R.
M-1321; 0.35 C, 2.8 Cr, 9 W, 0.3 V, bal Fe.
For extrusion rams and liners; hot work steel, oil hardened.

PWA-649.
M-1491; 0.03-0.10 C, 0.35 max Mn, 50.0-55.0 Ni 17.0-21.0 Cr, 2.8-3.0 Mo, bal Fe.
At R.T. 200,000 TS; 170,000 YS; 20 El; 25 RA; C 42 Rock.
At 1200°F; 165,000 TS; 140,000 YS; 20 El; 25 RA.
For aircraft and jet engine components, high temperature bolting.
Similar to alloy 718C; Corrosion and oxidation resistance.

PWA 651A.
M-1278; 23.5-26.5 Cr, 2-4 Mo, 5 max Fe, 10-15 Co, 6-8 W, bal Ni.
For jet engine components; heat and corrosion resistant.

PWA 652.
M-1491; 0.10 max C, 0.50 max Mn, 0.75 max Si, 18-21 Cr, 3.5-5.0 Mo, bal Ni.
Heat Treated: 188,000 TS; 115,000 YS; 28 El; 25 RA.
For jet engine and gas turbine buckets and discs, high temperature bolts.
Creep and heat resistant. Similar to Waspaloy.

PWA 653.
M-1491; 0.4-0.5 C, 10-12 W, 20-22 Cr, 1 max Ni, 0.6 max Fe, 1.5-2.0 Cb + Ta, bal Co.
Cast: 120,000 TS; 90,000 YS; C 38 Rock.
At 1800°F: 37,000 TS; 27,000 YS; 24 El.
For gas turbine engine components, turbine nozzle vanes.
High heat and oxidation resistant in the 1000-2000°F range.

PWA 656B.
M-1491; 0.05-0.09 C, 0.15 max Mn, 0.20 max Si, 14.0-15.5 Cr, 3.90-4.90 Mo, bal Ni.
Rolled: 205,000 TS; 140,000 YS; 17 El; 20 RA.
For turbine and jet engine components, combustion chambers.
Creep and oxidation resistant. Similar to U-700.

PWA 663.
M-1491; 0.1 C, 8 Cr, 10 Co, 6 Mo, 4.25 Ta, 6 Al, 1 Ti, Zr, B, bal Co.
At R.T.: 141,000 TS; 120,000 YS; 8 El.
At 1400°F: 138,000 TS; 117,000 YS; 4 El.
For high temperature applications, turbines, jet engines.
Heat resistant. Similar to B1900.

PWA 664.
M-1491; 0.12-0.17 C, 8-10 Cr, 11.5-13.5 W, 9-11 Co, 0.75-1.25 Cb, 4.7 5.2 Al, 2 Ti, 0.05 Zr, 0.01-0.02 B, bal Ni.
As Cast: 135,000 TS; 120,000 YS; 7 El.
At 1400°F: 135,000 TS; 122,000 YS; 3.5 El.
For cast turbine blades and vanes in aircraft gas turbines.
Precipitation hardening, high temperature strength.

PWA 786A.
  M-1278; 0.25 max C, 15.5-17.5 Cr, 3.5-5.0 Ni, 3.5-4.5 Cu, 0.2 Ti, bal Fe.
  For chemical plant equipment; heat and corrosion resistant.

PWA 1030.
  M-1491; 0.06 C, 0.08 Mn, 0.01 S, 0.004 P, 0.1 Si, 20.5 Cr, 14.2 Co, 4.2 Mo, 3 Ti, 1.5 Al, 0.03 Zr, 0.003 B, 0.05 Cu, bal Ni.
  Aged: 186,000 TS; 138,000 YS; 26 El; C 39 Rock.
  For jet engine turbine buckets and discs, high temperature bolts, missile systems.
  Precipitation hardened. Waspalloy. High temperature strength.

PYRAD see **AUBERT & DUVAL PYRAD.**

PYR-AIR-DIE.
  M-1432; C, alloy, bal Fe.
  For blanking and forming dies, cutter blades; air hardened, non-deforming.

PYR-AIR-DIE MODIFIED.
  M-1432; 0.55 C, 0.9 Mn, 0.25 Si, 3.5 Cr, 0.25 Ni, 0.25 V, 1.4 Mo, bal Fe.
  Air or oil hardening tool steel.

PYRAMID BABBITT.
  M-178; Sn, Sb, bal Pb.
  Cast: 17,800 TS; 8,870 Brin.
  For bearings, Babbitts; pouring temperature 875-1000°F.

PYRASTEEL-A.
  M-146; 0.3 C, 4-8 Ni, 8 Cr, bal Fe.
  For carburizing boxes, furnace parts and grids, heat and resistant parts; heat resistant.

PYRASTEEL-B.
  M-146; 0.3 C, 15 Ni, 14 Cr, bal Fe.
  For carburizing boxes, furnace parts; heat resistant.

PYRASTEEL-C.
  M-146; 0.3 C, 25 Ni, 20 Cr, bal Fe.
  Cast: 68,000 TS; 38,000 YS; 17 El; 160 Brin.
  For fornace parts, torch nozzles, trays; heat resistant.

PYRASTEEL-D.
  M-146; 0.3 C, 35 Ni, 25 Cr, bal Fe.
  Cast: 70,000 TS; 40,000 YS; 10 El; 12 RA; 170 Brin.
  For carburizing boxes, pots; heat resistant.

PYRASTEEL NO. 14.
  M-146; 0.2 C, 18 Cr, 0.5 Mo, bal Fe.
  At 70°F: 110,000 TS; 75,000 YS; 18 El; 32 RA; 215 Brin.
  At 1000°F: 70,000 TS; 49,000 YS; 24 El; 64 RA.
  For corrosion and heat resistant parts, oil refineries; corrosion and heat resistant.

PYRASTEEL NO. 18.
  M-146; 25-28 Ni, 16-18 Cr, 2.5 Si, bal Fe.
  For chain conveyors, lead pots; resistant to 1800°F.

PYRASTEEL NO. 20.
  M-146; 0.2 C, 17-18 Cr, 35-38 Ni, 2.5 Si, bal Fe.
  For heat resistant parts, furnace parts; corrosion and heat resistant.

PYRASTEEL NO. 2000.
  M-146; 25-27 Cr, 12-14 Ni, 1.5 Si, C, bal Fe.
  For castings, kilns, clinker coolers; heat and corrosion resistant to 2000°F.

PYRASTEEL SPECIAL.
  M-146; 14 Ni, 25-28 Cr, 2.5 Si, bal Fe.
  For salt pots; resistant to 2000°F.

PYRISTA.
  M-521; 0.09 C, 29.0 Cr, 1.8 Ni, bal Fe.
  Ferritic stainless steel bar.

PYROCAST.
  M-192; 1.75-2.0 C, 24-28 Cr, bal Fe.
  Cast: 70,000 TS; 70,000 YS; 0 El; 0 RA; 300 Brin.
  For heat treating and carburizing boxes, furnace parts; high heat resistance.

PYROCHROM 20.
  M-1342; 0.05 max C, 25 Cr, 20 Ni, bal Fe.
  Drawn wires: 60-75 kp/mm² TS; 30 min El.
  For electric heating elements.

PYROCHROM 30.
  M-1342; 0.05 max C, 20 Cr, 34 Ni, bal Fe.
  Drawn wires: 60-75 kp/mm² TS; 30 min El.
  For electric heating elements.

PYROCHROM 60.
  M-1342; 0.05 max C, 15 Cr, 64 Ni, bal Fe.
  Drawn wires: 60-75 kp/mm² TS; 30 min El.
  For electric heating elements.

PYROCHROM 80.
  M-1342; 0.05 max C, 18 Cr, 80 Ni, bal Fe.
  Drawn wires: 60-75 kp/mm² TS; 30 min YS.
  For electric heating elements.

PYRO DIE.
  M-38; 0.4 C, 1.0 Cr, 0.25 V, bal Fe.

PYRODIE.
  M-1724; 0.40 C, 1.0 Si, 5.0 Cr, 1.3 Mo, 1.0 V, bal Fe.

PYRODUR 7.
  M-1290; 0.30 C, 6.0 Cr, 2.25 Si, bal Fe.
  Cast, annealed: 71,000-100,000 TS; 6 min El.
  Elevated temperature applications; resistant to scaling to about 700°C.
  Werkstoff Nr. 1.4710; DIN G-X30CrSi6.

PYRODUR 8.
M-1290; 0.60 C, 13.0 Cr, 1.25 Si, bal Fe.
Cast, annealed: 71,000-100,000 TS; 4 min El.
Elevated temperature applications; resistant to scaling to about 800°C.
Werkstoff Nr. 1.4729; DIN G-X45CrSi13.

PYRODUR 9.
M-1290; 0.50 C, 17.0 Cr, 1.25 Si, bal Fe.
Cast, annealed: 71,000-100,000 TS; 2 min El.
Elevated temperature applications; resistant to scaling to about 900°C.
Werkstoff Nr. 1.4740; DIN G-X40CrSi17.

PYRODUR 9R.
M-1290; 1.5 C, 19.0 Cr, 1.25 Si, bal Fe.
Cast, annealed: 250-350 Brin.
Elevated temperature applications; resistant to scaling to about 900°C.
Werkstoff Nr. 1.4743; DIN G-X160CrSi18.

PYRODUR 10.
M-1290; 0.60 C, 22.5 Cr, 1.6 Si, bal Fe.
As cast: 200-280 Brin.
Elevated temperature applications; resistant to scaling to about 1050°C.
Werkstoff Nr. 1.4745; DIN G-X40CrSi23.

PYRODUR 12.
M-1290; 0.60 C, 29.0 Cr, 1.6 Si, bal Fe.
As cast: 220-300 Brin.
Elevated temperature applications; resistant to scaling to about 1180°C.
Werkstoff Nr. 1.4776; DIN G-X40CrSi29.

PYRODUR 12R.
M-1290; 1.40 C, 29.0 Cr, 1.60 Si, bal Fe.
Cast, annealed: 250-350 Brin.
Elevated temperature applications; resistant to scaling to about 1150°C.
Werkstoff Nr. 1.4777; DIN G-X130CrSi29.

PYRODUR 30-CN 10.
M-1290; 0.30 C, 22.0 Cr, 10.0 Ni, 1.25 Si, bal Fe.
As cast: 71,000-91,000 TS; 12 min El.
Elevated temperature operation: resistant to scaling to about 1050°C.
Werkstoff Nr. 1.4826; ASTM A297 Gr.HF. DIN G-X40CrNiSi229.

PYRODUR 30-CN 13.
M-1290; 0.30 C, 25.0 Cr, 12.5 Ni, 1.25 Si, bal Fe.
As cast: 71,000-92,000 TS; 10 min El.
Elevated temperature operations; resistant to scaling to about 1150°C.
Werkstoff Nr. 1.4837; DIN G-X35CrNiSi2512; ASTM A297; Gr.HH.

PYRODUR 40-CN5.
M-1290; 0.35 C, 27.0 Cr, 4.0 Ni, 1.25 Si, bal Fe.
As cast: 64,000-85,000 TS; 42,500 min YS; 4 min El.
Elevated temperature applications; resistant to scaling to about 1150°F.
Werkstoff 1.4823; DIN G-X40CrNiSi27 4. ASTM A297 Gr.HC.

PYRODUR 40-CN 20.
M-1290; 0.30 C, 25.0 Cr, 20.0 Ni, 1.25 Si, bal Fe.
As cast: 64,000-85,000 TS; 28,000 min YS; 8 min El.
Elevated temperature operations; resistant to scaling to about 1150°F.
Werksoff Nr. 1.4848; DIN G-X40CrNiSi2520; ASTM A297; Fr.HK.

PYRODUR 30-CN 38.
M-1290; 0.25 C, 37.0 Ni, 17.5 Cr, 1.25 Si, bal Fe.
As cast: 64,000-85,000 TS; 8 min El.
Elevated temperature operations; resistant to scaling to about 1120°C.
Werkstoff Nr. 1.4865; DIN G-X40NiCrSi3616. ASTM A297; Gr.HU.

PYRODUR 30 CN 38 NB.
M-1293; 0.10-0.20 C, 1.0-1.2 Si, 1.5 max Mn, 24.0-27.0 Cr, 19.0-21.0 Ni, bal Fe.
Castings for high temperature operation.
W.-Nr. 1.4849; DIN G-X 15 CrNiSi 25 20.

PYRODUR CO 50.
M-1290; 0.08 max C, 0.40 Si, 0.40 Mn, 27-29 Cr, 2.2 Mo, 48-50 Co, bal Fe.
Castings for high temperature operation.
W.-Nr. 2.4778; DIN G-CoCr 28.

PYRODUR CO 51.
M-1290; 0.08 max C, 0.40 Si, 0.40 Mn, 27-29 Cr, 2.2 Mo, 48-50 Co, Cb, bal Fe.
Castings for high temperature operations.
W.Nr. 1.4779; DIN G-CoCr 28 Nb.

PYROFIX-1.
M-1763; 0.25 max C, 2.0 max Mn, 1.0 max Si, 24.0-26.0 Cr, 19.0-22.0 Ni, bal Fe.
Austenitic stainless steel; for elevated temperature service.
AISI 310.

PYROFIX-2.
M-1763; 0.08 C, 2.0 max Mn, 0.75-1.5 Si, 17.0-20.0 Cr, 34.0-37.0 Ni, bal Fe.
Austenitic stainless steel; for high temperature equipment.
AISI 330.

PYROFIX-3.
M-1763; 0.20 max C, 2.0 max Mn, 1.0 max Si, 22.0-24.0 Cr, 12.0-15.0 Ni, bal Fe.
Austenitic stainless steel; for elevated temperature equipment.
AISI 309.

PYROFIX-5.
M-1763.
Austenitic stainless steel.

PYROFERAL.
M-Czech.; C, Cr, bal Fe.
For furnace parts and equipment; heat resistant.

PYR-OIL-DIE.
M-1432; C, alloy, bal Fe.
For dies, punches, crimpers, knives; oil hardened, non-deforming.

PYROMET see also **CARPENTER PYROMET.**

PYROMET 860.
M-1491; 0.05 C, 0.25 Mn, 0.1 Si, 13.0 Cr, 44.0 Ni, 4.0 Co, 6.0 mo, 3.0 Ti, 1.0 Al, 0.01 B, bal Fe.
For turbine engine parts.

PYROMIC.
M-108; 80 Ni, 20 Cr.
For electrical resistance; formerly "Pyromic No. 2."

PYRON 6.
M-1331; C, 2-3 Si, 6 Cr, bal Fe.
For oil refinery equipment; creep and heat resistant.

PYRON 7.
M-1331; C, 2-3 Si, 6 Cr, bal Fe.
For oil refinery equipment; creep and heat resistant.

PYRON 8.
M-1331; 0.12 max C, 2-3 Si, 6 Cr, bal Fe.
For oil refinery equipment; heat and creep resistant.

PYRON 8A1.
M-1331; 0.12 max C, 0.8 Si, 0.8 Al, 6.5 Cr, bal Fe.
For oil refinery equipment; heat and creep resistant.

PYRON 9.
M-1331; 0.12 max C, 2.2 Si, 13 Cr, bal Fe.
Annealed: 80,000 TS; 40,000 YS; 35 El; 65 RA; 160 Brin.
For valves, pumps, oil refinery equipment; heat and corrosion resistant.

PYRON 9A1.
M-1331; 0.12 max C, 1.2 Si, 1.0 Al, 13 Cr, bal Fe.
Annealed: 80,000 TS; 40,000 YS; 35 El; 65 RA; 160 Brin.
For valves, pumps, oil refinery equipment; heat and corrosion resistant.

PYRON 10.
M-1331; 0.12 max C, 2 Si, 18 Cr, bal Fe.
Annealed: 85,000 TS; 45,000 YS; 25 El; 50 RA; 160 Brin.
For furnace parts and equipment; heat and corrosion resistant.

PYRON 10A1.
M-1331; 0.12 max C, 1 Si, 1 Al, 18 Cr, bal Fe.
Annealed: 85,000 TS; 45,000 YS; 25 El; 50 RA; 160 Brin.
For oil refinery equipment; heat and corrosion resistant.

PYRON 12A1.
M-1331; 0.12 max C, 1.5 Si, 1.5 Al, 24 Cr, bal Fe.
Annealed: 85,000 TS; 50,000 YS; 30 El; 55 RA; 180 Brin.
For oil refinery and furnace equipment; heat resistant.

PYRON 100.
M-1777; 0.5 Mn, bal Fe.
Iron powder for production of high strength, low density P/M parts and bearings.

PYRON 2010.
M-1331; 0.15 C, 19.5 Cr, 9.5 Ni, bal Fe.
Annealed: 80,000 TS; 35,000 YS; 55 El; 75 RA; 150 Brin.
For chemical plant equipment, tanks, mixers; Type 302; stainless, austenitic.

PYRON 2419.
M-1331; 0.15 C, 2 Si, 24 Cr, 19 Ni, bal Fe.
Annealed: 100,000 TS; 45,000 YS; 50 El; 65 RA; 185 Brin.
For furnace parts, valves, pumps, turbine parts; Type 310; stainless, austenitic.

PYRON 2504.
M-1331; 0.2 C, 1.2 Si, 25 Cr, 4 Ni, bal Fe.
For furnace parts, valves, heat treating boxes; heat resistant.

PYRON 16.36.
M-1331; 0.12 C, 16 Cr, 25 Ni, bal Fe.
For furnace parts, chemical plant equipment; corrosion and heat resistant.

PYRON 20.13.
M-1331; 0.12 C, 20 Cr, 13 Ni, bal Fe.
Annealed: 80,000 TS; 35,000 YS; 50 El; 65 RA; 160 Brin.
For chemical plant equipment; stainless, austenitic.

PYRON 25.04 SPEZIAL.
M-1331; 0.12 C, 25 Cr, 4 Ni, bal Fe.
For furnace parts, valves, heat treating boxes; heat resistant.

PYRON AC-325.
M-1777.
Iron powder (95%-325 mesh); for chemical applications requiring high specific surface and reactivity.

PYRONEAL.
M-64; 0.55 C, 0.9 Cr, 0.7 Mo, 0.6 Si, 2.2 Ni, bal Fe.
For hot work dies, punches; oil or air hardened, hot work, press and upsetter dies.

PYRON D-63.
M-1777; 0.5 Mn, bal Fe.
Iron powder, more compressible than PYRON 100; particularly suited for manufacture of iron-copper-carbon P/M parts.

PYRON G85.
M-1331; 0.3 C, 2.2 Si, 6 Cr, bal Fe.
For oil refinery equipment; heat and creep resistant.

PYRON G95.
M-1331; 0.5 C, 1.5 Si, 17 Cr, bal Fe.
For furnace parts, chemical plant equipment; heat and corrosion resistant.

PYRON G105.
M-1331; 0.6 C, 1.5 Si, 22 Cr, bal Fe.
For furnace parts, heat treating boxes; heat resistant.

PYRON G115.
M-1331; 0.6 C, 195 Si, 29 Cr, bal Fe.
For furnace parts, heat treating boxes; heat resistant.

PYRON G2210.
M-1331; 0.4 C, 2 Si, 22 Cr, 9.5 Ni, bal Fe.
Cast: 85,000 TS; 45,000 YS; 35 El; 165 Brin.
For furnace parts, heat treating boxes, valves; corrosion and heat resistant.

PYRON G2519.
M-1331; 0.4 C, 2 Si, 25 Cr, 19 Ni, bal Fe.
Cast: 75,000 TS; 50,000 YS; 17 El; 170 Brin.
For pumps, valves, furnace parts, heat treating boxes; corrosion and heat resistant.

PYRON G27.04.
M-1331; 0.4 C, 1.3 Si, 27 Cr, 4 Ni, bal Fe.
Cast: 90,000 TS; 65,000 YS; 2 El; 212 Brin.
For furnace and turbine parts; heat resistant.

PYRON G55.24NI.
M-1331; 0.55 C, 1.3 Si, 55 Cr, 24 Ni, bal Fe.
For furnace parts; heat and oxidation resistant.

PYRON R-80.
M-1777.
Iron powder; low density, high green strength iron powder used in friction material formulations.

PYRON M-IRON.
M-1777; 1.0 Mo, bal Fe.
Iron powder used in low density P/M parts where high sintered hardness is required.

PYROPHORIC ALLOY.
M-U.S.; 10 Mn, 10 Sb, 20 Cr, 15 Ti, bal Fe.
For lighters for gas stoves, cigarette lighters; produces violent sparks when abrased.

PYROTEM.
M-64; 0.55 C, 2.1 Ni, 0.87 Cr, 0.73 Mo, bal Fe.
Annealed: 250 Brin.
Heat treated: 570 Brin.
For hot work dies, cold shear knives; oil hardened, shock resistant.

PYROTHERM 5 MO.
M-1342; 0.15 max C, 5 Cr, 0.5 Mo, bal Fe.
As tempered: 50-60 kp/mm^2 TS; 25 kp/mm^2 min YS; 23 min El.
Heat resistant rolled and wrought alloy.
For furnace engineering in cement, glass, metallurgical and petrochemical industries.

PYROTHERM 6.
M-1342; 0.12 max C, 6 Cr, bal Fe.
Ann: 55-70 kp/mm^2 TS; 35 kp/mm^2 min YS; 15 min El.
Heat resistant rolled and wrought alloy.
For furnace engineering in cement, glass, metallurgical and petrochemical industries.

PYROTHERM 14.
M-1342; 0.12 max C, 13 Cr, bal Fe.
Ann: 55-70 kp/mm^2 TS; kp/mm^2 min YS; 15 min El.
Heat resistant rolled and wrought alloy.
For furnace engineering in cement, glass, metallurgical, and petrochemical industries.

PYROTHERM 18.
M-1342; 0.12 max c, 18 Cr, bal Fe.
Ann: 55-70 kp/mm^2 TS; 35 kp/mm^2 min YS; 15 min El.
Heat resistant rolled and wrought alloy.
For furnace enginerring in cement, glass, metallurgical, and petrochemical industries.

PYROTHERM 28.
M-1342; 0.12 max C, 28 Cr, bal Fe.
Ann: 55-70 kp/mm^2 TS; 40 kp/mm^2 min YS; 12 min El.
Heat resistant rolled and wrought alloy.
For furnace engineering in cement, glass, metallurgical, and petrochemical industries.

PYROTHERM 18/8.
M-1342; 0.12 max C, 18 Cr, 9 Ni, bal Fe.
As quenched: 55-70 kp/mm^2 TS; 22 kp/mm^2 min YS; 40 min El.
Heat resistant rolled and wrought alloy.
For furnace engineering in cement, glass, metallurgical, and petrochemical industries.

PYROTHERM 18/8 TI.
M-1342; 0.15 max C, 18 Cr, 9 Ni, Ti = 5 x C, bal Fe.
As quenched: 55-70 kp/mm^2 TS; 22 kp/mm^2 min YS; 40 min El.
Heat resistant rolled and wrought alloy.
For furnace engineering in cement, glass, metallurgical, and petrochemical industries.

PYROTHERM 25/4.
M-1342; 0.20 C, 25 Cr, 4 Ni, bal Fe.
As quenched: 60-75 kp/mm^2 TS; 40 kp/mm^2 YS; 26 min El.
Heat resistant rolled and wrought alloy.
For furnace engineering in cement, glass, metallurgical, and petrochemical industries.

PYROTHERM 28/5.
M-1342; 0.20 C, 28 Cr, 5 Ni, bal Fe.
As quenched: 60-75 kp/mm² TS; 40 kp/mm² min YS; 26 min El.
Heat resistant rolled and wrought alloy.
For furnace engineering in cement, glass, metallurgical, and petrochemical industries.

PYROTHERM 15/16.
M-1342; 0.05 max C, 15 Cr, 64 Ni, bal Fe.
As quenched: 50-70 kp/mm² TS; 18 kp/mm² min YS; 30 min El.
Heat resistant rolled and wrought alloy.
For furnace engineering in cement, glass, metallurgical, and petrochemical industries.

PYROTHERM 18/36.
M--1342; 0.15 max C, 18 Cr, 36 Ni, bal Fe.
As quenched: 50-70 kp/mm² TS; kp/mm² min YS; 30 min El.
Heat resistant rolled and wrought alloy.
For furnace engineering in cement, glass, metallurgical, and petrochemical industries.

PYROTHERM 18/36 NB.
M-1342; 0.15 C, 18 Cr, 36 Ni, 1.2 Nb, bal Fe.
As quenched: 50-70 kp/mm² TS; 20 kp/mm² min YS; 30 Min El.
Heat resistant rolled and wrought alloy.
For furnace engineering in cement, glass, metallurgical and petrochemical industries.

PYROTHERM 20/15.
M-1342; 0.13 max C, 20 Cr, 15 Ni, bal Fe.
As quenched: 55-70 kp/mm² TS: 25 kp/mm² min YS; 40 min El.
Heat resistant rolled and wrought alloy.
For furnace engineering in cement, glass, metallurgical, and petrochemical industries.

PYROTHERM 20/15 MO.
M-1342; 0.13 max c, 20 Cr, 15 Ni, 1.2 Mo, bal Fe.
As quenched: 55-70 kp/mm² TS; 25 kp/mm² min YS; 40 min El.
Heat resistant rolled and wrought alloy.
For furnace engineering in cement, glass, metallurgical, and petrochemical industries.

PYROTHERM 20/15 MONB.
M-1342; 0.13 max C, 20 Cr, 15 Ni, 1.2 Mo, 1.2 Nb, bal Fe.
As quenched: 55-70 kp/mm² TS; 25 kp/mm² min YS; 40 min El.
Heat resistant rolled and wrought alloy.
For furnace engineering in cement, glass, metallurgical, and petrochemical industries.

PYROTHERM 20/33 ALTI.
M-1342; 0.05-0.10 C, 20 Cr, 33 Ni, 0.2 Al, 0.2 Ti, bal Fe.
As quenched: 50-70 kp/mm² TS; 20 kp/mm² min YS; 30 min El.
Heat resistant rolled and wrought alloy.
For furnace engineering in cement, glass, metallurgical, and petrochemical industries.

PYROTHERM 22/10.
M-1342; 0.20 max C, 21 Cr, 11 Ni, bal Fe.
As quenched: 60-75 kp/mm² TS; 30 kp/mm² min YS; 40 min El.
Heat resistant rolled and wrought alloy.
For furnace engineering in cement, glass, metallurgical, and petrochemical industries.

PYROTHERM 22/14.
M-1342; 0.10 max C, 22 Cr, 14 Ni, bal Fe.
As quenched: 55-70 kp/mm² TS; 22 kp/mm² min YS; 40 min El.
Heat resistant rolled and wrought alloy.
For furnace engineering in cement, glass, metallurgical, and petrochemical industries.

PYROTHERM 24/24 NB.
M-1342; 0.35 C, 24 Cr, 24 Ni, 1.5 Nb, bal Fe.
As quenched: 60-80 kp/mm² TS; 30 kp/mm² min YS; 20 min El.
Heat resistant rolled and wrought alloy.
For furnace engineering in cement, glass, metallurgical and petrochemical industries.

PYROTHERM 25/12.
M-1342; 0.20 max C, 23 Cr, 12 Ni, bal Fe.
As quenched: 60-75 kp/mm² TS; 30 kp/mm² min YS; 40 min El.
Heat resistant rolled and wrought alloy.
For furnace engineering in cement, glass, metallurgical, and petrochemical industries.

PYROTHERM 25/20.
M-1342; 0.15 max C, 25 Cr, 20 Ni, 1.8 Si, bal Fe.
As quenched: 55-75 kp/mm² TS; 30 kp/mm² min YS; 40 min El.
Heat resistant rolled and wrought alloy.
For furnace engineering in cement, glass, metallurgical, and petrochemical industries.

PYROTHERM 25/20 H.
M-1342; 0.40 C, 25 Cr, 20 Ni, bal Fe.
As quenched: 55-75 kp/mm² TS; 30 kp/mm² min YS; 20 min El.
Heat resistant rolled and wrought alloy.
For furnace engineering in cement, glass, metallurgical and petrochemical industries.

PYROTHERM 25/35 H.
M-1342; 0.45 C, 25 Cr, 35 Ni, bal Fe.
As quenched: 55-75 kp/mm² TS; 30 kp/mm² min YS; 12 min El.
Heat resistant rolled and wrought alloy.
For furnace engineering in cement, glass, metallurgical, and petrochemical industries.

PYROTHERM 28/48/5.
 M-1342; 0.15 max C, 28 Cr, 48 Ni, 5 W, bal Fe.
 As quenched: 55-80 kp/mm^2 TS; 25 kp/mm^2 min YS; 20 min El.
 Heat resistant rolled and wrought alloy.
 For furnace engineering in cement, glass, metallurgical and petrochemical industries.

PYROTHERM G 5 MO.
 M-1342; 0.10 C, 5 Cr, 0.5 Mo, bal Fe.
 Ann: 60-80 kp/mm^2 TS; 40 kp/mm^2 YS; 15 min El.
 Heat resistant cast alloy.
 For elevated temperature equipment.

PYROTHERM G 6.
 M-1342; 0.30 C, 6 Cr, bal Fe.
 Ann: 55-80 kp/mm^2 TS; 30 kp/mm^2 min YS; 3 min El.
 Heat resistant cast alloy.
 For elevated temperature equipment.

PYROTHERM G 13 MO.
 M-1342; 0.15 max C, 13 Cr, 0.4 Mo, bal Fe.
 Ann: 50-90 kp/mm^2 TS.
 Heat and corrosion resisting cast alloy.
 For elevated temperature equipment.

PYROTHERM G 18.
 M-1342; 0.40 C, 17 Cr, bal Fe.
 Ann: 50-70 kp/mm^2 TS.
 Heat and corrosion resisting cast alloy.
 For elevated temperature equipment.

PYROTHERM G 18 H.
 M-1342; 1.5 C, 18 Cr, bal Fe.
 Annealed heat and corrosion resisting cast alloy.
 For elevated temperature equipment.

PYROTHERM G 25.
 M-1342; 0.40 C, 25 Cr, bal Fe.
 Ann: 50-80 kp/mm^2 TS.
 Heat and corrosion resisting cast alloy.
 For elevated temperature equipment.

PYROTHERM G 25 H.
 M-1342; 1.3 C, 25 Cr, bal Fe.
 Heat and corrosion resisting cast alloy (as cast).
 For elevated temperature equipment.

PYROTHERM G 28.
 M-1342; 0.5 C, 28 Cr, bal Fe.
 As cast: 50-80 kp/mm^2 TS.
 Heat and corrosion resistant cast alloy.
 For elevated temperature equipment.

PYROTHERM G 28 H.
 M-1342; 1.3 C, 28 Cr, bal Fe.
 Heat and corrosion resisting cast alloy (as cast).
 For elevated temperature equipment.

PYROTHERM G 18/8.
 M-1342; 0.30 C, 18 Cr, 8 Ni, bal Fe.
 As cast: 45-65 kp/mm^2 TS; 21 kp/mm^2 min YS; 12 min El.
 Heat and corrosion resisting cast alloy.
 For elevated temperature equipment.

PYROTHERM G 25/4.
 M-1342; 0.40 C, 25 Cr, 4 Ni, bal Fe.
 As cast: 55-80 kp/mm^2 TS; 30 kp/mm^2 min YS; 2 min El.
 Heat and corrosion resistant cast alloy.
 For elevated temperature equipment.

PYROTHERM G 25/4 H.
 M-1342; 1.3 C, 25 Cr, 4 Ni, bal Fe.
 Heat and corrosion resistant cast alloy (as cast).
 For elevated temperature equipment.

PYROTHERM G 26/9.
 M-1342; 0.35 C, 26 Cr, 9 Ni, bal Fe.
 As cast: 45-65 kp/mm^2 TS; 21 kp/mm^2 min YS; 12 min El.
 Heat and corrosion resisting cast alloy.
 For elevated temperature equipment.

PYROTHERM G 28/5.
 M-1342; 0.40 C, 28 Cr, 5 Ni, bal Fe.
 As cast: 55-80 kp/mm^2 TS; 30 kp/mm^2 min YS; 2 min El.
 Heat and corrosion resisting cast alloy.
 For elevated temperature equipment.

PYROTHERM G 638.
 M-1342; 0.45 C, 26 Cr, 35 Ni, 16 Co, 5 W, 1 Nb, bal Fe.
 As cast: 45-65 kp/mm^2 TS; 25 kp/mm^2 min YS; 4 min El.
 Heat and corrosion resisting cast alloy.
 For elevated temperature equipment.

PYROTHERM G 16/13 NB.
 M-1342; 0.08 C, 16 Cr, 13 Ni, Nb=8xC, bal Fe.
 As cast: 40-60 kp/mm^2 TS; 18 kp/mm^2 min YS; 25 min El.
 Heat and corrosion resisting cast alloy.
 For elevated temperature equipment.

PYROTHERM G 17/53 W.
 M-1342; 0.45 C, 17 Cr, 53 Ni, 2 W, bal Fe.
 As cast: 45-65 kp/mm^2 TS; 25 kp/mm^2 min YS; 5 min El.
 Heat and corrosion resisting cast alloy.
 For elevated temperature equipment.

PYROTHERM G 18/36 H.
 M-1342; 0.40 C, 18 Cr, 36 Ni, bal Fe.
 As cast: 40-60 kp/mm^2 TS; 21 kp/mm^2 min YS; 8 min El.
 Heat and corrosion resisting cast alloy.
 For elevated temperature equipment.

PYROTHERM G 18/36 NB.
M-1342; 0.40 C, 18 Cr, 36 Ni, 1.2 Nb, bal Fe.
As cast: 45-65 kp/mm^2 TS; 24 kp/mm^2 min YS; 8 min El.
Heat and corrosion resisting cast alloy.
For elevated temperature equipment.

PYROTHERM G 20/15.
M-1342; 0.20 C, 20 Cr, 15 Ni, bal Fe.
As cast: 45-65 kp/mm^2 TS; 21 kp/mm^2 min YS; 14 min El.
Heat and corrosion resisting cast alloy.
For elevated temperature equipment.

PYROTHERM G 20/32 NB.
M-1342; 0.10 C, 20 Cr, 32 Ni, 1.1 Nb, bal Fe.
As cast: 45-65 kp/mm^2 TS; 18 kp/mm^2 min YS; 25 min El.
Heat and corrosion resisting cast alloy.
For elevated temperature equipment.

PYROTHERM G 22/10.
M-1342; 0.40 C, 22 Cr, 10 Ni, bal Fe.
As cast: 45-65 kp/mm^2 TS; 21 kp/mm^2 min YS; 12 min El.
Heat and corrosion resisting cast alloy.
For elevated temperature equipment.

PYROTHERM G 22/14.
M-1342; 0.15 C, 22 Cr, 14 Ni, bal Fe.
As cast: 40-60 kp/mm^2 TS; 20 kp/mm^2 min YS; 14 min El.
Heat and corrosion resisting cast alloy.
For elevated temperature equipment.

PYROTHERM G 22/14 H.
M-1342; 0.45 C, 22 Cr, 14 Ni, bal Fe.
At cast: 55-80 kp/mm^2 TS; 24 kp/mm^2 min YS; 4 min El.
Heat and corrosion resisting cast alloy.
For elevated temperature equipment.

PYROTHERM G 24/24 NB.
M-1342; 0.35 C, 24 Cr, 24 Ni, 1.5 Nb, bal Fe.
As cast: 50-70 kp/mm^2 TS; 25 kp/mm^2 min YS; 10 min El.
Heat and corrosion resisting cast alloy.
For elevated temperature equipment.

PYROTHERM G 25/12.
M-1342; 0.40 C, 25 Cr, 12 Ni, bal Fe.
As cast: 45-65 kp/mm^2 TS; 24 kp/mm^2 min YS; 10 min El.
Heat and corrosion resisting cast alloy.
For elevated temperature equipment.

PYROTHERM G 25/20.
M-1342; 0.15 C, 25 Cr, 20 Ni, bal Fe.
As cast: 45-65 kp/mm^2 TS; 21 kp/mm^2 min YS; 15 min El.
Heat and corrosion resisting cast alloy.
For elevated temperature equipment.

PYROTHERM G 25/20 H.
M-1342; 0.40 C, 25 Cr, 20 Ni, bal Fe.
As cast: 50-70 kp/mm^2 TS; 25 kp/mm^2 min YS; 10 min El.
Heat and corrosion resisting cast alloy.
For elevated temperature equipment.

PYROTHERM G 25/35 H.
M-1342; 0.45 C, 25 Cr, 35 Ni, bal Fe.
As cast: 50-70 kp/mm^2 TS; 25 kp/mm^2 min YS; 8 min El.
Heat and corrosion resisting cast alloy.
For elevated temperature equipment.

PYROTHERM G 25/35 NB.
M-1342; 0.45 C, 25 Cr, 35 Ni, 1.5 Nb, bal Fe.
As cast: 50-70 kp/mm^2 TS; 25 kp/mm^2 min YS; 8 min El.
Heat and corrosion resisting cast alloy.
For elevated temperature equipment.

PYROTHERM G 25/35 SO.
M-1342; 0.45 C, 25 Cr, 35 Ni, Mo, Al, bal Fe.
As cast: 50-70 kp/mm^2 TS; 25 kp/mm^2 min YS; 8 min El.
Heat and corrosion resisting cast alloy.
For elevated temperature equipment.

PYROTHERM G 26/32 MO.
M-1342; 0.40 C, 26 Cr, 32 Ni, 1.5 Mo, bal Fe.
As cast: 45-65 kp/mm^2 TS; 22 kp/mm^2 min YS; 8 min El.
Heat and corrosion resisting cast alloy.
For elevated temperature equipment.

PYROTHERM G 26/36 MO.
M-1342; 0.45 C, 25 Cr, 35 Ni, 1.5 Mo, bal Fe.
As cast: 50-70 kp/mm^2 TS; 25 kp/mm^2 YS; 8 min El.
Heat and corrosion resisting cast alloy.
For elevated temperature equipment.

PYROTHERM G 27/10 A1.
M-1342; 0.40 C, 27 Cr, 10 Ni, 0.7 Al, bal Fe.
As cast: 45-65 kp/mm^2 TS; 21 kp/mm^2 min YS; 12 min El.
Heat and corrosion resisting cast alloy.
For elevated temperature equipment.

PYROTHERM G28/14.
M-1342; 0.40 C, 28 Cr, 14 Ni, bal Fe.
As cast: 45-65 kp/mm^2 TS; 24 Kp/mm^2 min YS; 8 min El.
Heat and corrosion resisting cast alloy.
For elevated temperature equipment.

PYROTHERM G 28/15 W.
M-1342; 0.60 C, 28 Cr, 16 Ni, 2 W, bal Fe.
As cast: 45-65 kp/mm^2 TS; 25 kp/mm^2 min YS; 8 min El.
Heat and corrosion resisting cast alloy.
For elevated temperature equipment.

PYROTHERM G 28/20 AL.
M-1342; 0.40 C, 28 Cr, 20 Ni, 0.7 Al, bal Fe.
As cast: 50-70 kp/mm^2 TS; 25 kp/mm^2 min YS; 10 min El.
Heat and corrosion resisting cast alloy.
For elevated temperature equipment.

PYROTHERM G 28/20 MO.
M-1342; 0.40 C, 28 Cr, 20 Ni, 1.2 Mo, bal Fe.
As cast: 50-70 kp/mm^2 TS; 25 kp/mm^2 min YS; 10 min El.
Heat and corrosion resisting cast alloy.
For elevated temperature equipment.

PYROTHERM G 28/33 SI.
M-1342; 0.40 C, 28 Cr, 33 Ni, 2.5 Si, bal Fe.
As cast: 50-70 kp/mm^2 TS; 24 kp/mm^2 min YS; 5 min El.
Heat and corrosion resisting cast alloy.
For elevated temperature equipment.

PYROTHERM G 28/48 5.
M-1342; 0.40 C, 28 Cr, 48 Ni, 5 W, bal Fe.
As cast: 40-60 kp/mm^2 TS; 30 kp/mm^2 min YS; 3 min El.
Heat and corrosion resisting cast alloy.
For elevated temperature equipment.

PYROTHERM G 30/30.
M-1342; 0.55 C, 30 Cr, 30 Ni, bal Fe.
As cast: 45-65 kp/mm^2 TS; 22 kp/mm^2 min YS; 6 min El.
Heat and corrosion resisting cast alloy.
For elevated temperature equipment.

PYROTHERM G 50/50 NB.
M-1342; 0.05 max C, 49 Cr, 49 Ni, 1.5 Nb, bal Fe.
As cast: 55-75 kp/mm^2 TS; 30 kp/mm^2 min YS; 10 min El.
Heat and corrosion resisting cast alloy.
For elevated temperature equipment.

PYROTHERM G 60/40.
M-1342; 0.05 max C, 60 Cr, 39 Ni, bal Fe.
As cast: 75-95 kp/mm^2 TS.
Heat and corrosion resisting cast alloy.
For elevated temperature equipment.

PYROTHERM G UMCO 50.
M-1342; 0.20 C, 28 Cr, 50 Co, bal Fe.
As cast: 50-75 kp/mm^2 TS; 24 kp/mm^2 min YS; 6 min El.
Heat and corrosion equipment.
For elevated temperature equipment.

PYROTHERM G UMCO 51.
M-1342; 0.30 C, 28 Cr, 50 Co, 2 Nb, bal Fe.
As cast: 50-75 kp/mm^2 TS; 24 kp/mm^2 min YS; 3 min El.
Heat and corrosion resisting cast alloy.
For elevated temperature equipment.

PYROTHERM UMCO 50.
M-1342; 0.10 max c, 28 Cr, 50 Co, bal Fe.
As quenched: 70-90 kp/mm^2 TS; 25 kp/mm^2 min YS; 10 min El.
Heat resistant rolled and wrought alloy.
For furnace engineering in cement, glass, metallurgical anbd petrochemical industries.

PYROTHERM UMCO 51.
M-1342; 0.10 max C, 28 Cr, 50 Co, 2.2 Nb, bal Fe.
As quenched: 80-100 kp/mm^2 TS; 30 kp/mm^2 min YS; 6 min El.
Heat resistant rolled and wrought alloy.
For furnace engineering in cement, glass, metallurgical and petrochemical industries.

PYROTOOL A see CARPENTER PYROTOOL A.

PYROVAN.
M-73; 0.75 C, 1.0 Si, 5.25 Cr, 2.5 V, 1.1 Mo, bal Fe.
Hardened: C 55-63 Rock.
Hot work die steel with extra high carbon for higher hardness and wear resistance.
For forging dies, hot press and forming dies.
Air hardenable, good temperature resistance.

PYTHON.
M-1779; 0.8-1.2 C, 0.2 V, 0.25 Mn, 0.25 Si, bal Fe.
Water hardening tool steel; for blanking and forming dies, solid cold heading dies, machine parts. Typical heat treated hardness; Rc 56-60.
AISI W-2.

Q

Q-ALLOY A.
M-1005; 0.35-0.75 C, 15-19 Cr, 64-68 Ni, bal Fe.
Cast: 65,000 TS; 36,000 YS; 9 El; 176 Brin.
For lead pots, carburizing boxes, furnace fixtures, cyanide pots.
Heat and corrosion resistant.

Q-ALLOY B.
M-1005; 0.35-0.75 C, 58-62 Ni, 10-14 Cr, bal Fe.
Cast: 68,000 TS; 36,000 YS; 4 El; 185 Brin.
For furnace parts, carburizing and heat treating boxes, gas retorts, muffles.
Heat resistant. Max. operating temperature is 2000°F.

Q-ALLOY B-SI.
M-1005; 0.36-0.50 C, Si, 58-66 Ni, 16-23 Cr, bal Fe.
For heat and corrosion resistant parts; heat and corrosion resistant.

Q-ALLOY "C-1".
M-1005; 26-30 Cr, bal Fe.
Cast: 80,000 TS; 40,000 YS; 3 El; 4 RA; 180 Brin.
Heat treated: 40,000-55,000 TS; 30,000-45,000 YS; 0-1 El; 0-2 RA; 160-550 Brin.
Suited for acid mine water, nitric acid and bad sulfuric acid furnace conditions, heat treating boxes; corrosion and heat resisting.

Q-ALLOY C-1SC.
M-1005; 0.25 max C, 28 Cr, 3 max Ni, bal Fe.
For salt pots, furnace equipment; high corrosion resistance, brittle.

Q-ALLOY "C-2".
M-1005; 16-18 Cr, 0.4 C, bal Fe.
Cast: 60,000 TS; 55,000 YS; 12 El; 17 RA; 205 Brin.
Heat treated: 80,000 TS; 60,000 YS; 20 El; 35 RA.
Suited for acid mine water and nitric acid corrosion resisting parts; heat treating boxes; corrosion resisting; heat resisting.

Q-ALLOY "C-3".
M-1005; 28-30 Cr, high C, 3 max Ni, bal Fe.
Cast: 30,000-45,000 TS; 0 El; 0 RA; 300-600 Brin.
For rolling mill guides, wear plates for crusher parts, chute plates; heat and wear resisting.

Q-ALLOY CHROME C-1.
M-1005; 25-30 Cr, 3 max Ni, 0.2 C, bal Fe.
For corrosion resistant castings; resists mine water corrosion.

Q-ALLOY CHROME C-1A.
M-1005; 0.21-0.35 C, 24-30 Cr, bal Fe.
For heat and corrosion resistant parts; heat and corrosion resistant.

Q-ALLOY CHROME C-2.
M-1005; C, Cr, bal Fe.
For chemical equipment; resists nitric acid corrosion.

Q-ALLOY CHROME C-2A.
M-1005; 0.13-0.20 C, 12-15 Cr, bal Fe.
For corrosion resistant parts; corrosion resistant.

Q-ALLOY CHROME C-2B.
M-1005; 0.21-0.35 C, 12-15 Cr, bal Fe.
For corrosion resistant parts; corrosion resistant.

Q-ALLOY CHROME C-2C.
M-1005; 0.12 max C, 16-23 Cr, bal Fe.
For corrosion resistant parts; heat and corrosion resistant.

Q-ALLOY CHROME C-2D.
M-1005; 0.13-0.20 C, 16-23 Cr, bal Fe.
For corrosion resistant parts; heat and corrosion resistant.

Q-ALLOY CHROME C-2E.
M-1005; 0.21-0.35 C, 16-23 Cr, bal Fe.
For corrosion resistant parts; heat and corrosion resistant.

Q-ALLOY CHROME C-3.
M-1005; C, alloy, bal Fe.
500 Brin.
For mill guides; heat resisting to 2000°F.

Q-ALLOY CHROME C-3A.
M-1005; 0.81-1.10 C, 16-23 Cr, bal Fe.
For heat and corrosion resistant parts; heat and corrosion resistant.

Q-ALLOY CHROME CN1-A.
M-1005; 0.21-0.35 C, 24-30 Cr, 7-11 Ni, bal Fe.
For heat and corrosion resistant parts; heat and corrosion resistant.

Q-ALLOY CHROME CN1-B.
M-1005; 0.13-0.20 C, 24-30 Cr, 12-15 Ni, bal Fe.
For heat and corrosion resistant parts; heat and corrosion resistant.

Q-ALLOY CHROME CN1-S.
M-1005; 0.36-0.50 C, 24-30 Cr, 7-11 Ni, bal Fe.
For heat and corrosion resistant parts; heat and corrosion resistant.

Q-ALLOY CHROME CN2-A.
M-1005; 0.21-0.35 C, W, 16-23 Cr, 7-11 Ni, bal Fe.
For heat and corrosion resistant parts; heat and corrosion resistant.

Q-ALLOY CHROME J.
M-1005; 0.12 max C, 12-15 Cr, bal Fe.
For corrosion resistant parts; corrosion resistant.

Q-ALLOY CHROME K-1.
M-1005; Ni, Cr, bal Fe.
For furnace parts not subject to high sulfur fumes; high heat resistance to 2200°F.

Q-ALLOY CHROME KA-2.
M-1005; 0.13-0.20 C, 16-23 Cr, 7-11 Ni, bal Fe.
For heat and corrosion resistant parts; stainless.

Q-ALLOY CHROME KA-2H.
M-1005; 0.21-0.35 C, 16-23 Cr, 7-11 Ni, bal Fe.
For heat and corrosion resistant parts; stainless.

Q-ALLOY CHROME KA-2MO.
M-1005; 0.13-0.20 C, Mo, 16-23 Cr, 7-11 Ni, bal Fe.
For heat and corrosion resistant parts; heat and corrosion resistant.

Q-ALLOY CHROME KA-2S.
M-1005; 0.12 max C, 16-23 Cr, 7-11 Ni, bal Fe.
For stainless parts; stainless.

Q-ALLOY CHROME KA-4.
M-1005; 0.21-0.35 C, Mo, 16-23 Cr, 7-11 Ni, bal Fe.
For heat and corrosion resistant parts; heat and corrosion resistant.

Q-ALLOY KNC-3.
M-1005; 0.13-0.20 C, Si, 24-30 Cr, 16-23 Ni, bal Fe.
For heat and corrosion resistant parts; heat and corrosion resistant.

Q-ALLOY "C.N.-1".
M-1005; 24-26 Cr, 11-13 Ni, bal Fe.
Cast: 65,000-85,000 TS; 40,000 YS; 10-27 El; 15-45 RA; 180-220 Brin.
Heat treated: 50,000-65,000 TS; 35,000-50,000 YS; 0-11 El; 0-12 RA;
For furnace parts where temperature is not excessive, heat treating boxes; heat and corrosion resistant; max operating temperature 2100°F.

Q-ALLOY CN1-H.
M-1005; 0.13-0.20 C, 24-30 Cr, 7-11 Ni, bal Fe.
For furnace chains, muffles, retorts; heat and corrosion resistant.

Q-ALLOY CN-1H-MO.
M-1005; 0.13-0.20 C, Mo, 24-30 Cr, 7-11 Ni, bal Fe.
For heat and corrosion resistant parts; heat and corrosion resistant.

Q-ALLOY CN-1 MO.
M-1005; 24-26 Cr, 11-13 Ni, 1.5-3.0 Mo, bal Fe.
For pulp industry exposed to hot sulfite liquors and bleaches; corrosion and heat resistant.

Q-ALLOY "C.N.-2".
M-1005; 18-20 Cr, 8-10 Ni, bal Fe.
Cast: 60,000-80,000 TS; 35,000-40,000 YS; 10-50 El; 10-40 RA; 160-210 Brin.
Heat treated: 80,000-100,000 TS; 35,000-50,000 YS; 5-15 El; 10-20 RA.
For heat treating boxes, furnace parts; corrosion and heat resisting; max operating temperature 2000°F.

Q-ALLOY CN2-MO.
M-1005; 17-21 Cr, 7-9 Ni, 1-4 Mo, 0.2 C, bal Fe.
For chemical equipment; stainless, austenitic.

Q-ALLOY, GRADE A+.
M-1005; 66-68 Ni, 19-21 Cr, bal Fe.
Cast: 80,000-90,000 TS; 50,000-45,000 YS;
For carburizing boxes, retorts, solution pots, furnace parts; heat resistant to 2220°F.

Q-ALLOY X-2.
M-1005; C, 30 Ni, 30 Cr, bal Fe.
For high temperature applications; corrosion and heat resistant.

Q-ALLOY X-3.
M-1005; C, 20 Ni, 30 Cr, bal Fe.
For high temperature applications; corrosion and heat resistant.

Q-ALLOY X-4.
M-1005; C, 20 Ni, 25 Cr, bal Fe.
For high temperature applications; corrosion and heat resistant.

Q-ALLOY X-5.
M-1005; C, 5 Ni, 28 Cr, bal Fe.
For high temperature applications; corrosion and heat resistant.

Q-ALLOY X-6.
M-1005; C, 16 Ni, 28 Cr, bal Fe.
For high temperature applications; corrosion and heat resistant.

Q-ALLOY X-7.
M-1005; 0.2 C, 25 Cr, 12 Ni, bal Fe.
For furnace rails and rollers, oil still tube supports; austenitic, resist thermal shock and high stresses.

Q-ALLOY X-8.
M-1005; C, 10 Ni, 20 Cr, bal Fe.
For high temperature applications; corrosion and heat resistant.

QE 22A (STANDARD ALLOY).
M-43; 0.7 Zr, bal Mg.
T6-temper: 40,000 TS; 30,000 YS; 4 El.
At 400°F: 28,000 TS; 25,000 YS; 22 El.
For aircraft, jet engine and missile components; highest cast yield strength, age-hardenable.

QEZZA.
M-681; 2.0-3.0 Ag, 0.40-1.0 Zr, 1.75-2.25 Re, bal Mg.
Cast: 40,000 psi TS; 30,000 psi YS; 2 El.
High strength magnesium castings.

QT. 131 GRADE A.
M-1724; 0.20 max C, 0.50 max Si, 1.6 max Mn, 0.50 max Nb, bal Fe.
Quenched and tempered steel.

QT. 131 GRADE B.
M-1724; 0.20 max C, 0.10-0.50 Si, 1.80 max Mn, 0.50 max Nb, bal Fe.
Quenched and tempered steel.

QT. 445 GRADE A.
M-1724; 0.15-0.21 C, 0.80 max Si, 0.80-1.10 Mn, O, 18-0.40 Mo, 0.50- 0.80 Cr, 0.05-0.15 Zr, 0.0005-0.0025 B, bal Fe.
Quenched and tempered steel, for crane structural members, oil rigs, bridges.

QT. 445 GRADE B.
M-1724; 0.15-0.21 C, 0.90 max Si, 0.80-1.10 Mn, 0.25-0.60 Mo, 0.50-0.8 Cr, 0.05-0.15 Zr, 0.0005-0.0025 B, bal Fe.
Quenched and tempered steel, for crane structural members, oil rigs, bridges.

Q-TEMP 10B18Q.
M-604; 0.15-0.20 C, 0.80-1.10 Mn, 0.20-0.35 Si, 0.0005 min B, bal Fe.
Rolled: 70,400 TS; 38,700 YP; 25 El; 63 RA; B 72 Rock.
For fasteners, bolts, screws, clips. Tough, water hardening.

Q-TEMP 10B21Q.
M-604; 0.18-0.23 C, 0.80-1.10 Mn, 0.20-0.35 Si, 0.0005 min B, bal Fe.
Rolled: 72,000 TS; 39,000 YS; 24 El; 62 RA; B 73 Rock.
For fasteners, bolts, screws, clips. Tough, water hardening.

Q-TEMP 10B22Q.
M-604; 0.17-0.23 C, 1.0-1.3 Mn, 0.20-0.35 Si, 0.0005 min B, bal Fe.
Rolled: 75,000 TS; 40,000 YS; 22 El; 60 RA; B 76 Rock.
For fasteners, screws, bolts, nuts. Tough, water hardening.

Q-TEMP 10B23Q.
M-604; 0.17-0.23 C, 1.1-1.4 Mn, 0.20-0.35 Si, 0.0005 min B, bal Fe.
Rolled: 80,000 TS; 44,000 YS; 20 El; 56 RA; B 78 Rock.
For fasteners, screws, bolts, nuts. Tough, water hardening.

Q-TEMP 41B20 Q.
M-604; 0.18-0.23 C, 0.75-1.0 Mn, 0.10-0.35 Si, 0.25-0.40 Cr, 0.15-0.25 Mo, 0.0005 min B, bal Fe.
Hardenable (oil or water) to: 160 ksi min YS.
For bolts, fasteners.

Q-TEMP 41BV20 Q.
M-604; 0.18-0.23 C, 0.75-1.0 Mn, 0.10-0.35 Si, 0.25-0.40 Cr, 0.15-0.25 Mo, 0.03-0.08 V, 0.0005 min B, bal Fe.
Hardenable (oil or water) to: 160 ksi min YS.
For bolts, fasteners.

Q-TEMP 50B23 Q.
M-604; 0.20-0.25 C, 0.75-1.0 Mn, 0.10-0.35 Si, 0.25-0.40 Cr, 0.0005 min B, bal Fe.
Hardenable (oil or water) to: 165 ksi min YS; for bolts, fasteners.

Q-TEMP 60B20 Q.
M-604; 0.18-0.23 C, 0.75-1.0 Mn, 0.10-0.35 Si, 0.25-0.40 Cr, 0.03-0.08 V, 0.0005 min B, bal Fe.
Hardenable (oil or water) to: 160 ksi min YS.
For bolts, fasteners.

QUAKER.
M-1431; 0.4 C, 0.4 Mn, 4.5 Cr, 1 Mo, 0.8 W, bal Fe.
For tools, dies; air or oil hardened.

QUAKER LC2.
M-1628; 0.18 C, 0.65 Mn, 0.45 Si, 2.4 Ni, bal Fe.
Norm. Q.+T.: 70,000 psi TS; 40,000 psi YS; 24 El; 160 Brin.
Steel castings for low temperature operations, ASTM A 352 Gr. LC2.

QUAKER LC3.
M-1628; 0.10 C, 0.65 Mn, 0.45 Si, 3.5 Ni, bal Fe.
Norm. Q.+T.: 70,000 psi TS; 40,000 psi YS; 24 El; 170 Brin.
Nickel steel castings for low temperature operations.
ASTM A 352 Gr. LC3.

QUAKER LCA.
M-1628; 0.20 C, 0.60 Mn, 0.45 Si, bal Fe.
Norm. Q.+T.: 60,000 psi TS; 30,000 psi YS; 24 El; 150 Brin.
Carbon steel castings for low temperature operations.
ASTM A 352 Gr. LCA.

QUAKER LCB.
M-1628; 0.25 C, 0.70 Mn, 0.45 Si, bal Fe.
Norm. Q.+T.: 65,000 psi TS; 35,000 psi YS; 24 El; 160 Brin.
Carbon steel castings for low temperature operations.
ASTM A 352 Gr. LCB.

QUAKER LCC.
M-1628; 0.20 C, 0.85 Mn, 0.45 Si, bal Fe.
Norm. Q.+T.: 70,000 psi TS; 40,000 YS; 22 El; 160 Brin.
Carbon steel castings for low temperature operations.
ASTM A 352 Gr. LCC.

QUAKER Q05.
M-1628; 0.05 C, 0.10 Mn, 0.60 Si, bal Fe.
Cast, Ann: 50,000 psi TS; 28,000 psi YS; 30 El; 120 Brin.
Similar to AISI 1005.

QUAKER Q1CM.
M-1628; 0.15 C, 0.65 Mn, 0.45 Si, 1.25 Cr, 0.50 Mo, bal Fe.
Norm. & Temp.: 70,000 psi TS; 40,000 psi YS; 20 El; 150 Brin.
Alloy steel casting for elevated temperature service.
ASTM A 217 Gr. WC6; A 356 Gr. 6.

QUAKER Q2CM.
M-1628; 0.15 C, 0.60 Mn, 0.45 Si, 2.4 Cr, 1.0 Mo, bal Fe.
Norm. & Temp.: 70,000 psi TS; 40,000 psi YS; 20 El; 170 Brin.
Alloy steel casting for elevated temperature service.
ASTM A 217 Gr. WC9; A 356 Gr.10.

QUAKER Q5M.
M-1628; 0.15 C, 0.60 Mn, 0.45 Si, 5.5 Cr, 0.50 Mo, bal Fe.
Norm. & Temp.: 90,000 psi TS; 60,000 psi YS; 18 El; 200 Brin.
Alloy steel casting for elevated temperature service.
ASTM A 217 Gr. C5; SAE 60502.

QUAKER Q9M.
M-1628; 0.15 C, 0.50 Mn, 0.45 Si, 9.0 Cr, 1.0 Mo, bal Fe.
Norm. & Temp.: 90,000 psi TS; 60,000 psi YS; 18 El; 200 Brin.
Alloy steel casting for elevated temperature service.
ASTM A 217 Gr C12.

QUAKER Q41.
M-1628; 0.40 C, 0.70 Mn, 0.45 Si, 1.0 Cr, 0.25 Mo, bal Fe.
Norm. & Temp.: 100,000 psi TS; 70,000 psi YS; 15 El; 220 Brin.
Alloy steel casting; for structural service.
AISI 4140.

QUAKER Q60.
M-1628; 0.20 C, 0.60 Mn, 0.45 Si, bal Fe.
Norm. & Temp.: 60,000 psi TS; 30,000 psi YS; 24 El; 140 Brin.
Carbon steel casting.
ASTM A-27 Gr. 60-30.

QUAKER Q65.
M-1628; 0.20 C, 0.65 Mn, 0.45 Si, bal Fe.
Norm. & Temp.: 65,000 psi TS; 35,000 psi YS; 24 El; 140 Brin.
Carbon steel casting.
ASTM A-27 Gr. 65-35.

QUAKER Q70.
M-1628; 0.25 C, 0.70 Mn, 0.45 Si, bal Fe.
Norm. & Temp.: 70,000 psi TS; 36,000 psi YS; 22 El; 150 Brin.
Carbon steel casting.
ASTM A 27 Gr. 70-36.

QUAKER Q70-1.
M-1628; 0.20 C, 0.85 Mn, 0.45 Si, bal Fe.
Norm. & Temp.: 70,000 psi TS; 40,000 psi YS; 22 El; 160 Brin.
Carbon steel casting.
ASTM A-27 Gr. 70-40.

QUAKER Q80.
M-1628; 0.40 C, 0.70 Mn, 0.45 Si, bal Fe.
Norm. & Temp.: 80,000 psi TS; 40,000 psi YS; 17 El; 180 Brin.
Carbon steel casting.
ASTM A 148 Gr. 80-40.

QUAKER Q86.
M-1628; 0.15 C, 0.80 Mn, 0.45 Si, 0.60 Cr, 0.60 Ni, 0.20 Mo, bal Fe.
Norm. & Temp.: 80,000 psi TS; 50,000 psi YS; 22 El; 180 Brin.
Alloy steel casting; for structural service.
ASTM A 148 Gr. 80-50; similar to AISI 8615.

QUAKER Q90.
M-1628; 0.30 C, 0.80 Mn, 0.45 Si, 0.60 Cr, 0.60 Ni, 0.20 Mo, bal Fe.
Norm. & Temp.: 90,000 psi TS; 60,000 psi YS; 20 El; 190 Brin.
Alloy steel casting; for structural service.
ASTM A 148 Gr. 90-60; similar to AISI 8630.

QUAKER Q105.
M-1628; 0.30 C, 0.80 Mn, 0.45 Si, 0.60 Cr, 0.60 Ni, 0.20 Mo, bal Fe.
Norm. Q. + T.: 105,000 psi TS; 85,000 psi YS; 17 El; 230 Brin.
Alloy steel casting; for structural service.
ASTM A 148 Gr. 105-85; similar to AISI 8630.

QUAKER Q120.
M-1628; 0.30C, 0.80 Mn, 0.45 Si, 0.60 Cr, 0.60 Ni, 0.20 Mo, bal Fe.
Norm. Q. + T.: 120,000 psi; 95,000 psi YS; 14 El; 260 Brin.
Alloy steel casting; for structural service.
ASTM 148 Gr. 120-95; similar to AISI 8630.

QUAKER Q150.
M-1628; 0.30C, 0.80 Mn, 0.45 Si, 0.60 Cr, 0.60 Ni, 0.20 Mo, bal Fe.
Norm. Q. + T.: 150,000 psi TS; 125,000 psi YS; 9 El; 330 Brin.
Alloy steel casting; for structural service.
ASTM A 148 Gr. 150-125; similar to AISI 8630.

QUAKER Q4340.
M-1628; 0.40C, 0.70 Mn, 0.45 Si, 0.80 Cr, 2.0 Ni, 0.40 Mo, bal Fe.
Norm. & Temp.: 100,000 psi TS; 70,000 psi YS; 15 El; 220 Brin.
Alloy steel casting; for structural service.
Similar to AISI 4340.

QUAKER QCCM.
M-1628; 0.15 C, 0.65 Mn, 0.45 Si, 0.55 Cr, 0.50 Mo, bal Fe.
Norm. & Temp.: 70,000 psi TS; 40,000 psi YS; 22 El; 150 Brin.
Alloy steel casting for elevated temperature service.
ASTM A 356 Gr. 5.

QUAKER QCN.
M-1628; 0.03 C, 0.80 Mn, 0.40 Si, 0.30 Ni, 0.69 Cu, 1.0 Cb.
As cast: 60,000 psi TS; 32,000 psi YS; 20 El.
Cu-Ni casting.

QUAKER QHB.
M-1628; 0.08 C, 0.75 Mn, 0.40 Si, 66.0 Ni, 28.0 Mo, 0.35 V, bal Fe.
Sol. Ann.: 72,000 psi TS; 46,000 psi YS; 6 El.
High alloy, corrosion resistant casting.
Hastelloy B Type; ASTM A-296 N-12M.

QUAKER QHC.
M-1628; 0.08 C, 0.75 Mn, 0.40 Si, 16.5 Cr, 57.0 Ni, 17.0 Mo, 4.25 W, 0.30 V, bal Fe.
Sol. Ann.: 72,000 psi TS; 46,000 psi YS; 4 El.
High alloy, corrosion resistant casting.
Hastelloy C Type; ASTM A296 CW 12M.

QUAKER QINC.
M-1628; 0.05 C, 0.75 Mn, 0.75 Si, 15.5 Cr, 75.0 Ni, bal Fe.
As cast: 70,000 psi TS; 28,000 psi YS; 30 El.
Ni-Cr-Fe alloy casting; corrosion resistant.
ASTM A 296 Cy-40.

QUAKER QMC.
M-1628; 0.20 C, 0.65 Mn, 0.45 Si, 0.50 Mo, bal Fe.
Norm. & Temp.: 65,000 psi TS; 35,000 psi YS; 24 El; 140 Brin.
Alloy steel casting for elevated temperature service.
ASTM A 217 Gr. WC1; A 356 Gr.2.

QUAKER QML.
M-1628; 0.15 C, 1.0 Mn, 1.2 Si, 66.0 Ni, 30.0 Cu.
As cast: 65,000 psi TS; 32,000 psi YS; 25 El.
Monel casting; Monel 410. ASTM A296 M-35.

QUAKER QML-CB.
M-1628; 0.15 C, 1.0 Mn, 1.2 Si, 66.0 Ni, 3.0 Cu, 1.3 Cb.
As cast: 65,000 psi TS; 32,500 psi YS; 25 El.
Monel type casting; similar to Monel 411.

QUAKER QNI.
M-1628; 0.15 C, 1.0 Mn, 1.0 Si, 97.0 Ni, bal Fe.
As cast: 50,000 psi TS; 18,000 psi YS; 10 El.
High nickel casting, corrosion resistant.
ASTM A296 CZ100; Nickel 210.

QUALITAT-55.
M-1548; 3.0-5.0 Cu, 0.3-0.7 Si, 0.3-0.8 Mn, bal Al.
Heat Treated: 50,000-60,000 TS; 18-23 El; 100-110 Brin.
For rivets, hydraulic fittings, aircraft engine components.
Heat treatable, high fatigue strength.

QUALITAT-M.
M-1548; 0.85 Si, 0.6-0.9 Mn, 0.5-1.0 Mg, bal Al.
Heat Treated: 50,000-60,000 TS; 3-8 El; 95-100 Brin.
For general light structures, booms, scaffolds, transmission tires.
Good strength and fabricability. Heat treatable.

QUALITY STEEL.
M-237; 1.2-1.5 Cr, 4.5-5.0 Ni, 0.3-0.5 Mn, 0.3-0.35 C, bal Fe.
226,000 TS; 15 El; 45 RA.
For gears; air hardening.

QUARZAL.
M-Ger.; 15 Cu, 10 max Si, Fe, Ni, Cr, Mo, bal Al.
For bearings; wear resisting.

QUARZAL Q2.
M-Ger.; 2 Cu, 1.0 max Fe, 0.5 hardening agent, bal Al.
For bearings.

QUARZAL Q5.
M-Ger.; 5-15 Cu, 1.0 max Fe, 0.5 hardening agent, bal Al.
125 Brin.
For bearings.

QUARZAL Q8.
M-Ger.; 5-15 Cu, 1.0 max Fe, 0.5 hardening agent, bal Al.
For bearings.

QUARZAL Q12.
M-Ger.; 5-15 Cu, 1.0 max Fe, 0.5 hardening agent, bal Al.
125 Brin.
For bearings.

QUARZAL Q15.
M-Ger.; 5-15 Cu, 1.0 max Fe, 0.5 hardening agent, bal Al.
125 Brin.
For bearings.

QUEEN'S METAL.
M-Eng.; 88.5 Sn, 7.0 Sb, 3.5 Cu, 1.0 Zn.
For utensils; resists tarnishing.

QUEEN'S METAL "B".
M-Eng.; 50.5 Sn, 16.5 Sb, 16.5 Pb, 16.5 Zn.
For type metal.

QUEEN'S METAL "C".
M-Eng.; 87 Sn, 8.5 Sb, 3.5 Cu, 1 Zn.
For type metal.

QUEEN'S METAL "D".
M-Eng.; 73.36 Sn, 8.88 Sb, 8.88 Pb, 8.88 Zn.
For type metal.

QUEEN'S METAL "E".
M-Eng.; 88.5 Sn, 7 Sb, 3.5 Cu, 1 Bi.
For type metal.

QUICKSILVER SOLDER.
63-57 Ag, 21-25 Cu, 3.8-6.2 Sn, 10-12 Zn.
For silver solder; corrosion resistant.

R

R 2.
 M-912; 0.20 C, 0.25 Si, 0.50 Mn, bal Fe.
 Ht Tr: 410-540 N/mm² TS.
 Weldable, for forgings, tubing up to 350°C.
 W.-Nr. 1151.

R 3.
 M-912; 0.35 C, 0.25 Si, 0.65 Mn, bal Fe.
 Ht Tr: 500-650 N/mm² TS.
 For bolts and nuts for use up to 400°C.
 W.-Nr. 1181.

R 3F.
 M-912; 0.35 C, 0.25 Si, 0.55 Mn, bal Fe.
 Carbon steel for structural purposes.
 Fine grain steel.
 W.Nr. 1183.

R 4.
 M-912; 0.45 C, 0.25 Si, 0.65 Mn, bal Fe.
 Hardened: 540-690 N/mm² TS.
 For bolts and nuts resistant up to 400°C.
 W.-Nr. 1191.

R 4.
 M-1577; 0.07 C, 25.43 Cr, 4.19 Ni, 1.52 Mo, bal Fe.
 Corrosion resistant alloy.

R 4F.
 M-912; 0.45 C, 0.25 Si, 0.65 Mn, bal Fe.
 Carbon steel for structural purposes.
 Fine grain steel.
 W.-Nr. 1193.

R 5F.
 M-912; 0.53 C, 0.25 Si, 0.55 Mn, bal Fe.
 Carbon steel for structural purposes.
 Fine grain steel.
 W.Nr. 0213.

R6F2K8M5.
 M-USSR; 0.8 C, 5 Mo, 4 Cr, 2 V, 6 W, 8 Co, bal Fe.
 Hardened: C 64-67 Rock.
 For drills, reamers, broaches, lathe and planer tools.
 High-speed steel, high red-hardness.

R6F2K14M5.
 M-USSR; 0.8 C, 4 Cr, 2 V, 6 W, 5 Mo, 14 Co, bal Fe.
 Hardened: C 64-66 Rock.
 For lathe and planer tools, form cutters, reamers, drills, broaches.
 High-speed steel, high red-hardness.

R6K8F2M5.
 M-USSR; 0.8 C, 6 W, 4 Cr, 2 V, 5 Mo, 8 Co, bal Fe.
 Hardened: C 64-66 Rock.
 For lathe and planer tools, form cutters, reamers, broaches, drills.
 High-speed steel, high red-hardness.

R6K14F2M5.
 M-USSR; 0.8 C, 6 W, 4 Cr, 2 V, 5 Mo, 14 Co, bal Fe.
 Hardened: C 64-67 Rock.
 For form cutters, drills, reamers, broaches, lathe and planer tools.
 High-speed steel; high red-hardness.

R 7.
 M-912; 0.75 C, 0.25 Si, 0.70 Mn, bal Fe.
 Water hardenable construction steel.

R 7F.
 M-912; 0.70 C, 0.25 Si, 0.30 Mn, bal Fe.
 Carbon steel for structural purposes.
 Fine grain steel.
 W.Nr. 1249.

R9.
 M-USSR; 0.9 C, 8.5 W, 4 Cr, 2.1 V, bal Fe.
 Hardened: C 64-66 Rock.
 For lathe and planer tools, drills, reamers, broaches, form cutters.
 High-speed steel, high red-hardness.

R9K5 (EI 705).
 M-USSR; 0.8-0.9 C, 3.8-4.4 Cr, 5.0-6.0 Co, 9.0-10.5 W, 1.6-2.0 V, bal Fe.
 Hardened: C 64-67 Rock.
 For machining heat resistant steels, tools, cutters, planer and lathe tools, broaches, reamers, form tools.
 High-speed steel, high red-hardness.

R9K8F4M.
 M-USSR; 1.3 C, 8.8 W, 4.2 Cr, 3.6 V, 1 Mo, 8.1 Co, bal Fe.
 Hardened: C 65-67 Rock.
 For precision tools and cutters, reamers, broaches, drills.
 High-speed steel, high red-hardness.

R9K10.
 M-USSR; 0.95 C, 9.5 W, 4 Cr, 2.4 V, 10 Co, bal Fe.
 Hardened: C 65-67 Rock.
 For reamers, broaches, cut-off tools, form cutters, drills.
 High-speed steel, high red-hardness.

R9K30.
 M-USSR; 0.90 C, 4.0 Cr, 10.2 W, 1.6 V, 29.4 Co, bal Fe.
 Hardened: C 64-68 Rock.
 For heavy duty cutters, drills, reamers, broaches, lathe tools.
 High-speed steel, high red-hardness.

R10F5K5.
 M-USSR; 1.5 C, 10 W, 4 Cr, 4.5 V, 6 Co, bal Fe.
 Hardened: C 66-68 Rock.
 For cutters, reamers, broaches, drills, textile needles, form tools.
 High-speed steel, high red-hardness.

R15K5.
M-USSR; 0.8 C, 15 W, 3 Cr, 1.5 V, 5 Co, bal Fe.
Hardened: C 64-66 Rock.
For lathe and planer tools, form cutters, drills, reamers, broaches.
High-speed steel, high red-hardness.

R15K12.
M-USSR; 0.8 C, 15 W, 3 Cr, 1.5 V, 12 Co, bal Fe.
Hardened: C 64-67 Rock.
For lathe and planer tools, drills, reamers, broaches, form tools, hobs, taps.
High-speed steel, high red-hardness.

R18.
M-USSR; 0.7 C, 18 W, 4 Cr, 1 V, bal Fe.
Hardened: C 64-66 Rock.
For cutters, reamers, hobs, lathe and planer tools, form and milling cutters.
High red-hardness.
Type T1 high speed steel.

R18F4K8M.
M-USSR; 1.25-1.4 C, 4.4-5.0 Cr, 15.5-17 W, 3.2-3.8 V, 7.5-8.5 Co, 1.2-1.5 Mo, bal Fe.
Hardened: C 64-67 Rock.
For cutters, reamers, broaches, taps, hobs, textile needles.
High red-hardness.
High speed steel; abrasion resistant.

R18K8F4M.
M-USSR; 1.3 C, 17 W, 4 Cr, 4 V, 1.4 Mo, 8 Co, bal Fe.
Hardened: C 64-67 Rock.
For reamers, broaches, drills, hobs, taps, form cutters.
High-speed steel, high red-hardness.

R18K10.
M-USSR; 0.88 C, 4.0 Cr, 1.3 V, 18.3 W, 9.6 Co, bal Fe.
Hardened: C 64-67 Rock.
For cutters, reamers, broaches, drills, taps, hobs.
High-speed steel, high red-hardness.

R-41.
M-44; 19 Cr, 11 Co, 10 Mo, 3 Ti, 2 Al, bal Ni.
Wire, rod, ribbon for welding wire (Ren ae 41), and fastener stock.

R43 see HEPPENSTALL R43.

R-63.
M-44; 4 Mn, bal Ni.
Wire and ribbon for combustion engine electrodes.

R-235 see also ALLVAC R-235.

R 304 UD.
M-1541; 0.12 C, 1.0 Mn, 0.04 max P, 0.015 max S, 0.5 Si, 2.0 Cu, 7.2 Ni, 14.0 Cr, 0.40 Mo, bal Fe.
100,000 psi TS; 65 El.
Austenitic stainless steel for deep drawing purposes.

R 409 SR.
M-1541; 0.008 C, 0.50 Mn, 0.04 max P, 0.015 max S, 1.40 Si, 11.5 Cr, 0.25 Ti, bal Fe.
70,000 psi TS; 35 El.
Heat resistant stainless steel for automotive catalytic exhaust gas converter system.

R 410 L.
M-1541; 0.02 C, 0.45 Mn, 0.040 max P, 0.015 max S, 0.50 Si, 12.5 Cr, bal Fe.
63,000 psi TS; 36 El.
Heat resistant stainless steel for automotive exhaust gas system.

R 410 UL.
M-1541; 0.006 C, 0.45 Mn, 0.04 max P, 0.015 max S, 0.50 Si, 12.5 Cr, bal Fe.
60,000 psi TS; 40 El.
Heat resistant stainless steel for automotive exhaust gas system.

R 430 LT.
M-1541; 0.006 C, 0.90 Mn, 0.04 max P, 0.015 max S, 0.45 Si, 16.3 Cr, Ti = 20 x (C + N), bal Fe.
Corrosion resistant ferrite type stainless steel.
67,000 psi TS; 28 El.

R 434 LT (N)-1.
M-1541; 0.006 C, 0.40 Mn, 0.040 max P, 0.015 max S, 0.045 Si, 16.3 Cr, 0.95 Mo, Ti = 20 x (C + N), Cb = 20 x (C + N), bal Fe.
71,000 psi TS; 33 El.
Corrosion resistant ferrite type stainless steel.

R 434 LT (N)-2.
M-1541; 0.006 C, 0.45 Mn, 0.040 max P, 0.010 max S, 0.45 Si, 18.5 Cr 2.0 Mo, Ti = 20 x (C + N), Cb = 20 x (C + N), bal Fe.
77,000 psi TS; 32 El.
Corrosion resistant ferrite type stainless steel.

R448.
M-Eng.; C, 10-12 Cr, bal Fe.
For gas turbine parts; corrosion and heat resistant.

R2799.
M-108; 70 Fe, 30 Ni.
For compensating shunts for electrical equipment; temperature-sensitive, magnetic.

R 2800.
M-108; Ni-Fe.
Temperature compensating alloy.

RA 26-1.
M-1209; 0.02 C, 0.3 Mn, 0.3 Si, 26 Cr, 0.2 Ni, 1.0 Mo, 0.03 N, 0.05 Ti, bal Fe.
Ann: 75,000 TS; 25,000 YS; 30 El; 177 BHN.
Ferritic, titanium stabilized, heat and corrosion resistant alloy; weldable and machinable.
For recuperators, oil burner parts, furnace lining, combustion chambers, pressure vessels, Evaporator - Petrochemical - bleaching equipment, water purification facilities.

RA 309.
M-1209; 0.20 max C, 22-24 Cr, 12-15 Ni, bal Fe.
Annealed: 85,000-95,000 TS; 40,000-50,000 YS; 55-45 El; 150-185 Brin.
For furnace parts and equipment, refinery equipment; corrosion and heat resistant, austenitic.

RA 310.
M-1209; 0.25 max C, 24-26 Cr, 19-22 Ni, bal Fe.
Annealed: 95,000 TS; 45,000 YS; 50 El; 185 Brin.
For furnace parts and equipment, gas turbine and jet engine components; corrosion and heat resistant, austenitic.

RA 330.
M-1209; 0.05 C, 1.5 Mn, 1.25 Si, 19 Cr, 35 Ni, 43 Fe.
Room temp: 85,000 TS; 42,000 YS; 45 El.
1400°F: 34,000 TS; 18,800 YS; 64.8 El.
Austenitic, non-hardenable, heat and corrosion resistant alloy, weldable and machinable.
For high temp. applications; turbine parts, heat exchangers, heat treating equipment, radiant tubes, muffles.
AMS 5592; AMS 5716.

RA 330 HC.
M-1209; 0.40 C, 1.5 Mn, 1.25 Si, 19 Cr, 35 Ni, bal Fe.
Ann: 100,000 TS; 50,000 YS; 41 El; 55 RA; 175 BHN.
Austenitic, high carbon, heat and corrosion resistant alloy; weldable and machinable.
For furnace belt pins, high temperature bolting.

RA 333.
M-1209, M-1491; 0.05 C, 1.5 Mn, 1.25 Si, 25 Cr, 45 Ni, 3.0 Mo, 3.0 Co, 3.0 W, 18 Fe.
Room temp: 100,000 TS; 50,000 YS; 50 El.
1400°F: 57,200 TS; 28,300 YS; 44.8 El.
Austenitic, non-hardenable, heat and corrosion resistant alloy; weldable and machinable.
For high temp. application; turbine parts, radiant tubes, heat treating fixtures, chemical plant equipment.
ASM 5593; ASM-5517.

RA 446.
M-1209; 0.20 max C, 23-27 Cr, 0.25 max Ni, bal Fe.
Annealed: 80,000-90,000 TS; 50,000-60,000 YS; 25-20 El; 150-190 Brin.
For furnace parts, heat treat boxes, conveyor chains; heat resistant, ferritic.

RABW.
M-912; 0.45 C, 1.35 Cr, 0.25 Mo, 4.0 Ni, bal Fe.
Cold work tool steel; for cover dies, embossing dies, bending tools, shear blades.
W.-Nr. 2767; AFNOR 45 NCD 17.

RACO 5.
M-986; 0.10 C, 0.35 Mn, bal Fe.
For welding electrodes; for mild steel; AWS-E6030.

RACO 7.
M-986; 0.07 C, 0.35 Mn, 0.10 Si, 0.03 S, 0.03 P, bal Fe.
For welding electrodes for mild steel; E-6010.

RACO 13.
M-986; 0.07 C, 0.30 Mn, 0.10 Si, 0.03 S, 0.03 P, bal Fe.
For mild steel welding electrodes; E-6013.

RACO 20.
M-986; 0.08 C, 0.35 Mn, 0.10 Si, 0.03 S, 0.03 P, bal Fe.
For welding electrodes for mild steel; E-6020.

RACO 25.
M-986; 0.15 C, 1.0 Cr, 0.5 Mn, bal Fe.
Welded: 200 Brin.
For welding electrodes for mild steel; machinable.

RACO 45.
M-986; 0.25 C, 2.5 Cr, 1.0 Mn, bal Fe.
Welded: 450 Brin.
For hard surfacing electrodes; water resistant.

RACO 55.
M-986; 0.4 C, 3.0 Cr, 1.5 Mn, bal Fe.
Welded: 550 Brin.
For hard surfacing electrodes; water resistant.

RACO HD 64.
M-986; 0.08 C, 0.5 Mo, bal Fe.
For welding electrodes; for high tensile steel.

RACO NO. 8.
M-986; 0.07 C, bal Fe.
For welding electrodes; for mild steel.

RACO NO. 11.
M-986; C, bal Fe.
For A.C. welding electrodes; for mild steel.

RACO RED LABEL.
M-986; 0.1 C, 0.4 Mn, bal Fe.
For welding rods; gas welding.

RACO TYPE D.
M-986; 0.10 C, bal Fe.
For welding electrodes; E 4520 cored rod.

RACO TYPE M.
M-986; 0.10 C, 0.40 Mn, 0.01 Si, 0.03 S, 0.03 P, bal Fe.
For welding electrodes; copper coated.

RACOLLOY MANGANESE.
M-986; C, 14 Mn, 3.5 Ni, bal Fe.
For hard facing electrodes; for Mn steel, coated, austenitic.

RADAX SPEZIAL.
M-1340; 1.45 C, 1.40 Cr, 0.60 Mn, bal Fe.
For bearings, bushings, cutters; oil or water hardened, wear resistant.

RADAX W10.
M-1340; 1.2 C, 1.0 W, 0.1 V, 0.2 Cr, bal Fe.
For cutters, dies; water hardened, wear resistant.

RADECO.
M-1322; 1.3 C, 4.3 Cr, 0.85 Mo, 3.8 V, 12 W, bal Fe.
For forming and blanking dies, punches; high speed steel.

RADIKAL MO 5.
M-1309; 0.85 C, 6.7 W, 4.5 Cr, 2.0 V, 5.3 Mo, bal Fe.
Hardened: C 64-65 Rock.
For lathe and planer tools, hobs, taps, drills, reamers, broaches.
High-speed steel, high red-hardness.

RADIKAL MO 9.
M-1309; 0.85 C, 4 Cr, 9.2 Mo, 2 W, 1.3 V, bal Fe.
Hardened: C 64-65 Rock.
For milling cutters, lathe and planer tools, broaches, reamers, hobs, taps.
High-speed steel, high red-hardness.

RADIKAL MO 55.
M-1309; 0.85 C, 6.7 W, 4.5 Cr, 5.5 Co, 2.0 V, 5.3 Mo, bal Fe.
Hardened: C 64-66 Rock.
For lathe and planer tools, hobs, taps, reamers, drills, broaches.
High-speed steel, high red-hardness.

RADIKAL MO 60.
M-1309; 1.25 C, 6.7 W, 4.5 Cr, 3.4 V, 5.3 Mo, bal Fe.
Hardened: C 64-66 Rock.
For lathe and planer tools, reamers, broaches, hobs, taps, form cutters.
High-speed steel, high red-hardness.

RADIKAL MO 92.
M-1309; 1.1 C, 2.0 W, 4.2 Cr, 2.1 V, 9.2 Mo, bal Fe.
Hardened: C 64-65 Rock.
For reamers, broaches, taps, hobs, form cutters.
High-speed steel, high red-hardness.

RADIO ALLOY NO. 30.
M-61; 98 Cu, 2 Ni.
For electrical instruments, rheostats, voltage control relays; working temperature 350°C (max).

RADIO ALLOY NO. 60.
M-61; 95 Cu, 3.5 Ni.
For electrical instruments, resistors; working temperature 350°C (max).

RADIO ALLOY NO. 90.
M-61; 88 Cu, 12 Ni.
For electrical instruments, resistors; working temperature 400°C (max).

RADIO ALLOY NO. 180.
M-61; 78 cu, 22 Ni.
For electrical instruments, resistors; working temperature 400°C (max).

RADIOMETAL 36.
M-108; 36 Ni, bal Fe.
Soft magnetic alloy, high resistivity, low losses for transformers.

RADIOMETAL 50.
M-108; 50 Ni, 50 Fe.
Soft magnetic alloy, high initial permeability, high saturation for transformers and relays.

RADIO NOX A 182.
M-72; 0.04 C, 18 Cr, 12 Ni, Mo, B, bal Fe.
Austenitic stainless steel.

RADIUM A.
M-959; 4.5 Sb, 4.5 Cu, bal Sn.
Cast: 12,850 TS; 28 Brin.
For heavy duty bearings, bushings; Babbitt; M.P. 437-699°F.

RAE 1.
M-912; 0.15 C, 0.25 Si, 0.80 Mn, 1.05 Cr, 1.45 Ni, bal Fe.
Alloy carburizing steel; for gears, spline couplings, shafts, bolts.
W.-Nr. 5713.

RAE 3.
M-912; 0.14 C, 0.25 Si, 0.55 Mn, 0.75 Cr, 3.45 Ni, bal Fe.
Alloy carburizing steel; for automotive and tractor gears, spline couplings, universal couplings.
W.-Nr. 5752.

RAE 40 C.
M-Eng.; 0.55 Mg, 2.0 Cu, 2.5 Mn, 4.4 Ni, 0.27 Be, bal Al.
Wrought (at R.T.): 36,000 TS; 1.0 El.
Wrought (at 752°F): 4120 TS; 59.5 El.
For aircraft and jet engine parts.
Good resistance to loads at moderately elevated temperatures.

RAE 55.
M-Eng.; 0.71 Mg, 1.84 Cu, 2.02 Mn, 2.34 Ni, 0.2 Be, bal Al.
Wrought (at R.T.): 39,700 TS; 9.5 El.
Wrought (at 752°F): 4520 TS; 79.0 El.
For aircraft and jet engine parts.
Good resistance to loads at moderately elevated temperatures.

RAFFINAL.
M-249, M-493, M-541; 99.99 + Al.
Soft: 7000 TS; 3000 YS; 60 El; 14 Brin.
Hard: 16,000 TS; 15,000 YS; 6 El; 30 Brin.
For window frames, door handles, chemical plant equipment; corrosion resistant.

RAIL-ARC see **WEAR-O-MATIC RAIL-ARC.**

RAILENDER.
M-712; 0.35 C, 1.8 Cr, 0.12 V, bal Fe.
320-336 Brin.
For hard facing electrodes; for battered rail ends.

RAILROAD "A" BRONZE.
M-England; 0-20 Sn, 0-22 Zn, 0-20 Pb, 0-0.5 P, bal Cu.
For axle bearings, gears, piston rods, side valves; heavy duty.

RAILROAD BRONZE DURABLE "K".
73.5 Cu, 9.5 Sn, 9.5 Zn, 7.5 Pb.
For locomotive bearings; heavy duty.

RAILROAD BRONZE DUTCH "F".
85.25 Cu, 12.75 Sn, 2.0 Zn.
For eccentric straps for railroad; high strength.

RAILROAD BRONZE FRENCH "B".
M-French; 82 Cu, 10 Sn, 8 Zn.
For axle bearings; high strength.

RAILROAD BRONZE FRENCH COMMON "C".
M-French; 78 Cu, 20 Sn, 2 Zn.
For axle bearings; high strength.

RAILROAD BRONZE GEARING "H".
88.8 Cu, 8.5 Sn, 2.7 Zn.
For railroad gearing; tough.

RAILROAD BRONZE GERMAN "J".
81 Cu, 15 Zn, 4 Pb.
For locomotive bearings; heavy duty.

RAILROAD BRONZE HARD "E".
87.05 Cu, 7.88 Sn, 5.07 Zn.
For hard axle bearings, high strength.

RAILROAD BRONZE "I".
89 Cu, 2.4 Sn, 7.8 Zn.
For locomotive bearings; heavy duty.

RAILROAD BRONZE "L".
74.1 Cu, 3.7 Sn, 22.2 Zn.
For locomotive pistons and piston rods; high strength.

RAILROAD BRONZE LAFOND "D".
80 Cu, 18 Sn, 2 Zn.
For axle bearings; high strength.

RAILROAD BRONZE LAFOND "G".
84 Cu, 14 Sn, 2 Zn.
For eccentric straps for railroads; high strength.

RAILROAD BRONZE "M".
84.5 Cu, 10 Sn, 5 Pb, 0.5 P.
For locomotive side valves; high strength.

RAILROAD BRONZE "N".
84 Cu, 8.5 Sn, 5 Zn, 2.5 Pb.
For locomotive injectors; free-cutting.

RAILROAD BRONZE "O".
88.5 Cu, 10 Sn, 0.5 P.
For bronze parts; high strength.

RAILROAD BRONZE "P".
80 Cu, 5 Sn, 15 Pb.
For locomotive axle box bearings; heavy duty.

RAILROAD BRONZE "Q".
75 Cu, 5 Sn, 20 Pb.
For bearings, Babbitts; anti-friction.

RAILROAD ROD see **AIRCO RAILROAD ROD.**

RAIL STEEL.
M-Eng; 0.50-0.89 C, 0.60-0.90 Mn, 0.04 max. P, 0.15 min. Si, bal Fe.
For steel rails; wear resistant.

RAILWAY AXLE BOX.
M-657; 83 Cu, 7 Sn, 6 Zn, 4 Pb.
Cast: 30,000 TS; 12,000 YS; 18 El.
For castings, bearings; tough.

RAILWEAR.
M-844; 0.7 C, 13 Mn, bal Fe.
Cast: 350-380 Brin.
For hard surfacing electrodes; coated, for rail joints.

RAISED-LAY.
M-1066; precious metal tip on base metal blade.
For electrical bar contacts; laminated, composite.

RAKEL'S METAL.
88 Cu, 10 Ni, 1 Mn, 1 Zn.
For corrosion resistant parts; corrosion resistant.

RAMET 1.
M-329; Submicron dispersion-strengthened carbide.
Hardness: 91.5 Rock A;
Transverse rupture strength: 400,000 psi; Compressive strength: 525,000 psi; Young's modulus: 80 x 10^6 Psi;
For machining high strength alloys and most difficult-to-machine alloys.

RAMSOS 1.
M-1288; 0.30 C, 2.6 Cr, 0.3 V, 8.5 W, bal Fe.
For punches, shears, extrusion rams; hot work steel; 30WCrV3411.

RAMSOS 1C.
M-1288; 0.35 C, W, Cr, V, bal Fe.
For punches, shears, crimpers, upsetters; hot work steel; 35WCrV4012.

RAMSOS 1MO.
M-1288; 0.35 C, 5 Mo, 2.5 Cr, V, bal Fe.
For punches, shears, crimpers, upsetters; hot work steel; 35MoCrV179.

RAMSOS 1N.
M-1288; 0.25 C, W, Cr, Ni, V, bal Fe.
For punches, shears, crimpers; hot work steel; 25WCrNiV3610.

RAMSOS 2A.
M-1288; 0.60 C, Mo, Cr, W, V, bal Fe.
For punches, shears, crimpers; hot work steel; 60MoCrWV9515.

RAMSOS 2B.
M-1288; 0.60 C, W, Mo, Cr, V, bal Fe.
For punches, shears, crimpers; hot work steel; 60WMoCrV2726.

RAMSOS 3.
M-1288; 0.35 C, Cr, W, Mo, Si, V, bal Fe.
For extrusion rams, liners, punches; hot work steel; 35CrWMoSiV226.

RAMSOS 5A.
M-1288; 0.45 C, Cr, Ni, W, Si, Mn, bal Fe.
For hot work tools, punches, crimpers; hot work steel; 45CrNiWSiMn525.

RAMSOS 6MO.
M-1288; 0.30 C, Cr, Mo, V, bal Fe.
For hot work tools; hot work steel; 30CrMoV1328.

RAMSOS 6(S6).
M-1288; 0.30 C, 1.1 Cr, 0.18 V, 3.75 W, bal Fe.
For hot work tools, punches, crimpers; hot work steel; 30WCrV15.

RAMSOS 7.
M-1288; 0.45 C, 1.3 Cr, 0.45 Mo, 0.8 V, 0.45 W, bal Fe.
For hot work tools; hot work steel; 45CrVMoW58.

RAMSOS 7MO.
M-1288; 0.30 C, Cr, Mo, Mn, V, bal Fe.
For hot work tools; 30 CrMoMnV710; hot work steel.

RAMSOS 8.
M-1288; 0.40 C, 2 Cr, 1 Mn, 0.25 Mo, bal Fe.
For header tools, container jackets; 40CrMnMo7; hot work steel.

RAMSOS 9.
M-1288; 0.30 C, 2.5 Cr, 0.20 Mo, 0.15 V, bal Fe.
For extrusion press parts, mandrels, liners; 30CrMoV9; hot work steel.

RAMSOS 55.
M-1288; 0.55 C, 0.10-0.40 Si, 0.5-0.7 Mn, bal Fe.
For gears, shafts, axles, tools; D1N-C55WS; water hardened.

RAMSOS 80.
M-1288; 0.80 C, 0.10-0.40 Si, 0.5-0.7 Mn, bal Fe.
For tools, springs, cutters, drills; water hardened; Type W1.

RAMSOS G1.
M-1288; 0.55 C, 0.7 Cr, 0.18 Mo, 1.65 Ni, 0.2 V, bal Fe.
For hot work tools, punches, upsetters; 55NiCrMoV6; hot work steel.

RAMSOS G1M.
M-1288; 0.50 C, 0.8-1.1 Mn, 0.25-0.50 Si, bal Fe.
Heat treated; 115,000-135,000 TS.
For gears, shafts, punches, crimpers; 50Mn4; oil hardened.

RAMSOS G1 SPECIAL.
M-1288; 0.50-0.60 C, 1.2 Cr, 0.35 Mo, 1.6 Ni, 0.2 V, bal Fe.
For hot work tools, punches, crimpers; 56NiCrMoV7; hot work steel.

RAMSOS G3.
M-1288; 0.45 C, 0.25-0.50 Si, 0.3-0.8 Mn, bal Fe.
For gears, pinions, shafts, axles; water hardened; C45W3.

RAMSOS G4.
M-1288; 0.60 C, 0.25-0.50 Si, 0.3-0.8 Mn, bal Fe.
For crimpers, punches, tools, dies; C60W3; water hardened.

RAMSOS G5.
M-1288; 0.75 C, 0.25-0.50 Si, 0.3-0.8 Mn, bal Fe.
For crimpers, tools, springs, punches; C75W3; water hardened.

RAMSOS GE4.
M-1288; 1.0 C, 0.25-0.50 Si, 0.3-0.8 Mn, bal Fe.
For drills, taps, punches, tools; Type W1; water hardened; C100W2.

RAMSOS KSZ.
M-1288; 0.3 C, 0.55 Mn, 0.25 Si, 2.5 Cr, 0.15 V, bal Fe.
For extrusion press parts, mandrels, liners; 30CrMoV9; hot work steel.

RAMSOS MUK.
M-1288; 0.45 C, 0.9 Si, 0.3 Mn, 1 Cr, 0.2 V, 1.8 W, bal Fe.
For hot working tools, crimpers, upsetters; 45WCrV7; oil hardened.

RAMSOS O.
M-1288; 0.82 C, 4.1 Cr, 0.85 Mo, 1.6 V, 8.7 W, bal Fe.
For lathe and planer tools, reamers, broaches, hobs; high speed steel; 82WV3419.

RAMSOS O/2.
M-1288; 0.65 C, 3.75 Cr, 0.85 Mo, 0.7 V, 8.5 W, bal Fe.
For lathe and planer tools, drills, hobs, reamers; high speed steel; 65WMo348.

RAMSOS OK.
M-1288; 0.65 C, 3.75 Cr, 0.85 Mo, 8.5 W, 0.7 V, bal Fe.
For lathe and planer tools, hobs, broaches; high speed steel; 65WMo348.

RAMSOS OO.
M-1288; 0.95 C, 4 Cr, 1 V, Mo, W, bal Fe.
For cutters, drills, hobs, reamers; high speed steel; 95WMo1126.

RAMSOS S4.
M-1288; 0.30 C, 0.3 Mn, 2.35 Cr, 0.6 V, 4.25 W, bal Fe.
Heat treated: 198,000-242,000 TS.
For extrusion dies; rams and liners; hot work steel; 30WCrV179.

RAMSOS SGK.
M-1288; 0.30 C, Cr, Mo, Si, V, bal Fe.
For punches, upsetters, crimpers; 30CrMoSiV2011; hot work steel.

RAMSOS SUPER A.
M-1288; 0.40 C, W, Mo, Cr, V, bal Fe.
For tools, dies, upsetters, punches; 40WMoCrV1838; hot work steel.

RAMSOS SUPER K.
M-1288; 0.30 C, W, Co, Cr, Mo, V, bal Fe.
For crimpers, dies, punches, shears; 30WCoCrMoV2424; hot work steel.

RAMSOS SUPER WN.
M-1288; 0.55 C, Ni, W, Cr, Co, Si, V, bal Fe.
For hot work tools, dies; 55 NiWCrCoSiV484416; oil hardened.

RAMSOS VC.
M-1288; 0.40 C, Ni, Cr, Mo, bal Fe.
For gears, axles, crankshafts; 40NiCrMo176; oil hardened, shock resistant.

RAN 5.
M-912; 0.35 C, 0.25 Si, 0.35 Mn, 1.70 Cr, 3.75 Ni, bal Fe.
Deep hardening alloy steel for structural parts for tractors, motor bus and heavy equipment. W.-Nr. 5952.

RANALLOY TYPE A.
M-1128; C, Mo, Ni, Cr, bal Fe.
500 Brin.
For hard surfacing electrodes; wear resistant.

RANALLOY TYPE B.
M-1128; C, Mo, Ni, Cr, bal Fe.
530 Brin.
For hard surfacing electrodes; wear resistant.

RANALLOY TYPE C.
M-1128; C, Mo, Ni, Cr, bal Fe.
480 Brin.
For hard surfacing electrodes; wear resistant.

RANDALL.
M-1155; Cu-graphite.
For bearings; sintered.

RANDOLF METAL.
For dental alloy; corrosion resistant.

RAN FH.
M-912, M-501; 0.23-0.30 C, 0.7 Cr, 3.5 Ni, bal Fe.
For gears, shafts, crankshafts, bolts, studs; oil hardened, shock resistant.

RANITE 1.
M-1100.
Cobalt base (Co-Cr-W) electrode.
For non-magnetic, non-machinable hard-surfacing for resistance to extreme abrasion, corrosion, and high temperatures.
52 Rock C.

RANITE 5.
M-1100; 35% alloy, bal Fe.
Bare tube rod for gas application two-layer deposit for light to moderate impact and severe abrasion; 61 Rc.

RANITE 6.
M-1100.
Cobalt base (Co-Cr-W) electrode.
For non-magnetic, machinable build-up for shear blades, valves for good metal-to-metal wear, impact, and hot wear.
41 Rock C.

RANITE 12.
M-1100.
Cobalt base (Co-Cr-W) electrode.
For non-magnetic, machinable build-up for good abrasion resistance and medium impact.
45 Rock C.

RANITE A.
M-1100; 3% total C, Mn, Cr, bal Fe.
All position electrode, AC or DC; for build-up of worn parts as tractor rails, idlers, shovel buckets; 42-44 Rock C.

RANITE BU.
M-1100; 5% total of C, Mn, Cr, bal Fe.
All position electrode, AC or DC, straight or reverse polarity; for build-up of shafts, gears, idlers; machinable; 28-34 Rock C.

RANITE BX.
M-1100; 9% total of C, Mn, Cr, bal Fe.
All position electrode, AC or DC; for multipass build-up of dredge pump shells, cone crusher mantles and liners; 44-50 Rock C.

RANITE C.
M-1100; 14% total C, Cr, Mo, bal Fe.
All position electrode, AC or DC; for 1-2 passes hardsurfacing for extreme abrasion and medium impact are encountered; as plow shares, conveyor screws.
58-60 Rock C.

RANITE D.
M-1100; 14% total C, Mn, Cr, Mo, B, bal Fe.
All position electrode, AC or DC; for 1-2 passes for extreme abrasion resistance, as brick dies, drill collars, dredge teeth.
63-68 Rock C.

RANITE F.
M-1100; 18% total of C, Mn, Cr, Mo, bal Fe.
All position electrode, AC or DC; for hardfacing truck beds, earth moving equipment, guides and chutes; 52-57 Rock C.
Wear and abrasion resistant.

RANITE G.
M-1100; 10% total C, Mn, Cr, Ni, Mo, bal Fe.
All purpose electrode, AC or DC; for general purpose build-up and hardfacing.
40-54 Rock C.

RANITE J.
M-1100; 26% total C, Mn, Cr, Mo, bal Fe.
All position electrode, AC or DC; for 2 pass hardsurfacing at fast deposit rate.
Good wear resistance up to 800°F.
45-50 Rock C.

RANITE TYPE NO. 4.
M-1100; C, Si, Cr, Mo, Ti, B, bal Fe.
Welded: 520 Brin.
For hard facing electrodes; wear and shock resistant.

RANKIN 308-16.
M-1100; 0.07 C, 19.5 Cr, 9.5 Ni, 1.6 Mn, bal Fe.
AC-DC-16 coated electrode for welding 200-300 series stainless steels, to and including 308.
84,000-96,000 psi TS; 36-46 El.

RANKIN 308L.
M-1100; 0.04 C, 19.0 Cr, 9.5 Ni, 1.5 Mn, 0.3 Si, bal Fe.
AC-DC-16 coated electrode for welding 304 and 304L stainless steels.
80,000-90,000 psi TS; 36-46 El.

RANKIN 309-16.
M-1100; 0.10 C, 23.0 Cr, 13.0 Ni, 1.6 Mn, 0.5 Si, bal Fe.
AC-DC-16 coated electrode for welding dissimilar metals and cladding 18 to mild steels.
85,000-95,000 TS; 35-45 El.

RANKIN 310-16.
M-1100; 0.20 C, 26.0 Cr, 21.0 Ni, 1.8 Mn, 0.4 Si, bal Fe.
AC-DC-16 coated electrode for welding 310 and dissimilar metals.
85,000-95,000 psi TS; 35-45 El.

RANKIN 312-16.
M-1100; 0.15 C, 29.0 Cr, 9.5 Ni, 1.9 Mn, 0.5 Si, bal Fe.
AC-DC-16 coated electrode for welding high tensile to high temperature heat resisting alloys; and for dissimilar metals.
100,000-110,000 psi TS; 22-25 El.

RANKIN 316-16.
M-1100; 0.07 C, 18.0 Cr, 13.0 Ni, 2.25 Mo, 1.7 Mn, 0.3 Si, bal Fe.
AC-DC-16 coated electrode for welding type 316 stainless steel.
85,000-95,000 psi TS; 35-45 El.

RANKIN 316-16L.
M-1100; 0.04 max C, 18.0 Cr, 13.0 Ni, 2.25 Mo, 1.7 Mn, 0.50 Si, bal Fe.
AC-DC-16 coated electrode for welding type 316L stainless steel.
80,000-90,000 psi TS; 35-45 El.

RANMANG.
M-1100; 16% total Mn, Ni, bal Fe.
All position electrode, AC or DC; for welding and build-up of high manganese steel in rails, dredge shells, rolls.
As deposited: 75-85 Rb; cold worked: 45-50 Rc.
144,000 psi TS; 33 El.

RANMANG 2.
M-1100; 36% of total C, Mn, Cr, Ni, Si, bal Fe.
All position electrode, AC or DC; for joining and build-up of high manganese and other steels, on hammers, shovels, dragline pins and links.
As deposited: 17-22 Rc; work hardened: 43-48 Rc.

RANOMATIC A.
M-1100; C, 2.5 alloy, bal Fe.
Martensitic, non-work-hardening welding wire for overlay to give moderate abrasion and good impact resistance; 42-44 Rc.

RANOMATIC BU.
M-1100; 4% of total C, Cr, Mn, Mo, bal Fe.
Open-arc welding wire for machinable build-up on carbon and low alloy steel parts; 24-36 Rc.

RANOMATIC BX-2.
M-1100; 13.5% of total C, Mn, Cr, Mo, Cu, bal Fe.
Self-shielded, flux-cored wire for rebuilding and hardsurfacing carbon and manganese steel parts.
As deposited, third pass; 43-45 Rc; as cold worked 60 Rc.

RANOMATIC D.
M-1100; 14% of total C, Cr, Mo, B, bal Fe.
Open-arc welding wire for depositing an extremely hard, abrasion resistant coating; 63-68 Rc.

RANOMATIC F.
M-1100; 18% of total C, Cr, Mo, bal Fe.
Open-arc welding wire for depositing a hard, abrasion resistant coating; 55-60 Rc.

RANOMATIC H.
M-1100; 23% of total C, Mn, Cr, bal Fe.
High chromium, open-arc wire for two-pass coating for abrasion and corrosion resistance; 55 Rc.

RANOMATIC O.
M-1100; 23.5% of total C, Mn, Cr, Mo, bal Fe.
Self-shielded, flux-cored wire for multi-pass build-up of wear resistant coatings.
As deposited: 45-46 Rc; work hardened: 63 Rc.

RANSCO-ALLOY.
M-1164; 2.5-3.25 C, 1-2 Si, 12-15 Ni, 5-7 Cu, bal Fe.
Cast: 20,000-35,000 TS; 125-200 Brin.
For diesel engine cylinder liner; centrifugally cast.

RANTUNG.
M-1100.
Cast tungsten carbide with binder; electrode.
Coated for arc; uncoated for acetylene.
For extreme hard, wear resistant coating.

RAPIDE B.
M-1115; 0.65-0.80 C, 1 Co, 4-5 Cr, 17-19 W, 1-1.5 V, bal Fe.
For cutters, taps, reamers, hobs, broaches; high speed steel, oil hardened.

RAPIDE C.
M-1115; 0.75-0.90 C, 4-5 Cr, 17-19 W, 1.0-1.5 V, 0.5-1.0 Co, bal Fe.
For lathe and planer tools, milling cutters, taps; high speed steel, oil hardened.

RAPIDE J.
M-1115; 0.7-0.9 C, 4-5 Cr, 17-14 W, 9-11 Co, 2 V, bal Fe.
For lathe and planer tols, reamers, hobs, taps, broaches; high speed steel, oil hardened.

RAPID FINISHING.
M-261; C, alloy, bal Fe.
For fast finishing tools; water hardened.

RAPID SPEZIAL.
M-1345; 0.95 C, W, Mo, bal Fe.
For tools, dies; oil hardening.

RARE EARTH SILICIDE.
M-1038; 15-18 Ce, 28-33 Si, 30-35 total rare earths, bal Fe.
Ferro alloy for deoxidation, desulphurization and sulphide shape control of steels; also a source of cerium and rare earths for gray and cast iron.

RAUCHBERG.
M-Ger.; 66-75 Cu, 15-19 Pb, 10-5 Sn, 1-5 Sb.
For bearings; corrosion resistant.

RAUENLOY.
M-1138; C, Cr, Mo, V, bal Fe.
For leaf springs; tough, oil hardened.

RAVEN MST.
M-1182; 0.90 C, 1.6 Mn, bal Fe.
For punches, dies, crimpers, chisels, cutters; Type O2; non-deforming, oil hardened.

RAVEN NBS.
M-1182; 0.70 C, 0.75 Cr, 1.5 Ni, 0.25 Mo, bal Fe.
For punches, dies, rollers, mandrels; Type L6; oil hardened.

RAXA ATV.
M-1353; 1.45 C, 1.5 Cr, V, bal Fe.
For cutting tools, dies, mandrels, rolls, wear resistant; oil hardening.

RAXA BK 5.
M-1353; 1.0 C, 5.0 Cr, 1.0 Mo, 0.15 V, bal Fe.
For cutting tools, shear blades, threading rolls; wear resistant; air or oil hardening.

RAXA BK 12 G.
M-1353; 1.2 C, 12 Cr, V, bal Fe.
For wear parts in the cement and coal mining industry; air or oil hardening.
Special steel, casting-quality.

RAXA BK 60.
M-1353; 0.65 C, 14 Cr, 0.6 Mo, bal Fe.
For cutting tools, shear blades.

RAXA BK 100.
M-1353; 1.05 C, 13.5 Cr, 0.5 Mo, 0.3 Ni, bal Fe.
For cutting tools, shear blades, rolls, forming tools; wear resistant; air or oil hardening.

RAXA BK 110.
 M-1353; 0.95 C, 12 Cr, 1.0 Mo, 1.0 V, bal Fe.
 For shear blades, cutting tools, wear resistant; air or oil hardening.

RAXA BK 121.
 M-1353; 1.55 C, 12 Cr, 0.7 Mo, 1.0 V, bal Fe.
 For cutting tools, dies, forming rolls; high wear resistance; air or oil hardening.
 AISI D2.

RAXA BK 122.
 M-1353; 2.20 C, 12.5 Cr, 0.9 Mo, 2.2 V, bal Fe.
 For cutting tools, drawing dies, forming rolls, pressing tools; high wear resistance; air or oil hardening.
 Similar to AISI D7.

RAXA BK 211 G.
 M-1353; 1.9 C, 21 Cr, 0.7 Mo, 1.5 W, bal Fe.
 For rolling mill equipment and rolls; high wear resistance; special steel, casting-quality.

RAXA BKL.
 M-1353; 2.1 C, 12 Cr, 0.8 W, bal Fe.
 For cutting tools, shaping rolls, wood knives; high wear resistance; air or oil hardening.

RAXA BKLO.
 M-1353; 2.1 C, 1.0 Co, 13 Cr, 0.5 Mo, 0.7 W, bal Fe.
 For cutting tools for electric sheets; high wear resistance; air or oil hardening.

RAXA BKR.
 M-1353; 1.65 C, 12 Cr, 0.6 Mo, 0.5 W, bal Fe.
 For cutting tools, stamping tools, metal saws; high wear resistance; air or oil hardening.

RAXA BKS.
 M-1353; 2.1 C, 12 Cr, bal Fe.
 For cutting tools, shear blades, pressing tools; high wear resistance; air or oil hardening.

RAXA BKV.
 M-1353; 1.65 C, 12 Cr, 0.2 V, bal Fe.
 For stamping tools, wood cutters, metal saws, threading rolls; high wear resistance; oil hardening.

RAXA BKW.
 M-1353; 1.2 C, 12 Cr, 1.4 Mo, 1.7 V, 2.5 W, bal Fe.
 For high wear resistant cutting tools; air or oil hardening.

RAXA C 25.
 M-1353; 2.5 C, 23 Cr, 0.4 Ni, 0.3 V, bal Fe.
 For high wear resistant special steel casting tools for the cement and coal mining industry.
 W. Nr. 1.2288.

RAXA C 35.
 M-1353; 3.2 C, 25 Cr, 0.3 V, bal Fe.
 For extremely wear resistant special steel casting tools for the coal mining industry.

RAXA CM 4 G.
 M-1353; 1.3 C, 8 Co, 21 Cr, 6.5 Mo, bal Fe.
 For rolling mill equipment, twist rolls, guide rolls; special steel casting-quality; hot wear resistant.

RAXA CM 8 G.
 M-1353; 2.4 C, 1.5 Si, 24 Co, 17 Cr, 2.3 Mo, 6 Ni, 1.4 V, 2.8 W, bal Fe.
 For rolling mill equipment, twist rolls, guide rolls; special steel casting-quality; hot wear resistant.

RAXA CM 88.
 M-1353; 0.2 C, 10 Co, 9 Cr, 2.2 Mo, 7 W, bal Fe.
 For hot working tools in pressure casting moulds; hot wear resistant; oil hardening.

RAXA CM 89.
 M-1353; 0.45 C, 5 Co, 5 Cr, 3 Mo, 2 V, bal Fe.
 For hot working tools in pressure casting moulds; hot wear resistant; oil hardening.

RAXA CMD.
 M-1353; 0.40 C, 1.0 Si, 5 Cr, 1.2 Mo, 0.4 V, bal Fe.
 For forging dies, pressure casting moulds, extrusion tools, hot working tool steel, oil hardening.
 Similar to AISI H 11.

RAXA CMD/W.
 M-1353; 0.37 C, 1.0 Si, 5 Cr, 1.5 Mo, 0.25 V, 1.3 W, bal Fe.
 For pressure casting moulds, hot pressing and forging tools; hot working tool steel.
 Similar to AISI H 12.

RAXA CMG.
 M-1353; 2.6 C, 1.7 Si, 10 Co, 23 Cr, 8 Mo, bal Fe.
 For rolling mill equipment, twist rolls, special steel, casting-quality; hot wear resistant.

RAXA CMR.
 M-1353; 0.35 C, 5 Cr, 3 Mo, 0.7 V, bal Fe.
 For forging dies, hot drawing dies, mandrels; hot working tool steel; oil hardening.

RAXA CMS.
 M-1353; 0.4 C, 5 Cr, 1.4 Mo, 1.0 V, 0.12 S, bal Fe.
 For pressure casting moulds, recipient casings, punches; hot working tool steel; oil hardening with good machinability.

RAXA CMV.
 M-1353; 0.4 C, 1.0 Si, 5 Cr, 1.4 Mo, 1.0 V, bal Fe.
 For forging dies, pressure casting moulds and extrusion tools; hot working tool steel; oil hardening.

RAXA DBV.
M-1353; 0.45 C, 1.5 Si, 1.5 Cr, 0.15 V, bal Fe.
For drawing tools, cutting tools; cold working tool steel; oil hardening.

RAXA DBV/K.
M-1353; 0.6 C, 1.0 Si, 1.0 Mn, 1.2 Cr, 0.1 V, bal Fe.
For shear blades, stamping tools, forging dies, moulds; cold working tool steel; oil hardening.

RAXA DBV/W.
M-1353; 0.4 C, 1.5 Si, 1.5 Cr, 0.12 V, bal Fe.
For shear blades, drawing dies; cold working tool steel; water hardening.

RAXA GARANT 6 M.
M-1353; 0.88 C, 4 Cr, 5 Mo, 1.8 V, 6.5 W, bal Fe.
For twist drills, end mills, threading rolls; high speed steel.
AISI M2.

RAXA GARANT 6 MC.
M-1353; 0.9 C, 5 Co, 4 Cr, 5 Mo, 1.8 V, 6.5 W, bal Fe.
For twist drills, end mills, thread milling cutters; high speed steel.

RAXA GARANT 6 MCK.
M-1353; 1.1 C, 5 Co, 4 Cr, 4 Mo, 1.8 V, 6.75 W, bal Fe.
For milling cutters of all types; high speed steel.
AISI M 41.

RAXA GARANT 6 MK.
M-1353; 1.0 C, 4 Cr, 5 Mo, 1.8 V, 6.5 W, bal Fe.
For milling cutters of all types; high speed steel.
AISI M2.

RAXA GARANT 6 MV.
M-1353; 1.2 C, 4 Cr, 5 Mo, 3 V, 6.5 W, bal Fe.
For taps; high speed steel.
AISI M3 class 2.

RAXA GARANT 8 M.
M-1353; 0.8 C, 4 Cr, 8 Mo, 1.2 V, 1.8 W, bal Fe.
For twist drills; high speed steel.
AISI M1.

RAXA GARANT 8 M 34.
M-1353; 0.9 C, 8 Co, 4 Cr, 9 Mo, 2 V, 2 W, bal Fe.
For milling cutters of all types; high speed steel.
Similar to AISI M 34.

RAXA GARANT 8 M 42.
M-1353; 1.1 C, 8 Co, 4 Cr, 9.5 Mo, 1.2 V, 1.5 W, bal Fe.
For milling cutters of all types; high speed steel.
AISI M 42.

RAXA GARANT 8 MC.
M-1353; 0.95 C, 5 Co, 4 Cr, 8.5 Mo, 2 V, 2 W, bal Fe.
For end mills, twist drills; high speed steel.
Similar to AISI M 30.

RAXA GARANT 10.
M-1353; 0.9 C, 4 Cr, 1.0 Mo, 2.5 V, 12 W, bal Fe.
For profile cutters, broaches, taps; high speed steel.

RAXA GARANT 18.
M-1353; 0.74 C, 4.1 Cr, 1.1 V, 18.5 W, bal Fe.
For drills, hobs, reamers, taps, broaches; high speed steel. AISI T2.

RAXA GARANT 55.
M-1353; 0.79 C, 18 W, 4.7 Co, 1.5 V, 4.3 Cr, bal Fe.
For reamers, broaches, taps, form cutters; high speed steel.

RAXA GARANT 70.
M-1353; 1.3 C, 4.3 Cr, 0.85 Mo, 3.8 V, 12 W, bal Fe.
For engravers' tools, form cutters; high speed steel.

RAXA GARANT 100.
M-1353; 1.3 C, 5 Co, 4 Cr, 2 Mo, 3.75 V, 12 W, bal Fe.
For turning tools, milling cutters; high speed steel.

RAXA GARANT 200.
M-1353; 0.76 C, 10 Co, 4.2 Cr, 0.8 Mo, 1.8 V, 18 W, bal Fe.
For lathe and planer tools, reamers, broaches; high speed steel.

RAXA GARANT 300.
M-1353; 0.75 C, 15 Co, 4.3 Cr, 0.7 Mo, 1.2 V, 18 W, bal Fe.
For turning tools for roughing, planing tools; high speed steel.

RAXA GARANT 400.
M-1353; 1.25 C, 12 Co, 4.3 Cr, 5 Mo, 3.3 V, 9 W, bal Fe.
For turning tools of all types; high speed steel.

RAXA GARANT III.
M-1353; 1.0 C, 4 Cr, 2.7 Mo, 2.3 V, 2.8 W, bal Fe.
For circular metal saws, threading tools, band saws; high speed steel.

RAXA GF CR 12.
M-1353; 2.0 C, 12 Cr, 1.2 Ni, 0.5 V, bal Fe.
For rolling mill equipment, guides; special steel casting-quality; high wear resistant.

RAXA GFG.
M-1353; 0.3 C, 1.1 Cr, 0.5 Ni, bal Fe.
For rolling mill equipment, guides; special steel casting-quality; high wear resistant.

RAXA GL 2.
M-1353; 0.40 C, 1.5 Mn, 2.0 Cr, 0.2 Mo, bal Fe.
For forging dies, upsetting tools, extrusion discs; hot working tool steel; air or oil hardening.

RAXA GL 3.
M-1353; 0.55 C, 0.7 Cr, 0.3 Mo, 1.7 Ni, 0.1 V, bal Fe.
For forging dies, plastic moulds; hot working tool steel; oil hardening.

RAXA GL 4.
M-1353; 0.55 C, 1.1 Cr, 0.5 Mo, 1.7 Ni, 0.1 V, bal Fe.
For forging dies, extrusion punches; hot working tool steel; air or oil hardening.

RAXA GL 5.
M-1353; 0.6 C, 1.4 Cr, 0.8 Mo, 1.7 Ni, 0.1 V, bal Fe.
For forging dies, extrusion punches; hot working tool steel; air or oil hardening.

RAXA HDS.
M-1353; 0.45 C, 1.0 Si, 1.2 Cr, 0.2 V, 1.95 W, bal Fe.
For shear blades, pneumatic tools, cutting tools; cold work tool steel; oil hardening.

RAXA HSS.
M-1353; 1.05 C, 1.0 Mn, 1.0 Cr, 1.2 W, bal Fe.
For cutting tools, measuring tools; cold working tool steel; oil hardening.

RAXA HWA.
M-1353; 0.30 C, 2.65 Cr, 0.35 V, 8.5 W, bal Fe.
For extrusion rams and liners, mandrels; oil hardened, hot work steel.

RAXA HWA 10.
M-1353; 0.6 C, 4 Cr, 1.0 Mo, 0.7 V, 9 W, bal Fe.
For high stress hot working tools; hot working tool steel; air or oil hardening.

RAXA HWA CO.
M-1353; 0.45 C, 4.5 Co, 4.5 Cr, 0.5 Mo, 2.0 V, 4.5 W, bal Fe.
For high stress hot working tools; hot working tool steel; air or oil hardening.
AISI H19.

RAXA HWAS.
M-1353; 0.3 C, 2.0 Co, 2.5 Cr, 0.25 V, 8.5 W, bal Fe.
For tube extrusion tools, pressure casting moulds; hot working tool steel; air or oil hardening.

RAXA HWA SPEZIAL see **RAXE HWAS.**

RAXA KSM.
M-1353; 1.0 C, 0.1 V, bal Fe.
For cold impact tools, stamping tools; cold working tool steel; oil hardening.

RAXA MKS.
M-1353; 1.05 C, 1.1 Mn, 1.1 Cr, bal Fe.
For cutting tools, shaping rolls; cold working tool steel; oil hardening.

RAXA MSA.
M-1353; 0.9 C, 2.0 Mn, 0.15 V, bal Fe.
For cutting tools, measuring tools; cold working tool steel; oil hardening.

RAXA MWA.
M-1353; 0.3 C, 2.35 Cr, 0.6 V, 4.25 W, bal Fe.
For extrusion dies and rams, upsetters, headers; oil hardened, tough.

RAXA NWP.
M-1353; 0.5 C, 1.0 Cr, 3.25 Ni, bal Fe.
For stamping tools, cutlery punching tools; cold working tool steel; air or oil hardening.

RAXA NWP/K.
M-1353; 0.45 C, 1.4 Cr, 0.3 Mo, 4 Ni, bal Fe.
For high stressed stamping tools, billet shearing knives; cold working tool steel; air or oil hardening.

RAXA PDH.
M-1353; 0.27 C, 1.4 Cr, 1.3 Mo, 0.4 V, bal Fe.
For forging dies, hot working tool steel; oil hardening.
W. Nr. 1.2353.

RAXA PDL.
M-1353; 0.3 C, 0.8 Cr, 0.6 Mo, 2.5 Ni, 0.3 V, bal Fe.
For pilger mandrels 70-140 mm dia, mandrel bars, hot working tool steel; air or oil hardening.

RAXA PDM.
M-1353; 0.3 C, 0.4 Cr, 1.7 Mo, 4.5 Ni, 0.25 V, bal Fe.
For pilger mandrels up to 70 mm dia, mandrel bars; hot working tool steel; air or oil hardening.

RAXA PDR.
M-1353; 0.3 C, 3 Cr, 2.7 Mo, 0.5 V, bal Fe.
For forging dies, pressure casting mould tools; hot working tool steel; air or oil hardening.

RAXA PSV.
M-1353; 0.25 C, 0.8 Cr, 0.3 Mo, 1.5 Ni, 0.2 V, bal Fe.
For pilgrin mandrels 140-200 mm dia, mandrel bars; hot working tool steel; oil hardening.

RAXA PW 8.
M-1353; 0.8 C, 2.0 Cr, 1.3 W, bal Fe.
For pilger rolls,-hot work; special steel casting-quality.

RAXA PW 9.
M-1353; 0.9 C, 2.0 Cr, 1.0 W, bal Fe.
For pilger rolls,-hot work; special steel casting-quality.

RAXA PW 12.
M-1353; 1.2 C, 2.0 Cr, 1.3 W, bal Fe.
For pilger rolls, hot-work; special steel casting-quality.

RAXA ROV.
M-1353; 0.45 C, 4.5 Cr, 1.1 Mo, 12 Ni, 1.0 V, 9 W, bal Fe.
For extrusion dies, hot working tool steel.

RAXA RWS 5.
M-1353; 1.3 C, 1.0 Cr, 0.35 V, 5 W, bal Fe.
For cold drawing dies; cold working tool steel; water hardening.

RAXA RWS 5/W.
M-1353; 1.3 C, 5 W, bal Fe.
For cold drawing dies, turning knives; cold working tool steel; water hardening.

RAXA SKL.
M-1353; 1.0 C, 1.5 Cr, bal Fe.
For cold rolls, drawing dies, threading tools; cold working tool steel; oil or water hardening.

RAXA SL 35.
M-1353; 1.45 C, 3.5 V, bal Fe.
For high wear resistant cold extrusion and cold impact tools; cold working tool steel; water hardening.

RAXA SPS.
M-1353; 0.55 C, 1.7 Si, 0.3 Mo, 0.4 Ni, bal Fe.
For clamping tools, beating bars; cold working tool steel; oil hardening.

RAXA TWO 3.
M-1353; 0.45 C, 1.5 Cr, 0.5 Mo, 0.8 V, 0.5 W, bal Fe.
For forging tools; hot working tool steel; oil hardening.

RAXA WCNV.
M-1353; 0.5 C, 1.3 Si, 13 Cr, 13 Ni, 1.0 V, 2.0 W, bal Fe.
For extrusion dies; hot working tool steel.

RAXA WCNV/M.
M-1353; 0.35 C, 1.3 Si, 13 Cr, 6 Ni, 1.0 V, 2.8 W, bal Fe.
For extrusion dies, hot working tool steel.

RAXA WCR 8.
M-1353; 0.5 C, 1.0 Si, 8.5 Cr, 1.2 Mo, 1.2 W, bal Fe.
For wood chopping knives; cold working tool steel; oil or air hardening.

RAXA WF 6.
M-1353; 1.0 C, 1.2 Cr, 0.6 W, bal Fe.
For rolling mill equipment; special steel casting-quality.

RAXA WF 18.
M-1353; 2.0 C, 0.7 Co, 22 Cr, 1.9 Ni, 1.8 W, bal Fe.
For rolling mill equipment; special steel casting-quality.

RAXA WHW.
M-1353; 0.6 C, 1.0 Cr, 0.2 V, 2.0 W, bal Fe.
For shear blades, cutting tools; cold working tool steel; oil hardening.

RAXA ZRWS.
M-1353; 2.7 C, 1.5 Co, 22 Cr, 2.5 Ni, 0.3 V, 5 W, bal Fe.
For hot drawing dies; special steel casting-quality.

RAXIT 006 G.
M-1353; 0.1 C, 13 Cr, bal Fe.
For steam turbines, fittings, stainless steel casting-quality.
W. Nr. 1.4006.

RAXIT 008 G.
M-1353; 0.12 C, 13 Cr, 1.0 Ni, bal Fe.
For valve parts; stainless steel casting-quality.
W. Nr. 1.4008.

RAXIT 027 G.
M-1353; 0.2 C, 14 Cr, bal Fe.
For steam turbines, fittings, valves, glass moulds; stainless steel casting-quality.
W. Nr. 1.4027.

RAXIT 059 G.
M-1353; 0.22 C, 17 Cr, 1.5 Ni, bal Fe.
For fittings, pump parts, stainless steel casting-quality.
W. Nr. 1.4059.

RAXIT 085 G.
M-1353; 0.7 C, 29 Cr, bal Fe.
For chemical industry, mining; stainless steel casting-quality; wear resistant.
W. Nr. 1.4085.

RAXIT 086 G.
M-1353; 1.2 C, 29 Cr, bal Fe.
For chemical industry, mining; stainless steel casting-quality; wear resistant.
W. Nr. 4086.

RAXIT 136 G.
M-1353; 0.7 C, 29 Cr, 2 Mo, bal Fe.
For chemical industry, mining; stainless steel casting-quality; wear resistant.
W. Nr. 1.4136.

RAXIT 138 G.
M-1353; 1.2 C, 29 Cr, 2 Mo, bal Fe.
For chemical industry, mining; stainless steel casting-quality; wear resistant.

RAXIT 219 G.
M-1353; 0.25 C, 1.0 Cr, 0.25 Mo, bal Fe.
Special steel casting-quality; tough at sub zero temperatures.
W. Nr. 1.7219.

RAXIT 306 G.
M-1353; 0.03 C, 18 Cr, 11 Ni, bal Fe.
For chemical industry; stainless steel casting-quality.
W. Nr. 1.4306.

RAXIT 308 G.
M-1353; 0.07 C, 18 Cr, 10 Ni, bal Fe.
For chemical industry; stainless steel casting-quality.
W. Nr. 1.4308.

RAXIT 312 G.
M-1353; 0.1 C, 18 Cr, 9 Ni, bal Fe.
For water turbines, pump wheels; stainless steel casting-quality.
W. Nr. 1.4312.

RAXIT 313 G.
M-1353; 0.05 C, 13 Cr, 5 Ni, bal Fe.
For water turbines, pump-wheels; stainless steel casting-quality.
W. Nr. 1.4313.

RAXIT 340 G.
M-1353; 0.4 C, 27 Cr, 4 Ni, bal Fe.
For chemical industry, mining; stainless steel casting-quality; wear resistant.
W. Nr. 1.4340.

RAXIT 347 G.
M-1353; 0.08 C, 26 Cr, 7 Ni, bal Fe.
For chemical industry, fittings; stainless steel casting-quality.
W. Nr. 1.4347.

RAXIT 404 G.
M-1353; 0.03 C, 18 Cr, 2.3 Mo, 12 Ni, bal Fe.
For chemical industry; stainless steel casting-quality.
W. Nr. 1.4404.

RAXIT 408 G.
M-1353; 0.06 C, 18 Cr, 2.3 Mo, 11 Ni, bal Fe.
For chemical and textile industries; stainless steel casting-quality.
W. Nr. 1.4408.

RAXIT 410 G.
M-1353; 0.1 C, 18 Cr, 2.3 Mo, 10 Ni, bal Fe.
For chemical and textile industries; stainless steel casting-quality.
W. Nr. 1.4410.

RAXIT 437 G.
M-1353; 0.06 C, 18 Cr, 2.7 Mo, 12 Ni, bal Fe.
For chemical industry; stainless steel casting-quality.
W. Nr. 1.4437.

RAXIT 448 G.
M-1353; 0.06 C, 17 Cr, 4.5 Mo, 13 Ni, bal Fe.
For chemical and textile industries; stainless steel casting-quality.
W. Nr. 1.4448.

RAXIT 460 G.
M-1353; 0.08 C, 27 Cr, 1.6 Mo, 5 Ni, bal Fe.
For cellulose industry; stainless steel casting-quality.

RAXIT 464 G.
M-1353; 0.4 C, 27 Cr, 2.3 Mo, 5 Ni, bal Fe.
For chemical industry; stainless steel casting quality; wear resistant.
W. Nr. 1.4464.

RAXIT 500 G.
M-1353; 0.07 C, 20 Cr, 2 Cu, 3 Mo, 25 Ni, Nb, bal Fe.
For chemical industry; stainless steel casting-quality.
W. Nr. 1.4500.

RAXIT 542 G.
M-1353; 0.05 C, 17 Cr, 4 Cu, 4 Ni, Nb, bal Fe.
For water power plants; stainless steel casting-quality; wear resistant.
W. Nr. 1.4542.

RAXIT 552 G.
M-1353; 0.07 C, 18 Cr, 10 Ni, Nb, bal Fe.
For chemical industry; stainless steel casting-quality.
W. Nr. 1.4552.

RAXIT 581 G.
M-1353; 0.07 C, 18 Cr, 2.3 Mo, 10 Ni, + Nb, bal Fe.
For textile, cellulose and dye industries; stainless steel casting-quality.
W. Nr. 1.4581.

RAXIT 585 G.
M-1353; 0.07 C, 18 Cr, 2.3 Mo, 20 Ni, Nb, bal Fe.
For chemical industry; stainless steel casting-quality.
W. Nr. 1.4585.

RAXIT 588 G.
M-1353; 0.05.C, 18 Cr, 2 Cu, 3 Mo, 22 Ni, Nb, bal Fe.
For chemical industry; stainless steel casting-quality.
W. Nr. 1.4588.

RAXIT 593 G.
M-1353; 0.07 C, 25 Cr, 1.5 Cu, 2.5 Mo, 8 Ni, bal Fe.
For chemical industry, pump manufacturers; stainless steel casting-quality.

RAXIT 602 G.
M-1353; 0.1 C, 16 Cr, 17 Mo, 52 Ni, 4 W, bal Fe.
For chemical industry; stainless steel casting-quality.

RAXIT 619 G.
M-1353; 0.25 C, bal Fe.
Special steel casting-quality with high strength at elevated temperatures.
W. Nr. 1.0619.

RAXIT 810 G.
M-1353; 0.05 C, 28 Mo, bal Fe.
For chemical industry; stainless steel casting-quality.

RAXIT 902 G.
M-1353; 0.06 C, 18 Cr, 10 Ni, bal Fe.
Special stainles steel casting-quality, tough at sub zero temperatures.
W. Nr. 1.6902.

RAXIT 905 G.
M-1353; 0.10 C, 18 Cr, 10 Ni, Nb, bal Fe.
Special stainless steel casting-quality, tough at sub zero temperatures.
W. Nr. 1.6905.

RAXIT 948 G.
M-1353; 0.06 C, 18 Cr, 11 Ni, bal Fe.
For pressure vessels in nuclear reactors; stainless steel casting-quality with high strength at extremely high temperatures.

RAXIT 1815 SI.
M-1353; 0.03 C, 18 Cr, 15 Ni, 4 Si, bal Fe.
For chemical industry; stainless steel casting-quality.

RAXIT 5419.
M-1353; 0.22 C, 0.4 Mo, 0.7 Mn, bal Fe.
For turbine industry; special steel casting-quality with high strength at elevated temperatures.

RAXIT 7225.
M-1353; 0.42 C, 1.0 Cr, 0.25 Mo, 0.7 Mn, bal Fe.
For dredging and excavating industry; special steel casting-quality; oil hardening.

RAXOTHERM 150 G.
M-1353; 0.5 C, 1.5 Si, 15 Cr, 0.5 Ni, bal Fe.
For rolling mill equipment, wear parts; heat resisting casting-quality.

RAXOTHERM 250 G.
M-1353; 0.5 C, 1.0 Si, 25 Cr, bal Fe.
For pressure casting crucibles; heat resisting casting-quality.

RAXOTHERM 710 G.
M-1353; 0.3 C, 1.5 Si, 7 Cr, bal Fe.
For furnace industry; heat resisting up to 1300°F (704°C); heat resisting casting-quality.
W. Nr. 1.4710.

RAXOTHERM 729 G.
M-1353; 0.4 C, 1.5 Si, 13 Cr, bal Fe.
For furnace industry; heat resisting up to 1550°F (843°C); heat resisting casting-quality.
W. Nr. 1.4729.

RAXOTHERM 740 G.
M-1353; 0.4 C, 1.5 Si, 17 Cr, bal Fe.
For furnace industry; heat resisting up to 1650°F (900°C); heat resisting casting-quality.
W. Nr. 1.4740.

RAXOTHERM 743 G.
M-1353; 1.6 C, 1.5 Si, 18 Cr, bal Fe.
For furnace industry; wear resistant, heat resisting up to 1650°F (900°C); heat resisting casting-quality.
W. Nr. 1.4743.

RAXOTHERM 745 G.
M-1353; 0.4 C, 1.5 Si, 23 Cr, bal Fe.
For furnace industry; heat resisting up to 1900°F (1038°C); heat resisting casting-quality.
W. Nr. 1.4745.

RAXOTHERM 776 G.
M-1353; 0.4 C, 1.5 Si, 29 Cr, bal Fe.
For furnace industry; heat resisting up to 2100°F (1150°C); heat resisting casting-quality.
W. Nr. 1.4776.

RAXOTHERM 777 G.
M-1353; 1.3 C, 1.5 Si, 29 Cr, bal Fe.
For furnace industry; heat resisting up to 2000°F (1093°C); heat resisting and wear resisting casting-quality.
W. Nr. 1.4777.

RAXOTHERM 823 G.
M-1353; 0.4 C, 1.5 Si, 27 Cr, 4 Ni, bal Fe.
For furnace industry; heat resisting up to 2000°F (1093°C); heat resisting casting-quality.
W. Nr. 1.4823.

RAXOTHERM 825 G.
M-1353; 0.25 C, 1.5 Si, 18 Cr, 9 Ni, bal Fe.
For furnace industry; heat resisting up to 1650°F (900°C); heat resisting casting-quality.
W. Nr. 1.4825.

RAXOTHERM 826 G.
M-1353; 0.4 C, 1.5 Si, 22 Cr, 9 Ni, bal Fe.
For furnace industry; heat resisting up to 1750°F (954°C); heat resisting casting-quality.
W. Nr. 1.4826.

RAXOTHERM 832 G.
M-1353; 0.25 C, 1.5 Si, 20 Cr, 14 Ni, bal Fe.
For furnace industry; heat resistant to 1750°F (954°C); heat resisting casting-quality.
W. Nr. 1.4832.

RAXOTHERM 837 G.
M-1353; 0.4 C, 1.5 Si, 25 Cr, 12 Ni, bal Fe.
For furnace industry; heat resisting up to 1920°F (1050°C); heat resisting casting-quality.
W. Nr. 1.4837.

Section I: Alloy Data / 1147

RAXOTHERM 848 G.
M-1353; 0.4 C, 2 Si, 25 Cr, 21 Ni, bal Fe.
For furnace industry; heat resisting up to 2000°F (1093°C); heat resisting casting-quality.
W. Nr. 1.4848.

RAXOTHERM 849 G.
M-1353; 0.4 C, 1.5 Si, 18 Cr, 38 Ni, 1.5 Nb, bal Fe.
For furnace industry; heat resisting up to 2000°F (1093°C); heat resisting casting-quality.
W. Nr. 1.4849.

RAXOTHERM 857 G.
M-1353; 0.4 C, 2 Si, 25 Cr, 35 Ni, bal Fe.
For oil and gas installation; heat resisting casting-quality.
W. Nr. 1.4857.

RAXOTHERM 865 G.
M-1353; 0.4 C, 2 Si, 18 Cr, 36 Ni, bal Fe.
For furnace industry; heat resistant up to 2100°F (1150°C); heat resisting casting-quality.
W. Nr. 1.4865.

RAXOTHERM 879 G.
M-1353; 0.5 C, 1.0 Si, 28 Cr, 50 Ni, 5 W, bal Fe.
For furnace industry, hearth rolls; heat resisting up to 2100°F (1150°C); heat resisting casting-quality.
W. Nr. 1.4879.

RAXOTHERM CO 5 G.
M-1353; 0.1 C, 50 Co, 28 Cr, bal Fe.
For furnace industry; heat resisting up to 2150°F (1177°C); heat resisting casting-quality.
W. Nr. 1.4778.

RAXOTHERM CO 6 G.
M-1353; 0.35 C, 50 Co, 28 Cr, 2 Nb, bal Fe.
For furnace industry; heat resisting up to 2150°F (1177°C); heat resisting casting-quality.

RAXOTHERM CR NI 50/50.
M-1353; 0.07 C, 50 Cr, 50 Ni, + Nb.
For oil refining plants, power plants, high heat resisting; stainless special steel casting-quality.

RAXOTHERM WC MO/G.
M-1353; 0.48 C, 1.5 Cr, 0.8 Mo, 0.3 V, bal Fe.
For pressure casting industry; high temperature casting quality.

RAYMUR Cu-Ni.
For ornamental and corrosion resistant parts; corrosion resistant.

RAYO.
85 Ni, 15 Cr.
For resistance alloy; heat resistant.

RB 6.
M-912; 0.70 C, 0.20 Si, 0.25 Mn, bal Fe.
Carbon tool steel; for stamping and forming dies, woodworking tools.

RB 8.
M-912; 0.85 C, 0.20 Si, 0.25 Mn, bal Fe.
Carbon tool steel; for cutting and stamping tools, burring and trimming tools.
W.-Nr. 1625.

RB 10.
M-912; 1.05 C, 0.20 Si, 0.25 Mn, bal Fe.
Carbon tool steel; for hand tools as shears, numbering stamps.
W.-Nr. 1654.

RB 11.
M-912; 1.15 C, 0.20 Si, 0.25 Mn, bal Fe.
Carbon tool steel; for milling cutters, taps, broaches, reamers, shears.
W.-Nr. 1654.

RBFF.
M-912; 0.52 C, 0.25 Si, 1.75 Mn, 0.10 V, bal Fe.
For leaf springs.
W.-Nr. 0915.

R BRAND COPPER.
M-1778; Fire refined copper ingot.
For remelting copper for copper alloys.

RC.
M-1653; 0.70-0.80 C, 3.5-4.0 Cr, 17.0-19.0 W, 0.80-1.20 V, bal Fe.
Tungsten high speed steel.
UX 75 W 18 KU; AISI T1.

RC 2.
M-1653; 0.75-0.90 C, 3.5-4.5 Cr, 17.0-19.0 W, 1.7-2.2 V, bal Fe.
Tungsten high speed steel.
U X 82 WV 18 KU; AISI T2.

RCC.
M-912; 2.05 C, 11.5 Cr, bal Fe.
Cold work tool steel; for heavy duty stamping and punching, shear blades.
W.-Nr. 2080; similar to AISI D3.

R.C.C.B.
M-1118; 2.25 C, 13 Cr, bal Fe.
For tools, dies, non-deforming.

RCC EXTRA.
M-912; 2.05 C, 11.5 Cr, 0.70 W, bal Fe.
Cold work tool steel; for heavy duty stamping and cutting tools, punches, reamers, broaches.
W.-Nr. 2436; AFNOR Z 200 CW 12.

RCCM.
M-1488; 0.90 C, 1.90 Mn, 0.50 Cr, 0.10 V, bal Fe.
Cold work steel; oil hardening.
For punching and forming dies; reamers.
AFNOR 90 MCV 8.

RCC SPEZIAL.
M-912; 1.65 C, 11.50 Cr, 0.60 Mo, 0.10 V, 0.50 W, bal Fe.
Cold work tool steel; for blanking and trimming dies, thread rolling dies, reamers, plastic moulds.
W.-Nr. 2601; AFNOR Z 165 CDV 12.

RCC SUPRA.
M-912; 1.50 C, 12.0 Cr, 0.80 Mo, 0.85 V, bal Fe.
Cold work tool steel; for punching and stamping dies, thread rolling dies, reamers.
W.-Nr. 2379; Similar to AISI D2.

RCCV.
M-912; 2.10 C, 12.0 Cr, 1.15 Mo, 2.10 V, bal Fe.
Cold work tool steel; stamping dies for dynamo and transformer sheets, plastic moulds.
W.-Nr. 2378; similar to AISI D7.

RCK 5.
M-1653; 0.70-0.85 C, 3.5-4.5 Cr, 17.0-19.0 W, 1.0-1.5 V, 4.5-5.5 Co, bal Fe.
High speed tool steel.
W.-Nr. 1.3255; AISI T4.

RCK 10.
M-1653; 0.75-0.85 C, 3.5-4.5 Cr, 17.5-19.5 W, 1.5-2.0 V, 0.50-1.0 Mo, 9.0-10.0 Co, bal Fe.
High speed tool steel.
W.-Nr. 1.3265; similar to AISI T5.

RCW 1.
M-912; 0.58 C, 3.75 Cr, 0.70 Mo, 0.70 V, 9.0 W, bal Fe.
Hot work tool steel; for tube pressing mandrels up to 30 mm diameter.
W.-Nr. 2622; AFNOR Z 60 WC 09 04.

R.C.W. 2.
M-1118; 0.28 C, 9.5 W, 2.5 Cr, 0.1 V, bal Fe.
For hot work dies and tools; hot work steels.

RCW 2 H.
M-912; 0.32 C, 3.55 Cr, 0.35 V, 8.50 W, bal Fe.
Hot work steel; for hot extrusion dies, pressure casting moulds, forging dies.
W.-Nr. 2581; AISI H21.

R.C.W. 2VA.
M-1118; 0.6 C, 9.5 W, 2.5 Cr, 0.1 V, bal Fe.
For hot work dies and tools; hot work steels.

R.C.W.M.
M-1118; 0.5 C, 9.5 W, 2.5 Cr, 0.1 V, bal Fe.
For hot work dies and tools; hot work steels.

RD ALLOY.
M-US; Cu alloy.
For brazing of tools; age-hardenable.

RDC 1.
M-912; 0.36 C, 1.0 Si, 5.0 Cr, 1.40 Mo, 0.25 V, 1.35 W, bal Fe.
Hot work tool steel; pressing dies, forging dies.
W.-Nr. 2606; similar to AISI H 12.

RDC 2.
M-912; 0.38 C, 1.0 Si, 5.0 Cr, 1.25 Mo, 0.45 V, bal Fe.
Hot work tool steel; for pressure casting moulds for light metal; extrusion press tools.
W.-Nr. 2343; AISI H 11.

RDC 2V.
M-912; 0.40 C, 1.0 Si, 5.0 Cr, 1.35 Mo, 1.10 V, bal Fe.
Hot work tool steel; extruding and piercing mandrels, forging dies.
W.-Nr. 2344; similar to AISI H 13.

RDC 6.
M-912; 0.63 C, 1.10 Si, 5.25 Cr, 1.15 Mo, 0.30 V, bal Fe.
Hot or cold work tool steel; for trimming dies, punching dies, shears, plastic moulds.
W.-Nr. 2362; AFNOR Z60 CDV5.

RDC 8.
M-912; 0.81 C, 4.0 Cr, 4.25 Mo, 1.0 V, bal Fe.
Cold work tool steel; for coining and stamping dies, reamers, broaches.
W.-Nr. 2369; AFNOR 80 DCV 42.

R.D.S see **CARPENTER R.D.S.**

REACTAL.
M-419; 3.2 C, 2.2 Si, 1.5 Ni, 0.8 Cr, bal Fe.
Cast: 42,000-55,000 TS; 187 Brin.
For corrosion and heat resisting parts; corrosion and heat resistant castings.

READILY FUSIBLE ALLOY.
67 Zn, 33 Cu.
For brazing; high strength.

READY-MARK.
M-1172; 0.90 C, 1.2 Mn, 0.5 Cr, 0.2 V, 0.5 W, bal Fe.
AISI type 01 oil hardening tool steel.

RECIDAL.
M-Italy; Al alloy.
For screw machine products, fasteners; free-cutting.

RECIDAL 55.
M-Italy; 5.3-5.7 Cu, 0.2-0.3 Pb, 0.2-0.3 Cd, 0.4-0.6 Zn, 0.30 Fe, 0.25 Si, bal Al.
For screw machine products, fasteners. Free-cutting, corrosion resistant.

RECKHAMMER SSW.
M-Ger.; 0.90 C, 0.80 Cr, 0.30 Mn, bal Fe.
For bearings, liners, cutters, sleeves; water hardened, wear resistant.

RECN.
M-912; 0.18 C, 0.25 Si, 0.50 Mn, 1.95 Cr, 1.95 Ni, bal Fe.
Alloy carburizing steel; for highly stressed automotive and tractor gears and spline couplings.
W.-Nr. 5920.

RECO 1.
M-1188; 24 Ni, 13 Al, 4 Cu, bal Fe.
For permanent magnets, electrical and magnetic equipment.
High permeability.

RECO 2A.
M-Holland; 7 Al, 20 Ni, 20 Co, 7 Cu, 6.5 Ti, bal Fe.
For permanent magnets.

RECO 3A.
M-1188; 19 Ni, 12 Co, 10 Al, 6 Cu.
For permanent magnets, electrical and magnetic equipment.
High permeability.

RECO II A.
M-1188; C, alloy, bal Fe.
For permanent magnets; high permeability.

RECORD.
M-815; Sb, Sn, bal Pb.
For bearings; Babbitt.

RECORD 66.
M-336; 0.7 C, 5 W, 6.75 Mo, 4.25 Cr, 1.75 V, bal Fe.
For gear cutters, form cutters, cold or hot work dies; high speed steel.

RECORD-EMINENT.
M-336; 0.85 C, 18.5 W, 4.25 Cr, 2.5 V, 10.5 Co, bal Fe.
For tools, dies, cutters; high speed steel.

RECORD EMINENT-28.
M-387; 0.8 C, 4.5 Cr, 1 Mo, 10 Co, 18.5 W, 1.6 V, bal Fe.
For milling cutters, lathe and planer tools, reamers; high speed steel.

RECORD EXTRA.
M-336; 0.8 C, 18 W, 3.75 Cr, 1.25 V, 2.25 Co, bal Fe.
260 Brin.
For dies, tools for high speed production for turning and heavy cutting, sealing material; high speed steel.

RECORD MO 50.
M-1346; 0.85 C, Cr, V, Mo, W, bal Fe.
For lathe and planer tools, reamers, broaches, hobs; high speed steel.

RECORD STAR.
M-336; 0.7 C, 4 Cr, 14 W, 2 V, bal Fe.
For tools, cutters; high speed steel.

RECORD SUPERIOR.
M-336; 0.8 C, 18 W, 4 Cr, 1 V, bal Fe.
For tools, dies for boring, planing and slotting, taps, reamers, punches; high speed steel.

RECOVAC 100.
M-1631.
Soft magnetic high nickel content alloy with improved wear resistance for wear-resistant magnet heads and relay components.

REDALOY.
M-216; 55 Cu, 1 max Pb, 2.25 Al, bal Zn.
Extruded: 85,000 TS; 25,000 YS; 20 El; B 80 Rock.
Drawn: 90,000 TS; 35,000 YS; 15 El; B 85 Rock.
For hardware, plumbing, fixtures.
Corrosion resistance.

REDALOY NO. 1.
M-122; 53.75-54.25 Cu, 3.3-3.5 Pb bal Zn.
77,000-88,000 TS; 24,000-64,000 YS; 7-17 El; 11-22 RA; 70-100 Brin.
For screw machine parts; free cutting.

REDALOY NO. 2.
M-122; 57.5-58 Cu, 3.3-3.5 Pb, bal Zn.
63,000-70,000 TS; 23,000-56,000 YS; 13-32 El; 16-29 RA; 65-100 Brin.
For screw machine parts; free cutting.

REDALOY NO. 3.
M-122; 52-53.7 Cu, 0.4-0.6 Pb, bal Zn.
For pressure die castings; high strength.

REDALOY NO. 8.
M-122; 54 Cu, 3.4 Pb, bal Zn.
For forgings, hardware; free-cutting.

RED ANCHOR.
M-342; 0.95-1.1 C, bal Fe.
For general tools, chasers; water hardened.

RED ANCHOR DRILL ROD.
M-342, M-115; 0.95-1.10 C, bal Fe.
80,000 TS; 60,000 YS; 25 El; 50 RA.
For anvils, dental tools, spindles, precision shafts for motors; water hardened.

RED BRASS.
M-8; 85 Cu, 15 Zn.
Tubes and sheets; 42,000-50,000 TS; 20,000-40,000 YS; 43-4 El; 68 RA.
For salt water pipes, plumbing, hardware; corrosion resistant.

RED BRASS 24.
M-8; 85 Cu, 15 Zn.
Soft: 40,000 TS; 15,000 YS; 50 El.
Hard: 69,000 TS; 55,000 YS; 10 El.
For plumbing pipe, condenser tubes, radiator cores; red brass.

RED BRASS, 85%.
M-33, M-1789; 85 Cu, 15 Zn.
Annealed: 42,000 TS; 14,000 YS; 45 El.
Rolled: 70,000 TS; 58,000 YS; 5 El.
For plumbing, pipe, tubes, hardware; corrosion resisting.

RED BRASS 230.
M-8, M-279; 85 Cu, 15 Zn.
Ann: 39,000-47,000 TS; 8000-19,000 YS; 43-48 El.
Cold rolled: 44,000-90,000 TS; 23,000-80,000 YS; 2-39 El.
Golden color, stress-corrosion resistant brass.
For jewelry, pen caps, pencil ferrules, steam iron tubes.

RED CASTING BRASS.
M-165; 85 Cu, 5 Sn, 5 Pb, 5 Zn.
Cast: 27,000 TS; 18 El.
For castings, hardware; GM No. 4041 M.

RED CIRCLE.
M-905; 3.2 C, 1.5 Si, bal Fe.
For cast iron rolls; chilled.

RED CIRCLE.
M-750; C, alloy, bal Fe.
For maintenance and repair operations; oil hardening.

RED CUT COBALT.
M-115, M-34; 0.72 C, 17.25 W, 4 Cr, 4.5 Co, 1.0 V, 0.5 Mo, bal Fe.
For cutting and forming tools, lathe tools, planer tools, reamers, taps; applicable for deep cutting, scaly material.

RED CUT COBALT B.
M-34; 0.78 C, 18 W, 4 Cr, 8.75 Co, 1.85 V, 0.75 Mo, bal Fe.
For cutting and forming tools, reamers, taps; high speed steel.

RED CUT SUPERIOR.
M-115, M-34; 0.5-0.8 C, 0.25-0.40 Si, 17.5-18.5 W, 3.75-4.25 Cr, 0.95-1.15 V, bal Fe.
Annealed: 105,000 TS; 75,000 YS; 14 El; 17 RA; 255 Brin.
For tools, blanking dies, chasers, forming tools and dies, lathe and planer tools, taps; high speed steel; water and abrasion resistant.

RED CUT SUPERIOR TEMPER "J".
M-115; 0.50-0.55 C, 17.5-18.5 W, 3.75-4.25 Cr, 0.95-1.15 V, bal Fe.
For high speed cutting tools, blanking dies, forming tools, taps; wear and abrasion resistant.

RED CUT SUPERIOR TEMPER "P".
M-115; 0.71-0.75 C, 17.5-18.5 W, 3.75-4.25 Cr, 0.95-1.15 V, bal Fe.
For high speed cutting tools, blanking dies, forming tools, taps; water and abrasion resistant.

RED DEVIL.
M-712; 0.07 C, bal Fe.
Welded: 68,000 TS; 56,000 YS; 30 El; 50 RA.
For welding rods; flux coated.

RED DEVIL 75.
M-712; 0.1 C, 0.6 Mo, bal Fe.
Welded: 74,000 TS; 62,000 YS; 32 El; 65 RA.
For welding rods; flux coated.

RED DEVIL 85.
M-712; C, 0.9 Mo, bal Fe.
Welded: 80,000 TS; 72,000 YS; 24 El; 47 RA.
For welding rods; flux coated.

RED (EINHEITS) BRASS.
M-Germ.; 89-83 Cu, 5-12 Zn, 3-10 Pb, 2-5 Sn.
Hard: 80,000 TS; 71,000 YS; 135 El; 4 Brin.
Soft: 41,000 TS; 18,000 YS; 47 El; 52 Brin.
For tubing, hardware; Tombac, free-cutting.

RED FOX 34.
M-1724; 0.15 max C, 19.0 Cr, 11.0 Ni, 2.0 Si, bal Fe.
Good scaling resistance in air at temperatures up to 1000°C.
AISI 302 B.

RED FOX-309.
M-1724; 0.2 C, 23 Cr, 14 Ni, bal Fe.
Annealed: 75,000 TS; 30,000 YS; 40 El; 220 Brin.
For high temperature service, kilns, combustion chambers, salt pots, brazing fixtures.
AISI Type 309, austenitic. Heat and corrosion resistant.

RED FOX-310.
M-430, M-1724; 0.2 C, 25 Cr, 21 Ni, bal Fe.
Annealed: 87,000 TS; 36,000 YS; 54 El; B 85 Rock.
For high temperature service, furnace equipment, carburizing boxes, valves pumps.
Heat and corrosion resistant.
AISI Type 310, austenitic.

RED GOLD.
75 Au, 15 Cu.
For jewelry, ornaments; corrosion resistant.

REDHARD HS.
M-1082; 0.7 C, 18 W, 4 Cr, 2 V, bal Fe.
For cutting tools; high speed steel.

REDHARD NF.
M-1082; 45 Co, 32 Cr, 18 W, 2.5 C, 2.1 other elements.
Cast: 73,000 TS; 650 Brin.
For cutting tools, form tools; nonferrous, heat and abrasion resistant.

REDI H.
M-1322; 0.45 C, Si, Cr, V, bal Fe.
For gears, bolts, springs, crankshafts; oil hardened, shock resistant.

REDI-KOTE.
M-604; low C, Mn, bal Fe.
Steel sheet with remelted zinc coating.

RED INDIAN.
M-435; 0.35 C, 5 Cr, 4.5 W, 0.3 Mo, 0.3 V, 0.5 Co, 1 Si, bal Fe.
For die casting dies, extrusion dies and mandrels; hot work steel.

REDI SPEZIAL.
M-1322; 0.61 C, 1.18 Cr, 0.1 V, 0.75 Mn, bal Fe.
For springs, dies, tools; oil hardened, tough.

RED METAL.
70 Cu, 20 Zn, 6 Pb, 4 Sn.
For hardware; free-cutting.

RED RAY.
85 Ni, 15 Cr.
For heating elements; heat resistant.

RED SHADOW.
M-783; 0.40 C, 3.1 Cr, 0.25 Mn, 0.25 V, 10 W, bal Fe.
For tools, dies; hot work steel.

RED SHADOW.
M-289; 0.8 C, 4 Cr, 5 Mo, 6.5 W, 2 V, bal Fe.
For milling cutters, reamers, broaches, hobs, drills; high speed steel, oil hardened.

RED STAR TAP.
M-34; 1.1 C, bal Fe.
For taps, drills, cutters; highest quality carbon tool steel.

RED STAR TOOL.
M-34, M-115; 0.6-1.2 C, bal Fe.
For tools, drills, punches; standard carbon tool steel.

RED STAR TUNGSTEN.
M-115; 1.20 C, 0.30 Si, 0.25 Mn, 1.60 W, 0.70 Cr, 0.20 V, 0.25 Mo, bal Fe.
Oil or water hardening tool steel.
For blanking dies, taps, reamers, thread rolling dies, punches, finishing tools, gauges.
Modified AISI 07.

RED STAR TUNGSTEN 10 TEMPER.
M-115, M-34; 0.90-1.0 C, 0.35-0.50 Cr, 1.5-1.7 W, 0.06-0.10 V, bal Fe.
Annealed: 90,000 TS; 65,000 YS; 22 El; 42 RA; 190 Brin.
For taps, reamers, punches, dies, finishing tools, gages; water hardening.

RED STAR TUNGSTEN 12 TEMPER.
M-115, M-34; 1.25-1.15 C, 0.60-0.80 Cr, 1.5-1.7 W, 0.10-0.25 V, 0.25 Mo, bal Fe.
Heat treated: 180,000-145,000 TS; 154,000-210,000 YS; 1-13 El; 5-36 420-302 Brin.
For taps, reamers, punches, dies, finishing tools, gages; water or oil hardening.

RED STAR TUNGSTEN DRILL ROD.
M-342, M-115; 1.2 C, 1.6 W, 0.7 Cr, 0.25 Mo, 0.2 V, bal Fe.
For drill rod; water hardening.

RED STREAK see SIMONDS RED STREAK.

RED STREAK ALNICO.
M-373; 0.10 C, 17.5 Ni, 11 Al, 13 Co, bal Fe.
For permanent magnets; Alnico No.2.

RED TIP.
M-266; 61.5 Cu, 3 Pb, bal Zn.
Half hard: 62,000 TS; 53,000 YS; 25 El.
Annealed: 49,000 TS; 18,000 YS; 53 El.
For screw machine products; free-cutting.

RED TIP 201.
M-213; Zn, Pb, bal Cu.
For screw machine products; free-cutting.

RED, WHITE AND BLUE.
M-1112; 0.9-1.0 C, bal Fe.
For tools, taps, drills; water hardened.

RED-X-5.
M-130; 1-2 Cu, 0.2-0.5 Mg, 0.2-0.6 Mn, 4.5-6.5 Si, bal Al.
Permanent mold: 32,000 TS; 20,000 YS; 3.5 El; 65 Brin.
Sand cast: 26,000 TS; 16,000 YS; 2.5 El; 60 Brin.
For forgings and castings; age-hardenable.

RED-X-6.
M-906; 1-2 Cu, 0.2-0.5 Mg, 0.4 Mn, 6 Si, bal Fe.
For light alloy castings; age-hardenable.

RED-X-8.
M-130; 2 Cu, 0.5 Mg, 0.4 Mn, 8 Si, bal Al.
Cast: 33,000 TS; 20,000 YS; 3 El; 65 Brin.
Heat treated: 42,000 TS; 26,000 YS; 1 El; 95 Brin.
For light alloy parts, permanent mold castings; heat treatable.

RED-X-8-5.
M-130; 5 Cu, 0.25 Mg, 8 Si, 0.3 Mn, bal Al.
For light alloy parts, pistons; age-hardenable.

RED-X-10.
M-130; 9.5-10.5 Si, 2 Cu, 0.6 Mg, 0.6 Mn, bal Al.
Permanent mold: 40,000 TS; 30,000 YS; 0.5 El; 105 Brin.
For light alloy parts, pistons; high fluidity.

RED X-11.
M-130; 1-2 Cu, 10.5-11.8 Si, 0.4-1.0 Mg, 0.6 max Zn, 0.8 Mn, bal Al.
Cast-F: 34,000 TS; 23,000 YS; 2 El; 80 Brin.
T6-Temper: 45,000 TS; 35,000 YS; 2 El; 100 Brin.
T51-Temper: 39,000 TS; 31,000 YS; 1 El; 90 Brin.
For light alloy castings; high fluidity.

RED-X-13.
M-130; 2 Cu, 0.7 Mg, 0.75 Mn, 12 Si, bal Al.
Heat treated: 34,000-50,000 TS; 23,000-41,000 YS; 1.5-1.0 El; 85-110 Brin.
For permanent mold castings, crank cases, pistons; heat treatable.

RED-X-20.
M-130; 1.5 Cu, 20.5 Si, 0.6 Mg, 0.4 Mn, 0.7 Ni, bal Fe.
F temper: 25,000 TS; 1.0 El; 90 Brin.
T7 temper: 38,000 TS; 1.0 El; 110 Brin.
For permanent mold pistons; heat treatable.

REED BRASS.
M-U.S.; 69 Cu, 30 Zn, 1.0 Sn.
For pipes, tubes, fittings, condenser tubes; corrosion resistant.

REED BRASS.
M-1426; 67 Cu, 1 Sn, 32 Zn.

REEX.
M-Italy; 0.45-0.90 C, 11 Cr, 14 Ni, 1 Si, 2 W, bal Fe.
For valves; heat and corrosion resistant.

REFAX METAL.
M-1173; 3.2 C, 1 Ni, 0.8 Cr, 2 Si, bal Fe.
For valve bodies; cast iron.

REFLECTAL.
M-116; 1.2 Mg, bal Al.
Annealed: 21,000 TS; 8000 YS; 24 El.
For boats, fan blades, window frames; corrosion resistant.

REFLECTAL-05.
M-541; 0.2-1.0 Mg, bal Al.
R-Temper: 10,000-15,000 TS; 5000 YS; 20-40 El; 25 Brin.
1/2 Hard: 16,000-20,000 TS; 15,000 YS; 12-15 El; 30 Brin.
For costume jewelry, mirrors, reflectors, watch cases, window frames. Corrosion resistant.

REFLECTAL-050.
M-493; 0.5 Mg, bal Al.
Ann: 80 MPa TS; 25 El.
Hard: 180 MPa TS; 2 El.
Plate for decorative purposes.

REFLECTAL 5.
M-249, M-493; 0.3-1.0 Mg, bal Al.
Soft: 11,500 TS; 4500 YS; 35 El; 22 Brin.
1/4 hard temper: 13,500 TS; 13,000 YS; 20 El; 32 Brin..
For mirrors, jewelry, reflectors, watch cases; corrosion resistant.

REFLECTAL 20.
M-249, M-541, M-493; 1.5-2.5 Mg, bal Al.
1/2 H-temper: 31,500 TS; 27,300 YS; 5 El; 65 Brin.
H-temper: 35,500 TS; 31,500 YS; 3 El; 70 Brin.
For window frames, door handles, costume jewelry, high corrosion resistance.

REFLECTAL 74.
M-624, M-541; 0.5-1.0 Mg, 0.2-1.0 Si, bal Al.
TA-14 Temper: 20,000-28,400 TS; 11,400-21,500 YS; 20-35 El; 55-75 Brin.
TA-16 Temper: 43,000 TS; 36,000 YS; 5 El; 100 Brin.
For marine structures and hardware, architectural applications.
Heat treatable, corrosion resistant.

REFLECTAL-107.
M-249, M-493; 0.2-1.0 Si, 0.2-1.0 Mg, bal Al.
Heat treated: 28,500-42,500 TS; 21,500-35,500 YS; 15-5 El; 80-100 Brin.
For ornamental trim, window frames, door handles; age-hardenable, corrosion resistant.

REFOR-9-2.
M-1766; 0.44 C, 0.45 Mn, 3.10 Si, 9.0 Cr, bal Fe.
Martensitic steel for valves.
IHA F-322; DIN X 45 CrSi9; BS EN 52.

REFOR-14-14.
M-1766; 0.44 C, 1.0 max Mn, 2.0 max Si, 14.0 Cr, 14.0 Ni, 2.5 W, bal Fe.
Austenitic steel for valves.
IHA F-321; BS EN 54; AFNOR Z 45 CNWS 15-14.

REFOR-25-20.
M-1766; 0.20 max C, 2.0 max mn, 2.5 max Si, 25.0 Cr, 20.0 Ni, bal Fe.
Austenitic stainless steel for parts operating at high temperature.
IHA F-331; AISI 310; BS A-11.

REFOR-30.
M-1766; 0.20 max C, 1.0 max Mn, 2.0 max Si, 28.0 Cr, bal Fe.
Stainless steel for furnace equipment that operates at high temperatures.
AFNOR Z 15C27; similar to AISI 446.

REFORMEND.
M-677; 0.2 C, bal Fe.
For welding electrodes; for cast iron.

REFRACTALOY 26.
M-118, M-1491; 0.08 max C, 16-20 Cr, 35-39 Ni, 16-22 Co, 3 Mo, 3 Ti 0.3 max Al, 0.5-1.5 Si, 0.4-1.0 Mn, bal Fe.
Hardened: 170,000 TS; 100,000 YS; 18 El; 20 RA; 241-311 Brin.
For turbine blades and parts, bolts, springs; heat resistant to 1500°F; high creep strength.

REFRACTALOY 70.
M-118; 19-21 Ni, 28.5-31.5 Co, 19-21 Cr, 8 Mo, 4 W, 2 Mn, 0.8 Si, 0.08 max C, bal Fe.
Hardened: 132,000 TS; 87,000 YS; 3 El; 3 RA; 315 Brin.
For high temperature applications; resists heat over 1200°F.

REFRACTALOY 80.
M-118; 0.2 max C, 28.5-31.5 Co, 19.5-21.5 Cr, 19-21 Ni, 16 max Fe, 9-11 Mo, 4.5-5.5 W, 0.8 Mn, 0.8 Si.
Cast: 83,000 TS; 57,000 YS; 8 El; 11 Ra; 180 Brin.
Hardened: 100,000 TS; 84,000 YS; 3 El; 4 RA; 220 Brin.
For nozzle vanes, gas turbine parts; high ductility, creep strength and oxidation resistance.

REFRACTALOY B.
M-118; 25 Cr, 30 Ni, 8 Mo, bal Fe.
For combustion chambers, exhaust cones; high heat and corrosion resistant.

REFRACTORY.
M-Eng.; 0.05 C, 30 Co, 21 Ni, 14 Fe, 20 Cr, 8 Mo, 4 W.
For jet engine components; high strength, heat resistant.

REFRACTORY SOLDER.
M-Eng.; 50 Cu, 50 Zn.
For solder; brazing.

REGAL.
M-275; 1.35 C, 3.5 W, bal Fe.
For cutters, lathe tools, broaches; fast finishing steel.

REGAL.
M-1705; 1.55 C, 11.5 Cr, 1.0 V, 0.8 Mo, bal Fe.
High carbon, high chromium tool and die steel; air or oil hardening. AISI D2.

REGAL.
M-1339; 1.42 C, W, V, bal Fe.
For engravers' tools, forming and blanking dies; water or oil hardened.

REGAL SPEZIAL.
M-1339; 1.3 C, 5.0 W, bal Fe.
Cold work tool steel, for drawing dies.

REGEL-METALL.
M-Eng.; 83.3 Sn, 11.1 Sb, 5.6 Cu.
For bearings, bushings; anti-friction.

REGELMETALL WM 80.
80 Sn, Pb, Sb.
For bearings; antifriction metal.

REGENT.
M-1339; 0.90 C, 1.9 Mn, 0.1 V, bal Fe.
For blanking and forming dies; oil hardened, non-deforming.

REGENT.
M-1348; 2.1 C, 11.5 Cr, 0.30 Mn, bal Fe.
For forming and blanking dies, punches; oil hardened, non-deforming.

REGENT C.3.H.
M-1415; 0.15 C, bal Fe.
For plastic dies, hobbed dies; case hardened.

REGENT G.14.
M-1415; 1.5 C, 12 Cr, 0.25 V, 0.75 Mo, bal Fe.
For thread rolling dies, gages, broaches; non-deforming, air hardening.

REGENT G.22.
M-1415; 1.5 C, 1 Cr, 0.5 V, 0.2 Mo, 3 Ni, 0.5 Mn, bal Fe.
For shear blades, punches, dies; oil hardened, tough.

REGENT I.18.
M-1415; 2 C, 12 Cr, bal Fe.
For punches, blanking and forming dies; oil hardened, non-deforming.

REGENT J.7.
M-1415; 0.5 C, 1 Mn, 1 Cr, 0.35 Mo, bal Fe.
For hand chisels, sates, cutters; tough, air hardened, non-tempering.

REGENT J.15.
M-1415; 1.15 C, 1.5 W, 0.15 V, bal Fe.
For taps, drills, reamers, cutters, threading dies; water hardened, wear resistant.

REGENT L.5.
M-1415; 0.4 C, 0.5 Mn, 3.25 Ni, bal Fe.
For chisels, punches; tough, oil hardened.

REGENT N.1.
M-1415; 0.6 C, 2 Si, 0.25 V, 0.25 Mo, 0.74 Mn, bal Fe.
For punches, concrete breaker picks, chisels; oil hardened, tough.

REGENT N.1.H.
M-1415; 0.35 C, 1.25 Cr, 4.25 Ni, bal Fe.
For plastic mold dies; air or oil hardened.

REGENT N.3.C.
M-1415; 0.12 C, 1.25 Cr, 4 Ni, bal Fe.
For plastic mold dies; case hardened, tough.

REGENT N.13.
M-1415; 0.40 C, 1.25 Cr, 1.75 Ni, 0.20 Mo, bal Fe.
For plastic mold and die casting dies; oil hardened, tough.

REGENT R.16.
M-1415; 0.10 C, 13 Cr, bal Fe.
Annealed: 75,000 TS; 40,000 YS; 35 El; 70 RA; 155 Brin.
For multiple dies, valves, cutlery; corrosion resistant; Type 410.

REGENT Z.
M-1348; 2.1 C, 11.5 Cr, 0.70 W, bal Fe.
For forming and blanking dies, punches: oil hardened, non-deforming.

REGIN 3.
M-111; 0.48 C, 0.9 Si, 1.15 Cr, 0.25 Mo, 2.25 W, 0.15 V, bal Fe.
For chisels, dies; hot work steel AISI S1.

REGIN 711.
0.5 C, 1.15 Cr, 2.5 W, 0.2 V, bal Fe.
For shear blades, hot heading dies, chisels; hot work steel.

REGULAR STRAIGHT CARBON.
M-261; 0.7-1.3 C, bal Fe.
For tools, drills, taps; water hardening.

REGULIT K now **VEW K100.**

REGULIT KNL now **VEW K105.**

REGULIT KR now **VEW K107.**

REGULUS.
M-U.S.; 75 Pb, 25 Sb.
For acid valves, cocks, flanges, chemical apparatus; hard lead.

REGULUS METAL "A".
6-8 Sb, bal Pb.
For chemical plant equipment; hard lead.

REGULUS METAL "B".
8-10 Sb, bal Pb.
For chemical equipment; hard lead.

REGULUS METAL "C".
10-12 Sb, bal Pb.
For bearings; hard lead.

REGULUS METAL "D".
12 Sb, bal Pb.
For bearings; hard lead.

REGULUS VENUS.
M-Eng; 50 Cu, 50 Sb.

REICHSBAHN "B".
M-Ger.; 85 Cu, 11 Sn, 4 Zn.
For railroad bearings; heavy duty.

REICHSBAHN "C".
M-Ger.; 85 Cu, 9 Sn, 6 Zn.
For railroad valves, slides; high strength.

REICHSBAHN "D".
M-Ger.; 91 Cu, 5 Sn, 4 Zn.
For railroad pipe flanges; tough.

REICHSBAHN "E".
M-Ger.; 82 Cu, 10 Sn, 7 Zn, 1 Pb.
For machinery castings; free-cutting.

REICHSBAHN, EINHEITS RED BRASS "A".
M-Ger.; 85 Cu, 9 Sn, 6 Zn.
For railroad hardware; high strength.

REICHSBAHN "F".
M-Ger.; 85 Cu, 5 Sn, 8 Zn, 2 Pb.
For machinery castings; free-cutting.

REICHSBAHN "G".
M-Ger.; 86 Cu, 10 Sn, 4 Pb.
For railroad dynamo and cold rolled bearings; heavy duty.

REICHSBAHN "H".
M-Ger; 77 Cu, 8 Sn, 15 Pb.
For railroad heavy duty bearings; heavy duty.

REICHSBAHN "I".
M-Ger.; 79 Cu, 20 Sn, 1 Pb.
For car journal bearings; heavy duty.

REICHSBAHN "J".
M-Ger.; 85 Cu, 14 Sn, 1 Pb.
For railroad heavy duty bearing; heavy duty.

REICHSBAHN "K".
M-Ger.; 89 Cu, 10 Sn.
For worm wheels; tough.

REICH'S BRONZE.
M-Ger.; 85 Cu, 7.5 Fe, 0.6 Al, 0.5 Mn.
For strong corrosion resistant parts; corrosion resistant.

REIN-ALU 99.5.
M-324; 99.5 Al. (pure aluminium).

REINALUMINIUM.
M-116; 99 Al or 99.5 Al.
Annealed: 9000-16,000 TS; 40-30 El.
Hard: 29,000 TS; 25,000 YS; 4 El.
For chemical apparatus, tanks, dairy industry, electrical parts; commercially pure Al.

REINALUMINIUM 2S.
M-1140; 99.0-99.4 Al.
S-O Temper: 14,000 TS; 7100 YS; 35 El.
S-H Temper: 25,600 TS; 24,200 YS; 6 El.
For cooking utensils, food and chemical processing equipment: wrought, corrosion resistant.

REINING BKV see **RAXA BKV.**

REINNICKEL.
M-297; 99.0-99.7 Ni.
Annealed: 65,000 TS; 18,000 YS; 50 El; 90 Brin.
Hard rolled: 113,000 TS; 108,000 YS; 2 El; 200 Brin.
For chemical and food-handling equipment; pure Ni corrosion resistant.

REISHOLZ A22.
M-1335; 0.22 C, 0.25 Si, 0.45 Mn, bal Fe.
Annealed: 73,000 TS; 41,000 YS; 22 El; 58 RA; 140 Brin.
For bolts, fasteners, gears, cams, camshafts; case hardening steel.

REISHOLZ A22E.
M-1335; 0.28 C, Mn, bal Fe.
Hot rolled: 80,000 TS; 50,000 YS; 30 El; 56 RA; 165 Brin.
For bolts, fasteners, gears, machine tool parts; water hardened.

REISHOLZ A32E.
M-1335; 0.36 C, Mn, bal Fe.
Hot rolled: 85,000 TS; 54,000 YS; 30 El; 54 RA; 185 Brin.
For bolts, fasteners, machine tool parts, gears; water hardened.

REISHOLZ A35.
M-1335; 0.35 C, 0.25 Si, 0.45 Mn, bal Fe.
Hot rolled: 85,000 TS; 54,000 YS; 30 El; 54 RA; 185 Brin.
For gears, bolts, machine tool parts; water hardened.

REISHOLZ B44.
M-1335; 0.24 C, 1.15 Ni, 0.15 max Cr, bal Fe.
For gears, pinions, fasteners, shafts; oil hardened, shock resistant.

REISHOLZ B76.
M-1335; 0.35 C, 1.35 Ni, 0.6 max Cr, bal Fe.
For gears, pinions, fasteners, shafts, axles; oil hardened, shock resistant.

REISHOLZ B412.
M-1335; 0.24 C, Si, Mn, Ni, bal Fe.
For gears, pinions, shafts, bolts; oil hardened, shock resistant.

REISHOLZ B412 MO.
M-1335; 0.24 C, Si, Mn, Ni, Mo, bal Fe.
For gears, shafts, bolts, studs, crankshafts; oil hardened, shock resistant.

REISHOLZ BCD53T.
M-1335; 0.28 C, 1.25 Cr, 0.35 Mo, 1.95 Ni, bal Fe.
For gears, shafts, crankshafts, machine tool parts; oil hardened, shock resistant.

REISHOLZ BCD53W.
M-1335; 0.34 C, 1.45 Cr, 0.25 Mo, 1.55 Ni, bal Fe.
For gears, shafts, machine tool parts; oil hardened, shock resistant.

REISHOLZ BCD54.
M-1335; 0.25 C, 1 Cr, 0.18 Mo, 1.5 Ni, bal Fe.
For gears, shafts, crankshafts, machine tool parts; oil hardened, shock resistant.

REISHOLZ BCD66.
M-1335; 0.28 C, 1.15 Cr, 0.25 Mo, 1.15 Ni, bal Fe.
For gears, pinions, shafts, machine tool parts; oil hardened, shock resistant.

REISHOLZ BCD66T.
M-1335; 0.28 C, 1.05 Cr, 0.45 Mo, bal Fe.
For gears, pinions, shafts, machine tool parts; oil hardened, shock resistant.

REISHOLZ CE36.
M-1335; 0.36 C, 1.25 Mn, 1.15 Cr, bal Fe.
For forging dies, crankshafts, fasteners; oil hardened, tough.

REISHOLZ D39K.
M-1335; 0.17 C, Mo, V, bal Fe.
For gears, bolts, machine tool parts; case hardened.

REISHOLZ DC54.
M-1335; 0.24 C, 1.15 Cr, 0.25 Mo, 0.55 Mn, bal Fe.
For gears, bolts, machine tool parts; oil hardened, tough.

REISHOLZ DC84.
M-1335; 0.42 C, 1.1 Cr, 0.2 Mo, 0.65 Mn, bal Fe.
For gears, bolts, machine tool parts; oil hardened, tough.

REISHOLZ DE20.
M-1335; 0.20 C, 1.1 Mn, 0.25 Mo, 0.25 Si, bal Fe.
For gears, cams, camshafts, machine tool parts; case hardened, tough.

REISHOLZ DE46.
M-1335; 0.20 C, 0.25 Mo, 1.1 Mn, 0.25 Si, bal Fe.
For gears, cams, camshafts, machine tool parts; case hardened, tough.

REISHOLZ E36.
M-1335; 0.36 C, 0.25 Si, 1.25 Mn, bal Fe.
For gears, cams, camshafts, fasteners; water hardened.

REISHOLZ E46K.
M-1335; 0.20 C, 0.12 V, 1.45 Mn, 0.25 Si, bal Fe.
For gears, cams, camshafts, fasteners; case hardening steel, tough.

REISHOLZ E305.
M-1335; 0.30 C, 0.37 Si, 0.45 Mn, bal Fe.
Hot rolled: 80,000 TS; 50,000 YS; 30 El; 56 RA; 165 Brin.
For gears, bolts, machine tool parts; water hardened.

REISHOLZ EF18.
M-1335; 0.19 C, 0.50 Si, 1.15 Mn, bal Fe.
For gears, cams, machine tool parts; case hardened.

REISHOLZ EF46.
M-1335; 0.46 C, 0.9 Si, 0.9 Mn, bal Fe.
For gears, springs, punches, bolts, machine tool parts; water hardened.

REISHOLZ EF55W.
M-1335; 0.55 C, Cr, Mn, Si, bal Fe.
For gears, bolts, machine tool parts; oil hardened.

REISHOLZ EF60.
M-1335; 0.53 C, 0.8 Si, 1.05 Mn, bal Fe.
For gears, springs, bolts, fasteners; water hardened.

REISHOLZ M1.
M-1335; 0.12 C, 0.25 Si, 0.55 Mn, bal Fe.
Annealed: 65,000 TS; 48,000 YS; 28 El; 65 RA; 130 Brin.
For nails, rivets, screws, bolts, bushings; case hardened.

REISHOLZ M2.
M-1335; 0.22 max C, 0.20 Si, 0.30 Mn, bal Fe.
Annealaed: 73,000 TS; 60,000 YS; 22 El; 58 RA; 150 Brin.
For fan blades, bushings, gears, shafts, rivets; case hardened.

REISHOLZ M3.
M-1335; 0.25 max C, 0.20 Si, 0.30 Mn, bal Fe.
Hot rolled: 70,000 TS; 45,000 YS; 31 El; 58 RA; 145 Brin.
For gears, bolts, armature shafts, keys, brackets; water hardened.

REISHOLZ NI.
M-1335; 0.26 C, Cr, Mo, bal Fe.
For gears, bolts, machine tool parts; oil hardened, tough.

REISHOLZ NIA.
M-1335; 0.22 C, 0.25 Si, 0.60 Mn, 2.25 Cr, bal Fe.
For oil refinery equipment; creep and heat resistant.

REISHOLZ NIC.
M-1335; 0.22 C, 0.20 Si, 0.60 Mn, 1.5 Cr, bal Fe.
For oil refinery equipment, gears, shafts; heat and creep resistant.

REISHOLZ NIN.
M-1335; 0.24 C, 2.5 Cr, 0.25 Mo, 0.8 max Ni, bal Fe.
For oil refinery equipment; heat and creep resistant.

REISHOLZ NIT.
M-1335; 0.32 C, Cr, Ni, Mo, bal Fe.
For gears, pinions, shafts; oil hardened, shock resistant.

REISHOLZ S22.
M-1335; 0.22 C, 0.25 Si, 0.45 Mn, bal Fe.
Annealed; 75,000 TS; 45,000 YS; 22 El; 58 RA; 140 Brin.
For screws, bolts, camshafts, bushings; water hardened.

REISHOLZ S35.
M-1335; 0.35 C, 0.25 Si, 0.55 Mn, bal Fe.
Hot rolled; 85,000 TS; 54,000 YS; 30 El; 53 RA; 185 Brin.
For armature shafts, brackets, gears, bolts; water hardened.

REISHOLZ S45.
M-1335; 0.45 C, 0.25 Si, 0.65 Mn, bal Fe.
Hot rolled: 98,000 TS; 58,000 YS; 24 El; 45 RA; 212 Brin.
For axles, gears, bolts, bushings; water hardened.

REISHOLZ SK11H.
M-1335; 0.20 C, 0.25 Si, 0.60 Mn, 0.30 Mo, bal Fe.
Annealed: 75,000 TS; 42,000 YS; 20 El; 55 RA; 150 Brin.
For gears, bolts, fasteners, rivets; case hardened.

REISHOLZ SK11R.
M-1335; 0.15 C, 0.25 Si, 0.65 Mn, 0.30 Mo, bal Fe.
Annealed: 75,000 TS; 42,000 YS; 20 El; 55 RA; 150 Brin.
For gears, bolts, fasteners, camshafts; case hardened.

REISHOLZ SK12.
M-1335; 0.16 C, 1.05 Cr, 0.45 Mo, 0.65 Mn, bal Fe.
For camshafts, cams, bolts, gears; case hardened.

REISHOLZ SK12E.
M-1335; 0.24 C, 1.25 Cr, 0.45 Mo, 0.55 Mn, bal Fe.
For gears, bolts, camshafts, crankshafts; oil hardened, tough.

REISHOLZ SK12ES.
M-1335; 0.24 C, 1.35 Cr, 0.55 Mo, 0.20 V, bal Fe.
For die casting dies, gears, crankshafts; oil hardened, tough.

REISHOLZ SK12H.
M-1335; 0.20 C, 1.05 Cr, 0.45 Mo, 0.65 Mn, bal Fe.
For gears, bolts, camshafts, cams; case hardened.

REISHOLZ SK12R.
M-1335; 0.13 C, 0.85 Cr, 0.45 Mo, 0.55 Mn, bal Fe.
For gears, bolts, camshafts; case hardened.

REISHOLZ SK14R.
M-1335; 0.10 C, 1.1 Si, 1.8 Cr, 0.30 Mo, 0.30 V, bal Fe.
For camshafts, cams, plastic mold dies; case hardened.

REISHOLZ SK16.
M-1335; 0.10 C, 2.25 Cr, 1.1 Mo, 0.50 Mn, bal Fe.
For plastic mold dies; case hardened.

REISHOLZ SK20.
M-1335; 0.22 C, 11.5 Cr, 1 Mo, 0.5 Ni, 0.3 V, bal Fe.
Annealed: 90,000 TS; 40,000 YS; 25 El; B 92 Rock.
Hardened: 240,000 TS; 205,000 YS; 10 El; C 50 Rock.
For valves, bearings, cutlery, surgical instruments, pivots. Corrosion resistant, hardenable.

REISHOLZ SK20W.
M-1335; 0.20 C, 12 Cr, 1 Mo, 0.4 Ni, 0.3 V, 0.5 W, bal Fe.
Annealed: 95,000 TS; 40,000 YS; 25 El; B 92 Rock.
Hardened: 235,000 TS; 100,000 YS; 11 El; C 48 Rock.
For valves, bearings, cutlery, surgical instruments, gears. Corrosion resistant, hardenable.

REISHOLZ SK30.
M-1335; 1.4 C, Mn, bal Fe.
For bearings, cutters, engravers' tools; Type W1; water hardened.

REISHOLZ SK31H.
M-1335; 0.18 C, 0.65 Cr, 0.15 Mo, 0.85 Mn, bal Fe.
For gears, bolts, camshafts, cams; case hardened.

REISHOLZ SK31R.
M-1335; 0.15 C, 0.75 Cr, 0.15 Mo, 0.75 Mn, bal Fe.
For gears, bolts, camshafts, cams; case hardened.

REISHOLZ SK32.
M-1335; 0.16 C, 1.05 Cr, 0.25 Mo, 0.20 V, bal Fe.
For gears, bolts, camshafts, cams; case hardened.

REISHOLZ SK32R.
M-1335; 0.13 C, 1.05 Cr, 0.25 Mo, 0.55 Mn, bal Fe.
For gears, bolts, camshafts, cams; case hardened.

REISHOLZ ST 35.
M-1335; 0.17 max C, 0.35 max Si, 0.40 Max Mn, bal Fe.
Annealed: 70,000 TS; 40,000 YS; 25 El; 60 RA; 145 Brin.
For rivets, nails, bolts, nuts, gears, shafts; case hardened.

REISHOLZ ST 45.
M-1335; 0.22 max C, 0.35 max Si, 0.45 max Mn, bal Fe.
Annealed: 73,000 YS; 42,000 YS; 22 El; 58 RA; 150 Brin.
For fan blades, gears, bolts, rivets, shafts; case hardened.

REITHS ALLOY.
M-Ger.; 74.5 Cu, 11.6 Sn, 0.9 Pb, 4.9 Sb.
For bearings; heavy duty.

REKFORD EMINENT.
M-336; C, Mn, Cr, W, V, Mo, Co, bal Fe.
For tools, dies; high speed steel.

REKORD-25.
M-387; 0.70 C, 4.5 Cr, 18.5 W, 0.6 max Co, 1.2 V, bal Fe.
For lathe and planer tools, drills, reamers, hobs; high speed steel.

REKORD 1939.
M-387; 0.8 C, 0.2 Mo, 12.5 W, 0.8 Co, 2.5 V, 4 Cr, bal Fe.
For drills, reamers, broaches, taps, milling cutters; high speed steel.

REKORD BK03.
M-1346; 0.9 C, W, Co, V, 4 Cr, bal Fe.
For lathe and planer tools, reamers, broaches; high speed steel.

REKORD BK05.
M-1346; 0.79 C, 4.75 Co, 4.3 Cr, 0.75 Mo, 1.5 V, 18 W, bal Fe.
For lathe and planer tools, reamers, broaches; high speed steel.

REKORD BKO10.
M-1346; 0.76 C, 10 Co, 4.2 Cr, 0.8 Mo, 1.8 V, 18 W, bal Fe.
For lathe and planer tools, reamers, broaches, hobs; high speed steel.

REKORD BRKO.
M-1346; 0.80 C, Co, Cr, Mo, V, W, bal Fe.
For lathe and planer tools, reamers, broaches, taps; high speed steel.

REKORD BVK4.
M-1346; 1.35 C, Co, Mo, V, W, bal Fe.
For engravers' tools, forming dies, broaches, taps; high speed steel.

REKORD BVN.
M-1346; 1.3 C, 4.3 Cr, 0.85 Mo, 3.8 V, 12 W, bal Fe.
For engravers' tools, forming dies, broaches, taps; high speed steel.

REKORD BW018.
M-1346; 0.74 C, 4.1 Cr, 1.1 V, 18.5 W, bal Fe.
For lathe and planer tools, reamers, taps, hobs; high speed steel.

REKORD EXTRA-26.
M-387; 0.80 C, 1.2 Mo, 2.5 Co, 4.5 Cr, 18.5 W, 1.6 V, bal Fe.
For milling cutters, lathe and planer tools, reamers; high speed steel.

REKORD MO 90.
M-1346; 0.80 C, Cr, V, Mo, W, bal Fe.
For lathe and planer tools, reamers, taps, drills; high speed steel.

REKORD SELECT.
M-336; C, Mn, Cr, W, V, Mo, Cu, bal Fe.
For tools, dies.

REKORD SELECT-27.
M-387; 0.80 C, 1.2 Mo, 4.5 Cr, 18.5 W, 5.5 Co, 1.6 V, bal Fe.
For milling cutters, lathe and planer tools, drills; high speed steel.

REKORD SUPERIOR.
M-1346; 0.86 C, 4.1 Cr, 0.85 Mo, 2.5 V, 12 W, bal Fe.
For lathe and planer tools, milling cutters, drills; high speed steel.

RELIABRAZE 2417.
M-1073; paste of copper powder and vehicle for copper brazing.

RELIANCE.
M-370; 0.95 C, 0.2 Si, 0.35 Mn, bal Fe.
Cold drawn flats and squares.
AISI Type W1 water hardening tool steel.

RELIANCE.
M-907; 0.25-0.40 C, bal Fe.
For steel castings; water hardening.

RELIANCE.
M-1061; 0.07-0.13 C, 0.2-0.5 Mn, 3.0-3.5 Ni, 1.4-1.6 Cu, bal Fe.
Rolled: 95,000 TS; 75,000 YS; 30 El; 60 RA; 183 Brin.
For oil well sucker rods; for severe service.

RELIANITE.
M-1440; 3.2 C, Si, Mn, bal Fe.
For housings, gears, shafts; nodular iron, ductile.

RELIT.
M-Russian; WC + Co.
For cutting tools; hard sintered alloy.

RELY AWM1.
M-420; 89 Sn, 8 Sb, 3 Cu.
For bearings; Babbitt.

RELY C.M. 1.
M-420; 6 Cu, 84 Sn, 10 Sb.
For bearings; Babbitt.

RELY C.M.2.
M-420; 6 Cu, 72 Sn, 12 Pb, 10 Sb.
For bearings; Babbitt.

RELY C.M.3.
M-420; 3 Cu, 40 Sn, 42 Pb, 15 Sb.
For bearings; Babbitt.

RELY C.M.4.
M-420; 3 Cu, 31 Sn, 50 Pb, 16 Sb.
For bearings; Babbitt.

RELY C.M.5.
M-420; 1 Cu, 15 Sn, 70 Pb, 14 Sb.
For bearings; Babbitt.

RELY C.M.6.
M-420; 6 Sn, 80 Pb, 14 Sb.
For bearings; Babbitt.

RELY "YZ".
M-420; 91.5 Sn, 3.5 Sb, 4.3 Cu, 0.5 Ni,
For bearings.

REMA.
M-908; 0.05 C, 0.20 Mn, 0.10 Si, bal Fe.
For hobbing molds; deep cavity molds for plastic industry.

REMA.
M-1067; 0.05 C, 0.2 Mn, 0.1 Si, bal Fe.
For hobbing molds; case hardening steel.

REMALLOY.
M-670; 17 Mo, 12 Co, bal Fe.
Bar: 126,000 TS; 50,000 TrS; 60 Rock C.
10,000 Residual flux.
230 Coercive force.
For permanent magnets; precipitation hardened. Fabricated parts only.

REMALLOY.
M-908, M-1067; 0.10 C, 0.50 Mn, 0.6 Cr, 1.5 Ni, bal Fe.
For plastic dies, hobbing molds; case-hardening.

REMALLOY.
M-790; 0.7 C, 4 Cr, 18 W, 2 V, 12 Co, bal Fe.
For lathe and planer tools, reamers, broaches, taps, hobs; high speed steel, oil hardened.

REMALLOY 17.
M-22, M-7205; 17 Mo, 12 Co, 71 Fe, 3 Mn.
For permanent magnets; high coercive force.

REMALLOY 17.
M-373; 17 Mo, 12 Co, bal Fe.
Maximum energy product 1.10; residual flux density, Br. 10,000; coercive force Hc (Oersted) 250.
Rolled or cast permanent magnet.

REMALLOY 20.
M-373; 20 Mo, 12 Co, bal Fe.
Maximum energy product 1.25; residual flux density, Br. 8550; Coercive force Hc (Oersted) 335.
Hot rolled or cast permanent magnet.

REMALLOY 20.
M-205, M-373; 67.7 Fe, 12 Co, 20 Mo, 0.3 Mn.
For receiver magnets; permanent magnet. 355 coercive force.

REMANIT 1510.
M-1299; 0.15 max C, 11.5-13.5 Cr, bal Fe.
Heat treated: 120,000-135,000 TS; 110,000-117,000 YS; 16-15 El; 63-58 RA; 220-240 Brin.
For cutlery, valves, turbine blades; Type 410; corrosion resistant.

REMANIT 1510 H.
M-1299; 0.12 max C, 14-18 Cr, bal Fe.
Annealed: 70,000 TS; 40,000 YS; 30 El; 55 RA; 150 Brin.
Cold drawn: 130,000 TS; 120,000 YS; 2 El; 185 Brin.
For oil refinery equipment, bolts, oil burners; Type 430; stainless, ferritic.

REMANIT 1880SS.
M-1299; 0.10 max C, 16-18 Cr, 10-14 Ni, 2-3 Mo, bal Fe.
Annealed: 85,000-95,000 TS; 35,000-45,000 YS; 60-50 El; 75-60 RA; 150-190 Brin.
For chemical plant equipment, mixers, agitators, filters; Type 316; stainless, austenitic.

REMANIT 1880ST.
M-1299; 0.08 max C, 17-19 Cr, 8-11 Ni, Ti = 5 x C, bal Fe.
Normalized: 93,000 TS; 36,000 YS; 45 El; 60 RA; 165 Brin.
Annealed: 87,000 TS; 33,000 YS; 57 El; 73 RA; 155 Brin.
For chemical plant equipment; Type 321; austenitic, stainless.

REMANIT 4000 ETC.
M-1759.
See Werkstoff Nr. 1.4000 etc.
Stainless steels; 51 grades.

REM ARMS HD-1000 DENSITY 6.8.
M-1698; Iron, sintered.
As sintered: 23,000 TS; 12 El; Rf 40.
For low and medium stressed mechanical components; soft magnetic cores.
MPIF F-0000-S; ASTM B310-70, Type III; Class A.

REM ARMS HD-1000 DENSITY 7.0.
M-1698; Iron, sintered.
As sintered: 26,500 TS; 15 El; Rf; 50.
For low and medium stressed mechanical components; soft magnetic cores.
MPIF F-0000-S ASTM B310-70, Type IV, Class A.

REM ARMS HD-1000 DENSITY 7.2.
M-1698; Iron, sintered.
As sintered: 29,500 TS; 20 El; Rf 55.
For low and medium stressed mechanical components; soft magnetic cores.
MPIF F-0000-T; ASTM B310-70, Type IV, Class A.

REM ARMS HD-1000 DENSITY 7.4.
M-1698; Iron, sintered.
As sintered: 35,000 TS; 24 El; Rb 20.
For low and medium stressed mechanical components; soft magnetic cores.
MPIF F-0000-T ASTM B310-70, Type V, Class A.

REM ARMS HD-1000 DENSITY 7.6.
M-1698; Iron, sintered.
As sintered: 40,000 TS; 29 El; Rb 30.
For low and medium stressed mechanical components; soft magnet cores; this grade can be case hardened, plated and colored.
MPIF F-0000-U; ASTM B310-70, Type V, Class A.

REM ARMS HD-1003 DENSITY 6.8.
M-1698; Medium carbon steel, sintered.
As sintered: 32,000 TS; 9 El; Rb 12.
For medium stressed structural parts.
MPIF F-0000-S; ASTM B310-70; Type III, Class B.

REM ARMS HD-1003 DENSITY 7.2.
M-1698; Medium carbon steel, sintered.
As sintered: 43,000 TS; 11 El; Rb 37.
Can be heat treated to 50,000-80,000 psi.
For medium stressed structural parts.
MPIF F-0000-T; ASTM B310-70, Type IV, Class B.

REM ARMS HD-1007 DENSITY 6.8.
M-1698; High carbon steel, sintered.
As sintered: 46,000 TS; 4 El; Rb 44.
For highly stressed structural parts.
Heat treatable to 65,000-100,000 psi.
MPIF F-0008-S; ASTM B310-70, Type III, Class C.

REM ARMS HD-1007 DENSITY 7.2.
M-1698; High carbon steel, sintered.
As sintered: 55,000 TS; 5 El; Rb 58.
For highly stressed structural parts.
Heat treatable to 65,000-100,000 psi.
MPIF F-0008-T ASTM B310-70, Type IV, Class C.

REM ARMS HD-2020 DENSITY 6.8.
M-1698; Low nickel iron, sintered.
As sintered: 38,000 TS; 8 El; Rb 32.
Good strength and toughness; for small appliance parts.
MPIF FN-0200-S; ASTM B484-70, Type II, Grade 1, Class A.

REM ARMS HD-2020 DENSITY 7.0.
M-1698; Low nickel iron, sintered.
As sintered: 41,000 TS; 11 El; Rb 40.
Good strength and toughness; for small appliance parts.
MPIF FN-0200-S; ASTM B484-70, Type II, Grade 1, Class A.

REM ARMS HD-2020 DENSITY 7.2.
M-1698; Low nickel iron, sintered.
As sintered: 47,000 TS; 14 El; Rb 49.
Good strength and toughness; for gears, cams and small appliance parts.
MPIF FN-0200-T; ASTM B484-70, Type III, Grade 1, Class A.

REM ARMS HD-2020 DENSITY 7.4.
M-1698; Low nickel iron, sintered.
As sintered: 53,000 TS; 18 El; Rb 60.
Good strength and toughness; for gears, cams and small appliance parts.
MPIF FN-0200-T; ASTM B484-70, Type III, Grade 1, Class A.

REM ARMS HD-2027 DENSITY 6.8.
M-1698; Low nickel steel, sintered.
As sintered: 72,000 TS; 3 El; Rb 74.
Good strength and toughness; hardenable to Rc 40-45.
MPIF FN-0208-S; ASTM B484-70, Type II, Grade 1, Class C.

REM ARMS HD-2027 DENSITY 7.2.
M-1698; Low nickel steel, sintered.
As sintered: 91,000 TS; 4 El; Rb 86.
Good strength and toughness; hardenable to Rc 40-45.
MPIF FN-0208-T; ASTM B484-70 Type III, Grade 1, Class C.

REM ARMS HD-2040 DENSITY 6.8.
M-1698; Medium nickel iron, sintered.
As sintered: 37,000 TS; 5 El; Rb 35.
For case hardening.
MPIF FN-0400-T; ASTM B484-70, Type II, Grade 2, Class A.

REM ARMS HD-2040 DENSITY 7.2.
M-1698; Medium nickel iron, sintered.
As sintered: 49,000 TS; 9 El; Rb 58.
For case hardening.
MPIF FN-0400-T; ASTM B484-70, Type III, Grade 2, Class A.

REM ARMS HD-2047 DENSITY 6.8.
M-1698; Medium nickel steel, sintered.
As sintered: 74,000 TS; 2 El; Rb 80.
High strength, tough core; hardenable to 120,000-160,000 psi.
MPIF FN-0408-S; ASTM B484-70, Type II, Grade 2, Class C.

REM ARMS HD-2047 DENSITY 7.2.
M-1698; Medium nickel steel, sintered.
As sintered: 95,000 TS; 3.5 El; Rb 87.
High strength, tough core; hardenable to 120,000-160,000 psi.
MPIF FN-0408-T; ASTM B484-70, Type III, Grade 2, Class C.

REM ARMS HD-2070 DENSITY 6.8.
M-1698; High nickel iron, sintered.
As sintered: 65,000 TS; 4.5 El; Rb 64.
High strength, fair ductility; easily swaged or formed.
MPIF FN-0700-S; ASTM B484-70, Type II, Grade 3, Class A.

REM ARMS HD-2070 DENSITY 7.2.
M-1698; High nickel iron, sintered.
As sintered: 76,000 TS; 8 El; Rb 78.
High strength, fair ductility; easily swaged or formed.
MPIF FN-0700-T; ASTM B484-70, Type III, Grade 3, Class A.

REM ARMS HD-2076 DENSITY 6.8.
M-1698; High nickel steel, sintered.
As sintered: 94,000 TS; 3.5 El; Rb 86.
High strength; heat treatable.
MPIF FN-0705-S; ASTM B484-70, Type II, Grade 3, Class C.

REM ARMS HD-2076 DENSITY 7.2.
M-1698; High nickel steel, sintered.
As sintered: 105,000 TS; 4 El; Rc 20.
High strength; heat treatable.
MPIF FN-0705-T; ASTM B484-70, Type III, Grade 3, Class C.

REM ARMS HD-2108 DENSITY 7.4.
M-1698; Air hardening nickel steel, sintered.
As sintered: 150,000 TS; 4 El; Rc 40.
For jobs requiring strong, wear-resisting parts.

REM ARMS HD-3050 DENSITY 7.0.
M-1698; Remington alloy, sintered.
As sintered: 32,000 TS; 12 El; Rb 30.
High ductility by single pressing.

REM ARMS HD-3050 DENSITY 7.2.
M-1698; Remington alloy, sintered.
As sintered: 37,000 TS; 16 El; Rb 35.
High density and high ductility by single pressing.

REM ARMS HD-3050 DENSITY 7.4.
M-1698; Remington alloy, sintered.
As sintered: 47,000 TS; 25 El; Rb 50.
Good impact strength; high density and high ductility; for case hardening, plating, coloring.

REM ARMS HD-3050 DENSITY 7.6.
M-1698; Remington alloy, sintered.
As sintered: 53,000 TS; 30 El; Rb 60.
Good impact strength; high density and high ductility; for case hardening, plating, coloring.

REM ARMS HD-4058 DENSITY 6.6.
M-1698; Copper steel, sintered.
As sintered: 78,000 TS; 1 El; Rb 82.
Medium strength components.
ASTM B426-70, Grade 2, Type III.

REM ARMS HD-4058 DENSITY 6.8.
M-1698; Copper steel, sintered.
As sintered: 96,000 TS; 1 El; Rb 90.
Medium strength components.
ASTM B426-70, Grade 2, Type IV.

REM ARMS HD-4058 DENSITY 7.0.
M-1698; Copper steel, sintered.
As sintered: 105,000 TS; 2 El; Rb 96.
Medium strength components.
ASTM B426-70, Grade 2, Type IV.

REM ARMS HD-4058 DENSITY 7.2.
M-1698; Copper steel, sintered.
As sintered: 110,000 TS; 2 El; Rc 25.
For medium strength components.
ASTM B426-70, Grade 2, Type V.

REM ARMS HDM-1000 DENSITY 6.4.
M-1698; Iron-sintered.
Theoretical density, % 81; induction B max, kilogauss 9.1; remanence Br, kilogauss 7.4; coercive Hc Force, oersted 1.80; maximum permeability 2350; elec. Res., microhm-cm 12.
For pole pieces, cores, armatures, generator rotors, stator cores.
MPIF F-0000-P; ASTM B310-70, Type II, Class A.

REM ARMS HD-1000 DENSITY 6.8.
M-1698; Iron-sintered.
Theoretical density, % 86; induction B max, kilogauss 11.4; remanence Br kilogauss 9.6; coercive Hc Force, oersted 1.65; maximum permeability 2900; elec. res. microhm-cm 12.
For DC applications, generator rotors etc.
MPIF F-0000-R; ASTM B310, Type III, Class A.

REM ARMS HDM-1000 DENSITY 7.2.
M-1698; Iron-sintered.
Theoretical density, % 92; induction B max kilogauss 13.6; remanence Br, kilogauss 11.8; coercive Hc Force, oersted 1.60; maximum permeability 3700; elec. res., microhm-cm 12.
For DC applications, cores, rotors, stators.
MPIF F-0000-S; ASTM B310-70, Type IV, Class A.

REM ARMS HDM-1000 DENSITY 7.4.
M-1698; Iron-sintered.
Theoretical density, % 94; induction B max, kilogauss 14.7; remanence Br kilogauss 12.9; coercive Hc Force, oersted 1.50; maximum permeability 4200; elec res., microhm-cm 12.
For DC applications, pole pieces, stators.
MPIF F-0000-T; ASTM B310-70, Type V, Class A.

REM ARMS HDM-1000 DENSITY 7.6.
M-1698; Iron-sintered.
Theoretical density, % 97; induction B max, kilogauss 15.8; remanence Br kilogauss 14.0; coercive Hc Force, oersteds 1.35; maximum permeability 4750; elec. res., microhm-cm 12.
For DC applications, pole pieces, cores, stators.
MPIF F-0000-U; ASTM B310-70, Type V, Class A.

REM ARMS HDM-2500 DENSITY 6.8.
M-1698; Iron-nickel, sintered.
Theoretical density, % 83; induction B max, kilogauss 9.3; remanence Br, kilogauss 7.1; coercive Hc Force, oersted 0.26; elec. res., microhm-cm 43.
For torque motors, relays, missile components.

REM ARMS HDM-2500 DENSITY 7.1.
M-1698; Iron-nickel, sintered.
Theoretical density, % 87; induction B max, kilogauss 10.9; remanence Br kilogauss 8.0; coercive Hc Force, oersted 0.25; elec. res., microhm-cm 43.
For torque motors, relays, missile components.

REM ARMS HDM-2500 DENSITY 7.5.
M-1698; Iron-nickel, sintered.
Theoretical density, % 91; induction B max, kilogauss 12.7; remanence Br kilogauss 9.4; coercive Hc Force, oersted 0.24; maximum permeability 21,000; elec. res., microhm-cm 43.
Very low coercive force and high permeability; for torque motors, relays, missile components.

REM ARMS HDM-2500 DENSITY 7.7.
M-1698; Iron-nickel, sintered.
Theoretical density, % 94; induction B max, kilogauss 13.3; remanence Br kilogauss 9.5; coercive Hc Force, oersted 0.22; maximum permeability 22,000; elec. res., microhm-cm 43.
Very low coercive force and high permeability; for torque motors, relays, missile components.

REM ARMS HDM-3050 DENSITY 7.0.
M-1698; (Soft magnetic alloy, sintered).
Theoretical density, % 88; induction B max, kilogauss 11.6; remanence Br, kilogauss 6.7; coercive Hc Force, oersted 1.60; elec. res., microhm-cm 26.
For DC applications.

REM ARMS HDM-3050 DENSITY 7.2.
M-1698; (Soft magnetic alloy, sintered).
Theoretical density, % 90; induction B max, kilogauss 13.5; remanence Br, kilogauss 8.1; coercive Hc Force, oersted 1.50; elec. res., microhm-cm 26.
For DC applications.

REM ARMS HDM-3050 DENSITY 7.4.
M-1698; (Soft magnetic alloy, sintered).
Theoretical density, % 93; induction B max, kilogauss 13.8; remanence Br, kilogauss 10.2; coercive Hc Force, oersted 1.10; elec. res., microhm-cm 26.
For DC applications.

REM ARMS HDM-3050 DENSITY 7.6.
M-1698; (Soft magnetic alloy, sintered).
Theoretical density, % 96; induction B max, kilogauss 14.4; remanence Br, kilogauss 10.5; coercive Hc Force, oersted 1.10; elec. res., microhm-cm 26.
For DC applications.

REM ARMS HDM-5010 DENSITY 7.2.
  M-1698; Silicon-iron, sintered.
    Theoretical density, % 92; induction B max, kilogauss 13.6; remanence Br, kilogauss 10.3; coercive Hc Force, oersteds 1.5; maximum permeability 3450; elec. res., microhm-cm 22.
    For AC applications, solenoids, relays, armatures.

REM ARMS HDM-5020 DENSITY 7.1.
  M-1698; Silicon-iron, sintered.
    Theoretical density, % 92; induction B max, kilogauss 13.2; remanence Br, kilogauss 10.3; coercive Hc Force, oersted 1.20; elec. res., microhm-cm 39.
    For AC applications, relays, armatures, cores.

REM ARMS HDM-5030 DENSITY 6.8.
  M-1698; Silicon-iron, sintered.
    Theoretical density, % 89; induction B max, kilogauss 11.7; remanence Br, kilogauss 9.4; coercive Hc Force, oersted 1.30; elec. res., microhm-cm. 46.
    For AC applications, solenoids, pole pieces, stators.

REM ARMS HDM-5030 DENSITY 7.0.
  M-1698; Silicon-iron, sintered.
    Theoretical density, % 91; induction B max, kilogauss 13.1; remanence Br, kilogauss 10.9; coercive Hc Force, oersted 1.25; elec. res., microhm-cm 46.
    For AC applications, solenoids, armatures, cores.

REM ARMS HDM-5030 DENSITY 7.2.
  M-1698; Silicon-iron, sintered.
    Theoretical density, % 94; induction B max, kilogauss 13.9; remanence Br, kilogauss 11.8; coercive Hc Force, oersted 1.20; maximum permeability 4860; elec. res., microhm-cm 46.
    For AC applications, relays, actuator cores.

REM ARMS HDM-5030 DENSITY 7.4.
  M-1698; Silicon-iron, sintered.
    Theoretical density, % 96; induction B max, kilogauss 14.4; remanence Br, kilogauss 12.4; coercive Hc Force, oersted 1.00; elec. res., microhm-cm 46.
    For AC applications, electromagnetic devices.

REM ARMS HDM-5040 DENSITY 7.3.
  M-1698; Silicon-iron, sintered.
    Theoretical density, % 96; induction B max, kilogauss 14.2; remanence Br, kilogauss 10.8; coercive Hc Force, oersted 1.00; maximum permeability 5850; elec. res., microhm-cm 59.
    For AC applications, solenoids, pole pieces.

REM ARMS HDM-5050 DENSITY 7.2.
  M-1698; Silicon-iron, sintered.
    Theoretical density, % 96; induction B max, kilogauss 13.7; remanence Br, kilogauss 8.7; coercive Hc Force, oersted 0.85; maximum permeability 6400; elec. res., microhm-cm 68.
    For AC applications, relays, armatures, solenoids.

REM ARMS HDM-9800 DENSITY 7.8.
  M-1698 Nickel-Iron-Moly, sintered.
    Theoretical density, % 90; induction B max, kilogauss 7.2; remanence Br, kilogauss 4.8; coercive Hc Force, oersted 0.07; maximum permeability 37,000; elec. res., microhm-cm 55.
    For high permeability uses.

REM ARMS HDS-303L DENSITY 6.6.
  M-1698.
    303L-Stainless (Austenitic).
    As sintered: 62,000 TS; 4 El; Rb 63, Charpy 2 ft. lbs.
    Free machining, non-magnetic, non-hardenable.
    MPIF SS-303-R.

REM ARMS HDS-303L DENSITY 6.8.
  M-1698.
    303L-Stainless (Austenitic).
    As sintered: 69,000 TS; 12 El; Rb 70; Charpy 12 ft. lbs.
    Free machining, non-magnetic, non-hardenable.
    MPIF SS-303-R.

REM ARMS HDS-303L DENSITY 7.0.
  M-1698.
    303L-Stainless (Austenitic).
    As sintered: 75,000 TS; 16 El; Rb 75; Charpy 40 ft. lbs.
    Free machining, non-magnetic, non-hardenable.

REM ARMS HDS-304L DENSITY 6.6.
  M-1698.
    304L-Stainless (Austenitic).
    As sintered: 64,000 TS; 5 El; Rb 70; Charpy 3 ft. lbs.
    Non-magnetic, non-hardenable, stainless.
    ASTM B525-70, Type II, Grade I.

REM ARMS HDS-304L DENSITY 6.8.
  M-1698.
    304L-Stainless (Austenitic).
    As sintered: 71,000 TS; 12 El; Rb 75; Charpy 12 ft. lbs.
    Non-magnetic, non-hardenable, stainless.
    ASTM B525-70 Type II, Grade I.

REM ARMS HDS-304L DENSITY 7.0.
  M-1698.
    304L-Stainless (Austenitic).
    As sintered: 75,000 TS; 15 El; Rb 78; Charpy 42 ft. lbs.
    Non-magnetic, non-hardenable, stainless.
    ASTM B525-70 Type III, Grade I.

REM ARMS HDS-316L DENSITY 6.6.
M-1698.
316L-Stainless (Austenitic).
As sintered: 60,000 TS; 4 El; Rb 70; Charpy 4 ft. lbs.
Non-magnetic, non-hardenable, improved corrosion resistance.
MPIF SS-316-R; ASTM B525-70, Type II, Grade II.

REM ARMS HDS-316L DENSITY 6.8.
M-1698.
316L-Stainless (Austenitic).
As sintered: 68,000 TS; 12 El; Rb 74; Charpy 10 ft. lbs.
Non-magnetic, non-hardenable, improved corrosion resistance.
MPIF SS-316-R; ASTM B525-70, Type II, Grade II.

REM ARMS HDS-316L DENSITY 7.0.
M-1698.
316L-Stainless (Austenitic).
As sintered: 73,000 TS; 16 El; Rb 79; Charpy 45 ft. lbs.
Non-magnetic, non-hardenable, improved corrosion resistance.
ASTM B525-70, Type III, Grade II.

REM ARMS HDS-410 DENSITY 6.4.
M-1698.
Type 410 Stainless (Martensitic).
As sintered: 67,000 TS; 0 El; Rc 15; Charpy 1 ft. lb.
Hardenable to Rc 50 equivalent; fair corrosion resistance; ferromagnetic.
MPIF SS-410-P.

REM ARMS HDS-410 DENSITY 6.6.
M-1698.
Type 410 Stainless (Martensitic).
As sintered: 84,000 TS; 1 El; Rc 20; Charpy 1.5 ft. lb.
Hardenable to Rc 50 equivalent; fair corrosion resistance; ferromagnetic.

REM ARMS HDS-410 DENSITY 6.8.
M-1698.
Type 410 Stainless (Martensitic).
As sintered: 88,000 TS; 2 El; Rc 20; Charpy 1.5 ft. lb.
Hardenable to Rc 50 equivalent; fair corrosion resistance; ferromagnetic.

REM ARMS HDS-410 DENSITY 7.0.
M-1698.
Type 410 Stainless (Martensitic).
As sintered: 94,000 TS; 2 El; Rc 25; Charpy 2 ft. lb.
Hardenable to Rc 50 equivalent; fair corrosion resistance; ferromagnetic.

REM ARMS HDS-410 DENSITY 7.2.
M-1698.
Type 410 Stainless (Martensitic).
As sintered: 73,000 TS; 6 El; Rb 75; Charpy 3 ft. lb.
Hardenable to Rc 50 equivalent; fair corrosion resistance; ferromagnetic.

REM ARMS HDS 430 DENSITY 6.6.
M-1698.
Type 430 Stainless (Ferritic).
As sintered: 35,000 TS; 2 El; Rb 90.
Non-hardenable; fair corrosion resistance; ferromagnetic.

REM ARMS HDS-430 DENSITY 6.8.
M-1698.
Type 430 stainless (Ferritic).
As sintered: 50,000 TS; 2 El; Rb 85.
Non-hardenable; fair corrosion resistance; ferromagnetic.

REM ARMS HDS-430 DENSITY 7.0.
M-1698.
Type 430 Stainless (Ferritic).
As sintered: 60,000 TS; 4 El; Rb 70.
Non-hardenable; fair corrosion resistance; ferromagnetic.

REM ARMS HDS-9400 DENSITY 6.8.
M-1698.
High corrosion resistant stainless.
As sintered: 50,000 TS; 4 El; Rb 60.
Outstanding corrosion resistance to H_2SO_4, caustic solution brine, plating solutions.

REM ARMS HDS-9400 DENSITY 7.0.
M-1698.
High corrosion resistant stainless.
As sintered: 66,000 TS; 11 El; Rb 70.
Outstanding corrosion resistance to H_2SO_4, caustic solutions, brine, plating solutions.

REMENDUR.
M-22; 49 Co, 2-5 V, bal Fe.
Remanence-21,500 gausses.
Coercive force 20-60 oersteds.
Residual induction 16,000-21,500 gauss.
For magnetic devices, telephone switches.
Shows square hysteresis loop.
Non-directional properties, high curie temp.
Permanent magnet. High residual induction.

REMENDUR 27.
M-61; 2.7 V, 48.5 Co, bal Fe.
For self latching reed switches.
Saturation induction 20,000 gausses.
Hc = 27 oe, magnetically semihard.

REMENDUR 38.
M-61; 3.5 V, 48.5 Co, bal Fe.
Bias magnets for latching relays.
Saturation induction 20,000 gausses.
Hc = 38 oe, magnetically semihard.

REMENDUR 48.
M-61; 4.2 V, 48 Co, bal Fe.
Bias magnets.
Saturation induction 20,000 gausses.
Hc=48 oe, magnetically semihard.

REMIRAL.
M-Ger.; 0.5-2.0 Mg, bal Al.
H25 Temper: 30,000 TS; 25,000 YS.
For reflectors, automotive trim.
Bright aluminum sheets.
Corrosion resistant.

REMIRAL-100.
M-493; 1.0 Mg, bal Al.
Ann: 100 MPa TS; 25 El.
Hard: 200 MPa TS; 2 El.
For ornamental and decorative parts.

REMY AMN.
M-1339; 0.90-1.05 C, 1.0-1.2 Mn, 0.50-0.70 Cr, 0.50-0.70 W, 0.10 V, bal Fe.
Oil hardening, cold work tool steel.
W.-Nr. 1.2510; AISI O1.

REMY CE6.
M-1339; 0.15 C, 0.65 Cr, 0.5 Mn, bal Fe.
For gears, bolts, machine tool parts; case hardened.

REMY CE8.
M-1339; 0.16 C, 0.95 Cr, 1.15 Mn, bal Fe.
For gears, cams, camshafts; case hardened, tough.

REMY CEMO8.
M-1339; 0.22 C, 1.2 Cr, 0.25 Mo, 0.65 Mn, bal Fe.
For gears, bolts, machine tool parts; case hardened, tough.

REMY CEMO10.
M-1339; 0.24 C, 1.15 Cr, 0.25 Mo, 0.55 Mn, bal Fe.
For gears, bolts, machine tool parts; oil hardened, shock resistant.

REMY CEN4.
M-1339; 0.19 C, 1.25 Cr, 0.20 Mo, 3.75 Ni, bal Fe.
For gears, bolts, camshafts, cams; case hardened, shock resistant.

REMY CNLK.
M-1339; 0.45 C, 1.4 Cr, 0.7 Mo, 0.3 V, bal Fe.
For gears, bolts, machine tool parts; oil hardened, shock resistant.

REMY DS.
M-1339; 0.45 C, Cr, V, Si, bal Fe.
For springs, gears, crankshafts, bolts, studs; oil hardened, shock resistant.

REMY EPS.
M-1339; 0.30 C, 2.6 Cr, 8.5 W, 0.4 V, bal Fe.
Hot work tool steel; forging dies.
W.-Nr. 1.2581; Similar to AISI H21.

REMY EPS 33.
M-1339; 0.33 C, 3.0 Cr, 2.8 Mo, 0.5 V, bal Fe.
Hot work tool steel; pressing dies.
W.-Nr. 1.2365; similar to AISI H10.

REMY EPS 51.
M-1339; 0.38 C, 1.0 Si, 5.3 Cr, 1.1 Mo, 0.4 V, bal Fe.
Hot work tool steel; forging dies.
W.-Nr. 1.2343; Similar to AISI H11.

REMY EPS 51 V.
M-1339; 0.40 C, 1.0 Si, 5.2 Cr, 1.3 Mo, 1.0 V, bal Fe.
Hot work tool steel; extrusion dies, piercing mandrels, forging dies.
W.-Nr. 1.2344; Similar to AISI H13.

REMY EPS W 51.
M-1339; 0.36 C, 1.0 Si, 5.3 Cr, 1.5 Mo, 1.3 W, 0.30 V, bal Fe.
Air hardening, hot work tool steel.
W.-Nr. 1.2606; AISI H12.

REMY EXTRA 50.
M-1339; 0.45 C, 0.25-0.50 Si, 0.3-0.8 Mn, bal Fe.
Hot rolled: 98,000 TS; 59,000 YS; 24 El; 45 RA; 212 Brin.
For gears, bolts, machine tool parts; water hardened.

REMY EXTRA 60.
M-1339; 0.60 C, 0.25-0.50 Si, 0.3-0.8 Mn, bal Fe.
Heat treated: 160,000 TS; 113,000 YS; 12 El; 40 RA; 325 Brin.
For gears, rails, springs, machine tool parts; water hardened.

REMY EXTRA 70.
M-1339; 0.70 C, 0.25 max Si, 0.25 max Mn, bal Fe.
Heat treated: 175,000 TS; 128,000 YS; 12 El; 37 RA; 352 Brin.
For rails, punches, crimpers, axes; Type W1; water hardened.

REMY EXTRA 80.
M-1339; 0.85 C, 0.25 max Si, 0.25 max Mn, bal Fe.
Heat treated: 190,000 TS; 145,000 YS; 12 El; 35 RA; 390 Brin.
For tools, springs, cutters, drills, taps; Type W1; water hardened.

REMY EXTRA 100.
M-1339; 1.0 C, 0.25 max Si, 0.25 max Mn, bal Fe.
Annealed: 100,000 TS; 53,000 YS; 21 El; 42 RA; 200 Brin.
For drills, taps, cutters, springs; Type W1; water hardened.

REMY EXTRA 120.
M-1339; 1.15 C, 0.25 max Si, 0.25 max Mn, bal Fe.
Annealed: 110,000 TS; 56,000 YS; 18 El; 40 RA; 210 Brin.
For springs, taps, reamers, cutters, hobs; Type W1; water hardened.

REMY EXTRA 130.
M-1339; 1.3 C, 0.25 max Si, 0.25 max Mn, bal Fe.
For engravers' tools, taps, drills, cutters; Type W1; water hardened.

REMY EXTRA EDEL 7.
M-1339; 0.70 C, 0.25 max Si, 0.25 max Mn, bal Fe.
Heat treated: 175,000 TS; 128,000 YS; 12 El; 37 RA; 355 Brin.
For rails, tools, punches, axes, crimpers; Type W1; water hardened.

REMY EXTRA EDEL 8.
M-1339; 0.85 C, 0.25 max Si, 0.25 max Mn, bal Fe.
Heat treated: 190,000 TS; 145,000 YS; 10 El; 30 RA; 400 Brin.
For cutters, drills, taps, springs, reamers; Type W1; water hardened.

REMY EXTRA EDEL 10.
M-1339; 1.0 C, 0.25 max Si, 0.25 max Mn, bal Fe.
Annealed: 100,000 TS; 53,000 YS; 20 El; 40 RA; 200 Brin.
For springs, tools, drills, taps, broaches; Type W1; water hardened.

REMY EXTRA EDEL 10 SPEZIAL.
M-1339; 1.25 C, 1.15 Si, 1.2 Cr, 0.7 Mn, bal Fe.
For bearings, dies, cutters; water or oil hardened.

REMY EXTRA EDEL 12.
M-1339; 1.1 C, 0.25 max Si, 0.25 max Mn, bal Fe.
Annealed: 110,000 TS; 55,000 YS; 18 El; 40 RA; 210 Brin.
For springs, tools, cutters, taps, reamers, drills; Type W1; water hardened.

REMY KP.
M-1339; 0.50 C, 1.05 Cr, 3.25 Ni, bal Fe.
For gears, bolts, crankshafts; oil hardened, shock resistant.

REMY LDZ SPEZIAL.
M-1339; C, alloy, bal Fe.
For machine tool parts; oil hardened.

REMY ONLK.
M-1339; 0.45 C, 1.4 Cr, 0.7 Mo, 0.7 Mn, bal Fe.
For dies, gears, bolts, crankshafts; oil hardened, tough.

REMY P6R.
M-1339; 0.53 C, 0.90 Si, 0.90 Mn, bal Fe.
For gears, bolts, upsetters, shears, crimpers; water or oil hardened, tough.

REMY S30.
M-1339; 0.70 C, 1.7 Si, 0.70 Mn, bal Fe.
Heat treated: 340,000 TS; 283,000 YS; 5 El; 20 RA; 600 Brin.
For punches, chisels, springs, shear blades; oil hardened.

REMY SPEZIAL 1 GESTO.
M-1339; 1.05 C, 0.07 Mn, 0.25 Si, 1.1 Cr, bal Fe.
For bearings, liners, sleeves, cutters; water or oil hardened, wear resistant.

REMY SPEZIAL 2.
M-1339; 1.45 C, 0.60 Mn, 1.4 Cr, bal Fe.
For bearings, cutters, blanking and forming dies oil hardened, wear and abrasion resistant.

REMY SPEZIAL 3.
M-1339; 1.05 C, 1.0 Cr, 1.15 W, bal Fe.
For heading and blanking dies; oil hardened.

REMY SPEZIAL 4.
M-1339; 1.4 C, 0.30 Cr, 0.10 V, 0.30 Mn, bal Fe.
For blanking and forming dies, engravers tools; water hardened, wear resistant.

REMY SPEZIAL 6.
M-1339; 0.35 C, 1.05 Cr, 0.18 V, 1.85 W, 0.9 Si, bal Fe.
For forging and heading dies, die casting dies; oil hardened, tough.

REMY SPEZIAL 7.
M-1339; 0.45 C, 0.9 Si, 1.05 Cr, 0.18 V, 1.85 W, bal Fe.
For forging and heading dies; oil hardened, tough.

REMY SPEZIAL 8.
M-1339; 1.05 C, 1.2 Cr, bal Fe.
For cutters, bearings, liners, sleeves; oil or water hardened.

REMY SPEZIAL 12.
M-1339; 1.25 C, 1.15 Si, 0.70 Mn, 1.2 Cr, bal Fe.
For bearings, cold working tools, liners, sleeves; oil hardened.

REMY SPEZIAL 13.
M-1339; 1.05 C, 1.0 Cr, 0.30 Mn, 0.25 Si, bal Fe.
For bearings, cold working tools, liners, sleeves; water hardened, wear resistant.

REMY SPEZIAL 14.
M-1339; 0.50 C, 0.95 Mn, 1.05 Cr, 0.1 V, bal Fe.
For springs, gears, bolts, punches; oil hardened, shock resistant.

REMY SPEZIAL 14H.
M-1339; 0.58 C, 1.0 Cr, 0.09 V, 0.95 Mn, bal Fe.
For springs, gears, bolts, crankshafts; oil hardened, shock resistant.

REMY WP4.
M-1339; 0.30 C, 1.1 Cr, 0.18 V, 3.75 W, 1.0 Si, bal Fe.
For punches, forging dies, upsetters; oil hardened.

REMY WP12.
M-1339; 0.30 C, 2.4 Cr, 2 Co, 0.25 V, 8.5 W, bal Fe.
For extrusion rams, liners, punches; hot work steel, oil hardened.

REMY WPK45.
M-1339; 0.30 C, 2.35 Cr, 0.6 V, 4.2 W, bal Fe.
For extrusion rams and liners; oil hardened, tough.

REMY WZK SPEZIAL.
M-1339; 0.30 C, 0.35 V, 2.65 Cr, 8.5 W, bal Fe.
For extrusion rams and liners, punches, shears; hot work steel, oil hardened.

RENAL.
M-1359; 99.99 Al.
Soft: 13,000 TS; 5000 YS; 45 El; 23 Brin.
For electrical conductors; high electrical conductivity.

RENAULT ALLOY.
M-French; 88 Al, 2 Cu, 10 Zn.
For light alloy parts; non-hardenable.

RENAULT AT.2.
M-689; 0.8 C, 0.3 Ni, 5 Cr, 0.7 Mo, 18.5 W, 1.3 V, bal Fe.
For broaches, cutters, drills; high speed steel.

RENAULT AT.3.
M-689; 0.8 C, 0.3 Si, 0.3 Ni, 4.5 Cr, 6 Mo, 6.7 W, 1.6 V, bal Fe.
For tools, cutters, reamers, broaches; high speed steel.

RENAULT AT.4.
M-689; 0.8 C, 0.2 Ni, 5 Cr, 1.4 Mo, 19.5 W, 2 V, 5.2 Co, bal Fe.
For tools, cutters, hobs, reamers; high speed steel.

RENAULT AT.7.
M-689; 1.2 C, 0.35 Mn, 0.2 Ni, 1.35 Cr, bal Fe.
For tools; water hardening.

RENAULT AT.9.
M-689; 0.4 C, 0.4 Mn, 4.9 Ni, 0.4 Cr, 1.2 Mo, bal Fe.
For gears, shafts; tough.

RENAULT AT.12.
M-689; 1.9 C, 0.3 Mn, 0.7 Ni, 13 Cr, 0.3 Mo, bal Fe.
For tools, dies; nondeforming.

RENAULT AT.13.
M-689; 1.9 C, 0.4 Mn, 0.5 Ni, 12.5 Cr, 0.6 Mo, 0.8 W, 3.2 Co, bal Fe.
For tools, dies; nondeforming.

RENAULT AT.14.
M-689; 0.58 C, 0.67 Mn, 1.45 Ni, 0.67 Cr, 0.3 Mo, 0.12 V, bal Fe.
For forging dies, tools; oil hardening.

RENAULT AT.25.
M-689; 1.03 C, 1.82 Mn, 0.1 Ni, 0.13 V, 0.1 Cr, bal Fe.
For tools, dies; nondeforming.

RENAULT AT.26.
M-689; 0.8 C, 0.3 Mn, 0.1 Ni, bal Fe.
For tools, drills, hobs, springs, punches; water hardened.

RENAULT AT.35.
M-689; 0.3 C, 0.6 Mn, 0.6 Ni, 3.3 Cr, 0.5 Mo, 12 W, 0.1 V, 2.5 Co, bal Fe.
For dies, tools; hot work steel.

RENAULT G.
M-689; 0.45 C, 1.15 max Si, 0.7 Mn, 14 Ni, 14 Cr, 0.3 Mo, 2.25 W, bal Fe.
For exhaust valves; heat resistant.

RENAULT H.
M-689; 0.34 C, 0.4 Si, 0.75 Mn, 1.4 Ni, 0.92 Cr, bal Fe.
For gears, shafts, pinions; tough.

RENAULT P.
M-689; 1.25 C, 0.55 Mn, 0.45 Ni, 14 Cr, 1.2 Mo, 1.2 Co, bal Fe.
For valves; heat resistant.

RENAULT R.
M-689; 0.55 C, 0.35 max Si, 0.45 Mn, 0.4 Ni, 14 Cr, bal Fe.
For pump parts; corrosion resistant.

RENAULT S.
M-689; 0.14 C, 0.75 Mn, 1.4 Ni, 1 Cr, bal Fe.
For gears, shafts; case-hardening.

RENAULT WA.
M-689; 0.4 C, 0.7 Mn, 0.5 max Ni, 1 Cr, 0.3 Mo, bal Fe.
For gears, shafts; oil hardening.

RENAULT WB.
M-689; 0.35 C, 0.7 Mn, 0.4 max Ni, 1 Cr, 0.3 Mo, bal Fe.
For gears, shafts; oil hardening.

RENAULT WE.
M-689; 0.36 C, 0.7 Mn, 0.4 max Ni, 1 Cr, 0.3 Mo, bal Fe.
For gears, shafts; oil hardening.

RENAULT WZ.
M-689; 0.42 C, 0.7 Mn, 0.5 Ni, 1 Cr, 0.3 Mo, bal Fe.
For gears, shafts; oil hardening.

RENE 41.
M-1491 (and others); 0.12 C, 19 Cr, 11 Co, 10 Mo, 3 Ti, 1.5 Al, bal Ni.
Ht. Tr.: 160,000-206,000 TS; 120,000-154,000 YS; 14-18 El.
For jet engine components, after-burners, turbine wheels, bolts.
Age-hardenable, high heat and corrosion resistance.

RENE 41 see also **ALLVAC RENE 41, CRUCIBLE RENE 41.**

RENE 41 WIRE.
M-1507; 0.06-0.12 C, 18.0-20.0 Cr, 9.0-10.5 Mo, 5.0 Fe, 10.0-12.0 Co, 0.5-1.8 Al, 3.0-3.3 Ti, 0.010 max B, 53 Ni.
For high temperature springs in 1100-1500°F range.

RENE 62.
1.1-1.4 Al, 2.1-2.5 Cb, 2.35-2.63 Ti, 8.5-9.5 Mo, 13.5-16.5 Cr, 21.0-24.0 Fe, 0.02-0.08 C, 0.25 max Mn, 0.01 B, bal Ni.
At R.T.: 205,000 TS; 160,000 YS; 6 El.
At 1400°F: 145,000 TS; 117,000 YS; 20 El.
For turbine components, jet engines, turbine wheels and frames.
Precipitation hardening, good weldability.
High corrosion, heat and oxidation resistance.

RENE 63.
0.10 C, 14 Cr, 15 Co, 6 Mo, 3 W, 2.5 Ti, 3.8 Al, 0.015 B, bal Ni.
Ht. Tr.: 210,000 TS; 148,000 YS; 18 El.
At 1400°F: 155,000 TS; 138,000 YS; 4-9 El.
For jet engine and gas turbine parts, high temperature bolting.
High rupture strength. Corrosion, heat and oxidation resistant.

RENE 77.
M-1491; 15 Cr, 5 Mo, 3.5 Ti, 4.4 Al, 0.02 B, 0.06 C, 15 Co, 0.04 Z , bal Ni.
Cast: 125,000 TS; 110,000 YS; 8 RA.
At 1600°F: 85,000 TS; 75,000 YS; 12 RA.
For jet engine turbine blades.
High creep-rupture strength to 1800°F.
Resists hot sulfidation corrosion.

RENE 80.
M-1491; 14 Cr, 9.5 Co, 4 Mo, 4 W, 5 Ti, 3 Al, 0.17 C, 0.03 Zr, 0.015 B, bal Ni.
Cast alloy; for turbine blades.
Very good hot corrosion resistance.

RENE 85.
M-1491; 0.27 C, 9.3 Cr, 15.0 Co, 3.25 Mo, 5.35 W, 3.25 Ti, 5.25 Al, 0.015 B, 0.03 Zr, bal Ni.
Compressor disc alloy, cast.

RENE 95.
M-1491; 0.015 C, 14 Cr, 8.0 Co, 3.5 Mo, 3.5 W, 3.5 Cb, 2.5 Ti, 3.5 Al, 0.01 B, 0.05 Zr, bal Ni.
Turbine or compressor disc alloy, cast.

RENE 100.
M-1491; 4.0-4.4 Ti, 5.3-5.7 Al, 9-10 Cr, 14-16 Co, 2.7-3.3 Mo, 1 max Fe, 0.02 B, 0.2 C, 1.0 V, 0.03-0.09 Zr, bal Ni.
Cast: 147,000 TS; 123,000 YS; 9 El; C 30-44 Rock.
At 1500°F: 144,000 TS; 118,000 YS; 6 El; 7.2 RA.
For turbine blades operating to 1900°F.
Excellent high temperature rupture strength and long-time stability.

RENE 125.
M-1491; 0.10 C, 9.0 Cr, 10.0 Co, 2.0 Mo, 7.0 W, 2.6 Ti, 4.8 Al, 0.015 B, 0.05 Zr, 3.8 Ta, 1.6 Hf, bal Ni.
Turbine blade alloy.

RENNITE.
M-425; 0.9 C, 3.5 W, bal Fe.
For cutting tools; oil hardened.

RENNITE.
M-790; 0.7 C, 4 Cr, 18 W, 2 V, 12 Co, bal Fe.
For lathe and planer tools, hobs, taps, reamers, broaches; high speed steel, oil hardened.

RENYX, AL BASE.
M-125; 91.5 Al, 4 Ni, 4 Cu, 0.5 Si.
For die castings; corrosion resisting.

RENYX, ZN BASE.
M-125; 4 Al, 3 Cu, 92 Zn, 0.1 Mn.
For die castings; corrosion resisting.

REO EXTRA.
M-1705; 1.0 C, 0.35 Mn, 0.2 Si, 0.2 V, bal Fe.
Water hardening tool steel. AISI W2.

REO SPECIAL (W2, GRADE 1).
M-1705; 1.0 C, 0.35 Mn, 0.2 Si, bal Fe.
Water hardening tool steel, AISI W2.

REOSTENE.
M-Eng.; Ni, Fe.
For resistances; heat resistant.

REO (W2, GRADE 3).
M-1705; 1.0 C, 0.35 Mn, 0.2 Si, bal Fe.
Water hardening tool steel; AISI W2.

REPAROX XFC.
M-717.
Electrode for torch joining steel, cast iron, copper base alloys; bonding temp: 1400-1600°F; 65,000 psi TS.

REPUBLIC 15-5PH.
M-97; 0.07 C, 15.0 Cr, 4.5 Ni, 3.5 Cu, bal Fe.
Precipitation hardenable martensitic stainless.
Ht: 145-200 ksi TS; 125-185 ksi YS; 10-15 El.
For springs and high strength parts.

REPUBLIC 17-4PH.
M-97; 0.07 max C, 15.5-17.5 Cr, 3.0-5.0 Ni, 0.15-0.45 Cb + Ta, 3.0-5. Cu, bal Fe.
H900-temper: 200,000 TS; 180,000 YS; 14 El; 50 RA; 432 Brin.
H925-temper: 190,000 TS; 175,000 YS; 14 El; 54 RA; 409 Brin.
For valves, ball bearings, pump gears, fasteners; stainless, heat treatable, martensitic.

REPUBLIC 17-7PH.
M-97; 0.09 max C, 16.0-18.0 Cr, 6.5-7.75 Ni, 0.75-1.5 Al, bal Fe.
Condition A: 130,000 TS; 40,000 YS; 30 El; 180 Brin.
Condition RH950: 200,000 TS; 175,000 YS; 10 El; 30 RA; 440 Brin.
For conveyor chains, bearings, saw blades, pressure tanks, airframes; age hardenable, corrosion and heat resistant.

REPUBLIC 35.
M-97; 0.12 C, 0.75 Mn, 0.10 Si, Cb & V, bal Fe.
Rolled: 47,000 TS; 35,000 YS; 30 El.
Structural equipment, transportation equipment; weldable, tough.

REPUBLIC 50.
M-97; 0.12 max C, 0.5-1.0 Mn, 0.5-1.0 Cu, 0.5-1.1 Ni, 0.10 min Mo, bal Fe.
Rolled: 70,000 TS; 50,000 YS; 22 El.
For agricultural implements, mine and railroad cars, truck and trailer bodies; tough, good weldability.

REPUBLIC 60.
M-97; 0.15 C, 0.5-1.0 Mn, 0.3-1.0 Cu, 0.10 min Mo, 0.3 Cr, 0.4-1.1 Ni, bal Fe.
Rolled: 70,000 TS; 50,000 YS; 22 min El.
For welded structures, truck bodies, freight cars, mine cars; tough, good weldability.

REPUBLIC 70.
M-97; 0.20 C, 1.0 Mn, 0.25 Si, 1.15 Cu, 1.5 Ni, 0.22 Mo, bal Fe.
Rolled: 96,000 TS; 73,000 YS; 17 El.
For welded structures, truck trailers, mine cars; tough, good weldability.

REPUBLIC 80.
M-97; 0.20 C, 1.0 Mn, 1.0-1.5 Cu, 0.2-0.3 Mo, 1.2-1.75 Ni, Cb, V, bal Fe.
Rolled: 100,000 TS; 80,000 YS; 18 min El.
For transportation equipment, truck bodies and truck trailers, railroad hopper cars; tough, weldable.

REPUBLIC 98BV40.
M-97; 0.40-0.46 C, 0.75-1.0 Mn, 0.50-0.80 Si, 0.60-0.90 Ni, 0.80-1.05 Cr, 0.45-0.60 Mo, 0.01-0.06 V, 0.007 max B, bal Fe.
Ultra high strength steel.

REPUBLIC 100-AR.
M-97; 0.16 C, 1.40 Mn, 0.70 Cu, 1.40 Cr, 1.40 Ni, 0.45 Mo, 0.04 Al, bal Fe.
As rolled: 140 ksi TS; 100 ksi YS; 16-18 El.
High strength low alloy steel for structural purposes.

REPUBLIC 300M.
M-97; 0.38-0.43 C, 0.60-0.90 Mn, 1.5-1.8 Si, 1.65-2.0 Ni, 0.70-0.95 Cr, 0.30-0.50 Mo, 0.05-0.10 V, bal Fe.
Alloy ultra high strength steel; vacuum carbon deoxidized.

REPUBLIC 4330M.
M-97; 0.28-0.33 C, 0.80-1.1 Mn, 0.20-0.35 Si, 1.60-2.0 Ni, 0.75-0.95 Cr, 0.35-0.50 Mo, 0.05-0.10 V, bal Fe.
Alloy ultra high strength steel.

REPUBLIC 4350.
M-97; 0.48-0.53 C, 0.65-0.85 Mn, 0.20-0.35 Si, 1.65-2.0 Ni, 0.70-0.90 Cr, 0.20-0.30 Mo, bal Fe.
Alloy ultra high strength steel.

REPUBLIC 550.
M-97; 0.4-0.5 C, 0.6-0.7 Mn, 1.5-1.65 Cr, 0.9-1.2 Mo, 0.3 V, bal Fe.
For hot bolts and rivet headers; hot work steel.

REPUBLIC A-286.
M-97; 0.08 max C, 13.5-16.5 Cr, 24.0-27.0 Ni, 1.0-1.5 Mo, 1.9-2.35 Ti 0.01 B, 0.35 max Al, 0.1-0.5 V, bal Fe.
Condition A: 93,000 TS; 36,000 YS; 48 El; 70 RA; 143 Brin.
Condition STA: 145,000 TS; 100,000 YS; 24 El; 37 RA; 295 Brin.
For jet engine frames and casings, afterburner parts, fasteners; age hardenable, high heat resistance to 1200°F.

REPUBLIC A-440.
M-97; 0.25 C, 1.1-1.6 Mn, 0.20 min Cu, 0.30 Si, bal Fe.
As rolled: 75 ksi min TS; 50,000 min YS; 20 El.
High strength low alloy structural steel.

REPUBLIC A 441.
M-97; 0.22 max C, 1.2 Mn, 0.02 V, 0.2 Cu, bal Fe.
Rolled: 70,000 min TS; 50,000 min YP; 22 El.
For structural members, mine cars, booms, cranes, auto and bus bodies.
High strength steel and weldability.

REPUBLIC COR-TEN.
M-97; 0.12 max C, 0.2-0.5 Mn, 0.07-0.15 P, 0.25-0.75 Si, 0.25-0.55 Cu 0.65 max Ni, 0.5-1.25 Cr, bal Fe.
Hot rolled: 70,000 TS; 50,000 YS; 22 El.
For transport bodies; atmospheric corrosion resistant.

REPUBLIC D6AC.
M-97; 0.42-0.48 C, 0.60-0.90 Mn, 0.15-0.30 Si, 0.40-0.70 Ni, 0.90-1.20 Cr, 0.90-1.10 Mo, 0.05-0.10 V, bal Fe.
Alloy ultra high strength steel; vacuum carbon deoxidized.

REPUBLIC EXTRA SPECIAL TRANSFORMER.
M-97; Si, bal Fe.
For transformers; high permeability.

REPUBLIC H-11.
M-97; 0.38-0.43 C, 0.20-0.40 Mn, 0.80-1.0 Si, 4.75-5.25 Cr, 1.2-1.4 Mo, 0.4-0.6 V, bal Fe.
Alloy ultra high strength steel; also as hot work tool steel; vacuum carbon deoxidized.

REPUBLIC HP9-4-20.
M-97; 0.16-0.23 C, 0.2-0.4 Mn, 8.5-9.5 Ni, 0.65-0.85 Cr, 0.9-1.1 Mo, 0.06-0.12 V, 4.25-4.75 Co, bal Fe.
Heat treated: 200,000-250,000 TS; 190,000-220,000 TYS; 190,000 CYS 12-15 El; 55 RA.
For solid fuel rocket motor cases, submarine hulls, pressure vessels, aircraft and aerospace structures. Tough, shock resistant, good weldability.

REPUBLIC HP9-4-30.
M-97; 0.29-0.34 C, 0.15-0.35 Mn, 0.10 max Si, 0.01 max S & P, 7-8 Ni 0.9-1.1 Cr, 0.9-1.1 Mo, 0.006-0.12 V, 4-5 Co, bal Fe.
Plate: 230,000 TS; 210,000 YS; 15 El; 40 RA.
For solid fuel rockets, motor cases, submarine hulls, pressure vessels, aircraft structural components. Weldable and formable.
High strength and toughness. Heat treatable.

REPUBLIC LOCORE M22.
M-97; 3.5 Si, bal Fe.
For transformers; high permeability.

REPUBLIC LOCORE M27.
M-97; 2.4 Si, bal Fe.
For electrical equipment; high permeability.

REPUBLIC LOCORE M45.
M-97; 1 Si, bal Fe.
For electric motors; high permeability.

REPUBLIC LOCORE M47.
M-97; 0.50 Si, bal Fe.
For armatures; high permeability.

REPUBLIC M.
M-97; 0.25 max C, 1.1-1.6 Mn, 0.30 max Si, 0.20 min Cu, bal Fe.
Hot rolled: 75,000 TS; 50,000 YS; 20 El.
For earth moving equipment, trucks and trailers, cars; low alloy-high strength steel.

REPUBLIC NAX.
M-97; 0.08-0.15 C, 0.5-0.7 Mn, 0.6-0.9 Si, 0.5-0.6 Cr, 0.5-0.15 Zr, bal Fe.
Rolled: 70,000 TS; 50,000 YS; 150 Brin.
For railroad and mine cars, truck bodies; good weldability and toughness.

REPUBLIC PH13-8MO.
M-97; 0.05 max C, 0.10 max Mn, 12.75 Cr, 8.0 Ni, 1.2 Al, 2.2 Mo, 0.01 max N, bal Fe.
Ht: 170-225 ksi TS; 150-205 ksi YS; 12-16 El.
Precipitation hardening martensitic stainless.

REPUBLIC PH15-7MO.
M-97; 0.09 max C, 14.0-16.0 Cr, 6.5-7.75 Ni, 0.75-1.5 Al, 2.0-3.0 Mo, bal Fe.
Condition A: 150,000 TS; 65,000 YS; 25 El; 220 Brin.
Condition RH950: 225,000 TS; 200,000 YS; 2-5 El; 460 Brin.
For conveyor chains, bearings, pressure tanks, airframes; age hardenable, corrosion and heat resistant.

REPUBLIC SILICON, LOCORE M36.
M-97; 3.0 Si, bal Fe.
For electrical equipment; high permeability.

REPUBLIC SILICON, LOCORE M-43.
M-97; 0.25 Si, bal Fe.
For armatures, electrical equipment; high permeability.

REPUBLIC SPECIAL ALLOY TYPE "D".
M-97; 0.4 C, 1.5 Ni, 0.7 Cr, bal Fe.
Oil treated: 185,000-260,000 TS; 165,000-240,000 YS; 15-6 El; 45-17 RA.
For drive shafts, spindles; tough.

REPUBLIC SPECIAL TRANSFORMER.
M-97; 3.5 Si, bal Fe.
For transformers; high permeability.

REPUBLIC X-42-W.
M-97; 0.21 max C, 1.25 Mn, 0.30 Si, Cd, V, bal Fe.
Rolled: 60,000 TS; 42,000 YS; 24 El.
For structural work, transportation equipment, heavy containers. Weldable formable.

REPUBLIC X-45-W.
M-97; 0.22 max C, 1.25 Mn, 0.30 Si, Cb, V, bal Fe.
Rolled: 60,000 TS; 45,000 YS; 22 El.
For structural equipment, transportation equipment, heavy containers.
Weldable, formable.

REPUBLIC X-50-W.
M-97; 0.22 max C, 1.35 Mn, 0.30 Si, Cb, V, bal Fe.
Rolled: 65,000 TS; 50,000 YS; 22 El.
For truck and trailer bodies, railroad cars, ordnance equipment. Weldable formable.

REPUBLIC X-55-W.
M-97; 0.25 max C, 1.35 Mn, 0.30 Si, Cb, V, bal Fe.
Rolled: 70,000 TS; 55,000 YS; 20 El.
For structural work, cranes, booms, mine cars, bus and auto bodies. Weldable, formable.

REPUBLIC X-60-W.
M-97; 0.26 max C, 1.35 Mn, 0.30 Si, Cb, V, bal Fe.
Rolled: 75,000 TS; 60,000 YS; 18 El.
For structural parts, earth moving equipment, truck and tractor parts, auto and bus bodies. Weldable, formable.

REPUBLIC X-65-W.
M-97; 0.26 max C, 1.35 Mn, 0.30 Si, Cb, V, bal Fe.
Rolled: 80,000 TS; 65,000 YS; 16 El.
For structural work, railroad cars, bridge and building construction, cranes and booms. Weldable, formable.

REPUBLIC X-70-W.
M-97; 0.26 max C, 1.65 Mn, 0.30 Si, Cb, V, bal Fe.
Rolled: 85,000 TS; 70,000 YS; 14 El.
For structural equipment, earth moving equipment, railroad hopper cars, truck and trailer bodies. Weldable, formable.

RESISCO.
M-282; 91 Cu, 7 Al, 2 Ni.
Drawn: 100,000 TS; 85,000 YS; 11 El; 205 Brin.
For condensers, coolers; high corrosion resistant.

RESISCO (ALUMINUM-BRONZE).
M-286; 7 Al, bal Cu.
Ann: 68,100 psi TS; 27,500 psi YS; 60 El; 110 DPN (106 Brin).
For condenser tube.

RESISTAC.
M-129; 90 Cu, 9 Al, 1 Fe.
Cast: 75,000 TS; 37,000 YS; 15 El; 20 RA; 135 Brin.
Acid resisting castings and forgings; abrasion resisting.

RESISTAC.
M-129; 88 Cu, 10 Al, 2 Fe.
For chemical apparatus; Al-Bronze.

RESISTAC.
M-129; 89 Cu, 10 Al, 1 Fe.
Hot rolled: 100,000 TS; 60,000 YS; 6 El; 6 RA.
For equipment, chemical apparatus; Al-Bronze.

RESISTAC NO. 1.
M-129; 8-11 Al, 5 max Fe, 5 max Ni, bal Cu.
Cast or rolled: 65,000 TS; 28,000 YS; 20 El; 20 RA; 120 Brin.
For castings, gears; corrosion resistant.

RESISTAC NO. 2.
M-129; 8-11 Al, 5 max Fe, 5 max Ni, bal Cu.
Cast or rolled: 75,000 TS; 32,000 YS; 25 El; 25 RA; 150 Brin.
For castings, gears; corrosion resistant.

RESISTAC NO. 3.
M-129; 8-11 Al, 5 max Fe, 5 max Ni, bal Cu.
Cast: 90,000 TS; 42,000 YS; 15 El; 15 RA; 185 Brin.
Forged: 95,000 TS; 45,000 YS; 17 El; 17 RA; 195 Brin.
For castings, gears, pumps, impellers; corrosion resistant.

RESISTAL.
M-Eng.; 90 Cu, 9 Al, 1 Fe.
75,000 TS; 37,000 YS; 20 El.
For gears, slides; corrosion resistant.

RESISTALOY.
M-122; 59 Cu, 2 Al, 1 Ni, bal Zn.
Cold drawn: 112,750 TS; 85,000 YS; 8.5 El; 11.0 RA; 205 Brin.
For shafts, bearings, bolts, nuts, studs, high tensile forgings; resists sea water corrosion.

RESISTALOY NO. 12.
M-122; 59 Cu, 1 Ni, 2 Al, bal Zn.
Rolled: 95,000 TS; 58,000 YS; 190 Brin.
For propeller shafts, boat hardware, marine applications; nickel brass.

RESISTANCE NO. 1.
67 Ag, 33 Pt.
For resistances; heat resistant.

RESISTANCE NO. 2.
85 Cu, 12 Mn, 3 Fe.
For resistances; heat resistant.

RESISTANCE NO. 3.
57 Cu, 26 Zn, 18 Ni.
For resistances; corrosion resistant.

RESISTANCE NO. 4.
56 Cu, 26 Ni, 18 Zn, 1 Fe.
For resistances; corrosion resistant.

RESISTANCE, HIGH MAGNETIC.
70 Cr, 30 Ni.
For magnet; high permeability.

RESISTANCE, LUNGE.
M-Eng.; 87-84 Cu, 12-14 Mn, 1.8-1.9 Fe.
For electrical resistors.

RESISTIN-1.
M-Eng.; 1.8 Fe, 11.7 Mn, 86.5 Cu.
For resistance alloy.

RESISTIN-2.
M-Eng.; 3 Fe, 12 Mn, 85 Cu.
For resistance alloy.

RESISTO-CAST.
M-507; 3.5 C, 3.2 Si, 0.6 Mn, 0.067 S, 0.41 P, 0.29 Cr, 1.33 Ni, 0.2 Mo, 0.15 Cu, bal Fe.
For oxy-acetylene welding repair in cast iron castings.

RESISTO-LOY.
M-507; 3.5 C, 0.46 Si, 1.0 Mn, 29.3 Cr, 4.0 Mo, 0.3 Cu, bal Fe.
Coated, hardfacing electrode.
As arc welded: 100,000 psi approx.; 58-60 Rc.
For overlay or buildup on shovel teeth, dredging tools, plow shares, cement mill rings; valves for handling corrosive chemicals; good resistance to abrasion and corrosion.

RESOLUT 1.
M-1323; 2.1 C, 11.5 Cr, bal Fe.
For blanking and forming dies, punches; oil or air hardened, non-deforming.

RESOLUT 1E.
M-1323; 2.1 C, 11.5 Cr, 0.7 W, bal Fe.
For blanking, piercing, and forming dies, punches; oil or air hardened, non-deforming.

RESOLUT 1W.
M-1323; 1.65 C, 11.5 Cr, 0.1 V, bal Fe.
For blanking, piercing, and forming dies, punches; air hardened, non-deforming.

RESOLUT 2.
M-1323; 0.9 C, 1.9 Mn, 0.25 Si, 0.1 V, bal Fe.
For punches, cutters, crimpers, forming dies; oil hardened, non-deforming.

RESOLUT 3.
M-1323; 1.25 C, 1.15 Si, 0.7 Mn, 1.2 Cr, bal Fe.
For bearings, liners, cutters; water hardened, wear resistant.

RESOLUT 4.
M-1323; 1.45 C, 1.4 Cr, 0.6 Mn, bal Fe.
For bearings, bushings, liners, cutters; water hardened, wear resistant.

RETORT.
M-210; 0.30-0.35 C, 0.40-0.80 Si, 0.60-0.80 Mn, bal Fe.
For machinery parts, gears; castings.

REVALON.
M-238; 76 Cu, 22 Zn, 2 Al, 0.05 As.
Hard: 85,000 TS; 60,000 YS; 10 El.
Soft: 60,000 TS; 27,000 YS; 55 El.
For marine parts; hardware; tough.

REVERE 1100.
M-238; 99.0 min Al.
O-temper: 13,000 TS; 5000 YS; 45 El; 23 Brin.
H 18-temper: 24,000 TS; 22,000 YS; 15 El; 44 Brin.
For trim, housings, containers, chemical equipment; commercially pure aluminum.

REVERE 1145.
M-238; 99.45 min Al.
For wrappings, electrical condensers; foil.

REVERE 2014.
M-238; 0.8 Mn, 0.4 Mg, 0.8 Si, 4.4 Cu, bal Al.
O-temper: 27,000 TS; 14,000 YS; 18 El; 45 Brin.
T 4-temper: 62,000 TS; 44,000 YS; 20 El; 105 Brin.
T 6-temper: 70,000 TS; 60,000 YS; 13 El; 135 Brin.
For aircraft forgings, hardware, structural fittings; heat treatable, high strength.

REVERE 2017.
M-238; 0.5 Mn, 0.5 Mg, 4 Cu, bal Al.
O-temper: 26,000 TS; 10,000 YS; 22 El; 45 Brin.
T 4-temper: 62,000 TS; 40,000 YS; 22 El; 105 Brin.
For aircraft structures, fittings, screw machine products; heat treatable, good formability.

REVERE 2024.
M-238; 0.6 Mn, 1.5 Mg, 4.5 Cu, bal Al.
O-temper: 27,000 TS; 11,000 YS; 22 El; 47 Brin.
T 4-temper: 68,000 TS; 48,000 YS; 19 El; 120 Brin.
For aircraft structures; heat treatable.

REVERE 3003.
M-238; 1.2 Mn, bal Al.
O-temper: 16,000 TS; 6000 YS; 40 El; 28 Brin.
H 14-temper: 31,500 TS; 19,000 YS; 16 El; 40 Brin.
H 18-temper: 29,000 TS; 26,000 YS; 10 El; 55 Brin.
For structural work, tank cars, cooking utensils, heat exchangers; resists atmosphere corrosion.

REVERE 3004.
M-238; 1.2 Mn, 1 Mg, bal Al.
O-temper: 26,000 TS; 10,000 YS; 25 El; 45 Brin.
H-34 temper: 34,000 TS; 27,000 YS; 12 El; 63 Brin.
H 38-temper: 40,000 TS; 34,000 YS; 6 El; 77 Brin.
For fuel lines, fan blades, roofing, hydraulic tubing; high resistance to weather corrosion.

REVERE 3005.
M-238; 0.6 Si, 0.7 Fe, 0.3 Cu, 1.0-1.5 Mn, 0.2-0.6 Mg, 0.25 Zn, 0.10 Ti, bal Al.
Wrought aluminum alloy.
AA 3005.

REVERE 3150.
M-238; 99.45 min Al.
O-temper: 12,000 TS; 4000 YS.
H-16 temper: 18,000 TS; 16,000 YS.
For electrical conductors. Was Revere E.C.

REVERE 5005.
M-238; 0.8 Mg, bal Al.
O-temper: 18,000 TS; 6000 YS; 30 El.
H-16 temper: 26,000 TS; 25,000 YS; 5 El.
H-32 temper: 29,000 TS; 27,000 YS; 5 El.
For light alloy parts; strip and tube.

REVERE 5050.
M-238; 1.4 Mg, bal Al.
O-temper: 21,000 TS; 8000 YS; 24 El.
H-32 temper: 25,000 TS; 21,000 YS; 9 El.
H-38 temper: 32,000 TS; 29,000 YS; 6 El.
For light alloy parts; strip and tube.

REVERE 5052.
M-238; 2.5 Mg, 0.25 Cr, bal Al.
O-temper: 27,000 TS; 12,000 YS; 30 El; 45 Brin.
H 34-temper: 37,000 TS; 31,000 YS; 14 El; 67 Brin.
H 38-temper: 41,000 TS; 36,000 YS; 8 El; 85 Brin.
For camera cases, deck housings, gasoline tanks, aircraft components; resists sea water corrosion.

REVERE 6061.
M-238; 1 Mg, 0.25 Cr, 0.6 Si, 0.25 Cu, bal Al.
O-temper: 18,000 TS; 8000 YS; 30 El; 30 Brin.
T 4-temper: 35,000 TS; 21,000 YS; 25 El; 65 Brin.
T 6-temper: 45,000 TS; 40,000 YS; 17 El; 95 Brin.
For structures, marine equipment, engine baffles; heat treatable, good weldability.

REVERE 6062.
M-238; 0.25 Cu, 0.6 Si, 1.0 Mg, 0.06 Cr, bal Al.
O-temper: 18,000 TS; 8000 YS; 30 El.
T-4 temper: 35,000 TS; 21,000 YS; 25 El.
T-6 temper: 45,000 TS; 40,000 YS; 7 El.
For extrusions; age-hardened.

REVERE 6063.
M-238; 0.7 Mg, 0.4 Si, bal Al.
F-temper: 22,000 TS; 15,000 YS; 20 El; 42 Brin.
T 5-temper: 30,000 TS; 25,000 YS; 12 El; 65 Brin.
T 6-temper: 35,000 TS; 30,000 YS; 12 El; 73 Brin.
For light alloy parts, moldings, trim, architecture; heat treatable, good corrosion resistance.

REVERE 6101.
M-238; 0.4 max Fe, 0.3-0.6 Si, 0.4-0.8 Mg, bal Al.
Rolled: 29,000 TS; 25,000 YS.
For light alloy parts; corrosion resistant.

REVERE ALLOY NO. 102.
M-238; 99.96 Cu.
Hard: 50,000 TS; 48,000 YS; 5 El.
Soft: 32,000 YS; 8000 YS; 40 El.
For commercial products, deep drawn parts; oxygen-free. Was Revere 107.

REVERE ALLOY NO. 116.
M-238; 99.9 Cu, Ag.
Hard: 52,000 TS; 48,000 YS; 5 El; 83 Brin.
Soft: 33,000 TS; 7000 YS; 40 El; 75 Brin.
For automobile radiators; silver bearing copper. Was Revere 103.

REVERE ALLOY NO. 220A.
M-238; 90 Cu, 10 Zn.
Hard: 60,000 TS; 55,000 YS; 20 El; B 60 Brin.
Soft: 40,000 TS; 10,000 YS; 50 El; F 55 Brin.
For costume jewelry, screws, forgings; commercial bronze. Was Revere 121.

REVERE ALLOY NO. 226.
M-238; 87 Cu, bal Zn.
For jewelry, ornaments; Red Brass 87%. Was Revere 126.

REVERE ALLOY NO. 230.
M-238; 85 Cu, 15 Zn.
Hard: 57,000 TS; 52,000 YS; 23 El; B 75 Brin.
Soft: 40,000 TS; 10,000 YS; 55 El; F 55 Brin.
For jewelry, name plates, tubing, hardware; red brass. Was Revere 130.

REVERE ALLOY NO. 260.
M-238; 70 Cu, 30 Zn.
Hard: 70,000 TS; 52,000 YS; 30 El; B 80 Brin.
Soft: 48,000 TS; 16,000 YS; 65 El; F 65 Brin.
For cartridge cases, stampings; cartridge brass. Was Revere 160.

REVERE ALLOY NO. 268.
M-238; 66 Cu, 34 Zn.
Hard: 74,000 TS; 60,000 YS; 8 El; 80 B Brin.
Soft: 47,000 TS; 15,000 YS; 62 El; 64 F Brin.
For rivets, eyelets, electrical sockets; yellow brass. Was Revere 170.

REVERE ALLOY NO. 270.
M-238; 65 Cu, bal Zn.
Hard: 90,000 TS; 55,000 YS; 15 El.
Soft: 42,000 TS; 13,000 YS; 60 El.
For rivets, fasteners; high strength. Was Revere 170A.

REVERE ALLOY NO. 280.
M-238; 60 Cu, bal Zn.
Hard: 80,000 TS; 57,000 YS; 20 El.
Soft: 52,000 TS; 22,000 YS; 48 El.
For brazing; low M.P. Was Revere 182.

REVERE ALLOY NO. 280C.
M-238; 61.3 Cu, 0.25 Pb, bal Zn.
For stamped parts; yellow brass. Was Revere 238.

REVERE ALLOY NO. 310.
M-238; 5 Sn, 0.2 P, bal Cu.
Spring Temper: 91,000-105,000 TS; 4 El; B 95 Rock.
Soft Temper: 40,000-55,000 TS; 57-64 El.
For diaphragms, bellows, springs, fuse clips, lock washers, clutch discs.
High fatigue and corrosion resistance.
Grade A Phosphur Bronze. Was Revere 308.

REVERE ALLOY NO. 317.
M-238; 38 Zn, 2 Pb, bal Cu.
Soft: 52,000 TS; 20,000 YS; 45 El; 72 Brin.
For hardware, plumbing goods, forgings; extruded, forging brass. Was Revere 280.

REVERE ALLOY NO. 325.
M-238; 73 Cu, 24.5 Zn, 2.5 Pb.
Hard: 80,000 TS; 75,000 YS; 5 El.
Soft: 50,000 TS; 15,000 YS; 55 El.
For hardware, machinery parts; free-cutting. Was Revere 215.

REVERE ALLOY NO. 332.
M-238; 66.5 Cu, 1.6 Pb, bal Zn.
For bushings, screw machine products; tubing, free-machining. Was Revere 222.

REVERE ALLOY NO. 340.
M-238; 64.5 Cu, 1 Pb, bal Zn.
Hard: 90,000 TS; 66,000 YS; 5 El.
Soft: 47,000 TS; 12,000 YS; 60 El.
For hardware, bolts, gears; free-cutting. Was Revere 227.

REVERE ALLOY NO. 342.
M-238; 62 Cu, 1.75 Pb, bal Zn.
Hard: 80,000 TS; 53,000 YS; 15 El.
Soft: 50,000 TS; 17,000 YS; 49 El.
For rivets, gears, clock parts; free-cutting. Was Revere 247.

REVERE ALLOY NO. 344.
M-238; 63 Cu, 0.75 Pb, bal Zn.
Hard: 67,000 TS; 42,000 YS; 19 El.
Soft: 46,000 TS; 14,000 YS; 55 El.
For machined cold headed parts; free-cutting. Was Revere 250.

REVERE ALLOY NO. 362.
M-238; 61.5 Cu, 3.8 Pb, bal Zn.
Hard: 63,000 TS; 44,000 YS; 13 El.
Soft: 43,000 TS; 14,000 YS; 52 El.
For valve bonnets, screw machine products; free-cutting. Was Revere 252.

REVERE ALLOY NO. 365.
M-238; 59 Cu, 0.6 Pb, bal Zn.
Soft: 54,000 TS; 20,000 YS; 45 El.
For print rolls, tube headers; leaded Muntz metal. Was Revere 275.

REVERE ALLOY NO. 380.
M-238; 56.5 Cu, 41.25 Zn, 2.25 Pb.
Soft: 60,000 TS; 20,000 YS; 30 El; 116 Brin.
For architectural, lock parts; free-cutting. Was Revere 283.

REVERE ALLOY NO. 385.
M-238; 58 Cu, 2.75 Pb, bal Zn.
Soft: 60,000 TS; 20,000 YS; 30 El.
For architectural trim, hardware; free-cutting. Was Revere 277.

REVERE ALLOY NO. 411.
M-238; 90 Cu, 9.5 Zn, 0.5 Sn.
Hard: 65,000 TS; 57,000 YS; 5 El; 140 Brin.
Soft: 45,000 TS; 10,000 YS; 42 El; 100 Brin.
For bearings, lamp connections, weatherstrip; heavy duty. Was Revere 325.

REVERE ALLOY NO. 415.
M-238; 90 Cu, 2 Sn, bal Zn.
For weather strip; corrosion resistant. Was Revere 327.

REVERE ALLOY NO. 425.
M-238; 88 Cu, 2 Sn, bal Zn.
For jewelry; corrosion resistant. Was Revere 337.

REVERE ALLOY NO. 432.
M-238; 86 Cu, 0.5 Sn, bal Zn.
For jewelry; corrosion resistant. Was Revere 336.

REVERE ALLOY NO. 445.
M-238; 71 Cu, 0.05 P, 1 Sn, bal Zn.
Hard: 10,000 TS; 80,000 YS; 3 El.
Soft: 53,000 TS; 22,000 YS; 65 El.
For condenser tubes; Phosphorized Admiralty, Metal. Was Revere 362.

REVERE ALLOY NO. 464A.
M-238; 60 Cu, 39.25 Zn, 0.75 Sn.
Hard: 82,000 TS; 55,000 YS; 20 El.
Soft: 60,000 TS; 22,000 YS; 45 El.
For propeller and pump shafts, piston rods; Roman Bronze. Was Revere 380.

REVERE ALLOY NO. 464B.
M-238; 62 Cu, 0.75 Sn, bal Zn.
Hard: 10,000 TS; 70,000 YS; 4 El.
Soft: 58,000 TS; 21,000 YS; 36 El.
For marine applications; Hard Naval Brass. Was Revere 370.

REVERE ALLOY NO. 465A.
M-238; 60.5 Cu, 0.25 Pb, 0.07 Mn, 0.07 Sn, bal Zn.
For coal screens; high strength. Was Revere 443.

REVERE ALLOY NO. 482.
M-238; 60 Cu, 0.75 Pb, 0.75 Sn, bal Zn.
Hard: 86,000 TS; 59,000 YS; 10 El.
Soft: 63,000 TS; 26,000 YS; 35 El.
For tube headers; Leaded Naval Brass. Was Revere 387.

REVERE ALLOY NO. 485.
M-238; 60 Cu, 37.25 Zn, 2 Pb, 0.75 Sn.
Hard: 90,000 TS; 70,000 YS; 5 El.
Soft: 58,000 TS; 20,000 YS; 35 El.
For forgings, screw machine products; Leaded Naval Brass, free cutting. Was Revere 389.

REVERE ALLOY NO. 505.
M-238; 4 max Sn, 0.2 P, 98.75 Cu.
Spring Temper: 68,000-83,000 TS; 4 El; B 74-85 Rock.
Soft: 34,000-50,000 TS; 47-50 El; F 50-67 Rock.
For diaphragms, bellows, springs, fuse clips, lock washers, clutch discs.
High fatigue and corrosion resistance.
Grade E Phosphor Bronze. Was Revere 305.

REVERE ALLOY NO. 521.
M-238; 8 Sn, 0.2 P, bal Cu.
Spring Temper: 105,000-118,500 TS; 3 El; B 100 Rock.
Soft: 53,000-67,000 TS; 65-70 El; B 20-70 Rock.
For diaphragms, bellows, springs, fuse clips, lock washers, clutch discs.
High fatigue and corrosion resistance.
Grade D Phosphor Bronze. Was Revere 315.

REVERE ALLOY NO. 524.
M-238; 10 Sn, 0.2 P, bal Cu.
Hard Temper: 94,000-109,000 TS; 7 El; B 100 Rock.
Soft: 58,000-73,000 TS; 62-68 El; B 25-75 Rock.
For diaphragms, bellows, springs, fuse clips, lock washers, clutch discs.
High fatigue and corrosion resistance.
Grade D Phosphor Bronze. Was Revere 317.

REVERE ALLOY NO. 639.
M-238; 91 Cu, 7 Al, 2 Si.
Hard: 95,000 TS; 53,000 YS; 25 El.
Soft: 85,000 TS; 43,000 YS; 35 El.
For hardware, fittings, valve stems; Al bronze. Was Revere 436.

REVERE ALLOY NO. 675.
M-238; 58 Cu, 0.75 Fe, 0.5 Mn, 1.0 Sn, bal Zn.
Hard: 90,000 TS; 55,000 YS; 10 El; 185 Brin.
Soft: 65,000 TS; 27,000 YS; 35 El; 120 Brin.
For valve stems, clutch disks, pump parts; Mn bronze. Was Revere 454.

REVERE ALLOY NO. 677.
M-238; 56.7 Cu, 0.75 Pb, 1.75 Ni, 0.15 Mn, 1.2 Fe, 0.6 As, bal Zn.
Hard: 85,000 TS; 62,000 YS; 10 El.
Soft: 65,000 TS; 26,000 YS; 40 El.
For valve stems; Arsenical Bronze. Was Revere 469.

REVERE ALLOY NO. 700.
M-238; 55 Cu, 18 Ni, 27 Zn.
Hard: 100,000 TS; 85,000 YS; 3 El; 91 B Rock.
Soft: 60,000 TS; 27,000 YS; 40 El; 90 F Rock.
For springs, trim, plumbing fixtures; nickel silver 18% B. Was Revere 555.

REVERE ALLOY NO. 706.
M-128; 1.4 Fe, 0.3 Mn, 10 Ni, bal Cu.
For condenser and heat exchanger tubes; 90-10 Cupro-Nickel. Was Revere 508.

REVERE ALLOY NO. 735.
M-238; 72 Cu, 0.15 Mn, 18 Ni, bal Zn.
For hollow ware, cutlery, trim; Nickel Silver; deep drawing. Was Revere 536.

REVERE ALLOY NO. 752.
M-238; 65.5 Cu, 18 Ni, 16.5 Zn.
Hard: 85,000 TS; 74,000 YS; 3 El; 87 B Rock.
Soft: 58,000 TS; 25,000 YS; 40 El; 85 F Rock.
For marine and auto trim, hardware; nickel, silver 18% A. Was Revere 533.

REVERE ALLOY NO. 757.
M-238; 66 Cu, 12 Ni, bal Zn.
Hard: 95,000 TS; 80,000 YS; 3 El.
Soft: 53,000 TS; 17,000 YS; 46 El.
For jewelry, springs; 12% Nickel Silver. Was Revere 540.

REVERE ALLOY NO. 776.
M-238; 43 Cu, 13 Ni, bal Zn.
For elevator trim, springs; 13% Nickel Silver. Was Revere 560.

REVERE ALLOY NO. 788.
M-238; 65 Cu, 2 Pb, 10 Ni, 0.25 Mn, bal Zn.
For washers, jewelry, panels; Leaded 10% Nickel Silver. Was Revere 580.

REVERE ALLOY NO. 790.
M-238; 66 Cu, 2 Pb, 12 Ni, bal Zn.
Hard: 78,000 TS; 75,000 YS; 5 El.
Soft: 55,000 TS; 18,000 YS; 40 El.
For key blanks, hardware, stampings; nickel silver. Was Revere 575.

REVERE NO. 110.
M-238; 99.90+ Cu.
Hard: 51,000 TS; 46,000 YS; 5-15 El; Rockwell B58.
Soft: 30,000 TS; 45 El.
For downspouts, bus bars, gutters; electrolytic, tough pitch. Was Revere 100.

REVERE NO. 113.
M-238; 99.90 Cu, bal Ag.
Hard: 51,000 TS; 48,000 YS; 5-15 El; Rockwell B58. Soft 32,000 TS; 45 El.
For conductors; lake copper. Was Revere 101.

REVERE NO. 120.
M-238; 0.007 P, bal Cu.
For high conductivity tubing. Was Revere 112.

REVERE NO. 122.
M-238; 99.90 Cu + P.
Hard: 55,000 TS; 48,000 YS; 5-10 El; Rockwell B69.
Soft: 32,000 TS; 45 El.
For kettles, oil and water tubes, welded construction; phosphorized copper. Was Revere 111.

REVERE NO. 142.
M-238; 99.70 Cu + P, 0.3 As.
Hard: 60,000 TS; 55,000 YS; 5 El; 228 Brin.
Soft: 36,000 TS; 10,000 YS; 40 El; 63 Brin.
For tubes; arsenical copper. Was Revere 113.

REVERE NO. 145.
M-238; 0.5 Te, bal Cu.
1/2-H temper: 38,000 TS; 30,000 YS; 15 El.
H-temper: 44,000 TS; 38,000 YS; 10 El.
F.C. Copper. Was Revere 114.

REVERE NO. 210A.
M-238; 5 Zn, 95 Cu.
Hard: 55,000 TS; 39,000 YS; 5 El; 121 Brin.
Soft: 35,000 TS; 11,000 YS; 35 El.
For jewelry, primers; gilding bronze. Was Revere 119.

REVERE NO. 240.
M-238; 80 Cu, 20 Zn.
Hard: 85,000-98,000 TS; 55,000 YS; 5 El; 203 Brin.
Soft: 43,000-46,000 TS; 12,000 YS; 50 El; 57 Brin.
For bellows, jewelry, meters; red brass. Was Revere 140.

REVERE NO. 262.
M-238; 68 Cu, 32 Zn.
Hard: 90,000 TS; 64,000 YS; 3 El; 187 Brin.
Soft: 47,000 TS; 11,000 YS; 62 El; 62 Brin.
For drawn and spun parts; deep drawing and spinning brass. Was Revere 165.

REVERE NO. 280B.
M-238; 60 Cu, 40 Zn.
Hard: 88,000 TS; 80,000 YS; 12 El; 191 Brin.
Soft: 59,000 TS; 15,000 YS; 52 El; 64 Brin.
For architectural structures, panel sheets, condenser tubes; Muntz Metal. Was Revere 180.

REVERE NO. 330.
M-238; 67 Cu, 32.25 Zn, 0.75 Pb.
Hard: 50,000 TS; 5 El.
Soft: 44,000 TS; 45 El.
For hardware, screw machine products; leaded brass, free cutting. Was Revere 224.

REVERE NO. 342A.
M-238; 63 Cu, 35 Zn, 2 Pb.
Hard: 85,000 TS; 50,000 YS; 5 El; 150 Brin.
Soft: 45,000 TS; 10,000 YS; 50 El; 60 Brin.
For watch parts, locks; heavy leaded brass. Was Revere 235.

REVERE NO. 360.
M-238; 61.5 Cu, 35.5 Zn, 3.0 Pb.
Hard: 65,000 TS; 15 El; B 72 Rockwell.
Soft: 50,000 TS; 50 El; B 65 Rockwell.
For machining parts, screw machine products; free-cutting. Was Revere 240.

REVERE NO. 380B.
M-238; 56 Cu, 41.25 Zn, 2.75 Pb.
Hard: 70,000 TS; 55,000 YS; 10 El; 144 Brin.
Soft: 50,000 TS; 15,000 YS; 20 El; 64 Brin.
For builders hardware, architectural work; architectural bronze. Was Revere 285.

REVERE NO. 381.
M-238; 60 Cu, 39.25 Zn, 0.75 Sn.
Hard: 70,000 TS; 55,000 YS; 25 El; 140 Brin.
Soft: 54,000 TS; 15,000 YS; 50 El.
For hardware, condenser tubes; Roman Bronze.

REVERE NO. 422.
M-238; 87 Cu, 12 Zn, 1 Sn.
Hard: 80,000 TS; 4 El; 150 Brin.
Soft: 45,000 TS; 40 El; 100 Brin.
For chains; chain bronze. Was Revere 340.

REVERE NO. 443.
M-238; 71 Cu, 28 Zn, 1 Sn, 0.03 As.
Hard: 100,000 TS; 98,000 YS; 3 El; 215 Brin.
Soft: 53,000 TS; 18,000 YS; 60 El; 53 Brin.
For condenser tubes; Admiralty Metal; corrosion resistant. Was Revere 358.

REVERE NO. 464.
M-238; 60 Cu, 39 Zn, 0.25 Pb, 0.75 Sn.
Hard: 75,000 TS; 39,000 YS; 15 El; 150 Brin.
Soft: 54,000 TS; 15,000 YS; 45 El; 83 Brin.
For condenser heads, forgings; Naval Brass. Was Revere 386.

REVERE NO. 548R.
M-238; 62 Cu, 33 Zn, 5 Ni.
Hard: 85,000 TS; 60,000 YS; 2 El; 150 Brin.
Soft: 50,000 TS; 15,000 YS; 45 El; 70 Brin.
For jewelry, decorative trim; Nickel Silver, 5%.

REVERE NO. 608.
M-238; 95 Cu, 5 Al, 0.035 As.
Hard: 70,000 TS; 50,000 YS; 25 El; 150 Brin.
Soft: 57,000 TS; 10,000 YS; 55 El.
For hardware, bushings; Al bronze. Was Revere 429.

REVERE NO. 614.
M-238; 7 Al, 2 Fe, bal Cu.
Soft: 75,000 TS; 42,000 YS; 30 El.
Hard: 85,000 TS; 55,000 YS; 30 El.
For high strength plates for pressure vessels; Al-bronze, tough. Was Revere 430.

REVERE NO. 710.
M-238; 80 Cu, 20 Ni, 0.5 Fe, 0.7 max Mn.
Hard: 80,000 TS; 76,000 YS; 3 El; 152 Brin.
Soft: 49,000 TS; 14,000 YS; 35 El; 64 Brin.
For condenser tubes, corrosion resistant tanks; 20% Cupro-Nickel. Was Revere 505.

REVERE NO. 715.
M-238; 70 Cu, 30 Ni, 0.6 Fe, 1.0 max Mn.
Hard: 84,000 TS; 80,000 YS; 4 El; 187 Brin.
Soft: 49,000 TS; 18,000 YS; 50 El; 60 Brin.
For hardware, process equipment; 30% Cupro-Nickel. Was Revere 510.

REVERE NO. 745.
M-238; 65 Cu, 25 Zn, 10 Ni.
Hard: 90,000 TS; 65,000 YS; 3 El; 152 Brin.
Soft: 50,000 TS; 15,000 YS; 45 El; 70 Brin.
For stampings, jewelry; Nickel Silver, 10%. Was Revere 545.

REX.
M-1345; 0.56 C, 0.30 Cr, 0.18 Mo, 1.65 Ni, 0.10 V, bal Fe.
For forging and heading dies, punches; oil hardened, tough.

REX 49.
M-38; 1.1 C, 6.75 W, 3.75 Mo, 4.25 Cr, 2.0 V, 5.0 Co, bal Fe.
Hardened: C 64-69 Rock.
For special applications of twist drills, end mills, form cutters, lathe tools, reamers.
AISI Type M41 High speed tool steel.

REX 78.
M-1126, M-521; 0.07 C, 14 Cr, 18 Ni, 3.75 Mo, 0.65 Ti, 3.6 Cu, bal Fe.
Heat treated: 94,000 TS; 49,000 YS; 39 El; 57 RA.
For gas turbine disks and blades; corrosion and heat resistant.

REX 95.
M-38; 0.80 C, 14 W, 4.0 Cr, 2.0 V, 0.75 Mo, 5.25 Co, bal Fe.
Hardened: C 63-66 Rock.
For single point lathe tools, severe cutting operations, tools for stainless steels.
AISI Type T8 High speed tool steel.

REX 326.
M-1126; 0.25 C, 17 Cr, 17 Ni, 2.5 Mo, 7 Co, 1.8 Cb, 3 Mn, bal Fe.
For jet engine components; high heat resistance.

REX 326D.
M-114, M-1126; 0.45 Co, 14.3 C, 14.6 Ni, 2.2 W, 2.2 Mo, 9.5 Co, 2.8 Cb, bal Fe.
For valves, pump parts; corrosion and heat resistant.

REX 327.
M-1126; 0.20 C, 17 Cr, 17 Ni, 3 Mo, 7 Co, 0.8 Ti, 3 Cu, bal Fe.
For valves, pump parts; heat and corrosion resistant.

REX 337 A.
M-Eng.; C, Ni, Cr, bal Fe.
For high temperature applications; heat and corrosion resistant.

REX 400.
M-English; 76 Ni, 19 Cr, 0.6 Al, 2.1 Ti, bal Fe.
For high temp. applications; heat resistant.

REX 448.
M-531; 0.15 C, 0.75 Si, 0.75 Mn, 11.5 Cr, 0.75 Mo, 0.15 V, 0.45 Cb, bal Fe.
Annealed: 70,000 TS; 35,000 YS; 30 El; B 80 Rock.
For cutlery, surgical instruments, chemical plant equipment.
Corrosion and heat resistant.

REX 448.
M-1126; 0.10 C, 11 Cr, 0.8 Mo, 0.8 Ni, 0.5 Cb, 0.2 V, 1.0 Mn, bal Fe.
For aircraft and jet engine components; corrosion and heat resistant.

REX 467.
M-521; 0.20 C, 0.7 Si, 1.0 Mn, 9.5 Ni, 14.5 Cr, 2.5 Cu, 2 Mo, 0.8 Ti, bal Fe.
For turbine blades in jet engines; high creep resistance, stainless.

REX 467.
M-1126; 0.20 C, 14 Cr, 10 Ni, 2 Mo, 0.8 Ti, 2.5 Cu, bal Fe.
For valves, pump parts; corrosion and heat resistant.

REX 648C.
M-56; 0.18 max C, 0.10 max Si, 0.10 max Mn, 17.0-19.0 Ni, 0.20 max Cr, 4.6-5.2 Mo, 0.05-0.15 Al, 7.0-8.5 Co, 0.30-0.60 Ti, bal Fe.
Maraging steel.

REX AA.
M-38; 0.75 C, 0.3 Mn, 4.0 Cr, 1.15 V, 18 W, bal Fe.
Hardened: C 62-65 Rock.
For the lathe cutting tools, drills, end mills, reamers, slotting saws, gear cutters.
AISI Type T1 High speed tool steel.

REX AAA.
M-38; 0.7 C, 17-19 W, 0.5 Mo, 4 Cr, 1 V, 5 Co, bal Fe.
For cutters, reamers, drills, hobs, broaches, taps; Type T4; high speed steel.

REX-A-LITE.
M-1017; C, Cr, Mn, Si, bal Fe.
Welded: 570 Brin.
For hard facing electrodes; abrasion resistant.

REX-A-LITE 1.
M-1017; Co, Cr, W.
Welded: 550 Brin.
For hard facing electrodes; abrasion and corrosion resistant.

REX-A-LITE 6.
M-1017; Co, Cr, W.
Welded: 440 Brin.
For hard facing electrodes; shock and impact resistant.

REXARC.
M-1017; C, Mn, bal Fe.
Welded: 380 Brin.
For hard facing electrodes; wear resistant.

REXARC 1.
M-1017; Ni.
For welding electrodes for cast iron; machinable welds.

REXARC 30.
M-1017; C, Cr, Mn.
Welded: Rockwell C 30.
For hard facing electrodes; metallic coated.

REXARC 48.
M-1017; C, Cr, bal Fe.
Welded: Rockwell C 48.
For hard facing electrodes.

REXARC 50.
M-1017; C, Cr, bal Fe.
Welded: Rockwell C 50.
For hard facing electrodes; impact and abrasion resistant.

REXARC 55.
M-1017; C, Cr, bal Fe.
Welded: Rockwell C 55.
For hard facing electrodes; shock resistant.

REXARC 65.
M-1017; C, Cr, TiO, Mo, bal Fe.
Welded: Rockwell C 65.
For hard facing electrodes; abrasion resistant.

REXARC 520.
M-1017; C, Cr, Mo, bal Fe.
Welded: Rockwell C 54.
For hard facing electrodes; abrasion resistant.

REXARC MANGANESE M1.
M-1017; C, Mn, Ni, bal Fe.
Welded: Rockwell C 35.
For hard facing electrodes; build-up on Mn castings, wear resistant.

REXARC MANGANESE S-3.
M-1017; C, Mn, Ni, bal Fe.
Welded: Rockwell C 14-18.
For hard facing electrodes; used on 13 Mn steel, wear resistant.

REX BRONZE.
M-652; 80 Cu, 10 Zn, 10 Pb.
For castings.

REX CAST ALUMINUM.
M-844; Cu, bal Al.
For welding rod; for Al alloys.

REX CAST IRON.
M-844; 3.0-3.5 C, 3.0-3.5 Si, bal Fe.
For welding rod; for cast iron.

REXCOBALT.
M-French; C, alloy, bal Fe.
For tools.

REX COMPOSITON.
M-652; 78-86 Cu, 3-6 Sn, 2-6 Pb, 5-14 Zn.
For castings; free-cutting.

REX M-2.
M-38; 0.83 C, 6.4 W, 4.1 Cr, 1.9 V, 5.0 Mo, bal Fe.
For tools, cutters, broaches, chasers, hobs; high speed steel. AISI M2.

REX M2 HIGH CARBON.
M-38; 1.0 C, 6.4 W, 5.0 Mo, 4.15 Cr, 1.95 V, bal Fe.
Harden to C 62.5-66.5 Rock.
For boring tools, twist drills, milling cutters, threading tools, thread chasers; improved wear resistance.
AISI Type M2 High speed tool steel.

REX M2S.
M-38; 0.85 C, 6.4 W, 5.0 Mo, 4.15 Cr, 1.95 V, bal Fe.
Harden to C 61-67 Rock.
Free machining; for form tools, drills, end mills, milling cutters, hobs, broaches.
AISI Type M2 High speed tool steel.

REX M2S HIGH CARBON.
M-38; 1.0 C, 6.4 W, 5.0 Mo, 41.15 Cr, 1.95 V, 0.15 S, bal Fe.
Harden to: C 62-66 Rock.
Free machining type high speed steel.
For boring tools, drills, end mills, reamers, broaches, milling cutters, hobs.
AISI Type M2 High speed tool steel.

REX M-3.
M-38; 1 C, 4 Cr, 2.5 V, 5.7 W, 5.0 Mo, bal Fe.
For broaches, form tools, reamers, cut-off tools; high speed steel, oil hardened.

REX M3-1.
M-38; 1.05 C, 4 Cr, 2.4 V, 6.25 W, 6.25 Mo, bal Fe.
For lathe and planer tools, milling cutters, reamers; Type M3 high speed steel.

REX M3S-2.
M-38; 1.2 C, 0.3 Mn, 0.3 Si, 4 Cr, 6.25 W, 6.25 Mo, 0.15 S, bal Fe.
Hardened: 64-67 Rock C.
For form tools, milling cutters, hobs, cut-off tools, reamers, lathe tools.
High-speed steel, free-cutting.
Resistance to softening at high temperatures.

REX-M7.
M-38; 1.0 C, 0.3 Mn, 0.3 Si, 3.75 Cr, 1.75 W, 8.75 Mo, 2.0 V, bal Fe.
Hardened: 64-66 Rock C.
For twist drills, end mills, cutting tools, lathe tools, threading tools, hobs, reamers, taps, chasers, form cutters.
High speed steel. Improved wear resistance. Type M7 tool steel.

REX M33.
M-38; 0.90 C, 9.5 Mo, 1.75 W, 3.75 Cr, 1.25 V, 8.0 Co, bal Fe.
Hardened: C 63-68 Rock.
For tough machining and higher speed operations of drills, end mills, form cutters, lathe tools, taps, reamers.
AISI Type M33 High speed tool steel.

REX M42.
M-38; 1.1 C, 1.5 W, 9.5 Mo, 3.75 Cr, 1.15 V, 8.0 Co, bal Fe.
Hardened: C 64-69 Rock.
For special purpose drills, end mills, form cutters, reamers, lathe tools.
AISI Type M42 High speed tool steel.

REXOR.
M-750; 0.35 C, 0.7 Mn, 0.6 Cr, 0.35 Mo, 0.15 Si, 0.2 Ni, bal Fe.
Annealed: 107,000 TS; 80,000 YS; 22 El; 248 Brin.
For chisels, pneumatic tools; tough, wear resistant.

REX T15S.
M-38; 1.55 C, 4.0 Cr, 5.0 V, 12.25 W, 5.0 Co, 0.10 S, bal Fe.
High speed steel, free machining grade.
For cutting tools as boring tools, end mills, broaches, gear cutters.

REX-TMO.
M-38; 0.85 C, 3.75 Cr, 1.15 V, 1.70 W, 8.5 Mo, bal Fe.
For lathe and planer tools, milling cutters, hobs; high speed steel; Type M1.

REXTOX.
M-Eng.; Fe-Mn-Cu.
For resistance alloy.

REX-TUNG.
M-1017; WC.
For hard facing electrodes; mild steel tube filled with WC.

REX VM.
M-38; 0.98 C, 8.0 Mo, 4.0 Cr, 0.70 W, 1.95 V, bal Fe.
Harden to C 58-66 Rock.
For broaches, center drills, lathe tools, end mills, thread chasers, hobs.
AISI Type M10 High speed tool steel.

REX VM "PX" TEMPER.
M-38; 0.55 C, 4.0 Cr, 8.0 Mo, 2.0 V, bal Fe.
High speed steel, low carbon grade.
For extrusion dies, piercer points, valve extrusion die rings and inserts.

REX Z METAL.
M-566; 2.4 C, 0.75 Mn, 1.0 Si, 0.18 Cu, bal Fe.
Heat treated: 75,000 TS; 50,000 YS; 8 El; 200 Brin.
For chains, buckets, sprockets; pearlitic malleable iron.

REYNOLDS.
M-671.
All Reynolds Aluminum Alloys now identified by AA numbers.

REZISTAL KA2.
M-72; 0.15 max C, 0.40-0.65 Mn, 8-9.5 Ni, 17.5-19.0 Cr, bal Fe.
Heat treated: 85,000-95,000 TS; 30,000-40,000 YS; 55-60 El; 70-75 RA; 135-145 Brin.
For all welded construction which cannot receive the Strauss heat treatment after welding; radiators, lamps, food and cooking utensils; non-magnetic, austenitic.

REZISTAN.
M-1161; 3.2 C, Cr, Ni, Cu, Mo, Al, bal Fe.
For pumps, valves, combustion chambers; alloy cast iron, corrosion resistant.

RF-1.
M-1766; 0.18 C, 0.70 Mn, 0.25 Si, 0.50 Mo, bal Fe.
For tubing, vessels for elevated temperature.
DIN 15Mo3; ASTM A182-F1; AFNOR 13 CD2.

RF-2.
M-1766; 0.13 C, 0.50 Mn, 1.1 Cr, 0.50 Mo, bal Fe.
For equipment for elevated temperature.
ASTM A182-F12; AFNOR 10CD5.

RF-3.
M-1766; 0.13 C, 0.50 Mn, 0.40 Si, 2.3 Cr, 1.0 Mo, bal Fe.
Fittings for oil refineries and other high temperature operations.
ASTM A182-F22; AFNOR 10CD9.

RF-5.
M-1766; 0.11 C, 0.50 Mn, 0.25 Si, 5.0 Cr, 0.50 Mo, bal Fe.
Equipment for oil refineries and other high temperature operations.
Similar to SAE 51501; ASTM A182-F5/F5A; DIN 12CrMo195; BS625.

RFF.
M-912; 0.58 C, 1.75 Si, 0.75 Mn, 0.40 Cr, bal Fe.
For flat or coil springs.
W.-Nr. 0961.

R.F.F.
 M-1118; 0.5 C, 0.7 Mn, 1.9 Si, 0.2 Mo, bal Fe.
 For tools, dies, punches; tough.

RFFH.
 M-912; 0.65 C, 1.75 Si, 0.80 Mn, bal Fe.
 Steel for flat or spiral springs.
 W.-Nr. 0906.

RFFL.
 M-912; 0.63 C, 1.65 Si, 0.80 Mn, 0.25 Cr, bal Fe.
 Steel for flat or coil springs.
 W.-Nr. 0960.

RFFN.
 M-912; 0.51 C, 1.75 Si, 0.70 Mn, bal Fe.
 Water hardening spring steel.
 W.-Nr. 0903.

RFFO.
 M-912; 0.58 C, 1.75 Si, 0.80 Mn, bal Fe.
 Water hardening steel for flat or spiral springs.
 W.-Nr. 0904.

RFFW.
 M-912; 0.46 C, 1.80 Si, 0.70 Mn, bal Fe.
 For springs, laminated springs; water hardened.
 W.-Nr. 0902.

RFFWW.
 M-912; 0.38 C, 1.80 Si, 0.70 Mn, bal Fe.
 Water hardening spring steel.
 W.-Nr. 0970.

RG5.
 M-Ger.; 85 Cu, 5 Sn, 7 Zn, 3 Pb.
 For castings; pressure tight.

RG7H.
 M-1541.
 High magnetic induction, grain oriented silicon sheet: 0.012 inch thick.
 Max. permeability: 0.082 (H/M); resistivity: 45 micro-ohm-cm; core loss: (W15/50) 0.36 watts per pound.
 Grain oriented silicon steel; for motors and/or transformers.

RGS 1.
 M-912; 0.55 C, 0.85 Cr, 0.30 Mo, 1.65 Ni, 0.10 V, bal Fe.
 Hot work tool steel; for forging dies and metal extrusion tools.
 W.-Nr. 2713.

RGS 3.
 M-912; 0.58 C, 1.15 Cr, 0.35 Mo, 2.85 Ni, 0.10 V, bal Fe.
 Hot or cold work tool steel; for forging and embossing dies, shear blades, moulds.
 W.-Nr. 2743; AFNOR 60 NCD 12.

RGS 4.
 M-912; 0.56 C, 1.10 Cr, 0.50 Mo, 1.65 Ni, 0.18 V, bal Fe.
 Hot or cold work tool steel; for drop hammer dies, punches, extrusion dies.
 W. Nr. 2714; AFNOR 55 NCD 07 05.

RGS 6.
 M-912; 0.20 C, 3.35 Mo, 3.10 Ni, bal Fe.
 Hot work tool steel; for hot pressing dies.
 W.-Nr. 2777; AFNOR 20 DN 32-12.

RGT 0.
 M-912; 0.13 C, 19.5 Cr, 5 max Fe, 0.4 Ti, bal Ni.
 Room temp: 85 kg/mm^2 TS; 35 kg/mm^2 YS.
 Corrosion resistant and heat resistant alloy.
 For combustion chamber parts.
 W.-Nr. 2.4951.

RGT 1.
 M-912; 0.06 C, 15.0 Cr, 25.5 Ni, 1.30 Mo, 0.30 V, 2.10 Ti, bal Fe.
 Room temp: 110 kg/mm^2 TS; 75 kg/mm^2 YS; 20 El.
 Corrosion resisting and heat resisting alloy.
 Parts for gas and steam turbines to 750°C.
 W.-Nr. 1.4980.

RGT 3.
 M-912; 0.06 C, 20.0 Cr, 5.0 max Fe, 1.40 Al, 2.40 Ti, bal Ni.
 Room temp.: 120 kg/mm^2 TS; 75 kg/mm^2 YS; 20 El.
 Corrosion resisting and heat resisting alloy.
 Parts for gas turbines and transmissions, for use to 850°C.
 W.-Nr. 2.4952.

RGT 4.
 M-912; 0.06 C, 20.0 Cr, 4.5 Mo, 5.0 max Fe, 1.40 Al, 2.40 Ti, bal Ni.
 Room temp: 120 kg/mm^2 TS; 75 kg/mm^2 YS; 20 El.
 Corrosion resisting and heat resisting alloy.
 Parts for gas turbines and transmissions, for use to 850°C.
 W.-Nr. 2.4976.

RGT 5.
 M-912; 0.10 C, 22.0 Cr, 9.0 Mo, 1.5 Co, 0.6 W, 18.5 Fe, bal Ni.
 Room temp: 80 kg/mm^2 TS; 35 kg/mm^2 YS; 40 El.
 Corrosion resisting and heat resisting alloy.
 Parts for gas turbines and transmissions, for use to 850°C.
 W.-Nr. 2.4972.

RGT 6.
 M-912; 0.06 C, 15.0 Cr, 7.0 Fe, 0.7 Al, 2.5 Ti, 1.0 Nb + Ta, bal Ni.
 Room temp: 110 kg/mm^2 TS; 65 kg/mm^2 YS; 25 El.
 Corrosion resisting and heat resisting alloy, for parts operating up to 850°C.

RGT 8.
M-912; 0.08 C, 12.5 Cr, 43.5 Ni, 6.0 Mo, 0.3 Al, 3.0 Ti, bal Fe.
Room temp: 120 kg/mm^2 TS; 85 kg/mm^2 YS; 15 El.
Corrosion resisting and heat resisting alloy, for parts for jet power plants and gas turbines.
W.-Nr. 2.4975.

RGT 9.
M-912; 0.06 C, 15.0 Cr, 45.0 Ni, 4.0 Mo, 4.0 W, 1.0 Al, 3.0 Ti, bal Fe.
Room temp: 130 kg/mm^2 TS; 90 kg/mm^2 YS; 12 El.
Corrosion resisting and heat resisting alloy, for parts operating up to 850°C.

RGT 12.
M-912; 0.06 C, 20.0 Cr, 18.0 Co, 5.0 max Fe, 1.5 Al, 2.5 Ti, bal Ni.
Room temp: 125 kg/mm^2 TS; 80 kg/mm^2 YS; 25 El.
Corrosion resisting and heat resisting alloy, for parts for gas turbines and transmissions for use up to 900°C.
W.-Nr. 2.4969.

RGT 13.
M-912; 0.06 C, 20.0 Cr, 4.5 Mo, 18.0 Co, 5.0 max Fe, 1.5 Al, 2.5 Ti, bal Ni.
Room temp: 125 kg/mm^2 TS; 80 kg/mm^2 YS; 30 El.
Corrosion resisting and heat resisting alloy, for parts for gas turbines and transmissions operating up to 900°C.
W.-Nr. 2.4982.

RGT 14.
M-912; 0.10 C, 18.5 Cr, 4.0 Mo, 18.5 Co, 4.0 max Fe, 2.9 Al, 2.9 Ti, bal Ni.
Room temp: 130 kg/mm^2 TS; 80 kg/mm^2 YS; 15 El.
Corrosion resisting and heat resisting alloy, for parts for gas turbines and transmissions operating up to 950°C.
W.-Nr. 2.4983.

RGT 15.
M-912; 0.10 C, 19.0 Cr, 10.0 Mo, 11.0 Co, 5.0 max Fe, 1.5 Al, 3.0 Ti, bal Ni.
Room temp: 135 kg/mm^2 TS; 100 kg/mm^2 YS; 12 El.
Corrosion resisting and heat resisting alloy, for parts for gas turbines and transmissions operating up to 950°C. Also a tool steel.
W.-Nr. 2.4973.

RGT 16.
M-912; 0.15 C, 15.0 Cr, 5.0 Mo, 20.0 Co, 1.0 max Fe, 4.5 Al, 1.2 Ti, bal Ni.
Room temp: 120 kg/mm^2 TS; 80 kg/mm^2 YS; 16 El.
Corrosion resisting and heat resisting alloy, for parts for gas turbines and transmissions for operation up to 950°C.

RGT 18.
M-912; 0.15 C, 15.0 Cr, 4.0 Mo, 30.0 Co, 2.0 max Fe, 3.0 Al, 2.5 Ti, bal Fe.
Room temp: 115 kg/mm^2 TS; 75 kg/mm^2 YS; 15 El.
Corrosion resisting and heat resisting alloy, for parts for gas turbines and transmissions for operation up to 950°C.

RGT 32.
M-912; 0.12 C, 21.0 Cr, 20.0 Ni, 3.0 Mo, 20.0 Co, 2.5 W, 1.0 Nb + Ta, 0.15 N, bal Fe.
Room temp: 80 kg/mm^2 TS; 35 kg/mm^2 YS; 35 El.
Corrosion resisting and heat resisting alloy, for parts of gas turbines and transmissions for operation up to 850°C.
W.-Nr. 1.4971.

RGT 33.
M-912; 0.45 C, 20.0 Cr, 20.0 Ni, 4.0 Mo, 20.0 Co, 4.0 W, 4.0 Nb + Ta, bal Fe.
Room temp: 100 kg/mm^2 TS; 55 kg/mm^2 YS; 10 El.
Corrosion resisting and heat resisting alloy, for parts of gas turbines and transmissions for operation up to 850°C.
W.-Nr. 1.4977 and 1.4978.

RGT 35.
M-912; 0.38 C, 20.0 Cr, 20.0 Ni, 4.0 Mo, 42.0 Co, 4.0 W, 5.0 max Fe, 4.0 Nb + Ta, bal Fe.
Room temp: 100 kg/mm^2 TS; 50 kg/mm^2 YS; 25 El.
Corrosion resisting and heat resisting alloy, for parts for gas turbines and transmissions for operation up to 900°C.
W.-Nr. 2.4989.

RGT 36.
M-912; 0.10 C, 20.0 Cr, 10.0 Ni, 50.0 Co, 15.0 W, 3.0 max Fe.
Room temp: 85 kg/mm^2 TS; 35 kg/mm^2 YS; 35 El.
Corrosion resisting and heat resisting alloy, for parts of gas turbines and transmissions for operation up to 950°C.
W.-Nr. 2.4967.

RGT 101.
M-912; 0.06 C, 14.0 Cr, 26.0 Ni, 3.0 Mo, 1.7 Ti, bal Fe.
Room temp: 100 kg/mm^2 TS; 65 kg/mm^2 YS.
Corrosion resistant and heat resistant alloy for operation up to 750°C.

RGT 501.
M-912; 0.06 C, 15.0 Cr, 3.0 Mo, 3.0 W, 7.0 Fe, 3.0 Nb + Ta, bal Ni.
Room temp: 95 kg/mm^2 TS; 50 kg/mm^2 YS; 40 El.
Corrosion resisting and heat resisting alloy, for parts operating up to 700°C.

RGT 601.
 M-912; 0.05 C, 18.5 Cr, 53.0 Ni, 3.0 Mo, 0.6 Al, 0.8 Ti, 5.3 Nb + Ta, bal Fe.
 Room temp: 140 kg/mm^2 TS; 115 kg/mm^2 YS; 15 El.
 Corrosion resisting and heat resisting alloy, for parts operating 650-750°C.
 Also used as hot work tool steel.

R.H.1.
 M-1201; 0.8-0.9 C, 17-20 W, 3.5-5.0 Cr, 1.0 V, 3 Mo, bal Fe.
 620 Brin.
 For welding rod for high speed steel; heat and wear resistant.

RHEINDUR.
 M-1548; 3.5-5.0 Cu, 0.5-1.0 Si, 0.5-1.0 Mn, 0.5-1.2 Mg, bal Al.
 Heat Treated: 57,000-71,000 TS; 50,000-54,000 YS; 10-12 El; 130-140 Brin.
 For fasteners, aircraft and jet engine components.
 Age-hardenable. High strength.

RHEINROHR 3CA2.
 M-625; 0.15 C, 0.75 Cr, 0.15 Mo, bal Fe.
 For gears, shafts, cams, camshafts; case hardened, tough.

RHEINROHR 3DB22.
 M-625; 0.13 C, 1 Cr, 0.25 Mo, 0.2 V, bal Fe.
 For gears, shafts, cams, camshafts; case hardened, tough.

RHEINROHR 3HK5.
 M-625; 0.2 C, 0.5 Si, 2.0-2.5 Cr, 1 Mo, bal Fe.
 Rolled: 71,000-92,500 TS; 42,700 YS; 12-18 El.
 For superheater tubes, boiler ends, chemical tanks; heat resistant to 580°F.

RHEINROHR 9F2.
 M-625; 0.44 C, Cr, Si, Mn, bal Fe.
 For gears, shafts, crankshafts, bolts, studs; water or oil hardened.

RHEINROHR 18/12 MS.
 M-625; 0.10 C, 16-18 Cr, 10.5-13 Ni, 2.0-2.5 Mo, Ti = 4 x C, bal Fe.
 Rolled: 78,000-107,000 TS; 35,600 YS; 45-40 RA.
 For superheater tubes, boiler ends, chemical tanks; heat resistant to 850°C.

RHEINROHR 18/12S.
 M-625; 0.08 C, 16 Cr, 13 Ni, Cb, bal Fe.
 Rolled: 78,000-107,000 TS; 31,300 YS; 50-40 El.
 For superheater tubes, boiler ends, chemical tanks; heat resistant to 850°C.

RHEINROHR 230V.
 M-625; 0.12 C, 5 Ni, bal Fe.
 Rolled: 85,000-102,000 TS; 58,000 YS; 20 El.
 For refrigeration components, bolts, fasteners; ductile at low temperature.

RHEINROHR C73.
 M-625; 0.35 C, 0.27 Si, 0.50 Mn, bal Fe.
 Hot rolled: 85,000 TS; 54,000 YS; 30 El; 53 RA; 183 Brin.
 For gears, bolts, fasteners; water hardened.

RHEINROHR C1203.
 M-625; C, alloy, bal Fe.
 For machine tool parts; oil hardened, tough.

RHEINROHR CS65S.
 M-625; 0.12 C, 1.15 Si, 2.5 Cr, 0.45 Mo, bal Fe.
 Annealed: 65,000-85,000 TS; 36,000 YS; 21 El.
 For superheater tubes, boiler ends, drums; heat resistant to 650°F.

RHEINROHR CUNI 47.
 M-625; 0.20 C, Cu, Ni, bal Fe.
 For machine tool parts; case hardened.

RHEINROHR CUNI 52.
 M-625; 0.20 C, Cu, Ni, Mn, bal Fe.
 For machine tool parts.

RHEINROHR CUNI 52 MO.
 M-625; 0.20 C, Cu, Ni, Mn, Mo, bal Fe.
 For machine tool parts.

RHEINROHR CV18W.
 M-625; 0.2 C, 1 Si, 1.5-2.0 Cr, V, Mo, bal Fe.
 Rolled: 71,000-92,500 TS; 42,700 YS; 25-20 El.
 For superheater tubes, boiler ends, chemical tanks; heat resistant to 620°C.

RHEINROHR D33.
 M-625; 0.17 C, 1.05 Mn, 0.30 Si, bal Fe.
 For gears, bolts, machine tool parts; case hardened.

RHEINROHR D45.
 M-625; C, alloy, bal Fe.
 For machine tool parts; oil hardened, tough.

RHEINROHR D45V.
 M-625; 0.1 C, 1.5 Cr, V, bal Fe.
 Rolled: 64,000-85,000 TS; 42,700 YS; 25-18 El.
 For petroleum cracking and ammonia synthesis equipment; heat resistant to 550°C.

RHEINROHR D73.
 M-625; 0.40 C, 0.37 Si, 0.95 Mn, bal Fe.
 For gears, bolts, machine tool parts; water hardened.

RHEINROHR D83.
 M-625; 0.46 C, 0.35 Si, 0.80 Mn, bal Fe.
 For gears, bolts, machiine tool parts; water hardened.

RHEINROHR E33.
 M-625; 0.14 C, 1.05 Mn, 0.3 Si, bal Fe.
 For gears, shafts, machine tool parts; case hardening steel.

RHEINROHR E34SP.
M-625; 0.17 C, 1.05 Mn, 0.3 Si, bal Fe.
For gears, shafts, machine tool parts; case hardening steel.

RHEINROHR E35SP.
M-625; 0.19 C, 0.5 Si, 1.15 Mn, bal Fe.
For gears, machine tool parts; case hardening steel.

RHEINROHR E45.
M-625; 0.19 C, 1.15 Mn, 0.5 Si, bal Fe.
For gears, shafts, machine tool parts; case hardening steel.

RHEINROHR E45SP.
M-625; 0.22 C, 0.5 Si, 1.15 Mn, bal Fe.
For gears, shafts, machine tool parts; case hardening steel.

RHEINROHR E93.
M-625; 0.46 c, 0.96 Mn, 0.35 Si, bal Fe.
Hot rolled: 100,000 TS; 60,000 YS; 24 El; 45 RA; 215 Brin.
For gears, bolts, shafts, fasteners; water hardened.

RHEINROHR F1203.
M-625; C, alloy, bal Fe.
For machine tool parts; oil hardened.

RHEINROHR H-205.
M-625; C, bal Fe.
For boiler plates.

RHEINROHR H-215.
M-625; C, bal Fe.
For boiler plates.

RHEINROHR H-215A.
M-625; C, bal Fe.
For boiler plates.

RHEINROHR H-316.
M-625; C, bal Fe.
For boiler plates.

RHEINROHR H-326.
M-625; C, bal Fe.
For boiler plates.

RHEINROHR H-437.
M-625; C, bal Fe.
For boiler plates.

RHEINROHR HSB40.
M-625; C, bal Fe.
For machine tool parts; water hardened.

RHEINROHR HSB45.
M-625; C, bal Fe.
For machine tool parts; water hardened.

RHEINROHR HSB50.
M-625; C, bal Fe.
For machine tool parts; water hardened.

RHEINROHR HSB50R.
M-625; C, bal Fe.
For machine tool parts; water hardened.

RHEINROHR HSB55.
M-625; C, bal Fe.
For machine tool parts; water hardened.

RHEINROHR IKI.
M-625; 0.1 C, 0.5 Ti, bal Fe.
Rolled: 48,000-60,000 TS; 21,400 YS; 22 El.
For chromium impregnation; for stainless surfaces.

RHEINROHR MN47MO.
M-625; 0.19 C, Mn, Mo, bal Fe.
For gears, bolts, camshafts, cams; case hardened.

RHEINROHR N1K.
M-625; 0.16 C, 2.25 Cr, bal Fe.
For oil refinery equipment; heat and creep resistant.

RHEINROHR N8N.
M-625; 0.17 C, 2.7 Cr, 0.25 Mo, 0.15 V, bal Fe.
For camshafts, cams, oil refinery equipment; case hardened.

RHEINROHR N9.
M-625; 0.20 C, 3.25 Cr, 0.5 Mo, 0.5 V, bal Fe.
For die casting dies; oil hardened.

RHEINROHR N10.
M-625; 0.21 C, 3 Cr, 0.4 Mo, 0.8 V, 0.37 W, bal Fe.
For die casting dies; oil hardened.

RHEINROHR TH31.
M-625; 0.15 C, 0.30 Mo, 0.60 Mn, 0.25 Si, bal Fe.
For gears, bolts, machine tool parts; case hardening steel.

RHEINROHR TH32.
M-625; 0.13 C, 0.85 Cr, 0.45 Mo, bal Fe.
For machine tool parts, gears, pinions; case hardening steel.

RHEINROHR TS53.
M-625; 0.2 C, 0.5 Ti, bal Fe.
Rolled: 50,000-64,000 TS; 25,000 YS; 30-20 El.
For steam boilers; good creep strength.

RHEINROHR TS57.
M-625; C, alloy, bal Fe.
For machine tool parts; oil hardened.

RHEINROHR TSL.
M-625; C alloy, bal Fe.
For machine tool parts; oil hardened.

RHEINROHR TT.
M-625; 0.10 C, Mn, Al, bal Fe.
Rolled: 58,000-72,000 TS; 36,000 min YS; 22 min El.
For refrigeration equipment; ductile to -80°C.

RHEINROHR TTA23.
M-625; 0.15 C, 17 Mn, 3 Cr, bal Fe.
Rolled: 100,000-128,000 TS; 38,500 YS; 30 El.
For low temperature applications; ductile to -185°C.

RHENIUM.
M-133; 99.5 Re.
70,000 TS; 24 El.
For constituent of alloys and chemicals; electroplating.

RHENIUM.
M-1755; Re.
Purities: 99.999%, 99.99%, 99.9%, 99+ %.
Forms: Powder, wire, sheet, foil, sintered bar, tubing, arc melted buttons, single crystals.

RHEOTAN NO. 1.
M-Eng.; 84 Cu, 12 Mn, 4 Zn.
For resistances.

RHEOTAN NO. 2.
M-Eng.; 53 Cu, 18 Zn, 25 Ni, 5 Fe.
For resistances.

RHEOTAN NO. 3.
M-Eng.; 84 Cu, 2 Mn, 4 Zn, 12 Fe.
For resistances.

RHF 15.
M-912; 0.15 C, 0.20 max Si, 0.95 N, 1.35 Cr, 0.90 Mo, 0.25 V, bal Fe.
Q + T: 1080 N/mm^2 TS; 930 N/mm^2 YS; 10 El.
High strength low carbon steel; may be case hardened.
W.-Nr. 1.7735.

RHF 32.
M-912; 0.03 max C, 9.0 Co, 5.0 Mo, 18.5 Ni, 0.75 Ti, Al, B, Zr, bal Fe.
Hot work and corrosion resisting for special hot working of metals.
W.-Nr. 2709; AFNOR Z 2 NKD 18-09.

RHODIUM.
M-1755; Rh.
Purities: 99.999%, 99.99%, 99.9%.
Forms: Sponge, powder, wire, foil, sheet, rod, single crystal.

RHOTANIUM.
M-Eng.; 90-60 Au, 10-40 Pd.
For jewelry; corrosion resistant.

RH SPECIAL.
M-114; 0.55 C, 0.35 Mn, 0.9 Si, 5.0 Cr, 0.25 V, 1.25 W, 1.25 Mo, bal Fe.
AISI Type A8 air hardening tool steel.

RICHARD PLASTIC BABBITT.
M-Eng.; 82.43 Sn, 9.77 Sb, 8.1 Cu.
For Babbitt, bearings; antifriction.

RICHARDS.
M-Eng.; 71.5 Sn, 25 Zn, 3.5 Al.
For aluminum solder.

RICHARDS ALLOY.
M-Eng.; 96 Zn, 4 Al.
For die castings.

RICHARDS BRONZE.
M-Eng.; 42 Zn, 1 Fe, 1-2 Al, bal Cu.
For marine hardware; corrosion resistant, high strength.

RICHARDSON'S SPECULUM.
M-Eng.; 65 Cu, 30 Sn, 2 As, 2 Si, 0.7 Zn.
For reflectors; high polish.

RICH GOLD METAL.
90 Cu, 10 Zn.
For window screen wire; commercial bronze.

RICH LOW BRASS.
M-8; 15 Zn, Pb, Fe, bal Cu.
For hardware, fittings; red brass.

RICHLOY HI-TENSILE SOLDER NO. 3.
M-910; 20 Sn, 80 Pb.
Cast: 7183 TS; 47 El.
For soft solder; M.P. 416°F.

RICHLOY HI-TENSILE SOLDER NO. 4.
M-910; 30 Sn, 70 Pb.
Cast: 7883 TS; 42 El.
For soft solder; M.P. 390°F.

RICHLOY HI-TENSILE SOLDER NO. 5.
M-910; 40 Sn, 60 Pb.
Cast: 9000 TS; 36 El.
For soft solder; M.P. 370°F.

RICHLOY HI-TENSILE SOLDER NO. 6.
M-910; 50 Sn, 50 Pb.
Cast: 9412 TS; 34 El.
For soft solder; M.P. 356°F.

RICHLOY HI TENSILE SOLDER NO. 7.
M-910; 65 Sn, 35 Pb.
Cast: 9577 TS; 32 El.
For soft solder; M.P. 360°F.

RIGOR.
M-111; 1 C, 0.6 Mn, 5.25 Cr, 1.1 Mo, 2 V, bal Fe.
For dies; for cold work. AISI A2.

RIKEN 201.
M-1110; 5-7 Al, 0.1-0.5 Mn, 1.5 max Zn, bal Mg.
Rolled: 31,000-37,000 TS; 3-10 El.
For light parts; wrought alloy.

RIKEN 202.
M-1110; 8-11 Al, 0.1-0.5 Mn, 1.0 max Zn, bal Mg.
Rolled: 37,00-42,500 TS; 5-7 El.
For light parts; wrought alloy.

RIKEN 203.
 M-1110; 0.5 max Al, 0.5-2.5 Mn, 0.5 max Zn, bal Mg.
 Rolled: 25,500-30,000 TS; 2 El.
 For light parts; wrought alloy.

RIKEN 501.
 M-1110; 3.5-6.5 Al, 0.1-0.5 Mn, 2.5-3.5 Zn, bal Mg.
 Cast: 25,000 TS; 5 El.
 For light parts; sand casting.

RIKEN 502A.
 M-1110; 8-11 Al, 0.1-0.5 Mn, bal Mg.
 Cast: 21,300 TS.
 For light parts; sand casting.

RIKEN 502B.
 M-1110; 8-11 Al, 0.1-0.5 Mn, bal Mg.
 Cast: 28,500-30,000 TS.
 For light parts; sand casting.

RIKEN 601.
 M-1110; 8-11 Al, 0.1-1.0 Mn, 0.2-1.0 Zn, bal Mg.
 Die cast.
 For light die castings; high strength.

RIM-ROLL NO. 9.
 1.05 C, 0.5 Cr, bal Fe.
 For rolls; water hardened.

RIOK 5F5.
 M-USSR; 1.5 C, 10 W, 4 Cr, 4.5 V, 6 Co, bal Fe.
 Hardened: 65-68 Rock C.
 For broaches, reamers, knitting and textile needles, form cutters.
 High-speed steel, high red-hardness.

RI PREMIUM.
 M-101; WC.
 For cutting tools, for superalloys.
 Sintered carbide, abrasion resistant.

RIVER ACE 60.
 M-1541; 0.12 C, 1.23 Mn, 0.015 P, 0.016 S, 0.35 Si, 0.31 Ni, 0.05 Mo, 0.041 V, bal Fe.
 94,000 psi TS; 80,000 psi YS; 35 El.
 High strength low alloy quenched and tempered steel for weldable structures, penstock and pressure vessels.

RIVER ACE 60L.
 M-1541; 0.12 C, 1.37 Mn, 0.012 P, 0.007 S, 0.36 Si, 0.86 Ni, 0.128 Mo, 0.017 V, bal Fe.
 94,700 psi TS; 82,600 psi YS; 29 E.
 High strength low alloy quenched and tempered steel for low temperature pressure vessels and structures.

RIVER ACE 70.
 M-1541; 0.12 C, 1.04 Mn, 0.016 P, 0.011 S, 0.25 Si, 0.24 Cu, 0.58 Ni, 0.43 Cr, 0.28 Mo, 0.031 V, 0.0031 B, bal Fe.
 105,000 psi TS; 96,000 psi YS; 25 El.
 High strength low alloy quenched and tempered steel for weldable structures and pressure vessels.

RIVER ACE 70L.
 M-1541; 0.15 C, 1.12 Mn, 0.016 P, 0.006 S, 0.29 Si, 0.24 Cu, 0.08 Ni, 0.40 Cr, 0.29 Mo, 0.027 V, 0.0028 B, bal Fe.
 105,400 psi TS; 92,500 psi YS; 27 El.
 High strength low alloy quenched and tempered steel for low temperature pressure vessels and structures.

RIVER ACE 100.
 M-1541; 0.10 C, 0.54 Mn, 0.014 P, 0.007 S, 0.23 Si, 0.47 Cu, 1.54 Ni, 0.51 Cr, 0.66 Mo, 0.046 V, 0.0018 B, bal Fe.
 143,400 psi TS; 135,100 psi YS; 23 El.
 High strength low alloy quenched and tempered steel for weldable structures and construction equipment.

RIVER ACE K-O.
 M-1541; 0.15 C, 1.14 N, 0.012 P, 0.015 S, 0.27 Si, 0.21 Cu, 0.28 N, 0.45 Cr, 0.31 Mo, 0.015 V, 0.0022 B, bal Fe.
 123,000 psi TS; 114,000 psi YS; 24 El.
 High strength low alloy quenched and tempered steel for weldable structures and pressure vessels.

RIVER ACE K-OL.
 M-1541; 0.16 C, 0.88 N, 0.009 P, 0.010 S, 0.28 Si, 0.24 Cu, 1.0 Ni, 0.58 Cr, 0.53 Mo, 0.050 V, 0.0035 B, bal Fe.
 126,700 psi TS; 116,500 psi YS; 24 El.
 High strength low alloy quenched and tempered steel for low temperature pressure vessels and structures.

RIVER ACE K-O M.
 M-1541; 0.12 C, 0.97 N, 0.006 P, 0.008 S, 0.29 Si, 0.21 Cu, 1.08 Ni, 0.44 Cr, 0.36 Mo, 0.03 V, 0.003 B, bal Fe.
 122,300 psi TS; 112,400 psi YS; 24 El.
 High strength low alloy quenched and tempered steel for weldable structures; penstock and pressure vessel purposes.

RIVER-H360.
 M-1541; 0.16 C, 1.0 Mn, 0.016 P, 0.010 S, 0.24 Si, 0.66 Cr, 0.30 Mo, 0.032 V, 0.002 B, bal Fe.
 Brinell hardness: 380 min.
 Low alloy steel with minimum hardness for abrasion resistance applications.

RIVER TEN 41.
M-1541; 0.12 C, 0.52 N, 0.065 P, 0.020 S, 0.30 Cu, bal Fe.
62,300 TS; 42,800 psi YS; 28 El.
Low alloy steel with good atmosphere corrosion resistance.

RIVER TEN 45K.
M-1541; 0.12 C, 0.54 Mn, 0.010 P, 0.015 S, 0.27 Si, 0.25 C, 0.13 Ni, 0.28 Cr, bal Fe.
84,800 psi TS; 67,800 psi YS; 32 El.
Low alloy steel, resistant to sulphuric acid corrosion.

RIVER TEN 50.
M-1541; 0.10 C, 0.85 Mn, 0.018 P, 0.012 S, 0.38 Si, 0.30 Cu, 0.27 Ni, 0.42 Cr, 0.013 Cb, bal Fe.
77,100 psi TS; 57,300 psi YS; 25 El.
High strength low alloy steel, good atmosphere, corrosion resistance, for weldable structures.

RIVER TEN 50M.
M-1541; 0.08 C, 0.50 Mn, 0.086 P, 0.017 S, 0.30 Si, 0.40 Cu, 0.39 Ni, 0.72 Cr, 0.028 Cb, bal Fe.
81,600 psi TS; 65,600 psi YS; 30 El.
High strength low alloy steel with good atmospheric corrosion resistance.

RIVER TEN 53.
M-1541; 0.13 C, 0.67 Mn, 0.019 P, 0.008 S, 0.38 Si, 0.27 Cu, 0.34 Ni, 0.49 Cr, 0.035 Cb, bal Fe.
81,100 psi TS; 64,400 psi YS; 36 El.
High strength low alloy steel, good corrosion resistance, for weldable structures.

RIVER TEN 62.
M-1541; 0.13 C, 0.56 Mn, 0.022 P, 0.010 S, 0.50 Si, 0.33 Cu, 0.33 Ni, 0.49 Cr, 0.065 Mo, bal Fe.
93,200 psi TS; 79,100 psi YS; 29 El.
Quenched and tempered high strength low alloy steel, good atmosphere corrosion resistance, for weldable structures.

RIVER TEN R.
M-1541; 0.09 C, 0.43 Mn, 0.086 P, 0.009 S, 0.35 Si, 0.32 Cu, 0.32 Ni, 0.46 Cr, bal Fe.
73,800 psi TS; 57,700 psi YS; 37 El.
High strength low alloy steel with good atmosphere corrosion resistance.

RK5.
M-USSR; 0.7 C, 18 W, 4 Cr, 1.2 V, 0.4 Mo, 5 Co, bal Fe.
Hardened: 64-66 Rock C.
For lathe and planer tools, reamers, cut-off tools and drills, broaches, hobs.
High-speed steel, high red-hardness.

RK10.
M-USSR; 0.8 C, 18 W, 4 Cr, 1.5 V, 0.5 Mo, 10 Co, bal Fe.
Hardened: 64-68 Rock C.
For cutters, lathe and planer tools, reamers, hobs, form cutters.
High-speed steel, high red-hardness.

RK15.
M-USSR; 0.8 C, 18 W, 4 Cr, 1 V, 0.6 Mo, 14 Co, bal Fe.
Hardened: 64-66 Rock C.
For reamers, broaches, form cutters, lathe and planer tools, hobs, milling cutters.
High-speed steel, high red-hardness.

RKCM.
M-912; 1.0 C, 5.25 Cr, 1.10 Mo, 0.30 V, bal Fe.
Cold work tool steel; for punching, cutting and trimming tools.
W.-Nr. 2363; similar to AISI A2.

RM 210.
M-1578; 90 Cu, 10 Zn.
Known as 90/10 Gilding metal and as "commercial bronze".
For ornaments, imitation jewelry.
BS 2870 CZ.101; ASTM B36-220.

RM 215.
M-1578; 85 Cu, 15 Zn.
Known as 85/15 Gilding metal or as "red brass."
For ornamental work, vanity cases, piping for carrying liquids, diaphragms and bellows.
BS 2870 CZ.102; ASTM B36-230.

RM 220.
M-1578; 80 Cu, 20 Zn.
Known as 80/20 Gilding metal and as "low brass."
For architectural work, costume jewelry, strip for slide fasteners, diaphragms, bellows, wire for Fourdrinier wire cloth.
BS 2870 CZ.103; ASTM B36-240.

RM 230 (CARTRIDGE BRASS).
M-1578; 70 Cu, 30 Zn.
High ductility for deep drawing into cartridge cases, rivets, screws, lamp fixtures, pen and pencil parts, and many similar drawn or formed parts.
BS 2870 CZ.106; ASTM B36-260.

RM 421.
M-1578; 63 Cu, 1 Pb, bal Zn.
Called matrix and key brass. For key blanks, watch, clock and instrumental parts; free machining.

RM 433.
M-1578; 62 Cu, 2 Pb, bal Zn.
Strip is called "clock brass", wire is heading and drilling wire.
For mild cold work; free machining; clock and instrument work.
BS 2800 CZ.119; ASTM B121-353.

RM 444.
M-1578; 61 Cu, 1 Pb, bal Zn.
Free machining brass sheet called "engraving brass".
For ornamental work, plaques.

RM 445.
M-1578; 61 Cu, 1.5 Pb, bal Zn.
Sheet and strip, called "rule brass". For engraving, key blanks, watch and clock parts.

RM 656.
M-1578, M-1805; 95 Cu, 5 Sn, P.
Phosphor bronze sheet, various tempers; for springs and stampings in textile, paper and chemical plant operations.
BS 2870 Pb 1021; ASTM B 103 Al.

RM 848 (CUPRO-NICKEL).
M-1578; 75 Cu, 25 Ni.
Coinage alloy.
BS 2870 CN.104.

RMI 0.2% PD.
M-1229; 0.08 C, 0.30 Fe, 0.05 N, 0.02 Pd, 0.015 H, (sheet), (all values max), bal Ti.
R.T., wrought, min: 50,000 TS; 40,000 YS; 22 El.
At 800°F, typical: 26,000 TS; 15,000 YS; 25 El.
At 1000°F, typical: 19,000 TS; 11,000 YS; 32 El.
For corrosion resistance in chemical industry where media is mildly reducing or varies between oxidizing and reducing.

RMI 1 AL-8V-5FE.
M-1229; 0.05 C, 0.08 N, 0.0125 H (all max values), 0.25-0.50 O, 0.8-1.8 Al, 4.0-6.0 Fe, 7.5-8.5 V, bal Ti.
R.T. min: 210,000 TS; 200,000 YS; 6 El; (as heat treated).
For high strength fasteners, Alpha-beta-alloy.

RMI 3 AL-2.5V.
M-1229; 0.05 C, 0.02 N, 0.30 Fe, 0.12 O, 0.0125 H, (all max values) 2.5-3.5 Al, 2.0-3.0 V, bal Ti.
Cold worked, stress relieved tubing at room temp.: 125,000 TS; 105,000 YS; 10 El.
At 800°F: 55,000 TS; 40,000 YS; 22 El.
For aircraft hydraulic tubing and fittings. Alpha-beta alloy.

RMI 4AL-3MO-1V.
M-1229; 0.08 C, 0.25 Fe, 0.05 N, 0.015 H (all max values), 3.75-4.75 Al, 2.3-3.5 Mo, 0.5-1.5 V, bal Ti.
R.T. min: 125,000 TS; 115,000 Ys; 10 El.
At 1000°F: 65,000 TS; 55,000 YS; 35 El.
Plate, sheet, strip; Alpha-beta alloy.
For aircraft uses requiring high strength and elevated temperature stability.

RMI 5AL-2.5 SN.
M-1229; 0.08 C, 0.50 Fe, 0.05 N, 0.20 O, 0.175 H (billet) (all above values max), 4.0-6.0 Al, 2.0-3.0 Sn, bal Ti.
R.T. wrought, min: 120,000 TS; 115,000 YS; 10 El.
At 800°F, typical: 78,000 TS; 59,000 YS; 18 El.
At 1000°F, typical: 67,000 TS; 55,000 YS; 19 El.
For parts, requiring weldability, oxidation resistance, stability and strength at elevated temperatures. Alpha alloy.

RMI 5AL-2.5 SN ELI.
M-1229; 0.08 C, 0.15 Fe, 0.05 N, 0.12 O, 0.125 H (bar) (all above values max), 4.7-5.75 Al, 2.0-3.0 Sn, bal Ti.
R.T. Wrought, min: 105,000 TS; 100,000 YS; 1 El.
At -320°F typical: 180,000 TS; 168,000 YS; 16 El.
At -423°F, typical: 229,000 TS; 206,000 YS; 15 El.
For use in high pressure cryogenic vessels at temperatures below -320°F. Alpha alloy.

RMI 5AL-6SN-2ZR, 1MO-SI.
M-1229; 0.05 C, 0.03 N, 0.30 Fe, 0.15 O, 0.0125 H, (all above max values) 4.5-5.5 Al, 5.0-7.0 Sn, 1.5-2.5 Zr, 0.5-1.0 Mo, 0.15-0.35 Si, bal Ti.
At 900°F: 102,000 TS; 76,000 YS; 17 El.
For use where high creep strength and elevated temperature are required, as in jet engine components; Alpha alloy.

RMI 6AL-2CB-1TA-1MO.
M-1229; 0.05 C, 0.03 N, 0.25 Fe, 0.10 O, 0.0125 H, (all above values max) 5.5-6.5 Al, 1.5-2.5 Cb, 0.5-1.5 Ta, 0.5-1.5 Mo, bal Ti.
R.T. Wrought: 115,000-125,000 TS; 105,000-115,000 YS; 10 min El.
At 1000°F, typical: 70,000 TS; 55,000 YS; 15 El.
For parts requiring toughness, strength, weldability and resistance to sea water; Alpha alloy.

RMI 6AL-2SN-4ZR-2MO-SI.
M-1229; 0.10 C, 0.05 N, 0.15 O, 0.25 Fe, 0.010 H (all max values), 5.5-6.5 Al, 1.75-2.25 Sn, 3.5-4.5 Zr, 1.75-2.25 Mo, bal Ti.
R.T. min: 130,000 TS; 120,000 YS; 10 El.
At 1000°F: 75,000 TS; 60,000 YS; 25 El.
Alpha-beta alloy; good high temperature creep strength; for jet engine components.

RMI 6AL-4V.
M-1229; 0.08 C, 0.25 Fe, 0.05 N, 0.010 H (billet) (all above values max), 5.5-6.75 Al, 3.5-4.5 V, bal Ti.
R.T. min: 130,000 TS; 120,000 YS; 10 El.
At 800°F: 90,000 TS; 75,000 YS; 17 El.
For compressor blades, discs, rings for jet engines, aircraft components, pressure vessels. Alpha-beta alloy.

RMI 6AL-4V ELI.
M-1229; 0.08 C, 0.13 O, 0.010 H (Billet), 0.15 Fe (all max values), 5.5-6.5 Al, 3.5-4.5 V, bal Ti.
R.T. min: 130,000 TS; 120,000 YS; 10 El.
-320°F: 220,000 TS; 205,000 YS; 14 El.
-423°F: 265,000 TS; 250,000 YS; 6 El.
Extra low interstitials permit use for high pressure cryogenic vessels down to -320°F. Alpha-beta alloy.

RMI 6AL-6V-2SN.
M-1229; 0.05 C, 0.04 N, 0.125 H (all max values), 5.0-6.0 Al, 5.0-6.0 V, 1.5-2.5 Sn, 0.35-1.0 Fe, 0.35-1.0 Cu, bal Ti.
R.T. min: 150,000 TS; 140,000 YS; 8-20 El.
At 600°F: 135,000 TS; 117,000 YS.
For rocket engine cases, ordnance components, aircraft parts; Alpha-beta alloy.

RMI 7AL-4MO.
M-1229; 0.08 C, 0.25 Fe, 0.05 N, 0.010 H (billet) (all max values), 6.5-7.5 Al, 3.5-4.5 Mo, bal Ti.
R.T. min: 145,000 TS; 135,000 YS; 10 El.
At 1000°F: 105,000 TS; 88,000 YS; 20 El.
For aircraft and jet engines; Alpha-beta alloy.

RMI 8AL-1MO-1V.
M-1229; 0.08 C, 0.05 N, 0.0125 H (billet) (all max values), 7.5-8.5 Al, 0.75-1.25 Mo, 0.75-1.25 V, bal Ti.
R.T. min: 135,000 TS; 125,000 YS; 10 El.
At 1000°F: 80,000 TS; 60,000 YS; 20 El.
Alpha-beta alloy; good creep property.
For aircraft and jet engine parts.

RMI 8MN.
M-1229; 0.08 C, 0.05 N, 0.015 H (sheet) (all max values), 6.5-9.0 Mn, bal Ti.
R.T. min: 120,000 TS; 110,000 YS; 10 El.
At 800°F: 87,000 TS; 66,000 YS; 20 El.
Sheet and plate for aircraft skin and structural components, Alpha-beta alloy.

RMI 13V-11CR-3AL.
M-1229; 0.06 C, 0.05 N, 0.0175 H, 0.35 Fe, (all max values), 12.5-14.5 V, 10-12 Cr, 2.5-4.0 Al, bal Ti.
R.T. min: 130,000 TS; 125,000 YS; 10 El.
At 1000°F: 100,000 TS; 75,000 YS; 35 El.
Beta alloy; for high strength aircraft structures.

RMI 30.
M-1229; 0.08 C, 0.25 Fe, 0.05 N, 0.010 max H (billet), (all max values), bal Ti.
R.T. (wrought) min: 40,000 TS; 30,000 YS; 25 El.
800°F (typical): 20,000 TS; 13,000 YS; 26 El.
Commercially pure titanium, wrought.
For corrosion resistance in the chemical and marine industries; in airframe construction.

RMI 40.
M-1229; 0.08 C, 0.25 Fe, 0.05 N, 0.010 H (billet) (all max values) bal Ti.
R.T. wrought, min: 50,000 TS; 40,000 YS; 22 El.
800°F, typical: 26,000 TS; 15,000 YS; 25 El.
For corrosion resistance in the chemical and marine industries; in aircraft construction.

RMI 55.
M-1229; 0.08 C, 0.25 Fe, 0.05 N, 0.010 H, (billet) (all max values) bal Ti.
R.T. wrought, min: 65,000 TS; 55,000 YS; 20 El.
At 800°F, typiccal: 29,000 TS; 17,000 YS; 18 El.
At 1000°F, typical: 22,000 TS; 13,000 YS; 33 El.
For corrosion resistance in the chemical and marine industries and for aircraft construction.

RMI 70.
M-1229; 0.08 C, 0.40 Fe, 0.05 N, 0.010 H, (billet) (all max values), bal Ti.
R.T. wrought, min.: 80,000 TS; 70,000 YS; 15 El.
At 800°F, typical: 34,000 TS; 21,000 YS; 22 El.
At 1000°F, typical: 27,000 TS; 16,000 YS; 29 El.
For corrosion resistance in chemical and marine industries and for aircraft construction.

RMI TI 3AL-8V-4ZR-4MO.
M-1229; 0.05 C, 0.03 N, 3.0-4.0 Al, 7.5-8.5 V, 5.5-6.5 Cr, 3.5-4.5 M 3.5-4.5 Zr, 0.30 Fe, bal Ti.
Beta alloy; for heavy section, high strength forgings for airframe, cold headed rivets, high shear fasteners.

RMI TI-5AL-2SN-2ZR-4MO-4CR.
M-1229; 0.05 C, 0.03 N, 0.15 Fe, 0.13 O_2, 0.0100 H_2(billet), all max values, 4.5-5.5 Al, 1.75-2.25 Sn, 1.75-2.25 Zr, 3.5-4.5 Mo, 3.5-4.5 Cr, bal Ti.
STA, R.T.: 180,000 TS; 170,000 YS; 7 El; 15 RA.
For aircraft frames and jet engine components. Alpha-beta alloy.

RMI TI-6AL-2SN-2ZR-2MO-2CR-0.25 SI.
M-1229; 0.05 C, 0.25 Fe, 0.03 N, 0.14 O_2, 0.0100 H_2 (billet), all max values, 5.25-6.25 Al, 1.75-2.25 Sn, 1.75-2.25 Zr, 1.75-2.25 Mo, 1.75-2.25 Cr, 0.20-0.30 Si, bal Ti.
STA, R.T.: 170,000 TS; 160,000 YS; 10 El; 20 RA.
800°F: 130,000 TS; 102,000 YS; 21 El; 40 RA.
For air frame and jet engine parts requiring high strength and fracture toughness. Alpha-beta alloy.

RMI TI-6AL-2SN-4ZR-6MO.
M-1229; 0.04 C, 0.04 N, 0.15 Fe, 0.15 O_2, 0.0100 H_2 (billet) all max values, 5.5-6.5 Al, 1.75-2.25 Sn, 3.5-4.5 Zr, 5.5-6.5 Mo, bal Ti.
STA, R.T.: (min) 170,000 TS; 160,000 YS; 10 L-8T El; 20L-15T RA.
At 1000°F: 125,000 TS; 95,000 YS; 15 El, 50 RA.
For jet engine components requiring high tensile strength with intermediate creep strength. Alpha-beta alloy.

RMK 5.
M-1653; 0.80-0.90 C, 3.25-4.25 Cr, 4.5-5.5 Mo, 5.5-7.0 W, 1.8-2.2 V, 4.5-5.5 Co.
Mo-W-Co high speed tool steel.
W.-Nr. 1.3243.

R-MONEL see MONEL ALLOY R-405.

R.N.D.
M-1175; 0.08 C, 1-1.25 Mn, bal Fe.
Welded: 55,000 TS; 45,000 YS; 33-40 El; 60-80 RA.
For welding rod.

R-NICKEL.
M-303; 99.2 Ni.
For food handling equipment.

RNO 100.
M-912; 1.05 C, 0.40 Si, 0.30 Mn, 17.0 Cr, 0.50 Mo, bal Fe.
For corrosion resistant roller bearings.
DIN X 105 CrMo 17; Similar to AISI 440C.

RNOD CO.
M-912; 0.08 C, 10.5 Cr, 0.70 Ni, 6.50 Co, 0.35 V, 0.80 Mo, 0.25 Nb + Ta, bal Fe.
Room temp: 100-115 kg/mm² TS; 85 kg/mm² YS.
Corrosion resistant and heat resistant alloy for operation 500-600°C.
W.-Nr. 1.4911.

RNOD NI.
M-912; 0.10 C, 11.5 Cr, 2.6 Ni, 1.4 Mo, 0.20 V, 0.30 Nb, bal Fe.
Room temp: 95-115 kg/mm² TS; 80 kg/mm² YS.
Corrosion resistant and heat resistant alloy for operation 500-600°C.
W.-Nr. 4939.

RNO MOV.
M-912; 0.21 C, 0.20 Si, 0.70 Mn, 12.0 Cr, 1.0 Mo, 0.70 Ni, 0.30 Mo, bal Fe.
Ht: 785-930 N/mm² TS.
For parts requiring corrosion resistance and heat resistance to 580°C.
W.-Nr. 4922.

ROBERTS-AUSTEN (PURPLE GOLD).
M-Eng.; 79 Au, 21 Al.
For jewelry; corrosion resistant.

ROBINS C.I. 25.
M-460; 3.4 T.C., 0.5 C.C., 2.5 Si, bal Fe.
Cast: 28,000 TS; 170-190 Brin.
For gray iron castings; Tr.S. 2000; Tr. def. 0.25".

ROBINS C.I. 35G.
M-460; 3.3 T.C., 0.55 C.C., 2.3 Si, 0.3 Cr, bal Fe.
Cast: 35,000 TS; 190-230 Brin.
For cast iron gears; Tr.S. 2200; Tr. def. 0.30".

ROBINS C.I. 35SG.
M-460; 3.3 T.C., 0.65 C.C., 2.3 Si, 1.1 Ni, bal Fe.
Cast: 35,000 TS; 210-240 Brin.
For cast iron gears; Tr.S. 2300; Tr, def. 0.30".

ROBINS C.I. 50.
M-460; 3.1 T.C., 0.65 C.C., 2.4 Si, 0.75 Ni, 0.2 Cr, 0.5 Mo, bal Fe.
Cast: 50,000 TS; 230 Brin.
Heat treated: 70,000 TS; 270 Brin.
For cast iron castings, gears; Tr.S, 3000; Tr. def. 0.40".

ROBINS COPPER-FREE NI-RESIST.
M-460; 2.5-3.0 T.C., 2.0-2.5 Si, 13.5-14.5 Ni, bal Fe.
Cast: 25,000 TS; 120-140 Brin.
For Tr.S. 2100; Tr. def. 0.60".

ROBINS E-5.
M-460; 3.3-3.5 C, 2.0-2.2 Si, 0.55-0.70 Mn, 0.9-1.1 Ni, 0.2-0.3 Mo, 0.1-0.3 Cr, bal Fe.
Cast: 35,000 TS; 190-220 Brin.
For Al melting pots; cast iron.
For extrusions, forgings and pressings; corrosion resistant.

ROBINS HIGH CHROME NI-RESIST.
M-460; 2.5-3.0 T.C., 2.0-2.5 Si, 13.5-14.5 Ni, 5.5-6.5 Cu, 4-6 Cr, bal Fe.
Cast: 30,000 TS; 160-200 Brin.
For stoker gates, flue dampers; Tr.S. 2400; Tr. def. 0.45".

ROBINS NI-HARD.
M-460; 3.3-3.6 C, 0.7-1.0 Si, 0.7-1.0 Mn, 4.25-4.75 Ni, 1.25- 1.70 Cr, bal Fe.
Cast: 575-650 Brin.
For rolling mill guides, crusher segments, liners, rolls; cast iron.

ROBINS STANDARD NI-RESIST.
M-460; 2.5-3.0 T.C., 2.0-2.5 Si, 13.5-14.5 Ni, 5.5-6.5 Cu, 2-2.5 Cr, 0.9-1.1 Mn, bal Fe.
Cast: 25,000 TS; 130-160 Brin.
For castings; non-magnetic; corrosion resistant.

ROBUST.
M-501; 0.5 C, 2.25 W, 1.5 Cr, 0.2 V, bal Fe.
For tools, dies, hot work dies; hot work steel.

ROBUST.
M-617; 0.7 C, 18 W, 4 Cr, 1 V, bal Fe.
For tools, cutters; high speed steel.

ROBUST 35.
 M-912; 0.32-4.0 C, 1.2-1.5 Si, 1.2-1.5 Cr, 0.05-1.1 V, bal Fe.
 For tools, dies; hot work steel.

ROBUST 40.
 M-912; 0.4-0.5 C, 0.4-0.6 Mn, 1.2-1.5 Si, 1.2-1.5 Cr, 0.05-1.1 V, bal Fe.
 For tools, dies, punches; shock resistant.

ROBUST 50.
 M-501; 0.6 C, 1.18 Cr, 0.1 V, 0.8 Si, 0.75 Mn, bal Fe.
 For gears, springs, crankshafts, fasteners; oil hardened, shock resistant.

ROBUST M.
 M-912; 0.45 C, 1.1 Si, 1.2 Cr, 2.5 W, bal Fe.
 For tools, dies; tough.

ROC.
 M-1766; 1.90 C, 0.40 Mn, 0.50 max Si, 12.0 Cr, 0.70 W, bal Fe.
 High carbon-high chromium, air hardening tool steel; for cold work stamping dies.
 IHA F-521; DIN X210CrW12.

ROC EXTRA.
 M-1766; 1.55 C, 0.40 Mn, 0.50 max Si, 12.0 Cr, 0.80 Mo, 0.25 V, bal Fe.
 High carbon-high chromium, air hardening tool steel; for cold work stamping dies.
 DIN X165 CrMoV12; AFNOR Z160CD12.

ROCHETTE-UMCO50.
 M-1582; 0.05-0.12 C, 0.18 Ti, 0.6 Cb, 26-30 Cr, 47-52 Co, bal Fe.
 Wrought: 132,000 TS; 61,000 YS; 7 El; 6 RA; 350 Brin.
 For furnace baffles, burner tips, sintering grates, quench baskets.
 Corrosion, heat, thermal shock resistant.

ROCHLING ZK.
 M-501, M-912; 1.45 C, 0.25 Si, 0.60 Mn, 1.4 Cr, bal Fe.

ROCKALOY GRADE A.
 M-829; WC + Co.
 For cutting tools; carbide.

ROCKALOY NO. 110.
 M-829; WC + Co.
 For cutting tools; carbide.

ROCKALOY NO. 111.
 M-829; WC + Co.
 For cutting tools; carbides.

ROCKALOY NO. 112.
 M-829; WC + Co.
 For cutting tools; carbides.

ROCKET.
 M-688; C, alloy, bal Fe.
 For tools, dies; wear and abrasion resistant, air or oil hardening.

ROCKRITE.
 M-957; 1.0 C, 1.0 Cr, bal Fe.
 For bearings, liners; SAE-52100 tubing.

ROCKWELL.
 M-1705; 0.6 C, 0.7 Mn, 1.85 Si, 0.2 V, 0.45 Mo, bal Fe.
 Shock resisting tool steel; oil or water hardening. AISI S5.

ROCOLOY.
 M-U.S.; 0.39 C, 0.7 Mn, 1.1 Cr, 1.0 Si, 1.0 Co, 0.25 Mo, 0.15 V, bal Fe.
 Modified AISI 4137 with Co added.

ROCOLOY 270.
 M-US; 0.4 C, 1.3 Si, 1.35 Cr, 0.8 Ni, 0.5 Mo, 0.3 W, 0.15 V, 1.35 Co, bal Fe.
 Ht. Tr.; 325,000 TS; 270,000 YS; 7 El; C 55 Rock.
 For high strength, corrosion resistant parts, fasteners, rocket motor cases.
 Heat treatable. High strength.

RODAR.
 M-61; 29 Ni, 17 Co, 0.3 Mn, bal Fe.
 Annealed: 80,000 TS; 30,000 YS; Rockwell B82.
 Cold drawn: 90,000 TS; 50,500 YS; 250 Brin.
 For metal to glass seals; resists thermal shock.

RODDRAW 305.
 M-1663; 0.12 max C, 18 Cr, 11.5 Ni, bal Fe.
 Ann: 90,000 psi (620.5 MPa) TS; 38,000 psi (262 MPa) YS; 50 El; B 80 Rock.
 For deep drawn parts, aerosol tops, gaskets.
 AMS 5514.

RODFLEX 270.
 M-1663; 0.15 C, 16.0-18.0 Cr, 6.0-8.0 Ni, 2.0 max Mn, 1.0 max Si, 0.50 max Cu, 0.50 max Mo, bal Fe.
 Cold rolled: 270,000 psi min TS.
 "High Yield" Cr-Ni stainless steel strip.
 For springs, retainer rings, fasteners.

RODFLEX 290.
 M-1663; 0.15 max C, 16.0-18.0 Cr, 6.0-8.0 Ni, 2.0 max Mn, 1.0 max Si, 0.50 max Cu, 0.50 max Mo, bal Fe.
 Cold rolled: 290,000 psi min TS.
 "High Yield" Cr-Ni stainless strip.
 For springs, retainer rings, fasteners.

RODFOR.
 M-1766; 1.0 C, 0.30 Mn, 0.40 max Si, 1.5 Cr, bal Fe.
 Oil hardening drill rod; for drills, hardened shafts, drift pins.
 IHA F523-F131; AFNOR 100Cr6; AISI 52100.

RODINOX 301.
M-1663; 0.15 max C, 17 Cr, 7 Ni, bal Fe.
Ann: 100,00 psi (689.5 MPa) TS; 40,000 psi (275.8 MPa) YS; 60 El; B 85 Rock.
For name plates, helicopter blades, stamped parts.
MIL-S-5059.

RODINOX 302.
M-1663; 0.15 max C, 18 Cr, 9 Ni, bal Fe.
Ann: 90,000 psi (620.5 MPa) TS; 40,000 psi (275.8 MPa) YS; 60 El; B 85 Rock.
For laminated couplings, camera parts, diaphragms.
MIL-S-5059.

RODINOX 304.
M-1663; 0.08 max C, 19 Cr, 10 Ni, bal Fe.
Ann: 90,000 psi (620.5 MPa) TS; 40,000 psi (275.8 MPa) YS; 50 El; B 80 Rock.
For shim stock, deep drawing parts, watch bands.
MIL-S-5059.

RODINOX 304L.
M-1663; 0.03 max C, 19 Cr, 10 Ni, bal Fe.
Ann: 90,000 psi (620.5 MPa) TS; 40,000 psi (275.8 MPa) YS; 50 El; B 80 Rock.
For bellows, flexible metal hose, diaphragms.
AMS 5511.

RODINOX 410.
M-1663; 0.15 max C, 12 Cr, bal Fe.
Ann: 70,000 psi (482.6 MPa) TS; 45,000 psi (310.3 MPa) YS; 25 El; B 80 Rock.
Ht: 180,000 psi (1241 MPa) TS; 140,000 psi (965 MPa) YS; C 40 Rock; Ann: fasteners, valves;
HT: valves, shims, cutlery, springs.
AMS 5504.

RODINOX 430.
M-1663; 0.12 max C, 17 Cr, bal Fe.
Ann: 75,000 psi (517 MPa) TS; 45,000 psi (310 MPa) YS; 25 El; B 80 Rock.
For gaskets, trim, chemical handling; non-hardenable by heat treat.

RODMAG 80.
M-1663; 0.06 C, 79 Ni, 4.5 Mo, bal Fe.
For magnetic shielding of electrical components.
MIL-N-47037.

RODRESIST 316.
M-1663; 0.08 C, 18 Cr, 13.5 Ni, 2.5 Mo, bal Fe.
Ann: 90,000 psi (620.5 MPa) TS; 40,000 psi (275.8 MPa) YS; 50 El; B 80 Rock.
Bellows, surgical implants, diaphragms, marine applications.
AMS 5524.

RODRESIST 316L.
M-1663; 0.03 max C, 17 Cr, 13 Ni, 2.5 Mo, bal Fe.
Ann: 90,000 psi (620.5 MPa) TS; 40,000 psi (275.8 MPa) YS; 50 El; B 80 Rock.
For bellows, turbine heat exchangers, welded parts.
AMS 5507.

RODSEAL 29-17.
M-1663; 0.06 max C, 29 Ni, 17 Co, bal Fe.
Coef. Thermal Expan: 4.38×10^{-6} in/in/°F, 70-1000°F.
For glass seals and hermetic seals.
ASTM F-15.

RODSEAL 36.
M-1663; 0.04 C, 36 Ni, bal Fe.
Coef. Therm. Expan: 1.6×10^{-6} in/in/°F, 70-1000°F.
For precision instruments, bi-metals, optical equipment.

RODSEAL 42.
M-1663; 0.05 C, 41.5 Ni, bal Fe.
Coef. Therm. Expan: 4.8×10^{-6} in/in/°F, 70-1000°F.
For glass seals.
ASTM F-30.

RODTEMP 309.
M-1663; 0.08 max C, 23 Cr, 13.5 Ni, bal Fe.
Ann: 90,000 psi (620.5 MPa) TS; 45,000 psi (310.3 MPa) YS; 45 El; B 85 Rock.
Heat exchangers, furnace parts, and for high temperature operation.
AMS 5523.

RODTEMP 310.
M-1663; 0.08 max C, 25 Cr, 20.5 Ni, bal Fe.
Ann: 90,000 psi (620.5 MPa) TS; 45,000 psi (310.3 MPa) YS; 45 El; B 85 Rock.
Jet engine parts, oil refinery equipment, furnaces.
AMS 5521.

RODWELD 321.
M-1663; 0.08 max C, 18 Cr, 10.5 Ni, Ti = 5 x C min, bal Fe.
Ann: 95,000 psi (655 MPa) TS; 35,000 psi (241.3 MPa) YS; 45 El; B 80 Rock.
For jet engine insulation blankets, welded parts.
AMS 5510.

RODWELD 347.
M-1663; 0.08 max C, 18 Cr, 10.5 Ni, Cb-Ta = 10 x C min, bal Fe.
Ann: 95,000 psi (655 MPa) TS; 40,000 psi (275.8 MPa) YS; 45 El; B 85 Rock.
For jet engine seals, bellows, welded parts.
AMS 5512.

ROESCH.
M-Eng.; 50 Zn, 49 Sn, 0.7 Sb, 0.2 Cu.
For aluminum solder.

ROGER METAL.
M-700; C, Cr, Ni, bal Fe.
For stainless parts; corrosion resistant.

ROHMIUM.
Al alloy.
For light alloy parts.

ROHN.
50 Ni, 30 Cr, 17 Fe.
For resistances, heat and corrosion resistant parts; non-scaling, heat resistant.

ROHRENWERK HMS 42.
M-625; 0.18 C, bal Fe.
Rolled: 57,000-72,000 TS; 34,000-40,000 YS.
For gears, cams, camshafts; case hardening steel.

ROLEY 275.
M-1288; 0.95 C, W, Mo, 4 Cr, 1 V, bal Fe.
For lathe and planer tools, reamers, broaches, drills, hobs; high speed steel.

ROLEY 300.
M-1288; 0.82 C, 4.1 Cr, 0.85 Mo, 1.6 V, 8.7 W, bal Fe.
For lathe and planer tools, drills, hobs; high speed steel.

ROLEY BA.
M-1288; 0.15 max C, 18 Cr, 8.5 Ni, bal Fe.
Annealed: 80,000 TS; 35,000 YS; 55 El; 75 RA; 150 Brin.
Cold drawn: 180,000 TS; 150,000 YS; 10 El; 250 Brin.
For chemical plant equipment, tanks, mixers, filters; Type 302; stainless austenitic.

ROLEY BA88.
M-1288; 0.15 max C, 18 Cr, 8.5 Ni, 0.2 S, bal Fe.
For screw machine products, fasteners; Type 303; stainless, free-cutting.

ROLEY B SPEZIAL.
M-1288; 0.1 max C, 2 Si, 18 Cr, 8.5 Ni, bal Fe.
Annealed: 80,000 TS; 35,000 YS; 55 El; 75 RA; 150 Brin.
Cold drawn: 180,000 TS; 150,000 YS; 10 El; 250 Brin.
For chemical plant equipment, tanks, filters, mixers; Type 302; stainless austenitic.

ROLEY B SUPER.
M-1288; 0.1 max C, 2.Si, 2 Mo, 18 Cr, 8.5 Ni, bal Fe.
Annealed: 85,000 TS; 35,000 YS; 50 El; 65 RA; 160 Brin.
Cold drawn: 150,000 TS; 135,000 YS; 6 El; 300 Brin.
For acid resistant chemical plant equipment; type 316; stainless, austenitic.

ROLEY CM2.
M-1288; 0.35 C, 16.5 C, 1.15 Mo, bal Fe.
For oil refinery and chemical plant equipment; corrosion resistant.

ROLEY CME2.
M-1288; 0.20 C, 1.25 Mn, 1.15 Cr, bal Fe.
For gears, cams, camshafts; case hardened, tough.

ROLEY CME3.
M-1288; 0.15 C, 0.65 Cr, 0.5 Mn, bal Fe.
For gears, bolts, machine tool parts; case hardened.

ROLEY CN12.
M-1288; 0.10 max C, 12.5 Cr, 12 Ni, bal Fe.
For valves; corrosion and heat resistant.

ROLEY CNM.
M-1288; 0.07 max C, 17 Cr, 4.75 Mo, 13 Ni, bal Fe.
Annealed: 90,000 TS; 40,000 YS; 45 El; 60 RA; 170 Brin.
For acid resistant chemical plant equipment, tanks; Type 317; stainless, austenitic.

ROLEY CRP2.
M-1288; 0.40 C, Cr, Mn, Mo, bal Fe.
For machine tool parts, bolts, gears; oil hardened, tough.

ROLEY CT.
M-1288; 0.12 max C, 18 Cr, 9.5 Ni, Ti = 4 x C, bal Fe.
Annealed: 80,000 TS; 35,000 YS; 55 El; 75 RA; 150 Brin.
Cold drawn: 180,000 TS; 150,000 YS; 10 El; 250 Brin.
For chemical plant equipment, tanks; Type 321; stainless, austenitic.

ROLEY CT2N.
M-1288; 0.12 max C, 18 Cr, 9.5 Ni, Cb = 8 x C, bal Fe.
Annealed: 90,000 TS; 45,000 YS; 56 El; 65 RA; 160 Brin.
For welded chemical plant equipment; Type 347; stainless, austenitic.

ROLEY CU.
M-1288; 0.07 max C, 17.5 Cr, 2.0 Mo, 17.5 Ni, Ti = 7 x C, bal Fe.
For valves, chemical plant equipment, welded parts; austenitic, stainless.

ROLEY DSA.
M-1288; 1.42 C, W, V, bal Fe.
For cutters, engravers' tools; water hardened, wear resistant.

ROLEY DSA SPEZIAL.
M-1288; 1.3 C, 4.75 W, 0.2 max Cr, 0.3 Mn, bal Fe.
For engravers' tools, forming and blanking dies; water hardened, wear resistant.

ROLEY ECR15 see **VEW E410.**

ROLEY ECR20 see **VEW E400.**

ROLEY EFW see **VEW E920.**
M-1288; 0.15 C, 0.25 Si, 0.37 Mn, bal Fe.
Annealed: 70,000 TS; 40,000 YS; 25 El; 60 RA; 145 Brin.
For gears, bolts, fasteners, machine tool parts; case hardened.

ROLEY ERJ 15.
M-1288; 0.15 C, 1.55 Cr, 1.55 Mn, bal Fe.
For gears, cams, camshafts; case hardened, tough.

ROLEY ERJ 20.
M-1288; 0.18 C, 2 Cr, 2 Ni, 0.5 Mn, bal Fe.
For gears, cams, camshafts; case hardened, tough.

ROLEY ERJ 25.
M-1288; 0.13 C, 0.7 Cr, 2.5 Ni, 0.5 max Mn, bal Fe.
For gears, cams, camshafts; case hardened, tough.

ROLEY ERM 20 see **VEW E300.**

ROLEY ESPH.
M-1288; 1.3 C, 0.25 max Si, 0.25 max Mn, bal Fe.
For engraving tools, cutters, broaches; Type W1; water hardened.

ROLEY ESPZH.
M-1288; 1.0 C, 0.25 max Si, 0.25 max Mn, bal Fe.
Annealed: 100,000 TS; 53,000 YS; 21 El; 42 RA; 200 Brin.
For drills, taps, reamers, hobs, cutters; Type W1; water hardened.

ROLEY EXTRA HART.
M-1288; 1.0 C, 0.25 max Si, 0.25 max Mn, bal Fe.
Annealed: 100,000 TS; 53,000 YS; 21 El; 42 RA; 200 Brin.
For springs, drills, taps, reamers; Type W1; water hardened.

ROLEY EXTRA MH.
M-1288; 1.15 C, 0.25 max Si, 0.25 max Mn, bal Fe.
Annealed: 110,000 TS; 58,000 YS; 18 El; 38 RA; 210 Brin.
For springs, drills, reamers, taps; Type W1; water hardened.

ROLEY EXTRA SEHR ZAH.
M-1288; 0.70 C, 0.25 max Si, 0.25 max Mn, bal Fe.
Heat treated: 175,000 TS; 128,000 YS; 12 El; 37 RA; 355 Brin.
For springs, rails, cutters, tools, axes, hammers; Type W1; water hardened.

ROLEY EXTRA SPEZIAL MH.
M-1288; 1.1 C, 0.25 Max Si, 0.25 max Mn, bal Fe.
Annealed: 110,000 TS; 58,000 YS; 18 El; 40 RA; 210 Brin.
For springs, taps, reamers, hobs, broaches; Type W1; water hardened.

ROLEY EXTRA SPEZIAL SEHR ZAH.
M-1288; 0.70 C, 0.25 max Si, 0.25 max Mn, bal Fe.
Heat treated: 175,000 TS; 128,000 YS; 12 El; 37 RA; 355 Brin.
For springs, rails, hammers, tools, axes; Type W1; water hardened.

ROLEY EXTRA SPEZIAL ZAH.
M-1288; 0.85 C, 0.25 max Si, 0.25 max Mn, bal Fe.
Heat treated: 190,000 TS; 145,000 YS; 10 El; 30 RA; 400 Brin.
For springs, drills, cutters, taps, punches; Type W1; water hardened.

ROLEY EXTRA SPEZIAL ZH see **VEW K988.**

ROLEY EXTRA ZAH see **VEW K930.**

ROLEY EXTRA ZH see **VEW K990.**

ROLEY EZH see **VEW K990.**

ROLEY FAM see **VEW F180.**

ROLEY GW.
M-1288; 1.2 C, W, V, bal Fe.
For cutters, forming dies; water hardened.

ROLEY GWA.
M-1288; 1.05 C, 1.2 Cr, bal Fe.
For bearings, bushings, sleeves, liners; water hardened, wear resistant.

ROLEY GWK.
M-1288; 1.25 C, 1.15 Si, 0.7 Mn, 1.2 Cr, bal Fe.
For bearings, cutters, bushings, liners; water or oil hardened.

ROLEY GWV.
M-1288; 1.15 C, 0.65 Cr, 0.1 V, bal Fe.
For bearings, bushings, liners; water hardened.

ROLEY HIR35.
M-1288; 0.22-0.34 C, 0.7 Cr, 3.5 Ni, bal Fe.
For gears, bolts, machine tool parts; oil hardened, shock resistant.

ROLEY HIR45.
M-1288; 0.35 C, 0.6 Mn, 1.3 Cr, 4.5 Ni, bal Fe.
For gears, bolts, crankshafts; oil hardened, shock resistant.

ROLEY HMN15.
M-1288; 0.30 C, 1.35 Mn, 0.25 Si, bal Fe.
For gears, bolts, machine tool parts; water hardened, tough.

ROLEY HMV35 see **VEW V762.**

ROLEY HMV42 see **VEW V742**.

ROLEY HMV50 see **VEW F550**.

ROLEY HRIM20 see **VEW V145**.

ROLEY HRM15 see **VEW V340**.

ROLEY HRM35 see **VEW V330**.

ROLEY HRM40 see **VEW V320**.

ROLEY HRM45.
M-1288; 0.42 C, Cr, Mo, bal Fe.
For gears, bolts, machine tool parts; oil hardened, tough.

ROLEY HRMV25.
M-1288; 0.30 C, 2.5 Cr, 0.2 Mo, 0.15 V, bal Fe.
For forging dies; upsetters, header dies; oil hardened, tough.

ROLEY KGS.
M-1288; 1.0 C, 1.55 Cr, 0.35 Mn, bal Fe.
For bearings, cutters, bushings; oil or water hardened, wear resistant.

ROLEY KL3 see **VEW R100**.

ROLEY KLR see **VEW K505**.

ROLEY KLS see **VEW R100**.

ROLEY KSA.
M-1288; 0.90 C, 1.9 Mn, 0.1 V, 0.25 Si, bal Fe.
For punches, dies, cutters, shears; oil hardened, non-deforming.

ROLEY L1.
M-1288; 0.12 max C, 0.4 Si, 13 Cr, bal Fe.
Annealed: 75,000 TS; 40,000 YS; 35 El; 70 RA; 155 Brin.
For turbine blades, cutlery, valves; Type 410; corrosion resistant.

ROLEY L2.
M-1288; 0.12 max C, 13 Cr, 0.4 Si, bal Fe.
Annealed: 75,000 TS; 40,000 YS; 35 El; 70 RA; 155 Brin.
For turbine blades, cutlery, valves; Type 410; corrosion resistant.

ROLEY L3.
M-1288; 0.12 max C, 0.4 Si, 13 Cr, bal Fe.
Annealed: 75,000 TS; 40,000 YS; 35 El; 70 RA; 155 Brin.
For turbine blades, cutlery, valves, surgical instruments; Type 410; corrosion resistant.

ROLEY M1.
M-1288; 0.20 C, 0.4 Si, 13 Cr, bal Fe.
Annealed: 95,000 TS; 50,000 YS; 25 El; 55 RA; 195 Brin.
For turbine blades, valves, cutlery, surgical instruments; Type 420; corrosion resistant.

ROLEY M2.
M-1288; 0.22 C, 17 Cr, 1.5 Ni, bal Fe.
Annealed: 125,000 TS; 95,000 YS; 20 El; 55 RA; 260 Brin.
For pumps, marine hardware, valves; Type 431.

ROLEY M3.
M-1288; 0.4 C, 0.4 Si, 13 Cr, bal Fe.
Annealed: 100,000 TS; 55,000 YS; 20 El; 50 RA; 200 Brin.
For valves, cutlery, surgical instruments; Type 420; corrosion resistant.

ROLEY M3E.
M-1288; 0.90 C, 18 Cr, 1.15 Mo, 1.0 V, bal Fe.
Annealed: 107,000 TS; 62,000 YS; 18 El; 35 RA; 220 Brin.
Heat treated: 280,000 TS; 270,000 YS; 3 El; 15 RA; 555 Brin.
For cutlery, valves, ball bearings, surgical instruments; Type 440 B; corrosion resistant.

ROLEY M4.
M-1288; 0.12 C, 16.5 Cr, 0.25 Mo, 0.2 S, bal Fe.
For screw machine products, bolts, shafts, fasteners; Type 416; corrosion resistant, free-cutting.

ROLEY MA3 see **VEW V960**.

ROLEY MA4 see **VEW V945**.

ROLEY MA5 see **VEW V935**.

ROLEY MA6 see **VEW V920**.

ROLEY MO-1.
M-1288; 0.70 C, Mo, Cr, W, V, bal Fe.
For lathe and planer tools, reamers, drills, taps; high speed steel.

ROLEY MO-2.
M-1288; 0.85 C, W, Mo, bal Fe.
For lathe and planer tools, reamers, drills, taps; high speed steel.

ROLEY MO-3.
M-1288; 0.88 C, Co, W, Mo, Cr, V, bal Fe.
For lathe and planer tools, drills, high speed steel.

ROLEY MO-4.
M-1288; 1.15 C, Co, W, Mo, Cr, V, bal Fe.
For engravers' tools, form cutters, broaches, reamers; high speed steel.

ROLEY NMV see **VEW V304**.

ROLEY P1D.
M-1288; 0.46 C, Ni, Cr, W, bal Fe.
For forging and heading dies; upsetters; oil hardened, tough.

ROLEY P2.
M-1288; 0.50 C, 3.25 Ni, 1.05 Cr, 0.50 Mn, bal Fe.
For gears, bolts, crankshafts, fasteners; oil hardened, shock resistant.

ROLEY P3.
M-1288; 0.45 C, 0.70 Mn, 0.30 V, 1.4 Cr, 0.7 Mo, bal Fe.
For gears, bolts, machine tool parts; oil hardened, tough.

ROLEY P6.
M-1288; 1.1 C, 0.4 Cr, 0.3 Mn, 0.2 Si, bal Fe.
For bearings, cutters, bushings; water hardened, wear resistant.

ROLEY P7.
M-1288; 1.05 C, 1.0 Cr, 0.30 Mn, 0.25 Si, bal Fe.
For bearings, bushings, liners; water hardened, wear resistant.

ROLEY P8.
M-1288; 0.90 C, 0.80 Cr, 0.30 Mn, 0.30 Si, bal Fe.
For bearings, bushings, liners; water hardened, wear resistant.

ROLEY PRIMA SEHR ZAH.
M-1288; 0.45 C, 0.3-0.8 Mn, 0.25-0.50 Si, bal Fe.
Hot rolled: 98,000 TS; 60,000 YS; 24 El; 55 RA; 212 Brin.
For axles, gears, bolts, bushings, crankshafts; water hardened.

ROLEY PRIMA WEICH.
M-1288; 0.35 C, 0.25-0.50 Si, 0.25-0.80 Mn, bal Fe.
Hot rolled: 84,000 TS; 54,000 YS; 30 El; 53 RA; 183 Brin.
For gears, shafts, axles, bolts, screws; water hardened.

ROLEY PRIMA ZAH.
M-1288; 0.60 C, 0.25-0.50 Si, 0.25-0.80 Mn, bal Fe.
Heat treated: 160,000-115,000 TS; 113,000-77,000 YS; 12-23 El; 40-54 RA; 320-230 Brin.
For wheels, die blocks, rails, girders; water hardened.

ROLEY SAO.
M-1288; 0.90 C, 0.10 V, 1.9 Mn, bal Fe.
For punches, forming and blanking dies, shear blades; oil hardened, non-deforming.

ROLEY SF.
M-1288; 0.65 C, 1.7 Si, 0.70 Mn, bal Fe.
For springs, punches, upsetters; oil hardened, tough.

ROLEY SK.
M-1288; 0.55 C, 0.90 Si, 0.30 Mn, 1.05 Cr, 0.18 V, 1.85 W, bal Fe.
For cold heading and forming dies; oil hardened, tough.

ROLEY SP2.
M-1288; 0.70 C, 1.7 Si, 0.70 Mn, bal Fe.
For springs, chisels, punches, pneumatic tools; oil hardened, shock resistant.

ROLEY SP2W.
M-1288; 0.53 C, 0.80 Si, 1.05 Mn, bal Fe.
For crankshafts, bolts, gears, axles; water hardened, tough.

ROLEY SPEZIAL.
M-1288; 0.86 C, 4.1 Cr, 0.85 Mo, 2.5 V, 12 W, bal Fe.
For lathe and planer tools, drills; high speed steel.

ROLEY SUPER.
M-1288; 1.3 C, 4.3 Cr, 0.85 Mo, 3.8 V, 12 W, bal Fe.
For forming and blanking dies, cutters; high speed steel.

ROLEY SUPER 3.
M-1288; 0.74 C, 4.1 Cr, 1.1 V, 18.5 W, bal Fe.
For lathe and planer tools, reamers, broaches; high speed steel.

ROLEY SUPERIOR.
M-1288; 1.05 C, 1.0 Cr, 1.15 W, 0.90 Mn, bal Fe.
For heading and blanking dies; form tools; oil hardened, wear resistant.

ROLEY SUPER SPEZIAL.
M-1288; 0.65 C, W, Co, 4 Cr, V, Mo, bal Fe.
For lathe and planer tools, reamers, punches, taps; high speed steel.

ROLEY SVA.
M-1288; 1.0 C, 0.10 V, 0.25 Mn, 0.20 Si, bal Fe.
For blanking and forming dies; Type W2; water hardened.

ROLEY T70.
M-1288; 0.1 max C, 17.5 Cr, Ti = 7 x C, bal Fe.
Annealed: 80,000 TS; 50,000 YS; 25 El; 50 RA; 150 Brin.
For oil refinery equipment, dairy and food equipment; Type 430 Ti, corrosion resistant.

ROLEY TO1.
M-1288; 1.35 C, W, Co, bal Fe.
For engravers' tools, forming and blanking dies; oil hardened, wear resistant.

ROLEY TO2.
M-1288; 0.79 C, 4.7 Co, 4.3 Cr, 0.75 Mo, 1.5 V, 18 W, bal Fe.
For lathe and planer tools, cutters, reamers, drills; high speed steel.

ROLEY TO3.
M-1288; 0.86 C, 2.8 Co, 4.3 Cr, 0.85 Mo, 2.1 V, 12 W, bal Fe.
For lathe and planer tools, cutters, reamers, taps; high speed steel.

ROLEY TO4.
M-1288; 0.80 C, 4 Cr, Co, Mo, W, V, bal Fe.
For lathe and planer tools, broaches; high speed steel.

ROLEY TO SUPER.
 M-1288; 0.76 C, 10 Co, 4.2 Cr, 0.8 Mo, 1.8 V, 18 W, bal Fe.
 For lathe and planer tools, form cutters, drills; high speed steel.

ROLEY UDL.
 M-1288; 0.38 C, 1.05 Cr, 0.1 V, 0.95 Mn, bal Fe.
 For gears, bolts, springs, crankshafts; oil hardened, shock resistant.

ROLEY UDW.
 M-1288; 0.38 C, Si, Cr, V, bal Fe.
 For gears, bolts, springs; oil hardened, shock resistant.

ROLEY UDWM.
 M-1288; 0.61 C, 1.18 Cr, 0.10 V, 0.75 Mn, 0.85 Si, bal Fe.
 For springs, punches, crankshafts; oil hardened, shock resistant.

ROLEY UK.
 M-1288; 0.55 C, 0.90 Si, 1.05 Cr, 0.18 V, 1.85 W, bal Fe.
 For heading and blanking dies; cold work tools; oil hardened, tough.

ROLEY UKL.
 M-1288; 0.67 C, 1.3 Si, 0.50 Mn, 0.50 Cr, bal Fe.
 For punches, springs, pneumatic tools; oil hardened, tough.

ROLEY UKS.
 M-1288; 0.45 C, Si, Cr, V, bal Fe.
 For gears, bolts, crankshafts, springs; oil hardened, shock resistant.

ROLEY UKW(D).
 M-1288; 0.35 C, 0.90 Si, 1.05 Cr, 0.18 V, 1.85 W, bal Fe.
 For heading and forming dies, cold work tools; oil hardened, tough.

ROLEY WB10.
 M-1288; 0.25 C, 1.15 Cr, 0.25 Mo, 0.55 Mn, bal Fe.
 For gears, bolts, crankshafts, axles; oil hardened, tough.

ROLEY WB20.
 M-1288; 0.24 C, 1.35 Cr, 0.55 Mo, 0.20 V, bal Fe.
 For die casting dies, punches; oil hardened, tough.

ROLEY ZC10.
 M-1288; 0.41 C, 0.25 Si, 0.70 Mn, 1.1 Cr, bal Fe.
 For gears, bolts, machine tool parts; oil hardened, tough.

ROLEY ZCMO10.
 M-1288; 0.50 C, 1.0 Cr, 0.2 Mo, 0.65 Mn, 0.25 Si, bal Fe.
 For gears, bolts, machine tool parts; oil hardened, tough.

ROLEY ZIM EXTRA.
 M-1288; 1.4 C, 0.30 Cr, 0.10 V, 0.30 Mn, bal Fe.
 For engravers' tools, cutters, bearings; water or oil hardened, wear resistant.

ROLEY ZIM SPEZIAL.
 M-1288; 1.4 C, 0.30 Cr, 0.10 V, 0.30 Mn, bal Fe.
 For engravers' tools, cutters, form tools; oil hardened, wear resistant.

ROLLED AUGER.
 M-289; C, bal Fe.
 For tools; water hardened.

ROLLE MA-356.
 M-1501; 6.5-7.5 Si, 0.20-0.75 Mg, 0.03-0.25 Be, 0.20 Ti, 0.2 Cu, 0.1 Zn, bal Al.
 Cast: 40,000 TS; 30,000 YS; 5 El; 60 Brin.
 Heat treated: 50,000 TS; 40,000 YS; 2 El; 70 Brin.
 For gear cases, housings, crankcases, oil pans, airframe fittings; age-hardenable, high strength, corrosion resistant.

ROLLODUR 1.
 M-1247; 13 Cr, 10 W, 1.5 C, bal Fe.
 For milling cutters, tools; cast to shape.

ROLLODUR 2.
 M-1247; 13 Cr, 4 W, 1.5 C, bal Fe.
 For milling cutters, tools; cast to shape.

ROLLODUR 44-22.
 M-1247; 1.4 C, 4 W, Cr, Mo, Co, V, bal Fe.
 For cutting tools, milling cutters; cast to shape, abrasion resistant.

ROLLODUR 108-32.
 M-1247; 1.8 C, 10 W, Cr, Mo, Co, V, bal Fe.
 For cutting tools; abrasion resistant.

ROLLOY.
 M-170; 2.5-3.5 T.C., 4.5 Ni, 1.5 Cr, bal Fe.
 Cast: 570-650 Brin.
 For wear resistant castings; wear resistant.

ROLLSCHEN 1 CR 5 NI.
 M-1240; 0.2 C, 1 Cr, 0.5 Ni, bal Fe.
 Heat treated: 78,000-125,000 TS; 50,000 YS; 16 El; 55 RA; 170-250 Brin.
 For gears, shafts, pinions, cams, camshafts; tough, case-hardening steel.

ROLLSCHEN 1 CR 10 NI.
 M-1240; 0.2 C, 0.1 Cr, 1.0 Ni, bal Fe.
 Heat treated: 107,000-148,000 TS; 85,000 YS; 12 El; 45 RA; 210-250 Brin.
 For gears, shafts, cams, pinions, camshafts; case hardening.

ROLLSCHEN 1 CR 15 NI.
 M-1240; 0.2 C, 0.1 Cr, 1.5 Ni, bal Fe.
 Heat treated: 135,000-176,000 TS; 99,600 YS; 10 El; 40 RA; 220-260 Brin.
 For gears, shafts, pinions, camshafts; case hardening.

ROLLSCHEN 1 CR 20 NI.
M-1240; 0.2 C, 0.1 Cr, 2.0 Ni, bal Fe.
Heat treated: 162,000-205,000 TS; 114,000 YS; 7 El; 35 RA; 235-300 Brin.
For gears, shafts, camshafts; tough, case hardening.

ROLLSCHEN 2 MO 4.
M-1240; 0.2 C, 0.4 Mo, bal Fe.
Heat treated: 78,300-98,500 TS; 50,000 YS; 16 El; 60 RA.
For gears, shafts, pinions; case hardening.

ROLLSCHEN 3 CR 5 NI.
M-1240; 0.4 C, 0.3 Cr, 0.5 Ni, bal Fe.
Heat treated: 85,500-99,500 TS; 57,000 YS; 19 El; 55 RA; 210-230 Brin.
For gears, shafts, crankshafts; oil hardening, tough.

ROLLSCHEN 3 CR 10 MO 10.
M-1240; 0.3 C, 1.0 Cr, 1.0 Mo, bal Fe.
Heat treated: 121,000-142,000 TS; 107,000 YS; 15 El; 50 RA.
For gears, shafts, axles; oil hardening, tough.

ROLLSCHEN 3 CR 10 NI.
M-1240; 0.4 C, 0.3 Cr, 1.0 Ni, bal Fe.
Heat treated: 99,000-121,000 TS; 71,000 YS; 15 El; 55 RA; 230-250 Brin.
For gears, shafts, crankshafts; oil hardening, tough.

ROLLSCHEN 3 CR 15 NI.
M-1240; 0.4 C, 0.3 Cr, 1.5 Ni, bal Fe.
Heat treated: 107,000-128,000 TS; 85,000 YS; 14 El; 55 RA; 240-260 Brin.
For gears, shafts, crankshafts; oil hardening, tough.

ROLLSCHEN 3 CR 15 NI 12 MO AL.
M-1240; 0.3 C, 1.5 Cr, 0.8 Ni, 0.3 Mo, 1 Al, bal Fe.
Heat treated: 114,000-142,000 TS; 85,000 YS; 14 El; 40 RA; 250 Brin.
For gears, pinions, cams, shafts; nitriding steel.

ROLLSCHEN 3 CR 20 NI.
M-1240; 0.4 C, 0.3 Cr, 2 Ni, bal Fe.
Heat treated: 121,000-142,000 TS; 100,000 YS; 14 El; 50 RA; 260-280 Brin.
For gears, shafts, crankshafts; oil hardening, tough.

ROLLSCHEN 3 CR NI 10 MO.
M-1240; 0.3 Cr, Ni, 1 Mo, bal Fe.
Heat treated: 128,000-148,000 TS; 107,000 YS; 12 El; 50 RA; 230-260 Brin.
For gears, shafts, crankshafts; oil hardening, tough.

ROLLSCHEN 3 CR NI 15 MO.
M-1240; C, 0.3 Cr, Ni, 1.0 Mo, bal Fe.
Heat treated: 148,000-176,000 TS; 121,000 YS; 9 El; 45 RA; 245 Brin.
For gears, shafts, crankshafts; oil hardening, tough.

ROLLSCHEN 3 CR NI 20 MO.
M-1240; 0.3 Cr, Ni, 1.0 Mo, bal Fe.
Heat treated: 169,000-197,000 TS; 141,000 YS; 9 El; 40 RA; 260 Brin.
For gears, shafts, axles; oil hardening, tough.

ROLLSCHEN 4 CR 10 MO 4.
M-1240; 0.4 C, 1.0 Cr, 0.4 Mo, bal Fe.
Heat treated: 128,000-148,000 TS; 107,000 YS; 12 El; 45 RA.
For gears, shafts, axles; oil hardening, tough.

ROLLSCHEN 5 CR SI.
M-1240; 0.5 C, 0.7 Mn, 1.2 Si, bal Fe.
Heat treated: 220,000-250,000 TS; 185,000 YS; 4 El; 18 RA.
For springs; oil hardening.

ROLLSCHEN 5 MN SI.
M-1240; 0.5 C, 0.7 Mn, 1.2 Si, bal Fe.
Heat treated: 184,000-214,000 TS; 148,000 YS; 7 El; 20 RA.
For springs; oil hardening.

ROLLSCHEN 5 NI 12 CR 8 MO 3.
M-1240; 0.5 C, 1 Cr, 1.7 Ni, 0.3 Mo, bal Fe.
Heat treated: 190,000-200,000 TS.
For gears, shafts, crankshafts, axles; oil hardening, shock resistant.

ROLLSCHEN 5 NI 12 CR 8 MO 7.
M-1240; 0.5 C, 1 Cr, 1.7 Ni, 0.7 Mo, bal Fe.
Heat treated: 220,000-240,000 TS.
For gears, shafts, crankshafts, axles; oil hardening, shock resistant.

ROLLSCHEN WA EXTRA.
M-1240; 0.5 C, 0.7 Si, 1.5 Cr, 0.5 Mo, 0.3 V, bal Fe.
Heat treated: 156,000-184,000 TS.
For gears, shafts, cranshafts, axles; oil hardening, shock resistant.

ROLLSCHEN WAG46.
M-1240; 0.4 C, 1 Si, 5.5 Cr, 0.8 Mo, 0.5 V, bal Fe.
Heat treated: 185,000-205,000 TS.
For tools, dies, punches; oil hardening, shock resistant.

ROL-MAN.
M-180; 11-14 Mn, 1-1.3 C, bal Fe.
130,000-160,000 TS; 35-50 El; 180 Brin.
For woven wire screens, wear plates, pins, bushings, stokers; austenitic steel, non-magnetic.

ROMA BRONZE.
M-Eng; 59 Cu, 41 Zn, 0.4 Pb, 0.2 Al, traces Fe.
For hardware, nuts, bolts; free-cutting.

ROMANIUM.
M-Eng; 0.25 Cu, 1.75 Ni, 0.15 Sn, 0.17 W, 97.43 Al.
Hard rolled: 47,500 TS; 5 El.
Annealed: 34,000 TS; 19.4 El.
For light alloy parts.

ROMILLY BRASS.
M-England; 29 Zn, 0.2 Sn, 0.3 Pb, bal Cu.
For fittings, hardware; free-cutting, corrosion resistant.

RONENSIL.
M-501; C, 16 Mn, 7-9 Cr, 1 Mo, bal Fe.
For stainless steel parts; stainless.

RO NI 9.
M-1339; 0.10 max C, 9.0 Ni, bal Fe.
Steel that is tough at sub-zero temperature.
W.-Nr. 1.5662.

RONIA METAL.
Cu, Zn, Co, Mn, P.
For hardware; corrosion resistant.

ROOFLOY.
M-314; 0.02 Ca, 0.02 Mg, 0.02 Sn, bal Pb.
For roofing; rolled sheet.

RORSCHACH 23S.
M-1140; 3.5-4.5 Cu, 0.2-0.9 Si, 0.8-1.5 Mn, 0.8-1.5 Mg, bal Al.
For light alloy parts; age-hardened.

RORSCHACH 26S.
M-1140; 3.5-4.5 Cu, 1 max Si, 0.2-1.0 Mg, 0.2-1.0 Mn, bal Al.
For light alloy parts; age-hardened.

RORSCHACH 65S.
M-1140; 0.5-1.0 Si, 0.2-1.0 Mn, 0.15-0.40 Cr or Cu, 0.5-1.0 Mg, bal Al.
For light alloy parts.

RORSCHACH A4S.
M-1140; 1 max Mg, 0.5-1.0 Mn, bal Al.
For light alloy parts.

RORSCHACH A56S.
M-1140; 4-6 Mg, 0.4 max Mn or Cr, bal Al.
For light alloy parts; corrosion resistant.

RORSCHACH A57S.
M-1140; 2 max Mg, 0.5 max Mn or Cr, bal Al.
For light alloy parts; corrosion resistant.

RORSCHACH AR-A4S.
M-1140; 1 Mn, 0.7 Mg, bal Al.
O-temper: 17,000 TS; 5700 YS; 25 El; 40 Brin.
1/2 H-temper: 28,500 TS; 21,000 YS; 7 El; 60 Brin.
H-temper: 31,400 TS; 30,000 YS; 5 El; 70 Brin.
For roofing; good formability.

RORSCHACH AR-A57S.
M-1140; 1.5 Mg, bal Al.
O-temper: 18,500 TS; 7100 YS; 35 El; 30 Brin.
1/2 H-temper: 30,000 TS; 28,500 YS; 6 El; 60 Brin.
H-temper: 35,000 TS; 34,000 YS; 4 El; 70 Brin.
For chemical drums, architectural trim; good formability.

RORSHACK 3S.
M-1140; 1.0-2.0 Mn, bal Al.
Soft: 18,500 TS; 10,000 YS; 30 El; 25 Brin.
Hard: 33,500 TS; 31,500 YS; 4 El; 60 Brin.
For food, textile and marine equipment; corrosion resistant.

RORSHACK 4S.
M-1140; 0.5-1.5 Mg, 0.5-1.5 Mn, bal Al.
Soft: 28,500 TS; 10,000 YS; 17 El; 40 Brin.
Hard: 45,500 TS; 34,200 YS; 4 El; 90 Brin.
For transport and marine equipment; stronger than 3s.

RORSHACK 17S.
M-1140; 4 Cu, 0.5 Mg, 0.5 Mn, bal Al.
O-temper: 32,700 TS; 32,700 YS; 18 El; 45 Brin.
T6-temper: 64,000 TS; 42,700 YS; 17 El; 110 Brin.
For aircraft and transport equipment and parts, structures; age-hardenable.

RORSHACK 17S CLAD.
M-1140; 4 Cu, 0.5 Mg, 0.5 Mn, bal Al.
O-temper: 31,000 TS; 14,200 YS; 18 El; 45 Brin.
T6-temper: 59,800 TS; 39,800 YS; 17 El; 110 Brin.
RT-temper: 64,000 TS; 45,500 YS; 11 El; 115 Brin.
For aircraft structures, fuselage, wings; clad, age-hardenable.

RORSHACK 24S.
M-1140; 4.2 Cu, 1.5 Mg, 0.5 Mn, bal Al.
O-temper: 37,000 TS; 18,500 YS; 14 El; 45 Brin.
T-temper: 69,700 TS; 49,800 YS; 15 El; 110 Brin.
RT-temper: 71,100 TS; 59,800 YS; 9 El; 115 Brin.
For aircraft structures, fuselage, wings; age-hardenable.

RORSHACK 24S CLAD.
M-1140; 4.2 Cu, 1.5 Mg, 0.5 Mn, bal Al.
O-temper: 34,200 TS; 14,200 YS; 14 El.
T-temper: 66,800 TS; 45,500 YS; 15 El.
RT-temper: 68,300 TS; 47,000 YS; 9 El.
For aircraft structures, fuselage, wings; age-hardenable, clad.

RORSHACK 51S.
M-1140; 1 Si, 0.5-1.0 Mg, 0.2-1.0 Mn, bal Al.
O-temper: 18,500 TS; 11,400 YS; 22 El; 25 Brin.
W-temper: 38,400 TS; 24,400 YS; 10 El; 90 Brin.
T-temper: 47,000 TS; 38,400 YS; 11 El; 100 Brin.
For ship building and transportation industries; corrosion resistant.

RORSHACK 57S.
M-1140; 2.5 Mg, 0.25 Cr, bal Al.
O-temper: 31,300 TS; 19,900 YS; 25 El; 45 Brin.
H-temper: 45,500 TS; 39,800 YS; 4 El; 95 Brin.
For architectural trim, refrigerator trays; wrought.

RORSHACK 61S.
M-1140; 0.6 Mg, 1 Si, 0.25 Cr, bal Al.
O-temper: 19,900 TS; 11,400 YS; 20 El; 45 Brin.
W-temper: 42,700 TS; 35,500 YS; 17 El; 95 Brin.
RT-temper: 54,100 TS; 45,500 YS; 7 El; 105 Brin.
For structures; age-hardenable.

ROSE FUSIBLE ALLOY-1.
M-England; 50 Bi, 22 Sn, 28 Pb.
For safety plugs, fuses; M.P. 79°C.

ROSE FUSIBLE ALLOY-2.
M-England; 35 Bi, 30 Sn, 35 Pb.
For safety plugs, fire extinguishers; M.P. 98°C.

ROSEIN.
M-England; 40 Ni, 30 Al, 20 Sn, 10 Ag.
For jewelry, ornaments; corrosion resistant.

ROSE LABEL.
M-1.24; C, 0.27 Mn, bal Fe.
For tools, cutters; water hardened.

ROSENHAIN-ARCH-BUTT ALLOY.
72 Al, 25 Zn, 3 Cu.
For strong light alloy parts; non-hardenable.

RO SI.
M-912; 0.10 C, 0.25 Si, 0.40 Mn, bal Fe.
Carburizing steel for piston pins, camshafts, levers.
W.-Nr. 1121.

ROSIN.
M-Eng.; 40 Ni, 30 Al, 20 Sn, 10 Ag.
For jewelry and ornamental articles.

ROSS ALLOY.
M-Eng.; 68 Cu, 32 Sn.
For reflectors; "Rotguss."

ROSSLYN METAL.
M-1122; stainless-clad copper.
Rolled: 80,000-86,000 TS; 37,000-50,000 YS; 31-60 El.
For cooking utensils, kettles, heat exchangers; formerly "Amclad."

ROSTHORN-WILN BRASS.
M-Eng.; 31.9 Zn, bal Cu.
For condenser tubes, stampings, hardware; high ductility.

ROSTODUR R3.
M-1325; 0.20 C, 13 Cr, 1 Mo, bal Fe.
Annealed: 95,000 TS; 40,000 YS; 25 El; B 82 Rock.
Hardened: 235,000 TS; 200,000 YS; 10 El. C 50 Rock.
For valves, cutlery, bearings, surgical instruments.
Corrosion resistant, hardenable.

ROSTODUR R7.
M-1325; 0.15 C, 13 Cr, 1 Mo, bal Fe.
Annealed: 80,000 TS; 40,000 YS; 15 El; B 93 Rock.
For springs, chemical plant and oil refinery equipment, table flatware.
Corrosion resistant.

ROTAS.
M-498; 0.15-0.25 C, bal Fe.
For case-hardened parts; case-hardening steel.

ROTELLOY 2.
M-108; Fe-Co-V.
Soft magnetic alloy, high strength, for high speed generators.

ROTELLOY 3.
M-108; Fe-Co-V.
Soft magnetic alloy, high strength, for high speed generators.

ROTGUSS.
M-Germ.; 93-82 Cu, 4-10 Sn, 3-10 Zn, 1-2 Pb.
For cast fittings, valves, hardware; free-cutting.

ROTGUSS 4.
M-1420; Zn, bal Cu.
Cast: 26,000-29,000 TS; 15 El; 50 Brin.
For hardware; brass.

ROTGUSS 5.
M-1420; Zn, bal Cu.
Cast: 21,000-29,000 TS; 10 El; 60 Brin.
For hardware; brass.

ROTGUSS 8.
M-1420; Zn, bal Cu.
Cast: 21,000-29,000 TS; 8 El; 70 Brin.
For hardware; brass.

ROTGUSS 10.
M-1420; Zn, bal Cu.
Cast: 29,000-36,000 TS; 12 El; 70 Brin.
For hardware; brass.

ROTGUSS RED BRONZE.
M-Ger.; 4-10 Sn, 3-10 Zn, Pb, bal Cu.
For hardware, fittings; free-cutting, castings.

ROTOR-11.
M-1348; 1.2 C, 0.20 Cr, 0.10 V, 1.0 W, bal Fe.
For blanking and forming dies, headers; oil hardened, abrasion resistant.

ROTOR LDP.
M-1318; 1.05 C, 1.15 W, 0.90 Mn, 1.0 Cr, bal Fe.
For bearings, cutters, bushings; oil or water hardened, wear resistant.

ROTOR N.
M-1318; 1.05 C, 0.65 Cr, 0.10 V, 0.30 Mn, bal Fe.
For bearings, bushings, cutters; water hardened, tough.

ROTOR SPEZIAL N.
M-1318; 1.45 C, 1.4 Cr, 0.6 Mn, bal Fe.
For bearings, cutters, liners, forming dies; oil or water hardened.

ROTOXIT.
M-Eng.; 4 Si, bal Cu.
For chemical apparatus; corrosion resistant.

ROTPUNKT.
M-1331; 0.90 C, W, Mo, 4 Cr, 1 V, bal Fe.
For lathe and planer tools, hobs, taps, reamers; high speed steel.

ROTPUNKT DMO5.
M-1331; 0.7 C, W, Mo, 4 Cr, 1 V, bal Fe.
For lathe and planer tools, hobs, broaches; high speed steel.

ROTPUNKT DMO5CO.
M-1331; 0.7 C, W, Mo, 4 Cr, V, bal Fe.
For lathe and planer tools, drills, reamers, taps; high speed steel.

ROTPUNKT K5.
M-1331; 0.7 C, W, Mo, 4 Cr, V, bal Fe.
For lathe and planer tools, hobs, broaches; high speed steel.

ROTUNG D.
M-1246; W, Cu.
For electrical contacts; tungsten impregnated with copper.

ROTUNG P.
M-1246; W, Cu.
For electrical contacts; tungsten impregnated with copper.

ROULZ.
M-Ger.; 35-50 Cu, 25-30 Ni, 20-40 Ag.
For ornamental and corrosion resistant pens; corrosion resistant.

ROXO.
M-753; 0.4 C, 2.4 Si, bal Fe.
For castings; cast iron.

R-P.
M-40; 0.65 C, 0.8 Mn, 1.95 Si, bal Fe.
For tools; oil hardening.

RPG 3.
M-912; 0.30 C, 3.0 Cr, 2.85 Mo, 0.50 V, bal Fe.
Hot work tool steel; pressing dies, die inserts, punches, die casting dies.
W.-Nr. 2365; Similar to AISI H10.

RQ-100.
M-24; 0.12-0.21 C, 0.45-0.70 N, 0.20-0.35 S, 0.001-0.005 B, 0.45-0.60 Mo, 1.20-1.50 Ni, 0.85-1.20 Cr, bal Fe.
Roller quenched and tempered: 115,000-135,000 psi TS; 100,000 psi min YS; 18 El; 235-293 Brin.
Structural steel plate, to 2 1/2 inch thick.

RQ-100A.
M-24; 0.12-0.21 C, 0.45-0.70 N, 0.20-0.35 S, 0.001-0.005 B, 0.50-0.65 Mo, bal Fe.
Roller quenched and tempered: 115,000-135,000 psi TS; 100,000 psi YS; 18 El; 235-293 Brin.
Structural steel plate, to 1 1/4 inch thick.

RQ-100B.
M-24; 0.12-0.21 C, 0.45-0.70 Mn, 0.20-0.35 Si, 0.001-0.005 B, 0.45-0.60 Mo, 1.20-1.50 Ni, bal Fe.
Roller quenched and tempered: 115,000-135,000 psi TS; 100,000 psi min YS; 18 El; 235-293 Brin.
Structural steel plate, to 2 inch thick.

RQ-321.
M-24; 0.12-0.21 C, 0.45-0.70 Mn, 0.20-0.35 Si, 1.20-1.50 Ni, 0.85-1.20 Cr, 0.45-0.60 Mo, 0.001-0.005 B, bal Fe.
Q + T to 321 Brin min.
Abrasion resisting steel plate to 4 inch thick.

RQ-321A.
M-24; 0.12-0.21 C, 0.45-0.70 Mn, 0.20-0.3 Si, 0.50-0.65 Mo, 0.001-0.005 B, bal Fe.
Q + T to 321 Brin min.
Abrasion resisting steel plate to 1 1/4 inch thick.

RQ-321B.
M-24; 0.12-0.21 C, 0.45-0.70 Mn, 0.20-0.35 Si, 1.20-1.50 Ni, 0.45-0.60 Mo, 0.001-0.005 B, bal Fe.
Q + T to 321 Brin min.
Abrasion resisting steel plate to 2 inch thick.

RQ-340.
M-24; 0.12-0.21 C, 0.45-0.70 N, 0.20-0.35 Si, 1.20-1.50 Ni, 0.85-1.20 Cr, 0.45-0.60 Mo, 0.001-0.005 B, bal Fe.
Q + T to 340 Brin min.
Abrasion resisting steel plate to 2 inch thick.

RQ-340A.
M-24; 0.12-0.21 C, 0.45-0.70 Mn, 0.20-0.35 Si, 0.50-0.65 Mo, 0.001-0.005 B, bal Fe.
Q + T to 340 Brin min.
Abrasion resisting steel plate to 1 inch thick.

RQ-340B.
M-24; 0.12-0.21 C, 0.45-0.70 Mn, 0.20-0.35 Si, 1.20-1.50 Ni, 0.45-0.60 Mo, 0.001-0.005 B, bal Fe.
Q + T to 340 Brin min.
Abrasion resisting steel plate to 1 1/2 inch thick.

RQ-360.
M-24; 0.12-0.21 C, 0.45-0.70 Mn, 0.20-0.35 Si, 1.20-1.50 Ni, 0.85-1.20 Cr, 0.45-0.60 Mo, 0.001-0.005 B, bal Fe.
Q + T to 360 Brin min.
Abrasion resisting steel plate to 1 1/2 inch thick.

RQ-360A.
M-24; 0.12-0.21 C, 0.45-0.70 Mn, 0.20-0.35 Si, 0.50-0.65 Mo, 0.001-0.005 B, bal Fe.
Q + T to 360 Brin min.
Abrasion resisting steel plate to 1/2 inch thick.

RQ-360B.
M-24; 0.12-0.21 C, 0.45-0.70 Mn, 0.20-0.35 Si, 1.20-1.50 Ni, 0.45-0.60 Mo, 0.001-0.005 B, bal Fe.
Q + T to 360 Brin min.
Abrasion resisting steel plate to 1 inch thick.

RQ AR-321.
M-24; 0.25-0.32 C, 0.40-0.65 Mn, 0.20-0.35 Si, 0.80-1.15 Cr, 0.15-0.25 Mo, bal Fe.
Q + T to 321 Brin min.
Abrasion resisting steel plate to 3 inch thick.

RQ AR-340.
M-24; 0.25-0.32 C, 0.40-0.65 Mn, 0.20-0.35 S, 0.80-1.15 Cr, 0.15-0.25 Mo, bal Fe.
Q + T to 340 Brin min.
Abrasion resisting steel plate to 2 inch thick.

RQ AR-360.
M-24; 0.25-0.32 C, 0.40-0.65 Mn, 0.20-0.35 Si, 0.80-1.15 Cr, 0.15-0.25 Mo, bal Fe.
Q + T to 360 Brin min.
Abrasion resisting steel plate to 1 1/2 inch thick.

RQ AR-400.
M-24; 0.25-0.32 C, 0.40-0.65 Mn, 0.20-0.35 Si, 0.80-1.15 Cr, 0.15-0.25 Mo, bal Fe.
Q + T to 400 Brin min.
Abrasion resisting steel plate, 3/8 to 1 inch thick.

RQC-60.
M-24; 0.20 max C, 0.70-1.35 Mn, 0.15-0.50 Si, bal Fe.
As normalized: 70,000-90,000 psi TS; 50,000 min YS; 22 El.
Quenched and tempered: 80,000-100,000 psi TS; 60,000 min YS; 22 El.
For pressure vessels, weldable.

RQC-80.
M-24; 0.20 max C, 1.35 max Mn, 0.15-0.30 S, B, bal Fe.
Q + T: 95,000 psi min TS; 80,000 psi min YS; 18 min El.
Steel plates for structural applications; 3/16 inch to 1 1/2 inch thick.

RQC-90.
M-24; 0.20 max C, 1.35 max Mn, 0.15-0.30 S, 0.0005 min B, bal Fe.
Q + T: 100,000 psi min TS; 90,000 psi min YS; 18 min El.
Steel plates for structural purposes; 3/16 inch to 1 1/2 inch thick.

RQC-100.
M-24; 0.20 max C, 1.50 max Mn, 0.15-0.30 Si, 0.0005 min B, bal Fe.
Q + T: 110,000 psi min TS; 100,000 psi min YS; 18 min El.
Steel plates for structural purposes; 3/16 inch to 1 1/4 inch thick.

RQC-321.
M-24; 0.28 max C, 1.50 max Mn, 0.20-0.60 Si, 0.005 min B, bal Fe.
Q + T to 321 Brin min.
Abrasion resisting steel plate to 1 1/4 inch thick.

RQC-340.
M-24; 0.28 max C, 1.50 max Mn, 0.20-0.60 Si, 0.005 min B, bal Fe.
Q + T to 340 Brin min.
Abrasion resisting steel plate, 1/4 to 1 inch thick.

RR 5.
M-912; 0.55 C, 0.15 max Si, 0.40 Mn, bal Fe.
Carbon tool steel; stone working tools.
W.-Nr. 1820.

R.R. 50.
M-106; 0.8-2.0 Cu, 0.05-0.2 Mg, 1.5-2.8 Si, 0.8-1.4 Fe, 0.8-1.7 Ni, 0.05-0.3 Ti, bal Al.
Sand cast: 22,400-27,800 TS; 14,500-16,800 YS; 2-4 El; 60-70 Brin; 1.6 Izod.
General purpose sand or chill cast aluminum.
Gen. Eng. BS 1490 LM 23-P; BS 2L51.

RR.90-227.
M-1739; C-Cu-Fe, sintered.
Double press, double sintered.
Density: 70 gms/cc; UTS: 71 kg/mm^2 typical. Elon: 2-5%.
Gears, sprockets, levers.

R.R.250.
M-106; 4.5-5.5 Cu, 0.2-0.3 Mn, 0.8-1.2 Ni, 0.15-0.25 Ti, 0.1-0.4 Sb, 0.1-0.4 Co, bal Al.
Sand cast: 31,400-35,800 TS; 29,000 YS; 0-2 El; 90-100 Brin; 0.6 Izod.
Modified "Y" alloy for service up to 250°C.

R.R. 350.
M-106; 4.5-5.5 Cu, 0.2-0.3 Mn, 1.3-1.8 Ni, 0.15-0.25 Ti, 0.1-0.4 Sb, 0.1-0.4 Co, 0.1-0.3 Zr, bal Fe.
Sand cast: 31,400-35,800 TS; 29,000 YS; 0-2 El; 90-100 Brin; 0.6 Izod.
Modified "Y" alloy for service up to 350°C.

RRAC7.
M-426; 4.6 Sn, 1.6-2.0 Ni, 0.7-0.9 Mn, 0.4-0.8 Sb, 0.4-0.6 Si, 0.35-0.50 Mg, bal Al.
For bearings; for engine bearings.

RRAC9.
M-426; 55.5-70.0 Sn, 1.5-1.8 Ni, 0.6-0.9 Cu, 0.7-1.0 Mg, 0.15-0.35 Si 0.20-0.45 Fe, bal Al.
For bearings; connecting rods.

R.R.ALLOY C.242.
M-239; 0.3 C, 20 Cr, 10 Co, 10 Mo, 1.0 max Fe, bal Ni.
For high temperature applications; heat and corrosion resistant.

RS 115.
M-1633; 85 Cu, 5 Sn, 5 Pb, 5 Zn.
Bar: 32,000-43,000 TS; 15,000-24,000 YS; 20-35 El; 55-65 Brin.
For bearings, bushings, sleeves, liners.
Ounce metal, free machining.

RS 123.
M-1633; 81 Cu, 3 Sn, 7 Pb, 9 Zn.
Bar: 29,000-39,000 TS; 13,000-17,000 YS; 18-30 El; 50-60 Brin.
For bearings, bushings, liners, sleeves.
Good embeddability.

RS 194.
M-1633; 81 Cu, 19 Sn.
Bar: 40,000-48,000 TS; 150-170 Brin.
For bearings, bushings, linings.
Bridge bronze, corrosion resistant.

RS 197.
M-1633; 85 Cu, 14 Sn, 1 Zn.
Bar: 30,000-39,000 TS; 1-3 El.
For bearings, bushings, sleeves.
Corrosion resistant.

RS 198.
M-1633; 83 Cu, 14 Sn, 3 Zn.
Bar: 32,000 TS; 23,000 YS; 12 El; 75 Brin.
For bearings, bushings, sleeves.
Navy H, corrosion resistant.

RS 205.
M-1633; 89 Cu, 10 Sn, 2 Pb.
Bar: 35,000 TS; 21,000 YS; 12 El; 85 Brin.
For worm wheels, gears, shafts.
Gear Bronze, corrosion resistant, free-machining.

RS 206.
M-1633; 88 Cu, 10 Sn, 2 Pb.
Bar: 36,000-46,000 YS; 18,000-26,000 YS; 15-25 El; 65-80 Brin.
For bearings, bushings, liners.
Leaded Gun Metal, good embeddability.

RS 210.
M-1633; 88 Cu, 10 Sn, 2 Zn.
Bar: 40,000-50,000 TS; 20,000-25,000 YS; 20-35 El; 60-75 Brin.
For bearings, bushings, liners, gears.
Gun Metal, corrosion resistant.

RS 215.
M-1633; 87 Cu, 10 Sn, 1 Pb, 2 Zn.
Bar: 37,000-47,000 TS; 17,000-22,000 YS; 18-25 El.
For bearings, bushings, liners.
Commerical Gun Metal, corrosion resistant.

RS 225.
M-1633; 88 Cu, 8 Sn, 4 Zn.
Bar: 47,000-51,000 TS; 18,000-23,000 YS; 65-80 Brin; 25-35 El.
For bearings, bushings, gears.
Navy G Bronze, corrosion resistant.

RS 230.
M-1633; 87 Cu, 8 Sn, 1 Pb, 4 Zn.
Bar: 36,000-45,000 TS; 16,000-24,000 YS; 18-30 El; 60-75 Brin.
For gears, bushings, bearings. Leaded Navy G. Bronze.

RS 245.
M-1633; 88 Cu, 6 Sn, 1 Pb, 4 Zn.
Bar: 36,000-48,000 TS; 16,000-21,000 YS; 25-40 El; 60-72 Brin.
For bearings, bushings, liners.
Navy M. bronze, corrosion resistant.

RS 305.
M-1633; 80 Cu, 10 Sn, 10 Pb.
Bar: 27,000-37,000 TS; 15,000-22,000 YS; 10-20 El; 55-70 Brin.
For bearings, bushings, liners.
Corrosion resistant, good embeddability.

RS 311.
M-1633; 84 Cu, 8 Sn, 8 Pb.
Bar: 25,000-32,000 TS; 7-11 El.
For bearings, bushings, liners.
Corrosion resistant, good embeddability.

RS 315.
M-1633; 83 Cu, 7 Sn, 7 Pb, 3 Zn.
Bar: 30,000-38,000 TS; 17,000-21,000 YS; 12-20 El; 55-65 Brin.
For bearings, bushings, liners.
Bearing Bronze, corrosion resistant.

RS 319.
M-1633; 78 Cu, 6 Sn, 16 Pb.
Bar: 25,000-33,000 TS; 14,000-20,000 YS; 10-18 El; 50-60 Brin.
For bearings, bushings, liners.
Semi-Plastic Bronze, good embeddability.

RS 321.
M-1633; 71 Cu, 7 Sn, 21 Pb.
Bar: 24,000-31,000 TS; 15,000 YS; 15 El; 55 Brin.
For bearings, bushings, liners.
Hi-Leaded Tin Bronze.

RS 323.
M-1633; 70 Cu, 5 Sn, 25 Pb.
Bar: 23,000-30,000 TS; 11,000-16,000 YS; 7-16 El; 42-45 Brin.
For bearings, bushings, liners.
Hi-Leaded Tin Bronze.

RS 326.
M-1633; 85 Cu, 5 Sn, 9 Pb, 1 Zn.
Bar: 25,000-32,000 TS; 12,000-18,000 YS; 8-12 El; 60 Brin.
For bearings, bushings, liners.
Corrosion resistant.

RS 327.
M-1633; 75 Cu, 5 Sn, 20 Pb.
Bar: 22,000-30,000 TS; 7-10 El.
For bearings, bushings, liners.
Corrosion resistant. Good embeddability.

RS 410.
M-1633; 56 Cu, 2 Sn, 9 Pb, 20 Zn, 12 Ni.
Bar: 34,000-40,000 TS; 17,000-20,000 YS; 25 El; 50-60 Brin.
For bushings, bearings, liners.
Nickel Silver, corrosion resistant.

RS 415A.
M-1633; 87 Cu, 3 Fe, 9 Al.
Bar: 60,000-75,000 TS; 23,000-28,000 YS; 15-25 El; 90-125 Brin.
For gears, bushings, bearings, shafts.
Aluminum Bronze, corrosion resistant, tough.

RS 415B.
M-1633; 88 Cu, 1 Fe, 10 Al.
Bar: 70,000-85,000 TS; 25,000-30,000 YS; 20-35 El; 100-140 Brin.
For gears, bushings, bearings, shafts.
Aluminum Bronze, corrosion resistant, tough.

RS 415 C.
M-1633; 83 Cu, 2 Ni, 4 Fe, 11 Al.
Bar: 85,000-95,000 TS; 30,000-41,000 YS; 12-20 El; 160-195 Brin.
For bearings, liners, bushings, shafting.
Aluminum Bronze, corrosion resistant.

RS 415 D.
M-1633; 81 Cu, 4 Ni, 4 Fe, 11 Al.
Bar: 95,000 TS; 45,000 YS; 7 El; 195 Brin.
For bushings, bearings, liners, shafting.
Corrosion resistant. Aluminum Bronze.

RS 421.
M-1633; 58 Cu, 40 Zn, 1 Fe, 1 Al, 0.25 Mn.
Bar: 70,000-88,000 TS; 28,000-40,000 YS; 20-35 El; 90-120 Brin.
For bearings, bushings, sleeves, shafting, impellers.
Manganese Bronze, corrosion resistant.

RS 423.
M-1633; 64 Cu, 24 Zn, 3 Fe, 5 Al, 4 Mn.
Bar: 90,000-100,000 TS; 45,000-55,000 YS; 18-30 El; 160-200 Brin.
For bushings and bearings, shafting.
Manganese Bronze, corrosion resistant.

RS 424.
M-1633; 64 Cu, 24 Zn, 3 Fe, 5 Al, 4 Mn.
Bar: 11,000-120,000 TS; 65,000-90,000 YS; 12-20 El; 170-225 Brin.
For bushings and bearings, shafts.
Corrosion resistant. Manganese Bronze.

RS 500.
M-1633; 5 Zn, 1.5 Ni, 2.5 Fe, 1.5 Al, 3-5 Si, 0.25 Mn, bal Cu.
Bar: 45,000-65,000 TS, 20,000-35,000 YS; 15-40 El; 55-90 Brin.
For bushings and bearings, shafts.
Silicon Bronze, corrosion resistant.

R-S "A" METAL.
M-475; 3.2 C, Mn, Si, bal Fe.
For machinery castings, gears, abrasion resistant, cast iron.

R-S "H" METAL.
M-1475; C, alloy, bal Fe.
For valves; abrasion and corrosion resistant.

RS LEGIERUNG.
M-1548; 0.7-1.2 Si, 0.7-1.2 Mn, 0.8-1.2 Mg, bal Al.
Heat Treated: 40,000-51,000 TS; 28,000-43,000 YS; 5-10 El; 75-95 Brin.
For general structures, scaffolds, rails, booms, transmission towers.
Age-hardenable, good strength and fabricability.

RSM 200.
M-97; 0.03 max C, 0.10 max Mn, 17.0-19.0 Ni, 3.5-4.5 Mo, 7.0-8.0 Co, 0.05-0.15 Al, 0.10-0.25 Ti, bal Fe.
H.T.: 200 ksi min YS.
Maraging steel.

RSM 250.
M-97; 0.03 max C, 0.10 max Mn, 17.0-19.0 Ni, 4.6-5.1 Mo, 7.0-8.0 Co, 0.05-0.15 Al, 0.30-0.50 Ti, bal Fe.
H.T.: 250 ksi min YS.
Maraging steel.

RSM 300.
M-97; 0.03 max C, 0.10 max Mn, 0.10 max Si, 18.0-19.0 Ni, 8.5-9.5 Co, 4.7-5.2 Mo, 0.50-0.80 Ti, Al, bal Fe.
Maraging steel.

R.S.V.
M-1118; 0.7 C, 18 W, 4 Cr, 1 V, bal Fe.
For tools, dies, cutters, reamers; high speed steel.

R.T. 6.
M-1118; 0.7-0.8 C, bal Fe.
For tools, punches, springs; water hardened.

RT 6.
M-912; 0.70 C, 0.15 Si, 0.20 Mn, bal Fe.
Carbon tool steel; for hand tools as chisels, hammers, punches.

R.T. 7.
M-1118; 0.7-0.8 C, bal Fe.
For tools, punches, springs; water hardened.

R.T. 8.
M-1118; 0.9-1.0 C, bal Fe.
For tools, drills, springs; water hardened.

RT 8.
M-912; 0.85 C, 0.15 Si, 0.20 Mn, bal Fe.
Carbon tool steel for punching and stamping tools.
W.-Nr. 1525.

RT 10.
M-912; 1.05 C, 0.15 Si, 0.20 Mn, bal Fe.
Carbon tool steel for shear blades, punches, embossing dies, threading tools.
W.-Nr. 1545.

R.T. 10.
M-1118; 0.9-1.0 C, bal Fe.
For tools, taps, drills, punches; water hardened.

R.T. 10.
M-1118; 1.05-1.15 C, bal Fe.
For tools, taps drills, punches; water hardened.

RTC 14.
M-912; 1.0 C, 1.50 Cr, bal Fe.
Cold work tool steel; for drills, threading dies, lathe centers, reamers, drift pins.
W.-Nr. 2067; AFNOR 100 C 6.

R-THERM 20-12 MO.
M-1331; 0.20 C, 12 Cr, 1.2 Mo, bal Fe.
Annealed: 80,000 TS; 40,000 YS; 24 El; B 94 Rock.
Hardened: 220,000 TS; 180,000 YS; 12 El; C 48 Rock.
For gears, cutlery, surgical instruments, valves, bearings.
Corrosion resistant, hardenable.

R-THERM 2012 MOWV.
M-1331; 0.22 C, 11.5 Cr, 1 Mo, 0.5 Ni, 0.3 V, 0.5 W, bal Fe.
Annealed: 90,000 TS; 40,000 YS; 25 El; B 92 Rock.
Hardened: 235,000 TS; 200,000 YS; 10 El; C 50 Rock.
For valves, cutlery, bearings, surgical instruments.
Corrosion resistant, hardenable.

RTS.
M-912; 1.20 C, 0.65 Cr, 0.10 V, bal Fe.
Cold work tool steel; for taps, counterbores, broaches, punches.
W.-Nr. 2210; AFNOR 120 C 3.

RTS see JESSOP RTS.

RTW 1.
M-912; 1.20 C, 0.25 Cr, 0.10 V, 1.0 W, bal Fe.
Cold work tool steel; for drills, counter sinks, rotary files, milling cutters.
W.-Nr. 2516; AFNOR 120 W 12.

R.T.W. 2.
M-1118; 0.5 C, 1.5 Cr, 2.25 W, 0.2 V, bal Fe.
For tools, hot work dies; hot work steel.

RTW 2 H.
M-912; 0.45 C, 1.0 Si, 1.10 Cr, 0.18 V, 2.0 W, bal Fe.
Cold work tool steel; for punches, stamping tools, chisels, pneumatic tools.
W.-Nr. 2542; AFNOR 45 Wc 20.

R.T.W. 3.
M-1118; 1.3 C, 4.5 Cr, bal Fe.
For tools, dies; air or oil hardened.

RTWK.
M-912; 0.60 C, 0.60 Si, 1.10 Cr, 0.18 V, 2.0 W, bal Fe.
Cold work tool steel; for punches, cold shears, chisels.
W.-Nr. 2550; AFNOR 60 WC 20.

RUBBON see ALL-STATE NO. 55 RUBBON.

RUBEL BRONZE "D.W".
M-2; 40-55 Cu, 3-15 Ni, 1-3 Mn, 1-2 Fe, 0.5-3.0 Al, bal Zn.
Cast: 60,000 TS; 30 El; 90 Brin.
Hot pressed: 75,000 TS; 30 El; 100 Brin.
For printing rolls; corrosion resistant.

RUBEL BRONZE H-1.
M-2; 41 Cu, 48 Zn, 11 Mn + Ni + Fe + Al.
Cast: 99,000 TS; 8 El; 160 Brin.
Hot pressed: 105,000 TS; 12 El; 190 Brin.
For propellers, piston rings, valves, rotor covers, shrinking rings; high strength, non-corrosive.

RUBEL BRONZE H-2.
M-2; 50 Cu, 42 Zn, 8 Mn, + Ni + Fe + Al.
Cast: 82,000 TS; 8 El; 150 Brin.
Hot pressed: 105,000 TS; 15 El; 170 Brin.
For bearing bushings, piston rods, gears, valve stems; high strength, non-corrosive.

RUBEL BRONZE H-2 "W".
M-2; 40-55 Cu, 3.15 Ni, 1-3 Mn, 1-2 Fe, 0.5-3.0 Al, bal Zn.
Cast: 70,000 TS; 10 El; 140 Brin.
Hot pressed: 90,000 TS; 25 El; 160 Brin.
For bearings, bushings; corrosion resistant.

RUBEL BRONZE "L".
M-2; 40-55 Cu, 3-15 Ni, 1-3 Mn, 1-2 Fe, 0.5-3.0 Al, bal Zn.
Cast: 60,000 TS; 30 El; 90 Brin.
Hot pressed: 80,000 TS; 35 El; 115 Brin.
For bearings, bushings; corrosion resistant.

RUBEL BRONZE "L.W".
M-2; 40-55 Cu, 3-15 Ni, 1-3 Mn, 1-2 Fe, 0.5-3.0 Al, bal Zn.
Cash: 60,000 TS; 30 El; 120 Brin.
For bearings, bushings; corrosion resistant.

RUBEL BRONZE W-1.
M-2; 50 Cu, 41 Zn, 9 Mn + Ni + Fe + Al.
Cast: 70,000 TS; 20 El; 100 Brin.
Hot pressed: 85,000 TS; 25 El; 110 Brin.
For construction material for turbines, marine engines and locomotives; high strength, non-corrosive.

RUBEL BRONZE W-2.
M-2; 51 Cu, 43 Zn, 6 Mn + Ni + Fe + Al.
Cast: 70,000 TS; 18 El; 110 Brin.
Hot pressed: 85,000 TS; 22 El; 120 Brin.
For propellers, centrifugal drums, bearing bushings, steam and safety jackets; high strength, non-corrosive.

RUBEL BRONZE W-2.
M-2; 40-55 Cu, 3-15 Ni, 1-3 Mn, 1-2 Fe, 0.5-3.0 Al, bal Zn.
Cast: 65,000 TS; 15 El; 105 Brin.
Hot pressed: 75,000 TS; 20 El; 115 Brin.
For die forgings, rods, tubes, propellers; corrosion resistant.

RUBELIT.
M-2; 40-55 Cu, 1-3 Mn, 3-15 Ni, 1-2 Fe, 0.5-3.0 Al, bal Zn.
Hot pressed: 69,000-76,000 TS; 20-5 El; 130-190 Brin.
For bushings for bearings, piston bolts, valve guides; corrosion resistant.

RUBEL METAL.
M-Eng.; 50-55 Cu, 40 Zn, 4-5 Fe + Mn.
60,000 TS.
For fittings, hardware; tough with high ductility.

RUBIDIUM.
M-1755; Rb.
Purities: 99.9+ %, 99.5%.
Packaging: Glass ampules, steel containers (under vacuum inert gas, oil).

RUBIS BA1.
M-1323; 0.50 C, 1.05 Cr, 3.25 Ni, bal Fe.
For gears, bolts, forging dies, crankshafts; oil hardened, shock resistant.

RUBIS HKP.
M-1323; 1.0 C, 0.1 V, 0.25 Mn, 0.20 Si, bal Fe.
For blanking dies, bearings, cutters; Type W2; water hardened.

RUHRLIT KOBALT 5.
M-1311; 0.79 C, 4.7 Co, 4.3 Cr, 0.75 Mo, 1.5 V, 18 W, bal Fe.
For lathe and planer tools, reamers, taps, drills; high speed steel.

RUHRLIT KOBALT 10.
M-1311; 0.72 C, 10 Co, 4.2 Cr, 0.8 Mo, 1.8 V, 18 W, bal Fe.
For lathe and planer tools, hobs, milling cutters; high speed steel.

RUHRLIT KOBALT 135.
M-1311; 1.3 C, 4.3 Cr, 0.85 Mo, 3.8 V, 12 W, bal Fe.
For engraving tools, lathe and planer cutters, hobs; high speed steel.

RUHRLIT NH18.
M-1311; 0.74 C, 4.1 Cr, 1.1 V, 18.5 W, bal Fe.
For lathe and planer tools, reamers; high speed steel.

RUHRLIT NH260.
M-1311; 0.86 C, 4.1 Cr, 0.85 Mo, 2.5 V, 12 W, bal Fe.
For lathe and planer tools, reamers, broaches, drills, taps; high speed steel.

RUHRLIT NHMO.
M-1311; 0.85 C, 4.1 Cr, Mo, W, V, bal Fe.
For lathe and planer tools, reamers, hobs, drills; high speed steel.

RUHRLIT S32.
M-1311; 0.86 C, 2.8 Co, 4.3 Cr, 0.85 Mo, 2.1 V, 12 W, bal Fe.
For lathe and planer tools, reamers, broaches, drills; high speed steel.

RUHRLIT SS20.
M-1311; 0.82 C, 4.1 Cr, 0.85 Mo, 1.6 V, 8.7 W, bal Fe.
For lathe and planer tools, drills, reamers, hobs; high speed steel.

RUHRLIT SS150.
M-1311; 1.35 C, Co, Cr, V, W, bal Fe.
For reamers, broaches, milling cutters; high speed steel.

RUHRLIT SS315.
M-1311; 0.95 C, Cr, W, Mo, bal Fe.
For reamers, broaches, milling cutters; high speed steel.

RUHRLIT SS460.
M-1311; 1.3 C, 4.3 Cr, 0.85 Mo, 3.8 V, 12 W, bal Fe.
For engraving tools, milling cutters, broaches; high speed steel.

RUHRSTAHL MI.
M-1382; 0.17 max C, 0.30 Si, 0.30 Mn, bal Fe.
Annealed: 70,000 TS; 55,000 YS; 25 El; 60 RA; 140 Brin.
For gears, pinions, machine tool parts, rivets; case hardened.

RUHRSTAHL MII.
M-1382; 0.23 max C, 0.30 Si, 0.30 Mn, bal Fe.
Annealed: 75,000 TS; 60,000 YS; 22 El; 58 RA; 150 Brin.
For fan blades, bushings, bolts, nuts, rivets; water hardened.

RULE BRASS 238.
M-8; 35 Zn, 2.5 Pb, bal Cu.
Soft: 45,000 TS; 50 El.
Hard: 73,000 TS; 7 El.
For screw machine products, bolts, nuts, screws; free-cutting.

RUNE.
M-111; 1.4 C, 0.45 Cr, bal Fe.
For razors; water hardening.

RUPEE (SILVER).
91.6 Ag, 8.3 Cu.
For coinage; corrosion resistant.

R.U.S.
M-1118; 0.95 C, 1.2 Mn, 0.5 Cr, 0.5 W, bal Fe.
For tools, dies; non-deforming.

RUS.
M-912; 0.90 C, 2.0 Mn, 0.35 Cr, 0.20 Mo, bal Fe.
Cold work tool steel; for cutting and punching tools and dies, shear blades, reamers, moulds.
W.-Nr. 2842; Similar to AISI 02.

RUS 3.
M-912; 0.93 C, 1.1 Mn, 0.60 Cr, 0.10 V, 0.60 W, bal Fe.
Oil hardening cold work tool steel; for taps, milling cutters, punches, shears.
W.-Nr. 2510; Similar to AISI 01.

RUS 4.
M-912; 1.05 C, 0.90 Mn, 1.0 Cr, 1.15 W, bal Fe.
Cold work tool steel; for cutting and stamping sheet steel, hand taps, thread chasers.
W.-Nr. 2419.

RUSCAR.
M-405; Pb, Sn, Sb.
For antifriction bearings; Babbitt.

RUSELITE.
M-99, M-1476; 4 Cu, 2 Cr, 2 Mo, 0.2 Ti, bal Al.
Die cast: 40,000 TS; 6 El.
Annealed: 30,000 TS; 20 El.
For hardware, fixtures, housings, cases; corrosion resistant.

RUSELITE NO. 500.
M-99; Al, 4 Cu, 1.25-1.50 Cr, 0.2 Mo, 0.2 Ti.
For hardware, fixtures; corrosion resistant.

RUSSIAN-1.
M-USSR; 64 Cu, 18 Ni, 18 Zn, 0.3 Fe, 0.3 Pb.
For ornaments, corrosion resisting white metal parts; corrosion resistant.

RUSSIAN-2.
M-USSR; 1.3 Cu, 26.5 Zn, 72.2 Sn.
For bearings; anti-friction.

RUSSIAN-3.
M-USSR; 66-78 Cu, 21-34 Zn, 0-1 Sn.
For condenser tubes; high ductility.

RUSSIAN PACKING.
M-USSR; 98.5 Zn, 0.98 Sn, 0.32 Pb, 0.16 Fe.
For bearings; will not resist heat or live steam.

RUTHENIUM.
M-1755; Ru.
Purities; 99.999%, 99.99%, 99.9%.
Forms: Sponge, powder, rods, discs, single crystals.

RV 4.
M-1653; 1.3-1.5 C, 4.0-4.5 Cr, 1.0 max Mo, 11.5-12.5 W, 3.9-4.1 V, bal Fe.
High speed tool steel.
W.-Nr. 1.3302: UX 150 WV 1305 KU.

RV 5 K.
M-1653; 1.4-1.6 C, 4.0-5.0 Cr, 12.0-13.0 W, 4.7-5.3 V, 4.5-5.5 Co, bal Fe.
W-Co high speed tool steel.
W.-Nr. 1.3202; AISI T15.

RW 1.
M-912; 1.05 C, 0.25 Si, 0.35 N, 0.50 Cr, bal Fe.
For ball and roller bearings.
W.-Nr. 3501; DIN 105 Cr 2.

RW 3.
M-912; 1.00 C, 0.25 Si, 0.35 Mn, 1.50 Cr, bal Fe.
For ball and roller bearings.
W.-Nr. 3505; DIN 100 Cr 6.

RW 4.
M-912; 1.00 C, 0.60 Si, 1.10 Mn, 1.50 Cr, bal Fe.
For ball and roller bearings.
W.-Nr. 3520; DIN 100 CrMn 6.

RW 5.
M-912; 1.00 C, 0.30 Si, 0.70 Mn, 1.80 Cr, 0.30 Mo, bal Fe.
For ball and roller bearings.
W.-Nr. 3536.

RW 7.
M-912; 0.97 C, 0.50 Si, 1.0 Mn, 1.95 Cr, 0.55 Mo, bal Fe.
For ball and roller bearings.
W.-Nr. 3539; DIN 100 CrMnMo 8.

RW 50.
M-912; 0.83 C, 0.20 max Si, 0.35 max Mn, 4.20 Cr, 4.25 Mo, 1.0 V, bal Fe.
For roller bearings.
W.-Nr. 3551; Similar to AISI M 50.

RW 60.
M-912; 0.90 C, 0.30 Si, 0.30 Mn, 4.25 cr, 5.0 Mo, 1.85 V, 6.40 W, bal Fe.
For special roller bearigs; high speed steel.
W.-Nr. 3343; Similar to AISI M2.

RWA.
M-912; 0.30 C, 2.60 Cr, 0.60 V, 4.25 W, bal Fe.
Hot work tool steel; for pressure casting moulds, forging and pressing dies.
W.-Nr. 2567: AFNOR 30 WCV 40 12.

RXM15.
M-1541; 0.06 C, 1.20 Mn, 0.04 max P, 0.015 max S, 20.0 Cr, 3.2 Si, 14.0 Ni, 0.08 REM, bal Fe.
96,000 psi TS; 64 El.
Heat resistant stainless steel for automotive thermal reactor system.

RY-ALLOY.
M-240; 0.85-1.0 C, 1.0-1.2 Mn, 0.20 V, 0.50 W, 0.50 Cr, bal Fe.
Oil hardening tool steel.
AISI Type 01.

RYANITE.
M-464; 3.2 C, 2.4 Si, 1.5 Ni, 4 Cr, bal Fe.
Cast: 38,000-40,000 TS; 225-250 Brin.
Heat treated: 60,000-70,000 TS; 425-450 Brin.
For dies, brake drums, dies for sheet metal fabrication; cast iron.

RYANITE NO. 2.
M-464; 3.4 C, 1.5 Ni, 0.5 Cr, bal Fe.
For crankcases, cylinder blocks, cylinder barrels; cast iron.

RYCASE.
M-240; 0.14-0.20 C, 1.0-1.3 Mn, 0.04 max P, 0.08-0.13 S, 0.10 max Si, bal Fe.
HT: 82-119 ksi TS; 64-99 ksi YS; 19-22 El; 38-62 RA; 154-198 Brin.
For screw stock, carburized parts; free cutting carburizing steel.
AISI 1117.

RYCROME.
M-240; 0.40 C, 0.9 Mn, 0.9 Cr, 0.2 Mo, 0.2 Si, bal Fe.
HT: 105-135 ksi TS; 105 ksi YS; 16-20 El; 54-61 RA; 269-302 Brin.
For crankshafts, pinions, bolts, studs, machinery parts requiring moderate strength.
AISI 4140-4142.

RYCUT 20.
M-240; 0.20 C, 0.15-0.35 Pb, 0.5 Cr, 0.2 Mo, 0.6 Ni, 0.8 Mn, bal Fe.
Rolled: 92,000 TS; 65,000 YS; 24 El; 61 RA; 192 Brin.
For gears, spindles, cams, shafts, bushings; free-cutting, case hardening steel. AISI 86L20.

RYCUT 40 ANNEALED.
M-240; 0.40 C, 0.9 Cr, 0.2 Mo, 0.9 Mn, 0.15-0.35 Pb, bal Fe.
Ann: 91 ksi TS; 63 ksi YS; 27 El; 58 RA; 187 Brin.
Free machining steel; for gears, shafts, cams, studs, wrenches, arbors.
AISI 41L40.

RYCUT 50 ANN.
M-240; 0.50 C, 0.8 Mn, 0.9 Cr, 0.2 Mo, 0.15-0.35 Pb, bal Fe.
Ann: 103 ksi TS; 69 ksi YS; 23 El; 51 RA, 212 Brin.
Free cutting leaded steel; for shafts, collets, brake dies, gears, boring bars.
AISI 41L47-41L50.

RYCUT HT.
M-240; 0.47 C, 0.9 Mn, 1.0 Cr, 0.2 Mo, 0.15-0.35 Pb, bal Fe.
H.T.: 130 ksi TS; 110 ksi YS; 18 El; 52 RA.
For machinery parts, shafts, gears.
Free cutting.
AISI 41L42, 41L47, 41L50.

RYCUT RS ANN.
M-240; 0.50 C, 0.8 Mn, 0.9 Cr, 0.2 Mo, 0.06-0.10 S, bal Fe.
Ann: 103 ksi TS; 69 ksi YS; 23 El; 51 RA; 212 Brin.
For shafts, brake dies, gears, machinery parts; free cutting resulphurized 4150.

RYCUT RS HT.
M-240; 0.51 C, 1.0 Mn, 0.8 Cr, 0.2 O, 0.06-0.10 S, bal Fe.
H.T.: 130 ksi TS; 105 ksi YS; 16 El; 48 RA.
For shafts, brake dies, gears, machinery parts; free cutting resulphurized 4150.

RYERSON ABRASION RESISTING STEEL now AR-360.

RYERSON V.D see V.D. TOOL.

RYNALLOY.
M-1211; 2.5 max C, 20 Ni, 1.8 min Cr, 6 max Si, 1 max Mn, bal Fe.
For ball and socket joints in aircraft exhaust systems; anti-galling, for use up to 1800°F.

RYTENSE.
M-240; 0.40 C, 1.55 N, 0.040 max P, 0.08-0.13 S, 0.10 max Si, bal Fe.
Rolled: 99 ksi TS; 61 ksi YS; 25 El; 51 RA.
For gears, cams, arbors, rollers, bushings, machinery parts; free-cutting.
AISI 1141.

RYTENSE 44.
M-240; 0.47 C, 1.45 Mn, 0.04 max P, 0.24-0.33 S, bal Fe.
Rolled: 102 ksi TS; 57 ksi YS; 24 El; 48 RA.
For machinery parts, shafts, gears. Free cutting.
AISI 1144.

RYTENSE AA see **RYTENSE.**

R Z 1.
M-912; 1.05 C, 0.20 Si, 0.40 Mn, bal Fe.
For cold worked tension springs.
W.-Nr. 1274.

R Z 5.
M-Eng.; 3.5-5.0 Zn, 0.4-1.0 Zr, 0.75-1.75 rare metals, bal Mg.
Sand cast and precipitation treated: 200 MPa TS; 135 MPa YS; 4 El 65 Brin.
BS 2970 MAG 5.

RZF.
M-912; 0.67 C, 1.30 Si, 0.50 Mn, 0.50 Cr, bal Fe.
For valve springs, coil springs, torsion bar springs.
W.-Nr. 7103.

S

S 18/8.
M-1739; 18 Cr, 8 Ni, bal Fe, sintered.
Density: 6.7-7.1 gms/cc; UTS 44 kg/mm^2.
Elon: 10 min; Hardness: Rock B 30.
B.S.S. A520; AISI 316L.

S 18/8 3-020.
M-1739.
Austenitic 18-8 stainless steel, sintered.
Density: 6.7-7.1 gms/cc; Elon: 10 min.
UTS. 36.2-44 kg/mm^2; Hardness HV5-130.

S 333.
M-1290; 0.95-1.03 C, 3.8-4.5 Cr, 2.7-3.0 W, 2.2-2.5 V, 2.5-2.8 Mo, bal Fe.
High speed steel.
W.-Nr. 1.3333.

S 410.
M-1739; 0.10 C, 12 Cr, bal Fe, sintered.
Density: 6.7-7.1 gms/cc; UTS 50 kg/mm^2.
Elon: 7 min; Hardness Rock B 50.
B.S.S. A530; AISI 410.

S-590 see also **UNITEMP S-590.**

S-816 see also **UDIMET S-816, UNITEMP S-816.**

SA 22 see **CRUCIBLE SA 22.**

SA-32.
M-1008; 0.2 C, 20 Cr, 12 Ni, bal Fe.
For furnace parts; resists heat to 1400°F.

SABECO NO. 9.
M-966; 69-71 Cu, 8.5-9.5 Sn, 20-22 Pb.
25,100 TS; 9.5 El; 8.3 RA; 60 Brin.
For heavy load bearings; bearing bronze.

SABECO NO. 11.
M-966; 69-71 Cu, 10.5-11.5 Sn, 18-20 Pb.
26,500 TS; 6 El; 6.5 RA; 66 Brin.
For extra heavy pressure bearings; bearing bronze.

SABECO NO. 11 H G.
M-966; 69-71 Cu, 10.5-11.5 Sn, 18-20 Pb.
27,300 TS; 4.3 El; 3.5 RA; 78 Brin.
For worm wheels, clutch shifter shoes, heavy bearings; bearing bronze.

SABECO NO. 14.
M-966; 70 Cu, 14 Sn, 16 Pb.
Cast: 27,300 TS; 4.2 El; 3.5 RA; 78 Brin.
For bearings, bushings.

SABECO NO. 16.
M-966, M-221; 15-16.5 Sn, 13.5-14.5 Pb, bal Cu.
Cast: 32,250 TS; 1 El; 1.5 RA; 119 Brin.
For friction rings, heavy duty bearings; Babbitt.

SABEN.
M-618; 0.7 C, 14 W, 4 Cr, 1 V, bal Fe.
For punches, nail dies, drills, taps; high speed steel, oil hardened.

SABEN 6-5-2.
M-510; 0.83 C, 4.1 Cr, 5.0 Mo, 6.4 W, 1.9 V, bal Fe.
High speed tool steel.
B.S. 4659 Type BM 2; AISI M2.

SABEN EXTRA.
M-510; 0.75 C, 4.1 Cr, 18.0 W, 1.1 V, bal Fe.
High speed tool steel.
B.S. 4659 Type BT1; AISI T1.

SABEN HC.
M-510; 1.25 C, 4.5 Cr, 13.5 W, 0.3 Mo, 3.7 V, bal Fe.
W-V type high speed steel.

SABEN TENCO.
M-510; 0.75 C, 5.0 Cr, 0.60 Mo, 18.5 W, 1.75 V, 9.5 Co, bal Fe.
Cobalt high speed steel.
B.S. 4659 Type BT 5; AISI T5.

SABEN WUNDA.
M-510; 0.75 C, 4.25 Cr, 0.60 Mo, 18.5 W, 1.1 V, 5.5 Co, bal Fe.
Cobalt high speed steel.
B.S. 4659 Type BT4; AISI T4.

SABEX.
M-510; 0.28 C, 0.30 Si, 0.30 Mn, 2.25 Ni, 2.5 Cr, 9.5 W, 0.15 V, bal Fe.
Hot work tool steel.
B.S. 4659 Type BH21A.

SABRE.
M-435; 1.25 C, 0.30 Si, 0.30 Mn, 10 W, 4.25 Cr, 2.5 Mo, 4.3 V, 5.5 Co, bal Fe.
For tool bits, form tools, form and blank dies; high red hardness, high speed steel.

S. A. BRILLIANT.
M-333; 0.7 C, 18 W, 4 Cr, 1 V, bal Fe.
For drills, lathe and planer tools; high speed steel.

SABUS.
M-618; 0.7 C, 22 W, bal Fe.
For glass cutting knives, drills, milling cutters, saws; super high speed steel.

SA-CN.
M-333; 0.5 C, 2.5 W, bal Fe.
For bolt heading dies, hot work punches; hot work steel.

S. A. DURABIL.
M-333; 0.45 C, 1.25 Cr, 2.25 W, 0.2 V, bal Fe.
For hot work dies; impact resistant; hot work steel.

SAE EX1.
M-68; 0.15-0.21 C, 0.35-0.60 Mn, 0.3 Si, 4.8-5.3 Ni, 0.2-0.3 Mo, bal Fe.
Hardened: 200,000 TS; 155,000 YS; 15 El; 60 RA; C 43 Rock.
For heavy duty parts in earth moving equipment, and trucks, rotary and percussion rock bits, ring gears, transmissions.
High strength and toughness. Shock resistant. Heavy duty carburizing steel.

SAE EX2.
M-68; 0.65-0.75 C, 0.25-0.45 Mn, 0.20-0.35 Si, 0.70-1.0 Ni, 0.15-0.30 Cr, 0.08-0.15 Mo, bal Fe.
Heat treated: 60 min Rock C.
For stress-bearing applications, spindles, rolls, bearings.
Wear resistant, shock resistant.

SAFE 10NC6.
M-1757; 0.10 C, 0.75 Mn, 0.25 Si, 1.5 Ni, 1.0 Cr, bal Fe.
For bolts, shafts; case hardened for gears, pinions.

SAFE 14NC11.
M-1757; 0.14 C, 0.50 Mn, 0.25 Si, 2.75 Ni, 0.75 Cr, bal Fe.
Case hardened for large, highly stressed gears.

SAFE 16MC5.
M-1757; 0.16 C, 1.2 Mn, 0.30 Si, 1.0 Cr, bal Fe.
Low alloy-low carbon steel; usually case-hardened for gears, shafts, universal joints.

SAFE 16NC6.
M-1757; 0.16 C, 0.75 Mn, 0.25 Si, 1.5 Ni, 1.0 Cr, bal Fe.
For bolts, shafts; case hardened for gears, pinions.

SAFE 18CD4.
M-1757; 0.18 C, 0.75 Mn, 0.25 Si, 1.0 Cr, 0.25 Mo, bal Fe.
For shafts, bolts; carburized for gears.

SAFE 18NCD3.
M-1757; 0.18 C, 0.80 Mn, 0.25 Si, 1.5 Ni, 1.0 Cr, 0.25 Mo, bal Fe.
For cogwheels, pinions; usually case hardened. Similr to AISI 4320.

SAFE 20MC5.
M-1757; 0.20 C, 1.2 Mn, 0.30 Si, 1.2 Cr, bal Fe.
Low alloy-low carbon steel; usually case-hardened for sprockets, cardan joints.

SAFE 20NC6.
M-1757; 0.20 C, 0.75 Mn, 0.25 Si, 1.50 Ni, 1.0 Cr, bal Fe.
For bolts, shafts; case hardened for gears, cams.

SAFE 20NCD2.
M-1757; 0.20 C, 0.80 Mn, 0.25 Si, 0.55 Ni, 0.55 Cr, 0.20 Mo, bal Fe.
For bolts, flanges; case hardened for gears.
AISI 8720.

SAFE 25CD4.
M-1757; 0.25 C, 0.75 Mn, 0.25 Si, 1.0 Cr, 0.25 Mo, bal Fe.
For shafts, axles, pinions.

SAFE 30CAD6-12.
M-1757; 0.30 C, 0.70 Mn, 0.30 Si, 1.65 Cr, 0.35 Mo, 1.15 Al, bal Fe.
For nitrided highly stressed sliding parts.

SAFE 30CD4.
M-1757; 0.30 C, 0.75 Mn, 0.25 Si, 1.0 Cr, 0.25 Mo, bal Fe.
For shafts, bolts, pinions.

SAFE 30CD12.
M-1757; 0.30 C, 0.50 Mn, 0.25 Si, 3.0 Cr, 0.40 Mo, bal Fe.
For highly stressed mechanical parts.

SAFE 30NC11.
M-1757; 0.30 C, 0.50 Mn, 0.25 Si, 2.75 Ni, 0.75 Cr, bal Fe.
For cogwheels, crankshafts, gears.

SAFE 32C4.
M-1757; 0.32 C, 0.75 Mn, 0.30 Si, 1.0 Cr, bal Fe.
For arbors, axles, automotive parts.

SAFE 35CD4.
M-1757; 0.35 C, 0.75 Mn, 0.25 Si, 1.0 Cr, 0.25 Mo, bal Fe.
For crankshafts, shafts, pinions.

SAFE 35NC6.
M-1757; 0.35 C, 0.75 Mn, 0.25 Si, 1.5 Ni, 1.0 Cr, bal Fe.
For shafts, bolts, spindles, axles.

SAFE 35NCD6.
M-1757; 0.35 C, 0.80 Mn, 0.25 Si, 1.5 Ni, 1.5 Cr, 0.25 Mo, bal Fe.
For crankshafts, arbors, axles.

SAFE 35NCD16.
M-1757; 0.35 C, 0.45 Mn, 0.25 Si, 4.0 Ni, 1.8 Cr, 0.40 Mo, bal Fe.
Deep hardening steel for crankshafts, large, highly stressed components.

SAFE 38C2.
M-1757; 0.38 C, 0.75 Mn, 0.30 Si, 0.45 Cr, bal Fe.
For automotive parts.

SAFE 38C4.
M-1757; 0.38 C, 0.75 Mn, 0.30 Si, 1.0 Cr, bal Fe.
For shafts, axles, steering columns.

SAFE 40CAD6-12.
M-1757; 0.40 C, 0.70 Mn, 0.30 Si, 1.65 Cr, 0.35 Mo, 1.15 Al, bal Fe.
For nitrided highly stressed sliding parts.

SAFE 40NCD3.
M-1757; 0.40 C, 0.70 Mn, 0.25 Si, 0.85 Ni, 0.75 Cr, 0.25 Mo, bal Fe.
For gears, shafts, axles, bolts, flanges.

SAFE 42C2.
M-1757; 0.42 C, 0.75 Mn, 0.30 Si, 0.45 Cr, bal Fe.
For shafts, arbors.

SAFE 42C4.
M-1757; 0.42 C, 0.75 Mn, 0.30 Si, 1.0 Cr, bal Fe.
For axles, shafts, hand tools.

SAFE 42CD4.
M-1757; 0.42 C, 0.75 Mn, 0.25 Si, 1.0 Cr, 0.25 Mo, bal Fe.
For cogwheels, connecting rods, pinions.

SAFE 50CV4.
M-1757; 0.50 C, 0.90 Mn, 0.30 Si, 1.0 Cr, 0.15 V, bal Fe.
Chrome-vanadium steel for shafts, gears, axles, hand tools.

SAFE 100C6.
M-1757; 1.0 C, 0.30 Mn, 0.20 Si, 1.5 Cr, bal Fe.
For bearing rollers and balls.

SAFE CND8.
M-1757; 0.30 C, 0.45 Mn, 0.25 Si, 2.0 Ni, 2.0 Cr, 0.40 Mo, bal Fe.
For crankshafts, highly stressed shafting.

SAFLEX.
M-237; bimetal.
For thermostatic bimetal; active from 500-800°F.

SAGAMORE.
M-1779; 1.0 C, 5.0 Cr, 1.0 Mo, 0.3 V, bal Fe.
General purpose air hardening cold work die steel; good wear resistance and toughness.
For blanking and forming dies, punches, rolls and trim dies; usually at 58/62 Rc.
AISI A-2.

SAGAMORE EZ.
M-1779; 1.0 C, 5.0 Cr, 1.0 Mo, 0.3 V, 0.12 S, bal Fe.
Same as SAGAMORE except free-machining grade.

SA GREY LABEL.
M-333; 0.5 C, bal Fe.
For drop forge dies.

SAHIB.
M-502; C, bal Fe.
For tools.

SA HY-PRO.
M-333; 0.7 C, 18 W, 4 Cr, 1 V, 5 Co, bal Fe.
For cutting tools; high speed steel.

SA HY-PRO DIE STEEL.
M-333; 1.5 C, 13 Cr, bal Fe.
For dies; non-deforming.

SALAMANDRE D. 1.
M-1546; 1.0 C, 1.65 Cr, 0.5 Ni, 0.15 V, bal Fe.
Ht. Tr.: 200,000 TS; 185,000 YS; 3 El; 444 Brin.
For rollers, bearings, cams, bushings, sleeves, cutting tools, punches.
Cold work steel, oil hardening.

SALGE ANTIFRICTION METAL.
M-Eng.; 86 Zn, 9.9 Sn, 4.0 Cu, 1.1 Pb.
For bearings; anti-friction.

SA LILY BRAND.
M-333; C, alloy, bal Fe.
For brass working dies.

SALLIT'S SPECULUM.
65 Cu, 31 Sn, 4 Ni.
For metallic mirrors; corrosion resistant.

SAM ALLOY.
11 Mischmetal, 1.5 Cu, 1.25 Ni, 1 Mn, 0.3 Cr, 0.02 Ti, bal Al.
Cast: 15,000 TS; 0.5 El; 52 Brin.
Forged: 10.0 El; 62 Brin.
For aircraft engine components; high temperature resistance to 600°F.

SAMARIUM.
M-1755; Sa.
Purities: Special distilled grade 99.9%, 99.5%.
Forms: Ingot, lump, rod, filings, wire, sheet, foil, powder.

S. A. M. B.
0.2-1.0 Al, 56-59 Cu, 0.8-1.2 Fe, 0.2-2.0 Mn, 0.8-1.5 Sn, bal Zn.
75,000 TS; 35,000 YS; 20 El; 140 Brin.
For extrusions; corrosion resistant.

SAMLEGIERUNG.
13 Mn, 10.2 Si, 5.8 Al, 2.5 C, 0.3 Cu, bal Fe.

SAMPSON.
M-Eng.; 0.43 C, 0.43 Mn, 1.22 Ni, 0.43 Cr, bal Fe.
For tools, gears; oil hardened.

SAMPSON METAL.
88 Zn, 4 Cu, 8 Al.
For bearings; anti-friction.

SA-N-65.
M-333; 0.5 C, 0.9 Cr, 0.2 V, bal Fe.
For rivet sets; punches; shock resistant.

SA-N-88.
M-333; C, alloy, bal Fe.
For hobs, punches; tough.

SANBOLD CVS.
M-510; 0.50 C, 1.0 Cr, 0.15 V, bal Fe.
Chrome-vanadium steel.

SANBOLD NA 35.
M-510.
Ni-Fe high permeability alloy.
Max permeability μ max: 12,000-20,000; max flux density: 13,000 gauss.
For magnetic screens, relays in automatic telephones.

SANBOLD NA 47.
M-510; 40-50 Ni, bal Fe.
Max permeability μ max: 15,000-30,000; max flux density: 16,000 gauss.
For transformers and chokes.

SANBOLD NA 76.
M-510; 76-77 Ni, plus others, bal Fe.
Max permeability μ max: 50,000-150,000; max flux density: 8000 gauss.
Low hysteresis loss.
For relay devices, chokes, transformers.

SANCY.
M-1115; 1.5-2.0 C, 11-13 Cr, bal Fe.
For forming and blanking dies, punches; oil or air hardened, non-deforming.

SANDERSON 476.
M-618; 1.5 C, 11.5 Cr, 0.75 Mo, bal Fe.
For blanking and forming dies, punches; air hardened, non-deforming.

SANDERSON 476 HEAVY DUTY.
M-618; 1.85 C, 11-12 Cr, 0.33 V, 0.7 Mn, bal Fe.
For blanking and forming dies, threading dies; oil hardened, non-deforming.

SANDERSON 476 SPECIAL.
M-618; 2.25 C, 13 Cr, bal Fe.
For blanking and forming dies, punches; oil hardened, non-deforming.

SANDERSON D70 MAGNET STEEL.
M-618; 0.7 C, 6 W, bal Fe.
For permanent magnet.

SANDERSON EXTRA.
M-38; 1-1.1 C, 0.25 Mn, bal Fe.
For punches, dies, cutters, general tools; water hardening.

SANDERSON SPECIAL DRILL ROD.
M-38; 1.25 C, 0.25 Mn, bal Fe.
For tools, punches; drill rod.

SANDERSON XX.
M-903; 0.9 C, 1 Cr, bal Fe.
For burnishing tool; water hardening.

SANDOW.
M-173; 0.4 C, 0.9 Cr, 0.2 Mo, bal Fe.
Heat treated: 125,000 TS; 110,000 YS; 19 El; 47 RA; 293 Brin.
For axles, gears, crane hooks, mandrels; tough, oil hardened.

SANDUSKY 55.
M-295; 1.3 Fe, 20.1 Ni, 1.0 Mn, bal Cu.
Cast: 75,900 TS; 55,600 YS; 14 El; 140 Brin.
Tubes for marine propulsion shaft sleeves and for paper machine roll application.

SANDUSKY ALLOY 63.
M-295; 0.08 C, 2.0 max Mn, 2.0 max Si, 19.5-23.5 Cr, 7.0-11.0 Ni, 1.0-3.0 Mo, bal Fe.
Cast: 93,400 psi TS; 57,700 psi YS; 41 El; 182 BHN.
Rolls for paper machinery.

SANDUSKY ALLOY 70.
M-295; 0.03 C, 0.80 Mn, 0.50 Si, 11.9 Cr, 4.0 Ni, 1.5 Mo, bal Fe.
Cast: 122,600 psi TS; 99,500 psi YS; 20 El; 260 BHN.
Rolls for paper machinery.

SANDUSKY ALLOY 75.
M-295; 0.05 C, 0.75 Mn, 0.50 Si, 26.0 Cr, 6.8 Ni, bal Fe.
Cast: 107,200 psi TS; 57,800 psi YS; 13 El; 235 BHN.
Rolls for paper machinery.

SANDUSKY I N.
M-295; 85-87 Cu, 4-6 Sn, 3-5 Zn, 4-6 Pb, 0.5-1.0 Ni.
35,000-45,000 TS; 15,000 YS; 20-25 El; 60-70 Brin.
For rolls, cylinders, liners, shaft sleeves, bushings; centrifugal castings.

SANDVIK 1B.
M-101; 0.06 max C, bal Fe.
Normalized: 50,000 TS; 38,000 YS; 44 El; 67 RA; 100 Brin.
For flexible tubing, deep drawn parts, tin plate containers; ductile.

SANDVIK 1C27.
M-101, M-101a, M-101b; 0.08 max C, 0.3 Si, 0.3 Mn, 13.5 Cr, bal Fe.
Ferritic-martensitic chromium corrosion resistant steel; weldable.
For heat exchanger tubes.
W. Nr. 1.4000; AISI (405).

SANDVIK 1C34.
M-1018, M-101a, M-101b; 0.05 C, 0.3 Si, 0.3 Mn, 17.0 Cr, bal Fe.
Chromium stainless steel wire for cold heading; Werkstoff Nr. 4016; AISI 430.
Also used as MIG, TIG welding wire.

SANDVIK 1C345.
M-1018, M-101a, M-101b; 0.08 C, 0.5 Si, 0.5 Mn, 16.7 Cr, 1.8 Mo, 0.6 Ti, bal Fe.
Ferritic stabilized stainless tubing.
For heat exchangers, gas preheaters, evaporators.
DIN X8 Cr Mo Ti 17; Werkstoff Nr. 1.4523.

SANDVIK 1DT.
M-101; 0.07 max C, 0.03 Si, 0.15 Mn, bal Fe.
Normalized: 46,000 TS; 31,000 YS; 52 El; 79 RA; 90 Brin.
For lamp globes, housings, deep drawn parts; ductile.

SANDVIK 1DTR.
M-101; 0.07 max C, 0.03 Si, 0.15 Mn, bal Fe.
For magnetic purposes; low coercive force.

SANDVIK 1HS49.
M-101, M-101a, M-101b; 0.05 C, 0.5 Si, 1.9 Mn, 2.0 Ni, 0.55 Mo, bal Fe.
Wire as filler metal for welding low carbon, low alloy steels.

SANDVIK 1S15.
M-101; 0.05 max C, 1.5 Si, 0.1 Cu, bal Fe.
Annealed: 64,000 TS; 36 El.
For relays, small transformers and electric motors; dynamo iron, high permeability.

SANDVIK 1S35.
M-101; 0.06 C, 3.4 Si, bal Fe.
Annealed: 85,000 TS; 36 El.
For relay cores, transformers; magnetic iron, high permeability.

SANDVIK 1XR17.
M-101, M-101a, M-101b; 0.010 max C, 0.4 Si, 1.8 Mn, 0.015 max P, 0.015 max S, 20 Cr, 9.5 Ni, bal Fe.
Drawn wire for MIG, TIG, and submerged arc welding of 18-8 stainless steel.

SANDVIK 2 C-34.
M-101; 0.13 C, 17 Cr, bal Fe.
Annealed: 78,000 TS; 47,000 YS; 36 El; 66 RA; 165 Brin.
At 800°C: 6,720 TS; 90 El; 99 RA.
For use where material is to be subjected to hot bending or welding, superheater tubes; cutlery; resists corrosion of HNO_3 but not of H or H_2SO_4; stainless.

SANDVIK 2C344.
M-101, M-101a, M-101b; 0.10 C, 0.9 Si, 0.3 Mn, 17.5 Cr, 1.0 Al, bal Fe.
Ferritic, aluminum-alloyed stainless tubing.
For gas preheaters, furnace conveyor rollers, soot blower tubes; good scaling resistance.
Werkstoff Nr. 1.4742.

SANDVIK 2LS.
M-101; 0.10 C, 0.15 Si, 0.4 Mn, 0.2 Cr, 0.3 max Cu, bal Fe.
Normalized: 57,000 TS; 40,000 YS; 45 El; 70 RA; 110 Brin.
For machine parts; case hardening steel.

SANDVIK 2 N-3 CL.
M-101; 0.08-0.13 C, 0.35-0.60 Mn, 0.5-0.8 Cr, 2.75-3.5 Ni, bal Fe.
Annealed: 80,000-95,000 TS; 38,000-55,000 YS; 32-38 El; 60-70 RA; 160-195 Brin.
Hardened: 130,000-156,000 TS; 100,000-130,000 YS; 18-24 El; 58-65 R 285-350 Brin.
For gears, wheels, piston rods, cam shafts, steering arms, levers; case-hardening steel.

SANDVIK 2N9.
M-101, M-101a, M-101b; 0.08 C, 0.2 Si, 0.5 Mn, 0.02 max P, 0.02 max S, 9.3 Ni, bal Fe.
R.T.: 108,000 TS; 99,000 YS; 25 El.
-320°F: 149,000 TS; 135,000 YS; 25 El.
Weldable, for containers for low temperature operation, liquidified gases.

SANDVIK 2R12.
M-101; 0.020 max C, 0.10 max Si, 0.8 Mn, 18.5 Cr, 11 Ni, bal Fe.
For processing nuclear fuel; heat exchangers, piping nitric acid.

SANDVIK 2R16.
M-101, M-101a, M-101b; 0.020 max C, 0.3 Si, 1.0 Mn, 0.015 max P, 0.015 max S, 19.5 Cr, 10.5 Ni, bal Fe.
Drawn wire for MIG, TIG, and submerged arc welding of 18-8 stainless steel.
Werkstoff Nr. 4302/4316; AISI 304L.

SANDVIK 2R17.
M-101, M-101a, M-101b; 0.020 max C, 0.4 Si, 1.8 Mn, 0.015 max P, 0.015 max S, 20.5 Cr, 10.0 Ni, bal Fe.
Drawn wire for MIG, TIG and submerged arc welding of austenitic stainless steels.

SANDVIK 2R61.
M-101, M-101a, M-101b; 0.02 max C, 0.4 Si, 1.4 Mn, 0.015 P, 0.015 S, 18.5 Cr, 10.5 Ni, 2.7 Mo, bal Fe.
Drawn wire for MIG, TIG and submerged arc welding of stainless steel. AISI 316L.

SANDVIK 2R62.
M-101, M-101a, M-101b; 0.02 max C, 0.35 Si, 1.8 Mn, 0.015 max P, 0.0 max S, 19.5 Cr, 13.0 Ni, 2.2 Mo, bal Fe.
Drawn wire for MIG, TIG and submerged arc welding of stainless steel.
Similar to AISI 316/317.

SANDVIK 2R63.
M-101, M-101a, M-101b; 0.02 max C, 0.35 Si, 1.8 Mn, 0.015 max P, 0.0 max S, 1.0 Cr, 12.0 Ni, 2.8 Mo, bal Fe.
Drawn wire for MIG, TIG and submerged arc welding of 316 type stainless steel.
Werkstoff Nr. 4403/4430; AISI 316L.

SANDVIK 2 RE10.
M-101, M-101a, M-101b; 0.02 max C, 0.30 max Si, 1.75 Mn, 24.5 Cr, 20 Ni, 0.30 max Mo, bal Fe.
Austenitic stainless steel with special resistance to oxidizing conditions as in nitric acid; used in chemical plants.

SANDVIK 2RE69.
M-101; 0.02 max C, 0.40 max Si, 1.75 Mn, 25.0 Cr, 22.0 Ni, 2.1 Mo, 0.12 N, bal Fe.
For fertilizer industry; nitric acid production, Urea strippers.

SANDVIK 2RK65.
M-101, M-101a, M-101b; 0.02 max C, 1.8 Mn, 19.5 Cr, 25.0 Ni, 4.5 Mo 1.5 Cu, bal Fe.
Tube, ann: 71,000-107,000 TS; 31,000 min YS; 40 min El.
Austenitic stainless with high resistance to sulphuric acid, phosphoric acid; for heating coils, heat exchanger tubes, in chemical tanks.

SANDVIK 2RN65.
M-101, M-101a, M-101b; 0.020 max C, 0.45 Si, 1.8 Mn, 17.5 Cr, 24.0 4.7 Mo, bal Fe.
Tube, ann: 71,000-107,000 TS; 31,000 min YS; 40 min El; 165 VDH, (approx.).
Austenitic stainless steel tubes; high resistance to general corrosion, especially sulphuric acid and chlorides; for textile, pulp and paper industries.

SANDVIK 2S.
M-101; 0.10 C, 0.15 Si, 0.4 Mn, bal Fe.
Normalized: 57,000 TS; 39,000 YS; 45 El; 70 RA; 110 Brin.
For boiler and locomotive tubes; also case hardening steel.

SANDVIK 3HS32.
M-101, M-101a, M-101b; 0.15 C, 0.15 Si, 0.95 Mn, 1.35 Cr, 0.90 Mo, 0.25 V, bal Fe.
Tube: 135,000-164,000 TS; 114,000 YS; 10 El; 290-340 VDH.
Bainitic high strength steel tubes.
For mechanically loaded tubular construction.
Werkstoff Nr. 1.7734.

SANDVIK 3L7.
M-101; 0.12 C, 0.15 Si, 0.7 Mn, bal Fe.
Normalized: 64,000 TS; 44,000 YS; 42 El; 72 RA; 125 Brin.
For bus bodies; good welding properties.

SANDVIK 3LS.
M-101; 0.15 C, 0.15 Si, 0.5 Mn, bal Fe.
Normalized: 60,000 TS; 40,000 YS; 41 El; 65 RA; 115 Brin.
For machine parts, gears, shafts; case hardened steel.

SANDVIK 3 MO.
M-101, M-101a, M-101b; 0.15 C, 0.2 Si, 0.7 Mn, 0.02 max P, 0.02 max 0.33 Mo, bal Fe.
For tubes exposed to steam temperature. Yield point at 570°F: 31,500 psi min. Yield point at 930°F: 21,500 psi min.
DIN 15Mo3.

SANDVIK 3M01.
M-101; 0.15 C, 0.5 Mo, bal Fe.
Annealed: 68,000 TS; 47,000 YS; 38 El; 135 Brin.
For boiler and superheater tubes for steam; for steam temperatures up to 900°F.

SANDVIK 3R12.
M-101, M-101a, M-101b; 0.30 max C, 0.45 Si, 0.8 Mn, 18.5 Cr, 11.0 bal Fe.
1 % offset yield, 210°F: 24,200 psi min.
1 % offset yield, 1110°F: 15,600 psi min.
Low carbon austenitic stainless, weldable, for tubing for food and dairy plants, breweries, chemical and nuclear power plants.
Werkstoff Nr. 1.4306, 1.4316; AISI 304L.

SANDVIK 3R13.
M-101, M-101a, M-101b; 0.03 C, 0.45 Si, 0.85 Mn, 18.0 Cr, 12.5 Ni, bal Fe.
Austenitic stainless steel wire for cold-heading;
Werkstoff Nr. 4306; AISI 305.

SANDVIK 3R19.
M-101, M-101a, M-101b; 0.030 C, 0.45 Si, 0.8 Mn, 18.5 Cr, 11.0 Ni, 0.18 N, bal Fe.
0.2% offset yield, 68°F: 42,700 psi min.
0.2% offset yield, 750°F; 21,300 psi min.
Used largely for tubing carrying gas or fluid under pressure. Werkstoff Nr. 1.4311.

SANDVIK 3R60.
M-101, M-101a, M-101b; 0.030 max C, 0.6 Si, 1.7 Mn, 17.0 Cr, 13.6 N 2.8 Mo, bal Fe.
1% offset yield at 390°F: 21,300 psi min.
1% offset yield at 1110°F: 14,200 psi min.
Austenitic stainless tube steel, improved corrosion resistance, weldable.
Miscellaneous tubing applications where corrosion resistance is major requirement. Werkstoff Nr. 1.4435; AISI 316L.

SANDVIK 3R64.
M-101, M-101a, M-101b; 0.03 C, 0.45 Si, 1.8 Mn, 18.5 Cr, 14.5 Ni, 3 bal Fe.
Annealed: 85,000 TS; 45,000 YS; 60 El; B80 Rock.
For paper and pulp mill equipment, chemical plant equipment, acid tanks, digesters, evaporators.
Austenitic, stainless, Type 317L.

SANDVIK 3R67.
M-101, M-101a, M-101b; 0.025 max C, 0.4 Si, 1.8 Mn, 0.015 max P, 0.0 S, 19.0 Cr, 13.5 Ni, 3.3 Mo, bal Fe.
Drawn wire for MIG, TIG and submerged arc welding of Type 317 stainless steels. AISI 317.

SANDVIK 3RE13.
M-101, M-101a, M-101b; 0.025 max C, 0.4 Si, 1.8 Mn, 0.015 max P, 0.015 max S, 24.0 Cr, 13.5 Ni, bal Fe.
Wire for MIG, TIG and submerged arc welding of stainless steel. AISI 309.

SANDVIK 3RE60.
M-101, M-101a, M-101b; 0.30 max C, 1.65 Si, 1.5 Mn, 18.5 Cr, 4.7 Ni, 2.7 Mo, bal Fe.
Tube: 92,000-128,000 TS; 64,000 min YS; 30 El; 260 VDH.
Ferritic-austenitic stainless with high resistance to stress corrosion and intergranular corrosion; largely for heat exchanger tubes.

SANDVIK 3RN14.
M-101, M-101a, M-101b; 0.03 C, 0.45 Si, 0.85 Mn, 16.5 Cr, 18.0 Ni, bal Fe.
Austenitic stainless steel wire for cold-heading.

SANDVIK 3RS17.
M-101, M-101a, M-101b; 0.025 max C, 0.85 Si, 1.8 Mn, 20.6 Cr, 9.7 Ni, bal Fe.
Austenitic stainless steel wire for metallizing; also for welding using MIG methods.

SANDVIK 3RS61.
M-101, M-101a, M-101b; 0.025 max C, 0.85 Si, 1.7 Mn, 0.015 max P, 0. max S, 18.5 Cr, 10.5 Ni, 2.7 Mo, bal Fe.
Drawn wire for MIG welding of Type 316 stainless steel. Werkstoff Nr. 4403/4430.

SANDVIK 3RS62.
M-101, M-101a, M-101b; 0.025 max C, 0.85 Si, 1.8 Mn, 19.5 Cr, 13 Ni 2.3 Mo, bal Fe.
Austenitic stainless steel wire for metallizing; similar to AISI 316L. Also used for MIG welding.

SANDVIK 4C27A.
M-101; 0.20 C, 0.9 Ni, 1.0 Mn, 0.18 S, 12-14 Cr, 1.3 Mo, bal Fe.
Annealed: 95,000 TS; 50,000 YS; 25 El; 55 RA; 195 Brin.
Cold drawn: 105,000 TS; 85,000 YS; 17 El; 50 RA; 215 Brin.
For cutlery, valves, turbine blades, gages, gears; stainless, hardenable.

SANDVIK 4C54.
M-101, M-101a, M-101b; 0.18 C, 0.5 Si, 0.8 Mn, 27.0 Cr, bal Fe.
Ferritic chromium stainless steel.
Yield strength at 1110°F: 15,500 psi min.
Good resistance to scaling, sulphurous gases, oil-ash corrosion at elevate temperatures; for pyrometer protection tubes, soot-blowing tubes, heat exchangers.
Werkstoff Nr. 1.4083; AISI 446.

SANDVIK 4HSL31.
M-101, M-101a, M-101b; 0.18 C, 0.25 Si, 0.75 Mn, 1.5 Cr, 0.40 Mo, 0.03 Al, 0.004 B, bal Fe.
Tube: 121,000-142,000 TS; 100,000 YS; 15 El.
Bainitic high strength steel tube.
For mechanically loaded tubular construction.

SANDVIK 4L7.
M-101; 0.20 C, 0.7 Mn, bal Fe.
Normalized: 71,000 TS; 51,000 YS; 33 El; 67 RA; 140 Brin.
For bake oven tubes, boiler and superheater tubes for steam; steam temperature up to 850°F.

SANDVIK 4LM.
M-101; 0.20 C, 1.45 Mn, bal Fe.
Normalized: 80,000 TS; 48,000 YS; 28 El; 160 Brin.
For weldable tubes, engineering purposes; tough.

SANDVIK 4LS.
M-101; 0.20 C, 0.2 Si, 0.1 Mn, bal Fe.
Normalized: 65,000 TS; 46,000 YS; 38 El; 60 RA; 130 Brin.
For gears, pinions, shafts; case hardening steel.

SANDVIK 4N2C34.
M-101, M-101a, M-101b; 0.20 C, 0.3 Si, 0.5 Mn, 16.6 Cr, 2.2 Ni, bal Fe.
Martensitic stainless steel, hardenable to about 205,000 psi TS.
For stainless structural parts requiring high strength; good to about 900°F.
Werkstoff Nr. 1.4057; AISI 431.

SANDVIK 5LMV.
M-101, M-101a, M-101b; 0.25 C, 0.25 Si, 1.4 Mn, 0.10 V, bal Fe.
Normalized: 60,000-75,000 TS; 45,000 min YS; 22 El; 180-220 Brin.
Hardened: 78,000-95,000 TS; 70,000 min YS; 13 El; 240-280 Brin.
Mechanical tubing, weldable, for tubular load carrying structures. Werkstoff Nr. 1.5213.

SANDVIK 5R10.
M-101, M-101a, M-101b; 0.05 max C, 0.45 Si, 0.45 Mn, 18.5 Cr, 9.3 Ni, bal Fe.
Ann: 102,000 TS; 50,000 YS; 55 El.
Half hard: 156,000 TS; 128,000 YS; 17 El.
General purpose stainless steel for food, soft drink, brewery, dairy plant and for chemical plants.
Werkstoff Nr. 4301; AISI 304.

SANDVIK 5R60.
M-101, M-101a, M-101b; 0.05 max C, 0.6 Si, 1.7 Mn, 17.4 Cr, 13.4 Ni, 2.7 Mo, bal Fe.
Austenitic stainless with excellent corrosion resistance; for use in food and beverage plants, textile, chemical plants.
Werkstoff Nr. 4436; AISI 316.

SANDVIK 5RA50.
M-101, M-101a, M-101b; 0.05 max C, 0.45 Si, 1.8 Mn, 0.030 max P, 0.2 S, 18.0 Cr, 10.0 Ni, 0.5 Mo, bal Fe.
Austenitic stainless steel; for food and chemical processing equipment; free machining type.
Werkstoff Nr. 1.4305; AISI 303.

SANDVIK 6C27.
M-101; 0.30 C, 14 Cr, bal Fe.
Annealed: 104,000 TS; 74,000 YS; 28 El; 55 RA; 255 Brin.
For electric razors; stainless, martensitic, hardenable.

SANDVIK 6HS63.
M-101, M-101a, M-101b; 0.30 C, 0.25 Si, 0.55 Mn, 1.0 Cr, 3.3 Ni, 0.25 Mo, bal Fe.
Ann: (approx) 107,000 TS; 230 VDH.
Hardened: 156,000-199,000 TS; 128,000-149,000 YS; 10-12 El.
Steel tubing, hardenable.
For internal gears, female splined couplings.
Werkstoff Nr. 1.5755.

SANDVIK 6LM.
M-101; 0.3 C, 0.2 Si, 1.35 Mn, bal Fe.
Normalized: 92,000 TS; 62,000 YS; 32 El; 61 RA; 180 Brin.
For bicycle tubes, hollow blooms; water hardening, for cold drawing.

SANDVIK 6R42.
M-101, M-101a, M-101b; 0.06 max C, 0.4 Si, 1.3 Mn, 0.025 max P, 0.020 max S, 19.5 Cr, 9.5 Ni, 0.9 Nb, bal Fe.
Niobium stabilized wire for MIG, TIG, and submerged arc welding of austenitic stainless steels.
Werkstoff Nr. 4551. AISI 347.

SANDVIK 6R60.
M-101, M-101a, M-101b; 0.04 max C, 0.6 Si, 1.7 Mn, 17.5 Cr, 12.6 Ni, 2.8 Mo, bal Fe.
Austenitic stainless hard drawn wire for springs; 0.013-0.124 in dia. wire has 220,000-277,000 psi TS; very good corrosion resistance.
Werkstoff Nr. 4436; AISI 316.

SANDVIK 6R81.
M-101, M-101a, M-101b; 0.06 max C, 0.4 Si, 1.3 Mn, 0.025 max P, 0.020 max S, 19.0 Cr, 11.0 Ni, 2.7 Mo, bal Fe.
Niobium stabilized wire for MIG, TIG, and submerged arc welding of austenitic stainless steels.
Werkstoff Nr. 4576.

SANDVIK 6X10RN90.
M-101; 20 Cr, 30 Ni, 1.6 Al, 1.0 max Si, bal Fe.
Good creep strength at elevated temperatures.

SANDVIK 7C27MO2.
M-101; 0.35 C, 13.5 Cr, 0.4 Ni, 1.2 Mo, bal Fe.
Annealed: 95,000 TS; 50,000 YS; 25 El; 55 RA; 195 Brin.
For valves, cutlery, oil refinery equipment; corrosion resistant.

SANDVIK 7L.
M-101; 0.35 C, 0.2 Si, 0.5 Mn, bal Fe.
Normalized: 82,000 TS; 54,000 YS; 31 El; 50 RA; 170 Brin.
For mechanical parts, axles, hydraulic cylinders; water hardened.

SANDVIK 7R60.
M-101, M-101a, M-101b; 0.55 C, 0.6 Si, 1.7 Mn, 0.020 max P, 0.020 max S, 16.5 Cr, 13.5 Ni, 2.3 Mo, bal Fe.
1% proof stress at 750°F: 22,800 psi min.
1% proof stress at 1290°F: 18,500 psi min.
Austenitic stainless with molybdenum. Good hot strength and resistance to scaling; for stream pipes and superheater tubes.
ASTM A213 Gr. TP 316 H, etc.

SANDVIK 7RE10.
   M-101; 0.08 max C, 2.0 max Mn, 1.5 max Si, 25 Cr, 20 Ni, bal Fe.
   Austenitic stainless steel.
   AISI 310S.

SANDVIK 7RE12.
   M-101, M-101a, M-101b; 0.055 C, 0.3 Si, 1.5 Mn, 0.015 max P, 0.015 max S, 25.5 Cr, 21.0 Ni, bal Fe.
   Wire for MIG, TIG, and submerged arc welding of Type 310 stainless steel.
   Werkstoff Nr. 4842; AISI 310.

SANDVIK 8R30.
   M-101, M-101a, M-101b; 0.08 max C, 0.6 Si, 1.5 Mn, 0.020 max P, 0.02 max S, 17.5 Cr, 11.5 Ni, Ti = 5 x % C min, bal Fe.
   1% proof stress at 750°F: 23,500 psi min.
   1% proof stress at 1290°F: 17,800 psi min.
   Austenitic stabilized stainless for steam pipes and superheater tubes for temperatures up to about 800°C (1470°F).
   Werkstoff Nr. 4541.

SANDVIK 8R40.
   M-101, M-101a, M-101b; 0.08 max C, 0.45 Si, 1.8 Mn, 0.020 max P, 0.020 max S, 17.5 Cr, 11.5 Ni, Nb = 10 x %C min, bal Fe.
   1% proof stress at 750°F: 27,000 psi min.
   1% proof stress at 1290°F: 20,500 psi min.
   Austenitic stabilized stainless for superheater tubes for temperatures up to about 1290°F.
   Werkstoff Nr. 1.4550.

SANDVIK 8R41.
   M-101, M-101a, M-101b; 0.08 max C, 0.45 Si, 1.3 Mn, 0.020 max P, 0.020 max S, 16.5 Cr, 13.0 Ni, Nb = 10 x % C, bal Fe.
   1% proof stress at 750°F: 27,000 psi min.
   1% proof stress at 1290°F: 20,500 psi min.
   Austenitic stabilized stainless; mainly used for steam pipes and superheater tubes to temperatures of about 1290°F.
   Werkstoff Nr. 4961.

SANDVIK 8R42.
   M-101, M-101a, M-101b; 0.06 max C, 0.4 Si, 1.3 Mn, 19.5 Cr, 9.5 Ni, 0.9 Nb, bal Fe.
   Stabilized austenitic stainless steel wire for metallizing.

SANDVIK 8R70.
   M-101, M-101a, M-101b; 0.08 max C, 0.5 Si, 1.8 Mn, 17.0 Cr, 13.0 Ni, 2.2 Mo, Ti = 5 x % C min, bal Fe.
   Stabilized austenitic stainless steel; very good corrosion resistance even at elevated temperatures; for food and chemical industries.
   Werkstoff Nr. 1.4571.

SANDVIK 8R80.
   M-101; 0.08 max C, 17.5 Cr, 13.4 Ni, 2.45 Mo, Cb = 10 x C, bal Fe.
   Austenitic stainless steel.
   Similar to AISI 317.

SANDVIK 8R81.
   M-101, M-101a, M-101b; 0.06 max C, 0.45 Si, 1.3 Mn, 19 Cr, 11 Ni, 2.8 Mo, 0.9 Nb, bal Fe.
   Austenitic stainless steel wire for metallizing.

SANDVIK 10L.
   M-101; 0.5 C, 0.25 Si, 0.5 Mn, bal Fe.
   Normalized: 100,000 TS; 63,000 YS; 28 El; 41 RA; 205 Brin.
   For tubes, gears, shafts, water hardened.

SANDVIK 10RA50.
   M-101, M-101a, M-101b; 0.08 C, 0.65 Si, 1.6 Mn, 0.030 max P, 0.30 S, 17.5 Cr, 9.5 Ni, 0.5 Mo, bal Fe.
   Austenitic free-machining stainless steel.
   For shafts, pins, bolts, studs, screws, etc, that require much machining but must be stainless.

SANDVIK 10RE20.
   M-101; 0.08 C, 26.3 Cr, 5.0 Ni, bal Fe.
   Heat resisting steel.

SANDVIK 10RE21.
   M-101; 0.08 C, 0.45 Si, 0.45 Mn, 26.3 Cr, 5.0 Ni, 1.45 Mo, bal Fe.
   Annealed: 105,000 TS; 80,000 YS; 25 El; 230 Brin.
   For valves, valve fittings, pump parts.
   Type 329 heat resistant steel.

SANDVIK 11.
   M-101; 0.60 C, 0.20 Si, 0.30 Mn, bal Fe.
   Heat treated: 115,000-160,000 TS; 77,000-113,000 YS; 23-12 El; 54-40 RA; 229-320 Brin.
   For wheels, die blocks, girders, rails, discs; water hardened.

SANDVIK 11L.
   M-101; 0.6 C, 0.25 Si, 0.5 Mn, bal Fe.
   Normalized: 107,000 TS; 63,000 YS; 26 El; 37 RA; 225 Brin.
   For springs, typewriter parts; water hardened, for cold drawing.

SANDVIK 11R51.
   M-101a; 0.09 C, 8 Ni, 17 Cr, 1.1 Si, 1.25 Mn, 0.7 Mo, bal Fe.
   Rolled: 299,000 TS; 277,000 YS; 2-4 El.
   Tempered: 341,000 TS; 327,000 YS; 1-4 El.
   For springs for instruments; high fatigue life and tensile strength.

SANDVIK 12.
   M-101; 0.63 C, 0.25 Si, 0.35 Mn, bal Fe.
   Normalized: 114,000 TS; 64,000 YS; 24 El; 33 RA; 240 Brin.
   For tools, mechanical parts; water hardened.

SANDVIK 12C27.
M-101, M-101a, M-101b; 0.58 C, 0.35 Si, 0.35 Mn, 14.0 Cr, bal Fe.
Chromium stainless strip steel; hardenable.
For cutlery, surgical knives, food processing knives; in thin strips for razor blades.

SANDVIK 12R10.
M-101, M-101a, M-101b; 0.10 C, 0.45 Si, 0.45 Mn, 18 Cr, 9 Ni, bal Fe.
Hard drawn stainless steel wire for springs; .013-.125 in dia: 256,000-313,000 psi TS.
Werkstoff Nr. 4300; AISI 302.

SANDVIK 12R10HV.
M-101, M-101a, M-101b; 0.10 C, 0.45 Si, 0.45 Mn, 18 Cr, 9 Ni, bal Fe.
Vacuum melted austenitic hard drawn spring wire: 0.013-.125 in dia has 256,000-313,000 psi TS; very good fatigue properties.
Werkstoff Nr. 4300; AISI 302.

SANDVIK 12R11.
M-101; 0.10 C, 0.45 Si, 0.45 Mn, 18 Cr, 7.7 Ni, bal Fe.
Annealed: 110,000 TS; 40,000 YS; 60 El; B85 Rock.
Available in 8 other tempers.
For aircraft structural members, trailer bodies, wheel covers, diaphragms, springs.
Type 301 stainless, austenitic.

SANDVIK 12R72HV.
M-101, M-101a, M-101b; 0.10 C, 0.5 Si, 1.8 Mn, 0.020 max P, 0.010 max S, 15.0 Cr, 15.0 Ni, 1.2 Mo, 0.40 Ti, 0.006 B, bal Fe.
1% proof stress at 750°F: 27,700 psi min.
1% proof stress at 1200°F: 25,600 psi min.
Austenitic Mo-Ti stabilized Ni-Cr stainless for steam or superheater tubes or other parts at temperatures up to about 1380°F.

SANDVIK 12 W12-C 1.
M-101; 0.65 C, 5.5-6.0 W, bal Fe.
For magnetic parts; magnetic steel.

SANDVIK 13.
M-101; 0.70 C, 0.3 Si, 0.45 Mn, bal Fe.
Normalized: 117,000 TS; 64,000 YS; 23 El; 30 RA; 250 Brin.
For piston rings, umbrella ribs, clicker dies; water hardened.

SANDVIK 13LM.
M-101; 0.75 C, 0.65 Mn, bal Fe.
Annealed: 175 Brin.
For springs; oil or water hardened.

SANDVIK 13M.
M-101, M-101a, M-101b; 0.75 C, 0.2 Si, 0.65 Mn, bal Fe.
Bainite hardened strip steel; 192,000-228,000 TS; 171,000-199,000 YS; 6-8.5 El; 41-47 Rock C.
For stamped and lightly formed parts.
Werkstoff Nr. 0605/1.1248; AISI 1074.

SANDVIK 14BSA.
M-101, M-101a, M-101b; 0.68 C, 0.25 Si, 0.85 Mn, 0.015 max P, 0.15 S, bal Fe.
Free machining, water hardening, carbon steel; for shafts, pins, screws, studs, miscellaneous hardware.

SANDVIK 14N3.
M-101; 0.75 C, 2.6 Ni, bal Fe.
Heat treated: 199,000 TS.
For wood band saws; oil hardened.

SANDVIK 14NC.
M-101; 0.75 C, 0.15 Cr, 0.5 Ni, bal Fe.
Heat treated: 192,000 TS.
For wood band saws, chain saws; oil hardened.

SANDVIK 14P.
M-101; 0.75 C, 0.2 Pb, bal Fe.
Rolled: 120,000 TS.
For clock and watch parts; free-cutting.

SANDVIK 14S1C1.
M-101; 0.75 C, 0.35 Cr, 1.4 Si, bal Fe.
Heat treated: 127,000 TS.
For springs, friction saws, flapper valves; oil hardened.

SANDVIK 15RE10.
M-101, M-101a, M-101b; 0.12 C, 0.55 Si, 1.8 Mn, 24.5 Cr, 20.5 Ni, bal Fe.
0.2% offset yield at 1290°F: 17,000 psi min.
Austenitic stainless, weldable, good resistance to oxidation at elevated temperatures; for heat exchanger tubes, furnace parts.
Werkstoff Nr. 1.4845; AISI 310.

SANDVIK 15RE12.
M-101, M-101a, M-101b; 0.12 C, 0.45 Si, 1.8 Mn, 0.015 max P, 0.015 max S, 26.0 Cr, 21.0 Ni, bal Fe.
Wire for MIG, TIG, and submerged arc welding of Type 310 stainless steel.
Werkstoff Nr. 4842; AISI 310.

SANDVIK 15VDT.
M-101; 0.80 C, 0.1 V, bal Fe.
Drawn: 100,000 TS.
For wire for latch needles; water hardened.

SANDVIK 16CIV.
M-101; 0.90 C, 0.25 Si, 0.55 Mn, 0.6 Cr, 0.10 V, bal Fe.
Oil hardening cold work tool steel; for small drills, punches.

SANDVIK 17.
M-101; 1.0 C, 0.25 Si, 0.45 Mn, bal Fe.
For tools, springs, valves, doctor blades; water hardening.

SANDVIK 17AP.
M-101; 1.0 C, 0.2 Pb, 0.05 S, 0.025 P, bal Fe.
Annealed: 82,000 TS.
For watch parts, machinery parts; free-cutting, oil hardened.

SANDVIK 17C.
M-101; 1.0 C, 0.15 Cr, 0.3 Si, 0.3 Mn, bal Fe.
Annealed: 85,000 TS; 185 Brin.
For tools, saws, springs, piston skirt expanders; water hardening.

SANDVIK 17VDT.
M-101; 1.0 C, 0.1 V, bal Fe.
For needle wire; oil hardened.

SANDVIK 18C29.
M-101; 0.95 C, 13.5 Cr, bal Fe.
Air or oil hardening tool steel.
Corrosion resistant.

SANDVIK 18C283.
M-101; 1.05 C, 14 Cr, 0.5 Mo, 0.5 Co, 0.25 Cu, bal Fe.
Annealed: 125,000 TS; 250 Brin.
For knives, ball bearings, ball point pens; stainless, hardenable.

SANDVIK 18RM11.
M-101, M-101a, M-101b; 0.15 C, 0.4 Si, 6.0 Mn, 18.0 Cr, 8.0 Ni, bal Fe.
Austenitic stainless steel wire for metallizing and for welding. Werkstoff Nr. 4370.

SANDVIK 20.
M-101; 1.15 C, 0.25 Si, 0.35 Mn, bal Fe.
For needles, cutters, tools; water hardening.

SANDVIK 20P.
M-101; 1.15 C, 0.2 Pb, bal Fe.
For reeds in musical boxes; water hardened.

SANDVIK 20W.
M-101; 1.15 C, 0.15 W, bal Fe.
For watch mainsprings; oil hardened.

SANDVIK 20 WIW.
M-101; 1.15 C, 0.5 W, 1 V, bal Fe.
For tools, dies; water hardening.

SANDVIK 21C.
M-101; 1.25 C, 0.15 Cr, 0.2 Si, 0.35 Mn, bal Fe.
For razor blades, band saws; water hardening.

SANDVIK 21T10.
M-101; 1.25 C, 0.25 Cr, 1.75 W, 0.1 V, bal Fe. 210 Brin.
For tools, cutters, saw blades; fast finishing steel.

SANDVIK 21T10P.
M-101; 1.25 C, 0.25 Cr, 1.75 W, 0.1 V, 0.2 Pb, bal Fe.
For dental burs, saw blades; Ledloy.

SANDVIK 22C.
M-101, M-101a, M-101b; 1.29 C, 0.2 Si, 0.3 Mn, 0.15 Cr, bal Fe.
Strip steel, designed for metal band saws, hardenable.

SANDVIK 700.
M-101; 0.03 max C, 0.5 Si, 1.7 Mn, 21.0 Cr, 25.0 Ni, 4.5 Mo, 0.30 Cb, bal Fe.
For phosphoric acid; pulp + paper.

SANDVIK CORONA 62.
M-101, M-101a, M-101b; 0.80 C, 0.3 Si, 0.3 Mn, 4.0 Cr, 5.0 Mo, 6.3 W, 2.0 V, bal Fe.
High speed steel metal band saw steel. AISI M2.

SANDVIK CORONA 67.
M-101, M-101a, M-101b; 0.92 C, 0.3 Si, 0.3 Mn, 4.0 Cr, 2.0 Mo, 1.0 W, 2.25 V, bal Fe.
Air or oil hardening metal band strip steel.

SANDVIK GC 1025.
M-101.
Titanium coated carbide.

SANDVIK HARDFLEX 11L.
M-101, M-101a, M-101b; 0.55 C, 0.25 Si, 0.5 Mn, bal Fe.
Bainite hardened strip steel: 128,000-164,000 TS; 107,000-142,000 YS; 10-14 El; 27-36 Rock C.
For stamped and lightly formed parts.
Werkstoff Nr. 1.1210; AISI 1055.

SANDVIK HARDFLEX 13LM.
M-101, M-101a, M-101b; 0.65 C, 0.2 Si, 0.75 Mn, bal Fe.
Bainite hardened strip steel: 192,000-228,000 TS; 171,000-199,000 YS; 6-8.5 El; 41-47 Rock C.
For stamped and lightly formed parts.
Werkstoff Nr. 0603/1.1231; AISI 1065.

SANDVIK HT3.
M-101; 0.10 max C, 5 Cr, bal Fe.
Annealed: 70,000-72,000 TS; 28,000-30,000 YS; 28-36 El; 65-70 RA; 135-160 Brin.
For oil refinery equipment; Type 502; heat and creep resistant.

SANDVIK HT5.
M-101; 0.12 C, 0.9 Cr, 0.55 Mo, bal Fe.
Annealed: 74,000 TS; 46,000 YS; 30 El; 65 RA; 170 Brin.
For boiler and superheater tubes; for steam temperature up to 950°F.

SANDVIK HT7.
M-101; 0.08 C, 9 Cr, 1 Mo, 0.7 Si, bal Fe.
Annealed: 92,000 TS; 67,000 YS; 30 El; 73 RA; 195 Brin.
For tubes in oil refineries and superheaters.

SANDVIK HT8.
M-101; 0.12 C, 2.25 Cr, 1 Mo, bal Fe.
Annealed: 75,000 TS; 50,000 YS; 33 El; 76 RA; 170 Brin.
For boiler and superheater tubes; for steam temperature up to 1005°F.

SANDVIK HT9.
M-101, M-101a, M-101b; 0.20 C, 0.3 Si, 0.55 Mn, 0.02 max P, 0.02 max S, 12.0 Cr, 0.5 Ni, 1.0 Mo, 0.3 V, 0.5 W, bal Fe.
At 570°F: 57,000 min YS.
At 930°F: 43,000 min YS.
Ferritic chromium steel for superheater tubes for temperatures up to about 1200°F.
Werkstoff Nr. 1.4935.

SANDVIK HT91.
M-101, M-101a, M-101b; 0.20 C, 0.3 Si, 0.55 Mn, 0.02 max P, 0.02 max S, 12.0 Cr, 0.5 Ni, 1.0 Mo, 0.3 V, bal Fe.
0.2% proof stress at 570°F: 57,000 psi.
0.2% proof stress at 930°F: 43,000 psi.
Ferritic chromium steel for superheater tubes to operate up to about 1200°F.
Werkstoff Nr. 1.4922.

SANDVIK ON50HV.
M-101; 0.02 max C, 0.08 max Si, 0.45 Mn, 51 Ni, bal Fe.
Soft magnetic iron-nickel alloy.
Low coercive force; high permeability; coefficient of expansion same as lead glass; for reed relays.

SANDVIK SANICRO 30.
M-101, M-101a, M-101b; 0.030 max C, 0.55 Si, 0.55 Mn, 0.015 max P, 0.015 max S, 0.10 max Cu, 20.0 Cr, 34.0 Ni, 0.35 Ti, 0.30 Al, bal Fe.
Nickel-iron-chromium alloy tube, wire.
Good resistance to oxidation, stress corrosion, gases at elevated temperatures.
Werkstoff Nr. 1.4876; ASTM B163 Alloy 800.

SANDVIK SANICRO 31.
M-101, M-101a, M-101b; 0.04 C, 0.55 Si, 0.55 Mn, 0.015 max P, 0.015 max S, 0.10 max Cu, 21.0 Cr, 31.0 Ni, 0.35 Ti, 0.30 Al, bal Fe.
Nickel-iron-chromium alloy bar, tube, strip, wire.
High resistance to scaling in air, to carburization, to nitrogen absorption, to stress corrosion; used in petroleum industry.
Werkstoff Nr. 1.4876; ASTM B163, B407, B408, B409.

SANDVIK SANICRO 70.
M-101, M-101a, M-101b; 0.030 max C, 0.35 Si, 0.55 Mn, 0.015 max P, 0.010 max S, 16.0 Cr, 74 Ni, 9 Fe.
Austenitic nickel base alloy; good resistance to scaling, carburizing and steam at elevated temperatures.
Werkstoff Nr. 2.4816; ASTM B163, B166, B167.

SANDVIK SANICRO 71.
M-101, M-101a, M-101b; 0.55 C, 0.35 Si, 0.8 Mn, 0.015 max P, 0.015 max S, 0.10 max Cu, 16.0 Cr, 72.5 Ni, 9.5 Fe, 0.35 Ti, 0.25 Al.
Austenitic nickel base alloy; good resistance to attack by most gases at elevated temperatures.
Werkstoff Nr. 2.4816; ASTM B163, B166, B167.

SANDVIK SANICRO 72.
M-101, M-101a, M-101b; 0.030 max C, 0.20 Si, 3.0 Mn, 0.010 max P, 0.010 max S, 20.0 Cr, 72.5 Ni, 0.32 Ti, 1.0 max Fe, 2.5 Nb.
Wire for MIG, TIG, and submerged arc welding of nickel base alloys, stainless steels and carbon steels.

SANDVIK TITANIUM 9.
M-101; 0.15 max Fe, 0.035 max C, 0.01 max N, 0.09 max O, 0.005 max H, bal Ti.
Very high purity titanium; for parts and tubing for nuclear reactors, sea water, acids, plating equipment; also used in wire form for welding titanium.

SANDVIK TITANIUM 20.
M-101; 0.30 max Fe, 0.05 max C, 0.03 max N, 0.20 max O, 0.015 max H, bal Ti.
High purity titanium; for chemical industry, pulp and paper industry, electroplating racks, food industry, heat exchangers.

SANDVIK ZIRCALOY 2.
M-101, M-101a, M-101b; 1.20-1.70 Sn, Fe + Cr + Ni = 0.18-0.38, bal Zr (high purity).
For fuel element canning tubes in nuclear reactors. ASTM B353.

SANDVIK ZIRCALOY 4.
M-101, M-101a, M-101b; 1.20-1.70 Sn, Fe + Cr = 0.28-0.37, bal Zr (high purity).
For fuel element canning tubes in nuclear reactors, ASTM B353.

SANICRO see **SANDVIK SANICRO.**

SA N-N.
M-333; 0.5 C, 2 W, bal Fe.
For gripper dies, piercers; hot work steel.

SA NON PA-REIL.
M-333; 0.9 C, 1.2 Mn, bal Fe.
For dies, punches, upsetters, crimpers; Type O1; non-deforming.

S.A.P. 865.
M-555, M-249, M-1605; 13-14 Al_2O_2, bal Al.
Sintered: 49,000 TS; 35,000 YS; 8 El; 105 Brin.
At 930°F: 16,000 TS; 13,500 YS; 2 El.
For compressor blades, engine components, heat exchangers, pistons, cylinder heads.
High creep resistant. Sintered powder. Dispersion hardened.

S.A.P. 895.
M-249, M-1605, M-555; 10-11 Al_2O_2, bal Al.
At 70°F: 43,000 TS; 29,000 YS; 16 El; 100 Brin.
For compressor blades, engine components, heat exchangers, pistons, cylinder heads.
Sintered powder. Dispersion hardened.

S.A.P. 930.
M-249, M-1605, M-555; 7 Al_2O_2, bal Al.
At 70°F: 36,000 TS; 23,000 YS; 20 El; 90 Brin.
At 750°F: 12,100 TS; 11,200 YS; 5.7 El.
For compressor blades, engine components, heat exchangers, cylinder heads, pistons.
Strong high temperature material.
Sintered powder.
Dispersion hardened.

SAPHIR.
M-1120; 1.0-1.2 C, bal Fe.
For drills, taps, reamers, shears, punches; Type W1, water hardened.

SARATOGA.
M-1779; 0.90 C, 0.38 Si, 0.50 W, 1.2 Mn, 0.5 Cr, bal Fe.
Oil hardening, non-deforming tool steel, suitable for blanking and forming dies, punches, plug gages and trim dies; usually at Rc 56-60.
AISI 0-1.

SARDO.
M-344; 1.5 C, 13 Cr, bal Fe.
For tools, dies; non-deforming, air hardening.

SASHCHAIN.
92-95 Cu, 8-5 Sn.
For sash chain; bronze.

SATAN.
M-365; 0.23 C, 0.60 Mn, 1.25 Si, 10.0 Cr, 0.75 Ni, 1.0 V, 0.45 W, 0.45 Mo, bal Fe.
Hot work tool steel, chromium type.

SATCO.
M-88; 97.5 Pb, bal Ca, Sn, etc.
27 Brin.
For bearings, crosshead gibs, trailer and driving boxes; white metal, for high speeds and heavy loads.

SATMUMETAL.
M-108; 53-58 Ni, bal Fe.
Max. permeability 240,000.
Flux density 15,000, coercive force 0.025.
Initial permeability 65,000.
For small distribution transformers, instrument transformers, ground leakage protective devices, torroids.
Soft magnetic alloy.

S.A.T.T. SPECIAL.
M-333; 1.2 C, 0.75 Cr, 1.5 W, 0.2 Mo, bal Fe.
For tools, cutters; oil hardening.

SATURN.
M-114; 1.25 C, 3.5 W, bal Fe.
For wire drawing plates, dies, cutting tools, chasers, drills; fast finishing steel.

SAUREFEST ALLOYS see JUNKER.

SAUT-DU-TARN 18.CD.4F.
M-1119; 0.14-0.25 C, 0.80-1.2 Cr, 0.15-0.30 Mo, bal Fe.
Heat treated: 165,000-222,000 TS; 92,000 YS; 8 RA.
For gears, bolts, cams, camshafts; oil hardened, tough.

SAUT-DU-TARN 25.CD.4F.
M-1119; 0.22-0.30 C, 0.80-1.2 Cr, 0.15-0.30 Mo, bal Fe.
Heat treated: 186,000-242,000 TS; 107,000 YS; 3 RA.
For gears, bolts, crankshafts; oil hardened, tough.

SAUT-DU-TARN 30.NC.11F.
M-1119; 0.25-0.35 C, 0.60-0.90 Cr, 2.5-3.0 Ni, bal Fe.
Heat treated: 128,000-157,000 TS; 107,000 YS; 8 RA.
For gears, bolts, crankshafts; oil hardened, shock resistant.

SAUT-DU-TARN 35.CD.4F.
M-1119; 0.30-0.39 C, 0.80-1.2 Cr, 0.15-0.30 Mo, bal Fe.
Heat treated: 221,000-245,000 TS; 2 RA.
For gears, bolts, crankshafts; oil hardened, tough.

SAUT-DU-TARN 35.NC.15H.
M-1119; 0.30-0.38 C, 1.5-1.9 Cr, 3.5-4.0 Ni, bal Fe.
Heat treated: 242,000 TS; 5 RA.
For gears, bolts, crankshafts; oil hardened, shock resistant.

SAUT-DU-TARN 35.NCD.6F.
M-1119; 0.30-0.38 C, 0.8-1.2 Cr, 1.2-1.6 Ni, 0.15-0.30 Mo, bal Fe.
Heat treated: 165,000-200,000 TS; 144,000 YS.
For gears, bolts, crankshafts; oil hardened, shock resistant.

SAUT-DU-TARN 35.NCD.12H.
M-1119; 0.25-0.35 C, 0.70-1.2 Cr, 2.7-3.3 Ni, 0.25-0.45 Mo, bal Fe.
Heat treated: 150,000 TS; 127,000 YS.
For gears, bolts, crankshafts; oil hardened, shock resistant.

SAUT-DU-TARN 45.S7.
M-1119; 0.40-0.50 C, 1.6-2.1 Si, 0.4-0.8 Mn, bal Fe.
Heat treated: 192,000-242,000 TS; 150,000 YS; 3 El; 241 Brin.
For springs, torsion bars; oil hardened, tough.

SAUT-DU-TARN 55.S6.
M-1119; 0.55 C, 1.6-2.1 Si, 0.6-1.0 Mn, bal Fe.
Heat treated: 186,000-213,000 TS; 170,000 YS; 2 El; 270 Brin.
For springs, torsion bars; oil hardened, tough.

SAUT-DU-TARN NO. 1.
M-1119; 0.9-1.0 C, bal Fe.
For tools, drills, punches; water hardened.

SAUT-DU-TARN NO. 2.
M-1119; 1.1 C, bal Fe.
For tools, drills; water hardened.

SAVA-Y.
M-541, M-1220; 3.8-4.2 Cu, 1.3-1.7 Mg, 1.8-2.3 Ni, bal Al.
Cast: 36,000-47,000 TS; 32,700-42,700 YS; 0.5 El; 100-115 Brin.
For pistons, cylinder heads; age hardenable, heat resistant.

SAVILLE B 1.
M-486; 0.55 C, 0.6 Mn, bal Fe.
For rifle barrels, gear wheels, housings; water hardening.

SAVILLE BEST CAST STEEL.
M-486; 0.7-1.2 C, 0.3 Mn, bal Fe.
For blanking and forming dies, shear blades; water hardening.

SAVILLE CAST.
M-486; 0.7-1.2 C, 0.3 Mn, bal Fe.
For punches, shear blades, chisels, files; water hardening.

SAVILLE CAST STEEL WARRANTED.
M-486; 0.7-1.2 C, 0.3 Mn, bal Fe.
For wood working tools, punches, drills, taps; water hardening.

SAVILLE CROWN.
M-486; 0.44 C, 0.35 Mn, 1.3 Cr, 0.15 V, 2.3 W, bal Fe.
For punches, chipping chisels, pneumatic tools; tough, shock resistant.

SAVILLE F 2.
M-486; 0.57 C, 0.7 Cr, 0.2 V, bal Fe.
For die casting dies; hot die steel.

SAVILLE F.7.
M-486; 0.48 C, 1.5 Cr, 1.2 W, 0.12 V, 0.7 Ni, bal Fe.
For chisels, rivet snaps, beading tools; shock resistant.

SAVILLE G-0.
M-486; 0.33 C, 0.5 Mn, 3.5 Ni, 0.75 Cr, bal Fe.
For bearing caps, valve rockers, propeller shafts; shock resistant.

SAVILLE G-1 SPECIAL.
M-486; 0.32 C, 0.5 Mn, 4.1 Ni, 1.2 Cr, 0.25 Mo, bal Fe.
For gears, pinions, shafts; tough.

SAVILLE G-2.
M-486; 0.42 C, 1.3 Si, 13 Ni, 13 Cr, 2.5 W, bal Fe.
At 70°F: 106,000 TS; 31 El.
At 1100°F: 66,000 TS; 35 El.
At 1500°F: 32,000 TS; 55 El.
For aircraft engine exhaust valves; austenitic, resists shock at high temperature.

SAVILLE G-14.
M-486; 0.39 C, 4.1 Ni, 1.3 Cr, 0.3 Mo, bal Fe.
For chisels, cold punches, press tools; shock resistant.

SAVILLE G-18 B.
M-486; 0.4 C, 1 Si, 13 Ni, 13 Cr, 0.10 Co, 3 Nb, 2.5 W, 2 Mo, bal Fe.
At 70°F: 92,000 TS; 40 El.
At 1300°F: 56,000 TS; 12 El.
At 1500°F: 44,000 TS; 9 El.
For aircraft engine exhaust valves; austenitic, high hot strength.

SAVILLE H-27.
M-486; 0.37 C, 0.6 Mn, 3 Cr, 0.9 Mo, 0.2 V, bal Fe.
190,000 TS; 12 El; 435 Brin.
For crankshafts, air-screen shafts; shock resistant, nitriding steel.

SAVILLE H-30.
M-486; 0.6 C, 0.6 Mn, 0.6 Cr, bal Fe.
Heat treated: 110,000-130,000 TS; 80,000-90,000 YS; 15-12 El; 300-341 Brin.
For cylinder liners, gears, axles; water hardening.

SAVILLE H-42.
M-486; 1.65 C, 0.45 Mn, 13 Cr, 0.7 Mo, 0.3 V, bal Fe.
For burnishing rolls, shear blades, master hobs; air hardening, non-deforming.

SAVILLE H-44.
M-486; 1.0 C, 0.7 Mn, 1.45 Cr, bal Fe.
For press tools, shear blades, cold rolls; cold work steel.

SAVILLE J 23.
M-486; 0.36 C, 0.4 Si, 2.6 Cr, 1 V, 1.8 W, 4.3 Mo, bal Fe.
For die casting dies, swaging dies; hot die steel.

SAVILLE J-32.
M-486; 0.32 C, 4 Ni, 1.5 Cr, 5.8 W, 0.35 V, bal Fe.
For hot punches, stamping dies; hot work steel.

SAVILLE K 18.
M-486; 0.32 C, 0.95 Si, 5 Cr, 1.4 W, 1.7 Mo, bal Fe.
For die casting dies; hot die steel.

SAVILLE KSA.
M-486; 0.9 C, 1.85 Mn, 0.45 Cr, 0.45 W, bal Fe.
For gauges, taps, dies, chasers, cams, pawls; non-deforming, oil hardening.

SAVILLE O.K.
M-486; 0.44 C, 1.3 Cr, 2.3 W, 0.7 Si, 0.15 V, bal Fe.
For die and punches for hot work; resists heat checking.

SAVILLE PERMABRITE.
M-486; 0.10-0.14 C, Cr, Ni, bal Fe.
For stainless steel parts; austenitic super-stainless.

SAVILLE RUSTLESS IRON.
M-486; 0.15 C, Cr, Ni, bal Fe.
For stainless parts; stainless, dead soft type.

SAVILLE STAINLESS STEEL.
M-486; 0.28-0.32 C, Cr, Ni, bal Fe.
For stainless parts; stainless, hardening quality.

SAVILLE S.V.L.
M-486; 0.32 C, 4.1 Ni, 1.3 Cr, 0.2 Mo, bal Fe.
Hardened: 215,000 TS; 190,000 YS; 12 El; 477 Brin.
For plastic molding dies; shock resistant, air hardening.

SAVILLE WPS.
M-486; 2.3 C, 0.35 Mn, 13 Cr, bal Fe.
For gauges, shear blades, thread rollers; resists wear and corrosion, non-deforming.

SAVILLE XLD.
M-486; 0.82 C, 5 Cr, 0.6 Mo, 1.7 V, 22 W, 17 Co, bal Fe.
For turning, planing and boring tools; high speed steel.

SA/W.
M-459; C, Si, Mn, bal Fe.

SA-WHITE LABEL.
M-333; 0.9 C, bal Fe.
For taps, drills, forming and blanking dies; water hardening.

SAXONIA METAL.
M-England; 84.8 Zn, 5.3 Sn, 6 Cu, 3 Pb, 0.2 Al.
For bearings; will not resist heat or live steam.

SA-YELLOW LABEL.
M-333; 0.8 C, 0.9 Cr, 0.2 V, bal Fe.
For punches, dies, shear blades; oil hardening.

S-BABBITT.
M-815, M-314; 15 Sb, 1 As, 1 Sn, 0.5 Cu, bal Pb.
At 70°F: 10,350 TS; 2 El; 20 Brin.
At 392°F: 1260 TS; 95 El.
For bearings, liners; Babbitt.

SC.
M-459; C, Cr, Si, Mn, bal Fe.

SC-6.
M-1200; 0.50 C, 0.7 Mn, 1.5 Ni, 0.75 Cr, 0.5 Mo, bal Fe.
Normalized: 180,000 TS; 140,000 YS; 8 El; 12 RA; 240 Brin.
For gears, pinions, shafts, axles; oil hardened, shock resistant.

SC65.
M-1083; 6-7 Ag, bal Cu.
Forged: 61,000-67,000 TS; 5-12 El; 115-140 Brin.
For heavy duty seam welding wheels, 82-86% electrical conductivity.
BS 4577 and ISO 5182 alloy A/4/3.

SCANDIUM.
M-1755; Sc.
Purities: Special distilled 99.9%, 99.5%.
Forms: Ingot, lump, sheet, wire, foil.

SCASON.
M-1787; WC plus alloy.
For anti-skid and wear-resistant alloy for anti-skid studs for tires; and studded straps for snow and ice chains.

SCB OR SCB ALLOYS.
M-1528.
All alloys inactive; Stauffer no. longer supplies.

SCB-291 see **FANSTEEL SCB-291.**

SCB-885 see **COLUMBIUM D-36.**

SCB-990 see **FANSTEEL 80.**

SCEPTRE BRASS.
62 Cu, 36 Zn, 1.4 Fe, 1.1 Al, 0.07 Pb.
For ornamental parts, hardware; high strength.

SCH 3.
M-1488; 0.45 C, 9 Cr, 3 Si, bal Fe.
Ht. tr. (Q + T): 880 N/mm² UTS.
Valve steel.
AFNOR Z45CS9; W.-Nr. 1.4718.

SCH 4.
M-1488; 0.40 C, 9.5 Cr, 2.3 Si, 2.3 Mo, bal Fe.
Ht. Tr. (Q + T): N/mm² min UTS.
Valve steel.
AFNOR Z40 CSD 10.

SCHENK DGZNA14 (Z400).
M-1453; 3.5-4.3 Al, 0.6 max Cu, 0.02-0.05 Mg, bal Zn.
Die cast: 36,000 TS; 1.5 El; 70 Brin.
For ornaments, instrument housings, high strength.

SCHENK DGZNA14CUL (Z400).
M-1453; 3.5-4.3 Al, 0.6-1.0 Cu, 0.02-0.05 Mg, bal Zn.
Die cast: 38,000 TS; 2 El; 80 Brin.
For ornaments, instrument housings, high strength.

SCHENK DMGA1911.
M-1453; 7.6-10.0 Al, 0.1-1.0 Zn, 0.1-0.6 Mn, bal Mg.
Die cast: 22,000-31,000 TS; 1.0-0.2 RA; 55 Brin.
For instrument housings.

SCHENK GA1CUSI NO. 223.
M-1453; 4-7 Cu, 2-4 Si, bal Al.
Sand cast: 23,000-30,000 TS; 17,000-23,000 YS; 0.5-2 El; 75-100 Brin.
For aircraft castings; age-hardenable.

SCHENK GA1MG3(CU) NO. 241.
M-1453; 1.5-4.0 Mg, bal Al.
Sand cast: 20,000-27,000 TS; 11,000-14,500 YS; 6-2 El; 50-65 Brin.
For marine castings; corrosion resistant.

SCHENK GA1MG3HY31.
M-1453; 1.5-3.5 Mg, bal Al.
Sand cast: 19,900-27,000 TS; 11,400-14,300 YS; 8-3 El; 50-60 Brin.
For marine castings; corrosion resistant.

SCHENK GA1MG5HYD51.
M-1453; 4.5-5.5 Mg, 0.1-0.5 Mn, bal Al.
Sand cast: 22,800-28,000 TS; 12,000-14,000 YS; 5-2 El; 55-70 Brin.
For marine castings; corrosion resistant.

SCHENK GA1MGMN.
M-1453; 1.5-3.0 Mg, 0.6-1.3 Mn, 0.5 Fe, bal Al.
Sand cast: 20,000-27,000 TS; 11,400-14,300 YS; 8-3 El; 50-60 Brin.
For marine and chemical industry castings; corrosion resistant.

SCHENK GA1SI.
M-1453; 11.0-13.5 Si, 0.3-0.5 Mn, 0.15 Ti, 0.6 Fe, bal Al.
Sand cast: 24,200-31,000 TS; 11,400-12,800 YS; 8-4 El; 50-60 Brin.
For automotive castings; corrosion resistant.

SCHENK GA1SI5MG.
M-1453; 0.4-0.6 Si, 0.5-0.8 Mg, 0.2-0.6 Mn, bal Al.
Sand cast: 21,000-27,000 TS; 14,000-19,000 YS; 3-1 El; 60-70 Brin.
For light alloy parts.

SCHENK GA1SI6CU3 NO. 225.
M-1453; 5.5-7.0 Si, 2.0-4.0 Cu, 0.4-0.6 Mn, bal Al.
Sand cast: 23,000-30,000 TS; 11,000-23,000 YS; 3-1 El; 65-80 Brin.
For pressure tight castings; corrosion resistant.

SCHENK GA1SI9(CU) NO. 232.
M-1453; 7-11 Si, 0.2-0.5 Mn, bal Al.
Sand cast: 21,000-29,000 TS; 14,000-20,000 YS; 3-1 El; 65-85 Brin.
For pressure tight castings; corrosion resistant.

SCHENK GA1SI(CU) NO. 231.
M-1453; 11-13 Si, bal Al.
Sand cast: 15,000-31,000 TS; 11,000-14,500 YS; 4-1 El; 50-60 Brin.
For pressure proof castings; corrosion resistant.

SCHENK GA1SICU NO. 234.
M-1453; 5-6 Si, 1.2-1.6 Cu, 0.4-0.6 Mg, bal Al.
Sand cast: 23,100-31,000 TS; 14,000-20,000 YS; 3-1 El; 65-85 Brin.
For instrument housings; corrosion resistant.

SCHENK GA1SIMG.
M-1453; 9-10 Si, 0.25-0.40 Mg, 0.3-0.5 Mn, 0.15 Ti, 0.6 Fe, bal Al.
Sand cast: 20,000-39,000 TS; 12,700-15,700 YS; 5-2 El; 55-65 Brin.
For automotive castings; corrosion resistant.

SCHENK GA1SIMG(CU) NO. 233.
M-1453; 8.5-10.5 Si, 0.2-0.4 Mg, bal Al.
Sand cast: 27,000-35,000 TS; 13,000-16,000 YS; 5-2 El; 55-65 Brin.
For pressure tight castings; corrosion resistant.

SCHENK GA1ZNCU NO. 212.
M-1453; 3-5 Cu, 4-8 Zn, 1.2 Fe, 0.6 Mn, 1.5 max Si, bal Al.
Sand cast: 21,000-30,000 TS; 13,000-19,000 YS; 3-1 El; 65-80 Brin.
For light alloy parts.

SCHENK GDA1MG9.
M-1453; 6-10 Mg, 0.2-0.7 Mn, bal Al.
Die cast: 27,000-38,000 TS; 3-1 El; 65-85 Brin.
For marine castings; corrosion resistant.

SCHENK GDA1MGSI.
M-1453; 0.3-2.5 Mg, 1.5-5.0 Si, 1.5 max Mn, bal Al.
Die cast: 22,800-27,000 TS; 3-1 El; 55-70 Brin.
For marine castings, housings; corrosion resistant.

SCHENK GDA1SI7.
M-1453; 6-10 Si, 0.2-0.7 Mn, 0.5 max Mg, bal Al.
Die cast: 24,000-35,000 TS; 3-1 El; 55-75 Brin.
For thin wall castings; corrosion resistant.

SCHENK GDA1SI13.
M-1453; 11-13 Si, 0.2-0.7 Mn, 0.5 max Mg, bal Al.
Die cast: 28,000-36,000 TS; 3-1 El; 60-80 Brin.
For thin wall castings; corrosion resistant.

SCHENK GDA1SI(CU) NO. 231.
M-1453; 11-13 Si, 0.5 Mn, bal Al.
Die cast: 26,000-36,000 TS; 3-1 El; 55-75 Brin.
For pressure tight castings; corrosion resistant.

SCHENK GDA1SICU NO. 311.
M-1453; 5.0-6.5 Si, 2-3 Cu, 0.2-0.6 Mn, bal Al.
Die cast: 27,000-33,000 TS; 2.5-1.0 El; 55-75 Brin.
For instrument housings; corrosion resistant.

SCHENK GMGA13ZN.
M-1453; 2.6-3.4 Al, 0.6-1.4 Zn, 0.1-0.5 Mn, bal Mg.
Sand cast: 20,000-24,200 TS; 7000-8500 YS; 10-5 El; 40-50 Brin.
For light alloy parts; Type AZ31.

SCHENK GMGA14ZN.
M-1453; 3.3-4.1 Al, 2.3-3.1 Zn, 0.1-0.5 Mn, bal Mg.
Sand cast: 24,000-31,000 TS; 10,000-12,800 YS; 3 El; 45 Brin.
For light alloy parts; Type AZF.

SCHENK GMGA16ZN.
M-1453; 5.3-6.1 Al, 2.3-3.1 Zn, 0.1-0.5 Mn, bal Mg.
Sand cast: 18,600 TS; 11,500 YS; 1.5 El; 60 Brin.
For light alloy parts; Type AZG.

SCHENK GMGA19.
M-1453; 7.8-8.8 Al, 0.1-0.8 Zn, 0.1-0.5 Mn, bal Mg.
Sand cast: 24,000 TS; 13,000 YS; 4 El; 55 Brin.
For light alloy parts; Type A9V.

SCHIRM ZAH.
M-614; 0.85 C, 0.25 max Si, 0.25 max Mn, bal Fe.
Heat treated: 190,000 TS; 140,000 YS; 10 El; 30 RA; 400 Brin.
For springs, taps, reamers, broaches, drills; Type W1; water hardened.

SCHIRM ZH.
M-614; 1.0 C, 0.25 max Si, 0.25 max Mn, bal Fe.
Annealed: 100,000 TS; 53,000 YS; 21 El; 42 RA; 200 Brin.
For springs, taps, reamers, drills, lathe tools; Type W1; water hardened.

SCHMIDTCLEMENS PHM ETC see **MARKER** or **MAERKER.**

SCHMIDT LOCOMOTIVE BEARING.
M-Eng.; 86 Cu, 14 Sn.
For bearings; bronze.

SCHMIDT TSS.
M-196; 2.4-4.0 Mg, 0.4 max Mn, 0.3 max Cr, bal Al.
Soft: 28,000 TS; 13,000 YS; 30 El; 47 Brin.
Hard: 40,000 TS; 35,000 YS; 10 El; 73 Brin.
For aircraft tanks and fittings, fuel lines, marine parts; resists sea water corrosion.

SCHOELLER ARWA now **VEW N208.**

SCHOELLER ARWKN now **VEW N206.**

SCHOELLER MAO now **VEW A500.**

SCHOELLER MAO SUPERIOR now **VEW A604.**

SCHOELLER MAOY now **VEW A500.**

SCHOELLER MASD now **VEW A122.**

SCHOELLER MASN now **VEW A350.**

SCHOELLER MASO now **VEW A120.**

SCHOELLER MASO SUPERIOR now **VEW A200.**

SCHOELLER MASOY now **VEW A100.**

SCHOELLER MASOY SUPERIOR now **VEW A205.**

SCHOELLER MASWY now **VEW A305.**

SCHOELLER R1 now **VEW H525.**

SCHOELLER R9 now **VEW H550.**

SCHOLLER ARG now **VEW W501.**

SCHOLLER ARH HART now **VEW N540.**

SCHOLLER ARH ZAH now **VEW N324.**

SCHOLLER ARKS now **VEW N320.**

SCHOLLER ARL now **VEW N350.**

SCHOLLER ARW now **VEW N100.**

SCHOLLER ARWDT now **VEW N238.**

SCHOLLER ARWK now **VEW N200.**

SCHOLLER ARWKT now **VEW N205.**

SCHOLLER ARZ now **VEW N320.**

SCHOLLER BM25 now **VEW V920.**

SCHOLLER BM35 now **VEW V935.**

SCHOLLER BM45 now **VEW V945.**

SCHOLLER BM60 now **VEW V960.**

SCHOLLER CVS now **VEW K506.**

SCHOLLER DHS now **VEW S600.**

SCHOLLER DS now **VEW V320.**

SCHOLLER ECM now **VEW E300.**

SCHOLLER ED now **VEW E304.**

SCHOLLER ENC1 now **VEW E230.**

SCHOLLER ENC2 now **VEW E220.**

SCHOLLER ENC4 now **VEW E204.**

SCHOLLER EXTRA ZAH now **VEW K980.**

SCHOLLER EXTRA ZH now **VEW K990.**

SCHOLLER FCR now **VEW K245.**

SCHOLLER FSH now **VEW E920.**

SCHOLLER HNC1.5 now **VEW V224.**

SCHOLLER KHP now **VEW W326.**

SCHOLLER LES now **VEW E920.**

SCHOLLER LM1 now **VEW E200.**

SCHOLLER LM2 now VEW E216.

SCHOLLER MA0 now VEW A500.

SCHOLLER MA1 now VEW A505.

SCHOLLER MA3 now VEW A522.

SCHOLLER MAN now VEW A750.

SCHOLLER MASBW now VEW A102.

SCHOLLER MASW now VEW A300.

SCHOLLER MAT now VEW A700.

SCHOLLER MNF now VEW F180.

SCHOLLER MS now VEW K465.

SCHOLLER NSW now VEW W107.

SCHOLLER PRIMA ZAH now VEW K980.

SCHOLLER PRIMA ZH now VEW K990.

SCHOLLER S200H now VEW V330.

SCHOLLER S200W now VEW V340.

SCHOLLER SAWM now VEW B404.

SCHOLLER SC1 now VEW K240.

SCHOLLER SG now VEW W100.

SCHOLLER SGHV now VEW W100.

SCHOLLER SHM now VEW F124.

SCHOLLER SPZ now VEW K245.

SCHOLLER U3 now VEW K455.

SCHOLLER V230 now VEW K505.

SCHOLLER VC135 now VEW V510.

SCHOLLER VCMV now VEW V350.

SCHOLLER VNC1 now VEW V155.

SCHOLLER VNC2 now VEW V145.

SCHOLLER VPN now VEW V500.

SCHOLLER WKL now VEW K200.

SCHOMBERG ALLOY.
87 Zn, 10 Sn, 3 Cu.
For bearings; anti-friction.

SCHOMBERG BEARING.
59 Zn, 40 Sn, 0.4 Cu, 0.2 Pb, 0.2 Fe.
For bearings; anti-friction.

SCHULZ.
M-Ger.; 31 Co, 15 Cr, 35 W, 2 Fe, 10 Mo, 0.7 C.
For tools; corrosion and heat resistant.

SCHULZ ALLOY.
91 Zn, 6 Cu, 3 Al.
For die castings, ornamental parts; free-cutting.

SCHWABENSTAHL.
M-1383; C, alloy, bal Fe.
For machine tool parts; oil hardened.

SCHWARZPUNKT.
M-1331; 1.0 C, Cr, W, Mo, V, bal Fe.
For cutters, dies, reamers; oil hardened.

SCHWERMETALL.
M-1246; W alloy.
Sintered: 72,000-85,000 TS; 220-250 Brin.
For gyros, counterweights, radioactive shields; sintered, heavy alloy.

SCIMITARS.
M-135; 18 Ni, 30 Zn, bal Cu.
For domestic utensils, ornaments; nickel silver.

SCLERON.
M-116; 1.5 Cu, 4 Si, bal Al.
Heat treated: 58,000-72,000 TS; 42,000-50,000 YS; 15-10 El; 100-135 Brin.
For small structural parts; age-hardened.

SCLERON NO. 1.
M-320, M-116; 0.1 Li, 4 Cu, bal Al.
42,500 TS; 15-20 El.
For connecting rods, machine construction; similar to "Aeron."

SCN 1.
M-1653; 0.49-0.56 C, 1.2-1.5 Si, 0.70-0.90 Mn, 0.70-1.0 Cr, 0.50-0.70 Ni, bal Fe.
Silicon alloy steel. 52 SiCrNi 5.

SCOOTER.
M-478; 0.15 C, 0.3 S, bal Fe.
Drawn: 89,600 TS; 15 El; 45 RA.
For general purposes; free-cutting steel.

SCOOTER CASE.
M-478; 0.10 C, 1.0-1.5 Mn, 0.4 S, bal Fe.
For cams, gears, fasteners; case hardened.

SCOTT.
M-1705; 2.25 C, 0.35 Mn, 0.5 Si, 11.5 Cr, 0.2 V, 0.8 Mo, bal Fe.
High carbon, high chromium tool and die steel; air or oil hardening. AISI D4.

SCOTTS ACME.
0.4 C, bal Fe.
For shafting; water hardening.

SCR3.
M-1296; 0.03 max C, 1.8 Si, 1.5 Mn, 25.0 Ni, 25.0 Cr, 1.5 V, 0.3 Ti, bal Fe.
82,500 psi TS; 38,400 psi YS; 60 El.
Austenitic stainless steel resistant to stress corrosion cracking in high temperature and high pressure water containing chloride. Good formability and weldability.

SCREEN PLATES.
M-U.S.; 58 Cu, 41 Zn, 0.75 Sn, 0.25 Pb.
For screen plates; high strength.

SCREW BRONZE.
M-U.S.; 5 Zn, 1 Sn, 0.5 Pb, bal Cu.
For screws, bolts, nuts, hardware; free-cutting.

SCREW METAL.
M-U.S.; 60 Cu, 1.5 Pb, bal Zn.
For screws, bolts, studs; free-cutting.

SCREW NUT BRONZE.
M-U.S.; 86 Cu, 11 Sn, bal Zn.
For screws, nuts, bolts; high strength.

SCRIBE-IT.
M-1149; 0.95 C, 1.2 Mn, 0.25 Si, 0.5 Cr, 0.2 V, 0.5 W, bal Fe.
Oil hardening tool steel; AISI O1.

SC SPECIAL.
M-115; 0.50 C, 0.3 Mn, 3 Cr, 14 W, 0.5 V, 0.3 Si, bal Fe.
For hot work and extrusion dies; oil hardening, hot work steel.

SCV.5.
M-1744; 1.0 C, 0.5 Mn, 0.3 Si, 0.95 Mo, 5.0 Cr, 0.25 V, bal Fe.
5% Chromium cold work tool steel.

SCX SERIES.
M-1296; 0.12 C, 1.45 max Si, 1.6 max Mn, Cb, bal Fe.
0.024-0.091 in (0.6-2.3mm) thick cold rolled sheet or coil.
2 strength grades: SCX 50,60 (each number means min YP in ksi).
High strength low alloy steel for cold formed parts of automotive and appliances.

SCXD SERIES.
M-1296; 0.07 C, 0.4 Si, 2.9 Mn, Cb, bal Fe.
0.024-0.091 in (0.6-2.3mm) thick cold rolled sheet or coil.
3 strengths: SCXD 80,90,100 (80,90,100 ksi min TS respectively).
High strength low alloy steel for cold formed parts of automotive and appliances.
Good formability and spot-weldability.

SD 111.
M-1653; 1.0-1.10 C, 1.2 Mn, 0.70-1.0 Cr, bal Fe.
Tool steel, for special purposes.
U 100 CrMn 4; W.-Nr. 1.2127.

SDM.
M-1653; 0.80-0.95 C, 1.8-2.2 Mn, 0.10-0.20 V, bal Fe.
Oil hardening cold work tool steel.
W.-Nr. 1.2842; similar to AISI O2.

SD-NICKEL.
M-529; 99.9 Ni.
For anodes. Improves plating.

SDW.
M-1653; 1.0-1.1 C, 0.80-1.10 Mn, 0.90-1.10 Cr, 1.0-1.3 W, bal Fe.
Cold work tool steel.
U 100 WCr.

SEACO HIGH SPEED.
M-374; 0.7 C, 18 W, 4 Cr, 1 V, bal Fe.
For cutting tools; high speed steel.

SEALALLOY.
Bi alloy.
For sealing metal to glass.

SEALCOR see **ALL-STATE SEALCOR.**

SEALVAR.
M-1623; 29 Ni, 17 Co, bal Fe.
For glass to metal sealing, hermetic seals with the harder glasses and ceramics in electronic industries.
Controlled expansion.

SEAMLESS TUBING ALLOY.
M-U.S.; 60-61.5 Cu, 38.5-40 Zn.
For tubing; high strength.

SEARING BEARING.
M-Eng.; 86 Cu, 14 Sn.
For bearings; bronze.

SEARS HARD FACING.
M-374; C, 11-13.5 Mn, bal Fe.
Welded: 500-600 Brin.
For hard facing electrodes for build-up work; austenitic, work hardened.

SEA STEEL.
M-1005; 0.2 C, 18 Cr, 8 Ni, bal Fe.
For marine fittings, hardware, pulleys; stainless, austenitic.

SEA WATER ALLOY-1.
M-U.S.; 17 Ni, 5 Mn, 0.7 C, bal Fe.
For marine construction; stainless and corrosion resistant.

SEA WATER ALLOY-2.
M-U.S.; 1.2 Cr, 24 Ni, 0.6 Mn, 0.5 C, bal Fe.
For marine construction; stainless and corrosion resistant.

SEA WATER BRONZE.
M-England; 33 Ni, 16 Sn, 5.5 Zn, 1 Bi, bal Cu.
For marine parts, hardware; resists sea water corrosion.

SECAERO.
M-969; 1.0 C, 0.65 Mn, 5.0 Cr, 0.3 V, 1.1 Mo, bal Fe.
Air hardening tool steel, cold work type; AISI A2.

SECOBALT.
M-969; 0.75 C, 5 Cr, 11 Co, 1.5 V, 18 W, 0.8 Mo, bal Fe.
For cutters, tools; high speed steel.

SECODIE.
M-969; 1.35 C, 0.6 Cr, 0.35 V, 1.5 Mo, 1.2 Si, bal Fe.
For dies; non-deforming.

SECO II.
M-Sweden; TiC, WC.
For tools, dies; sintered carbides.

SECOLEO.
M-969; 0.75 C, 0.45 Mn, 0.2 Si, 1.1 Cr, 1.65 Ni, bal Fe.
Oil hardening tool and die steel.

SECOLOY.
M-969; 2.5 C, 33 Cr, 44 Co, 17 W, 2 Fe.
For tool bits, cutters, cast.

SECON 1% SILICON ALUMINUM UBG.
M-1524; 0.85-1.15 Si, very low Fe, Cu, Mg; bal Al.
For ultrasonic bonding.

SECON 422.
M-1524.
Platinum-Rhodium-Ruthenium alloy
180 ohms/C Ft.
Used in potentiometers.

SECON 430.
M-1524; Mo, bal Pd.
Wire: 215,000 TS.
For resistors, potentiometers; high electrical resistance.

SECON 436.
M-1524.
Tungsten-Platinum alloy.
400 ohms/C Ft.
For use in potentiometers and strain gages.

SECON 436D.
M-1524; W, Mo, bal Pt.
Wire: 300,000 TS.
For resistors, potentiometers; high electrical resistance.

SECON 445.
M-1524; Mo, bal Pd.
Wire: 250,000 TS.
For resistors, potentiometers; high electrical resistance.

SECON 449.
M-1524; Rh, bal Pt.
Wire: 240,000 TS.
For resistors, potentiometers; high electrical resistance.

SECON 486.
M-1524; Mo, bal Pd.
Wire: 250,000 TS.
For resistors, potentiometers; high electrical resistance.

SECOVAN.
M-969; 0.7 C, 4.5 Cr, 2.25 V, 18 W, 0.65 Mo, bal Fe.
For tools, cutters; high speed steel.

SECO XXX.
M-969; 0.5 C, 0.9 Mn, bal Fe.
For chisels, hammers; tough.

SECRETAN.
M-Eng.; 95-91 Cu, 9-5 Al, 1.5 Mg, 0.5 P.
For strong corrosion resistant parts; corrosion resistant.

SECURIT LDS.
M-1318; 0.90 C, 1.9 Mn, 0.1 V, bal Fe.
For punches, dies, crimpers, form tools; oil hardened, non-deforming.

SEFCO NO. 2.
M-970; 3.2 C, 2.5 Si, 1.5 Ni, bal Fe.
Cast: 66,000 TS; 230 Brin.
For gears, machinery parts; cast iron.

SEIBEL WSM.
M-1369; Al alloy.
For light alloy parts.

SEIFERT.
M-Eng.; 73 Sn, 21 Zn, 5 Pb, 0.5 P, 0.5 Sb.
For ornaments, bearings; anti-friction.

SEL 1.
0.08 C, 15 Cr, 26 Co, 4.5 Mo, 4.4 Al, 2.3 Ti, 0.015 B, 1.0 max Fe, bal Ni.
For gas turbines, high temperature parts, jet engine components.
Cast alloy, high heat and corrosion resistant.

SEL 15.
M-1491; 0.07 C, 11 Cr, 14.5 Co, 6.5 Mo, 1.5 W, 5.4 Al, 2.5 Ti, 0.5 max Fe, 0.4 Cb + Ta, 0.015 B, bal Ni.
For gas turbine components, high temperature parts.
Cast alloy, high heat and corrosion resistant.

SELECT B.
M-73; 1.0 C, 5.25 Cr, 0.3 V, 1.1 Mo, bal Fe.
For blanking and forming dies, punches, plastic dies; wear resistant, air hardening.

SELECT B.
M-1702; 1.0 C, 0.7 Mn, 0.3 Si, 5.25 Cr, 0.25 V, 1.1 Mo, bal Fe.
Air hardening tool steel, medium alloy, for cold work tools; AISI A2.
Type A2 air hardening tool steels.

SELENIUM COPPER-948.
M-8; 99.5 Cu, 0.5 Se.
Soft: 32,000 TS; 10,000 YS; 45 El.
Hard: 45,000 TS; 40,000 YS; 15 El.
For electrical conductors; hot work qualities.

SELF-HARDENING.
M-35, M-345; 1.6 C, 3.75 Cr, 9.5 W, bal Fe.

SELF-HARDENING.
M-38; 1.9 C, 4 Cr, 4.1 Mo, bal Fe.

SELF LUBE BRONZE see **KEYSTONE C-64 ETC.**

SELFLUBE IRON.
M-939; 95 Fe, 5 Cu.

SELFLUBE IRON.
M-989; 90 Fe, 10 Cu.

SELVA METAL.
M-Eng.; Zn, bal Cu.
For hardware; a high tension brass.

SEMALLOY NO. 100.
M-1729; 43 Bi, 22 Pb, 8 Sn, 5 Cd, 18 In, 4 Hg.
Liquidus 110°F; Solidus 100°F. (38-43°C).
Note: Semalloys No. 100 through No. 2400 are low melting alloys used for fusible elements in automatic sprinklers, solders, encapsulating electronic equipment, anchoring small assemblies, patterns, models, low melting castings, fusible cores, dental work.

SEMALLOY NO. 102.
M-1729; 45 Bi, 23 Pb, 11 Sn, 5 Cd, 16 In.
Liquidus 126°F; Solidus 117°F. (47-52°C).

SEMALLOY NO. 103.
M-1729; 49 Bi, 18 Pb, 11 Sn, 1 Cd, 21 In.
Liquidus 133°F; Solidus 129°F. (54-56°C).

SEMALLOY NO. 104.
M-1729; 49.3 Bi, 26.3 Pb, 13.2 Sn, 9.8 Cd, 1.4 Ga.
Liquidus 151°F; Solidus 149°F. (65-66°C).

SEMALLOY NO. 105.
M-1729; 48 Bi, 25 Pb, 13 Sn, 9 Cd, 5 In.
Liquidus 149°F; Solidus 134°F. (57-65°C).

SEMALLOY NO. 106.
M-1729; 48 Bi, 26 Pb, 13 Sn, 9 Cd, 4 In.
Liquidus 149°F; Solidus 142°F. (61-65°C).

SEMALLOY NO. 107.
M-1729; 43 Bi, 23 Pb, 12 Sn, 9 Cd, 13 Hg.
Liquidus 151°F; Solidus 131°F. (55-66°C).

SEMALLOY NO. 108.
M-1729; 49 Bi, 18 Pb, 15 Sn, 18 In.
Liquidus 156°F; Solidus 136°F. (58-69°C).

SEMALLOY NO. 110.
M-1729; 51 Bi, 28 Pb, 12 Sn, 9 Cd.
Liquidus 163°F; Solidus 158°F. (70-73°C).

SEMALLOY NO. 111.
M-1729; 50 Bi, 25 Pb, 12.5 Sn, 12.5 Cd.
Liquidus 165°F; Solidus 158°F. (70-74°C).

SEMALLOY NO. 112.
M-1729; 50 Bi, 24.95 Pb, 12.5 Sn, 12.5 Cd, 0.05 Ag.
Liquidus 165°F; Solidus 158°F. (70-74°C).

SEMALLOY NO. 113.
M-1729; 48 Bi, 20 Sn, 19 Pb, 13 Cd.
Liquidus 168°F; Solidus 158°F.

SEMALLOY NO. 114.
M-1729; 50 Bi, 35 Pb, 9 Sn, 6 Cd.
Liquidus 173°F; Solidus 158°F. (70-78°C).

SEMALLOY NO. 115.
M-1729; 42 Bi, 35 Pb, 13 Sn, 10 Cd.
Liquidus 176°F; Solidus 158°F. (70-80°C).

SEMALLOY NO. 116.
M-1729; 50 Bi, 39 Pb, 3 Sn, 8 Cd.
Liquidus 180°F; Solidus 170°F. (77-82°C).

SEMALLOY NO. 117.
M-1729; 50 Bi, 39 Pb, 1 Sn, 8 Cd, 2 In.
Liquidus 185°F; Solidus 178°F. (81-85°C).

SEMALLOY NO. 118.
M-1729; 43 Bi, 38 Pb, 10 Sn, 9 Cd.
Liquidus 190°F; Solidus 160°F. (71-78°C).

SEMALLOY NO. 119.
M-1729; 50 Bi, 39 Pb, 2 Sn, 8 Cd, 1 In.
Liquidus 192°F; Solidus 176°F. (80-89°C).

SEMALLOY NO. 120.
M-1729; 51 Bi, 31 Pb, 15 Sn, 1 Cd, 2 In.
Liquidus 192°F; Solidus 176°F. (80-89°C).

SEMALLOY NO. 121.
M-1729; 51 Bi, 40 Pb, 8 Cd, 1 In.
Liquidus 192°F; Solidus 188°F. (87-89°C).

SEMALLOY NO. 122.
M-1729; 40 Bi, 37 Pb, 13 Sn, 10 Cd.
Liquidus 185°F; Solidua 158°F. (70-85°C).

SEMALLOY NO. 123.
M-1729; 52 Bi, 32 Pb, 15 Sn, 1 Cd.
Liquidus 198°F; Solidus 181°F. (89-92°C).

SEMALLOY NO. 124.
M-1729; 51.5 Bi, 31.5 Pb, 15 Sn, 2 In.
Liquidus 200°F; Solidus 190°F. (88-93°C).

SEMALLOY NO. 125.
M-1729; 50 Bi, 39 Pb, 4 Sn, 7 Cd.
Liquidus 200°F; Solidus 165°F. (74-93°C).

SEMALLOY NO. 126.
M-1729; 45.3 Bi, 24.5 Sn, 17.9 Pb, 12.3 Cd.
Liquidus 190°F; Solidus 158°F. (70-88°C).

SEMALLOY NO. 127.
M-1729; 38.4 Bi, 30.8 Pb, 15.4 Sn, 15.4 Cd.
Liquidus 207°F; Solidus 158°F. (70-97°C).

SEMALLOY NO. 129.
M-1729; 50 Bi, 31 Pb, 19 Sn.
Liquidus 210°F; Solidus 200°F. (93-99°C).

SEMALLOY NO. 130.
M-1729; 41 Bi, 28 Pb, 22 Sn, 9 Cd.
Liquidus 215°F; Solidus 158°F. (70-102°C).

SEMALLOY NO. 131.
M-1729; 50 Bi, 25 Pb, 25 Sn.
Liquidus 239°F; Solidus 203°F. (95-115°C).

SEMALLOY NO. 132.
M-1729; 52 Bi, 31.7 Pb, 15.3 Sn, 1 In.
Liquidus 201°F; Solidus 195°F. (91-94°C).

SEMALLOY NO. 133.
M-1729; 38 Pb, 37 Bi, 25 Sn.
Liquidus 261°F; Solidus 199°F. (93-127°C).

SEMALLOY NO. 134.
M-1729; 56 Bi, 22 Pb, 22 Sn.
Liquidus 219°F; Solidus 203°F. (95-104°C).

SEMALLOY NO. 135.
M-1729; 51.6 Bi, 37.4 Sn, 6 In, 5 Pb.
Liquidus 264°F; Solidus 203°F. (95-129°C).

SEMALLOY NO. 136.
M-1729; 52 Bi, 38 Pb, 10 Sn.
Liquidus 221°F; Solidus 208°F. (98-105°C).

SEMALLOY NO. 137.
M-1729; 35 Bi, 35 Pb, 20 Sn, 10 Cd.
Liquidus 221°F; Solidus 158°F. (70-105°C).

SEMALLOY NO. 138.
M-1729; 45 Bi, 35 Pb, 20 Sn.
Liquidus 225°F; Solidus 205°F. (96-107°C).

SEMALLOY NO. 138A.
M-1729; 55 Bi, 39 Pb, 6 Sn.
Liquidus 226°F; Solidus 215°F. (102-108°C).

SEMALLOY NO. 139.
M-1729; 46 Bi, 34 Pb, 20 Sn.
Liquidus 227°F; Solidus 203°F. (95-108°C).

SEMALLOY NO. 140.
M-1729; 52 Bi, 41 Pb, 7 Sn.
Liquidus 234°F; Solidus 208°F. (98-112°C).

SEMALLOY NO. 141.
M-1729; 54 Bi, 44 Pb, 1 Sn, 1 Cd.
Liquidus 235°F; Solidus 219°F. (104-113°C).

SEMALLOY NO. 142.
M-1729; 40 Bi, 33 Pb, 14 Sn, 13 Cd.
Liquidus 235°F; Solidus 162°F. (92-113°C).

SEMALLOY NO. 143.
M-1729; 50 Bi, 30 Pb, 20 Sn.
Liquidus 239°F; Solidus 203°F. (95-104°C).

SEMALLOY NO. 144.
M-1729; 70 Pb, 25 Bi, 5 Sn.
Liquidus 240°F; Solidus 212°F. (100-116°C).

SEMALLOY NO. 145.
M-1729; 50 Bi, 25 Cd, 25 Sn.
Liquidus 235°F; Solidus 217°F. (103-113°C).

SEMALLOY NO. 146.
M-1729; 53 Bi, 42 Pb, 5 Sn.
Liquidus 243°F; Solidus 217°F. (103-117°C).

SEMALLOY NO. 147.
M-1729; 38 Bi, 26 Pb, 32 Sn, 3 Cd, 1 Sb.
Liquidus 244°F; Solidus 167°F. (75-118°C).

SEMALLOY NO. 148.
M-1729; 54 Bi, 43 Pb, 3 Sn.
Liquidus 246°F; Solidus 226°F. (108-119°C).

SEMALLOY NO. 149.
M-1729; 55 Bi, 44 Pb, 1 Sn.
Liquidus 248°F; Solidus 242°F. (117-120°C).

SEMALLOY NO. 150.
M-1729; 55 Bi, 44 Pb, 1 In.
Liquidus 250°F; Solidus 248°F. (120-121°C).

SEMALLOY NO. 151.
M-1729; 57 Bi, 41 Pb, 2 Cd.
Liquidus 250°F; Solidus 197°F. (92-121°C).

SEMALLOY NO. 152.
M-1729; 31 Bi, 46 Pb, 18 Sn, 5 Cd.
Liquidus 253°F; Solidus 158°F. (70-123°C).

SEMALLOY NO. 153.
M-1729; 52.98 Bi, 42.49 Pb, 4.53 Sn.
Liquidus 243°F; Solidus 217°F. (103-117°C).

SEMALLOY NO. 154.
M-1729; 58 Bi, 42 Pb.
Liquidus 259°F; Solidus 255°F.

SEMALLOY NO. 155.
M-1729; 50 Sn, 50 In.
Liquidus 257°F; Solidus 243°F. (117-125°C).

SEMALLOY NO. 159.
M-1729; 95 In, 5 Ga.
Liquidus 266°F; Solidus 256°F.

SEMALLOY NO. 160.
M-1729; 57 Bi, 2 Pb, 41 Sn.
Liquidus 271°F; Solidus 262°F. (128-133°C).

SEMALLOY NO. 161.
M-1729; 32 Bi, 34 Pb, 34 Sn.
Liquidus 271°F; Solidus 205°F. (96-133°C).

SEMALLOY NO. 162.
M-1729; 38 Bi, 31 Pb, 31 Sn.
Liquidus 275°F; Solidus 205°F. (96-135°C).

SEMALLOY NO. 163.
M-1729; 38.41 Bi, 30.77 Pb, 30.77 Sn, 0.5 Ag.
Melting range: 205-275°F; 96-135°C.

SEMALLOY NO. 164.
M-1729; 36.45 Bi, 31.75 Pb, 31.5 Sn, 0.25 Cd, 0.05 Ag.
Liquidus 277°F; Solidus 203°F. (95-136°C).

SEMALLOY NO. 164.
M-1729; 55 Bi, 5 Pb, 40 Sn.
Liquidus 277°F; Solidus 250°F.

SEMALLOY NO. 165.
M-1729; 36 Bi, 32 Pb, 31 Sn, 1 Ag.
Liquidus 277°F; Solidus 203°F. (95-136°C).

SEMALLOY NO. 166.
M-1729; 36 Bi, 32 Pb, 32 Sn.
Liquidus 279°F; Solidus 203°F. (95-137°C).

SEMALLOY NO. 167.
M-1729; 28.5 Bi, 43 Pb, 28.5 Sn.
Liquidus 279°F; Solidus 205°F. (96-137°C).

SEMALLOY NO. 169.
M-1729; 5 Bi, 32 Pb, 45 Sn, 18 Cd.
Liquidus 282°F; Solidus 270°F. (132-139°C).

SEMALLOY NO. 170.
M-1729; 31 Bi, 38 Pb, 31 Sn.
Liquidus 282°F; Solidus 205°F. (96-139°C).

SEMALLOY NO. 171.
M-1729; 33.33 Bi, 33.34 Pb, 33.33 Sn.
Melting range: 205-289°F; 96-143°C.

SEMALLOY NO. 172.
M-1729; 34 Bi, 33 Pb, 33 Sn.
Melting range: 205-289°F; 96-143°C.

SEMALLOY NO. 173.
M-1729; 58 Sn, 42 In.
Liquidus 292°F; Solidus 242°F.

SEMALLOY NO. 174.
M-1729; 21 Bi, 42 Pb, 37 Sn.
Liquidus 306°F; Solidus 248°F.

SEMALLOY NO. 175.
M-1729; 80 In, 15 Pb, 5 Ag.
Liquidus 315°F; Solidus 315°F.

SEMALLOY NO. 177.
M-1729; 45 Bi, 33 Pb, 22 Cd.
Liquidus 316°F; Solidus 197°F.

SEMALLOY NO. 178.
M-1729; 45 Bi, 55 Pb.
Liquidus 320°F; Solidus 253°F.

SEMALLOY NO. 179.
M-1729; 50 Sn, 25 Pb, 25 Cd.
Liquidus 320°F; Solidus 293°F.

SEMALLOY NO. 180.
M-1729; 70 Sn, 18 Pb, 12 In.
Liquidus 324°F; Solidus 324°F. (162°C).

SEMALLOY NO. 181.
M-1729; 16 Bi, 36 Pb, 48 Sn.
Liquidus 324°F; Solidus 284°F.

SEMALLOY NO. 182.
M-1729; 14 Bi, 43 Pb, 43 Sn.
Liquidus 325°F; Solidus 291°F.

SEMALLOY NO. 183.
M-1729; 50 Sn, 40 Pb, 10 Bi.
Liquidus 330°F; Solidus 248°F.

SEMALLOY NO. 184.
M-1729; 22 Bi, 51 Pb, 27 Sn.
Liquidus 338°F; Solidus 268°F.

SEMALLOY NO. 185.
M-1729; 40 Bi, 60 Sn.
Liquidus 338°F; Solidus 281°F.

SEMALLOY NO. 186.
M-1729; 20 Bi, 50 Pb, 30 Sn.
Liquidus 343°F; Solidus 266°F.

SEMALLOY NO. 187.
M-1729; 13 Bi, 47 Pb, 40 Sn.
Liquidus 349°F; Solidus 294°F.

SEMALLOY NO. 188.
M-1729; 61.5 Sn, 35.5 Pb, 3 Ag.
Liquidus 355°F; Solidus 355°F.

SEMALLOY NO. 189.
M-1729; 25 Bi, 15 Pb, 60 Sn.
Liquidus 356°F; Solidus 205°F.

SEMALLOY NO. 1001.
M-1729; 52 In, 48 Sn.
M.P. 243°F; 117°C.

SEMALLOY NO. 1002.
M-1729; 62.5 Ga, 21.5 In, 16 Sn.
M.P. 51°F; 11°C, eutectic alloy.

SEMALLOY NO. 1003.
M-1729; 75.5 Ga, 24.5 In.
M.P. 61°F; 16°C, eutectic alloy.

SEMALLOY NO. 1004.
M-1729; 82 Ga, 12 Sn, 6 Zn.
M.P. 63°F; 17°C, eutectic alloy.

SEMALLOY NO. 1005.
M-1729; 92 Ga, 8 Sn.
M.P. 68°F; 20°C, eutectic alloy.

SEMALLOY NO. 1006.
M-1729; 100 Cs.
M.P. 86°F; 30°C.

SEMALLOY NO. 1007.
M-1729; 55 In, 45 Pb.
M.P. 419°F; 215°C; eutectic alloy.

SEMALLOY NO. 1008.
M-1729; 100 Ga.
M.P. 86°F; 30°C.

SEMALLOY NO. 1009.
M-1729; 100 Rb.
M.P. 102°F; 30°C.

SEMALLOY NO. 1010.
M-1729; 93 Pb, 5 In, 2 Ag.
M.P. 541°F; 283°C; eutectic alloy.

SEMALLOY NO. 1011.
M-1729; 100 P.
M.P. 111°F; 44°C.

SEMALLOY NO. 1012.
M-1729; 95 Ga, 5 In.
M.P. 77°F; 25°C.

SEMALLOY NO. 1013.
M-1729; 90 In, 10 Ag.
Liquidus 459°F; Solidus 286°F. (141-237°C).

SEMALLOY NO. 1015.
M-1729; 45 Bi, 23 Pb, 19 In, 8 Sn, 5 Cd.
M.P. 117°F; 47°C.

SEMALLOY NO. 1040.
M-1729; 49 Bi, 21 In, 18 Pb, 12 Sn.
M.P. 136°F; 58°C; eutectic alloy.

SEMALLOY NO. 1050.
M-1729; 49 Bi, 21 In, 18 Pb, 1 Cd.
M.P. 136°F; 58°C; eutectic alloy.

SEMALLOY NO. 1060.
M-1729; 51 In, 33 Bi, 16 Sn.
M.P. 142°F; 61°C; eutectic alloy.

SEMALLOY NO. 1075.
M-1729; 62 In, 30 Bi, 8 Cd.
M.P. 144°F; 62°C; eutectic alloy.

SEMALLOY NO. 1090.
M-1729; 50 Bi, 27 Pb, 13 Sn, 10 Cd.
M.P. 158°F; 70°C; eutectic alloy.

SEMALLOY NO. 1100.
M-1729; 52 Bi, 26 Pb, 22 In.
M.P. 158°F; 70°C; eutectic alloy.

SEMALLOY NO. 1125.
M-1729; 66 In, 34 Bi.
M.P. 162°F; 72°C; eutectic alloy.

SEMALLOY NO. 1127.
M-1729; 48.5 Bi, 41.5 In, 10 Cd.
M.P. 172°F; 78°C; eutectic alloy.

SEMALLOY NO. 1150.
M-1729; 57 Bi, 17 Sn, 26 In.
M.P. 174°F; 79°C; eutectic alloy.

SEMALLOY NO. 1175.
M-1729; 54.02 Bi, 29.68 In, 16.38 Sn.
M.P. 178°F; 81°C.

SEMALLOY NO. 1220.
M-1729; 52 Bi, 40 Pb, 8 Cd.
M.P. 198°F; 92°C; eutectic alloy.

SEMALLOY NO. 1245.
M-1729; 44 In, 42 Sn, 14 Cd.
M.P. 200°F; 93°C; eutectic alloy.

SEMALLOY NO. 1270.
M-1729; 52.5 Bi, 32 Pb, 15.5 Sn.
M.P. 203°F; 95°C; eutectic alloy.

SEMALLOY NO. 1275.
M-1729; 52 Bi, 30 Pb, 18 Sn.
M.P. 205°F; 96°C; eutectic alloy.

SEMALLOY NO. 1280.
M-1729; 52 Bi, 32 Pb, 16 Sn.
M.P. 205°F; 96°C; eutectic alloy.

SEMALLOY NO. 1285.
M-1729; 46 Bi, 34 Sn, 20 Pb.
M.P. 212°F; 100°C; eutectic alloy.

SEMALLOY NO. 1290.
M-1729; 50 Bi, 28 Pb, 22 Sn.
Melting range: 205-229°F; 96-109°C.

SEMALLOY NO. 1330.
M-1729; 54 Bi, 26 Sn, 20 Cd.
M.P. 217°F; 103°C; eutectic alloy.

SEMALLOY NO. 1400.
M-1729; 67 Bi, 33 In.
M.P. 229°F; 109°C; eutectic alloy.

SEMALLOY NO. 1405.
M-1729; 100 S.
M.P. 235°F; 113°C.

SEMALLOY NO. 1510A.
M-1729; 74 In, 26 Cd.
M.P. 253°F; 123°C; eutectic alloy.

SEMALLOY NO. 1530.
M-1729; 55 Bi, 45 Pb.
M.P. 255°F; 124°C; eutectic alloy.

SEMALLOY NO. 1555.
M-1729; 46 Sn, 37 Te, 17 Cd.
M.P. 262°F; 128°C; eutectic alloy.

SEMALLOY NO. 1580.
M-1729; 56 Bi, 40 Sn, 4 Zn.
M.P. 266°F; 130°C; eutectic alloy.

SEMALLOY NO. 1590.
M-1729; 52 Sn, 48 In.
Liquidus 268°F; Solidus 243°F. (117-131°C).

SEMALLOY NO. 1592.
M-1729; 40 Bi, 40 Sn, 20 Pb.
Liquidus 266°F; Solidus 250°F. (121-130°C).

SEMALLOY NO. 1594.
M-1729; 70 In, 15 Sn, 9.6 Pb, 5.4 Cd.
M.P. 257°F; 125°C; eutectic alloy.

SEMALLOY NO. 1596.
M-1729; 58 Bi, 42 Pb.
Liquidus 261°F; Solidus 257°F. (125-127°C).

SEMALLOY NO. 1598.
M-1729; 95 In, 5 Bi.
Liquidus 302°F; Solidus 257°F. (125-150°C).

SEMALLOY NO. 1610.
M-1729; 37.5 Sn, 37.5 Pb, 25 In.
Liquidus 358°F; Solidus 273°F. (134-181°C).

SEMALLOY NO. 1630.
M-1729; 57 Bi, 1 Pb, 42 Sn.
M.P. 275°F; 135°C; eutectic alloy.

SEMALLOY NO. 1640.
M-1729; 55 Bi, 40 Sn, 5 Pb.
Melt range: 250-277°F; 121-136°C.

SEMALLOY NO. 1680.
M-1729; 58 Bi, 42 Sn.
M.P. 281°F; 138°C; eutectic alloy.

SEMALLOY NO. 1690.
M-1729; 51 Sn, 31 Pb, 18 Cd.
M.P. 288°F; 142°C; eutectic alloy.

SEMALLOY NO. 1710.
M-1729; 60 Bi, 40 Cd.
M.P. 291°F; 144°C; eutectic alloy.

SEMALLOY NO. 1715.
M-1729; 97 In, 3 Ag.
M.P. 290°F, 143°C.

SEMALLOY NO. 1725.
M-1729; 51.2 Sn, 30.6 Pb, 18.2 Cd.
M.P. 291°F; 144°C.

SEMALLOY NO. 1730.
M-1729; 58 Sn, 42 In.
Liquidus 292°F; Solidus 243°F. (117-144°C).

SEMALLOY NO. 1735.
M-1729; 70 Sn, 18 Pb, 12 Cu.
M.P. 324°F; 162°C.

SEMALLOY NO. 1740.
M-1729; 42 Pb, 37 Sn, 21 Bi.
Liquidus 306°F; Solidus 248°F. (120-152°C).

SEMALLOY NO. 1741.
M-1729; 99.4 In, 0.6 Ga.
M.P. 306°F; 152°C; eutectic alloy.

SEMALLOY NO. 1742.
M-1729; 99 In, 1 Cu.
M.P. 307°F; 153°C; eutectic alloy.

SEMALLOY NO. 1743.
M-1729; 99.6 In, 0.4 Ga.
M.P. 307°F; 153°C; eutectic alloy.

SEMALLOY NO. 1744.
M-1729; 99.5 In, 0.5 Ga.
M.P. 307°F; 153°C; eutectic alloy.

SEMALLOY NO. 1746.
M-1729; 50 Sn, 32 Pb, 18 Cd.
M.P. 295°F; 146°C; eutectic alloy.

SEMALLOY NO. 1760.
M-1729; 100% In. (Indium).
M.P. 313°F; 156°C.

SEMALLOY NO. 1765.
M-1729; 80 In, 15 Pb, 5 Ag.
M.P. 301°F; 149°C; eutectic alloy.

SEMALLOY NO. 1770.
M-1729; 45 Bi, 33 Pb, 22 Cd.
Liquidus 316°F; Solidus 198°F. (92-158°C).

SEMALLOY NO. 1775.
M-1729; 50 Sn, 25 Bi, 25 Pb.
Liquidus 320°F; Solidus 293°F. (145-160°C).

SEMALLOY NO. 1780.
M-1729; 55 Pb, 45 Bi.
Liquidus 320°F; Solidus 253°F. (123-160°C).

SEMALLOY NO. 1790.
M-1729; 50 Sn, 25 Pb, 25 Cd.
Liquidus 320°F; Solidus 293°F. (145-160°C).

SEMALLOY NO. 1810.
M-1729; 48 Sn, 36 Pb, 16 Bi.
Liquidus 324°F; Solidus 284°F. (140-162°C).

SEMALLOY NO. 1815.
M-1729; 95 In, 5 Ag.
Liquidus 324°F; Solidus 319°F. (159-162°C).

SEMALLOY NO. 1820.
M-1729; 43 Pb, 43 Sn, 14 Bi.
Liquidus 325°F; Solidus 291°F. (144-163°C).

SEMALLOY NO. 1830.
M-1729; 50 Sn, 40 Pb, 10 Bi.
Liquidus 332°F; Solidus 248°F. (120-167°C).

SEMALLOY NO. 1836.
M-1729; 48.8 Sn, 41 Pb, 10.2 Bi.
Liquidus 331°F; Solidus 288°F. (142-166°C).

SEMALLOY NO. 1840.
M-1729; 51 Pb, 27 Sn, 22 Bi.
Liquidus 338°F; Solidus 268°F. (131-170°C).

SEMALLOY NO. 1850.
M-1729; 60 Sn, 40 Bi.
Liquidus 338°F; Solidus 281°F. (138-170°C).

SEMALLOY NO. 1860.
M-1729; 50 Pb, 30 Sn, 20 Bi.
Liquidus 343°F; Solidus 266°F. (130-173°C).

SEMALLOY NO. 1870.
M-1729; 47 Pb, 40 Sn, 13 Bi.
Liquidus 349°F; Solidus 294°F. (146-176°C).

SEMALLOY NO. 1873.
M-1729; 68 Sn, 32 Cd.
M.P. 351°F; 177°C; eutectic alloy.

SEMALLOY NO. 1880.
M-1729; 62.5 Sn, 36 Pb, 1.5 Ag.
M.P. 355°F; 179°C; eutectic alloy.

SEMALLOY NO. 1890.
M-1729; 60 Sn, 25 Bi, 15 Pb.
Liquidus 356°F; Solidus 205°F. (96-180°C).

SEMALLOY NO. 1910.
M-1729; 62 Sn, 38 Pb.
M.P. 361°F; 183°C; eutectic alloy.

SEMALLOY NO. 1914.
M-1729; 63 Sn, 37 Pb.
M.P. 361°F; 183°C; eutectic alloy.

SEMALLOY NO. 1916.
M-1729; 65 Sn, 35 Pb.
Liquidus 364°F; Solidus 361°F. (183-184°C).

SEMALLOY NO. 1918.
M-1729; 65 Sn, 34 Pb, 1 Sb.
Liquidus 370°F; Solidus 363°F. (184-188°C).

SEMALLOY NO. 1920.
M-1729; 60 Sn, 40 Pb.
Liquidus 374°F; Solidus 361°F. (183-190°F).

SEMALLOY NO. 1923.
M-1729; 70 Sn, 30 Pb.
Liquidus 378°F; Solidus 361°F. (183-192°C).

SEMALLOY NO. 1927.
M-1729; 75 Sn, 25 Pb.
Liquidus 383°F; Solidus 361°F. (183-195°C).

SEMALLOY NO. 1930.
M-1729; 55.5 Pb, 40.5 Sn, 4 Bi.
Liquidus 388°F; Solidus 338°F. (170-198°C).

SEMALLOY NO. 1933.
M-1729; 91 Sn, 9 Zn.
M.P. 390°F; 199°C; eutectic alloy.

SEMALLOY NO. 1934.
M-1729; 92 Sn, 8 Zn.
Liquidus 392°F; Solidus 388°F. (198-200°C).

SEMALLOY NO. 1935.
M-1729; 55 Sn, 45 Pb.
Liquidus 392°F; Solidus 361°F. (183-200°C).

SEMALLOY NO. 1936.
M-1729; 80 Sn, 20 Pb.
Liquidus 390°F; Solidus 361°F. (183-199°C).

SEMALLOY NO. 1937.
M-1729; 50 Sn, 47 Pb, 3 Sb.
Liquidus 399°F; Solidus 365°F. (185-204°C).

SEMALLOY NO. 1938.
M-1729; 85 Sn, 15 Pb.
Liquidus 403°F; Solidus 361°F. (183-206°C).

SEMALLOY NO. 1940.
M-1729; 55 Pb, 44 Sn, 1 Ag.
Liquidus 410°F; Solidus 350°F. (177-210°C).

SEMALLOY NO. 1952.
M-1729; 90 Sn, 10 Pb.
Liquidus 415°F; Solidus 361°F. (183-213°C).

SEMALLOY NO. 1956.
M-1729; 50 In, 50 Pb.
Liquidus 408°F; Solidus 356°F. (180-209°C).

SEMALLOY NO. 1960.
M-1729; 50 Pb, 50 Sn.
Liquidus 421°F; Solidus 361°F. (183-216°C).

SEMALLOY NO. 1968.
M-1729; 52 Pb, 48 Sn.
Liquidus 424°F; Solidus 361°F. (183-218°C).

SEMALLOY NO. 1970.
M-1729; 55 Pb, 45 In.
Liquidus 428°F; Solidus 343°F. (173-220°C).

SEMALLOY NO. 1971.
M-1729; 100 Se (Selenium).
M.P. 428°F; 220°C.

SEMALLOY NO. 1978.
M-1729; 96.5 Sn, 3.5 Ag.
M.P. 430°F; 221°C.

SEMALLOY NO. 1980.
M-1729; 95.5 Sn, 3.5 Ag, 1 Cd.
Liquidus 430°F; Solidus 425°F. (218-221°C).

SEMALLOY NO. 1982.
M-1729; 95 Sn, 5 Pb.
Liquidus 432°F; Solidus 361°F. (183-222°C).

SEMALLOY NO. 1990.
M-1729; 97.5 Sn, 2.5 Ag.
Liquidus 438°F; Solidus 430°F. (221-226°C).

SEMALLOY NO. 1994.
M-1729; 55 Pb, 45 Sn.
Liquidus 441°F; Solidus 361°F. (183-227°C).

SEMALLOY NO. 2000.
M-1729; 48 Bi, 28.5 Pb, 14.5 Sn, 9 Sb.
Liquidus 440°F; Solidus 217°F. (103-227°C).

SEMALLOY NO. 2003.
M-1729; 61 Sn, 36 Pb, 3 Ag.
Liquidus 478°F; Solidus 354°F. (179-248°C).

SEMALLOY NO. 2006.
M-1729; 99.25 Sn, 0.75 Cu.
M.P. 440°F; 227°C; eutectic alloy.

SEMALLOY NO. 2010.
M-1729; 99 Sn, 1 Ga.
M.P. 442°F; 228°C; eutectic alloy.

SEMALLOY NO. 2012.
M-1729; 99 Sn, 1 Te.
Liquidus 842°F; Solidus 450°F. (232-450°C).

SEMALLOY NO. 2025.
M-1729; 58 Pb, 40 Sn, 2 Sb.
Liquidus 448°F; Solidus 365°F. (185-231°C).

SEMALLOY NO. 2030.
M-1729; 100% Sn. (Tin).
M.P. 450°F; 232°C.

SEMALLOY NO. 2040.
M-1729; 60 Pb, 37 Sn, 3 Ag.
Liquidus 450°F; Solidus 355°F. (179-232°C).

SEMALLOY NO. 2050.
M-1729; 68 Pb, 23 Sn, 9 Cd.
Liquidus 455°F; Solidus 293°F. (145-235°C).

SEMALLOY NO. 2053.
M-1729; 99 Sn, 1 Sb.
M.P. 456°F; 236°C; eutectic alloy.

SEMALLOY NO. 2055.
M-1729; 63 Pb, 35 Sn, 2 Sb.
Liquidus 468°F; Solidus 365°F. (185-242°C).

SEMALLOY NO. 2060.
M-1729; 60 Pb, 40 Sn.
Liquidus 455°F; Solidus 361°F. (183-235°C).

SEMALLOY NO. 2070.
M-1729; 62 Pb, 37 Sn, 1 As.
Liquidus 460°F; Solidus 361°F. (183-238°C).

SEMALLOY NO. 2080.
M-1729; 97 Sn, 3 Sb.
Liquidus 460°F; Solidus 450°F. (232-238°C).

SEMALLOY NO. 2090.
M-1729; 95 Sn, 5 Sb.
Liquidus 464°F; Solidus 452°F. (233-240°C).

SEMALLOY NO. 2092.
M-1729; 62 Pb, 38 Sn.
Liquidus 468°F; Solidus 361°F. (183-242°C).

SEMALLOY NO. 2093.
M-1729; 72 Sn, 28 Cd.
Liquidus 358°F; Solidus 351°F; (177-181°C).

SEMALLOY NO. 2094.
M-1729; 63.2 Pb, 35 Sn, 1.8 Sb.
Liquidus 470°F; Solidus 365°F. (185-243°C).

SEMALLOY NO. 2095.
M-1729; 60 Sn, 36 Pb, 4 Ag.
Liquidus 473°F; Solidus 354°F. (179-245°C).

SEMALLOY NO. 2096.
M-1729; 65 Pb, 35 Sn.
Liquidus 477°F; Solidus 361°F. (183-247°C).

SEMALLOY NO. 2097.
M-1729; 61.5 Sn, 35.5 Pb, 3 Ag.
Liquidus 478°F; Solidus 354°F. (179-248°C).

SEMALLOY NO. 2098.
M-1729; 83 Pb, 17 Cd.
M.P. 478°F; 248°C; eutectic alloy.

SEMALLOY NO. 2100.
M-1729; 95 Bi, 5 Sn.
Liquidus 483°F; Solidus 273°F. (134-251°C).

SEMALLOY NO. 2120.
M-1729; 85 Pb, 10 Sb, 5 Sn.
Liquidus 493°F; Solidus 464°F. (240-256°C).

SEMALLOY NO. 2125.
M-1729; 63 Pb, 34 Sn, 3 Zn.
Liquidus 493°F; Solidus 338°F. (170-256°C).

SEMALLOY NO. 2126.
M-1729; 82 Pb, 14.5 Bi, 2.5 Cd, 1 Sb.
Liquidus 498°F; Solidus 261°F. (127-259°C).

SEMALLOY NO. 2127.
M-1729; 75 Sn, 25 Zn.
Liquidus 572°F; Solidus 388°F. (198-300°C).

SEMALLOY NO. 2128.
M-1729; 50 Sn, 47 Pb, 3 Ag.
Liquidus 500°F; Solidus 354°F. (179-260°C).

SEMALLOY NO. 2129.
M-1729; 73.7 Pb, 25 Sn, 1.3 Sb.
Liquidus 504°F; Solidus 363°F. (184-262°C).

SEMALLOY NO. 2130.
M-1729; 90 Pb, 10 Sb.
Liquidus 500°F; Solidus 486°F. (252-260°C).

SEMALLOY NO. 2131.
M-1729; 82.5 Cd, 17.5 Zn.
M.P. 509°F; 265°C; eutectic alloy.

SEMALLOY NO. 2133.
M-1729; 91 Pb, 9 Sb.
Liquidus 509°F; Solidus 486°F. (252-265°C).

SEMALLOY NO. 2135.
M-1729; 75 Pb, 25 Sn.
Liquidus 514°F; Solidus 361°F. (183-268°C).

SEMALLOY NO. 2136.
M-1729; 79 Pb, 20 In, 1 Sb.
Liquidus 517°F; Solidus 363°F. (184-269°C).

SEMALLOY NO. 2137.
M-1729; 75 Pb, 15 Sb, 10 Sn.
Liquidus 514°F; Solidus 464°F. (240-268°C).

SEMALLOY NO. 2138.
M-1729; 75 Pb, 25 In.
Liquidus 508°F; Solidus 440°F. (227-264°C).

SEMALLOY NO. 2139.
M-1729; 80 Sn, 20 Zn.
Liquidus 518°F; Solidus 390°F. (199-270°C).

SEMALLOY NO. 2140.
M-1729; 100 Bi.
M.P. 520°F; 271°C.

SEMALLOY NO. 2153.
M-1729; 90 Sn, 10 Ag.
Liquidus 563°F; Solidus 430°F. (221-295°C).

SEMALLOY NO. 2160.
M-1729; 92.5 Pb, 5 Sn, 2.5 Ag.
M.P. 536°F; 280°C; eutectic alloy.

SEMALLOY NO. 2165.
M-1729; 81 Pb, 19 In.
Liquidus 536°F; Solidus 518°F. (270-280°C).

SEMALLOY NO. 2170.
M-1729; 92 Pb, 5 Sn, 3 Sb.
Liquidus 545°F; Solidus 463°F. (239-285°C).

SEMALLOY NO. 2172.
M-1729; 85 Pb, 11.5 Sb, 3.5 Sn.
M.P. 464°F; 240°C; eutectic alloy.

SEMALLOY NO. 2173.
M-1729; 94 Pb, 6 Sb.
Liquidus 545°F; Solidus 486°F. (252-285°C).

SEMALLOY NO. 2175.
M-1729; 92 Pb, 8 Sb.
Liquidus 520°F; Solidus 486°F. (252-271°C).

SEMALLOY NO. 2177.
M-1729; 93 Pb, 5 Sn, 2 Ag.
Liquidus 568°F; Solidus 530°F. (277-298°C).

SEMALLOY NO. 2178.
M-1729 60 Cd, 30 Zn, 10 Sn.
Liquidus 550°F; Solidus 315°F. (157-288°C).

SEMALLOY NO. 2180.
M-1729; 85 Pb, 15 Sn.
Liquidus 553°F; Solidus 437°F. (225-289°C).

SEMALLOY NO. 2187.
M-1729; 57 Pb, 40 Sn, 3 Ag.
Liquidus 543°F; Solidus 354°F. (179-284°C).

SEMALLOY NO. 2188.
M-1729; 90 Pb, 5 In, 5 Ag.
Liquidus 590°F; Solidus 554°F. (290-310°C).

SEMALLOY NO. 2190.
M-1729; 90 Pb, 5 Sn, 5 Ag.
M.P. 558°F; 292°C; eutectic alloy.

SEMALLOY NO. 2191.
M-1729; 90 Pb, 10 Ag.
Liquidus 842°F; Solidus 579°F. (304-450°C).

SEMALLOY NO. 2200.
M-1729; 95 Sn, 5 Ag.
Liquidus 473°F; Solidus 430°F. (221-245°C).

SEMALLOY NO. 2210.
M-1729; 95 Pb, 5 Sb.
Liquidus 563°F; Solidus 486°F. (252-295°C).

SEMALLOY NO. 2213.
M-1729; 90 Cd, 10 Zn.
Liquidus 570°F; Solidus 509°F. (265-299°C).

SEMALLOY NO. 2215.
M-1729; 96 Pb, 4 Sb.
Liquidus 570°F; Solidus 486°F. (252-299°C).

SEMALLOY NO. 2216.
M-1729; 97 Sn, 3 Cu.
Liquidus 572°F; Solidus 441°F. (227-300°C).

SEMALLOY NO. 2217.
M-1729; 90 Pb, 10 In.
Liquidus 572°F; Solidus 561°F. (294-300°C).

SEMALLOY NO. 2218.
M-1729; 90 Pb, 10 Sn.
Liquidus 573°F; Solidus 514°F. (268-301°C).

SEMALLOY NO. 2220.
M-1729; 60 Sn, 34 Pb, 6 Ag.
Liquidus 579°F; Solidus 351°F. (177-304°C).

SEMALLOY NO. 2230.
M-1729; 95.5 Pb, 2.5 Ag, 2 Sn.
Liquidus 580°F; Solidus 570°F. (299-304°C).

SEMALLOY NO. 2235.
M-1729; 92.5 Pb, 5 In, 2.5 Ag.
M.P. 585°F; 307°C; Autectic alloy.

SEMALLOY NO. 2240.
M-1729; 97.5 Pb, 1.5 Ag, 1 Se.
M.P. 588°F; 309°C; eutectic alloy.

SEMALLOY NO. 2245.
M-1729; 70 Sn, 30 Zn.
Liquidus 592°F; Solidus 390°F. (199-311°C).

SEMALLOY NO. 2253.
M-1729; 97.5 Pb, 1.75 Ag, 0.75 Sn.
M.P. 590°F; 310°C; eutectic alloy.

SEMALLOY NO. 2260.
M-1729; 70 Pb, 27 Sn, 3 Ag.
Liquidus 594°F; Solidus 354°F. (179-312°C).

SEMALLOY NO. 2261.
M-1729; 95 Pb, 5 In.
Liquidus 597°F; Solidus 594°F. (312-314°C).

SEMALLOY NO. 2262.
M-1729; 92 Pb, 8 In.
Liquidus 581°F; Solidus 572°F. (300-305°C).

SEMALLOY NO. 2264.
M-1729; 98 Pb, 1.2 Sb, 0.8 Ga.
M.P. 599°F; 315°C; eutectic alloy.

SEMALLOY NO. 2265.
M-1729; 78.4 Cd, 16.6 Zn, 5 Ag.
Liquidus 600°F; Solidus 480°F. (249-316°C).

SEMALLOY NO. 2269.
M-1729; 100% Pb. (Lead).
M.P. 621°F; 327°C.

SEMALLOY NO. 2270.
M-1729; 96.5 Pb, 3.5 Ag.
Liquidus 621°F; Solidus 579°F. (304-327°C).

SEMALLOY NO. 2275.
M-1729; 98 Pb, 2 Sb.
Liquidus 608°F; Solidus 572°F. (300-320°C).

SEMALLOY NO. 2278.
M-1729; 99 Pb, 1 Sb.
Liquidus 615°F; Solidus 594°F. (312-324°C).

SEMALLOY NO. 2284.
M-1729; 100% Cd. (Cadmium).
M.P. 610°F; 321°C.

SEMALLOY NO. 2287.
M-1729; 98.5 Pb, 1.5 Sb.
Liquidus 612°F; Solidus 583°F. (306-322°C).

SEMALLOY NO. 2290.
M-1729; 98 Sn, 2 As.
Liquidus 626°F; Solidus 448°F. (231-330°C).

SEMALLOY NO. 2291.
M-1729; 60 Zn, 40 Cd.
Liquidus 635°F; Solidus 509°F. (265-335°C).

SEMALLOY NO. 2292.
M-1729; 60 Sn, 40 Zn.
Liquidus 645°F; Solidus 390°F. (198-341°C).

SEMALLOY NO. 2294.
 M-1729; 99 Sn, 1 Ge.
 Liquidus 653°F; Solidus 450°F. (232-345°C).

SEMALLOY NO. 2295.
 M-1729; 83 Pb, 15 Sb, 1 Sn, 1 As.
 Liquidus 667°F; Solidus 594°F. (312-353°C).

SEMALLOY NO. 2300.
 M-1729; 95 Pb, 5 Ag.
 Liquidus 687°F; Solidus 579°F. (304-364°C).

SEMALLOY NO. 2306.
 M-1729; 70 Zn, 30 Sn.
 Liquidus 707°F; Solidus 388°F. (198-375°C).

SEMALLOY NO. 2310.
 M-1729; 99.5 Sn, 0.5 P.
 Liquidus 707°F; Solidus 450°F. (232-375°C).

SEMALLOY NO. 2313.
 M-1729; 95 Zn, 5 Al.
 M.P. 720°F; 382°C; eutectic alloy.

SEMALLOY NO. 2315.
 M-1729; 95 Cd, 5 Ag.
 Liquidus 740°F; Solidus 649°F. (343-393°C).

SEMALLOY NO. 2317.
 M-1729; 90 Zn, 10 Cd.
 Liquidus 750°F; Solidus 511°F. (266-399°C).

SEMALLOY NO. 2320.
 M-1729; 99 Pb, 1 Te.
 Liquidus 752°F; Solidus 621°F. (327-400°C).

SEMALLOY NO. 2330.
 M-1729; 94.5 Cd, 5.5 Ag.
 Liquidus 752°F; Solidus 649°F. (343-400°C).

SEMALLOY NO. 2340.
 M-1729; 98 Pb, 2 Zn.
 Liquidus 784°F; Solidus 604°F. (318-418°C).

SEMALLOY NO. 2345.
 M-1729; 100 Zn.
 M.P. 787°F; 419°C.

SEMALLOY NO. 2347.
 M-1729; 55 Ge, 45 Al.
 Melt point: 795°F; 424°C.

SEMALLOY NO. 2350.
 M-1729; 95 Sn, 5 As.
 Liquidus 797°F; Solidus 448°F. (231-425°C).

SEMALLOY NO. 2380.
 M-1729; 97 Sn, 3 P.
 Liquidus 932°F; Solidus 450°F. (232-500°C).

SEMALLOY NO. 2384.
 M-1729; 75 Al, 10 Si, 10 Zn, 4 Cu, 1 Fe.
 Liquidus 1040°F; Solidus 960°F. (516-560°C).

SEMALLOY NO. 2388.
 M-1729; 83 Mg, 5 Zn, 12 Al.
 Melting range: 1125-1145°F; 607-618°C.
 Note: SEMALLOY alloys, available in various forms, are used essentially in brazing operations, but also for fusable links for safety devices, special castings and other special applications.

SEMALLOY NO. 2389.
 M-1729; 100 Te (Tellurium).
 M.P. 842°F; 450°C.

SEMALLOY NO. 2391.
 M-1729; 95 Al, 5 Si.
 Liquidus 1165°F; Solidus 1071°F. (577-629°C).

SEMALLOY NO. 2395.
 M-1729; 86 Al, 10 Si, 4 Cu.
 Liquidus 1085°F; Solidus 970°F. (521-585°C).

SEMALLOY NO. 2400.
 M-1729; 100% Bi. (Bismuth).
 M.P. 520°F; 271°C.

SEMALLOY NO. 2401.
 M-1729; 92.5 Al, 7.5 Si.
 Liquidus 1135°F; Solidus 1071°F. (577-613°C).

SEMALLOY NO. 2403.
 M-1729; 88.3 Al, 11.7 Si.
 M.P. 1071°F; 577°C; eutectic alloy.

SEMALLOY NO. 2405.
 M-1729; 90 Al, 10 Si.
 Liquidus 1094°F; Solidus 1071°F. (577-590°C).

SEMALLOY NO. 2406.
 M-1729; 88 Al, 12 Si.
 Liquidus 1080°F; Solidus 1071°F. (577-582°C).

SEMALLOY NO. 2411.
 M-1729; 89 Mg, 9 Al, 2 Zn.
 Liquidus 1110°F; Solidus 770°F. (410-599°C).

SEMALLOY NO. 2414.
 M-1729; 100 Sb.
 M.P. 1168°F; 631°C.

SEMALLOY NO. 2420.
 M-1729; 45 Ag, 15 Cu, 16 Zn, 24 Cd.
 Melting range: 1125-1145°F; 607-618°C.

SEMALLOY NO. 2425.
 M-1729; 44 Ag, 27 Cu, 15 Cd, 13 Zn, 1 P.
 Liquidus 1205°F; Solidus 110°F. (599-652°C).

SEMALLOY NO. 2430.
 M-1729; 50 Ag, 15.5 Cu, 16.5 Zn, 18 Cd.
 Melting range: 1160-1175°F; 627-635°C.

SEMALLOY NO. 2435.
 M-1729; 51 Ag, 22 Cu, 19 Zn, 7 Cd, 1 Sn.
 Liquidus 1180°F; Solidus 1161°F. (627-638°C).

SEMALLOY NO. 2437.
 M-1729; 49 Ag, 23 Zn, 16 Cu, 7.5 Mn, 4.5 Ni.
 Liquidus 1300°F; Solidus 1161°F. (627-704°C).

SEMALLOY NO. 2440.
M-1729; 50 Ag, 15 Cu, 25 Zn, 10 Cd.
Melting range: 1160-1185°F; 627-640°C.

SEMALLOY NO. 2445.
M-1729; 41 Ag, 24 Cd, 18 Zn, 17 Cu.
Liquidus 1161°F; Solidus 1125°F. (607-627°C).

SEMALLOY NO. 2450.
M-1729; 45 Ag, 18 Cu, 21 Zn, 16 Cd.
Melting range: 1141-1190°F; 616-643°C.

SEMALLOY NO. 2454.
M-1729; 100 Mg.
M.P. 1202°F; 650°C.

SEMALLOY NO. 2460.
M-1729; 56 Ag, 22 Cu, 17 Zn, 5 Sn.
Melting range: 1145-1205°F; 618-627°C.

SEMALLOY NO. 2465.
M-1729; 81 Ag, 19 Ge.
M.P. 1206°F; 651°C.

SEMALLOY NO. 2467.
M-1729; 100 Al.
M.P. 1220°F; 660°C.

SEMALLOY NO. 2469.
M-1729; 99 Mg, 1 Mn.
Liquidus 1200°F; 649°C.

SEMALLOY NO. 2470.
M-1729; 61 Ag, 24 Cu, 15 In.
Melting range: 1166-1265°F; 630-685°C.

SEMALLOY NO. 2475.
M-1729; 61.5 Ag, 24 Cu, 14.5 In.
Liquidus 1305°F; Solidus 1155°F. (624-707°C).

SEMALLOY NO. 2477.
M-1729; 75 Cu, 17.75 Ag, 7.25 P.
M.P. 1189°F; 643°C; eutectic alloy.

SEMALLOY NO. 2480.
M-1729; 82.7 Cu, 10 Ag, 7.3 Pd.
Melting range: 1190-1265°F; 643-685°C.

SEMALLOY NO. 2485.
M-1729; 50 Ag, 15.5 Cu, 16 Cd, 3 Ni, 15.5 Zn.
Melting range: 1170-1270°F; 632-688°C.

SEMALLOY NO. 2490.
M-1729; 70 Ag, 30 In.
Melting range: 1130-1279°F; 610-693°C.

SEMALLOY NO. 2500.
M-1729; 35 Ag, 26 Cu, 21 Zn, 18 Cd.
Melting range: 1125-1295°F; 607-702°C.

SEMALLOY NO. 2503.
M-1729; 45 Ag, 30 Cu, 12 Zn, 13 Mn.
M.P. 1298°F; 703°C; eutectic alloy.

SEMALLOY NO. 2510.
M-1729; 80 Cu, 15 Ag, 5 Pd.
Melting range: 1185-1300°F; 640-704°C.

SEMALLOY NO. 2515.
M-1729; 60 Ag, 20 Cu, 7 Zn, 10 Cd, 3 Sn.
Melting range: 1270-1300°F; 688-704°C.

SEMALLOY NO. 2520.
M-1729; 30 Ag, 29 Cu, 21 Zn, 20 Cd.
Melting range: 1125-1300°F; 607-704°C.

SEMALLOY NO. 2521.
M-1729; 30 Ag, 27 Cu, 23 Zn, 20 Cd.
Liquidus 1310°F; Solidus 1125°F. (607-710°C).

SEMALLOY NO. 2523.
M-1729; 54.5 Ag, 15.5 Cu, 14 Zn, 15 Cd, 1 Sn.
Liquidus 1190°F; Solidus 1130°F. (610-643°C).

SEMALLOY NO. 2525.
M-1729; 45 Ag, 19 Zn, 18 Cu, 18 Cd.
Liquidus 1190°F; Solidus 1130°F. (610-643°C).

SEMALLOY NO. 2527.
M-1729; 40 Ag, 27 Cd, 18 Cu, 15 Zn.
Liquidus 1195°F; Solidus 1130°F. (610-646°C).

SEMALLOY NO. 2530.
M-1729; 60 Ag, 27 Cu, 13 In.
Melting range: 1175-1300°F; 635-704°C.

SEMALLOY NO. 2534.
M-1729; 100 Ba.
M.P. 1310°F; 710°C.

SEMALLOY NO. 2535.
M-1729; 66.7 Ag, 22.3 Cu, 11 Zn.
Melting range: 1280-1320°F; 693-716°C.

SEMALLOY NO. 2539.
M-1729; 40 Ag, 27 Cu, 18 Zn, 15 Cd.
Liquidus 1325°F; Solidus 1161°F; (627-718°C).

SEMALLOY NO. 2540.
M-1729; 96 Sn, 4 Se.
Liquidus 1310°F; Solidus 450°F. (232-710°C).

SEMALLOY NO. 2543.
M-1729; 45 Ag, 20.5 Cd, 17 Cu, 16.5 Zn, 0.5 Sn, 0.5 Pb.
Liquidus 1150°F; Solidus 1135°F. (613-621°C).

SEMALLOY NO. 2545.
M-1729; 50 Ag, 19 Cd, 16 Zn, 15 Cu.
Liquidus 1160°F; Solidus 1145°F. (618-627°C).

SEMALLOY NO. 2550.
M-1729; 60 Ag, 25 Cu, 15 Zn.
Melting range: 1245-1325°F; 674-718°C.

SEMALLOY NO. 2560.
M-1729; 60 Ag, 30 Cu, 10 Sn.
Melting range: 1115-1325°F; 602-718°C.

SEMALLOY NO. 2565.
M-1729; 65 Ag, 20 Cu, 15 Zn.
Melting range: 1240-1325°F; 671-718°C.

SEMALLOY NO. 2570.
M-1729; 87 Cu, 7 P, 6 Ag.
Melting range: 1190-1320°F; 643-716°C.

SEMALLOY NO. 2573.
M-1729; 60.5 Ag, 22.5 Cu, 7 Zn, 10 Cd.
Melting range: 1285-1335°F; 696-724°C.

SEMALLOY NO. 2577.
M-1729; 60 Ag, 20 Cu, 10 Cd, 10 Zn.
Melting range: 1285-1335°F; 696-724°C.

SEMALLOY NO. 2580.
M-1729; 50 Ag, 28 Cu, 22 Zn.
Melting range: 1250-1340°F; 677-727°C.

SEMALLOY NO. 2585.
M-1729; 50 Ag, 28 Zn, 20 Cu, 2 Ni.
Melting range: 1220-1301°F; 660-705°C.

SEMALLOY NO. 2590.
M-1729; 40 Ag, 30.5 Cu, 29.5 Zn.
Melting range: 1245-1340°F; 674-727°C.

SEMALLOY NO. 2600.
M-1729; 75 Ag, 25 Zn.
Melting range: 1300-1345°F; 704-729°C.

SEMALLOY NO. 2610.
M-1729; 57.3 Ag, 32.7 Cu, 7 Sn, 3 Mn.
Melting range: 1120-1345°F; 604-729°C.

SEMALLOY NO. 2617.
M-1729; 55 Ag, 31.5 Cu, 11.7 Zn, 1.8 Ni.
Melting range: 1300-1355°F; 704-735°C.

SEMALLOY NO. 2620.
M-1729; 60 Ag, 30 Cu, 10 Zn.
Melting range: 1245-1382°F; 674-750°C.

SEMALLOY NO. 2630.
M-1729; 45 Ag, 30 Cu, 25 Zn.
Melting range: 1250-1370°F; 677-743°C.

SEMALLOY NO. 2633.
M-1729; 61 Ag, 24 Cu, 15 Zn.
Melting range: 1260-1375°F; 682-745°C.

SEMALLOY NO. 2637.
M-1729; 98 Ag, 2 Li.
Melting range: 1274-1382°F; 690-750°C.

SEMALLOY NO. 2640.
M-1729; 34 Cu, 31.5 Ag, 15.5 Zn, 19 Cd.
Melting range: 1165-1390°F; 629-754°C.

SEMALLOY NO. 2650.
M-1729; 70 Ag, 20 Cu, 10 Zn.
Melting range: 1275-1360°F; 691-738°C.

SEMALLOY NO. 2653.
M-1729; 66.7 Ag, 27.7 Cu, 5.6 Zn.
Melting range: 1350-1390°F; 732-754°C.

SEMALLOY NO. 2655.
M-1729; 66.7 Ag, 28.25 Cu, 5.05 Zn.
Melting range: 1360-1395°F; 738-757°C.

SEMALLOY NO. 2657.
M-1729; 72.15 Ag, 22.8 Cu, 5.05 Zn.
Melting range: 1345-1400°F; 729-760°C.

SEMALLOY NO. 2660.
M-1729; 68 Ag, 27 Cu, 5 Zn.
Melting range: 1370-1400°F; 743-760°C.

SEMALLOY NO. 2663.
M-1729; 68 Ag, 26.6 Cu, 0.4 Zn, 5 Sn.
Melting range: 1370-1400°F; 743-760°C.

SEMALLOY NO. 2665.
M-1729; 40 Ag, 36 Cu, 24 Zn.
Melting range: 1229-1405°F; 665-763°C.

SEMALLOY NO. 2667.
M-1729; 48 Cu, 32 Zn, 20 Ag.
Melting range: 1370-1410°F; 743-766°C.

SEMALLOY NO. 2668.
M-1729; 34.2 Zn, 32.3 Cu, 28.5 Ag, 5 Mn.
Melting range: 1305-1385°F; 707-752°C.

SEMALLOY NO. 2670.
M-1729; 38 Cu, 32 Zn, 30 Ag.
Melting range: 1250-1410°F; 677-766°C.

SEMALLOY NO. 2675.
M-1729; 72 Ag, 28 Cu, 0.2 Li.
M.P. 1410°F; 760°C; eutectic alloy plus lithium.

SEMALLOY NO. 2680.
M-1729; 93 Cu, 7 P.
Melting range: 1305-1485°F; 707-807°C.

SEMALLOY NO. 2684.
M-1729; 100 St. (Strontium).
M.P. 1418°F; 770°C.

SEMALLOY NO. 2685.
M-1729; 75 Ag, 20 Cu, 5 Zn.
Melting range: 1350-1425°F; 732-744°C.

SEMALLOY NO. 2690.
M-1729; 50 Ag, 34 Cu, 16 Zn.
Melting range: 1270-1425°F; 688-724°C.

SEMALLOY NO. 2695.
M-1729; 40.15 Cu, 32.85 Zn, 27 Ag.
Melting range: 1350-1430°F; 732-777°C.

SEMALLOY NO. 2700.
M-1729; 40 Ag, 30 Cu, 28 Zn, 2 Ni.
Melting range: 1290-1435°F; 671-778°C.

SEMALLOY NO. 2710.
M-1729; 72 Ag, 28 Cu.
Melting range: 1435-1435°F; eutectic alloy.

SEMALLOY NO. 2720.
M-1729; 63 Ag, 27 Cu, 10 In.
Melting range: 1265-1346°F; 685-730°C.

SEMALLOY NO. 2727.
M-1729; 68 Ag, 27 Cu, 5 Sn.
Melting range: 1369-1400°F; 743-760°C.

SEMALLOY NO. 2735.
M-1729; 64.75 Zn, 27.25 Cu, 7.5 Sn, 0.5 Pb.
Melting range: 1385-1400°F; 752-782°C.

SEMALLOY NO. 2740.
M-1729; 91 Cu, 7 P, 2 Ag.
Melting range: 1185-1455°F; 643-791°C.

SEMALLOY NO. 2750.
M-1729; 71.15 Ag, 28.1 Cu, 0.75 Ni.
Melting range: 1436-1463°F; 780-795°C.

SEMALLOY NO. 2753.
M-1729; 75 Ag, 22 Cu, 3 Zn.
Melting range: 1365-1450°F; 741-788°C.

SEMALLOY NO. 2755.
M-1729; 68 Ag, 27 Cu, 5 Pd.
Melting range: 1480-1490°F; 804-810°C.

SEMALLOY NO. 2757.
M-1729; 92.75 Cu, 7.25 P.
Melting range: 1310-1456°F; 710-790°C.

SEMALLOY NO. 2758.
M-1729; 71.15 Ag, 28.1 Cu, 0.75 Ni.
Melting range: 1436-1463°F.

SEMALLOY NO. 2760.
M-1729; 50 Cu, 40 Ag, 10 Mn.
M.P. 1472°F; 800°C; eutectic alloy.

SEMALLOY NO. 2762.
M-1729; 71.5 Ag, 28 Cu, 0.5 Ni.
Melting range: 1435-1450°F; 779-788°C.

SEMALLOY NO. 2765.
M-1729; 80 Cu, 15 Ag, 5 P.
Melting range: 1190-1475°F.

SEMALLOY NO. 2770.
M-1729; 63 Ag, 28.5 Cu, 6 Sn, 2.5 Ni.
Melting range: 1275-1475°F; 690-802°C.

SEMALLOY NO. 2772.
M-1729; 75 Ag, 24.5 Cu, 0.5 Ni.
Melting range: 1435-1475°F; 779-802°C.

SEMALLOY NO. 2774.
M-1729; 88.75 Cu, 6.25 P, 5 Ag.
Melting range: 1190-1480°F.

SEMALLOY NO. 2775.
M-1729; 89 Cu, 5 Ag, 6 P.
Melting range: 1189-1480°F; 643-804°C.

SEMALLOY NO. 2776.
M-1729; 52.2 Ag, 38.4 Cu, 9.4 Zn.
Melting range: 1345-1435°F; 729-779°C.

SEMALLOY NO. 2778.
M-1729; 80 Ag, 16 Cu, 4 Zn.
Melting range: 1345-1490°F; 729-810°C.

SEMALLOY NO. 2780.
M-1729; 68.4 Ag, 26.6 Cu, 5 Pd.
Melting range: 1485-1490°F; 807-810°C.

SEMALLOY NO. 2783.
M-1729; 88 Cu, 6 Ag, 6 P.
Melting range: 1189-1480°F; 643-804°C.

SEMALLOY NO. 2800.
M-1729; 45 Cu, 30 Zn, 20 Ag, 5 Cd.
Liquidus 1500°F; Solidus 1140°F. (615-815°C).

SEMALLOY NO. 2805.
M-1729; 80 Cu, 15 Ag, 5 P.
Melting range: 1189-1450°F; 643-788°C.

SEMALLOY NO. 2810.
M-1729; 45 Cu, 35 Zn, 20 Ag.
Melting range: 1430-1500°F; 777-816°C.

SEMALLOY NO. 2813.
M-1729; 47.75 Cu, 32.8 Zn, 19.45 Ag.
Melting range: 1440-1500°F; 782-816°C.

SEMALLOY NO. 2817.
M-1729; 47 Cu, 35 Zn, 18 Ag.
Melting range: 1440-1500°F; 782-816°C.

SEMALLOY NO. 2819.
M-1729; 100 As.
M.P. 1503°F; 817°C.

SEMALLOY NO. 2845.
M-1729; 80 Cu, 20 Sn.
Melting range: 1470-1635°F; 798-890°C.

SEMALLOY NO. 2852.
M-1729; 95 Ag, 5 Al.
Melting range: 1440-1510°F; 782-821°C.

SEMALLOY NO. 2857.
M-1729; 60 Ag, 40 Cu.
Melting range: 1434-1535°F; 779-835°C.

SEMALLOY NO. 2858.
M-1729; 60 Ag, 20 Zn, 18 Cu, 2 Ni.
Melting range: 1434-1580°F; 779-860°C.

SEMALLOY NO. 2860.
M-1729; 77 Ag, 21 Cu, 2 Ni.
Melting range: 1435-1525°F; 779-829°C.

SEMALLOY NO. 2865.
M-1729; 60 Ag, 20 Zn, 15 Cu, 5 Ni.
Melting range: 1230-1545°F; 666-841°C.

SEMALLOY NO. 2874.
M-1729; 100 Ca.
M.P. 1562°F; 850°C.

SEMALLOY NO. 2890.
M-1729; 65 Ag, 28 Cu, 5 Mn, 2 Ni.
Melting range: 1385-1560°F; 752-849°C.

SEMALLOY NO. 2893.
M-1729; 87.75 Cu, 12 Ge, 0.25 Ni.
Melting range: 1508-1769°F; 820-965°C.

SEMALLOY NO. 2895.
M-1729; 58.5 Ag, 31.5 Cu, 10 Pd.
Melting range: 1515-1566°F; 824-852°C.

SEMALLOY NO. 2900.
M-1729; 53 Cu, 38 Zn, 38 Zn, 9 Ag.
Melting range: 1450-1565°F; 788-852°C.

SEMALLOY NO. 2910.
M-1729; 58 Ag, 32 Cu, 10 Pd.
Melting range: 1521-1565°F; 827-852°C.

SEMALLOY NO. 2915.
M-1729; 68.8 Ag, 26.7 Cu, 4.5 Ti.
Melting range: 1526-1562°F; 830-850°C.

SEMALLOY NO. 2920.
M-1729; 54 Ag, 40 Cu, 5 Zn, 1 Ni.
Melting range: 1325-1575°F; 718-857°C.

SEMALLOY NO. 2922.
M-1729; 52.5 Cu, 25 Ag, 22.5 Zn.
Melting range: 1250-1575°F; 677-857°C.

SEMALLOY NO. 2924.
M-1729; 40 Ag, 30 Cu, 25 Zn, 5 Ni.
Melting range: 1220-1580°F; 660-860°C.

SEMALLOY NO. 2926.
M-1729; 50 Cu, 40 Zn, 10 Ag.
Melting range: 1495-1590°F; 813-866°C.

SEMALLOY NO. 2928.
M-1729; 52.5 Cu, 47.5 Zn.
Melting range: 1570-1595°F; 854-868°C.

SEMALLOY NO. 2930.
M-1729; 96 Sn, 4 S.
Liquidus 1576°F; Solidus 450°F. (232-858°C).

SEMALLOY NO. 2940.
M-1729; 90 Ag, 10 Cu.
Melting range: 1434-1598°F; 779-870°C.

SEMALLOY NO. 2945.
M-1729; 52 Cu, 38 Zn, 10 Ag.
Melting range: 1470-1575°F; 799-857°C.

SEMALLOY NO. 2950.
M-1729; 58 Cu, 37 Zn, 5 Ag.
Melting range: 1575-1615°F; 841-879°C.

SEMALLOY NO. 2952.
M-1729; 56 Cu, 37 Zn, 7 Ag.
Melting range: 1575-1600°F; 857-871°C.

SEMALLOY NO. 2953.
M-1729; 90 Cu, 10 Sn.
Melting range: 1550-1830°F; 843-999°C.

SEMALLOY NO. 2954.
M-1729; 51.5 Cu, 45 Zn, 3.5 Sn.
Melting range: 1585-1610°F; 863-877°C.

SEMALLOY NO. 2955.
M-1729; 82 Ag, 9 Ga, 9 Pd.
Melting range: 1553-1616°F; 845-880°C.

SEMALLOY NO. 2956.
M-1729; 89 Ni, 11 P.
M.P. 1610°F; 880°C.

SEMALLOY NO. 2958.
M-1729; 57.5 Cu, 25 Ag, 17.5 Zn.
Melting range: 1255-1625°F; 880-885°C.

SEMALLOY NO. 2960.
M-1729; 62.5 Ag, 32.5 Cu, 5 Ni.
Melting range: 1435-1590°F; 779-866°C.

SEMALLOY NO. 2965.
M-1729; 56 Cu, 40 Zn, 4 Sn.
Melting range: 1590-1630°F; 866-888°C.

SEMALLOY NO. 2968.
M-1729; 77 Ni, 13 Cr, 10 P.
M.P. 1630°F; 888°C.

SEMALLOY NO. 2970.
M-1729; 56 Cu, 40 Zn, 1 Sn, 1 Fe, 1 Mn, 1 Ni.
Melting range: 1590-1630°F; 866-888°C.

SEMALLOY NO. 2973.
M-1729; 55 Cu, 44.75 Zn, 0.25 Mn.
Melting range: 1610-1635°F; 877-891°C.

SEMALLOY NO. 2975.
M-1729; 60 Cu, 39.2 Zn, 0.8 Sn.
Melting range: 1625-1652°F; 885-900°C.

SEMALLOY NO. 2980.
M-1729; 92.5 Ag, 7.3 Cu, 0.2 Li.
Melting range: 1400-1635°F; 760-891°C.

SEMALLOY NO. 2985.
M-1729; 51.5 Cu, 44 Zn, 4.5 Ag.
Melting range: 1410-1635°F; 766-891°C.

SEMALLOY NO. 2987.
M-1729; 92.5 Ag, 7.3 Cu, 0.2 Li.
Melting range: 1435-1634°F.

SEMALLOY NO. 2990.
M-1729; 92.5 Ag, 7.5 Cu.
Melting range: 1490-1634°F; 810-890°C.

SEMALLOY NO. 2993.
M-1729; 79.5 Au, 20.5 Cu.
Melting range: 1635-1635°F; eutectic alloy.

SEMALLOY NO. 2996.
M-1729; 56 Ag, 42 Cu, 2 Ni.
Melting range: 1420-1640°F; 771-893°C.

SEMALLOY NO. 2997.
M-1729; 76.85 Ni, 13 Cr, 10 P, 0.15 Mn.
Melting range: 1615-1640°F; 879-893°C.

SEMALLOY NO. 3000.
M-1729; 75 Au, 20 Cu, 5 Ag.
Melting range: 1625-1645°F; 885-895°C.

SEMALLOY NO. 3010.
M-1729; 52 Cu, 28 Cu, 20 Pd.
Melting range: 1614-1648°F; 879-898°C.

SEMALLOY NO. 3020.
M-1729; 95 Cu, 5 P.
Melting range: 1305-1650°F; 707-899°C.

SEMALLOY NO. 3025.
M-1729; 57 Cu, 42 Zn, 1 Sn.
Melting range: 1630-1650°F; 888-999°C.

SEMALLOY NO. 3027.
M-1729; 59 Cu, 40 Zn, 1 Sn.
Melting range: 1630-1650°F; 888-899°C.

SEMALLOY NO. 3030.
M-1729; 65 Ag, 20 Cu, 15 Pd.
Melting range: 1562-1652°F; 850-900°C.

SEMALLOY NO. 3043.
M-1729; 60 Cu, 40 Zn.
Melting range: 1645-1648°F; 895-898°C.

SEMALLOY NO. 3050.
M-1729; 60 Cu, 40 Ag.
Melting range: 1434-1670°F; 779-910°C.

SEMALLOY NO. 3063.
M-1729; 42 Ni, 34.5 Fe, 7.5 Co, 5 Cu.
Melting range: 1600-1675°F; 871-913°C.

SEMALLOY NO. 3071.
M-1729; 99.5 Ag, 0.5 Li.
Melting range: 1616-1697°F; 880-925°C.

SEMALLOY NO. 3073.
M-1729; 47 Cu, 42 Zn, 10.5 Ni, 0.5 Ag.
Melting range: 1685-1710°F; 818-932°C.

SEMALLOY NO. 3074.
M-1729; 48 Cu, 42 Zn, 10 Ni.
Melting range: 1690-1715°F; 921-935°C.

SEMALLOY NO. 3075.
M-1729; 100 Ge. (Germanium).
M.P. 1719°F; 937°C.

SEMALLOY NO. 3077.
M-1729; 71.5 Ti, 28.5 Ni.
M.P. 1728°F; 942°C.

SEMALLOY NO. 3080.
M-1729; 99 Ag, 1 As.
Melting range: 1670-1742°F; 910-950°C.

SEMALLOY NO. 3090.
M-1729; 54 Ag, 25 Pd, 21 Cu.
Melting range: 1654-1742°F; 900-950°C.

SEMALLOY NO. 3110.
M-1729; 70 Cu, 30 Ag.
Melting range: 1434-1742°F; 779-950°C.

SEMALLOY NO. 3120.
M-1729; 82 Au, 18 Ni.
M.P. 1742°F; 950°C.

SEMALLOY NO. 3123.
M-1729; 47 Cu, 43 Zn, 10 Ni.
Melting range: 1660-1750°F; 905-955°C.

SEMALLOY NO. 3127.
M-1729; 85 Ag, 15 Sn.
Melting range: 1335-1526°F; 724-830°C.

SEMALLOY NO. 3130.
M-1729; 100% Ag.
Melting point: 1761°F; 961°C.

SEMALLOY NO. 3140.
M-1729; 96 Ag, 4 Mn.
M.P. 1762°F; 961°C.

SEMALLOY NO. 3150.
M-1729; 62 W, 35 Ag, 3 Ni.
M.P. 1764°F; 962°C.

SEMALLOY NO. 3170.
M-1729; 85 Ag, 15 Mn.
Melting range: 1760-1780°F; 960-971°C.

SEMALLOY NO. 3190.
M-1729; 85 Cu, 8 Sn, 7 Ag.
Melting range: 1225-1805°F; 663-985°C.

SEMALLOY NO. 3201.
M-1729; 91.75 Fe, 4.5 Si, 3 B, 0.75 Ti.
Melting range: 1795-1820°F; 979-993°C.

SEMALLOY NO. 3202.
M-1729; 91.25 Ni, 4.5 Si, 2.9 B, 1.35 Fe.
Melting range: 1795-1820°F; 979-993°C.

SEMALLOY NO. 3203.
M-1729; 83.35 Ni, 6 Cr, 3 B, 2.5 Fe, 0.15 Mn, 5 Si.
Melting range: 1750-1825°F; 954-996°C.

SEMALLOY NO. 3204.
M-1729; 82 Ni, 7 Cr, 4.5 Si, 2.9 B, 3.6 Fe.
Melting range: 1780-1830°F; 971-999°C.

SEMALLOY NO. 3205.
M-1729; 80 Cu, 20 Ag.
Melting range: 1434-1796°F; 779-980°C.

SEMALLOY NO. 3206.
M-1729; 82.5 Ni, 7 Cu, 3 B, 4.5 Si, 3 Fe.
Melting range: 1780-1830°F; 971-999°C.

SEMALLOY NO. 3209.
M-1729; 72.5 Ni, 16 Cr, 3.5 B, 3 Fe, 5 Si.
Melting range: 1825-1840°F; 996-1005°C.

SEMALLOY NO. 3210.
M-1729; 95 Ag, 5 Pd.
Melting range: 1778-1850°F; 970-1010°C.

SEMALLOY NO. 3220.
M-1729; 60 Cu, 40 Au.
Melting range: 1805-1850°F.

SEMALLOY NO. 3225.
M-1729; 40 Au, 60 Cu.
Melting range: 1814-1850°F.

SEMALLOY NO. 3240.
M-1729; 62.5 Cu, 37.5 Au.
Melting range: 1814-1859°F.

SEMALLOY NO. 3241.
M-1729; 76.3 Ni, 7 Cr, 6 W, 4.5 Si, 3 Fe, 3.2 B.
Melting range: 1790-1900°F; 977-1038°C.

SEMALLOY NO. 3243.
M-1729; 63 Cu, 37 Au.
Melting range: 1815-1860°F; 990-1016°C.

SEMALLOY NO. 3247.
M-1729; 60 Mn, 40 Ni.
M.P. 1864°F; 1018°C; eutectic alloy.

SEMALLOY NO. 3250.
M-1729; 95.8 Cu, 3.1 Si, 1.1 Mn.
Melting range: 1742-1866°F; 950-1019°C.

SEMALLOY NO. 3252.
M-1729; 48.6 Ag, 22.5 Pd, 18.9 Cu, 10 Ni.
Melting range: 1796-1850°F; 980-1010°C.

SEMALLOY NO. 3255.
M-1729; 91.2 Cu, 7 Al, 1.8 Si.
Melting range: 1800-1840°F; 982-1004°C.

SEMALLOY NO. 3260.
M-1729; 65 Cu, 35 Au.
Melting range: 1832-1868°F.

SEMALLOY NO. 3263.
M-1729; 67 Mn, 16 Co, 16 Ni, 1 B.
Melting range: 1840-1870°F; 1005-1021°C.

SEMALLOY NO. 3267.
M-1729; 91.8 Ni, 4.5 Si, 3.5 B, 0.2 Mn.
Melting range: 1800-1875°F; 982-1024°C.

SEMALLOY NO. 3269.
M-1729; 65.5 Ni, 23 Mn, 7 Si, 4.5 Cu.
Melting range: 1800-1850°F; 982-1010°C.

SEMALLOY NO. 3270.
M-1729; 62 Cu, 35 Au, 3 Ni.
Melting range: 1787-1886°F; 975-1030°C.

SEMALLOY NO. 3272.
M-1729; 92.5 Ni, 4.5 Si, 3 B.
Melting range: 1800-1900°F; 982-1038°C.

SEMALLOY NO. 3280.
M-1729; 96 Pb, 4 S.
Liquidus 1868°F; Solidus 621°F. (327-1020°C).

SEMALLOY NO. 3290.
M-1729; 97 Cu, 3 Si.
Melting range: 1778-1877°F; 970-1025°C.

SEMALLOY NO. 3293.
M-1729; 89.35 Ni, 6.15 P, 4.5 Si.
Melting range: 1616-1880°F; 879-1027°C.

SEMALLOY NO. 3295.
M-1729; 92 Cu, 8 Sn.
Melting range: 1616-1868°F; 880-1020°C.

SEMALLOY NO. 3296.
M-1729; 93 Cu, 2.8-4.0 Si, 1.5 Sn, 1.5 Mn.
Melting range: 1780-1880°F; 971-1027°C.

SEMALLOY NO. 3311.
M-1729; 68 Mn, 32 Ni.
Melting range: 1859-1886°F; 1015-1030°C.

SEMALLOY NO. 3315.
M-1729; 74.85 Ni, 17 Mn, 8 Si, 0.15 C.
Melting range: 1850-1890°F; 1010-1032°C.

SEMALLOY NO. 3320.
M-1729; 70 Cu, 30 Au.
Melting range: 1859-1895°F.

SEMALLOY NO. 3322.
M-1729; 70.55 Ni, 16.5 Cr, 4.5 Si, 4 Fe, 3.85 B, 0.6 C.
Melting range: 1780-1900°F; 1010-1066°C.

SEMALLOY NO. 3324.
M-1729; 73.2 Ni, 13.5 Cr, 4.5 Fe, 4.5 Si, 3.5 B, 0.8 C.
Melting range: 1790-1900°F; 977-1038°C.

SEMALLOY NO. 3327.
M-1729; 91 Ni, 4.5 Si, 1.5 Fe, 3 B.
Melting range: 1800-1900°F; 982-1038°C.

SEMALLOY NO. 3328.
M-1729; 86 Cu, 11 Al, 3 Fe.
Melting range: 1880-1900°F; 1027-1038°C.

SEMALLOY NO. 3331.
M-1729; 92 Cu, 8 Al.
M.P. 1904°F; 1040°C.

SEMALLOY NO. 3332.
M-1729; 89 Cu, 10 Al, 1 Fe.
Melting range: 1904-1913°F; 1040-1045°C.

SEMALLOY NO. 3333.
M-1729; 95 Cu, 5 Sn.
Melting range: 1750-1920°F; 954-1049°C.

SEMALLOY NO. 3335.
M-1729; 97 Cu, 3 As.
Melting range: 1724-1922°F; 940-1055°C.

SEMALLOY NO. 3340.
M-1729; 70 Mn, 30 Ni.
Melting range: 1886-1922°F; 1030-1050°C.

SEMALLOY NO. 3341.
M-1729; 98.5 Cu, 1.5 Si.
Melting range: 1890-1940°F; 1032-1060°C.

SEMALLOY NO. 3343.
M-1729; 97 Cu, 3 Ge.
Melting range: 1894-1922°F; 1035-1050°C.

SEMALLOY NO. 3345.
M-1729; 95 Cu, 5 Al.
Melting range: 1920-1945°F; 1049-1063°C.

SEMALLOY NO. 3360.
M-1729; 94.6 Ni, 3.5 Si, 1.9 B.
Melting range: 1814-1931°F; 990-1055°C.

SEMALLOY NO. 3370.
M-1729; 81.35 Ni, 15 Cr, 3.5 B, 0.15 C.
M.P. 1931°F; 1055°C.

SEMALLOY NO. 3380.
M-1729; 95 Cu, 5 Ag.
Melting range: 1832-1940°F; 1000-1060°C.

SEMALLOY NO. 3400.
M-1729; 97 Cu, 3 Ga.
Melting range: 1922-1949°F; 1050-1065°C.

SEMALLOY NO. 3410.
M-1729; 72 Au, 22 Ni, 6 Cr.
Melting range: 1785-1950°F.

SEMALLOY NO. 3415.
M-1729; 93.8 Ni, 3.5 Si, 1.5 Fe, 1.2 B.
Melting range: 1800-1950°F; 932-1066°C.

SEMALLOY NO. 3420.
M-1729; 90 Ag, 10 Pd.
Melting range: 1835-1950°F; 1002-1066°C.

SEMALLOY NO. 3430.
M-1729; 71 Pt, 29 Sn.
M.P. 1958°F; 1070°C.

SEMALLOY NO. 3445.
M-1729; 74 Ni, 14 Cr, 4.5 Fe, 4.5 Si, 3 B.
Melting range: 1780-1900°F; 971-1038°C.

SEMALLOY NO. 3447.
M-1729; 74.5 Ni, 15 Cr, 4.5 Si, 3 Fe, 3 B.
Melting range: 1780-1970°F; 971-1077°C.

SEMALLOY NO. 3448.
M-1729; 98.7 Cu, 1.3 Sn.
Melting range: 1900-1970°F; 1035-1075°C.

SEMALLOY NO. 3449.
M-1729; 68.5 Ni, 12 W, 10 Cr, 3.5 Fe, 3.5 Si, 2.5 B.
Melting range: 1780-2000°F; 971-1093°C.

SEMALLOY NO. 3450.
M-1729; 99.7 Cu, 0.3 Si.
M.P. 1981°F; 1083°C.

SEMALLOY NO. 3460.
M-1729; 100 Cu.
Melting point: 1083°C; 1981°F.

SEMALLOY NO. 3470.
M-1729; 82 Cu, 18 Pd.
Melting range: 1985-2012°F; 1085-1100°C.

SEMALLOY NO. 3475.
M-1729; 71 Ni, 18 Cr, 11 B.
Melting range: 1975-2075°F; 1079-1135°C.

SEMALLOY NO. 3479.
M-1729; 61 Ni, 19 Cr, 10 Si, 10 Mn.
Melting range: 1975-2075°F; 1079-1135°C.

SEMALLOY NO. 3480.
M-1729; 81.35 Ni, 15 Cr, 3.5 B, 0.15 Mn.
Melting range: 1900-1990°F; 1038-1088°C.

SEMALLOY NO. 3490.
M-1729; 35 Cu, 30 Pd, 20 Ni, 15 Mn.
M.P. 1994°F; 1090°C.

SEMALLOY NO. 3500.
M-1729; 63 Ni, 16 W, 11.5 Cr, 3.75 Fe, 3.25 Si, 2.5 B.
Melting range: 1790-2020°F; 971-1104°C.

SEMALLOY NO. 3505.
M-1729; 55 Cu, 20 Pd, 15 Ni, 10 Mn.
M.P. 2020°F; 1104°C.

SEMALLOY NO. 3510.
M-1729; 75 Ag, 20 Pd, 5 Mn.
Melting range: 1960-2050°F; 1071-1121°C.

SEMALLOY NO. 3520.
M-1729; 70.85 Ni, 19 Cr, 10 Si, 0.15 C.
Melting range: 1975-2075°F; 1080-1135°C.

SEMALLOY NO. 3521.
M-1729; 79.75 Ni, 11 Cr, 3.5 Si, 3.5 Fe, 2.25 B.
Melting range: 1780-2120°F; 971-1160°C.

SEMALLOY NO. 3530.
M-1729; 71 Ni, 19 Cu, 10 Si.
Melting range: 1975-2075°F; 1079-1135°C.

SEMALLOY NO. 3540.
M-1729; 88.7 Cu, 10 Ni, 1.3 Fe.
Melting range: 2012-2102°F; 1100-1150°C.

SEMALLOY NO. 3550.
M-1729; 82.55 Ni, 10 Cr, 2.5 Si, 2.5 Fe, 2 B, 0.45 C.
Melting range: 1780-2120°F; 971-1160°C.

SEMALLOY NO. 3560.
M-1729; 73 Ag, 27 Pt.
Melting range: 1823-2120°F; 995-1160°C.

SEMALLOY NO. 3565.
M-1729; 50.8 Co, 19 Cr, 17 Ni, 8 Si, 4 W, 0.8 B, 0.4 C.
Melting range: 2025-2100°F; 1107-1149°C.

SEMALLOY NO. 3570.
M-1729; 80 Ag, 20 Pd.
Melting range: 1958-2102°F; 1070-1150°C.

SEMALLOY NO. 3580.
M-1729; 75 Cu, 25 Ni.
Melting range: 2102-2210°F; 1150-1210°C.

SEMALLOY NO. 3590.
M-1729; 65 Pd, 35 Co.
Melting range: 2246-2255°F; 1230-1235°C.

SEMALLOY NO. 3592.
M-1729; 79 Cu, 21 Ni.
Melting range: 2100-2190°F; 1149-1199°C.

SEMALLOY NO. 3600.
M-1729; 64 Ag, 33 Pd, 3 Mn.
Melting range: 2100-2250°F; 1149-1232°C.

SEMALLOY NO. 3610.
M-1729; 60 Pd, 40 Ni.
M.P. 2260°F; 1237°C.

SEMALLOY NO. 3615.
M-1729; 70 Cu, 30 Ni.
Melting range: 2140-2260°F; 1171-1238°C.

SEMALLOY NO. 3620.
M-1729; 69.4 Cu, 30 Ni, 0.6 Fe.
Melting range: 2138-2264°F; 1170-1240°C.

SEMALLOY NO. 3624.
M-1729; 100 Mn.
M.P. 2271°F; 1244°C.

SEMALLOY NO. 3625.
M-1729; 60 Cu, 40 Pt.
Melting range: 2165-2221°F; 1185-1216°C.

SEMALLOY NO. 3630.
M-1729; 60 Cu, 40 Pd.
Melting range: 2057-2102°F; 1125-1150°C.

SEMALLOY NO. 3640.
M-1729; 54 Pd, 36 Ni, 10 Cr.
Melting range: 2250-2300°F; 1232-1260°C.

SEMALLOY NO. 3644.
M-1729; 100 Be.
M.P. 2343°F; 1284°C.

SEMALLOY NO. 3650.
M-1729; 55 Cu, 45 Ni.
Melting range: 2246-2363°F; 1230-1295°C.

SEMALLOY NO. 3660.
M-1729; 70 Ni, 30 Pd.
Melting range: 2354-2408°F; 1290-1320°C.

SEMALLOY NO. 3670.
M-1729; 53.5 Ni, 46.5 Mo.
M.P. 2399°F; 1315°C.

SEMALLOY NO. 3680.
M-1729; 63 Co, 37 Mo.
M.P. 2440°F; 1340°C.

SEMALLOY NO. 3690.
M-1729; 67 Ni, 28.5 Cu, 2.5 Fe, 2 Mg.
Melting range: 2370-2460°F; 1299-1349°C.

SEMALLOY NO. 3700.
M-1729; 67 Ni, 29.25 Cu, 3 Al, 0.75 Sn.
Melting range: 2400-2460°F; 1316-1349°C.

SEMALLOY NO. 3710.
M-1729; 72.85 Ni, 17 Cr, 10 Fe, 0.15 C.
Melting range: 2540-2600°F; 1393-1427°C.

SEMALLOY NO. 3714.
M-1729; 100 Si.
M.P. 2574°F; 1412°C.

SEMALLOY NO. 3720.
M-1729; 94 Ni, 4.75 Al, 1 Sn, 0.25 C.
Melting range: 2615-2635°F; 1435-1446°C.

SEMALLOY NO. 3730.
M-1729; 64 Fe, 36 Ni.
Melting range: 2633-2642°F; 1445-1450°C.

SEMALLOY NO. 3740.
M-1729; 100 Ni. (Nickel).
Melting point; 1453°C; 2647°F.

SEMALLOY NO. 3744.
M-1729; 100 Co.
M.P. 2718°F; 1492°C.

SEMALLOY NO. 3750.
M-1729; 99.1 Fe, 0.45 Mn, 0.25 Si, 0.2 C.
M.P. 2760°F; 1516°C.

SEMALLOY NO. 3754.
M-1729; 100 Fe.
M.P. 2798°F; 1537°C.

SEMALLOY NO. 3760.
M-1729; 100 Pd. (Palladium).
Melting point: 1552°C; 2826°F.

SEMALLOY NO. 3765.
M-1729; 100 Ti. (Titanium).
Melting point: 1660°C; 3020°F.

SEMALLOY NO. 3774.
M-1729; 100 V. (Vanadium).
M.P. 3380°F; 1860°C.

SEMALLOY NO. 3775.
M-1729; 95 Pt, 5 Ir.
Melting range: 3216-3220°F; 1769-1771°C.

SEMALLOY NO. 3780.
M-1729; 100 Pt. (Platinum).
Melting point: 1773°C; 3216°F.

SEMALLOY NO. 3782.
M-1729; 90 Pt, 10 Ir.
Melting range: 3236-3272°F; 1780-1800°C.

SEMALLOY NO. 3784.
M-1729; 100 Cr.
M.P. 3362°F; 1850°C.

SEMALLOY NO. 3785.
M-1729; 85 Pt, 15 Ir.
Melting range: 3272-3308°F; 1800-1820°C.

SEMALLOY NO. 3786.
M-1729; 95 Pt, 5 Rh.
Melting range: 3290-3299°F; 1810-1815°C.

SEMALLOY NO. 3790.
M-1729; 90 Pt, 10 Rh.
Melting range: 3326-3362°F; 1830-1850°C.

SEMALLOY NO. 3794.
M-1729; 80 Pt, 20 Ir.
Melting range: 3326-3362°F; 1830-1850°C.

SEMALLOY NO. 3800.
M-1729; 100 Zr. (Zirconium).
Melting point: 1852°C; 3366°F.

SEMALLOY NO. 3805.
M-1729; 80 Pt, 20 Rh.
Melting range: 3425-3452°F; 1885-1900°C.

SEMALLOY NO. 3810.
M-1729; 80 Mo, 20 Ru.
M.P. 3452°F; 1900°C.

SEMALLOY NO. 3815.
M-1729; 70 Pt, 30 Rh.
Melting range: 3497-3506°F; 1925-1930°C.

SEMALLOY NO. 3820.
M-1729; 60 Pt, 40 Rh.
Melting range: 3515-3533°F; 1935-1945°C.

SEMALLOY NO. 3830.
M-1729; 100 Rh. (Rhodium).
Melting point: 1966°C; 3571°F.

SEMALLOY NO. 3840.
M-1729; 60 Pt, 40 Ir.
Melting range: 3632-3659°F; 2000-2015°C.

SEMALLOY NO. 3850.
M-1729; 90 Mo, 10 Si.
Melting range: 3848-3902°F; 2120-2150°C.

SEMALLOY NO. 3854.
M-1729; 100 Hf (Hafnium).
M.P. 4028°F; 2220°C.

SEMALLOY NO. 3856.
M-1729; 100 B. (Boron).
M.P. 3812°F; 2100°C.

SEMALLOY NO. 3860.
M-1729; 100 Ir. (Iridium).
Melting point: 2450°C; 4442°F.

SEMALLOY NO. 3870.
M-1729; 100 Nb. (Niobium or Columbium).
Melting point: 2468°C; 4474°F.

SEMALLOY NO. 3880.
M-1729; Tungsten carbide.
Melting point: 2480°C; 4496°F.

SEMALLOY NO. 3890.
M-1729; 100 Ru. (Ruthenium).
Melting point: 2500°C; 4532°F.

SEMALLOY NO. 3894.
M-1729; 100 Mo. (Molybdenum).
M.P. 4748°F; 2620°C.

SEMALLOY NO. 3900.
M-1729; 100 Ta. (Tantalum).
Melting point: 2996°C; 5425°F.

SEMALLOY NO. 3904.
M-1729; 100 Os. (Osmium).
M.P. 4892°F; 2700°C.

SEMALLOY NO. 3910.
M-1729; 100 Re. (Rhenium).
Melting point: 3180°C; 5756°F.

SEMALLOY NO. 3920.
M-1729; 100 W. (Tungsten).
M.P. 6116°F; 3380°C.

SEMALLOY NO. 3930.
M-1729; 100 C (Carbon).
M.P. 9032°F; 5000°C.

SEMALLOY NO. A902.
M-1729; 85 Pb, 15 Au.
M.P. 419°F; 215°C; eutectic alloy.

SEMALLOY NO. A903.
M-1729; 90 Sn, 10 Au.
M.P. 423°F; 217°C; eutectic alloy.

SEMALLOY NO. A904.
M-1729; 82 Bi, 18 Au.
M.P. 466°F; 241°C; eutectic alloy.

SEMALLOY NO. A905.
M-1729; 80 Au, 20 Sn.
M.P. 536°F; 280°C; eutectic alloy.

SEMALLOY NO. A906.
M-1729; 71 Sn, 29 Au.
Liquidus 541°F; Solidus 486°F. (252-283°C).

SEMALLOY NO. A907.
M-1729; 68 Pb, 32 Au.
Liquidus 568°F; Solidus 489°F. (254-298°C).

SEMALLOY NO. A908.
M-1729; 87 Cd, 13 Au.
M.P. 588°F; 309°C; eutectic alloy.

SEMALLOY NO. A909.
M-1729; 55 Sn, 43 Au.
Liquidus 644°F; Solidus 588°F. (309-340°C).

SEMALLOY NO. A910.
M-1729; 84 Au, 16 Ga.
M.P. 646°F; 341°C; eutectic alloy.

SEMALLOY NO. A911.
M-1729; 88 Au, 12 Ge.
M.P. 673°F; 356°C; eutectic alloy.

SEMALLOY NO. A912.
M-1729; 75 Au, 25 Sb.
M.P. 680°F; 360°C; eutectic alloy.

SEMALLOY NO. A913.
M-1729; 98 Au, 2 Si.
Liquidus 1724°F; Solidus 698°F. (370-940°C).

SEMALLOY NO. A914.
M-1729; 96.9 Au, 3.1 Si.
M.P. 698°F; 370°C; eutectic alloy.

SEMALLOY NO. A915.
M-1729; 83 Te, 17 Au.
M.P. 781°F; 416°C; eutectic.

SEMALLOY NO. A916.
M-1729; 58 Au, 42 Te.
M.P. 837°F; 447°C.

SEMALLOY NO. A917.
M-1729; 73 Au, 27 In.
M.P. 844°F; 451°C.

SEMALLOY NO. A918.
M-1729; 68 Au, 32 Ga.
M.P. 844°F; 451°C; eutectic alloy.

SEMALLOY NO. A919.
M-1729; 55 Sb, 45 Au.
M.P. 860°F; 460°C; eutectic alloy.

SEMALLOY NO. A920.
M-1729; 81 Au, 19 In.
Liquidus 910°F; Solidus 896°F. (480-488°C).

SEMALLOY NO. A921.
M-1729; 59 Au, 41 In.
M.P. 921°F; 494°C; eutectic alloy.

SEMALLOY NO. A922.
M-1729; 58 Cd, 42 Au.
M.P. 932°F; 500°C; eutectic alloy.

SEMALLOY NO. A923.
M-1729; 96 Au, 4 Al.
M.P. 977°F; 525°C; eutectic alloy.

SEMALLOY NO. A924.
M-1729; 53 Au, 47 Cd.
M.P. 1004°F; 540°C; eutectic alloy.

SEMALLOY NO. A925.
M-1729; 50 Au, 50 In.
Liquidus 1004°F; Solidus 921°F. (494-540°C).

SEMALLOY NO. A926.
M-1729; 91 Au, 9 Al.
M.P. 1056°F; 569°C; eutectic alloy.

SEMALLOY NO. A927.
M-1729; 61 Mg, 39 Au.
M.P. 1069°F; 576°C; eutectic alloy.

SEMALLOY NO. A928.
M-1729; 72 Au, 28 Cd.
M.P. 1134°F; 612°C; eutectic alloy.

SEMALLOY NO. A929.
M-1729; 69 Au, 31 Zn.
M.P. 1159°F; 626°C; eutectic alloy.

SEMALLOY NO. A930.
M-1729; 95 Al, 5 Au.
M.P. 1188°F; 642°C; eutectic alloy.

SEMALLOY NO. A931.
M-1729; 85 Au, 15 Zn.
M.P. 1188°F; 642°C.

SEMALLOY NO. A933.
M-1729; 78.5 Au, 21.5 Mg.
M.P. 1429°F; 776°C; eutectic alloy.

SEMALLOY NO. A934.
M-1729; 80.5 Au, 19.5 Mg.
M.P. 1436°F; 780°C; eutectic alloy.

SEMALLOY NO. A935.
M-1729; 89 Au, 11 Th.
M.P. 1490°F; 810°C; eutectic alloy.

SEMALLOY NO. A936.
M-1729; 94 Au, 6 Mg.
M.P. 1521°F; 827°C; eutectic alloy.

SEMALLOY NO. A961.
M-1729; 60 Au, 40 Cu.
Melting range: 1679-1724°F; 915-940°C.

SEMALLOY NO. A963.
M-1729; 75 Au, 25 Ag.
Melting range: 1886-1904°F.

SEMALLOY NO. A965.
M-1729; 80 Au, 20 Ag.
Melting range: 1900-1918°F; 1038-1048°C.

SEMALLOY NO. A966.
M-1729; 100 Au.
Melting point: 1945°F; 1063°C.

SEMALLOY NO. A971.
M-1729; 55 Au, 45 Pt.
Melting range: 2236-2786°F; 1280-1530°C.

SEMALLOY NO. A972.
M-1729; 92 Au, 8 Pd.
Melting range: 2174-2264°F; 1190-1240°C.

SEMALLOY NO. A973.
M-1729; 87 Au, 13 Pd.
Melting range: 2300-2381°F; 1260-1305°C.

SEMALLOY NO. A974.
M-1729; 75 Au, 25 Pt.
Melting range: 2210-2570°F; 1210-1410°C.

SEMALLOY NO. A975.
M-1729; 75 Au, 25 Pd.
Melting range: 2507-2552°F; 1375-1400°C.

SEMALLOY NO. A976.
M-1729; 65 Au, 35 Pd.
Melting range: 2590-2624°F; 1421-1440°C.

SEMALLOY NO. A977.
M-1729; 20 Au, 80 Pt.
Melting range: 2570-3002°F; 1410-1650°C.

SEMALLOY NO. A978.
M-1729; 91 Pt, 9 Au.
Melting range: 2912-3110°F; 1600-1710°C.

SEMALLOY NO. A2297.
M-1729; 60 Au, 20 Ag, 20 Sn.
M.P. 680°F; 360°C; eutectic alloy.

SEMALLOY NO. A2360.
M-1729; 75 Au, 25 In.
Liquidus 856°F; Solidus 844°F. (451-485°C).

SEMALLOY NO. A2370.
M-1729; 80 Au, 20 In.
Liquidus 905°F; Solidus 844°F. (451-485°C).

SEMALLOY NO. A2495.
M-1729; 38 Au, 26 Ag, 19 Cu, 16 Cd, 1 Zn.
Melting range: 1175-1300°F; 635-704°C.

SEMALLOY NO. A2531.
M-1729; 32 Ag, 25 Cu, 23 Au, 19 Cd, 1 Zn.
Melting range: 1200-1285°F; 649-696°C.

SEMALLOY NO. A2567.
M-1729; 31 Ag, 29 Au, 20 Cu, 19 Cd, 1 Zn.
Melting range: 1280-1400°F; 693-760°C.

SEMALLOY NO. A2820.
M-1729; 50 Au, 29.5 Ag, 20.5 Cu.
Melting range: 1460-1510°F; 793-821°C.

SEMALLOY NO. A2840.
M-1729; 49 Au, 27.5 Ag, 23.5 Cu.
Melting range: 1460-1515°F; 793-824°C.

SEMALLOY NO. A2850.
M-1729; 53 Au, 23.5 Ag, 23.5 Cu.
Melting range: 1480-1520°F; 804-877°C.

SEMALLOY NO. A2870.
M-1729; 58.3 Au, 24 Ag, 17.7 Cu.
Melting range: 1500-1550°F; 816-843°C.

SEMALLOY NO. A2880.
M-1729; 60 Au, 20 Ag, 20 Cu.
Melting range: 1535-1553°F; 835-845°C.

SEMALLOY NO. A2925.
M-1729; 42 Au, 32 Ag, 16 Cu, 10 Zn.
Melting range: 1335-1380°F; 724-749°C.

SEMALLOY NO. A2927.
M-1729; 48 Au, 29 Ag, 17 Cu, 6 Zn.
Melting range: 1350-1450°F; 732-788°C.

SEMALLOY NO. A3040.
M-1729; 60 Au, 37 Cu, 3 In.
Melting range: 1580-1652°F; 860-900°C.

SEMALLOY NO. A3060.
M-1729; 81.5 Au, 15.5 Cu, 3 Ni.
Melting range: 1652-1670°F.

SEMALLOY NO. A3067.
M-1729; 66.7 Au, 33.3 Cu.
Melting range: 1655-1685°F; 903-920°C.

SEMALLOY NO. A3070.
M-1729; 58.3 Au, 39.6 Cu, 2.1 Ag.
Melting range: 1660-1690°F; 905-921°C.

SEMALLOY NO. A3085.
M-1729; 81.5 Au, 16.5 Cu, 2 Ni.
Melting range: 1670-1697°F; 910-925°C.

SEMALLOY NO. A3160.
M-1729; 55.5 Cu, 41.7 Au, 2.8 Ag.
Melting range: 1695-1765°F; 924-963°C.

SEMALLOY NO. A3180.
M-1729; 50 Au, 50 Ag.
Melting range: 1742-1787°F; 950-975°C.

SEMALLOY NO. A3200.
M-1729; 94 Au, 6 Cu.
Melting range: 1769-1814°F; 965-990°C.

SEMALLOY NO. A3207.
M-1729; 35 Au, 65 Ag.
Melting range: 1796-1832°F; 980-1000°C.

SEMALLOY NO. A3208.
M-1729; 43 Au, 57 Ag.
Melting range: 1796-1832°F; 980-1000°C.

SEMALLOY NO. A3212.
M-1729; 50 Au, 40 Cu, 10 Ni.
Melting range: 1780-1839°F; 971-1004°C.

SEMALLOY NO. A3214.
M-1729; 50 Au, 50 Ag.
Melting range: 1832-1868°F; 1000-1020°C.

SEMALLOY NO. A3230.
M-1729; 77 Cu, 20 Au, 3 In.
Melting range: 1778-1859°F; 970-1015°C.

SEMALLOY NO. A3253.
M-1729; 55 Cu, 45 Au.
Melting range: 1749-1780°F; 954-971°C.

SEMALLOY NO. A3260.
M-1729; 73.8 Au, 26.2 Ni.
Melting range: 1796-1850°F; 980-1010°C.

SEMALLOY NO. A3265.
M-1729; 70 Au, 22 Ni, 8 Pd.
Melting range: 1841-1899°F; 1005-1037°C.

SEMALLOY NO. A3275.
M-1729; 60 Ag, 40 Au.
Melting range: 1814-1841°F; 990-1005°C.

SEMALLOY NO. A3297.
M-1729; 80 Cu, 20 Au.
Melting range: 1940-1976°F; 1060-1080°C.

SEMALLOY NO. A3310.
M-1729; 99 Au, 1 Ga.
Melting range: 1877-1886°F; 1025-1030°C.

SEMALLOY NO. A3330.
M-1729; 99.4 Au, 0.6 Sb.
Liquidus 1886°F; Solidus 680°F. (360-1030°C).

SEMALLOY NO. A3350.
M-1729; 75 Au, 20 Pd, 5 Mn.
Melting range: 1846-1962°F; 1008-1072°C.

SEMALLOY NO. A3440.
M-1729; 65 Au, 35 Ni.
Melting range: 1769-1967°F; 965-1075°C.

SEMALLOY NO. A3455.
M-1729; 90 Cu, 10 Au.
Melting range: 1945-1972°F; 1063-1078°C.

SEMALLOY NO. A3465.
M-1729; 98.5 Au, 1.5 Pd.
Melting range: 1967-2012°F; 1075-1100°C.

SEMALLOY NO. A3485.
M-1729; 50 Au, 25 Ni, 25 Pd.
Melting range: 2016-2050°F; 1102-1121°C.

SEMALLOY NO. A3575.
M-1729; 96.5 Au, 3.5 Pd.
Melting range: 2057-2102°F; 1125-1150°C.

SEMALLOY NO. A3585.
M-1729; 36 Ni, 34 Pd, 30 Au.
Melting range: 2075-2136°F; 1135-1169°C.

SEMALLOY NO. A3595.
M-1729; 94 Au, 6 Pd.
Melting range: 2102-2192°F; 1150-1200°C.

SEMALLOY NO. A3635.
M-1729; 91 Au, 9 Pd.
Melting range: 2192-2282°F; 1200-1250°C.

SEMALLOY NO. A3655.
M-1729; 82 Au, 18 Pd.
Melting range: 2417-2462°F; 1325-1350°C.

SEMALLOY NO. A3665.
M-1729; 74 Au, 26 Pd.
Melting range: 2534-2561°F; 1390-1405°C.

SEMALLOY NO. A3675.
M-1729; 62 Au, 38 Pd.
Melting range: 2606-2642°F; 1430-1450°C.

SEMALLOY NO. A3735.
M-1729; 60 Au, 40 Pd.
Melting range: 2642-2665°F; 1450-1463°C.

SEMALLOY NO. A3745.
M-1729; 60 Pd, 40 Au.
Melting range: 2723-2732°F; 1495-1500°C.

SEMALLOY NO. A3770.
M-1729; 75 Pt, 20 Pd, 5 Au.
Melting range: 2993-3083°F; 1645-1695°C.

SEMALLOY NO. A9231.
M-1729; 44 Ag, 38 Au, 18 Ge.
Liquidus 977°F; Solidus 959°F. (515-525°C).

SEMALLOY NO. A9251.
M-1729; 45 Ag, 38 Au, 17 Ge.
M.P. 977°F; 525°C; eutectic alloy.

SEMALLOY NO. A9524.
M-1729; 67 Au, 33 Cu.
Melting range: 1650-1684°F; 899-918°C.

SEMALLOY NO. A9540.
M-1729; 80 Au, 20 Cu.
M.P. 1632°F; 889°C; eutectic alloy.

SEMALLOY NO. A9570.
M-1729; 89 Au, 11 Mn.
M.P. 1760°F; 960°C.

SEMALLOY NO. A9580.
M-1729; 90 Au, 10 Mn.
Melting range: 1760-1787°F; 960-975°C.

SEMALLOY NO. A9585.
M-1729; 62 Cu, 38 Au.
Melting range: 1814-1859°F; 990-1015°C.

SEMALLOY NO. A9590.
M-1729; 60 Cu, 40 Au.
Melting range: 1796-1841°F; 980-1005°C.

SEMALLOY NO. A9610.
M-1729; 72 Au, 22 Ni, 6 Cr.
Melting range: 1785-1950°F; 974-1066°C.

SEMALLOY NO. A9614.
M-1729; 65 Cu, 35 Au.
Melting range: 1832-1868°F; 1000-1020°C.

SEMALLOY NO. A9615.
M-1729; 78 Cu, 20 Au, 2 In.
Melting range: 1787-1877°F; 975-1025°C.

SEMALLOY NO. A9640.
M-1729; 70 Cu, 30 Au.
Melting range: 1859-1895°F; 1015-1035°C.

SEMALLOY NO. A9650.
M-1729; 83 Au, 17 Ag.
Melting range: 1904-1922°F; 1040-1050°C.

SEMALLOY NO. A9670.
M-1729; 56 Au, 44 Mn.
M.P. 1963°F; 1073°C; eutectic alloy.

SEMALLOY NO. A9690.
M-1729; 54 Au, 46 Mn.
M.P. 1985°F; 1085°C; eutectic alloy.

SEMALLOY NO. A9700.
M-1729; 59 Au, 41 Mn.
M.P. 2003°F; 1095°C; eutectic alloy.

SEMALLOY NO. AG1.5.
M-1729; 97.5 Pb, 1.5 Ag, 1 Sn.
M.P. 588°F; 309°C; eutectic alloy.

SEMALLOY NO. AG2.5.
M-1729; 97.5 Pb, 2.5 Ag.
M.P. 580°F; 304°C; eutectic alloy.

SEMALLOY NO. AG5.5.
M-1729; 94.5 Pb, 5.5 Ag.
Liquidus 689°F; Solidus 579°F. (304-365°C).

SEMALLOY NO. PB65.
M-1729; 64.5 Pb, 34.5 Sn.
Liquidus 475°F; Solidus 360°F. (182-246°C).

SEMALLOY NO. PB70.
M-1729; 70 Pb, 30 Sn.
Liquidus 491°F; Solidus 361°F. (183-255°C).

SEMALLOY NO. PB80.
M-1729; 80 Pb, 20 Sn.
Liquidus 535°F; Solidus 361°F. (183-279°C).

SEMALLOY NO. SB5.
M-1729; 94.8 Sn, 5 Sb, 0.2 Pb.
Liquidus 464°F; Solidus 450°F. (232-240°C).

SEMALLOY NO. SN5.
M-1729; 95 Pb, 5 Sn.
Liquidus 594°F; Solidus 518°F. (270-312°C).

SEMALLOY NO. SN10.
M-1729; 87.9 Pb, 10 Zn, 2.1 Ag.
Liquidus 570°F; Solidus 514°F. (268-299°C).

SEMALLOY NO. SN20.
M-1729; 79 Pb, 20 Sn, 1 Sb.
Liquidus 530°F; Solidus 360°F. (182-277°C).

SEMALLOY NO. SN30.
M-1729; 68.4 Pb, 30 Sn, 1.6 Sb.
Liquidus 490°F; Solidus 360°F. (182-254°C).

SEMALLOY NO. SN35.
M-1729; 63 Pb, 35 Sn, 1.8 Sb, 0.2 Bi.
Liquidus 475°F; Solidus 360°F (182-246°C).

SEMALLOY NO. SN40.
M-1729; 59.4 Pb, 40 Sn.
Liquidus 460°F; Solidus 360°F. (182-238°C).

SEMALLOY NO. SN50.
M-1729; 50 Sn, 49.4 Pb.
Liquidus 420°F; Solidus 360°F. (182-216°C).

SEMALLOY NO. SN60.
M-1729; 60 Sn, 39.4 Pb.
Liquidus 375°F; Solidus 360°F. (182-191°C).
QQ S-571D.

SEMALLOY NO. SN62.
M-1729; 62 Sn, 35.4 Pb, 2 Ag.
Liquidus 372°F; Solidus 350°F. (177-189°C)9
QQ S-571D.

SEMALLOY NO. SN63.
M-1729; 63 Sn, 36.4 Pb.
M.P. 360°F; 182°C.
QQS-571D.

SEMALLOY NO. SN70.
M-1729; 70 Sn, 29.4 Pb.
Liquidus 380°F; Solidus 360°F. (182-193°C).
QQ S-591D.

SEMALLOY SN96.
M-1729; 96 Sn, 4 Ag.
Liquidus 432°F; Solidus 430°F. (221-222°C).

SEMDEX.
M-64; 0.60 C, 0.25 Si, 0.80 Mn, 0.25 Cr, bal Fe.
Water hardening tool steel for trimmer dies, sow blocks.

SEMINOLE HARD.
M-1779; 0.52 C, 1.3 Cr, 2 W, 0.25 V, 0.9 Si, bal Fe.
Oil hardening, shock resisting tool steel for moderate hot work; for shear blades, dies, punches, chisels, pneumatic tools.
AISI S-1.

SEMINOLE MEDIUM.
M-1779; 0.42 C, 1.3 Cr, 2 W, 0.25 Mo, 0.25 Si, bal Fe.
Oil hardening, shock resisting tool steel for moderate hot work; for shear blades, dies, punches, chisels, pneumatic tools.
Lower carbon version of AISI S-1.

SEMI-PLASTIC BRONZE.
M-Eng.; 79-75 Cu, 13.5-16.5 Pb, 7-9 Sn.
20,000 TS; 10 El.
For machine bearings; leaded bronze.

SEMI-STEEL.
M-Eng.; 3.0-3.5 total C, 0.5-0.81 combined C, 1.4-1.8 Si, bal Fe.
For frames, housings, cast gears; cast iron.

SEMPERAL.
M-971; 4 Cu, 1 Zn, 0.3 Mg, 0.6 Mn, 0.7 Fe, bal Al.
Soft: 36,000 TS; 22 El; 60 Brin.
Rolled: 71,000 TS; 2 El; 170 Brin.
For light alloy parts; heat treatable.

SENDUST.
M-198; 9.6 Si, 5.4 Al, 85 Co.
For cores in electromagnetic apparatus. Brittle, cast alloy.

SENDUST.
M-Japan; 9.6 Si, 5.4 Al, 85 Co.
For cores of electromagnetic apparatus; brittle, cast alloy.

SENECA.
M-972; 0.7 C, bal Fe.
For wire.

SENECA.
M-435; 0.35 C, 3.2 Cr, 9.5 W, 0.4 V, bal Fe.
Hot work tool steel.
Similar to AISI H21.

SENECA HOT DIE.
M-435; 0.28 C, 9.5 W, 3.2 Cr, 0.4 V, bal Fe.
Heat treat: 243,000 TS; 215,000 YS; 12 El; 37 RA; 50 Rock C.
For punches, hot dies, swaging dies; hotwork.

SERAING PISTON.
M-England; 89 Cu, 9 Zn, 2 Sn.
Cast: 35,000 TS; 16 El.
For piston rings; corrosion resistant.

SERMALLOY A_1.
M-1644; Fe, Co, Ni, Al.
Permanent magnet material.
Coercive force H_c (oersted) 2000 ± 100.
Energy product BH max (MG.Oe) 5.5 ± 0.5.
Remanence Br (gauss) 7400 ± 200.
Saturation field, H_s (oersted, approx.) 6000.

SERMALLOY A₂.
M-1644; Al-Ni-Co.
Permanent magnet material.
Coercive force H_c (oersted) 500.
Energy product BH max (MG.Oe) 5.
Remanence Br (gauss) 14,200 ± 200.
Saturation field H_s (oersted, approx.) 1800.

SERMALLOY A2.
M-1669; 24 Co, 7.5 Al, 12 Ni, 2 Cu, 0.2 Si, bal Fe.
Br: 14,300; H_c 500; BH max. 5,000,000.
For magnetometry.
High residual induction, Alnico type alloy. Permanent magnet.

SERMALLOY A₂ GD.
M-1644, M-1722; Al-Ni-Co.
Permanent magnet material.
Coercive force H_c (oersted) 600.
Energy product BH max (MG.Oe) 5.5 ± 0.5.
Remanence Br (gauss) 14,300.
Saturation field H_s (oersted, approx.) 2000.

SERMALLOY A2GD.
M-1669; 24 Co, 7.5 Al, 12 Ni, 2 Cu, 0.2 Si, bal Fe.
Br: 14,400 Hc 610; BH max 6,600,000.
For magnetometry.
Columnar Alnico type permanent magnet.

SERMALLOY A₂GDH.
M-1669; 24 Co, 7.5 Al, 12 Ni, 2 Cu, 0.2 Si, bal Fe.
Remanence 15,000 gauss, 280 Hc.
For hysteresis drives.
Alnico type permanent magnet.

SERMALLOY AL.
M-1669; 38 Co, 7.5 Al, 14 Ni, 3 Cu, 8 Ti, bal Fe.
Cast: 7400 Remanent induction, 2100 coercive force, 9482 magnetic saturation, 874°C. Curie temperature.
For magnetic and electrical equipment.
Permanent magnet. Magnetically anisotropic.

SEVA.
M-286; 70 Cu, 0.04 As, bal Zn.
Annealed: 40,300-47,000 TS; 18,000-24,000 YS; 60-75 El; 54-73 Brin.
For condensers, and coolers evaporators, juice heaters, vacuum pan tubes for the sugar industry.
Malleable and ductile brass; corrosion resistant.

SEVENTY-THIRTY.
M-Eng.; 70 Cu, 30 Ni.
For turbine blades, condenser tubes; corrosion resistant.

SEYCAST.
M-103; 99 Ni.
For anodes; controlled grain.

SEYMOUR 10A7.
M-103; 10 Ni, 18.5 Zn, 71.5 Cu.
Soft: 54,000 TS; 67 Brin.
1/2 H-temper: 70,000 TS; 144 Brin.
H-temper: 80,000 TS; 165 Brin.
For hollowware, flatware, zipper parts; nickel silver, good formability.

SEYMOUR 12A4.
M-103; 56.5 Cu, 12 Ni, 31.5 Zn.
1/4 H: 65,000 TS; 108 Brin.
Hard: 90,000 TS; 185 Brin.
Spring: 109,000 TS; 222 Brin.
For springs, clips: nickel silver, corrosion resistant.

SEYMOUR 18 AL.
M-103; 18 Ni, 17 Zn, 65 Cu.
Hard: 80,000-120,000 TS; 12-3 El; 185-210 Brin.
Soft: 50,000-55,000 TS; 45-40 El; 75-77 Brin.
For springs, bellows, hardware, welding rod; nickel silver, good formability.

SEYMOUR 265.
M-103; 65 Cu, 35 Zn.
For hardware, plumbing fixtures; yellow brass.

SEYMOUR 270.
M-103; 70 Cu, 30 Zn.
Hard: 76,000 TS; 62,000 YS; 10 El; 159 Brin.
Soft: 47,000 TS; 17,000 YS; 65 El; 62 Brin.
For tanks, cartridge cases, vessels; cartridge brass.

SEYMOUR 280.
M-103; 80 Cu, 20 Zn.
Hard: 91,000 TS; 55,000 YS; 10 El; 190 Brin.
Soft: 46,000 TS; 15,000 YS; 50 El; 57 Brin.
For hardware, fasteners; low brass.

SEYMOUR 285.
M-103; 85 Cu, 15 Zn.
Hard: 84,000 TS; 55,000 YS; 2 El; 170 Brin.
Soft: 43,000 TS; 15,000 YS; 42 El; 56 Brin.
For diaphragms, fasteners; red brass.

SEYMOUR 290.
M-103; 90 Cu, 10 Zn.
Hard: 72,000 TS; 47,000 YS; 6 El; 145 Brin.
Soft: 39,000 TS; 12,000 YS; 40 El; 54 Brin.
For hardware, fasteners; commercial brass.

SEYMOUR 430.
M-103; 5 Sn, 5 Pb, 0.25 P, bal Cu.
For bushings, gears, keys, bearings; leaded P-bronze.

SEYMOUR 971.
M-103; 3 Sn, 0.25 P, bal Cu.
1/2 H-temper: 60,000 TS; 125 Brin.
H-temper: 72,000 TS; 150 Brin.
Spring temper: 87,000 TS; 176 Brin.
For springs, clips, diaphragms, contacts; Phosphor Bronze Grade E.

SEYMOUR 1008.
M-103; 8 Ni, 61.5 Cu, bal Zn.
Soft: 60,000 TS; 40 El.
Hard: 110,000 TS; 7 El.
Spring: 130,000 TS; 2 El.
For cutlery, hardware, trim, appliances, hospital equipment; nickel silver corrosion resistant.

SEYMOUR 1009.
M-103; 61.5 Cu, 8 Ni, 2 Pb, bal Zn.
For keys, hardware, cutlery, trim; leaded nickel silver.

SEYMOUR 1011.
M-103; 65 Cu, 10 Ni, bal Zn.
Soft: 60,000 TS; 40 El.
Hard: 110,000 TS; 7 El.
Spring: 130,000 TS; 2 El.
For cutlery, hardware, trim, appliances, hospital equipment; nickel silver Gr. A, corrosion resistant.

SEYMOUR 1017.
M-103; 71.5 Cu, 10 Ni, bal Zn.
Soft: 55,000 TS.
Hard: 102,000 TS.
Spring: 120,000 TS.
For cutlery, hardware, trim, appliances, hospital equipment; nickel silver corrosion resistant.

SEYMOUR 1091.
M-103; 61.5 Cu, 10 Ni, 1 Pb, bal Zn.
Rolled: 90,000 TS.
For hardware, cutlery, trim, hospital equipment; leaded nickel silver.

SEYMOUR 1092.
M-103; 61.5 Cu, 10 Ni, 2 Pb, bal Zn.
For keys, hardware, cutlery; leaded nickel silver.

SEYMOUR 1095.
M-103; 61.5 Cu, 10 Ni, 0.5 Pb, bal Zn.
For hardware, trim, cutlery, hospital equipment; leaded nickel silver.

SEYMOUR 1211.
M-103; 65 Cu, 12 Ni, bal Zn.
Soft: 60,000 TS; 45 El.
Hard: 115,000 TS; 7 El.
Spring: 136,000 TS; 2 El.
For cutlery, hardware, trim, appliances, hospital equipment; nickel silver corrosion resistant.

SEYMOUR 1214.
M-103; 56.5 Cu, 12 Ni, bal Zn.
Soft: 60,000 TS; 47 El.
Hard: 103,000 TS; 7 El.
Spring: 117,000 TS; 2 El.
For cutlery, hardware, trim, appliances, hospital equipment; nickel silver corrosion resistant.

SEYMOUR 1291.
M-103; 61.5 Cu, 12 Ni, 1 Pb, bal Zn.
For hardware, trim, cutlery; leaded nickel silver.

SEYMOUR 1292.
M-103; 61.5 Cu, 12 Ni, 2 Pb, bal Zn.
For keys, hardware, cutlery, trim; leaded nickel silver.

SEYMOUR 1295.
M-103; 61.5 Cu, 12 Ni, 0.5 Pb, bal Zn.
For hardware, cutlery, hospital equipment; leaded nickel silver.

SEYMOUR 1511.
M-103; 65 Cu, 15 Ni, bal Zn.
Soft: 60,000 TS; 47 El.
Hard: 103,000 TS; 7 El.
Spring: 117,000 TS; 2 El.
For cutlery, hardware, trim, appliances, hospital equipment; nickel silver corrosion resistant.

SEYMOUR 1591.
M-103; 61.5 Cu, 15 Ni, 1 Pb, bal Zn.
For hardware, cutlery, hospital equipment; leaded nickel silver.

SEYMOUR 1811.
M-103; 65 Cu, 18 Ni, bal Zn.
Soft: 60,000 TS; 47 El.
Hard: 103,000 TS; 7 El.
Spring: 117,000 TS; 2 El.
For cutlery, hardware, trim, appliances, hospital equipment; nickel silver corrosion resistant.

SEYMOUR 1814.
M-103; 55 Cu, 18 Ni, bal Zn.
Soft: 70,000 TS; 45 El.
Hard: 123,000 TS; 6 El.
Spring: 145,000 TS; 2 El.
For cutlery, hardware, trim, appliances, hospital equipment; nickel silver corrosion resistant.

SEYMOUR 1815.
M-103; 65 Cu, 18 Ni, 0.15 Ag, bal Zn.
Soft: 60,000 TS; 47 El.
Hard: 103,000 TS; 7 El.
Spring: 117,000 TS; 2 El.
For cutlery, hardware, trim, appliances, hospital equipment; nickel silver corrosion resistant.

SEYMOUR 1817.
M-103; 72 Cu, 18 Ni, bal Zn.
For cutlery, hardware, trim, appliances, hospital equipment; nickel silver.

SEYMOUR 1891.
M-103; 61.5 Cu, 18 Ni, 1 Pb, bal Zn.
For hardware, cutlery, hospital equipment; leaded nickel silver.

SEYMOUR 1895.
M-103; 61.5 Cu, 18 Ni, 0.5 Pb, bal Zn.
For hardware, cutlery, hospital equipment; leaded nickel silver.

SEYMOUR 1897.
M-103; 65 Cu, 18 Ni, 0.15 Ag, 1 Pb, bal Zn.
For hardware, cutlery, hospital equipment; leaded nickel silver.

SEYMOUR 2265.
M-103; 66 Cu, bal Zn.
1/2 H-temper: 60,000 TS.
Hard temper: 73,000 TS.
Spring: 90,500 TS.
For electrical components, hardware, watch parts, jewelry; high brass.

SEYMOUR 2270.
M-103; 70 Cu, bal Zn.
1/2 H-temper: 62,000 TS; 121 Brin.
Hard temper: 76,000 TS; 159 Brin.
Spring: 95,400 TS; 190 Brin.
For electrical components, hardware, jewelry, watches, lamps; cartridge brass.

SEYMOUR 2280.
M-103; 80 Cu, bal Zn.
1/2 H-temper: 60,000 TS; 117 Brin.
Hard temper: 72,500 TS; 150 Brin.
Spring: 89,000 TS; 185 Brin.
For electrical components, hardware, jewelry, watches, lamps; low brass.

SEYMOUR 2285.
M-103; 85 Cu, bal Zn.
1/2 H-temper: 56,000 TS; 114 Brin.
Hard temper: 67,500 TS; 139 Brin.
Spring: 82,000 TS; 165 Brin.
For electrical components, hardware, jewelry, watches, lamps; rich low brass.

SEYMOUR 2287.
M-103; 87 Cu, bal Zn.
1/2 H-temper: 54,000 TS; 116 Brin.
Hard temper: 65,000 TS; 137 Brin.
Spring: 75,000 TS; 156 Brin.
For electrical components, hardware, jewelry, watches, lamps; Nugild.

SEYMOUR 2290.
M-103; 90 Cu, bal Zn.
1/2 H-temper: 52,000 TS; 104 Brin.
Hard temper: 61,500 TS; 125 Brin.
Spring: 73,000 TS; 153 Brin.
For electrical components, hardware, jewelry, watches, lamps; commercial bronze.

SEYMOUR 2295.
M-103; 95 Cu, bal Zn.
1/2 H-temper: 47,000 TS; 83 Brin.
Hard temper: 54,500 TS; 110 Brin.
Spring: 64,000 TS; 130 Brin.
For electrical parts, hardware, jewelry, watches, lamps; gilding bronze.

SEYMOUR 2401.
M-103; 90 Cu, 0.5 Sn, 9.5 Zn.
Soft: 38,000 TS; 12,000 YS; 45 El.
For weather stripping; brass, corrosion resistant.

SEYMOUR 2402.
M-103; 90 Cu, 2 Sn, 8 Zn.
Soft: 40,000 TS; 42 El.
For hardware, costume jewelry; corrosion resistant.

SEYMOUR 2405.
M-103; 90.5 Cu, 5 Sn, 4.5 Zn.
For hardware, costume jewelry; corrosion resistant.

SEYMOUR 2451.
M-103; 95 Cu, 1 Sn, 4 Zn.
For hardware, weather stripping; corrosion resistant.

SEYMOUR 2875.
M-103; 87.5 Cu, 2.25 Sn, 10.25 Zn.
Soft: 45,000 TS; 57 El.
Hard: 102,000 TS; 7 El.
For hardware, costume jewelry; corrosion resistant.

SEYMOUR 2876.
M-103; 87 Cu, bal Zn.
1/2 H-temper: 54,000 TS; 116 Brin.
Hard temper: 65,000 TS; 138 Brin.
Spring: 75,000 TS; 160 Brin.
For electrical components, hardware, watches, jewelry; high brass.

SEYMOUR 2877.
M-103; 86.5 Cu, 0.5 Sn, 13 Zn.
For hardware, chains, fasteners, costume jewelry; jeweler's bronze, corrosion resistant.

SEYMOUR 2963.
M-103; 63.5 Cu, 0.75 Pb, bal Zn.
1/4 H-temper: 60,000 TS; 50,000 YS; 23 El; 117 Brin.
Hard: 73,000 TS; 60,000 YS; 8 El; 150 Brin.
Spring: 90,500 TS; 185 Brin.
For hardware, hinges, watch backs, fasteners; low leaded brass.

SEYMOUR 2964.
M-103; 64 Cu, 2 Pb, bal Zn.
1/4 H-temper: 60,000 TS; 117 Brin.
Hard: 73,000 TS; 150 Brin.
Spring: 90,500 TS; 185 Brin.
For clock parts, gears, wheels, channel plates; high leaded brass, corrosion resistant.

SEYMOUR 2965.
M-103; 62 Cu, 2.5 Pb, bal Zn.
1/4 H-temper: 60,000 TS; 117 Brin.
Hard: 73,000 TS; 150 Brin.
Spring: 90,500 TS; 185 Brin.
For clock parts, gears, wheels, channel plates; extra high leaded brass, free-cutting.

SEYMOUR 5770.
M-103; 70 Cu, 30 Ni.
For process equipment; cupro-nickel, corrosion resistant.

SEYMOUR 5778.
M-103; 22 Ni, bal Cu.
For process equipment; cupro-nickel, corrosion resistant.

SEYMOUR 5780.
M-103; 80 Cu, 20 Ni.
For process equipment; cupro-nickel, corrosion resistant.

SEYMOUR 5788.
M-103; 10 Ni, bal Cu.
For process equipment; cupro-nickel, corrosion resistant.

SEYMOUR 5794.
M-103; 3.5 Ni, bal Cu.
For process equipment; cupro-nickel, corrosion resistant.

SEYMOUR 5798.
M-103; 2.0 Ni, bal Cu.
For process equipment; cupro-nickel, corrosion resistant.

SEYMOUR 8120.
M-103; 0.05 Ag, bal Cu.
For engraving plates; engraver's copper.

SEYMOUR 8121.
M-103; 99.9 Cu min, 0.04 oxygen.
1/4 H-temper: 38,000 TS; 30,000 YS; 25 El; 64 Brin.
1/2 H-temper: 42,000 TS; 36,000 YS; 14 El; 75 Brin.
Hard: 50,000 TS; 45,000 YS; 6 El; 83 Brin.
For gutters, radio parts, terminals, ball floats; tough pitch copper.

SEYMOUR 8122.
M-103; 99.92 min Cu.
1/4 H-temper: 38,000 TS; 30,000 YS; 35 El; 64 Brin.
Hard: 50,000 TS; 45,000 YS; 12 El; 83 Brin.
For bus bars and bus conductors, wave guides; OFHC copper.

SEYMOUR 9230.
M-103; 8 Sn, 0.30 P, bal Cu.
For springs, fuse clips, diaphragms; P-bronze.

SEYMOUR 9231.
M-103; 6 Sn, 0.30 P, bal Cu.
For instruments; special Fourdrinier wire, P-bronze.

SEYMOUR 9370.
M-103; 3 Sn, 0.23 P, bal Cu.
For springs, diaphragms, electrical contacts; Fourdrinier wire, P-bronze.

SEYMOUR 9430.
M-103; 5 Sn, 5 Pb, 0.15 P, bal Cu.
For bearings, bushings, liners; bearing bronze.

SEYMOUR 9444.
M-103; 4 Sn, 4 Pb, 4 Zn, 0.07 P, bal Cu.
For bearings, bushings; leaded bronze.

SEYMOUR 9494.
M-103; 5 Sn, 1 Pb, 0.07 P, bal Cu.
For bearings, liners, sleeves; leaded P-bronze.

SEYMOUR 9910.
M-103; 10 Sn, 0.10 P, bal Cu.
For springs, discs, diaphragms, bellows; P-bronze.

SEYMOUR 9928.
M-103; 8 Sn, 0.10 P, bal Cu.
For springs, thermostats, diaphragms, microphones; P-bronze.

SEYMOUR 9950.
M-103; 5 Sn, 0.15 P, bal Cu.
1/2 H-temper: 55,000-70,000 TS.
Hard temper: 72,000-87,000 TS.
For springs, switches, diaphragms, clutch discs, chemical hardware; P-bronze, 16% electrical conductivity.

SEYMOUR 9971.
M-103; 3 Sn, 0.10 P, bal Cu.
1/2 H-temper: 55,000-65,000 TS.
Hard: 67,000-77,000 TS.
For low duty springs, diaphragms, electrical contacts; P-bronze, 22% electrical conductivity.

SEYMOUR 9985.
M-103; 1.35 Sn, 0.07 P, bal Cu.
For low duty springs, diaphragms, electrical contacts; P-bronze.

SEYMOUR NO. 5.
M-103; 5 Ni, 34 Zn, 61 Cu.
Hard: 135,000 TS; 2 El.
Soft: 50,000 TS.
For screw machine parts; elec. cond. 12% (wire form).

SEYMOUR NO. 5A1.
M-103; 5 Ni, 27 Zn, bal Cu.
Hard: 100,000 TS; 3 El; 185 Brin.
For hardware, springs; nickel silver, corrosion resistant.

SEYMOUR NO. 10A1.
M-103; 10 Ni, 23 Zn, 67 Cu.
Hard: 109,000 TS; 1.5 El; 188 Brin.
Soft: 52,000 TS; 40 El; 70 Brin.
For hardware; elec. cond. 9% (sheet form); 10% "Nickel Silver."

SEYMOUR NO. 10X1.
M-103; 10 Ni, 28 Zn, 1 Pb, bal Cu.
Hard: 75,000 TS; 5 El; 176 Brin.
Soft: 54,000 TS; 30 El; 75 Brin.
For hardware, screw machine products; nickel silver, free-cutting, corrosion resistant.

SEYMOUR NO. 10X2.
M-103; 12 Ni, 25 Zn, 2 Pb, bal Cu.
Hard: 85,000 TS; 5 El; 172 Brin.
For marine hardware, screw machine products; nickel silver, free-cutting, corrosion resistant.

SEYMOUR NO. 12A1.
M-103; 12 Ni, 23 Zn, bal Cu.
Hard: 106,000 TS; 2 El; 210 Brin.
For marine hardware, fasteners; nickel silver, corrosion resistant.

SEYMOUR NO. 12X1.
M-103; 12 Ni, 1 Pb, 25 Zn, bal Cu,
Hard: 90,000 TS; 5 El; 172 Brin.
Soft: 56,000 TS; 35 El; 75 Brin.
For screw machine products, marine hardware; corrosion resistant, free-cutting.

SEYMOUR NO. 12X2.
M-103; 12 Ni, 26 Zn, 2 Pb, bal Cu,
Hard: 95,000 TS; 3 El; 195 Brin.
Soft: 57,000 TS; 45 El; 76 Brin.
For marine hardware, keys, fasteners; nickel silver, free-cutting, corrosion resistant.

SEYMOUR NO. 15A.
M-103; 15 Ni, 21 Zn, 64 Cu.
Hard: 93,000 TS; 5 El; 190 Brin.
Soft: 58,000 TS; 40 El; 75 Brin.
For hardware; corrosion resistant.

SEYMOUR NO. 15 AL.
M-103; 15 Ni, 65 Cu, bal Zn.
Hard: 97,000 TS; 195 Brin.
For hardware, marine parts; nickel silver, corrosion resistant.

SEYMOUR NO. 15B.
M-103; 15 Ni, 27 Zn, 58 Cu.
Hard: 100,000 TS; 2 El; 204 Brin.
Soft: 55,000 TS; 36 El; 77 Brin.
For hardware; corrosion resistant.

SEYMOUR NO. 15X1.
M-103; 15 Ni, 1 Pb, 23 Zn, bal Cu.
Hard: 85,000 TS; 10 El; 150 Brin.
Soft: 58,000 TS; 33 El; 80 Brin.
For marine hardware, screw machine products; corrosion resistant, free-cutting.

SEYMOUR NO. 15X2.
M-103; 15 Ni, 23 Zn, 2 Pb, bal Cu.
Hard: 92,000 TS; 4 El; 185 Brin.
Soft: 58,000 TS; 45 El; 75 Brin.
For hardware, screw machine products; nickel silver, free-cutting, corrosion resistant.

SEYMOUR NO. 18A3.
M-103; 18 Ni, 17 Zn, 65 Cu.
Hard: 100,000 TS; 3 El; 190 Brin.
Soft: 54,000 TS; 40 El; 78 Brin.
For hardware; corrosion resistant.

SEYMOUR NO. 18A4.
M-103; 18 Ni, 27 Zn, bal Cu.
Hard: 115,000 TS; 2 El; 210 Brin.
Soft: 65,000 TS; 45 El; 83 Brin.
For springs, marine hardware; nickel silver, corrosion resistant.

SEYMOUR NO. 18A7.
M-103; 18 Ni, 7 Zn, 75 Cu.
Hard: 85,000 TS; 4 El; 180 Brin.
Soft: 50,000 TS; 32 El; 65 Brin.
For hardware; corrosion resistant.

SEYMOUR NO. 18X1.
M-103; 18 Ni, 1 Pb, 19 Zn, 62 Cu.
Hard: 85,000 TS; 13 El; 150 Brin.
For screw machine products; 18% "Leaded Nickel Silver;" free cutting.

SEYMOUR NO. 25.
M-103; 25 Ni, 19 Zn, 56 Cu.
Hard: 110,000 TS; 4 El; 204 Brin.
Soft: 72,000 TS; 30 El; 110 Brin.
For hardware; corrosion resistant.

SEYMOUR NO. 287.
M-103; 13 Zn, bal Cu.
Hard: 65,000 TS; 137 Brin.
For jewelry, trim; "Nu-Gild."

SEYMOUR NO. 444.
M-103; 88 Cu, 4 Sn, 4 Pb, 4 Zn.
Hard: 60,000 TS; 20 El.
Soft: 48,000 TS.
For bearings, studs, shafts, gears, bolts; free-cutting, Grade B2 phosphor bronze.

SEYMOUR NO. 494.
M-103; 94 Cu, 5 Sn, 1 Pb.
Hard: 61,500 TS; 20 El.
Soft: 50,000 TS; 40 El.
For bolts, screws, bearings, gears, shafts; grade B1 phosphor bronze.

SEYMOUR NO. 780.
M-103; 20 Ni, 80 Cu.
Hard: 85,000 TS; 2 El; 165 Brin.
Soft: 50,000 TS; 30 El; 72 Brin.
For condenser tubes; corrosion resistant.

SEYMOUR NO. 785.
M-103; 15 Ni, 85 Cu.
Hard: 70,000 TS; 3 El; 160 Brin.
Soft: 45,000 TS; 30 El; 70 Brin.
For condenser tubes; corrosion resistant.

SEYMOUR NO. 895.
M-103; 89.5 Cu, 10.5 Sn, 0.10 min P.
For welding rod to build up bearings; Grade C arc welding.

SEYMOUR NO. 910.
M-103; 10 Sn, 0.2 P, bal Cu.
Soft: 65,000 TS; 65 El; 75 Brin.
Spring: 122,000 TS; 5 El; 240 Brin.
For springs, diaphragms, fasteners, cams, bellows, shafts; Gr. D. P-bronze.

SEYMOUR NO. 928.
M-103; 8 Sn, 0.25 P, bal Cu.
Hard: 112,000 TS; 38 El; 240 Brin.
Soft: 60,000 TS; 55 El; 75 Brin.
For springs, bellows, diaphragms, hardware, bolts; Grade C, P-bronze, tough.

SEYMOUR NO. 950.
M-103; 5 Sn, 0.25 P, bal Cu.
Hard: 100,000 TS; 1.5 El; 210 Brin.
Soft: 50,000 TS; 50 El; 75 Brin.
For springs, diaphragms, bolts, fasteners, hardware; Grade A, P-bronze, tough.

SEYMOUR NO. 9225.
M-103; 91.7 Cu, 8.25 Sn, 0.25 P.
For welding rod for bronze, brass and steel; Grade C arc welding.

SEYMOUR NO. 9522.
M-103; 95 Cu, 5 Sn, 0.22 min P.
For welding rod for general purpose; Grade A arc welding.

SEYMOURCO.
M-103; 34 Zn, 5 Ni, bal Cu.
For costume jewelry; nickel silver.

SEYMOURITE.
M-103; 64 Cu, 18 Ni, 18 Zn.
For ornaments, electrical equipment; high corrosion resistant.

SEYMOUR NICKEL SILVER GRADE A.
M-103; 23 Zn, 19 Ni, bal Cu.
100,000 TS.
For costume jewelry; corrosion resistant.

SF 67.
M-1724; 0.70 C, 13.0 Cr, bal Fe.
Razor blade strip.

SFC 2/61/0.
M-1739; Iron (some Cu-C-S, sintered).
Free machining grade of FC2/61/0.

S-G 9.
M-500; 0.26 C, 0.75 Mn, bal Fe.
Normalized: 70,000 TS; 36,000 YS; 22 El; 30 RA.
For general usage; case hardened.

S-G 13.
M-500; 0.25 C, 0.50 Mn, bal Fe.
Normalized: 60,000 TS; 30,000 YS; 24 El; 35 RA.
For general usage; case-hardened.

S-G 17.
M-500; 0.28 C, 0.6 Mn, bal Fe.
Normalized: 65,000 TS; 35,000 YS; 20 El; 30 RA.
For general usage; case hardened.

S-G 30.
M-1266; 6-8 Cu, 1.5-2.5 Si, 1.0-1.5 Zn, 0.2 Ti, bal Al.
Cast.
For general light alloy castings; modified Alcoa 212 alloy.

S-G 80-10-10.
M-1266; 9.25-10.5 Sn, 9-11 Pb, 0.75 max Zn, bal Cu.
Cast: 25,000-35,000 TS; 15,000-22,000 YS; 6-12 El; 50-60 Brin.
For bearings, acid resistant castings; corrosion resistant.

S-G 81-3-6-10.
M-1266; 2.5-3.5 Sn, 5-7 Pb, 9-11 Zn, bal Cu.
Cast: 28,000-36,000 TS; 13,000-16,000 YS; 15-28 El; 50-60 Brin.
For plumbing goods, pressure valves, fittings; free-cutting.

S-G 83-4-6-7.
M-1266; 3.5-4.2 Sn, 5.7-6.7 Pb, 6-8 Zn, bal Cu.
Cast: 30,000-35,000 TS; 12,000-16,000 YS; 15-25 El; 50-60 Brin.
For plumbing goods, valves, gas and water fittings; red brass, free-cutting.

S-G 83-7-7-3.
M-1266; 6.5-7.5 Sn, 6.5-8.0 Pb, 2.5-4.0 Zn, bal Cu.
Cast: 28,000-35,000 TS; 15,000-20,000 YS; 10-15 El; 55-65 Brin.
For bearings, bushings; free-cutting.

S-G 85-5-5-5.
M-1266; 4-6 Sn, 4-6 Pb, 4-6 Zn, bal Cu.
Cast: 30,000-40,000 TS; 15,000-25,000 YS; 15-33 El; 55-65 Brin.
For gears, valve bodies, plumbing; free-cutting bronze.

S-G 90.
M-1266; 3.7-4.7 Cu, 2.7-3.7 Si, 1 Fe, 0.5 Zn, 0.4 Ni, bal Al.
Cast.
For general light alloy castings; modified Alcoa 108 alloy.

SG-100 NICKEL.
M-1573; 99.93 Ni + Co, 0.18 Co, 0.004 Cu, 0.02 Fe, 0.0025 S, 0.007 C.
Annealed Strip: 50,000 TS; 12,000 YS; 45 El; 25 Rock 30 T.
For cathode sleeves and shields, heat exchanger fins, getter flags, passive cathodes, magnetostrictive transducers.
High purity nickel from powder metallurgy.

SG-101 NICKEL.
M-1573; 99.9 Ni + Co, 0.005 Cu, 0.01 Fe, 0.01 C, 0.0004 S.
High purity nickel strip.

S.G. IRON.
 M-494, M-1285; 3.3 C, 2.5 Si, 1.5 Ni, 0.05 Mg, bal Fe.
 Cast: 160-300 Brin.
 For high strength and tough castings; nodular cast iron.

SGM 4.
 M-912; 0.43 C, 4.25 Cr, 0.50 Mo, 2.0 V, 4.25 W, 4.25 Co, bal Fe.
 Hot work tool steel; for hot extruding dies, forging dies, brass casting molds.
 W.-Nr. 2678; AISI H19.

S.G.N1-RESIST D2C(MODIFIED).
 M-68; 2.2-2.6 C, 1.9-2.6 Si, 3.7-4.4 Mn, 21-24 Ni, 0.2 max Cr, 0.1 max P, bal Fe.
 For pumps, valves, compressors, turbo-expanders.
 Austenitic iron for subzero services. High notch toughness.

S.G.N1-RESIST D2M.
 M-68; 2.2-2.6 C, 1.9-2.6 Si, 3.7-4.4 Mn, 21-24 Ni, 0.06-0.12 Mg, 0.2 Cr, 0.1 max P, bal Ni.
 For pumps, valves, compressors, turbo-expanders, low temperature equipment
 Excellent cryogenic toughness and ductility.

SGS-TANTALUM.
 M-1624; 100 Ta + Yttrium.
 300% stronger than ordinary Ta at 1650°C.
 For electronic and electrical components, vacuum furnaces, aerospace and nuclear systems.
 Fine controlled grain. No oxygen embrittlement caused by grain coarsening to 2300°C.

S-G TYPE 12.
 M-1266; 6.0-8.5 Cu, 1.0-3.0 Si, 0.4 Mn, 0.8-1.5 Fe, 0.05 Mg, 0.2-2.2 Zn, 0.4 Ni, 0.2 Ti, bal Al.
 Sand cast: 19,000 TS.
 For low strength light weight castings.

S-G TYPE 13.
 M-1266; 0.6 Cu, 9.5-13.0 Si, 2.2 Fe, 0.1 Mg, 0.3 Mn, 0.5 Ni, bal Al.
 Die cast: 37,000 TS; 18,000 YS; 1.8 El.
 For meter cases, switch boxes, low stressed castings; thin wall castings.

S-G TYPE 43.
 M-1266; 0.1-0.6 Cu, 4.5-6.0 Si, 0.2 Mn, 0.8-2.0 Fe, 0.3 Zn, 0.3 Ni, 0.2 Ti, bal Al.
 Die cast: 30,000 TS; 14,000 YS; 5.0 El; 40 Brin.
 For marine castings, manifolds, cooking utensils; corrosion resistant.

S-G TYPE 81.
 M-1266; 6.0-8.5 Cu, 1.0-3.5 Si, 0.4 Mn, 2-3 Fe, 1.8-2.0 Zn, 0.2 Ti, 0.4 Ni, bal Al.
 Die cast: 33,000 TS; 25,000 YS; 1 El; 80 Brin.
 For instrument housings, cases; high fluidity.

S-G TYPE 85.
 M-1266; 3.0-4.5 Cu, 4.5-9.5 Si, 0.4 Mn, 1.3-2.3 Fe, 0.6-1.0 Zn, 0.4 0.2 Cr, bal Al.
 Die cast: 38,000 TS; 22,000 YS; 2.5 El; 85 Brin.
 For general die castings; high strength.

S-G TYPE 108.
 M-1266; 2.5-4.7 Cu, 2.5-4.5 Si, 0.4 Mn, 1.1 Fe, 0.3-1.0 Mg, 0.2 Ti, Ni, bal Al.
 Sand cast: 21,000 TS; 14,000 YS; 2.5 El; 60 Brin.
 For utility sand castings; good castability and machinability.

S-G TYPE 122.
 M-1266; 9.2-10.8 Cu, 1.0-1.7 Si, 0.2 Mn, 1.5 Fe, 0.2 Mg, 0.5 Zn, 0.3 Ni, 0.2 Ti, bal Al.
 Heat treated: 41,000-48,000 TS; 36,000-40,000 YS; 0.5 El; 115-140 Brin.
 For pistons, valve guides, bearings; age hardenable, wear resistant.

S-G TYPE 132.
 M-1266; 0.5-2.0 Cu, 11-13 Si, 0.6 Mn, 1.2 Fe, 1.2 Mg, 0.2 Zn, 0.5-3.0 Ni, 0.2 Ti, bal Al.
 Heat treated: 36,000-47,000 TS; 28,000-43,000 YS; 0.5 El; 105-125 Br
 For automotive engine pistons and sleeves; low expansion alloy.

S-G TYPE 138.
 M-1266; 9-11 Cu, 3.5-4.5 Si, 1.2 Fe, 0.3 Mn, 0.2 Mg, 0.5 Zn, bal Al.
 Permanent mold: 30,000 TS; 24,000 YS; 15 El; 90 Brin.
 For pistons, cylinder heads; similar to Alcoa 138.

S-G TYPE 142.
 M-1266; 3.5-4.5 Cu, 0.7 Si, 0.2 Mn, 0.9 Fe, 1.6 Mg, 0.5 Zn, 2.0 Ni, 0.2 Ti, bal Al.
 Heat treated: 27,000-47,000 TS; 18,000-42,000 YS; 0.5-1.0 El; 70-110 Brin.
 For pistons, cylinder heads; age-hardenable, strength at elevated temp.

S-G TYPE 195.
 M-1266; 3-5 Cu, 1.2-2.5 Si, 1.0 Fe, 0.5 Mn, 0.4 Zn, 0.2 Ti, bal Al.
 Heat treated: 32,000-40,000 TS; 16,000-34,000 YS; 8.5-1.5 El; 60-95 Brin.
 For crank cases, aircraft components; age hardenable.

S-G TYPE 212.
 M-1266; 6.0-8.5 Cu, 1-3 Si, 0.4 Mn, 0.8-1.5 Fe, 0.2-2.2 Zn, bal Al.
 Sand cast: 19,000-24,000 TS; 11,000-14,000 YS; 2.5-1.0 El; 50-65 Bri
 For general utility castings; similar to Alcoa 212.

S-G TYPE 214.
 M-1266; 0.1 Cu, 0.4 Si, 0.5 Fe, 3.25-4.5 Mg, 0.25 Zn, 0.2 Ti, bal Al.
 Sand cast: 25,000 TS; 12,000 YS; 9 El; 50 Brin.
 For light weight castings; similar to Alcoa 214.

S-G TYPE 218.
M-1266; 0.2 Cu, 0.3 Si, 0.3 Mn, 1.8 Fe, 7.5-8.5 Mg, 0.1 Zn, bal Al.
Die cast: 42,000 TS; 23,000 YS; 7.0 El; 75 Brin.
For airplane wheel flanges and brake shoes; similar to Alcoa 218.

S-G TYPE 220.
M-1266; 0.2 Cu, 0.2 Si, 9.5-10.6 Mg, 0.1 Zn, bal Al.
Heat treated: 46,000 TS; 25,000 YS; 14 El; 75 Brin.
For railway and bus equipment; similar to Alcoa 220; age hardenable.

S-G TYPE 355.
M-1266; 1.0-1.8 Cu, 4.5-5.5 Si, 0.5 Mg, 0.2 Ti, 0.2 Mn, bal Al.
Heat treated: 38,000-43,000 TS; 25,000-36,000 YS; 4-0.5 El; 80-90 Br
For aircraft engine castings, cylinder heads; similar to Alcoa 355; age hardenable.

S-G TYPE 356.
M-1266; 0.3 Cu, 0.5-7.5 Si, 0.4 Mn, 0.5 Fe, 0.3 Mg, bal Al.
Heat treated: 32,000-40,000 TS; 24,000-30,000 YS; 5-2 El; 70-90 Brin
For leakproof castings; similar to Alcoa 356; age hardenable.

S-G TYPE A108.
M-1266; 4-5 Cu, 5-6 Si, 1 Fe, 0.3 Mn, 0.2 Ti, 0.5 Zn, bal Al.
Permanent mold: 28,000 TS; 16,000 YS; 2 El; 65 Brin.
For utility castings; intricate shapes.

S-G TYPE AXS-679.
M-1266; 3-4 Cu, 4.5-9.5 Si, 1.3 Fe, 0.6 Zn, 0.5 Ni, 0.2 Cr, bal Al.
Die cast.
For general die castings; corrosion resistant.

S-G TYPE B195.
M-1266; 4.5 Cu, 2-3 Si, 0.05-4.0 Mn, 1.2 Fe, 0.5 Zn, 0.2 Ni, 0.2 Ti, bal Al.
Heat treated: 37,000-42,000 TS; 19,000-28,000 YS; 9-6 El; 75-90 Brin
For aircraft castings; age hardenable.

S-G TYPE C113.
M-1266; 6.0-8.5 Cu, 1.0-6.0 Si, 0.4 Mn, 1.5 Fe, 1.0-2.5 Zn, 0.2 Ti, 0.2 Ni, bal Al.
Permanent mold: 30,000 TS; 24,000 YS; 1 El; 70 Brin.
For utility castings; similar to Alcoa C113.

S-G TYPE SG 300.
M-1266; A1 alloy.
Cast: 32,000 min TS; 5 min El.
For light alloy castings.

S-G VALVE BRONZE.
M-1266; 5.5-6.5 Sn, 1-2 Pb, 3-5 Zn, bal Cu.
Cast: 34,000 max TS; 22 El.
For valve bodies; composition M alloy, free-cutting.

S-G YELLOW BRASS.
M-1266; 0.5-1.2 Sn, 3-5 Pb, 0.5 max Fe, 78-81 Cu, bal Zn.
Cast: 28,000-35,000 TS; 10,000-15,000 YS; 25-35 El; 25-35 Brin.
For plumbing fixtures, ferrules, battery terminals; free-cutting brass.

SHAKUDO (SHAKDO).
M-Japan; 96-94 Cu, 3.7-4.2 Au, 1.6-0.1 Ag, traces of Pb, Fe, As.
For ornamental parts; corrosion resistant.

SHARALLOY.
M-1071; 0.10 C, 0.40 Mn, Si, Cb, bal Fe.
HSLA hot rolled sheet and strip.
60,000 psi min TS; 45,000 psi min YS; 22 min El.
SAE 945 A, C.

SHARALLOY 45.
M-1071; 0.10 C, 0.40 Mn, Si, 0.01 min Cb, bal Fe.
HSLA hot rolled sheet and strip.
60,000 psi min TS; 45,000 psi min YS; 25 min El.
For automotive and structural applications.
SAE 945 X.

SHARALLOY 50.
M-1071; 0.16 C, 0.40 Mn, Si, 0.01 min Cb, bal Fe.
HSLA hot rolled sheet and strip.
65,000 psi min TS; 50,000 psi min YS; 22 min El.
For automotive and structural applications.
SAE 950 X.

SHARALLOY 55.
M-1071; 0.16 C, 0.50 Mn, Si, 0.01 min Cb, bal Fe.
HSLA hot rolled sheet and strip.
70,000 psi min TS; 55,000 psi min YS; 20 min El.
For automotive and structural applications.
SAE 955 X.

SHARALLOY 60.
M-1071; 0.16 C, 0.50 Mn, Si, 0.01 min Cb, bal Fe.
HSLA hot rolled sheet and strip.
75,000 psi min TS; 60,000 psi min YS; 18 min El.
For automotive and structural applications.
SAE 960 X.

SHARON 0126-8.
M-1071; 1.21-1.35 C, 0.10-0.25 Mn, 0.10-0.25 Si, 0.10-0.25 Cr, bal Fe.
Sheet or strip; usually for band saws.

SHARON 0127-5.
M-1071; 1.10-1.25 C, 0.30-0.50 Mn, 0.18-0.33 Si, 0.25-0.5 Cr, bal Fe.
Sheet and strip; for razor blades.

SHARON 0130-5.
M-1071; 1.20-1.50 C, 12.0-15.0 Mn, 0.55 max Si, bal Fe.
Sheet, plate or strip; for steel helmets.
(Hadfield Helmet Steel).

SHARON 0136-5.
M-1071; 0.90-1.10 C, 0.15-0.40 Mn, 0.20-0.35 Si, 0.55-0.70 Mo, bal Fe.
Sheet and strip; for hack saw blades.

SHARON 0137-5.
M-1071; 0.50-0.60 C, 1.30-1.65 Mn, 0.15-0.30 Si, 0.10-0.20 Mo, bal Fe.
Sheet, plate or strip; for hand shears.

SHARON 0141-8.
M-1071; 0.70-0.81 C, 0.30-0.45 Mn, 0.15-0.30 Si, 2.25-2.65 Ni, bal Fe.
Sheet, plate or strip; for large saw blades.

SHARON 0143-5.
M-1071; 0.08 max C, 0.30-0.45 Mn, 0.10-0.25 Si, Ti = 5 x C min, bal Fe.
Sheet, plate or strip; for nuclear applications.

SHARON 0166-5.
M-1071; 0.79-0.90 C, 0.20-0.35 Mn, 0.20-0.35 Si, 0.35-0.55 Cr, 0.60-0.90 Ni, bal Fe.
Typical use: saw bar.

SHARON 0168-5.
M-1071; 0.10-0.15 C, 0.50-0.70 Mn, 0.10-0.20 Si, 0.20-0.40 Cr, 0.20-0.40 Ni, bal Fe.
Typical use: needle bearing.

SHARON 0170-6.
M-1071; 0.95-1.10 C, 0.30-0.50 Mn, 0.10-0.25 Si, 0.40-0.60 Cr, 0.15-0.25 Mo, bal Fe.
Typical use: cutlery.

SHARON 0175-6.
M-1071; 0.70-0.81 C, 0.30-0.45 Mn, 0.15-0.30 Si, 2.25-2.65 Ni, 0.05-0.10 V, bal Fe.
Typical use: saw chain links.

SHARON 0181-5.
M-1071; 0.60-0.70 C, 0.30-0.50 Mn, 0.20-0.35 Si, 0.25-0.40 Cr, 1.2-1.5 Ni, 0.08-0.15 Mo, bal Fe.
Typical use: circular saws.

SHARON 0182-5.
M-1071; 0.61-0.72 C, 0.35-0.55 Mn, 0.20-0.35 Si, 0.40-0.60 Cr, 0.60-0.90 Ni, 0.10-0.20 Mo, bal Fe.
Typical use: saw chain links.

SHARON 0182-5EF.
M-1071; 0.62-0.71 C, 0.35-0.50 Mn, 0.20-0.30 Si, 0.60-0.90 Ni, 0.40-0.60 Cr, 0.10-0.20 Mo, bal Fe.
Hardenable to 58-63 Rc.
Typical use: chain saw cutters and drive links.

SHARON 0183-5.
M-1071; 0.69-0.80 C, 0.25-0.40 Mn, 0.15-0.30 Si, 0.10-0.25 Cr, 0.45-0.75 Ni, 0.10-0.20 Mo, bal Fe.
Typical use: saw bars.

SHARON 0184-5.
M-1071; 0.70-0.81 C, 0.35-0.55 Mn, 0.20-0.35 Si, 0.40-0.60 Cr, 0.60-0.90 Ni, 0.10-0.20 Mo, bal Fe.
Typical use: circular saws.

SHARON 0186-5.
M-1071; 0.64-0.75 C, 0.40-0.60 Mn, 0.20-0.35 Si, 0.30-0.50 Cr, 0.70-1. Ni, 0.08-0.15 Mo, bal Fe.
Typical use: circular saws.

SHARON 0188-5.
M-1071; 0.79-0.90 C, 0.25-0.40 Mn, 0.15-0.30 Si, 0.10-0.25 Cr, 0.60-0.90 Ni, 0.13-0.20 Mo, bal Fe.
Typical use: circular saws.

SHARON 0190-6.
M-1071; 0.28-0.33 C, 0.45-0.65 Mn, 0.45-0.75 Si, 1.0-1.5 Cr, 0.40-0.60 Mo, 0.20-0.30 V, bal Fe.
Typical use: AMS 6385 aircraft brakes.

SHARON 0191-6.
M-1071; 0.95-1.10 C, 0.20-0.40 Mn, 0.10-0.25 Si, 0.40-0.60 Cr, 0.10-0.20 Mo, 0.10-0.20 V, bal Fe.
Typical use: cutlery.

SHARON 3100 SERIES.
M-1071; 0.13-0.53 C, 0.40-0.90 Mn, 0.20-0.35 Si, 1.1-1.4 Ni, 0.55-0.90 Cr, bal Fe.
(Composition as desired). Oil hardenable.
Typical uses: axles, drive shafts, nuts & bolts, tubing, valve tips, washers.

SHARON 4000 SERIES.
M-1071; 0.09-0.50 C, 0.70-1.0 Mn, 0.20-0.35 Si, 0.15-0.30 Mo, bal Fe.
(Composition as desired). Oil hardenable.
Typical uses: springs (flat and coiled), mower blades, pruning shear blades, clutch plates.

SHARON 4100 SERIES A.
M-1071; 0.18-0.40 C, 0.49-0.90 Mn, 0.20-0.35 Si, 0.40-1.10 Cr, 0.08-0.25 Mo, bal Fe.
(Composition as desired). Oil hardenable.
Typical uses: seat belt hardware, fittings.

SHARON 4100 SERIES B.
M-1071; 0.38-0.65 C, 0.75-1.0 Mn, 0.20-0.35 Si, 0.70-1.10 Cr, 0.15-0.35 Mo, bal Fe.
(Composition as desired). Oil hardenable.
Typical uses: aircraft couplings, landing gear, trencher knives.

SHARON 4112.
M-1071; 0.10-0.15 C, 0.30-0.60 Mn, 0.15-0.30 Si, 0.30-0.50 Cr, 0.15-0.25 Mo, bal Fe.
Typical use: wheel bearing.

SHARON 5100 SERIES A.
M-1071; 0.13-0.33 C, 0.70-0.90 Mn, 0.20-0.35 Si, 0.70-1.10 Cr, bal Fe.
(Composition as desired). Oil hardenable.
Typical uses: fasteners, case hardened parts.

SHARON 5100 SERIES B.
M-1071; 0.30-0.51 C, 0.60-0.95 Mn, 0.20-0.35 Si, 0.70-1.15 Cr, bal Fe.
(Composition as desired). Oil hardenable.
Typical uses: transmission gears, roller chain, spline shafts.

SHARON 5100 SERIES C.
M-1071; 0.48-0.64 C, 0.70-1.0 Mn, 0.20-0.35 Si, 0.70-0.90 Cr, bal Fe.
(Composition as desired). Oil hardenable.
Typical uses: coil and flat springs, cutlery.

SHARON 6100 SERIES A.
M-1071; 0.16-0.21 C, 0.50-0.70 Mn, 0.20-0.35 Si, 0.50-0.70 Cr, 0.10-0.15 V, bal Fe.
For case hardened parts.

SHARON 6100 SERIES B.
M-1071; 0.48-0.53 C, 0.70-0.90 Mn, 0.20-0.35 Si, 0.80-1.10 Cr, 0.15 min V, bal Fe.
Typical uses: springs, circular saw blades, business machine parts.

SHARON 8600 SERIES A.
M-1071; 0.13-0.33 C, 0.70-0.90 Mn, 0.20-0.35 Si, 0.40-0.70 Ni, 0.40-0.60 Cr, 0.15-0.25 Mo, bal Fe.
(Composition as desired). Oil hardenable.
Typical uses: carburized gears, spline shafts, bearings and races.

SHARON 8600 SERIES B.
M-1071; 0.35-0.64 C, 0.75-1.00 Mn, 0.20-0.35 Si, 0.40-0.70 Ni, 0.40-0.60 Cr, 0.15-0.25 Mo, bal Fe.
(Composition as desired). Oil hardenable.
Typical uses: spline shafts, helical springs.

SHARON UT-1.
M-1071; 0.08 max C, 1.0 max Mn, 1.0 max Si, 10.5-11.75 Cr, Ti = 6 x 0.75 max Al, bal Fe.
Typical use: utility stainless.

SHARON UT-3.
M-1071; 0.08 max C, 1.0 max Mn, 1.0 max Si, 11.5-14.5 Cr, 0.10-0.30 A bal Fe.
Typical use: utility stainless.

SHARON UT-9.
M-1071; 0.03 max C, 1.0 max Mn, 1.0 max Si, 12.0-14.5 Cr, bal Fe.
Typical use: utility stainless.

SHARPALOY.
M-1172; 1.0 C, 1.75 W, 3.85 Cr, 1.1 V, 8.35 Mo, 3.85 Co, bal Fe.
High speed tool steel; cutting tools.

SHAWINIGAN "20".
M-443; 29 Ni, 19 Cr, 3 Mo, 4 Cu, 1 Si, 0.07 max C, bal Fe.
Annealed: 65,000-75,000 TS; 26,000-32,000 YS; 50-35 El; 60-40 RA; 120-150 Brin.
For sulfuric acid equipment; corrosion resistant to H_2SO_4.

SHAWINIGAN CA.
M-443; 0.20 max C, 12-14 Cr, 1.0 max Ni, bal Fe.
Annealed: 100,000 TS; 85,000 YS; 20 El; 35 RA; 200 Brin.
For valve seats; wear resisting, corrosion resisting.

SHAWINIGAN CF-SE.
M-443; 0.10 C, 18-20 Cr, 8-10 Ni, 0.25 Se, bal Fe.
Annealed: 80,000 TS; 40,000 YS; 45 El; 60 RA; 180 Brin.
For stainless parts; free-cutting, austenitic, stainless.

SHAWINIGAN HC.
M-443; 0.2-0.3 C, 26-30 Cr, 3 max Ni, bal Fe.
Cast: 112,000 TS; 75,000 YS; 25 El; 25 RA: 225 Brin.
For furnace parts; heat resisting to S atmosphere.

SHAWINIGAN HH.
M-443; 0.25-0.35 C, 24-26 Cr, 11-13 Ni, bal Fe.
Cast: 85,000 TS; 40,000 YS; 25 El; 45 RA; 180 Brin.
For furnace parts; heat resisting.

SHAWINIGAN HL.
M-443; 0.25-0.35 C, 20-22 Ni, 26-30 Cr, bal Fe.
Cast: 75,000 TS; 45,000 YS; 10 El; 180 Brin.
For furnace parts; heat resisting.

SHAWINIGAN HT.
M-443; 0.20-0.35 C, 35-40 Ni, 15-18 Cr, bal Fe.
Cast: 70,000 TS; 35,000 YS; 12 El; 15 RA.
For furnace parts; heat resisting to 2100°F.

SHAWINIGAN KA-2-MO.
M-443; 0.10 max C, 18-20 Cr, 8.0-10.0 Ni, 2.5-3.5 Mo, 0.5 Si, 0.4 Mn, bal Fe.
Annealed: 85,000 TS; 45,000 YS; 46 El; 60 RA; 180 Brin.
For apparatus in the sulfite industry and chemical plants with strong acid corrosion resistant.

SHAWINIGAN SSS.
M-443; C, 35 Ni, 15 Cr, bal Fe.
For heat and corrosion resistant castings and equipment; heat and corrosion resistant.

SHAWINIGAN X.
M-443; 0.2 C, 22 Cr, 14 Ni, W, bal Fe.
For jet engine parts; heat resistant.

SHEARCUT.
M-1431, M-1433; 0.55 C, 0.8 Mn, 0.25 Cr, 2 Si, 0.4 Mo, 0.3 V, bal Fe.
For shear blades, punches, jaws, chisels; Type S5; oil hardened.

SHEATHING BRONZE.
M-Eng.; 32.5 Ni, 16 Sn, 5.5 Zn, 1.5 Bi, bal Cu.
For sheathing, marine construction; resists water corrosion.

SHEEPBRIDGE STOKES MARK III.
M-494, M-1485; 0.3 C, 1.5 Ni, 0.9 Cr, bal Fe.
For centrifugal castings; tough.

SHEETWELD.
M-118; 0.2 C, bal Fe.
For welding electrodes; for light sheet metal.

SHEFFALLOY.
M-973; Pb-Sn.
For collapsible tubes.

SHEFFIELD.
M-261; 0.82-0.85 C, 0.3 Mn, 0.5-0.7 Cr, 0.18-0.22 Mo, bal Fe.
For coining and cutlery dies; water hardening.

SHEFFIELD HARD ALLOY.
M-Eng.; 46 Cu, 31 Sn, 20 Zn.
For ornamental parts; corrosion resistant.

SHEFFIELD NICKEL SILVER.
M-Eng.; 63-55 Cu, 17-37 Zn, 19-11 Ni, 0.3 Pb.
Used as base for plated tableware; nickel silver.

SHEFFIELD NO. 10.
M-207; 0.3 C, Ni, Mo, Cr, bal Fe.
Heat treated: 154,000 TS; 139,000 YS; 16 El; 55 RA; 300 Brin.
For gears, cams, arbors, shafts, axles, mandrels; preheat treated, tough.

SHEFFIELD NO. 11.
M-207; 0.4 C, bal Fe.
Annealed: 90,000 TS; 50,000 YS; 175 Brin.
For mandrels, springs, tongs, rivets, shackles; forging quality.

SHEFFIELD NO. 12.
M-207; C, alloy, bal Fe.
For gears, picks, lathe centers, shovel teeth; shock and fatigue resistant.

SHEFFIELD NO. 20.
M-207; C, alloy, bal Fe.
Heat treated: 154,000 TS; 139,000 YS; 16 El; 55 RA; 300 Brin.
For shafting, pump rods; oil hardened.

SHEFFIELD P.B.
M-207; C, alloy, bal Fe.
For press brake dies; oil hardening.

SHEFFIELD PLATE.
M-207; C, alloy, bal Fe.
Heat treated: 150,000 TS; 140,000 YS; 14 El; 55 RA; 365 Brin.
For chute liners, conveyors, baffle plates; work hardened, fatigue resistant.

SHEFFIELD PRESS BRAKE.
M-207; 0.53 C, 0.95 Mn, 0.31 Si, 1.0 Cr, 0.5 Ni, 0.24 V, 0.25 Mo, bal Fe.
Oil hardening steel for miscellaneous tools.

SHEF-LO-TEMP.
M-10, M-1428; 0.20 C, 0.7-1.35 Mn, 0.15-0.30 Si, bal Fe.
Heat treated: 70,000 TS; 50,000 YS; 21 El.
For structural members, pressure vessels, derricks, booms, bridges.
Tough, shock resistant.

SHEF-SUPER-LO-TEMP.
M-10, M-1428; 0.20 C, 1.2-1.45 Mn, 0.2-0.5 Si, bal Fe.
Heat treated: 80,000 TS; 60,000 YS; 21 El.
For mine and railroad cars, pressure vessels, derricks, bridges.
Wear resistant, shock resistant.

SHEF-TEN STEEL.
M-1428; 0.25 max C, 1.4 Mn, 0.2 min Cu, 0.4 max Ni, bal Fe.
Rolled: 70,000 TS; 40,000 YS; 20 El.
For mine and railroad cars, shovels, structures; good formability.

SHELLDIE.
M-55; 0.36 C, 0.6 Mn, 0.9 Si, 5 Cr, 1.85 Mo, 3.0 V, bal Fe.
Ht. Tr.; As required from 37 to 46 Rc.
For forging die inserts, punches, mandrels, heading dies.
Resists heat checking. AISI H-11.

SHEMTSCHUSHNY.
M-Eng.; 1.3-4.45 Fe, 11.1-13.8 Mn, 86.8-84.2 Cu, 0.6-0.5 Si.
For electrical resistances; resistance alloy.

SHIBU-ICHI.
M-Japan; 67.3 Cu, 32.1 Ag, 0.5 Fe, trace Au.
For ornaments; corrosion resistant.

SHIBU-ICHI.
M-Japan; 51.1 Cu, 48.9 Ag, 0.12 Au.
For ornaments, jewelry; corrosion resistant.

SHIELD-ARC 65T.
M-578; C, bal Fe.
Steel arc welding electrode.
As welded: 75,000-87,000 psi TS.
AWS E7010-G.

SHIELD-ARC 85.
M-578; 0.12 C, 0.60 Mn, 0.03 P, 0.04 S, 0.40 Si, 0.50 Mo, bal Fe.
Steel arc welding electrode.
AWS Class E7010-Al.

SHIELD-ARC 85P.
M-578; C, bal Fe.
Steel arc welding electrode.
AWS Class E7010-g.

SHIELD-ARC HYP.
M-578; C, bal Fe.
Steel arc welding electrode.

SHIELD-ARC X70.
M-578; C, bal Fe.
Steel arc welding electrode.
As welded: 92,000-93,000 psi TS.
AWS E8010-G.

SHIELD-O-MATIC.
M-844; 0.15 C, bal Fe.
For electrodes for welding; coated-arc.

SHIGA WHITE GOLD.
M-Japan; 90-60 Au, 5-20 Ni, 5-20 Cr.
For ornaments, jewelry; corrosion resistant.

SHIP BRAND ORANGE LABEL.
M-446; 0.07 C, 18-21 Cr, 8-11 Ni, bal Fe.
Rolled: 120,000 TS.
Heat treated: 190,000-290,000 TS.
For springs; corrosion resistant.

SHIP BRAND RED LABEL.
M-446; 0.35 C, 13.5 Cr, bal Fe.
Rolled: 100,000 TS.
For springs; corrosion resistant.

SHIP NAIL ALLOY.
M-Eng.; 50 Sn, 33 Pb, 17 Sb.
For bearings; anti-friction.

SHIP NAIL BRASS.
M-Eng.; 64 Cu, 25 Zn, 8.5 Pb, 2.5 Sn.
For ship nails, hardware; free-cutting.

SH-KH 15.
M-USSR; 0.95-1.10 C, 1.3-1.6 Cr, 0.2 Mn, 0.2 max Ni, bal Fe.
For ball bearings, races, pivots; water hardened.

SHOCK-DIE.
M-35; 0.50 C, 0.75 Mn, 1.40 Mo, 3.20 Cr, bal Fe.
Air hardening shock resisting tool steel.
For punches, blanking dies, chisels, gripper dies.
AISI S7.

SHOCK PROOF.
M-883; 3 C, 1.5 Si, bal Fe.
Cast: 53,000 TS.
For castings, gears; malleable iron.

SHOCK RESISTING.
M-261; C, alloy, bal Fe.
For tools, dies; shock resistant.

SHOCK RESISTING.
M-389; 0.6 C, 0.9 Mn, bal Fe.
For punches, shears; tough.

SHOCK-RITE.
M-1684; 0.65 C, 0.85 Mn, 2.0 Si, 0.25 Cr, 0.25 Mo, 0.2 V, bal Fe.
Oil or water hardening, shock resisting type tool steel; for shear blades, pneumatic tools. AISI S5.

SHOE NAIL BRASS.
M-U.S.; 63 Cu, 37 Zn.
For hardware, shoe nails; yellow brass.

SHOE TIP METAL.
M-Eng.; 88 Cu, 12 Zn.
For shoe tips; red brass.

SHOW CASE METAL.
M-Eng.; 58-59.5 Cu, 24-22.5 Zn, 18-8 Ni.
For show cases, architectural parts; corrosion resistant.

SHUNT STEEL.
M-1419; 70 Fe, 30 Ni.
For compensating shunts for electrical equipment; temperature-sensitive, magnetic.

SHURBOND 0.
M-1468; 92.8 Cu, 7.2 P.
Melt range: 1300-1450°F.
Economical, highly fluid, self-brazing on copper.

SHURBOND 02.
M-1468; 2 Ag, 91 Cu, 7 P.
Melt range: 1185-1270°F.
For brazing copper, brass and steel.

SHURBOND 06.
M-1468; 6 Ag, 86.5 Cu, 7.5 P.
Melt range: 1185-1350°F.
For brazing ferrous and non-ferrous alloys.
AWS BCuP-4.

SHURBOND 2.
M-1468; 91 Cu, 7 P, 2 Ag.
Melt range: 1190-1425°F.
For brazing copper and copper alloys.

SHURBOND 4.
M-1468; 10 Ag, 52 Cu, 38 Zn.
For silver solder; M.P. 1450-1565°F.

SHURBOND 5 BRAZE.
M-1468; 89 Cu, 6 P, 5 Ag.
Melt range: 1185-1485°F.
For brazing copper base alloys where poor fit-ups exist.

SHURBOND 5 SOLDER.
M-1468; 5 Ag, 16.5 Zn, 78.5 Cd.
Melt range: 500-725°F.
Solder for copper, brass and steel.

SHURBOND 5X.
M-1468; 5 Ag, 95 Cd.
Melt range: 620-725°F.
Solder for copper, brass and steel.

SHURBOND 6.
M-1468; 88 Cu, 6 P, 6 Ag.
Melt range: 1185-1480°F.
For brazing ferrous and non-ferrous alloys; ductile joints.

SHURBOND 6F.
M-1468; 86.75 Cu, 7.25 P, 6 Ag.
Melt range: 1190-1300°F.
Very fluid; for brazing copper base alloys with tight fits.

SHURBOND 13.
M-1468; 30 Ag, 38 Cu, 32 Zn.
For silver solder; M.P. 1370-1410°F.

SHURBOND 14.
M-1468; 40 Ag, 36 Cu, 24 Zn.
For silver solder; M.P. 1330-1445°F.

SHURBOND 15.
M-1468; 80 Cu, 5 P, 15 Ag.
Melt range: 1185-1460°F.
For brazing copper base alloys having wide gaps.
AWS-BCuP-5.

SHURBOND 20.
M-1468; 20 Ag, 45 Cu, 30 Zn, 5 Cd.
For silver solder; M.P. 1140-1500°F.

SHURBOND 28.
M-1468; 50 Ag, 28 Cu, 22 Zn.
For silver solder; M.P. 1250-1340°F.

SHURBOND 31.
M-1468; 34 Cu, 31.5 Ag, 19 Cd, 15.5 Zn.
Melt range: 1165-1390°F.
For torch and induction brazing ferrous and non-ferrous alloys.

SHURBOND 33.
M-1468; 60 Ag, 25 Cu, 15 Zn,
For silver solder; M.P. 1260-1325°F.

SHURBOND 35.
M-1468; 35 Ag, 26 Cu, 21 Zn, 18 Cd.
For silver solder; M.P. 1125-1295°F. Type BAg2.

SHURBOND 45.
M-1468; 45 Ag, 15 Cu, 16 Zn, 24 Cd.
For silver solder; M.P. 1125-1145°F; Type BAg1.

SHURBOND 50.
M-1468; 50 Ag, 15.5 Cu, 16.5 Zn, 18 Cd.
For silver solder; M.P. 1160-1175°F; Type BAg1a.

SHURBOND 54.
M-1468; 54 Ag, 40 Cu, 5 Zn, 1 Ni.
For silver solder; M.P. 1325-1575°F.

SHURBOND 56.
M-1468; 56 Ag, 22 Cu, 17 Zn, 5 Sn.
For silver solder; M.P. 1145-1205°F.

SHURBOND 60.
M-1468; 60 Ag, 30 Cu, 10 Sn.
For silver solder; M.P. 1115-1325°F.

SHURBOND 65.
M-1468; 65 Ag, 20 Cu, 15 Zn.
For silver solder; M.P. 1235-1310°F.

SHURBOND 70.
M-1468; 70 Ag, 20 Cu, 10 Zn.
For silver solder; M.P. 1275-1360°F.

SHURBOND 72.
M-1468; 72 Ag, 28 Cu.
For silver solder; M.P. 1435-1435°F.

SHURBOND 75.
M-1468; 75 Ag, 22 Cu, 3 Zn.
For silver solder; M.P. 1365-1450°F.

SHURBOND 79.
M-1468; 25 Ag, 52.5 Cu, 22.5 Zn.
For silver solder; M.P. 1500-1575°F.

SHURBOND 85.
M-1468; 85 Ag, 15 Mn.
Melt range: 1760-1778°F.
AMS 4766; ASTM B-260-52T BAg-Mn.

SHURBOND 120.
M-1468; 20 Ag, 45 Cu, 35 Zn.
For silver solder; M.P. 1430-1500°F.

SHURBOND 145.
M-1468; 45 Ag, 30 Cu, 25 Zn.
For silver solder; M.P. 1250-1370°F. Type BAg5.

SHURBOND 240.
M-1468; 40 Ag, 30 Cu, 28 Zn, 2 Ni.
For silver solder; M.P. 1240-1435°F; Type BAg4.

SHURBOND 250.
M-1468; 50 Ag, 34 Cu, 16 Zn.
For silver solder; M.P. 1270-1425°F; Type BAg6.

SHURBOND 254.
M-1468; 40 Ag, 30 Cu, 25 Zn, 5 Ni.
For silver solder; M.P. 1240-1560°F.

SHURBOND 350.
M-1468; 50 Ag, 15.5 Cu, 15.5 Zn, 3 Ni.
For silver solder; M.P. 1170-1270°F.

SHX SERIES.
M-1296; 0.18 max C, 0.55 max Si, 1.5 max Mn, (Cr, Mo, Cb, V, Ti added if necessary), bal Fe.
0.062-0.500 in (1.57-12.7 mm) thick hot rolled sheet or coil.
5 strength grades: SHX 50,55,60,70,80 (each number means min YP in ksi).
High strength low alloy steel for cold-worked parts of automotive and appliances.

SHXF SERIES.
M-1296; 0.12 max C, 0.55 max Si, 1.65 max Mn, (Mo, Cb, V, and Ti added if necessary), bal Fe.
0.062-0.186 in (1.57-4.76 mm) thick hot rolled sheet or coil.
6 strength grades: SHXF 50,60,70,80,100. (each number means min YP in ksi.
High strength low alloy steel for cold worked parts of automotive and appliances.
Good formability and weldability.

SIBLEY ALLOY.
M-Eng.; 67 Al, 33 Zn.
For strong light alloy parts; non-hardenable.

SIBLEY CASTING ALLOY.
M-Eng.; 80 Al, 20 Zn.
For strong light alloy castings; non-hardenable.

SICROMAL 8.
M-625; 6.3-6.8 Cr, 0.6-0.9 Al, 0.12 max C, 0.6-0.9 Si, bal Fe.
Air annealed: 64,000-85,500 TS; 35,000 YS; 20 El; 125-165 Brin.
For superheaters, recuperators, annealing ovens, pots, bolts; resists heat to 800°C.

SICROMAL 9.
M-625; 12.5-13.5 Cr, 1.0-1.3 Si, 0.8-1.1 Al, 0.12 max C, 1 max Mn, bal Fe.
Air annealed: 71,900-92,500 TS; 43,000 YS; 15 El; 140-180 Brin.
For annealing ovens, pyrometer tubes, superheaters; resists heat to 900°C.

SICROMAL 10.
M-625; 17.5-18.5 Cr, 0.8-1.1 Si, 0.8-1.1 Al, 0.12 max C, 1 max Mn, bal Fe.
Air annealed: 71,000-92,500 TS; 43,000 YS; 12 El; 140-180 Brin.
For superheaters, annealing ovens, pots, bolts; resists heat to 1000°C.

SICROMAL 12.
M-625; 23-25 Cr, 1.3-1.6 Al, 1 max Mn, 0.12 max C, 1.3-1.6 Si, bal Fe.
Air annealed: 71,000-92,500 TS; 43,000 YS; 10 El; 140-180 Brin.
For superheaters, pots, bolts, annealing ovens; resists heat to 1200°C.

SICROMAL 12H.
M-625; 0.15 C, 23 Cr, 1 Si, 2.5 Al, bal Fe.
Annealed: 92,000 TS; 57,000 YS; 10 El.
For resistances, heating elements; heat resistant to 1200°C.

SICROMAL 17/17 KMS.
M-625; 0.1 C, 0.3-0.5 Si, 2 max Mn, 17-18 Cr, 1.8-2.2 Cu, 17-18 Ni, 1.8-2.2 Mo, Ti = 7 x C, bal Fe.
Water quenched: 78,200-106,700 TS; 35,500 YS; 40 El; 155-210 Brin.
For sulfuric acid equipment; corrosion and acid resistant.

SICROMAL 17/17 KNS.
M-625; 0.7 max C, 17.5 Cr, 2 Mo, 17.5 Ni, 2 Cu, Ti = 7 x C, bal Fe.
For valves and pump parts, furnace equipment; corrosion and heat resistant.

SICROMAL 18/8.
M-625; 0.15 max C, 18 Cr, 8.5 Ni, bal Fe.
Annealed: 80,000 TS; 35,000 YS; 55 El; 75 RA; 150 Brin.
Cold drawn: 180,000 TS; 150,000 YS; 10 El; 250 Brin.
For chemical plant equipment, tanks, mixers, filters; Type 302; stainless, austenitic.

SICROMAL 18-8 MS.
M-625; 0.12 max C, 17.5-18.5 Cr, 10-11 Ni, 1.8-2.2 Mo, Ti = 4 x C, 2 max Mn, bal Fe.
Water quenched: 78,000-107,000 TS; 38,000 YS; 40 El; 155-210 Brin.
For chemical and textile industries; austenitic, stainless.

SICROMAL 18-8 S.
M-625; 0.12 max C, 17.5-18.5 Cr, 9-10 Ni, 2 max Mn, Ti = 4 x C, bal Fe.
Water quenched: 78,000-107,000 TS; 38,000 YS; 40 El; 155-210 Brin.
For chemical industries; austenitic, stainless.

SICROMAL 18/12 MS.
M-625; 0.10 max C, 2 max Mn, 16 Cr, 13 Ni, 2.0-2.5 Mo, 1 Ta/Nb, bal Fe.
Water quenched: 78,200-106,700 TS; 38,400 YS; 40 El; 150-210 Brin.
For stainless parts, oil refinery equipment; stainless, austenitic.

SICROMAL 18/12 S.
M-625; 0.10 max C, 2 max Mn, 18 Cr, 10 Ni, 1 Ta/Nb, bal Fe.
Water quenched: 78,200-106,700 TS; 38,400 YS; 40 El; 150-210 Brin.
For stainless parts, oil refinery equipment; stainless, austenitic.

SICROMAL 19 MS.
M-625; 0.10 max C, 16.5-17.5 Cr, 1.6-1.9 Mo, 1 max Mn, Ti = 7 x C, bal Fe.
Air annealed: 71,000-92,500 TS; 43,000 YS; 20 El; 140-180 Brin.
For chemical and textile industries; stainless, resists HNO_3.

SICROMAL 20-10.
M-625; 0.1-0.2 C, 19-20 Cr, 9-10 Ni, 1.8-2.3 Si, 2 max Mn, bal Fe.
Water quenched: 85,000-107,000 TS; 43,000 YS; 40 El; 165-210 Brin.
For furnace parts, crucibles, autoclaves; austenitic, resists heat to 1050°C.

SICROMAL 23-20.
M-625; 0.1-0.2 C, 23-25 Cr, 20-22 Ni, 1.8-2.3 Si, 2 max Mn, bal Fe.
Water quenched: 85,000-107,000 TS; 43,000 YS; 40 El; 165-210 Brin.
For recuperators, crucibles, autoclaves; austenitic, resists heat to 1200°C.

SICROMAL 85.
M-625; 0.15 C, 6.5 Cr, 2.0 Si, bal Fe.
Rolled: 78,000-107,000 TS; 56,900 YS; 18 El.
For steam boilers, superheater tubes, recuperators; heat resistant to 850°C.

SICROMAL 130.
M-625; Cr, Si, Al, bal Fe.
Annealed: 107,000 TS; 57,000 YS; 10 El.
For resistance wire; resists heat to 1250°C.

SICROMAL C 7A.
M-625; Cr, Si, Al, bal Fe.
Annealed: 78,000 TS; 57,000 YS; 10 El.
For resistance wire; resists heat to 1100°C.

SICROMAL CS 65 S.
M-625; 0.15 max C, 1.3-1.7 Si, 1 max Mn, 2.0-2.5 Cr, 0.4-0.5 Mo, bal Fe.
Normalized: 64,000-85,500 TS; 35,500 YS; 18 El; 125 Brin.
For oil refining equipment, steam boilers.

SICROMAL CV 18 W.
M-625; 0.1 C, 1.2-1.4 Si, 1 max Mn, 1.5-2.0 Cr, 0.2-0.3 Mo, 0.2-0.3 bal Fe.
Normalized: 64,000-85,500 TS; 42,700 YS; 20 El; 125-165 Brin.
For steam boilers.

SICROMAL D.
M-625; 0.15 max C, 0.3-0.5 Si, 1 max Mn, 4.5-5.5 Cr, 0.45-0.55 Mo, bal Fe.
Annealed: 64,000-85,300 TS; 42,600 YS; 20 El; 125-165 Brin.
For ammonia synthesis equipment; corrosion resistant.

SICROMAL D7.
M-625; 0.2 C, 7 Cr, Mo, bal Fe.
Rolled: 64,000 TS; 35,600 YS; 20 El.
For petroleum cracking equipment; heat resistant to 700°C.

SICROMAL D9.
M-625; 0.15 C, 9 Cr, 0.5 Mo, bal Fe.
Rolled: 64,000 TS; 35,600 YS; 20 El.
For hydrogenating and ammonia synthesis equipment; heat resistant to 750°C.

SICROMAL D12.
M-625; 0.12 max C, 12-13 Cr, 0.4 Si, bal Fe.
Rolled: 71,000-92,500 TS; 42,700 YS; 20 El.
For chemical plant equipment; corrosion resistant.

SICROMAL D16.
M-625; 0.12 C, 18 Cr, 0.4 Si, bal Fe.
Rolled: 64,000-85,000 TS; 42,700 YS; 20 El.
For nitric acid and soap making equipment; corrosion resistant.

SICROMAL D 16 S.
M-625; 0.10 max C, 17-18 Cr, 1 max Mn, Ti = 7 x C, bal Fe.
Air annealed: 64,000-85,500 TS; 43,000 YS; 20 El; 125-165 Brin.
For chemical industries; stainless, resists HNO_3.

SICROMAL D 45 V.
M-625; 0.12 max C, 0.4-0.8 Si, 1 max Mn, 1.4-1.8 Cr, 0.25-0.35 V, bal Fe.
Normalized: 64,000-85,300 TS; 42,670 YS; 18 El; 125-165 Brin.
For ammonia synthesis apparatus.

SICROMAL TS 57.
M-625; 0.1 max C, 0.7-1.1 Si, 1 max Mn, Ti = 5 x C, bal Fe.
Normalized: 48,300-68,300 TS; 28,500 YS; 20 El; 95-135 Brin.
For construction steel.

SICROMAL TSL.
M-625; 0.1 C, 0.7 Ti, bal Fe.
Rolled: 48,000-60,000 TS; 21,400 YS; 22 El.
For steam boilers and equipment in soap industries; resists caustic embrittlement.

SICROMO 5S.
M-376; 0.15 max C, 1-2 Si, 4-6 Cr, 0.45-0.65 Mo, bal Fe.
Annealed: 60,000 TS; 25,000 YS; 30 El; 50 RA; 163 Brin.
For high temperature tubing in oil refineries; heat resistant to 1500°F.

SICROMO 7.
M-376; 0.15 max C, 0.5-1.0 Si, 6-8 Cr, 0.45-0.65 Mo, bal Fe.
Annealed: 60,000 TS; 25,000 YS; 30 El; 50 RA; 179 Brin.
For high temperature tubing in oil refineries; corrosion resistant.

SICROMO 9M.
M-376; 0.15 max C, 0.5-1.0 Si, 7-9 Cr, 0.9-1.1 Mo, bal Fe.
Annealed: 60,000 TS; 25,000 YS; 30 El; 50 RA; 179 Brin.
For high temperature tubing in oil refineries; corrosion resistant.

SIDERAPHITE.
M-Eng.; 62 Fe, 23 Ni, 5 Cu, 5 Al, 4 W.
For acid resisting vessels and apparatus; stainless and corrosion resistant.

SIEMENS-HALSKE.
M-198; 90-95 Ni, 5-10 Ta.
For heat and corrosion resistant parts; stainless and corrosion resisting.

SIEMENS-HALSKE.
M-198; 70 Ni, 30 Ta.
For spark plug electrodes; stainless and corrosion resisting.

SIEMENS-HALSKE.
M-198; 48 Zn, 5 Sb, 47 Cd.
For solder; stainless and corrosion resisting.

SIFALUMIN NO. 14.
M-243; 99 Al.
Welded: 12,000 TS.
For welding rod for Al.

SIFALUMIN NO. 15.
M-243; 5 Si, bal Al.
Welded: 14,000 TS.
For welding rod for Duralumin.

SIFALUMIN NO. 16.
M-243; 10-12 Si, bal Al.
Welded: 20,000 TS.
For welding rod for Al.

SIFALUMIN NO. 27.
M-243; 5 Mg, 1 Mn, bal Al.
Welded: 34,000 TS.
For welding rod for Al alloys; MG5 Alloy.

SIFALUMIN NO. 36.
M-243; Al alloy.
For Al brazing rod; low temperature brazing.

SIFBRONZE NO. 1.
M-243; Sn, bal Cu.
Welded: 60,000 TS; 32,000 YS; 28 El; 35 RA; 117 Brin.
For welding rod; all purpose.

SIFBRONZE NO. 2.
M-243; 9 Ni, Sn, bal Cu.
Welded: 72,000 TS; 35,000 YS; 26 El; 33 RA; 126 Brin.
For welding rod; wear resistant.

SIFBRONZE NO. 3.
M-243; 15 Ni, bal Cu.
Welded: 66,000 TS; 36,000 YS; 24 El; 30 RA; 124 Brin.
For welding rod; corrosion resistant.

SIFBRONZE NO. 4.
M-243; Sn, bal Cu.
Welded: 60,000 TS; 32,000 YS; 28 El; 35 RA; 117 Brin.
For welding rod for Al-bronze; flux coated.

SIFBRONZE NO. 10.
M-243; Mn, Zn, bal Cu.
Welded: 60,000 TS; 32 El; 30 RA; 128 Brin.
For welding rod; resists sea water corrosion.

SIFBRONZE NO. 102 (FLUXCOATED).
M-243; 9 Ni, Sn, bal Cu.
For building up worn components.
Fluxcoated type of SIFBRONZE NO. 2.

SIFBRONZE NO. 104 (FLUXCOATED).
M-243; Sn, bal Cu.
For brazing all classes of ferrous metals.
Fluxcoated type of SIFBRONZE NO. 1.

SIFCOLOY.
M-889; 3.1 T.C., 1.14 Si, 0.62 Cu, 0.76 Mo, bal Fe.
Cast: 58,000 TS; 250 Brin.
For nozzle castings, lock gate valves; erosion resistant.

SIFCUPRON NO. 17.
M-243; 10-15 P, bal Cu.
For brazing alloy for Cu and brass; M.P. 750°C.

SIFCUPRON NO. 17.
M-243; 7-8 P, Cu.
Low melting Cu-P alloy, for brazing brass and copper.
BS. 1845 CP. 3.

SIFCUPRON NO. 17-2AG.
M-243; P, Ag, Cu.
Low melting brazing alloy; similar to SIFCUPRON NO. 17 plus 2% Ag. For brazing seams and fittings in copper hot water cylinders.
B.S. 1845 CP. 2.

SIF-PHOSPHOR BRONZE NO. 8.
M-243; 7 Sn, P, bal Cu.
For fusion welding of phosphor bronze castings; good color match.
BS.; 2901 C.11.

SIFSILCOPPER NO. 7.
M-243; Ag, bal Cu.
For welding rod for copper; easy flowing.

SIFSILCOPPER NO. 968.
M-243; 3 Si, Mn, Cu.
For fusion welding similar alloys; by T.I.G., carbon arc or Oxy-Acetylene process.
BS. 2901 C.9.

SIFSTEEL NO. 11.
M-243; 0.08 C, bal Fe.
Welded: 46,000 TS; 32,000 YS; 13 El; 20 RA.
For welding rod for steel; Cu coated.

SIFSTEEL NO. 12 (WEAR RESISTING).
M-243; C, Si, Mn, Cr, Fe.
Weld filler metal for building up worn surfaces of cams, rail points, steel pins.
BS. 1453 A.5.

SIFSTEEL NO. 19 SILICO MANGANESE.
M-243; C, Si, Mn, bal Fe.
Welded: 68,000 TS; 50,000 YS; 12.7 El; 23.4 RA.
For welding rod; high strength.

SIFSTEEL NO. 22.
M-243; C, 1.5 Mn, bal Fe.
For welding rod; tough.

SIFSTEEL NO. 22 (1 1/2 MN).
   M-243; C, 1 1/2 Mn, Fe.
   Low carbon alloy steel rod for welding pipe lines, pressure vessels.
   BS. 1453 A.2.

SIFSTEEL STAINLESS NO. 21.
   M-243; 0.25-0.50 C, 2-3.5 Si, 0.6-1.25 Mn, 24-28 Ni, 15.5-19 Cr, bal Fe.
   For welding rod for stainless steel; stainless.

SIFSTEEL STAINLESS NO. 22 (NB).
   M-243; 18/8 stainless plus Nb.
   Stabilized austenitic electrode for TIG, MIG, and Oxy-Acetylene welding.
   BS. 2901 347 S96.

SIGAL.
   M-Ger.; 87 Al, 13 Si.

SIGERON.
   M-506; 3.2 C, 2.2 Si, 1.5 Ni, bal Fe.
   Cast.
   Castings; nickel cast iron.

SIGMALUMIN.
   M-622; 0.8 Si, 3.8 Cu, 0.7 Mn, bal Al.
   Soft: 25,000-29,000 TS; 11,400-15,900 YS; 20-15 El. For light alloy parts; age-hardenable.

SIGMA ST 835-1030.
   M-72; 0.70 C, 1.30 Mn, 0.75 Si, bal Fe.
   For tools as screw drivers, pliers.
   W.Nr. 1.8861.

SIGMA ST 1325-1470.
   M-72; 0.48 C, 0.70 Mn, 1.80 Si, 0.40 Cr, bal Fe.
   For shock resisting tools.
   W. Nr. 1.8863.

SIL 1.
   M-U.S.; 0.45 C, 0.40 Mn, 3.3 Si, 8.5 Cr, bal Fe.
   1400°F: 20,000 psi TS.
   Exhaust valve steel.
   SAE HNV 3.

SIL-5.
   M-1013; 5 Ag, 37 Zn, 58 Cu.
   For silver solder; M.P. 1575-1600°F.

SIL-5C.
   M-1013; 5 Ag, 95 Cd.
   For silver solder; M.P. 626-740°F.

SIL-7T.
   M-1013; 7 Ag, 85 Cu, 8 Sn.
   For silver solder; M.P. 1225-1805°F.

SIL 9.
   M-1013; 9 Ag, 53 Cu, 38 Zn.
   For silver brazing; M.P. 1510-1575°F.

SIL 10.
   M-U.S.; 0.38 C, 1.0 Mn, 3.0 Si, 19.0 Cr, 8.0 Ni, bal Fe.
   1400°F: 41,200 psi TS; 28,400 psi YS; 18 El.
   Exhaust valve steel.
   SAE EV 5.

SIL 10 see also **SILCROME X10.**

SIL 10N.
   M-U.S.; 0.38 C, 1.0 Mn, 3.0 Si, 19.0 Cr, 8.0 Ni, 0.15-0.25 N, bal Fe.
   1400°F: 41,200 psi TS; 28,400 psi YS; 18 El.
   Exhaust valve steel.
   SAE EV 6.

SIL 20.
   M-1013; 20 Ag, 45 Cu, 35 Zn.
   For silver brazing; M.P. 1430-1500°F.

SIL 20C.
   M-1013; 20 Ag, 45 Cu, 30 Zn, 5 Cd.
   For silver brazing; M.P. 1140-1500°F.

SIL-25.
   M-1013; 25 Ag, 52.5 Cu, 22.5 Zn.
   For silver solder; M.P. 1500-1575°F.

SIL 30.
   M-1013; 30 Ag, 38 Cu, 32 Zn.
   For silver brazing; M.P. 1370-1410°F.

SIL-35.
   M-1013; 35 Ag, 32 Cu, 33 Zn.
   Melt range: 1260-1370°F; Braze: 1370-1550°F, 72,000 psi TS.
   For various brazing operations.

SIL-40.
   M-1013; 40 Ag, 30.5 Cr, 29.5 Zn.
   For silver solder; M.P. 1080-1340°F.

SIL-40N.
   M-1013; 40 Ag, bal Cu.
   For silver solder; M.P. 1220-1435°F; AWS-BAg 4.

SIL-40T.
   M-1013; 40 Ag, 30 Cu, 28 Zn, 2 Sn.
   Melt range: 1220-1320°F.
   For various brazing operations.

SIL 45.
   M-1013; 45 Ag, bal Cu.
   For silver solder; M.P. 1250-1370°F; AWS-BAg 5.

SIL-50.
   M-1013; 50 Ag, bal Cu.
   For silver solder; M.P. 1275-1425°F; AWS-BAg 6.

SIL-54N.
   M-1013; 54 Ag, bal Cu.
   For silver solder; M.P. 1225-1275°F.

SIL-56T.
M-1013; 56 Ag, bal Cu.
For silver solder; M.P. 1145-1205°F; AWS-BAg 7.

SIL-60.
M-1013; 60 Ag, 25 Cu, 15 Zn.
For silver solder; M.P. 1260-1325°F.

SIL-60T.
M-1013; 60 Ag, 30 Cu, 10 Sn.
Melt range: 1095-1325°F; Braze: 1325-1500°F.
70,000 psi TS; 36 El; 80 Rock 15T.
For various brazing operations.
Was SIL-6T.

SIL-65.
M-1013; 65 Ag, bal Cu.
For silver solder; M.P. 1280-1325°F; AWS-BAg 9.

SIL-70.
M-1013; 70 Ag, bal Cu.
For silver solder; M.P. 1335-1390°F; AWS-BAg 10.

SIL-72.
M-1013; 72 Ag, bal Cu.
For silver solder; M.P. 1435°F; AWS-BAg 8.

SIL-80.
M-1013; 80 Ag, 16 Cu, 4 Zn.
For silver solder; M.P. 1360-1490°F.

SIL-85M.
M-1013; 85 Ag, bal Cu.
For silver solder; M.P. 1745-1760°F.

SIL 746.
M-U.S.; 0.70 C, 6.3 Mn, 0.55 Si, 21.0 Cr, 1.9 max Ni, 0.23 N, bal Fe.
1400°F: 50,000 psi TS; 30,000 psi YS; 8 El.
Exhaust valve steel.
SAE EV 11.

SILAFOND BETA.
M-1176; 11.5-13.0 Si, 0.3 Mg, 0.5 Mn, bal Al.
Cast: 34,000-46,000 TS; 28,000-40,000 YS; 4-1 El; 80-110 Brin.
For light alloy parts; corrosion resistant.

SILAFONT 1.
M-1354, M-1420; 11.0-13.5 Si, 0.3-0.5 Mn, bal Al.
Sand cast: 25,000-32,000 TS; 11,400-12,000 YS; 4-8 El; 50-60 Brin.
Permanent mold: 28,000-37,000 TS; 12,000-16,000 YS; 3-7 El; 60-70 Brin.
For chemical industry, ship and auto construction; high corrosion resistance.

SILAFONT-1.
M-249, M-1605, M-1634; 12-13 Si, 0.3-0.6 Mn, bal Al.
Sand cast: 27,000 TS; 10,000 YS; 6 El; 55 Brin.
Chill cast: 33,000 TS; 14,000 YS; 5 El; 60 Brin.
For chemical industry, ship and auto construction; heat treatable.

SILAFONT-2.
M-249, M-1605, M-1634; 12-13 Si, 0.2-0.4 Mg, 0.3-0.6 Mn, bal Al.
Sand cast: 40,000 TS; 34,000 YS; 2.5 El; 95 Brin.
Chill cast: 43,000 TS; 38,000 YS; 2 El; 100 Brin.
For chemical industry, ship and auto construction; heat treatable.

SILAFONT 3.
M-1354, M-1420; 8.5-10.5 Si, 0.2-0.4 Mg, 0.5 max Mn, bal Al.
Sand cast: 45,000-54,000 TS; 31,000-37,000 YS; 3-7 El; 90-100 Brin.
Permanent mold: 48,000-58,000 TS; 35,000-40,000 YS; 4-8 El; 95-110 Brin.
For chemical industry, ship and auto construction; high corrosion resistance.

SILAFONT-3.
M-249, M-1605, M-1634; 9-10 Si, 0.2-0.4 Mg, 0.3-0.6 Mn, bal Al.
Sand cast: 38,000 TS; 30,000 YS; 7 El; 80 Brin.
Chill cast: 39,000 TS; 33,000 YS; 6 El; 85 Brin.
For chemical industry, ship and auto construction; heat treatable.

SILAFONT-4.
M-249, M-1605; 12-13 Si, 0.5-1.0 Cu, 0.3 Mn, bal Al.
Sand cast: 31,000 TS; 13,000 YS; 2-5 El; 55 Brin.
Permanent mold: 35,000 TS; 14,000 YS; 2-4 El; 65 Brin.
For instrument housings, cases, intricate castings; corrosion resistant, good castability.

SILAFONT-5.
M-1605; 9.0-9.8 Si, 0.2-0.35 Mg, 0.45-0.55 Co, bal Al.
As cast: 26,000-33,000 TS; 14,000 YS; 3-6 El; 50-60 Brin.
Heat treated: 42,000 TS; 37,000 YS; 2-5 El; 70-95 Brin.
For general sand and die castings, housings, instrument cases.
Impact and shock resistant.
Heat treatable.

SILAFONT-6.
M-1605; 8.5-11.0 Si, 0.2-0.35 Mg, bal Al.
Die cast: 40,000 TS; 26,000 YS; 1 El; 80 Brin.
For general die castings, instrument cases, housings.
Good strength and corrosion resistant.

SILAFONT-7.
M-1605; 12.8-13.4 Si, 0.9-1.2 Mg, 1.0-1.4 Cu, 0.9-1.2 Ni, bal Al.
Heat treated: 44,000 TS; 37,000 YS; 0.5 El; 135 Brin.
For engine pistons.
Sand and die casting alloy, good hot crack resistance. Heat treatable.

SILAFONT CUPRO.
M-1176; 0.8 Cu, 12.0-13.3 Si, 0.3 Mn, bal Al.
Cast: 31,000-37,000 TS; 17,000-20,000 YS; 3.0-1.5 El; 60-80 Brin.
For light alloy parts; corrosion resistant.

SILAFONT MN.
M-1176; 9 Si, 0.35 Mg, 0.5 Mn, bal Al.
Cast: 35,000-43,000 TS; 28,000-37,000 YS; 5.5-3.5 El; 80-95 Brin.
For light alloy parts; corrosion resistant.

SILAFONT NORMAL.
M-1176; 12.75-13.25 Si, bal Al.
Cast: 31,000-34,000 TS; 14,000-17,000 YS; 3.0-1.5 El; 60-80 Brin.
For light alloy parts; corrosion resistant.

SIL-AID.
M-1013; 35 Ag, 26 Cu, 21 Zn, 18 Cd.
Cast: 60,000 TS.
For silver brazing all metals; M.P. 1125-1300°F.

SILAL.
M-494, M-1485; 2.5 C, 1.0 max Al, 0.96 Mn, 5 Si, 0.16 P, 0.05 Cr, bal Fe.
Cast: 20,000 TS; 160 Brin.
For stove and furnace parts, fire bars; cast iron, scale resistant.

SILAL 53.
M-890; 2-4 Mg, 0.4 max Mn, 0.3 max Cr, bal Al.
Soft: 28,000 TS; 13,000 YS; 30 El; 47 Brin.
Hard: 40,000 ES; 35,000 YS; 10 El; 73 Brin.
For aircraft tanks and fittings, fuel lines, marine parts; resists sea water corrosion.

SILAL V.
M-890; 97.2 Al, 1.2 Mg, 0.8 Mn, 0.5 Si, 0.3 Ti.
For light alloy parts; rolling and extrusion.

SILANCA.
M-Eng.; 92.5-94.5 Ag, 0-2.5 Zn, 4-4.5 Sb, 1-3 Cd.
For silverware; corrosion resistant.

SILBEREISEN NO. 1.
M-403; 1.0-1.5 Si, 0.9-1.5 Mn, 0.07-0.09 P, 0.02-0.04 S, 2.8 C, bal Fe.
36,000-50,000 TS.
For steamship and locomotive cylinders, turbine housings, pump bodies, high pressure valves; high strength machine and centrifugal castings.

SILBEREISEN NO. 2.
M-403; 1.7-2.5 Si, 0.4-0.7 Mn, 0.07-0.09 P, 0.02-0.04 S, 2.8 C, bal Fe.
36,000-50,000 TS.
For steamship and locomotive cylinders, turbine housings, pump bodies, high pressure valves; malleable iron and chilled iron castings.

SILBEREISEN NO. 3.
M-403; 1.0-1.3 Si, 0.4-0.7 Mn, 0.3-0.8 P, 0.02-0.04 S, 2.8 C, bal Fe.
36,000-50,000 TS.
For steamship and locomotive cylinders, turbine housings, pump bodies, high pressure valves; heavy castings, hard iron (chilled) castings.

SILBEREISEN NO. 4.
M-403; 1.3-1.7 Si, 0.4-0.7 Mn, 0.3-0.8 P, 0.02-0.04 S, 2.8 C, bal Fe.
36,000-50,000 TS.
For steamship and locomotive cylinders, turbine housings, pump bodies, high pressure valves; ordinary machinery castings.

SILBEREISEN NO. 5.
M-403; 1.7-2.5 Si, 0.4-0.7 Mn, 0.3-0.8 P, 0.02-0.04 S, 2.8 C, bal Fe.
36,000-50,000 TS.
For steamship and locomotive cylinders, turbine housings, pump bodies, high pressure valves; thin sections, common machinery castings.

SILBEREISEN NO. 6.
M-403; 1.7-2.5 Si, 0.4-0.7 Mn, 2.8 max C, 1.2 Ni, 0.5-1.0 Cr, bal Fe.
For cast iron cylinders, pump bodies, acid resisting castings; cast iron.

SILBEREISEN NO. 7.
M-403; 1.7-2.5 Si, 0.4-0.7 Mn, 2.8 max C, 1-2 Ni, 0.5-1.0 Cr, bal Fe.
For cast iron cylinders, pump bodies, acid resisting castings; cast iron.

SILBERKREUZ 1.
M-1344; 0.86 C, 4.1 Cr, 0.85 Mo, 2.5 V, 12 W, bal Fe.
For lathe and planer tools, reamers, drills, hobs; high speed steel.

SILBERKREUZ 2.
M-1344; 0.82 C, 4.1 Cr, 0.85 Mo, 1.6 V, 8.7 W, bal Fe.
For lathe and planer tools, reamers, broaches, hobs; high speed steel.

SILBERKREUZ SPEZIAL.
M-1344; 0.74 C, 4.1 Cr, 1.1 V, 18.5 W, bal Fe.
For lathe and planer tools, drills, taps, reamers; high speed steel.

SILBERLOT.
M-Ger.; Brazing alloys.
Series of brazing alloys used in Germany.
German standard Nos.

SILBERPUNKT.
   M-1331; 0.74 C, 4.1 Cr, 1.1 V, 18 W, bal Fe.
   For lathe and planer tools, reamers, broaches, taps; high speed steel.

SIL-BOND 30.
   M-1013; 30 Ag, 27 Cu, 33 Zn, 20 Cd.
   Melt range: 1125-1310°F; Braze 1310-1550°F.
   66,000 psi TS; 35 El; 82 Rock 15T.
   For various brazing operations.

SIL-BOND 31.
   M-1013; 31.5 Ag, 34 Cu, 15.5 Zn, 19 Cd.
   For silver brazing; M.P. 1165-1390°F, fillet type.

SIL-BOND 35.
   M-1013; 35 Ag, 26 Cu, 21 Zn, 18 Cd.
   Melt range: 1125-1295°F; Braze: 1295-1550°F.
   65,000 psi TS; 28 El; 80 Rock 15T.
   For various silver brazing operations.

SIL-BOND 45.
   M-1013; 45 Ag, 15 Cu, 16 Zn, 24 Cd.
   Cast: 70,000 TS.
   For silver brazing; M.P. 1125-1145°F; general purpose.

SIL-BOND 50.
   M-1013; 50 Ag, 15.5 Cu, 16.5 Zn, 18 Cd.
   Cast: 65,000 TS.
   For silver brazing; M.P. 1160-1175°F; general purpose.

SIL-BOND 50N.
   M-1013; 50 Ag, 15.5 Cu, 15.5 Zn, 16 Cd, 3 Ni.
   For silver brazing for carbide tipped tools; M.P. 1195-1270°F, resists chloride corrosion.

SILBRASS.
   4-5 Si, 10-12 Zn, bal Cu.
   For corrosion resistant parts; corrosion resistant.

SILCAZ ALLOY 3.
   M-48; 35-40 Si, 9-11 Ca, 6-8 Al, 3-5 Zr, 9-11 Ti, 0.55-0.75 B, bal Fe.
   For addition agent in steel making; alloying agent.

SILCHROM 20.
   M-459; 0.8 C, 2 Si, 20 Cr, 1.5 Ni, bal Fe.
   For valves; heat and corrosion resistant.

SILCHROM 25.
   M-459; 0.45 C, 1.2 Si, 1.1 Mn, 23 Cr, 2.8 Mo, 5 Ni, bal Fe.
   For valves; heat and corrosion resistant.

SIL-CO.
   M-1013; 70 Ag, 20 Cu, 10 Zn.
   For silver solder; M.P. 1235-1360°F.

SIL-CON.
   M-1013; 50 Ag, 34 Cu, 16 Zn.
   For silver solder; M.P. 1260-1410°F.

SILCORO 60.
   M-1493; 60 Au, 20 Cu, 20 Ag.
   For brazing.
   M.P. 835-845°C.

SILCORO 75.
   M-1493; 75 Au, 20 Cu, 5 Ag.
   For brazing brasses.
   M.P. 1625-1643°F; corrosion resistant.

SILCOZ ALLOY.
   M-48; 35-40 Si, 9-11 Ca, 6-8 Al, 3-5 Zr, 9-11 Ti, 0.55-0.75 B.
   For boron additions to steel; increased hardenability.

SILCROME NO. 9.
   M-568; 0.5 C, 14 Ni, 14 Cr, 3 Si, 2.5 W, 0.5 Mo, bal Fe.
   For valve seats; heat and corrosion resistant.

SILCROME RA.
   M-1; 0.12 C, 16 Cr, 1.0 Si, 1.0 Cu, bal Fe.

SIL-EEN.
   M-1013; 40 Ag, 36 Cu, 24 Zn.
   For silver solder; M.P. 1330-1445°F.

SILESIA D610PW.
   M-1323; 0.30 C, 2.5 Cr, 0.2 Mo, 0.55 Mn, bal Fe.
   For gears, dies, forging dies; oil hardened, tough.

SILESIA E60.
   M-1323; 0.15 C, 0.65 Cr, 0.50 Mn, 0.25 Si, bal Fe.
   For gears, bolts, machine tool parts; case hardened.

SILESIA EK20.
   M-1323; 0.20 C, 0.25 Si, 1.25 Cr, 1.15 Mn, bal Fe.
   For gears, bolts, camshafts, cams; case hardening steel.

SILESIA F6.
   M-1323; 0.70 C, 1.7 Si, 0.70 Mn, bal Fe.
   For springs; oil hardened, tough.

SILESIA R2.
   M-1323; 1.3 C, 0.25 max Si, 0.25 max Mn, bal Fe.
   For engravers' tools, textile needles, reamers, drills; Type W1; water hardened.

SILESIA R3.
   M-1323; 1.15 C, 0.25 max Si, 0.25 max Mn, bal Fe.
   Annealed: 110,000 TS; 58,000 YS; 18 El; 38 RA; 210 Brin.
   For springs, drills, reamers, broaches; Type W1; water hardened.

SILESIA R4.
M-1323; 1.0 C, 0.25 max Si, 0.25 max Mn, bal Fe.
Annealed: 100,000 TS; 53,000 YS; 21 El; 42 RA; 200 Brin.
For springs, taps, drills, reamers, broaches; Type W1; water hardened.

SILESIA R5.
M-1323; 0.85 C, 0.25 max Si, 0.25 max Mn, bal Fe.
Heat treated: 190,000 TS; 145,000 YS; 10 El; 30 RA; 400 Brin.
For springs, taps, drills, cutters, reamers; Type W1; water hardened.

SILESIA R6.
M-1323; 0.70 C, 0.25 max Si, 0.25 max Mn, bal Fe.
Heat treated: 175,000 TS; 128,000 YS; 12 El; 37 RA; 355 Brin.
For rails, punches, springs, girders; Type W1; water hardened.

SILESIA RWA.
M-1323; 0.45 C, 1.4 Cr, 0.7 Mo, 0.3 V, bal Fe.
For forging and die casting dies; oil hardened.

SILESIA RWA EXTRA.
M-1323; 0.40 C, 0.25 Si, Cr, Mn, Mo, bal Fe.
For gears, bolts, crankshafts; oil hardened, tough.

SILESIA RWA SPEZIAL.
M-1323; 0.30 C, 1.1 Cr, 0.18 V, 3.75 W, 1.0 Si, bal Fe.
For header and forging dies, punches; oil hardened, tough.

SILESIA SC.
M-1323; 0.67 C, 1.3 Si, 0.5 Mn, 0.5 Cr, bal Fe.
For springs, chisels, punches; oil hardened, tough.

SILESIA SCN.
M-1323; 0.45 C, Ni, Cr, bal Fe.
For gears, bolts, shafts, axles; oil hardened, shock resistant.

SILESIA TCN SPEZIAL.
M-1323; 0.50 C, 1.05 Cr, 3.25 Ni, bal Fe.
For gears, bolts, crankshafts, axles; oil hardened, shock resistant.

SILESIA TCS.
M-1323; 0.61 C, 1.18 Cr, 0.10 V, 0.75 Mn, 0.85 Si, bal Fe.
For forging dies, punches, upsetters; oil hardened, tough.

SILESIA TCV.
M-1323; 1.15 C, 0.65 Cr, 10 V, bal Fe.
For bearings, cold work tools; oil hardened, wear resistant.

SILESIA TFE.
M-1323; 1.05 C, 1.0 Cr, 1.15 W, 0.90 Mn, bal Fe.
For bearings, sleeves, liners; water hardened, wear resistant.

SILESIA TFE24.
M-1323; 1.42 C, W, V, bal Fe.
For engravers' tools, form cutters; water hardened.

SILESIA TKL1.
M-1323; 1.0 C, 1.55 Cr, 0.35 Mn, bal Fe.
For bearings, liners, sleeves; oil or water hardened, wear resistant.

SILESIA TKL2.
M-1323; 1.05 C, 1.0 Cr, 0.30 Mn, bal Fe.
For bearings, liners, sleeves; water hardened, wear resistant.

SILESIA TKL3.
M-1323; 1.1 C, 0.4 Cr, 0.3 Mn, 0.2 Si, bal Fe.
For bearings, liners, cutters; water hardened, wear resistant.

SILESIA TSPE.
M-1323; 1.2 C, 0.1 V, 1.0 W, 0.28 Mn, bal Fe.
For blanking and heading dies; oil hardened, wear resistant.

SILESIA TV3.
M-1323; 1.0 C, 0.1 V, 0.25 Mn, 0.2 Si, bal Fe.
For bearings, heading dies; Type W2; water hardened.

SILESIA TWA.
M-1323; 0.30 C, 2.65 Cr, 0.35 V, 8.5 W, bal Fe.
For extrusion rams and liners, punches; oil hardened, hot work steel.

SILESIA TWA24.
M-1323; 0.30 C, 2.35 Cr, 0.6 V, 4.25 W, 0.30 Mn, bal Fe.
For extrusion rams, forging dies, punches; oil hardened, hot work steel.

SILESIA TWA SPEZIAL.
M-1323; 0.30 C, 2 Co, 2.4 Cr, 0.25 V, 8.5 W, bal Fe.
For extrusion rams and punches, shears; hot work steel, oil hardened.

SILESIA TZ.
M-1323; 1.4 C, 0.30 Cr, 0.10 V, 0.30 Mn, bal Fe.
For engravers' tools, blanking and forming dies; oil hardened, wear resistant.

SILESIA V201.
M-1323; 0.55 C, 0.7 Cr, 0.18 Mo, 1.65 Ni, 0.1 V, bal Fe.
For forging and heading dies, punches; oil hardened, tough.

SILESIA V201 SPEZIAL.
M-1323; 0.56 C, 0.9 Cr, 0.2 Mo, 0.1 V, 2.0 Ni, bal Fe.
For forging and die casting dies; oil hardened, tough.

SILESIA W35.
M-1323; 0.35 C, 0.25-0.50 Si, 0.30-0.80 Mn, bal Fe.
Hot rolled: 85,000 TS; 54,000 YS; 30 El; 54 RA; 185 Brin.
For gears, shafts, axles, bolts; water hardened.

SILESIA W45.
M-1323; 0.45 C, 0.25-0.50 Si, 0.30-0.80 Mn, bal Fe.
Hot rolled: 98,000 TS; 58,000 YS; 24 El; 45 RA; 212 Brin
For axles, gears, bolts, shafts, tie rods; water hardened.

SILESIA W60.
M-1323; 0.60 C, 0.25-0.50 Si, 0.30-0.80 Mn, bal Fe.
Heat treated: 160,000 TS; 113,000 YS; 12 El; 40 RA; 320 Brin.
For wheels, die blocks, girders, rails, bushings; water hardened.

SILESIA W75.
M-1323; 0.75 C, 0.25-0.50 Si, 0.30-0.80 Mn, bal Fe.
Heat treated: 180,000 TS; 135,000 YS; 12 El; 36 RA; 370 Brin.
For springs, rails, girders, clutch discs; Type W1; water hardened.

SILESIA W90.
M-1323; 0.90 C, 0.25-0.50 Si, 0.30-0.80 Mn, bal Fe.
Heat treated: 180,000 TS; 120,000 YS; 10 El; 30 RA; 380 Brin.
For springs, taps, reamers, drills, broaches; Type W1; water hardened.

SILESIA WG.
M-1323; 0.53 C, 0.90 Si, 0.90 Mn, bal Fe.
For punches, upsetters, bolts; water hardened.

SILESIA Z3.
M-1323; 0.35 C, 0.90 Si, 1.05 Cr, 0.18 V, 1.85 W, bal Fe.
For header and forging dies, punches; oil hardened, tough.

SILESIA Z4 SPEZIAL.
M-1323; 0.45 C, W, Cr, V, bal Fe.
For header and forging dies, die casting dies; oil hardened, tough.

SILESIA Z5.
M-1323; 0.55 C, 0.90 Si, 1.05 Cr, 0.18 V, 1.85 W, bal Fe.
For header and forging dies, punches; oil hardened, tough.

SILESIA Z6.
M-1323; 0.38 C, Si, Cr, V, bal Fe.
For gears, bolts, crankshafts; oil hardened, tough.

SILESIA Z7.
M-1323; 0.45 C, Si, Cr, V, bal Fe.
For springs, gears, bolts, crankshafts; oil hardened, shock resistant.

SIL-EX.
M-1013; 50 Ag, 15.5 Cu, 16.5 Zn, 18 Cd.
For silver brazing; M.P. 1160-1175°F.

SILEX-65.
M-1766; 0.65 C, 0.80 Mn, 1.80 Si, bal Fe.
Shock resistant tool steel; picks and shovels.

SILFERAL.
M-302; Si, bal Al.
Chill cast: 25,000-29,000 TS; 3-2 El; 65 Brin.
For cylinder heads, motorcycle cylinders; non-hardenable.

SIL-FIL.
M-1013; 50 Ag, 15.5 Cu, 15.5 Zn, 16 Cd, 3 Ni.
For silver solder; M.P. 1170-1270°F.

SILFLO "5" ETC see ALL-STATE SILFLO "5" ETC.

SIL-FOS.
M-63; 15 Ag, 80 Cu, 5 P.
M.P. 1185°F; Flow P. 1300°F.
For brazing copper and copper base alloys, where joints are wide.
AWS BCuP-5.

SIL-FOS-5.
M-63; 5 Ag, 89 Cu, 6 P.
M.P. 1185°F; Flow P. 1300°F.
For brazing copper and copper alloys.
AWS BCuP-3.

SIL-GON.
M-1013; Ag-Cu.
For silver solder; M.P. 1275-1425°F.

SILICAL.
M-Swed.; 11-14 Si, bal Al.
For exhaust valves; austenitic.

SILICO I.
M-72; 0.45 C, Cr, Si, bal Fe.
For gears, shafts, crankshafts, fasteners; water hardened.

SILICO II.
M-72; 0.45 C, Cr, Si, bal Fe.
For gears, bolts, shafts, fasteners; water hardened.

SILICO ALLOY (S5).
M-35; 0.58 C, 0.85 Mn, 1.95 Si, 0.45 Mo, 0.3 Cr, bal Fe.
Annealed: 108,000 TS; 64,000 YS; 27 El; 212 Brin.
For shear blades, pneumatic tools, mandrels, bending dies, pipe cutters, chisels, punches.
Oil or water hardened Type S5 tool steel. Shock resistant. Tough.

SILICO-CHROMIUM STEEL.
M-U.S.; 9-12 Cr, 5 Si, 1.2 C, bal Fe.
For valves, heat resistant parts; heat resistant.

SILICO-MANGANESE BRASS.
M-U.S.; 40 Zn, 1-2 Mn, 0.2-0.3 Fe, 0.05-1.0 Si, bal Cu.
For marine parts, architectural applications; corrosion resistant.

SILICO MANGANESE SPRING WIRE.
M-38; 0.5 C, 0.7 Mn, 2 Si, bal Fe.
For coil and flat springs, automatic springs; high fatigue resistance.

SILICO-MANGANESE STEEL.
M-U.S.; 0.45-1.5 Mn, 0.4-1.9 Si, bal Fe.
For springs.

SILICON.
M-1755; Si.
Purities: Semiconductor ("N" & "P" types), Hyper-Pure (99.9999%,) Optical (99.999%), 99.99%, 99.9%.
Forms: Polycrystalline, powders, float zoned & czochraiski single crystal granules, epitaxial, boules, deposited rods, chunks, etc.

SILICON COPPER E-1232.
M-1191; 2.75-3.25 Si, bal Cu.
Forged: 50,000-55,000 TS; 30,000-35,000 YS; 30-25 El; 95-120 Brin.
For electrical applications; forgings, high cond.

SILICON CORE IRON see CARPENTER SILICON CORE IRON.

SILICON CORE IRON "B-FM".
M-32; 0.05 max C, 0.40 max Mn, 2.25-2.75 Si, 0.08-0.15 P, bal Fe.
Hc from 10,000 gauss 0.65.
Br from 10,000 gauss 5600-7600.
Saturation at H-200, 20,600.
For solenoid switches, armatures, pole pieces, relays.
Free machining, high permeability.

SILICON-FERROCHROME.
M-U.S.; 55-45 Cr, 30-50 Fe, 1-17 Si, 2.5-7 C.
For steel making; Cr-addition agent.

SILICO NICKEL.
M-897; 96 Cu, 2.5 Si, 1.5 Mn.
For cathodes and filaments in electronic tubes; heat resistant.

SILICON-MANGANESE.
M-48; 68 Mn, 20 Si, 11 Fe, 0.7 C.
For steel making; Mn addition.

SILICON-NICKEL.
M-48; 80-40 Ni, 16-18 Si, 2.5-30 Fe.
For steel making; Ni addition.

SILICON RED BRASS.
M-33; 81.5 Cu, 14.5 Zn, 4 Si.
Ann: 85,000 TS; 43,000 YS; 25 El.
Drawn 7%: 100,000 TS; 57,000 YS; 21 El.
For high strength valve stems.

SILICON RED BRASS-6942.
M-8; 82 Cu, 17 Zn, 1.0 Si.
Soft: 55,000 TS; 20,000 YS; 60 El.
Hard: 90,000 TS; 50,000 YS; 8 El.
For resistance welding.

SILICON TIN BRONZE 5072.
M-8; 97.5 Cu, 1.75 Sn, 0.75 Si.
Forged: 45,000 TS; 14,000 YS; 40 El.
For electrical apparatus; corrosion resistant, forgeable.

SILICO-SPIEGEL.
M-48; 15-20 Mn, 8-15 Si, 0.15 P, C, bal Fe.
For steel making; Si addition.

SILIMAZ.
M-72; 0.6 C, 15.5 Si, 2.6 Mn, bal Fe.
Cast: 15,000 TS; C 50 Rock.
For pumps, valves, drains, agitators, fittings, castings.
Acid resistant. Brittle as cast. Sensitive to thermal shock.

SILKORRIT-1.
M-1323; 0.15 max C, 18 Cr, 8.5 Ni, bal Fe.
Annealed: 80,000 TS; 35,000 YS; 55 El; 75 RA; 150 Brin.
Cold drawn: 180,000 TS; 130,000 YS; 10 El; 250 Brin.
For chemical plant equipment, tanks, mixers, filters; Type 302; stainless austenitic.

SILKORRIT-2.
M-1323; 0.07 max C, 18 Cr, 9.5 Ni, bal Fe.
Annealed: 85,000 TS; 35,000 YS; 60 El; 75 RA; 150 Brin.
Cold drawn: 180,000 TS; 125,000 YS; 10 El; 330 Brin.
For chemical plant equipment, welded structures; Type 304; stainless, austenitic.

SILKORRIT-2T.
M-1323; 0.1 max C, 18 Cr, 8.5 Ni, bal Fe.
Annealed: 80,000 TS; 35,000 YS; 55 El; 75 RA; 150 Brin.
Cold drawn: 180,000 TS; 130,000 YS; 10 El; 250 Brin.
For chemical plant equipment, tanks, mixers, agitators; Type 302; stainless, austenitic.

SILKORRIT-2U.
M-1323; 0.12 max C, 18 Cr, 9.5 Ni, Ti = 4 x C, bal Fe.
Annealed: 85,000 TS; 35,000 YS; 55 El; 65 RA; 150 Brin.
Cold drawn: 95,000 TS; 60,000 YS; 40 El; 60 RA; 185 Brin.
For welded chemical plant equipment, tanks, mixers; Type 321; stainless, austenitic.

SILKORRIT-4.
M-1323; 0.07 max C, 18 Cr, 2 Mo, 10.5 Ni, bal Fe.
Annealed: 85,000 TS; 35,000 YS; 50 El; 65 RA; 160 Brin.
Cold drawn: 150,000 TS; 135,000 YS; 6 El; 300 Brin.
For acid resistant chemical plant equipment; Type 316; stainless, austenitic.

SILKORRIT-4T.
M-1323; 0.10 max C, 18 Cr, 2.2 Mo, 9.5 Ni, bal Fe.
Annealed: 85,000 TS; 35,000 YS; 50 El; 65 RA; 160 Brin.
Cold drawn: 150,000 TS; 135,000 YS; 6 El; 300 Brin.
For acid resistant chemical plant equipment; Type 316; stainless, austenitic.

SILKORRIT-4U.
M-1323; 0.12 max C, 18 Cr, 10.5 Ni, 2 Mo, Ti = 4 x C, bal Fe.
Annealed: 85,000 TS; 35,000 YS; 45 El; 60 RA; 160 Brin.
For welded acid resistant chemical plant equipment; Type 316 Ti; stainless, austenitic.

SILLMAN BRONZE.
M-Eng.; 9.7 Al, 3.9 Fe, bal Cu.
For jewelry, bullet shells; corrosion resistant, Al-bronze.

SIL-LO.
M-1013; 15 Ag, 80 Cu, 5 P.
For silver brazing nonferrous metals; M.P. 1185-1300°F.

SIL-LON.
M-1013; 30.5 Cu, 29.5 Zn, 40 Ag.
For silver brazing all metals; Cd free; M.P. 1250-1350°F.

SIL-LOY.
M-1013; 65 Ag, 20 Cu, 15 Zn.
For silver brazing; Cd free; M.P. 1265-1325°F.

SILMA.
M-1133; 85 Ag, 15 Mn, 0.5 max others.
For silver solder brazing; M.P. 1790°F; S.S. 27000; for high temperature service.

SILMA.
M-1178; 85 Ag, 15 Mn.
For solder; high temperature service; M.P. 1790°F.

SIL-MAG-M NO. 1.
M-398; 42-47 Si, 7.5-10.0 Mg, 1 Misch metal, bal Fe.
For cast iron inoculant; for ductile iron.

SIL-MAG-M NO. 2.
M-398; 60-65 Si, 16-20 Mg, 1 Misch metal, bal Fe.
For cast iron inoculant; for ductile iron.

SIL-MAG NO. 1.
M-398; 42-47 Si, 7.5-10.0 Mg, bal Fe.
For cast iron inoculant; for ductile iron.

SIL-MAG NO. 2.
M-398; 60-65 Si, 16-20 Mg, bal Fe.
For cast iron inoculant; for ductile iron.

SILMAL.
M-France; 1 Mg, 1.75 Si, bal Al.
Rolled: 38,000-43,000 TS; 25-30 El.
For light alloy parts; same as "Silumin."

SIL-MAN.
M-34, M-115; 0.55 C, 2.1 Si, 0.85 Mn, 0.25 Cr, 0.3 V, bal Fe.
205,000 TS; 175,000 YS; 8 El; 16 RA; 400 Brin.
For shock tools, punches, chisels, shear blades; shock resisting tool steel; silico-manganese steel.

SILMELEC.
M-France; 97.8 Al, 0.6 Mg, 0.6 Mn, 1.0 Si.
For electrical conductor wires; high electrical conductivity.

SILMET.
M-135; Ni-Cu-Zn alloys.
For deep stamping in electroplate trade; deep stamping and spinning.

SILNIC BRONZE (COPPER ALLOY NO. 647).
M-33; 1.6-2.2 Ni, 0.45-0.75 Si, bal Cu.
Aged: 90,000 TS; 80,000 YS; 12 El.
For fasteners, bolts, structural components. High strength, corrosion resistant.

SIL-NIK.
M-1013; 40 Ag, 30 Cu, 28 Zn, 2 Sn.
For silver solder; M.P. 1240-1435°F.

SIL NO. 1 see CARPENTER SIL NO. 1 also EMS-1.

SIL-O-ACID NO. 7.
M-481; Co, Si, Mg, bal Cu.
Cast: 52,000 TS; 21 El; 20 RA; 78 Brin.
For acid proof parts; resists acetic, carbolic, sulfuric and tartaric acid.

SIL-OID.
M-1013; 31.5 Ag, 34 Cu, 15.5 Zn, 19 Cd.
For silver brazing all metals; M.P. 1165-1390°F.

SIL-TEN S see USS SIL-TEN.

SIL-TEX.
M-1013; 60 Ag, 25 Cu, 15 Zn.
For silver brazing all metals; Cd free; M.P. 1230-1330°F.

SIL-TIN.
M-1013; 56 Ag, 22 Cu, 17 Zn, 5 Sn.
For silver solder; M.P. 1145-1205°F.

SIL-TITE.
M-1013; 45 Ag, 30 Cu, 25 Zn.
For silver brazing all metals; Cd free; M.P. 1240-1370°F.

SIL-TRON.
M-1013; 80 Ag, 16 Cu, 4 Zn.
For silver brazing; Cd free; M.P. 1375-1480°F.

SILUMIN.
M-116; 12-13.5 Si, bal Al.
Hard: 36,000 TS; 2 El.
For cylinders; wrought.

SILUMIN.
M-299, M-297; 11-14 Si, bal Al.
Annealed: 21,000 TS; 8500 YS; 20 El; 50 Brin.
For automotive castings, instruments, housings; corrosion resistant, easy to cast.

SILUMIN-BETA.
M-299, M-297; 12 Si, 0.3 Mg, 0.5 Mn, bal Al.
Sand cast: 25,000 TS; 14,000 YS; 2-4 El; 60 Brin.
For automotive castings; high fluidity, corrosion resistant.

SILUMIN-GAMMA.
M-297, M-299; 12 Si, 0.5 Mn, 0.3 Mg, bal Al.
Cast: 28,000 TS; 16,000 YS; 3 El; 60 Brin.
For automotive castings; good castability, corrosion resistant.

SIL-UTEC.
M-1013; 72 Ag, 28 Cu.
For silver solder; M.P. 1435°F.

SILVALOY.
M-897; 3 Si, 97 Ni.
Annealed: 75,000 TS; 55,000 YS; 25 El.
For cathodes and filaments in electronic tubes; heat resistant.

SILVALOY.
M-1432; C, alloy, bal Fe.
Heat treated: 155,000 TS; 132,000 YS; 17 El; 57 RA; 280 Brin.
For shafting, pump rods; preheat treated steel.

SILVALOY 5.
M-686; 5 Ag, 88.5 Cu, 6.5 P.
For brazing for joining copper; M.P. 1185-1280°F.

SILVALOY 6.
M-686; 6 Ag, 88 Cu, 6 P.
For brazing alloy.
M.P. 1190-1300°F.

SILVALOY 15.
M-686; 15 Ag, 5 P, 80 Cu.
For silver brazing for Cu alloys; M.P. 1185-1280°F.

SILVALOY 20.
M-686; 20 Ag, 45 Cu, 30 Zn, 5 Cd.
For silver brazing; M.P. 1130-1500°F.

SILVALOY 35.
M-686; 35 Ag, 26 Cu, 21 Zn, 18 Cd.
For silver brazing; M.P. 1125-1295°F.

SILVALOY 45.
M-686; 45 Ag, 15 Cu, 16 Zn, 24 Cd.
For silver brazing; M.P. 1145-1190°F.

SILVALOY 50.
M-686; 50 Ag, 15.5 Cu, 16.5 Zn, 18 Cd.
For silver brazing; M.P. 1160-1175°F.

SILVALOY 54.
M-686; 54 Ag, 40 Cu, 5 Zn, 1 Ni.
For brazing.
M.P. 1325-1575°F.

SILVALOY 60.
M-686; 60 Ag, 30 Cu, 10 Sn.
For atmosphere furnace brazing.
M.P. 600-725°C.

SILVALOY 105.
M-686; 45 Ag, 30 Cu, 12 Zn, 13 Mn.
For brazing; good strength to 700°F. M.P. 1298°F.

SILVALOY 250.
M-686; 40 Ag, 30 Cu, 28 Zn, 2 Ni.
For silver brazing for stainless steel; M.P. 1240-1435°F.

SILVALOY 254.
M-686; 40 Ag, 30 Cu, 25 Zn, 5 Ni.
For brazing steels; hardening and brazing in one operation. M.P. 1220-1580°F.

SILVALOY 301.
M-686; 72 Ag, 28 Cu.
For silver brazing; M.P. 1435°F.

SILVALOY 355.
M-686; 56 Ag, 22 Cu, 17 Zn, 5 Sn.
For brazing.
M.P. 1145-1205°F.

SILVALOY 377.
M-686; 5 Ag, 95 Cd.
Uses: Brazing alloy.
M.P. 640-740°F.

SILVALOY 503.
M-686; 50 Ag, 15.5 Cu, 15.5 Zn, 16 Cd, 3 Ni.
For silver brazing of carbide tipped tools; M.P. 1195-1270°F.

SILVALOY 850.
M-686; 85 Ag, 15 Mn.
For silver brazing; M.P. 1760-1778°F.

SILVALOY A-4.
M-686; 9 Ag, 53 Cu, 38 Zn.
For silver brazing; M.P. 1410-1565°F.

SILVALOY A-11.
M-686; 20 Ag, 45 Cu, 35 Zn.
For silver brazing; M.P. 1315-1500°F.

SILVALOY A-13.
M-686; 30 Ag, 38 Cu, 32 Zn.
For silver brazing; M.P. 1370-1410°F.

SILVALOY A-14.
M-686; 40 Ag, 36 Cu, 24 Zn.
For silver brazing; M.P. 1235-1415°F.

SILVALOY A-18.
M-686; 45 Ag, 30 Cu, 25 Zn.
For silver brazing; M.P. 1250-1370°F.

SILVALOY A-25.
M-686; 50 Ag, 34 Cu, 16 Zn.
For silver brazing; M.P. 1272-1425°F.

SILVALOY A-28.
M-686; 50 Ag, 28 Cu, 22 Zn.
For silver brazing; M.P. 1250-1340°F.

SILVALOY A-33.
M-686; 60 Ag, 25 Cu, 15 Zn.
For silver brazing; M.P. 1260-1325°F.

SILVALOY A-49.
M-686; 80 Ag, 16 Cu, 4 Zn.
For silver brazing; M.P. 1360-1490°F.

SILVALOY A-79.
M-686; 25 Ag, 52.5 Cu, 22.5 Zn.
For brazing alloy; M.P. 1250-1575°F.

SILVALOY AE-100.
M-686; 92.5 Ag, 7.3 Cu, 0.2 Li.
For atmosphere furnace brazing. M.P. 1435-1635°C.

SILVALOY AE-102.
M-686; 72 Ag, 27.8 Cu, 0.2 Li.
For brazing.
M.P. 1410°F. Eutectic alloy.

SILVALOY AT-600.
M-686; 98 Ag, 2 Li.
For brazing.
M.P. 1290-1400°F.

SILVALOY AT-601.
M-686; 97 Ag, 3 Li.
For brazing.
M.P. 1115-1235°F.

SILVALOY B-70.
M-686; 7 Ag, 85 Cu, 8 Sn.
For brazing.
M.P. 1225-1805°F.

SILVALOY EASY.
M-686; 65 Ag, 20 Cu, 15 Zn.
For silver solder; M.P. 1280-1325°F.

SILVALOY HARD.
M-686; 75 Ag, 22 Cu, 3 Zn.
For silver solder; M.P. 1365-1450°F.

SILVALOY I-401.
M-686; 5 Ag, 16.6 Zn, 78.4 Cd.
For brazing.
M.P. 480-600°F.

SILVALOY K-427.
M-686; 75 Ag, 25 Zn.
For brazing.
M.P. 1300-1345°F.

SILVALOY MEDIUM.
M-686; 70 Ag, 20 Cu, 10 Zn.
For silver solder; M.P. 1275-1360°F.

SILVALOY S-475.
M-686; 95 Ag, 5 Al.
For brazing.
M.P. 1550-1600°F.

SILVALOY T50.
M-686; 62.5 Ag, 32.5 Cu, 5 Ni.
For silver brazing for precipitation hardening steels; M.P. 1435-1590°F; good wetting.

SILVALOY T-51.
M-686; 75 Ag, 24.5 Cu, 0.5 Ni.
For brazing alloy.
M.P. 1435-1475°F.

SILVALOY T52.
M-686; 77 Ag, 21 Cu, 2 Ni.
For silver brazing for precipitation hardening steels; M.P. 1435-1525°F; good wetting.

SILVALOY T-53.
M-686; 71.5 Ag, 28 Cu, 5 Ni.
For brazing.
M.P. 1435-1450°F.

SILVALOY VTG 60.
M-686; 60 Ag, 30 Cu, 10 Sn.
For brazing filler metals for electronic applications.
M.P. 600-725°C.

SILVALOY VTG 301.
M-686; 72 Ag, 28 Cu.
For brazing filler metal for electronic applications.
M.P. 780°C.

SILVALOY VTG 303.
M-686; 50 Ag, 50 Cu.
Annealed wire: 60,8000 TS; 13 El.
For brazing filler metals for electronic applications.
M.P. 780-855°C.

SILVALOY VTG-AQ602.
M-686; 61.5 Ag, 24 Cu, 14.5 In.
Annealed wire: 58,700 TS; 37.5 El.
For brazing filler metals for electronic applications.
M.P. 630-705°C.

SILVANITE.
M-35; 1.30 C, 8.0 W, 4.0 Cr, 0.50 V, bal Fe.
Special purpose high speed tool steel; hardened (60-64 Rc), ready to use.
For woodworking tools, scrapers, paper knives.
Type F8.

SILVAZ.
M-48; 0.6 B, 38 Si, 6 Al, 10 Ti, 6 Zr, 10 V, bal Fe.
For deoxidizer; for steel making.

SILVAZ 3.
M-48; 0.5 B, 35-40 Si, 6 Al, 10 Ti, 6 Zr, 10 V, bal Fe.
For hardening constituent in steel; adding B to steel.

SILVEL.
67.5 Cu, 16 Zn, 6.5 Ni, 2.2 Fe.
Hard drawn: 130,000 TS; 2 El.
For tableware, plumbing fixtures, electric contact springs; German Silver.

SILVEL.
73 Cu, 12 Mn, 12 Sn, 1.8 Fe, 0.5 Pb, 0.3 Al.
Annealed: 73,000 TS; 32 El.
For tableware, plumbing fixtures, electric contact springs; German Silver.

SILVER.
M-1755; Ag.
Purities: Zone refined 99.9999%, 99.999%, 99.99%, 99.9%.
Forms: Rod, ingot, shot, sheet, powder, wire, foil, single crystals.

SILVER ALLOYED COPPER 110.
M-8; 1 Ag, 0.1 Mn, bal Cu.
For welding rod; oxy-acetylene for copper.

SILVER-BEARING COPPER.
M-1657; 25 oz/ton Ag, bal Cu.
Hard: 50,000 TS; 40,000 YS; 6 El; B 50 Rock.
Soft: 34,000 TS; 10,000 YS; 45 El; F 45 Rock.
For electrical and electronic components.
Electrical conductivity 100.5.
Superior elevated temperature properties.

SILVER BEARING COPPER 111.
M-8; 99.9 Cu, bal Ag.
Hard: 48,000 TS; 40,000 YS; 6 El; 83 Brin.
Soft: 33,000 TS; 10,000 YS; 45 El; 42 Brin.
For electrical apparatus; high conductivity.

SILVER BEARING COPPER-112.
M-8; 99.9 Cu, 10-15 oz Ag/ton.
Soft: 33,000 TS; 10,000 YS; 35 El.
Hard: 46,000 TS; 40,000 YS; 5 El.
For electrical apparatus; for high softening temp.

SILVER BEARING COPPER-114.
M-8; 99.9 Cu, 0.034-0.051 Ag.
Hard: 48,000 TS; 40,000 YS; 6 El; B 50 Rock.
Soft: 33,000 TS; 10,000 YS; 45 El; F 45 Rock.
For electrical and electronic components.
Elect. cond. 100.

SILVER BEARING COPPER, COPPER NO. 113.
M-8, M-33; 99.7 + Cu, 8 oz Ag per ton.
Sheet-annealed or cold rolled.
For core and fin stock for radiators, assemblies for tin dipping.
Higher softening temperature than pure copper.
ASTM B152.

SILVER BEARING COPPER, COPPER NO. 116.
M-8, M-33; 99.7 + Cu, 25 oz Ag per ton.
Sheet-annealed or cold rolled.
For core and fin stock for radiators, commutator segments, tin dipped assemblies.
Higher softening temperature than pure copper.
ASTM B152.

SILVER BRONZE.
M-Eng.; 13-23 Zn, 16-18 Ni, 0-2 Al, 0-2 Pb, 0.3 Si, bal Cu.
For bearings, bushings, hardware; corrosion resistant.

SILVER CHISEL.
M-661; C, alloy, bal Fe.
For welding electrode for chisels; tough, shock resistant.

SILVER, COINAGE U.S.A.
M-U.S.; 90 Ag, 10 Cu.
For coins; corrosion resistant.

SILVERCOTE BERYLLIUM COPPER.
M-1244; 2 Be, 0.5 max Co, bal Cu.
Heat treated: 200,000 TS; 450 Brin.
For springs, lead wires, pins; age hardenable, low creep and corrosion resistance.

SILVERCOTE BERYLLIUM COPPER NO. 10.
M-1244; 0.5 Be, 2.5 Co, bal Cu.
Cold drawn: 120,00-130,000 TS.
For lead wires, connectors; age hardenable, high fatigue resistance.

SILVER FOIL-1.
M-U.S.; 90 Sn, 10 Zn.
For foil for wrapping purposes.

SILVER FOIL-2.
M-U.S.; 98 Sn, 2 Cu.
For foil for wrapping purposes.

SILVER FOIL-3.
M-U.S.; 91 Sn, 8.3 Zn, 0.4 Pb.
For foil for wrapping purposes.

SILVERIN.
M-303; 67-70 Ni, 1-3 Mn and Fe, bal Cu.
Annealed: 74,000 TS; 40 El.
Heat treated: 146,000 TS; 20 El.
For chemical apparatus, dye vats, plating rolls, table ware; similar to "Monel"; corrosion resistant.

SILVERINE.
M-452; 80-71 Cu, 16-17 Ni, 1-8 Zn, 1-2.8 Sn, 1-2 Co, 1-1.5 Fe.
For ornamental and sanitary fittings; corrosion resistant.

SILVERITE.
M-Eng.; Cu, Ni.
For corrosion resistant parts; nickel silver.

SILVER LABEL.
M-343; 0.90 C, 0.5 Cr, 1.1 Mn, 0.5 W, bal Fe.
For hobs, taps, reamers, dies; non-deforming.

SILVER LABEL.
M-341; 0.6 C, 0.8 Mn, 0.45 Mo, 0.2 V, 1.85 Si, bal Fe.
For punching and shearing dies; oil or water hardening. AISI S5 tool steel.

SILVER LEAF.
M-U.S.; 8.25 Zn, 91 Sn, 0.35 Pb.
For foil for wrapping purposes.

SILVER MAG NICKEL.
M-237; 0.28 max Mg, 0.18 max Ni, 99.4 min Ag.
Annealed: 32,000-38,000 TS; 12-27 El; 75 Knoop.
Hardened: 68,000-74,000 TS; 2-6 El; 169 Knoop.
For springs, switches.
Vacuum melted, corrosion resistant.
Hardened by oxidation.

SILVER METAL-1.
M-Eng.; 67 Zn, 33 Ag.
For ornaments.

SILVER METAL-2.
M-Eng.; 53.3 Cu, 8.3 Zn, 31.6 Ni, 0.8 Fe, 1.2 Sn, 3.6 Pb, 0.3 Mn.
For corrosion resistant parts; corrosion resistant.

SILVER SOLDER AK NO. 5.
M-926; 35 Ag, 42 Cu, 11 Zn, 12 Cd.
For brazing; M.P. 1190-1180°F, easy flow.

SILVER SOLDER KA.
M-926; 53.5 Ag, 31.5 Cu, 15 Zn.
For brazing; M.P. 1280-1320°F, medium flow.

SILVER SOLDER KC NO. 4.
M-926; 54 Ag, 41.1 Cu, 3.9 Zn, 1 Ni.
For brazing; M.P. 1290-1400°F, hard flow.

SILVER SOLDER KL NO. 1.
M-926; 52 Ag, 16 Cu, 15.5 Zn, 16.5 Cd.
For brazing; M.P. 1150-1160°F, easy flow.

SILVER SOLDER MK NO. 3.
M-926; 75 Ag, 24 Zn, 1 Cd.
For brazing; M.P. 1380-1430°F, medium flow.

SILVER SOLDER NO. 38.
M-243.
Silver solder for brazing ferrous and non-ferrous alloys except aluminium alloys.

SILVER SOLDERS.
M-U.S.; 32 Ag, 23 Cu, 17 Zn, 18 Cd.
For solders, silver solders; corrosion resistant.

SILVER SOLDER SC NO. 1.
M-926; 63.3 Ag, 25.6 Cu, 11.1 Zn.
For brazing; M.P. 1280-1340°F, easy flow.

SILVER SOLDER SC NO. 4.
M-926; 63.8 Ag, 29.7 Cu, 5.71 Zn, 0.75 Ni.
For brazing; M.P. 1260-1430°F, medium flow.

SILVER STEEL.
M-510; 1.20 C, 0.40 Mn, 0.40 Cr, bal Fe.
Water hardening carbon tool steel.

SILVER STEEL.
M-1062; 1.0 C, bal Fe.
For saws; water hardened.

SILVER STRIPE HIGH SPEED.
M-1067; 0.7 C, 4 Cr, 18 W, 1.0 V, bal Fe.
For tools, cutters, boring and chasing tools; high speed steel.

SILVERTEM.
M-64; 0.35 C, 4.75 Cr, 1.25 W, 1.5 Mo, 1.0 Si, 0.2 Ag, bal Fe.
For hot work dies; prehardened die steel.

SILVERY PIG IRON 16%.
M-1038; 16.0 Si, 0.75 C, 1.0 Mn, bal Fe.
Ferro silicon for alloying in cast iron and some steels.

SILVERY PIG IRON 20%.
M-1038; 20.0 Si, 0.25 C, 1.0 Mn, bal Fe.
Ferrosilicon for alloying in cast iron and some steels.

SILVERY PIG IRON PULVERIZED.
M-1038; 15.01-15.50 Si, 0.80 C, 1.0 Mn, bal Fe.
For benefication of certain ores and minerals.

SILVIUM.
M-1384; Sb, bal Pb.
For grids for batteries; corrosion resistant.

SILVORE.
M-Eng.; 61.9 Cu, 19.5 Zn, 18.4 Ni, 0.3 Fe.
For show cases and architectural trim; corrosion resistant.

SILVUNG 240.
M-1246; W, Ag.
For contacts for interrupters; tungsten impregnated with Ag.

SILVUNG D.
M-1246; W, Ag. For contacts for voltage regulators; tungsten impregnate with Ag.

SILVUNG P.
M-1246; W, Ag.
For facing contacts; tungsten impregnated with Ag.

SIL-X.
M-1059; ferro silicon.
For steel deoxidizer; compounded with an oxidizing agent.

SIMALLOY 322.
M-373; 0.08 max C, 1.0 max Si, 17 Cr, 7 Ni, 0.70 Mo, bal Fe.
Modified AISI 301 austenitic stainless steel.

SIMALLOY 600.
M-373; 0.15 C, 1.0 Mn, 0.5 Si, 15.5 Cr, 72.0 min Ni, 8.0 Fe.
For heat exchangers, jet engine parts.
Good high temperature properties.
ASTM B-166, B-168; Inconel 600.

SIMALLOY 718 see INCONEL ALLOY 718.

SIMALLOY 722 see INCONEL ALLOY 722.

SIMALLOY 800 see INCOLOY 800.

SIMALLOY A-286 see A-286.

SIMALLOY HX.
M-373; 18.5 Fe, 22.0 Cr, 9.0 Mo, 1.5 Co, 0.6 W, 0.10 C, bal Ni.
High temperature alloy.
AMS 5536, 5754; AISI 680.

SIMALLOY L-605 see ALLOY L-605.

SIMALLOY N-155 see N-155.

SIMALLOY X 750.
M-373; 0.80 C, 1.0 Mn, 0.5 Si, 15.0 Cr, 70.0 min Ni, 1.0 Cb, 2.4 Ti, 0.7 Al, 6.5 Fe.
Age hardenable high temperature alloy.
ASTM A 637; Inconel X 750.

SIMBRAX.
85-97 Cu, 1.5 Fe, 1-3.5 Si, 2-7 Zn.
Soft: 45,000 TS; 20,000 YS; 50 El; 40 Brin.
Hard: 120,000 TS; 2 El; 250 Brin.
For pressure gauges; corrosion resistant.

SIMETEORA.
M-Eng.; 0.76 C, 0.17 Mn, 2.87 Cr, 16,86 W, bal Fe.
For tools, cutters, drills; high speed steel.

SIMGAL now ALCAN GB-50S.

SIMILARGENT.
M-Eng.; Cu, Ni.
For white metal parts; nickel silver.

SIMILOR.
M-Eng.; 89-80 Cu, 9-20 Zn, 0-7 Sn.
For hardware, pipe fittings; corrosion resistant.

SIMO.
M-1115; 0.3-0.4 C, 2.0-3.0 Si, 9-11 Cr, 0.8-1.3 Mo, bal Fe.
Rolled: 141,000 TS; 114,000 YS; 15 El.
For high temperature valves; corrosion and heat resistant.

SIMO.
M-1653; 0.50-0.65 C, 1.70-2.10 Si, 0.70-0.90 Mn, 0.25-0.45 Mo, bal Fe.
Shock resisting type tool steel.
U 58 SiMo 8 KU; AISI S5.

SIMOCH.
M-115; 0.46-0.50 C, 0.8-1.0 Si, 3.0-3.5 Cr, 1.4 Mo, 0.3 V, bal Fe.
For shear knives and blades, punches, rivet sets; air hardened, shock resistant.

SIMONDS now SIMONDS TEENAX 46.

SIMONDS.
M-373; 71 Ni, 2.8 Cr, 3.67 Si, bal Fe.
For radiant tubes, furnace muffles, conveyor belts, high temperature equipment.
Corrosion and oxidation resistant.
Heat resistant.

SIMONDS 17% COBALT MAGNET STEEL.
M-373; 17 Co, bal Fe.
For permanent magnets.

SIMONDS 30% NICKEL STEEL.
M-373; 0.1 C, 30 Ni, bal Fe.
For temperature compensator; controlled expansion.

SIMONDS 35% COBALT MAGNET STEEL.
M-373; 35 Co, bal Fe.
For permanent magnets.

SIMONDS 38-7FM.
M-373; 38 Ni, bal Fe.
For bourdon tubes, control devices, tuning forks.
Constant modulus, controlled expansion.

SIMONDS 40% COBALT MAGNET STEEL.
M-373; 40 Co, bal Fe.
For permanent magnets.

SIMONDS 42-43.
M-373; 42 Ni, bal Fe.
For glass to metal seals, sealing leads in electric light bulbs and radio tubes.
Controlled expansion.

SIMONDS 47.
M-373; 0.5 C, 1.4 Cr, 0.25 V, 2 W, 0.25 Mo, 0.8 Si, bal Fe.
For chisels; shock resistant.

SIMONDS 48-50.
M-373; 48 Ni, bal Fe.
For sealing leads in electric light bulbs and radio tubes, glass-to-metal seals.
Controlled expansion.

SIMONDS 73-3.5 CO.
M-373; 3.5 Cr, 1 C, bal Fe.
Maximum energy product 0.302; residual flux density, Br. 9700; coercive force Hc (Oersted) 81.
Rolled or cast permanent magnet.

SIMONDS 81-18.5 CO.
M-373; 8.5 Co, 3.75 Cr, 5 W, 0.75 C, bal Fe.
Maximum energy product 0.690; residual flux density, Br. 10700; coercive force Hc (Oersted) 160.
Rolled or cast permanent magnet.

SIMONDS 81C.
M-373; 0.7 C, 3 Cu, 4 Cr, 0.7 Mn, 20 Co, bal Fe.
Annealed: Rock C 25-35.
Hardened: 300,000 TS; Rock C 60-65.
For permanent magnets in hysteresis motors, speedometers, meters, gages.
Quench-hardened alloy. Non-directional magnetic properties.

SIMONDS 83-3 CO.
M-373; 3.25 Co, 4 Cr, 1 C, bal Fe.
Maximum energy product 0.382; Residual flux density. Br 9700; Coercive force Hc (Oersted) 81.
Rolled or cast permanent magnet.

SIMONDS 260 MBS.
M-373; 1.25 C, 0.30 Cr, bal Fe.
Cold rolled strip; as rolled or heat treated.

SIMONDS 1050 ROTARY RULE.
M-373; 0.50 C, 0.80 Mn, bal Fe.
Cold rolled strip; as rolled or heat treated.

SIMONDS 1064 RULE DIE.
M-373; 0.64 C, 0.60 Mn, bal Fe.
Cold rolled strip; as rolled or heat treated.

SIMONDS 1070 CIRCULAR SAW.
M-373; 0.70 C, 0.80 Mn, bal Fe.
Cold rolled strip; as rolled or heat treated.

SIMONDS 3500.
M-373; 38 Co, 3.8 Cr, 5 W, 0.75 C, bal Fe.
Maximum energy product, BH max 0.982; residual flux density, Br. 10400; coercive force, Hc (Oersted) 230.
Hot rolled or cast, 38-45 Rc; hardened 60-63 Rc.

SIMONDS 6150 EB BACKER.
M-373; 0.50 C, 0.80 Mn, 1.0 Cr, bal Fe.
Cold rolled strip; as rolled or heat treated.
Similar to AISI 6150.

SIMONDS AIRTRUE 51.
M-373; 1 C, 5 Cr, 0.25 V, 1 Mo, bal Fe.
For gages, knives, blanking and forming dies; Type A2; air hardened, non-deforming.

SIMONDS ALNICO NO. 2, 2C, 5 ETC see ALNICO 2, 2C, 5 ETC.

SIMONDS CCM.
M-373; 1.5 C, 12 Cr, 0.9 Mo, 0.9 V, bal Fe.
For tools, dies, cold work dies; air hardening, nondeforming.
AISI D2.

SIMONDS CHROMIUM MAGNET STEEL.
M-373; 3.5 Cr, 1 C, bal Fe.
For magnets; permanent.

SIMONDS CHROMIUM-MOLYBDENUM MAGNET STEEL.
M-373; 4 Cr, Mo, bal Fe.
For magnets; permanent.

SIMONDS ELINVAR see ELINVAR.

SIMONDS HIGH SPEED see SIMONDS RED STREAK.

SIMONDS INVAR 136.
M-373; 36 Ni, bal Fe.
Low expansion alloy; often used for bimetals.

SIMONDS MOLVA-T see MOLVA-T.

SIMONDS NO. 8 WOOD SAW.
M-373; 0.75 C, 2.50 Ni, bal Fe.
Oil hardening steel for wood-cutting saws.

SIMONDS NO. 116 PAPER KNIFE.
M-373; 1.18 C, 0.50 Mn, 0.50 Cr, 0.90 Mo, bal Fe.
Oil hardening tool steel for paper cutters, shears.

SIMONDS NO. 174 PLANER KNIFE.
M-373; 1.15 C, 8.5 Cr, 1.0 Si, 1.5 Mo, 1.5 W, 2.0 V, bal Fe.
Air hardening cold work tool steel for planer knives and blades.

SIMONDS NO. 176 CHIPPER KNIFE.
M-373; 0.48 C, 1.10 Si, 8.50 Cr, 1.50 Mo, 1.20 W, bal Fe.
Air hardening tool steel; for chipper knives.

SIMONDS NO. 12150.
M-373; 1.5 C, 12 Cr, 0.15 V, 0.8 Mo, bal Fe.
For punches, dies; abrasion and corrosion resistant.

SIMONDS PM DIE.
M-373; 0.35 C, 13.0 Cr, bal Fe.
Air hardenable tool steel for plastic mold dies.
AISI 420.

SIMONDS RED LABEL.
M-373; 0.8-1.2 C, 0.3 Mn, 0.25 Si, bal Fe.
For taps, drills, reamers, hobs, punches; Type W1; water hardened.

SIMONDS RED STREAK.
M-373; 0.75 C, 4 Cr, 1 V, 18 W, bal Fe.
For drills, reamers, cutting tools; high speed steel.

SIMONDS SLICER KNIFE.
M-373; 1.30 C, 1.0 Cr, bal Fe.
Oil hardening steel; for circular slicing knives.

SIMONDS S.T.M.
M-373; 0.8 C, 3.75 Cr, 1.1 V, 1.6 W, 8.6 Mo, bal Fe.
For cutting tools, drills, reamers; high speed steel.

SIMONDS SUPER INVAR TYPE A.
M-373; 31 Ni, 4.5 Co, bal Fe.
For bimetals.
Controlled expansion.

SIMONDS TEENAX 46.
M-373; 0.9 C, 0.5 Cr, 1.25 Mn, 0.15 V, 0.5 W, bal Fe.
For taps, punches, shear blades, chisels, upsetters; Type O1; oil hardened, non-deforming.

SIMONDS TUNGSTEN MAGNET STEEL.
M-373; 1 C, 3 W, bal Fe.
For permanent magnets.

SIMPLEX.
M-48; 0.01 max C, 5-7 Si, 63-66 Cr, bal Fe.
For making stainless steel; alloy additive.

SIN-CHU.
M-Japan; 66.5 Cu, 33.4 Zn, 0.1 Fe.
For hardware; brass.

SINTAG 852.
M-63; 85 Ag, 15 CdO.
For electric contacts.
Sintered, corrosion resistant.

SINTAG 903.
M-63; 90 Ag, 10 CdO.
For electric contacts.
Sintered, corrosion resistant.

SINTALOY B-901.
M-1427; 90 Cu, 10 Sn.
Sintered bronze.
As sintered: 30,000 TS; 2 El.
General purpose structural parts.

SINTALOY BB-901.
M-1427; 88 Cu, 10 Sn, 1.75 C.
Sintered bronze.
As sintered: 15,000 TS; 1 El.
General purpose structural parts. SAE 901.

SINTALOY CI-100 DENSITY 5.1-6.1.
M-1427; 99 Fe, 1 C.
Sintered steel, low density.
Sintered: 16,000-24,000 TS; 0.2 El; 10-50 Rb.
Low cost, quality structural parts.

SINTALOY CI-100 DENSITY 6.1-6.5.
M-1427; 99 Fe, 1 C.
Sintered steel, medium density.
As sintered: 19,000-30,000 TS; 0.3 El; 20-50 Rb.
Low cost, quality structural parts.
ASTM B310 Type 1 Class A.

SINTALOY CI-100-HT.
M-1427; 99 Fe, 1 C.
Sintered steel, heat treated.
HT: 30,000-40,000 TS; 10-35 Rc.
Improved strength and hardness.

SINTALOY CI-100-ST.
M-1427; 99 Fe, 1 C.
Sintered steel, steam treated.
As treated: 19,000-30,000 TS; 50-90 Rb.
Improved hardness, wear and corrosion resistance.

SINTALOY CX-140.
M-1427; 1.75 Ni, 0.5 Mo, 1.5 Cu, 0.6 C, bal Fe.
Sintered: 60,000 TS; 2-4 El; 60-80 Rb.
High strength, low alloy steel characteristics.

SINTALOY CX-140-HT.
M-1427; 1.75 Ni, 0.5 Mo, 1.5 Cu, 0.6 C, bal Fe.
Sintered and heat treated: 100,000 TS; 30-40 Rc.
Good strength, hardness, wear resistance.

SINTALOY EC-100.
M-1427; 100 Cu.
Sintered copper.
As sintered: 17,000-34,000 TS; 6-20 El.
Excellent electrical conductivity.

SINTALOY F-1003.
M-1427; 96 Fe, 3 Ni, 1 C.
Sintered: 60,000 min TS; 2-6 El; 60-90 Rb.
Good strength alloy.

SINTALOY F-1003-HT.
M-1427; 96 Fe, 3 Ni, 1 C.
Sintered and heat treated: 100,00-150,000 TS; 1-4 El; 20-45 Rc.
High strength alloy, good wear resistance.

SINTALOY HDM.
M-1427; 100 Fe.
Sintered iron.
As sintered: 25,000-40,000 TS; 2-15 El.
For magnetic applications.

SINTALOY HDS.
M-1427; 99 Fe, 1 C.
Sintered steel, high density.
As sintered: 45,000-70,000 TS; 2-12 El; 30-70 Rb.
Similar characteristics to annealed steel.

SINTALOY HDS-HT.
M-1427; 99 Fe, 1 C.
Sintered steel, heat treated, high density.
HT: 60,000-120,000 TS; 0.2 El; 20-50 Rc.
Hardened steel for wear parts.

SINTALOY L-1004.
M-1427; 95 Fe, 4 Cu, 1 C.
Sintered: 25,000-45,000 TS; 0.5 El; 20-60 Rb.
Moderate strength structural parts.

SINTALOY L-1007.
M-1427; 92 Fe, 7 Cu, 1 C.
Sintered: 30,000-50,000 TS; 0.5 El; 25-70 Rb.
Moderate strength structural parts.
ASTM B222.

SINTALOY LI-1010.
M-1427; 74 Fe, 25 Cu, 1 C.
Sintered: 55,000 TS; 0.5 El; 85-95 Rb.
Infiltrated alloy; good mechanical properties.
ASTM B303.

SINTALOY LI-1010 HT.
M-1427; 74 Fe, 25 Cu, 1 C.
Sintered and heat treated: 90,000 TS; 30-40 Rc.
Good strength, hardness and wear.

SINTALOY SS-316L.
M-1427; 16 Cr, 12 Ni, 2 Mo, bal Fe.
Sintered: 35,000-45,000 TS; 4-6 El; 55-70 Rb.
Austenitic stainless; good corrosion resistant properties.

SINTALOY SS-410L.
M-1427; 12.5 Cr, bal Fe.
Sintered: 34,000-49,000 TS; 5-8 El; 50-75 Rb.
Corrosion resistant; heat treatable.

SINTALOY SS-410L-HT.
M-1427; 12.5 Cr, C, bal Fe.
Sintered and heat treated: 100,000-140,000 TS; 1 El; 25-37 Rc.
High strength stainless steel.

SINTALOY Z-802.
M-1427; 80 Cu, 20 Zn.
Sintered: 29,000-36,0000 TS; 10-18 El.
For hardware, machine tool parts; yellow brass, powder metallurgy.

SINTALOY Z-802 PB.
M-1427; 78.5 Cu, 20 Zn, 1.5 Pb.
Sintered: 27,000-34,000 TS; 10-18 El.
For hardware, machine tool parts; leaded yellow brass.

SINTALOY Z-802 PB-P.
M-1427; 78.5 Cu, 19.7 Zn, 1.5 Pb, 0.3 P.
Sintered: 29,000-36,000 TS; 18-32 El.
For hardware, machine tool parts; powder metallurgy.

SINTALOY Z-901.
M-1427; 90 Cu, 10 Zn.
Sintered: 20,000-28,000 TS; 8-15 El.
For hardware, structural parts; high electrical conductivity.

SINTALOY Z-901P.
M-1427; 90 Cu, 9.5 Zn, 0.5 P.
Sintered: 25,000-30,000 TS; 18-25 El.
For hardware, structural parts; powder metallurgy.

SINTALOY Z-901 PB.
M-1427; 88.5 Cu, 10 Zn, 1.5 Pb.
Sintered: 18,000-27,000 TS; 8-15 El.
For hardware, structural parts; free-cutting, powder metallurgy.

SINTEEL M.
M-1020; steel impregnated with Cu.
Sintered: 100,000 TS.
Heat treated: 170,000 YS.
For machinery parts; sintered.

SINTEEL R.
M-1020; steel impregnated with Cu.
For machinery parts; sintered.

SINTEEL S.
M-1020; steel impregnated with Cu.
For machinery parts; sintered.

SINTERBRONZE.
M-1246; Sn, bal Cu.
Sintered.
For bearings; sintered.

SINTERLOY A.
M-799; 0.15 C, bal Fe.
For gears, cams, washers; sintered alloy.

SINTERLOY B.
M-799; 0.40 C, bal Fe.
For gears, cams, washers; sintered alloy.

SINTERLOY C.
M-799; 0.80 C, bal Fe.
For gears, cams; sintered alloy.

SINTHERM.
M-1246; TiC, Zr, B.
For evaporation boats.

SINTOX.
M-1379; Al_2O_3.
For cutting tools; sintered.

SINTREX D.
M-1710; 99.0 min Fe, other non-volatile 0.12 max.
Electrolytic iron powder.
100 mesh standard, annealed powder for sintering to soft magnetic iron parts.

SINTREX E.
M-1710; 99.0 min Fe, other non-volatile 0.12 max.
Electrolytic iron powder.
100 mesh sponge, annealed powder, for sintering to soft magnetic iron parts.

SINTREX F.
M-1710; 99.0 min Fe, other non-volatile 0.12 max.
Electrolytic iron powder.
200 mesh annealed powder for sintering into soft magnetic iron parts.

SIOUX H-HARDENED RINGS.
M-512; 0.76 C, 0.54 Mn, 1.38 Si, 3.8 Cr, 5.53 Mo, bal Fe.
Hardened: 200,000-225,000 TS; 11-9 El; 35-28 RA.
For valve rings, valve seat inserts; Rockwell C-47.

SIOUX SUPER CAST RINGS.
M-512; 0.45-0.55 Mo, 1.1-1.2 Si, 1.6 C, 0.6-0.7 Mn, 1.0-2.0 Ni, 0.55-0.65 Cr, bal Fe.
At 75°F: 75,000 TS; 46,000 YS; 35 El; 1.8 RA.
At 1000°F: 43,000 TS; 33,000 YS; 14 El; 13.3 RA.
For valve seat inserts, valve seat rings; heat and scale resistant.

SIRIUS-5 CAST.
M-1546; 0.30 C, 16.0 Cr, 12.0 Ni, 3.6 Si, 1.2 Mn, bal Fe.
For furnace parts, heat treating boxes, superheaters, carburizing boxes, salt pots.
Austenitic, stainless; useful to 1050-1100°C in slightly sulphur atmosphere.

SIRIUS-10 CAST.
M-1546; 0.25 C, 18.0 Cr, 9.5 Ni, 1.3 Si, bal Fe.
For furnace parts, superheaters, heat treating boxes.
Austenitic, stainless. Resists sulphur atmosphere.

SIRIUS 30.
M-1546; 0.19 C, 19.0 Cr, 7.0 Ni, 4.0 W, 1.5 Si, 1.5 Mn, bal Fe.
Heat treated: 107,000 TS; 43,000 YS; 30 El.
For turbine blades and discs, catalyzer equipment, jet engine parts.
Austenitic stainless steel. Heat resistant to 1000°C.

SIRIUS 35 see NICRAL C.

SIRIUS 36.
M-1488; 0.52 C, 21 Cr, 4 Ni, 9 Mn, 0.15 Nb, 0.35 N, bal Fe.
Sol. tr.: 1030 N/mm² min UTS.
Valve steel.
AFNOR Z52CMN 21-9 plus N; SAE EV 8.

SIRIUS 40.
M-1546; 0.32 C, 14.0 Cr, 14.0 Ni, 2.5 W, 1.5 Si, bal Fe.
Annealed: 114,000 TS; 64,000 YS; 40 El.
For valves, exhaust valves for automobile and aircraft engines.
Austenitic, stainless, heat resistant.

SIRIUS 345 see NICRAL H.

SIRIUS SUPER-HT.
M-1546; 0.20 C, 16 Cr, 14 Ni, 10 Co, 2.5 W, 2.0 Ti, 1.0 Si, bal Fe.
Heat treated: 100,000 TS; 72,000 YS; 20 El.
For gas turbine component parts.
Austenitic, stainless. Resists oxidation up to 950°C.

SIS.
Prefix used on some Swedish alloy standards.

SIS 22-40.
M-Sweden; 0.32 C, 2.4-2.6 Cr, 0.3 Mo, 0.3 V, 0.5 Ni, bal Fe.
This steel is usually nitrided; for gears, cams, pinions, camshafts.

SISERSKITE.
M-USSR; 57 Os, 8 Ru, 34 Rh + Ir, Pt, Pd, Au, Cu.
For fountain pen points; nautral osmiridium mined in U.S.S.R.

SISTAL.
M-115; 0.47 C, 0.90 Si, 0.30 Mn, 8.25 Cr, 1.2 W, 1.35 Mo, 0.30 V, bal Fe.
Air hardenable to about 58 Rc.
Hot work tool steel, for extrusion press, upsetting rams, pressure dies.

SITKIN NO. 100.
M-1727; 4.5-5.7 Sn, 2.75-4.0 Pb, 1.0-9.5 Zn, bal Cu.
For bearings, pump impellers; free cutting.

SITKIN NO. 101.
M-1727; 4.5-5.75 Sn, 2-3 Pb, 9.5-16 Zn, bal Cu.
For bushings, bearings, pump impellers; free-cutting.

SITKIN NO. 110.
M-1727; 4.5-5.5 Sn, 4.25-7.0 Pb, 2.0-7.0 Zn, bal Cu.
For pump impellers, bushings, bearings; free-cutting, red brass.

SITKIN NO. 115.
M-1727; 4.0-5.5 Sn, 4.25-7.0 Pb, 4.0-10.0 Zn, bal Cu.
Cast: 33,000-46,000 TS; 17,000-24,000 YS; 15-35 El; 12-32 RA; 55-65 Brin.
For hardware, plumbing; red brass, free-cutting.

SITKIN NO. 120.
M-1727; 3.0-4.5 Sn, 6-10 Pb, 5-10 Zn, bal Cu.
Cast: 30,000-38,000 TS; 12,000-17,000 YS; 15-27 El; 12-25 RA; 50-60 Brin.
For hardware, plumbing; red brass, free-cutting.

SITKIN NO. 123.
M-1727; 2-4 Sn, 5-8 Pb, 6-18 Zn, bal Cu.
Cast: 29,000-39,0000 TS; 13,000-17,000 YS; 18-30 El; 15-27 RA; 50-60 Brin.
For hardware, plumbing; plumbers brass, free-cutting.

SITKIN NO. 125.
M-1727; 2-4 Sn, 5-8 Pb, 6-18 Zn, bal Cu.
Cast: 30,000-40,000 TS; 12,000-16,000 YS; 20-35 El; 15-30 RA; 50-60 Brin.
For hardware, plumbing; plumbers brass, free-cutting.

SITKIN NO. 130.
M-1727; 1-3 Sn, 4-10 Pb, 8-20 Zn, bal Cu.
For hardware, plumbing; free-cutting.

SITKIN NO. 131.
M-1727; 1-2 Sn, 1-2 Pb, 2-8 Zn, bal Cu.
For hardware, plumbing; free-cutting.

SITKIN NO. 132.
M-1727; 1-3 Sn, 1-3 Pb, 86-88 Cu, bal Zn.
For hardware, plumbing; free-cutting.

SITKIN NO. 193.
M-1727; 21-23 Sn, 0.50 max others, bal Cu.
For bearings.

SITKIN NO. 194.
M-1727; 18-20 Sn, bal Cu.
For bearings.

SITKIN NO. 195.
M-1727; 16-18 Sn, bal Cu.
For bearings.

SITKIN NO. 196.
M-1727; 14-16 Sn, bal Cu.
For bearings.

SITKIN NO. 197.
M-1727; 13-15 Sn, 1.5 max Zn, 0.2 Pb, bal Cu.
For bearings.

SITKIN NO. 198.
M-1727; 12.5-14.5 Sn, 2.5-4.5 Zn, 1 max Pb, bal Cu.
For bearings.

SITKIN NO. 199.
M-1727; 12-14 Sn, bal Cu.
For bearings.

SITKIN NO. 200.
M-1727; 10.75-12.25 Sn, 0.75-1.25 Pb, 0.1 max Zn, bal Cu.
For bearings; free-cutting.

SITKIN NO. 201.
M-1727; 10.75-12.0 Sn, 0.8-1.2 Pb, 1-4 Zn, bal Cu.
For bearings.

SITKIN NO. 205.
M-1727; 9.75-12.0 Sn, bal Cu.
For bearings.

SITKIN NO. 206.
M-1727; 9-11 Sn, 1.0-2.5 Pb, bal Cu.
For bearings.

SITKIN NO. 210.
M-1727; 9.0-10.7 Sn, 1-7 Zn, 0.2 max Pb, bal Cu.
For bearings.

SITKIN NO. 215.
M-1727; 9.0-10.7 Sn, 1-7 Zn, 0.8-1.2 Pb, bal Cu.
For bearings.

SITKIN NO. 220.
M-1727; 9.0-10.7 Sn, 1.2-2.7 Pb, 1-5 Zn, bal Cu.
For bearings.

SITKIN NO. 221.
M-1727; 9.0-10.7 Sn, 2.7-4.2 Pb, 1-5 Zn, bal Cu.
For bearings.

SITKIN NO. 225.
M-1727; 7.5-9.7 Sn, 1-7 Zn, 0.2-0.4 Pb, bal Cu.
For bearings.

SITKIN NO. 230.
M-1727; 7.5-9.0 Sn, 0.80-1.2 Pb, 1-7 Zn, bal Cu.
Cast: 33,000-43,000 TS; 16,000-24,000 YS; 18-30 El; 15-30 RA; 60-75 Brin.
For bearings; Gun Metal.

SITKIN NO. 235.
M-1727; 7.5-9.0 Sn, 1.2-2.7 Pb, 1-6 Zn, bal Cu.
For bearings.

SITKIN NO. 240.
M-1727; 7.5-9.0 Sn, 2.7-4.2 Pb, 1-6 Zn, bal Cu.
For bearings.

SITKIN NO. 241.
M-1727; 7.5-9.0 Sn, 4.2-6.2 Pb, 1-6 Zn, bal Cu.
For bearings.

SITKIN NO. 242.
M-1727; 6-8 Sn, 0.5 max Zn, bal Cu.
For bearings.

SITKIN NO. 245.
M-1727; 5.7-7.5 Sn, 1-2 Pb, 1-7 Zn, bal Cu.
Cast: 34,000-42,000 TS; 16,000-21,000 YS; 20-35 El; 16-30 RA; 60-70 Brin.
For valves, pump parts; free-cutting.

SITKIN NO. 250.
M-1727; 5.7-7.5 Sn, 2.0-3.2 Pb, 2-6 Zn, bal Cu.
For bearings.

SITKIN NO. 251.
M-1727; 5.7-7.5 Sn, 3.2-5.2 Pb, 1-6 Zn, bal Cu.
For bearings.

SITKIN NO. 253.
 M-1727; 4.5-5.7 Sn, 1-2 Pb, 1-9 Zn, bal Cu.
 For bearings.

SITKIN NO. 255.
 M-1727; 4.0-5.7 Sn, 2.0-2.7 Pb, 1-4 Zn, bal Cu.
 For bearings.

SITKIN NO. 256.
 M-1727; 3.0-4.5 Sn, 1.7-3.0 Pb, 1-9 Zn, bal Cu.
 For bearings.

SITKIN NO. 257.
 M-1727; 3-4 Sn, 0.5-1.0 Pb, 85-90 Cu, bal Zn.
 For bearings.

SITKIN NO. 295.
 M-1727; 15-17 Sn, 4-6 Pb, 0.25-1.0 Zn, bal Cu.
 For bearings.

SITKIN NO. 296.
 M-1727; 12-14 Sn, 14-16 Pb, 0.5 max Zn, 0.5-1.0 Ni, bal Cu.
 For bearings.

SITKIN NO. 296.5.
 M-1727; 10.5-12.5 Sn, 7-10 Pb, bal Cu.
 For bearings.

SITKIN NO. 297.
 M-1727; 9.5-11.5 Sn, 4.2-6.0 Pb, 0.75-1.5 others, bal Cu.
 For bearings.

SITKIN NO. 298.
 M-1727 9-11 Sn, 4.2-6.2 Pb, 1-5 Zn, bal Cu.
 For bearings.

SITKIN NO. 299.
 M-1727; 9-11 Sn, 9-11 Pb, 0.15-0.75 Zn, bal Cu.
 For bearings.

SITKIN NO. 300.
 M-1727; 9.0-10.7 Sn, 8.5-11.7 Pb, 0.25 max Sb, bal Cu.
 For bearings.

SITKIN NO. 305.
 M-1727; 9.0-10.7 Sn, 8.5-21.7 Pb, 0.75-1.5 others, bal Cu.
 Cast: 27,000-37,000 TS; 15,000-22,000 YS; 6-12 El; 5-11 RA; 55-70 Brin.
 For bearings; bearing bronze.

SITKIN NO. 310.
 M-1727; 7.5-9.0 Sn, 7.2-8.7 Pb, 0.75-1.5 others, bal Cu.
 For bearings.

SITKIN NO. 311.
 M-1727; 7.5-9.0 Sn, 7.2-8.7 Pb, bal Cu.
 For bearings.

SITKIN NO. 312.
 M-1727; 7.5-9.0 Sn, 8.7-11.0 Pb, 0.75-1.5 others, 3.0 max Zn, bal Cu.
 For bearings.

SITKIN NO. 313.
 M-1727; 7.5-9.0 Sn, 12-16 Pb, 0.25-0.50 Ni, bal Cu.
 For bearings.

SITKIN NO. 314.
 M-1727; 7.5-9.0 Sn, 12-16 Pb, 0.25-0.75 others, bal Cu.
 For bearings.

SITKIN NO. 315.
 M-1727; 5.5-7.5 Sn, 6.0-10.5 Pb, 4 max Zn, 0.75-1.5 others, bal Cu.
 Cast: 30,000-38,000 TS; 17,000-21,000 YS; 12-20 El; 10-22 RA; 55-65 Brin.
 For bearings, bushings; bearing bronze.

SITKIN NO. 319.
 M-1727; 5.7-7.5 Sn, 12-16 Pb, 0.75-2.0 others, bal Cu.
 Cast: 25,000-30,000 TS; 14,000-20,000 YS; 10-18 El; 8-15 RA; 50-60 Brin.
 For bearings, bushings; bearing bronze.

SITKIN NO. 320.
 M-1727; 5.7-7.5 Sn, 16.7-29.7 Pb, 3.0 max Zn, bal Cu.
 For bearings.

SITKIN NO. 321.
 M-1727; 5.7-7.5 Sn, 16-29 Pb, 0.75-1.5 others, bal Cu.
 For bearings.

SITKIN NO. 322.
 M-1727; 4.0-5.7 Sn, 21-26 Pb, 1.5 max others, bal Cu.
 Cast: 23,000-30,000 TS; 11,000-15,000 YS; 7-16 El; 5-12 RA; 42-55 Brin.
 For bearings, bushings, bearing bronze.

SITKIN NO. 323.
 M-1727; 4-6 Sn, 21.7-26.0 Pb, 0.5 max Zn, bal Cu.
 For bearings.

SITKIN NO. 324.
 M-1727; 4.0-5.7 Sn, 26-32 Pb, 0.75-1.5 impurities, bal Cu.
 For bearings.

SITKIN NO. 325.
 M-1727; 4.5-5.7 Sn, 11.7-21.7 Pb, 3 max Zn, bal Cu.
 For bearings.

SITKIN NO. 326.
 M-1727; 3.5-5.5 Sn, 8-10 Pb, 4 max Zn, bal Cu.
 For bearings.

SITKIN NO. 400.
 M-1727; 0.5-1.7 Sn, 2-4 Pb, 70-74 Cu, bal Zn.
 Cast: 35,000-40,000 TS; 12,000-14,000 YS; 25-40 El; 20-40 RA; 40-55 Brin.
 For plumbing, hardware; leaded yellow brass, free-cutting.

SITKIN NO. 403.
M-1727; 66-70 Cu, 0.5-1.7 Sn, 2-4 Pb, bal Zn.
Cast: 30,000-38,000 TS; 11,000-15,000 YS; 20-35 El; 15-30 RA; 40-60 Brin.
For plumbing, hardware; yellow brass, free-cutting.

SITKIN NO. 405.
M-1727; 66 max Cu, 0.5-1.7 Sn, 2-4 Pb, bal Zn.
For hardware, screw machine products; free-cutting, yellow brass.

SITKIN NO. 405.1.
M-1727; 65 max Cu, 0.5-1.5 Sn, 0.5 max Pb, bal Zn.
For hardware, screw machine products; yellow brass.

SITKIN NO. 405.2.
M-1727; 65 max Cu, 0.5-1.5 Sn, 0.5-1.0 Pb, bal Zn.
For hardware, screw machine products; free-cutting, yellow brass.

SITKIN NO. 406.
M-1727; 66 max Cu, 0.5-1.0 Sn, 1.0-1.5 Pb, bal Zn.
For hardware, screw machine products; free-cutting, yellow brass.

SITKIN NO. 407.
M-1727; 66-72 Cu, 0.5 max Sn, 0.5 max Pb, bal Zn.
For hardware, screw machine products; yellow brass.

SITKIN NO. 407.5.
M-1727; 90-95 Cu, 0.20 max Sn, 0.50 max Pb, bal Zn.
For clips, fasteners, springs; good corrosion resistance.

SITKIN NO. 408.
M-1727; 83-86 Cu, 0.20 max Sn, 0.50 max Pb, bal Zn.
For clips, fuse parts, hardware; good corrosion resistance.

SITKIN NO. 409.
M-1727; 61 max Cu, 1.5-2.0 Pb, bal Zn.
For screw machine products, hardware; free-cutting, yellow brass.

SITKIN NO. 410.
M-1727; 5 max Sn, 10 max Pb, 9-14 Ni, bal Cu + Zn.
Cast: 30,000-40,000 TS; 15,000-20,000 YS; 10-25 El; 7-20 RA; 50-60 Brin.
Nickel silver, leaded.

SITKIN NO. 411.
M-1727; 5 max Sn, 10 max Pb, 14-18 Ni, bal Cu + Zn.
Cast: 35,000-45,000 TS; 17,000-24,000 YS; 15-30 El; 15-30 RA; 65-80 Brin.
For bearings; nickel silver, leaded.

SITKIN NO. 412.
M-1727; 5 max Sn, 10 max Pb, 18-22 Ni, bal Cu + Zn.
Cast: 40,000-60,000 TS; 17,000-30,000 YS; 15-25 El; 11-22 RA; 76-120 Brin.
For bearings; nickel silver, leaded.

SITKIN NO. 413.
M-1727; 5 max Sn, 10 max Pb, 22-27 Ni, bal Cu + Zn.
For bearings; nickel silver, leaded.

SITKIN NO. 414.
M-1727; 5 max Sn, 10 max Pb, 27-33 Ni, bal Cu + Zn.
For bearings; nickel silver, leaded.

SITKIN NO. 415.
M-1727; 1 max Sn, 0.1 max Pb, 8-13 Al, bal Cu + Zn.
Cast: 70,000-87,000 TS; 25,000-30,000 YS; 2-38 El; 20-36 RA; 110-140 Brin.
For hardware, propellers, gears; Al. Brinze.

SITKIN NO. 420.
M-1727; Zn, Mn, bal Cu.
Cast: 60,000-65,000 TS.
For hardware, propellers, gears; Mn. Bronze.

SITKIN NO. 421.
M-1727; 56-69 Cu, 0.75-2.0 Fe, 0.3 max Pb, 0.5 max Ni, bal Zn.
Cast: 65,000-68,000 TS; 28,000-40,000 YS; 20-35 El; 20-40 RA; 115 Brin.
For propellers, gears, hardware; Bronze.

SITKIN NO. 422.
M-1727; Zn, Mn, bal Cu.
Cast: 80,000-90,000 TS.
For hardware, gears, propellers; Mn. Bronze.

SITKIN NO. 423.
M-1727; Zn, Mn, bal Cu.
Cast: 90,000-100,000 TS.
For hardware, gears, propellers; Mn. Bronze.

SITKIN NO. 424.
M-1727; 60-68 Cu, 2-4 Fe, 1 max Ni, 0.1 max Pb, bal Zn.
Cast: 100,000-120,000 TS; 65,000-90,000 YS; 12-18 El; 5-18 RA; 190-235 Brin.
For hardware, gears, propellers; Mn. Bronze.

SITKIN NO. 500.
M-1727; 90 min Cu, 2 max Sn, 2.0-5.5 Si, 2.5 max Fe, 1.5 max Mn.
For bolts, gears, fasteners; Si. Bronze.

SIVYER 5% CR.
M-288; 0.25 max C, 4.5-6.5 Cr, 0.45-0.65 Mo, bal Fe.
Normalized: 100,000-110,000 TS; 75,000-90,000 YS; 22-18 El; 229-241 Brin.
For oil refinery and chemical plant equipment; Type 502; corrosion resistant.

SIVYER 20.
M-288; 0.07 max C, 29 Ni, 20 Cr, 2-3 Mo, 4 Cu, bal Fe.
Cast: 65,000-75,000 TS; 28,000-38,000 YS; 50-35 El; 50-40 RA; 120-150 Brin.
For chemical plant equipment, valves, mixers, tanks; resists mixed acids, austenitic.

SIVYER CHROME ELECTRIC STEEL.
M-288; 0.22 C, 5.0 Cr, 0.75 Mn, 0.40 Si, bal Fe.
Annealed: 95,000 TS; 55,000 YS; 18 El; 25 RA; 200 Brin.
Heat treated: 100,000-200,000 TS; 70,000-165,000 YS; 19-3 El; 35-4 RA; 200-550 Brin.
For miscellaneous castings; corrosion resistant.

SIVYER CHROME NICKEL ELECTRIC STEEL.
M-288; 0.40 C, 0.9 Cr, 3 Ni, bal Fe.
Annealed: 100,000 TS; 65,000 YS; 20 El; 35 RA; 200 Brin.
Heat treated: 115,000-210,000 TS; 85,000-173,000 YS; 18-6 El; 43-10 RA; 200-600 Brin.
For gears, shafts; oil hardening.

SIVYER "DYNAMO".
M-288; 0.12 C, 0.10 Mn, 0.20 Si, bal Fe.
Normalized: 55,000 TS; 35,000 YS; 30 El; 60 RA; 120 Brin.
For electrical parts; high magnetic permeability.

SIVYER HIGH CARBON STEEL.
M-288; 0.7 C, bal Fe.
Annealed: 80,000 TS; 40,000 YS; 20 El; 30 RA; 175 Brin.
Heat treated: 85,000-150,000 TS; 55,000-90,000 YS; 20-3 El; 35-4 RA; 175-400 Brin.
For tools, springs; water hardening.

SIVYER HIGH SILICON.
M-288; 0.09-0.12 C, 0.15 Mn, 1.75 Si, bal Fe.
For electrical equipment; high permeability.

SIVYER MANGANESE-CARBON ELECTRIC STEEL.
M-288; 0.32 C, 1.45 Mn, 0.40 Si, bal Fe.
Annealed: 95,000 TS; 55,000 YS; 22 El; 40 RA; 200 Brin.
Heat treated: 95,000 TS; 65,000 YS; 22-2.5 El; 50-3.5 RA; 200-450 Brin.
For gears, shafts; oil hardening.

SIVYER NITRO.
M-288; 0.30 C, 0.75 Mn, 0.40 Si, 1.5 Cr, 0.6 v, bal Fe.
For gears, pinions, shafts, cams; nitriding steel.

SIVYER NO. 60.
M-288; 18 Cr, 8 Ni, 0.50 Mn, 0.10 C, bal Fe.
Cast: 70,000-80,000 TS; 30,000-40,000 YS; 50-60 El; 50-70 RA; 135-150 Brin.
For stainless parts; corrosion resisting.

SIVYER NO. 62.
M-288; 23-25 Cr, 11-13 Ni, 1.0 Si, 0.5 Mn, 0.18 C, bal Fe.
Cast: 75,000-85,000 TS; 30,000-40,000 YS; 35-45 El; 35-45 RA.
For furnace parts, heat and corrosion resisting parts; heat and corrosion resisting.

SIVYER NO. 64.
M-288; 0.25 C, 0.50 Mn, 1.0 Si, 30 Cr, 9.5 Ni, bal Fe.
For corrosion and heat resistant parts; corrosion and heat resistant.

SIVYER NO. 66.
M-288; 12-14 Cr, 0.5 Mn, 1.0 Si, 0.10 C, bal Fe.
Cast: 75,000-170,000 TS; 45,000-130,00 YS; 8-25 El; 15-50 RA; 175-350 Brin.
For furnace parts, corrosion resisting parts; corrosion resisting.

SIVYER NO. 67.
M-288; 16-18 Cr, 0.25 C, 0.50 Mn, 1.0 Si, bal Fe.
Cast: 85,000-105,000 TS; 55,000-70,000 YS; 10-20 El; 15-35 RA; 160-190 Brin.
For furnace parts, corrosion resisting parts; corrosion resisting.

SIVYER NO. 70.
M-288; 15-17 Cr, 35-37 Ni, 1.0 Si, 0.45 C, bal Fe, 0.50 Mn.
Cast: 60,000-70,000 TS; 1-7 El; 1-8 RA; 200-225 Brin.
For furnace parts, heat and corrosion resisting parts; corrosion and heat resisting.

SIVYER "NULOY".
M-288; 0.28 C, 1.25 Mn, 0.40 Si, bal Fe.
For castings, gears; water hardenable.

SIVYER "POLE".
M-288; 0.18 C, 0.55 Mn, 0.25 Si, bal Fe.
For cast hardened parts; carburizing steel.

SIVYER VANADIUM ELECTRIC STEEL.
M-288; 0.45 C, 0.20 V, 0.75 Mn, 0.40 Si, bal Fe.
Annealed: 85,000 TS; 50,000 YS; 22 El; 40 RA; 180 Brin.
For shafts, gears, pinions; water hardening.

SIWI.
M-40; C, alloy, bal Fe.
For dies; hot die steel.

SIX EIGHTY ALLOY.
M-200; 65% Ag.
For dental amalgams.

SIXIX.
M-435; 0.8 C, 4 Cr, 6.5 W, 5 Mo, 2 V, bal Fe.
For cutters, tools; high speed steel.

SIXTY N.
M-604; 0.20 C, 1.35 Mn, 0.35 Si, 0.02 N, 0.04-0.11 V, bal Fe.
Normalized: 80 ksi TS; 60 ksi YS; 23 El.
Readily formed and welded.
For stressed structures at low temperatures, arctic and marine structures.
ASTM A633 Grade E.

SJ CALKING LEAD.
M-1747; 0.0005 Ag, 0.0003 Cd, 99.99 min Pb.
For plumbing applications.
Exceeds ASTM B29-55.

SJMFA-300.
M-1747; 0.030 Ca, 0.001 max Ag, bal Pb. (Bi, Sn, Sb, Cu, As, Zn, Cd, Ni, Fe 0.0005 max each).
Cast and aged: 3300-3700 psi TS; 40-50 El; melt range: 422°F (approx); 40-45 Rock R.
Cast grids for industrial stand-by batteries.

SJMFA-700.
M-1747; 0.070 Ca, 0.001 max Ag, bal Pb. (Bi, Sn, Sb, Cu, As, Zn, Cd, Ni, Fe 0.0005 max each).
Cast and aged: 5200-5600 psi TS; 30-45 El; melt range: 623°F (approx); 70-80 Rock R.
Cast lead products of moderate strength.

SJMFA-707.
M-1747; 0.70 Sn, 0.070 Cd, 0.001 max Ag, bal Pb. (Bi, Sb, Cu, As, Zn, Cd, Ni, Fe 0.0005 max each).
Cast and aged: 7000-7500 psi TS; 30-40 El; melt range: 619-633°F; 85-95 Rock R.
All purpose casting; battery grids.

SJMFA-903.
M-1747; 0.30 Sn, 0.10 Ca, 0.001 max Ag, bal Pb. (Bi, Sb, Cu, As, Zn, Cd, Ni, Fe 0.0005 max each).
Cast and aged: 6000-6500 psi TS; 25-35 El; melt range: 622-640°F; 80-85 Rock R.
Cast parts and grids for maintenance-free batteries.

SJMFA-905.
M-1747; 0.50 Sn, 0.10 Ca, 0.001 max Ag, bal Pb. (Bi, Sb, Cu, As, Zn, Cd, Ni, Fe 0.0005 max each).
Cast and aged: 6500-7000 psi TS; 25-35 El; melt range: 621-637°F; 85-90 Rock R.
All purpose casting; grids for storage batteries.

SJMFA-910.
M-1747; 1.0 Sn, 0.10 Ca, 0.001 max Ag, bal Pb. (Bi, Sb, Cu, As, Zn, Cd, Ni, Fe 0.0005 max each).
Cast and aged: 7500-8000 psi TS; 20-35 El; melt range: 617-630°F; 90-95 Rock R.
Cast grids for storage batteries.

SJMFA-1000.
M-1747; 0.10 Ca, 0.001 max Ag, bal Pb. (Bi, Sn, Sb, Cu, As, Zn, Cd, Ni, Fe 0.0005 max each).
Cast and aged: 5400-5800 psi TS; 30-45 El; melt range: 623-625°F; 65-80 Rock R.
Cast grids for maintenance-free batteries.

SK 42.
M-883; 0.27 C, 1.2 Mn, bal Fe.
Plate: 60,000 TS; 42,000 YS; 19 El.
For mine and railroad cars, booms, derricks, bridges, pressure vessels.
Good fabricability and weldability.

SK A45.
M-883; 0.22 C, 1.25 Mn, 0.10 Si, 0.02 min V, bal Fe.
Rolled: 65,000 TS; 45,000 YS; 19 El.
For railroad and mine cars, booms, bridges, derricks, pressure vessels.
Good fabricability and weldability.

SK A50.
M-883; 0.26 C, 1.3 Mn, 0.10 Si, 0.02 Min V, bal Fe.
Rolled: 70,000 TS; 50,000 YS; 18 El.
For railroad and mine cars, auto and bus bodies, bridges, derricks, pressure vessels.
Good fabricability and weldability.

SKEFKO NO. 1.
M-224; 0.95-1.05 C, 1.4-1.65 Cr, 0.25-0.35 Mn, bal Fe.
Heat treated: 90,000-160,000 TS; 60,000-140,000 YS; 20-6 El; 45-12 RA; 190-450 Brin.
For ball bearings; water hardening.

SKEFKO NO. 2.
M-224; 1.01-1.1 C, 0.4-0.6 Cr, 0.30-0.40 Mo, bal Fe.
For ball bearings; water hardening.

SKEFKO NO. 3.
M-224; 0.95-1.05 C, 1.0-1.25 Cr, 0.30-0.40 Mo, bal Fe.
For ball bearings; water hardening.

SKELTER PLATE.
M-15; 0.39 C, 2.3 Mn, 0.90 Cr, 0.34 Mo, bal Fe.
Ht Tr to 139,000 psi YS.

S.K.F. 100.
M-267; 0.7-1.0 C, bal Fe.
For stamping, cold heading and trimming dies, general tools; water hardened.

S.K.F. DRILL.
M-267; C, bal Fe.
For drills.

SKF NO. 2.
M-267; 0.87-0.97 C, 1.4-1.7 Mn, 1.4-1.7 Cr, bal Fe.
For trimming and stamping dies; oil hardened.

S.K.F. NO. 3 TUBING.
M-267; 1.0 C, 1 Cr, bal Fe.
For bearing races, bushings; water hardened.

SKF NO. 6.
M-267; 0.9 C, 1.1 Cr, 0.6 Mn, bal Fe.
For ball bearings, rollers, spindles; water hardened.

S.K.F. NO. 7.
M-267; 1.1 C, 0.6 Cr, bal Fe.
For balls, rollers, spindles; water hardened.

S.K.F. NO. 9.
M-267; 1.1 C, 0.9 Cr, bal Fe.
For balls, rollers, spindles; water hardened.

S.K.F. NO. 13.
M-267; 1.1 C, 1.0 Cr, bal Fe.
For balls, rollers, spindles; water hardened.

S.K.F. NO. 48 STEEL.
M-267; high C, high Cr, Mo, bal Fe.
For cold trimming, drawing, punching and blanking dies, gauges; air hardening steel.

SKF NO. 85.
M-267; C, alloy, bal Fe.
For chisels; water hardened.

S.K.F. NO. 711.
M-267; 0.5 C, 2.5 W, 1.2 Cr, 0.7 Si, 2 V, bal Fe.
For tools, shear blades, dies, cutters, chisels; good wear resistance.

S.K.F. STEEL NO. 1.
M-267; 1.0 C, 1.05 Mn, 0.45 Si, 1.05 Cr, bal Fe.
For ball bearings, races and balls; wear resistant.

S.K.F. STEEL NO. 3.
M-267; 1.0 C, 1.50 Cr, bal Fe.
For ball bearings, races and balls; wear resistant.

S.K.F. STEEL NO. 22.
M-267; 1.0 C, 0.35 Mo, 1.1 Cr, bal Fe.
For trimming and stamping dies, drills; wear resistant.

S.K.F. STEEL NO. 36.
M-267; 2.25 C, 11.5-12.5 Cr, 0.15 V, 1.2 Mn, 0.20 Si, bal Fe.
For cold trimming, drawing, punching, blanking dies, gages; formerly known as "VH-63 Steel."

SKF STEEL NO. 46.
M-267; C, Cr, W, V, bal Fe.
For tools, cutters, dies; non-deforming.

SK H8.
M-Japan; 0.65 C, 16.0 W, 4.3 Cr, 0.76 V, 2.64 Co, bal Fe.
Hardened: C 64-66 Rock.
For tools and cutters, lathe and planer tools, drills.
High-speed steel, high red-hardness.

SKHLF STEEL.
M-USSR; 0.2 C, 0.3 Si, 0.5 Mn, 0.55 Cr, 0.45 Ni, 0.68 Cu, bal Fe.
For welded structures; good weldability.

S.K. SILVER STEEL see **SILVER STEEL.**

SL250.
M-1522; 2.5 Ag, Cu, Pb.
Melt range: 302-305°C.
Max stress: 3.9 kgf/mm^2; 45 El.
For soft soldering; high temperature grade.

S.L.B.
M-802; 0.45 C, bal Fe.
For castings.

S-LESS STEEL (BREARLY).
M-Eng.; 13 Cr, 0.3 C, bal Fe.
For corrosion resisting parts; corrosion resistant.

SLICKER SOLDER.
M-U.S.; 66 Sn, 34 Pb.
Commercial lead solder; M.P. 356°F.

SLIDE VALVE BRONZE.
M-Eng.; 2.5 Sn, 9 Zn, bal Cu.
For slide valves; high strength.

SLIDEX.
M-1555; C 10-18 Ni, Zn, bal Cu.
Nickel silver alloy, for slide fasteners.

SLOCRODE.
M-494, M-1485; 3.2 C, 0.75-1.25 Ni, 0.5-0.75 Cr, bal Fe.
Oil treated: 36,000-40,000 TS; 200-500 Brin.
For castings for high duty purposes; cast iron.

SLT2N.
M-1296; 0.14 max C, 0.3 max Si, 0.7 max Mn, 2.1-2.5 Ni, bal Fe.
6-32 mm thick plate: 65,000-77,000 TS; 37,000 min YP.
2.5 Ni steel for low temperature service up to -70°C (-94°F) for pressure vessels, cryogenic equipment. High notch toughness.
Conforms to ASTM A203 Gr. A.

SLT3NA.
M-1296; 0.14 max C, 0.3 max Si, 0.7 max Mn, 3.25-3.75 Ni, bal Fe.
6-50 mm thick plate: 65,000-81,000 TS; 37,000 or 40,000 min YP.
3.5% Ni normalized steel for low temperature service to -101°C (-150°F).
High notch toughness.
Available in 2 strength grades: SLT3N-26A, SLT3N-28A.
Conforms to ASTM A203 Gr. D or E.

SLT3NB.
M-1296; 0.14 max C, 0.3 max Si, 0.9 max Mn, 3.25-3.75 Ni, 0.25 max Mo 0.5 max Cr, bal Fe.
6-50 mm thick plate: 79,000-100,000 TS; 64,000 min YP.
3.5% Ni quenched and tempered steel for low temperature service to -101°C (-150°F).
High notch toughness.

SLT9N-53.
M-1296; 0.12 max C, 0.15-0.30 Si, 0.9 max Mn, 8.5-9.5 Ni, bal Fe.
6-26 mm thick plate: 100,000-120,000 TS; 75,000 min YP.
9% Ni double normalized and tempered steel for low temperature service to -196°C(-320°F).
For LNG storage tanks.
Conforms to ASTM-A353.

SLT9N-60.
M-1296; 0.12 max C, 0.15-0.30 Si, 0.9 max Mn, 8.5-9.5 Ni, bal Fe.
6-50 mm thick plate: 100,000-120,000 TS; 85,000 min YP.
9% Ni quenched and tempered steel for low temperature service to -196°C (-320°F).
For LNG storage tanks.
Conforms to ASTM-A553 Type I.

SLUSH ALLOY.
M-91; 4.75 Al, 0.25 Cu, bal Zn.
For slush and permanent mold castings, primarily in making lighting fixtures.

S.L.V. STEEL.
M-56; M-521; 0.45 C, 8 Cr, 3.5 Si, bal Fe.
Oil treated: 145,600-156,800 TS; 100,000-112,000 YS; 20-15 El; 43-53 RA; 255-285 Brin.
For valves for automobile and aircraft engines; heat and corrosion resistant; Firth-Brown J-181.

SM-0005.
M-1186; 99.9 Cu.
Annealed: 30,000 TS; 6000 YS; 45 El; 70 RA; 50 Brin.
For electrical equipment; high conductivity.

SM-0010.
M-1186; 99.9 Cu.
Cold drawn: 61,000 TS; 57,000 YS; 3 El; 50 RA; 70 Brin.
For electrical equipment; high conductivity.

SM-0020.
M-1186; 99.9 Cu.
Wire: 61,000 TS; 57,000 YS; 3 El; 50 RA; 70 Brin.
Hard drawn: 37,000 TS; 29,000 YS; 10 El; 40 RA.
For electrical equipment; fire refined.

SM-0021.
M-1186; 99.8 Cu.
Annealed: 30,000 TS; 6000 YS; 45 El; 70 RA; 50 Brin.
Rolled: 39,000 TS; 33,000 YS; 20 El; 60 RA; 75 Brin.
For sheets, tubes; deoxidized copper.

SM 40.
M-106; 0.5-0.75 Mg, 4.8-5.7 Zn, 0.4-0.6 Cr, bal Al.
As sand cast: 215 MPa TS; 170 MPa YS; 4 El; 60 Brin.

S.M. 40 ETC see also JESSOP STEELS.

SM116.
M-137; 5.5-6.5 Mg, 11-13 Si, 1-2 Ni, bal Al.
44,000 TS; 36,000 YS; 10 El; 125 Brin.
For light alloy parts, die castings; hardenable, high temperature applications.

SM 200 now MAR-M-ALLOY 200.
M-1536; 0.15 C, 9 Cr, 12.5 W, 10 Co, 1 Cb, 5 Al, 2 Ti, 0.05 Zr, 0.015 B, 0.25 Fe, bal Ni.
At R.T.: 135,000 TS; 120,000 YS; 7 El.
At 1900°F: 60,000 TS; 50,000 YS; 5 El.
For cast gas-turbine blades, jet engine components.
Precipitation hardening. Good oxidation resistance, useful strength to 1900°F.

SM 211 now MAR-M-ALLOY 211.
M-1536; 0.15 C, 9.0 Cr, 10.0 Co, 2.5 Mo, 5.5 W, 2.7 Cb, 2.0 Ti, 5.0 Al, 0.015 B, 0.05 Zr, bal Ni.
For high temperature applications, fasteners, jet engine and gas turbine components.
High rupture strength, heat and corrosion resistant.

SM 237.
M-106; 0.2-0.4 Mg, 6.5-7.5 Si, 0.20 max Ti, bal Al.
As chill cast: 200 MPa TS; 95 MPa YS; 7 El.

SM 302 now MAR-M-ALLOY 302.
M-1536; 0.78-0.93 C, 20-23 Cr, 9-11 W, 8-10 Ta, 0.10-0.30 Zr, 1.5 max Fe, 0.010 max B, bal Co.
Cast: 140,000 TS; 100,000 YS; 2 El.
At 1800°F: 40,000 TS; 32,000 YS; 15 El.
For aircraft gas turbine guide vanes and buckets.
High oxidation resistance to 2000°F.

SM 322 now MAR-M-ALLOY 322.
M-1536; 0.9-1.1 C, 20-23 Cr, 8-10 W, 4-5 Ta, 2.0-2.5 Zr, 0.65-0.85 Ti, 1.5 max Fe, bal Co.
Cast: 121,000 TS; 91,000 YS; 3.2 El; 4.0 RA.
At 1400°F: 86,000 TS; 51,00 YS; 6.3 El; 9.3 RA.
For turbine blades and vanes.
High heat and oxidation resistant.
Vacuum melted and vacuum cast superalloy.

SM-0710.
M-1186; 1 Sn, bal Cu.
Annealed: 37,000 TS; 9000 YS; 45 El; 60 RA; 55 Brin.
Hard: 46,000 TS; 38,000 YS; 15 El; 40 RA; 95 Brin.
For sheets, wire, rod; phosphor bronze.

SM-0720.
M-1186; 2 Sn, bal Cu.
Annealed: 40,000 TS; 12,000 YS; 50 El; 60 RA; 60 Brin.
Hard: 51,000 TS; 44,000 YS; 15 El; 40 RA; 110 Brin.
For sheets; phosphor bronze.

SM-1063.
M-1186; 63 Cu, bal Zn.
Annealed: 47,000 TS; 14,000 YS; 60 El; 60 RA; 65 Brin.
Hardened: 60,000 TS; 45,000 YS; 25 El; 40 RA; 115 Brin.
For sheets, wire tubes; yellow brass.

SM-1067.
M-1186; 67 Cu, 23 Zn.
Drawn: 60,000 TS; 45,000 YS; 20 El; 45 RA.
For tubing; yellow brass.

SM-1072.
M-1186; 72 Cu, bal Zn.
Annealed: 45,000 TS; 14,000 YS; 65 El; 70 RA; 60 Brin.
For sheets, cartridge cases; cartridge brass.

SM-1076.
M-1186; 76 Cu, bal Zn.
Annealed: 43,000 TS; 11,000 YS; 50 El; 65 RA; 60 Brin.
For wire; ductile brass.

SM-1080.
M-1186; 80 Cu, bal Zn.
Annealed: 40,000 TS; 11,000 YS; 55 El; 75 RA; 55 Brin.
Rolled: 50,000 TS; 35,000 YS; 25 El; 65 RA; 110 Brin.
For sheets; low brass.

SM-1085.
M-1186; 85 Cu, bal Zn.
Annealed: 37,000 TS; 11,000 YS; 55 El; 75 RA; 55 Brin.
Rolled: 51,000 TS; 40,000 YS; 20 El; 65 RA; 110 Brin.
For wire, tube, rod; red brass.

SM-1092.
M-1186; 92 Cu, bal Zn.
Annealed: 34,000 TS; 11,000 YS; 50 El; 75 RA; 55 Brin.
Rolled: 45,000 TS; 33,000 YS; 20 El; 65 RA; 100 Brin.
For sheets; commercial bronze.

SM-1258.
M-1186; 1 Pb, 58 Cu, bal Zn.
Forged: 64,000 TS; 17,000 YS; 20 El; 20 RA; 110 Brin.
For architectural applications, trim, hardware; leaded brass, free-cutting.

SM-1361.
M-1186; 1.5 Pb, 61 Cu, bal Zn.
Annealed: 50,000 TS; 17,000 YS; 40 El; 40 RA; 80 Brin.
Cold drawn: 55,000 TS; 35,000 YS; 25 El; 30 RA; 100 Brin.
For hardware, bolts, nuts, screws, fasteners; leaded brass, free-cutting.

SM-1459.
M-1186; 2 Pb, 59 Cu, bal Zn.
Annealed: 53,000 TS; 17,000 YS; 30 El; 25 RA; 90 Brin.
Cold drawn: 68,000 TS; 57,000 YS; 8 El; 15 RA; 140 Brin.
For hardware, bolts, nuts, screws, fasteners; leaded brass, free-cutting.

SM-1658.
M-1186; 3 Pb, 59 Cu, bal Zn.
Cold drawn: 61,000 TS; 31,000 YS; 20 El; 15 RA; 110 Brin.
For screw machine products, fasteners; free-cutting, leaded brass.

SM-2060.
M-1186; 60 Cu, 0.5 Al, 0.5 Pb, 1 Sn, 2 Mn, 0.8 Fe, bal Zn.
Cold drawn: 68,000 TS; 43,000 YS; 20 El; 25 RA; 130 Brin.
For vessels, tanks, containers; brass.

SM-2265.
M-712; 2.2 Al, 2 Pb, 65 Cu, bal Zn.
Cast: 64,000 TS; 43,000 YS; 25 El; 30 RA; 120 Brin.
For welding rod; Al-brass.

SM-2266.
M-1186; 66 Cu, 4.5 Al, 2 Fe, 3.5 Mn, bal Zn.
Forged: 100,000 TS; 57,000 YS; 15 El; 20 RA; 180 Brin.
For rods, forgings; aluminum brass.

SM-2276.
M-1186; 76 Cu, 2 Al, bal Zn.
Hard: 61,000 TS; 40 El.
For tubes; Admiralty Brass.

SM-2359.
M-1186; 59 Cu, 2 Mn, bal Zn.
Drawn: 64,000 TS; 43,000 YS; 30 El; 35 RA; 120 Brin.
For wire, rod; manganese brass.

SM-2770.
M-1186; 70 Cu, 1 Sn, bal Zn.
Drawn: 57,000 TS; 30 El.
For tubes, condenser tubes; Admiralty Brass.

SM-3010.
 M-1186; 82 Cu, 10 Al, 4 Ni, 4 Fe.
 Forged: 92,000 TS; 50,000 YS; 15 El; 160 Brin.
 For rod, shapes, propellers; aluminum bronze.

SM-3080.
 M-1186; 85 Cu, 8 Al, 1 Ni, 2 Fe, 3 Mn, 1 Sn.
 Forged: 71,000 TS; 28,000 YS; 35 El; 40 RA; 120 Brin.
 For wire, rod, shapes; aluminum bronze.

SM-3090.
 M-1186; 91 Cu, 9 Al.
 Rolled: 71,000 TS; 28,000 YS; 30 El; 35 RA; 100 Brin.
 For sheets, rod; aluminum bronze.

SM-3099.
 M-1186; 81.5 Cu, 10 Al, 1.5 Ni, 3 Fe, 3 Mn, 1 Sn.
 Forged: 92,000 TS; 57,000 YS; 15 El; 20 RA; 160 Brin.
 For rods, shapes; aluminum bronze.

SM-3617.
 M-1186; 98 Cu, 1.7 Si, 0.3 Mn.
 Annealed: 38,000 TS; 11,000 YS; 35 El.
 Drawn: 50,000 TS; 28,000 YS; 15 El.
 For wire, rod; silicon bronze.

SM 3630.
 M-1186; 96 Cu, 3 Si, 1 Mn.
 Annealed: 57,000 TS; 17,000 YS; 50 El; 50 RA; 85 Brin.
 Hard: 71,000 TS; 43,000 YS; 20 El; 30 RA; 150 Brin.
 For sheets, rod; silicon bronze.

SM-3730.
 M-1186; 3 Sn, bal Cu.
 Annealed: 43,000 TS; 14,000 YS; 55 El; 60 RA; 70 Brin.
 Rolled: 54,000 TS; 46,000 YS; 30 El; 40 RA; 115 Brin.
 For sheets; phosphor bronze.

SM-3750.
 M-1186; 5 Sn, bal Cu.
 Annealed: 47,000 TS; 17,000 YS; 65 El; 65 RA; 70 Brin.
 Rolled: 63,000 TS; 54,000 YS; 25 El; 50 RA; 130 Brin.
 For sheets; phosphor bronze.

SM-3770.
 M-1186; 7 Sn, bal Cu.
 Rolled: 100,000 TS; 90,000 YS; 8 El; 35 RA: 200 Brin.
 Drawn: 135,000 TS.
 For sheets, wire; phosphor bronze.

SM-3790.
 M-1186; 9 Sn, bal Cu.
 Hard: 75,000 TS; 57,000 YS; 35 El; 50 RA; 170 Brin.
 For sheets, tubes; phosphor bronze.

SM-3840.
 M-1186; 4 Sn, 4 Pb, 4 Zn, bal Cu.
 Rolled: 57,000 TS; 48,000 YS; 20 El; 35 RA; 125 Brin.
 For sheets; free-cutting.

SM-4220.
 M-1186; 80 Cu, 20 Ni.
 Drawn: 60,000 TS; 43,000 YS; 5 El; 30 RA; 120 Brin.
 For tubes, condenser tubes; Cupro-Nickel.

SM-4230.
 M-1186; 70 Cu, 30 Ni.
 Drawn: 80,000 TS; 64,000 YS; 5 El; 30 RA; 155 Brin.
 For tubes, condenser tubes; Cupro-Nickel.

SM-4312.
 M-1186; 64 Cu, 12 Ni, bal Zn.
 Annealed: 57,000 TS; 21,000 YS; 35 El; 60 RA; 80 Brin.
 Hard: 80,000 TS; 64,000 YS; 8 El; 40 RA; 150 Brin.
 For sheets, tubes, wire, shapes; nickel silver.

SM-4318.
 M-1186; 60 Cu, 18 Ni, bal Zn.
 Annealed: 60,000 TS; 26,000 YS; 30 El; 50 RA; 90 Brin.
 Hard: 83,000 TS; 67,000 YS; 8 El; 35 RA; 150 Brin.
 For sheets, tubes, wires, shapes; nickel silver.

SM-5023.
 M-1186; 99.7 Al.
 Drawn: 27,000 TS.
 For wire.

SM-5025.
 M-1186; 99.5 Al.
 Drawn: 25,000 TS.
 For wire.

SM-5027.
 M-1186; 99.3 Al.
 Rolled: 18,000 TS; 16,000 YS; 4 El; 45 Brin.
 For sheets.

SM-5078.
 M-1186; 1.2 Mn, bal Al.
 Annealed: 14,000 TS; 6000 YS; 30 El; 35 Brin.
 Hard: 26,000 TS; 23,000 YS; 5 El; 60 Brin.
 For sheets, tube, shapes; 3 S alloy.

SM-5083.
 M-1186; 0.6 Mg, bal Al.
 Annealed: 14,000 TS; 6000 YS; 30 El; 35 Brin.
 Hard: 26,000 TS; 23,000 YS; 5 El; 60 Brin.
 For sheets, tubes, shapes; 63 S alloy.

SM-6307.
M-1186; 0.8 Mg, 0.6 Mn, 4.2 Cu, 0.5 Si, bal Al.
Age-hardened: 57,000 TS; 35,000 YS; 14 El; 120 Brin.
For sheets, tube, rod, shapes; A; 17 S alloy; age-hardenable.

SM-6315.
M-1186; 1.5 Mg, 0.6 Mn, 4.5 Cu, bal Al.
Age-hardened: 63,000 TS; 40,000 YS; 12 El; 130 Brin.
For sheets, tube, rod, shapes; A; 24 S alloy; age-hardenable.

SM-6508.
M-1186; 0.8 Mg, 0.7 Mn, 1 Ni, bal Al.
Age-hardened: 43,000 TS; 35,000 YS; 10 El; 100 Brin.
For sheets, tube, rod, shapes; age-hardenable.

SM-6525.
M-1186; 2.5 Mg, 0.25 Cr, bal Al.
Annealed: 26,000 TS; 11,000 YS; 18 El; 50 Brin.
Hard: 35,000 TS; 33,000 YS; 4 El; 90 Brin.
For sheets, tube, shapes; 52 S alloy.

SM-6536.
M-1186; 3.5 Mg, bal Al.
Annealed: 32,000 TS; 11,000 YS; 18 El; 60 Brin.
Hard: 43,000 TS; 35,000 YS; 6 El; 100 Brin.
For sheets, wire, shapes; A 214 alloy.

SM-6821.
M-1186; 1.5 Mg, 0.6 Mn, 4.5 Cu, bal Al.
Heat treated: 57,000 TS; 37,000 YS; 12 El; 130 Brin.
For sheets, aircraft structures; age-hardenable.

SM-6958.
M-1186; 2.5 Mg, 0.1 Mn, 1.6 Cu, 5.75 Zn, 0.2 Cr, 0.05 Ti, bal Al.
Age-hardened: 78,000 TS; 68,000 YS; 6 El; 155 Brin.
75 S alloy; age-hardenable.

S-M ALLOY.
M-244; 5 Si, bal Al.
Al welding rod for welding and soldering of Al; corrosion resistant.

SMENA NO. 1.
M-515; 20.6 W, 31.4 Cr, 33.5 Ni, 4.2 C, 1.6 Mn + Si, 8.7 Fe.
For cutting tools, drills, reamers, hobs; high speed steel.

SMENA NO. 2.
M-515; Co-Cr-W.
To replace Stellite for resurfacing; hard alloy, wear and corrosion resistant.

SM FERROCHROME (HIGH CARBON).
M-48; 60-65 Cr, 4-6 Si, 4-6 Mn, 4-6 C, bal Fe.
For Cr additions to steel or iron; Si and Mn protects the Cr.

SMG-2.
M-349.
Dental alloy, ceramic gold.
Melt range: 2035-2190°F; 150 Brin.
For all porcelain to gold restorations.

SMG-2 NEY CERAMIC SOLDER.
M-349.
Dental solder.
For use with SMG-2 and SMG-W casting gold.
Flows at 1925°F.

SMG-3.
M-349.
Dental alloy, ceramic gold.
Melt range: 2175-2310°F; 200 Brin.
For all porcelain to gold restorations; white gold color.

SMG-3 NEY CERAMIC SOLDER.
M-349.
Dental solder.
For use with SMG-3 casting golds.
Flows at 3100°F.

SMG-W.
M-349.
Dental alloy; white gold color.
Melt range: 2090-2230°F; 205 Brin.

SMG-Y.
M-349.
Dental alloy, ceramic gold.
Melt range: 2080-2285°F; 180 Brin.
Gold color; good bond strength.

SMG-YW NEY CERAMIC SOLDER.
M-349.
Dental solder.
For use with SMG-Y and SMG-W casting golds.
Flows at 2000°F.

SMITH NO. 10.
M-605; 37.5 Cr, 7.5 Al, bal Fe.
200-235 Brin.
For electrical resistors for heating furnaces, heating elements; heat resistant; El. Res. 1000 ohms; C.M.F. max operating temperature 2400°F.

SMITHWAY.
M-605; C, alloy, bal Fe.
For welding electrodes.

SMITTER LENIAN.
M-Eng.; 72 Cu, 13 Ni, 9.8 Zn, 2.3 Sn, 2 Fe, 1 Bi.
For corrosion resistant parts; corrosion resistant.

S.M.L. ALLOY (MONEL).
M-Eng.; 68 Ni, 28 Cu, 2.5 Fe, 2.5 Mn.
For parts requiring high corrosion resistance and strength at elevated temperatures; corrosion resistant.

SMO 5.
M-1290. 0.84-0.92 C, 3.8-4.5 Cr, 6.0-6.7 W, 1.7-2.0 V, 4.7-5.2 Mo, bal Fe.
High speed steel.
W.-Nr. 1.3343.

SMO 5 CO 5.
M-1290; 0.88-0.96 C, 3.8-4.5 Cr, 6.0-6.7 W, 1.7-2.0 V, 4.7-5.2 Mo, 4.5-5.0 Co, bal Fe.
High speed steel.
W.-Nr. 1.3243.

SMO 5 H.
M-1290; 0.95-1.05 C, 3.8-4.5 Cr, 6.0-6.7 W, 1.2-2.0 V, 4.7-5.2 Mo, bal Fe.
High speed steel.
W.-Nr. 1.3342.

SMO 5 V 3.
M-1290; 1.17-1.27 C, 3.8-4.5 Cr, 6.0-6.7 W, 2.7-3.2 V, 4.7-5.2 Mo, bal Fe.
High speed steel.
W.-Nr. 1.3344.

SMO 6 CO 5.
M-1290; 1.13-1.20 C, 3.8-4.5 Cr, 6.0-6.75 W, 3.0-3.5 V, 4.8-5.3 Mo, 4.5-5.0 Co, bal Fe.
High speed steel.
W.-Nr. 1.3244.

SMO 9.
M-1290; 0.78-0.86 C, 3.5-4.2 Cr, 1.5-2.0 W, 1.0-1.3 V, 8.0-9.2 Mo, bal Fe.
High speed steel.
W.-Nr. 1.3346.

SMO 9 CO 5.
M-1290; 0.90-0.98 C, 3.4-4.2 Cr, 1.5-2.0 W, 1.8-2.2 V, 8.0-9.2 Mo, 4.5-5.0 Co, bal Fe.
High speed steel.
W.-Nr. 1.3248.

SMO 9 V.
M-1290; 0.97-1.07 C, 3.5-4.2 Cr, 1.5-2.0 W, 1.8-2.2 V, 8.0-9.2 Mo, bal Fe.
High speed steel.
W.-Nr. 1.3348.

SMO 10 CO 8.
M-1290; 1.05-1.12 C, 3.6-4.4 Cr, 1.2-1.8 W, 1.0-1.3 V, 9.0-10.0 Mo, 7.5-8.5 Co, bal Fe.
High speed steel.
W.-Nr. 1.3247.

SMOOTHCOTE NO. 34 see **ALL-STATE NO. 34.**

SMOOTHCUT see also **ATMODIE SMOOTHCUT.**

S.M.S. ALLOY 1.
M-1093; Cu-W.
Bars: 135,000 TS; 130 Brin.
For resistance welding electrode; 35% conductivity, for stainless steel and brass.

S.M.S. ALLOY 10.
M-1093; Cu-W.
Bars: 160,000 TS; 205 Brin.
For resistance welding electrode; 28% conductivity, for inserts and facings.

S.M.S. ALLOY 20.
M-1093; Cu-W.
Bars: 170,000 TS; 228 Brin.
For resistance welding electrode; 27% conductivity, for heavy projection welds.

S.M.S. ALLOY 100W.
M-1093; Cu-W.
Bars: 200,000 TS; Rockwell A76.
For resistance welding electrode; 30% conductivity, for Red Brass.

S.M.S. ALLOY 101.
M-1093; Cu alloy.
Bars: 50,000-60,000 TS; 15,000-20,000 YS; 25-20 El; 100-116 Brin.
For resistance welding electrode; 80% conductivity, for coated metals.

S.M.S. ALLOY 103.
M-1093; Cu alloy.
Bars: 55,000-65,000 TS; 25,000-35,000 YS; 15 El; 116-137 Brin.
For resistance welding electrode; 75% conductivity, for carbon steel and brass.

S.M.S. ALLOY 116.
M-1093; Cu alloy.
Cast: 65,000-75,000 TS; 12,000-16,000 YS; 10-2 El; 116-165 Brin.
For resistance welding electrode.

S.M.S. ALLOY W-2.
M-1093; Cu alloy.
Cast: 90,000 TS; 68,000 YS; 1 El; 330 Brin.
Rolled: 140,000 TS; 85,000 YS; 2 El; 330 Brin.
For resistance welding electrode.

S.M.S. ALLOY W5.
M-1093; 0.4 Be, 2.6 Co, bal Cu.
Cast: 90,000 TS; 55,000 YS; 10-15 El.
Rolled: 125,000 TS; 65,000 YS; 8-15 El.
For springs, bellows, diaphragms; age-hardenable, corrosion resistant.

S.M. STEEL.
M-32; 2 Si, 0.8 Mn, 0.5-0.6 C, bal Fe.
For machinery parts, gears, shafts; water hardening.

S.M.Z. ALLOY.
M-48; 60-65 Si, 5-7 Mn, 5-7 Zr, bal Fe.
For graphitizing alloy for cast iron; ladle addition.

SN 5 1/2.
M-1739.
Nickel alloy steel, sintered.
Density: 6.6 gms/cc; UTS: 47 kg/mm² typical.
Hardenable to Rock C 40 approx.
B.S.S. A400; MPIE FN-0505-T.

SO 7.1002 (304 + SI).
M-106; 3.8-4.5 Cu, 1.0-1.5 Si, bal Al.
Sand cast: 40,000-45,000 TS; 27,000-29,000 YS; 4-6 El; 90-100 Brin; 3.5 Izod.
Sand or chill cast aluminum alloy, hardenable, good strength and less hot short.

SODERFORS NO. 05.
M-387; 0.06 C, 0.10 Si, 0.20 Mn, bal Fe.
For case hardened gears, cams, camshafts, plastic molds.
Case hardening steel, wear resistant.

SODERFORS NO. 1.
M-387; 0.6 C, 0.3 Mn, bal Fe.
For tools, punches; water hardening.

SODERFORS NO. 2.
M-387; 0.7 C, 0.2 Si, 0.3 Mn, bal Fe.
For tools, punches, dies; water hardening.

SODERFORS NO. 3.
M-387; 0.8 C, 0.2 Si, 0.3 Mn, bal Fe.
For tools, springs; water hardening.

SODERFORS NO. 3 1/2.
M-387; 0.9 C, 0.2 Si, 0.3 Mn, bal Fe.
For tools, springs, drills; water hardening.

SODERFORS NO. 4.
M-387; 1.0 C, 0.2 Si, 0.3 Mn, bal Fe.
For tools, drills, taps; water hardening.

SODERFORS NO. 5.
M-387; 1.1 C, 0.2 Si, 0.3 Mn, bal Fe.
For tools, drills, taps; water hardening.

SODERFORS NO. 6.
M-387; 1.2 C, 0.2 Si, 0.3 Mn, bal Fe.
For tools, cutters, drills; water hardening.

SODERFORS NO. 7.
M-387; 1.3 C, 0.2 Si, 0.3 Mn, bal Fe.
For tools, cutters; water hardening.

SODERFORS NO. 8.
M-387; 1.45 C, 0.2 Si, 0.3 Mn, bal Fe.
For tools, cutters; water hardening.

SODERFORS NO. 9.
M-387; 0.6-1.45 C, 0.2 Si, 0.3 Mn, 0.45 Cr, bal Fe.
For tools, punches; water hardening.

SODERFORS NO. 9A.
M-387; 0.6-1.45 C, 0.20 Si, 0.3 Mn, 0.25 Cr, bal Fe.
For tools, punches; water hardening.

SODERFORS NO. 9B.
M-387; 0.6-1.45 C, 0.2 Si, 0.3 Mn, 0.15 Cr, bal Fe.
For tools, springs, dies; water hardening.

SODERFORS NO. 10.
M-387; 0.6-1.45 C, 0.2 Si, 0.3 Mn, 1.0 Cr, bal Fe.
For tools, drills, springs; water hardening.

SODERFORS NO. 11.
M-387; 0.6-1.45 C, 0.2 Si, 0.3 Mn, 1.5 Cr, bal Fe.
For tools, drills, springs; water hardening.

SODERFORS NO. 11A.
M-387; 1.0 C, 2 Cr, bal Fe.
For tools, dies; water hardening.

SODERFORS NO. 12.
M-387; 0.8 C, 0.2 Si, 0.3 Mn, 0.45 Cr, 0.10 V, bal Fe.
For tools, drills, taps; water hardening.

SODERFORS NO. 12/0.8.
M-387; 0.75-0.85 C, 0.05-0.15 V, bal Fe.
Annealed: 88,000 TS; 52,000 YS; 20 El; 190 Brin.
For cutters, shears, slitters, gripper and forming dies, coining and embossing dies.
Type W2 water hardening steel, wear resistant.

SODERFORS 12/1.0.
M-387; 1.0 C, 0.2 Si, 0.3 Mn, 0.1 V, bal Fe.
Annealed: 100,000 TS; 55,000 YS; 20 El; 40 RA; 220 Brin.
For heading and forming dies, punches; Type W2; water hardened.

SODERFORS NO. 13.
M-387; 0.95 C, 0.2 Si, 0.3 Mn, 0.45 Cr, 0.10 V, bal Fe.
For tools, drills, cutters; water hardening.

SODERFORS NO. 14.
M-387; 1.05 C, 0.2 Si, 0.3 Mn, 0.45 Cr, 0.10 V, bal Fe.
For tools, drills, cutters; water hardening.

SODERFORS NO. 15.
M-387; 0.90-1.05 C, 1.35-1.65 Si, 0.6-0.9 Mn, 0.9-1.15 Cr, bal Fe.
For bearings, liners, cutters, bushings, cold heading tools.
Oil hardening, wear and shock resistant.

SODERFORS NO. 16.
M-387; 0.95 C, 0.2 Si, 1.25 Mn, 0.5 Cr, 0.5 W, 0.1 V, bal Fe.
For tools, dies, cutters; non-shrinking. AISI 01.

SODERFORS NO. 17.
M-387; 1.15 C, 0.2 Si, 0.3 Mn, 0.1 V, 0.5 W, bal Fe.
For drills, taps, hobs, broaches; water hardened.

SODERFORS NO. 17A.
M-387; 1.1 C, 0.25 Cr, 0.13 V, 1.1 W, bal Fe.
Annealed: 100,000 TS; 60,000 YS; 15 El; 210 Brin.
For punches, slitters, bearings, textile needles, broaches, reamers.
Type F1 tool steel. Water or oil hardening. Abrasion resistant.

SODERFORS NO. 18.
M-387; 0.45 C, 0.25 Si, 1.4 Cr, 2.2 W, 0.2 V, 0.2-0.3 Mo, bal Fe.
For chisels, punches, shock resisting.

SODERFORS NO. 19.
M-387; 1.15 C, 0.10 Si, 0.5 Cr, 1.5 W, bal Fe.
For tools, cutters; keen cutting edge.

SODERFORS NO. 20.
M-387; 1.3 C, 1.8 Cr, 4 W, 0.20 V, bal Fe.
For tools, dies; oil hardening.

SODERFORS NO. 21.
M-387; 1.3 C, 0.5 Cr, 5 W, bal Fe.
For tools, dies, cutters; oil hardening.

SODERFORS NO. 22.
M-387; 0.5 C, 3 Cr, 9 W, 0.3 V, bal Fe.
For die casting dies, extrusion mandrels and liners; hot work steel, oil hardened.

SODERFORS NO. 23.
M-387; 0.45 C, 3 Cr, 14 W, bal Fe.
For tools, dies; hot work steel.

SODERFORS NO. 24.
M-387; 0.45 C, 3.0 Cr, 0.3 V, 9.0 W, bal Fe.
For hot punches, extrusion dies; hot piercers and grippers.
Hot work tool steel, oil hardening, tough and shock resistant.

SODERFORS NO. 25.
M-387; 0.70 C, 4 Cr, 18 W, 1 V, 0.4 Co, bal Fe.
For drills, reamers, broaches, hobs, lathe cutters; high speed steel, oil hardened.

SODERFORS NO. 26.
M-387; 0.80 C, 5.5 Cr, 18 W, 1.5 V, 1.5 Mo, 2 Co, bal Fe.
For tools, cutters; high speed steel. AISI T2.

SODERFORS NO. 27.
M-387; 0.80 C, 5.5 Cr, 18 W, 1.5 V, 1.5 Mo, 5.5 Co, bal Fe.
For tools, cutters; high speed steel. AISI T4.

SODERFORS NO. 28.
M-387; 0.75 C, 5.5 Cr, 18 W, 1.5 V, 1.0 Mo, 10 Co, bal Fe.
For tools, cutters; high speed steel. AISI T5.

SODERFORS NO. 29.
M-387; 0.77-0.87 C, 3.5-4.5 Cr, 4.5-5.5 Mo, 1.7-2.1 V, 6.0-7.0 W, bal Fe.
Hardened: 64-67 Rock C.
For lathe and planer tools, form cutters, reamers, broaches, drills, hobs.
Type M2 high-speed steel, high red-hardness.

SODERFORS NO. 29A.
M-387; 0.82-0.90 C, 3.5-4.5 Cr, 2.8-3.6 Mo, 1.9-2.3 V, 6.0-7.0 W, bal Fe.
Hardened: 64-66 Rock C.
For lathe and planer tools, hobs, reamers, drills, taps.
High-speed steel, high red-hardness.

SODERFORS NO. 29B.
M-387; 1.45 C, 4 Cr, 5 Mo, 5 V, 6.5 W, bal Fe.
For textile needles, reamers, broaches, chasers, hobs.
High speed steel, high red-hardness.

SODERFORS NO. 30.
M-387; 1.1 C, 9.5 Co, 4.3 Cr, 5 Mo, 2.6 V, 6.5 W, bal Fe.
Hardened: 64-68 Rock C.
For broaches, reamers, twist drills, lathe and milling cutters, boring tools.
Type M36 high-speed steel, high red hardness.

SODERFORS NO. 31.
M-387; 1 C, 3 Cr, bal Fe.
For dies, tools; oil hardening.

SODERFORS NO. 32.
M-387; 0.65 C, 6 W, bal Fe.
For hot work tools; hot work steel.

SODERFORS NO. 41.
M-387; 0.5-1.3 C, 0.2 Si, 0.55 Mn, bal Fe.
For tools, dies, punches; water hardening.

SODERFORS NO. 42.
M-387; 0.45-0.70 C, 0.7 Si, 0.9 Mn, bal Fe.
For tools, dies, punches; water hardening.

SODERFORS NO. 43.
M-387; 0.60 C, 1.9 Si, 0.9 Mn, bal Fe.
For tools, springs; oil hardening.

SODERFORS NO. 46.
M-387; 0.5 C, 1.1 Cr, 0.15 V, bal Fe.
For gears, shafts; oil hardened, tough.

SODERFORS NO. 61.
M-387; 2.2 C, 13 Cr, 1 W-Mo-Ni-Co, bal Fe.
For tools, dies; non-deforming.

SODERFORS NO. 62.
M-387; 2.2 C, 12.5 Cr, bal Fe.
For dies; non-deforming. AISI D3.

SODERFORS 63.
M-387; 1.5 C, 12 Cr, 0.8 Mo, 0.2 V, bal Fe.
For shears, punches, blanking and forming dies; air-hardened, non-deforming.

SODERFORS NO. 65.
M-387; 0.95-1.05 C, 5.0-5.5 Cr, 1.0-1.2 Mo, 0.15-0.25 V, bal Fe.
Annealed: 103,000 TS; 51,000 YS; 26 El; C 18 Rock.
Ht. Tr.: 255,000 TS; 200,000 YS; 3 El; C 57 Rock.
For blanking and trimming dies, cutters, gauges, punches, burnishing tools, shear blades, broaches.
Type A2 air hardening steel, non-deforming.

SODERFORS NO. 67.
M-387; 0.35-0.42 C, 0.8-1.2 Si, 5.0-5.5 Cr, 1.2-1.6 Mo, 0.85-1.15 V, bal Fe.
Annealed: 98,000 TS; 74,000 YS; 28 El; 216 Brin.
Ht. Tr.: 290,000 TS; 228,000 YS; 3 El; C 55 Rock.
For forging and heading dies, compression tools, die casting and extrusion dies.
Type H13 hot work steel. High hot hardness. Shock resistant.

SODERFORS NO. 81.
M-387; 0.13 C, 0.7 Cr, 3 Ni, bal Fe.
For gears, shafts; case hardening.

SODERFORS NO. 82.
M-387; 0.3 C, 0.6 Cr, 2 Ni, bal Fe.
For shafts, gears, pinions; tough.

SODERFORS NO. 83.
M-387; 0.3 C, 0.8 Cr, 3 Ni, bal Fe.
For gears, pinions, shafts; tough.

SODERFORS NO. 84.
M-387; 0.6 C, 1.8 Ni, 0.7 Cr, 0.7 Mo, bal Fe.
For drop forging dies; oil hardened, tough.

SODERFORS NO. 85.
M-387; 0.5 C, 0.8 Cr, 3.25 Ni, 0.2 Mo, bal Fe.
For gears, dies, tools; tough.

SODERFORS NO. 214.
M-387; 1.05 C, 0.2 V, 0.2 Si, 0.3 Mn, bal Fe.
Annealed: 90,000 TS; 55,000 YS; 18 El; 200 Brin.
For shears, slitters, lathe and planer tools, drills, hobs.
Water hardening, wear resistant.

SODERFORS NO. 215.
M-387; 1.0 C, 1.3 Si, 0.8 Mn, 1.0 Cr, bal Fe.
For bearings, liners, bushings, cutters, punches, cold heading and drawing dies.
Oil hardening, abrasion resistant.

SODERFORS NO. 323.
M-387; 0.30 C, 0.30 Si, 4.8 Co, 1.5 Cr, 0.5 Mo, 0.10 V, 5.3 W, bal Fe.
For hot work tools, punches, hot shears.
Oil hardening, shock resistant.

SODERFORS NO. 364.
M-387; 1.45-1.65 C, 11-13 Cr, 0.7-0.9 Mo, 0.7-1.0 V, bal Fe.
Ht. Tr.: 278,000 TS; 214,000 YS; 1 El; C 57 Rock.
For blanking and drawing dies, broaches, hobs, shear blades, punches, gauges.
Type D2 air-hardening steel. Non-deforming, shock resistant.

SODERFORS NO. 423.
M-387; 1.15 C, 4.0 Cr, 5.0 Mo, 3.0 V, 6.0 W, bal Fe.
Hardened: 64-68 Rock C.
For drills, reamers, broaches, form cutters, lathe and planer tools, hobs.
Type M3 high-speed steel. High red hardness.

SODERFORS NO. 424.
M-387; 1.1 C, 5.3 Co, 4.3 Cr, 5.0 Mo, 2.6 V, 6.5 W, bal Fe.
Hardened: 64-68 Rock C.
For broaches, reamers, taps, chasers, drills, lathe and planer tools.
Type M35 high-speed steel, high red-hardness.

SODERFORS NO. 431.
M-387; 0.8 C, 4.0 Cr, 9.0 Mo, 1.2 V, 1.5 W, bal Fe.
Hardened: 64-66 Rock C.
For lathe and planer tools, form cutters, drills, hobs, reamers, gauges.
Type M1 high-speed steel, oil hardening. High red-hardness.

SODERFORS NO. 433.
M-387; 1.0 C, 4 Cr, 9 Mo, 2 V, 1.8 W, bal Fe.
Hardened: 64-67 Rock C.
For drills, end mills, thread rolling dies, chasers, punches, taps.
Type M-7 high-speed steel. High red-hardness.

SODERFORS NO. 434.
M-387; 1.2 C, 5 Co, 4 Cr, 9 Mo, 3 V, 1.8 W, bal Fe.
For reamers, broaches, drills, form cutters, chasers.
High-speed steel, good red-hardness.

SODERFORS NO. 481.
M-387; 0.8 C, 4.5 Cr, 8 Mo, 2 V, 9 W, bal Fe.
For chasers, taps, drills, reamers, broaches, lathe and planer tools.
High-speed steel, good red-hardness.

SODERFORS NO. 502.
M-387; 0.1 C, 14 Cr, bal Fe.
Annealed: 90,000 TS; 50,000 YS; 40 El; 50 RA; 160 Brin.
For cutlery, surgical instruments, chemical plant equipment; Type 410; corrosion resistant.

SODERFORS NO. 504.
M-387; 0.08 max C, 17.5 Cr, 12.0 Ni, 2.75 Mo, bal Fe.
Annealed: 80,000 TS; 30,000 YS; 60 El; B 80 Rock.
For chemical plant equipment, evaporators, acid tanks, digesters.
Corrosion resistant, austenitic, non-hardenable.

SODERFORS NO. 506.
M-387; 0.10 max C, 27 Cr, 5.3 Ni, 1.6 Mo, bal Fe.
For heat treating boxes, furnace parts, retorts, gas burners, combustion chambers.
Corrosion and heat resistant.

SODERFORS NO. 508.
M-387; 0.10 max C, 27 Cr, 5.3 Ni, 1.6 Mo, bal Fe.
For heat treating boxes, furnace parts, retorts, gas burners, combustion chambers.
Corrosion and heat resistant.

SODERFORS NO. 509.
M-387; 0.12 C, 12 Cr, 0.6 Ni, 0.5 Mo, bal Fe.
Annealed: 80,000 TS; 42,000 YS; 30 El; 160 Brin.
For chemical plant and oil refinery equipment, tableware, knives.
Heat and corrosion resistant.

SODERFORS NO. 510.
M-387; 0.15 C, 14 Cr, 0.3 Ni, bal Fe.
Annealed: 75,000 TS; 40,000 YS; 35 El; 70 RA; 155 Brin.
Cold drawn: 100,000 TS; 85,000 YS; 17 El; 60 RA; 205 Brin.
For cutlery, pump parts, valves; Type 410; corrosion resistant.

SODERFORS NO. 511.
M-387; 0.2 C, 14 Cr, 0.3 Ni, bal Fe.
For pump-shafts, cutlery, valves; corrosion resistant.

SODERFORS NO. 512.
M-387; 0.35 C, 14 Cr, 0.3 Ni, bal Fe.
Annealed: 95,000 TS; 50,000 YS; 25 El; 55 RA; 196 Brin.
Heat treated: 250,000 TS; 215,000 YS; 8 El; 25 RA; 512 Brin.
For cutlery, knives, surgical instruments, gears, shafts; Type 420; corrosion resistant.

SODERFORS NO. 514.
M-387; 0.53-0.58 C, 13.5-14.0 Cr, 0.3 Ni, bal Fe.
Annealed: 100,000 TS; 50,000 YS; B 90 Rock.
For surgical instruments, cutlery, knives, gears, shafts.
Corrosion and heat resistant.

SODERFORS NO. 516.
M-387; 0.20 C, 14 Cr, 0.3 Ni, 0.3 Mo, bal Fe.
Annealed: 95,000 TS; 50,000 YS; 25 El; B 92 Rock.
For cutlery, surgical instruments, gauges, needle valves, bearings.
Corrosion and heat resistant.

SODERFORS NO. 517.
M-387; 1.0 C, 14 Cr, 0.2 Mo, 0.6 Co, 0.1 V, bal Fe.
For bearings: corrosion resistant.

SODERFORS NO. 522.
M-387; 0.08 max C, 17.5 Cr, 0.5 Mn, bal Fe.
Annealed: 80,000 TS; 50,000 YS; 25 El; 50 RA; 150 Brin.
For household utensils, sinks, oil refinery equipment; stainless, ferritic; Type 430.

SODERFORS NO. 523.
M-387; 0.10 max C, 17 Cr, 0.35 Ni, 1.75 Mo, bal Fe.
Annealed: 90,000 TS; 50,000 YS; 20 El; 165 Brin.
For oil burners, hardware, marine fittings, food handling equipment.
Corrosion and heat resistant.

SODERFORS NO. 525.
M-387; 0.15-0.25 C, 16-18 Cr, 1.25-2.5 Ni, bal Fe.
Annealed: 125,000 TS; 90,000 YS; 20 El; 260 Brin.
For pump shafts, marine hardware, valve trim, aircraft components.
Type 431 corrosion and heat resistant.

SODERFORS NO. 526.
M-387; 0.15 C, 17.5 Cr, 1.8 Mo, bal Fe.
Annealed: 85,000 TS; 48,000 YS; 25 El; 160 Brin.
For oil burner parts, hardware, marine fixtures, food handling equipment.
Corrosion and heat resistant.

SODERFORS NO. 527.
M-387; 0.15 C, 17 Cr, 1.5 Ni, 0.5 Mo, 0.2 Si, bal Fe.
For cutlery, tableware; corrosion resistant, free-cutting.

SODERFORS NO. 532.
M-387; 0.2 C, 25 Cr, bal Fe.
For stainless parts; corrosion resistant.

SODERFORS NO. 541.
M-387; 0.41 C, 3 Si, 9 Cr, 0.3 Mo, bal Fe.
For valves; heat corrosion resistant.

SODERFORS NO. 552.
M-387; 0.03 max C, 17-20 Cr, 9-12 Ni, bal Fe.
Annealed: 80,000 TS; 30,000 YS; 60 El; B 74 Rock.
For architectural molding and trim, processing equipment, welded tanks, evaporators.
Type 304L stainless steel, austenitic.

SODERFORS NO. 553.
M-387; 0.05 C, 18 Cr, 9 Ni, bal Fe.
For chemical and textile plant equipment; Type 304; austenitic, stainless.

SODERFORS NO. 554.
M-387; 0.08 max C, 18 Cr, 9 Ni, bal Fe.
For stainless parts; austenitic, stainless.

SODERFORS NO. 555.
M-387; 0.13 C, 18 Cr, 9 Ni, bal Fe.
Annealed: 90,000 TS; 35,000 YS; 50 El; 16 RA; 140 Brin.
For chemical and plastic plant equipment; Type 302; stainless, austenitic.

SODERFORS NO. 556.
M-387; 0.10 max C, 18 Cr, 9 Ni, 0.2 S, bal Fe.
Rolled: 80,000-100,000 TS; 40,000-60,000 YS; 150-200 Brin.
For screw machine products, fasteners, machinery parts; Type 303; austenitic, free-cutting, stainless.

SODERFORS NO. 558.
M-387; 0.08 max C, 17-19 Cr, 9-13 Ni, Ti = 5 × C, bal Fe.
Annealed: 85,000 TS; 30,000 YS; 58 El; B 82 Rock.
For welded structures, chemical plant equipment, tanks, evaporators.
Type 321 stainless steel, austenitic, stabilized, welding grade.

SODERFORS NO. 558 CB.
M-387; 0.08 max C, 17-19 Cr, 9-13 Ni, 1.2 max Cb, bal Fe.
Annealed: 85,000 TS; 35,000 YS; 60 El; B 80 Rock.
For welded structures, chemical plant equipment, digesters.
Type 347 stainless steel, austenitic, stabilized, welding grade.

SODERFORS NO. 558 NB.
M-387; 0.08 max C, 17-19 Cr, 9-13 Ni, 1.2 max Cb, bal Fe.
Annealed: 90,000 TS; 40,000 YS; 55 El; B 82 Rock.
For welded structures, chemical plant equipment, digesters.
Type 347 stainless steel, austenitic, stabilized, welding grade.

SODERFORS NO. 558 TI.
M-387; 0.06 C, 18 Cr, 11 Ni, 0.4 Ti, bal Fe.
Annealed: 85,000 TS; 30,000 YS; 55 El; B 80 Rock.
For welded structures and tanks, chemical plant equipment, jet aircraft components, exhaust systems.
Type 321 stainless steel, austenitic, stabilized.

SODERFORS NO. 559.
M-387; 0.12 max C, 14 Cr, 13 Ni, bal Fe.
For valves; corrosion and heat resistant.

SODERFORS NO. 562.
M-387; 0.05 max C, 18 Cr, 10 Ni, 1.5 Mo, bal Fe.
For chemical plant equipment; acid resistant, austenitic.

SODERFORS NO. 562 EL.
M-387; 0.03 max C, 17-18 Cr, 12-14 Ni, 2.3-3.8 Mo, bal Fe.
Annealed: 85,000 TS; 35,000 YS; 50 El; B 82 Rock.
For chemical plant equipment, acid tanks, agitators, valve trim.
Type 316L stainless steel, austenitic, acid resistant.

SODERFORS NO. 562 N.
M-387; 0.06 max C, 17.5 Cr, 10.5-11.5 Ni, 2.25 Mo, bal Fe.
Annealed: 80,000 TS; 30,000 YS; 60 El; B 80 Rock.
For chemical plant equipment, agitators, digesters, evaporators.
Austenitic, stainless, Type 316.

SODERFORS NO. 563.
M-387; 0.08 max C, 18 Cr, 10 Ni, 1.5 Mo, bal Fe.
For chemical plant equipment; acid resistant, austenitic.

SODERFORS NO. 564.
M-387; 0.05 max C, 18 Cr, 12 Ni, 2.7 Mo, 1.2 Mn, bal Fe.
For chemical plant equipment; stainless, austenitic.

SODERFORS NO. 565.
M-387; 0.08 max C, 1.2 Mn, 18 Cr, 12 Ni, 2.7 Mo, bal Fe.
For stainless parts; austenitic, stainless.

SODERFORS NO. 566.
M-387; 0.10 max C, 27 Cr, 5 Ni, 1.6 Mo, bal Fe.
For chemical plant equipment, sulphuric acid tanks; Type 329; stainless steel.

SODERFORS NO. 571.
M-387; 0.45 C, 14 Cr, 14 Ni, 0.3 Mo, 2.2 W, bal Fe.
For engine valves; austenitic, stainless.

SODERFORS NO. 576.
M-387; 0.15 max C, 1.2 Si, 23 Cr, 21 Ni, bal Fe.
For furnace parts, heat treating boxes; heat resistant to 1100°C.

SODERFORS NO. 584.
M-387; 0.12 max C, 7.5-10.0 Mn, 17-19 Cr, 4-6 Ni, 0.25 max N, bal Fe.
Annealed: 100,000 TS; 50,000 YS; 60 El; B 90 Rock.
For dairy and chemical plant equipment springs, refrigerator trays, cabinets, architectural trim.
Type 202 stainless steel, austenitic.

SODERFORS ANVIL.
M-387; 0.8 C, 0.9 Mn, bal Fe.
For tools; water hardened.

SODERFORS ARROW.
M-387; High C, 1.2 Mn, 0.5 W, 0.5 Cr, 0.1 V, bal Fe.
For tools, dies, thread and block gages, master tools, taps; non-shrinking, non-warping.

SODERFORS CROWN "W".
M-387; 0.9 C, 0.3 Mn, 1.8 Cr, 4.0 W, 0.2 V, bal Fe.
For tools, for turning chilled iron rolls and all hard metals; non-warping.

SODERFORS DOUBLE CRANE.
M-387; 0.9-1.2 C, bal Fe.
For tools, drills, stone tools; water hardened.

SODERFORS DRILL.
M-387; 1.0-1.2 C, bal Fe.
For drills, tools, hollow and solid drills; water hardened.

SODING ADK.
M-1309; 2.1 C, 12.0 Cr, bal Fe.
For blanking and forming dies, punches, gauges; oil hardening, non-deforming.
DIN X210 Cr 12.

SODING ADK SPEZIAL.
M-1309; 2.1 C, 12.0 Cr, bal Fe.
For blanking and forming dies, gauges, punches; air hardened, non-deforming.

SODING ADP SPEZIAL.
M-1309; 1.7 C, 12.0 Cr, 0.6 Mo, 0.5 W, bal Fe.
For punches, crimping and forming dies, blanking and trimming dies, punches.
Air hardening, non-deforming. Cold work tool steel.
DIN X-165 CrMoV12; Werkstoff Nr.1.2601.

SODING MW.
M-1309; 0.45 C, 13.0 Cr, 13.0 Ni, 1.2 W, 1.2 V, bal Fe.
Austenitic type hot work tool steel for extrusion dies and tube presses handling heavy metals.
Hot work steel X50NiCrWV1313.

SODING MW 575.
M-1309; 0.35 C, 13.0 Cr, 5.75 Ni, 2.75 W, 0.70 V, bal Fe.
Alloy hot work steel; for extrusion dies.

SODING PERMANENT.
M-1309; 0.75 C, 4.5 Cr, 18.5 W, 1.2 V, bal Fe.
Hardenable to C 62-65 Rock.
For roughing and finishing cuts of all kinds, lathe tools, milling cutters, taps, drills, reamers.
High speed steel B18.

SODING PERMANENT 35.
M-1309; 0.8 C, 4.5 Cr, 18.5 W, 0.8 Mo, 1.7 V, 5.0 Co, bal Fe.
Hardenable to C 63-66 Rock.
For heavy and roughing cuts on lathes, planers, and slotters on cast iron, steel and cast steel including stainless grades.
High speed steel E18Co5.

SODING PERMANENT 70.
M-1309; 1.4 C, 4.5 Cr, 12.5 W, 1.0 Mo, 4.0 V, 5.0 Co, bal Fe.
Hardenable to C 63-66 Rock.
For form tools for automatics, lathe tools, relieving tools; medium and finishing cuts at high speed.
High speed steel EVA Co.

SODING PERMANENT 120.
M-1309; 1.35 C, 4.5 Cr, 9.0 W, 3.7 Mo, 3.75 V, 10.5 Co, bal Fe.
Hardenable to C 63-68 Rock.
For lathe tools, form tools for automatics, relieving tools and form cutters for planing and slotting operations.
High speed steel EW9Co10.

SODING RADIKAL MO5.
M-1309; 0.85 C, 4.5 Cr, 6.7 W, 5.3 Mo, 2.0 V, bal Fe.
Hardenable to C 62-66 Rock.
For milling cutters, twist drills, reamers, form cutters and lathe tools; for heavy and rough cuts; good toughness.
High speed steel D Mo5; AISI M2.

SODING RADIKAL MO9.
M-1309; 0.85 C, 4.2 Cr, 2.0 W, 9.2 Mo, 1.3 V, bal Fe.
Hardenable to C 62-66 Rock.
For twist drills, taps, reamers, milling cutters and tools for woodworking machines.
High speed steel BMo9; AISI M1.

SODING RADIKAL MO55.
M-1309; 0.85 C, 4.5 Cr, 6.7 W, 5.3 Mo, 2.0 V, 5.5 Co, bal Fe.
Hardenable to C 62-66 Rock.
For lathe, planer and slotter tools, milling cutters, twist drills; for difficult machining and drilling operations.
High speed steel E Mo5Co5.

SODING RADIKAL MO60.
M-1309; 1.25 C, 4.5 Cr, 6.7 W, 5.3 Mo, 3.4 V, bal Fe.
Hardenable to C 62-66 Rock.
For form tools on medium-sized lathes and automatics, milling cutters, twist drills, chasers, broaches, reamers.
High speed steel E Mo5V3.

SODING RADIKAL MO92.
M-1309; 1.1 C, 4.2 Cr, 2.0 W, 9.2 Mo, 2.1 V, bal Fe.
Hardenable to C 62-66 Rock.
For twist drills, reamers, lathe tools, milling cutters; and tools for woodworking machines.
High speed steel B Mo9V.

SODING SPH 7.
M-1309; 0.32 C, 2.9 Cr, 2.8 Mo, 0.5 V, bal Fe.
Oil or salt bath hardenable to C 40-51 Rock.
For extrusion ram heads for light and heavy metals, hot upsetting tools, pressure die casting dies for brass, light metals and zinc.
Hot work tool steel X32CrMoV33.

SODING SPH 8.
M-1309; 0.30 C, 2.8 Cr, 2.8 Mo, 2.6 Co, 0.5 V, bal Fe.
Oil hardenable to C 40-50 Rock.
For tools for metal extrusion and tube presses, extrusion dies for heavy metals, pressure die casting dies.
Hot work steel.

SODING SPH 9.
M-1309; 0.40 C, 4.0 Cr, 4.0 W, 2.0 V, 1.5 Co, bal Fe.
Hot work tool steel; for extrusion press dies, hot upsetting tools, sliding cores in dies.

SODING SPH 12.
M-1309; 0.50 C, 4.0 Cr, 12.0 Ni, 10.0 W, 1.5 Co, bal Fe.
Alloy hot work tool steel; for extrusion dies.

SODING W35.
M-1309; 0.4 C, 5.5 Cr, 1.4 Mo, 0.3 V, bal Fe.
Air or oil hardenable to C 40-54 Rock.
For impact and pressure forging dies, pressure die casting dies, extrusion dies, piercing and punching dies.
Hot work tool steel X38CrMoV51; AISI H11.

SODING W36.
M-1309; 0.37 C, 4.8 Cr, 1.5 Mo, 1.4 W, 0.9 Si, 0.6 Mn, 0.2 V, bal Fe.
Oil or air hardenable to C 40-54 Rock.
For pressure die casting dies for light metal, tin, zinc and lead alloys when long life is important.
For forging die inserts.
Hot work steel X37CrMoW51 AISI H12.

SODING W37.
M-1309; 0.40 C, 5.0 Cr, 1.5 Mo, 1.0 V, bal Fe.
Air or oil hardenable to C 40-54 Rock.
For pressure die casting dies for light metals, tin, zinc and lead alloys; also for forging die inserts, piercing mandrels and extrusion mandrels.
Hot work tool steel X40CrMoV51; AISI H13.

SODIUM.
M-1755; Na.
Purities: Distilled 99.99%; 99.95% (Nuclear Grade).
Packaging: Bottles, ampules, cylinders (under vacuum, inert gas, petroleum distillate).

SODIUM-LEAD ALLOY.
M-596; 98 Pb, 2 Na.
For non-ferrous alloy deoxidizer.

SODIUM-ZINC ALLOY.
M-596; 98 Zn, 2 Na.
For nonferrous alloy deoxidizer; brittle.

SOFT.
M-388; 0.5 C, bal Fe.
For tools, shafts, gears; water hardened.

SOFTITE.
M-206; galvanized steel.
For structural components.

SOFTWELD.
M-578; C, bal Fe.
Arc welding electrode.
AWS Class En: C1.

SOHO SELF-HARDENING.
M-340; 0.9 C, 4 Cr, 0.2 V, bal Fe.
For tools, dies; oil hardened.

SOLABRAZE C-6601.
M-1670; 43 Zr, 12 Ni, 2 Be, bal Ti.
For brazing titanium to itself or to ceramic parts in vacuum or oxygen-free inert gas.
Melting range 1470-1500°F. Braze at 1650°F.

SOLABRAZE C-6602.
M-1670; 45 Zr, 8 Ni, 2 Be, bal Ti.
For brazing titanium to itself or to ceramic parts in high vacuum or oxygen-free inert gas.
Melting point 1650°F. Braze at 1700°F.

SOLABRAZE C-6603.
M-1670; 47 Zr, 5 Be, bal Ti.
For brazing titanium to itself or to ceramic parts in vacuum or oxygen-free inert gas.
Melting range 1640-1660°F. Braze at 1725°F.

SOLABRAZE C-6604.
M-1670; 47 Zr, 5 Be, 5 Al, bal Ti.
For brazing titanium to itself or to ceramic parts in vacuum or oxygen-free inert gas.
Melting point 1700°F. Braze at 1750°F.

SOLAR.
M-96; 1.2 C, 1.6 Cr, 1.45 W, 0.15 V, bal Fe.
For tools, dies.

SOLAR.
M-1322; 1.05 C, 1.0 Cr, 1.15 W, 0.90 Mn, bal Fe.
For cutters, reamers, forming dies; water or oil hardened.

SOLAR see CARPENTER SOLAR.

SOLAR ECLIPSE.
M-891; C, alloy, bal Fe.
For cold hobbing dies; oil hardened.

SOLBISKYS ALLOY.
M-Ger.; 1.4 Zn, 0.5-1.0 Ni, 0-3 Cd, 0.3 Sn, bal Al.
For light alloy parts; non-hardenable.

SOLDERAL.
M-427; 5 Si, bal Al.
For Al solder, castings; non-corroding.

SOLDER HC.
M-870; Pb-Sn.
20,000 TS.
For solder for Al.

SOLDER NO. 1.
M-Eng.; 30-40 Sn, 2-3 Sb, bal Pb.
6890 TS; 117 El; 12 Brin.
For commercial lead solders, M.P. 401°.

SOLDER NO. 2.
M-Eng.; 50 Sn, 50 Pb.
6400 TS; 96 El; 11 Brin.
For commercial lead solders; M.P. 446°F.

SOLDERZIT.
M-1137; Pb, Sn.
For soft soldering; powdered soft solder and flux combination.

SOLDIER A26.
M-1415; 0.3 C, 9 W, 3.25 Cr, 0.2 V, bal Fe.
For rivet dies, extrusion and press dies, mandrels; hot work steel, oil hardened.

SOLDIER A26N.
M-1415; 0.25 C, 8.5 W, 3 Cr, 0.25 V, 2.5 Ni, bal Fe.
For rivet dies, extrusion and press dies, mandrels; hot work steel, oil hardened.

SOLDIER CHROME H.33.
M-1415; 0.95 C, 4 Cr, bal Fe.
For hot shears, upsetters, punches; hot work steel for nonferrous metals.

SOLDIER M.314.
M-1415; 0.35 C, 1 Cr, 3 Ni, 0.3 Mo, 0.5 Mn, bal Fe.
For die casting dies, stamping die blocks; hot work steel, oil hardened.

SOLDIER P.23.
M-1415; 0.4 C, 1 Si, 5 Cr, 4 W, 0.4 Mo, bal Fe.
For hot dies, punches, mandrels, gripper dies; hot work steel, air or oil hardened.

SOLDIER SCT C.16.B.
M-1415; 0.5 C, 1.5 Cr, 2 W, 0.2 V, 1 Si, bal Fe.
For shear blades, swaging dies; hot work steel, shock resistant.

SOLEIL 1.2.
M-1302; 0.15 min C, 12-14 Cr, bal Fe.
Annealed: 88,000 TS; 40,000 YS; 32 El; 68 RA; 170 Brin.
Heat treated: 256,000 TS; 190,000 YS; 6 El; 10 RA; 540 Brin.
For cutlery, valve trim, turbine blades; Type 420; stainless, hardenable.

SOLEIL 2B.
M-1546; 0.20 C, 13 Cr, bal Fe.
Annealed: 90,000 TS; 45,000 YS; 30 El; B 90 Rock.
Ht. Tr.; 225,000 TS; 160,000 YS; 8-16 El.
For cutlery, turbine blades, valves, pumps, furnace parts, pistons.
Type 420 stainless steel. Hardenable. Corrosion and heat resistant.

SOLEIL 3.
M-1302; 0.15 max C, 11.5-13.5 Cr, bal Fe.
Heat treated: 120,000-135,000 TS; 110,000-117,000 YS; 16-15 El; 63-58 RA; 220-240 Brin.
For cutlery, valves, turbine blades; Type 410; corrosion resistant.

SOLEIL 4S.
M-1546; 0.05 C, 13 Cr, 0.2 Al, bal Fe.
Annealed: 68,000 TS; 38,000 YS; 30 El.
For oil refinery equipment, cracking stills, tubes for the petroleum industry.
Ferritic stainless steel. Non-hardenable.

SOLEIL 16.
M-1546; 0.06 C, 16.5 Cr, bal Fe.
Annealed: 70,000 TS; 40,000 YS; 30 El; 140 Brin.
For food and chemical industries, nitric acid equipment, cooking utensils.
Ferritic stainless steel. Resists nitric acid, non-hardenable.

SOLEIL 17.
M-1546; 0.20 C, 17 Cr, bal Fe.
Annealed: 80,000 TS; 45,000 YS; 30 El; 130 Brin.
For chemical plant and oil refinery equipment, fixtures, pump parts.
Ferritic stainless steel, non-hardenable.

SOLEIL 18.
M-1302; 0.12 max C, 14-18 Cr, bal Fe.
Annealed: 70,000 TS; 40,000 YS; 30 El; 55 RA; 150 Brin.
Cold drawn: 130,000 TS; 120,000 YS; 2 El; 185 Brin.
For oil refinery equipment, bolts, kitchen sinks; Type 430; stainless, ferritic.

SOLEIL A1M.
M-1488; 0.07 max C, 1.0 max Mn, 1.0 max Si, 1.5-2.5 Ni, 11.0-13.5 Cr, 0.75 max Mo, bal Fe.
Stainless casting, ferritic type; 200-250 Brin.
Parts for steam turbines.
AFNOR Z4CN 13.2 M.

SOLEIL A2.
M-1488; 0.12 C, 12.5 Cr, bal Fe.
Martensitic stainless steel; weldable.
For kitchen equipment, knife handles, pump bodies, impellers.
AFNOR Z12C13; AISI 410.

SOLEIL A2M.
M-1488; 0.15 max C, 1.0 max Mn, 1.0 max Si, 11.5-14.0 Cr, bal Fe.
Stainless casting; ann: 180-220 Brin.
For petroleum and food industries.
AFNOR Z12C13M; ACI CA 15.

SOLEIL A2U.
M-1488; 0.15 max C, 13 Cr, 0.60 max Mo, 0.15 min S, bal Fe.
Q. & T.: 640-830 N/mm² UTS.
Free machining stainless steel.
AFNOR Z12CF13; similar to AISI 416.

SOLEIL A3.
M-1488; 0.12 max C, 0.5 Mn, 0.6 max Si, 13.0 Cr, bal Fe.
Martensitic type stainless steel; weldable.
AFNOR Z15C13; AWS ER 410.

SOLEIL A4.
M-1488; 0.20 C, 13.0 Cr, bal Fe.
Martensitic stainless steel; air or oil hardenable to 740-880 N/mm² UTS.
For pump bodies, valves, control rods.
AFNOR Z20C13; AISI 420.

SOLEIL A4M.
M-1488; 0.18-0.25 C, 1.0 max Mn, 1.0 max Si, 12.5-14.5 Cr, bal Fe.
Stainless casting; annealed: 200-250 Brin. Pumps and faucets for food industries.
AFNOR Z20C13M; similar to ACI CA-40.

SOLEIL A5.
M-1488; 0.30 C, 13.0 Cr, bal Fe.
Martensitic stainless steel; air or oil hardenable to 830-980 N/mm² UTS.
For axles, shafts, valves, pump parts.
AFNOR Z30C13; AISI 420.

SOLEIL A5M.
M-1488; 0.25-0.35 C, 1.0 max Mn, 1.0 max Si, 13.0-15.0 Cr, bal Fe.
Stainless casting; annealed: 250-300 Brin.
Faucets and pumps for paper industries.
AFNOR Z30C13M; similar to ACI CA-40.

SOLEIL A5U.
M-1488; 0.30 C, 13 Cr, 1.0 max Ni, 0.60 max Mo, 0.15 min S, bal Fe.
Q. & T.: 830-980 N/mm² UTS.
Free machining martensitic stainless steel.
AFNOR Z30CF13; AISI 420 F.

SOLEIL A6M.
M-1488; 0.35-0.45 C, 1.0 max Mn, 1.0 max Si, 13.0-15.0 Cr, bal Fe.
Stainless casting; ann: 260-300 Brin.
For pump parts in paper industries.
AFNOR Z38C13M; W.Nr. 1.4034.

SOLEIL A8.
M-1488; 0.70 C, 0.5 Mn, 0.6 max Si, 15.0 Cr, bal Fe.
Martensitic stainless.
AFNOR Z70C15; AISI 440A.

SOLEIL A21M.
M-1488; 0.08-0.15 C, 1.0 max Mn, 1.0 max Si, 0.50-1.50 Ni, 12.0-14.0 Cr, bal Fe.
Stainless casting; annealed: 170-290 Brin.
For pumps and faucets; food industries; cold dilute organic acids.
AFNOR Z12CN13M; W.Nr. 4008.

SOLEIL B2.
M-1488; 0.08 max C, 12.5 Cr, bal Fe.
Ferritic type stainless steel; not usually hardenable. Weldable.
440-640 N/mm² UTS.
For parts subject to water and steam.
AFNOR Z6C13.

SOLEIL B3.
M-1488; 0.08 max C, 12.5 Cr, 0.20 Al, bal Fe.
Ferritic type stainless steel; weldable; not hardenable by heat treating.
440-640 N/mm² UTS.
For use with water or steam; petroleum industries.
AFNOR Z6 CA13; AISI 405.

SOLEIL B4.
M-1488; 0.10 max C, 17.0 Cr, bal Fe.
Ferritic type stainless; not hardenable by heat treating.
440-640 N/mm² UTS.
Tableware forks and spoons, kitchenware, soap and dyeing industries.
AFNOR Z8C17; AISI 430.

SOLEIL B4U.
M-1488; 0.12 C, 17 Cr, 0.60 max Mo, 0.15 min S, bal Fe.
Wrought, ann: 440-640 N/mm² UTS.
Free machining ferritic type stainless steel.
AFNOR Z10CF17; AISI 430 F.

SOLEIL C1.
M-1488; 0.12 max C, 16.0 Cr, 3.0 Ni, 1.5 Cu, bal Fe.
Martensitic stainless steel; air or oil hardenable to 740-880 N/mm² UTS.
For pump parts in organic acid or neutral briny solutions.
AFNOR Z10CNU 17-04.

SOLEIL C2M.
M-1488; 0.15-0.25 C, 1.0 max Mn, 1.0 max Si, 1.5-3.0 Ni, 15.0-18.0 Cr, bal Fe.
Stainless casting; annealed: 275-310 Brin.
For parts subject to sea water or dilute organic acids.
AFNOR Z20Cn17.2M; similar to AISI 431.

SOLEIL C3M.
M-1488; 0.06 max C, 1.0 max Mn, 1.0 max Si, 3.5-5.0 Ni, 15.0-17.5 Cr, 3.0-4.5 Cu, 0.15 min Nb, bal Fe.
Stainless casting; Ht. Tr.; 365-410 Brin.
Good corrosion resistance at elevated temperature.
AFNOR Z4CNUNB 16.4M.

SOLEIL C4.
M-1488; 0.06 max C. 16.0 Cr, 5.0 Ni, 0.9 Mo, bal Fe.
Martensitic stainless; air or oil hardenable to 880-1080 N/mm² UTS.
Bars; good strength and corrosion resistance to sea water, organic acids.
AFNOR Z4CND 16-05; see also VIRGO 3.

SOLEIL C4M.
1488; 0.06 max C, 1.0 max Mn, 1.0 max Si, 4.0-5.50 Ni, 15.5-17.5 Cr, 0.50-1.50 Mo, bal Fe.
Stainless casting.
For water turbines, pumps and other parts in sea water.
AFNOR Z4 CND 17.4M.

SOLEIL C5M.
M-1488; 0.06 max C, 1.0 max Mn, 1.0 max Si, 3.5-5.0 Ni, 12.0-13.5 Cr, 0.4-0.7 Mo, bal Fe.
Stainless casting.
Parts for pumps, turbines, nuclear energy.
AFNOR Z4CND13.4M; ACI CA6NM.

SOLEIL D9.
M-1488; 1.10 C, 0.5 Mn, 0.6 max Si, 16.5 Cr, 0.5 Mo, bal Fe.
Martensitic stainless.
AFNOR Z110CD17; AISI 440C.

SOLEIL O.
M-1546; 0.44 C, 14 Cr, bal Fe.
Annealed: 95,000 TS; 50,000 YS; 25 El; B 92 Rock.
For cutlery, turbine blades, diesel engine pistons, surgical instruments.
Martensitic stainless steel. Hardenable.
Corrosion and heat resistant.

SOLEIL T1.
M-1546; 0.20 C, 0.4 Cb, 11 Cr, 0.75 Mo, 0.5 V, 0.3 Ni, bal Fe.
Annealed: 95,000 TS; 40,000 YS; 25 El; B 92 Rock.
Heat treated: 240,000 TS; 205,000 YS; 9 El; C 50 Rock.
For surgical instruments, hardware, knives, gears, shafts.
Corrosion resistant, hardenable.

SOLEIL T2.
M-1546; 0.20 C, 11 Cr, 1 Mo, 0.8 Ni, 0.35 V, 0.6 W, bal Fe.
Annealed: 95,000 TS; 40,000 YS; 25 El; B 92 Rock.
Hardened: 240,000 TS; 205,000 YS; El; C 50 Rock.
For valves, bearings, cutlery, surgical instruments.
Corrosion resistant, hardenable.

SOLID DRILL.
M-783; 1.05 C, 0.35 Mn, bal Fe.
For granite drills; water hardened.

SOLUMANG 75B.
M-1038; 75 Mn (pure electrolytic), 25 Al.
For use in steel melting.

SOLUMANG 90PF.
M-1038; 90.0 Mn (pure electrolytic), 10.0 flux (alkali halides).
For rapid addition of manganese in alloying and deoxidation of steel.

SO-LUMINUM.
55 Sn, 33 Zn, 11 Al, 1 Cu.
For aluminum solder.

SOLUSAL.
M-388, M-427; Cu, Si, bal Al.
34,000 TS; 16 El.
For airplane, automobile and marine parts to resist corrosion; light alloy.

SONA-METAL.
M-893; 0.2-0.7 Si, 3-4 Cu, 0.7 Sn, 0.5 Mg, 1.0 Fe, 0.2 Ni, bal Al.
36,000-50,000 TS; 4-7 El; 60-90 Brin.
For light alloy parts; age-hardenable.

SONDERLEGIERUNG.
M-116; 1 Mn, 0.2 Si, 0.5 Fe, bal Al.
Annealed: 19,000 TS; 14,000 YS; 30 El.
For deep drawing parts; corrosion resistant.

SONDERMESSING.
M-Ger.; Cu, alloy, bal Zn.
For nuts, bolts, hardware; special alloy high tensile brass.

SONDERSTAHL DAUER D-X.
M-1310; 2.1 C, 11.5 Cr, 0.35 Si, 0.30 Mn, bal Fe.
For blanking and forming dies, punches; oil hardened, non-deforming.

SONDERSTAHL MIME EXTRA ZH.
M-1310; 1.05 C, 0.25 Si, 1.0 Cr, 1.15 W, 0.90 Mn, bal Fe.
For blanking and forming dies; oil hardened, tough.

SONDERSTAHL TCK.
M-1310; 1.0 C, 1.1 Cr, 0.07 Mn, 0.25 Si, bal Fe.
For bearings, liners, sleeves, cutters; water hardened, wear resistant.

SONOSTON.
M-1675; 54.2 Mn, 37 Cu, 4.25 Al, 3 Fe, 1.5 Ni.
For marine propellers, diesel engines, pneumatic drills.
Dampens noise and vibrations; weldable.

SORCERY B1912 ETC see B1912 ETC.

SOREL'S ALLOY-1.
M-Eng.; 80 Zn, 10 Cu, 10 Fe.
For bearings.

SOREL'S ALLOY-2.
M-Eng.; 98 Zn, 1 Cu, 1 Fe.
For bearings.

SORMITE.
M-515; 30 Cr, 4.5 Si, 1.1 Mn, 3.2 C, 5.3 Ni, bal Fe.
To replace certain Stellite alloys, hard facing; hard alloy, wear and corrosion resistant.

SORMITE.
M-USSR; 1.75 C, 15.5 Cr, 2 Ni, 2 Si, bal Fe.
For blanking and forming dies, gages, punches; air hardened, non-deforming.

SOS6-6.
M-USSR; C, alloy, bal Fe.
Bearings for high speeds and pressures; anti-friction.

SOUDINOX 65.
M-1488; 0.02 max C, 7.0 Mn, 1.0 max Si, 25.5 Cr, 20.0 Ni, bal Fe.
Austenitic stainless steel; weldable grade.
AFNOR Z2CNM 25-20.

SOUDINOX 308L.
M-1488; 0.03 max C, 1.0 Mn, 1.0 max Si, 19.5 Cr, 10.75 Ni, bal Fe.
Austenitic stainless steel; weldable grade.
AFNOR Z2CN 20-10; low carbon modification of AISI 308.

SOUDINOX 308L TM.
M-1488; 0.025 max C, 1.75 Mn, 1.0 max Si, 19.5 Cr, 10.75 Ni, bal Fe.
Austenitic stainless steel; weldable grade.
AFNOR Z1CN 20-10; AWS ER 308L.

SOUDINOX 308W.
M-1488; 0.02 max C, 1.0 Mn, 1.0 max Si, 19.5 Cr, 11.0 Ni, bal Fe.
Austenitic stainless steel; weldable grade.
AFNOR Z1CN20-10; AWS ER308L.

SOUDINOX 309.
M-1488; 0.08 max C, 2.0 Mn, 1.0 max Si, 24.0 Cr, 13.5 Ni, bal Fe.
Austenitic stainless steel; weldable grade.
AFNOR Z10CN24-13; AWS ER309.

SOUDINOX 309L.
M-1488; 0.03 max C, 1.5 Mn, 1.0 max Si, 24.5 Cr, 13.0 Ni, bal Fe.
Austenitic stainless steel; weldable grade.
AFNOR Z2CN24-13; AWS ER309L.

SOUDINOX 310.
M-1488; 0.10 max C, 2.2 Mn, 1.0 max Si, 27.0 Cr, 21.0 Ni, bal Fe.
Austenitic stainless steel; weldable grade.
AFNOR Z12CN25-20; AWS ER310.

SOUDINOX 312.
M-1488; 0.15 max C, 2.0 Mn, 1.0 max Si, 30.0 Cr, 9.5 Ni, bal Fe.
Austenitic stainless steel, weldable grade.
AFNOR Z12CN30-10; AWS ER312.

SOUDINOX 316L.
M-1488; 0.03 max C, 1.2 Mn, 1.0 max Si, 17.5 Cr, 11.0 Ni, 2.6 Mo, bal Fe.
Austenitic stainless steel; weldable grade.
AFNOR Z2CND18.11; AISI 316L.

SOUDINOX 316L TM.
M-1488; 0.02 max C, 1.75 Mn, 1.0 max Si, 19.0 Cr, 13.25 Ni, 2.6 Mo, bal Fe.
Austenitic stainless steel; weldable grade.
AFNOR Z2CND19-13; AWS ER316L.

SOUDINOX 410.
M-1488; 0.05 max C, 0.5 Mn, 0.6 max Si, 13.0 Cr, bal Fe.
Chrome corrosion resisting steel, low carbon type; weldable.
For corrosion resistance and heat resistance.
AFNOR Z3C14.

SOUDINOX 1330.
M-1488; 0.03 max C, 1.5 Mn, 1.0 max Si, 20.5 Cr, 10.2 Ni, 2.9 Mo, bal Fe.
Austenitic stainless steel; weldable grade.
AFNOR Z2CND20-10; W. Nr. 1.4428.

SOUDINOX A.
M-1488; 0.03 max C, 1.0 Mn, 1.0 max Si, 20.5 Cr, 8.0 Ni, 2.5 Mo, 1.5 Cu, bal Fe.
Austenitic stainless steel; weldable grade.
AFNOR Z5CNDU21-08.

SOUDINOX B6.
M-1488; 0.02 max C, 1.2 Mn, 1.0 max Si, 20.0 Cr, 25.5 Ni, 4.3 Mo, 1.6 Cu, bal Fe.
Austenitic stainless steel; weldable grade.
AFNOR Z1NCDU25-20.

SOUDINOX S1.
M-1488; 0.02 max C, 6.5 Mn, 3.7 Si, 19.5 Cr, 14.0 Ni, bal Fe.
Austenitic stainless steel; weldable grade.
AFNOR Z2CNMS20-14.

SP.
M-1655; 0.70 C, 0.70 Mn, 1.70 Si, bal Fe.
Water hardenable to about 54 Rc for collets, chuck jaws, clamps, screw drivers.
W.-Nr. 1.2823.

SP-3.
M-513; 2.5 C, 1.0 max Si, 20.0 Cr, 40.0 Ni, 6.0 W, 12.0 Co, bal Fe.
For diesel engine valves.

SP5.
M-1687; 10 Sn, 5 Pb, 0.3 min P, bal Cu.
Leaded bronze; cast or wrought; 75 Brin.
AFNOR UE10 PB5.

SP10.
M-1687; 10 Sn, 9 Pb, bal Cu.
Leaded bronze; cast or wrought; 75 Brin.
AFNOR UE10 Pb8; SAE 64.

SP10.
M-1739.
Alloy steel, sintered.
Density: 5.9-6.2 gms/cc; UTS: 44 kg/mm² typical.
Elon: 1/2-2%; Hardness: Rock B 40.
SAE 866A; ASTM B426 Type 1 Grade 3.

SP15.
M-1687; 7.5 Sn, 15 Pb, bal Cu.
Copper-Tin-Lead alloy; cast or wrought. 45 Brin.
AFNOR U Pb15 E8.

SP20.
M-1739.
Cu-Ni-C Steel, sintered.
Density: 6.6-6.8 gms/cc; UTS: 39-47 kg/mm²; Elon: 1-3%; Hardness: HV5-200.
Hardenable to Rock C 35 approx.
For clutch transmission parts, hubs, levers.
B.S.S. A401; MPIE FN-0310-T.

SP22.
M-1687; 5 Sn, 18 Pb, bal Cu.
Copper-Lead-Tin alloy; cast or wrought; 40 Brin.
AFNOR UPb20 E5.

SP22.
M-1739.
Cu-Ni-C steel, sintered.
Density: 6.8-6.9 gms/cc; UTS: 63kg/mm²; Elon: 1%; Hardness: HV5-300.
High duty structural components.
B.S.S. A402; MPIE FN-0508-T.

SP24.
M-1739.
Nickel alloy steel, sintered.
Density: 6.8 gms/cc min; UTS: 39 kg/mm² typical; Elon: 4 min; Hardness: Rock B 40
Can be hardened or case hardened.
B.S.S. A400; MPIE FN-0550-T.

SP25.
M-1739.
Nickel alloy steel, sintered.
Density: 6.8-7.2 gms/cc; UTS: 41 kg/mm² typical.
Hardenable to Rock C 35 approx.
B.S.S. A401: MPIE FN-0310-T.

SP26.
M-1739.
Cu-Ni-C steel, sintered.
Density: 6.8 gms/cc; UTS: 47 kg/mm²; Elon: 1-3%; Hardness: HV5-200.
Gears for oil pumps, diesel engines.
B.S.S. A400; MPIE FN-0500-T.

SP 35X.
M-959; 35 Sn, bal Pb.
Block solder; melt range: 358-460°F.

SP 40X.
M-959; 40 Sn, bal Pb.
Block solder; melt range: 358-448°F.

SPA ALLOY.
M-47; 1.6 C, 28.0 Cr, 2.0 Ni, 2.0 Mo, 1.0 Cu, 2.0 max Si, 1.5 max M, bal Fe.
Cast alloy, hardenable to 600 Brin.
Very good abrasion resistance: excellent corrosion-erosion resistance in alumina slurry at pH 2.5 to pH 11.

SPANAL 320.
M-324; 2.5-5.0 Cu, 0.2-1.8 Mg, 0.3-1.5 Mn, 0.5-2.5 Pb, Sn, Cd, Bi, bal Al.
Annealed: 27,000 TS; 11,000 YS; 22 El; 47 Brin.
Heat treated: 72,000 TS; 57,000 YS; 130 Brin.
For screw machine products, aircraft parts; free-cutting, age-hardenable.

SPANASIL.
M-324; 0.8 Mg, 1.0 Si, 1.0 Pb, bal Al.
20-28 kp/mm² TS; 10-18 kp/mm² YS; 8-10 El.
Al Mg Si Pb.

SPANG CHALFANT 1.
M-735; 0.15 C, 2 Cr, 0.5 Mn, 0.5 Mo, bal Fe.
Rolled: 66,500 TS; 40,500 YS; 40 El; 131 Brin.
For gears, shafts, rolls; case hardened.

SPANG CHALFANT 2.
M-735; 0.1-0.2 C, 0.5 Mn, bal Fe.
Rolled: 62,400 TS; 36 El; 126 Brin.
For gears, shafts, rolls; case hardened.

SPANG CHALFANT 3.
M-735; 0.1-0.2 C, 0.45-0.65 Mo, bal Fe.
Rolled: 64,000 TS; 32,500 YS; 37 El; 126 Brin.
For gears, shafts, rolls; case hardened.

SPANVAL 320.
M-Eng.; C, alloy, bal Fe.
For engineering structures; strong.

SPARTA.
M-114; 1.05 C, 5 Cr, 0.9-1.25 Mo, 0.25 V, bal Fe.
For cutting tools, dies, punches; air hardened.

SPARTA 80.
M-114; 0.8 C, 0.65 Mn, 5.25 Cr, 1.1 Mo, 0.25 V, bal Fe.
For blanking and forming dies, punches; air hardening.

SPARTA CV.
M-114; 2.35 C, 5.25 Cr, 4 V, 1 Mo, 1 W, bal Fe.
Hardened: 65-67 Rock C.
For blanking and forming dies, lamination and coining dies; cold shear blades, plug gauges.
Type A7 cold work die steel. Abrasion resistant.

SPARTAN-5.
M-435; 0.5 C, 18 W, 4 Cr, 1 V, bal Fe.
For cutting tools, shear blades; high speed steel, hot work tools.

SPARTAN-7.
M-435; 0.7 C, 0.25 Mn, 18 W, 4 Cr, 1.2 V, bal Fe.
For hot shear blades, drills, seamers, tips, broaches; high speed steel.

SPEAR 3S.
M-1737; 0.80 C, 0.25 S, 0.30 Mn, 2.4 Ni, 0.20 Cr, bal Fe.
Plate or sheet for saw blades.

SPEAR "5-6-2".
M-1737; 0.83 C, 4.25 Cr, 6.0 W, 5.0 Mo, 2.0 V, bal Fe.
Moly-tungsten high speed steel for cutting tools.
AISI M-2.

SPEAR 50.
M-1737; 0.55 C, 0.60 Mn, bal Fe.
Oil or water hardening steel; for clipper tools, bolsters, pressure plates.

SPEAR 75.
M-1737; 0.75 C, 0.70 Mn, bal Fe.
Oil hardening steel; for press brakes, shear blades, drifts, machine parts.

SPEAR A.H.C.
M-1737; 0.50 C, 1.0 Si, 1.5 Cr, 2.0 W, 0.12 V, bal Fe.
Oil hardening steel for pneumatic tools, chisels and shock resisting tools and parts.
AISI S1.

SPEAR B.1.
M-1737; 0.32 C, 4.25 Ni, 1.25 Cr, 0.30 Mo, bal Fe.
Oil or air hardenable tool steel for plastic moulds.

SPEAR B.4.
M-1737; 0.32 C, 0.30 Si, 0.60 Mn, 2.1 Cr, 0.25 Mo, 0.15 V, bal Fe.
Alloy steel for nitriding.

SPEAR CRH.
M-1737; 0.75 C, 0.25 Si, 0.60 Mn, 0.40 Cr, bal Fe.
Plate or sheet for saw blades.

SPEAR CV.
M-1737; 0.78 C, 0.25 Si, 0.60 Mn, 0.45 Cr, 0.15 V, bal Fe.
Plate or sheet for saw blades.

SPEAR D.1.
M-1737; 0.50 C, 0.70 Mn, 1.0 Cr, 0.25 V, bal Fe.
Oil hardening steel for die casting dies, stamping and forging dies, hot shear blades.
AISI L2.

SPEAR D.5.
M-1737; 0.35 C, 1.0 Si, 5.75 Cr, 1.0 V, 1.5 Mo, bal Fe.
Oil hardening hot work tool steel; for die casting dies, gripping and forming tools, hot shear blades.
Chromium type hot work tool steel.
AISI H13.

SPEAR D.7.
M-1737; 0.32 C, 1.0 Si, 1.5 W, 5.0 Cr, 0.40 V, 1.5 Mo, bal Fe.
Air hardening tool steel for gripper dies, mandrels, header and trimmer dies.
Chromium type hot work tool steel.
AISI H12.

SPEAR D.9.
M-1737; 0.30 C, 9.5 W, 3.2 Cr, 0.30 V, bal Fe.
Tungsten type hot work tool steel for forging dies, header tools, extrusion dies.
AISI H21.

SPEAR D.12.
M-1737; 2.0 C, 13.0 Cr, bal Fe.
High carbon-high chrome type tool steel.
Oil hardenable; for abrasion and corrosion resisting tools and dies.
AISI D3.

SPEAR D.13.
M-1737; 1.9 C, 13.5 Cr, 1.3 Mo, bal Fe.
High carbon-high chrome type tool steel.
Air or oil hardening for heavy duty use, abrasion and corrosion resisting tools & dies.
Similar to AISI D3.

SPEAR D.14.
M-1737; 1.3 C, 13.0 Cr, 1.5 Mo, 2.75 Co, bal Fe.
High carbon-high chrome type tool steel.
Air or oil hardening for heavy duty use, abrasion and corrosion resisting tools and dies.
AISI D5.

SPEAR D.15.
M-1737; 1.0 C, 0.70 Mn, 5.0 Cr, 1.0 Mo, 0.25 V, bal Fe.
Air hardening tool steel for blanking and forming dies, shear blades, gauges.
AISI A2.

SPEAR D.17.
M-1737; 1.5 C, 12.0 Cr, 0.9 V, 0.9 Mo, bal Fe.
High carbon-high chromium type tool steel. Air or oil hardening for abrasion and corrosion resisting tools and dies.
AISI D2.

SPEAR D.X.
M-1737; 0.28 C, 0.85 Cr, 2.2 W, 0.25 V, 0.50 Mo, 2.25 Ni, bal Fe.
Oil hardening hot work tool steel for tools requiring maximum toughness.

SPEAR EXM.
M-1737; 0.73 C, 0.25 Si, 0.40 Mn, 1.4 Ni, 0.20 Cr, bal Fe.
Plate or sheet for saw blades.

SPEAR M.1.
M-1737; 0.80 C, 4.25 Cr, 1.25 W, 8.5 Mo, 1.15 V, bal Fe.
Moly high speed steel for cutting tools.
AISI M1.

SPEAR M.42.
M-1737; 1.05 C, 4.25 Cr, 1.0 W, 9.0 Mo, 1.1 V, 8.0 Co, bal Fe.
Cobalt-Moly high speed steel for cutting tools.
AISI M42.

SPEAR NO. 2. GRADE 1.
M-1737; 1.05 C, 0.20 Mn, bal Fe.
Water hardening tool steel; for taps, twist drills, lathe centers, shear blades.
AISI W1.

SPEAR NO. 2. GRADE 2.
M-1737; 0.95 Cu, 0.20 Mn, bal Fe.
Water hardening tool steel; for cold heading dies, punches, woodworking tools.
AISI W1.

SPEAR NO. 2. GRADE 3.
M-1737; 0.85 Cu, 0.20 Mn, bal Fe.
Water hardening tool steel; for pneumatic hammers and chisels, press tools, mason's tools.
AISI W1.

SPEAR NO. 2 VANADIUM.
M-1737; 1.0 C, 0.20 Mn, 0.25 V, bal Fe.
Water hardening tool steel; for blanking dies, punches, hand stamps, chuck jaws.

SPEAR N.S.
M-1737; 0.95 C, 1.2 Mn, 0.50 Cr, 0.50 W, 0.15 V, bal Fe.
Oil hardening tool steel for jigs and gauges, cold trimming dies, punches, shear blades.
AISI O1.

SPEAR P.S.
M-1737; 1.05 C, 0.60 Mn, 1.5 Cr, 0.25 Mo, bal Fe.
Oil hardening tool steel; for forming and blanking tools, gauges, taps, bushings.
AISI L4.

SPEAR TC.
M-1737; 0.45 C, 0.25 Si, 0.50 Mn, 1.5 Ni, 0.15 Cr, 0.30 Mo, bal Fe.
Plate or sheet for saw blades.

SPEAR VANCHIP.
M-1737; 0.68 C, 1.0 Si, 8.0 Cr, 1.5 Ni, 1.4 V, 1.4 Mo, bal Fe.
Air hardenable tool steel for chipper knives, veneer knives, shear blades.

SPECIAL.
0.7 C, 4 Cr, 1 V, 1.5 W, 9.5 Mo, bal Fe.
For cutting tools, drills; high speed steel.

SPECIAL.
M-344; C, W, bal Fe.
For dies; high duty shafts; hot dies; oil hardened.

SPECIAL.
M-388; 0.8 C, 0.2 V, 0.8 Cr, bal Fe.
For tools, cutters; oil hardened.

SPECIAL.
M-389; 0.7 C, 18 W, 4 Cr, 2 V, bal Fe.
For tools, cutters; high speed steel.

SPECIAL ALLOY.
M-389; 0.7 C, 1 Mn, 0.4 Cr, bal Fe.
For tools, dies; water hardened.

SPECIAL ALLOYS.
M-373; C, alloy, bal Fe.
For tools.

SPECIAL ALUMINUM.
M-U.S.; 10.5-10.7 Al, 3.1-4.25 Fe, 3.6 Ni, 0-1.6 Mn, bal Cu.
200-300 Brin.
For worm wheels, gears, trolley shoes; heat treatable.

SPECIAL ARDHO N.H.O.
M-1744; 0.35 C, 0.25 Si, 0.50 Mn, 3.2 Ni, 0.78 Cr, bal Fe.
3% Nickel-Chromium tool steel, for smiths' tools, punches, dies.

SPECIAL BB.
M-261; 0.28 C, 3.20 Cr, 0.30 V, 9.50 W, bal Fe.
Hardened: 243,000 TS; 215,000 YP; 12 El; 37 RA; 50 Rock C.
For die casting dies for brass and aluminum bronze, extrusion dies and liners, hot forging dies.
Oil or air hardening; AISI Type H21.

SPECIAL-BLOC.
M-1488; 0.55 C, 0.75 Mn, 0.55 Ni, 1.0 Cr, 0.45 Mo, 0.05 V, bal Fe.
Oil hardening hot work tool steel; for hot forming dies.
AFNOR 55 CNDV 4.

SPECIAL CARBON.
M-97; 0.65-1.4 C, 0.3 Mn, 0.25 Si, bal Fe.
For cold work dies; water hardening.

SPECIAL CARBON SSC.
M-40; 0.95 C, 0.35 Ni, V, bal Fe.
For tools, punches; water hardened.

SPECIAL CARBON WITH V.
M-73; 1.0 C, 0.2 Mn, 0.2 V, 0.10 max Cr, bal Fe.
For drills, tools; water hardening.

SPECIAL CHROME VANADIUM.
M-365; 0.6 C, 0.9-1.0 Cr, 0.2 V, bal Fe.
For tools, punches; oil hardened.

SPECIAL COLD PRESSING VANADIUM STEEL.
M-112; 1.1 C, 0.2 Si, 0.3 Mn, 0.2 V, bal Fe.
For cold heading dies; swaging dies; water hardened, shock resistant.

SPECIAL CONQUEROR.
M-1406; 0.85-1.1 C, bal Fe.
Heat treated: 190,000 TS; 145,000 YS; 10 El; 30 RA; 400 Brin.
For rock drills, punches, taps, shear blades; water hardened; Type W1.

SPECIAL D.
M-344; C, W, bal Fe.
For tools; oil hardened.

SPECIAL DRACO.
M-114; 0.7 C, 1.4 Cr, 0.2 V, bal Fe.
For tools, dies, drills; water hardened.

SPECIAL FA71.
M-1488; 0.40 C, 3.15 Cr, 1.0 Mo, 0.20 V, bal Fe.
Hot work tool steel for forging and hot forming dies; good thermal shock resistance.
AFNOR 40 CDV 12; similar to AISI S7.

SPECIAL FA82.
M-1488; 0.38 C, 1.0 Si, 5.25 Cr, 1.35 Mo, 0.5 V, bal Fe.
Hot work tool steel for forging and upsetting dies, and forming dies.
AFNOR Z38 CDV5; AISI H 11.

SPECIAL G.
M-782; 0.7-1.2 C, bal Fe.
For tools; water hardened.

SPECIAL HDS.
M-486; 0.28 C, 3.2 Cr, 0.3 V, 9.5 W, 0.3 Mo, bal Fe.
For casting dies, hot swaging dies; hot die steel.

SPECIAL HOT DIE.
M-1112; 9.5 W, 2.5 Cr, 0.10-0.15 V, bal Fe.
For cutting tools, dies; hot die steels.

SPECIAL HS-55.
M-24; 0.50 C, 4 Cr, 1 V, 18 W, bal Fe.
Air hardened: C 60-62 Rock.
Oil hardened: C 60-64 Rock.
For extrusion and forging dies, swaging and pressing dies, punches, hammer tools.
Type H26 Hot work steel, high red-hardness and toughness.

SPECIAL HS55 see **BETHLEHEM SPECIAL HS.**

SPECIAL K.
M-27, M-1405, M-1618; 1.9 C, 13.5 Cr, 0.65 Si, 0.3 Mn, bal Fe.
For punches, dies, swaging and forming tools; air or oil hardened; non-deforming.

SPECIAL K.
M-363; 2.2 C, 12 Cr, 0.6 Si, 0.3 Mn, bal Fe.
For punch and die work tools, cutting dies, plug gages; non-deforming.

SPECIAL K5.
M-118; 2 C, 5 Cr, bal Fe.
For tools, dies, forming and blanking dies; air hardened, non-deforming.

SPECIAL K5.
M-1618; 1.0 C, 0.7 Mn, 0.3 Si, 5.3 Cr, 0.15 V, 1.0 Mo, bal Fe.
Annealed: 103,000 TS; 51,000 YS; 26 El; C 18 Rock.
Ht. Tr.; 253,000 TS; 200,000 YS; 3 El; C 53 Rock.
For blanking and trimming dies, shear blades, broaches, cutters, punches.
Type A2 air hardening tool steel, non-deforming.

SPECIAL KMV now **VEW K110.**

SPECIAL KN.
M-1182; 1.5 C, 12 Cr, 1 Mo, bal Fe.
For blanking and forming dies, punches; Type D2; air hardened, non-deforming.

SPECIAL KR.
M-1182; 2.25 C, 12 Cr, bal Fe.
For lamination and blanking dies, punches; Type D3; oil hardened, non-deforming.

SPECIAL LOHYS.
M-1591; 0.75 Si, 99.25 Fe.
Initial permeability 400.
Max permeability 5350.
Coercive force 0.85.
For electrical equipment, magnetic instruments.
Soft magnetic material.

SPECIALLOY.
M-1704; 0.51 C, 0.93 Mn, 0.98 Cr, 0.52 Ni, 0.22 V, 0.25 Mo, bal Fe.
Low alloy medium carbon oil hardening tool steel.

SPECIALLOY 1026.
M-1236; 10 Fe, 90 Cu.
Low temperature introduction of iron to copper alloys.

SPECIALLOY 1029.
M-1236; 99.9 min Cu.
For iron and steel alloying; addition agent.

SPECIALLOY 1103.
M-1236; 2 Li, 98 Cu.
Deoxidizer for copper alloys.

SPECIALLOY 1105.
M-1236; 2 B, 98 Cu.
Deoxidizer for copper alloys.

SPECIALLOY 1124.
M-1236; 1 Cr, 99 Cu.
Conductivity copper casting alloy.

SPECIALLOY 1124CC.
M-1236; 1 Cr, 99 Cu.
Continous-cast extrusion or forging stock of CDA 182, RWMA Class 2 Alloy.

SPECIALLOY 1213CC.
M-1236; miscellaneous Al-Bronze alloys.
Extrusion and forging stock,-continuous cast billet.

SPECIALLOY 1224.
M-1236; 5 Cr, 95 Cu.
Master alloy for conductivity copper castings.

SPECIALLOY 1314.
M-1236; 10 Si, 90 Cu.
Silicon source for copper alloy addition.

SPECIALLOY 1327.
M-1236; 10 Co, 90 Cu.
Cobalt addition for copper alloys.

SPECIALLOY 1328.
M-1236; 1 Nb, 10 Ni, 89 Cu.
Cupro-nickel casting alloy.

SPECIALLOY 1328CC.
M-1236; 10 Ni, 90 Cu.
Continuous cast cupro-nickel extrusion and forging stock; CDA 706.

SPECIALLOY 1340.
M-1236; 13 Zr, 87 Cu.
Zirconium addition to copper alloys.

SPECIALLOY 1414.
M-1236; 15 Si, 85 Cu.
Silicon addition to copper alloys.

SPECIALLOY 1512.
M-1236; 20 Mg, 80 Cu.
Magnesium additions to copper alloys.

SPECIALLOY 1514.
M-1236; 20 Si, 80 Cu.
Silicon additions to copper alloys.

SPECIALLOY 1714.
M-1236; 30 Si, 70 Cu.
Silicon additions to copper alloys.

SPECIALLOY 1722.
M-1236; 30 Ti, 70 Cu.
Titanium additions to copper alloys.

SPECIALLOY 1725.
M-1236; 30 Mn, 70 Cu.
For brass alloying; addition agent.

SPECIALLOY 1726.
M-1236; 30 Fe, 70 Cu.
Iron additions for copper alloys.

SPECIALLOY 1727.
M-1236; 30 Co, 70 Cu.
Cobalt additions for copper alloys.

SPECIALLOY 1728.
M-1236; 1 Nb, 30 Ni, 69 Cu.
High strength weldable grade cupro-nickel casting alloy.

SPECIALLOY 1728CC.
M-1236; 30 Ni, 70 Cu.
Continuous cast cupro-nickel extrusion and forging stock. CDA 715.

SPECIALLOY 1733.
M-1236; 30 As, 70 Cu.
Arsenic additions to copper alloys.

SPECIALLOY 1826.
M-1236; 50 Fe, 50 Cu.
Iron additions for copper alloys.

SPECIALLOY 1827.
M-1236; 50 Co, 50 Cu.
Cobalt additions for copper alloys.

SPECIALLOY 1848.
M-1236; 50 Cd, 50 Cu.
Cadmium additions for copper alloys.

SPECIALLOY 1852.
M-1236; 50 Te, 50 Cu.
Tellurium additions for copper alloys.

SPECIALLOY 1928CC.
M-1236; 2.5 Cr, 30 Ni, 67.5 Cu.
Continuous cast, chromium modified, cupro-nickel extrusion and forging stock.
CDA 719.

SPECIALLOY 2029.
M-1236; 30 Cu, 70 Ni.
Copper plus nickel additions for iron alloys.

SPECIALLOY 2829.
M-1236; 50 Cu, 50 Ni.
Nickel addition to copper alloys; also Cu-Ni additions to iron alloys.

SPECIALLOY 2841.
M-1236; 68 Nb, 40 Ni.
Niobium (Columbium) additions to cupro-nickel alloys.

SPECIALLOY 2842.
M-1236; 50 Mo, 50 Ni.
Molybdenum additions to iron alloys.

SPECIALLOY 2929.
M-1236; 1 Si, 1 Mn, 2 Fe, 30 Cu, 66 Ni.
Corrosion resistant-high strength nickel copper casting alloy.

SPECIALLOY 2929B.
M-1236; 2 Si, 1 Mn, 2 Fe, 30 Cu, 65 Ni.
Medium hardness nickel copper casting alloy.

SPECIALLOY 2929C.
M-1236; 3 Si, 1 Mn, 2 Fe, 30 Cu, 64 Ni.
High hardness nickel copper casting alloy.

SPECIALLOY 2929D.
M-1236; 4 Si, 1 Mn, 2 Fe, 30 Cu, 63 Ni.
Extra high hardness nickel copper casting alloy.

SPECIALLOY 2929W.
M-1236; 2 Nb, 1 Mn, 2 Fe, 1 Si, 30 Cu, 64 Ni.
Weldable grade, corrosion resistant, high strength nickel copper casting alloy.

SPECIALLOY 2942.
M-1236; 25 Mo, 25 Fe, 50 Ni.
Molybdenum addition to iron alloys.

SPECIALLOY 3841.
M-1236; 60 Nb, 40 Fe.
Niobium additions to ferrous and non-ferrous alloys.

SPECIALLOY 3928.
M-1236; 8 Cr, 28 Cu, 56 Ni, bal Fe.
Nickel-copper-chromium additions to iron alloys.

SPECIALLOY 5025 MANGANINGOT TM.
M-1236; 99.5 min Mn.
Fused alloy elemental manganese addition for ferrous and non-ferrous alloy

SPECIALLOY 5813 ALUMANINGOT TM.
M-1236; 40 Al, 60 Mn.
High purity aluminum plus manganese addition to iron alloys.

SPECIAL MANGANESE NICKEL.
M-872; 0.4 C, 12-14 Mn + Ni, bal Fe.
For welding electrode, hard facing; shielded arc.

SPECIAL MD.
M-1488; 0.55 C, 1.0 Si, 1.10 Cr, 2.0 W, bal Fe.
Cold work tool steel; shock resisting type.
For chisels, rivet sets, pneumatic tools.
AFNOR 55 WCS 20; similar to AISI S1.

SPECIAL MS.
M-1496; 0.28 C, 0.5 Si, 0.3 Mn, 1.2 Cr, 1.4 Mo, 0.25 V, 2.3 W, 2.3 Co, bal Fe.
For die casting dies; extrusion press components.
Thermal shock resistance.

SPECIAL NO. 18.
M-336; 0.45 C, 1.4 Cr, 2.2 W, 0.2 V, 0.3 Mo, bal Fe.
180 Brin.
For pneumatic tools, shear blades, chisels, rivet sets; shock tool steel.

SPECIAL NO. 18S.
M-336; 0.45 C, 1.4 Cr, 2.2 W, 0.3 Mo, 1.0 Si, 0.2 V, bal Fe.
For shear blades, rivet busters, chisels; shock resistant.

SPECIAL NO. 19.
M-336; C, W, Mn, Cr, bal Fe.
200 Brin.
For taps, thread dies; oil hardened.

SPECIAL NO. 20.
M-336; C, Cr, W, Mn, V, bal Fe.
240 Brin.
For finishing tools; oil hardened.

SPECIAL OIL HARDENING.
M-389; 0.9 C, 1.2 Mn, bal Fe.
For tools, dies; non-deforming.

SPECIAL OIL HARDENING.
M-275; 0.9 C, 1.2 Mn, 0.5 Cr, 0.5 W, bal Fe.
For tools, dies, punches; non-deforming.

SPECIAL PISTON STEEL.
M-57; 1.05 C, 0.25 Mn, bal Fe.
For pistons; water hardened.

SPECIAL-PRESSE.
M-1488; 0.20 C, 3.15 Ni, 3.4 Mo, bal Fe.
Hot work tool steel for forging dies. AFNOR 20 DN 33-12.

SPECIAL-PURPOSE 0125-8.
M-1071; 1.18-1.31 C, 0.15-0.35 Mn, 0.25 Si, 0.10-0.25 Cr, bal Fe.
For razor blades, hack saw blades.

SPECIAL PURPOSE ALLOY.
M-1071; 1.18-1.31 C, 0.15-0.35 Mn, 0.25 Si, 0.10-0.25 Cr, bal Fe.
Hot rolled: 30-35 Rc; Cold rolled, ann: 20 Rc max. Hardenable to 65 R max.
For hack saw blades, razor blades, cutlery.

SPECIAL R-2.
M-261; M-486; 0.2 C, 13 Cr, bal Fe.
Hardened: 210,000-140,000 TS; 195,000-120,000 YS; 10-15 El; 461-300 Brin.
For plastic molding dies; corrosion resistant.

SPECIAL SHELL TURNING.
M-Eng.; 0.7 C, 18 W, 4 Cr, 1 V, bal Fe.
For high speed cutting; high speed steel.

SPECIAL U1.
M-1488; 0.90 C, 1.9 Mn, 0.50 Cr, 0.10 V, bal Fe.
Cold work tool steel; oil hardening.
For punching and forming dies, reamers.
AFNOR 90 MCV 8.

SPECIAL VANADIUM.
M-339; C, bal Fe.
For dies, cold punches, impact tools, shear blades; tough and shock resistant.

SPECIAL VANADIUM.
M-275; 0.8-1.6 C, 0.2 V, bal Fe.
For tools, dies; water hardened.

SPECIAL VANADIUM.
M-365; 0.8-1.1 C, 0.10-0.20 V, bal Fe.
For tools; water hardened.

SPECIAL VERY HARD.
M-363; 1.1 C, 2.8 W, bal Fe.
For fast finishing tools; oil or water hardened.

SPECULUM.
68.25 Cu, 31.75 Sn.
For reflectors, mirrors; Cu_4, Sn.

SPECULUM METAL.
M-U.S.; 68-66 Cu, 32-34 Sn.
Rolled: 4,000-6,000 TS; 0 El.
For telescope reflectors, mirrors; resists tarnish; approximately Cu_3Sn.

SPEDEX.
M-135; 8-20 Ni, 2 Pb, 25 Zn, bal Cu.
For keys, electroplaters, sundries, gas and electric meters, carburetors, screws; nickel silver; free machining.

SPEED ALLOY.
M-173; 0.35 C, 1.04 Cr, 0.002 B, 1.15 Mn, bal Fe.
Rolled: 105,000 TS; 65,000 YS; 20 El; 207 Brin.
For cams, gears, racks, yokes, platens, die casting dies. Oil hardening.

SPEED-ALLOY 66.
M-849; Flux cored weld wire.
Typical deposit analysis: 0.08 C, 1.50 Mn, 0.40 Si (using CO_2 shielded gas).
Welded: 77,000 TS; 65,000 YS; 21 El.
For welding mild steel. AWS E70T-G.

SPEED ALLOY 71.
M-849; Flux cored weld wire.
Typical deposit analysis: 0.08 C, 1.35 Mn, 0.50 Si (using CO_2 shielding gas).
Welded: 87,000 TS; 75,000 YS; 25 El.
For welding mild steel. AWS E70T-1.

SPEED-ALLOY 71-A1.
M-849.
Flux cored weld wire.
Typical deposit analysis: 0.085 C, 0.85 Mn, 0.55 Si, 0.55 Mo.
Welded: 85,000 TS; 74,000 YS; 22 El.
For welding carbon and 0.5% Mo steels. AWS E-70T-G.

SPEED-ALLOY 74.
M-849; Flux cored weld wire.
Typical deposit analysis: 0.09 C, 1.0 Mn, 0.60 Si.
Welded: 88,000 TS; 70,000 YS; 22 El.
For welding mild steel. AWS E70T-4.

SPEED-ALLOY 75.
M-849; Flux cored weld wire.
Typical deposit analysis: 0.08 C, 1.2 Mn, 0.55 Si (using CO_2 shielding gas).
Welded: 76,450 TS; 64,100 YS; 32 El.
For welding mild steel. AWS E70T-5.

SPEED-ALLOY 75-AL.
M-849; Flux cored low alloy weld wire.
Typical deposit analysis: 0.08 C, 0.85 Mn, 0.40 Si, 0.55 Mo.
Welded: 78,000 TS; 66,000 YS; 21 El.
For welding carbon 0.5 Mo steels.
AWS E70T-G.

SPEED-ALLOY 81-W.
M-849; Flux cored alloy weld wire.
Typical deposit analysis: 0.06 C, 1.05 Mn, 0.55 Si, 0.55 Cr, 0.60 Ni, 0.50 Cu, bal Fe.
Welded: 92,000 TS; 80,000 YS; 25 El.
For welding weathering steels such as ASTM A588, A242.
AWS E80T.

SPEED ALLOY 85.
M-849; Flux cored low alloy weld wire. Typical deposit analysis: 0.08 C 1.0 Mn, 0.55 Si, 1.30 Ni, 0.20 Mo.
Welded: 85,700 TS; 70,200 YS; 26 El.
For welding low alloy steels with a tensile strength of about 80,000 psi.
AWS E80T.

SPEED-ALLOY 85-B2L.
M-849; Flux cored low alloy weld wire.
Typical deposit analysis: 0.03 C, 0.90 Mn, 0.45 Si, 1.35 Cr, 0.50 Mo.
Welded: 85,000 TS; 72,000 YS; 17 El.
For welding Cr-Mo pipe and plate.

SPEED-ALLOY 85-C1.
 M-849; Flux cored low alloy weld wire.
 Typical deposit analysis: 0.08 C, 0.95 Mn, 0.50 Si, 2.60 Ni.
 Welded: 90,000 TS; 80,000 YS; 21 El.
 For welding 2 1/2% nickel steels; good low temperature impact properties.
 AWS E80T.

SPEED-ALLOY 85-C2.
 M-849; Flux cored alloy weld wire.
 Typical deposit analysis: 0.06 C, 1.0 Mn, 0.55 Si, 3.50 Ni.
 Welded: 95,000 TS; 85,000 YS; 28 El.
 For welding 3 1/2% nickel steels, good low temperature properties. AWS E80T.

SPEED-ALLOY 85-C3.
 M-849; Flux cored alloy weld wire.
 Typical deposit analysis: 0.05 C, 0.90 Mn, 0.55 Si, 0.90 Ni, 0.55 Mo.
 Welded: 86,000 TS; 72,000 YS; 25 El.
 For welding low alloy, high strength steels in the 80,000 psi tensile strength range.
 AWS E80T.

SPEED-ALLOY 91.
 M-849; Flux cored alloy weld wire.
 Typical deposit analysis: 0.08 C, 1.35 Mn, 0.55 Si, 1.30 Ni.
 Welded: 98,000 TS; 89,250 YS; 20 El.
 For welding low alloy, high strength steels in the 90,000 psi tensile strength range.

SPEED-ALLOY 91-B3.
 M-849; Flux cored alloy weld wire.
 Typical deposit analysis: 0.09 C, 0.90 Mn, 0.50 Si, 2.25 Cr, 1.0 Mo.
 Welded: 130,000 TS; 120,000 YS; 12 El.
 For welding Cr-Mo steel for high temperature operation.

SPEED-ALLOY 91-B3L.
 M-849; Flux cored alloy weld wire.
 Typical deposit analysis: 0.035 C, 0.80 Mn, 0.55 Si, 2.20 Cr, 1.0 Mo.
 Welded: 105,000 TS; 92,000 YS; 16 El.
 For welding Cr-Mo steel for high temperature operation.

SPEED-ALLOY 95.
 M-849; Flux cored alloy weld wire.
 Typical deposit analysis: 0.07 C, 1.30 Mn, 0.40 Si, 1.80 Ni, 0.25 Mo.
 Welded: 94,000 TS; 81,000 YS; 21 El.
 For welding low alloy, high strength steels in the 90,000 psi tensile strength range.
 AWS E90T.

SPEED-ALLOY 95-D1.
 M-849; Flux cored alloy weld wire.
 Typical deposit analysis: 0.07 C, 1.65 Mn, 0.40 Si, 0.45 Mo.
 Welded: 105,000 TS; 97,000 YS; 22 El.
 For welding Mn-Mo steels in the 90,000-100,000 psi tensile strength range. AWS E90T.

SPEED-ALLOY 105.
 M-849; Flux cored alloy weld wire.
 Typical deposit analysis: 0.07 C, 1.45 Mn, 0.55 Si, 2.10 Ni, 0.35 Mo.
 Welded: 108,500 TS; 96,000 YS; 22 El.
 For welding low alloy, high strength steels in the 100,000 psi tensile strength range.
 AWS E100T.

SPEED-ALLOY 105-D2.
 M-849; Flux cored alloy weld wire.
 Typical deposit analysis: 0.08 C, 1.90 Mn, 0.50 Si, 0.40 Mo.
 Welded: 107,500 TS; 86,750 YS; 25 El.
 For welding Mn-Mo steel for 100,000 min tensile strength.
 AWS E110T.

SPEED-ALLOY 111.
 M-849; Flux cored alloy weld wire.
 Typical deposit analysis: 0.07 C, 1.40 Mn, 0.40 Si, 2.30 Ni, 0.35 Mo.
 Welded: 117,000 TS; 105,000 YS; 21 El.
 For welding low alloy, high strength steels in the 110,000 psi tensile strength range.
 AWS 110T.

SPEED-ALLOY 115.
 M-849; Flux cored alloy weld wire.
 Typical deposit analysis; 0.08 C, 1.60 Mn, 0.40 Si, 2.30 Ni, 0.40 Mo.
 Welded: 118,000 TS; 108,000 YS; 21 El.
 For welding low alloy, high strength steels in the 110,000 psi tensile strength range.
 AWS E110T.

SPEED-ALLOY 121-H.
 M-849; Flux cored alloy weld wire.
 Typical deposit analysis: 0.18 C, 1.40 Mn, 0.40 Si, 0.45 Cr, 1.25 Ni, 0.30 Mo.
 Welded: 135,000 TS; 92,000 YS; 9 El.
 Designed for repair and reclamation of AISI 4130 castings; a heat treatable deposit.

SPEED-ALLOY 125.
 M-849; Flux cored alloy weld wire.
 Typical deposit analysis: 0.085 C, 1.60 Mn, 0.40 Si, 0.40 Cr, 2.20 Ni, 0.45 Mo.
 Welded: 130,000 TS; 120,000 YS; 19 El.
 For welding low alloy, high strength steels, in the 120,000-130,000 psi tensile strength range.
 AWS E110T.

SPEED-ALLOY 5025.
 M-849; Flux cored alloy weld wire.
 Typical deposit analysis: 0.05 C, 1.0 Mn, 0.55 Si, 5.10 Cr, 0.35 Ni, 0.55 Mo.
 Welded and stress relieved at 1350°F: 98,000 TS; 86,000 YS; 22 El.
 For welding 5% Cr-1% Cr-1% Mo plate and pipe.

SPEEDALOY.
 M-908, M-987, M-1067; 2.0 C, 32 Cr, 18 W, 48 Co.
 For dies, cutters, tool bits; cast alloy.

SPEED CASE.
 M-173, M-1024; 0.20 C, 1.25 Mn, 0.25 Si, 0.02 P, bal Fe.
 Rolled: 70,000 TS; 45,000 YS; 30 El; 54 RA.
 For carburized gears, cams, shafts, camshafts, jigs, figures.
 Free-machining, case-hardening.

SPEEDCASE X-1515.
 M-849; 0.15 C, Mn, bal Fe.
 For screw machine parts; free machining.

SPEED CASE X-1525.
 M-495; 0.20 C, bal Fe.
 Hot rolled: 70,000 TS; 52,000 YS; 37.5 El; 60 RA; 137 Brin.
 Cold rolled: 85,000 TS; 80,000 YS; 26 El; 48 RA; 156 Brin.
 For case hardened parts, gears, shafts, pinions; carburizing steel.

SPEED STAR see CARPENTER SPEED STAR.

SPEED TREAT.
 M-173; 0.42-0.53 C, 1.0-1.25 Mn, 0.20-0.30 S, 0.045 max P, 0.10 max S bal Fe.
 Hot rolled plate: 45,000 psi YS.
 Oil or water hardenable to 197,000 psi max YS.
 Free machining plate for structural parts.

SPEED TREAT.
 M-1024; 0.45 C, 1.25 Mn, bal Fe.
 Hot rolled: 90,000 TS; 60,000 YS; 23 El; 40 RA.
 Cold finished: 105,000-100,000 TS; 80,000-85,000 YS; 18-19 El; 30-37 RA.
 For gears, shafts, pinions, lead screws; X1545 analysis.

SPEED TREAT X-1535.
 M-495; 0.30-0.40 C, 0.25 S, bal Fe.
 For screw machine parts; free machining.

SPEED TREAT X-1545.
 M-849; 0.40-0.50 C, bal Fe.
 For screw machine parts; free machining.

SPEED TREAT X-1545.
 M-495; 0.40-0.50 C, 0.25 S, bal Fe.
 For screw machine parts; free machining.

SPELTER WIRE.
 M-Eng.; 64 Cu, 36 Zn, 0.4 Pb, 0.2 Fe.
 For brazing.

SPENARD.
 M-1744; 0.32 C, 0.30 Si, 0.50 Mn, 4.1 Ni, 1.3 Cr, 0.30 Mo, bal Fe.
 4 1/4% Nickel-Chromium tool steel, for punches, trimming dies, plastic moulds.

SPERRY ALLOY 37.
 M-701; 4.3 Si, 1.8 Mg, 3.5 Zn, 0.1 Cu, 0.2 Cr, 0.3 Fe, 0.4 Mn, bal Al.
 For die castings, gimbal rings; high yield strength.

SPEZIAL BRONZE.
 M-1386; Sn, bal Cu.
 For hardware; corrosion resistant.

SPEZIAL EINSATZSTAHL ES2.
 M-1352; 0.15 C, 0.10-0.40 Si, 0.25-0.80 Mn, bal Fe.
 Annealed: 70,000 TS; 55,000 YS; 25 El; 60 RA; 145 Brin.
 For gears, bolts, camshafts, cams, rivets; case hardened.

SPHERULITE.
 M-1257; 3.0 C, 2.5 Si, bal Fe.
 For housings, gears, shafts; cast iron.

SPHINX.
 M-1120; 0.8 C, 2 Mn, 0.2 V, bal Fe.
 For punches, dies, low temperature appliances; oil hardened.

SPHYNX.
 M-1120; 0.8 C, 2.0 Mn, 0.2 V, 0.5 W, bal Fe.
 For punches, crimpers, headers, cutters; oil hardened, non-deforming.

SPIAUTER.
 M-Eng.; 90 Zn, 8 Sb, 2 Cu.
 For bearings; hard zinc.

SPIEGELEISEN see also ELECTROMET SPEIGELEISEN.

SPIN 320.
 M-1709; 0.10 max C, 20.0-23.0 Cr, 8.0-10.0 Mo, 5.0 max Fe, 3.15-4.15 Cb (+Ta), 1.0 max Co, bal Ni.
 Cast: 75,000 psi min TS; 40,000 psi min YS; 18 min El; Rc 30 max hardness.
 For strength and oxidation resistance at elevated temperature (INCONEL 625).

SPIN 518-A.
 M-1709; 3.5 C, 2.2 Si, bal Fe.
 Ann: 65,000 psi TS; 45,000 psi YS; 12 El; 149-187 Brin.
 Ductile iron with ferrite matrix.
 ASTM A536 65-45-12.

SPIN 518-N.
 M-1709; 3.5 C, 2.2 Si, bal Fe.
 Normalized: 80,000 psi TS; 55,000 psi YS; 6 El; 202-269 Brin.
 Ductile iron with pearlite-ferrite matrix.
 ASTM A536 80-55-06.

SPIN 522.
M-1709; 3.5 C, 2.2 Si, 0.9 Ni, 0.5 Mo, bal Fe.
Normalized: 100,000 psi TS; 70,000 psi YS; 3 El; 241-285 Brin.
Ductile iron with pearlite matrix.
ASTM A536 100-70-03.

SPIN 775.
M-1709; 0.15 max C, 1.0 max Mn, 1.5 max Si, 11.5-14.0 Cr, 0.50 max Mo, 1.0 max Ni, bal Fe.
Centrifugally cast; heat treatable to 363-428 Brin and 160,000 psi TS (before tempering).
ACI CA-15.

SPIN 776.
M-1709; 0.32 C, 11.5-14.0 Cr, 1.0 max Mn, 1.5 max Si, 1.0 max Ni, 0.5 max Mo, bal Fe.
Centrifugally cast; heat treatable to about 500 Brin (before tempering).
ACI CA-40.

SPIN 803PH.
M-1748; 0.05 C, 16 Cr, 4 Ni, 3 Cu, 0.2 Cb, bal Fe.
Aged: 190,000 TS; 170,000 YS; 12 El; 35 RA; 415 Brin.
High strength stainless; good galling resistance. For landing gear bushings, sleeves, pressure vessels.

SPIN 902.
M-1709; 0.20-0.60 C, 2.0 max Mn, 2.0 max Si, 24.0-28.0 Cr, 18.0-22.0 Ni, 0.50 max Mo, bal Fe.
Centrifugally cast stainless steel.
Stress rupture: 100 hrs min at 4800 psi at 1800°F.
ACI CK-20.

SPIN 902 HK.
M-1748; 0.40 C, 25 Cr, 20 Ni, bal Fe.
As cast: 87,000 TS; 45,000 YS; 20 El; 16 RA.
Good creep strength and oxidation resistance to 2000°F; for furnace tubes, retorts.

SPIN 914.
M-1709; 0.15 max C, 1.0 max Mn, 1.5 max Si, 11.5-14.0 Cr, 1.0 max Ni, 1.0 max Mo, S, bal Fe.
Centrifugally cast; free machining grade of SPIN 775 (and ACI CA-15).

SPIN 935.
M-1709; 0.25-0.35 C, 1.0 max Mn, 1.0 max Si, 23.0-25.0 Cr, 23.0-25.0 Ni, 1.4-1.8 Cb, bal Fe.
Centrifugally cast alloy.
Stress rupture: 100 hrs min at 5500 psi at 1800°F.
(Inconel 519).

SPINNING SILVER.
M-Eng.; 67 Cu, 17 Zn, 16 Ni.
For spoons, forks, knives; spun and drawn.

SPOOLALLOY.
M-1263; C, alloy, bal Fe.
For heliarc welding wire; steel welding.

SPOOLARC ALUMINUM NO. 1100.
M-1713.
Continuous bare electrode wire, AA 1100 composition, for welding 1100 and 3003 aluminum.
AWS Class ER1100.

SPOOLARC ALUMINUM NO. 4043.
M-1713.
Continuous bare electrode wire, AA 4043 composition, for general welding of many aluminum alloys (5% Si).
AWS Class ER4043.

SPOOLARC ALUMINUM NO. 4047.
M-1713.
Continuous bare electrode wire, AA 4047 composition, for welding many aluminum alloys (12% Si). (Was No. 718).
AWS Class ER 4047; BA1Si-4.

SPOOLARC ALUMINUM NO. 5183.
M-1713.
Continuous bare electrode wire, AA 5183 composition, for welding many aluminum alloys containing magnesium.
AWS Class ER 5183.

SPOOLARC ALUMINUM NO. 5356.
M-1713.
Continuous bare electrode wire, AA 5356 composition, for welding many aluminum alloys containing magnesium.
AWS Class ER 5356.

SPOOLARC ALUMINUM NO. 5554.
M-1713.
Continuous bare electrode wire, AA 5554 composition, for welding many aluminum alloys contaning magnesium. Recommended for elevated temperature service.
AWS ER 5554, AA 5554.

SPOOLARC ALUMINUM NO. 5556.
M-1713.
Continuous bare electrode, AA 5556 composition for welding many aluminum alloys containing magnesium.
AWS Class ER 5556.

SPOOLARC DEOXIDIZED COPPER WELDING ELECTRODE.
M-1713.
As welded: up to 35,000 psi TS; 40-50 El; 25-40 Rock F.
For welding copper to itself and to mild steel or to overlay steel; with MIG equipment.
AWS A5.6-69 (E Cu).

SPOOLARC PHOSPHOR "BRONZEC" WELDING ELECTRODE.
M-1713.
As welded: up to 65,000 psi TS; up to 37,000 psi YS; 42-50 El; 80-100 Brin.
For welding or build-up on copper base alloys and cast iron; with MIG equipment.
AWS A5.6-69 (ECuSnC).

SPOOLARC SILICON BRONZE WELDING ELECTRODE.
M-1713.
As welded: up to 58,000 psi TS; up to 25,000 psi YS; 53-55 El; 80-100 Brin.
For welding silicon bronze, copper alloys and some iron base metals; with MIG equipment.
AWS A5.6-69 (ECuSi).

SPOON METAL.
Cu, Ni.
For spoons; nickel silver.

SPRABABBITT A.
M-853; 7.5 Sb, 3.5 Cu, bal Sn.
For wire for metal spraying; Babbitt.

SPRABRASS Y.
M-853; 34 Zn, bal Cu.
For wire for metal spraying; brass.

SPRABRONZE C.
M-853; 10 Sn, bal Cu.
For wire for metal spraying; bronze.

SPRABRONZE M.
M-853; 40 Zn, 58 Cu, 2 Sn, Fe, Mn.
For wire for metal spraying; bronze.

SPRABRONZE P.
M-853; 5 Sn, 0.25 P, bal Cu.
For wire for metal spraying; P-bronze.

SPRABRONZE T.
M-853; 39 Zn, 1 Sn, bal Cu.
For wire for metal spraying; Tobin bronze.

SPRAIRON A.
M-853; Fe.
For metal spraying; iron wire.

SPRAM 14112.
M-717.
Alloy powder for flame spray to develop surface for severe abrasive resistance; Matrix 59-62 Rc, Diamix C particles.

SPRAM 14493.
M-717.
Alloy powder for flame spray for ductile, unlimited buildup; resistant to cracking.
Rc 25-30.

SPRAM 14494.
M-717.
Alloy powder for flame spray; gives machinable, corrosion and impact resistant coating. All purpose.
Rc 35-40.

SPRAM 14495.
M-717.
Alloy powder for flame spray: machinable with carbide tools.
Rc 45-50.

SPRAM 14496.
M-717.
Alloy powder for flame spray; gives hard, smooth deposits; finish by grinding; use against seals, packing, sliding surfaces.
Rc 55-60.

SPRAM 14496 DIAMAX.
M-717.
Alloy powder for flame spray; best resistance to fine particle abrasion.
Requires grind finishing. Matrix: Rc 60-63, DIAMIX A particles.

SPRASTEEL 10.
M-853; 0.10 C, bal Fe.
For metal spraying; steel wire.

SPRASTEEL 25.
M-853; 0.25 C, bal Fe.
For metal spraying; steel wire.

SPRASTEEL 40.
M-853; 0.40 C, bal Fe.
For metal spraying; steel wire.

SPRASTEEL 80.
M-853; 0.80 C, bal Fe.
For metal spraying; steel wire.

SPRASTEEL 120.
M-853; 1.2 C, bal Fe.
For metal spraying; steel wire.

SPREEMETALL.
M-621; 43 Zn, 1.5 Mn, 0.5 Pb, bal Cu.
Cold rolled: 86,000 TS; 50,000 YS; 5 El; 140 Brin.
For sheets, bars, forgings; resists sea water corrosion.

SPRING BRONZE 0.8%.
M-1770; 85 Cu, 0.8 Sn, 14.2 Zn.
Sheet or strip, rolled.
CDA 434; ASTM 591.

SPRING BRONZE, 1%.
M-1770; 87.5 Cu, 1.0 Sn, 11.5 Zn.
Sheet or strip, rolled.
CDA 422; ASTM B591.

SPRING GOLD.
50 Cu, 25 Au, 25 Ag.
For jewelry, dental applications; corrosion resistant.

SPRING SILVER.
M-1066, M-237; 99.4 min Ag, 0.28 max Mg, 0.18 max Ni.
Air Hardened: 58,000 min TS; 5-15 El, 63-67 Rock. (30T).
For spring contacts, switches, relays.
Air hardens by internal oxidation; 70% elect. cond.
Good spring properties.

SPRING SILVER.
54.6 Cu, 27.3 Zn, 18 Ni.
For corrosion resistant springs; corrosion resistant.

SPRING STEEL.
0.9-1.1 C, 0.4-0.8 Mn, bal Fe.
For springs; oil hardening.

SPRING WIRE BRASS.
M-U.S.; 70-74 Cu, 30-26 Zn.
100,000 TS.
For springs.

SPS 245.
M-435; 0.40 C, 0.75 Mn, 0.60 Cr, 0.15 Mo, bal Fe.
Annealed: 100,000 TS; 68,000 YS; 30 El; 58 RA; 187 Brin.
For shafts, gears, bolts, studs; Izod-45.

S.P.S. SPECIAL.
M-96; 0.8 C, 2 W, 0.9 Cr, bal Fe.
For tools, cutters; water hardened.

SPZ.
M-91.
Special Purity Zinc.

SPZ.
M-1687; 7 Sn, 4 Zn, 6 Pb, bal Cu.
Leaded bronze, wrought; 75 Brin.
AFNOR UE7 Z5 Pbb.

SQUARE 50.
M-1089; 50 Ni, 50 Fe.
Grain oriented alloy processed for maximum B-H loop squareness.

SQUARE 80.
M-1089; 80 Ni/Fe/Mo.
High maximum flux density, high gain, high squareness ratio; soft magnetic alloy.

SR.
M-1290; 1.65-1.75 C, 0.25-0.40 Si, 0.20-0.40 Mn, 11.0-12.0 Cr, 0.07-0.12 V, bal Fe.
Cold work tool steel; W.-Nr. 1.2201.

SRA 100 (1144).
M-240; 0.45 C, 1.5 Mn, 0.04 max P, 0.24-0.33 S; 0.25 Si, bal Fe.
Cold drawn, strain hardened: 125 ksi TS (typical); 100 ksi min YS; 12 El; 34 RA.
For moderately stressed parts; free machining with minimum distortion.

SRMV.
M-1290; 1.5-1.6 C, 0.30-0.50 Si, 0.30-0.50 Mn, 11.5-12.5 Cr, 0.6-0.8 Mo, 0.90-1.10 V, bal Fe.
Cold work tool steel; W.-Nr. 1.2379.

SR-NICKEL.
M-303; 99.8 Ni.
For electroplating anodes; chemical equipment.

SR SPEZIAL.
M-1290; 1.55-1.75 C, 0.25-0.40 Si, 0.20-0.40 Mn, 11.0-12.0 Cr, 0.50-0.70 Mo, 0.40-0.60 W, 0.07-0.12 V, bal Fe.
Hot work tool steel; W.-Nr. 1.2601.

SR STEEL IIA-1121.
M-1161; C, Cr, Ni, bal Fe.
Cast: 112,000 TS; 44,000 YS; 23 El; 16 RA; 200 Brin.
Rolled: 95,000 TS; 75,000 YS; 50 El; 65 RA.
For pumps, valves, propellers, heat exchangers; stainless, austenitic.

SR STEEL IIA-1123.
M-1161; C, Cr, Ni, bal Fe.
For chemical plant equipment; stainless, austenitic.

SR STEEL IIA-1132.
M-1161; 0.12 max C, 13-15 Ni, 11-13 Cr, 6-7 Mo, 5.5-6.5 Cu, bal Fe.
Annealed: 95,000 TS; 75,000 YS; 50 El; 65 RA; 180 Brin.
At 1500°F: 32,500 TS.
For chemical plant and food processing equipment; stainless, austenitic.

SR STEEL UST.
M-1161; 0.1 C, 20 Ni, 20 Cr, 10 max Cu + Mo, bal Fe.
At 1600°F: 33,800 TS; 10 El; 17 RA.
For chemical plant equipment; austenitic, stainless.

SS-6.
M-1713; weld metal; 0.08 C, 0.59 Mn, 0.21 Si, bal Fe.
As welded: 76,000 psi TS; 69,000 psi YS; 26 El.
AC-DC, straight polarity electrode for high speed welding of horizontal fillets, especially in shipyards.
ABS Filler Grade No.H2.

SS 410 3.020.
M-1739.
Ferritic 410 stainless steel, sintered.
Density: 6.7-7.1 gms/cc; Elon: 7%.
UTS: 42.5-49 kg/mm^2; Hardness: HV5-170.

SS 511.
M-1655; 1.30 C, 0.30 Mn, 0.20 Si, 0.15 Cr, 5.0 W, bal Fe.
Hardenable to 63-65 Rc for drawing dies.
W.-Nr. 1.2453.

SS 931.
M-494, M-1485; 3.0 C, 2.2 Si, 1.3 Cr, bal Fe.
Cast: 260-310 Brin.
For valve seats, heat resistant castings; cast iron.

SSB.
4.5-5.5 Cu, bal Al.
For light alloy parts; super duralumin type.

SS CUT A.
M-373; 0.70 C, 0.75 Mn, 0.75 Si, 0.50 Mo, 17.5 Cr, bal Fe.
Martensitic chromium stainless steel for knives, cutlery.
AISI 440 A.

SS CUT B.
M-373; 0.90 C, 0.75 Mn, 0.75 Si, 0.50 Mo, 17.5 Cr, bal Fe.
Martensitic chromium stainless steel for knives, cutlery.
AISI 440 B.

SS CUT C.
M-373; 1.05 C, 0.75 Mn, 0.75 Si, 0.50 Mo, 17.5 Cr, bal Fe.
Martensitic chromium stainless steel for knives, cutlery.
AISI 440 C.

SS-DEMI-DUR.
M-1546; 0.45 C, 1.9 Si, bal Fe.
For chisels, punches.

SS EXTRA COLD HEADING.
M-908; 1.0-1.1 C, bal Fe.
For cold heading dies.

S/S FR. VIII.
M-1260; 80 Sn, 11.5 Sb, 5.5 Cu, 3 Pb.
Cast: 17,800 TS; 35 Brin.
For steam turbine and generator bearings; M.P. 355-675°F; wear resistant.

S. SIRIUS.
M-France; 0.25 C, 16 Ni, 17 Cr, 3 W, 12 Co, 2 Ti, bal Fe.
For valves; heat and corrosion resistant.

SSS-100 see ARMCO SSS 100.

ST-10 SOLDER.
M-815; 87.5 Pb, 10 Sn, 1.5 Ag, 0.5 Bi, 0.5 Sb.
Cast: 5335 TS; 11.5 El.
For solder; liquid at 554°F.

ST-15.
M-1073; 82.75 Pb, 15 Sn, 1.25 Ag, 0.5 Bi, 0.55 Sb.
For soft solder; M.P. 278-532°F.

ST-15 SOLDER.
M-815; 82.75 Pb, 15 Sn, 1.25 Ag, 0.5 Bi, 0.5 Sb.
Cast: 6180 TS; 11.5 El.
For solder; liquid st 532°F.

ST-20A SOLDER.
M-815; 76.75 Pb, 20 Sn, 1.25 Ag, 1.5 Bi, 0.5 Sb.
Cast: 6720 TS; 13 El.
For solder; liquid at 514°F.

ST-20B.
M-1073; 73.25 Pb, 20 Sn, 1.25 Ag, 5.0 Bi, 0.5 Sb.
For soft solder; M.P. 258-486°F.

ST-20B SOLDER.
M-815; 73.25 Pb, 20 Sn, 1.25 Ag, 5 Bi, 0.5 Sb.
Cast: 6810 TS; 27 El.
For solder; liquid at 486°F.

ST-20 SOLDER.
M-815; 77.75 Pb, 20 Sn, 1.25 Ag, 0.5 Bi, 0.5 Sb.
Cast: 6840 TS; 14 El.
For solder; liquid at 518°F.

ST-30.
M-1073; 67.75 Pb, 30 Sn, 1.25 Sb, 0.5 Bi, 0.5 Sb.
For soft solder; M.P. 248-478°F.

ST-30 SOLDER.
M-815; 67.75 Pb, 30 Sn, 1.25 Ag, 0.5 Bi, 0.5 Sb.
Cast: 7625 TS; 15 El.
For solder; liquid at 478°F.

ST. 37.161.
M-Swiss; 0.15 C, 0.06 S, 0.06 P, bal Fe.
Annealed: 70,000 TS; 40,000 YS; 25 El; 60 RA; 145 Brin.
For bridge construction; good weldability.

ST. 40.21.
M-Ger.; 0.18 C, bal Fe.
For welded sluice gates; excellent weldability.

ST-48.
M-895; 0.4 C, bal Fe.
For gears, shafts; water hardened.

ST-51.
M-516; 0.48-0.55 C, 1.35-1.65 Mn, 0.08-0.13 S, bal Fe.
Cold drawn, stress relieved: 140,000 TS min; 125,000 YS min: 285 Brin.
For machine parts, shafts.
Free machining, eliminates heat treatment.

ST-52.
M-Ger.; 0.12.0.15 C, 0.5-1.1 Si, 0.75-1.6 Mn, 0.25-1.0 Cu, 0-0.6 Cr, 0-0.25 Mo, bal Fe.
89,600 TS; 65,000 YS; 24 El.
For structural parts; excellent for welding.

ST. 52.
M-Ger.; 0.2 C, 0.15 Si, 0.5 Mn, 0.5 Cu, bal Fe.
For railway cars, bus and truck bodies; high strength structural steel.

ST. 60.
M-Swiss; 0.1 C, bal Fe.
Annealed: 64,000 TS; 41,000 YS; 28 El; 65 RA; 130 Brin.
For welded structures; good weldability.

ST 400.
M-1522; 4 Ag, Sn.
Melt range: 221-224°C.
Max stress: 5.5 kgf/mm^2; 48 El.
For soft soldering; corrosion resistant.

STA 8.
M-494; 3.2 C, 1.6-2.8 Si, 0.2 P, bal Fe.
Cast: 210-260 Brin.
For thin section castings; cast iron.

STABIL.
M-96; 0.8 C, 2.0 Mn, V, bal Fe.
Annealed: 107,000 TS; 215 Brin.
For blanking dies, threading tools, milling cutters, gages.
Oil hardening, shock resistant. Type 02 tool steel.

STABIL.
M-1322; 0.90 C, 1.9 Mn, 0.1 V, bal Fe.
For punches, cutters, forming dies; oil hardened, non-deforming.

STABIL SPECIAL.
M-96; 0.85 C, 1.7 Mn, 0.3 Cr, 0.2 V, bal Fe.
For tools, dies; non-deforming.

STABIL SPEZIAL.
M-1322; 1.05 C, 1.0 Cr, 1.15 W, 0.9 Mn, bal Fe.
For cutters, bearings, forming dies; oil or water hardened.

STABLE-ARC.
M-578; C, bal Fe.
Steel arc welding electrode.
AWS Class E 4510.

STABRITE.
M-550; Ni alloy.
For castings; corrosion resistant.

STACKPOLE CW335.
M-1069.
Silver-Tungsten, 35% Ag.
Density: 14.3 g/cc; Spec. Res.: 1 x 10^{-6} ohm/in; Hardness: 30T77; Strength: 110,000 psi.
Electrical contacts for circuit breaker applications.

STACKPOLE CW350.
M-1069; Silver-Tungsten, 50% Ag.
Density: 13.2 g/cc; Spec. Res.: 1 x 10^{-6} ohm/in; Hardness: 30T66; Strength: 85,000 psi.
Electrical contacts for circuit breaker applications.

STACKPOLE CW366.
M-1069; 99 min W.
Density: 18.0 g/cc; Spec. Res.: 3 x 10^{-6} ohm/in; Hardness: 30N55; Strength: 104,000 psi.
Heat sinks for semiconductor applications.

STACKPOLE FM26.
M-1069; Silver-Molybdenum, 52% Ag.
Density: 10.2 g/cc; Spec. Res.: 1 x 10^{-6} ohm/in; Hardness: 30T60; Strength: 90,000 psi.
Electrical contacts for circuit breaker applications.

STACKPOLE FM202.
M-1069; 99 min Mo.
Density: 9.7 g/cc; Spec. Res.: 3 x 10^{-6} ohm/in; Hardness: 30N31; Strength: 110,000 psi.
Heat sinks for semiconductor applications.

STACKPOLE SG124.
M-1069; Silver-graphite, 80% Ag.
Density: 5.5 g/cc; Spec. Res.: 4 x 10^{-6} ohm/in; Hardness: 15W44; Strength: 5,000 psi.
For sliding contacts for slip ring assemblies; brushes for electric motors.

STACKPOLE SG161.
M-1069; Silver-graphite, 50% Ag.
Density: 3.2 g/cc; Spec. Res.: 1 x 10^{-4} ohm/in; Hardness: 15W34; Strength: 3500 psi.
Brushes for electric motors.

STAFLO 761.
M-63; 20% Ni & 80% Lithobraze 925 (Lithobraze 925 is 92.5 Ag, 7.3 Cu, 0.2 Li).
Duplex brazing alloy for brazing honeycomb panels of stainless, helps control fillets.

STAG.
M-210; 1.1 C, bal Fe.
For tools, drills, taps; water hardened.

STAG 14 AIR HARDENING.
M-210, 0.67 C, 15 W, 3.5 Cr, 0.37 V, bal Fe.
For woodworking tools, end mills, drills; high speed steel.

STAG ALLENITE.
M-210; W C + Co.
For tipped cutting tools; fast cutting on cast iron; sintered carbide.

STAG ATHYWELD 60/1/13.
M-210; WC.
Rockwell C 52.
For hard facing brick making machines, mixer knives.
Wear and abrasion resistant.

STAG EXTRA SPECIAL.
M-210; 0.8 C, 19 W, 6 Co, 5 Cr, 1.5 V, 0.5 Mo, bal Fe.
For lathe and planer cutters, hobs, reamers, taps; high speed steel, oil hardened.

STAG MAJOR.
M-210; 0.8 C, 21 W, 11 Co, 5 Cr, 1.5 V, 0.5 Mo, bal Fe.
For drills, reamers, hobs, lathe and planer cutters; high speed steel, oil hardened.

STAG MO.
M-210; 0.8 C, 6.5 W, 4 Mo, 4 Cr, 1.5 V, bal Fe.
For tools, cutters, reamers, broaches, taps; high speed steel.

STAG MO562.
M-210; 0.85 C, 4.75 Cr, 5.0 Mo, 6.0 W, 2.0 V, bal Fe.
For drills, taps, reamers, form tools; high speed steel.

STAG SPECIAL.
M-210; 0.7 C, 19 W, 4 Cr, 1 V, bal Fe.
For planing and shaping tools, drills, reamers, hobs; high speed steel, oil hardened.

STAHLSCHMIDT C1V.
M-1294; 0.50 C, 1.0 Cr, 0.09 V, bal Fe.
For springs, gears, machine tool parts, bolts; oil hardened, shock resistant.

STAHLSCHMIDT C2V.
M-1294; 0.58 C, 1.0 Cr, 0.09 V, bal Fe.
For springs, gears, crankshafts; oil hardened, shock resistant.

STAHLSCHMIDT CE6.
M-1294; 0.15 C, 0.65 Cr, 0.5 Mn, bal Fe.
For gears, fasteners, machine tool parts; case hardened, shock resistant.

STAHLSCHMIDT CE8.
M-1294; 0.16 C, 0.95 Cr, 1.15 Mn, bal Fe.
For gears, cams, camshafts; case hardened, tough.

STAHLSCHMIDT CE10.
M-1294; 0.20 C, 1.25 Mn, 1.15 Cr, bal Fe.
For gears, cams, camshafts; case hardened, tough.

STAHLSCHMIDT CNL.
M-1294; 0.50 C, 1.05 Cr, 3.25 Ni, bal Fe.
For gears, bolts, crankshafts; oil hardened, shock resistant.

STAHLSCHMIDT E16.
M-1294; 0.15 C, 0.25 Si, 0.37 Mn, bal Fe.
Annealed: 70,000 TS; 40,000 YS; 25 El; 60 RA; 145 Brin.
For gears, bolts, machine tool parts; case hardening steel.

STAHLSCHMIDT E25.
M-1294; 0.22 C, 0.25 Si, 0.45 Mn, bal Fe.
Annealed: 75,000 TS; 42,000 YS; 22 El; 58 RA; 140 Brin.
For gears, shafts, fasteners, bolts, screws; water hardened.

STAHLSCHMIDT EXTRA.
M-1294; 0.82 C, 4.1 Cr, 0.85 Mo, 1.6 V, 8.7 W, bal Fe.
For lathe and planer tools, reamers, broaches; high speed steel.

STAHLSCHMIDT G500.
M-1294; 1.0 C, 0.20 Si, 0.25 Mn, 0.10 V, bal Fe.
Annealed: 100,000 TS; 53,000 YS; 21 El; 42 RA; 200 Brin.
For drills, taps, reamers, springs; Type W2; water hardened.

STAHLSCHMIDT HS23.
M-1294; 0.95 C, W, Mo, bal Fe.
For cutters, dies; oil hardened, tough.

STAHLSCHMIDT HS55.
M-1294; 0.85 C, W, Mo, Cr, V, bal Fe.
For cutters, dies; oil hardened, tough.

STAHLSCHMIDT HWG.
M-1294; 0.56 C, Ni, Cr, Mo, V, bal Fe.
For upsetters, extrusion dies; oil hardened, tough.

STAHLSCHMIDT HWGW.
M-1294; 0.55 C, 0.70 Cr, 0.18 Mo, 1.65 Ni, 0.1 V, bal Fe.
For forging dies, upsetters, shears; oil hardened, tough.

STAHLSCHMIDT MS5.
M-1294; 0.37 C, 1.25 Si, 1.25 Mn, bal Fe.
For gears, punches, machine tool parts; oil hardened, tough.

STAHLSCHMIDT MV7.
M-1294; 0.42 C, 1.75 Mn, 0.25 Si, 0.1 V, bal Fe.
For punches, gears, bolts, shafts; oil hardened, tough.

STAHLSCHMIDT NE15.
M-1294; 0.15 C, 1.55 Cr, 1.55 Ni, bal Fe.
For die casting dies, gears, cams; case hardened, tough.

STAHLSCHMIDT NE25.
M-1294; 0.13 C, 0.7 Cr, 2.5 Ni, bal Fe.
For gears, bolts, machine tool parts; case hardened, shock resistant.

STAHLSCHMIDT NE35.
M-1294; 0.13 C, 0.7 Cr, 3.5 Ni, bal Fe.
For gears, bolts, camshafts, cams; case hardened, shock resistant.

STAHLSCHMIDT NE45.
M-1294; 0.13 C, 1.1 Cr, 4.5 Ni, 0.5 Mn, bal Fe.
For gears, bolts, camshafts, cams; case hardened, shock resistant.

STAHLSCHMIDT NV15.
M-1294; 0.28-0.35 C, 0.5 Cr, 1.5 Ni, 0.6 Mn, bal Fe.
For bolts, gears, machine tool parts; oil hardened, shock resistant.

STAHLSCHMIDT NV25.
M-1294; 0.28-0.35 C, 0.6 Mn, 0.7 Cr, 2.5 Ni, bal Fe.
For gears, bolts, crankshafts, axles; oil hardened, shock resistant.

STAHLSCHMIDT NV35.
M-1294; 0.22-0.30 C, 0.6 Mn, 0.7 Cr, 3.5 Ni, bal Fe.
For gears, bolts, crankshafts, axles; oil hardened, shock resistant.

STAHLSCHMIDT NV45.
M-1294; 0.35 C, 4.5 Ni, 1.3 Cr, 0.6 Mn, bal Fe.
For gears, dies, crankshafts, bolts; oil hardened, shock resistant.

STAHLSCHMIDT PD EXTRA.
M-1294; 0.38 C, Si, Cr, V, bal Fe.
For gears, bolts, springs, crankshafts; oil hardened, shock resistant.

STAHLSCHMIDT RCM.
M-1294; 0.45 C, 1.4 Cr, 0.70 Mo, 0.30 V, 0.7 Mn, bal Fe.
For forging dies, punches, gears, crankshafts; oil hardened, tough.

STAHLSCHMIDT SC15.
M-1294; 0.67 C, 1.3 Si, 0.50 Cr, 0.5 Mn, bal Fe.
For shears, rivet sets, upsetters; oil hardened, tough.

STAHLSCHMIDT SCE.
M-1294; 0.15 C, 0.65 Cr, 0.5 Mn, bal Fe.
For gears, bolts, oil refinery equipment; case hardened, creep resistant.

STAHLSCHMIDT SCS.
M-1294; 1.05 C, 1.15 Cr, bal Fe.
For liners, sleeves, cutters, bearings; water or oil hardened, wear resistant.

STAHLSCHMIDT SE1.
M-1294; 0.20 C, 1.25 Mn, 1.25 Cr, bal Fe.
For gears, bolts, crankshafts, cams; case hardened, tough.

STAHLSCHMIDT SGK.
M-1294; 1.45 C, 1.4 Cr, 0.6 Mn, bal Fe.
For blanking and forming dies; water or oil hardened.

STAHLSCHMIDT SGS2.
M-1294; 1.3 C, 0.25 max Si, 0.25 max Mn, bal Fe.
For blanking and forming dies, cutters, reamers; Type W1; water hardened.

STAHLSCHMIDT SGS3.
M-1294; 1.15 C, 0.25 max Si, 0.25 max Mn, bal Fe.
Annealed: 110,000 TS; 58,000 YS; 18 El; 40 RA; 210 Brin.
For springs, drills, reamers, taps, broaches; Type W1; water hardened.

STAHLSCHMIDT SGS4.
M-1294; 1.0 C, 0.25 max Si, 0.25 max Mn, bal Fe.
Annealed: 100,000 TS; 53,000 YS; 21 El; 42 RA; 200 Brin.
For springs, tools, drills, taps, reamers; Type W1; water hardened.

STAHLSCHMIDT SGS5.
M-1294; 0.85 C, 0.25 max Si, 0.25 max Mn, bal Fe.
Heat treated: 185,000-125,000 TS; 140,000-85,000 YS; 14-24 El; 37-52 RA; 400-230 Brin.
For springs, drills, reamers, taps, cutters; Type W1; water hardened.

STAHLSCHMIDT SGS6.
M-1294; 0.70 C, 0.25 max Si, 0.25 max Mn, bal Fe.
Heat treated: 174,000-122,000 TS; 128,000-82,000 YS; 12-22 El; 37-52 RA; 352-240 Brin.
For wheels, die blocks, rails, springs; Type W1; water hardened.

STAHLSCHMIDT SNR.
M-1294; 1.65 C, 11.5 Cr, 0.10 V, bal Fe.
For blanking and forming dies, punches; air hardened, non-deforming.

STAHLSCHMIDT SNS EXTRA.
M-1294; 2.1 C, 11.5 Cr, 0.30 Mn, bal Fe.
For blanking and forming dies, punches; oil hardened, non-deforming.

STAHLSCHMIDT SNS SPEZIAL.
M-1294; 2.1 C, 11.5 Cr, 0.70 W, bal Fe.
For blanking and forming dies; oil hardened, non-deforming.

STAHLSCHMIDT SP EXTRA.
M-1294; 0.70 C, 1.7 Si, 0.70 Mn, bal Fe.
For punches, shears, upsetters; oil hardened, tough.

STAHLSCHMIDT SPEZIAL ZAH.
M-1294; 0.74 C, 4.1 Cr, 1.1 V, 18.5 W, bal Fe.
For lathe and planer tools, reamers, broaches, taps; high speed steel.

STAHLSCHMIDT SSF.
M-1294; 0.60 C, 0.25-0.50 Si, 0.30-0.80 Mn, bal Fe.
Heat treated: 160,000-115,000 TS; 113,000-77,000 YS; 12-23 El; 40-54 RA; 320-230 Brin.
For wheels, die blocks, rails, girders, springs; water hardened.

STAHLSCHMIDT SS II.
M-1294; 0.90 C, 1.9 Mn, 0.10 V, bal Fe.
For blanking and forming dies, cutters, punches; oil hardened, non-deforming.

STAHLSCHMIDT STANDARD.
M-1294; 0.79 C, 4.75 Co, 4.3 Cr, 0.75 Mo, 1.5 V, 18 W, bal Fe.
For lathe and planer tools, reamers, drills, hobs; high speed steel.

STAHLSCHMIDT STANDARD SPEZIAL.
M-1294; 0.76 C, 10 Co, 4.2 Cr, 0.8 Mo, 1.8 V, 18 W, bal Fe.
For lathe and planer tools, drills, reamers; high speed steel.

STAHLSCHMIDT STS3.
M-1294; 1.1 C, 0.25 max Si, 0.25 max Mn, bal Fe.
Heat treated: 200,000-140,000 TS; 125,000-90,000 YS; 10-20 El; 30-45 RA; 400-275 Brin.
For springs, reamers, drills, taps; Type W1; water hardened.

STAHLSCHMIDT STS4.
M-1294; 1.0 C, 0.25 max Si, 025 max Mn, bal Fe.
Heat treated: 185,000-130,000 TS; 120,000-80,000 YS; 10-20 El; 30-45 RA; 390-270 Brin.
For springs, hobs, cutters, reamers, taps; Type W1; water hardened.

STAHLSCHMIDT STS5.
M-1294; 0.85 C, 0.25 max Si, 0.25 max Mn, bal Fe.
Heat treated: 180,000-130,000 TS; 118,000-80,000 YS; 12-24 El; 32-47 RA; 370-260 Brin.
For springs, taps, reamers, drills; Type W1; water hardened.

STAHLSCHMIDT STS6.
M-1294; 0.70 C, 0.25 max Si, 0.25 max Mn, bal Fe.
Heat treated: 174,000-122,000 TS; 128,000-82,000 YS; 12-22 El; 37-52 RA; 352-240 Brin.
For wheels, die blocks, rails, springs; Type W1; water hardened.

STAHLSCHMIDT SUPERIOR.
M-1294; 0.86 C, 4.1 Cr, 0.85 Mo, 2.5 V, 12 W, bal Fe.
For lathe and planer tools, reamers, broaches; high speed steel.

STAHLSCHMIDT SZP 40.
M-1294; 0.30 C, 2.35 Cr, 0.6 V, 4.25 W, bal Fe.
For shear blades, pneumatic tools, punches; oil hardened, hot work steel.

STAHLSCHMIDT SZP 80.
M-1294; 0.30 C, 2.65 Cr, 0.35 V, 8.5 W, bal Fe.
For extrusion dies, rams and liners; oil hardened, hot work steel.

STAHLSCHMIDT SZP 120.
M-1294; 0.30 C, 2.0 Co, 0.25 V, 8.5 W, 2.4 Cr, bal Fe.
For extrusion dies, rams and liners, punches, shear blades; oil hardened, hot work steel.

STAHLSCHMIDT UE8.
M-1294; 0.15 C, Cr, Mo, bal Fe.
For gears, bolts, camshafts, cams; case hardened, tough.

STAHLSCHMIDT UE10.
M-1294; 0.20 C, Cr, Mo, bal Fe.
For gears, bolts, camshafts, cams; case hardened, tough.

STAHLSCHMIDT UV12.
M-1294; 0.25 C, 0.65 Mn, 1.0 Cr, 0.2 Mo, bal Fe.
For gears, bolts, camshafts; oil hardened, tough.

STAHLSCHMIDT UV13.
M-1294; 0.33 C, 1.0 Cr, 0.2 Mo, 0.65 Mn, bal Fe.
For gears, bolts, machine tool parts; oil hardened, tough.

STAHLSCHMIDT UV14.
M-1294; 0.42 C, 1.0 Cr, 0.2 Mo, bal Fe.
For gears, bolts, crankshafts, dies; oil hardened, tough.

STAHLSCHMIDT UV24.
M-1294; 0.50 C, 1.0 Cr, 0.2 Mo, 0.65 Mn, bal Fe.
For gears, bolts, crankshafts, fasteners; oil hardened, shock resistant.

STAHLSCHMIDT V35.
M-1294; 0.35 C, 0.55 Mn, 0.25 Si, bal Fe.
Hot rolled: 85,000 TS; 54,000 YS; 30 El; 53 RA; 183 Brin.
For gears, bolts, crankshafts, axles, screws; water hardened.

STAHLSCHMIDT V45.
M-1294; 0.45 C, 0.5 Mn, 0.25 Si, bal Fe.
Hot rolled: 98,000 TS; 59,000 YS; 24 El; 45 RA; 212 Brin.
For axles, gears, bolts, crankshafts; water hardened.

STAHLSCHMIDT V60.
M-1294; 0.61 C, 0.25 Si, 0.55 Mn, bal Fe.
Heat treated: 160,000-115,000 TS; 113,000-77,000 YS; 12-23 El; 40-54 RA; 320-230 Brin.
For girders, rails, springs, die blocks; water hardened.

STAINALOY.
M-1102; Cu, alloy.
For resistance and spot welding electrodes.

STAINLESS STEEL TYPE F.I.
M-521; 0.12 max C, 0.8 max Si, 1.0 max Mn, 11.5-13.5 Cr, bal Fe.
Martensitic type stainless steel.
B.S. 970 (PT4) 410S21; similar to AISI 403, 410.

STAINLESS W.
M-604; 0.06 C, 0.70 Mn, 1.25 Si, 7.0 Ni, 17.0 Cr, 1.1 Ti, 0.50 Al,
H.T.: 200 ksi TS; 170 ksi YS; 10 El.
Ultra high strength corrosion resistant steel.
For pump and valve parts, wind tunnels.

STAINLEND 8N12.
M-677; 0.03 C, 6 Mn, 0.5 Si, 16.5 Cr, 72.3 Ni, 0.2 Mo, 1.7 CbTa, 1.8 Fe.
For welding electrodes; Inconel, class ENiCrFe-3.

STAINLEND 14/75.
M-677; 0.04 C, 3 Mn, 0.6 Si, 15 Cr, 69 Ni, 1.3 Mo, 1.6 Cb, 8.9 Fe.
For welding electrodes; Inconel, class ENiCrFe-2.

STAINLEND 15/35 see **STAINLEND 330.**

STAINLEND 19/9 see **STAINLEND 347.**

STAINLEND 25/12 see **STAINLEND 309 CB.**

STAINLEND 308.
M-677; 19 Cr, 9 Ni, 0.07 C, bal Fe.
For welding rods; corrosion resistant.

STAINLEND 308HC.
M-677; 0.14 C, 1.7 Mn, 0.4 Si, 20 Cr, 9.7 Ni, bal Fe.
For welding electrodes; specially coated for welding austenitic CrNi stainless steel.

STAINLEND 308L.
M-677; 0.035 C, 1.8 Mn, 0.4 Si, 20 Cr, 9.8 Ni, bal Fe.
For welding electrodes; stainless steel type 308L.

STAINLEND 309.
M-677; 25 Cr, 12 Ni, 0.10 C, bal Fe.
For welding rods; corrosion and heat resistant.

STAINLEND 309 CB.
M-677; 0.08 C, 1.9 Mn, 0.5 Si, 23 Cr, 13 Ni, 0.8 CbTa, bal Fe.
For welding electrodes; Cb-stabilized stainless steel for root pass welding in Type 347 clad steel.

STAINLEND 309 MO.
M-677; 0.07 C, 1.8 Mn, 0.4 Si, 23 Cr, 12.8 Ni, 2.5 Mo, bal Fe.
For welding electrodes; Mo-bearing stainless steel for root pass welding of 316, 316L, 319 clad steel.

STAINLEND 310.
M-677; C, 25 Cr, 20 Ni, bal Fe.
For welding electrodes; E-310-16; stainless.

STAINLEND 310 CB.
M-677; 0.1 C, 2 Mn, 0.5 Si, 26 Cr, 21 Ni, 0.8 CbTa, bal Fe.
For welding electrodes; Cb-stabilized for root pass welds in type 347 clad plate.

STAINLEND 310 HC.
M-677; 0.22 C (also 0.3 C and 0.4 C), 1.8 Mn, 0.4 Si, 26.5 Cr, 21 Ni, bal Fe.
For welding electrodes; welding ACI types CK-20, HK, HL, HN castings, supplied in three carbon ranges.

STAINLEND 310 MO.
M-677; 0.1 C, 1.8 Mn, 0.4 Si, 25.5 Cr, 21.5 Ni, 2.3 Mo, bal Fe.
For welding electrodes; Mo-bearing stainless steel.

STAINLEND 312.
M-677; 0.12 C, 1.8 Mn, 0.5 Si, 29 Cr, 9.2 Ni, bal Fe.
For welding electrodes; for joining dissimilar metals.

STAINLEND 316.
M-677; 0.12 C, 1.8 Mn, 0.5 Si, 19 Cr, 9.2 Ni, bal Fe.
For welding electrodes; for joining dissimilar metals.

STAINLEND 316L.
M-677; 0.035 C, 1.8 Mn, 0.4 Si, 19.5 Cr, 12.5 Ni, 22 Mo, bal Fe.
For welding electrodes; stainless steel type 316L.

STAINLEND 317.
M-677; 0.35 C, 1.8 Mn, 0.4 Si, 19.5 Cr, 12.5 Ni, 2.2 Mo, bal Fe.
For welding electrodes; stainless steel type 316L.

STAINLEND 318.
M-677; 0.035 C, 1.8 Mn, 0.4 Si, 19.5 Cr, 12.5 Ni, 2.2 Mo, bal Fe.
For welding electrodes; stainless steel type 316 L.

STAINLEND 330.
M-677; 0.035 C, 1.8 Mn, 0.4 Si, 19.5 Cr, 12.5 Ni, 2.2 Mo, bal Fe.
For welding electrodes; stainless steel type 316 L.

STAINLEND 347.
M-677; 0.035 C, 1.8 Mn, 0.4 Si, 19.5 Cr, 12.5 Ni, 2.2 Mo, bal Fe.
For welding electrodes; stainless steel type 316 L.

STAINLEND 349.
M-677; 0.09 C, 1.8 Mn, 0.5 Si, 19.5 Cr, 9 Ni, 0.5 Mo, 0.9 Cb, 1.4 bal Fe.
For welding electrodes; superalloy electrode, stainless steel type 349.

STAINLEND 410.
M-677; 0.08 C, 0.9 Mn, 0.4 Si, 12.7 Cr, 0.2 Ni, 0.4 Mo, bal Fe.
For welding electrodes; martensitic stainless steel, type 410.

STAINLEND 430.
M-677; 0.08 C, 0.8 Mn, 0.4 Si, 16.5 Cr, 0.4 Ni, bal Fe.
For welding electrodes; ferritic stainless steel, type 430.

STAINLEND HCN see **STAINLEND 310.**

STAINLEND KMO see **STAINLEND 316.**

STAINLEND KMOCB see **STAINLEND 318.**

STAINLESS INVAR.
M-Japan; 5 Co, 31.5 Ni, 63.5 Fe.
For stainless and corrosion resistant parts and equipment; stainless, corrosion resistant.

STAINLESS INVAR.
M-U.S.; 36.5 Fe, 9.5 Cr, 54 Co.
For instruments; low expansion.

STAINLESS IRON (AMERICAN STAINLESS STEEL).
M-293; 0.12 max C, 8-60 Cr, bal Fe.
For hardware, tools, chemical equipment; stainless.

STAINLESS IRON C-1.
M-293; 0.12 max C, 15 max Cr, bal Fe.
70,000-190,000 TS; 35-16 El; 140-387 Brin.
For stampings, automobile trim moldings, machine parts, cheap cutlery; corrosion resistant.

STAINLESS IRON C-2.
M-293; 0.12 max C, 15-18 Cr, bal Fe.
90,000-75,000 TS; 28 El; 170 Brin.
For exterior trim for buildings and automobiles, cold rivets, nitric acid equipment, stampings; corrosion resistant.

STAINLESS IRON C-3.
M-293; 0.35 max C, 18-23 Cr, bal Fe.
100,000 TS; 22 El; 196 Brin.
For screw machine products, wire paper mill products, chemical apparatus; corrosion resistant.

STAINLESS IRON C-4.
M-293; 0.35 max C, 23-30 Cr, bal Fe.
95,000 TS; 18 El; 187 Brin.
For furnace parts, valve parts, conveyor chain bolts and nuts, heat resisting parts; corrosion and heat resistant.

STAINLESS IRON TYPE F.I.
M-521; 0.1 C, 11-13 Cr, 1.0 max Si, 1.0 max Mn, bal Fe.
Heat treated: 68,000-83,000 TS; 40,000-56,000 YS; 40-30 El; 60-50 R 150-180 Brin.
For turbine blades, shrouding strip, aircraft fittings; Type 403 and 410; corrosion resistant.

STAINLESS IRON TYPE FI 17.
M-521; 0.10 C, 0.2 Si, 0.2 Mn, 17 Cr, bal Fe.
Annealed: 75,000 TS; 48,000 YS; 32 El; 50 RA; 160 Brin.
For corrosion resistant parts; brittle welds, corrosion resistant.

STAINLESS S-20.
M-1262; 0.07 max C, 29 Ni, 20 Cr, 3-4 Mo, 4 Cu, bal Fe.
Cast: 65,000-75,000 TS; 28,000-38,000 YS; 50-35 El; 50-45 RA; 120-160 Brin.
For chemical plant equipment, valves, agitators, tanks; resists mixed acid austenitic.

STAINLESS STEEL A.
M-293; 0.12 min C, 8-16 Cr, bal Fe.
Hardened: 260,000 TS; 10 El; 512 Brin.
For cutlery, springs, balls, bearing races, surgical instruments, hypodermic needles; stainless and corrosion resistant.

STAINLESS STEEL A.S.S.
M-293; 8-60 Cr, 0.40 Mn, > 0.12 C, bal Fe.
Rolled: 80,000-250,000 TS; 24,000-150,000 YS; 10-15 El; 30-35 RA; 1 Brin.
For cutlery, surgical instruments, valve seats, pump shafts; corrosion and abrasion resisting.

STAINLESS STEEL B.
M-293; 0.12 min C, 16-18 Cr, bal Fe.
Hardened: 601 Brin.
For cutlery, bearing races, surgical instruments; stainless and corrosion resistant.

STAINLESS STEEL TYPE FAL.
M-521; 0.10 C, 0.4 Si, 0.6 Mn, 12 Cr, 4.5 Al, bal Fe.
Heat treated: 88,000 TS; 62,000 YS; 25 El.
For resistances, rheostats; maximum service temperature 700°C.

STAINLESS STEEL TYPE FG.
M-56, M-521; 0.25 C, 12-14 Cr, 0.4 Si, 0.3 Mn, bal Fe.
Heat treated: 90,000-112,000 TS; 56,000-78,500 YS; 30-20 El; 60-50 R 220-240 Brin.
For turbine blades, high pressure steam valves; Type 420; corrosion resistant.

STAINLESS STEEL TYPE FH.
M-521; 0.3 C, 12-14 Cr, bal Fe.
Heat treated: 230,000 TS; 450-550 Brin.
For cutlery, surgical instruments, knives, edge tools, rolls, springs, ball bearings and races; magnetic, stainless; Firth Brown I-162, Firth F.H. Stainless.

STAINLESS STEEL TYPE FHM.
M-521; 0.8 C, 0.2 Si, 0.2 Mn, 17 Cr, 0.5 Mo, bal Fe.
Annealed: 120,000 TS; 98,000 YS; 260 Brin.
For ball bearings; corrosion resistant.

STAIN-ROD NO. 308.
M-1096; 0.07 max C, 9 Ni, 19 Cr, bal Fe.
Welded: 85,000-95,000 TS; 40-50 El.
For stainless steel welding rod; austenitic, stainless.

STAINTIN 157PA.
M-717.
For soldering stainless steel.
M.P. 425°F.

STAINTRODE A.
M-717.
Electrode for AC-DC welding of 300 series stainless.
Tensile strength 90,000 psi.

STAINTRODE A-MO.
M-717.
Electrode for AC-DC welding of 303, 315, 316 and 329 stainless steels.
Tensile strength 80,000 psi.

STAINTRODE A-MO-L.
M-717.
Electrode for AC-DC welding of 316, 316L and 318 stainless steels.
Tensile strength 90,000 psi.

STAINTRODE B.
M-717.
For AC-DC metallic arc welding of 301, 302, 304, 305, 306 and 308 stainless steels.
85,000 psi TS.

STAINTRODE B-MO.
M-717.
Electrode for AC-DC welding of several 300 series stainless steels.
Tensile strength 90,000 psi.

STAINTRODE D.
M-717.
Electrode for AC-DC welding of type 310 stainless.
Tensile strength 95,000 psi.

STAINWELD 308-15.
M-578; 0.08 C, 18.0 Cr, 9.0 Ni, 2.5 Mn, 0.90 Si, 0.04 P, 0.03 S, bal Fe.
Stainless steel arc welding electrode.
AWS Class E308-15.

STAINWELD 308-16.
M-578; 0.08 C, 18.0 Cr, 9.0 Ni, 2.5 Mn, 0.90 Si, 0.04 P, 0.03 S, bal Fe.
Stainless steel arc welding electrode.
AWS Class E308-16.

STAINWELD 308L-16.
M-578; 0.08 C, 18.0 Cr, 9.0 Ni, 2.5 Mn, 0.90 Si, 0.04 P, 0.03 S, bal Fe.
Stainless steel welding electrode.
AWS Class E308L-16.

STAINWELD 309-16.
M-578; 0.08 C, 23.9 Cr, 12.4 Ni, bal Fe.
Stainless steel arc welding electrode.
AWS E309-16.

STAINWELD 310-15.
M-578; 0.20 C, 25.0 Cr, 20.0 Ni, 2.5 Mn, 0.75 Si, bal Fe.
Stainless steel arc welding electrode.
AWS Class E310-15.

STAINWELD 310-16.
M-578; 0.20 C, 25.0 Cr, 20.0 Ni, 2.5 Mn, 0.75 Si, bal Fe.
Stainless steel arc welding electrode.
AWS Class E310-16.

STAINWELD 316L-16.
M-578; 0.04 C, 17.0 Cr, 11.0 Ni, 2.0 Mo, 2.5 Mn, 0.90 Si, bal Fe.
Stainless Steel arc welding electrode.
AWS Class E316L-16.

STAINWELD 347-15.
M-578; 0.08 C, 18.0 Cr, 9.0 Ni, 2.5 Mn, 0.90 Si, bal Fe.
Stainless steel arc welding electrode.
AWS Class E347-15.

STAINWELD 347-16.
M-578; 0.08 C, 18.0 Cr, 9.0 Ni, 2.5 Mn, 0.90 Si, bal Fe.
Stainless steel arc welding rod.
AWS Class E347-16.

STALCREST.
M-1724; 0.17 max C, 0.40 max Si, 1.0 max Mn, 0.07-0.10 P, 0.70-1.0 Cr 0.25-0.55 Cu, 0.10 max V, 0.025 min Al, bal Fe.
Weldable steel with good weathering properties. Properties meet BS 4360 WR 50.

STALINITE.
M-USSR; C, 9.5 Cr, 11.5 Mn, bal Fe.
For machine parts, dredging, oil drilling, agricultural and kindred equipment; sintered alloy.

STALINIT NO. 2.
M-USSR; Ti-Cr-Fe.
For cutting tools; hard sintered alloy.

STALLOY.
M-Eng.; C, 3-4 Si, bal Fe.
For telephone and loud speaker diaphragms, electrical machinery, instruments; max permeability 8,000.

STAMINAL.
M-73; 0.55 C, 0.4 Cr, 2.7 Ni, 0.9 Mn, 0.15 V, 0.45 Mo, 1.0 Si, bal Fe.
For chisels, cold cutters, die blocks, punches; oil or air hardened; tough.

STANDARD.
M-336; 0.9 C, 0.3 Mn, bal Fe.
For general tools, drills, stone tools, dies; water hardened.

STANDARD.
M-275; 0.7-1.2 C, bal Fe.
For general tools, dies, punches; water hardened.

STANDARD ADMIRALTY BRONZE.
M-126; 88 Cu, 10 Sn, 2 Zn.
For bronze castings, gears, worm wheels, trolley wheels; tough, wear resistant.

STANDARD ALLOY CO. H R-5 M.
M-105; 50 Fe, 25 Cr, 20 Ni, 2.5-4.0 Mo, 0.40 Mn, 0.3 C.
Cast: 62,000 TS; 51,000 YS; 18 El; 13 RA; 190-210 Brin.
For furnace parts; heat and corrosion resistant.

STANDARD ALLOY CR-1.
M-105; 0.08 max C, 1.5 Mn, 2 Si, 18-21 Cr, 8-11 Ni, bal Fe.
Cast: 70,000 TS; 28,000 YS; 55 El; 150 Brin.
For chemical and food processing equipment, furnace parts; ACI-CF8; corrosion and heat resistant.

STANDARD ALLOY CR-2.
M-105; 0.08 max C, 1.5 Mn, 1.5 max Si, 18-21 Cr, 9-12 Ni, 2-3 Mo, bal Fe.
Cast: 75,000 TS; 30,000 YS; 50 El; 160 Brin.
For chemical plant equipment: ACI-CF8M; corrosion resistant.

STANDARD ALLOY CR-3.
M-105; 0.20 max C, 1.5 max Mn, 2.0 max Si, 18-21 Cr, 8-11 Ni, bal Fe.
Cast: 75,000 TS; 30,000 YS; 50 El; 160 Brin.
For chemical and food processing equipment; ACI-CF20; corrosion resistant.

STANDARD ALLOY "H.R.-1".
M-105; 20 Ni, 20 Cr, 0.4 C, 1.5 Si, 0.7 Mn, bal Fe.
For furnace parts; heat and corrosion resistant.

STANDARD ALLOY "H.R.-2".
M-105; 25 Ni, 20 Cr, 1.5 Si, 0.4 C, 0.7 Mn, bal Fe.
For furnace parts; heat and corrosion resistant.

STANDARD ALLOY "H.R.-3".
M-105; 36 Ni, 15 Cr, 0.4 C, 0.7 Mn, 1.5 Si, bal Fe.
Cast: 62,000 TS; 35,000 YS; 10 El; 11 RA; 190 Brin.
For lead pots, carburizing boxes, trays; heat resistant.

STANDARD ALLOY "H.R.-4".
M-105; 62 Ni, 14 Cr, 0.4 C, 0.7 Mn, 1.5 Si, bal Fe.
Cast: 65,000 TS; 35,000 YS; 5 El; 5 RA; 180 Brin.
For carburizing boxes, retorts; heat resistant.

STANDARD ALLOY "H.R.-5".
M-105; 20 Ni, 25 Cr, 0.4 C, 0.7 Mn, 1.7 Si, bal Fe.
73,000 TS; 46,000 YS; 17 El.
For furnace parts; heat resistant.

STANDARD ALLOY "H.R.-6".
M-105; 13 Ni, 25 Cr, 0.4 C, 0.7 Mn, 1.7 Si, bal Fe.
Cast: 85,000 TS; 50,000 YS; 15 El; 15 RA; 170 Brin.
For furnace rails, stack dampers; heat resistant.

STANDARD ALLOY "H.R.-7".
M-105; 8 Ni, 18 Cr, 0.3 C, 1.0 Mn, 1.0 Si, bal Fe.
Cast: 97,000 TS; 45,000 YS; 25 El; 20 RA.
For stainless castings; stainless.

STANDARD ALLOY "H.R.-8".
M-105; 28-30 Cr, Mn, 3 Ni, 0.4 Cr, 1.8 Si, bal Fe.
Cast: 50,000 TS; 40,000 YS; 1 El; 0 RA; 196 Brin.
For acid resisting equipment, grate bars; acid and corrosion resistant.

STANDARD ALLOY HR-9.
M-105; C, 18 Cr, bal Fe.
For corrosion resistant parts; corrosion resistant.

STANDARD ALLOY HR-10.
M-105; 0.21-0.35 C, 24-30 Cr, 7-11 Ni, Si, bal Fe.
For heat and corrosion resistant parts; heat and corrosion resistant.

STANDARD ALLOY HR-11.
M-105; 0.21-0.35 C, 16-23 Cr, 12-15 Ni, bal Fe.
For heat and corrosion resistant parts; heat and corrosion resistant.

STANDARD ALLOY K.A. 4.
M-105; 17-20 Cr, 8-10 Ni, 0.25 C, 2-4 Mo, bal Fe.
Annealed: 105,000 TS; 49,000 YS; 55 El; 62 RA.
For pipe fittings, pump casings, impellers, shafts; austenitic; will not respond to heat treatment.

STANDARD ALLOY, SPECIAL.
M-105; 20 Cr, 25 Ni, 0.40 C, bal Fe.
62,100 TS; 3 El; 177 Brin.
For furnace parts; heat and corrosion resistant.

STANDARD AR-1.
M-105; 0.2 C, 9 Ni, 19 Cr, 3 Si, bal Fe.
Cast.
For castings, valves and fittings for handling H_2SO_4; acid resisting.

STANDARD BEARING.
58 Pb, 26 Sn, 15 Sb, 1 Cu.
For bearings; anti-friction.

STANDARD CADMIUM.
M-U.S.; 92.5 Ag, 5.75 Cu, 1.75 Cd.
For solder, brazing; silver solder.

STANDARD CARBON.
M-73; 0.9-1.3 C, 0.2 Si, 0.25 Mn, bal Fe.
For tools, drills, taps; water hardening.

STANDARD CARBON WITH V.
M-73; 0.9-1.3 C, 0.2 V, bal Fe.
For drills, tools; water hardening.

STANDARD DRACO.
M-114; 0.7-1.2 C, 0.2 V, bal Fe.
For tools, dies, punches; water hardened.

STANDARD DRACO DV.
M-114; 1.0 C, 0.5 V, bal Fe.
For drills, taps, reamers, broaches; Type W3; oil or water hardened.

STANDARD G.
M-782; 0.7-1.4 C, bal Fe.
For chisels, tools; water hardened.

STANDARD GLYCO.
M-240; Sn, Sb, bal Pb.
For bearings; Babbitt.

STANDARD GOLD.
90 Au, 10 Cu.
For money, jewelry; corrosion resistant.

STANDARD GOLD (GREAT BRITAIN).
M-Eng.; 92 Au, 8 Cu.
For coinage; corrosion resistant.

STANDARD H.R. ALLOYS.
M-105; 16-25 Cr, 20-60 Ni, C, bal Fe.
For furnace parts; heat resisting.

STANDARD MISCO.
M-84; 46 Fe, 35 Ni, 15 Cr, 0.50 Mn, 0.60 C.
Cast: 75,000 TS; 70,000 YS; 3 El; 6 RA; 180 Brin.
Rolled: 100,000 TS; 81,000 YS; 42 El; 52 RA; 190 Brin.
For heat treating boxes, furnace parts, valves, stills; heat resisting; wear and corrosion resistant.

STANDARD NICKEL BRONZE.
M-126; 88 Cu, 5 Sn, 2 Zn, 5 Ni.
For valves, corrosion resistant parts; corrosion resistant.

STANDARD NO. 4 BABBITT.
M-428, M-88; Pb, Sb, Sn.
At 70°F: 14.3 Brin.
At 212°F: 6.4 Brin.
For bearings for line shafts; Babbitt, heavy loads.

STANDARD PHOSPHOR BRONZE.
M-U.S.; 80 Cu, 10 Sn, 10 Pb, up to 0.25 P.
Cast: 28,000 TS; 6-9 El; 55-65 Brin.
For high speed bearings; heavy duty.

STANDARD SILVER.
92.5 Ag, 7.5 Cu.
For coinage; corrosion resistant.

STANDARD SILVER STERLING.
90 Ag, 10 Cu.
For cutlery, ornaments, utensils.

STANDFEST SPEZIAL.
M-1307; 0.90 C, 0.25 Si, 1.9 Mn, 0.10 V, bal Fe.
For forming and blanking dies, punches; oil hardened, non-deforming.

STANELEC.
M-691; 0.25 Si, bal Fe.
For motors, armatures; high permeability.

STANFIRE.
M-896; 3.2 C, 1.5 Ni, 0.8 Cr, bal Fe.
For fire box grates; heat resisting cast iron.

STANNIOL.
96 Sn, 2.4 Pb, 1 Cu, 0.3 Ni, 0.1 Fe.
For bearings, wrapping foil.

STANNIOL.
98.9 Sn, 0.7 Pb, 0.33 Cu.
For bearings, wrapping foil.

STANNUM METAL.
M-77; 90 Sn, 6 Sb, 4 Cu, Cast.
For machine bearings; high tin Babbitt.

STANTUFONT.
M-1176; 0.5-1.0 Cu, 12.4-13.0 Si, 0.7-1.0 Mg, 2.0-2.4 Ni, bal Al.
Heat treated: 40,000-50,000 TS; 38,000-43,000 YS; 0.5-0.3 El; 120-130 Brin.
For light alloy parts; age-hardened.

STAR.
M-657; 78 Cu, 6 Sn, 10 Zn, 6 Pb.
For castings; free-cutting.

STAR BLUE CHIP.
M-57; 0.7 C, 14 W, 4 Cr, 2 V, bal Fe.
For cutters, drills, reamers, hobs, lathe and planer tools; high speed steel.

STAR J METAL see **HAYNES STELLITE STAR J.**

STARKAD.
M-111; 1.3 C, 0.25 Cr, 5.5 W, bal Fe.
For drawing dies; oil hardening.

STARLI.
M-1260; 85 Sn, 10 Sb, 5 Cu.
Cast: 19,900 TS; 28 Brin.
For engine bearings; M.P. 440-630°F; shock resistant.

STAR-MAX see **CARPENTER STAR MAX.**

STAR-MO.
M-57; 0.8 C, 6.5 W, 5 Mo, 4 Cr, 2 V, bal Fe.
For cutting tools; high speed steel.

STAR-MO F.M.
M-57; 0.83 C, 6.4 W, 5.0 Mo, 4.0 Cr, 2.0 V, 0.12 S, bal Fe.
For reamers, broaches, chasers, lathe and form tools, milling cutters; high speed steel, free-machining.

STAR-MO MOD.
 M-57; 0.65 C, 4.2 Cr, 6 Mo, 6 W, 2 V, bal Fe.
 For reamers, drills, lathe and planer tools; high speed steel.

STARO.
 M-1294; 0.40 C, 0.40 Si, 0.30 Mn, 13 Cr, bal Fe.
 Annealed: 100,000 TS; 55,000 YS; 20 El; 50 RA; 210 Brin.
 For valves, cutlery, surgical and dental instruments; Type 420; stainless.

STARRETT NO. 496.
 M-1143, M-1225; 0.90 C, 12.2 Mn, 0.5 Cr, 0.5 W, 0.2 V, bal Fe.
 For punches, crimpers, upsetters, headers; oil hardened, non-deforming.

STARRETT NO. 497.
 M-1225, M-1143; 0.95-1.05 C, 0.6 Mn, 0.4 Si, 5.0-5.5 Cr, 1.0 Mo, 0.2 V, bal Fe.
 For thread roller dies, master hubs, forming dies; air hardening, non-deforming.

STARRETT NO. 498.
 M-1225.
 Free machining low carbon steel (0.18C) as precision ground flat stock.

STAR-ZENITH see **CARPENTER STAR-ZENITH.**

STATO 3.
 M-111a; 0.15 C, 1.2 Cr, 0.25 Mo, bal Fe.
 For gears, shafts, machinery parts; case hardened.

STATO 5.
 M-111; 0.25 C, 0.65 Mn, 1.05 Cr, 0.2 Mo, bal Fe.
 For gears, shafts, bolts; tough.

STATO 8.
 M-111; 0.4 C, 0.65 Mn, 1.1 Cr, 0.25 Mo, bal Fe.
 For gears, shafts, bolts; tough.

STATO 23.
 M-111; 0.15 C, 0.8 Mn, 1 Cr, 0.35 Mo, bal Fe.
 For gears, shafts; case hardened.

STATOS EXTRA now **VEW K720.**

STATOS SPEZIAL now **VEW K100.**

STATOS SPEZIAL GW now **VEW K116.**

STATOS SPEZIAL W now **VEW K107.**

STATOS SUPERIOR now **VEW K465.**
 M-1314; 1.05 C, 1 Cr, 1.15 W, bal Fe.
 For cold working tools, headers, punches, bearings; water hardened, wear resistant.

STATOS V now **VEW K505.**
 M-1314; 1.45 C, 1.4 Cr, bal Fe.
 For punches, drawing dies, bearings, liners; water hardened.

STATUARY BRONZE.
 95-88 Cu, 1.4-10 Sn, 9.5 Zn, 0-6 Pb, trace P.
 For statuary, castings.

STATUARY BRONZE, AUGSBURG "A".
 89.43 Cu, 8.17 Sn, 1.07 Pb, 0.34 P, 0.19 Ni.
 For statuary; corrosion resistant.

STATUARY BRONZE, AUGSBURG "B".
 94.74 Cu, 1.64 Sn, 0.54 Zn 6.24 Pb. 0.71 Ni.
 For statuary; corrosion resistant.

STATUARY BRONZE, BACCHUS POTSDAM.
 89.34 Cu, 7.5 Sn, 1.63 Zn, 1.21 Pb, 0.18 P.
 For statuary; corrosion resistant.

STATUARY BRONZE, BAVARIA MUNICH.
 91.55 Cu, 1.77 Sn, 5.5 Zn, 1.3 Pb.
 For statuary; good castability.

STATUARY BRONZE, COLUMN VENDOME.
 89.2 Cu, 10.2 Sn, 0.5 Zn, 0.1 Pb.
 For statuary; good castability.

STATUARY BRONZE MELANCHTON, WITTENBERG.
 89.55 Cu, 2.99 Sn, 7.45 Zn.
 For statuary; good castability.

STATUARY BRONZE, MUNICH.
 92.88 Cu, 4.18 Sn, 0.44 Zn, 2.31 Pb, 0.15 P.
 For statuary; good castability.

STAVANGER N250.
 M-1304; 0.25 C, 23-30 Cr, bal Fe.
 Annealed: 85,000 TS; 55,000 YS; 25 El; B 83 Rock.
 For annealing boxes, oil burners, glass molds, furnace parts and fittings, exhaust manifolds.
 Type 446 stainless steel, heat and corrosion resistant.

STAVANGER N330.
 M-1304; 0.35 max C, 23-27 Cr, bal Fe.
 Annealed: 90,000 TS; 60,000 YS; 20 El; 45 RA; 180 Brin.
 For furnace parts, heat treating boxes; Type 446; heat resistant.

STAVANGER S1CHMO.
 M-1340; 0.12 C, 13 Cr, 1.1 Mo, bal Fe.
 Annealed: 80,000 TS; 35,000 YS; 25 El; B 92 Rock.
 For springs, shafts, table flatware, oil refinery and chemical plant equipment.
 Corrosion resistant.

STAVANGER S1CH-NI.
 M-1304; 0.15 max C, 13 Cr, 2 Ni, bal Fe.
 For oil refinery equipment, valves; corrosion resistant.

STAVANGER S1C SPEC.
M-1304; 0.08 C, 14 Cr, bal Fe.
Annealed: 75,000 TS; 36,000 YS; 30 El; B 80 Rock.
For flat springs, knives, tableware, chemical plant equipment.
Type 410 stainless steel, heat and corrosion resistant.

STAVANGER S23/20.
M-1304; 0.20 C, 2 Si, 25 Cr, 20 Ni, bal Fe.
Annealed: 100,000 TS; 50,000 YS; 45 El; B 90 Rock.
For furnace parts, annealing boxes, heat treating fixtures.
Type 314 stainless steel, austenitic, heat and corrosion resistant.

STAVANGER S128M SPECIAL.
M-1304; 0.06 max C, 18 Cr, 9 Ni, 1.5 Mo, bal Fe.
Annealed: 85,000 TS; 35,000 YS; 50 El; 65 RA; 160 Brin.
For acid resistant chemical plant equipment; Type 316; stainless.

STAVANGER S178.
M-1304; 0.08-0.20 C, 2 max Mn, 17-19 Cr, 8-10 Ni, bal Fe.
Annealed: 85,000 TS; 35,000 YS; 60 El; 70 RA; 150 Brin.
Cold drawn; 125,000 TS; 95,000 YS; 22 El; 55 RA; 277 Brin.
For oil refinery and chemical plant equipment; Type 302; stainless, austenitic.

STAVANGER S178 IISP.
M-1304; 0.08 C, 19 Cr, 11 Ni, bal Fe.
Annealed: 80,000 TS; 30,000 YS; 60 El; B 80 Rock.
For chemical plant equipment, digesters, evaporators, tanks.
Type 304 stainless steel, austenitic.

STAVANGER S178 SUPRA.
M-1304; 0.03 C, 18-20 Cr, 9-12 Ni, bal Fe.
Annealed: 77,000 TS; 30,000 YS; 60 El; B 75 Rock.
For chemical plant equipment, evaporators, digesters, tanks.
Type 304L stainless steel, austenitic.

STAVANGER S268M.
M-1304; 0.20 C, 23-28 Cr, 2.5-5.0 Ni, 1-2 Mo, bal Fe.
Normalized: 103,000 TS; 78,000 YS; 18 El; 45 RA; 235 Brin.
Annealed: 95,000 TS; 41,000 YS; 29 El; 60 RA; 225 Brin.
For valves, pumps, furnace parts: Type 327; heat resistant.

STAVANGER S278.
M-1304; 0.12 C, 17-19 Cr, 8-10 Ni, bal Fe.
Annealed: 90,000 TS; 40,000 YS; 50 El; B 58 Rock.
For chemical plant equipment, vessels, tanks, mixers, agitators.
Type 302 stainless steel, austenitic.

STAVANGER SC-140.
M-1304; 0.15 min C, 12-14 Cr, bal Fe.
Annealed: 88,000 TS; 40,000 YS; 32 El; 68 RA; 170 Brin.
For cutlery, valve trim, turbine blades; Type 420; stainless, hardenable.

STAVANGER SC140L.
M-1304; 0.20 C, 14 Cr, bal Fe.
Ht. Tr.; 225,000 TS; 200,000 YS; 12 El; C 48 Rock.
For cutlery, surgical instruments, scissors, knives, gears, shafts.
Type 420 stainless steel, hardenable. Heat and corrosion resistant.

STAVANGER SC140MO.
M-1304; 0.14 C, 14.5 Cr, 1 Mo, bal Fe.
Annealed: 80,000 TS; 35,000 YS; 25 El; B 92 Rock.
For springs, shafts, table flatware, oil refinery and chemical plant equipment.
Corrosion resistant, hardenable.

STAVANGER SCN192.
M-1304; 0.25 max C, 17 Cr, 2 Ni, bal Fe.
For oil refinery equipment; Type 341; corrosion and heat resistant.

STAVANGER SIC.
M-1304; 0.08 max C, 11.5-13 Cr, 0.1-0.3 Al, bal Fe.
Annealed: 71,000 TS; 42,600 YS; 22 El; 70 RA; 150 Brin.
Heat treated: 175,000 TS; 145,000 YS; 21 El; 64 RA; 352 Brin.
For oil refinery and chemical plant equipment; Type 405; corrosion resistant.

STAVANGER SIC16.
M-1304; 0.12 max C, 14-18 Cr, bal Fe.
Annealed: 70,000 TS; 40,000 YS; 30 El; 55 RA; 150 Brin.
Cold drawn: 130,000 TS; 120,000 YS; 2 El; 185 Brin.
For oil refinery equipment, bolts, kitchen sinks; Type 430; stainless, ferritic.

STAVANGER SIC18.
M-1304; 0.12 max C, 14-18 Cr, bal Fe.
Annealed: 70,000 TS; 40,000 YS; 30 El; 55 RA; 150 Brin.
Cold drawn: 130,000 TS; 120,000 YS; 2 El; 185 Brin.
For oil refinery equipment, bolts, kitchen sinks; Type 430; stainless, ferritic.

STAVANGER SICH.
M-1304; 0.15 max C, 11.5-13.5 Cr, 1 max Mn, bal Fe.
Annealed: 80,000 TS; 40,000 YS; 35 El; 70 RA; 155 Brin.
Cold drawn: 100,000 TS; 85,000 YS; 17 El; 60 RA; 205 Brin.
For valve parts, turbine blades, cutlery, knives; Type 410; corrosion resistant.

STAVANGER SN12-12.
M-1304; 0.08-0.16 C, 12 Cr, 12 Ni, bal Fe.
For valves; corrosion and heat resistant.

STAVANGER SN128IIISP.
M-1304; 0.06 C, 16-18 Cr, 10-14 Ni, 2-3 Mo, bal Fe.
Annealed: 80,000 TS; 30,000 YS; 60 El; B 78 Rock.
For chemical plant equipment, vessels, tanks, digesters, agitators.
Type 316 stainless steel, austenitic.

STAVANGER SN128M.
M-1304; 0.10 max C, 16-18 Cr, 10-14 Ni, 2-3 Mo, bal Fe.
Annealed: 85,000-95,000 TS; 35,000-45,000 YS; 60-50 El; 75-60 RA; 150-190 Brin.
For acid resistant chemical plant equipment; Type 316; stainless, austenitic.

STAVANGER SN128MIIISP.
M-1304; 0.08 C, 17 Cr, 12 Ni, 2 Mo, bal Fe.
Annealed: 80,000 TS; 30,000 YS; 60 El; B 78 Rock.
For chemical plant equipment, tanks, digesters, agitators, mixers.
Type 316 stainless steel. Austenitic.

STAVANGER SN128, SPEC.
M-1304; 0.10 max C, 16-18 Cr, 10-14 Ni, 1.7-2.7 Mo, Cb = 10 x C, bal Fe.
Annealed: 85,000-95,000 TS; 35,000-45,000 YS; 60-50 El; 75-60 RA; 150-190 Brin.
For acid resistant chemical plant equipment; Type 316 Cb; stainless, austenitic.

STAVANGER SN129.
M-1304; 0.08 max C, 18-20 Cr, 8-11 Ni, 2 max Mn, bal Fe.
Annealed: 90,000 TS; 45,000 YS; 60 El; 135 Brin.
Cold drawn: 180,000 TS; 150,000 YS; 10 El; 330 Brin.
For chemical plant equipment, welded structures; Type 304; stainless, austenitic.

STAVANGER SNW.
M-1304; 0.25 max C, 24-26 Cr, 19-22 Ni, bal Fe.
Annealed: 95,000 TS; 45,000 YS; 50 El; 65 RA; 185 Brin.
At 1200°F: 57,000 TS; 22,000 YS; 32 El; 45 RA.
For furnace parts and equipment, heat treating boxes; Type 310; austenitic, heat resistant.

STAVANGER SS286M.
M-1304; 0.12 C, 25-30 Cr, 3.5-6.0 Ni, 1-2 Mo, bal Fe.
Annealed: 105,000 TS; 80,000 YS; 25 El; 230 Brin.
For valves, valve fittings, pump parts, furnace equipment. Precipitation hardening.
Type 329 stainless steel, austenitic, corrosion and heat resistant.

STAYBLADE MAX.
M-521; 0.12 C, 18.5 Cr, 8.5 Ni, 0.8 Ti, 1.4 Al, bal Fe.
For turbine blades, boiler drums; austenitic; resists heat to 900°C.

STAYBLADE STEEL.
M-521; 0.22 C, 1 Si, 0.6 Mn, 20 Cr, 8.5 Ni, 1.2 Ti, bal Fe.
Heat treated: 95,000 TS; 58,000 YS; 30 El; 50 RA; 200 Brin.
For heat and corrosion resistant parts; heat and corrosion resistant.

STAYBOLT.
M-724; 0.05 C, 2 SiO_2, 0.15 Si, bal Fe.
For staybolts; wrought iron.

STAYBRITE 17/7.
M-521; 0.10 C, 0.6 Si, 1.3 Mn, 17.5 Cr, 7.5 Ni, bal Fe.
Soft Sheet: 108,000 TS; 38,000 YP; 55 El.
Hard Sheet: 180,000 TS; 146,000 YP; 10 El.
For springs, aircraft structural parts, diaphragms, household utensils, trim.
Type 301 stainless steel, nonmagnetic, austenitic.

STAYBRITE 317(L).
M-521; 0.03 C, 0.2-1.0 Si, 0.5-2.0 Mn, 18.0-19.5 Cr, 14-16 Ni, 3.0-4. Mo, bal Fe.
Ann: 72,000 min TS; 28,000 min YS; 40 min El.
Weldable without subsequent heat treatment.
Excellent corrosion resistant and resistance to pitting corrosion; for parts in textile, paper pulp and food industries. AISI 317L.

STAYBRITE D.D.Q.
M-521; 0.10 C, 12 Cr, 12 Ni, 0.3 Si, 0.8 Mn, bal Fe.
Annealed: 78,000-90,000 TS; 32,000-37,000 YS; 60-40 El; 60-40 RA; 130-150 Brin.
For domestic table and hollowware, decorative trim; austenitic, stainless.

STAYBRITE EMS.
M-521; 0.12 C, 0.8 Si, 1.6 Mn, 0.25 S, 18 Cr, 10 Ni, 0.28 Mo, 0.6 Ti, bal Fe.
Annealed: 82,000 TS; 36,000 YS; 53 El; 54 RA; 175 Brin.
For stainless parts; free-cutting, austenitic, stainless.

STAYBRITE FCB.
M-521; 0.12 C, 0.6 Si, 0.4 Mn, 18 Cr, 10 Ni, 1.2 Cb, bal Fe.
Annealed: 84,000 TS; 37,000 YS; 58 El; 65 RA; 175 Brin.
For stainless parts; austenitic, stainless.
Similar to AISI 347.

STAYBRITE F.C.B. (H) STEEL.
M-521; 0.04-0.09 C, 17.0-19.0 Cr, 9.0-12.0 Ni, Nb = 10 x C, bal Fe.
Stabilized austenitic stainless steel.
Similar to AISI 347.

STAYBRITE FCB (T).
M-521; 0.13 C, 0.6 Si, 1.2 Mn, 18 Cr, 12 Ni, 1.3 Cb, bal Fe.
Heat treated: 85,000 TS; 38,000 YS; 58 El; 65 RA; 175 Brin.
For stainless parts; stainless, austenitic.

STAYBRITE F.D.P.
M-521; 0.10 max C, 0.2-1.0 Si, 0.5-2.0 Mn, 17-19 Cr, 8.5-10.5 Ni, Ti x C min, bal Fe.
Ann: 78,000 min TS; 30,000 min YS; 40 min El; 180 VDH.
Weldable stainless; for welded assemblies; for operation at elevated temperatures.
BS: En.58B, S129, S520, S521; AISI 321.

STAYBRITE F.D.P. (H) STEEL.
M-521; 0.04-0.09 C, 17.0-19.0 Cr, 9.0-12.0 Ni, Ti = 5 x C min, 0.70 Mo, bal Fe.
Stabilized austenitic stainless steel.
Similar to AISI 321.

STAYBRITE F.D.P. (L).
M-521; 0.05 C, 0.8 Si, 0.8 Mn, 18 Cr, 11 Ni, Ti = 5 x C min, bal Fe.
Annealed: 90,000 TS; 40,000 YP; 50 El; 50 RA; 160 Brin.
For welded equipment in chemical, plastic and textile plants, tanks.
Type 321 stainless, austenitic, welding grade. Stabilized.

STAYBRITE F.M.B.
M-521; 0.07 C, 18 Cr, 8 Ni, 3 Mo, bal Fe.
Annealed: 90,000-112,000 TS; 36,000-42,500 YS; 60-40 El; 60-40 RA; 160-190 Brin.
For chemical apparatus to resist acetic and other acids at elevated temperatures; rust and acid resisting; non-magnetic; Firth Brown I-166.

STAYBRITE FMB (FC).
M-521; 0.08 max C, 2.0 max Mn, 1.0 max Si, 16.0-18.0 Cr, 10.0-14.0 Ni 0.20 P, 0.10 min S, 1.75-2.50 Mo, bal Fe.
Free machining austenitic stainless steel.

STAYBRITE F.M.B. (H) STEEL.
M-521; 0.04-0.09 C, 16.0-18.0 Cr, 10.0-13.0 Ni, 2.0-2.75 Mo, bal Fe.
Austenitic stainless steel.
Similar to AISI 316.

STAYBRITE F.M.B.(L).
M-521; 0.03 C, 0.4 Si, 1.5 Mn, 17.5 Cr, 12.0 Ni, 2.8 Mo, bal Fe.
Annealed: 88,000 TS; 35,000 YP; 50 El; 50 RA; 160 Brin.
For chemical, textile, plastic and pharmaceutical plant equipment.
Type 316L stainless steel, austenitic, nonmagnetic.

STAYBRITE FMB (L) CO.
M-521; 0.03 max C, 0.5-2.0 Mn, 0.2-1.0 Si, 16.5-18.0 Cr, 12.0-14.0 Ni, 2.25-3.0 Mo, Co, bal Fe.
Austenitic stainless steel.

STAYBRITE FMB(TI).
M-521; 0.07 C, 0.3 Si, 0.3 Mn, 18 Cr, 12-14 Ni, 3.25 Mo, Ti = 4 x min, bal Fe.
Annealed: 85,000 TS; 38,000 YS; 50 El; 50 RA; 170 Brin.
For chemical plant equipment, evaporators; austenitic, stainless.

STAYBRITE FML.
M-521; 0.07 C, 0.6 Si, 0.8 Mn, 18 Cr, 9.5 Ni, 1.25 Mo, bal Fe.
Annealed: 85,000 TS; 38,000 YS; 50 El; 50 RA; 170 Brin.
For stainless parts; austenitic, stainless.

STAYBRITE F.S.L. CO. TA.
M-521; 0.03 max C, 0.5-2.0 Mn, 0.2-1.0 Si, 18.0-19.5 Cr, 10.0-11.5 Ni, Co, Ta, bal Fe.
Austenitic stainless steel; for elevated temperature and corrosion resistant operations. Weldable.

STAYBRITE F.S.L.(L).
M-521; 0.03 C, 0.6 Si, 1.0 Mn, 18.5 Cr, 11.0 Ni, bal Fe.
Annealed: 85,000 TS; 36,000 YS; 60 El; 65 RA; 160 Brin.
For chemical, textile and pharmaceutical processing equipment, tanks, agitators, vessels, filters.
Type 304L stainless, austenitic, nonmagnetic.

STAYBRITE F.S.T.
M-521; 0.10 max C, 18 Cr, 8 Ni, 0.8 Mn, 0.6 Si, bal Fe.
Annealed: 83,000-101,000 TS; 34,000-41,000 YS; 60-40 El; 60-40 RA; Brin.
For vats, tanks, ship fittings, trim, reflectors; Type 302; austenitic, stainless.

STAYBRITE F.S.T.(FC).
M-521; 0.10 max C, 0.2-1.0 Si, 1.0-2.0 Mn, 0.15-0.30 S, 17-19 Cr, 8.5 10.5 Ni, 0.2-0.5 Mo, bal Fe.
Ann: 78,000 min TS; 30,000 min YS; 40 min El; 175 VDH.
Free machining austenitic stainless for machined parts as studs, cap screw shafts.
EN 58 AM; AISI 303.

STAYBRITE F.S.T.(L).
M-521; 0.06 C, 0.6 Si, 1.5 Mn, 18.5 Cr, 10 Ni, bal Fe.
Annealed: 90,000 TS; 36,000 YP; 60 El; 65 RA; 160 Brinell.
For architectural trim, kitchen equipment, chemical and textile processing equipment.
Type 304 stainless steel, nonmagnetic, austenitic.

STAYBRITE F.S.T. (L) (H) STEEL.
M-521; 0.04-0.09 C, 17.5-19.0 Cr, 8.0-11.0 Ni, bal Fe.
Austenitic stainless steel.
Weldable; for elevated temperature operation.

STEAM BRONZE-1.
M-U.S.; 85 Cu, 5 Zn, 5 Pb, 5 Sn.
26,000 TS; 16 El; 20 RA.
For valves and fittings, automotive parts; castings.

STEAM BRONZE-2.
M-U.S.; 88 Cu, 2 Pb, 8 Sn.
32,000 TS; 22 El; 25 RA.
For valves, fittings, automobile parts; free-cutting.

STEAM METAL.
M-126; 88 Cu, 6 Sn, 3 Pb, 3 Zn.
For injectors, valves, steam specialties; free-cutting.

STEAM METAL, AJAX.
M-123; 87 Cu, 6 Sn, 5 Zn, 2 Pb.

STEAM VALVE BRONZE.
Sn, Zn, Pb, bal Cu,
For steam valves; pressure tight castings.

STEB METAL.
M-496; Ag, Cu.
For enameling and electrical contacts; silver rolled into copper.

STEDMETAL.
M-518; 3.2 C, 2.5 Si, 1.5 Ni, 0.7 Mn, bal Fe.
For crushers, grinders, pulverizers; Ni-cast iron.

STEEL 1KH14ND.
M-USSR; 0.05 C, 13.5 Cr, 1.5 Ni, 1.37 Cu, bal Fe.
For stainless parts; stainless, austenitic.

STEEL 1KH18N9T.
M-USSR; C, Ni, Cr, Ti, bal Fe.
For stainless parts; stainless.

STEEL 15 CR 3.
M-Ger.; 0.12-0.18 C, 0.5 Mn, 0.20 Si, 0.5-0.8 Cr, bal Fe.
Heat treated: 85,000-120,000 TS.
For gears, shafts, machine tool parts; case hardening steel, water hardened.

STEEL 15 CR NI 6.
M-Ger.; 0.12-0.17 C, 0.5 Mn, 0.20 Si, 1.4-1.7 Ni, 1.4-1.7 Cr, bal Fe.
Heat treated: 140,000-185,000 TS.
For gears, cams, shafts, camshafts; case-hardening steel, oil hardened.

STEEL 16 MN CR 5.
M-Ger.; 0.14-0.19 C, 1.0-1.3 Mn, 0.25 Si, 0.8-1.1 Cr, bal Fe.
Heat treated; 115,000-155,000 TS.
For gears, shafts, machine tool parts, case hardening steel, oil hardened.

STEEL 18 CR NI 8.
M-Ger.; 0.15-0.20 C, 0.5 Mn, 0.20 Si, 1.8-2.1 Ni, 1.8-2.1 Cr, bal Fe.
Heat treated: 170,000-205,000 TS.
For gears, shafts, cams, pinions, camshafts; case hardening steel, oil hardened.

STEEL 20 MN CR 5.
M-Ger.; 0.17-0.22 C, 1.1-1.4 Mn, 0.20 Si, 1.0-1.3 Cr, bal Fe.
Heat treated: 127,000-170,000 TS.
For gears, pinions, cams, shafts, camshafts; case hardening steel, oil hardened.

STEEL 20-XGP.
M-USSR; C, alloy, bal Fe.
Rolled: 144,000 TS; 114,000 YS.
For machine tool parts; case hardened.

STEEL 20-XHM.
M-USSR; C, alloy, bal Fe.
For machine tool parts; case hardened.

STEEL 25 CR MO 4.
M-Ger.; 0.22-0.29 C, 0.7 Mn, 0.25 Si, 0.9-1.2 Cr, 0.15-0.25 Mo, bal Fe.
Heat treated: 115,000-135,000 TS.
For construction parts and equipment; oil hardened, shock resistant.

STEEL 30 CR MO V 9.
M-Ger.; 0.26-0.34 C, 0.6 Mn, 0.25 Si, 2.3-2.7 Cr, 0.15-0.25 Mo, 0.1-0.2 V, bal Fe.
Heat treated: 180,000-207,000 TS.
For construction equipment, gears, crankshafts, axles; oil hardened, shock resistant.

STEEL 30 CR NI MO 8.
M-Ger.; 0.26-0.34 C, 1.8 -2.1 Ni, 1.8-2.1 Cr, 0.25-0.35 Mo, bal Fe.
Heat treated: 177,000-207,000 TS.
For gears, shafts, crankshafts, axles; oil hardened, shock resistant.

STEEL 30 MN 5.
M-Ger.; 0.27-0.34 C, 1.2-1.5 Mn, 0.15-0.35 Si, bal Fe.
Heat treated: 115,000-135,000 TS.
For gears, shafts, machine tools, axles; oil hardened.

STEEL 34 CR 4.
M-Ger.; 0.30-0.37 C, 0.5-0.8 Mn, 0.15-0.35 Si, 0.9-1.2 Cr, bal Fe.
Heat treated: 127,000-150,000 TS.
For construction parts and equipment; oil hardened.

STEEL 34 CR MO 4.
M-Ger.; 0.30-0.37 C, 0.7 Mn, 0.25 Si, 0.9-1.2 Cr, 0.15-0.25 Mo, bal Fe.
Heat treated: 127,000-135,000 TS.
For construction parts and equipment; oil hardened, shock resistant.

STEEL 34 CR NI MO 6.
M-Ger.; 0.30-0.38 C, 1.4-1.7 Ni, 1.4-1.7 Cr, 0.15-0.25 Mo, bal Fe.
Heat treated: 157,000-185,000 TS.
For gears, shafts, crankshafts, axles; oil hardening, shock resistant.

STEEL 36 CR NI MO 4.
M-Ger.; 0.32-0.40 C, 0.9-1.2 Ni, 0.9-1.2 Cr, 0.15-0.25 Mo, bal Fe.
Heat treated: 142,000-170,000 TS.
For construction parts, gears, crankshafts; oil hardened, shock resistant.

STEEL 37 MN SI 5.
M-Ger.; 0.33-0.61 C, 1.1-1.4 Mn, 1.1-1.4 Si, bal Fe.
Heat treated: 127,000-150,000 TS.
For construction parts, machine tools; oil hardened.

STEEL 40 MN 4.
M-Ger.; 0.36-0.44 C, 0.8-1.1 Mn, 0.25-0.50 Si, bal Fe.
Heat treated: 115,000-135,000 TS.
For gears, shafts, machine tools, axles; oil hardening.

STEEL 41 CR 4.
M-Ger.; 0.38-0.44 C, 0.5-0.8 Mn, 0.15-0.35 Si, 0.9-1.2 Cr, bal Fe.
Heat treated: 127,000-150,000 TS.
For construction parts and equipment; oil hardened.

STEEL 42 CR MO 4.
M-Ger.; 0.38-0.45 C, 0.7 Mn, 0.25 Si, 0.9-1.2 Cr, 0.15-0.25 Mo, bal Fe.
Heat treated: 142,000-170,000 TS.
For construction parts and equipment; oil hardened, shock resistant.

STEEL 42 MN V 7.
M-Ger.; 0.38-0.45 C, 1.6-1.9 Mn, 0.15-0.35 Si, 0.07-0.12 V, bal Fe.
Heat treated: 142,000-170,000 TS.
For construction parts and equipment; oil hardened.

STEEL 50 CR MO 4.
M-Ger.; 0.46-0.54 C, 0.7 Mn, 0.25 Si, 0.9-1.2 Cr, 0.15-0.25 Mo, bal Fe.
Heat treated: 157,000-185,000 TS.
For construction parts and equipment; oil hardened, shock resistant.

STEEL 50 KHFA.
M-USSR; C, bal Fe.
For springs; water hardened.

STEEL 55C2.
M-USSR; C, alloy, bal Fe.
For machine tool parts.

STEELARC see **ALL-STATE STEELARC.**

STEEL EC30.
M-Ger.; 0.10-0.16 C, 0.3-0.5 Cr, 0.4-0.6 Mn, 0.35 max Si, bal Fe.
Heat treated: 78,000-100,000 TS.
For gears, pinions, shafts, cams; case hardening, water hardened.

STEEL EC60.
M-Ger.; 0.12-0.18 C, 0.6-0.9 Cr, 0.5 Mn, 0.35 max Si, bal Fe.
Heat treated: 100,000-127,000 TS.
For gears, pinions, shafts, cams; case hardening, water hardened.

STEEL ECMO 80.
M-Ger.; 0.13-0.17 C, 1.0-1.3 Cr, 0.2-0.3 Mo, 0.8-1.1 Mn, bal Fe.
Heat treated: 120,000-155,000 TS.
For gears, shafts, pinions, cams; case hardening, oil hardened.

STEEL ECMO 100.
M-Ger.; 0.18-0.23 C, 0.9-1.2 Mn, 1.1-1.4 Cr, 0.3 Mo, bal Fe.
Heat treated: 155,000-205,000 TS.
For gears, pinions, shafts, cams; case hardening, oil hardened.

STEEL ECN25.
M-Ger.; 0.10-0.17 C, 2.25-2.75 Ni, 0.55-0.95 Cr, bal Fe.
Heat treated: 115,000-155,000 TS.
For gears, pinions, shafts, cams; case hardening, shock resistant.

STEEL ECN35.
M-Ger.; 0.10-0.17 C, 3.25-3.75 Ni, 0.55-0.95 Cr, bal Fe.
Heat treated: 127,000-170,000 TS.
For gears, pinions, shafts, cams, crankshafts; case hardening, shock resistant.

STEEL ECN45.
M-Ger.; 0.10-0.17 C, 4.25-4.75 Ni, 0.90-1.35 Cr, bal Fe.
Heat treated: 170,000-200,000 TS.
For gears, pinions, shafts, cams, camshafts; case hardening, shock resistant.

STEEL EI-257.
M-USSR; 0.11 C, 13 Cr, 12 Ni, 2.5 W, 0.68 Mo, bal Fe.
For steam tubes, boilers.

STEEL EN15.
M-Ger.; 0.10-0.17 C, 1.25-1.75 Ni, 0.2 max Cr, bal Fe.
Heat treated: 85,000-115,000 TS.
For gears, pinions, shafts, cams; case hardening, shock resistant, water hardened.

STEELTECTIC.
M-717.
Electrode for AC-DC metallic arc pass over pass welding on mild steel without slag chipping; thin sections; 80,000 psi TS.

STEFANITE A2.
M-1188; carbide.
For cutting tools; for high speed finishing.

STEFANITE A3.
M-1188; carbide.
For cutting tools; for medium speed cutting.

STEFANITE A4.
M-1188; carbide.
For cutting tools; medium speed machining.

STEFANITE A6.
M-1188; carbide.
For cutting tools; for roughing and finishing.

STEFANITE A9.
M-1188; carbide.
For cutting tools; for low speed rough turning.

STEFANITE F6.
M-1188; carbide.
For cutting tools; for cutting cast iron.

STELCO CB45-60.
M-1124; 0.20 C, 1.2 Mn, 0.005 min Cb, bal Fe.
Rolled: CB45: 60,000 TS; 45,000 YS; 25 El. CB50: 65,000 TS; 50,000 YS; 22 El. CB55: 70,000 TS; 55,000 YS; 20 El. CB60: 75,000 TS; 60,000 YS; 18 El.
For mine and railroad cars, pressure vessels, derricks, booms, bridges.
Good fabricability and weldability.

STELCO COLUMBIUM.
M-1124; 0.20 C, 1.2 Mn, 0.005 min Cb, bal Fe.
Rolled: 60,000-75,000 TS; 45,000-60,000 YS; 18-25 El.
For bridges, pressure vessels, booms, mine cars, railroad and bus bodies.
High strength low alloy steel, good formability and weldability.

STELCOLOY 50.
M-1124; 0.15 max C, 1.35 max Mn, 0.15-0.30 Si, 0.20-0.50 Cu, 0.20-0.50 Ni, bal Fe.
A low alloy high-strength steel with emphasis on resistance to brittle fracture as well as atmosphere corrosion.
Rolled: 70,000 TS; 50,000 YS; 22 El.
For welded, riveted or bolted construction, especially welded bridges and buildings.

STELCOLOY 60.
M-1124; 0.20 C, 1.50 Mn, 0.15-0.30 Si, 0.20-0.50 Cu, 0.30-0.50 Cr, 0.25-0.50 Ni, 0.01-0.12 V, bal Fe.
Wrought: 80 ksi TS; 60 ksi YS; 18 El.
HSLA steel, improved atmosphere corrosion resistance.

STELCOLOY 60.
M-1124; 0.20 max C, 1.36 max Mn, 0.15-0.30 Si, 0.20-0.50 Cu, 6.25-6.50 Ni, 0.30-0.50 Cr, bal Fe.
Rolled: 80,000 TS; 60,000 YS; 18 El (8").
Low alloy high strength steel with emphasis on resistance to brittle fracture as well as atmosphere corrosion.
For welded, riveted or bolted construction, especially welded bridges and buildings.

STELCOLOY 70.
M-1124; 0.22 max C, 1.5 max Mn, 0.15-0.30 Si, 0.20-0.50 Cu, 0.25-0.50 Ni, 0.05 max Cb, 0.02-0.10 V, bal Fe.
A low alloy, high strength steel especially designed for the transportation industry.
Rolled: 90,000 TS; 70,000 YS; 14 El (8").

STELCO VANADIUM 441.
M-1124; 0.20 C, 1.2 Mn, 0.02 min V, bal Fe.
Rolled: 65,000 TS; 50,000 YS; 22 El.
For bridges, booms, pressure vessels, mine and railroad cars.
High strength low alloy steel, tough, good weldability and formability.

STELLAR.
M-1704; 0.34 C, 0.80 Mn, 0.50 Si, 0.90 Cr, 0.35 Mo, 0.33 Cu.
Low alloy, oil hardening tool steel.

STELLITE 1.
M-1580; 2.4 C, 33 Cr, 13 W, bal Co.
Cast: 95,000 TS; C 56 Rock.
For hard facing applications.
Heat, corrosion and wear resistant.

STELLITE 3.
M-1580; 2.4 C, 30 Cr, 13 W, bal Co.
Cast: 90,000 TS; 51-58 Rc.
For pump sleeves, rotary seal rings, wear pads, bearing sleeves.
High abrasion and corrosion resistant. DTD 4732.

STELLITE 4.
M-1580; 33 Cr, 14 W, 1 C, bal Co.
Cast: 130,000 TS; 45-49 Rc.
High temperature strength and wear resistant castings.

STELLITE 6.
M-1580; 26 Cr, 5 W, 1 C, bal Co.
Cast: 116,000 TS; 1 El; 400 Brin.
For weld deposits on steam and chemical valves, shear blades, tong bits.
Corrosion and heat resistant.
DTD 4733.

STELLITE 7.
M-1580; 26 Cr, 6 W, 0.4 C, bal Co.
Cast: 120,000 TS; 30-35 Rock C.
For gas turbine blades, brass casting dies, extrusion dies.
Resistant to thermal shock.
Excellent corrosion resistance and high temperature strength.
DTD 4736.

STELLITE 8.
M-1580; 0.2 C, 27 Cr, 6 Mo, 63.0 Co, 2 Ni.
Cast: 120,000 TS; 30-35 Rock C.
For gas turbine blades, brass casting dies, extrusion dies.
Cast alloy. High temperature strength and excellent corrosion resistance.

STELLITE 12.
M-1580; 1.8 C, 9 W, 29 Cr, bal Co.
Cast: 10,000 TS; 510 Brin.
For hard facing electrodes.
Good high temperature strength.
Heat and wear resistant.
DTD 4734.

STELLITE 12P.
M-1580; 31 Cr, 9 W, 1.4 C, bal Co.
Hardfacing alloy; as deposited 43-48 RC.

STELLITE 20.
M-1580; 33 Cr, 18 W, 2.5 C, bal Co.
Cast: 80,000 TS; 55-59 Rock C.
High abrasion and corrosion resistance.
DTD 4737.

STELLITE 100.
M-1580; 34 Cr, 19 W, 2 C, bal Co.
Cast: 56,000 TS; 61-66 Rc.
For tool bits, milling cutters, blades.

STELLITE 208.
M-1580; 49 Co, 26 Cr, 3 Mo, 20 Fe.
556 Mn/m^2; 10 El; 24-29 Rc; 210-245 DPH.
Casting alloy for gas turbine blades, brass casting dies and extrusion dies.

STELLITE 250.
M-1580; 28 Cr, 0.1 C, 20 Fe, bal Co.
As cast; 541 Mn/m^2 TS; 309 MN/m^2 YS; 8 El; 19-29 Rc.
Casting alloy for turbine blades, brass casting dies, extrusion dies.

STELLITE 251.
M-1580; 28 Cr, 0.3 C, 18 Fe, 2 Nb, bal Co.
As Cast: 618 MN/m^2 TS; 463 MN/m^2 YS; 4 El; 23-35 Rc.
Casting alloys for turbine blades, brass casting dies, extrusion dies.

STELLITE 306.
M-1580; 60 Co, 25 Cr, 2 W, 0.4 C, 5 Ni, 6 Nb.
As cast: 34-39 Rc; 345-380 DPH.
Hard facing alloy having good thermal and mechanical shock.

STELLITE 506.
M-1580; 55 Co, 35 Cr, 7.5 W, 1.6 C.
As cast: 39-43 Rc; 380-425 DPH.
For hard surfacing.

STELLITE ALLOY NO.40 see HAYNES ALLOY NO.40.

STELLITE ETC see also HAYNES STELLITE and DELORO STELLITE.

STELLITE F.
M-1580; 2.0 C, 0.30 Mn, 1.0 Si, 25.0 Cr, 22.0 Ni, 1.0 Fe, 0.60 Mo, 12.0 W, bal Co.
For hard facing of combustion engine valves to give enhanced resistance to corrosion and erosion; 40-45 Rock C equivalent.

STELLITE SF1.
M-1580; 45 Co, 19 Cr, 13 W, 1.3 C, 13 Ni, 3 Si, 2.5 B.
As cast: 54-58 Rock C.
For hard facing; wear and abrasion resistant.
DTD 4655 A.

STELLITE SF6.
M-1580; 50 Co, 19 Cr, 8 W, 0.7 C, 13 Ni, 3.0 Si, 1.7 B.
Cast: 43-46 Rock C.
For hard facing: wear and abrasion resistant; machinable.
DTD 4756.

STELLITE SF12.
M-1580; 52 Co, 19 Cr, 9 W, 1.0 C, 13 Ni, 2.5 Si, 1.5 B.
Cast: 48-50 Rock C.
For hard facing; wear and abrasion resistant; machinable with difficulty.
DTD 4757.

STELLITE SF20.
M-1580; 42 Co, 19 Cr, 15 W, 1.5 C, 13 Ni, 3 Si, 3 B.
Cast: 60-62 Rock C.
For hard facing, wear and abrasion resistant; not machinable.
DTD 4784.

STELLITE X40.
M-1580; 26 Cr, 7 W, 0.5 C, 10 Ni, bal Co.
Cast: 96,000 TS; 30-35 Rock C.
Corrosion resistant, high temperature strength and resistance to thermal shock.
DTD 4824.

STELLIT TTS NO. 1.
M-USSR; Co-W-Cr.
For cutting tools; hard sintered alloy.

STELLIT TTS NO. 2.
M-USSR; Co-W-Cr.
For cutting tools; hard sintered alloy.

STELLRAM.
M-1385; WC.
For cutting tools; sintered carbide.

STELMAX 45.
M-1124; 0.15 C, 1.5 Mn, 0.005 min Cb, 0.01 min V, bal Fe.
Wrought: 55 ksi TS; 45 ksi YS; 25 El.
HSLA steel; ASTM A715.

STELMAX 50.
M-1124; 0.15 C, 1.5 Mn, 0.005 min Cb, 0.01 min V, bal Fe.
Wrought: 60 ksi TS; 50 ksi YS; 24 El.
HSLA steel; ASTM A715.

STELMAX 60.
M-1124; 0.15 C, 1.5 Mn, 0.005 min Cb, 0.01 min V, bal Fe.
Wrought: 70 ksi TS; 60 ksi YS; 22 El.
HSLA steel; ASTM A715.

STELMAX 70.
M-1124; 0.15 C, 1.5 Mn, 0.005 min Cb, 0.01 min V, 0.02 N, bal Fe.
Wrought: 80 ksi TS; 70 ksi YS; 20 El.
HSLA steel; ASTM A715.

STELMAX 80.
M-1124; 0.15 C, 1.5 Mn, 0.005 min Cb, 0.01 min V, 0.02 N, bal Fe.
Wrought: 90 ksi TS; 80 ksi YS; 18 El.
HSLA steel; ASTM A715.

STEMALLOY.
M-264; Sn, alloy, bal Cu.
For valve stem and seats, globe valves; valve bronze, corrosion and wear resistant.

STENTOR see also **CARPENTER STENTOR.**

STEPHENSON-1.
M-Eng.; 84 Cu, 8.3 Zn, 2.9 Pb, 0.4 Fe.
30,000 TS; 17 El; 19 RA.
For piston rings; P.L. 16,000.

STEPHENSON-2.
M-Eng.; 31 Sn, 31 Fe, 19 Cu, 19 Zn.
For castings.

STEPHENSON LOCOMOTIVE BEARING.
M-Eng.; 79.5 Cu, 7.5 Sn, 5 Zn, 8 Pb.
For bearings for locomotives; heavy duty.

STEPHENSON PISTON RINGS.
M-Eng.; 84 Cu, 8.3 Sn, 2.9 Sn, 4.3 Pb, 0.4 Fe.
For piston rings; tough.

STERCON-1000.
M-57; 0.07 C, 19 Cr, 14 Ni, 4.3 Mo, 3 Ti, 1.3 Al, 1.0 Fe, bal Ni.
For high temperature applications; corrosion and heat resistant.

STEREOTYPE METAL.
60 Sn, 35 Pb, 5 Sb.
For type metal.

STEREOTYPE METAL.
82 Pb, 14.8 Sb, 3.2 Sn.
12,000 TS; 4 El; 22 Brin.
For type metal; M.P. 468°F.

STEREOTYPE METAL.
82 Pb, 6 Sn, 12 Sb.
For type.

STEREOTYPE METAL.
76 Pb, 4 Sn, 20 Sb.
For type.

STEREOTYPE METAL.
70 Pb, 7 Sn, 23 Sb.
For type.

STEREOTYPE METAL.
67 Pb, 17 Sn, 16 Sb.
For type.

STERLIN.
69 Cu, 18 Ni, 13 Zn, 0.8 Pb.
For base for plated tableware; corrosion resistant.

STERLINE.
68-68.5 Cu, 13.3-12.8 Zn, 18 Ni, 0.75-0.80 Fe, 0-0.8 Pb.
For tableware, architectural trim; corrosion resistant.

STERLING.
M-Eng.; 62 Sn, 15 Zn, 11 Al, 8.3 Pb, 2.5 Cu, 1.2 Sb.
For solder.

STERLING.
M-904; 0.1 C, 18 Cr, 8 Ni, bal Fe.
Cast: 75,000 TS; 30,000 YS; 25 El; 40 RA; 150 Brin.
For chemical plant equipment; Type 302; stainless, austenitic.

STERLING 18-8S.
M-57; 0.08 max C, 18-20 Cr, 8-10 Ni, bal Fe.
For stainless and heat resistant parts; stainless, malleable.

STERLING 304 + SI see **SO 7.1002 (304 + SI).**

STERLING "M" DRILL STEEL.
M-57; 1.25 C, 0.3 Mn, 0.25 Cr, bal Fe.
For drills, taps; water hardened.

STERLING METAL.
66 Cu, 33-27 Zn, 0.7 Fe, 0-2 Pb, Sn.
For hardware, fittings; free-cuttings.

STERLING SILVER.
M-63; 92.5 Ag, 7.5 Cu.
Sand cast: 30,000 TS; 18,000 YS; 41 El; 55 RA.
Chill cast: 24,000 TS; 19,000 YS; 7 El; 16 RA.
For silverware, coins, jewelry; corrosion resistant.

STERLING SILVER SOLDER.
80 Ag, 18 Zn, 2.5 Cu.
For silver solder; corrosion resistant.

STERLING STAINLESS 410.
M-57; 0.15 max C, 1 max Mn, 11.5-13.5 Cr, bal Fe.
Hardened: 190,000 TS; 145,000 YS; 15 El; 55 RA; 390 Brin.
For valve parts, pump rods, pistons, shafts, bolts; corrosion resistant, hardenable; Type 410.

STERLING STAINLESS 416.
M-57; 0.15 max C, 1.25 max Mn, 12-14 Cr, 0.07 min S, P or Se, 0.6 max Mo or Zr, bal Fe.
Hardened: 110,000 TS; 85,000 YS; 18 El; 55 RA; 230 Brin.
For screw machine products; corrosion resistant, hardenable, free-cutting; Type 416.

STERLING STAINLESS 420.
M-57; 0.15 min C, 1 max Mn, 12-14 Cr, bal Fe.
Hardened: 230,000 TS; 195,000 YS; 8 El; 25 RA; 500 Brin.
For cutlery, needles, optical mirrors; corrosion resistant, hardenable; Type 420.

STERLING STAINLESS 430.
M-57; 0.12 max C, 1 max Mn, 14-18 Cr, bal Fe.
Cold drawn: 85,000 TS; 70,000 YS; 20 El; 65 RA; 185 Brin.
For corrosion and heat resistant parts; non-hardenable, corrosion resistant; Type 430.

STERLING STAINLESS 440-A.
M-57; 0.60-0.75 C, 16-18 Cr, 0.75 max Mo, bal Fe.
Hardened: 260,000 TS; 240,000 YS; 5 El; 20 RA; 510 Brin.
For valve parts, knives, ball bearings; corrosion resistant, hardenable; Type 440-A.

STERLING STAINLESS 440-B.
M-57; 0.75-0.95 C, 16-18 Cr, 0.75 max Mo, bal Fe.
Hardened: 280,000 TS; 270,000 YS; 3 El; 15 RA; 555 Brin.
For valve parts, knives, ball bearings; corrosion resistant, hardenable; Type 440-B.

STERLING STAINLESS 440-BMO.
M-57; 0.85 C, 0.4 Mn, 17 Cr, 0.2 V, 2.5 Mo, bal Fe.
Hardened: 280,000 TS; 270,000 YS; 3 El; 15 RA; 590 Brin.
For oil refineries, valve parts; corrosion resistant, hardenable.

STERLING STAINLESS 440-BV.
M-57; 0.75-0.95 C, 0.3-0.5 Mn, 16-18 Cr, 0.4-0.6 Mo, 0.15-0.25 V, bal Fe.
Hardened: 280,000 TS; 270,000 YS; 3 El; 15 RA; 590 Brin.
For cutlery, knives; corrosion resistant, hardenable.

STERLING STAINLESS 440-CM.
M-57; 0.95-1.20 C, 1 max Mn, 16-18 Cr, 0.75 max Mo, bal Fe.
Hardened: 285,000 TS; 275,000 YS; 2 El; 10 RA; 580 Brin.
For bushings, ball races, valve parts, cutlery; corrosion resistant, hardenable; Type 440-C.

STERLING STAINLESS 440-CX.
M-57; 0.95-1.20 C, 1 max Mn, 16-18 Cr, 0.75 max Mo, bal Fe.
Hardened: 285,000 TS; 275,000 YS; 2 El; 10 RA; 610 Brin.
For ball bearings, cutlery; corrosion resistant, hardenable; Type 440-C.

STERLING STAINLESS A.
M-57; 0.35 C, 0.35 Mn, 12-15 Cr, bal Fe.
Oil treated: 240,000-150,000 TS; 200,000-125,000 YS; 4-12 El; 8-40 RA; 500-300 Brin.
Annealed: 100,000 TS; 65,000 YS; 27 El; 60 RA; 185 Brin.
For cutlery, sharp edged or pointed parts, scissors, surgical and dental instruments; stain resisting, tough and hard; max. operating temperature 1650°F.

STERLING STAINLESS B.
M-57; 16 Cr, 0.35 Mn, 0.65 C, bal Fe.
Oil treated: 270,000-130,000 TS; 245,000-100,000 YS; 2-12 El; 3-30 RA; 545-285 Brin.
Annealed: 95,000 TS; 54,000 YS; 27 El; 45 RA; 185 Brin.
For cutlery, surgical and dental instruments, ball bearings, valve seats; max. operating temperature 1650°F; requires heat treatment for stainless properties.

STERLING STAINLESS BHH.
M-57; 1.05 C, 17.5 Cr, bal Fe.
For ball races, valve parts, bearing surfaces, machine parts; heat and corrosion resistant; hard grade.

STERLING STAINLESS M.
M-57; 0.12 max C, 15-18 Cr, bal Fe.
Annealed: 100,000 TS; 70,000 YS; 25 El; 65 RA; 200 Brin.
For heat resisting parts, corrosion resisting parts; stain resisting without heat treatment; Iz-40.

STERLING STAINLESS MG.
M-57; 0.35 max C, 0.50 Mn, 18-23 Cr, bal Fe.
For shafts, pump rods, furnace parts, glass molds; high temperature resistance; corrosion resistant.

STERLING STAINLESS "NI-T".
M-57; 0.15 max C, 11-13 Cr, 1.2-2.5 Ni, bal Fe.
For gun barrels, pistons, bolts, pump rods; corrosion resistant, hardenable.

STERLING STAINLESS T.
M-57; 12.5 Cr, 0.35 Ni, 0.35 Mn, 0.12 C, bal Fe.
Oil treated: 185,000-120,000 TS; 160,000-100,000 YS; 17-22 El; 60-70 RA; 395-240 Brin.
Normalized: 145,000 TS; 125,000 YS; 20 El; 63 RA; 241 Brin.
For pump rods, machinery parts, turbine blades, shafts, pistons, gun barrels; resists stain without heat treatment; max operating temperature 1650°F.

STERLING "V".
M-57; 0.9 C, 0.35 Mn, 0.15 Cr, 0.15 V, bal Fe.
For punches, springs, shear blades, taps; water hardened.

STERLING XX.
M-57; 0.95 C, 0.30 Mn, 0.05 Cr, bal Fe.
For tools, dies; water hardened.

STERLITE.
M-201; 53 Cu, 25 Ni, 20 Zn, Mn.
53,650 TS; 27,250 YS; 33 El; 35 RA; 99 Brin.
For ship fittings, laundry and dairy machinery, surgical instruments; tough and corrosion resisting.

STERMET GR. 1.
M-1008; 65-70 Ni, 17-20 Cr, bal Fe.
For heat resistant castings, heat treating and carburizing pots; for heat shock.

STERMET GR. 2.
M-1008; 58-62 Ni, 10-14 Cr, bal Fe.
Cast: 60,000-75,000 TS; 30,000-40,000 YS; 2-5 El; 2-6 RA; 185 Brin.
For castings, rolls, plates, enameling fixtures; for heat shock; heat resistant.

STERMET GR. 3.
M-1008; 35-40 Ni, 16-20 Cr, bal Fe.
For castings, rolls, rails, furnace parts; heat resistant.

STERMET GR. 4.
M-1008; 33-37 Ni, 13-17 Cr, bal Fe.
Cast: 60,000-75,000 TS; 35,000-50,000 YS; 4-6 El; 4-7 RA; 185 Brin.
For furnace parts, conveyors, belts, burner parts; heat resistant, castings.

STERMET GR. 5.
M-1008; 8-10 Ni, 28-30 Cr, bal Fe.
For stoker parts, dampers, skid rails, castings; heat resistant.

STERMET GR. 6.
M-1008; 10-14 Ni, 24-26 Cr, bal Fe.
Cast: 80,000-100,000 TS; 40,000-55,000 YS; 10-18 El; 10-20 RA, 175 Brin.
For heat resistant castings; for severe S atmospheres and abrasive conditions.

STERMET GR. 7.
M-1008; 18-22 Ni, 23-27 Cr, bal Fe.
For castings, furnace parts; heat and corrosion resistant.

STERMET GR. 8.
M-1008; 8-12 Ni, 26-30 Cr, bal Fe.
For furnace parts, skid rails, castings; heat resistant.

STERMET GR. 9.
M-1008; 0-3 Ni, 27-30 Cr, bal Fe.
Cast: 40,000-55,000 TS; 35,000-45,000 YS; 1.0 El; 1.5 RA; 180 Brin.
For castings, salt pots; heat resistant.

STERMET GR. 10.
M-1008; 8-10 Ni, 18-20 Cr, bal Fe.
Cast: 80,000-95,000 TS; 30,000-45,000 YS; 40-60 El; 50-60 RA; 180 Brin.
For food and dairy equipment; heat and corrosion resistant.

STERRO METAL.
M-Eng.; 60-55 Cu, 38-42 Zn, 1.8-4.7 Fe.
Forged: 78,200 TS; 27,800 YS; 39 El.
Cast: 60,500 TS.
For hydraulic cylinders subjected to high pressures; pressure-tight castings.

STERVAC 1000.
M-57; 0.08 C, 19.5 Cr, 4.3 Mo, 13.5 Co, 1.3 Al, 3.1 Ti, 1.25 Fe, bal Ni.
At 70°F: 188,000 TS; 115,000 YS; 28 El; 25 RA.
At 1400°F: 117,000 TS; 99,000 YS; 28 El; 41 RA.
For turbo blades and rotor discs, high temperature bolts; heat resistant to 1500°F.

STERVAC 2000.
M-57; 0.15 C, 19 Cr, 9.75 Mo, 10 Co, 1.1 Al, 2.5 Ti, 1.25 Fe, 0.06 Zr, 0.005 Be, bal Ni.
At 70°F: 170,000 TS; 98,000 YS; 20 El; 22 RA.
At 1400°F: 120,000 TS; 92,000 YS; 20 El.
For jet engine and gas turbine buckets; heat resistant to 1500°F.

STERVAC 3000.
M-57; 0.08 C, 14 Cr, 33 Ni, 4 Mo, 0.25 Al, 0.25 Zr, 2 Ti, 6.5 W, bal Fe.
At 1200°F: 134,000 TS; 121,000 YS; 14 El; 20 RA.
For jet engine components; high heat resistance to 1350°F.

STERVAC 4000.
M-57; 0.09 C, 19 Cr, 52.5 Ni, 9.75 Mo, 11 Co, 1.65 Al, 3.15 Ti, 0.005 B, bal Fe.
At 1400°F: 120,000 TS; 110,000 YS; 8 El; 12 RA.
For jet engine components; useful up to 1650°F.

STERVAC 5000.
M-57; 0.15 max C, 3 Al, 3 Ti, 4 Mo, 18 Cr, 17 Co, 3 Fe, bal Ni.
At 70°F: 197,000 TS; 18 El; 22 RA.
At 1200°F: 175,000 TS; 8 El; 16 RA.
At 1800°F: 4,600 TS; 22 El; 41 RA.
For gas turbine components, bolts, valves; useful up to 1600°F, age-hardenable.

STIRLING CAST.
66.2 Cu, 33.11 Zn, 0.02 Pb, 0.66 Fe.
For hardware; high strength.

STITCHING WIRE.
74.5 Ni, 18 Cr, 7.5 Mn.
For electrical resistances; resistance alloy.

ST. JOE CHEMICAL LEAD.
M-1747; 0.05 Cu, 0.005 Ag, 0.0003 Cd, 0.0003 Ni, bal Pb.
Cast: 2400-2600 psi TS; 40-50 El; 5 Brin.
MP: 618°F; Density 0.41 lbs/in^3.
Chemical equipment; acid containers. Exceeds ASTM B29-55.

STM see **SIMONDS STM.**

STM.
M-1419; 0.15 max C, 11.5-14.0 Cr, 0.2-0.6 Mo, 0.6 max Ni, bal Fe.
Annealed: 80,000 TS; 38,000 YS; 25 El; B 92 Rock.
Cold Drawn: 100,000 TS; 80,000 YS; 15 El; B 96 Rock.
For springs, shafts, table flatware, oil refinery equipment. Corrosion resistant.

STMCO M-33.
M-373; 0.88 Co, 3.75 Cr, 1.75 W, 9.5 Mo, 1.2 V, 8.25 Co, bal Fe.
Co-Mo high speed steel for cutting tools; good red hardness.
AISI M33.

STOLLBERG BRASS.
M-Eng.; 32.8 Zn, 0.4 Sn, 2 Pb, bal Cu.
For forgings, hardware, fittings; good workability, free-cutting.

STONE BRONZE.
M-452; 58 Cu, 39 Zn, 1.5 Fe, 0.8 Al, 0.5 Mn, Sn.
For marine propellers, nuts, bolts; corrosion resistant.

STONES "608" SUPERHEAT BRONZE.
M-452; 7 Ni, bal Cu.
For parts exposed to superheated steam bucket wheels, pistons, locomotive slide valve, cocks; resists corrosion and erosion.

STONES ENGLISH GEAR BRONZE.
M-452; 89 Cu, 11 Sn, trace P.
Cast: 39,000-42,000 TS; 21,000-24,000 YS; 10-6 El; 9-7 RA; 80 Brin.
For gears, bearings; used for severe service; C.E.L. 16,000.

STONES GEAR BRONZE.
M-126; 11 Sn, 0.25 Pb, 0.25 P, bal Cu.
For gears, worm wheels; tough, wear resistant.

STONES Z-METAL.
M-452; Sn, Ni, bal Cu.
For fire box stays; heat resistant.

STONEWELL BABBITT.
M-565; Pb, Sb, Sn.
Cast: 10,650 TS; 7550 YS.
For bearings for heavy loads and pressures, machinery bearings; Babbitt metal.

STOODITE.
M-202; 4 C, 1.5 Si, 31 Cr, 4 Mn, bal Fe.
Weld: 600 Brin.
For hard facing rod for farm tools, crushers; abrasion resisting.

STOODY 1.
M-202; 2.5 C, 26 Cr, 10.5 W, 60 Co, 1 Fe.
Welded: 50,000 TS.
For hard facing electrode; abrasion, wear and heat resistant.

STOODY 6.
M-202; 1.2 C, 1 Si, 28 Cr, 5 W, 70 Co, 1 Fe.
Welded: 125,000 TS; 2 El; 400 Brin.
For hard facing electrodes; impact and corrosion resistant.

STOODY 31.
M-202; 2 C, 28 Cr, 4 Mo, bal Fe.
Coated electrode for out of position welding.
Hard facing for earth abrasion.

STOODY 33.
M-202; 2 C, 28 Cr, 4 Ni, bal Fe.
Coated electrode for hard facing.
For crack free deposits, earth abrasion hard facing.

STOODY 60.
M-202.
Nickel base hard facing powder.
As deposited: 50 Rc.

STOODY 63.
M-202.
Nickel base hard facing powder.
As deposited: 19 Rc.

STOODY 64.
M-202.
Nickle base hard facing powder.
As deposited: 40 Rc.

STOODY 65.
M-202.
Nickel base hard facing powder.
As deposited: 50 Rc.

STOODY 85.
M-202.
WC in STOODY 60 powder.
For severe earth abrasion and hard facing; matrix: 63 Rc.

STOODY 100.
M-202; 36 (Cr, Mo, Mn, C, Si), bal Fe.
Cast: 520-580 Brin.
For crushers, tool joints; hard facing electrode.

STOODY 100 HC.
M-202; 5 C, 28 Cr, 1 Mo, bal Fe.
Open arc wire electrode. Hard facing for earth abrasion.

STOODY 101.
M-202; 2 C, 28 Cr, 8 Ni, 1.5 W, 0.6 Mn, 1 Si, bal Fe.
Cast: 350-400 Brin.
For electrode for submerged arc; hard facing.

STOODY 102.
M-202; 0.3 C, 5 Cr, 1.5 W, 1.5 Mn, 0.75 Mn, 1 Si, bal Fe.
Cast: 520-550 Brin.
For electrode for submerged arc; hard facing.

STOODY 103.
M-202; 4 C, 30.5 Cr, 3.5 Mn, 1.2 Si, bal Fe.
560 Brin.
For welding electrodes; hard facing.

STOODY 104.
M-202; 3.2 (Mn, Mo, Si, C), bal Fe.
Cast: 270-320 Brin.
For tractor rollers, idlers, trunnions; hard facing electrode.

STOODY 105.
M-202; 8 (Cr, Mo, V, Mn, Si, C), bal Fe.
Forged: 200,000 TS; 14 El; 14 RA; 470 Brin.
For hard facing electrodes, rollers and idlers for tractors and shovels; abrasion and impact resistant.

STOODY 105B.
M-202; 0.25 C, 3 Cr, 1.5 Ni, 0.5 Mo, 0.3 V, 1.25 Mn, 0.5 Si, bal Fe.
Cast: 500-520 Brin.
For electrode for submerged arc; hard facing.

STOODY 106.
M-202; 13 (Ni, Mo, Si, Mn, C), bal Fe.
Cast: 140,000 TS; 400 Brin.
For hard facing electrode for steel mill roll necks, cable drums; abrasion and impact resistant.

STOODY 107.
M-202; 6.5 (Cr, Mn, Mo, Si, C), bal Fe.
Cast: 360-410 Brin.
For hard facing electrode.

STOODY 110.
M-202; 16 Mn, 16 Cr, bal Fe.
Open arc wire electrode.
For joining and rebuilding dissimilar metals, carbon and austenitic manganese steels.

STOODY 121.
M-202; 2.5 C, 15 Cr, 2 Mn, 1 Si, 0.25 Zr, bal Fe.
Cast: 560 Brin.
For electrode for submerged arc; hard facing.

STOODY 130.
M-202; 60 WC in 40 steel matrix.
For electrode for open arc; hard facing.

STOODY 131.
M-202; 2 C, 28 Cr, 4 Mo, bal Fe.
Open arc wire electrode.
Hard facing for earth abrasion.

STOODY 133.
M-202; 2 C, 28 Cr, 4 Ni, bal Fe.
Open arc wire electrode.
For crack free deposits, earth abrasion hard facing.

STOODY 1027.
M-202; 1.5 C, 4.5 Cr, 1.2 Mn, 1 Si, bal Fe.
For hard facing electrodes; impact resistant.

STOODY 2110.
M-202; 16 Mn, 16 Cr, bal Fe.
Coated electrode.
For joining and rebuilding dissimilar metals, carbon and austenitic manganese steels.

STOODY DYNAMANG.
M-202; 14 Mn, 4 Ni, 4 Cr, bal Fe.
Coated rod and open arc welding wire.
For joining and rebuilding austenitic manganese steel.

STOODY MANGANESE.
M-202; 0.7 C, 14 Mn, 5 Ni, bal Fe.
Cast: 82,000 TS; 60,000 YS; 15 El; 12 RA.
For welding electrodes for Mn steel; austenitic, wear resistant.

STOODY SELF-HARDENING.
M-202; 1 C, 0.1 Mo, 2 Mn, 6 Cr, 1.2 Si, bal Fe.
Weld: 550 Brin.
For welding rod, hard surfacing; hard-facing, tough; non-machinable.

STOODY SELF-HARDENING 21.
M-202; 3 C, 14 Cr, 2.5 Mn, 1.2 Si, bal Fe.
For welding electrodes, hard facing rod; severe abrasion resistance.

STOVALL.
72 Au, 16.5 Cu, 2 Pt, 13.5 Ni.
For dental alloy; corrosion resistant.

STRAIN TEMPERED 41L45.
M-516; 0.43-0.48 C, 0.75-1.0 Mn, 0.20-0.35 Si, 0.8-1.1 Cr, 0.15-0.25 Mo, 0.15-0.35 Pb, bal Fe.
Cold drawn, stress relieved.
150,000 TS min; 130,000 YS min; 10 El; 35 RA; 302 Brin.
For machine parts, shafts.
Eliminates heat treatment.

STRAIN TEMPERED 1045.
M-516; 0.43-0.50 C, 0.6-0.9 Mn, bal Fe.
Cold drawn, stress relieved. 125,000 TS min; 100,000 YS min.
For machine parts, shafts. Eliminates heat treatment.

STRAIN TEMPERED 1050.
M-516; 0.48-0.55 C, 0.6-0.9 Mn, bal Fe.
125,000 TS min, 100,000 YS min.
Cold drawn, stress relieved.
For machine parts, shafts.
Eliminates heat treatment.

STRAIN TEMPERED 1141.
M-516; 0.37-0.45 C, 1.35-1.65 Mn, 0.08-0.13 S, bal Fe.
Cold drawn, stress relieved.
125,000 TS min; 100,000 YS min.
For machine parts, shafts.
Free machining, eliminates heat treating.

STRAIN TEMPERED 1144.
M-516; 0.40-0.48 C, 1.35-1.65 Mn, 0.24-0.33 S, bal Fe.
Cold drawn, stress relieved.
125,000 TS min; 100,000 YS min.
For machine parts, shafts.
Free machining, eliminates heat treating.

STRAIN TEMPERED 1151.
M-516; 0.48-0.55 C, 0.70-1.0 Mn, 0.08-0.13 S, bal Fe.
Cold drawn, stress relieved.
125,000 TS min; 100,000 YS min.
For machine parts, shafts.
Free machining, eliminates heat treating.

STRAIN TEMPERED 4140.
M-516; 0.38-0.43 C, 0.75-1.0 Mn, 0.20-0.35 Si, 0.80-1.1 Cr, 0.15-0.25 Mo, bal Fe.
Cold drawn, stress relieved.
125,000 TS min; 105,000 YS min.
For machine parts, shafts.
Eliminates heat treatment.

S-TREATED.
M-64; 0.6 C, 0.3 Cr, bal Fe.
Prehardened: 201-235 Brin.
For die blocks; hot work, hammer press, short runs.

S-TREATED 14A60.
M-64; 0.60 C, 0.8 Mn, 0.25 Si, 0.25 Cr, bal Fe.
Water hardening tool steel; AISI W4.

STRENES B.
M-898; C, Cr, Ni, Mo, bal Fe.
For dies, bushings.

STRENES C.
M-898; 3 C, 2.2 Si, 0.9 Mn, 0.4 Mo, 0.4 Cr, bal Fe.
Cast: 50,000 TS; 190 Brin.
Heat treated: 75,000 TS; 550 Brin.
For pump impellers, fixtures, bushings; alloy cast iron.

STRENES D.
M-898; C, Cr, Ni, Mo, bal Fe.
For dies, bushings; heat resistant.

STRENES E.
M-898; 3 C, 1.5 Si, 14 Ni, 3 Cr, 6 Cu, bal Fe.

STRENICOR.
M-238; 3.5-5.0 Ni, 0.7-2.0 Si, 0.3-1.0 Fe, bal Cu.
Cast: 90,000 TS; 70,000 YS; 1 El; 8 RA.
Forged: 107,000 TS; 83,000 YS; 5 El; 7 RA.
For transformer terminal eye-bolts; corrosion resistant, no season cracking.

STRENLITE.
M-1124; 0.28 max C, 1.1-1.6 Mn, 0.31 max Si, 0.20 min Cu, bal Fe.
A high strength steel for construction of riveted or bolted bridges and buildings and for other structural purposes.
Rolled: 70,000 TS; 50,000 YS; 18 El.

STRESCON TM 195.
M-279; 97 Cu, 1.5 Fe, 0.8 Co, 0.6 Sn, 0.1 P.
Ann: 50,000-65,000 TS; 22,000-45,000 YS; 22-35 El.
Cold rolled: 60,000-100,000 TS; 35,000-87,000 YS; 3-24 El.
For electrical terminals, connectors, contact springs; high strength, high conductivity, corrosion resistant.

STRESSPROOF.
M-669, M-1124; 0.40-0.48 C, 1.35-1.65 Mn, 0.04 max P, 0.24-0.33 S, 0.15-0.30 Si, 0.25 max Cu, bal Fe.
Cold finished: 132,000 min TS; 100,000 min YS; 12 El; 83% machinability.
For bolts, shafts, studs; free machining.

STRIKING DIE.
M-275; 0.90 C, 0.30 Mn, 0.25 Si, bal Fe.
For hobs, punches, stamping dies, blanking and coining dies. Water hardening tool steel, AISI W1.

STROH.
M-899; 0.4 C, bal Fe.
For steel castings, gears; wear and shock resistant.

STROLOY 5C see **B&W STROLDY 5C.**

STRONGSET see **ALL-STATE NO 509 STRONGEST.**

STRONTIUM.
M-1755; Sr.
Purities: 99.95+%, 99.5-99.7%, 98%. Forms: Rod, lumps, wire.

STUDAL.
M-France; 97.7 Al, 1.0 Mg, 1.3 Mn.
For light alloy parts.

STUFFING BOX ALLOY.
M-Eng.; 61.5 Cu, 15.5 Ni, 11 Zn, 10 Pb, 2 Sn.
For stuffing boxes; corrosion resistant.

STYLUS METAL.
5 Sn, 20 Cu, 12 Sb, 63 Pb.

STYLUS METAL.
85 Pb, 5 Sn, 10 Sb.
For bearings; Babbitt.

STYRIA 779 now **VEW S302.**

STYRIA ACM now **VEW V800.**

STYRIA ACM-N now **VEW V810.**

STYRIA B100 now **VEW M112.**

STYRIA BCM now **VEW M-152.**

STYRIA BN3 now **VEW M120.**

STYRIA BN4 SPECIAL now **VEW M130.**

STYRIA BSP now **VEW P906.**

STYRIA CMG now **VEW W331.**

STYRIA CMH now **VEW V320.**

STYRIA CMM now **VEW V330.**

STYRIA CMS now **VEW V350.**

STYRIA CM SPEZIAL now **VEW V304.**

STYRIA CMW now **VEW V304.**

STYRIA CMZ now **VEW V320.**

STYRIA CMZH now **VEW V310.**

STYRIA CR35 now **VEW V510.**

STYRIA CR35H now **VEW V500.**

STYRIA CRD now **VEW K116.**

STYRIA CRD SPEZIAL now **VEW K105.**

STYRIA CRLS now **VEW K305.**

STYRIA CRS 815W now **VEW K301.**

STYRIA CR-V-SILBERSTAHL now **VEW K510.**

STYRIA CSF now **VEW F200.**

STYRIA CV5 now **VEW F550.**

STYRIA CVF now **VEW F550.**

STYRIA DGM now **VEW W322.**

STYRIA DPK now **VEW K247.**

STYRIA E60 now **VEW E525.**

STYRIA E80 now **VEW E304.**

STYRIA E100 now **VEW E300.**

STYRIA ECR now **VEW K204.**

STYRIA ECR2 now **VEW K201.**

STYRIA ECR EXTRA now **VEW K205.**

STYRIA EE now **VEW E204.**

STYRIA EM12 now **VEW K701.**

STYRIA ENC2 now **VEW E220.**

STYRIA ENC15 now **VEW E230.**

STYRIA ESPA now **VEW K721.**

STYRIA ESPAS now **VEW K722.**

STYRIA ESS now **VEW K978.**

Section I: Alloy Data / 1341

STYRIA EWP398 now **VEW W103**.

STYRIA EWPX now **VEW W100**.

STYRIA EWS1 now **VEW K405**.

STYRIA EXTRA HH now **VEW K993**.

STYRIA EXTRA MH now **VEW K992**.

STYRIA EXTRA ZH now **VEW K988**.

STYRIA EXTRA ZH SPEZIAL now **VEW K760**.

STYRIA EXTRA ZW now **VEW K976**.

STYRIA FLP now **VEW K607**.

STYRIA G3F now **VEW F912**.

STYRIA G4 now **VEW V969**.
 M-1291; 0.75 C, 0.25 Si, 0.55 Mn, bal Fe.
 For rails, punches, hammers, axes, springs; water hardened.

STYRIA G5 now **VEW V960**.

STYRIA G6 now **VEW V955**.

STYRIA G7 now **VEW V945**.

STYRIA G7S now **VEW V935**.

STYRIA G8 now **VEW V920**.

STYRIA G10 now **VEW V900**.

STYRIA GBK now **VEW K200**.

STYRIA HW7 now **VEW H160**.

STYRIA HW18 now **VEW H120**.

STYRIA HWDA now **VEW H100**.

STYRIA HWDN now **VEW H300**.

STYRIA KCN now **VEW V204**.

STYRIA KLR now **VEW K505**.

STYRIA KLS now **VEW R100**.

STYRIA KWSD now **VEW K331**.

STYRIA LH now **VEW M252**.

STYRIA LHP now **VEW W502**.

STYRIA P973 now **VEW K630**.
 M-1291, M-1114; 0.85 C, 0.6 Ni, bal Fe.
 For machine tool parts, cold heading dies, punches.
 Water hardening, wear resistant.

STYRIA PECO now **VEW K244**.

STYRIA PFM now **VEW K612**.

STYRIA RA now **VEW A505**.

STYRIA RA1 now **VEW A120**.

STYRIA RA1 SPEZIAL now **VEW A350**.

STYRIA RA1T now **VEW A300**.

STYRIA RA3 now **VEW A100**.

STYRIA RA3 EXTRA now **VEW A354**.

STYRIA RAF now **VEW H525**.

STYRIA RAFK now **VEW H550**.

STYRIA RAS (NB) now **VEW A750**.

STYRIA M-80 now **VEW E410**.

STYRIA M-90 now **VEW E400**.

STYRIA MAN now **VEW E930**.

STYRIA MCV now **VEW V564**.

STYRIA MNS now **VEW V762**.

STYRIA MNV now **VEW V742**.

STYRIA MP now **VEW K605**.
 M-1291, M-1114; 0.50 C, 1.0 Cr, 3.25 Ni, Mo, bal Fe.
 For cold working tools, upsetters, forging dies, gears, bolts; oil hardened, shock resistant.

STYRIA MSH now **VEW F105**.

STYRIA MSW now **VEW F100**.

STYRIAN BLUE LABEL.
 M-651; 0.9 C, 0.2 V, bal Fe.
 For cutting tools; water hardened.

STYRIA NCZ now **VEW V157**.

STYRIA NHP now **VEW W501**.

STYRIA NHP SPECIAL now **VEW W500**.

STYRIA NW3 now **VEW E200**.

STYRIA O5K now **VEW V953**.

STYRIA O7K now **VEW V946**.

STYRIA RAT now **VEW A700**.

STYRIA RAW now **VEW A500**.

STYRIA RAZ now **VEW A506**.

STYRIA RKH now **VEW N688**.

STYRIA RP1K now **VEW W326**.

STYRIA RS1 now **VEW N540**.

STYRIA RS2 now **VEW N530**.

STYRIA RS2M now **VEW N330**.

STYRIA RS2W now **VEW N320**.

STYRIA RS3 now **VEW N315**.

STYRIA RSK now **VEW N200**.

STYRIA RSMO now **VEW N242**.

STYRIA RSN now VEW N351.

STYRIA RSW now VEW N100.

STYRIA RSZ now VEW N310.

STYRIA S434 now VEW K760.

STYRIA SCV now VEW K510.

STYRIA SIMPLEX now VEW K608.

STYRIA SP5H now VEW K970.

STYRIA SP5W now VEW K960.

STYRIA SP6H now VEW K950.

STYRIA SP6W now VEW K945.

STYRIA SPAH now VEW E920.

STYRIA SPAW now VEW E900.

STYRIA SPG EXTRA now VEW W300.

STYRIA SPG EXTRA V now VEW W302.

STYRIA SPG SPECIAL W now VEW W304.

STYRIA SPMW now VEW V904.

STYRIA SS25K now VEW S610.

STYRIA SS39KN now VEW S203.

STYRIA SS 652 now VEW S600.

STYRIA SS 653 SPECIAL now VEW S607.
M-1291; 1.2 C, 6 W, 5 Mo, 3 V, 4 Cr, bal Fe.

STYRIA SS921 now VEW S401.

STYRIA SSF now VEW K240.

STYRIA SW1 now VEW K465.

STYRIA SWS now VEW K460.

STYRIA US-SILBERSTAHL now VEW K992.

STYRIA V66 now VEW H802.

STYRIA V101 now VEW H800.

STYRIA VCN now VEW V214.

STYRIA VK20N now VEW H730.

STYRIA VKS now VEW H700.

STYRIA VNC2 now VEW V145.

STYRIA VNC10 now VEW V165.

STYRIA VNC15 now VEW V155.

STYRIA WB now VEW M312.

STYRIA WCM now VEW D330.

STYRIA WCMV now VEW D240.

STYRIA WCMW now VEW V340.

STYRIA WD30 now VEW D500.

STYRIA WK5K now VEW W105.

STYRIA WKM 33 now VEW W320.

STYRIA WPS now VEW W706.

STYRIA W-SILBERSTAHL now VEW K405.

ST. Z1.
M-Swiss; 0.13 C, 0.05 S, 0.05 P, bal Fe.
Rolled: 52,000-64,000 TS; 34,000 YS; 25 El.
For bridge construction; good weldability.

SU-16.
M-1622; 11 W, 3 Mo, 2 Hf, 0.08 C, bal Cb.
At 78°F: 112,000 TS; 90,000 YS; 25 El; 43 RA.
At 1112°F: 90,500 TS; 64,500 YS; 15 El; 45 RA.
For jet engine and missile components.
High heat resistant. Good rupture strength.

SU-31 see COLUMBIUM SU-31.

SUBSTITUTE.
90 Cu, 6.5 Sn, 3 Zn, 0.5 Pb.

SUDAL.
M-380; Pb, Sn.
For Al solders.

SUDALON "A".
M-380; Pb, Sn.
For Al solders.

SUDALON "B".
M-380; Pb, Sn.
For Al solders.

SUDALON "C".
M-380; Pb, Sn.
For Al solders.

SUEDOIS-S1.
M-1546; 1.0 C, 0.4 Mn, 0.2 Si, bal Fe.
Hardened: 216,000 TS; 152,000 YP; 11 El; 600 Brin.
For precision tools, drills, cutters, chisels, punches, taps, springs, bumper bars, bearings.
Water hardening, wear resistant.

SUEDOIS-S2.
M-1546; 0.80 C, 0.7 Mn, 0.2 Si, bal Fe.
Hardened: 183,000 TS; 135,000 YP; 13 El; C 40 Rock.
For precision tools, drills, cutters, chisels, taps, punches.
Water hardening, wear resistant.

SUHLER-WHITE COPPER.
40 Cu, 32 Ni, 25 Zn, 2.6 Pb.
For ornamental parts; corrosion resistant.

SUJ2-N.
M-1589; 1.2 C, 1.0 Cr, 0.6 Mn, 0.4 Si, bal Fe.
Ht. Tr.: 225,000 TS; 210,000 YS; 444 Brin.
For ball bearings, balls and races, liners, bushings.
Water hardening. Similar to 52100 Steel.

SULFOR.
M-1766; 0.10 C, 1.10 Mn, 0.06 max Si, 0.05 max P, 0.30 S, 0.20 Pb, bal Fe.
Free machining, low carbon steel.
IHA F-212; AISI 12L14.

SULGERS ANTIFRICTION.
M-England; 83.6 Zn, 9.9 Sn, 4 Cu, 1.2 Pb.
For bearings; will not resist heat or live steam.

SULZER-2 13 CR CO MO.
M-1660; 0.14 C, 0.6 Si, 0.7 Mn, 13.0 Cr, 0.8 Mo, 10.0 Co, bal Fe.
Cast, heat tr.: 950-1150 N/mm^2 TS; 750 min N/mm^2 YS; 12 min El. (or stronger if desired.)
For cast pump impellers, wheels.
Corrosion resistant, high strength steel.
DIN: GX 14 Cr Co Mo 13 10.

SUMET BRONZE SM22.
M-71; 10 Pb, Sn, bal Cu.
For bearings; heavy duty.

SUMIALLOY E202.
M-1296; 0.20 C, 0.25 Si, 1.3 Mn, 0.025 P, 0.020 S, 0.60 Cr, 0.002 B, bal Fe.
25-100 mm diameter bar.
Case hardening steel with good hardenability; for automotive components as gears.

SUMISTRONG 60 Q-B.
M-1296; 0.18 max C, 0.50 max Si, 1.50 max Mn, 0.3 max Cu, 0.60 max Ni 0.20 max Cr, 0.30 max Mo, 0.15 max V, bal Fe.
85,000 min TS; 71,000 min YS; 20 min El.
High strength low alloy steel pipe for marine structural services. Good weldability.

SUMISTRONG 80 Q-C.
M-1296; 0.16 max C, 0.35 max Si, 1.20 max Mn, 0.3 max Cu, 0.4-1.2 Ni, 0.4-0.8 Cr, 0.3-0.6 Mo, 0.10 max V, bal Fe.
114,000-135,000 TS; 99,600 min YS; 18 min El.
High strength low alloy steel pipe for marine structural services. Good weldability.

SUMITEN 60 CLASS.
M-1296; 0.20 max C, 0.55 max Si, 0.9-1.6 Mn, 0.5 max Cu, 0.6 max Ni, 0.5 max Cr, 0.35 max Mo, 0.1 max V, bal Fe.
6-100 mm thick plate: 86,000-103,000 TS; 66,000 min YP.
High strength low alloy steel.
5 types: SUMITEN 60, 60 K, 60 W, 60 F, 60 LT.

SUMITEN 80 CLASS.
M-1296; 0.18 max C, 0.55 max Si, 1.2 max Mn, 0.15-0.50 Cu, 0-1.5 Ni, 0.3-1.3 Cr, 0.1-0.7 Mo, 0.1 max V, 0.006 max B, bal Fe.
6-100 mm thick plate: 114,000-136,000 TS; 100,000 min YP.
High strength low alloy steel for bridges, marine structures, penstocks, pressure vesssels.
4 types: SUMITEN 80, 80 W, 80.S, 80.SW.

SUMITEN 100 CLASS.
M-1296; 0.18 max C, 0.55 max Si, 1.5 max Mn, 0.15-0.50 Cu, 0.30-2.75 Ni, 0.3-1.2 Cr, 0.1-0.9 Mo, 0.1 max V, 0.006 max B, bal Fe.
6-50 mm thick plate: 134,000-164,000 TS; 128,000 min YP.
For heavy welded structures, bridges, pressure vessels.
2 types: SUMITEN 100, 100 W.

SUN BRONZE.
M-Eng.; 40-60 Co, 10 Al, 30-50 Cu.
For high temperature fittings; heat resistant.

SUN BRONZE-1.
M-Eng.; 95 Cu, 5 Al.
For pump rods, bushings, propeller blade bolts; corrosion resistant.

SUN BRONZE-2.
M-U.S.; 60-40 Co, 30-50 Cu, 10 Al.
For high temperature fittings; heat resistant.

SUNDEEL.
80 Au, 10 W, 7 Ni.
Used as dental alloy; corrosion resistant.

SUNRAY GOLD.
M-422; Ni, bal Cu.
For imitation gold, jewelry; corrosion resistant.

SUNSHINE METAL.
2.9 C, 2 Si, bal Fe.
For rubber tire molds; semi-steel cast iron.

SUPARD CHROME.
M-508; 0.4 C, 11.5-14 Cr, 1 max Ni, 0.5 max Mo, bal Fe.
Heat treated: 240,000 TS; 175,000 YS; 5 El; 7 RA; 450 Brin.
For pump liners, plunger sleeves, bushings; resists H_2SO_4.

SUPER.
M-783; 1.1 C, 3.75 Cr, 0.3 Mn, 3.25 V, 17.5 W, 0.8 Mo, bal Fe.
For lathe and planer tools; high speed steel.

SUPER 3 see HOWMET SUPER 3.

SUPER 6 see HOWMET SUPER 6.

SUPER-ALLOY.
M-335; 0.5 C, 0.75 Si, 1.15 Cr, 2.5 W, 0.2 V, bal Fe.
Oil hardening; for tools, dies.
AISI S1. Shock resisting tool steel.

SUPERALUMAG T.35.
M-770; 3 Zn, 1.5 Mg, 0.2 Cr, 0.4 Mn, bal Al.
Annealed: 26,000 TS; 10,000 YS; 20 El; 38 Brin.
Hardened: 50,000 TS; 35,000 YS; 16 El; 70 Brin.
For engine parts; corrosion resistant.

SUPERALUMAG T.45.
M-770; 6 Zn, 2 Mg, 0.3 Cr, 0.5 Mn, bal Al.
Annealed: 29,000 TS; 10,000 YS; 20 El; 50 Brin.
Hardened: 69,000 TS; 50,000 YS; 12 El; 150 Brin.
For aircraft parts; hardenable.

SUPERALUMAG T.60.
M-770; 8 Zn, 2 Mg, 1.5 Cu, 0.3 Cr, 0.5 Mn, bal Fe.
Annealed: 45,000 TS; 10,000 YS; 16 El; 55 Brin.
Hardened: 86,000 TS; 72,000 YS; 8 El; 180 Brin.
For aircraft structures; hardenable.

SUPER ANTINIT C now VEW L318.

SUPER AUSTENITE.
M-1749; 0.75 C, 4.5 Cr, 18.0 W, 0.5 Mo, 1.0 V, bal Fe.
"18-4-1" type high speed tool steel. For lathe and planing tools, drills, reamers, broaches, cores for pressure die casting.

SUPER AVIONAL.
M-249, M-624; 4.5 Cu, 0.8 Si, 1 Mg, 1.2 Mn, bal Al.
Aged: 68,000 TS; 48,000 YS; 15 El; 130 Brin.
For aircraft parts, structural members; age-hardenable.

SUPERB.
M-73; 0.7 C, 0.9 Cr, 0.2 V, bal Fe.
For tools, dies, chasers; water or oil hardened.

SUPERBASIQUE.
M-700; C, Cr, Ni, bal Fe.
For corrosion resistant parts; stainless, austenitic.

SUPERBRONZE.
M-Eng.; 57-69 Cu, 21-38 Zn, 3-3.2 Mn, 1.3-2 Fe, 1.2-5 Al.
For marine parts, propeller blades; corrosion resisting.

SUPER BUFFALOY.
M-962; High C, bal Fe.
Wire for screening abrasive material.

SUPERCASE.
M-406; 0.08-0.15 C, bal Fe.
Hot rolled: 60,000 TS; 33 El; 60 RA; 110 Brin.
Heat treated: 81,000 TS; 24 El; 69 RA; 217 Brin.
For camshafts, gears, bolts, transmission shafts, worms, sprockets; case hardening steel; free machining.

SUPER CAST 4.
M-1018; 4 Al, 0.02 Mg, bal Zn.
For die castings.

SUPER CAST ALLOY NO.6 see SUPER CAST 4.

SUPER CAST S49.
M-1018; 8.5 Si, 3.5 Cu, bal Al.
For aluminum die castings.

SUPER CAST S102.
M-1018; 10.5 Si, 2.5 Cu, bal Al.
For aluminum die castings.

SUPER CESCO.
M-356; 0.55 C, 1.0 Mn, 1.3 Mo, 0.35 V, 2.0 Si, bal Fe.
For tools; shock resistant.

SUPER CESCO DIAMOND now CESCO DIAMOND SPECIAL.
M-356; 0.50-0.65 C, 2.0 Si, 1.0 Mn, 1.25 Mo, 0.3 V, bal Fe.
Heat treated: 236,400-338,000 TS; 231,000-279,200 YS; 10.5-4 El; 31-7.7 RA; 612 Brin.
For chisels, concrete busters, drills, caulking and beading tools.

SUPERCHITONAL.
M-624; aluminum clad Superavional.
For aircraft structures.

SUPER CHLOR.
M-46; 4-5 Cr, 14.2-14.75 Si, 0.70-1.10 C, bal Fe.
Cast alloy for handling H_2SO_4, HNO_3 and HCl.

SUPER CHROME.
M-1705; 1.4 C, 0.3 Mn, 0.6 Si, 13.0 Cr, 0.5 Ni, 0.6 Mo, 3.3 Cu, bal Fe.
High carbon, high chromium tool and die steel; air or oil hardening. AISI D5.

SUPER COAT.
M-1540; Coated sheet steel.
Coated with 0.03-0.06 μ chromium and 0.014-0.2 μ CrO_3.
For beverage cans, crown caps, pails, dry cells, motor oil cans. Resists staining and corrosion.

SUPER COBALT.
M-1702; 0.80 C, 2.0 V, 9.0 Co, 18.0 W, 4.0 Cr, bal Fe.
T5 high speed steel tool bits; good red hardness and wear resistance.
(Electrite Super Cobalt).

SUPER-COR.
M-663; weld deposit: 0.052 C, 1.50 Mn, 0.49 Si, 0.05 Ni, 0.05 Cr, 0.01 Mo, 0.02 V, bal Fe.
As welded: 83,600 psi TS; 72,000 psi YS; 27 El.
Good impact strength.
AWS A5.20 Class E 70T-1.

SUPERDIE.
M-35; 2.20 C, 12 Cr, 0.8 W, 1.0 Si, bal Fe.
Annealed: 269 Brin.
For blanking, trimming dies, burnishing tools, extrusion and drawing dies; resistant to abrasive wear. AISI D3 tool steel.

SUPERDRAW.
M-210; 1.3 C, 0.3 Mn, 2.25 W, bal Fe.
For cold drawing dies; water hardened.

SUPER-DUCT F-3.
M-440; 0.4 C, 1.5 Ni, 0.8 Cr, 0.2 Mo, bal Fe.
For machinery parts; high strength and tough.

SUPERELSO.
M-1488; 0.18 max C, 0.90 Mn, 0.30 Si, 1.3 Cr, 0.2 Cu, 0.25 Mo, 1.5 Ni, V, bal Fe.
Plate: 114,000-135,000 TS; 97,000 YS; 15 min El.
For mine cars, chutes, bus and trailer bodies, derricks, structures, booms.
Low-alloy high strength structural steel, shock resistant.

SUPERELSO 70.
M-1488; 0.18 max C, 1.3 Mn, 0.50 max Si, 1.0 Ni, 1.0 Cr, 0.35 Mo, bal Fe.
Q + T: 780-930 N/mm^2 UTS; 685 N/mm^2 min YS.
Weldable: for civil construction.

SUPERELSO 600.
M-1488; 0.12 max C, 1.15 Mn, 0.50 max Si, 0.80 Ni, 1.75 Cr, 0.25 Mo, bal Fe.
Q + T: 700 N/mm^2 min UTS; 600 N/mm^2 min YS.
Weldable; for general civil construction.

SUPERFORM 30.
M-173; 0.03-0.08 C, 0.25-0.50 Mn, 0.01-0.06 Si, 0.05-0.12 Zr, 0.02-0.06 (typ) Al, bal Fe.
30,000 psi min YS.
Good cold-forming and welding.
For railroad car components, disc brake parts.

SUPERFORM 40.
M-173; 0.14-0.20 C, 0.40-0.90 Mn, 0.01-0.02 Si, 0.05-0.12 Zr, 0.02-0.06 (typ) Al, bal Fe.
40,000 psi min YS.
For wheel spiders, bumpers.

SUPER-GENUINE BABBITT see ADAMANT SUPER-GENUINE BABBITT.

SUPER HALLAMITE.
M-1202; 0.7 C, 18 W, 4 Cr, 1 V, 5 Co, bal Fe.
For hacksaw blades, saws, cutting tools; high speed steel, oil hardened.

SUPER HARDTEM.
M-64; 0.43 C, 2.6 Ni, 1.35 Cr, 0.55 Mo, 0.18 V, bal Fe.
Annealed: 250 Brin.
Heat treated: 470 Brin.
For die blocks; oil hardened, shock resistant.

SUPER-HIGH SPEED.
M-U.S.; 0.75-0.90 C, 17-22 W, 4.5-6 Cr, 1-2.5 V, 0.75-1.5 Mo, 8-15 Co, bal Fe.
For high speed tools; high speed steel; has great red hardness.

SUPER HI-PHY KIRKSITE.
M-88; 4 Al, 3 Cu, 0.01 Mg, 0.007 Be, bal Zn. (max .005 Pb, .002 Sn, .002 Cd, .05 Fe)
Sand cast: 40,000 TS; 3.0 El; 34 impact; 109 Brin.

SUPER-HOLFOS W.W.
M-169; 88.3 Cu, 11.5 Sn, 0.2 P.
Cent. cast: 45,000 TS; 27,000 YS; 8 El; 100 Brin.
Good wear and abrasion resistance.
For worm and helical gears, drive shafts, splined couplings.
BS 1400 PB2-C; BS 421; SAE 65.

SUPER IMPACTO.
M-1636, M-435; 0.12 C, 0.5 Mn, 1.5 Cr, 3.25 Ni, bal Fe.
Heat treated: 175,000 TS; 135,000 YS; 16 El; 59 RA; 341 Brin; 48 Izod.
For heavy duty gears, bearings, clutch dogs, broach holders, plastic molds, transmission shafts.
Tough and fatigue resistant.

SUPER INCOMPARABLE.
M-1409; 0.8 C, 19-20 W, 4.6 Cr, 1.6 V, 5.5 Co, Mo, bal Fe.
For continuous cutting tools for cast iron; high speed steel.

SUPER INMANITE.
M-1749; 0.80 C, 4.5 Cr, 22.0 W, 1.0 Mo, 1.5 V, 10.0 Co, bal Fe.
10% Cobalt-Tungsten high speed tool steel. For heavy duty machining operations.

SUPER-INVAR.
M-118; 31 Ni, 5 Co, bal Fe.
For instruments; zero expansion.

SUPER INVINCIBLE 5% COBALT.
M-1406; 0.75 C, 18 W, 4 Cr, 1.5 V, 5 Co, bal Fe.
For tools, cutters, reamers, hobs, chasers; high speed steel.

SUPER INVINCIBLE ADVANCE.
M-1406; 0.75 C, 19 W, 4.25 Cr, 1.75 V, 10 Co, bal Fe.
For lathe and turning tools, drills, broaches; high speed steel.

SUPER INVINCIBLE TB.
M-1406; 0.75 C, 14 W, 4 Cr, 2 V, 5 Co, bal Fe.
For shear blades, tools, cutters, punches; high speed steel.

SUPERIOR.
M-U.S.; 19.5 Cr, 0.5 Fe, 2.0 Mn, bal Ni.
For resistance; resistance alloy.

SUPERIOR.
M-901; 0.2 C, 18 Cr, 8 Ni, bal Fe.
For stainless steel parts; stainless.

SUPERIOR.
M-24; 0.7-1.3 C, 0.15-0.25 V, bal Fe.
For taps, reamers, cutters, punches, dies, broaches, chisels; tool steel, extra quality.

SUPERIOR 1.
M-140; 2.05-2.2 C, 11.5-13.5 Cr, 0.55-0.65 V, bal Fe.
For tools, cutters, blanking dies, shear knives, ceramic tools; non-deforming.

SUPERIOR 2.
M-140; 1.45-1.6 C, 11-12 Cr, 0.15-0.25 V, 0.70-0.80 Mo, bal Fe.
For tools, cutters, coining and forming and blanking dies; oil hardening.

SUPERIOR 3.
M-140; 1.5 C, 12 Cr, 0.87 V, 1.0 Mo, bal Fe.
For shear blades, blanking and forming dies; cold worked steel.

SUPERIOR ARK.
M-261; 0.7 C, 19 W, 4 Cr, 2 V, bal Fe.
For high speed cutting tools; high speed steel.

SUPERIOR CHROME.
M-275; C, Cr, bal Fe.
For tools, dies; oil hardened.

SUPERIOR COPPER STEEL.
M-967; C, alloy, Cu, bal Fe.
For replacing solid copper; Cu plated corrosion resistant steel.

SUPERIOR NO. 1.
M-1170; 3.00-3.25 T.C., 1.9-2.1 Si, 0.75-1.25 Ni, 0.3-0.6 Cr, 0.6-0.8 Mo, bal Fe.
Cast: 50,000-60,000 TS; 235-269 Brin.
For diesel pistons; cast iron.

SUPERIOR NO. 2.
M-1170; 3.00-3.25 T.C., 2.00-2.25 Si, 0.75 Ni, 0.3-0.5 Mo, 0.6-0.8 M bal Fe.
Cast: 40,000 TS; 196-235 Brin.
Tr. S. 3000; Tr. D. 0.30; cast iron.

SUPERIOR NO. 3.
M-1170; 3.00-3.25 T.C., 2.00-2.25 Si, 0.7-0.9 Mn, 0.75-1.00 Cu, bal Fe.
Cast: 38,000 TS; 187-217 Brin.
For machine tools; Tr. S. 2900; Tr. D. 0.30.

SUPERIOR NO. 4.
M-1170; 3.0-3.3 T.C., 2.0-2.2 Si, 0.6-0.8 Mn, 0.3 Cr, bal Fe.
Cast: 35,000 TS; 187-217 Brin.
For elevator sheaves; Tr. S. 2800; Tr. D. 0.28.

SUPERIOR NO. 5.
M-1170; 3.00-3.25 T.C., 2.1-2.4 Si, 0.6-0.8 Mn, 0.3-0.4 Mo, bal Fe.
Cast: 38,000 TS; 211-248 Brin.
For automotive manifold; cast iron.

SUPERIOR NO. 6.
M-1170; 3.0-3.3 T.C., 2.0-2.2 Si, 0.65-0.80 Mn, 0.3 Cr, 0.3 Mo, bal Fe.
Cast: 40,000 TS; 207-241 Brin.
For pressure castings; Tr. S. 3000; Tr. D. 0.30.

SUPERIOR NO. 7.
M-1170; 3.00-3.25 T.C., 2.0-2.2 Si, 0.6-0.9 Mn, 1.25-1.75 Ni, 0.4-0.6 Cr, 0.4-0.6 Mo, bal Fe.
Cast: 50,000-60,000 TS; 207-255 Brin.
For lathe beds, dies; suitable for flame hardening.

SUPERIOR NO. 8.
M-1170; 3.00-3.25 T.C., 1.8-2.0 Si, 0.7-0.9 Mn, 1 Ni, 0.25-0.35 Cr, 0.3-0.4 Mo, bal Fe.
Cast: 40,000-55,000 TS; 207-217 Brin.
For diesel heads; resists heat shock and wear.

SUPERIOR OIL HARDENING.
M-261; 0.9 C, 1.2 Mn, 0.5 Cr, 0.5 W, bal Fe.
For press tools, dies, gauges, punches; oil hardened, non-deforming.

SUPERIOR SHAFTING.
M-435; 0.35 C, 1.5 Mn, 0.25 Si, 2 Cr, 0.1 max S, bal Fe.
For shafting; free cutting.

SUPER JACKSBERG.
M-477; C, alloy, bal Fe.
For high speed tools and cutters; high speed steel.

SUPER KANTHAL.
M-1150; Mo-disilicide.
For heating elements; high oxidation resistance to 3000°F.

SUPER-KUT.
M-1067; 0.7 C, 4 Cr, 18 W, 0.5 Mo, 5 Cr, 1 V, bal Fe.
For tools, cutters, broaches, chasers; high speed steel.

SUPERLA.
M-336; 1.3 C, V, bal Fe.
For lathe and planer tools, reamers, taps, drills; Type W2; water hardened.

SUPER LA-LED.
M-669; 0.13 max C, 0.85-1.35 Mn, 0.15-0.35 Pb, 0.5 S, bal Fe.
Cold Drawn: 70,000 TS; 60,000 YS; 12 El; 150 Brin.
For screw machine products, bolts, fasteners, shafts.
Free-machining.

SUPER LION.
M-1415; 0.75 C, 19 W, 4.5 Cr, 1.25 V, 10 Co, 0.75 Mo, bal Fe.
For lathe and planer tools, drills, reamers; high speed steel.

SUPER LION EXTRA.
M-1415; 0.75 C, 20 W, 4 Cr, 1 V, 10 Co, 0.75 Mo, bal Fe.
For lathe and planer tools, hobs, drills, reamers; high speed steel.

SUPER LION SPECIAL.
M-1415; 0.80 C, 19 W, 4.5 Cr, 1.25 V, 15 Co, 0.75 Mo, bal Fe.
For lathe and planer tools, drills, reamers; high speed steel.

SUPER LMW EXTRA.
M-1779; 0.90 C, 3.75 Cr, 1.15 V, 1.65 W, 9.3 Mo, 8.25 Co, bal Fe.
Heavy duty cobalt bearing high speed steel used mainly for cutting tools requiring exceptional red hardness.
AISI M-33.

SUPER-LOY.
M-1085; 0.2 C, bal Fe.
For screws, filter cloth.

SUPERLOY.
M-1102; Cu alloy.
For resistance and spot welding electrodes.

SUPERLOY 30.
M-246; 0.25-0.35 C, 1.2-1.3 Mn, 0.75 max Si, bal Fe.
Annealed: 60,000 TS; 30,000 YS; 24 El; 35 RA; 150 Brin.
For castings, gears, shafts; water hardening.

SUPERLOY 42.
M-246; 0.25-0.35 C, 1.1-1.5 Mn, 0.75 max Si, bal Fe.
Annealed: 80,000 TS; 40,000 YS; 17 El; 25 RA; 175 Brin.
For castings, gears, shafts; water hardening.

SUPERLOY 53.
M-246; 0.30-0.40 C, 0.7-0.9 Mn, 0.3-0.5 Si, 0.6-0.7 Cr, 1.25-1.75 Ni, 0.25-0.35 Mo, bal Fe.
Annealed: 85,000 TS; 50,000 YS; 20 El; 30 RA; 190 Brin.
For castings, gears, shafts, crankshafts; tough, shock resistant.

SUPERLOY 64.
M-246; 0.35-0.45 C, 1.35-1.65 Mn, 0.3-5 Si, 0.7-0.9 Cr, 1.25-1.75 Ni, 0.3-0.4 Mo, bal Fe.
Annealed: 120,000 TS; 90,000 YS; 10 El; 15 RA; 260 Brin.
For castings, gears, shafts, crankshafts; tough, shock resistant.

SUPERLOY 530.
M-246; C, alloy, bal Fe.
For machine tool parts; high strength and shock resistant.

SUPERMAGALUMA.
M-622; 5.4 Mg, 0.15 Mn, bal Al.
Soft: 36,000-42,500 TS; 14,500-21,000 YS; 25-20 El.
Hard: 50,000-65,000 TS; 42,500-58,000 YS; 6-4 El.
For light alloy parts; heat treatable, corrosion resistant.

SUPERMAL.
M-170; 1.7 T.C., 0.3 Mn, 1.2 Si, bal Fe.
Heat treated: 75,000 TS; 55,000 YS; 8 El; 187 Brin.
For chains, buckets, conveyor equipment; pearlitic, wear resistant; malleable iron.

SUPERMALLOY.
M-670; 15.7 Fe, 5 Mo, 0.3 Mn, bal Ni.
For transformers, communication and radar equipment; magnetically soft. (Fabricated parts only).

SUPERMALLOY.
M-1089; 80 Ni/Fe/Mo.
Soft magnetic alloy processed for high initial permeability.

SUPERMALOY.
M-205; 79 Ni, 5 Mo, 15.7 Fe, 0.3 Mn.
For communication transformers, magnets; high permeability.

SUPER MANGANESE BRONZE.
M-129; 69 Cu, 20 Zn, 6.5 Al, 2.5 Fe, 2 Mn.
Cast: 110,000 TS; 200 Brin.
For propeller blades, marine parts; corrosion resistant.

SUPERMENDUR.
M-118, M-22, M-1089, M-1626; 49 Fe, 49 Co, 2 V.
For amplifiers, pulse and power transformers; high permeability. Soft magnet.

SUPERMENDUR.
M-108.
Soft magnetic alloy, high saturation, square loop, for special transformers.

SUPER MENDUR.
M-670; 49 Fe, 49 Co, 2 V.
Soft magnetic material, high permeability. (fabricated parts only)

SUPERMETONIC.
M-1783; 30 Ni, 2 Fe, 2 Mn, bal Cu.
Cu-Ni tubing for heat exchanger applications Cu Ni 30 Fe 2 Mn 2.

SUPER MO-CHIP HIGH SPEED.
M-57; 0.7 C, 4 Cr, 8 Co, 1.5 V, 8 Mo, bal Fe.
For tools, cutters, reamers; high speed steel.

SUPERMOLD.
M-73; 0.3 C, 1.7 Cr, 0.40 Mo, bal Fe.
Molds for plastics, die casting dies.
AISI P20.

SUPER MOTUNG.
M-114; 0.8 C, 3.75 Cr, 5 Co, 1.25 V, 1.5 W, 8.5 Mo, bal Fe.
For cutting tools, form tools, taps, broaches; high speed steel.

SUPER MOTUNG 33.
M-114; 0.88 C, 9.5 Mo, 1.75 W, 4.0 Cr, 1.25 V, 8.25 Co, bal Fe.
Hardened: 64-67 Rock C.
For cutters, drills, taps, end mills, lathe tools, milling cutters, chasers. Good red-hardness and wear resistance.
AISI-M33 High-speed steel.

SUPER MOTUNG 34.
M-114; 0.90 C, 8.5 Mo, 1.75 W, 4.0 Cr, 2.0 V, 8.25 Co, bal Fe.
Hardened: 64-67 Rock C.
For cutters, drills, taps, chasers, lathe tools, milling cutters, form tools.
Good cutting properties and red-hardness.
AISI-M34 High speed steel.

SUPERMU 10 replaced by **SUPERPERM 49.**

SUPERMU 40 replaced by **SUPERPERM 80.**

SUPERMUMETAL.
M-108; 14 Fe, 5 Cu, 4 Mo, bal Ni.
For electrical and magnetic equipment.
Vacuum melted. High magnetic permeability and low electrical losses.

SUPER NERVA.
M-807; 0.7 C, 18 W, 4 Cr, 1 V, bal Fe.
For tools, cutters; high speed steel.

SUPER NICKEL-701.
M-8; 70 Cu, 30 Ni.
Hard: 70,000 TS; 60,000 YS; 10 El.
For condenser tubes, corrosion resisting.

SUPER OILITE.
M-211; 18-22 Cu, bal Fe.
Sintered: 22,000 TS; 5.8-6.2 density.
Oilite bearing material.
ASTM B439-67 Grade 4; SAE 863.

SUPER OILITE 2-1.
M-211; 1.75-3.0 Cu, 0.5-1.0 C, bal Fe.
Sintered: 50,000 TS; 6.1-6.4 density; 15,000 fatigue strength.
ASTM B426-65, Type 11 Gr 1 Class C; SAE 864B.

SUPER OILITE 2-1 HARDENED.
M-211; 0.8 C, 2 Cu, 97 Fe.
Sintered and hardened: 3 ranges from 55,000-80,000 TS; 50,000-75,000 YS; 25-40 Rc; Density 6.1-7.2.
MPIF FC-0208-P or R or S; ASTM B-426 Grade 1 Type II or IV.

SUPER OILITE 5-1.
M-211; 4.0-6.0 Cu, 0.5-1.0 C, bal Fe.
Sintered: 50,000 TS; 6.1-6.4 density; 15,000 fatigue strength.
ASTM B426-65 Type 11 Gr 2, Class C; SAE 865B.

SUPER OILITE 9.
M-211; 7.0-11.0 Cu, 86.5 min Fe, 0.3 max C.
Sintered: 29,500-34,000 TS; 5.8-6.2 density; 17,000 shear; 13,000 fatigue.
ASTM B439-67 Gr 3; SAE 862.

SUPER OILITE 9-1.
M-211; 6.5-7.5 Cu, 89.0 min Fe, 0.6-1.0 C.
Sintered: 40,000-50,000 TS; 5.8-6.2 density.
ASTM B426-65 Gr 3 Cl.C; SAE 866 A; PMPMA FC-0710-N.

SUPER OILITE 16.
M-211; 18-22 Cu, 0.6-1.0 C, bal Fe.
Sintered: 40,000 TS; 5.8-6.2 density; Oilite bearing material, self lubricating bearings.

SUPER OILITE-M.
M-211; 0.5 C, 0.5 S, 2.0 Cu, 97 Fe.
Sintered: 40,000-50,000 TS; 36,000-45,000 YS; 1.0 El; Density 6.1-6.
Machinable grade Oilite bearing material.

SUPER PANTHER.
M-1; 0.8 C, 19 W, 4 Cr, 2 V, 8 Co, 0.7 Mo, bal Fe.
For lathe and planer tools, reamers, broaches, form cutters; high speed steel; Type T5.

SUPERPERM 49.
M-1089; 50 Ni/Fe.
Soft magnetic alloy. High initial permeability and high maximum flux.

SUPERPERM 80.
M-1089; 80 Ni/Fe/Mo.
Soft magnetic alloy processed for high permeability and low magnetizing force.

SUPER PERMALLOY "C".
M-269; 78 Ni, 13 Fe, 4 Mo, 5 Cu.
For telecommunications, magnetic amplifiers, wide band transformers, current transformers, computers.
High initial permeability with low losses.

SUPER PURITY.
M-28; 99.99 min Al.
O-temper: 8500 TS.
1/2 H-temper: 11,000 TS.
H-temper: 14,500 TS.
Extruded: 8000 TS.
For roofing flashings, reflectors; high corrosion resistance.

SUPER PYRONEAL.
M-64; 0.42-0.47 C, 0.3-0.6 Mn, 0.5-0.7 Si, 4.1-4.5 Ni, 1.25-1.65 Cr, 0.7-0.8 Mo, 0.12-0.16 V, bal Fe.
For press and hammer dies, upsetter and insert dies.
Oil hardening. High compressive strength.
Resists softening at high working temperatures.

SUPER PYROTEM.
M-64; 0.42-0.47 C, 0.30-0.60 Mn, 0.5-0.7 Si, 4.1-4.5 Ni, 1.25-1.65 Cr 0.7-0.8 Mo, 0.12-0.16 V, bal Fe.
For press and hammer dies, upsetter and insert dies.
Oil hardening; high compressive strength.
Resists softening at high working temperatures.

SUPER RADIOMETAL.
M-108; 50 Ni, 50 Fe.
Annealed: 58,000 TS; 19,000 YS; 27 El. (magnetic anneal).
For electrical and magnetic equipment, core material for audio and instrument transformers, magnetic shields.
Vacuum melted. Soft magnet. High magnetic permeability and low losses.

SUPER RAPID.
M-1405; 0.8 C, 4 Cr, 18 W, 1.2 V, bal Fe.
For lathe and planer tools, broaches, reamers, drills; high speed steel.

SUPER RAPID EXTRA now VEW S200.

SUPER-RAPID EXTRA.
M-363; 0.7 C, 18 W, Cr, 1 V, bal Fe.
For punches, dies, cutting tools; high speed steel.

SUPER RAPID EXTRA.
M-1182; 0.7 C, 18 W, 4 Cr, 1 V, bal Fe. For cutters, taps, broaches, drills; high speed steel.

SUPER RAPID EXTRA 500 now VEW S305.

SUPER RAPID EXTRA HVN now VEW S203.

SUPER RAPID EXTRA MO now VEW S600.

SUPER RAPID EXTRA N now VEW S205.

SUPER RAPID EXTRA V now VEW S201.

SUPER RUSTFREE SR STEEL.
M-1161; Cr, Ni, Mo, bal Fe.
Cold worked: 90,000-105,000 TS; 75,000-100,000 YS; 50-4 El; 65-28 RA; 156-410 Brin.
Annealed: 95,300 TS; 75,000 YS.
For pickling equipment, turbine compressors, gasoline cracking plants; corrosion resistant, "Super Rustfree Steels;" austenitic.

SUPER-SAFETY.
M-449; C, alloy, bal Fe.
For steel stamps.

SUPER SAMSON see CARPENTER SUPER SAMSON.

SUPER-SHOCK.
M-750; 0.5 C, 0.75 Si, 1.15 Cr, 2.5 W, 0.2 V, bal Fe.
For cold battering and chipping tools, caulking and beading tools; shock resistant, wear resistant.

SUPER SILICON CAST IRON NO. 9.
M-243; 3.5 C, 2.5 Si, bal Fe.
For welding rods for cast iron; cast iron.

SUPER SPECIAL EXPRESS.
M-1408; 0.7 C, 18 W, 4 Cr, 1 V, bal Fe.
For lathe and planer tools, hobs, reamers, taps; high speed steel.

SUPERSQUARE 80.
M-1089; 80 Ni/Fe/Mo.
Higher maximum flux density, high gain, high squareness ratio; soft magnetic alloy.

SUPERSTAR see CARPENTER SUPERSTAR.

SUPER STAR-MO M2-5.
M-57; 0.85 C, 6 W, 4.1 Cr, 2 V, 5 Mo, 5 Co, bal Fe.
For boring tools, broaches, hobs, milling cutters; high speed steel.

SUPER STAR ZENITH.
M-32; 0.8 C, 4 Cr, 2 V, 18 W, bal Fe.
For cutters, hobs, drills, taps, reamers; high speed steel; Type T2.

SUPERSTON 40.
M-452, M-13, M-539; 8 Al, 3 Fe, 12 Mn, 2 Ni, bal Cu.
Sand cast: 98,000 TS; 48,000 YS; 26 El; 28 RA; 185 Brin.
Centrifugal cast: 105,000 TS; 50,000 YS; 30 El; 30 RA; 190 Brin.
Heat treated: 125,000 TS; 75,000 YS; 12 El; 255 Brin.
For valves, pumps, propellers, gears, slides; corrosion and wear resistant.

SUPERSTON 60.
M-452, M-539; 11-14 Mn, 8.5-9.0 Al, 2-4 Fe, 1.5-4.5 Ni, 1.0 max Sn, 0.02 max Pb, bal Cu.
Sand cast: 107,500 TS; 56,000 YS; 10 El.
Chill cast: 127,000 TS; 83,000 YS; 5.5 El.
For marine propellers, shafts, hardware.
High strength, tough, corrosion resistant.

SUPERSTON 70.
M-1675, M-452; 72 Cu, 15 Mn, 3 Fe, 2.5 Ni, 7.5 Al.
Aluminum bronze casting for marine applications.

SUPERSTON L-189.
M-452; 8-12 Al, 4-6 Ni, 4-6 Fe, bal Cu.
Cast: 90,000 TS; 45,000 YS; 15-20 El; 180 Brin.
For engine parts; Al bronze, corrosion resistant.

SUPER STRENGTH 100.
M-1428; 0.15 C, 0.50 Mn, 0.35 Si, 1.65 Cr, 0.25 Cu, 0.50 Mo, 0.002 B, 0.07 Ti, bal Fe.
For mine cars, chutes, bus and trailer bodies, structures, derricks.
Low-alloy high-strength structural steel. Shock resistant.

SUPER STRENGTH MALLEABLE.
M-171; 3.0 C, 1.2 Si, 0.7 Mn, bal Fe.
Cast: 57,000 TS; 38,000 YS; 18 El; 120 Brin.
For machinery parts; malleable cast iron.

SUPER TERRIFIC.
M-1113; C, 18 W, 5 Co, 4 Cr, 1.3 V, bal Fe.
For tools, dies, cutters; high speed steel.

SUPER T-H24.
M-1799; 0.50 Fe, 0.50 Co, 0.06 Si, bal Al.
19.0 ksi UTS; 17.0 ksi YTS; 15 El; 61 min IACS.
For aircraft wire, high temperature wire.

SUPERTHERM.
M-47; 35 Ni, 26 Cr, 15 Co, 5 W, 1.6 Si, 0.5 C, bal Fe.
Cast: 75,000 TS; 48,000 YS; 5 El; 215 Brin.
For furnace parts, cement kilns, smelting and calcining equipment.
High heat resistance to 2300°F. Good hot ductility.

SUPERTOUGH G.
M-1724; 0.15-0.23 C, 0.35 max Si, 0.75-1.25 Mn, 0.25-1.25 Cr, 0.40- 0.60 Mo, 0.50-1.0 Ni, bal Fe.
High strength low alloy steel suitable for large forgings and fabricated pressure vessels. Weldable.

SUPER TUFF.
M-1705; 0.5 C, 0.7 Mn, 3.25 Cr, 1.4 Mo, bal Fe.
Shock resisting tool steel; air or oil hardening.
AISI S7.

SUPER UGIMAX.
M-1120; 14 Ni, 8 Al, 24-35 Co, 3 Cu, bal Fe.
13400 Br, 800 Hc, 8 BH max.
For magnets in loud speakers, microphones, measuring instruments.
Permanent magnet, high permeability.

SUPERWELD NO. 66.
M-607; C, bal Fe.
Deposited: 60,000-70,000 TS; 45,000-55,000 YS; 30-25 El.
For welding electrode for light and heavy gauge material.

SUPERWELD NO. 77.
M-607; 0.08 C, 0.45 Mn, 0.30 Si, bal Fe.
Deposited: 75,000 TS; 65,000 YS; 20 El.
AC-DC welding electrode for light and heavy gauge carbon steel; production and maintenance welding.
Conforms to AWS Class E-6012.

SUPERWELD NO. 77A.
M-607; 0.10 C, 0.35 Mn, 0.40 Si, bal Fe.
Deposited: 70,000 TS; 60,000 YS; 20 El.
AC-DC welding electrode for carbon sheet steel; improved weld appearance; easy clean-up.
Conforms to AWS Class E-6013.

SUPER-Y.
M-902; 3.3 C, 1.5 Ni, 0.5 Cr, bal Fe.
Annealed: 60,000 TS; 44,000 YS; 18 El; 143 Brin.
For castings; malleable iron.

SUPER-Z.
M-91.
Zn base alloy sheet and strip.
For hot forming, forging, extruding, vacuum pressure formed.

SUPER ZORITE.
M-184; 37-40 Ni, 17-21 Cr, bal Fe.
For beams, rails, trays, furnace parts; heat resisting to 1800°F.

SUPRA.
M-Eng.; 20 Si, 5 Cu, 2 Mn, 0.7 Fe, bal Al.
Cast: 21,000 TS; 0.5 El; 100-130 Brin.
For pistons; light weight.

SUPRAKOLBENMETALL.
20 Si, 5 Cu, 2 Mn, 0.7 Fe, bal Al.
For light alloy parts, pistons.

SUPRALOY.
M-1128; C, B, bal Fe.
640 Brin.
For hard surfacing electrodes; for metal to metal wear.

SUPRA REKORD I.
M-1307; 0.86 C, 2.8 Co, 4.3 Cr, 0.85 Mo, 2.1 V, 12 W, bal Fe.
For lathe and planer tools, drills, reamers, taps; high speed steel.

SUPRA REKORD II.
M-1307; 0.79 C, 4.7 Co, 4.3 Cr, 0.75 Mo, 1.5 V, 18 W, bal Fe.
For lathe and planer tools, reamers, broaches, taps; high speed steel.

SUPRA RECORD III.
M-1307; 0.76 C, 10 Co, 4.2 Cr, 0.8 Mo, 1.8 V, 18 W, bal Fe.
For lathe and planer tools, reamers, broaches, hobs; high speed steel.

SUPRA VENIVICI.
M-1307; 0.86 C, 4.1 Cr, 0.85 Mo, 2.5 V, 12 W, bal Fe.
For lathe and planer tools, milling cutters, hobs; high speed steel.

SUPRA VENIVICI EXTRA.
M-1307; 0.74 C, 4.1 Cr, 1.1 V, 18.5 W, bal Fe.
For lathe and planer tools, drills, taps, hobs; high speed steel.

SUPREMUS, (AISI T1).
M-69; 0.73 C, 18 W, 4 Cr, 1.1 V, bal Fe.
For high speed tools, cutters, planer tools, broaches, drills; high speed steel.

SUPREMUS EXTRA, (AISI T2).
M-69; 0.85 C, 0.7 Mo, 18.5 W, 4 Cr, 2 V, bal Fe.
For cutters, drills, lathe centers, broaches, reamers; high speed steel; Type T1.

SUPRIMPACTO.
M-435; 0.12 C, 5 Mn, 1.5 Cr, 3.75 Ni, bal Fe.
For plastic mold dies, carburized parts; machine cut cavities, case hardened.

SUPRIMPACTO.
M-435; 0.12 C, 1.5 Cr, 3.75 Ni, bal Fe.
Carburizing steel, deep hardening. For large gears.

SU-PYR-LOY.
M-1432; C, alloy, bal Fe.
Rolled: 153,000 TS; 120,000 YS; 241 Brin.
Heat treated: 264,000 TS; 210,000 YS; 530 Brin.
For shear blades, chisels, punches; water hardened, non-tempering.

SURA 13-CKN-30.
M-1189; 0.12 C, 0.75 Cr, 3 Ni, bal Fe.
For gears, shafts, camshafts; case hardened, shock resistant.

SURA 18-CKN-30.
M-1189; 0.18 C, 0.75 Cr, 3 Ni, bal Fe.
For gears, shafts; case hardened, shock resistant.

SURA 25-CKM-10.
M-1189; 0.25 C, 1.05 Cr, 0.2 Mo, bal Fe.
For machinery parts; water hardening.

SURA 30-CKMN-32.
M-1189; 0.30 C, 1.05 Cr, 3.25 Ni, 0.25 Mo, bal Fe.
For gears, shafts; oil hardened, shock resistant.

SURA 30-CMK-27.
M-1189; 0.30 C, 2.7 Cr, 0.5 Mo, bal Fe.
For machinery parts; oil hardened.

SURA 35-CKN-25.
M-1189; 0.35 C, 1.15 Cr, 2.6 Ni, bal Fe.
For shafts, gears, pinions; oil hardened, tough.

SURA 40-CKN-12.
M-1189; 0.4 C, 0.8 Mn, 0.8 Cr, 1.25 Ni, bal Fe.
For shafts, gears, pinions; oil hardened, tough.

SURA 40-CMK-10.
M-1189; 0.40 C, 1.1 Cr, 0.25 Mo, bal Fe.
For machinery parts; oil hardened.

SURA 50-CG-12.
M-1189; 0.5 C, 1.2 Mn, bal Fe.
For gears, shafts, machinery parts; oil hardened, tough.

SURA 50-CK-5.
M-1189; 0.5 C, 0.5 Cr, bal Fe.
For shafts; water hardening.

SURA-A30.
M-1189; 4.2 Si, bal Fe.
For transformers; silicon steel.

SURA-AT26.
M-1189; 4.2 Si, bal Fe.
For transformers; silicon steel.

SURA-C33.
M-1189; 3.5 Si, bal Fe.
For transformers; silicon steel.

SURA-C37.
M-1189; 3.5 Si, bal Fe.
For transformers, large electrical machines; silicon steel.

SURA-C43.
M-1189; 2.5 Si, bal Fe.
For tranformers, electrical machines; silicon steel.

SURA-C44.
M-1189; 2.1 Si, bal Fe.
For transformers, electrical machines; silicon steel.

SURA-CT28.
M-1189; 3.5 Si, bal Fe.
For tranformers; silicon steel.

SURA-CT30.
M-1189; 3.5 Si, bal Fe.
For machines and transformers for high frequencies; silicon steel.

SURA-D57.
M-1189; 1.7 Si, bal Fe.
For D.C, motors subject to heavy loads; silicon steel.

SURA-D59.
M-1189; 1.7 Si, bal Fe.
For electrical machines and tranformers; silicon steel.

SURA-D60.
M-1189; 1.0 Si, bal Fe.
For transformers, low frequency machines; silicon steel lamination.

SURA-D66.
M-1189; 1.3 Si, bal Fe.
For small electrical motors; silicon steel laminations.

SURA-D68.
M-1189; 1.0 Si, bal Fe.
For small electrical motors; silicon steel laminations.

SURA-D70.
M-1189; 1.0 Si, bal Fe.
For electrical machines; silicon steel.

SURA-D80.
M-1189; 0.5 Si, bal Fe.
For armatures, motors, generators; silicon steel laminations.

SURAHAMMAR 15-CMK-6.
M-1189; 0.15 C, 0.8 Cr, 0.6 Mo, bal Fe.
For machinery parts, gears; case hardened.

SUREWELD A.
M-844; 0.07 C, bal Fe.
Welded: 67,000 TS; 53,000 YS; 30 El.
For welding electrodes; coated.

SUREWELD B.
M-844; 0.06 C, bal Fe.
Welded: 70,000 TS; 55,000 YS; 30 El.
For welding electrodes; coated.

SUREWELD C.
M-844; C, bal Fe.
Welded: 67,000 TS; 25 El.
For welding electrodes; coated.

SUREWELD F.
M-844; 0.09 C, bal Fe.
Welded: 67,000 TS; 55,000 YS; 27 El; 35 RA.
For welding electrodes; coated.

SUREWELD H.
M-844; C, 12-15 alloy, bal Fe.
For hard facing electrodes; wear resistant.

SUREWELD H1.
M-844; C, 15 alloy, bal Fe.
Welded: 550-650 Brin.
For hard facing electrodes; impact resistance.

SUREWELD H2.
M-844; C, 18 alloy, bal Fe.
Welded: 500-575 Brin.
For hard facing electrodes; wear resistant.

SUREWELD MLY.
M-844; 0.06 C, bal Fe.
Welded: 80,000 TS; 62,000 YS; 25 El.
For welding electrodes; coated.

SUREWELD N.
M-844; 0.9 C, bal Fe.
Welded: 75,000 TS; 55,000 YS; 25 El; 20 RA.
For welding electrodes; coated.

SUREWELD NO. 450.
M-844; C, alloy, bal Fe.
For welding electrodes; coated, wear and shock resisting.

SUREWELD NO. 520.
M-844; C, 12-14 Mn, bal Fe.
600 Brin.
For Mn steel welding electrodes; hard surfacing.

SUREWELD NO. 550.
M-844; C, alloy, bal Fe.
600-675 Brin.
For hard surfacing electrodes; abrasion and wear resisting.

SUREWELD NO. 650.
M-844; C, alloy, bal Fe.
700-800 Brin. For hard-surfacing electrodes; severe abrasion resisting.

SUREWELD TD.
M-844; C, 2 alloy, bal Fe.
Welded: 600-780 Brin.
For hard facing electrodes for cutting tools and dies; wear resistant.

SURF-HARD.
M-1636; 0.40 C, 0.60 Mn, 1.6 Cr, 0.35 Mo, 1.05 Al, bal Fe.
Annealed: 87,000 TS; 54,000 YS; 20 El; 179 Brin.
For ejector pins for plastic molds, bushings, sleeves, shafts, gears.
Nitriding steel, wear and abrasion resistant, nitrided case.

SURGALOY.
M-1019; 0.20 max C, 18 Cr, 8 Ni, bal Fe.
For sutures for surgery; stainless.

SUSINI.
M-Japan; 1.5-4.5 Cu, 1-8 Mn, 0.5-1.5 Zn, bal Al.
For light alloy parts; non-hardenable.

SU-VENEER.
M-901; clad steel.
For corrosion resistant stampings and drawn parts; corrosion resistant.

SUWEFA 12M.
M-1307; 1.2 C, 0.4 Si, 12.5 Mn, bal Fe.
For wear plates, rails, crushers, rollers; wear and abrasion resistant. Work hardened.

SUWEFA A12.
M-1307; 0.10 max C, 12.5 Cr, 12 Ni, bal Fe.
For chemical plant equipment, valves; heat and corrosion resistant.

SUWEFA A18.
M-1307; 0.15 max C, 18 Cr, 9 Ni, bal Fe.
Annealed: 80,000 TS; 35,000 YS; 55 El; 75 RA; 150 Brin.
For chemical plant equipment, tanks, mixers; Type 302; stainless, austenitic.

SUWEFA A18Z.
M-1307; 0.12 max C, 18 Cr, 9.5 Ni, Ti = 4 x C, bal Fe.
Annealed: 85,000 TS; 35,000 YS; 55 El; 65 RA; 150 Brin.
For welded chemical plant equipment, tanks; Type 321; stainless, austenitic.

SUWEFA A18ZN.
M-1307; 0.12 max C, 18 Cr, 9.5 Ni, Cb = 8 x C, bal Fe.
Annealed: 90,000 TS; 45,000 YS; 56 El; 65 RA; 160 Brin.
For welded chemical plant equipment, tanks; vessels; Type 347; stainless, austenitic.

SUWEFA A182Z.
M-1307; 0.12 max C, 18 Cr, 10.5 Ni, 2 Mo, Ti = 4 x C, bal Fe.
Annealed: 85,000 TS; 35,000 YS; 55 El; 65 RA; 150 Brin.
For welded chemical plant equipment, tanks, vessels; Type 321; stainless, austenitic.

SUWEFA A182ZN.
M-1307; 0.12 max C, 18 Cr, 10.5 Ni, 2 Mo, Cb = 8 x C, bal Fe.
Annealed: 90,000 TS; 45,000 YS; 56 El; 65 RA; 160 Brin.
For welded chemical plant equipment, tanks, vessels; Type 347; stainless, austenitic.

SUWEFA A2182Z.
M-1307; 0.07 C, 17.5 Cr, 2 Mo, 17.5 Ni, 2 Cu, Ti = 7 x C, bal Fe.
For chemical plant equipment; heat and corrosion resistant.

SUWEFA AS18.
M-1307; 0.07 max C, 18 Cr, 9.5 Ni, bal Fe.
Annealed: 85,000 TS; 35,000 YS; 60 El; 70 RA; 150 Brin.
Cold drawn: 180,000 TS; 125,000 YS; 10 El; 330 Brin.
For welded chemical plant equipment, tanks, mixers; Type 304; stainless, austenitic.

SUWEFA AS175.
M-1307; 0.07 max C, 17 Cr, 4.75 Mo, 13 Ni, bal Fe.
Annealed: 90,000 TS; 40,000 YS; 55 El; 65 RA; 160 Brin.
For acid resistant chemical plant equipment, tanks; Type 317; stainless, austenitic.

SUWEFA AS182.
M-1307; 0.07 max C, 18 Cr, 10.5 Ni, Cb = 8 x C, bal Fe.
Annealed: 90,000 TS; 45,000 YS; 56 El; 65 RA; 160 Brin.
Cold drawn: 100,000 TS; 65,000 YS; 40 El; 60 RA; 205 Brin.
For welded chemical plant equipment, tanks, mixers; Type 347; stainless, austenitic.

SUWEFA BEM.
M-1307; 0.4 C, 0.38 Si, 0.95 Mn, bal Fe.
Hot rolled: 90,000 TS; 58,000 YS; 27 El; 50 RA; 200 Brin.
For gears, shafts, bolts, fasteners; water hardened.

SUWEFA BKM.
M-1307; 0.3 C, 0.25 Si, 1.35 Mn, bal Fe.
For cams, bolts, camshafts, machine tool parts; oil hardened, tough.

SUWEFA C1VV.
M-1307; 0.50 C, 1.0 Cr, 0.09 V, 0.85 Mn, bal Fe.
For springs, gears, crankshafts, bolts; oil hardened, shock resistant.

SUWEFA C12.
M-1307; 2.1 C, 0.30 Mn, 11.5 Cr, bal Fe.
For blanking and forming dies, punches; oil hardened, non-deforming.

SUWEFA C12CO.
M-1307; 1.65 C, 0.3 Mn, 11.5 Cr, Co, bal Fe.
For blanking and forming dies, punches; air hardened, non-deforming.

SUWEFA C12 SPEZIAL.
M-1307; 2.1 C, 0.3 Mn, 11.5 Cr, 0.7 W, bal Fe.
For blanking and forming dies, punches; oil hardened, non-deforming.

SUWEFA C45.
M-1307; 0.41 C, 1.0 Cr, 0.65 Mn, bal Fe.
For springs, gears, crankshafts, bolts, studs; oil hardened, shock resistant.

SUWEFA CG.
M-1307; 0.40 C, Cr, Mn, Mo, bal Fe.
For gears, bolts, machine tool parts; oil hardened, tough.

SUWEFA CIVV.
M-1307; 0.50 C, 1.0 Cr, 0.09 V, bal Fe.
For springs, gears, bolts, studs; oil hardened, tough.

SUWEFA CME.
M-1307; 0.15 C, Cr, Mo, bal Fe.
For gears, bolts, machine tool parts; case hardened, tough.

SUWEFA CMM.
M-1307; 0.20 C, Cr, Mo, bal Fe.
For gears, cams, machine tool parts; case hardened, tough.

SUWEFA CMV2.
M-1307; 0.25 C, 1.0 Cr, 0.2 Mo, 0.65 Mn, bal Fe.
For gears, bolts, machine tool parts; oil hardened, tough.

SUWEFA CMV3.
M-1307; 0.33 C, 1.0 Cr, 0.2 Mo, 0.65 Mn, bal Fe.
For gears, fasteners, machine tool parts; oil hardened, tough.

SUWEFA CMV4.
M-1307; 0.42 C, 1.0 Cr, 0.2 Mo, 0.65 Mn, bal Fe.
For gears, bolts, machine tool parts; oil hardened, tough.

SUWEFA CMV SPEZIAL.
M-1307; 0.50 C, Cr, Mo, bal Fe.
For gears, bolts, machine tool parts; oil hardened, tough.

SUWEFA CN1V.
M-1307; 0.28-0.35 C, 0.5 Cr, 1.5 Ni, bal Fe.
For gears, bolts, machine tool parts; oil hardened, tough.

SUWEFA CN2E.
M-1307; 0.13 C, 0.7 Cr, 2.5 Ni, bal Fe.
For gears, cams, camshafts; case hardened, tough.

SUWEFA CN2V.
M-1307; 0.28-0.35 C, 0.7 Cr, 2.5 Ni, bal Fe.
For gears, bolts, crankshafts; oil hardened, tough.

SUWEFA CN3E.
M-1307; 0.13 C, 0.7 Cr, 3.5 Ni, bal Fe.
For gears, cams, camshafts, machine tool parts; case hardened, shock resistant.

SUWEFA CN3V.
M-1307; 0.22-0.30 C, 0.7 Cr, 3.5 Ni, bal Fe.
For gears, bolts, shafts, studs, machine tool parts; oil hardened, shock resistant.

SUWEFA CN4E.
M-1307; 0.13 C, 1.1 Cr, 4.5 Ni, bal Fe.
For gears, bolts, cams, shafts, camshafts; case hardening steel.

SUWEFA CN4V.
M-1307; 0.35 C, 1.3 Cr, 3.5 Ni, bal Fe.
For gears, bolts, fasteners, crankshafts; oil hardened, shock resistant.

SUWEFA CO 300.
M-1307; 0.86 C, 2.8 Co, 4.3 Cr, 0.85 Mo, 2.1 V, 12 W, bal Fe.
For lathe and planer tools, hobs, reamers, drills, taps; high speed steel.

SUWEFA CO 500.
M-1307; 0.79 C, 4.75 Co, 4.3 Cr, 0.75 Mo, 1.5 V, 18 W, bal Fe.
For lathe and planer tools, reamers, broaches, taps; high speed steel.

SUWEFA CO 1000.
M-1307; 0.76 C, 10 Co, 4.2 Cr, 0.8 Mo, 1.8 V, 18 W, bal Fe.
For lathe and planer tools, milling cutters, hobs; high speed steel.

SUWEFA CSV4.
M-1307; 0.38 C, 0.85 Si, 1.0 Cr, 0.1 V, bal Fe.
For springs, gears, bolts, studs; oil hardened, shock resistant.

SUWEFA CSV5.
M-1307; 0.45 C, 1.0 Cr, 0.1 V, bal Fe.
For springs, gears, bolts, studs; oil hardened, shock resistant.

SUWEFA CSV6.
M-1307; 0.61 C, 1.18 Cr, 0.1 V, 0.75 Mn, bal Fe.
For springs, gears, bolts, studs; oil hardened, shock resistant.

SUWEFA CV.
M-1307; 0.41 C, 1.0 Cr, 0.65 Mn, bal Fe.
For gears, bolts, crankshafts; oil hardened, tough.

SUWEFA CV50.
M-1307; 0.41 C, 0.25 Sn, 0.65 Mn, 1.0 Cr, bal Fe.
For gears, bolts, studs, shafts, fasteners; oil hardened, shock resistant.

SUWEFA DMO 10.
M-1307; 0.24 C, 1.15 Cr, 0.25 Mo, bal Fe.
For gears, bolts, fasteners, machine tool parts; oil hardened, tough.

SUWEFA DMO 11.
M-1307; 0.15 C, 0.65 Mn, 0.30 Mo, bal Fe.
For gears, fasteners, machine tool parts; case hardening steel.

SUWEFA DMO 14.
M-1307; 0.13 C, 0.85 Cr, 0.45 Mo, bal Fe.
For gears, fasteners, machine tool parts; case hardening steel.

SUWEFA DMO 15.
M-1307; 0.24 C, 1.25 Cr, 0.45 Mo, bal Fe.
For gears, machine tool parts; oil hardened, tough.

SUWEFA DMO 20.
M-1307; 0.24 C, 1.35 Cr, 0.55 Mo, 0.20 V, bal Fe.
For gears, shafts, machine tool parts; oil hardened, tough.

SUWEFA E50.
M-1307; 0.35 C, 0.25 Si, 0.55 Mn, bal Fe.
For gears, bolts, machine tool parts; water hardened.

SUWEFA EC3.
M-1307; 0.15 C, 0.25 Si, 0.5 Mn, 0.65 Cr, bal Fe.
For gears, bolts, machine tool parts; case hardened.

SUWEFA ECM.
M-1307; 0.20 C, 0.25 Si, 1.25 Mn, 1.15 Cr, bal Fe.
For gears, cams, camshafts, fasteners; case hardened, shock resistant.

SUWEFA EH50.
M-1307; 0.15 C, 0.25 Si, 0.37 Mn, bal Fe.
Annealed: 70,000 TS; 40,000 YS; 25 El; 60 RA; 145 Brin.
For gears, fasteners, machine tool parts; case hardening steel.

SUWEFA EHC40.
M-1307; 0.13 C, 0.5 Cr, bal Fe.
For gears, bolts, oil refinery equipment; case hardening steel.

SUWEFA EHC70.
M-1307; 0.15 C, 0.65 Cr, 0.5 Mn, bal Fe.
For gears, shafts, oil refinery equipment; case hardening steel.

SUWEFA EHC90.
M-1307; 0.16 C, 0.95 Cr, 1.15 Mn, bal Fe.
For gears, bolts, camshafts, cams; case hardening steel.

SUWEFA ELCS.
M-1307; 1.45 C, 0.6 Mn, 1.4 Cr, bal Fe.
For bearings, liners, sleeves, bushings; water hardened, wear resistant.

SUWEFA EMC5.
M-1307; 0.16 C, 0.95 Cr, 1.15 Mn, bal Fe.
For gears, cams, camshafts, fasteners; case hardening steel, tough.

SUWEFA EMC5H.
M-1307; 0.20 C, 1.25 Mn, 1.15 Cr, bal Fe.
For gears, cams, camshafts, fasteners; case hardening steel, tough.

SUWEFA ENC14.
M-1307; 0.13 C, 0.7 Cr, 3.5 Ni, bal Fe.
For gears, cams, camshafts; case hardening steel, tough.

SUWEFA ENC18.
M-1307; 013 C, 1.1 Cr, 4.5 Ni, bal Fe.
For gears, bolts, machine tool parts; case hardening steel, tough.

SUWEFA EW3.
M-1307; 0.15 C, 0.5 Mn, 0.65 Cr, bal Fe.
For gears, cams, camshafts, machine tool parts; case hardening steel.

SUWEFA EW5H.
M-1307; 0.20 C, 1.15 Cr, 1.25 Mn, bal Fe.
For gears, cams, camshafts, fasteners; case hardening steel, tough.

SUWEFA EX15.
M-1307; 0.15 C, 1.55 Cr, 1.55 Ni, bal Fe.
For gears, cams, camshafts; case hardened, tough.

SUWEFA EX20.
M-1307; 0.18 C, 2 Cr, 2 Ni, bal Fe.
For gears, cams, camshafts; case hardened, tough.

SUWEFA EX25.
M-1307; 0.13 C, 0.7 Cr, 2.5 Ni, bal Fe.
For gears, bolts, machine tool parts; case hardened, shock resistant.

SUWEFA EXTRA.
M-1307; 1.0 C, 0.10 V, 0.25 Mn, bal Fe.
For cutters, taps, springs, drills, reamers; Type W2; water hardened.

SUWEFA EXTRA MH.
M-1307; 1.1 C, 0.25 max Si, 0.25 max Mn, bal Fe.
Annealed: 110,000 TS; 56,000 YS; 20 El; 40 RA; 210 Brin.
For springs, tools, cutters, taps, drills; Type W1; water hardened.

SUWEFA EXTRA WEICH.
M-1307; 0.70 C, 0.20 max Si, 0.25 max Mn, bal Fe.
Heat treated: 174,000 TS; 128,000 YS; 12 El; 37 RA; 355 Brin.
For springs, rails, punches, axes, hammers; Type W1; water hardened.

SUWEFA EXTRA ZAH.
M-1307; 0.70 C, 0.25 max Si, 0.25 max Mn, bal Fe.
Heat treated: 175,000 TS; 128,000, YS; 12 El; 37 RA; 355 Brin.
For springs, rails, axes, punches; Type W1; water hardened.

SUWEFA EXTRA ZH.
M-1307; 1.0 C, 0.25 max Si, 0.25 Mn, bal Fe.
Annealed: 100,000 TS; 53,000 YS; 21 El; 42 RA; 200 Brin.
For cutters, taps, springs, drills, broaches; Type W1; water hardened.

SUWEFA F11Z.
M-1307; 0.1 max C, 17.5 Cr, Ti = 7 x C, bal Fe.
Annealed: 80,000 TS; 50,000 YS; 25 El; 50 RA; 150 Brin.
For oil refinery equipment, sinks; Type 430 Ti; stainless.

SUWEFA F13.
M-1307; 0.12 max C, 13 Cr, bal Fe.
Annealed: 75,000 TS; 40,000 YS; 35 El; 70 RA; 155 Brin.
For turbine blades, valves, cutlery; Type 410; stainless.

SUWEFA F17.
M-1307; 0.8 C, 17 Cr, bal Fe.
Annealed: 107,000 TS; 62,000 YS; 18 El; 35 RA; 220 Brin.
For cutlery, valves, ball bearings, surgical instruments; Type 440B; corrosion resistant.

SUWEFA F17A.
M-1307; 0.12 C, 16.5 Cr, 0.25 Mo, 0.2 S, bal Fe.
Annealed: 80,000 TS; 50,000 YS; 25 El; 50 RA; 150 Brin.
For screw machine products, oil refinery equipment; Type 430F; corrosion resistant.

SUWEFA F171Z.
M-1307; 0.1 max C, 17 Cr, 1.8 Mo, Ti = 7 x C, bal Fe.
Annealed: 125,000 TS; 95,000 YS; 20 El; 55 RA; 260 Brin.
For welded chemical plant equipment, pump parts; corrosion resistant; Type 431 Ti.

SUWEFA FBV.
M-1307; 1.15 C, 0.65 Cr, 0.10 V, bal Fe.
For bearings, bushings, liners; water or oil hardened, wear resistant.

SUWEFA FPC4.
M-1307; 0.40 C, Cr, Mn, V, bal Fe.
For gears, bolts, crankshafts, fasteners; oil hardened, shock resistant.

SUWEFA GB1.
M-1307; 0.15 C, 0.25 Si, 0.37 Mn, bal Fe.
Annealed: 70,000 TS; 40,000 YS; 25 El; 60 RA; 145 Brin.
For gears, bolts, machine tool parts; case hardened.

SUWEFA GB2.
M-1307; 0.22 C, 0.25 Si, 0.45 Mn, bal Fe.
Annealed: 73,000 TS; 42,000 YS; 21 El; 57 RA; 145 Brin.
For machine tool parts, gears, fasteners; case hardened.

SUWEFA GB3.
M-1307; 0.35 C, 0.25 Si, 0.55 Mn, bal Fe.
Hot rolled: 85,000 TS; 54,000 YS; 30 El; 53 Ra; 185 Brin.
For gears, bolts, machine tool parts; water hardened.

SUWEFA GB4.
M-1307; 0.4 C, 0.25 Si, 0.65 Mn, bal Fe.
Hot rolled: 98,000 TS; 60,000 YS; 24 El; 45 RA; 212 Brin.
For gears, bolts, machine tool parts; water hardened.

SUWEFA GB5.
M-1307; 0.53 C, 0.25 Si, 0.65 Mn, bal Fe.
Normalized: 100,000 TS; 55,000 YS; 19 El; 26 RA; 200 Brin.
For gears, bolts, machine tool parts; water hardened.

SUWEFA GB5H.
M-1307; 0.56 C, 0.30 Si, 0.55 Mn, bal Fe.
Normalized: 102,000 TS; 56,000 YS; 18 El; 25 RA; 205 Brin.
For gears, bolts, machine tool parts; water hardened.

SUWEFA GB6.
M-1307; 0.61 C, 0.25 Si, 0.65 Mn, bal Fe.
Heat treated: 160,000 TS; 113,000 YS; 12 El; 40 RA; 320 Brin.
For rails, axes, hammers, springs; water hardened.

SUWEFA GB9.
M-1307; 0.90 C, 0.37 Si, 1.0 Mn, bal Fe.
For springs, cutters; Type W1; water hardened.

SUWEFA GB11.
M-1307; 0.15 C, 0.25 Si, 0.37 Mn, bal Fe.
Annealed: 70,000 TS; 40,000 YS; 25 El; 60 RA; 145 Brin.
For gears, bolts, machine tool parts; case hardened.

SUWEFA GB12.
M-1307; 0.22 C, 0.25 Si, 0.45 Mn, bal Fe.
Cold drawn: 78,000 TS: 68,000 YS; 20 El; 55 RA; 160 Brin.
For machine tool parts; case hardened.

SUWEFA GB13.
M-1307; 0.35 C, 0.25 max Si, 0.55 max Mn, bal Fe.
Hot rolled: 85,000 TS; 54,000 YS; 30 El; 53 RA; 185 Brin.
For gears, machine tool parts; water hardened.

SUWEFA GB14.
M-1307; 0.45 C, 0.25 Si, 0.65 Mn, bal Fe.
Hot rolled: 98,000 TS; 59,000 YS; 24 El; 45 RA; 212 Brin.
For gears, bolts, machine tool parts; water hardened.

SUWEFA GB16.
M-1307; 0.60 C, 0.38 Si, 0.65 Mn, bal Fe.
Heat treated: 160,000 TS; 113,000 YS; 12 El; 40 RA; 321 Brin.
For machine tool parts, axes, shafts; water hardened.

SUWEFA GB18.
M-1307; 0.75 C, 0.38 Si, 0.70 Mn, bal Fe.
Heat treated: 185,000 TS; 142,000 YS; 15 El; 40 RA; 390 Brin.
For rails, springs, axes, hammmers, punches; Type W1; water hardened.

SUWEFA GBS40.
M-1307; 0.45 C, Ni, Cr, bal Fe.
For gears, bolts, machine tool parts; oil hardened, tough.

SUWEFA GBS45.
M-1307; 0.45 C, Ni, Cr, W, bal Fe.
For gears, bolts, machine tool parts; oil hardened, tough.

SUWEFA GBS50.
M-1307; 0.50 C, 1.05 Cr, 3.25 Ni, 0.5 Mn, bal Fe.
For gears, bolts, machine tool parts; oil hardened, shock resistant.

SUWEFA GC2.
M-1307; 1.45 C, 1.4 Cr, 0.6 Mn, bal Fe.
For bearings, bushings, liners; water hardened, wear resistant.

SUWEFA GCS9.
M-1307; 0.90 C, 1.2 Cr, 1.15 Si, 0.7 Mn, bal Fe.
For bearings, bushings, liners; water hardened, wear resistant.

SUWEFA GCS12.
M-1307; 1.25 C, 1.2 Cr, 1.15 Si, 0.7 Mn, bal Fe.
For bearings, bushings, liners; water hardened, wear resistant.

SUWEFA GCV.
M-1307; 0.38 C, 1.2 Si, Cr, V, bal Fe.
For bolts, gears, machine tool parts; oil hardened, tough.

SUWEFA GDW.
M-1307; 0.40 C, Cr, Mn, Mo, bal Fe.
For gears, bolts, machine tool parts; oil hardened, shock resistant.

SUWEFA GF2.
M-1307; 1.3 C, 0.25 max Si, 0.25 max P, bal Fe.
For gears, shafts, machine tool parts, bolts; water hardened.

SUWEFA GF43.
M-1307; 0.45 C, 0.25-0.50 Si, 0.3-0.8 Mn, bal Fe.
For gears, shafts, fasteners, machine tool parts; water hardened.

SUWEFA GFD3.
M-1307; 0.46 C, 1.7 Mn, bal Fe.
For punches, crimpers, shear blades; oil hardened.

SUWEFA GFD4.
M-1307; 0.46 C, 1.7 Si, 0.65 Mn, bal Fe.
For upsetters, riveters, punches; shock resistant, oil hardened.

SUWEFA GFD5.
M-1307; 0.65 C, 1.75 Si, 0.70 Mn, bal Fe.
For springs; tough, shock resistant.

SUWEFA GFD5W.
M-1307; 0.55 C, 1.7 Si, 0.70 Mn, bal Fe.
For upsetters, springs, punches; tough, shock resistant.

SUWEFA GFD7.
M-1307; 0.65 C, 1.15 Si, 1 Mn, bal Fe.
For springs; shock resistant.

SUWEFA GFD7R.
M-1307; 0.7 C, 0.7 Mn, 1.7 Si, bal Fe.
For springs; shock resistant.

SUWEFA GFD8.
M-1307; 0.67 C, 1.3 Si, 0.5 Mn, 0.5 Cr, bal Fe.
For springs; shock resistant.

SUWEFA GHE SPEZIAL.
M-1307; 0.56 C, Ni, Cr, Mo, V, bal Fe.
For gears, bolts, machine tool parts; oil hardened, shock resistant.

SUWEFA GKLO.
M-1307; 1.0 C, 1.55 Cr, 0.35 Mn, 0.3 Si, bal Fe.
For bearings, liners, sleeves; water hardened.

SUWEFA GR.
M-1307; 1.05 C, 1.0 Cr, 1.15 W, 0.90 Mn, bal Fe.
For cutters, fast finishing tools; water hardened.

SUWEFA GR3.
M-1307; 1.42 C, W, V, bal Fe.
For fast finishing cutters, form tools; water hardened.

SUWEFA GS.
M-1307; 0.55 C, 0.70 Cr, 0.18 Mo, 1.65 Ni, 0.1 V, bal Fe.
For gears, bolts, crankshafts; oil hardened, shock resistant.

SUWEFA GS7.
M-1307; 0.80 C, 0.10-0.40 Si, 0.50-0.70 Mn, bal Fe.
Heat treated: 188,000 TS; 143,000 YS; 12 El; 35 RA; 390 Brin.
For springs, taps, drills, hobs, reamers; Type W1; water hardened.

SUWEFA GS8.
M-1307; 0.80 C, 0.10-0.40 Si, 0.50-0.70 Mn, bal Fe.
Heat treated: 188,000 TS; 143,000 YS; 12 El; 35 RA; 390 Brin.
For springs, cutters, drills, taps; Type W1; water hardened.

SUWEFA GSI.
M-1307; 1.15 C, 0.20 Si, 0.30 Mn, 0.65 Cr, 0.10 V, bal Fe.
For bearings, bushings, liners; oil hardened, wear resistant.

SUWEFA GSP5.
M-1307; 0.53 C, 0.90 Si, 0.90 Mn, bal Fe.
For gears, bolts, studs, fasteners; water hardened.

SUWEFA GSP6.
M-1307; 0.70 C, 1.7 Si, 0.70 Mn, bal Fe.
For springs; oil hardened, tough.

SUWEFA GSS.
M-1307; 0.90 C, 1.9 Mn, 0.1 V, bal Fe.
For punches, dies, upsetters, cutters; oil hardened, non-deforming.

SUWEFA GSS2.
M-1307; 0.90 C, 1.9 Mn, 0.1 V, bal Fe.
For punches, dies, upsetters, cutters; oil hardened, non-deforming.

SUWEFA GSS111.
M-1307; 1.05 C, 1.0 Cr, 1.15 W, 0.9 Mn, bal Fe.
For fast finishing tools, bearings, cutters; water hardened.

SUWEFA GVS6.
M-1307; 0.55 C, 0.10-0.40 Si, 0.50-0.70 Mn, bal Fe.
Normalized: 100,000 TS; 55,000 YS; 18 El; 26 RA; 200 Brin.
For gears, bolts, shafts, studs; water hardened.

SUWEFA GW7.
M-1307; 0.70 C, 0.25 max Si, 0.25 max Mn, bal Fe.
Heat treated: 175,000 TS; 128,000 YS; 12 El; 37 RA; 355 Brin.
For rails, springs, axes, hammers, punches; water hardened.

SUWEFA GW7E.
M-1307; 0.70 C, 0.25 max Si, 0.25 max Mn, bal Fe.
Heat treated: 175,000 TS; 128,000 YS; 12 El; 37 RA; 355 Brin.
For rails, springs, axes, hammers, punches; water hardened.

SUWEFA GW8.
M-1307; 0.85 C, 0.25 max Si, 0.25 max Mn, bal Fe.
Heat treated: 190,000 TS; 145,000 YS; 10 El; 30 RA; 400 Brin.
For springs, taps, cutters, drills; Type W1; water hardened.

SUWEFA GW8E.
M-1307; 0.85 C, 0.25 max Si, 0.25 max Mn, bal Fe.
Heat treated: 190,000 TS; 145,000 YS; 10 El; 30 RA; 400 Brin.
For springs, taps, cutters, drills; Type W1; water hardened.

SUWEFA GW10.
M-1307; 1.0 C, 0.25 max Si, 0.25 max Mn, bal Fe.
Annealed: 100,000 TS; 53,000 YS; 21 El; 42 RA; 200 Brin.
For springs, cutters, drills, taps, reamers; Type W1; water hardened.

SUWEFA GW10E.
M-1307; 1.0 C, 0.25 max Si, 0.25 max Mn, bal Fe.
Annealed: 100,000 TS; 53,000 YS; 21 El; 42 RA; 200 Brin.
For springs, cutters, drills, taps,reamers; Type W1; water hardened.

SUWEFA GW10V.
M-1307; 1.0 C, 0.20 Si, 0.25 Mn, 0.1 V, bal Fe.
For springs, cutters, drills, taps, reamers; Type W2; water hardened.

SUWEFA GW11.
M-u307; 1.15 C, 0.25 max Si, 0.25 max Mn, bal Fe.
Annealed: 110,000 TS; 56,000 YS; 18 El; 40 RA; 210 Brin.
For springs, taps, cutters, broaches; Type W1; water hardened.

SUWEFA GW11E.
M-1307; 1.1 C, 0.25 max Si, 0.25 max Mn, bal Fe.
Annealed: 110,000 TS; 56,000 YS; 20 El; 40 RA; 205 Brin.
For springs, taps, reamers, broaches; Type W1; water hardened.

SUWEFA GW13.
M-1307; 1.3 C, 0.25 max Si, 0.25 max Mn, bal Fe.
For engravers' tools, reamers, broaches, taps; Type W1; water hardened.

SUWEFA GW23.
M-1307; 0.15 C, 0.25-0.50 Si, 0.30-0.70 Mn, bal Fe.
Annealed: 70,000 TS; 40,000 YS; 25 El; 60 RA; 145 Brin.
For gears, bolts, machine tool parts; case hardened.

SUWEFA GW33.
M-1307; 0.35 C, 0.25-0.50 Si, 0.30-0.70 Mn, bal Fe.
Hot rolled: 85,000 TS; 54,000 YS; 30 El; 53 RA; 185 Brin.
For gears, bolts, machine tool parts; water hardened.

SUWEFA GW43.
M-1307; 0.45 C, 0.25-0.50 Si, 0.30-0.70 Mn, bal Fe.
Hot rolled: 98,000 TS; 60,000 YS; 24 El; 45 RA; 215 Brin.
For gears, bolts, machine tool parts; water hardened.

SUWEFA GW63.
M-1307; 0.60 C, 0.25-0.50 Si, 0.30-0.70 Mn, bal Fe.
Heat treated: 160,000 TS; 113,000 YS; 12 El; 40 RA; 32 Brin.
For gears, rails, springs, hammers; water hardened.

SUWEFA GW73.
M-1307; 0.67 C, 0.25-0.50 Si, 0.30-0.70 Mn, bal Fe.
Heat treated: 170,000 TS; 125,000 YS; 14 El; 38 RA; 340 Brin.
For rails, springs, axes, crimpers; water hardened.

SUWEFA GW93.
  M-1307; 0.90 C, 0.25-0.50 Si, 0.30-0.70 Mn, bal Fe.
  Heat treated: 190,000 TS; 145,000 YS; 10 El; 30 RA; 400 Brin.
  For springs, tools, cutters, drills; Type W1; water hardened.

SUWEFA GZB8F.
  M-1307; 0.12 max C, 0.7 Si, 0.3 Mn, 6.5 Cr, 0.75 Al, bal Fe.
  For oil refinery equipment; creep and heat resistant.

SWEFA GZB9F.
  M-1307; 0.12 max C, 1.15 Si, 0.3 Mn, 13 Cr, 0.95 Al, bal Fe.
  For oil refinery equipment; creep and heat resistant.

SUWEFA GZB10A.
  M-1307; 0.15 C, 2 Si, 19.5 Cr, 9.5 Ni, bal Fe.
  Annealed: 80,000 TS; 35,000 YS; 55 El; 75 RA; 150 Brin.
  For chemical plant equipment, tanks, mixers; Type 302; stainless, austenitic.

SUWEFA GZB10F.
  M-1307; 0.12 max C, 18 Cr, 0.95 Al, bal Fe.
  For oil refinery equipment; heat and creep resistant.

SUWEFA GZB11FA.
  M-1307; 0.2 C, 25 Cr, 1.2 Si, 4 Ni, bal Fe.
  Cast: 90,000 TS; 65,000 YS; 18 El; 210 Brin.
  For cylinder liners, valve seats and bodies; corrosion and heat resistant.

SUWEFA GZB12A.
  M-1307; 0.15 C, 24 Cr, 19 Ni, bal Fe.
  Annealed: 100,000 TS; 45,000 YS; 50 El; 65 RA; 185 Brin.
  For furnace parts, valves, pumps, turbines; Type 310; corrosion and heat resistant.

SUWEFA GZB12F.
  M-1307; 0.12 max C, 1.5 Al, 24 Cr, bal Fe.
  For oil refinery equipment; heat and creep resistant.

SUWEFA GZW.
  M-1307; 1.4 C, 0.1 V, 0.3 Cr, 0.3 Mn, bal Fe.
  For bearings, liners, cutters; water hardened, wear resistant.

SUWEFA H10.
  M-1307; 0.21 C, 3 Cr, 0.4 Mo, 0.8 V, 0.37 W, bal Fe.
  For upsetters, crimpers; oil hardened.

SUWEFA HE15.
  M-1307; 0.15 C, 0.25 Si, 0.37 Mn, bal Fe.
  Annealed: 70,000 TS; 40,000 YS; 25 El; 60 RA; 145 Brin.
  For gears, bolts, machine tool parts; case hardened.

SUWEFA HEC5.
  M-1307; 0.13 C, Cr, bal Fe.
  For machine tool parts.

SUWEFA HEC10.
  M-1307; 0.15 C, 0.25 Si, 0.5 Mn, 0.65 Cr, bal Fe.
  For gears, machine tool parts; case hardened.

SUWEFA HEN15.
  M-1307; 0.13 C, 0.2 Cr, 1.5 Ni, bal Fe.
  For gears, bolts, camshafts; case hardened, shock resistant.

SUWEFA HH.
  M-1307; 1.2 C, Mn, bal Fe.
  For cutters, wear parts; oil hardened.

SUWEFA HKC10.
  M-1307; 0.33 C, 1.0 Cr, 0.2 Mo, 0.65 Mn, bal Fe.
  For gears, bolts, crankshafts, fasteners; oil hardened, tough.

SUWEFA HSB2.
  M-1307; 0.75 C, 0.25 max Si, 0.25 max Mn, bal Fe.
  Heat treated: 180,000-130,000 TS; 135,000-90,000 YS; 10-20 El; 35-50 RA; 370-245 Brin.
  For springs, tools, cutters, rails, die blocks; Type W1; water hardened.

SUWEFA HSB3.
  M-1307; 0.85 C, 0.5 Cr, bal Fe.
  For bearings, bushings, cutters; water hardened.

SUWEFA HZCV.
  M-1307; 0.50 C, 0.95 Mn, 1.05 Cr, 0.1 V, bal Fe.
  For springs, gears, bolts, shafts; oil hardened, shock resistant.

SUWEFA KL3.
  M-1307; 1.0 C, 1.55 Cr, 0.35 Mn, bal Fe.
  For bearings, bushings, liners; water hardened, wear resistant.

SUWEFA KMC.
  M-1307; 0.20 C, 1.15 Cr, 1.25 Mn, bal Fe.
  For bearings, liners, races, bushings; case hardened, tough.

SUWEFA KZM.
  M-1307; 1.4 C, 0.30 Mn, 0.1 V, bal Fe.
  For engravers tools, punches, header dies; water hardened; Type W2.

SUWEFA LCN.
  M-1307; 0.50 C, 1.05 Cr, 3.25 Ni, bal Fe.
  For gears, shafts, countershafts, bolts, studs; oil hardened, shock resistant.

SUWEFA LCV.
  M-1307; 0.50 C, 1.05 Cr, 0.1 V, 0.95 Mn, bal Fe.
  For gears, springs, bolts, studs, shafts; oil hardened, shock resistant.

SUWEFA LCW1.
M-1307; 0.35 C, 0.9 Si, 1.05 Cr, 0.18 V, 1.85 W, bal Fe.
For header dies, upsetters, crimpers; oil hardened, tough.

SUWEFA LCW2.
M-1307; 0.35 C, 1.05 Cr, 0.18 V, 1.85 W, bal Fe.
For header dies, punches, crimpers, upsetters; oil hardened, tough.

SUWEFA LCW3.
M-1307; 0.55 C, 0.9 Si, 1.05 Cr, 0.18 V, 1.85 W, bal Fe.
For header dies, punches, shears, crimpers; oil hardened, tough.

SUWEFA LST.
M-1307; 0.45 C, Cr, V, Si, bal Fe.
For gears, springs, bolts, studs, shafts; oil hardened, shock resistant.

SUWEFA M13.
M-1307; 0.4 C, 0.4 Si, 13 Cr, bal Fe.
Annealed: 95,000 TS; 50,000 YS; 25 El; 55 RA; 195 Brin.
For valves, cutlery, surgical and dental instruments; Type 420; stainless.

SUWEFA M-M1.
M-1307; 0.90 C, 18 Cr, 1.15 Mo, 1.0 V, bal Fe.
Heat treated: 280,000 TS; 270,000 YS; 3 El; 15 RA; 555 Brin.
For cutlery, valves, ball bearings; hardenable, corrosion resistant.

SUWEFA MO1.
M-1307; 0.95 C, W, Mo, bal Fe.
For dies, tools, cutters; oil hardened.

SUWEFA MO-333.
M-1307; 0.95 C, Cr, W, Mo, bal Fe.
For lathe and planer tools, reamers, broaches, taps; high speed steel.

SUWEFA MO-560.
M-1307; 0.85 C, Cr, W, Mo, bal Fe.
For lathe and planer tools, drills, reamers, hobs; high speed steel.

SUWEFA MO-900.
M-1307; 0.80 C, Mo, Cr, W, V, bal Fe.
For lathe and planer tools, reamers, broaches, taps; high speed steel.

SUWEFA MOG.
M-1307; 0.45 C, 0.7 Mn, 1.4 Cr, 0.7 Mo, 0.3 V, bal Fe.
For forging dies, punches, shears; oil hardened, tough.

SUWEFA MVC20.
M-1307; 0.30 C, Cr, V, bal Fe.
For gears, bolts, crankshafts; oil hardened, tough.

SUWEFA MV EXTRA.
M-1307; 0.42 C, 1.75 Mn, 0.25 Si, 0.1 V, bal Fe.
For punches, upsetters, shears; oil hardened, tough.

SUWEFA NC6.
M-1307; 0.27 C, 0.6 Mn, 1.1 Al, 1.4 Cr, bal Fe.
For oil refinery equipment; heat and creep resistant.

SUWEFA NC6H.
M-1307; 0.34 C, 1.1 Al, 1.4 Cr, 0.6 Mn, bal Fe.
For oil refinery equipment; heat and creep resistant.

SUWEFA NCMO4.
M-1307; 0.32 C, 1.1 Cr, 1.1 Al, 0.18 Mo, bal Fe.
For oil refinery equipment; heat and creep resistant.

SUWEFA NCV9 SPEZIAL.
M-1307; 0.31 C, 0.6 Mn, 2.35 Cr, 0.18 Mo, 0.13 V, bal Fe.
For oil refinery equipment; heat and creep resistant.

SUWEFA NFKC.
M-1307; 1.42 C, W, V, bal Fe.
For forming and blanking dies, cutters; oil hardened, wear resistant.

SUWEFA NG.
M-1307; 0.55 C, 0.70 Cr, 0.18 Mo, 1.65 Ni, 0.1 V, bal Fe.
For forging and heading dies, upsetters; oil hardened, tough.

SUWEFA NG2 SUPRA.
M-1307; 0.56 C, 0.2 Mo, 1.8 Ni, 0.1 V, 0.8 Cr, bal Fe.
For forging and heading dies, upsetters; oil hardened, tough.

SUWEFA NI 2.
M-1307; 0.13 C, 0.2 Cr, 1.5 Ni, bal Fe.
For gears, bolts, camshafts; case hardened.

SUWEFA NIC2.
M-1307; 0.25-0.35 C, 0.7 Cr, 2.5 Ni, bal Fe.
For gears, bolts, fasteners, crankshafts; oil hardened, shock resistant.

SUWEFA NIC 3.
M-1307; 0.13 C, 2.5 Ni, 0.7 Cr, bal Fe.
For gears, bolts, crankshafts, cams; case hardened, shock resistant.

SUWEFA NIC 4.
M-1307; 0.13 C, 0.7 Cr, 3.5 Ni, bal Fe.
For gears, bolts, camshafts, cams; case hardened, shock resistant.

SUWEFA NIC 4H.
M-1307; 0.22-0.30 C, 0.7 Cr, 3.5 Ni, bal Fe.
For gears, bolts, crankshafts, fasteners; oil hardened, shock resistant.

SUWEFA NIC 5.
M-1307; 0.13 C, 1.1 Cr, 4.5 Ni, bal Fe.
For gears, bolts, camshafts, cams; case hardened, shock resistant.

SUWEFA NIC 5H.
M-1307; 0.35 C, 1.3 Cr, 4.5 Ni, bal Fe.
For gears, bolts, machine tool parts; oil hardened, shock resistant.

SUWEFA NSI.
M-1307; 0.70 C, 1.7 Si, 0.70 Mn, bal Fe.
For springs, pneumatic tools, chisels; oil hardened, tough.

SUWEFA PIL80.
M-1307; 0.28 C, Ni, Cr, Mo, V, bal Fe.
For forging and heading dies; oil hardened, tough.

SUWEFA PRIMA WEICH.
M-1307; 0.70 C, 0.25 max Si, 0.25 max Mn, bal Fe.
Heat treated: 175,000-122,000 TS; 128,000-82,000 YS; 12-22 El; 37-52 RA; 350-240 Brin.
For springs, rails, clutch discs, girders, rails; Type W1; water hardened.

SUWEFA PRIMA ZAH.
M-1307; 0.85 C, 0.25 max Si, 0.25 max Mn, bal Fe.
Heat treated: 190,000 TS; 145,000 YS; 10 El; 32 RA; 400 Brin.
For drills, taps, reamers, springs, hobs; Type W1; water hardened.

SUWEFA PRIMA ZH.
M-1307; 1.05 C, 0.25 max Si, 0.25 max Mn, bal Fe.
Heat treated: 200,000 TS; 125,000 YS; 8 El; 28 RA; 400 Brin.
For springs, drills, reamers, taps, broaches; Type W1; water hardened.

SUWEFA REA.
M-1307; 0.15 max C, 18 Cr, 8.5 Ni, bal Fe.
Annealed: 80,000 TS; 35,000 YS; 55 El; 75 RA; 150 Brin.
For chemical plant equipment, tanks; Type 302; stainless, austenitic.

SUWEFA REA2.
M-1307; 0.15 max C, 18 Cr, 8.5 Ni, bal Fe.
Annealed: 80,000 TS; 35,000 YS; 55 El; 75 RA; 150 Brin.
For chemical plant equipment, tanks, mixers; Type 302; stainless, austenitic.

SUWEFA REA2-K.
M-1307; 0.12 max C, 18 Cr, 9.5 Ni, Ti = 4 x C, bal Fe.
Annealed: 85,000 TS; 35,000 YS; 55 El; 65 RA; 150 Brin.
For welded chemical plant equipment, mixers, tanks; Type 321; stainless, austenitic.

SUWEFA REA2-KN.
M-1307; 0.12 max C, 18 Cr, 9.5 Ni, Cb = 8 x C, bal Fe.
Annealed: 90,000 TS; 45,000 YS; 56 El; 65 RA; 160 Brin.
For welded chemical plant equipment, tanks, vessels; Type 347; stainless, austenitic.

SUWEFA REA2-KS.
M-1307; 0.07 C, 18 Cr, 9.5 Ni, bal Fe.
Annealed: 85,000 TS; 35,000 YS; 60 El; 70 RA; 150 Brin.
For chemical plant equipment, tanks, vessels; Type 304; stainless, austenitic.

SUWEFA REA4-K.
M-1307; 0.12 max C, 18 Cr, 2 Mo, 10.5 Ni, Ti = 4 x C, bal Fe.
Annealed: 90,000 TS; 40,000 YS; 45 El; 60 RA; 170 Brin.
For welded acid resistant equipment, tanks, mixers; Type 316 Ti; stainless, austenitic.

SUWEFA REA4-KN.
M-1307; 0.12 max C, 18 Cr, 10.5 Ni, 2 Mo, Cb = 8 x C, bal Fe.
Annealed: 90,000 TS; 40,000 YS; 45 El; 60 RA; 70 Brin.
For welded acid resistant chemical plant equipment; Type 316 Cb; stainless, austenitic.

SUWEFA REA4-KS.
M-1307; 0.07 max C, 18 Cr, 10.5 Ni, 2 Mo, bal Fe.
Annealed: 85,000 TS; 35,000 YS; 50 El; 65 RA; 160 Brin.
Cold drawn: 150,000 TS; 135,000 YS; 6 El; 300 Brin.
For acid resistant chemical plant equipment; Type 316; stainless, austenitic.

SUWEFA REH.
M-1307; 0.4 C, 0.4 Si, 13 Cr, bal Fe.
Annealed: 100,000 TS; 55,000 YS; 20 El; 50 RA; 205 Brin.
For valves, cutlery, surgical and dental instruments; Type 420; stainless.

SUWEFA REHD.
M-1307; 0.20 C, 13 Cr, 1.15 Mo, bal Fe.
Annealed: 100,000 TS; 55,000 YS; 22 El; 52 RA; 200 Brin.
For valves, cutlery, chemical plant equipment; Type 420 Mo; stainless.

SUWEFA REH EXTRA.
M-1307; 0.90 C, Cr, V, bal Fe.
For bearings, blanking and forming dies; oil hardened, wear resistant.

SUWEFA REM.
M-1307; 0.20 C, 13 Cr, 0.4 Si, 0.3 Mn, bal Fe.
Annealed: 95,000 TS; 50,000 YS; 25 El; 55 RA; 196 Brin.
For surgical and dental instruments, valves, cutlery; Type 420; stainless.

SUWEFA RES.
 M-1307; 0.65 C, 1.3 Si, 0.5 Mn, 0.5 Cr, bal Fe.
 For springs; oil hardened, tough.

SUWEFA RESO.
 M-1307; 0.65 C, 1.15 Si, 1.0 Mn, bal Fe.
 For springs; oil hardened, tough.

SUWEFA REW.
 M-1307; 0.12 max C, 0.4 Si, 13 Cr, bal Fe.
 Annealed: 75,000 TS; 40,000 YS; 35 El; 70 RA; 155 Brin.
 For turbine blades, valves, cutlery, surgical instruments; Type 410; stainless.

SUWEFA REXH.
 M-1307; 0.22 C, 17 Cr, 1.5 Ni, bal Fe.
 Annealed: 125,000 TS; 95,000 YS; 20 El; 55 RA; 250 Brin.
 For pumps, marine hardware, valves; Type 431; heat resistant.

SUWEFA REXH EXTRA.
 M-1307; 0.22 C, 0.4 Si, 17 Cr, 1.5 Ni, bal Fe.
 Annealed: 125,000 TS; 95,000 YS; 20 El; 55 RA; 260 Brin.
 Cold drawn: 130,000 TS; 110,000 YS; 15 El; 35 RA; 270 Brin.
 For pumps, marine hardware, valves; Type 431; heat resistant.

SUWEFA REXN.
 M-1307; 0.1 max C, 17.5 Cr, Ti = 7 x C, bal Fe.
 Annealed: 80,000 TS; 50,000 YS; 25 El; 50 RA; 150 Brin.
 For oil refinery equipment, oil burner heaters; Type 430 Ti; stainless.

SUWEFA REXW.
 M-1307; 0.8 C, 18 Cr, bal Fe.
 Annealed: 110,000 TS; 65,000 YS; 18 El; 35 RA; 220 Brin.
 Heat treated: 280,000 TS; 270,000 YS; 3 El; 15 RA; 555 Brin.
 For bearings, cutlery, valves; stainless, hardenable.

SUWEFA SBH.
 M-1307; 1.25 C, 1.15 Si, 0.70 Mn, 1.2 Cr, bal Fe.
 For blanking and forming dies, bearings; oil hardened, wear resistant.

SUWEFA SBN.
 M-1307; 1.1 C, 0.40 Cr, 0.30 Mn, bal Fe.
 For bearings, cutters, liners; water or oil hardened.

SUWEFA SCHE.
 M-1307; 0.61 C, 0.85 Si, 0.75 Mn, 1.18 Cr, 0.10 V, bal Fe.
 For springs, gears, bolts, crankshafts; oil hardened, shock resistant.

SUWEFA SER.
 M-1307; 0.67 C, 1.3 Si, 0.5 Mn, 0.5 Cr, bal Fe.
 For springs, punches, pneumatic tools; oil hardened, tough.

SUWEFA SIW.
 M-1307; 1.2 C, 0.20 Cr, 0.10 V, 1.0 W, bal Fe.
 For bearings, sleeves, liners; water hardened, wear resistant.

SUWEFA SJI.
 M-1307; 0.46 C, 1.7 Si, 0.65 Mn, bal Fe.
 For springs, punches, chisels; oil hardened, tough.

SUWEFA SJII.
 M-1307; 0.55 C, 1.7 Si, 0.70 Mn, bal Fe.
 For springs, punches, chisels, pneumatic tools; oil hardened, tough.

SUWEFA SJIII.
 M-1307; 0.65 C, 1.7 Si, 0.70 Mn, bal Fe.
 For springs, punches, chisels, pneumatic tools; oil hardened, tough.

SUWEFA SJCV.
 M-1307; 1.15 C, 0.2 Si, 0.3 Mn, 0.65 Cr, 0.1 V, bal Fe.
 For bearings, sleeves, liners; water hardened.

SUWEFA T13.
 M-1307; 0.2 C, 0.4 Si, 13 Cr, bal Fe.
 Annealed: 95,000 TS; 50,000 YS; 25 El; 50 RA; 195 Brin.
 For valves, cutlery, surgical and dental instruments; Type 420; stainless.

SUWEFA T17.
 M-1307; 0.22 C, 0.4 Si, 17 Cr, 1.5 Ni, bal Fe.
 Annealed: 125,000 TS; 95,000 YS; 20 El; 50 RA; 260 Brin.
 For pumps, marine hardware, valves; Type 431; stainless.

SUWEFA T131.
 M-1307; 0.20 C, 0.4 Si, 13 Cr, 1.15 Mo, bal Fe.
 Annealed: 100,000 TS; 55,000 YS; 20 El; 45 RA; 200 Brin.
 For oil refinery equipment, valves, cutlery; Type 430 Mo; stainless.

SUWEFA T171.
 M-1307; 0.35 C, 1.65 Cr, 1.15 Mo, bal Fe.
 For oil refinery equipment, heat treating boxes; corrosion resistant.

SUWEFA TC4.
 M-1307; 0.33 C, 0.25 Si, 0.65 Mn, 1.0 Cr, bal Fe.
 For gears, bolts, crankshafts, axles; oil hardened, tough.

SUWEFA TC4 SPEZIAL.
 M-1307; 0.41 C, 1.0 Cr, 0.65 Mn, bal Fe.
 For gears, bolts, crankshafts, axles; oil hardened, tough.

SUWEFA TCMO4.
M-1307; 0.34 C, 1.05 Cr, 0.20 V, 0.65 Mn, bal Fe.
For gears, bolts, springs, crankshafts; oil hardened, shock resistant.

SUWEFA TCMO4H.
M-1307; 0.42 C, 1.05 Cr, 0.20 V, 0.65 Mn, bal Fe.
For gears, bolts, springs, crankshafts; oil hardened, shock resistant.

SUWEFA TCMO4W.
M-1307; 0.25 C, 1.0 Cr, 0.2 Mo, 0.65 Mn, bal Fe.
For gears, bolts, crankshafts; oil hardened, tough.

SUWEFA TCMO5.
M-1307; 0.50 C, 1.0 Cr, 0.2 Mo, 0.65 Mn, bal Fe.
For gears, bolts, crankshafts; oil hardened, tough.

SUWEFA TCV4.
M-1307; 0.50 C, 1.0 Cr, 0.09 V, bal Fe.
For springs, bolts, crankshafts; oil hardened, shock resistant.

SUWEFA TCV5.
M-1307; 0.58 C, 1.0 Cr, 0.09 V, bal Fe.
For springs, shafts, gears, crankshafts; oil hardened, shock resistant.

SUWEFA TCV9 SPEZIAL.
M-1307; 0.30 C, 2.5 Cr, 0.2 Mo, 0.15 V, bal Fe.
For gears, shafts, crankshafts, pinions, bolts; oil hardened, shock resistant.

SUWEFA TFH.
M-1307; 0.58 C, Cr, V, bal Fe.
For springs, bolts, crankshafts; oil hardened, shock resistant.

SUWEFA TM4.
M-1307; 0.40 C, 0.37 Si, 0.95 Mn, bal Fe.
Hot rolled: 92,000 TS; 58,000 YS; 27 El; 50 RA; 200 Brin.
For gears, bolts, shafts, axles, screws; water hardened.

SUWEFA TM5.
M-1307; 0.30 C, 0.25 Si, 1.35 Mn, bal Fe.
For gears, bolts, shafts, axles, crankshafts; water hardened.

SUWEFA TMCV4.
M-1307; 0.27 C, Mn, Cr, V, bal Fe.
For gears, bolts, machine tool parts; oil hardened, tough.

SUWEFA TMS4.
M-1307; 0.53 C, 0.80 Si, 1.05 Mn, bal Fe.
For gears, bolts, machine tool parts; water hardened.

SUWEFA TMS5.
M-1307; 0.37 C, 1.25 Si, 1.25 Mn, bal Fe.
Hot rolled: 97,000 TS; 59,000 YS; 25 El; 52 RA; 200 Brin.
For punches, chisels, pneumatic tools; oil hardened, shock resistant.

SUWEFA TMV7.
M-1307; 0.42 C, 1.75 Mn, 0.1 V, bal Fe.
For punches, chisels, pneumatic tools, gears; oil hardened, tough.

SUWEFA TNC6.
M-1307; 0.28-0.35 C, 0.5 Cr, 1.5 Ni, bal Fe.
For gears, bolts, crankshafts, fasteners; oil hardened, shock resistant.

SUWEFA TNC14H.
M-1307; 0.22-0.30 C, 0.7 Cr, 3.5 Ni, bal Fe.
For gears, bolts, crankshafts, fasteners; oil hardened, shock resistant.

SUWEFA TNC18.
M-1307; 0.35 C, 1.3 Cr, 4.5 Ni, bal Fe.
For gears, bolts, crankshafts, axles; oil hardened, shock resistant.

SUWEFA TX10.
M-1307; 0.36 C, 1.0 Ni, 1.0 Cr, 0.2 Mo, bal Fe.
For die casting and plastic mold dies; oil hardened, tough.

SUWEFA TX15.
M-1307; 0.35 C, 1.55 Ni, 1.55 Cr, 0.2 Mo, bal Fe.
For die casting and plastic mold dies; oil hardened, tough.

SUWEFA TX20.
M-1307; 0.30 C, 2.0 Cr, 0.3 Mo, 2.0 Ni, bal Fe.
For die casting and plastic mold dies; oil hardened, tough.

SUWEFA UCN.
M-1307; 0.19 C, 1.25 Cr, 0.20 Mo, 3.75 Ni, bal Fe.
For gears, bolts, camshafts, cams; oil hardened, tough.

SUWEFA UER.
M-1307; 0.40 C, 13 Cr, 0.30 Mn, bal Fe.
Annealed: 100,000 TS; 55,000 YS; 20 El; 50 RA; 200 Brin.
For valves, cutlery, surgical and dental instruments; Type 420; corrosion resistant.

SUWEFA UH50.
M-1307; 0.15 C, 0.25-0.50 Si, 0.30-0.80 Mn, bal Fe.
Annealed: 70,000 TS; 40,000 YS; 25 El; 60 RA; 143 Brin.
For gears, bolts, cams, screws, fan blades; case hardened.

SUWEFA UH60.
M-1307; 0.15 C, 0.65 Cr, 0.25 Si, 0.50 Mn, bal Fe.
For gears, bolts, fan blades; case hardened.

SUWEFA UMC.
M-1307; 0.20 C, 1.15 Cr, 1.25 Mn, 0.25 Si, bal Fe.
For gears, bolts, camshafts, cams; case hardened.

SUWEFA UNIVERSAL.
M-1307; 0.53 C, 0.90 Si, 0.90 Mn, bal Fe.
For springs, bolts, gears, machine tool parts; oil hardened, tough.

SUWEFA URV.
M-1307; 1.42 C, W, V, bal Fe.
For engravers' tools, cutters, forming dies; oil hardened, wear resistant.

SUWEFA USS.
M-1307; 0.90 C, 1.9 Mn, 0.10 V, bal Fe.
For punches, blanking and forming dies; oil hardening, non-deforming.

SUWEFA V200.
M-1307; 0.82 C, 4.1 Cr, 0.85 Mo, 1.6 V, 8.7 W, bal Fe.
For lathe and planer tools, reamers, broaches; high speed steel.

SUWEFA V300.
M-1307; 0.86 C, 4.1 Cr, 0.85 Mo, 2.5 V, 12 W, bal Fe.
For lathe and planer tools, broaches, taps, drills; high speed steel.

SUWEFA V400.
M-1307; 1.3 C, 4.3 Cr, 0.85 Mo, 3.8 V, 12 W, bal Fe.
For engravers' tools, form cutters, reamers, taps; high speed steel.

SUWEFA VC11.
M-1307; 1.1 C, 0.40 Cr, 0.30 Mn, bal Fe.
For bearings, liners, sleeves, cutters; water or oil hardened.

SUWEFA VC12.
M-1307; 2.1 C, 11.5 Cr, 0.30 Mn, bal Fe.
For blanking and forming dies; oil hardened, non-deforming.

SUWEFA VCN188.
M-1307; 0.45 C, Cr, Ni, W, bal Fe.
For forging and heading dies; oil hardened, tough.

SUWEFA VCS2.
M-1307; 0.45 C, Si, Cr, bal Fe.
For machine tool parts; oil hardened, tough.

SUWEFA VCS9.
M-1307; 0.45 C, Cr, Si, bal Fe.
For machine tool parts; oil hardened, tough.

SUWEFA VMS.
M-1307; 0.37 C, 1.25 Si, 1.25 Mn, bal Fe.
For machine tool parts; oil hardened, tough.

SUWEFA VRV.
M-1307; 1.3 C, 0.25 Si, 0.30 Mn, 0.20 max Cr, 4.75 W, bal Fe.
For cutters, engravers' tools, reamers; fast-finishing tool steel.

SUWEFA W6.
M-1307; 0.60 C, 0.38 Si, 0.65 Mn, bal Fe.
Heat treated: 160,000 TS; 115,000 YS; 12 El; 40 RA; 325 Brin.
For wheels, die blocks, girders, rails, springs; water hardened.

SUWEFA W18.
M-1307; 0.74 C, 4.1 Cr, 1.1 V, 18.5 W, bal Fe.
For lathe and planer tools, reamers, broaches, drills; high speed steel.

SUWEFA WCV4.
M-1307; 0.50 C, 1.0 Cr, 0.85 Mn, bal Fe.
For gears, bolts, crankshafts; oil hardened, tough.

SUWEFA WCV5.
M-1307; 0.58 C, 1.0 Cr, 0.09 V, 0.95 Mn, bal Fe.
For springs, gears, bolts, machine tool parts; oil hardened, shock resistant.

SUWEFA WCV9 SPEZIAL.
M-1307; 0.30 C, Cr, Mo, V, bal Fe.
For forging and heading dies, punches; oil hardened, tough.

SUWEFA WEL.
M-1307; 1.05 C, Cr, bal Fe.
For bearings, cutters, header dies; water hardened, wear resistant.

SUWEFA WES.
M-1307; 1.2 C, 0.10 V, 0.20 Cr, 0.28 Si, 1.0 W, bal Fe.
For header and blanking dies; oil hardened, tough.

SUWEFA WF8.
M-1307; 0.67 C, 1.3 Si, 0.5 Mn, 0.50 Cr, bal Fe.
For springs, punches, chisels, upsetters; oil hardened, tough.

SUWEFA WGKL.
M-1307; 1.0 C, 0.30 Si, 0.35 Mn, 1.55 Cr, bal Fe.
For bearings, cutters, blanking dies; oil hardened, abrasion resistant.

SUWEFA WKL.
M-1307; 1.0 C, 1.1 Cr, 0.07 Mn, 0.25 Si, bal Fe.
For bearings, header dies; water hardened, wear resistant.

SUWEFA WM13.
M-1307; 0.40 C, 0.40 Si, 0.30 Mn, 13 Cr, bal Fe.
Annealed: 100,000 TS; 55,000 YS; 25 El; 55 RA; 200 Brin.
For valves, cutlery, surgical and dental instruments; Type 420; stainless.

SUWEFA WMS.
M-1307; 0.30 C, 2.65 Cr, 0.33 V, 8.5 W, bal Fe.
For extrusion rams and liners, punches; hot work steel, oil hardened.

SUWEFA WMS5.
M-1307; 0.30 C, 2.35 Cr, 0.6 V, 4.25 W, 0.30 Mn, bal Fe.
For extrusion rams, upsetters, punches, shears; hot work steel, oil hardened.

SUWEFA WMS SPEZIAL.
M-1307; 0.30 C, 2.0 Co, 2.4 Cr, 0.25 V, 8.5 W, bal Fe.
For hot work tools, dies, punches; hot work steel, oil hardened.

SUWEFA WMV.
M-1307; 0.45 C, 1.35 Cr, 0.45 Mo, 0.8 V, 0.45 W, bal Fe.
For forging and heading dies; die casting dies; oil hardened, tough.

SUWEFA WRL.
M-1307; 0.90 C, 0.80 Cr, 0.30 Mn, 0.25 Si, bal Fe.
For bearings, cutters, sleeves, liners; water hardened, wear resistant.

SUWEFA WRL HART.
M-1307; 1.05 C, 1.0 Cr, 0.30 Mn, 0.25 Si, bal Fe.
For bearings, cutters, sleeves, liners; water hardened, wear resistant.

SUWEFA WS1.
M-1307; 1.20 C, W, bal Fe.
For cutters, dies; oil hardened, wear resistant.

SUWEFA W SPEZIAL 4.
M-1307; 0.45 C, 0.25-0.50 Si, 0.30 Mn, bal Fe.
Hot rolled: 98,000 TS; 58,000 YS; 24 El; 45 RA; 212 Brin.
For machine tool parts, gears; water hardened.

SUWEFA W SPEZIAL H.
M-1307; 0.90 C, 0.25-0.50 Si, 0.30-0.80 Mn, bal Fe.
Heat treated: 190,000 TS; 145,000 YS; 10 El; 30 RA; 400 Brin.
For springs, taps, cutters, reamers, drills; Type W1; water hardened.

SUWEFA W SPEZIAL MH.
M-1307; 0.75 C, 0.25-0.50 Si, 0.30-0.80 Mn, bal Fe.
Heat treated: 175,000 TS; 128,000 YS; 12 El; 37 RA; 352 Brin.
For springs, rails, clutch discs, girders; Type W1; water hardened.

SUWEFA W SPEZIAL WEICH.
M-1307; 0.35 C, 0.25-0.50 Si, 0.30-0.80 Mn, bal Fe.
Hot rolled: 85,000 TS; 54,000 YS; 30 El; 53 RA; 183 Brin.
For gears, bolts, shafts, axles; water hardened.

SUWEFA W SPEZIAL ZAH.
M-1307; 0.45 C, 0.25-0.50 Si, 0.30-0.80 Mn, bal Fe.
Hot rolled: 98,000 TS; 59,000 YS; 24 El; 45 RA; 212 Brin.
For axles, gears, bolts, tie rods; water hardened.

SUWEFA W SPEZIAL ZH.
M-1307; 0.60 C, 0.25-0.50 Si, 0.30-0.80 Mn, bal Fe.
Heat treated: 160,000 TS; 113,000 YS; 12 El; 40 RA; 320 Brin.
For wheels, die blocks, girders, springs; water hardened.

SUWEFA WT13.
M-1307; 0.20 C, 0.40 Si, 0.30 Mn, 13 Cr, bal Fe.
Annealed: 73,000 TS; 40,000 YS; 22 El; 58 RA; 140 Brin.
For turbine blades, valves, cutlery; Type 420; stainless.

SUWEFA WZK SPEZIAL.
M-1307; 0.30 C, 2.65 Cr, 0.35 V, 8.5 W, bal Fe.
For extrusion rams and liners, punches, upsetters; hot work steel, oil hardened.

SUWEFA ZGW.
M-1307; 1.65 C, 11.5 Cr, 0.10 V, bal Fe.
For forming and blanking dies; air hardened, non-deforming.

SUWEFA ZH12.
M-1307; 2.1 C, 11.5 Cr, 0.30 Mn, 0.25 Si, bal Fe.
For forming and blanking dies, punches; oil hardened, non-deforming.

SUWEFA ZH120.
M-1307; 2.1 C, 11.5 Cr, 0.30 Mn, 0.35 Si, bal Fe.
For blanking and forming dies, punches; oil hardened, non-deforming.

SV 2.
M-1290; 0.78-0.86 C, 3.8-4.5 Cr, 8.3-9.0 W, 1.4-1.7 V, 0.70-1.0 Mo, bal Fe.
High speed steel; W.-Nr. 1.3316.

SV 4.
M-1290; 1.2-1.35 C, 3.8-4.5 Cr, 11.5-12.5 W, 3.5-4.0 V, 0.70-1.0 Mo, bal Fe.
High speed steel; W.-Nr. 1.3302.

SV 4 CO 5.
M-1290; 1.30-1.45 C, 3.8-4.5 Cr, 11.5-12.5 W, 3.5-4.0 V, 4.5-5.0 Co, 0.70-1.0 Mo, bal Fe.
High speed steel; W.-Nr. 1.3202.

SVENSKA SM-0010.
M-1186; 99.99 Cu, 0.03 oxygen.
Annealed: 33,000 TS; 70,000 YS; 45 El.
1/2 H-temper: 40,000 TS; 36,000 YS; 15 El; 80 Brin.
For electrical terminals; tough pitch copper.

SVENSKA SM-1092.
M-1186; 8 Zn, 92 Cu.
Annealed: 36,000 TS; 10,000 YS; 45 El; 50 Brin.
H-temper: 64,000 TS; 56,000 YS; 5 El; 125 Brin.
For hardware; commercial bronze.

SVENSKA SM-1160.
M-1186; 60 Cu, 0.4 Pb, bal Zn.
Annealed: 52,000 TS; 18,000 YS; 40 El; 100 Brin.
1/4 H-temper: 57,000 TS; 43,000 YS; 20 El; 137 Brin.
For architectural trim, condenser tubes, hardware; Muntz Metal.

SVENSKA SM-1163.
M-1186; 63 Cu, 0.3 Pb, bal Zn.
Annealed: 51,000 TS; 18,000 YS; 50 El; 64 Brin.
H-temper: 86,000 TS; 65,000 YS; 5 El; 176 Brin.
For screw machine products; low leaded brass.

SVENSKA SM-1261.
M-1186; 61 Cu, 1 Pb, bal Zn.
Annealed: 52,000 TS; 18,000 YS; 40 El; 75 Brin.
H-temper: 70,000 TS; 45,000 YS; 20 El; 137 Brin.
For screw machine products; leaded brass.

SVENSKA SM-1263.
M-1186; 63 Cu, 1 Pb, bal Zn.
Annealed: 51,000 TS; 18,000 YS; 40 El; 69 Brin.
1/2 H-temper: 65,000 TS; 50,000 YS; 20 El; 132 Brin.
For screw machine products; leaded brass.

SVENSKA SM-1661.
M-1186; 61.5 Cu, 3 Pb, bal Zn.
1/2 H-temper: 49,000 TS; 18,000 YS; 40 El; 61 Brin.
H-temper: 68,000 TS; 52,000 YS; 20 El; 150 Brin.
For screw machine products; leaded brass.

SVENSKA SM-1956.
M-1186; 56 Cu, 0.5 Pb, 0.4 Al, bal Zn.
For architectural trim; brass.

SVENSKA SM-2276.
M-1186; 76 Cu, 2 Al, 0.05 As, bal Zn.
Annealed: 65,000 TS; 42,000 YS; 40 El; 107 Brin.
For condenser tubes; aluminum brass.

SVENSKA SM-2771.
M-1186; 71 Cu, 1 Sn, 0.4 As, bal Zn.
Annealed: 65,000 TS; 42,000 YS; 35 El; 107 Brin.
For condenser tubes, evaporator and heat exchanger tubes; Inhibited Admiralty.

SVENSKA SM-4210.
M-1186; 88.4 Cu, 10 Ni, 1.2 Fe, 0.4 Mn, bal Zn.
Cold drawn: 44,000-60,000 TS; 22,000-57,000 YS; 45-15 El; 64-120 Brin.
For condenser tubes; corrosion resistant.

SVERKER 1.
M-111; 2.C, 0.7 Mn, 13 Cr, 0.2 V, bal Fe.
For dies; for cold work, non-deforming.

SVERKER 2.
M-111; 1.5 C, 12 Cr, 0.08 Mo, 0.2 V, bal Fe.
For dies; for cold work, non-deforming.

SVERKER 3.
M-111; 2.05 C, 0.75 Mn, 13 Cr, 1.25 W, bal Fe.
For coldwork dies; air or oil hardening; non-deforming; AISI D6.

SVERKER 21.
M-111; 1.55 C, 12 Cr, 0.8 Mo, 0.9 V, bal Fe.
Hardened: 278,000 TS; 214,000 YS; 1 El; C 56 Rock.
For hobs, plastic molds, punches.
Type D2 air hardening steel, wear resistant.

SVM.
M-1618; 0.55 C, 0.75 Mn, 1.85 Si, 0.25 V, 0.50 Mo, bal Fe.
Ht. Tr.: 275,000 TS; 247,000 YS; 9 El; 514 Brin.
For pneumatic tools, shear blades, punches, chisels, caulking tools.
Type S5 tool steel; tough, shock resistant.

SV-RHF 10.
M-912; 0.38 C, 0.25 Si, 0.65 Mn, 0.80 Cr, 1.80 Ni, 0.35 Mo, 0.10 V, bal Fe.
Q + T: 1770 N/mm^2 TS; 1470 N/mm^2 YS; 6 El.
For highly stressed parts; vacuum melted.
W.-Nr. 1.6926.

SV-RHF 20.
M-912; 0.40 C, 0.90 Si, 0.30 Mn, 5.0 Cr, 1.30 Mo, 0.50 V, bal Fe.
Q + T: 1960 N/mm^2 TS; 1570 N/mm^2 YS; 6 El.
For highly stressed parts; vacuum melted.
W.-Nr. 1.7783.

SV-RHF 30.
M-912; 0.02 C, 7.50 Co, 4.80 Mo, 18.0 Ni, 0.40 Ti, Al, bal Fe.
Heat treated: 1720 N/mm^2 min TS; 1620 N/mm^2 min YS; 6 min El.
Vacuum melted, high strength alloy. Maraging steel.
W.-Nr. 1.6359, 1.2706; AFNOR Z 2 Ni Co Mo 18 85.

SV-RHF 32.
M-912; 0.01 C, 9.0 Co, 5.0 Mo, 18.5 Ni, 0.70 Ti, Al, bal Fe.
Ht Ts: 1960 N/mm^2 TS; 1910 N/mm^2 YS; 5 min El.
Vacuum melted, high strength alloy.
Maraging steel; W.-Nr. 1.6358.

SV-RHF 33.
M-912; 0.01 C, 12.0 Co, 4.0 Mo, 18.0 Ni, 1.60 Ti, Al, bal Fe.
Ht Ts: 2350 N/mm^2 TS; 2260 N/mm^2 YS; 4 min El.
Vacuum melted, high strength alloy.
Maraging steel. W.-Nr. 1.6356.

SW 9 CO 10.
M-1290; 1.2-1.35 C, 3.8-4.5 Cr, 9.5-11.0 W, 3.0-3.5 V, 3.5-4.0 Mo, 10.0-11.0 Co, bal Fe.
High speed steel; W.-Nr. 1.3207.

SW 12 CO 3.
M-1290; 0.77-0.85 C, 3.8-4.5 Cr, 11.5-12.5 W, 1.7-2.0 V, 0.70-1.0 Mo, 2.5-3.0 Co, bal Fe.
High speed steel; W.-Nr. 1.3211.

SW 12 V 2.
M-1290; 0.90-1.0 C, 3.8-4.5 Cr, 11.5-12.5 W, 2.3-2.6 V, 0.70-1.0 Mo, bal Fe.
High speed steel; W.-Nr. 1.3318.

SW-14.
M-1713; weld metal: 0.09 C, 0.50 Mn, 0.20 Si, 0.012 P, 0.021 S, bal Fe.
As welded: 73,500 psi TS; 60,000 psi YS; 24 El.
All-position, covered electrode, AC-DC, reverse polarity; for welding mild steel.
AWS Class E6011.

SW-15.
M-1713; weld metal: 0.10 C, 0.47 Mn, 0.45 Si, bal Fe.
As welded: 74,000 psi TS; 63,000 psi YS; 22 El.
Covered electrode, AC-DC, for general purpose welding and repair.
AWS E6013.

SW-15-IP.
M-1713; weld metal: 0.09 C, 0.50 Mn, 0.26 Si, bal Fe.
As welded: 73,000 psi TS; 61,000 psi YS; 25 El.
AC-DC electrode with an iron powder coating to improve deposition rate for general welding and repair.
AWS E7014.

SW 18.
M-1290; 0.70-0.78 C, 3.8-4.5 Cr, 17.5-18.5 W, 1.0-1.2 V, bal Fe.
High speed steel; W.-Nr. 1.3355.

SW 18 CO 5.
M-1290; 0.75-0.83 C, 3.8-4.5 Cr, 17.5-18.5 W, 1.4-1.7 V, 0.50-0.80 Mo, 4.5-5.0 Co, bal Fe.
High speed steel; W.-Nr. 1.3255.

SW 18 CO 10.
M-1290; 0.72-0.80 C, 3.8-4.5 Cr, 17.5-18.5 W, 1.4-1.7 V, 0.50-0.80 Mo 9.0-10.0 Co, bal Fe.
High speed steel; W.-Nr. 1.3265.

SW-44.
M-1713; weld metal: 0.09 C, 0.95 Mn, 0.83 Si, bal Fe.
As welded: 81,500 psi TS; 72,000 psi YS; 22 El.
High speed, heavy coated, iron-powder, AC-DC, electrode for high deposition.
AWS Class E7024.

SW-47.
M-1713; weld metal: 0.07 C, 0.88 Mn, 0.63 Si, bal Fe.
As welded: 81,500 psi TS; 67,000 psi YS; 33 El.
All-position, iron-powder, low-hydrogen, AC-DC, reverse polarity electrode for welding difficult to weld steels.
AWS Class E7018.

SW-610.
M-1713; weld metal: 0.10 C, 0.30 Mn, 0.010 P, 0.022 S, 0.20 Si, bal Fe.
As welded: 71,000 psi TS; 60,000 psi YS; 24 El.
Covered electrode for welding mild steel; all-position, DC reverse polarity.
AWS Class E6010.

SW-612.
M-1713; weld metal: 0.07 C, 0.42 Mn, 0.23 Si, 0.014 P, 0.028 S, bal Fe.
As welded: 71,500 psi TS; 61,500 psi YS; 24 El.
Covered electrode, AC-DC, straight polarity for welding mild steel with poor fit-up.
AWS Class E6012.

SWB 7SC5.
M-1331; 0.67 C, 1.3 Si, 0.5 Mn, 0.5 Cr, bal Fe.
For springs; oil hardened, good resiliency.

SWB-13C2.
M-1331; 0.13 C, Cr, bal Fe.
For gears, cams, machine tool parts; case hardening.

SWB 14N6.
M-1331; 0.13 C, 0.2 Cr, 1.5 Ni, bal Fe.
For gears, pinions, shafts, cams, camshafts; case hardened, shock resistant.

SWB 15C3.
M-1331; 0.15 C, 0.65 Cr, bal Fe.
For gears, pinions, shafts, cams, camshafts; case hardened, shock resistant.

SWB 15CN6.
M-1331; 0.15 C, 1.5 Cr, 1.5 Ni, bal Fe.
For gears, pinions, camshafts, cams, shafts; case hardened, shock resistant.

SWB 15MNC32.
M-1331; 0.15 C, bal Fe.
Cold drawn: 72,000 TS; 60,000 YS; 22 El; 58 RA; 145 Brin.
For gears, cams, camshafts, fasteners; case hardening steel.

SWB 16M5.
M-1331; 0.16 C, Si, 1.1 Mn, bal Fe.
For gears, cams, camshafts; case hardening.

SWB 16M8.
M-1331; 0.16 C, Si, 1.2 Mn, bal Fe.
For gears, cams, camshafts; case hardening.

SWB 16MC5.
M-1331; 0.16 C, 0.25 Si, 1.15 Mn, 0.95 Cr, bal Fe.
For gears, cams, camshafts, machine tool parts; case hardening.

SWB-18CN8.
M-1331; 0.18 C, 2 Cr, 2 Ni, bal Fe.
For gears, shafts, machine tool parts; case hardening.

SWB-20MC5.
M-1331; 0.2 C, 1.25 Mn, 1.15 Cr, bal Fe.
For gears, shafts, machine tool parts; case hardening.

SWB-22CV4.
M-1331; 0.22 C, 1.1 Cr, 0.2 V, bal Fe.
For gears, cams, camshafts; case hardening.

SWB-22M3.
M-1331; 0.22 C, 1 Mn, bal Fe.
For gears, cams, camshafts; case hardening.

SWB-24CMO5.
M-1331; 0.24 C, 1.15 Cr, 0.25 Mo, bal Fe.
For gears, bolts, fasteners, shafts; water or oil hardened, tough.

SWB-24CMO54.
M-1331; 0.24 C, 1.25 Cr, 0.45 Mo, bal Fe.
For gears, shafts, bolts, studs, fasteners; water or oil hardened, tough.

SWB-24CMOV55.
M-1331; 0.25 C, 1.35 Cr, 0.55 Mo, 0.2 V, bal Fe.
For gears, machine tool parts, bolts, shafts; oil hardened, shock resistant.

SWB-25CMO4.
M-1331; 0.25 C, 1.0 Cr, bal Fe.
For gears, shafts, bolts, studs; water hardened.

SWB-27MCV5.
M-1331; 0.27 C, Mn, Cr, V, bal Fe.
For gears, shafts, machine tool parts; water hardened, tough.

SWB-30M5.
M-1331; 0.30 C, 1.35 Mn, bal Fe.
For gears, shafts, machine tool parts; water or oil hardened.

SWB-30MOV9.
M-1331; 0.3 C, 2.5 Cr, 0.2 Mo, 0.15 V, bal Fe.
For gears, axles, machine tool parts, bolts; oil hardened, shock resistant.

SWB-31CMOV9.
M-1331; 0.31 C, 2.35 Cr, 0.18 Mo, 0.13 V, bal Fe.
For gears, shafts, machine tool parts, bolts; oil hardened, shock resistant.

SWB-32CAMO4.
M-1331; 0.32 C, 1.1 Al, 1.1 Cr, 0.18 Mo, bal Fe.
For oil refinery equipment, fasteners; creep resistant.

SWB-33CAN7.
M-1331; 0.33 C, 1.1 Al, 1.7 Cr, 1 Ni, bal Fe.
For oil refinery equipment, fasteners; creep resistant.

SWB-34C4.
M-1331; 0.33 C, 1 Cr, 0.65 Mn, bal Fe.
For gears, bolts, machine tool parts; water hardened.

SWB-34CA6.
M-1331; 0.34 C, 1.1 Al, 1 Cr, 0.65 Mn, bal Fe.
For oil refinery equipment, fasteners; creep resistant.

SWB-34CNMO6.
M-1331; 0.34 C, 0.55 Mn, 1.5 Cr, 0.2 Mo, 1.5 Ni, bal Fe.
For gears, shafts, machine tool parts, bolts; oil hardened, shock resistant.

SWB-35CMO4.
M-1331; 0.33 C, 0.65 Mn, 1 Cr, 0.2 Mo, bal Fe.
For gears, shafts, machine tool parts, bolts; oil hardened, shock resistant.

SWB-35MC72.
M-1331; 0.40 C, 1.5 Mn, Cr, bal Fe.
For gears, shafts, machine tool parts, crimpers; oil hardened, tough.

SWB-36C6.
M-1331; 0.36 C, 0.45 Mn, 1.6 Cr, bal Fe.
For gears, shafts, machine tool parts; water hardened.

SWB-36CNMO4.
M-1331; 0.36 C, 0.65 Mn, 1 Cr, 0.2 Mo, 1 Ni, bal Fe.
For gears, shafts, machine tool parts, bolts; oil hardened, shock resistant.

SWB-37MS5.
M-1331; 0.37 C, 1.25 Si, 1.25 Mn, bal Fe.
For machine tool parts, upsetters; oil hardened, shock resistant.

SWB-37MSG.
M-1331; 0.35 C, 1.25 Si, 1.25 Mn, bal Fe.
For machine tool parts, upsetters; oil hardened, shock resistant.

SWB-40M4.
 M-1331; 0.4 C, 0.95 Mn, bal Fe.
 Hot rolled: 95,000 TS; 60,000 YS; 25 El; 50 RA; 205 Brin.
 For gears, shafts, machine tool parts, bolts; water hardened.

SWB-41C4.
 M-1331; 0.41 C, 1.0 Cr, 0.65 Mn, bal Fe.
 For gears, shafts, machine tool parts; water hardened.

SWB-42CMO4.
 M-1331; 0.42 C, 0.65 Mn, 1 Cr, 0.2 Mo, bal Fe.
 For gears, shafts, crimpers, bolts, punches; oil hardened, tough.

SWB-42CV6.
 M-1331; 0.42 C, Cr, V, bal Fe.
 For gears, shafts, bolts, axes, fasteners; oil hardened, shock resistant.

SWB-42MV7.
 M-1331; 0.42 C, 1.75 Mn, 0.1 V, bal Fe.
 For punches, crimpers, upsetters; shock resistant.

SWB-46M7.
 M-1331; 0.46 C, 1.75 Mn, 0.1 V, bal Fe.
 For punches, crimpers, upsetters; shock resistant.

SWB-46MSG.
 M-1331; 0.46 C, 1.2 Si, 1.25 Mn, bal Fe.
 For punches, upsetters, riveters; oil hardened, shock resistant.

SWB-50CMO4.
 M-1331; 0.5 C, 0.65 Mn, 1 Cr, 0.2 Mo, bal Fe.
 For bolts, studs, machine tool parts; oil hardened, shock resistant.

SWB-50CV4.
 M-1331; 0.5 C, 0.85 Mn, 1 Cr, 0.09 V, bal Fe.
 For bolts, springs, shafts, crankshafts; oil hardened, shock resistant.

SWB-50CVG.
 M-1331; 0.5 C, 0.85 Mn, 1 Cr, 0.09 V, bal Fe.
 For bolts, fasteners, crankshafts, axles, springs; oil hardened, shock resistant.

SWB-53MS4.
 M-1331; 0.53 C, 0.8 Si, 1.05 Mn, bal Fe.
 For springs, bolts, shafts; tough, oil hardened.

SWB-58CV4.
 M-1331; 0.58 C, 0.95 Mn, 1 Cr, 0.09 V, bal Fe.
 For springs, crankshafts, axles; tough, oil hardened.

SWB-70WS7.
 M-1331; 0.70 C, 1.7 Si, 0.7 Mn, bal Fe.
 For dies, punches, upsetters, springs; oil hardened, shock resistant.

SWB-74NC2.
 M-1331; 0.75 C, Ni, Cr, bal Fe.
 For punches, upsetters, crimpers; oil hardened, shock resistant.

SWB-85C7.
 M-1331; 0.85 C, 1.75 Cr, 0.35 Mn, bal Fe.
 For bearings, liners, sleeves; water or oil hardened, wear resistant.

SWB-100C4.
 M-1331; 1.0 C, 1.1 Cr, 0.07 Mn, bal Fe.
 For bearings, liners, sleeves; water hardened, wear resistant.

SWB-100C6.
 M-1331; 1.0 C, 0.35 Mn, 1.55 Cr, bal Fe.
 For bearings, liners, races, sleeves; water hardened, wear resistant.

SWB-105C4.
 M-1331; 1.05 C, 0.25 Si, 0.3 Mn, 1 Cr, bal Fe.
 For bearings, liners, sleeves, cutters; water hardened, wear resistant.

SWB-110C2.
 M-1331; 1.1 C, 0.4 Cr, 0.3 Mn, bal Fe.
 For bearings, liners, races, sleeves; water hardened, wear resistant.

SWB-120C5.
 M-1331; 1.2 C, 0.3 Mn, 0.2 Si, bal Fe.
 For cutters, drills, reamers, hobs, taps; Type W1; water hardened.

SWB-130C1.
 M-1331; 1.25 C, 0.2 Si, 0.3 Mn, 1 Cr, bal Fe.
 For bearings, races, taps, reamers; water hardened, wear resistant.

SWB-140C3.
 M-1331; 1.4 C, 0.2 Si, 0.3 Mn, 1 Cr, bal Fe.
 For bearings, sleeves, liners, races; water hardened, wear resistant.

SWB-150C6.
 M-1331; 1.45 C, 0.25 Si, 0.6 Mn, 1.4 Cr, bal Fe.
 For bearings, sleeves, liners, races; water hardened, wear resistant.

SWB-200C8.
 M-1331; 2.0 C, 13 Cr, bal Fe.
 For blanking and forming dies; oil or air hardened, non-deforming.

SWB-A7.
 M-1331; 0.40 C, 0.25 Si, 0.25 Mn, bal Fe.
 Hot rolled: 92,000 TS; 58,000 YS; 27 El; 50 RA; 205 Brin.
 For gears, bolts, shafts, studs, fasteners; water hardened.

SWB-A8.
M-1331; 0.85 C, 0.25 Si, 0.25 Mn, bal Fe.
Heat treated: 190,000 TS; 145,000 YS; 12 El; 35 RA; 390 Brin.
For drills, hobs, springs, cutters; Type W1; water hardened.

SWB-A10.
M-1331; 1.0 C, 0.25 Si, 0.25 Mn, bal Fe.
Annealed: 100,000 TS; 53,000 YS; 21 El; 42 RA; 200 Brin.
For drills, taps, reamers, cutters, springs; Type W1; water hardened.

SWB-A10 SPEZIAL.
M-1331; 1.0 C, 0.2 Si, 0.25 Mn, 0.1 V, bal Fe.
Annealed: 100,000 TS; 53,000 YS; 21 El; 42 RA; 200 Brin.
For drills, taps, springs, reamers; Type W2; water hardened.

SWB-A11.
M-1331; 1.1 C, 0.25 Si, 0.25 Mn, bal Fe.
Annealed: 105,000 TS; 55,000 YS; 20 El; 40 RA; 200 Brin.
For drills, hobs, reamers, broaches; Type W1; water hardened.

SWB-A13.
M-1331; 1.3 C, 0.25 Si, 0.25 Mn, bal Fe.
For engravers' tools, taps, reamers; Type W1; water hardened.

SWB-CMO4.
M-1331; 0.33 C, 0.65 Mn, 1 Cr, 0.2 Mo, bal Fe.
For gears, shafts, crankshafts, machine tool parts, bolts; oil hardened, tough.

SWB-CNMO8.
M-1331; 0.30 C, 2 Cr, 0.3 Mo, 2 Ni, bal Fe.
For gears, bolts, machine tool parts; water or oil hardened, shock resistant.

SWEAT-ON PASTE see COLMONOY SWEAT-ON PASTE.

SWEDESTEEL "A".
M-367; 0.8 C, bal Fe.
For rock drills, concrete busters; water hardened.

SWEDESTEEL "B".
M-367; 0.8 C, 0.2 Mo, bal Fe.
For scarifier teeth, chisel blanks; water hardened.

SWEDESTEEL "C".
M-367; 0.5 C, 1 Cr, 1.5 Ni, bal Fe.
For shovel teeth; oil hardened.

SWEDESTEEL "D".
M-367; 0.5 C, 1.5 Cr, 0.2 V, bal Fe.
For diggings bars.

SWEDESTEEL "E".
M-367; 0.7 C, 18 W, 4 Cr, 1 V, bal Fe.
For tool bits, drills, taps; high speed steel.

SWEDISH STEEL NO. 4.
M-336; 0.9-1.0 C, bal Fe.
For embossing dies, wood and machine screw coldheader dies; water hardened.

SWEDISH STEEL NO. 9.
M-336; C, alloy, bal Fe.
For large cold-header dies; water hardened.

SWEDISH STEEL NO. 41.
M-336; C, high Si, high Mn, bal Fe.
For chipping chisels; retains keen cutting edge.

SWEDISH VANADIUM.
M-80; 0.7 C, 0.2 V, bal Fe.
For tools and dies, punches; water hardened.

SWED-OIL.
M-335; 0.9 C, 1.2 Mn, 0.5 Cr, 0.5 W, 0.2 V, bal Fe.
Oil hardening tool and die steel.

SWESCO E.
M-1026; 0.10 C, 17-20 Cr, bal Fe.
Heat treated: 75,000 TS; 40,000 YS; 50 El.
For heat treatment castings; resists oxidation to 1550°F.

SWESCO E-2.
M-1026; 0.20-0.30 C, 0.60-0.85 Mn, 0.35-0.50 Si, bal Fe.
Annealed: 72,000 TS; 42,000 YS; 29 El; 49 RA; 147 Brin.
Normalized: 78,500 TS; 44,500 YS; 30 El; 48 RA; 163 Brin.
For cast gears, shafts, machinery parts; water hardened.

SWESCO E-3.
M-1026; 0.30-0.35 C, 1 Mn, 0.30 Mo, 0.8 Cu, bal Fe.
Annealed: 97,000 TS; 58,500 YS; 24 El; 39 RA; 192 Brin.
Heat treated: 113,000 TS; 100,000 YS; 20 El; 55 RA; 245 Brin.
For cast gears, shafts, machinery parts; water hardened, tough.

SWESCO E-4.
M-1026; 0.45-0.50 C, 1 Mn, 0.30 Mo, 1.25 Cr, 0.9 Cu, bal Fe.
Cast: 107,000 TS; 78,250 YS; 19 El; 41 RA; 217 Brin.
For pumps, valves.

SWESCO E-5.
M-1026; 0.40-0.50 C, 0.70-0.90 Mn, 0.003 B, 0.3-0.5 Si, bal Fe.
Normalized: 98,500 TS; 49,500 YS; 22 El; 32 RA; 179 Brin.
For dies, gears, sprockets, runways; cast to shape.

SWESCO F.
M-1026; 0.40 C, 23-28 Cr, 9-13 Ni, bal Fe.
Cast: 70,000 TS; 45,000 YS; 15 El.
For heat and corrosion resistant castings; resists oxidation to 2100°F.

SWESCO G.
M-1026; 0.10 C, 35-37 Ni, 13-17 Cr, bal Fe.
Cast: 60,000 TS; 35,000 YS; 5 El.
For heat resistant castings; resists oxidation to 2100°F.

SWIFTWELD.
M-578; 0.2 C, bal Fe.
Cast: 45,000-55,000 TS; 5-10 El.
For welding electrodes; for mild steel.

SXZ 500.
M-1522; 5 Ag, Zn, Cd.
Melt range: 270-285°C.
Max stress: 20.5 kgf/mm^2; 8 El.
For soft soldering; high strength.

SYLCUM.
M-Eng.; 7.3 Cu, 0.5 Mn, 1.4 Ni, 0.5 Fe, 9 Si, bal Al.
For pistons.
Cast, corrosion resistant.

SYLVANIA A.
M-856; 0.5 Hf, 0.02 C, bal W.
Stress Relieved: 64,200 TS at 2000°F.
For gyros, high density materials.
Very little weldability.
Highest creep strength at 3200°F, brittle.
High ductile to brittle transition temp.

SYLVANIA MT-104.
M-856; 0.5 Ti, 0.08 Zr, 0.025 C, bal Mo.
Rolled: 145,000 TS; 117,000 YS; 21 El; 52 RA.
Recrystallized: 85,000 TS; 56,000 YS; 40 El; 34 RA.
Stress-Relieved: 133,000 TS; 100,000 YS; 31 El; 59 RA.
For heat engines, heat exchangers, nuclear reactors, radiation shields, rocket nozzles, vector controls.
Good high temperature properties. Powder metallurgy.

SYLVANIA NC73 now **SYLVANIA WN-103.**

SYLVANIA NF 22 now **SYLVANIA WN-102.**

SYLVANIA NO. 4.
M-856; 42 Ni, 5.6 Cr, 0.20 Mn, 0.25 Si, 0.15 max Al, bal Fe.
Annealed: 73,000 TS; 30,000 YS; 40 El; 107 Brin.
Cold drawn: 140,000 TS; 135,000 YS; 2 El; 210 Brin.
For glass to metal seals; controlled expansion.

SYLVANIA NR-106.
M-856; 8-10 Eu_2O_3 + Gd_2O_3 + Sm_2O_3, bal Ni.
R.T.: 56,000 TS; 28,000 YS; 15 El; 230 DPH.
At 1800°F 7000 TS; 5000 YS; 14 El.
For nuclear reactor control structures.
Neutron flux control. Used up to 2000°F.
Sintered powders. Dispersion alloy.

SYLVANIA NS55.
M-856; 0.0018 Na, 0.0049 K, 0.0003 Ca, 0.0008 Si, 0.0008 Cr, 0.0025 Fe, 0.0004 Al, 0.02 C, 0.00002 Cu, 0.0004 max Mg, bal Fe.
For high temperature filaments and instruments.
High purity iron.

SYLVANIA WN-102.
M-856; 95 W, 2.5 Ni, 2.5 Fe.
Sintered: 125,000 TS; 100,000 YS; 11 El; C 26-31 Rock.
For counterweights, radioactive shielding, electrical contacts, flywheels, governors, gyroscopes.
Sintered heavy alloy.

SYLVANIA WN-103.
M-856; 90 W, 7 Ni, 3 Cu.
Sintered: 110,000 TS; 95,000 YS; 8 El; C 20-25 Rock.
For counterweights, radioactive shielding, electrical contacts, gyros, flywheels, governors'.
Sintered heavy metal.

SYLVANITE F8.
M-735; 1.25 C, 4 Cr, 8 W, 0.5 V, bal Fe.
For woodworking knives, scrapers, woodworking tools.
High speed steel, good red-hardness.

SYLVAN STAR.
M-57; 1.05 C, 0.10 Cr, 0.2 V, bal Fe.
Annealed: 100,000 TS; 53,000 YP; 21 El; 197 Brin.
Ht. Tr.: 104,000-216,000 TS; 74,000-152,000 YP; 11-23 El; 32-50 RA; 200-600 Brin.
For cold work dies, tools, cutters, axes, chisels, punches.
Type W2 water hardening tool steel.

SYRACUSE GENUINE BABBITT.
M-565; Cu, Sb, Pb, bal Sn.
70,000 TS.
For bearings to resist shock and high temperature; Babbitt metal; shock resistant.

T

T-1-ALLOY.
M-190; 4 Cu, 0.2 Ti, bal Al.
Heat treated: 30,000-34,000 TS; 7-9 El.
For light alloy parts; age-hardenable.

"T-1" STEEL.
    M-604; 0.17 C, 0.80 Mn, 0.25 Si, 0.80 Ni, 0.55 Cr, 0.50 Mo, 0.30 Cu, B, V, bal Fe.
    W. Q. + T.: 125 ksi TS; 118 ksi YS; 21 El.
    May be machined, torch cut, formed, welded.
    For pressure vessels, earth moving equipment, heavy vehicles, cranes, low temperature parts subject only to atmosphere.
    ASTM A514, A517.

"T-1" STEEL 321.
    M-604; 0.17 C, 0.80 Mn, 0.25 Si, 0.80 Ni, 0.55 Cr, 0.50 Mo, B, V, Cu, bal Fe.
    W. Q. + T.: 321 Brin.
    Machine or torch cut, welded, mild forming.
    For heavy impact and abrasion; power shovel buckets, dump trucks, chutes.
    ASTM A678.

"T-1" STEEL 340.
    M-604; 0.17 C, 0.80 Mn, 0.25 Si, 0.80 Ni, 0.55 Cr, 0.50 Mo, B, V, Cu, bal Fe.
    W. Q. + T.: 340 Brin.
    Machine or torch cut, welded, mild forming.
    For heavy impact and abrasion; power shovel buckets, dump trucks, chutes.
    ASTM A678.

"T-1" STEEL 360.
    M-604; 0.17 C, 0.80 Mn, 0.25 Si, 0.80 Ni, 0.55 Cr, 0.50 Mo, B, V, Cu, bal Fe.
    W. Q. + T.: 360 Brin.
    Machine or torch cut, welded, mild forming.
    For heavy impact and abrasion; power shovel buckets, dump trucks, chutes.
    ASTM A678.

"T-1" TYPE A STEEL.
    M-604; 0.17 C, 0.80 Mn, 0.25 Si, 0.55 Cr, 0.20 Mo, B, V, bal Fe.
    W. Q. + T.: 125 ksi TS; 118 ksi YS; 21 El.
    May be machined, torch cut, formed, welded.
    For pressure vessels, earth moving equipment, heavy vehicles, cranes, low temperature parts subject only to atomsphere.
    ASTM A514, A517.

"T-1" TYPE A STEEL 321.
    M-604; 0.17 C, 0.80 Mn, 0.25 Si, 0.55 Cr, 0.20 Mo, B, V, bal Fe.
    W. Q. + T.: 321 Brin.
    Machine or torch cut, welded, mild forming.
    For heavy impact and abrasion; power shovel buckets, dump trucks, chutes.
    ASTM 678.

"T-1" TYPE A STEEL 340.
    M-604; 0.17 C, 0.80 Mn, 0.25 Si, 0.55 Cr, 0.20 Mo, B, V, bal Fe.
    W. Q. + T.: 340 Brin.
    Machine or torch cut, welded, mild forming.
    For heavy impact and abrasion; power shovel buckets, dump trucks, chutes.
    ASTM A678.

"T-1" TYPE A STEEL 360.
    M-604; 0.17 C, 0.80 Mn, 0.25 Si, 0.55 Cr, 0.20 Mo, B, V, bal Fe.
    W. Q. + T.: 360 Brin.
    Machine or torch cut, welded, mild forming.
    For heavy impact and abrasion; power shovel buckets, dump trucks, chutes.
    ASTM A678.

"T-1" TYPE B STEEL.
    M-604; 0.17 C, 1.15 Mn, 0.25 Si, 0.50 Ni, 0.55 Cr, 0.25 Mo, B, V, bal Fe.
    W. Q. + T.: 125 ksi TS; 118 ksi YS.
    May be machined, torch cut, formed, welded.
    For pressure vessels, earth moving equipment, heavy vehicles, cranes, low temperature parts subject only to atmosphere.
    ASTM A514, A517.

"T-1" TYPE B STEEL 321.
    M-604; 0.17 C, 1.15 Mn, 0.25 Si, 0.50 Ni, 0.55 Cr, 0.25 Mo, B, V, bal Fe.
    W. Q. + T.: 321 Brin.
    Machined, torch cut, welded, mild forming.
    For heavy impact and abrasion; power shovel buckets, dump trucks, chutes.
    ASTM A678.

"T-1" TYPE B STEEL 340.
    M-604; 0.17 C, 1.15 Mn, 0.25 Si, 0.50 Ni, 0.55 Cr, 0.25 Mo, B, V, bal Fe.
    W. Q. + T.: 340 Brin.
    Machined, torch cut, welded, mild forming.
    For heavy impact and abrasion; power shovel buckets, dump trucks, chutes.
    ASTM A678.

"T-1" TYPE B STEEL 360.
    M-604; 0.17 C, 1.15 Mn, 0.25 Si, 0.50 Ni, 0.55 Cr, 0.25 Mo, B, V, bal Fe.
    W. Q. + T.: 360 Brin.
    Machined, torch cut, welded, mild forming.
    For heavy impact and abrasion; power shovel buckets, dump trucks, chutes.
    ASTM A678.

T.2.
    M-1116; 0.35 C, 2.3 Cr, 3.0 W, 0.2 V, 0.3 Mo, 0.3 Mn, 0.3 Si, bal Fe.
    Oil hardening tool and die steel; for press and forging dies, shears, rivet hammers.
    AFNOR: 35 WCD 30.10.

T 2 EXTRA.
    M-96; 0.56 C, 0.9 Mn, 0.4 Si, 0.5 Cr, bal Fe.
    For shafts, dies, punches; water hardening.

T3.
    M-1116; 1.3 C, 4.5 Cr, bal Fe.
    For tools, dies; air or oil hardened.

T-3.
M-1769.
Sintered carbide.
230,000 TrS; Density: 13.2-13.5.
Hardness: 90.0-91.0 RA.
Industry code: C-5; ISO P-40.

T 4.
M-912; 0.45, 0.25, 0.70, bal Fe.
Carbon tool steel; for hand tools as axes, hatchets, knives, hand saws.
W.-Nr. 1730.

T 5.
M-912; 0.58 C, 0.25 Si, 0.70 Mn, bal Fe.
Carbon tool steel; for hand tools, for saws, hammers, screw drivers.
W.-Nr. 1740.

T5K12V.
M-USSR; 83 WC, 5 TiC, 12 Co.
For hard cutting tools, wear resistant dies.
Sintered carbide, wear resistant.

T 6.
M-912; 0.67 C, 0.25 Si, 0.70 Mn, bal Fe.
Carbon tool steel; for knives, hand saws, wood working tools.
W.-Nr. 1744.

T.6.
M-1116; 0.33 C, 3.0 Cr, 10.5 W, 0.4 V, 0.4 Mo, 0.3 Mn, 0.3 Si, bal Fe.
Air or oil hardening tool and die steel; for hot forging, thread rolling dies, extrusion dies for brass and bronze.
AFNOR: Z 30 WC 10.03; AISI H21/22.

T-6.
M-1769.
Sintered carbide.
175,000 TrS; Density: 11.0-11.2.
Hardness: 91.8-92.4 RA.

T 7.
M-912; 0.75 C, 0.25 Si, 0.70 Mn, bal Fe.
Carbon tool steel; for hand tools, planer blades, wood chisels, wood working tools.
W.-Nr. 1750.

T-8.
M-1769.
Sintered carbide.
140,000 TrS; Density: 5.8-6.0.
Hardness: 92.5-93.5 RA.
Industry code: C-8; ISO P-01.

T15K6.
M-USSR; WC, TiC, Co.
For cutting tools.
Sintered carbide, abrasion resistant.

T 18.
M-1290; 1.0-1.1 C, 0.15-0.30 Si, 1.0-1.2 Mn, 0.70-1.0 Cr, bal Fe.
Cold work tool steel.
W.-Nr. 1.2127.

T.35.
M-1116; 1.03 C, 0.7 Cr, 3.7 W, 0.3 V, 0.4 Mo, 0.55 Mn, 0.3 Si, bal Fe.
Oil hardening tool steels; for drills, boring tools, taps, threading tools.
AFNOR: 100 WC 35.03; AISI F2.

T-35.
M-1769.
Sintered carbide.
220,000 TrS; Density: 12.7-13.0.
Hardness: 90.0-91.0 RA.

T45.
M-1653; 0.30 max C, 1.0 Mn, bal Fe.
Steel, usually forgings requiring notch toughness.
Similar to ASTM A350 LF1.

T50.
M-1653; 0.18 max C, 0.70 max Mn, 2.30 Ni, bal Fe.
Nickel low carbon steel.
18 Ni9.

T-50.
M-1769.
Sintered carbide.
230,000 TrS; Density: 12.6-12.8.
Hardness; 91.5-92.2 RA.
Industry code: C-5-6; ISO P25,30.

T-51.
M-1769.
Sintered carbide.
230,000 TrS; Density: 12.25-12.45.
Hardness: 90.0-91.0 RA.
Industry code: C-5.

T51 see **HEPPENSTALL T51.**

T-56.
M-1769.
Sintered carbide.
280,000 TrS; Density: 13.15-13.30.
Hardness: 90.8-91.8 RA.
ISO P-35.

T-63.
M-1769.
Sintered carbide.
200,000 TrS; Density: 11.25-11.45.
Hardness: 91.7-92.3 RA.
Industry code: C-6.

T-64.
M-1769.
Sintered carbide.
220,000 TrS; Density: 12.25-12.45.
Hardness: 91.2-92.2 RA.
Industry code: C-6.

T-65.
M-1769.
Sintered carbide.
210,000 TrS; Density: 12.75-13.0.
Hardness: 91.7-92.3 RA.
Industry code: C-7.

T-76.
M-1769.
Sintered carbide.
200,000 TrS; Density: 12.27-12.4.
Hardness: 92.4-93.2 RA.
Industry code: C-7; ISO P-10, 20.

T-78.
M-1769.
Sintered carbide.
200,000 TrS; Density: 12.9-13.1.
Hardness: 93.0-94.0 RA.
ISO P-05.

T100.
M-1653.
0.18 C, 3.5 Ni, bal Fe.
Steel, usually forgings requiring notch toughness (18 Ni 14); ASTM A350-LF3.

T-111 see **WESTINGHOUSE T-111.**

T150.
M-1653; 0.4 C, 3.5 Ni, bal Fe.
Alloy steel for low temperature service.
Similar to ASTM A320 L9.

T200.
M-1653; 0.13 C, 9 Ni, bal Fe.
Nickel steel for low temperature service.
ASTM A353.

T-222 see **WESTINGHOUSE T-222.**

TA-10 W see **FANSTEEL 60 METAL.**

TA-135.
M-634; C, alloy, bal Fe.
For gear sectors, castings.

TAIFUN.
M-614; 1.05 C, 0.90 Mn, 1.0 Cr, 1.15 W, bal Fe.
For cold work tools, heading dies; oil hardened, tough.

TAIL SHAFT BRONZE.
M-508; 87 Cu, 6 Sn, 3 Pb, 4 Zn.
For bearings, pump liners; centrifugal casting.

TALABOT 10 NC 6.
M-1119; 0.07-0.12 C, 0.60-0.90 Mn, 1.20-1.60 Ni, 0.85-1.15 Cr, bal Fe.
Low alloy-low carbon structural steel.
AFNOR 10 NC 6; W.-Nr. 1.5713.

TALABOT 12 CD 4.
M-1119; 0.08-0.14 C, 0.50-0.80 Mn, 0.85-1.15 Cr, 0.15-0.30 Mo, bal Fe.
Low alloy-low carbon structural steel.
AFNOR 12 CD 4; W.-Nr. 1.7262.

TALABOT 14 NC 11.
M-1119; 0.11-0.17 C, 0.35-0.60 Mn, 2.5-3.0 Ni, 0.60-0.90 Cr, bal Fe.
Alloy steel for carburizing.
AFNOR 14 NC 11; W.-Nr. 1.5732.

TALABOT 15 ND 8.
M-1119; 0.13-0.18 C, 0.20-0.50 Mn, 1.80-2.30 Ni, 0.15-0.30 Mo, bal Fe.
Alloy structural steel; also for carburizing.
AFNOR 15 ND 8.

TALABOT 16 MC 5.
M-1119; 0.14-0.19 C, 1.0-1.3 Mn, 0.80-1.10 Cr, bal Fe.
Low alloy-low carbon structural steel.
AFNOR 16 MC 5; W.-Nr. 1.7131.

TALABOT 18 CD 4.
M-1119; 0.18-0.22 C, 0.60-0.90 Mn, 0.85-1.15 Cr, 0.15-0.30 Mo, bal Fe.
AFNOR 18 CD 4; W.-Nr. 1.7264.

TALABOT 20 MC 5.
M-1119; 0.17-0.22 C, 1.1-1.4 Mn, 1.0-1.3 Cr, bal Fe.
AFNOR 20 MC 5; W.-Nr. 1.7147.

TALABOT 25 CD 4.
M-1119; 0.22-0.29 C, 0.60-0.90 Mn, 0.85-1.15 Cr, 0.15-0.30 Mo, bal Fe.
AFNOR 25 CD 4; W.-Nr. 1.7218.

TALABOT 30 NC 11.
M-1119; 0.27-0.34 C, 0.35-0.60 Mn, 2.5-3.0 Ni, 0.60-0.90 Cr, bal Fe.
AFNOR 30 NC 11; W.-Nr. 1.5736.

TALABOT 32 C 4.
M-1119; 0.30-0.35 C, 0.60-0.90 Mn, 0.85-1.15 Cr, bal Fe.
AFNOR 32 C 4; W.-Nr. 1.7033.

TALABOT 35 CD 4.
M-1119; 0.30-0.37 C, 0.60-0.90 Mn, 0.85-1.15 Cr, 0.15-0.30 Mo, bal Fe.
Structural steel.
AFNOR 35 CD 4; W.-Nr. 1.7220.

TALABOT 35 NC 15.
M-1119; 0.30-0.38 C, 0.20-0.50 Mn, 3.50-4.0 Ni, 1.50-1.90 Cr, bal Fe.
Alloy structural steel.
AFNOR 35 NC 15.

TALABOT 35 NCD 14.
M-1119; 0.30-0.40 C, 0.20-0.50 Mn, 3.20-3.70 Ni, 1.20-1.60 Cr, 0.20- 0.40 Mo, bal Fe.
Alloy structural steel.
AFNOR 35 NCD 14. W.-Nr. 1.6746.

TALABOT 40 NC 17.
M-1119; 0.37-0.45 C, 0.15-0.55 Mn, 4.0-4.5 Ni, 1.5-2.0 Cr, bal Fe.
Deep hardening alloy constructional steel.
AFNOR 40 NC 17; W.-Nr. 1.5864.

TALABOT 45 C 4.
M-1119; 0.41-0.48 C, 0.60-0.90 C, 0.85-1.15 Cr, bal Fe.
Constructional steel.
AFNOR 45 C 4; Similar to AISI 5145.

TALABOT 50 CV 4.
M-1119; 0.47-0.55 C, 0.70-1.1 Mn, 0.85-1.15 Cr, 0.10-0.20 V, bal Fe.
Chrome-vanadium structural steel.
AFNOR 50 CV 4; AISI 6150.

TALABOT A.C.T. NO. 1.
M-1119; 0.5 C, 1.5 Cr, 2.2 W, 0.2 V, bal Fe.
For hot work dies and tools; hot work steel.

TALABOT A.C.T. NO. 2.
M-1119; 0.5 C, 1.5 Cr, 2.2 W, 0.2,V, bal Fe.
For hot work dies and tools; hot work steel.

TALABOT MNS.
M-1119; 0.5 C, 0.7 Mn, 1.9 Si, 0.2 Mo, bal Fe.
For tools, dies, punches; tough.

TALABOT SPECIAL OM.
M-1119; 0.95 C, 1.2 Mn, 0.5 Cr, 0.5 W, bal Fe.
For tools, dies, cutters; non-deforming.

TALIDE 101.
M-913; W, C, Co.
Sintered: 840 Brin.
For drawing dies, blast nozzles; Tr.S. 190,000; C.S. 810,000; sintered carbides.

TALIDE C-75.
M-913; WC.
Sintered: Rockwell A85.
For cutting tools; sintered carbide.

TALIDE C-80.
M-913; WC.
Sintered: Rockwell A87.
For cutting toools; sintered carbide.

TALIDE C-85.
M-913; WC.
Sintered: Rockwell A88.
For cutting tools; sintered carbide.

TALIDE C-88.
M-913; 9 Co, 91 Wc.
Sintered: A 91 Rock; 275,000 Tr.S.; 275,000 CS.
For extrusion and swaging dies, draw punches. Wear and abrasion resistant.

TALIDE C-89.
M-913; 85.4 W, 5.5 C, 10 Co.
Sintered: 150,000 TS; Rockwell A90.
For draw dies, cutting tools; sintered carbide.

TALIDE C-91.
M-913; 86.4 W, 5.6 C, 8 Co.
Sintered: 140,000 TS; Rockwell A 91.5.
For draw dies, cutting tools; sintered carbide.

TALIDE C93.
M-913; 86 W, 2.3 Ta, 5.8 C, 6 Co.
Sintered: 130,000 TS; 790 Brin.
For wear plates, guides, dies; heat resistant to 2000°F.

TALIDE C-95.
M-913; 91 W, 6 C, 3 Co.
Sintered: 810 Brin.
For precision boring and finishing cutters; Tr.S. 200,000; C.S. 775,000, sintered carbides.

TALIDE C-99.
M-913; WC.
Sintered: Rockwell A 91.
For cutting tools; sintered carbide.

TALIDE C-907.
M-913; Sintered carbide tool material.
For cutting tools on finish cutting operations, and for cold work dies for processing aluminum.

TALIDE C-7525.
M-913; 66 W, 5 Ta, 25 Co, 6 C.
Sintered: 210,000 TS; 650 Brin.
For cold working tools, blanking dies; sintered carbide.

TALIDE C-8020.
M-913; 5 C, 75 W, 20 Co.
Sintered: 190,000 TS; 680 Brin.
For punching and shearing tools; sintered carbides.

TALIDE C-8515.
M-913; 79.8 W, 5.2 C, 15 Co.
Sintered: 175,000 TS; 700 Brin.
For impact tools; sintered carbides.

TALIDE CT-20.
M-913; Sintered carbide tool material.
For dies for cold extrusion of non-ferrous metals.

TALIDE CT-28.
M-913; Sintered carbide tool material.
For cutting tools for flash trimming operations.

TALIDE S-88.
M-913; 75 W, 14 Ti, 5.6 Ta, 6.3 C, 9 Co.
Rockwell A 93.
For cutting tools, cold forming dies; sintered carbides.

TALIDE S-90.
M-913; 73 W, 8 Ti, 2 Ta, 7 C, 10 Co.
Rockwell A 91.
For cutting tools; sintered carbides.

TALIDE S-92.
M-913; 69.5 W, 9.6 Ti, 5.6 Ta, 7.3 C, 8 Co.
Rockwell A 92.
For cutting tools; sintered carbides.

TALIDE S-94.
M-913; 65.7 W, 12 Ti, 7.5 Ta, 7.8 C, 7 Co.
Sintered: 810 Brin.
For precision boring cutters; Tr.S. 180,000; C.S. 750,000; sintered carbides.

TALIDE S-880.
M-913; Sintered carbide tool material.
For cutting tools for rough cutting operations on steel.

TALIDE S-901.
M-913; Sintered carbide tool material.
For cutting tools for finish cutting operations on steel.

T-ALLOY.
M-140; 0.38 C, 3.5 Cr, 10.5 W, 0.4 V, bal Fe.
Hardened: C 48-52 Rock.
For brass forging dies, hot forming dies, hot punches, spreading punches, dummy blocks. Type H22 hot work tool steel.

T ALLOY "A".
M-140; 0.33 C, 3.5 Cr, 9.6 W, 0.5 V, bal Fe.
For tools, forming dies; hot work steel.

T-ALLOY B.
M-140; 0.5 C, 3.0 Cr, 15 W, 0.5 V, bal Fe.
For dies, punching and forming dies; hot work steels.

T-ALLOY C.
M-140; 0.25 C, 4 Cr, 15 W, 0.5 V, bal Fe.
For punching and forming dies; hot work steel.

TALMI GOLD-1.
M-Japan; 90 Cu, 8.9 Zn, 0.9 Au.
For cheap jewelry; Au welded on by rolling.

TALMI GOLD-2.
M-Japan; 86 Cu, 12 Zn, 1.1 Sn, 0.3 Fe.
For hardware, cheap jewelry; corrosion resistant.

TALON.
M-210; 0.9 C, bal Fe.
For tools.

TAM.
M-109; 12-15 Ti, 18-22 Al, 2-4 Si, 4.5 Cr.
For refining steel; grain refiner.

TAM CUPRO-TITANIUM.
M-109; 25-30 Ti, 0.2 max Fe, Si, C, bal Cu.
Used in adding Ti to Cu alloys; hardener.

TAM FOUNDRY FERRO TITANIUM.
M-109; 0.1 max C, 20-24 Si, 28-32 Ti, 0.5-1.5 Al, bal Fe.
For addition agent for alloy cast iron; inoculant.

TAM LOW CARBON FERRO TITANIUM 25% GRADE.
M-109; 0.1 max C, 3-4.5 Si, 26-29 Ti, 4.5-6.5 Al, bal Fe.
For addition agent for alloy steel; inoculant.

TAM LOW CARBON FERRO-TITANIUM (40% GRADE).
M-109; 38-42 Ti, 6-9 Al, 3-5 Si, 0.1 C, bal Fe.
For source of Ti for alloy steels; Ti-additions.

TAM MAGNESIUM-ZIRCONIUM K-50.
M-109; 50 Zr, 50 Mg.
Addition agent to Mg alloys; for alloying.

TAM MANGANESE-TITANIUM REGULAR GRADE.
M-109; 0.10 max C, 25-28 Ti, 8-10 Al, 1.5-3.5 Si, 20-20 Fe, bal Mn.
For steel deoxidizer; grain refiner.

TAM MANGANESE-TITANIUM SPECIAL GRADE.
M-109; 0.10 max C, 27-31 Ti, 8-10 Al, 2-4 Si, 2-5 Fe, bal Mn.
For steel deoxidizer; grain refiner.

TAM MED. CARBON FERRO-CARBON TITANIUM.
M-109; 17-21 Ti, 3-5 C, approx 2-3 Si, approx 1.3 Al, bal Fe.
For steel metallurgical applications; for extra low C rimming steel.

TAM METALLIC TITANIUM.
M-109; 98 Ti, 0.2 Fe, 0.2 Si, 0.2 C, 0.2 Al.
For making of special Titanium alloys; Ti-additions.

TAM MOLYBDENUM TITANIUM.
M-109; 48 Mo, 16 Ti, 2 Al, 7 Si, 27 Fe.
For ladle additions to cast iron; Mo-addition.

TAM NICKEL TITANIUM.
M-109; 62 Ni, 25 Ti, 8 Al, 1 Si, 3.5 Fe.
For ladle additions to steel; Ni-addition.

TAM ORIGINAL FERROCARBON TITANIUM.
M-109; 15-18 Ti, 6-8 C, 2.65 Si, 1.9 Al, bal Fe.
For deoxidizer in steel making.

TAM SILICO TITANIUM.
M-109; 44-47 Ti, 42-45 Si, 0.03-0.1 C, 1.5-3.0 Al, bal Fe.
For cast iron inoculant, steel deoxidizer; low M.P.

TANDEM.
78 Pb, 17 Sb, 5 Sn.
For bearings; anti-friction.

TANDEM B.M.
M-Eng.; 66 Zn, 21.5 Sn, 7 Cu, 4.8 Pb, 0.4 Al.
For bearings; will not resist heat or live steam.

TA-NI (TANTALUM-NICKEL).
M-53; Ni plus Ta.
115,000 TS.
For electronic tube parts, chemical applications; cheaper than pure Ta.

TANK BRAND.
0.3-0.5 C, bal Fe.
For machinery parts; water hardening.

TANNENBAUM.
M-1307; 0.55 C, 0.10-0.40 Si, 0.30-0.70 Mn, bal Fe.
Annealed: 100,000 TS; 55,000 YS; 15 El; 22 RA; 180 Brin.
For machine tool parts, bolts, gears; water hardened.

TANTALOY see also FANSTEEL 61 METAL.

TANTALUM.
M-1755; Ta.
Purities: Zone refined 99.995%, Premium grade 99.95%, 99.8% (Melting grade), 99.7% Sintering grade, Capacitor grades.
Forms: Powders, bars, sheets, foils, wire, pellets, rods, tubing, ingots, single crystals.

TANTALUM BRONZE.
M-99; 1.25 Mo, 0.20 Ta, 10 Al, bal Cu.
Sand cast: 68,700 TS; 44,000 YS; 8.5 El.
Drawn wire: 162,200 TS; 154,000 YS; 1.7 El.
For steam fittings, valves; corrosion resistant.

TANTUNG 144.
M-53; 43-48 Co, 25-30 Cr, 16-21 W, 2-4 C, 3-8 Cb, 1-3 Mn, 1 max Fe.
Cast cobalt base alloy.
Chill cast: 64 Rc; 275,000 TrS.
For cutting tools, wear and corrosion resistant applications; good toughness and excellent abrasion resistance.

TANTUNG G.
M-53; 45-50 Co, 27-32 Cr, 14-19 W, 2-4 C, 2-7 Cb, 1-3 Mo, 1 max Fe.
Cast cobalt base alloy.
Chill cast: 62 Rc; 300,000 TrS.
For cutting tools, wear and corrosion resistant applications; high toughness and good abrasion resistance.

TAPDIE.
M-35; 1.25 C, 1.40 W, 0.45 Cr, 0.20 V, bal Fe.
Oil or water hardening cold work tool steel.
For taps, threading tools, broaches, reamers.
AISI 07.

TARNAC.
Cu, Mn, Ni.
For electrical instruments; v. "Manganin."

TATA LA 55.
M-997; 0.14-0.18 C, 1.3-1.6 Mn, 0.10 max Nb, 0.18 max V, 0.50 max Si, bal Fe.
Rolled: 42 kgf/mm^2 min YS; 55 kgf/mm^2 min TS; 20 Min El.
For structural members in bridges and buildings, transportation, and earth moving equipment.

TATA LA 60.
M-997; 0.17-0.21 C, 1.4-1.6 Mn, 0.10 max Nb, 0.18 max V, 0.50 max Si, bal Fe.
Normalized: 45 kgf/mm^2 min YS; 60 kgf/mm^2 min TS; 20 min El.
For structural members in bridges and buildings, transportation, and earth moving equipment.

TATA SPL. "A" GRADE.
M-997; 0.05 max C, 0.20 max Mn, 0.015 max S, 0.025 max P, 0.05 max Si, Al killed, bal Fe.
Rolled and annealed.
Permeability at 1.5/2 oersteads-2000 min.
Total induction-200/250 oersteds-20,000 min Gauss.
Soft magnet yokes.
Rolled plates-Galvanizer's pot.

TATMO see ELECTRITE TATMO.

TATMO V see ELECTRITE TATMO V.

TAURUS BRONZE MARK I.
M-476; 87.7 Cu, 12 Sn, 0.3 P.
Sand cast: 24,700-40,300 TS; 15,700-24,700 YS; 25-5 El.
For high quality gear bronze; Iz-25-40.

TAURUS BRONZE MARK II.
M-476; 89.7 Cu, 10 Sn, 0.3 P.
Sand cast: 27,000-47,000 TS; 15,700-22,300 YS; 28-6 El.
For bearings requiring some plasticity; Iz-27-44.

TAURUS BRONZE MARK III.
M-476; 86 Cu, 14 Sn, trace P.
Sand cast: 29,000-33,200 TS; 24,700-29,000 YS; 3-1 El.
For bearings for turn tables, movable bridges, etc.; Iz-13-17.

TAURUS BRONZE MARK IV.
M-476; 88.25 Cu, 10.5 Sn, 0.25 P, 1.0 Ni.
Sand cast: 31,300-49,000 TS; 17,800-31,300 YS; 30-8 El.
For sand castings requiring high strength, toughness and wearing qualities; Iz-40-60.

TAURUS BRONZE MARK V.
M-476; 88 Cu, 11.2 Sn, 0.3 P, 0.5 Ni.
Centrifugal: 40,300-49,000 TS; 27,000-36,000 YS; 5-15 El.
For gear bronze; Iz-24-32.

TAURUS BRONZE MARK VI.
M-476; 88 Cu, 10 Sn, 2 Zn.
Sand cast: 31,300-44,700 TS; 17,800-22,300 YS; 10-25 El.
For general bronze castings; similar to "Admiralty Metal."

TAURUS BRONZE MARK VII.
M-476; 81.7 Cu, 12 Sn, 0.3 P, 1 Ni, 5 Pb.
Sand cast: 298000-40,000 TS; 22,300-31,000 YS; 4-10 El.
For cogging and rolling mill bearings, heavy duty bearings; Iz-40-50.

TAURUS BRONZE MARK VIII.
M-476; 80.7 Cu, 10 Sn, 0.3 P, 1 Ni, 8 Pb.
Sand cast: 31,300-37,800 TS; 22,300-27,000 YS; 6-12 El.
For bearings, running against hard shafts; Iz-80-90.

TAURUS BRONZE MARK IX.
M-476; 76.7 Cu, 10 Sn, 0.3 P, 1 Ni, 12 Pb.
Sand cast: 31,300-37,800 TS; 20,700-27,000 YS; 8-13 El.
For heavy duty plastic bronze bearings; Iz-60-80.

TAURUS BRONZE MARK X.
M-476; 84 Cu, 8 Sn, 4 Pb, 4 Zn.
Sand cast: 29,000-42,600 TS; 13,500-15,700 YS; 18-26 El.
For dense general purpose castings, pump bodies, etc.; Iz-90-120.

TAURUS BRONZE MARK XI.
M-476; 85 Cu, 5 Sn, 5 Pb, 5 Zn.
Sand cast: 29,000-40,300 TS; 13,500-15,700 YS; 20-30 El.
For castings not requiring hardness, bushes and bearings; Iz-70-120.

TAURUS BRONZE MARK XII.
M-476; 88 Cu, 6 Sn, 6 Zn.
Sand cast: 36,000-42,600 TS; 15,700-17,800 YS; 30-60 El.
For light, thin section, tough castings; Iz-120.

TAURUS MARK XIV.
M-476; 86 Cu, 7 Sn, 5 Zn, 2 Pb.
Cast: 31,300-40,300 TS; 13,500-17,900 YS; 12-8 El.
For general purpose castings; free-cutting.

TAURUS MARK XVI.
M-476; 90.75 Cu, 9 Sn, 0.5 Ni, 0.25 P.
Cast: 44,800-53,800 TS; 22,400-31,400 YS; 30-21 El.
For general purpose castings; corrosion resistant.

TAURUS MARK XIX.
M-476; 88.5 Cu, 10.5 Sn, 0.75 Ni, 0.25 P.
Centrifugal cast: 31,400-49,400 TS; 17,900-31,400 YS; 30-8 El.
For centrifugal castings; wear resistant.

TAURUS MARK XX.
M-476; 80.5 Cu, 4 Ni, 2 Mn, 4 Fe, 9.5 Al.
Centrifugal cast: 89,600-105,000 TS; 35,200-56,000 YS; 16-12 El.
For highly stressed bearings and structures; fatigue resistance.

TAURUS MARK XXI.
M-476; 55 Cu, 0.75 Mn, 1 Fe, 1 Al, bal Zn.
Sand cast: 67,000-78,500 TS; 25,900-31,400 YS; 28-20 El.
For castings; corrosion resistant.

TAURUS MARK XXII.
M-476; 65 Cu, 2 Mn, 2 Fe, 5 Al, bal Zn.
Sand cast: 85,000-94,200 TS; 40,300-49,300 YS; 18-15 El.
For castings; corrosion resistant.

TAURUS MARK XXIII.
M-476; 60 Cu, 2 Mn, 2 Fe, 5.5 Al, bal Zn.
Sand cast: 107,500-112,000 TS; 58,200-67,200 YS; 15-12 El.
For castings; corrosion resistant.

TAURUS MARK XXIV.
M-476; 73 Cu, 5 Sn, 17 Ni, 5 Pb.
Centrifugal cast: 48,800-56,500 TS; 35,840-44,800 YS; 2-1 El.
For bushings, castings; good machinability.

TAURUS MARK XXV.
M-476; 94.5 Cu, 1.5 Sn, 4 Ni.
Centrifugal cast: 38,200-49,300 TS; 20,300-26,800 YS; 42-25 El.
For slip rings for electrical machines; high conductivity.

TAURUS MARK XXVI.
M-476; 44 Cu, 11 Sn, 38 Ni, 7 Pb.
Centrifugal cast: 49,300-53,800 TS; 22,400-31,400 YS; 4-2 El.
For valve seats; for severe service.

TAURUS MARK XXX.
M-476; 88 Cu, 2 Sn, 10 Zn.
Sand cast: 26,900-31,300 TS; 35-30 El.
For brazing brass; high strength.

TAURUS MARK XXXI.
M-476; 59 Cu, 40 Zn, 1 Pb.
Sand cast: 40,300-44,800 TS; 15 El.
For valves, roll and ball bearing castings; free-cutting.

T.A.W see FANSTEEL 61 METAL.

TAWILCO.
M-1442; 3 C, alloy, bal Fe.
Cast: 50,000 TS; 45,000 YS; 0.1 El; 174-500 Brin.
For machinery parts, gears, housings; cast iron, heat and abrasion resistant.

TAYLOR 751.
M-1412; 3.3 C, 2.0 Si, Mn, alloy, bal Fe.
For brick mold die liners; cast iron, hard, wear resistant.

TAYLOR CRA.
M-1412; 3.3 C, Mn, 1.8 Si, alloy, bal Fe.
For castings for clay working industries; cast iron, hard, wear resistant.

TAYLOR NI-MO STEEL.
M-915; 0.2 C, 1.5 Ni, 0.2 Mo, bal Fe.
For hoist and drag chains.

TAYLOR TM ALLOY.
M-915; 0.18 -0.23 C, 0.4-0.7 Ni, 0.15-0.25 Mo, 0.4-0.6 Cr, bal Fe.
For chains; shock resistant.

TAYLOR-WHITE.
M-Eng.; 8.5 W, 3 Cr, 0.75-1.0 C, bal Fe.
For tools, dies, cutters; semi high speed steel.

TAZ-8.
M-1491; 8 Ta, 6 Cr, 6 Al, 4 W, 4 Mo, 2.5 V, 0.004 B, 1.0 Zr, 0.125 C, bal Ni.
At 1900°F: 56,000 TS.
For high speed aircraft nose cones, turbine buckets.
High stress rupture strength. Cast alloy.
Highest strength at 1900°F. Hardenable.

TAZ-8A.
M-1491; 8 Ta, 6 Cr, 6 Al, 4 Mo, 4 W, 2.5 Cb, 1 Zr, 0.125 C, 0.004 bal Ni.
As cast: 130,000 TS.
Rolled: 240,000 TS.
For low-stress components, stator vanes, turbine buckets, after- burner liners, transition ducts.
high oxidation resistance and high temperature strength.

TAZ-8B.
M-1491; 8 Ta, 6 Cr, 6 Al, 4 Mo, 4 W, 1.5 Cb, 1.0 Zr, 5 Co, 0.125 C, 0.004 B, bal Ni.
At R.T.: 150,000 TS; 3 El.
For turbine and jet engine buckets.
High creep and stress rupture strength.
High temperature strength. Heat and oxidation resistant.

TBS 600, TBS 1000 see **TIMKEN TBS 600, TBS 1000.**

T.C.12.
M-1116; 0.33 C, 12 Cr, 12 W, 1.0 V, 0.3 Mn, 0.5 Si, bal Fe.
Air or oil hardening tool and die steel; for extrusion dies, hot forging and threading dies; molds for brass and bronze castings.
AFNOR: Z 35 WCV 12.12; AISI H23.

T-C 20.
M-1022; 20 W, 80 Co.
For dies, cutlery, bearings; corrosion resistant.

T-C 25.
M-1022; 25 W, 75 Co.
For dies, cutlery, bearings; corrosion resistant.

T-C 30.
M-1022; 30 W, 70 Co.
For dies, cutlery, bearings; corrosion resistant.

T-C 35.
M-1022; 35 W, 65 Co.
For dies, cutlery, bearings; corrosion resistant.

T.C.610.
M-1116; 0.20 C, 10.0 Cr, 6.5 W, 2.0 Mo, 0.3 Mn, 0.4 Si, 10.0 Co, bal Fe.
Air or oil hardening tool and die steel; for hot extrusion dies; withstands hot contact for longer time.
AFNOR: Z 20 CKWD 10.10.07.02.

TCS (TERNE COATED STAINLESS STEEL).
M-806; Base: Type 304 (18 Cr, 8 Ni) stainless steel.
Coating: 80 Pb, 20 Sn. (Terne).
Soft: 80,000 TS; 30,000 YS.
For roofing, gutters, flashings, architectural brackets.
East to form and to solder.

TD12ZRE.
M-1120; 11.5 Mo, 6 Zr, 4.5 Sn, bal Ti.
Quenched and tempered: 1450 N/mm^2 TS; 1360 N/mm^2 YS; 4 El.
Metastable beta type alloy, Good properties in as-quenched condition; very strong after tempering.

TD NICKEL.
M-522, M 53, M 1491; ThO$_2$, 98 Ni.
At. R.T.: 65,000 TS; 50,000 YS; 23 El; 90 RA.
At 2400°F: 11,000 TS; 10,000 YS; 0.4 El; 5 RA.
At 1200°F: 32,000 TS; 30,000 YS; 10 El; 45 RA.
For aircraft gas turbines, afterburners, high temperature applications, fasteners.
Dispersion hardening alloy. Good characteristics at 900-1200°F.

TD NICR.
M-522, M-53, M-1491; 20 Cr, 2 ThO$_2$, bal Ni.
At R.T.: 125,000 TS; 80,000 YS; 18 El. (in 1").
At 1500°F: 35,000 TS; 30,000 YS; 7 El (in 1").
For high temperature applications, jet engines, aerospace structural components, furnace hardware.
Dispersion hardening alloy. High oxidation and sulfidation resistance.

TE.
M-U.S.; 0.1 C, 0.7 Mn, 20 Cr, 30 Ni, 4 Mo, 4 W, 1.9 Ta, 0.15 N$_2$, bal Fe.
For jet engine parts; high heat and oxidation resistant.

TEA LEAD.
M-Eng.; 98 Pb, 2 Sn.
For lead foil for packing tea.

TEC.
M-63; 5 Ag, 95 Cd.
For solder for copper, brass and steel; M.P. 640-740°F.

TEC see also **BRAZE 053 (TEC)**.

TECHALLOY 36 (INVAR).
M-1280; 0.04 C, 0.35 Mn, 0.12 Si, 36.0 Ni, bal Fe.
Very low thermal expansion rate.

TECHALLOY 294 (45-55).
M-1280; 43-45 Ni, 0.25 C, 2.0 Mn, 2.5 Fe, 1.0 Si, bal Cu.
For resistance wiring operating up to 110°F (595°C).

TECHALLOY 302.
M-1280; 0.2 max C, 18 Cr, 8 Ni, bal Fe.
Annealed: 90,000 TS; 40,000 YS; 50 El; 160 Brin.
Cold drawn: 180,000 TS; 150,000 YS; 10 El; 215 Brin.
For chemical plant equipment; stainless, austenitic.

TECHALLOY 304.
M-1280; 0.08 C, 18 Cr, 8 Ni, bal Fe.
Annealed: 85,000 TS; 35,000 YS; 60 El; 70 RA; 150 Brin.
Cold drawn: 180,000 TS; 125,000 YS; 10 El; 330 Brin.
For chemical and textile plant equipment, tanks, vessels; for welded structures, stainless, austentiic.

TECHALLOY 308.
M-1280; 0.08 C, 2 Mn, 19-21 Cr, 10-12 Ni, bal Fe.
Welding wire with enriched Cr-Ni content.

TECHALLOY 308L.
M-1280; 0.03 C, 2.0 Mn, 1.0 Si, 19-21 Cr, 10-12 Ni, bal Fe.
Low carbon austenitic stainless steel welding wire.

TECHALLOY 309.
M-1280; 0.1 C, 25 Cr, 12 Ni, bal Fe.
Annealed: 90,000 TS; 40,000 YS; 50 El; Rockwell B 83.
For refinery and chemical plant equipment, furnace parts; austenitic, corrosion and heat resistant.

TECHALLOY 309L.
M-1280; 0.03 C, 2.0 Mn, 1.0 Si, 19-21 Cr, 10-12 Ni, bal Fe.
Low carbon austenitic stainless steel welding wire for welding 309 stainless.

TECHALLOY 310.
M-1280; 0.25 max C, 25 Cr, 20 Ni, bal Fe.
Annealed: 100,000 TS; 45,000 YS; 50 El; 65 RA; 185 Brin.
For furnace parts, heat treat boxes, furnace linings; austenitic, corrosion and heat resistant.

TECHALLOY 314.
M-1280; 0.20 C, 25 Cr, 20 Ni, 2-3 Si, bal Fe.
Annealed: 100,000 TS; 50,000 YS; 45 El; 60 RA; 180 Brin.
For furnace parts, heat treat boxes, fixtures; heat resistant, austenitic.

TECHALLOY 316.
M-1280; 0.1 C, 18 Cr, 12 Ni, 2-3 Mo, bal Fe.
Annealed: 80,000 TS; 30,000 YS; 60 El; 80 RA; 135 Brin.
Cold drawn: 150,000 TS; 135,000 YS; 6 El; 300 Brin.
For chemical plant equipment, mixers, tanks, agitators; austenitic, acid resistant.

TECHALLOY 317.
M-1280; 0.1 C, 18 Cr, 12 Ni, 3-4 Mo, bal Fe.
Annealed: 85,000-90,000 TS; 40,000 YS; 50-45 El; 160 Brin.
For chemical plant equipment, mixers, tanks, agitators; austenitic, acid resistant.

TECHALLOY 330.
M-1280; 0.20 C, 15 Cr, 35 Ni, bal Fe.
Annealed: 80,000 TS; 50,000 YS; 30 El; 30 RA; 185 Brin.
For furnace parts, heat treat boxes; austenitic, heat resistant.

TECHALLOY 347.
M-1280; 0.08 C, 18 Cr, 10 Ni, Cb = 10 x C, bal Fe.
Annealed: 90,000 TS; 35,000 YS; 50 El; 65 RA; 110 Brin.
Cold drawn: 100,000 TS; 65,000 YS; 40 El; 60 RA; 210 Brin.
For chemical plant equipment, tanks, vessels; welded structures, austenitic, stainless.

TECHALLOY 430.
M-1280; 0.10 C, 14-18 Cr, bal Fe.
Annealed: 75,000 TS; 45,000 YS; 30 El; 65 RA; 155 Brin.
Cold drawn: 130,000 TS; 120,000 YS; 2 El; 185 Brin.
For screw machine products, pump and valve parts; corrosion resistant.

TECHALLOY A (80-20).
M-1280; 77-79 Ni, 0.15 C, 2.5 Mn, 1.0 Fe, 1.0 Si, 19-21 Cr.
For electrical resistance elements operating up to 2150°F (1175°C).

TECHALLOY C-20.
M-1280; 0.07 max C, 29 Cr, 20 Ni, 2 min Cu, 1 Si, 2 Mo, bal Fe.
Cast: 65,000 TS; 30,000 YS; 45 El; 50 RA; 130 Brin..
Rolled: 85,000 TS; 35,000 YS; 35 El; 50 RA; 180 Brin.
For chemical plant equipment, mixers, agitators; resists mixed acids, austenitic.

TECHALLOY C (62-16).
M-1280; 57 Ni, 0.15 C, 1.0 Mn, 1.0 Si, 14-18 Cr, bal Fe.
For electrical resistance elements operating up to 1700°F (925°C).

TECHALLOY D (35-19).
M-1280; 34-37 Ni, 0.15 C, 1.0 Mn 1-3 Si, 18-21 Cr, bal Fe.
For resistance wiring operating up to 1400-1600°F (760-870°C).

TECHALLOY GLASSEAL 29-17.
M-1280; 0.06 C, 29.0 Ni, 0.50 Mn, 17.0 Co, 0.50 Mg, 0.20 Si, bal Fe.
For sealing metal to hard glass.

TECHALLOY GLASSEAL 42.
M-1280; 0.03 C, 0.80 Mn, 0.30 Si, 42.0 Ni, bal Fe.
For sealing metal to 1075 glass, also 0120 and 0010 glasses, as sealed beam headlights, electronic tubes, industrial lamps.

TECHALLOY GLASSEAL 52.
M-1280; 0.05 C, 0.50 Mn, 0.30 Si, 50.5 Ni, bal Fe.
For sealing metal to soft glass.

TECHNALLOY A.
M-229; 1.2-1.3 C, bal Fe.
For rolls; for steel mills.

TECHNALLOY AX.
M-229; C, bal Fe.
For heavy duty rolls; for blooming mills.

TECHNALLOY B.
M-229; 1.8-2.0 C, bal Fe.
For rolls; for steel mills.

TECHNALLOY C.
M-229; 1.5-1.8 C, bal Fe.
For heavy duty rolls; for structural mills.

TECHNALLOY D.
M-229; 2.2-2.3 C, bal Fe.
For rolls; for steel mills.

TECHNALLOY E.
M-229; 2.4-2.6 C, bal Fe.
For rolls; for steel mills.

TECHNALLOY STANDARD.
M-229; 1.0 C, bal Fe.
For rolls; for steel mills.

TECHNICAL IRON.
M-Russia; 0.025 max C, 0.03 max Mn, 0.03 max Si, bal Fe.
For magnetic circuits, deflectors; high permeability.

TECH-TRONIC 32.
M-1280; 0.15 C, 11.0-14.0 Mn, 1.0 Si, 16.5-19.0 Cr, 0.20-0.45 N, 0.50 2.5 Ni, bal Fe.
Wire for springs, screens, cages, racks.

TECH-TRONIC 50.
M-1280; 0.06 C, 4.0-6.0 Mn, 1.0 Si, 20.5-23.5 Cr, 11.5-13.5 Ni, 1.5-3.0 Mo, 0.10-0.30 Cb, 0.10-0.30 V, 0.2-0.4 N, bal Fe.
Wire for AC power equipment, marine hardware, springs, fasteners.

TECO.
M-796; WC, VC, CrC, ThC, MoC, Co, Ni.
For tools, wearing parts, cutters; a series of refractory alloys.

TECO 38.
M-1054; WC, Co.
For cutting tools; sintered.

TECO 51.
M-1054; WC, Co.
For cutting tools; sintered.

TECO 469.
M-1054; WC + Co.
For cutting tools; cemented.

TECO 610 G.
M-1054; WC, Co.
For cutting tools; sintered.

TECO 946 X.
M-1054; WC, Co.
For cutting tools; sintered.

TECO A.
M-1054; WC, Co.
For cutting tools; sintered.

TECO A-1.
M-1054; WC, Co.
For cutting tools; sintered.

TECO-ATS1.
M-1054; 10 Mo, bal TiC.
For tools, bearings, seals, wear parts. Resists heat and wear, sintered carbide.

TECO B.
M-1054; WC, Co.
For cutting tools; sintered.

TECO C.
M-1054; WC, Co.
For cutting tools; sintered.

TECO CF.
M-1054; WC, Co.
For cutting tools; sintered.

TECO CHD.
M-1054; WC, Co.
For cutting tools; sintered.

TECO CMD.
M-1054; WC, Co.
For cutting tools; sintered.

TECO KT 13.
M-1054; WC, Co.
For cutting tools; sintered.

TECO TH.
M-1054; carbide.
For cold work dies; sintered.

TEC Z see BRAZE 056 (TEC Z).

T.E.D.
M-1116; 1.3 C, 4.3 Cr, 12 W, 0.9 Mo, 4 V, bal Fe.
Air or oil harden to 63-66 Rc.
For slotting tools, gear cutters, form cutters, punches.
AFNOR: Z 130 WVD 12.04.01.
Germany: S 12.1.4 (EV 4) W.Nr. 1.3302.

T.E.D.C.
M-1116; 0.80 C, 4.2 Cr, 12.5 W, 1.8 Mo, 1.8 V, 1.2 Co, bal Fe.
Air or oil harden to 63-66 Rc.
For lathe tools, twist drills, reamers, milling cutters, broaches.
AFNOR: Z 80 WDVK 12.02.02.

TEENAX 46 see SIMONDS TEENAX 46.

TEGO.
M-528; 78.83 Pb, 15-18 Sb, 1.5-3.0 Sn, 1-2 Cu, 0.3-0.8 As.
For bearings; antifriction.

TELASTIC MOLY see MOLY TELASTIC.

TELCALLOY 1.
M-108; Cu-Ni.
Electrical resistance alloy.
P-10 μ Ω cm for resistance.
(micro ohm/cm^3).

TELCALLOY 1.5.
M-108; Cu-Ni.
Electrical resistance alloy.
P-15 μ Ω cm for resistance.

TELCALLOY 2.
M-108; Cu-Ni.
Electrical resistance alloy.
P-20 μ Ω cm for resistance.

TELCALLOY 3.
M-108; Cu-Ni.
Electrical resistance alloy.
P-30 μ Ω cm. for resistance.

TELCALLOY 4.
M-108; Cu-Ni.
Electrical resistance alloy.
P-40 μ Ω cm. for resistance.

TELCON 15.
M-108; 36 Ni, bal Fe/Ni.
Thermostatic bimetal.
ASTM flexivity/deg F.: 10.3×10^{-6}.

TELCON 36/64.
M-108; 36 Fe, 64 Ni.
For pulse and radar transformers, relay cores; soft magnet, high permeability.

TELCON 41.
M-108; Ni/Ni, Cr, Fe.
Thermostatic bimetal.
ASTM flexivity/deg. F.: 4.6×10^{-6}.

TELCON 75.
M-108; 58 Ni, bal Fe/Ni, Mn, Fe.
Thermostatic bimetal.
ASTM flexivity/deg. F.: 7.2×10^{-6}.

TELCON 79.
M-108; 77 Ni, 14 Fe, 5 Cu, 4 Mo.
Soft magnetic alloy, high permcability at low field strengths for transformers, chokes.

TELCON 140.
M-108; 36 Ni, bal Fe/Ni, Mn, Fe.
Thermostatic bimetal.
ASTM flexivity/deg. F.: 15.5×10^{-6}.

TELCON 188.
M-108; Ni, Fe, Cr/Ni, Fe, Cr.
Thermostatic bimetal.
ASTM flexivity/deg. F.: 9.8×10^{-6}.

TELCON 200.
M-108; 36 Ni, bal Fe/Mn, Cu, Ni.
Thermostatic bimetal.
ASTM flexivity/deg.F.: 21.4×10^{-6}.

TELCON 400.
M-108; 42 Ni, bal Fe/Ni, Mn, Fe.
Thermostatic bimetal.
ASTM flexivity/deg.F.: 13.1×10^{-6}.

TELCON CU BE 250 see CU BE 250.

TELCON E140.
M-108; 36 Ni, bal Fe/Ni, Cr, Fe.
Thermostatic bimetal.
ASTM flexivity/deg.F. 14.4×10^{-6}.

TELCON E400.
M-108; 42 Ni, bal Fe/Ni, Cr, Fe.
Thermostatic bimetal.
ASTM flexivity/deg.F.: 11.7×10^{-6}.

TELCON MARAGING STEEL.
M-108.
For high strength applications and MIG and TIG welding.

TELCON R.2799.
M-108; 70 Fe, 30 Ni.
Rolled: 90,100 TS; 35,100 YS.
For magnetic shunts, speedometers, tachometers; soft magnet, Curie temperature at 60°C.

TELCON STAN.
M-108; Cu-Ni.
Electrical resistance alloy.
P-48 $\mu\,\Omega$ cm. for resistance. (micro ohm/cm^3).

TELCOSEAL.
M-108; 54 Fe, 29 Ni, 17 Co.
For sealing to borosilicate glasses in thermionic valve and cathode ray tube construction.
Coefficient of expansion same as glass.

TELCOSEAL 1.
M-108; 29 Ni, 17 Co, bal Fe.
For glass/ceramic-to-metal sealing.
Coefficient of expansion, 20-500°C: 6.2×10^{-6}.
For use with borosilicate glass.

TELCOSEAL 2.
M-108; 42 Ni, bal Fe.
For glass/ceramic-to-metal sealing.
Coefficient of expansion, 20-500°C: 8.0×10^{-6}.

TELCOSEAL 3.
M-108; 42 Ni, 6 Cr, bal Fe.
For glass/ceramic-to-metal sealing.
Coefficient of expansion, 20-500°C: 11.7×10^{-6}.

TELCOSEAL 3/2.
M-108; 47 Ni, 5 Cr, bal Fe.
For glass/ceramic-to-metal sealing.
Coefficient of expansion, 20-500°C: 10.7×10^{-6}.

TELCOSEAL 4.
M-108; 46 Ni, bal Fe.
For glass/ceramic-to-metal sealing.
Coefficient of expansion, 20-500°C: 8.4×10^{-6}.

TELCOSEAL 6.
M-108; 49 Ni, bal Fe.
For glass/ceramic-to-metal seal.
Coefficient of expansion, 20-500°C: 9.5×10^{-6}.

TELCOSEAL 6/2.
M-108; 48 Ni, bal Fe.
For glass/ceramic-to-metal sealing.
Coefficient of expansion, 20-500°C: 9.2×10^{-6}.

TELCOSEAL 6/3.
M-108; 50 Ni, bal Fe.
For glass/ceramic-to-metal sealing.
Coefficient of expansion, 20-500°C: 9.7×10^{-6}.

TELCOSEAL 6/4.
M-108; 51 Ni, bal Fe.
For glass/ceramic-to-metal sealing.
Coefficient of expansion, 20-500°C: 10.0×10^{-6}.

TELCUMAN.
M-108; Cu alloy.

TELECTAL.
M-299; 1.5 Si, 0.1 Li, bal Al.
43,000-47,000 TS; 6 El; 80-100 Brin.
For light alloy parts, electrical conductors; non-hardenable.

TELEDIUM.
M-458; 0.02-0.10 Te, bal Pb.
Cold rolled: 3800-5000 TS; 8 Brin.
For chemical plant equipment, cable sheathing, batteries, water supply pipes; corrosion resistant.

TELEGRAPH BRONZE.
M-Eng.; 80 Cu, 7.5 Pb, 7.5 Zn, 5 Sn.
For machine parts, switches; free-cutting.

TELEMET.
M-1; 0.25 max C, 16-23 Cr, 2 max Mn, 1 max Si, bal Fe.
Rolled: 60,000-75,000 TS; 40,000-45,000 YS; 35-30 El; 178-185 Brin.
For television picture tubes, metal-to-glass seals; heat and corrosion resistant.

TELLOY 4140.
M-67; 0.38-0.43 C, 0.75-1.0 Mn, 0.035 max P. 0.04-0.06 S, 0.010 min Te, 1.0 Cr, 0.20 Mo, bal Fe.
Free machining alloy steel.

TELLURIUM COPPER.
M-33; 0.5 Te, 0.008 P, bal Cu.
Ann: 33,000 TS; 8,000 YS; 3 El.
Drawn: 48,000 TS; 46,000 YS; 18 El.
For general use; high conductivity.

TELLURIUM COPPER 112.
M-141; 0.5 Te, bal Cu.
For welding torch and soldering iron tips; high conductivity.

TELLURIUM COPPER 145.
M-8; 99.49 Cu, 0.50 Te, 0.01 P.
Hard Rod: 48,000 TS; 40,000 YS; 12 El; B 50 Rock.
Soft Rod: 32,000 TS; 10,000 YS; 45 El; B 45 Rock.
For transformer and circuit breaker terminals, studs, bolts, current-carrying parts.
Elect. cond. 95%. Free-cutting. Not subject to hydrogen embrittlement.

TELLURIUM COPPER 1452.
M-8; 0.50 Te, bal Cu.
Hard: 48,000 TS; 40,000 YS; 12 El; 83 Brin.
Soft: 32,000 TS; 10,000 YS; 45 El; 42 Brin.
For electrical apparatus, screw machine parts; high conductivity, free-cutting.

TELLURIUM LEAD.
M-88; 0.06 Te, 0.06 Cu, bal Pb.
Rolled.
For pipes, chemical industry; greater corrosion resistance to H_2SO_4.

T E M see **TRIPLE EXTRA M**; **T.E.R.** see **TRIPLE EXTRA**; **T.E.D.** see **TRIPLE EXTRA D**.

T.E.M.
M-1116; 1.0 C, 4.2 Cr, 6.5 W, 5 Mo, 1.9 V, bal Fe.
Air or oil harden to 63-66 Rc.
For twist drills, reamers, lathe cutters, broaches.
AFNOR: Z 100 WDV 06.05.02.
US-AISI M2 (high carbon) high speed steel.

T.E.M.02.C.
M-1116; 1.0 C, 4.2 Cr, 6.5 W, 5 Mo, 1.9 V, 2 Co, bal Fe.
Air or oil harden to 63-67 Rc.
For drills, reamers, lathe cutting tools, taps, gear cutters.
AFNOR: Z 100 WDVK 06.05.02.02.

T.E.M.02.V.
M-1116; 1.08 C, 4.2 Cr, 6.3 W, 5.3 Mo, 2.4 V, bal Fe.
Air or oil harden to 62-67 Rc.
For drills, taps, threading form tools, lathe tools, hobs, broaches.
AFNOR: Z 110 WDV 06.05.02.
US-AISI M3(1).

T.E.M.03.V.
M-1116; 1.2 C, 4.0 Cr, 6.3 W, 5 Mo, 3 V, bal Fe.
Air or oil harden to 62-66.5 Rc.
For drills, reamers, gear cutters, broaches, lathe tools, thread chasers.
AFNOR: Z 120 WDV 06.05.03.
US-AISI M 3(2).
Germany: S 6.5.3(EMo5V3); W.-Nr. 1.3344.

T.E.M.05.C.
M-1116; 1.0 C, 4.3 Cr, 7.7 W, 5 Mo, 2.3 V, 5 Co, bal Fe.
Air or oil harden to 63-67 Rc.
For drills, taps, reamers, turning tools for machining hard metals.
AFNOR: Z 100 WDKV 07.05.05.02.
US-AISI M41.

T.E.M.08.C.
M-1116; 1.1 C, 4 Cr, 5 W, 6.7 Mo, 2 V, 8 Co, bal Fe.
Air or oil harden to 63-69 Rc.
For drills and lathe tools machining abrasive materials, for finishing cuts.
AFNOR: Z 110 KDWV 08.07.05.02.

T.E.M.60.
M-1116; 0.60 C, 4.0 Cr, 6.5 W, 5.0 Mo, 1.9 V, 0.3 Mn, 0.3 Si, bal Fe.
Air or oil hardening tool and die steel; for cold extrusion dies; for hot forming dies.
AFNOR: Z 60 WDCV.06.05.04.02; AISI H42.

T.E.M.92.
M-1116; 0.92 C, 4.2 Cr, 6.5 W, 5.0 Mo, 1.9 V, bal Fe.
Air or oil harden to 64-66 Rc.
For drills, broaches, reamers, lathe and threading tools; good wear resistance.
AFNOR Z 92 WDV 06.05.02. US-AISI: M2.
Germany: S 6.5.2 (DMo5); W. Nr. 1.3343.

T.E.M.C.
M-1116; 0.85 C, 4.3 Cr, 6.5 W, 5 Mo, 1.9 V, 5 Co, bal Fe.
Air or oil harden to 62-67 Rc.
For drills and lathe tools for special work, high temperature, high speeds special alloys.
AFNOR: Z 85 WDKV 06.05.05.02.
US-AISI M 35.
Germany: S 6.5.2.5 (EMo 5 Co 5); W.-Nr. 1.3243.

T.E.M.C V.
M-1116; 1.57 C, 4.3 Cr, 6.7 W, 5 Mo, 4.8 V, 5 Co, bal Fe.
Air or oil harden to 62-68 Rc.
For high speed finish cutting, particularly on hard and tough materials as stainless steels.
AFNOR: Z 150 WDKV 06.05.05.05.
US-AISI-M15.

T.E.M.C V.11.
M-1116; 1.65 C, 4.3 Cr, 6.8 W, 5.3 Mo, 4.8 V, 11 Co, bal Fe.
Air or oil harden to 62-69 Rc.
For tool bits for finishing and high speed machining of stainless steels, and non-metallics.
AFNOR: Z 160 KWDV 10.06.05.05.

T.E.M.P.
M-1116; 1.1 C, 4.3 Cr, 7.0 W, 4.0 Mo, 2.0 V, 5.0 Co, bal Fe.
Air or oil harden to 63-67 Rc.
For reamers, broaches, lathe tools, thread chasers.
AFNOR: Z 110 WDKV 07.04.05.02. US-AISI: M 41. Germany: S 7.4.2.5; W.Nr. 1.3246.

TEMP AIR 8.
M-1431; 0.6 C, 3.6 Cr, 1.75 V, 8.5 Mo, bal Fe.
For drills, cutters, extrusion dies; hot work steel, oil hardened.

TEMPALOY 630.
M-8; 82 Cu, 9.5 Al, 1.0 Mn, 5 Ni, 2.5 Fe.
Hard: 105,000 S; 60,000 YS; 12 El.
For propeller shafting; heat treatable.

TEMPALTO.
M-719; 56 Cu, 14 Ni, 30 Pb, P.
For metallic packing; for high temperatures.

TEMPERA 1.
M-1288; 0.12 max C, 0.8 Si, 0.8 Al, 6.5 Cr, bal Fe.
Annealed: 70,000 TS; 30,000 YS; 28 El; 65 RA; 150 Brin.
For oil refinery equipment; corrosion and creep resistant.

TEMPERA 2.
M-1288; 0.12 max C, 1.2 Si, 1 Al, 13 Cr, bal Fe.
Annealed: 75,000 TS; 40,000 YS; 35 El; 70 RA; 155 Brin.
For oil refinery equipment; corrosion and creep resistant.

TEMPERA 3.
M-1288; 0.12 max C, 1.5 Si, 1.5 Al, 24 Cr, bal Fe.
Annealed: 75,000-95,000 TS; 46,000-60,000 YS; 35-25 El; 65-46 RA; 160-200 Brin.
For furnace equipment, heat treat boxes; heat resistant.

TEMPERA 4.
M-1288; 0.15 C, 2 Si, 20 Cr, 9.5 Ni, bal Fe.
Annealed: 75,000 TS; 35,000 YS; 50 El; 65 RA; 160 Brin.
For chemical plant equipment; Type 302; stainless, austenitic.

TEMPERA 5.
M-1288; 0.15 C, 2 Si, 24 Cr, 19 Ni, bal Fe.
Annealed: 100,000 TS; 45,000 YS; 50 El; 65 RA; 185 Brin.
For furnace parts, heat treat boxes; Type 310; corrosion and heat resistant.

TEMPERA A.
M-1288; 0.12 max C, 1 Si, 1 Al, 18 Cr, bal Fe.
Annealed: 90,000 TS; 55,000 YS; 20 El; 40 RA; 160 Brin.
For chemical plant equipment, furnace parts; corrosion and creep resistant.

TEMPERA B.
M-1288; 0.2 C, 1.2 Si, 25 Cr, 4 Ni, bal Fe.
Annealed: 95,000 TS; 60,000 YS; 25 El; 45 RA; 180 Brin.
For furnace parts, heat treat boxes; heat and creep resistant.

TEMPERATURE COMPENSATOR 32 see **CARPENTER TEMPERATURE COMPENSATOR 32.**

TEMPERATURE COMPENSATOR ALLOY NO. 30.
M-Eng.; Fe-Ni.
For compensating shunts for electrical equipment; temperature-sensitive, magnetic.

TEMPERDIE.
M-1433; 0.40 C, 1 Mn, 1.5 Cr, 0.25 V, 1 Mo, bal Fe.
For hot header dies, punches, hot press dies; hot work steel oil hardened.

TEMPERIM.
M-566; 3.2 C, Si, Mn, other alloy elements, bal Fe.
Cast: 400 Brin.
For wear resistant parts; cast iron, abrasion resistant.

TEMPERITE.
M-917; Pb, Sn, Cd.
For temperature indicating alloys; low melting series of alloys.

TEMPER TOUGH.
M-40; 0.75 C, 0.6 Mn, 1.15 Si, 0.8 Cr, 0.15 V, 0.3 Mo, bal Fe.
Oil hardening tool steel; AISI L6.

TEMPLATE.
M-719; Zn alloy.
For general castings; white metal.

TEMPLESS.
M-1009; 0.9 C, alloy, bal Fe.
For punches, dies, forming and blanking dies, upsetters; no temper required.

TEMPLESS NO. 11.
M-1009; 0.3-0.37 C, 0.4 Mn, 0.7-0.9 Cr, 0.4-0.6 Mo, bal Fe.
Rolled: 130,000 TS; 16 El; 50 RA; 220 Brin.
For pneumatic tools, axes, chisels, caulking tools; shock resistant.

TEMPO.
M-1433; 0.90 C, 1.15 Mn, 0.5 Cr, 0.5 W, bal Fe.
For gauges, hobs, punches, master tools; oil hardened, non-deforming.

T.E.M.V.
M-1116; 1.28 C, 4.3 Cr, 5.6 W, 4.6 Mo, 4 V, bal Fe.
Air or oil harden to 63-66.5 Rc.
For drills, reamers, finish cutting lathe tools, by special temper for punches, files, chisels.
AFNOR: Z 130 WDV 06.05.04.
US-AISI M4.

T.E.M.V.11.
M-1116; 1.25 C, 4.3 Cr, 7.5 W, 6.3 Mo, 3 V, 11 Co, bal Fe.
Air or oil harden to 62-69 Rc.
For form tools, lathe tools, climb milling cutters for machining hard and tough steels.
AFNOR: Z 125 KWDV 11.08.06.03.
Germany: S 10.4.3.10 (EW9Co10); W.Nr. 1.3207.

TENAX METAL.
M-1322; 4.2-4.6 Al, 2.2-3.0 Cu, 1.2 Pb, 0.35 Fe, bal Zn.
For solder for aluminum ware; strong.

TENAX-N.
M-96; 0.45 C, 1.0 Cr, 2 W, bal Fe.
Annealed: 102,000 TS; 205 Brin.
For chisels, tools, punches, dies, hot shear blades.
Oil hardening, shock resistant.

TENAX NB.
M-96, M-1322; 0.55 C, 1 Si, 1.1 Cr, 1.9 W, 0.18 V, 0.3 Mn, bal Fe.
For punches, upsetters, riveters, dies, crimpers; oil hardened, shock resistant.

TENAZ-1.
M-1766; 0.38 C, 0.40 Mn, 1.0 Si, 1.15 Cr, 2.0 W, 0.15 V, bal Fe.
For chisels, punches, pneumatic tools.
IHA F-525; Din 35 WCrV7.

TENAZ-2.
M-1766; 0.50 C, 0.40 Mn, 1.0 Si, 1.15 Cr, 2.0 W, 0.15 V, bal Fe.
Shock resisting tool steel; for punches, pneumatic tools, chisels.
IHA F-524; similar to AISI S1.

TENELON.
M-604; 0.08-0.12 C, 17.0-18.5 Cr, 14.5-16.0 Mn, 0.30-1.0 Si, 0.75 max Ni, 0.35 min N, bal Fe.
Manganese austenitic steel; work-hardens.
For parts subject to extreme wear and abrasion.

TENCO.
M-618; C, 18 W, 9 Co, 1 Mo, 4 Cr, 2 V, bal Fe.
For tools, dies, cutters; high speed steel.

TENEM.
M-646; Sn, Be, bal Cu.
For corrosion and wear resistant parts; high strength.

TENIT 1720 SPEZIAL now **VEW W106.**

TENIT KL SPEZIAL.
M-1114; 0.55 C, 1 Cr, 0.18 V, 1.8 W, bal Fe.
For hot work tools and dies; hot work steel, oil hardened.

TENNESEAL.
M-604; low C, Mn, bal Fe.
Pre-formed, zinc-coated steel sheets for roofing and siding.

TENS-50.
M-1458; Si, Mg, Be, N, bal Al.
Sand cast: 46,000 TS; 36,000 YS; 5 El.
Permanent mold: 50,000 TS; 44,000 YS; 6 El.
For aircraft and missile components, gear cases; age-hardenable.

TENSIL-COR.
M-663; 0.04 C, 1.63 Mn, 0.5 Si, 0.5 Mo, bal Fe.
A CO_2 shielded, cored wire for single or multipass welding of high tensile low alloy steel.
As welded: 112,000 TS; 104,700 YS; 24 El.
Good toughness down to -75°F.
AWS E 70T-G.

TENSIL-COR.
M-663; weld deposit: 0.072 C, 1.63 Mn, 0.50 Si, 0.46 Mo, 0.03 V, 0.04 Cr, bal Fe.
As welded: 108,500 psi TS; 103,000 psi YS; 23 El.
Good impact strength.
AWS A5.20 class E70T-G.

TENSILE-FLEX.
M-1598.
Oxygen free, copper-based, precipitation-type alloy wire.
High strength, resistant to hydrogen embrittlement, long flex life, high electrical conductivity.

TENSILEND 70.
M-677; 0.06 C, 0.55 Mn, 0.35 Si, bal Fe.
Welded: 70,000-80,000 TS; 70,000 YS; 26 El; 60 RA; 180 Brin.
For welding electrodes for low alloy steel; low H_2.

TENSILEND 80.
M-677; 0.06 C, 0.8 Mn, 0.4 Si, 1.0 Ni, bal Fe.
For welding electrodes; class E8016-C3, 1% nickel steel.

TENSILEND 100.
M-677; 0.06 C, 0.55 Mn, 0.4 Si, 1.8 Ni, 0.6 Mo, 0.1 V, bal Fe.
Welded: 100,000-110,000 TS; 91,000 YS; 22 El; 58 RA; 230 Brin.
For welding electrodes for low alloy steel; low H_2.

TENSILEND 120.
M-677; 0.07 C, 0.9 Mn, 0.25 Si, 1.8 Ni, 0.8 Mo, 0.5 V, bal Fe.
Welded: 120,000-130,000 TS; 107,000 YS; 19 El; 52 RA; 260 Brin.
For welding rods for marine applications; low H_2.

TENSILIA.
M-Eng.; 0.48 C, 0.51 Mn, 0.18 V, bal Fe.
For tools, gears, pinions, shafts; water hardening.

TENSILITE.
M-129; 67-64 Cu, 24-29 Zn, 2.5-3.8 Mn, 3.1-4.4 Al, 0-1.2 Fe.
103,100 TS; 32,500 YS; 19 El; 23 RA.
For gears, worm wheels; tough.

TENSILITE.
M-129; 68.26 Cu, 21.04 Zn, 3.20 Mn, 0.04 Si, 0.14 Pb, 4.8 Al, 2.03 Fe.
For gears, worm wheels; wear resistant.

TENSILOY BORON.
M-1438; C, alloy, bal Fe.
For machinery parts, gears, pinions, shafts; preheat treated, abrasion and shock resistant.

TENSILOY EXTRA.
M-1438; 0.50 C, 1.3 Si, 0.3 Mo, 0.5 W, bal Fe.
For gears, crimpers, punches, dies, upsetters; oil hardened, shock resistant.

TENSILOY H.C.C.
M-1438; 1.5 C, 12 Cr, 1 V, 1 Mo, bal Fe.
For blanking and forming dies, punches; Type D2; air hardened, non-deforming.

TENSILOY H.S.S.
M-1438; 0.80 C, 4 Cr, 2 V, 18 W, 0.65 Mo, bal Fe.
For reamers, taps, broaches, drills, hobs, lathe tools; Type T2; high speed steel.

TENSILOY NITRIDING DIE.
M-1438; 0.80 C, 4 Cr, 1 V, 1.5 W, 8 Mo, bal Fe.
For reamers, broaches, dies, hobs; Type M1; high speed steel.

TENSILOY NON-TEMPERING.
M-1438; 0.33 C, 0.75 Cr, 0.75 Cu, 0.75 Mo, bal Fe.
For tools, dies; water hardened, non-tempering.

TENSILOY O.H. DIE.
M-1438; 0.90 C, 1.15 Mn, 0.5 Cr, 0.5 W, bal Fe.
For punches, dies, crimpers, cutting tools; oil hardened, non-deforming.

TENSITE.
M-897; 98 Ni, 2 Al.
Annealed: 57,000 TS; 18,000 YS; 45 El.
For cathodes and filaments in electronic tubes; heat resistant.

TEN STAR see **CARPENTER TEN STAR.**

TENUAL 50.
M-190; 6-8 Cu, 2.5 max Zn, 1.5 max Fe, 2.0 max Si, bal Al.
For crankcases, oil pans, cylinder heads.

TENUAL NO. 226.
M-190; 87 Min Al, 9.25-10.75 Cu, 0.90-1.50 Fe, 0.15-0.35 Mg.
Cast: 21,000-24,000 TS; 115-127 Brin.
For pistons, valve tappets, cam-shaft bearings, guides; same as SAE-34.

TENUAL 1.
M-190; 5 Si, bal Al.
Cast: 19,000 TS; 8000 YS; 8 El; 40 Brin.
For instrument cases, thin wall pressure tight castings; corrosion resistant.

TENUAL T.
M-190; 0.2 Ti, 1 Mn, bal Al.
For aircraft structures, cowling; corrosion resistant.

TENUAL T TITANITE.
M-190; 0.2 Ti, 1.5 Mn, bal Al.
For light alloy parts; corrosion resistant.

TENZALOY.
M-815; 0.8 Cu, 0.4 Mg, 8 Zn, bal Al.
Aged: 32,000 TS; 22,000 YS; 3 El; 65 Brin.
For light castings, housings, machinery parts; high strength, requires no heat treatment.

TEPAZ.
M-1318; 0.12 max C, 1.15 Si, 13 Cr, 0.95 Al, bal Fe.
For oil refinery equipment; corrosion and creep resistant.

TEPAZ 1020.
M-1318; 0.15 C, 2 Si, 19.5 Cr, 9.5 Ni, bal Fe.
Annealed: 80,000 TS; 35,000 YS; 55 El; 65 RA; 170 Brin.
For chemical plant equipment; Type 302; stainless, austenitic.

TEPAZ 2025.
M-1318; 0.15 C, 24 Cr, 19 Ni, bal Fe.
Annealed: 100,000 TS; 45,000 YS; 50 El; 65 RA; 180 Brin.
For furnace parts, heat treat boxes; heat resistant, austenitic; Type 310.

TERBIUM.
M-1755; Tb.
Purities: 99.9% (special distilled grade), 99.5+%.
Forms: Ingot, lump, sheet, foil, rod, wire, filings, sponge, single crystals.

TERMACID.
M-1245; 0.10-0.25 C, 27 Cr, 4.5 Ni, Mo, Mn, bal Fe.
For furnace parts, heat treating boxes; heat resistant to 2000°F.

TERMAFOND 245C.
M-1445; 3.75-4.25 Cu, 14-15 Si, 0.6-1.0 Mg, 3.7-4.3 Ni, 0.5-0.9 Co, bal Al.
Heat treated: 24,000-33,000 TS; 22,000-21,000 YS; 0.5-0.2 El; 115-140 Brin.
For light alloy parts; age-hardened, corrosion resistant.

TERMAFOND 245NT.
M-1445; 4.3-4.7 Cu, 14.5-15.5 Si, 0.7-1.1 Mg, 0.7 Mn, 0.2 Ti, 2.3-2.7 Ni, bal Al.
Heat treated: 21,000-28,500 TS; 20,000-27,000 YS; 0.5-2.0 El; 110-130 Brin.
For light alloy parts; age-hardened.

TERMAFOND C3.
M-1445; 2.9-3.2 Cu, 1.6 Fe, 0.7 Si, 0.6 Mg, 0.2 Ti, 0.6 Ni, bal Al.
Heat treated: 47,000-60,000 TS; 43,000-51,000 YS; 5-1 El; 115-150 Brin.
For light alloy parts; age-hardenable.

TERMAFOND C10.
M-1445; 9.3-10.7 Cu, 0.5-1.3 Fe, 0.2-0.4 Mg, bal Al.
Heat treated: 40,000-51,200 TS; 34,000-37,000 YS; 0.5-0.2 El; 125 150 Brin.
For pistons; age-hardenable.

TERMAFOND C12N.
M-1445; 11.0-12.5 Cu, 0.6-0.8 Fe, 0.1-0.2 Ti, 0.4-0.6 Ni, bal Al.
Heat treated: 40,000-65,000 TS; 24,000-27,000 YS; 1.5-1.0 El; 110-130 Brin.
For pistons; age-hardenable.

TERMAFOND C12T.
M-1445; 11.0-12.5 Cu, 0.2 Ti, 1 Fe, bal Al.
Heat treated: 38,000-42,7000 TS; 22,800-25,600 YS; 1.5-1.0 El; 110-125 Brin.
For pistons; age-hardenable.

TERMAFOND C46.
M-1445; 9.5-10.5 Cu, 0.8-1.2 Si, 0.3 Mg, 0.2 Ti, 1.3-1.7 Ni, bal Fe.
Heat treated: 45,00-54,000 TS; 37,000-48,000 YS; 1.0-0.5 El; 115-140 Brin.
For engine cylinder heads, pistons; age-hardenable.

TERMAFOND S10.
M-1445; 2.0-2.5 Cu, 9.5-10.5 Si, 1.4 Mg, 0.8-1.2 Ni, bal Al.
Heat treated: 36,00-50,000 TS; 34,000-42,000 YS; 0.5-0.3 El; 95-125 Brin.
For light alloy parts; for high temperature service.

TERMAFOND S121.
M-1445; 2.25 Cu, 0.8-1.2 Fe, 11-13 Si, 1.5 Mg, 0.5 Mn, bal Al.
Heat treated: 41,000-50,000 TS; 38,000-43,000 YS; 0.5-0.3 El; 120-135 Brin.
For light alloy parts; corrosion resistant.

TERMAFOND S122.
M-1445; 0.5-1.1 Cu, 13 Si, 1.2 Mg, 0.15 Ti, 2.2 Ni, bal Al.
Heat treated: 40,000-50,000 TS; 38,000-43,000 YS; 0.5-0.3 El; 120-130 Brin.
For light alloy parts; for high temperature service.

TERMAFOND Y.
M-1445; 3.8-4.2 Cu, 1.3-1.7 Mg, 1.8-2.3 Ni, bal Al.
Heat treated: 36,000-47,000 TS; 32,000-43,000 YS; 0.5-0.3 El; 95-115 Brin.
For engine cylinder heads, pistons; age-hardenable.

TERMAFOND YT.
M-1445; 4 Cu, 1.5 Mg, 2 Ni, bal Al.
For engine cylinders, pistons; age-hardenable.

TERNALLOY 5.
M-130; 1.6 Mg, 3 Zn, 0.3 Cr, 0.5 Mn, bal Al.
SC-F temper: 29,000 TS; 13,000 YS; 12 El; 50 Brin.
SC-T 51 temper: 30,000 TS; 18,000 YS; 7 El; 60 Brin.
PM-Age temper: 43,000 TS; 22,000 YS; 16 El; 70 Brin.
For castings; natural aging.

TERNALLOY 6.
M-906, M-130; 0.2-0.4 Mn, 3.4-3.9 Zn, 0.2-0.4 Cr, 0.6-2.0 Mg, bal Al.
Cast: 30,000 TS; 20,000 YS; 2 El; 70 Brin.
Aged: 35,000 TS; 23,000 YS; 4 El.
For light alloy castings; heat treatable.

TERNALLOY 7.
M-130; 0.2-0.4 Cr, 0.2 max Cu, 1.8-2.4 Mg, 0.2-0.4 Mn, 4-4.5 Zn, bal Al.
F-Temper: 30,000 TS; 19,000 YS; 5 El; 65 Brin.
T6-Temper: 44,000 TS; 40,000 YS; 1.5 El; 80 Brin.
For light alloy castings; responds to heat treatment.

TERNALLOY 8.
M-130; 0.2-0.4 Cr, 0.2 max Cu, 1.8-2.4 Mg, 0.2-0.4 Mn, 4.9 Zn, bal Al.
SC-F-Temper: 33,000 TS; 22,000 YS; 3 El; 70 Brin.
SC-T6-Temper: 57,000 TS; 55,000 YS; 100 Brin.
PM-6 Temper: 58,000 TS; 50,000 YS; 3 El; 110 Brin.
For light alloy castings; responds to heat treatment.

TERNE METAL.
M-88; 15-30 Sn, 85-70 Pb, up to 2.0 Sb.
For terne plate, roofing, gasoline and oil tanks; corrosion resistant, bearings.

TERNE ROOFING.
M-806; Base: copper bearing steel. Coating: 80 Pb, 20 Sn (Terne).
Soft: 45,000 TS.
For roofing, gutters, flashings, down drains, termite shields.
Easy to form and solder.

TERRA.
M-1422; C, alloy, bal Fe.
Hardenable to 60/61 Rock C.
For ball races, blanking and forming dies, cutlery dies, collets, drills, gauges, pistons, rings, taps, vise jaws; oil hardening, non-deforming.

TERRIFIC.
M-1113; C, 18 W, 4 Cr, 1 V, bal Fe.
For tools, dies, cutters; high speed steel.

TETMAJER ALUMINUM BRONZE.
  M-England; 4-10 Al, 1-3 Si, 0.7-1.0 Fe, bal Cu.
  For bushings, fittings; tough, high strength.

TETON.
  M-1779; 1.0 C, 1.3 Cr, bal Fe.
  For bearing races and balls, compression dies, machinery parts.
  Furnished vacuum melted for aircraft bearing components. Oil or water hardening.

TEW 15 CR 3 ETC.
  M-1759; see DIN 15 Cr 3 etc.
  Structural and case hardening grades.
  82 grades.

T G S see **CARPENTER T G S.**

T H-30.
  M-625; C, bal Fe.
  At 20°C.: 61,000 TS; 42,500 YS; 20 El.
  At 600°C.: 34,000 TS; 15,500 YS.
  For boiler and superheater tubes; good strength at high temperature; corrosion resistant.

T H-31.
  M-625; C, bal Fe.
  At 20°C: 71,700 TS; 44,000 YS; 18 El.
  At 600°C: 35,000 TS; 23,000 YS.
  For boiler tubes, superheater tubes; maintains strength at high temperature.

T H 60B.
  M-625; Cu-Ni.
  Rolled: 79,650 TS; 52,625 YS; 20 El.
  For thick walled, seamless high pressure drums; good strength at high temperatures.

T H 61B.
  M-625; Cu-Ni.
  Rolled: 85,340 TS; 55,470 YS; 23 El.
  For thick walled, seamless high pressure drums; good strength at high temperatures.

T H 62B.
  M-625; Cu-Ni.
  Rolled: 99,560 TS; 61,160 YS; 20 El.
  For thick walled, seamless high pressure drums; good strength at high temperatures.

THALASSAL.
  M-984; Si, 2.5 Mn, 2.3 Mg, 0.2 Sb, bal Al.
  Sand cast: 24,000-31,000 TS.
  For dirigible and airplane parts, marine parts; resists sea water corrosion.

THALLIUM.
  M-1755; Tl.
  Purities: Zone refined 99.9999%, 99.999%, 99.99%, 99+%.
  Forms: Rod, sticks, ingot, shot, wire, single crystals.

THERLO.
  M-44; 28.5-29.5 Ni, 16.5-17.5 Co, bal Fe.
  Sheet: 89,700 TS; 50,500 YP; 200-250 Brin.
  For electrical resistors, glass to metal seals, grid glow tubes.
  Heat resistant. Can be brazed or welded.

THERMAL B.
  M-750; 0.4 C, 3-4 Cr, 14 W, 0.6 V, bal Fe.
  For tools, cutters, hot shear blades; hot work steel.

THERMAL BB.
  M-750; 0.58 C, 3-4 Cr, 10-14 W, 0.4 V, bal Fe.
  For tools, cutters; high speed steel.

THERMALLOY see also **AMSCO THERMALLOY.**

THERMALLOY 4.
  M-9; C, Cr, Ni, bal Fe.
  For hard facing electrode; resists high temperature and thermal shock.

THERMALLOY 20.
  M-47; 1.0 max C, 20 Cr, 3 max Ni, bal Fe.
  Aged: 115,000 TS; 80,000 YS; 18 El.
  Cast: 75,000 TS; 60,000 YS; 2 El; 190 Brin.
  For furnace parts, grate bars, salt pots; ACI Type HC; heat resistant.

THERMALLOY 28.
  M-47; 0.2-0.4 C, 28 Cr, 3 Ni, bal Fe.
  Cast: 90,000 TS; 50,000 YS; 1 El; 1 RA; 240 Brin.
  Annealed: 67,000 TS; 48,000 YS; 4 El; 4 RA; 190 Brin.
  For salt pots, sintering bars, grate bars, dampers; Type HC; corrosion resistant to sulfur atmosphere.

THERMALLOY 30.
  M-47; 0.2-0.4 C, 18-21 Cr, 9-11 Ni, bal Fe.
  Cast: 93,000 TS; 50,000 YS; 32 El; 34 RA; 176 Brin.
  Annealed: 97,000 TS; 50,000 YS; 35 El; 38 RA; 173 Brin.
  For conveyor belts, oil stills, furnace parts; Type HF; heat resistant to 1600°F.

THERMALLOY 32.
  M-47; 0.40 C, 3.0 Ni, 29.0 Cr, 0.30 Mn, 2.0 Si, bal Fe.
  Cast heat resisting alloy.

THERMALLOY 34.
  M-47; 47;
  0.50 C, 5.0 Ni, 28.0 Cr, 0.50 Mn, 1.50 Si, bal Fe.
  Cast heat resisting alloy.

THERMALLOY 38.
  M-47; 0.40 C, 12.0 Ni, 28.0 Cr, 0.50 Mn, 1.50 Si, bal Fe.
  Cast heat resisting alloy.

THERMALLOY 40B (HH).
  M-47; 0.30 C, 12.0 Ni, 25.0 Cr, 0.50 Mn, 1.50 Si, bal Fe.
  Cast heat resisting alloy.
  ACI Grade HH Type II.

THERMALLOY 40E.
  M-47; 1.0 C, 12.0 Ni, 26.0 Cr, 0.50 Mn, 1.50 Si, bal Fe.
  Cast heat resisting alloy.

THERMALLOY 45.
  M-47; 0.40 C, 25.0 Ni, 20.0 Cr, 0.50 Mn, 1.50 Si, bal Fe.
  Cast heat resisting alloy.
  ACI Grade HN.

THERMALLOY 47 (HK).
  M-47; 0.40 C, 20.0 Ni, 26.0 Cr, 0.50 Mn, 1.50 Si, bal Fe.
  Cast heat resisting alloy.
  ACI Grade HK.

THERMALLOY 48.
  M-47; 0.40 C, 20.0 Ni, 30.0 Cr, 0.50 Mn, 1.50 Si, bal Fe.
  Cast heat resisting alloy.

THERMALLOY 50 CQ.
  M-47; 0.50 C, 35.0 Ni, 16.0 Cr, 0.50 Mn, 2.0 Si, 1.0 Cb, bal Fe.
  Cast heat resisting alloy.

THERMALLOY 50 (HT).
  M-47; 0.50 C, 35.0 Ni, 18.0 Cr, 0.50 Mn, 2.0 Si, bal Fe.
  Cast heat resisting alloy.
  ACI Grade HT.

THERMALLOY 58.
  M-47; 0.35-0.75 C, 17-19 Cr, 37-40 Ni, bal Fe.
  Cast: 72,000 TS; 40,000 YS; 4 El; 3 RA; 200 Brin.
  Annealed: 73,000 TS; 40,000 YS; 9 El; 9 RA; 200 Brin.
  For retorts, muffles, cyanide and lead pots; Type HU; heat resistant.

THERMALLOY 58 CQ.
  M-47; 0.50 C, 38.0 Ni, 18.0 Cr, 0.50 Mn, 2.0 Si, 1.0 Cb, bal Fe.
  Cast heat resisting alloy.

THERMALLOY 63 (HP).
  M-47; 0.35-0.75 C, 34-38 Ni, 25-28 Cr, 2.0 Si, bal Fe.
  Cast: 68,000 TS; 32,000 YS; 19 El; 18 RA; 170 Brin.
  Annealed: 78,000 TS; 46,000 YS; 6 El; 5 RA; 175 Brin.
  For calcining tubes, furnace shafts, heat treating boxes; heat resistant to 2100°F.
  ACI H.P.

THERMALLOY 72.
  M-47; 0.4-0.6 C, 58-62 Ni, 12 Cr, 0.50 Mn, 2.0 Si, bal Fe.
  Cast: 59,000-73,000 TS; 30,000-37,000 YS; 9-3 El; 9-4 RA; 166-198 Brin.
  For retorts, muffles, heat treating boxes; heat and corrosion resistant.

THERMALLOY 85.
  M-47; 0.35-0.75 C, 17-19 Cr, 65-85 Ni, bal Fe.
  Cast: 78,000-88,000 TS; 35,000-44,000 YS; 10-7 El; 15-8 RA; 190-220 Brin.
  For retorts, muffles, burner parts, heat treating boxes; Type HX; heat resistant to 2100°F in absence of sulfur.

THERMALLOY 85 EX.
  M-47; 0.50 C, 68.0 Ni, 17.0 Cr, 0.50 Mn, 2.0 Si, 2.0 Mo, bal Fe.
  Cast heat resisting alloy.

THERMALLOY HC-250.
  M-47; high C, 25 Cr, bal Fe.
  Cast and heat treated: 70,000-100,000 TS; 50,000-80,000 YS; 0.2-0.5 E 550-750 Brin.
  For castings, pumps, stokes, crusher rolls, grinding rolls; abrasion resistant, oxidation resistant to 1850°F.

THERMAX 10A.
  M-1299; 0.2 max C, 22-24 Cr, 12-15 Ni, 2 max Mn, 1 max Si, bal Fe.
  Annealed: 85,000-95,000 TS; 40,000-50,000 YS; 45-55 El, 150-185 Brin.
  For heat treating boxes, oil refinery and chemical plant equipment; Type 309; austenitic, heat resistant.

THERMAX 11A.
  M-1299; 0.25 max C, 24-26 Cr, 19-22 Ni, bal Fe.
  Annealed: 95,000 TS; 45,000 YS; 50 El; 65 RA; 180 Brin.
  At 1200°F: 57,000 TS; 22,000 YS; 32 El; 45 RA.
  For furnace parts, heat treating boxes, valves, pumps; Type 310; austenitic, heat resistant.

THERMAX 11F.
  M-1350; 0.6 C, 1.5 Si, 22 Cr, bal Fe.
  For valve and pump parts, furnace equipment; corrosion and heat resistant.

THERMAX 25.
  M-1485; 0.2-0.5 C, 2.0 max Si, 2.0 max Mn, 23-26 Cr, 18-22 Ni, 0.5 max Mo, bal Fe.
  At 800°C: 11.3 tons/sq.in 0.2% proof stress.
  Austenitic corrosion and heat resisting alloy.
  Cracking tubes in the gas and petro-chemical industry and for furnace parts; cast alloy.
  Similar to ASTM A297 HK.

THERMAX 30.
  M-1485; 0.20-0.35 C, 28-32 Cr, bal Fe.
  Annealed: 62,800 TS; 2 El; 200-240 Brin.
  For furnace parts, heat treating boxes; Brit. 1648 B, scale resistant to 1100°C.

THERMAX 37.
M-1485; 0.2-0.5 C, 2.0 max Si, 2.0 max Mn, 15-19 Cr, 33-37 Ni, bal Fe.
At 800°C: 15.4 tons/sq.in TS; 16 El.
Cast alloy.
Austenitic alloy for high temperature equipment as furnace parts, carburizing containers.

THERMAX 45.
M-1485; 0.30 max C, 2.5 max Si, 1.5 max Mn, 20 Cr, 45 Ni, bal Fe.
Cast: 70,000 TS; 6 El; 170-200 Br.
For furnace parts such as heat resisting trays and rollers; good resistanc to scaling and thermal shock.

THERMAX 60.
M-1485; 0.50 max C, 15-20 Cr, 60-65 Ni, bal Fe.
Annealed: 67,200 TS; 3 El; 180-210 Brin.
For furnace parts, muffles, heat treating equipment; corrosion and heat resistant.

THERMAX 70.
M-1485; 0.40-0.50 C, 1.2 Si, 1.5 Mn, 45-50 Ni, 26-29 Cr, 4.0-6.0 W, bal Fe.
Austenitic heat resisting cast alloy with good stress rupture properties and thermal fatigue resistance.

THERMAX 75.
M-1485; 0.10 C, 1.0 max Si, 1.0 max Mn, 5.0 max Fe, 0.4 Ti, 20 Cr, bal Ni.
For furnace castings and parts for gas turbines; resists scaling and thermal shock.

THERMAX 90.
M-1485; 0.10 C, 1.0 max Si, 1.0 max Mn, 0.9 Al, 1.6 Ti, 5.0 max Fe, 20 Cr, 18 Co, bal Ni.
For gas turbines and jet engine components; high creep properties in the range 1100-1650°F.

THERMAX 102.
M-1485; 0.25 max C, 0.50-1.50 Si, 1.0 max Mn, 10.0-14.0 Ni, 20.0-25.0 Cr, 2.5-3.5 W, bal Fe.
Cast, Ann: 35 tons/sq.in. TS; 10 El; 240 max Brin.
For furnace parts and heat treating trays.
B.S.1648 Grade E.

THERMAX 104.
M-1485; 0.25 max C, 0.50-1.50 Si, 0.50-1.5 Mn, 10.0-14.0 Ni, 20-25 Cr, 1.0 max W, bal Fe.
Cast, Ann: 35 tons/sq.in. TS; 10 El; 200-240 Brin.
For furnace parts and heat treating trays.
B.S. 1648 Grade E.

THERMAX 115.
M-1485; 0.08-0.15 C, 0.5-1.25 Mn, 0.04 max P, 0.04 max S, 10.0-12.5 Cr, 1.5 max Ni, 0.5-1.0 Mo, 0.1-0.5 V, 0.2-0.6 Nb, 1.0 max Si, bal Fe.
Cast, HT: 60 t.s.i. min UTS; 50 t.s.i. min YS; 10 El; 286-331 Brin.
Good creep resistance; for use to 650°C for gas turbine applications.
Modified AISI 410.

THERMAX 519.
M-1485; 0.25-0.35 C, 1.0 max Mn, 0.3 max Si, 0.030 max P, 0.040 max S 23-25 Cr, 23-25 Ni, 1.4-1.8 Nb, bal Fe.
36-38 t.s.i. UTS; 13-15 t.s.i. YS(0.1% P.S.); 18-25 El.
Heat resisting alloy for centrifugal cast tubes in the reformer and petrochemical industry; good stress rupture properties.

THERMAX 531.
M-1485; 0.35-0.50 C, 2.0 Si, 2.0 Mn, 23-26 Cr, 32-35 Ni, 1.0-1.5 W, bal Fe.
Stress to rupture at 1038°C: 2,060 psi in 1000 hrs.
Austenitic heat resisting cast alloy with good oxidation resistance, creep strength, and stress to rupture at elevated temperatures.

THERMAX 532.
M-1485; 0.08-0.12 C, 0.5-0.15 Mn, 0.3-1.0 Si, 29-33 Ni, 18-22 Cr, 0.5 max Mo, 0.5-0.6 Nb, bal Fe.
At 800°C: 12.2 tons/sq.in. TS; 6.2 tons/sq.in. YS; 56 El.
Austenitic heat resisting cast alloy.

THERMAX 533.
M-1485; 0.30-0.40 C, 1.5 max Mn, 2.0 max Si, 23-26 Cr, 33-37 Ni, 0.2-1.5 Nb, bal Fe.
At 800°C: 17.3 tons/sq.in. TS; 9.8 tons/sq.in. YS; 18 El.
Austenitic cast alloy for high temperature operation.

THERMAX 638.
M-1485; 0.4-0.55 C, 1.0 max Si, 1.0 max Mn, 34-36 Ni, 25-27 Cr, 14.16 Co, 4-5.5 W, 0.7-1.3 Nb, bal Fe.
At 800°C: 24.8 tons/sq.in. TS; 13.4 tons/sq.in. YS; 15 El.
Austenitic heat resisting cast alloy.

THERMAX 657.
M-1485; 0.1 max C, 1.0 max Fe, 0.5 max Si, 48-52 Cr, 0.16 max N_2, 1.4-1.7 Nb, 0.2 max C + N_2, bal Ni.
39-52 t.s.i. UTS; 16-27 t.s.i. YS (0.2 P.S.); 10-30 El.
High strength cast Ni-Cr-Nb alloy; resistant to ash products of low grade fuel; good creep-rupture properties.

THERMAX 4713 ETC.
M-1759; see Werkstoff Nr. 1.4713.
Heat resisting steels; 11 grades.

THERMCO 50.
M-1347; 0.1 C, 0.18 Ti, 0.6 Cb, 26-30 Cr, 47-52 Co, bal Fe.
Cast: 135,000 TS; 48,000 YS; 10 El; 10 RA; 250 Brin.
For furnace baffles, burner tips, sintering grates, quench baskets.
Corrosion, heat and thermal shock resistant.

THERMELAST 4002.
M-1631.
Constant modulus alloy on the basis of NiFeCr, hardenable through additions of Ti and Al, for balance springs, leaf springs, diaphragms.

THERMIMPHY.
M-1488; 0.12 C, 7.0 Cr, 0.5 Mo, 0.3 V, bal Fe.
Quenched and tempered: 590-730 N/mm² UTS; 440 N/mm² min YS; 15 El.
For high temperature bolts and parts.
AFNOR Z 12 CDV 7; ASTM A 182 Gr. F7.

THERMISILID.
M-72; special cast iron, 14-16 Si.
350-290 Brin.
For chemical equipment construction, acid plants; acid proof and heat resisting to H_2SO_4.

THERMISILID EXTRA.
M-Eng.; 14-18 Si, bal Fe.
Cast: 250 Brin.
For acid plants, explosives manufacturing; corrosion and acid resistant.

THERMIT.
M-528; 14-16 Sb, 5-7 Sn, 0.8-1.2 Cu, 0.7-1.5 Ni, 72-78.5 Pb, 0.3-0.8 As, 0.7-1.5 Cd.
29 Brin.
For bearings for pumps, railroads, engines; Standard German white metal; cast alloy.

THERMO.
M-260; C, alloy, bal Fe.
For tools, dies; oil hardening.

THERMO 50 see JUNKER THERMO 50.

THERMOCROM.
M-918; 3.2 C, 2.2 Si, 0.2 Mo, 0.8 Cr, bal Fe.
For piston rings; cast iron.

THERMODIT 101.
M-1488; 0.38 C, 1.0 Si, 5.25 Cr, 1.35 Mo, 0.5 V, bal Fe.
Hot work tool steel for forging, upsetting and hot forming dies.
AFNOR Z 38 CDV 5; AISI H 11.

THERMODUR 10 CR 20 32 NB.
M-1290; 0.10 C, 32 Ni, 20 Cr, 1 Nb, bal Fe.
Austenitic cast steel for elevated temperature service.
SEW 1.4859; G-X 10 NiCrNb 32 20.

THERMODUR 15 CN 13.
M-1290; 0.10-0.15 C, 0.50-1.0 Si, 1.0-1.5 Mn, 22.5-24.0 Cr, 12.0-13.0 Ni, bal Fe.
Cast: 70,000 min TS; 30,000 min YS; 30 min El.
Rupture strength: 185 psi, 10,000 hrs, 800°C.
Resistance to scaling in oxidizing and ashfree and sulfur free atmosphere to 1150°C.
DIN G-X15CrNiSi2512; ASTM A351-CH20.

THERMODUR 15 CN 15 T.
M-1290; 0.10-0.15 C, 0.50-1.0 Si, 1.0-1.5 Mn, 19-20 Cr, 14.5-15.5 Ni.
Cast: 64,000-85,000 TS; 28,000 min YS; 20 min El.
Rupture strength: 1850 psi, 10,000 hrs, 800°C.
Resistance to scaling in oxidizing and ashfree and sulfur free atmosphere to 1050°C.
DIN G-X15CrNiSi2014.

THERMODUR 15 CN 20.
M-1290; 0.10-0.15 C, 0.50-1.0 Si, 1.0-1.5 Mn, 23.5-25.0 Cr, 20-21 Ni, bal Fe.
Cast: 64,000-92,000 TS; 30,000 min YS; 30 min El.
Rupture strength: 2300 psi, 10,000 hrs, 800°C.
Resistance to scaling in oxidizing and ashfree and sulfur free atmosphere to 1150°C.
SEW-Nr.1.4849; DIN G-X15CrNiSi2520-ASTM A351-CK20.

THERMODUR 20 CMW 5.
M-1290; 0.20 C, 0.6 Mn, 2.3 Cr, 1.0 Mo, bal Fe.
Ferritic cast steel for elevated temperature service.
SEW 1.7379; GS-18 CrMo 9 10.

THERMODUR 30 CN 13.
M-1290; 0.25-0.30 C, 0.50-1.0 Si, 1.0-1.5 Mn, 24-25 Cr, 12-13 Ni, bal Fe.
Cast: 64,000-92,000 TS; 10 min El.
Rupture strength: 2400 psi, 10,000 hrs, 800°C.
Resistance to scaling in oxidizing and ashfree and sulfur free atmosphere to 1150°C.
SEW-Nr. 1.4837; DIN G-X35CrNiSi2512; ASTM A447-II.

THERMODUR 30 CN 24 24 NB.
M-1290; 0.30 C, 24 Cr, 24 Ni, 2 Si, Nb, bal Fe.
Austenitic cast steel for elevated temperature service.
SEW 1.4855; G-X 30 CrNiSiNb 24 24.

THERMODUR 30 CN 2535.
M-1290; 0.25-0.30 C, 0.50-1.0 Si, 1.0-1.5 Mn, 25.0-26.5 Cr, 34.0-35.0 Ni, bal Fe.
Cast: 64,000-92,000 TS; 35,000 min YS; 10 min El.
Rupture strength: 5100 psi, 10,000 hrs, 800°C.
Resistance to scaling in oxidizing and ashfree and sulfur free atmosphere to 1150°C.
SEW-Nr. 1.4857; DIN G-X30NiCrSi3525.

THERMODUR 40 CN 20.
M-1290; 0.20-0.50 C, 1.0-2.5 Si, 1.50 max Mn, 24.0-27.0 Cr, 19.0-21.0 Ni, bal Fe.
Austenitic cast steel for elevated temperature service.
SEW 1.4848; G-X 40 CrNiSi 25 20.

THERMODUR 40 CN CW 25.
M-1290; 0.35-0.40 C, 0.50-1.0 Si, 0.50-1.0 Mn, 25.0-26.5 Cr, 34.0-35.0 Ni, 14.0-15.0 Co, 5.0-5.5 W, bal Fe.
Cast: 71,000-100,000 TS; 38,000 min YS; 4 min El.
Rupture strength; 2800 psi, 10,000 hrs, 1000°C.
Resistance to scaling in oxidizing and ashfree and sulphur free atmosphere to 1200°C.
DIN G-X40NiCrCoW352515.

THERMODUR 50 CN 30 30.
M-1290; 0.50 C, 30 Ni, 30 Cr, bal Fe.
Austenitic cast steel for elevated temperature service.
SEW 1.4868; G-X 50 CrNi 30 30.

THERMODUR 50 CNW 30 50.
M-1290; 0.45-0.55 C, 0.50-1.0 Si, 0.50-1.0 Mn, 27.0-28.5 Cr, 46-48 Ni 4.5-5.0 W, bal Fe.
Cast: 57,000-85,000 TS; 35,000 min YS; 4 min El.
Rupture strength: 2400 psi, 10,000 hrs, 1000°C.
Resistance to scaling in oxidizing and ashfree and sulfur free atmosphere to 1150°C.
SEW-Nr. 2.4879; DIN G-X50NiCrW4828.

THERMODUR C18.8ES.
M-1290; 0.07 max C, 0.50-1.0 Si, 0.50-1.0 Mn, 18-19 Cr, 10-11 Ni, 2.0-2.2 Mo, bal Fe.
Cast: 64,000-92,000 TS; 35,000 min YS; 35 min El.
Rupture strength; 47,000 psi, 10,000 hrs, at 550°C.
Resistance to scaling in oxidizing and ashfree atmosphere without sulphur compounds: 750°C.
SEW-Nr. 1.4408; DIN G-X6CrNiMo1810; ASTM A351-CF8M.

THERMODUR C18 8ESS.
M-1290; 0.08 max C, 0.50-1.0 Si, 0.50-1.0 Mn, 17-18 Cr, 10.5-11.5 Ni, 2.0-2.2 Mo, Nb = 8 x %C, bal Fe.
Cast: 64,000-92,000 TS; 27,000 min YS; 20 min El.
Rupture strength: 48,000 psi, 10,000 hrs, 550°C.
Resistance to scaling in oxidizing and ashfree and sulphur free atmosphere to 750°C.
SEW-Nr. 1.4581; DIN G-X7CrNiMoNb1810.

THERMODUR C18.8S.
M-1290; 0.07 max C, 0.50-1.0 Si, 0.50-1.0 Mn, 18-19 Cr, 9-10 Ni, 0.75 max Mo, bal Fe.
Cast: 64,000-92,000 TS; 30,000 min YS; 35 min El.
Rupture strength: 30,000 psi, 10,000 hrs, at 500°C.
Resistance to scaling in oxidizing and ashfree atmosphere without sulphur compounds: 750°C.
SEW-Nr. 1.4308; DIN G-X6CrNi189; ASTM A351-CF8.

THERMODUR C18.8SS.
M-1290; 0.08 max C, 0.50-1.0 Si, 0.50-1.0 Mn, 18-19 Cr, 9.5-10.5 Ni, 0.75 max Mo, Nb = 8 x %C, bal Fe.
Cast: 64,000-92,000 TS; 30,000 min YS; 30 min El.
Rupture strength: 42,000 psi, 10,000 hrs, at 550°C.
Resistance to scaling in oxidizing and ashfree atmosphere without sulphur compounds: 750°C.
SEW-Nr. 1.4552; DIN G-X7CrNiNb189; ASTM A351-CF8C.

THERMODUR CMVW5.
M-1290; 0.15-0.20 C, 0.30-0.50 Si, 0.50-0.80 Mn, 1.2-1.5 Cr, 0.9-1.1 Mo, 0.20-0.30 V, bal Fe.
Cast: 85,000-112,000 TS; 64,000 min YS; 15 min El.
Resistance to scaling in oxidizing and ashfree atmosphere without sulphur compounds.
SEW-Nr. 1.7706; DIN GS-17CrMoV511.

THERMODUR CMW3.
M-1290; 0.15-0.20 C, 0.30-0.50 Si, 0.50-0.80 Mn, 1.0-1.5 Cr, 0.45-0.55 Mo, bal Fe.
Cast: 71,000-92,000 TS; 45,000 min YS; 20 min El.
Resistance to scaling in oxidizing and ashfree atmosphere without sulphur compounds: 570°C.
SEW-Nr. 1.7357; DIN GS-17CrMo55; ASTM A217-WC6.

THERMODUR CMW5.
M-1290; 0.10-0.15 C, 0.30-0.50 Si, 0.50-0.70 Mn, 2.0-2.5 Cr, 0.9-1.1 Mo, bal Fe.
Cast: 71,000-100,000 TS; 45,000 min YS; 20 min El.
Resistance to scaling in oxidizing and ashfree atmosphere without sulphur compounds: 580°C.
SEW-Nr. 1.7380; DIN GS-12CrMo910; ASTM A217-WC9.

THERMODUR CMW11.
M-1290; 0.20-0.26 C, 0.20-0.40 Si, 0.50-0.70 Mn, 11.3-12.2 Cr, 0.7-1.0 Ni, 1.0-1.2 Mo, 0.25-0.35 V, bal Fe.
Cast: 100,000-127,000 TS; 85,000 min YS; 15 min El.
Resistance to scaling in oxidizing and ashfree atmosphere without sulphur compounds.
SEW-Nr. 1.4931; DIN G-X22CrMoV121.

THERMODUR MW2.
M-1290; 0.18-0.23 C, 0.30-0.50 Si, 0.50-0.80 Mn, 0.30 max Cr, 0.35-0.4 Mo, bal Fe.
Cast: 64,000-85,000 TS; 35,000 min YS; 22 min El.
Resistance to scaling in oxidizing and ashfree atmosphere without sulphur compounds: 560°C.
SEW-Nr. 1.5419; DIN GS-22Mo4; ASTM A217-WC1.

THERMODUR MW 4.
M-1290; 0.20 C, 0.70 Mn, 0.6 Mo, bal Fe.
Low alloy cast steel.
GS-20 Mo 5.

THERMODUR P6M.
M-1290; 0.08-0.15 C, 0.30-0.50 Si, 0.40-0.70 Mn, 4.5-5.5 Cr, 0.45-0.55 Mo, bal Fe.
Cast: 92,000-120,000 TS; 60,000 min YS; 18 min El.
Resistance to scaling in oxidizing and ashfree atmosphere without sulphur compounds: 600°C.
SEW-Nr. 1.7363; DIN GS-12CrMo195; ASTM A217-C5.

THERMODUR P7M.
M-1290; 0.08-0.15 C, 0.30-0.50 Si, 0.50-0.80 Mn, 9.0-10.0 Cr, 1.1-1.2 Mo, bal Fe.
Cast: 92,000-120,000 TS; 60,000 min YS; 18 min El.
Resistance to scaling in oxidizing and ashfree atmosphere without sulphur compounds: 630°C.
SEW-Nr. 1.7389; DIN G-X12CrMo101; ASTM A217-C12.

THERMODUR W1.
M-1290; 0.18-0.23 C, 0.30-0.50 Si, 0.50-0.80 Mn, 0.30 max Cr, bal Fe.
Cast: 64,000-85,000 TS; 35,000 min YS; 22 min El.
Resistance to scaling in oxydizing and ashfree atmosphere without sulphur compounds: 560°C.
SEW-Nr. 1.0619; DIN GS-C25; ASTM A216-WCA.

THERMODUR W2.
M-1290; 0.27-0.30 C, 0.30-0.50 Si, 0.60-0.80 Mn, 0.30 max Cr, bal Fe.
Cast: 70,000 min TS; 40,000 min YS; 20 min El.
Resistance to scaling in oxidizing and ashfree atmosphere without sulphur compounds: 560°C.
DIN GS-C30; ASTM A216-WCB.

THERMODYNE.
M-815; Sb, Pb, bal Sn.
For bearings; Babbitt.

THERMOFLEX.
M-919.
For thermostatic and contact alloy; bi-metal.

THERMOFLUX.
M-1631.
Soft magnetic 30% Ni-Fe alloy for temperature compensation in permanent magnet systems.

THERMO-KANTHAL-N.
M-1150; 2-3 Si, bal Ni.
For thermocouples.
Max. operating temperature 2300°F.

THERMO-KANTHAL-P.
M-1150; 90 Ni, 10 Cr.
For thermocouples.
Max. operating temperature 2300°F.

THERMOLD 75.
M-114; 0.35 C, 1.0 Si, 0.3 Mn, 3.5 Cr, 0.75 V, 2.5 Mo, 2.0 Co, bal Fe.
Heat Treated: 296,000 TS; 247,000 YS; 9 El; 16 RA; 9 Charpy, C 59 Rock.
For aluminum die casting dies and inserts, cores, plungers, sleeves, slides, extrusion and forging dies.
Hot work steel. Air hardening. Good thermal fatigue resistance.

THERMOLD A replaced by **THERMOLD H11.**

THERMOLD AV replaced by **THERMOLD H13.**

THERMOLD B replaced by **THERMOLD H12.**

THERMOLD H10.
M-114; 0.40 C, 0.55 Mn, 1.0 Si, 3.25 Cr, 0.4 V, 2.5 Mo, bal Fe.
Heat treated: 233,000-301,000 TS; 201,000-260,000 YS; 8-10 El; 26-32 RA; Rock C 46-55.
For mandrels, dies, bolsters, dummy blocks, punches, gripper and header dies, hot shears, aluminum die casting dies.
Resists softening at high temperatures.
Type H10 hot work steel.

THERMOLD H11.
M-114; 0.35 C, 0.40 Mn, 1.0 Si, 5.0 Cr, 0.45 V, 1.5 Mo, bal Fe.
Heat Treated: 225,000-305,000 TS; 195,000-255,000 YS; 8-15 El; 29-50 RA; C 45-56 Rock.
For die casting dies, punches, piercing tools, mandrels, extrusion tooling, forging dies.
Type H11 hot work steel.
High hardenability. Good resistance to softening.

THERMOLD H12.
M-114; 0.35 C, 0.40 Mn, 1.0 Si, 5.0 Cr, 1.5 W, 0.4 V, 1.5 Mo, bal Fe.
Heat treated: 192,000-263,000 TS; 175,000-241,000 YS; 8-15 El; 25-43 RA; C 41-52 Rock.
For extrusion dies, dummy blocks, gripper and header dies, forging die inserts, punches, mandrels.
High hardenability, tough.
Type H12 hot work steel.

THERMOLD H13.
M-114; 0.35 C, 0.40 Mn, 1.0 Si, 5.0 Cr, 1.0 V, 1.5 Mo, bal Fe.
Heat treated: 189,000-258,000 TS; 170,000-238,000 YS; 8-18 El; 28-48 RA; C 41-51 Rock.
For die casting dies and inserts, cores, ejector pins, plungers, sleeves, extrusion and forging dies.
High hardenability and toughness.
Type H13 hot work steel. Air harden.

THERMOLD H19.
M-114; 0.40 C, 4.25 Cr, 4.25 W, 2.0 V, 4.25 Co, bal Fe.
Heat treated: 233,000 TS; 217,000 YS; 5.6 El; 17 RA; C 58 Rock.
For extrusion dies and inserts, dummy blocks, punches, mandrels, forging dies and inserts.
Good red hardness and shock resistance.
Type H19 hot work steel.

THERMOLD H21.
M-114; 0.35 C, 0.25 Mn, 0.35 Si, 3.5 Cr, 9.5 W, 0.5 V, bal Fe.
Heat treated: 242,500 TS; 220,500 YS; 7.4 El; 27 RA; C 54 Rock.
For mandrels, hot punches, and blanking dies, fly shear blades, extrusion and gripper dies, piercing points, dummy blocks. Resists softening.
Type H21 hot work steel.

THERMOLD H22.
M-114; 0.38 C, 0.25 Mn, 0.35 Si, 3.0 Cr, 11.0 W, 0.4 V, bal Fe.
Heat treated: 225,000 TS; 200,500 YS; 8.8 El; 30 RA; C 55 Rock.
For mandrels, hot blanking dies, hot punches, flying shear blades, extrusion dies, dummy blocks, gripper dies.
Shock resistant. Resists softening.
Type H22 hot work steel.

THERMOLD H23.
M-114; 0.32 C, 0.35 Mn, 0.50 Si, 12.0 Cr, 12.0 W, 1.0 V, bal Fe.
Heat treated: (at 1000°F) 167,500 TS; 135,000 YS; 7.4 El; 13.5 RA; C 34 Rock.
For extrusion dies, die casting dies, mandrels, hot punches, gripper dies.
High resistance to softening. Retains hot hardness.
Type H23 hot work steel.

THERMOLD H26.
M-114; 0.50 C, 0.25 Mn, 0.35 Si, 4.0 Cr, 18.0 W, 1.0 V, bal Fe.
Heat treated: 258,000 TS; 235,500 YS; 3.6 El; 6 RA; C 57 Rock.
For mandrels, hot blanking dies and punches, flying shear blades, extrusion and trim dies.
Maximum hot strength and resistance to softening.
Type H26 hot work steel.

THERMOLD J COLD WORK.
M-114; 0.50 C, 0.40 Mn, 1.1 Si, 5.0 Cr, 1.0 V, 1.4 Mo, 1.5 Ni, bal Fe.
For cold heading dies, die inserts, coining dies, forming rolls.
AISI Type A9 air hardening cold work tool steel.

THERMOLD J HOT WORK.
M-114; 0.50 C, 0.4 Mn, 1.0 Si, 5.0 Cr, 1.0 V, 1.4 Mo, 1.5 Ni, bal Fe.
For punches, piercing tools, forging dies, gripper dies, extrusion tooling, coining dies.
AISI Type A9 air hardening tool steel.

THERMOLD KL.
M-114; 0.35 C, 0.60 Mn, 1.5 Si, 7.25 Cr, 7.25 W, bal Fe.
Heat treated: 232,000 TS; 203,000 YS; 10.5 El; 34 RA.
For extrusion and gripper dies, forming and blanking dies, hot punches.
Good toughness and red hardness.
Hot work steel. Oil hardening.

THERMOLD Z.
M-114; 0.37 C, 0.8 Mn, 0.3 Si, 1.25 Cr, 0.15 V, 0.35 Mo, 0.001 B, bal Fe.
Hardenable to 35-54 Rc.
For plastic molds, zinc die casting molds.
AISI P20 Mold Steel.

THERMON 7.
M-1331; 0.45 C, 1.05 Cr, 0.20 V, 1.85 W, bal Fe.
For upsetters, heading dies, punches; oil hardened, tough.

THERMON 7H.
M-1331; 0.55 C, 1.05 Cr, 0.2 V, 1.85 W, bal Fe.
For upsetters, riveters, heading dies; oil hardened, tough.

THERMON 7W.
M-1331; 0.35 C, 1.05 Cr, 0.2 V, 1.85 W, bal Fe.
For upsetters, punches, heading dies; oil hardened, tough.

THERMON 12.
M-1331; 0.32 C, 0.30 Mn, 0.30 Si, 3 Cr, 2.8 Mo, 0.50 V, bal Fe.
For hot dies.
Hot work tool steel, tough, shock resistant.

THERMON 15.
M-1331; 0.30 C, 1.1 Cr, 0.18 V, 3.75 W, bal Fe.
For upsetters, punches, extrusion press parts; oil hardened, tough.

THERMON 17.
M-1331; 0.30 C, 2.35 Cr, 0.6 V, 4.25 W, bal Fe.
For extrusion press parts, rams, punches; hot work steel, oil hardened.

THERMON 17 EXTRA.
M-1331; 0.30 C, Cr, Co, V, W, bal Fe.
For extrusion press parts, rams, punches; hot work steel, oil hardened.

THERMON 21.
M-1331; 0.40 C, Cr, Mo, V, W, bal Fe.
For hot work tools, extrusion press parts; oil hardened, hot work steel.

THERMON 34.
M-1331; 0.30 C, 2.65 Cr, 0.35 V, 8.5 W, bal Fe.
For extrusion rams and liners, punches, upsetters; oil hardened, hot work steel.

THERMON 34 EXTRA.
M-1331; 0.30 C, 2 Co, 2.4 Cr, 0.25 V, 8.5 W, bal Fe.
For extrusion press parts, rams, punches; hot work steel, oil hardened.

THERMON 4630 ETC.
M-1759; see Werkstoff Nr. 1.4630.
High temperature alloys; 11 grades.

THERMON A.
M-1331; 0.45 C, Cr, Ni, W, bal Fe.
For hot work tools, dies, punches; hot work steel, oil hardened.

THERMONEAL 10.
M-64; 0.30-0.36 C, 0.2-0.4 Mn, 0.8-1.2 Si, 4.25-5.25 Cr, 1.75-2.00 Mo, bal Fe.
At 80°F: 217,000 TS; 184,000 YS; 13 El; 40 RA.
At 1000°F: 145,000 TS; 105,000 YS; 20 El; 65 RA.
For upsetting and forging dies, aluminum die casting dies, extrusion dies.
Type H11 hot work tool steel. Air hardening.

THERMONEAL 11.
M-64; 0.37-0.42 C, 5.0-5.5 Cr, 0.23-0.38 Mn, 0.85-1.1 Si, 1.0-1.2 Mo, 0.45-0.55 V, bal Fe.
At 80°F: 217,000 TS; 184,000 YS; 13 El; 40 RA.
At 1000°F: 145,000 TS; 105,000 YS; 20 El; 65 RA.
For die casting dies, upsetting and press forging dies, extrusion dies.
Type H11 hot work tool steel.

THERMONEAL 13.
M-64; 0.37-0.42 C, 5.0-5.5 Cr, 0.23-0.38 Mn, 1.0-1.3 Mo, 0.85-1.10 Si 0.90-1.0 V, bal Fe.
At 80°F: 217,000 TS; 184,000 YS; 13 El; 40 RA.
At 1000°F: 145,000 TS; 105,000 YS; 20 El; 65 RA.
For upsetting and press forge dies, extrusion dies.
Type H13 hot work steel.

THERMON RR.
M-1331; 0.20 C, 13 Cr, bal Fe.
Annealed: 95,000 TS; 50,000 YS; 25 El; 55 RA; 195 Brin.
Cold drawn: 105,000 TS; 85,000 YS; 17 El; 50 RA; 215 Brin.
For turbine blades, cutlery, valves, surgical instruments; Type 420; stainless.

THERMON SS.
M-1331; 0.65 C, 3.75 Cr, 0.85 Mo, 0.7 V, 8.5 W, bal Fe.
For cutting tools, drills, taps, hot work tools; high speed steel.

THERMON U.
M-1331; 0.45 C, 1.35 Cr, 0.45 Mo, 0.8 V, 0.45 W, bal Fe.
For upsetters, riveters, punches; oil hardened.

THERMOPERM.
M-72; 70 Fe, 30 Ni.
For compensating shunts for electrical equipment; temperature-sensitive, magnetic.

THERMOSIL.
M-1038.
61 Si (approx.), bal Fe.
Ferrosilicon used as graphitizing inoculant for gray and ductile iron when added to ladle; also as an additive.

THERMOTEM.
M-64; 0.33 C, 4.75 Cr, 1.87 Mo, bal Fe.
Annealed: 250 Brin.
Heat treated: 510 Brin.
For extrusion parts, dies, shear knives; oil hardened.

THERMOTEM 10.
M-64; 0.30-0.36 C, 4.25-5.25 Cr, 0.80-1.20 Si, 0.2-0.4 Mn, 1.75-2.00 Mo, bal Fe.
At 80°F: 217,000 TS; 184,000 YS; 13 El; 40 RA.
At 1000°F: 145,000 TS; 105,000 YS; 20 El; 65 RA.
For upsetting and forging dies, aluminum die casting dies, extrusion dies.
Type H11 hot work steel. Air hardening.

THERMOTEM-11.
M-64; 0.40 C, 1.0 Si, 5.25 Cr, 1.1 Mo, 0.5 V, bal Fe.
For hot work dies and punches; hot work steel, air hardened.

THERMOTEM-12.
M-64; 0.37 C, 14.0 Si, 5.0 Cr, 1.45 Mo, 1.25 W, 0.3 V, bal Fe.
For hot work dies and punches; hot work steel, air hardening.

THERMOTEM-13.
M-64; 0.40 C, 1.0 Si, 5.25 Cr, 1.1 Mo, 0.95 V, bal Fe.
For hot work dies and punches; hot work steel, air hardening.

THESSCO H12.
M-1522; 72 Ag, Cu.
Melt range: 780-780°C.
Yield point: 19.0 t.s.i.; 10 El.
Silver brazing, high quality joints in protective atmosphere.

THESSCO L3.
M-1522; 10 Ag, Cu, Zn.
Melt range: 840-855°C.
Yield point: 31.5 t.s.i.; 7 El.
For silver brazing, high melt point, high strength, low cost.

THESSCO L7.
M-1522; 14 Ag, Cu, Zn.
Melt range: 810-835°C.
Yield point: 33.5 t.s.i.; 3 El.
For silver brazing, high melt point, high strength, low cost.

THESSCO LX13.
M-1522; 20 Ag, Cu, Zn, Cd.
Melt range: 720-775°C.
Yield point: 20.5 t.s.i.; 9 El.
For silver brazing, high melt point.

THESSCO LX16.
M-1522; 23 Ag, Cu, Zn, Cd.
Melt range: 610-720°C.
Yield point: 17.0 t.s.i.; 18 El.
For silver brazing, general purpose, wide melt range.

THESSCO LX18.
M-1522; 25 Ag, Cu, Zn, Cd.
Melt range: 605-710°C.
Yield point: 20.0 t.s.i.; 14 El.
For silver brazing, general purpose, wide melt range.

THESSCO M1.
M-1522; 31 Ag, Cu, Zn.
Melt range: 715-755°C.
Yield point: 30.5 t.s.i.; 23 El.
For silver brazing, high melt point, high strength.

THESSCO MX0.
M-1522; 30 Ag, Cu, Zn, Cd.
Melt range: 605-680°C.
Yield point: 15 t.s.i.; 10 El.
For silver brazing, general purpose.

THESSCO MX4.
M-1522; 34 Ag, Cu, Zn, Cd.
Melt range: 610-665°C.
Yield point: 15.0 t.s.i.; 15 El.
For silver brazing, general purpose.

THESSCO MX8.
M-1522; 38 Ag, Cu, Zn, Cd.
Melt range: 605-650°C.
Yield point: 16.0 t.s.i.; 24 El.
For silver brazing, general purpose, low melt point.

THESSCO MX12.
M-1522; 42 Ag, Cu, Zn, Cd.
Melt range: 610-620°C.
Yield point: 17.0 t.s.i.; 27 El.
For silver brazing, lowest melting point.

THESSCO MX18.
M-1522; 48 Ag, Cu, Zn, Cd.
Melt range: 630-640°C.
Yield point: 19.5 t.s.i.; 43 El.
For silver brazing, extremely fluid.

THESSCONITE G.4.
M-1522; Ag, W.
Sintered: 15.0 density; 225 HV hardness; 43% elec. cond.
For heavy duty arc resistant contacts.

THESSCONITE G.7.
M-1522; Ag, W.
Sintered: 13.9 density; 165 HV hardness; 50% elec. cond.
For heavy duty arc resistant contacts.

THESSCONITE G.9.
M-1522; Ag, W.
Sintered: 13.5 density; 140 HV hardness; 55% elec. cond.
For heavy duty arc resistant contacts.

THESSCONITE GC.15.
M-1522; Ag, C.
Sintered: 10.2 density; 64 HV hardness; 85 elec. cond.
For sliding electrical contacts.

THESSCONITE GD.10.
M-1522; Ag, CdO.
Sintered: 10.0 density; 60 HV hardness; 82% elec. cond.
For anti-welding contacts.

THESSCONITE GN.1.
M-1522; Ag, Ni.
Sintered: 9.9 density, 90 HV hardness, 66% elec. cond.
For medium duty electrical contacts.

THESSCO PROOF SILVER.
M-1522; 99.99 Ag.
Melt point: 960°C.
Yield point: 2.9 t.s.i.; 45 El.
For brazing.

THESSCO REDIBRAZE.
M-1522; Bimetal of either GD.25 or GD.35 on silver.

THESSCO SILBRAZE.
M-1522; 1 Ag, Cu, Zn, Si.
Melt range: 885-895°C.
Yield point: 14.2 t.s.i.; 43 El.
For brazing; low cost silver bearing brass.

THETALOY.
M-1278; 0.38 C, 25 Cr, 7 W, 3 Mo, 12.5 Co, 6 Fe, 2.5 Mn, bal Ni.
For high temperature applications; corrosion and heat resistant.

THOR replaced by THERMOLD H23.

THORAN.
M-Germ.; tungsten carbide plus molybdenum carbide, 95.85 W, 3.94 C, Mo.
For hard cutting tools and dies, tips for high speed tools; cast alloy; W_2C + WC.

THOR DOPPELKREUZ EXTRA ZH.
M-1310; 0.6 C, 0.25-0.50 Si, 0.20-0.80 Mn, bal Fe.
For gears, rails, shafts, axles, crankshafts; water hardened.

THOR DOPPELKREUZ H.
M-1310; 0.9 C, 0.25-0.50 Si, 0.20-0.80 Mn, bal Fe.
Heat treated: 190,000 TS; 145,000 YS; 10 El; 30 RA; 400 Brin.
For drills, tools, springs, cutters; Type W1; water hardened.

THOR DOPPELKREUZ MH.
M-1310; 0.75 C, 0.25-0.50 Si, 0.3-0.8 Mn, bal Fe.
Heat treated: 185,000 TS; 140,000 YS; 14 El; 38 RA; 375 Brin.
For rails, springs, tools, hammers, axes; Type W1; water hardened.

THOR DOPPELKREUZ SEHR ZAH.
M-1310; 0.15 C, 0.15-0.35 Si, 0.25-0.50 Mn, bal Fe.
Cold drawn: 72,000 TS; 60,000 YS; 22 El; 58 RA; 145 Brin.
For gears, pinions, cams, bolts, camshafts; case hardening steel.

THOR DOPPELKREUZ ZAH.
M-1310; 0.35 C, 0.25-0.50 Si, 0.3-0.8 Mn, bal Fe.
Hot rolled: 85,000 TS; 54,000 YS; 30 El; 53 RA; 185 Brin.
For gears, bolts, fasteners, shafts; water hardened.

THOR DOPPELKREUZ ZH.
M-1310; 0.45 C, 0.25-0.50 Si, 0.3-0.8 Mn, bal Fe.
Hot rolled: 98,000 TS; 58,000 YS; 24 El; 45 RA; 212 Brin.
For gears, bolts, fasteners, shafts; water hardened.

THOR EXTRA EXTRA ZH.
M-1310; 1.1 C, 0.25 max Si, 0.25 max Mn, bal Fe.
Annealed: 100,000 TS; 53,000 YS; 21 El; 42 RA; 200 Brin.
For drills, taps, reamers, cutters, hobs; Type W1; water hardened.

THOR EXTRA SEHR ZAH.
M-1310; 0.75 C, 0.25 max Si, 0.25 max Mn, bal Fe.
Heat treated: 185,000 TS; 140,000 YS; 14 El; 38 RA; 375 Brin.
For springs, rails, axes, shafts; Type W1; water hardened.

THOR EXTRA ZAH.
M-1310; 0.85 C, 0.25 max Si, 0.25 max Mn, bal Fe.
Heat treated: 190,000 TS; 145,000 YS; 12 El; 35 RA; 390 Brin.
For springs, tools, drills, cutters; Type W1; water hardened.

THOR EXTRA ZH.
M-1310; 1.0 C, 0.25 max Si, 0.25 max Mn, bal Fe.
Annealed: 100,000 TS; 53,000 YS; 21 El; 42 RA; 200 Brin.
For drills, taps, cutters, springs, tools; Type W1; water hardened.

THORIUM.
M-1755; Th.
Purities: 99.99+%, Nuclear grade 99.8+%, commercial grade 98/99%.
Forms: Powder, crystal bar, sintered pellets, sheet, foil, ingot, rod.

THOR PRIMA EXTRA ZH.
M-1310; 1.15 C, 0.25 max Si, 0.25 max Mn, bal Fe.
Annealed: 110,000 TS; 58,000 YS; 18 El; 38 RA; 215 Brin.
For drills, taps, cutters, springs; Type W1; water hardened.

THOR PRIMA MH.
M-1310; 1.3 C, 0.25 max Mn, 0.25 max Si, bal Fe.
For cutting and engraving tools, taps; Type W1; water hardened.

THOR PRIMA SEHR ZAH.
M-1310; 0.7 C, 0.25 max Si, 0.25 max Mn, bal Fe.
Heat treated: 175,000 TS; 130,000 YS; 12 El; 36 RA; 355 Brin.
For rails, springs, tools, axes, hammers; Type W1; water hardened.

THOR PRIMA ZAH.
M-1310; 0.85 C, 0.25 max Si, 0.25 max Mn, bal Fe.
Heat treated: 190,000 TS; 145,000 YS; 10 El; 30 RA; 400 Brin.
For drills, taps, springs, tools, reamers; Type W1; water hardened.

THOR PRIMA ZH.
M-1310; 1.0 C, 0.25 max Si, 0.25 max Mn, bal Fe.
Annealed: 100,000 TS; 53,000 YS; 21 El; 42 RA; 200 Brin.
For drills, taps, hobs, saws, cutters, springs; Type W1; water hardened.

THREAD GAUGE STEEL.
M-32; 0.20 C, 1.3 Mn, 0.20 Si, bal Fe.
For thread guages; case hardening.

THREADING BRASS-223.
M-8; 65 Cu, 34.75 Zn, 0.25 Pb.
Soft: 45,000 TS; 17,000 YS; 55 El.
Hard: 73,000 TS; 60,000 YS; 10 El.
For drawing, forming, switch plates; high ductility.

THREE STAR.
M-85; 0.72 C, 4 Cr, 18.5 W, 1.2 V, 4 Co, 0.75 Mo, bal Fe.
For cutting tools; high speed steel.

THREE-TWENTY.
M-Eng.; 77 Al, 20 Zn, 3 Cu.
For light alloy parts; non-hardenable.

THURSTON BRASS.
M-U.S.; 44.5 Zn, 0.5 Sn, bal Cu.
For architectural trim; corrosion resistant.

THURSTON BUTTON.
M-Eng.; 33-37 Zn, 1.5-6 Sn, 12-15 Pb, bal Cu.
For ornamental and architectural parts; free-cutting.

THULIUM.
M-1755; Tm.
Purities: 99.9% (special distilled grade), 99.5+%.
Forms: Ingot, lump, turnings, sheet, rod, foil, wire, sponge.

THYRAPID 3202 ETC.
M-1759; see Werkstoff Nr. 1.3202 etc.
High speed tool steels; 10 grades.

THYRODUR 1525 ETC.
M-1759; see Werkstoff Nr. 1.1525 etc.
Cold work tool steels; 25 grades.

THYROPLAST 2162 ETC.
M-1759; see Werkstoff Nr. 1.2162 etc.
Plastic mould steels. 8 grades.

THYROTHERM 2323 ETC.
M-1759; see Werkstoff Nr. 1.2323 etc.
Hot work tool steels; 12 grades.

TI 3-2.5.
M-1798; 0.08 max C, 0.03 max N, 0.0125 max H, 0.25 max Fe, 92.0 min Ti, 0.175 max O, 2.5-3.5 Al; 2.0-3.0 V, others 0.40 max.
Alpha-beta alloy; weldable. Good strength.
May be strengthened by cold-working.

TI-3AL-2.5V.
M-1772; 3 Al, 2.5 V, bal Ti.
Cold drawn tube: 105,000-135,000 TS; 75,000-120,000 YS; 3-30 El.
For aerospace equipment, airplane components. High strength, ductility and fabricability. Used at temperatures up to 600°F.

TI-5AL-2.5SN.
M-144; 4.5-6.0 Al, 2.0-3.0 Sn, bal Ti.
Sheet: 125,000 TS; 120,000 YS; 15 El; 40 RA.
For airframes, high temperature applications; resists oxidation to 1200°F.

TI-5AL-2.5SN ELI.
M-144; 5 Al, 2.5 Sn, 0.25 max Fe, 0.12 max 0, bal Ti.
At -423°F: 229,000 TS; 206,000 YS; 10 El; 1.03 notch ratio.
Annealed: 100,000 TS; 90,000 YS; 10 El; C 36 Rock.
For high temperature applications, fasteners, airframes.
Extra low interstitials; improved notch strength at -423°F.

TI-6AL-2SN-4ZR-2MO.
M-144; 6 Al, 2 Sn, 4 Zr, 2 Mo, bal Ti.
Heat treated: 148,000 TS; 138,000 YS; 20 El; 46 RA.
Annealed: 130,000 TS; 120,000 YS; 10 El; 25 RA; C 36 Rock.
For high temperature components to 1050°F, compressor wheels, missiles space craft.
High temperature alloy. Alpha type. Good creep and tensile strength to 1050°F.

TI-6AL-2SN-4ZR-2MO-SI.
M-144; 6 Al, 2 Sn, 4 Zr, 2 Mo, 0.2 Si, bal Ti.
Heat treated: 167,000 TS; 151,000 YS; 7 El; 15 RA.
For engine discs.
For 800-1000°F service, good heat resistance.

TI-6AL-2SN-4ZR-6MO.
M-144; 5.5-6.5 Al, 1.8-2.2 Sn, 3.6-4.4 Zr, 5.5-6.5 Mo, bal Ti.
Heat treated: 160,000-195,000 psi TS; 150,000-180,000 psi YS; 9-12 El; 30-40 RA.
Good strength to 750°F.

TI-6AL-4V.
M-144; 5.5-6.75 Al, 3.5-4.5 V, bal Ti.
Annealed: 136,500 TS; 130,000 YS; 16 El; 40 RA; 320 Brin.
Heat treated: 137,000-168,000 TS; 128,500-157,000 YS; 9-16 El; 45 RA.
For high temperature applications; stable under stress to 950°F; heat treatable.

TI-6AL-4V ELI.
M-144; 6 Al, 4 V, 0.07 O_2, 0.15 Fe, bal Ti.
Annealed: 125,000 TS; 115,000 YS; 10 El; C 36 Rock. At -423°F: 263,000 TS.
For high temperature applications, jet engine components, fasteners, aircraft forgings.
Extra low interstitials. Improved notch strength at -423°F.

TI-6AL-6V-2SN.
M-144; 5.0-6.0 Al, 5.0-6.0 V, 1.5-2.5 Sn, 0.35-1.0 Fe, bal Ti.
Sheet heat treated: 213,000-219,000 TS; 200,000-213,000 YS; 4.0-0.5 E ; Rock C 46-48.
Annealed Bar: 165,000 TS; 152,000 YS; 15 El; 45 RA.
For high strength pressure vessels operating at ambient temperatures.
Heat treatable, alpha-beta type alloy.

TI-7AL-4MO.
M-144; 7 Al, 4 Mo, bal Ti.
Annealed: 140,000 TS; 130,000 YS; 10 El; 20 RA.
For aircraft gas turbine engines, compressor wheels, hub shafts; creep resistant, high temperature strength.

TI 8.
M-1697; Ti base, Mo, Ni sintered carbide.
175,000 TrS; 93 Rock A; density: 5.83.
For fast and general to light machining of steel, particularly where shock and scale are absent.

TI-8AL-1MO-1V.
M-144; 8 Al, 1 Mo, 1 V, bal Ti.
At 70°F: 165,000 TS; 148,000 YS; 17 El; 25 RA.
At 600°F: 128,000 TS; 110,000 YS; 16 El; 37 RA.
At 1000°F: 110,000 TS; 90,000 YS; 16 El; 33 RA.
For jet engine components to 900°F; excellent creep and rupture life.

TI-10V-2FE-3AL.
M-144; 2.6-3.4 Al, 9.0-11.0 V, 1.8-2.2 Fe, bal Ti.
Heat treated: 140,000-180,000 psi TS; 130,000-170,000 psi YS; 8-10 El 15-20 RA.
Good strength to 600°F.

TI-35A.
M-144; 0.08 max C, 0.05 max N, 0.15 max H, 0.25 max Fe, bal Ti.
Rolled: 35,000 TS; 25,000 YS; 22 El; 35 RA; 170 Brin.
For high temperature applications; commercially pure titanium.

TI-55A.
M-144; 0.20 max C, 0.015 max H, bal Ti.
Bar: 75,000 TS; 65,000 YS; 25 El.
For non-structural parts, fasteners.
Corrosion resistant.

TI-65A.
M-144; 0.08 max C, 0.20 max Fe, 0.0125 max H, 0.05 max N, bal Ti.
Annealed: 65,000 TS; 50,000 YS; 20 El; 35 RA; 225 Brin.
For aircraft structural members.
Corrosion resistant.

TI-75A.
M-144; 0.08 max C, 0.30 max Fe, 0.0125 max H, 0.05 max N, bal Ti.
Annealed: 70,000 TS; 60,000 YS; 15 El; 35 RA; 265 Brin.
For moderately stressed aircraft parts and structures.
Corrosion resistant.

TI-100A.
M-144; 0.08 max C, 0.05 max N, 0.15 max H, 0.25 max Fe, bal Ti.
Rolled: 80,000 TS; 70,000 YS; 15 El; 30 RA; 290 Brin.
For high temperature applications; commercially pure titanium.

TI-679.
M-282; 2.0-2.25 Al, 10.5-11.0 Sn, 4.0-5.0 Zr, 0.8-1.0 Mo, 0.15-0.21 Si, bal Ti.
Heat treated: 140,000 TS; 130,000 YS; 10 El; 20 RA.
For jet engine compressor wheels and blades. High stress stability at 900°F, good creep and short-time strength.

TIBELOY 637.
M-1734; 4.0 Al, 3.5 Pd, bal Ag.
For precious metal brazing.
Brazing temp.: 1770-1800°F.

TIBELOY 692.
M-1734; 8.0 Al, 4.0 Pd, bal Ag.
For precious metal brazing.
Brazing temp.: 1500-1550°F.

TIBON.
M-294; C, bal Fe.
For die blocks.

TICO.
M-Eng.; 27.5-30.4 Ni, 1.12 Mn, 1.1 Cu, bal Fe.
For electrical resistances; resistance alloy.

TICODE-12.
M-144; 0.8 Ni, 0.3 Mo, bal Ti.
70,000 psi (485 MPa) TS; 50,000 psi (345 MPa) YS; 18 El.
Excellent corrosion resistance and good thermal conductivity; good strength at 600°F.
For heat exchangers.

TI-CODE-12.
M-1758; 0.03 max N, 0.08 max C, 0.015 max H, 0.30 max Fe, 0.25 max O, 0.2-0.4 Mo, 0.6-0.9 Ni, 0.30 total other, bal Ti.
70,000 min UTS; 50,000 min YS; 18 El; 25 RA.
Improved corrosion resistance at higher temperatures, lower pH, and chloride containing environments.

TICONAL.
M-1127; 6-9 Al, 12-18 Ni, 0-2.5 Ti, 2-5 Cu, 20-30 Co, bal Fe.
For permanent magnet, sand and precision cast; similar to Alnico 5.

TICONAL 3A.
M-1127; 19 Ni, 12 Co, 10 Al, 6 Cu, bal Fe.
For permanent magnets in electrical and magnetic equipment.
High magnetic permeability.

TICONAL 600.
M-1127; 8 Al, 14 Ni, 24 Co, 3 Cu, bal Fe.
Cast: 5450 TrS.; C 50 Rock.
Br-1200, Hc-560, BH max 5,600,000.
For relays, motors, voltage regulators, magnetos, generators.
Similar to Alnico V permanent magnet.

TICONAL 700.
M-1127; 14 Ni, 8 Al, 24-35 Co, 3 Cu, bal Fe.
11,600 Br, 700 Hc, 5 BH max, Rock C 48.
Magnets for measuring instruments, electrical and magnetic equipment.
Permanent magnet, high magnetic permeability.

TICONAL 750.
M-1127; 14 Ni, 8 Al, 24-35 Co, 3 Cu, bal Fe.
11,000 Br, 750 Hc, 4.4 BH max., C 52 Rock.
For magnets in electrical apparatus and magnetic equipment.
Permanent magnet, high magnetic permeability.

TICONAL 800.
M-1127; 14 Ni, 8 Al, 24-35 Co, 3 Cu, bal Fe.
10,600 Br, 800 Hc, 4.3 BH max., C 52 Rock.
For magnets in electronic, electrical and magnetic equipment.
Same as Alcomax IV.
Permanent magnet. Thermal stability and resistant to demagnetizing fields.

TICONAL 1500.
M-1127; 35 Co, 13.5 Ni, 7.5 Al, 5 Ti, 3 Cu, bal Fe.
8500 Br, 1450 Hc, 5 BH max; Cast: 10,000 TS.
For magnets in electric clocks, electrical and magnetic equipment, magnetic separators, generators. Permanent magnet. Lowest losses after stabilization.
Same as Alnico VIII.

TICONAL 2000.
M-1127; 40 Co, 7.5 Ti, bal Fe.
Coercive force 2000.
For electric motors and magnetic equipment.
Permanent magnet, high magnetic permeability.

TICONAL C.
M-1127; 24 Co, 13.5 Ni, 8 Al, 3 Cu, 0.5 Cb, bal Fe.
For permanent magnets in electrical and magnetic equipment.
High magnetic permeability.

TICONAL D.
M-1127; 24 Co, 14 Ni, 8 Al, 3 Cu, 1.0 Ti, bal Fe.
For permanent magnets in electrical and magnetic equipment.
High magnetic permeability.

TICONAL E.
M-1127; 8 Al, 14 Ni, 24 Co, 3 Cu, bal Fe.
For permanent magnets; high permeability.

TICONAL E.
M-1188; 15 Ni, 8 Al, 25 Co, 3 Cu, bal Fe.
For permanent magnets; high permeability.

TICONAL F.
M-1127; 24 Co, 14 Ni, 8 Al, 3 Cu, 0.5 Ti, bal Fe.
For permanent magnets in electrical and magnetic equipment.
High magnetic permeability.

TICONAL-G.
M-1127; 28 Co, 13 Ni, 8 Al, 3 Cu, bal Fe.
For permanent magnets; Br. 12000, Hc 700.

TICONAL GG.
M-1127; 8 Al, 14 Ni, 24 Co, 3 Cu, bal Fe.
For permanent magnets; high permeability.

TICONAL GX.
M-1127; 24 Co, 14 Ni, 8 Al, 3 Cu, bal Fe.
For permanent magnets in metering devices, electrical and magnetic equipment.
High magnetic permeability. Strong permanent magnet.

TICONAL H.
M-1127; 24 Co, 13.5 Ni, 8 Al, 3 Cu, 2 Cb, bal Fe.
For permanent magnets in electrical and magnetic equipment.
High magnetic permeability.

TICONAL K.
M-1127; 6 Al, 18 Ni, 34 Co, 8 Ti, bal Fe.
For permanent magnets; high permeability.

TICONAL L.
M-1127; 19.5 Co, 14.5 Ni, 7 Al, 1.5 Cu, 0.8 Si, bal Fe.
For permanent magnets in electrical and magnetic equipment.
High magnetic permeability.

TICONAL S.
M-1127; 24 Co, 14 Ni, 8 Al, 3 Cu, bal Fe.
For permanent magnets in electrical and magnetic equipment.
High magnetic permeability.

TICONAL X.
M-1127; 34 Co, 14.5 Ni, 7.0 Al, 4.5 Cu, 5.0 Ti, bal Fe.
Max. energy product 13,400,000. 8800 Br, 1500 Hc, 5,300,000 (BH) max.
For permanent magnets, in electrical equipment. High magnetic permeability.

TICONAL XX.
M-1127; 7 Al, 15 Ni, 4 Cu, 5 Ti, 34 Co, bal Fe.
For permanent magnets; highly directional crystal oriented.

TICONIUM.
M-920; 35 Ni, 31 Co, 23 Cr, 6 Mo, 0.01 C, bal Fe.
Cast: 110,000 TS; 65,000 YS; 6 El; 350 Brin.
For bone surgery. Corrosion resistant.

TICUNI.
M-1493, M-856; 70 Ti, 15 Cu, 15 Ni.
For brazing titanium and stainless steel components.

TIERS ARGENT.
M-Eng.; 66 Al, 33 Ag.
For ornamental parts; corrosion resistant.

TIFICO.
M-294; C, alloy, bal Fe.
For engine crankshaft, die blocks, marine service parts; oil hardening.

TIGER BRONZE.
M-1794; 75 Cu, 6 Sn, 16 Pb, 3 Zn.
Sand cast: 20 ksi TS; 14 ksi YS; 7 El; 50 Brin. Similar to CDA 939.

TIGERLOY.
M-467; 0.3 C, 1.5 Ni, 0.2 Mo, bal Fe.
Heat treated: 100,000 TS; 60,000 YS; 25 El; 45 RA; 207 Brin.
For castings; impact resistant.

TIG-TECTIC 5-AH.
M-717.
Inert arc or torch overlays on gauges, blanking dies, forming dies.

TIG-TECTIC 5-HSS.
M-717.
Inert arc or torch high speed steel overlays.

TIG-TECTIC 5-HW.
M-717.
Inert arc or tooth overlays for hot work tools and dies.

TIG-TECTIC 5-OH.
M-717.
Inert arc or torch overlays for oil hardening steels.

TIG-TECTIC 5-WH.
M-717.
Inert arc or torch overlays for water hardening steels.

TIG-TECTIC 21.
M-717.
Inert arc overlays for aluminum.

TIG-TECTIC 23.
M-717.
Inert arc overlays for aluminum.

TIG-TECTIC 66.
M-717.
Inert arc electrode for medium carbon steel welding and repair.

TIG-TECTIC 182.
M-717.
Inert arc for joining copper base alloys and to steel.

TIG-TECTIC 224.
M-717.
Electrode for TIG welding gray cast iron. Machinable; 50,000 psi TS.

TIG-TECTIC 670.
M-717.
Inert arc welding of alloy and stainless steel.

TIG-TECTIC 680.
M-717.
Inert arc electrode for welding alloy steel; 120,000 TS.

TIG-TECTIC 1851.
M-717.
Aluminum bronze filler rod for inert arc joining and overlaying on bronze and other copper alloys.

TIG-TECTIC 6800.
M-717.
Inert arc electrode for corrosion resistant welds on nickel alloys.

TIG-TECTIC A.
M-717.
Columbium stabilized stainless electrode for inert arc welding stainless steel.

TIG-TECTIC A-MO.
M-717.
Mo bearing stainless electrode for inert arc welding stainless steel.

TIG-TECTIC B.
M-717.
For inert arc welding 18-8 stainless.

TIG-TECTIC D.
M-717.
25/20 stainless rod (Type 310) for inert arc welding 309-310 type stainless or dissimilar stainless.

TILOY.
M-294; Cr, Mo, C, bal Fe.
For die blocks; oil hardening.

TIMAXX 5.
M-271; 0.6-0.8 C, 13-15 Mn, 1.25-1.75 Si, 2.75-3.25 Ni, 4.0-4.5 Cr, bal Fe.
Cast: 100,000 TS; 40,000 YS; 40 El; 30 RA; 180 Brin.
For castings; wear and abrasion resistant to 800°F.

TIME-GRAPH.
M-1705; 1.45 C, 0.8 Mn, 1.2 Si, 0.2 Cr, 0.25 Mo, bal Fe.
Oil hardening tool steel. AISI 06.

TIMKEN 0.50 MO.
M-376; 0.08-0.25 C, 0.3-0.8 Mn, 0.1-0.5 Si, 0.45-0.65 Mo, bal Fe.
Annealed: 55,000 TS; 30,000 YS; 30 El; 150 Brin.
For cracking furnace tubes; to 1050°F.

TIMKEN 2 1/4 CR-1-MO.
M-376; 0.15 max C, 2-2.5 Cr, 0.9-1.1 Mo, bal Fe.
Annealed: 60,000 TS; 25,000 YS; 30 El; 50 RA; 163 Brin.
For high temperature tubing in oil refineries; resist heat to 1150°F.

TIMKEN 5% CR-MO.
M-376; 0.15 C, 0.50 Mn, 0.50 Si, 4.0-6.0 Cr, 0.4-0.6 Mo, bal Fe.
At 850°F: 66,600 TS; 26,000 YS; 39 El.
At 1400°F: 13,300 TS; 7250 YS; 65 El.
For still tubes, oil refinery tubing; age-hardenable, resists oil corrosion and heat to 1200°F.

TIMKEN 17-22 (A).
M-376; 0.41-0.48 C, 0.45-0.65 Mn, 0.55-0.70 Si, 1.0-1.5 Cr, 0.4-0.6 M , 0.2-0.3 V, bal Fe.
Normalized: 125,000 TS; 105,000 YS; 16 El; 50 RA; 250 Brin.
For high temperature bolting; air hardening.

TIMKEN 17-22 (A) S.
M-376; 0.28-0.33 C, 1.0-1.5 Cr, 0.4-0.6 Mo, 0.2-0.3 V, 0.55-0.75 Si, 0.45-0.65 Mn, bal Fe.
Normalized: 125,000 TS; 105,000 YS; 16 El; 50 RA; 250 Brin.
For bolting, flanges, valves; resists thermal cracking.

TIMKEN 17-22 "A" V.
M-376; 0.25 C, 0.8 Mn, 0.75 Si, 1.25 Cr, 0.5 Mo, 0.85 V, bal Fe.
For high temperature applications; good creep strength.

TIMKEN 4422.
M-376; 0.20-0.25 C, 0.7-0.9 Mn, 0.2-0.35 Si, 0.35-0.45 Mo, bal Fe.
For bearings, case hardening.

TIMKEN 4427.
M-376; 0.24-0.29 C, 0.35-0.45 Mo, 0.7-0.9 Mn, 0.3 Si, bal Fe.
For bearings, liners. Case hardening.

TIMKEN 4620.
M-376; 0.17-0.22 C, 0.5 Mn, 1.6-2.0 Ni, 0.2-0.3 Mo, bal Fe.
At 850°F: 72,400 TS; 52,500 YS; 37.5 El; 70.0 RA.
At 1200°F: 24,800 TS; 12,000 YS; 59.3 El; 79.9 RA.
For gears, pinions, shafts, high temperature steam service; case hardening steel; corrosion resistant.

TIMKEN 4718.
M-376; 0.16-0.21 C, 0.35-0.55 Cr, 0.90-1.2 Ni, 0.3-0.4 Mo, 0.7-0.9 Mn bal Fe.
Heat treated: 210,000 TS; 154,000 YS; 13.5 El; 50 RA; 470 Brin.
For gears, bolts, camshafts, cams; case hardened, shock resistant.

TIMKEN 52100.
M-376; 0.95-1.10 C, 0.25-0.45 Mn, 1.3-1.6 Cr, bal Fe.
Annealed: 94,500 TS; 62,000 YS; 27 El; 62 RA; 179 Brin.
Cold drawn: 107,000 TS; 87,000 YS; 17 El; 55 RA; 229 Brin.
For bearings, races, balls, bushings; water or oil hardenable.

TIMKEN CARBON.
M-376; 0.10-0.20 C, 0.3-0.6 Mn, bal Fe.
Annealed: 61,500-48,000 TS; 45,000-30,000 YS; 31-28 El; 70 RA; 132-125 Brin.
For oil refinery tubing, cracking furnace tubes, condenser tubes, heat exchanger tubes; for services to 900°F.

TIMKEN CBS-600.
M-376; 0.16-0.22 C, 0.9-1.2 Si, 1.25-1.65 Cr, 0.9-1.1 Mo, bal Fe.
Heat treated: 98,000-177,500 TS; 25,000-118,250 YS; 29-18 El; 72-45 RA; 197-341 Brin.
For jet engine and rocket components, anti-friction bearings; case hardening.

TIMKEN DM.
M-376; 0.15 max C, 0.30-0.60 Mn, 0.50-1.0 Si, 1.0-1.5 Cr, 0.45-0.65 Mo, bal Fe.
Annealed: 60,000 TS; 25,000 YS; 30 El; 50 RA; 163 Brin.
For high temperature tubing; heat resistant to 1150°F.

TIMKEN DM2.
M-376; 0.15 max C, 0.3-0.6 Mn, 0.5 max Si, 0.8-1.25 Cr, 0.45-0.65 Mo, bal Fe.
Annealed: 60,000 TS; 25,000 YS; 30 El; 163 Brin.
For superheaters, steam pipes, resists heat to 1150°F.

TIMKEN HS 250 (4741).
M-376; 0.37-0.44 C, 0.65-0.95 Mn, 0.4-0.6 Si, 0.8-1.1 Cr, 0.7-1.0 Ni, 0.3-0.4 Mo, bal Fe.
Heat treated: 187,000 TS; 182,000 YS; 14.5 El; 51 RA; 388 Brin.
For gears, shafts, bolts, crankshafts; oil hardened, shock resistant.

TIMKEN NITRIDING STEEL NO. 3.
M-376; 0.38-0.43 C, 0.5-0.7 Mn, 1.4-1.8 Cr, 0.3-0.4 Mo, 0.95-1.3 Al, bal Fe.
Nitriding steel; for gears, shafts, cylinders.

TIMKEN T-1 (TM-NO. 1).
M-376; 0.95-1.1 C, 1.2-1.5 Cr, 0.2-0.3 Mo, bal Fe.
For anti-friction bearings; deep hardening.

TIMKEN T-2 (TM-NO. 2).
M-376; 0.95-1.1 C, 1.05-1.35 Mn, 1.2-1.5 Cr, 0.45-0.55 Mo, bal Fe.
Heat treated.
For anti-friction bearings; deep hardening.

TIMKEN TBA.
M-376; 0.75 C, 1.2 Mn, 1.05 Cr, 1.45 Ni, 1.3 Mo, bal Fe.
An air hardening steel for large bearings and machinery parts.
Air hardenable to 56-60 Rock C in a 6" Rd.

TIMKEN TBS-9.
M-376; 0.95 C, 0.65 Mn, 0.5 Cr, 0.12 Mo, bal Fe.
A through hardening bearing steel.

TIMKEN TBS-600.
M-376; 0.95-1.1 C, 1.25-1.65 Cr, 0.25-0.35 Mo, 0.8-1.2 Si, bal Fe.
For jet engine and rocket components; anti-friction bearings; hard and abrasion resistant; for service up to 600°F.

TIMKEN TBS 1000.
M-376; 0.80 C, 0.50 Mn, 0.50 Si, 0.90-1.20 Cr, 5.0 Mo, 1.05 V, bal Fe.
For dies, cutters, punches, headers; oil or air hardening.

TIMKEN TDS-30.
M-376; 0.17-0.22 C, 0.45-0.65 Mn, 0.20-0.35 Si, 0.4-0.6 Cr, 1.65-2.0 Ni, 0.2-0.3 Mo, bal Fe.
Tubular drill steel, must be carburized to harden.

TIMKEN TDS-70.
M-376; 0.25-0.31 C, 0.8-1.2 Mn, 0.5-0.8 Si, 1.9-2.4 Cr, 0.10 max Ni, 0.25-0.35 Mo, bal Fe.
Tubular drill steel, air hardenable.

TIMKEN TDS-90.
M-376; 0.23-0.28 C, 0.4-0.6 Mn, 0.20-0.35 Si, 3.0-3.5 Cr, 0.45-0.6 Mo, bal Fe.
Tubular drill steel, air hardenable.

TIMKEN WHS-100.
M-376; 0.13-0.21 C, 1.0-1.3 Mn, 0.2-0.35 Si, 0.65-0.9 Cr, 0.4-0.7 Ni, 0.15-0.25 Mo, 0.03-0.08 V, 0.003 B, bal Fe.
A weldable 100,000 psi yield strength material.

TIMKEN WHS-130.
M-376; 0.20-0.27 C, 0.6-0.8 Mn, 0.2-0.35 Si, 0.7-0.9 Cr, 1.55-2.0 Ni, 0.2-0.3 Mo, bal Fe.
A weldable 130,000 psi minimum yield strength material.

TIN.
M-1755; Sn.
Purities: Zone refined 99.9999%, 99.999%, 99.99%, 99.9%.
Forms: Rod, bar, shot, powder, wire, sheet, foil, single crystals.

TI-NAMEL.
M-67; 0.06 C, 0.30 Mn, 0.12 max Cu, 0.05 Al, 0.30 Ti, bal Fe.
Hot rolled: 55,000 TS; 33,000 YS; 27 El; 121 Brin.
For sheets for enameling; specially prepared.

TIN BRASS 1% TIN.
 M-1789; 87.5 Cu, 11.5 Zn, 1.0 Sn.
 Ann: 45,000 TS; 19,000 YS; 45 El.
 Rolled hard: 73,000 TS; 68,000 YS; 9 El.
 For weather strips, springs, clips, switches, terminals.

TIN BRASS 2% TIN.
 M-1789; 88.5 Cu, 9.5 Zn, 2.0 Sn.
 Ann: 44,000 TS; 18,000 YS; 48 El.
 Rolled hard: 76,000 TS; 73,000 YS; 4 El.
 For weather strip, springs, clips, switches, terminals.

TINEA CLASSIC.
 M-921; 45 Sn, 42 Pb, 10 Sb, 3 Cu.
 For bearings; anti-friction metal.

TINEA DIESEL.
 M-921; 83 Sn, 11 Sb, 6 Cu.
 Cast: 26 Brin.
 For bearings; M.P. 380°C; antifriction.

TINEA DIESEL R.S.
 M-921; 76.5 Pb, 10 Sn, 13.5 Sb, 1 Cu.
 Cast: 26.7 Brin.
 For bearings; M.P. 400°C; antifriction.

TINEA MARINE EK.
 M-921; 72.5 Sn, 16 Pb, 7 Sb, 4.5 Cu.
 For bearings; M.P. 380°C; antifriction.

TINEA SUPER MARINE.
 M-921; 90 Sn, 6.5 Sb, 3.5 Cu.
 Cast: 23.7 Brin.
 For automobile bearings; M.P. 330°C; antifriction.

TIN FOIL.
 88 Sn, 8 Pb, 4 Cu, 0.5 Sb.
 For bearings, packing foil.

TINICOSIL NO. 10.
 M-122; Zn, Ni, bal Cu.
 For hardware; corrosion resistant.

TINICOSIL NO. 14.
 M-122; 42.5 Cu, 15 Ni, bal Zn.
 Rolled: 105,000 TS; 70,000 YS; 20 El; 215 Brin.
 For valve parts, instrument parts; nickel silver, corrosion resistant.

TINICOSIL NO. 15.
 M-122; 41.75-42.25 Cu, 14.75-15.25 Ni, bal Zn.
 Hard: 215,000 TS; 192,000 YS; 8 El; 10 RA; 140 Brin.
 Soft: 118,000 TS; 84,000 YS; 11 El; 10 RA; 122 Brin.
 For hot forgings, hardware; corrosion resistant.

TINICOSIL NO. 20.
 M-122; 46-47 Cu, 1.4-1.6 Pb, 15-16 Ni, bal Zn.
 For pressure die castings; corrosion resistant.

TINICOSIL NO. 21.
 M-122; 50 Cu, 10 Ni, 1 Pb, 1 Fe, 3 Mn, bal Zn.
 Extrusion: 140 Brin.
 For valve parts, instrument parts; extrusions, forgings, nickel silver, corrosion resistant.

TINICOSIL NO. 53.
 M-122; 46 Cu, 10 Ni, bal Zn.
 Hard: 105,000 TS; 80,000 YS; 20 El; 15 RA; 158 Brin.
 Forged: 92,000 TS; 42,000 YS; 26 El; 25 RA; 140 Brin.
 For forgings, hardware parts, plumbing fixtures; corrosion resistant.

TINICOSIL NO. 54.
 M-122; 46.5 Cu, 2.2 Pb, 10 Ni, 2 Mn, bal Zn.
 Cold drawn: 88,000 TS; 68,000 YS; 30 El; 20 RA; 150 Brin.
 For screw machine products, fishing tackle; leaded nickel silver, corrosion resistant.
 CA 798.

TINIDUR.
 M-72; 0.15 max C, 30 Ni, 15 Cr, 1.8 Ti, bal Fe.
 Rolled: 135,000-160,000 TS; 85,000-115,000 YS; 30-20 El; 55-30 RA.
 For gas turbine blades, jet engine components; heat and corrosion resistant.

TINIDUR 1650.
 M-72; 0.04 C, 14.7 Cr, 26 Ni, 2.3 Ti, 1 Mn, 0.15 Al, bal Fe.
 For jet engine components; heat and corrosion resistant.

TINIDUR 1875.
 M-72; 0.04 C, 14.7 Cr, 26 Ni, 2-3 Ti, 1 Mn, 0.15 Al, bal Fe.
 For jet engine components; heat and corrosion resistant.

TINMAN'S SOLDER.
 66.5 Sn, 33.5 Pb.
 For solder; soft.

TINSEL.
 60 Sn, 40 Pb.
 For tinsel for decorative purposes.

TIN WELD 1.
 M-717.
 Paint on solder paste combined with flux.
 All purpose; dissimilar metals.
 Bonding temp: 375°F; 7000 psi TS.

TIN WELD 3.
 M-717.
 Paint on solder paste containing neutral type flux for joining and tinning copper. Bonding temp: 360°F; 8000 psi TS.

TIOGA.
M-1779; 0.67 C, 0.65 Cr, 1.4 Ni, 0.2 Mo, bal Fe.
For lathe centers, clutch parts, cams, arbors; oil hardened, tough.
AISI L6.

TIONA.
M-114; 1.0 C, 0.85 Cr, 2.0 Mn, 0.85 Mo, bal Fe.
For blanking dies; air hardened.

TIPALOY 100.
M-1157.
Cd-Cu alloy electrode.
Wrought: 65,000 psi TS; 85% Elec. conductivity.
For spot welding light metal alloys, terne plate, galvanized and similar materials.

TIPALOY 130.
M-1157; Cu-Cr alloy.
Cast: 55,000 TS; 31,000 YS; 22 El; 125 Brin.
Wrought: 80,000 TS; 62,000 YS; 17 El; 165 Brin.
For resistance-welder electrodes; wear resistant.

TIPALOY 200.
M-1157; Co-Ag-Be, bal Cu.
Cast: 90,000 TS; 70,000 YS; 8-10 El; 190 Brin.
Wrought: 100,000 TS; 80,000 YS; 8-10 El; 210 Brin.
For resistance-welder electrodes; wear resistant.

TIPALOY T-1W.
M-1157.
Cu-W welding electrode.
Density: 12.6 g/cc; 63,000 psi TS; 35-42% Elec. conductivity.
Facings or electrodes for flash and butt welding electrodes and projection welding electrodes.

TIPALOY T-3W.
M-1157.
Cu-W welding electrode.
Density: 13.5 g/cc; 75,000 psi TS; 31-38 Elec. cond.
For projection welding electrodes.

TIPALOY T-3W53.
M-1157.
Cu alloy-W welding electrode.
Heat treated: 120,000 psi TS.
Density: 13.5 g/cc; 28-32% Elec. cond.
For high strength electrodes.

TIPALOY T-4.
M-1157.
Cu base electrode.
Cast: 110,000 psi TS; 20% Elec. cond.
Wrought: 170,000 psi TS; 23% Elec. cond.
Strong, wear resistant electrode.

TIPALOY T-5.
M-1157.
Al-Cu alloy electrode.
Cast: 70,00 psi TS; 15% Elec. cond.
For certain flash and butt welding operations.

TIPALOY T-5W.
M-1157.
Cu-W welding electrode.
Density: 13.8 g/cc; 85,000 psi TS; 28-33% Elec. cond.
For spot and projection welding electrodes.

TIPALOY T-10W.
M-1157.
Cu-W welding electrode.
Density: 14.3 g/cc; 90,000 psi TS; 28-32% Elec. conductivity.
For projection welding dies; for flash and butt welding dies.

TIPALOY T-10W53.
M-1157.
Cu alloy-W welding electrode.
Heat treated: 160,000 psi TS;
Density: 14.3 g/cc; 26-30% Elec. cond.
For high strength electrodes.

TIPALOY T20S.
M-1157.
Ag-W welding electrode.
70,000 psi TS; 49% Elec. cond.

TIPALOY T-20W.
M-1157.
Cu-W welding electrode.
Density: 14.7 g/cc; 95,000 psi TS; 27-31% Elec. cond.
For heavy duty projection welding.

TIPALOY T-30W.
M-1157.
Cu-W welding electrode.
Density: 14.6 g/cc; 100,000 psi TS; 26-30% Elec. cond.
For projection welding dies and cross wire welding.

TIPALOY T-35S.
M-1157.
Ag-W welding electrode.
50,000 psi TS; 51% Elec. cond.

TIPALOY-T-100M.
M-1157.
100% Mo welding electrode.
Density: 9.9 g/cc; 80,000 psi TS; 32% Elec. cond.
For welding or electro-brazing non-ferrous metals having relatively high electrical conductivity.

TIPALOY T-100W.
M-1157.
100% W welding electrode.
Density: 19.0 g/cc; 50,000 psi TS; 32% Elec. cond.
For spot welding copper and copper base alloys.

TIPALOY T-253.
M-1157; Ni-Be, bal Cu.
Cast: 80,000 psi TS; 40% Elec. cond.
Wrought: 100,000 psi TS; 40% Elec. cond.
Backing material for welding set-ups.

TIPALOY T-G12.
M-1157.
Ag-Wc welding electrode.
Density: 11.8 g/cc; 35,000 psi TS; 50-60% Elec. cond.
Good conductivity; -for light loads.

TIPALOY T-G13.
M-1157.
Ag-WC welding electrode.
Density: 12.2 g/cc; 40,000 psi TS; 45-55% Elec. cond.
Good conductivity; -for light loads.

TIPALOY T-G14.
M-1157.
Ag-WC welding electrode.
Density: 13.1 g/cc; 55,000 psi TS; 30-40% Elec. cond.

TIPALOY T-G17.
M-1157.
Ag-Mo welding electrode.
Density: 10.1 g/cc; 60,000 psi TS; 45-50% Elec. cond.

TIPALOY T-G18.
M-1157.
Ag-Mo welding electrode.
Density: 10.2 g/cc; 45,000 psi TS; 50-55% Elec. cond.

TIPALOY T-TC5.
M-1157.
Cu-W-WC welding electrode.
Density: 11.0 g/cc; 70,000 psi TS; 45-50% Elec. cond.
For projection welding dies; light loads but needs abrasion resistance.

TIPALOY T-TC10.
M-1157.
Cu-W-WC welding electrode.
Density: 11.5 g/cc; 75,000 psi TS; 40-45% Elec. cond.
For heavy duty projection welding and electro-forging.

TIPALOY T-TC20.
M-1157.
Cu-W-Wc welding electrode.
Density: 12.2 g/cc; 85,000 psi TS; 30-35% Elec. cond.
For electro-forging and upsetting.

TIPALOY T-TC53.
M-1157.
Cu alloy-W-WC welding electrode.
Heat treated: 150,000 psi TS.
Density: 12.3 g/cc; 18-23 Elec. cond.
For electro-forging and upsetting where hardness and abrasion resistance are required.

TI PD.
M-144; 0.15-0.20 Pd, bal Ti.
Plate: 62,000 TS; 46,000 YS; 27 El; 38 RA; 200 Brinell.
For chemical industry equipment, for special corrosion applications.
Alpha type. Resists seawater corrosion.

TIRFING 1.
M-111; 0.55 C, 1.75 Si, 0.75 Mn, 0.2 Cr, bal Fe.
For springs, chisels; oil hardening.

TIRFING 2.
M-111; 0.6 C, 1.5 Si, 0.5 Mn, 0.5 Cr, bal Fe.
For springs, punches; oil hardening.

TIRFING 3.
M-111; 0.5 C, 0.85 Mn, 1.05 Cr, 0.15 V, bal Fe.
For springs, dies; oil hardening.

TISCON.
M-997; 0.19-0.23 C, 0.75-0.85 Mn, bal Fe.
Twisted rolled bars: 49.5 kgf/mm^2 TS; 42.5 kgf/mm^2 YS; 14.5 El.
Reinforcing bars.

TISCO "TAPPI" CORROSION RESISTING STEEL.
M-271; 0.13-0.35 C, 16-23 Cr, 7-11 Ni, Mo, bal Fe.
Annealed: 70,000-75,000 TS; 60-50 El; 60-45 RA; 130-160 Brin.
For machinery parts in paper and pulp machinery; now Tisco No. 110.

TISCRAL.
M-997; 0.22 max C, 1.2-1.5 Mn, 0.08 max P, 0.50-0.80 Cr, 0.02-0.10 V, 0.005-0.015 Ti, 0.20-0.40 Al, bal Fe.
Rolled: 45 kgf/mm^2 YS; 62 kgf/mm^2 TS; 18 El; 217 BHN.
Wear and abrasion resistant steel.

TISSIERS METAL-1.
M-Eng.; 97 Cu, 2 Zn, 1 As.
For hardware; arsenical bronze; good conductivity.

TISSIERS METAL-2.
M-Eng.; 97 Cu, 2 Zn, 0.5 Sn.
For hardware; arsenical bronze; good conductivity.

TITAN.
M-32; 1.00 C, bal Fe.
For tools; water hardening.

TITAN 50.
M-1086.
TiC base sintered carbide.
Density: 5.82; Hardness: Rock A 90.5.
Transverse rupture strength: 275,000 psi (1940 N/mm^2).
High toughness cemented titanium carbide grade for general purpose and semi-rough machining of steel.
ISO P 20-P 30.

TITAN 60.
M-1086; TiC, Co.
For cutting tools in turning, boring and milling operations.
Sintered carbide. Wear and crater resistant.

TITAN 80.
M-1086; TiC, MoC, Ni.
Sintered: A 92.7 Rock; 200,000 min Tr.S.
For cutting tools in high speed machining.
Wear and crater resistant. Sintered carbide.

TITAN 100.
M-1086.
TiC base sintered carbide.
Density: 5.50; Hardness: Rock A 93.3.
Transverse rupture strength: 170,000 psi (1200 N/mm^2).
For high speed semi-finish and finish machining of alloy steels, tool steels, stainless steels and cast iron; for speeds exceeding 800 SFPM.
ISO P 01 - P 05.

TITANAL.
M-English; 12 Cu, 0.8 Mg, 0.5 Si, 0.5 Fe, bal Al, for pistons; cast.

TITANALOY.
M-1511; 0.12 Ti, 1.0 Cu, bal Zn.
Sheet: 24,000 TS; 10 El; 55 Brin.
For roofing, gutters, trim, housings, fuses, curtain walls; high creep resistance.

TITAN BEARING BRONZE CA 673.
M-122; 59.25 Cu, 1.1 Pb, 1.15 Si, 2.25 Mn, bal Zn.
For bushings, bearings, gears; free-cutting.

TITAN BRONZE "A".
M-122; 53.5 Cu, 46 Zn, 0.3 Fe, 0.3 Al.
Hot rolled: 75,000-85,000 TS; 40,000-48,000 YS; 25-30 El; 45-50 RA.
Cast: 50,000-55,000 TS; 30,000-35,000 YS; 15-20 El.
For gears, pinions, roller bearings, pump rods; non-corrosive.

TITAN BRONZE "B".
M-122; 60.5 Cu, 1.02 Sn, 38.5 Zn.
Annealed: 74,740 TS; 40,277 YS; 34 El; 37 RA; 109 Brin.
Hard drawn: 73,560 TS; 52,443 YS; 31 El; 44 RA.
For marine shafts, underwater fittings; non-corrosive.

TITAN BRONZE WELDING ROD.
M-122; 58.75 Cu, 0.75 Sn, bal Zn.
Cold drawn: 94,000 TS; 75,000 YS; 14 El; 80 RA; 174 Brin.
Welded: 52,000 TS; 40,000 YS; 15 El; 104 Brin.
For welding rod; M.P. 1615°F; non-forming.

TITANITE.
M-190, M-955, M-681; 4 Cu, 0.2 Ti, bal Al.
Heat treated: 30,000-37,000 TS; 15,000-24,000 YS; 6-5 El; 60-75 Brin.
For flywheel and axle housings, aircraft wheels, fittings; age-hardenable.

TITANIT FM.
M-1246; WC.
Sintered.
For cutting tools; sintered carbides.

TITANIT G 1.
M-1246; carbides.
For cutting tools; sintered carbide.

TITANIT G 2.
M-1246; carbide.
For cutting tools; sintered carbide.

TITANIT G 3.
M-1246; carbide.
For cutting tools; sintered carbide.

TITANIT G 4.
M-1246; carbide.
For cutting tools; sintered carbide.

TITANIT G 5.
M-1246; carbide.
For cutting tools: sintered carbide.

TITANIT G 6.
M-1246; carbide.
For cutting tools; sintered carbide.

TITANIT H 1.
M-1246; carbides.
For cutting tools; sintered carbide.

TITANIT SIT.
M-1246; WC.
Sintered.
For cutting tools; sintered carbides.

TITANIT S2T.
M-1246; WC.
Sintered.
For cutting tools; sintered carbides.

TITANIT S3T.
M-1246; WC.
Sintered.
For cutting tools; sintered carbides.

TITANIT S4T.
M-1246; WC.
Sintered.
For cutting tools; sintered carbides.

TITANIT U.
M-1246; carbides.
For cutting tools; sintered carbide.

TITANIT WZ1.
M-1246; WC.
Sintered.
For cutting tools; sintered carbides.

TITANIT WZ2.
M-1246; WC.
Sintered.
For cutting tools; sintered carbides.

TITANIUM.
M-1791.
Ti (C.P.).
Ann: 55,000 TS; 45,000 YS; 25 El.
Tube for chemical processing.
Corrosion resistant.

TITANIUM.
M-1755; Ti.
Purities: Zone refined 99.995%, special crystal grade 99.95% crystal bar and electro refined 99.9+%, commercial 99.5%.
Forms: Rods, sponge, granules, powder, wire, sheets, foils, ingot, tubing castings, single crystals.

TITANIUM 3 AL-2.5 V.
M-1084; 2.5-3.5 Al, 2.0-3.0 V, 0.12 max O, 0.045 max N, 0.05 max C, 0.30 max Fe. 0.015 max H, bal Ti.
Ann: 98,000 psi TS; 88,000 psi YS; 22 El.
75% cold worked: 160,000 psi TS; 148,000 psi YS; 5 El.
Good formability; used for honeycomb structures.

TITANIUM-3 ALUMINUM - 2.5 VANADIUM.
M-1791; 3 Al, 2.5 V, bal Ti.
Extruded and annealed: 95,000 psi TS; 80,000 psi YS; 10 El.
Aircraft hydraulic tube.
High strength to weight ratio.

TITANIUM 120.
M-282; Ti alloy.
Annealed: 67,200 TS; 20 El.
For aircraft frames; low density.

TITANIUM 130.
M-282; Ti alloy.
Annealed: 56,000-89,600 TS; 40,400 YS; 15 El.
For aircraft frames; low density.

TITANIUM 160.
M-282; Ti alloy.
Annealed: 89,600 TS; 67,200 YS; 12 El.
For aircraft frames; low density.

TITANIUM 314A.
M-282; 4 Al, 4 Mn, bal Ti.
Annealed: 139,000 TS; 127,000 YS; 10 El.
Hot rolled: 154,000 TS; 132,700 YS; 21 El; 33 RA.
For jet engine and guided missile components; high temperature applications.

TITANIUM 314C.
M-282; 2 Al, 2 Mn, bal Ti.
Annealed: 89,600 TS; 67,200 YS; 15 El.
Heat treated: 94,900 TS; 72,400 YS; 23 El; 47 RA.
For aircraft and missile components; high temperature use.

TITANIUM 317.
M-282; 5 Al, 2.5 Sn, bal Ti.
Annealed: 112,000 TS; 101,000 YS; 10 El.
For welded rings, compressor blades; weldable, good creep and fatigue strength to 800°F.

TITANIUM 318A.
M-282; 6 Al, 4 V, bal Ti.
Annealed: 139,000 TS; 130,000 YS; 10 El.
For jet engine components, airframe forgings; weldable, good hot strength.

TITANIUM 371.
M-282; Sn, Al, bal Ti.
Heat treated: 148,000 TS; 126,600 YS; 17 El; 19 RA.
For high temperature applications to 1000°F; heat treatable.

TITANIUM A-40.
M-1798; 0.02 C, 0.012 N, 0.005 H, 0.15 Fe, 0.150 O, bal Ti.
Alpha grade titanium; properties similar to austenitic stainless steel.

TITANIUM A-55.
M-1798; 0.03 C, 0.015 N, 0.006 H, 0.15 Fe, 0.180 O, bal Ti.
Alpha phase titanium; moderate strength with good formability.
Used in airframes, chemical and similar applications.

TITANIUM A-70.
M-1798; 0.035 C, 0.015 N, 0.006 H, 0.15 Fe, 0.250 O, bal Ti.
Alpha phase titanium; better strength than A-55.
For airframes and similar applications.

TITANIUM ALUMINUM BRONZE NO 1.
M-58; 90 Cu, 10 Al.
65,000-80,000 TS; 22,000-26,000 YS; 25-15 El; 24-16 RA; 90-100 Brin.
For gears; C.Y.P. 19,000.

TITANIUM ALUMINUM BRONZE NO 5.
M-58; 89 Cu, 10 Al, 1 Fe.
65,000-80,000 TS; 23,000-28,000 YS; 30-15 El; 29-21 RA; 92-100 Brin.
For gears: C.Y.P. 19,000.

TITANIUM BEARING STEEL.
M-67; 0.2 C, 0.3 Ti, bal Fe.
For jet engine parts; heat resistant; coated with porcelain enamel.

TITANIUM BRONZE.
M-92; 6 Ti, bal Cu.
Cast: 120,000 TS; 105,000 YS; 8 El.
For high strength castings; age-hardenable.

TITANIUM DISILICIDE.
M-109; 44.7 Ti, 51.8 Si, 1.5 Al, 2.1 Fe.
For specail alloy applications; cermet.

TITANIUM EP 20-2.
20 Al, 2 V, bal Ti.
At 70°F: 190,000 TS; 171,000 YS; 6 El; 380 Brin.
At 800°F: 150,000 TS; 130,000 YS; 12 El.
At 1100°F: 120,000 TS; 105,000 YS; 15 El.
For high temperature applications, aircraft and missile components; high temperature strength, weldable.

TITANIUM EP 90-10.
9.8-10.2 Cr, 0.005-0.05 Fe, bal Ti.
At 70°F: 201,000 TS; 175,000 YS; 6 El.
At 800°F: 150,000 TS; 121,000 YS; 12 El.
For pressure vessels, fasteners, clamps, airplane skins; heat treatable, high strength.

TITANIUM GR. NDA.
M-522; 97 min Ti, 0.3 max Fe, 0.2 N_2, 0.15 C.
For ladle additions to stainless steels; unsintered pellets.

TITANIUM STEEL.
M-U.S.; 0.3-9.0 Ti, 0.1-0.8 C, bal Fe.
For machinery parts, tools; water hardening.

TITAN MANGANESE BRONZE.
M-122; 59.2 Cu, 0.9 Fe, 0.9 Sn, 0.1 Mn, 0.1 Si, bal Zn.
Cold drawn: 89,000 TS; 58,700 YS; 21.5 El; 40 RA; 159 Brin.
For bolts, rods, forgings; corrosion resistant, high tensile.
CA 675.

TITAN MANGANESE BRONZE.
M-122; 59 Cu, 0.80 Pb, 0.90 Sn, 0.1 Mn, 0.9 Fe, bal Zn.
Rolled: 75,000 psi TS; 35 El.
For machine parts; free cutting.
CA 767.

TITAN MANGANESE BRONZE WELDING ROD.
M-122; 59.0 Cu, 0.8 Sn, 0.8 Fe, 0.1 Si, bal Zn.
Cold drawn; 96,000 TS; 79,000 YS; 16 El; 81 RA; 180 Brin.
Welded 53,000 TS; 38,000 YS; 12 El; 107 Brin.
For welding rod for cast iron; M.P. 1620°F; general purpose rod.
CA 681.

TITAN MANGANESE BRONZE WELDING ROD (NICKEL) CA 680.
M-122; 58.72 Cu, 0.30 Mn, 0.50 Fe, 0.50 Ni, 0.90 Sn, 0.12 Si, bal Zn.
General purpose manganese bronze rod with nickel; for brass, copper, bronze, nickel alloys, steel and cast iron.

TITAN MUNTZ METAL.
M-122; 61 Cu, bal Zn.
Cold drawn: 72,500 TS; 52,500 YS; 30 El; 66 RA; 135 Brin.
For forgings, rods; brass.

TITAN NAVAL BRONZE.
M-122; 60.25 Cu, 0.75 Sn, 39.0 Zn.
Annealed: 62,000 TS; 31,150 YS; 41 El; 95 Brin.
Cold drawn: 69,000 TS; 46,000 YS; 32 El; 46 RA; 137 Brin.
For ship construction, high strength forgings, corrosion resistant.

TITAN NAVAL BRONZE WELDING ROD.
M-122; 59 Cu, 0.8 Si, 0.75 Sn, bal Zn.
Cold Drawn: 90,000 TS; 70,000 YS; 16 El; 78 RA; 170 Brin.
Welded: 50,000 TS; 15 El; 104 Brin.
For welding rod; M.P. 1625°F.
CA 470.

TITAN NICKEL SILVER WELDING ROD CA773.
M-122; 48.0 Cu, 10 Ni, 0.10 Si, bal Zn.
Low fuming rod with nickel silver color and high mechanical properties.
M.P. 1680°F.

TITAN PRESSURE CASTING ALLOY.
M-122; 65.0 Cu, 0.10 Pb, 0.9 Si, bal Zn.
For pressure die castings; high strength.
CA 879.

TITAN SILICON BRONZE CA655.
M-122; 95.9 Cu, 0.80 Mn, 3.30 Si.
High copper rod or wire for welding or cold heading.

TITAN NO 4.
M-122; 65 Cu, 35 Zn.
For cold headed products, rivets, bolts; cold heading brass, good ductility.

TITAN NO 4 FORGE.
M-122; 59-59.5 Cu, 0.9-1.1 Pb, 0.5-0.7 Sn, bal Zn.
63,000-68,000 TS; 35,000-47,000 YS; 30-44 El; 20-32 RA; 86-107 Brin.
For screw machine parts; free cutting.

TITAN NO 6A.
M-122; 60.0-60.5 Cu, 0.65-0.85 Sn, bal Zn.
Hard: 63,000 TS; 36,000 YS; 42 El; 46 RA; 70 Brin.
Soft: 62,000 TS; 29,000 YS; 48 El; 53 RA; 72 Brin.
For hardware, marine parts; corrosion resistant.

TITAN NO 25.
M-122; 59.0-59.5 Cu, 0.90-1.10 Pb, 0.5-0.7 Sn, bal Zn.
Hard: 68,400 TS; 47,600 YS; 30 El; 20 RA; 99 Brin.
Soft: 66,000 TS; 35,050 YS; 38 El; 25 RA; 93 Brin.
For hardware, screw machine products; free-cutting.

TITAN NO 35.
M-122; 64.0 Cu, 1.00 Pb, bal Zn.
Hard: 45,500 TS; 24,000 YS; 54 El; 73 RA; 66 Brin.
Soft: 42,000 TS; 20,000 YS; 62 El: 74 RA; 60 Brin.
For hardware, screw machine products; free-cutting swaging brass.
CA 340.

TITAN NO 36.
M-122; 59 Cu, 2 Pb, bal Zn.
Extruded: 55,000 TS; 23,000 YS; 40 El; 80 Brin.
For hardware, screw machine products; forging brass, free-cutting.

TITAN NO 51.
M-122; 59 Cu, 0.7 Pb, 0.9 Sn, 0.9 Fe, 0.2 Mn, bal Zn.
Soft: 65,000 TS; 30,000 YS; 30 El; 116 Brin.
1/2 H-temper: 78,000 TS; 52,000 YS; 22 El; 150 Brin.
H-temper: 85,000 TS; 60,000 YS; 12 El; 165 Brin.
For screw machine products, marine hardware; leaded manganese bronze.

TITAN NO 52.
M-122; 57.25 Cu, 2.5 Pb, bal Zn.
For screw machine products; free-cutting.
CA 385.

TITAN NO 65.
M-122; 59 Cu, 0.9 Sn, 0.9 Fe, 0.2 Mn, bal Zn.
Soft: 65,000 TS; 35,000 YS; 30 El; 116 Brin.
1/2 H-temper: 78,000 TS; 52,000 YS; 22 El; 150 Brin.
H-temper 85,000 TS; 60,000 YS; 12 El; 165 Brin.
For propellers, marine hardware, bolts; corrosion resistant, manganese bronze.

TITAN NO 68.
M-122; 56-60 Cu, 0.8-1.3 Fe, 0.7-1.2 Al, 0.5 max Mn, bal Zn.
Cast: 80,000 TS; 40,000 YS; 30 El; 137 Brin.
For propellers, shafts, marine hardware; manganese bronze.

TITAN NO 73.
M-122; 1.9 Si, 6.9 Al, bal Cu.
Cast 85,000 TS; 43,000 YS; 30 El; 135 Brin.
For impellers, marine hardware; aluminium-silicon bronze.
CA 642.

TITAN NO 74.
M-122; 81.5 Cu, 4 Si, bal Zn.
Cold drawn: 90,000 TS; 45,000 YS: 25 El; 170 Brin.
For valve stems, line pole hardware; die castings and forgings.
CA 694.

TITAN NO 75.
M-122; 70 Cu, 30 Zn.
For cold headed products, rivets, bolts; cartridge brass high ductility.
CA 260.

TITAN NO 83.
M-122; 56 Cu, 0.5 Pb, bal Zn.
For marine hardware; high strength.

TITAN NO 92.
M-122; 58 Cu, 0.8 Si, 1.6 Al, 2.5 Mn, 0.3 Pb, bal Zn.
Cast: 85,000 TS; 45,000 YS; 18 El; 156 Brin.
For propellers, impellers, marine hardware; high manganese bronze, corrosion resistant.
CA 674.

TITAN NO 95.
M-122; 62.5 Cu, 0.60 Sn, bal Zn.
Soft: 56,000 TS; 24,000 YS; 37 El; 104 Brin.
1/2 H: 64,000 TS; 32,000 YS; 32 El; 121 Brin.
Hard: 70,000 TS; 46,000 YS; 24 El; 150 Brin.
For marine hardware, propeller shafts; corrosion resistant.

TITAN NO 96.
M-122; 90 Cu, 2 Pb, 1 Ni, bal Zn.
Soft: 38,000 TS; 12,000 YS; 45 El; 57 Brin.
1/2 H: 52,000 TS; 48,000 YS; 20 El; 107 Brin.
For marine hardware, screw machine products; free cutting, corrosion and wear resistant.

TITAN PENN BRONZE WELDING ROD.
M-122; 58.75 Cu, bal Zn.
Welded: 57,000 TS; 35,000 YS; 16 El; 107 Brin.
Cold drawn: 103,000 TS; 89,000 YS; 13 El; 83 RA; 207 Brin.
For welding rod; M.P. 1628°F; high strength; ductile.
W 16.

TITAN-SEEWASSER.
M-Ger.; 2-4 Mg, 1.2 Mn, 0-0.2 Sb or Ti, bal Al.
Sand cast: 20,000-27,000 TS; 11,400-14,300 YS; 3-8 El; 50-60 Brin.
For marine hardware, chemical plant equipment; resists salt water corrosion.

TITAN SWAGING BRASS.
M-122; 64-64.5 Cu, 0.70-1.0 Pb, bal Zn.
42,000-51,000 TS; 20,000-34,000 YS; 41-62 El; 71-74 RA; 58-84 Brin.
For swagings, spinnings, stampings; high strength.

TITUS.
M-114; 0.7 C, 0.75 Cr, 0.2 V, bal Fe.
Heat treated: 220,000-125,000 TS; 200,000-100,000 YS: 5-20 El; 5-53 RA; 487-320 Brin.
For punches and chisels, axes, ball races, dies, shear blades; "Cyclops Titus."

TIWZ-12.
M-1197, M-1246; TiC.
For gas turbines, jet engines, nuclear power plants; sintered, oxidation resistant.

TIZIRBE.
M-1493, M-856; 48 Ti, 48 Zr, 4 Be.
For brazing titanium stainless steel components. Corrosion resistant.

TIZIT.
M-1246; 40-80 W, 3-40 Fe, 4-15 Ti, 2-4 C, 1-5 Ce, 4 Cr.
Cast.
For hard cutting tools and dies: cast alloy.

TIZIT A.
M-1246; carbides.
For cutting tools, dies; sintered carbides.

TIZIT CR1.
M-1246; CrC, bal Ni.
For valves and pump parts; corrosion resistant.

TIZIT CR2.
M-1246; WC, bal Co.
For valves and pump parts; corrosion resistant.

TIZIT FM.
M-1246; WC, TiC, TaC, Co.
For cutting tools and dies; carbides, wear resistant.

TIZIT G1.
M-1246; WC, Co.
For cutting tools and dies; carbides, wear resistant.

TIZIT G2.
M-1246; WC, Co.
For cutting tools and dies; carbides, wear resistant.

TIZIT G3.
M-1246; WC, Co.
For cutting tools and dies; carbides, wear resistant.

TIZIT G4.
M-1246; WC, Co.
For cutting tools and dies; carbides, wear resistant.

TIZIT G5.
M-1246; WC, Co.
For cutting tools and dies; carbides, wear resistant.

TIZIT G6.
M-1246; WC, Co.
For cutting tools and dies; carbides, wear resistant.

TIZIT H1.
M-1246; WC, Co.
For cutting tools and dies; carbides, wear resistant.

TIZIT H2.
M-1246; WC, Co.
For cutting tools and dies; carbides, wear resistant.

TIZIT H3.
M-1246; WC, Co.
For cutting tools and dies; carbides, wear resistant.

TIZIT SIT.
M-1246; WC, TiC, TaC, Co.
For cutting tools and dies; carbides; wear resistant.

TIZIT S2T.
M-1246; WC, TiC, TaC, Co.
For cutting tools and dies; carbides, wear resistant.

TIZIT S3T.
M-1246; carbide.
For cutting tools; extreme wear resistance, sintered.

TIZIT S4T.
M-1246; WC, TiC, TaC, Co.
For cutting tools and dies; carbides, wear resistant.

TIZIT-U.
M-1246; carbide.
For cutting tools; wear resistant, sintered.

T-K ALLOY see CARPENTER T-K.

TL-ALLOY.
M-8; 3.2-3.8 Mn, bal Cu.
For tachometer drag cups.
Corrosion resistant, high electrical conductivity.

T-LOY 34.
M-634; 0.30-0.38 C, 0.8 Mn, 0.4 Si, 0.6-0.9 Ni, 0.4-0.6 Cr, 0.20 Mo, bal Fe.
Heat treateed: 90,000 TS; 60,000 YS; 20 El; 40 RA; 187 Brin.
For shovel bottoms, gears, shafts, racks; oil hardened, shock resistant.

T-LOY 42.
M-634; 0.38-0.46 C, 0.8 Mn, 0.9-1.1 Ni, 0.6-0.8 Cr, 0.25-0.35 Mo, bal Fe.
Normalized: 100,000 TS; 75,000 YS; 12 El; 25 RA; 250 Brin.
Oil hardened: 175,000 TS; 145,000 YS; 10 El; 25 RA; 250 Brin.
For tractor treads, wear parts, dies; tough, oil hardened.

TM-6 LOW CARBON.
M-275; 0.60 C, 4 Cr, 2 V, 6 W, 5 Mo, bal Fe.
For brass extrusion dies, hot forming dies, shear blades, cold heading inserts; AISI H42.

TMS-O.
M-1460; C, 11.0 min Mn, Si, bal Fe.
Austenitic manganese steel casting.
Similar to ASTM A 128.

TMS-20.
M-1460; C, 11.0 min Mn, Cr, Si, bal Fe.
Austenitic manganese steel casting.
Similar to ASTM A 128.

TMS-30.
M-1460; C, 11.0 min Mn, Cr, Si, bal Fe.
Austenitic manganese steel casting; more Cr than TMS-20.
Similar to ASTM A 128.

TMS-40.
M-1460; C, 11.0 min Mn, Cr, Si, bal Fe.
Austenitic manganese steel casting; more Cr than TMS-30.
Similar to ASTM A 128.

TMS-ASH.
M-1460; C, 11.0 min Mn, Mo, Si, bal Fe.
Austenitic manganese steel casting.
Similar to ASTM A 128.

TMS-MO.
M-1460; C 11.0 Mn, Mo, Si, bal Fe.
Austenitic manganese steel casting.
Similar to ASTM A 128.

TMS-NI.
M-1460; C, 11.0 min Mn, Ni, Si, bal Fe.
Austenitic manganese steel casting.
Similar to ASTM A 128 Gr.D.

TOBIN BRONZE 452 see **TOBIN BRONZE 4641.**

TOBIN BRONZE 470.
M-8; 59 Cu, 0.6 Sn, bal Zn.
For welding rod, for steel, cast iron and copper alloys; oxy-acetylene welding.

TOBIN BRONZE 4641.
M-8; 60 Cu, 0.75 Sn, bal Zn.
Hard: 63,000 TS; 35,000 YS; 35 El; 100 Brin.
Soft: 56,000 TS; 22,000 YS; 45 El; 59 Brin.
For piston rods, propeller shafts; corrosion resistant.

TOBIN BRONZE NO 1875.
58.2 Cu, 39.5 Zn, 2.3 Sn.
For piston rods, propeller shafts, bolts, nuts; corrosion resistant to sea water.

TOKUSHU-RHM5.
M-1647; 0.25-0.35 C, 12-14 Cr, 0.3-0.5 Mo, bal Fe.
Annealed: 75,000 TS; 40,000 YS; 25 El; B 78 Rock.
Heat treated: 200,000 TS; 150,000 YS; 10 El; C 45 Rock.
For oil refinery and chemical plant equipment, cutlery, tableware, knives.
Corrosion resistant, hardenable.

TOKUSHU-RHM7.
M-1647; 0.15-0.20 C, 11-12.5 Cr, 0.9-1.2 Mo, 0.7 max Ni, bal Fe.
Annealed: 70,000 TS; 35,000 YS; 30 El; B 80 Rock.
Heat treated: 190,000 YS; 145,000 YS; 15 El; C 40 Rock.
For chemical plant and oil refinery equipment.
Corrosion resistant, hardenable.

TOKUSHU-RHM8.
M-1647; 0.12-0.17 C, 11.0-12.5 Cr, 0.9-1.2 Mo, 0.6 max Ni, bal Fe.
Annealed: 70,000 TS; 35,000 YS; 30 El; B 80 Rock.
Heat treated: 135,000 TS; 105,000 YS; 10 El; C 29 Rock.
For chemical plant and oil refinery equipment.
Corrosion resistant.

TOKUSHU-RHM9.
M-1647; 0.20 max C, 11.5-14 Cr, 0.8-1.3 Mo, 1 max Ni, bal Fe.
Annealed: 70,000 TS; 35,000 YS; 30 El; b 80 rock.
Heat treated: 145,000 TS; 115,000 YS; 20 El; 300 Brin.
For oil refinery and chemical plant equipment.
Corrosion resistant, hardenable.

TOKUSHU-RHM10.
M-1647; 0.08-0.18 C, 12.0-13.5 Cr, 0.3-0.6 Mo, 0.6 max Ni, bal Fe.
Annealed: 70,000 TS; 35,000 YS; 30 El; B 80 Rock.
Heat treated: 180,000 TS; 140,000 YS; 15 El; 375 Brin.
For chemical plant and oil refinery equipment.
Corrosion resistant.

TOKUSHU-RHM11.
 M-1647; 0.1-0.2 C, 11.5-13.0 Cr, 0.6-1.5 Mo, 0.5 max Ni, bal Fe.
 Annealed: 75,000 TS; 38,000 YS; 28 El; B 82 Rock.
 Heat treated: 135,000 TS; 105,000 YS; 10 El; C 29 Rock.
 For oil refinery and chemical plant equipment. Corrosion resistant.

TOKUSHU-RHM37.
 M-1647; 0.08-0.18 C, 11.5-14.0 Cr, 0.3-0.6 Mo, 0.6 max Ni, bal Fe.
 Annealed: 70,000 TS; 35,000 YS; 30 El; Rock B 80.
 For oil refinery equipment, chemical and rubber processing apparatus. Corrosion resistant.

TOKUSHU-RHMV4.
 M-1647; 0.20-0.25 C, 11.0-12.5 Cr, 0.9-1.2 Mo, 0.4-0.8 Ni, 0.25-0.35 bal Fe.
 Annealed: 95,000 TS; 40,000 YS; 25 El; B 92 Rock.
 Hardened: 240,000 TS; 205,000 YS; 9 El;C 50 Rock.
 For valves, cutlery, bearings, surgical instruments.
 Corrosion resistant, hardenable.

TOKUSHU-RHMW1.
 M-1647; 0.20-0.25 C, 11.0-12.5 Cr, 0.9-1.25 Mo, 0.5-0.9 Ni, 0.25 V, 1.1 W, 0.05 Al, 0.25 Co, 0.04 Sn, 0.05 Ti, bal Fe.
 Annealed: 90,000 TS; 40,000 YS; 25 El; B 94 Rock.
 Hardened: 250,000 TS; 210,000 YS; 8 El; C 52 Rock.
 For valves, cutlery, surgical instruments, bearings.
 Corrosion resistant, hardenable.

TOKUSHU-RHMW2.
 M-1647; 0.18-0.23 C, 12-14 Cr, 0.75 -1.25 Mo, 0.5-1.0 Ni, 0.2-0.5 V, 0.8-1.2 W, bal Fe.
 Annealed: 85,000 TS; 40,000 YS; 25 El; B 94 Rock.
 Heat treated: 240,000 TS; 205,000 YS; 9 El; C 50 Rock.
 For surgical instruments, valves, bearings, cutlery, shafts.
 Corrosion resistant, hardenable.

TOLEDO.
 M-1704; 0.90 C, 1.2 Mn, 0.5 Cr, 0.2 V, 0.5 W, bal Fe.
 Oil hardening cold work tool steel.
 AISI O1.

TOLEDO.
 M-352; 1.47 C, 0.48 Mn, 4.16 W, bal Fe.
 For tools, dies; oil hardening.

TOLEDO 3/4% NICKEL.
 M-985; 0.50-0.60 C, 0.5-0.8 Ni, 0.5-0.8 Mn, bal Fe.
 For gears, bolts, cranskshafts, machine tool parts; oil or water hardened, tough.

TOLEDO 1% CHROME.
 M-985; 0.35-0.45 C, 0.85-1.15 Cr, 0.60-0.95 Mn, bal Fe.
 Heat treated: 102,000-133,000 TS; 72,000-92,000 YS; 22-18 El; 200-302 Brin.
 For gears, bolts, crankshafts, machine tool parts; Brit. En18, oil hardened.

TOLEDO 1% CHROME-MOLYBDENUM.
 M-985; 0.35-0.45 C, 0.9-1.2 Cr, 0.20-0.35 Mo, 0.5-0.8 Mn, bal Fe.
 For gears, bolts, crankshafts; Brit. En19A-B-C, oil hardened, tough.

TOLEDO 1% NI-CR.
 M-985; 0.30-0.40 C, 1.0-1.5 Ni, 0.45-0.75 Cr, bal Fe.
 Heat treated: 100,000-135,000 TS; 72,000-103,000 YS; 22-17 El; 200-32 Brin.
 For gears, bolts, machine tool parts; Brit. En 111, oil hardened, shock resistant.

TOLEDO 1% NI-CR-MO.
 M-985; 0.35-0.45 C, 1.2-1.6 Ni, 0.9-1.4 Cr, 0.10-0.20 Mo, bal Fe.
 Heat treated: 112,000-157,000 TS; 81,000-123,000 YS; 20-15 El; 223-37 Brin.
 For gears, bolts, crankshafts, machine tool parts; Brit. En110, oil hardened, shock resistant.

TOLEDO 3% CHROME MOLYBDENUM.
 M-985; 0.15-0.35 C, 2.5-3.5 Cr, 0.3-0.7 Mo, bal Fe.
 Heat treated: 101,000-224,000 TS; 72,000-180,000 YS; 22-10 El; 200-5 Brin.
 For oil refinery equipment; Brit. En29, creep resistant.

TOLEDO 3% NI-CR-MO.
 M-985; C, 3 Ni, Cr, Mo, bal Fe.
 For gears, bolts, camshafts, cams; Brit. S107, case hardened.

TOLEDO 13.
 M-985; 0.15-0.25 C, 0.15-0.35 Mo, 0.4-0.7 Ni, 1.4-1.8 Mn, bal Fe.
 Heat treated: 90,000 TS; 22 El; 180-230 Brin.
 For gears, cams, camshafts; Brin. En13, case hardened.

TOLEDO 15.
 M-985; 0.15-0.25 C, 0.4-0.6 Mn, bal Fe.
 Annealed: 73,000 TS; 41,000 YS; 22 El; 58 RA; 140 Brin.
 For nails, rivets, gears, cams, bushings; Brit. EN2C, case hardened.

TOLEDO 20.
M-985; 0.20-0.25 C, 0.05-0.35 Si, 0.6-1.0 Mn, bal Fe.
Cold drawn: 60,000 TS; 25 El; 140 Brin.
For gears, bolts, fasteners, brackets; Brit. EN3C water hardened.

TOLEDO 25.
M-985; 0.25-0.30 C, 0.05-0.35 Si, 1.0 max Mn, bal Fe.
Normalized: 80,000 TS; 25 El; 179 Brin.
For gears, bolts, shafts, keys, brackets; Brit. EN4, water hardened.

TOLEDO 28.
M-985; 0.25-0.40 C, 3.0-4.5 Ni, 0.75-1.25 Cr, 0.20-0.65 Mo, bal Fe.
Heat treated: 135,000-180,000 TS; 103,000-144,000 YS; 35-25 El; 270-415 Brin.
For gears, bolts, machine tool parts; Brit. EN 28, oil hardened, shock resistant.

TOLEDO 30.
M-985; 0.25-0.35 C, 0.05-0.35 Si, 0.60-1.0 Mn, bal Fe.
Hot rolled: 80,000 TS; 50,000 YS; 30 El; 56 RA; 163 Brin.
For armature shafts, gears, bolts, axles, screws; Brit. EN5, water hardened.

TOLEDO 40.
M-985; 0.35-0.45 C, 0.05-0.35 Si, 0.60-1.0 Mn, bal Fe.
Hot rolled: 91,000 TS; 58,000 YS; 27 El; 50 RA; 200 Brin.
For gears, shafts, bolts, axles, fasteners; Brit. EN8, water hardened.

TOLEDO 55.
M-985; 0.50-0.65 C, 0.05-0.35 Si, 0.60-0.80 Mn, bal Fe.
Cold drawn: 110,000-140,000 TS; 12 El; 223-302 Brin.
For wheels, die blocks, rails, girders, springs; Brit. EN9, water hardened.

TOLEDO 60 CARBON CHROME.
M-985; 0.5-0.7 C, 0.5-0.8 Cr, 0.5-0.8 Mn, bal Fe.
Heat treated: 123,000-155,000 TS; 15-21 El.
For axes, hammers, punches, die blocks; Brit. EN11, water or oil hardened.

TOLEDO 160.
M-985; 0.35-0.45 C, 1.5-2.0 Ni, 0.20-0.35 Mo, bal Fe.
Heat treated: 102,000-134,000 TS; 72,000-104,000 YS; 22-17 El; 200-300 Brin.
For gears, bolts, crankshafts, machine tool parts; Brit. EN160, oil hardened, shock resistant.

TOLEDO 206.
M-985; 0.12-0.17 C, 0.3-0.5 Cr, 0.3-0.5 Mn, bal Fe.
For gears, fasteners, machine tool parts; Brit. EN206, case hardened.

TOLEDO 207.
M-985; 0.16-0.21 C, 0.6-0.8 Cr, 0.6-0.8 Mn, bal Fe.
For gears, fasteners, machine tool parts; Brit. EN207, case hardened.

TOLEDO 325.
M-985; 0.22 C, 1.5-2.0 Ni, 0.4-0.6 Cr, 0.2-0.3 Mo, bal Fe.
Heat treated: 123,000 TS; 15 El.
For gears, bolts, cams, machine tool parts; Brit EN325, case hardened.

TOLEDO 351.
M-985; 0.20 max C, 0.6-1.0 Ni, 0.4-0.8 Cr, 0.1 max Mo, bal Fe.
Heat treated: 101,000 TS; 18 El.
For gears, bolts, cams, machine tool parts; Brit. EN351, case hardened, shock resistant.

TOLEDO 352.
M-985; 0.20 max C, 0.85-1.25 Ni, 0.6-1.0 Cr, 0.10 Max Mo, bal Fe.
Heat treated: 123,000 TS; 15 El.
For gears, bolts, cams, machine tool parts; Brit. EN352, case hardened, shock resistant.

TOLEDO 353.
M-985; 0.20 max C, 1.0-1.5 Ni, 0.75-1.25 Cr, 0.08-0.15 Mo, bal Fe.
Heat treated: 146,000 TS; 12 El.
For gears, bolts, cams, machine tool parts; Brit. EN353, case hardened, shock resistant.

TOLEDO 354.
M-985; 0.20 max C, 1.5-2.0 Ni, 0.75-1.25 Cr, 0.10-0.20 Mo, bal Fe.
Heat treated: 168,000 TS; 12 El.
For gears, bolts, cams, machine tool parts; Brit. EN354, case hardened, shock resistant.

TOLEDO 355.
M-985; 0.20 max C, 1.8-2.2 Ni, 1.4-1.7 Cr, 0.15-0.25 Mo, bal Fe.
Heat treated: 190,000 TS; 12 El.
For gears, bolts, machine tool parts; Brit. EN355, case hardened, shock resistant.

TOLEDO 361.
M-985; 0.13-0.17 C, 0.4-0.7 Ni, 0.55-0.85 Cr, 0.08-0.15 Mo, bal Fe.
Heat treated: 100,000 TS; 18 El.
For gears, bolts, machine tool parts; Brit. En361, case hardened, shock resistant.

TOLEDO 362.
M-985; 0.18-0.23 C, 0.4-0.7 Ni, 0.55-0.80 Cr, 0.08-0.15 Mo, bal Fe.
Heat treated: 123,000 TS; 15 El.
Heat gears, bolts, camshafts, cams; Brit. EN362, case hardened, shock resistant.

TOLEDO 363.
M-985; 0.22-0.26 C, 0.4-0.7 Ni, 0.55-0.80 Cr, 0.08-0.15 Mo, bal Fe.
Heat treated: 146,000 TS.
For gears, bolts, crankshafts, axles; Brit. EN363, case hardened, shock resistant.

TOLEDO ALLOY 131.
M-634; 0.22-0.28 C, 1.1 Cu, 0.06 V, 0.5 Si, 0.8 Mn, bal Fe.
Cast: 90,000-100,000 TS.
For machinery parts, gears, bolts; water hardened.

TOLEDO ALLOY 135.
M-634; 0.4-0.5 C, 0.8 Mn, 1 Cr, 1 Cu, 0.3 Mo, bal Fe.
Cast: 120,000 TS; 217 Brin.
For guides, bolts, gears, housings; water or oil hardened.

TOLEDO B.C.H.
M-985; 0.18 max C, 3.0-3.7 Ni, 0.6-1.1 Cr, 0.25 max Mo, bal Fe.
Heat treated: 123,000-156,000 TS; 13 El.
For gears, bolts, camshafts, cams; Brit. EN36, case hardened, shock resistant.

TOLEDO B.C.M.F.
M-985; 0.27-0.44 C, 2.3-2.8 Ni, 0.5-0.8 Cr, 0.4-0.7 Mo, bal Fe.
Heat treated: 123,000-224,000 TS; 92,000-180,000 YS; 18-10 El; 248-500 Brin.
For gears, bolts, shafts, axles, machine tool parts; Brit. EN25 and 26, oil hardened, shock resistant.

TOLEDO B.C.M.O.
M-985; 0.25-0.35 C, 3.0-3.7 Ni, 0.50-1.3 Cr, 0.2-0.65 Mo, bal Fe.
Heat treated: 123,000-157,000 TS; 92,000-123,000 YS; 18-15 El; 248-375 Brin.
For gears, bolts, machine tool parts, axles; Brit. EN27, oil hardened, shock resistant.

TOLEDO B.C.N.
M-985; 0.25-0.35 C, 0.5-1.0 Cr, 2.75-3.5 Ni, 0.65 max Mo, bal Fe.
Heat treated: 112,000-146,000 TS; 81,000-112,000 YS; 20-16 El; 223-340 Brin.
For gears, bolts, crankshafts, machine tool parts; Brit. EN23, oil hardened, shock resistant.

TOLEDO B.R.
M-985; 0.90-1.2 C, 0.30-0.75 Mn, 1.0-1.6 Cr, bal Fe.
For bearings, races, gauges, cutters; Brit. EN31, wear resistant, water hardened.

TOLEDO B.S.N.1.
M-985; 0.30-0.45 C, 0.6-1.0 Ni, 1.5 Mn, 0.10-0.35 Si, bal Fe.
Normalized: 90,000 TS; 22 El; 179-229 Brin.
For gears, bolts, crankshafts, machine tool parts; Brit. EN12, oil hardened, shock resistant.

TOLEDO B.S.N.3.
M-985; 0.25-0.35 C, 2.75-3.25 Ni, 0.3 max Cr, bal Fe.
Rolled: 100,000 TS; 70,000 YS; 22 El; 200-235 Brin.
For gears, bolts, crankshafts, axles; Brit. EN21, oil hardened, shock resistant.

TOLEDO B.S.N.35.
M-985; 0.35-0.45 C, 3.25-3.75 Ni, 0.5-0.8 Mn, 0.3 max Cr, bal Fe.
Rolled: 110,000-125,000 TS; 75,000-88,000 YS; 20-18 El; 220-302 Brin.
For gears, bolts, crankshafts, axles; Brit. EN22, oil hardened, shock resistant.

TOLEDO CARBON FILE.
M-985; 0.8 C, bal Fe.
For files; water hardened.

TOLEDO CARBON-MANGANESE.
M-985; 0.15-0.25 C, 1.3-1.7 Mn, 0.25 Cr, 0.40 max Ni, bal Fe.
Normalized: 80,000 TS; 20 El; 152-207 Brin.
For gears, cams, camshafts, fasteners; Brit. EN14A, water hardened.

TOLEDO CARBON SPRING.
M-985; 0.45-0.90 C, 0.10-0.40 Si, 0.55-0.90 Mn, bal Fe.
For laminated and coil springs; Brit. EN42, and 43, water hardened.

TOLEDO CHROME FILE.
M-985; 0.8 C, Cr, bal Fe.
For files; water hardened.

TOLEDO CHROME SPRING.
M-985; 0.45-0.60 C, 1.0-1.6 Si, 0.5-0.9 Mn, 0.55-1.4 Cr, bal Fe.
For springs; Brit. EN48, oil or water hardened.

TOLEDO CHROME VANADIUM VALVE SPRING.
M-985; 0.40-0.50 C, 1.0-1.5 Cr, 0.15 min V, 0.5-0.7 Mn, bal Fe.
For valve springs; Brit. EN50, oil hardened, tough.

TOLEDO EXTRA.
M-352; C, bal Fe.
For tools, fixtures, jigs; water hardened.

TOLEDO G 105.
M-985; 0.35-0.45 C, 1.3-1.6 Ni, 0.90-1.4 Cr, 0.20-0.35 Mo, bal Fe.
Heat treated: 112,000-224,000 TS; 81,000-179,000 YS; 20-8 El; 223-500 Brin.
For gears, bolts, crankshafts, machine tool parts; Brit. EN24, oil hardened, shock resistant.

TOLEDO G 110.
M-985; 0.26-0.34 C, 3.9-4.3 Ni, 1.1-1.4 Cr, 0.40 max Mo, bal Fe.
Heat treated: 224,000 TS; 180,000 YS; 10 El; 440-500 Brin.
For gears, bolts, crankshafts; Brit. EN30A, oil hardened, shock resistant.

TOLEDO HARD DRAWING CARBON SPRING.
M-985; 0.45-0.85 C, 1.0 max Mn, 0.30 max Si, bal Fe.
For springs; Brit. EN49, water hardened.

TOLEDO HIGH CARBON SPRING.
M-985; 0.90-1.2 C, 0.30 max Si, 0.45-0.70 Mn, bal Fe.
For laminated and coil springs; Brit. EN44, water hardened.

TOLEDO HIGH SPEED.
M-352; 0.7 C, 18 W, 4 Cr, 1 V, bal Fe.
For cutting tools, broaches; high speed steel.

TOLEDO H-O SILICOMANGANESE.
M-985; 0.50-0.60 C, 1.5-2.0 Si, 0.70-1.0 Mn, bal Fe.
For laminated springs, torsion bars; Brit. EN45 and 46, water or oil hardened, tough.

TOLEDO LANTERN.
M-985; 0.15 max C, 0.05-0.35 Si, 0.4-0.7 Mn, bal Fe.
Rolled: 70,000 TS; 20 El; 140 Brin.
For gears, bolts, rivets, nails, fan blades; Brit. EN32A, case hardened.

TOLEDO LOW ALLOY STEEL.
M-985; 0.25-0.45 C, 1.2-1.5 Mn, 0.50-1.0 Ni, 0.3-0.6 Cr, 0.15-0.25 Mo, bal Fe.
For gears, bolts, machine tool parts; Brit. EN100, oil hardened, shock resistant.

TOLEDO MANGANESE MOLYBDENUM.
M-985; 0.25-0.40 C, 1.3-1.8 Mn, 0.20-0.55 Mo, bal Fe.
Heat treated: 101,000-146,000 TS; 72,000-112,000 YS; 22-16 El; 200-34 Brin.
For gears, bolts, crankshafts, machine tool parts; Brit. EN 16 and 17, oil hardened, shock resistant.

TOLEDO-MOLYBDENUM BOLT STEEL.
M-985; 0.20-0.45 C, 0.5-1.5 Cr, 0.5-0.9 Mo, 0.4-0.7 Mn, bal Fe.
Heat treated: 123,000-146,000 TS; 92,000-112,000 YS; 18-16 El; 248-36 Brin.
For gears, bolts, machine tool parts; Brit. EN20A-B, oil hardened, tough.

TOLEDO NO 3.
M-634; 0.15 C, 1.5 Ni, 0.2 Mo, bal Fe.
Hardened: 110,000-115,000 TS; 70,000-80,000 YS; 26-24 El; 50-40 RA.
For mining equipment, carburized parts; case-hardening steel.

TOLEDO NO 4.
M-634; 0.7 C, 2 Si, 0.2 Mo, bal Fe.
90,000-115,000 TS; 60,000-75,000 YS; 207-228 Brin.
For coal mining equipment, pug mill parts, paddles, liners; abrasion resistant.

TOLEDO NO 6.
M-634; 0.9 C, 1.2 Mn, bal Fe.
For dies, trimming, stamping, forming and embossing; non-deforming, cast-to-shape.

TOLEDO NO 7.
M-634; 0.5 C, 1.2 Cr, 0.2 V, bal Fe.
90,000-95,000 TS; 60,000-67,000 YS; 25-22 El; 50-40 RA; 179-200 Brin.
For railway locomotive parts; tough, shock resistant.

TOLEDO NO 8.
M-634; 0.4 C, 1.5 Ni, 0.8 Cr, 0.2 Mo, bal Fe.
110,000-120,000 TS; 75,000-85,000 YS; 26-20 El; 50-40 RA; 228-241 Brin.
For automotive and aircraft parts, gears, short-run dies; oil hardening.

TOLEDO NON-SHRINKING.
M-352; 0.9 C, 1.2 Mn, 0.2 V, bal Fe.
For tools and dies; non-deforming.

TOLEDO N.T.R.1.
M-985; 0.10-0.30 C, 0.4 max Ni, 2.9-3.5 Cr, 0.4-0.7 Mo, bal Fe.
Heat treated: 100,000-135,000 TS; 72,000-103,000 YS; 22-17 El; 200- 277 Brin.
For gears, bolts, shafts, machine tool parts; Brit. EN40A, nitriding steel.

TOLEDO N.T.R.2.
M-985; 0.30-0.50 C, 0.4 max Ni, 2.5-3.5 Cr, 0.7-1.2 Mo, bal Fe.
Heat treated: 190,000 TS; 153,000 YS; 10 El; 375-444 Brin.
For gears, bolts, shafts, machine tool parts; Brit. EN40C nitriding steel.

TOLEDO N.T.R.3.
M-985; 0.25-0.45 C, 0.4 max Ni, 1.4-1.8 Cr, 0.10-0.25 Mo, bal Fe.
Heat treated: 100,000-123,000 TS; 72,000-92,000 YS; 20-17 El; 200-302 Brin.
For gears, bolts, shafts, machine tool parts; Brit. EN41, nitriding steel.

TOLEDO R.S.2.
M-985; 0.14-0.28 C, 1.5-2.0 Ni, 0.2-0.3 Mo, 0.3-0.6 Mn, bal Fe.
Heat treated: 123,000 TS; 15 El.
For machine tool parts, gears, bolts, cams; Brit. EN34 and 35, case hardened.

TOLEDO R.S.3.
M-985; 0.10-0.15 C, 2.75-3.5 Ni, 0.3 max C, bal Fe.
Rolled: 100,000 TS; 18 El.
For gears, bolts, camshafts, cams; Brit. EN33, case hardened.

TOLEDO R.S.5.
M-985; 0.16 max C, 4.5-5.5 Ni, 0.3 max Cr, bal Fe.
Rolled: 90,000 TS; 20 El.
For gears, cams, camshafts; Brit. EN37, case hardened.

TOLEDO S.82.
M-985; 0.12-0.18 C, 3.8-4.5 Ni, 1.0-1.4 Cr, 0.35 max Mo, bal Fe.
Heat treated: 190,000 TS; 12 El.
For gears, bolts, camshafts, cams; Brit. EN39, case hardened, shock resistant.

TOLEDO S.90.
M-985; 0.16 max C, 4.5-5.5 Ni, 0.15-0.35 Mo, 0.3 max Cr, bal Fe.
Heat treated: 146,000 TS; 13 El.
For gears, cams, camshafts; Brit. EN38, case hardened.

TOLEDO SCIMITAR.
M-352; C, W, Cr, bal Fe.
For tools, dies; oil hardened.

TOLEDO SILICO CHROME SPRING.
M-985; C, Si, Cr, bal Fe.
For spiral and coil springs; Brit. STA.2D, oil hardened, tough.

TOLEDO SPECIAL.
M-352; 0.7-1.2 C, bal Fe.
For general tools; water hardened.

TOLEDO SUPERIOR.
M-352; 0.7-1.2 C, bal Fe.
For general tools; water hardened.

TOLEDO T.P.C.
M-985; 0.45-0.55 C, 0.80-1.2 Cr, 0.15 min V, bal Fe.
For springs; Brit. EN47, water hardened.

TOLEDO X.B.P.
M-487; 0.65-1.25 C, bal Fe.
For drills, taps, reamers, hobs; water hardened.

TOMBAC-A.
M-France; 20 Zn, 80 Cu.
38,000 TS; 30 El.
For jewelry, bearings; corrosion resistant.

TOMBAC, ARCET.
82.3 Cu, 17.7 Zn.
For water pipes; red brass.

TOMBAC-B.
M-France; 17-20 Zn, 0.3 Sn, 80 Cu.
For jewelry, bearings; corrosion resistant.

TOMBAC, CAST.
87 Cu, 13 Zn.
For hardware; red brass.

TOMBAC, GALDEN.
87 Cu, 17.5 Zn. 0.5 Sn.
For pipes, hardware; red brass.

TOMBAC LUDENSCHEIDT.
82.3 Cu, 17.5 Zn.
For pipes; red brass.

TOMBAC, OKER.
85.3 Cu, 14.7 Zn.
For pipes; red brass.

TOMBAC, RED BRASS.
6 Zn, 85 Cu, 9 Sn.
Hard: 75,000 TS; 18,000 YS; 4 El; 135 Brin.
Soft: 42,000 TS; 7,000 YS; 43 El; 52 Brin.
For hardware, ornaments, tubing, marine parts; corrosion resistant.

TOMBAC, RED VIENNA.
M-Australia; 2 Zn, 98 Cu.
For jewelry, bearings; corrosion resistant.

TOMBAC, SHEET, PARIS.
92-84 Cu, 8-16 Zn.
For hardware, tubes; red brass.

TOMBASIL.
M-346; 4 Si, 13 Zn, bal Cu.
Sand cast: 65,000 TS; 30,000 YS; 20 El; 130 Brin.
Die cast: 90,000 TS; 50,000 YS; 25 El; 150 Brin.
Corrosion resistant bronze; for valve stems, brush holders, pump impellers structural castings.

TOMBASIL.
M-122; 81 Cu, 4 Si, bal Zn.
Rolled: 98,000 TS; 45,000 YS; 20 El; 22 RA; 185 Brin.
For gears, shafts, propellers; corrosion and wear resistant.

TONCAN ENAMELING IRON.
M-97; 0.03 max C, 0.04 max Mn, bal Fe.
Rolled: 45,000-50,000 TS; 30,000-33,000 YS; 32-30 El; 56-50 RA; 90 Brin.
For roofing and siding, pipes, locomotive boiler tubes; porcelain enameling.

TOOL.
M-364; 0.9 C, 0.2 V, bal Fe.
For drills, chisels, crow-bars; water or oil hardened.

TOOL-AGE see **MCKAY TOOL-AGE.**

TOOL-ALLOY see **MCKAY TOOL-ALLOY.**

TOOL-ARC AIR HARDENING.
M-1713; 1.0 C, 5 Cr, 1 Mo, 0.2 V, bal Fe.
For tool steel welding electrodes; for air hardening die steels.

TOOL-ARC HIGH SPEED.
M-1713; 0.7 C, 1.5 W, 9.5 Mo, 1 V, bal Fe.
For tool steel welding electrodes; for cutting tools.

TOOL-ARC HOT-WORK.
M-1713; 0.3 C, 5 Cr, 1.3 Mo, 1 W, bal Fe.
For tool steel welding electrodes; for hot die steels.

TOOL-ARC OIL HARDENING.
M-1713; 0.9 C, 1.2 Mn, 0.5 Cr, 0.5 W, bal Fe.
For tool steel welding electrodes; for non-deforming steel tools and dies.

TOOL-ARC WATER HARDENING.
M-1713; 0.9 C, 2 V, bal Fe.
For tool steel welding electrodes; for water hardening steel parts.

TOOLCRAFT.
M-175; 1.0 C, 1 Cr, bal Fe.
Welded: 600 Brin.
For hard facing electrodes, cutting tools; wear resistant.

TOOL-FORGE see **MCKAY TOOL-FORGE.**

TOOL-ROD NO 650.
M-1096; 0.5-0.7 C, 3.4 W, 6-8 Cr, 0.8 V, 0.4 Mn, bal Fe.
Welded; 500 Brin.
For hard facing electrode; for tool steels, for wear resistance.

TOOLTECTIC 6 HSS.
M-717.
For high speed steel overlays on tools by AC-DC Rev arc.
Hardness: C-62 Rock.

TOOLTECTIC 6HW.
M-717.
For hard overlays on hot working punches.
Hardness: C 52-55 Rock.

TOOLTECTIC 6SH.
M-717.
For hard overlays on steels.
Hardness: C 53-61 Rock.

TOOLTECTIC 6WH.
M-717.
For hard overlays on water hardening tools.
Hardness: C 56-59 Rock.

TOOLTECTIC 60H.
M-717.
For wear resistant overlays on steel machine parts
Hardness: C 59-62 Rock.

TOOLWELD A+O.
M-578; C, 5 Cr, bal Fe.
Welded: 650 Brin.
For hard surfacing electrodes; wear resistant, shielded arc.

TOPAL H.
M-2; 80 Cu, 20 Mn + Fe + Al.
Chill cast: 90,000 TS; 5 El; 170 Brin.
Hot pressed: 114,000 TS; 5 El; 200 Brin.
For shafts, bearings, dies, pump rods, steering nuts, gears; non-corrosive.

TOPAL "W".
M-2; 0.5 Mn, 1 Fe, 9 Al, bal Cu.
Cast: 50,000-60,000 TS; 30-15 El; 130-150 Brin.
Hot pressed: 60,000-80,000 TS; 30-15 El; 150-170 Brin.
For shafts, bearings, dies, gears; resists steam at 550°F.

TOPHAL X.
M-897; 35 Cr, 3 Al, 62 Fe.
Annealed: 85,000 TS; 57,000 YS; 20 El.
For heating elements; for induction furnaces.

TOPHAL Y.
M-897; 12 Cr, 5 Al, bal Fe.
Annealed: 80,000 TS; 65,000 YS; 15 El.
For heating elements; heat resistant.

TOPHAL Z.
M-897; 22 Cr, 5 Al, bal Fe.
Annealed: 85,000 TS; 65,000 YS; 20 El.
For heating elements; for electric furnaces.

TOPHEL.
M-61; 10 Cr, bal Ni.
Wire: 88,000 TS; 30,000 YS; 35 El.
For positive thermoelement for the standard Type K thermocouple.
Oxidation resistant and stabile.

TOPHET 30.
M-61; 70.0 Ni, 30.0 Cr.
For heating elements and furnace belts; max operating temperature 2300°F.
Resists green rot.

TOPHET "A".
M-61; 77-79 Ni, 19-20.5 Cr, 0.15 max C, 2 Mn.
Hard: 160,000 TS; 90,000 YS; 3 El; 45 RA.
Soft: 120,000 TS; 70,000 YS; 30 El; 45 RA; 190 Brin.
For resistance wire, heating elements, percolators, toasters, rheostats; tough to machine; max operating temperature 2100°F.

TOPHET A.
M-897; 80 Ni, 20 Cr.
Annealed; 105,000 TS; 70,000 YS; 40 El.
For heating elements; for high temperatures.

TOPHET A-CB.
M-61; 19.0 Cr, 1.3 Cb, 1.5 Si, bal Ni.
For high temperature furnace belts.

TOPHET "C".
M-61; 57-60 Ni, 14-18 Cr, 0.15 max C, bal Fe.
Hard: 100,000-140,000 TS; 85,000 YS; 100-3 El; 16 RA.
Soft: 100,000 TS; 60,000 YS; 25 El; 45 RA; 165 Brin.
For resistance wire, heating elements, percolators, toasters, rheostats; resists some common acids, max operating temperature 1850°F.

TOPHET C.
M-897; 60 Ni, 16 Cr, bal Fe.
Annealed: 105,000 TS; 70,000 YS; 30 El.
For heating elements; for domestic use.

TOPHET D.
M-61; 34 Ni, 20 Cr, 0.15 C, bal Fe.
Annealed: 75,000 TS; 20 El.
For moderate temperature service; used up to 1600°F.

TOPHET D-CB.
M-61; 35 Ni, 20 Cr, 1.5 Si, 1.25 Cb, bal Fe.
For high temperature furnace belts.

TOPHET E.
M-897; 75.5 Ni, 20 Cr, 2.5 Ti, 0.5 Si, 0.5 Mn, 0.5 Al.
Hardened: 280,000 TS; 210,000 YS; 2.5 El.
For springs, gas turbine accessories; age-hardenable.

TOPHET-F.
M-61; 77 Ni, 20 Cr, 3 Al.
Wire: 100,000-200,000 TS; 60,000-175,000 YS; 2-30 El.
For furnace belts in strongly reducing atmospheres at 1800-2100°F.
High heat and corrosion resistant.

TOPHET-F.
M-897; 36-38 Ni, 17-19 Cr, 2 Si, bal Fe.
For heating elements; useful operating temperature to 1000°C.

TOPHET-H.
M-897; 75-77 Ni, 19-21 Cr, 3.5 Al.
For heating elements; useful operating temperature to 1250°C.

TOPHET I-304.
M-897; 0.08 max C, 18-20 Cr, 8-10 Ni, bal Fe.
Annealed: 85,000 TS; 30,000 YS; 60 El; 70 RA; 160-140 Brin.
For tubing for chemical and food industries; stainless, austenitic.

TOPHET I-316.
M-897; 0.10 max C, 16-18 Cr, 10-14 Ni, 2.5 Mo, bal Fe.
Annealed: 80,000 TS; 30,000 YS; 50 El; 70 RA; 140-160 Brin.
For tubing for chemical and food industries; stainless, austenitic.

TOPHET M.
M-897; 8 Ni, 18 Cr, bal Fe.
290,000 TS; 1 El.
For magnetic recording wire; corrosion resist.

TOP NOTCH (AISI S1).
M-69; 0.50 C, 2.5 W, 1.25 Cr, 0.25 V, bal Fe.
For tools, chisels, track tools; shock resisting.

TORDAL.
M-297; 4 Cu, 0.5 Pb, 0.5 Bi, bal Al.
T3-temper: 55,000 TS; 43,000 YS; 15 El; 95 Brin.
T6-temper: 57,000 TS; 39,000 YS; 17 El; 97 Brin.
For screw machine products, machinery parts, fasteners; free-cutting.

TOREADOR.
M-489; C, alloy, bal Fe.
For high speed tools; high speed steel.

TOREADOR SUPRA.
M-489; C, alloy, bal Fe.
For high speed tools; high speed steel.

TORPEDO.
M-688; 0.9 C, 0.5 Cr, 1.3 Mn, 0.5 W, bal Fe.
For dies; punches, rivet sets; non-deforming, oil hardened.

TORPEDO BRONZE.
M-England; 0.5-1.5 Sn, 62-59 Cu, bal Zn.
For torpedo parts; corrosion resistant.

TOTO HS-A1.
M-1544; 0.15-0.18 C, 1.1-1.4 Mn, 0.35-0.60 Si, 0.20 min Cu, bal Fe.
Rolled: 70,000-81,000 TS; 46,000 min YP; 20% min El.
For agricultural equipment, structural members, buildings, bridges.
Structural steel.

TOTO HS-B.
M-1544; 0.19-0.26 C, 1.1-1.4 Mn, 0.40 max Si, 0.20 min Cu, bal Fe.
Rolled: 85,000-100,000 TS; 54,000 min YP; 17% min El.
For structural members, buildings, bridges, agricultural equipment.
Structural steel.

TOUCAS.
36 Cu, 29 Ni, 7.1 Fe, 7.1 Pb, 7.1 Sn, 7.1 Sb, 7.1 Zn.
For ornamental white metal parts; corrosion resistant.

TOUGH.
M-388; 0.6 C, bal Fe.
For tools, dies; water or oil hardened.

TOUGH DEVIL NO. 1.
M-712; 13 Mn, 3 Ni, C, bal Fe.
Cast; 135,000 TS; 45,000 YS; 80 El; 50 RA; 180-200 Brin.
For welding rods for high Mn steels; wear resistant, austenitic.

TOUGH DEVIL NO. 2.
M-712; 0.8 C, 14 Mn, 0.2 Mo, bal Fe, 520 Brin.
For hard surfacing electrodes; wear and abrasion resistant.

TOUGHDIE.
M-289; 0.45-0.55 C, 0.6-0.7 Mn, 0.25 Si, 3.0-3.25 Cr, 1.3-1.4 Mo, bal Fe.
AISI Type S7 shock resisting tool steel.

TOUGH HARD.
M-388; 0.9 C, 1.0 Mn, bal Fe.
For tools, dies; water or oil hardened.

TOUGH-ITE.
M-1036; 0.7 C, 18 W, 4 Cr, 1 V, bal Fe.
For cutting tool bits; high speed steel.

TOUGHITE NO. 5.
M-922; 0.7 C, 0.9 Cr, 0.2 V, bal Fe.
For tires and rollers in kilns, coolers and dryers; oil hardened.

TOUGH M.
M-24; 0.45 C, 0.55 Mn, 0.20 Si, 0.95 Cr, 0.20 V, bal Fe.
Low alloy, oil hardening tool steel, AISI L2.

TOUGH ONE.
M-1113; 1.00-1.15 C, bal Fe.
For tools, drills, taps, chasers; water hardened.

TOUGH TWO.
M-1113; 0.9-1.0 C, bal Fe.
For tools, drills, chasers; water hardened.

TOUGH THREE.
M-1113; 0.7-0.8 C, bal Fe.
For tools, springs, punches; water hardened.

TOURNAY METAL.
M-France; 82.5 Cu, 17.5 Zn.
For cheap jewelry; brass.

TOURUN LEONARD'S METAL.
90 Sn, 10 Cu.
For bearings; bronze, tough.

TPM.
M-U.S.; 0.05 C, 2.3 Mn, 0.05 Si, 16.0 Cr, 0.50 Co, 3.1 Ti, 0.05 Al, bal Ni.
For exhaust valves; SAE HEV 2.

T QUALITY.
M-1405; 0.7-1.1 C, 0.35 Mn, 0.35 Si, bal Fe.
Water hardening tool steel.
For masonry and wood working tools, files, blacksmith tools.
AISI W1.

TRABUK.
M-England; 5.5 Ni, 5 Sb, 2 Bi, bal Sn.
For food utensils; resists vegetable acids.

TRACKWEAR see **AIRCO TRACKWEAR.**

TRAFOPERM N3.
M-1631; 3% Si-Fe for pole pieces, relay components and measuring systems.

TRAKALOY.
M-1128; Ti, Mn, Ni, Cr.
300 Brin.
For hard facing electrodes; for rails and wheels.

TRANCE PHOSPHOR BRONZE.
M-103; 90 Cu, 5 Sn, 5 Pb, trace P.
For hardware; free cutting.

TRANCOR 3 W.
M-10; Si, bal Fe.
For transformer cores; oriented grains.

TRANCOR 4 W.
M-10; Si, bal Fe.
For transformer cores; oriented grains.

TRAN-COR A-5.
M-10; Si, bal Fe.
For audio transformer cores, low induction elect. equipment; for radio use.

TRAN-COR A-6.
M-10; Si, bal Fe.
For audio transformers, elect. equipment; low induction.

TRAN-COR M-14.
M-10; 4.5 Si, bal Fe.
For transformers; core loss 52; high permeability.

TRAN-COR M-15.
M-10; 4.0-4.4 Si, bal Fe.
For distribution transformers, rotating equipment; core loss 58; high permeability.

TRAN-COR M-17.
M-10; 4.0-4.4 Si, bal Fe.
For transformers, motors, magnetic cores; core loss 65; high permeability.

TRAN-COR M-19.
M-10; 3.0-3.4 Si, bal Fe.
For generators, transformers, magnetic cores; core loss 72; high permeability.

TRAN-COR M-22.
M-10; 3.2 Si, bal Fe.
For transformers, motors, magnetic cores; high permeability; core loss 82.

TRAN-COR M-27.
M-10; 2.8 Si, bal Fe.
For transformers, motors, magnetic cores; high permeability; core loss 101.

TRAN-COR M-36.
M-10; 1.0 Si, bal Fe.
For magnetic cores, rotating machines, elect. equipment; high permeability; core loss 117.

TRAN-COR M-43.
M-10; 0.5 Si, bal Fe.
For armatures, rotating machines, motors; high permeability; core loss 130.

TRAN-COR T.
M-10; Si, bal Fe.
For thin sheets for electric generators; high permeability.

TRAN-COR T-O.
M-10; Si, bal Fe.
For electric generators, armatures; high permeability.

TRAN-COR T-O-S.
M-10; Si, bal Fe.
For wound type transformers and reactors; high inductance, low core loss.

TRANSMISSION GLYCO.
M-240; Sn, Sb, bal Pb.
For bearings; Babbitt.

TRANSPARENT.
M-1318; 0.40 C, 13 Cr, 0.3 Mn, bal Fe.
Annealed: 95,000 TS; 50,000 YS; 25 El; 55 RA; 196 Brin.
For cutlery, valves, surgical and dental instruments; corrosion resistant; Type 420.

TRANSPARENT 3003.
M-1318; 0.20 C, 13 Cr, 0.3 Mn, 0.4 Si, bal Fe.
Cold drawn: 105,000 TS; 85,000 YS; 17 El; 50 RA; 215 Brin.
For cutlery, turbine blades, knives; Type 420; corrosion resistant.

TRANTINYL.
M-977; 0.7 C, 1 Cr, bal Fe.
For guide shoes, tools; abrasion resisting.

TREM BRONZE.
M-1194; 2 Sn, bal Cu.
Annealed: 90,000 TS; 35,000 YS; 50 El; 65 RA; 160 Brin.
For springs, clips, hardware, fasteners; tough, corrosion resistant.

TRENITE.
M-535; 0.1 C, 18 Cr, 8 Ni, bal Fe.
Annealed: 85,000 TS; 35,000 YS; 25 El; 45 RA; 150 Brin.
For chemical plant equipment, evaporators, tanks, Type 302; austenitic, stainless.

TRENITE-H.
M-535; 0.2 C, 18 Cr, 8 Ni, bal Fe.
45,000 TS; 0.4 El; 250 Brin.
For heat, wear, corrosion resistant parts; heat, wear, corrosion resistant.

TRENITE P.
M-535; 0.25 C, 20 Cr, 10 Ni.
Cast: 45,000 TS; 300 Brin.
For lead pots, heat treating equipment; resists heat to 1650°F.

TRENITE WX.
M-535; 0.3 C, 14 Cr, bal Fe.
500 Brin.
For coal grinders, hot coke car sides and mill liners; abrasion resistant.

T.R.I.
M-1398; 21 Sn, 1.8 Cu, 0.15 Si, 0.13 Fe, bal Al.
Cast: 15,500 TS; 4.8 El; 43 Brin.
For railroad bearings.

TRIALLOY.
M-909; Ni alloy.
For sanitary fittings and valves; corrosion resistant.

TRIANGLE 10.
M-1677; 65 Cu, 25 Zn, 10 Ni.
Hard: 115,000 TS; 5 El.
Soft: 55,000 TS; 45 El; 20,000 YS.
For hardware, costume jewelry, optical parts, holloware.
Nickel silver, elect. cond. 9%, corrosion resistant.

TRIANGLE 12.
M-1677; 65 Cu, 23 Zn, 12 Ni.
Hard: 115,000 TS; 3 El.
Soft: 58,000 TS; 45 El; 20,000 YS.
For slide fasteners, hardware, optical parts.
Elect. cond. 8%, nickel silver, corrosion resistant.

TRIANGLE 15.
M-1677; 65 Cu, 20 Zn, 15 Ni.
Hard: 103,000 TS; 5 El.
Soft: 58,000 TS; 45 El; 20,000 YS.
For camera goods, optical equipment, jewelry, etching stock.
Nickel silver, elect. cond. 7%, corrosion resistant.

TRIANGLE 18.
M-1677; 65 Cu, 17 Zn, 18 Ni.
Hard: 103,000 TS; 3 El; 90,000 YS.
Soft: 58,000 TS; 45 El; 25,000 YS.
For rivets, screws, zippers, hollow ware, truss wire, costume jewelry.
Elect. cond. 6%; nickel silver, corrosion resistant.

TRIANGLE 30.
M-1677; 98 Cu, 2 Ni.
Hard: 67,000 TS; 4 El.
Soft: 37,000 TS; 46 El.
For resistances.
Elect. cond. 34%.

TRIANGLE 60.
M-1677; 94 Cu, 6 Ni.
Hard: 73,000 TS; 4 El.
Soft: 37,000 TS; 46 El.
For resistances.
Elect. cond. 17.

TRIANGLE 70/30.
M-1677; 70 Cu, 30 Zn.
Hard: 110,000 TS; 7 El.
Soft: 50,000 TS; 50 El; 18,000 YS.
For fasteners, therminal plugs, plumbing goods, cartridge cases.
Elect. cond. 28%, high strength and ductility.

TRIANGLE 80.
M-1677; 1 Cd, 99 Cu.
Hard: 70,000 psi TS; 15 El. Soft: 42,000 psi TS; 45 El.
Good electrical conductivity.
CDA C16200.

TRIANGLE 85.
M-1677; 0.8 Cd, 99.2 Cu.
Hard: 69,000 psi TS; 15 El. Soft: 42,000 psi TS; 45 El.
Good electrical conductivity.
CDA C16200.

TRIANGLE 90.
M-1677; 89 Cu, 11 Ni.
Hard: 83,000 TS; 5 El.
Soft: 47,000 TS; 46 El; 16,000 YS; B 15 Rock.
For condensers, condenser plates, distiller tubes, ferrules, piping.
Elect. cond. 9%, corrosion resistant.

TRIANGLE 91/09.
M-1677; 91.5 Cu, 8 Zn, 0.5 Sn.
Hard: 77,000 TS; 6 El.
Soft: 44,000 TS; 40 El.
For metal hose braids.
Elect. cond. 39.

TRIANGLE 91X.
M-1677; 9.5 Zn, 0.5 Sn, bal Cu.
CDA C41100.

TRIANGLE 125.
M-1677; 90 Cu, 9 Zn, 1 Sn.
CDA C41300.

TRIANGLE 170.
M-1677; 70 Cu, 30 Zn, 0.02-0.05 P.
Hard: 110,000 TS; 7 El.
Soft: 50,000 TS; 50 El.
For condensers, heat exchangers, ferrules.
Elect. cond. 28%, inhibited.

TRIANGLE 171.
M-1677; 70 Cu, 30 Zn, 0.02-0.05 As.
Hard: 110,000 TS; 7 El.
Soft: 50,000 TS; 50 El.
For marine hardware, condensers, heat exchangers.
Elect. cond. 28% inhibited. Free from dezincification.

TRIANGLE 180.
M-1677; 23 Ni, 77 Cu.
CDA C71100.

TRIANGLE "X".
M-111; 1.0-1.2 C, bal Fe.
For rock drill pistons, punches, dies, broaches; water hardened.

TRIANGLE X-EXTRA.
M-111; 0.7 C, bal Fe.
For tools, drills, taps; water hardened.

TRIANGLE X-SPECIAL.
M-111; 0.9-1.2 C, bal Fe.
For tools, punches; water hardened.

TRIBALOY T-100.
M-167; 55 Co, 35 Mo, 10 Si.
Powder for powder metals or plasma spraying.
Two-phase composition for wear resistance.
Corrosion resistant; low co-efficient of friction.
Hardness of intermetallic: 1100 VHN (Vickers).
For valves, pistons, vanes, arbors.

TRIBALOY T-400.
M-167; 62 Co, 28 Mo, 8 Cr, 2 Si.
Rod or powder for plasma spray, powder metal, hard facing or remelt casting stock.
Wear resistant, corrosion resistant over wide temperature range.
For valves, tappet inserts, pump parts.

TRIBALOY T-700.
M-167; 50 Ni, 32 Mo, 15 Cr, 3 Si.
Rod or powder for plasma spray, powder metal, hard facing or remelt casting stock.
Resists mechanical wear and severe corrosion over wide temperature range.
For thrust washers, ball and roller bearings, sleeve bearings, valves.

TRIBALOY T-800.
M-167; 52 Co, 28 Mo, 17 Cr, 3 Si.
Rod or powder for plasma spray, powder metal, hard facing, or remelt casting stock.
Resists mechanical wear and severe corrosion over wide temperature range.
For valves, seals, piston rings, tappet inserts.

TRICENT.
M-68; 0.43 C, 1.6 Si, 0.8 Mn, 1.8 Ni, 0.85 Cr, 0.38 Mo, 0.08 V, bal Fe.
Heat treated: 290,000 TS; 250,000 YS; 10 El; 36 RA; 54 Brin.
For aircraft structures, landing gears; tough, shock resistant.

TRI-CLOVER.
M-909; 0.3-0.5 C, bal Fe.
For machinery parts; water hardening.

TRI-CORE.
M-1073; 40-60 Pb, bal Sn.
For solder; self fluxing soft solder.

TRI-LOK.
M-923; 0.4 C, bal Fe.
For gears, shafts; water hardened.

TRIMCO.
M-954; 0.6 C, 1.5 Ni, 0.8 Cr, bal Fe.
For pipe wrenches; oil hardening.

TRIMET 104.
M-63.
Easy-Flo 3 on both sides of copper in 1-4-1 ratio.
Melt point 1170°F (630°C); flow point 1270°F (690°C).

TRIMET 177.
M-63.
Easy-Flo 35 on both sides of copper in 1-2-1 ratio.
Melt point 1125°F (605°C); flow point 1295°F (700°C).
For brazing carbide tips.

TRIMET 201.
M-63.
BRAZE 403 on both sides of copper in 1-2-1 ratio.
Melt point 1220°F (660°C); flow point 1435°F (780°C).
For brazing carbide tips.

TRIMET 202.
M-63.
BRAZE 404 on both sides of copper in 1-2-1 ratio.
Melt point 1220°F (660°C); flow point 1580°F (860°C).
For brazing carbide tips.

TRIMET 258.
M-63.
Easy-Flo 3 on both sides of copper in 1-2-1 ratio.
Melt point 1170°F (630°C); flow point 1270°F (690°C).
For brazing carbide tips.

TRIMMAX.
M-64; 0.95 C, 3.75 Cr, 0.23 Mb, 0.16 V, bal Fe.
Annealed: 220 Brin.
Heat treated: 510 Brin.
For hot and cold trimmers; oil hardened.

TRIMMAX, H41.
M-64; 0.95 C, 0.30 Mn, 0.25 Si, 3.75 Cr, 0.17 V, 0.20 Mo, bal Fe.
Air hardening tool and die steel for cold work operations.

TRIMMER DIE.
M-80; C, alloy, bal Fe.
For tools; tough.

TRI-MO.
M-111; 1.5 C, 12 Cr, 2 V, 0.8 Mo, bal Fe.
Annealed: 102,000 TS; 14 El; 25 Brin.
For tools, dies, reamers, gages, forming and blanking dies, thread rolling dies.
Type D2 air hardening, non-deforming tool steel.

TRI-MO NO. 19.
M-341; 1.5 C, 12 Cr, 0.2 V, 0.9 Mo, bal Fe.
For forming and blanking dies; air hardened, non-deforming.

TRI-MO AIR-HARDENING.
M-350; C, Cr, Mo, bal Fe.
For heavy dies for severe requirements; air hardened.

TRIP STEELS MEANS TRANSFORMATION.
Induced plasticity. (Not an alloy).

TRIPLE A ALMAG 55.
M-1022; 10.0-11.5 Mg, 0.2-1.0 Be, 0.0001-0.05 B, bal Al.
Cast: 50,000-60,000 TS; 32,000-35,000 YS; 20-30 El; 80-100 Brin.
For light alloy castings, meters; sand or permanent mold.

TRIPLE CRESCENT.
M-1744; 1.25 C, 1.25 Cr, 4.3 W, 0.3 V, bal Fe.
4 1/4% Tungsten-chromium tool steel, for reamers, broaches, finishing tools, gauges.

TRIPLE CONQUEROR.
M-1406; 1.3 C, 4 W, 1.0 Cr, 0.3 Mo, bal Fe.
For drawing and extrusion dies, broaches; finishing steel, wear resistant.

TRIPLE E (X8176)-H24.
M-1799; 0.60 Fe, 0.06 Si, bal Al.
17.5 ksi UTS; 13.5 YTS; 15 El; 61 min IACS.
For building wire, magnet wire, automotive wire, communication cable.

TRIPLE ECLAIR.
M-1546; 0.78 C, 18.5 W, 1 V, 0.4 Mo, bal Fe.
For lathe and planer tools, hobs, milling cutters, taps, thread chasers, broaches, drills.
High-speed steel. Oil hardening. Good red-hardness.

TRIPLE EXPRESS 3X.
M-1546; 1.30 C, 10.5 W, 3.1 V, 0.5 Mo, bal Fe.
Hardened: C 64-68 Rock.
For finishing cutters, tools, deburrers, drills, broaches, milling cutters.
High speed steel, oil hardening, good red-hardness.

TRIPLE EXPRESS "A".
M-344; 0.7 C, 6 W, 6 Mo, 5 Co, 4 Cr, bal Fe.
For high speed lathe tools, twist drills, reamers; high speed steel.

TRIPLE EXPRESS "B".
M-344; 0.7 C, 8 Mo, 1 W, 4 Cr, 5 Co, bal Fe.
For high speed lathe tools, twist drills, reamers; high speed steel.

TRIPLE EXPRESS B.
  M-1546; 0.80 C, 18 W, 5 Co, 1.1 V, 0.9 Mo, bal Fe.
  Hardened: C 64-66 Rock.
  For turning tools, lathe and planer finishing cutters, hobs, reamers, broaches, taps, drills.
  High-speed steel, oil hardening; high red-hardness.

TRIPLE EXPRESS SUPER-4X.
  M-1546; 1.55 C, 13 W, 5 Co, 5 V, 5 Cr, 0.5 Mo, bal Fe.
  Hardened: C 64-68 Rock.
  For lathe and planer finishing tools, engraving tools, milling cutters, textile needles.
  High-speed steel, oil hardening.
  High red-hardness.

TRIPLE EXPRESS T.M.
  M-1546; 0.82 C, 6.0 W, 4.5 Mo, 1.4 V, bal Fe.
  Hardened: 64-66 Rock C.
  For lathe and planer tools, drills, saws, form cutters, broaches, milling cutters.
  High speed steel; oil hardening, high red-hardness.

TRIPLE EXPRESS V.
  M-1546; 0.85 C, 18.5 W, 1.9 V, 0.4 Mo, bal Fe.
  Hardened: 64-66 Rock C.
  For turning tools, lathe and planer cutters, drills, thread cutters, broaches, reamers, hobs.
  High speed steel; oil hardening, high red-hardness.

TRIPLE EXTRA 2 (T.E.R.2).
  M-1116; 0.82 C, 4.5 Cr, 18 W, 0.6 Mo, 2 V, bal Fe.
  Air or oil harden to 63-66 Rc.
  For cutting tools, lathe, gear cutters, broaches.
  AFNOR: Z 85 WV 18.02.
  US-AISI T2 high speed steel.
  Germany: W.nr. 1.3357.

TRIPLE EXTRA 60 (T.E.R.60).
  M-1116; 0.65 C, 4 Cr, 18 W, 0.4 Mo, 1 V, bal Fe.
  Air or oil harden to 60-63 Rc.
  For cutting tools, or cold work tools.
  AFNOR: Z 65 W 18.
  US-AISI T1 high speed steel.

TRIPLE EXTRA D (T.E.D.).
  M-1116; 0.80 C, 4.3 Cr, 12.5 W, 1.8 Mo, 1.8 V, bal Fe.
  Air or oil harden to 62-66 Rc.
  For lathe tools, drills, taps, reamers, cut-off tools.
  AFNOR: Z 80 WDV 12.02.02.
  US-AISI T7 High speed steel.
  Germany: S 12.1.2. (0); W. Nr. 1.3318.

TRIPLE EXTRA LION.
  M-1415; 0.75 C, 19 W, 4.25 Cr, 1.25 V, 5 Co, 0.75 Mo, bal Fe.
  For lathe and planer tools, drills, hobs, broaches; high speed steel.

TRIPLE EXTRA M (T.E.M.).
  M-1116; 0.83 C, 4.2 Cr, 6.5 W, 5 Mo, 1.9 V, bal Fe.
  Air or oil harden to 63-66 Rc.
  For twist drills, centre drills, reamers, taps, cutters, broaches.
  AFNOR: Z 85 WDV 06.05.02.
  US-AISI M2 high speed steel.
  Germany: S 6.5.2 (DMo5); W.nr. 1.3343.

TRIPLE EXTRA (T.E.R.).
  M-1116; 0.78 C, 4.5 Cr, 18 W, 0.6 Mo, 1.25 V, bal Fe.
  Air or oil harden to 63-66 Rc.
  For cutting tools, taps, thread chasers, turning tools.
  AFNOR: Z 80 W 18.
  US-AISI T1 high speed steel.
  Germany: S 18.01 (B-18); W.nr. 1.3355.

TRIPLE GRIFFIN.
  M-1112; 1.35 C, 0.4 Mn, 0.35 Cr, 2.75 W, 0.3 Si, bal Fe.
  For cold drawing dies, cutters; water hardened, hard case and tough core.

TRIPLE LIFE D-C.
  M-40; 1.2 C, 4 Cr, bal Fe.
  For dies, reamers; air hardened.

TRIPLE MERMAID.
  M-1737; 0.78 C, 4.25 Cr, 18.0 W, 1.25 V, 5.0 Co, bal Fe.
  Cobalt high speed steel for cutting tools.
  AISI T4.

TRIPLE VELOS.
  M-1744; 0.75 C, 18.0 W, 4.25 Cr, 1.1 V, 5.0 Co, bal Fe.
  Medium cobalt high speed steel.

TRIPLI-CAST.
  M-838; C, alloy, bal Fe.
  Rolled: 69,000-109,600 TS; 5-1.5 El; 132-205 Brin.
  For shafts; M.P. 1710°F.

TRI-PLY.
  M-114; Three-ply composite steel with steel center and Uniloy backing.
  For heat and corrosion resistant parts; stainless clad steels.

TRI-STEEL.
  M-67; 0.22 C, 1.25 Mn, 0.30 Si, 0.02 V, bal Fe.
  Rolled: 70,000 TS; 50,000 YS; 22 El.
  For railroad and agriculture equipment, mine cars; high-strength, low alloy construction steel.

TRI-TEN.
M-604; 0.22 max C, 1.25 max Mn, 0.2-0.6 Cu, 0.02 min V, bal Fe.
Rolled: 70,000 TS; 50,000 YS; 25 El; 140 Brin.
For cranes, shovels, derricks, mine cars, truck bodies.
Shock resistant. Good resistance to atmospheric corrosion.
ASTM A441.

TRI-TUNG see **UHB TRITUNG.**

TRIUMPH.
M-486; 0.68 C, 3.75 Cr, 0.6 Mo, 0.6 V, 14 W, bal Fe.
For slitting and hacksaw blades, punches; high speed steel.

TRIUMPH.
M-261; 0.68 C, 3.75 Cr, 0.6 V, 14.0 W, bal Fe.
Hardened: 64-66 Rock C.
For slitting saws, hacksaws, lathe centers, cold punches, reamers.
High speed steel for cutting operations under conditions of heavy vibration.

TRIUMPH 200.
M-261; 0.70 C, Cr, W, V, bal Fe.
For lathe and planer tools, reamers, hobs, drills, taps; high speed steel.

TRIUMPHATOR now **VEW K100.**

TRIUMPHATOR M now **VEW K116.**

TRIUMPHATOR W now **VEW K.**

TRIUMPH SUPERB.
M-261, M-486; 0.76 C, 4.25 Cr, 0.6 Mo, 1.4 V, 18 W, bal Fe.
For drills, broaches, reamers, hobs, gear cutters; high speed steel.

TRIUMPH SUPERB 1000.
M-486; 0.8 C, 4.75 Cr, 0.6 Mo, 1.6 V, 18.5 W, 5.7 Co, bal Fe.
For turning and boring tools, milling cutters, hobs; high speed steel.

TRIUMPH SUPERB DOUBLE 1000.
M-261, M-486; 0.82 C, 4.75 Cr, 0.6 Mo, 1.6 V, 20 W, 10.5 Co, bal Fe.
For tools, cutters, hobs; high speed steel.

TRIUMPH SUPERB EXTRA.
M-261, M-486; 0.78 C, 4.25 Cr, 0.6 Mo, 1.4 V, 22 W, bal Fe.
For form cutters, tools for chilled roll turning; high speed steel.

TRI-VAN.
M-111, 111a; 2.25 C, 12 Cr, bal Fe.
Hardened: 64-66 Rock C.
For blanking and drawing dies, punches, shears, slitters, lamination and coining dies, gauges.
Oil hardening, abrasion resistant. Type D4 die steel.

TRIVAN.
M-350; C, Cr, V, bal Fe.
For blanking and coining dies; oil hardened.

TROJAN.
M-435; 0.8 C, 18 W, 4 Cr, 2 V, 0.5 Mo, bal Fe.
For tools, dies, hobs, cutters; high speed steel.

TROJAN BABBITT.
M-428, M-88; Sb, Sn, Pb.
At 70°F: 27.4 Brin.
At 212°F: 11.2 Brin.
For bearings for steam and internal combustion engines; Babbitt, antifriction.

TROMALIT.
M-1120; 11 Co, 24 Ni, 11 Al, 3.5 Cu, bal Fe.
For electrical and magnetic equipment.
Permanent magnet; high permeability.

TROMALIT-III.
M-1120; 12 Ni, 10 Al, 4-20 Co, 2 Cu, bal Fe.
(Bakelite bonded).
3500 Br, 680 Hc, 0.8 BH max.
For magnets in speedometers and small battery motors.
Permanent magnet, high permeability.

TROMALIT-IV.
M-1120; 12 Ni, 10 Al, 4-20 Co, 2 Cu, bal Fe.
(Bakelite bonded).
3500 Br, 1000 Hc, 1 BH max.
For magnets in speedometers and in small battery motors.
Permanent magnet, high permeability.

TROMALIT-ALNI.
M-1120; 27 Ni, 13 Al, bal Fe.
For permanent magnets in electrical and magnetic equipment.
High magnetic permeability.

TROMALIT-ALNI090.
M-1120; 22 Ni, 12 Al, bal Fe.
For permanent magnets in electrical and magnetic equipment.
High magnetic permeability.

TROMALIT-ALNICO160.
M-1120; 24 Ni, 11 Al, 9.5 Co, 3.5 Cu, bal Fe.
For permanent magnets in electrical and magnetic equipment.
High magnetic permeability.

TRONAMANG.
M-1550; 100 Mn.
For an addition agent in the melting of steel and non-ferrous alloys.
Electrolytic manganese.

TRONAMANG 75.
M-1550; 75 Mn, 25 Al.
Briquettes used in alloying aluminum.

TROY.
M-1705; 1.25 C, 0.3 Mn, 0.35 Si, 0.4 Cr, 0.2 V, 1.4 W, bal Fe.
Oil hardening tool steel. AISI O7.

T.R.S.
M-750; C, Cr, Mo, W, bal Fe.
Rolled: 134,000 TS; 16 El; 37 RA; 269-550 Brin.
For tools, fixtures; general purpose tools, shock resistant.

TRS SPECIALITY.
M-750; low C, Cr, Mo, W, bal Fe.
Heat treated: 140,000 TS; 260-300 Brin.
For tools, dies, machine parts, gears, cams, pump shafts, clutches; tough, wear and fatigue resistant.

TRUALOY ALUMINUM.
M-925; 5 Si, bal Al.
Cast: 22,000 TS; 2 El; 60 Brin.
For light alloy castings; die or sand castings.

TRUALOY ALUMINUM BRONZE.
M-925; 90 Cu, bal Al.
Cast: 65,000 TS; 25,000 YS; 20 El; 115 Brin.
For sand castings.

TRUALLOY BEARING BRONZE.
M-925; Sn, Pb, bal Cu.
Cast: 25,000 TS; 12,000 YS; 8 El; 60-80 Brin.
For bearings; high compressive strength.

TRUALOY COPPER.
M-925; Cu.
For current conductors; high conductivity 90%.

TRU-CAST.
M-221; Be, bal Cu.
Cast: 125,000 TS; 110,000 YS; 200 Brin.
Hardened: 180,000 TS; 140,000 YS; 460 Brin.
For stamping and casting dies, molds; hardenable.

TRU-COR.
M-339; 0.9-1.05 C, 0.3-0.5 Mn, 0.15-0.30 Si, bal Fe.
For machine tool parts, drills, boring tools; type W1; water hardened.

TRU-COR DRILL ROD.
M-1703; 0.95-1.05 C, 0.3-0.5 Mn, bal Fe.
Water hardening drill rod; AISI W1.

TRUCOTE NO 101 F.C. ETC see **ALL-STATE NO 101 FC ETC.**

TRUDIE.
M-1683; 1.5 C, 12.0 Cr, 1.0 Mo, 1.0 V, bal Fe.
Air hardening cold work tool steel, high carbon-high chrome type.

TRUE-DIE.
M-750; C, alloy, bal Fe.
For dies; non-shrinking.

TRUDIE SPECIAL.
M-1683; 1.5 C, 12.0 Cr, 1.0 Mo, 3.0 Co, bal Fe.
Air hardening cold work tool steel.

TRUEWEAR.
M-69; 2.45 C, 12.25 Cr, 4.25 V, 1.1 Mo, bal Fe.
For dies, punches, pins, tools and forming rolls.
Extreme wear resistance.
AISI D7 cold work tool steel.

TRUFLEX A-1.
M-926; Ni-Cr-Fe.
For thermostatic bimetal; max temperature 300°F.

TRUFLEX A-2.
M-926; Ni-Cr-Fe.
For thermostatic bimetal; max temperature 300°F.

TRUFLEX B-1.
M-926; Alloy of 22 Ni, 3 Cr, 75 Fe bonded to alloy of 36 Ni, 64 Fe.
For thermostatic bimetal.
Differential expansion.
Max operating temperature 1200°F.

TRUFLEX B-2.
M-926; Ni-Cr-Fe.
For thermostatic bimetal; max temperature 1200°F.

TRUFLEX B3.
M-926; Ni-Cr-Fe.
For thermostats (0 to 650°F); bimetal max 1000°F.

TRUFLEX B12 1/2.
M-926; Ni-Cr-Fe.
For thermostatic bimetal; max temperature 1100°F.

TRUFLEX B-15.
M-926; Ni-Cr-Fe.
For thermostatic bimetal; max temperature 1100°F.

TRUFLEX B-20.
M-926; Ni-Cr-Fe.
For thermostatic bimetal; max temperature 1100°F.

TRUFLEX-25.
M-926; Ni-Cr-Fe.
For thermostatic bimetal; max temperature 1100°F.

TRUFLEX B-30.
M-926; Ni-Cr-Fe.
For thermostatic bimetal; max temperature 1100°F.

TRUFLEX B-35.
M-926; Ni-Cr-Fe.
For thermostatic bimetal; max temperature 1100°F.

TRUFLEX B-40.
M-926; Ni-Cr-Fe.
For thermostatic bimetal; max temperature 1100°F.

TRUFLEX B-47.
M-926; Ni-Cr-Fe.
For thermostatic bimetal; max temperature 1100°F.

TRUFLEX D.
M-926; Ni-Cr-Fe.
For thermostatic bimetal; max temperature 1200°F.

TRUFLEX E-1.
M-926; Ni-Cr-Fe.
For thermostatic bimetal; max temperature 1200°F.

TRUFLEX E-2.
M-926; Ni-Cr-Fe.
For thermostatic bimetal; max temperature 1200°F.

TRUFLEX E3.
M-926; Ni-Cr-Fe.
For thermostats (0 to 650°F); bimetal max 1000°F.

TRUFLEX E4.
M-926; Ni-Cr-Fe.
For thermostats (0 to 750°F); bimetal max 1000°F.

TRUFLEX E5.
M-926; Ni-Cr-Fe.
For thermostats (0 to 750°F); bimetal max 1000°F.

TRUFLEX F.
M-926; Ni-Cr-Fe.
For thermostatic bimetal; max temperature 1200°F.

TRUFLEX G-1.
M-926; Ni-Cr-Fe.
For thermostatic bimetal; max temperature 1200°F.

TRUFLEX G-2.
M-926; Ni-Cr-Fe.
For thermostatic bimetal; max temperature 1200°F.

TRUFLEX H.
M-926; Ni-Cr-Fe.
For thermostatic bimetal; max temperature 1200°F.

TRUFLEX J.
M-926; Ni-Cr-Fe.
For thermostatic bimetal; max temperature 500°F.

TRUFLEX J1.
M-926; Ni-Cr-Fe.
For thermostats (-50 to 400°F); bimetal max 625°F.

TRUFLEX J7.
M-926; Ni-Cr-Fe.
For thermostats; corrosion resistant, bimetal.

TRUFLEX K.
M-926; Ni-Cr-Fe.
For thermostatic bimetal; max temperature 1200°F.

TRUFLEX L.
M-926; Ni-Cr-Fe.
For thermostatic bimetal; max temperature 800°F.

TRUFLEX L1.
M-926; Ni-Cr-Fe.
For thermostats (-50 to 400°F); bimetal max 1000°F.

TRUFLEX M.
M-926; Ni-Cr-Fe.
For thermostatic bimetal; max temperature 1200°F.

TRUFLEX N.
M-926; Ni-Cr-Fe.
For thermostatic bimetal; max temperature 1000°F.

TRUFLEX NI.
M-926; Nickel bonded to Invar (36 Ni, bal Fe).
For thermostats.
Differential expansion bimetal. Low electrical resistance.

TRUFLEX O.
M-926; Ni-Cr-Fe.
For thermostatic bimetal; max temperature 400°F.

TRUFLEX P675R.
M-926; Ni-Cr-Fe.
For thermostats, thermal cutouts and circuit breakers; bimetal, high elec. resist.

TRUFLEX P850R.
M-926; Ni-Cr-Fe.
For thermal cutouts and circuit breakers; bimetal, high elec. resist.

TRUFLEX Q.
M-926; Ni-Cr-Fe.
For thermostatic bimetal; max temperature 1000°F.

TRUFLEX R.
M-926; Ni-Cr-Fe.
For thermostatic bimetal; max temperature 700°F.

TRUFLEX T.
M-926; Ni-Cr-Fe.
For thermostatic bimetal; max temperature 800°F.

TRUFORM.
M-69; 0.90 C, 1.2 Mn, 0.50 W, 0.50 Cr, bal Fe.
For dies, gages, tools, taps, reamers, screw dies; non-deforming tool steel. AISI 01.

TRUFORM SPECIAL.
M-261; C, alloy, bal Fe.
For tools, dies; non-deforming.

TRUFORM SPECIAL.
M-389; 0.9 C, 1.2 Mn, bal Fe.
For tools, dies; non-deforming.

TRU-GROUND.
M-69; C, alloy, bal Fe.

TRU-HEADERDIE.
M-69; 1.4 C, 3.5 V, bal Fe.

TRUMPET BRASS.
M-1426; 83 Cu, 1.5 Sn, bal Zn.

TRUMPET BRASS-435.
M-8; 81 Cu, 18 Zn, 1 Sn.
Soft: 48,000 TS; 60 El.
Hard: 80,000 TS; 10 El.
For Bourdon gauge tubes.

TRU-WEAR FM (AISI D7).
M-69; 2.2 C, 12.5 Cr, 4 V, 1.1 Mo, bal Fe.
For header and forming dies; cold work steel.

TRW 361.
M-1697; 90.0 WC, 10.0 Co.
400,000 TrS; A 89.0 Rock; Density 14.52.
Sintered carbide tool material.

TRW 362.
M-1697; 90.0 WC, 10.0 Co.
400.000 TrS; A 88.0 Rock; Density 14.53.
Sintered carbide tool material.

TRW 363.
M-1697; 88.5 WC, 0.5 TaC, 11.0 Co.
435,000 TrS; A 89.8 Rock; Density 14.48.
Sintered carbide tool material.

TRW 364.
M-1697; 87.0 WC, 13.0 Co.
400,000 TrS; A 88.5 Rock; Density 14.23.
Sintered carbide tool material.

TRW 366.
M-1697; 84.0 WC, 16.0 Co.
400,000 TrS; A 87.0 Rock; Density 13.90.
Sintered carbide tool material.

TRW 1800.
M-513, M-1491; 0.09 C, 13.0 Cr, 9.0 W, 1.5 Cb, 0.6 Ti, 6.0 Al, 0.07 B, 0.07 Zr, bal Ni.
High temperature alloy; for gas turbine blades, space vehicles.

TRW 1900.
M-513, M-1491; 10 Co, 0.11 C, 9 W, 1.5 Cb, 0.03 B, 0.1 Zr, 10.3 Cr Ni.
For jet engine and gas turbine components, space vehicles.
High heat and oxidation resistance.

TRW MOD 1900.
M-513; 0.13 C, 10.3 Cr, 10 Co, 9 W, 1.0 Ti, 6.3 Al, 1.5 Cb, 0.03 B, 0.10 Zr, 0.5 Hf, 0.5 Ta, 0.5 V, bal Ni.
High temperature alloy.

TRW VI A.
M-513, M-1491; 0.13 C, 6.1 Cr, 1.0 Ti, 5.4 Al, 2.0 Mo, 7.5 Co, 5.8 W, 0.5 Re, 0.43 Hf, 0.13 Zr, 0.02 B, 9.0 Ta, 0.5 Cb, bal Ni.
Cast, at 1600°F: 126,000 TS; 112,000 YS; 2.5 El.
For cast turbine blades.

T.S.B.
M-USSR; 0.7 C, 18 W, 4 Cr, 1 V, bal Fe.
For cutting tools, reamers; high speed steel.

T.S.R.
M-1116; 1.05 C, 0.85 Cr, 0.9 W, 0.15 Mo, 0.95 Mn, 0.25 Si, bal Fe.
Oil (or salt bath) hardening tool steel; for drills, cutting tools, forming tools, gages.
AFNOR: 100 MCW 04; AISI 01.

TSR BRAKE DIE.
M-1684; 0.50 C, 1.0 Mn, 1.2 Cr, 0.25 V, 0.3 Mo, bal Fe.
Oil hardening, low alloy tool steel.

T.S.S. 3 ALLOY.
M-196; 2.5 Mg, 0.3 Ti, 0.7 Si, bal Al.
Heat treated: 33,000 TS; 20,600 YS; 4 El; 5 RA; 75-90 Brin.
For furniture, interior light fixtures, wire, castings; resists sea water corrosion.

T.S.S. 5 ALLOY.
M-196; 5 Mg, 0.3 Ti, bal Al.
Sand cast: 27,000-33,000 TS; 14,000 YS; 5.5-12 El; 3-15 RA; 65 Brin.
For furniture, interior light fixtures, wire, castings; resists sea water corrosion.

T.S.S. 8 ALLOY.
M-196; 7.5 Mg, 0.3 Ti, bal Al.
Heat treated: 30,000-37,000 TS; 20,000 YS; 4.5-8 El; 2-20 RA; 80 Brin.
For furniture, interior light fixtures, wire, castings; resists sea water corrosion.

TT6.
M-1488; 0.85 C, 4.25 Cr, 5.0 Mo, 1.9 V, 6.0 W, bal Fe.
High speed steel for lathe tools, milling cutters, drills.
AFNOR Type 6-5-2; AISI M2. Was TRIPLE EXPRESS TM.

TT6 C5.
M-1488; 0.82 C, 4.25 Cr, 5.0 Mo, 1.85 V, 6.35 W, 4.75 Co, bal Fe.
High speed steel, good red hardness, for lathe tools, milling cutters, drills.
AFNOR Type 6-5-2-5.
Was TRIPLE EXTRA TM-K5.

TTV 1.
M-1495; 0.18 O, 0.03 N, 0.015 H, 0.20 Fe, 0.10 C, (all max), bal Ti.
Resistant to pitting and corrosion.

TTV 2.
M-1495; 0.25 O, 0.03 N, 0.015 H, 0.30 Fe, 0.10 C, (all max), bal Ti.
Resistant to pitting and corrosion.

TTV 3.
M-1495; 0.35 O, 0.03 N, 0.015 H, 0.30 Fe, 0.10 C, (all max), bal Ti.
Resistant to corrosion and pitting.

TTV 7.
M-1495; 0.25 O, 0.03 N, 0.015 H, 0.30 Fe, 0.10 C, 0.18 P, (all max), bal Ti.
Corrosion resistant.

TT7K12.
M-USSR; 4 TiC, 3 TaC, 81 WC, 12 Co.
For hard cutting tools, wear plates.
Sintered carbide, abrasion resistant.

TT7K15.
M-USSR; 78 WC, 4 TiC, 3 TaC, 15 Co.
For hard cutting tools, wear plates.
Sintered carbide, abrasion resistant.

TUBE-ALLOY see MCKAY TUBE-ALLOY.

TUBE BORIUM.
M-202; 60 WC, 40 steel.
For welding electrodes, hard facing rod; abrasion resistant.

TUBELOY.
M-314; 0.02 Ca, 0.02 Mg, 0.02 Sn, bal Pb.
Extruded: 40,000 TS; 15 El; 8 Brin.
For water service pipe; corrosion resistant.

TUBE MANDREL.
M-365; 1.15-1.35 C, 0.35 Cr, bal Fe.
For tools, mandrels, cutters; tough.

TUC-TUR-1.
M-Eng.; 63 Cu, 15 Ni, 22 Zn.
For machine parts, hardware; close grained.

TUC-TUR-2.
M-Eng.; 61-59 Cu, 13-18 Sn, 21-29 Zn, 0.3 Fe.
For fittings, machine parts.

TUCUNSIL.
M-1493, M-856; 10 Ti, Ag, bal Cu.
For brazing titanium and stainless steel components.

TUFALOY.
M-875; 3.3 C, 2.1 Si, bal Fe.
For rolls; cast iron.

TUFALOY.
M-164; 0.25-0.35 C, 1.25-1.35 Mn, bal Fe.
Normalized: 95,000 TS; 60,000 YS; 25 El; 55 RA; 180 Brin.
Heat treated: 160,000 TS; 110,000 YS; 12 El; 25 RA; 350 Brin.
For gears, shafts, high pressure castings; shock and wear resistant.

TUFANHARD 150.
M-1000; weld metal; 0.7 C, 14.0 Mn, 0.5 Si, 4.0 Cr, 4.0 Ni, bal Fe.
As welded: 14-18 Rc; work hardens to 45-50 Rc.
For build-up on manganese steel crusher rolls, dipper buckets, power shovel teeth.
DCEP or AC.

TUFANHARD 160.
M-1000; weld metal; 0.75 C, 4.0 Mn, 0.5 Si, 20.0 Cr, 9.0 Ni, bal Fe.
As welded: 15-20 Rc; work hardens to 50 Rc.
For overlay of railroad switches, frogs, railends.
DCEP or AC.

TUFANHARD 250.
M-1000; weld metal; 0.20 C, 0.12 Mn, 1.05 Cr, bal Fe.
As welded: 23-26 Rc.
For build-up on mild steel where good impact resistance is required.
DCEP or DCEN.

TUFANHARD 320.
M-1000; weld metal; 0.20 C, 1.0 Mn, 0.40 Si, 0.90 Cr, 0.30 Ni, 0.40 bal Fe.
As welded: 26-34 Rc.
For miscellaneous build-up.
DCEP or AC.

TUFANHARD 375.
M-1000; weld metal; 0.23 C, 0.69 Mn, 0.23 Si, 2.32 Cr, 0.18 Mo, bal
As welded: 29-40 Rc (depending on no. of passes).
For wear resistance under medium impact conditions.
DCEN or DCEP.

TUFANHARD 450.
M-1000; weld metal: 2.0 C, 1.9 Mn, 0.5 Si, 2.0 Cr, bal Fe.
As welded: 46-52 Rc.
For bulldozer blades, crusher jaws, bucket lips and teeth.
DCEP or AC.

TUFANHARD 550.
M-1000; weld metal; 3.5 C, 2.0 Mn, 1.5 Si, 12.0 Cr, 0.40 Mo, 0.60 V, bal Fe.
As welded: 47-54 Rc.
For overlay on hammer mill hammers, coke machinery, grader blades.
DCEP or AC.

TUFANHARD 580.
M-1000; weld metal; 6.0 C, 2.0 Mn, 2.0 Si, 26.0 Cr, 2.0 V, bal Fe.
As welded: 48-65 Rc.
For overlay on parts requiring abrasion and corrosion resistance.
DCEP or AC.

TUFANHARD 600.
M-1000; weld metal; 0.29 C, 0.84 Mn, 4.31 Cr, 0.28 Mo, bal Fe.
As welded: 52-58 Rc.
For overlay on parts subject to severe abrasion.
DCEN, DCEP, or AC.

TUF-COR.
M-663; weld deposit: 0.073 C, 1.40 Mn, 0.035 Ni, 0.05 Cr, 0.01 Mo, 0. 0.58 Si, bal Fe.
As welded: 89,000 psi TS; 78,000 psi YS; 25 El.
Good impact strength.
AWS A5.20 Class E70T-1.

TUFCUT.
M-750; C, alloy, bal Fe.
For tools; water hardened.

TUF-CUT.
M-1215; C, Mo, bal Fe.
For tools; water hardened.

TUFFALOY.
M-790; 0.8 C, 0.2 V, bal Fe.
For tools, dies, fixtures; water hardened.

TUFFALOY 1W3.
M-1092; Cu alloy.
Rolled: 135,000 TS; 130 Brin.
For resistance welding electrode; 35% conductivity, for stainless steel and brass.

TUFFALOY 10W3.
M-1092; 23 Cu-77W.
Bars: 160,000 TS; 205 Brin.
For resistance welding electrode; 25% conductivity, for inserts and facings.

TUFFALOY 20W3.
M-1092; Cu-W.
Bars: 170,000 TS; 228 Brin.
For resistance welding electrode; for heavy projection welding.

TUFFALOY 44.
M-1092; Cu alloy.
Cast: 90,000 TS; 60,000 YS; 1 El; 330 Brin.
Rolled: 140,000 TS; 85,000 YS; 2 El; 330 Brin.
For resistance welding electrode; for flash, butt and projection welds.

TUFFALOY 55.
M-1092; 0.4 Be, 2.6 Co, bal Cu.
Cast: 90,000 TS; 55,000 YS; 10-15 El.
Rolled: 125,000 TS; 65,000 YS; 8-15 El.
For springs, bellows, diaphragms; age-hardenable, corrosion resistant.

TUFFALOY 66.
M-1092; Cu, alloy.
Cast: 65,000-75,000 TS; 12,000-16,000 YS; 10-2 El; 116-165 Brin.
For resistance welding electrode; for high strength backing material.

TUFFALOY 77.
M-1092; Cu alloy.
Rolled: 55,000-65,000 TS; 25,000-35,000 YS; 15 El.
Cast: 45,000 TS; 20,000 YS; 12 El.
For resistance welding electrode; 75% conductivity, for steel and brass.

TUFFALOY 88.
M-1092; Cu alloy.
Bars: 50,000-60,000 TS; 15,000-20,000 YS; 25-20 El; 110-116 Brin.
For resistance welding electrode; for coated metals, terneplate, galvanized stock.

TUFFALOY 100W.
M-1092; Cu-W.
Bars: 200,000 TS; Rockwell A76.
For resistance welding electrode for nonferrous metals; 30% conductivity, for Red Brass and copper.

TUFF-CAST.
M-1029; 0.7 C, 18 W, 4 Cr, 1 V, bal Fe.
For cutting tools, hobs, drills; high speed steel.

TUFKUT.
M-1067; 0.5 C, 0.7 Mn, 1.5 Si, 0.35 Mo, bal Fe.
For tools, chisels; tough, shock resistant.

TUF-STUFF 224-C, 6230 (224 C).
M-266; 87.5 Cu, 9.75 Al, 2.75 Fe.
Forged: 60,000-107,000 TS; 35,000-82,000 YS; 35-15 El; 140-225 Brin.
Annealed: 75,000-90,000 TS; 40,000-52,000 YS; 25-15 El; 30 RA; 150 Brin.
For pickling equipment, piston rods, valve stems, gears, pumps, nuts, impellers, screws, valve seats in oil refineries; corrosion and wear resistant.

TUF-STUF 224C now **TUF-STUF 6181(224 C).**

TUF-STUF 224E-30 see **MUELLER 6181.**

TUF-STUF 224E-75 see **MUELLER 6180.**

TUF-STUF 224H now **TUF-STUF 6240(224 H).**

TUF-STUF 224K now **TUF-STUF 6300(224 K).**

TUF-STUF 6240 (224 H).
M-266; 86-87 Cu, 10.5-11.5 Al, 2-3 Fe.
Forged: 80,000 TS; 45,000 YS; 10 El; 170 Brin.
Heat treated: 90,000 TS; 70,000 YS; 3 El; 200 Brin.
For cams, rollers, gears, bearings, trolley shoes, shifter forks.
Aluminum bronze. Heat treatable. Wear and corrosion resistant.

TUF-STUF 6300 (224 K).
M-266; 78.0-82.0 Cu, 2.0-3.5 Fe, 4.5-5.5 Ni, 1.5 max Mn, 9.7-10.9 Al.
Forged: 90,000 TS; 50,000 YS; 20 El; 35 RA; 180-200 Brinell.
Heat treated: 105,000 TS; 65,000 YS; 15 El; 20 RA; 20-240 Brinell.
For valves, gears, worms, shafts, molds.
High strength, non-galling, wear resistant aluminum bronze. Heat treatable.

TUF-TIMBER.
M-32; 0.6-1.1 C, bal Fe.
For general forgings; water hardening.

TUFWEAR NO. 2.
M-927; C, Ni, Mn, bal Fe.
For sleeves, steel castings; for oil well drilling machines.

TUFWEAR NO. 3.
M-927; C, Ni, Cr, Mo, bal Fe.
For jaw clutches; for oil well drilling machines.

TUFWEAR NO. 4.
M-927; C, Ni, Mo, bal Fe.
For oil well equipment; tough.

TUF-WEAR 20.
M-927; 0.07 max C, 1.5 max Mn, 1.5 max Si, 19-22 Cr, 27.5-30.5 Ni, 2.0-3.0 Mo, 3.0-4.0 Cu, bal Fe.
Corrosion resistant steel casting.
ACI CN-7 M.

TUF-WEAR 30.
M-927; 0.04 max C, 1.0 max Mn, 1.0 max Si, 25-27 Cr, 4.75-6.0 Ni, 1.75-2.25 Mo, 2.75-3.25 Cu, bal Fe.
Corrosion resistant steel casting.
ACI CD-4 M Cu.

TULA.
Ag with a small amount of Cu and Pb.
For jewelry, silverware; corrosion resistant.

TULIPE 10.
M-1119; C, Mn, Si, bal Fe.
Heat treated: 50,000-92,000 TS; 30,000 YS; 20 El.
For punchings, bolts, rivets; water hardened, Afnor XC109.

TULIPE 12.
M-1119; C, Mn, Si, bal Fe.
Heat treated: 99,000-171,000 TS; 53,000 YS; 6 El.
For bolts, fasteners, forgings; water hardened, Afnor XC12G.

TULIPE 18.
M-1119; C, Mn, Si, bal Fe.
Heat treated: 78,000-107,000 TS; 54,000 YS; 16 El.
For bolts, fasteners, forgings; water hardened, Afnor XC18G.

TULIPE 18S.
M-1119; C, Mn, Si, bal Fe.
Heat treated: 74,000-103,000 TS; 50,000 YS; 16 El.
For bolts, fasteners, forgings; water hardened, Afnor XC18S.

TULIPE 32.
M-1119; C, Mn, Si, bal Fe.
Heat treated: 107,000-128,000 TS; 82,000 YS; 12 El.
For shunts, connecting rods, spindles; water hardened, Afnor XC32G.

TULIPE 35.
M-1119; C, Mn, Si, bal Fe.
Heat treated: 107,000-135,000 TS; 85,000 YS; 11 El.
For connecting rods, spindles; water hardened, Afnor XC35G.

TULIPE 38.
M-1119; C, Mn, Si, bal Fe.
Heat treated: 114,000-144,000 TS; 89,000 YS; 112 El.
For machinery parts, gears, bolts, shafts; water hardened, Afnor XC38G

TULIPE 42.
M-1119; C, Mn, bal Fe.
Heat treated: 128,000-157,000 TS; 101,000 YS; 9 El.
For bolts, fasteners; water hardened, Afnor XC42G.

TULIPE 48.
M-1119; C, Mn, Si, bal Fe.
Heat treated: 150,000 TS; 96,000 YS; 9 El.
For agriculture and machine tool parts; water hardened, Afnor XC48G.

TULIPE 55.
M-1119; C, Mn, Si, bal Fe.
Heat treated: 135,000-164,000 TS; 98,00 YS; 6 El.
For agriculture and machine tool parts; water hardened, Afnor XC55G.

TULIPE 65.
M-1119; 0.60-0.80 C, 0.65-0.90 Mn, 0.40 max Si, bal Fe.
Heat treated: 125,000 TS; 65,000 YS.
For agricultural equipment; water hardened, Afnor XC65G.

TULIPE 70.
M-1119; 0.67-0.75 C, 0.60-0.85 Mn, 0.40 max Si, bal Fe.
Heat treated: 128,000 TS; 70,000 YS.
For agricultural equipment; water hardened, Afnor XC70G.

TULIPE 80.
M-1119; 0.75-0.85 C, 0.50-0.75 Mn, 0.40 max Si, bal Fe.
Heat treated: 142,000 TS; 72,000 YS.
For tool equipment; water hardened, Afnor XC80G.

TULIPE EXTRA NO 3.
M-1119; 0.90 C, 0.40-0.60 Mn, bal Fe.
Water hardening tool steel.
AFNOR Type Y_3 90; W.-Nr. 1.1760; AISI W1.

TULIPE EXTRA NO 4.
M-1119; 0.75 C, 0.60-0.90 Mn, bal Fe.
Water hardening tool steel.
AFNOR Type Y_3 75; W.-Nr. 1.1750.

TULIPE EXTRA NO 5.
M-1119; 0.65 C, 0.60-0.90 Mn, bal Fe.
Water hardening tool steel.
AFNOR Type Y_3 65; W.-Nr. 1.1744.

TULIPE EXTRA NO 6.
M-1119; 0.55 C, 0.60-0.90 Mn, bal Fe.
Water hardening tool steel.
AFNOR Type Y_3 55; W.-Nr. 1.1735.

TULIPE EXTRA NO 7.
M-1119; 0.45 C, 0.60-0.90 Mn, bal Fe.
Water hardening tool steel.
AFNOR Type Y_3 45; W.-Nr. 1.1730.

TULIPE EXTRA NO 8.
M-1119; 0.35 C, 0.50-0.70 Mn, bal Fe.
Water hardening tool steel.
AFNOR Type Y_3 35; W.-Nr. 1.1720.

TULIPE EXTRA NO 10.
M-1119; 0.17 max C, 0.15-0.40 Si, 0.30-0.60 Mn, bal Fe.
Low carbon tool steel.
AFNOR Type Y_3 12; W.-Nr. 1.1705.

TUNCRO.
M-353; 0.50 C, 1.4 Cr, 0.25 V, 2 W, bal Fe.
For chisels, punches, pneumatic tools, hot work dies; hot work steel.

TUNGALOY.
M-336; 1.2 C, 0.6 Cr, 0.3 Mo, bal Fe.
For blanking and forming dies; oil or water hardened.

TUNGAY.
M-Eng.; Zn, Si, Al, Ni, bal Cu.
For corrosion resistant parts; alloyed by special process.

TUNGROD.
M-9; WC.
For hard facing electrodes; WC encased in a steel tube.

TUNGSIL.
M-237; 27-75 Ag, bal W.
For electrical contacts for circuit breakers; sintered.

TUNGSIT.
M-1046; W_2C + WC.
For hard facing and rebuilding bits; wear resist.

TUNGSITE.
M-1432; 0.33 C, 0.94 Cr, 0.48 Mo, 0.5 W, bal Fe.
Heat treated: 164,000 TS; 136,500 YS; 14 El; 43 RA; 275 Brin.
For punches, chisels, gauges, shear blades, gears; oil hardened, tough.

TUNGSTEEL.
M-1684; 0.35 C, 0.65 Mn, 0.35 Si, 0.75 Cr, 0.45 W, 0.40 Mo, bal Fe.
Oil or water hardening steel for shock resisting applications, as riveting hammers.

TUNGSTEN.
M-1755; W.
Purities: 99.999+%, 99.995% degassed, 99.99%, 99.95, commercial purity 99+%.
Forms: Powder, wires, sheets, foils, tubing, crucibles.

TUNGSTEN 3 RHENIUM.
M-65; 97 W, 3 Re.
For thermocouples to 5200°F.

TUNGSTEN 5 RHENIUM.
M-65; 95 W, 5 Rc.
For thermocouples to 5200°F.

TUNGSTEN 21.
M-1422; C, W, alloy, bal Fe.
Hardenable in oil or water to 54-56 Rock C.
Used for arbors, axles, boring bars, mandrels.

TUNGSTEN 26 RHENIUM.
M-65; 74 W, 26 Re.
For thermocouples to 5200°F.

TUNGSTEN 09991.
M-53; 1 Ni, 99 W.
Sintered: 28 Cr; Elect. cond.: 19% I.A.C.S.
For electrical contacts.

TUNGSTEN ALLOY 5H.
M-987; Sintered carbide tool material.
For cutting cast iron when cratering is a factor.

TUNGSTEN ALLOY 5S.
M-987; Sintered carbide tool material.
For fine finishing and precision boring of carbon and alloy steel.

TUNGSTEN ALLOY 8H.
M-987; Sintered carbide tool material.
High speed planing cast iron where cratering is a factor; also for heat and wear resistant parts as gages.

TUNGSTEN ALLOY 8T.
M-987; Sintered carbide tool material.
For finishing cuts on steel.

TUNGSTEN ALLOY 9.
M-987; Sintered carbide tool material.
For rough cutting cast iron, non-ferrous metals and non-metallics; wear resistant applications, light shock.

TUNGSTEN ALLOY 9A.
M-987; Sintered carbide tool materials.
For wear resistant applications including heavy shock.

TUNGSTEN ALLOY 9A15.
M-987; Sintered carbide tool material.
For wear resistant applications, including heavy shock.

TUNGSTEN ALLOY 9A20.
M-987; Sintered carbide tool material.
To withstand heavy impact.

TUNGSTEN ALLOY 9A25.
M-987; Sintered carbide tool material.
To withstand heavy impact.

TUNGSTEN ALLOY 9A30.
M-987; Sintered carbide tool material.
To withstand heavy impact.

TUNGSTEN ALLOY 9B.
M-987; Sintered carbide material.
Cutting tools for the fine finishing and precision boring of cast iron, non-ferrous metals and non-metallics.

TUNGSTEN ALLOY 9C.
M-987; Sintered carbide tool material.
Cutting tools for light finishing cast iron, non-ferrous metals and non-metallics.

TUNGSTEN ALLOY 9H.
M-987; Sintered carbide tool material.
Cutting tools for general purpose machining of cast iron, non-ferrous metals and non-metallics.

TUNGSTEN ALLOY 9K.
M-987; Sintered carbide tool material.
Cutting tools for general purpose machining of steel on non-rigid machines.

TUNGSTEN ALLOY 9M.
M-987; Sintered carbide tool material.
For wear resistant applications subject to only light shock.

TUNGSTEN ALLOY 9S.
M-987; Sintered carbide tool material.
For general purpose and heavy duty machining of carbon and alloy steels.

TUNGSTEN ALLOY 9T.
M-987; Sintered carbide tool material.
Cutting tools for semi-finish cuts on carbon and alloy steels.

TUNGSTEN ALLOY 10T.
M-987; Sintered carbide tool material.
Cutting tools for general purpose machining of carbon and alloy steels.

TUNGSTEN ALLOY 11C.
M-987; Sintered carbide tool material.
For light to medium impact applications in mining and rock drilling.

TUNGSTEN ALLOY 11H.
M-987; Sintered carbide tool material.
Cutting tools for planing steel when cratering is a factor.

TUNGSTEN ALLOY 11T.
M-987; Sintered carbide tool material.
Cutting tools for roughing operations on carbon and alloy steels.

TUNGSTEN ALLOY 12C.
M-987; Sintered carbide tool material.
Tools for mining and rock drilling applications.

TUNGSTEN ALLOY 12H.
M-987; Sintered carbide tool material.
Cutting tools where cratering and shock are factors as in shaping and planing.

TUNGSTEN ALLOY 13H.
M-987; Sintered carbide tool material.
Cutting tools for special applications, as: machining uranium, removal of hot flashwelds, extremely heavy roughing where cratering is a factor.

TUNGSTEN ALLOY 15C.
M-987; Sintered carbide tool material.
For percussion applications subject to medium to heavy impact.

TUNGSTEN ALLOY 16G.
M-987; Sintered carbide tool material.
Cutting tools for special applications, as: machining hot flash from steel welds.

TUNGSTEN ALLOY 20H.
M-987; Sintered carbide tool materials.
For parts and tools subject to heavy impact, galling conditions.

TUNGSTEN ALLOY 25H.
M-987; Sintered carbide tool material.
For parts and tools subject to very heavy impact, non-galling conditions.

TUNGSTEN ALLOY 30H.
M-987; Sintered carbide tool material.
For parts and tools subject to extra heavy impact, non-galling conditions.

TUNGSTEN BRASS.
M-England; 60 Cu, 2-4 W, 1-14 Ni, 0-3 Al, 0-0.2 Sn, bal Zn.
For hardware, fittings; corrosion resistant.

TUNGSTEN BRONZE.
M-England; 10 W, bal Cu.
For electrical contacts; wear resistant.

TUNGSTEN (COARSE GRAIN).
M-53; 99.9 W.
Swaged rod: 29 Rc; Elect. cond.: 31% I.A.C.S.
For electrical contacts.

TUNGSTEN-COPPER.
M-1698; 75 W, 25 C (or as ordered); sintered.
For use as EDM and ECM electrodes.

TUNGSTEN DIAMOND.
M-1405; 1.38 C, 4.5 W, 0.25 Si, 0.75 Cr, 0.35 Mn, bal Fe.
For form tools, chasers, reamers, hobs; oil hardened, wear resistant.

TUNGSTEN (FINE GRAIN).
M-53; 99.9 W.
Swaged rod: 41 Rc; Elect. cond.: 30% I.A.C.S.
For electrical contacts.

TUNGSTEN HOT WORK.
M-365; 0.43 C, 2.5 Cr, 0.1 V, 9 W, bal Fe.
For hot shearing, hot punches, hot drawing; hot work steel.

TUNGSTEN NICKELS.
M-44; 4 W, bal Ni.
Ribbons and strip for high strength electron tube cathodes.

TUNGSTEN-RHC.
M-1681; 4 Re, 0.35 Hf, 0.25 C, bal W.
At 3500°F: 75,000 min TS.
For components for electrical equipment and die casting machines, furnace heating elements, rocket engines.
High heat and oxidation resistance.

TUNGSTEN SPECIAL.
M-15; 0.34 C, 0.5 Mo, 0.5 W, 1.0 Cr, 0.8 Mn, bal Fe.
Heat treated: 167,000 TS; 137,000 YS; 13 El; 43 RA; 280 Brin.
For shafts, chuckjaws, chisels, punches, gears; shock resistant, oil or water hardened.

TUNGSTEN TAP.
M-387; 0.9 C, 3.5 W, bal Fe.
For tools, dies; oil hardened.

TUNGTEC 10112.
M-717.
Tungsten carbide and nickel base alloy powder for hard surfacing.

TUNGTIC TC1.
M-1667; 5 Mo, bal TiC.
Sintered: 120,000 Tr.S.; Rock A 93.5.
For tools, bearings, seals.
Resists heat and wear.

TUNGTIC TC2.
M-1667; 10 Mo, bal TiC.
Sintered: 170,000 Tr.S.; Rock A 92.5.
For tools, bearings, seals.
Resists heat and wear.

TUNGTUBE see AIRCO TUNGTUBE.

TUNGUM.
M-110; 0.99 Al, 82.48 Cu, 0.3 Fe, 0.72 Ni, 14.6 Zn, 0.76 Si.
Sand cast: 46,600 TS; 23,000 YS; 51 El; 98 Brin.
Chill cast: 49,300 TS; 22,000 YS; 67 El; 112 Brin.
Forged hard: 114,700 TS; 104,800 YS; 11 El; 218 Brin.
For compressor parts, pumps, centrifugal driers, chemical apparatus; resists acid corrosion and fatigue.

TUNGUM C.
M-110; 75 Cu, 3 Al, 2 Ni, 20 Zn.
Cast: 78,500 TS.
For castings in marine applications, hardware, shafts.
High strength and corrosion resistant.

TUNGWELD C.
M-578; WC alloy.
For hard facing electrode; abrasion resistant.

TUNGWELD F.
M-578; WC alloy.
For hard facing electrode; abrasion resistant.

TUNGWIN.
M-1510; 0.34 C, 0.78 Mn, 0.28 Si, 0.82 Cr, 0.40 Mo, 0.51 W, bal Fe.
Heat treated: 194,000-267,000 TS; 185,000-216,000 YS; 37-51 El; 9-15 RA; 395-498 Brin.
For arbors, dies, gears, shafts, spindles; oil or water hardened, non-tempering.

TURBADIUM BRONZE.
M-309; 50.96 Cu, 0.47 Sn, 0.30 Pb, 2.21 Mn, 1.75 Ni, 43 Zn, 1.36 Fe.
For propellers.

TURBIDE R34.
M-Eng.; TiC, Cr$_3$C$_2$, bal Ni.
For gas turbine blades; sintered, oxidation resistant.

TURBIDE R45.
TiC, Cr$_3$C$_2$, bal Ni.
Sintered: 670 Brin.
For gas turbine blades; sintered, oxidation resistant.

TURBINE BLADING-1.
82.1 Cu, 14.7 Ni, 2.5 Al, 0.7 Zn, 0.04 Si.
For turbine blades; corrosion resistant.

TURBINE BLADING-2.
78-81 Cu, 21-19 Ni, 0.75 max Fe.
For turbine blades; corrosion resistant.

TURBINE BRASS.
M-England; 22-32 Zn, 0.25 Pb, 0.4 Fe, bal Cu.
For turbine parts.

TURBINE BUSHING.
M-Eng.; 11 Zn, 15 Ni, 2 Sn, 10 Pb, bal Cu.
For turbine bearings and bushings; heavy duty.

TURBINE METAL.
M-129; 55 Cu, 35 Zn, Al, Mn, 3 Ni, bal Fe.
Cast: 80,000 TS; 40,000 YS; 20 El; 20 RA; 125-160 Brin.
For turbine runners; corrosion and erosion resisting.

TURBISTON BRONZE.
M-U.S.; 55 Cu, 41 Zn, 2 Ni, 1 Al, 0.84 Fe, 0.16 Mn.
For pistons; highly resistant to sea water; also called a brass.

TURBISTONS BRASS.
M-England; 41 Zn, 0.5 Fe, 2 Ni, 1 Al, 0.2 Mn, bal Cu.
For hardware, fittings; corrosion resistant, high strength.

TURBO.
M-2; 77 Cu, 10 Al, 1 Ni.
Cast: 78,000-98,000 TS; 15-8 El; 140-170 Brin.
Hot pressed: 98,000-121,000 TS; 20-10 El; 170-190 Brin.
For worm wheels, pumps, spindles, worm shafts; White alloy; resists acid corrosion.

TURBO GLYCO.
M-240; Sn, Sb, bal Pb.
For turbine bearings; Babbitt.

TURBO "S".
M-2; 8 Ni, 12 Al, bal Cu.
Cast: 86,000-69,000 TS; 8-15 El; 150-130 Brin.
Hot pressed: 100,800-127,700 TS; 15-5 El; 180-220 Brin.
For worm wheels, pumps, spindles, worm shafts; White alloy, resists acid corrosion; resists steam at 550°F.

TURBOTHERM 20CO-45 now **VEW L125.**

TURBOTHERM 20 MV now **VEW T550.**

TURBOTHERM 20 MVNB now **VEW T560.**

TURBOTHERM 20 MVW now **VEW T502.**

TURBOTHERM 1613MV now **VEW T250.**

TURBOTHERM 1613NB now **VEW T275.**

TURBOTHERM 1616M now **VEW T255.**

TUREX 4.
M-884; 0.3 C, 1.0 Si, 1.3 Cr, 0.2 V, 4.0 W, bal Fe.
For forging dies, hot shears, rivet sets; hot work steel, oil hardened.

TURNING-1.
58.5 Cu, 29 Zn, 12 Ni, 0-5 Pb.
For white metal parts; corrosion resistant.

TURNING-2.
65 Cu, 22 Zn, 12 Ni, 1 Pb.
For white metal parts; corrosion resistant.

TUSCA.
M-275; 0.50 C, 0.60 Mn, 1.3 Cr, 0.3 Mo, 0.25 V, bal Fe.
For punches, upsetters, riveting machines, plastic dies, jewelry dies, coining and silverware dies; oil hardened, shock resistant.

TUT-A-BRASE.
M-750; C, alloy, bal Fe.
For wear resistant plates; preheat treated, abrasion resistant.

TUTANIA, CAST.
M-Eng.; 92 Sn, 4.7 Sb, 2.5 Cu, 0.3 Pb.
For bearings, table and kitchenware; antifriction.

TUTANIA, ENGLISH.
M-Eng.; 80 Sn, 16 Sb, 2.7 Cu, 1.3 Zn.
For dishes, ornamental pieces; corrosion resistant.

TUTANIA, PLATE.
M-Eng.; 91-90 Sn, 8-6 Pb, 0.7-2.7 Cu, 0.3-1.3 Zn.
For household ware, platters, dishes; freecutting.

TUTENAG.
Commercial Zn.
For roofing, alloying.

TV2 ALLOY.
M-USSR; Ti alloy.
For high temperature applications; corrosion resistant.

T.V.K. 45.
 M-1116; 0.40 C, 4.5 Cr, 4.5 W, 2.2 V, 0.4 Mo, 0.3 Mn, 0.3 Si, 4.5 bal Fe.
 Air or oil hardening tool and die steel; for hot forging and thread rolling copper, brass and steel; hot punches; molds for brass.
 AFNOR: Z 35 WKCV 05.05.04.02; AISI H 19.

TWIN VAN 96.
 M-140; 0.96 C, 4.25 Cr, 2.1 V, 18.5 W, 0.65 Mo, bal Fe.
 High speed steel for cutting tools, tungsten type; AISI T2.

TWO-TO-ONE.
 66.7 Cu, 33.3 Zn.
 For drawn or spun brass parts; high ductility.

TYCROME now **RYCROME.**

TYLER.
 M-1705; 1.10 C, 0.3 Mn, 0.25 Si, 0.6 Cr, bal Fe.
 Water hardening tool steel; AISI W5.

TY-LOY.
 M-929; 0.7 C, 1.2 Cr, bal Fe.
 For wire screens; abrasion resistant.

TYPE 5 NICKEL-COBALT.
 M-580; Ni-Co.
 For anodes; electroplating.

TYPE 5 STEEL.
 M-1061; 0.18-0.23 C, 0.7-0.9 Mn, 1.6-2.0 Ni, 0.2-0.3 Mo, bal Fe.
 Rolled: 88,000 TS; 70,000 YS; 35 El; 65 RA; 174 Brin.
 For oil well sucker rods; case hardened.

TYPE F.
 M-55; 0.55 C, Cr, Ni, Mo steel.
 Prehardened to ordered hardness.
 Economy hot work steel for short runs.
 For hammer dies, sow blocks, and other hot work tools.

TYPE METAL.
 M-U.S.; 50 Pb, 28 Sb, 22 Bi.
 For type casting.

TYPE METAL, COMMON.
 M-U.S.; 60-56 Pb, 4.5-30 Sb, 10-40 Sn.
 For type metal.

TYPE METAL, ENGLISH.
 M-English; 78-55 Pb, 5-30 Sb, 2-35 Sn, 0-1 Cu.
 For type metal.

TYPE METAL, FRENCH.
 M-French; 78-55 Pb, 5-30 Sb, 2-35 Sn, 0-1 Cu.
 For type metal.

TYPE METAL, GERMAN.
 M-German; 78-55 Pb, 5-30 Sb, 2-35 Sn, 0-1 Cu.
 For type metal.

TYPE METAL, MONOTYPE.
 M-U.S.; 76 Pb, 16 Sb, 8 Sn.
 For type casting.

TYPE METAL, STANDARD.
 M-U.S.; 79 Pb, 16 Sb, 5 Sn.
 For type casting.

TYPE METAL, STANDARD.
 M-U.S.; 58 Pb, 15 Sb, 26 Sn, 1 Cu.
 For type casting.

TYPE METAL, STEREOTYPE.
 M-U.S.; 83.75 Pb, 11.75 Sb, 4 Sn, 0.4 Cu.
 For type casting.

TYPEWRITER METAL.
 57 Cu, 20 Ni, 20 Zn, 3 Al.
 For typewriter parts; corrosion resistant.

TYPLEX.
 M-289; 0.4 C, 4 Ni, 1.5 Cr, 0.8 Mo, 0.4 Mn, bal Fe.
 For hot forging dies, gauges; air or oil hardened, shock resistant.

TYRANN EXTRA now **VEW K466.**

TYSELEY ALLOY.
 M-Eng.; 3.5 Cu, 8.7 Al, 87 Zn, 0.3 Si.
 For castings, ornaments.

TZ 6.
 M-Eng; 5.0-6.0 Zn, 0.4-1.0 Zr, 1.5-2.3 Th, bal Mg.
 Sand cast and precipitation treated; 255 MPa TS; 155 MPa YS; 5 El 65 Brin.
 BS. 2970 MAG 9.

TZC.
 M-114, M-151; 1.25 Ti, 0.3 Zr, 0.15 C, bal Mo.
 Stress Relieved: 144,000 TS; 105,000 YS; 22 El; 36 RA.
 At 2000°F: 93,000 TS.
 At 2400°F: 60,000 TS.
 For aerospace equipment and components.
 High hot strength, heat treatable.

TZM.
 M-114, M-151; 0.5 Ti, 0.08 Zr, 0.015 C, bal Mo.
 Recrystallized: 80,000 TS; 55,000 YS; 20 El.
 At 2000°F: 73,000 TS; 150 DPH hard.
 At 2400°F: 53,500 TS.
 For heat engines, heat exchangers, nuclear reactors, radiation shields, extrusion dies, boring bars.
 High strength and hardness at high temperatures, heat and corrosion resistant.

U

U-10 ETC see **UNIMET U-10 ETC.**

U-20.
M-249; 13 Cu, 0.25 Mg, 1.5 Mn, 1.0 Ni, bal Al.
Heat treated: 40,000-43,000 TS; 0.5 El; 120-130 Brin.
For pistons; heat resistant.

U222.
M-1697.
Titanium carbide coated carbide.
Density: 14.95; 92.0 Rock A.
Very high resistance to abrasion; good shock resistance. For semi- roughing and finishing, particularly of austenitic stainless steel.

U225.
M-1697.
Titanium carbide coated carbide.
Density: 12.6; 91.3 Rock A.
For semi-roughing, including milling to general purpose machining for improvement in tool life.

U227.
M-1697.
Titanium carbide coated carbide.
Density 12.60; 92.6 Rock A.
High resistance to crater and abrasion; for light, high production machining.

U.B. DOUBLE BOLT.
M-111; 0.40-0.50 C, bal Fe.
Annealed: 2 El.
For tools, chisels; water hardened.

UC6.
M-1697; 93.5 Wc, 6.5 Ni.
A 91.2 Rock; Density 14.95.
Sintered carbide tool material.

U.C.A. 2-2.
M-111; 0.90 C, 1.6 Mn, 1.5 Cr, bal Fe.
640 Brin.
For blanking and forming dies, pressing dies; oil hardened, non-deforming.

UCHATINS BRONZE.
M-England; 8 Sn, 92 Cu.
For bearings, gears, worm wheels; wear resistant.

UCV-60.
M-176; 0.25 C, 1.50 Mn, 0.30 max Si, 0.25-0.40 Cu, 0.40-0.65 Cr, 0.02-0.10 V, bal Fe.
70,000 psi TS; 63,000 psi YS; 19 El.
High strength, low alloy steel.

UCV-65.
M-176; 0.23 C, 1.55 Mn, 0.30 Si, 0.015 V, bal Fe.
80,000 psi TS; 60,000 psi YS; 15 El.
High strength, low alloy steel.
ASTM 572.

UDDCO-12.
M-111; 0.20 C, 12 Cr, 1 Mo, 0.3 V, 0.5 W, bal Fe.
Annealed: 90,000 TS; 40,000 YS; 25 El; B 92 Rock.
Hardened: 240,000 TS; 205,000 YS; 9 El; C 50 Rock.
For cutlery, valves, bearings, surgical instruments, hardware.
Corrosion resistant, hardenable.

UDDEHOLM 5.75.
M-111; 0.35 C, 23-27 Cr, bal Fe.
Annealed: 90,000 TS; 60,000 YS; 20 El; 45 RA; 180 Brin.
For furnace parts, heat treating boxes; Type 446; heat resistant.

UDDEHOLM 6.31.26.
M-111; 0.15 min C, 12-14 Cr, bal Fe.
Annealed: 88,000 TS; 40,000 YS; 32 El; 68 RA; 170 Brin.
For cutlery, valve trim, turbine blades; Type 420; stainless, hardenable.

UDDEHOLM 63.
M-111; 0.08 max C, 17-19 Cr, 9-12 Ni, Cb = 10 x C, bal Fe.
Annealed: 85,000-95,000 TS; 35,000-45,000 YS; 55-50 El; 175 Brin.
For welded structures, chemical plant equipment; type 347, corrosion and heat resistant.

UDDEHOLM 121.
M-111; 0.08 max C, 11.5-13 Cr, 0.1-0.3 Al, bal Fe.
Annealed: 71,000 TS; 42,600 YS; 22 El; 70 Ra; 150 Brin.
Heat treated: 175,000 TS; 145,000 YS; 21 El; 64 RA; 352 Brin.
For oil refinery and chemical plant equipment; Type 405; corrosion resistant.

UDDEHOLMS STAINLESS 16.
M-111; 0.27 C, 13.5 Cr, 1.5 Mo, bal Fe.
Annealed: 80,000 TS; 40,000 YS; 24 El; B 94 Rock.
Hardened: 240,000 TS; 210,000 YS; 9 El; C 52 Rock.
For cutlery, surgical instruments, hardware, bearings, valves, gears.
Corrosion resistant, hardenable.

UDDEHOLMS STAINLESS 51.
M-111; 0.10 C, 14 Cr, 1 Mo, bal Fe.
Annealed: 80,000 TS; 38,000 YS; 25 El; B 93 Rock.
For springs, shafts, table flatware, oil refinery equipment.
Corrosion resistant.

UDDEHOLMS STAINLESS 851.
M-111; 0.20 C, 11.5 Cr, 1 Mo, 0.3 V, bal Fe.
Annealed: 95,000 TS; 40,000 YS; 25 El; B 92 Rock.
Hardened: 240,000 TS; 105,000 YS; 10 El; C 50 Rock.
For valves, bearings, cultery, gears, surgical instruments, shafts.
Corrosion resistant, hardenable.

UDIMAR B-250.
M-1317; 0.03 max C, 17.0-19.0 Ni, 7.0-8.5 Co, 4.6-5.1 Mo, 0.05-0.15 Al, 0.3-0.5 Ti, 0.002-0.006 B, 0.02 max Zr, bal Fe.
Hardened: 265-275 ksi TS; 255-265 ksi YS; 10-13 El, 50-60 RA; 48-50 Rock C.
Maraging steel.

UDIMAR B-300.
M-1317; 0.03 max C, 17.0-19.0 Ni, 8.5-9.5 Co, 4.7-5.2 Mo, 0.05-0.15 Al, 0.5-0.8 Ti, 0.002-0.006 B, 0.02 max Zr, bal Fe.
Hardened: 285-300 ksi TS; 270-280 ksi YS; 10-13 El; 50-60 RA; 51-55 Rock C.
Maraging steel.

UDIMET.
M-1317; 0.05 C, 18 Cr, 18 Fe, 3 Mo, 5.2 Cb, 1 Ti, 0.06 Al, 0.004 B, bal Ni.
Heat treated:
At 70°F: 190,000 TS; 165,000 YS; 20 El.
At 1200°F: 170,000 TS; 140,000 YS; 18 El.
For components of gas turbines, missile and booster assemblies; heat resistant; weldable; -423°F to +1300°F applications.

UDIMET 41.
M-1317; 0.09 C, 19 Cr, 11 Co, 10 Mo, 3 Ti, 1.5 Al, 5.0 max Fe, bal Ni.
Heat treated: 206,000 TS; 154,000 YS; 14 El; 400 Brin.
At 1200°F: 194,000 TS; 145,000 YS; 14 El.
At 1700°F: 58,000 TS; 50,000 YS; 26 El.
For jet engine components, afterburner parts, bolting; for severly stressed high temperature applications.

UDIMET 75.
M-1317; 0.08-0.15 C, 1.0 Mn, 1.0 Si, 18.0-21.0 Cr, 0.20-0.60 Ti, 5.0 Fe, bal Ni.
At 70°F: 108 ksi TS; 40 ksi YS; 42 El.
At 1200°F: 76 ksi TS; 30 ksi YS; 44 El.
Vacuum melted high temperature alloy.

UDIMET 80A.
M-1317; 0.10 C, 1.0 Mn, 1.0 Si, 18.0-21.0 Cr, 2.0 Co, 0.008 B, 1.0-1.8 Al, 1.8-2.7 Ti, 3.0 Fe, bal Ni.
At 70°F; 180 ksi TS; 113 ksi YS; 23 El.
At 1200°F: 146 ksi TS; 99 ksi YS; 18 El.
Vacuum melted, precipitation hardened, high temperature alloy.

UDIMET 90.
M-1317; 0.13 C, 1.0 Mn, 1.5 Si, 18.0-21.0 Cr, 15.0-21.0 Co, 0.010 B, 0.80-2.0 Al, 1.8-3.0 Ti, 3.0 Fe, bal Ni.
At 70°F: 180 ksi TS; 114 ksi YS; 27 El.
At 1200°F; 150 ksi TS; 100 ksi YS; 18 El.
Vacuum melted, precipitaiton hardened, high temperature alloy.

UDIMET 105.
M-1317; 0.20 C, 1.0 Mn, 1.0 Si, 13.5-15.75 Cr, 18.0-22.0 Co, 4.5-5.5 Mo, 0.010 B, 4.5-4.9 Al, 0.9-1.5 Ti, 2.0 Fe, bal Ni.
At 70°F: 166 ksi TS; 112 ksi YS; 12 El.
At 1600°F: 92 ksi TS; 67 ksi YS; 24 El.
Vacuum melted, precipitation hardened, high temperature alloy.

UDIMET 115.
M-1317; 0.20 C, 1.0 Mn, 1.0 Si, 14.0-16.0 Cr, 13.5-16.5 Co, 3.0-5.0 Mo, 4.5-5.5 Al, 3.5-4.5 Ti, 1.0 Fe, bal Ni.
At 70°F: 178 ksi TS; 120 ksi YS; 25 El.
At 1600°F: 120 ksi TS; 84 ksi YS; 18 El.
At 1800°F; 75 ksi TS; 38 ksi YS; 23 El.
Vacuum melted, precipitation hardened, high temperature alloy.

UDIMET 200.
M-1317; 0.10 max C, 0.35 Al, 2.5-3.0 Ti, 5.0-7.0 Mo, 11.0-14.0 Cr, 1. max Co, 40.0-45.0 Ni, 2.0 max Mn, 0.010-0.020 B, bal Fe.
At 70°F: 180 ksi TS; 130 ksi YS; 25 El.
At 1200°F: 124 ksi TS; 93 ksi YS; 10 El.
Vacuum melted, precipitation hardened, high temperature alloy.
Similar to Incoloy 901.

UDIMET 500.
M-1317, M-1491; 0.15 max C, 2.5-3.2 Al, 2.5-3.2 Ti, 3-5 Mo, 15-20 Cr, 13-20 Co, 4.0 max Fe, bal Ni.
At 70°F: 188,000 psi TS; 15 El; 22 RA.
At 1000°F: 185,000 psi TS; 11 El; 16 RA.
At 1800°F: 46,000 psi TS; 22 El; 41 RA.
For jet engine components; high heat resistance.

UDIMET 520.
M-1317, M 1491; 0.05 C, 19 Cr, 6 Mo, 3 Ti, 2 Al, 12 Co, 1 W, 0.005 B, bal Ni.
Heat treated: at 70°F: 190,000 TS;: 135,000 YS; 21 El.
At 1200°F: 170,000 TS; 115,000 YS; 17 El.
At 1800°F: 45,000 TS; 40,000 YS; 22 El.
For high temperature applications, jet engine parts; high heat resistance.

UDIMET 625.
M-1317; 0.05 C, 22 Cr, 3 Fe, 9 Mo, 4 Cb, 0.2 Ti, 0.2 Al, bal Ni.
Annealed:
At 70°F: 140,000 TS; 70,000 YS; 55 El.
At 1200°F: 103,000 TS; 60,000 YS; 80 El.
Good oxidation and corrosion resistance; high strength and toughness from cryogenic temperatures to 2000°F.

UDIMET 630.
M-1491; 0.04 max C, 0.2 max Mn, 0.2 max Si, 17.0 Cr, 1.0 max Co, 3.1 Mo, 3.0 W, 6.0 Cb, 1.1 Ti, 0.6 Al, 0.005 B, 17.5 Fe, bal Ni.
For parts operating up to 1000°F.

UDIMET 700.
M-1317, M 1491; 0.15 max C, 3.7-4.7 Al, 3-4 Ti, 4.5-5.7 Mo, 13-17 Cr, 17-20 Co, B, bal Ni.
Rolled 204,000 TS; 140,000 YS; 17 El; 20 RA.
Hr. Tr.: at 1200°F: 180,000 TS; 140,000 YS; 16 El.
Ht.Tr.: at 1400°F: 150,000 TS; 125,000 YS; 33 El.
Ht. Tr.: at 1800°F: 50,000 TS; 45,000 YS; 28 El.
For turbine blades, discs, combustion chambers; high creep and oxidation resistance.

UDIMET 710.
M-1317, M 1491; 0.13 C, 18 Cr, 15 Co, 3 Mo, 1.5 W, 5 Ti, 2.5 Al, 0.02 B, 0.08 Zr, bal Ni.
Heat treated (for optimum stress rupture properties):
At 1200°F: 190,000 TS; 130,000 YS; 8 El.
At 1600°F: 105,000 TS; 95,000 YS; 32 El.
At 1800°F: 60,000 TS; 40,000 YS; 30 El.
Land based gas turbine blades, high temperature application to 1800°F.

UDIMET 718.
M-1317; 0.05 C, 18 Cr, 18 Fe, 3 Mo, 5.2 Cb, 1 Ti, 0.06 Al, 0.004 B bal Ni.
Heat treated:
At 70°F: 190,000 TS; 165,000 YS; 20 El.
At 1200°F: 170,000 TS; 140,000 YS; 18 El.
For components of gas turbines, missiles and booster assemblies; heat resistant; weldable; -423°F to +1300°F applications.

UDIMET 901.
M-1317; 0.07 C, 13 Cr, 36 Fe, 5.8 Mo, 2.8 Ti, 0.3 Al, 0.015 B, bal Ni.
Heat treated:
At 70°F: 180,000 TS; 125,000 YS; 25 El.
At 1200°F: 140,000 TS; 110,000 YS; 22 El.
At 1400°F: 90,000 TS; 80,000 YS; 22 El.
For turbine discs and parts; for applications in the 1000-1400°F temperature range.

UDIMET A-286.
M-1317; 0.05 C, 0.2 Al, 2.0 Ti, 1.3 Mo, 15.0 Cr, 0.015 B, 26.0 Ni, bal Fe.
At 70°F: 146,000 psi TS; 105,000 psi YS.
At 1200°F: 104,000 psi TS; 88,000 psi YS.
Vacuum melted, precipitation hardened superalloy.

UDIMET D-979.
M-1317; 0.08 max C, 0.70-1.30 Al, 2.7-3.3 Ti, 3.0-4.5 Mo, 14.0-16.0 C 25.5-28.5 Fe, 0.008-0.016 B, 3.0-4.5 W, bal Ni.
At 70°F: 190,000 TS; 130,000 YS.
At 1200°F: 160,000 TS; 125,000 YS.
Vacuum melted, precipitation hardened superalloy.

UDIMET HX.
M-1317; 0.05-0.10 C, 1.0 max Mn, 1.0 max Si, 20.5-23.0 Cr, 0.5-2.5 Co, 0.2-1.0 W, 8.0-10.0 Mo, 17.0-20.0 Fe, bal Ni.
Vacuum melted, high temperature alloy.

UDIMET L-605.
M-1317; 0.10 C, 20.0 Cr, 10.0 Ni, 15.0 W, bal Co.
At 1200°F: 103,000 TS; 35,000 YS.
Vacuum melted high temperature alloy.

UDIMET M-252.
M-1317; 0.15 C, 1.0 Al, 2.5 Ti, 10.0 Mo, 19.0 Cr, 10.0 Co, 0.007 B, bal Ni.
At 70°F: 180,000 TS; 120,000 YS; at 1400°F: 130,000 TS; 105,000 YS.
Vacuum melted, precipitation hardened superalloy.
For turbine buckets, high temperature fasteners.

UDIMET N-155.
M-1317; 0.15 C, 3.0 Mo, 21.0 Cr, 20.0 Co, 10.0 Ni, 2.5 W, 1.5 Mn 1.0 Cb, 0.15 N, bal Fe.
At 70°F: 118,000 TS; 58,000 YS.
At 1200°F: 79,000 TS; 43,000 YS.
Vacuum melted high temperature alloy.

UDIMET NI-80 A.
M-1317; 0.06 C, 1.3 Al, 2.5 Ti, 19.5 Cr, 1.1 Co, bal Ni.
At 70°F: 145,000 psi TS; 90,000 psi YS.
At 1200°F: 115,000 psi TS; 80,000 psi YS.
Vacuum melted, precipitation hardened superalloy.

UDIMET S-816.
M-1517; 0.38 C, 4.0 Mo, 20.0 Cr, 4.0 Fe, 20.0 Ni, 4.0 W, 1.2 Mn, 4.0 Cb, bal Co.
At 70°F: 140,000 TS; 67,000 YS; at 1200°F: 112,000 TS; 44,000 YS.
Vacuum melted high temperature alloy.

UDIMET V-57.
M-1317; 0.05 C, 0.2 Al, 3.0 Ti, 1.3 Mo, 15 Cr, 0.015 B, 26.0 Ni, bal Fe.
At 70°F: 175,000 TS; 120,000 YS; at 1200°F: 130,000 TS; 110,000 YS.
Vacuum melted, precipitation hardened superalloy.

UDIMET WASPALOY.
M-1317; 0.07 C, 19 Cr, 13 Co, 4 Mo, 3 Ti, 1.3 Al, 0.005 B, 0.05 Zr, bal Ni.
Heat treated:
At 70°F: 190,000 TS; 130,000 YS; 25 El.
At 1200°F: 165,000 TS; 105,000 YS; 34 El.
At 1400°F: 125,000 TS; 95,000 YS; 28 El.
For gas turbine blading and discs; high temperature applications up to 1500°F where combined tensile and stress rupture properties are required.

U-E3-S.
M-French; Cu alloy.
100 Brin.
For aircraft construction.

UGICARB.
M-1787.
WC plus alloy.
For anti-skid tire studs and studded straps for snow and ice chains.

UGINARC 23 K.
M-1120; 0.02 max C, 9.8 Ni, 20.0 Cr, 0.4 max Si, 0.05 max Co, 0.015 max S, 0.020 max P, bal Fe.
Austenitic stainless steel for coated electrodes.
AWS E 308 L.

UGINARC 23 S1.
M-1120; 0.025 max C, 10.5 Ni, 20.0 Cr, 0.9 Si, 0.050 max Co, 0.015 max S, 0.025 max P, bal Fe.
Astenitic stainless steel wire with high Si for automatic MIG welding.
AWS Type ER 308 L Si.

UGINARC F 36.
M-1120; 0.12 C, 9.5 Ni, 30.0 Cr, 0.025 max S, 0.030 max P, bal Fe.
Stainless steel for welding.
AWS E 312 and ER 312.

UGINARC MKS.
M-1120; 0.020 max C, 11.3 Ni, 18.5 Cr, 2.8 Mo, bal Fe.
Austenitic stainless steel for coated electrodes.
Similar to AISI 316 L.

UGINARC MKS1.
M-1120; 0.025 max C, 12.5 Ni, 18.5 Cr, 2.7 Mo, 0.8 Si, bal Fe.
Austenitic stainless steel wire, high Si grade of AISI 316 L for MIG welding.

UGINARC R 27.
M-1120; 0.09 C, 13.0 Ni, 24.0 Cr, 0.015 max S, 0.030 max P, bal Fe.
Austenitic stainless steel for coated electrodes for welding with flux.
AWS E 309 and ER 309.

UGINARC R 31.
M-1120; 0.12 C, 21.0 Ni, 26.0 Cr, 0.020 max S, 0.030 max P, bal Fe.
Austenitic stainless steel for coated electrodes for welding with flux.
AWS E 310 and ER 310.

UGINE B21.
M-1120; 0.21 C, 0.70 Mn, B=40 ppm, bal Fe.
For improved hardenability. (Boron steel).
Note: B 21 F grade for cold heading.

UGINE B 38.
M-1120; 0.38 C, 0.70 Mn, B=40 ppm, bal Fe.
Boron steel for improved hardenability.
Note: B 38 F grade for cold heading.

UGINE C4.
M-1120; 0.40 C, Mn, bal Fe.
Water hardening steel; for jacks, press columns, shafts.
AISI 1040.

UGINE C5.
M-1120; 0.50 C, Mn, bal Fe.
Water hardening steel; for heat treated parts.
AISI 1050.

UGINE DY1.
M-1120; 0.08 C, 25.0 Cr, 0.2 N, bal Fe.
Ferritic type stainless steel; resistant to corrosion at high temperatures.
AISI Type 446.

UGINE F12U.
M-1120; 0.10 C, 13.0 Cr, 0.25 Mo, 0.30 S, bal Fe.
Free-machining martensitic stainless steel.
Similar to AISI 416.

UGINE F13.
M-1120; 0.13 C, 12.8 Cr, bal Fe.
Martensitic stainless steel.
For cutlery and tablewear; will take high finish.
AISI 410.

UGINE F13A.
M-1120; 0.13 C, 12.0 Cr, 0.040 N, bal Fe.
Martensitic stainless steel.
AISI Type 410 for turbines.

UGINE F13B.
M-1120; 0.15 C, 12.0 Cr, 0.35 max Cu, bal Fe.
Martensitic stainless steel.
For cutlery and tableware; will take high polish.

UGINE F13V.
M-1120; 0.13 C, 12.8 Cr, bal Fe.
Martensitic stainless steel.
For cold heading (bolts and nuts).
AISI 410.

UGINE F14PH.
M-1120; 0.06 max C, 4.5 Ni, 14.2 Cr, 1.4 Mo, 3.5 Cu, bal Fe.
Precipitation hardening stainless steel; Type 15-5 PH.

UGINE F17.
M-1120; 0.05 max c, 16.5 Cr, bal Fe.
Ferritic type stainless steel; non-hardenable by heat treating. For cold deformation and polishing.
AISI 430.

UGINE F17C.
M-1120; 0.12 C, 16.8 Cr, bal Fe.
Ferritic stainless steel; special grade for cutlery and tableware.

UGINE F17G.
M-1120; 0.06 C, 17.0 Cr, 0.015 max S, bal Fe.
Ferritic stainless steel; non-hardenable.
For cutlery and tableware.
AISI 430.

UGINE F17G1.
M-1120; 0.05 max C, 17.0 Cr, bal Fe.
Ferritic stainless steel; for wire-drawing.
AISI 430.

UGINE F17M.
M-1120; 0.05 C, 16.7 Cr, 1.0 Mo, bal Fe.
Ferritic stainless steel; improved corrosion resistance due to Mo.
AISI Type 434.

UGINE F17U.
M-1120; 0.09 C, 16.8 Cr, 0.25 S, bal Fe.
Free machining, ferritic stainless steel.

UGINE F1.
M-1120; 0.09 C, 13.0 Cr, bal Fe.
Martensitic stainless steel.
For petroleum industry.
Similar to AISI 410.

UGINE FID.
M-1120; 0.09 C, 13.0 Cr, bal Fe.
Martensitic stainless steel.
For tableware, easily polished to a mirror finish.
Similar to AISI 410.

UGINE G2.
M-1120; 0.14 C, 2.7 Ni, 0.7 Cr, Mn, bal Fe.
Carburizing steel; for case hardened gears and pinions, general macinery.

UGINE G4.
M-1120; 0.17 C, 3.0 Ni, 0.8 Cr, Mn, bal Fe.
Alloy case hardening steel; for gears, pinions, general machinery.

UGINE G12S.
M-1120; 0.13 C, 1.4 Ni, 1.0 Cr, 0.15 Mo, Mn, bal Fe.
Alloy carburizing steel; for case-hardened pinions.

UGINE G14S.
M-1120; 0.18 C, 1.3 Ni, 1.0 Cr, 0.20 Mo, Mn, bal Fe.
Case hardening steel; for large carburized gears, general machinery.

UGINE GB5.
M-1120; 0.19 C, 1.3 Ni, 1.0 Cr, 0.70 Mn, B=40 ppm, bal Fe.
Boron steel for improved hardenability.
For large case-hardened parts resistant to overload and fatigue.

UGINE GB14S.
M-1120; 0.20 C, 0.55 Ni, 0.50 Cr, 0.80 Mn, 0.20 Mo, B=40 ppm, bal Fe.
Boron steel for improved hardenability.
For general machinery and case hardened gears and pinions.

UGINE HD20.
M-1120; 0.21 C, 0.60 Ni, 0.50 Cr, 0.20 Mo, Mn, bal Fe.
Carburizing steel; pinions for trucks and tractors.
AISI 8620.

UGINE KI17.
M-1120; 1.0 C, 17.0 Cr, 0.5 Mo, bal Fe.
Martensitic stainless steel; for stainless ball bearings.
AISI 440 C.

UGINE KMT.
M-1120; 0.35 C, 1.0 Cr, 0.25 Mo, Mn, bal Fe.
Oil hardened steel; shafts and gears with high fatigue resistance.
AISI 4135.

UGINE KMX.
M-1120; 0.42 C, 1.0 Cr, 0.25 Mo, Mn, bal Fe.
Oil hardened steel; for petroleum industry.
ASTM A193 Gr. B7.

UGINE KMZ.
M-1120; 0.30 C, 3.0 Cr, 0.40 Mo, Mn, bal Fe.
Oil hardening steel, for large machinery parts.
Can be nitrided.

UGINE KN9.
M-1120; 0.36 C, 2.8 Ni, 2.8 Cr, 0.30 Mo, Mn, bal Fe.
Air or oil hardened steel; for large shafts and gears, for good fatigue and shock resistance.

UGINE KNA.
M-1120; 0.35 C, 3.8 Ni, 1.7 Cr, Mn, bal Fe.
Oil hardening steel; for large and highly stressed parts.

UGINE KNDMO.
M-1120; 0.33 C, 4.0 Ni, 1.8 Cr, 0.40 Mo, Mn, bal Fe.
Air or oil hardening steel; for large or highly stressed parts requiring high elastic limit and endurance.

UGINE KNHMO.
M-1120; 0.30 C, 4.0 Ni, 1.3 Cr, 0.40 Mo, Mn, bal Fe.
Oil hardening steel; for highly stressed parts, general machinery and aviation.

UGINE KNO.
M-1120; 0.35 C, 1.4 Ni, 1.0 Cr, 0.23 Mo, Mn, bal Fe.
Oil hardened steel; for shafts and gears for good fatigue and shock resistance.

UGINE MB20.
M-1120; 0.20 C, 1.20 Mn, B=40 ppm, bal Fe.
Boron steel for improved hardenability.
Note: MB20F for cold heading.

UGINE MB38.
M-1120; 0.38 c, 1.20 Mn, B=40 ppm, bal Fe.
Boron steel for improved hardenability.
Note: MB 38 F for cold heading.

UGINE NC16.
M-1120; 0.15 C, 1.3 Ni, 1.0 C, Mn, bal Fe.
Carburizing steel; automobile pinions.

UGINE NC20.
M-1120; 0.19 C, 1.3 Ni, 1.0 Cr, Mn, bal Fe.
Carburizing steel; for truck pinions.

UGINE NS20E.
M-1120; 0.1 C, 7.5 Ni, 17.4 Cr, bal Fe.
Austenitic stainless steel; work hardens rapidly; for cold-rolled springs.
AISI 301.

UGINE NS20R.
M-1120; 0.1 C, 8.5 Ni, 18.0 Cr, 0.9 Mn, bal Fe.
Austenitic stainless steel; for work-hardened springs.
AISI 302.

UGINE NS21A.
M-1120; 0.05 max C, 9.4 Ni, 18.5 Cr, bal Fe.
Austenitic stainless steel; wire drawing grade.
AISI 304.

UGINE NS21AS.
M-1120; 0.06 C, 8.9 Ni, 18.5 Cr, bal Fe.
Austenitic stainless steel; bars for machining.
AISI 304.

UGINE NS21NB.
M-1120; 0.06 C, 10.5 Ni, 17.8 Cr, Nb=10xC min, 1.0 max, B, bal Fe.
Stabilized austenitic stainless steel for welding and elevated temperature
Similar to AISI 347.

UGINE NS21R/NS20P.
M-1120; 0.07 C, 8.5 Ni, 18.5 Cr, bal Fe.
Austenitic stainless steel. For springs, cables, parts in polished wire.
AISI 302/304.

UGINE NS21T.
M-1120; 0.06 C, 10.0 Ni, 17.8 Cr, Ti=5xC min, 0.6 max, bal Fe.
Stabilized austenitic stainless steel for welding and high temperature.
AISI 321.

UGINE NS22S.
M-1120; 0.03 max C, 9.5 Ni, 18.5 Cr, bal Fe.
Austenitic stainless steel bars; low carbon for resistance to intergranular corrosion.
AISI 304 L.

UGINE NS22SV.
M-1120; 0.03 max C, 11.0 Ni, 18.5 Cr, bal Fe.
Austenitic stainless steel wire for cold heading of screws.
AISI 304 L.

UGINE NS24.
M-1120; 0.16 C, 13.5 Ni, 23.0 Cr, bal Fe.
Austenitic stainless steel, resistant to oxidation to 1050°C.
AISI 309.

UGINE NS30.
M-1120; 0.10 C, 20.0 Ni, 25.0 Cr, bal Fe.
Austenitic stainless steel, resistant to oxidation to 1100°C.
AISI 310.

UGINE NS30C.
M-1120; 0.06 C, 21.5 Ni, 25.0 Cr, 2.5 Si, bal Fe.
Austenitic stainless steel for elevated temperature operation.
Low carbon grade of AISI 314.

UGINE NS30Z.
M-1120; 0.10 C, 20.0 Ni, 25.0 Cr, 2.0 Si, bal Fe.
Austenitic stainless steel, for high temperature operation.
AISI 314.

UGINE NSCD.
M-1120; 0.030 max C, 16.5 Ni, 17.0 Cr, 5.25 Mo, 2.65 Cu, bal Fe.
Austentic stainless steel, excellent corrosion resistance in strong acid environment.

UGINE NSM21.
M-1120; 0.06 C, 11.0 Ni, 17.2 Cr, 2.25 Mo, bal Fe.
Austenitic stainless steel.
AISI 316.

UGINE NSM21S.
M-1120; 0.030 max C, 11.5 Ni, 17.2 Cr, 2.25 Mo, bal Fe.
Austenitic stainless steel; improved resistance to intergranular corrosion.
AISI 316L.

UGINE NSM21SV.
M-1120; 0.030 max C, 12.0 Ni, 17.3 Cr, 2.25 Mo, bal Fe.
Austenitic stainless steel wire for cold-heading of screws.
AISI 316L.

UGINE NSM22S.
M-1120; 0.03 C, 16-18 Cr, 10-14 Ni, 2-3 Mo, bal Fe.
Annealed: 85,000 TS; 35,000 YS; 50 El; B 82 Rock.
For chemical plant equipment, welded tanks, digesters, evaporators, valve trim.
Type 316L stainless steel, austenitic, acid resistant.

UGINE NSM23S.
M-1120; 0.030 max C, 14.5 Ni, 18.5 Cr, 3.25 Mo, bal Fe.
Austenitic stainless steel.
AISI 317L.

UGINE NSMC.
M-1120; 0.06 C, 11.0 Ni, 17.0 Cr, 2.25 Mo, Ti=5xC min, 0.6 max, B, bal Fe.
Stabilized austenitic stainless steel for chemical and mechanical resistance at high temperature.
Modified AISI 316.

UGINE NSV.
M-1120; 0.05 C, 12.5 Ni, 18.2 Cr, bal Fe.
Austenitic stainless steel for mass produced screws and bolts.
AISI 305.

UGINE NSV3.
M-1120; 0.020 max C, 9.5 Ni, 17.8 Cr, 3.5 Cu, bal Fe.
Low carbon austenitic stainless steel.
UNS S 30430.

UGINE NSZ1AV.
M-1120; 0.05 max C, 10.0 Ni, 18.7 Cr, bal Fe.
Austenitic stainless steel; wire for cold heading of screws.
AISI 304.

UGINE P12U.
M-1120; 0.30 C, 13.2 Cr, 0.25 Mo, 0.30 S, 1.25 max Mn, bal Fe.
Free-machining martensitic stainless steel.
AISI 420 F.

UGINE QA2D.
M-1120; 0.030 max C, 21.5 Ni, 33.5 Cr, 0.45 Ti, 0.3 Al, bal Fe.
Stabilized stainless steel for high temperature operation.
UNS N08800.

UGINE U12.
M-1120; 0.20 C, 13.0 Cr, bal Fe.
Martensitic stainless steel; for machinery parts which are in contact with steam, water, wine, beer, etc.
AISI 420.

UGINE U12A.
M-1120; 0.20 C, 13.0 Cr, bal Fe.
Martensitic stainless steel; very low sulphur grade, for turbines.
AISI Type 420.

UGINE U17N.
M-1120; 0.16 C, 2.7 Ni, 15.6 Cr, 0.15 Mo, bal Fe.
Martensitic stainless steel; high strength.
Similar to AISI 431.

UGINE UGV 182.
M-1120; 0.06 C, 18.0 Cr, 1.6 Mo, 1.5 Mn, 0.18 S, bal Fe.
Free machining ferritic stainless steel.
Improved corrosion resistance.
ASTM A582 XM34; UNS S 18200.

UGINE UM3.
M-1120; 0.18 C, 1.0 Cr, 0.22 Mo, Mn, bal Fe.
Case hardening steel; for bolts and nuts, parts for automobile steering gear.
Similar to AISI 4118.

UGINE UM5.
M-1120; 0.25 C, 1.0 Cr, 0.25 Mo, Mn, bal Fe.
For corbonitrided automobile pinions.

UGINE UM7.
M-1120; 0.35 C, 1 Cr, 0.25 Mo, bal Fe.
For shafts, axles, gears, pinions, crankshafts; SAE 4135; oil hardened.

UGINE YB4B.
M-1120; 0.20 C, 1.30 Mn, 1.20 Cr, B = 40 ppm, bal Fe.
Boron steel for improved hardenability.
Excellent formability; for case hardened gears and pinions for gear boxes.

UGINE YBB.
M-1120; 0.16 C, 1.10 Mn, 1.0 Cr, B = 40 ppm, bal Fe.
Boron steel for improved hardenability.
Excellent formability; for case hardened cams, and pinions for gear boxes.

UGINOX F 12 T.
M-1782; 0.05 max C, 11 Cr, 0.5 Ti, bal Fe.
Cold rolled: 380-580 N/mm^2 TS; 195 N/mm^2 min YS; 20 min El.
AFNOR Z 3 CT 11; W.Nr. 4512.

UGINOX F 17.
M-1782; 0.07 C, 17 Cr, bal Fe.
Hot or cold rolled: 460-610 N/mm^2 TS; 275 N/mm^2 min YS; 17 El.
AFNOR Z 8 C 17; AISI 430; W.Nr. 4016.

UGINOX F 17 MO.
 M-1782; 0.06 C, 17 Cr, 1 Mo, bal Fe.
 Cold rolled: 490-610 N/mm^2 TS; 310 N/mm^2 min YS; 17 min El.
 AFNOR Z 8 CD 17-01; AISI 434; W.Nr. 4113.

UGINOX FIA.
 M-1782; 0.03 C, 13 Cr, bal Fe.
 Hot or cold rolled: 430-580 N/mm^2 TS; 250 N/mm^2 min YS; 20 min El.
 AFNOR Z6C 13; AISI 410 S; W.Nr. 4000.

UGINOX NS 20.
 M-1782; 0.11 C, 17 Cr, 7.5 Ni, bal Fe.
 Hot or cold rolled: 650-850 N/mm^2 TS; 265 N/mm^2 min YS; 40 min El.
 AFNOR Z12 On 17-07; AISI 301; W.Nr. 4319.

UGINOX NS 20 E.
 M-1782; 0.10 C, 18 Cr, 7.5 Ni, bal Fe.
 Normally supplied as hard rolled sheet or strip.
 AFNOR Z12 CN 18-07; Similar to AISI 301; Type W.Nr. 4310.

UGINOX NS 20 P.
 M-1782; 0.09 max C, 18 Cr, 8.5 Ni, bal Fe.
 Cold rolled: 580-730 N/mm^2 TS; 245 N/mm^2 min YS; 40 min El.
 AFNOR Z 10 CN 18-09; AISI 302.

UGINOX NS 21 A.
 M-1782; 0.07 max C, 18 Cr, 9 Ni, bal Fe.
 Cold rolled: 560-710 N/mm^2 TS; 235 N/mm^2 min YS; 40 min El.
 AFNOR Z 6 CN 18-09; Similar to AISI 304; W.-Nr. 4301.

UGINOX NS 21 AN.
 M-1782; 0.06 max C, 18 Cr, 9.5 Ni, bal Fe.
 Cold rolled: 540-690 N/mm^2 TS; 210 N/mm^2 min YS; 40 min El.
 AFNOR Z 6 CN 18-09; BS 304 515.

UGINOX NS 21 AR.
 M-1782; 0.07 max C, 18 Cr, 9.5 Ni, bal Fe.
 Cold rolled: 540-690 N/mm^2 TS; 205 N/mm^2 min YS; 45 min El.
 AFNOR Z 6 CN 18-09; Similar to AISI 304; W.Nr. 4301.

UGINOX NS 21 AS.
 M-1782; 0.08 max C, 18 min Cr, 9.5 Ni, bal Fe.
 Cold rolled: 560-710 N/mm^2 TS; 215 N/mm^2 min YS; 40 min El.
 AFNOR Z 6 Cn 18-09; AISI 304; W.Nr. 4301.

UGINOX NS 21 C.
 M-1782; 0.08 max C, 18 Cr, 10 Ni, Ti = 5 X C min, bal Fe.
 Cold rolled: 540-690 N/mm^2 TS; 225 N/mm^2 min YS; 35 min El.
 AFNOR Z 6 CNT 18-10; AISI 321; W.Nr. 4541.

UGINOX NS 22 L.
 M-1782; 0.03 max C, 18 min Cr, 9.5 Ni, bal Fe.
 Cold rolled: 500-560 N/mm^2 TS; 205 N/mm^2 min YS; 40 min El.
 AFNOR Z 2 CN 18-10; AISI 304 L; W.Nr. 4306.

UGINOX NS 22 S.
 M-1782; 0.03 max C, 18 Cr, 10 Ni, bal Fe.
 Cold rolled: 500-650 N/mm^2 TS; 205 N/mm^2 min YS; 40 El.
 AFNOR Z 2 CN 18-10; Similar to AISI 304 L; Type W.Nr. 4306.

UGINOX NS 24.
 M-1782; 0.15 C, 23 Cr, 13.5 Ni, 1 max Mo, bal Fe.
 Cold rolled: 580-730 N/mm^2 TS; 270 N/mm^2 min YS; 30 min El.
 For elevated temperature operation.
 AFNOR Z 15 CN 24-13; AISI 309.

UGINOX NG 30.
 M-1782; 0.07 C, 25 Cr, 20 Ni, 1 max Mo, bal Fe.
 Cold rolled: 580-730 N/mm^2 TS; 250 N/mm^2 min YS; 35 min El.
 For high temperature operation.
 AFNOR Z 12 CN 25-20; AISI 310; W.Nr. 4845.

UGINOX NSCD.
 M-1782; 0.03 max C, 17.5 Cr, 16 Ni, 5 min Mo, 3 Cu, bal Fe.
 Cold rolled: 590-740 N/mm^2 TS; 255 N/mm^2 min YS; 35 min El.
 AFNOR Z 2 CNDU 17-16.

UGINOX NSM 21.
 M-1782; 0.07 max C, 17 Cr, 11.5 Ni, 2 min Mo, bal Fe.
 Cold rolled: 530-680 N/mm^2 TS; 225 N/mm^2 min YS; 40 min El.
 AFNOR Z 6 Cnd 17-11; AISI 316; Type W.Nr. 4401.

UGINOX NSM 21 S.
 M-1782; 0.03 max C, 17 Cr, 12 Ni, 2 min Mo, bal Fe.
 Cold rolled: 500-650 N/mm^2 TS; 215 N/mm^2 min YS; 40 min El.
 AFNOR Type Z 2 CND 17-12; AISI 316 L; Type W.Nr. 4404.

UGINOX NSM 22.
 M-1782; 0.05 max C, 17 Cr, 12.2 Ni, 2.6 Mo, bal Fe.
 Cold rolled: 520-670 N/mm^2 TS; 220 N/mm^2 min YS; 40 min El.
 AFNOR Type Z 6 CND 17-12; AISI 316; Type W.Nr. 4436.

UGINOX NSM 22 S.
 M-1782; 0.03 max C, 17.5 Cr, 13 Ni, 2.5 min Mo, bal Fe.
 Cold rolled: 500-650 N/mm^2 TS; 215 N/mm^2 min YS; 40 min El.
 AFNOR Type Z 2 CND 17-13; AISI 316 L; Type W.Nr. 4435.

UGINOX NSM 23 S.
M-1782; 0.03 max C, 18.5 Cr, 14.5 Ni, 3 min Mo, bal Fe.
Cold rolled: 520-670 N/mm² TS; 225 n/mm² min YS; 40 min El.
AFNOR Z 2 CND 19-15; AISI 317 L; Type W.Nr. 4438.

UGINOX NSMC.
M-1782; 0.08 max C, 17 Cr, 12 Ni, 2 min Mo, Ti = 5 X C min, bal Fe.
Cold rolled: 540-690 N/mm² TS; 235 N/mm² min YS; 35 min El.
AFNOR Z 6 CNDT 17-12; W.Nr. 4571.

UGINOX NSZ.
M-1782; 0.15 C, 20 Cr, 11 Ni, 2 Mo, bal Fe.
Hot or cold rolled: 590-740 N/mm² TS; 250 N/mm² min YS; 30 El.
For elevated temperature operation.
AFNOR Z 15 CNS 20-12; Type W.Nr. 4828.

UGINOX QA 2.
M-1782; 0.10 max C, 21 Cr, 32 Ni, 1 max Mo, 0.3 T; 0.3 Al, bal Fe.
Cold rolled: 520-670 N/mm² TS; 205 N/mm² min YS; 30 min El.
For high temperature equipment.
AFNOR Z 8 NC 32-21; W.Nr. 4876.

UGIPLUS GP KMT.
M-1120; 0.36 C, 1.0 Cr, 0.25 Mo, 0.070 S, Mn, bal Fe.
Heat-treated free-machining alloy steel.

UGIPLUS GP KNO.
M-1120; 0.35 C, 1.4 Ni, 1.0 Cr, 0.23 Mo, 0.070 S, Mn, bal Fe.
Ht.Tr free-machining alloy steel.

UGIPLUS GP KN 9.
M-1120; 0.35 C, 2.8 Ni, 2.8 Cr, 0.30 Mo, 0.070 S, Mn, bal Fe.
Heat-treated free-machining alloy steel.

UGIPLUS GP UM 7.
M-1120; 0.36 C, 1.0 Cr, 0.25 Mo, 0.070 S, Mn, bal Fe.
Free-machining alloy steel; oil hardenable.

U.H.B.
M-111; 0.8 C, 0.2 V, bal Fe.
For dies for cold heading, embossing, trimming and swaging; cold work steel.

UHB-3.
M-111, M-111a; 0.15 C, 0.70 Mn, 0.25 Si, bal Fe.
Annealed: 70,000 TS; 40,000 YS; 25 El; 60 RA: 145 Brin.
For screws, bolts, bushing, rivets; SAE 1015, case hardened.

UHB 3 M 15.
M-111; 0.12 C, 1.35 Mn, bal Fe.
For gears, shafts; case hardened.

UHB-3S.
M-111; 0.15 C, 0.25 Si, 0.70 Mn, bal Fe.
Annealed: 70,000 TS; 40,000 YS; 25 El; 60 RA; 145 Brin.
For gears cams, camshafts; case hardened.

UHB-7.
M-111, M-111a; 0.35 C, 0.35 Mn, 0.20 Si, bal Fe.
Hot rolled: 85,000 TS; 54,000 YS; 30 El; 53 RA; 185 Brin.
For gears, shafts, axles, bolts, screws; SAE1035, water hardened.

UHB 8 M 10.
M-111; 0.4 C, 1.25 Mn, bal Fe.
For gears, shafts; tough.

UHB-14.
M-111, M-111a; 0.70 C, 0.35 Mn, 0.20 Si, bal Fe.
Heat treated: 174,000-122,000 TS; 128,000-82,000 YS; 12-22 El; 37-52 RA: 352-240 Brin.
For die blocks, girders, rails, clutch discs; SAE 1070, water hardened.

UHB 15.
M-111; 0.75 C, bal Fe.
For hollow mine drills; oil hardening.

UHB 16.
M-111; 0.8 C, bal Fe.
For tools, mine drills; water hardening.

UHB 16 VA.
M-111; 0.8 C, 0.1 V, bal Fe.
For chisels.

UHB 19VA.
M-111; 1.0 C, 0.2 Si, 0.25 Mn, 0.1 V, bal Fe.
Heat treated: 120,000-215,000 TS; 85,000-150,000 YS; 20-10 El; 48-33 RA; 240-500 Brin.
For drills, taps, hobs, cutters, reamers; type W2; water hardened.

UHB-19 VA.
M-111a; 0.92 C, 0.1 V, 0.30 Mn, 0.25 Si, bal Fe.
For cold heading and coining dies; water hardened, cold work steel.

UHB 20.
M-111; 1.05 C, bal Fe.
For tools; water hardening.

UHB 20C15.
M-111; 1 C, 1.45 Cr, bal Fe.
For ball bearings; water hardening.

UHB 20 VA.
M-111; 1.05 C, 0.15 V, bal Fe.
For dies; for cold work.

UHB 21.
M-111; 1.05 C, bal Fe.
For solid mine drills; water hardening.

UHB 21C5.
M-111; 1.1 C, 0.2 Si, 0.3 Mn, 0.4 Cr, bal Fe.
Heat treated: 125,000-220,000 TS; 90,000-155,000 YS; 18-8 El; 45-30 RA; 245-560 Brin.
For bearings, liners, races, sleeves; water hardened.

UHB 21C10.
M-111; 1.05 C, 0.25 Si, 0.3 Mn, 1.0 Cr, bal Fe.
Heat treated: 225,000 TS; 160,000 YS; 7 El; 28 RA; 575 Brin.
For bearings, liners, races, sleeves; water hardened.

UHB-22.
M-111, M-111a; 1.1 C, 0.55 Mn, 0.20 Si, bal Fe.
Annealed: 100,000 TS; 53,000 YS; 21 El; 42 RA; 200 Brin.
For springs, cutters, drills, taps, hobs; type W1; water hardened.

UBH 24.
M-111; 1.2 C, bal Fe.
For tools; water or oil hardening.

UHB 24C5.
M-111; 1.15 C, 0.2 Si, 0.3 Mn, 0.65 Cr, 0.1 V, bal Fe.
Heat treated: 230,000 TS; 165,000 YS; 6 El; 25 RA; 580 Brin.
For cold header dies, punches, drills, cutters; water hardened.

UHB-25C.
M-111, M-111a; 1.25 C, 0.35 Mn, 0.20 Si, 0.17 Cr, bal Fe.
For bearings, pivots, bushings, liners; water hardened.

UHB 27C5.
M-111; 1.35 C, 0.3 Cr, bal Fe.
For files; water hardening.

UHB 30C5.
M-111; 1.4 C, 0.2 Si, 0.3 Mn, 0.3 Cr, 0.1 V, bal Fe.
For engravers tools, cutters, drills, taps; water hardened.

UHB 34L.
M-111; 0.03 max C, 17.0 Cr, 13.5 Ni, 4.3 Mo, bal Fe.
Austenitic stainless steel; for marine operations, pulp and paper.

UHB 46.
M-111, M-111a; 0.9 C, 0.3 Si, 1.1 Mn, 0.5 Cr, 0.5 W, 0.1 V, bal Fe.
For gauges, reamers, taps, hobs, knurling tools; type o1; non-deforming, oil hardened.

UHB-151.
M-111, M-111a; 1.0 C, 0.6 Mn, 0.25 Si, 5.25 Cr, 1.1 Mo, 0.2 V, bal Fe.
Annealed: 200-228 Brin.
For crimpers, gages, blanking and forming dies; Type A2, air hardened, nondeforming.

UHB-711.
M-111, M-111a; 0.5 C, 0.2 Mn, 0.75 Si, 1.25 Cr, 2.5 W, 0.2 V, bal Fe.
Annealed: 180-220 Brin.
For rivet sets, upsetters, punches, dies, type S1, oil hardened.

UHB-725LN.
M-111; 0.02 max C, 1.7 Mn, 25.0 Cr, 22.0 Ni, 2.1 Mo, 0.12 N, bal Fe.
Austenitic stainless; for chemical plants.

UHB-904L.
M-111; 0.02 max C, 20.0 Cr, 25.0 Ni, 4.5 Mo, 1.5 Cu, bal Fe.
Austenitic stainless steel; for chemical plant, pulp and paper.

UHB-AEB.
M-111; 1 C, 13.5 Cr, bal Fe.
For cutting, forming and blanking dies; air hardened, non-deforming.

UHB-BORE 2.
M-111; 1.1 C, 0.25 Cr, 1. W, 0.1 V, bal Fe.
For twist drills, taps; water hardened.

U.H.B. CALDUR.
M-111; 0.25 C, 12 Cr, 7 W, 4 Co, bal Fe.
Tempered at 600°F: C 50.5 Rock.
Tempered at 900°F: C 54 Rock.
Tempered at 1200°F: C, 45 Rock.
For die casting dies for brass and aluminum, forging and extrusion dies.
Hot work tool steel, Resists heat checking.
Good hardness and toughness.

UHB-CALIX.
M-111; 0.40 C, 1.3 Cr, 4 W, 0.25 V, bal Fe.
For upsetters, punches, extrusion dies and mandrels; hot work steel, oil hardened.

UHB-CALMAX.
M-111a; 0.28 C, 12 Cr, 7 W, 0.4 V, 9 Co, bal Fe.
For hot press dies, die casting dies, extrusion dies; air hardened.

U.H.B. CARBON TOOL STEEL.
M-111a; 1.05 C, 0.3 Mn, 0.25 Si, bal Fe.
For cold heading dies, punches, gauges; water hardening.

UHB-CASTOR 32.
M-111; 0.80 C, 4 Cr, 5 Mo, 6.5 W, 1.9 V, bal Fe.
For lathe and planer tools, hobs, reamers; high speed steel.

UHB COLOMO.
  M-111; 0.95 C, 1.1 Cr, 0.33 Mo, bal Fe.
  For tools, cutters, mining drills; oil hardened, tough.

UHB COLD HEADER DIE.
  M-111; 0.91-0.98 C, bal Fe.
  For cold heading dies; water hardened.

UHB COMPAX.
  M-111, M-111a; 0.13 C, 0.50 Mn, 0.30 Si, 1.4 Cr, 3.8 Ni, bal Fe.
  Annealed: 200-220 Brin, 100,000 TS.
  Ht. Tr.: 145,000 TS; 112,000 YP; 17 El; 290 Brin.
  For plastic molds, large-cavity dies for compression molding of plastics.
  Type P6 tool steel, case hardening.

"U.H.B." CROWN.
  M-111; 0.9 C, bal Fe.
  Annealed: 84,000 TS; 24 El.
  For cold heading dies, stamping dies, punches; water hardened.

UHB-DECOL LEDLOY.
  M-111; 0.97 C, 0.20 Si, 0.45 Mn, bal Fe.
  Annealed: 100,000 TS; 53,000 YS; 21 El; 42 RA; 200 Brin.
  For drills, hobs, reamers, taps; Type W1; water hardened.

UHB EXTRA.
  M-111, M-111a; 1.05 C, 0.25 Si, 0.3 Mn, bal Fe.
  For blanking dies, drills, hobs, reamers; Type W1; water hardened.

U.H.B. FINGAL.
  M-111; 0.5 C, 0.6 Cr, 0.5 W, 0.2 Mo, bal Fe.
  For pneumatic tools; oil or water hardening.

UHB FORMA.
  M-111; 0.05 max C, bal Fe.
  For plastic molding dies; case hardened.

UHB-FORMA-1.
  M-111; 0.05 max C, 0.10 max Si, 0.15 max Mn, bal Fe.
  Annealed: 58,000 TS; 45,000 YS; 30 El; 66 RA; 125 Brin.
  For plastic mold dies; case hardened.

U.H.B. HEADER.
  M-111; 0.9 C, 0.2 V, bal Fe.
  For tools, cold heading dies; water hardened.

UHB IMPAX.
  M-111, M-111a; 0.36 C, 1.4 Cr, 1.4 Ni, 0.2 Mo, bal Fe.
  Heat treated: 300-330 Brin.
  For molds, dies, tools, plastic molds, die casting tools for zinc.
  Prehardened, Type P20 tool steel.

UHB-MN-MO.
  M-111; 0.90 C, 1.4 Mn, 0.3 Si, 0.25 Mo, bal Fe.
  For tools, dies, punches; oil hardened, ground flat stock.

UHB-NICRO 33.
  M-111; 0.13 C, 0.75 Cr, 3 Ni, bal Fe.
  For gears, cams, camshafts; case hardened, oil hardened.

UHB-NICRO 43.
  M-111; 0.18 C, 0.75 Cr, 3 Ni, bal Fe.
  For gears, cams, camshafts; case hardened, oil hardened.

U.H.B. NO. 46 CROWN.
  M-111; 0.9 C, 1.2 Mn, 0.5 W, 0.5 Cr, bal Fe.
  Annealed: 104,000 TS; 21 El.
  For tools, dies, gauges, reamers; oil hardened; non-deforming.

UHB ORVAR-1.
  M-111; 0.38 C, 5.2 Cr, 1.3 Mo, 1.0 V, bal Fe.
  Ht. Tr.: 300,000 TS; 227,000 YS; 9 El; C 55 Rock.
  For hot dies, fasteners, extrusion and die casting dies, punches.
  Type H11 hot work tool steel, high toughness and wear resistance.

UHB ORVAR-2.
  M-111a, M-111; 0.35-0.40 C, 0.8-1.2 Si, 5.0-5.5 Cr, 1.2-1.5 Mo, 0.9-1.1 V, bal Fe.
  For die casting dies, punches, hot shears, forging dies; Type H13; hot work steel.

UHB-PILGRIM.
  M-111; 0.40 C, 0.50 Cr, 4 Ni, 1 Mo, 0.5 W, bal Fe.
  For forging dies, upsetters; hot work steel, oil hardened.

UHB PISTON.
  M-111; 1.05 C, 0.1 V, bal Fe.
  For pistons, drills, taps; water hardening.

UHB-PREGA.
  M-111; 0.45 C, 3 Cr, 0.45 Mo, 0.6 Mn, bal Fe.
  For drop forging dies, hobbing dies; hot work steel, oil hardened.

UHB PREMO.
  M-111, M-111a; 0.04 C, 4 Cr, 0.5 Mo, 0.1 Si, 0.1 Mn, bal Fe.
  For plastic mold dies, cold hubbed dies; case hardened, oil or water hardened.

UHB PREXI.
  M-111, M-111a; 0.15 C, 0.25 Si, 1.0 Mn, 1.2 Cr, 0.25 Mo, bal Fe.
  For die casting dies, plastic mold dies; case hardened, oil hardened.

U.H.B. REGIN 1.
  M-111; 0.42 C, 0.9 Si, 1 Cr, 2.5 W, bal Fe.
  For hot work tools, chisels; hot work steel.

UHB-RESISTO.
M-111a; 0.6 C, 1.85 Si, 0.7 Mn, 0.45 Mo, 0.2 V, bal Fe.
For shear blades, pneumatic chisels; shock resistant, oil hardened.

U.H.B. SPECIAL.
M-111; 0.35 C, 1.05 Si, 5.0 Cr, 0.4 V, 1.5 W, 1.65 Mo, bal Fe.
For blanking and forming dies; non-deforming; AISI H-12.

UHB STAINLESS 1.
M-111; 0.10 C, 13.5 Cr, bal Fe.
For corrosion resistant parts; corrosion resistant.

U.H.B. STAINLESS 1H.
M-111; 0.13-0.20 C, 12-15 Cr, bal Fe.
For corrosion resistant parts; corrosion resistant.

UHB STAINLESS 2.
M-111; 0.1 C, 17.5 Cr, 1 Si, bal Fe.
Annealed: 75,000 TS; 40,000 YS; 30 El; 50 RA; 140 Brin.
For chemical plant and oil refinery equipment; Type 430; corrosion resistant.

UHB STAINLESS 2M.
M-111; 0.08 C, 17 Cr, bal Fe.
Annealed: 80,000 TS; 45,000 YS; 30 El; 140 Brin.
For chemical plant equipment, kitchen sinks, auto trim.
Type 430 stainless steel, ductile, non-hardenable.

UHB STAINLESS 3.
M-111; 0.1 C, 18 Cr, 8 Ni, bal Fe.
Annealed: 85,000 TS; 35,000 YS; 60 El; 70 RA; 150 Brin.
For chemical plant equipment; Type 302; austenitic, stainless.

U.H.B. STAINLESS 3H.
M-111; 0.15 C, 18 Cr, 8.5 Ni, bal Fe.
For heat and corrosion resistant parts; stainless, heat and corrosion resistant.

UHB-STAINLESS 3L.
M-111; 0.08 C, 18 Cr, 10 Ni, bal Fe.
Annealed: 85,000 TS; 35,000 YS; 60 El; 70 RA; 150 Brin.
For chemical plant equipment, tanks, mixers; Type 304; stainless, austenitic.

UHB STAINLESS 3M.
M-111; 0.06 C, 19 Cr, 10 Ni, bal Fe.
Annealed: 85,000 TS; 40,000 YS; 60 El; 150 Brin.
For welded structures, chemical plant equipment, evaporators.
Type 304 stainless steel, austenitic.

UHB STAINLESS 3MM.
M-111; 0.07 max C, 9.5 Ni, 18 Cr, bal Fe.
Annealed: 85,000 TS; 35,000 YS; 60 El; 70 RA; 150 Brin.
Cold drawn: 180,000 TS; 125,000 YS; 10 El; 330 Brin.
For chemical plant equipment, tanks, mixers, agitators; Type 304; stainless, austenitic.

UHB STAINLESS 3 (P).
M-111; 0.15 max C, 18 Cr, 8.5 Ni, bal Fe.
Annealed: 80,000 TS; 35,000 YS; 55 El; 70 RA; 150 Brin.
Cold drawn: 180,000 TS; 150,000 YS; 10 El; 300 Brin.
For chemical plant equipment, tanks, mixers, agitators; Type 302; stainless, austenitic.

UHB STAINLESS 4.
M-111; 0.10 C, 18 Cr, 8 Ni, 1.3 Mo, bal Fe.
For stainless parts; stainless, austenitic.

U.H.B. STAINLESS 4H.
M-111; 0.13-0.20 C, Mo, 16-23 Cr, 7-11 Ni, bal Fe.
For heat and corrosion resistant parts; stainless, heat and corrosion resistant.

U.H.B. STAINLESS 4M.
M-111; 0.07 max C, Mo, 16-23 Cr, 7-11 Ni, bal Fe.
For heat and corrosion resistant parts; stainless, heat and corrosion resistant.

UHB STAINLESS 4MM.
M-111; 0.07 max C, 18 Cr, 2.1 Mo, 11.5 Ni, bal Fe.
Annealed: 85,000 TS; 35,000 YS; 50 El; 65 RA; 160 Brin.
For acid resistant chemical plant equipment, tanks; Type 316; stainless, austenitic.

UHB STAINLESS 5.
M-111; 0.20 C, 24 Cr, 1 Si, bal Fe.
For corrosion resistant parts; corrosion resistant.

UHB STAINLESS 6.
M-111; 0.35 C, 13.5 Cr, bal Fe.
For corrosion resistant parts; crrosion resistant.

U.H.B. STAINLESS 6H.
M-111; 0.36-0.50 C, 12-15 Cr, bal Fe.
For corrosion resistant parts; corrosion resistant.

UHB STAINLESS 7.
M-111; 0.95 C, 14 Cr, 1 Mn, bal Fe.
For corrosion resistant parts; corrosion resistant.

UHB STAINLESS 15.
M-111; 0.07 max C, 18.5 Cr, 20.5 Ni, bal Fe.
For chemical plant equipment; austenitic, heat and corrosion resistant.

UHB STAINLESS 16.
M-111; 0.27 C, 13.5 Cr, 1.5 Mo, bal Fe.
For surgical instruments, cutlery; hardenable, corrosion resistant.

UHB-STAINLESS 17H.
M-111; 0.25 C, 17.5 Cr, 12.5 Ni, W, Cb, bal Fe.
For chemical plant equipment; corrosion and heat resistant.

UHB STAINLESS 18.
M-111; 0.2 C, 18 Cr, 8 Ni, bal Fe.
Annealed: 80,000 TS; 35,000 YS; 55 El; 70 RA; 150 Brin.
For chemical plant equipment, tanks, filters, mixers; Type 302; stainless, austenitic.

UHB STAINLESS 20.
M-111; 0.2 C, 20 Cr, 10 Ni, bal Fe.
Annealed: 90,000 TS; 40,000 YS; 50 El; 60 RA; 170 Brin.
For chemical plant equipment; austenitic, stainless.

UHB STAINLESS 21.
M-111; 0.06 max C, 14 Cr, bal Fe.
For stainless parts; unhardenable, corrosion resistant parts.

UHB STAINLESS 22.
M-111; 0.18 C, 17 Cr, 1.5 Ni, bal Fe.
Heat treated: 135,000 TS; 112,000 YS; 15 El; 45 RA; 280 Brin.
For propeller shafts, pump spindles; Type 431; corrosion resistant.

UHB STAINLESS 24.
M-111; 0.06 max C, 17 Cr, 11.5 Ni, 3 Mo, bal Fe.
Annealed: 95,000 TS; 45,000 YS; 50 El; 60 RA; 190 Brin.
For chemical plant equipment, acid resistant parts; Type 317; stainless, austenitic.

UHB-STAINLESS 24L.
M-111; 0.08 C, 17.5 Cr, 13 Ni, 2.8 Mo, bal Fe.
Annealed: 85,000 TS; 35,000 YS; 50 El; 65 RA; 160 Brin.
For acid resistant chemical plant equipment; Type 316; stainless, austenitic.

UHB STAINLESS 25.
M-111; 0.07 max C, 23.5 Cr, 21.5 Ni, bal Fe.
Annealed: 95,000 TS; 45,000 YS; 55 El; 65 RA; 200 Brin.
For furnace parts, chemical plant equipment; Type 310; corrosion and heat resistant.

UHB STAINLESS 26.
M-111; 0.5 C, 14 Cr, bal Fe.
Annealed: 100,000 TS; 55,000 YS; 200 Brin.
For cutlery, valves, surgical instruments; Type 420; stainless, hardenable.

UHB-STAINLESS 27.
M-111; 0.12 C, 16 Cr, 13.5 Ni, W, bal Fe.
For chemical plant equipment; corrosion resistant.

UHB STAINLESS 31.
M-111; 0.18 C, 13.5 Cr, 0.7 Ni, bal Fe.
Annelaed: 95,000 TS; 50,000 YS; 25 El; 55 RA; 196 Brin.
For cutlery, surgical instruments, valves, turbine blades; Type 420; stainless, hardenable.

UHB STAINLESS 32.
M-111; 0.15 C, 16.5 Cr, 1 Ni, 0.5 Mo, bal Fe.
Heat treated: 130,000 TS; 110,000 YS; 16 El; 48 RA; 250 Brin.
For chemical plant equipment, pump parts; corrosion resistant, hardenable.

UHB STAINLESS 33.
M-111; 0.10 C, 14 Cr, 13 Ni, bal Fe.
Annealed: 75,000 TS; 30,000 YS; 35 El; 200 Brin.
For furnace parts, chemical plant equipment; stainless, austenitic.

UHB-STAINLESS 33MM.
M-111; 0.06 C, 12.5 Cr, 12.5 Ni, bal Fe.
For valves, pumps; corrosion and heat resistant.

UHB-STAINLESS 34.
M-111; 0.06 C, 17.4 Cr, 14 Ni, 4.5 Mo, Cb, bal Fe.
Annealed: 100,000 TS; 55,000 YS; 40 El; 55 RA; 200 Brin.
For welded chemical plant equipment, tanks; Type 317 Cb; stainless, austenitic.

UHB STAINLESS 35.
M-111; 0.20 max C, 24 Cr, 12 Ni, bal Fe.
Annealed: 90,000 TS; 40,000 YS; 50 El; 65 RA; 170 Brin.
For furnace parts and equipment, salt pots; Type 309; heat resistant.

UHB-STAINLESS 41.
M-111; 0.13 C, 13.5 Cr, bal Fe.
Annealed: 75,000 TS; 40,000 YS; 35 El; 70 RA; 155 Brin.
For valves, cutlery, turbine blades; Type 410; corrosion resistant.

UHB STAINLESS 43.
M-111; 0.08 C, 18 Cr, 9.5 Ni, 0.5 Mo, bal Fe.
Annealed: 85,000 TS; 55,000 YS; 60 El; 70 RA; 150 Brin.
Cold drawn: 120,000 TS; 95,000 YS; 25 El; 55 RA; 275 Brin.
For chemical plant equipment; stainless, austenitic.

UHB STAINLESS 44.
M-111; 0.13 C, 25 Cr, 4.5 Ni, 1.5 Mo, bal Fe.
Normalized: 103,000 TS; 78,000 YS; 18 El; 45 RA; 235 Brin.
For valves, pumps, chemical plant equipment; Type 329; corrosion resistant.

Section I: Alloy Data / 1451

UHB STAINLESS 45.
M-111; 0.13 C, 26 Cr, 4.5 Ni, bal Fe.
For chemical plant equipment; Type 327; corrosion resistant.

UHB STAINLESS 51.
M-111; 0.10 C, 14 Cr, 1 Mo, bal Fe.
For stainless parts; hardenable, corrosion resistant.

UHB STAINLESS 51 H.
M-111; 0.15 C, 14 Cr, Mo, bal Fe.
Annealed: 95,000 TS; 46,000 YS; 24 El; B 92 Rock.
Ht Tr.: 240,00 TS; 210,000 YS; 10 El; C 50 Rock.
For cutlery, surgical instruments, needle valves, gauges.
Type 420 stainless steel, hardenable.

UHB STAINLESS 52.
M-111; 0.10 C, 17 Cr, 1.5 Mo, bal Fe.
Annealed: 80,000 TS; 50,000 YS; 25 El; 50 RA; 150 Brin.
For oil refinery equipment, oil heaters; Type 430 Mo; corrosion resistant.

UHB STAINLESS 53.
M-111; 0.08 max C, 18 Cr, 9.5 Ni, Ti, bal Fe.
Annealed: 85,000 TS; 33,000 YS; 58 El; 75 RA; 150 Brin.
Cold drawn: 95,000 TS; 60,000 YS; 40 El; 60 RA; 185 Brin.
For welded structures, chemical plant equipment; Type 321; stainless, austenitic.

UHB-STAINLESS 54C.
M-111; 0.06 C, 17.5 Cr, 11.5 Ni, 2.2 Mo, Ti, bal Fe.
Annealed: 90,000 TS; 40,000 YS; 45 El; 60 RA; 180 Brin.
For welded acid resistant equipment; Type 316 Ti; stainless, austenitic.

UHB STAINLESS 55.
M-111; 0.20 max C, 24 Cr, 12 Ni, bal Fe.
Annealed: 85,000 TS; 35,000 YS; 60 El; 70 RA; 150 Brin.
For chemical plant equipment, tanks, filters, mixers; Type 304; stainless, austenitic.

UHB STAINLESS 62.
M-111; 0.1 max C, 17.5 Cr, Ti = 7 X C, bal Fe.
Annealed: 80,000 TS; 50,000 YS; 25 El; 50 RA; 150 Brin.
For welded oil refinery equipment, sinks; Type 430 Ti, austenitic, stainless.

UHB STAINLESS 63.
M-111; 0.12 max C, 18 Cr, 9.5 Ni, Cb = 8 x C, bal Fe.
Annealed: 90,000 TS; 45,000 YS; 56 El; 65 RA; 160 Brin.
For welded chemical plant equipment, tanks, mixers; Type 347; austenitic, stainless.

UHB STAINLESS 63 H.
M-111; 0.06 C, 16 Cr, 13 Ni, 0.6 Cb, bal Fe.
Annealed: 90,000 TS; 40,000 YS; 50 El; B 82 Rock.
For exhaust manifolds, superheaters, low temperature processing equipment.
Austenitic, non-hardenable.
Type 16-13 Cb stainless steel.

UHB-STAINLESS 64.
M-111; 0.06 C, 18 Cr, 18 Ni, 2 Mo, 2 Cb + Cu, bal Fe.
For chemical plant equipment; stainless, austenitic.

UHB-STAINLESS 72.
M-111; 0.13 C, 17 Cr, bal Fe.
Annealed: 80,000 TS; 50,000 YS; 25 El; 50 RA; 150 Brin.
For oil refinery equipment, oil burners and heaters; Type 430; corrosion resistant.

UHB STAINLESS 75.
M-111; 0.25 C, 25 Cr, bal Fe.
For corrosion and heat resistant parts; unhardenable, corrosion resistant.

UHB-STAINLESS 524.
M-111; 0.08 C, 17.5 Cr, 13 Ni, 2.8 Mo, Ti, bal Fe.
Annealed: 90,000 TS; 40,000 YS; 45 El; 60 RA; 180 Brin.
For acid resistant chemical plant equipment; Type 316 Ti; stainless, austenitic.

UHB-STAINLESS 525.
M-111; 0.10 C, 24 Cr, 18 Ni, Ti, bal Fe.
Annealed: 100,000 TS; 45,000 YS; 50 El; 65 RA; 185 Brin.
For furnace parts, valves, pumps; Type 310; stainless.

UHB-STAINLESS 624.
M-111; 0.06 C, 17.5 Cr, 13 Ni, 2.8 Mo, Cb, bal Fe.
Annealed: 90,000 TS; 40,000 YS; 45 El; 60 RA; 180 Brin.
For acid resistant chemical plant equipment; Type 316 Cb; stainless, austenitic.

UHB-STAINLESS 703.
M-111; 0.09 C, 17.5 Cr, 10.5 Ni, bal Fe.
Annealed: 85,000 TS; 35,000 YS; 60 El; 70 RA; 150 Brin.
For chemical plant equipment, tanks, mixers; Type 304; stainless, austenitic.

UHB-STAINLESS 716.
M-111; 0.35 C, 13.5 Cr, 1 Mo, bal Fe.
Annealed: 100,000 TS; 55,000 YS; 20 El; 50 RA; 210 Brin.
For valves, cutlery, surgical and dental instruments; Type 420; corrosion resistant.

UHB STAINLESS 731.
M-111; 0.20 C, 12-14 Cr, bal Fe.
Annealed: 95,000 TS; 45,000 YS; 25 El; B 92 Rock.
Ht Tr.: 250,000 TS; 215,000 YS; 8 El; C 52 Rock.
For cutlery, surgical instruments, gauges, needle valves.
Type 420 stainless steel, hardenable.

UHB STAINLESS 734.
M-111; 0.07 C, 18-20 Cr, 11-15 Ni, 3-4 Mo, bal Fe.
Annealed: 80,000 TS; 35,000 YS; 70 El; B 70 Rock.
Cold rolled: 150,000 TS; 125,000 YS; 15 El; C 30 Rock.
For paper and pulp equipment, textile and dye equipment, acid tanks, evaporators, digesters.
Type 317 stainless steel, austenitic, acid resistant.

UHB STANDARD.
M-111; 1.0 C, bal Fe.
For tools, drills, taps; water hardened.

UHB-STATO.
M-111; 0.35 C, 1.1 Cr, 0.20 Mo, bal Fe.
For gears, shafts, mandrels; oil hardened.

UHB-STATO 21.
M-111; 0.15 C, 0.25 Si, 0.60 Mn, 0.32 Mo, bal Fe.
For gears, cams, camshafts; case hardened, tough.

UHB-STATO 28.
M-111; 0.10 C, 2.5 Cr, 1.0 Mo, bal Fe.
For gears, cams, camshafts; case hardened, tough.

UHB STAVAX.
M-111a; 0.35 C, 0.45 Mn, 0.45 Si, 13.6 Cr, bal Fe.
Annealed: 210-220 Brin.
High strength, deep hardening plastic mold steel with ultra high polish potential; oil hardening, non-deforming.
AISI Type 420 corrosion resistant steel.

U.H.B. SUPER.
M-111; C, mo, W, bal Fe.
For chisels; shock resistant.

UHB TRIMO.
M-111a; 1.5 C, 0.3 Mn, 0.4 Si, 12.0 Cr, 0.8 Mo, 0.9 V, bal Fe.
Annealed: 200-210 Brin.
For dies, punches, shear blades, cold forming dies, exceptional wear resistance; air and oil hardening.
Type D2 cold work tool steel.

UHB TRITUNG.
M-111a, M-111; 2.0 C, 0.75 Mn, 0.3 Si, 13.0 Cr, 1.25 W, bal Fe.
Annealed: 230-260 Brin.
For cold and hot trimming dies; exceptional wear resistance; air and oil hardening non-deforming.
Type D6 cold work tool steel.

UHB "TRI-Z".
M-111a; 1.5 C, 12 Cr, 0.8 Mo, 0.9 V, bal Fe.
For blanking and drawing dies; thread rolling and cold forming dies, broaches; air hardened, low impurity count, non-deforming.

UHB UDDCO 1.
M-111a; 0.14 C, 5 Cr, bal Fe.
Annealed: 70,000 TS; 30,000 YS; 28 El; 65 RA; 160 Brin.
For petroleum and oil refinery equipment; Type 501; creep resistant.

UHB UDDCO 2.
M-111a; 0.14 C, 5 Cr, 0.5 Mo, bal Fe.
Annealed: 75,000 TS; 35,000 YS; 25 El; 60 RA; 170 Brin.
For petroleum and oil refinery equipment; Type 502; creep resistant.

UHB UDDCO 3.
M-111a; 0.06 C, 3 Cr, 0.5 Mo, bal Fe.
Annealed: 65,000 TS; 30,000 YS; 35 El; 70 RA; 150 Brin.
For petroleum and oil refinery equipment; Type 502; creep resistant.

UHB UDDCO 6.
M-111a; 0.18 C, 6 Cr, Si, bal Fe.
Annealed: 70,000 TS; 35,000 YS; 25 El; 60 RA; 180 Brin.
For petroleum and oil refinery equipment; Type 501; creep resistant.

UHB-VA.
M-111, M-111a; 1.05 C, 0.3 Mn, 0.25 Si, 0.2 V, bal Fe.
For drills, taps, reamers, blanking dies; Type W2; water hardened.

U.B.B. VALAND 1.
M-111; 0.3 C, 3 Cr, 1.75 Ni, 9.5 W, 0.3 V, bal Fe.
For hot work tools; hot work steel.

UHB VALAND 2.
M-111; 0.32 C, 2.5 Cr, 2.0 Co, 0.35 V, 8 W, bal Fe.
For hot work tools and dies; hot work steel.

UHB-WATER.
M-111, M 111a; 1.05 C, 0.30 Mn, 0.25 Si, bal Fe.
Annealed: 100,000 TS; 53,000 YS; 21 El; 42 RA; 200 Brim.
For blanking and cutting and forming dies; water hardened.

UHB-X1.
M-111a; alloy, bal Fe.
For dies; tools; oil hardened.

UKI.
M-1787.
WC plus alloy.
For anti-skid tire studs and studded straps for snow and ice chains.

ULBRASEAL 36.
M-1798; 36 Ni, bal Fe.
Low co-efficient of expansion.
Used in TV picture tubes as mask material, bimetallic thermostats.

ULBRASEAL 42.
M-1798; 41.0 Ni, 0.05 max C, 0.80 max Mn, 0.30 max Si, 0.10 max Al, 0.50 max Co, bal Fe.
Used for sealing of leads into light bulbs, automotive sealed beam headlights.

ULBRASEAL 42-8.
M-1798; 42.0 Ni, 0.07 C, 0.25 Mn, 0.30 max Si, 0.20 max Al, 5.6 Co, bal Fe.
Used in TV picture tubes.

ULBRASEAL 46.
M-1798; 46.0 Ni, 0.05 max C, 0.08 max Mn, 0.30 max Si, 0.10 max Al, 0.50 max Co, bal Fe.
Used for electrical resistors.

ULBRASEAL 52.
M-1798; 50.5 Ni, 0.05 max C, 0.60 Mn, 0.30 max Si, 0.10 max Al, 0.50 max Co, bal Fe.
Used for cores in mechanical rectifiers and magnetic amplifiers requiring rectangular hysteresis.

ULBRAVAR 29-17.
M-1798; 29.0 Ni, 0.06 C, 0.30 Mn, 0.20 Si, 17.0 Co, bal Fe.
Used for hermetic seals in vacuum tubes.
AMS 7728.

ULCONY.
65 Cu, 35 Pb.
For heavy duty bearings; for poor lubricating conditions.

ULTIMIUM N 112.
M-717.
Tungsten carbide containing electrodes for super hard overlays on steel; AC-DC; for surfacing conveyor screws, crusher liners, muller blades, wear plates.
Hardness C 68-72 Rock.

ULTIMO 4.
M-435; 0.45 C, 0.75 Mn, 1.25 Cr, 0.35 Mo, bal Fe.
Normalized: 110,000 TS; 85,000 YS; 25 El; 60 RA; 228 Brin.
For shafts, gears, arbors; heavy duty.

ULTIMO 6.
M-435; 0.55 C, 55 Mn, 0.80 Si, 1.0 Cr, 1.6 Ni, 0.75 Mo, bal Fe.
Oil hardening cold work tool steel; for shear blades, lead and zinc extrusion tools.

ULTIMO 200.
M-1636, M-435; 0.30 C, 0.8 Mn, 1.65 Cr, 0.4 Mo, bal Fe.
Annealed: 114,000 TS; 90,000 YS; 25 El; 229 Brin.
Hardened: 150,000 TS; 135,000 YS; 22 El; 302 Brin.
For heavy duty shafts, gears, crusher rolls, spindles, couplings, clamps.
Oil hardening, shock resisting.

ULTRA B.
M-912; 0.10 C, 65.0 Ni, 28.0 Mo, bal Fe.
Corrosion resistant alloy.
W.Nr. 2.4482.

ULTRA C.
M-912; 0.08 C, 16.5 Cr, 52.0 min Ni, 17.0 Mo, 4.0 W, bal Fe.
Corrosion resistant alloy.
W.Nr. 2.4537.

ULTRA CAPITAL.
M-1112; 0.75 C, 4 Cr, 18 W, 1 V, bal Fe.
For broaches, drills, reamers, hobs, lathe tools; high speed steel.

ULTRA CAPITAL 22.
M-1112; 0.75 C, 4.1 Cr, 22 W, 1.25 V, bal Fe.
For broaches, drills, reamers, milling cutters, hobs; high speed steel.

ULTRA CAPITAL PLUS ONE.
M-1112; 0.85 C, 4.25 Cr, 1.3 V, 5 Co, 18 W, bal Fe.
For lathe and planer tools, hobs, taps, reamers, drills; high speed steel, oil hardened.

ULTRA CAPITAL PLUS 2.
M-1112; 0.75 C, 4.2 Cr, 20 W, 1.5 V, bal Fe.
For broaches, reamers, drills, hobs, cutters, taps; high speed steel, oil hardened.

ULTRA-CAPITAL STEEL.
M-350, M-1112; 17 W, 3.4 Cr, 1.0 V, 0.7 C, bal Fe.
For turning tools and cutters; high speed steel.

ULTRADIE 1.
M-114; 2.25 C, 12 Cr, 0.8 Mo, 0.2 V, bal Fe.
For dies; for punching hard, brittle material, drills, reamers, shear blades, lamination dies; non-deforming: cold work die steel; AISI D3.

ULTRADIE 1M.
M-114; 2.25 C, 12 Cr, 2 V, 0.8 Mo, bal Fe.
For lamination and drawing dies; punches, gauges; Type D3; oil hardened, non-deforming.

ULTRADIE 2.
M-114; 1.5 C, 12 Cr, 0.8 Mo, bal Fe.
For tools, dies, reamers, taps; non-deforming.

ULTRADIE 3.
M-114; 1.5 C, 12 Cr, 1 V, 0.8 Mo, bal Fe.
For lamination and blanking dies, punches, shear knives; non-deforming: cold work die steel; AISI D2.

ULTRADUR.
M-1246; carbides.
For hard facing electrodes; sintered alloy.

ULTRAFORT 6355 ETC.
M-1759; see Werkstoff Nr. 1.6355.
High tensile steels; 5 grades.

ULTRALLOY.
M-1351; 48.5 Sn, 48.5 Zn, 2 Cu, 0.6 Si, 0.03 Ag.
Cast: 12,250 TS; 20 El; 23 Brin.
Solder for aluminum; M.P. 700-750°F.

ULTRALOY.
M-605; 37.5 Cr, 7.5 Al, bal Fe.
For furnace parts; see Smith No. 10.

ULTRALOY 10611.
M-717.

ULTRAPERM Z.
M-1631.
Soft magnetic high nickel content alloy with square hysteresis loop, e.g. for sensitive magnetic amplifiers.

ULTRA SUPERIOR 25.
M-1294; 1.3 C, 4.3 Cr, 0.85 Mo, 3.8 V, 12 W, bal Fe.
For forming and blanking dies, cutters; high speed steel.

ULTRA SUPERIOR 25Z.
M-1294; 0.86 C, 2.8 Co, 4.3 Cr, 0.85 Mo, 2.1 V, 12 W, bal Fe.
For lathe and planer tools, reamers, broaches; high speed steel.

ULTRATHERM 7.
M-1309; 0.25 C, 2.25 Si, 6.5 Cr, bal Fe.
Cast, ann: 95,000-130,000 TS; 4 El; 200-280 Brin.
Ferrite-pearlite, weldable.
For parts used at temperatures up to 850°C.
Din G-X30 CrSi 6.

ULTRATHERM 13.
M-1309; 0.45 C, 2.25 Si, 13 Cr, bal Fe.
Cast, ann: 95,000-135,000 TS; 4 El; 200-300 Brin.
Ferrite-pearlite; weldable.
For parts used at temperatures up to 900°C.
Din G-X45 CrSi 13.

ULTRATHERM 17.
M-1309; 0.45 C, 2.25 Si, 17 Cr, bal Fe.
Cast, ann: 95,000-135,000 TS; 2 El; 200-300 Brin.
Ferrite-carbide; weldable.
For parts used in oxidzing atmospheres at temperatures up to 950°C.
Din G-X40 CrSi 17.

ULTRATHERM 18-9.
M-1309; 0.18 C, 1.5 Si, 18.0 Cr, 9.0 Ni, bal Fe.
Cast, ann: 75,000-110,000 TS; 15 El; 130-200 Brin.
Austenitic, weldable.
For parts used at elevated temperatures.
Din G-X25 CrNiSi 18 9.

ULTRATHERM 19.
M-1309; 1.60 C, 1.50 Si, 18.0 Cr, bal Fe.
Cast,ann: 250-300 Brin.
Ferrite-carbide structure, machinability only fair.
For parts used at temperatures up to 950°C.
Din G-X 160 CrSi 18.

ULTRATHERM 20-40.
M-1309; 0.35 C, 1.35 Si, 17.5 Cr, 37.25 Ni, bal Fe.
Cast: 70,000 TS; 40,000 YS; 9 El; 170 Brin.
Aged: 73,000 TS; 43,000 YS; 5 El; 200 Brin.
For retorts, carburizing containers, rails, fixtures, furnace parts.
Corrosion and oxidation resistance.
Heat resistant, Type HU, austenitic.

ULTRATHERM 23.
M-1309; 0.45 C, 1.50 Si, 23.0 Cr, bal Fe.
Cast, ann: 200-300 Brin.
Ferrite-carbide structure, weldable.
For parts to be used in oxidizing atmospheres at temperature up to 1050°C.
Din G-X40 CrSi 22.

ULTRATHERM 23-11.
M-1309; 0.35 C, 1.75 Si, 22.0 Cr, 10.0 Ni, bal Fe.
Cast, ann: 75,000-110,000 TS; 12 El; 150-220 Brin.
Austenitic, weldable.
For parts used at elevated temperatures.
Din G-X40 CrNiSi, 229.

ULTRATHERM 23-15.
M-1309; 0.25 C, 1.75 Si, 20.0 Cr, 14 Ni, bal Fe.
Annealed: 90,000 TS; 40,000 YS; 50 El; B 83 Rock.
For heat treating equipment, heat exchangers, furnace parts and fixtures, salt pots, oil burners.
Non-hardenable. Heat resistant, austenitic.

ULTRATHERM 25-12.
M-1309; 0.35 C, 1.75 Si, 25.0 Cr, 12.0 Ni, bal Fe.
Ht. Tr.: 92,000 TS; 45,000 YS; 8 El; 200 Brin.
At 1400°F: 35,000 TS; 18,000 YS; 12 El.
For furnace parts, conveyor screws, salt pots, dampers, grate bars.
High corrosion and oxidation resistance.
Heat resistant, Type HH, austenitic.

ULTRATHERM 26-21.
M-1309; 0.35 C, 1.75 Si, 25.5 Cr, 20.0 Ni, bal Fe.
Cast: 75,000 TS; 50,000 YS; 17 El; 170 Brin.
Aged: 85,000 TS; 50,000 YS; 10 El; 190 Brin.
For salt pots, heat treating boxes, furnace parts, skids, rabble arms.
Corrosion and oxidation resistant. Heat resistant, austenitic.

ULTRATHERM 26-21 W.
M-1309; 0.15 C, 1.75 Si, 25.5 Cr, 20.0 Ni, bal Fe.
Annealed: 95,000 TS; 45,000 YS; 50 El; B 89 Rock.
For furnace parts and equipment, heat treating boxes and fixtures, boiler baffles, retorts, kiln linings.
High corrosion and oxidation resistance. Heat resistant, Type 310, austenitic.

UTRATHERM 27-15.
M-1309; 0.35 C, 1.75 Si, 26.5 Cr, 14 Ni, bal Fe.
Cast: 80,000 TS; 45,000 YS; 12 El; 180 Brin.
At 1400°F: 38,000 TS; 6 El.
For retorts, skids, brazing fixtures, lead pots, hearth plates, furnace parts.
Heat resistant. High corrosion and oxidation resistance.

ULTRATHERM 28-4H.
M-1309; 0.40 C, 1.5 Si, 27.0 Cr, 4.0 Ni, bal Fe.
Cast: 85,000 TS; 48,000 YS; 16 El; 190 Brin.
At 1400°F: 36,000 TS; 14 El.
For ore roasting furnaces, rabble arms, sintering bars, salt pots, furnace blowers, recuperators.
High oxidation resistance in sulphur atm. Heat resistant. Type HD.
Din G-X40 CrNi 27 4.

ULTRATHERM 30.
M-1309; 0.40 C, 1.5 Si, 28.5 Cr, bal Fe.
Cast: 110,000 TS; 75,000 YS; 19 El; 223 Brin.
For ore roasting furnaces, grate bars, salt pots, soot blowers, kiln parts.
Oxidation resistant to high sulphur atm. Heat resistant.
Din G-X40 CrSi 29.

ULTRATHERM 30 H.
M-1309; 1.30 C, 1.50 Si, 28.50 Cr, bal Fe.
Cast, ann: 250-350 Brin.
Ferrite, carbide structure.
For parts used in oxidizing atmospheres at temperatures up to 1100°C.
Din G-X130 CrSi 29.

ULTRAVAN see **ELECTRITE ULTRAVAN.**

UMCO-50.
M-1644; 48-52 Co, 0.05-0.12 C, 27-29 Cr, 0.5-1.0 Mn, 0.5-1.0 Si, bal Fe.
As cast, RT.: 80,000 TS; 45,600 YS; 8 El.
1830°F: 11,400 TS; 10,000 YS; 18 El.
Forged, RT.: 134,000 TS; 88,500 YS; 10 El.
1830°F: 11,400 TS; 8,600 YS; 18 El.
Excellent strength and corrosion resistance at elevated temperature; for furnaces, heat treat equipment, gas and pulverized coal burners.

UMCO-51.
M-1644; 48-52 Co, 0.25-0.40 C, 27-29 Cr, 2.0-2.2 Nb, 0.5-1.0 Mn, 0.5-1.0 Si, bal Fe.
As cast, RT.: 91,500 TS; 72,00 YS; 2.2 El.
1650°F: 30,400 TS; 26,200 YS; 13.8 El.
(properties vary with carbon content).
Excellent strength and corrosion resistance at elevated temperatures; for furnace parts, heat treat equipment, gas and pulverized coal burners.

UNA.
M-930; 0.1 C, bal Fe.
For welding rod.

UNA HIGH SPEED NO. 44.
M-930; 0.08 max C, 0.2-0.3 Mn, bal Fe.
For welding rod; flux coated for high speed.

UNA HIGH SPEED NO. 45.
M-930; 0.12-0.18 C, 0.4-0.6 Mn, bal Fe.
For welding rod; flux coated for high speed.

UNA HIGH TENSILE NO. 65.
M-930; 0.13-0.18 C, 0.4-0.6 Mn, bal Fe.
For welding rod; flux coated.

UNAMO 1.
M-1618; 0.8 C, 0.3 Mn, 0.3 Si, 4 Cr, 1 V, 1.5 W, 8 Mo, bal Fe.
Hardness: C 64-68 Rock.
For cutters, lathe and planer tools, drills, reamers, broaches, taps.
Type M-1 high speed steel, high red-hardness.

UNAMO 2.
M-1618; 0.8 C, 0.3 Si, 0.3 Mn, 4 Cr, 2 V, 6 W, 5 Mo, bal Fe.
Hardened: C 64-66 Rock.
For lathe and planer tools, reamers, broaches, chasers, hobs, taps, form cutters.
Type M-2 high-speed steel, high red-hardness.

UNAMO 7.
M-1618; 1.0 C, 0.3 Mn, 0.3 Si, 3.75 Cr, 2 V, 1.7 W, 8.75 Mo, bal Fe.
Hardened: C 64-67 Rock.
For broaches, chasers, reamers, hobs.
Type M-7 high speed steel, high red-hardness.

UNAMO 10.
M-1618; 0.85 C, 4 Cr, 2 V, 1.5 W, 8 Mo, bal Fe.
For twist drills, reamers, broaches, lathe and planer tools.
Type M-10 high speed steel, high red-hardness.

UNAMO 35.
M-1618; 0.80 C, 4 Cr, 2 V, 6 W, 5 Mo, 5 Co, bal Fe.
Hardened: C 65-67 Rock.
For reamers, broaches, lathe and planer tools, chasers, drills, hobs, reamers, form cutters.
Type M-35 high speed steel, high red-hardness.

UNA NO. 76.
M-930; 0.13-0.018 C, 0.4-0.6 Mn, bal Fe.
For welding rod; flux coated.

UNA NO. 156.
M-930; 0.13-0.18 C, bal Fe.
For welding rod; flux coated.

UNA NO. 158.
M-930; 0.08 C, 0.2-0.3 Mn, bal Fe.
For welding rod; flux coated.

UNA NO. 160.
M-930; 0.13-0.18 C, 0.4-0.6 Mn, bal Fe.
For welding rod; flux coated.

UNA NO. 200.
M-930; 0.07-0.12 C, 0.3-0.5 Mn, bal Fe.
For welding rod; flux coated.

UNA NO. 300.
M-930; 0.13-0.18 C, 0.4-0.6 Mn, bal Fe.
For welding rod; flux coated.

UNA NO. 325.
M-930; 0.85-1.10 C, 0.3-0.6 Mn, bal Fe.
For welding rod; hard and abrasion resisting.

UNA NO. 350.
M-930; 0.17 C, 4.75-5.25 Ni, bal Fe.
For welding rod.

UNA NO.425.
M-930; 0.08 C, 0.2-0.3 Mn, bal Fe.
Welding rod; protected arc.

UNA NO. 470.
M-930; Ni-Cu.
For welding rod; for machinable welds on cast iron.

UNA NO. 560.
M-930; Cu.
For welding rod for copper to steel; flux coated.

UNA NO. 601.
M-930; 0.5 Si, bal Al.
For welding rod; for Al.

UNA NO. 712.
M-930; 0.15-0.25 C, 0.3-0.6 Mn, 3.25-3.75 Ni, bal Fe.
For welding rod; flux coated.

UNA NO. 911.
M-930; 0.07 C, 0.5 Mn, 18-20 Cr, 8-10 Ni, bal Fe.
For welding rod; for 18-8 stainless steel.

UNA NO. 912.
M-930; 0.20 C, 2.0 Mn, 22-26 Cr, 12-14 Ni, bal Fe.
For welding rod; for stainless steel.

UNA NO. 914.
M-930; 0.10 C, 12-14 Cr, 0.25 Ni, bal Fe.
For welding rod; for corrosion resistant steel.

UNA NO. 916.
M-930; 0.10 C, 16-18 Cr, 0.25 Ni, bal Fe.
For welding rod; for stainless steel.

UNA NO. 917.
M-930; 0.10 C, 25-30 Cr, 0.25 Ni, bal Fe.
For welding rod; for stainless steel.

UNA NO. 918.
M-930; 0.15 C, 4-6 Cr, 0.4-0.6 Mo, bal Fe.
For welding rod; for 4-6 Cr steel.

UNA NO. 1500.
M-930; 0.08 C, bal Fe.
For welding rod.

UNA NO. 2150.
M-930; 0.03 C, 0.13 Mn, 0.46 Cu, 0.07 Mo, bal Fe.
For welding rod; for Cu-bearing steel.

UNA NO. 3175.
M-930; 0.08 C, 2 Ni, 1 Cu, bal Fe.
for welding rod; shielded coating.

UNA NO. 3200.
M-930; 0.10-0.14 C, 0.35-0.55 Mn, bal Fe.
For welding rod.

UNARAPID T-1.
M-1618; 0.7 C, 4 Cr, 1 V, 18 W, bal Fe.
Hardened: C 64-66 Rock.
For drills, saws, chasers, reamers, broaches, lathe and planer tools.
Type T-1 high speed steel, high red-hardness.

UNARAPID T-2.
M-1618; 0.8 C, 4 Cr, 2 V, 18 W, bal Fe.
Hardened: C 64-66 Rock.
For lathe and planer tools, forming tools, drills, broaches, reamers, taps.
Type T-2 high speed steel, high red-hardness.

UNARAPID T-4.
M-1618; 0.75 C, 4 Cr, 1 V, 18 W, 5 Co, bal Fe.
Hardened: C 64-66 Rock.
For lathe and planer tools, drills, chasers, reamers, broaches, hobs, taps
Type T-4 high speed steel, high red-hardness.

UNARAPID T-5.
M-1618; 0.8 C, 4 Cr, 2 V, 18 W, 8 Co, bal Fe.
Hardened: C 64-67 Rock.
For lathe and planer tools, form cutters, drills, reamers, broaches, hobs.
Type T-5 high speed steel, high red-hardness.

UNARAPID T-6.
M-1618; 0.8 C, 4.5 Cr, 1.5 V, 20 W, 12 Co, bal Fe.
Hardened: C 64-68 Rock.
For lathe and planer tools, chasers, form cutters, reamers, broaches, taps, hobs.
Type T-6 high speed steel, high red-hardness.

UNARAPID T-15.
M-1618; 1.5 C, 4.5 Cr, 4.75 V, 13.5 W, 0.5 Mo, 5 Co, bal Fe.
Hardened: C 64-66 Rock.
For textile needles, form cutters, broaches, reamers, taps.
Type T-15 high speed steel, high red-hardness.

UNAVAN 3 TYPE 1.
M-1618; 0.95-1.02 C, 4.5 Cr, 0.25 max Ni, 2.4-2.7 V, 5.8-6.3 W, 5.7- 6.2 Mo, 0.25 max Co, bal Fe.
Hardened: C 64-68 Rock.
For lathe and planer tools, form cutters, reamers, broaches, taps.
Type M-3 high speed steel, high red-hardness.

UNAVAN 4.
M-1618; 1.3 C, 4 Cr, 4 V, 5.5 W, 4.5 Mo, bal Fe.
Hardened: C 64-68 Rock.
For chasers, taps, reamers, form cutters, broaches, textile needles.
Type M-4 high speed steel, high red-hardness.

UNBREAKABLE METAL.
M-815; Al, Cu, bal Zn.
For lamp bases, toys, novelties; slush or permanent mold castings.

UNIBRAZE O PHOSPHOR-COPPER.
M-1711; 92.8 Cu, 7.2 P.
Melt range: 1300-1450°F; Braze: 1300-1450°F.
Filler metal for brazing copper in plumbing, air conditioning, electrical connections.
AWS A5.8-62T BCuP-2; ASTM B260-62T BCuP-2.

UNIBRAZE 2 PHOSPHOR-COPPER-SILVER.
M-1711; 91 Cu, 7 P, 2 Ag.
Melt range: 1190-1425°F; Braze: 1190-1425°F.
Filler metal for brazing copper in plumbing, air conditioning, electrical connections.
AWS A5.8-62T BCuP; ASTM B260-62T BCuP.

UNIBRAZE 5 PHOSPHOR-COPPER-SILVER.
M-1711; 89 Cu, 6 P,5 Ag.
Melt range: 1185-1485°F; Braze: 1185-1485°F.
For brazing copper base alloys where close fit-ups cannot be maintained.
AWS A5.8-62T BCuP-3; ASTM B260-62T BCuP-3.

UNIBRAZE 6 PHOSPHOR-COPPER-SILVER.
M-1711; 88 Cu, 6 P, 6 Ag.
Melt range: 1185-1480°F; Braze: 1185-1480°F.
For brazing ferrous and non-ferrous alloys.
AWS A5.8-62T BCuP-3; ASTM B260-62T BCuP-3.

UNIBRAZE 6F PHOSPHOR-COPPER-SILVER.
M-1711; 86.75 Cu, 7.25 P, 6 Ag.
Melt range: 1190-1300°F; Braze: 1190-1300°F.
For brazing ferrous aand non-ferrous alloys.
AWS A5.8-62T BCuP-4; ASTM B260-62T BCuP-4.

UNIBRAZE 15 PHOSPHOR-COPPER-SILVER.
M-1711; 80 Cu, 5 P, 15 Ag.
Melt range: 1185-1460°F; Braze: 1185-1460°F.
For brazing ferrous aand non-ferrous alloys.
AWS A5.8-62T BCuP-5; ASTM B260-62T BCuP-5.

UNIBRAZE 30.
M-1711; Copper cast iron rod.
For torch brazing and build up maintenance on cast iron.

UNIBRAZE 31.
M-1711; 31.5 Ag, 34 Cu, 15.5 Zn, 19 Cd.
Melt range: 1165-1390°F; Braze: 1165-1390°F.
For torch or induction brazing ferrous and non-ferrous metals.

UNIBRAZE 35.
M-1711; 35 Ag, 26 Cu, 21 Zn, 18 Cd.
Melt range: 1125-1295°F; Braze: 1125-1295°F.
For torch or induction brazing ferrous and non-ferrous metals where fit-up is poor.
AWS A5.8-62T BAg-2; ASTM B260-62T BAg-2; AMS 4768.

UNIBRAZE 40.
M-1711; Cast iron machinable electrode; AC-DC, straight or reverse polarity.
For repairing cast iron cylinder heads, motor blocks.

UNIBRAZE 45.
M-1711; 45 Ag, 15 cu, 16 Zn, 24 Cd.
Melt range: 1125-1145°F; Braze: 1125-1145°F.
For torch or induction brazing ferrous and non-ferrous metals where fit-up is good.
AWS A5.8 -62T BAg-1; ASTM B260-62T BAg-1; AMS 4769.

UNIBRAZE 50.
M-1711; 50 Ag, 15.5 Cu, 16.5 Zn, 18 Cd.
Melt range: 1160-1175°F; Braze: 1160-1175°F.
For torch or induction brazing ferrous and non-ferrous metals.
AWS A5.8-62T BAg-1a; ASTM B260-62T BAg-1a; AMS 4770B.

UNIBRAZE 60.
M-1711; cast iron non-machinable electrode; AC-DC, reverse polarity.
For repairing gears, motor housings, farm equipment, cams, levers, where machining is not necessary.

UNIBRAZE 80.
M-1711; Machinable cast iron electrode; AC-DC, straight or reverse polarity.
For repair work on cast iron and steel castings.

UNIBRAZE 110 (BARE,) UNIBRAZE 110 FC (FLUX COATED).
M-1711; Nickel silver brazing rod.
For maintenance repair and build up of steel tubing, bicycles, furniture, buildings.

UNIBRAZE 130 (BARE), UNIBRAZE 130 FC (FLUX COATED).
M-1711; Nickel silver brazing rod.
For repair and buildup of gears, valve seats, shafts.

UNIBRAZE 160.
M-1711; Low alloy steel electrode, AC-DC, reverse polarity.
For welding mild steel and low alloy steel to cast iron and higher carbon steel.

UNIBRAZE 210.
M-1711; Phosphor copper brazing rod.
Maintenance repair on copper tubing, wire, sheet.

UNIBRAZE 220.
M-1711; Silicon bronze electrode; AC-DC.
For surfacing steel, and for joining cast iron to steel and malleable iron.

UNIBRAZE 230.
M-1711; Phosphor copper silver brazing rod.
For brazing copper alloys.

UNIBRAZE 240.
M-1711; Phosphor bronze electrode; DC reverse polarity.
For welding and buildup on bronze castings, for joining steel, cast iron to bronze.

UNIBRAZE 250.
M-1711; All purpose stainless steel electrode, AC-DC, reverse polarity.
For maintenance repairs on steels, stainless, and clad steels.

UNIBRAZE 277.
M-1711; Nickel chrome high strength electrode, AC-DC.
For repairs of practically all steels.

UNIBRAZE 308 STAINLESS WELD WIRE.
M-1711; 0.07 max C, 20-22 Cr, 9.5-11.0 Ni, 1.0-2.0 Mn, 0.25-0.60 Si, bal Fe.
For MIG, TIG, manual and submerged arc welding AISI Types 201, 202, 204, 301, 302, 304, 305, 308, 410, 430 stainless.
AWS A5.9-62T; ASTM A371-627T.

UNIBRAZE 308 ELC STAINLESS WELD WIRE.
M-1711; 0.03 max C, 20-22 Cr, 9.5-11.0 Ni, 1.0-2.0 Mn, 0.25-0.60 Si, bal Fe.
For MIG, TIG, manual and submerged arc welding AISI 304L, 308L 321 & 347 stainless.
AWS A5.9-62T; ASTM A371-62T.

UNIBRAZE 309 STAINLESS WELD WIRE.
M-1711; 0.12 max C, 23-25 Cr, 13-14 Ni. 1.0-2.0 Mn, 0.25-0.60 Si, bal Fe.
For MIG, TIG, manual and submerged arc welding AISI 309 stainless and stainless to mild steel.
AWS A5.9-62T; ASTM A371-62T.

UNIBRAZE 310.
M-1711; Aluminum brazing rod.
For use on light gage aluminum sheet, tubing, furniture, trucks, trailers.

UNIBRAZE 310 STAINLESS WELD WIRE.
M-1711; 0.10-0.15 C, 26-28 Cr, 20.5-22.5 Ni, 1.25-2.5 Mn, 0.25-0.60 Si, bal Fe.
For MIG, TIG, manual and submerged arc welding AISI 310, 304 and some dissimiliar stainless and carbon or alloy steels.
AWS A5.9-62T; ASTM A371-62T; AMS 5694.

UNIBRAZE 312 STAINLESS WELD WIRE.
M-1711; 0.08-0.15 C, 28-31 Cr, 8.5-10.5 Ni, 1.0-2.0 Mn, 0.25-0.60 Si, bal Fe.
For MIG, TIG, manual and submerged arc welding stainless to mild steels and high strength steels.
AWS A5.9-62T; ASTM A371-62T; AMS 5784.

UNIBRAZE 316 STAINLESS WELD WIRE.
M-1711; 0.07 max C, 18-20 Cr, 12-14 Ni,1.0-2.0 Mn, 0.25-0.60 Si, 2.0-2.5 Mo, 1.0 max Cb, bal Fe.
For MIG, TIG, manual and submerged arc welding AISI 316 stainless steel.
AWS A5.9-62T; ASTM A371-62T; AMS 5690.

UNIBRAZE 316 ELC STAINLESS WELD WIRE.
M-1711; 0.03 max C, 18-20 Cr, 12-14 Ni, 1.0-2.25 Mn, 0.25-0.60 Si, 2.0-2.5 Mo, 1.0 max Cb, bal Fe.
For MIG, TIG, manual and submerged arc welding AISI 316 L stainless.
AWS A5.9-62T; ASTM A371-62T.

UNIBRAZE 330.
M-1711.
Aluminum brazing rod for build-up, and high strength joints; for castings also.

UNIBRAZE 345.
M-1711; Aluminum electrode-extruded coating, DC reverse polarity.
For miscellaneous maintenance repair of aluminum equipment.

UNIBRAZE 347 STAINLESS WELD WIRE.
M-1711; 0.07 max C, 19-21 Cr, 9.5-10.0 Ni, 1.2-2.0 Mn, 0.25-0.60 Si, Cb + Ta, bal Fe.
For MIG, TIG, manual and submerged arc welding AISI 321, 347, 30 and 304 L stainless.
AWS A5.9-62T; ASTM A371-62T.

UNIBRAZE 350.
M-1711; 50 Ag, 15.5 Cu, 15.5 Zn, 16 Cd, 3 Ni.
Melt range: 1195-1270°F; Braze: 1195-1270°F.
For torch or induction brazing ferrous and non-ferrous, including stainless steel; for brazing on carbide tips.
AWS A5.8-62T BAg-3; ASTM B260-62T BAg-3; AMS 4771.

UNIBRAZE 410 (BARE), UNIBRAZE 410 FC (FLUX COATED).
M-1711; Bronze brazing rod.
For joining and build-up on bronze, steel, cast iron, and miscellaneous alloys.

UNIBRAZE 410 STAINLESS WELD WIRE.
M-1711; 0.07-0.12 C, 12.0-13.5 Cr, 0.6 max Ni, 0.6 max Mn, 0.5 max Si 0.5 max Mo, bal Fe.
For MIG, TIG, manual and submerged arc welding of AISI 410, 403, 405, 414 & 416 stainless steels.
AWS A5.9-62T; ASTM A371-62T; AMS 5776.

UNIBRAZE 430 STAINLESS WELD WIRE.
M-1711; 0.01 max C, 15.5-17 Cr, 0.6 max Ni, 0.6 max Mn, 0.5 max Si, bal Fe.
For MIG, TIG, manual and submerged arc welding of AISI 430 stainless steel.
AWS A5.9-62T; ASTM A371-62T.

UNIBRAZE 533.
M-1711; Die cast rod, Zinc-base.
For joining zinc to itself or other alloys.

UNIBRAZE 555.
M-1711; Aluminum solder rod; for torch or other heat application; melts at 700°F.
For miscellaneous maintenance on aluminum base materials.

UNIBRAZE 610.
M-1711; Magnesium base rod.
For use on magnesium base materials.

UNIBRAZE 1000.
M-1711; High strength silver brazing rod, cadmium free.
For use on light gage materials.

UNIBRAZE 1010 (BARE), UNIBRAZE 1010 FC (FLUX COATED).
M-1711; low melting point silver brazing rod.
For electrical, air conditioning, heating and ventilating work; joins dissimilar metals.

UNIBRAZE 1110.
M-1711; lowest melting point silver brazing rod. General purpose repair and salvage.

UNIBRAZE 1550 (BARE), UNIBRAZE 1550 FC (FLUX COATED).
M-1711; premium silver brazing rod; cadmium free.
Particularly for joints that must be corrosion resistant.

UNIBRAZE 1640.
M-1711; silver brazing rod for tool tipping with carbide.

UNIBRAZE ALUMINUM BRONZE.
M-1711; 88.2 Cu, 1.5 Fe, 10.0 Al, 0.25 P.
MP.: 1967°F approx.
For shielded metal arc and gas metal arc braze braze welding and building up on ferrous and non-ferrous alloys.
AWS A5.6-57T Class A1-A2; ASTM B225-57T Class ECuAl-A2.

UNIBRAZE ALUMINUM WELDING ROD.
M-1711; Available in all aluminum alloy weldable grades.
For gas welding and inert tungsten gas (TIG) welding of aluminum alloys.
AWS A5-10; ASTM B285; Fed. QQ-R-566.

UNIBRAZE ALUMINUM WELDING WIRE.
M-1711; Available in all aluminum alloy weldable grades.
For inert gas metal arc (MIG) welding of aluminum alloys.

UNIBRAZE CUT ROD.
M-1711; AC-DC; reverse polarity on DC.
A fast working electrode without use of oxygen.
For cutting, beveling, gouging, piercing on most commercial metals.

UNIBRAZE DEOXIDIZED COPPER.
M-1711; 98.8 Cu, 0.75 Sn, 0.25 Si, 0.2 Mn.
MP: 1967°F approx.
For inert gas and oxyacetylene welding and build up on copper tanks and assemblies.
AWS A5.6-57T Class ECU; 5.7-57T Class RCu; ASTM B225-57T Class ECU; B259-57T Class RCu.

UNIBRAZE FABWELD.
M-1711; All purpose mild steel electrode; AC-DC, reverse polarity.
For maintenance and sheet metal welding.

UNIBRAZE GROOVING ROD.
M-1711; AC-DC; reverse polarity on DC.
For chamfering, routing, channeling, removing excess metal, gouging out old welds.

UNIBRAZE LOW FUMING BRONZE.
M-1711; 57.8 Cu, 40.3 Zn, 0.95 Sn, 0.85 Fe, 0.10 Si, 0.03 Mn.
MP: 1598°F. approx.
For oxyacetylene braze welding of steel, cast iron and copper alloys.
AWS A5.7-57T Class RCuZn-C; ASTM B259-57T Class RCuZn-

UNIBRAZE MAGNESIUM WELDING WIRE.
M-1711; available as AZ92A, AZ61A, or EZ33A.
Meets AWS, ASTM and Government Specs.

UNIBRAZE NAVAL BRONZE.
M-1711; 59 Cu, 40.4 Zn, 0.60 Sn.
MP: 1625°F. Approx.
For oxyacetylene braze welding and build-up on cast iron and steel.
AWS A5.7-57T Class RBCuZn-A; ASTM B259-57T Class RBCuZn-A.

UNIBRAZE NICKEL BRONZE.
M-1711; 59 Cu, 39 Zn, 0.9 Sn, 0.32 Ni, 0.42 Fe, 0.23 Mn, 0.7 Si.
MP: 1598°F. approx.
For oxyacetylene braze welding on malleable iron and steel.
AWS A5.7-57T Class RCuZu-B; ASTM B259-57T Class RCuZn-

UNIBRAZE NICKEL SILVER.
M-1711; 48.6 Cu, 41 Zn, 10.2 Ni, 0.15 Si, 0.02 P.
MP: 1706°F. approx.
For oxyacetylene braze welding iron and steel for color match.
AWS A5.7-57T Class RBCuZn-D; ASTM B259-57T Class RBCuZn-D.

UNIBRAZE NO. 716 ALUMINUM BRAZING WIRE.
M-1711.
Melting range: 970-1085°F.
For torch brazing aluminum.
AWS & ASTM: BA1Si-3.

UNIBRAZE NO. 718 ALUMINUM BRAZING WIRE.
M-1711; High silicon, aluminum base wire.
Melting range: 1070-1080°F.
For torch brazing aluminum.
AWS & ASTM: BA1Si-4.

UNIBRAZE PHOSPHOR BRONZE A.
M-1711; 94.8 Cu, 5.0 Sn, 0.20 P.
MP: 1922°F. Approx.
For inert gas and carbon arc welding of phosphor bronze and copper, and repair of castings.
AWS A5.13-56T Class RCuSn; ASTM A399-56T Class RCuSn.

UNIBRAZE PHOSPHOR BRONZE C.
M-1711; 91.9 Cu, 8.0 Sn, 0.1 P.
MP: 1880°F. approx.
For joining and repair of ferrous and non-ferrous parts, and for build-up.
AWS 5.6-57T Class ECuSn-C; ASTM B225-57T Class ECuSn-C.

UNIBRAZE PHOSPHOR BRONZE D.
M-1711; 89.75 Cu, 10.0 Sn, 0.25 P.
MP: 1832°F approx.
For inert gas and carbon arc braze welding of ferrous and non-ferrous alloys and for wear resistant surfacing and build-up.
AWS A5.13-56T Class RCuSn-D; ASTM A399-56T.

UNIBRAZE SILICON BRONZE.
M-1711; 95.8 Cu, 3.1 Si, 1.1 Mn.
MP: 1866°F. approx.
For inert gas, carbon-arc and oxyacetylene welding of copper base alloys and ferrous metals.
AWS A5.6-57T Class ECuSi; A5.7-57T Class RCuSi-A; ASTM B225-57T Class ECuSi; B259-57T Class RCuSi-A.

UNIBROACH.
M-114; 0.8 C, 4 Cr, 6.25 W, 6.25 Mo, 2.4 V, bal Fe.
For finishing tools, cutters, hobs, reamers; high speed steel.

UNICO.
M-344; 0.7 C, 0.4 Mn, 0.4 Si, 0.7 Cr, 1.4 W, bal Fe.
For pneumatic and flogging tools, shears; oil hardened, shock resistant.

UNICUT.
M-114; 1.0 C, 6.25 W, 4 Cr, 2.4 V, 6.25 Mo, bal Fe.
For lathe and planer tools, drills, taps, reamers, hobs; Type M3; high speed steel.

UNICUT-2.
M-114; 1.2 C, 6.0 Mo, 6.0 W, 4.0 Cr, 3.0 V, bal Fe.
Hardened: C64-68 Rock.
For cutters, drills, taps, end mills, reamers, hobs, form tools, lathe and planer tools, slitting saws, punches, drawing dies. Abrasion resistant, good red-hardness.
AISI M3, Type 2. High speed steel.

UNIDAL.
M-249; 4-6 Zn, 0.5-1.5 Mg, bal Al.
Extruded: 40,000 TS; 24,000 YS; 17 El; 85 Brin.
For architecture; heat treatable.

UNI-DIE.
M-35; 0.71 C, 2.0 Mn, 1.3 Mo, 1.0 Cr, bal Fe.
Air hardening cold work tool steel.
For blanking dies, punches, plug gages.
AISI A6.

UNIDUR-100.
M-493; 4.5 Zn, 1.3 Mg, bal Al.
Sol.Ht: 350 MPa TS; 10 El.
For welded construction of vehicles and equipment.

UNIFLUX 70 AND V 70.
M-1750; 0.07 C, 1.25-1.40 Mn, 0.45-0.55 Si, bal Fe.
Typical analysis and properties of deposited metal: 88,000 TS; 77,000 YS; 25 El; 65 RA.
Flux cored electrodes for shielded arc welding.
AWS A5.20 (69); ASME SFA 5.20.

UNIFLUX 75 AND V75.
M-1750; 0.07 C, 1.25-1.40 Mn, 0.45-0.55 Si, bal Fe.
Typical analysis and properties of deposited metal: 82,000-88,000 TS; 72,000-77,000 YS; 25 El; 65 RA.
Flux cored electrodes for shielded arc welding.
AWS A5.20 (69); ASME SFA 5.20.

UNIFLUX 90 AND V90.
M-1750; 0.07 C, 1.0 Mn, 0.45 Si, 2.40 Ni, bal Fe.
Typical analysis and properties of deposited metal: 85,000-90,000 TS; 78,000-82,000 YS; 26 El; 67 RA; good low temperature impact values.
Flux cored electrodes for shielded arc welding.

UNISTEEL 410.
M-1724; 0.25 max C, bal Fe.
Weldable, hot rolled reinforcing bar.

UNIFONT.
M-249, M-1605, M-1634; 4-6 Zn, 0.5-1.5 Mg, bal Al.
Sand cast: 28,000 TS; 20,000 YS; 4.5 El; 65 Brin.
Chill cast: 33,000 TS; 22,000 YS; 6 El; 75 Brin.
For architecture, ship and auto construction; not heat treatable.

UNIFONT 5.
M-1605; 0.8-1.0 Mg, 4.8-5.2 Zn, bal Al.
Sand cast & Aged: 37,000 TS; 24,000 YP; 3-6 El; 70 Brin.
Die cast and Aged: 40,000 TS; 31,000 YP; 4-8 El; 80 Brin.
For architecture, ship and auto components.
Self-hardening. Sand and die castings.

UNILOY 14 CMV replaced by **UNITEMP 14 CMV.**

UNILOY 14 HV replaced by **UNITEMP 14 HV.**

UNILOY 14 HW replaced by **UNITEMP 14 HW.**

UNILOY 14 MV replaced by **UNITEMP 14 MV.**

UNILOY 15-35 replaced by **UNITEMP 330.**

UNILOY 16-8 replaced by **UNILOY 301.**

UNILOY 16-18.
M-114; 0.08 max C, 15-17 Cr, 17-19 Ni, 2 max Mn, 1.0 max Si, bal Fe.
Wire 75,000 TS; 35,000 YS; 55 El.
For cold headed fasteners and parts.
Stainless, austenitic; excellent cold headability.

UNILOY 18-8 replaced by **UNILOY 302.**

UNILOY 18-8M replaced by **UNILOY 303.**

UNILOY 18-8S replaced by **UNILOY 304.**

UNILOY 18-8TI replaced by **UNILOY 321.**

UNILOY 18-12S replaced by **UNILOY 305.**

UNILOY 18-14S-MO replaced by **UNILOY 316.**

UNILOY 19-14 SM replaced by **UNILOY 317.**

UNILOY 19-9 DL replaced by **UNITEMP 19-9 DL.**

UNILOY 19-9 W MO replaced by **UNITEMP 19-9 W MO.**

UNILOY 19-9 WX replaced by **UNITEMP 19-9 WX.**

UNILOY 20-10S replaced by **UNILOY 308.**

UNILOY 20-25 replaced by **UNILOY 31.**

UNILOU 24-11.
M-114; 0.15-0.30 C, 0.5-0.8 Mn, 0.9-1.2 Si, 22-25 Cr, 10-12 Ni, bal Fe.
At 70°F.: 112,500 TS; 33 El; 236 Brin.
At 1200°F.: 67,500 TS; 2 El.
For heat resisting purposes; cannot be hardened by heat treatment; resists heat up to 2100°F.

UNILOY 25-12 replaced by **UNILOY 309.**

UNILOY 25-12S replaced by **UNILOY 309 S.**

UNILOY 25-20 replaced by **UNILOY 310.**

UNILOY 25-20H replaced by **UNILOY 314.**

UNILOY 36.
M-114; 35-36 Ni, 0.35 Mn, 0.12 Si, 0.04 C, bal Fe.
Bar: 68,000 TS; 36,000 YS; 44 El, 78 RA.
At -100°F: 87,000 TS; 60,000 YS; 42 El, 73 RA.
For thermostats, temperature control devices, bimetals, length standards.
Low coefficient expansion.

UNILOY 42 replaced by **UNISEAL 42.**

UNILOY 49 replaced by **UNISEAL 52.**

UNILOY 50 CR/50 NI.
M-114; 50 Cr, 1.0 Ti, 0.06 C, bal Ni.
Annealed: 125,000 TS; 72,000 YS; 18 El, 20 RA, C28 Rock.
For equipment in the pulp and paper industry, power production, waste incinerator, petroleum refining, fuel-oil fired heaters.
Excellent resistance to oil and acid corrosion.
Resists sulfidation.

UNILOY 17-4 MO.
M-114; 0.10 C, 17.0 Cr, 4.0 Ni, 3.0 Mo, bal Fe.
At R.T. 191,900 TS; 146,900 YS; 13.5 El.
At 1200°F: 45,240 TS; 29,230 YS; 42.0 El.
For knife blades, valves, springs, high strength structural applications up to 1000°F.
Precipitation hardening stainless.
Name changed to Unitemp 350.

UNILOY 201.
M-114; 0.15 max C, 16-18 Cr, 3.5-5.5 Ni, 5.5-7.5 Mn, 1.0 max Si, 0.25 max N, bal Fe.
Ann: 115,000 TS; 55,000 YS; 55 El; 90 Rock B.
Good corrosion resistance, nonmagnetic (Ann.), good resistance to scaling 1450°F; good impact strength.
For parts requiring corrosion resistance.
AISI Type 201 stainless steel; SAE 30201.

UNILOY 202.
M-114; 0.15 max C, 17-19 Cr, 4-6 Ni, 7.5-10.0 Mn, 1.0 max Si, 0.25 max N, bal Fe.
Ann: 105,000 TS; 55,000 YS; 55 El; 90 Rock B.
Good corrosion resistance, non-magnetic (Ann.), good resistance to scaling to 1500°F; good impact strength.
For parts requiring corrosion resistance.
AISI Type 202 stainless steel; SAE 30202.

UNILOY 301.
M-114; 0.15 max C, 16-18 Cr, 6-8 Ni, 2.0 max Mn, 1.0 max Si, bal Fe.
Ann: 110,000 TS; 40,000 YS; 60 El; 85 Rock B.
Corrosion resistant; work hardens rapidly to max of about 180,000 psi TS.
Cold worked materials for springs and strong, stainless stampings.
AISI Type 301 stainless steel; SAE 30301; AMS 5519.

UNILOY 302.
M-114; 0.15 max C, 17-19 Cr, 8-10 Ni, 2.0 max Mn, 1.0 max Si, bal Fe.
Ann: 85,000 TS; 35,000 YS; 60 El; 150 Brin.
Austenitic, non-magnetic (Ann.), corrosion resistant, ductile.
For stainless steel parts and assemblies, particulary stampings; food industry.
AISI Type 302 Stainless; SAE 30302.

UNILOY 303, 303 SE.
M-114; 0.15 max C, 17-19 Cr, 8-10 Ni, 2.0 max Mn, 1.0 max Si, 0.15 min S or Se, 0.6 max Mo, bal Fe.
Ann: 90,000 TS; 35,000 YS; 50 El; 160 Brin.
Corrosion resistant, austenitic, free machining.
For stainless parts made on automatic screw machines, screws, studs, nuts.
AISI Type 303; SAE 30303; AMS 5640.

UNILOY 303 MA.
M-114; 0.15 max C, 2 max Mn, 1 max Si, 0.11-0.16 S, 17-19 Cr, 8-10 Ni, 0.4-0.6 Mo, 0.6-1.0 Al, bal Fe.
Cold drawn; 100,000 TS; 60,000 YS; 40 El, 55 RA, 228 Brin.
Ann: 90,000 TS; 40,000 YS; 50 El, 60 RA, 160 Brin.
For stainless screw machine products, switch gears, pump and valve parts, fittings.
Free machining, stainless, austenitic, non-magnetic.
AMS 5638.

UNILOY 304.
M-114; 0.08 max C, 18-20 Cr, 8-10 Ni, 2.0 max Mn, 1.0 max Si, bal Fe.
Ann: 85,000 TS; 35,000 YS; 60 El; 150 Brin.
Austenitic, corrosion resistant.
For stainless stampings, structural parts, cold formed parts, rivets, beer barrels.
AISI Type 304 stainless; SAE 30304.

UNILOY 304 L.
M-114; 0.03 max C, 2.0 max Mn, 1.0 max Si, 18-20 Cr, 8-10 Ni, bal Fe.
Ann, R.T.: 85,000 TS; 37,000 YS; 57 El.
-320°F; 235,000 TS; 56,000 YS; 40 El.
1400°F; 29,000 TS; 11,000 YS; 36 El.
Austenitic, stainless, weldable.
For welded assemblies for cryogenic to room temperature to elevated temperature operation, tanks, etc.
AISI Type 3041; SAE 30304 L.

UNILOY 305.
M-114; 0.12 max C, 17-19 Cr, 10-13 Ni, 2.0 max Mn, 1.0 max Si, bal Fe.
Ann: 85,000 TS; 38,000 YS; 50 El; 8 Rock B.
Austenitic, corrosion resistant; less tendency to work harden.
For stainless stampings, formed parts.
AISI Type 305 stainless; SAE 30305.

UNILOY 308.
M-114; 0.08 max C, 2.0 max Mn, 1.0 max Si, 19-21 Cr, 10-12 Ni, bal Fe.
Wire, Ann: 115,000 TS; 80,000 YS; 40 El.
For weld rod, industrial furnaces, equipment for hot sulphite liquor.
AISI Type 308 stainless.

UNILOY 308L.
M-114; 0.03 max C, 2.0 max Mn, 1.0 max Si, 19-21 Cr, 10-12 Ni, bal Fe.
Main use is weld rod for welding austenitic and ferritic stainless steels.

UNILOY 309.
M-114; 0.20 max C, 2.0 max Mn, 1.0 max Si, 22-24 Cr, 12-15 Ni, bal Fe.
Ann, 70°F: 90,000 TS; 40,000 YS; 45 El; 160 Brin.
1600°F: 21,000 TS; 17,500 YS; 50 El.
Austenitic, corrosion resistant, good resistance to scaling to 1800°F, weldable.
For furnace parts, aircraft and jet engine parts, heat exchangers, chemical equipment.
AISI Type 309 Stainless; SAE 30309.

UNILOY 309 S.
M-114; 0.08 max C, 2.0 max Mn, 1.0 max Si, 22-24 Cr, 12-15 Ni, bal Fe.
Ann, 70°F: 90,000 TS; 40,000 YS; 45 El; 160 Brin.
1600°F: 21,000 TS; 17,500 YS; 50 El.
Austenitic, corrosion resistant, good resistance to scaling to 1800°F, preferred for welding.
For furnace parts, welded assemblies for aircraft and jet engine parts, heat exchangers.
AISI Type 309S; SAE 30309S; AMS 5650.

UNILOY 309S-CB.
M-114; 0.08 max C, 2.0 max Mn, 1.0 max Si, 22-24 Cr, 12-15 Ni, Cb + Ta = 10 x %C, bal Fe.
Low carbon, columbium stabilized, austenitic weld rod for welding Type 30 and similar metals for high temperature service.

UNILOY 310.
M-114; 0.25 max C, 2.0 max Mn, 1.5 max Si, 24-26 Cr, 19-22 Ni, bal Fe.
Ann, RT: 95,000 TS; 45,000 YS; 50 El; 89 Rock B.
1600°F: 25,000 TS; 17,000 YS; 33 El.
For furnace parts, aircraft and jet engine parts, heat exchangers, petroleum refining, weld rods, chemical equipment.
AISI Type 310 stainless; SAE 30310; AMS 5651, 5521.

UNILOY 310S.
M-114; 0.08 max C, 2.0 max Mn, 1.50 max Si, 24-26 Cr, 19-22 Ni, bal Fe.
Similar to Uniloy 310, but preferred for weld rods and some welded assemblies.
AISI Type 310S

UNILOY 314.
M-114; 0.25 max C, 2.0 max Mn, 1.5-3.0 Si, 23-26 Cr, 19-22 Ni, bal Fe.
Ann, RT: 95,000 TS; 45,000 YS; 50 El; 89 Rock B.
1600°F: 25,000 TS; 17,000 YS; 33 El.
For furnace parts, heat exchangers, oil refining; particularly for carburizing atmospheres as carburizing boxes and retorts.
AISI Type 314; SAE 30314; AMS 5652.

UNILOY 316.
M-114; 0.08 max C, 2.0 max Mn, 1.0 max Si, 16-18 Cr, 10-14 Ni, 2-3 Mo, bal Fe.
Ann: 80,000 TS; 30,000 YS; 60 El; 149 Brin.
Ann, CD: 90,000 TS; 60,000 YS; 45 El; 190 Brin.
Excellent corrosion resistance, weldable.
For photographic equipment, pulp and paper, pharmaceutical and textile equipment.
AISI Type 316; SAE 30316; AMS 5648.

UNILOY 316F.
M-114; 0.08 max C, 2.0 max Mn, 1.0 max Si, 0.10 min S, 0.10 min P, 17-19 Cr, 10-14 Ni, 1.75-3.0 Mo, bal Fe.
Free machining grade of Uniloy 316.
For stainless parts requiring appreciable machining.
AMS 5649

UNILOY 316L.
M-114; 0.03 max C, 2.0 max Mn, 1.0 max Si, 16-18 Cr, 10-14 Ni, 2.0-3. Mo, bal Fe.
Preferred for weld wire and welded assemblies for high temperature operation and for equipment for chemical plants.
AISI Type 316L; SAE 30316 L: AMS 5653.

UNILOY 317.
M-114; 0.08 max C, 2.0 max Mn, 1.0 max Si, 18-20 Cr, 11-15 Ni, 3.0-4. Mo, bal Fe.
Ann: 80,000 TS; 30,000 YS; 60 El; 149 Brin.
Austenitic, excellent corrosion resistance, maximum resistance to pitting.
For boat trim, chemical processing equipment, exhaust manifolds.
AISI Type 317; SAE 30317.

UNILOY 317L.
M-114; 0.03 max C, 2.0 max Mn, 1.0 max Si, 18-20 Cr, 11-15 Ni, 3.0-4. Mo, bal Fe.
For weld rod and for welded assemblies for corrosive conditions.

UNILOY 318.
M-114; 0.08 max C, 2.0 max Mn, 1.0 max Si, 17-19 Cr, 13-15 Ni, 2.0-3.0 Mo, Cb + Ta = 10 x %C, bal Fe.
Ann. RT: 80,000 TS; 30,000 YS; 60 El; 149 Brin.
1600°F: 24,000 TS; 16,000 YS; 39 El.
For weld wire, welded assemblies and for elevated temperature operation; furnace parts, heat exchangers, exhaust manifolds.

UNILOY 321.
M-114; 0.08 max C, 2.0 max Mn, 1.0 max Si, 17-19 Cr, 9.0-12.0 Ni, Ti = 5 x %C min, bal Fe.
Ann, RT: 85,000 TS; 35,000 YS; 55 El; 150 Brin.
1600°F: 17,000 TS; 13,500 YS; 57 El.
For parts and welded assemblies operating intermittently in 900-1600°F range.
Aircraft exhaust manifolds, jet engine parts, pressure vessels, heat treat equipment.
AISI Type 321; SAE 30321; AMS 5510, 5645.

UNILOY 325.
M-114; 0.4 C, 1.2 Si, 0.6 Mn, 7-10 Cr, 19-23 Ni, bal Fe.
Forged: 141,000-108,000 TS; 27-50 El; 248-167 Brin.
For applications where superheated steam, hot oils, sulphides, hot sulphur are encountered; non-corrosive and heat resistant; resists scaling up to 1500°F.

UNILOY 326.
M-114; 0.05 max C, 1.0 max Mn, 0.60 max Si, 25-27 Cr, 6-7 Ni, 0.25 max Ti, bal Fe.
About 30% austenitic but ferro magnetic.
Ann: 90,000-100,000 TS; 75,000 YS; 30 El.
Hardenable to about 180,000 TS by cold work; Resistant to corrosion and stress corrosion.
For cold headed bolts and cold headed fasteners for paper pulp, food and chemical industries.

UNILOY 330.
M-114; 0.20 max C, 2.0 max Mn, 1.5 max Si, 14.50-16.50 Cr, 34-37 Ni, bal Fe.
Ann. RT.: 85,000 TS; 42,000 YS; 43 El.
2000°F: 5100 TS; 3900 YS; 72 El.
Good strength and scaling resistance at high temperature.
For annealing fixtures, carburizing boxes, furnace parts operating up to 2100°F.

UNILOY 347.
M-114; 0.08 max C, 2.0 max Mn, 1.0 max Si, 17-19 Cr, 9-13 Ni, Cb + Ta = 10 x %C min, bal Fe.
Stabilized stainless for weld rod, welded assemblies, and for operation 900-1600°F.
For aircraft exhaust manifolds, fire walls, pressure vessels, high temperature chemical handling equipment.
AISI Type 347; SAE 30347; AMS 5512, 5646.

UNILOY 347F-SE.
M-114; 0.08 max C, 2.0 max Mn, 1.0 max Si, 0.10 min P, 0.030 max S, 17-19 Cr, 9-12 Ni, Cb + Ta = 10 x %C min, 0.15 min Se, bal Fe.
Ann. RT: 90,000 TS; 35,000 YS; 50 El, 160 Brin.
1600°F: 20,000 TS; 15,000 YS; 84 El.
Free machining, stabilized, austenitic, stainless.
For machined parts to operate intermittently to 900-1600°F range.
AMS 5642.

UNILOY 348.
M-114; 0.08 max C, 2.0 max Mn, 1.0 max Si, 17-19 Cr, 9-13 Ni, Cb + Ta = 10 x %C min, 0.10 max Ta, 0.20 max Co, bal Fe.
Stabilized, austenitic, weldable stainless, -similar to Uniloy 347 except restricted Tantalum and Cobalt content.
For nuclear applications.
AISI Type 348; SAE 30348; AMS 5646.

UNILOY 403.
M-114; 0.15 max C, 1.0 max Mn, 0.5 max Si, 11.5-13.0 Cr, bal Fe.
Ann: 75,000 TS; 40,000 YS; 35 El; 155 Brin.
Heat treatable, martensitic, weldable, stainless.
Hardenable to 100,000-195,000 psi TS.
For turbine wheels, valve parts, cutlery, sporting goods, rifle barrels.
AISI Type 403; SAE 51403; AMS 5613.

UNILOY 405.
M-114; 0.08 max C, 1.0 max Mn, 1.0 max Si, 11.5-14.5 Cr, 0.10-0.30 Al bal Fe.
Ann: CD.: 85,000 TS; 70,000 YS; 20 El, 185 Brin.
Weldable, not readily hardenable; good corrosion resistance.
Designed for welded assemblies that harden only slightly in the welded area.
AISI 405; SAE 51405.

UNILOY 410.
M-114; 0.15 max C, 1.0 max Mn, 1.0 max Si, 11.5-13.5 Cr, bal Fe.
Ann: 75,000 TS; 40,000 YS; 35 El; 155 Brin.
Hardenable to 100,000-190,000 TS; weldable, magnetic, martensitic, corrosion resistant.
For turbine blades, cutlery, pump parts, mining machinery, fishing poles.
AISI Type 410; SAE 51410; AMS 5504, 5613.

UNILOY 414.
M-114; 0.15 max C, 1.0 max Mn, 1.0 max Si, 11.5-13.5 Cr, 1.25-2.50 Ni, bal Fe.
Hardenable, martensitic, stainless.
Air or oil hardenable to approximately 200,000 TS max.
For beater bars, fasteners, gage parts, springs, mining equipment cutlery, shafts, valve seats.
AISI 414; SAE 51414; AMS 5615.

UNILOY 416.
M-114; 0.15 max C, 1.25 max Mn, 0.06 max P, 0.15 min S, 1.0 max Si, 12-14 Cr, 0.60 max Mo, bal Fe.
Ann, CD: 100,000 TS; 85,000 YS; 15 El; 215 Brin.
Ht. Tr.: 100,000-190,000 TS; 90,000-145,000 YS; 9-18 El; 26-42 Rock C.
Free machining, martensitic, corrosion resistant.
For shafts, valves, parts made on automatic screw machines, studs, pump parts.
AISI 416; SAE 51416; AMS 5610.

UNILOY 420.
M-114; 0.15 min C, 1.0 max Mn, 1.0 max Si, 12-14 Cr, bal Fe.
Air hardened: 110,000-250,000 TS; 70,000-200,000 YS; 22-7 El; 500-200 Brin.
Martensitic, magnetic, corrosion resistant.
For cutlery, hand tools, dental and surgical instruments, valve trim and parts, shafts.
AISI 420; SAE 51420; AMS 5506, 5621.

UNILOY 420 F.
M-114; 0.15 min C, 1.0 max Mn, 1.0 max Si, 0.04 max P, 0.15 min S, 12-14 Cr, 0.60 max Mo, bal Fe.
Martensitic, free-machining, air hardenable, corrosion resistant.
Air hardenable to 110,000-250,000 TS.
For shafts, hardware, screw machine parts, small turbine wheels, splined shafts.
AISI 420 F; SAE 51420 F; AMS 5620.

UNILOY 420 F-SE.
M-114; 0.15 min C, 1.0 max Mn, 1.0 max Si, 0.040 max P, 0.030 max S, 12-14 Cr, 0.60 max Mo, 0.15 min Se, bal Fe.
Same as Uniloy 420 F except that Selenium replaces Sulphur for free machining properties.

UNILOY 430.
M-114; 0.12 max C, 1.0 max Mn, 1.0 max Si, 14-18 Cr, bal Fe.
Ann: 80,000 TS; 55,000 YS; 30 El; 170 Brin.
Ferritic, magnetic, non-hardenable by heat treat, corrosion resistant.
For architectural and automotive trim and moldings, nitric acid tanks, table ware, plumbing supplies.
AISI 430; AMS 51430; AMS 5503, 5627.

UNILOY 430 F.
M-114; 0.12 max C, 1.25 max Mn, 1.0 max Si, 0.060 max P, 0.15 min S, 14-18 Cr, 0.60 max Mo, bal Fe.
Ann: 80,000 TS; 55,000 YS; 25 El; 170 Brin.
Ferritic, magnetic, non-hardenabe by heat treat, free machining, corrosion resistant.
For corrosion resistant studs, bolts, nuts, etc. being machined on automatic screw machines.
AISI 430 F; SAE 51430 F.

UNILOY 430 F-SE.
M-114; 0.12 max C, 1.25 max Mn, 1.0 max Si, 0.060 max P, 0.060 max S, 14-18 Cr, 0.15 min Se, bal Fe.
Same as Uniloy 430 F except that Selenium replaces Sulphur for free machining properties.
AISI 430 F-Se; SAE 51430 F-Se.

UNILOY 430 TI.
M-114; 0.12 max C, 1.0 max Mn, 1.0 max Si, 16-18 Cr, Ti = 6 x %C min, bal Fe.
Same as Uniloy 430 except that Titanium improves ductility of welds.

UNILOY 431.
M-114; 0.20 max C, 1.0 max Mn, 1.0 max Si, 15-17 Cr, 1.25-2.50 Ni, bal Fe.
Ann: 125,000 TS; 95,000 YS; 20 El; 260 Brin.
Oil hardened: 150,000-210,000 TS; 110,000-165,000 YS; 20-14 El; 31-45 Rock C.
Martensitic, weldable, corrosion resistant; good strength to 900°F.
Aircraft fittings, pump shafts, marine hardware, valve parts, conveyor parts.
AISI 431; SAE 51431; AMS 5628

UNILOY 434.
M-114; 0.12 max C, 1.0 max Mn, 1.0 max Si, 14-18 Cr, 0.75-1.25 Mo, bal Fe.
Same as Uniloy 430 except that added Molybdenum improves resistance to pitting from de-icing chemicals. AISI 434.

UNILOY 435.
M-114; 0.12 max C, 1.0 max Mn, 1.0 max 1.0 max Si, 14-18 Cr, 0.40-0.60 Cb + Ta, bal Fe.
Same as Uniloy 430 except that Cb + Ta improves surface smoothness when metal is stretch bent or deep drawn.

UNILOY 436.
M-114; 0.12 max C, 1.0 max Mn, 1.0 max Si, 14-18 Cr, 0.75-1.25 Mo, 0.40-0.60 Cb + Ta, bal Fe.
Similar to Uniloy 430 but produces smoother surfaces on deep drawing and improved resistance to pitting from de-icing chemicals. AISI 436.

UNILOY 440 A.
M-114; 0.60-0.75 C, 1.0 max Mn, 1.0 max Si, 16-18 Cr, 0.75 max Mo, bal Fe.
Ann: 115,000 TS; 90,000 YS; 12 El; 240 Brin.
Ht. Tr.: 260,000 TS; 240,000 YS; 5 El; 510 Brin.
Martensitic, magnetic, corrosion resistant.
For cutlery, bearings, valves, surgical and dental instruments.
AISI 440 A; SAE 51440 A; AMS 5631.

UNILOY 440 B.
M-114; 0.75-0.95 C, 1.0 max Mn, 1.0 max Si, 16-18 Cr, 0.75 max Mo, bal Fe.
Air or oil hardenable to 270,000-280,000 TS.
Holds high hardness to 900°F.
Corrosion resistant, magnetic.
For bearings, cutlery, spatula blades, food processing knives.
AISI 440 B; SAE 51440 B.

UNILOY 440 C.
M-114; 0.95-1.20 C, 1.0 max Mn, 1.0 max Si, 16-18 Cr, 0.75 max Mo, bal Fe.
Air or oil hardenable to 275,000-285,000 TS and 580 Brin; retains hardness after tempering to 900°F of 55-57 Rc; corrosion resistant.
For cutlery, bearings, nozzles, balls and seats for oil well pumps.
AISI 440 C; SAE 51440 C; AMS 5630.

UNILOY 440 F.
M-114; 0.95-1.20 C, 1.0 max Mn, 1.0 max Si, 0.040 max P, 0.05 min S, 16-18 Cr, 0.75 max Mo, bal Fe.
Free machining grade of Uniloy 440 C.
For parts requiring considerable machining before heat treating.
SAE 51440 F; AMS 5632.

UNILOY 440 F-SE.
M-114; 0.95-1.20 C, 1.0 max Mn, 1.0 max Si, 0.040 max P, 0.030 max S, 16-18 Cr, 0.75 max Mo, 0.10 min Se, bal Fe.
Similar to Uniloy 440 F except that Selenium replaces Sulphur for free machining properties.
SAE 51440 F-Se.

UNILOY 442.
M-114; 0.20 max C, 1.0 max Mn, 1.0 max Si, 18-23 Cr, bal Fe.
Ann, RT: 75,000 TS; 45,000 YS; 30 El; 160 Brin.
1600°F: 4500 TS; 3500 YS; 87 El.
Ferritic, non-hardenable stainless, weldable.
For annealing boxes, oil burner parts, nozzles, heating elements.
AISI 442.

UNILOY 443.
M-114; 0.20 max C, 1.0 max Mn, 1.0 max Si, 18-23 Cr, 0.90-1.25 Cr, bal Fe.
Same properties and characteristics as Uniloy 442, but more resistant to action of dilute sulphuric acid.

UNILOY 446.
M-114; 0.20 max C, 1.5 max Mn, 1.0 max Si, 23-27 Cr, 0.25 max N, bal Fe.
Ann, RT.: 80,000 TS; 50,000 YS; 25 El; 170 Brin.
1800°F: 2500 TS; 2000 YS; 135 El.
Ferritic, non-hardenable, stainless, weldable.
Oxidation resistant at high temperatures.
For annealing boxes, salt bath electrodes, oil burner parts, metal to glass seals.
AISI 446; SAE 51446.

UNILOY 501.
M-114; 0.10 min C, 4-6 Cr, bal Fe.
Annealed: 70,000 TS; 30,000 YS; 28 El; 65 RA; 160 Brin.
Heat treated: 175,000 TS; 135,000 YS; 15 El; 50 RA; 370 Brin.
For oil refinery equipment, fittings, grate bars; Type 501; corrosion resistant.

UNILOY 502.
M-114; 0.10 max C, 4-6 Cr, 0.5 Mo, bal Fe.
Annealed: 70,000 TS; 32,000 YS; 28 El; 65 RA; 160 Brin.
Heat treated: 180,000 TS; 140,000 YS; 13 El; 45 RA; 380 Brin.
For oil refinery equipment, fittings, grate bars; Type 502; corrosion resistant.

UNILOY 1409 replaced by **UNILOY 410.**

UNILOY 1409 AL.
M-114; 3.25-4.50 Al, 12-14 Cr, bal Fe.
For electrical resistance rod and wire, oxidation resistant to 1600°F.

UNILOY 1409 M replaced by **UNILOY 416.**

UNILOY 1409 NH replaced by **UNILOY 405.**

UNILOY 1409 NI replaced by **UNILOY 414.**

UNILOY 1409 TB replaced by **UNILOY 403.**

UNILOY 1415 NW replaced by **UNITEMP 1415 NW.**

UNILOY 1420 WM replaced by **UNITEMP 1420 WM.**

UNILOY 1430 MV replaced by **UNITEMP 1430 MV.**

UNILOY 1435 replaced by **UNILOY 420.**

UNILOY 1435 M replaced by **UNILOY 420 F.**

UNILOY 1809 replaced by **UNILOY 430.**

UNILOY 1809 M replaced by **UNILOY 430 F.**

UNILOY 1809 NI replaced by **UNILOY 431.**

UNILOY 1860 replaced by **UNILOY 440 A.**

UNILOY 1890 replaced by **UNILOY 440 B.**

UNILOY 2009 replaced by **UNILOY 442.**

UNILOY 2525 replaced by **UNILOY 446.**

UNILOY 15100 MO.
M-114; 1.05 C, 14.5 Cr, 4.0 Mo, bal Fe.
Martenitic, corrosion resistant, high strength and hardness to 1000°F.
For bearings, gears, shafts, loaded stress parts operating up to 1000°F.
AISI 618; modified 440 C + Mo.

UNILOY 18100 replaced by **UNILOY 440 C.**

UNILOY 18100 F SE replaced by **UNILOY 440 F SE.**

UNILOY 18110 BM replaced by **UNILOY 440 C.**

UNILOY A286 replaced by **UNITEMP A286.**

UNILOY D319.
M-114; 0.07 max C, 2.0 max Mn, 1.0 max Si, 17.5-19.5 Cr, 11-15 Ni, 2.25-3.0 Mo, bal Fe.
Austenitic, corrosion resistant.
For parts, and assemblies for chemical industry, certain improvements over Uniloy 316.

UNILOY EB 26-1.
M-114; 0.00k max C, 25.0-27.0 Cr, 0.40 max Mn, 0.75-1.25 Mo, 0.40 max Si, bal Fe.
Ann: 65,000-78,000 TS; 53,000-61,000 YS; 24-30 El; 83-86 Rock B.
Hardenable to about 116,000 TS; by cold work.
Corrosion and stress corrosion resistant, weldable, magnetic.
For marine equipment, chemical equipment, food and beverage equipment, pulp and paper.

UNILOY-HC.
M-114; 0.06 C, 15 Cr, 1.5 Co, 16 Mo, 3.75 W, 5.75 Fe. 0.2 V, bal Ni.
Heat Treated: 116,000 TS; 56,000 YS; 52 El, Rock B 95.
At 1600°F: 57,000 TS; 38,000 YS; 49 El.
For chemical processing equipment, process vessels, pump parts, valves, aircraft and aerospace components. Corrosion resistant. Resists repeated thermal shock at 1600 to 1800°F.

UNILOY HV replaced by **UNITEMP HV.**

UNILOY M-252 replaced by **UNITEMP M252.**

UNILOY N-155 replaced by **UNITEMP N-155.**

UNILOY S 590 replaced by **UNITEMP S 590.**

UNILOY S-817.
M-114; 0.32-0.42 C, 1-2 Mn, 19-21 Cr, 3.5-4.5 W, 19-21 Ni, 4 Mo, bal Co.
Rolled: 125,000-145,000 TS; 55,000-75,000 YS; 32-22 El; 31-20 RA.
For turbine rotors, buckets, shafts, bolts; age-hardenable, heat and corrosion resistant.

UNIMACH NO. 1.
M-114; 0.40 C, 0.40 Mn, 1.0 Si, 5 Cr, 1.4 Mo, 1.0 V, bal Fe.
Heat treated: 225,000-305,000 TS; 195,000-255,000 YS; 15.0-8.5 El; 50.2-29.5 RA; 445-560 Brin.
For die casting dies, hot work punches, hot shears, dummy blocks; hot work steel, air hardened, non-deforming.

UNIMAG 50.
M-114; 0.05 C, 0.05 Mn, 0.35 Si, 48.0 Ni, bal Fe.
After hydrogen anneal has high saturation flux of 15,000 gausses, and low hysteresis loss in DC, or in AC circuits with frequencies less than 400 cycles per second.
For laminated cores, magnetic shielding, relays, solenoid cores and electric motor rotors.

UNIMAR 250.
M-114; 0.02 C, 0.05 Mn, 0.05 Si, 18 Ni, 8 Co, 4.8 Mo, 0.4 Ti, 0.1 Al, bal Fe.
Ht. Tr.: 268,000 TS; 258,000 YS; 11 El, c52 Rock.
For jet engine, gas turbine and spacecraft components, pressure vessels, rocket motor cases.
Maraging steel, tough, ductile and shock resistant.
Ppt. hardened.

UNIMAR 300.
M-114; 0.02 C, 0.05 Mn, 0.05 Si, 18.5 Ni, 9 Co, 4.8 Mo, 0.6 Ti, 0.1 Al, bal Fe.
Annealed: 50,000 TS; 110,000 YS; 18 El; C32 Rock.
Ht. Tr.: 315,000 TS; 310,000 YS; 10 El, C55 Rock.
For aluminum die casting dies, pressure vessels, rocket motor cases, bolts and fasteners.
Maraging steel, vacuum melted. Precipitation hardened.

UNIMAR 350.
M-114; 0.02 C, 0.05 Mn, 0.05 Si, 17.5 Ni, 12 Co, 4.8 Mo, 1.5 Ti, 0.1 Al, bal Fe.
Ht. Tr. 360,000 TS; 350,000 YS; 10 El, C 58 Rock.
For jet engine, gas turbine and spacecraft components.
Maraging steel, tough, ductile and shock resistant.
Precipitation hardened.

UNIMATIC 6000.
M-717; mild steel electrode for AC-DC welding of mild steel; 68,000 TS.

UNIMET.
M-897; 1.57 Al, 1.35 W, 0.42 U, bal Ni.
For electron tube elements; heat resistant.

UNIMETAL.
M-391; Sn, Sb, bal Pb.
For welding of white metal; white metal.

UNIMET U-10.
M-1706; sintered carbide tool material.
For roughing cuts on cast iron, non-ferrous and non-metallic materials.

UNIMET U-20.
M-1706; sintered carbide tool material.
For general purpose roughing and finishing on cast iron, non-ferrous and non-metallic materials.

UNIMET U-20F.
M-1706; sintered carbide tool material.
For use on plastics, glass and highly abrasive exotic materials and for drilling plastic laminates.

UNIMET U-30.
M-1706; sintered carbide tool material.
For light roughing and average finishing of cast iron, non-ferrous and non-metallic materials. Medium high wear resistance and medium low shock resistance.

UNIMET U-40.
M-1706; sintered carbide tool material.
For precision boring of cast iron, non-ferrous and non-metallic materials.

UNIMET U-53.
M-1706; sintered carbide tool material.
For heavy duty roughing cuts on steel and steel alloys at extremely heavy feeds.

UNIMET U-60.
M-1706; sintered carbide tool material.
For general purpose machining and roughing on steel and steel alloys.

UNIMET U-70.
M-1706; sintered carbide tool material.
For average finishing and light roughing cuts on steel and steel alloys.

UNIMET U-73.
M-1706; sintered carbide tool material.
For light roughing and general finishing cuts on steel and steel alloys; fine feeds.

UNIMET U-80.
M-1706; sintered carbide tool material.
For precision boring of steel and steel alloys. High wear resistance coupled with better than average shock resistance. For precision turning, facing and high speed finishing.

UNIMET U-110.
M-1706; sintered carbide tool material.
For heavy duty and interrupted cutting of cast iron and non-ferrous matrials. Also for wear parts.

UNIMET U-130.
M-1706; sintered carbide tool material.
For shock resistant tools; withstands medium impact.

UNIMET U-135.
M-1706; sintered carbide tool material.
For fabrication of wear parts; resistant to impact, wear and abrasion.

UNIMET U-140.
M-1706; sintered carbide tool material.
To resist heavy impact.

UNION.
M-586; Sb, Pb, bal Sn.
For bearings; Babbitt.

UNION 20-CB.
M-1562; 0.07 max C, 20.0 Cr, 29.0 Ni, 2.0 min Mo, 3.0 min Cu, Cb = 8 x C min, 1.0 max Cb + Ta, bal Fe.
Annealed: 90,000 TS; 50,000 YS; 45 El; 65 RA; B 90 Rock.
For chemical plant equipment to resist sulphuric acid.
Stainless, austenitic, heat resistant.

UNIONALOY.
M-112; 3.2 C, 2.5 Si, 0.6 Mn, bal Fe.
For mill guides, tube mill plugs, hopper liners; resistant to abrasion; cast iron.

UNION FREECUT.
M-97; 0.13 C, 0.6-0.9 Mn, 0.07-0.12 P, 0.10-0.23 S, bal Fe.
Cold rolled: 80,000-100,000 TS; 70,000-80,000 YS; 20-10 El; 50-40 RA; 170-202 Brin.
For automatic screw machine parts; free machining steel.

UNION SUPERCUT.
M-97; 0.13 max C, 0.7-1.0 Mn, 0.08-0.12 P, 0.2-0.3 S, bal Fe.
Cold rolled: 80,000-95,000 TS; 70,000-80,000 YS; 20-10 El; 50-40 RA; 170-202 Brin.
For automatic screw machine parts; free machining steel.

UNISEAL 42.
M-114; 42 Ni, bal Fe.
For glass-to-metal and ceramic-to-metal seals. Controlled thermal expansion matching glasses.

UNISEAL 52.
M-114; 52 Ni, bal Fe.
For glass to metal and ceramic to metal seals. Controlled thermal expansion matching 0120 and 9010 glass.

UNISEAL 281 ON.
M-114; 0.10 C, 0.6 Mn, 0.4 Si, 0.4 Ni, 28.0 Cr, bal Fe.
For glass to metal seal.
Thermal expansion 77-1472°F: 6.48 in/in/°F x 10^{-6}.

UNISON.
M-1425; C, alloy, bal Fe.
For well bits, arbors, punches, crimpers; non-tempering, shock resistant.

UNISPAN 36.
M-114; 0.04 C, 0.35 Mn, 0.12 Si, 36.0 Ni, bal Fe.
Ann: 71,400 TS; 40,000 YS; 41 El; 80 Rock B.
Co-eff. thermal expansion: -1.07 to + 1.8 in/in°F x 10^{-6} from - to +500°F, depending on previous treatment of metal and temperature range measured.
For precision instruments.

UNISPAN LR 35.
M-114; 0.10 max C, 0.05 max Mn, 0.05 max Si, 36.5 max Ni, bal Fe.
Ann: 64,000-68,000 TS; 37,000-39,000 YS; 33-34 El; 74 Rock B.
Co/eff. thermal expansion: -8 to 212°F: 0.312 in/in°F x 10^{-6}.
For precision instruments.

UNITEMP 14-14 W.
M-114; 0.45 C, 13-15 Cr, 13-15 Ni, 1.75-3.0 W, 0.2-0.5 Mo, bal Fe.
Annealed: 110,000 TS; 48,500 YS; 38 El; 51 RA.
Aged: 130,000 TS; 68,500 YS; 28 El, 46 RA.
For parts requiring corrosion resistance to combustion products at elevated temperatures. Austenitic, stainless.

UNITEMP 14CMV.
M-114; 0.20 C, 1.0 Cr, 1.0 Mo, 0.10 V, 0.75 Si, 0.50 Mn, bal Fe.
Heat treated: 138,000 TS; 117,000 YS; 7 El; 45 RA.
For high temperature welded structures; good strength and weldability.

UNITEMP 14 HV.
M-114; 0.45 C, 1 Cr, 0.5 Mo, 0.3 V, bal Fe.
Ht. Tr.: 110,000-140,000 TS.
High temperature, high strength alloy.
For truck bodies, transportation equipment.
AISI 601.

UNITEMP 14 HW.
M-114; 0.45 C, 14 Cr, 14 Ni, 2.4 W; 0.35 Mo, bal Fe.
For parts for high temperature operation, heat resistant.

UNITEMP 14 MHV.
M-114; 0.25-0.30 C, 0.6-0.9 Mn, 0.7 Si, 1.0-1.5 Cr, 0.4-0.6 Mo, 0.75-0.95 V, bal Fe.
At R.T.: 150,000 TS; 140,000 YS; 16 El; 302 Brin.
At 1000°F: 86,000 TS; 74,000 YS; 28 E;, 82 RA.
For special bolting in steam turbines, hot dies, brake disks. Heat resistant. Useful to 1000°F.

UNITEMP 14 MV.
M-114; 0.28-0.33 C, 0.5 Mn, 0.6 Si, 1.0-1.5 Cr, 0.4-0.6 Mo, 0.2-0.3 V, bal Fe.
Heat treated: 167,000-298,000 TS; 164,000-272,000 YS; 16.5-8 El; 341-534 Brin.
For steam and gas turbine bolting up to 1050°F, brake disks; resists scaling and oxidation to 1200°F.

UNITEMP 19-9 DL.
M-114; 0.3 C, 19.2 Cr, 9 Ni, 1.25 W, 1.2 Mo, 0.4 Cb, 0.3 Ti, bal Fe.
Hot rolled: 118,000 TS; 69,000 YS; 55 El.
For jet engine components, exhaust manifolds; corrosion and heat resistant, austenitic.

UNITEMP 19-9 W-MO.
M-114; 0.08-0.12 C, 0.4-1.0 Mn, 18-20 Cr, 1.0-1.75 W, 8-10 Ni, 0.2-0. Mo, 0.3-0.6 Cb, 0.2-0.5 Ti, bal Fe.
For supercharger wheels, blades, casings, turbine blades; heat resistant.

UNITEMP 19-9 WX.
M-114; 0.07-0.13 C, 1-2 Mn, 19-22 Cr, 8-9.5 Ni, 0.5 Mo, 1.5 W, 1.2 Cb, 0.2 Ti, bal Fe.
For jet engine components, welding rod; heat and corrosion resistant.

UNITEMP 41.
M-114; 0.09 C, 19.0 Cr, 11.0 Co, 10.0 Mo, 3.1 Ti, 1.5 Al, 1.8 Fe, 0.005 B, bal Ni.
For highly stressed parts at high temperature; afterburner parts, turbine casings, nozzle parts, gas turbine parts, high temperature fasteners.
AISI 683 (Ren ae 41).

UNITEMP 212.
M-114, M-1491; 0.05-0.15 C, 15-17 Cr, 23-27 Ni, 4 Ti, 0.6 Cb, 0.1 B, 0.07 Zr, bal Fe.
At room temperature: 187,000 TS; 135,000 YS; 20 El; 32 RA.
At 1200°F: 145,000 TS; 123,000 YS; 15 El; 23 RA.
At 1400°F: 108,000 TS; 103,000 YS; 16 El; 25 RA.
For aircraft and guilded missile components; age-hardenable, stainless, good to 1400°F service.

UNITEMP 350.
M-114; 0.10 max C, 16.5-17.5 Cr, 4.0-4.5 Ni, 2.5-3.0 Mo, 0.50-0.75 Mn, 0.2-0.5 Si, bal Fe.
Heat treated: 200,000 TS; 118,000 YS; 10 El; C 43 Rock.
Annealed: 164,000 TS; 45,000 YS; 22 El; B 43 Rock.
For valves, springs, structures. Heat treatable, stainless.

UNITEMP 355.
M-114; 0.13 C, 0.75 Mn, 0.3 Si, 15,5 Cr, 4.25 Ni, 2.75 Mo, 0.10 N, bal Fe.
Sol. Tr. & Age: 180,000-220,000 TS; 170,000-185,000 YS; 12-19 El; 40-48 Rock C.
Corrosion resistant, good strength to 1000°F.
For jet engine parts, parts for high speed airplanes.
AISI 634 (AM-355).

UNITEMP 500.
M-114; 0.12 C, 17.5 Cr, 18.5 Co, 4.2 Mo, 3.1 Ti, 3 Al, 0.06 Zr, 0.005 B, 1.75 Fe, 0.03 Cu, bal Ni.
Heat treated: 195,000 TS; 125,000 yS; 31 El; 54 RA.
For jet engine components, turbine buckets, discs; corrosion and heat resistant.

UNITEMP 750.
M-114; 15.0 Cr, 73 Ni, 0.85 Cb, 2.5 Ti, 0.8 Al, 6.75 Fe.
Austenitic, age hardenable, nickel base alloy.
For springs operating up to 1300°F, for aircraft gas turbine parts.
AISI 688 (Inconel X).

UNITEMP 901.
M-114; 0.05 C, 12.5 Cr, 42.5 Ni, 6.0 Mo, 2.5 Ti, 0.20 Al, 0.015 B, bal Fe.
High temperature alloy.
For parts requiring high strength and corrosion resistance at 1000-1400°F.
AISI 601 or 602; (Incoloy 901).

UNITEMP 1415 NW.
M-114; 0.17 C, 0.4 Mn, 13.0 Cr, 2.0 Ni, 0.2 Mo, 2.95 W, bal Fe.
Hardenable to 140,000-180,000 TS.
Martensitic, corrosion resistant, good high temperature strength to 1200°F.
For compressor blades, turbine discs, hardware for jet engines and gas turbine engines.
AISI 615 (Greek Ascoloy).

UNITEMP 1420 WM.
M-114; 0.23 C, 0.75 Mn, 12 Cr, 0.8 Ni, 1.0 Mo, 1.0 W, 0.25 V, bal Fe.
Hardenable to 190,000-240,000 TS; good strength to 1000°F.
For parts for missiles, jet engines, gas turbines, aircraft, steam turbines; corrosion resistant.
AISI 616; AISI Type 422.

UNITEMP 1430 MV.
M-114; 0.25-0.35 C, 11-12 Cr, 0.5 max Ni, 2.5-3.0 Mo, 0.2 V, bal Fe.
Ht. Tr.: 125,000-210,000 TS; 85,000-120,000 YS; 12-20 El; 32-40 RA; 260-470 Brin.
For steam and gas turbine buckets, disks, bolts.
Heat resistant to 1200°F.

UNITEMP 1753.
M-114; 0.24 C, 16.2 Cr, 7.2 Co, 1.6 Mo, 8.4 W, 3.2 Ti, 1.9 Al, 0.008 B, 0.06 Zr, 9.5 Fe, bal Ni.
Heat treated: 195,000 TS; 130,000 YS; 20 El; 23 RA.
For high temperature applications, fasteners, turbine wheels; oxidation resistant to 1675°F.

UNITEMP A286.
M-114; 0.08 max C, 13-16 Cr, 24-27 Ni, 1-2 Mo, 1.7-2.2 Ti, 0.4 V, 0.35 max Al, bal Fe.
Rolled: 135,000-160,000 TS; 85,000-115,000 YS; 30-20 El; 50-30 RA.
For jet engine gas turbine buckets, discs and bolts; austenitic, corrosion and heat resistant.

UNITEMP AF2-1DA.
M-114; 0.33-0.37 C, 0.05-0.15 Zr, 0.015 B, 4.6 Al, 2.8-3.2 Ti, 11.5- 12.5 Cr, 5.8-6.2 W, 2.8-3.2 Mo, 9.5-10.5 Co, 1.3-1.7 Ta, bal Ni.
For gas turbine wheels and buckets.
High heat and oxidation resistance.

UNITEMP BHT.
M-114; 0.80 C, 0.3 Mn, 4 Cr, 4.25 Mo, 1 V, bal Fe.
For good strength at elevated temperature, bearings, races, gears, cams, and constant speed drive components; dies. AISI 613.

UNITEMP CMV.
M-114; 0.2 C, 1.0 Cr, 0.12 V, bal Fe.
For machine tool parts, gears, bolts; case hardened, tough.

UNITEMP D6C.
M-114; 0.45 C, 0.7 Mn, 0.3 Si, 1.1 Cr, 0.6 Ni, 1.0 Mo, 0.1 V, bal Fe.
Heat treated: 280,000 TS; 235,000 YS; 10 El; 38 RA.
For rocket motor cases; high strength and stability.

UNITEMP EME.
M-114; 0.1-0.2 C, 18-20 Cr, 11-13 Ni, 3.0-3.5 W, 0.85-1.2 Cb, 0.1-0.2 N, bal Fe.
Rolled: 135,000 TS; 100,000 YS; 20 El; 45 RA; 280 Brin.
For gas turbine components, high tempeature bolts; good strength and ductility to 1200°F, austenitic.

UNITEMP HV.
M-114; 0.45 C, 1.0 Cr, 0.55 Mo, 0.30 V, bal Fe.
Ht. Tr.: 167,000-298,000 TS; 164,000-272,000 YS; 8-17 El; 346-534 Brin.
Oil hardened; for die casting dies, high temperature bolts, brake discs.

UNITEMP HX.
M-114; 0.10 C, 0.65 Mn, 21.5 Cr, 1.5 Co, 9.0 Mo, 0.6 W, 18.5 Fe, bal Ni.
High temperature alloy for furnace parts and fixtures, jet engine tail pipes, afterburner components, shroud rings, turbine blades; forgeable, weldable, machinable.
AISI 680 (Hastelloy X).

UNITEMP L-605.
M-114; 0.12 C, 1.65 Mn, 20 Cr, 15 Ni, 15 W, 1.6 Fe, bal Co.
Annealed: 160,000 TS; 86,000 YS; 47 El.
At 1500°F: 55,300 TS; 44,900 YS; 16 El.
For afterburners, jet engine components; corrosion and heat resistant.

UNITEMP M252.
M-114; 0.15 C, 18-20 Cr, 9-11 Co, 9-11 Mo, 2.5 Ti, 0.8 Al, bal Ni.
At 70°F: 173,000 TS.
At 1000°F: 155,000 TS; 93,000 YS.
For jet engine and gas turbine bickets; high temperature strength, heat resistant.

UNITEMP M-308.
M-114; 0.08 C, 32.5 Ni, 13.75 Cr, 4.1 Mo, 6.5 W, 2.15 Ti, 0.3 Al, 0.004 B, 0.25 Zr, bal Fe.
For high temperature applications; heat and corrosion resistant.

UNITEMP N-155.
M-114; 0.10 C, 1.5 Mn, 20.7 Cr, 19.8 Ni, 19.5 Co, 3 Mo, 2.3 W, 1.2 Cb, 0.13 N, 2.0 Cu, bal Fe.
At 70°F: 119,000 TS; 57,000 YS; 43 El; 50 RA.
At 1200°F: 80,000 TS; 43,000 YS; 31 El; 42 RA.
For afterburners, engine tail pipes, combustion chambers; oxidation resistant to 1800°F.

UNITEMP S590.
M-114; 0.38-0.48 C, 19-22 Cr, 18-21 Ni, 18-21 Co, 4 Mo, 4 W, 4 Cb, bal Fe.
Rolled: 152,00 TS; 80,000 YS; 19 El; 25 RA; 300 Brin.
Cast: 84,000 TS; 38,000 YS; 4 El; 10 RA.
For jet engine tail pipes, after burners; austenitic, corrosion and heat resistant.

UNITEMP S-816.
M-114; 0.38 C, 1.35 Mn, 20 Cr, 20 Ni, 3.75 Mo, 4.2 W, 4.1 Cb, 5 max Fe, bal Co.
Heat treated: 140,000 TS; 55,000 YS; 35 El; 29 RA.
For jet engine components, turbine blades; high heat resistance.

UNITEMP VIRGO.
M-114; 0.60 C, 1.2 Si, 0.3 Mn, 5 Cr, 0.6 V, 5.25 Mo, bal Fe.
For punches, hot heading dies; hot work steel.

UNITEMP WASPALOY.
M-114; 0.07 C, 19.75 Cr, 13.50 Co, 4.45 Mo, 3.0 Ti, 1.4 Al, 0.75 Fe, 0.005 B, 0.04 Zr, bal Ni.
High strength at high temperature.
For turbine buckets for jet engines, discs, engine components.
AISI 685 (Waspaloy).

UNITO DRILL ROD.
M-339; C, bal Fe.
For drills, tools, water hardened.

UNIVAN.
M-11; C, Mn, V, bal Fe.
For fixtures, housings; tough.

UNIVAN.
M-112, M-1222; 0.28-0.32 C, 1.5 Ni, 0.12 V, 0.90-1.10 Mn, bal Fe.
Cast: 90,000 TS; 60,000 YS; 25 El; 50 RA; 175 Brin.
For locomotive frames, wheel centers, crossheads, coupling boxes, spindles, gears, pinions; castings to resist severe shocks and stresses.

UNIVAN C.
M-1122; 0.30-0.35 C, 1.3-1.7 Mn, 0.4-0.6 Cr, 0.27-0.33 Mo, bal Fe.
Cast: 85,000 TS; 55,000 YP; 22 El; 45 RA.
For locomotive frames, wheel centers, gears, axles, coupling boxes.
Wear resistant.

UNIVERSAL.
M-72; 0.45 C, 0.7 Si, 1.5 Cr, 0.5 Mo, 0.8 V, 0.5 W, bal Fe.
For tools, dies, punches; tough, oil hardening.

UNIVERSAL.
M-1331; 1.25 C, 1.15 Si, 0.70 Mn, bal Fe.
For cold work tools, punches; oil hardened, tough.

UNIVERSAL ALLOY NO. 10.
M-932; Cu alloy.
Welded: 65,000 TS; 60,000 YS; 20 El; 125 Brin.
For electrode tips, seam welding rolls; 85% conductivity.

UNIVERSAL NO. 10 BRASS.
M-1341; 57-62 Cu, 1-2 Sn, 1.5 max Pb, 0.4 Al, bal Zn.
Cast: 50,000 TS; 25,000 YS; 15 El; 83 Brin.
For bearings, gears, retainers, hardware; high strength, free cutting.

UNIVERSAL NO. 20, ALUMINUM BRONZE.
M-1341; 82 min Cu, 5 max Fe, 2 max Mn, bal Al.
Cast: 83,000 TS; 30,000 YS; 9 El; 150 Brin.
For gears, sprockets, bearings, pumps, marine parts; heat treatable, corrosion and wear resistant.

UNIVERSAL NO. XX, MANGANESE BRONZE.
M-1341; 60-68 Cu, 3.0-7.5 Al, 2-4 Fe, 2.5-5.0 Mn, 0.2 max Pb, bal Zn.
Cast: 95,000 TS; 60,000 YS; 5 El; 200 Brin.
For valve bodies, gears, boat fittings; resists sea water corrosion.

UNIVERSUM HARTE 1.
M-1306; 0.35 C, 0.90 Si, 1.05 Cr, 0.18 V, 1.85 W, bal Fe.
For shear blades, pneumatic tools, punches; oil hardened, shock resistant.

UNIVIT.
M-10; 0.1 C, bal Fe.
For steel for porcelain enameling.

UNIX.
M-1294; 1.05 C, 1.15 W, 0.90 Mn, bal Fe.
For cutters, blanking and forming dies; water or oil hardened.

UNMAGNETIZABLE WATCH WHEEL ALLOY.
M-Eng.; 65.75 Pt, 18 Ni, 18 Cu, 1.25 Cd.
For watch wheels; corrosion resistant.

UNNAMESSING MWU.
M-1371; Al alloy.
For light alloy parts.

URANIUM (DEPLETED AND NATURAL).
M-1755; U.
Purties: Dendritic 99.96+%, 99.7/8%.
Forms: Ingot, rod, powder, turnings, plate, sheet, foil.

URANIUM STEELS.
3 U, 0.2-0.7 C, Mn, Si, V, bal Fe.
For machinery parts; water hardening.

URANUS 4.
M-1546; 0.25 C, 20.0 Ni, 12.0 Cr, bal Fe.
Annealed: 85,300 TS; 49,800 YS; 30 El.
Forged: 107,000 TS; 71,000 YS; 25 El.
For pumps, turbines, valves.
Heat and corrosion resistant to 1000°C.

URANUS 11W.
M-614; 1.3 C, 0.25 max Si, 0.25 max Mn, bal Fe.
For engravers tools, reamers, broaches, hobs;
Type W1; water hardened.

URANUS 30.
M-1302; 0.2 C, 7 Ni, 4 Cr, 20 Cr, 1.7 Si, bal Fe.
For gas turbine components; heat and creep resistant.

URANUS 45 M.
M-1488; 0.07 max C, 2.0 max Mn, 1.5 max Si, 7.0-9.0 Ni, 20.0-22.0 Cr, 2.2-2.8 Mo, 0.50 max Cu, bal Fe.
Cast, ann: 600 N/mm^2 min UTS.
Austenitic stainless for pumps and piping for nuclear energy equipment.
AFNOR Z5 CND 20.8 M.

URANUS 50.
M-1488; 0.06 max C, 21.0 Cr, 7.0 Ni, 2.5 Mo, 1.5 Cu, bal Fe.
Wrought, annealed: 635 N/mm^2 min UTS.
Stainless and heat resistant steel; good resistance to stress corrosion.
AFNOR Z5 CNDU 21.08.

URANUS 50M.
M-1488; 0.07 max C, 2.0 max Mn, 1.5 max Si, 7.0-9.0 Ni, 20.0-22.0 Cr, 2.2-2.8 Mo, 1.0-2.0 Cr, bal Fe.
Cast, annealed: 600 N/mm^2 min UTS.
Austenitic stainless for pump and valve parts in chemical, food, petroleum fertilizer plants.
AFNOR Z5 CNDU 20.8 M.

URANUS 50T.
M-1302, M-1546; 0.03 C, 17 Cr, 13 Ni, 1.5 Cu, 2.5 Mo, bal Fe.
Annealed: 78,000 RS; 32,000 YS; 50 El; 80 Rock B.
For chemical and oil refinery plant equipment.
Stainless, heat resistant.

URANUS 50M.
M-1488; 0.07 max C, 2.0 max Mn, 1.5 max Si, 7.0-9.0 Ni, 20.0-22.0 Cr, 2.2-2.8 Mo, 1.0-2.0 Cr, bal Fe.
Cast, annealed: 600 N/mm^2 min UTS.
Austenitic stainless for pumpand valve parts in chemical, food, petroleum, fertilizer plants.
AFNOR Z5 CNDU 20.8 M.

URANUS 55 M.
M-1488; 0.05 max C, 2.0 max Mn, 1.5 max Si, 4.5-6.0 Ni, 25.0-27.0 Cr, 1.5-2.5 Mo, 2.5-3.5 Cu, bal Fe.
Cast,ann: 250-290 Brin.
Austenitic stainless for pumps, valves that also require abrasion resistance.
AFNOR Z3 CNUD 26.5 M; ACI CD 4 M Cu.

URANUS 65.
M-1488; 0.02 max C, 25.0 Cr, 20.0 Ni, Nb, bal Fe.
Wrought,ann: 490 N/mm^2 min UTS.
Stainless and heat resisting steel; good oxidation resistance at elevated temperature; resists hot nitric acid.
AFNOR Z2 CNNb 25.20; similar to AISI 310S.

URANUS 65M.
M-1488; 0.03 max C, 1.0 max Mn, 0.4 max Si, 19.0-22.0 Ni, 23.0-25.0 Cr, bal Fe.
Cast,ann: 490 N/mm^2 min UTS.
For nitric acid industries and atomic energy.
AFNOR Z2 CN 25.20 M.

URANUS 65.
M-1546; 0.03 max C, 25.5 Cr, 20.0 Ni, 0.25 Cb, bal Fe.
Annealed: 78,200 TS; 31,300 YS; 40 El.
For chemical equipment for hot concentrated nitric acid, oil burners, furnace parts, baffle plates.
Resists nitric acid, resists intergranular corrosion, austenitic. Oxidation resist.

URANUS 65M.
M-1546; 0.08 C, 25 Cr, 20 Ni, 1 Cb, bal Fe.
Annealed: 78,000 TS; 31,300 YS; 40 El.
For chemical equipment for hot concentrated nitric acid, oil burner parts, baffle plates.
Austenitic, corrosion and heat resistant.

URANUS B6.
M-1302; 0.03 C, 20 Cr, 25 Ni, 1.5 Cu, 4.5 Mo, bal Fe.
For chemical plant equipment, salt pots; corrosion resistant.

URANUS B6.
M-1488; 0.02 max C, 20.5 Cr, 25.5 Ni, 4.5 Mo, 1.5 Cu, bal Fe.
Wrought, ann: 550 N/mm^2 min UTS.
Stainless, austenitic grade for equipment for phosphoric and sulphuric acid plants; cellulose, paper and explosives industries.
AFNOR Z1 NCDU 25.20.

URANUS B6M.
  M-1488; 0.04 max C, 2.0 max Mn, 1.0 max Si, 24.0-27.0 Ni, 19.0-22.0 Cr, 4.0-4.8 Mo, 2.0-3.0 Cu, bal Fe.
  Cast,ann: 450 N/mm² min UTS.
  For chemical industries; phosphoric and sulphuric acids, petro-chemicals.
  AFNOR Z3 NCDU 25.20 M; similar to ACI CN-7M.

URANUS B 6 PM.
  M-1488; 0.03 max C, 2.0 max Mn, 1.0 max Si, 24.0-27.0 Ni, 19.0-22.0 Cr, 4.0-4.8 Mo, 2.0-3.0 Cu, 1.0-5.0 W, bal Fe.
  Resistant to phosphoric acid.
  AFNOR Z2 NCDUW 25.20 M.

URANUS B6SI.
  M-1546; 0.02 max C, 0.8 Mn, 2.7 Si, 25.0 Ni, 18.0 Cr, 3.8 Mo, 1.5 Cu, bal Fe.
  Annealed: 136,000 TS; 61,700 YS; 40 El.
  For chemical plant equipment, furnace parts, conveyor belts, high temperature equipment.
  Corrosion and heat resistant. Austenitic.

URANUS CH.
  M-1546; 0.2 C, 1.0 Si, 1.0 Mn, 22.5 Cr, 3.5 Ni, 2.5 Mo, 1.5 Cu, bal Fe.
  Heat treated: 121,000-135,000 TS; 75,000-100,000 YS; 25-35 El.
  For chemical and petroleum industries, oil refineries.
  Austenitic-ferritic, corrosion and heat resistant.

URANUS EXTRA.
  M-344; 1.0 C, 0.2 Si, 0.2 Mn, 0.1 V, bal Fe.
  For cold dies, punches, drills, taps, reamers; water hardened; Type W2.

URANUS F 1.
  M-344; 0.1 C, 2.5 Ni, 20 Cr, 1.5 Cu, bal Fe.
  For furnace parts, chemical plant equipment; heat and corrosion resistant.

URANUS R 3 see **ZCR157.**

URANUS R7 see **ATVS MO.**

URANUS S.
  M-1546; 0.02 max C, 0.8 Mn, 3.7 Si, 14.0 Ni, 17.5 Cr, bal Fe.
  Annealed: 132,300 TS; 55,200 YS; 32 El; 60 RA.
  For heat exchangers, chemical and atomic energy plants, nitric acid equipment, cryogenic equipment.
  Corrosion and heat resistant. Austenitic.

URANUS S1.
  M-1488; 0.015 C, 17.5 Cr, 14.5 Ni, 3.8 Si, bal Fe.
  Wrought,ann: 540 N/mm² min UTS.
  Stainless and heat resisting alloy; resistant to oxidizing nitric and sulphuric acids; also used in nuclear operations.
  AFNOR Z1 CNS 18.15.

URANUS SD.
  M-1546; 0.02 max C, 0.8 Mn, 3.7 Si, 15.0 Ni, 17.0 Cr, 2.5 Mo, bal Fe.
  Annealed: 143,300 TS; 60,000 YS; 35 El.
  For chemical plant equipment, digesters, acid containers.
  Heat and corrosion resistant. Austenitic.

URANUS SDM.
  M-1488; 0.03 max C, 2.0 max Mn, 3.0-4.0 Si, 15.0-17.0 Ni, 16.0-18.0 Cr, 2.0-3.0 Mo, bal Fe.
  Cast,ann: 450 N/mm² min UTS.
  Austenitic stainless steel.
  AFNOR Z2 CNSD 17.16 M.

URANUS SM.
  M-1488; 0.03 max C, 2.0 max Si, 3.5-4.5 Si, 13.0-15.0 Ni, 17.0-19.0 Cr, bal Fe.
  Cast, ann: 450 N/mm² min UTS.
  Resistant to various mixed acids, or chlorides.
  AFNOR Z2 CNS 18.14. M.

URBACH SK5.
  M-1320; 0.79 C, 4.75 Co, 4.3 Cr, 0.75 Mo, 1.55 V, 18 W, bal Fe.
  For lathe and planer tools, drills, reamers, hobs; high speed steel.

URBACH SK10.
  M-1320; 0.76 C, 10 Co, 4.2 Cr, 0.8 Mo, 1.8 V, 18 W, bal Fe.
  For lathe and planer tools, drills, broaches; high speed steel.

URBACH SS17.
  M-1320; 0.82 C, 4.1 Cr, 0.85 Mo, 1.6 V, 8.7 W, bal Fe.
  For lathe and planer tools, form cutters, drills; high speed steel.

URBACH SS17MO.
  M-1320; 0.95 C, 4 Cr, V, W, Mo, bal Fe.
  For lathe and planer tools, form cutters, reamers, taps; high speed steel.

URBACH SS25.
  M-1320; 0.86 C, 4.1 Cr, 0.85 Mo, 2.5 V, 12 W, bal Fe.
  For lathe and planer tools, drills, hobs, reamers; high speed steel.

URBACH SS35.
  M-1320; 0.85 C, W, Mo, 4 Cr, V, bal Fe.
  For lathe and planer tools, drills, reamers; high speed steel.

URBACH SS45.
  M-1320; 1.3 C, 4.3 Cr, 0.85 Mo, 3.8 V, 12 W, bal Fe.
  For blanking and forming dies, cutters; high speed steel.

URBACH U19.
  M-1320; 0.74 C, 4.3 Cr, 1.1 V, 18.5 W, bal Fe.
  For lathe and planer tools, reamers, broaches, taps; high speed steel.

URBACH V120.
M-1320; 0.74 C, 4.3 Cr, 1.1 V, 18.5 W, bal Fe.
For lathe and planer tools, reamers, taps, drills; high speed steel.

USALLOY.
M-203; 0.20 max C, 1.25 max Mn, 0.05 max S, 0.25 min Ni, 0.20 min Cu, (Ni + Cu + Cr = 1.25 min), bal Fe.
65,000 psi min TS; 45,000 psi min YS; 20 min El.
For corrosion resistant steel. Tee Head Bolts and nuts.

USAMET.
M-1712; 17 Cr, 7.2 Ni, 2 Mn, 1.4 Si, 0.10 C, bal Fe.
Hardened: 280,000 min TS; 265,000 YS.
For strip springs.

USCO 5.
M-585; 0.5 Cu, 0.6 Fe, 0.5 Zn, 5 Si, 0.4 Mn, bal Al.
Sand cast: 22,000 TS; 10,000 YS; 5 El; 47 Brin.
Die cast: 32,000 TS; 15,000 YS; 5 El.
For general castings; good corrosion resistance.

USCO 5-A.
M-585; 0.3 max Cu, 0.5 Fe, 0.3 max Zn, 5 Si, 0.3 max Mn, bal Al.
Sand cast: 19,000 TS; 9000 YS; 6 El; 40 Brin.
Die cast: 30,000 TS; 14,000 YS 7 El.
For food handling and chemical equipment; sand, permanent mold or die castings.

USCO 5-B.
M-585; 0.2 max Cu, 0.4 max Fe, 0.2 Zn, 7 Si, 0.3 Mg, 0.1 Mn, bal Al.
Heat treated: 34,000 TS; 30,000 YS; 2 El; 75 Brin.
For highly stressed parts; excellent corrosion resistance.

USCO 5-W.
M-585; 1.3 Cu, 0.6 max Fe, 0.3 Zn, 5 Si, 0.5 Mg, 0.4 max Mn, bal Al.
Sand cast: 24,000 TS; 21,000 YS; 4.5 El 55 Brin.
Heat treated: 37,000 TS; 29,000 YS; 2.5 El; 85 Brin.
For highly stressed parts; good corrosion resistance.

USCO 13.
M-585; 0.6 max Cu, 0.7 Fe, 0.5 max Zn, 12 Si, 0.05 Mg, 0.3 Mn, bal Al.
Sand cast: 26,000 TS; 12,000 YS; 8 El; 60 Brin.
Die cast: 35,000 TS; 16,000 YS; 3.5 El.
For pressure tight castings; corrosion resistant.

USCO 13A.
M-585; 0.7 Cu, 0.8 Fe, 0.1 Zn, 12 Si, 1 Mg, 0.1 Mn, 2.5 Ni, bal Al.
Permanent mold cast and heat treated: 47,000 TS; 33,000 YS; 0.5 El; 125 Brin.
For automotive and diesel pistons; high strength.

USCO A.
M-585; 6.5 Cu, 1 Fe, 1.4 Zn, 3.25 Si, 0.4 Mn, 0.05 Mg, bal Al.
Sand cast: 26,000 TS; 20,000 YS; 2 El; 74 Brin.
For general purpose castings; for low stress applications.

USCO A-3.
M-585; 7 Cu, 1 Fe, 1.5 Zn, 2 Si, 0.05 Mg, 0.4 Mn, 0.25 Ni, bal Al.
Sand cast: 26,000 TS; 23,000 YS; 1.5 El; 75 Brin.
For general purpose castings; for low stress applications.

USCO B-1.
M-585; 7 Cu, 0.9 Fe, 0.5 Zn, 5.5 Si, 0.3 Mg, 0.4 Mn, 0.3 Ni, bal Al.
Die cast: 28,000 TS; 26,000 YS; 1 El; 95 Brin.
Heat treated: 35,000 TS; 26,000 YS; 0.5 El; 100 Brin.
For automotive pistons; die and permanent mold castings.

USCO C-1.
M-585; 4.5 Cu, 0.8 Fe, 0.3 Zn 1.2 max Si, 0.03 max Mg, 0.4 Mn, bal Al.
Sand cast: 25,000 TS; 11,000 YS; 4 El; 65 Brin.
Heat treated: 36,000 TS; 26,000 YS; 4.5 El; 74 Brin.
For bus wheels, crankcases, transmission housings; for highly stressed parts.

USCO C-2.
M-585; 4.5 Cu, 0.8 Fe, 0.3 max Zn, 2.5 Si, 0.3 max Mn, bal Al.
Permanent mold cast: 27,000 TS; 15,000 YS; 2.5 El; 7 Brin.
Heat treated: 45,000 TS; 28,000 YS; 5 El; 90 Brin.
For automotive and aircraft parts; for highly stressed applications.

USCO D.
M-585; 4 Cu, 0.7 Fe, 0.6 Zn, 5 Si, 0.5 Mg, 0.3 Ni, bal Al.
Sand cast: 25,000 TS; 13,000 YS; 2 El; 70 Brin.
Die cast: 39,000 TS; 22,000 YS; 3 El.
For general purpose castings; heavy secton die castings.

USCO D-3.
M-585; 4.0 Cu, 0.7 Fe, 1 max Zn, 3 Si, 0.4 Mn, 0.3 Ni, bal Al.
Sand cast: 27,000 TS; 18,000 YS; 2.5 El; 66 Brin.
Heat treated: 38,500 TS; 29,000 YS; 3 El; 90 Brin.
For cylinder heads, manifold; heat treatable.

USCO D-4.
M-585; 2.9 Cu, 1 max Fe, 0.7 Zn, 5.25 Si, 0.15 max Mg, 0.5 Mn, 0.3 Ni, bal Al.
Sand cast: 28,500 TS; 16,300 YS; 2.7 El; 60 Brin.
Heat treated: 41,000 TS; 32,000 YS; 1.5 El; 80 Brin.
For general castings; age-hardenable.

USCO D-5.
M-585; 2.9 Cu, 1 max Fe, 0.7 Zn, 5.25 Si, 0.15 max Mg, 0.5 Mn, 0.3 Ni, bal Al.
Sand cast: 25,000 TS; 13,000 YS; 2 El; 70 Brin.
Aged permanent mold: 33,000 TS: 21,000 YS; 2 El; 79 Brin.
For general purpose castings; sand or permanent mold castings.

USCO-JOBBINS ALMAG 35 (535.2).
M-585; 6.2-7.5 Mg, 0.10-0.20 Mn, 0.10-0.20 Ti, bal Al.
As cast: 35,000 psi min TS; 18,000 psi min YS; 9.0 min El, 60 min Brin.
Weldable casting alloy with good strength and corrosion resistance; dimensionally stable.
AA 535.2 sand casting.

USCO-JOBBINS PRECEDENT 71A (771.2).
M-585; 6.5-7.5 Zn, 0.80-1.0 Mg, 0.06-0.20 Cr, 0.10-0.20 Ti, bal Al.
As cast: 37,000 psi TS; 23,000 psi YS; 10 El; 69 Brin.
Aged 3 weeks: 43,000 psi TS; 33,000 psi YS; 4.5 El; 93 Brin.
Sand casting alloy with high strength and good stability.
AA 771.2.

USCO K.
M-585; 3.5 Cu, 0.7 Fe, 0.5 Zn, 8.5 Si, 0.1 max Mg, 0.5 Mn, 0.5 max Ni, bal Al.
Sand cast: 25,000 TS; 14,000 YS; 2 El; 70 Brin.
Die cast: 43,000 TS; 27,000 YS; 2.5 El.
For general purpose castings; low shrinkage, very fluid.

USOC K-3.
M-585; 3.5 Cu, 1 max Fe, 1 max Zn, 6.5 Si, 0.2 Mg, 0.6 Mn, 0.3 max Ni, bal Al.
Sand cast: 28,000 TS; 16,000 YS; 3.5 El; 75 Brin.
Heat treated: 36,000 TS; 24,000 YS; 2 El; 80 Brin.
For highly stressed castings; heat treatable.

USCO K-5.
M-585; 2.5 Cu, 0.7 Fe, 0.35 Zn, 6.1 Si, 0.13 max Mg, 0.4 Mn, bal Al.
Sand cast: 27,000 TS; 15,500 YS; 3 El; 60 Brin.
Permanent mold cast: 32,000 TS; 17,500 YS; 3.5 El; 65 Brin.
For pressure tight castings; sand plaster or permanent mold castings.

USCO K-Z.
M-585; 2 max Zn, bal Al.
For die castings.

USCO L.
M-585; 4 Cu, 0.6 Fe, 0.1 Zn, 0.7 Si, 1.5 Mg, 0.1 Mn, 2 Ni, bal Al.
Heat treated: 44,000 TS; 0.5 El; 101 Brin.
Permanent mold: 47,000 TS; 42,000 YS; 0.5 El; 110 Brin.
For high temperature applications; high properties at elevated temperatures.

USCO R.
M-585; 0.1 max Cu, 0.5 Fe, 0.3 Zn, 0.3 max Si, 3.8 Mg, 0.6 max Mn, bal Al.
Sand cast: 25,000 TS; 12,000 YS; 9 El; 50 Brin.
For food processing equipment, fittings, hardware; corrosion resistant.

USCO S-1.
M-585; 9.5 Cu, 0.8 Fe, 0.5 max Zn, 4 Si, 0.25 Mg, 0.3 Mn, 0.1 Ni, bal Al.
Sand cast: 25,000 TS; 20,000 YS; 0.5 El; 80 Brin.
Permanent mold cast: 30,000 TS; 27,000 YS; 1 El; 90 Brin.
For high temperature applications; hard wearing surface.

USCO-S-22.
M-585; 8 Cu, 0.8 Fe, 0.2 Zn, 1.5 Si, 4 Mn, 0.1 Ni, bal Al.
Sand cast: 23,000 TS; 14,000 YS; 2 El; 74 Brin.
For general castings; low stress applications.

USCO SPECIAL K-9.
M-585; 11.0-12.0 Si, 1.0 max Zn, 0.50 max Cu, 0.25-0.40 Mg, 0.40-1.0 Fe, 0.30-0.40 Cr, 0.35 max Mn, bal Al.
Die cast: 23,000-30,000 psi TS; 17,400-23,400 psi YS; 2-5 El.
For clutch housings, farm implement parts, electrical equipment.

USCO T.
M-585; 0.8 Cu, 0.25 Fe, 8 Zn, 0.1 Si, 0.3 Mg, 0.15 Mn, bal Al.
Sand cast: 29,000 TS; 16,000 YS; 6.5 El; 60 Brin.
Aged: 35,000 TS; 25,000 YS; 4.5 El; 74 Brin.
For high strength castings; no heat treatment.

USCO Z-3.
M-585; Al, Mg.
For hardener for Zamac 3 alloys; hardener.

USCO Z-5.
M-585; Al, Mg.
Hardener.

USMAC 16.
M-1534; 0.10 C, 0.75 Mn, 0.65 Si, 12.5 Cr, 0.50 Ni, 0.20 S, bal Fe.
Cast and heat treated: 103,000-203,000 TS; 75,000-173,000 YS; 6-28 El; 9-62 RA; 185-415 Brin.
For chemical and food prcessing equipment, valve bodies, pump castings.
Type 416 stainless steel.

USMAC 40C.
M-1534; 1.05 C, 0.50 Mn, 0.50 Si, 17.5 Cr, 0.25 Ni, 0.50 Mo, bal Fe.
Cast: 110,000 TS; 70,000 YS; 10 El; 210 Brin.
For cutlery, valve parts, nozzles, pump parts.
Type 440 C stainless steel, wear and corrosion resistant.

USMAC 43.
M-1534; 0.25 C, 0.70 Mn, 0.60 Si, 21.0 Cr, 0.40 Ni, 1.00 Cu, bal Fe.
Cast: 85,000 TS; 55,000 YS; 5.0 El; 6.0 RA; 190 Brin.
For chemical plant and oil refinery equipment.
Type 443 stainless steel.

USMAC 436.
M-1534; 0.14 C, 0.50 Mn, 0.80 Si, 13.25 Cr, 1.8 Ni, 2.25 W, bal Fe.
Cast: 200,000 TS; 140,000 YS; 12.5 El; 40.0 RA; Rock C 43-44.
For jet engine and super charger parts, turbine blades and wheels.
Spec. AMS 5616. Ultra high strength.

USMAC ML.
M-1534; 0.18 C, 1.25 Mn, 1.0 Si, 31.75 Cu, 2.5 Fe, bal Ni.
Cast: 63,000 TS; 31,000 YS; 20 El; 135 Brin.
For chemical plant and plastic industry equipment.
Similar to Monel Metal. Corrosion resistant.

USO.
M-344; C, alloy, bal Fe.
For striking dies; water hardened.

USO 5S.
M-344; 0.1 max C, 4-6 Cr, 0.5 Mo, bal Fe.
Corrosion resistant.

USO 30CF.
M-344; 0.30 C, 3 Cr, 0.25 V, 8.8 W, bal Fe.
For hot dies and tools; hot work steel.

USO-AD.
M-344; 1.55 C, 11 Cr, 0.2 V, 0.7 Mo, bal Fe.
For blanking and trimming dies; oil hardened.

USO-HE.
M-344; 1.5 C, 0.5 Cr, 0.25 V, 1.6 W, bal Fe.
For dies, gages; oil hardened, non-deforming.

USO-NS.
M-344; 0.8 C, 0.2 Cr, 1.5 Mn, bal Fe.
For broaches, gages, blanking and forming dies; oil hardened, non-deforming.

USO VANADIUM.
M-344; 0.9 C, 0.2 V, bal Fe.
For heading and gripping dies, shear blades; water hardened.

USO TUNGSTEN.
M-344; 4 Cr, 0.8 V, 18 W, C, bal Fe.
For lathe tools, drills, milling cutters; high speed steel.

USO VANADIUM.
M-344; 0.8 C, 0.2 V, bal Fe.
For striking dies; water hardened.

U.S.S. 5 MO.
M-604; 0.10 min C, 4-6 Cr, 0.40-0.65 Mo, bal Fe.
Annealed: 70,000 TS; 30,000 YS; 28 El; 65 RA; 160 Brin.
Heat treated: 115,000 TS; 90,000 YS; 20 El; 60 RA; 240 Brin.
For petroleum industry, high temperature service; Type 501, scaling temperature 1150°F.

U.S.S. 5S MO.
M-604; 0.10 max C, 4-6 Cr, 0.40-0.65 Mo, bal Fe.
Annealed: 65,000 TS; 25,000 YS; 150 Brin.
For petroleum industry, high temperature service; Type 502, scaling temperature 1150°F.

USS 9% NICKEL STEEL.
M-604; 0.13 max C, 0.80 max Mn, 0.15-0.30 Si, 0.035 max P, 0.040 max S, bal Fe.
Minimum: 90,000 TS; 60,000 YS; 20 El.
For pressure vessels for low temperature operation; cryogenic alloy, good ductility and ease of fabrication. ASTM A353, A553.

U.S.S. 12.
M-604; 0.15 max C, 1.0 max Mn, 1.0 max Si, 11.5-13.0 Cr, bal Fe.
Annealed: 75,000 TS; 40,000 YS; 35 El; 70 RA; 155 Brin.
For turbine blades, pump rods, mine pumps, valve stems, bolts, coal screens, golf clubs; moderate corrosion and heat resistance.

U.S.S. 12 AL.
M-604; 0.08 max C, 11.5-14.5 Cr, 0.10-0.30 Al, bal Fe.
Annealed: 60,000 TS; 30,000 YS; 25 El; 163 Brin.
Drawn: 85,000 TS; 70,000 YS; 20 El; 60 RA; 185 Brin.
For oil and chemical industries, petroleum equipment; corrosion and oxidation resisting.

U.S.S. 12 F.M.
M-604; 0.15 max C, 0.15 min S, 0.6 max Mo or Zr, 1.25 Mn, bal Fe.
Annealed: 75,000 TS; 40,000 YS; 30 El; 60 RA; 155 Brin.
Heat treated: 110,000 TS; 85,000 YS; 18 El; 55 RA; 212 Brin.
For golf clubs, bolts, shafts, nuts, screws; good machinability, moderate corrosion and heat resistance; Type 416.

USS 12 MO.
M-604; 0.15 max C, 11.5-13.5 Cr, 0.40-0.60 Mo, bal Fe.
Martensitic stainless steel.
Modified AISI 410.

U.S.S. 12 TURBINE.
M-604; 0.15 max C, 11.5-13 Cr, bal Fe.
Annealed: 75,000 TS; 40,000 YS; 35 El; 70 RA; 155 Brin.
For turbine blading; Type 403; corrosion and heat resistant.

USS 12-2.
M-604; 0.15 max C, 11.5-13.5 Cr, 1.25-2.50 Ni, bal Fe.
Martensitic stainless steel.
For valve seats, gage parts, scissors.
AISI 414.

USS 12-12.
M-604; 0.12 max C, 11.5-13.5 Cr, 11.5-13.5 Ni, bal Fe.
Austenitic stainless steel.

USS 12-15.
M-604; 0.08 max C, 11.0-13.0 Cr, 14.0-16.0 Ni, bal Fe.
Austenitic stainless steel.
For cold headed parts.

USS-16-2.
M-604; 0.2 max C, 15-17 Cr, 1.25-2.5 Ni, 1 max Si, 1 max Mn, bal Fe.
For springs; corrosion resistant.

U.S.S. 17.
M-604; 0.12 max C, 1.0 max Mn, 1.0 max Si, 16-18 Cr, bal Fe.
Annealed: 75,000 TS; 40,000 YS; 30 El; 50 RA; 212 Brin.
For tanks for manufacture of HNO_3, auto tire covers, interior decoration; heat and corrosion resistant; Type 430.

USS 17-4-6.
M-604; 0.15 max C, 5.5-7.5 Mn, 3.5-5.5 Ni, 16-18 Cr, bal Fe.
Annealed: 115,000 TS; 55,000 YS; 55 El.
For stainless parts; stainless, alternate for Type 201.

USS 17-7.
M-604; 0.15 max C, 16-18 Cr, 6-8 Ni, bal Fe.
Annealed: 110,000 TS; 40,000 YS; 60 El; B85 Brin.
For aircraft, trailers, car construction; Stainless Type 301.

U.S.S. 17-FM.
M-604; 0.12 max C, 1.25 max Mn, 1.0 max Si, 16 Cr, 0.15 min S, 0.6 max Mo or Zr, bal Fe.
Annealed: 80,000 TS; 55,000 YS; 25 El; 60 Ra; 170 Brin.
Cold drawn: 90,000 TS; 80,000 YS; 15 El; 55 RA; 190 Brin.
For stainless parts; free-cutting, stainless.

USS 18-5-8.
M-604; 0.15 max C, 7.5-10 Mn, 4-6 Ni, 17-19 Cr, bal Fe.
Annealed: 105,000 TS; 55,000 YS; 55 El.
For stainless parts; austenitic, alternate for Type 202.

USS 18-8 CB.
M-604; 0.08 C, 2 max Mn, 1 max Si, 17-19 Cr, 9-12 Ni, Cb = 10 x C, bal Fe.
88,000-90,000 TS; 30,000 YS; 45-55 El; 70 RA; 165 Brin.
For engine manifolds, exhaust stacks, pressure vessels; stainless, austenitic.

U.S.S. 18-8 CB-TA.
M-604; 0.08 max C, 17-19. Cr, 9-12 Ni, Cb = 10 x C, 2 max Mn, 1 max Si, bal Fe.
Annealed: 95,000 TS; 40,000 YS; 45 El.
For exhaust stacks, airplane gasoline tanks, welded structures; resists intergranular corrosion; Type 347.

U.S.S. 18-8 F.M.
M-604; 0.15 max C, 2.0 max Mn, 1.0 max Si, 8-10 Ni, 17-19 Cr, 0.15 min S, 0.6 max Mo, bal Fe.
Annealed: 85,000 TS; 35,000 YS; 35 El; 50 RA; 217 Brin.
Drawn: 110,000 TS; 75,000 YS; 30 El; 50 RA; 240 Brin.
For bolts, nuts, shafts, couplings, piston rods, propellers; heat and corrosion resistant and good machinability, Type 303.

U.S.S. 18-8 FS.
M-604; 0.12 max C, 0.2 max Mn, 0.95 max Si, 17-19 Cr, 10-13 Ni, bal Fe.
Annealed: 85,000 TS; 38,000 YS; 50 El.
For food and dairy equipment, cold headed parts; corrosion resistant; Type 305.

USS 18-8-L.
M-604; 0.03 max C, 8-12 Ni, 18-20 Cr, bal Fe.
Annealed: 75,000 TS; 28,000 YS; 50 El; 140 Brin.
For welded stainless parts; resists intergranular corrosion after welding.

U.S.S. 18-8 MO.
M-604; 0.08 max C, 2.0 max Mn, 1.0 max Si, 10-14 Ni, 16-18 Cr, 2-3 Mo, bal Fe.
Annealed: 85,000 TS; 35,000 YS; 50 El; 55 RA; 150 Brin.
For textile industry, paper and pulp mill equipment, tubes; corrosion resistant; Type 316.

USS 18-8 MO CB TA.
M-604; 0.10 max C, 16.0-18.0 Cr, 10.0-14.0 Ni, 2.0-3.0 Mo, CbTa = 10 x C min, bal Fe.
Stainless steel.
Stabilized type AISI 316.

USS 18-8 MO-L.
M-604; 0.03 max C, 10-14 Ni, 16-18 Cr, 2-3 Mo, bal Fe.
Annealed: 75,000 TS; 32,000 YS; 50 El.
For welded stainless parts; resists integranular corrosion after welding.

U.S.S. 18-8 S.
M-604; 0.08 max C, 2.0 Si, 2.0 Mn, 8-12 Ni, 18-20 Cr, bal Fe.
Annealed: 85,000 TS; 35,000 YS; 50 El; 135 Brin.
For stainless parts, acid resisting apparatus, food-dairy, chemical and oil industries; corrosion resistant; minimized intergranular corrosion; Type 304.

U.S.S. 18-8 SI.
M-604; 0.15 max C, 2 max Mn, 2-3 Si, 17-19 Cr, 8-10 Ni, bal Fe.
Annealed: 90,000 TS; 40,000 YS; 50 El; 65 RA; Rockwell B85.
For furnace parts, air preheaters, still liners; oxidation and heat resistant.

U.S.S. 18-8 TI.
M-604; 0.08 max C, 2.0 max Mn, 1.0 max Si, 9-12 Ni, 17-19 Cr, 5 x C = Ti, bal Fe.
Annealed: 90,000 TS; 35,000 YS; 50 El; 60 RA; 202 Brin.
For tubes for stills, smoke stacks, gasoline tanks, oil tanks, aircraft exhaust; for heat and corrosion resistance (1000-1600°F) and when not practical to anneal after welding; Type 321.

U.S.S. 18-85.
M-604; 0.15 max C, 2.0 max Mn, 0.75 max Si, 8-10 Ni, 17-19 Cr, bal Fe.
Annealed: 90,000 TS; 40,00 YS; 50 El; 50 RA; 207 Brin.
For food containers, sinks, pipes, shafts bolts, railroad passenger cars, inside trim, architectural; heat and corrosion resistant; Type 302.

U.S.S. 19-9 MO.
M-604; 0.08 max C, 2 max Mn, 1 max Si, 18-20 Cr, 11-14 Ni, 3-4 Mo, bal Fe.
Annealed: 90,000 TS; 40,000 YS; 45 El; Rockwell B85.
For pulp and paper mill equipment, chemical industry; corrosion, heat and creep resistant.

U.S.S. 20-10 S.
M-604; 0.08 max C, 2.0 max Mn, 1.5 max Si, 10-12 Ni, 19-21 Cr, bal Fe.
Annealed: 80,000 TS; 35,000 YS; 35 El; 50 RA; 217 Brin.
For trim and structural puposes, furnace parts, weld rods; heat and corrosion resistance; Type 308.

U.S.S.21.
M-604; 0.20 max C, 18-23 Cr, 1.0 Si, 2.0 Mn, bal Fe.
Annealed: 75,000 TS; 40,000 YS; 22 El; 50 RA; 212 Brin.
For chemical apparatus, oil burner parts; corrosion and heat resistant.

U.S.S. 25-12.
M-604; 0.20 max C, 2 max Mn, 1.0 max Si, 12-15 Ni, 22-24 Cr, bal Fe.
Annealed: 90,000 TS; 45,000 YS; 45 El; 45 RA; 217 Brin.
For preheaters, furnace lining, furnace doors, annealing boxes; heat and corrosion resistant.

USS 25-12-S.
M-604; 0.08 max C, 12-15 Ni, 22-24 Cr, bal Fe.
Annealed: 90,000 TS; 45,000 YS; 45 El; 217 Brin.
For preheaters, furnace parts; corrosion and heat resistant.

USS 25-12SCB.
M-604; 0.08 max C, 22.0-24.0 Cr, 12.0-15.0 Ni, CbTa min 10 x C, (Ta 0.10 max), bal Fe.
Stainless and high temperature alloy; stabilized to improve welding and high temperature properties.
Modified AISI 309.

USS 25-12 SCBTA.
M-604; 0.08 max C, 22.0-24.0 Cr, 12.0-15.0 Ni, CbTa = 10 x C min, bal Fe.
Stainless and high temperature alloy; stabilized to improve welding and high temperature properties.
Modified AISI 309.

U.S.S. 25-20.
M-604; 0.25 max C, 2.0 max Mn, 1.5 Max Si, 24-26 Cr, 19-22 Ni, bal Fe.
Annealed: 90,000--110,000 TS; 40,000-60,000 YS; 35-50 El; 45-60 RA.
For corrosion and heat resisting parts; high heat resisting.

USS 25-20S.
M-604; 0.08 max C, 24.0-26.0 Cr, 19.0-22.0 Ni, 1.5 max Si, bal Fe.
Stainless and high temperature alloy.
For jet engineerings.
AISI 310S.

U.S.S. 25-20 SI.
M-604; 0.25 max C, 2 max Mn, 1.5-3.0 Si, 23-26 Cr, 19-22 Ni, bal Fe.
Annealed: 100,000 TS; 50,000 YS; 40 El; Rockwell B85.
For heat exchangers, furnace doors, carburizing boxes; oxidation and heat resistant.

U.S.S. 27.
M-604; 0.20 max C, 1.0 max Mn, 1.0 max Si, 23-27 Cr, bal Fe.
Annealed: 80,000 TS; 50,000 YS; 20 El; 50 RA; 217 Brin.
For furnace linings, furnace parts, gas preheaters, annealing boxes; heat resistant 1500-2000°F; Type 446.

USS 35 N-15 CR.
M-604; 0.08 max C, 17.0-20.0 Cr, 34.0-37.0 Ni, 0.75-1.50 Si, bal Fe.
For high temperature operation; as furnace parts.

USS 41.
M-604; 0.35 C, 1.50 Mn, 0.25 Si, bal Fe.
Steel wire and rod for cap screws and automatic wheel bolts. Can be cold-headed.

USS 1110.
M-604; 0.10 C, 0.45 Mn, 0.06 S, bal Fe.
Free-machining steel wire and rod.

USS 2090.
M-604; 0.22 C, 1.0 Mn, bal Fe.
Steel wire and rod for recessed-head screws.

USS 18-18-2.
M-640; 0.06 C, 18 Cr, 18 Ni, 2 Si, bal Fe.
Annealed: 80,000 TS; 36,000 YS; 54 El.
At 1000°F: 65,000 TS; 17,4000 YS; 47 El.
For the chemical, petroleum, electric power and food processing industries, tanks, kettles, pump parts.
Resists stress corrosion, austenitic, stainless.

USS AUSTENITIC MANGANESE STEEL see MANGANESE GR B.

USS B24.
M-604; 0.22 C, 1.0 Mn, 0.0005 min B, bal Fe.
Steel wire and rod for recessed head screws. Can be cold headed.

USS CHAR-PAC see CHAR-PAC.

U.S.S. COR-TEN.
M-604, M-377; 0.12 max C, 0.2-0.5 Mn, 0.07-0.15 P, 0.5 Ni, 0.25-0.75 Si, 0.2-0.5 Cu, bal Fe.
Rolled: 70,000-63,000 TS; 50,000-43,000 YS; 22 El.
For light weight construction, buses, trucks; corrosion resisting properties.

U.S.S. COR-TEN-B.
M-604; 0.10 max C, 0.10-0.30 Mn, 0.50-1.00 Si, 0.5-1.5 Cr, 0.3-0.5 Cu, bal Fe.
Normalized: 70,000 TS; 50,000 YS; 22 El.
For pipes where strength superior to low C steel as well as better resistance to corrosion is desired.

U.S.S. MAN-TEN.
M-377, M-604; 0.25 max C, 1.10-1.60 Mn, 0.045 max P, 0.05 max S, 0.15 min Si, 0.20 min Cu, bal Fe.
Rolled: 75,000-65,000 TS; 50,000-40,000 YS; 20 El.
For light weight construction, truck and trailer frames, crane masts; high strength, resists atmospheric corrosion.

U.S.S. MOTOR.
M-604; approximately 2.7 Si, bal Fe.
For silicon steel sheets for rotating electrical machinery; res. 40.

U.S.S. NO. 18-8 STABILIZED, see U.S.S. 18-8 TI.

USS PAR-TEN.
M-604; 0.12 max C, 0.75 max Mn, 0.07 max V, or 0.04 max Cb, 0.05 max S, 0.04 max P, 0.10 max Si, bal Fe.
Hot rolled sheet: 62,000 TS; 45,000 YP; 29 El.
For polished and plated components.
Good formability and weldability.

USS Q-CORE.
M-604.
Low core loss motor lamination sheet.

USS Q-TEMP 10B18Q, see Q-TEMP 10B18Q.

USS Q-TEMP 41BV20Q.
M-604; 0.18-0.23 C, 0.75-1.00 Mn, 0.25-0.40 Cr, 0.005 min B, 0.15-0.25 Mo, 0.03-0.08 V, bal Fe.
Heat treated: 189,000 TS; 156,000 YS; 15 El; 61 RA.
For fasteners, hardware, bolts, extruded parts. Readily cold headed and extruded.

USS RADIO TRANSFORMER 58.
M-604; approximately 3.8 Si, bal Fe.
For audio transformers; high permeability.

USS RADIO TRANSFORMER 65.
M-604; approximately 3.8 Si, bal Fe.
For audio transformers; high permeability.

U.S.S. SIL-TEN.
M-377; 0.40 max C, 0.6 min Mn, 0.2 min Si, 0.0-0.2 min Cu, bal Fe.
Hot rolled: 80,000-95,000 TS; 45,000 YS; 30-18 El.
For structural parts, mine cars, bus bodies; 0.20 min Cu added as ordered to increase corrosion resistance.

U.S.S TRANSFORMER 52.
M-604; 4.5 Si, bal Fe.
For transformers; sheets, high permeability.

U.S.S. TRANSFORMER 58.
M-604; approximately 4.3 Si, bal Fe.
For silicon steel sheets for transformers in rotating electrical equipment radio parts; res 60.

U.S.S. TRANSFORMER 65.
M-604; approximately 3.8 Si, bal Fe.
For silicon steel sheets for transformers in rotating electrical equipment radio parts; res. 59.

USS TRANSFORMER 66.
M-604; approximately 3.25 Si, bal Fe.
For high efficiency power and distribution transformers; Res. 50, watt loss 0.66.

U.S.S. TRANSFORMER 72.
M-604; approx 3.5 Si, bal Fe.
For silicon steel sheets for transformers, rotating machinery, electrical purposes; Res. 57.

USS TRI-TEN.
M-604; 0.22 max C, 1.25 max Mn, 0.2-0.6 Cu, 0.02 min V, bal Fe.
Rolled: 63,000-70,000 TS; 43,000-50,000 YS; 25 El.
For cranes, shovels, mine cars, truck bodies, derricks.
Good resistance to atmospheric corrosion, shock resistant.

USS TYPE 1 MOTOR LAMINATION SHEET.
M-604.
Low core loss lamination sheet.
Similar to AISI Type 1 and ASTM Type 1.

USS TYPE 2-S MOTOR LAMINATION SHEET.
M-604.
Low core loss sheet.
Similar to AISI and ASTM Type 2S.

U.S.S. VITRENAMEL.
M-604; 0.2 C, bal Fe.
For enameling purposes; sheets.

US ULTRA.
M-1618, M-1182; 0.36 C, 0.4 Mn, 1.1 Si, 5.0 Cr, 0.35 V, 1.3 Mo, bal Fe.
Ht. Tr.: 300,000 TS; 227,000 YS; 9 El; C 55 Rock.
For hot dies, fasteners, extrusion and die casting dies, bulkheads.
Type H-11 hot-work steel, high toughnes and wear resistance.

US ULTRA 2, now VEW W302.

US ULTRA-4, now VEW W304.

US ULTRA-4.
M-1618; M-1182; 0.36 C, 0.4 Mn, 1.0 Si, 5.0 Cr, 0.35 V, 1.3 W, 1.4 Mo, bal Fe.
Heat treated: 205,000 TS; 185,000 YS; 12 El; 42 RA; Rock C 44.
For die blocks, extrusion rams, shell piercing tools, upsetters, punches.
Type H12 hot work tool steel, air hardening. Resists heat checking.

UT 35.
M-1120; 0.08 max C, 0.05 max N, 0.0125 max H, 0.20 max O, 0.20 max Fe, bal Ti.
Ann,aged: 290-410 N/mm^2 TS; 195 N/mm^2 min YS; 22 min El.
For corrosion resistance and elevated temperature.

UT 35-02.
M-1120; 0.20 Pd, 0.08 max C, 0.05 max N, 0.0150 max H, 0.20 max O, 0.020 max Fe, bal Ti.
Ann,aged: 290-410 N/mm^2 TS; 195 N/mm^2 min YS; 22 min YS.
Improved corrosion resistance over unalloyed titanium.

UT 40.
M-1120; 0.08 max C, 0.06 max N, 0.0125 max H, 0.25 max O, 0.25 max Fe, bal Ti.
Ann,aged: 390-540 N/mm^2 TS; 275 N/mm^2 min YS; 20 min El.
For corrosion resistance and elevated temperature.

UT40R.
M-1120.
Same as UT 40 but specially designed as wire for use as rivets for aeronautical industry.

UT 50.
M-1120; 0.08 max C, 0.07 max N, 0.0125 max H, 0.35 max O, 0.25 max Fe, bal Ti.
Ann,aged: 490-640 N/mm^2 TS; 340 N/mm^2 min YS; 18 min El.
For aircraft structural parts.

UT60.
  M-1120; 0.10 max C, 0.07 max N, 0.125 max H, 0.40 max O, 0.35 max Fe, bal Ti.
  Ann,aged: 540-730 N/mm² TS; 440 N/mm² min YS; 15 min El.
  For structural, corrosion resisting parts.

UT 651A.
  M-1120; 6 Al, 5 Zr, 2 Sn, 1 Mo, 0.25 Si, bal Ti.
  Ht: 990 N/mm² TS; 850 N/mm² YS; 6 El.
  Heat treatable, high strength alloy.

UT 662.
  M-1120; 5.0-6.0 Al, 5.0-6.0 V, 1.5-2.5 Sn, 0.35-1.0 Cu, 0.04 max N, 0.015 max H, 0.20 O, 0.35-1.0 Fe, bal Ti.
  Ann: 1000 N/mm² TS; 930 Nmm² YS; 8 El.
  High strength, heat-treatable alloy.

UT 685.
  M-1120; 5.7-6.3 Al, 4.0-6.0 Zr, 0.25-0.75 Mo, 0.10-0.40 Si, 0.8 max C, 0.008 max H, 0.20 max O, 0.20 max Fe, bal Ti.
  Sol Tr, Age: 990 N/mm² TS; 850 N/mm² YS; 6 El.
  Aircraft structural parts, blades or disc for jet engine compressors up to 550°C.

UT 6242.
  M-1120; 5.5-6.5 Al, 1.8-2.2 Sn, 3.6-4.4 Zr, 1.8-2.2 Mo, 0.05 max C, 0.0125 max H, 0.12 O, 0.25 max Fe, bal Ti.
  Ann,Aged: 890 N/mm² TS; 820 N/mm² YS; 8 El.
  Weldable; for blades and discs of jet engine components.

UTA3V.
  M-1120; 2.5-3.5 Al, 2.0-3.0 V, 0.05 max C, 0.02 max N, 0.0125 max H, 0.25 O, 0.25 max Fe, bal Ti.
  Ann: 640 N/mm² TS; 550 N/mm² YS; 18 El.
  Weldable, ductile; for tubes, sheets, plates and wire.

UTA5E (L GRADE).
  M-1120; 4.5-5.5 Al, 2.0-3.0 Sn, 0.05 max C, 0.035 max N, 0.0125 max H, 0.12 max O, 0.25 max Fe, (Fe + O = 0.32 max), bal Ti.
  Ann: 700 N/mm² TS; 630 N/mm² YS; 10 El.
  For cryogenic parts and equipment.

UTA5E (NORMAL GRADE).
  M-1120; 4.5-5.5 Al, 2.0-3.0 Sn, 0.15 max C, 0.07 max N, 0.020 max H, 0.20 max O, 0.50 max Fe, bal Ti.
  Ann: 790 N/mm² TS; 760/mm² YS; 10 El.
  Weldable, not hardenable.

UT A6 V.
  M-1120; 5.5-6.75 Al, 3.5-4.5 V, 0.08 max C, 0.015 max H, 0.07 max N, 0.20 O, 0.30 max Fe, bal Ti.
  Bars, forgings, Ann: 900 N/mm² TS; 830 N/mm² YS; 10 El.
  Most widely used titanium alloy.

UT A 7 D.
  M-1120; 6.5-7.3 Al, 3.5-4.5 Mo, 0.08 max C, 0.5 max N, 0.0125 max H, 0.20 O, 0.25 max Fe, bal Ti.
  Quenched and aged: 1150 N/mm² TS; 1040 N/mm² YS; 8 El.
  High strength alloy,- usually as forgings.

UT A8 DV.
  M-1120; 7.3-8.5 Al, 0.75-1.25 V, 0.75-1.25 Mo, 0.08 max C, 0.05 N, 0.0060 max H, 0.12 max O, 0.30 max Fe, bal Ti.
  Ann: 1000 N/mm² TS; 950 N/mm² YS; 12 El.
  Forgings for aircraft.

UTALOY 41.
  M-1003; 0.40 C, 0.95 Mn, 0.55 Si, 1.10 Cr, 0.28 Mo, 0.05 max S, 0.05 max P, bal Fe.
  Cast, general engineering material.

UTALOY 86.
  M-1003; 0.37 C, 1.05 Mn, 0.35 Si, 0.75 Ni, 0.65 Cr, 0.25 Mo, bal Fe.
  Cast; general engineering material.

UTALOY 700.
  M-1003; 0.66 C, 1.10 Mn, 1.0 Si, 1.50 Cr, 0.50 Mo, 0.05 max S, 0.05 max P, bal Fe.
  Cast, furnished 350-550 Brin.
  For abrasion resistant applications.

UTC.
  M-1120; 2.0-3.0 Cu, 0.10 max C, 0.05 max N, 0.010 max H, 0.20 max O, 0.20 max Fe, bal Ti.
  Sheet, Sol. Tr., Aged: 690 N/mm² TS; 550 N/mm² min YS; 10 El.
  For aircraft structural parts.

UTEX.
  M-822; 0.95 C, 2.0 W, 0.6 Cr, bal Fe.
  For tools; oil hardened.

UTICA.
  M-1779; 1.25 C, 1.5 W, 0.4 Cr, 0.20 V, bal Fe.
  Heat treated: 444-740 Brin.
  For taps, reamers, tools, drills, broaches, punches, dies; non-deforming, non-shrinking steel.

UTILITY.
  M-688; 0.6-0.9 C, 0.25-0.60 Mn, bal Fe.
  For drills, chisels, blacksmith tools; water hardened.

UTILOY 12.
  M-1055; 0.12 max C, 11-13 Cr, 0.4 Mn, 0.9 Si, bal Fe.
  For castings; corrosion resistant.

UTILOY 12N.
  M-1055; 0.12 max C, 11-13 Cr, 1.7-2.5 Ni, bal Fe.
  For castings; corrosion resistant.

UTILOY 20.
M-1055; 29 Ni, 20 Cr, 2-3 Mo, 4 Cu, 0.07 max C, 1 Si, bal Fe.
Cast: 65,000-75,000 TS; 28,000-38,000 YS; 50-35 El; 50-40 RA; 120-150 Brin.
For chemical plant equipment; resists mixed acids.

UTILOY 46.
M-1055; 4-7 Cr, 0.25 max C, 0.5 Mo, 0.7 Mn, 0.4 Si, bal Fe.
For castings; corrosion resistant.

UTILOY 3085.
M-1055; 0.28-0.35 C, 0.8-1.1 Ni, bal Fe.
Cast: 82,000 TS; 45,000 YS; 22 El; 35 RA.
For pump parts; water or oil hardened.

UTILOY H.
M-1055; 0.28-0.32 C, 23-26 Cr, 11-13 Ni, 0.9-1.1 Si, bal Fe.
For castings; corrosion and heat resistant.

UTILOY NH.
M-1055; 35 Ni, 15 Cr, C, bal Fe.
For salt bath pots; heat and corrosion resistant.

UTILOY X.
M-1055; 0.15 max C, 18 Cr, 8 Ni, bal Fe.
For stainless castings; stainless.

UTILOY X-7.
M-1055; 0.08 max C, 1.5 max Mn, 2 max Si, 18-21 Cr, 8-11 Ni, bal Fe.
Cast: 78,000 TS; 43,000 YS; 45 El; 150 Brin.
Annealed: 78,000 TS; 38,000 YS; 55 El; 140 Brin.
For pumps, valves, autoclaves, mixers, kettles; ACI-CF8; stainless, austenitic.

UTILOY XX.
M-1055; 0.07 max C, 18 Cr, 8 Ni, 3.5-4.5 Mo, bal Fe.
For stainless castings; heat and corrosion resistant.

UTR.
M-1120.
Similar to UTC but as wire or bars designed to make rivets for aircraft.

UTTER.
M-111; 0.5 C, 0.9 Si, 0.9 Cr, bal Fe.
For springs; oil hardening.

UVW.
M-1655; 1.05 C, 1.0 Mn, 0.20 Si, 1.0 Cr, 1.20 W, bal Fe.
Cold work tool steel for knives, shears, punching dies for thin sheets.
W.-Nr. 1.2419.

UVW 2.
M-1655; 0.90 C, 2.0 Mn, 0.20 Si, 0.10 V, bal Fe.
Oil hardening cold work tool steel.
For stamping and punching dies.
W.-Nr. 1.2842; Similar to AISI 02.

V

V1.
M-1653; 0.15 C, 0.30-0.70 Mn, bal Fe.
Low carbon steel. Ital.
UNI C18; AISI 1015.

V2.
M-1653; 0.20 C, 0.40-0.80 Mn, bal Fe.
Low carbon steel. Ital.
UNI C20; AISI 1020.

V2AED.
M-Ger.; 0.15 C, 18 Cr, 8 Ni, W, Ti, Ta, bal Fe.
For turbine blading and nozzles; heat resistant.

V2M.
M-1653; 0.20 C, 0.30 Si, 1.5-2.0 Mn, bal Fe.
Manganese alloy carburizing steel. 20 Mn 8.

V3.
M-1653; 0.30 C, 0.65 Mn, bal Fe.
Carbon steel.
Ital. UNI C30; AISI 1030.

V3/1.
M-1653; 0.35 C, 0.75 Mn, bal Fe.
Medium carbon steel.
Ital. UNI C35; AISI 1035.

V4.
M-1653; 0.40 C, 0.75 Mn, bal Fe.
Medium carbon steel.
Ital. UNI C40; AISI 1040.

V4/1.
M-1653; 0.43 C, 0.75 Mn, 0.25 max Cr, 0.25 max Ni, bal Fe.
Medium carbon steel.
Ital. UNI C43; AISI 1042.

V5.
M-1653; 0.50 C, 0.40 max Si, 0.75 Mn, bal Fe.
Carbon structural steel.
Ital. UNI C50; AISI 1050.

V5/1.
M-1653; 0.48 C, 0.75 Mn, 0.25 max Cr, 0.25 max Ni, 0.25 max Cu, bal Fe.
Carbon steel.
Ital. UNI C48; similar to AISI 1050.

V6.
M-1653; 0.60 C, 0.15-0.40 Si, 0.70 Mn, bal Fe.
Carbon structural steel.
Ital. UNI C60; AISI 1060.

V7.
M-1653; 0.65-0.72 C, 0.15-0.40 Si, 0.60-0.90 Mn, bal Fe.
Carbon steel.
Ital. UNI C70; AISI 1070.

V13.
M-1653; 0.15 C, 0.45 Mn, 2.75 Ni, 0.75 Cr, bal Fe.
Ni-Cr carburizing steel.
15 Ni Cr 11.

V14.
M-1653; 0.18 C, 0.70 Mn, 0.90 Cr, 4.5 Ni, bal Fe.
High alloy carburizing steel.
(18 Ni Cr 18).

V17.
M-1653; 0.15 C, 0.85 Mn, 1.0 Ni, 0.80 Cr, bal Fe.
Alloy carburizing steel.
15 Cr Ni 4; similar to former SAE 3115.

V18.
M-1653; 0.18 -0.23 C, 0.8-1.10 Mn, 0.90-1.20 Ni, 0.90-1.20 Cr, bal Fe.
Ni-Cr carburizing steel.
20 Cr Ni 4.

V20M7K25.
M-USSR; 0.8 C, 18.9 W, 24.9 Co, 6.6 Mo, bal Fe.
For wear resistant cutting tools, wear plates.
Wear resistant, high red-hardness.

V22.
M-Eng.; W alloy.
For gyro and counterweight components; high heat resistant.

V 25/4 H.
M-1342; 1.3 C, 25 Cr, 4 Ni, bal Fe.
As cast: 300 HV 10.
For wear resistance against abrasion at elevated temperatures.

V 28 H.
M-1342; 1.3 C, 28.0 Cr, bal Fe.
As cast: 300 HV 10.
For wear resistance against abrasion.

V35.
M-1635; 0.32-0.38 C, 0.50-0.80 Mn, 2.0-2.5 Ni, 0.60-0.90 Cr, bal Fe.
Ni-Cr structural steel, deep hardening.
35 Ni Cr 9.

V-36.
M-1491; 0.27 C, 1.0 Mn, 0.40 Si, 25.0 Cr, 20.0 Ni, 4.0 Mo, 2.0 W, 2.0 Cb, 3.0 Fe, bal Co.
For high temperature sheets.

V41.
M-1653; 0.27-0.35 C, 0.50-0.80 Mn, 2.60-3.20 Ni, 0.60-1.0 Cr, 0.30-0.60 Mo, bal Fe.
Ni-Cr-Mo structural steel; deep hardening.
30 NiCrMo 12.

V44.
M-1653; 0.30-0.38 C, 0.30-0.80 Mn, 3.5-4.0 Ni, 1.5-1.8 Cr, 0.20-0.40 Mo, bal Fe.
Ni-Cr-Mo structural steel, deep hardening.
35 NiCrMo 15.

V-57 see **CARPENTER, LESCALLOY** and **UDIMET V-57.**

V64.
M-1653; 0.40 C, 1.35 Mn, bal Fe.
Manganese structural steel.
40 Mn 5.

V68.
M-1653; 0.95-1.10 C, 0.30-0.50 Mn, 1.40-1.65 Cr, bal Fe.
Chromium drill rod.
100 Cr 6; AISI 52100.

V 76 (WAS KLOSTER V-76).
M-335; 0.6 C, 1.85 Si, 0.7 Mn, 0.45 Mo, 0.2 V, bal Fe.
Oil or water hardening.
For tools, dies.
AISI S5. Shock resisting tool steel.

V175.
M-1744; 0.8 C, 18.0 W, 4.25 Cr, 1.8 V, bal Fe.
2% Vanadium high speed steel.

V-302.
M-1733; 0.15 max C, 18.0 Cr, 9.0 Ni, bal Fe.
Austenitic stainless steel.
AISI 302; DIN 1.4300.

V-303.
M-1733; 0.15 max C, 18.0 Cr, 9.0 Ni, 0.15 S, bal Fe.
Free cutting austenitic stainless steel.
AISI 303; DIN 1.4305.

V-304.
M-1733; 0.08 max C, 19.0 Cr, 9.5 Ni, bal Fe.
Austenitic stainless steel.
AISI 304; DIN 1.4301.

V-310.
M-1733; 0.20 max C, 25.0 Cr, 20.0 Ni, bal Fe.
Austenitic stainless steel.
AISI 310; DIN 1.4841.

V-316.
M-1733; 0.08 max C, 17.0 Cr, 12.0 Ni, 2.5 Mo, bal Fe.
Austenitic stainless steel.
AISI 316; DIN 1.4436.

V-347.
M-1733; 0.08 max C, 18.0 Cr, 11.0 Ni, Nb, bal Fe.
Stabilized Austenitic stainless steel.
AISI 347; DIN 1.4551.

V-416.
M-1733; 0.15 max C, 13.0 Cr, 0.15 S, bal Fe.
Free machining. Martensitic stainless.
AISI 416; DIN 1.4005.

VA-15.
M-1733. 0.15 C, 0.27 Si, 0.55 Mn, 1.20 Cr, 3.25 Ni, 0.12 Mo, bal Fe.
Alloy carburizing steel.
AISI 9315; DIN 14 NiCr 14.

VA-17.
M-1733; 0.17 C, 0.27 Si, 0.55 Mn, 1.2 Cr, 3.25 Ni, 0.12 Mo, bal Fe.
Alloy carburizing steel.
AISI 9317.

VA-ALLOY.
M-Eng.; 80.2 Al, 5.01 Cu, 13.7 Zn, 0.72 Fe, 0.2 V.
For light alloy parts; non-hardenable.

VACCUTHERM 5-32.
M-1307; 0.15 C, 11.5 Cr, 1.0 Mo, bal Fe.
Annealed: 80,000 TS; 40,000 YS; 40 El; B 92 Rock.
Heat treated: 200,000 TS; 160,000 YS; 15 El; C 45 Rock.
For chemical plant and oil refinery equipment, springs, table flatware.
Corrosion resistant, hardenable.

VACCUTHERM 5-32 H.
M-1307; 0.20 C, 11.5 Cr, 1.0 Mo, bal Fe.
Annealed: 95,000 TS; 40,000 YS; 25 El; 55 RA; B 92 Rock.
Heat treated: 240,000 TS; 210,000 YS; 8 El; 25 RA; C 50 Rock.
For gears, shafts, cutlery, knives, surgical instruments, stainless hardware, oil refinery equipment.
Corrosion resistant, hardenable.

VACCUTHERM 5-34.
M-1307; 0.20 C, 11.5 Cr, 1 Mo, 0.3 V, bal Fe.
Annealed: 95,000 TS; 40,000 YS; 25 El; B 92 Rock.
Heat treated: 245,000 TS; 210,000 YS; 10 El; C 50 Rock.
For gears, shafts, hardware, surgical instruments, oil refinery and equipment.
Corrosion resistant, hardenable.

VACCUTHERM 5-36.
M-1307; 0.20 C, 12 Cr, 1 Mo, 0.3 V, 0.5 W, bal Fe.
Annealed: 80,000 TS; 40,000 YS; 24 El; B 94 Rock.
Hardened: 240,000 TS; 205,000 YS; 9 El; C 50 Rock.
For cutlery, surgical instruments, valves, bearings, chemical plant and oil refinery equipment.
Corrosion resistant, hardenable.

VACCUTHERM 5-40 H.
M-1307; 0.18 C, 0.2 Cb, 11.5 Cr, 0.6 Mo, 0.6 Ni, 0.25 V, bal Fe.
Annealed: 85,000 TS; 40,000 YS; 24 El; B 95 Rock.
Hardened: 240,000 TS; 205,000 YS; 10 El; C 50 Rock.
For cutlery, surgical instruments, valves, bearings, pivots.
Corrosion resistant, hardenable.

VACODIL.
M-1631.
36 or 42% Ni Fe alloy with defined thermal expansion for temperature measuring and control devices.

VACODUR 16.
M-1631.
Soft magnetic Al-Fe alloy with high mechanical hardness for wear-resistant magnetic heads.

VACOFER.
M-1631.
Pure magnetic iron produced by powder metallurgy, for relay components, armatures, pole pieces and sintered shapes.

VACO FLEX.
M-1631.
Bimetal strips, each comprising 2 alloys with different thermal expansion.

VACOFLUX.
M-1631.
Soft magnetic, approx. 50% Co-Fe alloy for telephone receiver diaphragms, pole pieces, relay components.

VACOMAX.
M-1631.
Permanent magnet material on Co Fe basis with high energy product.

VACON.
M-1631; 28 Ni, 18 Co, bal Fe.
Sealing alloy for glass-to-metal and glass-to-ceramic seals.

VACOPLUS.
M-1631; 10 Cr, bal Ni.
For temperature measurement with thermocouples.

VACOVIT.
M-1631.
Ni-Fe, or Ni-Fe-Cr alloys for glass-to-metal seals.

VACOZET.
M-1631.
Magnetic semi-hard Co-Fe-Ni alloy for latching reed contacts and remanent magnetism relays.

VACRO 5.
M-111; 0.27 C, 1.15 Mn, 0.75 Cr, 0.15 V, bal Fe.
For gears, shafts; shock resistant.

VACRO 62.
M-111; 0.30 C, 2.5 Cr, 0.2 Mo, 0.15 V, bal Fe.
For die casting and plastic mold dies; oil hardened, tough.

VACROMIUM.
M-1631.
Ni-Cr and Ni-Cr-Fe alloys for electric heating applications.

VACRYFLUX.
M-1631.
High-field super-conductor based on Nb-Ti or Nb_3 Sn alloy.

VACUMET.
M-1631.
Reference materials for x-ray analysis and spectro-chemical emission analysis.

VACUMET WASPALLOY see CARPENTER WASPALLOY.

VACUMINUS.
M-1631.
95% Ni with additions, for temperature measurement with thermocouples.

V.A.G. 160.
M-Eng.; 2-5 Si, 0.6 Mg, 0.7 Mn, bal Al.
For light alloy parts; corrosion resistant.

VAL 1.
M-1653; 0.08-0.12 C, 1.0 max Si, 1.0 max Mn, 12.0-14.0 Cr, bal Fe.
Chromium stainless steel.
W.-Nr. 1.4006; similar to AISI 410.

VAL1-MP.
M-1653; 0.08-0.13 C, 1.0 max Si, 1.0 max Mn, 11.5-13.5 Cr, 0.40-0.60 Mo, 0.50-1.0 Ni, bal Fe.
Martensitic stainless steel.
W.-Nr. 1.4106.

VAL1-P.
M-1653; 0.08 max C, 1.0 max Si, 1.0 max Mn, 11.5-14.0 Cr, bal Fe.
Ferritic stainless steel.
W.-Nr. 1.4000.

VAL1-S.
M-1653; 0.16-0.25 C, 1.0 max Si, 1.0 max Mn, 12.0-14.0 Cr, 1.0 max Ni, bal Fe.
Martensitic stainless steel.
W.-Nr. 1.4021; similar to AISI 420.

VAL1-Z.
M-1653; 0.08-0.15 C, 1.0 max Si, 1.5 max Mn, 12.0-14.0 Cr, 0.60 max Mo, 1.0 max Ni, 0.15-0.25 S, bal Fe.
Free machining martensitic stainless steel.
W.-Nr. 1.4005; AISI 416.

VAL2.
M-1653; 0.28-0.35 C, 1.0 max Si, 1.0 max Mn, 12.0-14.0 Cr, 1.0 max Ni, bal Fe.
Martensitic stainless steel.
W.-Nr. 1.4028; X30 Cr 13; similar to AISI 420.

VAL2-CS.
M-1653; 0.36-0.45 C, 1.0 max Si, 1.0 max Mn, 12.5-14.5 Cr, 1.0 max Ni, bal Fe.
Martensitic stainless steel.
W.-Nr. 1.4034; X40 Cr 13.

VAL2-MV.
M-1653; 0.35-0.45 C, 1.0 max Si, 1.0 max Mn, 14.0-15.5 Cr, 0.40-0.60 Mo, bal Fe.
Martensitic stainless steel.
W.-Nr. 1.4140.

VAL4.
M-1653; 0.10-0.20 C, 1.0 max Si, 1.0 max Mn, 15.0-17.0 Cr, 1.5-2.5 Ni, bal Fe.
Martensitic stainless steel.
W.-Nr. 1.4057; X 18 CrNi 16; AISI 431.

VAL4-S.
M-1653; 0.10 max C, 1.0 max Si, 1.0 max Mn, 18.0-20.0 Cr, 1.5-2.5 Ni, bal Fe.
Martensitic stainless steel.
X 14 CrNi 19.

VAL 6.
M-1653; 0.50-0.65 C, 0.50-1.20 Si, 0.50 max Mn, 0.80-1.20 Cr, 2.0-2.5 W, bal Fe.
Shock resisting type tool steel.
U 58 WCr 9 KU; similar to AISI S1.

VAL 7.
M-1653; 1.0-1.3 C, 0.35 max Mn, 0.90-1.2 W, bal Fe.
Special purpose tool steel.
U 115 W 4 KU; AISI F1.

VAL 8.
M-1653; 0.40 C, 0.75 Si, 0.50 Mn, 1.0-1.2 Cr, 0.15-0.20 Mo, 2.0-2.1 W, bal Fe.
Hot work tool steel.
U 40 W 20 KU.

VAL 9.
M-1653; 0.23-0.33 C, 2.2-3.0 Cr, 8.0-95 W, 0.20-0.40 V, bal Fe.
Hot work tool steel.
W.-Nr. 1.2581; similar to AISI H21.

VAL 41/S.
M-1653; 0.45-0.60 C, 0.40 max Si, 0.60-0.90 Mn, 0.70-1.20 Cr, 0.40-0.6 Mo, 1.40-1.80 Ni, bal Fe.
Tool steel.
W.-Nr. 1.2714; U 52 NiCrMo 6 KU. Special purpose type, similar to AISI L6.

VAL 41/SD.
M-1653; 0.35-0.50 C, 0.40 max Si, 0.50-0.80 Mn, 1.50-1.80 Cr, 0.40-0.60 Mo, 3.3-4.0 Ni, bal Fe.
Hot work tool steel.
W.-Nr. 1.2766; U 42 NiCrMo 157 KU.

VAL 102.
M-1653; 0.30-0.38 C, 0.70-1.20 Si, 0.60 max Mn, 4.5-5.5 Cr, 1.0-1.5 Mo, 0.80-1.20 V, bal Fe.
Hot work tool steel.
W.-Nr. 1.2344; AISI H 13.

VAL 103.
M-1653; 0.30-0.38 C, 1.20 max Si, 0.60 Mn, 4.5-5.5 Cr, 1.0-1.5 Mo, 0.30-0.50 V, 1.0-1.6 W, bal Fe.
Hot work tool steel.
W.-Nr. 1.2608; AISI H12.

VAL 104.
M-1653; 0.30-0.38 C, 0.70-1.20 Si, 0.60 max Mn, 4.5-5.5 Cr, 1.0-1.5 Mo, 0.3-0.5 V, (1.0-1.6 W), bal Fe.
Hot work tool steel.
W.-Nr. 1.2343; AISI H11.

VALAND-1.
M-111; 0.3 C, 1.7 Ni, 9 W, 0.3 V, 3 Cr, bal Fe.
For die casting dies, hot working tools; hot work steel, oil hardened.

VALAND 2.
M-111; 0.30 C, 2 Co, 2.4 Cr, 0.25 V, 8.5 W, bal Fe.
For extrusion dies, rams and liners, punches; hot work steel, oil hardened.

VALENITE VC-1.
M-1668; Sintered carbide tool material.
Primarily for roughing cuts on cast iron; tough and abrasion resistant; good at withstanding shock of interrupted cuts.

VALENITE VC-2.
M-1668; Sintered carbide tool material.
General purpose grade machining cast iron and non-ferrous metals; more wear resistant than VC-1.

VALENITE VC-3.
M-1668; Sintered carbide tool material.
For finish machining on cast iron and on non-ferrous metals and non-metallics.
Withstands abrasion but not shock.

VALENITE VC-4.
M-1668; Sintered carbide tool material.
Hardest grade, good wear resistance; for finish machining and precision boring of cast iron and non-ferrous metals; low shock resistance.

VALENITE VC-6.
M-1668; Sintered carbide tool material.
For general purpose and light cuts on steel, where cratering is a problem.

VALENITE VC-7.
M-1668; Sintered carbide tool material.
High grade for finish cutting steel.

VALENITE VC-8.
M-1668; Sintered carbide tool material.
Very hard grade for fine finishing and precision boring steel.

VALENITE VC-9.
M-1668; Sintered carbide tool material.
Very hard wear resistant grade,-for finishing and wear application where no shock exists.

VALENITE VC-10.
M-1668; Sintered carbide tool material.
Excellent general purpose, wear grade,-for gage blocks, plug gages.

VALENITE VC-11.
M-1668; Sintered carbide tool material.
For wear applications where some shock is encountered, as planer tools.

VALENITE VC-12.
M-1668; Sintered carbide tool material.
High strength and shock resistant, for punches, blanking dies, lamination dies.

VALENITE VC-13.
M-1668; Sintered carbide tool material.
Good shock resistance; for blanking and punching dies.

VALENITE VC-14.
M-1668; Sintered carbide tool material.
For heavy shock resistance as in blanking dies, swaging, coining dies.

VALENITE VC-28.
M-1668; Sintered carbide tool material.
Premium grade for heavy duty and general purpose machining of cast iron and non-ferrous and non-metallic materials.
Good resistance to edge wear and chipping.

VALENITE VC-55.
M-1668; Sintered carbide tool material.
Premium grade for rough and general purpose machining of steel; heavy and interrupted cuts.

VALENITE VC-83.
M-1668; Sintered carbide tool material.
High titanium carbide content; finishing grade for light precision cuts.

VALENITE VC-125.
M-1668; Sintered carbide tool material.
For general purpose and roughing cuts on steel, including interrupted cuts.

VALENTINE STEEL.
0.75 C, 5.25 Cu, 1.2 Mn, 1.2 V, 12.5 W, bal Fe.

VALIMP.
  M-1653; 0.11 C, 12.5 Cr, 0.5 Mo, bal Fe.
  Annealed: 80,000 TS; 40,000 YS; 25 El; B 92 Rock.
  Hardened: 200,000 TS; 160,000 YS; 12 El; C 45 Rock.
  For springs, table flatware, oil refinery and chemical plant equipment.
  Corrosion resistant, hardenable.

VALIW.
  M-1653; 0.22 C, 12.5 Cr, 1.5 Mo, 0.75 Ni, 0.25 V, 1.0 W, bal Fe.
  Annealed: 98,000 TS; 42,000 YS; 22 El; B 95 Rock.
  Hardened: 245,000 TS; 210,000 YS; 8 El; C 50 Rock.
  For surgical instruments, knives, gears, shafts, hardware, pivots.
  Corrosion resistance, hardenable.

VALLINOX 301.
  M-1495; 0.08-0.15 C, 1.0 max Si, 2.0 max Mn, 16.-18.0 Cr, 6.5-8.5 Ni, bal Fe.
  Austenitic stainless steel; work-hardens rapidly.
  AFNOR Z 12 CN 17-08; AISI 301.

VALLINOX 304.
  M-1495; 0.07 max C, 1.0 max Si, 2.0 max Mn, 17.0-19.0 Cr, 8.0-10.0 Ni, bal Fe.
  Austenitic stainless steel.
  AFNOR Z 6 CN 18 09; AISI 304.

VALLINOX 316.
  M-1495; 0.07 max C, 1.0 max Si, 2.0 max Mn, 16.0-18.0 Cr, 10.0-12.0 Ni, 2.0-2.5 Mo, bal Fe.
  Austenitic stainless steel; for chemical equipment.
  AFNOR Z 6 CND 17-11; AISI 316.

VALLINOX 317 L.
  M-1495; 0.03 max C, 1.0 max Si, 2.0 max Mn, 18.0-20.0 Cr, 14.0-16.0 Ni, 3.0-4.0 Mo, bal Fe.
  Low carbon austenitic stainless steel.
  For welded stainless chemical equipment.
  AFNOR Z 2 CND 19-15; similar to AISI 317 L.

VALLINOX 410.
  M-1495; 0.08-0.15 C, 1.0 max Si, 1.0 max Mn, 11.5-13.5 Cr, 0.50 max Mo, bal Fe.
  Martensitic stainless steel.
  AFNOR Z 12 C 13; AISI 410.

VALLINOX 430.
  M-1495; 0.10 max C, 1.0 max Si, 1.0 max Mn, 16.0-18.0 Cr, 0.50 max Mo, bal Fe.
  Ferritic type stainless steel.
  AFNOR Z 8 C 17; AISI 430.

VALLINOX MO.
  M-1495; 0.10 max C, 1.0 max Si, 2.0 max Mn, 16.0-18. Cr, 11.0-13.0 Ni 2.0-2.5 Mo, Ti=5xC to 0.60, bal Fe.
  Stabilized austenitic stainless steel.
  AFNOR Z 8 CNDT 17-12. (Stabilized AISI 316).

VALLINOX MONB.
  M-1495; 0.08 max C, 1.0 max Si, 2.0 max Mn, 16.0-18.0 Cr, 11.0-13.0 N 2.0-2.5 Mo, Nb+Ta=10xC to 1.0, bal Fe.
  Stabilized austenitic stainless steel.
  AFNOR Z 8 CNDNb 17-12; (stabilized AISI 316).

VALLINOX MO T.B.C.
  M-1495; 0.03 max C, 1.0 max Si, 2.0 max Mn, 16.0-18.0 Cr, 11.0-13.0 Ni, 2.0-2.5 Mo, 0.10-0.20 N, bal Fe.
  Austenitic stainless steel with nitrogen.
  AFNOR Z 2 CND 17-12.

VALLINOX NB.
  M-1495; 0.08 max C, 1.0 max Si, 1.0 max Mn, 17.0-19.0 Cr, 10.0-12.0 Ni, Nb+Ta=10xC to 1.0, bal Fe.
  Stabilized austenitic stainless steel.
  AFNOR Z 8 CNNb 18-11; AISI 347.

VALLINOX T.B.C.
  M-1495; 0.03 max C, 1.0 max Si, 2.0 max Mn, 17.0-19.0 Cr, 9.0-11.0 Ni, 0.10-0.20 N, bal Fe.
  Austenitic stainless steel with nitrogen.
  AFNOR Z 2 CN 18-10.

VALLINOX TI.
  M-1495; 0.08 max C, 1.0 max Si, 2.0 max Mn, 17.0-19.0 Cr, 10.12 Ni, Ti=5xC to 0.60, bal Fe.
  Stabilized austenitic stainless steel.
  AFNOR Z 8 CNT 18-11; AISI 321.

VALVE BEARINGS.
  M-Eng.; 71 Sn, 24 Sb, 5 Cu.
  For bearings, valve packing; Babbitt.

VALVE BRONZE.
  M-129; 85-89 Cu, 2-10 Sn, 3-7 Zn, 3-6 Pb, Cast.
  For valves; corrosion resistant.

VALVE COPPER.
  M-234; 88 Cu, 4 Sn, 3 Zn, 3 Pb, 3 Ni.
  Cast: 30,000 TS; 15,000 YS; 15 El.
  For valves and pipe fittings.

VALVE STEEL.
  M-U.S.; 5.8 Si, 1.5 Ti, 1.5 V, bal Fe.
  For valves; heat resistant.

VALVE STEEL CHROME I.
  M-Eng.; 11-14 Cr, 0.4-1.2 C, 0.1-0.2 Si, bal Fe.
  For valves; heat resistant.

VALVE STEEL CHROME II.
  6.3 Cr, 0.5-1.0 C, 0.1-0.3 Si, bal Fe.
  For valves; heat resistant.

VALVE STEEL, TUNGSTEN.
  14 W, 3 Cr, 0.6 C, bal Fe.
  For valves; heat resistant.

VALVE STEEL VERY HARD.
  60 W, 26 Fe, 5 Ti, 4 Cr, 3 C, 2 Ce.
  For valves; heat resistant.

VAN-50.
M-173; 0.18 max C, 1.25 max Mn, 0.30 max Si, 0.02 min V or 0.01 Cb, 0.01 min Al, Ce, or Zr, bal Fe.
50,000 psi min YS.
Good cold-forming, weldability, impact toughness.
For bumpers, structural tubing, wheel spiders.
ASTM A-441 A-572.

VAN 60.
M-173; 0.18 max C, 1.40 max Mn, 0.40 max Si, 0.02 min V or 0.01 min Cb, 0.01 min Al, Ce or Zr, bal Fe.
Good cold-forming, weldability, impact toughness.
For automobile bumpers, agricultural equipment.

VAN 70.
M-173; 0.18 max C, 1.50 max Mn, 0.50 max Si, 0.02 min V or 0.01 min Cb, 0.01 min Al, Ce or Zr, bal Fe.
70,000 psi min YS.
Good cold-forming, weldability, impact toughness.
For automobile bumpers, wheel spiders.

VAN 80.
M-173; 0.18 max C, 1.60 max Mn, 0.60 max Si, 0.02 min Al, 0.05 min V, 0.005 min N, bal Fe.
80,000 psi min YS.
Good cold-forming, weldability, impact toughness.
For mobile cranes, transmission towers, truck frames.
ASTM A-656.

VANADIN 40.
M-1331; 1.3 C, 4.3 Cr, 0.85 Mo, 3.8 V, 12 W, bal Fe.
For engravers' tools, heading and blanking dies; high speed steel.

VANADIN 40 CO.
M-1331; 1.35 C, Cr, Co, Mo, V, W, bal Fe.
For engravers' tools, heading and blanking dies; high speed steel.

VANADIUM.
M-1755; V.
Purities: zone refined 99.99+%, dendritic, 99.9%, 99.7-99.8%, 99.3-99.5%, alumothermic 98/99%, 90+%.
Forms: Granules, ingot, sheet, rod, foil, wire, powders, single crystals.

VANADIUM 6-6-2 see VASCO M2.

VANADIUM 40.
M-1331; 1.3 C, 4.3 Cr, 0.85 Mo, 3.8 V, 12 W, bal Fe.
For blanking and forming dies, engravers' tools; high speed steel.

VANADIUM 40 CO.
M-1331; 1.35 C, Cr, Co, Mo, V, W, bal Fe.
For blanking and forming dies, engravers' tools; high speed steel.

VANADIUM BRASS.
M-U.S.; 70 Cu, 29.5 Zn, 0.5 V.
For condenser tubes, sheets, hardware.

VANADIUM BRONZE.
M-U.S.; 38.5 Zn, 0.5 V, bal Cu.
For pipes; high strength.

VANADIUM EXTRA.
M-35; 1.06 C, 0.30 Mn, 0.25 Si, 0.20 V, bal Fe.
Water hardening tool steel.
For trim dies, lathe centers, mandrels.
Type W2-2.

VANADIUM GRAINAL NO. 1.
M-1038; 25 V, 15 Ti, 10 Al, 0.2 B, bal Fe.
Used in steel manufacturing; increases hardenability.

VANADIUM HOT WORK.
M-364; 0.5 C, 0.5 Cr, 1 V, bal Fe.
For hot work tools; hot work steel.

VANADIUM PERMENDUR.
M-22, M-205; 49 Fe, 49 Co, 2 V.
For electrical apparatus working at high flux density; high permeability at high flux density.

VANADIUMSTAHL EXTRA ZH.
M-1310; 1.15 C, 0.65 Cr, 0.10 V, bal Fe.
For blanking and forming dies; oil or water hardened.

VANADIUM STANDARD.
M-35; 1.06 C, 0.30 Mn, 0.25 Si, 0.2 V, bal Fe.
For mandrels, cold chisels, blacksmith tools, jigs.
Type W2-3 water hardening tool steel.

VANADIUM STRIKING DIE.
M-275; 0.90 C, 0.25 V, bal Fe.
For drills, reamers, hobs, taps, striking dies; Type W2; water hardened.

VANADIUM TOOL.
M-Eng.; 0.9 C, 1 Cr, 0.2 V, bal Fe.
For tools; oil hardening.

VANADIUM TYPE "H".
M-115; 0.66-0.75 C, 0.15-0.25 V, 0.70-0.90 Cr, bal Fe.
Oil treated: 255,000 TS; 222,000 YS; 6 El; 8 RA; 460 Brin.
For punches, dies, tools, axes, hatchets, caulking tools, cold cutters, swedges; shock resistant.

VANALIUM.
80.2 Al, 5.1 Cu, 13.7 Zn, 0.72 Fe, 0.20 V.
Cast: 22,000 TS; 16,000 YS; 8 El.
For cast parts, for aircraft and automotive engines; resists corrosion and erosion.

VAN ALLEN.
M-Eng.; 64 Au, 18.75 Ag, 9 Cu, 8 Pd, 0.25 Al.
For dental alloy; corrosion resistant.

VANASIL.
M-1477; 21-23 Si, 1-1.5 Cu, 2-2.5 Ni, 0.75-1.25 Mg, 0.1 V, 0.15 Ti, bal Al.
Cast: 20,000-36,000 TS; 0.5 El; 90-150 Brin.
For pistons, cylinder sleeves and liners; low coefficient of friction.

VANASIL 77.
M-739; 21-23 Si, 2.0-2.5 Ni, 0.90-1.20 Cu, 0.75-1.25 Mg, bal Al.
T 6 at 72°F: 31,000 psi UTS; 300°F: 26,000 psi UTS; 600°F: 12,000 psi UTS.
Coefficient of thermal expansion, 72-300°F: 9.05×10^{-6}; 300-600°F: 10.45×10^{-6}.
For pistons, compressor blades; good dimensional stability and elevated temperature strength.

VAN CHIP.
M-57; 1.15 C, 6 W, 4.1 Cr, 3 V, 5.75 Mo, bal Fe.
For plane and lathe tools, reamers, broaches, hobs, taps, drills; high speed steel, Type M3.

VANCO 5.
M-487; 1.5 C, 12 W, 5 Cr, 5 V, 6 Co, bal Fe.
For taps, shear blades, boring and turning tools; high speed steel.

VANCRO.
M-908; 0.45 C, 0.7 Mn, 1.0 Cr, 0.15 V, bal Fe.
For dies, cold work tools; water hardening.

VANCRO.
M-1067; 0.45 C, 1 Cr, 0.15 V, bal Fe.
For dies, tools; tough, water hardening.

VANCRO.
M-1653; 0.95-1.05 C, 4.75-5.25 Cr, 1.1-1.3 Mo, 0.10-0.20 V, bal Fe.
Air hardening cold work tool steel.
W.-Nr. 1.2363; AISI A2.

VAN-CRO 12.
M-1684; 1.5 C, 12.0 Cr, 1.0 V, 1.0 Mo, bal Fe.
Air or oil hardening cold work tool steel, high carbon-high chromium type; for shear blades, blanking and trimming dies, gages. AISI D2.

VANCROM.
M-1425; C, alloy, bal Fe.
For gears, machine parts, pins, spindles; oil hardened, tough.

VAN CUT.
M-115; 1.0 C, 6 W, 4 Cr, 2.5 V, 5.5 Mo, bal Fe.
For broaches, form tools, chasers; high speed steel.

VANDERLOY.
M-1500; 99.9 Fe.
For electroformed molds and dies; electrolytic iron.

VANICK.
M-79; 2.5 C, 2.5 Si, 0.5 Ni, 0.50 Mn, trace V, bal Fe.
Cast: 50,000 TS.
For general castings; high test gray iron.

VANIDUR.
M-Ger.; 0.10 C, 18.0 Cr, 10.0 Ni, 1.0 V, 0.6 Ti, bal Fe.
Annealed: 85,000 TS; 30,000 YS; 55 El; B 80 Rock.
For welded structures, and tanks, chemical and dairy plant equipment.
Corrosion and heat resistant, austenitic, stabilized.

VANITE.
M-35; 0.82 C, 18 W, 4 Cr, 2 V, 0.6 Mo, bal Fe.
For tools, cutters, reamers, broaches, hobs, milling cutters.
Type T2 high speed steel.

VAN-LOM.
M-115; 0.85-0.89 C, 3.8-4.2 Cr, 8.0-8.5 Mo, 2.0 V, bal Fe.
For tools, cutters, chasers, reamers, drills, broaches, lathe and planer tools.
AISI M1.

VAN SNAP.
M-1409; low C, Cr, V, M, Mo, bal Fe.
For rivet snaps; oil hardened, non-tempering.

VAPOPERM 70.
M-1631.
Soft magnetic 75% Ni-Fe alloy for transformer cores, leakage current protective switches, and high-grade relays. (The same alloy is supplied under the Trade Mark MUMETALL to all countries except the Commonwealth countries, Eire, France and the USA).

VARIOPERM.
M-Eng.; 70 Fe, 30 Ni.
For compensating shunts for electrical equipment; temperature-sensitive, magnetic.

VAROSS.
M-1068; 35 Ni, bal Fe.
For springs; low expansion, corrosion resistant, non-magnetic.

VASCO DIE.
M-115; 0.82 C, 1.0 Si, 0.3 Mn, 7.75 Cr, 2.5 V, 1.55 Mo, bal Fe.
Ht. Tr.: 224,000-349,000 TS; 190,000-279,000 YS; 6.8-3.8 El; 22.7- 11.5 RA; 47-60 Rock C.
For trim dies, thread roll dies, coining dies, plastic molds, slitter knives, shear blades, punches, back up rolls, extrusion dies.

VASCOJET 90.
M-1723; 0.12-0.18 C, 0.20 max Si, 0.80-1.10 Mn, 1.25-1.50 Cr, 0.80-1.0 Mo, 0.20-0.30 V, bal Fe.
Air or oil hardenable.
Hardened: 140,000-184,000 TS; 112,000 min YS; 10 min El.
High strength, low carbon-low alloy steel.
Can be welded; can be case hardened for improved wear resistance. AFNOR 15 CDV 6.

VASCOJET 1000.
M-115; 0.40 C, 5 Cr, 1.3 Mo, 0.5 V, bal Fe.
Heat treated: 136,000-310,000 TS; 101,000-241,000 YS; 16-3 El; 42-7 RA; 290-570 Brin.
For die blocks, fasteners; air hardened, thermal stability.

VASCOJET 1000 VM.
M-115; 0.37-0.43 C, 4.75-5.25 Cr, 1.3 Mo, 0.5 V, 0.9 Si, 0.3 Mn, bal Fe.
Ht. Tr.: 136,000-311,000 TS; 101,000-240,000 YS; 16-9 El; 42-29 RA 290-570 Brin.
For aircraft landing gear components, engine mounts, fatigue resistant, high strength.

VASCO JET X5 now VASCOJET 1000 VM. 5 TEMPER.

VASCOLOY see VR/WESSON.

VASCOLOY RAMET see VR/WESSON.

VASCOLOY VR-75 see VR/WESSON VR-75.

VASCO M-2.
M-115; 0.80-0.85 C, 6.0-6.75 W, 4.75-5.25 Mo, 1.75-2.05 V, 3.9-4.4 Cr, bal Fe.
High speed tool steel, Mo-W type.
For cutting tools, taps, broaches, hobs, milling cutters; AISI M2.

VASCO M-2 DRILL ROD.
M-342; 0.82 C, 4.25 Cr, 1.9 V, 6.4 W, 5 Mo, bal Fe.
For tools, cutters; high speed steel.

VASCO M7.
M-115; 0.99 C, 1.75 W, 3.75 Cr, 2.05 V, 8.75 Mo, bal Fe.
Ht. Tr. Rock C 22-26.
For drills, end mills, reamers, hobs, lathe and planer tools, slitting saws, blanking and trimming dies, chasers.
Type M-7 high speed steel.

VASCO M10.
M-115; 0.85 C, 4 Cr, 2 V, 8 Mo, bal Fe.
For cutting tools, lathe and planer tools; high speed steel.

VASCO M50 VM.
M-115; 0.80 C, 0.30 Mn, 0.30 Si, 4.0 Cr, 1.0 V, 4.0 Mo, bal Fe.
Annealed: 202 Brin.
For gears, valve parts, elevated temperature bearings; cutting tools. AISI M50.

VASCOMAX 250, VM.
M-115; 0.03 max C, 17-19 Ni, 7.0-8.5 Co, 4.6-5.1 Mo, 0.3-0.5 Ti, 0.05-0.15 Al, bal Fe.
Annealed: 140,000 TS; 95,000 YS; 17% El; 75% RA; 290 Brinell.
Heat treated: 275,000 TS; 268,000 YS; 10% El; 48% RA; 520 Brinell.
For missile and aircraft components, rocket cases. Maraging steel, high strength, tough and ductile.

VASCOMAX 250W, VM.
M-115; 0.03 max C, 0.10 max Si, 0.10 max Mn, 17.5-18.5 Ni, 7.5-8.5 Co 4.0-5.0 Mo, 0.05-0.15 Al, 0.40-0.55 Ti, bal Fe.
For filler metal in MIG and TIG welding of 18% Nickel Maraging Steel.

VASCOMAX 300 VM.
M-115; 0.03 max C, 4.8 Mo, 18.0 Ni, 9.0 Co, 0.1 Al, 0.6 Ti, bal Fe.
Annealed: 150,000 TS; 110,000 YS; 18 El; 72 RA; Rock C 30-32.
Heat treated: 294,000-315,000 TS; 290,000-310,000 YS; 9-12 El; 35-56 RA; Rock C 54-55.
For missiles and aircraft components. High tensile strength and toughness good corrosion resistance, high notch to smooth tensile ratios.

VASCOMAX 300 W, VM.
M-115; 0.03 max C, 0.10 max Si, 0.10 max Mn, 17.5-18.5 Ni, 9.5-10.5 Co, 4.0-5.0 Mo, 0.05-0.15 Al, 0.55-0.75 Ti, bal Fe.
For filler metal in MIG and TIG welding of 18% nickel maraging steel.

VASCOMAX 350, VM.
M-115; 0.03 max C, 18 Ni, 8.5 Co, 5 Mo, 0.7 Ti, 0.10 Al, bal Fe.
Annealed: 165,000 TS; 120,000 YS; 18 El; C 35 Rock.
Heat treated: 340,000-360,000 YS; 350,000-370,000 TS; 10 El; 50 RA; C 58 Rock.
For shear blades, gripper dies, punches, extrusion tools, torsion bars, collets, missile and aircraft components.
Maraging steel, durable and tough. Consumable vacuum melted.

VASCO NON-SHRINKABLE see COLONIAL NO. 6.

VASCO SUPREME.
M-115; 1.5 C, 12.5 W, 5 V, 5 Co, bal Fe.
For circular form tools, broaches, cutting tools; high speed steel, free-cutting.

VASCO TUF.
M-115; 0.50 C, 0.90 Si, 0.30 Mn, 7.75 Cr, 1.4 V, 1.35 Mo, bal Fe.
Ht. Tr.: 234,000-314,000 TS; 198,000-236,000 YS; 9.4-8.0 El; 22-34 RA; C 48-56 Rock.
For chipper knives, shear blades, heavy duty slitters, punches, gripper dies, plastic molds.

VASCO-VJ AND VASCO VJVM now **VASCOJET 1000 VM.**

VASCO-WEAR.
M-115; 1.12 C, 1.2 Si, 0.3 Mn, 1.1 W, 7.75 Cr, 1.6 Mo, 2.4 V, bal Fe.
Air hardened to Rockwell C 62-64.
Has high compressive strength and excellent toughness.
For cold finishing rolls, cold forming rolls, cold extrusion dies, trim dies, blanking dies, shear blades.

VAUCHERS ALLOY.
75 Zn, 18 Sn, 4.5 Pb, 2.5 Sb.
For bearings.

VB-15.
M-1733; 0.15 C, 0.27 Si, 0.80 Mn, 0.50 Cr, 0.55 Ni, 0.20 Mo, bal Fe.
Alloy carburizing steel; AISI 8615.

VB-17.
M-1733; 0.17 C, 0.27 Si, 0.80 Mn, 0.50 Cr, 0.55 Ni, 0.20 Mo, bal Fe.
Alloy carburizing steel; AISI 8617.

VB-20.
M-1733; 0.20 C, 0.27 Si, 0.80 Mn, 0.50 Cr, 0.55 Ni, 0.20 Mo, bal Fe.
Alloy carburizing steel; AISI 8620.

VB-30.
M-1733; 0.30 C, 0.27 Si, 0.80 Mn, 0.50 Cr, 0.55 Ni, 0.20 Mo, bal Fe.
Hardenable alloy steel; AISI 8630.

VB-40.
M-1733; 0.40 C, 0.27 Si, 0.87 Mn, 0.50 Cr, 0.55 Ni, 0.20 Mo, bal Fe.
Hardenable alloy steel; AISI 8640.

VB-50.
M-1733; 0.50 C, 0.27 Si, 0.87 Mn, 0.50 Cr, 0.55 Ni, 0.20 Mo, bal Fe.
Hardenable alloy steel; AISI 8650.

VB-60.
M-1733; 0.60 C, 0.27 Si, 0.87 Mn, 0.50 Cr, 0.55 Ni, 0.20 Mo, bal Fe.
Hardenable alloy steel; AISI 8660.

VC see **VALENITE VC.**

VC2.
M-1653; 0.15 C, 0.85 Mn, 0.85 Cr, bal Fe.
Chromium carburizing steel.
(15 Mn Cr 5); similar to former SAE 5115.

VC-12.
M-1744; 0.80 C, 20.5 W, 4.25 Cr, 1.6 V, 12.0 Co, bal Fe.
Cobalt super high speed steel for heavy duty.

VC-13.
M-1733; 0.85 C, 1.8 Cr, 0.2 Mo, 0.12 V, bal Fe.
Cold work tool steel; AISI L-3; DIN 1.2067.

VC-52.
M-1733; 1.05 C, 0.27 Si, 0.35 Mn, 1.45 Cr, bal Fe.
Steel for bearings.
AISI 5210D; W.-Nr. 1.3505.

VC-130.
M-1733; 1.8 C, 12.5 Cr, 0.2 V, bal Fe.
Cold work tool steel; similar to AISI D-3; DIN 1.2080.

VC-131.
M-1733; 2.0 C, 12.5 Cr, 1.0 W, 0.2 V, bal Fe.
Cold work tool steel; DIN 1.2436.

VC 135.
M-912; 0.34 C, 0.25 Si, 0.85 Mn, 1.0 Cr, bal Fe.
Heat treatable steel for axles, crankshafts.
W.-Nr. 7033; Similar to AISI 5135.

VC-140.
M-1733; 0.15 max C, 12.5 Cr, bal Fe.
Martensitic type stainless steel.
AISI 410; DIN 1.4024.

VC-150.
M-1733; 0.35 C, 13.0 Cr, bal Fe.
Martensitic type stainless steel.
AiSI 420; DIN 1.4034.

VCA ALLOY.
13 V, 11 Cr, 3 Al, bal Ti.
Annealed: 135,000 TS; 131,000 YS; 21 El.
Aged: 221,000 TS; 207,500 YS; 4 El.
For missile cases, fasteners, welded pressure vessels; recommended for service from -65°F to 600°F.

VCD2.
M-1653; 0.18 C, 0.60-0.90 Cr, 0.10-0.20 Mo, bal Fe.
Low carbon-low alloy steel for carburizing; also for elevated temperature service.
18 Cr Mo 3.

VCD2/S.
M-1653; 0.15 C, 0.5 Mn, 0.7 Si, 1.2 Cr, 0.5 Mo, bal Fe.
Low alloy steel for valves and parts for high temperature service.
Similar to ASTM A182-F11.

VCD3.
M-1653.
Low carbon, low alloy carburizing steel.
18 Cr Mo V.

VCD4.
M-1653; 0.15 max C, 2.2 Cr, 1 Mo, bal Fe.
Steel for elevated service.

VCD5.
M-1653; 0.15 C, 5.0 Cr, 0.5 Mo, bal Fe.
Steel for high temperature equipment.
12 Cr Mo 20.

VCD6.
M-1653; 0.18 C, 2.7-3.0 Cr, 0.25 Mo, 0.15 V, bal Fe.
Alloy steel, often used for high pressure hydrogenation vessels.
18 Cr Mo V 10.

VCD7.
M-1653; 0.15 C, 9.0 Cr, 1.0 Mo, bal Fe.
Steel for high temperature equipment.

VCM.
M-1733; 0.32 C, 2.85 Cr, 2.9 Mo, 0.5 V, bal Fe.
Hot work tool steel.
DIN 1.2365.

VCMV.
M-912; 0.30 C, 0.25 Si, 0.55 Mn, 2.40 Cr, 0.20 Mo, 0.15 V, bal Fe.
Deep hardening alloy steel for structural part on heavy equipment; crankshafts, bolts.
W.-Nr. 7707.

VCO.
M-1733; 0.55 C, 0.95 Cr, 3.25 Ni, 0.3 Mo, bal Fe.
Hot work tool steel.
DIN 1.2721.

VC06.
M-USSR; Co, VC, bal WC.
For hard cutting tools and wear parts.
Sintered carbides, abrasion resistant.

VC08.
M-USSR; Co, VC, bal WC.
For hard cutting tools and wear parts.
Sintered carbides, abrasion resistant.

VC015.
M-USSR; Co, VC, bal WC.
For hard cutting tools and wear parts.
Sintered carbides, abrasion resistant.

VD-2.
M-1733; 1.5 C, 12.0 Cr, 0.9 Mo, 0.9 V, bal Fe.
Cold work tool steel.
AISI D2; DIN 1.2379.

VD3.
M-1653; 0.18-0.23 C, 0.70-0.90 Mn, 0.40-0.70 Ni, 0.40-0.60 Cr, 0.15-0.25 Mo, bal Fe.
Ni-Cr-Mo carburizing steel.
20 NiCrMo 2; AISI 8620.

VD4.
M-1653; 0.18 C, 0.75 Mn, 1.30 Ni, 0.80 Cr, 0.20 Mo, bal Fe.
Alloy carburizing steel.
18 NiCrMo 5.

VD 4 A.
M-1488; 1.30 C, 4.5 Cr, 4.5 Mo, 4.0 V, 5.5 W, bal Fe.
High carbon, high speed steel for reamers, broaches, lathe finishing tools, extrusion tooling.
AFNOR Type 6-5-4; AISI M4.

VD5.
M-1653; 0.15-0.21 C, 0.55 Mn, 1.50-1.80 Ni, 0.40-0.70 Cr, 0.20-0.30 Mo, bal Fe.
Ni-Cr-Mo carburizing steel.
18 NiCrMo 7; AISI 4320.

VD40.
M-1653; 0.37-0.43 C, 0.50-0.80 Mn, 1.60-1.90 Ni, 0.60-0.90 Cr, 0.20 -0.30 Mo, bal Fe.
Ni-Cr-Mo structural steel.
40 NiCrMo 7; AISI 4340.

VD45.
M-1653; 0.34-0.42 C, 0.50-0.80 Mn, 0.70-1.0 Ni, 0.70-1.0 Cr, 0.15-0.25 Mo, bal Fe.
Ni-Cr-Mo structural steel.
38 NiCrMo 4.

VDC see LATROBE VDC.

VDC DRILL ROD.
M-1702; 0.40 C, 5.0 Cr, 1.0 Mo, 1.0 Si, 1.0 V, bal Fe.
H13 type hot work tool steel in drill rod form; air hardenable.
(Latrobe VDC).

VD CHISEL.
M-240; 0.90 C, 0.4 Mn, 0.25 Si, bal Fe.
AISI Type W1 Water hardening tool steel.

V.D. TOOL.
M-240; 0.95-1.05 C, 0.40 Mn, 0.15-0.18 V, 0.25 Si, bal Fe.
For punches, dies, forming tools, shear blades, mandrels; water hardened; W2.

VECTOLITE.
M-38; 30 Fe_2O_3, 44 Fe_3O_4, 26 Co_2O_3.

VEDAL.
M-Eng.; Si, Mg, bal Al.
For light alloy parts; non-hardenable.

VEGA see also CARPENTER VEGA.

VEGA 12.
M-1546; 0.10 C, 12.5 Cr, bal Fe.
Heat treated: 87,000 TS; 64,000 YS; 18 El.
Annealed: 64,000 TS; 37,000 YS; 32 El.
For turbine parts, cutlery, oil refinery equipment, valves, pivots.
Martensitic stainless steel, Hardenable.

VEGA 12VS.
M-1546; 0.20 C, 12.5 Cr, 1.1 Mo, 1.1 W, 0.25 V, bal Fe.
Heat treated: 114,000-143,000 TS; 85,000-114,000 YS; 12-17 El.
For gas and steam turbine blades and rotors, surgical instruments.
Martensitic stainless steel. Useful to 650°C, hardenable.

VEGA 15.
M-1546; 0.05 C, 12.5 Cr, bal Fe.
Annealed: 78,000 TS; 57,000 YS; 30 El.
For oil refinery and food industry equipment, tanks, vessels, trim.
Ferritic stainless steel. Not hardenable.

VELINVAR.
M-Japan; 55-63 Co, 7-13 V, bal Fe.
For instruments, chronometers; low coefficient of expansion.

VELOCITAS SPEZIAL HI.
M-614; 0.48 C, Cr, Si, bal Fe.
For gears, pinions, bolts, shafts; oil hardened, tough.

VELOCITAS SPEZIAL HII.
M-614; 0.45 C, Cr, Si, V, bal Fe.
For gears, shafts, crankshafts, fasteners; oil hardened, tough.

VELODAL now **WIELAND A42**

VELOS.
M-1744; 0.85 C, 6.4 W, 5.2 Mo, 4.2 Cr, 1.9 V, bal Fe.
Tungsten-Molybdenum high speed steel.

VELOS 42.
M-1744; 1.05 C, 1.6 W, 9.5 Mo, 4.0 Cr, 1.2 V, 8.5 Co, bal Fe.
8% Cobalt-Moly high speed steel.

VELOS UR.
M-1744; 0.75 C, 18.0 W, 4.25 Cr, 1.1 V, bal Fe.
18% Tungsten high speed steel.

VELVET.
M-Eng.; C, W, bal Fe.
For tools, drills, taps; water hardened.

VELVET.
M-588; 10 Sn, 2 Pb, bal Cu.
For bushings, Babbitts, bars; hard.

VELVETOUCH.
M-933; Cu-Pb-Sn-graphite.
For clutch and brake discs, linings, bearings, facings; sintered.

VENANGO replaced by **CYCLOPS S2.**

VENANGO SPECIAL SHOCK RESISTING STEEL.
M-114; 0.65 C, 0.5 Mn, 1.1 Si, 0.2 V, 0.5 Mo, bal Fe.
Oil harden: 250,000-340,000 TS; 7-10 El; 10-14 Izod (notched); 50-59 Rc.
Good strength, hardness, shock resistance.
For power shock tools, driver bits, knock out pins.

VENIVICI.
M-1307; 0.82 C, 4 Cr, 0.85 Mo, 1.6 V, 8.7 W, bal Fe.
For lathe and planer tools, reamers, broaches, taps; high speed steel.

VENTOS 4718 ETC.
M-1759; see Werkstoff Nr. 1.4718.
Valve steels; 9 grades.

VENTURELOY II.
M-1430; Ag alloy.
For electrical contacts; 75% electrical conductivity.

VEP.
M-1733; 0.08 max C, 5.0 Cr, 0.55 Mo, bal Fe.
Hot work or mold steel.
AISI P-4; DIN 1.2341.

VERALOY GR. V.
M-1289; WC.
For cutting tools; sintered carbide.

VERIBEST DRILL ROD.
M-822; 0.9 C, 1.1 Mn, 0.6 Cr, 0.9 W, 0.24 V, bal Fe.
For dies, tools, punches; oil hardening drill rod.

VERILITE-1.
M-274; 0.3 Ni, 2.5 Cu, 0.7 Fe, 0.4 Si, bal Al.
Cast: 16,000 TS; 4 El; 4 RA.
For aircraft cylinder heads; age-hardenable.

VERILITE-2.
M-274; 1.5 Ni, 1.0 Cu, 1.5 Cr, 0.5 Mn, bal Al.
For aircraft cylinder heads; age-hardenable.

VERNALLOY.
M-934; 0.2 C, 20 Cr, 10 Ni, bal Fe.
For heat resistant parts; heat resistant.

VERNICON.
M-303.
Resistance alloy (Cuni 44).

VERNISIL 12.
M-303; 12-13 Ni, 62-64 Cu, bal Zn.
For hardware, cutlery; nickel silver, corrosion resistant.

VERNISIL 15.
M-303; 14-16 Ni, 62-64 Cu, bal Zn.
For hardware, cutlery; nickel silver, corrosion resistant.

VERNISIL 18.
M-303; 18 Ni, 65 Cu, bal Zn.
For deep drawn parts, wash basins, fittings; nickel silver, corrosion resistant.

VERNISIL 25.
M-303; 25 Ni, 15 Zn, bal Cu.
Annealed: 65,000 TS; 40,000 YS; 42 El; 130 Brin.
For bathroom fixtures, marine hardware; nickel silver, corrosion resistant.

VERNISIL 183F.
M-303; 17-19 Ni, 54-56 Cu, bal Zn.
For hardware, cutlery; nickel silver, corrosion resistant.

VERONICA C33.
M-1352; 0.95 C, W, Mo, Cr, V, bal Fe.
For lathe and planer tools, hobs, drills, taps; high speed steel.

VERONICA C65.
M-1352; 0.85 C, 4 Cr, W, Mo, V, bal Fe.
For lathe and planer tools, hobs, reamers; high speed steel.

VERONICA EXTRA.
M-1352; 0.74 C, 4 Cr, 1.1 V, 18.5 W, bal Fe.
For lathe and planer tools, reamers, broaches, taps; high speed steel.

VERONICA HOCHLEISTUNG 1000.
M-1352; 0.86 C, 4.1 Cr, 0.85 Mo, 2.5 V, 12 W, bal Fe.
For lathe and planer tools, reamers, hobs, drills; high speed steel.

VERONICA HOCHLEISTUNG GOLD.
M-1352; 0.76 C, 4.2 Cr, 0.8 Mo, 1.8 V, 18 W, bal Fe.
For lathe and planer tools, hobs, broaches, taps; high speed steel.

VERONICA HOCHLEISTUNG K3.
M-1352; 0.86 C, 2.8 Co, 4.3 Cr, 0.85 Mo, 2.1 V, 12 W, bal Fe.
For lathe and planer tools, reamers, taps, drills; high speed steel.

VERONICA HOCHLEISTUNG SILBER.
M-1352; 0.79 C, 4.7 Co, 4.3 Cr, 1.5 V, 18 W, bal Fe.
For lathe and planer tools, reamers, broaches, hobs; high speed steel.

VERONICAL HOCHLEISTUNG SILBER 4V.
M-1352; 1.35 C, Co, Cr, V, W, bal Fe.
For engravers' tools, blanking dies; high speed steel.

VERONICA HOCHLEISTUNG VS10.
M-1352; 1.3 C, 4.3 Cr, 0.85 Mo, 3.8 V, 12 W, bal Fe.
For engravers' tools, blanking and forming dies; high speed steel.

VERONICA SPEZIAL.
M-1352; 0.82 C, 4.1 Cr, 0.85 Mo, 1.6 V, 8.7 W, bal Fe.
For lathe and planer tools, reamers, hobs, drills; high speed steel.

VEROTEC 19666.
M-717.
Alloy powder for metal spraying thick build-ups; will accept final coat.

VERSALLOY.
M-435; 0.11 C, 17.0 Cr, 8.0 Ni, bal Fe.
Austenitic stainless steel.

VERSASTEEL.
M-38; 1.0 C, 0.3 Mn, 2.0 Si, 4.25 Cr, 1.15 V, 0.3 W, 2.5 Mo, bal Fe.
Air hardening tool steel.

VERTOMAR 1.
M-677; 0.015 max C, bal Fe. (ingot iron).
For electroslag welding wire; extra low carbon.

VERTOMAR 2M.
M-677; 0.11-.17 C, 1.75-2.1 Mn, bal Fe.
For electroslag welding wire; 2% manganese steel.

VERTOMAR 2MM.
M-677; 0.11-.17 C, 1.75 -2.1 Mn, 0.4-.6 Mo, bal Fe.
For electroslag welding wire; 2% manganese 0.5% molybdenum steel.

VERTOMAR 6.
M-677; 0.06 max C, 0.15 max Mn, bal Fe.
For electroslag welding wire; 0.06 max carbon.

VERTOMAR 8.
M-677; 0.10 max C, 0.35-.65 Mn, 0.1-2 Si, bal Fe.
For electroslag welding wire; silicon killed.

VERTOMAR 10.
M-677; 0.07-.13 C, 0.4-.7 Mn, bal Fe.
For electroslag welding wire; 0.10 max carbon.

VERTOMAR 15.
M-677; 0.13-.19 C, 0.95-1.3 Mn, 0.15-.3 Si, bal Fe.
For electroslag welding wire; 1% manganese 0.2 C.

VERTOMAR 60.
M-677; 0.55-.66 C, 0.9-1.25 Mn, 0.1-.2 Si, bal Fe.
For electroslag welding wire; medium carbon steel.

VERTOMAR 410 NI.
M-677; 0.04 C, 0.6 Mn, 0.4 Si, 12 Cr, 4 Ni, 0.8 Mo, bal Fe.
For electroslag welding wire; welding CA6NM castings.

VERTOMAR CM30.
M-677; 0.28-.33 C, 0.4-.6 Mn, 0.2-.35 Si, 0.8-1.1 Cr, 0.15-.25 Mo, bal Fe.
For electroslag welding wire; AISI type 4130 steel.

VERTOMAR CNM20.
M-677; 0.18-.23 C, 0.7-.9 Mn, 0.2-.35 Si 0.4-.6 Cr, 0.4-0.7 Ni, 0.15-.25 Mo, bal Fe.
For electroslag welding wire; AISI type 8620.

VERTOMAR CV50.
M-677; 0.48-0.53 C, 0.7-.9 Mn, 0.2-.35 Si, 0.8-1.1 Cr, 0.15 min V, bal Fe.
For electroslag welding wire; AISI type 6150.

VERY BEST.
  M-336; 1.05 C, 0.3 Mn, 0.5 Cr, 0.1 V, bal Fe.
  Annealed: 185 Brin.
  For dies and tools, drawing and forming dies, taps, drills, mandrels; tough and hard; water hardened.

VES.
  M-1307; 1.0 C, 0.1 V, 0.25 Mn, 0.20 Si, bal Fe.
  For drills, taps, springs, tools, cutters; Type W2; water hardened.

VESTALIN.
  28 Ni, C, bal Fe.
  For electrical resistances; resistance alloy.

VESUVIUS.
  M-521; 0.10 C, 30 Cr, 1.7 Ni, bal Fe.
  Oil treated: 78,000-101,000 TS; 58,000-85,000 YS; 25-15 El; 40-30 RA; 175-225 Brin.
  For furnace parts, fire bars, stokers, grids; heat and corrosion resistant; Firth Brown J-182.

VET-3.
  M-1733; 0.70 C, bal Fe.
  Water hardening tool steel.
  AISI W 1; DIN 1.1620.

VETD.
  M-1733; 1.0 C, 0.10 V, bal Fe.
  Water hardening tool steel.
  AISI W 2; DIN 1.1640.

VEW A100.
  M-1802; 0.07 max C, 17.5 Cr, 12.5 Ni, 2.8 Mo, bal Fe.
  Austenitic stainless steel for chemical equipment; improved corrosion resistance.
  W. Nr. 1.4436; AISI 316.

VEW A100G.
  M-1802; 0.07 max C, 17.5 Cr, 2.75 Mo, 12.5 Ni, bal Fe.
  Austenitic stainless steel casting for corrosion resistant containers to 300°C.
  W. Nr. 1.4437.

VEW A100R.
  M-1802; 0.07 max C, 17.5 Cr, 2.75 Mo, 13.0 Ni, bal Fe.
  Austenitic stainless steel for corrosion resistant parts and assemblies; to 350°C.
  W. Nr. 1.4436; AISI 316.

VEW A102.
  M-1802; 0.07 max C, 17.0 Cr, 13.0 Ni, 4.75 Mo, bal Fe.
  Austenitic stainless steel; extra good corrosion resistance; for chemical equipment.
  W. Nr. 1.4449; similar to AISI 317.

VEW A102R similar to **VEW A102.**

VEW A114 similar to **VEW A120.**

VEW A120.
  M-1802; 0.07 max C, 18 Cr, 10.5 Ni, 2.0 Mo, bal Fe.
  Austenitic stainless steel; extra good corrosion resistance; for chemical plant equipment.
  W. Nr. 1.4401; AISI 316.

VEW A120 G.
  M-1802; 0.07 max C, 2.0 max Si, 1.5 max Mn, 18.5 Cr, 2.25 Mo, 11.0 Ni, bal Fe.
  Austenitic stainless casting; for pumps, centrifuges.
  W. Nr. 1.4408.

VEW A120R similar to **VEW A120.**

VEW A121 similar to **VEW A120G.**

VEW A122.
  M-1802; 0.06 max C, 18 Cr, 8.5 Ni, 1.6 Mo, bal Fe.
  Austenitic stainless steel; for fittings for chemical plant equipment.
  Similar to W. Nr. 1.4420.

VEW A128.
  M-1802; 0.12 max C, 2.0 max Si, 1.5 max Mn, 18.5 Cr, 2.25 Mo, 10.0 Ni, bal Fe.
  Austenitic stainless casting; for pumps, stirring gear, filters.
  W. Nr. 1.4410.

VEW A200.
  M-1802; 0.03 max C, 18 Cr, 12 Ni, 2.5 Mo, bal Fe.
  Austenitic stainless steel, weldable; for chemical plant equipment.
  W. Nr. 1.4404; AISI 316 L.

VEW A205.
  M-1802; 0.03 max C, 18 Cr, 13.5 Ni, 2.8 Mo, bal Fe.
  Austenitic stainless steel, weldable, for chemical plant equipment.
  W.Nr. 1.4435; AISI 316L.

VEW A220.
  M-1802; 0.03 max C, 17.5 Cr, 2.75 Mo, 13.5 Ni, bal Fe.
  Austenitic stainless steel, weldable, for chemical plant equipment.
  W.Nr. 1.4435; AISI 316L.

VEW-A300.
  M-1802; 0.10 max C, 17.5 Cr, 10.5 Ni, 2 Mo, T=5 x C, bal Fe.
  Austenitic stainless steel, weldable; for tanks and other chemical plant equipment.
  W.Nr. 1.4571.

VEW A305.
  M-1802; 0.10 max C, 17.5 Cr, 13.5 Ni, 3.0 Mo, Ti=5 x C, bal Fe.
  Stabilized austenitic stainless steel, weldable, for chemical plant equipment.

VEW A350.
 M-1802; 0.10 max C, 18 Cr, 12 Ni, 2.2 Mo, Nb/Ta=8 x C, bal Fe.
 Stabilized austenitic stainless steel; weldable; for chemical plant equipment.
 W.Nr. 1.4580.

VEW A350 G.
 M-1802; 0.08 C, 1.5 max Si, 1.5 max Mn, 18.5 Cr, 2.25 Mo, 11.5 Ni, Nb=8 x C, bal Fe.
 Austenitic stainless casting; weldable; for armatures, pumps; housings.
 W.Nr. 1.4581; ACI CF-8M.

VEW A354G.
 M-1802; 0.10 max C, 1.0 max Si, 2.0 max Mn, 17.5 Cr, 2.75 Mo, 13 Ni, Nb=8 x C, bal Fe.
 Austenitic stainless casting; weldable; for containers in chemical industries.
 W.Nr. 1.4583.

VEW A400.
 M-1802; 0.04 max C, 17.5 Cr, 13.5 Ni, 4.5 Mo, 0.10-0.20 N, bal Fe.
 Austenitic stainless steel.
 W.Nr. 1.4439.

VEW A405.
 M-1802; 0.02 C, 25 Cr, 22 Ni, Mo, N, bal Fe.
 Austenitic stainless steel; good high temperature properties.
 W.Nr. 1.4466; X2CrNiMoN25 22.

VEW A410.
 M-1802; 0.03 max C, 17.5 Cr, 2.75 Mo, 13.0 Ni, bal Fe.
 Austenitic stainless steel; weldable; for parts and equipment in chemical and textile industries.
 W.Nr. 1.4429.

VEW A500.
 M-1802; 0.07 max C, 18 Cr, 9 Ni, bal Fe.
 Austenitic stainless steel; parts for food, beverage, textile industries.
 W.Nr. 1.4301; AISI 304.

VEW A500G.
 M-1802; 0.07 max C, 2.0 max Si, 1.5 max Mn, 18.5 Cr, 10.0 Ni, bal Fe.
 Austenitic stainless steel casting; for armatures, housings, pumps.
 W.Nr. 1.4308.

VEW A505.
 M-1802; 0.12 max C, 18 Cr, 9 Ni, bal Fe.
 Austenitic stainless steel.
 W.Nr. 1.4300; AISI 302.

VEW 505G.
 M-1802; 0.10 C, 18 Cr, 8.5 Ni, bal Fe.
 Austenitic stainless steel casting; for chemical plant equipment.
 W.Nr. 1.4312.

VEW A506.
 M-1802; 0.12 max C, 18 Cr, 9 Ni, 0.20 S, bal Fe.
 Free machining austenitic stainless steel.
 For stainless parts made by automatic screw machines.
 W.Nr. 1.4305; AISI 303.

VEW A511.
 M-1802; 0.07 max C, 19.0 Cr, 11.0 Ni, bal Fe.
 Austenitic stainless steel; for chemical industry.
 W.Nr. 1.4303; AISI 305.

VEW A520.
 M-1802; 0.12 max C, 17 Cr, 7 Ni, bal Fe.
 Austenitic stainless steel; work-hardens.
 For springs, trailer bodies; wheelcovers.
 W.Nr. 1.4310; AISI 301.

VEW A522.
 M-1802; 0.10 max C, 12.5 Cr, 12 Ni, bal Fe.
 Austenitic stainless; for valves, pump parts.
 W.Nr. 1.4307.

VEW A600.
 M-1802; 0.03 max C, 19 Cr, 11 Ni, bal Fe.
 For equipment requiring welding as tanks for food and beverage processing.
 W.Nr. 1.4306; AISI 304L.

VEW A604.
 M-1802; 0.03 max C, 18.5 Cr, 11.0 Ni, bal Fe.
 Austenitic stainless steel; weldable; for containers for food and beverage processing.
 W.Nr. 1.4306; AISI 304L.

VEW A610.
 M-1802; 0.02 max C, 18 Cr, 15 Ni, S, bal Fe.
 Free machining austenitic stainless steel.
 W.Nr. 1.4361.

VEW A700.
 M-1802; 0.10 max C, 18 Cr, 10 Ni, Ti=5 x C, bal Fe.
 Stabilized austenitic stainless steel; for welded equipment for food, textile equipment.
 W.Nr. 1.4541; AISI 321.

VEW A750.
 M-1802; 0.10 max C, 18 Cr, 10 Ni, Nb=8 x C, bal Fe.
 For welded chemical plant equipment.
 W.Nr. 1.4550; AISI 347.

VEW A750G.
 M-1802; 0.08 max C, 1.5 max Si, 1.5 max N, 18.5 Cr, 10 Ni, Nb=8 x C bal Fe.
 Austenitic stainless steel casting; weldable; for equipment for food, paper and textile industries.
 W.Nr. 1.4552.

VEW A900.
 M-1802; 0.10 max C, 27 Cr, 4.5 Ni, 1.5 Mo, bal Fe.
 Stainless steel for chemical plant equipment.
 W.Nr. 1.4460; AISI 329.

VEW A955.
M-1802; 0.07 max C, 17.5 Cr, 22.5 Ni, 3.5 Mo, 0.7 Nb, 1.8 Cu, bal Fe.
Austenitic stainless; for parts for chemical, petroleum and dye plants.
W.Nr. 1.4586.

VEW A960.
M-1802; 0.07 max C, 18.0 Cr, 20.0 Ni, 2.0 Mo, Nb=8 x C, 2.0 Cu, bal Fe.
Austenitic stainless steel; weldable.
For parts and equipment in chemical plants.
W.Nr 1.4505.

VEW A962.
M-1802; 0.02 C, 25 Ni, 20 Cr, 5 Mo, Cu, bal Fe.
Austenitic stainless steel; excellent corrosion resistance.
X2NiCrMoCu25 20 5.

VEW B110.
M-1802; 0.85 C, 0.65 Mn, 0.40 Si, 0.40 Cr, bal Fe.
Cold work tool steel.
For cutting tools, gages, machine knives.
W.-Nr. 1.2004.

VEW B112.
M-1802; 0.75 C, 0.60 Mn, 0.35 Cr, bal Fe.
Cold work tool steel for small tools as punches, mandrels, stamping tools.
W.Nr. 1.2003.

VEW B114.
M-1802; 1.25 C, 0.35 Mn, 0.35 Cr, bal Fe.
Water hardening tool steel for punches, cold chisels, mandrels.
W.Nr. 1.2002.

VEW B116.
M-1802; 0.85 C, 0.40 Mn, 0.40 Cr, bal Fe.
Cold work tool steel; for gauges, knives.
W.Nr. 1.2004.

VEW B304.
M-1802; 0.73 C, 0.50 Cr, 0.35 Mo, 0.25 V, 0.60 W, bal Fe.
Cold work tool steel; for paper shears, wood cutting saws.
W.Nr. 1.2604.

VEW B306.
M-1802; 1.15 C, 0.20 Cr, 2.0 W, bal Fe.
Cold work tool steel; for metal cutting saws and hack saw blades.
W.-Nr. 1.2442.

VEW B400.
M-1802; 0.80 C, 0.40 Mn, 0.50 Cr, 0.20 V, bal Fe.
Cold work tool steel; for paper shears, saws.
W.Nr. 1.2235.

VEW B404.
M-1802; 1.15 C, 0.65 Cr, 0.10 V, bal Fe.
Cold work tool steel; for cutters, saws, bearings.

VEW B535.
M-1802; 0.75 C, 0.40 Mn, 0.25 Cr, 0.55 Ni, bal Fe.
Cold work tool steel; water hardening; for hand tools.
W.Nr. 1.2703.

VEW B908.
M-1802; 0.78 C, 0.70 Mn, bal Fe.
Water hardening tool steel; for mandrels, collets.
W.Nr. 1.1750.

VEW D220.
M-1802; 0.20 C, 1.3 Cr, 1.1 Mo, 0.3 V, bal Fe.
Oil hardening; for bolts and nuts resistant to elevated temperatures to 530°C.
W.Nr. 1.8070.

VEW D230.
M-1802; 0.22 C, 1.3 Cr, 0.85 Mo, 0.25 V, bal Fe.
Oil hardening; for parts resistant to elevated temperature.
W.Nr. 1.7703.

VEW D240.
M-1802; 0.24 C, 1.40 Cr, 0.55 Mo, 0.2 V, bal Fe.
Oil hardening; for bolts and nuts resistant to elevated temperatures to 530°C.
W.Nr. 1.7733.

VEW D310.
M-1802; 0.10 C, 5.3 Cr, 0.55 Mo, bal Fe.
Air or oil hardening; for tubes for petroleum: distilling, and for hydrogenation plants.
W.Nr. 1.7362.

VEW D320.
M-1802; 0.10 C, 2.3 Cr, 1.0 Mo, bal Fe.
Oil hardening; for steam boiler and super heater tubes up to 530°C.
W.Nr. 1.7380.

VEW D330.
M-1802; 0.13 C, 0.85 Cr, 0.45 Mo, bal Fe.
Oil hardening; for steam boilers and superheater tubes to 530°C.
W.Nr. 1.7335.

VEW D330G.
M-1802; 0.22 C, 0.70 Mn, 1.0 Cr, 0.45 Mo, bal Fe.
Oil or water hardening; for elevated temperature operation.
W.Nr. 1.7354.

VEW D500.
M-1802; 0.15 C, 0.65 Mo, 0.30 Mo, bal Fe.
Oil hardening; for flanges resistant to elevated temperature to 530°F.
W.Nr. 1.5415.

VEW D502.
M-1802; 0.22 C, 0.60 Mn, 0.30 max Cr, 0.40 Mo, bal Fe.
For thick-walled, high pressure tubes, or small forgings, for elevated temperature operation.
W.Nr. 1.5419.

VEW E110.
M-1802; 0.17 C, 1.65 Cr, 0.30 Mo, 1.55 Ni, bal Fe.
Deep hardening carburizing steel; high core strength.
For plate wheels, driving pinions and highly stressed gears and cog wheels.
W.Nr. 1.6587.

VEW E115.
M-1802; 0.20 C, 0.75 N, 0.50 Cr, 0.60 Ni, 0.20 Mo, bal Fe.
Low alloy carburizing steel.
For gears, pinions, arbors, bushings.
W.Nr. 1.6523; AISI 8620.

VEW E154.
M-1802; 0.20 C, 0.50 Ni, 0.50 Cr, 0.20 Mo, bal Fe.
Carburizing steel.

VEW E200.
M-1802; 0.13 C, 3.5 Ni, 0.7 Cr, bal Fe.
Alloy carburizing steel; deep hardening.
For cams, camshafts, gears, universal joints.
W.Nr. 1.5752.

VEW E204.
M-1802; 0.13 C, 1.1 Cr, 4.5 Ni, bal Fe.
Alloy carburizing steel; deep hardening.
For gears, cams, high stressed gear wheels.
W.Nr. 1.5860.

VEW E220.
M-1802; 0.18 C, 2.0 Ni, 2.0 Cr, 0.1 Mo, bal Fe.
Alloy carburizing steel; deep hardening, high core strength.
For gears, shafts, heavy bolts, high stressed parts.
W.Nr. 1.5920.

VEW E224.
M-1802; 0.18 C, 0.5 Mn, 2 Cr, 2 Ni, bal Fe.
Alloy carburizing steel; high core strength.
For gears, cams, cog wheels.
W.Nr. 1.5920.

VEW E230.
M-1802; 0.15 C, 1.55 Ni, 1.55 Cr, bal Fe.
Alloy carburizing steel.
For gears, cams, camshafts, chain wheels.
W.Nr. 1.5919.

AVEW E234.
M-1802; 0.13 C, 1.4 Ni, 0.75 Cr, bal Fe.
Alloy carburizing steel.
For cams, camshafts, pinions, bearings.
W.Nr. 1.5713.

VEW E300.
M-1802; 0.20 C, 1.25 Cr, 0.25 Mo, bal Fe.
Alloy carburizing steel.
For cams, pinions, gears, bearings.
W.Nr. 1.7264.

VEW E320.
M-1802; 0.20 C, 0.75 Mn, 0.40 Cr, 0.45 Mo, bal Fe.
Carburizing steel; for small gears, cams, arbors, bushings.
W.Nr. 1.7321.

VEW E321.
M-1802; 0.20 C, 0.75 Mn, 0.035 P, 0.02-0.035 S, 0.40 Cr, 0.45 Mo, bal Fe.
Carburizing steel.
W.Nr. 1.7323.

VEW E400.
M-1802; 0.20 C, 1.25 Mn, 1.15 Cr, bal Fe.
Alloy carburizing steel.
For cams, pinions, bearings.
W.Nr. 1.7147.

VEW E401.
M-1802; 0.20 C, 1.25 Mn, 1.10 Cr, bal Fe.
Alloy carburizing steel.
For bearings, gears, cams.
W.Nr. 1.7149.

VEW E406.
M-1802; 0.18 C, 1.15 Mn, 1.0 Cr, bal Fe.
Alloy carburizing steel.
For cams, bearings, pinions.
W.Nr. 1.7168.

VEW E410.
M-1802; 0.15 C, 1.15 Mn, 1.0 Cr, bal Fe.
Low alloy carburizing steel.
For bearings, cams, pinions, arbors.
W.Nr. 1.7131.

VEW E411.
M-1802; 0.18 C, 1.15 Mn, 1.0 Cr, bal Fe.
Low alloy carburizing steel.
For pinions, shafts, bearings, cams.
W.Nr. 1.7139.

VEW E416.
M-1802; 0.16 C, 1.1 Mn, 1.05 Cr, B, bal Fe.
Low alloy carburizing steel.
For gears, shafts, universal joints.
W.Nr. 1.7160.

VEW E502.
M-1802; 0.20 C, 0.90 Mn, 1.0 Cr, 0.1 V, bal Fe.
Low alloy carburizing steel.
For pinions, gears, cams, bearings.
W.Nr. 1.7510.

VEW E525.
M-1802; 0.15 C, 0.5 Mn, 0.65 Cr, bal Fe.
Low alloy carburizing steel.
For piston pins, roller bearings, cams.
W.Nr. 1.7015.

VEW E900.
M-1802; 0.13 C, 0.25 Si, 0.37 Mn, bal Fe.
Carburizing steel, unalloyed.
For light loaded cams, small machine parts.
W.Nr. 1.1121; AISI 1010.

VEW E920.
M-1802; 0.15 C, 0.25 Si, 0.37 Mn, bal Fe.
Carburizing steel; unalloyed.
For light loaded cams, small machine parts.
W.Nr. 1.1141; AISI 1015.

VEW F14.
M-1802; 0.45 C, 1.65 Si, 0.70 Mn, bal Fe.
Si-Mn spring steel; oil hardening.
Laminated springs, or elliptical springs for vehicles.
W.Nr. 1.0902.

VEW F100.
M-1802; 0.50 C, 1.7 Si, 0.7 Mn, bal Fe.
Si-Mn spring steel; oil hardening.
Laminated springs for rail vehicles, bumper springs; also shock resisting tools.
W.Nr. 1.0903; similar to AISI 9255.

VEW F105.
M-1802; 0.65 C, 1.7 Si, 0.70 Mn, bal Fe.
Si-Mn spring steel; oil hardening.
Laminated or helical springs for automotive.
W.Nr. 1.0906; similar to AISI 9260.

VEW F108.
M-1802; 0.60 C, 1.7 S, 0.70 Mn, bal Fe.
Si-Mn spring steel; oil hardening.
For laminated or helical springs for vehicles.
W.Nr. 1.0909; similar to AISI 9260.

VEW F110.
M-1802; 0.55 C, 1.7 Si, 0.70 Mn, bal Fe.
Si-Mn spring steel; oil hardening.
Laminated or helical springs for automotive.
W.Nr. 1.0904; similar to AISI 9255.

VEW F114.
M-1802; 0.46 C, 1.65 Si, 0.65 Mn, bal Fe.
For laminated or elliptical springs.
W.Nr. 1.0902.

VEW F124.
M-1802; 0.65 C, 1.15 Si, 1.0 Mn, bal Fe.
Si-Mn spring steel; oil hardening.

VEW F128.
M-1802; 0.60 C, 1.65 Si, 0.70 Mn, 0.30 Ni, bal Fe.
Si-Mn spring steel; oil hardening.
Laminated springs, plate and spiral springs.
W.Nr. 1.0961.

VEW F180.
M-1802; 0.46 C, 1.8 Mn, bal Fe.
Water or oil hardening spring steel.
W.Nr. 1.0913.

VEW F200.
M-1802; 0.67 C, 1.30 Si, 0.50 Mn, 0.50 Cr, bal Fe.
Oil hardening spring steel.
For valve springs, helical springs.
W.Nr. 1.7103.

VEW F300.
M-1802; 0.55 C, 0.85 Mn, 0.75 Cr, bal Fe.
Helical springs, torsion bar springs; oil hardening.
W.Nr. 1.7176; AISI 5155.

VEW F500.
M-1802; 0.52 C, 0.90 Mn, 1.05 Cr, 0.20 Mo, 0.10 V, bal Fe.
Oil hardening spring steel.
W.Nr. 1.7701; AISI 4150.

VEW F550.
M-1802; 0.50 C, 0.90 Mn, 1.05 Cr, 0.15 V, bal Fe.
Chrome-vanadium spring steel.
For coiled springs; oil hardening.
W.Nr. 1.8159; AISI 6150.

VEW H100.
M-1802; 0.12 max C, 1.5 Si, 24.0 Cr, 1.5 Al, bal Fe.
Stainless and temperature resisting steel; for furnace parts.
W.Nr. 1.4762.

VEW H102.
M-1802; 0.20 C, 1.5 Mn, 25 Cr, 0.25 N, bal Fe.
Stainless and oxidation resisting steel for elevated temperature operations.
AISI 446.

VEW H103.
M-1802; 0.45 C, 2.0 Si, 23 Cr, bal Fe.
Stainless cast steel for elevated temperature operations.
W.Nr. 1.4745.

VEW H120.
M-1802; 0.12 max C, 18 Cr, 1.0 Si, 1.0 Al, bal Fe.
Stainless and temperature resisting steels for furnace parts, oil refinery equipment.
W.Nr. 1.4742.

VEW H120G.
M-1802; 0.45 C, 2.0 Si, 1.0 max Mn, 17 Cr, bal Fe.
Stainlss cast steel for elevated temperature operations.
W.Nr. 1.4740.

VEW H140.
M-1802; 0.12 max C, 1.2 Si, 13.0 Cr, 1.0 Al, bal Fe.
Stainless and temperature resisting steel; for furnace parts, oil refining equipment.
W.Nr. 1.4724.

VEW H160.
M-1802; 0.12 max C, 6.5 Cr, 0.8 Al, bal Fe.
Heat resisting steel; for oil refinery equipment.
W.Nr. 1.4713.

VEW H160G.
M-1802; 0.30 C, 2.0 Si, 1.0 max Mn, 7.0 Cr, bal Fe.
Cast steel for elevated temperature operation as burner parts, tempering furnaces.
W.Nr. 1.4710.

VEW H161.
M-1802; 0.30 C, 1.75 Si, 1.0 max Mn, 6-8 Cr, bal Fe.
Cast steel for elevated temperature operation.
W.Nr. 1.4710.

VEW H300.
M-1802; 0.20 C, 1.1 Si, 25 Cr, 4 Ni, bal Fe.
Heat resisting steel for furnace parts.
W.Nr. 1.4821.

VEW H300G.
M-1802; 0.40 C, 1.5 Si, 1.5 max Mn, 27 Cr, 4.5 Ni, bal Fe.
Stainless casting for elevated temperature operation, as furnace parts.
W.Nr. 1.4823.

VEW H301.
M-1802; 0.40 C, 1.5 Si, 1.5 max Mn, 27 Cr, 4.5 Ni, bal Fe.
Stainless casting for high temperature parts.
W.Nr. 1.4823.

VEW H500.
M-1802; 0.15 max C, 1.5 Si, 16 Cr, 35 Ni, bal Fe.
Heat resisting alloy; for furnace parts and heat treating equipment.
W.Nr. 1.4876; similar to AISI 330.

VEW H520.
M-1802; 0.12 max C, 2.0 Si, 21 Cr, 34 Ni, bal Fe.
Heat resisting alloy; for furnace parts and heat treating equipment.
W.Nr. 1.4864.

VEW H520G.
M-1802; 0.30 C, 2.0 Si, 1.5 max Mn, 18 Cr, 37 Ni, bal Fe.
Stainless casting for high temperature parts as furnace parts.
W.Nr. 1.4865.

VEW H521 similar to **VEW H520G.**

VEW H522.
M-1802; 0.15 max C, 2 max Si, 25 Cr, 20 S, bal Fe.
For gas turbines, furnace equipment.
W.Nr. 1.4845; similar to AISI 310.

VEW H525.
M-1802; 0.20 max C, 2 Si, 25 Cr, 20 Ni, bal Fe.
For furnace equipment, heat treat equipment.
W.Nr. 1.4841; similar to AISI 314.

VEW H527.
M-1802; 0.15 C, 1.5 Si, 1.5 max Mn, 25 Cr, 20 Ni, bal Fe.
Stainless casting for elevated temperature operation.
W.Nr. 1.4849.

VEW H529.
M-1802; 0.35 C, 2.0 Si, 1.5 max Mn, 25 Cr, 20 Ni, bal Fe.
Stainless casting for elevated temperature operation.
W.Nr. 1.4848; ACI HK.

VEW H532.
M-1802; 0.25 C, 2.0 Mn, 1.5 Si, 25 Cr, 20 Ni, bal Fe.
Stainless steel; for high temperature operations.
AISI 310.

VEW H537.
M-1802; 0.4 C, 2 Si, 25 Cr, 13 Ni, bal Fe.
Stainless casting; for furnace parts, heat treat boxes, pump bodies.
W.Nr. 1.4837; ACI HI.

VEW H539.
M-1802; 0.35 C, 2.0 S, 1.5 max Mn, 26 Cr, 14 Ni, bal Fe.
Stainless casting for elevated temperature operation; as furnace parts.
W.Nr. 1.4846.

VEW H550.
M-1802; 0.15 C, 2.0 Si, 20 Cr, 12 Ni, bal Fe.
Heat resisting steel; for furnace parts, oil refinery equipment.
W.Nr. 1.4828; AISI 309, 305.

VEW H551.
M-1802; 0.4 C, 2 Si, 22 Cr, 9.5 Ni, bal Fe.
Stainless steel casting for elevated temperature operation.
W.Nr. 1.4826; ACI HF.

VEW H566.
M-1802; 0.08 C, 2.0 Mn, 1.0 Si, 23.0 Cr, 13.0 Ni, bal Fe.
Stainless steel for high temperature operations.
AISI 309S.

VEW H700.
M-1802; 0.45 C, 9 Cr, 3 Si, bal Fe.
For exhaust valves in automotive engines.
W.Nr. 1.4718; similar to SAE HNV 3.

VEW H710.
M-1802; 0.40 C, 2.5 Si, 10 Cr, 1.0 Mo, bal Fe.
For exhaust valves in automotive engines.
W.Nr. 1.4731.

VEW H730.
M-1802; 0.80 C, 2.0 Si, 21.0 Cr, 1.5 Ni, bal Fe.
For exhaust valves in automotive engines.
W.Nr. 1.4747; similar to SAE HNV6.

VEW H800.
M-1802; 0.45 C, 2.5 Si, 19 Cr, 10 Ni, 1.2 W, bal Fe.
For exhaust valves in automotive engines.
W.Nr. 1.4873; similar to SAE EV 5.

VEW H850.
M-1802; 0.53 C, 9.0 Mn, 21.0 Cr, 4 Ni, 0.06 S, 0.42 N, bal Fe.
For exhaust valves in heavy-duty engines.
W.Nr. 1.4871; similar to SAE EV 8.

VEW K100.
M-1802; 2.1 C, 12.0 Cr, bal Fe.
Cold work tool steel; air or oil hardening.
For punches, blanking and forming dies, trimming dies.
W.-Nr. 1.2080; similar to AISI D3.

VEW K102.
M-1802; 2.9 C, 12.0 Cr, bal Fe.
Cold work tool steel; air or oil hardening.
For trimming and coining dies, thread rollers.
W.Nr. 1.2086.

VEW K103.
M-1802; 2.0 C, 12.0 Cr, 0.5 Mo, 0.5 V, 1.0 W, bal Fe.
Cold work tool steel; air or oil hardening.
For punching and trimming dies, coining dies, broaches, thread rolling dies.
W.Nr. 1.2600; modified AISI D3.

VEW K105.
M-1802; 1.65 C, 11.5 Cr, 0.6 Mo, 0.5 W, 0.3 V, bal Fe.
Cold work tool steel; air or oil hardening.
For broaches, forming dies, punches.
W.-Nr. 1.2601.

VEW K107.
M-1802; 2.1 C, 12.0 Cr, 0.7 W, V, bal Fe.
Cold work tool steel; air or oil hardening.
Heavy duty punching and trimming dies, reamers, broaches.
W.-Nr. 1.2436; similar to AISI D3.

VEW K110.
M-1802; 1.5 C, 12.0 Cr, 1.0 V, 0.7 Mo, bal Fe.
Cold work tool steel; air or oil hardening.
For punching and forming dies, trimming dies, thread rolling dies.
W.-Nr. 1.2379; AISI D2.

VEW K116.
M-1802; 1.65 C, 11.5 Cr, 0.1 V, bal Fe.
Cold work tool steel; air or oil hardening.
For punching, forming and trimming dies, thread rolling dies.
W.-Nr. 1.2201.

VEW K200.
M-1802; 1.05 C, 1.5 Cr, bal Fe.
Cold work tool steel; water or oil hardening.
For reamers, lathe centers, drills, bearings.
W.-Nr. 1.2067.

VEW K201.
M-1802; 0.85 C, 1.75 Cr, 0.35 Mn, bal Fe.
Cold work tool steel; water or oil hardening.
For bushings, liners, hand tools.
W.-Nr. 1.2064.

VEW K205.
M-1802; 1.4 C, 0.8 Cr, bal Fe.
Cold work tool steel; water or oil hardening.
For files, needle files, precision tools.
W.-Nr. 1.2008.

VEW K240.
M-1802; 0.90 C, 0.70 Mn, 1.2 Si, 1.2 Cr, bal Fe.
Cold work tool steel; oil hardening.
For punches, shear blades, engravers' tools, bushings.
W.-Nr. 1.2108.

VEW K243.
M-1802; 0.6 C, 1.7 Si, 0.8 Cr, bal Fe.
Cold work tool steel, shock resisting type.
For chisels, rivet sets, staking tools.

VEW K244.
M-1802; 0.67 C, 1.3 Si, 0.5 Mn, 0.5 Cr, bal Fe.
Cold work tool steel; shock resisting type.
For pneumatic chisels, rivet sets, staking tools.

VEW K245.
M-1802; 0.67 C, 1.5 Si, 0.7 Mn, Cr, bal Fe.
Cold work tool steel; shock resisting type.
For chisels, staking tools, rivet sets.
W. Nr. 1.2101.

VEW K300.
M-1802; 0.50 C, 0.6 Mn, 8.5 Cr, 1.2 Mo, 1.2 W, bal Fe.
Cold work tool steel; oil hardening.
Shear blades for metal sheet.
W.-Nr. 1.2631.

VEW K301.
M-1802; 0.50 C, 8.0 Cr, 1.6 Mo, 1.4 W, V, bal Fe.
Cold work tool steel; oil hardening.
Special tools.
W.-Nr. 1.2631.

VEW K305.
M-1802; 1.0 C, 5.0 Cr, 1.0 Mo, 0.15 V, bal Fe.
Cold work tool steel; air hardening.
For forming, blanking, embossing.
W.-Nr. 1.2363; AISI A2.

VEW K306.
M-1802; 0.47 C, 1.0 Si, 5.0 Cr, 1.35 Mo, 1.4 V, bal Fe.
Cold work tool steel; air hardening.
W.-Nr. 1.2345.

VEW K310.
M-1802; 0.80 C, 2.0 Cr, bal Fe.
Cold work tool steel.
For cold rolls, cams, press discs.
W.-Nr. 1.2327.

VEW K311.
M-1802; 0.45 C, 0.90 Mn, 1.8 Cr, 0.25 Mo, bal Fe.
Cold work tool steel.
For chisels, center punches.
W.-Nr. 1.2328.

VEW K400.
M-1802; 1.4 C, 0.35 Cr, 3.3 W, 0.25 V, bal Fe.
Cold work tool steel; oil or water hardening.
For engravers' tools; fast finishing tools.
W.-Nr. 1.2562.

VEW K405.
M-1802; 1.2 C, 0.20 Cr, 1.0 W, 0.10 V, bal Fe.
Cold work tool steel; water hardening.
For fast finishing tools for short runs; rotary files, counter bores.
W.-Nr. 1.2516.

VEW K450.
M-1802; 0.45 C, 1.0 Cr, 2.0 W, 0.15 V, bal Fe.
Cold work tool steel; oil hardening.
For sheet shears, pneumatic chisels; shock resisting type.
W.-Nr. 1.2542; similar to AISI S1.

VEW K451.
M-1802; 0.8 C, 2.6 W, 0.5 Mo, bal Fe.
Cold work tool steels.
For punches, header dies, upsetters.

VEW K455.
M-1802; 0.55 C, 1.0 Cr, 1.8 W, 0.18 V, bal Fe.
Hot or cold work tool steel; oil hardening.
For header dies, upsetters, chisels.
W.-Nr. 1.2550.

VEW K457.
M-1802; 1.10 C, 1.2 Cr, 1.3 W, 0.20 V, bal Fe.
Cold work tool steel; oil hardening.
Cutting tools for leather, plastic and other non-metallic materials.
W.-Nr. 1.2519.

VEW K458.
M-1802; 1.10 C, 1.2 Cr, 1.3 W, 0.20 V, bal Fe.
Cold work tool steel; oil hardening.
Cutting tools for wood, leather, plastics.
W.-Nr. 1.2519.

VEW-K460.
M-1802; 1.0 C, 1.1 Mn, 0.6 Cr, 0.6 W, 0.1 V, bal Fe.
Cold work tool steel; oil hardening.
For reamers, hand taps, pipe threading tools.
W.-Nr. 1.2510; AISI 01.

VEW K465.
M-1802; 1.05 C, 0.90 Mn, 1.0 Cr, 1.15 W, bal Fe.
Cold work tool steel; oil hardening.
For cold forming and cold heading dies; punches.
W.-Nr. 1.2419.

VEW K466.
M-1802; 0.35 C, 1.05 Cr, 1.85 W, 0.18 V, bal Fe.
Cold work tool steel; oil hardening.
For cold header dies, punches for thin sheet, pneumatic tools.
W.-Nr. 1.2542.

VEW K467.
M-1802; 1.2 C, 1.0 W, bal Fe.
Cold work tool steel; water hardening.
For center drills, counterbores, reamers.
W.-Nr. 1.2414.

VEW K505.
M-1802; 1.45 C, 0.60 Mn, 1.4 Cr, bal Fe.
Cold work tool steel; oil hardening.
For blanking and drawing dies, punches, bushings; abrasion resistant.
W.-Nr. 1.2063.

VEW K506.
M-1802; 0.45 C, Mn, Cr, Mo or V, bal Fe.
Cold work tool steel; oil hardening.
For cold heading and forming.
Similar to W.-Nr. 1.2241.

VEW K508.
M-1802; 0.50 C, 0.90 Mn, 1.05 Cr, 0.10 V, bal Fe.
Cold work tool steel; oil hardening.
For carbide tool shanks, screwdrivers, hatchets, various hand tools.
W.-Nr. 1.2241; similar to AISI 6150.

VEW K510.
M-1802; 1.15 C, 0.65 Cr, 0.10 V, bal Fe.
Cold work tool steel; oil or water hardening.
For drills, reamers, countersinks, scraping tools, punches.
W.-Nr. 1.2210.

VEW K511.
M-1802; 0.32 C, 0.50 Mn, 0.55 Cr, 0.10 V, bal Fe.
Cold work tool steel; water hardening.
For hand tools as screwdrivers and wrenches.
W.-Nr. 1.2208.

VEW K600.
M-1802; 0.45 C, 4.0 Ni, 1.3 Cr, 0.30 Mo, bal Fe.
Cold work tool steel; oil or air hardening.
For embossing dies, forming dies, shear blades.
W.-Nr. 1.2767.

VEW K605.
M-1802; 0.50 C, 3.25 Ni, 1.05 Cr, Mo, bal Fe.
Cold work tool steel; air or oil hardening.
For cold heading dies, shear blades.
W.-Nr. 1.2721.

VEW K618.
M-1802; 0.7 C, 1.3 Mn, 1 Si, 1 Cr, 1 Ni, bal Fe.
Heat treated: 227,000 TS; 350 Brin.
For machine tool parts; oil hardened.

VEW K630.
M-1802; 0.85 C, 0.7 Ni, V, bal Fe.
Cold work tool steel.
For cold heading tools.

VEW K700.
M-1802; 1.2 C, 12.5 Mn, bal Fe.
Austenitic manganese steel.
For wear and abrasion resisting parts.
W.-Nr. 1.3401.

VEW K701.
M-1802; 1.2 C, 12.5 Mn, bal Fe.
Austenitic manganese steel.
For wear and abrasion resisting parts.
W.-Nr. 1.3401.

VEW K707.
M-1802; 1.2 C, 12.5 Mn, bal Fe.
Manganese abrasion resisting steel casting, for rails, crushers, shovel teeth, drag lines.
W.-Nr. 1.3401.

VEW K708 similar to **K707**; **VEW K710** similar to **K707**.

VEW K720.
M-1802; 0.90 C, 2.0 Mn, 0.1 V, bal Fe.
Cold work tool steel; oil hardening.
For punches, trimming dies, small shears.
W.-Nr. 1.2842; similar to AISI 02.

VEW K722.
M-1802; 0.60 C, 0.90 S, 1.0 Mn, bal Fe.
Cold work tool steel; oil hardening.
Stamping or heading dies, rivet sets.
W.-Nr. 1.2826.

VEW K724.
M-1802; 0.50 C, 1.8 Mn, bal Fe.
Oil or water hardenable.
W.Nr. 1.0913.

VEW K760.
M-1802; 1.0 C, 0.25 Mn, 0.20 Si, 0.10 V, bal Fe.
Cold work tool steel; water hardening.
For drills, reamers, punches.
W.-Nr. 1.2833; AISI W1.

VEW K765.
M-1802; 1.45 C, 3.25 V, bal Fe.
Cold work tool steel; water hardening.
For deep drawing tools and dies, cold heading dies.
W.-Nr. 1.2838.

VEW K935.
M-1802; 0.35 C, 0.65 Mn, bal Fe.
Carbon tool steel; water hardening.
For screwdrivers, wrenches.

VEW K945.
M-1802; 0.45 C, 0.65 Mn, bal Fe.
Carbon tool steel; water hardening.
For hand tools; axes, hammers, screwdrivers.
W.-Nr. 1.1730.

VEW K950.
M-1802; 0.53 C, 0.38 Si, 0.55 Mn, bal Fe.
Cold work tool steel; water hardening.

VEW K960.
M-1802; 0.60 C, 0.38 Si, 0.65 Mn, bal Fe.
Cold work tool steel; water hardening.
For tool shanks, punches, dies.
W.-Nr. 1.1740.

VEW K970.
M-1802; 0.70 C, 0.38 Si, 0.70 Mn, bal Fe.
Cold work tool steel; water hardening.
For hand saws, knives.
W.-Nr. 1.1744; AISI W1.

VEW K980.
M-1802; 0.85 C, 0.20 Si, 0.20 Mn, bal Fe.
Hot or cold work tool steel; water hardening.
Cold: reamers, taps, drills.
Hot: forming or punching dies.
W.-Nr. 1.1625; AISI W1.

VEW K985.
M-1802; 0.90 C, 0.3-0.5 Si, 0.3-0.8 Mn, bal Fe.
Carbon tool steel; water hardening.
For taps, reamers, drills, woodworking tools.
W.-Nr. 1.1830; AISI W1.

VEW K990.
M-1802; 1.0 C, bal Fe.
Carbon tool steel; water hardening.
For drills, reamers, taps, punches.
W.-Nr. 1.1545 or 1.1645; AISI W1.

VEW K991.
M-1802; 1.05 C, bal Fe.
Carbon tool steel; water hardening.
For drills, reamers, taps, punches.
AISI W1.

VEW K995.
M-1802; 1.3 C, bal Fe.
Cold work tool steel; water hardening.
For files.
W.-Nr. 1.1663.

VEW K996.
M-1802; 1.25 C, 0.35 Mn, 0.35 Cr, bal Fe.
Cold work tool steel; water hardening.
For drawing dies, mandrels, reamers, punches.
W.-Nr. 1.2002.

VEW L125.
M-1802; 0.45 max C, 20 Cr, 20 Ni, 4 Mo, 4 W, 4.5 max Nb, 5.0 max Fe bal Co.
High temperature alloy; for components of gas turbines.
W.-Nr. 2.4989.

VEW L208.
M-1802; 2.4 C, 31 Cr, 18 W, 45 Co, bal Fe.
For hard facing electrodes; heat and abrasion resistant.

VEW L216.
M-1802; 2.0 C, 32 Cr, 14 W, 50 Co.
For corrosion resistant castings.

VEW L219.
M-1802; 1.2 C, 25 Cr, 4 W, 65 Co, bal Fe.
Hard facing electrodes; for dies, exhaust valves.

VEW L300.
M-1802; 0.08-0.15 C, 20 Cr, 0.40 Ti, 0.50 max Cu, 5.0 max Fe, 1.0 max Mn, bal Ni.
High temperature alloy.
W.Nr. 1.4951.

VEW L312.
M-1802; 0.10 max C, 27 Cr, 4.5 Ni, 1.6 Mo, bal Fe.
Good strength and oxidation resistance at elevated temperatures.
W.Nr. 1.4460.

VEW L314.
M-1802; 0.05 max C, 21.5 Cr, 42 Ni, 3.0 Mo, 0.80 Ti, 2.5 Cu, bal Fe.
For high temperature operations.
W.Nr. 1.4858.

VEW L318.
M-1802; 60 Ni, 17 Mo, 16 Cr, bal Fe.
For boilers, tanks, chemical plant equipment; heat and corrosion resistant.
W.Nr. 2.4811.

VEW M100.
M-1802; 0.20 C, 1.25 Mn, 1.15 Cr, bal Fe.
To be carburized and hardened for plastic molding dies, or structural purposes.
Good core strength.
W.-Nr. 1.2162.

VEW M112.
M-1802; 0.20 C, 1.0 Mn, 1.2 Cr, Mo, bal Fe.
To be carburized and hardened for plastic molding dies, or structural purposes.
Good core strength.
W.-Nr. 1.2160.

VEW M120.
M-1802; 0.15 C, 0.75 Cr, 3.4 Ni, bal Fe.
To be carburized and hardened for plastic mold dies, or structural purposes.
W.-Nr. 1.2735.

VEW M130.
M-1802; 0.20 C, 1.2 Cr, 4.4 Ni, Mo or W, bal Fe.
To be carburized and hardened for plastic mold dies, or for structural purposes.
Good core strength.
W.-Nr. 1.2764.

VEW M150.
M-1802; 0.05 C, 4.0 Cr, 0.5 Mo, bal Fe.
To be carburized and hardened for plastic mold dies, or for structural purposes.
Designed to be hobbed before heat treat.
W. Nr. 1.2341.

VEW M152.
M-1802; 0.06 C, 4.0 Cr, 0.5 Mo, bal Fe.
To be hobbed, carburized and hardened for plastic mold dies or die casting dies.
W.-Nr. 1.2341.

VEW M200.
M-1802; 0.40 C, 0.40 Si, 1.50 Mn, 1.9 Cr, 0.20 Mo, bal Fe.
Plastic mold or die-casting die steel; oil hardenable.
W.-Nr. 1.2312.

VEW M210.
M-1802; 0.40 C, 0.40 Si, 1.5 Mn, 1.9 Cr, 0.2 Mo, bal Fe.
Plastic mold or die casting die steel; oil hardenable.
W.-Nr. 1.2312; similar to AISI P20.

VEW M252.
M-1802; 0.35 C, 0.35 Si, 0.6 Mn, 1.3 Cr, 4.5 Ni, bal Fe.
Plastic mold or die casting die steel; oil hardening.

VEW M300.
M-1802; 0.37 C, 16.5 Cr, 1.15 Mo, bal Fe.
Stainless plastic mold or die cast die steel.
Air or oil hardening.
W.-Nr. 1.2316.

VEW M310.
M-1802; 0.40 C, 0.40 Mn, 13.0 Cr, bal Fe.
Stainless plastic mold or die casting die steel; air or oil hardening.
W. Nr. 1.2083; similar to AISI 420.

VEW M312.
M-1802; 0.40 C, 13.0 Cr, Ni, Mo, Mn, bal Fe.
Stainless plastic mold or die casting die steel; air or oil hardening.
W. Nr. 1.2083; similar to AISI 420.

VEW N100.
M-1802; 0.08-0.12 C, 12-14 Cr, bal Fe.
Corrosion resistant steel.
For water or steam turbine blades.
W. Nr. 1.4006; similar to AISI 410.

VEW N104.
M-1802; 0.08 C, 12-14 Cr, bal Fe.
Corrosion resistant steel.
For structural parts in water and steam.
W.Nr. 1.4000.

VEW N106.
M-1802; 0.08 max C, 13-15 Cr, bal Fe.
Corrosion resisting steel.
For tableware, building fittings.
W. Nr. 1.4001.

VEW N108.
M-1802; 0.06 C, 13 Cr, 0.2 Al, bal Fe.
Ferritic type corrosion resisting steel.
Weldable,-for annealing boxes, furnace parts, tableware.
W.Nr. 1.4002; AISI 405.

VEW N200.
M-1802; 0.10 max C, 17 Cr, bal Fe.
Ferritic type corrosion resisting steel.
For tableware, building fittings.
W.Nr. 1.4016; AISI 430.

VEW N205.
M-1802; 0.10 max C, 17.5 Cr, Ti = 7xC, bal Fe.
Stabilized ferritic corrosion resisting steel.
For welded oil refinery and dairy equipment.
W.Nr. 1.4510; modified AISI 430.

VEW N238.
M-1802; 0.10 max C, 17.5 Cr, 1.75 Mo, 1.0 max Ni, Ti = 7xC, bal Fe.
Ferritic type stainless, weldable.
For corrosion resistant equipment.
W. Nr. 1.4523.

VEW N242.
M-1802; 0.10 max C, 18.5 Cr, 2.0 Mo, Ti = 7xC, bal Fe.
For welded assemblies with good corrosion and heat resisting properties.
W.Nr. 1.4523.

VEW N310.
M-1802; 0.14 C, 17 Cr, 0.25 Mo, 0.20 S, bal Fe.
Free machining, ferritic type stainless.
For stainless bolts, and fasteners.
W.Nr. 1.4104; similar to AISI 430 F.

VEW N315.
M-1802; 0.15 C, 13 Cr, bal Fe.
Martensitic stainless steel for turbine blades, cutlery, tableware.
W.Nr. 1.4024; AISI 410-420.

VEW N316.
M-1802; 0.15 max C, 12-13 Cr, 0.20 S, bal Fe.
Martensitic stainless for hardenable threaded parts.
W.Nr. 1.4005; AISI 416.

VEW N320.
M-1802; 0.20 C, 13 Cr, bal Fe.
Martensitic stainless for cutlery, surgical instruments, dental tools.
W.Nr. 1.4021; AISI 420.

VEW N324.
M-1802; 0.20 C, 0.4 Si, 13 Cr, bal Fe.
Martensitic stainless for valves, cutlery, surgical and dental tools.
AISI 420.

VEW N330.
M-1802; 0.20 C, 13 Cr, 1.15 Mo, bal Fe.
Martensitic stainless steel.
For turbine blades, valve cones.
W. Nr. 1.4120.

VEW N335.
M-1802; 0.35 C, 16.5 Cr, 1.15 Mo, bal Fe.
Martensitic stainless steel.
For cutlery, high temperature valves and fittings, arbors, spindles, bolts.
W. Nr. 1.4122.

VEW N350.
M-1802; 0.20 C, 17 Cr, 2 Ni, bal Fe.
Martensitic stainless steel.
For marine hardware.
W.Nr. 1.4057; similar to AISI 431.

VEW N350G.
M-1802; 0.23 C, 17 Cr, 1.5 Ni, bal Fe.
Corrosion resistant steel castings; for valve guides, springfaces, spindles.
W.Nr. 1.4059.

VEW N351.
M-1802; 0.20 C, 18 Cr, 2 Ni, bal Fe.
Stainless and heat resisting steel.
For marine equipment.
W.Nr. 1.4057; similar to AISI 431.

VEW N352.
M-1802; 0.20 C, 1.0 Mn, 1.0 Si, 16.0 Cr, 2.0 Ni, bal Fe.
Martensitic stainless steel; good hardenability.
AISI 431.

VEW N358 similar to **VEW N359**.

VEW N359.
M-1802; 0.10 C, 13 Cr, 1.0 Ni, bal Fe.
Corrosion resistant steel casting; for pump parts, valves, rotors.
W.Nr. 1.4008; ACI CA-15.

VEW N400.
M-1802; 0.07 max C, 12.5 Cr, 4.25 Ni, bal Fe.
Corrosion resistant casting.
For hydraulic turbine equipment.
W.Nr. 1.4313.

VEW N530.
M-1802; 0.25-0.30 C, 13 Cr, bal Fe.
Martensitic stainless steel.
For valves, springs, cutlery, surgical and dental instruments.
W. Nr. 1.4028; AISI 420.

VEW N540.
M-1802; 0.40 C, 13 Cr, bal Fe.
Hardenable martensitic stainless.
For cutlery, springs, surgical instruments.
W. Nr. 1.4034; similar to AISI 420.

VEW N555.
M-1802; 0.55 C, 14 Cr, 0.55 Mo, bal Fe.
Hardenable martensitic stainless steel.
For cutlery, shears, cutting and forming tools.
W. Nr. 1.4110.

VEW N685.
M-1802; 0.80-0.90 C, 16-18 Cr, 1.0 Mo, 0.1 V, bal Fe.
Hardenable martensitic stainless steel.
For ball or roller bearings, cutlery, shears, surgical equipment.
W. Nr. 1.4112; similar to AISI 440B.

VEW N688.
M-1802; 1.0 C, 16 Cr, 0.8 Mo, 2 Co, bal Fe.
Hardenable martensitic stainless steel.
For ball or roller bearings, cutlery, shears, surgical and dental equipment.
W.Nr. 1.4535; similar to AISI 440C.

VEW N690.
M-1802; 1.05 C, 17.5 Cr, 1.25 Mo, 1.5 Co, 0.1 V, bal Fe.
High carbon martensitic stainless.
For ball or roller bearings, valves, pump parts.
W.Nr. 4528; similar to AISI 440C.

VEW N691.
M-1802; 1.10 C, 2.0 max Si, 1.0 max Mn, 28 Cr, 2.25 Mo, bal Fe.
Stainless steel casting; for paper and photo industries.
W. Nr. 1.4138.

VEW N692.
M-1802; 0.90 C, 16.5 Cr, 0.5 Mo, 0.25 V, 1.4 Co, bal Fe.
Martensitic stainless steel.
For cutlery, knife blades.
W.Nr. 1.4535; similar to AISI 440B.

VEW N693.
M-1802; 1.10 C, 2.0 max Si, 1.0 max Mn, 28 Cr, bal Fe.
Corrosion resistant casting; good resistance to heat and abrasion.
W. Nr. 1.4086.

VEW N695.
M-1802; 1.05 C, 17 Cr, 0.5 Mo, bal Fe.
High carbon martensitic stainless steel.
For ball or roller bearings, nozzles, valve seats.
W.Nr. 1.4125; AISI 440C.

VEW N700.
M-1802; 0.07 max C, 16.5 Cr, 4 Ni, 4 Cu, 0.30 Nb, bal Fe.
Precipitation hardenable stainless steel.
W.Nr. 1.4542; similar to 17-4 PH.

VEW N702.
M-1802; 0.07 C, 16.5 Cr, 4 Ni, 4 Cu, 0.3 Nb, bal Fe.
Corrosion resistant steel.
W.Nr. 1.4542; similar to 17-4 PH.

VEW P505.
M-1802; 0.05 max C, 23 Cr, 15 Ni, 1.5 Mo, 0.35 N, bal Fe.
Stainless steel.
W. Nr. 1.3951.

VEW P530.
M-1802; 0.08 max C, 18 Mn, 13 Cr, 2.5 Ni, 0.15 Ni, bal Fe.
Non-magnetizable stainless steel for welded construction.
W. Nr. 1.3949.

VEW P550.
M-1802; 0.50 C, 18 Mn, 5.0 Cr, bal Fe.
Non-magnetizable steel; for electrical equipment.
W. Nr. 1.3813.

VEW P600.
M-1802; 0.10 max C, 0.25 Si, 0.50 Mn, 9.0 Ni, bal Fe.
Steel tough at sub-zero temperatures.
For tanks and containers for cryogenic operations.
W. Nr. 1.5662.

VEW P602.
M-1802; 0.12 C, 5.0 Ni, bal Fe.
Oil hardenable steel for low temperature operations.

VEW P752.
M-1802; 0.05 max C, 26 Ni, 14 Al, bal Fe.
Cast permanent magnet.
W. Nr. 1.3728.

VEW P754.
M-1802; 0.05 C, 15 Co, 20 Ni, 9.5 Al, 3.5 Cu, bal Fe.
For cast permanent magnet.
W. Nr. 1.3743; ALNICO 160.

VEW P756.
M-1802; 0.05 C, 15 Co, 22 Ni, 10 Al, 3 Cu, bal Fe.
For cast permanent magnet.
W. Nr. 1.3745; ALNICO 190.

VEW P758.
M-1802; 0.05 C, 24 Co, 14 Ni, 8 Al, 3 Cu, Ti, bal Fe.
For permanent magnets.
W. Nr. 1.3761.

VEW P760.
M-1802; 0.05 C, 24 Co, 21 Ni, 8 Co, 3 Cu, bal Fe.
Permanent magnet.
W. Nr. 1.3760; ALNICO 400 or 500.

VEW P764.
M-1802; 0.05 C, 32 Co, 15 Ni, 7 Al, 4.5 Cu, 5 Ti, bal Fe.
Permanent magnet.
W. Nr. 1.3758; ALNICO 350.

VEW P804.
M-1802; 0.05 max C, 42 Ni, bal Fe.
Low thermal expansion; for instruments.
W. Nr. 1.3917.

VEW P906.
M-1802; 0.08 max C, 0.15 max Si, 0.50 max Mn, bal Fe.
Soft magnetic metal; for magnet cores, solenoids.
W. Nr. 1.1009.

VEW R100.
M-1802; 1.0 Cr, 0.35 Mn, 1.55 Cr, bal Fe.
Oil or water hardening steel for ball or roller bearings, bushings.
W. Nr. 1.3505; AISI 52100.

VEW R102.
M-1802.
Similar to VEW R100.

VEW R104.
M-1802.
Similar to VEW R100.

VEW R110.
M-1802; 1.0 C, 0.60 Si, 1.1 Mn, 1.55 Cr, bal Fe.
Oil or water hardening steel for ball or roller bearings, bushings: for larger sections than VEW R100.
W. Nr. 1.3520; similar to AISI 52100.

VEW S200.
M-1802; 0.74 C, 4.1 Cr, 18.0 W, 1.1 V, bal Fe.
High speed steel; for lathe and planer tools.
W. Nr. 1.3355; S18-0-1; AISI T1.

VEW S201.
M-1802; 0.80 C, 4.0 Cr, 18 W, 1.7 V, Mo, bal Fe.
For lathe tools, drills, reamers, broaches.
W. Nr. 1.3357; AISI T2.

VEW S203.
M-1802; 0.86 C, 4.1 C, 12 W, 2.5 V, 0.85 V, 0.85 Mo, bal Fe.
For lathe and planer tools, broaches, taps, milling cutters.

VEW S205.
M-1802; 0.82 C, 4.1 Cr, 9 W, 1.6 V, 0.85 Mo, bal Fe.
High speed steel; for drills, milling cutters, taps, lathe tools.

VEW S300.
M-1802; 0.76 C, 4.2 Cr, 18 W, 1.8 V, 0.8 Mo, 10 Co, bal Fe.
Cobalt-tungsten high speed steel for lathe tools, milling cutters, threading tools.
W. Nr. 1.3256; S18-1-2-10; similar to AISI T5.

VEW S302.
M-1802; 0.60 C, 4 Cr, 18 W, 1.3 V, 128 Co, bal Fe.
Cobalt-tungsten high speed steel.
For lathe and planer tools, milling cutters, hot shears.
Good red hardness.

VEW S305.
M-1802; 0.80 C, 4.3 Cr, 18 W, 1.6 V, 0.75 Mo, 4.7 Co, bal Fe.
For lathe and planer tools, milling cutters, threading tools.
Cobalt-tungsten high speed steel; good red hardness.
AISI T4; W. Nr. 1.3255.

VEW S307.
M-1802; 1.5 C, 5 Cr, 12 W, 5 V, 5 Co, bal Fe.
High carbon high speed steel.
For special lathe and threading tools.
Good wear resistance.
Similar to AISI T15.

VEW S308.
M-1802; 1.35 C, 4.3 Cr, 12 W, 4 V, 5 Co, bal Fe.
High carbon high speed steel.
For lathe tools, form tools, threading tools.
Good wear resistance; good red hardness.
W. Nr. 1.3202.

VEW S400.
M-1802; 1.0 C, 4.0 Cr, 9.0 Mo, 2 V, 2 W, bal Fe.
High speed steel; for milling cutters.
W.Nr. 1.3348; AISI M7.

VEW S401.
M-1802; 0.80 C, 4.0 Cr, 2 W, 1 V, 9 Mo, bal Fe
Molybdenum high speed steel.
For drills, reamers, form cutters, lathe tools.
1.3346; AISI M1.

VEW S500.
M-1802; 1.05 C, 3.7 Cr, 1.5 W, 9.5 Mo, 1.2 V, 8 Co, bal Fe.
Co-Mo high speed steel.
For twist drills, reamers, threading cutters, lathe tools, milling cutters.
For difficult machining; good red hardness.
W.-Nr. 1.3247; AISI M42.

VEW S600.
M-1802; 0.85 C, 4.1 Cr, 6.0 W, 5.0 Mo, 2 V, bal Fe.
Mo-W high speed steel.
For drills, reamers, milling cutters, lathe tools.
W.Nr. 1.3343; AISI M2.

VEW S601, S602 similar to **VEW S600.**

VEW S604.
M-1802; 1.0 C, 4 Cr, 6.3 W, 1.8 V, 5 Mo, bal Fe.
For twist drills, taps, broaches, lathe tools, milling cutters.
W.Nr. 1.3342; AISI M2.

VEW S607.
M-1802; 1.2 C, 4 Cr, 6 W, 3 V, 5 Mo, bal Fe.
High speed steel.
For lathe tools, thread cutters, milling cutters, broaches.
W.Nr. 1.3344; AISI M3 Class 2.

VEW S609 similar to VEW S610.

VEW S610.
M-1802; 0.95 C, 4.3 Cr, 2.8 W, 2.4 V, 2.8 Mo, bal Fe.
For twist drills, milling cutters, broaches.
W.Nr. 1.3333.

VEW S700.
M-1802; 1.3 C, 4 Cr, 10 W, 3.5 V, 3.5 Mo, 10.5 Co, bal Fe.
High speed steel.
For lathe tools, threading cutters.
W.Nr 1.3207.

VEW S705.
M-1802; 0.80 C, 4 Cr, 6.5 W, 5.0 Mo, 2 V, 5 Co, bal Fe.
High speed steel.
For lathe and planer tools, milling cutters.
W.Nr. 1.3243; AISI M41.

VEW T200.
M-1802; 0.08 max C, 1.0-2.0 Mn, 15.0 Cr, 1.25 Mo, 25.0 Ni, 2.0 Ti, 0.006 B, Al, V, bal Fe.
Austenitic temperature resisting alloy.
For parts for gas turbine engines.
W.Nr. 1.4980; AISI 660.

VEW T240.
M-1802; 0.15 max C, 16.0 Cr, 13.5 Ni, 2.75 W, 0.50 Ti, bal Fe.
Austenitic temperature resisting alloy.
For parts for steam and gas turbine engines.
W.Nr. 1.4962.

VEW T245.
M-1802; 0.10 max C, 16.5 Cr, 16.5 Ni, 3.0 W, Nb/Ta=10xC, bal Fe.
Austenitic temperature resisting alloy.
Blades for steam turbine, turbo blades, rotor wheels.
W.Nr. 1.4945.

VEW T250.
M-1802; 0.10 max C, 16.5 Cr, 13.5 Ni, 1.3 Mo, Nb/Ta=10xC, bal Fe.
Austenitic temperature resisting alloy.
Parts for steam turbine plants.
W.Nr. 1.4988.

VEW T255.
M-1802; 0.10 max C, 16.0 Cr, 16.0 Ni, 2.0 Mo, Nb/Ta=10xC, bal Fe.
Austenitic temperature resisting alloy.
Parts for steam and gas turbine engines.
W.Nr. 1.4981.

VEW T270.
M-1802; 0.08 max C, 2.0 max Mn, 18.0 Cr, 11.0 Ni, bal Fe.
Austenitic temperature resisting alloy.
For elevated temperature pipe lines.
W.-Nr. 1.4948.

VEW T275.
M-1802; 0.10 max C, 16.0 Cr, 13.0 Ni, Nb/Ta=10xC, bal Fe.
Austenitic temperature resisting alloy.
Parts for steam and gas turbines.
W. Nr. 1.4961.

VEW T502.
M-1802; 0.20 C, 12 Cr, 1.0 Mo, 0.5 Ni, 0.5 W, 0.3 V, bal Fe.
Heat resisting alloy; hardenable; for elevated temperature hardware.
W. Nr. 1.4935; similar to AISI 616.

VEW T550.
M-1802; 0.20 C, 12.0 Cr, 1.0 Mo, 0.5 Ni, 0.3 V, bal Fe.
Martensitic temperature resisting alloy.
For engine components; heat resistant to 550°C.
W. Nr. 1.4922.

VEW T552.
M-1802; 0.12 C, 12 Cr, 1.8 Mo, 2.5 Ni, 0.3 V, 0.3 N, bal Fe.
Heat resisting alloy; hardenable; for elevated temperature hardware.
W.Nr. 1.4939.

VEW T558.
M-1802; 0.20 C, 11 Cr, 1.0 Mo, 0.5 Ni, 0.3 V, bal Fe.
Corrosion resistance at elevated temperature; hardenable; hot piping gas or liquids.
W. Nr. 1.4922.

VEW T560.
M-1802; 0.20 C, 10.5 Cr, 0.8 Mo, 0.5 Ni, 0.2 V, 0.35 Nb, B, bal Fe.
Corrosion resistance at elevated temperature; hardenable; weldable.
W. Nr. 1.4913.

VEW T602.
M-1802; 0.20 C, 12 Cr, 1.0 Mo, 0.8 max Ni, bal Fe.
Temperature resisting alloy; hardenable. Parts for thermal power plants.
W. Nr. 1.4921.

VEW-V110.
M-1802; 0.32 C, 1.25 Cr, 0.45 Mo, 3.4 Ni, bal Fe.
Alloy steel for heavy automotive equipment; deep hardening; oil hardening.
W. Nr. 1.6746.

VEW V130.
 M-1802; 0.40 C, 1.1 C, 1.5 Ni, 0.3 Mo, V, bal Fe.
 Alloy structural steel; for automotive forgings and couplings; oil hardening.
 W.Nr. 1.6565; similar to AISI 4340.

VEW V145.
 M-1802; 0.30 C, 2.0 Ni, 2.0 Cr, 0.3 Mo, bal Fe.
 Alloy steel for automotive shafts, crankshafts, gears; oil hardening.
 W.Nr. 1.6580.

VEW V155.
 M-1802; 0.34 C, 1.5 Ni, 1.5 Cr, 0.2 Mo, bal Fe.
 Alloy steel for automotive parts as shafts, connecting rods, gears; oil hardening.
 W.Nr. 1.6582.

VEW-V157.
 M-1802; 0.40 C, 1.0 Cr, 2.0 Ni, 0.25 Mo, bal Fe.
 Alloy steel for automotive parts as axles, gears, crankshafts, spline couplings.
 Oil hardening; similar to AISI 4340.

VEW V165.
 M-1802; 0.36 C, 1.0 Ni, 1.0 Cr, 0.2 Mo, bal Fe.
 Alloy steel for automotive parts as shafts, axles, spline couplings; oil hardening.
 W.Nr. 1.6511.

VEW V174.
 M-1802; 0.36 C, 0.85 Ni, 0.60 C, 0.15 Mo, bal Fe.
 For structural parts; oil hardening.
 W.Nr. 1.6506; similar to AISI 8637.

VEW V204.
 M-1802; 0.35 C, 0.6 Mn, 1.3 Cr, 4.5 Ni, bal Fe.
 Alloy structural steel; deep hardening; for heavy shafts on earth moving equipment.
 W.Nr. 1.5864.

VEW V214.
 M-1802; 0.26 C, 0.70 Cr, 3.5 Ni, bal Fe.
 Alloy structural steel; deep hardening, for diesel crankshafts, drive couplings.
 W.Nr. 1.5755.

VEW V228.
 M-1802; 0.40 C, 0.8 Mn, 0.65 Cr, 1.25 Ni, bal Fe.
 Alloy structural steel, for automotive and machine parts; oil hardening.
 W.Nr. 1.5711; similar to old SAE 3140.

VEW V304.
 M-1802; 0.31 C, 2.35 Cr, 0.18 Mo, 0.15 V, bal Fe.
 Alloy steel, deep hardening; - for shafts, axles, large machine parts; may also be nitrided.
 W.Nr. 1.8515.

VEW V310.
 M-1802; 0.50 C, 1.0 Cr, 0.20 Mo, V, bal Fe.
 Alloy steel for shafts, arbors, bushings, axles; oil hardening.
 W.Nr.. 1.7228; AISI 4150.

VEW V320.
 M-1802; 0.42 C, 1.0 C, 0.20 Mo, bal Fe.
 Alloy steel for cog wheels, connecting rods, spline couplings.
 W.Nr. 1.7223; AISI 4142.

VEW V322 similar to **VEW V320.**

VEW V330.
 M-1802; 0.33 C, 1.0 Cr, 0.2 Mo, bal Fe.
 Alloy steel for bolts, shafts, arbors; weldable; oil hardening.
 W.Nr. 1.7220; AISI 4130, 4135.

VEW V340.
 M-1802; 0.25 C, 0.65 Mn, 1.0 Cr, 0.20 Mo, bal Fe.
 For shafts, bolts, lever arms.
 W.Nr. 1.7218; similar to AISI 4130.

VEW V340G.
 M-1802; 0.25 C, 0.65 Mn, 1.0 Cr, 0.20 Mo, bal Fe.
 Alloy steel casting.
 W.Nr. 1.7218.

VEW V350.
 M-1802; 0.30 C, 0.55 Mn, 2.50 Cr, 0.2 Mo, 0.15 V, bal Fe.
 For bolts, crankshafts, die casting dies.
 W.Nr. 1.7707.

VEW V500.
 M-1802; 0.41 C, 0.65 Mn, 1.0 Cr, bal Fe.
 For bolts, shafts, machine tool parts; oil hardening.
 W.Nr. 1.7035; similar to AISI 5140.

VEW V510.
 M-1802; 0.33 C, 0.70 Mn, 1.0 Cr, bal Fe.
 For bolts, shafts, machine parts; oil or water hardening.
 W.Nr. 1.7033; AISI 5132.

VEW V520.
 M-1802; 0.46 C, 0.65 Mn, 0.50 Cr, bal Fe.
 For shafts, bolts, machine parts; oil or water hardening.
 W.Nr. 1.7006.

VEW V560.
 M-1802; 0.60 C, 0.85 Mn, 1.0 Cr, bal Fe.
 Oil hardening steel for shafts.
 W.Nr. 1.8161; similar to AISI 5160.

VEW V622.
 M-1802; 0.15 C, 8.5 Ni, bal Fe.
 Corrosion resisting steel; for rifle barrels.
 W.Nr. 1.5662.

VEW V734.
M-1802; 0.15 C, 2.0 Mn, bal Fe.
Water or oil hardening steel; may be carburized before hardening.
W.Nr. 1.5074.

VEW V742.
M-1802; 0.42 C, 0.25 Si, 1.75 Mn, bal Fe.
For axles, chain wheels, bolts, shafts.
W.Nr. 1.5223; AISI 1340.

VEW V762.
M-1802; 0.37 C, 1.25 Si, 1.25 Mn, bal Fe.
For crankshafts, axles, shock resisting tools and parts; oil or water hardening.
W.Nr. 1.5122.

VEW V800.
M-1802; 0.38 C, 1.1 Cr, 1.1 Al, 0.25 Mo, bal Fe.
Nitriding steel.
W.Nr. 1.8509.

VEW V810.
M-1802; 0.32 C, 1.1 Cr, 1.1 Al, 0.20 Mo, bal Fe.
Nitriding steel.
W.Nr. 1.8507.

VEW V820.
M-1802; 0.35 C, 0.55 Mn, 1.8 Cr, 0.2 Mo, 1.1 Al, 1.0 Ni, bal Fe.
Nitriding steel for large cross section parts.
W.Nr. 1.8550.

VEW V918.
M-1802; 0.25 max C, 0.60 max Si, 0.40 Mn, bal Fe.
Carbon steel; water hardening; for small fasteners.
W.Nr. 1.0443.

VEW V920.
M-1802; 0.22 C, 0.25 Si, 0.45 Mn, bal Fe.
For small fasteners; water hardening.
W.Nr 1.1151; AISI 1023.

VEW V922.
M-1802; 0.20 C, 0.45 Si, 1.45 Mn, bal Fe.
For high pressure vessels and tubes for low temperature operation, down to -100°C.
W.Nr. 1.1169.

VEW V923.
M-1802; 0.30 C, 0.40 Si, 0.40 Mn, bal Fe.
Carbon steel; water hardening; for small fasteners.
W.Nr 1.0551.

VEW V924 similar to **V923.**

VEW V930.
M-1802; 0.30 C, 0.25 Si, 1.35 Mn, bal Fe.
For shafts, bolts, fasteners; water or oil hardening.
W.Nr. 1.1165.

VEW V935.
M-1802; 0.35 C, 0.25 Si, 0.65 Mn, bal Fe.
For bolts, shafts, fasteners; water hardening.
W.Nr. 1.1181; AISI 1035.

VEW V936.
M-1802; 0.40 C, 0.40 Si, 0.40 Mn, bal Fe.
Carbon steel; water hardening; for small fasteners.
W.Nr. 1.0553.

VEW V937 similar to **V936.**

VEW V940.
M-1802; 0.40 C, 0.37 Si, 0.95 Mn, bal Fe.
For bolts, shafts, machine parts; water hardening.
W.Nr. 1.1157.

VEW V943.
M-1802; 0.40 C, 0.65 Mn, bal Fe.
Carbon steel; for shafts, axles, bolts.
W.Nr. 1.1186; AISI 1040.

VEW V945.
M-1802; 0.45 C, 0.25 Si, 0.65 Mn, bal Fe.
For bolts, shafts, gears; water hardening.
W.Nr. 1.1191; AISI 1042.

VEW V946.
M-1802; 0.45 C, 0.35 Si, 0.68 Mn, bal Fe.
Designed particularly for shafts, axles, to be flame or induction hardened on the surface or localized area.
W.Nr. 1.1193; AISI 1045.

VEW V953.
M-1802; 0.56 C, 0.30 Si, 0.55 Mn, bal Fe.
For gears, shafts, worms, camshafts to be induction or flame hardened.
W.Nr. 1.1213; similar to AISI 1055.

VEW V955.
M-1802; 0.45 C, 0.25 Si, 0.65 Mn, bal Fe.
For shafts, connectors, automotive parts; water hardening.
W.Nr. 1.1203; AISI 1045.

VEW V960.
M-1802; 0.61 C, 0.25 Si, 0.65 Mn, bal Fe.
For axles, shafts, pins; water hardening.
W.Nr. 1.1221; AISI 1060.

VEW V969.
M-1802; 0.75 C, 0.25 Si, 0.55 Mn, bal Fe.
For shafts, pins, hand tools, springs.
W.Nr. 1.1231; AISI 1078.

VEW W100.
M-1802; 0.32 C, 8.7 W, 2.7 Cr, 0.35 V, bal Fe.
Hot work tool steel; oil hardening.
For pressure casting molds, hot extrusion dies.
W.-Nr. 1.2581; similar to AISI H21.

VEW-W103.
M-1802; 0.30 C, 8.5 W, 2.5 Cr, 1.8 Ni, bal Fe.
Hot work tool steel; oil hardening.
For pressure casting molds, hot extrusion dies.
W.-Nr. 1.2759.

VEW W105.
M-1802; 0.30 C, 2.35 Cr, 4.25 W, 0.6 V, bal Fe.
Hot work tool steel for pressure casting molds, cores, dies for non-ferrous metals.
W.-Nr. 1.2567.

VEW W106.
M-1802; 0.30 C, 1.1 Cr, 3.75 W, 0.18 V, bal Fe.
Hot work tool steel; oil hardening.
For extrusion dies, rams, pressing mandrels for non-ferrous metals.

VEW W108.
M-1802; 0.45 C, 4.5 Cr, 4.5 W, 4.5 Co, 0.50 Mo, 2.0 V, bal Fe.
Hot work tool steel; oil hardening.
For hot extruding dies, mandrels, pressure casting molds for brass.
W.R. 1.2678; AISI H19.

VEW W300.
M-1802; 0.35 C, 5.0 Cr, 1.3 Mo, 0.5 V, 1.0 Si, bal Fe.
Hot work tool steel; oil hardening.
Pressure casting molds for light metal, extrusion press tools, forging dies.
W.Nr. 1.2343; AISI H11.

VEW W301.
M-1802; 0.38 C, 5.0 Cr, 1.3 Mo, 0.5 V, bal Fe.
Hot work die steel; oil hardening.
Pressure casting molds, forging dies.
W.Nr. 1.7783; similar to AISI H11.

VEW W302.
M-1802; 0.38 C, 5.2 Cr, 1.3 Mo, 1.0 V, bal Fe.
Hot work tool steel; oil hardening.
For extrusion press rams and liners, die casting dies, forging dies.
W.-Nr. 1.2344; similar to AISI H13.

VEW W304.
M-1802; 0.35 C, 5.0 Cr, 1.4 W, 1.4 Mo, 0.4 V, bal Fe.
Hot work tool steel; oil hardening.
For extrusion rams and liners, die casting dies, forging dies.
W.Nr. 1.2606; similar to AISI H12.

VEW W320.
M-1802; 0.32 C, 3.0 Cr, 2.8 Mo, 0.5 V, bal Fe.
Hot work tool steel; oil hardening.
For die casting dies, heading dies, rams, molds.
W.-Nr. 1.2365.

VEW W321.
M-1802; 0.45 C, 4.5 Cr, 3.0 Mo, 2.0 V, 4.5 Co, bal Fe.
Hot work tool steel; oil hardening.
For die casting dies, extrusion dies, forging dies.
W.-Nr. 1.2889.

VEW W322.
M-1802; 0.30 C, 1.9 Cr, 2.6 Ni, 1.0 W, 2.6 Mo, bal Fe.
Hot work tool steel; oil hardening.
For die casting dies, hot forming dies.

VEW W326.
M-1802; 0.45 C, 1.4 Cr, 0.70 Mo, 0.30 V, bal Fe.
Hot work tool steel, oil hardening.
For pressing punches, heading dies, shears.
W.Nr. 1.2323.

VEW W327.
M-1802; 0.45 C, 1.4 Cr, 0.8 V, 0.5 Mo, 0.5 W, bal Fe.
Hot work tool steel; for light upsetting tools, shear knives, pressing punches.
W.Nr. 1.2603.

VEW W329.
M-1802; 0.20 C, 2.4 Cr, 0.35 Mo, bal Fe.
Hot work tool steel; oil hardening.
For pressure casting; may be case-hardened.
W.Nr. 1.2313.

VEW W500.
M-1802; 0.56 C, 1.0 Cr, 1.7 Ni, 0.5 Mo, 0.1 V, bal Fe.
Hot work tool steel; oil hardening.
For forging and upsetting dies.
W.-Nr. 1.2714.

VEW W501.
M-1802; 0.55 C, 0.7 Cr, 1.65 Ni, 0.18 Mo, 0.1 V, bal Fe.
Hot work tool steel; oil hardening.
Forging dies, extrusion dies.
W.-Nr. 1.2713.

VEW W502.
M-1802; 0.35 C, 1.35 Cr, 4.5 Ni, 0.4 Mo, bal Fe.
Hot work tool steel; oil hardening.
For pressing dies, roll rings.
W.-Nr. 1.2766.

VEW W600.
M-1802; 0.21 C, 3.3 Mo, 3.0 Ni, bal Fe.
Hot work tool steel.
W.Nr. 1.2777.

VEW W701.
M-1802; 0.50 C, 1.4 Si, 0.6 Mn, 4.0 Cr, 11.5 Ni, 0.7 Mo, 12.0 W, 1. Co, 1.0 V, bal Fe.
Hot work tool steel; air or oil hardening.
For extrusion presses.
W.-Nr. 1.2758.

VEW W705.
M-1802; 0.15 C, 10.0 Cr, 5.0 Mo, 10.0 Co, 0.5 V, bal Fe.
Hot work tool steel; may be case hardened.
W.Nr. 1.2886.

VEW W720.
M-1802; 0.03 max C, 17.5 Ni, 9.0 Co, 5.0 Mo, Al, Ti, bal Fe.
To be hubbed before heat treating.
For plastic mold or die cast dies.
W.Nr. 1.6358.

VEW W725.
M-1802; 0.03 max C, 18 Ni, 12.5 Co, 4 Mo, 1.8 Ti, bal Fe.
To be hubbed before heat treating.
For plastic mold or die cast dies.
W.Nr. 1.6356.

VEW Z904.
M-1802; 0.13 max C, 0.90 Mn, 0.10 P, 0.22 S, bal Fe.
Free-machining carbon steel for automatic screw machine operations.
W.Nr 1.0711; similar to AISI 1212.

VEW Z906.
M-1802; 0.14 max C, 1.1 Mn, 0.10 P, 0.24-0.32 S, bal Fe.
Free-machining carbon steel for automatic screw machine operations.
W.Nr. 1.0715; similar to AISI 1213.

VEW Z908.
M-1802; 0.15 max C, 1.25 Mn, 0.10 P, 0.36 S, bal Fe.
Free-machining carbon steel for automatic screw machine operations.
W.Nr. 1.0736; DIN 9 SM 36; similar to old SAE B1113.

VEW Z952.
M-1802; 0.15 max C, 1.25 Mn, 0.10 P, 0.36 S, 0.22 Pb, bal Fe.
Leaded, free-machining steel.
W.Nr. 1.0737; similar to AISI 12L14.

VEW Z980.
M-1802; 0.10 C, 0.70 Mn, 0.06 P, 0.20 S, bal Fe.
Free-machining steel, for case hardening.
W.Nr. 1.0721; similar to AISI 1212.

VEW Z982.
M-1802; 0.36 C, 0.70 Mn, 0.06 max P, 0.20 S, bal Fe.
Free machining, heat treatable steel.
W.Nr. 1.0726.

VEW Z984.
M-1802; 0.36 C, 0.70 Mn, 0.07 P, 0.20 S, 0.22 Pb, bal Fe.
Leaded, free-machining steel.
W.Nr. 1.0756.

VEW Z986.
M-1802; 0.46 C, 0.70 Mn, 0.06 max P, 0.20 S, bal Fe.
Free-machining, heat treatable steel.
W.Nr. 1.0727.

VEW Z988.
M-1802; 0.60 C, 0.70 Mn, 0.06 max P, 0.20 S, bal Fe.
Free-machining, heat treatable steel.
W.Nr. 1.0728.

VFC.
M-1733; 0.15 C, 0.25 Si, 0.75 Mn, bal Fe.
Low carbon steel; AISI 1015; DIN CK 15.

VIADUCT 15.
M-1112; 0.45 C, 1. Cr, 2 W, 0.3 V, 0.55 Si, 0.45 Mn, bal Fe.
Annealed: 220 Brin.
For pneumatic chisels, shear blades, cutters; tough, oil hardened.

VIAG.
M-935; 4.5-5.0 Cu, bal Al.
For light alloy parts; age-hardenable.

VIBRALLOY.
M-205, M-670; 9 Mo, 38-42 Ni, bal Fe.
For instruments, diaphragms, mechanical filters, vibrating reeds; constant modulus, high permeability.

VIBRALOY.
M-936; 0.9 C, 0.8 Cr, bal Fe.
For vibrating screens; abrasion resistant.

VIBRESIST.
M-435; 1.0 C, 1.25 V, 0.3 Mo, bal Fe.
For mining drills; oil hardened, hollow drill.

VIBRO.
M-140; 0.5 C, 1.4 Cr, 0.3 V, 1.9 W, bal Fe.
For tools, dies; hot work steel.

VICALLOY.
M-61; 10 V, 52 Co, bal Fe.
For hysteresis motors, magnetic clutches, spec. record tape, magnetic memory, magnets.
Saturation induction 10,000 gausses.
$H_c = 200$ oe, magnetically semihard.

VICALLOY.
M-115; 9.0 V, 38.1 Fe, 0.70 Mn, bal Co.
Ductile Co-Fe-V permanent magnet material.
9000 Br, 300 H_c, 1.0×10^6 BH max.
Designed for magnetic recording tapes.

VICALLOY 1.
M-108; Co-Fe-V.
Permanent magnet alloy for rotor assemblies, etc.

VICALLOY 2.
M-108; Co-Fe-V.
Permanent magnet alloy for rotor assemblies, etc.

VICALLOY I.
M-22; 10 V, 52 Co, 0.3 Mn, 1.9 W, 0.6 Si, bal Fe.
For permanent magnets, cold workable. 9000 Br; 300 H_c; 5500 Bo; 60 Rock C.

VICALLOY I.
M-670; 10 V, 52 Co, bal Fe.
Permanent magnet material.
7500 Br.: 250 H_c; 0.80 (BH) max (MGO).

VICALLOY VS30.
M-72; 30 Co, 15 Cr, bal Fe.
Coercive force-25, remanence 18,000.
For electrical and magnetic equipment.
Permanent magnet. High magnetic permeability.

VI CHROME.
M-694; 2.0 C, 0.3 Mn, 12-14 Cr, 0.2 V, bal Fe.
For dies; non-deforming, oil hardening.
AISI D4.

VI-CHROME "W".
M-694; 2.0 C, 13.0 Cr, 1.2 W, bal Fe.
Air or oil hardening cold work tool and die steel; chromium type; AISI D6.

VICTOR BRONZE.
M-U.S.; 39 Zn, 1.5 Al, 1 Fe, 0.03 V, bal Cu.
For pipes; corrosion resistant.

VICTOR DRILL ROD.
M-38; 1.05 C, bal Fe.
For shafts, rollers, pins, dowels, push rods; water hardened.

VICTORIA ALUMINUM.
M-Eng.; Cu, Zn, Si, Fe, bal Al.
For light alloy parts; v. "Partinium."

VICTOR METAL.
50 Cu, 35 Zn, 15 Ni.
For cast fittings; corrosion resisting.

VICTORY.
M-688; 0.8 C, 4 Cr, 2 V, 6 W, 5 Mo, bal Fe.
For drills, hobs, taps, broaches, reamers; Type M2; high speed steel.

VICTORY COBALT.
M-115; 0.85 C, 6 W, 5 Mo, 4 Cr, 2 V, 8.5 Co, bal Fe.
For cutting tools; high speed steel.

VICTRIX SPECIAL.
M-96; 0.15 C, 1 Cr, 3.65 Ni, 0.35 Mo, bal Fe.
For gears, shafts; case hardening.

VICULOY NO. 1.
M-1058; 2.0-2.2 Be, bal Cu.
Cast: 85,000-90,000 TS; 60,000-70,000 YS; 4-12 El; 9 RA; 200-220 Brin.
For gears, shafting welding dies and welding electrodes; corrosion resistant, age hardening.

VICULOY NO. 2.
M-1058; 1.6-1.8 Be, bal Cu.
Cast: 45,000-50,000 TS; 30,000-35,000 YS; 17-22 El; 40 RA; 120-140 Brin.
For circuit breakers, contacts, electrode jaws; corrosion resistant, age hardenable.

VICULOY NO. 3.
M-1058; 1.8-2.2 Be, bal Cu.
Heat treated: 160,000-170,000 TS; 120,000 YS; 2-8 El; 4 RA; 370-400 Brin.
For gears, die molds, rocker arms, non-sparking tools; corrosion resistant, age-hardenable.

VIENNESE ORNAMENTS.
M-Austria; 55 Cu, 25 Zn, 20 Ni.
For brass ornaments; corrosion resistant.

VIENNESE SHEET.
M-Austria; 60 Cu, 20 Zn, 20 Ni.
For ornaments, hardware; corrosion resistant.

VIENNESE TABLEWARE.
50 Cu, 25 Zn, 25 Ni.
For cutlery, ornaments; corrosion resistant.

VIGILANT.
M-1406; 0.5-1.0 C, bal Fe.
For chisels, hard tools, springs; water hardened.

VIKING.
M-140; 1.05 C, 1.05 Cr, 0.45 Mo, bal Fe.
For mandrels, rams; oil hardened.

VIKMANSHYTTAN VH94.
M-1447; 0.15 max C, 11.5-13.5 Cr, 1.25-2.50 Ni, bal Fe.
For chemical plant and oil refinery equipment; corrosion resistant.

VIKMANSHYTTAN VH217.
M-1447; 0.19-0.25 C, 13 Cr, bal Fe.
Annealed: 95,000 TS; 50,000 YS; 25 El; 55 RA; 195 Brin.
For valves, cutlery, surgical and dental instruments; Type 420; stainless.

VIKMANSHYTTAN VH273.
M-1447; 0.08 max C, 13 Cr, bal Fe.
Annealed: 75,000 TS; 40,000 YS; 35 El; 70 RA; 155 Brin.
For valves, cutlery, turbine blades; Type 403; stainless.

VIKMANSHYTTAN VH274.
M-1447; 0.10 C, 18 Cr, 8 Ni, bal Fe.
Annealed: 80,000 TS; 35,000 YS; 55 El; 75 RA; 150 Brin.
For chemical plant equipment, tanks, mixers; Type 302; stainless, austenitic.

VIKMANSHYTTAN VH276.
M-1447; 0.07-0.12 C, 18 Cr, 9 Ni, 1.5 Mo, bal Fe.
Annealed: 85,000 TS; 35,000 YS; 50 El; 65 RA; 160 Brin.
For acid resistant equipment, tanks, mixers; Type 316; stainless, austenitic.

VIKMANSHYTTAN VH280.
M-1447; 0.14 C, 13 Cr, 1 Mo, bal Fe.
For chemical plant and oil refinery equipment; corrosion resistant.

VIKMANSHYTTAN VH295.
M-1447; 0.75-0.95 C, 16-18 Cr, 0.75 Max Mo, bal Fe.
Annealed: 207,000 TS; 62,000 YS; 18 El; 35 RA; 220 Brin.
Heat treated; 280,000 TS; 270,000 YS; 3 El; 15 RA; 555 Brin.
For cutlery, valves, ball bearings; Type 440 B; corrosion resistant.

VIKMANSHYTTAN VH376.
M-1447; 0.06 max C, 18 Cr, 9 Ni, 1.5 Mo, bal Fe.
Annealed: 85,000 TS; 35,000 YS; 50 El; 65 RA; 160 Brin.
For acid resistant equipment, tanks, mixers; Type 316; stainless, austenitic.

VIKMANSHYTTAN VH399.
M-1447; 0.10 max C, 17 Cr, 1.5 Mo, bal Fe.
Annealed: 80,000 TS; 50,000 YS; 25 El; 50 RA; 150 Brin.
For oil refinery equipment, oil burners and heaters; corrosion resistant.

VIKMANSHYTTAN VH406.
M-1447; 0.13-0.18 C, 13 Cr, bal Fe.
Annealed: 95,000 TS; 50,000 YS; 25 El; 55 RA; 195 Brin.
For valves, cutlery, surgical and dental instruments; Type 420; stainless.

VIKMANSHYTTAN VH407.
M-1447; 0.12 max C, 20 Cr, 20 Ni, bal Fe.
For furnace parts and equipment; corrosion and heat resistant.

VIKMANSHYTTAN VH417.
M-1447; 0.15 min C, 12-14 Cr, bal Fe.
Annealed: 95,000 TS; 50,000 YS; 25 El; 55 RA; 195 Brin.
For valves, cutlery, surgical and dental instruments; Type 420; stainless.

VIKMANSHYTTAN WH605.
M-1447; 0.20 C, 12-14 Cr, bal Fe.
Annealed: 95,000 TS; 50,000 YS; 25 El; 196 Brin.
Hardened: 250,000 TS; 190,000 YS; 10 El; C 48 rock.
For cutlery, surgical instruments, valves, gears, shafts.
Type 420 stainless steel, hardenable.

VIKMANSHYTTAN WH620.
M-1447; 0.10 C, 14-18 Cr, bal Fe.
Annealed: 75,000 TS; 45,000 YS; 28 El; 140 Brin.
Cold rolled: 110,000 TS; 90,000 YS; 10 El; 195 Brin.
For automotive trim, kitchen sinks, oil burners, fasteners.
Type 430 stainless steel, nonhardenable.

VIKMANSHYTTAN WH627.
M-1447; 0.30 C, 23-30 Cr, bal Fe.
Annealed: 85,0000 TS; 55,000 YS; 25 El; B 83 Rock.
For oil burners, heat treating equipment, valves and fittings, furnace parts.
Type 446 stainless steel. High heat and corrosion resistant.

VIKMANSHYTTAN WH638.
M-1447; 0.06 C, 19 Cr, 10 Ni, bal Fe.
Annealed: 85,000 TS; 35,000 YS; 60 El; 150 Brin.
For chemical and dairy plant equipment, agitators, tanks, digesters.
Type 304 stainless steel, austenitic.

VIKMANSHYTTAN WH639.
M-1447; 0.06 C, 18-20 Cr, 9-11 Ni, bal Fe.
Annealed: 85,000 TS; 35,000 YS; 60 El; 150 Brin.
For chemical plant equipment, digesters, agitators. vessels.
Type 304 stainless steel, austenitic.

VIKMANSHYTTAN WH640.
M-1447; 0.06 C, 17-19 Cr, 9-12 Ni, 0.5 Cb, bal Fe.
Annealed: 85,000 TS; 35,000 YS; 60 El; B 83 Rock.
For welded chemical plant equipment, tanks, vessels, agitators.
Type 347 stainless steel, austenitic.

VIKMANSHYTTAN WH642.
M-1447; 0.12 C, 17-19 Cr, 8-10 Ni, bal Fe.
Annealed: 90,000 TS; 40,000 YS; 50 El; B 85 Rock.
For chemical plant equipment, tanks, vessels, digesters, trim, agitators.
Type 302 stainless steel, austenitic.

VIKMANSHYTTAN WH643.
M-1447; 0.06 C, 18 Cr, 11 Ni, 0.4 Ti, bal Fe.
Annealed: 85,000 TS; 35,000 YS; 55 El; B 80 Rock.
For welded chemical plant equipment, tanks, vessels, agitators, mixers.
Type 321 stainless steel, austenitic, stabilized.

VIKMANSHYTTAN WH645.
M-1447; 0.05 C, 18 Cr, 8 Ni, 2 Mo, bal Fe.
Annealed: 85,000 TS; 35,000 YS; 50 El; B 80 Rock.
For acid mixers, chemical plant equipment, agitators, tanks, valve trim.
Type 18-8 Mo steel, austenitic.

VIKMANSHYTTAN WH649.
M-1447; 0.08 C, 18 Cr, 12 Ni, 3 Mo, bal Fe.
Annealed: 80,000 TS; 40,000 YS; 40 El; B 85 Rock.
For dairy and chemical plant equipment, mixers, digesters, tanks.
Type 18-12 Mo stainless steel, austenitic, acid resistant.

VIKMANSHYTTAN WH650.
M-1447; 0.06 C, 17 Cr, 12 Ni, 2 Mo, bal Fe.
Annealed: 80,000 TS; 30,000 YS; 60 El; B 80 Rock.
For textile, pharmaceutical and chemical plant equipment.
Type 316-319 stainless steel, austenitic.

VIKMANSHYTTAN WH651.
M-1447; 0.08 C, 17 Cr, 13 Ni, 3 Mo, 0.8 Cb, bal Fe.
Annealed: 85,000 TS; 40,000 YS; 50 El; B 82 Rock.
For chemical plant equipment, mixers, agitators, welded tanks.
Type 318 stainless steel, austenitic, welding grade.

VIKMANSHYTTAN WH662.
M-1447; 0.15 C, 18 Cr, 8 Ni, bal Fe.
Annealed: 80,000 TS; 35,000 YS; 60 El; B 80 Rock.
For chemical and pharmaceutical plant equipment, tanks, digesters, mixers, agitators.
Type 302 stainless steel, austenitic.

VIKMANSHYTTAN WH662D.
M-1447; 0.06 C, 19 Cr, 11 Ni, bal Fe.
Annealed: 70,000 TS; 30,000 YS; 60 El; B 80 Rock.
For chemical plant equipment, mixers, tanks, vessels, digesters.
Type 304 stainless steel, austenitic.

VILLARES N 0721.
M-1733; 0.09 C, 0.70 Mn, 0.07 max P, 0.18-0.26 S, bal Fe.
Free cutting steel, similar to Werkstoff Nr. 1.0721.

VILLARES N 0726.
M-1733; 0.36 C, 0.70 Mn, 0.07 max P, 0.15-0.25 S, bal Fe.
Free cutting steel, similar to Werkstoff Nr. 1.0726.

VILLARES S-1112.
M-1733; 0.13 max C, 0.85 Mn, 0.07-0.12 P, 0.16-0.23 S, bal Fe.
Free cutting steel, similar to SAE 1212.

VILLARES S 1117.
M-1733; 0.17 C, 1.15 Mn, 0.04 max P, 0.08-0.13 S, bal Fe.
Free cutting steel, similar to SAE 1117.

VILLARES S 1118.
M-1733; 0.17 C, 1.45 Mn, 0.04 max P, 0.08-0.13 S, bal Fe.
Free cutting steel, similar to SAE 1118.

VILLARES S 1120.
M-1733; 0.21 C, 0.85 Mn, 0.04 max P, 0.08-0.13 S, bal Fe.
Free cutting steel.

VILLARES S 1126.
M-1733; 0.26 C, 0.85 Mn, 0.04 max P, 0.08-0.13 S, bal Fe.
Free cutting steel.

VILLARES S 1137.
M-1733; 0.36 C, 1.50 Mn, 0.04 max P, 0.08-0.13 S, bal Fe.
Free cutting steel, similar to SAE 1137.

VILLARES S 1141.
M-1733; 0.41 C, 1.50 Mn, 0.04 max P, 0.08-0.13 S, bal Fe.
Free cutting steel, similar to SAE 1141.

VILLARES S 1144.
M-1733; 0.44 C, 1.50 Mn, 0.04 max P, 0.24-0.33 S, bal Fe.
Free cutting steel, similar to SAE 1144.

VILLARES S 1146.
M-1733; 0.46 C, 0.85 Mn, 0.04 max P, 0.08-0.13 S, bal Fe.
Free cutting steel, similar to SAE 1146.

VINCO.
M-140; 0.70 C, 0.25 Mn, 3.75-4.25 Cr, 1.0 V, 17.5-18.5 W, bal Fe.
For cutters, hot nut dies, shear knives; Type T1; hot work steel.

VINCO HOT WORK.
M-140; 0.50 C, 18.0 W, 4.0 Cr, 1.0 V, bal Fe.
Hot work die steel; for hot nut dies, hot extrusion dies, press die insert gripper dies, punches, shear knives. Type H-26.

VINCO HW same as **VINCO HOT WORK.**

VINERTIA.
M-41; Co, Cr, Mo, V.
Cast: 90,000 TS; 50,000 YS; 2.5 El; 200 Brin.
Rolled: 190,000 TS; 175,000 YS; 9 El; 330 Brin.
For dental and surgical tools; corrosion resistant.

VIOLA AT13.
M-1452; 0.15 C, 23 Cr, 14 Ni, 0.15 S, bal Fe.
Annealed: 90,000 TS; 40,000 YS; 50 El; B 83 Rock.
For furnace parts, skids, incinerators, heat treating boxes, permanent molds. Free-machining.
Type 309S stainless steel, corrosion and heat resistant.

VIOLA AT15.
M-1452; 0.20 C, 25 Cr, 21 Ni, bal Fe.
Annealed: 95,000 TS; 45,000 YS; 50 El; B 89 Rock.
For furnace parts, skids, heat treating fixtures, pumps, valves.
Free machining. Type 310S stainless steel, corrosion and heat resistant.

VIOLA ICS.
M-1452; 0.10 C, 14-18 Cr, bal Fe.
Annealed: 70,000 TS: 40,000 YS: 30 El; 140 Brin.
Cold Rolled: 120,000 TS; 110,000 YS; 5 El; 195 Brin.
For oil burners, oil refinery equipment, heaters, kitchen sinks, furnace parts, septic tanks, fasteners.
Type 430 stainless steel, non-hardenable, ferritic.

VIOLA IN.
M-1452; 0.12 C, 18 Cr, 9 Ni, bal Fe.
Annealed: 90,000 TS; 40,000 YS; 50 El; B 85 Rock.
For food processing and chemical plant equipment, tanks, valve trim.
Type 302 stainless steel, austenitic, non-hardenable.

VIOLA INC.
M-1452; 0.06 C, 18 Cr, 12 Ni, 0.6 Cb, bal Fe.
Annealed: 85,000 TS; 35,000 YS; 60 El; B 80 Rock.
For welded structures, tanks, mixers, agitators, chemical plant equipment.
Type 347 stainless steel, stabilized, austenitic, welding grade.

VIOLA IND.
M-1452; 0.10 C, 16-18 Cr, 6-8 Ni, bal Fe.
Annealed: 110,000 TS; 40,000 YS; 60 El; B 85 Rock.
For aircraft and railroad car structures, trailer bodies, springs, automotive and architectural trim.
Type 301 stainless steel, austenitic, good ductility.

VIOLA INF25.
M-1452; 0.18 C, 24 Cr, bal Fe.
Annealed: 88,000 TS; 56,000 YS; 24 El; B 83 Rock.
For heat treating boxes, fixtures and furnace parts, oil burners, conveyor exhaust manifolds.
Type 446 stainless steel, corrosion and heat resistant.

VIOLA INF30.
M-1452; 0.20 C, 25 Cr, bal Fe.
Annealed: 90,000 TS; 58,000 YS; 23 El; B 85 Rock.
For heat treating boxes, fixtures and furnace parts, oil burners, conveyor exhaust manifolds.
Type 446 stainless steel, corrosion and heat resistant.

VIOLA INI.
M-1452; 0.06 C, 19 Cr, 10 Ni, bal Fe.
Annealed: 85,000 TS; 35,000 YS; 60 El; B 80 Rock.
For welded structures, chemical and pharmaceutical plant equipment, tanks, mixers.
Type 304 stainless steel, austenitic, non-magnetic.

VIOLA INI/BC.
M-1452; 0.05 C, 20 Cr, 12 Ni, bal Fe.
Annealed: 88,000 TS; 38,000 YS; 58 El; B 82 rock.
For chemical and pharmaceutical plant equipment, welded structures, tanks, mixers.
Type 304 stainless steel, austenitic, non-magnetic.

VIOLA INMI/-BC.
M-1452; 0.8 C, 18 Cr, 2 Mo, 8 Ni, bal Fe.
Annealed: 80,000 TS; 30,000 YS; 60 El; B 80 Rock.
For chemical plant equipment, mixers, digesters, valve trim.
Type 18-8 Mo stainless steel. Acid resistant, austenitic.

VIOLA INMS.
M-1452; 0.06 C, 18 Cr, 12 Ni, 2 Mo, bal Fe.
Annealed: 85,000 TS; 35,000 YS; 50 El; B 80 Rock.
For chemical plant equipment, agitators, mixers, digesters, valve trim.
Type 18-12 Mo stainless steel, acid resistant, austenitic.

VIOLA INS.
M-1452; 0.06 C, 18 Cr, 11 Ni, 0.3 Ti, bal Fe.
Annealed: 85,000 TS; 30,000 YS; 55 El; B 80 Rock.
For welded structures, exhaust systems, engine manifolds, refinery equipment, radiant superheaters.
Type 321 stainless steel, austenitic, welding grade.

VIOLA MIC.
M-1452; 0.12 C, 12-14 Cr, bal Fe.
Annealed: 95,000 TS; 50,000 YS; 25 El; B 92 Rock.
For cutlery, surgical instruments, gears, shafts, springs, bearings.
Type 420 stainless steel, hardenable.

VIRGINIA SILVER.
Ni, Zn, bal Cu.
For ornaments; Nickel silver.

VIRGINIA STEEL.
C, bal Fe.
For construction work.

VIRGO see also UNITEMP VIRGO.

VIRGO 7A.
M-1588; 0.22 C, 11.5 Cr, 1.1 Mo, 0.4 V, 0.4 Ni, bal Fe.
Heat treated: 100,000 TS; 72,000 YS; 10 El.
For turbine blades, bolts; corrosion resistant.

VIRGO 14 SB 1 see ICL 164 FLUAGE.

VIRGO 15SB.
M-1588; 0.06 C, 0.4 Si, 1.6 Mn, 18 Cr, 12 Ni, 0.4 Ti, bal Fe.
Annealed: 85,000 TS; 35,000 YS; 60 El; 70 RA; 150 Brin.
Cold drawn: 180,000 TS; 125,000 YS; 10 El; 330 Brin.
For chemical plant equipment; Type 304; stainless, austenitic.

VIRGO 17SB.
M-1588; 0.08 C, 17 Cr, 13 Ni, 2.5 Mo, 0.5 Ti or Cb, bal Fe.
Annealed: 80,000 TS; 30,000 YS; 60 El; 80 RA; 135 Brin.
Cold drawn: 150,000 TS; 135,000 YS; 6 El; 320 Brin.
For acid resistant equipment, mixers, evaporators; Type 316 Ti; stainless, austenitic.

VIRGO 38.
M-1488; 0.06 max C, 1.0 max Mn, 1.0 max Si, 4.0-5.5 Ni, 15.5-17.5 Cr, bal Fe.
Stainless casting.
Parts for water turbines.
AFNOR Z 6 CN 16 5M.

VIRGO 39 (CASTING).
M-1488; 0.06 max C, 1.0 max Mn, 1.0 max Si, 4.0-5.5 Ni, 15.5-17.5 Cr, 0.5-1.5 Mo, bal Fe.
Stainless casting.
For water turbines, pumps and other parts used in seawater.
AFNOR Z 4 CND 17.4 M.

VIRGO 39 (WROUGHT).
M-1802; 0.06 max C, 16.0 Cr, 5.0 Ni, 0.9 Mo, bal Fe.
Martensitic stainless; air or oil hardenable to 880-1080 N/mm^2 UTS.
Plate; good strength and corrosion resistance to sea water, organic acids.
AFNOR Z4 CND 16-05; see also SOLEIL C4.

VIRGO 94.
M-1588; 0.1 C, 16-20 Cr, 40-48 Ni, 1.9-2.6 Ti, 20-24 Co.
Cold drawn: 127,000 TS; 85,000 YS; 20 El.
For turbine components, valve and pump parts.
Austenitic, oxidation annd heat resistant to 1150°C.

VIRGO 104.
M-1488; 0.06 max C, 1.0 max Mn, 1.0 max Si, 3.5-5.0 Ni, 12.0-13.5 Cr, 0.4-0.7 Mo, bal Fe.
Stainless casting.
Parts for pumps, turbines, nuclear energy.
AFNOR Z4 CND 13.4 M; ACI CA 6NM.

VIRGO 105D.
M-1488; 0.07 max C, 1.0 max Mn, 1.0 max Si, 1.5-2.5 Ni, 11.0-13.5 Cr, 0.75 max Mo, bal Fe.
Stainless casting, ferritic type; 200-250 Brin.
Parts for steam turbines.
AFNOR Z 4 CN 13.2 M.

VIRGO 120.
M-1301, M-1488; 0.25 max C, 12-14 Cr, bal Fe.

VIRGO L 16.
M-1488, M-533; 0.50 C, 14.5-17.0 Si, 0.50 Mn, bal Fe.

VIRILLIUM.
M-Eng.; 67.9 Co, 24.1 Cr, 1.4 Ni, 5.3 Mo, bal Fe.
Cast: 96,000-98,000 TS; 64,000-69,000 YS; 10-11 El; 295 Brin.
For dental alloy; corrosion resistant.

VISCOTHERM 4.
M-1317; Cr, Co, Mo, bal Ni.

VISCOTHERM 5.
M-1317; Cr, Co, Mo, bal Ni.

VISCOTHERM 6.
M-1317; Cr, Co, Mo, bal Ni.

VISCOTHERM 7.
M-1317; Cr, Co, Mo, bal Ni.

VISCOTHERM 1612.
M-1317; Cr, Co, Mo, bal Ni.

VISCOUNT 20.
M-73; 0.40 C, 5 Cr, 1 V, 1 Si, 0.3 Mn, 1.2 Mo, S, bal Fe.
For white metal extrusion dies; Type H13; hot work steel.

VISCOUNT 44.
M-73, M-1702; 0.40 C, 5 Cr, 1 V, 1 Si, 0.3 Mn, 1.2 Mo, S, bal Fe.
Heat treated: 420-460 Brin.
For die casting dies, extrusion tools, forging die blocks; prehardened, air-hardened, resists heat checking.
AISI H13.

VISTA.
M-1705; 0.23 C, 0.6 Mn, 1.25 Si, 10.0 Cr, 0.75 Ni, 1.0 V, 0.45 W, 1.2 Mo, 0.10 N, bal Fe.
Air or oil hardening hot work tool and die steel, chromium type.

VISTA 11.
M-1705 ; 0.40 C, 0.3 Mn, 0.9 Si, 5.0 Cr, 0.5 V, 1.3 Mo, bal Fe.
Chromium type hot work tool and die steel.
AISI H 11

VISTA 12.
M-1705; 0.32 C, 0.3 Mn, 0.9 Si, 5.0 Cr, 0.25 V, 1.25 W, 1.50 Mo, bal Fe.
Chromium type hot work tool and die steel.
AISI H 12.

VISTA 13.
M-1705; 0.40 C, 0.35 Mn, 1.0 Si, 5.0 C, 1.0 V, 1.2 Mo, bal Fe.
Chromium type hot work tool and die steel.
AISI H 13.

VISTA 21.
M-1705; 0.30 C, 0.3 Mn, 3.0 Cr, 0.5 V, 9.0 W, bal Fe.
Hot work tool and die steel, tungsten type.
AISI H21.

VISTA 24.
M-1705; 0.42 C, 3.5 Cr, 0.3 V, 14.0 W, bal Fe.
Hot work tool and die steel, tungsten type.
AISI H 24.

VISTA 26.
M-1705; 0.52 C, 3.75 Cr, 0.9 V, 17.5 W, bal Fe.
Hot work tool and die steel, tungsten type.
AISI H 26.

VISTA 43.
M-1705; 0.50 C, 4.0 Cr, 1.95 V, 8.0 Mo, bal Fe.
Hot work tool and die steel, molybdenum type.
AISI W 43.

VISTO.
M-344; C, Cr, V, W, bal Fe.
For tools, cutters; high speed steel.

VISTO.
M-614; 0.7 C, 18 w, 4 Cr, 1 V, bal Fe.
For cutters, tools; high speed steel.

VISTO A50.
M-614; 0.56 C, Ni, Cr, Mo, V, bal Fe.
For gears, shafts, crankshafts, bolts, fasteners; oil hardened, shock resistant.

VISTO ABC III.
M-614; 0.95 C, W, Mo, bal Fe.
For cutters, dies, bearings, liners; water hardened, wear resistant.

VISTO DM1.
M-614; 0.45 C, 1.4 Cr, 0.7 Mo, 0.3 V, bal Fe.
For gears, bolts, crankshafts; oil hardened, tough.

VISTO KD6.
M-614; 0.86 C, 4.1 Cr, 0.85 Mo, 2.5 V, 12 W, bal Fe.
For lathe and planer tools, drills, reamers, taps; high speed steel.

VISTO KD13.
M-614; 0.86 C, 2.8 Co, 4.3 Cr, 0.85 Mo, 2.1 V, 12 W, bal Fe.
For lathe and planer tools, drills, reamers, hobs; high speed steel.

VISTO KD15.
M-614; 0.79 C, 4.7 Co, 4.3 Cr, 0.75 Mo, 1.5 V, 18 W, bal Fe.
For lathe and planer tools, broaches; high speed steel.

VISTO KD16.
M-614; 0.76 C, 10 Co, 4.2 Cr, 0.8 Mo, 1.8 V, 18 W, bal Fe.
For lathe and planer tools, form cutters; high speed steel.

VISTO KD22.
M-614; 1.3 C, 4.3 Cr, 0.85 Mo, 3.8 V, 12 W, bal Fe.
For engravers' tools, blanking and forming dies; high speed steel.

VISTO KD25.
M-614; 1.35 C, Cr, W, Co, V, bal Fe.
For engravers' tools, blanking and forming dies; high speed steel.

VISTO PL.
M-614; 0.45 C, W, Cr, V, bal Fe.
For punches, crimpers, upsetters; hot work steel, oil hardened.

VISTO REKORD.
M-614; 0.74 C, 4.1 Cr, 1.1 V, 18.5 W, bal Fe.
For lathe and planer tools, reamers, broaches, hobs; high speed steel.

VISTO SUPERIOR.
M-344; 0.8 C, 4 Cr, 0.8 Mo, 1.6 V, 8.7 W, bal Fe.
For shear blades, dies, drills, reamers, broaches, taps; high speed steel, oil hardened.

VISTO W5.
M-614; 0.30 C, 2.3 Cr, 0.6 V, 4.2 W, bal Fe.
For punches, riveters, extrusion tools; hot work steel, oil hardened.

VISTO W9.
M-614; 0.30 C, 2.65 Cr, 0.35 V, 8.5 W, bal Fe.
For extrusion rams and liners, punches; hot work steel, oil hardened.

VISTO W11C.
M-614; 0.30 C, 2 Co, 2.4 Cr, 0.25 V, 8.5 W, bal Fe.
For extrusion tools, punches, shears; hot work steel, oil hardened.

VISTO WC5.
M-614; 0.40 C, Cr, W, V, Si, bal Fe.
For hot work punches and shears; hot work steel, oil hardened.

VISTO WC SPEZIAL.
M-614; 0.38 C, Cr, Si, Mo, V, bal Fe.
For upsetters, riveters, punches, shears; hot work steel, oil hardened.

VISTO Z652.
M-614; 0.85 C, Cr, V, W, Mo, bal Fe.
For lathe and planer tools, drills, taps, hobs; high speed steel.

VITAL.
M-Ger.; 0.6 Si, 0.9 Cu, 1 Zn, bal Al.
60,000 TS; 5 El.
For light alloy parts; non-hardenable.

VITALLIUM.
M-937; 0.5 max C, 0.6 max Si, 0.75 max Mn, 5-7 Mo, 28-32 Cr, bal Co.
Cast: 97,000-130,000 YS; 58,000-80,000 YS; 15-4 El; 20-2 RA; 250-45 Brin.
For orthopedic and dental appliances; corrosion resistant.

VITALU.
M-Eng.; 5 Si, bal Al.
For light alloy parts; die castings.

VITA M-1.
M-1396; 0.7 C, 8.5 Mo, 4 Cr, 1 V, 1.5 W, bal Fe.
For lathe and planer tools; high speed steel.

VITA M-2.
M-1396; 0.7 C, 5 Mo, 4 Cr, 2 V, 6 W, bal Fe.
For lathe and planer tools; high speed steel.

VITA M-3.
M-1396; 0.7 C, 6 Mo, 4 Cr, 2.4 V, 6 W, bal Fe.
For lathe and planer tools; high speed steel.

VITA M-4.
M-1396; 0.7 C, 4.5 Mo, 4.5 Cr, 4 V, 5.5 W, bal Fe.
For lathe and planer tools; high speed steel.

VITA M-6.
M-1396; 0.7 C, 5 Mo, 4 W, 12 Co, 4.5 Cr, 1.5 V, bal Fe.
For lathe and planer tools; high speed steel.

VITA M-8.
M-1396; 0.7 C, 4.5 Mo, 4.2 Cr, 1.5 V, 5.5 W, 1.2 Cb, bal Fe.
For lathe and planer tools; high speed steel.

VITA M-10.
M-1396; 0.7 C, 8 Mo, 4 Cr, 2 V, bal Fe.
For lathe and planer tools; high speed steel.

VITA M-15.
M-1396; 0.7 C, 3 Mo, 6.7 W, 5 V, 5 Co, bal Fe.
For lathe and planer tools; high speed steel.

VITA M-20.
M-1396; 0.7 C, 8 Mo, 4 Cr, 1 V, 2.5 Co, bal Fe.
For lathe and planer tools; high speed steel.

VITA M-30.
M-1396; 0.7 C, 8 Mo, 4 Cr, 1 V, 2 W, 5 Co, bal Fe.
For lathe and planer tools; high speed steel.

VITA M-34.
M-1396; 0.7 C, 8.5 Mo, 4 Cr, 2 V, 2 W, 8 Co, bal Fe.
For lathe and planer tools; high speed steel.

VITA M-36.
M-1396; 0.7 C, 6 Mo, 4 Cr, 2 V, 6 W, 8 Co, bal Fe.
For lathe and planer tools; high speed steel.

VITA M-40.
M-1396; 0.7 C, 8 Mo, 4 Cr, 1.5 V, 8 Co, bal Fe.
For lathe and planer tools; high speed steel.

VITA M-50.
M-1396; 0.7 C, 4.2 Mo, 4 Cr, 1 V, bal Fe.
For lathe and planer tools; high speed steel.

VITA M-52.
M-1396; 0.7 C, 4.2 Mo, 4 Cr, 2 V, bal Fe.
For lathe and planer tools; high speed steel.

VITA M-54.
M-1396; 0.7 C, 4.2 Mo, 4 Cr, 3 V, bal Fe.
For lathe and planer tools; high speed steel.

VITA M-56.
M-1396; 0.7 C, 4.2 Mo, 4 Cr, 4 V, bal Fe.
For lathe and planer tools; high speed steel.

VITA T-1.
M-1396; 0.7 C, 18 W, 4 Cr, 1 V, bal Fe.
For lathe and planer tools, reamers, taps; high speed steel.

VITA T-2.
M-1396; 0.7 C, 18 W, 4 Cr, 2 V, bal Fe.
For lathe and planer tools; high speed steel.

VITA T-3.
M-1396; 0.7 C, 18 W, 4 Cr, 3.2 V, bal Fe.
For lathe and planer tools; high speed steel.

VITA T-4.
M-1396; 0.7 C, 18 W, 4 Cr, 1 V, 5 Co, bal Fe.
For lathe and planer tools; high speed steel.

VITA T-5.
M-1396; 0.7 C, 18 W, 4 Cr, 2 V, 8 Co, bal Fe.
For lathe and planer tools; high speed steel.

VITA T-6.
M-1396; 0.7 C, 22 W, 4.5 Cr, 1.5 V, 12 Co, bal Fe.
For lathe and planer tools; high speed steel.

VITA T-7.
M-1396; 0.7 C, 14 W, 4 Cr, 2 V, bal Fe.
For lathe and planer tools; high speed steel.

VITA T-8.
M-1396; 0.7 C, 14 W, 4 Cr, 2 V, 5 Co, bal Fe.
For lathe and planer tools; high speed steel.

VITA T-12.
M-1396; 0.7 C, 14 W, 4 Cr, 3 V, bal Fe.
For lathe and planer tools; high speed steel.

VITA T-15.
M-1396; 0.7 C, 13 W, 4.2 Cr, 5 V, 5 Co, bal Fe.
For lathe and planer tools; high speed steel.

VITA T-20.
M-1396; 0.7 C, 18.5 W, 4 Cr, 4 V, bal Fe.
For lathe and planer tools; high speed steel.

VITINOX 18 CU.
M-1488; 0.06 max C, 18 Cr, 10 Ni, 3.5 Cu, bal Fe.
Wrought ann: 490 N/mm² min UTS.
Free machining austenitic stainless steel.
AFNOR Z4 CNU 18.9.

VITINOX 18X.
M-1488; 0.12 max C, 18 Cr, 9 Ni, 0.60 max Mo, 0.15 min S, bal Fe.
Wrought ann: 510 N/mm² min UTS.
Free machining austenitic stainless steel.
AFNOR Z10 CNF 18.9.

VITRENAMEL 1.
M-604.
Low C, Mn, bal Fe.
For porcelain enameled steel sheets for stoves, kitchenware.

VITRENAMEL 2.
M-604.
Low C, Mn, bal Fe.
For porcelain enameled steel sheets for stoves, kitchenware.

VITRIFORM.
M-1761.
Sheet steel with colored porcelain coating; for interior walls.

VITRIX.
M-1322; 0.35 C, 1.3 Cr, 4.5 Ni, 0.6 Mn, bal Fe.
For gears, bolts, crankshafts, forging dies; oil hardened, shock resistant.

VIVAL.
M-807; 0.5 Si, 1 Mg, 0.5 Mn, bal Al.
Rolled: 36,000 TS; 34,000 YS; 5 El.
For light alloy parts.

VK2.
M-USSR; 2 Co, bal WC.
Sintered: Rock A 90-93.
For hard cutting tools, dies.
Sintered carbide of tungsten.

VK3.
M-USSR; WC, 3 Co.
Sintered: Rock A 92.
For cutting tools, dies.
Cemented carbides.

VK4.
M-USSR; WC, 4 Co.
Sintered: Rock A 92.
For cutting tools, dies.
Sintered carbide.

VK6.
M-USSR; 6 Co, bal WC.
Sintered: Rock A 92.
For hard cutting tools, dies.
Sintered carbide of tungsten.

VK6M.
M-USSR; 6 Co, bal WC.
Sintered: Rock A 92.
For hard cutting tools.
Sintered tungsten carbide, low porosity.

VK8.
M-USSR; 8 Co, bal WC.
Sintered: Rock A 91.5.
For hard cutting tools, dies.
Cemented carbides of tungsten.

VK8V.
M-USSR; 8 Co, bal WC.
Sintered: Rock A 91.8.
For hard cutting tools, dies.
Sintered carbide.

VK-10E.
M-1733; 1.30 C, 4.2 Cr, 4.5 Mo, 8.0 W, 2.7 V, 10.0 Co, bal Fe.
High speed steel.
DIN 1.3207.

VK15.
M-USSR; 15 Co, bal WC.
Sintered: Rock A 90.
For hard cutting tools, dies.
Sintered tungsten carbide.

VK20.
M-USSR; 20 Co, bal WC.
Sintered: Rock A 86-88.
For hard cutting tools, dies.
Sintered tungsten carbide.

VK25.
M-USSR; 25 Co, bal WC.
Sintered: Rock A 86-88.
For hard cutting tools, dies.
Sintered tungsten carbide.

V-KUT.
M-35; 0.71 C, 0.25 Mn, 0.28 Si, 0.80 Cr, 0.20 V, bal Fe.
Water or oil hardening tool steel; for rock drills, shear blades, hand stamps.

VL3.
M-1653; 0.47-0.55 C, 0.70-0.90 Mn, 0.80-1.20 Cr, 0.10-0.20 V, bal Fe.
Cr-V structural steel.
50 CrV 4; AISI 6150.

VL7-45U.
M-USSR; 0.16 C, 20 Cr, 8 W, 0.06 B, 25 Fe, 46 Ni.
Nickel-iron base superalloy.
For nozzle guide vanes.

VL-30.
M-1733; 0.30 C, 0.27 Si, 0.50 Mn, 0.95 Cr, 0.20 Mo, bal Fe.
Hardenable alloy steel.
AISI 4130; DIN 25 CrMo 4.

VL-40.
M-1733; 0.40 C, 0.27 Si, 0.87 Mn, 0.95 Cr, 0.20 Mo, bal Fe.
Hardenable alloy steel.
AISI 4140; DIN 42 CrMo 4.

VLM.
M-1779; 0.85 C, 8 Mo, 4 Cr, 2 V, bal Fe.
Hardened: C 63-66 Rock.
For tools, cutters, reamers, twist drills, milling cutters.
Type M-10 high-speed steel.

V L W 1.
M-116; 5 Cu, 0.4 Si, bal Al.
For light alloy parts; similar to "Lautal."

V L W 2.
M-116; 4 Cu, 0.1 Li, bal Al.
For light alloy parts; similar to "Scleron."

V L W 3.
M-116; 11-14 Si, bal Al.
For light alloy parts; similar to "Silumin."

V L W 6.
M-116; 1.3 Mn, 2.2 Mg, 0.7 Si, bal Al.
For marine hardware; resists sea water.

V L W 14.
M-116; 5 Cu, 0.4 Si, bal Al.
For light alloy parts; same as "Lautal."

V L W 17.
M-116; 4-5 Cu, bal Al.
For light alloy parts; same as "Duralumin."

V L W 19.
M-116; 0.8 Mg, 0.8 Mn, 1.4 Si, bal Al.
For light alloy parts; same as "Pantal."

V L W 23.
M-116; 0.1 Li, 4 Cu, bal Al.
For light alloy parts; same as "Scleron."

V L W 31.
M-116; 11-14 Si, bal Al.
For light alloy parts; same as "Silumin."

V L W 41.
M-116; 1-2 Mn, bal Al.
For light alloy parts; same as "Aluman."

V L W 61.
M-116; 1.3 Mn, 2.2 Mg, 0.2 Sb, 0.7 Si, bal Al.
For marine hardware: same as "K.S. Seewasser."

V L W 63.
M-116; 7.5 Mg, 0.3 Mn, bal Al.
For marine instruments; same as "B.S. Seewasser."

V L W 99.
M-116; Al.
For cable; same as pure Al.

V L W LEICHTMETALLE.
M-116; Al alloys.
For light alloy parts for airplanes, dirigible and automobiles; formerly known as "Leichstahl."

VLX 25-20.
M-1495; 0.10 C, 25 Cr, 20 Ni, bal Fe.
Oxidation resistant up to 1050°C.
Furnace tubes, refinery piping, burners.
Similar to AISI 310.

VLX MOBTBC 3.
M-1495; 0.03 C, 17 Cr, 13 Ni, 2.7 Mo, bal Fe.
Good resistance to organic acids and chlorinated media.
For petrochemical, textile, paper mill and pharmacy industries.
AISI 316L.

VLX SA.
M-1495; 0.03 C, 25 Cr, 20 Ni, Nb, bal Fe.
Intergranular corrosion resistance to nitric media.
For chemical and nuclear industries.
Similar to AISI 310.

VLX SL.
M-1495; 0.03 C, 20 Cr, 25 Ni, 5.5 (Mo + Cu), Nb, bal Fe.
Intergranular corrosion resistance to sulphuric, phosphoric and hydrochloric media.
For chemical, petrochemical industries, bleaching.

VLX ST.
M-1495; 0.04 C, 12 Cr, Ti, bal Fe.
Resistant to stress corrosion by Cl ions and to oxidation up to 700°C.
For heat exchangers, exhaust pipes.

VM-20.
M-1733; 0.20 C, 0.27 Si, 0.55 Mn, 0.50 Cr, 1.8 Ni, 0.25 Mo, bal Fe.
Alloy carburizing steel.
AISI 4320.

VM-40.
M-1733; 0.40 C, 0.27 Si, 0.70 Mn, 0.80 Cr, 1.80 Ni, 0.25 Mo, bal Fe.
Hardenable alloy steel.
AISI 4340.

VM 118.
M-912; 0.20 C, 0.25 Si, 1.45 Mn, bal Fe.
Low carbon manganese steel for construction purposes.
W.-Nr. 0499.

V-MANG.
M-9; C, 12-14 Mn, Mo, bal Fe.
For welding rod; for Mn steel.

VMARILITE.
M-Eng.; Al-Mg.
For light alloy parts.

VMC.
M-1653; 0.37-0.44 C, 0.50-0.80 Mn, 0.90-1.20 Cr, bal Fe.
Chromium structural steel. 40 Cr 4; similar to AISI 5140.

VMC1.
M-1653; 0.25 C, 0.30 Si, 1.5-2.0 Mn, 0.60-0.90 Cr, bal Fe.
Mn-Cr structural steel. (25 Mn Cr 8).

VMC2.
M-1653; 0.32-0.39 C, 0.80-1.10 Mn, 1.0-1.3 Cr, bal Fe.
Cr-Mn structural steel. 35 CrMn 5; similar to AISI 5135.

VMK.
M-1342; 2.5 C, 28 Cr, 3 Ni, 2 W, bal Fe.
As cast: 55 Rock C.
For wear resistance against abrasion.

VML.
M-1733; 0.55 C, 1.0 Cr, 0.45 Mo, 0.07 V, 0.85 Mn, bal Fe.
Hot work tool steel.

VMO.
M-1733; 0.57 C, 0.9 Cr, 1.55 Ni, 0.45 Mo, 0.10 V, bal Fe.
Hot work tool steel; DIN 1.2713.

VMS 135.
M-912; 0.37 C, 1.25 Si, 1.25 Mn, bal Fe.
Heat treatable steel; for shafts and arbors.
W.-Nr. 5122.

"V.M." STEEL.
0.7-1 Cr, 0.35-0.85 Mo, 0.4 C, $\leq$ 0.17 V, 0.4-0.6 Mn, bal Fe.
For gears, shafts, blacksmith tools; oil hardening.

VN-32.
M-1733; 0.32 C, 0.30 Si, 0.40 Mn, 0.60 Cr, 0.10 V, bal Fe.
Wr.-N. 1.2208; DIN 31 CrV 3.

VN-50.
M-1733; 0.50 C, 0.27 Si, 0.80 Mn, 0.95 Cr, 0.15 V, bal Fe.
Steel for springs.
AISI 6150; DIN 50 CrV 4.

VND.
M-1733; 0.95 C, 0.5 Cr, 0.55 W, 0.12 V, 1.25 Mn, bal Fe.
Oil hardening tool steel.
AISI 01; DIN 1.2419.

VNT see ARMCO VNT.

VOELKLINGEN DOCO.
M-1421; 1.6 C, 13 Cr, 1 Mo, 1 V, 2 Co, bal Fe.
For blanking and forming dies, punches; air hardened, non-deforming.

VOELKLINGEN NGSA.
M-1421; 0.45 C, 1.5 Cr, 0.4 W, 0.5 Mo, 0.8 V, bal Fe.
For punches, shear blades; hot work steel, oil hardened.

VOLKLINGEN NH4.
M-1421; 0.16 C, 23 Cr, 14 Ni, bal Fe.
Annealed: 75,000 TS; 30,000 YS; 40 El; 220 Brin.
For heat exchangers, combustion chambers, salt pots, oil burners.
Type 309 stainless steel, austenitic, heat and corrosion resistant.

VOLKLINGEN NH8G.
M-1421; 0.15 C, 24 Cr, 13 Ni, bal Fe.
Annealed: 75,000 TS; 32,000 YS; 210 Brin.
For furnace parts, salt pots, heat exchangers, combustion chambers.
Type 309 stainless steel, corrosion and heat resistant.

VOLKLINGEN NH22.
M-1421; 0.20 C, 19-22 Cr, 19-22 Ni, bal Fe.
Annealed: 95,000 TS; 45,000 YS; 50 El; B 90 Rock.
For furnace parts and equipment, valves, pumps, boiler baffles, turbine and jet engine components.
Type 310 stainless steel, heat and corrosion resistant.

VOELKLINGEN RCC.
M-1421; 2 C, 13 Cr, W, bal Fe.
For blanking and forming dies; oil or air hardened, non-deforming.

VOELKLINGEN RCW1.
M-1421; 0.60 C, 3.8 Cr, 10 W, 1 Mo, 0.8 V, bal Fe.
For shears, punches, upsetters; hot work steel, oil hardened.

VOELKLINGEN RCW2.
M-1421; 0.30 C, 3 Cr, 10 W, 0.3 Mo, 0.4 V, 2 Ni, bal Fe.
For extrusion dies, liners, punches; hot work steel, oil hardened.

VOLKLINGEN RNO.
M-1421; 0.20 C, 12-14 Cr, bal Fe.
Annealed: 95,000 TS; 50,000 YS; 25 El; B 92 Rock.
For pump and valve parts, surgical instruments, cutlery, gauges.
Type 420 stainless steel, heat and corrosion resistant. Hardenable.

VOLKLINGEN RNOF.
M-1421; 0.25 C, 12-14 Cr, bal Fe.
Annealed: 98,000 TS; 52,000 YS; 20 El; B 95 Rock.
For cutlery, surgical instruments, pump and valve parts.
Type 420 stainless steel, heat and corrosion resistant. Hardenable.

VOLKLINGEN RNOW.
M-1421; 0.12 C, 12-14 Cr, bal Fe.
Annealed: 80,000 TS; 40,000 YS; 30 El; B 85 Rock.
For furnace parts, oil refinery equipment, springs, tableware.
Type 410 stainless steel, corrosion and heat resistant.

VOLKLINGEN RNOWW.
M-1421; 0.10 C, 14-18 Cr, bal Fe.
Annealed: 75,000 TS; 45,000 YS; 30 El; B 80 Rock.
For automotive trim, kitchen sinks, fasteners, heat treating boxes.
Type 430 stainless steel, corrosion resistant, non-hardenable.

VOELKLINGEN RT10 EXTRA.
M-1421; 1.C, 0.15 Si, 0.30 Mn, bal Fe.
Heat treated: 185,000 TS; 120,000 YS; 10 El; 30 RA; 390 Brin.
For springs, taps, drills, reamers; Type W1; water hardened.

VOELKLINGEN RTWK.
M-1421; C, alloy, bal Fe.
For machine tool parts; oil hardened.

VOELKLINGEN RUS.
M-1421; 0.90 C, 2 Mn, 0.3 Si, bal Fe.
For punches, form dies, rolls; oil hardened, non-deforming.

VOELKLINGEN RUS4.
M-1421; C, alloy, bal Fe.
For machine tool parts; oil hardened.

VOELKLINGEN RWA.
M-1421; 0.30 C, 2.5 Cr, 4.5 W, 0.3 Mo, 0.5 V, bal Fe.
For extrusion press liners, mandrels, dies; hot work steel, oil hardened.

VOELKLINGEN RWS.
M-1421; 0.40 C, 3.3 Cr, 3 W, 0.5 Mo, 0.3 Co, bal Fe.
For pneumatic shears, punches, chisels; hot work steel, oil hardened.

VOELKLINGEN RWS2.
M-1421; 0.25 C, 1.3 Cr, 4.5 W, 0.5 Mo, 0.3 V, bal Fe.
For extrusion press tools, liners; hot work steel, oil hardened.

VOELKLINGEN SGM2.
M-1421; 0.30 C, 2.5 Cr, 8 W, 0.5 V, 2.5 Co, bal Fe.
For extrusion press tools; hot work steel, oil hardened.

VOELKLINGEN SGM5.
M-1421; 0.35 C, 5 Cr, 4 W, 0.4 Mo, 0.4 V, 0.5 Co, bal Fe.
For extrusion press tools; hot work steel, oil hardened.

VOELKLINGEN T7.
M-1421; C, alloy, bal Fe.
For forging and die casting dies; hot work steel.

VOIZIT.
M-Ger.; 96.5 Fe, 3.5 Graphite.
Sintered: 25-50 Brin.
For antifriction metal, bearings; pressed powders, 30-40% porosity.

VOKAR.
M-USSR; 8-15 C, 78-86 W, Mn, bal Fe.
For cutting tools; sintered hard alloy.

VOLCANO.
M-688; 0.7 C, 18 W, 4 Cr, 1 V, 5 Co, bal Fe.
For high temperature die work; high speed steel.

VOLCO NO. 5.
M-998; 63 Cu, 32 Zn, 5 Ni.
Hard: 80,000 TS; 7 El.
Soft 50,000 TS; 50 El.
For plumbing fixtures, jewelry; 12% cond.

VOLCO NO. 8.
M-998; 65 Cu, 27 Zn, 8 Ni.
Hard: 82,000 TS; 7 El.
Soft: 53,000 TS; 45 El.
For hardware; 9% conductivity.

VOLCO NO. 10.
M-998; 85 Cu, 15 Zn.
Hard: 69,000 TS; 7 El.
Soft: 40,000 TS; 45 El.
For jewelry, hardware; 37% cond.

VOLCO NO. 11.
M-998; 65 Cu, 25 Zn, 10 Ni.
Hard: 88,000 TS; 7 El.
Soft: 55,000 TS; 42 El.
For hardware; 8.4% conductivity.

VOLCO NO. 12.
M-998; 65 Cu, 23 Zn, 12 Ni.
Hard: 83,000 TS; 6 El.
Soft: 54,000 TS; 41 El.
For hardware; 6.3% conductivity.

VOLCO NO. 14.
M-998; 65 Cu, 21 Zn, 14 Ni.
Hard: 84,000 TS; 5 El.
Soft: 45,000 TS; 40 El.
For hardware; 6.2% conductivity.

VOLCO NO. 30.
M-998; 90 Cu, 10 Zn.
Hard: 62,000 TS; 6 El.
Soft: 57,000 TS; 4 El.
For screen wire, jewelry, screw stock; 43% cond.

VOLCO NO. 65.
M-998; 65 Cu, 35 Zn.
Hard: 73,000 TS; 10 El.
Soft: 45,000 TS; 60 El.
For rivets, screws, eyelets, reflectors; brass.

VOLCO NO. 68.
M-998; 70 Cu, 30 Zn.
Hard: 105,000 TS; 1 El.
Soft: 49,000 YS; 48 El.
For cartridges, clips; brass.

VOLCO NO. 80.
M-998; 80 Cu, 20 Zn.
Hard: 73,000 TS; 5 El.
Soft: 43,000 TS; 8 El.
For flexible hose, stampings; 32.5% conductivity.

VOLCO NO. 95.
M-998; 95 Cu, 5 Zn.
Hard: 55,000 TS; 5 El.
Soft: 35,000 TS; 38 El.
For jewelry, coins, stampings; 56% cond.

VOLOMIT.
WC + MoC.
For cutting tools, dies; similar to "Stellite."

VOLTAL.
M-1355; 4.7 Cu, 2-4 Si, 2.5 max Zn, 1.1 max Fe, bal Al.
For light alloy parts; age-hardenable.

VOLUMIT.
Tungsten carbide plus molybdenum.
For hard cutting tools and dies; sintered alloy.

VOLVIC.
M-1115; 0.2-0.3 C, 2.5-3.5 Cr, 8-10 W, bal Fe.
For rivet sets, extrusion press parts, punches; hot work steel, oil hardened.

VOLVIT.
M-462; 9 Sn, 91 Cu.
For bearings.

VPC.
M-1733; 0.36 C, 5.0 Cr, 1.3 Mo, 0.35 V, 1.1 Si, bal Fe.
Hot work tool steel.
AISI H 11; DIN 1.2343.

VPCW.
M-1733; 0.38 C, 5.0 Cr, 1.35 Mo, 1.15 W, 0.25 V, 0.90 Si, bal Fe.
Hot work tool steel.
AISI H 12; DIN 1.2606.

VPE.
M-1733; 1.0 C, 3.75 Cr, bal Fe.
Tool steel.

"V" PERMANDUR.
M-269; 49.3 Fe, 48.4 Co, 2.3 V.
For telecommunications, diaphragms in telephone receivers; high permeability.

VR-15.
M-1733; 0.15 C, 0.27 Si, 0.80 Mn, 0.80 Cr, bal Fe.
Alloy carburizing steel.
AISI 5115; DIN 15 Cr 3.

VR-30.
M-1733; 0.30 C, 0.27 Si, 0.80 Mn, 0.95 Cr, bal Fe.
Hardenable alloy steel.
AISI 5130.

VR-35.
M-1733; 0.35 C, 0.27 Si, 0.70 Mn, 0.90 Cr, bal Fe.
Hardenable alloy steel.
AISI 5135; DIN 34 Cr 4.

VR-40.
M-1733; 0.40 C, 0.27 Si, 0.80 Mn, 0.80 Cr, bal Fe.
Hardenable alloy steel.
AISI 5140; DIN 41 Cr 4.

VR-50.
M-1733; 0.50 C, 0.27 Si, 0.80 Mn, 0.80 Cr, bal Fe.
Spring steel.
AISI 5150.

VR-60.
M-1733; 0.60 C, 0.27 Si, 0.87 Mn, 0.80 Cr, bal Fe.
Spring steel.
AISI 5160.

VR/WESSON 2A1.
M-53, M-329; 86 Wc, 14 Co.
Sintered: 88.5 RA, 375,000 TrS, 14.1 density.
For cutting tools requiring medium shock resistance; and for light impact applications.
C-11, C-12 type.

VR/WESSON 2A3.
M-329, M-53; 89.0 WC, 11.0 Co.
Sintered: 325,000 TrS; A 89.3 Rock.
Density: 14.40.
For heavy roughing cuts and interrupted machining of ferrous and non-ferrous metals, and for non-metallics. Excellent shock resistance.

VR/WESSON 2A5.
M-329, M-53; 94 WC, 6 Co.
Sintered: 250,000 TrS; A 91.8 Rock.
Density: 14.90.
For general purpose machining, including hardened ferrous and non-ferrous. Very good wear resistance.

/R/WESSON 2A6.
M-329, M-53; 92 WC, 8 Co.
Sintered: 300,000 TrS; A 89.5 Rock.
Density: 14.70.
For heavy roughing and interrupted cuts, also for large wire drawing dies. Good shock resistance.
C-10 Type.

VR/WESSON 2A7.
M-329, M-53; 95.5 WC, 4.5 Co.
Sintered: 225,000 TrS; A 92.2 Rock.
Density: 15.10.
For fine finishing of ferrous and non-ferrous alloys. Excellent wear resistance.
C-3, 3-4 Type.

VR/WESSON 2A68.
M-329, M-53; 94 WC, 6 Co.
Sintered: 275,000 TS; A 90.8 Rock.
Density: 14.85.
For rough machining and interrupted cuts on ferrous and non-ferrous metals.
Good shock and wear resistance.

VR/WESSON 26.
M-329, M-53; 83 WC, 10 TiC 7 Co.
Sintered; 225,000 TrS; A 91.8 Rock.
Density: 12.40.
For general purpose and finish machining of steel and non-ferrous metals. Good shock and wear resistance.
C-6 Type.

VR/WESSON 630.
M-53; M-329.
TiC-base coating on cemented carbide.
For cutting tools, excellent abrasion resistance, good toughness and wear resistance.
For finishing and general purpose machining.
C-2 and C-3 type applications.

VR/WESSON 650.
M-53; M-329.
TiC-base coating on alloyed cemented carbide.
For cutting tools, excellent shock and wear resistance to cutting temperature, especially for milling and interrupted cuts.

VR/WESSON 660.
M-53; M-329.
TiC-base coating on alloyed cemented carbide.
For cutting tools, good shock and wear resistance.

VR/WESSON 670.
M-53; M-329.
TiC-base coating on alloyed cemented carbide.
For cutting tools, mainly semi-finishing and high speed finish machining of steel.

VR/WESSON RAMET I.
M-53; M-329; 89.5 WC, 0.5 Cr_3C_2, 10 Co.
Sintered: 91.5 RA; 375,000 TrS; 14.5 density.
For machining in problem areas, particularly slow speeds; also for dies.
C-0, C-1, C-9, C-10, C-11 type.

VR/WESSON VR-13.
M-329, M-53; 90 WC, 10 Co.
Sintered: 350,000 TrS; A 88.8 Rock.
Density: 14.50.
Cutting tools for mining, quarrying and percussion drilling.
C-13 Type.

VR/WESSON VR 14.
M-329, M-53; 88.5 WC, 11.5 Co.
Sintered: 350,000 TrS; A 88.2 Rock.
Density: 14.40.
Cutting tools for mining, quarrying and percussion drilling.
C-13 Type.

VR/WESSON VR-15.
M-329, M-53; 87 WC, 13 Co.
Sintered: 375,000 TrS; A 87.5 Rock.
Density: 14.30.
Cuttiing tools for mining, quarrying and percussion drilling.
C-14 Type.

VR/WESSON VR-54.
M-329, M-53; 93 WC, 7 Co.
Sintered: 275,000 TrS; A 91.8 Rock.
Density: 14.60.
For milling, broaching, hobbing, cast ferrous and non-ferrous alloys. Very good shock resistance.

VR/WESSON VR-65.
M-53; M-329; TiC, Mo_2C, Ni.
Sintered: 92.3 RA; 150,000 TrS; 5.85 density.
For high speed fine finishing and boring.
C-4, C-8 types.

VR/WESSON VR-71.
M-329, M-53; 66 WC, 18 TiC, 10 TaC + CbC, 6 Co.
Sintered: 200,000 TrS; A 92,5 Rock.
Density: 10.90.
For high speed finishing and precision boring of steel. Excellent wear resistance.
High resistance to cutting temperature.
C-7 type.

VR/WESSON VR-73.
M-329, M-53; 71.5 WC, 12 TiC, 10 TaC + CbC, 6.5 Co.
Sintered: 200,000 TrS; A 92.0 Rock.
Density: 11.90.
For finish machining of steel. Very good wear resistance; resistance to high cutting temperature.
C-7 Type.

VR/WESSON VR-75.
M-329, M-53; 74.5 WC, 8.0 TiC, 10 TaC + CbC, 7.5 Co.
Sintered: 250,000 TrS; A 91.5 Rock.
Density: 12.70.
For general purpose machining steel.
Good shock and wear resistance; resistance to high cutting temperature.

VR/WESSON VR-77.
M-329, M-53; 73.5 WC, 8 TiC, 10 TaC + CbC, 8.5 Co.
Sintered: 265,000 TrS; A 91.3 Rock.
Density: 12.60.
For heavy roughing and interrupted cuts on steel.
Very good shock resistance.

VR/WESSON VR-82.
M-53, M-329; 92 WC, 2.5 TaC/CbC, 5.5 Co.
Sintered: 92.5 RA; 250,000 TrS; 14.95 density.
For straight machining of high temperature alloys where chips are tough. Fair shock resistance; good wear resistance.
C-2, C-3, C-9 types.

VR/WESSON VR-85.
M-53, M-329.
Sintered: 91.5 Rd, 260,000 TrS; 14.9 density.
For straight machining of high temperature alloys where the chips are tough.

VR/WESSON VR-87.
M-329, M-53; 55 WC, 28 TaC, 17 Co.
Sintered: 350,000 TrS; A 85.0 Rock.
Density: 13.50.
Good hot hardness, for machining hot flash from heavy weld in steel pipe. Excellent resistance to shock and to cutting temperatures.

VR/WESSON VR-89.
M-329, M-53; 71 WC, 18 TaC, 11 Co.
Sintered: 275,000 TrS; A 89.9 Rock.
Density: 14.15.
For turning and milling tough alloy steels as high manganese steels; excellent shock resistance.

VR/WESSON VR-97.
M-53, M-329.
Al_2O_3, ceramic.
Sintered: 93.5 RA; 100,000 TrS; 3.98 density.
For cutting tools, continuous cuts at extremely high speeds, and hard metals, up to 60 Rc, at lower speeds.

VR/WESSON VR-100.
M-53, M-329.
Al_2O_3, TiC ceramic.
Hot pressed: 95.0 RA; 115,000 TrS; 4.3 density.
For cutting tools, turning and boring cast irons and steels up to 70 Rc, non-metallic materials, and milling various cast irons and steels at high speeds.

VR/WESSON WH.
M-329, M-53; 78 WC, 13 TiC, 2 TaC, 7 Co.
Sintered: 220,000 TrS; A 92.0 Rock.
Density: 11.75.
For general purpose and finish machining of steel. Good wear resistance.
C-7 Type.

VR/WESSON WM.
M-329, M-53; 75 WC, 13 TiC, 2 TaC + CbC, 10 Co.
Sintered: 250,000 TrS; A 91.0 Rock.
Density: 11.50.
For roughing and general machining steel. Good shock resistance.

VR/WESSON WS.
M-329, M-53; 83.5 WC, 4.0 TiC, 2.0 TaC + CbC, 10.5 Co.
Sintered: 275,000 TrS; A 90.5 Rock.
Density: 13.20.
For rough machining steel; interrupted cuts. Excellent shock resistance.

V.S.4 STEEL.
M-210; 0.75-1.05 C, 0.30 Mn, 0.25 V, bal Fe.
For cold heading dies, hammers, pistons; Type W2; water hardened.

VS-30.
M-72; 15 Cr, 52 Co, bal Fe.
Remanence 18,000; coercive force 25.
For hysteresis motors, electro and magnetic devices.
Vicalloy type permanent magnet.

VS-60.
M-1733; 0.60 C, 2.0 Si, 0.87 Mn, bal Fe.
Spring steel.
AISI 9260; DIN 65 Si 7.

VSC.
M-1655; 2.25 C, 0.30 Mn, 0.30 Si, 14.5 Cr, 1.0 Mo, 0.10 V, bal Fe.
Oil or air hardenable to 64-66 Rc.
For cold forming and precision rollers.

V.S.M see CARPENTER V.S.M.

VT-1-O.
M-USSR.
Unalloyed titanium.

VT-5.
M-USSR; 4.0-5.5 Al, bal Ti.

VT-5-1.
M-USSR; 4.0-5.5 Al, 2.0-3.0 Sn, bal Ti.
Annealed: 125,000 TS; 120,000 YS; 15 El.
For compressor blades, engine cowlings, support rings.
Alpha alloy. Corrosion resistant.

VT-6.
M-USSR; 5.0-6.5 Al, 3.5-4.5 V, bal Ti.
Annealed: 150,000 TS; 140,000 YS; 15 El; C 30 Rock.
For jet engine components, airframe parts, fasteners.
Alpha-beta alloy, good hot strength.

VT-6S.
M-USSR; 4.5 Al, 3.5 V, bal Ti.
Alpha-beta alloy.

VT-20.
M-1733; 0.20 C, 0.25 Si, 0.45 Mn, bal Fe.
Low carbon steel; AISI 1020; DIN ck22.

VT-21.
M-1733; 0.20 C, 0.25 Si, 0.75 Mn, bal Fe.
Low carbon steel.
AISI 1021.

VT-30.
M-1733; 0.30 C, 0.25 Si, 0.75 Mn, bal Fe.
Carbon steel.
AISI 1030.

VT-38.
M-1733; 0.38 C, 0.25 Si, 0.75 Mn, bal Fe.
Carbon steel.
AISI 1038; DIN CK35.

VT-40.
M-1733; 0.40 C, 0.25 Si, 0.75 Mn, bal Fe.
Carbon steel.
AISI 1040; DIN ck40.

VT-45.
M-1733; 0.45 C, 0.25 Si, 0.75 Mn, bal Fe.
Carbon steel.
AISI 1045.

VT-50.
M-1733; 0.50 C, 0.25 Si, 0.75 Mn, bal Fe.
Carbon steel.
AISI 1050; DIN ck50.

VT-60.
M-1733; 0.60 C, 0.25 Si, 0.75 Mn, bal Fe.
Carbon steel.
AISI 1060; DIN ck60.

VT-70.
M-1733; 0.70 C, 0.25 Si, 0.75 Mn, bal Fe.
Carbon steel.
AISI 1070; DIN ck69.

VT-80.
M-1733; 0.80 C, 0.25 Si, 0.75 Mn, bal Fe.
Carbon steel.
AISI 1080.

VT-95.
M-1733; 0.95 C, 0.25 Si, 0.40 Mn, bal Fe.
Carbon steel.
AISI 1095; DIN ck101.

VTG 129.
M-686; 59.01-61.0 Pd, bal Ni.
Uses: Brazing electron tubes.
M.P. 2260°F.

VTG 238.
M-686; 79.5-80.5 Au, 0.05 max Ag, bal Cu.
For brazing electron tubes.
M.P. 1630°F.

VTG 255.
M-686; 81.6-82.5 Au, bal Ni.
For brazing electron tubes.
M.P. 1742°F.

VTG 260.
M-686; 34.5-35.5 Au, 0.05 max Ag, bal Cu.
For brazing electron tubes.
M.P. 1832-1870°F.

VTG 261.
M-686; 74.5-75.5 Au, 4.5-5.5 Ag, bal Cu.
For brazing electron tubes.
M.P. 1625-1640°F.

VTG 301.
M-686; 71-73 Ag, bal Cu.
For brazing of electron tubes.
M.P. 1435°F.

VTG 428.
M-686; 4.5-5.5 Pd, bal Ag.
For brazing electron tubes.
M.P. 1780-1850°F.

VTG 431.
M-686; 9.5-10.5 Pd, bal Ag.
For brazing electron tubes.
M.P. 1835-1950°F.

VTG 447.
M-686; 19.5-20.5 Pd, bal Ag.
For brazing electron tubes.
M.P. 1960-2150°F.

VTG 478.
M-686; 4.5-5.5 Pd, 67.69 Ag, bal Cu.
For brazing electron tubes.
M.P. 1480-1490°F.

VTG 490.
M-686; 14.5-15.5 Pd, 64.66 Ag, bal Cu.
For brazing electron tubes.
M.P. 1565-1650°F.

VTG 491.
M-686; 9.5-10.5 Pd, 57.59 Ag, bal Cu.
For brazing electon tubes.
M.P. 1520-1565°F.

VTG 492.
M-686; 24.5-25.5 Pd, 53-56 Ag, bal Cu.
For brazing electron tubes.
M.P. 1650-1740°F.

VTG T51.
M-686; 74-76 Ag, 0.25-0.75 Ni, bal Cu.
For brazing electron tubes.
M.P. 1435-1475°F.

V-TOOL.
M-347; 0.9-1.0 C, 0.25-0.40 Mn, 0.2-0.3 V, bal Fe.
For taps, reamers, drills, hobs, broaches; water hardened.

VUL-BRO.
M-275; 1.0 C, 4 Cr, 2.7 V, 6 W, 5 Mo, bal Fe.
For lathe and planer tools, forming tools; Type M3; high speed steel.

VULCAN.
M-275; 0.7-1.2 C, 0.2 V, bal Fe.
For tags, milling cutters, reamers, drills, priming tools; water hardened.

VULCAN.
M-1742; 87% Sn content White Metal.
High speed thin wall linings; Loco/Diesel engines; aircraft applications.

VULCAN 4.
M-275; 0.9 C, 4 Cr, 0.75 V, 0.45 Mo, bal Fe.
For hot work tools and dies, hot shears; hot work steel.

VULCAN 4 HW.
M-275; 0.9 C, 4 Cr, 0.4 Mo, 0.75 V, bal Fe.
For drawing and forming dies, hot bolt and rivet gripper dies; hot work steel.

VULCAN 6-HW.
M-275; 0.60 C, 4 Cr, 0.75 V, 0.45 Mo, bal Fe.
For hot work tools, gripper dies; hot work steel, oil hardened.

VULCAN NO. 14.
M-275; C, alloy, bal Fe.
For tools, dies; oil hardening.

VULCAN NO. 50.
M-275; C, alloy, bal Fe.
For tools, dies; oil hardening.

VULCAN NO. 212.
M-275; 2.1-2.2 C, 0.3 Mn, 12-12.5 Cr, bal Fe.
For tools, dies, cutters; oil hardening, non-deforming.

VULCAN NO. 2300.
M-275; C, alloy, bal Fe.
For tools, dies; oil hardening.

VULCAN NO. 4870.
M-275; 0.5 C, 2.0 Si, 0.8 Mn, 0.25 V, 0.25 Cr, bal Fe.
For cold punches, shears, dies, chisels; shock resisting.

VULCAN A-24.
M-275; 0.3 C, 9 W, 2.75 Cr, bal Fe.
For die casting dies; oil hardening.

VULCAN A-41.
M-275; 0.45 C, 1.6 Cr, 0.25 V, 1.1 Mo, bal Fe.
For hot bolt and rivet headers, hot extrusion dies; hot work steel.

VULCAN A-42.
M-275; 0.25 C, 2.75 Cr, 10 W, 2 Ni, bal Fe.
For brass forging and die casting dies, extrusion and gripping dies, hot punches.
AISI Type H21 modified.

VULCAN ALIDIE.
M-275; 1.55 C, 12 Cr, 0.75 Mo, 0.25 V, bal Fe.
For drawing and blanking dies, shear blades, cold swaging dies; air or oil hardening.

VULCAN AUTO.
M-275; 0.3-0.55 C, 1.0 Cr, 0.2 V, 0.8 Mn, 0.25 Si, bal Fe.
For drive shafts, spindles, clutch dogs, plastic dies, punches, arbors; water or oil hardening; tough and fatigue resistant.

VULCAN CARBON VANADIUM.
M-275; 0.7-1.4 C, 0.2 V, bal Fe.
For tools, dies, drills; water hardening.

VULCAN CK.
M-275; C, alloy, bal Fe.
For tools; water hardening.

VULCAN DRAWING DIE.
M-275; C, alloy, bal Fe.
For tools, dies; oil hardening.

VULCAN EXTRA.
M-275; 0.7-1.0 C, bal Fe.
For lathe tools, milling cutters, twist drills, taps, reamers, punches, dies, shear blades; water hardened.

VULCAN EXTRA VANADIUM.
M-275; 0.7-1.2 C, 0.2 V, bal Fe.
For tools, dies; water hardening.

VULCAN FORT PITT.
M-275; 0.7-1.2 C, 0.3 Mn, 0.25 Si, bal Fe.
Annealed: 100,000 TS; 53,000 YS; 21 El; 197 Brin.
Ht. Tr.: 200,000 TS; 138,000 YS; 12 El; 390 Brin.
For tools, cutters, drills, reamers, die blocks, chasers, chisels, shear blades, forming dies. Type W1 water hardening tool steel. Wear resistant.

VULCAN-KIDD BFW.
M-275; 0.45 C, 0.35 Mn, 1.5 Si, 1.6 Cr, 0.25 V, bal Fe.
For coining and embossing dies, punches, stamps, rivet sets, upsetters.
Oil hardening, high toughness, shock resistant.

VULCAN "K.R.".
M-275; 1.1 C, 0.3 Mn, 0.35 Si, 0.6 Cr, bal Fe.
For brake drum drawing dies, automotive rim rolls, dies and punches for deep drawing steel; wear and abrasion resistant; AISI W5.

VULCAN METAL.
M-Eng.; 80.5 Cu, 11.7 Al, 4.4 Fe, 0.25 Zn, 1.5 Ni, 0.7 Sn, 0.7 Cr.
For corrosion resistant structural parts; corrosion resistant.

VULCAN NON-SHRINKABLE.
M-275; 0.9 C, 1.6 Mn, 0.3 Cr, bal Fe.
For plug and ring gages, intricate dies; non-shrinking and non-deforming, oil hardening.

VULCAN OIL HARD.
M-275; 0.9 C, 1.2 Mn, 0.5 Cr, 0.5 W, 0.15 V, bal Fe.
For arbors, bushings, dies, punches, cutters; Oil hardening, tough and shock resisting; AISI O1.

VULCAN PLASTIC DIE.
M-275; 0.08 max C, 0.3 max Mn, 0.1 max Si, bal Fe.
Water hardening steel for cavity dies, and plastic mold dies; usually case hardened; also for machine gaskets.

VULCAN Q.A.
M-275; 0.5 C, 1.2 Cr, 0.3 V, 2.2 W, bal Fe.
For hot punches, dies, shear blades, chisels, rivet sets, swaging and forming dies; hot work steel; AISI S1.

VULCAN REGAL.
M-275; 1.35 C, 3.5 W, bal Fe.
For cutters, tools, shears; finishing steel.

VULCAN REGAL NO.2.
M-275; 1.3 C, 5.5 W, 0.6 Cr, bal Fe.
For finishing tools, dies; keen cutting edge.

VULCAN RMK.
M-275; 0.52 C, 5 Cr, 1.1 Si, 1 V, 1.5 Mo, 1.5 Ni, bal Fe.
For hot and cold work tools, punches; Type A9, hot work steel, air hardened.

VULCAN SILICO-MANGANESE.
M-275; 0.9 C, 2 Si, 1.1 Mn, bal Fe.
For tools, dies; oil hardening.

VULCAN SPECIAL.
M-275; 1.0-1.2 C, bal Fe.
For milling cutters, lathe and planer tools, dies and punches; for excessive requirements.

VULCAN SPECIAL VANADIUM.
M-275; 0.85-1.05 C, 0.3 Mn, 0.3 Si, 0.3 V, bal Fe.
For taps, blanking dies, cold heading dies, reamers, chisels, shear blades pneumatic rivet sets. Water hardening tool steel. AISI W2.

VULCAN STAINLESS IRON C-2.
M-275; 0.12 max C, 15 max Cr, bal Fe.
For corrosion resistant parts; corrosion resistant.

VULCAN STAINLESS TYPE A.
M-275; 0.12 min C, 9-16 Cr, bal Fe.
For corrosion resistant parts; corrosion resistant.

VULCAN SUPER.
M-275; 0.8 C, 19 W, 4 Cr, 2 V, 0.7 Mo, bal Fe.
For cutting tools for very hard metal; high speed steel.

VULCAN TCM.
M-275; 0.35 C, 5 Cr, 1.0-1.5 Mo, 1.0 Si, 0.25 V, 1.25 W, bal Fe.
For die casting dies, hot punches; hot work steel.

VULCAN TM-5.
M-275; 0.7-0.8 C, 4 Cr, 1 V, 8-9 Mo, 1.5-1.6 W, bal Fe.
For broaches, chasers, forming dies, lathe and planer cutters, taps; high speed steel.

VULCAN TM-6.
M-275; 0.79-0.86 C, 4.15 Cr, 6.4 W, 5.0 Mo, 1.9 V, bal Fe.
For tools, cutters, broaches, drills, hobs; high speed steel.

VULCAN TOOL.
M-275; 0.7-1.2 V, bal Fe.
For tools, cutters; water hardening.

VULCAN VAIRLOY.
M-275; 1.0 C, 2.0 Mn, 0.9 Cr, 0.9 Mo, bal Fe.
For air hardening, non-deforming blanking or bending dies, gauges, punches, cams, shaving dies.
AISI Type A4.

VULCAN VULDIE.
M-275; 1 C, 5.25 Cr, 1.15 Mo, 0.25 V, bal Fe.
For cold work tools; air hardening, non-deforming, wear resistant.

VULCAN WOLFRAM COBALT.
M-275; 0.72 C, 18 W, 4 Cr, 1 V, 5 Co, 0.5 Mo, bal Fe.
For cutting tools, reamers, drills, hobs; high speed steel.

VULCAST.
M-275; 0.35 C, 1.5 Mo, 1 V, 5 Cr, bal Fe.
Heat treated: 215,000 TS; 184,000 YS; 13 El; 40 RA.
For upsetters, punches, forming dies; hot work steel; Type H.

VULC-IRON.
M-938; 3.3 C, 2.5 Si, bal Fe.
For die casting equipment; cast iron.

VULKAN.
M-1331; 0.56 C, Ni, Cr, Mo, V, bal Fe.
For gears, crimpers, crankshafts, bolts, studs; oil hardened, shock resistant.

VULKAN.
M-1796; 0.80 C, 4.25 Cr, 0.6 Mo, 18.0 W, 1.6 V, 5 Co, bal Fe.
High speed steel; for rough machining.

VULKAN 1.
M-1288; 2.1 C, 11.5 Cr, 0.7 W, bal Fe.
For blanking, piercing and forming dies, punches; oil or air hardened, non-deforming.

VULKAN 2B.
M-1288; 1.65 C, 11.5 Cr, 0.1 V, bal Fe.
For blanking, piercing and forming dies, punches; air hardened, non-deforming.

VULKAN 2MC.
M-1288; 1.5 C, Co, Mo, Cr, bal Fe.
For blanking and piercing dies, punches; air hardened, non-deforming.

VULKAN 2MW.
M-1288; 1.65 C, Cr, Mo, W, V, bal Fe.
For blanking and piercing dies, punches; air hardened, non-deforming.

VULKAN 3.
M-1288; 2.1 C, 11.5 Cr, bal Fe.
For blanking, piercing and forming dies, punches; oil or air hardened, non-deforming.

VULKAN 3MW.
M-1288; 1.5 C, Cr, Mo, V, bal Fe.
For blanking and piercing dies; air hardened, non-deforming.

VULKAN 3V.
M-1288; 2.34 C, Cr, V, bal Fe.
For blanking, piercing, forming dies, punches; oil hardened, wear resistant.

VULKAN 4.
M-1288; 1.35 C, Cr, bal Fe.
For bearings, liners, cutters, engravers' tools; water hardened, wear resistant.

VULKAN 55.
M-1331; 0.55 C, 0.25-0.50 Si, 0.30-0.80 Mn, bal Fe.
Annealed: 100,000 TS; 55,000 YS; 14 El; 20 RA; 175 Brin.
For axes, axles, shafts, gears, crankshafts; water hardened.

VULKAN 65.
M-1331; 0.60 C, 0.25-0.50 Si, 0.3-0.8 Mn, bal Fe.
Heat treated: 160,000 TS; 113,000 YS; 12 El; 40 RA; 325 Brin.
For rails, hammers, tools, shafts, crankshafts; water hardened.

VULKAN CM.
M-1331; 0.40 C, Cr, Mn, Mo, bal Fe.
For gears, bolts, fasteners, shafts, punches; water hardened.

VULKAN MS.
M-1331; 0.53 C, 0.90 Si, 0.90 Mn, bal Fe.
Annealed: 98,000 TS; 55,000 YS; 15 El; 22 RA; 180 Brin.
For punches, cripers, gears, shafts; water hardened.

VULKAN N.
M-1331; 0.55 C, 0.70 Cr, 0.18 Mo, 1.65 Ni, 0.1 V, bal Fe.
For gears, shafts, crankshafts, bolts, studs; oil hardened, shock resistant.

VULKAN SPEZIAL.
M-1331; 0.35 C, 1.35 Cr, 0.25 Mo, 3.9 Ni, bal Fe.
For gears, bolts, crankshafts; oil hardened, shock resistant.

VUL-MAX.
M-275; C, alloy, bal Fe.
For hot work tools and dies; hot work steel, oil hardened.

VUL-MO.
M-275; high C, 3.25-4.25 Cr, 1.25-2 W, 0.75-1.5 V, 7.5-9.5 Mo, (Co), bal Fe.
For reamers, lathe tools, drills, broaches, dies, hot work dies, cutting tools; high speed steel; wear resistant.

VULMOLD.
M-275; 0.10 max C, 0.3 Mn, 1.4 Cr, 0.52 Ni, 0.25 Mo, bal Fe.
For plastic dies, hubbed cavities; carburized grade.

VV.
M-1653; 0.10 C, 0.30-0.70 Mn, bal Fe.
Low carbon steel.
Ital. UNI C10; AISI 1010.

VV-35.
M-1733; 0.35 C, 19.0 Cr, 8.0 Ni, 3.0 Si, bal Fe.
Valve steel; SAE EV-5.

VV-45.
M-1733; 0.45 C, 9.5 Cr, 3.0 Si, bal Fe.
Valve steel.
SAE HNV-3; DIN 1.4718.

VV-53.
M-1733; 0.53 C, 21.0 Cr, 4.0 Ni, 9.0 Mn, 0.42 N, bal Fe.
Valve steel.
SAE EV-8.

VV-73.
M-1733; 0.73 C, 21.0 Cr, 1.7 Ni, 6.3 Mn, 0.2 Ni, bal Fe.
Valve steel.
SAE EV-11.

VV-80.
M-1733; 0.80 C, 19.5 Cr, 1.4 Ni, 2.0 Si, bal Fe.
Valve steel.
SAE HNV-6; DIN 1.4747.

VW-1.
M-1733; 1.15 C, 0.2 Cr, 1.1 W, 0.12 V, bal Fe.
Oil hardening tool steel. For cold work.
Similar to AISI O7; DIN 1.2516.

VW-3.
M-1733; 0.45 C, 1.0 Cr, 0.2 Mo, 1.9 W, 0.12 V, 1.0 Si, bal Fe.
Shock resistant tool steel.
Similar to AISI S 1; DIN 1.2542.

VW-9.
M-1733; 0.30 C, 2.75 Cr, 9.0 W, 0.35 V, bal Fe.
Hot work tool steel.
AISI H-20; DIN 1.2581.

VWK-5.
M-1733; 0.80 C, 4.25 Cr, 14.3 W, 2.15 V, 5.0 Co, bal Fe.
High speed steel.
AISI T-8; DIN 1.3251.

VWK-10.
M-1733; 0.75 C, 4.25 Cr, 0.9 Mo, 18.0 W, 1.6 V, 10.0 Co, bal Fe.
High speed steel.
AISI T5; DIN 1.3255.

VWM-1.
M-1733; 0.80 C, 4.25 Cr, 9.0 Mo, 2.0 W, 1.2 V, bal Fe.
High speed steel.
AISI M 1; DIN 1.3346.

VWM-2.
M-1733; 0.85 C, 4.25 Cr, 5.0 Mo, 6.4 W, 1.85 V, bal Fe.
High speed steel.
AISI M2; DIN 1.3343.

VWM-7.
M-1733; 1.0 C, 4.0 Cr, 8.7 Mo, 1.9 W, 2.1 V, bal Fe.
High speed steel.
AISI M7; DIN 1.3348.

VW MAGNESIUM.
M-Eng.; 8 Al, 0.6 Zn, 0.3 Mn, 0.005 Be, bal Mg.
Permanent mold: 20,000-21,500 TS; 14,000-18,500 YS; 3.25 El; 54-56 Brin.
Die cast: 21,500-24,000 TS; 17,000-20,000 YS; 2 El; 60-64 Brin.
For light alloy parts.

VZH 36-L1.
M-USSR; 10 Cr, 5 Al, 8 W, 4 Mo, 0.3 B, bal Ni.
Cast nickel-base superalloy.

VZH 36-L2.
M-USSR; 0.06 max C, 19-22 Cr, 2.3-2.7 Ti, 3.5-4.0 Al, 0.03 max B, max Fe, bal Ni.
Cast nickel-base superalloy.
For automotive turbine blades.

VZHL8.
M-USSR; 0.1-0.2 C, 4-17 Cr, 1.8-2.5 Ti, 2.5-3.5 Al, 4.5-6.0 Mo, 0.06 B, 8-12 Fe, bal Ni.
Cast nickel-base superalloy.
For nozzle guide vanes.

VZH 98.
M-USSR; 0.10 max C, 25.5 Cr, 15.0 W, bal Ni.
High temperature alloy for tailpipes, after burner liners, combustion cans.

VZKG.
M-Czechoslovakia; 0.20 C, 11 Cr, Mo, V, W, bal Fe.
Annealed: 95,000 TS; 40,000 YS; 25 El; B 92 Rock.
Hardened: 240,000 TS; 205,000 YS; 8 El; C 50 Rock.
For valves, cutlery, gears, bearings, surgical instruments.
Corrosion resistant, hardenable.

W

W-0.33.
85 Al, 14 Cu, 1 Mn.
For pistons, light alloy castings; non-hardenable.

W 2.
M-England; 0.7 C, 3.3 Cr, 0.8 Si, 16 W, 4 Co, 0.4 Ni, 0.6 Mo, bal Fe.
For cutters, taps, drills; high speed steel.

W. 3 TUNGSTEN.
M-344; C, W, bal Fe.
For hot dies; heat resistant.

W4X see FINKL W4X.

W-5.
M-69; 1.15 C, 0.25 Mn, 0.25 Si, 0.50 Cr, bal Fe.
AISI Type W5 water hardening tool steel.

W 6 ALLOY.
M-121; 0.5 Mn, 2.0 Si, bal Ni.
Annealed: 81,000 TS; 24,000 YS; 56 El; 77 RA; 130 Brin.
For spark plug electrodes; resists engine fuels.

W6 AND 4.
M-1113; C, 5.5 W, 4 Mo, 4 Cr, 1.5 V, bal Fe.
For tools, cutters; high speed steel.

W.9 TUNGSTEN.
M-344; C, W, bal Fe.
For hot dies; heat resistant.

W-25% RE same as WAH CHANG W-25 RE.

W 311.
M-1290; 0.35-0.45 C, 0.20-0.40 Si, 1.30-1.60 Mn, 1.80-2.10 Cr, 0.15-0.25 Mo, bal Fe.
Hot work tool steel.
Werkstoff Nr. 1.2311.

W 367.
M-1290; 0.37-0.43 C, 0.30-0.50 Si, 0.30-0.60 Mn, 4.7-5.2 Cr, 2.7-3.3 Mo, 0.80-1.0 V, bal Fe.
Deep hardening structural steel.
Werkstoff Nr. 1.2367.

W 376.
M-1290; 0.90-1.0 C, 0.20-0.40 Si, 0.20-0.40 Mn, 11.0-12.0 Cr, 0.85-0.9 Mo, 0.85-0.95 V, bal Fe.
Cold work tool steel.
Werkstoff Nr. 1.2376.

W-545 see also WESTINGHOUSE W-545.

W-545.
M-1491; 0.08 max C, 1.5 Mn, 0.4 Si, 13.5 Cr, 26 Ni, 1.5 Mo, 2.85 Ti Al, 0.08 B, bal Fe.
For gas turbine parts, high temperature bolting.

W 565.
M-1290; 0.35-0.45 C, 0.80-1.10 Si, 0.30-0.50 Mn, 5.1-5.3 Cr, 3.5-4.0 W, 0.15-0.20 V, bal Fe.
Hot work tool steel.
Werkstoff Nr. 1.2565.

W 622.
M-1290; 0.55-0.65 C, 0.20-0.40 Si, 0.20-0.40 Mn, 3.0-4.1 Cr, 0.85-0.95 8.5-9.5 W, 0.60-0.80 V, bal Fe.
Hot work tool steel.
Werkstoff Nr. 1.2622.

W-722 see NICKELVAC W-722.

WA-1.
M-1707; 94.0 WC; 6 Co.
Sintered: 290,000 TrS; A 91.0 Rock.
Density: 14.95; Abrasion resistance: 75%.
For roughing cuts on cast iron and non-ferrous materials: burr blanks, lathe cutters, masonry drills.
Code C-1.

WA-2.
M-1707; 94 WC, 6 Co.
Sintered: 270,000 TrS; A 92.0 Rock.
Density: 14.95; Abrasion resistance: 100%.
For general purpose machining cast iron and non-ferrous materials.
Code C-2 or C-3.

WA-3.
M-1707; 95.7 WC, 4.3 Co.
Sintered: 250,000 TrS; A 92.6 Rock.
Density: 15.1; Abrasion resistance: 110%.
Good general purpose machining cast iron and non-ferrous materials.
Code C-3.

WA-4.
M-1707; 97 WC, 3 Co.
Sintered: 240,000 TrS; A 92.8 Rock.
Density: 15.2; Abrasion resistance: 140%.
Hardest grade of tungsten carbide; used for sand blast nozzles, wear parts, precision boring tools; brittle.
Code C-4.

WA-5.
M-1707; 72 WC, 8 Tic, 11.5 TaC, 8.5 Co.
Sintered: 300,000 TrS; A 91.2 Rock.
Density: 12.45; Abrasion resistance: 35%.
General purpose roughing cuts on steel, premium steel cutting grade.
Code C-5.

WA-7.
M-1707; 76.5 WC, 12.0 Tic, 4.0 TaC, 7.5 Co.
Sintered: 225,000 TrS; A 92 Rock.
Density: 11.70; Abrasion resistance: 41%.
For light roughing cuts or heavy finishing cuts on steel.
Code C-7.

WA-8.
M-1707; 75 WC, 16 TiC, 4 TaC, 5 Co.
Sintered: 150,000 TrS; A 93.0 Rock.
Density: 11.30; Abrasion resistance: 53%.
For precision boring or turning of steel.
Code C-8.

WA-10.
M-1707; 91 WC, 9.0 Co.
Sintered: 325,000 TrS; A 89.2 Rock.
Density: 14.65; Abrasion resistance: 39%.
Essentially for wear applications with light shock, as light blanking dies, hard mandrels.
Code C-10.

WA-12.
M-1707; 87 WC, 13.0 Co.
Sintered: 375,000 TrS; A 88.2 Rock.
Density: 14.20; Abrasion resistance: 21%.
For impact applications as blanking and forming dies, rock bits, coal mining tools.
Code C-12.

WA-14.
M-1707; 75 WC, 25 Co.
Sintered: 400,000 TrS; A 84.5 Rock.
Density: 13.05.
Strong material that can be drilled with carbide drills; cold header tooling.
Code C-14.

WA-35.
M-1707; 94 WC, 6 Co.
Sintered: 260,000 TrS; A 92.6 Rock.
Density: 14.95.
For light finishing cuts on cast iron and non-ferrous materials. Code C-3.

WA-41.
M-1707; 92 WC, 8 Co.
Sintered: 310,000 TrS; A 91.0 Rock.
Density: 14.75.
For rough machining on cast iron; also for light percussion tools as vibrating masonry drills, percussion bits, etc.
Code C-1.

WA-47.
M-1707; 78 WC, 8 Tic, 6 Tac, 8 Co.
Sintered: 250,000 TrS; A 91.9 Rock.
Density: 12.40.
Premium steel cutting grade; for heavy cuts on railroad wheels; milling cutter blades, trepanning tools.
Code C-7.

WA-54.
M-1707.
Sintered; 375,000 TrS; A 90.7 Rock.
Density: 13.5.
Very heavy roughing cuts on steel.
Code C-5.

WA-57.
M-1707.
Sintered: 300,000 TrS; A 91.8 Rock.
For medium to heavy cuts in steel; resists cutting edge deformation.
Code C-5.

WA-59.
M-1707; 9 WC, 1 Tic, 8 Co.
Sintered: 300,000 TrS; A 91.8 Rock.
Density; 14.75; Abrasion resistance 85%.
For roughing cuts on cast iron and non-ferrous materials; for general machining, including stainless.
Code C-1.

WA-68.
M-1707; 79.5 WC, 8.5 Tic, 6.0 TaC, 6.0 Co.
Sintered: 200,000 TrS; A 93.0 Rock.
Density: 12.6.
High hardness; for finishing cuts on steel and special alloys; precision boring.
Code C-7.

WA-69.
M-1707; 91 WC, 4 TaC, 5 Co.
Sintered: 250,000 TrS; A 91.9 Rock.
Density: 14.9.
For general machining of cast iron rolls, brake drums.
Code C-2 or C-3.

WA-110.
M-1707.
Sintered: 375,000 TrS; A 91.7 Rock.
For end mills, drills, and for machining high tensile material and super alloys.
Code C-1-C-2.

WA-114.
M-1707; WC base.
Sintered: 400,000 TrS; A 90.0 Rock.
For light blanking dies, coal and rock drilling tools, impact and cold extrusion, shock resistant wear parts.
Code C-11 or C-12.

WA-119.
M-1707; WC base.
Sintered: 500,000 TrS; A 88.0 Rock.
High strength and toughness; for cold header punches and dies, blanking an stamping dies, coining punches, impact tooling.
Code C-12.

WA-510.
M-1707.
Sintered: A 91.0 Rock.
Non-magnetic.
Main use is wear parts for magnetic tape handling equipment.

WA-800.
M-1707; TiC base.
Sintered: 225,000 TrS; A 92.7 Rock.
Finish machining pearlitic irons and steels.
Code C-7 or C-8.

WABCOLOY.
M-384; 2.6 C, 2.5 Si, 0.6 Mn, 1.1 Ni, 0.9 Mo, bal Fe.
Cast: 55,000 TS; 0 El; 255-321 Brin.
For cylinders of steam driven compressors, crankshafts; cast iron; wear resistant.

WAGNER & GUHRS ALUMINUM.
80 Sn, 20 Zn.
For aluminum solder; soft.

WAGNERS FORMULA.
50-66 Cu, 19-31 Zn, 13-18 Ni.
For base for plated tableware; corrosion resistant.

WAH CHANG 129Y.
M-1537; 9-11 W, 9-11 Hf, 0.5 Ta, 0.1-0.4 Y, 0.5 W, bal Cb.
Recrystallized: 89,500 TS; 75,700 YS; 25 El.
At 1200°F: 65,000 TS; 41,600 YS; 17 El.
At 3000°F: 11,000 TS; 9900 YS; 75 Min El.
For space vehicles, missiles, rocket components.
High temperature refractory alloy. Good weldability and fabricability.

WAH CHANG C-120.
M-1537; 5 Mo, 1 Zr, 15 W, bal Cb.
At 70°F: 116,000 TS; 103,000 YS; 4 El.
At 2500°F: 27,000 TS; 20,000 YS; 50 El.
For space vehicles, nuclear reactors.
Good coombination of density, strength and oxidation resistant at high temperatures.

WAH CHANG C129.
M-1537; 10 W, 10 Hf, bal Cb.
Recrystallized: 89,000 TS; 70,000 YS; 23 El.
At 1200°F: 60,000 TS; 37,000 YS; 17 El.
At 3000°F: 10,000 TS; 9500 YS; 75 El.
For space vehicles, missiles, rocket components.
High temperature refractory alloy.
Good weldability and fabricability.

WAH CHANG CB-33 ZR.
M-1537; 33 Zr, bal Cb.
Uses: superconductors.

WAH CHANG W-25RE.
M-1537; 24-26 Re, bal W.
Sintered: 177,000 TS; 175,000 YS; 33 RA.
Recrystallized: 178,000 TS; 164,000 YS; 1-2 El.
For electronic tubes, cathodes, heat exchangers, furnace parts.
High temperature properties, heat resistant.

WAH CHANG WC-1ZR.
M-1537; 0.8-1.2 Zr, bal Cb.
Recrystallized: 42,000 TS; 23,000 YS; 42 El.
For thermal barriers, high temperature parts.
High heat and oxidation resistant.

WAH CHANG WC103.
M-1537; 1.0 Ti, 10.0 Hf, 0.7 Zr, 0.5 Ta, 0.5 W, bal Cb.
Stress Relieved:
At 70°F: 93,500 TS; 88,000 YS; 9 El.
At 2000°F; 26,400 TS; 18,200 YS; 63 El.
Recrystallized:
At 70°F: 61,000 TS; 42,000 YS; 25-30 El.
At 2500°F: 13,000 TS; 10,500 YS; 70 min El.
For space vehicles, nuclear reactors.
Good combination of strength, density, and oxidation resistance at high temperatures.

WAH CHANG WC-3015.
M-1596, M-1537; 28-30 Hf, 13-16 W, 0.4 Ta, 1-2 Zr, 0.07-0.30 C, bal Cb.
Extruded: 140,000-147,000 TS; 135,000-140,000 YS; 5-20 El.
For turbines, machine guns, high temperature components.
Resists high temperature oxidation.
Develops its own surface coating.

WAH CHANG XB-88.
M-1537, M-118; 28 W, 2 Hf, 0.067 C, bal Cb.
For gas turbine blades and buckets.
Low resistance to shock, good fatigue properties.
Heat resistant.

WALCALOY NO. 1.
M-963.
General purpose 18-8 stainless wire for use with Wirespray gun for build-up on 18-8 stainless and similar steels.

WALCALOY NO. 2.
M-963.
Martensitic stainless wire for use with Wirespray gun for build-up on carbon, alloy and 400 series stainless for wear and low shrink requirements.

WALCOLOY NO. 4.
M-963.
An 18-10 (316 type) stainless (with Mo) wire for use with Wirespray gun for build-up with improved corrosion resistance.

WALCOLOY NO. 5.
M-963.
Austenitic stainless wire for use with Wirespray gun for build-up of good work-hardening and wear surface.

WALCO METAL.
M-88; Pb alloy.
For anode for Cr plating, tank lining: resists chromic acid.

W-AL-CO TYPE 1.
M-939; 5 Si, bal Al.
For arc welding electrodes; for Al.

W-AL-CO TYPE 2.
M-939; 5 Si, bal Al.
For gas welding rod; for Al.

W-AL-CO TYPE E-2S.
M-939; 5 Si, bal Al.
For welding rod; for Al arc welding.

W-AL-CO TYPE G-2S.
M-939; 5 Si, bal Al.
For welding rod; for Al gas welding.

WALDE-LOY 100 METAL.
M-940; 87.5 Pb, 5.5 Sn, 5.3 Sb, 1.6 Cu, 0.10 As.
For storage battery cable terminals and lugs; resists H_2SO_4.

WALLEX NO. 1.
M-963; 30.0 Cr, 12.5 W, 2.25 C, 1.25 Si, bal Co, (others 6.0 max).
Melt point: 2355°F; 1290°C.
Bare rods or castings: 50-55 Rc; excellent corrosion resistance and low coefficient of friction gives good metal to metal wear.

WALLEX NO. 6.
M-963; 0.90-1.4 C, 25-31 Cr, 3.5-6.0 W, 0.75-1.75 Si, 6.5 max others, 55 min Co.
Cast: 39-44 Rock C.
For hard facing electrode; corrosion, heat and wear resistant.

WALLEX NO. 40.
M-963; 24.0 Ni, 16.0 Cr, 7.0 W, 2.0 Fe, 1.5 Si, 0.50 C, 2.0 B, bal Co.
Melt point: 2080°F; 1140°C.
Atomized powder for Spraywelder, Fusewelder.
Deposited metal 41-46 Rc.

WALLEX NO. 50.
M-963; 18.0 Ni, 19.0 Cr, 10.0 W, 2.75 Si, 1.0 Fe, 0.80 C, 3.5 B, bal Co.
Melt point: 2050°F; 1120°C.
Bare rods, atomized powder, castings; deposited metal 56-61 Rc.

WALLEX NO. 55.
M-963; Wallex No. 50 with tungsten carbide particles added.
Cast: C 55-60 Rock; M.P. 2050°F (approx).
Atomized and crushed powder for hard facing areas requiring extreme abrasion resistance; good resistance to heat, galling and corrosion.
Applied by Spraywelder.

WALLEX NO. 505.
M-963; Wallex No. 50 with tungsten carbide particles added.
Cast: C 55-60 Rock; M.P. 2050°F (approx).
Atomized and crushed powder for hard facing areas requiring extreme abrasion resistance; good resistance to heat, galling and corrosion.
Applied by Fusewelder Torch.

W-ALLOY.
82 Al, 12 Cu, 4.5 Zn, 1 W.
For light alloy castings; leak proof, hard.

WALLRAM H1.
M-1349; WC.
For cutting tools, dies; sintered carbides.

WALLRAM H1P.
M-1692; 94 WC, 6 Co.
For cutters and dies.
Wear and abrasion resistant.
Sintered carbides.

WALLRAM Z10.
M-1692; MoC, WC, Ni, Co, V.
For extrusion dies and nibs.
Sintered carbides. High wear and abrasion resistant.

WALMANG NO. 3.
M-963; 0.75 C, 3.5 Ni, 0.6 Si, 14 Mn, bal Fe.
Weld metal: 55 Rock C.
Electrode for hard facing: abrasion impact resisting deposits.

WALMANG NO. 8.
M-963; 0.75 C, 1 Mo, 0.6 Si, 14 Mn, bal Fe.
Weld metal: 50 Rock C.
Electrode wire for hard facing; abrasion and impact resisting deposits.

WALRAMITE.
Tungsten carbide W_2C + WC.
For cutting tools, dies; sintered.

WANDO.
M-114; 0.95 C, 1.2 Mn, 0.5 Cr, 0.2 V, 0.5 W, bal Fe.
For dies, gauges, taps, reamers, punches, slitting saws; Type 01; non-deforming.

WAR BABBITT.
M-993; Sn, Pb, Cu.
Cast: 10,750 TS; 5,630 YS; 2 El; 22 Brin.
For bearings; Babbitt.

WAR BRONZE.
M-Ger.; 2.19 Al, 4.85 Cu, 0.92 Pb, 0.15 Sn, 0.03 Fe, bal Zn.
For shell fuses; cast alloy.

WARDLOWS.
M-392; 0.7-1.4 C, bal Fe.
For tools; water hardened.

WARDLOWS TOUGH.
M-392; 0.9 C, 1.2 Mn, bal Fe.
For tools, dies; tough, oil hardened.

WARDLOWS UUT.
M-1113; 0.5 C, 1.5 Cr, 2.25 W, 0.25 V, bal Fe.
For hot work dies; hot work steel.

WARDLOWS WSV.
M-1113; 0.9 C, 0.1 V, bal Fe.
For tools, taps, drills, punches; water hardened.

WARMAN 5.
M-277; 0.13-0.20 C, 5 Cr, bal Fe.
For corrosion resistant parts; corrosion resistant.

WARMAN 5 M.
M-277; 0.13-0.20 C, 5 Cr, bal Fe.
For corrosion resistant parts; corrosion resistant.

WARMAN 6.
M-277; 0.15 C, 5-7 Cr, bal Fe.
Cast: 90,000 TS: 65,000 YS; 15 El; 35 RA.
For corrosion resisting parts; corrosion resisting.

WARMAN 6 M.
M-277; 5-7 Cr, 0.5-0.7 Mo, 0.18 min C, bal Fe.
Heat treated: 125,000 TS; 100,000 YS; 18 El; 42 RA.
For corrosion resisting parts; corrosion resistant.

WARMAN CALOXO 8-18.
M-277; 7-10 Cr, 17-20 Ni, 0.16 min C, bal Fe.
For stainless, corrosion resisting parts; corrosion resistant, stainless.

WARMAN 13.
M-277; 11-14 Cr, 0.12 min C, bal Fe.
Heat treated: 90,000 TS: 60,000 YS; 24 El; 50 RA.
For corrosion resisting parts; corrosion resistant.

WARMAN 13M.
M-277; 0.12 max C, 11.5-14.0 Cr, 0.5-0.7 Mo, bal Fe.
Heat treated: 95,000 TS; 65,000 YS; 25 El; 54 RA.
For corrosion resisting castings; corrosion resistant.

WARMAN CALDURO 13-2.
M-277; 0.20 C, 12.5-15.0 Cr, 1.5-2.5 Ni, bal Fe.
Cast: 80,000-110,000 TS; 60,000-90,000 YS; 25-10 El; 40-30 RA.
For corrosion resisting parts; corrosion and abrasion resisting; formerly "Nirosta Calduro KM-1."

WARMAN CALMAR 18-8.
M-277; 0.16 C, 17-20 Cr, 7-10 Ni, bal Fe.
Cast: 70,000 TS; 30,000 YS; 50 El; 180 Brin.
For stainless parts; corrosion resisting; formerly Nirosta Calmar KA2.

WARMAN CALMAR 18-8 M.
M-277; 17-20 Cr, 7-10 Ni, 3-4.5 Mo, 0.16 min C, bal Fe.
For stainless, heat and corrosion resisting parts; stainless, corrosion and heat resistant.

WARMAN COLOXO 15-25.
M-277; 14-17 Cr, 23-27 Ni, 3.0-4.5 Mo, 0.20 min C, bal Fe.
For heat and corrosion resisting parts; heat and corrosion resistant.

WARMAN CALOXO 15-25 M.
M-277; 14-17 Cr, 23-27 Ni, 3.0-4.5 Mo, 0.20 min C, bal Fe.
For heat and corrosion resisting parts; heat and corrosion resistant.

WARMAN CALOXO 15-35.
M-277; 14-17 Cr, 33-37 Ni, 0.20 min C, bal Fe.
For heat and corrosion resisting parts; heat and corrosion resistant.

WARMAN CALOXO 18-2.
M-277; 16-20 Cr, 1.5-2.5 Ni, 0.20 min C, bal Fe.
For corrosion resisting parts; corrosion resistant.

WARMAN CALOXO 25-20.
M-277; 0.25 C, 23-27 Cr, 19-21 Ni, bal Fe.
Cast: 75,000 TS, 35,000 YS; 25 El; 200 Brin.
For furnace and heat resisting parts; corrosion and abrasion resisting; formerly "Nirosta Caloxo KNC-3."

WARMAN CALOXO 25-20 M.
M-277; 23-27 Cr, 17-21 Ni, 3.0-4.5 Mo, 0.20 min C, bal Fe.
For heat and corrosion resisting parts; heat and corrosion resistant.

WARMANS CALOXO 18.
M-277; 0.16 C, 16-20 Cr, bal Fe.
Cast: 90,000 TS; 45,000 YS; 15 El; 30 RA.
For corrosion resisting parts; corrosion resisting; formerly "Warman Chrome Stainless."

WARMBRONZE.
M-1426; 88 Cu, 2 Sn, 10 Zn.
1/4 H-temper: 53,000 TS; 37 El; 107 Brin.
H-temper: 95,000 TS; 3 El; 172 Brin.
For hardware, plumbing; corrosion resistant.

WARMPRESSTAHL M43W.
M-1310; 0.30 C, 2.65 Cr, 0.35 V, 8.5 W, bal Fe.
For extrusion dies, rams and liners; oil hardened, hot work steel.

WARM WORKED 321.
M-1724; 0.08 max C, 17.0-19.0 Cr, 9.0-11.0 Ni, Ti = 5 X C/0.70, bal Fe.
High proof stress version of 321, obtained by controlled low temperature hot working.

WARNES METAL.
37 Sn, 26 Ni, 26 Bi, 11 Co.
Used as substitute for Ag in making ornamental articles; corrosion resistant.

WARPLIS; DRILL ROD.
M-370; 0.90 C, 0.5 Cr, 1.1 Mn, 0.5 W, 0.15 V, bal Fe.
Oil hardenable to 64 Rc max.
For tools, dies, gages, jigs.
Type O1 oil hardening tool steel.

WARRANTED BEST.
M-261; 0.7-1.0 C, bal Fe.
For tools; water hardened.

WARRANTED BEST CAST STEEL.
M-1408; high C, bal Fe.
For taps, reamers, broaches; water hardened, general purpose.

WARRANTED BEST DOUBLE SHEAR STEEL.
M-1408; high C, bal Fe.
For dies, machine knives; water hardened.

WARRANTED C 16.A.
M-1415; 0.5 C, 1 Si, 1.5 Cr, bal Fe.
For pneumatic tools, chisels, beading tools; oil hardened, shock resistant

WARRANTED CAST STEEL.
M-1112; 1.0 C, bal Fe.
For tools, taps, reamers; water hardened.

WASHER BRASS.
M-U.S.; 62 Cu, 38 Zn.
For washers, hardware; yellow brass.

WASHINGTON. (AISI W1 GR. 1).
M-166, M-69; 0.6-1.4 C, bal Fe.
For tools, general tools, broaches, cold heading dies, cutters; water hardened.

WASHINGTON SPECIAL.
M-69; 0.60-1.40 C, 0.2 V, 0.35 Mn, 0.25 Si, bal Fe.
For drills, reamers, springs, punches, taps, cutters; AISI W2; water hardened.

WASPALLOY.
M-44; 20 Cr, 13 Co, 4 Mo, 3 Ti, 1 Al, bal Ni.
Welding wire.

WASPALOY see also **ALLVAC, CARPENTER; CRUCIBLE, UDIMET** and **UNITEMP WASPALOY.**

WASPALOY.
M-1, M-1214, M-1317; 59 Ni, 1.95 Cr, 13.5 Co, 4.2 Mo, 3.0 Ti, 1.2 0.07 C, 0.7 Mn, 2 max Fe.
Heat treated: 188,000 TS; 115,000 YS; 28 El; 25 RA; 375 Brin.
At 1400°F; 117,000 TS; 99,000 YS; 28 El; 41 RA.
For jet engine turbine buckets and discs, high temperature bolts; heat resistant, high strength, high stress rupture strength up to 1400°F.

WATCH ALLOY-1.
70 Pd, 25 Cu, 4 Ag, 1 Ni.
For watch cases; corrosion resistant.

WATCH ALLOY-2.
50 Cu, 47.2 Ni, 2.8 Cd.
For watch cases; corrosion resistant.

WATCH ALLOY-3.
37.5 Au, 27 Cu, 23 Ag, 12.5 Pd.
For watch cases; corrosion resistant.

WATCH CASE BEZEL.
60-63 Cu, 24-21 Zn, 16 Ni.
For watch case bezel; corrosion resistant.

WATCH CASE METAL.
55-65 Cu, 30-16 Zn, 10-28 Ni.
For watch cases; corrosion resistant.

WATCHMAKERS ALLOY.
59 Cu, 40 Zn, 1.2 Pb.
For watch parts; free-cutting.

WATCH NICKEL.
M-942; 12 Ni, 64 Cu, 1 Pb, 23 Zn.

WATERCRAT.
M-1149; 1.05 C, 0.35 Mn, 0.2 Si, 0.5 Cr, bal Fe.
For tools, dies, shear blades, punches; precision ground flat stock.

WATERDIE EXTRA.
M-35; 1.0 C, 0.35 Mo, 0.25 Si, 0.50 Cr, bal Fe.
Water hardening tool steel.
For bushings, forming dies, burnishing rolls.
Type W5-2.

WATERDIE STANDARD.
M-35; 1.0 C, 0.35 Mn, 0.25 Si, 0.50 Cr, bal Fe.
Water hardening tool steel.
For automotive tools, wear plates, mandrels.
Type W5-3.

WAUKESHA B.
M-941; 0.12 C, 1.0 Mn, 1.0 Si, 1.0 Cr, 26.6-33.0 Mo, 2.5 Co, 0.60 V, 6.0 Fe, bal Ni.
Cast corrosion resistant alloy.
ACI N-12 M.

WAUKESHA C.
M-941; 0.12 C, 1.0 Mn, 1.5 Si, 15.0-20.0 Cr, 16.0-20.0 Mo, 2.5 Co, 0.40 V, 7.5 Fe, 5.25 W, bal Ni.
Cast corrosion resistant alloy.
ACI CW-12 M.

WAUKESHA NO. 0.
M-941; 47-53 Cu, 4-6 Pb, 10-13 Sn, 28-32 Ni, 2.5-3.5 Sb.
Cast: 60,000 TS; 60,000 YS; 1 El; 230 Brin.
For dairy and food processing equipment; corrosion and wear resistant.

WAUKESHA NO. 1.
M-941; 48-54 Cu, 5-7 Pb, 1.5-2.5 Sn, 20-30 Ni, 16-20 Zn.
Cast: 47,000 TS; 25,000 YS; 26-30 El; 105 Brin.
For dairy and food processing equipment; corrosion resistant.

WAUKESHA NO. 3.
M-941; 61-67 Cu, 28-32 Ni, 2-3 Fe, 1-2 Si, 1-2 Mn.
Cast: 105,000 TS; 65,000 YS; 2-5 El; 270 Brin.
For dairy and food processing equipment; corrosion resistant.

WAUKESHA NO. 4.
M-941; 54-60 Cu, 4-6 Sn, 23-27 Ni, 3 Fe, 6 Zn, 1 Si, 2-3 Mn.
Cast: 77,000 TS; 67,000 YS; 1-3 El; 240 Brin.
For dairy and food processing equipment; corrosion resistant.

WAUKESHA NO. 7.
M-941; 54-60 Cu, 4-7 Pb, 3 Sn, 23-27 Ni, 3 Fe, 4-7 Zn.
Cast: 48,000 TS; 27,000 YS; 18-22 El; 104 Brin.

WAUKESHA NO. 11.
M-941; 61-67 Cu, 23-27 Ni, 6-9 Fe, 1-2 Mn, 0.5 max Si.
Cast: 62,000-80,000 TS; 50,000 YS; 16-9 El; 158-164 Brin.
For dairy and food processing equipment; corrosion resistant.

WAUKESHA NO. 23.
M-941; 3.5-4.5 Pb, 7-9 Sn, 78-82 Ni, 6-9 Zn, 0.05-0.20 C.
Cast: 75,000 TS; 50,000 YS; 10 El; 165-190 Brin.
For dairy and food processing equipment, bearings; corrosion and galling resistant.

WAUKESHA NO. 23C.
M-941; 3.0-4.5 Pb, 7-9 Sn, 6-9 Zn, 0.05-0.20 C, 1.5-2.5 Mn, bal Ni.
Cast: 75,000 TS; 65,000 YS; 10 El; 200-230 Brinell.
For dairy and food processing equipment, bearings, bushings, pumps.
Non-galling, corrosion resistant.

WAUKESHA NO. 54C.
M-941; 8 Sn, 8 Zn, 6 Ag, 0.2 C, 2 Mn, bal Ni.
Cast: 85,000 TS; 60,000 YS; 10 El; 220 Brin.
For pumps, valve parts, bushings, bearings, dairy and food processing equipment.
High wear and corrosion resistance. Non-galling.

WAUKESHA NO. 88.
M-941; 0.05 max C, 3-5 Sn, 3.0-4.5 Bi, 2.5-3.5 Mo, 11-14 Cr, 2 max Fe, bal Ni.
Cast: 40,000 TS; 25,000 YS; 6 El; 7 RA; 150 Brin.
For food and chemical industry equipment; non-galling against stainless steel.

WAUKESHA NO. 118.
M-941; 63-69 Cu, 3.5-4.5 Pb, 2.5-3.5 Sn, 18-21 Ni, 3 Fe, 3-5 Zn.
Cast: 52,000 TS; 30,000 YS; 20 El; 123 Brin.
For dairy and food processing equipment; corrosion resistant.

WAUKESHA NO. 120.
M-941; 58-61 Cu, 5.5-6.5 Sn, 3.5-4.5 Pb, 3-5 Zn, 23-26 Ni, 1.5-2.5 Fe.
Cast: 60,000 TS; 45,000 YS; 5 El; 160 Brin.
For food and dairy.

WAUSAU.
M-1106; 3.1 T.C., 0.6 C.C., 1.2 Mo, 2.2 Si, 0.8 Mn, 0.5 Ni, 0.15 Cr, bal Fe.
Cast: 250-270 Brin.
For valve seats; alloy cast iron.

WAZ-16.
M-1491; 16 W, 7 Al, 2 Mo, 2 Cb, 0.5 Zr, 0.2 C, bal Ni.
Improved strength at 2200°F (1205°C).

WAZ-20.
M-1491; 0.15 C, 18.5 W, 6.2 Al, 1.5 Zr, bal Ni.
Directionally solidified. Cast alloy for jet engine discs & vanes.

WAZ-D.
M-US; 0.06 C, 16.5 W, 7.0 Al, 0.8 Zr, 4.3 Fe, 2.0 Y_2O_3, bal Ni.
Good sress rupture at 2200°F (1200°C).

WBD 200.
M-61; 99.5 min Ni.
Commercially pure nickel.
ASTM B160.

WBD 205.
M-61; 99.5 min Ni.
Commercially pure nickel for electronic applications.
ASTM F-9.

WBD 400.
M-61; 66 Ni, 31 Cu, 1 Fe.
For corrosion resistance marine and chemical processing.
Fed: QQN 281 (class A).

WBD 600.
M-61; 76.0 Ni, 16.0 Cr, 7.0 Fe.
For high temperature heat resistant applications.
ASTM B-166.

WBD WELD 55.
M-61; 55 Ni, bal Fe.
Core wire for electrodes for welding cast iron.
AWS 5.15 E NiFe-1.

WBD WELD 60.
M-61; 65 Ni, 27 Cu, 3.5 Mn, 2 Ti, 1 Si.
For welding nickel-copper alloys.
AWS 5.14 ERNiCu-7.

WBD WELD 61.
M-61; 96 Ni, 3 Ti.
For arc welding pure nickel.
AWS A5.11 ENi-1 and AWS A5.14 ERNi-3.

WBD WELD 62.
M-61; 74 Ni, 16 Cr, 7.5 Fe, 2 Cb.
For welding nickel base alloys.
AWS A5.14 ERNiCrFe-5.

WBD WELD 67.
M-61; 67 Cu, 31 Ni, 0.75 Mn.
For welding copper-nickel alloys.
AWS A5.7 RCuNi, and A5.6 ECuNi.

WBD WELD 82.
M-61; 72 Ni, 20 Cr, 3 Mn, 2.5 Cb.
For welding nickel base alloys.
Corrosion and heat resistant overlays on steel.
AWS A5.14 ERNiCr-3.

WBD WELD 92.
  M-61; 71 Ni, 16 Cr, 7 Fe, 3 Ti.
  For welding dissimilar alloys.
  AWS A5.14 ERNiCrFe-6.

W BRAND see **DARWIN W BRAND.**

WBZ 6 BRONZE.
  M-Ger.; 94 Cu, 6 Sn.
  Cast: 35,600 TS; 15,600 YS; 32 El; 50 Brin.
  Annealed: 36,900 TS; 15,650 YS; 40 El; 51 Brin.

WC 10.
  M-1740; 0.08-0.12 C, 0.30-0.60 Si, 0.30-0.50 Mn, 0.40 max Ni, 0.25 max Cr, 0.15 max Mo, 0.30 max Cu, bal Fe.
  B.S. 1617 Gr. A.

WC 18.
  M-1740; 0.16-0.20 C, 0.30-0.60 Si, 0.70-0.90 Mn, 0.40 max Ni, 0.25 max Cr, 0.15 max Mo, 0.30 max Cu, bal Fe.
  Easily machined; weldable.
  B.S. 1617 Gr. B; B.S. 592 Gr. A.

WC 20.
  M-1740; 0.25 max C, 0.50 max Si, 0.90 max Mn, 0.40 max Ni, 0.25 max Cr, 0.40 max Cu, bal Fe. (Total Ni-Cr-Mo-Cu max 0.80%).
  D.G.S. 8081A Admiralty general purpose castings.

WC 25.
  M-1740; 0.22-0.30 C, 0.30-0.60 Si, 0.60-0.80 Mn, 0.40 max Ni, 0.25 max Cr, 0.15 max Mo, 0.30 max Cu, bal Fe.
  B.S. 592 Gr. B.

WC 33.
  M-1740; 0.31-0.35 C, 0.30-0.60 Si, 0.60-0.90 Mn, 0.40 max Ni, 0.25 max Cr, 0.15 max Mo, 0.30 max Cu, bal Fe.
  ASTM A358-68 Grade 1.

WC 40.
  M-1740; 0.36-0.45 C, 0.30-0.60 Si, 0.70-1.0 Mn, bal Fe.
  Yield Sf. 19 tons/sq.in/min; UTS 35 tons/sq.in/min.
  B.S. 592 Gr. C.

WC 48.
  M-1740; 0.46-0.50 C, 0.30-0.50 Si, 0.60-1.0 Mn, 0.40 max Ni, 0.25 max Cr, 0.15 max Mo, 0.30 max Cu, bal Fe.
  B.S. 1760 Gr. A.

WC 55.
  M-1740; 0.51-0.60 C, 0.30-0.50 Si, 0.60-1.0 Mn, 0.40 max Ni, 0.25 max Cr, 0.15 max Mo, 0.30 max Cu, bal Fe.
  B.S. 1760 Gr. B.

WC-103 see **COLUMBIUM C-103.**

WC-3015 see **WAH-CHANG WC-3015.**

WC-3015.
  M-1596; 28-30 Hf, 1-2 Zr, 13-16 W, 0-4 Ta, 0.07-0.33 C, bal Cb.
  Extruded: 147,000 TS; 140,000 YS; 5-20 El.
  For turbine components.
  Good creep and stress-rupture properties.
  High oxidation and heat resistance.

WCC.
  M-115; 0.38-0.43 C, 4.0-4.5 Cr, 4.0-4.5 W, 4.0-4.5 Co, 2.1 V, 0.4 Mo, bal Fe.
  For punches, dies, extrusion dies, permanent molds; hot work steel, wear and thermal fatigue resistant.

W.D.D.
  M-73; 0.5 C, 3-6 W, bal Fe.
  For tools, dies; hot work steel.

W DIE STEEL.
  M-1409; 0.50 C, 2.2 W, 1 Cr, 0.35 V, 0.5 Mn, bal Fe.
  For punches, piercing dies, chisels; oil or water hardened, tough.

WE-II.
  M-206; 0.5 Si, bal Fe.
  Annealed: 45,000 TS; 25,000 YS; 25 El.
  For armatures, electrical equipment; high permeability.

WE III.
  M-206; 1.0 Si, bal Fe.
  Sheet: 60,000 TS; 40,000 YS; 15 El.
  For electrical equipment, chokes, radio transformers; high permeability.

WE IV.
  M-206; 1.0-3.5 Si, bal Fe.
  Sheet: 50,000-65,000 TS; 32,000-53,000 YS; 22-12 El.
  For electrical equipment, motors, radio transformers; high permeability.

WE V.
  M-206; 1.0-3.5 Si, bal Fe.
  Sheet: 50,000-65,000 TS; 32-000-53,000 YS; 12-22 El.
  For electrical equipment, motors, radio transformers; high permeability.

WE VI.
  M-206; 1.03-3.5 Si, bal Fe.
  Sheet: 50,000-65,000 TS; 32,000-53,000 YS; 12-22 El.
  For electrical equipment, motors, radio transformers; high permeability.

WE VII.
  M-206; 3.0-4.2 Si, bal Fe.
  Sheet: 67,000-70,000 TS; 55,000-65,000 YS; 8-4 El.
  For power and distribution transformers; high permeability.

WE VIII.
M-206; 3.0-4.2 Si, bal Fe.
Sheet: 67,000-70,000 TS; 55,000-65,000 YS; 8-4 El.
For power and distribution transformers; high permeability.

WE IX.
M-206; 3.0-4.2 Si, bal Fe.
Sheet: 67,000-70,000 TS; 55,000-65,000 YS; 8-4 El.
For power and distribution transformers; high permeability.

WE-X.
M-206; 3.0-4.2 Si, bal Fe.
Rolled: 70,000 TS; 65,000 YS; 4 El.
For motors, electrical equipment; high permeability.

WEARALOY.
M-655; 0.5 C, 1.5 Ni, 0.9 Cr, 0.2 Mo, bal Fe.
For steam shovels, dipper teeth, sheaves; wear and abrasion resistant.

WEAR-ARC see WEAR-ARC 40.

WEAR-ARC 3 IP.
M-1713; weld metal: 0.20 C, 0.90 Mn, 2.3 Cr, 1.1 Mo, 0.70 Si, bal Fe
AC-DC, reverse polarity, covered electrode for build-up to resist wear, impact and compressive loads. 101,000 psi TS; 24 El; 29 Rc.

WEAR-ARC 4 IP.
M-1713; weld metal: 0.45 C, 0.90 Mn, 1.30 Si, 2.2 Cr, 1.0 Mo, bal Fe.
AC-DC, straight or reverse polarity, covered electrode for hard surfacing. 54-56 Rc.

WEAR-ARC 5 IP.
M-1713; weld metal: 0.65 C, 1.0 Mn, 0.80 Si, 5.75 Cr, 0.65 Mo, bal Fe.
AC-DC, reverse polarity, covered electrode for hard surfacing. 58-60 Rc.

WEAR-ARC 6 IP.
M-1713; weld metal: 3.0 C, 0.80 Mn, 6.5 Cr, 1.8 Si, bal Fe.
AC-DC, straight or reverse polarity, covered electrode for build-up having high abrasion resistance but light impact. 56-59 Rc.

WEAR-ARC 12 IP.
M-1713; weld metal: 3.5 C, 2.7 Mn, 13.0 Cr, 1.8 Si, 1.1 Mo, bal Fe.
AC-DC, straight or reverse polarity, covered electrode for build-up for heavy impact and good abrasion resistance. 54-56 Rc.

WEAR-ARC 40.
M-1713; weld metal: 4.5 C, 0.30 Mn, 30.0 Cr, 1.8 Si, bal Fe.
AC-DC, reverse polarity, covered electrode for build-up having extreme abrasion resistance with medium impact, even at elevated temperatures. 57 Rc as welded.

WEAR-ARC NICKEL MANGANESE.
M-1713; weld metal: 0.60 C, 14.0 Mn, 0.55 Si, 4.0 Ni, bal Fe.
AC-DC, reverse polarity, covered electrode for welding or build-up on high manganese steel equipment, as bucket teeth.
Soft, austenitic, as welded; work hardens to about 48 Rc.

WEAR-ARC WH.
M-1713; 0.5 C, 4 Mn, 19 Cr, 9.5 Ni, bal Fe.
Welded: 127,000 TS; 19 El; 320 Brin.
For hard facing electrode; austenitic, stainless.

WEAR DEVIL A.
M-712; 0.65 C, 3.7 Cr, 0.2 V, 0.5 Mo, bal Fe.
550 Brin.
For hard facing welding electrodes; abrasion resisting.

WEAR DEVIL B.
M-712; 1 C, 5 Cr, 1.7 Mo, bal Fe.
500 Brin.
For hard facing welding electrodes; abrasion and erosion resistant.

WEAR-EX.
M-1432; C, alloy, bal Fe.
For hammer mills, mine cars, crushers, pulverizers; tough and abrasion resistant.

WEAREX.
M-487; 0.7 C, 14.5 W, 4 Cr, 1 V, bal Fe.
For lathe and planer tools, wood working tools, milling cutters; high speed steel.

WEAR-FLAME 40.
M-692; 4.5 C, 5 Mn, 30 Cr, 1.8 Si, bal Fe.
Weld: 600 Brin.
For hard facing electrodes.
Wear and abrasion resistant.

WEARGARD.
M-1709; 3.6 C, 3.2 Ni, 1.3 Cr, 0.2 Mo, bal Fe.
As Cast: 555 Brin.
Abrasion resistant cast iron.
Resin bonded waffle patterns for overlays.

WEAR GARD.
M-1748; 3.6 C, 3.2 Ni, 1.3 Cr, bal Fe.
As cast: 50,000 TS, 600 Brin.
For abrasion resisting liners.

WEARITE 4-11.
M-266, M-341; 10.0-11.2 Al, 3.0-4. Fe, 0.5 max others, bal Cu.
Plate: 100,000 TS; 50,000 YS; 7 El; 210 Brin.
For slides, cams, bushings, liners, bearings, chuck jaws, lathe beds.
Extruded aluminum bronze, tough.

WEARITE 4-13.
M-266, M-341; 12.5-13.6 Al, 3.5-5.0 Fe, bal Cu.
Plate: 105,000 TS; 65,000 YS; 1.5 El; 313 Brin.
For slides, cams, bushings, liners, bearings, chuck jaws, lathe beds.
Extruded aluminum bronze, tough.

WEARLOY.
M-571; 0.7 C, 1.5 Ni, 0.6 Cr, bal Fe.
For brake drums; for buses and trucks.

WEARMANG.
M-435; 0.30 C, 1.5 Mn, 0.25 Si, 0.20 Mo, bal Fe.
Rolled: 125,200 TS; 94,000 YS; 20 El; 285 Brin.
For wear plates, mining equipment; high wear and abrasion resistance.

WEAR-O-MATIC 3.
M-1713; weld metal: 0.07 C, 2.0 Mn, 2.0 Si, 0.50 Cr, 0.50 Mo, bal Fe.
Semi-automatic, open-arc, DC, straight or reverse polarity electrode for machinable build-up, usually prior to hard surfacing. 30-36 Rc.

WEAR-O-MATIC 6.
M-692; 0.65 C, 2.6 Mn, 3.0 Cr, 0.2 Si, 0.5 Mo, bal Fe.
Weld: C 48 Rock.
For hard surfacing conveyor buckets, dragline and power shovel lips and sides, scraper blades.
Severe impact and abrasion resistant.
Good compressive strength and high hardness.

WEAR-O-MATIC 6.
M-1713; weld metal: 0.65 C, 2.6 Mn, 3.0 Cr, 0.20 Si, 0.50 Mo, bal Fe.
Semi-automatic, open-arc, DC, straight or reverse polarity electrode for hard surfacing.
High impact resistance; 48 Rc.

WEAR-O-MATIC 7.
M-692; 0.4 C, 3.0 Mn, 5.3 Cr, 1.2 Si, 1.5 Mo, 1.4 W, bal Fe.
Weld: C 55 Rock.
For hard surfacing cable sheaves, crane wheels, pinch and pipe forming rolls.
Resists severe abrasion and compression.

WEAR-O-MATIC 12.
M-1713; weld metal: 2.3 C, 0.30 Mn, 17.0 Cr, 1.0 Si, 0.80 Mo, bal Fe.
Semi-automatic, open-arc, DC, reverse polarity electrode for hard surface build-up for heavy impact and severe abrasion resistance.
As deposited: 50 Rc.

WEAR-O-MATIC 12A.
M-692; 3.0 C, 4.0 Mn, 14.5 Cr, 1.3 Si, 0.8 Mo, bal Fe.
Weld: C 48 Rock.
For hard surfacing crushing equipment, muller tires, steel mill pinch roll scraper blades.
Submerged arc weld wire.
Resists severe abrasion and impact.

WEAR-O-MATIC 12B.
M-692; 2.3 C, 0.3 Mn, 17.0 Cr, 1.0 Si, 0.8 Mo, bal Fe.
Weld: C 50 Rock.
For hard surfacing heavy rock handling equipment, crushers, impactors, power shovel and dragline bucket parts. Open arc wire.
Heavy impact and abrasion resistance.

WEAR-O-MATIC 14.
M-692; 0.25 C, 1.3 Mn, 5.0 Ni, 0.3 Si, 4.8 Mo, 0.5 W, bal Fe.
Weld: C 48 Rock.
For hard surfacing steel mill rolls, blast furnace bells and hoppers, roll necks, bearings, journals.
Resists heat and abrasion.

WEAR-O-MATIC 15.
M-692; 4.0 C, 0.3 Mn, 5.5 Cr, 0.6 Si, 5.0 Mo, bal Fe.
Weld: C 60 Rockwell.
For hard surfacing pug mill knives and augers, cement pump screws, conveyors.
Wear and abrasion resistant.

WEAR-O-MATIC 15.
M-1713; weld metal: 4.0 C, 0.30 Mn, 5.5 Cr, 0.60 Si, 5.0 Mo, bal Fe.
Semi-automatic, open-arc, DC, straight or reverse polarity electrode for hard surfacing for extreme abrasion resistance. 60 Rc.

WEAR-O-MATIC 16.
M-692; 0.05 C, 1.6 Mn, 0.3 Si, 6.2 Ni, 5.3 Mo, bal Fe.
Weld: C 42-44 Rock.
For hard surfacing of blast furnace bells and hoppers, steel mill twist rolls, pinch rolls.
For heat and abrasion resistance without heat checking.

WEAR-O-MATIC 40.
M-692; 4.0 C, 1.5 Mn, 27.0 Cr, 1.0 Mo, 1.5 Si, bal Fe.
Welded: C 58 Rockwell.
For hard surfacing compression type crusher parts and hammer mills, mill guides.
Wear and abrasion resistant.

WEAR-O-MATIC 40.
M-1713; weld metal: 4.0 C, 1.5 Mn, 27.0 Cr, 1.0 Mo, 1.5 Si, bal Fe.
Semi-automatic, open-arc, DC, reverse polarity electrode for build-up for severe abrasion and compression. 58 Rc.

WEAR-O-MATIC 420.
M-692; 0.30 C, 1.5 Mn, 14 Cr, 1.0 Si, bal Fe.
Welded: C 52 Rock.
For hard surfacing paper mill rolls, pipe forming rolls, hot bar mill guides.
Corrosion and abrasion resistance.

WEAR-O-MATIC BR.
M-1713; weld metal: 0.12 C, 0.37 Si, 1.6 Mn, 2.5 Cr, 0.55 Mo, bal Fe.
Semi-automatic, open-arc, DC, reverse polarity electrode designed for repair of railroad freight car bolster bowls using 98 Argon - 2 Oxygen. 35-40 Rc.

WEAR-O-MATIC NICKEL MANGANESE.
M-1713; weld metal: 0.30 C, 13.5 Mn, 3.9 Ni, 0.60 Si, bal Fe.
Semi-automatic, open-arc, DC, reverse polarity electrode for build-up on austenitic manganese steel as in bucket teeth.
Soft as welded: work hardens to about 48 Rc.

WEAR-O-MATIC RAIL-ARC.
M-1713; weld metal: 0.06 C, 3.3 Mn, 0.50 Si, 17.8 Cr, 8.1 Ni, bal Fe.
Semi-automatic, open-arc, DC, reverse polarity electrode for multipass build-up on all weldable carbon and austenitic manganese steel rails.
Soft as deposited; work hardens to 35 Rc.

WEAR-O-MATIC SUPER WH.
M-1713; weld metal: 1.1 C, 15.0 Mn, 0.65 Si, 17.0 Cr, 1.4 Ni, bal Fe.
Semi-automatic, open-arc, DC, reverse polarity electrode for build-up having severe impact resistance with some abrasion.
As welded: 30 Rc; work hardens to 47-49 Rc.

WEAR-O-MATIC WH.
M-1713; weld metal: 0.38 C, 4.23 Mn, 0.47 Si, 20.2 Cr, 9.65 Ni, bal Fe.
Semi-automatic, open-arc, DC, reverse polarity electrode for weld or build-up for tough surface.
As welded: 36 Rc; work hardens to about 47 Rc.

WEAR-O-MATIC WH.
M-692; 0.38 C, 4.23 Mn, 0.47 Si, 20.2 Cr, 9.65 Ni, bal Fe.
Weld: 103,000 TS; 70,000 YS; 36 El.
For welding manganese steel to carbon steel, C 18 Rock (C 47 rock after work hardening).
Tough, resilient, work hardenable.

WEAR-PROOFT MAC HEMPITE.
M-229; 0.45 C, 1.25 Mn, 0.35 Mo, bal Fe.
Hardened: 700 Brin.
For gears, pinions, mill couplings, boxes, cams; wear and shock resistant.

WEARWELD.
M-578; 0.37 C, 2.2 Mn, 0.15 Si, 3.3 Cr, bal Fe.
Hard surfacing arc welding electrode for resistance to metal-to-metal wear.

WEARWELL.
M-1124; 0.30-0.35 C, 1.30-1.65 Mn, 0.15-0.30 Si, bal Fe.
Primarily for riveted and bolted applications where resistance to abrasion is important.

WEBERT ALLOY.
M-8; 14 Zn, 4 Si, Mn, bal Cu.
For pressure die castings.

WEFAHUTTE HECN.
M-1380; 0.13 C, 0.7 Cr, 2.5 Ni, 0.5 Max Mn, bal Fe.
For gears, bolts, plastic mold dies, shafts; case hardened, shock resistant.

WEFAHUTTE HECN15.
M-1380; 0.15 C, 1.55 Ni, 1.55 Cr, bal Fe.
For camshafts, cams, bolts, gears; case hardened, shock resistant.

WEFAHUTTE HECN35.
M-1380; 0.13 C, 0.7 Cr, 3.5 Ni, 0.5 Max Mn, bal Fe.
For gears, bolts, plastic mold dies, shafts; case hardened, shock resistant.

WEFAHUTTE HK60.
M-1380; 0.30 C, 0.25 Si, 1.35 Mn, bal Fe.
For punches, gears, crankshafts; water or oil hardened.

WEFAHUTTE HK75.
M-1380; 0.37 C, 1.2 Si, 1.25 Mn, bal Fe.
For punches, upsetters, crimpers; oil hardened, shock resistant.

WEFAHUTTE HKCN15W/H.
M-1380; 0.28-0.35 C, 0.5 Cr, 1.5 Ni, bal Fe.
For gears, bolts, crankshafts, fasteners; oil hardened, shock resistant.

WEFAHUTTE HKCN25W/H.
M-1380; 0.18-0.35 C, 0.7 Cr, 2.5 Ni, bal Fe.
For gears, bolts, machine tool parts; oil hardened, shock resistant.

WEFAHUTTE HKCN35W/H.
M-1380; 0.22-0.30 C, 0.7 Cr, 3.5 Ni, 0.6 Mn, bal Fe.
For gears, bolts, crankshafts; oil hardened, shock resistant.

WEFAHUTTE HKCNX25.
M-1380; 0.36 C, 1.0 Ni, 1.0 Cr, bal Fe.
For gears, bolts, machine tool parts; oil hardened, tough.

WEFAHUTTE HLMF.
M-1380; 0.46 C, Mn, bal Fe.
For machine tool parts, gears, shafts; water hardened.

WEFAHUTTE HLSFH.
M-1380; 0.65 C, 1.7 Si, 0.70 Mn, bal Fe.
For springs, punches, crimpers; oil hardened, shock resistant.

WEFAHUTTE HLSFW.
M-1380; 0.55 C, 1.7 Si, 0.70 Mn, bal Fe.
For springs, punches, upsetters; oil hardened, shock resistant.

WEFAHUTTE HUK10.
M-1380; 0.34 C, 0.1 V, 0.65 Mn, bal Fe.
For gears, bolts, shafts, fasteners; water hardened.

WEFAHUTTE KWF.
M-1380; 0.46 C, 1.7 Si, 0.65 Mn, bal Fe.
For springs, rivet sets, punches; oil hardened, tough.

WEGNER.
80 Sn, 20 Zn.
For bearings; anti-friction.

WEHRALLOY NO. 1.
M-599; 0.07 max C, 16-23 Cr, 7-11 Ni, bal Fe.
For heat and corrosion resistant parts; stainless, heat and corrosion resistant.

WEHRALLOY NO. 2.
M-599; 0.12 max C, 16-23 Cr, 7-11 Ni, bal Fe.
For heat and corrosion resistant parts; stainless, heat and corrosion resistant.

WEHRALLOY NO. 3.
M-599; 0.21-0.35 C, 24-30 Cr, 12-15 Ni, bal Fe.
For heat and corrosion resistant parts; stainless, heat and corrosion resistant.

WEHRALLOY NO. 4.
M-599; 0.13-0.35 C, 24-30 Cr, 7-11 Ni, bal Fe.
For heat and corrosion resistant parts; stainless, heat and corrosion resistant.

WEHRALLOY NO. 5.
M-599; C, Cr, Ni, bal Fe.
For heat and corrosion resistant parts; stainless, heat and corrosion resistant.

WEHRALLOY NO. 6.
M-599; 0.13-0.35 C, Si, 24-30 Cr, bal Fe.
For heat and corrosion resistant parts; heat and corrosion resistant.

WEHRALLOY NO. 7.
M-599; 0.21-0.50 C, Si, 12-23 Cr, bal Fe.
For heat and corrosion resistant parts; heat and corrosion resistant.

WEHRALLOY NO. 8.
M-599; 0.21-0.50 C, Si, 24-30 Ni, 16-23 Cr, bal Fe.
For heat and corrosion resistant parts; heat and corrosion resistant.

WEHRALLOY NO. 9.
M-599; 0.36-0.50 C, Si, 31-39 Ni, 16-23 Cr, bal Fe.
For heat and corrosion resistant parts; heat and corrosion resistant.

WEHRALLOY NO. 10.
M-599; 0.21-0.50 C, 58-66 Ni, 16-23 Cr, bal Fe.
For heat and corrosion resistant parts; heat and corrosion resistant.

WEHRALLOY NO. 12.
M-599; 0.12 max C, 12-15 Cr, bal Fe.
For corrosion resistant parts; corrosion resistant.

WEHRALLOY NO. 13.
M-599; 0.13-0.20 C, 12-15 Cr, bal Fe.
For corrosion resistant parts; corrosion resistant.

WEHRALLOY NO. 14.
M-599; 0.21-0.35 C, 12-15 Cr, bal Fe.
For corrosion resistant parts; corrosion resistant.

WEHRALLOY NO. 16.
M-599; 0.21-0.35 C, 16-23 Cr, bal Fe.
For heat and corrosion resistant parts; heat and corrosion resistant.

WEHRALLOY NO. 17.
M-599; 0.81-1.10 C, 16-23 Cr, bal Fe.
For heat and corrosion resistant parts; heat and corrosion resistant.

WEHRALLOY NO. 18.
M-599; 0.13-0.20 C, 24-30 Cr, bal Fe.
For heat and corrosion resistant parts; heat and corrosion resistant.

WEHRALLOY NO. 19.
M-599; 0.13-0.20 C, Si, 24-30 Cr, bal Fe.
For heat and corrosion resistant parts; heat and corrosion resistant.

WEHRALLOY NO. 21.
M-599; 0.21-0.35 C, 24-30 Cr, bal Fe.
For heat and corrosion resistant parts; heat and corrosion resistant.

WEHRALLOY NO. 22.
M-599; 0.13-0.20 C, 16-23 Cr, 7-11 Ni, bal Fe.
For heat and corrosion resistant parts; stainless, heat and corrosion resistant.

WEHRALLOY NO. 23.
M-599; 0.21-0.35 C, 16-23 Cr, 7-11 Ni, bal Fe.
For heat and corrosion resistant parts; stainless, heat and corrosion resistant.

WEHRALLOY NO. 24.
M-599; 0.21-0.35 C, 16-23 Cr, 7-11 Ni, bal Fe.
For heat and corrosion resistant parts; stainless, heat and corrosion resistant.

WEHRALLOY NO. 26.
M-599; 0.36-0.50 C, Si, 31-39 Ni, 16-23 Cr, bal Fe.
For heat and corrosion resistant parts; heat and corrosion resistant.

WEHRALLOY NO. 27.
  M-599; 0.36-0.50 C, Si, 31-39 Ni, 16-23 Cr, bal Fe.
  For heat and corrosion resistant parts; heat and corrosion resistant.

WEHRALLOY NO. 28.
  M-599; 0.51-0.80 C, Si, 67-75 Ni, 16-23 Cr, bal Fe.
  For heat and corrosion resistant parts; heat and corrosion resistant.

WEHRALLOY NO. 29.
  M-599; 0.21-0.35 C, 5-7 Cr, bal Fe.
  For corrosion resistant parts; corrosion resistant.

WEHRALLOY NO. 31.
  M-599; 0.13-0.20 C, 24-30 Cr, 12-15 Ni, bal Fe.
  For heat and corrosion resistant parts; heat and corrosion resistant.

WEHRALLOY NO. 43.
  M-599; 0.13-0.20 C, 24-30 Cr, 12-15 Ni, bal Fe.
  For heat and corrosion resistant parts; heat and corrosion resistant.

WEHRALLOY STAINLESS 18-8.
  M-599; 0.13-0.20 C, 18 Cr, 8 Ni, bal Fe.
  For stainless parts; stainless.

WEHRALLOY STAINLESS 18-8 SPECIAL.
  M-599; 0.07 max C, 18 Cr, 8 Ni, bal Fe.
  For stainless parts; stainless.

WEIDRIUM.
  M-Ger.; Ni, Cu, Zn, Sn, Fe, Mg.

WEIGER.
  77-80 Ag, 18-20 Cu, 2-5 Pt.
  For silver solder; corrosion resistant.

WEIGHTS.
  90 Cu, 8 Sn, 2 Zn.
  For weights; corrosion resistant.

WEINGARTNER EW 1540 EXTRA.
  M-1731; 1.0 C, bal Fe.
  Water hardening tool steel.
  Werkstoff Nr. 1.1540; AISI W1 tool steel.

WEINGARTNER EW 2343.
  M-1731; 0.38 C, 1.0 Si, 0.4 Mn, 5.3 Cr, 1.1 Mo, 0.4 V, bal Fe.
  Air or oil hardening hot work tool steel.
  Werkstoff Nr. 1.2343; similar to AISI H11.

WEINGARTNER EW 2601.
  M-1731; 1.65 V, 0.3 Si, 0.3 Mn, 12.0 C, 0.6 Mo, bal Fe.
  Air or oil hardening cold work tool steel.
  Werkstoff Nr. 1.2601; AISI D2.

WEINGARTNER EW 3343.
  M-1731; 0.78-0.86 C, 4.0 Cr, 5.0 Mo, 1.75 V, 6.0 W, bal Fe.
  High speed tool steel.
  Werkstoff Nr. 1.3343; AISI M2.

WEINGARTNER EW 4024.
  M-1731; 0.12-0.17 C, 1.0 max Si, 1.0 max Mn, 13.0 Cr, bal Fe.
  Air or oil hardenable martensitic stainless steel.
  Werkstoff Nr. 1.4024; similar to AISI 410 stainless steel.

WEINGARTNER EW 4301.
  M-1731; 0.07 max C, 1.0 max Si, 2.0 max Mn, 18.0 Cr, 10.0 Ni, bal Fe.
  Austenitic stainless steel.
  Werkstoff Nr. 1.4301; AISI 304 Stainless.

WEINGARTNER EW 4704.
  M-1731; 0.45 C, 0.4 Si, 0.45 Mn, 2.65 Cr, bal Fe.
  Oil hardening valve steel.
  Werkstoff Nr. 1.4704; DIN X 45 SiCr4.

WEINGARTNER EW 4841.
  M-1731; 0.20 max C, 2.05 Si, 2.0 max Mn, 25 Cr, 20 Ni, bal Fe.
  Austenitic high temperature alloy.
  Werkstoff Nr. 1.4841; similar to AISI 310 stainless steel.

WEINGARTNER EW 5919.
  M-1731; 0.15 C, 0.4 Mn, 1.55 Cr, 1.55 Ni, bal Fe.
  Oil hardenable carburizing steel.
  Werkstoff Nr. 1.5919; DIN 15 CrNi6.

WEINGARTNER EW 8159.
  M-1731; 0.50 C, 0.25 Si, 0.95 Mn, 1.05 Cr, 0.10 V, bal Fe.
  Oil hardening steel; for springs, shafts.
  Werkstoff Nr. 1.8159; DIN 50 CrV4; AISI 6150.

WEINGARTNER EW 8507.
  M-1731; 0.34 C, 0.6 Mn, 1.0 Cr, 0.2 Mo, 1.0 Al, bal Fe.
  Nitriding steel.
  Werkstoff Nr. 1.8507.

WEIRITE.
  M-836; Sn coated soft steel.
  For cans, roofing; tin plate, electrolytic.

WEIRZIN.
  M-836; Zn coated steel.
  For structural sheets; electrolytic plate.

WEISSPUNKT.
  M-1331; 0.90 C, Mo, Cr, V, W, Bal Fe.
  For lathe and planer tools, reamers, broaches; high speed steel.

WEISSPUNKT BM09.
  M-1331; C, 4 Cr, W, Mo, V, bal Fe.
  For lathe and planter tools, reamers; high speed steel.

WELCHS ALLOY.
  52 Sn, 48 Ag.
  For dental alloy, solder; corrosion resistant.

WELCON 2H.
M-1539; 0.15 C, 0.5 Si, 1.2 Mn, bal Fe.
Heat treated: 85,000-100,000 TS; 71,000 min YP; 16 min El.
For pressure vessels, bridges, mine cars, cranes, truck bodies.
Good fabricability and weldability.
Constructional steel plates, tough.

WELCON 2H-100.
M-1539; 0.18 C, 0.45 Si, 0.60 Mo, 0.10 V, bal Fe.
Plate: 138,000-164,000 TS; 128,000 YP.
For pressure vessels, bridges, bus and trailer bodies, derricks, booms.
Wear and shock resistant.

WELCON-2H CR.
M-1539; 0.11 C, 0.35 Si, 1.01 Mn, 0.014 P, 0.008 Si, 0.14 Ni, 0.50 Cr, 0.34 Mo, bal Fe.
29 mm thick plate: 97,000 TS; 86,100 YS; 45 El.
High strength low alloy steel for structural purposes.

WELCON 2H SUPER.
M-1539; 0.08-0.16 C, 0.6-1.2 Mn, 0.55 Si, 0.5 Cr, 0.4 Mo, 1.0 Ni, bal Fe.
Plate: 100,000-114,000 TS; 90,000 min YS; 18 El.
For pressure vessels, bridges, bus and trailer bodies, cranes, mine cars.
Wear and shock resistant. Heat treated by mill.

WELCON 2H ULTRA.
M-1539; 0.08-0.16 C, 0.60-1.2 Mn, 0.006 B, 0.8 Cr, 0.7 Mo, 0.15-0.50 Cu, 1.5 Ni, 0.1 V, bal Fe.
Plate: 114,000-135,000 TS; 100,000 min YS; 18-22 El.
For pressure vessels, bridges, mine cars, power shovels, cranes, trucks, trailers.
Shock and wear resistant. Heat treated by mill.

WELCON 50.
M-1539; 0.18 C, 1.35 max Mn, 0.55 max Si, bal Fe.
Rolled plate: 71,000-82,000 TS; 47,000 min YP; 26 min El.
For ship plate, oil tankers, thermal power generating equipment, mine cars, bus bodies.
Constructional steel plate, tough.

WELDBEST ALBRONZE 100.
M-1230; Al, bal Cu.
Welded: 100 Brin.
For welding electrode for Cu alloys; all purpose, Al-bronze.

WELDBEST ALBRONZE 250.
M-1230; Al, Bal Cu.
Welded: 250 Brin.
For welding electrode for Cu alloys; all purpose, Al-bronze.

WELDEX.
M-543; 2.2 C, (comb. C = 0.08) 0.25 Mn, 2.4 Si, 0.03 P, 0.01 S, bal Fe.
Malleable iron. 65,000-68,000 TS; 47,000-51,000 YS; 16-12 El.

WELDFIL 525 ATOMIZED STEEL POWDER.
M-1781; 1.8 Ni, 0.5 Mo, 0.2 Mn, 0.02 C, (0.5 H_2-loss), bal Fe.
For one-pass S.A.W. of plates to 2 in thick.

WELDMACO.
M-943; 0.2 C, bal Fe.
For welding rod.

WELDOLOY NO. 685.
M-944; 95 Cd, 5 Ag.
For solder on steel, bronze and cast iron; flows at 750°F.

WELDOLOY NO. 690.
M-944; 8 Si, bal Al.
Cast: 35,000 TS.
For harder solder for Al; flows at 930°F.

WELMET NO. 1.
M-946; 0.15 max C, 25 Cr, 35 Ni, 5 Mo, bal Fe.
For pipes, valves, fittings, steam nozzles; heat and corrosion resistant in SO_2 atmosphere.

WELMET 20.
M-946; 0.07 max C, 29 Ni, 20 Cr, 3-4 Mo, 4 Cu, bal Fe.
Cast: 65,000-75,000 TS; 28,000-38,000 YS; 50-35 El; 50-45 RA; 120-160 Brin.
For chemical plant equipment, valves, agitators, mixers; resists mixed acids, austenitic.

WEL-MET BRONZE.
M-1162; 84-86 Cu, 7-8 Sn, 5-6 Pb, 1-2 graphite.
Sintered: 12,000 TS; 5 El; 30 Brin.
For bearings; self-lubricating, sintered, oil impregnated.

WEL-MET STEEL.
M-1162; 89-92 Fe, 7-10 Cu, 0.5-2.0 graphite.
Sintered: 27,000 TS; 1-2 El; 60-80 Brin.
For bearings; sintered, self-lubricating.

WEL-TEN 50.
M-1540; 0.18 max C, 0.9-1.5 Mn, 0.25-0.45 Si, bal Fe.
Rolled: 71,000-82,000 TS; 47,000 min YS; 20 min El.
For pressure vessels, bridges, penstocks, construction machinery.
Constructional steel. Good weldability.

WEL-TEN 55.
M-1540; 0.18 max C, 1.2-1.5 Mn, 0.35-0.55 Si, bal Fe.
Normalized: 78,000-90,000 TS; 51,000 min YP; 20% min El.
For pressure vessels, bridges, penstocks, construction machinery, booms, derricks.
Readily fabricated, tough, construction steel. Good weldability.

WEL-TEN 60.
M-1540; 0.16 max C, 1.3 max Mn, 0.55 max Si, 0.6 Max Ni, 0.4 max Cr, 0.15 max V, bal Fe.
Heat treated: 85,000 min TS; 65,000 min YP; 16 min El.
For railroad and mine cars, pressure vessels, agricultural equipment, derricks.
Good weldability, tough. High-strength low-alloy constructional steel.

WEL-TEN 60.
M-1732; 0.13 C, 0.35 Si, 1.24 Mn, 0.012 P, 0.008 S, 0.04 V, bal Fe.
6-50 mm plate, Q&T: 92,900 psi TS; 82,200 psi YS; 30 El.
High strength steel for welded structures of oil storage tanks, bridges, earth moving and off shore structures. (Composition may vary with thickne on all WEL-TEN alloys).

WEL-TEN 60 CF.
M-1732; 0.07 C, 0.26 Si, 1.30 Mn, 0.015 P, 0.005 S, 0.21 Cr, 0.15 M V, bal Fe.
6-50 mm plate, Q&T: 92,000 psi TS; 81,400 psi YS; 27 El.
High strength steel for welded structures as spherical pressure vessels; low susceptibility to weld cracking.

WEL-TEN 60H.
M-1540; 0.18 max C, 0.15-0.75 Si, 1.0-1.5 Mn, 1.0 max Ni, 0.15 max Nb + V, bal Fe.
Normalized: 85,000-102,000 TS; 64,000 min YP; 20 min El.
For pressure vessels, bridges, penstocks, construction machinery.
Good weldability and formability.

WEL-TEN 60-LT.
M-1540; 0.16 max C, 0.90-1.4 Mn, 0.60 max Ni, 0.30 max Cr, 0.12 V, bal Fe.
Palte: 85,300-102,000 TS; 64,500-71,100 YS.
For pressure vessels, cryogenic equipment.
High notch toughness. Good weldability.

WEL-TEN 62.
M-1540; 0.18 max C, 0.15-0.55 Si, 0.9-1.4 Mn, 0.6 max Ni, 0.3 max Cr, 0.12 max V, bal Fe.
Quenched and Tempered: 88,000-107,000 TS; 71,000 min YS; 19 min El.
For pressure vessels, penstocks, bridges, construction machinery.
Good formability and weldability.

WEL-TEN 62.
M-1732; 0.14 C, 0.46 Si, 1.18 Mn, 0.017 P, 0.008 S, 0.15 Cr, 0.04 V, bal Fe.
6-50 mm plate, Q&T: 89,900 psi TS; 78,200 psi YS; 29 El.
High strength steel for pressure vessels.

WEL-TEN 62 CF.
M-1732; 0.07 C, 0.28 Si, 1.36 Mn, 0.014 P, 0.003 S, 0.25 Cr, 0.22 Mo, 0.03 V, bal Fe.
6-50 mm plate, Q&T: 93,900 psi TS; 82,500 psi YS; 26 El.
High strength steel for welded structures as spherical pressure vessels; low susceptibility to weld cracking.

WEL-TEN 70.
M-1732; 0.11 C, 0.45 Si, 1.0 Mn, 0.010 P, 0.003 S, 0.02 Cu, 0.90 Ni, 0.30 Cr, 0.40 Mo, 0.04 V, bal Fe.
50 mm thick: 112,800 TS; 103,000 YS; 44 El.
High strength low alloy steel for structural purposes.

WEL-TEN 80.
M-1540; 0.18 C, 0.6-1.2 Mn, 0.15-0.35 Si, 0.006 B, 0.4-0.8 Cr, 0.15-0.50 Cu, 0.6 Mo, 1.5 Ni, 0.1 V, bal Fe.
Plate: 114,000-135,000 TS; 100,000 min YS; 18-20 El; (Quenched and tempered).
For pressure vessels, bridges, mine cars, power shovels, cranes, trucks, trailers.
Shock and wear resistant. Good weldability.

WEL-TEN 80.
M-1732; 0.11 C, 0.21 Si, 0.85 Mn, 0.015 P, 0.006 S, 0.22 Cu, 0.97 Ni, 0.53 Cr, 0.43 Mo, 0.05 V, 0.0008 B, bal Fe.
6-100 mm plate, Q&T: 119,000 psi TS; 112,000 psi YS; 23 El.
High strength steel for welded structures as long span bridge, penstock, earth moving equipment.

WEL-TEN 80 C.
M-1732; 0.12 C, 0.26 Si, 0.86 Mn, 0.010 P, 0.004 S, 0.29 Cu, 0.74 Cr, 0.42 Mo, 0.04 V, 0.0008 B, bal Fe.
6-50 mm plate, Q&T: 118,000 psi TS; 110,000 psi YS; 23 El.
High strength steel for welded structures as pressure vessels, earth moving equipment.

WEL-TEN 80C.
M-1540; 0.18 max C, 0.15-0.35 Si, 0.6-1.2 Mn, 0.15-0.50 Cu, 0.7-1.3 Cr, 0.6 max Mo, 0.0006 max B, bal Fe.
Quenched and Tempered: 114,000-135,000 TS; 100,000 min YP.
For pressure vessels, bridges, penstocks, construction machinery, bulldozers, power shovels.
Good formability and weldability.

WEL-TEN 80 C-LT.
M-1540; 0.18 max C, 0.60-1.20 Mn, 0.7-1.3 Cr, 0.6 max Mo, 0.15-0.50 V, 0.006 max B, bal Fe.
Plate: 113,800-135,100 TS; 99,600 min YS. For pressure vessels, cryogenic equipment.
High notch toughness, good weldability.

WEL-TEN 80 E.
M-1732; 0.19 C, 0.35 Si, 1.43 Mn, 0.016 P, 0.010 S, 0.0008 B, bal Fe.
6-20 mm plate, Q&T: 121,000 psi TS; 110,000 psi YS; 36 El.
High strength steel for welded structures as earth moving equipment.

WEL-TEN 80P.
M-1732; 0.10 C, 0.31 Si, 1.46 Mn, 0.010 P, 0.010 S, 0.07 Cu, 0.59 Mo, 0.003 B, 0.049 Cb, bal Fe.
25 mm thick plate: 116,000 TS; 111,000 YS; 14.3 El.
High strength low alloy steel for structural purposes.

WEL-TEN 80S.
M-1732; 0.08 C, 0.54 Si, 1.36 Mn, 0.016 P, 0.0008 S, 0.14 Cu, 0.97 Ni, 1.03 Cr, 0.56 Mo, 0.05 V, 0.015 Ti, bal Fe.
25 mm thick plate: 118,000 TS; 108,000 YS; 26 El.
High strength low alloy steel for structural purposes.

WEL-TEN 100N.
M-1540; 0.18 max C, 1.5 max Ni, 0.15-0.50 Cu, 0.4-0.8 Cr, 0.6 max Mo, 0.10 max V, 0.6-1.2 Mn, bal Fe.
Quenched and Tempered Plate: 140,000-163,000 TS; 130,000 YS; 15 El.
For heavy duty welded structures and equipment, pressure vessels, bridges.
Conforms to ASTM-A300 Cl.1.
Good weldability, high strength.

WENDT-SONIS see also CQ AND CY.

WESGO 35 AU-65 CU.
M-1493; 35 Au, 65 Cu.
For brazing Cu, Kovar and nickel brass; M.P. 970-1005°C.

WESGO 40 AU-60 CU.
M-1493; 40 Au, 60 Cu.
For brazing Cu, Kovar and nickel brass.

WESGO 50 AU-50 CU.
M-1493; 50 Au, 50 Cu. For brazing Cu, Kovar, nickel brass; M.P. 930-950°C.

WESGO DECARBONIZED.
M-1493; 28.1 Cu, bal Ag.
For brazing; M.P. 780°C.

WESSEL'S SILVER.
66-51 Cu, 19-32 Ni, 12.5-17 Zn, 0-0.5 Fe, 0-2 Ag.
For ornamental parts, architectural trim; corrosion resistant.

WESSLING.
M-1293; 0.38 C, 17.6 Si, 0.8 Mn, bal Fe.
Cast: 15,000 TS; C 50 Rock.
For pumps, valves, drains, fittings, agitators, castings.
Acid resistant. Brittle as cast.

WESSONMETAL 26.
M-1498; 83 WC, 10 TiC, 7 Co.
Sintered: 225,000 TrS; 91.7 Rock A.
For cutting tools, dies; high hardness and strength; wear and crater resistant.

WESSONMETAL GR. 900.
M-1498; TiC, MoC, Ni.
Sintered: A 92.2-92.7 Rock; 200,000 TrS.
For cutters in high speed finish machining.
Excellent wear resistance with moderate shock resistance.

WESSONMETAL GR. GA.
M-1498; 95.5 WC, 4.5 Co.
Sintered: A 92.5 Rock; 210,000 TrS.
For cutting tools to machine cast iron and non-ferrous materials.
Extreme abrasion resistant.

WESSONMETAL GR. GF.
M-1498; 97 WC; 3 Co.
Sintered: A 92.7 Rock; 195,000 TrS.
For cutting tools to machine cast iron and non-ferrous materials.
High wear resistant. Hard and strong.

WESSONMETAL GR. GI.
M-1498; 94 WC, 6 Co.
Sintered: A 91.5-92.5 Rock; 250,000 TrS.
For cutting tools to machine cast iron, non-ferrous materials.
High strength and hardness. Abrasion resistant.

WESSONMETAL GR. GS.
M-1498; 93 WC, 7 Co.
Sintered: A 91.2 Rock; 260,000 TrS.
For cutting tools to machine cast iron and non-ferrous materials.
Shock resistant, high strength.

WESSONMETAL GR. HR.
M-1498; W, TiC, TaC, CbC, Co.
Sintered: A 91.2 Rock; 300,000 TrS.
For cutting tools for roughing and interrupted cuts.
High heat resistant.

WESSONMETAL GR. HV.
M-1498; 83 WC, 13 TiC, 1 Mo, 3 Co.
Sintered: A 94.0 Rock; 185,000 TrS.
For cutting tools for high speed machining.
Extremely high hardness.

WESSONMETAL GR. M.
M-1498; 83.5 WC, 3.5 TiC, 13 Co.
Sintered: A 90.2 Rock; 290,000 TrS.
For cutting tools in heavy or interrupted cuts on steel or cast iron.
High strength and hardness.

WESSONMETAL GR. WH.
M-1498; 80 WC, 13 TiC, 7 Co.
Sintered: A 92.4 Rock; 212,000 TrS.
For cutting tools in medium and light machining of steel at high speeds.
High crater resistance.

WESSONMETAL GR. WM.
M-1498; 77.25 WC, 12.75 TiC, 10 Co.
Sintered: A 91.2 Rock; 225,000 TrS.
For general purpose cutters.

WESSONMETAL GR. WP.
M-1498; 87 WC, 13 Co.
Sintered: A 88.5-89.5 Rock; 312,000 TrS.
For cutting tools and mining tools.
High impact resistance.

WESSONMETAL GR. WS.
M-1498; 83 WC, 4 TiC, 13 Co.
Sintered: A 90.5 Rock; 280,000 TrS.
For cutting tools in heavy, intermittent machining of tough steel.
High shock resistance.

WEST 60 STANCAST.
M-465; C, alloy, bal Fe.
Cast: 60,000 TS; 35,000 YS; 32 El; 54 RA.
For annealing boxes, pots, fittings, autoclaves; heat resistant.

WEST 75 HY-CAST.
M-465; C, alloy, bal Fe.
Cast: 75,000 TS; 42,000 YS; 28 El; 45 RA.
For anvil blocks, car wheels, pump impellers; oil hardening.

WEST 90 DURACAST.
M-465; C, alloy, bal Fe.
Cast: 90,000 TS; 58,000 YS; 22 El; 40 RA.
For gears, pipe fittings, clutch jaws; oil hardening.

WEST 110 SUPERCAST.
M-465; C, alloy, bal Fe.
Cast: 110,000 TS; 90,000 YS; 15 El; 25 RA.
For die blocks, car wheels, skip buckets; oil hardening.

WEST NO. 1.
M-465; 0.28-0.20 C, 0.6-0.8 Mn, 0.3-0.4 Si, bal Fe.
Annealed: 38,000-65,000 TS; 112-149 Brin.
For machinery parts, shafts; S.A.E. 1225.

WEST NO. 2.
M-465; 0.4-0.5 C, 0.6-0.8 Mn, 0.3-0.4 Si, bal Fe.
Annealed: 45,000-80,000 TS; 156-196 Brin.
For machinery parts, gears, shafts, axles; tough; S.A.E. 1245.

WEST NO. 3.
M-465; 0.15-0.20 C, 0.3-0.5 Mn, 2.0-3.0 Si, bal Fe.
70,000 TS; 170 Brin.
For heat treating boxes; cannot be welded.

WEST NO. 3A.
M-465; 0.15-0.20 C, 0.3-0.5 Mn, 1.5-2.0 Si, bal Fe.
Special: 70,000 TS; 156 Brin.
For heat resistant parts; heat resistant.

WEST NO. 4.
M-465; 0.20 max C, 0.3-0.5 Mn, 0.3-0.4 Si, bal Fe.
Annealed: 30,000-55,000 TS; 103-131 Brin.
For electrical and magnetic parts; S.A.E. 1215.

WEST NO. 4A.
M-465; 0.20 max C, 0.6-0.8 Mn, 0.3-0.4 Si, bal Fe.
Annealed: 35,000-60,000 TS; 103-137 Brin.
For machinery parts, case-hardened parts; carburizing steel; S.A.E. 1215.

WEST NO. 5.
M-465; 0.28-0.35 C, 0.6-0.8 Mn, 0.3-0.4 Si, bal Fe, Ti.
Annealed: 42,000-75,000 TS; 137-156 Brin.
For machinery parts; S.A.E. 1230.

WEST NO. 6.
M-465; 0.30-0.40 C, 1.25-1.50 Mn, 0.30-0.40 Si, 0.12 V, bal Fe.
Annealed: 60,000-95,000 TS; 126-228 Brin.
For cylinders, housings, valves; tough, close grained.

WEST NO. 6A (HY-CAST).
M-465; 0.3-0.40 C, 0.8-1.1 Mn, 0.30-0.4 Si, bal Fe.
Annealed: 55,000-90,000 TS; 170-208 Brin.
For shafts, gears, axles; water hardening.

WEST NO. 7 (DURACAST).
M-465; 0.3-0.4 C, 0.6-0.8 Mn, 0.3-0.4 Si, 0.15-0.20 Mo, bal Fe.
Annealed: 50,000-80,000 TS; 170-277 Brin.
For tools, dies; tough.

WEST NO. 7A (DURACAST).
M-465; 0.3-0.4 C, 0.6-0.8 Mn, 0.3-0.4 Si, 0.3-0.4 Mo, bal Fe.
Heat treated: 60,000-90,000 TS; 170-302 Brin.
For tools, dies; tough.

WEST NO. 8 (DURACAST).
M-465; 0.4-0.5 C, 0.6-0.8 Mn, 0.3-0.4 Si, 0.15-0.18 V, bal Fe.
Heat treated: 60,000-90,000 TS; 170-402 Brin.
For tools, dies, shafts, gears, anvils; tough.

WEST NO. 9 (DURACAST).
M-465; 0.6-0.8 C, 0.6-0.8 Mn, 0.3-0.4 Si, 0.15-0.18 V, bal Fe.
Heat treated: 80,000-110,000 TS; 228-600 Brin.
For tools, dies, punches; water hardened.

WEST NO. 9A (CUMLOY).
M-465; 0.3-0.4 C, 1.25-1.50 Mn, 0.3-0.4 Si, 0.2-0.3 Mo, 0.12 V, bal Fe.
Heat treated: 70,000-100,000 TS; 196-514 Brin.
For shafts, gears, machinery parts; tough.

WEST NO. 10 (CUMLOY).
M-465; 0.4-0.5 C, 0.6-0.8 Mn, 0.3-0.4 Si, 0.8-1.1 Cr, 0.2-0.3 Mo, bal Fe.
Heat treated: 60,000-100,000 TS; 196-514 Brin.
For gears, shafts, machinery parts; S.A.E. 4145; resists wear; cannot be welded.

WEST NO. 11 (CUMLOY).
M-465; 0.3-0.4 C, 0.8-1.1 Mn, 0.3-0.4 Si, 0.8-1.1 Cr, 0.2-0.3 Mo, bal Fe.
Heat treated: 75,000-110,000 TS; 196-514 Brin.
For high strength forgings; S.A.E. 4153; wear resistant.

WEST NO. 12 (CUMLOY).
M-465; 0.2-0.3 C, 0.6-0.8 Mn, 0.7-0.8 Si, 0.9-1.25 Cr, 0.4-0.6 Mo, bal Fe.
Heat treated: 70,000-90,000 TS; 170-402 Brin.
For machinery parts; tough.

WEST NO. 13 (CUMLOY).
M-465; 0.35-0.45 C, 1.00-1.25 Mn, 0.7-0.8 Si, 0.8-1.1 Cr, 0.2-0.3 Mo, bal Fe.
Heat treated: 80,000-120,000 TS; 217-600 Brin.
For dies; cannot be welded.

WEST NO. 14.
M-465; 0.5-0.6 C, 0.6-0.8 Mn, 0.3-0.4 Si, 1.0-1.25 Ni, 0.4-0.5 Mo, bal Fe.
Heat treated: 80,000-120,000 TS; 217-512 Brin.
For dies, blocks; Cumloy.

WEST NO. 14A.
M-465; 0.3-0.4 C, 0.6-0.8 Mn, 0.3-0.4 Si, 1.25-1.50 Ni, 0.3-0.4 Mo, bal Fe.
Heat treated: 70,000-100,000 TS; 196-514 Brin.
For cams; Cumloy.

WESTA 2.
M-1322; 1.3 C, 0.25 max Si, 0.25 max Mn, bal Fe.
For engravers' tools, drills, cutters, reamers; Type W1; water hardened.

WESTA 3.
M-1322; 1.15 C, 0.25 max Si, 0.25 max Mn, bal Fe.
Annealed: 110,000 TS; 58,000 YS; 18 El; 38 RA; 210 Brin.
For reamers, taps, form and milling cutters, drills; Type W1; water hardened.

WESTA 003A.
M-1322; 1.15 C, 0.25 max Si, 0.25 max Mn, bal Fe.
Annealed: 110,000 TS; 58,000 YS; 18 El; 38 RA; 210 Brin.
For drills, taps, reamers, broaches; Type W1; water hardened.

WESTA 4.
M-1322; 1.0 C, 0.25 max Si, 0.25 max Mn, bal Fe.
Heat treated: 120,000-215,000 TS; 85,000-150,000 YS; 20-10 El; 48-33 RA; 240-550 Brin.
For taps, drills, reamers, hobs, Type W1; water hardened.

WESTA 5.
M-1322; 0.85 C, 0.25 max Si, 0.25 max Mn, bal Fe.
Heat treated: 130,000-190,000 TS; 88,000-145,000 YS; 20-12 El; 50-35 RA; 255-390 Brin.
For tools, cutters, springs; Type W1; water hardened.

WESTA 6.
M-1322; 0.70 C, 0.25 max Si, 0.25 max Mn, bal Fe.
Heat treated: 122,000-175,000 TS; 82-000-128,000 YS; 22-12 El; 52-37 RA; 240-350 Brin.
For punches, rails, crimpers, springs; Type W1; water hardened.

WESTA 212.
M-1322; 0.3 C, 2.6 Cr, 0.35 V, 8.5 W, bal Fe.
For extrusion rams, liners and dies; hot work steel, oil hardened.

WESTA 212D.
M-1322; 0.3 C, 2.3 Cr, 0.6 V, 4.25 W, bal Fe.
For hot and cold work tools and dies; oil hardened, tough.

WESTA 301.
M-1322; 0.3 C, 2 Co, 2.4 Cr, 0.25 V, 8.5 W, bal Fe.
For extrusion rams and liners, punches, shears; hot work steel, oil hardened.

WESTA 425.
M-1322; 0.3 C, 1.1 Cr, 0.18 V, 3.75 W, 1.0 Si, bal Fe.
For cold work tools and dies, upsetters; oil hardened, tough.

WESTA 465M.
M-1322; 0.90 C, 1.9 Mn, 0.1 V, bal Fe.
For punches, crimpers, blanking dies; oil hardened, non-deforming.

WESTA 546.
M-1322; 0.45 C, 1.3 Cr, 0.45 Mo, 0.8 V, 0.45 W, bal Fe.
For cold work tools and dies; oil hardened, tough.

WESTA 725.
M-1322; 0.65 C, 3.7 Cr, 0.85 Mo, 0.7 V, 8.5 W, bal Fe.
For lathe and planer tools, drills, reamers, hobs; high speed steel, oil hardened.

WESTA 2002.
M-1322; 2.1 C, 11.5 Cr, bal Fe.
For blanking and forming dies, punches; oil hardened, non-deforming.

WESTA 2002 SPEZIAL.
M-1322; 2.1 C, 11.5 Cr, 0.7 W, bal Fe.
For blanking and forming dies, gages, punches; oil hardened, non-deforming.

WESTA 2002W.
M-1322; 1.65 C, 11.5 Cr, 0.1 V, bal Fe.
For blanking and forming dies, gages, punches; air hardened, non-deforming.

WESTA AK SPEZIAL.
M-1322; 0.4 C, 13 Cr, 0.4 Si, 0.3 Mn, bal Fe.
Annealed: 95,000 TS; 50,000 YS; 20 El; 50 RA; 200 Brin.
Cold drawn: 110,000 TS; 90,000 YS; 15 El; 45 RA; 240 Brin.
For turbine blades, cutlery, gauges; Type 420; corrosion resistant.

WESTA AK2 SPEZIAL.
M-1322; 0.2 C, 0.4 Si, 0.3 Mn, 13 Cr, bal Fe.
Annealed: 95,000 TS; 50,000 YS; 25 El; 55 RA; 195 Brin.
Cold drawn: 105,00 TS; 85,000 YS; 17 El; 50 RA; 215 Brin.
For turbine blades, surgical instruments; Type 420; corrosion resistant.

WESTA AL 14.
M-1322; 0.33 C, 1.1 Al, 1.7 Cr, 1.0 Ni, bal Fe.
For oil refinery equipment; creep resistant.

WESTA AL16.
M-1322; 0.27 C, 1.1 Al, 1.4 Cr, bal Fe.
For oil refinery equipment; creep resistant.

WESTA BE.
M-1322; 0.13 C, 0.7 Cr, 2.5 Ni, bal Fe.
For gears, fasteners, camshafts, cams, machine tool parts; case hardened, shock resistant.

WESTA BOW.
M-1322; 0.28 C, 0.7 Cr, 2.5 Ni, bal Fe.
For gears, bolts, fasteners, machine tool parts; oil hardened, shock resistant.

WESTA CA.
M-1322; 0.13 C, 0.5 Cr, bal Fe.
For gears, bolts, machine tool parts; case hardened.

WESTA CE.
M-1322; 0.15 C, 0.5 Mn, 0.65 Cr, bal Fe.
For gears, bolts, machine tool parts; case hardened.

WESTA CE SPEZIAL.
M-1322; 0.15 C, 0.65 Cr, 0.5 Mn, bal Fe.
For gears, bolts, machine tool parts; case hardened.

WESTA CE2 SPEZIAL.
M-1322; 0.2 C, 1.15 Cr, 1.25 Mn, bal Fe.
For gears, cams, camshafts, fasteners; case hardened, tough.

WESTA CM1.
M-1322; 0.15, C, 1.2 Cr, 0.2 Mo, bal Fe.
For gears, bolts, machine tool parts; case hardened, tough.

WESTA CM2.
M-1322; 0.20 C, 1.2 Cr, 0.2 Mo, bal Fe.
For gears, bolts, camshafts, fasteners; case hardened, tough.

WESTA CM3.
M-1322; 0.25 C, 1.0 Cr, 0.2 Mo, 0.65 Mn, bal Fe.
For gears, bolts, machine tool parts; oil hardened, tough.

WESTA CM4.
M-1322; 0.33 C, 0.65 Mn, 1.0 Cr, 0.2 Mo, bal Fe.
For gears, bolts, machine tool parts; oil hardened, tough.

WESTA CM5.
M-1322; 0.42 C, 0.65 Mn, 1.0 Cr, 0.2 Mo, bal Fe.
For gears, bolts, crankshafts, studs; oil hardened, tough.

WESTA CNBD.
M-1322; C, alloy, bal Fe.
For machine tool parts; oil hardened.

WESTA CNE.
M-1322; 0.13 C, 2.5 Ni, 0.7 Cr, bal Fe.
For gears, bolts, cams, camshafts; case hardened, shock resistant.

WESTA CNH SPEZIAL.
M-1322; 0.50 C, 1 Cr, 3.5 Ni, 0.5 Mn, bal Fe.
For gears, bolts, crankshafts, axles, shafts; oil hardened, shock resistant.

WESTA CNL.
M-1322; 0.35 C, 0.6 Mn, 1.3 Cr, 4.5 Ni, bal Fe.
For gears, bolts, crankshafts, machine tool parts; oil hardened, shock resistant.

WESTA CNL EXTRA.
M-1322; 0.50 C, 1.05 Cr, 3.25 Ni, bal Fe.
For gears, bolts, crankshafts, axles; oil hardened, shock resistant.

WESTA CNS 95.
M-1322; 0.22-0.30 C, 0.7 Cr, 3.5 Ni, bal Fe.
For gears, bolts, shafts, studs, fasteners; oil hardened, shock resistant.

WESTA CNSH.
M-1322; 0.22-0.30 C, 0.7 Cr, 3.5 Ni, bal Fe.
For gears, bolts, crankshafts; oil hardened, shock resistant.

WESTA CN SPEZIAL.
M-1322; 0.28-0.35 C, 0.7 Cr, 2.5 Ni, bal Fe.
For gears, bolts, crankshafts, machine tool parts; oil hardened, shock resistant.

WESTA CNS SPEZIAL.
M-1322; 0.22-0.30 C, 0.7 Cr, 3.5 Ni, bal Fe.
For gears, bolts, crankshafts, machine tool parts; oil hardened, shock resistant.

WESTA CR.
M-1322; 0.90 C, 0.8 Cr, 0.3 Mn, 0.25 Si, bal Fe.
For bearings, cutters, springs, liners, bushings; water hardened.

WESTA CR1.
M-1322; 1.0 C, 1.1 Cr, 0.07 Mn, 0.25 Si, bal Fe.
For bearings, liners, bushings; water hardened, wear resistant.

WESTA CR1/W.
M-1322; 0.90 C, 0.25 Si, 0.3 Mn, 0.8 Cr, bal Fe.
For bearings, liners, sleeves, bushings; water hardened, wear resistant.

WESTA CR2.
M-1322; 0.85 C, 0.25 Si, 0.35 Mn, 1.75 Cr, bal Fe.
For bearings, bushings, liners; water hardened, wear resistant.

WESTA CRK.
M-1322; 1.05 C, Cr, bal Fe.
For bearings, liners, bushings; water hardened, wear resistant.

WESTA CVM.
M-1322; 0.50 C, 0.65 Mn, 1.0 Cr, 0.2 Mo, bal Fe.
For gears, bolts, crankshafts, axles; oil hardened, shock resistant.

WESTA CX(2).
M-1322; 0.38 C, Si, Cr, V, bal Fe.
For gears, bolts, machine tool parts; oil hardened, shock resistant.

WESTA DS SPEZIAL.
M-1322; 1.15 C, 0.65 Cr, 1 V, bal Fe.
For bearings, cutters, liners; oil or water hardened.

WESTA E.
M-1322; 0.55 C, 0.10-0.40 Si, 0.5-0.7 Mn, bal Fe.
For gears, bolts, shafts, fasteners; water hardened.

WESTA ECN.
M-1322; 0.13 C, 0.35 max Si, 0.5 Mn, 0.7 Cr, 2.5 Ni, bal Fe.
For gears, bolts, camshafts, cams; case hardened, shock resistant.

WESTA EK.
M-1322; 1.45 C, 1.4 Cr, 0.6 Mn, bal Fe.
For bearings, liners, sleeves, bushings; water hardened, wear resistant.

WESTA EKA.
M-1322; 1.45 C, 1.4 Cr, 0.6 Mn, bal Fe.
For bearings, liners, sleeves, bushings; water hardened, wear resistant.

WESTA ES EXTRA.
M-1322; 0.53 C, 0.9 Si, 0.9 Mn, bal Fe.
For machine tool parts, gears; water hardened.

WESTA EXTRA.
M-1322; 1.3 C, 0.3 Mn, 0.2 max Cr, 4.7 W, bal Fe.
For cutters; water hardened.

WESTA EXTRA ZH.
M-1322; 1.0 C, 0.25 max Si, 0.25 max Mn, bal Fe.
Annealed: 100,000 TS; 53,000 YS; 21 El; 42 RA; 200 Brin.
For springs, drills, taps, reamers; Type W1; water hardened.

WESTA EZH.
M-1322; 1.0 C, 0.25 max Mn, 0.25 max S, bal Fe.
For springs, tools, drills, taps, reamers; Type W1; water hardened.

WESTA EZH SPEZIAL.
M-1322; 1.0 C, 0.1 V, 0.5 Mn, 0.20 Si, bal Fe.
For drills, taps, cutters, springs; Type W2; water hardened.

WESTA F.
M-1322; 1.1 C, 0.25 max Si, 0.25 max Mn, bal Fe.
Annealed: 110,000 TS; 56,000 YS; 18 El; 40 RA; 200 Brin.
For drills, taps, cutters, hobs, springs; Type W1; water hardened.

WESTA KZA.
M-1322; 1.45 C, 0.25 Si, 0.60 Mn, 1.4 Cr, bal Fe.
For bearings, bushings, cutters; water hardened, wear resistant.

WESTA MK.
M-1322; 0.76 C, 4.75 Co, 4.3 Cr, 0.75 Mo, 1.5 V, 18 W, bal Fe.
For lathe and planer tools, drills; high speed steel.

WESTA ORI.
M-1322; 1.25 C, 1.15 Si, 0.7 Mn, 1.2 Cr, bal Fe.
For cutters, forming dies; oil hardened.

WESTA R.
M-1322; 1.05 C, 1.0 Cr, 1.15 W, 0.90 Mn, bal Fe.
For cutters, tools; water or oil hardened.

WESTA RCR1.
M-1322; 1.4 C, 0.30 Cr, 0.10 V, 0.30 Mn, bal Fe.
For forming and blanking dies, cutters; oil or water hardened, wear resistant.

WESTA R-SPEZIAL.
M-1322; 1.42 C, W, V, bal Fe.
For bearings, cutters, forming dies; oil hardened, wear resistant.

WESTA SC.
M-1322; 0.67 C, 1.3 Si, 0.50 Mn, 0.50 Cr, bal Fe.
For springs, chisels, upsetters; oil hardened, shock resistant.

WESTA SP.
M-1322; 1.2 C, 0.20 Cr, 0.10 V, 1.0 W, bal Fe.
For cutters, reamers, forming dies; water or oil hardened.

WESTA T3.
M-1322; 1.1 C, 0.25 max Si, 0.25 max Mn, bal Fe.
Heat treated: 200,000 TS; 125,000 YS; 8 El; 27 RA; 400 Brin.
For springs, reamers, drills, taps; water hardened; Type W1.

WESTA T4.
M-1322; 1.0 C, 0.25 max Si, 0.25 max Mn, bal Fe.
Heat treated: 190,000 TS; 120,000 YS; 10 El; 30 RA; 390 Brin.
For drills, taps, reamers, broaches; Type W1; water hardened.

WESTA T5 EXTRA.
M-1322; 0.75 C, 0.25-0.50 Si, 0.30-0,80 Mn, bal Fe.
Heat treated: 175,000 TS; 130,000 YS; 12 El; 37 RA; 360 Brin.
For springs, clutch discs, girders, rails; Type W1; water hardened.

WESTA T5 HART.
M-1322; 0.61 C, 0.25-0.65 Si, 0.65 Mn, bal Fe.
Heat treated: 160,000 TS; 112,000 YS; 12 El; 40 RA; 330 Brin.
For wheels, die blocks, rails, girders; water hardened.

WESTA T5W.
M-1322; 0.60 C, 0.25-0.50 Si, 0.30-0.80 Mn, bal Fe.
Heat treated: 160,000 TS; 112,000 YS; 12 El; 40 RA; 330 Brin.
For wheels, die blocks, rails, girders; water hardened.

WESTA T6H.
M-1322; 0.45 C, 0.25 Si, 0.65 Mn, bal Fe.
Hot rolled: 98,000 TS; 58,000 YS; 24 El; 45 RA; 212 Brin.
For axles, gears, bolts, crankshafts; water hardened.

WESTA T6H EXTRA.
M-1322; 0.45 C, 0.25-0.50 Si, 0.30-0.80 Mn, bal Fe.
Hot rolled: 98,000 TS; 58,000 YS; 24 El; 45 RA; 212 Brin.
For axles, gears, bolts, crankshafts; water hardened.

WESTA T6W.
M-1322; 0.35 C, 0.25 Si, 0.55 Mn, bal Fe.
Hot rolled: 85,000 TS; 54,000 YS; 30 El; 53 RA; 180 Brin.
For gears, shafts, axles, bolts; water hardened.

WESTA T6W EXTRA.
M-1322; 0.35 C, 0.25-0.50 Si, 0.30-0.80 Mn, bal Fe.
Hot rolled: 85,000 TS; 54,000 YS; 30 El; 53 RA; 180 Brin.
For gears, shafts, axles, bolts; water hardened.

WESTA T7.
M-1322; 0.22 C, 0.25 Si, 0.45 Mn, bal Fe.
Annealed: 73,000 TS; 40,000 YS; 22 El; 58 RA; 140 Brin.
For fan blades, bolts, screws, gears; water hardened.

WESTA TBM1.
M-1322; 0.55 C, 0.60 Mn, 0.70 Cr, 0.18 Mo, 1.65 Ni, 0.1 V, bal Fe.
For forging and heading dies, shear blades; oil hardened, tough.

WESTA TBM1 EXTRA.
M-1322; 0.56 C, 0.85 Cr, 0.20 Mo, 2.0 Ni, 0.1 V, bal Fe.
For forging and heading dies; oil hardened, tough.

WESTA TEJ.
M-1322; 0.13 C, 4.5 Ni, 1.1 Cr, bal Fe.
For gears, bolts, machine tool parts; case hardened, shock resistant.

WESTA TEM.
M-1322; 0.13 C, 0.7 Cr, 3.5 Ni, bal Fe.
For gears, bolts, machine tool parts; case hardened, shock resistant.

WESTA TE SPEZIAL.
M-1322; 0.13 C, 0.7 Cr, 3.5 Ni, bal Fe.
For gears, cams, camshafts, fasteners; case hardening steel.

WESTA TH.
M-1322; 1.05 C, 1.0 Cr, 0.3 Mn, 0.25 Si, bal Fe.
For bearings, liners, sleeves; water hardened, wear resistant.

WESTA TPA.
M-1322; 0.35 C, 1.3 Cr, 4.5 Ni, bal Fe.
For gears, bolts, crankshafts, axles; oil hardened, shock resistant.

WESTA TY 1 W.
M-1322; C, alloy, bal Fe.
For machine tool parts; oil hardened, tough.

WESTA TY 2 W.
M-1322; 0.13 C, 0.2 Cr, 1.5 Ni, bal Fe.
For gears, bolts, shafts, cams, camshafts; case hardened, shock resistant.

WESTA VCN.
M-1322; 0.28-0.35 C, 0.7 Cr, 2.5 Ni, bal Fe.
For gears, bolts, crankshafts, fasteners; oil hardened, shock resistant.

WESTA W8.
M-1322; 0.15 C, 0.25 Si, 0.37 Mn, bal Fe.
For gears, bolts, machine tool parts; case hardened.

WESTA WM4.
M-1322; 0.30 C, 1.0 Si, 1.1 Cr, 0.18 V, 3.75 W, bal Fe.
For pneumatic tools, chisels, upsetters, punches; oil hardened, tough.

WESTA WP.
M-1322; 0.45 C, 1.4 Cr, 0.70 Mo, 0.30 V, 0.7 Mn, bal Fe.
For forging and die casting dies; oil hardened, tough.

WESTEECO NO. 4.
M-468; 0.30 C, Ni, Cr, Mo, bal Fe.
Heat treated: 110,000 TS; 80,000 YS; 20 El; 38 RA; 230 Brin.
For gears, shafts, machinery parts; oil hardened, tough.

WESTEECO NO. 5.
M-468; 0.15 C, 55-60 Ni, 15-18 Cr, bal Fe.
For heat resisting castings; heat resistant.

WESTEECO NO. 6.
M-468; 0.25 C, 1.0-1.25 Ni, 0.25 Mo, bal Fe.
Heat treated: 95,000 TS; 70,000 YS; 24 El; 42 RA; 195-215 Brin.
For lift arms for dump trucks; oil hardened, shock resistant.

WESTEECO NO. 35-15.
M-468; C, 35 Cr, 15 Ni, bal Fe.
For furnace parts, heat resisting castings; heat resistant.

WESTFALIA RF 1.
M-1655; 0.12 C, 13.0 Cr, bal Fe.
Martensitic, corrosion resistant steel for pump shafts, armature shafts, spindles.
Werkstoff Nr. 1.4001; X 7 Cr 14; AISI 410.

WESTFALIA RF 2.
M-1655; 0.15 C, 13.0 Cr, bal Fe.
Martensitic stainless steel; for valves, surgical instruments, armatures.
Werkstoff Nr. 1.4024; similar to AISI 410.

WESTFALIA RF 3.
M-1655; 0.20 C, 13.0 Cr, bal Fe.
Martensitic stainless steel; for instrument shafts, valves, surgical instruments, cutlery.
Werkstoff Nr. 1.4021; similar to AISI 420.

WESTFALIA RF 4.
M-1655; 0.40 C, 13.0 Cr, bal Fe.
Martensitic stainless steel; for pump parts, instrument shafts and gears, cutlery.
Werkstoff Nr. 1.4034; similar to AISI 420.

WESTFALIA RF 5.
M-1655; 0.20 C, 13.0 Cr, 1.2 Mo, bal Fe.
Martensitic stainless steel; for blades and parts for steam turbines.
Werkstoff Nr. 1.4120; similar to AISI 420.

WESTFALIA RFS 1.
M-1655; 0.08 C, 16.5 Cr, bal Fe.
Ferritic type stainless steel; for lightly loaded structural parts in wate or steam. Not hardenable by heat treat.
Werkstoff Nr. 1.4016; AISI 430.

WESTFALIA RFS 2.
M-1655; 0.10 C, 17.0 Cr, Ti, bal Fe.
Stabilized ferritic type stainless steel; for welded corrosion resistant parts.
Werkstoff Nr. 1.4510.

WESTFALIA RFS 3.
M-1655; 0.15 C, 16.5 Cr, 0.25 Mo, S, bal Fe.
Free-machining martensitic stainless steel for screws, nuts, bolts.
Werkstoff Nr. 1.4104.

WESTFALIA RFS 4.
M-1655; 0.22 C, 17.0 Cr, 1.5 Ni, bal Fe.
Ht Tr: 80-95 kp/mm^2 TS; 60 kp/mm^2 YS.
For strong shafts, arbors, pump parts to resist corrosion.
Werkstoff Nr. 1.4057.

WESTFALIA RFS H1.
M-1655; 1.10 C, 15.0 Cr, 0.50 Mo, V, bal Fe.
Martensitic stainless steel; hardenable to about 59 Rc.
For stainless ball and roller bearings, valves, spray nozzles.
Werkstoff Nr. 1.4111.

WESTFALIA RFS H2.
M-1655; 0.90 C, 18.0 Cr, 1.0 Mo, 0.10 V, bal Fe.
Martensitic stainless steel; hardenable to about 57 Rc.
For stainless parts subject to high wear as ball and roller bearings.
Werkstoff Nr. 1.4112.

WESTFALIA RFS H3.
M-1655; 0.90 C, 16.5 Cr, 0.50 Mo, 1.5 Co, 0.25 V, bal Fe.
Martensitic stainless steel; hardenable to about 60 Rc.
For ball and roller bearings, valves, cutlery.
Werkstoff Nr. 1.4535.

WESTFALIA RFS H4.
M-1655; 0.35 C, 16.5 Cr, 1.2 Mo, bal Fe.
Martensitic stainless steel; hardenable to about 450 Brin.
For cutlery; for arbors, spindles operating at elevated temperatures.
Werkstoff Nr. 1.4122.

WESTFALIA S B 1.
M-1655; 0.15 C, 18.0 Cr, 8.0 Ni, bal Fe.
Austenitic stainless steel.
For food and dairy equipment.
Werkstoff Nr. 1.4300; AISI 302.

WESTFALIA S B 2.
M-1655; 0.07 C, 18.0 Cr, 10.0 Ni, bal Fe.
Austenitic stainless steel.
For food and dairy equipment.
Werkstoff Nr. 1.4301; AISI 304.

WESTFALIA S B 3.
M-1655; 0.10 C, 18.0 Cr, 10.0 Ni, Ti, bal Fe.
Stabilized austenitic stainless steel.
Apparatus and parts for photo, paper, soap and textile industries.
Werkstoff Nr. 1.4541.

WESTFALIA S B 4.
M-1655; 0.10 C, 18.0 Cr, 9.0 Ni, Nb, bal Fe.
Stabilized austenitic stainless steel.
Apparatus and parts for photo, paper, soap, and textile industries.
Werkstoff Nr. 1.4550.

WESTFALIA S B 5.
M-1655; 0.07 C, 18.0 Cr, 11.5 Ni, 2.0 Mo, bal Fe.
Austenitic stainless steel.
Equipment for chemical industry, photo, rubber, die and textile industries
Werkstoff Nr. 1.4401; AISI 316.

WESTFALIA S B 6.
M-1655; 0.10 C, 18.0 Cr, 10.0 Ni, 2.0 Mo, Ti, bal Fe.
Titanium stabilized austenitic stainless steel for welded assemblies in chemical textile, die and photo industries.
Werkstoff Nr. 1.4571.

WESTFALIA S B 7.
M-1655; 0.10 C, 18.0 Cr, 10.0 Ni, 2.0 Mo, Nb, bal Fe.
Niobium stabilized austenitic stainless steel for welded assemblies in photo, chemical and textile industries.
Werkstoff Nr. 1.4580.

WESTFALIA S B 9.
M-1655; 0.10 C, 27.0 Cr, 1.5 Mo, 5.0 Ni, bal Fe.
Ferrite-austenite stainless.
65-80 kp/mm^2 TS; 50 Kp/mm^2 min YS.
Stainless parts for high chemical and mechanical stress.
Werkstoff Nr. 1.4460.

WESTFALISCHE CDM.
M-1345; 1.1 C, 0.30 Mn, 0.4 Cr, bal Fe.
For cutters, bearings, tools; water hardened, wear resistant.

WESTFALISCHE CEZ.
M-1345; 1.0 C, 0.07 Mn, 1.1 Cr, bal Fe.
For bearings, cutters, bushings; water hardened, wear resistant.

WESTFALISCHE CFSS.
M-1345; 1.05 C, Cr, bal Fe.
For bearings, bushings, liners; water hardened, wear resistant.

WESTFALISCHE EH10.
M-1345; 1.0 C, 0.25 max Si, 0.25 max Mn, bal Fe.
Annealed: 100,000 TS; 53,000 YS; 21 El; 42 RA; 200 Brin.
For tools, cutters, drills, taps, reamers; Type W1; water hardened.

WESTFALISCHE EM07.
M-1345; 0.70 C, 0.25 max Si, 0.25 max Mn, bal Fe.
Heat treated: 175,000 TS; 130,000 YS; 12 El; 36 RA; 355 Brin.
For springs, tools, punches, rails, cutters, axes; Type W1; water hardened.

WESTFALISCHE EPS.
M-1345; 0.30 C, 2.65 Cr, 0.35 V, 8.5 W, bal Fe.
For punches, shears, extrusion rams and liners; hot work steel, oil hardened.

WESTFALISCHE EWP970.
M-1345; 0.30 C, 2 Co, 2.4 Cr, 0.25 V, 8.5 W, bal Fe.
For punches, shears, extrusion rams and liners; hot work steel, oil hardened.

WESTFALISCHE EZ13.
M-1345; 1.3 C, 0.25 max Si, 0.25 max Mn, bal Fe.
For engravers' tools, forming dies, reamers; Type W1; water hardened.

WESTFALISCHE HSS.
M-1345; 1.65 C, 11.5 Cr, 0.1 V, 0.30 Mn, bal Fe.
For blanking and forming dies, punches; air hardened, nondeforming.

WESTFALISCHE ISS.
M-1345; 0.90 C, 1.90 Mn, 0.1 V, bal Fe.
For punches, dies, shears, cutters; oil hardened, nondeforming.

WESTFALISCHE OPS.
M-1345; 0.45 C, Si, Cr, V, bal Fe.
For springs, gears, bolts, fasteners; oil hardened, tough.

WESTFALISCHE OWS.
M-1345; 0.38 C, Si, Cr, V, bal Fe.
For gears, bolts, crankshafts, fasteners; oil hardened, tough.

WESTFALISCHE PM06.
M-1345; 0.60 C, 0.25-0.50 Si, 0.30-0.80 Mn, bal Fe.
Heat treated: 160,000-115,000 TS; 113,000-77,000 YS; 12-23 El; 40-54 RA; 320-230 Brin.
For wheels, die blocks, rails, girders; water hardened.

WESTFALISCHE PM45.
M-1345; 0.45 C, 0.25-0.50 Si, 0.30-0.80 Mn, bal Fe.
Hot rolled: 98,000 TS; 59,000 YS; 24 El; 45 RA; 212 Brin.
For axles, gears, bolts, bushings, crankshafts; water hardening.

WESTFALISCHE PW10.
M-1345; 1.1 C, 0.25 max Si, 0.25 max Mn, bal Fe.
For springs, taps, cutters, hobs, reamers; Type W1; water hardened.

WESTFALISCHE RZ5.
M-1345; 0.50 C, 0.95 Mn, 1.05 Cr, 0.1 V, bal Fe.
For springs, gears, bolts, crankshafts; oil hardened, shock resistant.

WESTFALISCHE SP07.
M-1345; 0.75 C, 0.25-0.50 Si, 0.30-0.80 Mn, bal Fe.
Heat treated: 185,000 TS; 140,000 YS; 14 El; 38 RA; 370 Brin.
For springs, rails, clutch discs, girders; Type W1; water hardened.

WESTFALISCHE SS09.
M-1345; 0.85 C, 0.25 max Si, 0.25 max Mn, bal Fe.
Heat treated: 190,000 TS; 145,000 YS; 10 El; 30 RA; 400 Brin.
For springs, drills, taps, reamers, hammers; Type W1; water hardened.

WESTFALISCHE SS10.
M-1345; 1.0 C, 0.25 max Si, 0.25 max Mn, bal Fe.
Annealed: 100,000 TS; 53,000 YS; 21 EL; 42 RA; 200 Brin.
For springs, taps, reamers, drills, hobs; Type W1; water hardened.

WESTFALISCHE SS BLAU.
M-1345; 0.86 C, 4.1 Cr, 0.85 Mo, 2.5 V, 12 W, bal Fe.
For drills, taps, reamers, broaches; high speed steel.

WESTFALISCHE SS GELB.
M-1345; 0.86 C, 2.8 Co, 4.3 Cr, 0.85 Mo, 2.1 V, 12 W, bal Fe.
For broaches, taps, cutters, drills, reamers; high speed steel.

WESTFALISCHE SS GOLD.
M-1345; 0.76 C, 10 Co, 4.2 Cr, 0.85 Mo, 1.8 V, 18 W, bal Fe.
For cutters, taps, drills, hobs, reamers; high speed steel.

WESTFALISCHE SS GRUN.
M-1345; 0.82 C, 4.1 Cr, 0.85 Mo, 1.6 V, 8.7 W, bal Fe.
For reamers, drills, broaches, taps, cutters; high speed steel.

WESTFALISCHE SS KUPFER.
M-1345; 0.74 C, 4.1 Cr, 1.1 V, 18.5 W, bal Fe.
For reamers, drills, broaches, hobs, taps; high speed steel.

WESTFALISCHE SS ROT.
M-1345; 1.3 C, 4.3 Cr, 0.85 Mo, 3.8 V, 12 W, bal Fe.
For blanking and forming dies, engravers' tools; high speed steel.

WESTFALISCHE SS SCHWARTZ.
M-1345; 1.35 C, 4 Cr, V, W, Mo, bal Fe.
For blanking and forming dies, reamers, taps; high speed steel.

WESTFALISCHE SS SILBER.
M-1345; 0.79 C, 4.75 Co, 4.3 Cr, 0.75 Mo, 1.5 V, 18 W, bal Fe.
For taps, chasers, drills, hobs, reamers; high speed steel.

WESTFALISCHE SS WEISS.
M-1345; 0.85 C, 4 Cr, V, W, Mo, bal Fe.
For lathe and planer tools, reamers; high speed steel.

WESTFALISCHE SZ09.
M-1345; 0.85 C, 0.25 max Si, 0.25 max Mn, bal Fe.
Heat treated: 190,000 TS; 145,000 YS; 10 El; 30 RA; 400 Brin.
For springs, taps, reamers, drills; Type W1; water hardened.

WESTFALISCHE TSS.
M-1345; 1.15 C, 0.65 Cr, 0.10 V, bal Fe.
For blanking and forming dies, bearings; water hardened, wear resistant.

WESTFALISCHE USS.
M-1345; 1.25 C, 1.15 Si, 0.70 Mn, 1.2 Cr, bal Fe.
For blanking and forming dies, bearings, liners; oil hardened.

WESTIG 31 CR V 3.
M-1730; 0.28-0.35 C, 0.25-0.40 Si, 0.40-0.60 Mn, 0.40-0.70 Cr, 0.07-0.12 V, bal Fe.
Cold work tool steel; for screwdrivers, wrenches.
Werkstoff Nr. 1.2208.

WESTIG 50 CR V 4.
M-1730; 0.47-0.55 C, 0.70-1.10 Mn, 0.90-1.20 Cr, 0.10-0.20 V, bal Fe.
Cr-V spring steel.
Werkstoff Nr. 1.8159; similar to AISI 6150.

WESTIG 55 NI CR MO V 6.
M-1730; 0.55 C, 0.60-0.80 Cr, 1.50-1.80 Ni, 0.25-0.35 Mo, 0.07-0.12 V bal Fe.
Cold work tool steel; forging and stamping dies.
Werkstoff Nr. 1.2713.

WESTIG 100 K6 now WESTIG 100 CR 6; WESTIG SSRM5 now WESTIG S 6-5-2; WESTIG SSR SPEZIAL now WESTIG S 18-0-1.

WESTIG 105 CR 4.
M-1730; 1.0-1.1 C, 0.90-1.15 Cr, bal Fe.
Ball bearing steel.
Werkstoff Nr. 1.3503; similar to AISI 52100.

WESTIG 110 W CR V 5.
M-1730; 1.10 C, 0.15-0.30 Si, 0.20-0.40 Mn, 1.10-1.30 Cr, 1.20-1.40 W, 0.15-0.25 V, bal Fe.
Cold work tool steel; for leather and rubber machine knives, staybolt taps.
Werkstoff Nr. 1.2519.

WESTIG 115 CR V 3.
M-1730; 1.10-1.25 C, 0.15-0.30 Si, 0.20-0.40 Mn, 0.50-0.80 Cr, 0.07-0.12 V, bal Fe.
Cold work tool steel; for drills, taps, reamers, scraping tools.
Werkstoff Nr. 1.2210.

WESTIG 115 W 8.
M-1730; 1.15 C, 0.15-0.30 Si, 0.20-0.40 Mn, 0.15-0.25 Cr, 1.80-2.10 W, bal Fe.
Cold work tool steel; for metal-cutting saws and hack saw blades.
Werkstoff Nr. 1.2442.

WESTIG 120 W 4.
M-1730; 1.20 C, 0.15-0.30 Si, 0.20-0.35 Mn, 0.90-1.10 W, bal Fe.
Cold work tool steel; for taps, center drills, twist drills.
Werkstoff Nr. 1.2414.

WESTIG 120 WV 4.
M-1730; 1.20 C, 0.15-0.30 Si, 0.20-0.35 Mn, 0.15-0.25 Cr, 0.90-1.10 W, 0.07-0.12 V, bal Fe.
Cold work tool steel; for twist drills, rotary files, countersinks.
Werkstoff Nr. 1.2516.

WESTIG 125 CR 1.
M-1730; 1.25 C, 0.15-0.30 Si, 0.25-0.40 Mn, 0.30-0.40 Cr, bal Fe.
Cold work tool steel; for mandrels, punches, taps, drawing dies.
Werkstoff Nr. 1.2002.

WESTIG 140 CR 3.
M-1730; 1.35-1.50 C, 0.15-0.30 Si, 0.25-0.40 Mn, 0.40-0.70 Cr, bal Fe.
Cold work tool steel; for files, burnishing tools, cutters.
Werkstoff Nr. 1.2008.

WESTIG 4006.
M-1730; 0.08-0.12 C, 12.0-14.0 Cr, bal Fe.
Chrome stainless steel.
Werkstoff Nr. 1.4006. (Was CHRONIFER F 14).

WESTIG 4016.
M-1730; 0.10 max C, 15.5-17.5 Cr, bal Fe.
Ferritic stainless steel, non-hardenable.
Werkstoff Nr. 1.4016; similar to AISI 430. (Was CHRONIFER F-17).

WESTIG 4021.
M-1730; 0.17-0.22 C, 12.0-14.0 Cr, bal Fe.
Martensitic stainless steel.
Werkstoff Nr. 1.4021; similar to AISI 420. (Was CHRONIFER V-13).

WESTIG 4301.
M-1730; 0.07 max C, 18.5 Cr, 9.5 Ni, bal Fe.
Austenitic stainless steel.
Werkstoff Nr. 1.4301; AISI 304. (Was CHRONIFER SPEZIAL SUPRA).

WESTIG 4305.
M-1730; 0.15 max C, 0.15-0.35 S, 18.0 Cr, 9.0 Ni, bal Fe.
Free machining austenitic stainless steel.
Werkstoff Nr. 1.4305; AISI 303. (Was CHRONIFER SPEZIAL D).

WESTIG 4401.
M-1730; 0.07 max C, 17.5 Cr, 12.0 Ni, 2.0-2.5 Mo, bal Fe.
Austenitic stainless steel.
Werkstoff Nr. 1.4401; similar to AISI 316. (Was CHRONIFER SPEZIAL D SUPRA).

WESTIG 4541.
M-1730; 0.10 max C, 18.0 Cr, 10 Ni, Ti = 5 X C, bal Fe.
Stabilized austenitic stainless steel.
Werkstoff Nr. 1.4541; AISI 321. (Was CHRONIFER SPEZIAL EXTRA).

WESTIG A 60.
M-1730; 0.60 C, 0.50-0.70 Mn, 0.070 max P, 0.15-0.25 S, bal Fe.
Free machining medium-high carbon steel.
Werkstoff Nr. 1.0729.

WESTIG A 60 PB.
M-1730; 0.60 C, 0.50-0.70 Mn, 0.070 max P, 0.15-0.25 S, Pb, bal Fe.
Free machining steel.

WESTIG C 75 W 3.
M-1730; 0.72-0.82 C, 0.15-0.40 Si, 0.60-0.80 Mn, bal Fe.
Carbon tool steel; for mandrels, collet chucks.
Werkstoff Nr. 1.1750.

WESTIG C 105 W 2.
M-1730; 1.05 C, 0.10-0.30 Si, 0.10-0.35 Mn, bal Fe.
Carbon tool steel; for stone tools, embossing tools, scythes.
Werkstoff Nr. 1.1645.

WESTIG C 110 W 2.
M-1730; 1.10 C, 0.30 max Si, 0.35 max Mn, bal Fe.
Carbon tool steel; for wood and leather working tools.
Werkstoff Nr. 1.1654.

WESTIG C 125 W 2.
M-1730; 1.20-1.35 C, 0.10-0.30 Si, 0.10-0.35 Mn, bal Fe.
Carbon tool steel; for files, scrapers, hard knives.
Werkstoff Nr. 1.1663.

WESTIG MK 63.
M-1730; 0.60-0.64 C, 0.35-0.55 Mn, 0.007 max N, bal Fe.
Structural steel.
Werkstoff Nr. 1.1222.

WESTIG N 100.
M-1730; 1.0 C, 0.15-0.30 Si, 0.15-0.30 Mn, 0.025 max S, 0.025 max P, 0.10-0.15 V, bal Fe.
Wire, for needles.
Werkstoff Nr. 1.2833.

WESTIG S 2-9-2.
M-1730; 1.0 C, 3.8 Cr, 8.6 Mo, 2.0 V, 1.8 W, bal Fe.
High speed steel; for milling tools, drills, lathe tools.
Werkstoff Nr. 1.3348; similar to AISI M1.

WESTIG S 2-10-1-8.
M-1730; 1.08 C, 4.0 Cr, 9.5 Mo, 1.2 V, 1.5 W, 8.0 Co, bal Fe.
High speed steel; for engraving milling cutters, lathe tools for hard materials.
Werkstoff Nr. 1.3247; similar to AISI M42.

WESTIG S 3-3-2.
M-1730; 1.0 C, 4.2 Cr, 2.7 Mo, 2.4 V, 2.85 W, bal Fe.
High speed steel; for milling cutters, twist drills, broaches.
Werkstoff Nr. 1.3333.

WESTIG S 6-5-2.
M-1730; 0.90 C, 4.2 Cr, 5.0 Mo, 1.85 V, 6.4 W, bal Fe.
High speed steel; for twist drills, lathe tools, milling cutters.
Werkstoff Nr. 1.3343; similar to AISI M2.

WESTIG S 6-5-2-5.
M-1730; 0.92 C, 4.1 Cr, 5.0 Mo, 1.9 V, 8.4 W, 4.8 Co, bal Fe.
High speed steel; for highly stressed twist drills.
Werkstoff Nr. 1.3243.

WESTIG S 6-5-3.
M-1730; 1.2 C, 4.2 Cr, 5.0 Mo, 3.0 V, 6.4 W, bal Fe.
High speed steel; for heavy duty milling cutters, broaches, reamers.
Werkstoff Nr. 1.3344.

WESTIG S 7-4-2-5.
M-1730; 1.10 C, 4.2 Cr, 3.8 Mo, 6.8 W, 1.8 V, 5.0 Co, bal Fe.
High speed steel; for twist drills, taps, reamers, milling cutters.
Werkstoff Nr. 1.3246.

WESTIG S 12-1-2.
M-1730; 0.95 C, 4.2 Cr, 0.85 Mo, 2.2 V, 12.0 W, bal Fe.
High speed steel; for lathe tools, milling cutters.
Werkstoff Nr. 1.3318.

WESTIG S 18-0-1.
M-1730; 0.75 C, 4.2 Cr, 1.0 V, 18.0 W, bal Fe.
High speed steel; for lathe tools, milling cutters, drills.
Werkstoff Nr. 1.3355; similar to AISI T1.

WESTIG X 165 CR MO V 12.
M-1730; 1.55-1.75 C, 11.0-12.0 Cr, 0.50-0.70 Mo, 0.40-0.60 W, 0.10-0.50 V, bal Fe.
Cold work tool steel; for coining dies, sheet metal punching dies.
Werkstoff Nr. 1.2601.

WESTINGHOUSE 52 NI-FE.
M-118; 52 Ni, bal Fe.
For semiconductors and hermetically sealed electronic components.
Controlled expansion.

WESTINGHOUSE B-33.
M-118; 4 V, bal Cb.
At 73°F: 81,000 TS; 595,000 YS; 32 El.
At 2000°F: 33,000 TS; 30,300 YS; 34 El.
At -148°F: 91,700 TS; 74,200 YS; 31 El.
For space vehicles, nuclear reactors.
Good combination of density, strength and oxidation resistance at high temperatures.

WESTINGHOUSE B-66.
M-118; 5 V, 5 Mo, 1 Zr, bal Cb.
At -148°F: 128,000 TS; 108,000 YS; 12 El.
At 73°F: 115,000 TS; 91,000 YS; 14 El.
At 2000°F: 65,000 TS; 58,000 YS; 28 El.
For space vehicles, nuclear reactors.
Good combination of density, strength and oxidation resistance at high temperatures.

WESTINGHOUSE B-77.
M-118; 5 V, 10 W, 1 Zr, bal Cb.
At -148°F: 159,000 TS; 137,000 YS; 20 El.
At 73°F: 132,000 TS; 106,500 YS; 18 El.
At 2400°F: 30,000 TS; 27,000 YS; 34 El.
For space vehicles, nuclear reactors.
Good combination of density, strength, and oxidation resistance at high temperatures.

WESTINGHOUSE HI-120.
M-118; Ti, bal Cb.
The ductility of the alloy windings enable the magnet to cycle from full strength to zero field without damage.
Operates in liquid helium at -452°F.
Superconducting magnet.

WESTINGHOUSE NC155.
M-118; 5 Mo, 5 V, bal Cb.
For high temperature applications, space vehicles; heat resistant.

WESTINGHOUSE SR26.
M-118; 0.05 max C, 18 Cr, 37 Ni, 20 Co, 3.5 Mo, 4.2 Ti, 0.2 Al, 0.05 B, 0.05 Zr, bal Fe.
For high temperature applications; heat resistant.

WESTINGHOUSE T-111.
M-118; 7.0-9.0 W, 2.0-2.8 Hf, 0.003 C, bal Ta.
Stress Relieved Sheet: 150,000 TS; 144,800 YS; 9 El.
At 2000°F: 92,100 TS; 67,500 YS; 8 El.
At 3000°F: 16,300 TS; 14,100 YS; 52 El.
For aerospace structural components, containers, for high temperature liquid metals in nuclear reactors.
Superior strength at 2000-3500°F. Good ductility for forming.

WESTINGHOUSE T-222.
M-118; 0.01 C, 9.64 W, 2.4 Hf, bal Ta.
At -320°F: 184,600 TS; 175,000 YS; 28 El.
At 75°F: 110,600 TS; 105,000 YS; 30 El.
At 3000°F: 24,200 TS; 24,000 YS; 26 El.
For high temperature, re-entry space vehicles, rocket reaction chambers, nozzle parts, liquid metal systems.
High temperature strength and good ductility at low temperatures, good welding.

WESTINGHOUSE W545.
M-118; 0.05 C, 26 Ni, 12.5 Cr, 1.5 Mo, 0.08 B, 2.6 Ti, 0.2 Al, bal Fe.
Bar: 162,000 TS; 115,000 YS; 20 El; 25 RA; 350 Brin.
Sheet: 142,000 TS; 108,000 YS; 8 El; 320 Brin.
For jet engine components, gas turbine discs; age-hardenable, heat resistant, high tensile and creep strength.

WESTINGHOUSE XB-88.
M-118; 28 W, 2 Hf, 0.067 C, bal Cb.
For gas turbine buckets and blades.
High heat and corrosion resistant, good stress-rupture properties.

WF see FINKL WF.

WF 11 see CRUCIBLE WF 11.

WF-11.
M-1491; 0.10 C, 1.5 Mn, 0.50 Si, 20.0 Cr, 10.0 Ni, 15.0 W, bal Co.
For jet engines parts, sheets.

WF-31.
M-1491; 0.15 C, 1.42 Mn, 20.0 Cr, 10.0 Ni, 2.6 Mo, 10.7 W, 1.0 Ti, bal Co.
For high temperature operation.

WFF.
M-1113; 1.3 C, 4.5 W, bal Fe.
For tools, cutters; fast finishing steel.

WHEEL BRAND BABBITT.
83 Sn, 14 Sb, 3 Cu.
For bearings, bushings; anti-friction.

WHEEL BRASS.
M-U.S.; 68 Cu, 30 Zn, 2 Pb.
For wheels, hardware; free-cutting.

WHEELERITE.
M-1210; 3.0-3.5 T.C., 1.5-2.5 Si, 0.05-0.15 Mg, bal Fe.
Cast: 102,000 TS; 80,000 YS; 9 El; 241 Brin.
Annealed: 67,000 TS; 46,000 YS; 26 El; 207 Brin.
For gears, housings, machine tool castings; ductile cast iron.

WHEELING METAL.
M-451; 0.5 C, bal Fe.
For rolls; water hardening.

WHEELING SUPER STEEL.
M-451; 0.7 C, bal Fe.

WHELCO NO. 5.
M-119; 16-23 Cr, 0.12 max C, bal Fe.
For corrosion resistant parts; corrosion resistant.

WHELCO OIL HARDENING.
M-119; 0.9 C, 1.2 Mn, Cr, V, bal Fe.
For tools and dies, punches; oil hardened.

WHISTLE LAFOND'S.
80-81 Cu, 17-18 Sn, 0-2 Zn, 0-2 Sb.
For steam whistles; bronze.

WHIT-ALLOY.
M-1077; Al alloy.
For body chuck; non-hardenable.

WHITE ALLOY-1.
64-5 Cu, 32 Sn, 3.5 As.
For metallic mirrors; high polish.

WHITE ALLOY-2.
53-49 Cu, 23-25 Zn, 22-25 Ni, 2-2.4 Fe.
For corrosion resistant white metal parts; corrosion resistant.

WHITE BENEDICT.
60.4 Cu, 18.1 Zn, 16.4 Ni, 0.7 Sn, 4.4 Pb.
For restaurant white metal ware; corrosion resistant.

WHITE BRASS.
M-U.S.; 66 Zn, 34 Cu.
For ornamental parts, castings.

WHITE BRASS.
M-U.S.; 65 Cu, 32-33 Zn, 3-2 Sn.
For ornamental parts.

WHITE BUTTON ALLOY.
48.8-53 Cu, 24.4-23.0 Zn, 24.4-22 Ni, 2.4-2.0 Fe.
For white buttons; corrosion resistant.

WHITE CAST IRON NO. 1.
M-U.S.; 2.5 combined C, 0.20 Mn, 0.75 Si, bal Fe.
Cast: 29,000 TS; 225 Brin.
For cylinders, housings; very hard and brittle.

WHITE COPPER.
70 Cu, 18 Zn, 12 Ni.
For corrosion resistant white metal parts; corrosion resistant.

WHITE END.
M-259; 0.7 C, 1.5 Cr, 0.2 V, bal Fe.
For rolls; oil hardening.

WHITE GOLD-1.
90 Au, 10 Pd.
For jewelry; corrosion resistant.

WHITE GOLD-2.
85-75 Au, 10-8 Ni, 2-9 Zn.
For jewelry; corrosion resistant.

WHITE GOLD, 10 CARAT.
41.6 Au, 25 Cu, 25 Ni, 8.3 Zn.
For jewelry; corrosion resistant.

WHITE GOLD, 14 CARAT "A".
58.3 Au, 17 Cu, 17 Ni, 7.6 Zn.
For jewelry; corrosion resistant.

WHITE GOLD, 14 CARAT "B".
59 Au, 25.5 Cu, 12.3 Ni, 3.2 Zn.
For jewelry; corrosion resistant.

WHITE GOLD, 18 CARAT.
75 Au, 18.5 Ag, 1 Cu, 5.5 Zn.
For jewelry; corrosion resistant.

WHITE GOLD SOLDER, 10 CARAT.
30 Au, 55 Ag, 1 Cu, 12 Zn, 2 Cd.
For gold solders; corrosion resistant.

WHITE GOLD SOLDER, 14 CARAT.
50 Au, 36 Ag, 1 Cu, 11 Zn, 2 Cd.
For gold solder; corrosion resistant.

WHITE GOLD SOLDER, 18 CARAT "A".
60.8 Au, 14 Ag, 4 Cu, 6 Ni, 5.2 Zn.
For gold solder; corrosion resistant.

WHITE GOLD SOLDER, 18 CARAT "B".
61 Au, 13.5 Ag, 1.5 Cu, 7 Ni, 17 Zn.
For gold solder; corrosion resistant.

WHITE GOLD SOLDER, 18 CARAT "C".
82 Au, 10 Ni, 6 Zn, 2 Cd.
For gold solder; corrosion resistant.

WHITE LABEL.
M-341; 1.55 C, 11.5 Cr, 0.8 Mo, 0.9 V, bal Fe.
For blanking, punching, forming and shearing dies, wear plates, thread rollers and drawing dies.
AISI D2 Cold work tool steel.

WHITE LABEL.
M-343; 0.75-0.90 C, bal Fe.
For blacksmith tools depending on temper, chisels, rock drills; "Hellers Electrical Tool."

WHITE LABEL.
M-363; 0.7 C, 18 W, 4 Cr, 1 V, bal Fe.
For cutting tools, hobs; high speed steel.

WHITE LABEL.
M-365; 0.95-1.30 C (as desired).
AISI Type W1 water hardening tool steel.

WHITE LABEL.
M-694; 1.05 C, 0.3 Mn, 0.25 Si, 0.2 V, bal Fe.
AISI Type W2 water hardening tool steel.

WHITE LABEL S.
M-341; 2.05 C, 11.75 Cr, 0.6 V, bal Fe.
For dies, tools; oil hardening, non-deforming.

WHITE LABEL STYRIAN-1.
M-651; 0.77 C, 0.17 Mn, 4.34 Cr, 20.7 W, bal Fe.
For tools, cutters, drills; high speed steel.

WHITE LABEL STYRIAN-2 NEW RAPID.
M-651; 0.70 C, 0.14 Mn, 4.63 Cr, 15.42 W, 1.0 V, bal Fe.
For tools, cutters, reamers; high speed steel.

WHITELEY.
45-55 Au, 30-35 Pd, 15-20 Pt.
For dental alloy; corrosion resistant.

WHITELIGHT ALUMINUM.
M-659; There are 13 aluminum alloys extruded; see AA listing for properties.

WHITELIGHT AZ10.
M-659; 0.1-1.5 Al, 0.20 min Mn, 0.2-0.6 Zn, bal Mg.
Extruded: 33,000 TS; 20,000 YS; 5-10 El.
Low cost, good machinability, weldability, workability.
For light weight machined parts.

WHITELIGHT AZ31B.
M-659; 2.5-3.5 Al, 0.7-1.5 Zn, bal Mg.
Extruded: 38,000 TS; 26,000 YS; 15 El; 49 Brin.
Rolled: 42,000 TS; 32,000 YS; 15 El; 73 Brin.
For light alloy parts; wrought.

WHITELIGHT AZ51.
M-659; 4.1-5.5 Al, 0.15 min Mn, 0.4-1.3 Zn, bal Mg.
Extruded: 38,000-41,000 TS; 23,000-28,000 YS; 10-11 El.
Good structural properties, high endurance limit, good weldability, machinablity, and corrosion resistance.

WHITELIGHT AZ61A.
M-659; 5.8-7.2 Al, 0.15 min Mn, 0.4-1.5 Zn, bal Mg.
Extruded: 45,000 TS; 32,000 YS; 16 El; 60 Brin.
For light alloy parts; wrought.

WHITELIGHT AZ80A.
M-659; 8.5 Al, 0.5 Zn, bal Mg.
Extruded: 49,000 TS; 36,000 YS; 11 El; 60 Brin.
For light alloy parts; wrought.

WHITELIGHT M1A.
M-659; 1.2 Mn, bal Mg.
Extruded: 32,000-39,000 TS; 26,000-29,000 YS; 11-10 El; 44 Brin.
For light alloy parts; wrought.

WHITELIGHT ZK20A.
M-659; 2.3 Zn, 0.55 Zr, bal Mg.
Extruded: 38,000 TS; 28,000 YS; 4 El.
For light weight structures.

WHITELIGHT ZK30A.
M-659; 3 Zn, 0.7 Zr, bal Mg.
Extruded: 44,000 TS; 34,000 YS.
For light weight structures.

WHITELIGHT ZK60A.
M-659; 5.7 Zn, 0.55 Zr, bal Mg.
Extruded: 49,000 TS; 38,000 YS; 14 El; 75 Brin.
T5-temper: 52,000 TS; 43,000 YS; 11 El; 82 Brin.
For light weight structures; age-hardened.

WHITE METAL-1.
77 Pb, 15 Sb, 5 Sn, 2.3 Cu.
For bearings, bushings; antifriction.

WHITE METAL-2.
53-49 Sn, 33-34 Pb, 11-14 Sb, 2.4-3.3 Cu, 0.1 Zn.
For bearings, bushings; antifriction.

WHITE METAL, DUTCH.
81.5 Sn, 8.8 Sb, 9.6 Cu.
For bearings, domestic ware; Babbitt.

WHITE METAL, HANOVER.
86.8 Sn, 7.6 Sb, 5.6 Cu.
For bearings, domestic ware; Babbitt.

WHITE NICKEL.
64-55 Cu, 18 Ni, 0.35 Fe, bal Zn.
For trimmings, control brackets, levers, fittings, plumbing; S.A.E. Spec 42.

WHITE NICKEL ALLOY.
65 Cu, 32.25 Ni, 2.75 Al.
For strong heat and corrosion resistant parts; corrosion resistant.
For bearings, bushings; antifriction.

WHITE TOMBASIL.
M-346; Zn, Mn, Ni, Pb, bal Cu.
Cast: 46,000-90,000 TS; 18,000-45,000 YS; 50-7 El; 100-180 Brin.
Die cast and permanent mold; low melting temperature white alloy; corrosion resistant; for building hardware, marine hardware, ornamental, drain covers, pumps, swimming pool fixtures.

WHITEX ZINC.
M-659; Zn.
For trim, moulding.

WHITWORTH.
M-Eng.; C, alloy, bal Fe.
For armor plate.

WHP.
M-1113; 1.5 C, 11.5 Cr, 0.75 Mo, bal Fe.
For dies, tools; non-deforming.

WI-52.
M-1512, M-1491; 20-22 Cr, 10-12 W, 0.4-0.5 C, 1.0-2.5 Fe, 1 max Ni, 1.5-2.5 Cb + Ta, bal Co.
At 70°F: 125,000 TS; 85,000 YS; 5 El; 5 RA; 380 Brin.
At 1500°F: 75,000 TS; 52,000 YS; 8 El; 12.5 RA.
For turbine nozzle vanes in jet engines; high heat resistance, investment cast.

WI 765.
M-1290; 0.60-0.75 C, 1.0 max Si, 1.0 max Mn, 13.0-15.0 Cr, 0.50-0.60 bal Fe.
Martensitic stainless steel.
Werkstoff Nr. 1.4109.

WICROMAL.
M-1359; 0.5-1.5 Mn, 0.3 max Cr, bal Al.
Soft: 16,000 TS; 6000 YS; 40 El.
Hard: 29,000 TS; 27,000 YS; 10 El.
For cooking utensils, heat exchangers, tanks, furniture; good forming and welding properties.

WIDALOX.
M-72; Al$_2$O$_3$ + additives.
For cutting tools.
Ceramic, high hardness.

WIDDER.
M-Ger.; 10-25 Sn, bal Pb.
For soft solder.

WIDIA-1.
M-371; 84 W, 3 C, 13 Co.
Sintered: 215,000 TS.
For tools, dies, tool bits, lathe and planer tool cutters, shaper tools, reamers, twist drills; sintered alloy.

WIDIA-2.
M-72; 87.4 W, 5.68 C, 6.1 Co.
Sintered: 250,000 TS.
For tools, dies, tool bits, lathe and planer tool cutters, shaper tools, reamers, twist drills; sintered alloy.

WIDIA F 1.
M-72; 69 WC, 25 TiC, 8 Co.
For cutting tools, dies; sintered.

WIDIA F 2.
M-72; 34.5 WC; 60 TiC, 5.5 Co.
For cutting tools, dies; sintered.

WIDIA G1.
M-72; 94 WC, 6 Co.
For cutting tools for cast iron; cemented.

WIDIA G2.
M-72; 89 WC, 11 Co.
For cutting tools for cast iron; cemented.

WIDIA G3.
M-72; 85 WC, 15 Co.
For cutting tools for cast iron; cemented.

WIDIA H.
M-371, M-72; WC + Co.
For machining hard materials, chilled castings; glass; cemented carbide.

WIDIA H1.
M-72; 94 WC, 6 Co.
For cutting tools for cast iron; cemented.

WIDIA H2.
M-72; 91.5 WC, 1 TaC, 7 Co, 0.5 VC.
For cutting tools, dies; sintered.

WIDIA N.
M-371, M-72; 88 W, 5 C, 6 Co.
For machining gray cast iron, brass, non-ferrous; best combination of hardness and toughness.

WIDIA S.
M-371, M-72; WC + Co.
For machining light work, used for light tools as broaches, reamers, forming tools; cemented carbide.

WIDIA S1.
M-72; 78 WC, 16 TiC, 8 Co.
For cutting tools for steel; cemented.

WIDIA S2.
M-72; 78 WC, 14 TiC, 8 Co.
For cutting tools for steel; cemented.

WIDIA S3.
M-72; 88 WC, 5 TiC, 7 Co.
For cutting tools for steel; cemented.

WIDIA TH10.
M-72; 91 WC, 3 (TiC + TaC), 6 Co.
For cutting tools, and dies, high temperature components.
Cemented carbides. Hard and abrasion resistant.

WIDIA TT25.
M-72; 20 TiC + TaC, 9 Co, bal WC.
For high temperature applications, cutters, dies.
Oxidation and corrosion resistant, wear and abrasion resistant.

WIDIA TT40.
M-72; 12 TiC + TaC, 13 Co, bal WC.
For cutting tools and drills, dies, wear parts.
Cemented carbides, hard and abrasion resistant.

WIDIA TT50.
M-72; 15 TiC + TaC, 17 Co, bal WC.
For cutting tools and drills, dies, wear parts.
Cemented carbides. Hard and abrasion resistant.

WIDIA X.
M-371, M-72; WC, Ti, C, Co.
For machining alloy steel of over 275 Brinell; cemented carbide.

WIDIA XX.
M-371, M-72; WC, Ti, C, Co.
For machining mild and soft steels up to 275 Brinell; cemented carbide.

WIDIE.
M-Ger.; 30 Sn, bal Pb.
For soft solder.

WIEGOLD now **WIELAND-S10.**

WIELAND-65.
M-764; 2.4 Fe, 0.03 P, 0.12 Zn, bal Cu.
Soft: 320-380 MPa TS; 200 MPa YS; 27 El; 85 HB.
Hard: 410-480 MPa TS; 370 MPa YS; 8 El; 130 HB.
Connector and switch parts, lead frames for integrated circuits in semi-conductor engineering.
High electrical and thermal conductivity, good temperature resistance, very good cold-forming and excellent soldering properties.

WIELAND-A05.
M-764; 99.5 min Al.
Extruded only: 75 MPa TS; 30 MPa YS; 25 El; 19 HB.
For construction of containers, vehicles and machinery.

WIELAND-A08.
M-764; 99.8 min Al.
Extruded only: 65 MPa TS; 30 MPa YS; 27 El; 18 HB.
For construction of containers, vehicles and machinery.

WIELAND-A11.
M-764; 1 Mg, bal Al.
Extruded only: 120 MPa TS; 50 MPa YS; 18 El; 40 HB.
For construction of containers, railroad cars, and chemical process equipment.

WIELAND-A13.
M-764; 3 Mg, bal Al.
For ship-building, vehicle, construction, chemical and food, optical, precision mechanics.
Resistant to corrosion.
Extruded only: 170 MPa TS; 90 MPa YS; 17 El; 48 HB.

WIELAND-A15.
M-764; 5 Mg, bal Al.
Extruded only: 265 MPa TS; 135 MPa YS; 15 El; 60 HB.
For chemical industry, optical, water and land craft, building trade.
Corrosion resistant.

WIELAND-A22.
M-764; 0.5 Mg, 0.5 Si, bal Al.
Naturally age-hardened: 140 MPa TS; 70 MPa YS; 23 El; 45 HB.
Artificially age-hardened: 225 MPa TS; 170 MPa YS; 20 El; 75 HB.
For building industry; age-hardenable.

WIELAND-A25.
M-764; 1 Mg, 0.5 Si, bal Al.
Naturally age-hardened: 160 MPa TS; 80 MPa YS; 23 El; 47 HB.
Artificially age-hardened: 255 MPa TS; 230 MPa YS; 13 El; 75 HB.
For building industry; age-hardenable.

WIELAND-A28.
M-764; 1 Mg, 0.8 Si, bal Al.
Naturally age-hardened: 210 MPa TS; 110 MPa YS; 16 El; 70 HB.
Artificially age-hardened: 285 MPa TS; 250 MPa YS; 12 El; 92 HB.
For building industry, construction of machinery and chemical process equipment.

WIELAND-A30.
M-764; 0.6 Mg, 0.9 Si, 0.25 Mn, bal Al.
Naturally age-hardened: 210 MPa TS; 110 MPa YS; 16 El; 70 HB.
Artificially age-hardened: 285 MPa TS; 250 MPa YS; 12 El; 92 HB.
For building industry, chemical process equipment.
Age-hardenable material.

WIELAND-A32.
M-764; 1 Mg, 1 Si, 1 Mn, bal Al.
Naturally age-hardened: 250 MPa TS; 190 MPa YS; 15 El; 68 HB.
Artificially age-hardened: 340 MPa TS; 310 MPa YS; 12 El; 100 HB.
Age-hardenable material for construction.

WIELAND-A42.
M-764; 4.5 Zn, 1.0 Mg, bal Al.
Naturally age-hardened: 340 MPa TS; 230 MPa YS; 17 El; 90 HB.
Artificially age-hardened: 420 MPa TS; 360 MPa YS; 12 El; 110 HB.
For vehicles, machinery and chemical process equipment; no loss of material strength after welding.

WIELAND-A61.
M-764; 1.3 Mn, 0.5 Fe, bal Al.
Extruded only: 110 MPa TS; 60 MPa YS; 30 El; 29 HB.
Tubing for heat exchangers and irrigating plants; containers, chemical process equipment.
Corrosion resistant.

WIELAND-A90.
M-764; 4 Cu, 0.5 Mg, 0.5 Mn, bal Al.
Similar to AA 2014.

WIELAND-B06.
M-764; 6 Sn, bal Cu.
Soft: 360 MPa TS; 180 MPa YS; 55 El; 90 HB.
Hard: 490 MPa TS; 420 MPa YS; 24 El; 155 HB.
Tin bronze; for paper and chemical industry.
Good spring properties; resistant to corrosion; solderable.
Similar to CDA and UNS C 90200.

WIELAND-B08.
M-764; 8 Sn, bal Cu.
Soft: 390 MPa TS; 180 MPa YS; 65 El; 90 HB.
Hard: 550 MPa TS; 510 MPa YS; 26 El; 165 HB.
For slide bearings; resistant to wear.
Similar to CDA and UNS C 90200.

WIELAND-B09.
M-764; 8 Sn, bal Cu.
Soft: 390 MPa TS; 180 MPa YS; 65 El; 90 HB.
Hard: 550 MPa TS; 510 MPa YS; 26 El; 165 HB.
For slide bearings; resistant to wear and to corrosion.
Similar to CDA and UNS C 90200.

WIELAND-B10.
M-764; 8 Sn, bal Cu.
Used only for Bourdon tubes.

WIELAND-B14.
M-764; 4 Sn, bal Cu.
For connectors, switches and similar electrical components.
Good cold-forming, soldering properties.
Spring properties; resistant to corrosion.

WIELAND-B15.
M-764; 5 Sn, bal Cu.
For connectors, switches and similar electrical components.
Good cold-forming, solderable properties.
Spring properties; resistant to corrosion.

WIELAND-B16.
M-764; 6 Sn, bal Cu.
Soft: 375 MPa TS; 180 MPa YS; 55 El; 85 HB.
Hard: 530 MPa TS; 490 MPa YS; 25 El; 160 HB.
For electrical switches, connectors, relays, springs; resistant to stress corrosion.

WIELAND-B18.
M-764; 8 Sn, bal Cu.
Soft: 390 MPa TS; 180 MPa YS; 65 El; 90 HB.
Hard: 550 MPa TS; 490 MPa YS; 25 El; 180 HB.
Relay springs, parts for electrical, ship-building, paper industry.
Solderable; resistant to stress corrosion.

WIELAND-B66.
M-764; 6 Sn, 6 Zn, bal Cu.
Soft: 400 MPa TS; 195 MPa YS; 60 El; 100 HB.
Spring hard: 670 MPa TS; 620 MPa YS; 14 El; 200 HB.
Leaf springs of switches and electrical equipment, diaphragms.

WIELAND-B92.
M-764; 5 Sn, 2 Ni, 10 Zn, bal Cu.
Nickel-tin bronze.

WIELAND-B98.
M-764; 4 Sn, 4 Zn, bal Cu.
Soft: 345 MPa TS; 140 MPa YS; 50 El; 80 HB.
Hard: 490 MPa TS; 430 MPa YS; 12 El; 145 HB.
For diaphragms, leaf springs of switches and electrical equipment.

WIELAND EK2 now **WIELAND-S20..**
M-764; 76 Cu, 2 Al, 0.03 P, bal Zn.
For condenser tubes; resists sea water corrosion.

WIELAND FW6 now **WIELAND-B06 AND B16.**

WIELAND FW8 now **WIELAND B09 AND B19.**
M-764; 7.5-9.0 Sn, 0.4 max P, 0.3 max Zn, bal Cu.
Soft: 57,000 TS; 60 El; 85 Brin.
Hard: 80,000 TS; 18 El; 155 Brin.
For paper and chemical industries.
Corrosion resistant.

WIELAND-G05.
M-764; 5 Sn, 5 Zn, 4.5 Pb, bal Cu.
Cast: 250 MPa min TS; 110 MPa YS (approx); 13 El; 65 HB.
For bearing bushes and bushings.
Similar to CDA and UNS C 83600.

WIELAND-G07.
M-764; 6.5 Sn, 6 Pb, 3.8 Zn, bal Cu.
Cast: 320 MPa min TS; 170 MPa YS; 30 El; 85 HB.
For bearing bushes and bushings.

WIELAND-G10.
M-764; 9.5 Sn, 1.0 Ni, 0.7 Pb, bal Cu.
Cast: 350 MPa min TS; 210 MPa YS; 18 El; 100 HB.
For bearing bushes and bushings.

WIELAND-G12.
M-764; 11.5 Sn, 1.0 Ni, 0.7 Pb, bal Cu.
Cast: 350 MPa TS; 230 MPa YS; 15 El; 105 HB.
For bearing bushes and bushings.

WIELAND-G14.
M-764; 13.5 Sn, 1.0 Ni, 0.7 Pb, bal Cu.
Cast: 330 MPa TS; 230 MPa YS; 8 El; 120 HB.
For bearing bushes and bushings.

WIELAND-G20.
M-764; 9.5 Sn, 5 Pb, bal Cu.
Cast: 300 MPa TS; 180 MPa YS; 16 El; 95 HB.
For bearing bushes and bushings.

WIELAND-G21.
M-764; 10 Sn, 10 Pb, bal Cu.
Cast: 290 MPa TS; 180 MPa YS; 16 El; 85 HB.
For bearing bushes and bushings.

WIELAND-G22.
M-764; 8 Sn, 15 Pb, bal Cu.
Cast: 260 MPa TS; 160 MPa YS; 15 El; 75 HB.
For bearing bushes and bushings.

WIELAND-G27.
M-764; 4.5 Sn, 20.5 Pb, bal Cu.
Cast: 200 MPa TS; 120 MPa YS; 14 El; 55 HB.
For bearing bushes and bushings.

WIELAND-G90.
M-764; 10 Sn, 2 Zn, bal Cu.
Similar to CDA and UNS C 90500.

WIELAND-G91.
M-764; 12 Sn, 1.5 Pb, bal Cu.
Similar to CDA and UNS C 92500.

WIELAND-G92.
M-764; 7 Pb, 7 Sn, bal Cu.
Similar to CDA and UNS C 93400.

WIELAND-G93.
M-764; 7.5 Sn, 7.5 Pb, 1.2 Zn, 1.3 Ni, bal Cu.
Similar to CDA and UNS C 93400.

WIELAND-G94.
M-764; 9.5 Sn, 5 Pb, 4 Ni, bal Cu.

WIELAND-G95.
M-764; 11.5 Sn, 2 Ni, bal Cu.
Similar to CDA and UNS C 91700.

WIELAND-G96.
M-764; 13.5 Sn, 3 Ni, bal Cu.

WIELAND HLOS now WIELAND-K10.

WIELAND-K10.
M-764; 99.9 Cu, oxygen-free.
Electrical and electronic industry, vacuum technology (especially suitable). Very high purity, high electrical conductivity.

WIELAND-K12.
M-764; 99.9 Cu, oxygen-free.
Electrical industry; high electrical conductivity.

WIELAND-K20.
M-764; 99.90 min Cu, 0.015-0.040 P.
For heating and sanitary installations, refrigeration and air-conditioning equipment.

WIELAND-K21.
M-764; 99.90 min Cu, 0.015-0.040 P.
For heat exchangers in refrigeration and air-conditioning plants. Very good cold-forming properties.

WIELAND-K30.
M-764; 99.90 min Cu, oxygen-containing.
Electrical industry; high electrical conductivity.

WIELAND-K60.
M-764; 0.3-1.2 Cr, bal Cu.
Age-hardened: 360 MPa TS; 290 MPa YS, 18 El; 120 HB.
Cold work-hardened: 520 MPa TS; 470 MPa YS, 8 El; 155 HB.
Electrical engineering, welding electrodes.
High electrical conductivity, high strength, good tempering properties.

WIELAND-L05.
M-764; 5 Ni, 1.0 Fe, bal Cu.
Soft: 260 MPa TS; 100 MPa YS; 45 El; 65 HB.
Tubing for heat exchangers, chemical process equipment.
Resistant to corrosion.

WIELAND-L10.
M-764; 10 Ni, 1.0 Fe, bal Cu.
Soft: 330 MPa TS; 120 MPa YS; 35 El; 70 HB.
Tubing for heat exchangers, chemical process equipment, ship building industry.
Resistant to corrosion and salt water.

WIELAND-L30.
M-764; 30 Ni, 0.5 Fe, 1.0 Mn, bal Cu.
Soft: 420 MPa TS; 160 MPa YS; 35 El; 95 HB.
Tubing for heat exchangers and conduits in ships, sea water desalination plants.
Very resistant to corrosion and to salt water.

WIELAND-L49.
M-764; 89 Cu, 9 Ni, 2 Sn.
Soft: 390 MPa TS; 180 MPa YS; 40 El; 85 HB.
Hard: 530 MPa TS; 520 MPa YS; 5 El; 150 HB.
Contact elements, connectors, relays.
Good spring qualities, resistant to tarnishing, corrosion, stress corrosion; solderable.

WIELAND-M05.
M-764; 5 Zn, 95 Cu.
Soft: 250 MPa TS; 100 MPa YS; 45 El; 65 HB.
Hard: 330 MPa TS; 300 MPa YS; 10 El; 110 HB.
For jewelry, art objects, watch and clock making industry, electrical industry.
Resistant to stress corrosion, good for enameling.

WIELAND-M10.
M-764; 10 Zn, 90 Cu.
Soft: 270 MPa TS; 100 MPa YS; 45 El; 65 HB.
Hard: 360 MPa TS; 330 MPa YS; 10 El; 115 HB.
For jewelry, art objects, watch and clock making industry, electrical industry.
Resistant to stress corrosion.

WIELAND-M15.
M-764; 15 Zn, 85 Cu.
Soft: 290 MPa TS; 110 MPa YS; 45 El; 65 HB.
Hard: 420 MPa TS; 390 MPa YS; 12 El; 120 HB.
For jewelry, art objects, electrical industry.

WIELAND-M20.
M-764; 20 Zn, 80 Cu.
Soft: 310 MPa TS; 120 MPa YS; 50 El; 70 HB.
Hard: 420 MPa TS; 350 MPa YS; 20 El; 125 HB.
For jewelry, art objects, Bourdon tubes, electric equipment for automotive corrugation tubes.

WIELAND-M28.
M-764; 28 Zn, 72 Cu.
Soft: 330 MPa TS; 110 MPa YS; 55 El; 70 HB.
Hard: 450 MPa TS; 410 MPa YS; 20 El; 130 HB.
For metal goods and musical instruments.
Very good cold forming properties.

WIELAND-M30.
M-764; 30 Zn, 70 Cu.
Soft: 320 MPa TS; 130 MPa YS; 55 El; 70 HB.
Hard: 450 MPa TS; 410 MPa YS; 20 El; 130 HB.
For jewelry, dials, contact springs; very good cold forming properties, good etching.

WIELAND-M32.
M-764; 31 Zn, 69 Cu.
Soft: 320 MPa TS; 130 MPa YS; 55 El; 70 HB.
Hard: 450 MPa TS; 410 MPa YS; 20 El; 130 HB.
For jewelry, dials, contact springs.
Very good cold-forming properties.

WIELAND-M33.
M-764; 33 Zn, 67 Cu.
Soft: 330 MPa TS; 130 MPa YS; 55 El; 70 HB.
Hard: 460 MPa TS; 410 MPa YS; 20 El; 135 HB.
For musical instruments, watch and clock parts, electrical goods.
Good cold-forming properties and mechanical polishing.

WIELAND-M36.
M-764; 36 Zn, 64 Cu.
Soft: 340 MPa TS; 130 MPa YS; 55 El; 70 HB.
Hard: 490 MPa TS; 470 MPa YS; 13 El; 145 HB.
Musical instruments, watch and clock parts.
Good for deep-drawing, forming and spinning.

WIELAND-M37.
M-764; 37 Zn, 63 Cu.
Soft: 340 MPa TS; 130 MPa YS; 55 El; 70 HB.
Hard: 490 MPa TS; 470 MPa YS; 13 El; 145 HB.
Metal goods, electrical parts; good spring properties.

WIELAND-M38.
M-764; 37 Zn, 63 Cu.
Soft: 340 MPa TS; 130 MPa YS; 55 El; 70 HB.
Hard: 490 MPa TS; 470 MPa YS; 13 El; 145 HB.
For musical instruments, contact springs.
Good cold-forming properties; good spring properties.

WIELAND-N12.
M-764; 64 Cu, 12 Ni, bal Zn.
Soft: 380 MPa TS; 160 MPa YS; 45 El; 85 HB.
Hard: 510 MPa TS; 470 MPa YS; 13 El; 155 HB.
For tableware, optical and precision instruments.
Can be deep-drawn; resistant to tarnishing and corrosion.

WIELAND-N17.
M-764; 55 Cu, 18 Ni, bal Zn.
Soft: 440 MPa TS; 200 MPa YS; 45 El; 95 HB.
Hard: 580 MPa TS; 550 MPa YS; 15 El; 170 HB.
For relay springs; good punching and soldering properties; resistant to tarnishing.

WIELAND-N18.
M-764; 61 Cu, 18 Ni, bal Zn.
Soft: 410 MPa TS; 170 MPa YS; 40 El; 90 HB.
Hard: 550 MPa TS; 520 MPa YS; 10 El; 170 HB.
For cutlery, relay springs, ornaments.
Soft temper grade can be deep-drawn.
Solderable; resistant to tarnishing.

WIELAND-N30.
M-764; 46 Cu, 10 Ni, 1 Pb, bal Zn.
Hard: 660 MPa TS; 490 MPa YS; 16 El; 180 HB.
Machined parts for watches, clocks, optical and precision instruments.
Good machining and hot forming properties.

WIELAND-N32.
M-764; 57 Cu, 12 Ni, 1 Pb, bal Zn.
Hard: 530 MPa TS; 480 MPa YS; 20 El; 150 HB.
Machined parts for watches, clocks, optical and precision instruments.
Good machining and hot forming properties.

WIELAND-N37.
M-764; 60 Cu, 17.5 Ni, 1.0 Pb, bal Zn.
Hard: 550 MPa TS; 510 MPa YS; 10 El; 160 HB.
For optical and precision instrument parts, drawing instruments.
Good machining; resistant to tarnishing.

WIELAND-N90.
M-764; 62 Cu, 23 Ni, 2 Sn, bal Zn.
Used in optical industry.

WIELAND PK1 now WIELAND-S28.

WIELAND-S10.
M-764; 1 Sn, 1 Ni, 11 Zn, bal Cu.
Soft: 290 MPa TS; 120 MPa YS; 40 El; 70 HB.
Hard: 420 MPa TS; 340 MPa YS; 15 El; 125 HB.
For art objects; color similar to gold.

WIELAND-S20.
M-764; 76 Cu, 2 Al, 0.03 As, bal Zn.
Soft: 370 MPa TS; 130 MPa YS; 55 El; 80 HB.
1/2 hard: 410 MPa TS; 180 MPa YS; 45 El; 90 HB.
Tubes for evaporators, condensers, sea-water desalination plants, ship equipment.
Salt water resistant.

WIELAND-S23.
M-764; 73 Cu. 3.5 Al, 0.5 Co, bal Zn.
Soft: 560-600 MPa TS; 430 MPa max YS; 40 El; 125 HB.
Hard: 790-840 MPa TS; 670 MPa min YS; 3 min El; 210 HB.
Cold work-hardened special brass alloy.
Spring contacts for electrical engineering.

WIELAND-S28.
M-764; 71 Cu, 1.2 Sn, 0.03 As, bal Zn.
Soft: 340 MPa TS; 120 MPa YS; 55 El; 70 HB.
1/2 hard: 390 MPa TS; 190 MPa YS; 45 El; 90 HB.
Tubes for condensers, heat exchangers in land plants (not sea water). Corrosion resistant.

WIELAND-S31.
M-764; 68 Cu, 1 Si, 0.2 Pb, bal Zn.
Drawn: 370-570 MPa TS; 130-470 MPa YS; 15 El; 160 HB.
For slide bearings, valve guides, fasteners, bolts. Corrosion resistant.

WIELAND-S37.
M-764; 60 Cu, 1 Al, 1 Mn, 0.3 Pb, bal Zn.
Soft: 420 MPa TS; 170 MPa YS; 32 El; 110 HB.
Hard: 570 MPa TS; 370 MPa YS; 18 El; 145 HB.
For slide bearings and parts subject to heavy wear.

WIELAND-S40.
M-764; 58 Cu, 2 Mn, 1.5 Al, 0.7 Pb, bal Zn.
Extruded: 640 MPa TS; 340 MPa YS; 18 El; 155 HB.
Drawn: 690 MPa TS; 390 MPa YS; 15 El; 170 HB.
For slide bearings and parts subject to heavy wear.

WIELAND-S91.
M-640; 60 Cu, 0.7 Pb, 0.7 Sn, bal Zn.

WIELAND SOMS68 now **WIELAND-S31.**

WIELAND T now **WIELAND-M20.**

WIELAND T5 now **WIELAND-M15.**

WIELAND T15 now **WIELAND-M05.**

WIELAND-U20.
M-764; 10 Al, 2 Fe, 2 Mn, bal Cu.
Extruded: 670 MPa TS; 260 MPa YS; 20 El; 150 HB.
Drawn: 730 MPa TS; 460 MPa YS; 16 El; 175 HB.
For bushes, worm wheels, spindle nuts, pump parts in ships and sea-water desalination plants.
High strength, good temperature stability, resistant to both corrosion and to wear.

WIELAND-U90.
M-764; 10 Al, 3 Fe, 2 Ni, bal Cu.
Similar to WIELAND-U20.

WIELAND WB206 now **WIELAND-S37.**

WIELAND WB681 now **WIELAND-S31.**

WIELAND WB800 now **WIELAND-B09.**

WIELAND WBT II now **WIELAND-W20.**

WIELAND WS 100 now **WIELAND-N30.**

WIELAND WS 120 now **WIELAND-N32.**

WIELAND WS 125 now **WIELAND-N12.**

WIELAND WS 175 now **WIELAND-N37.**

WIELAND WS 180 now **WIELAND-N18.**

WIELAND-Z10.
M-764; 63 Cu, 0.3 Pb, bal Zn.
Soft: 340 MPa TS; 130 MPa YS; 55 El; 75 HB.
Hard: 410 MPa TS; 310 MPa YS; 12 El; 140 HB.
For jewelry, connector parts; for spinning and embossing; good brazing and welding.

WIELAND-Z11.
M-764; 63 Cu, 1.0 Pb, bal Zn.
Soft: 330 MPa TS; 120 MPa YS; 55 El; 70 HB.
Hard: 480 MPa TS; 450 MPa YS; 12 El; 140 HB.
Carbon brush holders, electric terminals.
Good machining and embossing.

WIELAND-Z20.
M-764; 61 Cu, 0.2 Pb, bal Zn.
Soft: 360 MPa TS; 140 MPa YS; 45 El; 75 HB.
Hard: 490 MPa TS; 430 MPa YS; 18 El; 140 HB.
For fixtures, hardware and lock parts, watch and clock casings.
Suitable for embossing and hot stamping.

WIELAND-Z21.
M-764; 61 Cu, 1.7 Pb, bal Zn.
Soft: 350 MPa TS; 120 MPa YS; 50 El; 75 HB.
Hard: 480 MPa TS; 450 MPa YS; 10 El; 140 HB.
For watch and clock parts, valve parts, taps and fittings, brass rules.
Good machining and riveting.

WIELAND-Z23.
M-764; 61 Cu, 3 Pb, bal Zn.
Soft: 360 MPa TS; 200 MPa YS; 37 El; 85 HB.
Hard: 480 MPa TS; 440 MPa YS; 12 El; 140 HB.
Turned parts for watch, clock, electrical work, screws, bolts, nuts.
Excellent for machining on automatic lathes.

WIELAND-Z30.
M-764; 59 Cu, 1.7 Pb, bal Zn.
Soft: 380 MPa TS; 160 MPa YS; 45 El; 84 HB.
Hard: 530 MPa TS; 490 MPa YS; 12 El; 150 HB.
For side plates and movements for watches and clocks, instruments.
Good hot-forming and machining properties.

WIELAND-Z31.
M-764; 58 Cu, 2 Pb, bal Zn.
Hard: 530 MPa TS; 460 MPa YS; 13 El; 145 HB.
For lock parts, electric terminals, hardware, movements and side plates for clocks, watches.

WIELAND-Z33.
M-764; 58 Cu, 3 Pb, bal Zn.
Hard: 540 MPa TS; 480 MPa YS; 12 El; 150 HB.
Machined parts (gears) for clocks, watches and precision instruments; screws, nuts, bolts.
Excellent for machining on automatic lathes.

WIELAND-Z34.
M-764; 58 Cu, 4 Pb, bal Zn.
For machining on automatic lathes.

WIELAND-Z40.
M-764; 55 Cu, 2 Pb, bal Zn.
Extruded only: 580 MPa TS; 250 MPa YS; 16 El; 130 HB.
For extruded sections and shapes; good machining; not suitable for cold-forming.

WIGGIN ALLOY C263.
M-86; 20 Cr, 20 Co, 5.9 Mo, 2.2 Ti, 0.5 Al, bal Ni.
For gas turbine jet pipes, thrust reversers, noise suppressors, flame tubes.
Creep resistant for service to 850°C.

WIG METAL.
M-270; tungsten carbide.
For hard tools, dies, drill bits; high wear resistance.

WIKMANS CRU-1.
M-1190; 1.2 C, 0.20 Si, 0.30 Mn, bal Fe.
For springs, cutters, drills, reamers, broaches; Type W1; water hardened.

WIKMAN'S CRU-2.
M-1190; 1 C, 0.20 Si, 0.30 Mn, bal Fe.
Annealed: 100,000 TS; 53,000 YS; 21 El; 42 RA; 200 Brin.
For springs, cutters, reamers, drills, taps, hobs; Type W1; water hardened.

WIKMANS CRU-02.
M-1190; 1.5 C, 0.30 Si, 0.30 Mn, bal Fe.
For engravers' tools, blanking tools; water hardened.

WIKMANS CRU-03.
M-1190; 1.4 C, 0.30 Si, 0.30 Mn, bal Fe.
For engravers' tools, blanking and forming dies; Type W1; water hardened.

WIKMANS CRU-3.
M-1190; 0.80 C, 0.2 Si, 0.3 Mn, bal Fe.
Heat treated: 130,000-188,000 TS; 87,000-145,000 YS; 21-12 El; 50-35 RA; 235-390 Brin.
For drills, punches, reamers, hobs; Type W1; water hardened.

WIKMANS CRU-4.
M-1190; 0.70 C, 0.20 Si, 0.30 Mn, bal Fe.
Heat treated: 122,000-174,000 TS; 82,000-128,000 YS; 22-12 El; 52-37 RA; 240-355 Brin.
For wheels, die blocks, girders, rails, springs; Type W1; water hardened.

WIKMANS CRU-EX12.
M-1190; 0.70 C, 4.5 Cr, 18.5 W, 0.6 max Co, bal Fe.
For drills, reamers, broaches, lathe and planer tools; high speed steel.

WIKMANS CRU-EX20.
M-1190; 0.80 C, 1.2 Mo, 4.5 Cr, 18.5 W, 2.5 Co, 1.6 V, bal Fe.
For milling cutters, planer and shaper tools, broaches; high speed steel.

WIKMANS CRU-EX22.
M-1190; 0.80 C, 1.2 Mo, 4.5 Cr, 18.5 W, 5.5 Co, 1.6 V, bal Fe.
For milling cutters, hobs, planer and shaper tools; high speed steel.

WIKMANS CRU-EX26.
M-1190; 0.80 C, 1 Mo, 18.5 W, 4.5 Cr, 10 Co, 1.6 V, bal Fe.
For milling cutters, planer and shaper cutters; high speed steel.

WIKMAN'S CRU-EX49.
M-1190; 0.8 C, 4 Cr, 5 Mo, 6.2 W, 2 V, bal Fe.
For lathe and planer tools, hobs, reamers; high speed steel.

WIKMANS CRU-X42.
M-1190; 0.8 C, 4 Cr, 3 Mo, 6.2 W, 2 V, bal Fe.
For lathe and planer tools, hobs, drills, taps; high speed steel.

WIKMANS VH7.
M-1190; 0.70 C, 0.20 Si, 0.30 Mn, bal Fe.
Heat treated: 122,000-174,000 TS; 82,000-128,000 YS; 22-12 El; 52-37 RA; 240-355 Brin.
For wheels, die blocks, girders, rails, springs; Type W1; water hardened

WIKMANS VH8.
M-1190; 0.80 C, 0.2 Si, 0.3 Mn, bal Fe.
Heat treated: 130,000-188,000 TS; 87,000-145,000 YS; 21-12 El; 50-35 RA; 235-390 Brin.
For drills, punches, reamers, hobs; Type W1; water hardened.

WIKMANS VH10.
M-1190; 1 C, 0.20 Si, 0.30 Mn, bal Fe.
Annealed: 100,000 TS; 53,000 YS; 21 El; 42 RA; 200 Brin.
For springs, cutters, reamers, drills, taps, hobs; Type W1; water hardened.

WIKMANS VH10CV.
M-1190; 1 C, 0.20 Si, 0.30 Mn, 0.50 Cr, 0.1 V, bal Fe.
For bearings, heading and blanking dies; oil or water hardened.

WIKMAN'S VH10KA.
M-1190; 1.0-1.5 C, 0.20 Si, 0.30 Mn, 1 Cr, bal Fe.
For cutting tools, files, bearings; water hardened, wear resistant.

WIKMANS VH11.
M-1190; 1.1 C, 0.2 Si, 0.3 Mn, bal Fe.
Annealed: 100,000 TS; 53,000 YS; 21 El; 42 RA; 200 Brin.
For springs, cutters, drills, reamers, hobs; Type W1; water hardened.

WIKMANS VH12.
M-1190; 1.2 C, 0.20 Si, 0.30 Mn, bal Fe.
For springs, header dies, punches, cutters, drills; Type W1; water hardened.

WIKMANS VH13C.
M-1190; 1.35 C, 0.30 Cr, 0.25 Mn, 0.25 Si, bal Fe.
For files, razors, finishing tools; water hardened.

WIKMANS VH33.
M-1190; 0.95 C, 1.5 Si, 0.75 Mn, 1.0 Cr, bal Fe.
For drawing and forming dies, punches; oil hardened, shock resistant.

WIKMANS VH-35.
M-1190; 0.40 C, 1.15 Mn, 1.15 Cr, bal Fe.
For machinery parts, gears; oil hardened.

WIKMANS VH54.
M-1190; 0.30 C, 1 Cr, 3.5 Ni, 5.5 W, 0.25 Mn, 0.20 Si, bal Fe.
For mandrels, extrusion press dies and liners; hot work steel, oil hardened.

WIKMANS VH60.
M-1190; 1.25 C, 2 Cr, 0.15 V, 0.7 Si, 0.3 Mn, bal Fe.
For finishing tools, cutters; oil or water hardened.

WIKMANS VH63.
M-1190; 2.2 C, 1.2 W, 13 Cr, 0.75 Mn, 0.3 Si, bal Fe.
For forming and drawing dies, cold work tools; oil or air hardened, non-deforming.

WIKMANS VH-68.
M-1190; 0.30 C, 0.55 Mn, 1.25 Cr, 4.25 Ni, bal Fe.
For gears, shafts; tough, oil hardened.

WIKMANS VH78.
M-1190; 1.1 C, 0.20 Si, 0.30 Mn, 0.30 Cr, 1 W, 0.1 V, bal Fe.
For twist drills, punches, reamers, cutting tools; oil or water hardened.

WIKMANS VH94.
M-1190; 0.15 min C, 12-14 Cr, bal Fe.
Annealed: 88,000 TS; 40,000 YS; 32 El; 68 RA; 170 Brin.
For cutlery, valve trim, turbine blades; Type 420; stainless, hardenable.

WIKMANS VH214.
M-1190; 0.15 min C, 12-14 Cr, bal Fe.
Annealed: 88,000 TS; 40,000 YS; 32 El; 68 RA; 170 Brin.
For cutlery, valve trim, pump parts; Type 420; stainless, hardenable.

WIKMANS VH223.
M-1190; 1.8 C, 12 Cr, 0.6 Mo, 3.25 Co, bal Fe.
For cutting and cold work tools; air hardened, non-deforming.

WIKMANS VH225.
M-1190; 0.45 C, 0.9 Si, 0.3 Mn, 1.2 Cr, 0.25 Mo, 2.2 W, 0.15 V, bal Fe.
For pneumatic tools, chisels, shears, punches; hot work steel, oil hardened.

WIKMANS VH233.
M-1190; 0.80 C, 0.20 Si, 0.30 Mn, 0.1 V, bal Fe.
Heat treated: 190,000 TS; 145,000 YS; 12 El; 32 RA; 400 Brin.
For blanking and forming dies, punches; water hardened.

WIKMANS VH236.
M-1190; 0.55 C, 1.5 Cr, 3 Ni, 0.4 Mn, bal Fe.
For bakelite dies, cold work tools; oil hardened, shock resistant.

WIKMANS VH238.
M-1190; 0.90 C, 1.2 Mn, 0.5 Cr, 0.5 W, 0.1 V, bal Fe.
For drawing and forming dies, punches; oil hardened, non-deforming.

WIKMANS VH239.
M-1190; 0.15 min C, 12-14 Cr, bal Fe.
Annealed: 88,000 TS; 40,000 YS; 32 El; 68 RA; 170 Brin.
For cutlery, valve trim, turbine blades, Type 420; stainless, hardenable.

WIKMANS VH240.
M-1190; 0.35 C, 23-27 Cr, bal Fe.
Annealed: 90,000 TS; 60,000 YS; 20 El; 45 RA; 180 Brin.
For furnace parts, heat treating boxes; Type 446; heat resistant.

WIKMANS VH244.
M-1190; 1.35 C, 0.30 Cr, 6.0 W, 0.30 Mn, bal Fe.
For drawing and forming dies, finishing tools; oil hardened, high hardness.

WIKMANS VH266.
M-1190; 0.55 C, 1 Cr, 3 Ni, 0.3 Mo, 0.4 Mn, 0.3 Si, bal Fe.
For drop forging dies, cold work tools; oil or air hardened, shock resistant.

WIKMANS VH267.
M-1190; 0.40 C, 1.3 Si, 5 Cr, 0.75 Mo, 5 W, bal Fe.
For Al die casting dies, punches; oil or air hardened, hot work steel.

WIKMANS VH273.
M-1190; 0.12 max C, 11.5-13 Cr, 0.1-0.3 Al, bal Fe.
Annealed: 71,000 TS; 42,600 YS; 22 El; 70 RA; 150 Brin.
Heat treated: 175,000 TS; 145,000 YS; 21 El; 64 RA; 352 Brin.
For oil refinery and chemical plant equipment; Type 405; corrosion resistant.

WIKMANS VH274.
M-1190; 0.10 max C, 2 max Mn, 17-19 Cr, 8-10 Ni, bal Fe.
Annealed: 85,000 TS; 35,000 YS; 60 El; 70 RA; 150 Brin.
Cold drawn: 125,000 TS; 95,000 YS; 22 El; 55 RA; 277 Brin.
For oil refinery and chemical plant equipment; Type 304; stainless, austenitic.

WIKMANS VH276.
M-1190; 0.10 max C, 18 Cr, 10 Ni, 1.3 Mo, bal Fe.
Annealed: 85,000-95,000 TS; 35,000-45,000 YS; 60-50 El; 75-60 RA; 150-190 Brin.
For acid resistant chemical plant equipment; Type 316; stainless, austenitic.

WIKMANS VH277.
M-1190; 0.65 C, 1.4 Ni, 0.3 Mo, 0.7 Cr, 0.1 V, bal Fe.
For forging and cold working dies; oil hardened, shock resistant.

WIKMANS VH282.
M-1190; 0.4 C, 15 Cr, 15 Ni, 1.5 Si, 2.1 W, bal Fe.
For valves for large diesel engines; corrosion and heat resistant.

WIKMANS VH300.
M-1190; 0.30 C, 3 Cr, 1.7 Ni, 9 W, 0.3 V, 0.3 Mn, bal Fe.
For hot work tools, Cu die casting dies; hot work steel, oil hardened.

WIKMANS VH-312.
M-1190; 0.25 C, 0.65 Mn, 1.05 Cr, 0.20 Mo, bal Fe.
For gears, shafts, machinery parts; oil or water hardened, tough.

WIKMANS VH-313.
M-1190; 0.3 C, 0.55 Mn, 1 Cr, 3.2 Ni, 0.25 Mo, bal Fe.
For gears, shafts, axles, mandrels, machinery parts; oil hardened, shock resistant.

WIKMANS VH320.
M-1190; 1.5 C, 12 Cr, 0.80 Mo, 0.2 V, bal Fe.
For drawing and blanking dies, cold work tools; air hardened, non-deforming.

WIKMANS VH322.
M-1190; 0.4 C, 12 Cr, 1 Mo, 2.5 Si, bal Fe.
For values for large diesel engines; corrosion and heat resistant.

WIKMANS VH 341.
M-1190; 0.35 C, 0.55 Mn, 1.15 Cr, 2.6 Ni, bal Fe.
For gears, shafts; tough, oil hardened.

WIKMANS VH357.
M-1190; 1.2 C, 0.20 Si, 0.30 Mn, 0.55 W, 0.1 V, bal Fe.
For drills, taps, cutters, heading dies; oil or water hardened.

WIKMANS VH-364.
M-1190; 0.12 C, 0.55 Mn, 0.75 Cr, 3.0 Ni, bal Fe.
For gears, camshafts, crankshafts; case hardened.

WIKMANS VH376.
M-1190; 0.08 max C, 18 Cr, 9 Ni, 1.3 Mo, bal Fe.
Annealed: 85,000 TS; 35,000 YS; 50 El; 65 RA; 160 Brin.
For chemical plant and oil refinery equipment; Type 316; stainless, austenitic.

WIKMANS VH388.
M-1190; 1 C, 0.2 Si, 0.6 Mn, 5.2 Cr, 1.1 Mo, 0.2 V, bal Fe.
For forming and drawing dies, punches, shears; oil or air hardened, non-deforming.

WIKMANS VH-392.
M-1190; 0.20 C, 0.55 Mn, 1.8 Ni, 0.25 Mo, bal Fe.
For gears, pinions, machinery parts; case hardened, shock resistant.

WIKMANS VH407.
M-1190; 0.12 max C, 20 Cr, 19 Ni, bal Fe.
For heat treat boxes, furnace parts and equipment; stainless, austenitic.

WIKMANS VH410.
M-1190; 0.35 C, 1 Si, 5 Cr, 1.5 Mo, 0.4 V, bal Fe.
For extrusion dies for Al; oil or air hardened, non-deforming.

WIKMANS VH417.
M-1190; 0.30-0.36 C, 13 Cr, bal Fe.
Annealed: 100,000 TS; 55,000 YS; 25 El; 55 RA; 195 Brin.
For surgical instruments, valves, cutlery; Type 420; corrosion resistant.

WIKMANS VH602.
M-1190; 0.09-0.13 C, 12 Cr, 0.6 Ni, 0.5 Mo, bal Fe.
For turbine blades, valves; corrosion resistant.

WIKMANS VH604.
M-1190; 0.12 max C, 15.5 Cr, bal Fe.
Annealed: 80,000 TS; 50,000 YS; 25 El; 50 RA; 150 Brin.
For hardware, kitchen sinks, fasteners; Type 430; corrosion resistant.

WIKMANS VH611.
M-1190; 0.53-0.58 C, 14 Cr, bal Fe.
For butchers' knives, surgical instruments; corrosion resistant, hardenable.

WIKMANS VH620.
M-1190; 0.10 max C, 18 Cr, bal Fe.
Annealed: 80,000 TS; 50,000 YS; 25 El; 50 RA; 150 Brin.
For oil refinery equipment, soot blowers and bolts; corrosion and heat resistant.

WIKMANS VH650.
M-1190; 0.07 max C, 18 Cr, 10 Ni, 2.6 Mo, bal Fe.
Annealed: 85,000 TS; 35,000 YS; 50 El; 65 RA; 160 Brin.
For acid resistant chemical plant equipment; Type 316; stainless, austenitic.

WIKMANS VH-CN2.
M-1190; 0.25 C, 0.70 Cr, 3.0 Ni, bal Fe.
For gears, shafts, machinery parts; shock resistant.

WIKMANS VH-CN3.
M-1190; 0.18 C, 0.3 Si, 0.55 Mn, 0.75 Cr, 3.0 Ni, bal Fe.
For gears, camshafts, axles, machinery parts; case hardened, shock resistant.

WIKMANS VH-EX1939.
M-1190; 0.80 C, 0.2 Mo, 4 Cr, 12.5 W, 0.8 Co, 2.5 V, bal Fe.
For drills, planer and shaper tools, hobs, reamers; high speed steel.

WIKMANS VH-FS.
M-1190; 0.55 C, 1.75 Si, 0.75 Mn, bal Fe.
For springs; oil hardened.

WIKMANS VH VULC.
M-1190; 0.15 min C, 12-14 Cr, bal Fe.
Annealed: 88,000 TS; 40,000 YS; 32 El; 68 RA; 170 Brin.
For cutlery, valve trim, pump parts; Type 420; stainless, hardenable.

WIKMANS WH602.
M-1190; 0.08 C, 14.5 Cr, 1.0 Mo, bal Fe.
Annealed: 80,000 TS; 35,000 YS; 30 El; B 92 Rock.
For springs, table flatware, oil refinery and chemical plant equipment.
Corrosion resistant.

WILCO AMPLEX.
M-237.
For thermal cutouts in circuit breakers; bi-metal, temperature range 0-350°F.

WILCO C-5.
M-237; Ag alloy.
For electrical contacts; sintered.

WILCO DA.
M-237; Pd, bal Pt.
For electrical relays, voltage regulators; low contact resistance.

WILCO HIGHFLEX.
M-237; 18 Ni, 11 Cr, 0.25 Cr, bal Fe, and 36 Ni + Co, 0.12 C, bal Fe.
For temperature indicating instruments, contact making devices; bi-metal, temperature range 0-350°F.

WILCO HIGHFLEX 45.
M-237; 19.4 Ni, 0.55 C, 0.9 Mn, 2.2 Cr, 0.1 Co, bal Fe and 36 Ni + Co, 0.12 C, bal Fe.
For temperature indicating instruments, contact making devices; bi-metal, temperature range 0-350°F.

WILCO HIGH HEAT.
M-237.
For thermostats, temperature controls; bi-metal, temperature range 300-650°F.

WILCO HIGHHEAT 47.
M-237.
For thermostats, temperature controls; bi-metal, temperature range 300-650°F.

WILCO H T CONSTANT.
M-237.
For thermostats, temperature controls; bi-metal, temperature range 250-600°F.

WILCOLOY.
M-237; WC.
For telegraph relays; tungsten carbide.

WILCOLOY-B.
M-237; Be, bal Cu.
Heat treated: 190,000 TS; 150,000 YS; 400 Brin.
For contact backing, spring arms, circuit vibrators; age-hardened, maximum service temperature to 700°F.

WILCOLOY-C.
M-237; Cr, Si, bal Cu.
Heat treated: 76,000 TS; 61,000 YS; 144 Brin.
For contact backing, circuit breakers; heat treatable, high electrical conductivity.

WILCOLOY-E.
M-237; Cd, bal Cu.
Cold drawn: 64,000 TS; 55,000 YS; 137 Brin.
For contact backing springs, reeds, relays; high electrical conductivity and strength.

WILCOLOY T-1.
M-237; Co, Be, bal Cu.
Heat treated: 120,000 TS; 90,000 YS; 222 Brin.
For contact backing, spring arms, reeds, switches; age-hardenable.

WILCO NO. 3.
M-237; Au, bal Pt.
For electric contacts; low corrosion resistant.

WILCO NO. 4.
M-237; Au, bal Pt.
For electric contacts; free from atm. films.

WILCO NO. 6.
M-237.
For electrical contacts; resists oxidation and corrosion.

WILCO NO. 7.
M-237; Au, bal Pt.
For electric contacts.

WILCO NO. 10.
M-237; Al alloy.
For electric contacts; hard, wear resistant.

WILCO NO. 14.
M-237; Pt alloy.
For electric contacts; arc-resisting.

WILCO NO. 19.
M-237; Pt, bal Ag.
For electric contacts; low contact resistance.

WILCO NO. 20.
M-237; 80 Ag, 20 Cd.
For relay contacts; high conductivity.

WILCO NO. 28.
M-237; Zn, bal Ag.
For electric contacts; resists film formation.

WILCO NO. 29.
M-237; Pd, bal Ag.
For electric contacts; for relays.

WILCO NO. 31.
M-237; Ag alloy.
For electric contacts; high electrical conductivity.

WILCO NO. 33.
M-237; Au, bal Ag.
For electric contacts; corrosion resistant.

WILCO NO. 34.
M-237; Pd, bal Pt.
For electric contacts; for low current applications.

WILCO NO. 35.
M-237; Pd, bal Ag.
For electric contacts.

WILCO NO. 38.
M-237; Ag, Cu.
For electric contacts; high conductivity.

WILCO NO. 39.
M-237; Ag alloy.
For electric contacts; for voltage regulators.

WILCO NO. 40.
M-237; Ag alloy.
For electric contacts.

WILCO NO. 43.
M-237; Ru, bal Pt.
For electric contacts.

WILCO NO. 50.
M-237; 33 Rh, 67 Os.
For voltage regulators; corrosion resistant.

WILCO P-5.
M-237; Ir, bal Pt.
For electrical contacts, relays, thermostats; hard, wear resistant.

WILCO P-6.
M-237; Ru, bal Pt.
For electrical contacts, thermostats, relays; hard, wear resistant.

WILCO P-10.
M-237; Ir, bal Pt.
For electrical contacts, thermostats, relays; hard, wear resistant.

WILCO P-15.
M-237; Ir, bal Pt.
For electrical contacts, thermostats, relays; hard, wear resistant.

WILCO P-20.
M-237; Ir, bal Pt.
For electrical contacts, thermostats, relays; hard, wear resistant.

WILCO P-21.
M-237; Ir, bal Pt.
For electrical contacts, thermostats, relays; hard, wear resistant.

WILCO P-34.
M-237; Ag, bal Pd.
For electrical contacts, thermostats, relays; low surface contact resistance.

WILCO P-35.
M-237; Pd, bal Ag.
For electrical contacts, relays, radio vibrators; resists sulfide formation.

WILCO P-43.
M-237; Ru, bal Pt.
For electrical contacts, thermostats, relays; hard, wear resistant.

WILCO P-50.
M-237; Ir, Ru, Os, Rh, Pd, bal Pt.
For voltage regulators, electrical contacts, speed governors; high corrosion and wear resistance.

WILCO P-53.
M-237; Ru, bal Pt.
For electrical contacts, thermostats, relays; hard, wear resistant.

WILCO P-54.
M-237; Ag, bal Pd.
For electrical contacts, thermostats, relays; high hardness and wear resistance.

WILCO P-58.
M-237; Ru, bal Pd.
For electrical contacts, thermostats, relays; hard, and wear resistant.

WILCO P-60.
M-237; Pd, bal Pt.
For electrical contacts, relays, magnetos; wear and impact resistant.

WILCO P-61.
M-237; Ag, Ni, bal Pd.
For electrical contacts, thermostats, relays; low surface contact resistance.

WILCO P-62.
M-237; Ag, Ni, bal Pd.
For electrical contacts, voltage regulators; low surface contact resistance.

WILCO P-63.
M-237; Pd, Ni, bal Ag.
For electrical contacts, relays, voltage regulators; low surface contact resistance.

WILCO P-68.
M-237; Ag, Ni, bal Pd.
For electrical contacts, thermostats, relays; low surface contact resistance.

WILCO P-150.
M-237; Ir, Ru, Os, Rh, Pd, bal Pt.
For electrical contacts, voltage regulators, speed controls; oxidation and wear resistant.

WILCO SCOFLEX.
M-237.
For temperature indicating instruments, contact making devices; bi-metal, temperature range 0-350°F.

WILCO SILVER SOLDER NO. 2.
M-237; 65 Ag, 25 Cu, 10 Zn.
For silver solder; corrosion resistant.

WILCO SILVER SOLDER NO. 18.
M-237; 49.5 Ag, 34.5 Cu, 16 Zn.
For silver solder; corrosion resistant.

WILCO SILVER SOLDER NO. 25.
M-237; 22 Ag, 44 Cu, 34 Zn.
For silver solder; corrosion resistant.

WILCO SILVER SOLDER NO. 27.
M-237; 40 Ag, 36 Cu, 24 Zn.
For silver solder; corrosion resistant.

WILCO STANDARD.
M-237.
For temperature indicating instruments, contact making devices; bi-metal, temperature range 0-300°F.

WILCO THERMO-METAL CIRFLEX.
M-237; bimetal.
Thermostatic bimetal; operating temperature 50-300°F.

WILCO THERMO-METAL MIDFLEX 46.
M-237; bimetal.
Thermostatic bimetal; operating temperature 150-450°F.

WILCO THERMOMETAL RUFLEX.
M-237; bimetal.
Thermometal for steam traps, gas pilot controls; bimetal.

WILCO W METAL.
M-237.
For thermostatic bimetals.

WILLEYS 6A now **EX-CELL-O 6A.**

WILLEYS 6AX now **EX-CELL-O 6AX.**

WILLEYS 8A now **EX-CELL-O 8A.**

WILLEYS 10A now **EX-CELL-O 10A.**

WILLEYS 509 now **EX-CELL-O 509.**

WILLEYS 606 now **EX-CELL-O 606.**

WILLEYS E-3 now **EX-CELL-O E3.**

WILLEYS E-5 now **EX-CELL-O E5.**

WILLEYS E-6 now **EX-CELL-O E6.**

WILLEYS E-8 now **EX-CELL-O E8.**

WILLEYS E9 now **EX-CELL-O E9.**

WILLEYS E16 now **EX-CELL-O E16.**

WILLEYS E-25 now **EX-CELL-O E25.**

WILMOTT'S ALUMINUM.
86 Sn, 14 Bi.
For aluminum solder.

WILRICH 142.
M-947; 0.4 C, 13 Cr, 13 Ni, 3 W, 1.5 Si, bal Fe.
For supercharger buckets; high heat resistant.

WILRICH 300.
M-947; bimetal.
For hard facing electrodes; corrosion resistant.

WILRICH 301.
M-947; Ni-Cr.
Cast: 300-400 Brin.
For hard facing electrodes; corrosion resistant.

WILRICH 350.
M-947; Ni-Cr-Cu.
For hard facing electrodes; corrosion resistant.

WILRICH 600.
M-947; Cr, W.
Cast: 620 Brin.
For hard facing electrodes; corrosion resistant.

WILRICH 625.
M-947; Cr, W, Co.
Cast: 650 Brin.
For hard facing electrodes; heat, abrasion and corrosion resisting.

WILSON.
M-662; C, bal Fe.
For welding rod.

WILSON.
C, alloy, bal Fe.
For armor plate.

WILSON NO. 10LC.
M-662; 0.17 max C, 4.5-5.2 Ni, bal Fe.
For welding rod; wear resistant.

WILSON NO. 10SA.
M-662; 0.15 C, 5 Ni, bal Fe.
For welding rod; shielded.

WILSON NO. 512.
M-662; C, bal Fe.
For welding electrodes; mild steel.

WILSON NO. 520.
M-237; C, bal Fe.
For welding electrodes.

WILSON NO. 520.
M-662; C, bal Fe.
For welding rod for mild steel; shielded, arc.

WILSON NO. 520A.
M-662; C, bal Fe.
For welding rod Cr-Mo steel; shielded, arc.

WILSON NO. 575.
M-662; high Ni core.
For welding electrodes for cast iron; coated.

WIMET 3F.
M-1006; 3 Co, bal WC (micro grain).
Sintered: 1950 VDH hardness.
For special wire dies.

WIMET 15F.
M-1006; 15 Co, bal WC (micro grain).
Sintered: 1400 VDH.
For forming, blanking, cold extrusion.

WIMET 10F.
M-1006; 10 Co, bal WC (micro grain).
Sintered: 1650 VDH hardness.
For cutting aero space materials.
Special forming and extrusion tools.

WIMET 90B.
M-1006; 9 Co, bal WC.
Sintered: 1260 VDH.
For rock drilling and general mining applications.

WIMET BP1.
M-1006; 16 Co, bal WC.
Sintered: 1150 VDH.
For pressing and blanking tools, forming tools, heavy duty dies.

WIMET CM.
M-1006; 9.5 Co, bal WC.
Sintered: 1400 VDH.
For wood working tools.
ISO K30.

WIMET CT.
M-1006; 9 Co, bal WC.
Sintered: 1300 VDH.
For rock drilling; rotary tools and percussive drills for abrasive rocks.

WIMET CW540.
M-1006.
TiCN coated carbide for steel cutting.

WIMET CW620.
M-1006.
TiCN coated carbide for cutting cast iron.

WIMET CXT.
M-1006; 9 Co, bal WC.
Sintered: 1225 VDH.
For tools for rock drilling.

WIMET G.
M-1006; 11 Co, bal WC.
Sintered: 1325 VDH.
For dies, forming tools, cutting tools for wood and plastics.
ISO K40 (Am. Code C-1).

WIMET GW520.
M-1006.
TiCN coated carbide for steel cutting.

WIMET H.
M-1006; 6 Co, bal WC.
Sintered: 1750 VDH.
For cutting iron at high speeds; glass, non-ferrous metals.
ISO K10.

WIMET N.
M-1006; 6 Co, bal WC.
Sintered: 1575 VDH.
For turning and milling cast iron and non-ferrous metals; wire drawing dies, wear parts.
ISO K20 (Am. Code C-2).

WIMET R11.
M-1006; 11 Co, bal WC.
Sintered: 1140 VDH.
Heavy duty percussive rock drilling.

WIMET T.
M-1006; 20 Co, bal WC.
Sintered: 1050 VDH.
For cold heading and extrusion dies.

WIMET TT.
M-1006; 25 Co, bal WC.
Sintered: 950 VDH.
For cold heading and extrusion dies.

WIMET X L 2.
M-1006; Co, TiC, TaC, WC.
Sintered: 1575 VDH.
For steel cutting at high speeds and light loads.
ISO P20.

WIMET XL2B.
M-1006; 19 TiC, 15 (Ta+Nb*) C, 10 Co, bal WC. (*Nb=Cb).
Sintered: 1525 VDH.
Steel turning at medium to high speeds, semi-finish and finishing cuts.
ISO P10 (Am. Code C-7).

WIMET XL3.
M-1006; 9 TiC, 12 (Ta+Nb*) C, 9 Co, bal WC. (*Nb=Cb).
Sintered: 1450 VDH.
General purpose steel cutting with good resistance to shock and impact.
ISO P30 (Am. Code C-6).

WIMET XL35.
M-1006; 10 Tic, 2 (Ta+Nb*) C, 10 Co, bal WC. (*Nb=Cb).
Sintered: 1550 VDH.
For turning and milling steel having interrupted cuts.
ISO P25 (Am. Code C-6).

WIMET XL45.
M-1006; 4 TiC, 8 (Ta+Nb*) C; 11 Co, bal WC. (*Nb=Cb).
Sintered: 1500 VDH.
For rough turning and planing operations on steel.

WINCHESTER DRAWING DIE.
1.54 C, 0.25 Mn, 5.11 W, bal Fe.
For tools, dies; oil hardening.

WINDSOR.
M-69; 0.98 C, 5.25 Cr, 0.25 V, 1.0 Mo, bal Fe.
For dies; air hardened. AISI A2.

WINDSOR (CAST TO SHAPE).
M-69; 1.0 C, 0.45 Mn, 0.25 Si, 5.25 Cr, 0.25 V, 1.1 Mo, bal Fe.
For heavy duty production loads; machinery cams, etc.
Air hardening tool steel.

WINNS SUPERHEAT.
62 Cu, 35 Zn, 2 Sn, 0.4 Pb, 0.13 As, 0.2 Fe.
For valves, fittings for super heated steam.

WINSTON.
M-1705; 1.0 C, 0.6 Mn, 5.25 Cr, 0.25 V, 1.1 Mo, bal Fe.
Medium alloy air hardening tool steel.
AISI A2.

WINSTON V.
M-1705; 2.4 C, 5.25 Cr, 4.25 V, 1.1 Mo, bal Fe.
Air hardening tool steel. AISI A7.

WIPING SOLDER.
M-U.S.; 40 Sn, 60 Pb.
For commercial lead solder; M.P. 446°F.

WIPLA METAL.
M-Ger.; 0.2 C, 18 Cr, 8 Ni, bal Fe.
For machinery parts; non-rusting.

WIPTAM.
M-72; 45.5 Co, 28.3 Cr, 24.4 Ni, 0.1 C, 1.1 Si, 0.7 Mn.
Rolled: 90,000 TS; 74,000 YS; 1.2 El; 378 Brin.
For dental alloy, dentures.
Corrosion and wear resistant.

WIRE BRASS.
72-65 Cu, 27-35 Zn, 0.17 Sn, 0.28 Pb.
For wire applications; high ductility.

WIRESPRAY ALUMINUM.
M-963.
Aluminum wire for Wirespray gun build-up, mainly on aluminum parts.

WIRESPRAY ALUMINUM S.
M-963.
Silicon-Aluminum wire for Wirespray gun coating and build-up.

WIRESPRAY BABBITT "A".
M-963.
High tin, lead free Babbitt wire for spray build-up of bearings.

WIRESPRAY BRASS "Y".
M-963.
Yellow brass wire for spray build-up and repair of brass parts.

WIRESPRAY BRONZE "A".
M-963.
Widely used bronze (containing Al) wire for spray build-up and repair.
Machinable, produces good finish.

WIRESPRAY BRONZE "C".
M-963.
Commercial bronze wire for spray build-up and repair for light wear applications.

WIRESPRAY BRONZE "M".
M-963.
Manganese bronze wire for spray build-up and repair on manganese bronze parts.

WIRESPRAY BRONZE "P".
M-963.
Phosphor bronze wire for spray build-up and repair of phosphor bronze parts.

WIRESPRAY BRONZE "T".
M-963.
Tobin bronze wire for spray build-up and repair.

WIRESPRAY COPPER.
M-963.
High purity copper wire for spray build-up. For electrical applications.

WIRESPRAY CUPRO-NICKEL.
M-963; 70 Cu, 30 Ni.
Wire for spray build-up and repair.

WIRESPRAY MOLYBDENUM.
M-963.
Molybdenum wire for spray coating and build-up. Usually used as thin undercoat between base metal and final coating material.

WIRESPRAY MONEL.
M-963.
Monel wire for spray build-up and repair. Good corrosion protection where wear is not severe.

WIRESPRAY NICHROME 62-16.
M-963.
Ni-Cr-Fe high temperature wire for spray build-up. Used alone or in combination with and where aluminum is the top coating.

WIRESPRAY NICHROME 80-20.
M-963.
Ni-Cr alloy wire used usually for spray undercoat for ceramics. Heat and oxidation resistant.

WIRESPRAY NICKEL.
M-963.
Nickel wire for spray build-up and repair.

WIRESPRAY NO. 3 LO-SHRINK.
M-963.
High chrome stainless wire for metallizing gun; for producing heavy, machinable, stainless overlays.

WIRESPRAY NO. 10.
M-963.
A general purpose carbon steel wire for metallizing gun; deposits are machinable.

WIRESPRAY NO. 25.
M-963.
Carbon steel wire for metallizing gun; deposits harder than No. 10 but still machinable.

WIRESPRAY NO. 80.
M-963.
A low shrink steel wire for metallizing gun; too hard to machine except with carbide; excellent wear resistance.

WIRESPRAY TIN.
M-963.
Tin wire for spray build-up or repair.
Used for dairy equipment, decorative purposes, and to prepare surfaces for soldered joints.

WIRESPRAY ZINC.
M-963.
Zinc wire for spray build-up for corrosion protection in outdoor atmospheres.

WISIL.
M-72; 66.2 Co, 27.0 Cr, 4.5 Mo, 0.35 C, 0.4 Si, 1.0 Mn.
Cast: 123,000 TS; 87,000 YS; 10 El; 362 Brin.
For dental alloy; corrosion resistant.

WISSCO NO. 1.
M-120; 0.12 max C, 12-14 Cr, bal Fe.
130,000-233,000 TS.
For stainless wires; stainless, can be forged and gas welded.

WISSCO NO. 1 M.
M-120; 0.12 C, 12-15 Cr, S, Se or Mo, bal Fe.
For stainless wire; free machining.

WISSCO NO. 2.
M-120; 0.12 min C, 12-15 Cr, bal Fe.
For stainless wires; stainless, oil hardened.

WISSCO NO. 3.
M-120; 0.12 max C, 16-18 Cr, bal Fe.
For stainless wires.

WISSCO NO. 4.
M-120; 18 Cr, 8 Ni, 0.20 C, bal Fe.
170,000-300,000 TS.
For stainless wires; stainless, can be forged and welded.

WISSCO NO. 4 M.
M-120; 0.08-0.20 C, 17-19 Cr, 7-9 Ni, P, S or Se, bal Fe.
For stainless wire; free machining.

WISSCO NO. 4 S.
M-120; 0.11 max C, 17-19 Cr, 7-9 Ni, bal Fe.
For welding rods; stainless.

WISSCO NO. 5 A.
M-120; 0.20 max C, 22-26 Cr, 11-13 Ni, bal Fe.
For heat resistant wire; heat resistant.

WISSCO NO. 5 C.
M-120; 27 Cr, 0.30 C, bal Fe.
For high heat resistant wires; stainless.

WISSCO NO. 5 C N.
M-120; 0.25 max C, 19-21 Cr, 24-26 Ni, Mn, Si, bal Fe.
For high heat resistant wires, springs for high temperature service; stainless.

WISSLER HIGH SPEED.
M-Eng.; 15-50 Ni or Co, 15-35 Cr, 15-40 W, 0.5-2.5 B, 0.75-2.5 C.
For cutting tools, dies; cast nonferrous.

WIZARD.
M-289; 0.45 C, 0.3 Mn, 0.9 Cr, 1 W, 0.2 Mo, bal Fe.
For pneumatic tools, hand chisels, punches, crimpers; shock resistant, oil or water hardened.

WIZARD.
M-498; C, alloy, bal Fe.
For high speed tools; high speed steel.

WIZARD.
M-783; 0.45 C, 0.8 Cr, 0.85 W, bal Fe.
For pneumatic chisels, shock tools; tough.

WK 249.
M-1290; 0.40-0.50 C, 1.30-1.60 Si, 0.50-0.70 Mn, 1.30-1.60 Cr, 0.07-0.12 V, bal Fe.
Cold work tool steel.
Werkstoff Nr. 1.2249.

WK 500.
M-1290; 0.36-0.42 C, 0.90-1.20 Si, 0.30-0.50 Mn, 4.8-5.8 Cr, 0.80-1.40 Mo, 0.25-0.50 V, bal Fe.
Hot work tool steel.
Werkstoff Nr. 1.2343.

WK 575.
M-1290; 0.28-0.35 C, 0.20-0.40 Si, 0.20-0.40 Mn, 2.7-3.2 Cr, 2.6-3.0 Mo, 0.40-0.70 V, bal Fe.
Hot work tool steel.
Werkstoff Nr. 1.2365.

WK 600.
M-1290; 0.32-0.40 C, 0.90-1.20 Si, 0.30-0.60 Mn, 5.0-5.6 Cr, 1.3-1.6 Mo, 1.2-1.4 W, 0.15-0.40 V, bal Fe.
Hot work tool steel.
Werkstoff Nr. 1.2606.

WKE 4.
M-52; 1.25 C, 4.1 Cr, 3.1 Mo, 3.1 V, 9.0 W, 9.0 Co, bal Fe.
Cobalt high speed steel for lathe tools, drills, reamers, milling cutters. Good red hardness.

WKE-4 see FAGERSTA WKE-4.

WKE 45.
M-52; 1.40 C, 4.2 Cr, 3.5 Mo, 3.5 V, 9.0 W, 11.0 Co, bal Fe.
Cobalt high speed steel for lathe tools, milling cutters, reamers. Good abrasion resistance and red hardness.

W.K.Z. HOT WORK.
M-363; 0.3 C, 9.5 W, 2.8 Cr, 4 V, bal Fe.
For hot work tools, dies; hot work steel.
AISI H 21.

WKZV.
M-1290; 0.25-0.35 C, 0.15-0.30 Si, 0.20-0.40 Mn, 2.5-2.8 Cr, 8.0-9.0 W, 0.30-0.40 V, bal Fe.
Hot work tool steel.
Werkstoff Nr. 1.2581.

WM 80F.
M-Ger.; 80 Sn, 11 Sb, 9 Cu.
For bearings; anti-friction.

WM 344.
M-1290; 0.37-0.42 C, 0.90-1.20 Si, 0.30-0.50 Mn, 5.0-5.5 Cr, 1.20-1.50 Mo, 0.90-1.10 V, bal Fe.
Hot work tool steel.
Werkstoff Nr. 1.2344.

WM 362.
M-1290; 0.60-0.65 C, 1.0-1.2 Si, 0.30-0.50 Mn, 5.0-5.5 Cr, 1.0-1.3 Mo, 0.25-0.35 V, bal Fe.
Cold work tool steel.
Werkstoff Nr. 1.2362.

WM 631.
M-1290; 0.45-0.55 C, 0.80-1.0 Si, 0.40-0.60 Mn, 8.0-9.0 Cr, 1.1-1.3 Mo, 1.1-1.3 W, bal Fe.
Cold work tool steel.
Werkstoff Nr. 1.2631.

W.M. 802.
M-1177; 80 Sn, 12 Sb, 6 Cu, 2 Pb.
For bearings; white metal.

W.M. 810.
M-1177; 80 Sn, 10 Sb, 10 Cu.
For bearings; white metal.

W.M. 855.
M-1177; 85 Sn, 10 Sb, 5 Cu, 0.35 max Pb.
For bearings; white metal.

W.M. 903.
M-1177; 90 Sn, 7 Sb, 3 Cu, 0.35 max Pb.
For bearings; white metal.

W.M. 1735.
M-1177; 10 Sn, 15 Sb, 1.5 Cu, 73.5 Pb.
For bearings; white metal.

WN 4.
M-1290; 0.25-0.35 C, 0.80-1.1 Si, 0.30-0.50 Mn, 0.90-1.2 Cr, 3.5-4.0 W, 0.15-0.20 V, bal Fe.
Hot work tool steel.
Werkstoff Nr. 1.2564.

WN 5.
M-1290; 0.25-0.35 C, 0.15-0.30 Si, 0.20-0.40 Mn, 2.2-2.5 Cr, 4.0-4.5 0.50-0.70 V, bal Fe.
Hot work tool steel.
Werkstoff Nr. 1.2567.

WN-102 see **SYLVANIA WN-102.**

WN-103 see **SYLVANIA WN-103.**

WNS.
M-1113; 0.95 C, 1.2 Mn, 0.5 W, 0.5 Cr, bal Fe.
For tools, dies; non-deforming.

W O 0.
M-1655; 0.15 C, 0.25 Si, 0.40 Mn, bal Fe.
Carbon steel for case-hardened tools.
Werkstoff Nr. 1.1705.

W O 2.
M-1655; 0.35 C, 0.40 Si, 0.50 Mn, bal Fe.
Carbon tool steel.
Werkstoff Nr. 1.1720.

W O 3.
M-1655; 0.45 C, 0.35 Si, 0.70 Mn, bal Fe.
Carbon tool steel.
Werkstoff Nr. 1.1730.

W O 4.
M-1655; 0.60 C, 0.40 Si, 0.70 Mn, bal Fe.
Carbon tool steel.
Werkstoff Nr. 1.1740.

W O 5.
M-1655; 0.75 C, 0.40 Si, 0.70 Mn, bal Fe.
Carbon tool steel.
Werkstoff Nr. 1.1740.

W O 7 EXTRA.
M-1655; 0.70 C, 0.25 Si, 0.25 Mn, bal Fe.
Carbon tool steel.
Werkstoff Nr. 1.1520.

W O 7 PRIMA.
M-1655; 0.70 C, 0.30 Si, 0.35 Mn, bal Fe.
Carbon tool steel.
Werkstoff Nr. 1.1620.

W O 9 EXTRA.
M-1655; 0.85 C, 0.25 Si, 0.25 Mn, bal Fe.
Carbon tool steel.
Werkstoff Nr. 1.1530.

W O 9 PRIMA.
M-1655; 0.85 C, 0.30 Si, 0.35 Mn, bal Fe.
Carbon tool steel.
Werkstoff Nr. 1.1630.

W O 10 EXTRA.
M-1655; 1.0 C, 0.25 Si, 0.25 Mn, bal Fe.
Carbon tool steel.
Werkstoff Nr. 1.1540.

W O 10 PRIMA.
M-1655; 1.0 C, 0.30 Si, 0.35 Mn, bal Fe.
Carbon tool steel.
Werkstoff Nr. 1.1640.

W O 11 EXTRA.
M-1655; 1.10 C, 0.25 Si, 0.25 Mn, bal Fe.
Carbon tool steel.
Werkstoff Nr. 1.1550.

W O 11 PRIMA.
M-1655; 1.15 C, 0.30 Si, 0.35 Mn, bal Fe.
Carbon tool steel.
Werkstoff Nr. 1.1650.

W O 13 PRIMA.
M-1655; 1.30 C, 0.30 Si, 0.35 Mn, bal Fe.
Carbon tool steel.
Werkstoff Nr. 1.1660.

W O 120.
M-1655; 1.40 C, 0.30 Si, 0.30 Mn, 0.30 Cr, 0.10 V, bal Fe.
Cold work tool steel for punches, stamps, centering punch.
Werkstoff Nr. 1.2206.

WODURIT.
M-1331; 1.42 C, W, V, bal Fe.
For blanking and forming dies, punches; oil hardened, wear resistant.

WOLFF CC5.
M-1340; 1.1 C, 0.4 Cr, 0.3 Mn, bal Fe.
For bearings, liners, sleeves, cutters; water hardened, wear resistant.

WOLFF CC16.
M-1340; 1.0 C, 0.35 Mn, 1.55 Cr, 0.3 Si, bal Fe.
For bearings, liners, bushings; water or oil hardened, wear resistant.

WOLFF CM SPEZIAL.
M-1340; 0.45 C, 0.7 Mn, 1.4 Cr, 0.7 Mo, 0.3 V, bal Fe.
For gears, bolts, crankshafts, fasteners; oil hardened, tough.

WOLFF CMV12.
M-1340; 0.50 C, 0.95 Mn, 1.05 Cr, 0.1 V, bal Fe.
For gears, springs, bolts, shafts; oil hardened, shock resistant.

WOLFF CR-08.
M-1340; 0.90 C, 0.25 Si, 0.30 Mn, 0.80 Cr, bal Fe.
For bearings, liners, bushings; water hardened, wear resistant.

WOLFF CR-10.
M-1340; 1.05 C, 0.25 Si, 0.30 Mn, 1.2 Cr, bal Fe.
For bearings, liners, bushings; water hardened, wear resistant.

WOLFF CR-12.
M-1340; 2.1 C, 11.5 Cr, 0.35 Si, 0.30 Mn, bal Fe.
For blanking and forming dies, punches; oil hardened, non-deforming.

WOLFF CSC 12.
M-1340; 1.25 C, 1.15 Si, 0.70 Mn, 1.2 Cr, bal Fe.
For bearings, bushings, liners, sleeves; water hardened.

WOLFF CV 60.
M-1340; 0.55 C, 0.10-0.40 Si, 0.50-0.70 Mn, bal Fe.
For gears, pinions, machine tool parts; water hardened.

WOLFF EC 15.
M-1340; 0.15 C, 0.25 Si, 0.25 Mn, bal Fe.
Annealed: 70,000 TS; 40,000 YS; 25 El; 60 RA.
For gears, bolts, machine tool parts; case hardened.

WOLFF EW.
M-1340; 0.15 C, 0.15-0.35 Si, 0.25-0.50 Mn, bal Fe.
For gears, pinions, machine tool parts; case hardened.

WOLFF EXTRA HART.
M-1340; 1.0 C, 0.25 max Si, 0.25 max Mn, bal Fe.
For drills, taps, reamers, broaches, springs; Type W1; water hardened.

WOLFF EXTRA ZAH.
M-1340; 0.85 C, 0.25 max Si, 0.25 max Mn, bal Fe.
Heat treated: 190,000 TS; 145,000 YS; 10 El; 30 RA; 400 Brin.
For springs, tools, drills, taps; Type W1; water hardened.

WOLFF MCVII.
M-1340; 0.58 C, 1.0 Cr, 0.09 V, 0.95 Mn, bal Fe.
For springs, gears, bolts, shafts; oil hardened, shock resistant.

WOLFF MS 10.
M-1340; 0.53 C, 0.90 Si, 0.90 Mn, bal Fe.
For punches, crimpers, bolts, gears; water hardened.

WOLFF MS 15.
M-1340; 0.70 C, 1.7 Si, 0.70 Mn, bal Fe.
For springs; oil hardened.

WOLFF SCRV 15.
M-1340; 0.45 C, Si, Cr, V, bal Fe.
For springs, gears, bolts, shafts; oil hardened, shock resistant.

WOLFF SVC 13H.
M-1340; 0.61 C, 1.18 Cr, 0.10 V, 0.75 Mn, bal Fe.
For springs, crankshafts, punches; oil hardened, shock resistant.

WOLFRAM.
M-275; 0.58 C, 18 W, 4 Cr, 1 V, bal Fe.
For punches, shear blades, hot forming and brass extrusion dies.
Type H26 hot work tool steel.

WOLFRAM.
M-1406; 0.55 C, 2.75 W, 1.25 Cr, 0.25 V, bal Fe.
For hot stamping dies; shock resistant, oil hardened.

WOLFRAM 184.
M-1326; 0.74 C, 4.1 Cr, 1.1 V, 18.5 W, bal Fe.
For lathe and planer tools; reamers; high speed steel.

WOLFRAMANT.
M-459; 0.7 C, 18 W, 4 Cr, 1 V, 5 Co, bal Fe.
For cutters, fluters, hobs; high speed steel.

WOLFRAMANT.
M-1246; WC, Co.
For inserts for boring crowns; sintered carbides.

WOLFRAMANT.
M-1340; 1.42 C, W, V, bal Fe.
For engraving tools, fast finishing cutters; Type W2; water hardened.

WOLFRAMINIUM.
98 Al, 1.4 Sb, 0.36 Cu, 1.0 Sn, 0.05 W.
Cast: 20,000 TS; 6 El.
Rolled: 48,000 TS; 6 El.
Annealed: 24,000 TS; 16 El.
For motor car body work; non-tarnishing.

WOLFRAM LOW CARBON.
M-275; 0.50 C, 4 Cr, 1 V, 18 W, bal Fe.
For drills, reamers, taps, broaches, end mills; Type H26; high speed steel.

WOLFRAMSTAHL, EXTRA SPEZIAL.
M-1310; 1.42 C, W, V, bal Fe.
For engravers' tools, cutters, blanking dies; oil hardened, wear resistant.

WOLVA.
M-73; 0.9 C, 0.9 W, 0.2 Cr, bal Fe.
For punches, dies, reamers; water hardened.

WOLVERINE ALLOY 01.
M-1521; 99.9 Cu, 0.025 P.
Ann: 34,000 TS; 45 El.
Ductile copper with good corrosion resistance; used in air conditioners, plumbing tubes, and heat exchanger tubes.

WOLVERINE ALLOY 02.
M-1521; 99.4 Cu, 0.3 As, 0.025 P.
Ann: 34,000 TS; 45 El.
For heat exchanger applications.

WOLVERINE ALLOY 07.
M-1521; 99.9 Cu, 0.007 P.
Ann: 34,000 TS; 45 El.
High conductivity copper for electrical conductivity applications such as waveguide and bus tube.

WOLVERINE ALLOY 08.
M-1521; 98.7 Cu, 0.025 P, 1.0 Fe.
Ann: 40,000 TS; 40 El.
For air conditioners, plumbing tube and heat exchangers.

WOLVERINE ALLOY 21.
M-1521; 66.0 Cu, 0.5 Pb, bal Zn.
Cold drawn: 68,000 TS; 29 El.
Improved machinable brass used in plumbing brass goods, pump cylinders.

WOLVERINE ALLOY 22.
M-1521; 66.5 Cu, 1.5 Pb, 32.0 Zn.
Ann: 52,000 TS; 50 El.
Hard drawn: 75,000 TS; 7 El.
For applications requiring extensive machining; free cutting material.

WOLVERINE ALLOY 24.
M-1521; 70 Cu, 30 Zn.
Ann: 52,000 TS; 55 El.
Hard drawn: 78,000 TS; 8 El.
Highly ductile brass for miscellaneous fabrications; plumbing brass goods and pump cylinders.

WOLVERINE ALLOY 26.
M-1521; 71 Cu, 28 Zn, 1 Sn, As, Sb or P = 0.02.
Ann: 53,000 TS; 65 El.
Used as heat exchanger tube for steam condensers in circulating sea water or brackish water.

WOLVERINE ALLOY 28.
M-1521; 85 Cu, 15 Zn.
Ann: 44,000 TS; 45 El.
Condenser and heat exchanger alloy for use in fresh waters, water pipe.

WOLVERINE ALLOY 29.
M-1521; 90 Cu, 10 Zn.
Ann: 38,000 TS; 50 El.
Drawn: 52,000 TS; 20 El.
For use in rotating bands, marine hardware.

WOLVERINE ALLOY 30.
M-1521; 76 Cu, 22 Zn, 2 Al, As or P = 0.05%.
Ann: 60,000 TS; 55 El.
For condenser and heat exchanger applications in sea water; resistant to impingement attack.

WOLVERINE ALLOY 32.
M-1521; 95 Cu, 5 Al, 0.02 As.
Ann: 52,000 TS; 60 El.
For condenser and heat exchanger tube applications, especially in sea water, brackish waters and polluted sea waters.

WOLVERINE ALLOY 40.
M-1521; 99.0 min Al, 0.4 Fe, 0.15 Si, 0.1 Cu.
O Temper: 13,000 TS; 45 El.
H-18: 24,000 TS; 15 El.
For refrigeration tube.

WOLVERINE ALLOY 41.
M-1521; 97.0 min Al, 1.25 Mn, 0.5 Fe, 0.3 Si.
O Temper: 16,000 TS; 40 El.
H-18: 29,000 TS; 10 El.
Air conditioning and refrigeration tubing, condensers and heat exchangers in sulphur dioxide and hydrogen sulphide environments.

WOLVERINE ALLOY 42.
M-1521; 0.4 Si + Fe, 2.5 Mg, 0.25 Cr, bal Al.
O Temper: 28,000 TS; 30 El.
Good fatigue resistance; hydraulic lines.

WOLVERINE ALLOY 43.
M-1521; 0.8 Mg, 0.5 Si, 0.25 Fe, bal Al.
O Temper: 13,000 TS.
T6 Temper: 35,000 TS; 12 El.
Heat treatable high strength alloy used for furniture tube.

WOLVERINE ALLOY 44.
M-1521; 99.45 min Al.
O Temper: 12,000 TS; 35 El.
H-19 Temper: 27,000 TS; 5 El.
For electrical connectors and bus tubing.

WOLVERINE ALLOY 45.
M-1521; 0.68 Si, 0.11 Mn, 1.1 Mg, 0.3 Cu, 0.2 Cr, bal Al.
O Temper: 18,000 TS; 25 El.
T6 Temper: 45,000 TS; 12 El.
Heat treatable alloy, high strength structural applications.

WOLVERINE ALLOY 51.
M-1521; 68.7 Cu, 30.0 Ni, 0.7 Mn, 0.5 Fe.
Ann: 60,000 TS; 45 El.
Condenser and heat exchanger tube alloy in sea or brackish waters, resistant to impingement attack.

WOLVERINE ALLOY 52.
M-1521; 64.0 Ni, 1.0 Mn, 1.0 Fe, 34.0 Cu.
Ann: 78,000 TS; 43 El.
Heat exchanger tube alloy having high strength and overall excellent corrosion resistance.
Also known by Trade Name "Cu-Nel".

WOLVERINE ALLOY 53.
M-1521; 89 Cu, 10 Ni, 1 Fe.
Ann: 44,000 TS; 42 El.
Condenser and heat exchanger tube alloy in sea or brackish waters, resistant to impingement attack.

WONICO.
M-U.S.; 80 W, 13 Ni, 5 Co.
For alloy for sealing in glass; coefficient of expansion 55 x 10^{-7}.

WOOD FUSIBLE ALLOY-1.
M-U.S.; 50 Bi, 12.5 Sn, 25 Pb, 12.5 Cd.
For safety plugs, fuses; M.P. 71°C.

WOOD FUSIBLE ALLOY-2.
M-U.S.; 52.5 Bi, 31.5 Pb, 16 Sn.
For safety plugs, fuses; M.P. 98°C.

WORKWEAR 14.
M-1422; C, 14 Mn, bal Fe.
Plate: 145,000-155,000 TS; 73-75 El; 54-56 RA; 200 Brin as rolled, over 500 Brin, as work hardened.
For chute liners, crushers, shovel and dragline buckets, quarry and mine skips.
Austenitic, wear and abrasion resistant, shock and impact resistant.

WORTHITE.
M-463; 0.07 max C, 1.0 max Mn, 2.5-3.5 Si, 18.0-20.0 Cr, 22.0-25.0 Ni 2.5-3.0 Mo, 1.5-2.0 Cu, bal Fe.
Wrought: 80,000 psi min TS; 35,000 psi min YS; 40 El; 137-183 Brin.
Stainless steel for pumps requiring great resistance to corrosion.

WORTLE DIE.
M-355; 1.2 C, 3 W, 0.6 Mn, bal Fe.
For tools, dies; oil hardening.

WOTAN AKS-98.
M-1313; Al, bal Cu.
20°C: 71,100 TS; 34,100 YS; 44 El.
At 600°C: 8500 TS; 2850 YS; 2 El.
For autoclaves, chemical equipment, pumps, valves; Al-bronze, corrosion resistant, tough.

WOTAN COBALT 3.
M-1287; 0.86 C, 2.8 Co, 4.3 Cr, 0.85 Mo, 2.1 V, 12 W, bal Fe.
For lathe and planer tools, reamers, broaches, drills; high speed steel; 91WCoV3811.

WOTAN COBALT 5.
M-1287; 0.79 C, 4.75 Co, 4.3 Cr, 0.75 Mo, 1.5 V, 18 W, bal Fe.
For reamers, hobs, drills, taps, broaches; high speed steel; 79WCo7419.

WOTAN COBALT 10.
M-1287; 0.76 C, 10 Co, 4 Cr, 0.8 Mo, 1.8 V, 18 W, bal Fe.
For lathe and planer tools, taps, reamers, hobs; high speed steel; 76WCoV7240.

WOTAN COBALT 513.
M-1287; 0.80 C, Cr, Co, W, V, Mo, bal Fe.
For lathe and planer tools, taps, reamers, hobs; high speed steel; 80WCoCrVMo50.22.

WOTAN DRILLING.
M-1287; 1.3 C, 4.3 Cr, 0.85 Mo, 3.8 V, 12 W, bal Fe.
For reamers, drills, lathe and planer tools, broaches; high speed steel; 130WV3838.

WOTAN DRILLING CO.
M-1287; 1.35 C, Cr, W, V, bal Fe.
For cutters, form tools; high speed steel; 135WCo4619.

WOTAN EXTRA F.
M-1287; 0.74 C, 4.1 Cr, 1.1 V, 18.5 W, bal Fe.
For lathe and planer tools, drills, hobs, reamers; high speed steel; 74WV74.

WOTAN III/3.
M-1287; 0.95 C, 4 Cr, 1 V, Mo, W, bal Fe.
For lathe and planer tools, reamers, hobs, taps; high speed steel; 95WMo1126.

WOTAN ILLING.
M-1287; 0.82 C, 4.1 Cr, 0.85 Mo, 1.6 V, 8.7 W, bal Fe.
For lathe and planer tools, reamers, drills, hobs; high speed steel; 82WV3419.

WOTAN MO 5.
M-1287; 0.85 C, Cr, W, Mo, bal Fe.
For lathe and planer tools, hobs, broaches, taps; high speed steel; 82WV3419.

WOTAN MO 5 CO.
M-1287; C, Cr, W, Mo, V, bal Fe.
For cutters, hobs, broaches, drills; high speed steel.

WOTAN MO 5 V 3.
M-1287; C, Cr, W, Mo, V, bal Fe.
For cutters, hobs, broaches, drills; high speed steel.

WOTAN MO 9.
M-1287; C, Cr, W, Mo, V, bal Fe.
For cutters, hobs, broaches, drills; high speed steel.

WOTAN ZWILLING.
M-1287; 0.86 C, 4.1 Cr, 0.85 Mo, 2.5 V, 12 W, bal Fe.
For lathe and planer tools, broaches, reamers, hobs, taps; high speed steel; 86WV3826.

WPS.
M-1655; 0.55 C, 0.30 Si, 0.60 Mn, 0.70 Cr, 0.30 Mo, 1.7 Ni, 0.10 V, bal Fe.
Hot work tool steel.
For forging and pressing dies.
Werkstoff Nr. 1.2713.

WPS 1.
M-1655; 0.55 C, 0.30 Si, 0.70 Mn, 1.0 Cr, 0.50 Mo, 1.7 Ni, 0.10 V, bal Fe.
Hot work tool steel; forging dies.
Werkstoff Nr. 1.2714.

WPS 2.
M-1655; 0.45 C, 0.30 Si, 0.70 Mn, 1.50 Cr, 0.70 Mo, 0.30 V, bal Fe.
Hot work tool steel.
Molds for casting or forming light metals, lead, tin and zinc.
Werkstoff Nr. 1.2323.

WS 353.
M-1290; 0.24-0.30 C, 0.40-0.60 Si, 0.30-0.70 Mn, 1.30-1.50 Cr, 1.10-1.40 Mo, 0.35-0.45 V, bal Fe.
Oil hardening structural steel.
Werkstoff Nr. 1.2353.

WS 718.
M-1290; 0.50-0.57 C, 0.15-0.30 Si, 0.40-0.50 Mn, 0.50-0.70 Cr, 2.5-3.0 Ni, bal Fe.
Cold work tool steel.
Werkstoff Nr. 1.2718.

W. TAP.
M-73; 1.25 C, 1.5 W, 0.4 Cr, 0.3 Mo, bal Fe.
For taps, drills, reamers; oil hardened.

WUEST NO. 1.
50 Zn, 30 Al, 20 Cu.
For aluminum solder.

WUEST NO. 2.
65 Zn, 20 Al, 15 Cu.
For aluminum solder.

WUNDUS.
M-618; 0.7 C, 6 Co, 22 W, bal Fe.
For milling cutters, twist drills, turning, planing, shaping and slotting tools; for cutting hardest metals.

WWS.
M-912; 0.60 C, 0.90 Si, 1.0 Mn, bal Fe.
Hot or cold work tool steel; for hammer dies, trimming tools, punches; shock resistant.
Werkstoff Nr. 1.2826; AFNOR 60 MS 4.

WYCLIFFE BLACK HEART MALLEABLE IRON.
M-752; Meets British Standard B.S. 310/1958, Grade B 22/14.
Castings for agriculture, automobile, shipbuilding, textile, commercial vehicle.

WYNDALOY.
M-1064; 60 Cu, 20 Ni, 20 Mn.
Hardened: 200,000 TS; 170,000 YS; 475 Brin.
Annealed: 98,000 TS; 80,000 YS; 140 Brin.
For valves, pistons, pump rods, screws, bolts, nuts, corrosion resistant, hardenable, non-magnetic.

WYNITE HIGH DUTY NI-CR CAST IRON.
M-752; Meets British Standard BS 1452, Grade 14 or Grade 17.
Cast: 16/18 tons/sq.in.TS.
For pumps, jigs.

WZ-1.
M-1197; TiC.
For gas turbine parts, jet engine and nuclear power plants; sintered, high heat resistant.

WZ-1B.
M-1197, M-1246; 60 TiC, 32 Ni, 8 Cr.
For gas turbines, jet engines, nuclear power plants; sintered, oxidation resistant.

WZ-1C.
M-1246; 50 TiC, 40 Ni, 10 Cr.
For gas turbine blades; sintered, oxidation resistant.

WZ-1D.
M-1246; 35 TiC, 52 Ni, 13 Cr.
For high temperature applications; sintered cermet.

WZ-2.
M-1246; 60 TiC, 28 Co, 12 Cr.
For gas turbine blades; sintered, oxidation resistant.

WZ3.
M-Austria; 50 TiC, 10 TaC, 32 Ni, 8 Cr.
Sintered: 1070 Vickers.
For jet engine components; sintered, carbides.

WZ-12A.
M-1246; 75 Ti, 1.5 Ni, 5 Co, 5 Cr.
For gas turbine blades; sintered, oxidation resistant.

WZ-12B.
M-1197, M-1246; 60 TiC, 24 Ni, 8 Co, 8 Cr.
For gas turbines, jet engines, nuclear power plants; sintered, oxidation resistant.

WZ-12C.
M-1197, M-1246; 50 TiC, 30 Ni, 10 Co, 10 Cr.
For gas turbines, jet engines, nuclear power plants; sintered, oxidation resistant.

WZM.
M-US; 25 W, 0.1 Zr, 0.03 C, bal Mo.

X

X see BETHLEHEM X.

X-2.
M-115; 0.24 C, 0.89 Si, 0.32 Mn, 4.9 Cr, 0.54 V, 1.3 Mo, bal Fe.
Annealed: 85,000 TS; 35,000 YS; 12 El; 50 RA; 512 Brin.
For gears and shafts, strength structures; air or oil hardened, hot work steel.

X 8 NICOCRTI55-20-20.
M-Ger.; 0.10 C, 1.5 max Si, 1 max Mn, 0.8-1.8 Al, 5 max Fe, 1.8-2.7 Ti, 18-21 Cr, 15-21 Co, bal Ni.
At 20°C: 155,000 TS; 90,000 YS; 39 El; 20 RA.
At 800°C: 78,000 TS; 57,000 YS; 7 El; 4 RA.
For jet engine and gas turbine components.
High heat, corrosion and oxidation resistant.
Nimonic 90. Heat treatable.

X 8 NICRALTI 75-20.
  M-Germany; 0.04 C, 21 Cr, 2.5 Ti, 0.7 Al, 0.6 Mn, bal Ni.
  At 20°C: 132,000 TS; 80,000 YS; 45 El; 36 RA.
  At 800°C: 62,000 TS; 53,000 YS; 8 El; 10 RA.
  For gas turbine blades, jet engine components.
  High heat, corrosion and oxidation resistant.
  Nimonic 80A. Creep resistant.

X17C.
  M-1653; 0.10 max C, 1.0 max Si, 1.0 max Mn, 16.0-18.0 Cr, 0.50 max Ni, bal Fe.
  Ferritic stainless steel.
  Werkstoff Nr. 1.4016; X 8 Cr 17; AISI 430.

X17Z.
  M-1653; 0.12 max C, 1.0 max Si, 1.5 max Mn, 16.0-18.0 Cr, 0.60 max Mo, 0.50 max Ni, 0.15-0.25 S, bal Fe.
  Free machining ferritic type stainless steel.
  Werkstoff Nr. 1.4104; AISI 430 F.

X-40.
  M-1317, M-1490; 0.5 C, 0.5 Mn, 0.5 Si, 22 Cr, 10 Ni, 7.5 W, 1.5 Fe, bal Co.
  High temperature alloy.
  (Haynes Stellite Alloy 31).

X40COCRNIW 45-20.
  M-Ger.; 0.4 C, 1.2 Mn, 20 Cr, 20 Ni, 43 Co, 4 W, 3 Fe, 4 Cb.
  Ht. Tr.: 140,000 TS; 67,000 YS; 31 El; 300 Brin.
  For turbine blades, rotors, hardware, jet engine components.
  High heat and oxidation resistance to 1500°F.

X-41.
  0.5 C, 0.5 Mn, 25 Cr, 8 Ni, 7.5 W, 1.75 Cr, 2 B, 1 Fe, bal Co.
  For high temperature applications; heat resistant.

X-42-W ETC see REPUBLIC X-42-W ETC.

X-45.
  M-31; 0.25 C, 25 Cr, 10 Ni, 7 W, 1.5 Fe, bal Co.
  High temperature alloy.

X-45.
  M-1491; 0.25 C, 1.0 max Mn, 25.5 Cr, 10.5 Ni, 7.5 W, 0.010 B, 2.0 max Fe, bal Co.
  Cobalt base superalloy; casting.
  For marine gas turbines, nozzle vanes.

X-45 W ETC see REPUBLIC X-45 W ETC.

X-50.
  M-31, M-60; 0.75 C, 22 Cr, 20 Ni, 12 W, 2.5 Fe, bal Co.
  Cobalt base superalloy.

X-50.
  M-31, m-60; 0.76 C, 0.6 Mn, 22 Cr, 20 Ni, 40 Co, 12 W, 1 Fe.
  For high temperature applications; heat resistant.

X-63.
  M-31, M-60; 0.40 C, 22.0 Cr, 10.0 Ni, 58.5 Co, 6.0 Mo, 1.25 Al, 2.0 Fe.
  Cobalt base superalloy; casting.

X-76.
  M-U.S.; 0.5-2.0 Cu, 0.4-0.8 Mn, 1.4-3.0 Mg, 5-8 Zn, bal Al.
  For supercharger casings, diffusers, impellers; age-hardenable.

X-110 same as COLUMBIUM D-43.

X-750 see INCONEL ALLOY X-750; and NICKELVAC X-750.

X-782.
  M-US; 2.0 C, 0.5 Mn, 0.5 Si, 26.0 Cr, 4.0 Fe, 8.7 W, bal Ni.
  For engine valves.

X-1900 same as RENE 85.

X-ALLOY.
  M-Eng.; 3.6 Cu, 0.6 Mg, 0.7 Ni, 1.2 Fe, 0.7 Si, bal Al.
  For pistons; cast.

XALOY 100.
  M-1473; 3.5 C, B, Ni, Si, Mn, bal Fe.
  For extruder and injection molding barrels, cylinder liners, pump parts; hard, wear resistant white cast iron.
  Was DI-HARD.

XALOY 101.
  M-1473; 3.5 C, B, Ni, Cr, Si, Mn, bal Fe.
  For extruder and injection molding barrels, cylinder liners, pump parts; hard, wear resistant white cast iron.

XALOY 306.
  M-1473; 7 Cr, 40 Ni, B, Si, Mn, bal Co.
  For extruder and injection molding barrels, cylinder liners, pump parts; wear and corrosion resistant.

XALOY 800.
  M-1473; WC dispersion in Ni-Si-B alloy matrix. Wear and corrosion resistant bore coating for extruder and injection molding barrels, cylinder liners, pump parts.

XANTAL A.
  M-541; 87.2 Cu, 9 Al, 3 Fe, 0.8 Mn.
  Sand cast: 71,000-85,000 TS; 25-15 El.
  For castings, propellers; corrosion resistant.

XANTAL B.
  M-541; 81 Cu, 11 Al, 4 Fe, 4 Ni.
  Cast: 101,000 TS; 1-2 El.
  Heat treated: 114,000 TS; 2-4 El.
  For castings, marine hardware; corrosion resistant.

XANTAL M.
M-541; 89.5 Cu, 9 Al, 1.5 Ni.
Sand cast: 64,000-78,000 TS; 28-18 El.
Tough.

XANTAL S.
M-541; 90.4 Cu, 7.8 Al, 0.5 Ni, 0.3 Mn, 1.0 Zn.
Annealed: 71,000 TS; 50 El.
Forged: 65,000 TS; 50 El.
For marine hardware castings; corrosion resistant.

X A R 15.
M-836; 0.10-0.21 C, 0.6-1.1 Mn, 0.4-0.8 Si, 0.0025 B, 0.4-0.8 Cr, 0.18-0.28 Mo, 0.05-0.15 Zr, bal Fe.
Ht. Tr. Plate: 360 Brin. min; 190,000 TS; 175,000 YS; 16 El; 55 RA.
For bridges, booms, dipper sticks, penstocks, chutes, wear plates, truck bodies, fan blades, floor plates.
Quenched and tempered at mill. Tough. Wear and abrasion resistant.

X A R 30.
M-836; 0.30 max C, 0.6-1.1 Mn, 0.4-0.8 Si, 0.0025 B, 0.4-0.8 Cr, 0.18-0.28 Mo, 0.05-0.15 Zr, bal Fe.
Ht. Tr. Plate: 360 min Brin; 190,000 TS; 175,000 YS; 16 El; 55 RA.
For bridges, booms, dipper sticks, penstocks, chutes, wear plates, truck bodies, fan blades, floor plates.
Quenched and tempered at mill. Tough. Wear and abrasion resistant.

X-B.
M-1005; 33-37 Ni, 13-17 Cr, bal Fe.
Cast: 64,000 TS; 36,000 YS; 5 El; 196 Brin.
For castings, furnace parts; resists heat.

XB see CARPENTER C-XB VALVE STEEL; and FIRTH-BRWN XB.

XB-88 see WESTINGHOUSE XB-88; WAH CHANG XB-88; and COLUMBIUM XB-88.

XB, CAST.
M-513; 1.45 C, 0.4 Mn, 2.25 Si, 20.0 Cr, 1.3 Ni, bal Fe.
For diesel engine valves.

XB-STEEL.
M-Italy; 0.80 C, 19.5 Cr, 1.45 Ni, 2.0 Si, 0.4 Mn, bal Fe.
For exhaust valves operating up to 750°C.
Heat and corrosion resistant.

XCR.
M-U.S. 23.75 Cr, 4.75 Ni, 1.0 max Si, 2.75 Mo, 1 Mn, 0.45 C, bal Fe.
For exhaust valves; heat resistant.

XCV GRADE 140.
M-Eng.; 1.15 C, 0.24 Mn, 0.3 Cr, 0.15 V, bal Fe.
For tools, dies, drills, taps; water hardened.

XH-80.
M-USSR; 0.1 C, 20 Cr, 80 Ni.
Resistance wire for electric toasters, heaters.

X-ITE.
M-1005; 1.12 Mn, 38 Ni, 18 Cr, 2.0 Si, 0.5 C, bal Fe.
65,000-75,000 TS; 53,000-56,000 YS; 2.5 El; 4.0 RA.
For furnace parts, hearth plates, skid and roller rails; heat, corrosion and abrasion resistant.

X-ITE CB.
M-1005; 0.35-0.75 C, 17-21 Cr, 37-41 Ni, Cb, bal Fe.
Cast: 70,000 TS; 40,000 YS; 10 El; 180 Brin.
For furnace parts, heat treating boxes, salt pots, retorts, burners.
Heat and corrosion resistant.

XL CHISEL.
M-73; 0.55 C, 2.1 W, 1.5 Cr, 0.2 V, bal Fe.
For chisels, punches; oil hardening.

XL CUT.
M-1724; 0.09 max C, 0.02 max Si, 0.97 Mn, 0.05 P, 0.30 S, bal Fe.
Low carbon resulphurized grade.
For automatic screw machines.

XL CUT PB.
M-1724; 0.09 max C, 0.02 max Si, 0.97 Mn, 0.06 P, 0.30 S, 0.15 min Pb, bal Fe.
Low carbon resulphurized, leaded steel.
For automatic screw machines.

XL CUT SPB.
M-1724; 0.08 max C, 0.01 max Si, 0.97 Mn, 0.09 P, 0.30 S, 0.20 min Pb, 0.008 N, bal Fe.
Low carbon resulphurized, leaded steel.
For automatic screw machines.

XLO.
M-435; 0.6 C, 0.75 Cr, 0.3 Mo, 1.75 Ni, bal Fe.
For drop forge die blocks; oil hardening.

XL PLATE.
M-370; 0.20 C, Mn, bal Fe.
Stress relieved.
AISI C-1020.

XL-STEEL.
M-73; 0.45 C, 0.20 Si, 2.25 W, 1.30 Cr, 0.25 V, bal Fe.
For chisel steel, pneumatic hammers, shock tools; heavy duty, tough.

X-SUPERMAL.
M-170; 1-1.6 T.C., 1.1-1.3 Si, 0.3 Mn, bal Fe.
Heat treated: 60,000-70,000 TS; 50,000-60,000 YS; 8-6 El; 179-201 Brin.
For chainlinks, buckets, conveyor equipment; wear resistant.

XTRA TOUGH see DARWIN EXTRA TOUGH.

XUPER 16 XFC.
M-717.
Electrode for torch brazing all steels; bonding temp: 1400-1600°F; 100,000 psi TS.

XUPER 18 XFC.
M-717.
Electrode for metallic arc joining copper base alloys, steel, cast iron. Bonding temp: 1400-1600°F; 70,000 psi TS.

XUPER 146 XFC.
M-717.
Bronze type alloy for torch brazing ferrous and copper base alloys; bonding temp: 1400-1600°F; 65,000 psi TS.

XUPER 185 XFC.
M-717.
Bronze electrode for torch build-up; machinable and wear resistant; bonding temp: 1400-1600°F; 130 Brin, work hardens to 200 Brin.

XUPER 680 CGS.
M-717.
Electrode for AC-DC metallic arc welding all steels, dies, tools; 120 psi TS.

XUPER 1020 XFC.
M-717.
Cadmium free silver solder type for brazing stainless steel, copper alloys. Bonding temp: 1050°F; 85,000 psi TS.

XUPER 2100.
M-717.
For DC metallic arc welding of aluminum alloys containing magnesium; 30-35,000 psi TS.

XUPER 2240.
M-717.
For AC-DC metallic arc to penetrate contaminated cast iron surfaces; nodular deposits; machinable; 55-60,000 psi TS.

XUPER 9080.
M-717.
Electrode for AC-DC metallic arc build-up to give high hardness, corrosion resistance and resistance to high temperature impact; Rc 30, work hardens to Rc 45-50.

XUPERBOND.
M-717.
Alloy powder for metal spraying base coat on all metals except pure copper.
Maximum adherence.

XUPER BRONZTEC 19868.
M-717.
High hardness aluminum bronze powder for metal spraying.

XUPER DIAMAX 10999.
M-717.
Powder for spray coating that has good abrasion resistance.
RA 80-85.

XUPER DRIL-TEC 8800.
M-717.
Flux coated electrode containing hard particles for torch build-up on metals; bonding temperature 1400-1600°F. Matrix 200 Brin, particles 89-91. Good abrasion resistance.

XUPER ELASTODUR R8811.
M-717.
Electrode for torch build-up; has hard particles in matrix for abrasion resistance.

XUPER EUTECBOR 9000.
M-717.
Electrode for torch build-up on metals for high hardness and corrosion resistance.
Bonding temp: 1800°F; Rc 55-62.

XUPER EXOTRODE.
M-717.
Electrode for AC-DC to chamfer, gouge, cut and pierce all metals.

XUPER MODULTEC 2250.
M-717.
For AC-DC nodular graphite deposits on nodular cast iron for improved ductility; machinable; 55-60,000 psi TS.

XUPER NUCLEOTEC 2222.
M-717.
Electrode for AC-DC metallic arc welding all steels and nickel alloys; massive sections; good shock resistance; 100,000 psi TS.

XX.
M-24; 0.9-1.3 C, bal Fe.
For gripper and header dies, punches; Type W1; cold heading.

XX METAL.
M-174; C, alloy, bal Fe.
For machinery parts; oil hardening.

XX SUPERIOR.
M-Eng.; C, Cr, bal Fe.
For tools; water hardened.

XYRON 2-23.
M-717.
For machinable welds on cast iron; AC-DC reverse; for pump housings, pressure chambers.
Strength: 55,000-60,000 psi.

XYRON 2-24.
M-717.
For machinable welds on cast iron; AC-DC; strength: 50,000 psi.

XYRON 2-25.
M-717.
For machinable welds on cast iron, Meehanite and Ni-Resist; AC-DC.
Strength: 56,000 psi.

XYRON 244.
M-717.
Electrode for machinable welds and repairs on gray and alloyed castings; can join cast iron to steel; AC-DC.
Strength 53,000 psi.

Y

YALE BRONZE.
M-U.S.; 7-8 Zn, 0.5-1.5 Sn, 0.7-1.5 Pb, bal Cu.
For screw machine products, bolts, bushings; free-cutting.

Y ALLOY.
M-106; 3.5-4.5 Cu, 1.2-1.7 Mg, 1.8-2.3 Ni, bal Al.
Sand cast: 31,000-36,000 TS; 28,000-30,000 YS; 0-1 El; 95-105 Brin; 0.6 Izod.
Sand or chill cast aluminum for high temperature parts as automobile pistons.
Gen. Eng. BS 1490: LM 14-WP, 16-W, 16-WP.

Y-ALLOY.
M-137; 3.5-4.5 Cu, 1.2-1.7 Mg, 1.8-2.3 Ni, 0-0.2 Ti, bal Al.
34,000 TS; 28,000 YS; 100 Brin.
For pistons, cylinder heads; hardenable.

Y-ALLOY.
M-1176; 3.8-4.2 Cu, 1.3-1.7 Mg, 1.8-2.3 Ni, bal Al.
For pistons, cylinder heads; high temperature properties.

YAW-TEN 41.
M-1540; 0.18 max C, 0.35 max Si, 1.2 max Mn, 0.25-0.5 Cu, 0.4-0.65 Cr, bal Fe.
Normalized: 74,000 TS; 31,000 YP.
Rolled: 58,000 TS; 34,000 YP.
For buildings, bridges, industrial machinery.
Good fabricability and weldability.
Resists atmosphere corrosion.

YAW-TEN 50.
M-1540; 0.12 max C, 0.15-0.35 Si, 0.90 max Mn, 0.06-0.12 P, 0.25-0.50 Cu, 0.15 max Ti, bal Fe.
Annealed: 67,000 min TS; 50,000 min YP.
Rolled: 71,000 min TS; 57,000 min YP.
For bridges, structures, mine cars, bus bodies, railroad cars, ships, farm implements.
Good weldability and impact resistance.
Resists atmospheric corrosion.

YAW-TEN 60.
M-1540; 0.16 max C, 0.15-0.55 Si, 0.8-1.4 Mn, 0.25-0.5 Cu, 0.4-0.65 Cr, 0.15 max Ti, bal Fe.
Heat Treated: 82,000-100,000 TS; 63,000-67,000 YP.
For buildings, bridges, industrial machinery.
Good fabricability and weldability.
Resists atmospheric corrosion.

YAW-TEN M.
M-1540; 0.12 max C, 0.15-0.35 Si, 0.90 max Mn, 0.25-0.50 S, 0.5-1.0 Cr, 0.15 max Ti, bal Fe.
Rolled: 64,000 min TS; 47,000 min YP.
For air preheaters, flues, stacks, gas exhaust systems.
For applications exposed to corrosive attack from sulphuric acid gases and other corrosive chemicals.

YB-TEN.
M-531; 0.12 max S, 0.75 max Mn, 0.04 max P, 0.05 max S, 0.10 max Si, 0.01 min Cb or V, bal Fe.
Sheet: 62,000 TS; 45,000 YP; 28 El.
For automobile bumpers.
High strength low alloy steel. Good weldability and fabricability.

YC135A.
M-U.S.; 82 Sn, 13 Sb, 5 Cu.
Die casting alloy.
As cast: 10,000 psi TS; 11 Elon.

YCB.
M-U.S.; 15 Cr, 25 Ni, 4 Mo, 1 Si, 2 Cb, bal Fe.
For supercharger wheels for jet engine; heat resistant.

YD NICRAL.
M-1317; 16 Cr, 4.8 Al, 1.0 Y_2O_3, bal Ni.
Yttrium oxide dispersion nickel-chromium alloy, for high temperature operation.
At 70°F: 170 ksi TS, 12 El.
At 2000°F: 11 ksi TS, 10-20 El.
Good high temperature stress rupture strength.

YD-NI.
M-US; 16 Cr, 4 Al, 1.5 Y_2O_3, bal Ni.
Cast superalloy.

YELLOW BRASS, 65%.
M-33; 65 Cu, 35 Zn.
Ann: 50,000 TS; 60 El.
Drawn 84%: 128,000 TS; 3 El.
For springs, pins, rivets, screws.
CDA 270.

YELLOW BRASS 66%.
M-33; 66 Cu, 34 Zn.
Annealed: 49,000 TS; 17,000 YS; 57 El.
Rolled: 74,000 TS; 40,000 YS; 8 El.
For ornaments, lamp fixtures, automobile parts; deep drawing.

YELLOW BRASS 59.
M-8; 34 Zn, bal Cu.
Soft: 45,000 TS; 60 El; 52 Brin.
Hard: 76,000 TS; 4 El; 153 Brin.
For bead chains, eyelets, grommets, lamp fixtures; yellow brass.

YELLOW BRASS-61.
M-8; 63 Cu, 37 Zn.
Soft: 46,000 TS; 17,000 YS; 60 El.
Hard: 65,000 TS; 50,000 YS; 20 El.
For pins, screws, rivets; yellow brass, high strength.

YELLOW BRASS 268.
M-279; 66 Cu, 34 Zn.
Ann: 44,000-61,000 TS; 13,000-34,000 YS; 35-65 El.
Cold Rolled: 49,000-99,000 TS; 24,000-76,000 YS; 3-50 El.
General purpose yellow brass.
For formed parts, electrical hardware, hinges.

YELLOW BRASS 274.
M-8; 63 Cu, 37 Zn.
Hard: 65,000 TS; 57,000 YS; 20 El; B 75 Rock.
Soft: 46,000 TS; 17,000 YS; 60 El; B 20 Rock.
For cold headed screws, bolts, rivets, studs and accessories, novelties, fixtures, appliances. Elect. cond. 26. High strength.

YELLOW GOLD-1.
M-U.S.; 53 Au, 25 Ag, 22 Cu.
Cast: 205 Brin.
For jewelry; corrosion resistant.

YELLOW GOLD-2.
M-U.S.; 50 Au, 25 Ag, 25 Cu.
Heat treated: 159 Brin.
For jewelry; corrosion resistant.

YELLOW LABEL.
M-341; 0.9 C, 1.2 Mn, 0.5 Cr, 0.5 W, 0.2 V, bal Fe.
For tools, dies, punches; non-deforming, oil hardening.

YELLOW LABEL.
M-343; 0.8-1.1 C, 0.15-0.25 V, bal Fe.
For tools, dies, boring and shaping tools, milling cutters; "Hellers Alloy Die."

YELLOW LABEL CAST.
M-261; 0.7-1.2 C, bal Fe.
For tools and dies, drills, taps, punches; water hardened.

YELLOW TUBE BRASS-218.
M-8; 66.5 Cu, 33 Zn, 0.5 Pb.
Soft: 45,000 TS; 17,000 YS; 55 El.
Hard: 73,000 TS; 60,000 YS; 10 El.
For plumbing goods, flashlight shells; tubes.

YIDOR.
M-111; 1.45 C, 1.4 Cr, 0.60 Mn, bal Fe.
For blanking and forming dies, cutters; water or oil hardened, abrasion resistant.

Y-LEGIERUNG-2.
M-1548; 4.5 Cu, 1.5 Mg, 2.0 Ni, bal Al.
Heat treated: 54,000-58,000 TS; 31,000-36,000 YS; 16-20 El; 100-120 Brin.
For aircraft engine cylinder heads, pistons, jet engine impellers.

YND-30.
M-1540; 0.14 max C, 1.0-1.5 Mn, bal Fe.
Plate: 42,700 min YS; 61,200 min TS.
For storage tanks, transportable containers for liquified petroleum gas, pressure vessels.
High notch toughness for cryonics temperatures. Good weldability.

YND-33.
M-1540; 0.14 C, 1.0-1.5 Mn, 0.035 max S & P, bal Fe.
Plate: 64,000 min TS; 46,900 min YS.
For storage tanks and transportable containers for liquified petroleum gas chemical industry apparatus, refrigerated ships.
Low temperature steel, high notch toughness.

YND-37.
M-1540; 0.14 C, 1.0-1.5 Mn, 0.70 max Ni, bal Fe.
Plate: 71,000 min TS; 52,600 min YS.
For storage tanks and transportable containers for liquified petroleum gas chemical industry apparatus, refrigerated ships.
High notch toughness at low temperature.

YND-58.
M-1540; 0.14 max C, 2.0-2.7 Ni, 0.50 max Cr, 0.55 max Mo, bal Fe.
Plate 96-700-16,000 TS; 82,500 min YS.
For storage tanks and transportable containers for liquified petroleum gas chemical equipment, pressure vessels. High notch toughness, low temperature steel, good weldability.

YOCOMITE.
M-1151; high C, 3 Ni, 0.75 Cr, bal Fe.
50,000 TS; 450 Brin.
For dies, cast iron.

YO-FLEX 60.
M-531; 0.13 C, 0.45 Mn, 0.01 P, 0.025 S, bal Fe.
60,000 psi YS.
For high strength hot dipped galvanized sheet steel applications.

YO-FLEX 70.
M-531; 0.13 C, 0.45 Mn, 0.01 P, 0.025 S, bal Fe.
70,000 psi YS.
For high strength hot dipped galvanized sheet steel applications.

YO-FLEX 80.
M-531; 0.13 C, 0.45 Mn, 0.01 P, 0.025 S, bal Fe.
80,000 psi YS.
For high strength hot dipped galvanized sheet steel applications.

YO-FLEX 90.
M-531; 0.13 C, 0.45 Mn, 0.01 P, 0.025 S, bal Fe.
90,000 psi YS.
For high strength hot dipped galvanized sheet steel applications.

YO-LEAD TYPE A.
M-531; 0.13 C, 1.0 Mn, 0.06 P, 0.30 S, 0.25 Pb, bal Fe.
For free machining bar applications.

YO-LEAD TYPE B.
M-531; 0.15 C, 1.2 Mn, 0.6 P, 0.40 S, 0.25 Pb, bal Fe.
For free machining bar applications.

YOLOY ACR replaced by **YOLOY YSW-50 STEEL.**

YOLOY C.
M-531; 0.10 max C, 0.30 max Cr, 0.60 max Mn, 0.25-0.50 Cu, bal Fe.
Rolled: 60,000 TS; 45,000 YS; 30 El.
For caskets, tanks, truck trailers, trim; good welding and forming.

YOLOY HS.
M-531; 0.15 max C, 0.75 max Mn, 0.10 max P, 0.30 max Si, 0.75-1.25 Cu, bal Fe.
Rolled: 70,000 TS; 50,000 YS; 22 El; 62 RA; 85 Brin.
For trucks, trailers, cars, bodies, bridges; good weldability and formability.

YOLOY HSX.
M-531; 0.15 max C, 1.0 max Mn, 0.75 Cu, 1.0 Ni, bal Fe.
Rolled: 65,000 TS; 45,000 YS; 25 El.
For railroad cars, trucks, bridges, mine equipment; good weldability and formability.

YOLOY S.
M-531; 0.12 max C, 0.75-1.25 Cu, 1.65-2.0 Ni, 0.6 max Mn, bal Fe.
Rolled: 60,000 TS; 45,000 YS; 25 El.
For transportation equipment, mine cars, truck bodies; good weldability, precipitation hardenable.

YOLOY T-50.
M-531; 0.10 C, 0.45 Mn, 0.01 P, 0.025 S, 0.40 Si, 0.25 Cu, 0.55 Cr, Al, 0.10 Ti, bal Fe.
50,000 psi YS.
Hot rolled and cold rolled sheet for high strength weathering applications.

YOLOY T-60.
M-531; 0.10 C, 0.45 Mn, 0.01 P, 0.025 S, 0.40 Si, 0.25 Cu, 0.55 Cr, Al, 0.10 Ti, bal Fe.
60,000 psi YS.
Hot rolled and cold rolled sheet for high strength weathering steel applications.

YOLOY T-70.
M-531; 0.10 C, 0.45 Mn, 0.01 P, 0.025 S, 0.40 Si, 0.25 Cu, 0.55 Cr, Al, 0.20 Ti, bal Fe.
70,000 psi YS.
Hot rolled and cold rolled sheet for high strength weathering steel applications.

YOLOY T-80.
M-531; 0.10 C, 0.45 Mn, 0.01 P, 0.025 S, 0.40 Si, 0.25 Cu, 0.55 Cr, Al, 0.20 Ti, bal Fe.
80,000 psi TS.
Hot rolled and cold rolled sheet for high strength weathering steel applications.

YOLOY YSW 42.
M-531; 0.16 C, 0.75-1.0 Mn, 0.02 V, or 0.02 Cb, bal Fe.
Plate: 62,000 min TS; 42,000 min YP; 19 min El.
For trucks, crane booms, mine cars, bridges, pressure vessels, trailers.
Low-alloy high-strength steel. High weldability.

YOLOY YSW 45.
M-531; 0.16 C, 0.75-1.0 Mn, 0.02 V, or 0.02 Cb, bal Fe.
Plate: 60,000 min TS; 45,000 min YP; 19 min El.
For trucks, crane booms, bridges, mine cars, trailers, pressure vessels.
Low-alloy high-strength steel. High weldability.

YOLOY YSW 50 STEEL.
M-531; 0.22 max C, 1.25 max Mn, 0.01 min Cb or V, bal Fe.
Rolled: 65,000 TS; 50,000 YS; 22 El.
For structurals, auto and truck frames, R.R. construction; high strength low alloy steel.

YOLOY YSW 55.
M-531; 0.16 C, 1.0 Mn, 0.02 V, or 0.02 Cb, bal Fe.
Plate: 70,000 TS; 55,000 YP; 17 El (minimum).
For bridges, trucks, crane booms, mine cars, pressure vessels, trailers.
Low-alloy high-strength steel. High weldability.

YOLOY YSW 60.
M-531; 0.16 C, 0.75-1.0 Mn, 0.02 V, or 0.02 Cb, bal Fe.
Plate: 75,000 min TS; 60,000 min YP; 16 min El.
For mine cars, trailers, trucks, crane booms, bridges, power shovels.
Low-alloy high-strength steel.
Good weldability.

YOLOY YSW 65.
M-531; 0.16 C, 1.0 Mn, 0.02 V, or 0.02 Cb, bal Fe.
Plate: 80,000 min TS; 65,000 min YP; 15 min El.
For power shovels, crane booms, railroad cars, pressure vessels, mine cars.
Low-alloy high-strength steel. Good weldability.

YOLOY YSW 70.
M-531; 0.16 C, 1.0 Mn, 0.02 V, bal Fe.
Plate: 85,000 min TS; 70,000 min YP; 12 min El.
For mine cars, crane booms, buildings, pressure vessels, bridges, power shovels.
Low-alloy high-strength steel. High weldability.

YOLOY YSW-A441.
M-531; 0.22 max C, 1.25 max Mn, 0.20 min Cu, 0.02 min V, bal Fe.
Sheet: 60,000 TS; 45,000 YP; 25 El (minimum).
Bar: 67,000 TS; 46,000 YP; 19 El (minimum).
For railroad cars, bridges, buildings, crane booms, power shovels, mining machinery.
High-strength low alloy steel.

YO-MAN.
M-531; 0.25 max C, 1.1-1.6 Mn, 0.3 max Si, 0.20 min Cu, bal Fe.
Plate: 70,000 min TS; 50,000 min YP; 20 El.
For shovels, structural members, shovel blades.
High strength low alloy steel.

YO-NAMEL.
M-531; 0.005 C, 0.30 Mn, 0.003 P, 0.020 S, bal Fe.
For porcelain enameling.

YO-OH TYPE 1.
M-531; 0.09 C, 0.85 Mn, 0.10 P, 0.30 S, bal Fe.
For free machining bar applications.

YO-OH TYPE 2.
M-531; 0.09 C, 0.90 Mn, 0.06 P, 0.32 S, bal Fe.
For free machining bar applications.

YORCALBRO/ALUMBRO.
M-1560; 76 Cu, 22 Zn, 2 Al, 0.04 As.
For tubes in steam condensers and heat exchangers, oil refineries, sea water pipe lines.
Resists corrosion and erosion of sea water.
Aluminum brass.

YORCALBRO (ALUMINIUM BRASS).
M-286; 0.04 As, 2.0 Al, 22.0 Zn, bal Cu.
Ann: 47,800 psi TS; 23,200 psi YS; 70 El; 70-75 DPN (68-73 Brin).
Tube for condensers, and seawater pipelines.

YOR CASTAN.
M-1560; 88 Cu, 12 Sn.
As Drawn: 106,000-121,000 TS; 10-15 El; 194-242 Brin.
For condenser and heat exchanger tubes.
Resists sea water corrosion.

YORCORON.
M-286, M-1560; 31 Ni, 2 Fe, 2 Mn, bal Cu.
Plates and tubes.
Annealed: 62,700-71,700 TS; 22,400-36,000 YS; 40-50 El; 97-116 Brin.
Hard: 98,600-110,000 TS; 90,000-107,000 YS; 4-8 El; 175-213 Brin.
Very resistant to abrasion by such as sand.

YORCUNIC/KUNIFER 10.
M-1560; 10 Ni, 2 Fe, 1 Mn, bal Cu.
For evaporator tubes in sugar industry, feed water heaters.
Resists sea water corrosion.

YORCUNIFE/KUNIFER 5.
M-1560; 5.5 Ni, 1.2 Fe, 0.5 Mn, bal Cu.
For sea water pipelines.
Resists sea water corrosion.

YORK.
M-1705; 0.75 C, 2.0 Mn 0.3 Si, 1.0 Cr, 1.35 Mo, bal Fe.
Air hardening tool steel. AISI A6.

YORKSHIRE 70/30.
M-1560, M-286; 30 Ni, 0.7 Fe, 0.8 Mn, bal Cu.
For condenser tubes, evaporator tubes, feed water heater tubes, heat exchangers.
Resists sea water corrosion.
Cupro-nickel.

YOUNGSTOWN GALVANNEALED.
M-531; Zn coated low carbon steel.
For roofing, siding, culverts, fence wire.

Y-PHOSPHOR BRONZE NO 207A.
M-423; 70 Cu, 29 Zn, 1 P.
For wire for weaving purposes; great ductility.

YS-T50.
M-531; 0.09 C, 0.45 Mn, 0.01 P, 0.025 S, 0.04 Al, 0.10 Ti, bal Fe.
Hot rolled sheet and plate, cold rolled sheet: 50,000 psi YS.
For high strength steel applications.

YS-T60.
M-531; 0.09 C, 0.45 Mn, 0.01 P, 0.025 S, 0.04 Al, 0.15 Ti, bal Fe.
Hot rolled sheet and plate, cold rolled sheet: 60,000 psi YS.
For high strength steel applications.

YS-T70.
M-531; 0.09 C, 0.45 Mn, 0.01 P, 0.025 S, 0.04 Al, 0.20 Ti, bal Fe.
Hot rolled sheet and plate, cold rolled sheet: 70,000 psi YS.
For high strength steel applications.

YS-T80.
 M-531; 0.09 C, 0.45 Mn, 0.01 P, 0.025 S, 0.04 Al, 0.25 Ti, bal Fe.
 Hot rolled sheet and plate, cold rolled sheet: 80,000 psi YS.
 For high strength steel applications.

YS-T90.
 M-531; 0.09 C, 0.90 Mn, 0.01 P, 0.025 S, 0.04 Al, 0.25 Ti, bal Fe.
 Hot rolled sheet: 90,000 psi YS.
 For high strength steel applications.

YS-T100.
 M-531; 0.09 C, 0.90 Mn, 0.01 P, 0.025 S, 0.04 Al, 0.25 Ti, bal Fe.
 100,000 psi YS.
 For high strength cold rolled steel sheet applications.

YS-T120.
 M-531; 0.09 C, 0.90 Mn, 0.01 P, 0.025 S, 0.04 Al, 0.25 Ti, bal Fe.
 120,000 psi YS.
 For high strength cold rolled steel sheet applications.

YS-T140.
 M-531; 0.09 C, 0.90 Mn, 0.01 P, 0.025 S, 0.04 Al, 0.25 Ti, bal Fe.
 140,000 psi YS.
 For high strength cold rolled steel sheet applications.

YSW 42 ETC see YOLOY YSW 42 ETC.

YSW-50 F.
 M-531; 0.09 C, 0.90 Mn, 0.01 P, 0.025 S, 0.04 Cb, bal Fe.
 Hot rolled and cold rolled high strength steel sheet: 50,000 psi YS.

YSW-55 F.
 M-531; 0.09 C, 0.90 Mn, 0.010 P, 0.025 S, 0.05 Cb, bal Fe.
 Hot rolled and cold rolled high strength steel sheet: 55,000 psi YS.

YSW-60 F.
 M-531; 0.09 C, 0.90 Mn, 0.01 P, 0.025 S, 0.06 Cb, bal Fe.
 Hot rolled and cold rolled high strength steel sheet: 60,000 psi YS.

YSW-65 F.
 M-531; 0.09 C, 1.0 Mn, 0.01 P, 0.025 S, 0.07 Cb, bal Fe.
 Hot rolled and cold rolled high strength steel sheet: 65,000 psi YS.

YTTERBIUM.
 M-1755; Yb.
 Purities: 99.9% (Special distilled grade), 99.5 + %.
 Forms: Ingot, lump, sheet, foil, wire, turnings, rod.

YTTRIUM.
 M-1755; Y.
 Purities: 99.9%, 99.5%, 90%, 70%.
 Forms: Ingot, sponge, lump, sheet, foil, rod, wire, powder, turnings.

YUNDK 24.
 M-USSR; 14 Ni, 24 Co, 9 Al, 3 Cu, bal Fe.
 Sintered: Rockwell C 45.
 For electrical and magnetic equipment.
 Permanent magnet. Sintered alloy.
 High external energy and permeability.

Y W A STEEL.
 M-210; 0.28 C, 0.45 Si, 1.3 Cr, 3.5 Ni, 5.8 W, 0.25 V, bal Fe.
 For extrusion dies, hot punches.
 High temperature and abrasion resistance, tough.

Z

Z-1R.
 M-1134; 0.8-1.0 Al, 0.3-0.4 Cu, bal Zn.
 Rolled: 31,000-42,500 TS; 40-70 Brin.
 For structures.

Z-2.
 M-949; 4 Al, 3 Cu, 0.03 Mg, bal Zn.
 Die cast: 50,000 TS; 4-8 El; 82-87 Brin.
 For gears, pumps, motor frames, hardware; die castings.

Z-3.
 M-106; 6-8 Cu, 2-4 Si, 2-4 Zn, bal Al.

Z-3.
 M-949; 4 Al, 0.04 Mag, bal Zn.
 Die cast: 37,000 TS; 4-6 El; 60-65 Brin.
 For gears, frames, hardware, die castings.

Z-3 ALLOY.
 M-137; 6.0-8.0 Cu, 2-4 Zn, 2-4 Si, 0.7-1.1 Fe, bal Al.
 Cast: 23,000-28,000 TS; 1-2 El; 90 Brin.
 For crankcases, gear boxes, flywheel housings, brake shoes; die and sand cast.

Z 5 Z.
 M-Eng.; 3.5-5.5 Zn, 0.4-1.0 Zr, bal Mg.
 Sand cast and precipitation treated: 230 MPa TS; 145 MPa YS; 5 El Brin.
 Higher strength sand or chill casting.
 BS 2970 MAG 4.

Z 178.
 M-1290; 0.07 max C, 1.0 max Si, 2.0 max Mn, 17.0-20.0 Cr, 8.5-10.0 Ni, bal Fe.
 Austenitic stainless steel.
 Werkstoff Nr. 1.4301; similar to AISI 304.

Z 178 E.
M-1290; 0.10 max C, 1.0 max Si, 2.0 max Mn, 17.0-19.0 Cr, 9.0-11.5 Ni, Ti = 5 X C min, bal Fe.
Stabilized austenitic stainless steel.
Werkstoff Nr. 1.4541; similar to AISI 321.

Z 178 H.
M-1290; 0.12 max C, 1.0 max Si, 2.0 max Mn, 17.0-19.0 Cr, 8.0-10.0 Ni, bal Fe.
Austenitic stainless steel.
Werkstoff Nr. 1.4300; AISI 302.

Z A.
M-580; Zn, Al.
For electroplating anode.

ZADUR 5 BR.
M-1324; 1.42 C, W, V, bal Fe.
For engravers' tools, bearings, cutters; water or oil hardened.

ZADUR B 25.
M-1324; C, Cr, Ni, V, bal Fe.
For molds, dies.

ZADUR B25S.
M-1324; C, Cr, Ni, V, bal Fe.
For headers, upsetters, punches; cold work steel.

ZADUR BC1.
M-1324; 1.05 C, 1.0 Cr, 0.25 Si, 0.30 Mn, bal Fe.
For bearings, liners, sleeves, forming dies; water hardened.

ZADUR BC12.
M-1324; 2.1 C, 11.5 Cr, bal Fe.
For blanking and forming dies; oil or air hardened, nondeforming.

ZADUR BC12W.
M-1324; 1.65 C, 11.5 Cr, 0.1 V, bal Fe.
For blanking and forming dies, punches; air hardened, nondeforming.

ZADUR BC13.
M-1324; 2.1 C, 11.5 Cr, 0.7 W, bal Fe.
For blanking and forming dies, punches; oil or air hardened, nondeforming.

ZADUR BC13M EXTRA.
M-1324; C, Cr, Mn, bal Fe.
For upsetters, headers, dies, punches; cold work steel.

ZADUR BC13W.
M-1324; 1.65 C, 11.5 Cr, 0.1 V, 0.2 Mo, bal Fe.
For blanking and forming dies, punches; air hardened, nondeforming.

ZADUR BCN.
M-1324; 0.45 C, Ni, Cr, W, bal Fe.
For upsetters, heading dies, crimpers; oil hardened, tough.

ZADUR BDK.
M-1324; 0.45 C. 1.05 Cr, 0.2 V, 1.85 W, bal Fe.
For cold heading dies, headers, punches; oil hardened, tough.

ZADUR BDK EXTRA.
M-1324; 0.45 C, 0.9 Si, 1.05 Cr, 0.2 V, 1.85 W, bal Fe.
For cold heading dies, headers, punches; oil hardened, tough.

ZADUR BF2.
M-1324; 1.05 C, 1.25 Mn, 1.15 Cr, bal Fe.
For bearings, liners, sleeves; water hardened, water resistant.

ZADUR BF2 EXTRA.
M-1324; 0.9 C, 1.9 Mn, 0.1 V, bal Fe.
For punches, dies, upsetters, shears, tools; oil hardened, nondeforming.

ZADUR BFC.
M-1324; 1.05 C, 1.0 Cr, 1.15 W, 0.9 Mn, bal Fe.
For cutting tools, bearings, liners; oil or water hardened, wear resistant

ZADUR BGM.
M-1324; 0.34 C, 1.55 Cr, 1.55 Ni, 0.2 Mo, bal Fe.
For gears, shafts, crankshafts, bolts, fasteners; oil hardened, shock resistant.

ZADUR BGS.
M-1324; 0.45 C, Cr, Ni, bal Fe.
For gears, bolts, fasteners; oil hardened, shock resistant.

ZADUR BKM.
M-1324; 1.0 C, 0.1 V, 0.25 Mn, 0.25 Si, bal Fe.
For cutters, hobs, drills, taps, reamers; Type W2; water hardened.

ZADUR BLS.
M-1324; 0.55 C, 1.05 Cr, 0.18 V, 1.85 W, bal Fe.
For cold heading dies, punches, headers; oil hardened, tough.

ZADUR BMR.
M-1324; 0.40 C, Cr, Mo, bal Fe.
For gears, pinions, shafts, bolts; oil hardened, tough.

ZADUR BP5.
M-1324; 0.45 C, Si, Cr, V, bal Fe.
For springs, bolts, gears; oil hardened, tough.

ZADUR BPK.
M-1324; 0.40 C, Cr, V, Mn, bal Fe.
For gears, bolts, crankshafts; oil hardened, tough.

ZADUR BRK.
M-1324; 0.50 C, 1.05 Cr, 3.25 Ni, 0.50 Mn, bal Fe.
For gears, bolts, crankshafts, fasteners; oil hardened, shock resistant.

ZADUR BST1.
M-1324; C, Cr, W, Ni, V, Co, bal Fe.

ZADUR BWF2.
M-1324; C, Cr, Ni, V, bal Fe.
For tools, dies; oil hardened, tough.

ZADUR BWM.
M-1324; 0.3 C, 2.65 Cr, 0.35 V, 8.5 W, bal Fe.
For extrusion press dies and rams, upsetters; hot work steel, oil hardened

ZADUR BWM5.
M-1324; 0.3 C, 2.35 Cr, 0.6 V, 4.25 W, bal Fe.
For extrusion press dies and rams, upsetters; hot work steel, oil hardened

ZADUR BWM100.
M-1324; 0.3 C, 1.1 Cr, 1 Si, 0.18 V, 3,75 W, bal Fe.
For header dies, upsetters, crimpers; oil hardened, tough.

ZADUR BWMCO.
M-1324; 0.30 C, 2 Co, 2.4 Cr, 0.25 V, 8.5 W, bal Fe.
For extrusion press dies and rams, upsetters; oil hardened, tough.

ZADUR BWM EXTRA.
M-1324; C, Cr, W, V, Ni, bal Fe.
For header and upsetter dies, shears, punches; hot work steel, oil hardened.

ZADUR CCK.
M-1324; 1.45 C, 1.4 Cr, 0.60 Mn, bal Fe.
For bearings, bushings, liners, sleeves; water hardened, wear resistant.

ZADUR CCW.
M-1324; C, Cr, Ni, W, V, bal Fe.
Oil hardened; tough.

ZADUR D0.
M-1324; 0.28 C, Ni, Mo, bal Fe.
For gears, bolts, machine tool parts; oil hardened, tough.

ZADUR D1.
M-1324; 0.28 C, Ni, Cr, Mo, V, bal Fe.
For header dies, machine tool parts; oil hardened, tough.

ZADUR D3.
M-1324; 0.26 C, Ni, Cr, Mo, bal Fe.
For machine tool parts, gears, bolts; oil hardened, tough.

ZADUR DS10.
M-1324; 0.65 C, 3.75 Cr, 0.85 Mo, 8.5 W, 0.7 V, bal Fe.
For lathe and planer tools, reamers, drills, taps; high speed steel.

ZADUR EDH2.
M-1324; 0.42 C, 1.0 Cr, 0.2 Mo, bal Fe.
For gears, bolts, shafts, fasteners; oil hardened, tough.

ZADUR EDH3.
M-1324; 0.40 C, Cr, Mo, bal Fe.
For gears, bolts, shafts, fasteners; oil hardened, tough.

ZADUR EDH11.
M-1324; 0.37 C, 1.25 Si, 1.25 Mn, bal Fe.
For gears, shafts, crankshafts; oil hardened, tough.

ZADUR EDHV.
M-1324; 0.50 C, 0.85 Mn, 1.0 Cr, 0.09 V, bal Fe.
For gears, springs, bolts, shafts; oil hardened, shock resistant.

ZADUR EH.
M-1324; 0.15 C, 0.65 Cr, 0.5 Mn, bal Fe.
For gears, machine tool parts, shafts; case hardening steel.

ZADUR EHC125.
M-1324; 0.16 C, 1.15 Mn, 0.95 Cr, bal Fe.
For gears, bolts, fasteners, cams; case hardening steel, tough.

ZADUR EHC135.
M-1324; 0.20 C, 1.15 Mn, 0.95 Cr, bal Fe.
For bolts, gears, camshafts, cams, pinions; case hardening steel.

ZADUR FE501.
M-1324; 0.7 C, V, W, Cr, bal Fe.
For lathe and planer tools, drills, reamers, hobs; high speed steel.

ZADUR FE501R.
M-1324; 0.7 C, 4 Cr, V, W, bal Fe.
For lathe and planer tools, drills, reamers, taps; high speed steel.

ZADUR FE601.
M-1324; 0.86 C, 4.1 Cr, 0.85 Mo, 2.5 V, 12 W, bal Fe.
For lathe and planer tools, reamers, drills, taps; high speed steel.

ZADUR FEX.
M-1324; 0.74 C, 4.1 Cr, 1.1 V, 18.5 W, bal Fe.
For lathe and planer tools, reamers, drills, taps; high speed steel.

ZADUR FN.
M-1324; 0.82 C, 4.1 Cr, 0.85 Mo, 1.6 V, 8.7 W, bal Fe.
For lathe and planer tools, reamers, broaches, taps; high speed steel.

ZADUR GC15.
M-1324; 0.25 C, 14.5 Cr, 1 max Ni, bal Fe.
For oil refinery equipment; creep and heat resistant.

ZADUR GC17.
M-1324; 0.25 C, 17 Cr, 1.8 Ni, bal Fe.
Annealed: 125,000 TS; 95,000 YS; 20 El; 55 RA; 260 Brin.
Cold drawn: 130,000 TS; 110,000 YS; 15 El; 35 RA; 270 Brin.
For pumps, marine hardware, cutlery; Type 431; stainless and heat resistant.

ZADUR GC30H.
M-1324; 1.2 C, 1.3 Si, 29 Cr, bal Fe.
For wear plates, grates, grids, liners, valves; heat and abrasion resistant.

ZADUR GC30W.
M-1324; 0.40 C, 1.3 Si, 29 Cr, bal Fe.
Cast: 90,000 TS; 65,000 YS; 2 El; 212 Brin.
For cylinder liners, valve seats and bodies, bushings; heat and abrasion resistant.

ZADUR GCM30H.
M-1324; 1.2 C, 1.3 Si, 29 Cr, 2 Mo, bal Fe.
For cylinder liners, valve seats and bodies, bushings; heat and abrasion resistant.

ZADUR GD2.
M-1324; 0.7 C, Cr, W, V, bal Fe.
For lathe and planer tools, drills, reamers; high speed steel.

ZADUR GD15.
M-1324; 0.38 C, Si, Cr, V, bal Fe.
For springs, bolts, gears, machine tool parts; oil hardened, shock resistant.

ZADUR GD22.
M-1324; 0.35 C, 1.05 Cr, 0.18 V, 1.85 W, bal Fe.
For upsetter dies, header dies, crimpers, punches; oil hardened, tough.

ZADUR GRMN.
M-1324; 0.15 C, 18 Cr, 8.5 Ni, bal Fe.
Annealed: 80,000 TS; 35,000 YS; 55 El; 75 RA; 150 Brin.
For chemical plant equipment, tanks; Type 302; stainless, austenitic.

ZADUR GRMP.
M-1324; 0.15 C, 18 Cr, 9.5 Ni, 2 Mo, bal Fe.
Annealed: 85,000 TS; 35,000 YS; 50 El; 65 RA; 160 Brin.
For acid resistant chemical plant equipment, tanks, mixers, filters; Type 316; stainless, austenitic.

ZADUR HDL.
M-1324; C, W, V, bal Fe.
For machine tool parts; oil hardened.

ZADUR MSA1.
M-1324; 0.53 C, 0.9 Si, 0.9 Mn, bal Fe.
For gears, bolts, shafts, fasteners; water hardened.

ZADUR MSA2.
M-1324; 0.55 C, 0.7 C, O.18 Mo, 1.65 Ni, 0.10 V, bal Fe.
For gears, bolts, crankshafts, fasteners; oil hardened, shock resistant.

ZADUR MSA3.
M-1324; 0.56 C, 0.8 Cr, 0.2 Mo, 1.75 Ni, 0.1 V, bal Fe.
For gears, bolts, crankshafts, fasteners; oil hardened, shock resistant.

ZADUR RBM.
M-1324; 0.45 C, 1.4 Cr, 0.7 Mo, 0.3 V, bal Fe.
For forging and heading dies, upsetters, die casting dies; oil hardened, tough.

ZADUR RMA.
M-1324; 0.90 C, 1.15 Mo, 1.0 V, 18 Cr, bal Fe.
For cutlery, valves, ball bearings; hardenable, corrosion resistant.

ZADUR RMC.
M-1324; 0.20 C, 13 Cr, 1.15 Mo, bal Fe.
Annealed: 95,000 TS; 50,000 YS; 25 El; 55 RA; 195 Brin.
For turbine blades, cutlery, valves; hardenable, corrosion resistant.

ZADUR RME.
M-1324; 0.35 C, 16.5 Cr, 1.15 Mo, bal Fe.
For oil refinery equipment; corrosion and heat resistant.

ZADUR RMH.
M-1324; 0.40 C, 13 Cr, 0.30 Mn, bal Fe.
Annealed: 100,000 TS; 55,000 YS; 20 El; 50 RA; 200 Brin.
For turbine blades, valves, cutlery, surgical instruments; Type 420; stainless.

ZADUR RMI.
M-1324; 0.1 max C, 2 Si, 18 Cr, 8.5 Ni, bal Fe.
Annealed: 85,000 TS; 35,000 YS; 60 El; 70 RA; 150 Brin.
For chemical plant equipment, tanks, agitators; Type 304; stainless, austenitic.

ZADUR RMK.
M-1324; 0.12 max C, 18 Cr, 9.5 Ni, Ti = 4 x C, bal Fe.
Annealed: 85,000 TS; 35,000 YS; 55 El; 65 RA; 150 Brin.
For welded chemical plant equipment; Type 321; stainless, austenitic.

ZADUR RML.
M-1324; 0.12 max C, 18 Cr, 10.5 Ni, 2 Mo, Ti = 4 x C, bal Fe.
Annealed: 85,000 TS; 35,000 YS; 50 El; 65 RA; 160 Brin.
For welded acid resistant equipment; Type 316 Ti; stainless, austenitic.

ZADUR RMM.
M-1324; 0.2 C, 13 Cr, 0.4 Si, bal Fe.
Annealed; 95,000 TS; 50,000 YS; 25 El; 55 RA; 195 Brin.
For turbine blades, valves, cutlery, surgical instruments; Type 420 stainless, hardenable.

ZADUR RMN.
M-1324; 0.15 max C, 18 Cr, 8.5 Ni, bal Fe.
Annealed: 80,000 TS; 35,000 YS; 55 El; 75 RA; 150 Brin.
For chemical plant equipment, tanks, mixers, filters; Type 302; stainless austenitic.

ZADUR RMO.
M-1324; 0.22 C, 17 Cr, 1.5 Ni, 0.4 Si, bal Fe.
Annealed: 125,000 TS; 90,000 YS; 20 El; 55 RA; 260 Brin.
For pumps, marine hardware, valves; Type 431; corrosion resistant.

ZADUR RMP.
M-1324; 0.1 max C, 18 Cr, 2 Mo, 9.5 Ni, bal Fe.
Annealed: 85,000 TS; 35,000 YS; 50 El; 65 RA; 160 Brin.
For acid resistant chemical plant equipment; Type 316; stainless, austenitic.

ZADUR RMR.
M-1324; 0.07 max C, 18 Cr, 9.5 Ni, bal Fe.
Annealed: 85,000 TS; 35,000 YS; 60 El; 70 RA; 150 Brin.
For chemical plant equipment, tanks, mixers; Type 304; stainless, austenitic.

ZADUR RMT.
M-1324; 0.07 max C, 18 Cr, 10.5 Ni, 2 Mo, bal Fe.
Annealed: 85,000 TS; 35,000 YS; 50 El; 65 RA; 160 Brin.
For acid resistant chemical plant equipment; Type 316; stainless, austenitic.

ZADUR RMW.
M-1324; 0.12 max C, 0.4 Si, 13 Cr, bal Fe.
Annealed: 75,000 TS; 40,000 YS; 35 El; 70 RA; 155 Brin.
For turbine blades, valves, cutlery; Type 410; stainless.

ZADUR SO EXTRA.
M-1324; 1.3 C, 4.3 Cr, 0.85 Mo, 3.8 V, 12 W, bal Fe.
For blanking and forming dies; high speed steel.

ZADUR SSB50N.
M-1324; 0.80 C, W, Co, Cr, V, Mo, bal Fe.
For lathe and planer tools, reamers, broaches; high speed steel.

ZADUR SS C03.
M-1324; 0.86 C, 2.8 Co, 4.3 Cr, 0.85 Mo, 2.1 V, 12 W, bal Fe.
For lathe and planer tools, reamers, drills; high speed steel.

ZADUR SS CO5.
M-1324; 0.79 C, 4.75 Co, 4.3 Cr, 0.75 Mo, 1.5 V, 18 W, bal Fe.
For lathe and planer tools, cutters, reamers; high speed steel.

ZADUR SS CO10.
M-1324; 0.76 C, 10 Co, 4.2 Cr, 0.8 Mo, 1.8 V, 18 W, bal Fe.
For lathe and planer tools, form cutters, hobs; high speed steel.

ZADUR SS CO15.
M-1324; 0.7 C, 4 Cr, W, V, Co, bal Fe.
For lathe and planer tools, drills, taps, hobs; high speed steel.

ZADUR SS CO SPEZIAL.
M-1324; 1.35 C, Co, W, Mo, Cr, V, bal Fe.
For blanking and forming dies, engravers' tools; high speed steel.

ZADUR ZRS.
M-1324; 1.4 C, 0.30 Cr, 0.10 V, bal Fe.
For blanking and heading dies, engravers' tools; water or oil hardened, wear resistant.

Z-ALLOY.
M-18; 1 Pt, bal Ag.

Z A M.
M-580; 0.5 Al, 0.25 Hg, bal Zn.
For anodes for Zn plating; acid resistant.

ZAMA.
M-Italy; Al, Mg, bal Zn.
For die castings, instrument cases; corrosion resistant.

ZAMAK ALPHA.
M-1134; 4 Al, 0.5-1 Cu, 0.3 Mg, bal Zn.
Rolled: 44,000-71,000 TS; 75-140 Brin.
For die castings; rolled or cast.

ZAMAK BETA.
M-1134; 10 Al, 0.7 Cu, 0.03 Mg, bal Zn.
Rolled: 45,000-78,000 TS; 75-160 Brin.
For housings, hardware.

ZAMAK ETA.
M-1134; 10 Al, 0.3 Cu, bal Zn.
Rolled: 31,000-45,000 TS; 55-65 Brin.
For housings, hardware.

ZAMAK ETA H.
M-1134; 10 Al, 0.3 Cu, 0.01 Mg, bal Zn.
Rolled: 56,700-70,000 TS; 100-130 Brin.
For housings, hardware.

ZAMAK GAMMA.
M-Ger.; 0.8 Al, 0.4 Cu, 0-0.2 Mg, bal Zn.
For castings or drawn wire.

ZAMAK NO. 3.
M-91; 3.9-4.3 Al, 0.025-0.050 Mg, 0.10 max Cu, bal Zn.
Die Cast: 41,000 TS; 10 El; 82 Brin.
For gears, frames, hardware, pumps, die castings.

ZAMAK NO. 5.
M-91; 3.9-4.3 Al, 0.75-1.25 Cu, 0.03-0.04 Mg, bal Zn.
Die cast: 47,000 TS; 7 El; 91 Brin.
For die castings, gears, hardware.

ZAMAK NO. 7.
M-91; 3.9-4.3 Al, 0.10 max Cu, 0.010-0.020 Mg, 0.010-0.020 Ni, bal Zn.
For die castings, housings; improved castability.

ZAMAK Z 400.
M-299; 3.7-4.3 Al, 0-0.6 Cu, 0.02-0.05 Mg, bal Zn.
Die cast: 36,000-43,000 TS; 4-1.5 El; 70-80 Brin.
For housings, cases, ornaments, general die castings; impact resistant, shrinks on aging.

ZAMAK Z 410.
M-299; 3.7-4.3 Al, 0.6-1.0 Cu, 0.02-0.05 Mg, bal Zn.
Die cast: 39,000-46,000 TS; 5-2 El; 80-90 Brin.
For housings, cases, ornaments, general die castings; good corrosion resistance.

ZAMAK Z 430.
M-299; Zn alloy.
For mold inserts, tools, dies.

ZAMIUM.
60 Ni, 40 Cr, Mn, W.
For heat resistant parts; corrosion and heat resistant.

ZAPP 13.
M-1332; 0.08 max C, 13.0 Cr, bal Fe.
Ferritic type stainless steel.
Ann: 130-180 Brin.
Werkstoff Nr. 4000.

ZAPP 13A1.
M-1332; 0.08 max C, 13.0 Cr, 0.10-0.30 Al, bal Fe.
Ferritic type stainless steel.
Ann: 130-180 Brin.
Werkstoff Nr. 4002; AISI 405.

ZAPP 13B.
M-1332; 0.08 max C, 14.0 Cr, bal Fe.
Ferritic type stainless steel.
Ann: 130-180 Brin.
Werkstoff Nr. 4001.

ZAPP 17.
M-1332; 0.10 max C, 17.0 Cr, bal Fe.
Ferritic type stainless steel.
Ann: 130-170 Brin.
Werkstoff Nr. 4016; AISI 430.

ZAPP 17 F.
M-1332; 0.07 max C, 17.0 Cr, 1.0 Mo, bal Fe.
Ferritic type stainless steel.
Ann: 130-180 Brin.
Werkstoff Nr. 4113; similar to AISI 434.

ZAPP 17 NB.
M-1332; 0.10 max C, 17.0 Cr, Nb = 12 X C min; bal Fe.
Ferritic type stainless steel.
Ann: 130-170 Brin.
Werkstoff Nr. 4511.

ZAPP 17 T.
M-1332; 0.10 max C, 17.0 Cr, Ti = 7 X C min; bal Fe.
Ferritic type stainless steel.
Ann: 130-170 Brin.
Werkstoff Nr. 4510.

ZAPP 17 U.
M-1332; 0.15 max C, 17.0 Cr, 0.25 Mo, S, bal Fe.
Ferritic type stainless steel.
Ann: 190-230 Brin.
Werkstoff Nr. 4104; AISI 430 F.

ZAPP 26 NL.
M-1332; 0.10 max C, 26.0 Cr, 1.5 Mo, 4.5 Ni, bal Fe.
Ferritic type stainless steel.
Ann: 190-230 Brin.
Werkstoff Nr. 4460.

ZAPP 58 M.
M-1332; 0.10 max C, 18.0 Cr, 5.5 Ni, 8.5 Mn, N, bal Fe.
Manganese austenitic stainless.
Werkstoff Nr. 4371; AISI 202.

ZAPP 80 ELC.
M-1332; 0.03 max C, 18.0 Cr, 11.0 Ni, bal Fe.
Low carbon austenitic stainless steel.
Solution annealed: 130-180 Brin.
Werkstoff Nr. 4306; AISI 304 L.

ZAPP 80 FH.
M-1332; 0.15 max C, 17.0 Cr, 7.5 Ni, bal Fe.
Austenitic type stainless steel.
Solution treated: 170-210 Brin.
Work-hardens rapidly.
Werkstoff Nr. 4310; AISI 301.

ZAPP 80 K.
M-1332; 0.12 max C, 18.0 Cr, 9.0 Ni, bal Fe.
Austenitic type stainless steel.
Solution treated: 130-180 Brin.
Werkstoff Nr. 4300; AISI 302.

ZAPP 80 NB.
M-1332; 0.10 max C, 18.0 Cr, 10.5 Ni, Nb/Ta, bal Fe.
Stabilized austenitic stainless steel.
For welded structures and elevated temperatures.
Werkstoff Nr. 4550; AISI 347.

ZAPP 82 NB.
M-1332; 0.10 max C, 17.5 Cr, 2.3 Mo, 12.0 Ni, Nb/Ta, bal Fe.
Stabilized austenitic stainless steel.
For welded chemical equipment.
Werkstoff Nr. 4580.

ZAPP 80P.
M-1332; 0.07 max C, 18 Cr, 9.5 Ni, bal Fe.
Annealed: 85,000 TS; 35,000 YS; 60 El; 70 RA; 150 Brin.
Cold drawn: 180,000 TS; 125,000 YS; 10 El; 330 Brin.
For chemical plant equipment, welded structures, tanks; Type 304; stainless, austenitic.
W. Nr. 4301.

ZAPP 80T.
M-1332; 0.12 max C, 18 Cr, 9.5 Ni, Ti = 4 x C, bal Fe.
Annealed: 85,000 TS; 35,000 YS; 55 El; 65 RA; 150 Brin.
Cold drawn: 95,000 TS; 60,000 YS; 40 El; 60 RA; 185 Brin.
For welded structures, chemical plant equipment; Type 321; stainless, austenitic.
W. Nr. 4541.

ZAPP 80U.
M-1332; 0.12 C, 9.5 Ni, 18 Cr, 0.2 S, bal Fe.
Annealed: 80,000 TS; 35,000 YS; 40 El; 55 RA; 160 Brin.
For screw machine products, bolts; Type 302; stainless, free-cutting.
W. Nr. 4305.

ZAPP 82P.
M-1332; 0.07 max C, 18 Cr, 10.5 Ni, 2 Mo, bal Fe.
Annealed: 80,000 TS; 35,000 YS; 55 El; 70 RA; 150 Brin.
Cold drawn: 140,000 TS; 130,000 YS; 7 El; 280 Brin.
For acid resistant equipment, mixers, agitators; Type 316L; stainless, austenitic.
W. Nr. 4401.

ZAPP 82 T.
M-1332; 0.10 max C, 17.5 Cr, 2.3 Mo, 12.0 Ni, Ti, bal Fe.
Stabilized austenitic stainless steel.
Werkstoff Nr. 4571; similar to AISI 316.

ZAPP 83 ELC.
M-1332; 0.03 max C, 18.0 Cr, 2.8 Mo, 13.5 Ni, bal Fe.
Low carbon austenitic steel; for chemical equipment.
Werkstoff Nr. 4435; AISI 316 L.

ZAPP 83 NB.
M-1332; 0.10 max C, 17.5 Cr, 2.8 Mo, 13.0 Ni, Nb/Ta, bal Fe.
Stabilized austenitic stainless steel.
For welded chemical equipment.
Werkstoff Nr. 4583.

ZAPP 83 P.
M-1332; 0.07 max C, 18.0 Cr, 2.8 Mo, 13.0 Ni, bal Fe.
Austenitic stainless steel.
For chemical equipment.
Werkstoff Nr. 4436; AISI 316.

ZAPP 83 T.
M-1332; 0.10 max C, 17.5 Cr, 2.8 Mo, 13.0 Ni, Ti, bal Fe.
Stabilized austenitic stainless steel; for welded chemical equipment.
Werkstoff Nr. 4573.

ZAPP 120.
M-1332; 0.08 max C, 13.0 Cr, 13 Ni, bal Fe.
For chemical plant equipment; corrosion resistant.
W. Nr. 4307.

ZAPP 135 P.
M-1332; 0.07 max C, 17.0 Cr, 4.5 Mo, 13.5 Ni, bal Fe.
Austenitic stainless steel; for chemical equipment.
Werkstoff Nr. 4449; AISI 317.

ZAPP 182 RNB.
M-1332; 0.07 max C, 18.0 Cr, 2.3 Mo, 20.0 Ni, Cu/Ti/Ta, bal Fe.
Austenitic stainless steel.
Werkstoff Nr. 4505.

ZAPP 223 RNB.
M-1332; 0.07 max C, 17.0-18.0 Cr, 3.0-3.5 Mo, 22.0-23.0 Ni, Nb, bal Fe.
Austenitic stainless steel.
Werkstoff Nr. 4586.

ZAPP 252 T.
M-1332; 0.06 max C, 25.0 Cr, 2.2 Mo, 25.0 Ni, N, bal Fe.
Austenitic stainless steel.
Werkstoff Nr. 4577.

ZAPP AC14.
M-1332; 0.34 C, 1.1 Al, 1.4 Cr, bal Fe.
For oil refinery equipment; heat and creep resistant.

ZAPP BC115.
M-1332; 1.55-1.75 C, 0.25-0.40 Si, 0.2-0.4 Mn, 1.2-1.4 Co, 11.0-12.0 Cr, 0.5-0.6 Mo, bal Fe.
Cold work tool steel.
Werkstoff Nr. 1.2880.

ZAPP BC115S.
 M-1332; 2.0-2.25 C, 0.2-0.4 Si, 0.2-0.4 Mn, 0.8-1.1 Co, 11.0-12.5 Cr, 0.3-0.5 Mo, 0.6-0.8 W, bal Fe.
 Cold work tool steel.
 Werkstoff Nr. 1.2884.

ZAPP BC120.
 M-1332; 1.65 C, 11.5 Cr, Co, bal Fe.
 For blanking and forming dies, punches; air hardening, nondeforming.

ZAPP BCVW85.
 M-1332; 0.3 C, 2.4 Cr, 0.25 V, 8.5 W, 2 Co, bal Fe.
 For hot work tools, punches, shears, extrusion rams; oil hardened, tough.

ZAPP C3.
 M-1332; 1.4 C, 0.3 Cr, 0.1 V, bal Fe.
 For blanking and forming dies, engravers' tools; water hardened, wear resistant.

ZAPP C8.
 M-1332; 0.90 C, 0.8 Cr, 0.3 Mn, bal Fe.
 For bearings, cutters, punches; water hardened, wear resistant.

ZAPP C10.
 M-1332; 1.05 C, 0.3 Mn, 1.0 Cr, bal Fe.
 For bearings, cutters, punches; water hardened, wear resistant.

ZAPP C14.
 M-1332; 1.05 C, 1.2 Cr, 0.3 Mn, bal Fe.
 For bearings, cutters, punches; water hardened, wear resistant.

ZAPP C16.
 M-1332; 1.0 C, 1.55 Cr, 0.35 Mn, bal Fe.
 For bearings, liners, sleeves; water hardened, wear resistant.
 W. Nr. 1.2067.

ZAPP C120.
 M-1332; 2.1 C, 11.5 Cr, 0.70 W, bal Fe.
 For blanking and forming dies, gauges, punches; oil or air hardened, non-deforming.

ZAPP C130W.
 M-1332; 0.20 C, 13 Cr, 0.3 Mn, 0.4 Si, bal Fe.
 Annealed: 95,000 TS; 50,000 YS; 25 El; 55 RA; 196 Brin.
 For valve trim, turbine blades, surgical instruments; Type 420; corrosion resistant.

ZAPP C135M.
 M-1332; 0.4 C, 0.4 Si, 0.3 Mn, 13 Cr, bal Fe.
 Annealed: 100,000 TS; 55,000 YS; 22 El; 50 RA; 200 Brin.
 For valves, turbine blades, cutlery, knives; Type 420; stainless, hardenable.

ZAPP CLWV8.
 M-1332; 0.45 C, 1.35 Cr, 0.8 V, 0.45 Mo, 0.45 W, bal Fe.
 For forging dies, upsetters, header dies; oil hardened, tough.

ZAPP CM12.
 M-1332; 1.0-1.1 C, 0.15-0.30 Si, 1.0-1.2 Mn, 0.70-1.0 Cr, bal Fe.
 Cold work tool steel.
 Werkstoff Nr. 1.2127.

ZAPP CN14.
 M-1332; 0.40-0.50 C, 0.15-0.35 Si, 0.5-0.8 Mn, 1.2-1.5 Cr, 1.5-1.8 Ni, bal Fe.
 Cold work tool steel.
 Werkstoff Nr. 1.2710.

ZAPP CN15.
 M-1332; 0.12-0.17 C, 0.4-0.6 Mn, 1.4-1.7 Ni, 1.4-1.7 Cr, bal Fe.
 Alloy carburizing steel.
 Werkstoff Nr. 1.2712.

ZAPP CN35.
 M-1332; 0.50 C, 1.05 Cr, 3.25 Ni, 0.50 Mn, bal Fe.
 For gears, bolts, machine tool parts; oil hardened, shock resistant.

ZAPP CN58.
 M-1332; 0.10-0.17 C, 0.20-0.35 Si, 0.3-0.5 Mn, 0.65-0.85 Cr, 3.3-3.6 Ni, bal Fe.
 Cold work tool steel.
 Werkstoff Nr. 1.2735.

ZAPP CN59.
 M-1332; 0.10-0.17 C, 0.2-0.3 Si, 0.3-0.5 Mn, 0.9-1.20 Cr, 4.2-4.7 Ni, bal Fe.
 Cold work tool steel.
 Werkstoff Nr. 1.2745.

ZAPP CNL3.
 M-1332; 0.55 C, 0.7 Cr, 0.18 Mo, 1.65 Ni, 0.1 V, bal Fe.
 For gears, bolts, crankshafts; oil hardened, shock resistant.

ZAPP CNL5.
 M-1332; 0.57 C, 0.8 Cr, 0.2 Mo, 1.75 Ni, 0.1 V, bal Fe.
 For gears, bolts, crankshafts; oil hardened, shock resistant.

ZAPP CVL6.
 M-1332; 0.45 C, 0.7 Mn, 1.4 Cr, 0.7 Mo, 0.3 V, bal Fe.
 For gears, springs, bolts, shafts; oil hardened, tough.

ZAPP CVL10.
 M-1332; 0.38-0.42 C, 0.9-1.2 Si, 0.3-0.5 Mn, 4.8-5.8 Cr, 0.8-1.4 Mo, 0.25-0.50 V, bal Fe.
 Hot work tool steel.
 Werkstoff Nr. 1.2343.

ZAPP CVL 10 EXTRA.
 M-1332; 0.32-0.40 C, 0.9-1.2 Si, 0.3-0.5 Mn, 5.0-5.6 Cr, 1.3-1.6 Mo, 1.2-1.4 W, 0.15-0.40 V, bal Fe.
 Hot work tool steel.
 Werkstoff Nr. 1.2606.

ZAPP CVL10V.
M-1332; 0.37-0.42 C, 0.9-1.2 Si, 0.3-0.5 Mn, 5.0-5.5 Cr, 1.2-1.5 Mo, 0.9-1.1 V, bal Fe.
Hot work tool steel.
Werkstoff Nr. 1.2344.

ZAPP CVL 30.
M-1332; 0.28-0.35 C, 0.2-0.4 Si, 0.2-0.4 Mn, 2.7-3.2 Cr, 2.0-3.0 Mo, 0.4-0.7 V, bal Fe.
Hot work tool steel.
Werkstoff Nr. 1.2365.

ZAPP CVW35.
M-1332; 1.42 C, W, V, bal Fe.
For cutters, forming dies, tools; water or oil hardened.

ZAPP CVW40.
M-1332; 0.30 C, 0.18 V, 3.75 W, 1.1 Cr, bal Fe.
For extrusion dies, liners, rams, upsetters; oil hardened, tough.

ZAPP CVW45.
M-1332; 0.30 C, 2.35 Cr, 0.6 V, 4.25 W, bal Fe.
For extrusion dies, liners, rams, upsetters; oil hardened, tough.

ZAPP CVW50.
M-1332; 1.25-1.35 C, 0.2-0.3 Si, 0.2-0.4 Mn, 0.20 max Cr, 4.7-5.2 W, bal Fe.
Cold work tool steel.
Werkstoff Nr. 1.2453.

ZAPP CVW85.
M-1332; 0.30 C, 0.35 V, 8.5 W, 2.65 Cr, bal Fe.
For extrusion dies, rams, liners; oil hardened, tough.

ZAPP CVW95.
M-1332; 0.65 C, 0.85 Mo, 8.5 W, 3.75 Cr, 0.70 V, bal Fe.
For extrusion dies, rams, liners, cutters; high speed steel.

ZAPP CWN130.
M-1332; 0.45 C, Ni, Cr, W, bal Fe.
For forging dies, upsetters; oil hardened, tough.

ZAPP GUSS 1.4740.
M-1332; 0.30-0.60 C, 1.0-2.5 Si, 1.0 max Mn, 16.0-18.0 Cr, bal Fe.
Heatproof steel castings.
DIN G-X40 CrSi 17.

ZAPP GUSS 1.4776.
M-1332; 0.30-0.60 C, 1.0-2.5 Si, 1.0 max Mn, 27.0-30.0 Cr, bal Fe.
Heatproof steel castings.
DIN G-X40 CrSi 29.

ZAPP GUSS 1.4777.
M-1332; 1.2-1.4 C, 1.0-2.5 Si, 1.0 max Mn, 27.0-30.0 Cr, bal Fe.
Heatproof steel castings.
DIN G-X130 CrSi 29.

ZAPP GUSS 1.4823.
M-1332; 0.30-0.50 C, 1.0-2.0 Si, 1.5 max Mn, 26.0-28.0 Cr, 3.5-5.5 Ni, bal Fe.
Heatproof steel castings.
DIN G-X40 CrNiSi 27 4.

ZAPP GUSS 1.4826.
M-1332; 0.30-0.50 C, 1.0-2.5 Si, 1.5 max Mn, 21.0-23.0 Cr, 9.0-11.0 Ni, bal Fe.
Heatproof steel castings.
DIN G-X40 CrNiSi 22 9.

ZAPP GUSS 1.4837.
M-1332; 0.20-0.50 C, 1.0-2.5 Si, 1.5 max Mn, 24.0-26.0 Cr, 11.0-14.0 Ni, bal Fe.
Heatproof steel castings.
DIN G-X35 CrNiSi 25 12.

ZAPP GUSS 1.4846.
M-1332; 0.20-0.50 C, 1.0-2.5 Si, 1.5 max Mn, 25.0-28.0 Cr, 13.0-16.0 Ni, bal Fe.
DIN G-X40 CrNiSi 26 14.

ZAPP GUSS 1.4848.
M-1332; 0.20-0.50 C, 1.0-2.5 Si, 1.5 max Mn, 24.0-27.0 Cr, 19.0-21.0 Ni, bal Fe.
Heatproof steel castings.
DIN G-X40 CrNiSi 25 20; AISI 310.

ZAPP GUSS 1.4865.
M-1332; 0.20-0.50 C, 1.0-2.5 Si, 1.5 max Mn, 16.0-19.0 Cr, 36.0-39.0 Ni, bal Fe.
Heatproof steel castings.
DIN G-X40 CrNiSi 36 16.

ZAPP K4 EXTRA.
M-1332; 1.0 C, 0.25 max Si, 0.25 max Mn, bal Fe.
Annealed: 110,000 TS; 56,000 YS; 18 El; 40 RA; 210 Brin.
For springs, tools, cutters, taps, reamers; Type W1; water hardened.
W. Nr. 1.1543.

ZAPP K4 PRIMA.
M-1332; 1.0 C, 0.25 max Si, 0.25 max Mn, bal Fe.
Annealed: 110,000 TS; 56,000 YS; 18 El; 40 RA; 210 Brin.
For springs, tools, cutters, taps, reamers; Type W1; water hardened.
W. Nr. 1.1645.

ZAPP K5 EXTRA.
M-1332; 0.85 C, 0.25 max Si, 0.25 max Mn, bal Fe.
Heat treated: 190,000 TS; 145,000 YS; 10 El; 30 RA; 400 Brin.
For springs, tools, cutters, drills, taps, hobs; Type W1; water hardened.
W. Nr. 1.1525.

ZAPP K5 PRIMA.
M-1332; 0.85 C, 0.25 max Si, 0.25 max Mn, bal Fe.
Heat treated: 190,000 TS; 145,000 YS; 10 El; 30 RA; 400 Brin.
For springs, tools, cutters, drills, taps, hobs; Type W1; water hardened.
W. Nr. 1.1625.

ZAPP K6 EXTRA.
M-1332; 0.70 C, 0.25 max Si, 0.25 max Mn, bal Fe.
Heat treated: 175,000 TS; 128,000 YS; 12 El; 37 RA; 355 Brin.
For rails, springs, hammers, axes, punches; Type W1; water hardened.

ZAPP K6 PRIMA.
M-1332; 0.70 C, 0.25 max Si, 0.25 max Mn, bal Fe.
Heat treated: 175,000 TS; 128,000 YS; 12 El; 37 RA; 355 Brin.
For rails, hammers, springs, axes, punches; Type W1; water hardened.

ZAPP K 10.
M-1332; 0.10 C, 13.0 Cr, bal Fe.
Martensitic type stainless steel.
Heat treated: 170-210 Brin. (700-750°C temper).
Werkstoff Nr. 4006; AISI 403; 410.

ZAPP K 15.
M-1332; 0.15 C, 13.0 Cr, bal Fe.
Martensitic type stainless steel.
Heat treated: 190-240 Brin. (700-750°C temper).
Werkstoff Nr. 4024; similar to AISI 410.

ZAPP K15.
M-1332; 0.15 C, 0.3-0.5 Si, 0.25-0.50 Mn, bal Fe.
Cold drawn: 72,000 TS; 60,000 YS; 22 El; 58 RA; 145 Brin.
For gears, shafts, machine tool parts; case hardening steel.

ZAPP K 15 L.
M-1332; 0.15 C, 13.0 Cr, 1.2 Mo, bal Fe.
Martensitic type stainless steel.
Heat treated: 220-260 Brin. (600-700°C temper).
Werkstoff Nr. 4119.

ZAPP K 20.
M-1332; 0.20 C, 13.0 Cr, bal Fe.
Martensitic type stainless steel.
Heat treated: 180-250 Brin. (700-750°C temper).
Werkstoff Nr. 4021; AISI 420.

ZAPP K20.
M-1332; 0.2 C, 0.4 Si, 13 Cr, bal Fe.
Annealed: 95,000 TS; 50,000 YS; 25 El; 55 RA; 195 Brin.
Cold drawn: 105,000 TS; 85,000 YS; 17 El; 50 RA; 215 Brin.
For valves, cutlery, turbine blades, surgical instruments; Type 420; stainless.

ZAPP K20L.
M-1332; 0.2 C, 1.15 Mo, 13 Cr, bal Fe.
Annealed: 95,000 TS; 50,000 YS; 25 El; 55 RA; 195 Brin.
For valves, cutlery, pumps, turbine blades; Type 420 Mo; stainless.
W.Nr. 4120.

ZAPP K20N.
M-1332; 0.22 C, 17 Cr, 1.5 Ni, bal Fe.
Annealed: 125,000 TS; 95,000 YS; 20 El; 55 RA; 260 Brin.
For pumps, marine hardware, valves; Type 431; heat anad corrosion resistant.
W. Nr. 4057.

ZAPP K35L.
M-1332; 0.35 C, 1.15 Mo, 16.5 Cr, bal Fe.
Annealed: 130,000 TS; 95,000 YS; 18 El; 50 RA; 270 Brin.
For oil refinery equipment, marine hardware; heat and corrosion resistant.
W. Nr. 4122.

ZAPP K 40.
M-1332; 0.45 C, 13.0 Cr, bal Fe.
Martensitic type stainless steel.
Heat treatable to 50 Rc min.
Werkstoff Nr. 4034.

ZAPP K45.
M-1332; 0.45 C, 0.3-0.5 Si, 0.3-0.8 Mn, bal Fe.
Hot rolled: 98,000 TS; 60,000 YS; 24 El; 45 RA; 212 Brin.
For gears, bolts, machine tool parts; water hardened.

ZAPP K 50.
M-1332; 0.50 C, 13.5 Cr, 0.5 Mo, bal Fe.
Martensitic type stainless steel.
Ann: 180-225 Brin; hardenable to 50 Rc.
Werkstoff Nr. 4110.

ZAPP K55S.
M-1332; 0.55 C, 0.1-0.4 Si, 0.5-0.7 Mn, bal Fe.
Annealed; 100,000 TS; 55,000 YS; 15 El; 22 RA; 180 Brin.
For gears, bolts, machine tool parts; water hardened.

ZAPP K60.
M-1332; 0.60 C, 0.10-0.40 Si, 0.5-0.7 Mn, bal Fe.
Heat treated: 160,000 TS; 115,000 YS; 12 El; 40 RA; 325 Brin.
For gears, shafts, fasteners; water hardened.

ZAPP K 90 B.
M-1332; 0.90 C, 17.0 Cr, 0.5 Mo, Co, V, bal Fe.
Martensitic type stainless steel.
Werkstoff Nr. 4535.

ZAPP K 90 L.
M-1332; 0.90 C, 18.0 Cr, 1.2 Mo, V, bal Fe.
Martensitic type stainless steel.
Hardenable to 52 Rc min.
Werkstoff Nr. 4112; AISI 440 B.

ZAPP LC12.
M-1332; 0.90-1.1 C, 0.15-0.30 Si, 0.20-0.40 Mn, 1.1-1.3 Cr, 0.20-0.40 bal Fe.
Cold work tool steel.
Werkstoff Nr. 1.2303.

ZAPP LC40.
M-1332; 0.07 max C, 0.20 max Si, 0.20 max Mn, 3.5-4.0 Cr, 0.30-0.60 Mo, bal Fe.
Cold work tool steel.
Werkstoff Nr. 1.2341.

ZAPP LC115S.
M-1332; 0.90-1.0 C, 0.2-0.4 Si, 0.2-0.4 Mn, 11.0-12.0 Cr, 0.85-0.95 Mo, 0.85-0.95 V, bal Fe.
Cold work tool steel.
Werkstoff Nr. 1.2376.

ZAPP LC 120.
M-1332; 1.65 C, Cr, Mo, V, bal Fe.
For blanking and piercing dies, punches; air hardened.

ZAPP LC120S.
M-1332; 1.5-1.6 C, 0.3-0.5 Si, 0.3-0.5 Mn, 11.5-12.5 Cr, 0.6-0.8 Mo, 0.9-1.1 V, bal Fe.
Cold work tool steel.
Werkstoff Nr. 1.2379.

ZAPP LCN.
M-1332; 0.55 C, 0.7 Cr, 0.18 Mo, 1.65 Ni, 0.1 V, bal Fe.
For header and forging dies; oil hardened, tough.

ZAPP LCN35E.
M-1332; 0.16-0.22 C, 0.15-0.30 Si, 0.3-0.5 Mn, 1.1-1.4 Cr, 3.8-4.3 Ni, 0.15-0.25 Mo, bal Fe.
Cold work tool steel.
W.-Nr. 1.2764.

ZAPP LCN40.
M-1332; 0.32-0.38 C, 0.15-0.30 Si, 0.4-0.6 Mn, 1.2-1.5 Cr, 0.2-0.4 Mo, 3.8-4.3 Ni, bal Fe.
Hot work tool steel.
W.-Nr. 1.2766.

ZAPP LCN45.
M-1332; 0.40-0.50 C, 0.15-0.30 Si, 0.3-0.5 Mn, 1.2-1.5 Cr, 3.8-4.3 Ni, 0.15-0.35 Mo, bal Fe.
Cold work tool steel.
W.-Nr. 1.2767 Mo.

ZAPP LCN EXTRA.
M-1332; 0.50 C, 0.8 Cr, 0.2 Mo, 1.8 Ni, 0.1 V, bal Fe.
For header and forging dies, punches; oil hardened, tough.

ZAPP LCN SUPRA.
M-1332; 0.50-0.60 C, 0.15-0.35 Si, 0.6-0.8 Mn, 0.9-1.2 Cr, 0.7-0.9 Mo, 1.5-1.8 Ni, 0.7-0.12 V, bal Fe.
Hot work tool steel.
W.-Nr. 1.2744.

ZAPP LCW 50.
M-1332; 0.40-0.50 C, 0.3-0.5 Si, 0.3-0.5 Mn, 4.0-5.0 Cr, 4.0-5.0 Co, 0.4-0.6 Mo, 4.0-5.0 W, 1.8-2.1 V, bal Fe.
Hot work tool steel.
W.-Nr. 1.2678.

ZAPP LVC25.
M-1332; 0.26-0.34 C, 0.15-0.35 Si, 0.40-0.70 Mn, 2.3-2.7 Cr, 0.15-0.25 Mo, 0.1-0.2 V, bal Fe.
W.-Nr. 1.2307.

ZAPP LVC50.
M-1332; 0.90-1.05 C, 0.20-0.40 Si, 0.40-0.70 Mn, 4.8-5.5 Cr, 0.9-1.25 Mo, 0.1-0.3 V, bal Fe.
Cold work tool steel.
W.-Nr. 1.2363.

ZAPP M 180.
M-1332; 0.15 max C, 12.0 Cr, 0.5 Mo, 2.0 Ni, 18.0 Mn, bal Fe.
Manganese austenitic stainless steel.
W.-Nr. 4451.

ZAPP MC13E.
M-1332; 0.20 C, 1.25 Mn, 1.15 Cr, bal Fe.
For gears, cams, camshafts; case hardening steel.

ZAPP MCL3.
M-1332; 0.40 C, Cr, Mn, Mo, bal Fe.
For gears, bolts, crankshafts, fasteners; oil hardened, tough.

ZAPP MCW10.
M-1332; 1.05 C, 1.0 Cr, 1.15 W, 0.90 Mn, bal Fe.
For bearings, cutters, form tools; water hardened.

ZAPP MCWV.
M-1332; 0.90-1.05 C, 0.15-0.35 Si, 1.0-1.2 Mn, 0.5-0.7 Cr, 0.5-0.7 W, 0.05-0.15 V, bal Fe.
Cold work tool steel.
Werkstoff Nr. 1.2510.

ZAPP MS10.
M-1332; 0.53 C, 0.90 Si, 0.90 Mn, bal Fe.
For gears, bolts, crankshafts, shears, punches; oil hardened, tough.

ZAPP MVC12.
M-1332; 0.61 C, 0.85 Si, 0.75 Mn, 1.18 Cr, 0.1 V, bal Fe.
For springs, dies, upsetters, shears; oil hardened, shock resistant.

ZAPP P300.
M-1332; 0.50-0.57 C, 0.15-0.30 Si, 0.4-0.5 Mn, 0.5-0.7 Cr, 2.5-3.0 Ni, bal Fe.
Cold work tool steel.
Werkstoff Nr. 1.2718.

ZAPP SC5.
M-1332; 0.67 C, 0.5 Cr, 0.50 Mn, 1.3 Si, bal Fe.
For springs, shears, chisels, punches; oil hardened, tough.

ZAPP SC10.
M-1332; 0.90 C, 1.2 Cr, 1.15 Si, 0.70 Mn, bal Fe.
For blanking and forming dies; oil hardened, wear resistant.

ZAPP SC12.
M-1332; 1.25 C, 1.15 Si, 0.70 Mn, 1.2 Cr, bal Fe.
For blanking and forming dies; oil hardened, wear resistant.

ZAPP SCW20H.
M-1332; 0.55 C, 0.90 Si, 1.05 Cr, 0.20 V, 1.85 W, bal Fe.
For forging and die casting dies; oil hardened, tough.

ZAPP SCW 20M.
M-1332; 0.45 C, W, Cr, V, bal Fe.
For extrusion liners and rams, upsetters; hot work tools, oil hardened.

ZAPP SSB30.
M-1332; 0.86 C, 2.8 Co, 4.3 Cr, 0.85 Mo, 2.1 V, 12 W, bal Fe.
For lathe and planer tools, reamers, drills, taps; high speed steel.

ZAPP SSB50.
M-1332; 0.79 C, 4.75 Co, 4.3 Cr, 0.75 Mo, 1.55 V, 18 W, bal Fe.
For lathe and planer tools, reamers, broaches, taps; high speed steel.

ZAPP SSB 100.
M-1332; 0.76 C, 10 Co, 4.2 Cr, 0.8 Mo, 1.8 V, 18 W, bal Fe.
For lathe and planer tools, form tools; high speed steel.

ZAPP SSL25.
M-1332; 0.95 C, 4 Cr, W, Mo, V, bal Fe.
For lathe and planer tools, reamers, broaches; high speed steel.

ZAPP SSV25.
M-1332; 0.86 C, 4.1 Cr, 0.85 Mo, 2.5 V, 12 W, bal Fe.
For lathe and planer tools, drills, reamers, taps; high speed steel.

ZAPP SSV50.
M-1332; 1.3 C, 4.3 Cr, 0.85 Mo, 3.8 V, 12 W, bal Fe.
For blanking and forming dies, cutters, reamers; high speed steel.

ZAPP SSVB50.
M-1332; 1.35 C, 4 Cr, W, Co, V, Mo, bal Fe.
For blanking and forming dies; high speed steel.

ZAPP SSW85.
M-1332; 0.82 C, 4.1 Cr, 0.85 Mo, 1.6 V, 8.7 W, bal Fe.
For lathe and planer tools, reamers, broaches; high speed steel.

ZAPP SSW180.
M-1332; 0.74 C, 4.3 Cr, 1.1 V, 18.5 W, bal Fe.
For lathe and planer tools, drills, reamers, taps; high speed steel.

ZAPP SSWL50.
M-1332; 0.85 C, 4 Cr, W, Mo, V, bal Fe.
For lathe and planer tools, reamers, drills, taps; high speed steel.

ZAPP SVC14H.
M-1332; 0.45 C, Si, Cr, V, bal Fe.
For springs, bolts, gears, crankshafts; oil hardened, shock resistant.

ZAPP SVC14W.
M-1332; 0.45 C, Si, Cr, V, bal Fe.
For springs, bolts, gears, crankshafts; oil hardened, shock resistant.

ZAPP SVC25.
M-1332; 0.38 C, Si, Cr, V, bal Fe.
For springs, bolts, gears, crankshafts; oil hardened, shock resistant.

ZAPP VC4.
M-1332; 0.75-0.85 C, 0.25-0.40 Si, 0.30-0.50 Mn, 0.40-0.70 Cr, 0.15-0. V, bal Fe.
Cold work tool steel.
Werkstoff Nr. 1.2235.

ZAPP VC5.
M-1332; 1.15 C, 0.65 Cr, 0.10 V, 0.30 Mn, bal Fe.
For bearings, blanking and forming dies; oil hardened, wear resistant.

ZAPP VC10.
M-1332; 0.50 C, 1.05 Cr, 0.1 V, 0.95 Mn, bal Fe.
For gears, bolts, crankshafts; oil hardened, shock resistant.

ZAPP VC12.
M-1332; 0.55-0.62 C, 0.15-0.35 Si, 0.80-1.10 Mn, 0.90-1.2 Cr, 0.07-0.1 bal Fe.
Cold work tool steel.
Werkstoff Nr. 1.2242.

ZAPP VC120.
M-1332; 1.65 C, 11.5 Cr, 0.10 V, bal Fe.
For blanking and forming dies, punches; air hardened, non-deforming.

ZAPP VC120S.
M-1332; 2.1-2.3 C, 0.2-0.4 Si, 0.2-0.4 Mn, 11.5-12.5 Cr, 0.85-0.95 Mo, 2.1-2.3 V, bal Fe.
Cold work tool steel.
Werkstoff Nr. 1.2378.

ZAPP VK2 SUPRA.
M-1332; 1.4-1.5 C, 0.2-0.35 Si, 0.3-0.5 Mn, 3.0-3.5 V, bal Fe.
Cold work tool steel.
Werkstoff Nr. 1.2838.

ZAPP VK4 EXTRA.
M-1332; 1.0 C, 0.10 V, 0.25 Mn, 0.20 Si, bal Fe.
For heading and blanking dies, bearings, cutters; Type W2; water hardened.

ZAPP VK4 SUPRA.
M-1332; 0.90-1.0 C, 0.2-0.35 Si, 0.3-0.5 Mn, 0.35-0.45 V, bal Fe.
Cold work tool steel.
W.-Nr. 1.2835.

ZAPP VM20.
M-1332; 0.90 C, 1.9 Mn, 0.10 V, 0.25 Si, bal Fe.
For punches, forming dies, shear blades, upsetters; oil hardened, non-deforming.

ZAPP VW10.
M-1332; 1.2 C, 0.2 Cr, 0.1 V, 1.0 W, bal Fe.
For forming and blanking dies; water hardened, tough.

ZAPP WC120.
M-1332; 2.1 C, 11.5 Cr, 0.70 W, bal Fe.
For blanking and forming dies, punches; oil hardened, non-deforming.

ZAPP WCN45.
M-1332; 0.40-0.50 C, 0.15-0.30 Si, 0.3-0.5 Mn, 1.2-1.5 Cr, 3.8-4.3 Ni, 0.50 W, bal Fe.
Cold work tool steel.
Werkstoff Nr. 1.2767W.

ZCR 157 (URANUS R3).
M-1488; 0.07 C, 15.0 Cr, 7.0 Ni, 2.2 Mo, 1.15 Al, bal Fe.
Precipitation hardenable.
Stainless: 1460-1760 N/mm^2 UTS max; 7 El.
For mechanical structures in off-shore drilling equipment.
AFNOR Z8CNDA.

ZCR 173.
M-1488; 0.06 max C, 1.0 max Mn, 1.0 max Si, 3.5-5.0 Ni, 15.0-17.0 Cr, 2.3-3.3 Cu, Nb, bal Fe.
Stainless casting, Ht. Tr.: 365-420 Brin.
Good corrosion resistance at elevated temperature.
AFNOR Z4CNV 16.4.

ZCR 174.
M-1488; 0.06 max C, 1.0 max Mn, 1.0 max Si, 3.5-5.0 Ni, 15.0-17.5 Cr, 3.0-4.5 Cu, 0.15 min Nb, bal Fe.
Stainless casting, Ht. Tr.: 365-410 Brin.
Good corrosion resistance at elevated temperature.
AFNOR Z4CNVN6 16.4M.

ZE 10 A see MAGNESIUM ZE 10 A.

ZE41A.
M-Various foundries; 4.2 Zn, 1.2 RE(Ce, etc), 0.7 Zr, bal Mg.
T5-temper: 28,000-30,000 TS; 16,000-20,000 YS; 4-5 El.
At 600°F: 11,000 TS; 8,000 YS; 43 El.
Magnesium sand castings with good pressure tightness and elevated temperature properties.
For parts operating at 350-600°F.
ASTM B80-57T; British RZ5.

ZE41A (STANDARD ALLOY).
M-43; 0.75-1.75 rare earths, 3.5-5.0 Zn, 0.4-1.0 Zr, bal Mg.
Heat treated: 30,000-32,500 TS; 20,000-22,000 YS; 5-3 El; 65-75 Brin.
For structures up to 400°F service; age hardenable, pressure tight.

ZEDABRONZE ZA4.
M-1687; 23 Zn, 5 Al, 1 Ni, 2 Fe, 2.5 Mn, bal Cu.
Zinc-Manganese bronze, cast or wrought; 160-210 Brin.
AFNOR Z23/A4; SAE 430A.

ZEDABRONZE ZA7.
M-1687; 19-25 Zn, 5-5.5 Al, 3 Ni, 4-5 Mn, bal Cu.
Zinc-Manganese bronze, cast or wrought; 200-240 Brin.
SAE 430 B; ASTM B 138; 670.

ZEDABRONZE ZA9.
M-1687; 18-21 Zn, 6.5 Al, 3.5 Fe, 4.5-5.5 Mn, bal Cu.
Zinc-Manganese bronze, cast or wrought; 225-275 Brin.
AFNOR UZ19 A6.

ZELCO.
83 Zn, 15 Al, 2 Cu.
For aluminum solder.

ZELNICKER BABBITTS.
M-950; Pb, Sb, bal Sn.
For bearings; Babbitt.

ZENITE.
M-1472; 2.85-3.15 C, 1.6-1.8 Si, 0.8 Mn, 0.65-0.85 Ni, 0.15-0.35 Cu, max Cr, 0.2-0.4 Mo, bal Fe.
Cast: 45,000 min TS; 210-260 Brin.
For gears, shafts, housings.
High strength alloy cast iron.

ZENITH HIGH SPEED.
0.61 C, 0.12 Mn, 3.39 Cr, 19.2 W, bal Fe.
For tools, high speed cutting tools; high speed steel.

ZENITH SPECIAL.
M-1472; 2.00-2.30 Si, 3.20-3.50 C, 0.12 max S, 0.15 max P, 0.50-0.80 Mn, 0.2-0.4 Ni, 0.15-0.35 Cr, 0.5-0.7 Mo, bal Fe.
Cast: 40,000 min TS; 212-262 Brin.
For gears, shafts, housings.
High strength alloy cast iron.

ZENO.
M-1472; 3.3 C, Si, Mn, alloy, bal Fe.
For gears, shafts, machinery housings; cast iron.

ZEPHYR.
M-349.
Special NEY alloy plate gold.
Neutral color, non-oxidizing.
Fusing temp.: 2000°F.

ZERGAL-X3.
M-Italy; 5.1-6.1 Zn, 2.1-2.9 Mg, 1.2-2.2 Cu, 0.18-0.40 Zr, 0.1-0.3 Mn, 0.2 max Ti, bal Al.
O Temper: 40,000 TS; 24,000 YS; 10 El.
T6 Temper: 80,000 TS; 72,000 YS; 7 El.
For aircraft structures, mobile equipment, hydraulic systems.
Age-hardenable, high strength.

ZERON METAL.
C, 0.8-1.0 Mn, bal Fe.
Heat treated: 70,000-80,000 TS; 55,000-60,000 YS; 16-14 El; 170-180 Brin.
For wrenches, hardware, fittings; heat treated malleable cast iron.

ZETONIA.
M-Ger.; Pb, Sb, bal Sn.
For bearings, bushings; white antifriction.

ZEUS.
M-Ger.; 80 Cu, 20 Ag.
For safety fuses.

ZEUS EV37T.
M-72; 0.12 C, 0.3-0.5 Mn, 0.04 P, 0.035 S, 0.05 Ti, bal Fe.
For electrodes; welding.

ZEUS GV37 T.
M-72; 0.12 C, 0.05 Si, 0.5 Mn, 0.05 Ti, bal Fe.
For electrodes; welding.

ZEVESCAL.
M-1166; C, Cr, Mo, bal Fe.
Cast: 700 Brin.
For gravel mixers, pug mill paddles; abrasion resistant.

ZEVESCAL W.
M-1166; 1.8-2.4 C, 15-17 Cr, 1.5-2.0 Mo, bal Fe.
Annealed: 120,000 TS; 95,000 YS; 2 El; 2 RA; 200 Brin.
For chemical plant equipment; wear and abrasion resistant.

Z65 PLATINUM.
M-1765; 99.95 Pt, 600 ppm Zirconia.
Ann: 186.2 MPa UTS; 40 El; 60 HV.
Higher strength and creep resistance of platinum at higher temperatures.

Z65 RHODIUM-PLATINUM.
M-1765; 89.95 Pt, 10 Rh, 600 ppm Zirconia.
Ann: 344.7 MPa UTS; 30 El; 110 Hv.
Higher strength and creep resistance at elevated temperatures.

Z-GUSS.
M-302; 6 Al, 3 Zn, bal Mg.
33,000-40,000 TS; 3-1 El; 100 Brin.
For light alloys; age-hardenable.

ZH62A.
M-Various foundries; 5.7 Zn, 0.7 Zr, 1.8 Th, bal Mg.
T5-temper: 35,000-40,000 TS; 22,000-25,000 YS; 4-6 El. At 500°F: 14,000 TS; 10,000 YS; 30 El.
Magnesium sand castings with good pressure tightness and freedom from porosity.
For parts operating at 300-500°F.
ASTM B80-69; AMS 4438; QQ-M-56; SAE 508.

ZH62A (STANDARD ALLOY).
M-43; 1.4-2.2 Th, 4.8-6.2 Zn, 0.5 min Zr, bal Mg.
Aged: 43,500 TS; 26,500 YS; 4 El; 75 Brin.
At 500°F: 14,000 TS; 10,000 YS; 30 El.
For airframe castings, gear boxes, engine components; fatigue resistant to 400°F.

ZHS3.
M-USSR; 0.14 C, 15 Cr, 2.0 Ti, 2.0 Al, 5.5 W, 0.02 max B, bal Ni.
Cast nickel-base superalloy for nozzle guide vanes.

ZHS-3.
M-USSR; 0.16 C, 15 Cr, 4 Mo, 5 W, 2 Ti, 2 Al, 0.02 B, bal Ni.
For gas turbine engines, buckets, and discs.
Heat and oxidation resistant.

ZHS-6.
M-USSR; 0.15 C, 12.5 Cr, 4.7 Mo, 7.0 W, 2.6 Ti, 5.0 Al, 0.01 B, bal Ni.
For gas turbine engines, buckets and discs.
Heat and oxidation resistant.

ZHS-6K.
M-USSR; 0.17 C, 11.5 Cr, 5 Co, 4 Mo, 5 W, 2.7 Ti, 5.5 Al, 0.01 B, bal Ni.
For gas turbine engines, buckets and discs.
Heat and oxidation resistant.

ZI-184.
M-U.S.; 0.8-1.0 C, 7-9 Cr, 4-5 W, 1.1-1.5 V, bal Fe.
For tools, dies; oil hardening.

ZICRAL.
M-1792; Al alloy.
For light alloy parts; non-hardenable.

ZICRAL A-28GU.
M-French; Cu, Mg, Zn, bal Al.
For light alloy parts; heat treatable.

ZIERAL.
M-462; 0.6-1.4 Mg, 0.6-1.6 Si, 0.6-1.0 Mn, 0.3 max Cr, bal Al.
Annealed: 21,000 TS; 8000 YS; 24 El.
For window frames, gutters, fan blades, boats; good welding and forming properties.

ZILLOY 15.
M-91; 1.05 Cu, 0.01 Mg, bal Zn.
Hot rolled: 30,000-40,000 TS; 30-15 El; 130 Brin.
Cold drawn: 36,000-47,000 TS; 25-15 El; 150 Brin.
For corrugated roofing, weatherstrips; corrosion resistant.

ZILLOY 25.
M-91.
Zinc base alloy.

ZILLOY 25.
M-91; 0.12-0.16 Ti, 0.50-0.70 Cu, 0.30 max Pb, bal Zn.
29,000-40,000 TS; 14-26 El; 60-76 Rock 15T.
For roofing, gutters, trim, housings, fuses, curtain walls, high creep resistance.

ZILLOY 40 see ZILLOY 45.

ZILLOY 45.
M-91; 0.65-0.85 Cu, 0.10 max Pb, bal Zn.
Hot rolled: 22,000-30,000 TS; 41-21 El; 60 Rock 15 T.
Cold Rolled: 26,000-35,000 TS; 37-25 El; 65 Rock 15 T.
For weather strips, name plates, ferrules, drawn parts; corrosion resistant.

ZIMAL.
M-137; 4 Al, 3 Cu, bal Zn.
Cast-Die: 45,000 TS; 36,000 YS; 1 El; 100 Brin.
For carburetors, gears, corrosion resistant.

ZIMALIUM.
M-Ger.; 3.7-7.1 Mg, 2.8-4.5 Zn, bal Al.

ZINC.
M-1755; Zn.
Purities: Zinc refined 99.9999%, 99.9995%, 99.999%, 99.99%.
Forms: Rod, ingot, shot, powder, sheet, foil, wire, splatters, semicircular bar, single crystals.

ZINC BABBITT.
M-Eng.; 69 Zn, 26 Sn, 5 Cu, 3 Sb.
For bearings; anti-friction.

ZINC DURALUMIN ("E" ALLOY).
M-Eng.; 20 Zn, 2.5 Cu, 0.5 Mg, 0.5 Mn, bal Al.
Heat treated: 91,000 TS.
For strong light alloy parts; non-hardenable.

ZINC-RICH PRIMED.
M-604; low C, Mn, bal Fe.
Prepainted steel sheet.

ZINELL.
M-975; 0.4 C, bal Fe.
For forgings; water hardening.

ZINKALIUM.
M-Eng.; 0.8-8.3 Mg, 0.8-8.3 Zn, bal Al.
For light alloy parts; non-hardenable.

ZINKAL M.
M-1134; 1 max Mn, bal Zn.
Rolled: 21,300-35,500 TS; 25-40 Brin.
For structures.

ZINNAL.
M-204; Al sheet coated on both sides with Sn and rolled so as to form a firmly welded whole.
For instrument cases; corrosion resistant.

ZINNBRONZE SNBZ4.
M-297; 4 Sn, 0.3 P, bal Cu.
Hard: 71,600-92,000 TS; 20-35 El; 140-170 Brin.
Soft: 46,000-60,000 TS; 55-65 El; 70-80 Brin.
For paper and chemical plant equipment; corrosion resistant, P-bronze.

ZINNBRONZE SNBZ6.
M-297; 6 Sn, 0.3 P, bal Cu.
Hard: 71,600-92,000 TS; 20-35 El; 150-175 Brin.
Soft: 50,000-64,000 TS; 50-65 El; 75-85 Brin.
For paper and chemical plant equipment; corrosion resistant, P-bronze.

ZINNBRONZE SNBZ8.
M-297; 8 Sn, 0.2 P, bal Cu.
Hard: 78,000-100,000 TS; 20-35 El; 160-190 Brin.
Soft: 58,000-72,000 TS; 45-60 El; 80-90 Brin.
For paper and chemical plant equipment; corrosion resistant, P-bronze.

ZIP.
M-783; 0.65 C, 4 Cr, 7.5 Co, 1.65 V, 17.5 W, 0.85 Mo, bal Fe.
For lathe, planer and shaper tools; high speed steel.

ZIP.
M-289; 0.75 C, 19 W, 7 Co, 4.5 Cr, 2 V, 1 Mo, bal Fe.
For reamers, broaches, lathe and planer cutters, hobs; Type T5; high speed steel.

ZIPPALLOY.
M-103; 87 Cu, bal Zn.
1/2 H-temper: 54,000 TS; 116 Brin.
Hard temper: 65,000 TS; 137 Brin.
Spring: 75,000 TS; 160 Brin.
For electrical components, hardware, jewelry, watches, lamps; high brass, corrosion resistant.

ZIRCAL.
M-France; 7-8.5 Zn, 2.5 Mg, 1.5 Cu, 0.25 Cr, bal Al.
Rolled: 78,000-92,000 TS; 64,000-71,000 YS.
Wire: 97,000 TS; 92,000 YS; 6 El; 180 Brin.
For light alloy parts; age-hardenable.

ZIRCALLOY-2.
M-1791; 1.5 Sn, 0.12 O, 0.13 Fe, 0.10 Cr, 0.05 Ni, bal Zr.
Extruded and drawn: 65,000 TS; 47,000 YS; 25 El.
Low neutron cross section.
For pressure tubes in nuclear reactors.

ZIRCALOY-2.
M-118; M-101; 1.5 Sn, 0.12 Fe, 0.10 Cr, 0.05 Ni, bal Zr.
At 70°F: 76,000 TS; 49,000 YS; 28 RA; 185 Brin.
At 500°F: 42,000 TS; 24,000 YS; 42 RA.
For nuclear reactors; low neutron absorption.

ZIRCALOY 3.
M-118; 0.05 max C, 0.2-0.3 Sn, 0.2-0.3 Fe, bal Zr.
R.T.: 61,100 TS; 44,200 YS; 36.5 RA; 165 Brin.
At 500°F: 26,400 TS; 16,700 YS; 43.0 RA.
For high temperature applications; good strength to 750°F.

ZIRCALOY-4.
M-1791; 1.5 Sn, 0.12 O, 0.21 Fe, 0.10 Cr, bal Zr.
Extruded and drawn: 65,000 TS; 47,000 YS; 25 El.
Low neutron cross section.
For pressure tubes in nuclear reactors.

ZIRCALOY-4.
M-1235, M-118, M-114; 1.5 Sn, 0.2 Fe, 0.1 Cr, bal Zr.
Bar: 80,000-115,000 TS; 40,000-105,000 YS; 3-20 El; B 82-100 Rock.
For reactors, fuel sheathing, structural components in pressurized water reactors.
Corrosion resisting, high strength.

ZIRCONIUM.
M-1495; 99.5 min Zr.
Resistant to acids. For chemical industries.

ZIRCONIUM.
M-1791; 4 Hf, bal Zr.
Ann: 60,000 TS; 35,000 YS; 20 El.
Corrosion resistant.
Tube for chemical processing.

ZIRCONIUM.
M-1755; Zr.
Purities: zone refined 99.995+%, crystal bar 99.95%, reactor grade 99.6+% commercial grade.
Forms: sponge, rod, bar, powder, wire, plate, sheet, foil, tubing, single crystals.

ZIRCONIUM-2.5 NIOBIUM.
M-1791; 2.6 Nb, 0.11 O, bal Zr.
Extruded and drawn: 100,000 TS; 70,000 YS; 12 El.
Low neutron cross section.
For pressure tubes in nuclear reactors.

ZIRCONIUM-AJR.
M-83; 1.0 Cu, 1.5 Mo, ba Zr.
For nuclear reactors.
Resists CO_2 corrosion at high temperatures.

ZIRCONIUM-ATR.
M-83; 0.5 Cu, 0.5 Mo, bal Zr.
At 600°F: 45,000 TS; 42,000 YS; 24 El; B 84 Rock.
For nuclear reactors, fuel sheathing, reactor structural components.
Resists CO_2 corrosion at high temperatures.

ZIRCONIUM COPPER 150.
M-8; 99.83 Cu, 0.17 Zr.
Ht. Tr.: 60,000 TS; 50,000 YS; 12 El; B 75 Rock.
For commutators, collector rings, canned motor windings, rectifier bases.
Heat treatable, corrosion resistant.
93% electrical conductivity.

ZIRCONIUM-COPPER 992.
M-141; 0.13 Zr, bal Cu.
Annealed: 30,000 TS; 10,000 YS; 45 El; 40 Brin.
Aged: 60,000 TS; 55,000 YS; 15 El; 116 Brin.
For resistance welding electrodes, grid wires, contacts; heat treatable, high conductivity.

ZIRCONIUM-COPPER N-4 (NIPPERT ALLOY N-4).
M-1402; 0.25 Zr, bal Cu.
At 75°F: 53,000 TS; 50,500 YS; 10 El; 54 RA; 115 Brin.
At 550°F: 42,000 TS; 40,500 YS; 8 El; 37 RA; 63 Brin.
For commutators; high strength and conductivity at high temperature.
Same as "AMZIRC".

ZIRCONIUM DISILICIDE.
M-109; 52.7 Zr, 35.5 Si, 8 Al, 1.5 Fe.
For special alloy applications; cermet.

ZIRCONIUM GR. 11.
M-1561; 0.2 Fe + Cr, 0.01 N, 99.5 Min Zr + Hf.
165 max Brin.
For chemical process applications.
Corrosion resistant.

ZIRCONIUM GR. 12.
M-1561; 0.05 Fe + Cr, 0.01 N, 99.5 min Zr + Hf.
150 Brin.
For chemical plant process equipment.
High corrosion resistant.

ZIRCONIUM GR. 21.
M-1609; 0.17 Fe + Cr, 0.007 N, 0.02 Hf, 99.5 min Zr.
Annealed: 65,000 TS; 38,000 YS; 19 El; 31 RA; Rock B 76.
For atomic reactor components, fuel sheathing, reactor structures.
Thermal neutron absorption 0.18-0.20 barns.

ZIRCONIUM GR. 32.
M-1561, M-1609; 1.5 Sn, 0.14 Fe, 0.10 Cr, 0.05 Ni, bal Zr.
Annealed: 70,000 TS; 43,000 YS; 28 El; 150 Brin.
At 1000°F: 18,000 TS; 10,000 YS; 45 El.
Cold Reduced 60%: 110,000 TS; 102,000 YS; 35 RA.
For nuclear reactors, fuel sheathing, boiling and pressurized water reactors.
Resists irradiated water corrosion.

ZIRCONIUM GR. 34.
M-1561; 1.5 Sn, 0.21 Fe, 0.10 Cr, bal Zr.
Annealed: 70,000 TS; 43,000 YS; 28 El.
At 500°F: 35,000 TS; 120 DPH.
For nuclear reactors, fuel sheathing, structural components.
Resists irradiated water corrosion.

ZIRCONIUM STEEL.
M-Eng.; 0.1-0.6 Zr, 0.2-0.6 C, Mn, Si, P, S, bal Fe.
For machinery parts, shafts; water hardened.

ZIRKONAL.
M-Ger.; 0.5 Si, 15 Cu, 8 Mn, bal Al.
Rolled: 170-190 Brin.
For light alloy parts; not hardenable.

ZIRTEN.
M-1539; 0.16 max C, 0.35-0.65 Si, 0.3-1.2 Mn, 0.12 max P, 0.25-0.55 Cu, 0.5 max Ni, 0.4-0.8 Cr, 0.15 max Zr, bal Fe.
Plate: 70,000 min TS; 50,000 min YP; 18 min El.
For railroad and mine cars, bridges, booms, chutes.
Corrosion resisting to atmosphere.
High strength low alloy steel plates.

ZIRTUNG.
M-856; W-Zr.
For welding rod; for inert gas shielded arc welding.

ZISIUM.
M-Eng.; 82-83 Al, 1-3 Cu, 15 Zn, 0-1 Sn.
For strong light alloy parts; non-hardenable.

ZISKON.
60 Al, 40 Zn.
For light alloy parts; non-hardenable.

ZISKON.
25-33 Al, 75-67 Zn.
For castings; non-hardenable.

ZIVAN 45.
M-289; 0.45 C, 0.75 Mn, 0.95 Cr, 0.16 V, bal Fe.
For machinery parts; oil hardening.

ZIVCO.
M-289; 0.95-1.05 C, 0.5 Mn, 0.20 Si, bal Fe.
For sledges, picks, drills, taps; Type W1; water hardened.

ZIVCO VANADIUM.
M-289; 1.0-1.1 C, 0.15-0.25 V, bal Fe.
AISI Type W2 water hardening tool steel.

ZIVS COBALT.
M-289; 0.7 C, 19 W, 4.5 Cr, 2 V, 5 Co, bal Fe.
For broaches, reamers, gear cutters, hobs, taps; Type T4; high speed steel.

ZIV EXTRA.
M-289; 1.1 C, 0.3 Mn, 0.2 Si, 0.2 V, bal Fe.
For forming dies, shear blades, reamers, taps; water hardened; Type W2.

ZIVS REGULAR.
M-289; 0.95 C, 0.20 Si, 0.25 Mn, bal Fe.
For cold chisels, shear knives, punches, drills; water hardened; Type W1.

ZIV'S SOLID DRILL.
M-289; C, bal Fe.
For mining drills; water hardened.

ZIV SPECIAL.
M-289; 1.1 C, 0.3 Mn, 0.2 V, 0.2 Cr, bal Fe.
For tools, taps, reamers, shear blades; water hardened.

ZIV SUPER.
M-289; 0.7 C, 18 W, 4 Cr, 1 V, bal Fe.
Hardened: C 64-66 Rock.
For drills, hobs, reamers, broaches, taps, lathe and planer tools.
Type T1 high-speed steel.

ZK-NICKEL.
M-303; Mn, Cu, bal Ni.
For spark plug electrodes.

ZK 51.
M-France; 4.5 Zn, 0.7 Zr, bal Mg.
Cast: 34,000-38,000 TS; 23,000-26,000 YS; 6-8 El.
AMS 4443; SAE 509.

ZK51A.
M-Various foundries; 4.5 Zn, 0.7 Zr, bal Mg.
T5-Temper: 34,000-40,000 TS; 20,000-24,000 YS; 5-8 El.
At 600°F: 8,000 TS; 6,000 YS; 16 El.
Magnesium sand castings with good corrosion resistance; recommended for simple, highly stressed parts of uniform cross section.
ASTM B80-69; AMS 4443; SAE 509; QQ-M-56.

ZK51A (STANDARD ALLOY).
M-43; 3.6-5.5 Zn, 0.55 min Zr, 0.15 max Mn, bal Mg.
Cast: 33,000 TS; 19,000 YS; 11 El; 60 Brin.
For airframe structures, landing wheels, high yield strength casting alloy.

ZK60A.
M-43; 5.7 Zn, 0.5 Zr, bal Mg.
F-temper: 43,000-49,000 TS; 31,000-38,000 YS; 5-14 El; 75 Brin.
T5-temper: 45,000-53,000 TS; 36,000-44,000 YS; 4-11 El; 82 Brin.
Extrusions for highly stressed parts, primarily in aircraft and military; also used as forgings.
ASTM B107-69, B91-68; AMS 4352, 4362.

ZK61A.
M-43, others; 6.0 Zn, 0.8 Zr, bal Mg.
Casting alloy.

Z-METAL.
M-209, M-400; 2.0-2.6 T.C, 0.3-0.8 CC, 0.9-1.1 Si, 0.75-1.25 Mn, bal Fe.
Cast: 75,000-85,000 TS; 50,000-55,000 YS; 18-12 El; 155-225 Brin.
For gears, sprockets, wrenches, bolts, pipe clamps, tool bits, brake drums, air drills; a malleable cast iron; resistance to wear and corrosion; spheroidized.

Z-METAL.
M-904, M-775; 2.0-2.6 C, 1.1 Si, 0.7-1.2 Mn, bal Fe.
Cast: 70,000-90,000 TS; 40,000-60,000 YS; 8-18 El; 155-225 Brin.
For castings, camshafts, gears, nozzles; spheroidized cast iron.

Z METAL M-10.
M-275; 2.2 T.C., 1.0 Si, 1.0 Mn, bal Fe.
Heat treated: 65,000-85,000 TS; 45,000-55,000 YS; 10-18 El; 163-212 Brin.
For camshafts, axles, gears, machinery housings; pearlitic malleable iron.

Z-METAL M-15.
M-275; 2.2 T.C., 1.0 Si, 0.75-1.0 Mn, bal Fe.
Heat treated: 75,000-95,000 TS; 50,000-60,000 YS; 8-15 El; 170-223 Brin.
For camshafts, axles, gears, machinery parts, housings; pearlitic malleable iron.

Z-METAL M-16.
M-275; 2.2 T.C., 1.0 Si, 0.75-1.0 Mn, bal Fe.
Heat treated: 85,000-100,000 TS; 55,000-65,000 YS; 12-16 El; 201-235 Brin.
For gears, axles, housings, camshafts; pearlitic malleable iron.

Z-METAL M-17.
M-775; 2.0-2.6 T.C., 0.3-0.8 C, C., 0.9-1.1 Si, 0.75-1.2 Mn, bal Fe.
Heat treated: 80,000-100,000 TS; 60,000-75,000 YS; 6-14 El; 197-241 Brin.
For gears, crankshafts, axle housings, hammer heads; pearlitic malleable iron.

Z-METAL M-50.
M-775; 2.0-2.6 T.C., 0.3-0.8 C.C., 0.9-1.1 Si, 0.75-1.2 Mn, bal Fe.
Heat treated: 70,000-90,000 TS; 45,000-60,000 YS; 3-9 El; 179-223 Brin.
For gears, crankshafts, axle housings, shears; pearlitic malleable iron.

Z-METAL P7.
M-275; 3.0 C, 2.0 Si, Mn, bal Fe.
Cast: 100,000 TS; 85,000 YS; 3 El; 227 Brin.
For castings, housings; pearlitic, malleable iron.

ZN AL-1 ALLOY.
M-Ger.; 0.7-0.9 Al, 0.35-0.5 Cu, bal Zn.
For conductors for power transmission; high conductivity.

ZN-FE ALLOY.
M-Ger.; 0.13 Fe, bal Zn.
Rolled: 23,000-25,000 TS; 25-40 Brin.
For conductors for power transmission.

ZOLLNER Z132.
11-13 Si, 0.5-2.75 Cu, 0.7-1.3 Mg, 0.15 max Ti, 1.3 max Fe, bal Al.
Cast: 31,000 min TS; 90-120 Brin.
For pistons; age hardenable.

ZORITE.
M-184; 35 Ni, 17-15 Cr, 1.75 Mn, 0.5 C, 1.0 Si, bal Fe.
Cast: 65,000-55,000 TS; 45,000-40,000 YS; 2-5 El; 4-6 RA; 190-180 Brin.
For furnaces, heaters, electrical resistances; heat and corrosion resisting; max. working temperature 1900°F.

ZRE 1.
M-Eng.; 0.8-3.0 Zn, 0.4-1.0 Zr, 2.5-4.0 R. E. Metals, bal Mg.
Sand cast and precipation treated.
140 MPa TS; 95 MPa YS; 3 El; 50 Brin.
BS 2970 MAG 6.

ZT 1.
M-Eng.; 1.7-2.5 Zn, 0.4-1.0 Zr, 2.5-4.0 Th, bal Mg.
Sand cast and precipitation treated: 185 MPa TS; 85 MPa YS; 5 El; 50 Brin.
BS 2970 MAG 8.

Z.T.M.
M-73; 0.35 C, 0.50 Si, 0.8 Mn, 0.80 Cr, 0.32 Mo, bal Fe.
For die casting dies; hot work steel.

ZUNIT 13A.
M-1332; 0.12 max C, 1.2 Si, 1.0 Al, 13 Cr, bal Fe.
Annealed: 75,000 TS; 40,000 YS; 35 El; 70 RA; 155 Brin.
For oil refinery equipment, valves; creep and heat resistant.

ZUNIT 17 see ZAPP GUSS 1.4740.

ZUNIT 18A.
M-1332; 0.12 max C, 1 Si, 1 Al, 18 Cr, bal Fe.
Annealed: 80,000 TS; 50,000 YS; 25 El; 50 RA; 150 Brin.
For valves, oil refinery equipment; heat and creep resistant.

ZUNIT 24A.
M-1332; 0.12 max C, 1.5 Al, 24 Cr, bal Fe.
Annealed: 85,000 TS; 50,000 YS; 30 El; 55 RA; 180 Brin.
For oil refinery equipment, valves, pumps; creep and heat resistant.

ZUNIT 25N.
M-1332; 0.2 C, 1.2 Si, 25 Cr, 4 Ni, bal Fe.
Cast: 90,000 TS; 65,000 YS; 2 El; 212 Brin.
Heat treated: 97,000 TS; 65,000 YS; 18 El; 210 Brin.
For cylinder liners, bushings, valve seats and bodies; corrosion and heat resistant; Type CC-20.

ZUNIT 28 see ZAPP GUSS 1.4776.

ZUNIT 28K see ZAPP GUSS 1.4777.

ZUNIT 28N see ZAPP GUSS 1.4873.

ZUNIT-GUSS 17.
M-1332; 0.5 C, 1.5 Si, 17 Cr, bal Fe.
For furnace parts, heat treating boxes, retorts; corrosion and heat resistant.

ZUNIT-GUSS 28.
M-1332; 0.6 C, 1.5 Si, 29 Cr, bal Fe.
For grates, baffles, support skids, furnace parts; heat resistant.

ZUNIT-GUSS 28K.
M-1332; 1.3 C, 1.5 Si, 29 Cr, bal Fe.
For dies, crusher parts; heat resistant.

ZUNIT-GUSS 28N.
M-1332; 0.4 C, 1.3 Si, 27 Cr, 4 Ni, bal Fe.
Cast: 70,000 TS; 65,000 YS; 2 El; 190 Brin.
Aged: 115,000 TS; 80,000 YS; 18 El.
For furnace parts, salt pots, soot blowers, grate bars; Type HC; heat resistant.

ZUNIT-GUSS N10.
M-1332; 0.4 C, 1.3 Si, 27 Cr, 4 Ni, bal Fe.
Cast: 70,000 TS; 65,000 YS; 2 El; 190 Brin.
Aged: 115,000 TS; 80,000 YS; 18 El.
For furnace parts, salt pots, soot blowers, grate bars; Type HC; heat resistant.

ZUNIT-GUSS N11.
M-1332; 0.4 C, 2 Si, 28 Cr, 11 Ni, bal Fe.
Cast: 70,000 TS; 65,000 YS; 2 El; 190 Brin.
Aged: 115,000 TS; 80,000 YS; 15 El.
For furnace parts, salt pots, baffles, grids; Type HE; corrosion and heat resistant.

ZUNIT-GUSS N15.
M-1332; 0.4 C, 2 Si, 26 Cr, 14 Ni, bal Fe.
Cast: 75,000 TS; 47,000 YS; 17 El; 25 RA; 200 Brin.
For heat treating boxes, furnace parts, fixtures, grates; Type HH; corrosion and heat resistant.

ZUNIT-GUSS N20.
M-1332; 0.4 C, 2 Si, 26 Cr, 4 Ni, bal Fe.
Cast: 70,000 TS; 65,000 YS; 2 El; 190 Brin.
Aged: 115,000 TS; 80,000 YS; 15 El.
For furnace parts, salt pots, baffles, blowers; Type HC; corrosion and heat resistant.

ZUNIT-GUSS N30.
M-1332; 0.5 C, 1.8 Si, 25 Cr, 30 Ni, bal Fe.
For furnace parts, heat treating boxes, retorts; corrosion and heat resistant.

ZUNIT N8.
M-1332; 0.15 max C, 1.0 max Si, 20 max Mn, 17.0-19.0 Cr, 9.0-11.0 Ni 0.7 max Ti, bal Fe.
Austenitic stainless; heat resisting.
W.-Nr. 1.4878.

ZUNIT N10 see ZAPP GUSS 1.4826.

ZUNIT N10.
M-1332; 0.15 C, 19.5 Cr, 9.5 Ni, 2 Si, bal Fe.
Cast: 85,000 TS; 45,000 YS; 35 El; 165 Brin.
For heat treating boxes, baskets, burner tips, conveyors; Type HF; corrosion and heat resistant.

ZUNIT N11 see ZAPP GUSS 1.4837.

ZUNIT N15 see ZAPP GUSS 1.4846.

ZUNIT N20 see ZAPP GUSS 1.4848.

ZUNIT N20.
M-1332; 0.15 C, 2 Si, 24 Cr, 19 Ni, bal Fe.
Cast: 75,000 TS; 50,000 YS; 17 El; 170 Brin.
Aged: 85,000 TS; 50,000 YS; 10 El; 190 Brin.
For furnace parts, retorts, rabble arms, dampers; Type HK; corrosion and heat resistant.

ZUNIT N30 see ZAPP GUSS 1.4865.

ZUNIT ST.
M-1332; 0.10 max C, 1.0 Si, 1.0 max Mn, Ti = 5 x C, bal Fe.
Ferritic low carbon rolling and forging steel; heat resisting.
W.-Nr. 1.5310.

ZURCO GALVACOTE.
M-290; 0.2 C, bal Fe.
For structures; galvanized sheets.

ZURCO NO. 18.
M-290; 16-18 Cr, 0.1 C, bal Fe.
Annealed: 80,000 TS; 55,000 YS.
For non-corrosive parts; rustless steel; for continuous heat up to 1200°F.

ZURCO NO. 25.
M-290; 25-30 Cr, 0.1 C, bal Fe.
For non-corrosive parts; rustless steel; for continuous heat up to 2200°F.

ZURCO RUSTLESS IRON.
M-290; 16-18 Cr, C, bal Fe.
For non-corrosive parts; rustless.

ZURCO RUSTLESS STEEL.
M-290; 0.10 C, 12-14 Cr, bal Fe.
Annealed: 75,000 TS; 44,000 YS.
For corrosion resisting parts; rustless, max operating temperature 1000°F.

ZURCO SILVATEX.
M-290; 0.2 C, bal Fe.
For structures; coated sheets.

ZURCO STEEL "A".
M-290; 0.08-0.12 C, bal Fe.
For cold rolled or annealed pickled sheets; deep drawing.

ZURCO STEEL "AAA".
M-290; 0.08-0.12 C, bal Fe.
For annealed pickled sheets for Ni and Cr plating; extra deep drawing.

ZURCO STEEL "B".
M-290; 0.08-0.12 C, bal Fe.
For furniture sheets, filing cabinets, lockers; deep drawing.

ZURCO STEEL "C".
M-290; 0.08-0.12 C, bal Fe.
For sheets; deep drawing.

ZYKLON 5.
M-1340; 0.30 C, 2.35 Cr, 0.6 V, 4.25 W, bal Fe.
For upsetters, extrusion press parts; oil hardened, hot work steel.

ZYKLON 5SN.
M-1340; 0.30 C, 1.1 Cr, 0.18 V, 3.75 W, bal Fe.
For upsetters, extrusion press parts, rams; hot work steel, oil hardened.

ZYKLON 10.
M-1340; 0.30 C, 2.65 Cr, 0.35 V, 8.5 W, bal Fe.
For extrusion press parts, rams, and liners; hot work steel, oil hardened.

ZYKLON 12C.
M-1340; 0.3 C, 2 Co, 2.4 Cr, 0.25 V, 8.5 W, bal Fe.
For extrusion press parts, hot punches and shears; oil hardened, hot work steel.

ZYKLON 122.
M-1340; 0.35 C, 1.05 Cr, 0.18 V, 1.85 W, bal Fe.
For punches, shears, upsetters, cold headers; oil hardened, hot work steel.

ZYKLON CMV211.
M-1340; 0.40 C, Cr, Mn, Mo, bal Fe.
For gears, bolts, crankshafts; oil hardened.

ZYKLON CSM155.
M-1340; 0.67 C, 1.3 Si, 0.5 Mn, 0.5 Cr, bal Fe.
For springs; oil hardened.

ZYKLON MNV15.
M-1340; 0.55 C, 0.7 Cr, 0.18 Mo, 1.65 Ni, 0.1 V, bal Fe.
For gears, pinions, bolts, studs, crankshafts; oil hardened, shock resistant.

ZYKLON MNU35.
M-1340; 0.56 C, 0.7 Cr, 0.18 Mo, 1.65 Ni, 0.1 V, bal Fe.
For gears, pinions, bolts, studs, crankshafts; oil hardened, shock resistant.

SECTION II

ALPHABETICAL LIST OF MANUFACTURERS

A

M-1099	ABEX Corp., Mahwah, NJ 07430 (formerly American Brake Shoe Co.).
M-1794	ABEX Corp., Engineered Products Div., Meadville Plant, Meadville, PA 16335.
M-1191	Accurate Brass Co., Bristol, CT 06010.
M-608	Achorn Steel Co., Cambridge, MA (3).
M-281	Acieral Co. of America, New York, NY (3).
M-1043	Aciéries & Forges de Firminy, Paris, France (4) (now part of Creusot-Loire, M-1448).
M-1116	Aciéries de Champagnole, La Courneuve, France.
M-1487	Aciéries du Forez, St-Etienne (Loire), France.
M-1645	Aciéries et Forges d'Anor, Paris 11e, France.
M-455	Aciéries S. A. Bedel, F-42 Saint Etienne, France (3).
M-285	Acipco Steel Products, Div. of American Cast Iron Pipe Co., Birmingham, AL 35202.
M-694	Ackerlind Steel Co., Inc., Long Island City, NY 11104.
M-1022	Acme Aluminum Alloys, Inc., Dayton, OH (3).
M-1102	Acme Electric Welder Co., Los Angeles, CA 90058.
M-654	Acme Foundry & Machine Co., Coffeyville, KS 67337.
M-1142	Acme Foundry Co., Detroit, MI (3).
M-191	Acme-Newport Steel Co. (2) (part of Interlake, Inc.).
M-568	Acme Steel Co., Chicago, IL (2) (part of Interlake, Inc.).
M-209	Acme Steel & Malleable Iron Works, Div. of Buffalo Brake Beam Co., Buffalo, NY 14207.
M-1733	Aços Villares, São Paulo, Brazil.
M-1611	Adalet Mfg. Co., Cleveland, OH (2).
M-1086	Adamas Carbide Corp., Kenilworth, NJ 07033.
M-1756	Adams Alloy Rod Div. Adams Hardfacing Co., Wakita, OK 73771.
M-656	Adams & Co., J. D., Indianapolis, IN (2).
M-350	Adams & Osgood Steel Co., Boston, MA (3).
M-655	Adirondack Steel Casting Co., Watervliet, NY 12189.
M-1462	Advance Aluminum Casting Corp., Chicago, IL (3).
M-898	Advance Foundry Co., Dayton, OH 45401.
M-1746	Advanced Material Div., Armco Steel Corp., Baltimore, MD 21204.
M-1241	Aetna Standard Engineering Co., Ellwood City, PA 16117.
M-732	A.F.C., Société des, Paris, France.
M-1302	AFY, Société des, Paris, France.
M-783	Agawam Tool Co., Springfield, MA (3).
M-1175	Agil Chemie, Berlin, Germany.
M-661	Agile Div. Nagle/Sybron Corp., Rochester, NY (3).
M-1695	Aiken Industries, Inc., Carbide Division, Irwin, PA 15642.
M-663	Airco Vacuum Metals, Berkeley, CA 94710.
M-663	Airco Welding Products, New Providence, NJ 07974.
M-123	Ajax Metal Div. H. Kramer & Co. (4).
M-1673	Ajax Steel & Forge Co.
M-96a	Ajax Tool Sales, New York, NY (3).
M-1058	Akron Bronze & Aluminum Inc., Akron, OH 44309.
M-404	Aktiebolaget Skandinaviska Armaturfabriken, Stockholm, Sweden.
M-882	Aktiebolaget Svenska Metallverken, Stockholm, Sweden.
M-666	Aladdin Welding Products Inc., Grand Rapids, MI 49507.
M-678	Alais, Fonderies et Forges, Compagnie des Mines, Temaris, France.
M-620	Alais, Forges et Camargue, Paris, France.
M-559	Alamo Iron Works, San Antonio, TX 78237 (2).
M-537	Alan Wood Steel Co., Conshohocken, PA 19428 (2).
M-1272	Alar, Ltd., London, England (3).
M-1628	Albright & Wilson Ltd., London, England (2) (3).
M-1276	Alcaloy Inc., Cleveland, OH (3).
M-1625	Alcan Aluminium Corp., see Aluminium Goods.
M-643	Alcan Booth Industries Ltd., Alcan House, London, W1X 6DP England.
M-1775	Alcan Metal Powders, Div. of Alcan Aluminium Corp., Elizabeth, NJ 07207.

(1) Company reported to be out of business.
(2) Does not produce trade name alloys at present; listed for information only.
(3) Old listing; company name and/or address may be inaccurate.
(4) Merged, for information contact new corporate name.

M-1715	Alcoa of Great Britain, Droitwich, Worcestershire, WR9 7BG England.	M-1073	Alpha Metals Inc., Jersey City, NJ 07304.
M-1378	Alco Products, Schenectady, NY (3).	M-1314	Alpine Montan Co. (4) (merged into Vereinigte Edelstahlwerke, M-1802).
M-595	Alemite Div. Stewart Warner Corp., Chicago, IL 60614.	M-672	Alsia, Société de, Paris, France.
M-658	Alfa Romeo, Milan, Italy.	M-1779	AL Tech Specialty Steel Corp., Dunkirk, NY 14048.
M-1253	Algoma Steel Corp. Ltd., Sault Ste. Marie, Ontario, Canada.	M-1040	Alten Foundry & Machine Works, Lancaster, OH 43130.
M-124	Allan & Son, A., Carlstad, NJ 07072.	M-1216	Alter Co. (2) (alloys now produced by Alloy Metal Products, M-1795).
M-1179	Allard, Soc. Anon. Usines et Aciéries, Mont-sur-Marchienne, Belgium.	M-622	Aluminium Belge, S.A., Luttich, Belgium.
M-1	Allegheny Ludlum Steel Corp., Pittsburgh, PA 15222.	M-871	Aluminium Français, Paris 8, France.
M-210	Edgar Allen Balfour Steels Ltd., Sheffield, S9 1QY England.	M-1354	Aluminium G.m.b.H., Rheinfelden/ Baden, West Germany.
M-242	Allen Co., L. B., Shiller Park, IL 60176.	M-1625	Aluminium Goods, Div. of Alcan Canada Products Ltd., Toronto 154 Canada.
M-688	Allen Mfg. Co., Hartford, CT 06101.		
M-1249	Allen-Sherman-Hoff Pump Co., Wynnewood, PA (3).	M-249	Aluminium Industrie Aktiengesellschaft, Lausanne-Ouchy, Switzerland.
M-210a	Allen Steel Co., Edgar, New York, NY.		
M-621	Allgemeine Elektrizitäts-Gesellschaft, Berlin 1000 West Germany.	M-1634	Aluminium Industries, AG., Chippis, Switzerland.
M-2	Allgemeines Deutsches Metallwerk, G.m.b.H., Berlin, Germany.	M-1784	l'Aluminium Pechiney, Soc. de Vente de la, 75008 Paris, France.
M-1761	Alliance Wall Corp., Alliance, OH 44601.	M-1268	Aluminium, T. I., Ltd. (4) (now part of British Aluminium, M-28).
M-125	Allied Die Casting Corp., New York, NY (3).	M-644	Aluminium Union, Ltd., London, England (3).
M-953	Allied Process Corp., New York, NY (1).	M-493	Aluminium Walzwerke Singen G.m.b.H., 7700 Singen/Hohentweil, West Germany.
M-1227	Allied Products Corp., Chicago, IL 60650.		
M-1438	Allied Steel & Chemical Co., New York, NY (3).	M-1140	Aluminiumwerke A.G., Rorschach, Switzerland.
M-642	Allis-Chalmers Mfg. Co., Milwaukee, WI 53201.	M-302	Aluminiumwerke Maulbronn, Maulbronn, Württ, Germany.
M-1743	Allmetal Screw Products Co., Inc., Garden City, NY 11530.	M-555	Aluminiumwerke Nürnberg, G.m.b.H., Nürnberg, West Germany.
M-3	Alloy Cast Steel Co., Marion, OH 43302 (2).	M-982	Aluminium Wire & Cable Co., Port Tennant Works, Swansea SA1 8PS Glamorgan, England.
M-1005	Alloy Engineering & Casting Co., Champaign, IL 61820.		
M-775	Alloy Foundries Div. of the Eastern Co., Naugatuck, CT 06770.	M-1548	Aluminium Zentral, 4000 Düsseldorf 1, Germany (2) (this is an association, no alloys, information only).
M-1795	Alloy Metal Products, Inc., Davenport, IA 52808 (took over Alter Co. Alloys).	M-4	Aluminum Co. of America, Pittsburgh, PA 15219.
M-1734	Alloy Metals Inc., Troy, MI 48084.		
M-248	Alloy Metal Wire Works (now H. K. Porter Co., Inc., Prospect Park, PA 19076).	M-219	Aluminum Co. of Canada, Ltd., Montreal, Canada.
		M-6	Aluminum Industries Inc., Chicago, IL (3).
M-692	Alloy Rods Co., see Chemetron Corp., M-1713.	M-1233	Aluminum Smelting & Refining Co., Maple Heights, OH 44137.
M-1713	Alloy Rods Co., see Chemetron Corp., M-1713.	M-1279	Aluminum Smelting & Refining Co., Maple Heights, OH 44137.
M-126	Alloys & Products, Inc., New York, NY (3).	M-1158	Aluminum Solder Corp. (ALSOCO), New York, NY (3).
M-1265	Alloys & Products, Inc., New York, NY (3).	M-750	Amalgamated Steel Div. Allied Steel and Tractor Products Inc., Solon, OH 44139.
M-541	Alluminio S.A., Milan, Italy.		
M-1490	Allvac Metals (now Teledyne Allvac, Monroe, NC 28110).		
M-464	Allyne-Ryan Foundry Co., Cleveland OH (3).	M-1778	AMAX Copper Div. AMAX Corp., New York, NY 10017.
M-1023	ALOYCO Inc. (now Walworth Co., Linden, NJ 07036).	M-558	AMAX Lead & Zinc, Clayton, MO 63105.

Section II: Alphabetical List of Manufacturers

M-1708	AMAX Metal Powders (alloys transferred to PYRON Corp., M-1777).
M-151	AMAX Molybdenum Div. AMAX Corp., Greenwich, CT 06830 (was Climax Molybdenum Div.).
M-127	Ambolt Machine Tool Co., New York, NY (3).
M-1315	Ambo-Stahl-Gesellschaft, 5000 Koln 1, West Germany.
M-250	American Abrasive Metals Co., Irvington, NJ 07111.
M-1638	American Alloys Corp., Kansas City, MO 64120.
M-1456	American Art Alloys Inc., Kokomo, IN (3).
M-761	American Art Metals Inc., New York, NY (3).
M-7	American Boron Products Div. Continental Copper & Steel Industries, Buffalo, NY.
M-8	American Brass Co. (now known as Anaconda Co., Brass Div., Waterbury, CT 06720).
M-1483	American Brazing Alloys Co., Pelham, NY (3).
M-877	American Bridge Co. (4), see U.S. Steel Corp., M-604.
M-1564	American Carbide Corp., Union City, NJ (3).
M-285	American Cast Iron Pipe Co. (now ACIPCO Steel Products), Birmingham, AL 35202.
M-653	American Chain & Cable Co. Inc. (2).
M-1122	American Clad Metals Inc., Pawtucket, RI 02862.
M-956	American Crucible Products Co., Lorain, OH 44052.
M-276	American Cutting Alloys Inc., New York, NY (3).
M-1197	American Electro Metals Corp., Yonkers, NY (3).
M-1200	American Hoist & Derrick Co., St. Paul, MN 55107.
M-128	American Injector Co., Detroit, MI (3).
M-1639	American Light Alloys Co., Little Falls, NJ (3).
M-524	American Machine & Foundry (AMF Inc.), White Plains, NY 10604.
M-313	American Magnesium Co., Cleveland, OH (1).
M-129	American Manganese Bronze Co., Philadelphia, PA (3).
M-197	American Manganese Mfg. Co., Dunbar, PA (1).
M-558	American Metal Climax, Div. of AMAX Corp., Greenwich, CT 06830.
M-1508	American Metal Climax, Div. of AMAX Corp., Greenwich, CT 06830.
M-1213	American Metallurgical Products Co., Pittsburgh, PA 15239.
M-1440	American Meter Co., Reliance Foundry, Pittsburgh, PA (3).
M-538	American Nickeloid Co., Peru, IL 61354.
M-1153	American Non-Gran Bronze Co., Berwyn, PA (3).
M-686	American Platinum & Silver Div. Engelhard Minerals & Chemicals Corp., Carteret, NJ 07008.
M-1550	American Potash & Chemical Co., see Kerr-McGee Chemical Corp., M-1550.
M-684	American Radiator Co., New York, NY (2).
M-39	American Saw & Mfg. Co. (2).
M-11	American Sheet & Tin Plate Div. (4), see U.S Steel Corp., M-604.
M-1595	American Silver Co., Flushing, NY.
M-314	American Smelting & Refining Co. (ASARCO), New York, NY 10005.
M-1088	American Solder Co., Philadelphia, PA (3).
M-293	American Stainless Steel Co., Pittsburgh, PA (3).
M-12	American Steel & Wire Div. (4), see U.S Steel Corp., M-604.
M-551	American Steel Co., Ellwood City, PA (2).
M-991	American Steel Foundries, Chicago, IL 60601 (2).
M-1057	American Tank & Fabricating Co., Cleveland, OH 44111.
M-252	Amesbury Brass & Foundry Co., Amesbury, MA (3).
M-524	AMF Inc., White Plains, NY 10604.
M-13	Ampco Metal Div. Ampco-Pittsburgh Corp., Milwaukee, WI 53201.
M-211	Amplex Div. Chrysler Corp., Detroit, MI 48231.
M-9	AMSCO Div. ABEX Corp., Chicago Heights, IL 60411.
M-8	Anaconda Co., Brass Division, Waterbury, CT 06720.
M-1468	Anchor Alloys Inc., Brooklyn, NY 11222.
M-342	Anchor Drawn Steel Co. (4), (see Teledyne Vasco, M-115).
M-489	Andrews, Ltd., Sheffield, S4 7WZ England.
M-641	Andrews & Co., Ltd., Thomas, Royds Works, Sheffield, S4 7WZ England.
M-352	Andrews Toledo Ltd., London, England (1).
M-606	Ansoldo, Società Anonima, Rome, Italy.
M-639	Antaciron, Inc., Wellsville, NY (3).
M-579	A. O. Smith Co., Milwaukee, WI 53201 (2).
M-1716	A. O. Smith-Inland Inc., Milwaukee, WI 53201 (1).
M-564	Apex Bronze Foundry Co., Oakland, CA (3).
M-130	Apex International Alloys Inc., Des Plaines, IL 60018 (formerly Apex Smelting).
M-485	Apex Steel Co., Sheffield, England (1).
M-687	Apollo Metals Inc., Chicago, IL 60638.
M-131	Apollo Steel Co., Apollo, PA (3).
M-1075	Apothecaries Hall Co., Waterbury, CT (3).

M-448b	Appleby-Frodingham Steel Co. (4) (now part of British Steel Corp., M-1724).	M-1755	Atomergic Chemicals Co., Plainview, NY 11803.
M-1763	ARBED, (see TradeARBED).	M-1769	Atrax Cemented Carbides, Wallace Murray Corp., McKeesport, PA 15134.
M-14	Arcade Malleable Iron Co., Worcester, MA (3).	M-1115	Aubert et Duval, 92202 Neuilly-sur-Seine, France.
M-677	Arcos Corp., Philadelphia, PA 19143.	M-936	Audubon Metalwove Belt Corp., Philadelphia, PA 19134 (3).
M-1074	ARC Products Div. Chemetron Corp. (formerly ALL-STATE Welding Alloys), see Chemetron Corp. M-1713.	M-603	Aurora Metal Co., Montgomery, IL 60538.
M-581	Ardal, Ltd., Worle, Weston-Super-Mare, England (3).	M-937	Austenal Div. Howmet Corp., Dover, NJ 07801.
M-563	Ariston Metal Co., Jersey City, NJ (2) (3).	M-1152	Auto Specialties Mfg. Co., St. Joseph, MI 49085.
M-1807	Armco Ltd., Letchworth, Herts, SG6 ING England (same alloys as Armco, M-10).	M-1653	AVEC S.P.A Acciaieria Valbruna di Ernesto Gresele, 36100 Vicenza, Italy.
M-10	Armco Steel Corp., Middletown, OH 45042.	M-16	Avesta Jernverks Aktiebolag, Avesta, Sweden.
M-132	Armstrong Bros. Tool Co., Chicago, IL 60646.	M-703	Axelson Mfg. Co., Los Angeles, CA (3).
M-670	Arnold Engineering Co., Marengo, IL 60152.		**B**
M-1012	Arpocalloy Co., Kansas City, MO 64139.	M-447	Babcock & Wilcox Co., Power Generation Group, Barberton, OH 44203.
M-1662	Arwood Corp., Rockleigh, NJ 07647 (2).		
M-314	ASARCO and Federated Metals Corp., Div. of American Smelting & Refining Co., New York, NY 10005.	M-17	Babcock & Wilcox, Tubular Products Div., Beaver Falls, PA 15010.
M-815	ASARCO and Federated Metals Corp., Div. of American Smelting & Refining Co., New York, NY 10005.	M-704	Bachite Development Corp. (3).
		M-1535	Badell Co., Inc., Hammond, IN 46327.
		M-134	Baker, Perkins & Co. Ltd., Peterborough, England (3).
M-87	Ashby, Ltd., Morris, London, England (3).	M-309	Baldwin-Lima-Hamilton Corp., Eddystone Div., Philadelphia, PA.
M-1256	Ashton Ltd., N.C., Huddersfield, HD1 6RE England.	M-1510	Baldwin Steel Co., Jersey City, NJ 07305.
M-1637	Associated Electrical Industries Ltd., Manchester, England.	M-814	Baldwins Ltd., London, England (now part of British Steel Corp., M-1724).
M-15	Associated Steel Corp., Cleveland, OH 44128.	M-1112	Balfour & Co., Ltd., Arthur, Sheffield, England (part of Balfour-Darwins Ltd., Sheffield, England).
M-674	Ateliers de la Gironde, Paris, France.		
M-1470	Athenia Steel Div. National Standard Co., Clifton, NJ (2).	M-1725	Balfour-Darwins Ltd., Sheffield, England (now part of Edgar Allen Balfour Steels Ltd., Sheffield, S9 1QY, England, M-210).
M-637	Athens Foundry Div. Ingersoll-Rand Co., Athens, PA 18810.		
M-1062	Atkins & Co., E.C., Indianapolis, IN (3).		
M-492	Atkinson & Sons Ltd., E., Sheffield, S18 6NS England.	M-412	Ballard & Co., F.J., Tipton, England (2).
		M-19	Barber Asphalt Co., Philadelphia, PA (2).
M-567	Atkinson Co., Rochester, NY (3).		
M-695	Atlantic Casting & Engineering Co., Clifton, NJ 07012.	M-705	Barium Stainless Steel Corp., Canton, OH (3).
M-569	Atlantic Steel Casting Co., Chester, PA 19016.	M-135	Barker & Allen Ltd., Birmingham, England.
M-353	Atlantic Steel Corp., Astoria, NY 11106.	M-20	Barronia Metals, Ltd., London, England.
M-591	Atlantic Zinc Works, New York, NY (3).		
M-1474	Atlas Brass Foundry, Los Angeles, CA 90021.	M-619	Barrow Haematite Steel Co., Ltd., Barrow-in-Furness, Lancashire, England.
M-159	Atlas Foundry Co., Irvington, NJ (1).		
M-1049	Atlas Metal & Alloys Co. Ltd., Chicago, IL (3).	M-758	Bartlett Hayward Div. Koppers Co. (4), see Koppers Co., M-399.
M-696	Atlas Pattern & Model Works, Brooklyn, NY 11237.	M-1407	Barworth Flocton Steel Works Ltd., Ecclesfield, S30 3XH England.
M-1636	Atlas Steels Div. Rio Algom Corp., Cleveland, OH (3).	M-21	Batterium Metal & Vislok, Ltd., Harborough, England.
M-435	Atlas Steels Ltd., Welland, Ontario, Canada.	M-254	Bausch Machine Tool Co., Springfield, MA (2).

M-1406	Beardshaw & Co. Ltd., Joseph, Sheffield, S3 8DJ England.	M-321	Birmetals Ltd., Birmabright Works, Birmingham, B32 3BX England.
M-1258	Beardsley & Piper, Div. of Pettibone Corp., Chicago, IL 60639.	M-325	Birmetals Ltd., Birmabright Works, Birmingham, B32 3BX England.
M-720	Bearings Div. NL Industries, Toledo, OH 43614 (formerly Bunting Brass & Bronze).	M-137	Birmingham Aluminium (Casting) Co., Smethwick, Birmingham, England.
M-431	Bearium Metals Corp., Rochester, NY 14624.	M-1551	Birmingham Battery & Metal Co., Ltd., Birmingham, 29 England.
M-1183	Beaumont Birch Co., Philadelphia, PA 19102.	M-1772	Bishop Tube Co., Damascus Tube Div., Frazer, PA 19355.
M-706	Becker Bros. Carbon Corp., Cicero, IL 60650.	M-334	Bisset Steel Co., Cleveland, OH 44103.
M-613	Becker, Stahlwerk, AG., Krefeld, Rhineland, Germany.	M-1095	Blackalloy Co. of America, Paterson, NJ 07509.
M-544	Beckett Bronze Co., Muncie, IN (2).	M-1254	Black-Clawson Co., Middletown, OH 45042.
M-455	Bedel & Cie, see Aciéries SA Bedel, Saint Etienne, France.	M-861	Black Drill Co., Inc., Cleveland, OH 44117.
M-1429	Bedford Tool & Forge Co., Bedford, OH 44146.	M-26	Blackor Co., Los Angeles, CA (3).
M-707	Belle City Malleable Co., Racine, WI (1).	M-1640	Blackstown Corp., and Jamestown Malleable Iron Div. Blackstown Corp., Jamestown, NY 14701.
M-22	Bell Telephone Laboratories, Murray Hill, NJ (2).	M-1643	Blackstown Corp., and Jamestown Malleable Iron Div. Blackstown Corp., Jamestown, NY 14701.
M-1255	Belmont Smelting & Refining Works, Brooklyn, NY 11207.	M-413	Blackwell's Metallurgical Works, Lancashire, England (3).
M-1694	Bendix Steel Co., Cleveland, OH 44128 (3).	M-1222	Blaw-Knox Co. (4) (merged with Duraloy to form Duraloy-Blaw-Knox, M-45).
M-823	Benecke, Inc., Alexander, New York, NY (3).	M-143	Blaw-Knox Food & Chemical Div (4) (see Duraloy-Blaw-Knox, M-45).
M-1146	Benedict-Miller Inc., Lyndhurst, NJ 07071.	M-382	Bleiwerk Goslar, Goslar & Harz, Germany.
M-1290	Bergische Stahl-Industrie, 5630 Remscheid 1, West Germany.	M-516	Bliss & Laughlin Steel Co., Harvey, IL 60426.
M-708	Bergmann Elektrizitäwerke, Berlin, 1000, West Germany.	M-1435	Bochlet Gebruder, Kapfenberg, Steiermark, Austria.
M-1260	Bergsoe & Son, Paul, Glostrup, Denmark.	M-1331	Bochum Stahlwerk AG., Bochum, Germany.
M-1261	Bergsoe & Son, Paul, Lanskrona, Sweden.	M-1343	Bochumer Verein A, Bochum, Germany.
M-1128	Bergstrom Alloys Corp., New York, NY (3).	M-1184	Bofors AB., Bofors, Sweden.
M-1281	Berry Metal Co., Harmony, PA 16037.	M-27	Bohler Bros., Vienna, Austria (4) (merged to form Vereinigte Edelstahlwerke, Vienna, Austria, M-1802).
M-1435	Berted Foundry Co., Columbiana, OH (3).		
M-646	Beryllium Corp. of America (now Kawecki Beryllium Industries Inc., New York, NY 10017).	M-138	Bohn Aluminum & Brass Div., Southfield, MI 48075.
M-23	Bethlehem Foundry & Machine Co., Bethlehem, PA (2).	M-345	Boker & Co., H., New York, NY (1).
		M-1678	Boller Industries Div. of Boller Development Corp., Marion, IN 46952 (3).
M-24	Bethlehem Steel Corp., Bethlehem, PA 18016.	M-575	Bolton & Sons Ltd., Thomas, Stoke-on-Trent, ST10 2HF England.
M-1042	Biad Powder Metallurgy Co., Pittsburgh, PA (2) (3).	M-1674	Bolton-Emerson Inc., Lawrence, MA (2).
M-698	Bidault, Paris, France.	M-139	Bonney-Floyd Co., Columbus, OH (2).
M-410	Billington & Newton Ltd., Longport, Staffordshire, England (3).	M-975	Bonney Tool & Forge, Div. Gulf & Western, Allentown, PA 18105.
M-1177	Billiton Maatschappu, N.V., The Hague, Holland.	M-1649	Bonpertuis, Forges et Aciéries de, 38140 Rives-sur-Fure, France.
M-25	Binney Castings Co., Toledo, OH 43607.	M-338	Booth Aluminium Ltd., James (4) (merged to form Alcan Booth, M-643).
M-136	Birdsboro Corp., Birdsboro, PA 19508.		
M-577	Birkett, Billington & Newton Ltd., Hanley, England (3).	M-1284	Borolite Corp., Niagara Falls, NY (2) (3).

M-304   Bosch Metallwerk, Robert, A.G., Stuttgart, Germany.
M-632   Bound Brook Bearing Corp. (now GKN Powder Met. Inc., Bound Brook Div., Worcester, MA 01604).
M-1434 Bowsteel Distributors Corp., Linden, NJ (2).
M-336   Boyd-Wagner Co., Chicago, IL (3).
M-1154 Bradley & Foster Ltd., Staffordshire, England.
M-846   Bradley Laboratories, London, England.
M-140   Braeburn Alloy Steel (now CCS Braeburn Alloy Steel, Braeburn, PA 15068).
M-597   Braun-Steeples Co. Ltd., San Francisco, CA (3).
M-1449 Breda Co., Rome, Italy.
M-1451 Bresciana Metallurgica, Rome, Italy.
M-1394 Brevets Berthelmy, Société des, Paris, France.
M-141   Bridgeport Brass Co. (Div. of National Distillers & Chemical Corp.), Norwalk, CT 06856.
M-1389 Bridgeport Brass Co., Riverside, CA, see Bridgeport Brass, M-141.
M-509   Bridgeport Deoxidized Bronze Corp., Bridgeport, CT 06605 (2).
M-1579 Bridgeport Rolling Mills Co., Bridgeport, CT 06601.
M-424   Brighton Electric Steel Casting Co., Beaver Falls, PA 15010.
M-1572 Bristol Brass Co., Bridgeport, CT 06010.
M-28   British Aluminium Co. Ltd., Chalfont Technological Center, Buckinghamshire, SL9 OQB England.
M-416   British Driver-Harris Co. Ltd., Manchester, England (2).
M-978   British General Electric Co. Ltd., London, England.
M-711   British Insulated Callender's Cables, Ltd., Prescot, England.
M-699   British Metal Corp., Ltd., London EC 2, England.
M-396   British Non-Ferrous Research Association, London, England (2) (information only).
M-667   British Oxygen Co., Ltd., London, England (3).
M-715   British Piston Ring Co., Coventry, England.
M-478   British Rolling Mills, Ltd., Tipton, Staffordshire, England.
M-504   British Standards Institution, London, N1 9ND England (2) (not a producer; information only).
M-1724 British Steel Corp., Sheffield, S3 8AZ England.
M-306   British Thomson Houston Co., Ltd., Rugby, Warwickshire, England.
M-961   Brockhouse Casting Co., Wolverhampton, England (3).
M-593   Bronze Die Casting Co., Pittsburgh, PA 15233.

M-1687 Le Bronze Industrial, Rene Loiseau & Cie, 93001 Bobigny, France.
M-1395 Brooks & Perkins Inc., Livonia, MI 48150.
M-996   Brookside Metal Co., Ltd., Watford, Herts, WO2 4NF England (2).
M-759   Brown Alloy Works, Detroit, MI (3).
M-1172 Brown & Sharp Mfg. Co., Centerdale, RI 02911.
M-472   Brown, Bayley's Steel Works (2) (4) (now part of Dunford Hadfields Ltd., M-62).
M-1126 Brown-Firth Co., see Firth Brown Ltd., M-56.
M-476   Brown Foundries Co., David, Penistone, Nr. Sheffield, England.
M-213   Brown-Wales Co., Cambridge, MA 02146 (distributer for Horace Potts, M-307).
M-385   Brukskoncernen A.B., Fagersta, Sweden.
M-1037 Brush Wellman Inc., Cleveland, OH 44110 (formerly Brush Beryllium).
M-256   Buckeye Brass & Mfg. Co., Cleveland, OH 44103.
M-142   Buffalo Bronze Die Casting Co., Buffalo, NY (1).
M-962   Buffalo Wire Works Co. Inc., Buffalo, NY 14240.
M-143   Buflovak Equipment (now Blaw-Knox Food & Chemical Div. Duraloy-Blaw-Knox) (2).
M-719   Bull's Metal & Marine Ltd., Glasgow, Scotland.
M-401   Bundy Tubing Div. of Bundy Corp., Warren, MI 48089.
M-720   Bunting Brass & Bronze (now Bearings Div. NL Industries, Toledo, OH 43614).
M-721   Burden Iron Co., Troy, NY (1).
M-1467 Burdett Oxygen Co. (now Burdox Inc., Cleveland, OH 44114).
M-29   Burgess-Parr Inc., Freeport IL (1) (alloys now produced by Stainless Foundry, M-1262).
M-722   Burndy Corp., Norwalk, CT 06856.
M-1415 Burys & Co. Ltd., Sheffield, England (3).
M-1355 Busch-Jäger Ludenscheider Metallwerke, Ludenscheid, Germany.
M-292   Byers Co., A.M., Ambridge, PA (1).
M-723   Byrwill Co., Cincinnati, OH (3).

C

M-214   Cadman Mfg. Co., A.W., Pittsburgh, PA 15222.
M-725   Callite Tungsten Corp. (now GTE Sylvania, Towanda, PA 18848).
M-511   Calloy Ltd., Avonmouth, England.
M-726   Caloriz Corp. of Great Britain Ltd., London, England (2).
M-30   Calorizing Co., Pittsburgh, PA(3).
M-1166 Calumet Steel Castings Corp., Hammond, IN 46320.

Section II: Alphabetical List of Manufacturers / 1615

M-727	Cambridge Wire Cloth Co., Cambridge, MD 21613.
M-490	Cameron & Son Ltd., Sheffield, England (1).
M-1264	Cameron & Son Ltd., Sheffield, England (1).
M-1116	Campagnole, Aciéries de, 93123 Courneuve, France.
M-570	Canada Electric Steel Castings Ltd., Orillia, Ontario, Canada.
M-1530	Canadian Quebec Metallurgical Corp., Quebec, Canada.
M-1491	Cannon-Muskegon Corp., Muskegon, MI 49443.
M-354	Cannon-Stein Steel Co., Syracuse, NY (1).
M-1654	Capito & Klein.
M-1771	Capitol Castings Div. Midland Ross Corp., Phoenix, AZ 85001.
M-1557	Capper Pass & Son, Ltd., Bristol, England (3).
M-1695	Carbidie, Div. of Aiken Industries, Inc., Irwin, PA 15642.
M-1609	Carborundum Co., Niagara Falls, NY 14302.
M-1793	Carlson, Inc., G.O., Thorndale, PA 19372.
M-1696	Carmet Materials Div., Madison Heights, MI 48071.
M-251	Carobronze Ltd., London W4 England.
M-1587	Carondelet Foundry Co., St. Louis, MO 63110.
M-1330	Carp & Hones, 4150 Krefeld 1, West Germany (3).
M-32	Carpenter Technology Corp., Reading, PA 19603.
M-1410	Carr & Co., Ltd., Richard, Sheffield, S6 1LL England.
M-405	Carr Ltd., Charles, Birmingham, England (3).
M-600	Castalloy Co., Inc., Natick, MA.
M-904	Castings Corp., Brewster, NY 10509.
M-1091	Castings Development Co., Philadelphia, PA (3).
M-648	Castolin Welding Alloys Inc., New York, NY (3).
M-140	CCS Braeburn Alloy Steel, Braeburn, PA 15068.
M-1792	Cegedur Pechiney, 75361 Paris Cedex 08, France.
M-215	Central Brass & Aluminum Foundry Co., Cincinnati, OH 45203.
M-736	Central Engineering & Supply Co., Passaic, NJ 07055.
M-592	Central Foundry Co., Joplin, MO.
M-690	Central Foundry, Div. of General Motors, Saginaw, MI 48605.
M-515	Central Institute of Metals, Leningrad, USSR.
M-649	Central Iron & Steel Co., Harrisburg, PA (3).
M-845	Central Pattern & Foundry Co., Chicago, IL (3).
M-1709	Centrifugal Products Inc., Long Beach, CA 90801.
M-1770	Century Brass Products, Waterbury, CT 06720.
M-216	Cerro Copper Products, Div. of the Cerro-Marmon Corp., East St. Louis, IL 62202 (2), see Cerro Metal Products.
M-122	Cerro Metal Products, Div. of the Marmon Group, Bellefonte, PA 16823.
M-1600	Certanium Alloys & Research Co.
M-1748	Certified Alloy Products Inc., Long Beach, CA 90801.
M-582	Chace Co., W.M., Detroit, MI 48209.
M-566	Chain Belt Co., Milwaukee, WI.
M-835	Chain Div., C.M., Columbus McKinnon Corp., Tonawanda, NY 14150.
M-474	Chambersburg Engineering Co., Chambersburg, PA 17201.
M-1116	Champagnole, Aciéries de, 93123 Courneuve, France.
M-712	Champion Rivet Co., Cleveland, OH (3).
M-1683	Champion Steel Co., Orwell, OH 44076.
M-1021	Chance Vought Aircraft Div. United Aircraft Corp., Stratford, CT (2).
M-145	Chapman Valve Mfg. Co., Indian Orchard, MA (3).
M-33	Chase Brass & Copper Co., Cleveland OH 44122.
M-1790	Chase Brass & Copper Co. Inc, Williams County Div., Montpelier, OH 43543.
M-1789	Chase Brass & Copper Co., Sheet Div., Cleveland, OH 44122.
M-1791	Chase Nuclear, Div. Chase Brass & Copper Co., Inc., Cleveland, OH 44122.
M-257	Chateaugay Ore & Iron Co., Standish, NY (3).
M-1117	Chatillon, see Chiers-Chatillon.
M-520	Chatillon & Sons, John, Kew Gardens, NY 11415.
M-76	Chemalloy Electronics Corp., Santee, CA 92071.
M-1713	Chemetron Welding Products, Chicago, IL 60601.
M-725	Chemical & Metallurgical Div., GTE Sylvania, Towanda, PA 18848.
M-856	Chemical & Metallurgical Div., GTE Sylvania, Towanda, PA 18848.
M-1187	Chicago Development Corp., Ashland, VA 23005.
M-810	Chicago Hardware Foundry, North Chicago, IL (3).
M-902	Chicago Malleable Casting Co., Chicago, IL (3).
M-146	Chicago Steel Foundry, Chicago, IL (3).
M-1117	Chiers-Chatillon, 75009 Paris, France.
M-1492	Chromalloy American Corp., West Nyack, NY 10994.
M-147	Chrome Alloys Mfg. Co., Oakland, CA (3).
M-148	Chrome Steel Corp., Carteret, NJ (1).
M-980	Chromium Corp. of America, Cleveland, OH 44105.

M-1059	Chromium Mining & Smelting Corp. Ltd., Montreal, Canada.	M-530	Colvilles Ltd., Glasgow, Scotland (4) (now part of British Steel Corp., M-1724).
M-680	Chrysler Corp., Detroit, MI 48288 (2) (information only).	M-1741	Comalco Ltd., Melbourne, Australia.
M-211	Chrysler Corp., Amplex Div., Detroit, MI 48231.	M-1679	Cominco Trail, British Columbia, Canada.
M-807	Cie Français des Métaux, Paris, France.	M-1723	Commentrenne, 75429 Paris Cedex 09 France.
M-867	Cinaudagraph Corp., Stamford, CT (3).	M-1248	Commerce Pattern Co., New York, NY (3).
M-437	Cincinnati Steel Castings Co., Cincinnati, OH (3).	M-152	Commercial Alloys Co., San Francisco, CA (3).
M-298	Cindal Aluminium Ltd., Birmingham, England (3).	M-1808	Commonwealth Aircraft Corp., Melbourne, 3001, Australia.
M-149	Cleveland Automatic Machinery Co., Cleveland, OH (2).	M-623	Compagnie des Alliages Speciaux d'Aluminium, Asnières, France.
M-480	Cleveland Brass Corp., Cleveland OH (2).	M-1546	Compagnie des Ateliers et Forges de la Loire, F-75, Paris 9e, France.
M-738	Cleveland Refractory Metals Div., Chase Brass & Copper Co., Solon, OH 44139.	M-1117	Compagnie des Forges de Chatillon, see Chiers-Chatillon.
M-1681	Cleveland Refractory Metals Div., Chase Brass & Copper Co., Solon, OH 44139.	M-678	Compagnie des Mines, Fonderies et Forges d'Alais, Temaris, France.
M-612	Cleveland Twist Drill Co., Cleveland, OH (2).	M-258	Compagnie d'Orleans, Paris, France (3).
M-739	Clevite Bearing Div., Gould Inc., (now Gould, Inc., Engine Parts Div., Cleveland, OH 44110).	M-921	Compagnie Français de l'Etain, Paris 8e, France.
M-150	Clifford Industries Ltd., Charles, Birmingham, England (2).	M-780	Compound Electro Metals Ltd., London, England (3).
M-151	Climax Molybdenum Co., Greenwich, CT 06830 (part of AMAX Inc.).	M-734	Compressed Industrial Gases Co., Chicago, IL (3).
M-1569	Clyde Alloy Steel Co. (now part of British Steel Corp., M-1724).	M-1676	Connecticut Metals Corp., Hampden, CT (3).
M-1416	Clyde Works, Sheffield, see Osborn Steels, M-233.	M-636	Consolidated Ashcroft Hancock Co. Inc., Bridgeport, CT (3).
M-265	CMW Inc., Indianapolis, IN 46206 (formerly Mallory Metallurgical).	M-920	Consolidated Car Heating Co., Albany, NY (3).
M-218	Coan Ltd., R.W., London, England (3).	M-746	Constrictor Ltd., London, England (1).
M-1060	Coast Metals Inc., Little Ferry, NJ 07643.	M-451	Continental Foundry & Machine Co., East Chicago, IN (3).
M-1722	Cobalt Information Center, Columbus, OH (1) (2).	M-676	Continental Industries Corp., New York, NY (3).
M-503	Cochrane Corp., Philadelphia, PA (3).	M-631	Continental Steel Corp. (now Penn-Dixie Corp., Kokomo, IN 46901).
M-1450	Cogne Co., Rome, Italy.	M-1405	Cook & Co. Ltd., George H., Sheffield, S3 8RE England.
M-1295	Cohn Corp., Sigmund, Mount Vernon, NY 10553.	M-36	Cooper Alloy Corp., Hillside, NJ 07205.
M-657	Cohn Ltd., A., London, England (3).	M-1011	Cooper Metallurgical Corp., Cleveland, OH 44102 (formerly Cooper Products Co.).
M-740	Cold Metals Products Co., Youngstown, OH (3).	M-693	Copper Development Association, Potters Bar, Hertfordshire, England (2) (information only).
M-744	Colonial Alloys Co., Philadelphia, PA 19129.	M-1657	Copper Range Co., New York, NY 10020.
M-1393	Colonial Metals Co., Columbia, PA 17512.	M-1677	Copper Rod & Brass Products Div. Triangle Conduit & Cable Co., New Brunswick, NJ 08903.
M-34	Colonial Steel Div. Teledyne Vasco (4), see Teledyne Vasco, M-115.	M-682	Copperweld Steel Co., Warren, OH 44400 (2).
M-1529	Colorado Fuel & Iron Co., Denver, CO.	M-1217	Cop-Sil-Loy Inc., Hollywood, CA (3).
M-1392	Columbia Bronze Corp., Freeport, NY (3).	M-1506	Corning Glass Co., Corning, NY 14830 (2).
M-35	Columbia Tool Steel Co., Chicago Heights, IL 60411.	M-917	Cornish Wire Co. (now General Cable Corp., Cornish Wire Products, Williamstown, MA 01267).
M-835	Columbus-McKinnon Chain Corp. (now C.M. Chain Div. Columbus McKinnon Corp., Tonawanda, NY 15219).		

Section II: Alphabetical List of Manufacturers / 1617

M-1437 Coulter Steel & Forge Co., Emeryville, CA 94662.
M-337   Craine-Schrage Steel Div. Detroit Steel Co. (1).
M-748   Crane Co., Chicago, IL 60623.
M-1109 Crane Co., Torrey S., Plantsville, CT 06479.
M-749   Crescent Tool Co., Jamestown, NY.
M-1488 Creusot-Loire, 75428 Paris Cedex 09 France.
M-1052 Crobalt Inc., Ann Arbor, MI (3).
M-37   Cronite Foundry Co. Ltd., Tottingham, London N 15, England.
M-355   Crowley Inc., John A., New York, NY (3).
M-356   Crucible Electric Steel Co., Homestead, PA (4).
M-38   Crucible Specialty Metals Div. Colt Industries, Syracuse, NY 13201.
M-153   Crucible Steel Castings Co., Cleveland OH 44102.
M-1199 Crucible Steel Castings Co., Lansdowne, PA (2).
M-449   Cunningham Co., M.E., Pittsburgh, PA 15233.
M-154   Curtis Bay Copper & Iron Works (1).
M-1502 Curtis-Wright Corp., Quehanna, PA (3).
M-308   Cutler Hammer Inc., Milwaukee, WI 53216.
M-155   Cuyahoga Steel & Wire Div. Hoover Ball & Bearing Co. (2).
M-347   CWC Castings Div. of Textron Inc., Muskegon, MI 49443.

D

M-1773 D.A.B. Industries Inc., Troy, MI 48084.
M-1545 Daido Steel Co. Ltd., New York, NY 10017.
M-754   Damascus Steel Casting Co., New Brighton, PA 15066.
M-1016 Darbyshire Steel Co. Inc., El Paso, TX 79912.
M-40   Darwin & Milner Inc., Cleveland, OH (3).
M-487   Darwin's Ltd., Sheffield, England (now part of Balfour-Darwins, M-1725).
M-985   Darwin-Toledo Ltd. (Toledo Steel Works), Sheffield, England (3).
M-1019 Davis & Geck Inc., Pearl River, NY 10965.
M-1097 Dee Div. of Handy & Harmon (1).
M-584   Degefors Iron & Steel Works, Degefors, Sweden.
M-756   Delloy Metals, Philadelphia, PA (3).
M-41   Deloro Stellite, Belleville, Ontario, Canada.
M-1580 Deloro Stellite Ltd., Swindon, Wiltshire, SN3 40A England.
M-432   Delsteel Inc., Wilmington, DE 19899.
M-156   Delta Metal (BW) Ltd., West Bromwich, West Midlands, B70 9ER England.

M-179   Delta Metal (BW) Ltd., West Bromwich, West Midlands, B70 9ER England.
M-755   Denman & Davis Co., North Bergen, NJ (3).
M-157   Detroit Gray Iron Foundry-Detroit Alloy Steel Co., Detroit, MI (3).
M-1079 Detroit Mold Engineering Co., Newark, NJ (3).
M-1105 Detroit Tap & Tool Co., Warren, MI 48090.
M-1300 Deutro G.m.b.H., Berlin, Germany.
M-1356 Deutsch Delta Metallgesellschaft, Düsseldorf-Grafenberg, Germany.
M-459   Deutsche Edelstahlwerke G.m.b.H. (4) (merged to form Thyssen Edelstahlwerke A.G., M-1759).
M-1133 Deutsche Gold und Silber Scheide, Hanau, West Germany.
M-895   Deutsche Industrie, Nemen, Germany.
M-1334 Deutsche Mannesmannrohren-Werke AG., 4000 Düsseldorf 1, West Germany (2) (all DIN or W.Nr. grades).
M-981   Deutsche Messingwerke, Berlin, Germany.
M-1489 Dewrance & Co., Ltd., Skelmersdale, Lancashire, England (3).
M-1726 Diamond Metal Alloys, Houston, TX 77001.
M-819   Die Casting Appliance Corp., Ltd., London, England (3).
M-760   Die Castings Ltd., Birmingham, England.
M-1803 Diederichs, Karl, Stahl-,Walz-und Hammerwerk, D-5630 Remscheid 11, West Germany.
M-1357 Diehl Metall-Guss & Presswork, Hrch., GmbH., Nürnberg, West Germany (2).
M-822   Diehl Steel Co., Cincinnati, OH (3).
M-505   Dinorm Co., Berlin, Germany (2).
M-594   Dirigold Corp., Kokomo, IN (3).
M-1399 Dirilyte Co. of America, Kokomo, IN 46901.
M-357   Disston Inc., Pittsburgh, PA 15219 (2).
M-1121 Division Lead Co., Summit, IL 60501.
M-1427 Dixon Sintaloy Inc., Stamford, CT 06906.
M-1616 D-M-E Corp., Madison Heights, MI 48071.
M-1424 DoAll Co., Des Plaines, IL 60016.
M-762   Dodge Foundry & Machine Co., Philadelphia, PA 19135.
M-158   Doehler-Jarvis Div. NL Industries, Toledo, OH 43691 (2).
M-855   Doelger & Kirsten Inc., Milwaukee, WI 53210.
M-1348 Dohlen-Stahl Gusstahl-Handels, GmbH., Wetter/Ruhr, Germany.
M-1033 Dominion Foundries & Steel Ltd., Hamilton, Ontario, L8N 3J5 Canada.
M-1239 Dominion Magnesium Ltd., Toronto, Ontario, Canada.

1618 / Woldman's Engineering Alloys

M-763	Dominion Wheel & Foundry Co., Toronto, Canada.	M-501	Edelstahlwerk Buderus or Rochling Buderus (4) (merged to form Rochling-Burbach, M-912).
M-1223	Donegal Steel Foundry Co., Marietta, OH (3).	M-1324	Edelstahlwerke Düsseldorf-Heerdt GmbH. & Co., 4000 Düsseldorf-Heerdt, West Germany (3).
M-470	Dorman, Long & Co. (4) (now part of British Steel Corp., M-1724).	M-617	Edelstahlwerk Rochling A.G., Berlin, Germany.
M-614	Dorrenberg Edelstahl, 5252 Runderoth, West Germany.	M-1326	Edelstahl Witten A.G. (4) (merged to form Thyssen Edelstahlwerk A.G., M-1759).
M-222	Dortmund-Hoerder Huttenverein A.G., Dortmund, West Germany (1).	M-210	Edgar Allen Balfour Steels Ltd., Sheffield, S9 1QY England.
M-43	Dow Chemical Co., Midland, MI 48640.	M-433	Edgcomb Metals Co., Philadelphia, PA 19114 (2).
M-924	Dravo-Doyle Co., Pittsburgh, PA 15222.	M-607	Edgcomb Metals Co., Tulsa, OK 74103.
M-61	Driver Co., Wilbur B., Newark, NJ 07104.	M-999	Edgcomb Steel Co., Hillside, NJ (2).
M-44	Driver Harris Co., Harrison, NJ 07029.	M-161	Egal Metal Products Co., Baltimore, MD (3).
M-358	Duke Steel Co., Inc., New York, NY (3).	M-549	Egge Co., E.N., Los Angeles, CA (3).
M-1413	Dunford & Elliott Ltd. (4) (merged into Dunford-Hadfields, M-62).	M-1003	Eimco Foundry Div. Envirotech Corp., Salt Lake City, UT 84110.
M-62	Dunford Hadfields Ltd., Sheffield, S9 1TZ England.	M-557	Eisenwerke Neubrandenburg, G.m.b.H., Berlin, Germany.
M-522	DuPont de Nemours & Co., E.I., Wilmington, DE 19898 (2).	M-796	Eisler Electric Co., Union City, NJ (3).
M-1563	DuPont de Nemours & Co., E.I., Wilmington, DE 19898 (2).	M-1760	Ekatit Stahl Erich K. Tittel, 7120 Bietigheim-Bissinger, West Germany.
M-993	Duquesne Smelting Corp., Pittsburgh, PA (3).	M-359	Ekstrand & Tholand Co., New York, NY (3).
M-45	Duraloy Blaw-Knox Inc., Scottdale, PA 15683.	M-1141	Ekstrom, Carlson & Co., Rockford, IL 61110 (2).
M-767	Duralumin, Société du, Paris, France.	M-851	Electric Auto-Lite Co., Woodstock, IL (3).
M-769	Duraweld Metal Products Corp., Long Island City, NY (3).	M-673	Electro-Cables, Société de, Paris, France.
M-457	Durener Metallwerke, Duren, Germany.	M-1210	Electro Foundry Co., Troutdale, OR (3).
M-46	Duriron Co., Inc., Dayton, OH 45401.	M-1690	Electro Hercules Steel Co., Cleveland, OH 44102.
M-160	Dursar Corp., Newark, NJ (3).	M-778	Electroloy Co. Inc., Bridgeport, CT 06605.
M-948	Dusenberry & Stracken, Inc., New York, NY (2) (3).	M-1736	Electrometal-Acos Finos-SA, C. Postal 944-Campinos, SP. Brazil.
M-1324	Düsseldorf-Heerdt GmbH. & Co., KG., 4000 Düsseldorf-Heerdt, West Germany.	M-1020	Electro Metal Corp., Yonkers, NY (3).
M-757	Duval et Foulain, Paris, France.	M-1642	Electronic Specialties Co., H. & S. Metals Co., Pomona, CA (3).
M-773	Dymonhard Corp. of America, New York, NY (2) (3).	M-777	Electro Refractories & Abrasives Div. FERRO Corp., Buffalo, NY 14218 (2).
M-1218	Dyn-Metal Ltd., London, England (3).	M-388	Electro-Steel Co., Pittsburgh, PA (3).

E

M-1275	E. & E. Kaye Ltd, Middlesex, EN3 4SS England (now part of Pechiney).	M-779	Elesco Smelting Co., Chicago, IL 60623 (2).
M-1444	E. & E. Kaye Ltd, Middlesex, EN3 4SS England (now part of Pechiney).	M-1712	Elgiloy Co., Div. of American Gage & Machine Co., Elgin, IL 60120.
M-1014	Eastern Stainless Steel Co. Div. of Eastmet Corp., Baltimore, MD 21203.	M-1087	Elgin Watch Co., Chicago, IL 60606.
M-1710	Easton Metal Powder Inc., Easton, PA 18042 (3).	M-1358	Elisental Drahtwerk, Neuenrade/Westf., Germany (2).
M-947	Eaton Corp., Cleveland, OH 44114.	M-175	E.M.F. Electric Co. Ltd., Victoria, Australia. M-1201.
M-947	Eaton, Yale & Towne (now Eaton Corp.).	M-691	Empire Sheet & Tin Plate Co., Mansfield, OH (3).
M-1035	Eccles & Davis Machinery Co., Los Angeles, CA (3).	M-162	Empire Steel Castings Inc., Reading, PA 19603.
M-1768	Echevarria, S.A. (Aceros HEVA), Apartado 46, Bilboa, Spain.	M-49	Empire Steel Corp., Syracuse, NY (3).
M-776	Eclipse-Pioneer Foundries (Bendix Foundries), Teterboro, NJ (2).	M-1061	Emsco Derrick & Equipment Co., Los Angeles, CA (3).

Section II: Alphabetical List of Manufacturers / 1619

M-1083 Enfield Rolling Mills Ltd., Enfield, Middlesex, EN3 7QB England.
M-18 Engelhard Industries Div. Engelhard Mineral & Chemical Corp., Carteret, NJ 07008.
M-50 English Steel Corp. Ltd. (4) (merged into British Steel Corp., M-1724).
M-1414 English Steel Rolling Mills Corp. Ltd. (4) (merged into British Steel Corp., M-1724).
M-1359 Erbsloh, J. & A., Wuppertal-Barmen, Germany.
M-1333 Ergst, Stahlwerke, Kommanditgesellschaft, 5840 Schwerte 1, West Germany.
M-1593 Erie Forge & Steel Co., Erie, PA (3).
M-51 Erie Malleable Iron Co., Erie, PA 16501 (2).
M-1425 Erie Steel Co., Erie, PA (3).
M-1336 Erkenzweig & Schwemann, Hagen, Germany.
M-1024 Escaut & Meuse, Sté d', Paris 17e, France.
M-438 ESCO Corp., Portland, OR 97210 (formerly Electric Steel Foundry).
M-921 Etain, Compagnie Française de l', Paris 8e, France.
M-1787 Eurotungstene, 65X-38041 Grenoble Cedex, France (part of Pechiney).
M-717 Eutectic Corp., Flushing, NY 11358.
M-259 Evans Steel Co., H. D., Boston, MA (3).
M-589 Everard Tap & Die Corp. (2).
M-965 Excelite Co., Woodbridge, NJ (3).
M-976 Ex-Cell-O Corp., Tool & Abrasive Products Div., Detroit, MI 48232 (formerly Willeys Carbide).
M-1384 Exide Power Systems Div. ESB Inc., Philadelphia, PA (2).

F

M-724 F.A.C.A., Levallois-Perret (92) France.
M-52 Fagersta Bruks Aktiebolag, Fagersta, Sweden.
M-52 Fagersta Inc., West Caldwell, NJ 07006 (agent).
M-994 Fahralloy Co., Harvey, IL 60426.
M-886 Fairbanks, Morse & Co., Beloit, WI 53511.
M-491 Fairley & Sons, Jas., Ltd., Birmingham, England (3).
M-1464 Fairmount Foundry Inc., Hamburg, PA 19526.
M-627 Faitout Iron & Steel Co., Newark, NJ.
M-1312 Fakirstahl Hoffmans & Co., Edelstahlwerk, 5630 Remscheid 14, West Germany.
M-862 Falk Corp., Milwaukee, WI 53201.
M-53 Fansteel Inc., North Chicago, IL 60064.
M-78 Farbenindustrie Aktiengesellschaft, I. G. - Abtielung Elektronmetall, Frankfurt a.M., Germany.
M-163 Farrell Co., Div of USM Corp., Ansonia, CT 06401.
M-220 Farrell-Cheek Steel Co., Sandusky, OH 44870.
M-253 Farrelloy Co., Div. American Solder & Flux Co., Inc., Paoli, PA 19301 (2).
M-1145 Federal Foundries & Steel Co., Ltd., London, England.
M-709 Federal-Mogul Corp., Detroit, MI 48235.
M-815 Federated Metals Div. American Smelting & Refining Co., see ASARCO.
M-1360 Felten & Guillaume Carlswerke, A.G., Köln-Mülheim, West Germany.
M-1283 Feltrina Società Metallurgica, Milan, Italy.
M-104a Ferner Co., Inc., R. Y., Malden, MA (3).
M-54 Ferranti Ltd., Hollinwood, Lancashire, OL9 7JS England.
M-660 Ferrous Metals Corp., New York, NY (2).
M-1232 Ferroxcube Corp. Div. of North American Phillips Co., Saugerties, NY 12477.
M-1717 Ferry-Capitain, Usines De Bussy, B.P. 33-52 Joinville, France.
M-425 Field Co., B. A., Newark, NJ (3).
M-718 Fillmore Foundry Inc., Buffalo, NY (3).
M-55 Finkl & Sons Co., A., Chicago, IL 60614.
M-587 Finn Metal Works, John, San Francisco, CA (3).
M-1043 Firminy, Aciéries & Forges de (4) (now part of Creusot-Loire, M-1488).
M-56 Firth Brown Ltd., Sheffield, S4 7US England.
M-1699 Firth-Loach (now Howmet Corp., Carbide Div., McKeesport, PA 15134).
M-57 Firth-Sterling (now Teledyne Firth-Sterling, McKeesport, PA 15134).
M-521 Firth-Vickers Special Steels Ltd., Sheffield, S9 2FU England.
M-1659 Fisher AG., Georg, Germany (3).
M-378 Fisher Scientific Co., Pittsburgh, PA 15219.
M-894 Fitzsimmons Steel Co., Youngstown, OH 44501.
M-1409 Flockton, Tompkin & Co., Ltd., Sheffield, England (3).
M-806 Follansbee Steel Co., Follansbee, WV 26037.
M-752 Follsain-Wycliffe Foundries, Ltd., Lutterworth, Nr. Rugby, England.
M-679 Fonderie de Precision S.A., 92003 Nanterre, Paris, France.
M-804 Foote Bros. Gear & Machine Corp., Chicago, IL (3).
M-1038 Foote Mineral Co. (Ferroalloys Div.), Exton, PA 19341.
M-747 Ford Motor Co., Chemical-Metallurgical Products Div., Dearborn, MI 48121.
M-1649 Forges et Aciéries de Bonpertuis, 38140 Rives-sur-Fure, France.
M-439 Forging & Casting Div. Allegheny-Ludlum (4), see Allegheny-Ludlum, M-1.
M-1766 Forjas Alavesas S.A., 22 Vitoria, Spain.

M-164   Fort Pitt Steel Casting Co., Div. of Conval Corp., McKeesport, PA 15134.
M-104   Fourchambault et Decazeville, Soc. Anon. de Commentry (4) (now part of Creusot-Loire, M-1488).
M-430   Fox & Co., Samual (4) (now part of British Steel Corp., M-1724).
M-571   Frank Foundries Corp., Moline, IL 61265.
M-360   Frasse & Co., Peter A., Lake Success, NY (2).
M-72   Fried. Krupp Huttenwerke, Bochum, West Germany.
M-403   Friedrich Wilhelms-Hütte, Mülheim, Ruhr, West Germany. .
M-1285 Fromson Co., Inc., Rockville, CT 06066.
M-58   Frontier Bronze Corp., Niagara Falls, NY 14304.
M-1361 Fuchs, Otto, 5882 Meinerzhagen, West Germany.
M-1538 Fugi Iron & Steel Co., Ltd., Chou-Ku, Tokyo, Japan.
M-1589 Fujikoshi Steel Industry Co., Ltd., Tokyo, Japan.
M-808   Fuller & Basche Co., New York, NY (3).
M-1101 Fulton Gold Refineries Corp., New York, NY (3).
M-809   Fulton Iron Works Co., St. Louis, MO (2).
M-1243 Fusion Inc., Willoughby, OH 44094.

G

M-407   Gabriel & Co., Ltd., A.B., Row, Birmingham, England (3).
M-409   Gallimore & Sons, Wm., Ltd., Sheffield, England (1).
M-1104 Galv-Weld, Inc., Dayton, OH (3).
M-1205 Gardiner Metal Co., Chicago, IL 60632 (2).
M-287   Garford Engineering Co., Garfield, IN (3).
M-550   Garrett Brass & Foundry Co., Garrett, IN (3).
M-782   Gautier & Co., D. G., New York, NY (3).
M-729   Gayer Co., A., Paris, France.
M-27   Gebr. Bohler & Co. (4) (merged into Vereinigte Edelstahlwerke, M-1802).
M-1435 Gebruder Bochlet, Kapfenberg, Steiermark, Austria.
M-1004 General Aircraft Equipment Co. (1).
M-59   General Alloys Co., (1) (business and alloys taken over by Alloy Engineering & Casting Co., M-1005).
M-917   General Cable Corp., Cornish Wire Products, Williamstown, MA 01267.
M-1111 General Cerium Co. (1), see Ronson Metals, M-1497.
M-1520 General Communications Co., Boston, MA (2).
M-31   General Electric Co.: Carboloy Systems Dept., Detroit, MI 48232, Metallurgical Products Dept., Detroit, MI 48232, Magnetic Materials Section, Edmore, MI 48829. M-60.
M-831   General Electric Ltd., England (2).
M-1478 General Malleable Corp., Waukesha, WI (3).
M-1063 General Metals Powder Co., Akron, OH 44305.
M-165   General Motors Corp. and Central Foundry Div. of General Motors, Saginaw, MI 48605.
M-1286 General Motors Corp. and Central Foundry Div. of General Motors, Saginaw, MI 48605.
M-690   General Motors Corp. and Central Foundry Div. of General Motors, Saginaw, MI 48605.
M-926   General Plate Div. Metals & Controls Corp., Attleboro, MA 02703.
M-1034 General Steel Industries Inc., St. Louis, MO 63105.
M-1513 General Thermoelectric Corp., Princeton, NJ (3).
M-797   General Tool & Die Co., East Orange, NJ (3).
M-1251 Georgia Iron Works, Grovetown, GA 30813.
M-1323 Georgsmarienwerke Selesiastahl, G.m.b.H., Georgsmarienhütte, Germany.
M-1362 Gerhardi & Co., Ludenscheid, Germany.
M-854   G.H.R. Foundry Div. Dayton Malleable Iron Co., Dayton, OH (2).
M-650   Gibson Electric Co., Delmont, PA 15626.
M-897   Gilby-Fodor S.A., Rueil-Maimaison (Seine-et-Oise), France.
M-1477 Gillett & Eaton Inc., Lake City, MN (3).
M-817   Gilmore & Co., F. F., Needham Heights, MA 02194.
M-1298 Gilson, Ltd., Brussels, Belgium.
M-609   Gimo-Osterby Bruks, A.B., Gimo, Sweden.
M-284   Giulini Werke A.G., Rohrsbach, Germany.
M-632   GKN Powder Met Inc., Bound Brook Div., The Presmet Div., Worcester, MA 01604 (formerly Bound Brook Bearing Corp.).
M-818   Glacier Metal Co., Ltd., Alpertone, England (3).
M-586   Glacier Metal Co., Richmond, VA (3).
M-1700 Glader Co., K. C., Niles, IL 60648.
M-1728 Glidden Metals, Chemical/Metallurgical Div. of SCM Corp., Cleveland, OH 44113.
M-1030 Globe Steel Tubes Co., Milwaukee, WI (2).
M-1576 Glyco-Metallwerke Daelen & Loos, G.m.b.H., Germany (3).

M-528 Goldschmidt, Th., A.G., Abteilung Metalle, Essen, Germany.
M-1108 Goldsmith Bros., Chicago, IL (part of NL Industries).
M-1178 Gold und Silber Scheide Anstalt, Hanau, Germany.
M-458 Goodlass Wall & Lead Industries, Ltd., London, England (3).
M-820 Gorham Tool Industries Inc., Detroit, MI 48238.
M-1363 Göttingen Aluminiumwerke G.m.b.H., Göttingen, Germany.
M-739 Gould Inc., Engine Parts Div., Cleveland, OH 44110.
M-816 Grammer, Dempsey & Hudson Co., Newark, NJ (3).
M-633 Granite City Steel Co., Granite City, IL (4), see National Steel, M-836.
M-805 Grant & West Ltd., London, England (3).
M-247 Graphite Metallizing Corp., Yonkers, NY 10702.
M-821 Graphitized Alloy Corp., New York, NY (3).
M-596 Grasselli Chemical Co., Cleveland, OH (3).
M-1009 Grayborn Steel Co., New York, NY (3).
M-766 Great Lakes Steel Co., Div. National Steel Corp., see National Steel, M-836.
M-1067 Great Western Steel Co., Div. Hoyland Steel Co., Los Angeles, CA.
M-1267 Grede Foundries Inc., Milwaukee, WI 53204.
M-1194 Greenleaf Corp., Saegertown, PA 16433.
M-1635 Green River Steel Co., Owensboro, KY 43201 (Div. of Jessop Steel Co.).
M-949 Grey Mfg. Co., C. M., East Orange, NJ (3).
M-380 Griesogen Griesheimer Autogen Verkaufs, G.m.b.H., Greisheim a.M., Germany.
M-1316 Grimm, Gustav, Stahl-und Hammerwerk, 5630 Remscheid-Haddenbach, West Germany.
M-1404 Guertler, William, G.m.b.H., Berlin, Germany.
M-1431 Gulf Steel Corp., Dallas, TX (3).
M-223 Gunite Div., Kelsey Hayes Co., Rockford, IL (2).
M-1329 Guronitwerke Vervoort, G.m.b.H., Düsseldorf, West Germany.
M-1293 Gusstahl-Handels, G.m.b.H., van Dohlen-Stahl, Wetter, Ruhr, Germany.

H

M-825 Hackett Brass Foundry, Detroit, MI 48214.
M-62 Hadfields Ltd. (now Dunford Hadfields Ltd., Sheffield, S9 1TZ England).
M-841 Hadstrom Pattern Works, Oscar W., Chicago, IL (3).

M-1318 Haeckerstahl Willy Haecker, 5609 Huckeswagen-Rhld, West Germany.
M-1652 Hagener Gusstahlwerl Remy (now Remystahl), see M-1339.
M-260 Halcomb Steel Div. Crucible Steel (4), see Crucible Specialty Metals, M-38.
M-1202 Hallamshire Steel Co., Sheffield, 3 England.
M-1411 Hall & Pickles Ltd. (4) (now part of Osborn Steels, M-233).
M-1514 Hamilton Die Cast Inc., Hamilton, OH 45011.
M-850 Hamilton Foundry Div. Hamilton Allied Corp., Hamilton, OH 45011.
M-1084 Hamilton Precision Metals (now Hamilton Technology, Inc., Lancaster, PA 17604).
M-697 Hammond & Irving Inc., Auburn, NY 13201.
M-992 Hammond Brass Works, Hammond, IN (3).
M-361 Hand, Edward A., Philadelphia, PA (3).
M-827 Handcock Valve Div. Manning, Maxwell & Moore Co., Bridgeport, CT.
M-1328 Handler, G.m.b.H., Düsseldorf, West Germany.
M-63 Handy & Harmon, New York, NY 10022.
M-1269 Hanford Foundry Co., San Bernardino, CA 92404.
M-1574 Hanna Furnace Corp., Buffalo, NY 14224 (subsidiary of National Steel Co.).
M-833 Hansell-Elcock Co., Chicago, IL (3).
M-316 Hans-Heinrich Hütte, G.m.b.H., Langelsheim a.Harz, Germany.
M-580 Hanson-Van Winkle-Munning Co., Indianapolis, IN 46236.
M-1471 Hard Alloys Ltd., High Wycombe, Buckinghamshire, England.
M-502 Hardenite Steel Co. Ltd., Sheffield, S4 7WZ England.
M-166 Hardite Metals Inc., New York, NY (3).
M-799 Hardy & Co., Charles, New York, NY (3).
M-828 Harnischfeger Corp. (2) (alloy welding rod business transferred to Chemetron, M-1713).
M-1156 Harris, Arthur & Co., Chicago, IL 60607.
M-310 Harrison, Fischer & Co., Ltd., Sheffield, S3 7WJ England.
M-1235 Harvey Aluminum Inc. (2) (now Martin Marietta Aluminum).
M-830 Harville Machine Inc. (2).
M-497 Hassall & Sons, Wm., Manchester, England (3).
M-362 Hawkridge Bros Co., Malden, MA 02148.
M-167 Haynes Stellite (now Stellite Div. Cabot Corp., Kokomo, IN 46901).
M-1553 Haywood Foundries, Ltd., London, England (3).

M-411	Haywoods NCA Metal Ltd., London, England (3).
M-1270	Haywood Tyler of Canada, Ltd., Kitchener, Ontario, Canada.
M-1554	Heckford Ltd., Arthur E., Birmingham, England.
M-841	Hedstrom Corp., Oscar W., Chicago, IL (3).
M-1131	Heinkel-Herth Co., Berlin, Germany.
M-217	Hellefors Bruks Aktiebolag, Hellefors, Filipstad, Sweden.
M-343	Heller Bros. Co., Newark, NJ, see Heller Tool Div., M-1221.
M-1221	Heller Tool Div., Wallace Murray Corp., Newcomerstown, OH 43832.
M-488	Hemmings & Co., Rotherham, England (3).
M-1287	Henckels Zwillingswerke, G.A., Solingen, Germany.
M-1391	Henning Bros. & Smith Inc., Brooklyn, NY 11237.
M-1188	Henricot, S.A. Usines, Emile, Court-St.-Etienne, Belgium.
M-64	Heppenstall Co., Pittsburgh, PA 15201.
M-296	Heraeus-Vacuumschmelze (now Vacuumschmelze, M-1631).
M-1806	Herbert-Cutanit Ltd., Warrington, Lancashire, WA4 3JX England.
M-482	Herbert Small Tools & Equipment Ltd., see Herbert-Cutanit Ltd.
M-1181	Hetzel & Co., Nürnberg, Germany.
M-204	Hetzel, Vereinigte Silberhammerwerke, Nürnberg, Germany.
M-959	Hewitt Metals Corp., Detroit, MI 48208.
M-1044	Hewson Co., John, New York, NY (3).
M-344	Hidalgo Steel Co. Inc., New York, NY (3).
M-562	Hiertz Metal Co., Theodore, St. Louis, MO (3).
M-426	High Duty Alloys Ltd., Slough, Buckinghamshire, SLI 4PA England.
M-168	Hills-McCanna Co., Creston, IA 50801.
M-1000	Hobart Bros. Co., Troy, OH 45373.
M-340	Hobson, Houghton & Co., New York, NY (3).
M-1350	Hochfrequenz-Tiegelstahl, G.m.b.H., Bochum, Germany (2).
M-1781	Hoeganaes Corp. (subsidiary of Interlake Inc.), Riverton, NJ 08077.
M-1526	Hoeganaes Metal Powders Ltd., Birmingham, England.
M-1767	Hoesch Huttenwerke AG., 4600 Dortmund 1, West Germany.
M-798	Hoesch-Kohn-Nevessen (now Hoesch Huttenwerke, M-1767).
M-1338	Hoffmann & Co., Köln, Germany (2).
M-1325	Hoffman Elektrogusstahlwerk, Alb., Eschweiler, Germany.
M-224	Hofors Steel Works, Hofors, Sweden.
M-638	Hoganas-Billesholms Aktiebolaget, Hoganas, Sweden (2).
M-844	Hollup Corp., Chicago, IL.
M-169	Holroyd & Co. (now Holcroft Castings & Forgings Ltd., Rochdale, OL12 OLL England).
M-1282	Holt Equipment Co., Independence, OR (3).
M-1303	Holzer, Jacob (merged into Creusot-Loire, M-1488).
M-1800	Homogeneous Metals Inc., Herkimer, NY 13350.
M-1364	Honsel-Werke AG., Leichtmetallwerke, 5778 Meschede, West Germany.
M-995	Hopkinsons Ltd., Huddersfield, HD2 2UR England.
M-1377	Horbach & Schmitz G.m.b.H., 5000 Köln, Hansaring 49-51, West Germany.
M-610	Horndal Jernverks, A.B., Horndal, Sweden.
M-65	Hoskins Mfg. Co., Detroit, MI 48208.
M-363	Houghton & Richards Inc., Boston, MA (3).
M-1797	Hover, Chr., & Sohn, Edelstahlwerk, 5251 Berghausen uber Engelskirchen, West Germany.
M-1306	Hover, Gebruder, Edelstahlwerk, 5251 Kaiserau uber Engelskirchen, West Germany.
M-1045	Howard Foundry Co., Chicago, IL (1).
M-1699	Howmet Corp., Carbide Div., McKeesport, PA 15134 (formerly Firth-Loach Products).
M-908	Hoyland Steel Co., New York, NY (3).
M-428	Hoyt Metal Co., New York, NY (3).
M-317	Hoyt Metal Co. of London Ltd., Putney, London, SW15 2NX England.
M-225	Hubbard Steel Co., East Chicago, IN (3).
M-890	Hueck, Edward, Ludenscheid, Germany.
M-1796	Hufnagel G.m.b.H., 8500 Nürnberg, West Germany.
M-784	Hughes & Co., Ltd., F. A., London, England (3).
M-1499	Huntington Alloys Inc., Huntington, WV 25720.
M-664	Hunt-Spiller Div. Power Products Inc. (2).
M-892	Hurbenium Co. of America, Detroit, MI (3).
M-1271	Husqvarna Vapenfabrik Aktiebolag, Stockholm, Sweden.
M-194	Hybnickel Alloys Co., Wilmington, DE (3).
M-905	Hyde Park Foundry & Machine Co., Hyde Park, PA 15641.
M-42	Hydril Co., Los Angeles, CA (2).
M-1505	Hydrometals Inc., Dallas, TX 75206.
M-847	Hytensil Aluminum Co., Chicago, IL (3).

I

M-1346	Idealstahl Breidenbach KG, 5270 Gummersbach 31, Germany.

M-346	Illingworth Steel Co., Philadelphia, PA (3).	M-1751	Interlake Inc., Steel Division, Chicago, IL 60627.
M-1226	Illinois Tool Works, Chicago, IL 60631 (2).	M-5	International Alloys Ltd., Aylesbury, Buckinghamshire, England.
M-226	Illinois Zinc Co., Chicago, IL (3).	M-2a	International Development Corp., New York, NY (3).
M-29	Illium Corp. (1) (alloys now produced by Stainless Foundry, M-1262).	M-1661	International Lead-Zinc Research Association Inc., New York, NY 10017.
M-1452	Ilssa-Viola SpA, I-20159 Milan, Italy.		
M-1805	IMI Rod & Wire, Birmingham, England.	M-68	International Nickel Co., Inc., New York, NY 10004.
M-1578	IMI Rolled Metals, Birmingham, B6 7BA England.	M-1598	International Wire Products Co., Wyckoff, NJ 07481.
M-282	Imperial Aluminium Chemical Industries (Metals Div.) (now IMI (Knoch) Ltd., Birmingham, England, M-1578).	M-441	Interstate Foundry & Machine Co., Johnson City, TN (2).
		M-1764	IPM Corp., Columbus, OH 43207.
		M-1621	Isaacson Iron Works.
M-1547	Imperial Aluminium Co., Ltd., Birmingham, England (2), (alloys now supplied by Alcoa of Great Britain, M-1715).	M-789	Isabellenhuette, Berlin, Germany.
		M-1745	Iscar Ltd., Nahariya, Israel, or James Bellandi, Elmhurst IL 60126, or M. Baumgarten, Iscar Metals Inc., Raritan Center, Edison, NJ 08817.
M-786	Imperial Brass Mfg. Co., Chicago, IL (3).		
M-469	Imperial Chemical Industries (2) (alloy business transferred to IMI, M-1578, or Alcoa of Great Britain, M-1715).	M-1570	Ishikawajima-Harima Heavy Industries Co. Ltd., New York, NY 10048.
		M-471	Isteg Steel Products Co., Westminster, England (3).
M-1578	Imperial Metal Industries (now IMI Ltd., Birmingham, B6 7BA England).	M-1397	Istituto Sperimentali Metalli Leggeri, Milan, Italy.
M-351	Imperial Smelting Co. Ltd., St. James Square, London, SW 1 England.	M-269	ITT Components Group Europe, Harlow, Essex, United Kingdom.
M-771	Imphy, Société Metallurgique d', Paris 7ᵉ, France (now part of Creusot-Loire, M-1488).		

J

M-1076	Indiana General, Valparaiso, IN 46383.	
M-1041	Indium Corp. of America, Utica, NY 13502.	M-477	Jackman, Joseph, & Co., Ltd., Sheffield, England (1).
M-1482	Induction Steel Castings Co. Inc., East Detroit, MI 48021.	M-1032	Jackson Iron & Steel Co., Jackson, OH 45640.
M-1689	Industrial Alloy Steel Co., Cleveland, OH (2) (3).	M-1297	Jadot, Usines de, Brussels, Belgium.
		M-1643	Jamestown Malleable Iron Div., Blackstone Corp., Jamestown, NY 14701.
M-400	Industrial Furnace Corp., Buffalo, NY (3).		
M-1171	Industrial Overlay Metals Corp., New York, NY (3).	M-1423	Jamison Steel Corp., San Francisco, CA (3).
M-440	Industrial Steels Inc., Cambridge, MA (3).	M-508	Janney Cylinder Co., Philadelphia, PA 19136.
M-499	Industrial Steels Ltd., Sheffield (4) (merged into British Steel Corp., M-1724).	M-1577	Japan Metal Industry Co. Ltd., Kawasaki, Japan.
		M-1419	Japan Special Steel Co., Tokyo, Japan.
M-1585	Industria Magenti Fermanenti E. Affini (Impia), Milan, Italy.	M-1542	Japan Steel & Tube Co. Ltd., Tokyo, Japan.
M-1219	Industria Nazionale Alluminio, Mori (Trento), E. Bolzano, Italy.	M-1539	Japan Steel Works Ltd., New York, NY 10017.
M-829	Industries Trading Co., New York, NY (3).	M-170	Jeffrey Mfg. Div., Dresser Industries Inc., Columbus, OH 43216.
M-1224	Ingersoll-Rand Co., Phillipsburg, NJ 08865.	M-791	Jelenko & Co., J. F., New Rochelle, NY 10801.
M-328	Ingersoll Steel Co., New Castle, IN 47362.	M-231	Jelliff Corp., C. O., Southport, CT 06490.
M-1606	Inland Electronics Products Corp., Pasadena, CA (3).	M-261	Jessop-Saville Ltd., Sheffield, 9 England (3).
M-67	Inland Steel Co., Chicago, IL 60603.		
M-1749	Inman & Co. Ltd., T., Sheffield, S4 7YA England.	M-486	Jessop-Saville Ltd., Sheffield, 9 England (3).
M-322	INSILCO Corp., Meriden, CT 06450.	M-69	Jessop Steel Co., Washington, PA 15301.

M-1103	Jobbins, Inc., W. F., Aurora, IL (1) (bought out by U.S. Reduction, M-585).	M-792	Kellogg & Co., M. W. (now Pullman Kellogg Div. of Pullman Inc., E. Houston, TX 77046).
M-1588	Jodots Freres, Ets, Boloeil, Belgium.	M-900	Kelly Foundry & Machine Co., Elkins, WV 26241.
M-332	Johnson & Co., A., New York, NY (3) (agent for M-16).	M-171	Kencroft Malleable Co., Buffalo, NY (3).
M-172	Johnson Bronze Co., New Castle, PA (2).	M-793	Kennametal Inc., Latrobe, PA 15650.
		M-1590	Kensington Steel Co., Chicago, IL (2).
M-393	Johnson Mfg. Co., Princeton, IA 52768.	M-1039	Keokuk Electro-Metals Co., Keokuk, IA (3).
M-200	Johnson Matthey Metals Ltd., Southgate, London, N14 6ET England.	M-1129	Kirkling & Co., Whittier, CA (2).
		M-1550	Kerr-McGee Chemical Corp., Oklahoma City, OK 73125.
M-1620	Johnston Steel & Wire Co., Worcester, MA 01607.	M-1125	Kester Solder Co., Chicago, IL (2).
M-1195	Jonas & Culver Ltd., Sheffield, 9 England.	M-989	Keystone Carbon Co., St. Marys, PA 15857.
M-173	Jones & Laughlin Steel Co., Pittsburgh, PA 15263.	M-339	Kidd Drawn Steel Co., Aliquippa, PA (3).
M-1555	Jones & Rooke Ltd., Birmingham, England.	M-1352	Kind & Co., Edelstahlwerk, 5276 Wiehl 2-Bielstein, Germany.
M-422	Jones, Rd., Ltd., Birmingham, England (3).	M-1601	King Materials Laboratory, Victor, Palo Alto, CA (2).
M-1780	Jorgensen, Earle M., Los Angeles, CA 90054.	M-1641	King Materials Laboratory, Victor, Palo Alto, CA (2).
M-891	Joselin & Co., C., New York, NY (3).	M-174	Kinite Corp., Milwaukee, WI (3).
M-560	Joslyn Stainless Steels Co., Ft. Wayne, IN 46801.	M-546	Kinney Iron Works, Los Angeles, CA (3).
M-1347	Junker G.m.b.H., Otto, 5101 Lammersdorf uber Aachen, West Germany.	M-794	Kirk & Son Inc., Morris P., Los Angeles, CA (taken over by NL Industries, M-88).
		M-1159	Kling Bros. Engineering Works, Chicago, IL (3).

K

		M-1650	Klockner-Werke AG., Duisberg, Germany (2).
M-1310	Kabel Stahlwerke, C., Pouplier, Hagen-Kabel, Germany.	M-611	Klosters A.B., Langshyttan, Sweden (2).
M-517	Kahl-Holt Co., Baltimore, MD (3).	M-355	Kloster Steel Corp., Chicago, IL 60607.
M-429	Kahl Iron Foundry, Detroit, MI 48216.	M-1193	Knapp Mills, Inc., Long Island City, NY (3).
M-1147	Kaiser Aluminum & Chemical Corp., Oakland, CA 94643.	M-1518	Knowsley Cast Metal Co., Ltd., Manchester, England (3).
M-1148	Kaiser Steel Corp., Oakland, CA 94643.	M-645	Knutange, Sté Metallurgique de, Paris, France (2).
M-262	Kalif Corp., Emeryville, CA (3).		
M-70	Kanthal A.B., Hallstammer, Sweden.	M-1604	Kobe Steel Ltd., New York, NY 10022.
M-1150	Kanthal Corp., Bethel, CT 06801.	M-75	Koch Light Alloys Ltd., Willesden, London, NW 10 England.
M-1313	Kanz Metallwerke, Hans, Zurich-Albisrieden, Switzerland.	M-1583	Koerver & Nehring G.m.b.H., 4150 Krefield, West Germany.
M-1132	Kapfenberg, Steiermark, Austria.		
M-1701	Kasle Steel Co., Detroit, MI (2).	M-1702	Koncor Industries, Div. of Latrobe Steel Co. (4), see Latrobe Steel Co., M-73.
M-1541	Kawasaki Steel Corp., New York, NY 10017, or Fukiai-Ku, Kobe 651 Japan.	M-399	Koppers Co. Inc., Metal Products Div., Baltimore, MD 21203.
M-646	Kawecki Berylco Industries Inc., New York, NY 10017.	M-346	Kramer & Co., Chicago, IL 60608.
M-1622	Kawecki Chemical Co., see Kawecki Berylco Industries, M-646.	M-324	Kreidler Werke G.m.b.H., 7000 Stuttgart 40, West Germany.
M-572	Kay-Brunner Steel Products Inc., Alhambra, CA (2).	M-772	Kries & Sons Co., Henry A., Baltimore, MD (3).
M-1275	Kaye, E. & E., Ltd., Enfield, Middlesex, EN3 4SS England M-1444 (now part of Pechiney).	M-1135	Kropp Forge Co., Chicago, IL 60650.
		M-72	Krupp Huttenwerke, Fried., 4630 Bochum, West Germany.
M-510	Kayser-Ellison (now Sanderson Kayser Ltd., Sheffield, S9 2SD England).	M-1327	Kuhbier & Sohn, C., Zweigbetrieb, 5885 Schalksmühle 2, West Germany.
M-731	Keasby & Matteson Co., Ambler, PA (3).	M-1517	Kulite Tungsten Corp., Ridgefield, NJ 07657.

L

M-263	LaBour Pump Co., Elkhart, IN 46514.
M-552	LaClede Brass (or Steel) Works, St. Louis, MO 63102.
M-1527	Ladish Co., Cudahy, WI 53110.
M-857	Lake & Elliot Ltd., Braintree, England.
M-888	Lake City Malleable Co., Cleveland, OH (2).
M-800	Lake Erie Engineering Corp., Buffalo, NY (3).
M-1480	Lakeside Malleable Castings Co., Racine, WI (3).
M-1165	Lancaster Steel Co. Inc., New York, NY (3).
M-1169	Langley Alloys Ltd., Langley, Slough, Buckinghamshire, SL3 6EA England.
M-527	Lanz, Heinrich, A.G., Mannheim, Germany (3).
M-669	La Salle Steel Co., Chicago, IL 60680.
M-73	Latrobe Steel Co., Latrobe, PA 15650.
M-801	Lavin & Sons Inc., R., Chicago, IL 60623.
M-624	Lavorazione Leghe Leggere, Porto Marghera (Venezia), E. Ferrara, Milan, Italy.
M-1047	Lazalloys Ltd., Bently Works, Doncaster, Lancashire, England (2).
M-1408	Leadbeater & Scott Ltd., Sheffield, England (1).
M-74	Lebanon Steel Foundry, Lebanon, PA 17042.
M-1687	Le Bronze Industrial, Rene Loiseau & Cie, 93001 Bobigny, France.
M-1586	Leghe E. Accini Speciali (LEAS), Milan, Italy.
M-300	Lehigh Babbitt Co., Allentown, PA (3).
M-688	Lehigh Steel Corp., New York, NY (3).
M-1196	Le Moyne Steel Co., Pittsburgh, PA (3).
M-1459	Lesjefors Aktiebolag, Lesjefors, Sweden.
M-227	Levett, Walker M., New York, NY (3).
M-1396	LeVita Metal Alloy Co., Detroit, MI (3).
M-540	Lewin Metals Corp., East St. Louis, IL (3).
M-1762	Ley's Malleable Castings Co. Ltd., Derby, DE3 8LY England.
M-573	LFM Mfg. Co., Atchison, KS (3).
M-75	Lightalloys Ltd. (now Koch Lightalloys Ltd., Willesden, London NW 10 England).
M-1481	Light Alloys Products Co. Ltd., Munworth, England.
M-1031	Light Metals Inc., Indianapolis, IN 46205.
M-951	Light Metal Works, Tokyo, Japan.
M-578	Lincoln Electric Co., Cleveland, OH 44117.
M-605	Lindberg Hevi Duty Div. Solar Basic Industries, Chicago, IL 60612.
M-843	Linde Div. Union Carbide Corp, Metals Division, New York, NY 10017.
M-1319	Lindenberg Edelstahlwerk (4) (merged into Bergische Stahl- Industrie, M-1290).
M-228	Link-Belt Div. F.M.C. Corp., Chicago, IL.
M-1050	Lithaloys Corp., New York, NY (3).
M-1479	Little Bros. Foundry Co., Port Huron, MI (3).
M-1244	Little Falls Alloys Inc., Paterson, NJ 07501.
M-1466	Lobdell Co., Wilmington, DE (3).
M-573	Locomotive Finished Materials Co., Atchison, KS (3).
M-802	Logan Iron & Steel Co., Philadelphia, PA (3).
M-1311	Lohmann, Fried., G.m.b.H., Gusstahlfabrik, Walz-und Hammerwerke, 5821 Herbede (Ruhr), West Germany.
M-1546	Loire, Compagnie des Ateliers et Forges de la, F-75, Paris 9e France.
M-1118	Longwy, Société des Aciéries, Mont-Saint-Martin, France.
M-1495	Lorraine-Escault (now Vallourec, 75764 Paris Cedex 16, France).
M-1457	Lovsted & Co. Inc., C. M., Seattle, WA (2).
M-803	Lowmoor Best Yorkshire Iron Ltd., Lowmoor England (3).
M-1137	L & R Mfg. Co., Arlington, NJ (3).
M-1085	Ludlow-Saylor Wire Cloth Div. General Steel Industries, St. Louis, MO 63114.
M-1422	Ludlow Steel Corp., Bedford, OH 44146.
M-176	Lukens Steel Co., Coatesville, PA 19320.
M-77	Lumen Bearing Co., Buffalo, NY (3).
M-264	Lunkenheimer Co., Cincinnati, OH 45214.
M-534	Lunn & Co., E., Glasgow, Scotland.
M-1443	Luria Steel & Trading Co., New York, NY (2).
M-523	Lynchburg Foundry Co., Lynchburg, VA (2).
M-456	Lyon, Conklin & Co., Baltimore, MD (2).

M

M-1680	Macauley Foundry Co., Berkeley, CA 94710.
M-323	Machinenbau A.G., Golzern-Grimma, Germany.
M-1218a	Machinery & Machine Supplies Co. Inc. New York, N.Y. (3).
M-305	Machlett Laboratories, Stamford, CT (2).
M-177	Mackenzie's Sons Co. Inc., Duncan, Trenton, NJ (3).
M-229	Mackintosh-Hemphill Div. Gulf & Western Mfg. Co., Pittsburgh, PA 15203.
M-1461	Madison Foundry Co., Cleveland, OH (3).
M-1051	Madison-Kipp Corp., Madison, WI 53704.

M-1721	Magnacast Corp., Arlington Heights, IL 60005 (formerly Masten Corp.).
M-556	Magnesium Castings & Products Ltd., Slough, England (3).
M-78a	Magnesium Development Corp., Newark, NJ (3).
M-1130	Magnesium Elektron Ltd., St. James Square, London, SW 1 England.
M-1089	Magnetic Metals Co., Camden, NJ 08101.
M-1626	Magnetics, Div. of Spang Industries Inc., Butler, PA 16001.
M-436	Magnolia Anti-Friction Metal Co., London SW 1 England.
M-178	Magnolia Metal Corp., Auburn, NE 68305.
M-958	Magnus Metal Corp. (2) (now Bearings Div. NL Industries, St. Louis, MO).
M-1581	Magotteaux, Les Fonderies, Vaux-sous-Chevrement, (Liege), Belgium.
M-1365	Mahle G.m.b.H., D-700 Stuttgart 50, West Germany.
M-1292	Main Metal Ltd., London, WC 2 England (3).
M-859	Major Engineering Co., Tulsa, OK 74101.
M-1066	Makepeace Co., D. E., Attleboro, MA 02703 (Div. of Engelhard).
M-79	Malleable Iron Fitting Co., Branford, CT (3).
M-265	Mallory Metallurgical (now CMW Inc., Indianapolis, IN 46206) (Div. of P. R. Mallory & Co. Inc.).
M-1229	Mallory-Sharon (now RMI Co., Niles, OH 44446).
M-221	Manco Products Co., Melvindale, MI 48122.
M-179	Manganese Bronze & Brass Co. Ltd., Ipswich, England (4) (part of Delta Metal, M-156).
M-1504	Manganese Chemicals Corp. (now Chemetals Div. Diamond Shamrock, Cleveland, OH 44114).
M-180	Manganese Steel Forge Co., Philadelphia, PA 19134.
M-1381	Mannesmann-Hüttenwerk A.G., Duisberg, Germany.
M-1334	Mannesmannrohren-Werke A.G., 4000 Düsseldorf 1, West Germany.
M-230	Manning, Maxwell & Moore, Div. of Dresser Industries, Muskegon, MI 49443.
M-1509	Manoir A., Pitres (Eure), Usines du, Paris, France.
M-893	Manos Ltd., Zurich, Switzerland.
M-1068	Manross Div. Associated Spring Co., Bristol, CT 06010.
M-1691	Mansfield Brass & Aluminum Corp., Mansfield, OH (2).
M-1496	Marathon Specialty Steels Inc., New York, NY 10001
M-1671	Markey Bronze Corp., Delta, OH 43515.
M-1096	Marquette Corp., Minneapolis, MN (3).
M-1180	Marrel Freres, S.A., F-42800 Rive-de-Gier, France.
M-1149	Marshall Steel Co., La Grange, IL 60525.
M-753	Marsh Bros. & Co. Ltd., Sheffield, England.
M-348	Martinel Steel Co., Ltd., Sheffield, England (3).
M-1536	Martin Metals Co., Martin-Marietta Corp., Baltimore, MD 21227 (also Wheeling, IL 60090).
M-323	Maschinenbau A.G., Golzern-Grimma, Germany.
M-1277	Massey-Harris Ltd., Toronto, Ontario, Canada.
M-467	Massillon Steel Casting Co., Massillon, OH 44646.
M-1721	Maston Corp. (now Magnacast Corp., Arlington Heights, IL 60005).
M-1627	Materials Research Corp. (2).
M-454	Mather & Platt Ltd., Manchester, England.
M-1511	Mathieson-Heglar Co., La Salle, IL 61301.
M-1765	Matthey Bishop, Malvern, PA 19355.
M-302	Maulbronn, Aluminiumwerke, Maulbronn-Württ, Germany.
M-315	Maywood Chemical Works, Maywood, NJ (3).
M-212	McCallum-Hatch Bronze Co. Inc., Buffalo, NY (3).
M-1107	McCauley Alloy Sales Co., New York, NY (3).
M-80	McDonald & Co., P.F., Boston, MA (3).
M-710	McGean Chemical Co. Inc., Cleveland, OH 44109.
M-181	McGill Mfg. Co. Inc., Valparaiso, IN (2).
M-81	McInnes Steel Co., Corry, PA (2).
M-849	McKay Co. (now Teledyne McKay, Pittsburgh, PA 15219).
M-914	McKechnie Metals Ltd., Walsall, WS9 8DN England.
M-1592	McLouth Steel Corp., Detroit, MI 48209.
M-182	McPhail & Sons, Wm., Glasgow, Scotland (3).
M-1208	McQuay-Norris Mfg. Co., St. Louis, MO (3).
M-1441	Medart Engineering & Equipment Co., St. Louis, MO (2).
M-1167	Meech Foundry Inc., Cleveland, OH 44125.
M-82	Meehanite Worldwide, Div. of Meehanite Metal Corp., Chattanooga, TN 37401.
M-1556	Meigh Castings Co. Ltd., Cheltenham, England.
M-1349	Mentah Co., Vorgtlinder, West Berlin, Germany.
M-519	Merco Nordstrom Valve Co., Pittsburgh, PA (3).
M-1436	Meridian Steel Co. Inc., Jericho, NY 11753.

M-852	Mesta Machine Co., Pittsburgh, PA 15230.	M-1212	Miller Co., Meriden, CT 06450.
M-1065	Metal & Alloy Div. Silver Creek Precision Co. (1).	M-743	Miller Steel Co., Newark, NJ (3).
		M-187	Mills & Co. Ltd., William, Wednesbury, Staffordshire, WS10 0JY England.
M-529	Metal & Thermit Corp., New York, NY (3).	M-365	Milne & Co., A., Atlanta, GA 30318.
M-1559	Metal Alloys Ltd., South Wales (2).	M-1476	Milwaukee Die Casting Co., Milwaukee, WI (2).
M-913	Metal Carbides Corp., Youngstown, OH 44512.	M-475	Milwaukee Steel Foundry Co., Milwaukee, WI (3).
M-968	Metal Castings Ltd., Worcester, England (3).	M-858	Minimax Co., Div. of Graham Chemical Corp., Chicago, IL 60660.
M-320	Metallbank, A.G., (Heddernheimer Copper Works), Heddernheim, Germany.	M-733	Mining & Chemical Products Ltd., Alberton, Wembly, England.
M-299	Metallgesellschaft A.G., Frankfurt/Main, Hesse 6000, West Germany.	M-1752	3M Company, Industrial Electrical Products Div., 3M Center, St. Paul, MN 55101.
M-402	Metallhüttenwerke Schaefer und Schael, A.G., Breslau, Germany.	M-415	Miralite Ltd., Mortlake, England (1).
M-381	Metallo-Chemische Fabrik, Dr. Leopold Rostosky, Berlin, Germany.	M-988	Mir-o-Col Alloy Co., Los Angeles, CA (3).
M-813	Metalloy Products Co., Newport Beach, CA (3).	M-536	Mitchell Co. Inc., Robert, Montreal, Canada.
M-31	Metallurgical Products Dept., General Electric Corp., Detroit, MI 48232.	M-1685	Mitsubishi Metal Products of Japan, Tokyo, Japan.
M-971	Metallwerke, A.G., Vienna, Austria.	M-1543	Mitsubishi Steel Mfg. Co. Ltd., New York, NY 10017, or Tokyo, Japan.
M-1246	Metallwerk Plansee Gesellschaft, Reutte-Tirol, Austria.	M-1523	Molecu Wire Corp., Farmington, NJ 07727.
M-183	Metal Sales Corp., Jersey City, NJ (3).		
M-926	Metals & Controls and General Plate Div. of, Attleboro, MA 02703.	M-860	Moltrop Steel Products Co., Beaver Falls, PA (2).
M-1562	Metals & Controls and General Plate Div. of, Attleboro, MA 02703.	M-837	Molybdenum Co., N. O., Reutte-Tyrol, Austria.
M-1753	Metal Specialties Inc., E. Fairfield, CT 06430.	M-640	Molybdenum Corp. of America (now Molycorp., Inc., White Plains, NY 10604).
M-853	METCO Inc., Westbury, Long Island, NY 11590.		
M-1552	Metro-Cutanit Ltd., Warrington, Lancashire, England.	M-863	Monarch Alloy Co., Ravenna, OH (3).
		M-495	Monarch Steel Corp., Indianapolis, IN (3).
M-83	Metropolitan-Vickers Electrical Co. Ltd., Manchester, England.	M-86	Mond Nickel Co. Ltd., London, England.
M-450	Micheville, Sté des Aciéries de, Paris, France.	M-1445	Montecatini Settore Alluminio, Milan, Italy.
M-184	Michiana Products Corp., Michigan City, IN (3).	M-839	Moraine Mfg., Inc., Dayton, OH 45439.
M-1174	Michigan Powdered Metal Products Inc., Livonia, MI 48150.	M-1519	Morgan Crucible Co. Ltd., London, England (3).
M-545	Michigan Smelting & Refining Co., Detroit, MI (3).	M-87	Morris Ashby Ltd., London, England (3).
M-105	Michigan Standard Alloy Inc., Benton Harbor, MI 49022.	M-1250	Morris Machine Works, Baldwinsville, NY (2).
M-84	Michigan Steel Casting Co., Detroit, MI.	M-601	Motor Castings Co., Milwaukee, WI 53214.
M-751	Michigan Tool Co., Detroit, MI (3).	M-934	Mt. Vernon Furnace & Mfg. Co., Mt. Vernon, IL (3).
M-312	Michigan Valve & Foundry Co., Detroit, MI (2).	M-266	Mueller Brass Co., Port Huron, MI 48060.
M-185	Midland Motor Cylinder Co., Smithwick, England (3).		
M-85	Midvale-Heppenstall Co., Philadelphia, PA 19140.	M-188	Muntz & Co., P. H., Ltd., Birmingham, England (3).
M-1607	Midwest Materials Co., Cleveland, OH (2) (3).	M-526	Murex Ltd., Rainham, Essex, England.
M-186	Millbury Steel Foundry Co., Millbury, MA (3).	M-1531	Muskegon Piston Ring Co., Muskegon, MI 49443.
M-1144	Miller Co., Meriden, CT 06450.	M-1245	Myrens Verkstec A/S, Oslo, Norway.

N

M-679	Nanterre, Fonderie de Precision, Paris, France.
M-952	Napraloy Co., Jersey City, NJ (3).
M-548	Nassau Smelting & Refining Co. Inc., Staten Island, NY (2).
M-1463	National Airoil Burner Co., Philadelphia, PA 19134.
M-414	National Alloys Ltd., London, England (3).
M-189	National Alloy Steel Div. Blaw-Knox (2) (alloys are now made at Duraloy Blaw-Knox, M-45).
M-1688	National Alloy Tool Co., Cleveland, OH (2) (3).
M-539	National Bearing Div. ABEX Corp., Meadville, PA 16335.
M-864	National Broach & Machine, Detroit, MI 48213.
M-190	National Bronze & Aluminum Foundry, Cleveland, OH (3).
M-910	National Cable & Metal Co., Glendale, CA (3).
M-89	National Castings Div. Midland-Ross Corp., Cleveland, OH.
M-232	National Erie Corp., Erie, PA (3).
M-483	National Forge Co., Irvine, PA 16329.
M-88	National Lead (now NL Industries, Metals Div., Highstown, NJ 08520).
M-311	National Physical Laboratory, Teddington, England.
M-1624	National Research Corp., Metals Div., Newton, MA 02158.
M-1507	National Standard Co., Niles, MI 49120.
M-836	National Steel Co., Pittsburgh, PA 15219.
M-1446	National Tool Div., National Cleveland Corp., Cleveland, OH (3).
M-90	National Tube Div. U.S. Steel (4), see U.S. Steel Corp., M-604.
M-1458	Navan Inc., Anaheim, CA 92803.
M-865	Nesaloy Products Inc., New York, NY (3).
M-1386	Neubrandenburg Eisenwerke, G.m.b.H., Berlin, Germany.
M-1366	Neumayer Kabel u. Metallwerke A.G., Nürnberg, Germany.
M-1203	Newcomer Products Inc., Latrobe, PA 15650.
M-1720	New England Brass, Taunton, MA 02781.
M-973	New England Collapsible Tube Co., New London, CT (3).
M-442	New England High Carbon Wire Co., Millbury, MA (2).
M-91	New Jersey Zinc Co., Bethlehem, PA 18018.
M-191	Newport Steel Co. (4) (part of Interlake Inc., M-1751).
M-349	Ney Co., J. M., Bloomfield, CT 06002.
M-92	Niagara Falls Smelting & Refining Div. Continental Copper & Steel Industries Inc., Buffalo, NY (3).
M-1693	Niborium Industries Inc., Providence, RI 02905.
M-93	Nicralium Co., Jackson, MI (3).
M-716	Nihon Jyokiko Seikosho Goshi, Koisha, Japan.
M-1402	Nippert Co., Delaware, OH 43015.
M-1612	Nippon Kokon K.K., New York, NY 10017, or Tokyo, Japan.
M-1651	Nippon Stainless Steel Co. Ltd., C/O Sumitomo Metal Ind., Ltd., New York, NY 10017, or Tokyo, Japan.
M-1732	Nippon Steel U.S.A. Inc., New York, NY 10022.
M-386	Nitralloy Corp., New York, NY (3).
M-88	NL Industries Inc., Metal Division, Highstown, NJ 08520.
M-427	Non-Corrodal Alloys Ltd., London, England (3).
M-576	Non-Ferrous Castings Co. Ltd., Smethwick, England.
M-1608	Noranda Metal Industries Ltd., Montreal, Quebec, H3C 2Y4 Canada.
M-1238	Nordisk Aluminium Industry, A/S, Holmestrand, Norway.
M-1237	Norsk Aluminium Co. A/S, Hoyanger, Norway.
M-1139	North American Philips Co. Inc., New York, NY 10017.
M-1401	North American Philips Co. Inc., New York, NY 10017.
M-207	North American Steel Co., Cleveland, OH (3).
M-683	Northeast Metals Co., Philadelphia, PA (3).
M-643	Northern Aluminium, see Alcan Booth, M-643.
M-542	Northfield Iron Co., Northfield, MN (3).
M-532	Norton Co., Metals Div., Newton, MA 02164.
M-868	Novo Pump & Machine Div. American Marsh Pumps Inc., Lansing, MI 48905.
M-1629	Nuclear Metals Inc., Concord, MA 01742.
M-555	Nürnberg, Aluminiumwerke, G.m.b.H., Nürnberg, Germany.
M-1454	Nusite Steel Process Co., Warsaw, IN (3).
M-1448	Nyby, Granges, AB., S-644 80 Torshalla, Sweden.

O

M-1094	Oakes Bronze & Aluminum Co., Warren, OH 44482.
M-1420	Oederlin & Co. Ltd., Baden, Switzerland.
M-94	Ohio Brass Co., Mansfield, OH 44902.
M-398	Ohio Ferro Alloys Corp., Canton, OH 44711.

M-1163	Ohio Stainless & Commercial Steel Co., Cleveland, OH (3).	M-1804	Pechiney Ugine Kuhlmann Corp., New York, NY 10022 (parent company of many French metal companies).
M-788	Oil Well Supply Co., Houston, TX (3).		
M-927	Oklahoma Steel Castings Co., Tulsa, OK 74101.	M-876	Peckovers, Ltd., Toronto, Canada.
		M-341	Peninsular Steel Co., Detroit, MI 48205.
M-866	Oldham & Co., F. B., Buffalo, NY (3).	M-1433	Pennsylvania Steel Corp., Detroit, MI (3).
M-1007	Olds Alloys Co., South Gate, CA (3).		
M-279	Olin Brass Div. Olin Corp., East Alton, IL 62024.	M-466	Pennsylvania Steel Foundry & Machine Co., Hamburg, PA (2).
M-1566	Olin Corp., Stamford, CT 06904.	M-1252	Penrold Div. Brush Wellman Corp., Reading, PA.
M-1719	Olin Metals Research Labs., New Haven, CT (2).		
		M-1599	Perfect Circle Div. Dana Corp., Richmond, IN (2).
M-714	Oman Non-Friction Metal Co., El Paso, TX (3).		
		M-1615	Perfection Mica Co., Bensonville, IL 60106.
M-297a	Ore & Chemical Corp., New York, NY 10016, or Houston, TX 77092 (part of VDM, M-297).		
		M-1025	Permanente Magnesium Inc., Oakland, CA (1).
M-1603	Oregon Steel Mills, Portland, OR 97203.	M-1010	Permanent Magnet Association, London, England (1).
M-258	Orléans, Compagnie de, Paris, France.		
M-233	Osborn Steel Co. Ltd., Sheffield, S30 3ZU England.	M-880	Permo Inc., Chicago, IL (3).
		M-881	Permold Inc., Medina, OH 44256 (2).
M-1028	Oscap Mfg. Co., New York, NY (2) (3).	M-878	Perry Barr Metal Co., Ltd., Birmingham, England.
M-462	Osnabrück Kupfer und Drahtwerke, Osnabrück, Germany.		
		M-728	Perry Equipment Corp., Hainesport, NJ (2).
M-301	Osram, G.m.b.H., Kommandit-Ges., Berlin, Germany (2).		
		M-1053	Peterson Steels Inc., Union, NJ 07083.
M-1655	Ossenberg & Cie., W., Edelstahlwerke, Altena, Westf. 7, West Germany.	M-1623	Pfizer Inc., New York, NY 18042.
		M-319	Phelps Dodge Co., New York, NY 10022.
M-330	Ostermann & Co., Metallwerke, KG., Köln NRW 5000, West Germany.		
		M-1228	Philadelphia Bronze & Brass Corp., see Ampco Metal, M-13.
M-331	Otis Elevator Co., New York, NY 10001.		
M-840	Otis Steel Div. Jones & Laughlin Steel Co. (4), see Jones & Laughlin, M-173.	M-1549	Philips Co., N. V., Eindhoven, Netherlands.
		M-1127	Philips Electronic & Associated Industries Ltd., 11/12 Hanover Square, London, W 1 England.
M-1117a	Oxium S.A., 75002 Paris, France (dealer for M-1117).		
M-484	Oxley & Co., William, Rotherham, England (3).		
		M-685	Phillips & Co., C. E., Detroit, MI 48208.
		M-1658	Phoenix-Rheinrohr, AG. (now part of Mannesmannrohren, M-1334).
	P		
		M-1594	Phoenix Steel Corp., Claymont, DE (2).
M-192	Pacific Foundry Co., San Francisco, CA (3).	M-1742	Phosphor Bronze Co. Ltd., Birmingham, B12 0NA England.
M-553	Pacific Metal Co., Portland, OR (3).		
M-1263	Pacific Welding Alloys Mfg. Co., Los Angeles, CA (3).	M-1072	Pickands, Mather & Co., Cleveland, OH 44114.
M-872	Page Steel & Wire Div. Am. Chain & Cable, Monesson, PA (3).	M-326	Pioneer Alloy Products Co., Cleveland, OH.
M-367	Paragon Steel Co., Rutherford, NJ (3).	M-1403	Pioneer Aluminum Products Inc., Los Angeles, CA.
M-879	Paraloy Co., Chicago, IL (3).		
M-1185	Paris & d'Outreau, Soc. des Aciéries de, Paris 17ᵉ, France.	M-1078	Pipe Machinery Corp., Cleveland, OH (3).
M-234	Parker Appliance Co., Cleveland, OH (3).	M-543	Pittsburgh Brass Mfg. Co., Irwin, PA 15642.
M-874	Parker-Kalon Corp., Clifton, NJ.	M-235	Pittsburgh Crucible Steel (4), see Crucible Specialty Steel, M-38.
M-1160	Parker Pen Co., Janesville, WI 53545.		
M-635	Parkersburg Steel Co., Parkersburg, WV (2).	M-626	Pittsburgh Metallurgical Co. Inc., Niagara Falls, NY (3).
M-1738	Parkin, F. M., (Sheffield) Ltd., Sheffield, S8 0YW England.	M-1206	Pittsburgh Metal Purifying Corp., Saxonburg, PA (3).
M-944	Park Sales Co., New York, NY (3).	M-369	Pittsburgh Rolls Div. (4), see Duraloy Blaw-Knox, M-45.
M-368	Patriarche & Bell, New York, NY (3).		
M-832	Paulson & Sons, Thomas, Inc., Brooklyn, NY (2).	M-236	Pittsburgh Steel Foundry Corp., Glassport, PA (3).

M-370	Pittsburgh Tool Steel Wire (now Teledyne Pittsburgh Tool Steel Wire, Monaca, PA 15061).
M-193	P.L. & M. Co., Los Angeles, CA (3).
M-1246	Plansee, Metallwerk Gesellschaft M.B.H., Reutte-Tirol, Austria.
M-1090	Plastic Metals Div. National Radiator Co., Johnstown, PA (3).
M-1321	Plattawerke G.m.b.H., West Berlin, Germany.
M-1801	Plessey Inc., Metals Div., Melville, NY 11746.
M-1344	Plettenberger Gusstahlfabrik, 5970 Plettenberg 2, West Germany.
M-1215	Plew Tool Co., Columbia City, IN 46725.
M-547	Plouff Metallographic Institute, Boston, MA.
M-318	Plykrome Corp., New York, NY (3).
M-1322	poldihütte, Westa-Westdeutsche Edelstahlhandels-Gesellschaft, Düsseldorf, West Germany.
M-96	Poldi Steel Works, Prague, Czechoslovakia.
M-1567	Polymer Corp. Ltd., Sarnia, Ontario, Canada.
M-884	Pompey, Société Nouvelle des Aciéries de, Neuilly-sur- Seine 92202 France.
M-1648	Pont-St-Martin, Officine Metallurgiche d', France.
M-525	Poro Metals Ltd., London, England (3).
M-1590	Portec Inc. Casting Div., Kingsbury, IN 46345 (2) (formerly Kensington Steel).
M-248	Porter, H. K., Co. Inc., Alloy Metal Wire Works, Prospect Park, PA 19076, (formerly Alloy Metal Wire Co.).
M-1342	Pose-Marre Edelstahlwerke G.m.b.H., 4006 Erkrath bei Düsseldorf, West Germany.
M-307	Potts, Horace T., Co., Philadelphia, PA 19134.
M-1568	Powder Alloys Corp., Clifton, NJ (3).
M-1098	Powder Metal Products Co., Franklin Park, IL (3).
M-885	Powder Metals Inc., Long Island City, NY (3).
M-940	Pratt & Co., J. M., Cincinnati, OH (3).
M-1278	Pratt & Whitney Cutting Tool & Gage Operation, West Hartford, CT 06101.
M-838	Precious Metals Research Works Inc., Mt. Vernon, NY 10553.
M-887	Precision Casting Co., Cleveland, OH 44111, or Fayetteville, NY 13066.
M-1619	Precision Cast Parts Corp., Portland, OR 97206.
M-1703	Precision-Kidd Inc., West Aliquippa, PA 15001.
M-602	Prentiss & Co., George W., Holyoke, MA (3).
M-1168	Pressco Casting & Mfg. Corp., Chesterton, IN (3).
M-506	Prescott Co., Menominee, MI (3).
M-583	Pressed Steel Car Co., Chicago, IL (3).
M-371	Prosser & Sons, Thos., New York, NY (3).
M-561	Puget Sound Metal Works, Tacoma, WA (3).
M-792	Pullman Kellogg, Div. of Pullman Inc., E. Houston, TX 77046.
M-372	Purdy Co. Inc., A. R., Lyndhurst, NJ (3).
M-1432	Pyramid Steel Co., Cleveland, OH.
M-1777	PYRON Corp., Niagara Falls, NY 14304 (formerly AMAX Metal Powders).

Q

M-630	Q. & C. Co., New York, NY (3).
M-1628	Quaker Alloy Casting Co., Myerstown, PA 17067.
M-1390	Quint Alloy Corp., Dearborn, MI (2).

R

M-1155	Randall Graphite Products Co., Chicago, IL (3).
M-1100	Rankin Mfg. Co., San Diego, CA 92121.
M-1164	Rasmussen Mfg. Co., Hollydale, CA (3).
M-812	Ratcliff Metals Ltd., J. F., Birmingham, England (3).
M-1182	Raven Steel & Tool Co., Valley Stream, NY (3).
M-1465	RCA Corporate Standards Engineering, Camden, NJ 08101 (2) (information only).
M-1229	Reactive Metals (now RMI Co., Niles, OH 44446).
M-291	Reading Iron Co., Reading, PA (3).
M-1082	Redhard Metals Inc., Hatboro, PA (3).
M-1242	Reeves International Inc., New York, NY (2).
M-689	Regie Nationale des Usines Renault, St. Michael de Maurienne, 55 bd Charonne, Paris 11e, France.
M-986	Reid-Avery Co., Baltimore, MD 21222.
M-1353	Reining, Heinrich, G.m.b.H., 4000 Düsseldorf 1, West Germany.
M-1335	Reisholz, Stahl und Röhrenwerke, G.m.b.H., 4000 Düsseldorf 16, West Germany.
M-1440	Reliance Foundry, American Meter Co., Pittsburgh, PA (3).
M-907	Reliance Steel Casting Co., Pittsburgh, PA (3).
M-420	Rely Metal Works, Cape Town, South Africa.
M-1299	Remanit, G.m.b.H., Berlin, Germany.
M-255	Rem-Cru Titanium Inc., Midland, PA 15159.
M-1698	Remington Arms Co. Inc., Ilion, NY 13357.
M-1339	Remystahl, D-58 Hagen, F.R. Germany (see also M-1652).
M-689	Renault, Regie Nationale des Usines, St. Michael de Maurienne, Paris 11e, France.

Section II: Alphabetical List of Manufacturers / 1631

M-1633	Renewal Services Inc., Philadelphia, PA 19123.	M-1207	Royalloy Inc., New York, NY (2) (3).
M-1273	Renfrew Foundries Ltd., Hillingdon, Glasgow, Scotland.	M-1475	R-S Products Corp., Philadelphia, PA (3).
M-790	Rennie Tool Co. Ltd., Manchester, England (3).	M-1382	Ruhrstahl A.G., Henrichshütte, Hattingen/Ruhr, Germany.
M-97	Republic Steel Corp., Cleveland, OH 44101.	M-99	Ruselite Corp., Milwaukee, WI (3).
M-507	Resisto-Loy Co. Inc., Grand Rapids, MI 49507.	M-100	Rustless Iron Div. Armco Steel Corp (4), see Armco Steel, M-10.
M-238	Revere Copper & Brass Inc., Rome, NY 13440.	M-1211	Ryan Aeronautical (now Teledyne Ryan Aeronautical, San Diego, CA 92112).
M-1525	Rex Buckeye Co., Cleveland, OH 44144.	M-1672	Ryanite Tips Inc., St. Paul, MN (3).
M-671	Reynolds Metals Co., Richmond, VA (2).	M-434	Ryer Inc. Ltd., Los Angeles, CA (3).
M-1558	Reynolds Tube Co., Tyseley, Birmingham, 11 England (2).	M-240	Ryerson, Joseph T. & Son, Inc., Chicago, IL 60680.
M-625	Rheinische Röhrenwerke Aktiengesellschaft, Mülheim, Germany.		**S**
		M-1757	SAFE/Société des Aciers Fins de l'EST, 92106 Boulogne-Billancourt, France.
M-990	Rhodes Metaline Co., R. W., Long Island City, NY (3).	M-966	Saginaw Bearing Co., Saginaw, MI 48605.
M-1367	Rieger Gebr., Aalen, Germany.	M-1646	Sambra & Meuse, Aciéries de, France.
M-419	Rieter & Co., J. J., New York, NY (3).	M-618	Sanderson Kayser Ltd., Sheffield, S9 2SD England.
M-1110	Riken Metal Mfg. Co., Ube, Japan.	M-903	Sanderson Steel Div. (4), see Crucible Specialty Steel, M-38.
M-911	Riverside Foundry & Galvanizing Co., Kalamazoo, MI (2).	M-295	Sandusky Foundry & Machine Co., Sandusky, OH 44870.
M-98	Riverside Metals Corp., Riverside, NJ 08075 (2).	M-101	Sandvikens Jernverks Aktiebolag, Sandviken, Sweden.
M-1229	RMI Co., Niles, OH 44446 (formerly Reactive Metals).	M-101a	Sandvik Inc., Fairlawn, NJ 07410, or Scranton, PA 18501.
M-195	Robbins & Meyers, Inc., Springfield, OH 45501.	M-101b	Sandvikstahl G.m.b.H., Düsseldorf, West Germany.
M-1288	Robert-Leyer-Pritzkow & Co., Solingen-Ohligs, Germany.	M-1591	Sankey & Sons Ltd., London, England.
M-460	Robins Conveyors Inc., Passaic, NJ (3).	M-241	Sargent & Co., New Haven, CT (2).
M-1582	Rochette, Laminoire de la, Chaudfontaine, Belgium.	M-1494	Saut-du-Tarn, Société Nouvelle du, 75008 Paris, France.
M-501	Rochling Buderus (merged with Rochlingstahl, M-912, to form Rochling-Burbach, M-912) (all M-501 numbers dropped).	M-1119	Saut-du-Tarn, Société Nouvelle du, 75008 Paris, France.
		M-486	Saville & Co. Ltd., J. J., Sheffield, England (3).
M-912	Rochling-Burbach G.m.b.H., 6620 Volklingen-Saar, West Germany.	M-1516	Scaife Co., Oakmont, PA (2).
M-912	Rochlingstahl (now Rochling-Burbach).	M-1575	Scandinavian Metal Co., Sweden.
M-1123	Rochlingstahl Steel Works, Wetzlar, Germany.	M-1453	Schenk Leichtgusswerke, Maulbronn, Germany.
M-1663	Rodney Metals (now Teledyne Rodney, New Bedford, MA 02742).	M-1305	Schmidt & Clemens, Edelstah! ~erke, 5251 Kaiserau Bez, Köln uber Engelskirchen, West Germany.
M-1209	Rolled Alloys Inc., Detroit, MI 48211.	M-196	Schmidt Co., Karl, Neckarsulm, Germany (3).
M-1501	Rolle Mfg. Co., Lansdale, PA (3).		
M-1240	Rollschen Eisenwerke, AG., Gesellschaft der Ludv. von, Gerlafingen, Germany.	M-615	Schmidt Stahlwerke, Rudolf, Vienna, Austria.
M-1247	Rollschen Eisenwerke, AG., Ludv. von, Soluthurn, Switzerland.	M-1368	Schmole, RuG., Metallwerke Menden, Krs. Iserlohn, Westf. Germany.
M-239	Rolls-Royce Mfg. Co., Derby, England.	M-533	Schneider & Cie (4) (merged into Creusot-Loire, M-1488).
M-1497	Ronson Metals Corp., Newark, NJ 07105.	M-1301	Schneider, Virgo (4) (merged into Creusot-Loire, M-1488).
M-1140	Rorschach, Aluminiumwerke, Rorschach, Switzerland.	M-616	Schoeller-Bleckmann (4) (merged into Vereinigte Edelstahlwerke, M-1802).
M-713	Rossell & Co. Ltd., H., Sheffield, 4 England.	M-1308	Schoeller-Bleckmann (4) (merged into Vereinigte Edelstahlwerke, M-1802).
M-82a	Ross-Meehan Foundries, Chattanooga, TN 37401.		

M-1718	Schulte Eisenhandlung, Heinr. Aug., G.m.b.H., 4600 Dortmund, Germany.	M-1017	Sight Feed Generator Co., Richmond, IN (3).
M-1383	Schwabenstahl Hüttenwerke, G.m.b.H., Wasseralfingen (Württ), Germany.	M-1295	Sigmund Cohn Corp., Mount Vernon, NY 10553.
M-481	Scientific Alloys Inc., Los Angeles, CA (3).	M-461	Silicum Pistons Ltd., London, England (3).
M-1561	Scientific Products Inc., Detroit, MI 48211.	M-373	Simonds Steel Div. Wallace-Murray Corp., Lockport, NY 14094.
M-408	Scott & Co. Ltd., A. C., Manchester, England.	M-1015	Simonds Worden White Co., Dayton, OH (3).
M-102	Scovill Mfg. Co., Waterbury, CT (2) (sold metal business to Century Brass Products, Inc., M-1770).	M-875	Simpson Bros. Machine Works, Portsmouth, OH 45662.
M-1048	Scraw Alloys Ltd., Banstead, Surrey, England (2).	M-493	Singen, Aluminium Walzwerke, G.m.b.H., Singen/Hohentwell, West Germany.
M-969	Seaboard Steel Co. of America, New York, NY (3).	M-1469	Sintercast Div. of Chromalloy American Corp., West Nyack, NY 10994.
M-1524	Secon Metals Corp., White Plains, NY 10606.	M-1739	Sintered Products Ltd., Sutton-in-Ashfield, Nottinghamshire, NG17 5LL England.
M-1369	Seibel, W., AG, Mettmann, Rhld, Germany.	M-1379	Sintox Corp. of America, Allentown, PA (3).
M-590	Seitzinger's Inc., Atlanta, GA 30318.	M-512	Sioux Tools Inc., Sioux City, IA 51102.
M-1729	Semi-Alloys, Mt. Vernon, NY 10550.	M-1080	S.I.P.I. Metals Corp., Chicago, IL 60622.
M-972	Seneca Wire & Mfg. Co., Fostoria, OH 44830.	M-1727	Sitkin Smelting & Refining Co., Lewistown, PA 17044.
M-1644	SERMAG (Société d'Etudes et de Recherches Magnetiques), 38 St-Martin-d'Heres, France.	M-288	Sivyer Steel Casting Co., Milwaukee, WI 53202 (3).
M-1669	SERMAG (Société d'Etudes et de Recherches Magnetiques), 38 St-Martin-d'Heres, France.	M-404	Skandinaviska Armaturfabriken AG., Stockholm, Sweden.
M-970	Sessions Foundry Co., Bristol, CT (3).	M-267	S.K.F. Industries Inc., King of Prussia, PA 19406.
M-103	Seymour Products Co., Seymour, CT 06483 (3).	M-651	Skoda Works, National Corporation, 316 000 Plzen, Czechoslovakia.
M-453	Shanks & Co. Ltd., Barrhead, Scotland.	M-1716	Smith, A. O. - Inland Inc., Powder Metallurgy Div., Milwaukee, WI (2).
M-1071	Sharon Steel Corp., Sharon, PA 16146.	M-579	Smith Co., A. O., Milwaukee, WI 53201 (2).
M-443	Shawinigan Chemicals, Ltd., Montreal, Canada.	M-244	S-M Metal Works, London, England (3).
M-417	Shaw, J., Son & Greenhalgh, Ltd., Albert Works, Huddersfield, England.	M-1093	SMS Corp., Detroit, MI (3).
M-1485	Sheepbridge Alloy Castings Ltd., Sutton-in-Ashfield, Nottinghamshire, NG17 5LL England.	M-870	Snowbar, J. L., New York, NY (3).
		M-1176	Società Alluminio Veneto per Azioni, Porto Marghera, Milan, Italy.
M-494	Sheepbridge Engineering Ltd., Sheepbridge Works, Chesterfield, Derbyshire, England.	M-1220	Società Alluminio Veneto per Azioni, Porto Marghera, Venezia, Italy.
M-906	Sheepbridge Equipment Ltd., Chesterfield, England.	M-606	Società Anonima Ansaldo, Rome, Italy.
M-494	Sheepbridge Stokes Centrifugal Casting Co., see Sheepbridge Engineering.	M-768	Società Metallurgica Italia, Firenze, Italy.
M-1522	Sheffield Smelting Co. Ltd., Sheffield, S4 7WD England.	M-672	Société Alsia, Paris, France.
		M-104	Société Anon. de Commentry Fourchambault et Decazeville (4) (merged into Creusot Loire, M-1488).
M-1428	Sheffield Steel Div. Armco Steel Co., Houston, TX (1), see Armco Steel, M-10.	M-1180	Société Anon. des Etablissements Marrel Frères, B.P. 46, F-42 Rive-de-Gier, France.
M-974	Shenango-Penn Div. Shenango Furnace Co. (2).		
M-1002	Sherman & Co., New York, NY (3).	M-1119	Société Anon. des Hauts-Fourneaux (now Société Nouvelle du Saut-du-Tarn, 75008 Paris, France).
M-1373	Sherritt Gordon Mines Ltd., Toronto M5L 1B1 Ontario, Canada.		
M-198	Siemans & Halske AG., Berlin, Germany.	M-1179	Société Anon. Usines et Aciéries Allard, Mont-sur- Marchienne, Belgium.
M-199	Siemens-Schuckert AG., Berlin, Germany, or London, England.	M-1118	Société des Aciéries de Longwy, Mont-Saint-Martin, France.

M-1723	Société Commentryenne des Aciers (now Commentrenne, 75429, Paris Cedex 09-B.P. 381-09 France).	M-889	Spuck Iron & Foundry Co., St. Louis, MO (2).
M-1488	Société des Forges et Ateliers du Creusot (now Creusot- Loire, 75428 Paris Cedex 09 France).	M-1069	Stackpole Carbon Co., St. Marys, PA 15857.
		M-1754	Stackpole Magnet Division, Kane, PA 16735.
M-1644	Société d'Etudes et de Recherches Magnetiques, (SERMAG) B.P. 17926 38 St-Martin-d'Heres, France (3).	M-1335	Stahl-und Rohrenwerk Reisholz G.m.b.H., 4000 Düsseldorf 16, West Germany.
M-767	Société du Duralumin, Paris, France (3).	M-613	Stahlwerk Becker AG., Krefeld (Rheinland), Germany.
M-673	Société Electro-Cables, Paris, France.		
M-1043	Société et Forges de Firminy (4) (now part of Creusot-Loire , M-1488).	M-1331	Stahlwerke Bochum AG., 4630 Bochum, West Germany.
M-1786	Société Francais d'Electrometallurgie (SOFREM) 75361 Paris Cedex 08 France (part of Pechiney).	M-1310	Stahlwerke Kabel, C., Pouplier, Hagen-Kabel, Germany.
		M-1337	Stahlwerke R & H Plate, 5880 Ludenscheid-Platehof, West Germany.
M-1785	Société Metallurgique de Gerzat (S.M.G.) 63360 Gerzat, France (part of Pechiney).	M-912	Stahlwerke Rochling-Burbach G.m.b.H., 6620 Volklingen, Saar, West Germany.
M-884	Société Nouvelle des Aciéries de Pompey, Neuilly-sur- Seine 92202 France.	M-1307	Stahlwerke Sudwestfalen AG., 5900 Siegen 1, West Germany.
		M-1294	Stahlwerk Stahlschmidt & Co. KG., 4000 Düsseldorf-Heerdt, West Germany.
M-647	Société des Sondures, Castolin, South America.		
M-1784	Société de Vente de l'Aluminium Pechiney, 75008 Paris, France.	M-1730	Stahlwerk Westig G.m.b.H., 4750 Unna, West Germany.
M-1494	Société Nouvelle du Saut-du-Tarn, 75008 Paris, France.	M-1262	Stainless Foundry & Engineering Inc., Milwaukee, WI 53209.
M-1309	Soding, J. C., & Halbach, 58 Hagen/Westf., West Germany.	M-105	Standard Alloy Co. Inc., or Michigan Standard Alloy Inc., Benton Harbor, MI 49022.
M-1670	Solar Div. International Harvester Co., San Diego, CA 92138.	M-896	Standard Brake Shoe & Foundry Co., Pine Bluff, AR (2).
M-1136	Soldering Specialties Co., Summit, NJ (3).	M-1439	Standard Pressed Steel Co., Jenkintown, PA 19046.
M-960	Somers Thin Strip Inc., Waterbury, CT 06720 (4) (now part of Ohio Brass, M-279).	M-1602	Standard Steel Div., Titanium Metals Corp., Burnham, PA 17009.
M-647	Sonduras, Société des, Castolin, South America.	M-269	Standard Telephones & Cables (now ITT Components Group Eurpoe, Edinburgh Way, Harlow, Essex, England).
M-1266	Sonken-Galamba Corp., Kansas City, KS (3).		
M-1788	Sorcery Metals, Delray Beach, FL 33444.	M-1370	Standard-Werke, G.m.b.H., Werl/Westf., Germany (2).
M-1274	Southern Forge Ltd., England (now part of Imperial Aluminium, M-1547).	M-574	Stanley Steel Div. of the Stanley Works, New Britain, CT 06050.
M-268	Southern Malleable Iron Co., East St. Louis, IL (3).	M-1225	Starrett Co., L. S., Athol, MA 01331.
		M-1528	Stauffer Metals Div., Stauffer Chemical Corp., Westport, CT 06880 (2).
M-665	Southern Metals Co., St. Louis, MO (3).		
M-1799	Southwire Co., Carrollton, GA 30117.	M-1304	Stavenger Co., Oslo, Norway.
M-479	Sowers Mfg. Co., Buffalo, NY (3).	M-700	Sté de Produits Metallurgiques, Paris, France.
M-735	Spang Chalfont Div. National Supply Co., Pittsburgh, PA (3).		
		M-1024	Sté d'Escaut & Meuse, Paris 17e, France.
M-1737	Spear & Jackson (Industrial) Ltd., Sheffield, S4 7UR England.	M-675	Sté des Laminoirs et Tréfileries du Havre (now part of Cegedur Pechiney, M-1792).
M-1236	Specialloy Inc., Chicago, IL 60632.		
M-1317	Special Metals Corp., New Hartford, NY 13413.		
		M-518	Stedman Foundry & Machine Co., Aurora, IN 47001.
M-1704	Specialty Steel Co. of America, Warrensville Heights, OH (3).	M-395	Steel & Tubes Div. Republic Steel, see Republic Steel Corp., M-97 (4).
M-1744	Spenser & Co. Ltd., Walter, (now Spencer Clark Metal Industries Ltd., Sheffield, S6 3AF England).		
		M-742	Steel Co., Ltd., Motherwell, England (3).
M-701	Sperry Gyroscope Co., Great Neck, NY 11020.	M-1714	Steel Co. of Australia (2).

M-1124	Steel Co. of Canada. Ltd., Hamilton 23 Montreal, Canada.	M-1296	Sumitomo Metal America Inc., New York, NY 10017.
M-826	Steel Co. of Scotland (4) (now part of British Steel Corp. M-1724).	M-1515	Sumitoms Trading Co., Osaka, Japan.
M-448a	Steel, Peech & Tozer, Sheffield (4) (now part of British Steel Corp., M-1724).	M-1018	Superior Die Casting Corp., Cleveland, OH 44110.
M-607	Steel Sales Corp (now Edgcomb Metals, Tulsa, OK 74103).	M-1070	Superior Flux & Mfg. Co., Cleveland, OH 44143 (2).
M-270	Steinvertriebs Aktiengesellschaft, Berlin, Germany.	M-1170	Superior Foundry Co., Cleveland, OH (3).
M-1114	Steirische Gusstahlwerke (merged into Vereinigte Edelstahlwerke, A-1011 Wien, Austria, M-1802).	M-967	Superior Metal Co., Chicago, IL (3).
		M-635	Superior Sheet Steel Div. Parkersburg Steel Co., Parkersburg, WV (2).
M-1291	Steirische Gusstahlwerke (merged into Vereinigte Edelstahlwerke, A-1011 Wien, Austria, M-1802).	M-901	Superior Steel Corp., Pittsburgh, PA (3).
		M-945	Superior Tube Co., Norristown, PA 19404.
M-167	Stellite Div. Cabot Corp., Kokomo, In 46901.	M-406	Supersteels Inc., Cleveland, OH (3).
		M-916	Super Tool Co., Elk Rapids, MI 49629 (2).
M-1008	Sterling Alloys Inc., Woburn, MA (3).	M-1189	Surahammer Bruks, A.B., Surahammer, Sweden.
M-106	Sterling Metals Ltd., Gypsy Lane, Nuneaton, England.	M-107	Sutcliff, Speakman & Co., Leigh, Lancaster, England.
M-201	Sterlite Foundry & Mfg. Co., Auburn, IN (3).	M-1186	Svenska Metallverken A.B., Stockholm, Sweden.
M-514	Stewart-Warner, Die Casting Div., Chicago, IL (2).	M-333	Swedish American Steel Corp., Brooklyn, NY (3).
M-1486	St. Jacques, Société des Usines, Montlucon, France.	M-1026	Swedish Crucible Steel Co., Detroit, MI (3).
M-1656	St. Joseph Lead Co., Monaca, PA.	M-375	Swedish Iron & Steel Corp., Middlesex, NJ 08846 (2).
M-1684	St. Lawrence Steel Co., Cleveland, OH 44128 (3).	M-785	Swedish Steel Mills, A.A., New York, NY (3).
M-452	Stone Manganese Marine; J. Stone & Co., London, SE 17, England.	M-1605	Swiss Aluminium Ltd., Neuhausen, Rhf., Switzerland.
M-1675	Stone Manganese Marine; J. Stone & Co., London, SE 17, England.	M-66	Swiss Laboratory Inc., Akron, OH (2).
M-202	Stoody Co., Industry, CA 91749.	M-418	Sybry, Searls & Co., Ltd., Cannon Steel Works, Sheffield, England (3).
M-283	Strasser Co., E. Rorschach, Switzerland.	M-1632	Sydney Steel Corp., Sydney, Nova Scotia, Canada (2).
M-899	Stroh Process Steel Co. (now SPS Industries Inc., Allison Park, PA 15101).	M-856	Sylvania Electric Products (now GTE Sylvania, Towanda, PA 18848).
M-1081	Strong, Carlisle & Hammond Co., Cleveland, OH.	M-500	Symington-Gould Corp., Depew, NY (3).
M-208	Studebaker Chemical Co., Elyria, OH 44035.		**T**
M-374	Stulz Sickles Steel Co., Elizabeth, NJ 07207.	M-390	Tacony Steel Co., Philadelphia, PA (1).
M-928	Stupakoff Co., Pittsburgh, PA (2).	M-997	Tata Iron & Steel Co., Calcutta 16 India.
M-795	Stupakoff Co., Pittsburgh, PA (2).	M-1412	Taylor & Co., Robert, Labert, Scotland.
M-1114	Styria-Stahl Steirische Gusstahlwerke AG., Vienna (4) (now merged into Vereinigte Edelstahlwerke, A-1011 Wien, Austria).	M-915	Taylor Chain Co., S. G., Hammond, IN 46320.
		M-496	Taylor Ltd., Samuel, Birmingham, England (3).
M-1291	Styria-Stahl Steirische Gusstahlwerke AG., Vienna (4) (now merged into Vereinigte Edelstahlwerke, A-1011 Wien, Austria).	M-271	Taylor-Wharton Div. of Harsho Corp., Easton, PA (2).
		M-1442	Taylor-Wilson Mfg. Co., Pittsburgh, PA 15222.
M-1307	Sudwestfalen Stahlwerke AG., 5900 Siegen 1, West Germany.	M-1280	Techalloy Co. Inc., Rahns, PA 19426.
M-243	Suffolk Iron Foundry Ltd., Stowmarket, England.	M-108	Telcon Metals Ltd., Crawley, Sussex, England.
M-1660	Sulzer Brothers Ltd., Winterthur, Switzerland, Sulzer Bros. Inc., New York, NY 10006.	M-1490	Teledyne Allvac, Monroe, NC 28110.
		M-57	Teledyne Firth-Sterling, McKeesport, PA 15134.
M-71	Sumet Corp., Buffalo, NY (3).	M-849	Teledyne McKay, Pittsburgh, PA 15219.

Section II: Alphabetical List of Manufacturers / 1635

M-95	Teledyne Ohiocast, Springfield, OH 45501.
M-370	Teledyne Pittsburgh Tool Steel Co., Monaca, PA 15061.
M-1663	Teledyne Rodney Metals, New Bedford, MA 02742.
M-1211	Teledyne Ryan Aeronautical, San Diego, CA 92112.
M-115	Teledyne Vasco, Latrobe, PA 15650.
M-1537	Teledyne Wah Chang, Albany, OR 97321.
M-1596	Teledyne Wah Chang, Albany, OR 97321.
M-377	Tennessee Coal, Iron & Railroad (4), see U.S. Steel Corp., M-604.
M-444	Texas Electric Steel Casting Co., Houston, TX (2).
M-421	Thermit Ltd., Edmonton, England (2).
M-1417	Thomas & Baldwins (4) (now part of British Steel Corp., M-1724).
M-1192	Thomas & Skinner Inc., Indianapolis, IN 46205.
M-1460	Thomas Foundries Inc., Birmingham, AL 35201.
M-445	Thomas Steel Co., Warren, OH (2).
M-1617	Thompson Wire Co., Chicago, IL (3).
M-1400	Thomson Industries Inc., Manhasset, NY 11030.
M-1234	Thyssen Co., August, Berlin, Germany.
M-1759	Thyssen Edelstahlwerke AG., 4150 Krefeld 1, West Germany.
M-1759	Thyssen Edelstahlwerke A.G., 4150 Krefeld 1, West Germany (new conglomerate).
M-1268	T. I. Aluminium Ltd. (now part of British Aluminium, M-28).
M-1705	Time Steel Service Inc., Cleveland, OH 44124.
M-144	Timet, a Div. of Titanium Metals Corp. of America, Pittsburgh, PA 15230.
M-1758	Timet, a Div. of Titanium Metals Corp. of America, Pittsburgh, PA 15230.
M-376	Timken Co., Canton, OH 44706.
M-702	Timken-Detroit Axle Co., Detroit, MI (3).
M-1398	Tin Research Institute, Greenford, England.
M-1157	Tipaloy Inc., Detroit, MI 48211.
M-955	Titanite Alloys Corp., Cleveland, OH (3).
M-109	Titanium Alloy Mfg., Div. NL Industries, Niagara Falls, NY (3).
M-294	Titusville Forge Co., Titusville, PA (3).
M-787	T. L. M. Co., Paris, France (3).
M-1647	Tokoshu Seiko Ltd., Tokyo, Japan.
M-1614	Tokyo Keikiseizosho Ltd., Tokyo, Japan.
M-985	Toledo Steel Works, or Darwin Toledo Steel, Sheffield, England (1).
M-1782	Les Toles Inoxydables et Specials, Ugine-Guegnon, 92307 Levallois-Perret, France (part of Pechiney).
M-1001	Tonawanda Iron Corp., North Tonawanda, NY.
M-1667	Toshiba Tungaloy Co., Ltd., New York, NY 10017.
M-1544	Toto Steel Co. Ltd., Tokyo, Japan.
M-1036	Toughite Process Co., Chicago, IL (3).
M-1763	TradeARBED Inc., Luxembourg, B.P. 1802.
M-1664	Transformer Steels Ltd.
M-834	Transleteur & Co., Berlin, Germany.
M-979	Tréfileries du Havre, Paris, France.
M-770	Tréfileries & Laminoirs du Havre, Antony (Seine), France.
M-1783	Tréfimétaux, 92115 Clichy, France (part of Pechiney).
M-535	Trenite Foundry Corp., Trenton, NJ (3).
M-1056	Trethaway Associates, New York, NY (3).
M-1677	Triangle Conduit & Cable Co., Copper Rod & Brass Products Div., New Brunswick, NJ 08903.
M-909	Tri-Clover Div. Ladish Co. (4), see Ladish Co., M-1527.
M-923	Tri-Lok Co., Pittsburgh, PA (3).
M-954	Trimout Mfg. Co., Boston, MA (3).
M-925	True Alloys Inc., Detroit, MI (3).
M-513	TRW Metals Div., Minerva, OH 44657.
M-1706	TRW United Greenfield Div., Northbrook, IL 60062 (formerly Unimet Carbides).
M-1697	TRW Wendt-Sonis Division, TRW Inc., Rogers, AR 72756.
M-957	Tube Reducing Corp., Wallington, NJ (3).
M-1029	Tuff-Hard Corp., Detroit, MI (3).
M-737	Tullis, Ltd., D. & J., London, England (3).
M-1046	Tungsit Electro-Metals Works Ltd., London, England (3).
M-987	Tungsten Alloy Mfg. Co., Inc., Harrison, NJ 07029.
M-1385	Tungsten & Molybdenum Ltd., Nyon, Switzerland.
M-1054	Tungsten Electric Corp., Union City, NJ (3).
M-598	Tungsten Widia Tool Corp., New York, NY (3).
M-110	Tungum Alloy Co., Ltd., Cheltenham, England.
M-929	Tyler Inc., W. S., Mentor, OH 44060.

U

M-111	Uddeholms, A.B., Uddeholm, Sweden.
M-111a	Uddeholm Steel Corp., Totowa, NJ 07512.
M-1120	Ugine Aciers (Steel Div.) 75361 Paris Cedex 08 France.
M-1120	Ugine Aciers (Titanium Div.) 75361 Paris Cedex 08 France.
M-1798	Ulbrich Stainless Steels & Special Metals Inc., North Haven, CT 06473.
M-1231	Ullman & Associates, A. E., New York, NY (3).
M-1351	Ultraloy Corp., Chicago, IL (3).

M-272	Ultralumin Leichtmetall A.G., Germany (1).	M-1371	Unna Messingwerk, A.G., Unna, Westf., Germany.
M-1618	U.N. Alloy Steel Corp., Boston, MA 02027.	M-379	Upson Nut Div. or Tool Steel Div. Republic Steel (2), see Republic Steel, M-97.
M-930	Una Welding Inc., Cleveland, OH (3).		
M-745	Unexcelled Mfg. Co., New York, NY (3).	M-1320	Urbach & Co., Carl, 5809 Huckeswagen, West Germany.
M-1711	Unibraze Corp., Covington, OH 45318.		
M-634	Unicast Corp., Div. Midland Ross Corp., Toledo, OH.	M-1597	Urich Foundry Co., Erie, PA (2).
		M-1188	Usines Emile Henricot, S.A., Court-Saint-Etienne, Belgium.
M-1750	Unicore Inc., North Haven, CT 06473.		
M-391	Unimetal Co., Franklin, PA (3).	M-1495	Usines Vallouric (now Vallouric 75764 Paris Cedex 16 France).
M-1706	Unimet Carbides (now TRW United Greenfield Div., Northbrook, IL 60062).		
		M-1686	Usines Vallouric (now Vallouric 75764 Paris Cedex 16 France).
M-588	Union Bronze Co., Reading, PA (3).	M-1534	U.S. Magnet & Alloy Corp., Bloomfield, NJ (3).
M-48	Union Carbide Corp., Danbury, CT 06810.		
		M-585	U.S. Reduction Co., East Chicago, IN 46312.
M-397	Union Drawn Steel Div. (4), see Republic Steel, M-97.		
		M-1138	U.S. Spring & Bumper Co., Los Angeles, CA (3).
M-931	Union Metal Mfg. Co., Canton, OH (2).		
M-1533	Union Minière du Haut-Katanga, Central Workshop, Jadotville, Katanga, Congo.	M-1055	Utility Electric Steel Foundry Co., Los Angeles, CA (3).

V

M-112	Union Steel Casting Co., Pittsburgh, PA, Div. Duraloy Blaw-Knox (2) (their castings made at other divisions).	M-1630	Vacuum Metals Corp., Cambridge, MA (3).
		M-1214	Vacuum Metals Div. Crucible Steel (4), see Crucible Specialty Steel, M-38.
M-1610	Union Steel Corp., Union, NJ (2).		
M-397	United Alloy Steel Corp., Canton, OH (4), see Republic Steel Corp., M-97.	M-1631	Vacuumschmelze G.m.b.H., 6450 Hanau 1, West Germany.
M-565	United American Metals Corp., Chicago, IL 60612.	M-1653	Valbruna Ernesto Gresele, AVEC SPA Acciaieria, 36100 Vicenza, Italy.
M-273	United Lead Co., see NL Industries, M-88.	M-1668	Venite Metals Div. Valenite Corp., Madison Heights, MI 48071.
M-1776	United States Bronze Powders Inc., Flemington, NJ 08822.	M-1495	Vallourec, 75764 Paris Cedex 16, France.
		M-1686	Vallourec, 75764 Paris Cedex 16, France.
M-245	United States Graphite Corp. (now Wickes Engineered Materials, a Div. of Wickes Engineering, Saginaw, MI 48601).	M-394	Vanadium Corp. of America (4) (now part of Foote Minerals, M-1038.
		M-1500	Van der Horst Corp., (Terrel, TX) Los Angeles, CA 94103.
M-558	United States Metals Refining Co., Carteret, NJ.	M-297	VDM, abbreviation for Vereinigte Deutsche Metallwerke, M-297.
M-203	United States Pipe & Foundry Co., Birmingham, AL 35202.	M-1176	Veneto per Azioni, Soc. Alluminio, Porto Marghera, Milan, Italy.
M-604	United States Steel Corp., Pittsburgh, PA 15320	M-1220	Veneto per Azioni, Soc. Alluminio, Porto Marghera, Venezia, Italy.
M-964	United States Steel Supply Co. (4), see United States Steel, M-604.	M-1430	Venture Corp., Newark, NJ (3).
		M-1289	Veraloy Products Ltd., Birmingham, England.
M-448	United Steel Companies Ltd. (4) (now part of British Steel Corp., M-1724).		
		M-935	Vereinigte Aluminiumwerke, Bohn, Germany.
M-327	United Steel Div. (4), see Republic Steel Corp., M-97.		
		M-133	Vereinigte Chemische Fabriken, Leopoldshall, Germany.
M-1013	United Wire & Supply Co., Cranston, RI 02910.		
		M-297	Vereinigte Deutsche Metallwerke, D 5980 Werdolhl, West Germany (or Ore & Chemical Corp., New York, NY, 10016).
M-932	Universal Alloys Inc., Newark, NJ (3).		
M-1341	Universal Castings Corp., Chicago, IL 60638.		
M-114	Universal Cyclops Specialty Steel Div., Pittsburgh, PA 15228.	M-303	Vereinigte Deutsch Nickel-Werke AG., 5840 Schwerte, West Germany.
M-811	Universal Power Corp., Cleveland, OH (3).	M-1802	Vereinigte Edelstahlwerke, A-1011 Wien, Wildpretmarkt 2, Austria (took over Bohler, Schoeller-Bleckmann, Styria and Alpine).
M-1161	Uniworld Corp. of America, Cleveland, OH 44120.		

M-116	Vereinigte Leichtmetallwerke, G.m.b.H., Hannover-Linden, Germany.	M-942	Waltham Precision Instruments, Waltham, MA 02154.
M-1735	Vereinigte Osterreichische Eisen und Stahlwerke (VOEST) A.G., Linz-Austria.	M-1023	Walworth Co., Aloyco Plant, Linden, NJ 07036 (formerly ALOYCO Inc.).
		M-392	Wardlow, S. & C., Nutley, NJ (3).
M-204	Vereinigte Silberhammerwerke Hetzel, Nüremburg, Germany.	M-1113	Wardlows Ltd., Sheffield, S4 7LJ England.
M-383	Vereinigte Stahlwerke, Düsseldorf (1).	M-389	Ward's Sons Co., Edgar T., Newark, NJ (3).
M-274	Verilite Metals Co., New York, NY (3).		
M-1372	Vesevorder Metallwerke G.m.b.H., Werdohl/Westf, Germany.	M-277	Warman Steel Casting Co., Huntington Park, CA (3).
M-117	Vickers-Armstrong Co. (4) (now part of British Steel Corp., M-1724).	M-774	Washburn Wire Co., New York, NY (2).
		M-246	Washington Iron Works, Seattle, WA 98134.
M-1204	Victor Equipment Co. (4) (now part of Stellite Div. Colt Industries, M-167).	M-1484	Washington Steel Co., Washington, PA 15301.
M-423	Victor Mfg. Co., London, England (3).		
M-1447	Vikmanshyttan AG., Vikmanshyttan, Sweden.	M-1426	Waterbury Rolling Mills Inc., Waterbury, CT 06720.
M-1418	Vimetal S.A., Berne, Switzerland (2).	M-1173	Watsontown Foundry, Watsontown, PA 17777.
M-1571	Vitra Chemical Co., Chattanooga, TN (2).	M-941	Waukesha Foundry Co., Waukesha, WI 53186.
M-1421	Voelkingen, Forges et Aciéries de, Voelklingen (Sarre), France.	M-1106	Wausau Motor Parts Co., Schofield, WI.
M-1735	Voest, Linz, Austria or Voest International Inc., New York, NY 10022.	M-278	Weatherly Foundry & Mfg. Co., Weatherly, PA 18255.
		M-446	Webb Wire Works, New Brunswick, NJ (3).
M-998	Volco Brass & Copper Co., Kenilworth, NJ 07033.	M-599	Wehr Steel Co., Milwaukee, WI 53210.
M-329	VR/Wesson or Vascoloy-Ramet, Waukegan, IL (4), (now part of Fansteel, M-53).	M-1731	Weingartner & Co., KG., Emil, Edelstahlgrosshandel, 2000 Hamburg 54, West Germany.
M-1666	VR/Wesson or Vascoloy-Ramet, Waukegan, IL (4), (now part of Fansteel, M-53).	M-628	Weirton Steel Co. Div. of National Steel Corp., Weirton, WV 26026 (same alloys as Great Lakes Steel).
M-1257	Vulcan Foundry Co., Oakland, CA 94601.	M-1388	Welded Carbide Co. Inc., Clifton, NJ 07013.
M-922	Vulcan Iron Works, Wilkes-Barre, PA 18701.	M-939	Welding Alloy Mfg. Co., Malden, MA (3).
M-275	Vulcan-Kidd Div. H. K. Porter Co., Aliquippa, PA 15001 (3).	M-765	Welding Equipment & Supply Co., Detroit, MI 48212.
M-938	Vulcan Mold & Iron Co., Latrobe, PA 15650.	M-943	Welding Material Co., New Orleans, LA (3).
		M-1092	Welding Sales & Engineering Co., Detroit, MI (3).

W

		M-1230	Weldwire Co., Inc., King of Prussia, PA 19406.
M-1565	Wabash Alloys Inc., Wabash, IN 46992 (2).	M-946	Welland Electric Steel Foundry, Welland, Ontario, Canada.
M-1537	Wah Chang (now Teledyne Wah Chang, Albany, OR 97321).	M-681	Wellman Bronze & Aluminum (now Wellman Dynamics Corp., Creston, IA 50801).
M-1596	Wah Chang (now Teledyne Wah Chang, Albany, OR 97321).		
M-1512	Waimet Alloys Co., Dearborn, MI (3).	M-933	Wellman Corp., S. K., Bedford, OH 44146.
M-824	Wakefield Corp., Wakefield, MA 01880.		
M-1027	Walker Metal Products Ltd., Walkerville, Ontario, Canada.	M-1387	Wells Mfg. Co., Skokie, IL 60067.
		M-1162	Wel-Met Co., Kent, OH (3).
M-963	Wall-Colmonoy Corp., Detroit, MI 48203.	M-1697	Wendt-Sonis (now TRW Wendt-Sonis, Rogers, AR 72756).
M-1692	Wallram Hartmetall G.m.b.H., Essen, Germany.	M-1498	Wesson Co., Ferndale, MI (3).
		M-1322	Westa-Westdeutsche Edelstahlhandelsgesellschaft, Von Poldi-hütte, Düsseldorf, West Germany.
M-1707	Walmet Cemented Carbides, Div. of GTE Sylvania, Royal Oak, MI 48068.		
M-1740	Walsingham Steel Co., Ltd., Walsingham, Bishop Auckland, Co. Durham, DL13 3HX England.	M-1198	Western Alloyed Steel Casting Co., Minneapolis, MN (2).

M-279	Western Brass Metals Div. (now Olin Brass Div. Olin Corp., East Alton, IL 62024).
M-1774	Western Cold Drawn Steel Div. of Stanadyne, Inc., Elyria, OH 44035.
M-468	Western Crucible Steel Casting Co., Minneapolis, MN (3).
M-205	Western Electric Co., Chicago, IL (2) (information only).
M-1493	Western Gold & Platinum Co., Belmont, CA 94002.
M-1380	Westfalenhütte Dortmund A.G., Dortmund, Germany.
M-1373	Westfalische Kupfer und Messingwerke A.G., Lundenscheid, NRW 5880, West Germany.
M-1375	Westfalische Leichtmetallwerke G.m.b.H., Nachrodt, Krs. Altena, Westf., Germany.
M-1345	Westfalische Stahlgesellschaft, 5970 Plettenberg, West Germany.
M-384	Westinghouse Air Brake Co., Wilmerding, PA.
M-118	Westinghouse Electric Corp., Specialty Metals Div., Blairsville, PA 15717.
M-465	West Steel Casting Co., Cleveland, OH (3).
M-206	Wheeling-Pittsburgh Steel Corp., Pittsburgh, PA 15230.
M-883	Wheeling-Pittsburgh Steel Corp., Pittsburgh, PA 15230.
M-119	Wheelock, Lovejoy & Co., Div. of the Metalsource Corp., Cleveland, OH 44122.
M-652	Whipple & Choate Co., Bridgeport, CT (3).
M-659	White Metal Rolling & Stamping Corp., Brooklyn, NY 11222.
M-1077	Whiton Machine Co., New London, CT (3).
M-1665	Whittaker Corp., Nuclear Metals Div., W. Concord, MA 01781.
M-245	Wickes Engineered Materials, Saginaw, MI 48601 (formerly United States Graphite Corp.).
M-1006	Wickman Wimet, see Wimet Ltd.
M-120	Wickwire Spencer Steel Co., New York, NY (3).
M-764	Wieland-Werke AG., D-7900 Ulm, West Germany.
M-121	Wiggin & Co. Ltd., Henry, Hereford, HRA 9SL England.
M-1190	Wikmanshytte Bruks, AB., Wikmanshyttan, Sweden.
M-919	Wilkinsen, Wm., Shustoke, England (3).
M-1584	Willan Ltd., G. L., Sheffield, England.
M-976	Willeys Carbide, see Ex-Cell-O.
M-741	Williams & Sons, E. A., Carlstad, NJ 07072.
M-781	Williamson Bros. Inc., Bridgeport, CT (3).
M-918	Willworthy Piston Ring Ltd., London, England (3).
M-1374	Wilms A.G., G. Robert, Remscheid-Hasten, Germany.
M-237	Wilson Co., H. A., Engelhard Minerals & Chemicals Corp., Newark, NJ 07114.
M-662	Wilson Welder & Metals Co., New York, NY (3).
M-1006	Wimet Ltd., Coventry, CV4 9AD England (formerly Wickman Wimet).
M-554	Wing & Co., J. T., Detroit, MI (3).
M-848	Winn, W. Martin, Ltd., S. Staffordshire, England.
M-869	Wintershall, A.G., Heringer, Germany.
M-1613	Wisconsin Steel Div. International Harvester Co., Chicago, IL 60611.
M-1326	Witten Edelstahl (4) (now part of Thyssen Edelstahlwerke, M-1759).
M-1340	Wolff Handelsgesellschaft, Otto, 5000 Köln 1, West Germany.
M-1521	Wolverine Tube Div. Calumet & Hecla, Universal Oil Products, Allen Park, MI 48101.
M-842	Woodworkers Tool Works, Chicago, IL 60606.
M-280	Workington Iron & Steel Co. (4) (now part of British Steel Corp., M-1724).
M-463	Worthington Pump Inc., Mountainside, NJ 07092.
M-498	Worthington Steel & Annealing Co., Sheffield, England (3).
M-730	Wrought Bearing Metals Inc., New York, NY (3).
M-1376	Wutoschingen Aluminiumwerke G.m.b.H., Wutoschingen, Germany.
M-1259	W. W. Alloys, Detroit, MI (1) (formerly part of Fansteel, M-53).
M-629	Wyckoff Steel Div. AMCO-Pittsburgh Corp. (2).
M-1503	Wyman-Gordon Co., Worcester, MA 01601 (2).
M-1064	Wyndale Mfg. Co., Indianapolis IN (3).

X

M-1473	Xaloy Inc., New Brunswick, NJ 08903.

Y

M-1540	Yawata Iron & Steel Co., Ltd., Tokyo, Japan.
M-1151	Yocum & Son Inc., James, Philadelphia, PA (3).
M-1143	York Machine & Supply Co., York, PA (3).
M-286	Yorkshire Imperial Metals Ltd., Leeds, LS1 1RD England.
M-1560	Yorkshire Imperial Metals Ltd., Leeds, LS1 1RD England.
M-977	Youngstown Alloy Castings Co., Youngstown, OH (3).
M-873	Youngstown Foundry & Machine Co., Youngstown, OH (3).

M-531	Youngstown Steel, Youngstown Sheet & Tube Co., Youngstown, OH 44501.	M-950	Zelnicker, W. A., St. Louis, MO (3).
		M-1472	Zenith Foundry Co., West Allis, WI (3).
		M-1134	Zinkberatungsstelle, G.m.b.H., Berlin, Germany.
		M-289	Ziv Steel & Wire Co., Detroit, MI (3).
M-1332	Zapp, Robert, Edelstahl, 4000 Düsseldorf, West Germany.	M-290	Zurbach Steel Co., Salem, NH 03079.

Z

SECTION III

NUMERICAL LIST OF MANUFACTURERS

M-1	Allegheny Ludlum Steel Corp., Pittsburgh, PA 15222.	M-23	Bethlehem Foundry & Machine Co., Bethlehem, PA (2).
M-2	Allgemeines Deutsches Metallwerk, G.m.b.H., Berlin, Germany.	M-24	Bethlehem Steel Corp., Bethlehem, PA 18016.
M-2a	International Development Corp., New York, NY (3).	M-25	Binney Castings Co., Toledo, OH 43607.
		M-26	Blackor Co., Los Angeles, CA (3).
M-3	Alloy Cast Steel Co., Marion, OH 43302 (2).	M-27	Bohler Bros., Vienna, Austria (4) (merged to form Vereinigte Edelstahlwerke, Vienna, Austria, M-1802).
M-4	Aluminum Co. of America, Pittsburgh, PA 15219.		
M-5	International Alloys Ltd., Aylesbury, Buckinghamshire, England.	M-27	Gebr. Bohler & Co. (4) (merged into Vereinigte Edelstahlwerke, M-1802).
M-6	Aluminum Industries Inc., Chicago, IL (3).	M-28	British Aluminium Co. Ltd., Chalfont Technological Center, Buckinghamshire, SL9 0QB England.
M-7	American Boron Products Div. Continental Copper & Steel Industries, Buffalo, NY.	M-29	Burgess-Parr Inc., Freeport IL (1) (alloys now produced by Stainless Foundry, M-1262).
M-8	American Brass Co. (now known as Anaconda Co., Brass Div., Waterbury, CT 06720).	M-29	Illium Corp. (1) (alloys now produced by Stainless Foundry, M-1262).
M-8	Anaconda Co., Brass Division, Waterbury, CT 06720.	M-30	Calorizing Co., Pittsburgh, PA(3).
		M-31	General Electric Co.: Carboloy Systems Dept., Detroit, MI 48232, Metallurgical Products Dept., Detroit, MI 48232, Magnetic Materials Section, Edmore, MI 48829. M-60.
M-9	AMSCO Div. ABEX Corp., Chicago Heights, IL 60411.		
M-10	Armco Steel Corp., Middletown, OH 45042.		
M-11	American Sheet & Tin Plate Div. (4), see U.S Steel Corp., M-604.	M-31	Metallurgical Products Dept., General Electric Corp., Detroit, MI 48232.
M-12	American Steel & Wire Div. (4), see U.S Steel Corp., M-604.	M-32	Carpenter Technology Corp., Reading, PA 19603.
M-13	Ampco Metal Div. Ampco-Pittsburgh Corp., Milwaukee, WI 53201.	M-33	Chase Brass & Copper Co., Cleveland OH 44122.
M-14	Arcade Malleable Iron Co., Worcester, MA (3).	M-34	Colonial Steel Div. Teledyne Vasco (4), see Teledyne Vasco, M-115.
M-15	Associated Steel Corp., Cleveland, OH 44128.	M-35	Columbia Tool Steel Co., Chicago Heights, IL 60411.
		M-36	Cooper Alloy Corp., Hillside, NJ 07205.
M-16	Avesta Jernverks Aktiebolag, Avesta, Sweden.	M-37	Cronite Foundry Co. Ltd., Tottingham, London N 15, England.
M-17	Babcock & Wilcox, Tubular Products Div., Beaver Falls, PA 15010.	M-38	Crucible Specialty Metals Div. Colt Industries, Syracuse, NY 13201.
M-18	Engelhard Industries Div. Engelhard Mineral & Chemical Corp., Carteret, NJ 07008.	M-39	American Saw & Mfg. Co. (2).
		M-40	Darwin & Milner Inc., Cleveland, OH (3).
M-19	Barber Asphalt Co., Philadelphia, PA (2).	M-41	Deloro Stellite, Belleville, Ontario, Canada.
M-20	Barronia Metals, Ltd., London, England.	M-42	Hydril Co., Los Angeles, CA (2).
		M-43	Dow Chemical Co., Midland, MI 48640.
M-21	Batterium Metal & Vislok, Ltd., Harborough, England.	M-44	Driver Harris Co., Harrison, NJ 07029.
		M-45	Duraloy Blaw-Knox Inc., Scottdale, PA 15683.
M-22	Bell Telephone Laboratories, Murray Hill, NJ (2).	M-46	Duriron Co., Inc., Dayton, OH 45401.

(1) Company reported to be out of business.
(2) Does not produce trade name alloys at present; listed for information only.
(3) Old listing; company name and/or address may be inaccurate.
(4) Merged; for information contact new corporate name.

M-48	Union Carbide Corp., Danbury, CT 06810.	M-79	Malleable Iron Fitting Co., Branford, CT (3).
M-49	Empire Steel Corp., Syracuse, NY (3).	M-80	McDonald & Co., P.F., Boston, MA (3).
M-50	English Steel Corp. Ltd. (4) (merged into British Steel Corp., M-1724).	M-81	McInnes Steel Co., Corry, PA (2).
M-51	Erie Malleable Iron Co., Erie, PA 16501 (2).	M-82	Meehanite Worldwide, Div. of Meehanite Metal Corp., Chattanooga, TN 37401.
M-52	Fagersta Bruks Aktiebolag, Fagersta, Sweden.	M-82a	Ross-Meehan Foundries, Chattanooga, TN 37401.
M-52	Fagersta Inc., West Caldwell, NJ 07006 (agent).	M-83	Metropolitan-Vickers Electrical Co. Ltd., Manchester, England.
M-53	Fansteel Inc., North Chicago, IL 60064.	M-84	Michigan Steel Casting Co., Detroit, MI.
M-54	Ferranti Ltd., Hollinwood, Lancashire, OL9 7JS England.	M-85	Midvale-Heppenstall Co., Philadelphia, PA 19140.
M-55	Finkl & Sons Co., A., Chicago, IL 60614.	M-86	Mond Nickel Co. Ltd., London, England.
M-56	Firth Brown Ltd., Sheffield, S4 7US England.	M-87	Ashby, Ltd., Morris, London, England (3).
M-57	Firth-Sterling (now Teledyne Firth-Sterling, McKeesport, PA 15134).	M-87	Morris Ashby Ltd., London, England (3).
M-57	Teledyne Firth-Sterling, McKeesport, PA 15134.	M-88	National Lead (now NL Industries, Metals Div., Highstown, NJ 08520).
M-58	Frontier Bronze Corp., Niagara Falls, NY 14304.	M-88	NL Industries Inc., Metal Division, Highstown, NJ 08520.
M-59	General Alloys Co., (1) (business and alloys taken over by Alloy Engineering & Casting Co., M-1005).	M-89	National Castings Div. Midland-Ross Corp., Cleveland, OH.
M-61	Driver Co., Wilbur B., Newark, NJ 07104.	M-90	National Tube Div. U.S. Steel (4), see U.S. Steel Corp., M-604.
M-62	Dunford Hadfields Ltd., Sheffield, S9 1TZ England.	M-91	New Jersey Zinc Co., Bethlehem, PA 18018.
M-62	Hadfields Ltd. (now Dunford Hadfields Ltd., Sheffield, S9 1TZ England).	M-92	Niagara Falls Smelting & Refining Div. Continental Copper & Steel Industries Inc., Buffalo, NY (3).
M-63	Handy & Harmon, New York, NY 10022.	M-93	Nicralium Co., Jackson, MI (3).
M-64	Heppenstall Co., Pittsburgh, PA 15201.	M-94	Ohio Brass Co., Mansfield, OH 44902.
M-65	Hoskins Mfg. Co., Detroit, MI 48208.	M-95	Teledyne Ohiocast, Springfield, OH 45501.
M-66	Swiss Laboratory Inc., Akron, OH (2).	M-96	Poldi Steel Works, Prague, Czechoslovakia.
M-67	Inland Steel Co., Chicago, IL 60603.	M-96a	Ajax Tool Sales, New York, NY (3).
M-68	International Nickel Co., Inc., New York, NY 10004.	M-97	Republic Steel Corp., Cleveland, OH 44101.
M-69	Jessop Steel Co., Washington, PA 15301.	M-98	Riverside Metals Corp., Riverside, NJ 08075 (2).
M-70	Kanthal A.B., Hallstammer, Sweden.	M-99	Ruselite Corp., Milwaukee, WI (3).
M-71	Sumet Corp., Buffalo, NY (3).	M-100	Rustless Iron Div. Armco Steel Corp (4), see Armco Steel, M-10.
M-72	Fried. Krupp Huttenwerke, Bochum, West Germany.	M-101	Sandvikens Jernverks Aktiebolag, Sandviken, Sweden.
M-72	Krupp Huttenwerke, Fried., 4630 Bochum, West Germany.	M-101a	Sandvik Inc., Fairlawn, NJ 07410, or Scranton, PA 18501.
M-73	Latrobe Steel Co., Latrobe, PA 15650.	M-101b	Sandvikstahl G.m.b.H., Düsseldorf, West Germany.
M-74	Lebanon Steel Foundry, Lebanon, PA 17042.	M-102	Scovill Mfg. Co., Waterbury, CT (2) (sold metal business to Century Brass Products, Inc., M-1770).
M-75	Koch Light Alloys Ltd., Willesden, London, NW 10 England.	M-103	Seymour Products Co., Seymour, CT 06483 (3).
M-75	Lightalloys Ltd. (now Koch Lightalloys Ltd., Willesden, London NW 10 England).	M-104	Fourchambault et Decazeville, Soc. Anon. de Commentry (4) (now part of Creusot-Loire, M-1488).
M-76	Chemalloy Electronics Corp., Santee, CA 92071.		
M-77	Lumen Bearing Co., Buffalo, NY (3).		
M-78	Farbenindustrie Aktiengesellschaft, I. G. - Abteilung Elektronmetall, Frankfurt a.M., Germany.		
M-78a	Magnesium Development Corp., Newark, NJ (3).		

M-104	Société Anon. de Commentry Fourchambault et Decazeville (4) (merged into Creusot Loire, M-1488).	M-134	Baker, Perkins & Co. Ltd., Peterborough, England (3).
M-104a	Ferner Co., Inc., R. Y., Malden, MA (3).	M-135	Barker & Allen Ltd., Birmingham, England.
M-105	Michigan Standard Alloy Inc., Benton Harbor, MI 49022.	M-136	Birdsboro Corp., Birdsboro, PA 19508.
		M-137	Birmingham Aluminium (Casting) Co., Smethwick, Birmingham, England.
M-105	Standard Alloy Co. Inc., or Michigan Standard Alloy Inc., Benton Harbor, MI 49022.	M-138	Bohn Aluminum & Brass Div., Southfield, MI 48075.
M-106	Sterling Metals Ltd., Gypsy Lane, Nuneaton, England.	M-139	Bonney-Floyd Co., Columbus, OH (2).
		M-140	Braeburn Alloy Steel (now CCS Braeburn Alloy Steel, Braeburn, PA 15068).
M-107	Sutcliff, Speakman & Co., Leigh, Lancaster, England.		
M-108	Telcon Metals Ltd., Crawley, Sussex, England.	M-140	CCS Braeburn Alloy Steel, Braeburn, PA 15068.
M-109	Titanium Alloy Mfg., Div. NL Industries, Niagara Falls, NY (3).	M-141	Bridgeport Brass Co. (Div. of National Distillers & Chemical Corp.), Norwalk, CT 06856.
M-110	Tungum Alloy Co., Ltd., Cheltenham, England.	M-142	Buffalo Bronze Die Casting Co., Buffalo, NY (1).
M-111	Uddeholms, A.B., Uddeholm, Sweden.	M-143	Blaw-Knox Food & Chemical Div (4) (see Duraloy-Blaw-Knox, M-45).
M-111a	Uddeholm Steel Corp., Totowa, NJ 07512.		
M-112	Union Steel Casting Co., Pittsburgh, PA, Div. Duraloy Blaw-Knox (2) (their castings made at other divisions).	M-143	Buflovak Equipment (now Blaw-Knox Food & Chemical Div. Duraloy-Blaw-Knox) (2).
		M-144	Timet, a Div. of Titanium Metals Corp. of America, Pittsburgh, PA 15230.
M-114	Universal Cyclops Specialty Steel Div., Pittsburgh, PA 15228.	M-145	Chapman Valve Mfg. Co., Indian Orchard, MA (3).
M-115	Teledyne Vasco, Latrobe, PA 15650.	M-146	Chicago Steel Foundry, Chicago, IL (3).
M-116	Vereinigte Leichtmetallwerke, G.m.b.H., Hannover-Linden, Germany.	M-147	Chrome Alloys Mfg. Co., Oakland, CA (3).
M-117	Vickers-Armstrong Co. (4) (now part of British Steel Corp., M-1724).	M-148	Chrome Steel Corp., Carteret, NJ (1).
		M-149	Cleveland Automatic Machinery Co., Cleveland, OH (2).
M-118	Westinghouse Electric Corp., Specialty Metals Div., Blairsville, PA 15717.	M-150	Clifford Industries Ltd., Charles, Birmingham, England (2).
M-119	Wheelock, Lovejoy & Co., Div. of the Metalsource Corp., Cleveland, OH 44122.	M-151	AMAX Molybdenum Div. AMAX Corp., Greenwich, CT 06830 (was Climax Molybdenum Div.).
M-120	Wickwire Spencer Steel Co., New York, NY (3).		
M-121	Wiggin & Co. Ltd., Henry, Hereford, HRA 9SL England.	M-151	Climax Molybdenum Co., Greenwich, CT 06830 (part of AMAX Inc.).
		M-152	Commercial Alloys Co., San Francisco, CA (3).
M-122	Cerro Metal Products, Div. of the Marmon Group, Bellefonte, PA 16823.	M-153	Crucible Steel Castings Co., Cleveland OH 44102.
M-123	Ajax Metal Div. H. Kramer & Co. (4).	M-154	Curtis Bay Copper & Iron Works (1).
M-124	Allan & Son, A., Carlstad, NJ 07072.	M-155	Cuyahoga Steel & Wire Div. Hoover Ball & Bearing Co. (2).
M-125	Allied Die Casting Corp., New York, NY (3).		
M-126	Alloys & Products, Inc., New York, NY (3).	M-156	Delta Metal (BW) Ltd., West Bromwich, West Midlands, B70 9ER England.
M-127	Ambolt Machine Tool Co., New York, NY (3).	M-157	Detroit Gray Iron Foundry-Detroit Alloy Steel Co., Detroit, MI (3).
M-128	American Injector Co., Detroit, MI (3).	M-158	Doehler-Jarvis Div. NL Industries, Toledo, OH 43691 (2).
M-129	American Manganese Bronze Co., Philadelphia, PA (3).	M-159	Atlas Foundry Co., Irvington, NJ (1).
M-130	Apex International Alloys Inc., Des Plaines, IL 60018 (formerly Apex Smelting).	M-160	Dursar Corp., Newark, NJ (3).
		M-161	Egal Metal Products Co., Baltimore, MD (3).
M-131	Apollo Steel Co., Apollo, PA (3).		
M-132	Armstrong Bros. Tool Co., Chicago, IL 60646.	M-162	Empire Steel Castings Inc., Reading, PA 19603.
M-133	Vereinigte Chemische Fabriken, Leopoldshall, Germany.	M-163	Farrell Co., Div of USM Corp., Ansonia, CT 06401.

M-164	Fort Pitt Steel Casting Co., Div. of Conval Corp., McKeesport, PA 15134.	M-194	Hybnickel Alloys Co., Wilmington, DE (3).
M-165	General Motors Corp. and Central Foundry Div. of General Motors, Saginaw, MI 48605.	M-195	Robbins & Meyers, Inc., Springfield, OH 45501.
M-166	Hardite Metals Inc., New York, NY (3).	M-196	Schmidt Co., Karl, Neckarsulm, Germany (3).
M-167	Haynes Stellite (now Stellite Div. Cabot Corp., Kokomo, IN 46901).	M-197	American Manganese Mfg. Co., Dunbar, PA (1).
M-167	Stellite Div. Cabot Corp., Kokomo, In 46901.	M-198	Siemans & Halske AG., Berlin, Germany.
M-168	Hills-McCanna Co., Creston, IA 50801.	M-199	Siemens-Schuckert AG., Berlin, Germany, or London, England.
M-169	Holroyd & Co. (now Holcroft Castings & Forgings Ltd., Rochdale, OL12 OLL England).	M-200	Johnson Matthey Metals Ltd., Southgate, London, N14 6ET England.
M-170	Jeffrey Mfg. Div., Dresser Industries Inc., Columbus, OH 43216.	M-201	Sterlite Foundry & Mfg. Co., Auburn, IN (3).
M-171	Kencroft Malleable Co., Buffalo, NY (3).	M-202	Stoody Co., Industry, CA 91749.
M-172	Johnson Bronze Co., New Castle, PA (2).	M-203	United States Pipe & Foundry Co., Birmingham, AL 35202.
M-173	Jones & Laughlin Steel Co., Pittsburgh, PA 15263.	M-204	Hetzel, Vereinigte Silberhammerwerke, Nürnberg, Germany.
M-174	Kinite Corp., Milwaukee, WI (3).	M-204	Vereinigte Silberhammerwerke Hetzel, Nüremburg, Germany.
M-175	E.M.F. Electric Co. Ltd., Victoria, Australia. M-1201.	M-205	Western Electric Co., Chicago, IL (2) (information only).
M-176	Lukens Steel Co., Coatesville, PA 19320.	M-206	Wheeling-Pittsburgh Steel Corp., Pittsburgh, PA 15230.
M-177	Mackenzie's Sons Co. Inc., Duncan, Trenton, NJ (3).	M-207	North American Steel Co., Cleveland, OH (3).
M-178	Magnolia Metal Corp., Auburn, NE 68305.	M-208	Studebaker Chemical Co., Elyria, OH 44035.
M-179	Delta Metal (BW) Ltd., West Bromwich, West Midlands, B70 9ER England.	M-209	Acme Steel & Malleable Iron Works, Div. of Buffalo Brake Beam Co., Buffalo, NY 14207.
M-179	Manganese Bronze & Brass Co. Ltd., Ipswich, England (4) (part of Delta Metal, M-156).	M-210	Edgar Allen Balfour Steels Ltd., Sheffield, S9 1QY England.
M-180	Manganese Steel Forge Co., Philadelphia, PA 19134.	M-210	Edgar Allen Balfour Steels Ltd., Sheffield, S9 1QY England.
M-181	McGill Mfg. Co. Inc., Valparaiso, IN (2).	M-210a	Allen Steel Co., Edgar, New York, NY.
M-182	McPhail & Sons, Wm., Glasgow, Scotland (3).	M-211	Amplex Div. Chrysler Corp., Detroit, MI 48231.
M-183	Metal Sales Corp., Jersey City, NJ (3).	M-211	Chrysler Corp., Amplex Div., Detroit, MI 48231.
M-184	Michiana Products Corp., Michigan City, IN (3).	M-212	McCallum-Hatch Bronze Co. Inc., Buffalo, NY (3).
M-185	Midland Motor Cylinder Co., Smithwick, England (3).	M-213	Brown-Wales Co., Cambridge, MA 02146 (distributer for Horace Potts, M-307).
M-186	Millbury Steel Foundry Co., Millbury, MA (3).	M-214	Cadman Mfg. Co., A.W., Pittsburgh, PA 15222.
M-187	Mills & Co. Ltd., William, Wednesbury, Staffordshire, WS10 0JY England.	M-215	Central Brass & Aluminum Foundry Co., Cincinnati, OH 45203.
M-188	Muntz & Co., P. H., Ltd., Birmingham, England (3).	M-216	Cerro Copper Products, Div. of the Cerro-Marmon Corp., East St. Louis, IL 62202 (2), see Cerro Metal Products.
M-189	National Alloy Steel Div. Blaw-Knox (2) (alloys are now made at Duraloy Blaw-Knox, M-45).		
M-190	National Bronze & Aluminum Foundry, Cleveland, OH (3).	M-217	Hellefors Bruks Aktiebolag, Hellefors, Filipstad, Sweden.
M-191	Acme-Newport Steel Co. (2) (part of Interlake, Inc.).	M-218	Coan Ltd., R.W., London, England (3).
M-191	Newport Steel Co. (4) (part of Interlake Inc., M-1751).	M-219	Aluminum Co. of Canada, Ltd., Montreal, Canada.
M-192	Pacific Foundry Co., San Francisco, CA (3).	M-220	Farrell-Cheek Steel Co., Sandusky, OH 44870.
M-193	P.L. & M. Co., Los Angeles, CA (3).		

M-221 Manco Products Co., Melvindale, MI 48122.
M-222 Dortmund-Hoerder Huttenverein A.G., Dortmund, West Germany (1).
M-223 Gunite Div., Kelsey Hayes Co., Rockford, IL (2).
M-224 Hofors Steel Works, Hofors, Sweden.
M-225 Hubbard Steel Co., East Chicago, IN (3).
M-226 Illinois Zinc Co., Chicago, IL (3).
M-227 Levett, Walker M., New York, NY (3).
M-228 Link-Belt Div. F.M.C. Corp., Chicago, IL.
M-229 Mackintosh-Hemphill Div. Gulf & Western Mfg. Co., Pittsburgh, PA 15203.
M-230 Manning, Maxwell & Moore, Div. of Dresser Industries, Muskegon, MI 49443.
M-231 Jelliff Corp., C. O., Southport, CT 06490.
M-232 National Erie Corp., Erie, PA (3).
M-233 Osborn Steel Co. Ltd., Sheffield, S30 3ZU England.
M-234 Parker Appliance Co., Cleveland, OH (3).
M-235 Pittsburgh Crucible Steel (4), see Crucible Specialty Steel, M-38.
M-236 Pittsburgh Steel Foundry Corp., Glassport, PA (3).
M-237 Wilson Co., H. A., Engelhard Minerals & Chemicals Corp., Newark, NJ 07114.
M-238 Revere Copper & Brass Inc., Rome, NY 13440.
M-239 Rolls-Royce Mfg. Co., Derby, England.
M-240 Ryerson, Joseph T. & Son, Inc., Chicago, IL 60680.
M-241 Sargent & Co., New Haven, CT (2).
M-242 Allen Co., L. B., Shiller Park, IL 60176.
M-243 Suffolk Iron Foundry Ltd., Stowmarket, England.
M-244 S-M Metal Works, London, England (3).
M-245 United States Graphite Corp. (now Wickes Engineered Materials, a Div. of Wickes Engineering, Saginaw, MI 48601).
M-245 Wickes Engineered Materials, Saginaw, MI 48601 (formerly United States Graphite Corp.).
M-246 Washington Iron Works, Seattle, WA 98134.
M-247 Graphite Metallizing Corp., Yonkers, NY 10702.
M-248 Alloy Metal Wire Works (now H. K. Porter Co., Inc., Prospect Park, PA 19076).
M-248 Porter, H. K., Co. Inc., Alloy Metal Wire Works, Prospect Park, PA 19076, (formerly Alloy Metal Wire Co.).
M-249 Aluminium Industrie Aktiengesellschaft, Lausanne-Ouchy, Switzerland.

M-250 American Abrasive Metals Co., Irvington, NJ 07111.
M-251 Carobronze Ltd., London W4 England.
M-252 Amesbury Brass & Foundry Co., Amesbury, MA (3).
M-253 Farrelloy Co., Div. American Solder & Flux Co., Inc., Paoli, PA 19301 (2).
M-254 Bausch Machine Tool Co., Springfield, MA (2).
M-255 Rem-Cru Titanium Inc., Midland, PA 15159.
M-256 Buckeye Brass & Mfg. Co., Cleveland, OH 44103.
M-257 Chateaugay Ore & Iron Co., Standish, NY (3).
M-258 Compagnie d'Orleans, Paris, France (3).
M-258 Orléans, Compagnie de, Paris, France.
M-259 Evans Steel Co., H. D., Boston, MA (3).
M-260 Halcomb Steel Div. Crucible Steel (4), see Crucible Specialty Metals, M-38.
M-261 Jessop-Saville Ltd., Sheffield, 9 England (3).
M-262 Kalif Corp., Emeryville, CA (3).
M-263 LaBour Pump Co., Elkhart, IN 46514.
M-264 Lunkenheimer Co., Cincinnati, OH 45214.
M-265 CMW Inc., Indianapolis, IN 46206 (formerly Mallory Metallurgical).
M-265 Mallory Metallurgical (now CMW Inc., Indianapolis, IN 46206) (Div. of P. R. Mallory & Co. Inc.).
M-266 Mueller Brass Co., Port Huron, MI 48060.
M-267 S.K.F. Industries Inc., King of Prussia, PA 19406.
M-268 Southern Malleable Iron Co., East St. Louis, IL (3).
M-269 ITT Components Group Europe, Harlow, Essex, United Kingdom.
M-269 Standard Telephones & Cables (now ITT Components Group Eurpoe, Edinburgh Way, Harlow, Essex, England).
M-270 Steinvertriebs Aktiengesellschaft, Berlin, Germany.
M-271 Taylor-Wharton Div. of Harsho Corp., Easton, PA (2).
M-272 Ultralumin Leichtmetall A.G., Germany (1).
M-273 United Lead Co., see NL Industries, M-88.
M-274 Verilite Metals Co., New York, NY (3).
M-275 Vulcan-Kidd Div. H. K. Porter Co., Aliquippa, PA 15001 (3).
M-276 American Cutting Alloys Inc., New York, NY (3).
M-277 Warman Steel Casting Co., Huntington Park, CA (3).
M-278 Weatherly Foundry & Mfg. Co., Weatherly, PA 18255.
M-279 Olin Brass Div. Olin Corp., East Alton, IL 62024.

M-279 Western Brass Metals Div. (now Olin Brass Div. Olin Corp., East Alton, IL 62024).
M-280 Workington Iron & Steel Co. (4) (now part of British Steel Corp., M-1724).
M-281 Acieral Co. of America, New York, NY (3).
M-282 Imperial Aluminium Chemical Industries (Metals Div.) (now IMI (Knoch) Ltd., Birmingham, England, M-1578).
M-283 Strasser Co., E. Rorschach, Switzerland.
M-284 Giulini Werke A.G., Rohrsbach, Germany.
M-285 Acipco Steel Products, Div. of American Cast Iron Pipe Co., Birmingham, AL 35202.
M-285 American Cast Iron Pipe Co. (now ACIPCO Steel Products), Birmingham, AL 35202.
M-286 Yorkshire Imperial Metals Ltd., Leeds, LS1 1RD England.
M-287 Garford Engineering Co., Garfield, IN (3).
M-288 Sivyer Steel Casting Co., Milwaukee, WI 53202 (3).
M-289 Ziv Steel & Wire Co., Detroit, MI (3).
M-290 Zurbach Steel Co., Salem, NH 03079.
M-291 Reading Iron Co., Reading, PA (3).
M-292 Byers Co., A.M., Ambridge, PA (1).
M-293 American Stainless Steel Co., Pittsburgh, PA (3).
M-294 Titusville Forge Co., Titusville, PA (3).
M-295 Sandusky Foundry & Machine Co., Sandusky, OH 44870.
M-296 Heraeus-Vacuumschmelze (now Vacuumschmelze, M-1631).
M-297 VDM, abbreviation for Vereinigte Deutsche Metallwerke, M-297.
M-297 Vereinigte Deutsche Metallwerke, D 5980 Werdolhl, West Germany (or Ore & Chemical Corp., New York, NY, 10016).
M-297a Ore & Chemical Corp., New York, NY 10016, or Houston, TX 77092 (part of VDM, M-297).
M-298 Cindal Aluminium Ltd., Birmingham, England (3).
M-299 Metallgesellschaft A.G., Frankfort/Main, Hesse 6000, West Germany.
M-300 Lehigh Babbitt Co., Allentown, PA (3).
M-301 Osram, G.m.b.H., Kommandit-Ges., Berlin, Germany (2).
M-302 Aluminiumwerke Maulbronn, Maulbronn, Württ, Germany.
M-302 Maulbronn, Aluminiumwerke, Maulbronn-Württ, Germany.
M-303 Vereinigte Deutsch Nickel-Werke AG., 5840 Schwerte, West Germany.
M-304 Bosch Metallwerk, Robert, A.G., Stuttgart, Germany.
M-305 Machlett Laboratories, Stamford, CT (2).
M-306 British Thomson Houston Co., Ltd., Rugby, Warwickshire, England.
M-307 Potts, Horace T., Co., Philadelphia, PA 19134.
M-308 Cutler Hammer Inc., Milwaukee, WI 53216.
M-309 Baldwin-Lima-Hamilton Corp., Eddystone Div., Philadelphia, PA.
M-310 Harrison, Fischer & Co., Ltd., Sheffield, S3 7WJ England.
M-311 National Physical Laboratory, Teddington, England.
M-312 Michigan Valve & Foundry Co., Detroit, MI (2).
M-313 American Magnesium Co., Cleveland, OH (1).
M-314 American Smelting & Refining Co. (ASARCO), New York, NY 10005.
M-314 ASARCO and Federated Metals Corp., Div. of American Smelting & Refining Co., New York, NY 10005.
M-315 Maywood Chemical Works, Maywood, NJ (3).
M-316 Hans-Heinrich Hütte, G.m.b.H., Langelsheim a.Harz, Germany.
M-317 Hoyt Metal Co. of London Ltd., Putney, London, SW15 2NX England.
M-318 Plykrome Corp., New York, NY (3).
M-319 Phelps Dodge Co., New York, NY 10022.
M-320 Metallbank, A.G., (Heddernheimer Copper Works), Heddernheim, Germany.
M-321 Birmetals Ltd., Birmabright Works, Birmingham, B32 3BX England.
M-322 INSILCO Corp., Meriden, CT 06450.
M-323 Machinenbau A.G., Golzern-Grimma, Germany.
M-323 Maschinenbau A.G., Golzern-Grimma, Germany.
M-324 Kreidler Werke G.m.b.H., 7000 Stuttgart 40, West Germany.
M-325 Birmetals Ltd., Birmabright Works, Birmingham, B32 3BX England.
M-326 Pioneer Alloy Products Co., Cleveland, OH.
M-327 United Steel Div. (4), see Republic Steel Corp., M-97.
M-328 Ingersoll Steel Co., New Castle, IN 47362.
M-329 VR/Wesson or Vascoloy-Ramet, Waukegan, IL (4), (now part of Fansteel, M-53).
M-330 Ostermann & Co., Metallwerke, KG., Köln NRW 5000, West Germany.
M-331 Otis Elevator Co., New York, NY 10001.
M-332 Johnson & Co., A., New York, NY (3) (agent for M-16).
M-333 Swedish American Steel Corp., Brooklyn, NY (3).
M-334 Bisset Steel Co., Cleveland, OH 44103.
M-336 Boyd-Wagner Co., Chicago, IL (3).
M-337 Craine-Schrage Steel Div. Detroit Steel Co. (1).

Section III: Numerical List of Manufacturers / 1647

M-338	Booth Aluminium Ltd., James (4) (merged to form Alcan Booth, M-643).	M-374	Stulz Sickles Steel Co., Elizabeth, NJ 07207.
M-339	Kidd Drawn Steel Co., Aliquippa, PA (3).	M-375	Swedish Iron & Steel Corp., Middlesex, NJ 08846 (2).
M-340	Hobson, Houghton & Co., New York, NY (3).	M-376	Timken Co., Canton, OH 44706.
M-341	Peninsular Steel Co., Detroit, MI 48205.	M-377	Tennessee Coal, Iron & Railroad (4), see U.S. Steel Corp., M-604.
M-342	Anchor Drawn Steel Co. (4), (see Teledyne Vasco, M-115).	M-378	Fisher Scientific Co., Pittsburgh, PA 15219.
M-343	Heller Bros. Co., Newark, NJ, see Heller Tool Div., M-1221.	M-379	Upson Nut Div. or Tool Steel Div. Republic Steel (2), see Republic Steel, M-97.
M-344	Hidalgo Steel Co. Inc., New York, NY (3).	M-380	Griesogen Griesheimer Autogen Verkaufs, G.m.b.H., Greisheim a.M., Germany.
M-345	Boker & Co., H., New York, NY (1).		
M-346	Illingworth Steel Co., Philadelphia, PA (3).	M-381	Metallo-Chemische Fabrik, Dr. Leopold Rostosky, Berlin, Germany.
M-346	Kramer & Co., Chicago, IL 60608.	M-382	Bleiwerk Goslar, Goslar & Harz, Germany.
M-347	CWC Castings Div. of Textron Inc., Muskegon, MI 49443.	M-383	Vereinigte Stahlwerke, Düsseldorf (1).
M-348	Martinel Steel Co., Ltd., Sheffield, England (3).	M-384	Westinghouse Air Brake Co., Wilmerding, PA.
M-349	Ney Co., J. M., Bloomfield, CT 06002.	M-385	Brukskoncernen A.B., Fagersta, Sweden.
M-350	Adams & Osgood Steel Co., Boston, MA (3).	M-386	Nitralloy Corp., New York, NY (3).
M-351	Imperial Smelting Co. Ltd., St. James Square, London, SW 1 England.	M-388	Electro-Steel Co., Pittsburgh, PA (3).
		M-389	Ward's Sons Co., Edgar T., Newark, NJ (3).
M-352	Andrews Toledo Ltd., London, England (1).	M-390	Tacony Steel Co., Philadelphia, PA (1).
M-353	Atlantic Steel Corp., Astoria, NY 11106.	M-391	Unimetal Co., Franklin, PA (3).
M-354	Cannon-Stein Steel Co., Syracuse, NY (1).	M-392	Wardlow, S. & C., Nutley, NJ (3).
		M-393	Johnson Mfg. Co., Princeton, IA 52768.
M-355	Crowley Inc., John A., New York, NY (3).	M-394	Vanadium Corp. of America (4) (now part of Foote Minerals, M-1038.
M-355	Kloster Steel Corp., Chicago, IL 60607.	M-395	Steel & Tubes Div. Republic Steel, see Republic Steel Corp., M-97 (4).
M-356	Crucible Electric Steel Co., Homestead, PA (3).	M-396	British Non-Ferrous Research Association, London, England (2) (information only).
M-357	Disston Inc., Pittsburgh, PA 15219 (2).		
M-358	Duke Steel Co., Inc., New York, NY (3).	M-397	Union Drawn Steel Div. (4), see Republic Steel, M-97.
M-359	Ekstrand & Tholand Co., New York, NY (3).	M-397	United Alloy Steel Corp., Canton, OH (4), see Republic Steel Corp., M-97.
M-360	Frasse & Co., Peter A., Lake Success, NY (2).	M-398	Ohio Ferro Alloys Corp., Canton, OH 44711.
M-361	Hand, Edward A., Philadelphia, PA (3).	M-399	Koppers Co. Inc., Metal Products Div., Baltimore, MD 21203.
M-362	Hawkridge Bros Co., Malden, MA 02148.	M-400	Industrial Furnace Corp., Buffalo, NY (3).
M-363	Houghton & Richards Inc., Boston, MA (3).	M-401	Bundy Tubing Div. of Bundy Corp., Warren, MI 48089.
M-365	Milne & Co., A., Atlanta, GA 30318.	M-402	Metallhüttenwerke Schaefer und Schael, A.G., Breslau, Germany.
M-367	Paragon Steel Co., Rutherford, NJ (3).		
M-368	Patriarche & Bell, New York, NY (3).	M-403	Friedrich Wilhelms-Hütte, Mülheim, Ruhr, West Germany. .
M-369	Pittsburgh Rolls Div. (4), see Duraloy Blaw-Knox, M-45.	M-404	Aktiebolaget Skandinaviska Armaturfabriken, Stockholm, Sweden.
M-370	Pittsburgh Tool Steel Wire (now Teledyne Pittsburgh Tool Steel Wire, Monaca, PA 15061).	M-404	Skandinaviska Armaturfabriken AG., Stockholm, Sweden.
M-370	Teledyne Pittsburgh Tool Steel Co., Monaca, PA 15061.	M-405	Carr Ltd., Charles, Birmingham, England (3).
M-371	Prosser & Sons, Thos., New York, NY (3).	M-406	Supersteels Inc., Cleveland, OH (3).
M-372	Purdy Co. Inc., A. R., Lyndhurst, NJ (3).	M-407	Gabriel & Co., Ltd., A.B., Row, Birmingham, England (3).
M-373	Simonds Steel Div. Wallace-Murray Corp., Lockport, NY 14094.		

M-408	Scott & Co. Ltd., A. C., Manchester, England.	M-444	Texas Electric Steel Casting Co., Houston, TX (2).
M-409	Gallimore & Sons, Wm., Ltd., Sheffield, England (1).	M-445	Thomas Steel Co., Warren, OH (2).
M-410	Billington & Newton Ltd., Longport, Staffordshire, England (3).	M-446	Webb Wire Works, New Brunswick, NJ (3).
M-411	Haywoods NCA Metal Ltd., London, England (3).	M-447	Babcock & Wilcox Co., Power Generation Group, Barberton, OH 44203.
M-412	Ballard & Co., F.J., Tipton, England (2).	M-448	United Steel Companies Ltd. (4) (now part of British Steel Corp., M-1724).
M-413	Blackwell's Metallurgical Works, Lancashire, England (3).	M-448a	Steel, Peech & Tozer, Sheffield (4) (now part of British Steel Corp., M-1724).
M-414	National Alloys Ltd., London, England (3).	M-448b	Appleby-Frodingham Steel Co. (4) (now part of British Steel Corp., M-1724).
M-415	Miralite Ltd., Mortlake, England (1).	M-449	Cunningham Co., M.E., Pittsburgh, PA 15233.
M-416	British Driver-Harris Co. Ltd., Manchester, England (2).	M-450	Micheville, Sté des Aciéries de, Paris, France.
M-417	Shaw, J., Son & Greenhalgh, Ltd., Albert Works, Huddersfield, England.	M-451	Continental Foundry & Machine Co., East Chicago, IN (3).
M-418	Sybry, Searls & Co., Ltd., Cannon Steel Works, Sheffield, England (3).	M-452	Stone Manganese Marine; J. Stone & Co., London, SE 17, England.
M-419	Rieter & Co., J. J., New York, NY (3).	M-453	Shanks & Co. Ltd., Barrhead, Scotland.
M-420	Rely Metal Works, Cape Town, South Africa.	M-454	Mather & Platt Ltd., Manchester, England.
M-421	Thermit Ltd., Edmonton, England (2).	M-455	Aciéries S. A. Bedel, F-42 Saint Etienne, France (3).
M-422	Jones, Rd., Ltd., Birmingham, England (3).	M-455	Bedel & Cie, see Aciéries SA Bedel, Saint Etienne, France.
M-423	Victor Mfg. Co., London, England (3).		
M-424	Brighton Electric Steel Casting Co., Beaver Falls, PA 15010.	M-456	Lyon, Conklin & Co., Baltimore, MD (2).
M-425	Field Co., B. A., Newark, NJ (3).	M-457	Durener Metallwerke, Duren, Germany.
M-426	High Duty Alloys Ltd., Slough, Buckinghamshire, SLI 4PA England.	M-458	Goodlass Wall & Lead Industries, Ltd., London, England (3).
M-427	Non-Corrodal Alloys Ltd., London, England (3).	M-459	Deutsche Edelstahlwerke G.m.b.H. (4) (merged to form Thyssen Edelstahlwerke A.G., M-1759).
M-428	Hoyt Metal Co., New York, NY (3).		
M-429	Kahl Iron Foundry, Detroit, MI 48216.		
M-430	Fox & Co., Samual (4) (now part of British Steel Corp., M-1724).	M-460	Robins Conveyors Inc., Passaic, NJ (3).
		M-461	Silicum Pistons Ltd., London, England (3).
M-431	Bearium Metals Corp., Rochester, NY 14624.	M-462	Osnabrück Kupfer und Drahtwerke, Osnabrück, Germany.
M-432	Delsteel Inc., Wilmington, DE 19899.		
M-433	Edgcomb Metals Co., Philadelphia, PA 19114 (2).	M-463	Worthington Pump Inc., Mountainside, NJ 07092.
M-434	Ryer Inc. Ltd., Los Angeles, CA (3).	M-464	Allyne-Ryan Foundry Co., Cleveland OH (3).
M-435	Atlas Steels Ltd., Welland, Ontario, Canada.	M-465	West Steel Casting Co., Cleveland, OH (3).
M-436	Magnolia Anti-Friction Metal Co., London SW 1 England.	M-466	Pennsylvania Steel Foundry & Machine Co., Hamburg, PA (2).
M-437	Cincinnati Steel Castings Co., Cincinnati, OH (3).	M-467	Massillon Steel Casting Co., Massillon, OH 44646.
M-438	ESCO Corp., Portland, OR 97210 (formerly Electric Steel Foundry).	M-468	Western Crucible Steel Casting Co., Minneapolis, MN (3).
M-439	Forging & Casting Div. Allegheny-Ludlum (4), see Allegheny-Ludlum, M-1.	M-469	Imperial Chemical Industries (2) (alloy business transferred to IMI, M-1578, or Alcoa of Great Britain, M-1715).
M-440	Industrial Steels Inc., Cambridge, MA (3).	M-470	Dorman, Long & Co. (4) (now part of British Steel Corp., M-1724).
M-441	Interstate Foundry & Machine Co., Johnson City, TN (2).	M-471	Isteg Steel Products Co., Westminster, England (3).
M-442	New England High Carbon Wire Co., Millbury, MA (2).	M-472	Brown, Bayley's Steel Works (2) (4) (now part of Dunford Hadfields Ltd., M-62).
M-443	Shawinigan Chemicals, Ltd., Montreal, Canada.		

Section III: Numerical List of Manufacturers / 1649

M-474 Chambersburg Engineering Co., Chambersburg, PA 17201.
M-475 Milwaukee Steel Foundry Co., Milwaukee, WI (3).
M-476 Brown Foundries Co., David, Penistone, Nr. Sheffield, England.
M-477 Jackman, Joseph, & Co., Ltd., Sheffield, England (1).
M-478 British Rolling Mills, Ltd., Tipton, Staffordshire, England.
M-479 Sowers Mfg. Co., Buffalo, NY (3).
M-480 Cleveland Brass Corp., Cleveland OH (2).
M-481 Scientific Alloys Inc., Los Angeles, CA (3).
M-482 Herbert Small Tools & Equipment Ltd., see Herbert-Cutanit Ltd.
M-483 National Forge Co., Irvine, PA 16329.
M-484 Oxley & Co., William, Rotherham, England (3).
M-485 Apex Steel Co., Sheffield, England (1).
M-486 Jessop-Saville Ltd., Sheffield, 9 England (3).
M-486 Saville & Co. Ltd., J. J., Sheffield, England (3).
M-487 Darwin's Ltd., Sheffield, England (now part of Balfour-Darwins, M-1725).
M-488 Hemmings & Co., Rotherham, England (3).
M-489 Andrews, Ltd., Sheffield, S4 7WZ England.
M-490 Cameron & Son Ltd., Sheffield, England (1).
M-491 Fairley & Sons, Jas., Ltd., Birmingham, England (3).
M-492 Atkinson & Sons Ltd., E., Sheffield, S18 6NS England.
M-493 Aluminium Walzwerke Singen G.m.b.H., 7700 Singen/Hohentweil, West Germany.
M-493 Singen, Aluminium Walzwerke, G.m.b.H., Singen/Hohentwell, West Germany.
M-494 Sheepbridge Engineering Ltd., Sheepbridge Works, Chesterfield, Derbyshire, England.
M-494 Sheepbridge Stokes Centrifugal Casting Co., see Sheepbridge Engineering.
M-495 Monarch Steel Corp., Indianapolis, IN (3).
M-496 Taylor Ltd., Samuel, Birmingham, England (3).
M-497 Hassall & Sons, Wm., Manchester, England (3).
M-498 Worthington Steel & Annealing Co., Sheffield, England (3).
M-499 Industrial Steels Ltd., Sheffield (4) (merged into British Steel Corp., M-1724).
M-500 Symington-Gould Corp., Depew, NY (3).
M-501 Edelstahlwerk Buderus or Rochling Buderus (4) (merged to form Rochling-Burbach, M-912).
M-501 Rochling Buderus (merged with Rochlingstahl, M-912, to form Rochling-Burbach, M-912) (all M-501 numbers dropped).
M-502 Hardenite Steel Co. Ltd., Sheffield, S4 7WZ England.
M-503 Cochrane Corp., Philadelphia, PA (3).
M-504 British Standards Institution, London, N1 9ND England (2) (not a producer; information only).
M-505 Dinorm Co., Berlin, Germany (2).
M-506 Prescott Co., Menominee, MI (3).
M-507 Resisto-Loy Co. Inc., Grand Rapids, MI 49507.
M-508 Janney Cylinder Co., Philadelphia, PA 19136.
M-509 Bridgeport Deoxidized Bronze Corp., Bridgeport, CT 06605 (2).
M-510 Kayser-Ellison (now Sanderson Kayser Ltd., Sheffield, S9 2SD England).
M-511 Calloy Ltd., Avonmouth, England.
M-512 Sioux Tools Inc., Sioux City, IA 51102.
M-513 TRW Metals Div., Minerva, OH 44657.
M-514 Stewart-Warner, Die Casting Div., Chicago, IL (2).
M-515 Central Institute of Metals, Leningrad, USSR.
M-516 Bliss & Laughlin Steel Co., Harvey, IL 60426.
M-517 Kahl-Holt Co., Baltimore, MD (3).
M-518 Stedman Foundry & Machine Co., Aurora, IN 47001.
M-519 Merco Nordstrom Valve Co., Pittsburgh, PA (3).
M-520 Chatillon & Sons, John, Kew Gardens, NY 11415.
M-521 Firth-Vickers Special Steels Ltd., Sheffield, S9 2FU England.
M-522 DuPont de Nemours & Co., E.I., Wilmington, DE 19898 (2).
M-523 Lynchburg Foundry Co., Lynchburg, VA (2).
M-524 American Machine & Foundry (AMF Inc.), White Plains, NY 10604.
M-524 AMF Inc., White Plains, NY 10604.
M-525 Poro Metals Ltd., London, England (3).
M-526 Murex Ltd., Rainham, Essex, England.
M-527 Lanz, Heinrich, A.G., Mannheim, Germany (3).
M-528 Goldschmidt, Th., A.G., Abteilung Metalle, Essen, Germany.
M-529 Metal & Thermit Corp., New York, NY (3).
M-530 Colvilles Ltd., Glasgow, Scotland (4) (now part of British Steel Corp., M-1724).
M-531 Youngstown Steel, Youngstown Sheet & Tube Co., Youngstown, OH 44501.
M-532 Norton Co., Metals Div., Newton, MA 02164.
M-533 Schneider & Cie (4) (merged into Creusot-Loire, M-1488).
M-534 Lunn & Co., E., Glasgow, Scotland.
M-535 Trenite Foundry Corp., Trenton, NJ (3).

M-536   Mitchell Co. Inc., Robert, Montreal, Canada.
M-537   Alan Wood Steel Co., Conshohocken, PA 19428 (2).
M-538   American Nickeloid Co., Peru, IL 61354.
M-539   National Bearing Div. ABEX Corp., Meadville, PA 16335.
M-540   Lewin Metals Corp., East St. Louis, IL (3).
M-541   Alluminio S.A., Milan, Italy.
M-542   Northfield Iron Co., Northfield, MN (3).
M-543   Pittsburgh Brass Mfg. Co., Irwin, PA 15642.
M-544   Beckett Bronze Co., Muncie, IN (2).
M-545   Michigan Smelting & Refining Co., Detroit, MI (3).
M-546   Kinney Iron Works, Los Angeles, CA (3).
M-547   Plouff Metallographic Institute, Boston, MA.
M-548   Nassau Smelting & Refining Co. Inc., Staten Island, NY (2).
M-549   Egge Co., E.N., Los Angeles, CA (3).
M-550   Garrett Brass & Foundry Co., Garrett, IN (3).
M-551   American Steel Co., Ellwood City, PA (2).
M-552   LaClede Brass (or Steel) Works, St. Louis, MO 63102.
M-553   Pacific Metal Co., Portland, OR (3).
M-554   Wing & Co., J. T., Detroit, MI (3).
M-555   Aluminiumwerke Nürnberg, G.m.b.H., Nürnberg, West Germany.
M-555   Nürnberg, Aluminiumwerke, G.m.b.H., Nürnberg, Germany.
M-556   Magnesium Castings & Products Ltd., Slough, England (3).
M-557   Eisenwerke Neubrandenburg, G.m.b.H., Berlin, Germany.
M-558   AMAX Lead & Zinc, Clayton, MO 63105.
M-558   American Metal Climax, Div. of AMAX Corp., Greenwich, CT 06830.
M-558   United States Metals Refining Co., Carteret, NJ.
M-559   Alamo Iron Works, San Antonio, TX 78237 (2).
M-560   Joslyn Stainless Steels Co., Ft. Wayne, IN 46801.
M-561   Puget Sound Metal Works, Tacoma, WA (3).
M-562   Hiertz Metal Co., Theodore, St. Louis, MO (3).
M-563   Ariston Metal Co., Jersey City, NJ (2) (3).
M-564   Apex Bronze Foundry Co., Oakland, CA (3).
M-565   United American Metals Corp., Chicago, IL 60612.
M-566   Chain Belt Co., Milwaukee, WI.
M-567   Atkinson Co., Rochester, NY (3).
M-568   Acme Steel Co., Chicago, IL (2) (part of Interlake, Inc.).
M-569   Atlantic Steel Casting Co., Chester, PA 19016.
M-570   Canada Electric Steel Castings Ltd., Orillia, Ontario, Canada.
M-571   Frank Foundries Corp., Moline, IL 61265.
M-572   Kay-Brunner Steel Products Inc., Alhambra, CA (2).
M-573   LFM Mfg. Co., Atchison, KS (3).
M-573   Locomotive Finished Materials Co., Atchison, KS (3).
M-574   Stanley Steel Div. of the Stanley Works, New Britain, CT 06050.
M-575   Bolton & Sons Ltd., Thomas, Stoke-on-Trent, ST10 2HF England.
M-576   Non-Ferrous Castings Co. Ltd., Smethwick, England.
M-577   Birkett, Billington & Newton Ltd., Hanley, England (3).
M-578   Lincoln Electric Co., Cleveland, OH 44117.
M-579   A. O. Smith Co., Milwaukee, WI 53201 (2).
M-579   Smith Co., A. O., Milwaukee, WI 53201 (2).
M-580   Hanson-Van Winkle-Munning Co., Indianapolis, IN 46236.
M-581   Ardal, Ltd., Worle, Weston-Super-Mare, England (3).
M-582   Chace Co., W.M., Detroit, MI 48209.
M-583   Pressed Steel Car Co., Chicago, IL (3).
M-584   Degefors Iron & Steel Works, Degefors, Sweden.
M-585   U.S. Reduction Co., East Chicago, IN 46312.
M-586   Glacier Metal Co., Richmond, VA (3).
M-587   Finn Metal Works, John, San Francisco, CA (3).
M-588   Union Bronze Co., Reading, PA (3).
M-589   Everard Tap & Die Corp. (2).
M-590   Seitzinger's Inc., Atlanta, GA 30318.
M-591   Atlantic Zinc Works, New York, NY (3).
M-592   Central Foundry Co., Joplin, MO.
M-593   Bronze Die Casting Co., Pittsburgh, PA 15233.
M-594   Dirigold Corp., Kokomo, IN (3).
M-595   Alemite Div. Stewart Warner Corp., Chicago, IL 60614.
M-596   Grasselli Chemical Co., Cleveland, OH (3).
M-597   Braun-Steeples Co. Ltd., San Francisco, CA (3).
M-598   Tungsten Widia Tool Corp., New York, NY (3).
M-599   Wehr Steel Co., Milwaukee, WI 53210.
M-600   Castalloy Co., Inc., Natick, MA.
M-601   Motor Castings Co., Milwaukee, WI 53214.
M-602   Prentiss & Co., George W., Holyoke, MA (3).
M-603   Aurora Metal Co., Montgomery, IL 60538.
M-604   United States Steel Corp., Pittsburgh, PA 15320

M-605	Lindberg Hevi Duty Div. Solar Basic Industries, Chicago, IL 60612.	M-633	Granite City Steel Co., Granite City, IL (4), see National Steel, M-836.
M-606	Ansoldo, Società Anonima, Rome, Italy.	M-634	Unicast Corp., Div. Midland Ross Corp., Toledo, OH.
M-606	Società Anonima Ansaldo, Rome, Italy.	M-635	Parkersburg Steel Co., Parkersburg, WV (2).
M-607	Edgcomb Metals Co., Tulsa, OK 74103.		
M-607	Steel Sales Corp (now Edgcomb Metals, Tulsa, OK 74103).	M-635	Superior Sheet Steel Div. Parkersburg Steel Co., Parkersburg, WV (2).
M-608	Achorn Steel Co., Cambridge, MA (3).	M-636	Consolidated Ashcroft Hancock Co. Inc., Bridgeport, CT (3).
M-609	Gimo-Osterby Bruks, A.B., Gimo, Sweden.	M-637	Athens Foundry Div. Ingersoll-Rand Co., Athens, PA 18810.
M-610	Horndal Jernverks, A.B., Horndal, Sweden.	M-638	Hoganas-Billesholms Aktiebolaget, Hoganas, Sweden (2).
M-611	Klosters A.B., Langshyttan, Sweden (2).		
M-612	Cleveland Twist Drill Co., Cleveland, OH (2).	M-639	Antaciron, Inc., Wellsville, NY (3).
M-613	Becker, Stahlwerk, AG., Krefeld, Rhineland, Germany.	M-640	Molybdenum Corp. of America (now Molycorp., Inc., White Plains, NY 10604).
M-613	Stahlwerk Becker AG., Krefeld (Rheinland), Germany.	M-641	Andrews & Co., Ltd., Thomas, Royds Works, Sheffield, S4 7WZ England.
M-614	Dorrenberg Edelstahl, 5252 Runderoth, West Germany.	M-642	Allis-Chalmers Mfg. Co., Milwaukee, WI 53201.
M-615	Schmidt Stahlwerke, Rudolf, Vienna, Austria.	M-643	Alcan Booth Industries Ltd., Alcan House, London, W1X 6DP England.
M-616	Schoeller-Bleckmann (4) (merged into Vereinigte Edelstahlwerke, M-1802).	M-643	Northern Aluminium, see Alcan Booth, M-643.
M-617	Edelstahlwerk Rochling A.G., Berlin, Germany.	M-644	Aluminium Union, Ltd., London, England (3).
M-618	Sanderson Kayser Ltd., Sheffield, S9 2SD England.	M-645	Knutange, Sté Metallurgique de, Paris, France (2).
M-619	Barrow Haematite Steel Co., Ltd., Barrow-in-Furness, Lancashire, England.	M-646	Beryllium Corp. of America (now Kawecki Beryllium Industries Inc., New York, NY 10017).
M-620	Alais, Forges et Camargue, Paris, France.	M-646	Kawecki Berylco Industries Inc., New York, NY 10017.
M-621	Allgemeine Elektrizitäts-Gesellschaft, Berlin 1000 West Germany.	M-647	Société des Sondures, Castolin, South America.
M-622	Aluminium Belge, S.A., Luttich, Belgium.	M-647	Sonduras, Société des, Castolin, South America.
M-623	Compagnie des Alliages Speciaux d'Aluminium, Asnières, France.	M-648	Castolin Welding Alloys Inc., New York, NY (3).
M-624	Lavorazione Leghe Leggere, Porto Marghera (Venezia), E. Ferrara, Milan, Italy.	M-649	Central Iron & Steel Co., Harrisburg, PA (3).
M-625	Rheinische Röhrenwerke Aktiengesellschaft, Mülheim, Germany.	M-650	Gibson Electric Co., Delmont, PA 15626.
		M-651	Skoda Works, National Corporation, 316 000 Plzen, Czechoslovakia.
M-626	Pittsburgh Metallurgical Co. Inc., Niagara Falls, NY (3).	M-652	Whipple & Choate Co., Bridgeport, CT (3).
M-627	Faitout Iron & Steel Co., Newark, NJ.	M-653	American Chain & Cable Co. Inc. (2).
M-628	Weirton Steel Co. Div. of National Steel Corp., Weirton, WV 26026 (same alloys as Great Lakes Steel).	M-654	Acme Foundry & Machine Co., Coffeyville, KS 67337.
		M-655	Adirondack Steel Casting Co., Watervliet, NY 12189.
M-629	Wyckoff Steel Div. AMCO-Pittsburgh Corp. (2).	M-656	Adams & Co., J. D., Indianapolis, IN (2).
M-630	Q. & C. Co., New York, NY (3).		
M-631	Continental Steel Corp. (now Penn-Dixie Corp., Kokomo, IN 46901).	M-657	Cohn Ltd., A., London, England (3).
		M-658	Alfa Romeo, Milan, Italy.
M-632	Bound Brook Bearing Corp. (now GKN Powder Met. Inc., Bound Brook Div., Worcester, MA 01604).	M-659	White Metal Rolling & Stamping Corp., Brooklyn, NY 11222.
		M-660	Ferrous Metals Corp., New York, NY (2).
M-632	GKN Powder Met Inc., Bound Brook Div., The Presmet Div., Worcester, MA 01604 (formerly Bound Brook Bearing Corp.).	M-661	Agile Div. Nagle/Sybron Corp., Rochester, NY (3).

M-662	Wilson Welder & Metals Co., New York, NY (3).	M-690	General Motors Corp. and Central Foundry Div. of General Motors, Saginaw, MI 48605.
M-663	Airco Vacuum Metals, Berkeley, CA 94710.	M-691	Empire Sheet & Tin Plate Co., Mansfield, OH (3).
M-663	Airco Welding Products, New Providence, NJ 07974.	M-692	Alloy Rods Co., see Chemetron Corp., M-1713.
M-664	Hunt-Spiller Div. Power Products Inc. (2).	M-693	Copper Development Association, Potters Bar, Hertfordshire, England (2) (information only).
M-665	Southern Metals Co., St. Louis, MO (3).	M-694	Ackerlind Steel Co., Inc., Long Island City, NY 11104.
M-666	Aladdin Welding Products Inc., Grand Rapids, MI 49507.	M-695	Atlantic Casting & Engineering Co., Clifton, NJ 07012.
M-667	British Oxygen Co., Ltd., London, England (3).	M-696	Atlas Pattern & Model Works, Brooklyn, NY 11237.
M-669	La Salle Steel Co., Chicago, IL 60680.	M-697	Hammond & Irving Inc., Auburn, NY 13201.
M-670	Arnold Engineering Co., Marengo, IL 60152.	M-698	Bidault, Paris, France.
M-671	Reynolds Metals Co., Richmond, VA (2).	M-699	British Metal Corp., Ltd., London EC 2, England.
M-672	Alsia, Société de, Paris, France.	M-700	Sté de Produits Metallurgiques, Paris, France.
M-672	Société Alsia, Paris, France.	M-701	Sperry Gyroscope Co., Great Neck, NY 11020.
M-673	Electro-Cables, Société de, Paris, France.	M-702	Timken-Detroit Axle Co., Detroit, MI (3).
M-673	Société Electro-Cables, Paris, France.	M-703	Axelson Mfg. Co., Los Angeles, CA (3).
M-674	Ateliers de la Gironde, Paris, France.	M-704	Bachite Development Corp. (3).
M-675	Sté des Laminoirs et Tréfileries du Havre (now part of Cegedur Pechiney, M-1792).	M-705	Barium Stainless Steel Corp., Canton, OH (3).
M-676	Continental Industries Corp., New York, NY (3).	M-706	Becker Bros. Carbon Corp., Cicero, IL 60650.
M-677	Arcos Corp., Philadelphia, PA 19143.	M-707	Belle City Malleable Co., Racine, WI (1).
M-678	Alais, Fonderies et Forges, Compagnie des Mines, Temaris, France.	M-708	Bergmann Elektrizitäwerke, Berlin, 1000, West Germany.
M-678	Compagnie des Mines, Fonderies et Forges d'Alais, Temaris, France.	M-709	Federal-Mogul Corp., Detroit, MI 48235.
M-679	Fonderie de Precision S.A., 92003 Nanterre, Paris, France.	M-710	McGean Chemical Co. Inc., Cleveland, OH 44109.
M-679	Nanterre, Fonderie de Precision, Paris, France.	M-711	British Insulated Callender's Cables, Ltd., Prescot, England.
M-680	Chrysler Corp., Detroit, MI 48288 (2) (information only).	M-712	Champion Rivet Co., Cleveland, OH (3).
M-681	Wellman Bronze & Aluminum (now Wellman Dynamics Corp., Creston, IA 50801).	M-713	Rossell & Co. Ltd., H., Sheffield, 4 England.
M-682	Copperweld Steel Co., Warren, OH 44400 (2).	M-714	Oman Non-Friction Metal Co., El Paso, TX (3).
M-683	Northeast Metals Co., Philadelphia, PA (3).	M-715	British Piston Ring Co., Coventry, England.
M-684	American Radiator Co., New York, NY (2).	M-716	Nihon Jyokiko Seikosho Goshi, Koisha, Japan.
M-685	Phillips & Co., C. E., Detroit, MI 48208.	M-717	Eutectic Corp., Flushing, NY 11358.
M-686	American Platinum & Silver Div. Engelhard Minerals & Chemicals Corp., Carteret, NJ 07008.	M-718	Fillmore Foundry Inc., Buffalo, NY (3).
M-687	Apollo Metals Inc., Chicago, IL 60638.	M-719	Bull's Metal & Marine Ltd., Glasgow, Scotland.
M-688	Allen Mfg. Co., Hartford, CT 06101.	M-720	Bearings Div. NL Industries, Toledo, OH 43614 (formerly Bunting Brass & Bronze).
M-688	Lehigh Steel Corp., New York, NY (3).		
M-689	Regie Nationale des Usines Renault, St. Michael de Maurienne, 55 bd Charonne, Paris 11e, France.	M-720	Bunting Brass & Bronze (now Bearings Div. NL Industries, Toledo, OH 43614).
M-689	Renault, Regie Nationale des Usines, St. Michael de Maurienne, Paris 11e, France.	M-721	Burden Iron Co., Troy, NY (1).
		M-722	Burndy Corp., Norwalk, CT 06856.
M-690	Central Foundry, Div. of General Motors, Saginaw, MI 48605.	M-723	Byrwill Co., Cincinnati, OH (3).

Section III: Numerical List of Manufacturers / 1653

M-724	F.A.C.A., Levallois-Perret (92) France.	M-758	Bartlett Hayward Div. Koppers Co. (4), see Koppers Co., M-399.
M-725	Callite Tungsten Corp. (now GTE Sylvania, Towanda, PA 18848).	M-759	Brown Alloy Works, Detroit, MI (3).
M-725	Chemical & Metallurgical Div., GTE Sylvania, Towanda, PA 18848.	M-760	Die Castings Ltd., Birmingham, England.
M-726	Caloriz Corp. of Great Britain Ltd., London, England (2).	M-761	American Art Metals Inc., New York, NY (3).
M-727	Cambridge Wire Cloth Co., Cambridge, MD 21613.	M-762	Dodge Foundry & Machine Co., Philadelphia, PA 19135.
M-728	Perry Equipment Corp., Hainesport, NJ (2).	M-763	Dominion Wheel & Foundry Co., Toronto, Canada.
M-729	Gayer Co., A., Paris, France.	M-764	Wieland-Werke AG., D-7900 Ulm, West Germany.
M-730	Wrought Bearing Metals Inc., New York, NY (3).	M-765	Welding Equipment & Supply Co., Detroit, MI 48212.
M-731	Keasby & Matteson Co., Ambler, PA (3).	M-766	Great Lakes Steel Co., Div. National Steel Corp., see National Steel, M-836.
M-732	A.F.C., Société des, Paris, France.		
M-733	Mining & Chemical Products Ltd., Alberton, Wembly, England.	M-767	Duralumin, Société du, Paris, France.
M-734	Compressed Industrial Gases Co., Chicago, IL (3).	M-767	Société du Duralumin, Paris, France (3).
M-735	Spang Chalfont Div. National Supply Co., Pittsburgh, PA (3).	M-768	Società Metallurgica Italia, Firenze, Italy.
M-736	Central Engineering & Supply Co., Passaic, NJ 07055.	M-769	Duraweld Metal Products Corp., Long Island City, NY (3).
M-737	Tullis, Ltd., D. & J., London, England (3).	M-770	Tréfileries & Laminoirs du Havre, Antony (Seine), France.
M-738	Cleveland Refractory Metals Div., Chase Brass & Copper Co., Solon, OH 44139.	M-771	Imphy, Société Metallurgique d', Paris 7e, France (now part of Creusot-Loire, M-1488).
M-739	Clevite Bearing Div., Gould Inc., (now Gould, Inc., Engine Parts Div., Cleveland, OH 44110).	M-772	Kries & Sons Co., Henry A., Baltimore, MD (3).
M-739	Gould Inc., Engine Parts Div., Cleveland, OH 44110.	M-773	Dymonhard Corp. of America, New York, NY (2) (3).
M-740	Cold Metals Products Co., Youngstown, OH (3).	M-774	Washburn Wire Co., New York, NY (2).
M-741	Williams & Sons, E. A., Carlstad, NJ 07072.	M-775	Alloy Foundries Div. of the Eastern Co., Naugatuck, CT 06770.
M-742	Steel Co., Ltd., Motherwell, England (3).	M-776	Eclipse-Pioneer Foundries (Bendix Foundries), Teterboro, NJ (2).
M-743	Miller Steel Co., Newark, NJ (3).	M-777	Electro Refractories & Abrasives Div. FERRO Corp., Buffalo, NY 14218 (2).
M-744	Colonial Alloys Co., Philadelphia, PA 19129.	M-778	Electroloy Co. Inc., Bridgeport, CT 06605.
M-745	Unexcelled Mfg. Co., New York, NY (3).	M-779	Elesco Smelting Co., Chicago, IL 60623 (2).
M-746	Constrictor Ltd., London, England (1).		
M-747	Ford Motor Co., Chemical-Metallurgical Products Div., Dearborn, MI 48121.	M-780	Compound Electro Metals Ltd., London, England (3).
M-748	Crane Co., Chicago, IL 60623.	M-781	Williamson Bros. Inc., Bridgeport, CT (3).
M-749	Crescent Tool Co., Jamestown, NY.	M-782	Gautier & Co., D. G., New York, NY (3).
M-750	Amalgamated Steel Div. Allied Steel and Tractor Products Inc., Solon, OH 44139.	M-783	Agawam Tool Co., Springfield, MA (3).
		M-784	Hughes & Co., Ltd., F. A., London, England (3).
M-751	Michigan Tool Co., Detroit, MI (3).	M-785	Swedish Steel Mills, A.A., New York, NY (3).
M-752	Follsain-Wycliffe Foundries, Ltd., Lutterworth, Nr. Rugby, England.	M-786	Imperial Brass Mfg. Co., Chicago, IL (3).
M-753	Marsh Bros. & Co. Ltd., Sheffield, England.	M-787	T. L. M. Co., Paris, France (3).
M-754	Damascus Steel Casting Co., New Brighton, PA 15066.	M-788	Oil Well Supply Co., Houston, TX (3).
		M-789	Isabellenhuette, Berlin, Germany.
M-755	Denman & Davis Co., North Bergen, NJ (3).	M-790	Rennie Tool Co. Ltd., Manchester, England (3).
M-756	Delloy Metals, Philadelphia, PA (3).	M-791	Jelenko & Co., J. F., New Rochelle, NY 10801.
M-757	Duval et Poulain, Paris, France.		

M-792 Kellogg & Co., M. W. (now Pullman Kellogg Div. of Pullman Inc., E. Houston, TX 77046).
M-792 Pullman Kellogg, Div. of Pullman Inc., E. Houston, TX 77046.
M-793 Kennametal Inc., Latrobe, PA 15650.
M-794 Kirk & Son Inc., Morris P., Los Angeles, CA (taken over by NL Industries, M-88).
M-795 Stupakoff Co., Pittsburgh, PA (2).
M-796 Eisler Electric Co., Union City, NJ (3).
M-797 General Tool & Die Co., East Orange, NJ (3).
M-798 Hoesch-Kohn-Nevessen (now Hoesch Huttenwerke, M-1767).
M-799 Hardy & Co., Charles, New York, NY (3).
M-800 Lake Erie Engineering Corp., Buffalo, NY (3).
M-801 Lavin & Sons Inc., R., Chicago, IL 60623.
M-802 Logan Iron & Steel Co., Philadelphia, PA (3).
M-803 Lowmoor Best Yorkshire Iron Ltd., Lowmoor England (3).
M-804 Foote Bros. Gear & Machine Corp., Chicago, IL (3).
M-805 Grant & West Ltd., London, England (3).
M-806 Follansbee Steel Co., Follansbee, WV 26037.
M-807 Cie Français des Métaux, Paris, France.
M-808 Fuller & Basche Co., New York, NY (3).
M-809 Fulton Iron Works Co., St. Louis, MO (2).
M-810 Chicago Hardware Foundry, North Chicago, IL (3).
M-811 Universal Power Corp., Cleveland, OH (3).
M-812 Ratcliff Metals Ltd., J. F., Birmingham, England (3).
M-813 Metalloy Products Co., Newport Beach, CA (3).
M-814 Baldwins Ltd., London, England (now part of British Steel Corp., M-1724).
M-815 ASARCO and Federated Metals Corp., Div. of American Smelting & Refining Co., New York, NY 10005.
M-815 Federated Metals Div. American Smelting & Refining Co., see ASARCO.
M-816 Grammer, Dempsey & Hudson Co., Newark, NJ (3).
M-817 Gilmore & Co., F. F., Needham Heights, MA 02194.
M-818 Glacier Metal Co., Ltd., Alpertone, England (3).
M-819 Die Casting Appliance Corp., Ltd., London, England (3).
M-820 Gorham Tool Industries Inc., Detroit, MI 48238.
M-821 Graphitized Alloy Corp., New York, NY (3).
M-822 Diehl Steel Co., Cincinnati, OH (3).
M-823 Benecke, Inc., Alexander, New York, NY (3).
M-824 Wakefield Corp., Wakefield, MA 01880.
M-825 Hackett Brass Foundry, Detroit, MI 48214.
M-826 Steel Co. of Scotland (4) (now part of British Steel Corp. M-1724).
M-827 Handcock Valve Div. Manning, Maxwell & Moore Co., Bridgeport, CT.
M-828 Harnischfeger Corp. (2) (alloy welding rod business transferred to Chemetron, M-1713).
M-829 Industries Trading Co., New York, NY (3).
M-830 Harville Machine Inc. (2).
M-831 General Electric Ltd., England (2).
M-832 Paulson & Sons, Thomas, Inc., Brooklyn, NY (2).
M-833 Hansell-Elcock Co., Chicago, IL (3).
M-834 Transleteur & Co., Berlin, Germany.
M-835 Chain Div., C.M., Columbus McKinnon Corp., Tonawanda, NY 14150.
M-835 Columbus-McKinnon Chain Corp. (now C.M. Chain Div. Columbus McKinnon Corp., Tonawanda, NY 15219).
M-836 National Steel Co., Pittsburgh, PA 15219.
M-837 Molybdenum Co., N. O., Reutte-Tyrol, Austria.
M-838 Precious Metals Research Works Inc., Mt. Vernon, NY 10553.
M-839 Moraine Mfg., Inc., Dayton, OH 45439.
M-840 Otis Steel Div. Jones & Laughlin Steel Co. (4), see Jones & Laughlin, M-173.
M-841 Hadstrom Pattern Works, Oscar W., Chicago, IL (3).
M-841 Hedstrom Corp., Oscar W., Chicago, IL (3).
M-842 Woodworkers Tool Works, Chicago, IL 60606.
M-843 Linde Div. Union Carbide Corp, Metals Division, New York, NY 10017.
M-844 Hollup Corp., Chicago, IL.
M-845 Central Pattern & Foundry Co., Chicago, IL (3).
M-846 Bradley Laboratories, London, England.
M-847 Hytensil Aluminum Co., Chicago, IL (3).
M-848 Winn, W. Martin, Ltd., S. Staffordshire, England.
M-849 McKay Co. (now Teledyne McKay, Pittsburgh, PA 15219).
M-849 Teledyne McKay, Pittsburgh, PA 15219.
M-850 Hamilton Foundry Div. Hamilton Allied Corp., Hamilton, OH 45011.
M-851 Electric Auto-Lite Co., Woodstock, IL (3).
M-852 Mesta Machine Co., Pittsburgh, PA 15230.
M-853 METCO Inc., Westbury, Long Island, NY 11590.
M-854 G.H.R. Foundry Div. Dayton Malleable Iron Co., Dayton, OH (2).

M-855	Doelger & Kirsten Inc., Milwaukee, WI 53210.	M-891	Joselin & Co., C., New York, NY (3).
M-856	Chemical & Metallurgical Div., GTE Sylvania, Towanda, PA 18848.	M-892	Hurbenium Co. of America, Detroit, MI (3).
M-856	Sylvania Electric Products (now GTE Sylvania, Towanda, PA 18848).	M-893	Manos Ltd., Zurich, Switzerland.
		M-894	Fitzsimmons Steel Co., Youngstown, OH 44501.
M-857	Lake & Elliot Ltd., Braintree, England.	M-895	Deutsche Industrie, Nemen, Germany.
M-858	Minimax Co., Div. of Graham Chemical Corp., Chicago, IL 60660.	M-896	Standard Brake Shoe & Foundry Co., Pine Bluff, AR (2).
M-859	Major Engineering Co., Tulsa, OK 74101.	M-897	Gilby-Fodor S.A., Rueil-Maimaison (Seine-et-Oise), France.
M-860	Moltrop Steel Products Co., Beaver Falls, PA (2).	M-898	Advance Foundry Co., Dayton, OH 45401.
M-861	Black Drill Co., Inc., Cleveland, OH 44117.	M-899	Stroh Process Steel Co. (now SPS Industries Inc., Allison Park, PA 15101).
M-862	Falk Corp., Milwaukee, WI 53201.		
M-863	Monarch Alloy Co., Ravenna, OH (3).	M-900	Kelly Foundry & Machine Co., Elkins, WV 26241.
M-864	National Broach & Machine, Detroit, MI 48213.	M-901	Superior Steel Corp., Pittsburgh, PA (3).
M-865	Nesaloy Products Inc., New York, NY (3).	M-902	Chicago Malleable Casting Co., Chicago, IL (3).
M-866	Oldham & Co., F. B., Buffalo, NY (3).	M-903	Sanderson Steel Div. (4), see Crucible Specialty Steel, M-38.
M-867	Cinaudagraph Corp., Stamford, CT (3).		
M-868	Novo Pump & Machine Div. American Marsh Pumps Inc., Lansing, MI 48905.	M-904	Castings Corp., Brewster, NY 10509.
		M-905	Hyde Park Foundry & Machine Co., Hyde Park, PA 15641.
M-869	Wintershall, A.G., Heringer, Germany.	M-906	Sheepbridge Equipment Ltd., Chesterfield, England.
M-870	Snowbar, J. L., New York, NY (3).		
M-871	Aluminium Français, Paris 8, France.	M-907	Reliance Steel Casting Co., Pittsburgh, PA (3).
M-872	Page Steel & Wire Div. Am. Chain & Cable, Monesson, PA (3).		
		M-908	Hoyland Steel Co., New York, NY (3).
M-873	Youngstown Foundry & Machine Co., Youngstown, OH (3).	M-909	Tri-Clover Div. Ladish Co. (4), see Ladish Co., M-1527.
M-874	Parker-Kalon Corp., Clifton, NJ.	M-910	National Cable & Metal Co., Glendale, CA (3).
M-875	Simpson Bros. Machine Works, Portsmouth, OH 45662.		
		M-911	Riverside Foundry & Galvanizing Co., Kalamazoo, MI (2).
M-876	Peckovers, Ltd., Toronto, Canada.		
M-877	American Bridge Co. (4), see U.S. Steel Corp., M-604.	M-912	Rochling-Burbach G.m.b.H., 6620 Volklingen-Saar, West Germany.
M-878	Perry Barr Metal Co., Ltd., Birmingham, England.	M-912	Rochlingstahl (now Rochling-Burbach).
		M-912	Stahlwerke Rochling-Burbach G.m.b.H., 6620 Volklingen, Saar, West Germany.
M-879	Paraloy Co., Chicago, IL (3).		
M-880	Permo Inc., Chicago, IL (3).	M-913	Metal Carbides Corp., Youngstown, OH 44512.
M-881	Permold Inc., Medina, OH 44256 (2).		
M-882	Aktiebolaget Svenska Metallverken, Stockholm, Sweden.	M-914	McKechnie Metals Ltd., Walsall, WS9 8DN England.
M-883	Wheeling-Pittsburgh Steel Corp., Pittsburgh, PA 15230.	M-915	Taylor Chain Co., S. G., Hammond, IN 46320.
M-884	Pompey, Société Nouvelle des Aciéries de, Neuilly-sur- Seine 92202 France.	M-916	Super Tool Co., Elk Rapids, MI 49629 (2).
M-884	Société Nouvelle des Aciéries de Pompey, Neuilly-sur- Seine 92202 France.	M-917	Cornish Wire Co. (now General Cable Corp., Cornish Wire Products, Williamstown, MA 01267).
M-885	Powder Metals Inc., Long Island City, NY (3).	M-917	General Cable Corp., Cornish Wire Products, Williamstown, MA 01267.
M-886	Fairbanks, Morse & Co., Beloit, WI 53511.	M-918	Willworthy Piston Ring Ltd., London, England (3).
M-887	Precision Casting Co., Cleveland, OH 44111, or Fayetteville, NY 13066.	M-919	Wilkinsen, Wm., Shustoke, England (3).
		M-920	Consolidated Car Heating Co., Albany, NY (3).
M-888	Lake City Malleable Co., Cleveland, OH (2).		
M-889	Spuck Iron & Foundry Co., St. Louis, MO (2).	M-921	Compagnie Français de l'Etain, Paris 8e, France.
M-890	Hueck, Edward, Ludenscheid, Germany.	M-921	Etain, Compagnie Française de l', Paris 8^e, France.

M-922	Vulcan Iron Works, Wilkes-Barre, PA 18701.
M-923	Tri-Lok Co., Pittsburgh, PA (3).
M-924	Dravo-Doyle Co., Pittsburgh, PA 15222.
M-925	True Alloys Inc., Detroit, MI (3).
M-926	General Plate Div. Metals & Controls Corp., Attleboro, MA 02703.
M-926	Metals & Controls and General Plate Div. of, Attleboro, MA 02703.
M-927	Oklahoma Steel Castings Co., Tulsa, OK 74101.
M-928	Stupakoff Co., Pittsburgh, PA (2).
M-929	Tyler Inc., W. S., Mentor, OH 44060.
M-930	Una Welding Inc., Cleveland, OH (3).
M-931	Union Metal Mfg. Co., Canton, OH (2).
M-932	Universal Alloys Inc., Newark, NJ (3).
M-933	Wellman Corp., S. K., Bedford, OH 44146.
M-934	Mt. Vernon Furnace & Mfg. Co., Mt. Vernon, IL (3).
M-935	Vereinigte Aluminiumwerke, Bohn, Germany.
M-936	Audubon Metalwove Belt Corp., Philadelphia, PA 19134 (3).
M-937	Austenal Div. Howmet Corp., Dover, NJ 07801.
M-938	Vulcan Mold & Iron Co., Latrobe, PA 15650.
M-939	Welding Alloy Mfg. Co., Malden, MA (3).
M-940	Pratt & Co., J. M., Cincinnati, OH (3).
M-941	Waukesha Foundry Co., Waukesha, WI 53186.
M-942	Waltham Precision Instruments, Waltham, MA 02154.
M-943	Welding Material Co., New Orleans, LA (3).
M-944	Park Sales Co., New York, NY (3).
M-945	Superior Tube Co., Norristown, PA 19404.
M-946	Welland Electric Steel Foundry, Welland, Ontario, Canada.
M-947	Eaton Corp., Cleveland, OH 44114.
M-947	Eaton, Yale & Towne (now Eaton Corp.).
M-948	Dusenberry & Stracken, Inc., New York, NY (2) (3).
M-949	Grey Mfg. Co., C. M., East Orange, NJ (3).
M-950	Zelnicker, W. A., St. Louis, MO (3).
M-951	Light Metal Works, Tokyo, Japan.
M-952	Napraloy Co., Jersey City, NJ (3).
M-953	Allied Process Corp., New York, NY (1).
M-954	Trimout Mfg. Co., Boston, MA (3).
M-955	Titanite Alloys Corp., Cleveland, OH (3).
M-956	American Crucible Products Co., Lorain, OH 44052.
M-957	Tube Reducing Corp., Wallington, NJ (3).
M-958	Magnus Metal Corp. (2) (now Bearings Div. NL Industries, St. Louis, MO).
M-959	Hewitt Metals Corp., Detroit, MI 48208.
M-960	Somers Thin Strip Inc., Waterbury, CT 06720 (4) (now part of Ohio Brass, M-279).
M-961	Brockhouse Casting Co., Wolverhampton, England (3).
M-962	Buffalo Wire Works Co. Inc., Buffalo, NY 14240.
M-963	Wall-Colmonoy Corp., Detroit, MI 48203.
M-964	United States Steel Supply Co. (4), see United States Steel, M-604.
M-965	Excelite Co., Woodbridge, NJ (3).
M-966	Saginaw Bearing Co., Saginaw, MI 48605.
M-967	Superior Metal Co., Chicago, IL (3).
M-968	Metal Castings Ltd., Worcester, England (3).
M-969	Seaboard Steel Co. of America, New York, NY (3).
M-970	Sessions Foundry Co., Bristol, CT (3).
M-971	Metallwerke, A.G., Vienna, Austria.
M-972	Seneca Wire & Mfg. Co., Fostoria, OH 44830.
M-973	New England Collapsible Tube Co., New London, CT (3).
M-974	Shenango-Penn Div. Shenango Furnace Co. (2).
M-975	Bonney Tool & Forge, Div. Gulf & Western, Allentown, PA 18105.
M-976	Ex-Cell-O Corp., Tool & Abrasive Products Div., Detroit, MI 48232 (formerly Willeys Carbide).
M-976	Willeys Carbide, see Ex-Cell-O.
M-977	Youngstown Alloy Castings Co., Youngstown, OH (3).
M-978	British General Electric Co. Ltd., London, England.
M-979	Tréfileries du Havre, Paris, France.
M-980	Chromium Corp. of America, Cleveland, OH 44105.
M-981	Deutsche Messingwerke, Berlin, Germany.
M-982	Aluminium Wire & Cable Co., Port Tennant Works, Swansea SA1 8PS Glamorgan, England.
M-985	Darwin-Toledo Ltd. (Toledo Steel Works), Sheffield, England (3).
M-985	Toledo Steel Works, or Darwin Toledo Steel, Sheffield, England (1).
M-986	Reid-Avery Co., Baltimore, MD 21222.
M-987	Tungsten Alloy Mfg. Co., Inc., Harrison, NJ 07029.
M-988	Mir-o-Col Alloy Co., Los Angeles, CA (3).
M-989	Keystone Carbon Co., St. Marys, PA 15857.
M-990	Rhodes Metaline Co., R. W., Long Island City, NY (3).
M-991	American Steel Foundries, Chicago, IL 60601 (3).
M-992	Hammond Brass Works, Hammond, IN (3).
M-993	Duquesne Smelting Corp., Pittsburgh, PA (3).

Section III: Numerical List of Manufacturers / 1657

M-994	Fahralloy Co., Harvey, IL 60426.	M-1028	Oscap Mfg. Co., New York, NY (2) (3).
M-995	Hopkinsons Ltd., Huddersfield, HD2 2UR England.	M-1029	Tuff-Hard Corp., Detroit, MI (3).
M-996	Brookside Metal Co., Ltd., Watford, Herts, WO2 4NF England (2).	M-1030	Globe Steel Tubes Co., Milwaukee, WI (2).
M-997	Tata Iron & Steel Co., Calcutta 16 India.	M-1031	Light Metals Inc., Indianapolis, IN 46205.
M-998	Volco Brass & Copper Co., Kenilworth, NJ 07033.	M-1032	Jackson Iron & Steel Co., Jackson, OH 45640.
M-999	Edgcomb Steel Co., Hillside, NJ (2).	M-1033	Dominion Foundries & Steel Ltd., Hamilton, Ontario, L8N 3J5 Canada.
M-1000	Hobart Bros. Co., Troy, OH 45373.	M-1034	General Steel Industries Inc., St. Louis, MO 63105.
M-1001	Tonawanda Iron Corp., North Tonawada, NY.	M-1035	Eccles & Davis Machinery Co., Los Angeles, CA (3).
M-1002	Sherman & Co., New York, NY (3).	M-1036	Toughite Process Co., Chicago, IL (3).
M-1003	Eimco Foundry Div. Envirotech Corp., Salt Lake City, UT 84110.	M-1037	Brush Wellman Inc., Cleveland, OH 44110 (formerly Brush Beryllium).
M-1004	General Aircraft Equipment Co. (1).	M-1038	Foote Mineral Co. (Ferroalloys Div.), Exton, PA 19341.
M-1005	Alloy Engineering & Casting Co., Champaign, IL 61820.	M-1039	Keokuk Electro-Metals Co., Keokuk, IA (3).
M-1006	Wickman Wimet, see Wimet Ltd.	M-1040	Alten Foundry & Machine Works, Lancaster, OH 43130.
M-1006	Wimet Ltd., Coventry, CV4 9AD England (formerly Wickman Wimet).	M-1041	Indium Corp. of America, Utica, NY 13502.
M-1007	Olds Alloys Co., South Gate, CA (3).	M-1042	Biad Powder Metallurgy Co., Pittsburgh, PA (2) (3).
M-1008	Sterling Alloys Inc., Woburn, MA (3).	M-1043	Aciéries & Forges de Firminy, Paris, France (4) (now part of Creusot-Loire, M-1448).
M-1009	Grayborn Steel Co., New York, NY (3).		
M-1010	Permanent Magnet Association, London, England (1).		
M-1011	Cooper Metallurgical Corp., Cleveland, OH 44102 (formerly Cooper Products Co.).	M-1043	Firminy, Aciéries & Forges de (4) (now part of Creusot-Loire, M-1488).
M-1012	Arpocalloy Co., Kansas City, MO 64139.	M-1043	Société et Forges de Firminy (4) (now part of Creusot-Loire , M-1488).
M-1013	United Wire & Supply Co., Cranston, RI 02910.	M-1044	Hewson Co., John, New York, NY (3).
M-1014	Eastern Stainless Steel Co. Div. of Eastmet Corp., Baltimore, MD 21203.	M-1045	Howard Foundry Co., Chicago, IL (1).
M-1015	Simonds Worden White Co., Dayton, OH (3).	M-1046	Tungsit Electro-Metals Works Ltd., London, England (3).
M-1016	Darbyshire Steel Co. Inc., El Paso, TX 79912.	M-1047	Lazalloys Ltd., Bently Works, Doncaster, Lancashire, England (2).
M-1017	Sight Feed Generator Co., Richmond, IN (3).	M-1048	Scraw Alloys Ltd., Banstead, Surrey, England (2).
M-1018	Superior Die Casting Corp., Cleveland, OH 44110.	M-1049	Atlas Metal & Alloys Co. Ltd., Chicago, IL (3).
M-1019	Davis & Geck Inc., Pearl River, NY 10965.	M-1050	Lithaloys Corp., New York, NY (3).
M-1020	Electro Metal Corp., Yonkers, NY (3).	M-1051	Madison-Kipp Corp., Madison, WI 53704.
M-1021	Chance Vought Aircraft Div. United Aircraft Corp., Stratford, CT (2).	M-1052	Crobalt Inc., Ann Arbor, MI (3).
M-1022	Acme Aluminum Alloys, Inc., Dayton, OH (3).	M-1053	Peterson Steels Inc., Union, NJ 07083.
M-1023	ALOYCO Inc. (now Walworth Co., Linden, NJ 07036).	M-1054	Tungsten Electric Corp., Union City, NJ (3).
M-1023	Walworth Co., Aloyco Plant, Linden, NJ 07036 (formerly ALOYCO Inc.).	M-1055	Utility Electric Steel Foundry Co., Los Angeles, CA (3).
M-1024	Escaut & Meuse, Sté d', Paris 17°, France.	M-1056	Trethaway Associates, New York, NY (3).
M-1024	Sté d'Escaut & Meuse, Paris 17°, France.	M-1057	American Tank & Fabricating Co., Cleveland, OH 44111.
M-1025	Permanente Magnesium Inc., Oakland, CA (1).	M-1058	Akron Bronze & Aluminum Inc., Akron, OH 44309.
M-1026	Swedish Crucible Steel Co., Detroit, MI (3).	M-1059	Chromium Mining & Smelting Corp. Ltd., Montreal, Canada.
M-1027	Walker Metal Products Ltd., Walkerville, Ontario, Canada.	M-1060	Coast Metals Inc., Little Ferry, NJ 07643.

M-1061 Emsco Derrick & Equipment Co., Los Angeles, CA (3).
M-1062 Atkins & Co., E.C., Indianapolis, IN (3).
M-1063 General Metals Powder Co., Akron, OH 44305.
M-1064 Wyndale Mfg. Co., Indianapolis IN (3).
M-1065 Metal & Alloy Div. Silver Creek Precision Co. (1).
M-1066 Makepeace Co., D. E., Attleboro, MA 02703 (Div. of Engelhard).
M-1067 Great Western Steel Co., Div. Hoyland Steel Co., Los Angeles, CA.
M-1068 Manross Div. Associated Spring Co., Bristol, CT 06010.
M-1069 Stackpole Carbon Co., St. Marys, PA 15857.
M-1070 Superior Flux & Mfg. Co., Cleveland, OH 44143 (2).
M-1071 Sharon Steel Corp., Sharon, PA 16146.
M-1072 Pickands, Mather & Co., Cleveland, OH 44114.
M-1073 Alpha Metals Inc., Jersey City, NJ 07304.
M-1074 ARC Products Div. Chemetron Corp. (formerly ALL-STATE Welding Alloys), see Chemetron Corp. M-1713.
M-1075 Apothecaries Hall Co., Waterbury, CT (3).
M-1076 Indiana General, Valparaiso, IN 46383.
M-1077 Whiton Machine Co., New London, CT (3).
M-1078 Pipe Machinery Corp., Cleveland, OH (3).
M-1079 Detroit Mold Engineering Co., Newark, NJ (3).
M-1080 S.I.P.I. Metals Corp., Chicago, IL 60622.
M-1081 Strong, Carlisle & Hammond Co., Cleveland, OH.
M-1082 Redhard Metals Inc., Hatboro, PA (3).
M-1083 Enfield Rolling Mills Ltd., Enfield, Middlesex, EN3 7QB England.
M-1084 Hamilton Precision Metals (now Hamilton Technology, Inc., Lancaster, PA 17604).
M-1085 Ludlow-Saylor Wire Cloth Div. General Steel Industries, St. Louis, MO 63114.
M-1086 Adamas Carbide Corp., Kenilworth, NJ 07033.
M-1087 Elgin Watch Co., Chicago, IL 60606.
M-1088 American Solder Co., Philadelphia, PA (3).
M-1089 Magnetic Metals Co., Camden, NJ 08101.
M-1090 Plastic Metals Div. National Radiator Co., Johnstown, PA (3).
M-1091 Castings Development Co., Philadelphia, PA (3).
M-1092 Welding Sales & Engineering Co., Detroit, MI (3).
M-1093 SMS Corp., Detroit, MI (3).
M-1094 Oakes Bronze & Aluminum Co., Warren, OH 44482.
M-1095 Blackalloy Co. of America, Paterson, NJ 07509.
M-1096 Marquette Corp., Minneapolis, MN (3).
M-1097 Dee Div. of Handy & Harmon (1).
M-1098 Powder Metal Products Co., Franklin Park, IL (3).
M-1099 ABEX Corp., Mahwah, NJ 07430 (formerly American Brake Shoe Co.).
M-1100 Rankin Mfg. Co., San Diego, CA 92121.
M-1101 Fulton Gold Refineries Corp., New York, NY (3).
M-1102 Acme Electric Welder Co., Los Angeles, CA 90058.
M-1103 Jobbins, Inc., W. F., Aurora, IL (1) (bought out by U.S. Reduction, M-585).
M-1104 Galv-Weld, Inc., Dayton, OH (3).
M-1105 Detroit Tap & Tool Co., Warren, MI 48090.
M-1106 Wausau Motor Parts Co., Schofield, WI.
M-1107 McCauley Alloy Sales Co., New York, NY (3).
M-1108 Goldsmith Bros., Chicago, IL (part of NL Industries).
M-1109 Crane Co., Torrey S., Plantsville, CT 06479.
M-1110 Riken Metal Mfg. Co., Ube, Japan.
M-1111 General Cerium Co. (1), see Ronson Metals, M-1497.
M-1112 Balfour & Co., Ltd., Arthur, Sheffield, England (part of Balfour-Darwins Ltd., Sheffield, England).
M-1113 Wardlows Ltd., Sheffield, S4 7LJ England.
M-1114 Steirische Gusstahlwerke (merged into Vereinigte Edelstahlwerke, A-1011 Wien, Austria, M-1802).
M-1114 Styria-Stahl Steirische Gusstahlwerke AG., Vienna (4) (now merged into Vereinigte Edelstahlwerke, A-1011 Wien, Austria).
M-1115 Aubert et Duval, 92202 Neuilly-sur-Seine, France.
M-1116 Aciéries de Champagnole, La Courneuve, France.
M-1116 Campagnole, Aciéries de, 93123 Courneuve, France.
M-1116 Champagnole, Aciéries de, 93123 Courneuve, France.
M-1117 Chatillon, see Chiers-Chatillon.
M-1117 Chiers-Chatillon, 75009 Paris, France.
M-1117 Compagnie des Forges de Chatillon, see Chiers-Chatillon.
M-1117a Oxium S.A., 75002 Paris, France (dealer for M-1117).
M-1118 Longwy, Société des Aciéries, Mont-Saint-Martin, France.
M-1118 Société des Aciéries de Longwy, Mont-Saint-Martin, France.
M-1119 Saut-du-Tarn, Société Nouvelle du, 75008 Paris, France.
M-1119 Société Anon. des Hauts-Fourneaux (now Société Nouvelle du Saut-du-Tarn, 75008 Paris, France).
M-1120 Ugine Aciers (Steel Div.) 75361 Paris Cedex 08 France.

M-1120	Ugine Aciers (Titanium Div.) 75361 Paris Cedex 08 France.	M-1156	Harris, Arthur & Co., Chicago, IL 60607.
M-1121	Division Lead Co., Summit, IL 60501.	M-1157	Tipaloy Inc., Detroit, MI 48211.
M-1122	American Clad Metals Inc., Pawtucket, RI 02862.	M-1158	Aluminum Solder Corp. (ALSOCO), New York, NY (3).
M-1123	Rochlingstahl Steel Works, Wetzlar, Germany.	M-1159	Kling Bros. Engineering Works, Chicago, IL (3).
M-1124	Steel Co. of Canada. Ltd., Hamilton 23 Montreal, Canada.	M-1160	Parker Pen Co., Janesville, WI 53545.
M-1125	Kester Solder Co., Chicago, IL (2).	M-1161	Uniworld Corp. of America, Cleveland, OH 44120.
M-1126	Brown-Firth Co., see Firth Brown Ltd., M-56.	M-1162	Wel-Met Co., Kent, OH (3).
M-1127	Philips Electronic & Associated Industries Ltd., 11/12 Hanover Square, London, W 1 England.	M-1163	Ohio Stainless & Commercial Steel Co., Cleveland, OH (3).
		M-1164	Rasmussen Mfg. Co., Hollydale, CA (3).
		M-1165	Lancaster Steel Co. Inc., New York, NY (3).
M-1128	Bergstrom Alloys Corp., New York, NY (3).	M-1166	Calumet Steel Castings Corp., Hammond, IN 46320.
M-1129	Kirkling & Co., Whittier, CA (2).	M-1167	Meech Foundry Inc., Cleveland, OH 44125.
M-1130	Magnesium Elektron Ltd., St. James Square, London, SW 1 England.	M-1168	Pressco Casting & Mfg. Corp., Chesterton, IN (3).
M-1131	Heinkel-Herth Co., Berlin, Germany.	M-1169	Langley Alloys Ltd., Langley, Slough, Buckinghamshire, SL3 6EA England.
M-1132	Kapfenberg, Steiermark, Austria.	
M-1133	Deutsche Gold und Silber Scheide, Hanau, West Germany.	M-1170	Superior Foundry Co., Cleveland, OH (3).
M-1134	Zinkberatungsstelle, G.m.b.H., Berlin, Germany.	M-1171	Industrial Overlay Metals Corp., New York, NY (3).
M-1135	Kropp Forge Co., Chicago, IL 60650.	M-1172	Brown & Sharp Mfg. Co., Centerdale, RI 02911.
M-1136	Soldering Specialties Co., Summit, NJ (3).	M-1173	Watsontown Foundry, Watsontown, PA 17777.
M-1137	L & R Mfg. Co., Arlington, NJ (3).	
M-1138	U.S. Spring & Bumper Co., Los Angeles, CA (3).	M-1174	Michigan Powdered Metal Products Inc., Livonia, MI 48150.
M-1139	North American Philips Co. Inc., New York, NY 10017.	M-1175	Agil Chemie, Berlin, Germany.
		M-1176	Società Alluminio Veneto per Azioni, Porto Marghera, Milan, Italy.
M-1140	Aluminiumwerke A.G., Rorschach, Switzerland.	M-1176	Veneto per Azioni, Soc. Alluminio, Porto Marghera, Milan, Italy.
M-1140	Rorschach, Aluminiumwerke, Rorschach, Switzerland.	M-1177	Billiton Maatschappu, N.V., The Hague, Holland.
M-1141	Ekstrom, Carlson & Co., Rockford, IL 61110 (2).	M-1178	Gold und Silber Scheide Anstalt, Hanau, Germany.
M-1142	Acme Foundry Co., Detroit, MI (3).	M-1179	Allard, Soc. Anon. Usines et Aciéries, Mont-sur-Marchienne, Belgium.
M-1143	York Machine & Supply Co., York, PA (3).	M-1179	Société Anon. Usines et Aciéries Allard, Mont-sur- Marchienne, Belgium.
M-1144	Miller Co., Meriden, CT 06450.	
M-1145	Federal Foundries & Steel Co., Ltd., London, England.	M-1180	Marrel Freres, S.A., F-42800 Rive-de-Gier, France.
M-1146	Benedict-Miller Inc., Lyndhurst, NJ 07071.	M-1180	Société Anon. des Etablissements Marrel Frères, B.P. 46, F-42 Rive-de-Gier, France.
M-1147	Kaiser Aluminum & Chemical Corp., Oakland, CA 94643.	
M-1148	Kaiser Steel Corp., Oakland, CA 94643.	M-1181	Hetzel & Co., Nürnberg, Germany.
M-1149	Marshall Steel Co., La Grange, IL 60525.	M-1182	Raven Steel & Tool Co., Valley Stream, NY (3).
M-1150	Kanthal Corp., Bethel, CT 06801.	M-1183	Beaumont Birch Co., Philadelphia, PA 19102.
M-1151	Yocum & Son Inc., James, Philadelphia, PA (3).	M-1184	Bofors AB., Bofors, Sweden.
M-1152	Auto Specialties Mfg. Co., St. Joseph, MI 49085.	M-1185	Paris & d'Outreau, Soc. des Aciéries de, Paris 17°, France.
M-1153	American Non-Gran Bronze Co., Berwyn, PA (3).	M-1186	Svenska Metallverken A.B., Stockholm, Sweden.
M-1154	Bradley & Foster Ltd., Staffordshire, England.	M-1187	Chicago Development Corp., Ashland, VA 23005.
M-1155	Randall Graphite Products Co., Chicago, IL (3).		

M-1188	Henricot, S.A. Usines, Emile, Court-St.-Etienne, Belgium.	M-1221	Heller Tool Div., Wallace Murray Corp., Newcomerstown, OH 43832.
M-1188	Usines Emile Henricot, S.A., Court-Saint-Etienne, Belgium.	M-1222	Blaw-Knox Co. (4) (merged with Duraloy to form Duraloy-Blaw-Knox, M-45).
M-1189	Surahammer Bruks, A.B., Surahammer, Sweden.	M-1223	Donegal Steel Foundry Co., Marietta, OH (3).
M-1190	Wikmanshytte Bruks, AB., Wikmanshyttan, Sweden.	M-1224	Ingersoll-Rand Co., Phillipsburg, NJ 08865.
M-1191	Accurate Brass Co., Bristol, CT 06010.	M-1225	Starrett Co., L. S., Athol, MA 01331.
M-1192	Thomas & Skinner Inc., Indianapolis, IN 46205.	M-1226	Illinois Tool Works, Chicago, IL 60631 (2).
M-1193	Knapp Mills, Inc., Long Island City, NY (3).	M-1227	Allied Products Corp., Chicago, IL 60650.
M-1194	Greenleaf Corp., Saegertown, PA 16433.	M-1228	Philadelphia Bronze & Brass Corp., see Ampco Metal, M-13.
M-1195	Jonas & Culver Ltd., Sheffield, 9 England.	M-1229	Mallory-Sharon (now RMI Co., Niles, OH 44446).
M-1196	Le Moyne Steel Co., Pittsburgh, PA (3).	M-1229	Reactive Metals (now RMI Co., Niles, OH 44446).
M-1197	American Electro Metals Corp., Yonkers, NY (3).	M-1229	RMI Co., Niles, OH 44446 (formerly Reactive Metals).
M-1198	Western Alloyed Steel Casting Co., Minneapolis, MN (2).	M-1230	Weldwire Co., Inc., King of Prussia, PA 19406.
M-1199	Crucible Steel Castings Co., Lansdowne, PA (2).	M-1231	Ullman & Associates, A. E., New York, NY (3).
M-1200	American Hoist & Derrick Co., St. Paul, MN 55107.	M-1232	Ferroxcube Corp. Div. of North American Phillips Co., Saugerties, NY 12477.
M-1202	Hallamshire Steel Co., Sheffield, 3 England.		
M-1203	Newcomer Products Inc., Latrobe, PA 15650.	M-1233	Aluminum Smelting & Refining Co., Maple Heights, OH 44137.
M-1204	Victor Equipment Co. (4) (now part of Stellite Div. Colt Industries, M-167).	M-1234	Thyssen Co., August, Berlin, Germany.
		M-1235	Harvey Aluminum Inc. (2) (now Martin Marietta Aluminum).
M-1205	Gardiner Metal Co., Chicago, IL 60632 (2).	M-1236	Specialloy Inc., Chicago, IL 60632.
M-1206	Pittsburgh Metal Purifying Corp., Saxonburg, PA (2).	M-1237	Norsk Aluminium Co. A/S, Hoyanger, Norway.
M-1207	Royalloy Inc., New York, NY (2) (3).	M-1238	Nordisk Aluminium Industry, A/S, Holmestrand, Norway.
M-1208	McQuay-Norris Mfg. Co., St. Louis, MO (3).		
M-1209	Rolled Alloys Inc., Detroit, MI 48211.	M-1239	Dominion Magnesium Ltd., Toronto, Ontario, Canada.
M-1210	Electro Foundry Co., Troutdale, OR (3).	M-1240	Rollschen Eisenwerke, AG., Gesellschaft der Ludv. von, Gerlafingen, Germany.
M-1211	Ryan Aeronautical (now Teledyne Ryan Aeronautical, San Diego, CA 92112).		
M-1211	Teledyne Ryan Aeronautical, San Diego, CA 92112.	M-1241	Aetna Standard Engineering Co., Ellwood City, PA 16117.
M-1212	Miller Co., Meriden, CT 06450.	M-1242	Reeves International Inc., New York, NY (2).
M-1213	American Metallurgical Products Co., Pittsburgh, PA 15239.	M-1243	Fusion Inc., Willoughby, OH 44094.
M-1214	Vacuum Metals Div. Crucible Steel (4), see Crucible Specialty Steel, M-38.	M-1244	Little Falls Alloys Inc., Paterson, NJ 07501.
M-1215	Plew Tool Co., Columbia City, IN 46725.	M-1245	Myrens Verkstec A/S, Oslo, Norway.
M-1216	Alter Co. (2) (alloys now produced by Alloy Metal Products, M-1795).	M-1246	Metallwerk Plansee Gesellschaft, Reutte-Tirol, Austria.
M-1217	Cop-Sil-Loy Inc., Hollywood, CA (3).	M-1246	Plansee, Metallwerk Gesellschaft M.B.H., Reutte-Tirol, Austria.
M-1218	Dyn-Metal Ltd., London, England (3).		
M-1218a	Machinery & Machine Supplies Co. Inc., New York, N.Y. (3).	M-1247	Rollschen Eisenwerke, AG., Ludv. von, Soluthurn, Switzerland.
M-1219	Industria Nazionale Alluminio, Mori (Trento), E. Bolzano, Italy.	M-1248	Commerce Pattern Co., New York, NY (3).
M-1220	Società Alluminio Veneto per Azioni, Porto Marghera, Venezia, Italy.	M-1249	Allen-Sherman-Hoff Pump Co., Wynnewood, PA (3).
M-1220	Veneto per Azioni, Soc. Alluminio, Porto Marghera, Venezia, Italy.	M-1250	Morris Machine Works, Baldwinsville, NY (2).

M-1251	Georgia Iron Works, Grovetown, GA 30813.	M-1282	Holt Equipment Co., Independence, OR (3).
M-1252	Penrold Div. Brush Wellman Corp., Reading, PA.	M-1283	Feltrina Società Metallurgica, Milan, Italy.
M-1253	Algoma Steel Corp. Ltd., Sault Ste. Marie, Ontario, Canada.	M-1284	Borolite Corp., Niagara Falls, NY (2) (3).
M-1254	Black-Clawson Co., Middletown, OH 45042.	M-1285	Fromson Co., Inc., Rockville, CT 06066.
M-1255	Belmont Smelting & Refining Works, Brooklyn, NY 11207.	M-1286	General Motors Corp. and Central Foundry Div. of General Motors, Saginaw, MI 48605.
M-1256	Ashton Ltd., N.C., Huddersfield, HD1 6RE England.	M-1287	Henckels Zwillingwerke, G.A., Solingen, Germany.
M-1257	Vulcan Foundry Co., Oakland, CA 94601.	M-1288	Robert-Leyer-Pritzkow & Co., Solingen-Ohligs, Germany.
M-1258	Beardsley & Piper, Div. of Pettibone Corp., Chicago, IL 60639.	M-1289	Veraloy Products Ltd., Birmingham, England.
M-1259	W. W. Alloys, Detroit, MI (1) (formerly part of Fansteel, M-53).	M-1290	Bergische Stahl-Industrie, 5630 Remscheid 1, West Germany.
M-1260	Bergsoe & Son, Paul, Glostrup, Denmark.	M-1291	Steirische Gusstahlwerke (merged into Vereinigte Edelstahlwerke, A-1011 Wien, Austria, M-1802).
M-1261	Bergsoe & Son, Paul, Lanskrona, Sweden.	M-1291	Styria-Stahl Steirische Gusstahlwerke AG., Vienna (4) (now merged into Vereinigte Edelstahlwerke, A-1011 Wien, Austria).
M-1262	Stainless Foundry & Engineering Inc., Milwaukee, WI 53209.		
M-1263	Pacific Welding Alloys Mfg. Co., Los Angeles, CA (3).	M-1292	Main Metal Ltd., London, WC 2 England (3).
M-1264	Cameron & Son Ltd., Sheffield, England (1).	M-1293	Gusstahl-Handels, G.m.b.H., van Dohlen-Stahl, Wetter, Ruhr, Germany.
M-1265	Alloys & Products, Inc., New York, NY (3).		
M-1266	Sonken-Galamba Corp., Kansas City, KS (3).	M-1294	Stahlwerk Stahlschmidt & Co. KG., 4000 Düsseldorf-Heerdt, West Germany.
M-1267	Grede Foundries Inc., Milwaukee, WI 53204.	M-1295	Cohn Corp., Sigmund, Mount Vernon, NY 10553.
M-1268	Aluminium, T. I., Ltd. (4) (now part of British Aluminium, M-28).	M-1295	Sigmund Cohn Corp., Mount Vernon, NY 10553.
M-1268	T. I. Aluminium Ltd. (now part of British Aluminium, M-28).	M-1296	Sumitomo Metal America Inc., New York, NY 10017.
M-1269	Hanford Foundry Co., San Bernardino, CA 92404.	M-1297	Jadot, Usines de, Brussels, Belgium.
M-1270	Haywood Tyler of Canada, Ltd., Kitchener, Ontario, Canada.	M-1298	Gilson, Ltd., Brussels, Belgium.
		M-1299	Remanit, G.m.b.H., Berlin, Germany.
M-1271	Husqvarna Vapenfabrik Aktiebolag, Stockholm, Sweden.	M-1300	Deutro G.m.b.H., Berlin, Germany.
		M-1301	Schneider, Virgo (4) (merged into Creusot-Loire, M-1488).
M-1272	Alar, Ltd., London, England (3).		
M-1273	Renfrew Foundries Ltd., Hillingdon, Glasgow, Scotland.	M-1302	AFY, Société des, Paris, France.
		M-1303	Holzer, Jacob (merged into Creusot-Loire, M-1488).
M-1274	Southern Forge Ltd., England (now part of Imperial Aluminium, M-1547).		
		M-1304	Stavenger Co., Oslo, Norway.
M-1275	E. & E. Kaye Ltd, Middlesex, EN3 4SS England (now part of Pechiney).	M-1305	Schmidt & Clemens, Edelstahlwerke, 5251 Kaiserau Bez, Köln uber Engelskirchen, West Germany.
M-1275	Kaye, E. & E., Ltd., Enfield, Middlesex, EN3 4SS England M-1444 (now part of Pechiney).	M-1306	Hover, Gebruder, Edelstahlwerk, 5251 Kaiserau uber Engelskirchen, West Germany.
M-1276	Alcaloy Inc., Cleveland, OH (3).		
M-1277	Massey-Harris Ltd., Toronto, Ontario, Canada.	M-1307	Stahlwerke Sudwestfalen AG., 5900 Siegen 1, West Germany.
M-1278	Pratt & Whitney Cutting Tool & Gage Operation, West Hartford, CT 06101.	M-1307	Sudwestfalen Stahlwerke AG., 5900 Siegen 1, West Germany.
M-1279	Aluminum Smelting & Refining Co., Maple Heights, OH 44137.	M-1308	Schoeller-Bleckmann (4) (merged into Vereinigte Edelstahlwerke, M-1802).
M-1280	Techalloy Co. Inc., Rahns, PA 19426.	M-1309	Soding, J. C., & Halbach, 58 Hagen/Westf., West Germany.
M-1281	Berry Metal Co., Harmony, PA 16037.		

M-1310	Kabel Stahlwerke, C., Pouplier, Hagen-Kabel, Germany.	M-1331	Stahlwerke Bochum AG., 4630 Bochum, West Germany.
M-1310	Stahlwerke Kabel, C., Pouplier, Hagen-Kabel, Germany.	M-1332	Zapp, Robert, Edelstahl, 4000 Düsseldorf, West Germany.
M-1311	Lohmann, Fried., G.m.b.H., Gusstahlfabrik, Walz-und Hammerwerke, 5821 Herbede (Ruhr), West Germany.	M-1333	Ergst, Stahlwerke, Kommanditgesellschaft, 5840 Schwerte 1, West Germany.
		M-1334	Deutsche Mannesmannrohren-Werke AG., 4000 Düsseldorf 1, West Germany (2) (all DIN or W.Nr. grades).
M-1312	Fakirstahl Hoffmans & Co., Edelstahlwerk, 5630 Remscheid 14, West Germany.		
M-1313	Kanz Metallwerke, Hans, Zurich-Albisrieden, Switzerland.	M-1334	Mannesmannrohren-Werke A.G., 4000 Düsseldorf 1, West Germany.
M-1314	Alpine Montan Co. (4) (merged into Vereinigte Edelstahlwerke, M-1802).	M-1335	Reisholz, Stahl und Röhrenwerke, G.m.b.H., 4000 Düsseldorf 16, West Germany.
M-1315	Ambo-Stahl-Gesellschaft, 5000 Koln 1, West Germany.	M-1335	Stahl-und Rohrenwerk Reisholz G.m.b.H., 4000 Düsseldorf 16, West Germany.
M-1316	Grimm, Gustav, Stahl-und Hammerwerk, 5630 Remscheid-Haddenbach, West Germany.	M-1336	Erkenzweig & Schwemann, Hagen, Germany.
M-1317	Special Metals Corp., New Hartford, NY 13413.	M-1337	Stahlwerke R & H Plate, 5880 Ludenscheid-Platehof, West Germany.
M-1318	Haeckerstahl Willy Haecker, 5609 Huckeswagen-Rhld, West Germany.	M-1338	Hoffmann & Co., Köln, Germany (2).
M-1319	Lindenberg Edelstahlwerk (4) (merged into Bergische Stahl- Industrie, M-1290).	M-1339	Remystahl, D-58 Hagen, F.R. Germany (see also M-1652).
		M-1340	Wolff Handelgesellschaft, Otto, 5000 Köln 1, West Germany.
M-1320	Urbach & Co., Carl, 5809 Huckeswagen, West Germany.	M-1341	Universal Castings Corp., Chicago, IL 60638.
M-1321	Plattawerke G.m.b.H., West Berlin, Germany.	M-1342	Pose-Marre Edelstahlwerke G.m.b.H., 4006 Erkrath bei Düsseldorf, West Germany.
M-1322	poldihütte, Westa-Westdeutsche Edelstahlhandels-Gesellschaft, Düsseldorf, West Germany.	M-1343	Bochumer Verein A, Bochum, Germany.
M-1322	Westa-Westdeutsche Edelstahlhandelsgesellschaft, Von Poldi-hütte, Düsseldorf, West Germany.	M-1344	Plettenberger Gusstahlfabrik, 5970 Plettenberg 2, West Germany.
		M-1345	Westfalische Stahlgesellschaft, 5970 Plettenberg, West Germany.
M-1323	Georgsmarienwerke Selesiastahl, G.m.b.H., Georgsmarienhütte, Germany.	M-1346	Idealstahl Breidenbach KG, 5270 Gummersbach 31, Germany.
M-1324	Düsseldorf-Heerdt GmbH. & Co., KG., 4000 Düsseldorf-Heerdt, West Germany.	M-1347	Junker G.m.b.H., Otto, 5101 Lammersdorf uber Aachen, West Germany.
		M-1348	Dohlen-Stahl Gusstahl-Handels, GmbH., Wetter/Ruhr, Germany.
M-1324	Edelstahlwerke Düsseldorf-Heerdt GmbH. & Co., 4000 Düsseldorf-Heerdt, West Germany (3).	M-1349	Mentah Co., Vorgtlinder, West Berlin, Germany.
M-1325	Hoffman Elektrogusstahlwerk, Alb., Eschweiler, Germany.	M-1350	Hochfrequenz-Tiegelstahl, G.m.b.H., Bochum, Germany (2).
M-1326	Edelstahl Witten A.G. (4) (merged to form Thyssen Edelstahlwerk A.G., M-1759).	M-1351	Ultraloy Corp., Chicago, IL (3).
		M-1352	Kind & Co., Edelstahlwerk, 5276 Wiehl 2-Bielstein, Germany.
M-1326	Witten Edelstahl (4) (now part of Thyssen Edelstahlwerke, M-1759).	M-1353	Reining, Heinrich, G.m.b.H., 4000 Düsseldorf 1, West Germany.
M-1327	Kuhbier & Sohn, C., Zweigbetrieb, 5885 Schalksmühle 2, West Germany.	M-1354	Aluminium G.m.b.H., Rheinfelden/Baden, West Germany.
M-1328	Handler, G.m.b.H., Düsseldorf, West Germany.	M-1355	Busch-Jäger Ludenscheider Metallwerke, Ludenscheid, Germany.
M-1329	Guronitwerke Vervoort, G.m.b.H., Düsseldorf, West Germany.	M-1356	Deutsch Delta Metallgesellschaft, Düsseldorf-Grafenberg, Germany.
M-1330	Carp & Hones, 4150 Krefeld 1, West Germany (3).	M-1357	Diehl Metall-Guss & Presswork, Hrch., GmbH., Nürnberg, West Germany (2).
M-1331	Bochum Stahlwerk AG., Bochum, Germany.	M-1358	Elisental Drahtwerk, Neuenrade/Westf., Germany (2).

M-1359 Erbsloh, J. & A., Wuppertal-Barmen, Germany.
M-1360 Felten & Guillaume Carlswerke, A.G., Köln-Mülheim, West Germany.
M-1361 Fuchs, Otto, 5882 Meinerzhagen, West Germany.
M-1362 Gerhardi & Co., Ludenscheid, Germany.
M-1363 Göttingen Aluminiumwerke G.m.b.H., Göttingen, Germany.
M-1364 Honsel-Werke AG., Leichtmetallwerke, 5778 Meschede, West Germany.
M-1365 Mahle G.m.b.H., D-700 Stuttgart 50, West Germany.
M-1366 Neumayer Kabel u. Metallwerke A.G., Nürnberg, Germany.
M-1367 Rieger Gebr., Aalen, Germany.
M-1368 Schmole, RuG., Metallwerke Menden, Krs. Iserlohn, Westf., Germany.
M-1369 Seibel, W., AG, Mettmann, Rhld, Germany.
M-1370 Standard-Werke, G.m.b.H., Werl/Westf., Germany (2).
M-1371 Unna Messingwerk, A.G., Unna, Westf., Germany.
M-1372 Vesevorder Metallwerke G.m.b.H., Werdohl/Westf, Germany.
M-1373 Sherritt Gordon Mines Ltd., Toronto M5L 1B1 Ontario, Canada.
M-1373 Westfalische Kupfer und Messingwerke A.G., Lundenscheid, NRW 5880, West Germany.
M-1374 Wilms A.G., G. Robert, Remscheid-Hasten, Germany.
M-1375 Westfalische Leichtmetallwerke G.m.b.H., Nachrodt, Krs. Altena, Westf., Germany.
M-1376 Wutoschingen Aluminiumwerke G.m.b.H., Wutoschingen, Germany.
M-1377 Horbach & Schmitz G.m.b.H., 5000 Köln, Hansaring 49-51, West Germany.
M-1378 Alco Products, Schenectady, NY (3).
M-1379 Sintox Corp. of America, Allentown, PA (3).
M-1380 Westfalenhütte Dortmund A.G., Dortmund, Germany.
M-1381 Mannesmann-Hüttenwerk A.G., Duisberg, Germany.
M-1382 Ruhrstahl A.G., Henrichshütte, Hattingen/Ruhr, Germany.
M-1383 Schwabenstahl Hüttenwerke, G.m.b.H., Wasseralfingen (Württ), Germany.
M-1384 Exide Power Systems Div. ESB Inc., Philadelphia, PA (2).
M-1385 Tungsten & Molybdenum Ltd., Nyon, Switzerland.
M-1386 Neubrandenburg Eisenwerke, G.m.b.H., Berlin, Germany.
M-1387 Wells Mfg. Co., Skokie, IL 60067.
M-1388 Welded Carbide Co. Inc., Clifton, NJ 07013.
M-1389 Bridgeport Brass Co., Riverside, CA, see Bridgeport Brass, M-141.
M-1390 Quint Alloy Corp., Dearborn, MI (2).
M-1391 Henning Bros. & Smith Inc., Brooklyn, NY 11237.
M-1392 Columbia Bronze Corp., Freeport, NY (3).
M-1393 Colonial Metals Co., Columbia, PA 17512.
M-1394 Brevets Berthelmy, Société des, Paris, France.
M-1395 Brooks & Perkins Inc., Livonia, MI 48150.
M-1396 LeVita Metal Alloy Co., Detroit, MI (3).
M-1397 Istituto Sperimentali Metalli Leggeri, Milan, Italy.
M-1398 Tin Research Institute, Greenford, England.
M-1399 Dirilyte Co. of America, Kokomo, IN 46901.
M-1400 Thomson Industries Inc., Manhasset, NY 11030.
M-1401 North American Philips Co. Inc., New York, NY 10017.
M-1402 Nippert Co., Delaware, OH 43015.
M-1403 Pioneer Aluminum Products Inc., Los Angeles, CA.
M-1404 Guertler, William, G.m.b.H., Berlin, Germany.
M-1405 Cook & Co. Ltd., George H., Sheffield, S3 8RE England.
M-1406 Beardshaw & Co. Ltd., Joseph, Sheffield, S3 8DJ England.
M-1407 Barworth Flocton Steel Works Ltd., Ecclesfield, S30 3XH England.
M-1408 Leadbeater & Scott Ltd., Sheffield, England (1).
M-1409 Flockton, Tompkin & Co., Ltd., Sheffield, England (3).
M-1410 Carr & Co., Ltd., Richard, Sheffield, S6 1LL England.
M-1411 Hall & Pickles Ltd. (4) (now part of Osborn Steels, M-233).
M-1412 Taylor & Co., Robert, Labert, Scotland.
M-1413 Dunford & Elliott Ltd. (4) (merged into Dunford-Hadfields, M-62).
M-1414 English Steel Rolling Mills Corp. Ltd. (4) (merged into British Steel Corp., M-1724).
M-1415 Burys & Co. Ltd., Sheffield, England (3).
M-1416 Clyde Works, Sheffield, see Osborn Steels, M-233.
M-1417 Thomas & Baldwins (4) (now part of British Steel Corp., M-1724).
M-1418 Vimetal S.A., Berne, Switzerland (2).
M-1419 Japan Special Steel Co., Tokyo, Japan.
M-1420 Oederlin & Co. Ltd., Baden, Switzerland.
M-1421 Voelkingen, Forges et Aciéries de, Voelklingen (Sarre), France.
M-1422 Ludlow Steel Corp., Bedford, OH 44146.
M-1423 Jamison Steel Corp., San Francisco, CA (3).
M-1424 DoAll Co., Des Plaines, IL 60016.
M-1425 Erie Steel Co., Erie, PA (3).

M-1426	Waterbury Rolling Mills Inc., Waterbury, CT 06720.	M-1460	Thomas Foundries Inc., Birmingham, AL 35201.
M-1427	Dixon Sintaloy Inc., Stamford, CT 06906.	M-1461	Madison Foundry Co., Cleveland, OH (3).
M-1428	Sheffield Steel Div. Armco Steel Co., Houston, TX (1), see Armco Steel, M-10.	M-1462	Advance Aluminum Casting Corp., Chicago, IL (3).
M-1429	Bedford Tool & Forge Co., Bedford, OH 44146.	M-1463	National Airoil Burner Co., Philadelphia, PA 19134.
M-1430	Venture Corp., Newark, NJ (3).	M-1464	Fairmount Foundry Inc., Hamburg, PA 19526.
M-1431	Gulf Steel Corp., Dallas, TX (3).	M-1465	RCA Corporate Standards Engineering, Camden, NJ 08101 (2) (information only).
M-1432	Pyramid Steel Co., Cleveland, OH.		
M-1433	Pennsylvania Steel Corp., Detroit, MI (3).	M-1466	Lobdell Co., Wilmington, DE (3).
M-1434	Bowsteel Distributors Corp., Linden, NJ (2).	M-1467	Burdett Oxygen Co. (now Burdox Inc., Cleveland, OH 44114).
M-1435	Berted Foundry Co., Columbiana, OH (3).	M-1468	Anchor Alloys Inc., Brooklyn, NY 11222.
M-1435	Bochlet Gebruder, Kapfenberg, Steiermark, Austria.	M-1469	Sintercast Div. of Chromalloy American Corp., West Nyack, NY 10994.
M-1435	Gebruder Bochlet, Kapfenberg, Steiermark, Austria.	M-1470	Athenia Steel Div. National Standard Co., Clifton, NJ (2).
M-1436	Meridian Steel Co. Inc., Jericho, NY 11753.	M-1471	Hard Alloys Ltd., High Wycombe, Buckinghamshire, England.
M-1437	Coulter Steel & Forge Co., Emeryville, CA 94662.	M-1472	Zenith Foundry Co., West Allis, WI (3).
M-1438	Allied Steel & Chemical Co., New York, NY (3).	M-1473	Xaloy Inc., New Brunswick, NJ 08903.
		M-1474	Atlas Brass Foundry, Los Angeles, CA 90021.
M-1439	Standard Pressed Steel Co., Jenkintown, PA 19046.	M-1475	R-S Products Corp., Philadelphia, PA (3).
M-1440	American Meter Co., Reliance Foundry, Pittsburgh, PA (3).	M-1476	Milwaukee Die Casting Co., Milwaukee, WI (2).
M-1440	Reliance Foundry, American Meter Co., Pittsburgh, PA (3).	M-1477	Gillett & Eaton Inc., Lake City, MN (3).
M-1441	Medart Engineering & Equipment Co., St. Louis, MO (2).	M-1478	General Malleable Corp., Waukesha, WI (3).
M-1442	Taylor-Wilson Mfg. Co., Pittsburgh, PA 15222.	M-1479	Little Bros. Foundry Co., Port Huron, MI (3).
M-1443	Luria Steel & Trading Co., New York, NY (2).	M-1480	Lakeside Malleable Castings Co., Racine, WI (3).
M-1444	E. & E. Kaye Ltd, Middlesex, EN3 4SS England (now part of Pechiney).	M-1481	Light Alloys Products Co. Ltd., Munworth, England.
M-1445	Montecatini Settore Alluminio, Milan, Italy.	M-1482	Induction Steel Castings Co. Inc., East Detroit, MI 48021.
M-1446	National Tool Div., National Cleveland Corp., Cleveland, OH (3).	M-1483	American Brazing Alloys Co., Pelham, NY (3).
M-1447	Vikmanshyttan AG., Vikmanshyttan, Sweden.	M-1484	Washington Steel Co., Washington, PA 15301.
M-1448	Nyby, Granges, AB., S-644 80 Torshalla, Sweden.	M-1485	Sheepbridge Alloy Castings Ltd., Sutton-in-Ashfield, Nottinghamshire, NG17 5LL England.
M-1449	Breda Co., Rome, Italy.		
M-1450	Cogne Co., Rome, Italy.	M-1486	St. Jacques, Société des Usines, Montlucon, France.
M-1451	Bresciana Metallurgica, Rome, Italy.		
M-1452	Ilssa-Viola SpA, I-20159 Milan, Italy.	M-1487	Aciéries du Forez, St-Etienne (Loire), France.
M-1453	Schenk Leichtgusswerke, Maulbronn, Germany.	M-1488	Creusot-Loire, 75428 Paris Cedex 09 France.
M-1454	Nusite Steel Process Co., Warsaw, IN (3).	M-1488	Société des Forges et Ateliers du Creusot (now Creusot- Loire, 75428 Paris Cedex 09 France).
M-1456	American Art Alloys Inc., Kokomo, IN (3).		
M-1457	Lovsted & Co. Inc., C. M., Seattle, WA (2).	M-1489	Dewrance & Co., Ltd., Skelmersdale, Lancashire, England (3).
M-1458	Navan Inc., Anaheim, CA 92803.	M-1490	Allvac Metals (now Teledyne Allvac, Monroe, NC 28110).
M-1459	Lesjefors Aktiebolag, Lesjefors, Sweden.		

M-1490	Teledyne Allvac, Monroe, NC 28110.	M-1523	Molecu Wire Corp., Farmington, NJ 07727.
M-1491	Cannon-Muskegon Corp., Muskegon, MI 49443.	M-1524	Secon Metals Corp., White Plains, NY 10606.
M-1492	Chromalloy American Corp., West Nyack, NY 10994.	M-1525	Rex Buckeye Co., Cleveland, OH 44144.
M-1493	Western Gold & Platinum Co., Belmont, CA 94002.	M-1526	Hoeganaes Metal Powders Ltd., Birmingham, England.
M-1494	Saut-du-Tarn, Société Nouvelle du, 75008 Paris, France.	M-1527	Ladish Co., Cudahy, WI 53110.
M-1494	Société Nouvelle du Saut-du-Tarn, 75008 Paris, France.	M-1528	Stauffer Metals Div., Stauffer Chemical Corp., Westport, CT 06880 (2).
M-1495	Lorraine-Escault (now Vallourec, 75764 Paris Cedex 16, France).	M-1529	Colorado Fuel & Iron Co., Denver, CO.
		M-1530	Canadian Quebec Metallurgical Corp., Quebec, Canada.
M-1495	Usines Vallouric (now Vallouric 75764 Paris Cedex 16 France).	M-1531	Muskegon Piston Ring Co., Muskegon, MI 49443.
M-1495	Vallourec, 75764 Paris Cedex 16, France.	M-1533	Union Minière du Haut-Katanga, Central Workshop, Jadotville, Katanga, Congo.
M-1496	Marathon Specialty Steels Inc., New York, NY 10001		
M-1497	Ronson Metals Corp., Newark, NJ 07105.	M-1534	U.S. Magnet & Alloy Corp., Bloomfield, NJ (3).
M-1498	Wesson Co., Ferndale, MI (3).	M-1535	Badell Co., Inc., Hammond, IN 46327.
M-1499	Huntington Alloys Inc., Huntington, WV 25720.	M-1536	Martin Metals Co., Martin-Marietta Corp., Baltimore, MD 21227 (also Wheeling, IL 60090).
M-1500	Van der Horst Corp., (Terrel, TX) Los Angeles, CA 94103.		
		M-1537	Teledyne Wah Chang, Albany, OR 97321.
M-1501	Rolle Mfg. Co., Lansdale, PA (3).		
M-1502	Curtis-Wright Corp., Quehanna, PA (3).	M-1537	Wah Chang (now Teledyne Wah Chang, Albany, OR 97321).
M-1503	Wyman-Gordon Co., Worcester, MA 01601 (2).	M-1538	Fugi Iron & Steel Co., Ltd., Chou-Ku, Tokyo, Japan.
M-1504	Manganese Chemicals Corp. (now Chemetals Div. Diamond Shamrock, Cleveland, OH 44114).	M-1539	Japan Steel Works Ltd., New York, NY 10017.
		M-1540	Yawata Iron & Steel Co., Ltd., Tokyo, Japan.
M-1505	Hydrometals Inc., Dallas, TX 75206.		
M-1506	Corning Glass Co., Corning, NY 14830 (2).	M-1541	Kawasaki Steel Corp., New York, NY 10017, or Fukiai-Ku, Kobe 651 Japan.
M-1507	National Standard Co., Niles, MI 49120.		
M-1508	American Metal Climax, Div. of AMAX Corp., Greenwich, CT 06830.	M-1542	Japan Steel & Tube Co. Ltd., Tokyo, Japan.
M-1509	Manoir A., Pitres (Eure), Usines du, Paris, France.	M-1543	Mitsubishi Steel Mfg. Co. Ltd., New York, NY 10017, or Tokyo, Japan.
M-1510	Baldwin Steel Co., Jersey City, NJ 07305.	M-1544	Toto Steel Co. Ltd., Tokyo, Japan.
		M-1545	Daido Steel Co. Ltd., New York, NY 10017.
M-1511	Mathieson-Heglar Co., La Salle, IL 61301.		
		M-1546	Compagnie des Ateliers et Forges de la Loire, F-75, Paris 9ᵉ, France.
M-1512	Waimet Alloys Co., Dearborn, MI (3).		
M-1513	General Thermoelectric Corp., Princeton, NJ (3).	M-1546	Loire, Compagnie des Ateliers et Forges de la, F-75, Paris 9e France.
M-1514	Hamilton Die Cast Inc., Hamilton, OH 45011.	M-1547	Imperial Aluminium Co., Ltd., Birmingham, England (2), (alloys now supplied by Alcoa of Great Britain, M-1715).
M-1515	Sumitoms Trading Co., Osaka, Japan.		
M-1516	Scaife Co., Oakmont, PA (2).		
M-1517	Kulite Tungsten Corp., Ridgefield, NJ 07657.	M-1548	Aluminium Zentral, 4000 Düsseldorf 1, Germany (2) (this is an association, no alloys, information only).
M-1518	Knowsley Cast Metal Co., Ltd., Manchester, England (3).		
		M-1549	Philips Co., N. V., Eindhover, Netherlands.
M-1519	Morgan Crucible Co. Ltd., London, England (3).		
		M-1550	American Potash & Chemical Co., see Kerr-McGee Chemical Corp., M-1550.
M-1520	General Communications Co., Boston, MA (2).		
		M-1550	Kerr-McGee Chemical Corp., Oklahoma City, OK 73125.
M-1521	Wolverine Tube Div. Calumet & Hecla, Universal Oil Products, Allen Park, MI 48101.		
		M-1551	Birmingham Battery & Metal Co., Ltd., Birmingham, 29 England.
M-1522	Sheffield Smelting Co. Ltd., Sheffield, S4 7WD England.	M-1552	Metro-Cutanit Ltd., Warrington, Lancashire, England.

M-1553 Haywood Foundries, Ltd., London, England (3).
M-1554 Heckford Ltd., Arthur E., Birmingham, England.
M-1555 Jones & Rooke Ltd., Birmingham, England.
M-1556 Meigh Castings Co. Ltd., Cheltenham, England.
M-1557 Capper Pass & Son, Ltd., Bristol, England (3).
M-1558 Reynolds Tube Co., Tyseley, Birmingham, 11 England (2).
M-1559 Metal Alloys Ltd., South Wales (2).
M-1560 Yorkshire Imperial Metals Ltd., Leeds, LS1 1RD England.
M-1561 Scientific Products Inc., Detroit, MI 48211.
M-1562 Metals & Controls and General Plate Div. of, Attleboro, MA 02703.
M-1563 DuPont de Nemours & Co., E.I., Wilmington, DE 19898 (2).
M-1564 American Carbide Corp., Union City, NJ (3).
M-1565 Wabash Alloys Inc., Wabash, IN 46992 (2).
M-1566 Olin Corp., Stamford, CT 06904.
M-1567 Polymer Corp. Ltd., Sarnia, Ontario, Canada.
M-1568 Powder Alloys Corp., Clifton, NJ (3).
M-1569 Clyde Alloy Steel Co. (now part of British Steel Corp., M-1724).
M-1570 Ishikawajima-Harima Heavy Industries Co. Ltd., New York, NY 10048.
M-1571 Vitra Chemical Co., Chattanooga, TN (2).
M-1572 Bristol Brass Co., Bridgeport, CT 06010.
M-1574 Hanna Furnace Corp., Buffalo, NY 14224 (subsidiary of National Steel Co.).
M-1575 Scandinavian Metal Co., Sweden.
M-1576 Glyco-Metallwerke Daelen & Loos, G.m.b.H., Germany (3).
M-1577 Japan Metal Industry Co. Ltd., Kawasaki, Japan.
M-1578 IMI Rolled Metals, Birmingham, B6 7BA England.
M-1578 Imperial Metal Industries (now IMI Ltd., Birmingham, B6 7BA England).
M-1579 Bridgeport Rolling Mills Co., Bridgeport, CT 06601.
M-1580 Deloro Stellite Ltd., Swindon, Wiltshire, SN3 40A England.
M-1581 Magotteaux, Les Fonderies, Vaux-sous-Chevrement, (Liege), Belgium.
M-1582 Rochette, Laminoire de la, Chaudfontaine, Belgium.
M-1583 Koerver & Nehring G.m.b.H., 4150 Krefield, West Germany.
M-1584 Willan Ltd., G. L., Sheffield, England.
M-1585 Industria Magenti Fermanenti E. Affini (Impia), Milan, Italy.
M-1586 Leghe E. Accini Speciali (LEAS), Milan, Italy.

M-1587 Carondelet Foundry Co., St. Louis, MO 63110.
M-1588 Jodots Freres, Ets, Boloeil, Belgium.
M-1589 Fujikoshi Steel Industry Co., Ltd., Tokyo, Japan.
M-1590 Kensington Steel Co., Chicago, IL (2).
M-1590 Portec Inc. Casting Div., Kingsbury, IN 46345 (2) (formerly Kensington Steel).
M-1591 Sankey & Sons Ltd., London, England.
M-1592 McLouth Steel Corp., Detroit, MI 48209.
M-1593 Erie Forge & Steel Co., Erie, PA (3).
M-1594 Phoenix Steel Corp., Claymont, DE (2).
M-1595 American Silver Co., Flushing, NY.
M-1596 Teledyne Wah Chang, Albany, OR 97321.
M-1596 Wah Chang (now Teledyne Wah Chang, Albany, OR 97321).
M-1597 Urich Foundry Co., Erie, PA (2).
M-1598 International Wire Products Co., Wyckoff, NJ 07481.
M-1599 Perfect Circle Div. Dana Corp., Richmond, IN (2).
M-1600 Certanium Alloys & Research Co.
M-1601 King Materials Laboratory, Victor, Palo Alto, CA (2).
M-1602 Standard Steel Div., Titanium Metals Corp., Burnham, PA 17009.
M-1603 Oregon Steel Mills, Portland, OR 97203.
M-1604 Kobe Steel Ltd., New York, NY 10022.
M-1605 Swiss Aluminium Ltd., Neuhausen, Rhf., Switzerland.
M-1606 Inland Electronics Products Corp., Pasadena, CA (3).
M-1607 Midwest Materials Co., Cleveland, OH (2) (3).
M-1608 Noranda Metal Industries Ltd., Montreal, Quebec, H3C 2Y4 Canada.
M-1609 Carborundum Co., Niagara Falls, NY 14302.
M-1610 Union Steel Corp., Union, NJ (2).
M-1611 Adalet Mfg. Co., Cleveland, OH (2).
M-1612 Nippon Kokon K.K., New York, NY 10017, or Tokyo, Japan.
M-1613 Wisconsin Steel Div. International Harvester Co., Chicago, IL 60611.
M-1614 Tokyo Keikiseizosho Ltd., Tokyo, Japan.
M-1615 Perfection Mica Co., Bensonville, IL 60106.
M-1616 D-M-E Corp., Madison Heights, MI 48071.
M-1617 Thompson Wire Co., Chicago, IL (3).
M-1618 U.N. Alloy Steel Corp., Boston, MA 02027.
M-1619 Precision Cast Parts Corp., Portland, OR 97206.
M-1620 Johnston Steel & Wire Co., Worcester, MA 01607.
M-1621 Isaacson Iron Works.
M-1622 Kawecki Chemical Co., see Kawecki Berylco Industries, M-646.
M-1623 Pfizer Inc., New York, NY 18042.

M-1624	National Research Corp., Metals Div., Newton, MA 02158.	M-1649	Forges et Aciéries de Bonpertuis, 38140 Rives-sur-Fure, France.
M-1625	Alcan Aluminium Corp., see Aluminium Goods.	M-1650	Klockner-Werke AG., Duisberg, Germany (2).
M-1625	Aluminium Goods, Div. of Alcan Canada Products Ltd., Toronto 154 Canada.	M-1651	Nippon Stainless Steel Co. Ltd., C/O Sumitomo Metal Ind., Ltd., New York, NY 10017, or Tokyo, Japan.
M-1626	Magnetics, Div. of Spang Industries Inc., Butler, PA 16001.	M-1652	Hagener Gusstahlwerl Remy (now Remystahl), see M-1339.
M-1627	Materials Research Corp. (2).	M-1653	AVEC S.P.A Acciaieria Valbruna di Ernesto Gresele, 36100 Vicenza, Italy.
M-1628	Albright & Wilson Ltd., London, England (2) (3).
M-1628	Quaker Alloy Casting Co., Myerstown, PA 17067.	M-1653	Valbruna Ernesto Gresele, AVEC SPA Acciaieria, 36100 Vicenza, Italy.
M-1629	Nuclear Metals Inc., Concord, MA 01742.	M-1654	Capito & Klein.
		M-1655	Ossenberg & Cie., W., Edelstahlwerke, Altena, Westf. 7, West Germany.
M-1630	Vacuum Metals Corp., Cambridge, MA (3).	M-1656	St. Joseph Lead Co., Monaca, PA.
		M-1657	Copper Range Co., New York, NY 10020.
M-1631	Vacuumschmelze G.m.b.H., 6450 Hanau 1, West Germany.	M-1658	Phoenix-Rheinrohr, AG. (now part of Mannesmannrohren, M-1334).
M-1632	Sydney Steel Corp., Sydney, Nova Scotia, Canada (2).	M-1659	Fisher AG., Georg, Germany (3).
		M-1660	Sulzer Brothers Ltd., Winterthur, Switzerland, Sulzer Bros. Inc., New York, NY 10006.
M-1633	Renewal Services Inc., Philadelphia, PA 19123.
M-1634	Aluminium Industries, AG., Chippis, Switzerland.	M-1661	International Lead-Zinc Research Association Inc., New York, NY 10017.
M-1635	Green River Steel Co., Owensboro, KY 43201 (Div. of Jessop Steel Co.).
M-1636	Atlas Steels Div. Rio Algom Corp., Cleveland, OH (3).	M-1662	Arwood Corp., Rockleigh, NJ 07647 (2).
		M-1663	Rodney Metals (now Teledyne Rodney, New Bedford, MA 02742).
M-1637	Associated Electrical Industries Ltd., Manchester, England.
M-1638	American Alloys Corp., Kansas City, MO 64120.	M-1663	Teledyne Rodney Metals, New Bedford, MA 02742.
M-1639	American Light Alloys Co., Little Falls, NJ (3).	M-1664	Transformer Steels Ltd.
		M-1665	Whittaker Corp., Nuclear Metals Div., W. Concord, MA 01781.
M-1640	Blackstown Corp., and Jamestown Malleable Iron Div. Blackstown Corp., Jamestown, NY 14701.	M-1666	VR/Wesson or Vascoloy-Ramet, Waukegan, IL (4), (now part of Fansteel, M-53).
M-1641	King Materials Laboratory, Victor, Palo Alto, CA (2).	M-1667	Toshiba Tungaloy Co., Ltd., New York, NY 10017.
M-1642	Electronic Specialties Co., H. & S. Metals Co., Pomona, CA (3).	M-1668	Venite Metals Div. Valenite Corp., Madison Heights, MI 48071.
M-1643	Blackstown Corp., and Jamestown Malleable Iron Div. Blackstown Corp., Jamestown, NY 14701.	M-1669	SERMAG (Société d'Etudes et de Recherches Magnetiques), 38 St-Martin-d'Heres, France.
M-1643	Jamestown Malleable Iron Div., Blackstone Corp., Jamestown, NY 14701.	M-1670	Solar Div. International Harvester Co., San Diego, CA 92138.
M-1644	SERMAG (Société d'Etudes et de Recherches Magnetiques), 38 St-Martin-d'Heres, France.	M-1671	Markey Bronze Corp., Delta, OH 43515.
		M-1672	Ryanite Tips Inc., St. Paul, MN (3).
		M-1673	Ajax Steel & Forge Co.
M-1644	Société d'Etudes et de Recherches Magnetiques, (SERMAG) B.P. 17926 38 St-Martin-d'Heres, France (3).	M-1674	Bolton-Emerson Inc., Lawrence, MA (2).
		M-1675	Stone Manganese Marine; J. Stone & Co., London, SE 17, England.
M-1645	Aciéries et Forges d'Anor, Paris 11e, France.	M-1676	Connecticut Metals Corp., Hampden, CT (3).
M-1646	Sambra & Meuse, Aciéries de, France.	M-1677	Copper Rod & Brass Products Div. Triangle Conduit & Cable Co., New Brunswick, NJ 08903.
M-1647	Tokoshu Seiko Ltd., Tokyo, Japan.
M-1648	Pont-St-Martin, Officine Metallurgiche d', France.
M-1649	Bonpertuis, Forges et Aciéries de, 38140 Rives-sur-Fure, France.	M-1677	Triangle Conduit & Cable Co., Copper Rod & Brass Products Div., New Brunswick, NJ 08903.

1668 / Woldman's Engineering Alloys

M-1678 Boller Industries Div. of Boller Development Corp., Marion, IN 46952 (3).
M-1679 Cominco Trail, British Columbia, Canada.
M-1680 Macauley Foundry Co., Berkeley, CA 94710.
M-1681 Cleveland Refractory Metals Div., Chase Brass & Copper Co., Solon, OH 44139.
M-1683 Champion Steel Co., Orwell, OH 44076.
M-1684 St. Lawrence Steel Co., Cleveland, OH 44128 (3).
M-1685 Mitsubishi Metal Products of Japan, Tokyo, Japan.
M-1686 Usines Vallouric (now Vallouric 75764 Paris Cedex 16 France).
M-1686 Vallourec, 75764 Paris Cedex 16, France.
M-1687 Le Bronze Industrial, Rene Loiseau & Cie, 93001 Bobigny, France.
M-1687 Le Bronze Industrial, Rene Loiseau & Cie, 93001 Bobigny, France.
M-1688 National Alloy Tool Co., Cleveland, OH (2) (3).
M-1689 Industrial Alloy Steel Co., Cleveland, OH (2) (3).
M-1690 Electro Hercules Steel Co., Cleveland, OH 44102.
M-1691 Mansfield Brass & Aluminum Corp., Mansfield, OH (2).
M-1692 Wallram Hartmetall G.m.b.H., Essen, Germany.
M-1693 Niborium Industries Inc., Providence, RI 02905.
M-1694 Bendix Steel Co., Cleveland, OH 44128 (3).
M-1695 Aiken Industries, Inc., Carbide Division, Irwin, PA 15642.
M-1695 Carbidie, Div. of Aiken Industries, Inc., Irwin, PA 15642.
M-1696 Carmet Materials Div., Madison Heights, MI 48071.
M-1697 TRW Wendt-Sonis Division, TRW Inc., Rogers, AR 72756.
M-1697 Wendt-Sonis (now TRW Wendt-Sonis, Rogers, AR 72756).
M-1698 Remington Arms Co. Inc., Ilion, NY 13357.
M-1699 Firth-Loach (now Howmet Corp., Carbide Div., McKeesport, PA 15134).
M-1699 Howmet Corp., Carbide Div., McKeesport, PA 15134 (formerly Firth-Loach Products).
M-1700 Glader Co., K. C., Niles, IL 60648.
M-1701 Kasle Steel Co., Detroit, MI (2).
M-1702 Koncor Industries, Div. of Latrobe Steel Co. (4), see Latrobe Steel Co., M-73.
M-1703 Precision-Kidd Inc., West Aliquippa, PA 15001.
M-1704 Specialty Steel Co. of America, Warrensville Heights, OH (3).
M-1705 Time Steel Service Inc., Cleveland, OH 44124.
M-1706 TRW United Greenfield Div., Northbrook, IL 60062 (formerly Unimet Carbides).
M-1706 Unimet Carbides (now TRW United Greenfield Div., Northbrook, IL 60062).
M-1707 Walmet Cemented Carbides, Div. of GTE Sylvania, Royal Oak, MI 48068.
M-1708 AMAX Metal Powders (alloys transferred to PYRON Corp., M-1777).
M-1709 Centrifugal Products Inc., Long Beach, CA 90801.
M-1710 Easton Metal Powder Inc., Easton, PA 18042 (3).
M-1711 Unibraze Corp., Covington, OH 45318.
M-1712 Elgiloy Co., Div. of American Gage & Machine Co., Elgin, IL 60120.
M-1713 Alloy Rods Co., see Chemetron Corp., M-1713.
M-1713 Chemetron Welding Products, Chicago, IL 60601.
M-1714 Steel Co. of Australia (2).
M-1715 Alcoa of Great Britain, Droitwich, Worcestershire, WR9 7BG England.
M-1716 A. O. Smith-Inland Inc., Milwaukee, WI 53201 (1).
M-1716 Smith, A. O. - Inland Inc., Powder Metallurgy Div., Milwaukee, WI (2).
M-1717 Ferry-Capitain, Usines De Bussy, B.P. 33-52 Joinville, France.
M-1718 Schulte Eisenhandlung, Heinr. Aug., G.m.b.H., 4600 Dortmund, Germany.
M-1719 Olin Metals Research Labs., New Haven, CT (2).
M-1720 New England Brass, Taunton, MA 02781.
M-1721 Magnacast Corp., Arlington Heights, IL 60005 (formerly Masten Corp.).
M-1721 Maston Corp. (now Magnacast Corp., Arlington Heights, IL 60005).
M-1722 Cobalt Information Center, Columbus, OH (1) (2).
M-1723 Commentrenne, 75429 Paris Cedex 09 France.
M-1723 Société Commentryenne des Aciers (now Commentrenne, 75429, Paris Cedex 09-B.P. 381-09 France).
M-1724 British Steel Corp., Sheffield, S3 8AZ England.
M-1725 Balfour-Darwins Ltd., Sheffield, England (now part of Edgar Allen Balfour Steels Ltd., Sheffield, S9 1QY, England, M-210).
M-1726 Diamond Metal Alloys, Houston, TX 77001.
M-1727 Sitkin Smelting & Refining Co., Lewistown, PA 17044.
M-1728 Glidden Metals, Chemical/Metallurgical Div. of SCM Corp., Cleveland, OH 44113.
M-1729 Semi-Alloys, Mt. Vernon, NY 10550.
M-1730 Stahlwerk Westig G.m.b.H., 4750 Unna, West Germany.

M-1731 Weingartner & Co., KG., Emil, Edelstahlgrosshandel, 2000 Hamburg 54, West Germany.
M-1732 Nippon Steel U.S.A. Inc., New York, NY 10022.
M-1733 Aços Villares, São Paulo, Brazil.
M-1734 Alloy Metals Inc., Troy, MI 48084.
M-1735 Vereinigte Osterreichische Eisen und Stahlwerke (VOEST) A.G., Linz-Austria.
M-1735 Voest, Linz, Austria or Voest International Inc., New York, NY 10022.
M-1736 Electrometal-Acos Finos-SA, C. Postal 944-Campinos, SP. Brazil.
M-1737 Spear & Jackson (Industrial) Ltd., Sheffield, S4 7UR England.
M-1738 Parkin, F. M., (Sheffield) Ltd., Sheffield, S8 0YW England.
M-1739 Sintered Products Ltd., Sutton-in-Ashfield, Nottinghamshire, NG17 5LL England.
M-1740 Walsingham Steel Co., Ltd., Walsingham, Bishop Auckland, Co. Durham, DL13 3HX England.
M-1741 Comalco Ltd., Melbourne, Australia.
M-1742 Phosphor Bronze Co. Ltd., Birmingham, B12 0NA England.
M-1743 Allmetal Screw Products Co., Inc., Garden City, NY 11530.
M-1744 Spenser & Co. Ltd., Walter, (now Spencer Clark Metal Industries Ltd., Sheffield, S6 3AF England).
M-1745 Iscar Ltd., Nahariya, Israel, or James Bellandi, Elmhurst IL 60126, or M. Baumgarten, Iscar Metals Inc., Raritan Center, Edison, NJ 08817.
M-1746 Advanced Material Div., Armco Steel Corp., Baltimore, MD 21204.
M-1748 Certified Alloy Products Inc., Long Beach, CA 90801.
M-1749 Inman & Co. Ltd., T., Sheffield, S4 7YA England.
M-1750 Unicore Inc., North Haven, CT 06473.
M-1751 Interlake Inc., Steel Division, Chicago, IL 60627.
M-1752 3M Company, Industrial Electrical Products Div., 3M Center, St. Paul, MN 55101.
M-1753 Metal Specialties Inc., E. Fairfield, CT 06430.
M-1754 Stackpole Magnet Division, Kane, PA 16735.
M-1755 Atomergic Chemicals Co., Plainview, NY 11803.
M-1756 Adams Alloy Rod Div. Adams Hardfacing Co., Wakita, OK 73771.
M-1757 SAFE/Société des Aciers Fins de l'EST, 92106 Boulogne-Billancourt, France.
M-1758 Timet, a Div. of Titanium Metals Corp. of America, Pittsburgh, PA 15230.
M-1759 Thyssen Edelstahlwerke AG., 4150 Krefeld 1, West Germany.
M-1759 Thyssen Edelstahlwerke A.G., 4150 Krefeld 1, West Germany (new conglomerate).
M-1760 Ekatit Stahl Erich K. Tittel, 7120 Bietigheim-Bissinger, West Germany.
M-1761 Alliance Wall Corp., Alliance, OH 44601.
M-1762 Ley's Malleable Castings Co. Ltd., Derby, DE3 8LY England.
M-1763 ARBED, (see TradeARBED).
M-1763 TradeARBED Inc., Luxembourg, B.P. 1802.
M-1764 IPM Corp., Columbus, OH 43207.
M-1765 Matthey Bishop, Malvern, PA 19355.
M-1766 Forjas Alavesas S.A., 22 Vitoria, Spain.
M-1767 Hoesch Huttenwerke AG., 4600 Dortmund 1, West Germany.
M-1768 Echevarria, S.A. (Aceros HEVA), Apartado 46, Bilboa, Spain.
M-1769 Atrax Cemented Carbides, Wallace Murray Corp., McKeesport, PA 15134.
M-1770 Century Brass Products, Waterbury, CT 06720.
M-1771 Capitol Castings Div. Midland Ross Corp., Phoenix, AZ 85001.
M-1772 Bishop Tube Co., Damascus Tube Div., Frazer, PA 19355.
M-1773 D.A.B. Industries Inc., Troy, MI 48084.
M-1774 Western Cold Drawn Steel Div. of Stanadyne, Inc., Elyria, OH 44035.
M-1775 Alcan Metal Powders, Div. of Alcan Aluminium Corp., Elizabeth, NJ 07207.
M-1776 United States Bronze Powders Inc., Flemington, NJ 08822.
M-1777 PYRON Corp., Niagara Falls, NY 14304 (formerly AMAX Metal Powders).
M-1778 AMAX Copper Div. AMAX Corp., New York, NY 10017.
M-1779 AL Tech Specialty Steel Corp., Dunkirk, NY 14048.
M-1780 Jorgensen, Earle M., Los Angeles, CA 90054.
M-1781 Hoeganaes Corp. (subsidiary of Interlake Inc.), Riverton, NJ 08077.
M-1782 Les Toles Inoxydables et Specials, Ugine-Guegnon, 92307 Levallois-Perret, France (part of Pechiney).
M-1783 Tréfimétaux, 92115 Clichy, France (part of Pechiney).
M-1784 l'Aluminium Pechiney, Soc. de Vente de la, 75008 Paris, France.
M-1784 Société de Vente de l'Aluminium Pechiney, 75008 Paris, France.
M-1785 Société Metallurgique de Gerzat (S.M.G.) 63360 Gerzat, France (part of Pechiney).
M-1786 Société Francais d'Electrometallurgie (SOFREM) 75361 Paris Cedex 08 France (part of Pechiney).
M-1787 Eurotungstene, 65X-38041 Grenoble Cedex, France (part of Pechiney).
M-1788 Sorcery Metals, Delray Beach, FL 33444.

M-1789	Chase Brass & Copper Co., Sheet Div., Cleveland, OH 44122.	M-1800	Homogeneous Metals Inc., Herkimer, NY 13350.
M-1790	Chase Brass & Copper Co. Inc, Williams County Div., Montpelier, OH 43543.	M-1801	Plessey Inc., Metals Div., Melville, NY 11746.
M-1791	Chase Nuclear, Div. Chase Brass & Copper Co., Inc., Cleveland, OH 44122.	M-1802	Vereinigte Edelstahlwerke, A-1011 Wien, Wildpretmarkt 2, Austria (took over Bohler, Schoeller-Bleckmann, Styria and Alpine).
M-1792	Cegedur Pechiney, 75361 Paris Cedex 08, France.	M-1803	Diederichs, Karl, Stahl-,Walz-und Hammerwerk, D-5630 Remscheid 11, West Germany.
M-1793	Carlson, Inc., G.O., Thorndale, PA 19372.		
M-1794	ABEX Corp., Engineered Products Div., Meadville Plant, Meadville, PA 16335.	M-1804	Pechiney Ugine Kuhlmann Corp., New York, NY 10022 (parent company of many French metal companies).
M-1795	Alloy Metal Products, Inc., Davenport, IA 52808 (took over Alter Co. Alloys).	M-1805	IMI Rod & Wire, Birmingham, England.
M-1796	Hufnagel G.m.b.H., 8500 Nürnberg, West Germany.	M-1806	Herbert-Cutanit Ltd., Warrington, Lancashire, WA4 3JX England.
M-1797	Hover, Chr., & Sohn, Edelstahlwerk, 5251 Berghausen uber Engelskirchen, West Germany.	M-1807	Armco Ltd., Letchworth, Herts, SG6 ING England (same alloys as Armco, M-10).
M-1798	Ulbrich Stainless Steels & Special Metals Inc., North Haven, CT 06473.	M-1808	Commonwealth Aircraft Corp., Melbourne, 3001, Australia.
M-1799	Southwire Co., Carrollton, GA 30117.		

SECTION IV

OBSOLETE ALLOYS

Name	Code	Name	Code	Name	Code
000 Extra	M-459	Accoloy CNC-4B	M-1005	Adalloy No. 3	M-655
000 Spezial 31	M-459	Accoloy CNC-4D	M-1005	Adamant	M-1410
4 Best (Frasse)	M-360	Accoloy CNC-5A	M-1005	Adamant XL	M-56
11A45	M-64	Accoloy CNC-5B	M-1005	Adamas 5X	M-1086
13 Cr. Mo 44	M-459	Accoloy CNC-5C	M-1005	Adamas 387	M-1086
14 Ni. Cr 74	M-459	Accoloy NC-5	M-1005	Adamas 484	M-1086
15 Cr. Ni 6	M-459	Acculoy 280 C	M-1538	Adamas 619	M-1086
16 Mn. Cr 5	M-459	Accumet	M-38	Adamas CC	M-1086
18 Cr. Ni. 8	M-459	Ace Oil Hardening	M-307	Adamas D	M-1086
20 Mn. Cr. 5	M-459	Acibel	M-280	Adamas DD	M-1086
24 Cr. Mo. 5	M-459	Acier A3	M-1043	Adams HFA	M-656
30 Cr. Ni. Mo. 8	M-459	Acier ACM	M-1043	Adams HFB	M-656
34 Cr. 4	M-459	Acier ACMA	M-1043	Admiralty Alloy-442	M-8
34 Cr. Ni. Mo. 6	M-459	Acier AF No. 1	M-1043	Admiralty Bronze	M-8
36 Cr. Ni. Mo. 4	M-459	Acier AFS	M-1043	Admiralty Metal	M-33
37 Mn. Si. 5	M-459	Acier AM3	M-1043	Adnic	M-102
42 Cr. Mo. 4	M-459	Acier AS	M-1043	Aegis	M-327
47 Cr. 4	M-459	Acier BLN	M-1043	Aeral	M-187
50 Cr. Mo. 4	M-459	Acier BTR	M-1043	Aero-12-3	M-38
50 Cr. V 4	M-459	Acier CMYO	M-1043	Aerolite	M-28
100 Cr. 6	M-459	Acier CNW	M-1043	A.E. Supernickel	M-282
100 Cr. Mn 6	M-459	Acier CTN2	M-1043	Aeterna 600 Metal	M-953
"202" Nickel	M-1499	Acier CTN6	M-1043	Afcoloy 13	M-159
"204" Nickel	M-1499	Acier Diabolique		Afcoloy 17H	M-159
"330" Nickel	M-1499	Satan No. 2	M-1043	Afcoloy 18-8	M-159
No. 000	M-111	Acier Fam.	M-1043	Afcoloy 18-8 Cb	M-159
No. 1 Standard	M-179	Acier FFV No. 3	M-1043	Afcoloy 18-8 L	M-159
No. 1 TA Quality	M-179	Acier M13AFY	M-1043	Afcoloy 18-8 Mo	M-159
No. 2 B	M-260	Acier NC2	M-1043	Afcoloy 18-8 Se	M-159
No. 3 C	M-69	Acier Triple Satan	M-1043	Afcoloy 18-8 Ti	M-159
No. 4 Vasco Finish	M-115	Acier VDLD	M-1043	Afcoloy 20	M-159
No. 11V	M-32	Acier VDLDM	M-1043	Afcoloy 25-12	M-159
No. 35 Monel	M-68	Acimet C-1	M-480	Afcoloy 25-20	M-159
No. 426 Alloy	M-31	Acimet C-2	M-480	Afcoloy 28	M-159
No. 812	M-57	Acimet C-3	M-480	Afcoloy 35	M-159
No. 1015-65	M-669	Acimet C-4	M-480	Afcoloy 35L	M-159
No. 1020-90 Steel	M-397	Acimet Hard Lead		Afcoloy 60	M-159
		Alloy	M-480	Afcoloy 60L	M-159
A		Ack Die Steel	M-480	Afcoloy 1025	M-159
		Acme	M-248	Afcoloy 1040	M-159
A71NB Quality	M-179	Acme Silcrome 12	M-568	Afcoloy 2140	M-159
AA-1	M-586	Acme Silcrome CC	M-568	Afcoloy 3140	M-159
AB164 Quality	M-179	Acme Silcrome KA2	M-568	Afcoloy 4025	M-159
AB197 Quality	M-179	Acme Silcrome		Afcoloy 4140	M-159
Abradur 500	M-884	KA2S	M-568	Afcoloy 4340	M-159
Abrasist	M-1460	Acme Silcrome L-12	M-568	Afcomet 20	M-159
Abrasocote 10	M-828	Acme Silcrome		Afcomet 20 Cr	M-159
Ac-amsco No. 217	M-607	M-17	M-568	Afcomet 30	M-159
Ac-amsco No. 459	M-607	Acmite	M-35	Afcomet 30 Cr	M-159
Ac-coated 18-8	M-607	Acmite-L	M-35	Afcomet 40	M-159
Accolloy	M-653	Acorn	M-375	Afcomet 40 Cr	M-159
Accoloy CN-5	M-1005	AC-Weld	M-607	Afcomet 50	M-159

1671

Afcomet 50 Cr	M-159	Airco No. 19	M-663	Airco Stainless Ply	M-663
Afcomet 150-12	M-159	Airco No. 23 A		Airco Timang	M-663
Afcomet 250-12	M-159	Silicon Copper		Airco Trae-Rod 27	M-663
Afcomet 250-25	M-159	Rod	M-663	Airco Tungserts	M-663
Affreeze-III	M-448	Airco No. 25	M-663	Airdi 110	M-38
Agathon Armature	M-97	Airco No. 26	M-663	Airdi 225	M-38
Agathon Bit Steel	M-97	Airco No. 28	M-663	Airex	M-115
Agathon Dynamo	M-97	Airco No. 41	M-663	Airex	M-35
Agathon Electric	M-97	Airco No. 48	M-663	Air-Hard	M-32
Agathon Nickel		Airco No. 53	M-663	Air-Hardening	M-97
Alloy	M-97	Airco No. 55	M-663	Air-Hardening	M-260
Agathon Special		Airco No. 58	M-663	Airloy	M-1
Electric	M-97	Airco No. 62	M-663	Airomat 33 Self-	
Airco	M-662	Airco No. 63	M-663	Hardening	M-353
Airco 4-6 Cr-0.5 Mo		Airco No. 72	M-663	Airpro	M-24
Lime	M-663	Airco No. 76	M-663	Airque 4	M-140
Airco 9 Cr-1 Mo		Airco No. 78	M-663	Airque Special	M-140
Lime	M-663	Airco No. 78E	M-663	Air-Tough	M-32
Airco 12 Cr Lime	M-663	Airco No. 81	M-663	Air-Wear	M-32
Airco 16 Cr Lime	M-663	Airco No. 82	M-663	Aitch Metal	M-413
Airco 18-8 3.5 Mo		Airco No. 83	M-663	Ajax	M-123
Lime Type	M-663	Airco No. 84	M-663	Ajax Acid Resisting	
Airco 18-8 3.5 Mo		Airco No. 85	M-663	Bronze	M-123
Titania Type	M-663	Airco No. 86	M-663	Ajax Aluminum	
Airco 18-8 Cb-Mo		Airco No. 87	M-663	Bronze-1	M-123
Lime Type	M-663	Airco No. 90	M-663	Ajax Aluminum	
Airco 18-8 MoCb		Airco No. 91	M-663	Bronze-2	M-123
Titania Type	M-663	Airco No. 93	M-663	Ajax Anti-Acid	
Airco 18-8 Mo Lime	M-663	Airco No. 94	M-663	Metal	M-123
Airco 19-9 Cb		Airco No. 190	M-663	Ajax Bull	M-123
Titania	M-663	Airco No. 230	M-663	Ajax Extra	
Airco 19-9 Lime	M-663	Airco No. 312	M-663	Manganese	
Airco 19-9 Lime		Airco No. 315	M-663	Bronze	M-123
Type	M-663	Airco No. 382	M-663	Ajax Gear Bronze	
Airco 19-9 Titania		Airco No. 383	M-663	Gr. M	M-123
Type	M-663	Airco No. 387	M-663	Ajax Gear Bronze	
Airco 25-12	M-663	Airco No. 388	M-663	Gr. S	M-123
Airco 25-12 Lime		Airco No. 393	M-663	Ajax Gear Bronze	
Type	M-663	Airco No. 394	M-663	Gr. "X"	M-123
Airco 25-12 Titania		Airco No. 395	M-663	Aluminum 4543	M-4,
Type	M-663	Airco No. 396	M-663		M-671
Airco 25-20 Cb		Airco No. 397	M-663	Allegheny 46M	M-1
Titania	M-663	Airco No. 398	M-663	Ajax Golden Glow	
Airco 25-20 Titania	M-663	Airco S-10	M-663	Brass	M-123
Airco 28 Cr Lime	M-663	Airco S-15	M-663	Ajax Hamilton	
Airco 308	M-663	Airco S-16	M-663	Gear Bronze	M-123
Airco 410	M-663	Airco S-20	M-663	Ajax Honest Solder	M-123
Airco 4130	M-663	Airco S-65	M-663	Ajax Hydraulic	
Airco 5554	M-663	Airco S-308	M-663	Metal	M-123
Airco 6150	M-663	Airco S-309	M-663	Ajax Manganese	
Airco A715	M-663	Airco S-310	M-663	Bronze	M-123
Airco A725	M-663	Airco S-316	M-663	Ajax Manganese	
Airco Electrode	M-663	Airco S-347	M-663	Bronze Gr. A	M-123
Airco KA2S	M-663	Airco Self-		Ajax Manganese	
Airco KA2SMo	M-663	Hardening		Bronze SX	M-123
Aircolite	M-663	Electric	M-663	Ajax Manganese	
Aircomatic A309		Airco Self-		Bronze XX	M-123
ELC	M-663	Hardening Gas	M-663	Ajax Nickalloy	
Aircomatic A666	M-663	Airco Semi-		Bronze Grade 12	M-123
Airco Nickel		Automatic Air-		Ajax Nickalloy	
Manganese	M-663	Mang	M-663	Bronze Grade 21	M-123
Airco No. 5	M-663	Aircosil 105	M-663	Ajax Nickalloy	
Airco No. 6	M-663	Aircosil Q	M-663	Bronze, Grade 23	M-123
Airco No. 11	M-663	Airco Silicon Bronze	M-663		

Section IV: Obsolete Alloys / 1673

Ajax Nickel Aluminum Bronze	M-123	Alcoa 83	M-4	Alcoloy	M-279
		Alcoa 85	M-4	Alcoloy	M-279
		Alcoa 93	M-4	Alcomax-I	M-1010
Ajax Nickel Plastic Bronze	M-123	Alcoa 100	M-4	Alcrosil 5	M-376
		Alcoa 101	M-4	Alcrosil No. 3	M-376
Ajax Phosphor Bronze	M-123	Alcoa 105	M-4	Alculoy	M-158
		Alcoa 109	M-4	Alculoy 12	M-158
Ajax Plastic Bronze-A	M-123	Alcoa 112	M-4	Alculoy 75	M-158
		Alcoa 113	M-4	Alcumite	M-46
Ajax Plastic Bronze-B	M-123	Alcoa 132	M-4	Alcunic	M-102
		Alcoa 144	M-4	Alcunic "G"	M-102
Ajax Plastic Bronze Gr. A	M-123	Alcoa 145	M-4	Aldecor	M-97
		Alcoa 152	M-4	Aldivan	M-73
Ajax Plastic Bronze Gr. B.	M-123	Alcoa 170	M-4	Aldural	M-338
		Alcoa 172	M-4	Aldural E	M-338
Ajax Plastic Bronze Gr. C	M-123	Alcoa 196	M-4	Aldural-G	M-338
		Alcoa 212	M-4	Aldural K	M-338
Ajax Pressure Bronze Grade M	M-123	Alcoa 213	M-4	Aldural-L	M-338
		Alcoa 216	M-4	Aldural-LC	M-338
Ajax Pressure Bronze Grade S	M-123	Alcoa 231	M-4	Aldural Q	M-338
		Alcoa 304	M-4	Aldural S	M-338
Ajax Pressure Bronze Grade X	M-123	Alcoa 314	M-4	Aldural Z	M-338
		Alcoa 315	M-4	Alfenol	M-38
Ajax Special Valve Stem Bronze	M-123	Alcoa 348	M-4	Alfenol-16	M-32
		Alcoa 406	M-4	Alferium	M-533
Ajax Structural Bronze Grade S	M-123	Alcoa 505	M-4	Algo-Loy 1315	M-1253
		Alcoa 645	M-4	Algo-Loy 1317	M-1253
Ajax Structural Bronze, Grade Z	M-123	Alcoa 1130	M-4	Algoma 1315	M-1253
		Alcoa 1260	M-4	Algoma 1317	M-1253
Ajax Super Manganese Bronze	M-123	Alcoa 6062	M-4	Alhead	M-1
		Alcoa A100	M-4	Alkrohmit 140	M-27
		Alcoa A 213	M-4	Alkumag 302	M-324
Ajax Tool Steel	M-39	Alcoa A 254	M-4	Alkumag 303	M-324
Ajax Wiping Solder	M-123	Alcoa A334	M-4	Alkumag 311	M-324
Akro 6	M-1308	Alcoa A355	M-4	Al-L-24	M-28
A-L 18	M-1	Alcoa AM100A	M-4	Allegheny 12	M-1
A-L 20	M-1	Alcoa AM262	M-4	Allegheny 16-16-1	M-1
A-L 25	M-1	Alcoa AM266	M-4	Allegheny 18-8C	M-1
A-L-602	M-1	Alcoa AM-C88S	M-4	Allegheny 18-8SM	M-1
Alakron	M-187	Alcoa AMX269	M-4	Allegheny 19-9DX	M-1
Alamo	M-325	Alcoa AZ31	M-4	Allegheny 19-10M	M-1
Albanoid	M-121	Alcoa AZ63A	M-4	Allegheny 19-10SM	M-1
Albany	M-1	Alcoa AZ81A	M-4	Allegheny 20-10	M-1
Albertson Special	M-512	Alcoa AZ91C	M-4	Allegheny 20-10S	M-1
Albor Die	M-694	Alcoa B-17S	M-4	Allegheny 20-25S	M-1
Alcan 117	M-643	Alcoa B50S	M-4	Allegheny 20-25SM	M-1
Alcan GB162	M-643	Alcoa B 105	M-4	Allegheny 21	M-1
Alcan GB-B110	M-643	Alcoa B-113	M-4	Allegheny 22	M-1
Alcan GB-C125	M-643	Alcoa B 213	M-4	Allegheny 22 SP	M-1
Alchrome 6	M-61	Alcoa B 355	M-4	Allegheny 25-12	M-1
Alcoa 5 S	M-4	Alcoa C-17S	M-4	Allegheny 25-12S	M-1
Alcoa 6	M-4	Alcoa C 214	M-4	Allegheny 25-20	M-1
Alcoa 8	M-4	Alcoa D132	M-4	Allegheny 25-20S	M-1
Alcoa 8 S	M-4	Alcoa D-195	M-4	Allegheny 28	M-1
Alcoa 12	M-4	Alcoa EK41A	M-4	Allegheny 28-4	M-1
Alcoa 15 S	M-4	Alcoa EZ33A	M-4	Allegheny 29-9	M-1
Alcoa 27 S	M-4	Alcoa Mg 99.8	M-4	Allegheny 29-9S	M-1
Alcoa 45	M-4	Alcoa MIA	M-4	Allegheny 33 FM	M-1
Alcoa 47	M-4	Alcoa No. 89 E-2	M-4	Allegheny 33 NH Non-Hardening	M-1
Alcoa 51	M-4	Alcoa TA54	M-4		
Alcoa 70 S	M-4	Alcoa TA54A	M-4	Allegheny 33 Turbine	M-1
Alcoa 79	M-4	Alcoa XA75S	M-4		
Alcoa 81	M-4	Alcoa XA80S	M-4	Allegheny 33 W	M-1
Alcoa 82	M-4	Alcoa ZK60A	M-4	Allegheny 34	M-1

Name	Ref	Name	Ref	Name	Ref
Allegheny 46 Tungsten	M-1	Allegheny Metal 16-1	M-1	Allegheny Metal "B-Ti"	M-1
Allegheny 66 W	M-1	Allegheny Metal 17	M-1	Allegheny Metal "C"	M-1
Allegheny 67	M-1	Allegheny Metal 17EZ	M-1	Allegheny Metal L-12	M-1
Allegheny 216L	M-1	Allegheny Metal 17 Pluramelt	M-1	Allegheny Metalclad Steel	M-1
Allegheny 303 EZ	M-1	Allegheny Metal 18-8 Cast	M-1	Allegheny Metal FM Free Machining	M-1
Allegheny 416EZ	M-1	Allegheny Metal 18-8C Cast	M-1	Allegheny Metal (Type 345)	M-1
Allegheny 433	M-11	Allegheny Metal 18-8C Pluramelt	M-1	Allegheny Metal (Type 346)	M-1
Allegheny 4750	M-1	Allegheny Metal 18-8FS	M-1	Allegheny MF-1	M-1
Allegheny AF-71	M-1	Allegheny Metal 18-8M	M-1	Allegheny Moly Iron	M-1
Allegheny AF-183	M-1	Allegheny Metal 18-8MC	M-1	Allegheny No. 33	M-1
Allegheny ALX	M-1	Allegheny Metal 18-8M Cast	M-1	Allegheny No. 44	M-1
Allegheny AM 357	M-1	Allegheny Metal 18-8M Special	M-1	Allegheny No. 46	M-1
Allegheny AM-367	M-1	Allegheny Metal 18-8 Pluramelt	M-1	Allegheny No. 55	M-1
Allegheny Armature	M-1	Allegheny Metal 18-8S	M-1	Allegheny No. 66	M-1
Allegheny Audi Transformer "A"	M-1	Allegheny Metal 18-8S Cast	M-1	Allegheny R41	M-1
Allegheny Audio Transformer "B"	M-1	Allegheny Metal 18-8S Pluramelt	M-1	Allegheny Radio Transformer	M-1
Allegheny D979	M-1	Allegheny Metal 18-11C	M-1	Allegheny S-816+B	M-1
Allegheny Dynamo	M-1	Allegheny Metal 25-2	M-1	Allegheny Super Dynamo	M-1
Allegheny Dynamo Special	M-1	Allegheny Metal 25-2 Cast	M-1	Allegheny Transformer A	M-1
Allegheny Electrical Sheet Steel	M-1	Allegheny Metal 25-12C	M-1	Allegheny Transformer AA	M-1
Allegheny Electric Metal	M-1	Allegheny Metal 25-12 Cast	M-1	Allegheny Transformer B	M-1
Allegheny Field	M-1	Allegheny Metal 25-20C Cast	M-1	Allegheny Transformer C	M-1
Allegheny G-192	M-1	Allegheny Metal 25-20H	M-1	Allegheny TRC	M-1
Allegheny Ludlum 21-12N	M-1	Allegheny Metal 25-20H Cast	M-1	Allegheny Type 203 EZ	M-1
Allegheny-Ludlum 88	M-1	Allegheny Metal 25-20+Si	M-1	Allegheny V-36	M-1
Allegheny Ludlum 129	M-1	Allegheny Metal 28 Pluramelt	M-1	Alloy 4D	M-28
Allegheny Ludlum 418	M-1	Allegheny Metal 42	M-1	Alloy 8J	M-61
Allegheny Ludlum 419	M-1	Allegheny Metal 46SM	M-1	Alloy 9Al-30Mn	M-747
Allegheny Ludlum 901	M-1	Allegheny Metal "A"	M-1	Alloy 11A45	M-64
Allegheny Ludlum 1251	M-1	Allegheny Metal "A-Mo"	M-1	Alloy 61	M-167
Allegheny Ludlum-LMW	M-1	Allegheny Metal "A-Ti"	M-1	Alloy 89MC	M-69
Allegheny M-17	M-1	Allegheny Metal "B"	M-1	Alloy 150S	M-1147
Allegheny M-252	M-1	Allegheny Metal B(SP)	M-1	Alloy 422-19	M-167
Allegheny Metal 12-2	M-1			Alloy 936	M-1295
Allegheny Metal 12 Cast	M-1			Alloy 2129	M-108
Allegheny Metal 12EZ	M-1			Alloy 6059	M-167
Allegheny Metal 12-NH	M-1			Alloy Broachrite	M-669
Allegheny Metal 12 Pluramelt	M-1			Alloy CH42	M-694
Allegheny Metal 12-TB	M-1			Alloy Cross Cuts	M-69
Allegheny Metal 12-W	M-1			Alloyed Genuine Wrought Iron	M-292
Allegheny Metal 15-35	M-1			Alloy Gas Welding 2320	M-12
				Alloy Gas Welding A-2317	M-12
				Alloy Gas Welding A-2330	M-12
				Alloy Hollow Drill	M-24

Alloy	Code	Alloy	Code	Alloy	Code
Alloy HTB-1	M-1	Alnico	M-60	Alpine HRMV25	M-1314
Alloy Hubbing Die	M-24	Alnico I	M-31	Alpine MAC	M-1314
Alloy No. 6	M-85	Alnico I	M-38	Alpine NALH	M-1314
Alloy No. 20	M-508	Alnico II	M-38	Alpine NALJ	M-1314
Alloy No. 471	M-65	Alnico 2	M-60	Alpine NALM	M-1314
Alloy No. 484	M-65	Alnico 3	M-60	Alpine NALW	M-1314
Alloy No. 548	M-31	Alnico III	M-38	Alpine PRIMA-H	M-1314
Alloy No. 1270	M-607	Alnico 4	M-60	Alpine PRIMA ZAH	M-1314
Alloy Shoe Die	M-38	Alnico IV	M-38	Alpine WMC 5	M-1314
Alloy Trimmer	M-357	Alnico 4 Sintered	M-1076	Alpine WMW	M-1314
Alloy-W5	M-121	Alnico 5	M-60	Alpine WMWS	M-1314
Alloy-W6	M-121	Alnico V	M-38	Alray C	M-248
Alloy-W7	M-121	Alnico 5 CB	M-1192	Alray D	M-248
Alloy-W9	M-121	Alnico 6	M-60	Alronze-619	M-279
Alloy Wood Chisel	M-38	Alnico VI	M-38	Alronze Gr M	M-279
Alloy X40	M-60	Alnico 7	M-60	Alronze R 619	M-279
Alloy X40	M-31	Alnico 7-5 Oriented	M-1076	Alronze Standard	M-279
Alloy X41	M-31	Alnico 7-5 Unoriented	M-1076	Alsichrom-1(CA30)	M-1309
Alloy X50	M-31	Alnico VII A	M-1076	Alsichrom-2(CAF)	M-1309
Alloy X63	M-31	Alnico 7C	M-1192	Alsiloy	M-158
All-State HS-1	M-1713	Alnico VII-S	M-1076	Alsiloy 1	M-158
All-State No. 1	M-1074	Alnico 8	M-60	Alsiloy 3	M-158
All State No. 2	M-1074	Alnico VIII	M-31	Alsiloy 5	M-158
All-State No. 22	M-1713	Alnico 9	M-60	Alsiloy 9	M-158
All-State No. 32	M-1074	Alnico 12	M-60	Alsiloy 10	M-158
All-State No. 35	M-1074	Alnico XII	M-1076	Alsiloy S-1	M-158
All-State No. 61	M-1713	Alniloy	M-158	Alsiloy S-5	M-158
All-State No. 120	M-1074	Aloi	M-575	Alsteel	M-345
Allvac 18-15B	M-1490	Aloyco 18-8 Mo	M-1023	Altemp D979	M-1
Allvac-N	M-1490	Aloyco 35	M-1023	Altemp L605	M-1
Almanite-C	M-82	Aloyco 37	M-1023	Altemp M-252	M-1
Almanite S	M-82	Alpha	M-33	Altemp S-590	M-1
Almar 15(280HT)	M-1	Alpha	M-539	Altemp S-816	M-1
Almet 1-8	M-248	Alpha Chisel	M-85	Altemp V-36	M-1
Almet 12	M-248	Alpha Crusher	M-539	Alto	M-34
Almet 12 M	M-248	Alpha No. 2	M-85	Alumac No. 13	M-Eng.
Almet 17	M-248	Alpha Tool	M-85	Alumac No. 83	M-Eng.
Almet 17-7PH	M-248	Alpine AHD	M-1314	Alumac No. 85	M-Eng.
Almet 17Cu	M-248	Alpine AHKW	M-1314	Alumagnese 10	M-1444
Almet 18-8 Cb	M-248	Alpine ANP3	M-1314	Alumagnese 20	M-1275
Almet 18-8M	M-248	Alpine BEC60	M-1314	Alumagnese 20	M-1444
Almet 18-8S	M-248	Alpine BEC80	M-1314	Alumagnese 35	M-1275
Almet 18-8Ti	M-248	Alpine BLH	M-1314	Alumagnese 35	M-1444
Almet 18-12SMo	M-248	Alpine BR55	M-1314	Alumagnese 50	M-1275
Almet 20-10	M-248	Alpine-EC3	M-1314	Alumagnese 50	M-1444
Almet 21	M-248	Alpine EJR35	M-1314	Alumend	M-677
Almet 25-12	M-248	Alpine EJR45	M-1314	Aluminum 2EC	M-1147
Almet 25-20	M-248	Alpine EJRA	M-1314	Aluminum 3L-11	M-86, M-28
Almet 28	M-248	Alpine EL1	M-1314		
Almet 35-15	M-248	Alpine ERJ15	M-1314	Aluminum 4L-11	M-28
Almet 405	M-248	Alpine ERJ20	M-1314	Aluminum 50S	M-1147
Almet 420	M-248	Alpine Extra M	M-1314	Aluminum 55EC	M-1147
Almet 434	M-248	Alpine Extra Spezial ZH II	M-1314	Aluminum-100	M-297
Almet 442	M-248			Aluminum 218SP	M-671
Almet 446	M-248	Alpine EZH	M-27	Aluminum 310	M-4
Almet 1235	M-248	Alpine HIR15 W/H	M-1314	Aluminum 344	M-4
Almo No. 1	M-38	Alpine HIR 25	M-1314	Aluminum 357	M-4
Almo No. 1 C.H.	M-38	Alpine HIRA 35 W/H	M-1314	Aluminum 359	M-4
Almo No. 2	M-38			Aluminum 360	M-4
Almo No. 2 C.H.	M-38	Alpine HIRA W/H	M-1314	Aluminum 363	M-1147
Almo No. 3	M-38	Alpine HMV25	M-1314	Aluminum 385	M-4
Almo No. 3 C.H.	M-38	Alpine HMV60	M-1314	Aluminum 390	M-4, M-671
Alni	M-271	Alpine HRM45	M-1314		
Alnic	M-60				

Alloy	Ref	Alloy	Ref	Alloy	Ref
Aluminum 613	M-671, M-4, M-1147	Aluminum-A750	M-4, M-1147	Aluminum-SC114A+Mg	M-158
Aluminum 718	M-4, M-1147	Aluminum Alloy No. 119	M-249	Aluminum SG-81	M-158
Aluminum-1145	M-4, M-671	Aluminum Alloy No. 119 M	M-249	Aluminum Silicon Bronze	M-33
Aluminum 1180	M-671	Aluminum ASCR	M-114m	Aluminum-SN122A	M-158
Aluminum 1188	M-671	Aluminum-B750	M-4, M-1147	Aluminum X250	M-4
Aluminum 1199	M-671	Aluminum BA46	M-28	Aluminum X357	M-1147
Aluminum 1230	M-671	Aluminum Brass	M-102	Aluminum X2219	M-4
Aluminum 1235	M-671	Aluminum Brass	M-8	Aluminum XA140	M-4
Aluminum 2020	M-4	Aluminum Bronze	M-8	Aluminum XC65S	M-4
Aluminum 2021	M-4	Aluminum Bronze	M-129	Alunyte	M-543
Aluminum 2219	M-4, M-671	Aluminum Bronze	M-285	Alva Alloy Hollow Drill Steel	M-38
Aluminum 2364	M-1147	Aluminum C58	M-4	Alva Duplex Forging	M-38
Aluminum 2393	M-1147	Aluminum-C612	M-671, M-4	Alva Special	M-38
Aluminum 2618	M-1147	Aluminum C872	M-4	AM 53S	M-4
Aluminum 3002	M-4	Aluminum C914	M-4	AM 57S	M-4
Aluminum 3005	M-4	Aluminum C-960	M-4	AM 58S	M-4
Aluminum 3005	M-238	Aluminum C989	M-4	AM 59S	M-4
Aluminum 3105	M-4, M-671	Aluminum-CG86	M-158	AM 61S	M-4
		Aluminum CH-70	M-4	AM 67S	M-4
		Aluminum Chromium	M-394	AM 88S	M-4
Aluminum 4245	M-4	Aluminum D132	M-4	AM 112	M-4
Aluminum 4643	M-4	Aluminum-D612	M-671, M-1147	AM 230	M-4
Aluminum 5039	M-1147			AM 241	M-4
Aluminum 5080	M-4			AM 242	M-4
Aluminum 5180	M-4	Aluminum DTD 424	M-28	AM 244	M-4
Aluminum 5205	M-4, M-671	Aluminum DTD 428	M-28	AM 246	M-4
Aluminum 5257	M-4	Aluminum ECH17	M-4	AM 403	M-4
Aluminum 5557	M-4	Aluminum-F214	M-671	AM 555	M-4
Aluminum 5652	M-4, M-671	Aluminum Genuine	M-565	AM 740	M-4
Aluminum 5757	M-4	Aluminum HP356	M-1147	AM 764	M-4
Aluminum 6003	M-671	Aluminum Iron Bronze	M-8	Amanimphy	M-771
Aluminum 6051	M-1147	Aluminum K-155	M-1147	Ambo AST	M-1315
Aluminum 6071	M-4	Aluminum K 183	M-1147	Ambo BWG	M-1315
Aluminum 6253	M-671	Aluminum K 186	M-1147	Ambo CM04E	M-1315
Aluminum 6563	M-4	Aluminum-L214	M-671	Ambo CM05E	M-1315
Aluminum 6951	M-4	Aluminum LAC 10	M-28	Ambo CMV	M-1315
Aluminum 7002	M-671, M-4	Aluminum LAC112A Type 1	M-28	Ambo CMV25	M-1315
Aluminum 7006	M-4	Aluminum LAC 112A Type 2	M-28	Ambo CMV30	M-1315
Aluminum 7007	M-4	Aluminum LAC 113B	M-28	Ambo CMV50	M-1315
Aluminum 7038	M-1147			Ambo CSJ	M-1315
Aluminum 7080	M-4			Ambo DCN5	M-1315
Aluminum 7106	M-4	Aluminum Manganese	M-394	Ambo DCN5E	M-1315
Aluminum 7275	M-1147	Aluminum Molybdenum	M-394	Ambo DCNV	M-1315
Aluminum 7279	M-1147	Aluminum No. 2EC	M-4	Ambo DCZ	M-1315
Aluminum 8002	M-4	Aluminum No. 16	M-514	Ambo DM2	M-1315
Aluminum 8013	M-1147	Aluminum No. 37	M-514	Ambo DM3	M-1315
Aluminum 8212	M-1147	Aluminum No. 46	M-514	Ambo DM4	M-1315
Aluminum-A108	M-671	Aluminum-S12A	M-158	Ambo DM5	M-1315
Aluminum-A132	M-1147	Aluminum S457	M-4	Ambo DM6	M-1315
Aluminum-A214	M-671	Aluminum-SC84A	M-158	Ambo HHCA8	M-1315
Aluminum A357	M-4	Aluminum SC84A+Mg	M-158	Ambo HHCA10	M-1315
Aluminum-A360	M-4			Ambo HHCA12	M-1315
Aluminum-A380	M-4, M-1147			Ambo HHCN100	M-1315
Aluminum-A390	M-4, M-671	Aluminum-SC114A	M-158	Ambo HHCN120	M-1315
				Ambo HHCN120S	M-1315
				Ambo MS1N	M-1315
				Ambo MSD	M-1315
				Ambo NR2A	M-1315
				Ambo NR2AE	M-1315
				Ambo NR2AS	M-1315

Ambo NR2ATE	M-1315	American Type 23	M-991	Ampco-Trade 21	M-13
Ambo NR4AE	M-1315	American Type 31	M-991	Ampco-Trade 22	M-13
Ambo NR4AS	M-1315	American Valve		AMS	M-459
Ambo WMA	M-1315	Spring	M-12	Amsco 60	M-9
Ambrac 815	M-8	American XB	M-430	Amsco Air	
Ambrac A	M-8	American XCR	M-430	Hardening	M-9
Ambrac "B"	M-8	Amer-Led A	M-12	Amsco AW-	
Ambralloy	M-292	Amer-Led B	M-12	Thermalloy 4	M-9
Ambraloy 929	M-8	Amerspring	M-12	Amsco Chromeface	M-9
Am C52S	M-4	Amerstitch	M-12	Amsco Co-Mang	M-9
AMC 74S	M-4	Amerstrip	M-12	Amsco Dieweld	M-9
Amco OFHC	M-1508	Ampco	M-681	Amsco F-8N	M-9
Amercut 1020	M-12	Ampco 18-13	M-13	Amsco HF20	M-9
Amercut 4640	M-12	Ampco 18-33	M-13	Amsco HF-40	M-9
Amercut B-17	M-12	Ampco Beryllium		Amsco HF-60	M-9
Amercut B-24	M-12	Copper	M-13	Amsco Machine-	
Amercut B-28	M-12	Ampcoloy 3W1	M-31	Face	M-9
Amercut B-1111	M-12	Ampcoloy 3W10	M-13	Amsco Mo-Mang	M-9
Amercut B-1112	M-12	Ampcoloy 3W20	M-13	Amsco Nickel	
Amercut C-1115	M-12	Ampcoloy 3W-86	M-13	Manganese	M-9
Amercut C-1117	M-12	Ampcoloy 10W-86	M-13	Amsco Ni-Hard	M-9
Amercut C-1118	M-12	Ampcoloy 30	M-13	Amsco Ni-Mn Steel	M-9
Amercut C-1120	M-12	Ampcoloy 31	M-13	Amsco No. 217	M-9
Amercut C-1132	M-12	Ampcoloy 32	M-13	Amsco No. 459	M-9
Amercut C-1137	M-12	Ampcoloy 34	M-13	Amsco Railface	M-9
Amercut C-1141	M-12	Ampcoloy 35	M-13	Amsco Resistwear	M-9
Amercut X1020	M-12	Ampcoloy 38	M-13	Amsco SA-33	M-9
Amerhead 1330	M-12	Ampcoloy 40	M-13	Amsco SA-Build Up	M-9
Amerhead 1335	M-12	Ampcoloy 42	M-13	Amsco Toolface	M-9
Amerhead 4037	M-12	Ampcoloy 44	M-13	Amsco Trackrail	M-9
Amerhead 4140	M-12	Ampcoloy 49	M-13	Amsco V-Mang	M-9
Amerhead 4615	M-12	Ampcoloy 71	M-13	AMS Extra	M-459
Amerhead 8620	M-12	Ampcoloy 80	M-13	Amtite	M-558a
Amerhead A-2317	M-12	Ampcoloy 83	M-13	Amvilloy 1000	M-265
Amerhead A-2330	M-12	Ampcoloy 83-20	M-13	Amvilloy 2000	M-265
Amerhead A-3115	M-12	Ampcoloy 84	M-13	Anaconda 24	M-8
Amerhead A-3120	M-12	Ampcoloy 84-20	M-13	Anaconda 32	M-8
Amerhead A-3135	M-12	Ampcoloy 86	M-13	Anaconda 67	M-8
Amerhead C-1018	M-12	Ampcoloy 86-20	M-13	Anaconda 85	M-8
Amerhead NE 9437	M-12	Ampcoloy 91	M-13	Anaconda 205	M-8
American 10	M-991	Ampcoloy 95	M-13	Anaconda 211	M-8
American Black		Ampcoloy 95-20	M-13	Anaconda 624	M-8
Heart Malleable		Ampcoloy 97-20	M-13	Anaconda 719	M-8
Iron	M-113, M-24	Ampcoloy 98	M-13	Anaconda	
		Ampcoloy 342	M-13	Beryllium Copper	M-8
American Ca-14	M-991	Ampcoloy 382	M-13	Anaconda Bronze	
American CF-7	M-991	Ampcoloy 405	M-13	No. 14	M-8
American CF-7C	M-991	Ampcoloy 501	M-13	Anaconda Electro	
American CF-7MC	M-991	Ampcoloy 502	M-13	Sheet Copper	M-8
American CK-25	M-991	Ampcoloy 531	M-13	Anaconda Phosphor	
American CN-7	M-991	Ampcoloy 561	M-13	Bronze Gr. E.	M-8
American CT-7M	M-991	Ampcoloy A-3	M-13	Anaconda Special	
American Hylastic	M-991	Ampcoloy A-323	M-13	Phosphor Bronze	M-8
American Marine	M-565	Ampcoloy E-2	M-13	Anaconda Super	
American Marine		Ampcoloy E-3	M-13	Nickel	M-8
Genuine	M-565	Ampcoloy E-117	M-13	Anchor Chaser Die	M-342
American Piston		Ampcoloy E-123	M-13	Anchor Special Tap	M-342
Ring Standard		Ampcoloy E-133	M-13	Anfriloy	M-681
Iron	M-399	Ampco Metal	M-285	Anhyster A	M-771
American Super-		Ampco Trodaloy	M-13	Anhyster B	M-771
Tens	M-12	Ampco-Trade 12	M-13	Anhyster C	M-104
American Type 12	M-991	Ampco-Trade 16	M-13	Anhyster D	M-104
American Type 20	M-991	Ampco-Trade 18	M-13	Anhyster M	M-104
American Type 22	M-991	Ampco-Trade 20	M-13		

Name	Code	Name	Code	Name	Code
Anhyster Mau Molybdene	M-104	Antitherm FB 30 S	M-27	Argeste 13PX	M-1333
Anibal	M-104	Antitherm FB30SG	M-27	Argeste 17LT	M-1333
Anka "H"	M-472	Antitherm FB95	M-27	Argeste 17Nb	M-1333
Anka Steel	M-472	Antitherm FB 105	M-27	Argeste 17 PX	M-1333
Annite	M-38, M-334	Antoxyd	M-96	Argeste 17UZ	M-1333
		APM M-256	M-4	Argeste 58M	M-1333
		APM M-257	M-4	Argeste 80K	M-1333
Annite No. 1	M-24	APM M-430	M-4	Argeste 80 Mn	M-1333
Annite No. 2	M-24	APM M-470	M-4	Argeste 80 PX	M-1333
Anode	M-578	APW 4	M-686	Argeste 80 PXLC	M-1333
Anoxin 1	M-501	APW-301	M-686	Argeste 80 UZ	M-1333
Anoxin-1F	M-912	APW-A4	M-686	Argeste 80 X	M-1333
Anoxin 2	M-501	APW-A11	M-686	Argeste 82 PX	M-1333
Anoxin 2G	M-501	APW-A14	M-686	Argeste 82 PXLC	M-1333
Anoxin 3	M-501	APW-A25	M-686	Argeste 82 SG	M-1333
Anoxin 4	M-501	APW-A28	M-686	Argeste 82 X	M-1333
Anoxin 4G	M-501	APW-A33	M-686	Argeste 83 Nb	M-1333
Anoxin 5	M-501	APW-A49	M-686	Argeste 83 PX	M-1333
Anoxin 5G	M-501	APW-A850	M-686	Argeste 83 PXLC	M-1333
Anoxin 8	M-501	APW-B211	M-686	Argeste 83 T	M-1333
Anoxin 20	M-501	APW-B-217	M-686	Argeste 83 X	M-1333
Anoxin 22	M-501	APW-C-250	M-686	Argeste 120	M-1333
Anoxin 40	M-501	APW-Easy	M-686	Argeste 135PX	M-1333
Anoxin 44	M-501	APW-Hard	M-686	Argeste 180K	M-1333
Anoxin 66	M-501	APW-L451	M-686	Argeste 180M	M-1333
Anoxin 70	M-912	APW-Medium	M-686	Argeste 182 RX	M-1333
Anoxin 77	M-912	APW No. 18	M-686	Argeste K 15L	M-1333
Anoxin 200	M-912	APW No. 201	M-686	Argeste K 50	M-1333
Anoxin G	M-501	APW No. 205	M-686	Argeste K 100	M-1333
Anoxin Ultra-C	M-501	Aquatough 70	M-1724	Argus	M-548
Antag Chisel	M-85	Aquatough 100	M-1724	Aristocrat	M-77
Antimonial Admiralty	M-33	Aquatough 100	M-1569	Aristoloy 4-6Cr-Mo	M-682
		Aquila	M-96	Aristoloy 7-9Cr-Mo	M-682
Antinit 1KB2	M-27	ARA	M-114	Aristoloy 302	M-682
Antinit 1KB4	M-27	Arc 2233 A2	M-771	Aristoloy 303	M-682
Antinit AS2(H)	M-27	Arc 2233 A8	M-771	Aristoloy 304	M-682
Antinit AS2H	M-27	Arc 2233 B6	M-771	Aristoloy 305	M-682
Antinit AS2K	M-27	Arc 2266 A2	M-771	Aristoloy 306	M-682
Antinit AS4K	M-27	Arc 2266 A6	M-771	Aristoloy 307	M-682
Antinit ASA	M-27	Arc 2266 B6	M-771	Aristoloy 308	M-682
Antinit KW5	M-27	Arc 2266 Cb	M-104	Aristoloy 309	M-682
Antinit KW10 Spezial	M-27	Arc 2266 TRS	M-104	Aristoloy 310	M-682
		Arc 2702 A2	M-771	Aristoloy 316	M-682
Antinit KW10Z	M-27	Arc 2702 B	M-771	Aristoloy 317	M-682
Antinit KW15M	M-27	Arc 2702 B2	M-771	Aristoloy 321	M-682
Antinit KW40M	M-27	Arc 2702 B6	M-771	Aristoloy 347	M-682
Antinit KW50	M-27	Arc 2702 B8	M-771	Aristoloy 410	M-682
Antinit KW50M	M-27	Arc 2702 Cb	M-104	Aristoloy 414	M-682
Antinit KWG	M-27	Arc 2702 M	M-104	Aristoloy 416	M-682
Antinit KWMZ	M-27	Arc 2702 SP	M-104	Aristoloy 420	M-682
Antinit KWZ	M-27	Arc 2702 T	M-771	Aristoloy 430	M-682
Antinit RS30	M-27	Arc 2702 TRS	M-104	Aristoloy 446	M-682
Antinit RS 30M	M-27	Arc 6015	M-104	Aristoloy 4027 Leaded	M-682
Antinit SAS2 Extra	M-27	Arc 7915	M-104		
Antinit SAS5	M-27	Arcast-67	M-1662	Aristoloy 4140	M-682
Antinit SAS5G	M-27	Architectural Bronze C-94	M-179	Armature Electric	M-806
Antinit-Sil18	M-27			Armco 4-79 Ni	M-10
Antinit SKWA	M-27	Arcoloy	M-684	Armco 12T	M-10
Antinit SKWN	M-27	Ardent	M-1411	Armco 13	M-10
Antiox-1	M-1117	Ardoloy	M-575	Armco 15-35	M-10
Antiox-2	M-1117	Ardoloy Gr. 1A179	M-306	Armco 15 Type 425	M-10
Antiox-3	M-1117	Ardoloy Gr. 2	M-306	Armco 16-6	M-10
Antitherm FB7	M-27	Ardoloy Gr. 3	M-306	Armco 16-6S	M-10
Antitherm FB30G	M-27	Argeste 13 A1	M-1333	Armco 17 Cu	M-10

Armco 17FM	M-10	Armco Ti-6Al-2Sn-4Zr-6Mo	M-10	A.S. Special Hobbing Iron	M-694
Armco 17-14 CuMo	M-10	Armco Ti-6Al-4V	M-10	Astralloy Gr2	M-24, M-176
Armco 18-12 MoCb	M-10	Armco Ti-6Al-6V-2Sn	M-10	A.S. Tri-ack	M-694
Armco 18-20 SMo	M-10	Armco Ti-8Al-1Mo-1V	M-10	A.S. Tri-Mo	M-694
Armco 19-9	M-10	Armco Ti 40	M-10	Astroloy	M-31
Armco 19-9 Type 305	M-10	Armco Ti 55	M-10	A.S. White Label	M-694
Armco 19-9 Type 306	M-10	Armco Ti 70	M-10	Aterite (3 grades)	M-19
Armco 21	M-10	Armco Trancor 52	M-10	ATG-M	M-104
Armco 21-4N	M-10	Armco Tran-Cor 58	M-10	ATGS	M-French
Armco 25-20 Si	M-10	Armco Tran-Cor 65	M-10	Atha Champion	M-38
Armco 48 Ni	M-10	Armco Tran-Cor 72	M-10	Atha Chrome Roll	M-38
Armco 48 Ni-R	M-10	Armco Tran-Cor A-6	M-10	Atha No. 2500	M-38
Armco 317 ELC	M-10	Armco Tran-Cor M 14,	M-10	Atha No. 2600	M-38
Armco 430 ELC	M-10	Armco Tran-Cor T	M-10	Atha No. 2600	M-38
Armco Armature	M-10	Armco Tran-Cor X	M-10	Atha Rim Roll	M-38
Armco Beta 3	M-10	Armco Tran-Cor XX	M-10	Atlantaloy No. 12	M-695
Armco Electric	M-10	Armco Tran-Cor XXX	M-10	Atlantaloy No. 20	M-695
Armco Electromagnet Iron	M-10	Armco Type 410Se	M-10	Atlantaloy No. 21	M-695
Armco Field Grade	M-10	Armco X	M-10	Atlantaloy No. 22	M-695
Armco High Strength 55Y	M-10	Armco Zincgrip Paintgrip	M-10	Atlantaloy No. 23	M-695
Armco High Strength No. 1	M-10	Armorarc-B	M-692	Atlantaloy No. 24	M-695
Armco High Strength No. 2	M-10	Armorloy A-5	M-849	Atlantaloy No. 42	M-695
Armco High Strength No. 3	M-10	Arrestite	M-97	Atlantaloy No. 60	M-695
Armco High Strength No. 4	M-10	Arrow	M-73	Atlantic	M-353
Armco High Strength No. 6	M-10	Arrow (Latrobe)	M-73	Atlantic 33C	M-353
Armco High Strength No. 7	M-10	Arsenical Copper	M-33	Atlantic 44	M-353
Armco High Tensile	M-10	A.R. Stainless	M-153	Atlantic Die	M-435
Armco H T 50 Y	M-10	Art Die	M-35	Atlas 10	M-13
Armco Ingot Iron	M-10	A.S. Ack-Low	M-694	Atlas 20	M-13
Armco Intermediate Transformer	M-10	A.S. Bearcat	M-694	Atlas 89-A-3	M-13
Armco Oriented M-7	M-10	A.S. Blue Label	M-694	Atlas 89-A-207	M-13
Armco Oriented M-8	M-10	A.S. Brake & Die Steel	M-694	Atlas 89-A-303	M-13
Armco Oriented T	M-10	Ascoloy No. 44	M-1	Atlas 89-A-318	M-13
Armco Oriented TG	M-10	Ascoloy No. 55	M-1	Atlas 89-A-323	M-13
Armco Oriented T-S	M-10	Ascoloy No. 66	M-1	Atlas 89-E-103	M-13
Armco Radio 1	M-10	A.S. Cromat V	M-694	Atlas 89-E-106	M-13
Armco Radio 2	M-10	Asculoy	M-158	Atlas 89-E-108	M-13
Armco Radio 3	M-10	A.S. Duramold B	M-694	Atlas 89-E-117	M-13
Armco Radio 4	M-10	A.S. Hobbing Iron	M-694	Atlas 89-E-123	M-13
Armco Radio 5	M-10	A.S. Hollow Die	M-694	Atlas 89-E-133	M-13
Armco Radio 6	M-10	A.S. Hot Die	M-694	Atlas 89-E-135	M-13
Armco R A Type 434A	M-10	A.S. Lustre-Die	M-694	Atlas 89-E-203	M-13
Armco Special Electric	M-10	A.S. Nickel-Chrome Hobbing Steel	M-694	Atlas 90	M-13
Armco Ti-5Al-2 1/2 Sn	M-10	A.S. No. 15	M-694	Atlas B	M-1
Armco Ti-6Al-2Sn-4Zr-6Mo	M-10	A.S. No. 27	M-694	Atlas Double Extra	M-435
		A.S. No. 42	M-694	Atlas Hollow-Drill	M-435
		A.S. No. 46 Green Label	M-694	Atlas Iron No. 1	M-285
		A.S. No. 66	M-694	Atlas Iron No. 2	M-285
		A.S. No. 85	M-694	Atlas Iron No. 3	M-285
		A.S. No. 121	M-694	Atlas No. 50	M-76, M-1
		A.S. No. 670	M-694	Atlas No. 89 E-1	M-13
		A. Special	M-57	Atlas No. 89 E-3	M-13
				Atlas No. 89 E-4	M-13
				Atlas No. 89 E-5	M-13
				Atlas No. A-2	M-13
				Atlas Refined	M-435
				Atlas XXX	M-435
				Atrix 1	M-501
				Atrix 3R	M-501
				Atrix 4R	M-501
				Atrix 5R	M-501
				Atrix 10A	M-501

Atrix 10E	M-501	Auto Sumus	M-1337	AWX-45	M-537
Atrix 10S	M-501	Automotive Die	M-140	AWX-50	M-537
Atrix 10SS	M-501	Autovalve steel	M-32	AWX-55	M-537
Atrix 12	M-501	Autoxyd GuB	M-96	Axle Steel	M-618
Atrix 15A	M-501	Autoxyd 2-GuB	M-96	Axlo	M-499
Atrix 15E	M-501	Avesta	M-332	AZ31	M-325
Atrix 15S	M-501	Avesta 100	M-16	AZ31A	M-43
Atrix 15SS	M-501	Avesta 249	M-332	Azowit 25	M-1326
Atrix 20	M-501	Avesta 249 EH	M-16	Azowit 25S	M-1326
Atrix N10	M-501	Avesta 249H	M-332	Azowit 30M	M-1326
ATS	M-459	Avesta 249M	M-16	Azowit 30ML	M-1326
ATS2	M-459	Avesta 249S	M-16	Azowit 32M	M-1326
ATS6	M-459	Avesta 254 EV	M-16	Azowit 35	M-1326
ATS-103	M-459	Avesta 254 EVT	M-16	Azowit 35M	M-1326
ATS-113	M-459	Avesta 358	M-16	Azowit 35ML	M-1326
ATS-360	M-459	Avesta 393 H	M-16	Azowit 38M	M-1326
ATS-390	M-459	Avesta 393 S	M-16	Azowit 40	M-1326
Atsina	M-76	Avesta 739	M-16		
ATV-3	M-104	Avesta 739H	M-16	**B**	
ATV-R	M-104	Avesta 739HH	M-16		
ATV-S	M-104	Avesta 739S	M-16	B-24 Bronze	M-399
Aubert & Duval		Avesta 739SH	M-16	B-36	M-399
CNS3	M-1115	Avesta 831E	M-16	"B.50" Quality	M-179
Aubert & Duval		Avesta 831 H	M-16	B.A. 20	M-28
NC36	M-1115	Avesta 832	M-16	B.A. 23	M-28
Aubert & Duval		Avesta 832C	M-16	B.A. 24-MS	M-28
R12S	M-1115	Avesta 832 EMV	M-16	B.A. 29	M-28
Aubert & Duval		Avesta 832MVT	M-16	BA-31	M-28
VCD	M-1115	Avesta 832 P	M-16	BA 32	M-28
Audioloy	M-38	Avesta 832S	M-16	BA 33	M-28
Aufriloy	M-681	Avesta 832 SKE	M-16	B.A. 34	M-28
Auger	M-24	Avesta 832 SKTE	M-16	BA 35	M-28
Aur-O-Met 11	M-603	Avesta 832 SV/NB	M-16	B.A. 37	M-28
Aur-O-Met 12	M-603	Avesta 832 SVT	M-16	BA 40	M-28
Aur-O-Met 15	M-603	Avesta 832 T	M-16	B.A./40D Alloy	M-28
Aur-O-Met 52	M-603	Avesta HN	M-16	B.A./40J	M-28
Aur-O-Met 55	M-603	Avesta ST37	M-16	B.A. 40M	M-28
Aur-O-Met 57	M-603	AW.4V	M-50	BA 41	M-28
Aur-o-met 115	M-603	"A.W. 70-90" Type		BA 42	M-28
Aur-o-met 141	M-603	A	M-537	BA 45	M-28
Aur-o-met 305	M-603	A.W. 70-90 Type B	M-537	BA 46	M-28
Aur-o-met 305	M-603	AWCO-07	M-982	B.A./50 Alloy	M-28
Aur-o-met 315	M-603	AWCO-21	M-982	B.A. 160A Alloy	M-28
Aur-o-met 403	M-603	AWCO-24	M-982	BA 301	M-28
Aur-o-met 424	M-603	AWCO-25	M-982	BA 303	M-28
Aur-O-Met X-10	M-603	AWCO-27	M-982	BA 306	M-28
Aur-O-Met X-14	M-603	AWCO-28	M-982	BA 352	M-28
Autocrat	M-565	AWCO-31	M-982	BA 353	M-28
Autocrat Bushing		AWCO-35	M-982	BA 701	M-28
Bronze	M-565	AWCO-40	M-982	BA 704	M-28
Auto Die	M-260,	AWCO-45	M-982	BA 751	M-28
	M-38	AWCO-60	M-982	B.A.C. Brightray	M-121
Auto Extra N.C.	M-27	AWCO-301	M-982	Bain Alloy	M-90
Auto Extra NC		AWCO-303	M-982	Bain Bolt Steel	M-604
W/H	M-501,	AWCO-304	M-982	Bain Flange Steel	M-604
	M-912	AWCO-EP	M-982	Bain Steel	M-113
Auto First Quality		AWCO-SP	M-982	Bain Tube Steel	M-604
N.C.	M-27	AWCO-SP12	M-982	Bakadie	M-140
Auto NC W/h	M-501,	AWCO-SP16	M-982	Bakedie No. 2	M-140
	M-912	A.W. Dynalloy	M-537	Bal-Cut 1035	M-516
Auto No. 1	M-38	AW Dynalloy 50	M-537	Bal-Cut Steels	M-516
Auto Prima NC		A.W. Special	M-57	Bal-Cut X-1314	M-516
W/H	M-27	AW-Ten	M-537	Bal-Cut X-1335	M-516
Auto Spezial 2	M-1337	AW V-Steel	M-537	Baldwins P25	M-1417

Baldwins P25-12	M-1417	B & W No. 1300	M-17, M-447	Bergstahl AM17	M-1290
Baldwins P25-22	M-1417			Bergstahl AM17-13E	M-1290
Baldwins PEC	M-1417	B & W No. 1400	M-17, M-447	Bergstahl AM 18-8	M-1290
Baldwins PEH	M-1417			Bergstahl AM 858	M-1290
Baldwins PEL	M-1417	B & W No. 1500	M-17, M-447	Bergstahl AMG10	M-1290
Baldwins PET	M-1417			Bergstahl C1H	M-1290
Baldwins PKH	M-1417	B & W No. 5202	M-447	Bergstahl C2H	M-1290
Baldwins PKI	M-1417	Barbarite	M-19	Bergstahl CM1	M-1290
Baldwins PKL	M-1417	Baros	M-104	Bergstahl CM1B	M-1290
Baldwins PKM	M-1417	B-Arrow 1733	M-267	Bergstahl CM1H	M-1290
Baldwins PKN	M-1417	Baush A-5 Casting Metal	M-254	Bergstahl CM3	M-1290
Baldwins PMH	M-1417	Baush Duralumin Grade A	M-254	Bergstahl CMVW2	M-1290
Baldwins PML	M-1417			Bergstahl CMW1	M-1290
Baldwins PTL	M-1417	Baush Duralumin Grade B	M-254	Bergstahl CMW2	M-1290
Baltoc	M-260			Bergstahl CMW3	M-1290
Band File	M-38	Baxtron-DBA	M-522	Bergstahl CMW4	M-1290
B & K AZ-31A	M-1395	Baxtron-DBW	M-522	Bergstahl CN5	M-1290
B & W 12-14 Cr	M-447	B.A./"Y" Alloy	M-28	Bergstahl CN7	M-1290
B & W 16-3-3	M-17	B-Carbon Vanadium	M-76	Bergstahl CN8	M-1290
B & W 441	M-447			Bergstahl CN10	M-1290
B & W Alloy No. 1000	M-17, M-447	B.C. Hotwork	M-57	Bergstahl CN15	M-1290
		BC-No. 7	M-50	Bergstahl CN20	M-1290
		B.C. No. 8	M-50	Bergstahl CN35	M-1290
B & W Alloy No. 1100	M-17, M-447	BC-No. 8V	M-50	Bergstahl CN38	M-1290
		BC-No. 9	M-50	Bergstahl CV1	M-1290
		B.C. No. 10	M-50	Bergstahl CV1H	M-1290
B. & W. Nirosta KA-2	M-447	BC-No. 10V	M-1724	Bergstahl CV2	M-1290
		BC-No. 11	M-50	Bergstahl E2	M-1290
B. & W. Nirosta KA-2S	M-447	B.C. No. 12	M-50	Bergstahl E12	M-1290
		BC-No. 12	M-50	Bergstahl E16	M-1290
B & W No. 400	M-447	B.C. No. 14	M-50	Bergstahl E28	M-1290
B & W No. 401	M-447	B-D Cu	M-8	Bergstahl MNA1	M-1290
B & W No. 402	M-447	Bearium 82	M-431	Bergstahl MNA2	M-1290
B & W No. 420	M-447	Bearium B-6	M-431	Bergstahl MVW2	M-1290
B & W No. 445	M-447	Beaver	M-34	Bergstahl MVW4	M-1290
B & W No. 450	M-447	Beaver	M-253	Bergstahl MW2	M-1290
B & W No. 602	M-447	Beaver Babbitt	M-565	Bergstahl P6M	M-1290
B & W No. 603	M-447	Beaver D-2	M-52	Bergstahl P7M	M-1290
B & W No. 604	M-447	Beaver Die	M-115	Bergstahl SP120	M-1290
B & W No. 610	M-447	Beckett Metal	M-544	Bergstahl VCN170	M-1290
B & W No. 640	M-447	Belectric No. 0	M-707	Bergstahl VMS	M-1290
B & W No. 640S	M-447	Belectric No. 1	M-707	Berrydur	M-296
B & W No. 650	M-17, M-447	Belectric No. 2	M-707	Berrydur Contracid	M-296
		Belectric No. 3	M-707	Berrydur-Cu.	M-296
B & W No. 690	M-447	Belectromal	M-707	Berrydur-Cu.	M-296
B & W No. 692	M-447	B-Elite	M-459	Berylco 5	M-646
B & W No. 700	M-17, M-447	B-Elite HS	M-459	Berylco 50C	M-646
		B-Elite KVA	M-27	Berylco 225C	M-646
B & W No. 701	M-17, M-447	B-Elite KVA	M-459	Beryllium Copper 175	M-8
		Belmalloy	M-707		
B & W No. 900	M-447, M-17	Bemal-1	M-286	Beryl-Trode	M-13
		Bemal-2	M-286	Best	M-459
B & W No. 902	M-17, M-447	Benedict Metal	M-98	Best Cast E.S.C.-BC	M-50
		Benedict Nickel	M-8		
B & W No. 910	M-17, M-447	Benedict Nickel 812	M-8	Best (Disston)	M-357
		Beraloy 1	M-61	Beta Chisel Steel	M-85
B & W No. 912	M-17, M-447	Beraloy A	M-61	Bethadur 301	M-24
		Beraloy B	M-61	Bethadur 301X	M-24
B & W No. 950	M-17, M-447	Beraloy C	M-61	Bethadur 302B	M-24
		Beraloy D	M-61	Bethadur 304	M-24
B & W No. 951	M-447	Beraloy No. 25	M-61	Bethadur 305	M-24
B & W No. 1200	M-17, M-447	Bergstahl AM16-13N	M-1290	Bethadur 306	M-24
				Bethadur 307	M-24

Name	Ref	Name	Ref	Name	Ref
Bethadur 308	M-24	Bethlehem Hollow Drill	M-24	Birmabright BB 7	M-325
Bethadur 309	M-24			Birmametal BMB 761	M-321
Bethadur 310	M-24	Bethlehem H.V.	M-24		
Bethadur 316	M-24	Bethlehem Moco	M-24	Birmetal 230	M-321
Bethadur 317	M-24	Bethlehem Mokut	M-24	Birmetal-477	M-321
Bethadur 320	M-24	Bethlehem Moly	M-24	Birmetal-AM503	M-321
Bethadur 321	M-24	Bethlehem No. 1 Permanent Magnet	M-24	Birmetal-AZ31	M-321
Bethadur 345	M-24			Birmetal-ZW6	M-321
Bethadur 346	M-24			Bisco 18-4-1	M-24
Bethadur 347	M-24	Bethlehem No. 5	M-24	Bisco Airpro	M-334
Bethadur 348	M-24	Bethlehem No. 6 Nickel Steel	M-24	Bisco Annite No. 2	M-334
Bethadur 403	M-24			Bisco Best	M-334
Bethadur 405	M-24	Bethlehem No. 7	M-24	Bisco Die	M-334
Bethadur 410	M-24	Bethlehem No. 235	M-24	Bisco Tool Steel Tubing	M-334
Bethadur 414	M-24	Bethlehem No. 300	M-24		
Bethadur 425	M-24	Bethlehem Piston	M-24	Bisco Tool Steel Tubing	M-334
Bethadur 431	M-24	Bethlehem Pivot Steel	M-24		
Bethadur 440B	M-24			Bismo M-2	M-33
Bethadur 440 C	M-24	Bethlehem Shank Steel	M-24	Bismo M-3	M-24; M-334
Bethadur 501	M-24				
Bethadur 501C	M-24	Bethlehem Silvery Mayari Iron	M-24	Bit & Jar Steel	M-85
Bethadur 501D	M-24			B.K.L.	M-430
Bethadur No. 3	M-24	Bethlehem Special Gear Steel	M-24	BKS	M-459
Bethadur No. 302	M-24			BKS-3	M-459
Bethadur No. 430	M-24	Bethlehem Stainless Type A	M-24	BKS Extra	M-459
Bethadur No. 440	M-24			BKS Special	M-459
Bethadur No. 440-A	M-24	Bethlehem Stone Dressing	M-24	Black Jiant	M-24
Bethadur No. 442	M-24			Blackor	M-578
Bethadur No. 486	M-24	Bethlehem Superior Hollow Drill	M-24	Blanco F	M-1314
Bethalon 416	M-24			Blaw-Knox C-1	M-1222, M-112
Bethalon 430F	M-24	Bethlehem Tough	M-24		
Bethalon D	M-24	Bethlehem Tough M	M-24	Blaw-Knox C-1-0	M-1222, M-112
Bethalon No. 303	M-4				
Bethanized Products	M-24	Bethlehem XLC	M-24	Blaw-Knox C-1-1	M-1222, M-112
		Bethlehem XXX	M-24		
Bethlehem 5% Cr Air Hardening	M-24	Bethlehem XXX Special	M-24	Blaw-Knox C-1-2	M-1222, M-112
Bethlehem 6-6	M-24	BF 4-6 CrMo	M-139	Blaw-Knox C-1-3	M-112, M-1222
Bethlehem 6 Ni	M-24	BF 302 Stainless	M-139		
Bethlehem 33A	M-24	BF 316 Stainless	M-139	Blaw-Knox C-1-4	M-112, M-1222
Bethlehem 33B	M-24	BF 410 Stainless	M-139		
Bethlehem 33C	M-24	BF-CrMo	M-139	Blaw-Knox C-1-5	M-112, M-1222
Bethlehem 67 TAP	M-24	B-F H.T. Low Carbon	M-139		
Bethlehem 88-80	M-24			Blaw-Knox C-1-6	M-112, M-1222
Bethlehem 445	M-24	BF-MM	M-139		
Bethlehem Air Die	M-24	B-F N.C.M	M-139	Blaw-Knox C-2	M-1222
Bethlehem Alloy Hollow Drill Steel	M-24	BFS	M-24	Blaw-Knox C-2-A	M-112, M-1222
		"B-F" Tool	M-97		
Bethlehem Auger Drill	M-24	B-F Unique No. 20	M-97	Blaw-Knox C-3-A	M-112, M-1222
		BHS	M-459		
Bethlehem B-7	M-24	BIH	M-459	Blaw-Knox C-4	M-112, M-1222
Bethlehem B-7a	M-24	Bilame A	M-771		
Bethlehem B-14	M-24	Bilame AS	M-771	Blaw-Knox C-5	M-112, M-1222
Bethlehem BA-H	M-24	Bilame BC	M-771		
Bethlehem Bearing Cr-Mo	M-24	Binney Heat Resisting Alloy	M-25	Blaw-Knox C-6	M-112, M-122
Bethlehem BFS	M-24	Binney No. 71	M-25	Blaw-Knox C-6-M	M-112, M-1222
Bethlehem Broaching Steel	M-24	Binney No. 73	M-25		
		Birdsboro 50	M-136	Blue Anchor Drill Rod	M-342
Bethlehem Carbon	M-24	Birdsboro DA	M-136		
Bethlehem Cr-Mo-V-High V	M-24	Birdsboro Metal	M-136	Blue Chip Superior	M-57
		Birdsboro No. 26	M-136	Blue Devil 85	M-712
Bethlehem Extra Special	M-24	Birmabright BB 1-X	M-325	Blue Devil 100	M-712
				Blue Label	M-389

Blue Seal	M-510	Bohler CNW	M-27	Bohler Fox DUR	
BM	M-459	Bohler COH	M-27	600	M-27
B Monel	M-61, M-68	Bohler COK	M-27	Bohler Fox FB 30 S	M-27
		Bohler COM	M-27	Bohler Fox GFW	M-27
BMS	M-459	Bohler Crucible		Bohler Fox MSU	M-27
BMSC	M-459	Quality	M-27	Bohler Fox SAS 4	M-27
BMS Extra	M-459	Bohler CS	M-27	Bohler Fox SAS 8	M-27
BMSW	M-459	Bohler CSI	M-27	Bohler-Fox SFW	M-27
Bofors CR 33	M-1184	Bohler CV	M-27	Bohler Fox SpE	M-27
Bofors DR83	M-1184	Bohler DC 7	M-27	Bohler Fox UMZ	M-27
Bofors EO-43	M-1184	Bohler DCM 10	M-27	Bohler Fox WKZ	M-27
Bofors HRO564	M-1184	Bohler DCM 12	M-27	Bohler FSB	M-27
Bofors IRO-743	M-1184	Bohler DCM 15	M-27	Bohler G55	M-27
Bofors NN3R	M-1184	Bohler DCM54	M-27	Bohler GA	M-27
Bofors NR25	M-1184	Bohler DCM195	M-27	Bohler GCM0	M-27
Bofors NR45	M-1184	Bohler DCMV 7	M-27	Bohler GCN4	M-27
Bofors NROP23	M-1184	Bohler DCMV 20	M-27	Bohler Geodurit SH	M-27
Bofors R34	M-1184	Bohler DCMV 30	M-27	Bohler GMC	M-27
B.O.H.	M-345	Bohler DCMVW12	M-27	Bohler GMCA	M-27
Bohler	M-27	Bohler DM06	M-27	Bohler GMME	M-27
Bohler 2M	M-27	Bohler D Special	M-27	Bohler GMNE	M-27
Bohler 2NCMO	M-27	Bohler E2	M-27	Bohler GSF	M-27
Bohler 3NIMo	M-27	Bohler EB30	M-27	Bohler GSI	M-27
Bohler 3NM	M-27	Bohler EB80	M-27	Bohler H VII/VI	M-27
Bohler 3WKZ	M-27	Bohler EB95	M-27	Bohler Hard	M-27
Bohler 701	M-27	Bohler EB100	M-27	Bohler Hard Core	M-27
Bohler 751	M-27	Bohler EBK	M-27	Bohler Hart	M-27
Bohler 851	M-27	Bohler EBKW	M-27	Bohler HH	M-27
Bohler AC1	M-27	Bohler ECN 100	M-27	Bohler HSB	M-27
Bohler AC3	M-27	Bohler EMC	M-27	Bohler IMP	M-27
Bohler ACE	M-27	Bohler ENA	M-27	Bohler IN 1 A	M-27
Bohler ACo	M-27	Bohler EPB Extra	M-27	Bohler IN5A	M-27
Bohler AS-4	M-27	Bohler EPB Extra		Bohler IN8	M-27
Bohler AS-8	M-27	M	M-27	Bohler IN 10	M-27
Bohler Auto MS	M-27	Bohler EPB Prima	M-27	Bohler INOIL	M-27
Bohler Auto Special PA	M-27	Bohler ES	M-27	Bohler INSN	M-27
		Bohler ES2	M-27	Bohler Invar Steel	M-27
Bohler AWP	M-27	Bohler ESK	M-27	Bohlerit WC+Co, TiC+WC+Co	M-27
Bohler AZH	M-27	Bohler Extra H	M-27		
Bohler Bats	M-27	Bohler Extra Hard	M-27	Bohler Jar Steel	M-27
Bohler B-Elite	M-27	Bohler Extra K 5	M-27	Bohler K	M-27
Bohler B-Elite 18	M-27	Bohler Extra MG	M-27	Bohler K-3	M-27
Bohler B-Elite U	M-27	Bohler Extra Mittel-Hart	M-27	Bohler K-3 S	M-27
Bohler BH	M-27			Bohler K-100	M-27
Bohler BHC	M-27	Bohler Extra Rapid 300	M-27	Bohler K-100/1	M-27
Bohler BH-Extra	M-27			Bohler K-100 S	M-27
Bohler BHS	M-27	Bohler Extra Soft	M-27	Bohler K100 W/S	M-27
Bohler BHW	M-27	Bohler Extra Tough	M-27	Bohler KHMU	M-27
Bohler Blade Steel	M-27	Bohler Extra Weich	M-27	Bohler KK	M-27
Bohler BM	M-27	Bohler F 120	M-27	Bohler KMC	M-27
Bohler Boreas	M-27	Bohler F 145	M-27	Bohler Knife Steel	M-27
Bohler Brownie	M-27	Bohler FB30G	M-27	Bohler KP	M-27
Bohler Brownie Extra	M-27	Bohler FB30S	M-27	Bohler KPV	M-27
		Bohler FF	M-27	Bohler KR Special	M-27
Bohler BS2	M-27	Bohler FF	M-27	Bohler KW5	M-27
Bohler B-Special	M-27	Bohler FFB	M-27	Bohler KW-60	M-27
Bohler BW	M-27	Bohler FFBG	M-27	Bohler KW60-1	M-27
Bohler BW-VII	M-27	Bohler FFG	M-27	Bohler Lightning	M-27
Bohler BW-IX	M-27	Bohler First Quality	M-27	Bohler M751	M-27
Bohler BW-XII	M-27	Bohler FM Extra	M-27	Bohler M851	M-27
Bohler C/2	M-27	Bohler Fox A 7	M-27	Bohler Magnet Steel	M-27
Bohler CC Special	M-27	Bohler Fox DCMS	M-27	Bohler Martin Steel	M-27
Bohler CM Extra	M-27	Bohler Fox DMO	M-27	Bohler ME	M-27
Bohler CNME	M-27				

Bohler ME-6	M-27	Bohler Spezial Extra Hart	M-27	Bohler ZM	M-27
Bohler Middle Hard	M-27	Bohler Spezial ZAH	M-27	Bohler ZNM	M-27
Bohler Middle Hard 115	M-27	Bohler SPI	M-27	Bohler ZNM4	M-27
Bohler Mittel-Hart	M-27	Bohler SPU	M-27	Bohler ZRH	M-27
Bohler ML	M-27	Bohler SPV	M-27	Bohler ZRW	M-27
Bohler Molette	M-27	Bohler SSW	M-27	Bohler ZSV	M-27
Bohler MPA	M-27	Bohler Super Rapid	M-27	Bohnalite 2S	M-138
Bohler MPD Extra	M-27	Bohler Super Rapid Extra 214	M-27	Bohnalite 3S	M-138
Bohler MS45	M-27			Bohnalite 11S	M-138
Bohler MS60	M-27	Bohler Super Rapid Extra HV	M-27	Bohnalite 14S	M-138
Bohler MS70	M-27			Bohnalite 17S	M-138
Bohler MS85	M-27	Bohler SVM	M-27	Bohnalite 18S	M-138
Bohler MS90	M-27	Bohler SW	M-27	Bohnalite 24S	M-138
Bohler MSI	M-27	Bohler Tough	M-27	Bohnalite 25S	M-138
Bohler MS Steel	M-27	Bohler TW	M-27	Bohnalite 32S	M-138
Bohler MST	M-27	Bohler TWV	M-27	Bohnalite 53S	M-138
Bohler MYD	M-27	Bohler TWW	M-27	Bohnalite 61S	M-138
Bohler My Extra W	M-27	Bohler UF100	M-27	Bohnalite 63S	M-138
Bohler NAB	M-27	Bohler UM 1	M-27	Bohnalite 75S	M-138
Bohler Needle Die Steel	M-27	Bohler UM 2	M-27	Bohnalite A51S	M-138
		Bohler UM2M	M-27	Bohnalite B	M-138
Bohler NH	M-27	Bohler UMB	M-27	Bohnalite C	M-138
Bohler NI	M-27	Bohler US	M-27	Bohnalite E	M-138
Bohler NIP29	M-27	Bohler US 25	M-27	Bohnalite F	M-138
Bohler NIP36	M-27	Bohler USK	M-27	Bohnalite I	M-138
Bohler NIP50	M-27	Bohler V444D	M-27	Bohnalite J	M-138
Bohler No. 3 NW	M-27	Bohler VB 135	M-27	Bohnalite J-2	M-138
Bohler No. 4	M-27	Bohler VB150	M-27	Bohnalite K	M-138
Bohler No. 5 NM	M-27	Bohler VBS 135	M-27	Bohnalite L	M-138
Bohler No. 16	M-27	Bohler VBV140	M-27	Bohnalite L-2	M-138
Bohler No. 36 N	M-27	Bohler VCL240	M-27	Bohnalite L-3	M-138
Bohler No. 90	M-27	Bohler VCN400W	M-27	Bohnalite L-4	M-138
Bohler No. 711	M-27	Bohler VSI	M-27	Bohnalite L-6	M-138
Bohler NW	M-27	Bohler VSK	M-27	Bohnalite M-4	M-138
Bohler OFH70	M-27	Bohler W50	M-27	Bohnalite M-4A	M-138
Bohler PA2	M-27	Bohler W100	M-27	Bohnalite M-10	M-138
Bohler Panther	M-27	Bohler W 150	M-27	Bohnalite O	M-138
Bohler PAZ	M-27	Bohler WACE	M-27	Bohnalite O-2	M-138
Bohler PNA	M-27	Bohler WACV	M-27	Bohnalite R-1	M-138
Bohler Prima H	M-27	Bohler WB	M-27	Bohnalite S	M-138
Bohler Prima Hart	M-27	Bohler WD3	M-27	Bohnalite S-3	M-138
Bohler Prima Mittelhart 115	M-27	Bohler WD6	M-27	Bohnalite S-17	M-138
		Bohler WD15	M-27	Bohnalite S-25	M-138
Bohler PT 15	M-27	Bohler WD17	M-27	Bohnalite S-43	M-138
Bohler PV 35	M-27	Bohler WEICH	M-27	Bohnalite S-51	M-138
Bohler Rapid Steel	M-27	Bohler WFO	M-27	Bohnalite S-53	M-138
Bohler Remanenceless Steel	M-27	Bohler WH	M-27	Bohnalite U	M-138
		Bohler WKD	M-27	Bohnalite W	M-138
Bohler SAF	M-27	Bohler WKW2	M-27	Bohnalite W-3	M-138
Bohler SAS4MN	M-27	Bohler WKW2M	M-27	Bohnalite W-4	M-138
Bohler SC Extra	M-27	Bohler WKW6	M-27	Bohnalite W-5	M-138
Bohler SCV	M-27	Bohler WKW8	M-27	Bohnalite W-6	M-138
Bohler SK3	M-27	Bohler WKZ	M-27	Bohnalite X-1 S	M-138
Bohler SKVL	M-27	Bohler WKZ100	M-27	Bohnalite X-2	M-138
Bohler SMF	M-27	Bohler WM2	M-27	Bohnalite X-3	M-138
Bohler Special	M-27	Bohler WON	M-27	Bohnalite X-3 S	M-138
Bohler Special Extra MG	M-27	Bohler WPD	M-27	Bohnalite X-4	M-138
		Bohler WPZ	M-27	Bohnalite X-5	M-138
Bohler Special KRM	M-27	Bohler WV	M-27	Bohnalite X-6	M-138
		Bohler Z-II	M-27	Bohnalite X-7	M-138
Bohler Special KV	M-27	Bohler ZCS	M-27	Bohnalite X-11 S	M-138
Bohler Special W-43	M-27	Bohler ZE	M-27	Bohnalite Y	M-138
		Bohler ZK	M-27	Bohnalite Y-2	M-138
				Bohn Alloy 12-N	M-138

Section IV: Obsolete Alloys / 1685

Alloy	Ref	Alloy	Ref	Alloy	Ref
Bohn Alloy 16-N	M-138	Bohnolloy 70	M-138	Bolton 1% Tin Bronze	M-575
Bohn Alloy 20-N	M-138	Bohnolloy 70D	M-138	Bolton 1.5% Tin Phosphor Bronze	M-575
Bohn Alloy 25-N	M-138	Bohnolloy 74	M-138		
Bohn Alloy 60	M-138	Bohnolloy 75	M-138		
Bohn Alloy 66	M-138	Bohnolloy 79	M-138	Bolton 2.5% Tin Phosphor Bronze	M-575
Bohn Alloy 70	M-138	Bohnolloy 80	M-138		
Bohn Alloy 70 B	M-138	Bohnolloy 80C	M-138	Bolton No. 10 Phosphor Bronze	M-575
Bohn Alloy 76	M-138	Bohnolloy 82	M-138		
Bohn Alloy 78 B	M-138	Bohnolloy 82D	M-138	Bolton No. 11 Phosphor Bronze	M-575
Bohn Alloy 80	M-138	Bohnolloy 83	M-138		
Bohn Alloy 81	M-138	Bohnolloy 84	M-138	Bolton No. 12 Phosphor Bronze	M-575
Bohn Alloy 83	M-138	Bohnolloy 85	M-138		
Bohn Alloy 83 B	M-138	Bohnolloy 86	M-138	Bolton No. 14 Phosphor Bronze	M-575
Bohn Alloy 85	M-138	Bohnolloy 86D	M-138		
Bohn Alloy 85 B	M-138	Bohnolloy 87	M-138	Bolton No. 15 Phosphor Bronze	M-575
Bohn Alloy 88C	M-138	Bohnolloy 88	M-138		
Bohn Alloy 88 G	M-138	Bohnolloy 88C	M-138	Bolton No. 16 Phosphor Bronze	M-575
Bohn Alloy 88 M	M-138	Bohnolloy 89	M-138		
Bohn Alloy A	M-138, M-645	Bohnolloy 91	M-138	Boltons "Special" Bearing Metal	M-575
		Bohnolloy No. 2	M-138		
Bohn Alloy AX	M-138, M-645	Bohnolloy No. 4	M-138	Bonded Carbide	M-140
		Bohnolloy No. 5	M-138	Bonney-Floyd 4-6 Cr-Mo	M-139
Bohn Alloy C	M-138	Bohnolloy No. 6	M-138		
Bohn Alloy X	M-138	Bohnolloy No. 9	M-138	Bonney-Floyd Nirresist	M-139
Bohn Alloy XX	M-138	Bohnolloy No. 17 B	M-138		
Bohn Alloy XXA	M-138	Bohnolloy No. 45	M-138	Bonney-Floyd Stainless	M-139
Bohn Alloy XXX	M-138, M-645	Bohnolloy No. 66	M-138		
		Bohnolloy No. 70 A	M-138	Bonney-Floyd Stainless "N"	M-139
Bohn Alloy No. 90	M-138	Bohnolloy No. 70 C	M-138		
Bohn Alloy No. 90A	M-138	Bohnolloy No. 80 B	M-138	Bonney-Floyd Stainless "O"	M-139
		Bohnolloy No. 152	M-138		
Bohn Alloy No. 90H	M-138	Bohnolloy R-53	M-138	Bonney-Floyd Stainless "S"	M-139
		Bohnolloy R-54A	M-138		
Bohn Alloy No. 90M	M-138	Bohnolloy R54B	M-138	Booth 20S	M-338
		Bohnolloy R54C	M-138	Booth 20SA	M-338
Bohn Alloy No. 90S	M-138	Bohnolloy R-56C	M-138	Booth 40D	M-28
Bohn Alloy No. 90V	M-138	Bohnolloy R-57	M-138	Booth 76A	M-338
		Bohnolloy R-58	M-138	Booth A1	M-338
Bohn No. 7	M-138	Bohnolloy R-58E	M-138	Booth AD	M-338
Bohn No. 10-90	M-138	Bohnolloy R-59	M-138	Booth ADA	M-338
Bohn No. 15/85	M-138	Bohnolloy R-59B	M-138	Booth ALM	M-338
Bohn No. 20-80	M-138	Bohnolloy R-59C	M-138	Booth B1	M-338
Bohn No. 25-75	M-138	Bohnolloy R-59T	M-138	Booth B2C	M-338
Bohn No. 30-70 M	138	Bohnolloy R-60	M-138	Booth B76	M-338
		Bohnolloy R60M	M-138	Booth Cap Copper	M-338
Bohn No. 35-65	M-138	Bohnolloy R64R	M-138	Booth J5K	M-338
Bohn No. 40-60	M-138	Bohnolloy R65	M-138	Booth MG3	M-338
Bohn No. 45/55	M-138	Bohnolloy R66	M-138	Booth MG5	M-338
Bohn No. 50-50	M-138	Bohnolloy R68	M-138	Booth MG7	M-338
Bohn No. 60-40	M-138	Bohnolloy R69	M-138	Booth MV3	M-338
Bohn No. 95/5	M-138	Bohnolloy R-90	M-138	Booth T2	M-338
Bohn No. 97.5/2.5	M-138	Bohnolloy R-90A	M-138	Booth YDA	M-338
Bohn No. 100	M-138	Bohnolloy R-90 AA	M-138	Booth YDB	M-338
Bohn No. 100C	M-138	Bohnolloy R-90H	M-138	Bora	M-459
Bohn No. R-45	M-138	Bohnolloy R-90M	M-138	Bora 5	M-459
Bohn No. R-47	M-138	Bohnolloy R-90S	M-138	Bora 12	M-459
Bohnolloy-12	M-138	Bohnolloy R-90V	M-138	Bora 318	M-459
Bohnolloy 53	M-138	Bohnolloy R93	M-138	Bora Special	M-4
Bohnolloy 56	M-138	Bohnolloy W-90 VI	M-138	Bora Special M	M-459
Bohnolloy 57	M-138	Bokebit	M-345	Borcoloy Cr. 6	M-1004
Bohnolloy 58	M-138	Boker Power Chisel	M-345	Borcoloy Gr. 5	M-1004
Bohnolloy 58	M-138	Boker Special No. 847	M-345	Borcoloy Gr. 7	M-1004
Bohnolloy 59	M-138			Boron Deoxidized Copper 109	M-8
Bohnolloy 68	M-138	Boker Super Cobalt	M-345		

Boron-T	M-50	British Aluminum		B.S.C. "H.T.C."	M-1724	
Borotec 10009	M-717	No. 4	M-28	B.S.C. "H.T.C.N."	M-1724	
Bortam	M-109	British Aluminum		B.S.C. Med. "C."	M-1724	
Bosch (Cu-11)	M-304	No. 6	M-28	B.S.C. M.I.C.	M-1724	
Bound Brook	M-632	British Aluminum		B.S.C. M.I.C.4	M-1724	
Bourne Fuller Air Hardening	M-379, M-97	No. 6A	M-28	B.S.C. M.I.C.8	M-1724	
		British Aluminum No. 12	M-28	B.S.C. "N.C."	M-1724	
Bourne-Fuller "H-C"	M-379, M-97			B.S.C. "Nilex"	M-1724	
		Brittania Metal-1	M-332	B.S.C. "S.H.N.C."	M-1724	
		Brittania Metal-2	M-332	B.S.C. "Si-Cr"	M-1724	
Bow and Arrow	M-233	Brittania Metal-3	M-332	B.S.C. "Si-Mn"	M-1724	
B-Quality	M-618	Broaching	M-24	B.S.C. Super C 12	M-1724	
Braeburn-BDM	M-140	Bronzend	M-677	B.S.C. Super "C.H.N.C."	M-1724	
Braeburn Cobalt	M-140	Bronzend E	M-677			
Braeburn Extra	M-140	Bronzend H	M-677	B.S.C. Super "S.H.N.C."	M-1724	
Braeburn High Speed	M-140	Bronzend P	M-677			
		Bronzochrom No. 185	M-717	B.S.C. Supertough "A"	M-1724	
Braeburn M-7	M-140					
Braeburn M-10	M-140	Bronzochrom No. 186	M-717	B.S.C. Supertough "B"	M-1724	
Braeburn S.O.D.	M-140					
Braeburn Special	M-140	Bronzochrom No. 187	M-717	B.S.C. Supertough "C"	M-1724	
Braeburn Stainless	M-140					
Braeburn Standard	M-140	Bronzstox	M-214	B.S.C. Supertough "C-20"	M-1724	
Braeburn T-15	M-140	Broo-Zinc	M-996			
Brae-Cast	M-140	Brown & Sharpe Water Hardening	M-1172	B.S.C. Super "V.N.C.A"	M-1724	
Braemow	M-140					
Braemow Special	M-140	Brown Bailey BB4K-TI	M-472	B.S.C. "V.A.65"	M-1724	
Braetuf	M-140			B.S.C. V.A.P.	M-1724	
Braevac 718	M-140	Brown Bayley BBH	M-472	B.S.C. "V.C.M."	M-1724	
Brake Die	M-140	Brown Bailey QS	M-472	B.S.C. "V.H.R.D."	M-1724	
Brass, Common High	M-8	Brown-Bayleys BB 2 K	M-472	B.S.C. "V.N.C.A."	M-1724	
				B.S.C. "V.N.C.G."	M-1724	
Brass, Deep Drawing	M-8	Brown-Bayleys No. 33	M-472	BSEM 558	M-106	
				BSEM 558	M-106	
Brassoid	M-538	Brown Baileys QSS	M-472	BSF	M-459	
Brass Tin	M-538	B.R.S.	M-499	BSF; DMOC	M-459	
Brastil	M-158	Brush 240-C	M-1252	BSI 30 VMS	M-1290	
Braze 251	M-63	Brymill BRM-2	M-478	BSI AM 12	M-1290	
Braze 752	M-63	Brymill BRM-4	M-478	BSI AM 16.13 S	M-1290	
Braze-Clad	M-771	Brymill CH-1	M-478	BSI AM 16.13 SS	M-1290	
Braze AE	M-63	Brymill CH-10	M-478	BSI AM 16.16 ES	M-1290	
Brearley A	M-472	Brymill X	M-478	BSI AM 16.16 ESS	M-1290	
Brearley B	M-472	BS 1	M-459	BSI AM G 10	M-1290	
Brearley B.B.H.	M-472	BS 4	M-459	BSI CM 1B	M-1290	
Brearley C	M-472	B.S.C. 3 NS	M-1724	BSI GS-38D	M-1290	
Brearley "K"	M-472	B.S.C. 3 1/2 NS	M-1724	BSI GS-45D	M-1290	
Brearley SI	M-472	B.S.C. 5 CC	M-1724	BSI GS-C 25	M-1290	
Bright Extruded Bronze	M-33	B.S.C. "A"	M-1724	BSI Ni 2	M-1290	
		B.S.C. "A-31"	M-1724	BSI Ni 3	M-1290	
Brightray A.	M-121	B.S.C. A.W.	M-1724	BSI SHF	M-1290	
Brightray A	M-121	B.S.C. "B.C.T.B."	M-1724	BSO Spezial	M-459	
Brightray N	M-121	B.S.C. C 12	M-1724	BSW	M-459	
Brightray-H	M-121	B.S.C. C.D.	M-1724	BSW, DMOCN	M-459	
Brightway H	M-121	B.S.C. "C.H.-2N"	M-1724	Buderus RCW2	M-501	
Brilliant AXL	M-52	B.S.C. "C.H.-3N"	M-1724	Buderus RSZ	M-501	
Brilliant WKE	M-52	B.S.C. "C.H.-5 N"	M-1724	Buderus RSZ Especial	M-501	
Brilliant WKE Extra	M-52	B.S.C. "C.H.M.S."	M-1724			
		B.S.C. "C.H.N.C."	M-1724	Buderus RT9	M-501	
Brimco Bronze	M-1579	B.S.C. "C.H.N.M."	M-1724	Buderus RT12	M-501	
Brimcolloy-100	M-1579	B.S.C. "C.O.M.O."	M-1724	Buderus RTC20	M-501	
Brimcolloy-300	M-1579	B.S.C. C.V.M.	M-1724	Buderus RVS	M-501	
Brimcolloy-400	M-1579	B.S.C. CVM2	M-1724	Buderus T76	M-501	
		B.S.C. CVM3	M-1724	Buderus TRW2	M-501	
		B.S.C. H.S.M.	M-1724	Bunting 102	M-720	

Alloy	Code	Alloy	Code	Alloy	Code
Bunting 200	M-720	Callinite ST-3	M-725	Carbon-Ford	M-327
Bunting 202	M-720	Callinite TC-1	M-725	Carbon Magnet	M-60
Bunting No. 51	M-720	Callinite TC-2	M-725	Carbon-Molybdenum Steel	M-176
Bunting No. 78	M-720	Callinite TC-3	M-725		
Bunting No. 116	M-720	Calmalloy No. 1	M-31		
Bunting No. 122	M-720	Calmalloy No. 2	M-31	Carbon Moly Steel	M-139
Bunting No. 143	M-720	Calmaloy	M-60	Carbon Shoe Die	M-38
Bunting No. 158	M-720	Calmolloy	M-725	Carbon Tool Double Extra	M-85
Bunting No. 161	M-720	Calorite 1	M-60		
Bunting No. 162	M-720	Calorite 2	M-60	Carbon Vanadium	M-604
Bunting No. 170	M-720	Calorized 1% Mo Steel	M-30	Carbon-Vanadium	M-85
Bunting No. 178	M-720			Carbo Tool	M-334
Bunting No. 183	M-720	Calorized DM Steel	M-30	Carend	M-677
Bunting No. 188	M-720	Calorized Steel	M-30	Carilloy Carbon-Manganese	M-604
Burden Best	M-721	Calumet	M-38		
Burgess-Parr No. 85 Alloy	M-29	Calumetal NE	M-1166	Carilloy Carbon-Manganese	M-604
		Calumet Box and Pin	M-38		
Burndy No. 111	M-722			Carilloy Carbon-Manganese	M-604
Burndy No. 113	M-722	Calumet BR	M-1166		
Burndy No. 113LM	M-722	Calumet GH	M-1166	Carilloy Carbon-Manganese-Copper	M-604
Burndy No. 202	M-722	Calumet GR	M-1166		
Burndy No. 206	M-722	Cambriloy 2	M-727		
Burndy No. 302	M-722	Cambriloy 4	M-727	Carilloy Carbon-Molybdenum	M-604
Burndy No. 308	M-722	Cambriloy 5	M-727		
Burndy No. 309	M-722	Cambriloy 17	M-727	Carilloy Carbon-Molybdenum	M-604
Burndy No. 314	M-722	Cambriloy 25-12	M-727		
Burndy No. 316	M-722	Cambriloy 25-20	M-727	Carilloy Carbon-Molybdenum	M-604
Bush Brand Bearing Metal	M-575	Cambriloy A.	M-727		
		Camite	M-149	Carilloy Carbon-Molybdenum	M-604
Bush Hammer	M-38	Camloy	M-490		
Buster Brand	M-35	"CA" Nickel	M-121	Carilloy Carbon-Molybdenum	M-604
BW XII	M-459	Canon	M-1043		
B-Y Hot Work	M-114	Canon Superior	M-1043	Carilloy Carbon-Molybdenum	M-604
Byloy GR. W2	M-292	Canzler Brass	M-728		
		Canzler No. 1	M-728	Carilloy Carbon-Vanadium	M-604
C		Canzler No. 4	M-728		
		Capaloy	M-1028	Carilloy Carbon-Vanadium	M-604
C-55	M-64	Carbidie CD-355	M-1695		
"C63" Quality	M-179	Carboloy	M-31	Carilloy Carbon-Vanadium	M-604
"C90" Quality	M-179	Carboloy 77A	M-31		
"C92" Quality	M-179	Carboloy 78C	M-31	Carilloy Carbon-Vanadium	M-604
C-207	M-31	Carboloy 119A	M-31		
CA Double Diamond	M-38	Carboloy 608	M-31	Carilloy Carbon-Vanadium	M-604
		Carboloy 715	M-31		
Calido	M-47	Carboloy 831	M-31	Carilloy Carbon-Vanadium	M-604
Calido (Elalco)	M-44	Carboloy 831A	M-31		
Calido (Elalco)	M-47	Carboloy 906	M-31	Carilloy Chromium-Manganese-Silicon	M-604
Calite B29	M-30	Carboloy 958	M-31		
Calite BL	M-30	Carboloy 1078A	M-31		
Calite BL-28	M-30	Carboloy 1570	M-31	Carilloy Chromium-Manganese-Silicon	M-604
Calite C	M-30	Carboloy A-44	M-31		
Calite D	M-30	Carboloy Gr. 160	M-31		
Calite F	M-30	Carboloy Gr. 330	M-31	Carilloy Chromium-Molybdenum	M-604
Calite N2	M-30	Carboloy Gr. M-252	M-31		
Calite NCT3	M-30	Carbon	M-44	Carilloy Chromium-Vanadium	M-604
Calite Nirosta KA2	M-30	Carbon	M-90		
Calite Nirosta KA4	M-30	Carbon 1/2 Mo	M-90	Carilloy High-Carbon Low-Chromium	M-604
Calite R	M-30	Carbon Chisel	M-334		
Calite S	M-30	Carbon Cold Header No. V	M-73		
Calite S28	M-30			Carilloy High-Chromium-Molybdenum	M-604
Calliflex	M-725	Carbon Cold Header with Mo	M-73		
Callinite	M-725				
Callinite ST-1	M-725	Carbon Cold Header with V	M-73	Carilloy High Chromium Nickel	M-604
Callinite ST-2	M-725				

Carilloy High Nickel	M-604	
Carilloy High Nickel	M-604	
Carilloy High Nickel	M-604	
Carilloy High Nickel	M-604	
Carilloy High Nickel	M-604	
Carilloy High Nickel	M-604	
Carilloy Low Nickel Chromium	M-604	
Carilloy Low Nickel Chromium	M-604	
Carilloy Low Nickel Chromium	M-604	
Carilloy Low Nickel Chromium	M-604	
Carilloy Low Nickel Chromium	M-604	
Carilloy Manganese-Molybdenum	M-604	
Carilloy Manganese-Molybdenum	M-604	
Carilloy Manganese-Molybdenum	M-604	
Carilloy Manganese-Molybdenum	M-604	
Carilloy Manganese-Molybdenum	M-604	
Carilloy Manganese-Vanadium	M-604	
Carilloy Medium Nickel	M-604	
Carilloy Medium Nickel	M-604	
Carilloy Nickel-Chromium-Molybdenum	M-604	
Carilloy Nickel-Chromium-Molybdenum	M-604	
Carilloy Nickel-Chromium-Molybdenum	M-604	
Carilloy Nickel-Chromium-Molybdenum-Vanadium	M-604	
Carilloy Silicon-Vanadium	M-604	
Carmet ACA-1	M-1	
Carmet CA-22	M-1	
Carnegie-Illinois Spec. No. 2	M-604	
Carnegie-Illinois Spec. No. 3	M-604	
Carnegie-Illinois Spec. No. 4	M-604	
Carnegie-Illinois Spec. No. 5	M-604	
Carnegie-Illinois Spec. No. 6	M-604	
Carnegie-Illinois Spec. No. 7	M-604	
Carnegie-Illinois Spec. No. 13	M-604	
Carnegie-Illinois Spec. No. 19	M-604	
Caroga	M-1	
Caroph Carbon	M-85	
Carpaloy No. 1	M-32	
Carpaloy No. 2	M-32	
Carpaloy No. 3	M-32	
Carpenter 7Mo	M-32	
Carpenter 12-Alfenol	M-32	
Carpenter 16-25-6	M-32	
Carpenter 45-5	M-32	
Carpenter 49	M-32	
Carpenter B	M-32	
Carpenter C	M-32	
Carpenter Chrome Bearing Steel	M-32	
Carpenter Chrome Magnet	M-32	
Carpenter Chrome Steel No. 12S	M-32	
Carpenter DD	M-32	
Carpenter G.V.L.	M-32	
Carpenter H-9 Extra	M-32	
Carpenter High Expansion Alloy	M-32	
Carpenter High Nickel	M-32	
Carpenter Hi Wear 64	M-32	
Carpenter Hy-RA49	M-32	
Carpenter Illium R	M-32	
Carpenter Jason Steel No. 12-324	M-32	
Carpenter J. Y. Steel No. 656	M-32	
Carpenter KR Type B	M-32	
Carpenter Low Expansion 49	M-32	
Carpenter MEL-TROL K-W	M-32	
Carpenter Midas Steel No. 9-961	M-32	
Carpenter Nimark II	M-32	
Carpenter No. 1 Turbine Blade	M-32	
Carpenter No. 2-317	M-32	
Carpenter No. 2-408	M-32	
Carpenter No. 2-720	M-32	
Carpenter No. 2 Samson	M-32	
Carpenter No. 3-314	M-32	
Carpenter No. 3-317	M-32	
Carpenter No. 3-427	M-32	
Carpenter No. 3-547	M-32	
Carpenter No. 3-720	M-32	
Carpenter No. 4	M-32	
Carpenter No. 4-317	M-32	
Carpenter No. 4-408	M-32	
Carpenter No. 5-427	M-32	
Carpenter No. 5-720	M-32	
Carpenter No. 5-876	M-32	
Carpenter No. 11 Extra	M-32	
Carpenter No. 11 Extra Vanadium	M-32	
Carpenter No. 11 Nitro	M-32	
Carpenter No. 11 Special Vanadium	M-32	
Carpenter No. 30	M-32	
Carpenter No. 37-7FM	M-32	
Carpenter No. 111	M-32	
Carpenter No. 200	M-32	
Carpenter No. 314	M-32	
Carpenter No. 426	M-32	
Carpenter No. 427	M-32	
Carpenter No. 436	M-32	
Carpenter No. 478	M-32	
Carpenter No. 482	M-32	
Carpenter No. 484-FM	M-32	
Carpenter No. 492	M-32	
Carpenter No. 500	M-32	
Carpenter No. 500 Extra Special	M-32	
Carpenter No. 547	M-32	
Carpenter No. 709 Type 1	M-32	
Carpenter No. 720	M-32	
Carpenter No. 871	M-32	
Carpenter No. 872	M-32	
Carpenter No. 874	M-32	
Carpenter Permanent Magnet Type A	M-32	
Carpenter Permanent Magnet Type B	M-32	
Carpenter Permanent Magnet Type C	M-32	
Carpenter Permanent Magnet Type D	M-32	
Carpenter Permanent Magnet Type E	M-32	
Carpenter Permanent Magnet Type F	M-32	
Carpenter Permanent Magnet Type G	M-32	
Carpenter Samson No. 1	M-32	
Carpenter Samson No. 2	M-32	

Section IV: Obsolete Alloys / 1689

Alloy	Code
Carpenter Samson No. 3	M-32
Carpenter Samson No. 4A	M-32
Carpenter Samson No. 4B	M-32
Carpenter Samson No. 4C	M-32
Carpenter Samson No. 5	M-32
Carpenter Sil No. 10	M-32
Carpenter Special	M-32
Carpenter Stainless 20 Cb	M-32
Carpenter Stainless 304+B	M-32
Carpenter Stainless No. 1 Jr.	M-32
Carpenter Stainless No. 2-B	M-32
Carpenter Stainless No. 2 FM	M-32
Carpenter Stainless No. 4A	M-32
Carpenter Stainless No. 4B	M-32
Carpenter Stainless No. 4C	M-32
Carpenter Stainless No. 4 Cb	M-32
Carpenter Stainless No. 4 Ti	M-32
Carpenter Stainless No. 6	M-32
Carpenter Stainless No. 6-20	M-32
Carpenter Stainless No. 6 FM	M-32
Carpenter Stainless No. 7 (Type 329)	M-32
Carpenter Stainless No. 8	M-32
Carpenter Stainless No. 8 Se	M-32
Carpenter Stainless No. 12	M-32
Carpenter Stainless No. 347 FM	M-32
Carpenter Stainless No. D-1	M-32
Carpenter Stainless Steel No. 1	M-32
Carpenter Stainless Steel No. 3	M-32
Carpenter Stainless Steel No. 4	M-32
Carpenter Stainless Steel No. 7	M-32
Carpenter Stainless Steel No. 20	M-32
Carpenter Ten Star	M-32
Carpenter TGS	M-32
Carpenter Titanium 6-4	M-32
Carpenter Titanium 6-6-2	M-32
Carpenter Type A	M-32
Carpenter Type D	M-32
Carpenter V-Chrome Steel	M-32
Carrilloy Nickel Steel	M-604
Carrilloy Nickel-Vanadium-Molybdenum	M-604
Carrilloy Structural Manganese	M-604
Carrilloy Structural Manganese	M-604
Carrilloy Structural Nickel	M-604
Carrilloy Structural Nickel	M-604
Carrilloy Structural Nickel	M-604
Carrilloy Structural Nickel	M-604
Carrilloy Structural Steel	M-604
Carrilloy Structural Steel	M-604
Carvan	M-24
Casona	M-1411
Cassilloy	M-1569
Castdie	M-35
Catawba	M-69
Cathaloy-1	M-945
Cathaloy-2	M-945
Cathaloy A-30	M-945
Cathaloy A32	M-945
Cathermalite	M-69
Causal Metal	M-264
Cb-10W-5Zr	M-167
Cb-752	M-167
Cb-753	M-167
C.C.A.	M-38
C.C. Chisel	M-260
C.C.S. Die	M-38
C.C.S. Die Casting	M-362
CDC Alloy 50	M-1187
CDC Manganese Alloy No. 730	M-1187
C.D.C. No. 595	M-1187
CDC VAN-AD	M-1187
Cecolloy	M-474
Cecolloy C	M-474
C.E.C. Smooth Cut	M-35
Celero	M-357
Celfor	M-618
Cellini	M-24
Cello-Vanadium	M-81
Celsit	M-27
Celsit K	M-27
Celsit SEO	M-27
Celsit V	M-27
Celsit V-300	M-459
Celsit V-400	M-459
Celsit VEO	M-459
Cenco	M-215
Central Pure Iron	M-97
Centra Steel	M-690
Ceralloy 100	M-1497
Ceralloy 100F	M-1497
Ceramiseal	M-61
C.F.S. No. 1	M-73
Chace No. 307	M-582
Chace No. 720	M-582
Chace No. 772	M-582
Chamax O	M-1116
Chamet Bronze B	M-33
Chamet Bronze, Type A	M-33
Champagnole C7	M-1116
Champagnole C.8.	M-1116
Champagnole C.8.V.M.	M-1116
Champagnole C9	M-1116
Champagnole C10	M-1116
Champagnole C.12	M-1116
Champagnole Cr. 3	M-1116
Champagnole CRED	M-1116
Champagnole T1	M-1116
Champagnole T5	M-1116
Champagnole T.E.D.C. Steel	M-1116
Champaloy No. 2	M-38
Champion No. 48-1	M-712
Champion No. 58-12	M-712
Champion No. 64-6	M-712
Champion Non-Changeable	M-38
Champion Tool	M-38
Channeler	M-377
Channeller	M-24
Chase 5 Re-50 Mo.	M-33
Chase 75W-25 Re	M-33
Chase 149 Bronze	M-33
Chase "444" Bronze	M-33
Chase Alloy No. 58	M-33
Chase Bright Extruded Bronze	M-33
Chase Cupro-Nickel 80-20	M-33
Chase High Strength Tank Brass	M-33
Chase Nickel Aluminum Bronze	M-33
Chase Nickel Silver (12%)	M-33
Chase Nickel Silver 20%	M-33
Chase Telnic Bronze	M-33
Chase Valve Stem Bronze	M-33
Chatillon 3100	M-1117

Chatillon 3308	M-1117	Chroman "Bo."	M-296	Chrome Vanadium	
Chatillon 3333	M-1117	Chroman "C."	M-296	Tool	M-140
Chatillon 3400	M-1117	Chroman "Co."	M-296	Chromeweld 4-6	M-578
Chatillon 3408	M-1117	Chroman "D."	M-296	Chromic-A	M-98
Chatillon 5250	M-1117	Chroman "E."	M-296	Chromic-C	M-98
Chatillon 5650	M-1117	Chromang	M-677	Chromium Boron	M-394
Chatillon 5654	M-1117	Chromax	M-162	Chromium Carbide	
Chatillon 5655	M-1117	Chromax (Cast)	M-44	Gr. CR-1	M-57
Chatillon 5755	M-1117	Chrome	M-1405	Chromium Carbide	
Chatillon 5815	M-1117	Chrome 3 1/2%	M-1076	Gr. CR-2	M-57
Chatillon 5820	M-1117	Chrome 6%	M-1076	Chromium Carbide	
Chatillon 5830	M-1117	Chrome Ball Steel	M-260	Gr. CR-3	M-57
Chatillon MF No. 1	M-1117	Chrome Bearing		Chromium Magnet	M-60
Chatillon MN Si	M-1117	Steel	M-260	Chromium	
Chatillon TSM No.		Chrome-Copper-		Permalloy	M-22
2	M-1117	Nickel Steel	M-176	Chromoloy	M-31
Chatillon W 18	M-1117	Chrome Drill Rod	M-38	Chronifer Spezial	M-1730
Checkno	M-334	Chrome Firminy		Chronit 216U	M-1309
Checkno No. 1	M-24	No. 0	M-1043	Chronit 1603	M-1309
Checkno No. 1	M-334	Chrome Gummer		Chronit 1810MN	M-1309
Checkno No. 2	M-24	File	M-38	Chronit 1811N	M-1309
Checkno No. 2	M-334	Chrome Hot Die	M-373	Chronit CR150H	M-27
Checkno No. 3	M-24	Chromel B	M-65	Chronit CR150W	M-27
Checkno No. 3	M-334	Chromel R	M-65	Chronit CSFG	M-27
Chemalloy A28C	M-47	Chrome Magnet	M-115	Chronit Spezial KG	M-27
Chemalloy A28F	M-47	Chrome Magnet		Chronitherm 20	M-1309
Chemalloy A28M	M-47	M-31	M-260	Chronitherm	
Chemalloy A32	M-47	Chrome Magnet		20/Spez.	M-1309
Chemalloy A32M	M-47	No. 154	M-260	Chronitherm 30	M-1309
Chemalloy A38	M-47	Chrome Manganese		Chronitherm	
Chemalloy A38M	M-47	Steel	M-176	30/Spez.	M-1309
Chemalloy A45	M-47	Chrome Moly Roll	M-260	Chronitherm 60	M-1309
Chemalloy A45N	M-47	Chromend 8/18	M-677	Chronitherm	
Chemalloy A50N	M-47	Chromend 12C	M-677	60/Spez.	M-1309
Chemalloy A52N	M-47	Chromend 13/60	M-677	Chronitherm 80	M-1309
Chemalloy F6	M-47	Chromend 15/85	M-677	Chronitherm	
Chemalloy F8	M-47	Chromend 16/7	M-677	80/Spez.	M-1309
Chemalloy F12	M-47	Chromend 20/80	M-677	Cimet, Cast	M-44
Chemalloy F12F	M-47	Chromend 25/3Mo	M-677	Cimet (Malleable)	M-44
Chemalloy F20	M-47	Chromend 28/3 Mo	M-677	Circle L 1	M-74
Chemalloy F32F	M-47	Chromend HN	M-677	Circle L 2	M-74
Chemalloy H1	M-47	Chromend KS	M-677	Circle L 4	M-74
Chemalloy H2	M-47	Chromend Special	M-677	Circle L 6	M-74
Chemalloy H3	M-47	Chromend W	M-677	Circle L 11	M-74
Chemalloy H4	M-47	Chrome Nickel		Circle L 14	M-74
Chemalloy N1	M-47	Skate Blade	M-38	Circle L 16	M-74
Chemalloy N2	M-47	Chrome Vanadium		Circle L 18	M-74
Chemalloy N3	M-47	Chisel	M-38	Circle L-19-9DL	M-74
Chemalloy N4	M-47	Chrome Vanadium		Circle L 22Ag	M-74
Chemalloy N5	M-47	D	M-140	Circle L 23M	M-74
Chevis	M-1405	Chrome Vanadium		Circle L 23XM	M-74
Chevre No. 2 1/2	M-1117	Die	M-85	Circle L 24	M-74
Chevre No. 3	M-1117	Chrome Vanadium		Circle L 25M	M-74
Chimo	M-57	Die Casting	M-275	Circle L 30Cb	M-74
Chippaway	M-69	Chrome Vanadium		Circle L 30XM	M-74
Chipper Knife	M-357	G	M-140	Circle L 32XMC	M-74
Choice No. 1	M-115	Chrome Vanadium		Circle L 34S	M-74
Choice No. 2	M-115	H	M-140	Circle L 40	M-74
Chromalloy	M-114	Chrome Vanadium		Circle L 42	M-74
Chromaloid	M-538	K	M-140	Circle L 43	M-74
Chroman Alloy	M-296	Chrome Vanadium		Circle L 44	M-74
Chroman "Ao."	M-296	Screw Driver	M-38	Circle L 45	M-74
Chroman "B."	M-296	Chrome Vanadium		Circle L 47	M-74
Chroman B2 Mo	M-296	Spring Wire	M-260	Circle L 48	M-74

Section IV: Obsolete Alloys / 1691

Name	Code	Name	Code	Name	Code
Circle L 106	M-74	Climax	M-44	Colmonoy Overlay	M-963
Circle L 119	M-74	Climax Machinery	M-38	Colmonoy Special	
Circle L 155	M-74	Climelt		No. 1	M-963
Circle L 230	M-74	Molybdenum		Colmonoy Special	
Circle L 330	M-74	0.5% Titanium	M-151	No. 4	M-963
Circle L 430	M-74	Clydall 5 Special	M-1569	Colmonoy WCR	
Circle L 446	M-74	Clydall 12 Special	M-1569	100	M-963
Circle L-C	M-74	Cly-Die	M-233	Colmonoy WCR	
Circle L-CD	M-74	Clydmo	M-1569	200	M-963
Circle L-D	M-74	CMC	M-459	Colmonoy WCR	
Circle L No. 7	M-74	CMP(3)	M-533	300	M-963
Circle L No. 21	M-74	CMS	M-459	Colmonoy WCR	
Circle L No. 30	M-74	C.M. Tap	M-115	400	M-963
Circle N No. 25	M-74	C.N. Die	M-345	Colona	M-34
Circular Saw Bit	M-38	CNK2	M-459	Colonial 795F	M-115
Circular Saw Plate		C N M	M-200	Colonial Bit Steel	M-34
55	M-38	C.N.S.	M-69	Colonial C-2	M-34,
Circular Saw Plate		CNS2H	M-459		M-115
B	M-38	CNS-TH	M-459	Colonial C-2F	M-34,
Circular Saw Plate		Coast Metal	M-1060		M-115
C	M-38	Coast No. 6	M-1060	Colonial FMS	M-34,
CK 15	M-459	Coast No. 11	M-1060		M-115
CK 22	M-459	Coast No. 106	M-1060	Colonial Gripper	M-34
CK 35	M-459	Coast No. 109	M-1060	Colonial H-H No. 3	M-34
CK 45	M-459	Cobaflux	M-24	Colonial High	
CK 60	M-459	Cobaflux "B"	M-24	Speed	M-34
Clad R-303	M-671	Cobaflux Magnet A	M-24	Colonial Jar Steel	M-34
Clarite HW	M-35	Cobalt I	M-459	Colonial No. 3	M-34
Clarite HW50	M-35	Cobalt II	M-459	Colonial No. 4	M-34,
Clarite-L	M-35	Cobalt III	M-459		M-115
Clarite M	M-35	Cobalt III	M-459	Colonial No. 14	
Cletaloy 10-BC	M-738	Cobalt III, NX	M-459	Special	M-34
Cletaloy 10-CC	M-738	Cobalt 17%	M-1076	Colonial No. 35	M-34
Cletaloy 10-CSS	M-738	Cobalt 36%	M-1076	Colonial No. 36	M-34
Cletaloy 10-CT	M-738	Cobalt 125	M-459	Colonial No. 100	M-1393
Cletaloy 10-CTA	M-738	Cobalt 160	M-459	Colonial No. 101	M-1393
Cletaloy 20-BC	M-738	Cobalt 200	M-459	Colonial No. 110	M-1393
Cletaloy 20-CC	M-738	Cobalt 300	M-459	Colonial No. 115	M-1393
Cletaloy 20-CT	M-738	Cobalt Aluminum		Colonial No. 120	M-1393
Cletaloy 20-CTA	M-738	Bronze	M-1228	Colonial No. 123	M-1393
Cletaloy 30-BC	M-738	Cobalt Chrome-FM	M-73	Colonial No. 125	M-1393
Cletaloy 30-CC	M-738	Cobalt High Speed	M-81	Colonial No. 130	M-1393
Cletaloy 30-CT	M-738	Cobalt Magnet 37%	M-73	Colonial No. 131	M-1393
Cletaloy 30-CTA	M-738	Cobalt Magnet		Colonial No. 132	M-1393
Cletaloy CT-86	M-738	Steel (17%)	M-260	Colonial No. 193	M-1393
Cletaloy TA	M-738	Cobalt Magnet		Colonial No. 194	M-1393
Cletaloy TS	M-738	Steel (35%)	M-260	Colonial No. 195	M-1393
Clevite	M-739	Cobalt Major	M-69	Colonial No. 197	M-1393
Clevite 77	M-739	Cobalt Special	M-459	Colonial No. 198	M-1393
Clevite 90-10	M-739	Cobalt Steel	M-140	Colonial No. 200	M-1393
Clevite 100	M-739	Cobenium	M-61	Colonial No. 201	M-1393
Clevite 112	M-739	Cobite	M-35	Colonial No. 206	M-1393
Clevite 153	M-739	Co-Co	M-34	Colonial No. 210	M-1393
Clevite 250	M-739	Co-Elinvar	M-104	Colonial No. 215	M-1393
Clevite No. 8	M-739	Colhed	M-115	Colonial No. 220	M-1393
Clevite No. 10	M-739	Colmonoy No. 3	M-963	Colonial No. 221	M-1393
Clevite No. 25	M-739	Colmonoy No. 7	M-963	Colonial No. 225	M-1393
Clevite No. 35	M-739	Colmonoy No. 9	M-963	Colonial No. 230	M-1393
Clicker Die	M-357	Colmonoy No. 10	M-963	Colonial No. 235	M-1393
Clicker Die Welding		Colmonoy No. 221	M-963	Colonial No. 240	M-1393
Rod	M-38	Colmonoy No. 233	M-963	Colonial No. 241	M-1393
Clicking Die	M-357	Colmonoy No. 300	M-963	Colonial No. 245	M-1393
Clicking Die	M-357	Colmonoy No.		Colonial No. 250	M-1393
Climax	M-38	HC240	M-963	Colonial No. 251	M-1393

Colonial No. 253	M-1393	Colonial Stainless		Compo 59-E	M-632
Colonial No. 255	M-1393	Iron 797	M-34	Compo 60-Y	M-632
Colonial No. 256	M-1393	Colonial Stainless		Compo 61-A	M-632
Colonial No. 257	M-1393	Iron C N C	M-34	Compo 62-E	M-632
Colonial No. 295	M-1393	Colonial Stainless		Compo 63-H	M-632
Colonial No. 296	M-1393	Iron F M S	M-34	Compo 65-R-2	M-632
Colonial No. 296.5	M-1393	Colonial Stainless		Compo 66-H	M-632
Colonial No. 297	M-1393	Steel "A"	M-34	Compo 66-Q	M-632
Colonial No. 298	M-1393	Colonial Stainless		Compo 66-R	M-632
Colonial No. 299	M-1393	Steel "B"	M-34	Compo 70-H	M-632
Colonial No. 300	M-1393	Colonial Stainless		Compo 70-Q	M-632
Colonial No. 301	M-744	Steel "I"	M-34	Compo 70-R	M-632
Colonial No. 305	M-1393	Colonial Stainless		Composite	M-345
Colonial No. 310	M-1393	Steel "N"	M-34	Cond-Al	M-60
Colonial No. 311	M-1393	Colonial Stainless		Congo Hot Work	M-140
Colonial No. 312	M-1393	Steel "U"	M-34	Con-Pac	M-604
Colonial No. 313	M-1393	Colorado Water		Conservaloy	M-1
Colonial No. 314	M-1393	Hardening	M-618	Consil 983	M-63
Colonial No. 315	M-1393	Colt Hot	M-55	Constant	M-1114
Colonial No. 319	M-1393	Coltuf 28	M-530	Constant	M-85
Colonial No. 320	M-1393	Columbia CCB	M-35	Constantin	M-44
Colonial No. 321	M-1393	Columbia Double		Contracid	M-296
Colonial No. 322	M-1393	Special	M-35	Contracid-B2M	M-296
Colonial No. 323	M-1393	Columbia Electrex	M-35	Contracid B 2.5 M	M-296
Colonial No. 324	M-1393	Columbia Extra		Contracid B 4 M	M-296
Colonial No. 325	M-1393	Headerdie	M-35	Contracid B 6 M	M-296
Colonial No. 326	M-1393	Columbia Extra		Contracid B 7 M	M-296
Colonial No. 400	M-1393	Vanadium	M-35	Contracid B7Mo	M-296
Colonial No. 403	M-1393	Columbia Special	M-35	Contracid B 10 W	M-296
Colonial No. 405	M-1393	Columbia Special		Contracid B W M	
Colonial No. 405.1	M-1393	Wire Drawing	M-35	C	M-296
Colonial No. 405.2	M-1393	Columbia Spring	M-35	Cook H.D.Z.	M-1405
Colonial No. 406	M-1393	Columbia Standard	M-35	Cook I.L.O.	M-1405
Colonial No. 407	M-1393	Columbia Standard		Cook M.Y.	M-1405
Colonial No. 407.5	M-1393	Die Block	M-35	Cook W.Z.	M-1405
Colonial No. 408	M-1393	Columbia		Cooper Alloy 5	M-36
Colonial No. 409	M-1393	Vanadium		Cooper Alloy 14	M-36
Colonial No. 410	M-1393	Standard	M-35	Cooper Alloy 14A	M-36
Colonial No. 410	M-34,	Columbium 10W-5		Cooper Alloy 15	M-36
	M-115	Zr	M-1537	Cooper Alloy 15A	M-36
Colonial No. 410F	M-34,	Columbium SU-16	M-1528	Cooper Alloy 16C	M-36
	M-115	Columbium SU-31	M-1528	Cooper Alloy 16 C3	M-36
Colonial No. 411	M-1393	Columbus K 03	M-1305	Cooper Alloy 16 H	M-36
Colonial No. 412	M-1393	Columbus K 011	M-1305	Cooper Alloy 17C	M-36
Colonial No. 413	M-1393	Columbus K 055	M-1305	Cooper Alloy 17G	M-36
Colonial No. 414	M-1393	Columbus K 0109	M-1305	Cooper Alloy	
Colonial No. 415	M-1393	Columbus MO	M-1305	17M-ELC	M-36
Colonial No. 420	M-1393	Columbus SS10	M-1306	Cooper Alloy 17 Mo	M-36
Colonial No. 421	M-1393	Columbus SS11	M-1305	Cooper Alloy 17 Se	M-36
Colonial No. 422	M-1393	Columbus SS13	M-1305	Cooper Alloy	
Colonial No. 423	M-1393	Columbus SS19	M-1305	17-S-Mo	M-36
Colonial No. 424	M-1393	Colville Stainless		Cooper Alloy 18	M-36
Colonial No. 430	M-34,	Iron	M-530	Cooper Alloy 18-8	M-36
	M-115	Co Major	M-69	Cooper Alloy 19	M-36
Colonial No. 500	M-1393	Combarloy	M-575	Cooper Alloy 19AH	M-36
Colonial No. 610	M-34,	Comet	M-44	Cooper Alloy 19	
	M-115	Commercial Drill		AM	M-36
Colonial No. 610F	M-34,	Rod	M-1	Cooper Alloy 19	
	M-115	Como	M-40	AMO	M-36
Colonial No. 795	M-34,	Comokut	M-24,	Cooper Alloy 19B	M-36
	M-115		M-341	Cooper Alloy 20-10	M-36
		Comol-17	M-38	Cooper Alloy 21A	M-36
Colonial Stainless		Comol-20	M-60	Cooper Alloy 21 B	M-36
Iron 430	M-34	Compo	M-632	Cooper Alloy 21C	M-36

Section IV: Obsolete Alloys / 1693

Name	Code	Name	Code	Name	Code
Cooper Alloy 21 D	M-36	Cormet-A	M-1506	Corrochrom-18T	M-1309
Cooper Alloy 21 E	M-36	Cornix-1	M-501	Corrochroni-170	M-1309
Cooper Alloy 22M	M-36	Cornix-10	M-912	Corrochroni-187F	M-1309
Cooper Alloy 22PM	M-36	Cornix-40	M-912	Corrochroni-188C	M-1309
Cooper Alloy 22S Cb	M-36	Corona 1	M-101	Corrochroni-188T	M-1309
		Corona 2	M-101	Corrochroni-189E	M-1309
Cooper Alloy 22 SMO	M-36	Corona 3	M-101	Corrochroni-189N	M-1309
		Coronze	M-279	Corrochroni-189S	M-1309
Cooper Alloy 22 W	M-36	Corosoloy	M-964	Corrochroni-1310K	M-1309
Cooper Alloy 25	M-36	Corresist 13	M-1342	Corrochroni-1810A	M-1309
Cooper Alloy 25-20	M-36	Corresist 13H	M-1342	Corrochroni-1810M	M-1309
Cooper Alloy 25-20S	M-36	Corresist-13HMo	M-1342	Corrochroni-1810U	M-1309
Cooper Alloy 31	M-36	Corresist 17	M-1342	Corrochroni-1812	M-1309
Cooper Alloy 58	M-36	Corresist 17H	M-1342	Corrochroni-1812C	M-1309
Cooper Alloy 60	M-36	Corresist 17M	M-1342	Corrodur 14	M-1290
Cooper Alloy 531	M-36	Corresist 18/8	M-1342	Corrodur 16-13 N	M-1290
Cooper Alloy KA-4	M-36	Corresist 18/8M	M-1342	Corrodur 17-13 E3S	M-1290
Cooper Alloy P.H.20	M-36	Corresist 18/8 MW	M-1342	Corrodur 17-13 E5S	M-1290
		Corresist 18/8 N	M-1342	Corrodur 18-8	M-1290
Cooper Alloy S-21W	M-36	Corresist 18/8 Nb	M-1342	Corrodur 18-8 E	M-1290
Cooper Alloy S-23	M-36	Corresist 18/8 S	M-1342	Corrodur 18-8 ES	M-1290
Cooper Alloy V2B	M-36	Corresist 18/8 Ti	M-1342	Corrodur 18-8 ESS	M-1290
Copel-X	M-65	Corresist 18/8 W	M-1342	Corrodur 18-8 S	M-1290
Copnic	M-60	Corresist 18/10 MoNb	M-1342	Corrodur 18-8 SS	M-1290
Coppco 75	M-682			Corrodur 20-25 E	M-1290
Coppco 110	M-682	Corresist 18/10 MoS	M-1342	Corrodur 24-7 E	M-1290
Coppco 120	M-682			Corrodur 28-4	M-1290
Coppco 200	M-682	Corresist 18/10 MoTi	M-1342	Corrodur 30	M-1290
Coppco ACE	M-682			Corrodur 30E	M-1290
Coppco ACE A Temper	M-682	Corresist 18/20 MoCu	M-1342	Corrodur 30EH	M-1290
				Corrodur 30H	M-1290
Coppco ACE B Temper	M-682	Corresist 25/25 MoTi	M-1342	Corrodur C 14 N	M-1290
				Corrodur C 17.13 E3S	M-1290
Coppco ACE-C Temper	M-682	Corresist 28	M-1342		
		Corresist 28/5	M-1342	Corrodur C 18	M-1290
Coppco Alnico	M-271	Corresist 28H	M-1342	Corrodur C 18.8 ESSD	M-1290
Coppco Cro-Tung	M-682	Corresist 28 MH	M-1342		
Coppco Extra	M-682	Corresist G14	M-1342	Corrodur C 18.8 O	M-1290
Coppco Fast Finishing	M-682	Corresist G17	M-1342	Corrodur C 18.8 S	M-1290
		Corresist G18/8	M-1342	Corrodur C 18.8 SS	M-1290
Coppco Hot Work No. 1	M-682	Corresist G18/8 M	M-1342	Corrodur C 18.8 SSO	M-1290
		Corresist G18/8 Nb	M-1342		
Coppco Shock	M-682	Corresist G18/8 S	M-1342	Corrodur C 18 E	M-1290
Coppco Special	M-682	Corresist G18/10 MoNb	M-1342	Corrodur C 22.7	M-1290
Coppco Standard	M-682			Corrodur C 24.7 E	M-1290
Coppco Universal	M-682	Corresist G18/10 MoS	M-1342	Corrodur C 30 EH	M-1290
Copper Bearing Low-Metalloid	M-377			Corrodur C 30 H	M-1290
		Corresist G18/20 MoCu	M-1342	Corrodur N 15	M-1290
Copper Bearing Steel	M-90			Corrodur Sp 120	M-1290
		Corresist G27/4	M-1342	Corronel 210	M-121
Copper-Nickel (4%)	M-121	Corresist G28	M-1342	Corronel-220	M-121
Copper-Nickel 10%	M-297	Corresist G28/5	M-1342	Corronel-230	M-121
Copper-Nickel 15%	M-297	Corresist G28H	M-1342	Corronel B	M-121
Copper-Nickel 20%	M-121	Corresist G28HM	M-1342	Corronil	M-121
Copperoid	M-538	Corresist G28Mo	M-1342	Corsonite B	M-280
Copper-Silicon Steel	M-747	Corrochrom-13G	M-1309	Corsonite C	M-280
Copper Steel	M-377	Corrochrom-13M	M-1309	Corsonite D	M-280
Copper Steel	M-377	Corrochrom-13T	M-1309	Corsonite E	M-280
Copper Steel	M-377	Corrochrom-13W	M-1309	Corsonite F	M-280
Coppro Nitriding G	M-682	Corrochrom-17A	M-1309	Cor-Ten A	M-604
Coppro Nitriding G-Modified	M-682	Corrochrom-17F	M-1309	Cor-Ten B	M-604
		Corrochrom-17N	M-1309	Corvic Bronze	M-33
Coppro Nitriding H	M-682	Corrochrom-17T	M-1309	Cosmos	M-77
Cop-R-Loy	M-206	Corrochrom-18E	M-1309	Covan	M-140

Covandur	M-32	Crane No. 4 Carbon-Molybdenum	M-748	Creusot 4DF02/3	M-1488
CQ-2	M-1697			Creusot 4D02	M-1488
CQ-3	M-1697			Creusot 4PF0V	M-1488
CQ-4	M-1697	Crane No. 5 Chrome-Molybdenum	M-748	Creusot 4.5DF1	M-1488
CQ-12	M-1697			Creusot 4.5DF01	M-1488
CQ-13	M-1697			Creusot 5D6/7	M-1488
CQ-14	M-1697	Crane No. 44 Nickel Alloy	M-748	Creusot 5D8	M-1488
CQ-15	M-1697			Creusot 5FA0	M-1488
CQ-16	M-1697	Crane No. 49 Nickel Alloy	M-748	Creusot 5F0P	M-1488
CQ-23	M-1697			Creusot 5F0V	M-1488
Cramp Alloy No. 79	M-309	Crane Special Brass	M-748	Creusot 12FFB	M-1488
Cramp Alloy No. 100	M-309	Crane Steam Brass	M-748	Creusot 12FF0	M-1488
		Crescent Double Special	M-38	Creusot 13FF	M-1488
Cramp Alloy No. 119	M-309			Creusot 13FFCo	M-1488
		Crescent Extra	M-38	Creusot 14FF	M-1488
Cramp Alloy No. 250	M-309	Crescent Hot Work No. 2	M-38	Creusot A 1B	M-1488
				Creusot A 1H	M-1488
Cramp Alloy No. 252	M-309	Crescent Rim Roll	M-38	Creusot A 2	M-1488
		Crescent Special	M-38	Creusot A 2B	M-1488
Cramp Alloy No. 257	M-309	Crescent Special Carbon Tool	M-38	Creusot A 3/4	M-1488
				Creusot A 3/4B	M-1488
Cramp Alloy No. 261	M-309	Crescent Special Wire Die	M-38	Creusot A 3B	M-1488
				Creusot A3H	M-1488
Cramp Alloy No. 263	M-309	Crescent Steel	M-38	Creusot A 4/5B	M-1488
		Crescent Tool	M-38	Creusot A 4/5H	M-1488
Cramp Alloy No. 265	M-309	Creusabro 41	M-1488	Creusot A 4H	M-1488
		Creusabro 43	M-1488	Creusot A 5	M-1488
Cramp Alloy No. 268	M-309	Creusabro MLD	M-1488	Creusot A 5/6H	M-1488
		Creuselso 22	M-1488	Creusot A 5H	M-1488
Cramp Alloy No. 271	M-309	Creuselso 26	M-1488	Creusot A 6/7	M-1488
		Creuselso 31	M-1488	Creusot A 6/7B	M-1488
Cramp Alloy No. 274	M-309	Creuselso 38	M-1488	Creusot A 6B	M-1488
		Creuselso 42	M-1488	Creusot A 6H	M-1488
Cramp Alloy No. 276	M-309	Creuselso 50	M-1488	Creusot A 7B	M-1488
		Creusot 0.5DF03	M-1488	Creusot A 8B	M-1488
Cramp Alloy No. 278	M-309	Creusot 0.5DF06	M-1488	Creusot A 9	M-1488
		Creusot 0.5 F08	M-1488	Creusot A 9B	M-1488
Cramp's Elfur Iron No. 4	M-309	Creusot 1F3	M-1488	Creusot A 9H	M-1488
		Creusot 1F3/4	M-1488	Creusot A 10	M-1488
Cramp's Elfur Iron No. 5	M-309	Creusot 1F4/5	M-1488	Creusot A 10B	M-1488
		Creusot 1F5/6	M-1488	Creusot A 10HC	M-1488
Cramp's Elfur Iron No. 6	M-309	Creusot 1F03	M-1488	Creusot A 11	M-1488
		Creusot 1F04	M-1488	Creusot A 11B	M-1488
Cramp's Elfur Iron No. 7	M-309	Creusot 1F05/6	M-1488	Creusot A 11C	M-1488
		Creusot 1F07	M-1488	Creusot A 11HC	M-1488
Cramp's Elfur Iron No. 8 (Ni-Resist)	M-309	Creusot 1F08	M-1488	Creusot AM 6/7S	M-1488
		Creusot 1FV2	M-1488	Creusot AM 8/9S	M-1488
Cramp's Elfur Iron No. 9	M-309	Creusot 1.1F08	M-1488	Creusot AM 9/10S	M-1488
		Creusot 1.5DF0	M-1488	Creusot AM 10/11S	M-1488
Cramp's Elfur Iron No. 11	M-309	Creusot 1.5DF0V4S	M-1488	Creusot AM 11S	M-1488
		Creusot 2D8	M-1488	Creusot A 0	M-1488
Crane 5 Cr-Mo	M-748	Creusot 2D9	M-1488	Creusot A 0/1H	M-1488
Crane 18-8 Mo	M-748	Creusot 2D10	M-1488	Creusot A 0B	M-1488
Crane Carbon Cast Steel	M-748	Creusot 2D11	M-1488	Creusot A/0H	M-1488
		Creusot 2.2F0	M-1488	Creusot A 00	M-1488
Crane Cast Manganese Bronze	M-748	Creusot 3D8	M-1488	Creusot B 2/3	M-1488
		Creusot 3DF02	M-1488	Creusot B 3/4	M-1488
		Creusot 3DF05	M-1488	Creusot B 4/5	M-1488
		Creusot 3DF08	M-1488	Creusot B 5/6	M-1488
Crane Hardened Stainless Steel	M-748	Creusot 3F0	M-1488	Creusot B 6	M-1488
		Creusot 3FP	M-1488	Creusot BD 5/6	M-1488
Craneloy 20	M-748	Creusot 3.5DF 3/4	M-1488	Creusot BD 8/9	M-1488
Crane No. 3 Nickel Alloy	M-748	Creusot 3.5DF 7/8	M-1488	Creusot BD 9/10	M-1488
		Creusot 4DF02	M-1488	Creusot BF 1/2	M-1488

Section IV: Obsolete Alloys / 1695

Creusot BF 4/5	M-1488	Croloy 18-8		Crown	M-34
Creusot C 1/2	M-1488	Stabilized	M-17	Crown W 20	M-136
Creusot C 4/5	M-1488	Croloy 19-9DL	M-17	C.R.U.	M-267
Creusot CF	M-1488	Croloy 25-12	M-17	Crucast CA-15	M-38
Creusot CMP	M-1488	Cromal	M-342	Crucast CB30	M-38
Creusot CMV	M-1488	Cromaloy	M-85	Crucast CF-8	M-38
Creusot DF3	M-1488	Croman	M-115	Crucast CF-8C	M-38
Creusot DF 4/5	M-1488	Cromansil	M-604	Crucast CF-8M	M-38
Creusot DF5	M-1488	Cromaz-H	M-104	Crucast CF16FA	M-38
Creusot DF6	M-1488	Cromaz-N	M-104	Crucast CF20	M-38
Creusot DF6S	M-1488	Crome	M-24	Crucast HH	M-38
Creusot DF7	M-1488	Cromex 1	M-115	Crucast HK	M-38
Creusot DF 7/8	M-1488	Cromex 2	M-115	Crucast HT	M-38
Creusot DF8	M-1488	Cromic A	M-602	Crucia	M-327
Creusot DF03	M-1488	Cromic C	M-602	Crucia Steel	M-327
Creusot DMF	M-1488	Cromic D	M-602	Crucible 13% Mn	M-38
Creusot D0F1	M-1488	Cromimphy 1	M-771	Crucible 17 Cr-4 Ni	M-38
Creusot F04	M-1488	Cromimphy 1 Bis	M-104	Crucible 52 CB Var	
Creusot F06S	M-1488	Cromimphy 1 Bis		Bearing Steel	M-38
Creusot F06/7S	M-1488	Mo	M-771	Crucible 56	M-38
Creusot F07S	M-1488	Cromimphy 2	M-771	Crucible 201	M-38
Creusot F08S	M-1488	Cromimphy 2C	M-104	Crucible 202	M-38
Creusot F0V	M-1488	Cromimphy 4	M-771	Crucible 302B	M-38
Creusot FP	M-1488	Cromimphy 4 Bis	M-104	Crucible 309B	M-38
Creusot FR	M-1488	Cromimphy 33	M-771	Crucible 311	M-38
Creusot LM0	M-1488	Cromimphy 2100	M-771	Crucible 312	M-38
Creusot M	M-1488	Cromimphy 2200	M-771	Crucible 314	M-38
Creusot MF 6/7S	M-1488	Cromimphy A5	M-771	Crucible 317	M-38
Creusot ML	M-1488	Cromimphy A8	M-771	Crucible 318	M-38
Creusot MLD	M-1488	Cromimphy A8Mo	M-771	Crucible 325	M-38
Creusot M07	M-1488	Cromimphy A10M	M-771	Crucible 329	M-38
Creusot MV6/7S	M-1488	Cromimphy A15	M-771	Crucible 330	M-38
Creusot P1	M-1488	Cromimphy A15Mo	M-771	Crucible 405	M-38
Creusot P2	M-1488	Cromimphy A18Mo	M-771	Crucible 406	M-38
Creusot S1SA	M-1488	Cromimphy A20Mo	M-771	Crucible 422M	M-38
Creusot S2V	M-1488	Cromimphy A1007		Crucible 1383	M-38
Creusot S3	M-1488	Mo	M-771	Crucible A-110 AT	M-38
Creusot S3V	M-1488	Cromimphy A2100	M-771	Crucible B 110 Mo	M-38
Creusot S03	M-1488	Cromimphy A2200	M-771	Crucible B-120 VCA	M-38
Creusot SS5	M-1488	Cromimphy A3100	M-771	Crucible C 105 VA	M-38
Creusot SS5V	M-1488	Cromimphy A3200	M-771	Crucible C 110M	M-38
Creusot SS10	M-1488	Cromimphy No. 1	M-104	Crucible C 115	
Creusot SS14V	M-1488	Cromimphy No. 2	M-104	AMoV	M-38
Creusot SS.15	M-1488	Cromimphy No. 4	M-104	Crucible C-120AM	M-38
Creusot SV4	M-1488	Cromimphy No. 6	M-104	Crucible C 130AM	M-38
C.R.M.	M-162	Cromin D	M-61	Crucible C 130AMo	M-38
CrMo Bearing	M-24	Cromino 2	M-1315	Crucible C-135 AMo	M-38
CRM Special	M-459	Cromo	M-334	Crucible CCS	M-38
Crocar FM	M-115	Cromoco	M-57	Crucible CMC	M-38
Crofer 118	M-287	Cromodie	M-56	Crucible Collet	M-38
Croloy	M-357	Cromodie-W	M-56	Crucible Copper	
Croloy 1	M-17	Cromova	M-334	Bond	M-38
Croloy 1 3/4	M-17	Cromo WV	M-24	Crucible CSA39	M-38
Croloy 2 Si	M-17	Cromva	M-24	Crucible CSM No. 2	M-38
Croloy 5Cb	M-17	Cromva	M-334	Crucible CVM	M-38
Croloy 5M	M-17	Cronimo	M-38	Crucible D 6	M-38
Croloy 5Ti	M-17	Cronitung	M-260	Crucible D-319	M-38
Croloy 8M	M-17	Cro-Sil No. 10	M-38	Crucible Diamond	
Croloy 9	M-17	Cro-Sil No. 14	M-38	Brand	M-38
Croloy 15-15N	M-17	Croterite IV	M-179	Crucible HNM	M-38
Croloy 16-8-2	M-17	Croterite V	M-179	Crucible HRB	M-38
Croloy 16-13-3	M-17	Crotorite	M-179	Crucible HTX	M-38
Croloy 16-13-8	M-17	Crotung	M-151	Crucible Jail Bar	
Croloy 18-8S	M-17	Crow	M-1	Steel	M-38

Crucible M-252	M-38	Cumanite	M-9	D-2	M-338
Crucible M-308	M-38	Cunico	M-31	D-2 Alloy	M-28
Crucible NCR124MOD	M-38	Cunife II	M-60	D-2 Disston	M-357
		Cupaloy	M-118	D2-SA-Tool	M-357
Crucible Nitrard No. 1	M-76	Cupro-Arc "C"	M-692	D6-AC	M-115, M-1527
		Cupro-Nickel 10%	M-33		
Crucible Orthopedic Steel	M-38	Cupro-Nickel 20%	M-33	D-6-Co	M-357
		Curoloy	M-136	"D7" Quality	M-179
Crucible PHV	M-38	Cutlery	M-472	D 9 Mo	M-357
Crucible Self-Tem	M-38	Cutter Alloy	M-35	D9-VA	M-357
Crucible Stainless Iron	M-38	Cut Trode	M-717	D-10	M-989
		Cuyo	M-155	D 22 S	M-459
Crucible Stainless Iron No. 12	M-38	CV1 Extra	M-459	D29	M-57
		CV 30	M-459	D-29 Chisel Steel	M-357
Crucible Stainless Iron No. 18	M-38	CV 60	M-459	D-29 Knife	M-357
		CV 70	M-459	D. 70	M-50
Crucible Stainless Iron No. 24	M-38	CV 110	M-459	DA8	M-158
		CV 120	M-459	DA-99	M-158
Crucible Stainless Steel	M-38	CVFS	M-459	DA-105	M-158
		C.V.S.	M-499	Dacar	M-69
Crucible Stainless Steel No. 12	M-38	C.V.S.S.	M-499	Dalzell Steel	M-530
		CWM, S 431	M-459	Damascite	M-211
Crucible Stainless Steel No. 18	M-38	CY-2	M-1697	Damascus	M-539
		CY-5	M-1697	Damaxine Bronze	M-179
Crucible Stainless Steel No. 24	M-38	CY-12	M-1697	Dauphinox-TPMo	M-1649
		CY-14	M-1697	DBL	M-1
Crucible TXCR	M-38	CY-16	M-1697	DBL-2	M-1
Crucible UHS 260	M-38	CY-17	M-1697	DCN 2W	M-459
Crucible Versasteel	M-38	CY-31	M-1697	DCN 3W	M-459
Crucible Waspaloy	M-38	C.Y. Alloy	M-752	DCN 60	M-459
C.R.U. Composite	M-267	Cycloid Gear Steel	M-32	DCN 80	M-459
C.R.U. Excelsior	M-267	Cyclone 56	M-50, M-1414	DCNA	M-459
				DCN Extra	M-459
C.R.U. Excelsior XXII	M-267	Cyclone Extra	M-50, M-1414	DCNO	M-459
C.R.U. Excelsior Extra	M-267	Cyclone Supercut	M-50, M-1414	DCNR	M-459
				DCN Spezial	M-459
C.R.U. Excelsior Extra XII	M-267	Cyclops Ajax	M-114	Deco	M-96
		Cyclops B4	M-114	Decora	M-96
Crusca 12B	M-38	Cyclops B7	M-114	Decora W	M-96
Crusca Cold Hubbing	M-38	Cyclops B8	M-114	Deefour	M-1097
		Cyclops B42	M-114	Deefourteen	M-1097
Crusca Flame Hardening	M-38	Cyclops K	M-114	Deelite	M-1097
		Cyclops KL	M-114	Deeone	M-1097
Crusca New Process Hollow Drill	M-38	Cyclops KM	M-114	Deepep	M-1097
		Cyclops KR	M-114	Deeseven	M-1097
Crusca Solid Drill	M-38	Cyclops KS	M-114	Deesix	M-1097
Crusca "V.J." Extra	M-38	Cyclops No. 17 Metal	M-39, M-114	Deethree	M-1097
Crusca "V.J." Special	M-38			Deetwo	M-1097
				Defiheat	M-100
Crusher Bearing Bronze	M-508	Cyclops No. 67	M-114	Defirust	M-100
		Cyclops R.B.C.	M-114	Defirust No. 410	M-100
CRWXI	M-459	Cyclops Special	M-114	Defirust No. 416	M-100
CS	M-459	Cyclops Stainless Gr. A	M-114	Defirust Machining	M-100
CSA	M-38			Defirust N	M-100
CSA	M-459	Cyclops Stainless Gr. B	M-114	Defirust Special	M-100
CSA-G40.8	M-1253			Defirust, Turbine	M-100
C.S.M.	M-260	Cyclops Standard	M-114	Defistain	M-100
CSP	M-459	Cyclops W9	M-114	Defistain, Machining	M-100
CSV1	M-459	C.Y.W. Choice Die	M-114		
C.T.c. Ferry	M-121			Defistain, Special	M-100
Cube-Alloy	M-63	**D**		Defistain, Special	M-100
Cube Injection Mold	M-108			Delcondex 10	M-338
				Delcondex 12	M-338
Cufenloy-40	M-319	D-1 Disston	M-357	Delcondex 20	M-338

Section IV: Obsolete Alloys / 1697

Delcondex 30	M-338	D-H Manganese		Dilatherm 85	M-27
Delcondex 31	M-338	Alloy No. 720	M-44	Dilatherm 90	M-27
Delcondex 32	M-338	D-H Nirosta	M-44	Dilphy	M-771
Delconion	M-338	D-H No. 11	M-44	Dilver	M-104
Deloro 1300K	M-41	D-H No. 14	M-44	Dilver O	M-104
Delsteel Alloy S.T.	M-432	D-H No. 33	M-44	Dilver P	M-104
Delta II	M-764	D-H No. 95	M-44	Dilver Pl	M-771
Delta III	M-764	D-H No. 111	M-44	Dilver T	M-771
Delta IV	M-764	D-H No. 129	M-44	Dimondite	M-57
Delta Antifriction		D-H No. 146	M-44	Disc Alloy	M-1490
Metal	M-450,	D-H No. 245	M-44	Discaloy 24	M-118
	M-156	D-H No. 399	M-44	Disque E	M-884
Delta Bronze I	M-156	D-H No. 446	M-44	Disque H	M-884
Delta Bronze II	M-156	DH No. 499	M-44	Dissteel 812	M-357
Delta Bronze III	M-156	D-H No. 599	M-44	Disston 6-N-6	M-357
Delta Bronze V	M-156	D-H No. 799	M-44	Disston 6N6-M2	M-357
Delta Bronze VI	M-156	D-H No. 899	M-44	Disston 66	M-357
Delta Bronze VII	M-156	D-H No. 999	M-44	Disston HRW	M-140
Delta Bronze VIII	M-156	D-H Nirosta	M-44	Disston No. 712	M-357
Delta Bronze IX	M-156	D-H OHM Alloy	M-44	Disston No. 819	M-357
Delta Bronze IXa	M-156	D-H Special Alloy		Disston No. 821	M-357
Delta Bronze No.		Steel	M-1016	Disston No. 825	M-357
IV	M-156	D-H Tool Steel	M-1016	Disston No. 826	M-357
Delta Bronze No.		Diabolique Satan		Disston No. 827	M-357
VII	M-156	No. 2	M-1043	Disston No. 841	M-357
Delta Bronze No.		Dialoy	M-349	Disston No. 842	M-357
VIII	M-156	Diamant	M-1114	Dixtampo	M-156
Delta Bronze No.		Diamant 3	M-459	Dixtrudo	M-156
IX	M-156	Diamant 5	M-459	DLP Copper 120	M-8
Delta Bronze No.		Diamant Bronze	M-German	DM2-45	M-376
IXa	M-156	Diamant Extra	M-459	DM-35	M-376
Delta High Tensile		Diamant Extra	M-459	DM-45	M-376
Brass HT3	M-156	Diamant Extra		DMOC-Weich	M-459
Deltamax	M-1	Duro	M-616	DN-1	M-459
Delta Propeller		Diamant Hochhart	M-1308	DN-3	M-459
Bronze	M-156	Diamant R	M-459	DN 5	M-459
Delta Propeller		Diamant R	M-459	DN 15	M-459
Bronze	M-156	Diamant Special	M-1114	DN 25	M-459
Deltoid	M-156	Diamond E	M-999	DN 33	M-459
Demmler "D"	M-57	Diamond Extra	M-373	DN 36	M-459
Denscast	M-44	Diamond S	M-373	DN 42	M-459
Dense Alloy 112		Dibronze	M-13	"D" Nickel	M-1499
Type E	M-725	Dica A	M-69	D.N.S.	M-1724
Dense Alloy 112		Dica B Tool Room	M-69	Dodge D-1	M-762
Type P	M-725	Dica C	M-69	Dodge D-2	M-762
Dense Alloy 112		Dica D	M-69	Dodge D-2 A	M-762
Type Y	M-725	Die-Blanking	M-85	Dodge D-2 B	M-762
Densified Cru-Die	M-38	Die Block (A)	M-85	Dodge D-2 C	M-762
Deoxidized Copper	M-33	Die Block (B)	M-85	Dodge D3	M-762
Devils Iron	M-1	Die Block "C"	M-85	Dodge D-4	M-762
Dew-AMS Extra	M-459	Die Block "E"	M-85	Dodge D-5	M-762
Deward	M-1	Die Casting	M-140	Dodge D-6	M-762
Dew CMS	M-459	Die Casting Alloy		Dodge D-7	M-762
Dew D12L	M-459	624	M-8	Dodge D-8	M-762
Dew E38Mo	M-459	Die Casting Alloy		Dodge D-9	M-762
Dew E38V	M-459	1026	M-8	Dodge D10	M-762
Dew E38W	M-459	Diehard BB	M-56	Dodge D11	M-762
Dew ECR	M-459	Diehard HCD	M-56	Dodge D-12	M-762
Dew GSE	M-459	Diehard LC	M-56	Dodge D14	M-762
Dew Spezial-MS	M-459	Diemac	M-1680	Dodge DTS 30W8	M-762
Dew Spezial-W5	M-459	Die Steel	M-373	Dodge DTS 40	M-762
Dew WM559	M-459	Die Steel No. 13200	M-373	Dodge DTS 60	M-762
DG Cyclone	M-1414	Dieweld	M-607	Dodge DTS 100	M-762
D.G. Cyclone	M-1724	Dilatherm 15	M-27	Dodge DTS 150	M-762

Dodge DTS 165	M-762	Dowmetal J1	M-43	Dunelt 65	M-1413
Dodge DX	M-762	Dowmetal JK31A	M-43	Dunelt 67	M-1413
Dodge DZ	M-762	Dowmetal K	M-43	Dunelt 69	M-1413
Do-Di	M-158	Dowmetal L	M-43	Dunelt 82M	M-1413
Dofascoloy No. 1	M-1033	Dowmetal M (MIA)	M-43	Dunelt 83	M-1413
Do-It	M-289	Dowmetal P	M-43	Dunelt 84	M-1413
Doler Alumin No. 99	M-158	Dowmetal RC	M-43	Dunelt 38	M-1413
Doler Alumin No. 133	M-158	Dowmetal T	M-43	Dunelt 90	M-1413
		Dowmetal X	M-43	Dunelt 100	M-1413
Doler Alumin No. 308	M-158	Dow Solder A	M-43	Dunelt 101	M-1413
		Dow Solder B	M-43	Dunelt 128	M-1413
Doler Brass	M-158	Draco 3460	M-1117	Dunelt 129	M-1413
Doler Brass 1	M-158	Dragon	M-1	Dunkirk EZ	M-1
Doler Brass 2	M-158	Drawalloy	M-765	Duozinc	M-522
Doler Brass 3	M-158	Drawalloy 240	M-765	Duplex C.H. No. 1	M-38
Doler Brass 4	M-158	Drawalloy 340	M-765	Duplex C.H. No. 2	M-38
Doler Brass 5	M-158	Dreadnaught	M-362	Duplex C.H. No. 3	M-38
Doler-Mag	M-158	Dreadnought	M-38	Duplex Gear No. 1	M-38
Doler Mag No. 2	M-158	Drill	M-377	Duplex Gear No. 2	M-38
Doler Mag No. 6	M-158	Dril-Tec 88	M-717	Duplex Gear No. 3	M-38
Doler Mag No. 10	M-158	Ducol Steel	M-530	DUR 500	M-459
Doler No. 112	M-158	Ducol W 21	M-530	Durachrome	M-513
Doler Zinc	M-158	Ducol W 25	M-530	Duralite 35	M-541
Doler Zinc 2	M-158	Ducol W 30	M-530	Duraloy "18-8"	M-45
Doler-Zinc 3	M-158	Ductalloy 60	M-1099	Duraloy 18-8S	M-45
Doler-Zinc 5	M-158	Ductalloy 80	M-1099	Duraloy 18-8 SMO	M-45
Doler-Zinc 8	M-158	Ductalloy A50	M-1099	Duraloy 25-20 Mo	M-45
Domestic	M-69	Ductillite	M-206	Duraloy "A"	M-45
DOS Spezial	M-1291	Dullray	M-121	Duraloy "B"	M-45
Double Chrome		Dumet	M-104	Duraloy "C"	M-45
Vanadium	M-1	Dumost No. 1	M-97	Duraloy CB	M-45
Double Extra	M-85	Dumost No. 2	M-97	Duraloy CC	M-45
Double Mushet	M-233	Dumost No. 3	M-97	Duraloy CE30	M-45
Double Satan	M-1043	Dumost No. 16	M-97	Duraloy CF	M-45
Double Six	M-1	Dundie	M-1	Duraloy CF-7	M-45
Double Special	M-38	Dunelt 3	M-1413	Duraloy CF-7C	M-45
Double Special	M-57	Dunelt 4	M-1413	Duraloy CF-7M	M-45
Double Special	M-362	Dunelt 5	M-1413	Duraloy CF-7Se	M-45
Double Special	M-618	Dunelt 6	M-1413	Duraloy CF-10	M-45
Double Special D.S.	M-35	Dunelt 7	M-1413	Duraloy CF-10M	M-45
Double Super		Dunelt 10	M-1413	Duraloy CF-16	M-45
Hydra	M-1411	Dunelt 14	M-1413	Duraloy CF-16M	M-45
Dow HZ32A	M-43	Dunelt 15	M-1413	Duraloy CG-7	M-45
Dowmetal 251X	M-43	Dunelt 16	M-1413	Duraloy CG-7C	M-45
Dowmetal A	M-43	Dunelt 17	M-1413	Duraloy CG-7M	M-45
Dowmetal AM60X1	M-43	Dunelt 18	M-1413	Duraloy CG-10	M-45
Dowmetal AZ 63A	M-43	Dunelt 20	M-1413	Duraloy CG-16	M-45
Dowmetal AZ81A	M-43	Dunelt 21	M-1413	Duraloy CG-16Se	M-45
Dowmetal AZ91B	M-43	Dunelt 26	M-1413	Duraloy CH-10	M-45
Dowmetal B	M-43	Dunelt 33	M-1413	Duraloy CH-20	M-45
Dowmetal CM62	M-43	Dunelt 34	M-1413	Duraloy CK25	M-45
Dowmetal D	M-43	Dunelt 36	M-1413	Duraloy CM-25	M-45
Dowmetal E	M-43	Dunelt 39	M-1413	Duraloy CT-7	M-45
Dowmetal EK30A	M-43	Dunelt 40	M-113	Duraloy DF-20	M-45
Dowmetal EK31A	M-43	Dunelt 41	M-1413	Duraloy "H"	M-45
Dowmetal EK41A	M-43	Dunelt 50	M-1413	Duraloy H-2	M-45
Dowmetal EX	M-43	Dunelt 51	M-1413	Duraloy HA	M-45
Dowmetal EZ33A	M-43	Dunelt 53	M-1413	Duraloy HCA	M-45
Dowmetal F	M-43	Dunelt 55	M-1413	Duraloy HCN	M-45
Dowmetal FS-1	M-43	Dunelt 60	M-1413	Duraloy HE	M-45
Dowmetal G	M-43	Dunelt 61	M-1413	Duraloy HF	M-45
Dowmetal H	M-43	Dunelt 62	M-1413	Duraloy HL	M-45
Dowmetal HK32	M-43	Dunelt 63	M-1413	Duraloy HN	M-45
		Dunelt 64	M-1413	Duraloy-HOM	M-45

Section IV: Obsolete Alloys / 1699

Name	Code	Name	Code	Name	Code
Duraloy HP	M-45	Durco D-6	M-46	Dymonhard No. 95	M-773
Duraloy HT	M-45	Durco D-7	M-46	Dymonhard No. 96	M-773
Duraloy HW	M-45	Durco D-10	M-46	Dynalloy II	M-537
Duraloy HX	M-45	Durco D-12	M-46	Dynalloy 6731 (603)	M-266
Duraloy MH	M-45	Durco D-18	M-46	Dynavar	M-1084
Duraloy "N"	M-45	Durco D-28	M-46	Dyn-El	M-537
Duraloy No. 15-35	M-45	Durco D-181	M-46	DZ	M-459
Duraloy NSMO	M-45	Durco KA-2-Mo	M-46		
Duraloy R	M-45	Durehete	M-1724	**E**	
Duraloy "X"	M-45	Durehete 1050	M-1724		
Duralumin "B"	M-338	Durex Iron No. 76	M-839	E20M	M-459
Duralumin-C	M-338	Durface	M-717	E 21 S	M-459
Duralumin E	M-338	Durichlor	M-46	E 22 S	M-459
Duralumin G	M-338	Durimet	M-46	E 724Z	M-459
Duralamin H	M-338	Durimet-20	M-46	E-52100 Wire	M-12
Duralumin HX	M-338	Durimet A	M-46	E-Alloy	M-158
Duralumin J	M-338	Durimet B	M-46	Eastern ES-18-8FM	M-1014
Duralumin K	M-338	Durimet D	M-46	Eastern ES-18-11	M-1014
Duralumin KC	M-338	Durimet K	M-46	Eastern	
Duralumin L	M-338	Durimet L	M-46	ES-18-12-3Mo	M-1014
Duralumin LC	M-338	Durimet T	M-46	Eastern ES-18-12-3MoCb	M-1014
Duralumin M	M-338	Durinval	M-104	Eastern ES-18-12-3Mo-L	M-1014
Duralumin Q	M-338	Duriron	M-46		
Duralumin S	M-338	Durite	M-35	Eastern ES-18-12-4Mo	M-1014
Duralumin, Vickers	M-338	Duro	M-81, M-261		
Duralumin X	M-338				
Duramold A	M-24	Duro 6	M-459	Eastern No. 20	M-1014
Duramold C	M-24	Duro 70	M-459	Eastern No. 41	M-1014
Duramold Ni-Cr	M-24	Duro Antifriction	M-709	Eastern No. 155	M-1014
Duranickel K	M-68	Durochrome	M-513	Eastern No. 200	M-1014
Duranickel R	M-68	Durode	M-575	Eastern No. 286	M-1014
Duraperm	M-1084	Durode CC	M-575	Eastern No. 530	M-1014
Duraspan	M-45	Durode XH	M-575	Eastern No. 600	M-1014
Durax D	M-459	Duro Gloss C-1	M-69	Eastern No. 605	M-1014
Durax Extra Hart	M-459	Duro Gloss C-2	M-69	Eastern No. 702	M-1014
Durax H	M-459	Duro Gloss C-3	M-69	Eastern No. 718	M-1014
Durax N	M-459	Duro Gloss C-4	M-69	Eastern No. 750	M-1014
Durax Special	M-459	Duro Gloss C-12	M-69	Eastern No. 801 n	M-1014
Durax W2	M-459	Duro Gloss Free Machining	M-69	Eastern No. 1030	M-1014
Durax W3	M-459	Duro Gloss S	M-69	"EB" Silver Brazing Alloy	M-63
Durax W5	M-459	Duro High Speed	M-81	Eclair	M-1043
Durbar	M-142	Duron	M-459	Eclair 33	M-1043
Durbar Hard	M-142	Duval AU	M-1115	Eclair 44	M-1043
Durbar Standard	M-142	Duval DU-MUV	M-1115	Eclair 55	M-1043
Durcilium-F	M-1444	Duval Extra F	M-1115	Eclair 88	M-1043
Durcilium-M	M-1444	Duval SF2	M-1115	Eclair 99	M-1043
Durcilium-R	M-1444	Duval U 7	M-1115	Eclipsaloy 1	M-776
Durcilium-W	M-1444	Duval U 8	M-1115	Eclipsaloy 2	M-776
Durcilium-Y	M-1444	Dux 4	M-1724	Eclipsaloy 4	M-776
Durcilium-Z	M-1444	Dux 4	M-1414	Eclipsaloy 5	M-776
Durco	M-46	Dux 10	M-50, M-1414	Eclipsaloy 6	M-776
Durco 18-8	M-46			Eclipsaloy 12	M-776
Durco 18-8-S	M-46	Dux 12	M-50, M-1414	Eclipsaloy 13	M-776
Durco 18-8-S-Cb	M-46			Eclipsaloy 15	M-776
Durco 18-8-S-Mo	M-46	Dycast	M-73	Eclipsaloy 15A	M-776
Durco 18-8-S-Mo-Cb	M-46	Dy-Krome Special	M-604	Eclipsaloy 15B	M-776
Durco 25-12	M-46	Dymonhard No. 55	M-773	Eclipsaloy 15C	M-776
Durco 25-12-S	M-46	Dymonhard No. 65	M-773	Eclipsaloy 34	M-776
Durco 26-10	M-46	Dymonhard No. 90	M-773	Eclipsaloy 35	M-776
Durco 26-12	M-46	Dymonhard No. 91	M-773	Eclipsaloy 38	M-776
Durco 28-12	M-46	Dymonhard No. 92	M-773	Eclipsaloy 38B	M-776
Durco 30-15	M-46	Dymonhard No. 93	M-773	Eclipsaloy 56	M-776
Druco 181	M-46	Dymonhard No. 94	M-773	Eclipsaloy 57	M-776

Name	Code	Name	Code	Name	Code
Eclipsaloy 68	M-776	Edelstahl DMW	M-459	Edelstahl WKL	M-459
Eclipsaloy 122	M-776	Edelstahl DMZ	M-459	Edelstahl WKL3	M-459
Eclipsaloy 130	M-776	Edelstahl DN2	M-459	Edelstahl WM559	M-459
Eclipsaloy 304	M-776	Edelstahl DN19CR	M-459	Edelstahl WS1 Extra	M-459
Eclipsaloy 305A	M-776	Edelstahl DN20	M-459	Edelstahl WSP	M-459
Eclipsaloy 320	M-776	Edelstahl DVCN	M-459	Edelstahl WSPS	M-459
Eclipsaloy 322	M-776	Edelstahl E15Z	M-459	Edelstahl WVK135	M-459
Eclipsaloy 322C	M-776	Edelstahl E20M	M-459	Edelstahl ZNK2	M-459
Eclipsaloy 323	M-776	Edelstahl E22Z	M-459	Edelweiss AKRO 6	M-1308
Eclipsaloy 324	M-776	Edelstahl E38	M-459	Edelweiss Alsicro 8	M-1308
Eclipsaloy 517	M-776	Edelstahl E612	M-459	Edelweiss R 1	M-1308
Eclipsaloy 520	M-776	Edelstahl E975	M-459	Edelweiss RIG	M-1308
Eclipse Bronze	M-241	Edelstahl ECM	M-459	Edelweiss R 9	M-1308
Econo	M-140	Edelstahl ECMH	M-459	Edelweiss R9G	M-1308
Econoloy	M-964	Edelstahl ECR	M-459	Edelweiss R16	M-1308
Economo No. 20	M-119	Edelstahl EMOC	M-459	Edelweiss R25	M-1308
Economo No. 50	M-119	Edelstahl EMOCh	M-459	Edelweiss R25N	M-1308
Economy Bronze	M-8	Edelstahl ES2	M-459	Edelweiss R25NG	M-1308
Edelstahl	M-German	Edelstahl ES6	M-459	Edgar Allen Double Shear Temper	M-210
Edelstahl 000-Special-31	M-459	Edelstahl EW Extra	M-459	Edgar Allen Drift Steel	M-210
Edelstahl ATS1	M-459	Edelstahl GSV	M-459		
Edelstahl ATS5	M-459	Edelstahl KF15	M-459		
Edelstahl Best	M-459	Edelstahl KM80	M-459	Edgar Allen Drill Steel	M-210
Edelstahl BMS Special	M-459	Edelstahl KS	M-459	Edgar Allen Dunter-Pick Steel	M-210
		Edelstahl KSN	M-459		
Edelstahl BMSW	M-459	Edelstahl KSS Supra	M-459		
Edelstahl BSC	M-459	Edelstahl KW	M-459	Edgar Allen Feather Steel	M-210
Edelstahl BSV	M-459	Edelstahl LEMSI	M-459		
Edelstahl BSZ	M-459	Edelstahl M882	M-459	Edgar Allen Granite-Chisel Steel	M-210
Edelstahl BVT90	M-459	Edelstahl MC60	M-459		
Edelstahl CMC	M-459	Edelstahl MC Spezial	M-459		
Edelstahl CN Super	M-459	Edelstahl Mo20	M-459	Edgar Allen Imp. C.S.	M-210
Edelstahl CNW3	M-459	Edelstahl Mo-20S	M-459	Edgar Allen Iron Fibered Steel	M-210
Edelstahl CNWZ	M-459	Edelstahl Mo325	M-459		
Edelstahl CR25	M-459	Edelstahl MSJ	M-459		
Edelstahl CRM	M-459	Edelstahl MSO	M-459	Edgar Allen Miners Drill Steel	M-210
Edelstahl CRS	M-459	Edelstahl MTS1	M-459		
Edelstahl CRZ	M-459	Edelstahl MTS2	M-459	Edgar Allen No. 4 Hot Die	M-210
Edelstahl CSV2	M-459	Edelstahl-MTS5	M-459		
Edelstahl CSV3	M-459	Edelstahl-MTS6	M-459	Edgar Allen Plug Steel	M-210
Edelstahl CV1 Extra	M-459	Edelstahl NPU	M-459	Edgar Allen Red Label	M-210
Edelstahl CWH	M-459	Edelstahl NW2	M-459		
Edelstahl CWM	M-459	Edelstahl PKD	M-459	Edgar Allen Safe-Plate	M-210
Edelstahl D10S	M-459	Edelstahl PW2	M-459		
Edelstahl D15S	M-459	Edelstahl PW10	M-459	Edgar Allen Shear Temper	M-210
Edelstahl D20S	M-459	Edelstahl RS2	M-459		
Edelstahl D20V	M-459	Edelstahl SAK3	M-459		
Edelstahl D22S	M-459	Edelstahl SCH	M-459	Edgar Allen Silver Steel	M-210
Edelstahl DCN1	M-459	Edelstahl SCSP	M-459		
Edelstahl DCN2	M-459	Edelstahl SK	M-459	Edgar Allen Tool Class "E"	M-210
Edelstahl DCN3	M-459	Edelstahl SPG15	M-459		
Edelstahl DCN4	M-459	Edelstahl SS2	M-459	Edgar Allen Tool Class "F"	M-210
Edelstahl DCN36	M-459	Edelstahl SS3V	M-459		
Edelstahl DCN42	M-459	Edelstahl SS6	M-459	Edgar Allen Tool Class "H"	M-210
Edelstahl DCNR	M-459	Edelstahl SSC	M-459		
Edelstahl DM	M-459	Edelstahl TOS	M-459	Edgar Allen Tool Class "P"	M-210
Edelstahl DM18	M-459	Edelstahl VSK	M-459		
Edelstahl DMF	M-459	Edelstahl W15A1	M-459	Edgar Allen Turners Safe Plate	M-210
Edelstahl DMH	M-459	Edelstahl W20S	M-459		
Edelstahl DMN	M-459	Edelstahl WC2	M-459		
Edelstahl DMS	M-459	Edelstahl WE5	M-459	Edgar Allen Welding Steel	M-210
Edelstahl DMV	M-459	Edelstahl WE Extra	M-459		

Name	Code
"EH" Silver Brazing Alloy	M-63
E.H.W. No. 2	M-73
E.H.W. No. 3	M-73
E.H.W. No. 3	M-73
"E.H.W." Steel	M-73
E.I.S. 14	M-64
E.I.S. 15	M-64
E.I.S. 31 "Hotkut"	M-64
E.I.S. 42 "Kleenkut"	M-64
Elastic No. 3	M-349
Elastique Spring	M-327
Elastuf 44	M-307
Elastuf B-W	M-213
Elastuf C.H.	M-307
Elastuf Chro-Moly	M-213, M-307
Elastuf Chro-Moly (Modified)	M-307
Elastuf JJ	M-307
Elastuf Media Precision Finish	M-213
Elastuf Media Steel	M-213, M-307
Elastuf SAE 2335	M-213, M-307
Elastuf SAE 3140	M-213, M-307
Elastuf Stainfree	M-307
Elastuf Type A	M-307
Elastuf Type A Steel No. 9	M-213, M-307
Elastuf Type A Steel No. 14	M-213, M-307
Elastuf Type K	M-307
Elcomet "F"	M-263
Elcomet "L"	M-263
Elcomet "M"	M-263
Electrex	M-35
Electric Railway Babbitt	M-317
Electric Spindle	M-38
Electrite Co-6XL	M-73
Electrite Co-12	M-73
Electrite Cobalt XL	M-73
Electrite Cromolon	M-73
Electrite Double Six	M-73
Electrite Dynavan XL	M-73
Electrite HV-6	M-73
Electrite MHV-6	M-73
Electrite MV-4	M-73
Electrite No. 2	M-73
Electrite No. 5XL	M-73
Electrite No. 7XL	M-73
Electrite No. 19XL	M-73
Electrite Steel	M-73
Electrite Super Cobalt XL	M-73
Electrite Super CoMo	M-73
Electrite Tatmo XL	M-73
Electrite TNW-XL	M-73
Electrite U	M-73
Electrite UB	M-73
Electrite UB 4	M-73
Electrite Ultravan XL	M-73
Electrite Vanadium XL	M-73
Electroloy Match Plate Plate Aluminum	M-777
Electroloy Pattern Aluminum	M-777
Electrolytic Tough Pitch Copper	M-33
Electrometall	M-1130
Electron A Z	M-302
Electron AZ-102	M-106, M-78
Electron "A.Z.D."	M-78, M-217
Electron "A.Z.F."	M-78, M-217
Electron CM-Si	M-78
Electron No. 23	M-78
Electron "V-1"	M-78, M-302
Electron Z-1-B	M-78
Electron "Z-3"	M-78
Electro Sheet Copper Foil	M-8
Electrunite Boiler Tubing	M-97
Elektron 3Z3	M-1130
Elektron A4	M-1130
Elektron A6	M-338
Elektron A7	M-137
Elektron A 8	M-338
Elektron A8 HP	M-106
Elektron A9	M-78
Elektron A10	M-78
Elektron A11V	M-137
Elektron AM6	M-137
Elektron AM 537	M-784, M-1130
Elektron AZ31	M-338
Elektron AZ 551	M-137
Elektron AZ-855	M-338
Elektron AZF	M-137
Elektron AZG	M-137, M-784, M-1130
Elektron M2	M-338
Elektron MCZ	M-106
Elektron MG5	M-106
Elektron-MSR	M-1130
Elektron-MT2	M-1130
Elektron RZ5	M-106
Elektron TZ6	M-106
Elektron TZ6	M-1130
Elektron Z5Z	M-106
Elektron Z5Z	M-1130
Elektron ZRE2	M-1130, M-137, M-452
Elektron ZRE3	M-1130, M-137, M-452
Elektron ZREO	M-106
Elektron ZREI	M-106
Elektron ZTI	M-106
Elektron ZTI	M-1045
Elektron ZTY	M-1130
Elektron ZWI	M-338
Elektron ZWI	M-1130
Elektron ZW3	M-1130
Elektron ZW6	M-1130
Elephant Brand Phosphor-Copper	M-450
Elephant Brand Phosphorized Antifriction Metal	M-450
Elfur Iron No. 1	M-309
Elfur Iron No. 2	M-309
Elfur Iron No. 3	M-309
Elkaloy A	M-265
Elkaloy D	M-265
Elkaloy D110	M-265
Elkaloy D-112	M-265
Elkaloy D-120	M-265
Elkaloy D-144	M-265
Elkaloy D-146	M-265
Elkaloy DX	M-265
Elkonite 1W	M-265
Elkonite 5W-53	M-265
Elkonite 3W3	M-265
Elkonite 3W53	M-265
Elkonite 5-S	M-265
Elkonite 5W3	M-265
Elkonite 5W-53	M-265
Elkonite 10W3	M-265
Elkonite 10-W-53	M-265
Elkonite 20S	M-265
Elkonite 20W3	M-265
Elkonite 30W3	M-265
Elkonite 35S	M-265
Elkonite 50S	M-265
Elkonite 100M	M-265
Elkonite 100W	M-265
Elkonite 1062	M-265
Elkonite 2052	M-265
Elkonite 2110	M-265
Elkonite 2165	M-265
Elkonite 2173	M-265
Elkonite 2650	M-265
Elkonite 2665	M-265
Elkonite 2673	M-265
Elkonite 3042	M-265
Elkonite 3135	M-265
Elkonite 3150	M-265
Elkonite 3165	M-265
Elkonite 4050	M-265
Elkonite 4055	M-265
Elkonite 7130	M-265
Elkonite 7150	M-265
Elkonite 7160	M-265

Elkonite G12	M-265	empire 2511	M-162	English Steel	
Elkonite G13	M-265	Empire 2810	M-162	AW.4P	M-1414
Elkonite G14	M-265	Empire 2815	M-162	English Steel AW.5	M-1414
Elkonite G17	M-265	Empire CF-3	M-162	English Steel	
Elkonite G18	M-265	Empire CF-3M	M-162	B.4CR	M-1414
Elkonite IW3	M-265	Empire CF-16FA	M-162	English Steel BC	M-1414
Elkonite OW3	M-265	Empire CF-30	M-162	English Steel	
Elkonite TC5	M-265	Empire "D"	M-162	BC.8V	M-1414
Elkonite TC10	M-265	Empire Machinery		English Steel	
Elkonite TC-20	M-265	Steel	M-38	BC.10V	M-1414
Elkonite TC-53	M-265	Empire No. 5	M-162	English Steel BCTA	M-1414
Elkonium 12	M-265	Empire No. 11	M-162	English Steel CD	M-1414
Elkonium 15	M-265	Empire No. 13	M-162	English Steel CV.4	M-1414
Elkonium 21	M-265	Empire No. 16A	M-162	English Steel CVM	M-1414
Elkonium 24	M-265	Empire No. 18	M-162	English Steel	
Elkonium 26	M-265	Empire No. 18-8	M-162	HSM/W.9A	M-1414
Elkonium 31	M-265	Empire No. 18B	M-162	English Steel HW	M-1414
Elkonium 32	M-265	Empire No. 18C	M-162	English Steel M1C	M-1414
Elkonium 35	M-265	Empire No. 18Cu	M-162	English Steel	
Elkonium 37	M-265	Empire No. 23	M-162	M!C.3	M-1414
Elkonium 38	M-265	Empire No. 24-12	M-162	English Steel PN.1	M-1414
Elkonium 39	M-265	Empire No. 25-5	M-162	English Steel T.3N	M-1414
Elkonium 41	M-265	Empire No. 30	M-162	English Steel VAP	M-1414
Elkonium 42	M-265	Empire No. 35-15	M-162	English Steel	
Elkonium 44	M-265	Empire No. 60-20	M-162	VAP.2V	M-1414
Elkonium 45	M-265	Empire No. 75-20	M-162	English Steel W.3	M-1414
Elkonium 46	M-265	Empire Resista	M-162	English Steel XX.9	
Elkonium 47	M-265	Empire SHA	M-162	Cr	M-1414
Elkonium 49	M-265	Empire Steel	M-38	Engraver Plate	M-38
Elkonium 60	M-265	Enduria	M-24	Engraver's Brass	M-102
Elkonium 73	M-265	Enduro 4-6 Chrome	M-97	E No. 25	M-69
Elkonium 75	M-265	Enduro 4-6 Cr-Mo	M-97	Enorm S	M-1339
Elkonium 117	M-265	Enduro 4-6 Cr-W	M-97	Entecrod 192	M-717
Elkonium 305	M-265	Enduro 18-8 HT	M-97	ERA 59	M-62
Elkonium 323	M-265	Enduro 18-8 Mo	M-97	ERA 60 M	M-62
Elkonium 401	M-265	Enduro 18.8 Ti	M-97	ERA 131	M-62
Elkonium 404	M-265	Enduro 19-9	M-97	ERA 156	M-62
Elkonium 460	M-265	Enduro 19-9S	M-97	ERA 165	M-62
Elkonium No. 13	M-265	Enduro 19-9S Ti	M-97	ERA 171	M-62
Elkonium No. 25	M-265	Enduro 20-10	M-97	ERA 1414	M-62
Elkonium No. 57	M-265	Enduro 20-23	M-97	ERA 7393	M-62
Elkro	M-357	Enduro 20-25	M-97	E.R.A. "A.T.V."	M-Eng.
Eltun Die Steel	M-357	Enduro A	M-97	ERA A.T.V.	M-62
Elvandi	M-115	Enduro Free		ERA Boron 1	M-62
Elverite A	M-17	Cutting (F.C.)	M-97	ERA Boron 2	M-62
Elverite "B"	M-447	Enduro KA2	M-97	ERA Boron 3	M-62
Elverite C	M-447	Enduro K.A.-2F.M.	M-97	ERA Boron 4	M-62
Elverite D	M-447	Enduro KA2Mo	M-97	ERA C.R.2	M-62
EM IV	M-459	Enduro KA2S	M-97	ERA C.R.3	M-62
Em 15	M-1	Enduro KA2SMo	M-97	ERA CR4	M-62
Embeebush Metal	M-179	Enduro KA2STi	M-97	ERA CR4S	M-62
EMOCW	M-459	Enduro KA2Ti	M-97	ERA C.R. Boron	M-62
EMP 300 M	M-1716	Enduro KA28	M-97	ERA C.R.I.	M-62
EMP 400 MS	M-1716	Enduro KM1	M-97	ERA C.R.I.L.	M-62
EMP 4600	M-1716	Enduro KNC3	M-97	ERA C.R.I.S.	M-62
EMP 8600	M-1716	Enduro NCT3	M-97	ERA C.R.I.S. (Cb)	M-62
EMP 9400	M-1716	Enduro S	M-97	ERA H.R.-1	M-62
Empire 12	M-162	Enduro S15	M-97	ERA HR2	M-62
Empire 18	M-162	Enduro SFC	M-97	ERA HR3	M-62
Empire 22 C	M-162	Endweldur	M-653	ERA H.R. 4	M-62
Empire 30 H	M-162	Endweldur 85	M-653	ERA H.R. 5	M-62
Empire 46	M-162	Endweldur 125	M-653	ERA H.R. 6	M-62
Empire 1214	M-162	Engineering	M-472	ERA HR7	M-62
Empire 2010	M-162	English Steel AW	M-1414	ERA H.R.I.S.	M-62

Ergste 17	M-1333	ES-18-12 No	M-1014	E.S.C. W12	M-50	
Ergste 17L	M-1333	ES-19-9	M-1014	E.S.C. XX9Cr	M-50	
Ergste 17LT	M-1333	ES-19-9L	M-1014	ES Extra	M-459	
Ergste 17T	M-1333	ES 19-12 MoL	M-1014	ES Prima	M-27	
Ergste 17U	M-1333	ES-21	M-1014	E-Steel	M-173	
Ergste 25L	M-1333	ES-25-12	M-1014	Eternos 10	M-1339	
Ergste 80K	M-1333	ES-25-20	M-1014	Eternos 11	M-1339	
Ergste 80P	M-1333	ES-25-20Si	M-1014	Eternos 23	M-1339	
Ergste 80S	M-1333	ES-27	M-1014	Etiral	M-104	
Ergste 80T	M-1333	ES-A286	M-1014	Eureka	M-35	
Ergste 80U	M-1333	E.S.C. 5CC	M-50	Eureka 40-A A.H.		
Ergste 80X	M-1333	Escalloy 20	M-1014	Bare Rod	M-765	
Ergste 82P	M-1333	E.S.C. AW5	M-50	Eureka 50 Bronze	M-765	
Ergste 82RT	M-1333	E.S.C. B4CR	M-50	Eureka 332 Heat		
Ergste 82S	M-1333	E.S.C. BCTA	M-50	Resisting	M-765	
Ergste 82T	M-1333	E.S.C. CHC	M-50	Eureka 400P	M-765	
Ergste 82X	M-1333	E.S.C. CP	M-50	Eureka 500 and 580		
Ergste 120	M-1333	E.S.C. CRC	M-50	High Alloy		
Ergste 135P	M-1333	E.S.C. CV4	M-50	Electrodes	M-765	
Ergste 182RT	M-1333	E.S.C. CVM2	M-50	Eureka 1000-A		
Ergste CN20	M-1333	E.S.C. CVM3	M-50	H.W. Bare Rod	M-765	
Ergste CN30	M-1333	E.S.C. DSCO	M-50	Eureka Exp-10 and		
Ergste CN60	M-1333	E.S.C. HCRS	M-50	Exp-20 Surfacing		
Ergste CN80	M-1333	E.S.C. HSM/W9A	M-50	Electrodes	M-765	
Ergste E110S	M-1333	E.S.C. HW	M-50	Eureka High Speed	M-765	
Ergste ESCV110	M-1333	E.S.C. HYKRO	M-50	Eurekamold Bare		
Ergste ESWV	M-1333	E.S.C. Immaculate		Rod	M-765	
Ergste K13	M-1333	No. 1	M-50	Eureka No. 50 C	M-765	
Ergste K20	M-1333	E.S.C. Immaculate		Eutallite-6	M-717	
Ergste K20L	M-1333	No. 2	M-50	Eutallite Universal		
Ergste K20N	M-1333	E.S.C. Immaculate		10092	M-717	
Ergste K40	M-1333	No. 3	M-50	Eutecbor 9	M-717	
Ergste K90	M-1333	E.S.C. Immaculate		Eutec Chrom 9	M-717	
Ergste K90L	M-1333	No. 4	M-50	Eutec Rod 14	M-717	
Erm 3A	M-1083	E.S.C. Immaculate		Eutecrod 19	M-717	
Ermalite	M-51	No. 6	M-50	Eutecrod 80	M-717	
Erm ALW	M-1083	E.S.C. Immaculate		Eutecrod 80 FC	M-717	
Erm CCS	M-1083	No. 7	M-50	Eutecrod 91	M-717	
Erm HSM	M-1083	E.S.C. M1C	M-50	Eutecrod 155	M-717	
Erm NS	M-1083	E.S.C. M1C3	M-50	Eutecrod 158-B	M-717	
Erm SC65	M-1083	E.S.C. M1C4	M-50	Eutecrod 186 FC	M-717	
Erodur E 2	M-1290	E.S.C. M1C8	M-50	Eutec Rod 189	M-717	
Erodur E-C1 H	M-1290	ESCO 5	M-438	Eutecrod 800	M-717	
Erodur E-VMS	M-1290	ESCO Nirosta		Eutec Rod 1700	M-717	
Erodur MNA 1 U	M-1290	KNC-3	M-438	Eutecrod 1803	M-717	
Erodur MNA 2	M-1290	ESCO No. 16-N	M-438	Eutec Rod 1850FC	M-717	
Erzberg Weich	M-1314	ESCO No. 20	M-748	Eutecrod 1909	M-717	
Erzberg ZAH	M-1314	ESCO No. 40	M-438	Eutec-Tinweld I	M-717	
Erzberg Zahhart	M-1314	ESCO No. 41W	M-438	Eutec-Tinweld II	M-717	
ES3	M-459	ESCO No. 43	M-438	Eutec-Tinweld III	M-717	
ES4	M-459	ESCO No. 45	M-438	Eutec-Tinweld IV	M-717	
ES5	M-459	ESCO Stainless No.		Eutec Trode 24	M-717	
ES6	M-459	33	M-438	Eutectrode 24/49	M-717	
ES-12	M-1014	ESCO Stainless No.		Eutectrode 65	M-717	
ES-17	M-1014	34	M-438	Eutectrode 300	M-717	
ES 17-7	M-1014	ESCO Stainless No.		Eutectrode 691	M-717	
ES-18-8	M-1014	35	M-438	Eutec Trode 2100	M-717	
ES-18-8 LC	M-1014	ESCO Stainless No.		Eutectrode N 4	M-717	
ES 18-8Si	M-1014	49	M-438	Eutectrode N 40	M-717	
ES-18-10 CB	M-1014	E.S.C. PN1	M-50	Eutectrode N 61	M-717	
ES-18-10Cb-LO-Ta	M-1014	E.S.C. T3N	M-50	Eutectrode N 6800	M-717	
ES-18-10 Ti	M-1014	E.S.C. VAGS	M-50	E.V. 300	M-752	
ES-18-12	M-1014	E.S.C. W3	M-50	E-Van	M-85	
ES-18-12 MoL	M-1014	E.S.C. W5	M-50	Evansteel "O"	M-146	

Everard	M-589	Extrude Die	M-115	Fagersta FB-H13	M-52
Everbrite No. 82	M-154	Extrusion Die	M-73	Fagersta FB-M1	M-52
Everbrite No. 90	M-154	E-Z-Cut 20	M-240	Fagersta FB-M2	M-52
Everbrite No. 92	M-154	EZ Head	M-12	Fagersta FB-M3	
Everdur No. 50	M-8	E-Z Stainless Steel	M-73	(Class 2)	M-52
Everdur No. 1026	M-8			Fagersta FB-M7	M-52
Evershync	M-233	**F**		Fagersta FB-M10	M-52
EVHI Alloy	M-752			Fagersta FB-O1	M-52
EW	M-459	F-8 Iron	M-399	Fagersta FB-O6	
EWP	M-1114	F-16	M-399	Graphitic	M-52
Excel Hollow Drill	M-34	F-17 Trimetal	M-739	Fagersta FB-S1	M-52
Excello	M-345	F-23	M-739	Fagersta FB-S1	
Excelo	M-32	F-37	M-399	Mirycal	M-52
Excelsior	M-248	FA-3	M-1699	Fagersta FB-S4	M-52
Excelsior	M-98	FA-4	M-1699	Fagersta FB-S5	
Exelloy	M-748	FA-5	M-1699	Tufcut	M-52
Exl-Die Smoothcut	M-35	FA-6	M-1699	Fagersta FB-S7 Tuf	
Exlo	M-394	FA-7	M-1699	Die Air	
Exocut	M-1	FA-8	M-1699	Hardening Shock	
Exohard	M-1	FA-9	M-1699	Resisting	M-52
Exp. 99	M-106	FA-62	M-1699	Fagersta FB-T1	M-52
Extra	M-1116	Fagersta-16	M-52	Fagersta FB-T4	M-52
Extra	M-32	Fagersta-20	M-52	Fagersta FB-W1	
Extra 214	M-27	Fagersta-24	M-52	Cold Header Die	
Extra Best Cast	M-389	Fagersta A540	M-52	Steel	M-52
Extra Carbon		Fagersta B85	M-52	Fagersta FB-W1	
Vanadium	M-1	Fagersta B 103	M-52	Extra Carbon	M-52
Extra Cost	M-618	Fagersta B 805	M-52	Fagersta FB-W2	M-52
Extrad	M-85	Fagersta C46	M-52	Fagersta High	
Extra D	M-96	Fagersta C50	M-52	Production	M-385
Extra (Disston)	M-357	Fagersta C71	M-52	Fagersta Hollow	
Extra D.R.	M-73	Fagersta C143	M-52	Drill	M-385
Extra Duro MX4	M-616	Fagersta C182	M-52	Fagersta K291	M-52
Extra Extra	M-459	Fagersta C265	M-52	Fagersta K333	M-52
Extra Extra		Fagersta C345	M-52	Fagersta K336	M-52
E.S.C.-XX	M-50	Fagersta C424	M-52	Fagersta K669	M-52
Extra H	M-73	Fagersta C550	M-52	Fagersta K825	M-52
Extra Header Die	M-35	Fagersta C642	M-52	Fagersta K845	M-52
Extra L	M-115	Fagersta C643	M-52	Fagersta L97	M-52
Extra "L"	M-34	Fagersta Circular		Fagersta L441	M-52
Extra No. 5	M-360	Plates	M-385	Fagersta L536	M-52
Extra Punch	M-72	Fagersta Cold		Fagersta Overcoat	
Extra Rapid 300	M-27	Header	M-385	Axe	M-608
Extra Rapid 300A	M-27	Fagersta Cutlery		Fagersta P10	M-52
Extra Special	M-35	Steel	M-385	Fagersta P15	M-52
Extra Special	M-74	Fagersta D65	M-52	Fagersta Pavement	
Extra Special	M-260	Fagersta D66	M-52	Breaker	M-385
Extra Special Alloy	M-73	Fagersta D110	M-52	Fagersta Polheim	
Extra Special High		Fagersta D249	M-52	Wire	M-385
Speed	M-24	Fagersta D366	M-52	Fagersta Q5	M-52
Extra Special		Fagersta FB-A-2	M-52	Fagersta Q10	M-52
(Potts)	M-307	Fagersta FB-A-4	M-52	Fagersta QRO-45	
Extra Special Power		Fagersta FB-A-6	M-52	Hot Work	M-52
Transformer		Fagersta FB-		Fagersta R140	M-52
52-58	M-806	Brilliant Die	M-52	Fagersta R350	M-52
Extra Tenaz Duro	M-27	Fagersta FB-D2	M-52	Fagersta R740	M-52
Extra Tenaz Duro	M-616	Fagersta FB-D2		Fagersta RO 8155	
Extra Tough and		High Production	M-52	Hot Work	M-52
Hard	M-27	Fagersta FB-D3		Fagersta RRJ10	M-52
Extra Tough No. 7	M-69	High Production		Fagersta RRJ11	M-52
Extra Tough No. 7		"H"	M-52	Fagersta RRJ11	M-572
H.C.	M-69	Fagersta FB-D6	M-52	Fagersta RRJ14	M-52
Extra Valtool	M-357	Fagersta FB-H11	M-52	Fagersta RRM20	M-52
Extra Warranted	M-260	Fagersta FB-H12	M-52	Fagersta RRM22	M-52

Fagersta RRM23	M-52	Fahrite N-2	M-95	Fasaloy No. 106	M-53
Fagersta RRM24	M-52	Fahrite N-3-35	M-95	Fasaloy No. 136	M-53
Fagersta RRM28	M-52	Fahrite N-3-35-N	M-95	Fasaloy No. 138	M-53
Fagersta RRNJ30	M-52	Fahrite N-3-40	M-95	Fasaloy No. 139	M-53
Fagersta RRNJ30.32	M-52	Fahrite N-5	M-95	Fastbor Hollow Drill	M-908
Fagersta RRNJ31	M-52	Fahrite N-6	M-95	Fastell BH	M-53
Fagersta RRNJ32	M-52	Fahrite N-6 Mo	M-95	Fastell E	M-53
Fagersta RRNJ33	M-52	Fahrite N-7	M-95	Fastell N	M-53
Fagersta RRNJ35	M-52	Fahrite N-7A	M-95	Fastell NL	M-53
Fagersta RRNJ36	M-52	Fahrite N-7B	M-95	Fastell UM	M-53
Fagersta RRNJ38	M-52	Fahrite N-9	M-95	Fastell UP	M-53
Fagersta RRNJ40	M-52	Fahrite N-9A	M-95	Fastell UR	M-53
Fagersta RRNJ41	M-52	Fahrite N-10	M-95	Fastell No. 31	M-53
Fagersta RRNJ42	M-52	Fahrite N-11	M-95	Fastell No. 35	M-53
Fagersta RRNJ44	M-52	Fahrite N-17	M-95	Fastell No. 41	M-53
Fagersta RRNJ45	M-52	Fahrite N-31	M-95	Fastell No. 42	M-53
Fagersta RRNJ46	M-52	Fanite	M-53	Fastell No. 51	M-53
Fagersta RRNJ47	M-52	Fansteel 17	M-53	Fastell No. 72	M-53
Fagersta RRNJ50	M-52	Fansteel 21	M-53	Fastell No. 130	M-53
Fagersta RRNJ51	M-52	Fansteel 42	M-53	Fastell No. 142	M-53
Fagersta RRNJ52	M-52	Fansteel 60	M-53	Fastell 2900	M-53
Fagersta RRNJ59	M-52	Fansteel 77	M-53	Fast Finishing	M-32
Fagersta RRO4	M-52	Fansteel 80	M-53	Fast Finishing	M-365
Fagersta RRO6	M-52	Fansteel 82	M-53	Fast Finishing (Disston)	M-357
Fagersta RRS70	M-52	Fansteel 85	M-53	Fatigue-Proof	M-669
Fagersta RRS70 (72)	M-52	Fansteel 99	M-53	Faultless "A" Babbitt	M-317
Fagersta RRS71 (7.3)	M-52	Fansteel 222	M-53	Favorito SC1	M-616
		Fansteel 601	M-53	FB-3	M-1699
Fagersta RRS72	M-52	Fansteel 602	M-53	FB-4	M-1699
Fagersta RRS73	M-52	Fansteel 603	M-53	FB-5	M-1699
Fagersta RRS74	M-52	Fansteel 606	M-53	FB-6	M-1699
Fagersta RRT80	M-52	Fansteel 607	M-53	FC-5XI Spec.	M-439
Fagersta RRT83	M-52	Fansteel BL2	M-53	FC-5XIV	M-1
Fagersta RRT85	M-52	Fansteel No. 1	M-53	FC-85A	M-220
Fagersta RRV60	M-52	Fansteel No. 2	M-53	FC-85AC	M-220
Fagersta RRV61	M-52	Fansteel SCB-291	M-53	FC-85-AKN	M-220
Fagersta RRV61	M-52	Fanweld	M-53	FC-85-AN	M-220
Fagersta RRV62	M-52	Farco	M-253	FC-85-AV	M-220
Fagersta RRV62	M-52	Farm-Alloy	M-507	FC-85-BN	M-220
Fagersta RRV64	M-52	Farrell-Cheek F-85	M-220	FC-85-HN	M-220
Fagersta RRV64	M-52	Farrell-Cheek F-85-ACM	M-220	FC-85 T	M-220
Fagersta RRV66	M-52	Farrell-Cheek F-85-AKN	M-220	FC-440C	M-1
Fagersta S310	M-52	Farrell-Cheek FC-1020	M-220	FC-1025	M-220
Fagersta W221	M-52	Farrell F-85-B	M-220	FC-1025V	M-220
Fagersta W401	M-52	Farrell-F-85-BKN	M-220	FC 1040	M-220
Fagersta W406	M-52	Farrell F-85-HCM	M-220	FC-1040 N	M-220
Fahrite C-1	M-95	Farrell F-85-L	M-220	FC-1040 Q	M-220
Fahrite C-2	M-95	Farrell's "85" Alloy (A)	M-220	FC-1084	M-1
Fahrite C-7	M-95	Farrell's Hard Edge	M-220	FC-8740	M-439
Fahrite C-7M	M-95	Fasaloy	M-53	FC-AH	M-439
Fahrite C-8	M-95	Fasaloy 72	M-53	FC Air Hardening Cast to Shape	M-1
Fahrite C-8-16	M-95	Fasaloy 99	M-53	FC Air Hardening Forging	M-1
Fahrite C-8-16M	M-95	Fasaloy No. 3	M-53	FC-ALX	M-1
Fahrite C-8A	M-95	Fasaloy No. 4	M-53	FC-ALX No. 6	M-439
Fahrite C-8AM	M-95	Fasaloy No. 5	M-53	F.C.C. 4	M-439, M-1
Fahrite C-12-20	M-95	Fasaloy No. 6	M-53		
Fahrite C-38	M-95	Fasaloy No. 7	M-53	F.C.C. 6	M-439, M-1
Fahrite C-41	M-95	Fasaloy No. 12	M-53		
Fahrite C-316	M-95	Fasaloy No. 14	M-53		
Fahrite C-317	M-95	Fasaloy No. 24	M-53		
Fahrite C-S	M-95				
Fahrite N-1A	M-95				

F.C.C. 7	M-439, M-1	FC-Roloy No. 2	M-1	Federaloy HF-24	M-709
F.C.C. 11	M-439, M-1	FC Spezial	M-459	Federaloy RS-25	M-709
		FCW Special	M-459	Federaloy RS-27	M-709
		FD-3	M-1699	Federaloy RS-29	M-709
FCC 16B Cast-to-Shape	M-1	FD-4	M-1699	Fedol	M-375
		FD-5	M-1699	Fenicoloy	M-1523
F.C.C. 17	M-439, M-1	Federal High Speed	M-375	Feralloy	M-104
		Federal-Mogul		Feran	M-798
F.C.C. 21	M-439, M-1	B-101	M-709	Ferimphy	M-771
		Federal-Mogul		Fermo Special	M-499
F.C.C. 25	M-439	B-103	M-709	Fernichrome	M-60
F.C.C. 26	M-439	Federal-Mogul		Fernico	M-60
F.C.C. 27	M-439	B-104	M-709	Fernico II	M-60
F.C.C. 34M	M-439	Federal-Mogul		Fernico 5	M-31
F.C.C. 39	M-439	B-109	M-709	Fernite 17A	M-439
F.C.C. 41M	M-439	Federal-Mogul		Fernite 17B	M-439
F.C.C. 44	M-439	B-10025	M-709	Fernite 24	M-439
F.C.C. 46M	M-439	Federal-Mogul		Fernite No. 1	M-439
F.C.C. 47M	M-439	B-10105	M-709	Fernite No. 2	M-439
F.C.C. 49	M-439	Federal-Mogul		Fernite No. 3	M-439
F.C.C. 53	M-439	B-10325	M-709	Fernite No. 4	M-439
FCC 168	M-1	Federal-Mogul		Fernite No. 6	M-439
F.C.C. Air Hardening	M-1	B-10425	M-709	Fernite No. 7	M-439
		Federal-Mogul		Fernite No. 8	M-439
F.C.C. Air Hardening No. 48M	M-439	B-10925	M-709	Fernite No. 9	M-439
		Federal-Mogul		Fernite SC	M-439
		B-11125	M-709	Ferrimag 1	M-38
FC Cast Tool Steel	M-1	Federal-Mogul		Ferrimag 5	M-38
F.C.C. Oil Hardening	M-439	CS-50	M-709	Ferro-Titanium 18-25	M-394
FC-CMS	M-439	Federal-Mogul Duro	M-709	Ferro-Titanium	
F.C.C. No. 3 1/2	M-439, M-1	Federal-Mogul F-1	M-709	High Carbon	M-394
		Federal-Mogul F-2	M-709	Ferro-Titanium Low	
		Federal-Mogul F-3	M-709	Carbon	M-394
F.C.C. No. 22	M-439, M-1	Federal-Mogul F-5	M-709	Ferrotrode N2B	M-717
		Federal-Mogul F-6	M-709	Ferrozoid	M-121
FC-CNS	M-439	Federal-Mogul F-8	M-709	Festel Metal	M-167
F.C.C.-P.R.K.-33 (Cast-to-shape)	M-439	Federal-Mogul F-11	M-709	Festel Stellite	M-167
		Federal-Mogul F-13	M-709	F Extra	M-459
F.C.C. R3	M-1, M-439	Federal-Mogul F-15	M-709	FH-2	M-1699
		Federal-Mogul F-16	M-709	FH-3	M-1699
F.C.C. TMS	M-439, M-1	Federal-Mogul F-18	M-709	FH-4	M-1699
		Federal-Mogul F-19	M-709	FH-5	M-1699
F.C.C. TS	M-439, M-1	Federal-mogul F-20	M-709	FH-6	M-1699
		Federal-Mogul F-22	M-709	Fibro	M-499
FC-CTS	M-1	Federal-Mogul F-23	M-709	Fibrtough	M-435
FC-CV	M-1	Federal-Mogul F-26	M-709	Filnic F	M-44
FC-EZ	M-439	Federal-Mogul F-27	M-709	Filnic H	M-44
FC-FH	M-1	Federal-Mogul F-28	M-709	Filnic I	M-44
FC-Flamhard	M-1	Federal-Mogul F-32	M-709	Filnic J	M-44
FCF Special	M-459	Federal-Mogul F-33	M-709	Filnic K	M-44
FC-NCI	M-439	Federal-Mogul F-34	M-709	Filnic M	M-44
FC No. 1	M-439	Federal-Mogul		Filnic T	M-44
FC No. 2	M-439	L-100	M-709	Finkl Cold-Hot	M-55
FC No. 5	M-439	Federal-Mogul No. 407	M-709	Finkl FI-Carbo	M-55
FC No. 5XI	M-439			Finkl FX	M-55
FC No. 10	M-439	Federal-Mogul No. 408	M-709	Finkl Grade F	M-55
FC No. 14	M-439			Finkl Mo-Cro-Ni	M-55
FC No. 19	M-439	Federal-Mogul Special "B"	M-709	Finkl Type R	M-55
FC No. 20	M-439			Firedie 13	
FC No. 23	M-439	Federaloy AF-6	M-709	Smoothcut	M-35
FC No. 38M	M-439	Federaloy B-30	M-709	Firminy A.C.M	M-1043
FC No. 55	M-439	Federaloy B-35	M-709	Firminy A.D.F.	M-1043
FC No. 66	M-439	Federaloy B-40	M-709	Firminy A.D.F.M.	M-1043
FC-Roloy	M-1	Federaloy F-12	M-709		

Section IV: Obsolete Alloys / 1707

Firminy A.L.S.	M-1043	Firminy V.D.L.D.	M-1043	Firth-Brown CR3T	M-56
Firminy A.M.3	M-1043	Firth	M-56	Firth-Brown CR4T	M-56
Firminy B.L.N.	M-1043	Firth Best Tool	M-57	Firth-Brown CRM4	M-56
Firminy B.O.L.	M-1043	Firth-Brown 20C	M-56	Firth-Brown CRM5	M-56
Firminy B.T.R.	M-1043	Firth-Brown 30C	M-56	Firth-Brown CRM6	M-56
Firminy C.L.D.	M-1043	Firth-Brown 40C	M-56	Firth-Brown CRM0	M-56
Firminy C.M.Y.1	M-1043	Firth-Brown 50C	M-56	Firth-Brown CRV3	M-56
Firminy C.M.Y.2	M-1043	Firth-Brown 50CT	M-56	Firth-Brown CRWC	M-56
Firminy C.M.Y.4	M-1043	Firth-Brown 60CT	M-56	Firth-Brown CVW3	M-56
Firminy C.M.Y.5	M-1043	Firth-Brown 65CF	M-56	Firth-Brown D-60	
Firminy C.M.Y.6	M-1043	Firth-Brown 65CT	M-56	(Atlas A.T.M.N.)	M-56
Firminy C.M.Y.7	M-1043	Firth-Brown A-1	M-56	Firth-Brown D-62	
Firminy C.M.Y.16	M-1043	Firth-Brown A-2	M-56	(Firth F.N.35)	
Firminy C.N.W.	M-1043	Firth-Brown A-3	M-56	(Atlas A.T.N.X.)	M-56
Firminy C.R.V.	M-1043	Firth-Brown A-4	M-56	Firth-Brown D-63	
Firminy C.T.	M-1043	Firth-Brown A-5	M-56	(Firth F.N.5)	M-56
Firminy C.T.N.2	M-1043	Firth-Brown A-6	M-56	Firth-Brown D-65	
Firminy C.T.N.6	M-1043	Firth-Brown A-7	M-56	(Atlas A.T.C.V.)	M-56
Firminy C.T.N.C.4	M-1043	Firth-Brown ANC2	M-56	Firth-Brown E-80	
Firminy C.T.N.C.5	M-1043	Firth-Brown B-20	M-56	(Duratlas)	M-56
Firminy C.T.N.M.	M-1043	Firth-Brown B-21	M-56	Firth-Brown E-82	
Firminy C.T.N.V.	M-1043	Firth-Brown B-22	M-56	(Firth Bolt Steel)	M-56
Firminy C.V.A.1	M-1043	Firth-Brown B-23	M-56	Firth-Brown E-83	
Firminy C.V.A.2	M-1043	Firth-Brown B-24	M-56	(Atlas N.C.100)	M-56
Firminy C.V.A.3	M-1043	Firth-Brown B-25		Firth-Brown Extra	
Firminy C.V.A.4	M-1043	(Atlas S.P. Steel)	M-56	CS	M-56
Firminy C.V.A.6	M-1043	Firth-Brown B-26	M-56	Firth-Brown F60C	M-56
Firminy C.V.A.7	M-1043	Firth-Brown B-27	M-56	Firth-Brown F65CH	M-56
Firminy C.W.2	M-1043	Firth-Brown B-28		Firth-Brown F95C	M-56
Firminy C.W.3	M-1043	(Firth Roll Steel)		Firth-Brown F-103	
Firminy F.C.K.	M-1043	(Atlas Die Ring		(Atlas N.C.C.H.)	M-56
Firminy F.G.O.	M-1043	Steel)	M-56	Firth-Brown F543	M-56
Firminy G.R.N.	M-1043	Firth-Brown B-29		Firth-Brown FCMO	M-56
Firminy ICN001	M-1043	(Firth Shoe &		Firth-Brown FCVA	M-56
Firminy ICN64T	M-1043	Die Steel)	M-56	Firth-Brown FCW5	M-56
Firminy ICN164	M-1043	Firth-Brown B-30		Firth-Brown FN15	M-56
Firminy ICN472	M-1043	(Firth S.H.B.)		Firth-Brown FN50	M-56
Firminy ICN472T	M-1043	(Atlas Ball Race		Firth-Brown FNCF	M-56
Firminy ICN583	M-1043	Steel)	M-56	Firth-Brown FPRF	M-56
Firminy L.F.	M-1043	Firth-Brown Best		Firth-Brown G-120	
Firminy M.12A.F.Y.	M-1043	C.S.	M-56	(Nitralloy Grade	
Firminy M.14A.F.Y.	M-1043	Firth-Brown BRB	M-56	1)	M-56
Firminy M.A.S.	M-1043	Firth-Brown C-40		Firth-Brown G-121	M-56
Firminy M.C.R.1	M-1043	(Atlas A.T.C.20)	M-56	Firth-Brown G-122	M-56
Firminy M.C.R.2	M-1043	Firth-Brown C-42		Firth-Brown G-123	
Firminy M.C.R.3	M-1043	(Atlas A.T.C.40)	M-56	(Nitralloy Grade	
Firminy M.C.R.4	M-1043	Firth-Brown C-44		7)	M-56
Firminy M.C.R.5	M-1043	(Firth F.T.50)		Firth-Brown GK3	M-56
Firminy M.L.	M-1043	(Atlas A.T.C.55)	M-56	Firth-Brown GK5	M-56
Firminy N.2	M-1043	Firth-Brown C-45		Firth-Brown GK7	M-56
Firminy N.3	M-1043	(Altas A.T.C.75)	M-56	Firth-Brown H-140	M-56
Firminy N.5	M-1043	Firth-Brown Cast		Firth-Brown H-141	M-56
Firminy N.C.1	M-1043	Steel	M-56	Firth-Brown H-142	M-56
Firminy N.C.2	M-1043	Firth-Brown Cast		Firth-Brown H-143	M-56
Firminy N.C.3	M-1043	Steel Warranted	M-56	Firth-Brown H-144	M-56
Firminy N.C.4	M-1043	Firth-Brown CCR 1	M-56	Firth-Brown H-145	M-56
Firminy N.C.M.2	M-1043	Firth-Brown CCRF	M-56	Firth-Brown	
Firminy N.C.M.4	M-1043	Firth-Brown CCW	M-56	Hacksaw	M-56
Firminy P.F.	M-1043	Firth-Brown CDD	M-56	Firth-Brown HDBS	M-56
Firminy P.F.3	M-1043	Firth-Brown		Firth-Brown I-163	
Firminy S.A.M.	M-1043	Chromva-W	M-56	(Saitie)	M-56
Firminy T.R.D.	M-1043	Firth-Brown CMCR	M-56	Firth-Brown J-183	M-56
Firminy T.S.W.	M-1043	Firth-Brown Crit	M-56	Firth-Brown J-185	M-56
Firminy V.D.L.	M-1043	Firth-Brown CR2T	M-56		

Name	Ref
Firth-Brown L-223 (Firth "T" Brand)	M-56
Firth-Brown L-225 (Firth Textile Spindle)	M-56
Firth-Brown LCMO	M-56
Firth-Brown M13F	M-56
Firth-Brown M-241 (Firth T.M.S.)	M-56
Firth-Brown M-242 (Firth M.C.T.)	M-56
Firth-Brown M-243 (Trutap)	M-56
Firth-Brown M-244 (Atlas Self-Hardening Chisel)	M-56
Firth-Brown M-245 (Firth Extra W.C.S.)	M-56
Firth-Brown M-246 (Firth W.C.S.)	M-56
Firth-Brown M-247 (Firth T.D.C.)	M-56
Firth-Brown M-248 (Firth Hacksaw Steel)	M-56
Firth-Brown M-249 (Firth Extra Superior)	M-56
Firth-Brown M-250 (Wolfram C.S.)	M-56
Firth-Brown M-251 (Lascut)	M-56
Firth-Brown M-252 (Firth 5C.S.)	M-56
Firth-Brown M-253 (Firth C.Y.W.)	M-56
Firth-Brown M-255 (Firth Extra Hard Sheet)	M-56
Firth-Brown M-256 (Firth C.C.W.)	M-56
Firth-Brown M-258 (Stanslog)	M-56
Firth-Brown M-259 (Firth B.R.B.)	M-56
Firth-Brown M-260 (C.D.D.)	M-56
Firth-Brown M-262 (Dylasta)	M-56
Firth-Brown MCMT	M-56
Firth-Brown MCT	M-56
Firth-Brown MM0F	M-56
Firth-Brown MN1F	M-56
Firth-Brown MN3F	M-56
Firth-Brown MO1F	M-56
Firth-Brown MOVA (1)	M-56
Firth-Brown MOVA (2)	M-56
Firth-Brown NCM2	M-56
Firth-Brown NCM7	M-56
Firth-Brown NCMCH	M-56
Firth-Brown NCMF	M-56
Firth-Brown NCMV	M-56
Firth-Brown NCR1	M-56
Firth-Brown NCR3	M-56
Firth-Brown NIMCH	M-56
Firth-Brown NMCW	M-56
Firth-Brown NMM2	M-56
Firth-Brown O5CF	M-56
Firth-Brown O-275 (Firth N.M.C.)	M-56
Firth-Brown P-280 (Foundry A.A.)	M-56
Firth-Brown P-281 (Tungsten Magnet) (Atlas Tungsten Magnet)	M-56
Firth-Brown P-282 (Tungsten Chrome Magnet)	M-56
Firth-Brown P-283 (Cobalt Magnet)	M-56
Firth-Brown Q-300	M-56
Firth-Brown Q-301	M-56
Firth-Brown Q-302	M-56
Firth-Brown Q-303	M-56
Firth-Brown Q-304	M-56
Firth-Brown Q-305	M-56
Firth-Brown Q-306	M-56
Firth-Brown Q-307	M-56
Firth-Brown Q-308	M-56
Firth-Brown R.320	M-56
Firth-Brown R.321	M-56
Firth-Brown R.322	M-56
Firth-Brown R.323	M-56
Firth-Brown RBD	M-56
Firth-Brown RBD-E	M-56
Firth-Brown S-340 (Foundry A)	M-56
Firth-Brown S-341 (Foundry B Mild)	M-56
Firth-Brown S-342 (Foundry B)	M-56
Firth-Brown S-343 (Foundry C)	M-56
Firth-Brown S-344 (Foundry D)	M-56
Firth-Brown S-345 (Foundry E)	M-56
Firth-Brown SHC1	M-56
Firth-Brown SP	M-56
Firth-Brown Standard	M-56
Firth-Brown T-360 (Foundry "A, Mn")	M-56
Firth-Brown T-361 (Foundry "C-M")	M-56
Firth-Brown T-362 (Foundry N.C.)	M-56
Firth-Brown T-363 (Foundry N.C.Mo)	M-56
Firth-Brown T-364 (Foundry C-Cr)	M-56
Firth-Brown T-365 (Foundry A-Mo)	M-56
Firth-Brown T-366 (Foundry B, Mo)	M-56
Firth-Brown T-367 (Foundry A, Mn, Mo)	M-56
Firth-Brown T-368 (Adamant T.M.)	M-56
Firth-Brown T-369 (Adamant A.)	M-56
Firth-Brown T-370 (Adamant G-R)	M-56
Firth-Brown T-371 (Adamant H.P.)	M-56
Firth-Brown T-372 (Adamant Extra)	M-56
Firth-Brown T-373 (Foundry Pilger Roll)	M-56
Firth-Brown TDC	M-56
Firth-Brown TMS	M-56
Firth-Brown Treble Extra	M-56
Firth-Brown "U" Brand	M-56
Firth-CHQ	M-57
Firth "Diehard"	M-56
Firth "F-65"	M-56
Firth FCN-5 Steel	M-56
Firth FCNC Steel	M-56
Firth FN-3 Steel	M-56
Firth FNC-3	M-56
Firth FNC Steel	M-56
Firth FT-30 Steel	M-56
Firth FT-40 Steel	M-56
Firthite AAV	M-57
Firthite BDC	M-57
Firthite DC-1	M-57
Firthite DC-2	M-57
Firthite DC-3	M-57
Firthite DC-4	M-57
Firthite DCX	M-57
Firthite Gr. HAX	M-57
Firthite Gr. HE	M-57
Firthite H6	M-57
Firthite H8	M-57
Firthite H-13	M-57
Firthite H-15	M-57
Firthite HAX	M-57
Firthite HC	M-57
Firthite HD	M-57
Firthite HD-3	M-57
Firthite ND	M-57
Firthite ND-20	M-57
Firthite ND-25	M-57
Firthite NHA	M-57

Alloy	Code	Alloy	Code	Alloy	Code
Firthite T-16	M-57	Firth-Vickers		Follansbee Special	
Firthite T-41H	M-57	Immaculate No. 6	M-521	Motor No. 3	M-806
Firthite T-83	M-57	Firth-Vickers		Follansbee Special	
Firthite T-89M	M-57	Stainless FTV	M-521	Power	
Firthite T-90	M-57	Five Star	M-85	Transformer No.	
Firthite TA	M-57	Fixamper	M-771	6	M-806
Firthite XDL	M-57	Fixinvar	M-104	Forez 3ASR	M-1487
Firthite XS	M-57	Flecto Iron	M-94	Forez CH1	M-1487
Firth L.T.	M-57	Fleetweld	M-578	Forez CH3	M-1487
Firth L.T.L.	M-57	Fleetweld 11 HT	M-578	Forez CX50	M-1487
Firth M-10-V	M-57	Fleetweld No. 8	M-578	Forez CX75	M-1487
Firth "Nonvar"		Fleetweld No. 9	M-578	Forez CX100	M-1487
Cast Steel	M-56	Fleetweld No. 9 HT	M-578	Forez CX125	M-1487
Firthob	M-56	Fleetweld No. 10	M-578	Forez E3	M-1487
Firth Psusnap	M-56	Fleetweld No. 11	M-578	Forez M06	M-1487
Firth "RBD" Steel	M-56	Flexible Hack Band	M-357	Forez MT18	M-1487
Firth RT	M-56	Flexograin		Forez N15	M-1487
Firth Sterling Best	M-57	Phosphor Bronze	M-98	Forez NID	M-1487
Firth Sterling CR-1	M-57	Flex-o-loy	M-1579	Forez NVM	M-1487
Firth Sterling CR-2	M-57	Flexo Steel	M-32	Forez SM	M-1487
Firth Sterling CR-3	M-57	Fliegw. 1420	M-96	Forez SMX	M-1487
Firth Sterling Extra	M-57	Flintuff	M-95	Forez TES	M-1487
Firth Sterling		Flintmetal	M-1250	Forez TFR	M-1487
Meteor	M-57	Fluginox 51	M-1120	Forge Die C-30	M-115
Firth Sterling		Fluginox 60	M-1120	Formdie Smoothcut	M-35
Nirosta 19-9	M-57	Fluginox 62	M-1120	Formex 1	M-115
Firth Sterling		Fluginox 65	M-1120	Formex 2	M-115
Nirosta F.C.	M-57	Flylight No. 5	M-1045	Formite No. 1	M-35
Firth Sterling		Flylite No. 4	M-1045	Formite No. 2	M-35
Nirosta "K2A"	M-57	Flylite No.8	M-1045	Formite No. 3	M-35
Firth Sterling		Flylite No.9	M-1045	Fornanc Special	M-1724
Nirosta KA2	M-57	FM-2	M-1699	Fortuna NCS 5	M-1307
Firth Sterling		FM-3	M-1699	Fourdrinier Brass	M-33
Nirosta KA2S	M-57	FM-5	M-1699	Four Star	M-85
Firth Sterling		FM-6	M-1699	Fox CN 16/13 Co	M-430
Special	M-57	Folder Die Steel	M-81	Fox No. 001	M-430
Firth Sterling		Follansbee		Fox No. 002	
Stainless Type		Armature Electric		"Mancase"	M-430
FC	M-57	No. 1	M-806	Fox No. 012 (512)	M-430
Firth Sterling Type		Follansbee Extra		Fox No. 013 (513)	M-430
440-C	M-57	Special Power		Fox No. 014 (514)	
Firth-Vickers 326	M-521	Transformer No.		M 430	
Firth-Vickers 337	M-521	7	M-806	Fox No. 015 (515)	M-430
Firth-Vickers 507	M-521	Follansbee		Fox No. 017 (517)	M-430
Firth-Vickers F17	M-521	Improved Electric		Fox No. 020 (520)	M-430
Firth-Vickers		No. 2	M-806	Fox No. 030 (530)	
FCB(T)	M-521	Follansbee Radio		Ni-Cr Steel	M-430
Firth-Vickers F.I.	M-521	A-1 No. 13	M-806	Fox No. 031 (531)	M-430
Firth-Vickers		Follansbee Radio		Fox No. 040 (540)	M-430
F.M.B. 3T	M-521	A-2 No. 12	M-806	Fox No. 120	M-430
Firth-Vickers FSL	M-521	Follansbee Radio		Fox No. 130	M-430
Firth-Vickers		A-3 No. 11	M-806	Fox No. 131	M-430
F.S.M.1	M-521	Follansbee Radio B		Fox No. 135	M-430
Firth-Vickers F.V.		No. 10	M-806	Fox No. 140 (640)	M-430
507	M-521	Follansbee Radio C		Fox No. 143 (Du-	
Firth-Vickers F.V.		No. 9	M-806	Nic 143)	M-430
702	M-521	Follansbee Radio D		Fox No. 144 (644)	M-430
Firth-Vickers H.R.		No. 8	M-806	Fox No. 145 (645)	M-430
2610	M-521	Follansbee Regular		Fox No. 160 (660)	M-430
Firth-Vickers H.R.		Power		Fox No. 171 (671)	M-430
Crown 1 Steel	M-521	Transformer No.		Fox No. 172 (672)	M-430
Firth-Vickers H.R.		5	M-806	Fox No. 174 (674)	M-430
Crown Max Steel	M-521	Follansbee Special		Fox No. 190 (690)	M-430
		Dynamo No. 4	M-806	Fox No. 195	M-430

Fox No. 200	M-430	Fox SF320	M-430	Furodit NH12	M-912
Fox No. 207 (707)	M-430	Fox SF321	M-430	Furodit No. 8	M-501
Fox No. 208 (708)	M-430	Fox SF 347	M-430	Furodit "S"	M-501
Fox No. 209 (709)	M-430	Fox SF 610	M-430	Furodit "SS"	M-501
Fox No. 213 (713)	M-430	Fox SF 620	M-430	Furodit "Z"	M-501
Fox No. 525	M-430	Fox SF620	M-430	Furodit Z Special	M-501
Fox No. 540	M-430	Fox SF835	M-430	Fultalloy	M-809
Fox No. 721 Nitralloy	M-430	Fox SF920	M-430		
Fox No. 723 Nitralloy	M-430	"F" Quality	M-179	**G**	
		Framdie	M-35		
Fox No. 725 Nitralloy	M-430	Franz Mayer No. 2	M-27	G-97	M-249
		Frary Metal, Cast	M-88	Galahad A	M-62
Fox No. 727 Nitralloy	M-430	Frary Metal, Hard	M-88	Galahad A.C.	M-62
		Frary Metal, Medium	M-88	Galahad A.F.C.	M-62
Fox No. 755	M-430	Frasse Grade A	M-360	Galahad B	M-62
Fox No. 757	M-430	Frasse Grade B	M-360	Galahad B.F.C.	M-62
Fox No. 769	M-430	Frasse Grade C	M-360	Galahad C	M-62
Fox No. 1037	M-430	Frasse Grade H	M-360	Galahad C.F.C.	M-62
Fox No. 1038	M-430	Frasse Temper "A"	M-360	Galahad D	M-62
Fox No. 1049	M-430	Frasse Temper "B"	M-360	Galahad D.F.C.	M-62
Fox No. 1282	M-430	Frasse Temper "C"	M-360	Galahad E	M-62
Fox RF 10	M-430	Free Cutting Brass,		Galahad F	M-62
Fox RF 11	M-430	Scovill 276	M-102	Galvamatt	M-814
Fox RF 12	M-430	Frontier Alloy No.		Gaman H	M-38
Fox RF 21	M-430	24	M-58	Gaman L	M-38
Fox RF 31	M-430	Frontier No. 1	M-58	Gaman R	M-38
Fox RF 32	M-430	Frontier No. 12	M-58	Gamma Columbium	M-1
Fox RF 36	M-430	Frontier No. 32	M-58	Gannaloy	M-U.S.
Fox RF XR3	M-430	Frontier No. 33	M-58	Gas Engine Babbitt	M-317
Fox SF6	M-430	Frontier No. 37	M-58	G.C.C. No. 112	M-1111
Fox SF10	M-430	Frontier No. 39	M-58	G.C.C. Cerium Metal	M-1111
Fox SF 10	M-430	Frontier No. 40	M-58	G.C.C. Didymium	
Fox SF 11	M-430	Frontier No. 40X	M-58	Metal	M-1111
Fox SF11	M-430	Frontier No. 88	M-58	G.C.C. Mischmetal	M-1111
Fox SF12	M-430	FS 2-5	M-57	G.C.C. Pure Cerium	
Fox SF 12	M-430	F Spezial	M-459	Metal	M-1111
Fox SF 13	M-430	FT-2	M-1699	G.E. 33	M-60
Fox SF13	M-430	FT-3	M-1699	G.E. Alloy 76	M-31
Fox SF 14	M-430	FT-4	M-1699	G.E. Alloy DCM	M-31
Fox SF 17	M-430	FT-5	M-1699	G.E. Alloy F48	M-31
Fox SF17	M-430	FT-6	M-1699	G.E. Alloy F50	M-31
Fox SF 18	M-430	FT-7	M-1699	G.E. Alloy I-1360	M-31
Fox SF 20	M-430	FT-8	M-1699	G.E. Alloy J-406	M-31
Fox SF20	M-430	FT-9	M-1699	G.E. Alloy J-1300	M-31
Fox SF 21	M-430	FT-21	M-1699	G.E. Alloy J-1500	M-31
Fox SF 22	M-430	FT-35	M-1699	G.E. Alloy J-1570	M-31
Fox SF22	M-430	FT-42	M-1699	G.E. Alloy J-1650	M-31
Fox SF23	M-430	FT-42	M-1699	G.E. Alloy M-813	M-31
Fox SF25	M-430	FT-62	M-1699	G.E. Alloy ML-1700	M-31
Fox SF 25	M-430	Fuchsal	M-1361	G.E. Alloy P-6	M-31
Fox SF 26	M-430	Fuchsdur	M-1361	G.E. Alloy Rene 41	M-31
Fox SF 35	M-430	Fuchsman	M-1361	G.E. B-129	M-31
Fox SF35	M-430	Furious	M-1411	G.E. Co. No. 1	
Fox SF 50	M-430	Furodit 7G	M-912	Silver Solder	M-60
Fox SF80T	M-430	Furodit 8G	M-501	G.E. Co. No. 2	
Fox SF301	M-430	Furodit 9G	M-501	Silver Solder	M-60
Fox SF 302	M-430	Furodit 10	M-501	G.E. Co. No. 3	
Fox SF 304	M-430	Furodit 10G	M-501	Silver Solder	M-60
Fox SF304 ELC	M-430	Furodit 12G	M-501	G.E. Electrode	
Fox SF 316	M-430	Furodit 25	M-501	Type W-85	M-60
Fox SF316 Ti	M-430	Furodit 30	M-501	Gecor	M-60
Fox SF 317	M-430	Furodit N6	M-501	Gemma	M-771
Fox SF318	M-430	Furodit NH11	M-501	Genessee 110	M-500

Section IV: Obsolete Alloys / 1711

Name	Code	Name	Code	Name	Code
Genessee 130	M-500	Gold Label	M-389	Gripmore No. 1	M-24
Genessee 190	M-500	Gold Tip	M-669	Gripmore No.1-V	M-334
Genuine "A" Babbitt Metal	M-317	Goliath	M-459	Gripmore No. 2	M-334
		Goliath G	M-459	Gripmore Tool	M-334
Genuine Wrought Iron	M-292	Goliath GNV	M-459	GS	M-459
		Goliath Special M	M-459	GSC	M-459
German Silver Extra White	M-102	Goliath Spezial	M-459	GSE	M-459
		Goliath Spezial M	M-459	GSM	M-459
German Silver, Fifths	M-102	Goliath V	M-459	Gun Iron	M-664
		Gollet Steel	M-24	Gunite A	M-223
German Silver Firsts "A-1" Best	M-102	Gordon	M-73	Gunite B-1	M-223
		Gordon Die Steel	M-73	Gunite C	M-223
German Silver, Firsts "A" Special	M-102	Gorham Imperial	M-820	Gunite D	M-223
		Gorham Imperial 9	M-820	Gunite E	M-223
German Silver Foundry Alloy	M-102	Gorham M-40-C	M-820	Gunite E-1	M-223
		Gorham M-40-T	M-820	Gunite F	M-223
German Silver, Fourths	M-102	Gorham M-40-U	M-820	Gunite K	M-223
		Gorham Molybdenum	M-820	Gunite R	M-223
German Silver, Seconds	M-102			Gusstahl 3W	M-1314
		Gormet	M-820	Guy Alloy	M-31
German Silver, Special Thirds	M-102	Gormet FMC	M-820	Gyro	M-140
		Government Bronze	M-102	Gyromet	M-265
German Silver, Thirds	M-102	Government Bronze (Scovill No. 99 Alloy)	M-102	Gyromet 1100	M-265
German Silver White	M-102			**H**	
Gibraltar	M-345	Grade "A" Phosphor Bronze 351 Alloy	M-8	H-11	M-1602
Gigant	M-912			H 12	M-1602
Gigant 5	M-501	Grade H	M-360	Hacksaw Steel	M-38
Gigant 10	M-501	Grade X	M-85	Hadfield Manganese Steel	M-604
Gigant 11	M-912	Grainloy	M-136		
Gigant 12	M-912	Granada Tool	M-38	Hadfield Silicon Steel	M-62
Gigant 33	M-912	Grandios 3VN	M-1314		
Gigant 44	M-912	Grandios 5VN	M-1314	Hadfields Cr-Mn Steel	M-139
Gigant 50	M-501	Granit 0	M-1319		
Gigant 55	M-501	Granit 1	M-1319	Hadara	M-62
Gigant 60	M-501	Granit 2	M-1319	Halco	M-260, M-38
Gigant 66	M-912	Granit 3	M-1319		
Gigant 70	M-501	Granit IV	M-1319	Halcomb 440	M-362
Gigant 77	M-501	Granit V	M-1319	Halcomb 777	M-38
Gigant 88	M-501	Granit 30	M-1319	Halcomb 999	M-260
Gigant BST	M-501	Granite	M-85	Halcomb 1370	M-38
Gigant Duplo	M-501	Graph-Al	M-336	Halcomb Brake Die	M-38
Gigant M5	M-501	Graph M.N.S.	M-823, M-376	Halcomb C.C.S.	M-260
Gigant N	M-912			Halcomb Chrome-Moly Hog Knife	M-38
Gigant UNO	M-912	Graphidox No. 4	M-394		
Gilding Metal	M-33	Graphite Nitralloy	M-24	Halcomb CSM No. 2	M-38
Gilding Metal, Scovill Alloy 110	M-102	Graph-Sil	M-376		
		Graph-Tung	M-376	Halcomb Double Special W	M-260
Gilmore Tool	M-334	Gray Cut Cobalt	M-115		
Glidden J-8100	M-1728	Gray Label Stayput	M-343	Halcomb First Quality	M-38
Glidden J-8102	M-1728	Green Label	M-373		
Glidden J-8103	M-1728	Green Label	M-389	Halcomb FM-2 Free Machining	M-260, M-38
Glidden J-8104	M-1728	Gridaloy M	M-1523		
Glidden J-8202	M-1728	Gridaloy P	M-1523	Halcomb Grade A Stainless	M-260, M-38
Glidden J-8207	M-1728	Gridnic A	M-44		
Glidden J-8300	M-1728	Gridnic B	M-44		
Glidden J-8400	M-1728	Gridnic C	M-44	Halcomb Grade B High Carbon Stainless	M-38, M-260
Glidden J-8600	M-1728	Gridnic D	M-44		
Glidden J-8950	M-1728	Gridnic E	M-44		
Globeiron	M-1030	Gridnic F	M-44		
Glowray	M-121	Gridnic T	M-44	Halcomb Grade B Stainless	M-260
GM-02 Die	M-35	Gripmore	M-334		
"G" Nickel	M-68	Gripmore No. 1	M-334		

Name	Ref	Name	Ref	Name	Ref
Halcomb High Speed	M-38	Hansa Spezial K6	M-1308	Haynes 4560	M-167
Halcomb Hot work	M-260	Hansa Spezial T17	M-1308	Haynes 4561 Alloy	M-167
Halcomb L.C.T.	M-260	Hansa Spezial T5D	M-1308	Haynes 5260	M-167
Halcomb L.C.T. No. 2	M-260	Harchrome	M-828	Haynes 5261 Alloy	M-167
Halcomb Lo Chro Stud	M-260, M-38	Harchrome 5 Cr	M-828	Haynes 5461 Alloy	M-167
		Harchrome 16 Cr	M-828	Haynes Alloy 88	M-167
		Harcote 20	M-828	Haynes Alloy Cb-752	M-167
		Harcote 35	M-828		
Halcomb Nu-Die	M-38	Harcote 45	M-828	Haynes Alloy HE 1049	M-167
Halcomb Pyro	M-260	Harcote 55	M-828		
Halcomb Special	M-260	Hardface 200	M-712	Haynes Alloy No. 13	M-167
Halcomb "S.R.B."	M-38	Hardface 400	M-712		
Halcomb SS Tool Steel	M-38	Hardface 600	M-712	Haynes Alloy No. 36	M-167
		Hardfacer No. 11	M-1000		
		Hardfacer No. 112	M-1000	Haynes Alloy No. 56	M-167
Halcomb Stainless No. 16	M-260, M-38	Hard Kote "H."	M-139		
		Hard Kote "T."	M-139	Haynes Alloy No. 88	M-167
		Hard Surface Welding A-6120	M-12	Haynes Alloy No. 99	M-167
Halcomb Stainless No. 18	M-260, M-38	Hardware Bronze	M-33		
		Hardweld	M-578	Haynes Alloy No. 151	M-167
		Hardweld 50	M-578		
Halcomb Stainless No. 24	M-260, M-38	Hardweld 100	M-578	Haynes Alloy No. 152	M-167
		Harmomang A	M-828		
		Harmomang B	M-828	Haynes Alloy No. 294	M-167
Halcomb Stainless Iron No. 12	M-260, M-38	Harnichrome	M-828		
		Harnimang A	M-828	Haynes Alloy No. 713C	M-167
		Harnimang B	M-828		
Halcot	M-260	Harnimolly	M-828	Haynes Metal	M-167
Haldi	M-260	Harstain 18-8	M-828	Haynes Metal, Hard	M-167
Haldi No. 2	M-38	Harstain 18-8-2 Mo	M-828		
Hallside	M-826	Harstain 18-8-3 Mo	M-828	Haynes Metal, Soft	M-167
Halmo	M-38	Harstain 18-8 Cb	M-828	Haynes Nickel-Manganese	M-167
Handy 94 Cu-6 Ag Alloy	M-63	Harstain 25-12	M-828		
		Harstain 25-20	M-828	Haynes No. 1000	M-167
Handy EB	M-63	Harten A	M-828	Haynes Stellite Alloy No. 3	M-167
Handy ER	M-63	Harten B	M-828		
Handy ET	M-63	Hartop Brown	M-828		
Handy Hi-Temp 90	M-63	Hartop Green	M-828	Haynes-Stellite No. 1	M-167
Handy Hi-Temp 095	M-63	Hartop Red	M-828		
		Hartop Yellow	M-828	Haynes-Stellite No. 2	M-167
Handy Hi-Temp 710	M-63	Hartung	M-828	Haynes Stellite No. 7	M-167
		Harvill No. 1	M-830		
Handy Hi-Temp 720	M-63	Harvill No. 2	M-830	Haynes Stellite No. 8	M-167
		Harvill No. 3	M-830		
Handy Hi-Temp 721 (Low Carbon)	M-63	Harvill No. 4	M-830	Haynes Stellite No. 23	M-167
		Hastelloy A	M-167		
Handy Hi-Temp 820	M-63	Hastelloy Alloy 500	M-167	Haynes-Stellite No. 27	M-167
		Hastelloy Alloy 700	M-167		
Handy Hi-Temp 910	M-63	Hastelloy Alloy D	M-167	Haynes Stellite No. 30	M-167
		Hastelloy CHF	M-167		
Handy Hi-Temp 930	M-63	Hastelloy F	M-167	Haynes Stellite No. 88	M-167
		Hastelloy R-235	M-167		
Handy RSNI	M-63	Havoc	M-373	Haynes Stillite Alloy No. 4	M-167
Handy Silver Solder Hard No. 1	M-63	Hawk 777	M-362		
		Hawk 977	M-362	Hazel Bronze	M-575
Handy TR No. 1	M-63	Hawk 3110	M-362	HB 5	M-459
Hansa E special K5	M-616	Hawk Cold Header Die	M-362	HC 5	M-459
Hansa E special K10	M-616			HC 8	M-459
		Hawk "H"	M-362	HC 9	M-459
Hansa E special T17 Extra	M-616	Hawk Impacto	M-362	HC Pick Steel	M-38
		Hawk Prefak	M-362	HC Poker Bar Steel	M-38
Hansa Spezial	M-1308	Hawk Special	M-362	HC Quality	M-179
Hansa Spezial 325D	M-1308	Haynes 42	M-167	H.D.A. 8151	M-167
Hansa Spezial K3	M-1308				

Section IV: Obsolete Alloys / 1713

Headmore	M-334	Hecla 317	M-62	Henricot HL3X	M-1188
Heat-Resisting		Hecla A.T.G.	M-62	Henricot HL5C	M-1188
Steels, No. 4	M-69	Hecla A.T.V.	M-62	Henricot HL6	M-1188
Heavy Alloy	M-831	Hecla Bronze	M-62	Henricot HL27	M-1188
Heavy Tungsten		Hecla CH31	M-62	Henricot HM	M-1188
Alloy Class 2	M-793	Hecla D17	M-62	Henricot HMES	M-1188
Heavy Tungsten		Hecla EM20	M-62	Henricot HMH	M-1188
Alloy Class 3	M-793	Hecla EM35	M-62	Henricot HMVX	M-1188
Heavy Tungsten		Hecla HCT-4	M-62	Henricot HMX	M-1188
-W2	M-793	Hecla-HGT4	M-62	Henricot HN	M-1188
Heavy		Hecla MM20	M-62	Henricot HN50	M-1188
Tungsten-W5	M-793	Hecla MM35	M-62	Henricot HNF	M-1188
Hecla 10	M-62	Hecla S55	M-62	Henricot HO	M-1188
Hecla 18	M-62	Hecla SNS	M-62	Henricot HP2	M-1188
Hecla 34	M-62	Helbimphy	M-104	Henricot HP93	M-1188
Hecla 35	M-62	Helbimphy Special	M-104	Henricot HPB	M-1188
Hecla 36	M-62	Helco H-40	M-833	Henricot HPC	M-1188
Hecla 37	M-62	Helco HD-50	M-833	Henricot HPNR	M-1188
Hecla 40	M-62	Helco HD-60	M-833	Henricot HPO	M-1188
Hecla 41	M-62	Helco N-140	M-833	Henricot HPT	M-1188
Hecla 42	M-62	Helios D	M-140	Henricot HRI	M-1188
Hecla 46	M-62	Helios G	M-140	Henricot HR2	M-1188
Hecla 66	M-66	Helios H	M-140	Henricot HR2F	M-1188
Hecla 67	M-62	Hellers K	M-140	Henricot HR2L	M-1188
Hecla 67B	M-62	Hellers Hollow Drill	M-343	Henricot HR2X	M-1188
Hecla 70	M-62	Henricot BMES	M-1188	Henricot HR3	M-1188
Hecla 73	M-62	Henricot BO	M-1188	Henricot HR3L	M-1188
Hecla 78	M-62	Henricot FAMO	M-1188	Henricot HR4	M-1188
Hecla 98	M-62	Henricot FC	M-1188	Henricot HR4LA	M-1188
Hecla 100	M-62	Henricot FC2	M-1188	Henricot HR4LB	M-1188
Hecla 104	M-62	Henricot FP1	M-1188	Henricot HR4LX	M-1188
Hecla 105	M-62	Henricot FP2	M-1188	Henricot HR5	M-1188
Hecla 110	M-62	Henricot FPNC	M-1188	Henricot HR5II	M-1188
Hecla 115	M-62	Henricot H1F	M-1188	Henricot HR5III	M-1188
Hecla 116	M-62	Henricot HA0	M-1188	Henricot HR5VI	M-1188
Hecla 120	M-62	Henricot HA1	M-1188	Henricot HR6	M-1188
Hecla 138	M-62	Henricot HA2	M-1188	Henricot HR6P	M-1188
Hecla 142	M-62	Henricot HA3	M-1188	Henricot HR8S	M-1188
Hecla 143	M-62	Henricot HA4	M-1188	Henricot HRC	M-1188
Hecla 143B	M-62	Henricot HA5	M-1188	Henricot HRM	M-1188
Hecla 146	M-62	Henricot HBA	M-1188	Henricot HRM2M1	M-1188
Hecla 146B	M-62	Henricot HBE	M-1188	Henricot HRM2M2	M-1188
Hecla 147	M-62	Henricot HBNO	M-1188	Henricot HT1	M-1188
Hecla 150	M-62	Henricot HD2	M-1188	Henricot HT2	M-1188
Hecla 151	M-62	Henricot HDK	M-1188	Henricot HT3	M-1188
Hecla 151B	M-62	Henricot HDKW	M-1188	Henricot HT3S	M-1188
Hecla 152	M-62	Henricot HDZ	M-1188	Henricot HT6	M-1188
Hecla 153	M-62	Henricot HEB	M-1188	Henricot HTL	M-1188
Hecla 157	M-62	Henricot HEM3	M-1188	Henricot HTY	M-1188
Hecla 159	M-62	Henricot HF	M-1188	Henricot HV1S	M-1188
Hecla 163	M-62	Henricot HF1	M-1188	Henricot HV2S	M-1188
Hecla 174	M-62	Henricot HF2	M-1188	Henricot HV3	M-1188
Hecla 181	M-62	Henricot HF3	M-1188	Henricot HV3P	M-1188
Hecla 182	M-62	Henricot HF4	M-1188	Henricot HV4	M-1188
Hecla 183	M-62	Henricot HIF	M-1188	Henricot HV5	M-1188
Hecla 184	M-62	Henricot HKZ	M-1188	Henricot HV6	M-1188
Hecla 185	M-62	Henricot HL	M-1188	Henricot HV6S	M-1188
Hecla 191	M-62	Henricot HL1	M-1188	Henricot HV7	M-1188
Hecla 192	M-62	Henricot HL1A	M-1188	Henricot HV8	M-1188
Hecla 193	M-62	Henricot HL1X	M-1188	Henricot HVI	M-1188
Hecla 196	M-62	Henricot HL2	M-1188	Henricot HVIR	M-1188
Hecla 197	M-62	Henricot HL2X	M-1188	Henricot HVIX	M-1188
Hecla 206	M-62	Henricot HL3	M-1188	Henricot HW2	M-1188
Hecla 207	M-62	Henricot HL3C	M-1188	Henricot HW2E	M-1188

Henricot HW3	M-1188	High Speed Mo		Holfos 10% Lead	
Henricot HWC	M-1188	Extra	M-85	Bronze	M-169
Henricot HWD	M-1188	High Tensile Alloy	M-836	Holfos 20% Lead	
Henricot HWF	M-1188	Hi-Gloss "C"	M-69	Bronze	M-169
Henricot HWF1	M-1188	Hi-Gloss "DD"	M-69	Holfos AB1	M-169
Henricot HWF2	M-1188	Hi-Gloss Free		Holfos AB2	M-169
Henricot HWIE	M-1188	Machining	M-69	Holfos Bronze, Chill	
Henricot HWMO4	M-1188	Hilo, Modified	M-61	Bar Grade	M-169
Henricot HWMO7	M-1188	Hi-Mo-V	M-57	Holfos G2	M-169
Henricot HWS	M-1188	Hi-Nickel Alloy	M-373	Holfos G3-WP	M-169
Henricot HWT	M-1188	Hioloy "CU"	M-95	Holfos HTB 2	M-169
Henricot Ni-Al-O	M-1188	Hioloy O-3	M-95	Holfos JHR42	M-169
Henricot Ni-Al-1	M-1188	Hioloy-0-2	M-95	Holfos LB3	M-169
Henricot Ni-Al-Co-2	M-1188	Hioloy-0-4	M-95	Holfos LG3	M-169
Henricot Ni-Al-Co-2B	M-1188	Hioloy-0-6	M-95	Holfos PB3	M-169
Henricot Ni-Al-Co-6	M-1188	Hioloy-0-7	M-95	Holfos Super W.W. Bronze	M-169
Heppenstall 3N24	M-64	Hiperco	M-118	Holfos W.W.	
Heppenstall 3N25	M-64	Hiperco 27	M-118	Bronze	M-169
Heppenstall C45	M-64	Hiperco 35	M-118	Hollow Drill	M-307
Heppenstall C-55	M-64	Hiperco-50	M-118	Hoskin 10	M-65
Heppenstall X-60	M-64	Hipernik	M-118	Hoskins 46	M-65
Herculite	M-85	Hipernik V	M-118	Hoskins 502	M-65
Herculoy	M-238	Hipernom V	M-118	Hoskins 651	M-65
Herculoy 418	M-238	Hi-Proof 304	M-1724	Hoskins 667	M-65
Herculoy 419	M-238	Hi-Proof 304L	M-1724	Hoskins 670	M-65
Hermes 35	M-1308	Hi-Proof 316	M-1724	Hoskins 785	M-65
Hermes 45	M-1308	Hi-Proof 316L	M-1724	Hoskins 815-R	M-65
Hermes 60	M-1308	Hi-Proof 347	M-1724	Hoskins 827	M-65
Hewitt Mill	M-959	Hitem Iron No. 5	M-23	Hoskins No. 835	M-65
HG-Nickel	M-121	Hitem Iron No. 7	M-23	Hot Die "C"	M-140
HGT3	M-62	Hi-Temp 30	M-63	Hot Die (Chrome)	M-39
HI-10	M-98	Hi-Temp AF-183	M-1	Hot Die No. 2	M-140
HIA	M-1	Hitensiloy Bronze	M-122	Hot Die Steel	M-115
Hi-Alloy	M-43	Hi-Test	M-1441	Hot Die (Tungsten)	M-39
Hi-Cobalt	M-1076	Hi-Thoria	M-60	Hot Header No. 3	M-34
Hicore 75	M-1724	Hi TM 900	M-1439	Hotspur	M-472
Hicore 90	M-1724	HM	M-459	Howal	M-1364
Hi-C-Super Hy-Tuf	M-38	H-M-Blue Chip	M-57	Howard Alloy No. 1	M-1045
Hidalgo	M-344	"H" Monel	M-1499	Howard Alloy No. 3	M-1045
Hidurax 1/12 A	M-1169	HMS Steels	M-459	Howard Alloy No. 5	M-1045
Hidurax 1A	M-1169	Hoballoy	M-38	Howard Alloy No. 6	M-1045
Hidurax 3	M-1169	Hobart Hi-Carbon No. 40-HC	M-1000	Howard Alloy No. 7	M-1045
Hidurax 4/16A	M-1169	Hobart Hobronze	M-1000	Howard Alloy No. 8	M-1045
Hidurax 4/17A	M-1169	Hobart Manganick	M-1000	Howe Brown Extra	M-38
Hidurax 5	M-1169	Hobart Manganol	M-1000	Howe Brown	
Hidurax 5/2A	M-1169	Hobart No. 13	M-1000	Special	M-38
Hidurax 5/27A	M-1169	Hobart No. 55	M-1000	Hoyt Marine A	M-317
Hidurax 21A	M-1169	Hobart No. 77	M-1000	Hoyt Marine B	M-317
Hidurax 26A	M-1169	Hobart No. 90-PL	M-1000	Hoyt Metal	M-317
Hidurel 7	M-1169	Hobart No. 111	M-1000	Hoyt No. 1 Phosphor Bronze	M-317
Hidurit 10	M-1169	Hobart No. 111HT	M-1000	Hoyt No. 4A	M-317
Hidurit 11	M-1169	Hobart No. 885	M-1000	Hoyt No. 11 D	M-317
Hidurit 12	M-1169	Hobart No. 912	M-1000	Hoyt No. 156B	M-317
High Double Extra	M-85	Hobart No. 914	M-1000	H P A Nickel	M-121
High Dynamic	M-376	Hobart No. 932	M-1000	H P B Nickel	M-121
High Graphitic Iron	M-508	Hobart Softcast A	M-1000	HPM-Nickel	M-121
High Leaded Brass	M-33	Hobart Softcast C	M-1000	HPW-Nickel	M-121
High Speed	M-370	Hobart Strongcast	M-1000	HR Crown Max	M-56
High Speed Brass	M-256	Hobart Sulkote	M-1000	"HR" Monel	M-68
High Speed No. 2	M-210	Hobart Toolfacer	M-1000	H.R.S.	M-499
High Speed Extra Drill	M-85	Hobbing Die Steel	M-24	HRW Die Steel	M-357
		Hodur	M-1364		

HS-23-61	M-60	Hydril Bearing		Immadium IV	M-179
HS Prima	M-459	Metal	M-42	Immadium V	M-179
"HS" Quality	M-179	Hyflux Alnico V	M-1076	Immadium VI	M-179
HST 100	M-430	Hy-glo Hyglo	M-73	Immadium B.B.	M-179
HST 120	M-430	Hykro	M-50	Impacto	M-38
HST 140	M-430	Hykro V	M-50	Imperial	M-210
HTB-3	M-1	Hymu-400	M-32	Imperial	M-24
HTC Nickel	M-121	Hynico -I	M-200	Imperial Major	M-210
HT Silver Fuse	M-1243	Hynico-II	M-200	Imperial Malleable	
H.T.M. Steel	M-240	Hypar	M-114	Stainless Steel	M-210
H.T.S.	M-499	Hyperm-702	M-1309	Imperial Permanent	
Hubbing Die	M-24	Hyperm -766	M-1309	Magnet	M-210
Hubertus	M-96	Hypro	M-334	Imperial Stainless	
Hunt-Spiller High		Hypro A	M-24	Steel	M-210
Carbon Gun Iron	M-664	Hypro A	M-334	Imperial Turning	
Hunt-Spiller No.		Hypro B	M-24	Finishing	M-201
101	M-664	Hypro B	M-334	Imphram	M-104
Hunt-Spiller No.		Hyspeed Nickel		Imphy 1W1	M-771
101A	M-664	Bronze	M-129	Imphy 4	M-771
Hunt-Spiller No.		Hytempite		Imphy 6	M-771
104	M-664	(Commercial No.		Imphy 30CD12SP	M-771
Huron V	M-1	1)	M-412	Imphy 31 BIS SP	M-771
HY-140	M-604	Hy-ten "A" No. 1-5	M-119	Imphy 32SP	M-771
HY-150 Steel	M-604	Hy-ten A No. 1X	M-119	Imphy 34R	M-771
Hychrome 5616	M-73	Hy-ten A2	M-119	Imphy 34 SP	M-771
Hyco-Span	M-12	Hy-Ten Alloy		Imphy 38MS5SP	M-771
Hydra	M-1411	Chisel Steel	M-119	Imphy 41SP	M-771
Hydra-E	M-1411	Hy-Ten "A" No. 1	M-119	Imphy 301	M-771
Hydra-HD	M-1411	Hy-Ten "B" No. 3	M-119	Imphy 302	M-771
Hydra Husky	M-1411	Hy-ten "B" No. 4	M-219	Imphy 305	M-771
Hydra-M	M-1411	Hy-Ten "B" No. 5	M-119	Imphy 308	M-771
Hydra Multico	M-1411	Hy-ten B No. 340	M-119	Imphy 309	M-771
Hydra Vantage	M-1411	Hy-ten B No. 350	M-119	Imphy 310	M-771
Hydra-VK	M-1411	Hy-ten "B" Temper		Imphy 317	M-771
Hydra-Z	M-1411	No. 3-X-40	M-119	Imphy 347	M-771
Hydrex 188	M-1411	Hy-Ten SAE 2315	M-119	Imphy 431	M-771
Hydrex 367	M-1411	Hy-Ten SAE 3140	M-119	Imphy 505	M-104
Hydrex 368	M-1411	Hy-Ten SAE 4150	M-119	Imphy 766	M-104
Hydrex 369	M-1411	Hy-Ten SAE 4615	M-119	Imphy 1640	M-104
Hydrex 372	M-1411	Hy-Ten SAE 6145	M-119	Imphy 1691	M-104
Hydrex 375	M-1411	Hy-Tenso M-50	M-775	Imphy 3693	M-104
Hydrex 399	M-1411			Imphy ADR	M-104
Hydrex 1212	M-1411	**I**		Imphy A.F.T.	M-104
Hydrex CTY	M-1411			Imphy AMF	M-104
Hydrex ENGG	M-1411	IA-IA	M-345	Imphy ARC	M-104
Hydrex ENGG-FM	M-1411	"IAT" Quality	M-179	Imphy ARC 098	M-104
Hydrex HCI	M-1411	Ideal	M-115	Imphy ARC 164	M-104
Hydrex HR 356	M-1411	Ideal (Elalco)	M-44	Imphy ARC 1047	M-104
Hydrex HR 393	M-1411	Idealoy	M-681	Imphy ARC 2266S	M-104
Hydrex HR 2520	M-1411	IK1	M-459	Imphy ARC	
Hydrex HR 6015	M-1411	Illinois Chrome-		2266SP	M-104
Hydrex HRV	M-1411	Nickel-Moly	M-604	Imphy ARC 2266T	M-104
Hydrex MC1	M-1411	Illinois Nickel Steel	M-240	Imphy ARC 2702A	M-104
Hydrex MC1-FM	M-1411	Illium D	M-1262	Imphy ARC 2702B	M-104
Hydrex Modified		Illium S	M-1262	Imphy ARC 2702S	M-104
WIM	M-1411	Illium-R	M-1262	Imphy ATE	M-104
Hydrex S80	M-1411	Immaculate 4 W	M-521	Imphy ATG	M-104
Hydrex Special		Immaculate 5	M-50,	Imphy ATG-B	M-771
HR2520	M-1411		M-521	Imphy ATG-R	M-771
Hydrex W	M-1411	Immaculate 9	M-50,	Imphy ATV-1	M-104
Hydrex W-FM	M-1411		M-521	Imphy BCM	M-771
Hydrex WIM	M-1411			Imphy BTE	M-104
Hydrex XB (HR		Immadium I	M-179	Imphy BTE-ATE	M-771
329)	M-1411	Immadium II	M-179	Imphy BTE-CTE	M-771
		Immadium III	M-179		

Imphy BTE-NTE	M-771	Imphy N.C.-4	M-104	Inconel Alloy 718C	M-1499
Imphy B.Y.- 1	M-104	Imphy N.C.M.	M-104	Inconel Alloy 806	M-68
Imphy B.Y. 2	M-104	Imphy N C M	M-771	Inconel B	M-68
Imphy CCA 1007	M-771	Imphy N.D.	M-104	Inconel (Cast)	M-1499
Imphy CCR	M-104	Imphy N.F.	M-104	Inconel (Cast-Weldable)	M-1499
Imphy C.C.R.-Co.	M-104	Imphy N.F.C. - 2	M-104	Inco PDRL-102	M-68
Imphychrome	M-104	Imphy N.F.C. - 3	M-104	Incut 4140	M-67
Imphy CMO	M-771	Imphy N.F.C. No. 1	M-104	Incut 4142	M-67
Imphycorroye	M-104	Imphy NFCO	M-771	Indalloy	M-1076
Imphy CRB	M-771	Imphy N.M.F.	M-104	Indar 5003	M-265
Imphy C.T.E.	M-104	Imphy NMHG	M-771	Indar 6003	M-265
Imphy CVC	M-104	Imphynusable	M-104	Indar 6005	M-265
Imphy "D."	M-104	Imphy NY	M-104	Indar 6012	M-265
Imphy "D.D."	M-104	Imphy R 23	M-771	Indar 6015	M-265
Imphy "D.D.D."	M-104	Imphy R24	M-771	Indar 6016	M-265
Imphy "D.T."	M-104	Imphyrapide	M-104	Indar 6022	M-265
Imphy ERA HR1	M-104	Imphyrapide Ultra	M-104	Indar 6027	M-265
Imphy ERA HR2	M-104	Imphy Rapide Ultra	M-771, M-772	Indar 6028	M-265
Imphy "F.F."	M-104			Indar 6030	M-265
Imphy HPM	M-771	Imphy RCA 33	M-104	Indar 6031	M-265
Imphy ICR	M-771	Imphy RCA 33	M-771	Indar 6051	M-265
Imphy IR	M-771	Imphy RCA 44	M-104	Indar 6053	M-265
Imphy I.R.R.-S.	M-104	Imphy RCA 44	M-771	Indar 7004	M-265
Imphy I.R.U.	M-104	Imphy RES	M-104	Indar 8001	M-265
Imphy IRU	M-771	Imphy RNCO	M-104	Indco	M-499
Imphy ISO	M-771	Imphy RNCI	M-104	Independence	M-618
Imphy I.T.R.	M-104	Imphy RNC 2	M-104	Index	M-96
Imphy ITR	M-771	Imphy RNC 3	M-104	Indox V	M-1076
Imphy IVO	M-771	Imphy RNC 30	M-104	Indox VI-A	M-1076
Imphy I.W.3.	M-104	Imphy RNC 44	M-104	Indura Drawing Die	M-307
Imphy M 550	M-104	Imphy RWS	M-104	Indus	M-499
Imphy M-1628	M-104	Imphysil	M-771	Indus N.C.C.H. No. 1	M-499
Imphy MA-42	M-104	Imphysta	M-104		
Imphy M.C.T. 1	M-104	Imphy "T"	M-104	Indus N.C.C.H. No. 2	M-499
Imphy M.C.T. 2	M-104	Imphy Triple Rapide	M-772		
Imphy MCT2S	M-771			Indus N.C.C.H. No. 3	M-499
Imphy M.C.T. 3	M-104	Imphy "T.T."	M-104		
Imphy MCT4	M-771	Imphy "T.T.T."	M-104	Indus N.C.C.H. No. 4	M-499
Imphy MCT-19	M-771	Imphy Tungstene	M-772		
Imphy MCTO	M-771	Imphytungstene	M-104	Indus N.C.C.H. No. 5	M-499
Imphy "M.D."	M-104	Imphy UM Co 50	M-771		
Imphy MIP1	M-771	Imphy VY2-SP	M-771	Indus Nickel Molybdenum	M-499
Imphy MOS	M-104	Imphy VYP	M-771		
Imphy "M.T."	M-104	Imphy ZCR 716C		Indus "ONC"	M-499
Imphy MWS	M-104	Improved Electric	M-806	Infiloy	M-1708
Imphy N5CM	M-104	Improvite	M-346	Ingaclad	M-328
Imphy N. 7C.M.	M-104	Inco 101	M-68	Ingaclad Type 304	M-328
Imphy N42	M-771	Inco 739	M-68	Ingaclad Type 309	M-328
Imphy N48	M-771	Inco Alloy 739	M-68	Ingaclad Type 316	M-328
Imphy N54	M-771	Incoloy Alloy 810	M-68	Ingaclad Type 317	M-328
Imphy N58	M-771	Inco MM	M-68	Ingaclad Type 347	M-328
Imphy N.A.	M-104	Inconel 42	M-68	Ingersall	M-328
Imphynaturel	M-104	Inconel 52	M-68	Ingersoll	M-328
Imphy NB	M-104	Inconel 604	M-68	Ingersoll DBL	M-328
Imphy NC3HO	M-771	Inconel 700	M-68	Ingersoll IC-4C	M-328
Imphy N.C.3H.-1	M-104	Inconel 713C	M-68	Ingersoll IC-125	M-328
Imphy NC3H1	M-77	Inconel 739	M-68	Ingersoll ICN-K	M-328
Imphy NC3H1R.	M-771	Inconel 806	M-68	Ingersoll ICNM-9M	M-328
Imphy NC-3H2	M-104	Inconel Alloy 610	M-68	Ingersoll ICV-35	M-328
Imphy NC3H2R.	M-771	Inconel Alloy 611	M-68	Ingersoll ICV-100	M-328
Imphy NC3H8R.	M-771	Inconel Alloy 705	M-68	Ingersoll IW-120	M-328
Imphy NC3H10R	M-771	Inconel Alloy 713	M-68	Ingersoll IWM-65	M-328
Imphy NC3H16R	M-771	Inconel Alloy 717C	M-1499	InKomo	M-618
Imphy NC3H20R	M-771				

InKromstahl IK1	M-1234	Isocast 35	M-162	Janney No. 20	M-508
InKromstahl IK 3	M-38	Isocast 60-20A	M-162	J. C. Dairy Metal	M-508
InKromstahl IK25	M-1234	Isocast B-75	M-162	Jelliff Alloy 70	M-231
InKromstahl IK85	M-1234	Isocast CF-12M	M-162	Jelliff Alloy 1000	M-231
Inland Hi-Steel	M-67	Iso-cast M	M-162	Jelliff Alloy D	M-231
Inland Zinc-Alloy Steel	M-67	Isocast N	M-162	Jessco	M-69
		Isocast WC1	M-162	Jessco A	M-69
Innerberg HG	M-27	Isocast WC2	M-162	Jessco B	M-69
Innershield NR-202	M-578	Isocast WC3	M-162	Jessop 181 Nickel Steel	M-69
Inor-8	M-1490	Isocast WC5	M-162		
Inox 431	M-884	Isocast WC6	M-162	Jessop 317Cb	M-69
Inoxesco 1	M-1495	Isocast WC9	M-162	Jessop 318	M-69
Inoxesco 2	M-1495	Isodisc-1	M-27, M-1618	Jessop 420	M-69
Inoxesco 13	M-1495			Jessop 430	M-69
Insuluminum	M-60	Isodisc-2	M-27, M-1618	Jessop A-3C	M-69
Intermediate Manganese Abrasion Steel	M-337			Jessop Alloy 104	M-69
		Isodisc-4	M-27, M-1618	Jessop Armat	M-69
				Jessop Chrome Magnet Steel	M-69
Intermediate Manganese Medium Carbon	M-377	Isoloy 302	M-98		
		Isoloy 304	M-98	Jessop CNS No. 2	M-69
		Isoloy 316	M-98	Jessop Coal and Mine Bit Steel	M-69
Intermediate Manganese Rail Steel	M-377	"IX" Quality	M-179		
				Jessop Cobalt Magnet Steel	M-69
Intra	M-345	**J**		Jessop Composite "R."	M-69
Invar	M-44	J-4 Chisel	M-694		
Invar Free-Machining	M-73	Jalcase 1	M-173	Jessop DB No. 1	M-69
		Jalcase 2	M-173	Jessop Dril-It Drill	M-69
Invarod	M-32	Jalcase 3	M-173	Jessop E-9	M-69
Invaro No. 2	M-57	Jalcase 4	M-173	Jessop E-25CV	M-69
Invarstahl	M-303	Jalcase 5	M-173	Jessop E-83	M-69
Inventor	M-472	Jalcase 6	M-173	Jessop ET No. 4	M-69
Invincible Drill Rod	M-1703	Jalcase 7	M-173	Jessop ET No. 6	M-69
Iron "8-J"	M-664	Jalcase 8	M-173	Jessop E.T. No. 7	M-69
Ironite	M-174	Jalcase 9	M-173	Jessop H-4	M-69
Ironweld	M-607	Jalcase 10	M-173	Jessop H-9	M-69
Iroquis Special	M-1	Jalcold	M-173	Jessop Heat-Resisting No. 5	M-69
Irrubigo 16W	M-1305	Jalloy 1	M-173		
Irrubigo 17A	M-1305	Jalloy 3	M-173	Jessop High Speed Steel	M-69
Irrubigo 17W	M-1305	Jalloy 7	M-173		
Irrubigo Extra W	M-1305	Jalloy 280	M-173	Jessop Hot Working Die	M-69
Irrubigo GS18	M-1305	J & L 1113 Bessemer - Selenium Leaded	M-173		
Irrubigo GS30	M-1305			Jessop "J" Hot Work	M-69
Irrubigo MC18	M-1305				
Irrubigo S14	M-1305	J & L -B11L13	M-173	Jessop J.J.	M-69
Irrubigo S18	M-1305	J & L Correct Balance	M-173	Jessop K-2	M-69, M-261
Irrubigo S30	M-1305				
Irrubigo S30M	M-1305	J.&L. E-15	M-173	Jessop K-6	M-69, M-261
Irrubigo SN5	M-1305	J.&L. E-23	M-173		
Irrubigo SN8	M-1305	J.&L. E-33	M-173	Jessop M.D. Drill	M-69
Irrubigo SN8S	M-1305	"J" Hot Working Die	M-69	Jessop New Process	M-69
Irrubigo SN9	M-1305			Jessop No. 2B	M-69
Irrubigo SN9S	M-1305	J & L Type 303SM	M-173	Jessop No. 3C	M-69
Irrubigo SN12	M-1305	J & L Type 304SM	M-173	Jessop No. 5	M-69
Irrubigo SN18	M-1305	J & L Type 316SM	M-173	Jessop No. 5B	M-69
Isocast 1 EL	M-162	J & L Type 416SM	M-173	Jessop No. 7	M-69
Isocast 1L	M-162	J & L Type A	M-173	Jessop No. 8	M-69
Isocast 6	M-162	J & L Type A-Selenium	M-173	Jessop No. 9	M-69
Isocast 7A	M-162			Jessop No. 10	M-69
Isocast 24-12B	M-162	J & L Type B	M-173	Jessop No. 11	M-69
Isocast 30A	M-162	Janney No. 2 Bronze	M-508	Jessop No. 12	M-69
Isocast 30-20	M-162			Jessop No. 14	M-69
Isocast 31	M-162	Janney No. 8 Bronze	M-508	Jessop No. 18	M-69
Isocast 32	M-162			Jessop No. 27	M-69

Name	Ref
Jessop No. 27-V	M-69
Jessop No. 29-V	M-69
Jessop No. 51	M-69
Jessop No. 67	M-69
Jessop No. 96	M-69
Jessop No. 96C	M-69
Jessop No. 96K	M-69
Jessop No. 96-KC	M-69
Jessop No. 157	M-69
Jessop No. 202	M-69
Jessop No. 271	M-69
Jessop No. 276	M-69
Jessop No. 773	M-69
Jessop No. 774	M-69
Jessop Non-Magnetic Steel	M-69
Jessop Oil Bit Drill	M-69
Jessop R-22	M-261
Jessop Rapid Finishing Steel	M-69
Jessop RT	M-69
Jessop Steel No. 5	M-69
Jessop Superior Oil Hardening	M-69
Jessop T and V	M-69
Jessop TMC	M-69
Jessop Tungsten Magnet Steel	M-69
Jessop Type R.	M-69
Jetalloy 1570	M-60, M-31
Jethete-M. 140	M-430
Jethete M. 160	M-1724
Jewell Alloy	M-171
Jewell Alloy 42	M-131
Jewell Alloy S	M-131
Jewell Alloy V	M-131
Jewell Steel	M-171
Jewell Super Strength Malleable	M-171
Jobbins 3-6-6 Supreme	M-1103
Jobbins 3-6 Supreme	M-1103
Jobbins 4-8 Supreme	M-1103
Jobbins Almag 56	M-1103
Johnson Bronze	M-172
Johnson Bronze No. 12	M-172
Johnson Bronze No. 19	M-172
Johnson Bronze No. 25	M-172
Johnson Bronze No. 27	M-172
Johnson Bronze No. 40	M-172
Johnson Bronze No. 44	M-172
Johnson Bronze No. 51	M-172
Johnson Bronze No. 55	M-172
Johnson Bronze No. 66	M-172
Johnson Bronze No. 71	M-172
Johnson Bronze No. 72	M-172
Johnson Bronze Babbitt No. 11	M-172
Johnson Bronze Babbitt No. 97	M-172
Johnson Bronze Graphited Bearings	M-172
Johnson Bronze No. LX	M-172
Johnson No. 9 Babbitt Alloy	M-172
Joslyn Stainless Type 202	M-560
Joslyn Stainless Type 302B	M-560
Joslyn Stainless Type 302 Cu	M-560
Joslyn Stainless Type 305	M-560
Joslyn Stainless Type 305H	M-560
Joslyn Stainless Type 305 MH	M-560
Joslyn Stainless Type 308 FM	M-560
Joslyn Stainless Type 430 F Se	M-560
Joslyn Stainless Type 434	M-560
Joslyn Stainless Type 434 FM	M-560
J.S.B.	M-172
"J-S" Steel	M-57
Junker 951	M-1347
Junker A 17 MK Nb	M-1347
Junker A 25 M Nb	M-1347
Junker A 25 MK Nb	M-1347
Junker A 56 M	M-1347

K

Name	Ref
Kaisaloy No. 1 Standard	M-1148
Kaisaloy No. 2	M-1148
Kaisaloy No. 3	M-1148
Kaiser 2 S	M-1147
Kaiser 3 S	M-1147
Kaiser 24 S	M-1147
Kaiser 52 S	M-1147
Kaiser 61 S	M-1147
Kaiser 75 S	M-1147
Kaiser Mo-B Steel	M-1148
Kaiser RP-1	M-1147
Kaiser RP-S	M-1147
Kalif Metal	M-42, M-262
Kalloy	M-472
Kalloy	M-572
Kanthal-DL	M-1150
Kanthal DR.	M-1150
KAPO	M-335
Kasle KA-2	M-1701
Kasle KA-6	M-1701
Kasle KA8-2H	M-1701
Kasle KA-9	M-1701
Kasle KD-2	M-1701
Kasle KD-3	M-1701
Kasle KD-5	M-1701
Kasle KD-6	M-1701
Kasle KD-7	M-1701
Kasle KF-2	M-1701
Kasle KH-11	M-1701
Kasle KH-12	M-1701
Kasle KH-13	M-1701
Kasle KH-19	M-1701
Kasle KH-21	M-1701
Kasle KH-26	M-1701
Kasle KHD	M-1701
Kasle KM-1	M-1701
Kasle KM-2	M-1701
Kasle KM-2C	M-1701
Kasle KM-3	M-1701
Kasle KM-3II	M-1701
Kasle KM-4	M-1701
Kasle KM-7	M-1701
Kasle KM-10	M-1701
Kasle KM-35	M-1701
Kasle KM-42	M-1701
Kasle KM-50	M-1701
Kasle KNL	M-1701
Kasle KO-1	M-1701
Kasle KO-2	M-1701
Kasle KO-6	M-1701
Kasle KS-1	M-1701
Kasle KT-1	M-1701
Kasle KT-5	M-1701
Kasle KT-15	M-1701
Kasle KW-1	M-1701
Kasle KW-2	M-1701
Kasle KW-5	M-1701
Kawasaki-HTP52W	M-1541
Kawasaki-KO	M-1541
Kawasaki NTK-M7	M-1577
Kawasaki-QT60A	M-1541
Kawasaki-QT60B	M-1541
Kay-Brunner Niresist	M-572
K.E. 25	M-510
K.E. 37D	M-510
K.E. 40 AM	M-510
K.E. 41	M-510
K.E. 43	M-510
K.E. 127	M-510
K.E. 144	M-510
K.E. 156	M-510
K.E. 160	M-510
K.E. 200	M-510
K.E. 226	M-510

Alloy	Designation	Alloy	Designation	Alloy	Designation
K.E. 232	M-510	Kennametal K138	M-793	K.L.S. Hot Work	M-335
K.E. 241	M-510	Kennametal K139A	M-793	"K" Monel	M-1499
K.E. 275	M-510	Kennametal K140A	M-793	K Monel 44	M-68
K.E. 287	M-510	Kennametal K141A	M-793	K Monel 64	M-68
K.E. 332	M-510	Kennametal K150A	M-793	K-O	M-1541
K.E. 339	M-510	Kennametal K151A	M-793	Kobalt	M-1315
K.E. 354	M-510	Kennametal K152B	M-793	Kobalt 32V	M-1326
K.E. 484	M-510	Kennametal K501	M-793	Kobalt 98M	M-1326
K.E. 581	M-510	Kennametal KE 5	M-793	Kobalt - 125	M-459
K.E. 621	M-510	Kennametal KE7	M-793	Kobalt - 160	M-459
K.E. 637	M-510	Kennametal KH	M-793	Kobalt - 200	M-459
K.E. 660	M-510	Kennametal KM	M-793	Kobalt - 300	M-459
K.E. 708	M-510	Kennametal KS	M-793	Komalp 3 Herz M	M-1314
K.E. 753	M-510	Kennametal KWH	M-793	Komalp 400	M-1314
K.E. 795	M-510	Ken P-2	M-1700	Komalp Extra	
K.E. 798	M-510	Ken P-4	M-1700	Spezial	M-1314
K.E. 881	M-510	Ken P-6	M-1700	Koppers F-37	M-399
K.E. 933	M-510	Ken Prehard No. 2	M-1700	Koppers F-95	M-399
K.E. 954	M-510	Ken S-1	M-1700	Koppers K-6	M-399
K.E. 961	M-510	Ken S-2	M-1700	Kosmos	M-912
K.E. 965	M-510	Ken S-4	M-1700	Kossil	M-618
K.E. 1029	M-510	Ken S-5	M-1700	Kovar A	M-118
K.E. 1036	M-510	Ken T-2	M-1700	"K.R." Alloy	M-32
K E-2258	M-510	Ken T-4	M-1700	Kramer "5" Nickel	
K.E. 2301	M-510	Ken T-5	M-1700	Silver	M-346
K.E.A. 147	M-510	Ken T-6	M-1700	Kramer 81-2-2-15	M-346
K.E.A. 172	M-510	Ken T-8	M-1700	Kramer 83-3-3-11	
K.E.A. 203	M-510	Ken T-15	M-1700	Alloy	M-346
K.E.A. 207	M-510	Ken W-2	M-1700	Kramer 87-1-3-9	
K.E.A. 231	M-510	Kentanium K151	M-793	Alloy	M-346
K.E.A. 508	M-510	Kentanium K151B	M-793	Kramer 88-3-2-7	
K.E. Extra Quality	M-510	Kentanium K151C	M-793	Alloy	M-346
K.E. H22	M-510	Kentanium K152A	M-793	Kramer Alloy No.	
K.E. KHRS	M-510	Kentanium K152C	M-793	22	M-346
Kelly-Iron	M-900	Kentanium K153B	M-793	Kramer Alloy No.	
Kelock 788	M-510	Kentanium K161B	M-793	33	M-346
Kelock 1014	M-510	Kentanium K162C	M-793	Kramer Alloy No.	
Kelock A72	M-510	Kentanium K163B	M-793	44	M-346
Ken A-7	M-1700	Kentanium K164B	M-793	Kramer Alloy No.	
Ken Air A-4	M-1700	Kentanium K173B	M-793	88	M-346
Ken Chrome D-3	M-1700	Kentanium K183A	M-793	Kramer "B"	
Ken Chrome D-4	M-1700	Kentanium K184B	M-793	Manganese	
Ken-Chrome D-7	M-1700	Kerau	M-618	Bronze	M-346
Ken F-2	M-1700	Kerau Wunda	M-618	Kramer "K-MAG"	M-346
Ken H-11	M-1700	Kerus	M-618	Kramer Nickel	
Ken H 12	M-1700	K E X 369	M-510	Silver	M-346
Ken H 21	M-1700	Keystone	M-539	Kramer "O"	
Ken H 24	M-1700	Keystone-A	M-357	Aluminum	
Ken H 26	M-1700	Keystone Alloy		Bronze	M-346
Ken H 43	M-1700	Chisel	M-357	Kramer Special	
Ken L-2	M-1700	Keystone B	M-140	"M" Nickel Silver	M-346
Ken L-6	M-1700	Keystone Copper		"KR" Monel	M-1499
Ken M-1	M-1700	Steel	M-604	Kromal	M-69
Ken M-3	M-1700	KF 15	M-459	Kromarc 55	M-118
Ken M-4	M-1700	King Cobalt	M-69	Kromarc-58	M-118
Ken M-41	M-1700	Kissock Steel	M-139	Kromar D 70	M-1724
Kennametal 3410	M-793	Kloster D-C-66	M-335	Kromore	M-44
Kennametal K3H	M-793	Kloster Hollow		Kromover	M-1074
Kennametal K4	M-793	Drill	M-335	Krosil	M-38
Kennametal K12	M-793	Kloster Hot Work		Krosil Chisel	M-362
Kennametal K12T	M-793	Steel	M-335	Krovan	M-362
Kennametal K18	M-793	Kloster KLS-44	M-335	Krovan	M-38
Kennametal K20	M-793	Kloster Solid Drill	M-335	KSS Supra	M-459
Kennametal K25	M-793	Kloster V-76	M-335	Kubax	M-912

L

Name	Code
Kuplus	M-448
Kuplus Chrome	M-448
Kuplus High Tensile	M-448
Kynal-PA15	M-282
Kynal PA16	M-282
Kynal PA17	M-282
L-10 Alloy	M-28
LA-685	M-1520
La Belle 1089	M-38
La Belle "C" & "H"	M-38
La Belle CCV	M-38
La Belle Chromonic	M-38
La Belle CSD	M-38
La Belle Eagle	M-38
La Belle Extra 2 1/2	M-38
La Belle Extra 3A	M-38
La Belle Extra 3 1/2	M-38
La Belle Hot Header	M-38
La Belle N-183 Mix	M-38
La Belle Nickle Moly	M-38
La Belle No. 24 Die	M-38
La Belle No. 89 Die	M-38
La Belle No. 211 Die	M-38
La Belle Selftem	M-38
La Belle Semi-Hot	M-38
La Belle Silicon Punch No. 1	M-38
La Belle Special Hot Die	M-38
La Belle Stamp	M-38
La Belle Striking Die	M-38
La Belle TNU	M-38
La Belle Type 3	M-38
La-Bour G-60	M-263
La-Bour R-50	M-263
La-Bour R-50 (Modified)	M-263
Lamite Grade 1	M-115
Lamite Grade 3	M-115
Lamite Grade 6	M-115
Ladish D-6	M-38
Lancaster Silexite 55	M-1165
Langalloy 2V	M-1169
Langalloy 4R	M-1169
Langalloy 5R	M-1169
Langalloy 6R	M-1169
Langalloy 6V	M-1169
Langalloy 10V	M-1169
Langalloy 26V	M-1169
L-Aquila	M-96
L-Aquila Extra D	M-96
Larport	M-1411
LaSalle 1020-90	M-669
Latrobe	M-115
Latrobe 9T	M-115
Latrobe 10T	M-115
Latrobe 11T	M-115
Latrobe 12T	M-115
Latrobe AGT	M-73
Latrobe BR-4FM	M-73
Latrobe CC	M-73
Latrobe CEM	M-73
Latrobe EUB-4	M-73
Latrobe EUB-4M	M-73
Latrobe "Extra"	M-73
Latrobe GSN-FM	M-73
Latrobe L.C.X.	M-73
Latrobe L.P.O.	M-73
Latrobe MCH	M-73
Latrobe MCL	M-73
Latrobe No. 426	M-73
Latrobe O.K.V.	M-73
Latrobe Special	M-73
Latrobe Standard	M-73
Latrobe VDC-MF	M-73
Lavin No. 230	M-801
Lavin No. 300	M-801
Lavin No. 310	M-801
Lavin No. 320	M-801
Lavin No. 384	M-801
Lavin No. 420	M-801
Lavin No. 483	M-801
Lavin No. 598	M-801
Lavin No. 599	M-801
Lavin No. 685	M-801
Lawn Mower	M-357
L-Boz	M-96
L.C.Electrite Double Six M-2	M-73
"L.C." Iron	M-664
LCT	M-38
LCT No. 2	M-38
Leaded Copper 123	M-8
Leaded Nickel Commercial Bronze	M-33
Leaded Nickel Silver 12%	M-33
Leaded Nickel Silver 12%-796	M-8
Leaded Nickel Silver 18%-789	M-8
Leaded Red Brass	M-33
Leaded TS4140 Modified	M-669
Leantin	M-77
Lebanon 442	M-74
Lebanon Grade 22	M-74
Lebanon HR-A	M-74
Lectri-Led 4140	M-176
Leda	M-56
Ledaloyl	M-172
Ledloy	M-240
Ledloy 170	M-240
Ledloy 5120	M-682
Ledloy 8620	M-67
Ledloy-A	M-682
Ledloy C-1120	M-67
Ledo	M-38
Led-o-loy	M-1579
Ledurit Z IV	M-459
Legierung 600	M-324
Legierung 800	M-324
Lehigh Die and Tool	M-24
Lehigh S	M-24
Lems 1	M-459
Lescalloy	M-688
Lescalloy 422 Stainless	M-73
Lescalloy 600	M-73
Lescalloy 718	M-73
Lescalloy 901	M-73
Lescalloy 5616	M-73
Lescalloy A-286	M-73
Lescalloy BG66	M-73
Lescalloy D-979	M-73
Lescalloy Linco	M-73
Lescalloy Super Nitralloy	M-73
Lescalloy V-57	M-73
Lescalloy Waspalloy	M-73
Lescalloy X-750	M-73
Lesco "12-1-1/2"	M-73
Lesco "18-8"	M-73
Lesco 18-8 S	M-73
Lesco "21-12"	M-73
Lesco 25-12	M-73
Lesco "25-20"	M-73
Lesco BG41	M-73
Lesco BG43	M-73
Lesco C.E.M.	M-73
Lesco Crobalt Stainless	M-73
Lesco E.H.W.	M-73
Lesco E.S.A.	M-73
Lesco E.V.S.	M-73
Lesco Extra	M-73
Lesco "H"	M-73
Lesco "H.H."	M-73
Lesco "L"	M-73
Lesco L-410	M-73
Lesco "L.M."	M-73
Lesco "L.M.S"	M-73
Lesco L.S.-416	M-73
Lesco M-430	M-73
Lesco No. 18-8 Mo	M-73
Lesco No. 18-8 Se	M-73
Lesco Special	M-73
Lesco Standard	M-73
Lesco W.W.S.H.	M-73
Lewis Iron	M-240
Lewis Special Iron	M-240
Liberty	M-260
Liberty	M-362
Light Leaded Brass	M-33
Lindenberg 55SWC	M-1319
Lindenberg A3V	M-1319
Lindenberg A18V80	M-1319

Section IV: Obsolete Alloys / 1721

Lindenberg A18V150Co	M-1319	Loxley	M-618	Maerker Irrubigo 1 Mo	M-1305
Lindenberg A300	M-1319	LS	M-459	Maerker Irrubigo 1 S	M-1305
Lindenberg A600	M-1319	LSC	M-1	Maerker Irrubigo 2 Mo	M-1305
Lindenberg BFH	M-1319	LSD	M-210	Maerker Irrubigo 2 Mo	M-1305
Lindenberg BLA 702	M-1319	L.T. Forging	M-57	Maerker Irrubigo 3	M-1305
Lindenberg C615	M-1319	Lubeco	M-77	Maerker Irrubigo 6M	M-1305
Lindenberg CCN	M-1319	Lubrik	M-543	Maerker Irrubigo 12	M-1305
Lindenberg CCNW	M-1319	Lucero	M-44	Maerker Irrubigo 17h	M-1305
Lindenberg Extra 2	M-1319	Ludlum No. 545	M-1	Maerker Irrubigo 25	M-1305
Lindenberg Extra 3	M-1319	Ludlum S.S.V.	M-1	Maerker Irrubigo 42	M-1305
Lindenberg Extra 4	M-1319	Lukens 10L45	M-176	Maerker Irrubigo 188	M-1305
Lindenberg Extra 5	M-1319	Lukens AAR-M128 Gr. A	M-176	Maerker Irrubigo 188A	M-1305
Lindenberg Extra 6	M-1319	Lukens AAR-M128 Gr. B	M-176	Maerker Irrubigo 188EL	M-1305
Lindenberg HSO	M-1319	Lukens Carbon-Molybdenum Steel	M-176	Maerker Irrubigo 188N	M-1305
Lindenberg HW3MH	M-1319	Lukens Chrome-Nickel Steel	M-176	Maerker Irrubigo 188 S	M-1305
Lindenberg HW4ZH	M-1319	Lukens Inconel-Clad Steel	M-176	Maerker Irrubigo 188 SS	M-1305
Lindenberg HW5Z	M-1319	Lukens L-XX	M-176	Maerker Irrubigo 199	M-1305
Lindenberg IL15	M-1319	Lukens M7	M-176	Maerker Irrubigo 199 EL	M-1305
Lindenberg IL 25	M-1319	Lukens M36	M-176	Maerker Irrubigo 199 N	M-1305
Lindenberg IL 35	M-1319	Lukens M113	M-176	Maerker Irrubigo 199 S	M-1305
Lindenberg IL 45	M-1319	Lukens M-131	M-176	Maerker Irrubigo 199 SS	M-1305
Lindenberg IL45Co	M-1319	Lukens M283	M-176	Maerker Irrubigo 200 EL	M-1305
Lindenberg IL232	M-1319	Lukens M284	M-176	Maerker Irrubigo 200 N	M-1305
Lindenberg K	M-1319	Lukens M573 Gr. 65	M-176	Maerker Irrubigo 200 SS	M-1305
Lindenberg NWNH2	M-1319	Lukens M573 Gr. 70	M-176	Maerker Irrubigo 1818	M-1305
Lindenberg RS	M-1319	Lukens MP	M-176	Maerker Irrubigo Extra M	M-1305
Lindenberg T18V	M-1319	Lukens Nickel Steel	M-176	Maerker Irrubigo MC 18 W	M-1305
Lindenberg WLH	M-1319	Lumen Alloy No. 3	M-77	Magalloy PX-4	M-138
Lion Special	M-69	Lumen Alloy No. 4 A	M-77	Magalloy RX-1	M-138
Lithobraze 846	M-63	Lumen Alloy No. 7	M-77	Magalloy RX-2	M-138
L.L.C.T.	M-260	Lumen Alloy No. 42	M-77	Magalloy RX-3	M-138
L.L.L. Extra	M-267	Lumen No. 27	M-77	Magalloy RX-9	M-138
L.L.L. No. 44	M-267	Lumen Metal-1	M-77	Magalloy RX-13	M-138
L.L.L. No. 82	M-267	Lumen Metal-2	M-77	Magalloy X-5	M-138
L.L.L. No. 83	M-267	Lundenheimer A-7-S	M-264	Magalloy X-6	M-138
L.L.L. Standard	M-267	Lusterite-440 B	M-73	Magalloy X-7	M-138
"L.M." Steel	M-327	Lusterite 440 C	M-73	Magalloy X-8	M-138
L-Nickel	M-68	LXX-5T	M-1	Magalloy X-10	M-138
Lo Cro 4-6	M-38	LXX Hil	M-435	Magalloy X-11	M-138
Lo Cro 4-6 MO	M-38	Lyncast	M-523	Magalloy X-12	M-138
Lo Cro 4-6 W	M-38	Lynite	M-4	Magalloy Z-104	M-138
Lo Cro Type 501	M-38	Lyonore	M-456	Magalloy Z-107	M-138
Lo Cro Type 502	M-38				
Lodex	M-60				
Lo-Flo No. S-0	M-1243	**M**			
Lo-Flo No. S-1	M-1243				
Lo-Flo No. S-4	M-1243	M-3 Dreadnaught	M-362		
Lohete	M-345	M-3 Silectron	M-1		
Lomu	M-32	M-10 Air Hardening	M-81		
Lorraine A48S	M-1495	Macalloy	M-34		
Lo-Tin Silver Fuse	M-1243	Macalloy No. 1	M-115		
Lotur	M-357	Macalloy No. 2	M-115		
Low Carbon GFS	M-370	Maerker Irrubigo	M-1305		
Low Carbon Ontario	M-1	Maerker Irrubigo 1	M-1305		
Lowscore	M-472				
Low-Tin Commercial Bronze	M-33				

Magalloy Z-110	M-138	Mallory D581	M-265	M and A No. 36	M-1065
Magan-Nickel AlNi	M-303	Mallory D582	M-265	M and A No. 37	M-1065
Magan-Nickel NiC 1.5	M-303	Mallory Elkon Bronze	M-265	M and A No. 38	M-1065
Magan-Nickel NiC 4	M-303	Mallory Ha Tungsten	M-265	M and A No. 40	M-1065
Magan-Nickel NiC 5	M-303	Mallory HB Metal	M-265	Manganend 13	M-677
Magan-Nickel ZKNi	M-303	Mallory Manganese Aluminum Bronze	M-265	Manganese Boron	M-394
Magic	M-69			Manganese Carbon Alloy	M-669
Magnesium AZ53	M-43			Manganese Metal	M-394
Magnesium AZ91A	M-43	Mallory Manganese Bronze	M-265	Manganese Molybdenum Steel	M-176
Magnesium AZ91C	M-43				
Magnesium EK-30	M-43	Mallory MK-Tungsten	M-265	Manganese Nickel	M-68
Magnesium EK-30A	M-43			Manganese-Nickel (2%)	M-121
Magnesium EK31A	M-43	M-Alloy	M-575		
Magnesium EK31D	M-43	M-Alloy	M-575	Manganese-Nickel (5%)	M-121
Magnesium-EK41A	M-43	M-Alloy	M-1405		
Magnesium-EX31A	M-43	Malta GR	M-69	Manganese Nickel No. 484 Alloy	M-65
Magnesium HM11A	M-43	Malta JC Gr. CR	M-69		
Magnesium KIXI	M-43	Malta JC Gr. MF	M-69	Manganese Screw Stock	M-669
Magnesium LA91	M-1395	Malta JC Gr. SF	M-69		
Magnesium-LA91A	M-1395	Malta SR	M-69	Manganese Steel Medium	M-24
Magnesium LA141	M-1395	Malta SS	M-69		
Magnesium-LA141A	M-1395	Mammut	M-1114	Mangano	M-73
Magnesium-LA142	M-1395	Mammut	M-1291	Manganoid	M-170
Magnesium-LAZ933A	M-1395	Mammut Spezial XX	M-1291	Mangano Special	M-73
				Manganweld A	M-578
Magnesium ZE10A	M-43	Manaurite 12	M-884	Manganweld B	M-578
Magnesium ZK60B	M-43	Manaurite 50w	M-884	Mang-Arc	M-692
Magnesium ZM41	M-43	M and A Alloy No. 13	M-1065	Manga-Tone	M-507
Magnet C	M-459			Mangdie	M-1724
Magnet CM	M-459	M and A Alloy No. 24	M-1065	Mangonic 2	M-121
Magnet W	M-459			Mangonic 3	M-121
Magnet WH	M-459	M and A Alloy No. 27	M-1065	Mangonic 5	M-121
Magno	M-44			Mangrid D	M-61
Magnolia Bearing Bronze	M-178	M and A Alloy No. 33	M-1065	Mangrid E	M-61
				Mang-Trode	M-13
Magno-Nickel	M-68	M and A No. 1	M-1065	Manmo	M-115
Magnox-A12	M-321	M and A No. 2	M-1065	Manmo 6 Temper	M-115
Magnox ZA	M-321	M and A No. 3	M-1065	Manmo 7 Temper	M-115
Mahle 136	M-1365	M and A No. 4	M-1065	Marathon	M-45
Mallory 22	M-265	M and A No. 5	M-1065	Marathon	M-260
Mallory 53	M-265	M and A No. 6	M-1065	Markana Metal	M-303
Mallory 53-Z	M-265	M and A No. 7	M-1065	Marker 2x Columbus	M-1305
Mallory 84	M-265	M and A No. 8	M-1065		
Mallory 125	M-265	M and A No. 10	M-1065	Marker 2x Columbus-Hart	M-1305
Mallory 333	M-265	M and A No. 11	M-1065		
Mallory 333	M-265	M and A No. 14	M-1065	Marker 3x Rose	M-1305
Mallory 1000 Gyromet	M-265	M and A No. 15	M-1065	Marker 8A	M-1305
		M and A No. 16	M-1065	Marker 8F	M-1305
Mallory 2960	M-265	M and A No. 17	M-1065	Marker 9F	M-1305
Mallory 2985	M-265	M and A No. 18	M-1065	Marker 10A	M-1305
Mallory D-51	M-265	M and A No. 19	M-1065	Marker 10AX	M-1305
Mallory D-52	M-265	M and A No. 20	M-1065	Marker 10F	M-1305
Mallory D-53	M-265	M and A No. 21	M-1065	Marker 11A	M-1305
Mallory D53X	M-265	M and A No. 22	M-1065	Marker 11AN	M-1305
Mallory D-58F	M-265	M and A No. 23	M-1065	Marker 11AX5	M-1305
Mallory D-59	M-265	M and A No. 24	M-1065	Marker 11F	M-1305
Mallory D-154X	M-265	M and A No. 25	M-1065	Marker 12A	M-1305
Mallory D-157	M-265	M and A No. 26	M-1065	Marker 12A30	M-1305
Mallory D-258F	M-265	M and A No. 28	M-1065	Marker 12 AC 15	M-1305
Mallory D-511	M-265	M and A No. 32	M-1065	Marker 12AX	M-1305
Mallory D-558F	M-265	M and A No. 34	M-1065	Marker 12 AXN	M-1305
		M and A No. 35	M-1065	Marker 12 AXS	M-1305

Marker 12F	M-1305	Marker S & C SN 6		Markey M-24	M-1671
Marker 210 M	M-1305	M	M-1305	Markey M-25	M-1671
Marker 476 M		Marker S & C SN 8	M-1305	Markey M-34	M-1671
Extra B	M-1305	Marker S & C SN 8		Marog	M-34
Marker 476 M		EL	M-1305	Marques D1S	M-884
Spezial	M-1305	Marker S & C SN 8		Marques D1SS	M-884
Marker CVC	M-1305	N	M-1305	Marques D2S	M-884
Marker D 606 G	M-1305	Marker S & C SN 8		Marques D2SS	M-884
Marker D 607 G	M-1305	S	M-1305	Marques D3S	M-884
Marker DC 10	M-1305	Marker S & C SN 9	M-1305	Marques D3SS	M-884
Marker DCMC	M-1305	Marker S & C SN 9		Marwe 13P	M-1334
Marker ED 5 SG	M-1305	EL	M-1305	Marwe 14D	M-1334
Marker ED 12 G	M-1305	Marker S & C SN 9		Marwe 17L	M-1334
Marker Euzonit 60	M-1305	N	M-1305	Marwe 126E	M-1334
Marker Euzonit 70	M-1305	Marker S & C SN 9		Marwe 127M	M-1334
Marker Extra-Hart	M-1305	S	M-1305	Marwe 134A	M-1334
Marker Extra-		Marker S & C SN		Marwe 176M	M-1334
Mittelhart	M-1305	10	M-1305	Marwe 213ESV	M-1334
Marker Extra-Sehr		Marker S & C SN		Marwe 215E	M-1334
Zah	M-1305	10 M	M-1305	Marwe 215ESV	M-1334
Marker Extra-Zah	M-1305	Marker S & C SN		Marwe 220EV	M-1334
Marker Extra-		12	M-1305	Marwe 220 MH	M-1334
Zahhart	M-1305	Marker S & C SN		Marwe 221M	M-1334
Marker H7F	M-1305	18	M-1305	Marwe 228E	M-1334
Marker H8A	M-1305	Marker S & C SN		Marwe 251CBH	M-1334
Marker H9A	M-1305	25	M-1305	Marwe 251DV	M-1334
Marker H13F	M-1305	Marker S & C SN		Marwe 253DV	M-1334
Marker H14A	M-1305	26 M	M-1305	Marwe 262EB	M-1334
Marker H18F	M-1305,	Marker S & C SN		Marwe 291DV	M-1334
	M-1309	42	M-1305	Marwe 310CW	M-1334
Marker H20A	M-1305	Marker S & C SN		Marwe 320ES	M-1334
Marker H24F	M-1305	99 N	M-1305	Marwe 321ES	M-1334
Marker H24 FS	M-1305	Marker Spezial M		Marwedur A61	M-1334
Marker H40A	M-1305	Extra	M-1305	Marwedur A63	M-1334
Marker HM 3	M-1305	Marker Spezial		Marwedur AN11	M-1334
Marker HOV D	M-1305	M-Mittelhart	M-1305	Marwedur AN21	M-1334
Marker KO 15	M-1305	Marker Spezial		Marwenit 12M	M-1334
Marker MHM	M-1305	M-Sehr Zah	M-1305	Marwenit A10S	M-1334
Marker MW 3	M-1305	Marker Spezial		Marwenit A20S	M-1334
Marker MW 5	M-1305	M-Zah	M-1305	Marwenit AN10	M-1334
Marker Polyten	M-1305	Marker Spezial		Marwenit AN20	M-1334
Marker R 18	M-1305	M-Zahhart	M-1305	Marwenit AT10	M-1334
Marker S 7 N	M-1305	Marker SS 11	M-1305	Marwenit AT20	M-1334
Marker S & C S 14	M-1305	Marker SS 13	M-1305	Marwenit AX10	M-1334
Marker S & C S 14		Marker TAM	M-1305	Marwenit AX20	M-1334
S	M-1305	Marker Vg 508	M-1305	Marwenit F10	M-1334
Marker S & C S 15	M-1305	Marker W 18 K	M-1305	Marwenit F15	M-1334
Marker S & C S 17		Marker WAGT		Marwenit F20	M-1334
M	M-1305	Extra	M-1305	Marwenit FT13	M-1334
Marker S & C S 18	M-1305	Marker WL 4	M-1305	Marwenit FT23	M-1334
Marker S & C S 30	M-1305	Marker WMC	M-1305	Marwenit M10	M-1334
Marker S & C S 30		Marker ZES Spezial	M-1305	Marwenit M 11	M-1334
H	M-1305	Markey M-8	M-1671	Marwenit M 12	M-1334
Marker S & C S 30		Markey M-9	M-1691	Marwetherm A61	M-1334
M	M-1305	Markey M-11	M-1671	Marwetherm A63	M-1334
Marker S & C S 30		Markey M-12	M-1671	Marwetherm F90	M-1334
MH	M-1305	Markey M-13	M-1671	Marwetherm F95	M-1334
Marker S & C SG		Markey M-14	M-1671	Marwetherm F105	M-1334
25	M-1305	Markey M-18	M-1671	Marwetherm F120	M-1334
Marker S & C SG		Markey M-19	M-1671	MAS	M-459
25 M	M-1305	Markey M-20	M-1671	Masiloy I TM 669	M-279
Marker S & C SN 5	M-1305	Markey M-21	M-1671	Masiloy II TM 672	M-279
Marker S & C SN 5		Markey M-22	M-1671	Mastalloy 1011	M-1721
M	M-1305	Markey M-23	M-1671	Mastalloy 1201	M-1721

Mastalloy 1941	M-1721	McKay 5 Cr Mo	M-849	McKay AM 363	M-849
Mastalloy 1971	M-1721	McKay 9 Cr Mo	M-849	McKay Hardalloy 1	M-849
Mastalloy 2001	M-1721	McKay 10	M-849	McKay Hardalloy 6	M-849
Mastalloy 2061	M-1721	McKay 11	M-849	McKay Tool & Die	M-849
Mastalloy 2062	M-1721	McKay 12 Cr	M-849	McKenna K4	M-793
Mastalloy 2351	M-1721	McKay 14	M-849	M.C. Mold and	
Mastalloy 2961	M-1721	McKay 15-35	M-849	Cavity Steel	M-115
Matreloy	M-1627	McKay 15-60	M-849	Medium Leaded	
Matrix	M-60	McKay 15-D	M-849	Brass	M-33
Maxel 1	M-38	McKay 16-25-6	M-849	Medium Leaded	
Maxel 2	M-38	McKay 16 Cr	M-849	Naval Brass	M-33
Maxel 3	M-38	McKay 16 FBS	M-849	Meehanite A	M-82
Maxel 7	M-38	McKay 17-4 Cu Mo	M-849	Meehanite B	M-82
Maxel No. 4 Collet	M-38	McKay 17-7 PH	M-849	Meehanite C	M-82
Maxeloy	M-38	McKay 18	M-849	Meehanite CB	M-82
Maxel Shank Steel	M-38	McKay 18-8	M-849	Meehanite CB3	M-82
Maxilvry "A.W."	M-210	McKay 18-8 Cb	M-849	Meehanite D	M-82
Maximum G	M-96	McKay 18-8 Cb		Meehanite E	M-82
Maximum P	M-96	ELC	M-849	Meehanite GA	M-82
Maximum Special		McKay 18-8 ELC	M-849	Meehanite GAH	M-82
G3	M-96	McKay 18-8 Mo	M-849	Meehanite GB	M-82
Maximum Steel	M-360	McKay 18-8 Mo		Meehanite GC	M-82
Max-Wear	M-38	(317)	M-849	Meehanite GD	M-82
Mayari A	M-24	McKay 18-8 Mo-		Meehanite GE	M-82
Mayari B	M-24	ELC	M-849	Meehanite GM	M-82
Mayari Engine Bolt		McKay 18-8 WMo	M-849	Meehanite HA	M-82
Steel	M-24	McKay 18 Cr	M-849	Meehanite HB	M-82
Mayari Iron	M-24	McKay 19	M-849	Meehanite HC	M-82
Mayari Pig Iron	M-24	McKay 20	M-849	Meehanite HD	M-82
Mayari Staybolt		McKay 20-80	M-849	Meehanite HS-100	M-82
Steel	M-24	McKay 20 (320)	M-849	Meehanite KC	M-82
Mayari Superheater		McKay 20-H	M-849	Meehanite SC	M-82
Bot Steel	M-24	McKay 20-SP	M-849	Meehanite SF-60	M-82
Mayor	M-616	McKay 21	M-849	Meehanite SH-100	M-82
MC III V	M-459	McKay 24	M-849	Meehanite SP-80	M-82
MC 40	M-459	McKay 25-12	M-849	Meehanite WA	M-82
MC 100	M-459	McKay 25-12 Cb	M-849	Meehanite WAH	M-82
McGill No. 2	M-181	McKay 25-20	M-849	Meehanite WB	M-82
McGill No. 3	M-181	McKay 25-20 Cb	M-849	Meehanite WBC	M-82
McGill No. 4	M-181	McKay 25-20 Mo	M-849	Meehanite WEC	M-82
McGill Brass		McKay 28 Cr	M-849	Meehanite WH	M-82
Pressure Casting		McKay 29-9	M-849	Megaperm 4510	M-296
Alloy	M-181	McKay 29-9 Mo	M-849	Megaperm 6510	M-296
McGill Metal	M-181	McKay 116	M-849	Megapyr II German	
McInnes Folder Die	M-81	McKay 116 HV	M-849	Melloid A	M-719
McInnes HC-HC	M-81	McKay 117	M-849	Melloid AA	M-719
McInnes High		McKay 349	M-849	Melloid AAA	M-719
Carbon High		McKay 350	M-849	Melloid B	M-719
Chrome	M-81	McKay 360	M-849	Melloid C	M-719
McInnes Machine	M-81	McKay 380	M-849	Melloid D	M-719
McInnes Record A	M-81	McKay 430	M-849	Melloid E	M-719
McInnes Special		McKay 442	M-819	Mercury	M-114
Cr-Ni Steel	M-81	McKay 446	M-849	Mesmeric	M-56
McInnes Special		McKay 711	M-849	Mesoloy	M-1523
Tool Steel	M-81	McKay 714	M-849	Metal Band Saw	M-38
McInnes Standard		McKay 715	M-849	Metaloy No. I	M-1065
Va High Speed	M-81	McKay 720	M-849	Metaloy No. II	M-1065
McInnes "V"	M-81	McKay 720-H	M-849	Metaloy No. III	M-1065
McInnes Vanadium		McKay 724	M-849	Metamoid	M-289
Crucible	M-81	McKay 6010 IP	M-849	Metite	M-60
McK-Alloy	M-849	McKay 6011 IP	M-849	Mezzo Steel	M-612
McKay 2 Cr Mo	M-849	McKay 6020	M-849	Mg-9Y-1 Zn-0.5 Zr	M-43
McKay 3	M-849	McKay 13018	M-849	MGN	M-104
McKay 4-6 Cr	M-849	McKay AM 363	M-849	Miarmi Iron	M-854

Section IV: Obsolete Alloys / 1725

Michigan Metal	M-766	Midvaloy Extra		MMM	
Microlim	M-50	High Speed	M-85	Dreadnought	M-260
Midcyl	M-185	Midvaloy Finishing	M-85	"MNS" Nickel	M-121
Mid-Max	M-85	Midvaloy GTA	M-85	Mo 10	M-459
Midvale Extra		Midvaloy H.C.	M-85	Mo-20	M-459
Grade	M-85	Midvaloy "HR-1"	M-85	Mo-30	M-459
Midvale Extra Tool	M-85	Midvaloy "HY-X"	M-85	Mo-325	M-459
Midvale N.D.	M-85	Midvaloy "K.A. 2"	M-85	Mo-325 X	M-459
Midvale No. 44	M-85	Midvaloy "K.A.2 Mo"	M-85	Mo-1225 X	M-459
Midvale Non-Tempering	M-85	Midvaloy "K.A.2 S"	M-85	Mocarb	M-140
Midvale Special	M-85	Midvaloy "N" Metal	M-85	Modulvar	M-771
Midvale Stainless Iron	M-85	Midvaloy No. 3-3-4	M-85	Mogul	M-709
Midvaloy 13	M-85	Midvaloy No. 21-00 Cu	M-85	Moheco	M-601
Midvaloy 77F	M-85	Midvaloy No. 26	M-85	Moil Point	M-38
Midvaloy 77W	M-85	Midvaloy Nut Piercer	M-85	Moil Point	M-85
Midvaloy 0500	M-85	Midvaloy Special Finishing	M-85	Molite	M-35
Midvaloy 976	M-85	Midvaloy Stainless FC	M-85	Molite 5	M-35
Midvaloy 1225	M-85	Midvaloy Stainless Iron C-1	M-85	Molite 8	M-35
Midvaloy 1300-15	M-85	Midvaloy Stainless Iron C-2	M-85	Molite 9	M-35
Midvaloy 1300-40	M-85	Midvaloy Stainless Iron C-3	M-85	Molite HW10	M-35
Midvaloy 1700	M-85	Midvaloy Stainless Iron C4	M-85	Molite HW60	M-35
Midvaloy 1700-30	M-85	Midvaloy Stainless Steel 7	M-85	Molite M-1	M-35
Midvaloy 1735	M-85	Midvaloy Stainless Steel A	M-85	Molite M2	M-35
Midvaloy 1760	M-85	Midvaloy Stainless Steel B	M-85	Molite Smoothcut	M-35
Midvaloy 1808-7	M-85	Midvaloy V-2-A	M-85	Molybdenum-0.05 Zr	M-151
Midvaloy 1808-7 Cb	M-85	Mild Self Hardening	M-38	Molybdenum Metal	M-394
Midvaloy 1808-7 Mo	M-85	Millard	M-8	Molybdenum Self-Hardening	M-114
Midvaloy 1808-7S	M-85	Mills-LM1M	M-187	Molybdie	M-55
Midvaloy 1808-20	M-85	Mills-LM7P	M-187	Molybdie Grade FG	M-55
Midvaloy 1808 C	M-85	Mills-LM10W	M-187	Molybdie Grade FM	M-55
Midvaloy 1808 Se	M-85	Mills-LM11W	M-187	Molybdie Grade FS	M-55
Midvaloy 2025	M-85	Minimum	M-459	Molybdie Grade FX	M-55
Midvaloy 2060	M-85	Minimum Extra	M-459	Molybdie Grade WG	M-55
Midvaloy 2100	M-85	Minimum V Special	M-459	Molybdie Ni-Mo	M-55
Midvaloy 2512-7 M	M-85	Min-OX-Grade A	M-25	Molybdie Type C	M-55
Midvaloy 2512-10 M	M-85	Miralite	M-415	Molybdie Type R	M-55
Midvaloy 2512-10 MC	M-85	Misco "A."	M-81	Molycut 562	M-56
Midvaloy 2512-16 M	M-85	Misco B	M-81	Moly High Speed	M-81
Midvaloy 2512-16 Se	M-85	Mishima	M-31	Molyite	M-35
Midvaloy 2512-20	M-85	Mitia Gr. A	M-56	Molymet	M-61
Midvaloy 2512 B	M-85	Mitia Gr. B	M-56	Moly Permalloy	M-1
Midvaloy 2602	M-85	Mitia Gr. C	M-56	Mo-Max	M-612
Midvaloy 2700	M-85	Mitia Gr. TA	M-56	Monaca Drill Rod	M-370
Midvaloy 2802 M 85		Mitia Gr. TA5	M-56	Monaca Special	M-370
Midvaloy 2803	M-85	Mitia Gr. TE	M-56	Monarch	M-34
Midvaloy 3030	M-85	Mitia Gr. TE10	M-56	Monarch 2	M-435
Midvaloy "A"	M-85	Mitifine	M-123	Mond "70"	M-86
Midvaloy Aero-Valve	M-85	Mixend	M-677	Mond Nickel	M-86
Midvaloy "A.M.F." Alloy	M-85	MK Special	M-96	Monel 43	M-68
Midvaloy ATV-1	M-85	MM Dreadnought	M-260	Monel 50	M-68
Midvaloy "B"	M-85			Monel-140	M-68
Midvaloy BTG	M-85			Monel 326	M-68
Midvaloy EME	M-85			Monel 402	M-1499
				Monel 403	M-1499
				Monel Alloy 406	M-68
				Monel Alloy 410	M-68
				Monel Alloy 411	M-68
				Monel Alloy 474	M-68
				Monel Alloy 501	M-68
				Monel Alloy 505	M-68

Name	Ref	Name	Ref	Name	Ref
Monel Alloy 506	M-68	MTS4-Cr	M-459	Murex Stainless No. 7	M-529
Monel Alloy 507	M-68	Mueller 25S	M-266	Murex Stainless No. 15	M-529
Monel C	M-121	Mueller 205	M-266	Murex Surfacing	M-529
Monel (Cast)	M-1499	Mueller 206	M-266	Murex Type F	M-529
Monel Cast, Weldable Grade	M-1499	Mueller 224F	M-266	Murex Vertex	M-529
Monel Filler Metal 64	M-1499	Mueller 224N	M-266	Mustang	M-529
Monel G	M-121	Mueller 240	M-266	Mustang Special	M-69
Monel Welding Electrode	M-897, M-1499	Mueller 260	M-266	Mutemp	M-1417
		Mueller 303	M-266	Muvar	M-1084
		Mueller 571	M-266	My No. 3	M-27
Money (Wrought)	M-1499	Mueller 601L	M-266		
Monimax	M-1	Mueller 701	M-266	**N**	
Monix 20	M-501	Mueller 702	M-266		
Monox	M-602	Mueller 710	M-266		
Moonestone Bronze	M-575	Mueller 784	M-266	NA-4	M-189
Moore-Jones Bearing	M-Eng.	Mueller 786	M-266	N.A.-4A	M-189
		Mueller 801	M-266	NA-4CB	M-189
Mo Rapid Extra 3A	M-27	Mueller 803	M-266	NA-4-M	M-189
Morrison's Ductile Bronze	M-409	Mueller 4700	M-266	NA-6A	M-189
		Mueller 4701	M-266	NADA C	M-548
Mo-Star	M-85	Mueller 6140	M-266	NADA NT	M-548
Mostar	M-884	Mueller 6550	M-266	NADA NT-2	M-548
Mo-Steel No. 1	M-151	Mueller 6800	M-266	NADA NT-3	M-548
Mo-Steel No. 2	M-151	Mueller 6801	M-266	Nail Die Steel	M-24
Mo-Steel No. 3	M-151	Mueller 6810	M-266	National Tube 1 Cr-1/2 Mo	M-90
Mo TAP	M-73	Mueller 7986	M-266	National Tube 1 1/4 Cr-1/2 Mo	M-90
Mo-Tiger	M-24	Mueller 8360	M-266	National Tube 2 Cr-1/2 Mo	M-90
Motor 03A	M-1410	Mueller 8386	M-266	National Tube 2 1/4 Cr-1 Mo	M-90
Motor 05S	M-1410	Mueller 8420	M-266	National Tube 2 1/2 Cr-1/2 Mo-3/4 Si	M-90
Motor 07S	M-1410	Mueller 8440	M-266		
Motor 21SW	M-1410	Mueller 9030	M-266		
Motor 25B	M-1410	Mueller 9050	M-266		
Motor 25S	M-1410	Mueller 9220	M-266	National Tube 2 1/2 Nickel Steel	M-90
Mo-Tung 6	M-114	Mueller 9230	M-266	National Tube 3 Cr-1 Mo	M-90
Mo-Tung 54	M-114	Mueller 9340	M-266	National Tube 3 1/2 Ni	M-90
Moulimphy	M-104	Mueller 9370	M-266	National Tube 5 Cr-1/2 Mo-1 1/2 Si	M-90
Mova	M-56	Mueller 9380	M-266		
Mo-VAN	M-38	Mueller A51S	M-266		
Mow	M-459	Mueller Bronze Blue Tip	M-266	National Tube 5 Ni	M-90
M.P.M.-3 C-Steel	M-38	Mu-Hole Punch	M-85	National Tube 7 Cr-1/2 Mo	M-90
MSA	M-459	Multimold	M-24	National Tube 8 Cr-1 Mo	M-90
MSM 3A1-2 1/2 V	M-1229	Multole	M-85	National Tube 9 Cr-1 Mo	M-90
MSM 185	M-1229	Muraloy	M-552	National Tube 9 Ni	M-90
MSO	M-459	Murex Carbon-Moly	M-529	National Tube C-1118	M-90
MSR-A	M-106	Murex Chromansil	M-529		
MSR-B	M-106	Murex Chrome-Copper	M-529		
"M.S." Steel	M-151			N-A-X AC9111	M-766
MST 2Al-2Fe	M-1229	Murex Chrome-Moly	M-529	N-A-X AC9115	M-766
MST 2.5Al-16 V	M-1229	Murex Cresta	M-529	N-A-XTRA 55	M-766
MST 3Mn Complex	M-1229	Murex Hardex 20	M-529	N-A-XTRA 60	M-766
MST 4Al-4Mn	M-1229	Murex Hardex 60	M-529	N-A-XTRA 65	M-766
MST 5 Al-2.5Sn	M-1229	Murex Manganese Steel	M-529	N-A-XTRA 70	M-766
MST 6Al-4V	M-1229			N-A-XTRA 75	M-766
MST 7Al-4Mo	M-1229	Murex Nickel Steel	M-529	N-A-XTRA 125	M-766
MST 8Mn	M-1229	Murex No. 5 Chrome	M-529	N-A-XTRA 150	M-766
MST-40	M-1229	Murex No. 14 Chrome	M-529		
MST-55	M-1229	Murex Rolex	M-529		
MST-70	M-1229	Murex Special A	M-529		
MST 431	M-1229	Murex Stainless	M-529		
MST 821	M-1229				
MST 881	M-1229				
MST Gr. III	M-1229				

Section IV: Obsolete Alloys / 1727

NBD No. 36 Bronze	M-539	Nichrofy 424	M-1043	Nickimphy	M-104
		Nichrome I	M-44	Nickkal Steel	M-69
NBD Universal Babbitt	M-539	Nichrome II	M-44	Nickonomy	M-345
		Nichrome II	M-416	Nicloy 5	M-17
N.B.M. Silver Babbitt	M-539	Nichrome III	M-44	Ni-Copper	M-60
		Nichrome IV	M-44	Nicosel	M-56
N.C.C. Alloy	M-121	Nichrome, Cast	M-44	Nicovar	M-32
N.C. Heat Resisting	M-153	Nichrome S	M-44	Nicral A	M-104
N.C.M.	M-240	Nichrome Type A	M-44	Nicral B	M-104
N.C.M.	M-499	Nichrome Type B	M-44	Nicral DC	M-104
NCMV	M-1724	Nichrome, Wrought	M-44	Nicral DS	M-104
N.C.O.H.	M-499	Nichro-Zink	M-158	Nicral F	M-104
NCR-238	M-38	Nickahl Steel	M-69	Nicral H9	M-104
Necroni Steel	M-232	Nickel 41	M-68	Nicral J	M-104
N.D. Forging Steel	M-57	Nickel 51	M-68	Nicral KM	M-104
NEATRO FM	M-115	Nickel 202	M-68	Nicralloy-A	M-1411
Needle Wire	M-38	Nickel 204	M-68	Nicralloy-B	M-1411
Neo-Baros	M-104	Nickel 210	M-68	Nicral M	M-104
Neuf Eclairs	M-1043	Nickel 213	M-68	Nicral-O	M-771
Neusilber 4D	M-303	Nickel 224	M-68	Nicral P	M-104
Neusilber PDS	M-303	Nickel 225	M-1499	Nicral S	M-104
Neu-Tec-Tronic 157 BN	M-717	Nickel 305	M-68	Nicral S-Cu	M-104
		Nickel Alloy, Grade A	M-33	Nicral V	M-104
Neutralloy	M-23			Nicral W	M-104
Nevastain 25-12	M-1	Nickel-AT	M-121	Nicral X	M-104
New Bide	M-73	Nickel Cast Iron No. 2	M-68	Ni Cr Hobbing	M-24
New Capital	M-1114			Nicrobraz 45	M-963
Newco	M-442	Nickel Cast Iron No. 4	M-68	Nicrobraz-60	M-963
Newcomer C-2	M-1203			Nicrobraz 180	M-963
Newcomer C-3	M-1203	Nickel-Chromium-Iron Alloy 37/18	M-121	Nicrobraz 220	M-963
Newcomer C-4	M-1203			Nicrobraz 230	M-963
Newcomer C-5	M-1203	Nickelchromweld	M-578	Nicrobraz WG	M-963
Newcomer S-2	M-1203	Nickel-Clad Steel	M-176	Nicroex	M-35
Newcomer S-4	M-1203	Nickelend	M-677	Nicroma 120	M-1315
Newcomer S-6	M-1203	Nickelend 811	M-677	Nicroma 130	M-1315
New Kathode	M-578	Nickel "GFA"	M-121	Nicroma 150	M-1315
New K.S.	M-38	Nickel Grade E	M-68	Nicroman	M-357
New Lightweld	M-578	Nickel Grade T	M-68	Nicromaz-B	M-104
Neydium No. 8	M-349	Nickel "HPA"	M-121	Nigy	M-104
Neydium No. 9	M-349	Nickel "HPB"	M-121	Nikalium	M-179
Ney-Oro "A"	M-349	Nickel Leaded Commercial Bronze	M-33	Nikro	M-34
Ney Oro "A-W"	M-349			Nikrome	M-240
Ney Oro "B"	M-349			Nikrothal-2	M-70, M-1150
Ney Oro "B-W"	M-349	Nickel "O"	M-121		
Ney Oro "E"	M-349	Nickeloid	M-182	Nikrothal-L	M-70, M-1150
Ney Oro Elastic	M-349	Nickel Oreide	M-303		
Ney-Oro Elastic No. 2	M-349	Nickel Silver 10%-752	M-8	Nilcor	M-1470
				Nilo 501	M-121
Ney-Oro Elastic No. 3	M-349	Nickel Silver 18%	M-33	Nilomag 772	M-121
		Nickel Silver 18% B	M-33	Nilomag 800	M-121
Ney Oro "G-W"	M-349	Nickel Silver Grade A	M-98	Nilomag 801	M-121
Ney's No. 125	M-349			Nilstain 308	M-61
NiAG	M-266	Nickel Silver Grade B	M-98	Nilstain 309	M-61
Niagara	M-38			Nilstain 309 Cb	M-61
Niagra	M-373	Nickel Silver Grade B	M-98	Nilstain 310	M-61
Ni-Bral	M-1228			Nilstain 314	M-61
NIC Alloy	M-1026	Nickel Tin	M-538	Nilstain 316	M-61
Nicaloi	M-1	Nickelvac-F	M-1490	Nilstain 317	M-61
Nicaloy	M-60	Nickelvac-Inor-8	M-1490	Nilstain 321	M-61
Nicast	M-1713	Nickelvac L-605	M-1490	Nilstain 325	M-61
Nichranel-c	M-1687	Nickelvac-L-605	M-1490	Nilstain 347	M-61
Nichrofy 152	M-1043	Nickelvac N-155	M-1490	Nilstain 416	M-61
Nichrofy 345	M-1043	Nickelvac Super-X	M-1490	Nilstain C-20	M-61
Nichrofy 345 SP	M-1043	Nickend 1	M-677	Nilstain X	M-61

Name	Code	Name	Code	Name	Code
Nimar 110	M-1724	Nitralloy No. 630	M-115, M-386	Non-Shrink	M-233, M-334
Nimar 125	M-50, M-1724	Nitralloy Special Sulfur	M-97	Non-shrink Die	M-353
Nimark-II	M-32			Nonspall	M-1437
Ni-MoC	M-1119	Nitriding Steel 125 Type H	M-38	Non-Stretch Steel	M-669
Nimocast 258	M-121			Non-Temp Chisel	M-85
Nimocast-PD 16	M-121	Nitriding Steel 135 Type 6	M-38	Non-Wair	M-1724
Nimocast-PK 36	M-121	Nitriding Steel 230	M-38	Non-Warp Steel	M-669
Nimonic 95	M-121	Nitriding Steel BM	M-38	Nonwear	M-1437
Nimonic C75	M-1485	Nitriding Steel N	M-38	Noral	M-643
Nimonic C.B.	M-1485	Nitrodur 65	M-459	Noral 1S	M-643
Nimonic C.C.	M-1485	Nitrodur 65A	M-459	Noral 1SC	M-643
Nimonic DS	M-121	Nitrodur 80	M-459	Noral 2S	M-643
Nimonic DT	M-121	Nitrodur 81	M-459	Noral 3 S	M-643
Nimonic M4VC	M-121	Nitrodur 85	M-459	Noral 4 S	M-643
Nimonic M6VC	M-121	Nitrodur 90	M-459	Noral 10 S	M-643
Nimonic M14V	M-121	Nitrodur 91	M-459	Noral 11 S	M-643
Nimonic M15V	M-121	Nittany	M-122	Noral 13 S	M-643
Nimonic M17V	M-121	Nituf	M-140	Noral 15 S	M-643
Nimoply	M-121	Nitutec 10020	M-717	Noral 16 S	M-643
Nioloy	M-95	Nivaflex	M-1631	Noral 17 S	M-643
Ni-O-Nel	M-68, M-1280	Nivarox	M-296	Noral 19 S	M-643
		Nivarox CT	M-296	Noral 22 S	M-643
Ni-O-Nel Alloy 826	M-68	Nivarox M	M-296	Noral 24 S	M-643
Niperm 50	M-27	Nivarox N	M-296	Noral 25 S	M-643
Ni-Resist Low Expansion	M-68	Nivarox W	M-296	Noral 26 S	M-643
		"NM" Steel	M-151, M-117	Noral 28 S	M-643
Ni-Rod 55	M-68			Noral 31 S	M-643
Niron 46	M-61	No-Co-Ro	M-97	Noral 33 S	M-643
Nisiloy	M-1499	Noduloy Type 7 C	M-394	Noral 38 S	M-643
Nisimaz	M-104, M-771	Noduloy Type 8 C	M-394	Noral 42 S	M-643
		Noduloy Type 18 C	M-394	Noral 50 S	M-643
Ni-Span-D	M-121	Nogroth	M-162	Noral 51 S	M-643
"Ni" Stainless Steel	M-73	N.O.H.	M-499	Noral 54 S	M-643
Nital	M-140	No-Kor-0-12	M-435	Noral 55 S	M-643
Nit CA1	M-1315	No-Kor-0-12-F	M-435	Noral 57 S	M-643
Nit CA2	M-1315	No-Kor-0-14	M-435	Noral 58 S	M-643
Nit CMA	M-1315	No-Kor-0-18	M-435	Noral 62 S	M-643
Nit CMNA	M-1315	No-Kor-0-18-2	M-435	Noral 65 S	M-643
Nitralloy 125, Type H	M-1	No-Kor-0-18-8	M-435	Noral 80 S	M-643
		No-Kor-0-18-8F	M-435	Noral 100	M-643
Nitralloy Alamo	M-1	No-Kor-0-18-8M	M-435	Noral 111	M-643
Nitralloy CM	M-56	No-Kor-0-18-8S	M-435	Noral 115	M-643
Nitralloy EZ 5D5	M-1	No-Kor-0-18-H-60	M-435	Noral 123	M-643
Nitralloy G (135)	M-38	No-Kor-0-25-20	M-435	Noral 124	M-643
Nitralloy G Gr. 69	M-1	No-Kor-O-H	M-435	Noral 125	M-643
Nitralloy GK3	M-56	No-kor-O-HR	M-435	Noral 126	M-643
Nitralloy GK7	M-56	No-Kor-O-KA2-HS	M-435	Noral 127	M-643
Nitralloy Gr. 3	M-56	Nonabrade A.H.	M-1437	Noral 135	M-643
Nitralloy Gr. 5	M-56	Nonabrade Cobalt	M-1437	Noral 140	M-643
Nitralloy H (125)	M-38	Nonabrade O.H.	M-1437	Noral 150	M-643
Nitralloy H70	M-1	Nonbrit	M-50	Noral 155	M-643
Nitralloy Hi-Carbon	M-97	Nonchange A.H.	M-1437	Noral 156	M-643
Nitralloy "I"	M-97	Noncheck	M-1437	Noral 158	M-643
Nitralloy "L"	M-115	Nondistort O.H.	M-1437	Noral 160	M-643
Nitralloy LK1	M-56	Nonerode	M-1437	Noral 161	M-643
Nitralloy LK7	M-56	Nonexcelled Tungsten	M-1437	Noral 162	M-643
Nitralloy No. 115	M-115, M-386			Noral 165	M-643
		Non Magnetic	M-69	Noral 172	M-643
Nitralloy No. 125	M-115, M-386	Nonretreat	M-1437	Noral 211	M-643
		Non-Scuff	M-1437	Noral 218	M-643
Nitralloy No. 225	M-115, M-386	Nonshock Tungsten	M-1437	Noral 222	M-643
		Nonshrink	M-1437	Noral 224	M-643
				Noral 225	M-643

Section IV: Obsolete Alloys / 1729

Noral 226	M-643	NS-2 1/4 Cr-1 Mo		NYBY 1708 Mo Nb	M-1448
Noral 232	M-643	GMA	M-1507	NYBY 1708 MoT	M-1448
Noral 234	M-643	NS-2 1/4 Cr-1 Mo		NYBY 1708 Nb	M-1448
Noral 235	M-643	SA	M-1507	NYBY 1708 T	M-1448
Noral 236	M-643	NS-2 1/4 Nickel	M-1507	NYBY 1710	M-1448
Noral 238	M-643	NS-3 1/2 Nickel	M-1507	NYBY 1710 Si	M-1448
Noral 240	M-643	NS-25	M-1507	NYBY 2520	M-1448
Noral 242	M-643	NS-60M	M-1507	Nymphe	M-104
Noral 244	M-643	NS-61N	M-1507		
Noral 250	M-643	NS-62I	M-1507	**O**	
Noral 252	M-643	NS-67C	M-1507		
Noral 260	M-643	NS-82I	M-1507	O.B. Alloy No. 3	M-94
Noral 305	M-643	NS-92I	M-1507	O.B. Alloy No. 4 B	M-94
Noral 307	M-643	NS-355	M-1507	O.B. Alloy No. 5	M-94
Noral 320	M-643	Nu-Gild (Scovill		O.B. Alloy No. 5 A	M-94
Noral 322	M-643	Alloy 226)	M-102	O.B. Alloy No. 17	M-94
Noral 330	M-643	NW 1	M-459	O.B. Alloy No. 29	M-94
Noral 350	M-643	NW 3	M-459	Octanium	M-1160
Noral 450	M-643	NW 5	M-459	O Diamanthart D	M-96
Noral 465	M-643	NYBY 12-12	M-1448	Oerstit 30	M-459
Noral 510	M-643	NYBY 12-12 EL	M-1448	Oerstit 40	M-459
Noral 8000 S	M-643	NYBY 14-12	M-1448	Oerstit 50	M-459
Noral A56S	M-643	NYBY 16-13 LNb	M-1448	Oerstit 60	M-459
Noral B26S	M-643	NYBY 16-14		Oerstit 70	M-459
Noral B51S	M-643	LMoNbV	M-1448	Oerstit 90W	M-459
Noral B54S	M-643	NYBY 16-16		Oerstit 120	M-459
Noral B75S	M-643	LMoNb	M-1448	Oerstit 120 Cu	M-459
Noral B116	M-643	NYBY 17-7	M-1448	Oerstit 120 K	M-459
Noral B320	M-643	NYBY 17-12 UL	M-1448	Oerstit 130	M-459
Noral D50S	M-643	NYBY 18-5-9	M-1448	Oerstit 160	M-459
Noral D57S	M-643	NYBY 18-8	M-1448	Oerstit 190 K	M-459
Noral H10	M-643	NYBY 18-8 EMo	M-1448	Oerstit 250	M-459
Noral M57S	M-643	NYBY 18-8 H	M-1448	Oerstit 260	M-459
Noral M75S	M-643	NYBY 18-8 LMo	M-1448	Oerstit 360	M-459
Noral C77S	M-643	NYBY 18-8 LMoNb	M-1448	Oerstit 400	M-459
Noral NA17S		NYBY 18-8 T	M-1448	Oerstit 450	M-459
Alclad	M-643	NYBY 18-8 UMo	M-1448	Oerstit 450 K	M-459
Noral NA22S		NYBY 18-12 LMo	M-1448	Oerstit 700	M-459
Alclad	M-643	NYBY 18-12		Oerstit 800	M-459
Noral NA24S		LMoNb	M-1448	Oerstit 900 Cp	M-459
Alclad	M-643	NYBY 18-12 Mo	M-1448	Oerstit 1000	M-459
Noral NA26ST		NYBY 18-14 LMo	M-1448	Oerstit Spezial	M-459
Alclad	M-643	NYBY 20-12 L	M-1448	O Extra	M-96
Noranda-113	M-1608	NYBY 20-20	M-1448	O Extra D	M-96
Noranda-820	M-1608	NYBY 23-14	M-1448	Ohio 8-16 M	M-95
Noranda-3770	M-1608	NYBY 23-14 L Nb	M-1448	Ohio Air Die	M-34,
Nordic	M-291	NYBY 25-21	M-1448		M-115
Normalloy	M-394,	NYBY 25-24	M-1448	Ohio C-1	M-95
	M-604	NYBY 25-24 EMo	M-1448	Ohio C-2	M-95
No-Tin Silver Fuse	M-1243	NYBY 27-4	M-1448	Ohio C-3	M-95
Novantiox	M-1117	NYBY 27-5 Mo	M-1448	Ohio C-3-10	M-95
Novite	M-868	NYBY 0908 Mo	M-1448	Ohio C-8	M-95
Novo 2	M-345	NYBY 1408	M-1448	Ohio C-8-7	M-95
Novo High Speed	M-345	NYBY 1408 Al	M-1448	Ohio C-8-7 M	M-95
Novo Special	M-345	NYBY 1410	M-1448	Ohio C-8-20	M-95
Novo Steel	M-345	NYBY 1410 Mo	M-1448	Ohio C-8A	M-95
Novo Superior	M-345	NYBY 1415	M-1448	Ohio C-8C	M-95
Novo Superior		NYBY 1415 Mo	M-1448	Ohio C-8-Cb	M-95
Vanadium	M-345	NYBY 1420	M-1448	Ohio C-8M	M-95
No-Wear	M-725	NYBY 1425	M-1448	Ohio C-10	M-95
NS-1 1/4 Cr-1/2		NYBY 1435	M-1448	Ohio C-12	M-95
Mo GMA	M-1507	NYBY 1706	M-1448	Ohio C-12-40	M-95
NS-1 1/4 Cr-1/2		NYBY 1708	M-1448	Ohio C-12P	M-95
Mo SA	M-1507	NYBY 1708 Mo	M-1448	Ohio C-18	M-95

Ohio C-20	M-95	Ohio 0-23	M-95	Orlo	M-96
Ohio C-41	M-95	Ohio 0-24	M-95	Ormulu	M-77
Ohio C-63	M-95	Ohio 0-27	M-95	Ormulu, Large	M-77
Ohio C-73	M-95	Ohio 0-28	M-95	Ormulu, Small	M-77
Ohio C-316	M-95	Ohio 0-29	M-95	Orthosil	M-1192
Ohio C-317	M-95	Ohio 0-41-45	M-95	Osborn-EWC	M-233
Ohio Die FM	M-115	Ohio 0-43-45	M-95	Osemund	M-1337
Ohio HC-28	M-95	Ohmalon	M-61	Otisel K2	M-331
Ohioloy	M-95	Ohmaloy	M-1	Otisel K3	M-331
Ohioloy 1045	M-95	Ohmax	M-44	Otisel K7	M-331
Ohioloy 1670	M-95	Ohmsnit 110	M-27	Otisel K8	M-331
Ohioloy 4130	M-95	Oil Engine Babbitt	M-317	Otisel K9	M-331
Ohioloy 4140	M-95	Oilgraph EZ	M-1	Otisel 0-2	M-331
Ohioloy 4330	M-95	Oiltemp	M-24	Otisel 0-3	M-331
Ohioloy 8035	M-95	Oilway	M-345	Otisel 0-4	M-331
Ohioloy 8730	M-95	Old HYCC	M-38	Otisel 0-5	M-331
Ohioloy 8740	M-95	Olin-0629	M-279	Otisel 0-6	M-331
Ohioloy C5	M-95	Olin-0629M	M-279	Otisel 0-7	M-331
Ohioloy C12	M-95	Olin KO-1	M-1642,	Otisel 0-8	M-331
Ohioloy C16	M-95		M-279	Otisel 0-9	M-331
Ohioloy C355	M-95	Olin No. 605	M-279	Otisel 0-10	M-331
Ohioloy C411	M-95	Olympic Bronze		Otisel O-D	M-331
Ohioloy CE30	M-95	Type A	M-33	Otisel O-H	M-331
Ohioloy LC3	M-95	Olympic Bronze		Otisel O-M	M-331
Ohioloy Ni	M-95	Type B	M-33	Ottawa 60	M-1
Ohioloy WC6	M-95	Olympic Bronze		Oxweld No. 2	M-843
Ohioloy WC9	M-95	Type C	M-33	Oxweld No. 21 H.S.	M-843
Ohioloy WCA	M-95	Olympic Bronze			
Ohioloy WCB	M-95	Type D	M-33	**P**	
Ohioloy WCI (or LCI)	M-95	Olympic Bronze Type G	M-33	P2	M-1114
Ohio N-3A	M-95	Olympic FM	M-73	P3	M-1114
Ohio N-3K	M-95	OMC-7 Al-4 Mo	M-1532	P4	M-1114
Ohio N-3M	M-95	OMC-40	M-1532	P5	M-1114
Ohio N-3S	M-95	OMC-166A (cast)	M-1532	P6	M-1114
Ohio N-53	M-95	Omega Brand		PA 22 Nickel	M-121
Ohio N-88	M-95	Beryllium Copper		PA 23 Nickel	M-121
Ohio O-1A	M-95	Alloy No. 32	M-98	P & A Enameling	M-691
Ohio O-1W	M-95	Omega Brand		Pandex	M-73
Ohio 0-2	M-95	Nickel Silver		P & H 70 LA	M-828
Ohio 0-2W	M-95	Alloy No. 3	M-98	P & H 70 LA-2	M-1713
Ohio 0-4	M-95	Omega Brand		P & H 70 LB	M-828
Ohio 0-5	M-95	Nickel Silver		P & H 75 LP	M-1713
Ohio 0-5-20	M-95	Alloy No. 7	M-98	P & H 80 LE	M-1713
Ohio 0-5-B	M-95	Omega Brand		P & H 90 LE	M-828
Ohio 0-5M	M-95	Nickel Silver No. 6	M-98	P & H 90 LH No. 2	M-828
Ohio 0-6	M-95			P & H AC-1	M-828
Ohio 0-9	M-95	Omega Brand		P & H AC-3	M-828
Ohio 0-10	M-95	Phosphor Bronze No. 30	M-98	P & H AW-2B	M-1713
Ohio 0-11	M-95			P & H AW-3C	M-828
Ohio 0-11-15	M-95	Omega Brand		P & H AW-4	M-828
Ohio 0-11A	M-95	Phosphor Bronze No. 47	M-98	P & H DH-2	M-828
Ohio 0-12	M-95			P & H FW	M-828
Ohio 0-12A	M-95	One Star	M-85	P & H No. 21	M-1713
Ohio 0-12B	M-95	On-Plus	M-341	P & H No. 7	M-828
Ohio 0-14	M-95	Oranium Bronze "M"	M-8	P & H SM	M-828
Ohio 0-15-45	M-95			P & H Smootharc	M-828
Ohio 0-15E	M-95	Ordix Extra	M-501, M-912	Pantanax	M-459
Ohio 0-17	M-95			Panther 5N	M-1291, M-1114
Ohio 0-19	M-95	Oregon 60W-40Mo	M-1532		
Ohio 0-20	M-95	Oregon 85W-15Mo	M-1532	Panther 12	M-1
Ohio 0-21	M-95	Oreide .5%	M-102	Panther 750 Special	M-1114, M-1291
Ohio 0-21-20	M-95	Oreide Scovill Alloy 321	M-102		
Ohio 0-22	M-95			Panther Extra	M-1

Section IV: Obsolete Alloys / 1731

Name	Code	Name	Code	Name	Code
Panther N	M-1291, M-1114	Perdonal	M-764	Philadelphia Bronze 77	M-1228
Panther Special	M-1114	Permabraze 630	M-63	Philadelphia Bronze 85	M-1228
Panther XX	M-1	Permadur-ADS	M-1309	Philadelphia Bronze 137	M-1228
Paragon	M-38	Permadur-ADWS	M-1309	Philadelphia Bronze 137 T	M-1228
Paragon A	M-38	Permafly	M-1043	Philadelphia Bronze 404	M-1228
Paragon Oil Hardening	M-38	Permalloy 70	M-61	Philadelphia Bronze 4010	M-1228
Park A Hollo Drill	M-38	Permanent 10	M-1309	Philadelphia Bronze 4040	M-1228
Park Alloy Clipper Blade	M-38	Permanent 17	M-1309	Philo Brand Ferrochrome	M-398
Park Orthopedic Steel	M-38	Permanent 40	M-1309	Phoenix	M-35, M-1308
Park Silver	M-38	Permanent 50	M-1309	Phoenix 430A	M-1594
Parsons 2 S A	M-179	Permanent 80	M-1309	Phoenix 430B	M-1594
Parson's 2 S.A.	M-179	Permanente	M-1025	Phoenix-A387 Gr.D	M-1594
Parson's Manganese Bronze	M-179	Permanit 38	M-27	Phoenix CM-70	M-448
Parson's Manganese Bronze	M-179	Permanit 50	M-27	Phoenix CM-70 (SPT 539)	M-1724
Parson's Mota Metal	M-179	Permanit 70	M-27	Phoenix CM-80F (SPT 566)	M-1724
Parson's Silver Bronze	M-179	Permanit 95	M-27	Phoenix CM-90F (SPT 567)	M-1724
Parson's Star	M-179	Permanit 110	M-27	Phoenix Hansa	M-616
Parson's Star	M-179	Permanit 140	M-27	Phoenix Rapid Machining Steel	M-448
Parson's White Brass	M-179	Permanit 180K	M-27	Phosnic Bronze	M-33
Patina Steel	M-383	Permanit 200	M-27	Phosphor Bronze 4% 903 Gr. 1	M-8
Patina Steel Beizerei	M-383	Permanit 250	M-27	Phosphor Bronze 310	M-8
Patina Steel Rauchgas	M-383	Permanit 400K	M-27	Phosphor Bronze 316	M-8
PDCP	M-319	Permanit 600	M-27	Phosphor Bronze, American No. 1	M-98
Peerless 56	M-38	Permant	M-56	Phosphor Bronze, American No. 2	M-98
Peerless B	M-38	Permax	M-104	Phosphor Bronze, English No. 1	M-98
Peerless C	M-38	Permenor-4801	M-296	Phosphor Bronze, English No. 2	M-98
Peerless Cold Header	M-38	Permet PF-1	M-38	Phosphor Bronze Grade A	M-33, M-102
Peerless D	M-38	Permite	M-240	Phosphor Bronze Grade C	M-33, M-102
Peerless Extra	M-38	Permite No. 2001	M-6	Phosphor Bronze Grade E-1	M-33
Peerless J	M-38	Permold	M-881	Phosphor Bronze, Locomotive Bearing	M-98
Peerless LCT-2	M-38	Permold 3-5	M-881	Phosphor Bronze No. 30	M-98
Peerless W Co	M-38	Permold 7.3	M-881	Phosphor Bronze, Russian	M-98
Penco 57HW	M-341	Permold 15.3	M-881	Phosphorized Admiralty Metal	M-102
Penco 70	M-341	Permold 15.5	M-881	PHV Die	M-38
Penco 71	M-341	Permold 18.3	M-881		
Penco No. 3	M-341	Permold 40	M-881		
Penco No. 4	M-341	Permold 40E	M-881		
Penco No. 20	M-341	Permold 42N	M-881		
Penco WH	M-341	Permold 45	M-881		
Pencoyd	M-877	Permold 75.3	M-881		
Pennaloy "A"	M-466	Permold 90-10	M-881		
Pennaloy "B"	M-466	Permold 90-10S	M-881		
Pennaloy "C"	M-466	Permold 95.5	M-881		
Pennaloy "D"	M-466	Permold 103.7N	M-881		
Pennaloy Heat Resistant	M-466	Permold 110.5	M-881		
Pennaloy Heat Resistant	M-466	Permold 113.5	M-881		
Pentanax S	M-459	Permold 120.5	M-881		
Pequot	M-1	Permold 135W	M-881		
Percussion Cap Brass	M-33	Permold H.P.7.3	M-881		
		Permold H.P.15.5	M-881		
		Permold H.P.40	M-881		
		Permold H.P.42	M-881		
		Permold X-12	M-881		
		Permold Y	M-881		
		Perunal	M-493		
		PH No. 2	M-38		
		Phenix-HD32	M-616		
		Phenix-HD40	M-1308		
		Phenix-HD50	M-616		
		Phenix-HD301	M-616		
		Philadelphia Bronze 40	M-1228		
		Philadelphia Bronze 41	M-1228		

Piladuc	M-531	Plate UP24	M-1337	Poldi 425 D	M-96
Piston	M-140	Plate US Mo Co	M-1337	Poldi 702	M-96
Pitho	M-618	Plate UW280	M-1337	Poldi 702 M	M-96
Pivot Drill Rod	M-38	Plate WCN	M-1337	Poldi 895	M-96
Placovar	M-1084	Plate WKL	M-1337	Poldi AK	M-96
Planeweld No. 1	M-578	Plate WL30	M-1337	Poldi AK1B-GuB	M-96
Planeweld No. 2	M-578	Plate WP32	M-1337	Poldi AK1BS	M-96
Plasdie	M-35	Plate WV30	M-1337	Poldi AK1BS-GuB	M-96
Plasmold	M-56	Plate ZM13	M-1337	Poldi AK1B Special	M-96
Plastalloy	M-357	Platikut	M-357	Poldi AK1U	M-96
Plastic CSM No. 2	M-38	Platinite	M-104	Poldi AK 4	M-96
Plastiform	M-1724	Platinite + Cr	M-771	Poldi AK5M	M-96
Plastiron	M-357	Plumbite	M-123	Poldi AKC-GuB	M-96
Plate AJ 30/4	M-1337	Pluralloy 3 1/2 Ni	M-849	Poldi AKL	M-96
Plate AJ30N	M-1337	Pluralloy 80	M-849	Poldi AKM 2	M-96
Plate AK14	M-1337	Pluralloy 90	M-849	Poldi AKM 3	M-96
Plate AK20	M-1337	Pluralloy 110	M-849	Poldi AKMF	M-96
Plate AR 40/2	M-1337	Plutair-675	M-1410	Poldi AKMF 1	M-96
Plate AR404	M-1337	Pluteous 1000	M-1410	Poldi AKMF 2	M-96
Plate BP 16 E	M-1337	Pluteous 1001	M-1410	Poldi AKRV	M-96
Plate BP 18 E	M-1337	Pluteous 1001/31 C	M-1410	Poldi AKRVB	M-96
Plate BV 36	M-1337	Pluteous 1003	M-1410	Poldi AKRV-GuB	M-96
Plate Extra Extra P10	M-1337	Pluteous 1009	M-1410	Poldi AKS-GuB	M-96
		Pluto	M-616	Poldi AKS2-GuB	M-96
Plate Extra Spezial MH	M-1337	Plutoil 708	M-1410	Poldi AK Special	M-96
		Plutoil 723	M-1410	Poldi AKV Extra-GuB	M-96
Plate Extra Spezial ZAH	M-1337	Plutonic-12S	M-1410		
		Plutonic-157	M-1410	Poldi AKV Extra H-GuB	M-96
Plate Extra Spezial ZH	M-1337	Plutonic-163	M-1410		
		Plutonic-164	M-1410	Poldi AKV-GuB	M-96
Plate K 14	M-1337	Plutonic-165	M-1410	Poldi AKVH-GuB	M-96
Plate KP260W	M-1337	Plutonic-167	M-1410	Poldi AKX 6	M-96
Plate KS 7	M-1337	Plutonic-451	M-1410	Poldi AKX 8	M-96
Plate KS 12	M-1337	Plutonic-601	M-1410	Poldi AKX 12	M-96
Plate LM 22	M-1337	Plutonic-606	M-1410	Poldi AKX 12 Special	M-96
Plate MS4	M-1337	Ply Steel-Hard	M-34		
Plate MS7	M-1337	PM-20	M-884	Poldi AKX 20 GuB	M-96
Plate MS8	M-1337	P.M.G. Alloy	M-117, M-319	Poldi AKX 28 GuB	M-96
Plate NN26M	M-1337			Poldi AKX 30 GuB	M-96
Plate NR 40/D	M-1337	P.M.G. Grade 8	M-319	Poldi AKX 30-M GuB	M-96
Plate ON28	M-1337	P.M.G. Grade 30	M-319		
Plate PA 188	M-1337	P.M.G. Grade 77	M-319	Poldi AKX Extra	M-96
Plate PA 188 S	M-1337	P.M.G. No. 1	M-319	Poldi AKX Extra GuB	M-96
Plate PGS 1	M-1337	P.M.G. No. 3	M-319		
Plate Platit Extra	M-1337	P.M.G. No. 4	M-319	Poldi AKX GuB	M-96
Plate PM 512 Co	M-1337	P.M.G. No. 6	M-319	Poldi AKX Special	M-96
Plate PM 1291	M-1337	P.M.G. No. 10	M-319	Poldi AKX Special 1 GuB	M-96
Plate PMCO 1310	M-1337	P.M.G. No. 13	M-319		
Plate PP36J	M-1337	P.M.G. No. 14	M-319	Poldi AKX Special 1M GuB	M-96
Plate PR97K	M-1337	P.M.G. No. 95	M-319		
Plate PU 27	M-1337	PMM No. 96	M-319	Poldi AKX Special 20 GuB	M-96
Plate PU 83	M-1337	Pneu-Die	M-357		
Plate PV 30	M-1337	Pneutough	M-1724	Poldi AKX Special F GuB	M-96
Plate PV 36	M-1337	Pnusnap OH	M-56		
Plate PW 32	M-1337	Pnusnap WH	M-56	Poldi AKX Special GuB	M-96
Plate PW 83	M-1337	Polar	M-1490		
Plate PW 97	M-1337	Polaris	M-357	Poldi AKX Special M GuB	M-96
Plate PW 100 Extra	M-1337	Poldi 00	M-96		
Plate PW 134	M-1337	Poldi 000	M-96	Poldi AL 12	M-96
Plate PZ 190	M-1337	Poldi 000 Extra	M-96	Poldi AL 30	M-96
Plate PZ 194	M-1337	Poldi 000 Extra 6	M-96	Poldi AM 40	M-96
Platers Metal	M-102	Poldi 1 Extra Hart	M-96	Poldi AMA	M-96
Plate SVZ 203	M-1337	Poldi 5 HN	M-96	Poldi AMD	M-96
Plate TV27	M-1337	Poldi 301	M-96	Poldi BE	M-96

Section IV: Obsolete Alloys / 1733

Poldi BO-2	M-96	Poldi L-BO-4	M-96	Poldi T5 Extra	M-96
Poldi Bo-3	M-96	Poldi L-CE	M-96	Poldi T6H Extra T	M-96
Poldi BO4-GuB	M-96	Poldi L-CM 2	M-96	Poldi TBM	M-96
Poldi BOZ-GuB	M-96	Poldi L-CM 3	M-96	Poldi TBM Extra	M-96
Poldi BZ	M-96	Poldi L-CM 4	M-96	Poldi TEI	M-96
Poldi C	M-96	Poldi L-CM4W	M-96	Poldi TEIN	M-96
Poldi CA	M-96	Poldi L-CM-5	M-96	Poldi TEM	M-96
Poldi CKVM	M-96	Poldi L-CM6L	M-96	Poldi TEMW	M-96
Poldi CM3 Extra GuB	M-96	Poldi L-CNI	M-96	Poldi TE Special	M-96
Poldi CM3-GuB	M-96	Poldi L-CNLW	M-96	Poldi TI Special	M-96
Poldi CM 4	M-96	Poldi L-CNSW	M-96	Poldi TPA	M-96
Poldi CM 5	M-96	Poldi LDH 2	M-96	Poldi TPE	M-96
Poldi CM 6	M-96	Poldi LDH 4	M-96	Poldi TY1W	M-96
Poldi CM6L	M-96	Poldi LDH 5	M-96	Poldi TY3H	M-96
Poldi CNB 4	M-96	Poldi LDH 5 Special	M-96	Poldi TY3M	M-96
Poldi CNC	M-96	Poldi LDM 1	M-96	Poldi TY3W	M-96
Poldi CNE	M-96	Poldi LDM 2	M-96	Poldi TY4M	M-96
Poldi CNI	M-96	Poldi LDM 3	M-96	Poldi TY5M	M-96
Poldi CNL-GuB	M-96	Poldi LDM 3 Special	M-96	Poldi TY5W	M-96
Poldi CN Special	M-96	Poldi LDM Special	M-96	Poldi UG2-GuB	M-96
Poldi C.R.4 Regular	M-96	Poldi L-ECN	M-96	Poldi UG3-GuB	M-96
Poldi CRK	M-96	Poldi LG2M-GuB	M-96	Poldi UG4 Extra-GuB	M-96
Poldi CV1	M-96	Poldi LG2V-GuB	M-96	Poldi UG4-GuB	M-96
Poldi CVM	M-96	Poldi LG4-GuB	M-96	Poldi UG5 Extra-GuB	M-96
Poldi CVM 1	M-96	Poldi LG4M-GuB	M-96	Poldi UG5-GuB	M-96
Poldi CVM 2	M-96	Poldi LG4V-GuB	M-96	Poldi VCN	M-96
Poldi C.V. Special	M-96	Poldi L-M 10	M-96	Poldi VCN 45	M-96
Poldi CZ	M-96	Poldi L-M 15	M-96	Poldi VCNW	M-96
Poldi CZ Special	M-96	Poldi L-TEI	M-96	Poldi WO 1	M-96
Poldi DIN 1662	M-96	Poldi L-TEIN	M-96	Poldi WO 2	M-96
Poldi DS	M-96	Poldi L-TEM	M-96	Poldi WO 3	M-96
Poldi ECN	M-96	Poldi L-TEMW	M-96	Poldi WO3H Special	M-96
Poldi ESL	M-96	Poldi L-TY1W	M-96	Poldi WO 3 Special	M-96
Poldi ESW 2	M-96	Poldi L-TY3M	M-96	Poldi ZR-GuB	M-96
Poldi EX	M-96	Poldi L-TY3W	M-96	Poldi ZY	M-96
Poldi Extra C3-GuB	M-96	Poldi L-TY5W	M-96	Pompey 8CN2	M-884
Poldi Extra TH	M-96	Poldi L-VCN	M-96	Pompey 8NK	M-884
Poldi EZ	M-96	Poldi L-VCNW	M-96	Pompey 8SRX	M-884
Poldi EZH2	M-96	Poldi Maximum	M-96	Pompey 9KX	M-884
Poldi F	M-96	Poldi MK-H	M-96	Pompey 10 CNK3	M-884
Poldi FS 148	M-96	Poldi ML Special	M-96	Pompey 10K2	M-884
Poldi F.S. Extra Best	M-96	Poldi N 125	M-96	Pompey 10WK2	M-884
Poldi FS Special	M-96	Poldi N 136	M-96	Pompey 11	M-884
Poldi FSX	M-96	Poldi NIS	M-96	Pompey 11WK1	M-884
Poldi FZHX	M-96	Poldi NIS 2	M-96	Pompey 12KV	M-884
Poldi GS1	M-96	Poldi NISC	M-96	Pompey 12W1	M-884
Poldi GS2	M-96	Poldi NIT 2	M-96	Pompey 14CNK3	M-884
Poldi GS4	M-96	Poldi NIT 4	M-96	Pompey 15N3	M-884
Poldi HD 50	M-96	Poldi O Special	M-96	Pompey 20N2	M-884
Poldi HS 2	M-96	Poldi R	M-96	Pompey 20ND2	M-884
Poldi HS-GuB	M-96	Poldi RCR	M-96	Pompey 20NK3	M-884
Poldi HS Special	M-96	Poldi REDI-H	M-96	Pompey 32NK3	M-884
Poldi HS Special-GuB	M-96	Poldi S 2	M-96	Pompey 35CP	M-884
Poldi IHN	M-96	Poldi S 3	M-96	Pompey 35NKD1	M-884
Poldi Kaptor-D	M-96	Poldi S 5	M-96	Pompey 40NKD	M-884
Poldi KNO	M-96	Poldi SCH 1	M-96	Pompey 45CP	M-884
Poldi KNO Extra	M-96	Poldi SCW	M-96	Pompey Antichoc-1	M-884
Poldi L-ADVD	M-96	Poldi S.P. Special	M-96	Pompey Antichoc-2	M-884
Poldi L-AK5M	M-96	Poldi SR 2	M-96	Pompey Antichoc-H	M-884
Poldi L-AKX 8	M-96	Poldi SR 3	M-96	Pompey Antichoc-HD	M-884
Poldi L-BO-3	M-96	Poldi SR 5	M-96		
		Poldi SR 6	M-96		

Name	Code
Pompey APS5	M-884
Pompey APS6	M-884
Pompey APS10A	M-884
Pompey APS10C	M-884
Pompey APS10M4	M-884
Pompey APS20	M-884
Pompey APS20C	M-884
Pompey APS20M	M-884
Pompey APS1001	M-884
Pompey C1K	M-884
Pompey C2K	M-884
Pompey CKD1	M-884
Pompey CNKD	M-884
Pompey CNKD1	M-884
Pompey Crostar	M-884
Pompey D3	M-884
Pompey D48	M-884
Pompey DC1	M-884
Pompey EF10X	M-884
Pompey EF12	M-884
Pompey Etan	M-884
Pompey Eth	M-884
Pompey Eth	M-884
Pompey F8	M-884
Pompey GK	M-884
Pompey Hardinox 3	M-884
Pompey K74	M-884
Pompey L7	M-884
Pompey L12	M-884
Pompey L13	M-884
Pompey LCKD	M-884
Pompey LK	M-884
Pompey LRT	M-884
Pompey Manganese-Chrome	M-884
Pompey MAS	M-884
Pompey MC7	M-884
Pompey MC8	M-884
Pompey MC12	M-884
Pompey MKND	M-884
Pompey MKND-1	M-884
Pompey Modal-1	M-884
Pompey Modal-2	M-884
Pompey Modal-3	M-884
Pompey Modal-4	M-884
Pompey Nitral 3	M-884
Pompey PM17	M-884
Pompey PM18	M-884
Pompey PM20	M-884
Pompey PM35	M-884
Pompey PP	M-884
Pompey PSK	M-884
Pompey RMKR	M-884
Pompey RP	M-884
Pompey RS4E	M-884
Pompey S8	M-884
Pompey S8R	M-884
Pompey TRK	M-884
Pompey Turex 4	M-884
Pompey Turex 9	M-884
Pompton D.R.	M-1
Potomac 2V	M-1
Potomac 4V	M-1
Potts Ground Flat Stock	M-307
Potts Special Drill Rod	M-307
Potts Superior	M-307
Powdiron 51-1	M-632
Powdiron 55 P	M-632
Powdiron 56-L	M-632
Powdiron 59-FM	M-632
Powdiron 59-I	M-632
Powdiron 59-IC	M-632
Powdiron 59-PC	M-632
Powdiron 59-T	M-632
Powdiron 61-IC	M-632
Powdiron 61-P	M-632
Powdiron TCU	M-632
Power	M-178
Precision A-74	M-887
Precision ZN-4	M-887
Premabraze 101	M-63
Premabraze 128	M-63
Premet	M-1
Premier	M-248
Premier 3 1/2 Ni Weld	M-12
Premier AW Weld	M-12
Premier "C"	M-248
Premier EA Weld	M-12
Premier EB Weld	M-12
Premier EP Weld	M-12
Premier GA Weld	M-12
Premier GS Weld	M-12
Premier HC Weld	M-12
Premier HS Weld	M-12
Premier MB Spring	M-12
Premier Nickel Chrome	M-248
Premier Spring	M-12
Press E-Z	M-69
Pressurdie 5	M-140
Pressurdie No. 7	M-140
Presto	M-1339
Presto	M-32
Presto B	M-1339
Presto Steel No. 9-266	M-32
Prima 2H	M-1337
Prima 3MH	M-1337
Prima 6 SEHR ZAH	M-1337
Prima 65	M-1319
Prima 85	M-1319
Prima 90	M-1319
Prima 95	M-1319
Prima 100	M-1319
Prima H	M-1309
Prima MH	M-1309
Prima MH 100	M-27
Prima MH 115	M-27
Prima Zahweich	M-1314
Prima ZH I	M-1314
Prima ZH II	M-1314
Prima ZH III	M-1314
P.R.K (Potts)	M-307
Promet 6CR	M-956
Promet 18	M-956
Promet 22	M-956
Promet 86P	M-956
Promet 100	M-956
Promet Bronze 53	M-956
Promet Bronze 53L	M-956
Promet Bronze 53N	M-956
Promet No. 1	M-956
Promet No. 100-LT	M-956
PSP	M-459
PSU	M-459
Pure Ore A.D. 95	M-335
Pure Ore Cobalt	M-335
Pure Ore Cold Heading	M-335
Pure Ore D-52	M-335
Pure Ore D-C-33	M-335
Pure Ore D-C-66	M-335
Pure Ore Die Casting	M-335
Pure Ore Drill Rod	M-335
Pure Ore E-N-97	M-335
Pure Ore Finishing	M-335
Pure Ore High Speed	M-335
Pure Ore Kapo	M-335
Pure Ore KLS-44	M-335
Pure Ore Moly 6-6	M-335
Pure Ore No. 10	M-335
Pure Ore No. 14	M-335
Pure Ore No. 25 Special Alloy	M-335
Pure Ore Prior	M-335
Pure Ore Prior Extra	M-335
Pure Ore Production	M-335
Pure Ore Production Hi-Cr, Hi-C	M-335
Pure Ore Quarry Tool Steel	M-335
Pure Ore Rolmo	M-335
Pure Ore Solid Drill	M-335
Pure Ore V-995	M-335
Pure Ore Z-457 Die Cast Steel	M-335
Purple Cut Steel	M-1
Purple Label Special	M-69
PW2	M-459
P.X.D.	M-73
Pyral	M-104
Pyralloy	M-472
Pyramid Special	M-433
Pyrista	M-500, M-521
Pyrodie	M-64
Pyrodur P-CN 15	M-1290
Pyrodur P-CN 15H	M-1290
Pyrodur P-CN 15T	M-1290
Pyrodur P-CN 15TW	M-1290

Section IV: Obsolete Alloys / 1735

Pyrodur P-CN 15W	M-1290	Ran	M-501	Raxa GCN12	M-1353
Pyrodur P-CN 20H	M-1290	Ran 1 W/H	M-501	Raxa GCN12X	M-1353
Pyrodur P-CN 20W	M-1290	Ran 2 W/H	M-501	Raxa GCV9	M-1353
Pyrodur P-CN 35	M-1290	Ran 3 W/H	M-501	Raxa GCV11	M-1353
Pyrodur P-CN 35W	M-1290	Ran 6	M-501	Raxa GL1	M-1353
Pyrodur P-CN 60	M-1290	Ranalloy Type A	M-964	Raxa HDS/H	M-1353
Pyrodur P-CN 80	M-1290	Ranalloy Type B	M-964	Raxa HM	M-1353
Pyrodur P-CN BH	M-1290	Ranalloy Type C	M-964	Raxa KEV	M-1353
Pyrodur P-CN BS	M-1290	Ranger D.R.	M-69	Raxa KLW	M-1353
Pyrodur P-CN BW	M-1290	Ranite C-X	M-1100	Raxa MN 12	M-1353
Pyromic No. 1	M-108	Ranite Type B	M-1100	Raxa MWH	M-1353
Pyros	M-104	Raonel 600	M-1209	Raxa MWV	M-1353
Pyrotherm 25	M-1342	Rapide L	M-1115	Raxa PCO	M-1353
Pyrotherm 25/20 Nb	M-1342	Rapid Extra	M-1114	Raxa R10	M-1353
Pyrotherm G14	M-1342	Rapid Extra Mo	M-27	Raxa R15	M-1353
Pyrotherm G17	M-1342	Rapid Finish	M-69	Raxa R60	M-1353
Pyrotherm G 18/36 H	M-1342	Rapidit M	M-459	Raxa R70	M-1353
Pyrotherm G 20/33	M-1342	Rapidit MA	M-459	Raxa REC60	M-1353
Pyrotherm G 20/33 H	M-1342	Rapidit W	M-459	Raxa RZM	M-1353
Pyrotherm G 20/33 Nb	M-1342	Rapidit WA	M-459	Raxa Spezial 1	M-1353
Pyrotherm G 25/35	M-1342	Rapid Panther	M-1114	Raxa Spezial 2	M-1353
Pyrotherm G 25 H	M-1342	Rapid Special BNX	M-459	Raxa Spezial 3	M-1353
Pyrotherm G 26/14	M-1342	Rapid Spezial	M-459	Raxa Spezial 4	M-1353
Pyrotough	M-1724	Rapid Spezial BN	M-459	Raxa Spezial 5	M-1353
		Raxa 129	M-1353	Raxa Spezial 6	M-1353
		Raxa C10	M-1353	Raxa SSWR	M-1353
Q		Raxa CMER	M-1353	Raxa UW10	M-1353
		Raxa CNV3	M-1353	Raxa UW13	M-1353
		Raxa CNV5	M-1353	Raxa VS6	M-1353
		Raxa CNZ3	M-1353	Raxa WCH	M-1353
		Raxa CNZ5	M-1353	Raxa WCNU	M-1353
		Raxa CRO	M-1353	Raxa WP6	M-1353
Quality Carbon Drill Rod	M-1702	Raxa CRW	M-1353	Raxa WS1	M-1353
Quatre Eclairs	M-1043	Raxa CWV5/W	M-1353	Raxa WSM	M-1353
Quatre Eclairs Special	M-1043	Raxa Extra 1	M-1353	Raxa WZR	M-1353
Quinze Eclairs	M-1043	Raxa Extra 2	M-1353	Raxa ZHW	M-1353
		Raxa Extra 3	M-1353	Raxa ZVW	M-1353
		Raxa Extra 4	M-1353	R.B. Chisel	M-140
		Raxa Extra 5	M-1353	R.B. Special	M-140
R		Raxa Extra-Extra 1	M-1353	RCA-N9	M-1465
		Raxa Extra-Extra 2	M-1353	RCA-N91	M-1465
R-303	M-671	Raxa Extra-Extra 3	M-1353	RCA-N97	M-1465
RA-321	M-1209	Raxa Extra-Extra 4	M-1353	RCA-N100	M-1465
RA 430	M-1209	Raxa Extra-Extra 5	M-1353	RCA No. 91	M-1465
RA 600	M-1209	Raxa Extra-Extra 6	M-1353	RCA No. 97	M-1465
Raco HD6	M-986	Raxa G15K	M-1353	R.C.F.	M-200
Raco HD30	M-986	Raxa G17K	M-1353	R Cupro-Nickel (70/30)	M-121
Raco HD-82	M-986	Raxa G28NK	M-1353	R Cupro-Nickel (80/20)	M-121
Radeco 2	M-96	Raxa G30K	M-1353	R Cupro-Nickel (90/10)	M-121
Radeco K	M-96	Raxa G30MK	M-1353	R.D.X.	M-240
Radiametal	M-1	Raxa G188K	M-1353	Ready-Flow	M-686
Radianite	M-73	Raxa G188SK	M-1353	Readyweld	M-578
Radikal Extra	M-1309	Raxa G199K	M-1353	Record Select	M-336
Radikal Extra M	M-1309	Raxa G199SK M 1353		Re-Cro	M-85
Radikal Mo6	M-1309			Redalloy	M-33
Radikal Mo95	M-1309	Raxa Garant 33	M-1353	Red Arrow	M-73
Radikal Mo98	M-1309	Raxa Garant 50	M-1353	Red Arrow 48 S	M-73
Radiohm	M-44	Raxa Garant II	M-1353	Red Arrow 101	M-73
Radio Metal "A"	M-108	Raxa GBS	M-1353	Red Arrow 202	M-73
Raffinal	M-764	Raxa GC8	M-1353	Red Arrow 222	M-73
Railife	M-211	Raxa GC9	M-1353	Red Arrow 303	M-73
Rajah	M-114	Raxa GC10	M-1353		
Ram Brand Swedish Iron	M-335	Raxa GC 11	M-1353		
		Raxa GCN10	M-1353		

Red Arrow 333	M-73	Remanit 1510 Al	M-459	Remanit G Hard	M-459
Red Arrow Welding Rod	M-73	Remanit 1515	M-459	Remanit G Soft	M-459
		Remanit 1515 Mo	M-459	Remanit HB	M-459
Red Brass	M-33	Remanit 1520	M-459	Remanit HC	M-459
Red Chip	M-57	Remanit 1530	M-459	Re-Mo 50/50	M-33
Red Cut Superior FM	M-115	Remanit 1530-40-50	M-1299	Remount	M-1405
		Remanit 1530F	M-459	Remy MS70	M-1339
Red Cut Superior Temper "M"	M-115	Remanit 1540	M-459	Remy MS75	M-1339
		Remanit 1540 Mo	M-459	Remy NV50	M-1339
Red Cut Superior Temper "O"	M-115	Remanit 1610	M-1299	Remy WPN	M-1339
		Remanit 1610	M-459	Remy WS1	M-1339
Red Fox 10	M-430	Remanit 1610S	M-459	Renny D150 Meisselstahl	M-1339
Red Fox 11	M-430	Remanit 1610 ST	M-459		
Red Fox 12	M-430	Remanit 1620	M-1299	Renown	M-73
Red Fox 21	M-430	Remanit 1620	M-459	Republic 6-H-W	M-97
Red Fox 30	M-430	Remanit 1690	M-459	Republic 07	M-97
Red Fox 31	M-430	Remanit 1690 V	M-459	Republic 65	M-97
Red Fox 32	M-430	Remanit 1710	M-459	Republic A 2	M-97
Red Fox 33	M-430	Remanit 1710 A	M-459	Republic A 4	M-97
Red Fox 65/15	M-430	Remanit 1710A	M-1299	Republic Acme	M-97
Red Fox 80/20	M-430	Remanit 1710S	M-459	Republic D2	M-97
Red Fox 135	M-430	Remanit 1710 ST	M-459	Republic D3	M-97
Red Fox 326	M-430	Remanit 1740	M-459	Republic Double Strength, Grade 1	M-97
Red Label	M-389	Remanit 1740	M-1299		
Red Label Extra	M-373	Remanit 1790	M-1299	Republic Double Strength, Grade 1-A	M-97
Red Label Peerless	M-343	Remanit 1790	M-459		
Red Label (Wards)	M-389	Remanit 1790C	M-459		
Red Sabre	M-24	Remanit 1790C	M-1299	Republic Extra Special	M-379
Red Seal Die Steel	M-57	Remanit 1790H	M-459		
Red Star Die	M-34	Remanit 1790 V	M-459	Republic Fast Finishing	M-97
Red Star Die	M-115	Remanit 1800M	M-459		
Red Star Tap	M-115	Remanit 1810	M-459	Republic H12	M-97
Red Star Vanadium	M-34	Remanit 1810 SST	M-459	Republic H13	M-97
Red Tiger	M-24	Remanit 1810 SSW	M-459	Republic H14	M-97
Refiner Cast E.S.C.-RCS	M-50	Remanit 1813 SSW	M-459	Republic H21	M-97
		Remanit 1860 M	M-459	Republic H21M	M-97
Regular	M-73	Remanit 1880	M-459	Republic H24	M-97
Regular Bolt Die	M-85	Remanit 1880	M-1299	Republic H26	M-97
Regular Carbon	M-85	Remanit 1880 A	M-459	Republic H42	M-97
Regular Power Transformer No. 72	M-806	Remanit 1880 FH	M-459	Republic HP9-4-25	M-97
		Remanit 1880 S	M-459	Republic HP9-4-45	M-97
		Remanit 1880 SEW	M-459	Republic L2	M-97
Regular-SS	M-73	Remanit 1880 SS	M-459	Republic M1	M-97
Regular Stainless-420	M-73	Remanit 1880 SSEW	M-459	Republic M2	M-97
				Republic M3	M-97
Regular Tool	M-38	Remanit 1880 SST	M-459	Republic M4	M-97
Regular Tool	M-85	Remanit 1880 SSW	M-459	Republic M7	M-97
Reining MSM	M-1353	Remanit 1880 SW	M-459	Republic M10	M-97
Rein-Nickel GNi	M-303	Remanit 1880 SZT	M-459	Republic M43	M-97
Relleum	M-266	Remanit 1880 T	M-459	Republic O1	M-97
Rema	M-52	Remanit 1990 S-Cr	M-459	Republic Plastic Die	M-97
Rema Hobbing Die	M-52, M-385	Remanit 1990 SS-Cr	M-459		
				Republic RS-6Al-4V	M-97
Remanit	M-459	Remanit 1990SW	M-459	Republic RS-40	M-97
Remanit 8A	M-459	Remanit 2525 SST	M-459	Republic RS-55	M-97
Remanit 11A	M-459	Remanit 2804 G	M-459	Republic RS-70	M-97
Remanit 0327	M-459	Remanit 2810	M-459	Republic RS-100	M-97
Remanit 1200 M	M-459	Remanit 2810	M-1299	Republic RS-110	M-97
Remanit 1212	M-459	Remanit 2810 Mo	M-459	Republic RS-110A	M-97
Remanit 1218S	M-459	Remanit 2890	M-459	Republic RS-110B	M-97
Remanit 1218 ST	M-459	Remanit 2890 Mo	M-459	Republic RS-110C	M-97
Remanit 1510 (14 Cr)	M-1299	Remanit ATS 1	M-459	Republic RS-120	M-97
		Remanit G Feder Hard	M-459	Republic RS-120A	M-97
Remanit 1510 AB	M-459			Republic RS-130	M-97

Republic RS-135	M-97	Revere Alloy No. 428	M-238	Rexite Grade CR-7	M-38
Republic RS-140	M-97			Rexite Grade CR-8	M-38
Republic S1	M-97	Revere Alloy No. 432	M-238	Rexite Grade UC-14	M-38
Republic S3	M-97			Rexite Grade UC-16	M-38
Republic S4	M-97	Revere Alloy No. 440	M-238	Rexite UC-25	M-38
Republic Special Auto Type A	M-97	Revere Alloy No. 444	M-238	Rex LA	M-38
				Rex MM	M-38
Republic T1	M-97			Rex MMM	M-38
Republic T2	M-97, M-379	Revere Alloy No. 447	M-238	Rex Supercut	M-38
				Rex Supervan	M-38
Republic T4	M-97, M-379	Revere Alloy No. 456	M-238	Rex TMO-5	M-38
				Rex TMO-8	M-38
Republic Tool Steel	M-97, M-379	Revere Alloy No. 467	M-238	Rex TMO-S	M-38
		Revere Alloy No. 530	M-238	Rex VM Dreadnaught	M-38
Republic UA-6	M-97			Rex VM-S	M-38
Republic UA-8	M-97	Revere Alloy No. 562	M-238	Rexweld 27	M-38
Republic Unique Alloy	M-97, M-379			Rexweld 33W	M-38
		Revere Alloy No. 574	M-238	Rexweld 54	M-38
				Rexweld 57	M-38
Republic W1	M-97	Revere Alloy No. 576	M-238	Rexweld 64	M-38
Republic W2	M-97			Rexweld 66	M-38
Republic W5	M-97	Revere Alloy No. 577	M-238	Rexweld A	M-38
Republic XX Superior	M-379			Rexweld B	M-38
		Revere Alloy No. 578	M-238	Rexweld C	M-38
Republic YA-4	M-97			Rexweld VT	M-38
Resilia	M-24	Revere FS	M-238	Reynolds 2S	M-671
Resistvar 1	M-1084	Revere FS-1	M-238	Reynolds 3S	M-671
Resource	M-1411	Revere J-1	M-238	Reynolds 13	M-671
Resulphurized Stock	M-377	Revere JS	M-238	Reynolds 14S	M-671
		Revere JS-1	M-238	Reynolds 17S	M-671
Revere Alloy No. 770	M-238	Revere M	M-238	Reynolds 18S	M-671
		Revere No. 158	M-238	Reynolds 24S	M-671
Revere	M-238	Revere No. 234	M-238	Reynolds 25S	M-671
Revere 53S	M-238	Revere No. 260	M-238	Reynolds 32S	M-671
Revere 5051	M-238	Revere No. 535	M-238	Reynolds 43	M-671
Revere Alloy 464	M-238	Revere O-1	M-238	Reynolds 52S	M-671
Revere Alloy No. 99	M-238	Revere Phosphor Bronze A	M-238	Reynolds 56S	M-671
Revere Alloy No. 115	M-238			Reynolds 63S	M-671
		Revere Phosphor Bronze B	M-238	Reynolds 85	M-671
Revere Alloy No. 116	M-238			Reynolds 108	M-671
		Revere Phosphor Bronze C	M-238	Reynolds 112	M-671
Revere Alloy No. 122	M-238			Reynolds 113	M-671
		Revere Phosphor Bronze D	M-238	Reynolds 122	M-671
Revere Alloy No. 150	M-238			Reynolds 138	M-671
		Revere Phosphor Bronze E	M-238	Reynolds 142	M-671
Revere Alloy No. 173	M-238			Reynolds 152	M-671
		Rex 3-V	M-38	Reynolds 195	M-671
Revere Alloy No. 232	M-238	Rex 4-V	M-38	Reynolds 212	M-671
		Rex 18-8	M-38	Reynolds 214	M-671
Revere Alloy No. 233	M-238	Rex 440	M-38	Reynolds 218	M-671
		Rex 539	M-56	Reynolds 220	M-671
Revere Alloy No. 254	M-238	Rex 939	M-38	Reynolds 319	M-671
		Rex 1059	M-38	Reynolds 333	M-671
Revere Alloy No. 309	M-238	Rex 1092	M-38	Reynolds 356	M-671
		Rex "A"	M-38	Reynolds 360	M-671
Revere Alloy No. 342	M-238	Rex AA-OX	M-38	Reynolds 380	M-671
		Rex AA-PX	M-38	Reynolds 406	M-671
Revere Alloy No. 350	M-238	Rexalloy	M-38	Reynolds 645	M-671
		Rexalloy 33	M-38	Reynolds 750	M-671
Revere Alloy No. 385	M-238	Rexalloy 108	M-38	Reynolds A17S	M-671
		Rexalloy A	M-38	Reynolds A51S	M-671
Revere Alloy No. 411	M-238	Rexaloy	M-1017	Reynolds A108	M-671
		Rex Champion	M-38	Reynolds A132	M-671

Reynolds A214	M-671	Rezistal		Rita Hot-Forging	
Reynolds AS No. 1	M-671	KA-2S-20-10	M-38	Die No. 2	M-354
Reynolds AS No. 2	M-671	Rezistal KA2SCb	M-38	Rita Hot-Forging	
Reynolds B195	M-671	Rezistal KA-2S-Mo	M-38	No. 1	M-354
Reynolds B214	M-671	Rezistal KA-2S-Mo-T	M-38	Rita Nochange Oil Hardening	M-354
Reynolds C113	M-671	Rezistal KA-2ST	M-38	Rita Oil Hardening	
Reynolds F214	M-671	Rezistal KA-2ST Special 19-9	M-38	Drill Rod	M-354
Reynolds J51S	M-671	Rezistal KA-2T	M-38	Rita Special Drill Rod	M-354
Reynolds Metal	M-671	Rezistal NCR 124	M-38	Rita Special Tool	M-354
Reynolds No. 2364	M-671	Rezistal NCR-238	M-260, M-38	Rita Standard Drill Rod	M-354
Reynolds No. 2393	M-671	Rezistal No. 2 CW	M-38	Rita Standard Tool	M-354
Reynolds R317	M-671	Rezistal Safweld	M-362	Rita T.C.V. Alloy	M-354
Rezistal 2 C	M-38	Rezistal Stainless 122	M-38	Rita Three Point	M-354
Rezistal 3	M-38	Rezistal Stainless 162	M-38	Rita Vanadium Drill Rod	M-354
Rezistal 3C	M-38	Rezistal Stainless B-80	M-38	Rivaloy No. 24	M-98
Rezistal 4	M-38	Rezistal Stainless B-100	M-38	Riverside 27	M-98
Rezistal 7	M-38	Rezistal Stainless BM	M-38	Riverside 61	M-98
Rezistal 301	M-38	Rezistal Stainless Iron 12	M-38, M-260	Riverside 62	M-98
Rezistal 302	M-38			Riverside 63	M-98
Rezistal 302 B	M-38			Riverside 64	M-98
Rezistal 303	M-38			Riverside 66	M-98
Rezistal 304	M-38			Riverside 70	M-98
Rezistal 305	M-38			Riverside 72-28	M-911
Rezistal 308	M-38			Riverside 74-26	M-911
Rezistal 309	M-38	Rezistal Stainless Iron 17	M-38	Riverside 78	M-98
Rezistal 309 B	M-38	Rezistal Stainless Iron 20	M-38	Riverside 78-22	M-911
Rezistal 310	M-38			Riverside 80-20	M-911
Rezistal 311	M-38	Rezistal Stainless Iron 27	M-38	Riverside 80-20 Hard	M-911
Rezistal 314	M-38	Rezistal Stainless Iron F.M. 2	M-38, M-260	Riverside 109	M-98
Rezistal 316	M-38			Riverside 110	M-98
Rezistal 317	M-38			Riverside 111	M-98
Rezistal 318	M-38	Rezistal Stainless Steel Grade A	M-38	Riverside No. 1	M-98
Rezistal 321	M-38			Riverside No. 2	M-98
Rezistal 325	M-38	Rezistal Stainless Steel Grade B	M-38	Riverside No. 3	M-98
Rezistal 329	M-38			Riverside No. 4	M-98
Rezistal 347	M-38	Rezistal Turbine	M-38	Riverside No. 5	M-98
Rezistal 403	M-38	Rezistal VT	M-38	Riverside No. 6	M-98
Rezistal 405	M-38	Rezistal WH	M-38	Riverside No. 7	M-98
Rezistal 406	M-38	Rezistal X-40	M-38	Riverside No. 8	M-98
Rezistal 410	M-38	"RH" Monel	M-1499	Riverside No. 9	M-98
Rezistal 414	M-38	Rheinrohr-MV12	M-1658	Riverside No. 10	M-98
Rezistal 416	M-38	Rheinrohr-MVW12	M-1658	Riverside No. 11	M-98
Rezistal 420	M-38	Rhometal	M-108	Riverside No. 12	M-98
Rezistal 430	M-38	Rita Alloy	M-354	Riverside No. 13	M-98
Rezistal 431	M-38	Rita Alloy No. 2	M-354	Riverside No. 14	M-98
Rezistal 440A	M-38	Rita Alloy No. 4	M-354	Riverside No. 15	M-98
Rezistal 440B	M-38	Rita Alloy No. 5	M-354	Riverside No. 16	M-98
Rezistal 440BM	M-38	Rita Alloy No. 7	M-354	Riverside No. 17	M-98
Rezistal 440C	M-38	Rita Car-Van Die	M-354	Riverside No. 18	M-98
Rezistal 442	M-38	Rita Chrome Vanadium	M-354	Riverside No. 20	M-98
Rezistal 446	M-38			Riverside No. 21	M-98
Rezistal 2600	M-38			Riverside No. 23	M-98
Rezistal 2600-S	M-38	Rita Cobalt High Speed	M-354	Riverside No. 30	M-98
Rezistal FM-18-8	M-38			Riverside No. 31	M-98
Rezistal KA-2	M-38	Rita Electric Tool	M-354	Riverside No. 32	M-98
Rezistal KA-2-19-9	M-38	Rita Endur	M-354	Riverside No. 33	M-98
Rezistal KA2-20-10	M-38	Rita Extra Special	M-354	Riverside No. 39	M-98
Rezistal "K.A.2H."	M-38	Rita Extra Tool	M-354	Riverside No. 40	M-98
Rezistal "K.A.2M."	M-38	Rita High Speed	M-354	Riverside No. 44	M-98
Rezistal K A 2 Mo	M-38				
Rezistal KA-2-Mo-T	M-38				
Rezistal "K.A.2S."	M-38				
Rezistal KA-2S-19-9	M-38				

Alloy	Code	Alloy	Code	Alloy	Code
Riverside No. 47	M-98	Rochling MC15	M-912, M-501	Rochling R 2/3	M-912, M-501
Riverside No. 80	M-98	Rochling MC20	M-501, M-912	Rochling R5	M-501, M-912
Riverside No. 101	M-98	Rochling MFR	M-501, M-912	Rochling R6	M-501, M-912
Riverside No. 102	M-98	Rochling MM	M-501, M-912	Rochling R 6/7	M-501, M-912
Riverside No. 105	M-98	Rochling Mo 15	M-501, M-912	Rochling R8	M-501, M-912
Riverside No. 108	M-98	Rochling Mo 20	M-501, M-912	Rochling RA2	M-501, M-912
Riverside No. 120	M-98	Rochling Mo 25	M-501, M-912	Rochling RA2W	M-501, M-912
Riverside No. 205	M-98	Rochling Mo 35	M-501, M-912	Rochling RAB1	M-501, M-912
Riverside No. 209	M-98	Rochling Mo 40	M-501, M-912	Rochling RAE2	M-501, M-912
Riverside No. 209D	M-98	Rochling Mo 50	M-501, M-912	Rochling RAE3B	M-501, M-912
Riverside No. 330	M-98	Rochling Mo 230	M-501, M-912	Rochling RAE5	M-501, M-912
Riverside No. 400	M-98	Rochling Mo 240	M-912, M-501	Rochling RB9	M-501, M-912
Riverside No. 401	M-98	Rochling NG5A	M-501, M-912	Rochling RBF	M-501, M-912
Riverside No. 402	M-98	Rochling NGS	M-501, M-912	Rochling RBH	M-501, M-912
Riverside No. 403	M-98	Rochling NH4	M-501, M-912	Rochling RCCK	M-501, M-912
Riverside No. 404	M-98	Rochling NH4G	M-501, M-912	Rochling RCCW	M-501, M-912
Riverside No. 405	M-98	Rochling NH8	M-501, M-912	Rochling RCNH	M-501, M-912
Riverside No. 409	M-98	Rochling NH11G	M-501, M-912	Rochling RCW2	M-501, M-912
Riverside Grade A Phosphor Bronze	M-98	Rochling NH22G	M-501, M-912	Rochling RCW3	M-501, M-912
Riverside Grade E Phosphor Bronze	M-98	Rochling NH25G	M-501, M-912	Rochling RE2	M-501, M-912
Riverside Nickel Silver No. 7	M-98	Rochling NH40G	M-501, M-912	Rochling RE2C	M-501, M-912
Riverside Nickel Silver No. 18	M-98	Rochling NH60	M-501, M-912	Rochling RE2CW	M-501, M-912
Riverside Phosphor Bronze No. 30	M-98	Rochling NH80	M-501, M-912	Rochling RE2W	M-501, M-912
Riverside Phosphor Bronze No. 47	M-98	Rochling OCE12	M-501, M-912	Rochling RECN1	M-501, M-912
Riverside Phosphor Bronze No. 209	M-98	Rochling OCE34	M-501, M-912	Rochling RECNW	M-501, M-912
R-M-20	M-82a	Rochling OCE57	M-501, M-912	Rochling REN2	M-501
RM Special Sehr Hart	M-27	Rochling PG1	M-501, M-912	Rochling RF6	M-501, M-912
RNC Carbimphy	M-771	Rochling PG3	M-501, M-912	Rochling RFA	M-501, M-912
RNC Superimphy	M-771	Rochling Prima-H	M-501	Rochling RFAW	M-501, M-912
Rocan	M-238	Rochling Prima-MH	M-501	Rochling RG6	M-501, M-912
Roch 2	M-501	Rochling Prima-ZAH	M-501	Rochling RG7	M-501, M-912
Roch 3	M-501	Rochling Prima-ZH	M-501		
Roch 4	M-501	Rochling R1	M-501, M-912	Rochling RG8	M-501, M-912
Roch 5	M-501				
Roch 6	M-501				
Roch 7	M-501				
Roch 8	M-501				
Rochling D Spezial	M-501, M-912				
Rochling ECMo100	M-501, M-912				
Rochling ECN180	M-501, M-912				
Rochling ECV150	M-501, M-912				
Rochling ES114	M-501, M-912				
Rochling F1620	M-912, M-501				
Rochling HOWG	M-501, M-912				
Rochling K	M-501, M-912				
Rochling M10C	M-501, M-912				

Alloy	Spec
Rochling RG10	M-501, M-912
Rochling RG11	M-501, M-912
Rochling RG12	M-501, M-912
Rochling RGS2	M-501, M-912
Rochling RH15	M-501, M-912
Rochling RHB9	M-501, M-912
Rochling RKMV	M-501, M-912
Rochling RLB	M-501, M-912
Rochling RM2	M-501, M-912
Rochling RM 2/3	M-501, M-912
Rochling RM3	M-501, M-912
Rochling RM4	M-501, M-912
Rochling RM5	M-501, M-912
Rochling RM6	M-501, M-912
Rochling RM 6/7	M-501, M-912
Rochling RNO	M-501, M-912
Rochling RNO18	M-501, M-912
Rochling RNO25	M-912
Rochling RNO30EG	M-501, M-912
Rochling RNO30G	M-501, M-912
Rochling RNOA17	M-501, M-912
Rochling RNOB	M-501, M-912
Rochling RNOC	M-501, M-912
Rochling RNOF	M-501, M-912
Rochling RNOG	M-501, M-912
Rochling RNOH	M-501, M-912
Rochling RNOM	M-501, M-912
Rochling RNOMO	M-912
Rochling RNOMOWV	M-912
Rochling RNOS	M-501, M-912
Rochling RNOSG	M-501, M-912
Rochling RNO Spezial G	M-501, M-912
Rochling RNOT	M-501, M-912
Rochling RNOW	M-501, M-912
Rochling RNOW-W	M-501, M-912
Rochling RNOX	M-501, M-912
Rochling ROT 40	M-501, M912
Rochling ROT 60	M-501, M-912
Rochling RPDW	M-501, M-912
Rochling RPDZ	M-501, M-912
Rochling RPDZW	M-501, M-912
Rochling RR3	M-501, M-912
Rochling RR4	M-501, M-912
Rochling RR6	M-501, M-912
Rochling RR7	M-501, M-912
Rochling RR8	M-501, M-912
Rochling RSH	M-501, M-912
Rochling RT7 Extra	M-501, M-912
Rochling RT9	M-501, M-912
Rochling RT10EE	M-501, M-912
Rochling RT11 Cr	M-501, M-912
Rochling RT11 Extra	M-501, M-912
Rochling RT12 Extra	M-501, M-912
Rochling RT 14 Cr	M-501, M-912
Rochling RT14 Extra	M-501, M-912
Rochling RTC20	M-501, M-912
Rochling RT Extra	M-501, 912
Rochling RTK10	M-501, M-912
Rochling RTW2	M-501, M-912
Rochling RTW3	M-501, M-912
Rochling RVE	M-501, M-912
Rochling RVE42	M-501, M-912
Rochling RVE Extra	M-501, M-912
Rochling RW2	M-501, M-912
Rochling RW3A	M-501, M-912
Rochling RWS2	M-501, M-912
Rochling SGM	M-501, M-912
Rochling SGM2	M-501, M-912
Rochlingstahl 30	M-501
Rochlingstahl 32	M-501
Rochlingstahl 34	M-501
Rochlingstahl BP21	M-501
Rochlingstahl BP22	M-501
Rochlingstahl BP25	M-501
Rochlingstahl BP29	M-501
Rochlingstahl BP29 Extra	M-501
Rochlingstahl B.S. 2	M-912
Rochlingstahl D Special	M-912
Rochlingstahl ES	M-912
Rochlingstahl FVC	M-912
Rochlingstahl FVM45	M-912
Rochlingstahl MFR	M-912
Rochlingstahl NGS	M-912
Rochlingstahl PG3	M-912
Rochlingstahl PWD	M-912
Rochlingstahl PWD6	M-912
Rochlingstahl PWD13	M-912
Rochlingstahl PWDN	M-912
Rochlingstahl RAB1	M-912
Rochlingstahl RAN	M-912
Rochlingstahl RB7	M-912
Rochlingstahl RCNW	M-912
Rochlingstahl RCW2	M-912
Rochlingstahl RKMV	M-912
Rochlingstahl RLB	M-912
Rochlingstahl R.R. 7	M-912
Rochlingstahl RSH	M-912
Rochlingstahl RSV	M-912
Rochlingstahl RSZ	M-912
Rochlingstahl RSZ Special	M-912
Rochlingstahl RT7	M-912
Rochlingstahl RT7Cr	M-912
Rochlingstahl RT11	M-912
Rochlingstahl RT11 Cr	M-912

Section IV: Obsolete Alloys / 1741

Alloy	Code
Rochlingstahl RT14 Cr	M-912
Rochlingstahl RTC14 Mo	M-912
Rochlingstahl RTC 20	M-912
Rochlingstahl RTK10	M-912
Rochlingstahl RTW2	M-912
Rochlingstahl RTW3 Special	M-912
Rochlingstahl RUS2	M-912
Rochlingstahl RWM	M-912
Rochlingstahl RWS	M-912
Rochlingstahl RWS2	M-912
Rochlingstahl RWS4	M-912
Rochlingstahl S	M-912
Rochlingstahl SGM4	M-912
Rochlingstahl SKSV	M-912
Rochlingstahl T76	M-912
Rochlingstahl WNC	M-912
Rochlingstahl ZK	M-912
Rochling VC140	M-501, M-912
Rochling VC170	M-501, M-912
Rochling VM80	M-501, M-912
Rochling VM 90	M-501, M-912
Rochling VM100	M-501, M-912
Rochling VM125	M-501, M-912
Rochling VME	M-501, M-912
Rochling VMS140	M-501, M-912
Rochling VMS160	M-501, M-912
Rochling VMV125	M-501, M912
Rochling VMV140	M-501, M-912
Rochling WP3	M-501, M-912
Rodney Type R215	M-1663
Rokbore	M-430
Roley ERJ 35	M-1228
Roley ERJ 45	M-1228
Roley ERM 15	M-1228
Roley HIR15 W/R	M-1228
Roley HIR25	M-1228
Roley NALH	M-1228
Roley NALJ	M-1228
Roley NALM	M-1228
Roley NALW	M-1228
Rollo	M-50
Roman Bronze	M-238
Roncan	M-238
Ronensil	M-912
Rose 3X	M-1305
RS 1 Spezial	M-459
RS 2 Spezial	M-459
RS 3	M-459
RS 4	M-459
R.T. Steel	M-57
Ruinella	M-92
Rustless 5	M-100
Rustless 8-20	M-100
Rustless 12	M-100
Rustless 12-2	M-100
Rustless 12 Al	M-100
Rustless 12 FM	M-100
Rustless 12 T	M-100
Rustless 13 C 35	M-100
Rustless 13HC	M-100
Rustless 13HC35	M-100
Rustless 15	M-100
Rustless 16-2	M-100
Rustless 16-6	M-100
Rustless 17	M-100
Rustless 17-7	M-100
Rustless 17-C-60	M-100
Rustless 17-C-80	M-100
Rustless 17-C-100	M-100
Rustless 17FM	M-100
Rustless 17HC60	M-100
Rustless 17HC90	M-100
Rustless 18-8	M-100
Rustless 18-8 Cb	M-100
Rustless 18-8FM (SELENIUM)	M-100
Rustless 18-8FM (SULPHUR)	M-100
Rustless 18-8 Ti	M-100
Rustless 18-12-3 Mo	M-100
Rustless 18-12-4 Mo	M-100
Rustless 18-12 MD	M-100
Rustless 20-10	M-100
Rustless 21	M-100
Rustless 25-12	M-100
Rustless 25-20	M-100
Rustless 27	M-100
Rustless 29-9	M-100
Rustless RR-11	M-100
R.V.M.	M-1724
Ry-Alloy Drill	M-240
Ry-Alloy Flat	M-240
Ry-Arm	M-240
Ry-AX	M-240
RYCO 44	M-240
Rycut 47	M-240
RY-DIE	M-240
Ryerson Carbon	M-240
Ryerson H.T.M.	M-240
Ryerson Shock Tool	M-240
Ryolite 4-Point	M-240
Ryolite "B.F.D."	M-240
Ryolite Chromium Hot Work	M-240
Ryolite Diamond B	M-240
Ryolite Special High Speed Tool Steel	M-240
Ryolite Tungsten Hot Work	M-240
Ryolite X	M-240
Ryolite XX	M-240
Ryolite XXX	M-240

S

Alloy	Code
S-495 Alloy	M-1
S-497 Alloy	M-1
S-588 Alloy	M-1
S-590 Alloy	M-1
S-816 Alloy	M-1
SA	M-459
SA-50	M-725
SA200	M-459
SA300	M-459
SA500	M-459
SA900	M-459
Sabeco Metal	M-221
SABECO No 5	M-966
SA-Best Extra Standard	M-357
SAE 300 (7Q5)	M-106
Sagamore V	M-1
Saint Juery Extra	M-1119
Salvo	M-85
Samson	M-32
Samson Extra	M-32
Samson No. 2-547	M-32
Samson No. 3-427	M-32
Samson No. 3-547	M-32
Samson No. 4-408	M-32
Samson No. 5-317	M-32
Samson No. 5-720	M-32
Samson No. 158	M-32
Samson No. 436	M-32
Samson No. 500	M-32
Sanbron AA	M-618
Sanbron BB	M-618
Sanbron CC	M-618
Sanbron DD	M-618
Sanbron EE	M-618
Sanbron FF	M-618
Sanbron H.S. No. 1	M-618
Sanbron H.S. No. 2	M-618
Sanderson	M-38
Sanderson 3890	M-618
Sanderson 4379 Non-stain	M-618
Sanderson B. 883	M-618
Sanderson Chrome-Vanadium	M-618
Sanderson Double Special	M-618
Sanderson Extra 3 1/2	M-618
Sanderson Extra 4 1/2	M-618
Sanderson N.2832	M-618
Sanderson N.2834	M-618

Sanderson N.3603	M-618	Sandvik 3C27	M-101	Sandvik 17T4	M-101
Sanderson N.5366	M-618	Sandvik 3C27Mo2	M-101	Sandvik 17V	M-101
Sanderson NK2237	M-618	Sandvik 3C341	M-101	Sandvik 18C2	M-101
Sanderson NK2833	M-618	Sandvik 3LC2	M-101	Sandvik 19	M-101
Sanderson NK8053	M-618	Sandvik 3MC2	M-101	Sandvik 21C1	M-101
Sanderson NK8055	M-618	Sandvik 3N3	M-101	Sandvik 22C1	M-101
Sanderson NK8073	M-618	Sandvik 3N4C2	M-101	Sandvik 23C	M-101
Sanderson No. 1	M-618	Sandvik 3N5	M-101	Sandvik 23C1	M-101
Sanderson No. 2	M-618	Sandvik 3R9	M-101	Sandvik 24	M-101
Sanderson Special	M-38	Sandvik 3R14	M-101	Sandvik 40T13	M-101
Sanderson Special	M-618	Sandvik 3R14Mo	M-101	Sandvik 178Co	M-101
Sanderson SS8085	M-618	Sandvik 3S	M-101	Sandvik 282	M-101
Sanderson Tool	M-38	Sandvik 4C27	M-101	Sandvik 286	M-101
Sanderson Tungsten Tap	M-38	Sandvik 4C48	M-101	Sandvik 389	M-101
		Sandvik 4C48P3	M-101	Sandvik HT1	M-101
S + J Special	M-38	Sandvik 4C48S	M-101	Sandvik HT2	M-101
Sandusky No. 2	M-295	Sandvik 4C58	M-101	Sandvik No. 2M.F.X.	M-101
Sandusky No. 3	M-295	Sandvik 4N2C36	M-101		
Sandusky No. 4	M-295	Sandvik 4N3C2	M-101	Sandvik OR2	M-101
Sandusky No. 5	M-295	Sandvik 4N3C2Mo	M-101	Sandvik OR2N10	M-101
Sandusky No. 6	M-295	Sandvik 5A	M-101	Sandvik OR3	M-101
Sandusky No. 8	M-295	Sandvik 5C2Mo	M-101	Sandvik OR11	M-101
Sandusky No. 9	M-295	Sandvik 5L	M-101	Sandvik OR12	M-101
Sandusky No. 10	M-295	Sandvik 6C283	M-101	Sandvik OR18	M-101
Sandusky SA-351	M-295	Sandvik 6L	M-101	Sandvik OR18M	M-101
Sandvik 1C10S1	M-101	Sandvik 6N3C2	M-101	Sandvik OR19	M-101
Sandvik 1C36	M-101	Sandvik 6N3C2Mo	M-101	Sandvik OR20	M-101
Sandvik 1HT1	M-101	Sandvik 6N4C2	M-101	Sandvik OR23	M-101
Sandvik 1HT2	M-101	Sandvik 6T8	M-101	Sandvik OR26	M-101
Sandvik 1R2	M-101	Sandvik 7	M-101	Sandvik OR27	M-101
Sandvik 1R4	M-101	Sandvik 7BS	M-101	Sandvik OR28	M-101
Sandvik 1R8	M-101	Sandvik 7C27	M-101	Sans Rival	M-1119
Sandvik 1R10	M-101	Sandvik 7N3C3	M-101	SA/O	M-459
Sandvik 1R41	M-101	Sandvik 8	M-101	Saranac	M-38
Sandvik 1R42	M-101	Sandvik 8C2W4V	M-101	SATCO	M-958
Sandvik 1R43	M-101	Sandvik 8L	M-101	Saubron Hot Die	M-618
Sandvik 1R44	M-101	Sandvik 8LM	M-101	SA/W	M-459
Sandvik 1R45	M-101	Sandvik 8M	M-101	S. B. Steel	M-38
Sandvik 1R46	M-101	Sandvik 8MC2Mo	M-101	S.C.C.-210	M-1295
Sandvik 1S30	M-101	Sandvik 8N	M-101	S.C.C.-213	M-1295
Sandvik 2C27	M-101	Sandvik 8N1C2	M-101	S.C.C.-213M	M-1295
Sandvik 2C34A	M-101	Sandvik 8N3	M-101	S.C.C.-531	M-1295
Sandvik 2C40	M-101	Sandvik 9L	M-101	S.C.C.-925C	M-1295
Sandvik 2C341	M-101	Sandvik 10C2V	M-101	S.C.C.-C43	M-1295
Sandvik 2HT3	M-101	Sandvik 10C27	M-101	S.C.C.-W4	M-1295
Sandvik 2LC1	M-101	Sandvik 10LS2	M-101	Schmidtclemens, 62 alloys discontinued or name changed to Marker or Maerker	M-1305
Sandvik 2MF	M-101	Sandvik 10T14	M-101		
Sandvik 2N5	M-101	Sandvik 11S2	M-101		
Sandvik 2N36	M-101	Sandvik 12C	M-101		
Sandvik 2R1	M-101	Sandvik 12C331	M-101		
Sandvik 2R2	M-101	Sandvik 12L7	M-101		
Sandvik 2R2A	M-101	Sandvik 12LM	M-101	Schneider 0.5 F08	M-533
Sandvik 2R2N8	M-101	Sandvik 13W12C1	M-101	Schneider 1	M-1301
Sandvik 2R2N10	M-101	Sandvik 14LM	M-101	Schneider 1A	M-1301
Sandvik 2R3	M-101	Sandvik 14V	M-101	Schneider 1F 3/4	M-533
Sandvik 2R3N11	M-101	Sandvik 15	M-101	Schneider 1F 4/5	M-533
Sandvik 2R6	M-101	Sandvik 15D	M-101	Schneider 1 (Ni)	M-1301
Sandvik 2R14	M-101	Sandvik 16C2	M-101	Schneider 1S	M-1301
Sandvik 2R15	M-101	Sandvik 16C6	M-101	Schneider 1.1 F08	M-533
Sandvik 2R24	M-101	Sandvik 16LC	M-101	Schneider 1.3 FOV	M-533
Sandvik 2R25	M-101	Sandvik 16N2C1	M-101	Schneider 1.4DF3	M-533
Sandvik 2S35	M-101	Sandvik 16T2	M-101	Schneider 1.4DF5	M-533
Sandvik 2SE	M-101	Sandvik 17C1	M-101	Schneider 1.4DF6	M-533
Sandvik 3C2Mo1	M-101	Sandvik 17C3	M-101	Schneider 1.4DF7C	M-533

Alloy	Code	Alloy	Code	Alloy	Code
Schneider 1.4DF8C	M-533	Scholler AS4W	M-1308	Scovill 325	
Schneider 2	M-1301	Scholler ASA	M-1308	Lancashire Brass	M-102
Schneider 2DFO3	M-533	Scholler BRU1	M-1308	Scovill 342 High	
Schneider 2.2 FO	M-533	Scholler DSB	M-1308	Leaded Brass-65%	M-102
Schneider 3	M-1301	Scholler DW	M-1308	Scovill 350	M-102
Schneider 3FDO1	M-533	Scholler EC1	M-1308	Scovill 403 Gilding	
Schneider 3FO	M-533	Scholler EN1.5	M-1308	Bronze	M-102
Schneider 4	M-62	Scholler ENC3	M-1308	Scovill 405 Penny	
Schneider 5.1B	M-1301	Scholler Extra MH	M-1308	Bronze	M-102
Schneider 11	M-1301	Scholler Extra MH	M-1308	Scovill 434 Spring	
Schneider 11DS	M-1301	Scholler Extra		Bronze 0.8%	M-102
Schneider 13	M-1301	Weich	M-1308	Scovill 651 Low	
Schneider 13 FF	M-533	Scholler FLA	M-1308	Silicon Bronze	M-102
Schneider 14	M-1301	Scholler HCN2	M-1308	Scovill 706 Cupro	
Schneider 14S	M-1301	Scholler HCN3	M-1308	Nickel 10%	M-102
Schneider 14S-Ti	M-1301	Scholler HCN4	M-1308	Scovill 735 Nickel	
Schneider 15	M-1301	Scholler HLDH	M-1308	Silver 72-18	M-102
Schneider 31	M-1301	Scholler HLDW	M-1308	Scovill 745	M-102
Schneider 60	M-1301	Scholler HMC	M-1308	Scovill 788	M-102
Schneider 61	M-1301	Scholler KLI	M-1308	Scovill 810	
Schneider 63	M-1301	Scholler KW10	M-1308	Aluminum	
Schneider A8B	M-533	Scholler KW10		Bronze 5%	M-102
Schneider A9/10B	M-533	Spec.	M-1308	Scovill Admiralty	
Schneider A10/11B	M-533	Scholler KW20	M-1308	445	M-102
Schneider A11B	M-533	Scholler KW40	M-1308	Scovill Alloy 169	
Schneider C1	M-533	Scholler KW80	M-1308	Yellow Brass, 65%	
Schneider C2-C2B	M-533	Scholler KW100	M-1308	270	M-102
Schneider C4	M-533	Scholler KWA	M-1308	Scovill Alloy 280	
Schneider C5	M-533	Scholler KWB	M-1308	Muntz Metal	M-102
Schneider C6-C6S	M-533	Scholler KWZ	M-1308	Scovill Alloy 314	M-102
Schneider C.9	M-533	Scholler LM3 W/H	M-1308	Scovill Alloy 320	M-102
Schneider C.10	M-533	Scholler MACDW	M-1308	Scovill Alloy 323	
Schneider C.12	M-533	Scholler MAG	M-1308	Tin Brass (1.75%)	M-102
Schneider CMP	M-533	Scholler MAS	M-1308	Scovill Alloy 360	M-102
Schneider DFO	M-533	Scholler MASWS	M-1308	Scovill Alloy 362	
Schneider FLO6	M-533	Scholler OF56	M-1308	(Admiralty)	M-102
Schneider FLOV	M-533	Scholler PM	M-1308	Scovill Alloy 413	
Schneider FO3	M-533	Scholler Prima	M-1308	Tin Brass (1%)	M-102
Schneider FO4	M-533	Scholler Prima		Scovill Alloy 415	
Schneider FO 5/6	M-533	Weich	M-1308	Tin Brass (2%)	M-102
Schneider FO 6/7 S	M-533	Scholler S200	M-1308	Scovill Alloy 419	
Schneider FO7C	M-533	Scholler SAS2	M-1308	Scovill Alloy 430	
Schneider FO8C	M-533	Scholler SAS4	M-1308	Spring Bronze	M-102
Schneider FOV	M-533	Scholler SAS4MN	M-1308	Scovill Alloy 445	M-102
Schneider F.P.	M-533	Scholler SAS5	M-1308	Scovill Alloy 464	M-102
Schneider F.P.B.	M-533	Scholler SAST2	M-1308	Scovill Alloy 482	M-102
Schneider FR	M-533	Scholler SAST4	M-1308	Scovill Alloy 485	M-102
Schneider M.L.	M-533	Scholler SC2	M-1308	Scovill Alloy 510	M-102
Schneider M.L.S.	M-533	Scholler SGH	M-1308	Scoville Alloy 510	M-102
Schneider MO7	M-533	Scholler SKVL	M-1308	Scovill Alloy 521	M-102
Schneider P2	M-533	Scholler SNK3	M-1308	Scovill Alloy 525	M-102
Schneider S.3	M-533	Scholler Tamb	M-1308	Scovill Alloy 630	
Schneider S.3V	M-533	Scholler V132S	M-1308	Cupro Nickel	
Schneider S.I.S.	M-533	Scholler VCV230	M-1308	(15%)	M-102
Schneider S.I.S.B.	M-533	Scholler WU	M-1308	Scovill Alloy 675	M-102
Schneider		Schweistahl 5H	M-1314	Scovill Alloy 710	M-102
S.I.S.B.-S.I.	M-533	Scleron No. 2	M-320,	Scovill Alloy 715	M-102
Schneider SO3-M2	M-533		M-116	Scovill Alloy 743	
Schneider SS4	M-533	Scleron No. 3	M-320,	Nickel Silver (8%)	M-102
Schneider S.S.8	M-533		M-116	Scovill Alloy 752	M-102
Scholler 1116S	M-1308	Scovill 7% Nickel-		Scovill Alloy 754	
Scholler AMC	M-1308	Silver 743	M-102	Nickel Silver	
Scholler AS2	M-1308	Scovill 322 High		(15%)	M-102
Scholler AS2W	M-1308	Leaded Brass	M-102		

Alloy	Code
Scovill Alloy 757 Nickel Silver (12%)	M-102
Scovill Alloy 770 Nickel Silver (18%) (B)	M-102
Scovill Alloy 787 Aluminum Brass Type B	M-102
Scovill Aluminum Brass Type A	M-102
Scovill Cartridge Brass, 260	M-102
Scovill Commercial Bronze, Alloy 220	M-102
Scovill Extra High Leaded Brass 356	M-102
Scovill Forging Brass, Alloy 377	M-102
Scovill Hardware Bronze	M-102
Scovill High Brass, 270	M-102
Scovill Leaded Muntz Metal 380	M-102
Scovill Leaded Nickel Silver	M-102
Scovill Leaded Rod 360	M-102
Scovill Low Brass, Alloy 240	M-102
Scovill Low Leaded Brass (64%) Alloy 330	M-102
Scovill Low Leaded Brass (67%) Alloy 335	M-102
Scovill Matrix Brass 340	M-102
Scovill Muntz 280	M-102
Scovill Naval Brass 464	M-102
Scovill Nickel Silver (5%)	M-102
Scovill No. 1032	M-102
Scovill No. 3105	M-102
Scovill Red Brass, 85% Alloy 230	M-102
Scovill Tin Brass (5%) 419	M-102
S.C.S.S.	M-499
S.D.S	M-430
S.D.T.-T.C.O.	M-1119
Seaco	M-374
Seaco "70-20"	M-374
Seaco Blue Diamond	M-374
Seaco Chrome Vanadium	M-374
Seaco Exert	M-374
Seaco Hot Die	M-374
Seaco No-Label	M-374
Seaco Non-Shrinkable	M-374
Seaco O.H. No. 4 Non-Shrink	M-374
Seaco Red Diamond	M-374
Seaco Standard	M-374
Seaco Yellow Diamond	M-374
Sealmet 1	M-1
Sealmet 4	M-1
Sealmet HC-1	M-1
Sealmet HC4	M-1
Sealvac A	M-38, M-1630
Searcher	M-1411
Searcher A.1	M-1411
Segment Steel	M-357
Select	M-73
Select B-M	M-73
Select M	M-73
Self-Hardening	M-38
Self-Hardening	M-210
Self-Hardening	M-345
Self-Hardening	M-389
Semalloy No. 2507	M-1729
Semalloy No. 3265	M-1729
Semi-Alloys No. 157	M-1729
Semi-Alloys No. 158	M-1729
Semi High	M-357
Seminole	M-1
Seminole Steel	M-1
SFS	M-1114
SH 50	M-64
Shaperite	M-669
Sharon 18-8	M-600, M-1071
Sharon 18-8S	M-600
Sharon 21	M-600
Sharon RA	M-600, M-1071
Sharon-Sharp KN-44	M-1071
Sharon Stainless 12	M-600, M-1071
Sharon Stainless 17	M-600
Sharon Stainless 18-8	M-1071
Sharon Stainless 27	M-600
Shawinigan HR-1	M-443
Shawinigan HR-2	M-433
Shawinigan HR-3	M-443
Shawinigan KA2	M-443
Shawinigan Nirosta	M-443
Sheartough	M-1724
Sheffield-AR	M-1428
Sheffield High Strength-A	M-1428
Sheffield Hi-Strength-B	M-1428, M-10
Sheffield SSS-100	M-1428, M-10
Sheffield SSS-100A	M-1428, M-10
Sheffield SSS-100A-AR	M-1428, M-10
Sheffield SSS-100AR	M-1428, M-10
Sheffield SSS-100B	M-1428, M-10
Sheffield SX4	M-1552
Sheffield SX5	M-1552
Sheffield SX25	M-1552
Shellex	M-55
Shenango (46 alloys all cancelled)	M-974
Shield-Arc 100	M-578
Shield-Arc LH-70	M-578
Ship Brand Blue Label	M-446
Shock Steel	M-69
Shoe Die	M-1
Sicrimphy 1	M-771
Sicrimphy 2	M-771
Sicrimphy 3	M-104
Sicromal No. 8	M-383
Sicromal No. 9	M-383
Sicromal No. 10	M-383
Sicromal No. 12	M-383
Sicromo 1	M-376
Sicromo 2	M-376
Sicromo 2 1/2	M-376
Sicromo 3	M-376
Sicromo 5	M-376
Sicromo 5M	M-376
Sicromo 5MS	M-376
Sicromo 7M	M-376
Sicromo 9	M-376
Sifalumin No. 28	M-243
Sifalumin No. 29	M-243
Sifbrass No. 6	M-243
Sifbronze	M-243
Sifbronze No. 5	M-243
S.I.F. No. 23	M-243
S.I.F. No. 24	M-243
Sifonil No. 10	M-243
Sifonite No. 18	M-243
Sifsteel AMS No. 13	M-243
Sifsteel No. 26	M-243
Sifsteel No. 34	M-243
Sifsteel Stainless No. 26	M-243
Sifsteel W.R. No. 12	M-243
Silblock 25	M-394
Silblock 50	M-394
Silbrax	M-677
Silcalfa	M-658
Silchrome	M-1
Silchrome	M-1
Silchrome Wire	M-1
Silchrom I	M-459
Silchrom II	M-459
Silchrom V	M-459
Silchrom IX	M-459

Name	Code	Name	Code	Name	Code
Silcrome 1	M-1	Silver Fox 13	M-1724	Simonds Green Streak	M-373
Silchrome 12	M-1	Silver Fox 14	M-1724	Simonds H-12	M-373
Silcrome 12-2	M-1	Silver Fox 15	M-1724	Simonds H-13	M-373
Silcrome 12-EZ	M-1	Silver Fox 17	M-1724	Simonds Molva	M-373
Silcrome 17	M-1	Silver Fox 18	M-1724	Simonds No. 864	M-373
Silcrome 21	M-1	Silver Fox 19	M-430	Simonds No. 3400	M-373
Silcrome 28	M-1	Silver Fox 20	M-430	Simonds Non-Shrinking	M-373
Silcrome 46 M	M-1	Silver Fox 21	M-430	Simonds O.H.D. No. 106	M-373
Silcrome CC	M-1	Silver Fox 22	M-430	Simonds Red Label Special	M-373
Silcrome H-17	M-1	Silver Fox 23	M-430	Simonds Red Streak Air Hardening	M-373
Silcrome H-17-EZ	M-1	Silver Fox 24	M-1724	Simonds Red Streak Oil Hardening	M-373
Silcrome KA-2	M-76	Silver Fox 25	M-430	Simonds Super Cobalt	M-373
Silcrome KA-2B	M-1	Silver Fox 26	M-430	Simonds "T.A.S."	M-373
Silcrome KA2-EZ	M-1	Silver Fox 35	M-430	Simonds Tunco	M-373
Silcrome KA2M	M-1	Silver Fox 320	M-430	Simplex "CH"	M-38
Silcrome KA2S	M-1	Silver Fox 321	M-1724	Simplex Forging	M-38
Silcrome KA2T	M-1	Silver Fox 610	M-1724	"S" Inconel	M-68
Silcrome KM-1	M-1	Silver Fox 611	M-1724	Sinimax	M-1
Silcrome L-12	M-1	Silver Fox 612	M-1724	Sintaloy N-6418	M-1427
Silcrome M-17	M-1	Silver Fox 613	M-1724	Sintaloy Z-703	M-1427
Silcrome RA-EZ	M-1	Silver Fox 620	M-430	Sintaloy Z-703P	M-1427
Silcrome X	M-1	Silver Fox 920	M-430	Sintaloy Z-851	M-1427
Silcrome X9	M-1	Silver Hardtem	M-64	Sintramant	M-1246
Silcrome X10	M-1	Silver-Magnesium-Nickel	M-63	S.I.P. No. 25	M-243
Silcrome X142	M-1	Silveroid	M-121	Sirius-7	M-1303
Silcrome XCR	M-1	Silver Solder 154	M-926	Sirius HT	M-1303
Silcry	M-771	Silver Solder BH-1	M-926	Sironze	M-61
Silectron	M-1	Silver Solder CH-1	M-926	Sisco	M-375
Silectron	M-602	Silver Solder CLC	M-926	Sisco C.N.V.	M-375
Silfram	M-202	Silver Solder CM-3	M-926	Sisco C.V.R.	M-375
Silicarb	M-24	Silver Solder KH-7	M-926	Sisco C.V.S.	M-375
Silico-manganese	M-394	Silver Solder KH-105	M-926	Sisco Extra No. 2	M-375
Silico manganese Spring	M-12	Silver Solder LH-3	M-926	Sisco Oil Hardening Chisel	M-375
Silicon Aluminum Vanadium	M-394	Silver Solder LM-1	M-926	Sisco Standard No. 3	M-375
Silicon Copper	M-8	Silver Solder SB-2	M-926	Sisco V.A.S.	M-375
Silicon Monel Metal 2.5%	M-68	Silver Solder SC-2	M-926	Sivan	M-38
Silicon Nickel A	M-61	Silver Solder SH-2	M-926	Sivar 48	M-459
Silicon Nickel B	M-61	Silver Solder SH-7	M-926	Six-max	M-435
Silicon Steel-M15	M-604, M-1	Silver Solder SM-1	M-926	SK-1	M-459
		Silver Star	M-57	SK-3	M-459
Silicon Structural Steel	M-24	Silver Tip	M-669	SKF No. 50	M-267
Silicon Vanadium	M-394	Simagal 200	M-324	SKF No. 99	M-267
Silimo	M-24	Simanal	M-398	SKF No. 100	M-267
Silmalar	M-982	Simonds 74	M-373	SKF No. 114	M-267
Silmalec	M-338	Simonds 139	M-373	SKF No. 685	M-267
Silmanal	M-38, M-1076	Simonds 142	M-373	SKF Special	M-267
		Simonds 149	M-373	S-M	M-32
Silmo	M-376	Simonds 168	M-373	SM 100	M-38
Sil-Trode	M-13	Simonds 12225	M-373	SMC-155	M-38
Silumin	M-315	Simonds Blue Label	M-373	S Metal	M-399
Silvaloy 40	M-686	Simonds Blue Streak	M-373	Smith SW-15	M-579
Silvaloy 45M	M-686	Simonds Chromium Magnet Steel	M-373	Smith SW-17	M-579
Silvan Star	M-57	Simonds CR-32	M-373	"S" Monel	M-1499
Silverbond	M-69	Simonds Diamond "S"	M-373		
Silver Die No. 1	M-57	Simonds D.N.V.	M-373		
Silver Die No. 2	M-57	Simonds Frontier	M-373		
Silver Fox 6	M-1724	Simonds Frontier Special	M-373		
Silver Fox 10	M-1724				
Silver Fox 11	M-1724				
Silver Fox 12	M-1724				

Smootharc	M-828	Soding GCr30	M-1309	Soding UNB	M-1309
Smootharc AP	M-828	Soding GCr30Si	M-1309	Soding Universal	M-1309
Smootharc CM-50	M-828	Soding GCr30W	M-1309	Soding W15	M-1309
Smootharc CM-50-1	M-828	Soding GMN	M-1309	Soding WM	M-1309
Smootharc CM-50-2	M-828	Soding GMN Extra	M-1309	Soding ZWAT	M-1309
Smootharc DH	M-828	Soding GNRSN	M-1309	Soldaloy 2	M-265
Smootharc DH-2	M-828	Soding GONC	M-1309	Soldaloy No. 1	M-265
Smootharc DH-3	M-828	Soding GONF	M-1309	Soldaloy No. 8	M-265
Smootharc FR	M-828	Soding H	M-1309	Soleil-1	M-1043
Smootharc FW	M-828	Soding HBC/Ni	M-1309	Soleil-2	M-1043
Smootharc Harcast	M-828	Soding HID	M-1309	Soleil-4	M-1043
Smootharc Harcote	M-828	Soding HID2	M-1309	Soleil-30	M-1043
Smootharc Harmang	M-828	Soding HID3	M-1309	Soleil No. 2	M-1043
Smootharc Harni	M-828	Soding HID-4	M-1309	Solid Drill	M-307
Smootharc Har-Ten	M-828	Soding HPM	M-1309	Spare's Manganese Bronze	M-129
Smootharc HCPF	M-828	Soding KLR	M-1309	Sparkaloy	M-61
Smootharc LG	M-828	Soding LKW	M-1309	Sparkonite 1	M-265
Smootharc Litecote	M-828	Soding LMD	M-1309	Sparkonite 2A	M-265
Smootharc PF	M-828	Soding LME	M-1309	Sparkonite 3	M-265
Smootharc WC	M-828	Soding MSW	M-1309	Sparkonite 7	M-265
Smooth-Cut	M-35	Soding NR17T	M-1309	Sparkonite 10	M-265
Smoothole	M-85	Soding NR18T	M-1309	Spark Plug Metal	M-68
Smoothole Granite Drill	M-85	Soding NRSA	M-1309	Special	M-24
		Soding NRSB 1320	M-1309	Special	M-69
S.N.C.	M-73	Soding NRSB 1340	M-1309	Special 8 T	M-115
S.N.C.G.	M-499	Soding NRSE	M-1309	Special 10 T	M-115
S-Nickel	M-68	Soding NRSF	M-1309	Special Alloy (Disston)	M-357
"S" No. 2	M-81	Soding NRSG	M-1309		
"S" No. 4	M-81	Soding NRSM	M-1309	Special A.S.V.	M-57
Sobu	M-233	Soding NRSN	M-1309	Special Bolt Die	M-85
S.O.B.V. Alloy	M-233	Soding NRST	M-1309	Special Carbon	M-85
S.O.D. Die Steel	M-140	Soding NRSW	M-1309	Special Carbon	M-73
Soding ADP	M-1309	Soding OA 708	M-1309	Special Chrome Vanadium	M-210
Soding ADT	M-1309	Soding OA 1310	M-1309		
Soding CNF	M-1309	Soding OA 1810	M-1309	Special Double Crown	M-530
Soding CNL	M-1309	Soding OA 2415	M-1309		
Soding CNL Extra	M-1309	Soding ON 16 36	M-1309	Special Dynamo	M-806
Soding CNL Spezial	M-1309	Soding ON 20 12	M-1309	Special High Carbon 51-V	M-69
Soding CNM	M-1309	Soding ON 25 4	M-1309		
Soding CNP	M-1309	Soding ON 25 20	M-1309	Special "H.M." Oil Hardening Steel	M-69
Soding CNU Extra	M-1309	Soding ON 189	M-1309		
Soding CNU Spezial	M-1309	Soding ONC	M-1309	Special KCO	M-27, M-1618
		Soding ONIA	M-1309		
Soding Corrochrom 13T	M-1309	Soding PK35	M-1309	Special KRM	M-27, M-1618
		Soding Prima Weich	M-1309		
Soding CRWC	M-1309	Soding Prima ZAH	M-1309	Specialloy 1214	M-1236
Soding CSI	M-1309	Soding Prima ZH	M-1309	Specialloy 2929A	M-1236
Soding CSJ	M-1309	Soding SCHM	M-1309	Specialloy 3028	M-1236
Soding Extra Hart	M-1309	Soding SCHM Extra	M-1309	Specialloy 3028A	M-1236
Soding Extra MH	M-1309			Special M	M-1
Soding Extra Spezial 35	M-1309	Soding Special Diamant	M-1309	Special M-O High Speed	M-97
Soding Extra Weich	M-1309	Soding Special MH	M-1309	Special Motor	M-806
Soding Extra ZAH	M-1309	Soding Spezial Weich	M-1309	Special No. 0	M-389
Soding Extra ZH	M-1309			Special No. 13	M-1
Soding GCNS	M-1309	Soding Spezial ZAH	M-1309	Special Oil Hardening	M-57
Soding GCNV	M-1309	Soding Spezial ZH	M-1309		
Soding GCr6	M-1309	Soding SPG	M-1309	Special Oil Hardening	M-69
Soding GCr17Si	M-1309	Soding SPH2	M-1309		
Soding GCr23	M-1309	Soding SPH3	M-1309	Special Oilway	M-345
Soding GCr27Ni	M-1309	Soding SPHX	M-1309	Special Power Transformer No. 65	M-806
Soding GCr28	M-1309	Soding SPM	M-1309		
Soding GCr28Ni	M-1309	Soding STA	M-1309		

Section IV: Obsolete Alloys / 1747

Special Processed	M-69	Stainex No. 1	M-35	Stauffer Sta 880	M-1528
Special Punch	M-85	Stainex No. 2	M-35	Stauffer Sta 900	M-1528
Special Punch	M-73	Stainex No. 3	M-35	Staybrite 254	M-521
Special Punch &		Stainless Iron	M-248	Staybrite FG	M-521
Chisel Steel	M-81	Stainless Iron 2 Fm	M-38	Staybrite FH	M-521
Special SEHR Hart	M-27	Stainless Iron "N"	M-73	Staybrite FJ	M-521
Special SEHR Hart		Stainless Iron No.		Staybrite F.M.B.3	
A	M-27	16	M-38	T	M-521
Special Sifbronze		Stainless Iron No.		Staybrite FMB	
22% Nickel	M-243	18	M-38	(V4A)	M-1420
Special Tool	M-38	Stainless Iron No.		Staybrite F.M.S.	M-521
Special V	M-140	24	M-38	Staybrite F.M.X.	M-521
Special Vanadium	M-38	Stainless M.S. No.		Staybrite FSL	M-521
Special Vanadium	M-57	2	M-32	Staybrite FSM (2)	M-521
Special Vanadium	M-35	Stainless No. 20	M-46	Staybrite Steel	M-56
Special W2	M-459	Stainless Steel	M-61	Steel-Cote Nitra-	
Special W5	M-459	Stainless Steel		Alloy	M-1243
Special W10D1	M-35	Grade A	M-362	Steel-Cote No. G	M-1243
Special Wire		Stainless Steel		Steel-Cote No. L	M-1243
Drawing Die	M-35	Grade B	M-362	Steel-Cote No. T	M-1243
Special Wortle Die	M-1	Stainless Steel		Steel-Cote No. W	M-1243
Specification		Grade H.C.	M-362	Steel-Cote No. Z	M-1243
55-Alloy Steel	M-38	Stainless Steel		Steelmet 100	M-265
Speed-Alloy 73	M-849	(Thomas Steel		Steelmet 101	M-265
Speed-Cut	M-115	Co)	M-445	Steelmet 302	M-265
Speedicut 14	M-56	Stainless TX	M-57	Steelmet 302	M-265
Speedicut Leda	M-56	Stainless Steel Type		Steelmet 600	M-265
Speedicut		F.T.V.	M-521	Steel-Oilite	M-211
Maximum	M-56	Stainless "X"	M-115	Steel Oilite A	M-211
Speedicut		Staintec 10670	M-717	Steel Oilite AH	M-211
Maximum-18	M-56	Stainweld A	M-578	Steelton L.P. Iron	M-24
Speedicut Sixleda	M-56	Stainweld A-5	M-578	Steierische No. 779	M-1114
Speedicut Superleda	M-56	Stainweld A-5 Cb	M-578	Stelcoloy	M-1124
Speed Treat X-1535	M-894	Stainweld A-7	M-578	Stelcoloy G	M-1124
Spezial No. 5	M-459	Stainweld A-7 Cb	M-578	Stelcoloy S	M-1124
Spezial W	M-459	Stainweld B	M-578	Stellite 33	M-167
Spezial WSF	M-459	Stainweld B-Cb	M-578	Stellite 34	M-167
Spinzwel	M-121	Stainweld C	M-578	Stellite 50	M-167
Spring Brass	M-33	Stainweld D	M-578	Stellite Alloy No. 4	M-167
Spring Oreide	M-102	Staminal	M-85	Stellite Malleable	M-167
S.R. 4 Regular		Standard Alnico	M-271	Stellite No. 2	M-167
(Poldi)	M-96	Standard Bearing	M-24	Stellite No. 100	M-167
SS 1-6	M-459	Standard Carbon	M-97	Stellite No. 2400	M-167
SS 1 Extra Extra	M-459	Standard Diehard	M-56	Stentor	M-501,
SS 2	M-459	Standard (Disston)	M-357		M-912
SS 2 Extra Extra	M-459	Standard (Halcomb)	M-260,	Sterling	M-57
SS 3	M-459		M-38	Sterling 17-12 Mo	M-57
SS 3 Extra Extra	M-459	Standard Tool	M-38	Sterling 18-8FC	M-57
SS 4	M-459	Standard (Wards)	M-389	Sterling 18-9Ti	M-57
SS 4 Extra Extra	M-459	Stanley, Grade A	M-574	Sterling 300	M-106
SS 5	M-459	Stanley, Grade B	M-574	Sterling 305/3051	M-106
SS 5 Extra Extra	M-459	Stanley, Grade C	M-574	Sterling 356	M-106
SS 6	M-459	Stanley, Grade D	M-574	Sterling A. 356	M-106
SS 6 Extra Extra	M-459	Stanley, Grade E	M-574	Sterling D.R.	M-57
"SS 15" Quality	M-179	Star-Boron	M-32	Sterling Nitrard No.	
SSB	M-459	Star Columbium		1	M-57
SSF	M-459	M-8		Sterling Stainless	
SSM	M-459	Star ETD	M-27	302	M-57
S. S. Tool	M-38	Star Moly	M-32	Sterling Stainless	
Stabil	M-360	Star Mo-M2	M-57	303	M-57
Stackpole FW-41	M-1069	Starrett No. 495	M-1225	Sterling Stainless	
Sta-Gloss "A"	M-69	Stauffer 90 Ta-10W	M-1528	304	M-57
Sta-Gloss B	M-69	Stauffer SCb-291	M-1528	Sterling Stainless	
Sta-Gloss C	M-69	Stauffer SM291	M-1528	316	M-57

Sterling Stainless 321	M-57	Strainfree Elastuf Penn	M-307	Styria KS45	M-1291
Sterling Stainless 347	M-57	Straus Metal	M-76	Styria KWS	M-1291
Sterling Stainless 403	M-57	Strenes A	M-898	Styria LBS	M-1291
Sterling Stainless 414	M-57	Strenes AA	M-898	Styria M1H	M-1291
Sterling Stainless 431	M-57	Stressite Copper	M-211	Styria M3MD	M-1291
Sterling Stainless BH	M-57	Stressite Iron	M-211	Styria MN2	M-1291
Sterling Stainless BHHX	M-57	Stress-Proof No. 1	M-669	Styria MNC	M-1291
Sterling Stainless H	M-57	Stress-Proof No. 2	M-669	Styria MSS	M-1291
Sterling Stainless SB	M-57	Stress-Proof No. 3	M-669	Styria NH2H	M-1291
Sterling TX	M-57	Striking Die	M-38	Styria NH2W	M-1291
Sterling "V"	M-57	Stronger-Than-Steel	M-603	Styria NH3	M-1291
Sterling V (tap)	M-57	Structural Silicon (Sil-Ten)	M-604	Styria NHM	M-1291
Steropes	M-114	Structural Silicon Steel	M-377	Styria NW1	M-1291
Stewart Alloy No. 3	M-514	Sturdy Forty	M-669	Styria NW2	M-1291
Stewart Alloy No. 10	M-514	Styria 2	M-1291	Styria O6K	M-1291
Stewart Alloy No. 10R	M-514	Styria 3	M-1291	Styria OM Extra	M-1291
Stewart Alloy No. 12	M-514	Styria 4	M-1291	Styria OMS	M-1291
Stewart Alloy No. 13X	M-514	Styria 5	M-1291	Styria P2	M-1291
Stewart Alloy No. 16	M-514	Styria 6	M-1291	Styria P3	M-1291
Stewart Alloy No. 18	M-514	Styria 1818	M-1291	Styria P4	M-1291
Stewart Alloy No. 35	M-514	Styria 2028	M-1291	Styria P5	M-1291
Stewart Lumite No. 5	M-514	Styria A-1	M-1291	Styria P5M	M-1291
Stewart Lumite No. 10	M-514	Styria A-2	M-1291	Styria P6	M-1291
Stewart Magnesium R	M-514	Styria A-3	M-1291	Styria PK	M-1114
Stewart White Brass No. 3	M-514	Styria BHM	M-1291	Styria PS2	M-1291, M-1114
Stewart White Brass No. 5	M-514	Styria CE	M-1291	Styria PS3	M-1291, M-1114
Stewart White Brass No. 19	M-514	Styria CM Extra	M-1291	Styria PS4	M-1291, M-1114
Stewart Zn No. 2	M-514	Styria CM Extra K	M-1291	Styria RA1 Extra	M-1291, M-1114
Stewart Zn No. 3	M-514	Styria CMHH	M-1291	Styria RA22	M-1291
Stewart Zn No. 5	M-514	Styria CV3	M-1291	Styria RCP	M-1291, M-1114
Stoodex	M-202	Styria CV3M	M-1291	Styria RK15	M-1291, M-1114
Stoodite 45	M-202	Styria CVZ	M-1291	Styria RKW	M-1291
Stoodite 54	M-202	Styria DPKH	M-1291	Styria RS3N	M-1291, M-1114
Stoodite 63	M-202	Styria E3D	M-1291	Styria RS3S	M-1291, M-1114
Stoodite K	M-202	Styria EHH	M-1291	Styria RS3T	M-1291, M-1114
Stoody 120	M-202	Styria EMH	M-1291	Styria RS20N	M-1291
Stoody 122	M-202	Styria EN1W	M-1291	Styria RSV	M-1291, M-1114
Stoody AC	M-202	Styria EN3M	M-1291	Styria SKL	M-1291, M-1114
Stoody Manganese Nickel	M-202	Styria EN3W	M-1291	Styria Spezial CN	M-1291, M-1114
Strain Tempered Steel	M-516	Styria EN5M	M-1291	Styria SPMK	M-1114, M-129
		Styria EN5W	M-1291	Styria SS39K	M-1291, M-1114
		Styria EN36	M-1114	Styria SS50KN	M-1114, M-1291
		Styria ENC10	M-1291	Styria SS1200 Ultra	M-1291
		Styria EWP	M-1291	Styria S-Special	M-1291
		Styria EWP 970	M-1291	Styria SX3 Tannen	M-1291, M-1114
		Styria EWS	M-1291	Styria Tenit-KL	M-1291
		Styria EWS Spezial	M-1291	Styria VCN15	M-1291, M-1114
		Styria Extra Extra	M-1291		
		Styria EZH	M-1291		
		Styria EZH Spezial	M-1291		
		Styria EZW	M-1291		
		Styria G2	M-1291		
		Styria G3	M-1291		
		Styria G4F	M-1114		
		Styria G5F	M-1291		
		Styria G6S	M-1291		
		Styria G7W	M-1291, M-1114		
		Styria GWR	M-1291		
		Styria HWD	M-1291		
		Styria KCR	M-1291		

Styria VCV2	M-1291, M-1114	Super Hy-Tuf	M-38	Supertough D	M-1414
Styria VKSN	M-1291, M-1114	Super Hy-Tuf High Carbon	M-38	Supertough D	M-50
		Superimphy	M-104	Super Tricent	M-24
Styria VMS	M-1291, M-1114	Superior	M-140	Super Tyr	M-50
		Superior	M-248	Super Tyr	M-1414
Styria WCMD	M-1291, M-1114	Superior 5	M-140	Super Van Dreadnought	M-38, M-362
Styria WCMDW	M-1114, M-1291	Superior A Nickel Chrome	M-248	Super VNCA	M-50, M-1414
Styria WKM	M-1114, M-1291	Superior Ark High Speed	M-69, M-261	Superweld No. 55	M-607
Styria WPN Extra	M-1291	Superior Oil	M-694	Super WH Meehanite	M-82
Sub66	M-618	Superior Tool	M-38	Super X Leaded	
Sumet Bronze SM-2	M-71	Superior Tool Steel	M-81	Nickel Silver 8%	M-279
Sumet Bronze SM-4	M-71	Superior X-3012	M-945	Super X Nickel	
Sumet Bronze SM-10	M-1065	Super Kinite	M-345	Silver 12 Alloy No. 176	M-279
		Superkore BB	M-604		
Sumet Bronze SM-12	M-1065	Super Leda	M-56	Super X Nickel Silver 18 Alloy	
		Super LMW	M-1	163	M-279
Sumet Bronze SM-14	M-1065	Super LMW Special	M-1	Super X Nickel	
		Superloy	M-136	Silver 18 Alloy	
Sumet Bronze SM-16	M-1065	Superloy	M-507	164	M-279
		Superloy K2Mo	M-246	Super X Nickel	
Sumet Bronze SM-18	M-1065	Superloy No. 4	M-246	Silver 18 Alloy	
		Superloy No. 7	M-246	168	M-279
Sumiten 70R	M-1296	Superloy No. 10	M-246	Super X Phosphor	
Sumitomo ESD	M-1296	Superloy		Bronze A Alloy	
Super A	M-1243	Manganese Steel	M-246	No. 106	M-279
Super Alloy	M-328	Supermetal	M-635	Super X Phosphor	
Super A Meehanite	M-82	Super Motung		Bronze Gr. C	
Super Antinit B	M-27	Special	M-114	Alloy 113	M-279
Super-Ascoloy	M-1	Super-Mumetal 50	M-108	Super-X-Phosphor	
Super AW 23	M-50, M-1414	Supernickimphy	M-104	Bronze Grade D	M-279
		Super Nilvar	M-44	Super X Phosphor	
Superbolt	M-376	Super Rapid Extra 214A	M-27	Bronze Gr. E Alloy 101	M-279
Super Bronzochrom 10186	M-717	Super Rapid Extra 214N	M-27	Supraloy	M-964
Super C 12	M-1414	Super Rapid Extra		Supreme	M-73
Super CHNC	M-50	500N	M-27	Supremus T-1	M-69
Super CHNC	M-50, M-1414	Super Rapid Extra A	M-27	Surfaceweld	M-578
Super Chrome	M-345	Super Rapid Extra		Suttonite H.S.S.	M-765
Super Chrome		HVA	M-27	SVEA Metal	M-602
Vanadium	M-618	Super Sirius	M-1303	Switch Copper	M-238
Super Cyclone	M-50, M-1414	Super Sirius HT	M-1303	Sylvaloy	M-61
Super DBL	M-1	Super Squaremu 79	M-1089		
Super-De Lavaud	M-203	Super Steel Star	M-32	**T**	
Super-De Lavaud Metal	M-203	Supertemp	M-24	Ta-8W-2Hf	M-53
		Super Tiger	M-24	Ta-12.5W	M-53
Super Dreadnought	M-260, M-38	Super-TM-2	M-376	Tack Die	M-38
		Supertough B.20	M-50, M-1414	Tack Die	M-357
Super Duralumin	M-338			Tacony Alloy Die	M-390
Super Eclair	M-1043	Supertough B.25	M-50, M-1414	Tacony Alloy Steel	M-390
Superelso 52	M-1488			Tacony Alloy	
Superelso 60	M-1488	Supertough B.35	M-50, M-1414	Trimmer	M-390
Superfine IX	M-900			Tacony Bit	M-390
Superfine MM	M-900	Supertough C.20	M-50, M-1414	Tacony Carbon	M-390
Superfine Mold	M-900			Tacony Chisel and	
Superfine MX	M-900	Supertough C.40	M-50, M-1414	Drill	M-390
Superfine XXX	M-900			Tacony Gol Chisel	M-390
Super HI-Mo	M-57	Supertough C.40M	M-50, M-1414	Tacony Jar	M-390
Super Hydra	M-1411			Tacony Regular Die	M-390

Name	Code	Name	Code	Name	Code
Tacony Regular Trimmer	M-390	Tata T.H.S. 6	M-997	Thermalloy K	M-47
Talabot F.C. 2	M-1119	Tata TS	M-997	Thermalloy L	M-459
Talisman	M-178	Tata W6F	M-997	Thermanit A	M-459
T-Alloy "N"	M-140	Ta-W (Tantalum-Tungsten)	M-53	Thermanit C	M-459
T-Alloy Tool	M-140	T.C. Ferry	M-121	Thermanit D	M-459
Tally-Ho 100	M-430	TC Quality	M-1405	Thermanit G	M-459
Tally-Ho 200	M-430	Tec 4	M-63	Thermanit J	M-459
Tally-Ho 300	M-430	Tech-Aloy	M-349	Thermanit K	M-459
Tally-Ho 400	M-430	Teco Grade 23	M-916	Thermanit L	M-459
Tamco	M-109	Teco Grade 68	M-916	Thermanit M	M-459
Tanne 3M	M-27	Teco Grade 607	M-916	Thermanit P.33	M-459
Tannenbaum	M-1308	Teco Grade 610	M-916	Thermanit X	M-459
Tantaloy	M-53, M-329	Teco Grade 625	M-916	Thermax 8	M-459
		Teco Grade 946	M-916	Thermax 8A	M-459
Tantiron	M-23	Teco Grade 946X	M-916	Thermax 8AF	M-459
Tantung	M-53, M-329	Teco Grade 964	M-916	Thermax 8 F	M-459
		Teco Grade K-65	M-916	Thermax 8FAL	M-459
Tantung 148	M-53, M-329	Teco Grade KT-13	M-916	Thermax 9	M-459
		Teco Grade T-13	M-916	Thermax 9 AM	M-459
Tantung 162 A	M-53, M-329	TEC-Z	M-63	Thermax 9F	M-459
		Teelo	M-618	Thermax 9FAL	M-459
Tantung 166A	M-53, M-329	Telcon Bronze	M-108	Thermax 9S	M-459
		Telcon Cu-Be Mold	M-108	Thermax 10	M-459
Tantung 171	M-53, M-329	Telemet	M-1	Thermax 10 A	M-459
		Tellurium Copper Type B	M-33	Thermax 10 AM	M-459
Tantung G2	M-53, M-329	Telphy	M-771	Thermax 10AU	M-459
		Temcross	M-328	Thermax 10F	M-459
Tap Steel Special	M-85	Tempalloy	M-159	Thermax 10FAL	M-459
TA Quality	M-1405	Temp Alloy	M-451	Thermax 11	M-459
Tarpon	M-1	Tempaloy	M-8	Thermax 11A	M-459
Tata BB	M-997	Tempaloy 841	M-8	Thermax 11AS G	M-459
Tata CRDY	M-997	Tempaloy "A"	M-8	Thermax 11AST	M-459
Tata CRW	M-997	Tempaloy B	M-8	Thermax 11 FH	M-459
Tata DYH	M-997	Tempest	M-1411	Thermax 11 FN	M-459
Tata G-30-N	M-997	Templex	M-748	Thermax 12	M-459
Tata G-35-CRN	M-997	Tenaille No. 2 1/2	M-1117	Thermax 12A	M-459
Tata G-45-N	M-997	Tenax NF	M-96	Thermax 12FAL	M-459
Tata HC-12-CR	M-997	Tenit 1720	M-1114	Thermax 12F.-Cr	M-459
Tata HCRH	M-997	Tenit KL	M-1114	Thermax 12FH G	M-459
Tata HCRM	M-997	Tenit W	M-1114	Thermax 12 FN G	M-459
Tata HD	M-997	Tennax	M-389	Thermax 13 Co-G	M-459
Tata HMV	M-997	Tennessee Special	M-377	Thermax F	M-1485
Tata LC-12-CR	M-997	Tensiloy CT	M-1438	Thermax W	M-1485
Tata LSB	M-997	Tensite	M-61	Thermenol	M-38
Tata NCMA	M-997	Texalloy Heat Resistant	M-444	Thermodic HW	M-24
Tata NCMB	M-997	Tex Non-Shock	M-618	Thermodie	M-24
Tata NCRDY	M-997	Thermalloy 18-8	M-47	Thermo Die	M-38
Tata NDY	M-997	Thermalloy 38E	M-47	Thermo Die	M-260
Tata NIMN	M-997	Thermalloy 40A2	M-47	Thermo Die	M-362
Tata P-15-N	M-997	Thermalloy 43	M-47	Thermodur 10CN5050	M-1290
Tata P-35-N	M-997	Thermalloy 47D	M-47	Thermodur 10CN6040	M-1290
Tata SS	M-997	Thermalloy 58B	M-47	Thermodur 15CN1775	M-1290
Tata T.C.S.7	M-997	Thermalloy 72M	M-47	Thermodur 15CN2035	M-1290
Tata T.C.S.8	M-997	Thermalloy 85E	M-47	Thermodur 15CN2535	M-1290
Tata T.C.S.9	M-997	Thermalloy "A"	M-47		
Tata T.C.S.10	M-997	Thermalloy "B"	M-47		
Tata T.C.S.11	M-997	Thermalloy "C"	M-47	Thermodur 30CN15T	M-1290
Tata T.C.S.12	M-997	Thermalloy "D"	M-47	Thermodur 30CN20	M-1290
Tata T.H.S. 1	M-997	Thermalloy "E"	M-47	Thermodur 30CN2035	M-1290
Tata T.H.S. 2	M-997	Thermalloy H	M-47		
Tata T.H.S. 3	M-997	Thermalloy J	M-47		
Tata T.H.S. 4	M-997				
Tata T.H.S. 5	M-997				

Section IV: Obsolete Alloys / 1751

Alloy	Code
Thermodur 30CN2055	M-1290
Thermodur CMVW11	M-1290
Thermodur CMW3K	M-1290
Thermodur CW11	M-1290
Thermodur MVW2	M-1290
Thermodur T-C 16.13 S	M-1290
Thermodur T-C 16.13 SS	M-1290
Thermodur T-C 16.16 ES	M-1290
Thermodur T-C 16.16 ESS	M-1290
Thermodur T-C 18.8	M-1290
Thermodur T-C 18.8 ES	M-1290
Thermodur T-C 18.8 ESS	M-1290
Thermodur T-C 18.8 S	M-1290
Thermodur T-C 18.8 SS	M-1290
Thermodur T-CN 13W	M-1290
Thermodur T-CN 15 TW	M-1290
Thermodur T-CN 20	M-1290
Thermodur T-CN 20 W	M-1290
Thermodur T-CN 35 W	M-1290
Thermodur T-CN 60	M-1290
Thermodur T-CN 80	M-1290
Thermo-Electric Nickel Babbitt	M-143
Thermo-Lectric Copper-Hardened Babbitt	M-143
Thesscal A	M-1522
Thesscal P12	M-1522
Thessco 625	M-1522
Thessco E5	M-1522
Thessco Gd. 25	M-1522
Thessco Gd. 35	M-1522
Thessco HT5	M-1522
Thessco HXO	M-1522
Thessco MX18 Plus	M-1522
Thessconite HW.10	M-1522
Thessconite U.3	M-1522
Thessconite U.4	M-1522
Thessconite U.8	M-1522
Thessco P.G.S.	M-1522
Thessco Phosphalloy No. 1	M-1522
Thessco Phosphalloy No. 2	M-1522
Thessco Phosphalloy No. 5	M-1522
Thessco SX4	M-1552
Thessco SX25	M-1552
Thessco Trimetal	M-1552
Thessco Type 7	M-1552
Thin Fin	M-33
Thomastrip	M-445
Thredwell	M-307
Ti 01	M-1268
Ti-0.15 Pd	M-144
Ti.03	M-1268
Ti-3Al-8Mo-8V-2Fe	M-144
Ti.04	M-1268
Ti-4Al-3Mo-1V	M-144
Ti.05	M-1268
Ti-5Al,2.5Sn,2.5V, 1.3Cb,1.3Ta	M-144
Ti-5Al-4Fe-Cr	M-144
Ti-5Al-5Sn-5Zr	M-144
Ti-5Al-6Sn-2Zr-Mo-0.25Si	M-144
Ti-6Al-2Cb-1Ta-0.8Mo	M-144
Ti-6.5Al-2Cb-1Ta-1.2Mo	M-144
Ti-6.5Al-3Mo-1V	M-144
Ti.07	M-1268
Ti-7Al-2Cb-1Ta	M-144
Ti-7Al-2.5Mo	M-144
TI-7Al-12Zr	M-144
Ti-8Al-2Cb-1Ta	M-144
Ti-8Al-10V	M-97
Ti.11	M-1268
Ti.22	M-1268
Ti.33	M-1268
Ti.40	M-1268
Ti.44	M-1268
Ti-45A	M-144
Ti-50A	M-144
Ti.53	M-1268
Ti.55	M-1268
Ti.56	M-1268
Ti.66	M-1268
Ti.77	M-1268
Ti.88	M-1268
Ti.111	M-1268
Ti-140A	M-144
Ti-150B	M-144
Ti 155A	M-144
Ti-222	M-1268
Ti-223	M-1268
Ti-225	M-1268
Ti 227	M-1268
Ti 441	M-1268
Ti-444	M-1268
Ti-445	M-1268
Ti 551	M-1268
Ti-554	M-1268
Ti 663	M-1268
Ti-666	M-1268
Ti-667	M-1268
Ti 775	M-1268
Ti 886	M-1268
Ti-Brush 40	M-739
Ti-Brush 50	M-739
Ti-Brush 65A	M-739
Ti-Brush 120-AM	M-739
Tiger	M-334
Tiger Brand	M-24
Tiger Special	M-24
Tiger Special	M-334
Tiger Van	M-334
Tigervan	M-24
Tig-Tectic 2-24	M-717
Tig-Tectic 21FC	M-717
Tig-Tectic 183	M-717
Tig-Tectic 222	M-717
Timang 11-7	M-271
Timang 13-3	M-271
Timang (Tisco No. 4)	M-271
Timken 2Cr-1/2Mo	M-376
Timken 4-6% Cr	M-376
Timken 4-6% Cr W	M-376
Timken 5% Cr-Mo Plus Ti	M-376
Timken 16-13-3	M-376
Timken 16-15-6	M-376
Timken 16-25-6	M-376
Timken 17-22	M-376
Timken 17-22-V	M-376
Timken 18-8Cb (347TP)	M-376
Timken 18-8-S	M-376
Timken 18-8Ti (321TP)	M-376
Timken 25-12 (309TP)	M-376
Timken 25-20 (310TP)	M-376
Timken 35-15 (330 TP)	M-376
Timken 301	M-376
Timken 302	M-376
Timken 303	M-376
Timken 304	M-376
Timken 309	M-376
Timken 310	M-376
Timken 316	M-376
Timken 4140 Bolting	M-376
Timken 4520	M-376
Timken 91140	M-376
Timken CNM-40	M-376
Timken DM-15	M-376
Timken Fine Grained Carbon-Molybdenum	M-376
Timken Iron-Chromium (Type 430)	M-376
Timken Krupp	M-376
Timken MM-9	M-376

Timken Nitriding No. 2M	M-376	Timken Type TP 420 Stainless	M-376	Tisco No. 22	M-271
				Tisco No. 23	M-271
Timken Nitriding Steel No. 1	M-376	Timken Type TP 430 Stainless	M-376	Tisco No. 23 X	M-271
				Tisco No. 40	M-271
Timken Nitriding Steel No. 2	M-376	Timken Type TP 443 Stainless	M-376	Tisco No. 41	M-271
				Tisco No. 50	M-271
Timken No. 1 Analysis	M-376	Timken Type TP 446	M-376	Tisco No. 51	M-271
				Tisco No. 53	M-271
Timken No. 2 Analysis	M-376	Tin-Bearing Commercial Bronze	M-33	Tisco No. 65	M-271
				Tisco No. 70	M-271
Timken QV	M-376			Tisco No. 71	M-271
Timken Special Roll Steel	M-376	Tinite	M-1	Tisco No. 72	M-271
		Tirano Extra	M-616	Tisco No. 80	M-271
Timken TP-304L	M-376	Tirano Prima	M-616	Tisco No. 81	M-271
Timken TP-316	M-376	Tisco 5	M-271	Tisco No. 90	M-271
Timken TP-321H	M-376	Tisco 17	M-271	Tiscor	M-997
Timken TW	M-376	Tisco 23 W	M-271	Tiscrom	M-997
Timken Type 302B Stainless	M-376	Tisco 52	M-271	Tiska Nirosta KA-2	M-271
		Tisco 54	M-271	Tiska Nirosta KA-2-H	M-271
Timken Type 308 Stainless	M-376	Tisco 60	M-271		
		Tisco 90 High Strength	M-271	Tiska Nirosta KA-2-H-Mo	M-271
Timken Type 321 Stainless	M-376				
		Tisco 90 Regular	M-271	Tiska Nirosta KA-2-Mo	M-271
Timken Type 347 Stainless	M-376	Tisco 101	M-271		
		Tisco 101A	M-271	Tiska Nirosta KA-2-S	M-271
Timken Type 403 Stainless	M-376	Tisco 102	M-271		
		Tisco 102A	M-271	Tiska Nirosta KA-2-S-Mo	M-271
Timken Type 405 Stainless	M-376	Tisco 103	M-271		
		Tisco 103A	M-271	Tiska Nirosta KNC-3	M-271
Timken Type 406 Stainless	M-376	Tisco 104	M-271		
		Tisco 104A	M-271	Titan Bronze	M-122
Timken Type 410 Stainless	M-376	Tisco 105	M-271	Titania	M-1579
		Tisco 105	M-271	"Titanic" Carbon Steel	M-233
Timken Type 414 Stainless	M-376	Tisco 105A	M-271		
		Tisco 106	M-271	Titanit	M-459
Timken Type 416 Stainless	M-376	Tisco 106A	M-271	Titanite	M-1279
		Tisco 107	M-271	Titanit Ti 1	M-459
Timken Type 420 Stainless	M-376	Tisco 108	M-271	Titanit S Ti 1	M-459
		Tisco 109	M-271	Titanit S Ti 2	M-459
Timken Type 430 Stainless	M-376	Tisco 110	M-271	Titanit S Ti 3	M-459
		Tisco 120	M-271	Titanit S Ti 4	M-459
Timken Type 430F Stainless	M-376	Tisco 130	M-271	Titanium Ti-175A	M-144
		Tisco 131	M-271	Titanium Type TI-75A	M-144
Timken Type 431 Stainless	M-376	Tisco 132	M-271		
		Tisco 133	M-271	Titanium Type TI-100A	M-144
Timken Type 440 A Stainless	M-376	Tisco 150	M-271		
		Tisco 150X	M-271	Titanium Type TI-125A	M-144
Timken Type 440 B Stainless	M-376	Tisco 150-Y	M-271		
		Tisco 160	M-271	Titanium Type TI-150A	M-144
Timken Type 440 C Stainless	M-376	Tisco High Chrome Iron	M-271		
				"Titan" Manganese Steel	M-233
Timken Type 442 Stainless	M-376	Tisco Low Chrome Iron	M-271		
				Titan P-1	M-122
Timken Type 443 Stainless	M-376	Tisco Medium Chrome Iron	M-271	Ti-Tax	M-371
				Titem	M-294
Timken Type 446 Stainless	M-376	Tisco Nirosta 28-11	M-271	Tivan	M-294
		Tisco No. 1	M-271	TM	M-38
Timken Type 501 Stainless	M-376	Tisco No. 2	M-271	TMC	M-69
		Tisco No. 6	M-271	TMO Dreadnought	M-260, M-38
Timken Type 502	M-376	Tisco No. 17	M-271		
Timken Type TP 405 Stainless	M-376	Tisco No. 18	M-271		
		Tisco No. 19	M-271	TM Quality	M-1405
Timken Type TP 410 Stainless	M-376	Tisco No. 20	M-271	Tombasil A	M-123
		Tisco No. 21	M-271	Toncan Iron	M-97

Name	Code	Name	Code	Name	Code
Tool-Acc Work Hardening	M-692	Triunefador	M-616	Two-Tone H.C.	M-507
		Triunefador Z	M-616	Two-Tone T.H.	M-507
Tool Holder	M-85	Trodaloy No. 1	M-60	Tycon	M-32
Tool-N-Die No. 10-WH	M-1000	Trodaloy No. 7	M-60	Tylerite	M-929
		Trojan	M-956	Type SM-1	M-725
Tool-N-Die No. 15-AH	M-1000	Trojan Shank	M-435	Type SM-2	M-725
		Trojan Steel	M-32	Type SM-3	M-725
Tool-N-Die No. 16-HS	M-1000	Trojan Steel No. 8	M-307		
		Trojan Steel No. 9 1/2	M-307	**U**	
Tool-N-Die No. 24-HA	M-1000	"Tropic" Hot Die Steel	M-233	Ucar-75	M-280
Tool-N-Die No. 34-DA	M-1000	Troxeit Grade 85	M-53	Udimet 600	M-1317
		Troxeit Grade 95	M-53	Ugine 2	M-1120
Tool-N-Die No. 70-OH	M-1000	Troxeit Grade 100	M-53	Ugine 2PICS	M-1120
		T.T.Q. High Speed	M-69	Ugine 3-E3	M-1120
Tool-N-Die No. 71-OH	M-1000	Tube Brass	M-8	Ugine 4-E4	M-1120
		Tufanhard 400	M-1000	Ugine 5-E5	M-1120
Tool-N-Die No. 72-HW	M-1000	Tuffaloy 20W	M-663	Ugine AGT	M-1120
		Tuffaloy 53	M-663	Ugine BUL	M-1120
Tool-N-Die No. 73-HW	M-1000	Tuffaloy 55	M-663	Ugine C2	M-1120
		Tuffaloy 55S	M-663	Ugine C7	M-1120
Tooltec 10675	M-717	Tuftest	M-1441	Ugine C500	M-1120
Tooltectic 623	M-717	Tuftork	M-669	Ugine C600	M-1120
Toolweld	M-578	Tung-Alloy	M-507	Ugine C700	M-1120
Toolweld 55	M-578	Tungo	M-34	Ugine C800	M-1120
Toolweld 60	M-578	Tungo	M-115	Ugine CH 110	M-1120
Tophel II	M-61	Tungsten 6%	M-1076	Ugine CHOC	M-1120
Tophet B	M-61	Tungsten Hack Saw	M-38	Ugine CU	M-1120
Tophet M-5	M-61	Tungsten Hack Saw	M-69	Ugine D36	M-1120
Tophet X	M-61	Tungsten Hack Saw	M-357	Ugine D100	M-1120
Tormanc	M-1724	Tungsten Magnet	M-1	Ugine DS-TF	M-1120
Tormanc Major	M-1724	Tungsten Magnet	M-60	Ugine E1	M-1120
Tormanc Major	M-430	Tungsten Magnet	M-73	Ugine F15	M-1120
Tormanc Special	M-430	Tungsten Magnet M-5	M-38	Ugine FC	M-1120
Tormanc Special	M-1724			Ugine FJ	M-1120
Tormol	M-1724	Tungsten Magnet M-6	M-38	Ugine FIA	M-1120
Tormol	M-430			Ugine FIB	M-1120
TP	M-459	Tungsten Metal	M-394	Ugine FN	M-1120
T.P.A.	M-513	Tungsten Tap	M-38	Ugine FN2	M-1120
Tranelec A	M-691	Tungsten Tap Steel	M-57	Ugine G1	M-1120
Tranelec B	M-691	Turbadium Bronze	M-179	Ugine G1T	M-1120
Tranelec C	M-691	Turbaloy	M-31	Ugine G5	M-1120
Transweld	M-578	Turbaloy 13	M-31	Ugine G5T	M-1120
Tri-Ack	M-115	Turbine Alloy	M-115	Ugine G11S	M-1120
Tri-Alloy	M-77	Turbotherm 12	M-27	Ugine IK	M-1120
Tricrank	M-294	Turbotherm 13CO-10	M-27	Ugine K2C	M-1120
Trident	M-73			Ugine K2D	M-1120
Trimet	M-725	Turbotherm 15 M	M-27	Ugine K2F	M-1120
Triple Alloy Punch	M-38	Turbotherm 20CO-20S	M-27	Ugine K12	M-1120
Triple Die	M-57			Ugine KMD	M-1120
Triple Eclair	M-1043	Turbotherm 20 M	M-27	Ugine KNAMO	M-1120
Triple Extra	M-85	Turbotherm 35CO-20	M-27	Ugine KNB	M-1120
Triple Extra Die	M-85			Ugine KNE	M-1120
Triple Mushet	M-233	Turbotherm 1810Nb	M-27	Ugine KNH	M-1120
Triple Satan	M-1043			Ugine KNMO	M-1120
Triplex	M-748	Tusco	M-327	Ugine KOR	M-1120
Tri-Ten E	M-604	Twin Six	M-1	Ugine KORS	M-1120
Tritex	M-669	Twin-Mo	M-345	Ugine KS	M-1120
Tritex Broachite	M-669	Twin Mo-Co	M-345	Ugine KS2	M-1120
Tritex No. 1	M-669	Twin Mo-VA 3	M-345	Ugine KT	M-1120
Tritex No. 2	M-669	Twoscore	M-472	Ugine KTD	M-1120
Tritex Plus	M-669	Two Star	M-85	Ugine KTH	M-1120
Triton	M-140	Two Star Special	M-85	Ugine KVR	M-1120
Triton A	M-140			Ugine KW	M-1120

Ugine M51S	M-1120	Ugine TP	M-1120	Universal	M-1309
Ugine MA	M-1120	Ugine UM6	M-1120	Universal UC101	M-1341
Ugine MARS	M-1120	Ugine VR	M-1120	Uranium B	M-73
Ugine MBA	M-1120	Ugine W	M-1120	Uranium R	M-764
Ugine MFC	M-1120	Ugine WF	M-1120	Uranus 3	M-1303
Ugine Mn12	M-1120	Ugine XLX	M-1120	Uranus 10	M-1303
Ugine MP14	M-1120	Uginium	M-1120	Uranus 10M	M-1303
Ugine MS	M-1120	Ulcometal	M-88	Uranus 10SI	M-1303
Ugine MSS	M-1120	Ulmal	M-982	Uranus 10ST	M-1303
Ugine MTC	M-1120	Ulminium S	M-764	Uranus 15	M-1303
Ugine N2C	M-1120	U-Loy	M-97	Uranus 50	M-1303
Ugine N3C	M-1120	Ultimium N 113	M-717	USS 1/2% Cr-1/2%	
Ugine N5C	M-1120	Ultimium N 114		Mo	M-604
Ugine NC	M-1120	Express	M-717	U.S.S. 4/6 Cr-Mo	M-604
Ugine NC	M-1120	Ultra-cut	M-517	U.S.S. 4/6 Cr-Mo	
Ugine NK8	M-1120	Ultralumin	M-272	Steel	M-90
Ugine NS10	M-1120	U.M.A. (Agathon		U.S.S. 4/6 Cr-Mo	
Ugine NS20	M-1120	No. 1)	M-97	Steel	M-90
Ugine NS20C	M-1120	U.M.A. (Agathon)		U.S.S. 4/6 Cr-Mo-Ti	M-604
Ugine NS20CT	M-1120	No. 2	M-97	U.S.S. 4/6 Cr-Mo-Ti	
Ugine NS20S	M-1120	U.M.A. (Agathon)		Steel	M-90
Ugine NS21S	M-1120	No. 3	M-97	U.S.S. 4/6 Cr-Ti	M-604
Ugine NS22	M-1120	U.M.A. (Agathon)		USS 5	M-604
Ugine NS80C	M-1120	No. 4	M-97	USS 5S	M-604
Ugine NS95	M-1120	U.M.A. (Agathon)		U.S.S. 5W	M-604
Ugine NS190	M-1120	No. 5	M-97	U.S.S. 12 Al	M-604
Ugine NSF	M-1120	U.M.A. (Agathon)		U.S.S. 12 High	
Ugine NSM	M-1120	Spring	M-97	Carbon	M-604
Ugine NSM20	M-1120	Uni-Die Smoothcut	M-35	USS 12 MoV	M-604
Ugine NSM22	M-1120	Uniform Oil		USS 12 Ni-5 Cr-3	
Ugine NSMB	M-1120	Hardening	M-69	Mo	M-604
Ugine NSMC		Uniloy 1	M-114	U.S.S. 12 W	M-604
(Cu,Ti)	M-1120	Uniloy 2	M-114	U.S.S. 12 Z	M-604
Ugine ORH	M-1120	Uniloy 4-6 Cr	M-114	USS 15-7AMV	M-604
Ugine P	M-1120	Uniloy 18-8B	M-114	U.S.S. 16-6	M-604
Ugine PCV	M-1120	Uniloy 19-9Cb	M-114	U.S.S. 17-5 MnV	M-604
Ugine PM	M-1120	Uniloy 19-9Cb-M	M-114	U.S.S. 17 High-	
Ugine PM1	M-1120	Uniloy 19-9 DX	M-114	Carbon	M-604
Ugine PM2	M-1120	Uniloy 430MR	M-114	U.S.S. 17 W	M-604
Ugine PMV	M-1120	Uniloy 435	M-114	U.S.S. 18-8 Cold	
Ugine PY	M-1120	Uniloy 1409Cb	M-114	Rolled	M-604
Ugine QA2	M-1120	Uniloy 1416MV	M-114	U.S.S. 18-8	
Ugine QA6	M-1120	Uniloy 1422MV	M-114	Columbium	
Ugine QA8	M-1120	Uniloy 1430W	M-114	Bearing Type 345	M-604
Ugine QA18	M-1120	Uniloy 2009 U	M-114	U.S.S. 18-8	
Ugine QMS	M-1120	Uniloy 2825	M-114	Columbium	
Ugine R	M-1120	Uniloy 18100-F	M-114	Bearing Type 346	M-604
Ugine R12	M-1120	Uniloy Special No.		U.S.S. 18-8	
Ugine R18	M-1120	12-12	M-114	Molybdenum	
Ugine RC	M-1120	Unimach No. 2	M-114	Bearing	M-11
Ugine RD	M-1120	Unimach UCX2	M-114	U.S.S. 18-8 S-Mo	M-12
Ugine RDS	M-1120	Union Hymo	M-397	U.S.S. 18-8 S MO	M-604
Ugine ROC	M-1120	Union Hymo Steel	M-397	U.S.S. 18-8	
Ugine ROV	M-1120	Union-Larssen Steel		Stabilized	M-90
Ugine RS	M-1120	"Resista"	M-383,	U.S.S. 18-8	
Ugine RSA	M-1120		M-222	Stabilized Type	
Ugine RSE	M-1120	Union Maxcut	M-397	320	M-11
Ugine RSK	M-1120	Union McQuaid-		U.S.S. 18-8	
Ugine RSO	M-1120	Ehn	M-397	Stabilized Type	
Ugine RSS	M-1120	Union Metal	M-931	321	M-11
Ugine S	M-1120	Union Multicut	M-397	U.S.S. 18-8 Type	
Ugine S12	M-1120	Union Special		302	M-604
Ugine Special 400	M-1120	Carburizing Steel	M-397	U.S.S. 18-8 Type	
Ugine Special 460	M-1120	United Magnet	M-327	304	M-604

Section IV: Obsolete Alloys / 1755

Alloy	Ref
U.S.S. 18-8 Type 306	M-604
U.S.S. 18-8 Type 348	M-604
U.S.S. 18-10 Cb	M-12
U.S.S. 18-12	M-604
U.S.S. 19-9	M-604
U.S.S. 19-9S	M-604
U.S.S. 19-9 Ti	M-604
U.S.S. 20-10	M-12
U.S.S. 20-10S	M-604
U.S.S. 25-12 Ti	M-604
U.S.S. 25-20	M-604
USS 304 LN	M-604
U.S.S. 9260	M-12
USS Airsteel X-200	M-604
U.S.S. Air-Ten	M-604
USS Amer-Led Bessemer	M-12
USS Amer-Led Gr. A	M-12
USS Amer-Led Gr. B	M-12
U.S.S. Armature	M-604
U.S.S. C-1210 Cu	M-604
U.S.S. Copper Steel	M-377
USS Cu-Ni-Mo	M-604
U.S.S. Dynamo	M-604
U.S.S. Electrical	M-604
U.S.S. Field	M-604
USS Free-Machining Steel, Leaded	M-604
U.S.S. Lynore	M-604
U.S.S. Manganese-Nickel-Copper	M-604
USS Maraging Steel Gr. 200	M-604
USS Maraging Steel Gr. 250	M-604
USS Maraging Steel Gr. 300	M-604
USS "MX"	M-604
USS MX-Bess	M-12
USS MXI	M-604
USS MX-OH	M-12
U.S.S. Open Hearth Iron	M-604
U.S.S. Pole	M-604
U.S.S. Radio Transformer 52	M-604
U.S.S. Radio Transformer 72	M-604
U.S.S. Strux	M-604
U.S.S. Super-Kore A	M-604
U.S.S. Super-Kore AA	M-604
U.S.S. Super-Kore B	M-604
U.S.S. Super-Kore C	M-604
USS Superkore CC	M-604
Utaloy	M-1003
Utilitas	M-166, M-69
Utility Bearing Metal	M-575

V

Alloy	Ref
V-5 Alloy	M-394
V-7 Alloy	M-394
V 45 Steel	M-24
V 50 Steel	M-24
V 55 Steel	M-24
V 60 Steel	M-24
V 65 Steel	M-24
V444D	M-459
Vac-Arc AGT	M-73
Vac-Arc BG41	M-73
Vac-Arc Regent	M-73
Vac-Melt A	M-602
Vac-Melt AA	M-602
Vac-Melt B7M	M-602
Vac-Melt C	M-602
Va-Cro	M-85
Valiant	M-1411
Valimphy	M-104
Valimphy 550	M-771
Valimphy 600	M-771
Valmax 3/8	M-430
Valray 1	M-121
Valutap Die Steel	M-115
Valve-Loy	M-507
Vampire	M-1411
Vanadium 4	M-1326
Vanadium Castdie	M-35
Vanadium Firedie	M-35
Vanadium Grainal No. 6	M-394
Vanadium Metal	M-394
Vanadium Permandur	M-1
Vanadium Potts Best	M-307
Vanadium Premier	M-50
Vancoram Alsifer	M-394
Vancoram Aluminum Iron Vanadium	M-394
Vancoram Aluminum Vanadium	M-394
Vancoram Copper Aluminum Vanadium	M-394
Vancoram Copper Manganese Vanadium	M-394
Vancoram Copper Nickel Vanadium	M-394
Vancoram Ferro Chromium High Carbon	M-394
Vancoram Ferrovanadium	M-394
Vancoram Ferro Vanadium Crucible Grade B	M-394
Vancoram Ferro Vanadium Special High Vanadium	M-394
Vancoram High Carbon Chromium Metal	M-394
Vancoram Manganese Vanadium	M-394
Vancoram Nickel Vanadium	M-394
Vancoram Titanium 70% Cr	M-394
Vancorum Ferro Vanadium Open Hearth Grade A	M-394
Vancorum Ferro Vanadium Primos Grade C	M-394
Van Die Car	M-289
Vanguard	M-1411
Vanimoloy	M-1387
Van Lom FM	M-115
Vanquish	M-1411
Vantro-S	M-115
Vap	M-50
Vap.2V	M-50
Var Steel Gr. 250	M-1602
Var Steel Gr. 280	M-1602
Vasco	M-115
Vasco 6-6-2 Drill Rod	M-342
Vasco 8N2 FM	M-115
Vasco 14-4 CVM	M-115
Vasco 4650-A	M-115
Vasco Chromold CVM	M-115
Vasco CM	M-115
Vasco Greek Ascoloy	M-115
Vascojet 1000-5 Temper	M-115
Vascojet M-A	M-115
Vascoloy	M-115
Vascoloy Ramet 2A8	M-329
Vascoloy Ramet EH	M-329
Vascoloy Ramet Grade 2A	M-329
Vascoloy Ramet Grade 2A9	M-329
Vascoloy Ramet Grade 3150	M-329
Vascoloy Ramet Grade A	M-329, M-115
Vascoloy Ramet Grade AA	M-329, M-115

Vascoloy Ramet Grade AAA	M-329, M-115	Vasco Special 14 T	M-115	Vibrac Steel	M-117
		Vasco Stainless 302	M-115	Vibrac V.30	M-50, M-1414
		Vasco Stainless 410	M-115		
Vascoloy Ramet Grade AM	M-329	Vasco Stainless 416	M-115	Vibrac V.45	M-50, M-1414
		Vasco Stainless 420	M-115		
Vascoloy Ramet Grade AR	M-329	Vasco Stainless 430	M-115	Vickers B.C.T.	M-117, M-50
		Vasco Stainless 440	M-115		
Vascoloy Ramet Grade AT	M-329, M-115	Vasco Stainless 440C	M-115	Victoralloy	M-1204
				Victor Hi-Chrome A	M-694
		Vasco Stainless 441	M-115	Victor Hi-Chrome B	M-694
Vascoloy Ramet Grade AW	M-329, M-115	Vasco Supreme	M-115	Victorieux Saint Juery	M-1119
		Vasco Supreme A	M-115		
		Vasco Vanadium	M-115	Victorite-1	M-1204
Vascoloy Ramet Grade B	M-329, M-115	Vasco Vanadium Type "BB"	M-115	Victorite-6	M-1204
				Victorite 12	M-1204
		Vasco Vanadium Type "D"	M-115	Victortube	M-1204
Vascoloy Ramet Grade C	M-329, M-115			Victor Tungsmooth	M-1204
		Vasco Vanadium Type "G"	M-115	Victrix Extra	M-96
				Victrix Special 32	M-96
Vascoloy Ramet Grade CC	M-329, M-115	Vasco Vanadium Type "K"	M-115	Viking Extra	M-140
				Violet Label	M-694
		Vasco Vanadium Type "N"	M-115	Viper	M-1411
Vascoloy Ramet Grade D	M-329, M-115			Virgo 1	M-533
		Vasco WCC27202	M-115	Virgo 1.0	M-1488
		Vatool	M-357	Virgo 1B	M-533
Vascoloy Ramet Grade DD	M-329, M-115	Vedos	M-1411	Virgo 1S	M-1301
		Vega 1	M-1303	Virgo 1SA	M-1488
		Vega 2	M-1303	Virgo 2	M-533
Vascoloy Ramet Grade DT	M-329	Vega 3	M-1303	Virgo 2.0	M-1488
		Vega 10	M-1303	Virgo 2/3	M-533
Vascoloy Ramet Grade DW	M-329	Vega 16	M-1303	Virgo 2S	M-1301
		Vega 20	M-1303	Virgo 3	M-1301
Vascoloy Ramet Grade E	M-329	Vega 100	M-1303	Virgo 3S	M-1301
		Velodur	M-764	Virgo 4	M-1301, M-1488
Vascoloy Ramet Grade EE	M-329	Velvet Antifriction Metal	M-914	Virgo 5	M-1301, M-1488
Vascoloy Ramet Grade EM	M-329	Ventos 4631	M-1326		
		Ventos 4718	M-1326	Virgo 6	M-1488, M-533
Vascoloy Ramet Grade M	M-329	Ventos 4721	M-1326		
		Ventos 4732	M-1326	Virgo 7	M-1301, M-1488
		Ventos 4747	M-1326		
Vascoloy Ramet Grade S	M-329, M-115	Ventos 4748	M-1326	Virgo-11DS	M-533
		Ventos 4780	M-1326	Virgo 11S	M-533, M-1488
Vascoloy Ramet Grade SS	M-329	Ventos 4790	M-1326		
		Ventos 4871	M-1326	Virgo 11SA	M-533, M-1488
Vascoloy Ramet Grade X	M-329, M-115	Ventos 4873	M-1326		
		Ventos 4875	M-1326	Virgo 11SB	M-533, M-1488
		Ventos 4971	M-1326		
Vascoloy Ramet Grade XX	M-329, M-115	Veresta	M-459	Virgo 11SS	M-1488
		Veresta Special	M-459	Virgo 11SSB	M-533, M-1488
		Veresta-V	M-459		
Vascoloy X-5	M-115	Veriloy	M-44	Virgo 12	M-533
Vasco Marvel 721	M-115	Vernial	M-303	Virgo 13	M-533
Vasco Matrix II-CVM	M-115	Vernidur	M-303	Virgo 14	M-533
		Vernikorr	M-303	Virgo 14S	M-533
Vasco MC	M-115	Vernimag M	M-303	Virgo 14SB	M-533, M-1488
Vasco Momarc CVM	M-115	Versatile	M-1411		
		Very Best	M-334	Virgo 14SR	M-1488
Vasco Special	M-115	Very Hard	M-338	Virgo 14SR1	M-533, M-1488
Vasco Special 7 T	M-115	VH-63	M-267		
Vasco Special 9 T	M-115	V.H. Atlas	M-267	Virgo 14SR2	M-533, M-1488
Vasco Special 11 T	M-115	"V" High Speed Steel	M-81		
Vasco Special 12 T	M-115			Virgo 14SS	M-533, M-1488
Vasco Special 13 T	M-115	V.H. No. 54	M-267		

Section IV: Obsolete Alloys / 1757

Alloy	Code	Alloy	Code	Alloy	Code
Virgo 14SSB	M-533, M-1488	Virgo 87B	M-1301, M-1488	Waraloy 20X	M-959
Virgo 14SSR	M-1301, M-1488	Virgo 92	M-1301, M-1488	Waraloy 30X	M-959
Virgo 14SSR1	M-533, M-1488	Virgo 106	M-1301, M-1488	Warranted 50-50	M-959
				Washconite	M-246
Virgo 14SSR2	M-533, M-1488	Virgo 112	M-1301, M-1488	Washcote	M-828
				Waterbury A5	M-1426
Virgo 15	M-533	Vital	M-1411	Waterbury A10	M-1426
Virgo 15S	M-533, M-1488	Vital-X	M-1411	Waterbury A15	M-1426
		Vizor	M-1411	Waterbury A18	M-1426
Virgo 15SA	M-1488	VM Dreadnaught	M-362	Waterbury B-12-S	M-1426
Virgo 17	M-533	"V" Moly High		Waterbury B-18-S	M-1426
Virgo 17S	M-533, M-1488	Speed	M-81	Waterbury K-8	M-1426
		VS 4	M-210	Waterbury K-12	M-1426
Virgo 17SA	M-1488	VS Spezial Extra	M-459	Waterbury PBA	M-1426
Virgo 17SR	M-1301, M-1488	V-Star	M-85	Waterbury PBC	M-1426
		V-Star Special	M-85	Waterbury PBD	M-1426
Virgo 18S	M-1301, M-1488	V T-Steel	M-383, M-222	Waukesha No. 8	M-941
				Waukesha No. 13	M-941
Virgo 18SB	M-1301, M-1488	V T-Steel (V T-Stahl)	M-383, M-222	Waukesha No. 18	M-941
				Waukesha No. 20	M-941
Virgo 19S	M-1301, M-1488	Vulcan A-42	M-275	Waukesha No. 21	M-941
		Vulcan CWD	M-275	Waukesha No. 22	M-941
Virgo 19SB	M-533	Vulcan Extra Drill	M-275	Waukesha No. 35	M-941
Virgo 19SR	M-1301, M-1488	Vulcan Heavy Duty	M-275	Waukesha No. 50	M-941
		Vulcan Self-		WCV	M-459
Virgo 19SR1	M-1301, M-1488	Hardening	M-275	WCVH	M-459
		Vulcan Superior	M-275	WE	M-459
Virgo 19SR2	M-1301, M-1488			WE5	M-459
		W		Wear-Arc 12	M-692
Virgo 20	M-533			Wear-Arc 28	M-692
Virgo 23	M-1488			Wear-Arc 300	M-692
Virgo 31	M-1488	W4 Alloy	M-121	Wear-Arc 400	M-692
Virgo 31S	M-1301, M-1488	W4 Alloy	M-459	Wear-Arc 500	M-692
		W5 Alloy	M-121	Wear-Arc 600	M-692
Virgo 32	M-1488	W5 Alloy	M-121	Wear-Arc Chrome-Boride	M-1713
Virgo 41	M-1488, M-533	W7 Alloy	M-121	Wear-Arc "FP"	M-692
		W8 Alloy	M-121	Wear-Arc Super WH	M-1713
Virgo 41B	M-1488, M-533	W20S	M-459	Wear Devil C-1	M-712
		W22V	M-459	Wear Devil C-2	M-712
Virgo 41S	M-533, M-1301, M-1488	W65	M-459	Wear Devil C-3	M-712
		W85	M-459	Wearpact	M-991
		W-90 (Edelstahl)	M-459	Weartuf Steel	M-307
Virgo 42	M-1488, M-533	W95-Cr	M-459	Webb Blue Label	M-446
		W100	M-459	Webbite (Tam)	M-109
Virgo 43	M-1488, M-533	WA-6	M-1707	WE Extra	M-459
		WA-9	M-1707	Weiralead	M-628
Virgo 51	M-1301, M-1488	WA-11	M-1707	Weircoloy	M-628
		WA-13	M-1707	Weldanka	M-472
Virgo 52	M-1488	WA-40	M-1707	Weld-Arc	M-712
Virgo 60	M-533	WA-63	M-1707	Weld-Fast	M-238
Virgo 60A	M-1488	WA-301	M-764	Weldrawn 18-8	M-945
Virgo 61	M-533, M-1488	WA-870	M-1707	Wellcast A	M-681
		Wabik Metal	M-691	Well-Cast E	M-681
Virgo 61SS	M-533, M-1488	Walcoloy No. 1	M-963	Well-Cast J	M-681
		Walcoloy No. 2	M-963	Well-Cast K	M-681
Virgo 63	M-1488	Walcoloy No. 5	M-963	Wellcast No. 3S	M-681
Virgo 70	M-533	Wallex	M-963	Wellcast No. 4A	M-681
Virgo 74	M-533, M-1488	Wallex No. 1	M-963	Wellcast No. 5S	M-681
		Wallex No. 2	M-963	Wellcast No. 13A	M-681
Virgo 84	M-1488	W Alloy No. 6	M-963	Wellcast No. 16S	M-681
Virgo 86	M-1301, M-1488	W Alloy No. 6	M-963	Wellcast No. 17S	M-681
		Wapresta	M-1308	Wellcast No. 65	M-681

Wellcast No. 73	M-681	Western Cupro-Nickel-7%-#142	M-279	Western Nickel Silver 66-10, #180	M-279
Wellcast No. 78	M-681	Western Cupro-Nickel-9%	M-279	Western Nickel Silver 70-12, #176	M-279
Wellcast No. 80	M-681	Western Cupronickel 1290	M-279	
Wellcast No. 85	M-681			
Wellcast No. 88	M-681			
Wellcast No. 99C	M-681	Western Deoxidized Copper #52	M-279	Western Nickel Silver 72-18, #163	M-279
Wellcast No. 116	M-681	Western Gilding Alloy No. 5	M-279	Western No. 30	M-279
Well-Cast P	M-681			
Well-Cast R	M-681			
Well-Cast T	M-681	Western High Brass	M-279	Western OFHC #53	M-279
Well-Cast V	M-681	Western High Conductivity Bronze #86	M-279	Western Phosphor Bronze 1.25%, #101	M-279
Well-Cast W	M-681			
Well-Cast X	M-681			
Well-Cast Y	M-681			
Well-Cast Z	M-681	Western High Leaded Brass #64	M-279	Western Phosphor Bronze 5%, #106	M-279
Western Alloy 34	M-279			
Western Alloy 98	M-279			
Western Alloy No. 10	M-279	Western High Leaded Brass 64	M-279	Western Phosphor Bronze 8%, #113	M-279
Western Alloy No. 13	M-279	Western High Leaded Brass #68	M-279	Western Red Brass #15	M-279
Western Alloy No. 15	M-279	Western Jewelry Bronze #13	M-279	Western Silver Bearing Copper #51	M-279
Western Alloy No. 51	M-279	Western Leaded Bearing Bronze #116	M-279	Westrohr 17L	M-1334
Western Alloy No. 52	M-279			Westrohr N5B	M-1334
				Westrohr N8	M-1334
				Westrohr N9	M-1334
Western Alloy No. 53	M-279	Western Leaded Nickel Silver 8%, #130	M-279	Westrohr N10	M-1334
Western Alloy No. 55	M-279	Western Leaded Nickel Silver 10%, #131	M-279	Wexite Grade Cr-6	M-38
				Whelco A	M-119
Western Alloy No. 62	M-279			Whelco Alloy Tap	M-119
				Whelco B	M-119
Western Alloy No. 63	M-279	Western Leaded Nickel Silver 12%, #133	M-279	Whelco Die and Tool	M-119
Western Alloy No. 68	M-279			Whelco Finishing	M-119
		Western Low Brass 20	M-279	Whelco High Speed	M-119
Western Alloy No. 86	M-279			Whelco Hot Die	M-119
		Western Lubaloy #84	M-279	Whelco Hot Work	M-119
Western Alloy No. 116	M-279			Whelco M	M-119
		Western Lubaloy-X #80	M-279	Whelco No. 1	M-119
Western Alloy No. 133	M-279			Whelco No. 2	M-119
		Western Lubronze #83	M-279	Whelco No. 3	M-119
Western Alloy No. 178	M-279			Whelco No. 4	M-119
		Western Manganese Brass #277	M-279	Whelco No. 6	M-119
Western Alloy No. 180	M-279			Whelco No. 7	M-119
				Whelco No. 8	M-119
Western Alloy No. 222	M-279	Western Medium Leaded Brass #62	M-279	Whelco No. 9	M-119
				Whelco Piston	M-119
Western Aluminum Bronze	M-279	Western Medium Leaded Brass #63	M-279	Whelco Standard	M-119
				Whelco Std	M-119
Western Bearing Bronze	M-279			Whelco Superior	M-119
		Western Muntz Metal	M-279	Whiz	M-307
Western Best Quality Brass	M-279			WI-52	M-167
		Western Nickel Silver 55-18, #168	M-279	WI-301	M-764
Western Brazing Brass #340	M-279			Widia "M-68"	M-371
				Wieland AlBz4	M-764
Western Bronze No. 222	M-279	Western Nickel Silver 56-12, #178	M-279	Wieland AlBz9	M-764
				Wieland Fw	M-764
Western Copper #55	M-279			Wieland Fw2	M-764
		Western Nickel Silver 65-18, #164	M-279	Wieland Fw444	M-764
Western Cupro Nickel-7%	M-279			Wieland Tf	M-764
				Wieland WB444	M-764
				Wieland Ws80	M-764

Section IV: Obsolete Alloys / 1759

Wieland Ws131	M-764	Wironit 770	M-1326	Witten C 38	M-1326
Wiesilber	M-764	Wironit 785	M-1326	Witten C 40	M-1326
Wilco H.T. Special	M-237	Wironit 795	M-1326	Witten C 46	M-1326
Willeys 710	M-976	Wironit 798	M-1326	Witten CKZ	M-1326
Willeys 945	M-976	Wironit 801	M-1326	Witten CN 15	M-1326
Willeys E-4.5	M-976	Wironit 803	M-1326	Witten CN 15 L	M-1326
Willeys E-10	M-976	Wironit 805	M-1326	Witten CN 18	M-1326
Willeys E-12	M-976	Wironit 807	M-1326	Witten CN 18 L	M-1326
Willeys E-13	M-976	Wironit 807L	M-1326	Witten CN41	M-1326
Willeys E-18	M-976	Wironit 808	M-1326	Witten CN80	M-1326
Willeys W9C	M-976	Wironit 812	M-1326	Witten CNE2	M-1326
Willeys W12C	M-976	Wironit 812A	M-1326	Witten CNE 3	M-1326
Willeys X3	M-976	Wironit 812E	M-1326	Witten CNE4	M-1326
Willeys X620	M-976	Wironit 812EL	M-1326	Witten CNL	M-1326
Wilmil	M-187	Wironit 812EN	M-1326	Witten CNO 17	M-1326
Wilmil M	M-187	Wironit 812L	M-1326	Witten CNO 18 F	M-1326
Wilson 109	M-663	Wironit 815	M-1326	Witten CNO 36	M-1326
Wilson 512	M-663	Wironit 818	M-1326	Witten CNO 40	M-1326
Wilson 520 and		Wironit 2001	M-1326	Witten CNOV 30	M-1326
520A	M-663	Wironit 2003	M-1326	Witten CNOV 30 L	M-1326
Wilson 575	M-663	Wironit 2004	M-1326	Witten CNOV 37 L	
Wimet 110B	M-1006	Wironit 2005	M-1326	VA	M-1326
Wimet F	M-1006	Wironit 2007	M-1326	Witten CNV 1	M-1326
Wimet FS	M-1006	Wironit 2008	M-1326	Witten CNV 2	M-1326
Wimet HH	M-1006	Wironit 2010	M-1326	Witten CNV 3	M-1326
Wimet NC	M-1006	Wironit 2010 E	M-1326	Witten CNV 4	M-1326
Wimet NH	M-1006	Wironit 2010 EN	M-1326	Witten CO 15	M-1326
Wimet S58	M-1006	Wironit 2012 E	M-1326	Witten CO 17	M-1326
Wimet TTX	M-1006	Wironit 2012 EN	M-1326	Witten CO 20 AF	M-1326
Wimet X	M-1006	Wironit 2020	M-1326	Witten CO 20 Al	M-1326
Wimet X8	M-1006	Wironit 2107	M-1326	Witten CO 20 L	M-1326
Wimet XL1	M-1006	Wironit 2307	M-1326	Witten CO 23	M-1326
Wimet XL2	M-1006	Wironit 4113	M-1326	Witten CO23 L	M-1326
Wimet XL4	M-1006	Wironit 4460	M-1326	Witten CO 25 F	M-1326
Wimet XX	M-1006	Wironit B	M-1326	Witten CO 32	M-1326
Wimet XX7	M-1006	Wironit C	M-1326	Witten CO 38	M-1326
Wire Drawing Alloy	M-35	Wissco No. 4 H T	M-120	Witten CO 46	M-1326
Wironit 140	M-1326	Witan 2	M-1326	Witten COL 30	M-1326
Wironit 150	M-1326	Witherm	M-1326	Witten COV 13 L	M-1326
Wironit 151	M-1326	Witherm 8	M-1326	Witten COV 30	M-1326
Wironit 151 L	M-1326	Witherm 10	M-1326	Witten COV 30 L	M-1326
Wironit 599	M-1326	Witherm 15	M-1326	Witten CRL	M-1326
Wironit 599 Al	M-1326	Witherm 20	M-1326	Witten CRL25	M-1326
Wironit 600	M-1326	Witherm 25	M-1326	Witten CRW 1	M-1326
Wironit 600 A	M-1326	Witherm 29	M-1326	Witten CRW 2	M-1326
Wironit 601	M-1326	Witherm 32	M-1326	Witten CRW 3	M-1326
Wironit 602	M-1326	Witherm 36	M-1326	Witten CSP	M-1326
Wironit 610	M-1326	Witherm 48	M-1326	Witten CV 48	M-1326
Wironit 610 E	M-1326	Witherm 48 L	M-1326	Witten CV58	M-1326
Wironit 610 EN	M-1326	Witherm 52	M-1326	Witten D3	M-1326
Wironit 615	M-1326	Witherm 61	M-1326	Witten D4	M-1326
Wironit 620	M-1326	Witten 5H	M-1326	Witten D6	M-1326
Wironit 632	M-1326	Witten 5SC5	M-1326	Witten D 6 G	M-1326
Wironit 632 L	M-1326	Witten 7H	M-1326	Witten D 7 G	M-1326
Wironit 645	M-1326	Witten 2260	M-1326	Witten D 7 S	M-1326
Wironit 696	M-1326	Witten AZ 35 M	M-1326	Witten D9	M-1326
Wironit 706E	M-1326	Witten AZ 38 M	M-1326	Witten D 9 S	M-1326
Wironit 710A	M-1326	Witten BCNO 18	M-1326	Witten D 1414	M-1326
Wironit 715	M-1326	Witten BCNO 22	M-1326	Witten D2414	M-1326
Wironit 720	M-1326	Witten BCO 23	M-1326	Witten D 2417	M-1326
Wironit 735	M-1326	Witten BCO 24	M-1326	Witten D2420	M-1326
Wironit 745	M-1326	Witten BL Extra	M-1326	Witten D3508	M-1326
Wironit 755	M-1326	Witten C 33	M-1326	Witten D4412S	M-1326
Wironit 765	M-1326	Witten C 37	M-1326	Witten D 4412 S	M-1326

Alloy	Ref	Alloy	Ref	Alloy	Ref
Witten D 4415	M-1326	Witten HBL	M-1326	Witten SS 115 C	M-1326
Witten D 4519 S	M-1326	Witten HCE	M-1326	Witten SS 116	M-1326
Witten D 5325	M-1326	Witten HFC 5 LVA	M-1326	Witten SS 117 M	M-1326
Witten D 5509	M-1326	Witten HFC 16	M-1326	Witten SS 120	M-1326
Witten D 5512	M-1326	Witten HFN 18	M-1326	Witten SS 125	M-1326
Witten D5514	M-1326	Witten HSO	M-1326	Witten SS 131	M-1326
Witten D 5514 S	M-1326	Witten K 1 L	M-1326	Witten SS 213 C	M-1326
Witten D5515	M-1326	Witten K 2 L	M-1326	Witten SSO	M-1326
Witten D 5515 S	M-1326	Witten K 3 L	M-1326	Witten SW	M-1326
Witten D 5517	M-1326	Witten Kobalt 5	M-1326	Witten TL65	M-1326
Witten D 5520	M-1326	Witten Kobalt 10	M-1326	Witten TL80	M-1326
Witten D 5535	M-1326	Witten Kobalt 54 V	M-1326	Witten TL90	M-1326
Witten D 5607	M-1326	Witten Kobalt 55 H	M-1326	Witten UAM	M-1326
Witten D5610	M-1326	Witten Kobalt 55 M	M-1326	Witten UAM Extra	M-1326
Witten D 8512	M-1326	Witten Kobalt 105 M	M-1326	Witten UAM Special	M-1326
Witten D 8514	M-1326	Witten Kobalt 810 M	M-1326	Witten UM 30	M-1326
Witten D 8516	M-1326	Witten KV 32 COL	M-1326	Witten UM 40	M-1326
Witten D 8518	M-1326	Witten LDM	M-1326	Witten UM 45	M-1326
Witten D 8518 L	M-1326	Witten M 28	M-1326	Witten UM 50	M-1326
Witten D 8518 W	M-1326	Witten M32	M-1326	Witten UM 60	M-1326
Witten DA 1525 LVA	M-1326	Witten M-36	M-1326	Witten UM 70	M-1326
Witten DA 1525 VU	M-1326	Witten M 41	M-1326	Witten UM 80	M-1326
Witten DA 1573	M-1326	Witten MC 14	M-1326	Witten UM 85	M-1326
Witten DA 1613	M-1326	Witten MC 18	M-1326	Witten UM 90	M-1326
Witten DA 1613 V	M-1326	Witten Mo 5	M-1326	Witten UM 100	M-1326
Witten DA 1616	M-1326	Witten Mo 5H	M-1326	Witten UM 110	M-1326
Witten DA 1616 B	M-1326	Witten Mo 9	M-1326	Witten UM 115	M-1326
Witten DA 1616 L	M-1326	Witten Mo 9V	M-1326	Witten UM 125	M-1326
Witten DA 1616 W	M-1326	Witten Mo 53V	M-1326	Witten UMK	M-1326
Witten DA 1713	M-1326	Witten MS 5	M-1326	Witten Vanadium 4	M-1326
Witten DA 1713 W	M-1326	Witten MS 5 P(HT)	M-1326	Witten VM 03	M-1326
Witten DA 1811	M-1326	Witten MS 33	M-1326	Witten VM6	M-1326
Witten DA 2019	M-1326	Witten MS53	M-1326	Witten VM 23	M-1326
Witten DA 2019 L	M-1326	Witten MSL	M-1326	Witten VS12C	M-1326
Witten DA 2020	M-1326	Witten MV40	M-1326	Witten W	M-1326
Witten DA 2040	M-1326	Witten NE1	M-1326	Witten W 01 E	M-1326
Witten DA 2040 L	M-1326	Witten PCVWL	M-1326	Witten W1W7	M-1326
Witten DA 2060 Ti	M-1326	Witten PO	M-1326	Witten W1W8	M-1326
Witten DA 2060 TiL	M-1326	Witten PO43	M-1326	Witten W1ZM	M-1326
Witten DA 2080	M-1326	Witten POL	M-1326	Witten W5S	M-1326
Witten DA 2080 L	M-1326	Witten PW	M-1326	Witten W 10 E	M-1326
Witten DA 2080 Ti	M-1326	Witten PWH	M-1326	Witten W 10 L	M-1326
Witten DA 2080 TiL	M-1326	Witten PW Spezial	M-1326	Witten W20E	M-1326
Witten DCM12	M-1326	Witten PWV	M-1326	Witten W22	M-1326
Witten DCMV30	M-1326	Witten RMH	M-1326	Witten W 25 E	M-1326
Witten DM	M-1326	Witten RMH 735	M-1326	Witten W 30 E	M-1326
Witten D Mo	M-1326	Witten S	M-1326	Witten W 30 L	M-1326
Witten DMW	M-1326	Witten S2	M-1326	Witten W 34	M-1326
Witten EM 3	M-1326	Witten S 48 L	M-1326	Witten W 40 E	M-1326
Witten EM 15	M-1326	Witten S 87 NL	M-1326	Witten W 40 L	M-1326
Witten EML	M-1326	Witten S 1525 LVA	M-1326	Witten W 44	M-1326
Witten EMW	M-1326	Witten SC67	M-1326	Witten W 50 E	M-1326
Witten G 65	M-1326	Witten SCR	M-1326	Witten W 53 Special	M-1326
Witten G 75	M-1326	Witten SCS	M-1326	Witten W 60 E	M-1326
Witten G 110	M-1326	Witten SO8	M-1326	Witten W 66	M-1326
Witten GB	M-1326	Witten SOZ	M-1326	Witten W 66 Special	M-1326
Witten GBA	M-1326	Witten SS 65 M	M-1326	Witten W 77	M-1326
Witten GKM	M-1326	Witten SS 105	M-1326	Witten W84	M-1326
Witten GLHA	M-1326	Witten SS 110	M-1326	Witten W 88 A	M-1326
Witten GSO	M-1326	Witten SS 115	M-1326	Witten W 99	M-1326
				Witten WE	M-1326

Section IV: Obsolete Alloys / 1761

Witten WHL	M-1326	WW-10W3	M-1259	Youngstown Special Motor	M-531
Witten WIB	M-1326	WW-20W3	M-1259	Youngstown Transformer	M-531
Witten WIB 100	M-1326	WW-100W	M-1259	Youngstown Transformer Extra Special	M-531
Witten WIW 9	M-1326	WW105F	M-1259	Youngstown Transformer Special	M-531
Witten WIW 10	M-1326	WW110F	M-1259	Yuma Chrome Magnet Steel	M-1
Witten WIZ 7	M-1326	WW-120	M-1259		
Witten WIZ 8	M-1326	WW-125	M-1259		
Witten WIZ 9	M-1326	WW140F	M-1259		
Witten WIZ 11	M-1326	WW-150	M-1259		
Witten WIZ 11 L	M-1326	WW-175	M-1259		
Witten WIZ 12	M-1326	WW175A	M-1259		
Witten WM45	M-1326	WW175B	M-1259	**Z**	
Witten WN14P	M-1326	WW-190	M-1259		
Witten Wolfram 184	M-1326	WW-200	M-1259		
		WW-230	M-1259	Zal-Die	M-233
Witten WS 45	M-1326	WW-275	M-1259	Zamak No. 2	M-91
Witten ZR 3	M-1326	WW-325	M-1259	Zamak No. 6	M-91
Witten ZR 5	M-1326	WW-375	M-1259	Zapp 28	M-1332
Witten ZW	M-1326	Wyckoff 12L14X	M-629	Zapp 28L	M-1332
WKL	M-459	Wyckoff 1215	M-629	Zapp 28N	M-1332
WM 559	M-459	Wyckoff WXB-1	M-629	Zapp 28W	M-1332
WMD	M-1701	Wyckoff WXB-2	M-629	Zapp 42MNV7	M-1332
WMD Extra	M-1701	Wyckoff WXB-3	M-629	Zapp 80	M-1332
W.M.S.	M-499	Wyco 100-44	M-629	Zapp 80X	M-1332
WOCO	M-24	Wyco 100-50	M-629	Zapp 82	M-1332
WOCO	M-334	Wyco 125-44	M-629	Zapp 82RT	M-1332
Woco	M-334	Wyco Cold Finished Steel	M-629	Zapp 82S	M-1332
Wolfram 12V	M-1326			Zapp 82X	M-1332
Wolfram 33	M-1326			Zapp 135T	M-1332
Wolfram 94	M-1326	**X**		Zapp 182RT	M-1332
Wompco	M-463			Zapp C5	M-1332
Wood Band Saw	M-38	X13RG	M-1115	Zapp K2 Prima	M-1332
Wood Saw	M-357	X-63	M-60	Zapp K3 Extra	M-1332
Wortle Die	M-38	XAloy 1525	M-1473	Zapp K3 Prima	M-1332
Wortle Die No. 1	M-38	XB75S	M-4	Zapp K5 Extra	M-1332
Wortle Die No. 4	M-38	XC-60 Steel	M-766	Zapp K5 Prima	M-1332
Wortle Die No. 5	M-38	X-CEL Super X	M-959	Zapp K60G	M-1332
Wortle Die Steel	M-73	X.D.H. Steel	M-57	Zapp K90G	M-1332
Wortle Drawing Die	M-38	X.D.M. Steel	M-57	Zapp MC9	M-1332
Wortle No. 4 Drawing Die	M-362	XL Metal	M-399	Zephor Bronze	M-1579
		Xtrocut-250	M-240	Zilloy	M-91
WRM Alloy A-5	M-1426	X Wiping Solder	M-959	Zilloy 202	M-91
WRM Alloy A-10	M-1426	Xyron 2-26	M-717	Zilloy 203	M-91
WRM Alloy A-12	M-1426			Zilloy 204	M-91
WRM Alloy A-15	M-1426			Zinc A	M-91
WRM Alloy A-18	M-1426	**Y**		Zincdie	M-35
WRM Alloy B-12-S	M-1426			Zip Zip	M-166
WRM Alloy B-18-S	M-1426	Yellow End	M-510	Zirmet	M-1038
WRM Alloy K-8	M-1426	Yoloy 45-W	M-531	Zivs Block Tested Hollow Drill	M-289
WRM Alloy K-12	M-1426	Yoloy "E" ACR	M-531	Z.N.	M-140
Wrought Iron 4D	M-292	Yoloy E-HS	M-531	Zncube	M-60
WSCR	M-459	Yoloy "E" HSX	M-531	Zodiac	M-121
WSI EX	M-459	Yoloy M Gr. A	M-531	ZOVM	M-459
WT Drill Rod	M-73	Yoloy M Gr. B	M-531	Z.T. Die Steel	M-73
WW-1	M-1259	Yorcalnic	M-948	Zunit 6	M-1332
WW-1W3	M-1259	Youngstown Armature	M-531	Zunit 13	M-1332
WW-2	M-1259			Zunit 18	M-1332
WW-3	M-1259	Youngstown Electrical	M-531	Zunit-Guss 22	M-1332
WW-5	M-1259			ZW.1	M-325
WW-6	M-1259	Youngstown Field	M-531	ZW.3	M-325
WW-7	M-1259	Youngstown Special Dynamo	M-531	ZW.6	M-325
WW-8	M-1259				

Chemical Composition Limits (a,b)

(Only composition limits which are identical to those listed herein or are registered with The Aluminum Association should be designated as "AA" alloys.)

Natural Impurity Limits for Wrought Unalloyed Aluminum

AA number	Registered date	Si	Fe	Cu	Mn	Mg	Cr	Ni	Zn	Ga	V		Ti	Others(c) Each	Others(c) Total	Aluminum min(d,e)
1030	...	0.35	0.6	0.10	0.05	0.05	...	...	0.10	...	0.05	...	0.03	0.03	...	99.30(e)
1035	8/22/78	0.35	0.6	0.10	0.05	0.05	...	...	0.10	...	0.05	...	0.03	0.03	...	99.35(e)
1040	8/22/73	0.30	0.50	0.10	0.05	0.05	...	...	0.10	...	0.05	...	0.03	0.03	...	99.40(e)
1045	8/22/73	0.30	0.45	0.10	0.05	0.05	...	...	0.05	...	0.05	...	0.03	0.03	...	99.45(e)
1050	...	0.25	0.40	0.05	0.05	0.05	...	...	0.05	...	0.05	...	0.03	0.03	...	99.50(e)
1055	8/22/73	0.25	0.40	0.05	0.05	0.05	...	...	0.05	...	0.05	...	0.03	0.03	...	99.55(e)
1060	...	0.25	0.35	0.05	0.03	0.03	...	...	0.05	...	0.05	...	0.03	0.03	...	99.60(e)
1065	8/22/73	0.25	0.30	0.05	0.03	0.03	...	...	0.05	...	0.05	...	0.03	0.03	...	99.65(e)
1070	...	0.20	0.25	0.04	0.03	0.03	...	...	0.04	...	0.05	...	0.03	0.03	...	99.70(e)
1075	...	0.20	0.20	0.04	0.03	0.03	...	...	0.04	...	0.05	...	0.03	0.03	...	99.75(e)
1080	...	0.15	0.15	0.03	0.02	0.02	...	...	0.03	0.03	0.05	...	0.03	0.02	...	99.80(e)
1085	...	0.10	0.12	0.03	0.02	0.02	...	...	0.03	0.03	0.05	...	0.02	0.01	...	99.85(e)
1090	...	0.07	0.07	0.02	0.01	0.01	...	...	0.03	0.03	0.05	...	0.01	0.01	...	99.90(e)
1095	...	0.030	0.040	0.010	0.010	0.010	...	...	0.010	...	...	...	0.005	0.005	...	99.95(d)

Registered Compositions

AA number	Registered By	Registered Date	Si	Fe	Cu	Mn	Mg	Cr	Ni	Zn	Ga	V		Ti	Others(c) Each	Others(c) Total	Aluminum min(d,e)
1100	...	...	1.0 Si + Fe	...	0.05-0.20	0.05	...	...	...	0.10	...	...	(t)	...	0.05	0.15	99.00(e)
1200	AATD	9/23/66	1.0 Si + Fe	...	0.05	0.05	...	...	...	0.10	...	...	...	0.05	0.05	0.15	99.00(e)
1230(j)	...	...	0.7 Si + Fe	...	0.10	0.05	0.05	...	...	0.10	...	0.05	...	0.03	0.03	...	99.30(e)
1135	Alcoa	4/23/57	0.65 Si + Fe	...	0.05-0.20	0.04	0.05	...	...	0.10	...	0.05	...	0.03	0.03	...	99.35(e)
1235(g)	...	...	0.65 Si + Fe	...	0.05	0.05	0.05	...	...	0.10	...	0.05	...	0.03	0.03	...	99.35(e)
1435	Kaiser	3/5/58	0.15	0.30-0.50	0.02	0.05	0.05	...	...	0.10	...	0.05	...	0.03	0.03	...	99.35(e)
1145(g)	...	...	0.55 Si + Fe	...	0.05	0.05	0.05	...	...	0.05	...	0.05	...	0.03	0.03	...	99.45(e)
1345(v)	Alcoa	10/8/56	0.30	0.40	0.10	0.05	0.05	...	...	0.05	...	0.05	...	0.03	0.03	...	99.45(e)
1250	Alcoa	1/3/56	0.20	0.40	0.10	0.01	0.01	0.01	...	0.05	...	...	(ff)	...	0.03	...	99.50(e)
1350(w)	AATD	1/24/75	0.10	0.04	0.05	0.01	...	0.01	...	0.05	0.03	...	0.05 B, (ff)	...	0.03	0.10	99.50(e)
1170	Alcoa	1/16/58	0.30 Si + Fe	...	0.03	0.03	0.02	0.03	...	0.04	...	0.05	...	0.03	0.03	...	99.70(e)
1175(h)	...	...	0.15 Si + Fe	...	0.10	0.02	0.02	...	...	0.04	0.03	0.05	...	0.02	0.02	...	99.75(e)
1180(i)	...	...	0.09	0.09	0.01	0.02	0.02	...	...	0.03	0.03	0.05	...	0.02	0.02	...	99.80(e)

Registered Compositions (continued)

AA number	Registered By	Date	Si	Fe	Cu	Mn	Mg	Cr	Ni	Zn	Ga	V	Ti		Others(c) Each	Others(c) Total	Aluminum min(d,e)
1185	...	...	0.15 Si + Fe		0.01	0.02	0.02	...	...	0.03	0.03	0.05	0.02		0.01	...	99.85(e)
1285	...	...	0.08(k)	0.08(k)	0.02	0.01	0.01	...	...	0.03	0.03	0.05	0.02		0.01	...	99.85(e)
1188(i)	...	...	0.06	0.06	0.005	0.01	0.01	...	...	0.03	0.03	0.05	0.01		0.01	...	99.88(e)
1199(i)	Alcoa	3/12/56	0.006	0.006	0.006	0.002	0.006	...	...	0.006	0.005	0.005	0.002		0.002	...	99.99(d)
2011	...	...	0.40	0.7	5.0-6.0	...	...	...	...	0.30	...	...	...		0.05	0.15	Remainder
2014	...	...	0.50-1.2	0.7	3.9-5.0	0.40-1.2	0.20-0.8	0.10	...	0.25	...	...	0.15	0.20 Zr + Ti	0.05	0.15	Remainder
2214	...	...	0.50-1.2	0.30	3.9-5.0	0.40-1.2	0.20-0.8	0.10	...	0.25	...	...	0.15	0.20 Zr + Ti	0.05	0.15	Remainder
2017	...	...	0.20-0.8	0.7	3.5-4.5	0.40-1.0	0.40-0.8	0.10	...	0.25	...	...	0.15	0.20 Zr + Ti	0.05	0.15	Remainder
2117	...	...	0.8	0.7	2.2-3.0	0.20	0.20-0.50	0.10	...	0.25	...	...	...		0.05	0.15	Remainder
2018	...	...	0.9	1.0	3.5-4.5	0.20	0.45-0.9	0.10	1.7-2.3	0.25	...	...	...		0.05	0.15	Remainder
2218	...	...	0.9	1.0	3.5-4.5	0.20	1.2-1.8	0.10	1.7-2.3	0.25	...	...	...		0.05	0.15	Remainder
2618	...	...	0.10-0.25	0.9-1.3	1.9-2.7	...	1.3-1.8	...	0.9-1.2	0.10	...	...	0.04-0.10		0.05	0.15	Remainder
2219	Alcoa	8/13/54	0.20	0.30	5.8-6.8	0.20-0.40	0.02	...	...	0.10	...	0.05-0.15	0.02-0.10	(II)	0.05	0.15	Remainder
2319(u)	Alcoa	6/5/58	0.20	0.30	5.8-6.8	0.20-0.40	0.02	...	...	0.10	...	0.05-0.15	0.10-0.20	(II, t)	0.05	0.15	Remainder
2419	Kaiser	10/12/72	0.15	0.18	5.8-6.8	0.20-0.40	0.02	...	...	0.10	...	0.05-0.15	0.02-0.10	(II)	0.05	0.15	Remainder
2024	...	...	0.50	0.50	3.8-4.9	0.30-0.9	1.2-1.8	0.10	...	0.25	...	...	0.15	0.20 Zr + Ti	0.05	0.15	Remainder
2124	AATD	10/2/70	0.20	0.30	3.8-4.9	0.30-0.9	1.2-1.8	0.10	...	0.25	...	...	0.15	0.20 Zr + Ti	0.05	0.15	Remainder
2025	...	...	0.50-1.2	1.0	3.9-5.0	0.40-1.2	0.05	0.10	...	0.25	...	...	...		0.05	0.15	Remainder
2036	Reynolds	8/13/70	0.50	0.50	2.2-3.0	0.10-0.40	0.30-0.6	0.10	...	0.25	...	...	0.15		0.05	0.15	Remainder
2048	Reynolds	8/2/72	0.15	0.20	2.8-3.8	0.20-0.6	1.2-1.8	...	...	0.25	...	...	0.10		0.05	0.15	Remainder
3002	Alcoa	7/3/61	0.08	0.10	0.15	0.10-0.25	0.05-0.20	...	...	0.05	...	0.05	0.03		0.03	0.10	Remainder
3102	Alcoa	3/1/72	0.40	0.7	0.10	0.05-0.40	...	...	...	0.30	...	...	0.10		0.05	0.15	Remainder
3003	...	...	0.6	0.7	0.05-0.20	1.0-1.5	...	...	...	0.10	...	...	...		0.05	0.15	Remainder
3303	Alcoa	5/31/74	0.6	0.7	0.05-0.20	1.0-1.5	...	...	...	0.30	...	...	...		0.05	0.15	Remainder
3004	...	...	0.30	0.7	0.25	1.0-1.5	0.8-1.3	...	...	0.25	...	...	...		0.05	0.15	Remainder
3005	...	...	0.6	0.7	0.30	1.0-1.5	0.20-0.6	0.10	...	0.25	...	...	0.10		0.05	0.15	Remainder
3105	Alcoa	5/27/60	0.6	0.7	0.30	0.30-0.8	0.20-0.8	0.20	...	0.40	...	...	0.10		0.05	0.15	Remainder
3006	Alumax	8/24/73	0.50	0.7	0.10-0.30	0.50-0.8	0.30-0.6	0.20	...	0.15-0.40	...	...	0.10		0.05	0.15	Remainder
3007	Nat. Alum.	3/17/76	0.50	0.7	0.10-0.30	0.30-0.8	0.6	...	...	0.40	...	...	...		0.05	0.15	Remainder
4032	AATD	6/11/68	3.5-4.5	0.35	0.05-0.15	0.03	0.05-0.15	...	0.50-1.3	0.15	...	...	0.02		0.05	0.15	Remainder
4004(hh)	Reynolds	10/5/71	9.0-10.5	0.8	0.25	0.10	1.0-2.0	...	...	0.20	...	...	...	0.8-1.4 Cd	0.05	0.15	Remainder
X4104	Reynolds	2/26/74	9.0-10.5	0.8	0.25	0.10	1.0-2.0	...	...	0.20	...	...	...	0.02-0.20 Bi	0.05	0.15	Remainder
X4005(f)	Reynolds	10/5/71	9.5-11.0	0.8	0.25	0.10	0.20-1.0	...	...	0.20	...	...	...		0.05	0.15	Remainder
4032	...	...	11.0-13.5	1.0	0.50-1.3	...	0.8-1.3	0.10	0.50-1.3	0.25	...	...	...		0.05	0.15	Remainder
4043(u)	...	...	4.5-6.0	0.8	0.30	0.05	0.05	...	...	0.10	...	...	0.20	(t)	0.05	0.15	Remainder
4343(o)	...	...	6.8-8.2	0.8	0.25	0.10	...	...	...	0.20	...	...	...		0.05	0.15	Remainder
4543	...	...	5.0-70	0.50	0.10	0.05	0.10-0.40	0.05	...	0.10	...	...	...		0.05	0.15	Remainder
4643(u)	Alcoa	8/14/63	3.6-4.6	0.8	0.10	0.05	0.10-0.30	...	...	0.10	...	...	0.10	(t)	0.05	0.15	Remainder
4044(bb)	Reynolds	7/15/69	7.8-9.2	0.8	0.25	0.10	...	...	...	0.20	...	...	0.15		0.05	0.15	Remainder
4545(o)	...	...	9.0-11.0	0.8	0.30	0.05	0.05	...	...	0.10	...	...	0.20		0.05	0.15	Remainder

Registered Compositions (continued)

AA number	Registered By	Date	Si	Fe	Cu	Mn	Mg	Cr	Ni	Zn	Ga	V	Ti	Others (c) Each	Others (c) Total	Aluminum min (d,e)
4145(o,u)	Alcoa	4/30/57	9.3-10.7	0.8	3.3-4.7	0.15	0.15	0.15	...	0.20	...	...	...	(t)	0.05	0.15	Remainder
4047(o,u)	...	...	11.0-13.0	0.8	0.30	0.15	0.10	...	...	0.20	...	...	...	(t)	0.05	0.15	Remainder
5005	...	...	0.30	0.7	0.20	0.20	0.50-1.1	0.10	...	0.25	...	...	...	...	0.05	0.15	Remainder
5205	AATD	5/29/67	0.15	0.7	0.03-0.10	0.10	0.6-1.0	0.10	...	0.05	...	...	...	...	0.05	0.15	Remainder
5010	Alumax	10/3/61	0.40	0.7	0.25	0.10-0.30	0.20-0.6	0.15	...	0.30	...	...	0.10	...	0.05	0.15	Remainder
X5020	Alcoa	7/12/72	0.30	0.7	1.3-1.9	0.10-0.50	2.4-3.2	0.20	...	0.20	...	...	0.10	...	0.05	0.15	Remainder
5040	Alcoa	2/24/61	0.30	0.7	0.25	0.9-1.4	1.0-1.5	0.10-0.30	...	0.25	...	...	...	...	0.05	0.15	Remainder
5050	...	...	0.40	0.7	0.20	0.10	1.1-1.8	0.10	...	0.25	...	...	...	...	0.05	0.15	Remainder
5051	Alumax	3/1/67	0.40	0.7	0.25	0.20	1.7-2.2	0.10	...	0.25	...	...	0.10	...	0.05	0.15	Remainder
5151	Alcoa	9/25/70	0.20	0.35	0.15	0.10	1.5-2.1	0.10	...	0.15	...	...	0.10	...	0.05	0.15	Remainder
5052	...	...	0.25	0.40	0.10	0.10	2.2-2.8	0.15-0.35	...	0.10	...	...	...	...	0.05	0.15	Remainder
5252	Kaiser	2/24/61	0.08	0.10	0.10	0.10	2.2-2.8	...	...	0.05	...	0.05	...	...	0.03	0.10	Remainder
5352	Alcoa	9/23/71	0.45 Si + Fe		0.10	0.10	2.2-2.8	0.10	...	0.10	...	...	0.10	...	0.05	0.15	Remainder
5652	...	...	0.40 Si + Fe		0.04	0.01	2.2-2.8	0.15-0.35	...	0.10	...	...	...	...	0.05	0.15	Remainder
5154(n)	...	...	0.25	0.40	0.10	0.10	3.1-3.9	0.15-0.35	...	0.20	...	...	0.20	...	0.05	0.15	Remainder
5254	Alcoa	7/8/57	0.45 Si + Fe		0.05	0.01	3.1-3.9	0.15-0.35	...	0.20	...	...	0.05	...	0.05	0.15	Remainder
5554(u)	Alcoa	3/5/58	0.25	0.40	0.10	0.50-1.0	2.4-3.0	0.05-0.20	...	0.25	...	...	0.20	(t)	0.05	0.15	Remainder
5654(u)	AATD	5/29/68	0.45 Si + Fe		0.05	0.50-1.0	2.4-3.0	0.05-0.20	...	0.25	...	...	0.05-0.20	(t)	0.05	0.15	Remainder
5056	...	...	0.30	0.40	0.10	0.01	3.1-3.9	0.15-0.35	...	0.20	...	...	0.05-0.15	...	0.05	0.15	Remainder
5356(u)	...	...	0.25	0.40	0.10	0.05-0.20	4.5-5.6	0.05-0.20	...	0.10	...	...	0.06-0.20	...	0.05	0.15	Remainder
5456	Alcoa	10/4/56	0.25	0.40	0.10	0.50-1.0	4.7-5.5	0.05-0.20	...	0.25	...	...	0.20	(t)	0.05	0.15	Remainder
5556(u)	Alcoa	10/9/56	0.25	0.40	0.10	0.50-1.0	4.7-5.5	0.05-0.20	...	0.25	...	...	0.05-0.20	...	0.05	0.15	Remainder
5357	...	...	0.12	0.17	0.20	0.15-0.45	0.8-1.2	...	...	0.05	...	...	...	...	0.05	0.15	Remainder
5457	Alcoa	12/24/57	0.08	0.10	0.20	0.15-0.45	0.8-1.2	...	...	0.05	...	0.05	...	...	0.03	0.10	Remainder
5657	Reynolds	2/26/60	0.08	0.10	0.10	0.03	0.6-1.0	...	...	0.05	0.03	0.05	...	...	0.02	0.05	Remainder
5082	Alcoa	8/14/63	0.20	0.35	0.15	0.15	4.0-5.0	0.15	...	0.25	...	...	0.10	...	0.05	0.15	Remainder
5182	Alcoa	11/10/67	0.20	0.35	0.15	0.20-0.50	4.0-5.0	0.10	...	0.25	...	...	0.10	...	0.05	0.15	Remainder
5083	...	...	0.40	0.40	0.10	0.40-1.0	4.0-4.9	0.05-0.25	...	0.25	...	...	0.15	...	0.05	0.15	Remainder
5183(u)	Kaiser	6/7/57	0.40	0.40	0.10	0.50-1.0	4.3-5.2	0.05-0.25	...	0.25	...	...	0.15	(t)	0.05	0.15	Remainder
X5085	Alcoa	7/12/72	...	0.40	0.15	0.20	5.8-6.8	0.20	...	0.20	...	...	0.10	...	0.05	0.15	Remainder
5086	...	...	0.30	0.50	0.10	0.20-0.7	3.5-4.5	0.05-0.25	...	0.25	...	...	0.15	...	0.05	0.15	Remainder
X5087(u)	Alcoa	7/10/74	0.40	0.7	0.10	0.50-1.0	4.7-5.5	0.05-0.20	...	0.25	...	...	0.05-0.20	(t, jj)	0.05	0.15	Remainder
X5090	Conalco	3/19/70	0.20	0.35	0.25	0.35	6.0-8.0	0.05-0.30	...	0.20	...	...	0.02	(gg)	0.05	0.15	Remainder
6101(p)	Reynolds, Revere	7/8/55	0.30-0.7	0.50	0.10	0.03	0.35-0.8	0.03	...	0.10	...	...	...	0.06 B	0.03	0.10	Remainder
6201(z)	Kaiser	9/7/60	0.50-0.9	0.50	0.10	0.03	0.6-0.9	0.03	...	0.10	...	...	...	0.06 B	0.03	0.10	Remainder
6301	Kaiser	4/27/70	0.50-0.9	0.7	0.10	0.15	0.6-0.9	0.10	...	0.25	...	...	0.15	...	0.05	0.15	Remainder
6003(q)	...	...	0.35-1.0	0.6	0.10	0.8	0.8-1.5	0.35	...	0.20	...	...	0.10	...	0.05	0.15	Remainder
6004	Kaiser	2/9/73	0.30-0.6	0.10-0.30	0.10	0.20-0.6	0.40-0.7	...	...	0.05	...	...	...	...	0.05	0.15	Remainder

Registered Compositions (continued)

AA number	Registered By	Date	Si	Fe	Cu	Mn	Mg	Cr	Ni	Zn	Ga	V		Ti	Others(c) Each	Others(c) Total	Aluminum min (d,e)
6005	AATD	12/20/62	0.6-0.9	0.35	0.10	0.10	0.40-0.6	0.10		0.10				0.10	0.05	0.15	Remainder
6105	Revere	11/23/65	0.6-1.0	0.35	0.10	0.10	0.45-0.8	0.10		0.10				0.10	0.05	0.15	Remainder
6205	Conalco	3/19/70	0.6-0.9	0.7	0.20	0.05-0.15	0.40-0.6	0.05-0.15		0.25			0.05-0.15 Zr	0.15	0.05	0.15	Remainder
6006	Kaiser	10/20/71	0.20-0.6	0.35	0.15-0.30	0.05-0.20	0.45-0.9	0.10		0.10				0.10	0.05	0.15	Remainder
6007	Conalco	4/4/75	0.9-1.4	0.7	0.20	0.05-0.25	0.6-0.9	0.05-0.25		0.25			0.05-0.20 Zr	0.15	0.05	0.15	Remainder
6011			0.6-1.2	1.0	0.40-0.9	0.8	0.6-1.2	0.30		1.5				0.20	0.05	0.15	Remainder
6151			0.6-1.2	1.0	0.35	0.20	0.45-0.8	0.15-0.35	0.20	0.25				0.15	0.05	0.15	Remainder
6351	Kaiser	12/16/58	0.7-1.3	0.50	0.10	0.40-0.8	0.40-0.8			0.20				0.20	0.05	0.15	Remainder
6951			0.20-0.50	0.8	0.15-0.40	0.10	0.40-0.8			0.20					0.05	0.15	Remainder
6053			(r)	0.35	0.10		1.1-1.4	0.15-0.35		0.10					0.05	0.15	Remainder
6253(x)			(r)	0.50	0.10		1.0-1.5	0.04-0.35		1.6-2.4					0.05	0.15	Remainder
6061			0.40-0.8	0.7	0.15-0.40	0.15	0.8-1.2	0.04-0.35		0.25				0.15	0.05	0.15	Remainder
6261	Alcan	4/23/68	0.40-0.7	0.40	0.15-0.40	0.20-0.35	0.7-1.0	0.10		0.20				0.10	0.05	0.15	Remainder
6162	Reynolds	3/26/59	0.40-0.8	0.50	0.20	0.10	0.7-1.1	0.10		0.25				0.10	0.05	0.15	Remainder
6262	Alcoa	1/14/60	0.40-0.8	0.7	0.15-0.40	0.15	0.8-1.2	0.04-0.14		0.25			(aa)	0.15	0.05	0.15	Remainder
6063			0.20-0.6	0.35	0.10	0.10	0.45-0.9	0.10		0.10				0.10	0.05	0.15	Remainder
6463	Alcoa	4/15/57	0.20-0.6	0.15	0.20	0.05	0.45-0.9			0.05					0.05	0.15	Remainder
6763	National	12/4/72	0.20-0.6	0.08	0.04-0.16	0.03	0.45-0.9			0.03					0.03	0.10	Remainder
6066			0.9-1.8	0.50	0.7-1.2	0.6-1.1	0.8-1.4	0.40		0.25				0.20	0.05	0.15	Remainder
6070	Alcoa	1/18/62	1.0-1.7	0.50	0.15-0.40	0.40-1.0	0.50-1.2	0.10		0.25		0.05		0.15	0.05	0.15	Remainder
7001			0.35	0.40	1.6-2.6	0.20	2.6-3.4	0.18-0.35		6.8-8.0				0.20	0.05	0.15	Remainder
7004	Alcan	3/19/64	0.25	0.35	0.05	0.20-0.7	1.0-2.0	0.05		3.8-4.6			0.10-0.20 Zr	0.05	0.05	0.15	Remainder
7104	Alcan	3/19/64	0.25	0.40	0.03		0.50-0.9			3.6-4.4				0.10	0.05	0.15	Remainder
7005(dd)	Alcoa	8/13/62	0.35	0.40	0.10	0.20-0.7	1.0-1.8	0.06-0.20		4.0-5.0			0.08-0.20 Zr	0.01-0.06	0.05	0.15	Remainder
7008(ee)	Alcoa	11/15/68	0.10	0.10	0.05	0.05	0.7-1.4	0.12-0.25		4.5-5.5				0.05	0.05	0.10	Remainder
7011(ee)	Reynolds	12/2/68	0.15	0.20	0.05	0.10-0.30	1.0-1.6	0.05-0.20		4.0-5.5				0.05	0.05	0.15	Remainder
7013(f)	Alcoa	1/29/76	0.6	0.7	0.10	1.0-1.5				1.5-2.0					0.05	0.15	Remainder
X7016	Reynolds	6/29/72	0.10	0.12	0.45-1.0	0.03	0.8-1.4			4.0-5.0		0.05		0.03	0.03	0.10	Remainder
X7116	Reynolds	6/12/75	0.15	0.30	0.50-1.1	0.05	0.8-1.4			4.2-5.2	0.03	0.05		0.05	0.03	0.10	Remainder
X7029	Reynolds	12/8/75	0.10	0.12	0.50-0.9	0.03	1.3-2.0			4.2-5.2		0.05		0.03	0.03	0.10	Remainder
7039	Kaiser	7/16/62	0.30	0.40	0.10	0.10-0.40	2.3-3.3	0.15-0.25		3.5-4.5				0.10	0.05	0.15	Remainder
X7046	Alcoa	5/16/73	0.20	0.40	0.25	0.05-0.30	1.0-1.6	0.06-0.20		6.6-7.6			0.06-0.18 Zr	0.06	0.05	0.15	Remainder
7049	Kaiser	5/10/68	0.25	0.35	1.2-1.9	0.20	2.0-2.9	0.10-0.22		7.2-8.2				0.10	0.05	0.15	Remainder
7149	Kaiser	10/28/75	0.15	0.20	1.2-1.9	0.20	2.0-2.9	0.10-0.22		7.2-8.2				0.10	0.05	0.15	Remainder
7050	Alcoa	2/1/71	0.12	0.15	2.0-2.6	0.10	1.9-2.6	0.04		5.7-6.7			0.08-0.15 Zr	0.06	0.05	0.15	Remainder
7070(f)	Alcan	12/20/72	0.15	0.25	0.05					1.3-1.8					0.05	0.15	Remainder
7072(y)			0.7 Si + Fe		0.10	0.10	0.10			0.8-1.3					0.05	0.15	Remainder
7472	Alcoa	12/19/60	0.25	0.6	0.05	0.05	0.9-1.5			1.3-1.9					0.05	0.15	Remainder
7075			0.40	0.50	1.2-2.0	0.30	2.1-2.9	0.18-0.28		5.1-6.1			0.25 Zr + Ti	0.20	0.05	0.15	Remainder
7175	Alcoa	11/8/57	0.15	0.20	1.2-2.0	0.10	2.1-2.9	0.18-0.28		5.1-6.1				0.10	0.05	0.15	Remainder

Registered Compositions (continued)

AA number	Registered By	Date	Si	Fe	Cu	Mn	Mg	Cr	Ni	Zn	Ga	V	Ti		Others(c) Each	Total	Aluminum min (d,e)
7475	Alcoa	9/15/69	0.10	0.12	1.2-1.9	0.06	1.9-2.6	0.18-0.25	...	5.2-6.2	...	...	0.06	...	0.05	0.15	Remainder
7076	...	...	0.40	0.6	0.30-1.0	0.30-0.8	1.2-2.0	...	...	7.0-8.0	...	...	0.20	...	0.05	0.15	Remainder
7277(n)	...	...	0.50	0.7	0.8-1.7	...	1.7-2.3	0.18-0.35	...	3.7-4.3	...	...	0.10	...	0.05	0.15	Remainder
7178(n)	...	...	0.40	0.50	1.6-2.4	0.30	2.4-3.1	0.18-0.35	...	6.3-7.3	...	...	0.20	...	0.05	0.15	Remainder
7079	...	...	0.30	0.40	0.40-0.8	0.10-0.30	2.9-3.7	0.10-0.25	...	3.8-4.8	...	...	0.10	...	0.05	0.15	Remainder
7179	Alcoa	11/8/57	0.15	0.20	0.40-0.8	0.10-0.30	2.9-3.7	0.10-0.25	...	3.8-4.8	...	...	0.10	...	0.05	0.15	Remainder
8001	Alcoa	9/5/57	0.17	0.45-0.7	0.15	0.6	0.7	...	0.9-1.3	0.05	...	...	...	...	0.05	0.15	Remainder
8112(n)	...	...	1.0	1.0	0.40	0.005	...	0.20	...	1.0	...	...	0.20	(s)	0.05	0.15	Remainder
8020	Conalco	6/13/73	0.10	0.10	0.005	...	...	...	...	...	...	...	...	(kk)	0.03	0.10	Remainder
X8030	A.E.I.	9/29/75	0.10	0.30-0.8	0.15-0.30	...	0.05	...	...	...	...	0.05	...	0.001-0.04 B	0.03	0.10	Remainder
X8130	Reynolds	3/31/76	0.15(mm)	0.40-1.0(mm)	0.05-0.15	...	...	...	...	0.05	...	...	...	...	0.03	0.15	Remainder
8040	Conalco	11/15/62	1.0 Si + Fe		0.20	0.05	...	...	...	0.10	...	...	...	0.10-0.30 Zr	0.05	0.15	Remainder
8076	Alcoa	7/24/72	...	0.6-0.9	0.04	...	0.08-0.22	...	...	0.20	...	...	...	0.04 B	0.03	0.10	Remainder
X8176	Southwire	1/21/76	0.03-0.15	0.40-1.0	...	...	...	...	...	0.05	0.03	...	...	...	0.05	0.15	Remainder
X8077	Alcan	5/20/75	0.10	0.10-0.40	0.05	...	0.10-0.30	...	...	0.10	...	...	...	(m)	0.03	0.10	Remainder
8079(g)	Reynolds	1/9/69	0.05-0.30	0.7-1.3	0.05	...	...	...	...	0.10	...	...	...	...	0.05	0.15	Remainder
8280(n)	...	...	1.0-2.0	0.7	0.7-1.3	0.10	...	...	0.20-0.7	0.05	...	...	0.10	5.5-7.0 Sn	0.05	0.15	Remainder
8081	Alcoa	2/8/65	0.7	0.7	0.7-1.3	0.10	...	...	...	0.05	...	...	0.10	18.0-22.0 Sn	0.05	0.15	Remainder

(a) Composition in percent maximum unless shown as a range or a minimum. Standard limits for alloying elements and impurities are expressed to the following places: less than 1/1000 percent, 0.000X; 1/1000 to 1/100 percent, 0.00X; 1/100 to 1/10 percent unalloyed aluminum made by a refining process, 0.0XX; 1/100 to 1/10 percent alloys and unalloyed aluminum not made by a refining process, 0.0X; 1/10 through 1/2 percent, 0.XX; over 1/2 percent 0.X, X.X, etc. (b) For purposes of determining conformance to these limits, an observed value or a calculated value obtained from analysis is rounded off to the nearest unit in the last right-hand place of figures used in expressing the specified limit, in accordance with the following: American National Standard Rules for Rounding Off Numerical Values (ANSI Z25.1). When the figure next beyond the last figure or place to be retained is less than 5, the figure in the last place retained should be kept unchanged. When the figure next beyond the last figure or place to be retained is greater than 5, the figure in the last place to be retained should be increased by 1. When the figure next beyond the last figure or place to be retained is 5 and (1) there are no figures, or only zeroes, beyond this 5, if the figure in the last place to be retained is odd, it should be increased by 1; if even, it should be kept unchanged; (2) if the 5 next beyond the figure in the last place to be retained is followed by any figures other than zero, the figure in the last place retained should be increased by 1, whether odd or even. (c) Analysis is regularly made only for the elements for which specific limits are shown, except for unalloyed aluminum. If, however, the presence of other elements is suspected to be, or in the course of routine analysis is indicated to be in excess of the specified limits, further analysis is made to determine that these other elements are not in excess of the amount specified. (d) The aluminum content for unalloyed aluminum made by a refining process is the difference between 100.00 percent and the sum of all other metallic elements present in amounts of 0.0010 percent or more each, expressed to the third decimal before determining the sum, which is rounded to the second decimal before subtracting. (e) The aluminum content for unalloyed aluminum not made by a refining process is the difference between 100.00 percent and the sum of all other metallic elements present in amounts of 0.010 percent or more each, expressed to the second decimal before determining the sum. (f) Cladding alloy. (g) Foil. (h) Cladding on clad 1100 and clad 3003 reflector sheet. (i) Capacitor alloy. (j) Cladding on alclad 2014. (r) Silicon 45-65 percent of magnesium. (s) Boron, cobalt 0.001 max. each; cadmium 0.003 max.; lithium 0.008 max. (t) Beryllium alloy. (o) Brazing alloy. (p) Bus conductor. (q) Cladding on alclad 2024. (k) Silicon plus iron, 0.14 max. (l) Lead, bismuth 0.20-0.6 each. (m) Boron 0.05 max.; zirconium 0.02-0.08. (n) Consider as original 0.0008 max. for welding electrode only. (u) Mechanical wire. (v) Welding electrode. (w) Formerly designated EC. (x) Cladding on alclad 5056. (y) Cladding on alclad 2219, alclad 3003, alclad 3004, alclad 5050, alclad 5155, alclad 6061, alclad 7075, alclad 7475 and alclad 7178. (z) Conductor alloy. (aa) Lead, bismuth 0.40-0.7 each. (bb) Cladding on brazing sheet. (cc) Cadmium 0.05-0.20; tin 0.03-0.08. (dd) Extruded products. (ee) High strength cladding for sheet and plate products. (ff) Vanadium plus titanium, 0.02 max. (gg) Beryllium 0.001-0.02; boron 0.001-0.05. (hh) Cladding on X7, X8, X13 and X14 brazing sheet. (ii) These designations may be used for registration of new compositions after all unregistered numbers in the same series are used. Thereafter, the appropriate number with the oldest cancellation date shall be the next number re-used. (jj) Cobalt 0.30-0.7. (kk) Bismuth 0.10-0.50; tin 0.10-0.25. (ll) Zirconium 0.10-0.25. (mm) Silicon plus iron, 1.0 max.

Chemical Composition Limits (a, b)

(Only composition limits which are identical to those listed herein or are registered with The Aluminum Association should be designated as "AA" alloys.)

Registered alloys in the form of XXX.0 castings, XXX.1 ingot and XXX.2 ingot

AA number	Former designation	Registered By	Date	Product(c)	Si	Fe	Cu	Mn	Mg	Cr	Ni	Zn	Sn	Ti	Others(d) Each	Others(d) Total	Al min (e)
100.1(f)	...	AATL(g)	6/30/70	Ingot	0.15	0.6-0.8	0.10	(h)	...	(h)	...	0.05	...	(h)	0.03(h)	0.10	99.00
130.1(f)	...	AATD	6/30/70	Ingot	(i)	(i)	0.10	(h)	...	(h)	...	0.05	...	(h)	0.03(h)	0.10	99.30
150.1(f)	...	AATD	6/30/70	Ingot	(j)	(j)	0.05	(h)	...	(h)	...	0.05	...	(h)	0.03(h)	0.10	99.50
160.1	...	Alcoa	1/28/76	Ingot	0.10(j)	0.25(j)	...	(h)	...	(h)	...	0.05	...	(h)	0.03(h)	0.10	99.60
170.1(f)	...	AATD	6/30/70	Ingot	(k)	(k)	...	(h)	...	(h)	...	...	...	(h)	0.03(h)	0.10	99.70
201.0	...	Conalco	4/17/68	S	0.10	0.15	4.0-5.2	0.20-0.50	0.15-0.55	...	...	...	...	0.15-0.35	0.05(l)	0.10	Remainder
201.2	...	Conalco	4/17/68	Ingot	0.10	0.10	4.0-5.2	0.20-0.50	0.20-0.55	...	...	...	...	0.15-0.35	0.05(l)	0.10	Remainder
A201.0	...	Conalco	10/9/70	S	0.05	0.10	4.0-5.0	0.20-0.40	0.15-0.35	...	...	...	...	0.15-0.35	0.03(l)	0.10	Remainder
A201.2	...	Conalco	10/9/70	Ingot	0.05	0.07	4.0-5.0	0.20-0.40	0.20-0.35	...	...	...	...	0.15-0.35	0.03(l)	0.10	Remainder
202.0	...	Conalco	4/17/68	S	0.10	0.15	4.0-5.2	0.20-0.8	0.15-0.55	...	...	...	...	0.15-0.35	0.05(l)	0.10	Remainder
202.2	...	Conalco	4/17/68	Ingot	0.10	0.10	4.0-5.2	0.20-0.8	0.20-0.55	...	...	...	...	0.15-0.35	0.05(l)	0.10	Remainder
203.0	Hiduminium 350	M&A Co.	12/2/72	S	0.30	0.50	4.5-5.5	0.20-0.30	0.10	...	1.3-1.7	0.10	...	0.15-0.25(m)	0.05(n)	0.20	Remainder
203.2	Hiduminium 350	M&A Co.	12/2/72	Ingot	0.20	0.35	4.8-5.2	0.20-0.30	0.10	...	1.3-1.7	0.10	...	0.15-0.25(m)	0.05(n)	0.20	Remainder
204.0	A-U5GT	Howmet	10/1/74	S&P	0.20	0.35	4.2-5.0	0.10	0.15-0.35	0.20-0.5	0.05	0.05	0.05	0.15-0.30	0.05	0.15	Remainder
204.2	A-U5GT	Howmet	10/1/74	Ingot	0.15	0.10-0.20	4.2-4.9	0.05	0.20-0.35	0.20-0.5	0.03	0.05	0.05	0.15-0.25	0.05	0.15	Remainder
X206.0	...	Trialco	4/23/76	S&P	0.10	0.15	4.2-5.0	0.20-0.50	0.15-0.35	...	0.05	0.10	0.05	0.15-0.30	0.05	0.15	Remainder
X206.2	...	Trialco	4/23/76	Ingot	0.10	0.10	4.2-5.0	0.20-0.50	0.20-0.35	...	0.03	0.05	...	0.15-0.25	0.05	0.15	Remainder
XA206.0	...	Trialco	4/23/76	S&P	0.05	0.10	4.2-5.0	0.20-0.50	0.15-0.35	...	0.05	0.10	0.05	0.15-0.30	0.05	0.15	Remainder
XA206.2	...	Trialco	4/23/76	Ingot	0.05	0.07	4.2-5.0	0.20-0.50	0.20-0.35	...	0.03	0.05	...	0.15-0.25	0.05	0.15	Remainder
208.0	...	AATD	...	S	2.5-3.5	1.2	3.5-4.5	0.50	0.10	...	0.35	1.0	...	0.25	...	0.50	Remainder
208.1	108	AATD	...	Ingot	2.5-3.5	0.9	3.5-4.5	0.50	0.10	...	0.35	1.0	...	0.25	...	0.50	Remainder
208.2	108	AATD	...	Ingot	2.5-3.5	0.8	3.5-4.5	0.30	0.03	...	...	0.20	...	0.20	...	0.30	Remainder
213.0	C113	AATD	...	S&P	1.0-3.0	1.2	6.0-8.0	0.6	0.10	...	0.35	2.5	...	0.25	...	0.50	Remainder
213.1	C113	AATD	...	Ingot	1.0-3.0	0.9	6.0-8.0	0.6	0.10	...	0.35	2.5	...	0.25	...	0.50	Remainder
222.0	122	AATD	...	S&P	2.0	1.5	9.2-10.7	0.50	0.15-0.35	...	0.50	0.8	...	0.25	...	0.35	Remainder
222.1	122	AATD	...	Ingot	2.0	1.2	9.2-10.7	0.50	0.20-0.35	...	0.50	0.8	...	0.25	...	0.35	Remainder
224.0	...	Alcoa	4/2/69	S&P	0.06	0.10	4.5-5.5	0.20-0.50	...	...	...	...	...	0.35	0.03(o)	0.10	Remainder
224.2	...	Alcoa	4/2/69	Ingot	0.02	0.04	4.5-5.5	0.20-0.50	...	...	...	...	...	0.25	0.03(o)	0.10	Remainder
238.0	138	AATD	...	P	3.5-4.5	1.5	9.0-11.0	0.6	0.15-0.35	...	1.0	1.5	...	0.25	...	0.50	Remainder
238.1	138	AATD	...	Ingot	3.5-4.5	1.2	9.0-11.0	0.6	0.20-0.35	...	1.0	1.5	...	0.25	...	0.50	Remainder
238.2	138	AATD	...	Ingot	3.5-4.5	1.2	9.5-10.5	0.50	0.20-0.35	...	0.50	0.50	...	0.20	...	0.50	Remainder
A240.0	A140	Alcoa	...	S	0.50	0.50	7.0-9.0	0.30-0.7	5.5-6.5	...	0.30-0.7	0.10	...	0.20	0.05	0.15	Remainder
A240.1	A140	Alcoa	...	Ingot	0.50	0.40	7.0-9.0	0.30-0.7	5.6-6.5	...	0.30-0.7	0.10	...	0.20	0.05	0.15	Remainder
242.0	142	AATD	...	S&P	0.7	1.0	3.5-4.5	0.35	1.2-1.8	0.25	1.7-2.3	0.35	...	0.25	0.05	0.15	Remainder
242.1	142	AATD	...	Ingot	0.7	0.8	3.5-4.5	0.35	1.3-1.8	0.25	1.7-2.3	0.35	...	0.25	0.05	0.15	Remainder

Registered alloys in the form of XXX.0 castings, XXX.1 ingot and XXX.2 ingot (continued)

AA number	Former designation	Registered By	Date	Product(e)	Si	Fe	Cu	Mn	Mg	Cr	Ni	Zn	Sn	Ti	Others(d) Each	Others(d) Total	Al min (e)
242.2	142	AATD	...	Ingot	0.6	0.6	3.5-4.5	0.10	1.3-1.8	...	1.7-2.3	0.10	...	0.20	0.05	0.15	Remainder
A242.0	A142	AATD	...	S	0.6	0.8	3.7-4.5	0.10	1.2-1.7	0.15-0.25	1.8-2.3	0.10	...	0.07-0.20	0.05	0.15	Remainder
A242.1	A142	AATD	...	Ingot	0.6	0.6	3.7-4.5	0.10	1.3-1.7	0.15-0.25	1.8-2.3	0.10	...	0.07-0.20	0.05	0.15	Remainder
A242.2	A142	AATD	...	Ingot	0.35	0.6	3.7-4.5	0.10	1.3-1.7	0.15-0.25	1.8-2.3	0.10	...	0.07-0.20	0.05	0.15	Remainder
249.0	X149	Alcoa	...	P	0.05	0.10	3.8-4.6	0.25-0.50	0.25-0.50	...	...	2.5-3.5	...	0.02-0.35	0.03	0.10	Remainder
249.2	X149	Alcoa	...	Ingot	0.05	0.07	3.8-4.6	0.25-0.50	0.30-0.50	...	...	2.5-3.5	...	0.02-0.12	0.03	0.10	Remainder
295.0	195	AATD	...	S	0.7-1.5	1.0	4.0-5.0	0.35	0.03	...	...	0.35	...	0.25	0.05	0.15	Remainder
295.1	195	AATD	...	Ingot	0.7-1.5	0.8	4.0-5.0	0.35	0.03	...	...	0.35	...	0.25	0.05	0.15	Remainder
295.2	195	AATD	...	Ingot	0.7-1.2	0.8	4.0-5.0	0.30	0.03	...	...	0.30	...	0.20	0.05	0.15	Remainder
B295.0	B195	AATD	...	P	2.0-3.0	1.2	4.0-5.0	0.35	0.05	...	0.35	0.50	...	0.25	...	0.35	Remainder
B295.1	B195	AATD	...	Ingot	2.0-3.0	0.9	4.0-5.0	0.35	0.05	...	0.35	0.50	...	0.25	...	0.35	Remainder
B295.2	B195	AATD	...	Ingot	2.0-3.0	0.8	4.0-5.0	0.30	0.03	...	...	0.30	...	0.20	0.05	0.15	Remainder
305.0	...	Reynolds	11/5/74	S&P	4.5-5.5	0.6	1.0-1.5	0.50	0.10	0.25	...	0.35	...	0.25	0.05	0.15	Remainder
305.2	...	Reynolds	9/24/73	Ingot	4.5-5.5	0.14-0.25	1.0-1.5	0.05	...	...	...	0.05	...	0.20	0.05	0.15	Remainder
A305.0	...	Reynolds	11/5/74	S&P	4.5-5.5	0.20	1.0-1.5	0.10	0.10	...	...	0.10	...	0.20	0.05	0.15	Remainder
A305.1	...	Kaiser	6/4/74	Ingot	4.5-5.5	0.15	1.0-1.5	0.05	...	...	...	0.05	...	0.20	0.05	0.15	Remainder
A305.2	...	Reynolds	9/24/73	Ingot	4.5-5.5	0.13	1.0-1.5	0.05	...	...	...	0.05	...	0.20	0.05	0.15	Remainder
308.0	A108	AATD	...	S&P	5.0-6.0	1.0	4.0-5.0	0.50	0.10	...	...	1.0	...	0.25	...	0.50	Remainder
308.1	A108	AATD	...	Ingot	5.0-6.0	0.8	4.0-5.0	0.50	0.10	...	...	1.0	...	0.25	...	0.50	Remainder
308.2	A108	AATD	...	Ingot	5.0-6.0	0.8	4.0-5.0	0.30	0.10	...	...	0.50	...	0.20	...	0.50	Remainder
319.0	319, AllCast	AATD	...	S&P	5.5-6.5	1.0	3.0-4.0	0.50	0.10	...	0.35	1.0	...	0.25	...	0.50	Remainder
319.1	319, AllCast	AATD	...	Ingot	5.5-6.5	0.8	3.0-4.0	0.50	0.10	...	0.35	1.0	...	0.25	...	0.50	Remainder
319.2	319, AllCast	AATD	...	Ingot	5.5-6.5	0.6	3.0-4.0	0.10	0.10	...	0.10	0.10	...	0.20	...	0.20	Remainder
A319.0	...	AATD	8/28/70	S&P	5.5-6.5	1.0	3.0-4.0	0.50	0.10	...	0.35	3.0	...	0.25	...	0.50	Remainder
A319.1	...	AATD	8/28/70	Ingot	5.5-6.5	0.8	3.0-4.0	0.50	0.10	...	0.35	3.0	...	0.25	...	0.50	Remainder
324.0	324	Alcoa	...	P	7.0-8.0	1.2	0.40-0.6	0.50	0.40-0.7	...	0.30	1.0	...	0.20	0.15	0.20	Remainder
324.1	324	Alcoa	...	Ingot	7.0-8.0	0.9	0.40-0.6	0.50	0.45-0.7	...	0.30	1.0	...	0.20	0.15	0.20	Remainder
324.2	324	Alcoa	1/26/72	Ingot	7.0-8.0	0.6	0.40-0.6	0.10	0.45-0.7	...	0.10	0.10	...	0.20	0.05	0.15	Remainder
328.0	Red X-8	AATD	...	S	7.5-8.5	1.0	1.0-2.0	0.20-0.6	0.20-0.6	0.35	0.25	1.5	...	0.25	...	0.50	Remainder
328.1	Red X-8	AATD	...	Ingot	7.5-8.5	0.8	1.0-2.0	0.20-0.6	0.25-0.6	0.35	0.25	1.5	...	0.25	0.05	0.50	Remainder
A332.0	A132	AATD	...	P	11.0-13.0	1.2	0.50-1.5	0.35	0.7-1.3	...	2.0-3.0	0.35	...	0.25	0.05	...	Remainder
A332.1	A132	AATD	...	Ingot	11.0-13.0	0.9	0.50-1.5	0.35	0.8-1.3	...	2.0-3.0	0.35	...	0.25	0.05	0.15	Remainder
A332.2	A132	AATD	...	Ingot	11.0-13.0	0.9	0.50-1.5	0.10	0.9-1.3	...	2.0-3.0	0.10	...	0.20	...	0.50	Remainder
F332.0	F132	AATD	...	P	8.5-10.5	1.2	2.0-4.0	0.50	0.50-1.5	...	0.50	1.0	...	0.25	...	0.50	Remainder
F332.1	F132	AATD	...	Ingot	8.5-10.5	0.9	2.0-4.0	0.50	0.6-1.5	...	0.50	1.0	...	0.25	...	0.50	Remainder
F332.2	F132	AATD	...	Ingot	8.5-10.0	0.6	2.0-4.0	0.10	0.9-1.3	...	0.10	0.10	...	0.20	...	0.30	Remainder
Z332.0	Z132	AATD	...	P	11.0-13.0	1.2	1.5-3.0	0.50	0.50-1.5	...	0.50-1.5	1.0	...	0.25	...	0.50	Remainder
Z332.1	Z132	AATD	...	Ingot	11.0-13.0	0.9	1.5-3.0	0.50	0.6-1.5	...	0.50-1.5	1.0	...	0.25	...	0.50	Remainder

Registered alloys in the form of XXX.0 castings, XXX.1 ingot and XXX.2 ingot (continued)

AA number	Former designation	Registered By	Date	Prod-uct(e)	Si	Fe	Cu	Mn	Mg	Cr	Ni	Zn	Sn	Ti	Others(d) Each	Others(d) Total	Al min (e)
333.0	333	AATD	...	P	8.0-10.0	1.0	3.0-4.0	0.50	0.05-0.50	...	0.50	1.0	...	0.25	...	0.50	Remainder
333.1	333	AATD	...	Ingot	8.0-10.0	0.8	3.0-4.0	0.50	0.10-0.50	...	0.50	1.0	...	0.25	...	0.50	Remainder
A333.0	...	AATD	8/28/70	P	8.0-10.0	1.0	3.0-4.0	0.50	0.05-0.50	...	0.50	3.0	...	0.25	...	0.50	Remainder
A333.1	...	AATD	8/28/70	Ingot	8.0-10.0	0.8	3.0-4.0	0.50	0.10-0.50	...	...	3.0	...	0.25	...	0.50	Remainder
343.0	X443Z	MRCI	10/27/72	D	6.7-7.7	1.2	0.50-0.9	0.50	0.10	0.10	...	1.2-2.0	0.5	...	0.10	0.35	Remainder
343.1	X443Z	MRCI	10/27/72	Ingot	6.7-7.7	0.50-0.9	0.50-0.9	0.50	0.10	0.10	...	1.2-1.9	0.5	...	0.10	0.35	Remainder
354.0	354	AATD	...	P	8.6-9.4	0.20	1.6-2.0	0.10	0.40-0.6	...	...	0.10	...	0.20	0.05	0.15	Remainder
354.1	354	AATD	...	Ingot	8.6-9.4	0.15	1.6-2.0	0.10	0.45-0.6	...	...	0.10	...	0.20	0.05	0.15	Remainder
355.0	355	AATD	...	S&P	4.5-5.5	0.6(p)	1.0-1.5	0.50(p)	0.40-0.6	0.25	...	0.35	...	0.25	0.05	0.15	Remainder
355.1	355	AATD	...	Ingot	4.5-5.5	0.50(p)	1.0-1.5	0.50(p)	0.45-0.6	0.25	...	0.35	...	0.25	0.05	0.15	Remainder
355.2	355	AATD	...	Ingot	4.5-5.5	0.14-0.25	1.0-1.5	0.05	0.50-0.6	...	...	0.05	...	0.20	0.05	0.15	Remainder
C355.0	C355	AATD	...	S&P	4.5-5.5	0.20	1.0-1.5	0.10	0.40-0.6	...	...	0.10	...	0.20	0.05	0.15	Remainder
C355.1	...	Kaiser	6/4/74	Ingot	4.5-5.5	0.15	1.0-1.5	0.10	0.40-0.6	...	...	0.10	...	0.20	0.05	0.15	Remainder
C355.2	C355	AATD	...	Ingot	4.5-5.5	0.13	1.0-1.5	0.05	0.50-0.6	...	...	0.05	...	0.20	0.05	0.15	Remainder
356.0	356	AATD	...	S&P	6.5-7.5	0.6	0.25	0.35	0.20-0.40	...	...	0.35	...	0.25	0.05	0.15	Remainder
356.1	356	AATD	...	Ingot	6.5-7.5	0.50	0.25	0.35	0.25-0.40	...	...	0.35	...	0.25	0.05	0.15	Remainder
356.2	356	AATD	...	Ingot	6.5-7.5	0.13-0.25	0.10	0.05	0.30-0.40	...	...	0.05	...	0.20	0.05	0.15	Remainder
A356.0	A356	AATD	...	S&P	6.5-7.5	0.20	0.20	0.10	0.20-0.40	...	...	0.10	...	0.20	0.05	0.15	Remainder
A356.1	...	Kaiser	6/4/74	Ingot	6.5-7.5	0.15	0.20	0.05	0.30-0.40	...	...	0.05	...	0.20	0.05	0.15	Remainder
A356.2	A356	AATD	...	Ingot	6.5-7.5	0.12	0.10	0.05	0.17-0.25	...	...	0.05	...	0.20	0.05	0.15	Remainder
F356.0	...	Reynolds	10/20/71	S&P	6.5-7.5	0.20	0.20	0.10	0.17-0.25	...	...	0.10	...	0.20	0.05	0.15	Remainder
F356.2	...	Reynolds	10/20/71	Ingot	6.5-7.5	0.12	0.10	0.05	0.45-0.6	...	...	0.05	...	0.20	0.05	0.15	Remainder
357.0	357	AATD	...	S&P	6.5-7.5	0.15	0.05	0.03	0.45-0.6	...	...	0.05	...	0.20	0.05	0.15	Remainder
357.1	357	AATD	...	Ingot	6.5-7.5	0.12	0.05	0.03	0.40-0.7	...	...	0.05	...	0.20	0.05	0.15	Remainder
A357.0	A357	AATD	...	S&P	6.5-7.5	0.20	0.20	0.10	0.45-0.7	...	...	0.10	...	0.10-0.20	0.05(q)	0.15	Remainder
A357.2	A357	AATD	...	Ingot	6.5-7.5	0.12	0.10	0.05	0.40-0.6	0.20	...	0.05	...	0.10-0.20	0.03(q)	0.10	Remainder
B358.0	Tens-50	AATD	...	S&P	7.6-8.6	0.30	0.20	0.20	0.45-0.6	0.05	...	0.20	...	0.10-0.20	0.05(r)	0.15	Remainder
B358.2	Tens-50	AATD	...	Ingot	7.6-8.6	0.20	0.10	0.10	0.50-0.7	...	...	0.10	...	0.12-0.20	0.05(s)	0.15	Remainder
359.0	359	AATD	...	S&P	8.5-9.5	0.20	0.20	0.10	0.55-0.7	...	...	0.10	...	0.20	0.05	0.15	Remainder
359.1	359	AATD	...	Ingot	8.5-9.5	0.12	0.10	0.10	0.40-0.6	...	...	0.10	...	0.20	0.05	0.15	Remainder
360.0(t)	360	AATD	...	D	9.0-10.0	2.0	0.6	0.35	0.45-0.6	...	0.50	0.50	0.15	...	...	0.25	Remainder
360.2	360	AATD	...	Ingot	9.0-10.0	0.7-1.1	0.10	0.10	0.40-0.6	...	0.10	0.10	0.10	...	...	0.20	Remainder
A360.0(t)	A360	AATD	...	D	9.0-10.0	1.3	0.6	0.35	0.45-0.6	...	0.50	0.50	0.15	...	...	0.25	Remainder
A360.1(t)	A360	AATD	...	Ingot	9.0-10.0	1.0	0.6	0.35	0.45-0.6	...	0.50	0.40	0.15	...	...	0.25	Remainder
A360.2	A360	AATD	...	Ingot	9.0-10.0	0.6	0.10	0.05	0.15-0.40	...	...	0.05	...	...	0.05	0.15	Remainder
363.0	363	Kaiser	1/16/70	S&P	4.5-6.0	1.1	2.5-3.5	(u)	0.20-0.40	(u)	0.25	3.0-4.5	0.25	0.20	(v)	0.30	Remainder
363.1	363	Kaiser	1/16/60	Ingot	4.5-6.0	0.8	2.5-3.5	(u)	0.20-0.40	(u)	0.25	3.0-4.5	0.25	0.20	(v)	0.30	Remainder
364.0	364	AATD	...	D	7.5-9.5	1.5	0.20	0.10	0.25-0.50	...	0.15	0.15	...	...	0.05(w)	0.15	Remainder

Registered alloys in the form of XXX.0 castings, XXX.1 ingot and XXX.2 ingot (continued)

AA number	Former designation	Registered By	Registered Date	Product(e)	Si	Fe	Cu	Mn	Mg	Cr	Ni	Zn	Sn	Ti	Others(d) Each	Others(d) Total	Al min (e)
364.2	364	AATD	...	Ingot	7.5-9.5	0.7-1.1	0.20	0.10	0.25-0.40	0.25-0.50	0.15	0.15	0.15	...	0.05(w)	0.15	Remainder
380.0(t)	380	AATD	...	D	7.5-9.5	2.0	3.0-4.0	0.50	0.10	...	0.50	3.0	0.35	...	...	0.50	Remainder
380.2	380	AATD	...	Ingot	7.5-9.5	0.7-1.1	3.0-4.0	0.10	0.10	...	0.10	0.10	0.10	...	...	0.20	Remainder
A380.0(t)	A380	AATD	...	D	7.5-9.5	1.3	3.0-4.0	0.50	0.10	...	0.50	3.0	0.35	...	...	0.50	Remainder
A380.1(t)	A380	AATD	...	Ingot	7.5-9.5	1.0	3.0-4.0	0.50	0.10	...	0.50	2.9	0.35	...	...	0.50	Remainder
A380.2	A380	AATD	...	Ingot	7.5-9.5	0.6	3.0-4.0	0.10	0.10	...	0.10	0.10	...	...	0.05	0.15	Remainder
B380.0	A380	AATD	...	D	7.5-9.5	1.3	3.0-4.0	0.50	0.10	...	0.50	1.0	0.35	...	...	0.50	Remainder
B380.1	A380	AATD	...	Ingot	7.5-9.5	1.0	3.0-4.0	0.50	0.10	...	0.50	0.9	0.35	...	...	0.50	Remainder
383.0	...	AATD	...	D	9.5-11.5	1.3	2.0-3.0	0.50	0.10	...	0.30	3.0	0.15	...	...	0.50	Remainder
383.1	...	AATD	...	Ingot	9.5-11.5	0.6-1.0	2.0-3.0	0.50	0.10	...	0.30	2.9	0.15	...	...	0.50	Remainder
383.2	...	AATD	...	Ingot	9.5-11.5	0.6-1.0	2.0-3.0	0.10	0.10	...	0.10	0.10	0.10	...	...	0.20	Remainder
384.0	384	AATD	...	D	10.5-12.0	1.3	3.0-4.5	0.50	0.10	...	0.50	3.0	0.35	...	...	0.50	Remainder
384.1	384	AATD	...	Ingot	10.5-12.0	1.0	3.0-4.5	0.50	0.10	...	0.50	2.9	0.35	...	...	0.50	Remainder
384.2	384	AATD	...	Ingot	10.5-12.0	0.6-1.0	3.0-4.5	0.10	0.10	...	0.10	0.10	0.10	...	...	0.20	Remainder
A384.0	384	AATD	...	D	10.5-12.0	1.3	3.0-4.5	0.50	0.10	...	0.50	1.0	0.35	...	...	0.50	Remainder
A384.1	384	AATD	...	Ingot	10.5-12.0	1.0	3.0-4.5	0.50	0.10	...	0.50	0.9	0.35	...	...	0.50	Remainder
B384.0	384	Alcoa	1/21/70	D	11.0-13.0	2.0	2.0-4.0	0.50	0.30	...	0.50	3.0	0.30	...	...	0.50	Remainder
B384.1	384	Alcoa	1/21/70	Ingot	11.0-13.0	0.7-1.1	2.0-4.0	0.10	0.30	...	0.50	2.9	0.30	...	...	0.20	Remainder
390.0	390	AATD	...	D	16.0-18.0	1.3	4.0-5.0	0.10	0.45-0.65	...	...	0.10	...	0.20	0.10	0.20	Remainder
390.2	390	AATD	...	Ingot	16.0-18.0	0.6-1.0	4.0-5.0	0.10	0.50-0.65	...	...	0.10	...	0.20	0.10	0.20	Remainder
A390.0	A390	AATD	...	S&P	16.0-18.0	0.50	4.0-5.0	0.10	0.45-0.65	...	...	0.10	...	0.20	0.10	0.20	Remainder
A390.1	A390	AATD	...	Ingot	16.0-18.0	0.40	4.0-5.0	0.10	0.50-0.65	...	...	0.10	...	0.20	0.10	0.20	Remainder
392.0	392	AATD	...	D	18.0-20.0	1.5	0.40-0.8	0.20-0.6	0.8-1.2	...	0.50	0.50	0.30	0.20	0.15	0.50	Remainder
392.1	392	AATD	...	Ingot	18.0-20.0	1.1	0.40-0.8	0.20-0.6	0.9-1.2	...	0.50	0.40	0.30	0.20	0.15	0.50	Remainder
393.0	Vanasil	USCO	...	SP&D	21.0-23.0	1.3	0.7-1.1	0.10	0.7-1.3	...	2.0-2.5	0.10	...	0.10-0.20	0.05(x)	0.15	Remainder
393.1	Vanasil	USCO	...	Ingot	21.0-23.0	1.0	0.7-1.1	0.10	0.8-1.3	...	2.0-2.5	0.10	...	0.10-0.20	0.05(x)	0.15	Remainder
393.2	Vanasil	USCO	...	Ingot	21.0-23.0	0.8	0.7-1.1	0.10	0.8-1.3	...	2.0-2.5	0.10	...	0.10-0.20	0.05(x)	0.15	Remainder
408.2(y)	...	Reynolds	9/24/73	Ingot	8.5-9.5	0.6-1.3	0.10	0.10	...	...	...	0.10	0.15	...	0.10	0.20	Remainder
409.2(y)	...	Reynolds	9/24/73	Ingot	9.0-10.0	0.6-1.3	0.10	0.10	...	...	...	0.10	0.10	...	0.10	0.20	Remainder
411.2(y)	...	Reynolds	9/24/73	Ingot	10.0-12.0	0.6-1.3	0.20	0.10	...	...	...	0.10	0.15	...	0.10	0.20	Remainder
413.0(t)	13	AATD	...	D	11.0-13.0	2.0	1.0	0.35	0.10	...	0.50	0.50	0.15	...	...	0.25	Remainder
413.2	13	AATD	...	Ingot	11.0-13.0	0.7-1.1	0.10	0.10	0.07	...	0.10	0.10	0.05	...	...	0.20	Remainder
A413.0(t)	A13	AATD	...	D	11.0-13.0	1.3	1.0	0.35	0.10	...	0.50	0.50	0.15	...	...	0.25	Remainder
A413.1(t)	A13	AATD	...	Ingot	11.0-13.0	1.0	1.0	0.35	0.10	...	0.50	0.40	0.15	...	...	0.25	Remainder
A413.2	A13	AATD	...	Ingot	11.0-13.0	0.6	0.10	0.05	0.05	...	0.05	0.05	0.05	...	...	0.10	Remainder
443.0	43	AATD	...	S	4.5-6.0	0.8	0.6	0.50	0.05	0.25	...	0.50	...	0.25	...	0.35	Remainder
443.1	43	AATD	...	Ingot	4.5-6.0	0.6	0.6	0.50	0.05	0.25	...	0.50	...	0.25	...	0.35	Remainder
443.2	43	AATD	...	Ingot	4.5-6.0	0.6	0.10	0.10	0.05	...	...	0.10	...	0.20	0.05	0.15	Remainder

Aluminum Association/1771

Registered alloys in the form of XXX.0 castings, XXX.1 ingot and XXX.2 ingot (continued)

AA number	Former designation	Registered By	Registered Date	Product(e)	Si	Fe	Cu	Mn	Mg	Cr	Ni	Zn	Sn	Ti	Others(d) Each	Others(d) Total	Al min (e)
A143.0	43(0.30 max Cu)	AATD	...	S	4.5-6.0	0.8	0.30	0.50	0.05	0.25	...	0.50	...	0.25	...	0.35	Remainder
A443.1	43(0.30 max Cu)	AATD	...	Ingot	4.5-6.0	0.6	0.30	0.50	0.05	0.25	...	0.50	...	0.25	...	0.35	Remainder
B443.0	43(0.15 max Cu)	AATD	...	S&P	4.5-6.0	0.8	0.15	0.35	0.05	...	...	0.35	...	0.25	0.05	0.15	Remainder
B443.1	43(0.15 max Cu)	AATD	...	Ingot	4.5-6.0	0.6	0.15	0.35	0.05	...	...	0.35	...	0.25	0.05	0.15	Remainder
C443.0	A43	AATD	...	D	4.5-6.0	2.0	0.6	0.35	0.10	...	0.50	0.50	0.15	...	...	0.25	Remainder
C443.1	A43	AATD	...	Ingot	4.5-6.0	1.0	0.6	0.35	0.10	...	0.50	0.40	0.15	...	...	0.25	Remainder
C443.2	A43	Reynolds	11/5/74	Ingot	4.5-6.0	0.7-1.1	0.10	0.10	0.05	...	...	0.10	...	0.25	0.05	0.15	Remainder
444.0	...	Reynolds	9/24/73	S&P	6.5-7.5	0.6	0.25	0.35	0.10	...	...	0.35	...	0.25	0.05	0.15	Remainder
444.2	...	Reynolds	9/24/73	Ingot	6.5-7.5	0.13-0.25	0.10	0.05	0.05	...	...	0.05	...	0.20	0.05	0.15	Remainder
A444.0	A344	AATD	...	P	6.5-7.5	0.20	0.10	0.10	0.05	...	...	0.10	...	0.20	0.05	0.15	Remainder
A444.1	...	Kaiser	6/4/74	Ingot	6.5-7.5	0.15	0.10	0.10	0.05	...	...	0.05	...	0.20	0.05	0.15	Remainder
A444.2	A344	AATD	...	Ingot	6.5-7.5	0.12	0.05	0.05	0.05	...	...	0.05	...	0.20	0.05	0.15	Remainder
B444.2(y)	...	Reynolds	9/24/73	Ingot	6.5-7.5	0.6-1.3	0.10	0.10	...	...	...	0.10	...	...	0.10	0.20	Remainder
514.0	214	AATD	...	S	0.35	0.50	0.15	0.35	3.5-4.5	...	...	0.15	...	0.25	0.05	0.15	Remainder
514.1	214	AATD	...	Ingot	0.35	0.40	0.15	0.35	3.6-4.5	...	...	0.15	...	0.25	0.05	0.15	Remainder
514.2	214	AATD	...	Ingot	0.30	0.30	0.10	0.10	3.6-4.5	...	...	0.10	...	0.20	0.05	0.15	Remainder
A514.0	A214	AATD	...	P	0.30	0.40	0.10	0.30	3.5-4.5	...	...	1.4-2.2	...	0.20	0.05	0.15	Remainder
A514.2	A214	AATD	...	Ingot	0.30	0.30	0.10	0.10	3.6-4.5	...	...	1.4-2.2	...	0.20	0.05	0.15	Remainder
B514.0	B214	AATD	...	S	1.4-2.2	0.6	0.35	0.8	3.5-4.5	...	...	0.35	...	0.25	0.05	0.15	Remainder
B514.2	B214	AATD	...	Ingot	1.4-2.2	0.30	0.10	0.10	3.6-4.5	0.25	...	0.10	...	0.20	0.05	0.15	Remainder
F514.0	F214	AATD	...	S	0.30-0.7	0.50	0.15	0.35	3.5-4.5	...	...	0.15	...	0.25	0.05	0.15	Remainder
F514.1	F214	AATD	...	Ingot	0.30-0.7	0.40	0.15	0.35	3.6-4.5	...	...	0.15	...	0.25	0.05	0.15	Remainder
F514.2	F214	AATD	...	Ingot	0.30-0.7	0.30	0.10	0.10	3.6-4.5	...	...	0.10	...	0.20	0.05	0.15	Remainder
L514.0	L214	Reynolds	1/2/70	D	0.50-1.0	1.3	0.20	0.40-0.6	2.5-4.0	...	...	0.10	...	...	0.05	0.15	Remainder
L514.2	L214	Reynolds	1/2/70	Ingot	0.50-1.0	0.6-1.0	0.10	0.40-0.6	2.7-4.0	...	...	0.05	...	0.25	0.05	0.15	Remainder
518.0	218	AATD	...	D	0.35	1.8	0.25	0.35	7.5-8.5	...	0.15	0.15	0.15	...	...	0.25	Remainder
518.1	218	AATD	...	Ingot	0.35	1.0	0.25	0.35	7.8-8.5	...	0.15	0.15	0.15	...	...	0.25	Remainder
518.2	218	AATD	...	Ingot	0.25	0.7	0.10	0.10	7.6-8.5	...	0.05	0.10	0.05	...	...	0.10	Remainder
520.0	220	AATD	...	S	0.25	0.30	0.25	0.15	9.5-10.6	...	...	0.15	...	0.25	0.05	0.15	Remainder
520.2	220	AATD	...	Ingot	0.15	0.20	0.20	0.10	9.6-10.6	...	...	0.10	...	0.20	0.05	0.15	Remainder
535.0	Almag 35	AATD	...	S	0.15	0.15	0.05	0.10-0.25	6.2-7.5	...	...	...	...	0.10-0.25	0.05(z)	0.15	Remainder
535.2	Almag 35	AATD	...	Ingot	0.10	0.10	0.05	0.10-0.25	6.6-7.5	...	...	...	...	0.10-0.25	0.05(z)	0.15	Remainder
A535.0	A218	AATD	...	S	0.20	0.20	0.10	0.10-0.25	6.5-7.5	...	...	...	...	0.25	0.05	0.15	Remainder
A535.1	A218	AATD	...	Ingot	0.20	0.15	0.10	0.10-0.25	6.6-7.5	...	...	...	...	0.25	0.05	0.15	Remainder
B535.0	B218	AATD	...	S	0.15	0.15	0.05	0.05	6.5-7.5	...	...	...	...	0.10-0.25	0.05	0.15	Remainder
B535.2	B218	AATD	...	Ingot	0.10	0.12	0.05	0.05	6.6-7.5	...	...	...	...	0.10-0.25	0.05	0.15	Remainder
705.0	603, Ternalloy 5	AATD	...	S&P	0.20	0.8	0.20	0.40-0.6	1.4-1.8	0.20-0.40	...	2.7-3.3	...	0.25	0.05	0.15	Remainder
705.1	603, Ternalloy 5	AATD	...	Ingot	0.20	0.6	0.20	0.40-0.6	1.5-1.8	0.20-0.40	...	2.7-3.3	...	0.25	0.05	0.15	Remainder

Registered alloys in the form of XXX.0 castings, XXX.1 ingot and XXX.2 ingot (continued)

AA number	Former designation	Registered By	Registered Date	Product(e)	Si	Fe	Cu	Mn	Mg	Cr	Ni	Zn	Sn	Ti	Others(d) Each	Others(d) Total	Al min (e)
707.0	607, Ternalloy 7	AATD	...	S&P	0.20	0.8	0.20	0.40-0.6	1.8-2.4	0.20-0.40	...	4.0-4.5	...	0.25	0.05	0.15	Remainder
707.1	607, Ternalloy 7	AATD	...	Ingot	0.20	0.6	0.20	0.40-0.6	1.9-2.4	0.20-0.40	...	4.0-4.5	...	0.25	0.05	0.15	Remainder
A712.0	A612	AATD	...	S	0.15	0.50	0.35-0.65	0.05	0.6-0.8	...	...	6.0-7.0	...	0.25	0.05	0.15	Remainder
A712.1	A612	AATD	...	Ingot	0.15	0.40	0.35-0.65	0.05	0.65-0.8	...	...	6.0-7.0	...	0.25	0.05	0.15	Remainder
C712.0	C612	Alcoa	...	P	0.30	0.7-1.4	0.35-0.65	0.05	0.25-0.45	...	...	6.0-7.0	...	0.20	0.05	0.15	Remainder
C712.1	C612	Alcoa	...	Ingot	0.30	0.7-1.1	0.35-0.65	0.05	0.30-0.45	...	...	6.0-7.0	...	0.20	0.05	0.15	Remainder
D712.0	D612, 40E	AATD	...	S	0.30	0.50	0.25	0.10	0.50-0.65	0.40-0.6	...	5.0-6.5	...	0.15-0.25	0.05	0.20	Remainder
D712.2	D612, 40E	AATD	...	Ingot	0.15	0.40	0.25	0.10	0.50-0.65	0.40-0.6	...	5.0-6.5	...	0.15-0.25	0.05	0.20	Remainder
713.0	613, Tenzaloy	AATD	...	S&P	0.25	1.1	0.40-1.0	0.6	0.20-0.50	0.35	0.15	7.0-8.0	...	0.25	0.10	0.25	Remainder
713.1	613, Tenzaloy	AATD	...	Ingot	0.25	0.8	0.40-1.0	0.6	0.25-0.50	0.35	0.15	7.0-8.0	...	0.25	0.10	0.25	Remainder
771.0	Precedent 71A	USCO	...	S	0.15	0.15	0.10	0.10	0.8-1.0	0.06-0.20	...	6.5-7.5	...	0.10-0.20	0.05	0.15	Remainder
771.2	Precedent 71A	USCO	...	Ingot	0.10	0.10	0.10	0.10	0.85-1.0	0.06-0.20	...	6.5-7.5	...	0.10-0.20	0.05	0.15	Remainder
B771.0	Precedent 71B	USCO	...	S	0.15	0.15	0.10	0.10	0.6-0.8	0.06-0.20	...	6.0-7.0	...	0.10-0.20	0.05	0.15	Remainder
B771.2	Precedent 71B	USCO	...	Ingot	0.10	0.10	0.10	0.10	0.65-0.8	0.06-0.20	...	6.0-7.0	...	0.10-0.20	0.05	0.15	Remainder
850.0	750	AATD	...	S&P	0.7	0.7	0.7-1.3	0.10	0.10	...	0.7-1.3	...	5.5-7.0	0.20	...	0.30	Remainder
850.1	750	AATD	...	Ingot	0.7	0.50	0.7-1.3	0.10	0.10	...	0.7-1.3	...	5.5-7.0	0.20	...	0.30	Remainder
A850.0	A750	AATD	...	S&P	2.0-3.0	0.7	0.7-1.3	0.10	0.10	...	0.30-0.7	...	5.5-7.0	0.20	...	0.30	Remainder
A850.1	A750	AATD	...	Ingot	2.0-3.0	0.50	0.7-1.3	0.10	0.10	...	0.30-0.7	...	5.5-7.0	0.20	...	0.30	Remainder
B850.0	B750	AATD	...	S&P	0.40	0.7	1.7-2.3	0.10	0.6-0.9	...	0.9-1.5	...	5.5-7.0	0.20	...	0.30	Remainder
B850.1	B750	AATD	...	Ingot	0.40	0.50	1.7-2.3	0.10	0.7-0.9	...	0.9-1.5	...	5.5-7.0	0.20	...	0.30	Remainder
XC850.0	XC750	Alcoa	...	S&P	5.5-6.5	0.7	3.0-4.0	0.50	...	...	...	...	5.5-7.0	0.20	...	0.30	Remainder
XC850.2	XC750	Alcoa	...	Ingot	5.5-6.5	0.50	3.0-4.0	0.10	...	...	...	...	5.5-7.0	0.20	...	0.30	Remainder

(a) Composition in percent maximum unless shown as a range or a minimum. Standard limits for alloying elements and impurities are expressed to the following places: Less than 1/100 percent, 0.000X; 1/1000 to 1/100 percent, 0.00X; 1/100 to 1/10 percent alloys and unalloyed aluminum not made by a refining process, 0.0X; 1/10 through 1/2 percent, 0.XX; over 1/2 percent, 0.X, X.X, etc. (Magnesium percent for some alloys are exceptions to this rule.) (b) For purposes of determining conformance to these limits, an observed value or a calculated value obtained from analysis is rounded off to the nearest unit in the last right-hand place of figures used in expressing the specified limit, in accordance with the following: American National Standard Rules for Rounding Off Numerical Values (ANSI Z25.1). When the figure next beyond the last figure or place to be retained is less than 5, the figure in the last place retained should be kept unchanged. When the figure next beyond the last figure or place to be retained is greater than 5, the figure in the last place retained should be increased by 1. When the figure next beyond the last figure or place to be retained is 5 and (1) there are no figures, or only zeroes, beyond this 5, the figure in the last place to be retained is, if even, it should be kept unchanged; (2) if the 5 next beyond the figure in the last place to be retained is followed by any figures other than zero, the figure in the last place retained should be increased by 1, whether odd or even. (c) D = die casting; P = permanent mold; S = sand. (d) Analysis is regularly made only for the elements for which specific limits are shown, except for minimum purities of 99.00 percent. If, however, the presence of other elements is suspected to be, or in the course of routine analysis is indicated to be in excess of the specified limits, further analysis is made to determine that these other elements are not in excess of the amount specified. (e) The aluminum content for unalloyed aluminum not made by a refining process is the difference between 100.00 percent and the sum of all other metallic elements present in amounts of 0.010 percent or more each, expressed to the second decimal before determining the sum. (f) Rated minimum conductivity characteristic is based on established relations between electrical conductivity and metal composition. (g) Aluminum Association Technical The rating of ingot metal for minimum conductivity characteristic is based on established relations between electrical conductivity and metal composition. Division. (h) Manganese + chromium + titanium + vanadium 0.025 max. (i) Iron/silicon ratio 2.5 min. (j) Iron/silicon ratio 1.5 min. (l) Silver 0.40-1.0. (m) Titanium + zirconium 0.50 max. (n) Antimony 0.20-0.30; cobalt 0.20-0.30; zirconium 0.10-0.30. (o) Vanadium 0.05-0.15; zirconium 0.10-0.25. (p) If iron exceeds 0.45, manganese content shall not be less than one-half iron content. (q) Beryllium 0.04-0.07. (r) Beryllium 0.10-0.30. (s) A360.1, A380.1 and A413.1 ingot is used to produce 360.0 and A360.0; 380.0 and A380.0; 413.0 and A413.0 castings, respectively. (u) Manganese anc chromium 0.8 max. total. (v) Lead 0.25 max. (w) Beryllium 0.02-0.04. (x) Vanadium 0.08-0.15. (y) B444.2, 408.2, 409.2 and 411.2 are used to coat steel. (z) Beryllium 0.003-0.007, boron 0.002 max.

Wrought Alloys

Coppers

Copper alloy no.	Desig- nation	Description	Cu (incl. Ag) (% min)	Ag % min	Ag troy oz min	As	Sb	P	Te	Other named elements
C10100(a)	OFE	Oxygen free electronic	99.99(b)	...	...	(c)	(c)	.003	.0010	(c)
C10200(a)	OF	Oxygen free	99.95	...	...	...	...	...	...	...
C10300	OFXLP	...	99.95(d)	...	...	...	...	.001-.005	...	...
C10400(a)	OFS	Oxygen free with Ag	99.95	.027	8	...	...	...	...	...
C10500(a)	OFS	Oxygen free with Ag	99.95	.034	10	...	...	...	...	...
C10700(a)	OFS	Oxygen free with Ag	99.95	.085	25	...	...	...	...	...
C10800	OFLP	...	99.95(d)	...	...	...	...	.005-.012	...	...
C10920	...	...	99.90	...	...	...	...	...	...	.02 O
C10930	...	...	99.90	.044	13	...	...	...	...	.02 O
C10940	...	...	99.90	.085	25	...	...	...	...	.02 O
C11000(a)	ETP	Electrolytic tough pitch	99.90	...	...	...	...	...	...	...
	FRHC	Fire-refined high conductivity	99.90	...	...	...	...	...	...	...
	CRTP	Chemically refined tough pitch	99.90	...	...	...	...	...	...	...
C11100(a)	...	Electrolytic tough pitch anneal resistant	99.90	...	...	...	...	...	...	(e)
C11300(a)(f)	STP	Tough pitch with Ag	99.90	.027	8	...	...	...	...	...
C11400(a)(f)	STP	Tough pitch with Ag	99.90	.034	10	...	...	...	...	...
C11500(a)(f)	STP	Tough pitch with Ag	99.90	.054	16	...	...	...	...	...
C11600(a)(f)	STP	Tough pitch with Ag	99.90	.085	25	...	...	...	...	...
C11700	...	...	99.9(g)	...	...	...	...	.04	...	.004-.02 B
C12000	DLP	Phosphorus deoxidized, low residual phosphorus	99.90	...	...	...	...	.004-.012	...	...
C12100	...	...	99.90	.014	4	...	...	.005-.012	...	...
C12200(h)	DHP	Phosphorus deoxidized, high residual phosphorus	99.9	...	...	...	...	.015-.040	...	...
C12300	...	...	99.90	.014	4	...	...	.015-.040	...	...
C12500(i)	FRTP	Fire refined tough pitch	99.88	...	...	.012	.003	...	.025(j)	.050 Ni, .003 Bi, .004 Pb
C12700(i)	FRSTP	Fire refined tough pitch with Ag	99.88	.027	8	.012	.003	...	.025(j)	.050 Ni, .003 Bi, .004 Pb
C12800(i)	FRSTP	Fire refined tough pitch with Ag	99.88	.034	10	.012	.003	...	.025(j)	.050 Ni, .003 Bi, .004 Pb
C12900(i)	FRSTP	Fire refined tough pitch with Ag	99.88	.054	16	.012	.003	...	.025(j)	.050 Ni, .003 Bi, .004 Pb
C13000(i)	FRSTP	Fire refined tough pitch with Ag	99.88	.085	25	.012	.003	...	.025(j)	.050 Ni, .003 Bi, .004 Pb
C14200	DPA	Phosphorus deoxidized arsenical	99.4	...	...	.15-.50	...	.015-.040	...	...
C14300	...	Cadmium copper, deoxidized	99.90(k)	...	...	...	...	...	...	.05-.15 Cd
C14310	...	...	99.90(k)	...	...	...	...	...	...	.10-.30 Cd
C14500(l)	DPTE	Phosphorus deoxidized tellurium bearing	99.90(m)	...	...	...	...	.004-.012(n)	.40-.6	...
C14700	...	Sulfur bearing	99.90(o)	...	...	...	...	...	...	.20-.50 S
C14710	...	...	99.90(o)(p)	...	...	...	...	.010-.030	...	.20-.50 S, .10 Pb
C14720	...	...	99.50(o)(p)	...	...	...	...	.010-.030	...	.05-.15 S, .05 Pb
C15000	...	Zirconium copper	99.80	...	...	...	...	...	...	.10-.20 Zr
C15500	...	...	99.75	.027-.10	8-30	...	...	.040-.080	...	.08-.13 Mg
C15710	...	...	99.80(q)	...	...	...	...	...	...	.15-.25 Al$_2$O$_3$, .01 Fe, .01 Pb, .04 O
C15720	...	...	99.80(q)	...	...	...	...	...	...	.35-.45 Al$_2$O$_3$, .01 Fe, .01 Pb, .04 O
C15735	...	...	99.80(q)	...	...	...	...	...	...	.65-.75 Al$_2$O$_3$, .01 Fe, .01 Pb, .04 O

High Copper Alloys

Copper alloy no.	Previous trade name	Cu (incl. Ag) + elements with specific limits (r) (% min)	Composition, per cent maximum (unless shown as a range or minimum)									Other named elements
			Fe	Sn	Ni	Co	Cr	Si	Be	Pb	Cd	
C16200	Cadmium copper	99.8	.02	...	...	...	...	...	...	...	.7-1.2	...
C16500	...	99.8	.02	.50-.7	...	...	...	...	...	...	.6-1.0	...
C17000	Beryllium copper	99.5(b)	(s)	...	(s)	(s)	...	...	1.60-1.79	...	...	...
C17200	Beryllium copper	99.5(b)	(s)	...	(s)	(s)	...	...	1.80-2.00	...	...	...
C17300	...	99.5(b)	(s)	...	(s)	(s)	...	...	1.80-2.00	.20-.6	...	...
C17500	Beryllium copper	99.5(b)	.10	...	...	2.4-2.7	...	...	.40-.17	...	...	...
C17600	...	99.5(b)	.10	...	...	1.4-1.7	...	...	.25-.50	...	...	.9-1.1 Ag
C17700	...	99.5(b)	.10	...	...	2.4-2.7	...	...	.40-.7	...	...	.40-.6 Te
C18200	Chromium copper	99.5	.10	...	...	...	.6-1.2	.10	...	.05	...	...
C18400	Chromium copper	99.8	.15	...	...	...	.40-1.2	.10	...	...	...	.005 As, .005 Ca, .05 Li, .05 P, .7 Zn
C18500	Chromium copper	99.8	...	...	...	...	.40-1.0	...	...	.015	...	.04 P, .08-.12 Ag
C18700	...	99.9	...	...	...	...	...	...	...	.8-1.5	...	...
C18900	...	99.9	...	.6-.9	...	...	...	.15-.40	...	.02	...	.05 P, .01 Al, .10-.30 Mn, .10 Zn
C19000	...	99.5	.10	...	.9-1.3	...	...	...	...	.05	...	.8 Zn, .15-.35 P
C19100	...	99.5	.20	...	.9-1.3	...	...	...	...	.10	...	.50 Zn, .35-.6 Te, .15-.35 P

Copper alloy no.	Composition, per cent maximum (unless shown as a range or minimum)							
	Cu	Fe	Sn	Zn	Al	Pb	P	Co
C19200	98.7 min (t)	.8-1.2	...	...	...	...	.01-.04	...
C19400	97.0 min (t)	2.1-2.6	...	.05-.20	...	.03	.015-.15	...
C19500	96.0 min (t)	1.3-1.8	.40-.7	.20	.02	.02	.08-.12	.6-1.1

Copper alloy no.	Cu (incl. Ag) + elements with specific limits (r) (% min)	Composition, per cent maximum (unless shown as a range or minimum)						
		Fe	Sn	Zn	Al	Pb	P	Co
C19600	99.7 min	.9-1.2	...	.35	...	...	.25-.35	...

Copper–Zinc Alloys (Brasses)

Copper alloy no.	Previous trade name	Composition, per cent maximum (unless shown as a range or minimum)					
		Cu	Pb	Fe	Zn(t)	P	Other named elements
C20500	...	97.0-98.0	.02	.05	Rem.	...	...
C21000	Gilding, 95%	94.0-96.0	.05	.05	Rem.	...	...
C22000	Commercial bronze, 90%	89.0-91.0	.05	.05	Rem.	...	...
C22600	Jewelry bronze, 87½%	86.0-89.0	.05	.05	Rem.	...	...
C23000	Red brass, 85%	84.0-86.0	.05	.05	Rem.	...	...
C23030	...	83.5-85.5	.05	.05	Rem.	...	.20-.40 Si
C23400	...	81.0-84.0	.05	.05	Rem.	...	...
C24000	Low brass, 80%	78.5-81.5	.05	.05	Rem.	...	...
C25000	...	74.0-76.0	.05	.05	Rem.	...	...
C26000	Cartridge brass, 70%	68.5-71.5	.07	.05	Rem.	...	...
C26100	...	68.5-71.5	.05	.05	Rem.	.02-.05	...
C26200	...	67.0-70.0	.07	.05	Rem.	...	...
C26800	Yellow brass, 66%	64.0-68.5	.15	.05	Rem.	...	...
C27000	Yellow brass, 65%	63.0-68.5	.10	.07	Rem.	...	...
C27200	...	62.0-65.0	.07	.07	Rem.	...	...
C27400	Yellow brass, 63%	61.0-64.0	.10	.05	Rem.	...	...
C28000	Muntz metal, 60%	59.0-63.0	.30	.07	Rem.	...	...
C28200	...	58.0-61.0	.03	.05	Rem.	.12-.22	.005 Al(u), .05 Sn

Copper–Zinc–Lead Alloys (Leaded Brasses)

Copper alloy no.	Previous trade name	Composition, per cent maximum (unless shown as a range or minimum)					
		Cu	Pb	Fe	Sn	Zn(t)	Other named elements
C31400	Leaded commercial bronze	87.5-90.5	1.3-2.5	.10	...	Rem.	.7 Ni
C31600	Leaded commercial bronze (nickel bearing)	87.5-90.5	1.3-2.5	.10	...	Rem.	.7-1.2 Ni, .04-.10 P
C32000	Leaded red brass	83.5-86.5	1.5-2.2	.10	...	Rem.	.25 Ni
C33000	Low leaded brass (tube)	65.0-68.0	.20-.8(v)	.07	...	Rem.	...
C33100	...	65.0-68.0	.7-1.2	.06	...	Rem.	...
C33200	High leaded brass (tube)	65.0-68.0	1.3-2.0	.07	...	Rem.	...
C33500	Low leaded brass	62.5-66.5	.30-.8	.10	...	Rem.	...
C34000	Medium leaded brass, 64½%	62.5-66.5	.8-1.4	.10	...	Rem.	...
C34200	High leaded brass, 64½%	62.5-66.5	1.5-2.5	.10	...	Rem.	...
C34400	...	62.0-66.0	.50-1.0	.10	...	Rem.	...
C34500	...	62.0-64.0	1.5-2.8	.10	...	Rem.	...
C34700	...	62.5-64.5	1.0-1.8	.10	...	Rem.	...
C34800	...	61.5-63.5	.40-.8	.10	...	Rem.	...
C34900	...	61.0-64.0	.10-.50	.10	...	Rem.	...
C35000	Medium leaded brass, 62%	59.0-64.0(w)	.8-1.4	.10	...	Rem.	...
C35300	High leaded brass, 62%	59.0-64.5(w)	1.3-2.3	.10	...	Rem.	...
C35600	Extra high leaded brass	59.0-64.5(x)	2.0-3.0	.10	...	Rem.	...
C36000	Free cutting brass	60.0-63.0	2.5-3.7	.35	...	Rem.	...
C36200	...	60.0-63.0	3.5-4.5	.15	...	Rem.	...
C36500	Leaded muntz metal, uninhibited	58.0-61.0	.40-.9	.15	.25	Rem.	...
C36600	Leaded muntz metal, arsenical	58.0-61.0	.40-.9	.15	.25	Rem.	.02-.10 As
C36700	Leaded muntz metal, antimonial	58.0-61.0	.40-.9	.15	.25	Rem.	.02-.10 Sb
C36800	Leaded muntz metal, phosphorized	58.0-61.0	.40-.9	.15	.25	Rem.	.02-.10 P
C37000	Free cutting muntz, metal	59.0-62.0	.9-1.4	.15	...	Rem.	...
C37100	...	58.0-62.0	.6-1.2	.15	...	Rem.	...
C37700	Forging brass	58.0-61.0	1.5-2.5	.30	...	Rem.	...
C37800	...	57.0-60.0	1.0-2.5	.30	...	Rem.	...
C38000	...	55.0-60.0	1.5-2.5	.35	.30	Rem.	.50 Al
C38500	Architectural bronze	55.0-60.0	2.0-3.8	.35	...	Rem.	...
C38590	...	56.5-60.0	2.0-3.5	.35	...	Rem.	...
C38600	...	56.0-59.0	2.5-4.5	.35	...	Rem.	.02 Sb

Copper–Zinc–Tin Alloys (Tin Brasses)

Copper alloy no.	Previous trade name	Composition, per cent maximum (unless shown as a range or minimum)								Other named elements
		Cu	Pb	Fe	Sn	Zn(t)	P	As	Sb	
C40500	...	94.0-96.0	.05	.05	.7-1.3	Rem.	...	...	...	...
C40800	...	94.0-96.0	.05	.05	1.8-2.2	Rem.	...	...	...	...
C41000	...	91.0-93.0	.05	.05	2.0-2.8	Rem.	...	...	...	...
C41100	...	89.0-92.0	.10	.05	.30-.7	Rem.	...	...	...	...
C41300	...	89.0-93.0	.10	.05	.7-1.3	Rem.	...	...	...	...
C41500	...	89.0-93.0	.10	.05	1.5-2.2	Rem.	...	...	...	...
C42000	...	88.0-91.0	...	...	1.5-2.0	Rem.	.25	...	...	...
C42100	...	87.5-89.0	.05	.05	2.2-3.0	Rem.	.35	...	...	.15-.35 Mn
C42200	...	86.0-89.0	.05	.05	.8-1.4	Rem.	.35	...	...	...
C42500	...	87.0-90.0	.05	.05	1.5-3.0	Rem.	.35	...	...	...
C43000	...	84.0-87.0	.10	.05	1.7-2.7	Rem.	...	...	...	...
C43200	...	85.0-88.0	.05	.05	.40-.6	Rem.	.35	...	...	...
C43400	...	84.0-86.0	.05	.05	.50-1.0	Rem.	...	...	...	...
C43500	...	79.0-83.0	.10	.05	.6-1.2	Rem.	...	...	...	...
C43600	...	80.0-83.0	.05	.05	.20-.50	Rem.	...	...	...	...
C44300	Admiralty, arsenical	70.0-73.0	.07	.06	.9-1.2(y)	Rem.	...	.02-.10	...	...
C44400	Admiralty, antimomial	70.0-73.0	.07	.06	.9-1.2(y)	Rem.	...	...	.02-.10	...
C44500	Admiralty, phosphorized	70.0-73.0	.07	.06	.9-1.2(y)	Rem.	.02-.10	...	...	...
C46200	Naval brass, 63½%	62.0-65.0	.20	.10	.50-1.0	Rem.	...	...	...	...
C46400	Naval brass, uninhibited	59.0-62.0	.20	.10	.50-1.0	Rem.	...	...	...	...
C46500	Naval brass, arsenical	59.0-62.0	.20	.10	.50-1.0	Rem.	...	.02-.10	...	...
C46600	Naval brass, antimonial	59.0-62.0	.20	.10	.50-1.0	Rem.	...	...	.02-.10	...
C46700	Naval brass, phosphorized	59.0-62.0	.20	.10	.50-1.0	Rem.	.02-.10	...	...	...
C47000	Naval brass welding and brazing rod	57.0-61.0	.05	...	.25-1.0	Rem.	...	...	...	.01 Al
C47600	...	86.0-88.0	1.8-2.2	.05	1.8-2.2	Rem.	.03-.07	...	...	.05-.15 Mn
C48200	Naval brass, medium leaded	59.0-62.0	.40-1.0	.10	.50-1.0	Rem.	...	...	...	...
C48500	Naval brass, high leaded	59.0-62.0	1.3-2.2	.10	.50-1.0	Rem.	...	...	...	...

Copper–Tin Alloys (Phosphor Bronzes)

Copper alloy no.	Previous trade name	Cu + Sn + P (% min)	Composition, per cent maximum (unless shown as a range or minimum)					
			Pb	Fe	Sn	Zn	P	Al
C50100	...	99.5	.05	.05	.50-.8	...	.01-.05	...
C50200	...	99.5	.05	.10	1.0-1.5	...	.04	...
C50500	Phosphor bronze, 1.25% E	99.5	.05	.10	1.0-1.7	.30	.03-.35	...
C50700	...	99.5	.05	.10	1.5-2.0	...	.30	...
C50800	...	99.5	.05	.10	2.6-3.4	...	.01-.07	...
C50900	...	99.5	.05	.10	2.5-3.8	.30	.03-.30	...
C51000	Phosphor bronze, 5% A	99.5	.05	.10	4.2-5.8	.30	.03-.35	...
C51100	...	99.5	.05	.10	3.5-4.9	.30	.03-.35	...
C51800	Phosphor bronze	99.5	.02	...	4.0-6.0	...	.10-.35	.01
C51900	...	99.5	.05	.10	5.0-7.0	.30	.03-.35	...
C52100	Phosphor bronze, 8% C	99.5	.05	.10	7.0-9.0	.20	.03-.35	...
C52400	Phosphor bronze, 10% D	99.5	.05	.10	9.0-11.0	.20	.03-.35	...

Copper–Tin–Lead Alloys (Leaded Phosphor Bronzes)

Copper alloy no.	Previous trade name	Cu + Sn + P + Pb (% min)	Composition, per cent maximum (unless shown as a range or minimum)				
			Pb	Fe	Sn	Zn	P
C53200	Phosphor bronze B	99.5	2.5-4.0	.10	4.0-5.5	.20	.03-.35
C53400	Phosphor bronze B-1	99.5	.8-1.2	.10	3.5-5.8	.30	.03-.35
C54400	Phosphor bronze B-2	99.5(z)	3.5-4.5	.10	3.5-4.5	1.5-4.5	.01-.50
C54800	...	99.5(z)	4.0-6.0	.10	4.0-6.0	.30	.03-.35

Copper–Aluminum Alloys (Aluminum Bronzes)

Composition, per cent maximum (unless shown as a range or minimum)

Copper alloy no.	Cu + elements with specific limits (r) (% min)	Cu (incl. Ag)	Pb	Fe	Sn	Zn	Al	As	Mn	Si	Ni (incl. Co)	Co	P
C60600	99.5	92.0-96.0	...	.50	...	...	4.0-7.0	...	...	...	...	...	...
C60700	99.5	94.6-96.0	.01	...	1.7-2.0	...	2.3-2.9	...	...	...	...	...	...
C60800	99.5	92.5-94.8	.10	.10-1	...	...	5.0-6.5	.02-.35	...	...	...	...	...
C61000	99.5	90.0-93.0	.02	.50	...	.20	6.0-8.5	...	...	.10	...	...	...
C61300	99.5	86.5-93.8	...	3.5	.20-.50	...	6.0-8.0	...	.50	...	.50	...	...
C61400	99.5	88.0-92.5	.01	1.5-3.5	...	.20	6.0-8.0	...	1.0	...	...	...	.015
C61500	99.5	89.0-90.5	.015	...	...	...	77.7-8.3	...	...	...	1.8-2.2	...	...
C61800	99.5	86.9-91.0	.02	.50-1.5	...	.02	8.5-11.0	...	...	.10	...	...	...
C61900	99.5	83.6-88.5	.02	3.0-4.5	.6	.8	8.5-10.0	...	...	...	...	...	...
C62200	99.5	83.2-86.0	.02	3.0-4.2	...	.02	11.0-12.0	...	...	.10	...	...	...
C62300	99.5	82.2-89.5	...	2.0-4.0	.6	...	8.5-11.0	...	.50	.25	1.0	...	...
C62400	99.5	82.8-88.0	...	2.0-4.5	.20	...	10.0-11.5	...	.30	.25	...	...	...
C62500	99.5	79.0-84.0	...	3.5-5.0	...	...	12.5-13.5	...	2.0	...	...	...	...
C63000	99.5	78.0-85.0	...	2.0-4.0	.20	.30	9.0-11.0	...	1.5	.25	4.0-5.5	...	...
C63200	99.5	75.9-84.5	.02	3.0-5.0(aa)	...	...	8.5-9.5	...	3.5	.10	4.0-5.5(aa)	...	...
C63400	99.5	94.9-97.1	.05	.15	.20	.50	2.6-3.2	.15	...	.25-.45	.15	...	...
C63600	99.5	93.2-96.3	.05	.15	.20	.50	3.0-4.0	.15	...	.7-1.3	.15	...	...
C63800	99.5	93.0-95.7	.05	.10	...	.8	2.5-3.1	...	.10	.10	.10	.25-.55	...
C64200	99.5	88.2-92.2	.05	.30	.20	.50	6.3-7.6	.15	.10	1.5-2.2	.25	...	...
C64210	99.5	88.2-92.2	.05	.30	.20	.50	6.3-7.0	.15	.10	1.5-2.0	.25	...	...
C64400	99.5	88.8-91.5	.03	.05	.10	.20	3.5-4.5	...	...	.8-1.3	4.2-5.0	...	...

Copper–Silicon Alloys (Silicon Bronzes)

Composition, per cent maximum (unless shown as a range or minimum)

Copper alloy no.	Previous trade name	Cu + elements with specific limits (r) (% min)	Cu (incl. Ag) (% min)	Pb	Fe	Sn	Zn	Al	Mn	Si	Ni (incl. Co)
C64700	...	99.5	97.0	.10	.10	...	.50	...	...	.40-.8	1.6-2.2
C64900	...	99.5	96.2	.05	.10	1.2-1.6	.20	.10	...	.8-1.2	.10
C65100	Low silicon bronze B.	99.5	96.0	.05	.8	...	1.5	...	.7	.8-2.0	...
C65300	...	99.7	97.4	.05	.8	...	...	...	...	2.0-2.6	...
C65500	High silicon bronze A	99.5	94.8	.05	.8	...	1.5	...	.50-1.3	2.8-3.8	.6
C65600	...	99.5	94.0	.02	.50	1.5	1.5	.01	1.5	2.8-4.0	...
C65800	...	99.5	94.8	.05	.25	...	...	.01	.50-1.3	2.8-3.8	...
C66100	...	99.5	94.0	.20-.8	.25	...	1.5	...	1.5	2.8-3.5	...

Miscellaneous Copper–Zinc Alloys

Copper alloy no.	Previous trade name	Cu (incl. Ag)	Pb	Fe	Sn	Zn(t)	Ni (incl. Co)	Al	Mn	Si	Other named elements
						Composition, per cent maximum (unless shown as a range or minimum)					
C66400	...	Rem.(t)	.015	1.3-1.7(bb)	.05	11.0-12.0	.05	.05	.05	.05	.02 P, .30-.7 Co (bb), .05 Ag
C66700	Manganese brass	68.5-71.5	.07	.10	...	Rem.	...	...	.8-1.5	...	...
C66800	...	60.0-63.0	.50	.35	.30	Rem.	.25	.25	2.0-3.5	.50-1.5	...
C66900	...	62.5-64.5	.05	.25	...	Rem.	...	...	11.5-12.5	...	...
C67000	Manganese bronze B	63.0-68.0	.20	2.0-4.0	.50	Rem.	...	3.0-6.0	2.5-5.0	...	...
C67300	...	58.0-63.0	.40-3.0	.50	.30	Rem.	.25	.25	2.0-3.5	.50-1.5	...
C67400	...	57.0-60.0	.50	.35	.30	Rem.	.25	.50-2.0	2.0-3.5	.50-1.5	...
C67410	...	55.5-59.0	.8	1.0	.50	Rem.	2.0	1.3-2.3	1.0-2.4	.7-1.3	...
C67500	Manganese bronze A	57.0-60.0	.20	.8-2.0	.50-1.5	Rem.	...	.25	.05-.50	...	...
C67600	...	57.0-60.0	.50-1.0	.40-1.3	.50-1.5	Rem.	...	...	.05-.50	...	...
C67700	...	55.5-58.0	.50-1.0	.7-1.5	...	Rem.	1.5-2.3	...	.05-.30	...	.40-.8 As
C67800	...	56.0-59.0	.30	.7-1.5	.20	Rem.	...	.50-1.5	.20-.6	...	...
C67810	...	56.5-59.5	1.0	1.0	.50	Rem.	1.5	.40-1.6	.40-1.8	.6	...
C68000	Bronze, low fuming (nickel)	56.0-60.0	.05	.25-1.25	.75-1.10	Rem.	.20-.8	.01	.01-.50	.04-.15	...
C68100	Bronze, low fuming	56.0-60.0	.05	.25-1.25	.75-1.10	Rem.	...	.01	.01-.50	.04-.15	...
C68200	...	58.0-60.0	...	.6-1.0	...	Rem.	...	...	.6-1.0	.07-.15	...
C68700	Aluminum brass, arsenical	76.0-79.0	.07	.06	...	Rem.	...	1.8-2.5	...	...	.02-.10 As
C68800	...	Rem.(t)	.05	.05	...	21.3-24.1(cc)	...	3.0-3.8(cc)	...	...	.25-.55 Co
C69000	...	72.0-74.5	.025	.05	...	Rem.	.50-.8	3.3-3.5	...	...	...
C69400	Silicon red brass	80.0-83.0	.30	.20	...	Rem.	...	...	...	3.5-4.5	...
C69430	...	80.0-83.0	.30	.20	...	Rem.	...	...	...	3.5-4.5	.03-.06 As
C69440	...	80.0-83.0	.30	.20	...	Rem.	...	...	...	3.5-4.5	.03-.06 Sb
C69450	...	80.0-83.0	.30	.20	...	Rem.	...	...	.40	3.5-4.5	.03-.06 P
C69700	...	75.0-80.0	.50-1.5	.20	...	Rem.	...	...	.40	2.5-3.5	...
C69710	...	75.0-80.0	.50-1.5	.20	...	Rem.	...	...	.40	2.5-3.5	.03-.06 As
C69720	...	75.0-80.0	.50-1.5	.20	...	Rem.	...	...	.40	2.5-3.5	.03-.06 Sb
C69730	...	75.0-80.0	.50-1.5	.20	...	Rem.	...	...	.40	2.5-3.5	.03-.06 P
C69800	...	66.0-70.0	.8	.4	...	Rem.	.50	...	...	.7-1.3	...
C69900	Incramute	99.5(dd)	.02	.10	...	.14	.10	1.4-2.3	40.0-48.0	...	.20 Co, .05 C, .05 Ag, .05 Cd, .01 As
C69910	...	Rem. (q)	.01	1.0-1.4	...	3.0-5.0	...	.25-.8	28.0-32.0	...	...

Copper – Nickel Alloys

Copper alloy no.	Previous trade name	Cu + elements with specific limits (r) (% min)	Cu (incl. Ag) (% min)	Pb	Fe	Zn	Ni (incl. Co)	Mn	Other named elements
C70100	...	99.7	...	...	.05	.25	3.0-4.0	.50	...
C70200	...	99.7	...	.05	.10	...	2.0-3.0	.40	...
C70400	Copper nickel, 5%	99.5	...	.05	1.3-1.7	1.0	4.8-6.2	.30-.8	...
C70500	Copper nickel, 7%	99.5	...	.05	.10	.20	5.8-7.8	.15	...
C70600	Copper nickel, 10%	99.5	86.5	.05(ee)	1.0-1.8	1.0(ee)	9.0-11.0	1.0	(ee)
C70690	...	99.5	89.0	.001	.005	.001	9.0-11.0	.001	(ff)
C70700	...	99.5	...	...	.05	...	9.5-10.5	.50	...
C70800	Copper nickel, 11%	99.5	...	.05	.10	.20	10.5-12.5	.15	...
C70900	...	99.5	...	.05	.6	1.0	13.5-16.5	.6	...
C71000	Copper nickel, 20%	99.5	...	.05	1.0	1.0	19.0-23.0	1.0	...
C71100	...	99.5	...	.05	.10	.20	22.0-24.0	.15	...
C71300	...	99.5	...	.05	.20	1.0	23.5-26.5	1.0	...
C71500	Copper nickel, 30%	99.5	65.0	.05(ee)	.40-1.0	1.0(ee)	29.0-33.0	1.0	(ee)
C71580	...	99.5	65.0	.05	.5	.05	29.0-33.0	.30	(gg)
C71590	...	99.5	67.0	.001	.005	.001	29.0-33.0	.001	(ff)
C71700	...	99.5	...	...	.40-1.0	...	29.0-33.0	...	.30-.7 Be
C71900	...	99.5	...	...	.50	...	28.0-32.0	.20-1.0	2.4-3.2 Cr, .02-.25 Zr, .01-.20 Ti, .01 C, .25 Si
C72200	...	99.5	...	(ee)	.7-1.0	(ee)	15.0-18.0	1.0	(ee), .30-.7 Cr, .03 Si, .03 Ti
C72500	...	99.8	...	.05	.6	.50	8.5-10.5	.20	1.8-2.8 Sn

Copper – Nickel – Zinc Alloys (Nickel Silvers)

Copper alloy no.	Previous trade name	Cu	Pb	Fe	Zn(t)	Ni (incl. Co)	Mn	Other named elements
C73200	...	70.0 min	.05	.6	3.0-6.0	19.0-23.0	1.0	...
C73500	...	70.5-73.5	.10	.25	Rem.	16.5-19.5	.50	...
C73800	Nickel silver, 70-12	68.5-71.5	.05	.25	Rem.	11.0-13.0	.50	...
C74000	...	69.0-73.5	.10	.25	Rem.	9.0-11.0	.50	...
C74300	...	63.0-66.0	.10	.25	Rem.	7.0-9.0	.50	...
C74500	Nickel silver, 65-10	63.5-66.5	.10(hh)	.25	Rem.	9.0-11.0	.50	...
C75200	Nickel silver, 65-18	63.0-66.5	.10	.25	Rem.	16.5-19.5	.50	...
C75400	Nickel silver, 65-15	63.5-66.5	.10	.25	Rem.	14.0-16.0	.50	...
C75700	Nickel silver, 65-12	63.5-66.5	.05	.25	Rem.	11.0-13.0	.50	...
C75900	...	60.0-63.0	.10	.25	Rem.	17.0-19.0	.50	...
C76000	...	60.0-63.0	.10	.25	Rem.	7.0-9.0	.50	...
C76100	...	59.0-63.0	.10	.25	Rem.	9.0-11.0	.50	...
C76200	...	57.0-61.0	.10	.25	Rem.	11.0-13.5	.50	...
C76300	...	60.0-64.0	.50-2.0	.50	Rem.	17.0-19.0	.50	...
C76400	...	58.5-61.5	.05	.25	Rem.	16.5-19.5	.50	...
C76600	...	55.0-58.0	.10	.25	Rem.	11.0-13.5	.50	...
C76700	Nickel silver, 56.5-15	55.0-58.0	...	...	Rem.	14.0-16.0	.50	...
C77000	Nickel silver, 55-18	53.5-56.5	.10	.25	Rem.	16.5-19.5	.50	...
C77300	...	46.0-50.0	.05	...	Rem.	9.0-11.0	...	.01 Al, .25 P, .04-.25 Si
C77400	...	54.0-47.0	.20	...	Rem.	9.0-11.0	...	...
C77600	Nickel silver, 43.5-13	42.0-45.0	.25	.20	Rem.	12.0-14.0	.25	.15 Sn
C78200	...	63.0-67.0	1.5-2.5	.35	Rem.	7.0-9.0	.50	...
C78800	...	63.0-67.0	1.5-2.0	.25	Rem.	9.0-11.0	.50	...
C79000	...	63.0-67.0	1.5-2.2	.35	Rem.	9.0-11.0	.50	...
C79200	...	59.0-66.5	.8-1.4	.25	Rem.	11.0-13.0	.50	...
C79300	...	55.0-59.0	.50-2.0	.50	Rem.	11.0-13.0	.50	...
C79600	Leaded nickel silver, 10%	43.5-46.5	.8-1.2	...	Rem.	9.0-11.0	1.5-2.5	...
C79800	...	45.5-48.5	1.5-2.5	.25	Rem.	9.0-11.0	1.5-2.5	...
C79900	...	47.5-50.5	1.0-1.5	.25	Rem.	6.5-8.5	.50	...

Cast Alloys
Coppers

Copper alloy no.	Cu (incl. Ag) (% min)	Ag % min	Ag troy oz min	B
C80100	99.95	...	...	...
C80300	99.95	.034	10	...
C80500	99.75	.034	10	.02
C80700	99.75	...	...	.02
C80900	99.70	.034	10	...
C81100	99.70	...	...	...

Composition, per cent maximum (unless shown as a range or minimum)

High Copper Alloys

Composition, per cent maximum (unless shown as a range or minimum)

Copper alloy no.	Cu(t)(dd)	Ag	Be	Co	Si	Ni	Fe	Al	Sn	Pb	Zn	Cr
C81300	98.5 min	...	.02-.10	.6-1.0	...	...	...	...	...	...	...	...
C81400	98.5 min	...	.02-.10	...	...	...	...	...	...	...	...	.6-1.0
C81500	98.0 min	...	...	...	.15	...	.10	.10	.10	.02	.10	.40-1.5
C81700	94.2 min	.8-1.2	.30-.55	.25-1.5	...	.25-1.5	...	...	...	...	...	...
C81800	95.6 min	.8-1.2	.30-.55	1.4-1.7	...	...	...	...	...	...	...	...
C82000	95.0 min	...	.45-.8	2.4-2.7(ii)	.15	.20	.10	.10	.10	.02	.10	.10
C82100	95.5 min	...	.35-.8	.25-1.5	...	.25-1.5	...	...	...	...	...	...
C82200	96.5 min	...	.35-.8	...	...	1.0-2.0	...	...	...	...	...	...
C82400	96.4 min	...	1.65-1.75	.20-.40	...	.10	.20	.15	.10	.02	.10	.10
C82500	95.5 min	...	1.90-2.15	.35-.7(ii)	.20-.35	.20	.25	.15	.10	.02	.10	.10
C82600	95.2 min	...	2.25-2.45	.35-.7	.20-.35	.20	.25	.15	.10	.02	.10	.10
C82700	94.6 min	...	2.35-2.55	...	.15	1.0-1.5	.25	.15	.10	.02	.10	.10
C82800	94.8 min	...	2.50-2.75	.35-.7(ii)	.20-.35	.20	.25	.15	.10	.02	.10	.10

Copper–Tin–Zinc and Copper–Tin–Zinc–Lead Alloys
(Red Brasses and Leaded Red Brasses)

Composition, per cent maximum (unless shown as a range or minimum)

Copper alloy no.	Cu	Sn	Pb	Zn	Fe	Sb	Ni (incl. Co)	S	P(kk)	Al	Si
C83300	92.0-94.0	1.0-2.0	1.0-2.0	2.0-6.0	...	...	...	...	...	...	...
C83400	88.0-92.0	.20	.50	8.0-12.0	...	...	...	...	...	...	...
C83500	86.0-88.0(jj)	5.5-6.5	3.5-5.5	1.0-2.5	.25	.25	.50-1.0	.08	.03	.005	.005
C83600	84.0-86.0(jj)	4.0-6.0	4.0-6.0	4.0-6.0	.30	.25	1.0(jj)	.08	.05	.005	.005
C83800	82.0-83.8(jj)	3.3-4.2	5.0-7.0	5.0-8.0	.30	.25	1.0(jj)	.08	.03	.005	.005

Semi-red Brasses and Leaded Semi-red Brasses

Composition, per cent maximum (unless shown as a range or minimum)

Copper alloy no.	Cu	Sn	Pb	Zn	Fe	Sb	Ni(jj) (incl. Co)	S	P(kk)	Al	Si
C84200	78.0-82.0	4.0-6.0	2.0-3.0	1.0-16.0	.40	.25	.8	.08	.05	.005	.005
C84400	78.0-82.0(jj)	2.3-3.5	6.0-8.0	7.0-10.0	.40	.25	1.0	.08	.02	.005	.005
C84500	77.0-79.0(jj)	2.0-4.0	6.0-7.5	10.0-14.0	.40	.25	1.0	.08	.02	.005	.005
C84800	75.0-77.0(jj)	2.0-3.0	5.5-7.0	13.0-17.0	.40	.25	1.0	.08	.02	.005	.005

Yellow Brasses and Leaded Yellow Brasses

Composition, per cent maximum (unless shown as a range or minimum)

Copper alloy no.	Cu	Sn	Pb	Zn(t)	Fe	Sb	Ni (incl. Co)	Mn	As	S	P	Al	Si
C85200	70.0-74.0	.7-2.0	1.5-3.8	20.0-27.0	.6	.20	1.0	...	...	.05	.02	.005	.05
C85400	65.0-70.0	.50-1.5	1.5-3.8	24.0-32.0	.7	...	1.0	...	...	...	...	.35	.05
C85500	59.0-63.0	.20	.20	Rem.	.20	...	.20	.20	...	...	...	...	...
C85700	58.0-64.0	.50-1.5	.8-1.5	32.0-40.0	.7	...	1.0	...	...	...	...	.55	.05
C85800	57.0 min	1.5	1.5	31.0-41.0	.50	.05	.50	.25	.05	.05	.01	.55	.25

Manganese and Leaded Manganese Bronze Alloys

Composition, per cent maximum (unless shown as a range or minimum)

Copper alloy no.	Cu	Sn	Pb	Zn	Fe	Ni (incl. Co)	Al	Mn
C86100	66.0-68.0	.20	.20	Rem.	2.0-4.0	...	4.5-5.5	2.5-5.0
C86200	60.0-66.0	.20	.20	22.0-28.0	2.0-4.0	1.0	3.0-4.9	2.5-5.0
C86300	60.0-66.0	.20	.20	22.0-28.0	2.0-4.0	1.0	5.0-7.5	2.5-5.0
C86400	56.0-62.0	.50-1.5	.50-1.5	34.0-42.0	.40-2.0	1.0	.50-1.5	.10-1.0
C86500	55.0-60.0	1.0	.40	36.0-42.0	.40-2.0	1.0	.50-1.5	.10-1.5
C86700	55.0-60.0	1.5	.5-1.5	30.0-38.0	1.0-3.0	1.0	1.0-3.0	1.0-3.5
C86800	53.5-57.0	1.0	.20	Rem.	1.0-2.5	2.5-4.0	2.0	2.5-4.0

Copper–Zinc–Silicon Alloys (Silicon Bronzes and Silicon Brasses)

Composition, per cent maximum (unless shown as a range or minimum)

Copper alloy no.	Cu(dd)	Sn	Pb	Zn	Fe	Al	Si	Mn	Mg	As	Sb	Ni (incl. Co)	S	P
C87200	89.0 min	1.0	.50	5.0	2.5	1.5	1.0-5.0	1.5	...	...	...	...	...	...
C87400	79.0 min	...	1.0	12.0-16.0	...	.8	2.5-4.0	...	...	...	...	...	...	...
C87410	79.0 min	...	1.0	12.0-16.0	...	.8	2.5-4.0	...	...	.03-.06	...	...	...	...
C87420	79.0 min	...	1.0	12.0-16.0	...	.8	2.5-4.0	...	...	...	.03-.06	...	...	...
C87430	79.0 min	...	1.0	12.0-16.0	...	.8	2.5-4.0	...	...	...	...	...	...	.03-.06
C87500	79.0 min	...	.50	12.0-16.0	...	.50	3.0-5.0	...	...	...	...	...	...	...
C87510	79.0 min	...	.50	12.0-16.0	...	.50	3.0-5.0	...	...	.03-.06	...	...	...	...
C87520	79.0 min	...	.50	12.0-16.0	...	.50	3.0-5.0	...	...	...	.03-.06	...	...	...
C87530	79.0 min	...	.50	12.0-16.0	...	.50	3.0-5.0	...	...	...	...	...	...	.03-.06
C87600	88.0 min	...	.50	4.0-7.0	...	...	3.5-5.5	...	...	...	...	...	...	...
C87800	80.0 min	.25	.15	12.0-16.0	.15	.15	3.8-4.2	.15	.01	.05	.05	.20	.05	.01
C87900	63.0 min	.25	.25	30.0-36.0	.40	.15	.8-1.2	.15	...	.05	.05	.50	.05	.01

Copper–Tin Alloys (Tin Bronzes)

Composition, per cent maximum (unless shown as a range or minimum)

Copper alloy no.	Cu	Sn	Pb	Zn	Fe	Sb	Ni (incl. Co)	S	P(kk)	Al	Si
C90200	91.0-94.0(ll)	6.0-8.0	.30	.50	.20	.20	.50	.05	.05	.005	.005
C90300	86.0-89.0(jj)	7.5-9.0	.30	3.0-5.0	.20	.20	1.0(jj)	.05	.05	.005	.005
C90500	86.0-89.0(jj)	9.0-11.0	.30	1.0-3.0	.20	.20	1.0(jj)	.05	.05	.005	.005
C90700	88.0-90.0(mm)	10.0-12.0	.50	.50	.15	.20	.50	.05	.30	.005	.005
C90800	85.0-89.0(mm)	11.0-13.0	.25	.25	.15	.20	.50	.05	.30	.005	.005
C90900	86.0-89.0(dd)	12.0-14.0	.25	.25	.15	.20	.50	.05	.05	.005	.005
C91000	84.0-86.0	14.0-16.0	.20	1.5	.10	.20	.8	.05	.05	.005	.005
C91100	82.0-85.0	15.0-17.0	.25	.25	.25	.20	.50	.05	1.0	.005	.005
C91300	79.0-82.0	18.0-20.0	.25	.25	.25	.20	.50	.05	1.0	.005	.005
C91600	86.0-89.0(mm)	9.7-10.8	.25	.25	.20	.20	1.2-2.0	.05	.30	.005	.005
C91700	84.0-87.0(mm)	11.3-12.5	.25	.25	.20	.20	1.2-2.0	.05	.30	.005	.005

Copper Development Association/1783

Copper–Tin–Lead Alloys (Leaded Tin Bronzes)

Copper alloy no.	Composition, per cent maximum (unless shown as a range or minimum)										
	Cu	Sn	Pb	Zn	Fe	Sb	Ni (incl. Co)	S	P(kk)	Al	Si
C92200	86.0-90.0(jj)	5.5-6.5	1.0-2.0	3.0-5.0	.25	.25	1.0(jj)	.05	.05	.005	.005
C92300	85.0-89.0(jj)	7.5-9.0	.30-1.0	2.5-5.0	.25	.25	1.0(jj)	.05	.05	.005	.005
C92400	86.0-89.0(jj)	9.0-11.0	1.0-2.5	1.0-3.0	.25	.25	1.0(jj)	.05	.05	.005	.005
C92500	85.0-88.0	10.0-12.0	1.0-1.5	.50	.30	.25	.8-1.5	.05	.30	.005	.005
C92600	86.0-88.5	9.3-10.5	.8-1.2	1.3-2.5	.20	.25	.7	.05	.03	.005	.005
C92700	86.0-89.0	9.0-11.0	1.0-2.5	.7	.20	.25	1.0	.05	.25	.005	.005
C92800	78.0-82.0	15.0-17.0	4.0-6.0	.8	.20	.25	.8	.05	.05	.005	.005
C92900	82.0-86.0(mm)	9.0-11.0	2.0-3.2	.25	.20	.25	2.8-4.0	.05	.50	.005	.005

Copper–Tin–Lead (High–Leaded Tin Bronzes)

Copper alloy no.	Composition, per cent maximum (unless shown as a range or minimum)										
	Cu	Sn	Pb	Zn	Fe	Sb	Ni (incl. Co)	S	P(kk)	Al	Si
C93200	81.0-85.0(jj)	6.3-7.5	6.0-8.0	2.0-4.0	.20	.35	1.0(jj)	.08	.15	.005	.005
C93400	82.0-85.0	7.0-9.0	7.0-9.0	.8	.20	.50	1.0	.08	.50	.005	.005
C93500	83.0-86.0(jj)	4.3-6.0	8.0-10.0	2.0	.20	.30	1.0(jj)	.08	.05	.005	.005
C93700	78.0-82.0(jj)	9.0-11.0	8.0-11.0	.8	.15(nn)	.55	1.0(jj)	.08	.15	.005	.005
C93800	75.0-79.0(jj)	6.3-7.5	13.0-16.0	.8	.15	.8	1.0(jj)	.08	.05	.005	.005
C93900	76.5-79.5(jj)	5.0-7.0	14.0-18.0	1.5	.40	.50	.8(jj)	.08	1.5	.005	.005
C94000	69.0-72.0(jj)	12.0-14.0	14.0-16.0	.50	.25	.50	.50-1.0(jj)	.08(oo)	.05	.005	.005
C94100	65.0-75.0(jj)	4.5-6.5	15.0-22.0	3.0	.25	.8	.8	.08	.05	.005	.005
C94300	68.5-73.5(jj)	4.5-6.0	22.0-25.0	.8	.15	.8	1.0(jj)	.08	.05	.005	.005
C94400	Rem.(jj)(ll)	7.0-9.0	9.0-12.0	.8	.15	.8	1.0	.08	.50	.005	.005
C94500	Rem.(jj)(ll)	6.0-8.0	16.0-22.0	1.2	.15	.8	1.0	.08	.05	.005	.005

Copper–Tin–Nickel Alloys (Nickel–Tin Bronzes)

Copper alloy no.	Composition, per cent maximum (unless shown as a range or minimum)										
	Cu	Sn	Pb	Zn	Fe	Sb	Ni (incl. Co)	Mn	S	P	Al	Si
C94700	85.0-90.0	4.5-6.0	.10(pp)	1.0-2.5	.25	.15	4.5-6.0	.20	.05	.05	.005	.005
C94800	84.0-89.0	4.5-6.0	.30-1.0	1.0-2.5	.25	.15	4.5-6.0	.20	.05	.05	.005	.005
C94900	79.0-81.0	4.0-6.0	4.0-6.0	4.0-6.0	.30	.25	4.0-6.0	.10	.08	.05	.005	.005

Copper–Aluminum–Iron and Copper–Aluminum–Iron–Nickel Alloys (Aluminum Bronzes)

Copper alloy no.	Composition, per cent maximum (unless shown as a range or minimum)						
	Cu	Pb	Fe	Ni (incl. Co)	Al	Mn	Si
C95200	86.0 min (ll)	...	2.5-4.0	...	8.5-9.5	...	...
C95300	86.0 min (ll)	...	.8-1.5	...	9.0-11.0	...	...
C95400	83.0 min (dd)	...	3.0-5.0	2.5	10.0-11.5	.50	...
C95410	83.0 min (dd)	...	3.0-5.0	1.5-2.5	10.0-11.5	.50	...
C95500	78.0 min (dd)	...	3.0-5.0	3.0-5.5	10.0-11.5	3.5	...
C95600	88.0 min (ll)	...	...	.25	6.0-8.0	...	1.8-3.3
C95700	71.0 min (ll)	.03	2.0-4.0	1.5-3.0	7.0-8.5	11.0-14.0	.10
C95800	79.0 min (dd)	.03	3.5-4.5 (aa)	4.0-5.0 (aa)	8.5-9.5	.8-1.5	.10

Copper−Nickel−Iron Alloys (Copper−Nickels)

Copper alloy no.	Cu	Pb	Fe	Ni (incl. Co)	Mn	Si	Nb	C	Be
				Composition, per cent maximum (unless shown as a range or minimum)					
C96200	84.5-87.0	.03 (qq)	1.0-1.8	9.0-11.0	1.5	.30	1.0	.15	...
C96300	Rem.	.03 (qq)	.40-1.0	18.0-22.0	1.0	.7	1.0	...	...
C96400	65.0-69.0	.03 (qq)	.25-1.5	28.0-32.0	1.5	.50	.50-1.5	.15	...
C96600	Rem.	.01	.8-1.1	29.0-33.0	1.0	.15	...	...	.40-.7

Copper−Nickel−Zinc Alloys (Nickel Silvers)

Copper alloy no.	Cu	Sn	Pb	Zn	Fe	Sb	Ni (incl. Co)	S	P	Al	Mn	Si
	Composition, per cent maximum (unless shown as a range or minimum)											
C97300	53.0-58.0	1.5-3.0	8.0-11.0	17.0-25.0	1.5	.35	11.0-14.0	.08	.05	.005	.50	.15
C97400	58.0-61.0	2.5-3.5	4.5-5.5	Rem.	1.5	...	15.5-17.0	...	...	...	.50	...
C97600	63.0-67.0	3.5-4.5	3.0-5.0	3.0-9.0	1.5	.25	19.0-21.5	.08	.05	.005	1.0	.15
C97800	64.0-67.0	4.0-5.5	1.0-4.0	1.0-4.0	1.5	.20	24.0-27.0	.08	.05	.005	1.0	.15

Copper−Lead Alloys (Leaded Coppers)

Copper alloy no.	Cu	Sn	Pb	Ag	Zn	P	Fe	Total other elements
	Composition, per cent maximum (unless shown as a range or minimum)							
C98200	73.0-79.0	.50	21.0-27.0	...	...	...	.35	.45
C98400	67.0-74.0	.25	25.0-32.0	1.5	.10	.02	.35	.15
C98600	60.0-70.0	.50	30.0-40.0	1.5	...	...	.35	.30
C98800	56.5-62.5	.25	37.5-42.5	5.5	.10	.02	.35	.30

Special Alloys

Copper alloy no.	Other designations	Cu	Sn	Pb	Ni	Fe	Al	Co	Si	Mn	Other named elements
		Composition, per cent maximum (unless shown as a range or minimum)									
C99300	Incramet 800	Rem.	.05	.02	13.5-16.5	.40-1.0	10.7-11.5	1.0-2.0	.02	...	...
C99400	...	Rem.	...	.25	1.0-3.5	1.0-3.0	.50-2.0	...	.50-2.0	.50	.50-5.0 Zn
C99500	...	Rem.	...	.25	3.5-5.5	3.0-5.0	.50-2.0	...	.50-2.0	.50	.50-2.0 Zn
C99600	Incramute 1	Rem. (dd)	.10	.02	.20	.20	1.0-2.8	.20	.10	39.0-45.0	.20 Zn, .05 C
C99700	...	54.0 min (dd)	1.0	2.0	4.0-6.0	1.0	.50-3.0	...	...	11.0-15.0	19.0-25.0 Zn
C99750	...	55.0-61.0	.50-2.5	...	5.0 (aa)	1.0 (aa)	.25-3.0	...	...	17.0-23.0	17.0-23.0 Zn

(a) These are high conductivity coppers which have in the annealed condition a minimum conductivity of 100% IACS. (b) The value of Cu is exclusive of Ag. (c) The total of the seven following elements, Se, Te, Bi, As, Sb, Sn and Mn not to exceed 40 ppm (.0040%). Hg, max, 1 ppm (.0001%); Zn, max, 1 ppm (.0001%); Cd, max, 1 ppm (.0001%); S, max, 18 ppm (.0018%); Pb max, 10 ppm (.0010%); Se, max, 10 ppm (.0010%); Bi, max, 10 ppm (.0010%); Oxygen, max, 10 ppm (.0010%). (d) Includes P. (e) Small amounts of Cd or other elements may be added by agreement to improve resistance to softening at elevated temperatures. (f) This includes low resistance lake copper and electrolytic copper to which Ag is added. (g) Includes B. (h) This includes oxygen-free copper which contains P in an amount agreed upon. (i) This includes high resistance lake copper. (j) Te + Se. (k) Includes Cd. Deoxidized with lithium or other suitable elements as agreed upon. (l) This includes oxygen-free tellurium bearing copper which contains P in an amount agreed upon. (m) Includes Te. (n) Other deoxidizers may be used as agreed upon, in which case P need not be present. (o) Includes Ag, S, P and Pb. (p) Includes oxygen-free or deoxidized grades with deoxidizers (such as phosphorus, boron, lithium or other) in an amount agreed upon. (q) Total named elements shall be 99.8% minimum. (r) Specific limits are defined as any numerical values, whether maximum only, minimum only or ranges. (s) Ni + Co, .20% min.; Ni + Fe + Co, .6% max. (t) These specification limits do not preclude the possible presence of other unnamed elements. However, analysis shall regularly be made only for the minor elements listed in the table plus all major elements except one. The major element which is not analyzed shall be determined by difference between the sum of those elements analyzed and 100 per cent. By agreement between producer and consumer, analysis may be required and limits established for elements not specified. (u) Al + Si, .005% max. (v) For tube over 5 inches O.D., the Pb may be less than .20%. (w) Copper, 61.0% min for rod. (x) Copper, 60.0% min for rod. (y) For flat products, the minimum tin content may be .8%. (z) Includes Zn. (aa) Iron content shall not exceed nickel content. (bb) Fe + Co shall be 1.8-2.3%. (cc) Al + Zn shall be 25.1-27.1%. (dd) Total named elements shall be 99.5% minimum. (ee) When the product is for subsequent welding applications and so specified by the purchaser, Zn shall be .50% max., Pb .02% max., P .02% max, Sulfur .02% max, and Carbon .05% max. (ff) The following maximum limits shall apply: .03 C, .02 Si, .003 S, .002 Al, .001 P, .0005 Hg, .001 Ti, .001 Sb, .001 As, .001 Bi, and .001 Sn. (gg) The following maximum limits shall apply: .07 C, .15 Si, .024 S, .05 Al, and .03 P. (hh) Pb, .05% maximum for rod and wire. (ii) Nickel plus cobalt. (jj) In determining copper minimum, copper may be calculated as Cu + Ni. (kk) For continuous castings, phosphorus shall be 1.5% maximum. (ll) Total named elements shall be 99.0% minimum. (mm) Cu + Sn + Pb + Ni + P shall be 99.5% min. (nn) The iron shall be .35% max when used for steel backed bearings. (oo) For continuous castings, sulfur shall be .25% max. (pp) The mechanical properties of copper alloy no. C94700 (heat treated) may not be attained if the lead content exceeds .01%. (qq) For welding grades, lead may not exceed .01%.

Standard Alloy Steels

Cast or Heat Chemical Ranges and Limits for Bars, Blooms, Billets and Slabs

Steel designation, AISI or SAE	UNS number	Chemical composition, per cent							
		C	Mn	P max	S max	Si	Ni	Cr	Mo
1330	G13300	0.28-0.33	1.60-1.90	0.035	0.040	0.15-0.30	...	...	...
1335	G13350	0.33-0.38	1.60-1.90	0.035	0.040	0.15-0.30	...	...	...
1340	G13400	0.38-0.43	1.60-1.90	0.035	0.040	0.15-0.30	...	...	...
1345	G13450	0.43-0.48	1.60-1.90	0.035	0.040	0.15-0.30	...	...	...
4023	G40230	0.20-0.25	0.70-0.90	0.035	0.040	0.15-0.30	...	...	0.20-0.30
4024	G20240	0.20-0.25	0.70-0.90	0.035	0.035-0.050	0.15-0.30	...	...	0.20-0.30
4027	G40270	0.25-0.30	0.70-0.90	0.035	0.040	0.15-0.30	...	...	0.20-0.30
4028	G40280	0.25-0.30	0.70-0.90	0.035	0.035-0.50	0.15-0.30	...	...	0.20-0.30
4037	G40370	0.35-0.40	0.70-0.90	0.035	0.040	0.15-0.30	...	...	0.20-0.30
4047	G40470	0.45-0.50	0.70-0.90	0.035	0.040	0.15-0.30	...	...	0.20-0.30
4118	G41180	0.18-0.23	0.70-0.90	0.035	0.040	0.15-0.30	...	0.40-0.60	0.08-0.15
4130	G41300	0.28-0.33	0.40-0.60	0.035	0.040	0.15-0.30	...	0.80-1.10	0.15-0.25
4137	G41370	0.35-0.40	0.70-0.90	0.035	0.040	0.15-0.30	...	0.80-1.10	0.15-0.25
4140	G41400	0.38-0.43	0.75-1.00	0.035	0.040	0.15-0.30	...	0.80-1.10	0.15-0.25
4142	G41420	0.40-0.45	0.75-1.00	0.035	0.040	0.15-0.30	...	0.80-1.10	0.15-0.25
4145	G41450	0.43-0.48	0.75-1.00	0.035	0.040	0.15-0.30	...	0.80-1.10	0.15-0.25
4147	G41470	0.45-0.50	0.75-1.00	0.035	0.040	0.15-0.30	...	0.80-1.10	0.15-0.25
4150	G41500	0.48-0.53	0.75-1.00	0.035	0.040	0.15-0.30	...	0.80-1.10	0.15-0.25
4161	G41610	0.56-0.64	0.75-1.00	0.035	0.040	0.15-0.30	...	0.70-0.90	0.25-0.35
4320	G43200	0.17-0.22	0.45-0.65	0.035	0.040	0.15-0.30	1.65-2.00	0.40-0.60	0.20-0.30
4340	G43400	0.38-0.43	0.60-0.80	0.035	0.040	0.15-0.30	1.65-2.00	0.70-0.90	0.20-0.30
E4340	G43406	0.38-0.43	0.65-0.85	0.025	0.025	0.15-0.30	1.65-2.00	0.70-0.90	0.20-0.30
4615	G46150	0.13-0.18	0.45-0.65	0.035	0.040	0.15-0.30	1.65-2.00	...	0.20-0.30
4620	G46200	0.17-0.22	0.45-0.65	0.035	0.040	0.15-0.30	1.65-2.00	...	0.20-0.30
4626	G46260	0.24-0.29	0.45-0.65	0.035	0.040	0.15-0.30	0.70-1.00	...	0.15-0.25
4720	G47200	0.17-0.22	0.50-0.70	0.035	0.040	0.15-0.30	0.90-1.20	0.35-0.55	0.15-0.25
4815	G48150	0.13-0.18	0.40-0.60	0.035	0.040	0.15-0.30	3.25-3.75	...	0.20-0.30
4817	G48170	0.15-0.20	0.40-0.60	0.035	0.040	0.15-0.30	3.25-3.75	...	0.20-0.30
4820	G48200	0.18-0.23	0.50-0.70	0.035	0.040	0.15-0.30	3.25-3.75	...	0.20-0.30
5117	G51170	0.15-0.20	0.70-0.90	0.035	0.040	0.15-0.30	...	0.70-0.90	...
5120	G51200	0.17-0.22	0.70-0.90	0.035	0.040	0.15-0.30	...	0.70-0.90	...
5130	G51300	0.28-0.33	0.70-0.90	0.035	0.040	0.15-0.30	...	0.80-1.10	...
5132	G51320	0.30-0.35	0.60-0.80	0.035	0.040	0.15-0.30	...	0.75-1.00	...
5135	G51350	0.33-0.38	0.60-0.80	0.035	0.040	0.15-0.30	...	0.80-1.05	...
5140	G51400	0.38-0.43	0.70-0.90	0.035	0.040	0.15-0.30	...	0.70-0.90	...
5150	G51500	0.48-0.53	0.70-0.90	0.035	0.040	0.15-0.30	...	0.70-0.90	...
5155	G51550	0.51-0.59	0.70-0.90	0.035	0.040	0.15-0.30	...	0.70-0.90	...
5160	G51600	0.56-0.64	0.75-1.00	0.035	0.040	0.15-0.30	...	0.70-0.90	...
E51100	G51986	0.98-1.10	0.25-0.45	0.025	0.025	0.15-0.30	...	0.90-1.15	...
E52100	G52986	0.98-1.10	0.25-0.45	0.025	0.025	0.15-0.30	...	1.30-1.60	...
6118	G61180	0.16-0.21	0.50-0.70	0.035	0.040	0.15-0.30	...	0.50-0.70	0.10-0.15 V
6150	G61500	0.48-0.53	0.70-0.90	0.035	0.040	0.15-0.30	...	0.80-1.10	0.15 min V
8615	G86150	0.13-0.18	0.70-0.90	0.035	0.040	0.15-0.30	0.40-0.70	0.40-0.60	0.15-0.25
8617	G86170	0.15-0.20	0.70-0.90	0.035	0.040	0.15-0.30	0.40-0.70	0.40-0.60	0.15-0.25
8620	G86200	0.18-0.23	0.70-0.90	0.035	0.040	0.15-0.30	0.40-0.70	0.40-0.60	0.15-0.25
8622	G86220	0.20-0.25	0.70-0.90	0.035	0.040	0.15-0.30	0.40-0.70	0.40-0.60	0.15-0.25
8625	G86250	0.23-0.28	0.70-0.90	0.035	0.040	0.15-0.30	0.40-0.70	0.40-0.60	0.15-0.25
8627	G86270	0.25-0.30	0.70-0.90	0.035	0.040	0.15-0.30	0.40-0.70	0.40-0.60	0.15-0.25
8630	G86300	0.28-0.33	0.70-0.90	0.035	0.040	0.15-0.30	0.40-0.70	0.40-0.60	0.15-0.25
8637	G86370	0.35-0.40	0.75-1.00	0.035	0.040	0.15-0.30	0.40-0.70	0.40-p.60	0.15-0.25
8640	G86400	0.38-0.43	0.75-1.00	0.035	0.040	0.15-0.30	0.40-0.70	0.40-0.60	0.15-0.25
8642	G86420	0.40-0.45	0.75-1.00	0.035	0.040	0.15-0.30	0.40-0.70	0.40-0.60	0.15-0.25
8645	G86450	0.43-0.48	0.75-1.00	0.035	0.040	0.15-0.30	0.40-0.70	0.40-0.60	0.15-0.25
8655	G86550	0.51-0.59	0.75-1.00	0.035	0.040	0.15-0.30	0.40-0.70	0.40-0.60	0.15-0.25
8720	G87200	0.18-0.23	0.70-0.90	0.035	0.040	0.15-0.30	0.40-0.70	0.40-0.60	0.20-0.30
8740	G87400	0.38-0.43	0.75-1.00	0.035	0.040	0.15-0.30	0.40-0.70	0.40-0.60	0.20-0.30
8822	G88220	0.20-0.25	0.75-1.00	0.035	0.040	0.15-0.30	0.40-0.70	0.40-0.60	0.30-0.40
9260	G92600	0.56-0.64	0.75-1.00	0.035	0.040	1.80-2.20	...	...	...

Standard Boron Steels (a)

Cast or Heat Chemical Ranges and Limits for Bars, Blooms, Billets and Slabs

Steel designation, AISI or SAE	UNS number	Chemical composition, per cent							
		C	Mn	P max	S max	Si	Ni	Cr	Mo
50B44	G50441	0.43-0.48	0.75-1.00	0.035	0.040	0.15-0.30	...	0.40-0.60	...
50B46	G50461	0.44-0.49	0.75-1.00	0.035	0.040	0.15-0.30	...	0.20-0.35	...
50B50	G50501	0.48-0.53	0.75-1.00	0.035	0.040	0.15-0.30	...	0.40-0.60	...
50B60	G50601	0.56-0.64	0.75-1.00	0.035	0.040	0.15-0.30	...	0.40-0.60	...
51B60	G51601	0.56-0.64	0.75-1.00	0.035	0.040	0.15-0.30	...	0.70-0.90	...
81B45	G81451	0.43-0.48	0.75-1.00	0.035	0.040	0.15-0.30	0.20-0.40	0.35-0.55	0.08-0.15
94B17	G94171	0.15-0.20	0.75-1.00	0.035	0.040	0.15-0.30	0.30-0.60	0.30--0.50	0.08-0.15
94B30	G94301	0.28-0.33	0.75-1.00	0.035	0.040	0.15-0.30	0.30-0.60	0.30-0.50	0.08-0.15

(a) These steels can be expected to contain 0.0005 to 0.003 per cent boron.

Note 1. Grades shown in the above list with prefix letter E are normally made by the basic electric furnace process. All others are normally manufactured by the basic open hearth or basic oxygen process but may be manufactured by the basic electric furnace process. **Note 2.** If electric furnace practice is specified or required for grades other than those designated (i.e., E4340) the limits for phosphorus and sulfur are 0.025 per cent respectively and the prefix E is added. **Note 3.** For acid electric and acid open hearth steels, the limits for phosphorus and sulfur are 0.050 per cent, respectively. **Note 4.** In the case of certain qualities, the foregoing standard steels are ordinarily furnished to lower phosphorus and lower sulfur maxima as hereinafter indicated. **Note 5.** Small quanities of certain elements are present in alloy steels which are not specified or required. These elements are considered as incidental and may be present to the following maximum amounts: copper, 0.35 per cent; nickel, 0.25 per cent; chromium, 0.20 per cent; and molybdenum, 0.06 per cent. **Note 6.** Standard steels can be produced with a lead range of 0.15-0.35 per cent. Such steels are identified by inserting the letter "L" between the second and third numerals of the AISI number, e.g., 41L40. Lead is generally reported as a range of 0.15-0.35 per cent. **Note 7.** Where minimum and maximum sulfur content is shown, it is indicative of resulfurized steel. **Note 8.** Standard alloy steels, which are generally fine grain, may be produced with a boron treatment addition to improve hardenability. Such steels can be expected to contain 0.0005 to 0.003 per cent boron. These steels are identified by inserting the letter "B" between the second and third numerals of the AISI number, e.g., 50B46.

Standard Nonresulfurized Carbon Steels (a)

Cast or Heat Chemical Ranges and Limits for Bars, Blooms, Billets and Slabs

Steel designation, AISI or SAE	UNS number	Chemical composition, per cent			
		C	Mn	P max	S max
1005(b)	G10050	0.06 max	0.35 max	0.040	0.050
1006(b)	G10060	0.08 max	0.25 max	0.040	0.050
1008	G10080	0.10 max	0.30-0.50	0.040	0.050
1010	G10100	0.08-0.13	0.30-0.60	0.040	0.050
1012	G10120	0.10-0.15	0.30-0.60	0.040	0.050
1015	G10150	0.13-0.18	0.30-0.60	0.040	0.050
1016	G10160	0.13-0.18	0.60-0.90	0.040	0.050
1017	G20170	0.15-0.20	0.30-0.60	0.040	0.050
1018	G10180	0.15-0.20	0.60-0.90	0.040	0.050
1019	G10190	0.15-0.20	0.70-1.00	0.040	0.050
1020	G10200	0.18-0.23	0.30-0.60	0.040	0.050
1021	G10210	0.18-0.23	0.60-0.90	0.040	0.050
1022	G10220	0.18-0.23	0.70-1.00	0.040	0.050
1023	G10230	0.20-0.25	0.30-0.60	0.040	0.050
1025	G10250	0.22-0.28	0.30-0.60	0.040	0.050
1026	G10260	0.22-0.28	0.60-0.90	0.040	0.050
1029	G10290	0.25-0.31	0.60-0.90	0.040	0.050
1030	G10300	0.28-0.34	0.60-0.90	0.040	0.050
1035	G10350	0.32-0.38	0.60-0.90	0.040	0.050
1037	G10370	0.32-0.38	0.70-1.00	0.040	0.050
1038	G10380	0.35-0.42	0.60-0.90	0.040	0.050
1039	G10390	0.37-0.44	0.70-1.00	0.040	0.050
1040	G10400	0.37-0.44	0.60-0.90	0.040	0.050
1042	G10420	0.40-0.47	0.60-0.90	0.040	0.050
1043	G10430	0.40-0.47	0.70-1.00	0.040	0.050
1044	G10440	0.43-0.50	0.30-0.60	0.040	0.050
1045	G10450	0.43-0.50	0.60-0.90	0.040	0.050
1046	G10460	0.43-0.50	0.70-1.00	0.040	0.050
1049	G10490	0.46-0.53	0.60-0.90	0.040	0.050
1050	G10500	0.48-0.55	0.60-0.90	0.040	0.050
1053	G10530	0.48-0.55	0.70-1.00	0.040	0.050
1055	G10550	0.50-0.60	0.60-0.90	0.040	0.050
1059(b)	G10590	0.55-0.65	0.50-0.80	0.040	0.050
1060	G10600	0.55-0.65	0.60-0.90	0.040	0.050
1070	G10700	0.65-0.75	0.60-0.90	0.040	0.050
1078	G10780	0.72-0.85	0.30-0.60	0.040	0.050
1080	G10800	0.75-0.88	0.60-0.90	0.040	0.050
1084	G10840	0.80-0.93	0.60-0.90	0.040	0.050
1086(b)	G10860	0.80-0.93	0.30-0.50	0.040	0.050
1090	G10900	0.85-0.98	0.60-0.90	0.040	0.050
1095	G10950	0.90-1.03	0.30-0.50	0.040	0.050

(a) Manganese 1.00 per cent maximum. (b) Standard steel grades for wire rods and wire only.
Note. In the case of certain qualities, the foregoing standard steels are ordinarily furnished to lower phosphorus and lower sulfur maxima as hereinafter indicated. **Copper.** When copper is required, 0.20 per cent minimum is generally specified. **Lead.** Standard carbon steels can be produced with a lead range of 0.15 to 0.35 per cent, to improve machinability. Such steels are identified by inserting the letter "L" between the second and third numerals of the AISI number, e.g., 10L45. Lead is generally reported as a range of 0.15 to 0.35 per cent. **Boron.** Standard killed carbon steels, which are generally fine grain, may be produced with a boron treatment addition to improve hardenability. Such steels can be expected to contain 0.005 to 0.003 per cent boron. These steels are identified by inserting the letter "B" between the second and third numerals of the AISI number, e.g., 10B46.

Standard Resulfurized Carbon Steels

Cast or Heat Chemical Ranges and Limits for Bars, Blooms, Billets and Slabs

Steel designation, AISI or SAE	UNS number	Chemical composition, per cent			
		C	Mn	P max	S
1110	G11100	0.08-0.13	0.30-0.60	0.040	0.08-0.13
1117	G11170	0.14-0.20	1.00-1.30	0.040	0.08-0.13
1118	G11180	0.14-0.20	1.30-1.60	0.040	0.08-0.13
1137	G11370	0.32-0.39	1.35-1.65	0.040	0.08-0.13
1139	G11390	0.35-0.43	1.35-1.65	0.040	0.13-0.20
1140	G11400	0.37-0.44	0.70-1.00	0.040	0.08-0.13
1141	G11410	0.37-0.45	1.35-1.65	0.040	0.08-0.13
1144	G11440	0.40-0.48	1.35-1.65	0.040	0.24-0.33
1146	G11460	0.42-0.49	0.70-1.00	0.040	0.08-0.13
1151	G11510	0.48-0.55	0.70-1.00	0.040	0.08-0.13

Lead. Standard carbon steels can be produced with a lead range of 0.15 to 0.35 per cent, to improve machinability. Such steels are identified by inserting the letter "L" between the second and third numerals of the AISI number, e.g., 10L45. Lead is generally reported as a range of 0.15 to 0.35 per cent.

Standard Rephosphorized and Resulfurized Carbon Steels

Cast or Heat Chemical Ranges and Limits for Bars, Blooms, Billets and Slabs

Steel designation, AISI or SAE	UNS number	Chemical composition, per cent				
		C	Mn	P	S	Pb
1211	G12110	0.13 max	0.60-0.90	0.07-0.12	0.10-0.15	...
1212	G12120	0.13 max	0.70-1.00	0.07-0.12	0.16-0.23	...
1213	G12130	0.13 max	0.70-1.00	0.07-0.12	0.24-0.33	...
1215	G12150	0.09 max	0.75-1.05	0.04-0.09	0.26-0.35	...
12L14	G12144	0.15 max	0.85-1.15	0.04-0.09	0.26-0.35	0.15-0.35

Silicon. It is not common practice to produce the 12XX series of steels to specified limits for silicon because of its adverse effect on machinability. **Lead.** Standard carbon steel scan be produced with a lead range of 0.15 to 0.35 per cent to improve machinability. Such steel is identified by inserting the letter "L" between the second and third numerals of the AISI steel designation, e.g., 12L15.

Standard Nonresulfurized Carbon Steels (a)

Cast or Heat Chemical Ranges and Limits for Bars, Blooms, Billets and Slabs

Steel designation, AISI or SAE	UNS number	Chemical composition, per cent			
		C	Mn	P max	S max
1513	G15130	0.10-0.16	1.10-1.40	0.040	0.050
1522	G15220	0.18-0.24	1.10-1.40	0.040	0.050
1524	G15240	0.19-0.25	1.35-1.65	0.040	0.050
1526	G15260	0.22-0.29	1.10-1.40	0.040	0.050
1527	G15270	0.22-0.29	1.20-1.50	0.040	0.050
1541	G15410	0.36-0.44	1.35-1.65	0.040	0.050
1548	G15480	0.44-0.52	1.10-1.40	0.040	0.050
1551	G15510	0.45-0.56	0.85-1.15	0.040	0.050
1552	G15520	0.47-0.55	1.20-1.50	0.040	0.050
1561	G15610	0.55-0.65	0.75-1.05	0.040	0.050
1566	G15660	0.60-0.71	0.85-1.15	0.040	0.050

(a) Manganese maximum over 1.00 per cent.

Note. In the case of certain qualities, the foregoing standard steels are ordinarily furnished to lower phosphorus and lower sulfur maxima as hereinafter indicated. **Copper.** When copper is required, 0.20 per cent minimum is generally specified. **Lead.** Standard carbon steels can be produced with a lead range of 0.15 to 0.35 per cent, to improve machinability. Such steels are identified by inserting the letter "L" between the second and third numerals of the AISI number, e.g., 10L45. Lead is generally reported as a range of 0.15 to 0.35 per cent. **Boron.** Standard killed carbon steels, which are generally fine grain, may be produced with a boron treatment addition to improve hardenability. Such steels can be expected to contain 0.005 to 0.003 per cent boron. These steels are identified by inserting the letter "B" between the second and third numerals of the AISI number, e.g., 10B46.

Assumed Densities by Silicon and Aluminum Content
(ASTM A34)

Silicon-aluminum factor, % Si + 1.7 (% Al)	Assumed density, grams per cubic centimeter (a)
0.00-0.65	7.85
0.66-1.40	7.80
1.41-2.15	7.75
2.16-2.95	7.70
2.96-3.70	7.65

(a) Kilograms/m³ = (grams/cubic centimeter) X 1000.

Maximum Core Losses for Flat Rolled, Nonoriented Fully Processed Electrical Steel (a)
(ASTM A677)

ASTM type	Former AISI type	Thickness in.	Thickness mm	Maximum core loss at 15 kG (1.5 T) W/lb 60 Hz	Maximum core loss at 15 kG (1.5 T) W/kg 50 Hz
36F145	M-15	.014	.36	1.45	2.53
47F168	M-15	.0185	.47	1.68	2.93
36F158	M-19	.014	.36	1.58	2.75
47F174	M-19	.0185	.47	1.74	3.03
64F208	M-19	.025	.64	2.08	3.62
36F168	M-22	.014	.36	1.68	2.93
47F185	M-22	.0185	.47	1.85	3.22
64F218	M-22	.025	.64	2.18	3.80
36F180	M-27	.014	.36	1.80	3.13
47F190	M-27	.0185	.47	1.90	3.31
64F225	M-27	.025	.64	2.25	3.92
36F190	M-36	.014	.36	1.90	3.31
47F205	M-36	.0185	.47	2.05	3.57
64F240	M-36	.025	.64	2.40	4.18
47F230	M-43	.0185	.47	2.30	4.01
64F270	M-43	.025	.64	2.70	4.70
47F305	M-45	.0185	.47	3.05	5.31
64F360	M-45	.025	.64	3.60	6.27
47F400	M-47	.0185	.47	4.00	6.96
64F490	M-47	.025	.64	4.90	8.53
47F475	...	.0185	.47	4.75	8.27
64F610	...	.025	.64	6.10	10.62

(a) Tests are made on Epstein specimens not annealed after shear ng, and consisting of strips of which half are cut parallel and half are cut transverse to the rolling direction.

Maximum Core Losses for
Flat Rolled Nonoriented Semiprocessed Electrical Steel (a)
(ASTM A683)

ASTM type	Former AISI type	Thickness in.	Thickness mm	Maximum core loss at 15 kG (1.5 T) W/lb 60 Hz	Maximum core loss at 15 kG (1.5 T) W/kg 50 Hz
47S178	M-27	.0185	.47	1.78	3.10
64S194	M-27	.025	.64	1.94	3.38
47S188	M-36	.0185	.47	1.88	3.27
64S213	M-36	.025	.64	2.13	3.71
47S200	M-43	.0185	.47	2.00	3.48
64S230	M-43	.025	.64	2.30	4.01
47S250	M-45	.0185	.47	2.50	4.35
64S280	M-45	.025	.64	2.80	4.88
47S300	. . .	.0185	.47	3.00	5.22
64S350	. . .	.025	.64	3.50	6.10

(a) Core loss is determined after a quality evaluation anneal. This anneal is performed on Epstein specimens consisting of strips of which half are cut parallel and half are cut transverse to the rolling direction. Customarily, the carbon level is reduced to 0.005% or less and the anneal requires using a suitable atmosphere and a soak temperature of 1550 °F (845 °C) for approimately one hour.

Note. Above table does not apply to "full hard" material.

Maximum Core Losses for
Flat Rolled Grain-oriented Fully Processed Electrical Steel (a)

ASTM type	Former AISI type	Thickness in.	Thickness mm	Maximum core loss at 15 kG (1.5 T) W/lb 60 Hz	Maximum core loss at 15 kG (1.5 T) W/kg 50 Hz
ASTM A665					
27G053	M-4	0.0106	0.27	0.53	0.89
30G058	M-5	0.0118	0.30	0.58	0.97
35G066	M-6	0.0138	0.35	0.66	1.11
ASTM A725					
27H076	M-4	0.0106	0.27	0.76	1.27
30H083	M-5	0.0118	0.30	0.83	1.39
35H094	M-6	0.0138	0.35	0.94	1.57

(a) Tests are made on Epstein specimens cut parallel to the direction of rolling and stress-relief annealed in a manner to ensure magnetic characteristics are like those inherent in the materials from which the specimens were taken. Practices vary, but anneals at temperatures in the range of 1450 °F (790 °C) to 1550 °F (845 °C) for times of about one hour in atmospheres having dew points not greater than 0 °F (-20 °C) and comprised to mixtures of pure nitrogen and pure hydrogen usually produce satisfactory results.

Alloy Steel Plate Compositions

Open Hearth and Basic Oxygen

AISI no.	C	Mn	P max	S max	Si(a)	Ni	Cr	Mo	V
1330	0.27-0.34	1.50-1.90	0.035	0.040	0.15-0.30	...	...	...	...
1335	0.32-0.39	1.50-1.90	0.035	0.040	0.15-0.30	...	...	...	...
1340	0.36-0.44	1.50-1.90	0.035	0.040	0.15-0.30	...	...	...	..
1345	0.41-0.49	1.50-1.90	0.035	0.040	0.15-0.30	...	...	...	...
4118	0.17-0.23	0.60-0.90	0.035	0.040	0.15-0.30	...	0.40-0.65	0.08-0.15	...
4130	0.27-0.34	0.35-0.60	0.035	0.040	0.15-0.30	...	0.80-1.15	0.15-0.25	...
4135	0.32-0.39	0.65-0.95	0.035	0.040	0.15-0.30	...	0.80-1.15	0.15-0.25	...
4137	0.33-0.40	0.65-0.95	0.035	0.040	0.15-0.30	...	0.80-1.15	0.15-0.25	...
4140	0.36-0.44	0.70-1.00	0.035	0.040	0.15-0.30	...	0.80-1.15	0.15-0.25	...
4142	0.38-0.46	0.70-1.00	0.035	0.040	0.15-0.30	...	0.80-1.15	0.15-0.25	...
4145	0.41-0.49	0.70-1.00	0.035	0.040	0.15-0.30	...	0.80-1.15	0.15-0.25	...
4340	0.36-0.44	0.55-0.80	0.035	0.040	0.15-0.30	1.65-2.00	0.60-0.90	0.20-0.30	...
E4340	0.37-0.44	0.60-0.85	0.025	0.025	0.15-0.30	1.65-2.00	0.65-0.90	0.20-0.30	...
4615	0.12-0.18	0.40-0.65	0.035	0.040	0.15-0.30	1.65-2.00	...	0.20-0.30	...
4617	0.15-0.21	0.40-0.65	0.035	0.040	0.15-0.30	1.65-2.00	...	0.20-0.30	...
4620	0.16-0.22	0.40-0.65	0.035	0.040	0.15-0.30	1.65-2.00	...	0.20-0.30	...
5160	0.54-0.65	0.70-1.00	0.035	0.040	0.15-0.30	...	0.60-0.90	...	...
6150	0.46-0.54	0.60-0.90	0.035	0.040	0.15-0.30	...	0.80-1.15	...	0.15 min
8615	0.12-0.18	0.60-0.90	0.035	0.040	0.15-0.30	0.40-0.70	0.35-0.60	0.15-0.25	...
8617	0.15-0.21	0.60-0.90	0.035	0.040	0.15-0.30	0.40-0.70	0.35-0.60	0.15-0.25	...
8620	0.17-0.23	0.60-0.90	0.035	0.040	0.15-0.30	0.40-0.70	0.35-0.60	0.15-0.25	...
8622	0.19-0.25	0.60-0.90	0.035	0.040	0.15-0.30	0.40-0.70	0.35-0.60	0.15-0.25	...
8625	0.22-0.29	0.60-0.90	0.035	0.040	0.15-0.30	0.40-0.70	0.35-0.60	0.15-0.25	...
8627	0.24-0.31	0.60-0.90	0.035	0.040	0.15-0.30	0.40-0.70	0.35-0.60	0.15-0.25	...
8630	0.27-0.34	0.60-0.90	0.035	0.040	0.15-0.30	0.40-0.70	0.35-0.060	0.15-0.25	...
8637	0.33-0.40	0.70-1.00	0.035	0.040	0.15-0.30	0.40-0.70	0.35-0.60	0.15-0.25	...
8640	0.36-0.44	0.70-1.00	0.035	0.040	0.15-0.30	0.40-0.70	0.35-0.60	0.15-0.25	...
8655	0.49-0.60	0.70-1.00	0.035	0.040	0.15-0.30	0.40-0.70	0.35-0.60	0.15-0.25	...
8742	0.38-0.46	0.70-1.00	0.035	0.040	0.15-0.30	0.40-0.70	0.35-0.60	0.20-0.30	...

(a)Silicon available in ranges of 0.10-0.20 per cent, 0.20-0.35 per cent, and 0.35 per cent maximum (when carbon deoxidized) when so specified by the purchaser.
Note 1. Small quantities of certain elements not required may be found. These elements are to be considered as incidental and are acceptable to the following maximum amounts: copper to 0.35 per cent, nickel to 0.25 per cent, chromium to 0.20 per cent, and molybdenum to 0.06 per cent.
Note 2. When electric furnace steel is ordered, the carbon range is restricted 0.01 per cent, manganese 0.05 per cent, chromium 0.05 per cent up to 1.25 per cent, incl., and 0.10 per cent over 1.25 per cent. The maximum phosphorus and sulphur is 0.025 per cent each. **Note 3.** Boron or lead may be added to the above compositions.

Alloy Steel for Plates

ASTM Specification Designations

Desig-nation	Grade	Thickness, in.	C	Mn	P max	S max	Si	Cr	Ni	Mo	V	Ti	Zr	Cu	B	Co	Al
A 202	A	...	0.17 max	1.05-1.40	0.035	0.04	0.60-0.90	0.35-0.60	...	...	...	...	...	...	...	...	...
	B	...	0.25 max	1.05-1.40	0.035	0.04	0.60-0.90	0.35-0.60	...	...	...	...	...	...	...	...	...
A 203	A	Up to 2, incl.	0.17 max	0.70 max	0.035	0.04	0.15-0.30	...	2.10-2.50	...	...	...	...	...	...	...	...
		Over 2 to 4, incl.	0.20 max	0.80 max	0.035	0.04	0.15-0.30	...	2.10-2.50	...	...	...	...	...	...	...	...
		Over 4 to 6, incl.	0.23 max	0.80 max	0.035	0.04	0.15-0.30	...	2.10-2.50	...	...	...	...	...	...	...	...
	B	Up to 2, incl.	0.21 max	0.70 max	0.035	0.04	0.15-0.30	...	2.10-2.50	...	...	...	...	...	...	...	...
		Over 2 to 4, incl.	0.24 max	0.80 max	0.035	0.04	0.15-0.30	...	2.10-2.50	...	...	...	...	...	...	...	...
		Over 4 to 6, incl.	0.25 max	0.80 max	0.035	0.04	0.15-0.30	...	2.10-2.50	...	...	...	...	...	...	...	...
	D	Up to 2, incl.	0.17 max	0.70 max	0.035	0.04	0.15-0.30	...	3.25-3.75	...	...	...	...	...	...	...	...
		Over 2 to 4, incl.	0.20 max	0.80 max	0.035	0.04	0.15-0.30	...	3.25-3.75	...	...	...	...	...	...	...	...
	E	Up to 2, incl.	0.20 max	0.70 max	0.035	0.04	0.15-0.30	...	3.25-3.75	...	...	...	...	...	...	...	...
		Over 2 to 4, incl.	0.23 max	0.80 max	0.035	0.04	0.15-0.30	...	3.25-3.75	...	...	...	...	...	...	...	...
A 204	A	1 and under	0.18 max	0.90 max	0.035	0.04	0.15-0.30	...	...	0.45-0.60	...	...	...	...	...	...	...
		Over 1 to 2, incl.	0.21 max	0.90 max	0.035	0.04	0.15-0.30	...	...	0.45-0.60	...	...	...	...	...	...	...
		Over 2 to 4, incl.	0.23 max	0.90 max	0.035	0.04	0.15-0.30	...	...	0.45-0.60	...	...	...	...	...	...	...
		Over 4 to 6, incl.	0.25 max	0.90 max	0.035	0.04	0.15-0.30	...	...	0.45-0.60	...	...	...	...	...	...	...
	B	1 and under	0.20 max	0.90 max	0.035	0.04	0.15-0.30	...	...	0.45-0.60	...	...	...	...	...	...	...
		Over 1 to 2, incl.	0.23 max	0.90 max	0.035	0.04	0.15-0.30	...	...	0.45-0.60	...	...	...	...	...	...	...
		Over 2 to 4, incl.	0.25 max	0.90 max	0.035	0.04	0.15-0.30	...	...	0.45-0.60	...	...	...	...	...	...	...
		Over 4 to 6, incl.	0.27 max	0.90 max	0.035	0.04	0.15-0.30	...	...	0.45-0.60	...	...	...	...	...	...	...
	C	1 and under	0.23 max	0.90 max	0.035	0.04	0.15-0.30	...	...	0.45-0.60	...	...	...	...	...	...	...
		Over 1 to 2, incl.	0.26 max	0.90 max	0.035	0.04	0.15-0.30	...	...	0.45-0.60	...	...	...	...	...	...	...
		Over 2 to 4, incl.	0.28 max	0.90 max	0.035	0.04	0.15-0.30	...	...	0.45-0.60	...	...	...	...	...	...	...
A 225	A	...	0.18 max	1.45 max	0.035	0.04	0.15-0.30	...	...	...	0.09-0.14	...	...	...	...	...	...
	B	...	0.20 max	1.45 max	0.035	0.04	0.15-0.30	...	...	...	0.09-0.14	...	...	...	...	...	...
A 302	A	1 and under	0.23 max	0.95-1.30	0.035	0.040	0.15-0.30	...	...	0.45-0.60	...	...	...	...	...	...	...
		Over 1 to 2, incl.	0.23 max	0.95-1.30	0.035	0.040	0.15-0.30	...	...	0.45-0.60	...	...	...	...	...	...	...
		Over 2	0.25 max	0.95-1.30	0.035	0.040	0.15-0.30	...	...	0.45-0.60	...	...	...	...	...	...	...
	B	1 and under	0.20 max	1.15-1.50	0.035	0.040	0.15-0.30	...	...	0.45-0.60	...	...	...	...	...	...	...
		Over 1 to 2, incl.	0.23 max	1.15-1.50	0.035	0.040	0.15-0.30	...	...	0.45-0.60	...	...	...	...	...	...	...
		Over 2	0.25 max	1.15-1.50	0.035	0.040	0.15-0.30	...	...	0.45-0.60	...	...	...	...	...	...	...
	C	1 and under	0.20 max	1.15-1.50	0.035	0.040	0.15-0.30	...	0.40-0.70	0.45-0.60	...	...	...	...	...	...	...
		Over 1 to 2, incl.	0.23 max	1.15-1.50	0.035	0.040	0.15-0.30	...	0.40-0.70	0.45-0.60	...	...	...	...	...	...	...
		Over 2	0.25 max	1.15-1.50	0.035	0.040	0.15-0.30	...	0.40-0.70	0.45-0.60	...	...	...	...	...	...	...
	D	1 and under	0.20 max	1.15-1.50	0.035	0.040	0.15-0.30	...	0.70-1.00	0.45-0.60	...	...	...	...	...	...	...
		Over 1 to 2, incl.	0.23 max	1.15-1.50	0.035	0.040	0.15-0.30	...	0.70-1.00	0.45-0.60	...	...	...	...	...	...	...
		Over 2	0.25 max	1.15-1.50	0.035	0.040	0.15-0.30	...	0.70-1.00	0.45-0.60	...	...	...	...	...	...	...
A 353	...	...	0.13 max	0.90 max	0.035	0.040	0.15-0.30	...	8.50-9.50	...	...	...	...	...	...	...	...

ASTM Specification Designations (continued)

Designation	Grade	Thickness, in.	C	Mn	P max	S max	Si	Cr	Ni	Mo	V	Ti	Zr	Cu	B	Co	Al
A 387																	
Group I	A	...	0.21 max	0.55-0.80	0.035	0.040	0.15-0.30	0.50-0.80	...	0.45-0.60	...	...	...	...	...	...	...
	B	...	0.17 max	0.40-0.65	0.035	0.040	0.15-0.30	0.80-1.15	...	0.45-0.60	...	...	...	...	...	...	...
	C	...	0.17 max	0.40-0.65	0.035	0.040	0.50-0.80	1.00-1.50	...	0.45-0.65	...	...	...	...	...	...	...
Group II	D	...	0.15 max	0.30-0.60	0.035	0.035	0.15-0.30	2.00-2.50	...	0.90-1.10	...	...	...	...	...	...	...
	E	...	0.15 max	0.30-0.60	0.035	0.035	0.15-0.30	2.75-3.25	...	0.90-1.10	...	...	...	...	...	...	...
A 514 and A 517	A	...	0.15-0.21	0.80-1.10	0.035	S max for A 514:0.04	0.40-0.80	0.50-0.80	...	0.18-0.28	...	...	...	...	0.0025 max	...	...
	B-514	...	0.12-0.21	0.70-1.00	0.035	S max for A 517:0.040	0.20-0.35	0.40-0.65	...	0.15-0.25	0.03-0.08	0.01-0.03	0.05-0.15	...	0.0005-0.005	...	...
	B-517	...	0.15-0.21	0.70-1.00	0.035		0.20-0.35	0.40-0.65	...	0.15-0.25	0.03-0.08	0.01-0.03	...	...	0.0005-0.005	...	...
	C	...	0.10-0.20	1.10-1.50	0.035		0.15-0.30	...	...	0.20-0.30	...	...	...	...	0.001-0.005	...	...
	D	...	0.13-0.20	0.40-0.70	0.035		0.20-0.35	0.85-1.20	...	0.15-0.25	(a)	0.04-0.10	...	0.20-0.40	0.0015-0.005	...	...
	E	...	0.12-0.20	0.40-0.70	0.035		0.20-0.35	1.40-2.00	...	0.40-0.60	(a)	0.04-0.10	...	0.20-0.40	0.0015-0.005	...	...
	F	...	0.10-0.20	0.60-1.00	0.035		0.15-0.35	0.40-0.65	0.70-1.00	0.40-0.60	0.03-0.08	...	...	0.15-0.50	0.002-0.006	...	...
	G	...	0.15-0.21	0.80-1.10	0.035		0.50-0.90	0.50-0.90	...	0.40-0.60	...	...	...	...	0.0025 max	...	...
	H-514	...	0.12-0.21	0.95-1.30	0.035		0.20-0.35	0.40-0.65	0.30-0.70	0.20-0.30	0.03-0.08	...	...	...	0.0005-0.005	...	...
	H-517	...	0.12-0.21	0.95-1.30	0.035		0.20-0.35	0.40-0.65	0.30-0.70	0.20-0.30	0.03-0.08	...	0.05-0.15	...	0.0005 min	...	...
	J	...	0.12-0.21	0.45-0.70	0.035		0.20-0.35	...	...	0.50-0.65	...	...	...	...	0.001-0.005	...	...
	K	...	0.10-0.20	1.10-0.50	0.035		0.15-0.30	...	...	0.45-0.55	...	...	...	...	0.001-0.005	...	...
	L	...	0.13-0.20	0.40-0.70	0.035		0.20-0.35	1.15-1.65	...	0.25-0.40	(a)	0.04-0.10	...	0.20-0.40	0.0015-0.005	...	...
	M	...	0.12-0.21	0.45-0.70	0.035		0.20-0.35	...	1.20-1.50	0.45-0.60	...	...	...	...	0.001-0.005	...	...
	N-514 only	...	0.15-0.21	0.80-1.10	0.035		0.40-0.90	0.50-0.80	...	0.25 max	...	...	0.05-0.15	...	0.0005-0.0025	...	...
	P	...	0.12-0.21	0.45-0.70	0.035		0.20-0.35	0.85-1.20	1.20-1.50	0.45-0.60	...	...	...	...	0.001-0.005	...	...
A 533	A	...	0.25 max	1.15-1.50	0.035	0.040	0.15-0.30	...	...	0.45-0.60	...	...	...	...	...	...	...
	B	...	0.25 max	1.15-1.50	0.035	0.040	0.15-0.30	...	0.40-0.70	0.45-0.60	...	...	...	...	...	...	...
	C	...	0.25 max	1.15-1.50	0.035	0.040	0.15-0.30	...	0.70-1.00	0.45-0.60	...	...	...	...	...	...	...
	D	...	0.25 max	1.15-1.50	0.035	0.040	0.15-0.30	...	0.20-0.40	0.45-0.60	(b)	...	...	...	...	...	...
A 538	A	...	0.03 max	0.10 max	0.010	0.010	0.10 max	17.0-19.0	17.0-19.0	4.0-4.5	...	0.10-0.25	...	...	...	7.0-8.5	0.05-0.15
	B	...	0.03 max	0.10 max	0.010	0.010	0.10 max	17.0-19.0	17.0-19.0	4.6-5.1	...	0.30-0.50	...	...	...	7.0-8.5	0.05-0.15
	C	...	0.03 max	0.10 max	0.010	0.010	0.10 max	18.0-19.0	18.0-19.0	4.6-5.2	...	0.55-0.80	...	...	...	8.0-9.5	0.05-0.15
A 542	...	...	0.15 max	0.30-0.60	0.035	0.035	0.15-0.30	2.00-2.50	...	0.90-1.10	...	...	...	...	...	...	...
A 543	A	Up to 4, incl.	0.23 max	0.40 max	0.035	0.040	0.20-0.35	1.50-2.00	2.60-3.25	0.45-0.60	0.03 max(b)	...	...	...	...	...	...
		Over 4	0.23 max	0.40 max	0.035	0.040	0.20-0.35	1.50-2.00	3.00-4.00	0.45-0.60	0.03 max(b)	...	...	...	...	...	...
	B	Up to 4, incl.	0.23 max	0.40 max	0.020	0.020	0.20-0.35	1.50-2.00	2.60-3.25	0.45-0.60	0.03 max(b)	...	...	...	...	...	...
		Over 4	0.23 max	0.40 max	0.020	0.020	0.20-0.35	1.50-2.00	3.00-4.00	0.45-0.60	0.03 max(b)	...	...	...	...	...	...
A 553	A	...	0.13 max	0.90 max	0.035	0.040	0.15-0.30	...	8.50-9.50	...	...	...	...	...	...	...	...
	B	...	0.13 max	0.90 max	0.035	0.040	0.15-0.30	...	7.50-8.50	...	...	...	...	...	...	...	...
A 562	...	...	0.12 max	1.20 max	0.04	0.05	0.15-0.50	...	...	...	...	4 x C min	...	0.15 max	...	...	...

(a) May be substituted for part or all of titanium content on a one for one basis. (b) Vanadium content of 0.05 per cent max shall be furnished when specified.
Note: ASME may or may not have specifications with similar chemical composition ranges and limits. Refer to ASME publications.

Standard Types of Stainless and Heat Resisting Steels

Chemical Ranges and Limits of Cast or Heat (Formerly Ladle) Analysis

		Chemical composition, per cent (maximum unless otherwise shown)								
Type number	UNS number	C	Mn	P	S	Si	Cr	Ni	Mo	Other elements
201	S20100	0.15	5.50-7.50	0.060	0.030	1.00	16.00-18.00	3.50-5.50	...	N 0.25
202	S20200	0.15	7.50-10.00	0.060	0.030	1.00	17.00-19.00	4.00-6.00	...	N 0.25
205	S20500	0.12-0.25	14.00-15.50	0.060	0.030	1.00	16.50-18.00	1.00-1.75	...	N 0.32-0.40
301	S30100	0.15	2.00	0.045	0.030	1.00	16.00-18.00	6.00-8.00	...	...
302	S30200	0.15	2.00	0.045	0.030	1.00	17.00-19.00	8.00-10.00	...	...
302B	S30215	0.15	2.00	0.045	0.030	2.00-3.00	17.00-19.00	8.00-10.00	...	...
303	S30300	0.15	2.00	0.20	0.15 min	1.00	17.00-19.00	8.00-10.00	0.60(a)	...
303Se	S30323	0.15	2.00	0.20	0.060	1.00	17.00-19.00	8.00-10.00	...	Se 0.15 min
304	S30400	0.08	2.00	0.045	0.030	1.00	18.00-20.00	8.00-10.50	...	...
304L	S30403	0.030	2.00	0.045	0.030	1.00	18.00-20.00	8.00-12.00	...	...
	S30430	0.08	2.00	0.045	0.030	1.00	17.00-19.00	8.00-10.00	...	Cu 3.00-4.00
304N	S30451	0.08	2.00	0.045	0.030	1.00	18.00-20.00	8.00-10.50	...	N 0.10-0.16
305	S30500	0.12	2.00	0.045	0.030	1.00	17.00-19.00	10.50-13.00	...	...
308	S30800	0.08	2.00	0.045	0.030	1.00	19.00-21.00	10.00-12.00	...	...
309	S30900	0.20	2.00	0.045	0.030	1.00	22.00-24.00	12.00-15.00	...	...
309S	S30908	0.08	2.00	0.045	0.030	1.00	22.00-24.00	12.00-15.00	...	...
310	S31000	0.25	2.00	0.045	0.030	1.50	24.00-26.00	19.00-22.00	...	...
310S	S31008	0.08	2.00	0.045	0.030	1.50	24.00-26.00	19.00-22.00	...	...
314	S31400	0.25	2.00	0.045	0.030	1.50-3.00	23.00-26.00	19.00-22.00	...	...
316	S31600	0.08	2.00	0.045	0.030	1.00	16.00-18.00	10.00-14.00	2.00-3.00	...
316F	S31620	0.08	2.00	0.20	0.10 min	1.00	16.00-18.00	10.00-14.00	1.75-2.50	...
316L	S31603	0.030	2.00	0.045	0.030	1.00	16.00-18.00	10.00-14.00	2.00-3.00	...
316N	S31651	0.08	2.00	0.045	0.030	1.00	16.00-18.00	10.00-14.00	2.00-3.00	N 0.10-0.16
317	S31700	0.08	2.00	0.045	0.030	1.00	18.00-20.00	11.00-15.00	3.00-4.00	...
317L	S31703	0.030	2.00	0.045	0.030	1.00	18.00-20.00	11.00-15.00	3.00-4.00	...
321	S32100	0.08	2.00	0.045	0.030	1.00	17.00-19.00	9.00-12.00	...	Ti 5XC min
329	S32900	0.10	2.00	0.040	0.030	1.00	25.00-30.00	3.00-6.00	1.00-2.00	...
330	N08330	0.08	2.00	0.040	0.030	0.75-1.50	17.00-20.00	34.00-37.00	...	...
347	S34700	0.08	2.00	0.045	0.030	1.00	17.00-19.00	9.00-13.00	...	Cb+Ta 10 X C min
348	S34800	0.08	2.00	0.045	0.030	1.00	17.00-19.00	9.00-13.00	...	Cb+Ta 10 X C min Ta 0.10 max Co 0.20 max

Chemical Ranges and Limits of Cast or Heat (Formerly Ladle) Analysis (continued)

Type number	UNS number	C	Mn	P	S	Si	Cr	Ni	Mo	Other elements
384	S38400	0.08	2.00	0.045	0.030	1.00	15.00-17.00	17.00-19.00	...	...
403	S40300	0.15	1.00	0.040	0.030	0.50	11.50-13.00	...	...	...
405	S40500	0.08	1.00	0.040	0.030	1.00	11.50-14.50	...	...	Al 0.10-0.30
409	S40900	0.08	1.00	0.045	0.045	1.00	10.50-11.75	...	...	Ti 6 X C min 0.75 max
410	S41000	0.15	1.00	0.040	0.030	1.00	11.50-13.50	...	...	...
414	S41400	0.15	1.00	0.040	0.030	1.00	11.50-13.50	1.25-2.50	...	...
416	S41600	0.15	1.25	0.060	0.15 min	1.00	12.00-14.00	...	0.60(a)	...
416Se	S41623	0.15	1.25	0.060	0.060	1.00	12.00-14.00	...	...	Se 0.15 min
420	S42000	Over 0.15	1.00	0.040	0.030	1.00	12.00-14.00	...	...	...
420F	S42020	Over 0.15	1.25	0.060	0.15 min	1.00	12.00-14.00	...	0.60(a)	...
422	S42200	0.20-0.25	1.00	0.025	0.025	0.75	11.00-13.00	0.50-1.00	0.75-1.25	V 0.15-0.30 W 0.75-1.25
429	S42900	0.12	1.00	0.040	0.030	1.00	14.00-16.00	...	...	...
430	S43000	0.12	1.00	0.040	0.030	1.00	16.00-18.00	...	...	...
430F	S43020	0.12	1.25	0.060	0.15 min	1.00	16.00-18.00	...	0.60(a)	...
430FSe	S43023	0.12	1.25	0.060	0.060	1.00	16.00-18.00	...	...	Se 0.15 min
431	S43100	0.20	1.00	0.040	0.030	1.00	15.00-17.00	1.25-2.50	...	...
434	S43400	0.12	1.00	0.040	0.030	1.00	16.00-18.00	...	0.75-1.25	...
436	S43600	0.12	1.00	0.040	0.030	1.00	16.00-18.00	...	0.75-1.25	Cb+Ta 5 XC min – 0.70 max
440A	S44002	0.60-0.75	1.00	0.040	0.030	1.00	16.00-18.00	...	0.75	...
440B	S44003	0.75-0.95	1.00	0.040	0.030	1.00	16.00-18.00	...	0.75	...
440C	S44004	0.95-1.20	1.00	0.040	0.030	1.00	16.00-18.00	...	0.75	...
442	S44200	0.20	1.00	0.040	0.030	1.00	18.00-23.00	...	...	...
446	S44600	0.20	1.50	0.040	0.030	1.00	23.00-27.00	...	...	N 0.25
501	S50100	Over 0.10	1.00	0.040	0.030	1.00	4.00-6.00	...	0.40-0.65	...
502	S50200	0.10	1.00	0.040	0.030	1.00	4.00-6.00	...	0.40-0.65	...
	S13800	0.05	0.10	0.01	0.008	0.10	12.25-13.25	7.50-8.50	2.00-2.50	Al 0.90-1.35 N 0.010
	S15500	0.07	1.00	0.040	0.030	1.00	14.00-15.50	3.50-5.50	...	Cu 2.50-4.50 Cb+Ta 0.15-0.45
	S17400	0.07	1.00	0.040	0.030	1.00	15.50-17.50	3.00-5.00	...	Cu 3.00-5.00 Cb+Ta 0.15-0.45
	S17700	0.09	1.00	0.040	0.040	1.00	16.00-18.00	6.50-7.75	...	Al 0.75-1.50

(a) May be added at manufacturer's option.

Note: For some tube-making processes it is necessary that the nickel content of several of the austenitic grades be slightly higher than shown in the above table. The producer should be consulted for the appropriate nickel ranges for such grades.

Identification and Type Classification of Tool Steels with Typical Applications

High-speed Tool Steels
Symbol M, Molybdenum Types

The high-speed steels listed under the letter symbol M generally are considered to have molybdenum as the principal alloying element, although several contain an equal or a slightly greater amount of such elements as tungsten or cobalt. Those types with higher carbon and vanadium contents generally offer improved abrasion resistance, but machinability and grindability may be adversely affected. The series beginning with M41 is characterized by the capability of attaining exceptionally high hardness in heat treatment.

Typical Applications (a)

Broaches	Hobs	Reamers
Chasers	Lathe tools	Routers
Cheeking tools	Milling cutters	Saws
Drills	Planer tools	Taps
End mills	Punches	Woodworking tools

(a) The producer should be consulted for specific recommendations and availability from warehouse stocks.

In addition to the above cutting tools, some of these high speed steels are successfully used for such cold work applications as cold header die inserts, thread rolling dies, punches, and blanking dies. For such applications in order to increase toughness, the high-speed steels frequently are hardened from a lower temperature than used for cutting tools.

It should be recognized that because of differences in properties, availability, and economic considerations, all of these steels should not be considered for each application.

Type	UNS	\multicolumn{8}{c}{Identifying elements, per cent}
		C	Mn	Si	W	Mo	Cr	V	Co
M1	T11301	.85(a)	...	...	1.50	8.50	4.00	1.00	...
M2	T11302	.85; 1.00(a)	...	...	6.00	5.00	4.00	2.00	...
M3 Class 1	T11313	1.05	...	...	6.00	5.00	4.00	2.40	...
M3 Class 2	T11323	1.20	...	...	6.00	5.00	4.00	3.00	...
M4	T11304	1.30	...	...	5.50	4.50	4.00	4.00	...
M6	T11306	.80	...	...	4.00	5.00	4.00	1.50	12.00
M7	T11307	1.00	...	...	1.75	8.75	4.00	2.00	...
M10	T11310	.85; 1.00(a)	...	...	...	8.00	4.00	2.00	...
M30	T11330	.80	...	...	2.00	8.00	4.00	1.25	5.00
M33	T11333	.90	...	...	1.50	9.50	4.00	1.15	8.00
M34	T11334	.90	...	...	2.00	8.00	4.00	2.00	8.00
M36	T11336	.80	...	...	6.00	5.00	4.00	2.00	8.00
M41	T11341	1.10	...	...	6.75	3.75	4.25	2.00	5.00
M42	T11342	1.10	...	...	1.50	9.50	3.75	1.15	8.00
M43	T11343	1.20	...	...	2.75	8.00	3.75	1.60	8.25
M44	T11344	1.15	...	...	5.25	6.25	4.25	2.00	12.00
M46	T11346	1.25	...	...	2.00	8.25	4.00	3.20	8.25
M47	T11347	1.10	...	...	1.50	9.50	3.75	1.25	5.00

(a) Other carbon contents may be available.
Note: Some of the types can be produced with a sulphur addition to improve machinability.

High-speed Tool Steels (continued)

Symbol T, Tungsten Types

The high-speed steels listed under the letter symbol T have tungsten as the principal alloying element. They are more resistant to decarburization in heat treatment than the molybdenum types, and usually are hardened from higher temperatures.

Typical Applications (a)

Type	UNS	C	Mn	Si	W	Mo	Cr	V	Co
T1	T12001	.75(b)	...	...	18.00	...	4.00	1.00	...
T2	T12002	.80	...	...	18.00	...	4.00	2.00	...
T4	T12004	.75	...	...	18.00	...	4.00	1.00	5.00
T5	T12005	.80	...	...	18.00	...	4.00	2.00	8.00
T6	T12006	.80	...	...	20.00	...	4.50	1.50	12.00
T8	T12008	.75	...	...	14.00	...	4.00	2.00	5.00
T15	T12015	1.50	...	...	12.00	...	4.00	5.00	5.00

(a) The producer should be consulted for specific recommendations and availability from warehouse stocks. (b) Other carbon contents may be available.
Note: Some of the types can be produced with a sulphur addition to improve machinability.

Hot Work Tool Steels

Symbol H
H1-H19, incl., Chromium Types

These hot work steels are referred to as chromium types, although there may be other significant alloying elements present. The types containing molybdenum, H10, H11, H12 and H13, are the most widely used of all the hot work steels, and are characterized by high hardenability and excellent toughness. They offer the best selection from warehouse stocks.

The chromium-tungsten type H14 and the chromium-tungsten-cobalt type H19 offer greater resistance to softening, but are less ductile in service, and are not as readily available from stock.

Typical Hot Work Applications (a)

Bolsters	Forging dies	Mandrels
Die casting dies	Gripper dies	Punches
Dummy blocks	Hot header dies	Piercing tools
Extrusion dies		

(a) The producer should be consulted for specific recommendations and availability from warehouse stocks.

In addition to the above hot work tooling, some of these hot work steels are used for cold work applications requiring exceptional toughness at relatively high hardness levels. Such applications may include coining dies, forming rolls, and header die casings.

Type	UNS	C	Mn	Si	W	Mo	Cr	V	Co
H10	T20810	.40	...	...	...	2.50	3.25	.40	...
H11	T20811	.35	...	...	...	1.50	5.00	.40	...
H12	T20812	.35	...	...	1.50	1.50	5.00	.40	...
H13	T20813	.35	...	...	...	1.50	5.00	1.00	...
H14	T20814	.40	...	...	5.00	...	5.00	...	...
H19	T20819	.40	...	...	4.25	...	4.25	2.00	4.25

Note: Some of the types can be produced with a sulphur addition to improve machinability.

Hot Work Tool Steels (continued)
Symbol H
H20-H-39, incl., Tungsten Types

The tungsten types are intended for those hot work applications where resistance to the softening effect of elevated temperature is of greatest importance and a lesser degree of toughness can be tolerated.

Typical Hot Work Applications (a)

Hot blanking dies	Gripper dies	Punches
Dummy blocks	Mandrels	Hot shear blades
Extrusion dies	Piercer points	Hot trim dies

Type	UNS	\multicolumn{8}{c}{Identifying elements, per cent}
		C	Mn	Si	W	Mo	Cr	V	Co
H21	T20821	.35	...	...	9.00	...	3.50	.50	...
H22	T20822	.35	...	...	11.00	...	2.00	.40	...
H23	T20823	.30	...	...	12.00	...	12.00	1.00	...
H24	T20824	.45	...	...	15.00	...	3.00	.50	...
H25	T20825	.25	...	...	15.00	...	4.00	.50	...
H26	T20826	.50	...	...	18.00	...	4.00	1.00	...

(a) The producer should be consulted for specific recommendations and availability from warehouse stocks.

H40-H59, incl., Molybdenum Types

The molybdenum types are low-carbon modifications of molybdenum high-speed steels. They offer excellent resistance to the softening effect of elevated temperature, but, like the tungsten types, should be restricted to those applications where less ductility is acceptable. They generally are not readily available except on a mill delivery basis.

Typical Applications (a)

Type	UNS	\multicolumn{8}{c}{Identifying elements, per cent}
		C	Mn	Si	W	Mo	Cr	V	Co
H42	T20842	.60	...	...	6.00	5.00	4.00	2.00	...

(a) The producer should be consulted for specific recommendations and availability from warehouse stocks.

Cold Work Tool Steels

Symbol D, High-carbon High-chromium Types

The cold work tool steels listed under the letter symbol D are all characterized by high-carbon content, from 1.00 to 2.35 per cent, and nominally 12 per cent chromium. The types containing molybdenum are air hardening and therefore offer a high degree of dimensional stability in heat treatment. The D series exhibit high wear resistance, which increases with increasing carbon and vanadium content.

Typical Applications (a)

Blanking dies	Gages	Rolls
Burnishing tools	Knurls	Shear knives
Cold forming dies	Lamination dies	Slitter knives
Drawing dies	Punches	Thread rolling dies

(a) The producer should be consulted for specific recommendations and availability from warehouse stocks.

The final selection is dependent upon differences in properties, availability, production requirements, and economic considerations.

Type	UNS	\multicolumn{9}{c}{Identifying elements, per cent}								
		C	Mn	Si	W	Mo	Cr	V	Co	Ni
D2	T30402	1.50	...	...	...	1.00	12.00	1.00	...	...
D3	T30403	2.25	...	...	...	...	12.00	...	...	...
D4	T30404	2.25	...	...	...	1.00	12.00	...	...	...
D5	T30405	1.50	...	...	...	1.00	12.00	...	3.00	...
D7	T30407	2.35	...	...	...	1.00	12.00	4.00	...	...

Note: Some of the types can be produced with a sulphur addition to improve machinability.

Symbol A, Medium Alloy Air Hardening Types

The cold work tool steels listed under the letter symbol A cover a wide range of carbon and alloy contents, but all have high hardenability and exhibit a high degree of dimensional stability in heat treatment. The low-carbon types A8 and A9 offer greater shock resistance than the other steels in this group, but are lower in their resistance to wear. Type A7, which has high carbon and vanadium, exhibits maximum abrasion resistance, but should be restricted to applications where toughness is not a prime consideration.

Typical Applications (a)

Blanking dies	Knurls	Rolls
Coining dies	Mandrels	Shear knives
Cold forming tools	Master hubs	Slitter knives
Forming dies	Molds	Spindles

(a) The producer should be consulted for specific recommendations and availability from warehouse stocks.

It should be recognized that because of the wide differences in properties and availability, all of these steels should not be considered for each application.

Symbol A, Medium Alloy Air Hardening Types (continued)

Type	UNS	\multicolumn{9}{c	}{Identifying elements, per cent}
		C	Mn	Si	W	Mo	Cr	V	Co	Ni
A2	T30102	1.00	...	...	...	1.00	5.00	...	...	...
A3	T30103	1.25	...	...	...	1.00	5.00	1.00	...	...
A4	T30104	1.00	2.00	...	...	1.00	1.00	...	...	...
A6	T30106	.70	2.00	...	...	1.25	1.00	...	...	...
A7	T30107	2.25	...	...	1.00(a)	1.00	5.25	4.75	...	...
A8	T30108	.55	...	...	1.25	1.25	5.00	...	...	...
A9	T30109	.50	...	...	...	1.40	5.00	1.00	...	1.50
A10(b)	T30110	1.35	1.80	1.25	...	1.50	...	...	...	1.80

(a) Optional. (b) Contains free graphite in the microstructure to improve machinability.
Note: Some of the types can be produced with a sulphur addition to improve machinability.

Symbol O, Oil Hardening Types

The cold work tool steels listed under the letter symbol O are all low alloy types that must be oil quenched in heat treatment. Sizes over approximate 2 to 2½ inches (50.8 to 63.5 millimeters) may exhibit lower hardness in the interior. They generally are available from warehouse stocks and are widely used.

Typical Applications (a)

Blanking dies	Drawing dies	Machine ways
Cams	Forming dies	Plastic mold dies
Coining dies	Gages	Shear blades
		Trim dies

(a) The producer should be consulted for specific recommendations and availability from warehouse stocks.

Type	UNS	\multicolumn{9}{c	}{Identifying elements, per cent}
		C	Mn	Si	W	Mo	Cr	V	Co	Ni
01	T31501	.90	1.00	...	.50	...	.50	...	...	...
02	T31502	.90	1.60	...	...	...	...	...	...	...
06(a)	T31506	1.45	.80	1.00	...	.25	...	...	...	...
07	T31507	1.20	...	...	1.75	...	.75	...	...	...

(a) Contains free graphite in the microstructure to improve machinability.

Shock Resisting Tool Steels
Symbol S

The shock resisting tool steels under the letter symbol S. There is considerable variation in their alloy content and hardenability, but all are intended for applications requiring high toughness and resistance to shock loading.

Typical Applications (a)

Chipping chisels	Knock out pins	Screw driver blades
Circular pipe cutters	Mandrels	Shear blades
Drift pins	Punches	Stamps
Grippers	Rivet sets	Track tools
Hand chisels		

(a) The producer should be consulted for specific recommendations and availability from warehouse stocks.

Type	UNS	Identifying elements, per cent									
		C	Mn	Si	W	Mo	Cr	V	Co	Ni	Al
S1	T41901	.50	...	...	2.50	...	1.50	...	...	...	...
S2	T41902	.50	...	1.00	...	.50	...	...	...	...	...
S5	T41905	.55	.80	2.00	...	.40	...	...	...	...	...
S6	T41906	.45	1.40	2.25	...	.40	1.50	...	...	...	...
S7	T41907	.50	...	...	...	1.40	3.25	...	...	...	...

Mold Steels
Symbol P

The steels listed under the letter symbol P generally are intended for mold applications. The types P2-P6 are very low in carbon content, and usually are supplied at very low hardness to facilitate cold hobbing of the impression. They are then carburized to develop the required surface properties for injection and compression molds for plastics.

The types P20 and P21 usually are supplied in the prehardened condition, so that the cavity can be machined and the mold placed directly in service. They may be used for plastic molds, zinc die casting dies, and holder blocks.(a)

(a) The producer should be consulted for specific recommendations and availability from warehouse stocks.

Type	UNS	Identifying elements, per cent									
		C	Mn	Si	W	Mo	Cr	V	Co	Ni	Al
P2	T51602	.07	...	...	...	.20	2.00	...	...	.50	...
P3	T51603	.10	...	...	...	...	.60	...	...	1.25	...
P4	T51604	.07	...	...	...	.75	5.00	...	...	...	...
P5	T51605	.10	...	...	...	...	2.25	...	...	...	...
P6	T51606	.10	...	...	...	...	1.50	...	...	3.50	...
P20	T51620	.35	...	...	...	.40	1.70	...	...	...	...
P21	T51621	.20	...	...	...	...	...	...	...	4.00	1.20

American Iron and Steel Institute/1803

Special Purpose Tool Steels
Symbol L, Low Alloy Types

The tool steels listed under the letter symbol L cover a wide range of alloy content and physical properties, and are intended for special applications. Type L6 and the low-carbon versions of L2 generally are used for machine parts and in applications where toughness is an important consideration.

Typical Applications (a)

Low-carbon types		High-carbon types
Arbors	Machine tool parts	Arbors
Brake dies	Pinions	Dies
Cams	Punches	Drills
Chucks	Rivet sets	Gages
Collets	Rolls	Knife edges
Drift pins	Spindles	Knurls
Jigs	Wrenches	Rolls
		Taps

(a) The producer should be consulted for specific recommendations and availability from warehouse stocks.

			Identifying elements, per cent							
Type	UNS	C	Mn	Si	W	Mo	Cr	V	Co	Ni
L2	T61202	.50-1.10(b)	...	...	...	...	1.00	.20	...	...
L6	T61206	.70	...	...	...	.25(a)	.75	...	...	1.50

(a) Optional. (b) Various carbon contents are available.

Water Hardening Tool Steels
Symbol W

The tool steels listed under the letter symbol W are essentially carbon steels and are among the least expensive of tool steels. They must be water quenched to attain the necessary hardness, and except in very small sizes, will harden with a hard case and a soft core. They may be used for a wide variety of tools, as indicated below, but their limitations must be recognized, and discussion of the application with the producer is strongly recommended.

Typical Applications (a)

Axes	Countersinks	Jewelers dies
Blanking dies	Drills	Reamers
Cold heading dies	Files	Taps
Cold striking dies	Forming dies	Woodworking tools

(a) The producer should be consulted for specific recommendations and availability from warehouse stocks.

			Identifying elements, per cent							
Type	UNS	C	Mn	Si	W	Mo	Cr	V	Co	Ni
W1	T72301	.60-1.40(a)	...	...	...	...	...	...	...	...
W2	T72302	.60-1.40(a)	...	...	...	...	...	.25	...	...
W5	T72305	1.10	...	...	...	...	.50	...	...	...

(a) Various carbon contents are available.

Nominal Composition of Tool Steel Types Formerly Listed by AISI

Type	C	Mn	Si	W	Mo	Cr	V	Co	Cb
High-speed Tool Steels									
M8	.80	...	...	5.00	5.00	4.00	1.50	...	1.25
M15	1.50	...	...	6.50	3.50	4.00	5.00	5.00	...
M35	.80	...	...	6.00	5.00	4.00	2.00	5.00	...
M45	1.25	...	...	8.00	5.00	4.25	1.60	5.50	...
T3	1.05	...	...	18.00	...	4.00	3.00	...	...
T7	.75	...	...	14.00	...	4.00	2.00	...	...
T9	1.20	...	...	18.00	...	4.00	4.00	...	...
Hot Work Tool Steels									
H15	.40	...	...	...	5.00	5.00	...	...	...
H16	.55	...	...	7.00	...	7.00	...	...	...
H20	.35	...	...	9.00	...	2.00	...	...	...
H41	.65	...	...	1.50	8.00	4.00	1.00	...	...
H43	.55	...	...	...	8.00	4.00	2.00	...	...
Cold Work Tool Steels									
D1	1.00	...	...	...	1.00	12.00	...	...	...
D6	...	...	...	...	...	...	...	...	...
A5	1.00	3.00	...	...	1.00	1.00	...	...	...
Shock Resisting Tool Steels									
S3	.50	...	...	1.00	...	.74	...	...	...
S4	.55	.80	2.00	...	...	...	...	...	...
Mold Steels									
P1	.10	...	...	...	...	...	...	...	...
Special Purpose Tool Steels									
L1	1.00	...	...	...	...	1.25	...	...	...
L3	1.00	...	...	...	...	1.50	.20	...	...
L4	1.00	.60	...	...	...	1.50	.25	...	...
L5	1.00	1.00	...	...	.25	1.00	...	...	...
L7	1.00	.35	...	...	.40	1.40	...	...	...
F1	1.00	...	...	1.25	...	...	...	...	...
F2	1.25	...	...	3.50	...	...	...	...	...
F3	1.25	...	...	3.50	...	.75	...	...	...
Water Hardening Tool Steels									
W3	1.00	...	...	...	...	...	.50	...	...
W4	.60/1.40(a)	...	...	...	...	.25	...	...	...
W6	1.00	...	...	...	...	.25	.25	...	...
W7	1.00	...	...	...	...	.50	.20	...	...

(a) Various carbon contents may be available.

Standard Designations and Chemical Composition Ranges for Heat- and Corrosion-resistant Castings(a)

Cast alloy designation	Wrought alloy type(b)	C	Mn max	Si max	P max	S max	Cr	Ni	Other elements
CA-15	410	0.15 max	1.00	1.50	0.04	0.04	11.5-14	1 max	Mo 0.5 max(c)
CA-40	420	0.20-0.40	1.00	1.50	0.04	0.04	11.5-14	1 max	Mo 0.5 max(c)
CB-30	431	0.30 max	1.00	1.50	0.04	0.04	18-22	2 max	...
CB-7Cu	17-4PH	0.07 max	1.00	1.00	0.04	0.04	15.5-17	3.6-4.6	Cu 2.3-3.3
CC-50	446	0.50 max	1.00	1.50	0.04	0.04	26-30	4 max	...
CD-4MCu	...	0.040 max	1.00	1.00	0.04	0.04	25-27	4.75-6.00	Mo 1.75-2.25, Cu 2.75-3.25
CE-30	...	0.30 max	1.50	2.00	0.04	0.04	26-30	8-11	...
CF-3	304L	0.03 max	1.50	2.00	0.04	0.04	17-21	8-12	...
CF-8	304	0.08 max	1.50	2.00	0.04	0.04	18-21	8-11	...
CF-20	302	0.20 max	1.50	2.00	0.04	0.04	18-21	8-11	...
CF-3M	316L	0.03 max	1.50	1.50	0.04	0.04	17-21	9-13	Mo 2.0-3.0
CF-8M	D319(316)	0.08 max	1.50	2.00	0.04	0.04	18-21	9-12	Mo 2.0-3.0
CF-8C	347	0.08 max	1.50	2.00	0.04	0.04	18-21	9-12	Cb 8xC min, 1.0 max
CF-16F	303	0.16 max	1.50	2.00	0.17	0.04	18-21	9-12	Mo 1.5 max, Se 0.20-0.35
CG-8M	317	0.08 max	1.50	1.50	0.04	0.04	18-21	9-13	Mo 3.0-4.0
CH-20	309	0.20 max	1.50	2.00	0.04	0.04	22-26	12-15	...
CK-20	310	0.20 max	1.50	2.00	0.04	0.04	23-27	19-22	...
CN-7M	...	0.07 max	1.50	(d)	0.04	0.04	18-22	21-31	Mo-Cu (d)
CY-40	...	0.40 max	1.50	3.00	0.015	0.015	14-17	Bal.	Fe 11.0 max
CZ-100	...	1.00 max	1.50	2.00	0.015	0.015	...	95 min	Fe 1.50 max
M-35	...	0.35 max	1.50	2.00	0.015	0.015	...	Bal.	Cu 26-33, Fe 3.50 max
HA	...	0.20 max	0.35-0.65	1.00	0.04	0.04	8-10	...	Mo 0.90-1.20
HC	446	0.50 max	1.00	2.00	0.04	0.04	26-30	4 max	Mo 0.5 max(c)
HD	327	0.50 max	1.50	2.00	0.04	0.04	26-30	4-7	Mo 0.5 max(c)
HE	...	0.20-0.50	2.00	2.00	0.04	0.04	26-30	8-11	Mo 0.5 max(c)
HF	302B	0.20-0.40	2.00	2.00	0.04	0.04	19-23	9-12	Mo 0.5 max(c)
HH	309	0.20-0.50	2.00	2.00	0.04	0.04	24-28	11-14	Mo 0.5 max(c), N 0.2 max
HI	...	0.20-0.50	2.00	2.00	0.04	0.04	26-30	14-18	Mo 0.5 max(c)
HK	310	0.20-0.60	2.00	2.00	0.04	0.04	24-28	18-22	Mo 0.5 max(c)
HL	...	0.20-0.60	2.00	2.00	0.04	0.04	28-32	18-22	Mo 0.5 max(c)
HN	...	0.20-0.50	2.00	2.00	0.04	0.04	19-23	23-27	Mo 0.5 max(c)
HT	330	0.35-0.75	2.00	2.50	0.04	0.04	13-17	33-37	Mo 0.5 max(c)
HU	...	0.35-0.75	2.00	2.50	0.04	0.04	17-21	37-41	Mo 0.5 max(c)
HW	...	0.35-0.75	2.00	2.50	0.04	0.04	10-14	58-62	Mo 0.5 max(c)
HX	...	0.35-0.75	2.00	2.50	0.04	0.04	15-19	64-68	Mo 0.5 max(c)

(a)Most of the standard grades listed are covered for general applications by American Society for Testing and Materials specifications A 296 and A 297. ASTM specifications A 217, A 351, A 362, A 447, A 448, A 451 and A 452 also apply to some of the grades. (b)Wrought alloy type numbers are listed only for the convenience of those who want to determine corresponding wrought and cast grades. Because the cast alloy chemical composition ranges *are not the same* as the wrought composition ranges, buyers should use cast alloy designations for proper identification of castings. (c)Molybdenum not intentionally added. (d)There are several proprietary alloy compositions falling within the stated chromium and nickel ranges, and containing varying amounts of silicon, molybdenum and copper.

Note. Designations with the initial letter "C" indicate alloys generally used to resist corrosive attack at temperatures less than 1200F. Designations with the initial letter "H" indicate alloys generally used under conditions where the metal temperature is in excess of 1200F. The second letter represents the nominal chromium-nickel type; the nickel content increasing in amount from "A" to "Z". For example, "F" stands for the 19% Cr-9% Ni, "K" for the 25% Cr-20% Ni, and "W" for the 12% Cr-60% Ni alloy types. Numerals following the letters indicate the *maximum* carbon content of the corrosion resistant alloys; carbon content may also be designated in the heat resistant grades by following the letters with a numeral to indicate the *midpoint* of a ±0.05% carbon range. If special elements are included in the composition they are indicated by the addition of a letter to the symbol. Thus, "CF-8M" is an alloy for corrosion resistant service, of the molybdenum-containing 19% Cr-9% Ni type with a maximum carbon content of 0.08%.

Carbon and Alloy Cast Steels

ASTM A 27-73	Mild to medium-strength carbon-steel castings for general application.
ASTM A 148-73	High-strength steel castings for structural purposes.
ASTM A 216-75	Carbon steel castings suitable for fusion welding for high temperature service.
ASTM A 217-75	Martensitic stainless steel and alloy steel castings for pressure containing parts suitable for high temperature service.
ASTM A 352-76	Ferritic steel castings for pressure containing parts suitable for low temperature service.
ASTM A 356-75	Heavy-walled carbon and low-alloy steel castings for steam turbines.
ASTM 389-74a	Alloy steel castings specially heat treated for pressure containing parts suitable for high temperature service.
ASTM A 486-74	Steel castings for highway bridges.
ASTM A 487-76	Steel castings suitable for pressure vessels.
ASTM A 643-75	Steel castings, heavy-walled, carbon and alloy, for pressure vessels.
SAE J435a	1975 automotive steel castings.
AAR M 201-66	Steel castings.
ABS	American Bureau of Shipping, Steel Castings – 1964 Rules Edition – Machinery and Hull Castings.

Specification and heat treatment			Mechanical properties (minimum unless range is given)					Other tests: bend, impact hardness	Chemical composition (maximum percent unless range is given)								
Specification	Grade	Heat treatment	Tensile strength ksi	Tensile strength MPa	Yield strength ksi	Yield strength MPa	Elongation in 2″, %	Reduction of area, %		C	Mn	P	S	Si	Ni	Cr	Mo	Other elements
ASTM A 27-73	N-1	A, N, NT or QT	...	...	...	...	...	...	...	.25(a)	.75(a)	.05	.06	.80	.50(b)	.40(b)	.20(b)	Cu .30 (b)
	N-2	A, N, NT or QT	...	...	...	...	...	...	...	.35(a)	.60(a)	.05	.06	.80	.50(b)	.40(b)	.20(b)	Cu .30(b)
	U-60-30	...	60	414	30	207	22	30	...	.25(a)	.75(a)	.05	.06	.80	.50(b)	.40(b)	.20(b)	Cu .30(b)
	60-30	A, N, NT or QT	60	414	30	207	24	35	...	.30(a)	.60(a)	.05	.06	.80	.50(b)	.40(b)	.20(b)	Cu .30(b)
	65-35	A, N, NT or QT	65	448	35	241	24	35	...	.30(a)	.70(a)	.05	.06	.80	.50(b)	.40(b)	.20(b)	Cu .30(b)
	70-36	A, N, NT or QT	70	483	36	248	22	30	...	.35(a)	.70(a)	.05	.06	.80	.50(b)	.40(b)	.20(b)	Cu .30(b)
	70-40(c)	A, N, NT or QT	70	483	40	276	22	30	...	.25(a)	1.20(a)	.05	.06	.80	.50(b)	.40(b)	.20(b)	Cu .30(b)
ASTM A 148-73	80-40	A, N, NT or QT	80	552	40	276	18	30	...	(d)	...	.05	.06	...	...	...	...	...
	80-50	A, N, NT or QT	80	552	50	345	22	35	...	(d)	...	.05	.06	...	...	...	...	...
	90-60	A, N, NT or QT	90	621	60	414	20	40	...	(d)	...	.05	.06	...	...	...	...	...
	105-85	A, N, NT or QT	105	724	85	586	17	35	...	(d)	...	.05	.06	...	...	...	...	...
	120-95	A, N, NT or QT	120	827	95	655	14	30	...	(d)	...	.05	.06	...	...	...	...	...
	150-125	A, N, NT or QT	150	1034	125	862	9	22	...	(d)	...	.05	.06	...	...	...	...	...
	175-145	A, N, NT or QT	175	1207	145	1000	6	12	...	(d)	...	.05	.06	...	...	...	...	...
ASTM A 216-75	WCA	A, N, or NT	60-85	415-585	30	205	24	35	Bend – degrees 90(e)	.25(f)	.70(f)	.04	.045	.60	.50(g)	.40(g)	.25(g)	Cu .50(g), V .03
	WCB	A, N, or NT	70-95	485-655	36	250	22	35	90(e)	.30	1.00	.04	.045	.60	.50(g)	.40(g)	.25(g)	Cu .50(g), V .03
	WCC	A, N, or NT	70-95	485-655	40	275	22	35	90(e)	.25(h)	1.20(h)	.04	.045	.60	.50(g)	.40(g)	.25(g)	Cu .50(g), V .03
ASTM A 217-75	WC1	NT, 1100 F (595 C)	65-90	450-620	35	240	24	35	90(i)	.25	.50-.80	.04	.045	.60	.50(j)(k)	.35(j)	.45-.65	Cu .50(j)(k), W .10(j)(k)
	WC4	NT 1100 F (595 C)	70-95	485-655	40	275	20	35	90(i)	.20	.50-.80	.04	.045	.60	.70-1.10	.50-.80	.45-.65	Cu .50(j)(l), W .10(j)(l)
	WC5	NT 1100 F (595 C)	70-95	485-655	40	275	20	35	90(i)	.20	.40-.70	.04	.045	.60	.60-1.00	.50-.90	.90-1.20	Cu .50(j)(l), W .10(j)(l)
	WC6	NT 1100 F (595 C)	70-95	485-655	40	275	20	35	90(i)	.20	.50-.80	.04	.045	.60	.50(j)(k)	1.00-1.50	.45-.65	Cu .50(j)(k), W .10(j)(k)
	WC9	NT 1250 F (675 C)	70-95	485-655	40	275	20	35	90(i)	.18	.40-.70	.04	.045	.60	.50(j)(k)	2.00-2.75	90-1.20	Cu .50(j)(k), W .10(j)(k)
	C5	NT 1250 F (675 C)	90-115	620-795	60	415	18	35	90(i)	.20	.40-.70	.04	.045	.75	.50(j)(k)	4.00-6.50	.45-.65	Cu .50(j)(k), W .10(j)(k)
	C12	NT 1250 F (675 C)	90-115	620-795	60	415	18	35	90(i)	.20	.35-.65	.04	.045	1.00	.50(j)(k)	8.00-10.00	90-1.20	Cu .50(j)(k), W .10(j)(k)
	CA-15	NT 1100 F (595 C)	90-115	620-795	65	450	18	30	...	.15	1.00	.04	.04	1.50	1.00	11.5-14.0	.50	Cu .50(j)(k), W .10(j)(k)
ASTM A 352-76	LCA	NT or QT 1100 F (590 C)	60-85	415-585	30	205	24	35	Impact (b) ft.-lb. @ °F J @ °C 13-25 18-32	.25(o)	.70(o)	.04	.045	.60	...	...	...	...
	LCB	NT or QT 1100 F (590 C)	65-90	450-620	35	240	24	35	13-50 18-46	.30	1.00	.04	.045	.60	...	...	...	...
	LCC	NT or QT 1100 F (590 C)	70-95	485-655	40	275	22	35	15-50 20-46	.25(o)	1.20(o)	.04	.045	.60	...	...	.45-.65	...
	LC1	NT or QT 1100 F (590 C)	65-90	450-620	35	240	24	35	13-75 18-59	.25	.50-.80	.04	.045	.60	...	...	.45-.65	...
	LC2	NT or QT 1100 F (590 C)	70-95	485-655	40	275	24	35	15-100 20-73	.25	.50-.80	.04	.045	.60	2.00-3.00	...	...	...
	LC2-1	NT or QT 1100 F (590 C)	105-130	725-895	80	550	18	30	30-100 41-73	.22	.55-.75	.04	.045	.50	2.50-3.50	1.35-1.85	.30-.60	...
	LC3	NT or QT 1100 F (590 C)	70-95	485-655	40	275	24	35	15-150 20-101	.15	.50-.80	.04	.045	.60	3.00-4.00	...	...	...
	LC4	NT or QT 1050 F (570 C)	70-95	485-655	40	275	24	35	15-175 20-115	.15	.50-.80	.04	.045	.60	4.00-5.00	...	...	...
ASTM A 356-75(p)	1	NT 1100 F (595 C)	70	485	36	250	20	35	...	.35(a)	.70(a)	.035	.030	.60	...	...	...	...
	2	NT 1100 F (595 C)	65	450	35	240	22	35	...	.25(a)	.70(a)	.035	.030	.60	...	.40-.70	.45-.65	...
	5	NT 1100 F (595 C)	70	485	40	275	22	35	...	.25(a)	.70(a)	.035	.030	.60	...	.40-.70	.40-.60	...
	6	NT 1100 F (595 C)	70	485	45	310	22	35	...	.20	.50-.80	.035	.030	.60	...	1.00-1.50	.45-.65	...

Carbon and Alloy Cast Steels (continued)

Specification and heat treatment			Mechanical properties (minimum unless range is given)						Other tests: bend, impact hardness	Chemical composition (maximum percent unless range is given)								
Specification	Grade	Heat treatment	Tensile strength ksi	Tensile strength MPa	Yield strength ksi	Yield strength MPa	Elongation in 2", %	Reduction of area, %		C	Mn	P	S	Si	Ni	Cr	Mo	Other elements
ASTM A 389-74a	8	NT 1100 F (595 C)	80	550	50	345	18	45	...	.13-.20	.50-.90	.035	.030	.20-.60	...	1.00-1.50	.90-1.20	V .05-.15
	9	NT 1100 F (595 C)	85	585	60	415	15	45	...	.13-.20	.50-.90	.035	.030	.20-.60	...	1.00-1.50	.90-1.20	V .20-.35
	10	NT 1100 F (595 C)	85	585	55	380	20	35	...	.20	.50-.80	.035	.030	.60	...	2.00-2.75	.90-1.20	...
ASTM A 389-74a	C23	NT 1350 F (730 C) max NT 1250 F (675 C) min 1h/in.(r)	70	483	40	276	18	35	...	.20	.30-.80	.04	.045	.60	...	1.00-1.50	.45-.65	V .15-.25
	C24	NT 1350 F (730 C) max NT 1250 F (677 C) min 12h(r)	80	552	50	345	15	35	...	.20	.30-.80	.04	.045	.60	...	.80-1.25	.90-1.20	V .15-.25
									Impact(s)(t) ft.-lb. @ °F J @ °C									
ASTM A 486-74	70	N, NT, or QT	70	483	36	248	22	30	25-70 34-21	.35	.90	.05	.06	.80	...	...	...	...
	90	N, NT, or QT	90	621	60	414	20	40	25-70 34-21	.35	(u)	.05	.06	(u)	...	...	...	(u)
	120	QT	120	827	95	655	14	30	30-70 41-21	.35	(u)	.05	.06	(u)	...	...	...	(u)
ASTM A 487-76	1N	NT 1100 F (595 C)	85-110	585-760	55	380	22	40	...	.30	1.00	.04	.045	.80	.50(v)	.35(v)	Mo + W .25(v)	V .04-.12, Cu .50(v)
	2N	NT 1100 F (595 C)	85-110	585-760	53	365	22	35	...	.30	1.00-1.40	.04	.045	.80	.50(v)	.35(v)	.10-.30	V .30, Cu .50(v), W .10(v)
	4N	NT 1100 F (595 C)	90-115	620-795	60	415	20	40	...	.30	1.00	.04	.045	.80	.40-.80	.40-.80	.15-.30	V .03, Cu .50(w), W .10(w)
	6N	NT 1100 F (595 C)	115	795	80	550	18	30	...	.38	1.30-1.70	.04	.045	.80	.40-.80	.40-.80	.30-.40	V .03, Cu .50(w), W .10(v)
	8N	NT 1250 F (680 C)	85-110	585-760	55	380	20	35	...	.20	.50-.90	.04	.045	.80	(w)	2.00-2.75	.90-1.10	V .03, Cu .50(w), W .10(v)
	9N	NT 1100 F (595 C)	90	620	60	415	20	35	...	.33	.60-1.00	.04	.045	.80	.50(v)	.75-1.10	.15-.30	V .03, Cu .50(v), W .10(v)
	10N	NT 1100 F (595 C)	100	690	70	485	18	35	...	.30	1.00	.04	.045	.80	1.4-2.0	.55-.90	.20-.40	V .03, Cu .50(w), W .10(w)
	11N	NT 1100 F (595 C)	70-95	485-655	40	275	20	35	...	.20	.50-.80	.04	.045	.60	.70-1.10	.50-.80	.45-.65	V .03, Cu .50(x), W .10(x)
	12N	NT 1100 F (595 C)	70-95	485-655	40	275	20	35	...	.20	.40-.70	.04	.045	.60	.60-1.00	.50-.90	.90-1.20	V .03, Cu .50(x), W .10(x)
	13N	NT 1100 F (595 C)	90-115	620-795	60	415	18	35	...	.30	.80-1.10	.04	.045	.60	1.40-1.75	.40(y)	.20-.30	V .03, Cu .50(y), W .10(y)
	A, AN	NT 1100 F (595 C)	60-85	415-585	30	205	24	35	...	.25(z)	.75(z)	.04	.045	.60	.50(v)	.40(y)	.25(v)	V .03, Cu .50(v), W .10(v)
	B, BN, C, CN(aa)	NT 1100 F (595 C)	70-95	485-655	36	248	22	35	...	.30	1.00	.04	.045	.80	.50(v)	.40(y)	.25(v)	V .03, Cu .50(v), W .10(v)
	DN	NT 1100 F (595 C)	70-95	485-655	40	275	17	25	...	.25(bb)	1.20(bb)	.04	.045	.60	.50(v)	.40(y)	.25(v)	V .03, Cu .35(v), W .10(v)
	CA 6NM	NT 1100 F (595 C)	80	550	40	275	15	35	...	.40-.50	.50-.90	.04	.045	.60	3.5-4.5	11.5-14.0	.4-1.0	V .03, Cu .50(x), W .10(x)
	CA 15M	NT 1100 F (595 C)	110-135	760-930	80	550	15	35	...	.06	1.00	.03	.04	.65	1.00	11.5-14.0	.15-1.0	
	CA 15a	NT 1100 F (595 C)	90-115	620-793	65	448	18	30	...	.15	1.00	.04	.04	1.50	1.00	11.5-14.0	.50	
	1Q	QT 1100 F (595 C)	140-170	965-1170	110-130	760-896	10	25	...	.30	1.00	.04	.045	.80	.50(w)	.35(w)	Mo + W .25	V .04-.2, Cu .50(v), W .10(v)
	2Q	QT 1100 F (595 C)	90-115	620-795	65	450	22	45	...	.30	1.00	.04	.045	.80	.50(w)	.35(w)	.10-.30	V .03, Cu .50(w), W .10(w)
	4Q	QT 1100 F (595 C)	90-115	620-795	65	450	22	40	...	.30	1.00-1.40	.04	.045	.80	.50(w)	.40-.80	.15-.30	V .03, Cu .50(w), W .10(w)
	4QA	QT 1100 F (595 C)	105-130	725-895	85	585	17	35	...	.30	1.00	.04	.045	.80	.40-.80	.40-.80	.15-.30	V .03, Cu .50(w), W .10(x)
	6Q	QT 1100 F (595 C)	115	795	95	655	15	35	...	.38	1.30-1.70	.04	.045	.80	.40-.80	.40-.80	.30-.40	V .03, Cu .50(w), W .10(y)
	7Q(cc)	QT 1100 F (595 C)	115	795	100	690	12	30	...	.20	.60-1.00	.04	.045	.80	.70-1.00	.40-.80	.40-.60	V .03-.10, B .002-.006, Cu .15-.50, W .10(w)
	8Q	QT 1250 F (680 C)	105	725	85	585	17	30	...	.20	.50-.90	.04	.045	.80	.50(w)	2.00-2.75	.90-1.10	V .03, Cu .50(w), W .10(w)
	9Q	QT 1100 F (595 C)	105	725	85	585	16	35	...	.33	.60-1.00	.04	.045	.80	...	.75-1.10	.15-.30	V .03, Cu .50(w), W .10(v)
	10Q	QT 1100 F (595 C)	125	860	100	690	15	35	...	.30	1.00	.04	.045	.80	1.40-2.00	.55-.90	.20-.40	V .03, Cu .50(w), W .10(w)
	11Q	QT 1100 F (595 C)	105-130	725-895	85	585	17	35	...	.20	.50-.80	.04	.045	.60	.70-1.10	.50-.80	.45-.65	V .03, Cu .50(x), W .10(x)
	12Q	QT 1100 F (595 C)	105-130	725-895	85	585	15	35	...	.20	.40-.70	.04	.045	.60	.60-1.00	.50-.90	.90-1.20	V .03, Cu .50(w), W .10(x)
	13Q	QT 1100 F (595 C)	105-130	725-895	85	585	17	35	...	.30	.80-1.10	.04	.045	.60	1.40-1.75	.40(y)	.20-.30	V .03, Cu .50(y), W .10(y)
	14Q	QT 1100 F (595 C)	120-145	825-1000	95	655	14	30	...	.55	.80-1.10	.04	.045	.60	1.40-1.75	.40(y)	.20-.30	V .03, Cu .50(y), W .10(y)
	AQ	QT 1100 F (595 C)	70-90	485-620	30	205	24	35	...	.25(z)	.75(z)	.04	.045	.60	.50(w)	.40(y)	.25(w)	V .03, Cu .50(w), W .10(y)
	BQ	QT 1100 F (595 C)	80-105	550-725	36	248	22	35	...	.30	1.00	.04	.045	.60	.50(w)	.40(y)	.25(w)	V .03, Cu .50(w), W .10(w)
	CQ	QT 1100 F (595 C)	80-105	550-725	40	275	22	35	...	.25(bb)	1.20(bb)	.04	.045	.60	.50(w)	.40(y)	.25(w)	V .03, Cu .50(w), W .10(w)

Carbon and Alloy Cast Steels (continued)

Specification and heat treatment			Mechanical properties (minimum unless range is given)						Chemical composition (maximum percent unless range is given)							
Specification	Grade	Heat treatment	Tensile strength ksi	Tensile strength MPa	Yield strength ksi	Yield strength MPa	Elongation in 2", %	Reduction of area, %	Other tests: bend, impact hardness	C	Mn	P	S	Si	Ni	Cr	Mo	Other elements

Specification	Grade	Heat treatment	ksi	MPa	ksi	MPa	%	%		C	Mn	P	S	Si	Ni	Cr	Mo	Other elements
ASTM A 643-75	A1	N or QT 1100 F (595 C)(dd)	70	485	40	275	22	35	(ee)	.25(ff)	1.20(ff)	.045	.045	.60	.50(gg)	.40(gg)	.25(gg)	V .03(gg), Cu .50(gg), W .10(gg)
	B1	NT or QT 1150 F (620 C)(dd)	80-110	550-760	50	345	22	35	(ee)	.25	1.15-1.50	.035	.035	.60	.45-1.00	.40(gg)	.45-.60	V .03(gg), Cu .50(gg), W .10(gg)
	C1	NT or QT 1150 F (620 C)(dd)	85-115	585-795	55	380	20	35	(ee)	.20	.40-.80	.035	.035	.60	.50(gg)	2.00-2.75	.90-1.20	V .03(gg), Cu .50(gg), W .10(gg)
	C2	NT or QT 1150 F (620 C)(dd)	95-125	655-860	75	515	18	35	(ee)	.20	.40-.80	.035	.035	.60	.50(gg)	2.00-2.75	.90-1.20	V .03(gg), Cu .50(gg), W .10(gg)
	C3	NT or QT 1150 F (620 C)(dd)	105-135	725-930	85	585	15	30	(ee)	.20	.40-.80	.035	.035	.60	.50(gg)	2.00-2.75	.90-1.20	V .03(gg), Cu .50(gg), W .10(gg)
	D1	NT or QT 1150 F (620 C)(dd)	105-135	725-930	85	585	15	30	(ee)	.20	.40-.70	.020	.020	.60	2.75-3.90	1.50-2.00	.40-.60	V .03(gg), Cu .50(gg), W .10(gg)
	D2	NT or QT 1150 F (620 C)(dd)	115-145	795-1000	100	690	13	30	(ee)	.20	.40-.70	.020	.020	.60	2.75-3.90	1.50-2.00	.40-.60	V .03(hh), Cu .50(hh), W .10(hh)
SAE Automotive J435a	0022	A, N, or NT	…	…	…	…	…	…	HBN 187 max	.12-.22	.50-.90	.050	.06	.60	…	…	…	…
	0025	A, N, or NT	60	414	30	207	22	30	187 max	.25(ii)	.75(ii)	.050	.06	.80	…	…	…	…
	0030	A, N, NT, or QT	65	448	35	241	24	35	131-187	.30(ii)	.70(ii)	.050	.06	.80	…	…	…	…
	0050A	N or NT	85	586	45	310	16	24	170-229	.40-.50	.50-.90	.050	.06	.80	…	…	…	…
	0050B	QT	100	689	70	483	10	15	207-255	.40-.50	.50-.90	.050	.06	.80	…	…	…	…
	080	A, N, NT, or QT	80	552	50	345	22	35	163-207	…	…	.050	.06	…	…	…	…	…
	090	NT or NQT	90	621	60	414	20	40	187-241	…	…	.050	.06	…	…	…	…	…
	0105(jj)	NQT	105	724	85	586	17	35	217-248	…	…	.050	.06	…	…	…	…	…
	0120(jj)	NQT	120	827	95	655	14	30	248-311	…	…	.050	.06	…	…	…	…	…
	0150(jj)	NQT	150	965	125	862	9	22	311-363	…	…	.050	.06	…	…	…	…	…
	0175(jj)	NQT	175	1206	145	1000	6	12	363-415	…	…	.050	.06	…	…	…	…	…
AAR M 201-66	A	Unannealed	60	414	30	207	22	30	…	(kk)	.85	.05	Basic	…	…	…	…	…
	A	A or N	60	414	30	207	26	38	…	(kk)	.85	.05	.05	…	…	…	…	…
	B	A or N	70	483	38	262	24	36	…	(kk)	.85	.05	Acid	…	…	…	…	…
	C	NT or QT	90	621	60	414	22	45	(ll)	.35	(kk)	.05	.06	…	…	…	…	…
	D	QT	105	724	85	586	17	35	…	(kk)	…	.05	…	…	…	…	…	…
	E	QT	120	827	100	689	14	30	…	(kk)	…	.05	…	…	…	…	…	…
ABS	1	A or NT	60	414	30	207	24	35	Bend – degrees 120	…	…	…	…	…	…	…	…	…
	2	A or NT	70	483	36	248	22	30	90	…	…	…	…	…	…	…	…	…
	Hull	A or NT	60	414	30	207	24	35	120	…	…	…	…	…	…	…	…	…

(a) For each reduction of 0.01 percent carbon below the maximum specified, an increase of 0.04 percent manganese above the maximum specified will be permitted to a maximum of 1.40 percent for grades 70-40 and 1.00 percent for the other grades. (b) Maximum content of unspecified elements. (c) Grade 70-40 may be used to meet the requirement of Grade 70-36 when agreed upon by the manufacturer and the purchaser. (d) Alloying elements shall be selected by the manufacturer unless otherwise specified. (e) Test not required unless stipulated by customer as per ASTM A-703. (f) For each reduction of 0.01 percent below the specified maximum carbon content, an increase of 0.04 percent manganese above the specified maximum will be permitted up to a maximum of 1.10 percent. (g) It is recognized that residual elements are unavoidable in steel, and, in the interest of uniform welding, the restrictions shown shall be complied with. Report of analysis shall be made for these residual elements only when Supplementary Requirements SI is specified on the order. (h) For each reduction of 0.01 percent below the specified maximum carbon content an increase of 0.04 percent manganese above the specified maximum will be permitted to a maximum of 1.4 percent. (i) Test not required unless stipulated by customer as per ASTM A-703. (j) Maximum unspecified element. It is recognized that residual elements are unavoidable in steel, and in the interest of uniform welding, the restrictions shown shall be complied with. Report of analysis shall be made for these residual elements only when Supplementary Requirement SI is specified in the order. (k) The total maximum content of unspecified elements = 1.00 percent. (l) The total maximum content of unspecified elements = 0.60 percent. (m) The test temperature shall be as specified by the customer. When a test temperature other than that listed in this table is used, the lowest temperature at which the material met the impact requirements shall be stamped on a raised pad located immediately ahead of the material symbol, for example, 25LCB. (n) Lateral expansion to be reported if stipulated by the customer. (o) For each reduction of 0.01 percent C below the maximum specified, an increase of 0.04 percent Mn above the maximum specified will be permitted up to a maximum of 1.10 percent Mn (Grade CA) and 1.40 percent Mn (Grade LCC). (p) 1: Deoxidation shall be by manganese and silicon. Furnace or ladle deoxidation with other agents is permissible with the approval of the purchaser. 2: In no case, for grades other than Grade 1, shall the amount of aluminum added, directly or as a constituent of other deoxidizers, exceed ½ lb/ton (0.25 g/kg) of steel. 3. The purchaser may specify that no aluminum be added. 4: Vacuum deoxidation is acceptable. The specific method shall be subject to approval by the purchaser. (q) For each 0.01 percent reduction in carbon below the maximum specified, an increase of 0.04 percentage points of manganese over the maximum specified for that element will be permitted up to 1.00. (r) Refer to original specification for additional information. (s) Values apply only to sections up to 2 inches. (t) Refer to original specification for additional impact testing information. (u) The manganese, silicon, and other alloying elements which are added to obtain the mechanical properties specified shall be selected by the manufacturer. (v) Unspecified element, maximum content as residual element, total residual elements = 1 percent maximum. (w) Unspecified element, maximum content as residual elements = .60 percent. (x) Unspecified element, maximum content as residual elements, total residual elements = .50 percent. (y) Unspecified element, maximum content as residual element, total residual elements = .75 percent. (z) For each reduction of 0.01 percent below the specified maximum carbon content, an increase of 0.04 percent manganese above the specified maximum will be permitted up to a maximum of 1.1 percent. (aa) Yield strength at 0.2 percent offset or total extension under load of 0.005 in./in. of gage length to 90 ksi (620 MPa) yield strength and 0.006>90 ksi yield strength. (bb) For each reduction of 0.01 percent below the specified maximum carbon content, an increase of 0.04 percent manganese above the specified maximum will be permitted up to a maximum of 1.4 percent. (cc) Maximum thickness 2½ in. (63.5 mm). (dd) Refer to original specification for additional heat treatment information. (ee) After machining test castings to the hydrostatic test pressure requirements stated on the drawing or purchase order. (ff) For each 0.01 percent reduction in carbon below the maximum specified, an increase of 0.04 percent manganese over the maximum specified will be permitted up to 1.40 percent. (gg) Unspecified element, maximum content as residual elements, total residual elements = 1 percent maximum. (hh) Unspecified element, maximum content as residual elements, total residual elements = .70 percent maximum. (ii) For each reduction of 0.01 percent C below the maximum specified, an increase of .04 percent Mn above the maximum specified will be permitted to a maximum of 1.0 percent Mn. (jj) Hardenability requirements when specified. (kk) Alloying elements shall be selected by the manufacturer unless otherwise specified. (ll) Hardenability requirement of Rc = 40 maximum at 10/16 in. (15.9 mm).

High-alloy Cast Steels

Specification		
ASTM A128-75a	Austenitic manganese steel castings.	
ASTM A743-77	Corrosion-resistant, iron-chromium, iron-chromium-nickel, and nickel base alloy castings for general application.	
ASTM A744-77	Corrosion-resistant iron-chromium, iron-chromium-nickel, and nickel base alloy castings for sever service.	
ASTM A297-76	Heat-resistant iron-chromium and iron-chromium-nickel alloy castings for general application.	
ASTM A351-76	Austenitic steel castings for high temperature service.	
ASTM A447-74	Chrom-nickel-iron alloy castings (25-12 class) for high-temperature service.	
ASTM 448-50	Nickel-chrom-iron alloy castings (35-15 class) for high temperature service.	
ASTM A494-76	Nickel and nickel alloy castings.	
ASTM A560-74	Chrom-nickel alloy castings.	
Military	MIL-S-16993A[1] steel castings (12 percent chromium).	
Military	MIL-S-867A December 1951 steel castings corrosion resisting austenitic.	

Specification and heat treatment			Mechanical properties (minimum unless range is given)						Chemical composition (maximum percent unless range is given)							
Specification	Grade	Heat treatment	Tensile strength ksi	MPa	Yield strength ksi	MPa	Elongation in 2″, %	Reduction of area, %	Other tests; bend, impact hardness	C	Mn	Si	P	S	Ni	Cr	Mo	Other elements
ASTM A-128-75a	A(a/b)	Q 1800 F (1000 C)	70	485	30(d)	205(d)	35	...	Bend test(c)	1.05-1.35	11.0	1.00	.07	...	...	...	...	...
	B-1(a/b)	Q 1800 F (1000 C)	70	485	28	195	35	...	Bend test(c)	.9 - 1.05	11.5-14.0	1.00	.07	...	...	...	...	...
	B-2(a/b)	Q 1800 F (1000 C)	70	485	30	205	30	...	Bend test(c)	1.05-1.2	11.5-14.0	1.00	.07	...	...	...	...	...
	B-3(a/b)	Q 1800 F (1000 C)	70	485	30	205	30	...	Bend test(c)	1.12-1.28	11.5-14.0	1.00	.07	...	...	...	...	...
	B-4(a/b)	Q 1800 F (1000 C)	70	485	30	205	30	...	Bend test(c)	1.2 - 1.35	11.5-14.0	1.00	.07	...	...	...	...	...
	C(a/b)	Q 1800 F (1000 C)	70	485	30	205	25	...	Bend test(c)	1.05-1.35	11.5-14.0	1.00	.07	...	...	1.5-2.5	...	...
	D(a/b)	Q 1800 F (1000 C)	70	485	30	205	30	...	Bend test(c)	.7 - 1.3	11.5-14.0	1.00	.07	...	3.0-4.0	...	...	...
	E-1(a/b)	Q 1800 F (1000 C)	65	450	28	195	30	...	Bend test(c)	.7 - 1.3	11.5-14.0	1.00	.07	...	...	...	.9-1.2	...
	E-2(a/b)	Q 1800 F (1000 C)	80	550	40	275	10	...	Bend test(c)	1.05-1.45	11.5-14.0	1.00	.07	...	...	...	1.8-2.1	...
	F(a/b)	Q 1800 F (1000 C)	90	620	65	450	18	30	Bend test(c)	1.05-1.35	6.0-8.0	1.00	.07	...	...	...	0.9-1.2	...
ASTM A-743-77	CF-8	SQ 1900 F (1040 C)	90	620	65	450	18	30	(e)	0.08	1.50	2.00	0.04(f)	0.04(f)	8.0-11.0	18.0-21.0	...	...
	CG-12	SQ 1900 F (1040 C)	65	450	30	205	...	...	(e)	0.12	1.50	2.00	0.04(f)	0.04(f)	10.0-13.0	20.0-23.0	...	...
	CF-20	SQ 1900 F (1040 C)	55	380	...	...	...	...	(e)	0.20	1.50	2.00	0.04(f)	0.04(f)	8.0-11.0	18.0-21.0	...	...
	CF-8M	SQ 1900 F (1040 C)	100	690	70	485	15	25	(e)	0.08	1.50	2.00	0.04(f)	0.04(f)	9.0-12.0	18.0-21.0	2.0-3.0(f)	...
	CF-8C	SQ 1900 F (1040 C)	70	485	30	205	35	...	(e)	0.08	1.50	2.00	0.04(f)	0.04(f)	9.0-12.00	18.0-21.0	...	Cb(g)
	CF-16F(h)	SQ 1900 F (1040 C)	70	485	30	205	30	...	(e)	0.16	1.50	2.00	0.04(f)	0.04(f)	9.0-12.00	18.0-21.0	(f)	...
	CH-20(i)	SQ 2000 F (1090 C)	70	485	30	205	30	...	(e)	0.20	1.50	2.00	0.04(f)	0.04(f)	12.0-15.0	22.0-26.0	...	...
	CK-20	SQ 2000 F (1093 C)	65	450	28	195	30	...	(e)	0.20	2.00	2.00	0.04(f)	0.04(f)	19.0-22.0	23.0-27.0	...	...
	CE-30	SQ 2000 F (1093 C)	80	550	40	275	10	...	(e)	0.30	1.50	2.00	0.04(f)	0.04(f)	8.0-11.0	26.0-30.0	...	...
	CA-15	NT or A	90	620	65	450	18	30	HBN-241 max(e)	0.15	1.00	1.50	0.04(f)	0.04(f)	1.0	11.5-14.0	0.50(f)	...
	CA-15M	NT or A(j)	90	620	65	450	18	30	HBN-241 max(e)	0.15	1.00	0.65	0.04(f)	0.04(f)	1.0	11.5-14.0	0.15-1.00(f)	...
	CB-30	N or A(g)	65	450	30	205	...	...	HBN-241 max(e)	0.30	1.00	1.50	0.04(f)	0.04(f)	2.00	18.0-21.0	...	...
	CC-50	N or A(j)	55	380	...	...	...	...	HBN-241 max(e)	0.50	1.00	1.50	0.04(f)	0.04(f)	4.00	26.0-30.0	...	...
	CA-40	NT or A(j)	100	690	70	485	15	25	HBN-269 max(e)	0.2-0.4	1.00	1.50	0.04(f)	0.04(f)	1.0	11.5-14.0	0.5(f)	...
	CF-3	As cast or SQ(j)	70	485	30	205	35	...	(e)	0.03(l)	1.50	2.00	0.04(f)	0.04(f)	8.0-12.0	17.0-21.0	...	...
	CF-3M	As cast or SQ(j)	70	485	30	205	30	...	(e)	0.03(l)	1.50	1.50	0.04(f)	0.04(f)	9.0-13.0	17.0-21.0	2.0-3.0(f)	...
	CG-8M	SQ 1900 F (1040 C)	75	520	35	240	25	...	(e)	0.08	1.50	1.50	0.04(f)	0.04(f)	9.0-13.0	18.0-21.0	3.0-4.0(f)	...
	CN-7M	SQ 2050 F (1120 C)	62	485	25	170	35	...	(e)	0.07	1.50	1.50	0.04(f)	0.04(f)	27.5-30.5	19.0-22.0	2.0-3.0(f)	Cu 3.0-4.0
	CN-7MS	SQ 2050 F (1120 C)	70	485	30	205	35	...	(e)	0.07	1.00	2.50-3.50	0.03(f)	0.03(f)	22.0-25.0	18.0-20.0	2.5-3.0(f)	Cu 1.5-2.0
	CW-12M	(j)	72	495	46	320	4.0	...	(e)	0.12	1.00	1.50	0.04(f)	0.03(f)	Remainder	15.50-20.00(f)	16.00-20.00(f)	W 5.25, V .40, Co 2.50, Fe 7.5
	CY-40	As cast	70	485	28	195	30	...	(e)	0.40	1.50	3.00	.03(f)	0.03(f)	Remainder	14.00-17.00	...	Fe 11.00
	CZ-100	As cast	50	345	18	125	10	...	(e)	1.00	1.50	2.00	.03(f)	0.03(f)	Remainder	...	...	Cu 1.25, Fe 3.00
	M-35	As cast	65	450	30	205	25	...	(e)	0.35	1.00	1.00	.03(f)	.03(f)	Remainder	...	...	Cu 26.0-33.0, Fe 3.50
	N-12M	(j)	72	495	46	320	6	...	(e)	0.12	1.00	1.00	0.04(f)	0.03(f)	Remainder	1.00	26.0-33.0(f)	V 0.60, Co 2.50, Fe 6.00
	CA-6NM	NT 1100 F (590 C)	110	760	80	550	15	...	HBN-285 max(e)	0.06	1.00	1.00	0.04(f)	0.03(f)	3.5-4.5	11.5-14.0	0.40-1.0(f)	...
	CD4MCu	SQ 1900 F (1040 C)	100	689	70	483	16	...	(e)	0.04	1.00	1.00	0.04(f)	0.04(f)	4.75-6.00	24.5-26.5	1.75-2.25(f)	Cu 2.75-3.25
	CA-6N	NT 1500 F (815 C)	140	965	135	930	15	...	(e)	0.06	0.50	1.00	0.02(f)	0.02(f)	6.0-8.0	10.5-12.5	...	...
ASTM 744-77	CF-8	SQ 1900 F (1040 C)	70(m)	485(m)	30(m)	205(m)	35	...	(m)	0.08	1.50	2.00	0.04	0.04	8.0-11.0	18.0-21.0	...	...
	CF-8M	SQ 1900 F (1040 C)	70	485	30	205	30	...	(m)	0.08	1.50	2.00	0.04	0.04	9.0-12.0	18.0-21.0	2.0-3.0	...

Steel Founders' Society/1811

High-alloy Cast Steels (continued)

Specification		Specification and heat treatment		Mechanical properties (minimum unless range is given)						Chemical composition (maximum percent unless range is given)								
		Grade	Heat treatment	Tensile strength ksi	MPa	Yield strength ksi	MPa	Elongation in 2", %	Reduction of area, %	Other tests; bend, impact hardness	C	Mn	P	S	Si	Ni	Cr	Mo	Other elements
ASTM A-297-76		CF-8C	SQ 1900 F (1040 C)	70	485	30	205	30	...	(n)	0.08	1.50	0.04	0.04	2.00	9.0-12.0	18.0-21.0	...	Cb(o)
		CF-3	SQ 1900 F (1040 C)	70	485	30	205	35	...	(n)	0.03(p)	1.50	0.04	0.04	2.00	8.0-12.0	17.0-21.0	...	...
		CF-3M	SQ 1900 F (1040 C)	70	485	30	205	30	...	(n)	0.03(p)	1.50	0.04	0.04	1.50	9.0-13.0	17.0-21.0	2.0-3.0	...
		CG-8M	SQ 1900 F (1040 C)	75	520	35	240	25	...	(n)	0.08	1.50	0.04	0.04	1.50	9.0-13.0	18.0-21.0	3.0-4.0	...
		CN-7M	SQ 2050 F (1120 C)	62	425	25	170	35	...	(n)	0.07	1.50	0.04	0.04	1.50	27.5-30.5	19.0-22.0	2.0-3.0	Cu 3.0-4.0
		CN-7MS	SQ 2050 F (1120 C)	70	485	30	205	35	...	(n)	0.07	1.00	0.04	0.03	2.50-3.50	22.0-25.0	18.0-20.0	2.5-3.0	Cu 1.5-2.0
		CW-12M	(q)	72	495	46	315	4.0	...	(n)	0.12	1.00	0.04	0.03	1.50	Remainder	15.50-20.00	16.00-20.00	W 5.25, V .40, Fe 7.50
		CY-40	As cast	70	485	28	195	30	...	(j)	0.40	1.50	0.03	0.03	3.00	Remainder	14.00-17.00	...	Fe 11.00
		CZ-100	As cast	50	345	18	125	10	...	(n)	1.00	1.50	0.03	0.03	2.00	Remainder	...	...	Cu 1.25 max, Fe 3.00
		M-35	As cast	65	450	30(r)	205(r)	25	...	(n)	0.35	1.50	0.03	0.03	2.00	Remainder	...	...	Cu 26.0-33.0, Fe 3.50
		N-12M	(q)	72	495	46	315	6	...	(n)	0.12	1.00	0.04	0.03	1.00	Remainder	1.00	26.0-33.0	V .60, Fe 6.00
		CD-4MCu	SQ 1900 F (1040 C)(s)	100	690	70	485	16	...	(n)	0.04	1.00	0.04	0.04	1.00	4.75-6.00	24.5-26.5	1.75-2.25	Cu 2.75-3.25
ASTM A-297-76		HC	...	55	380	...	...	...	...	...	.50	1.00	.04	.04	2.00	4.00	26-30	.50(t)	...
		HD	...	75	515	35	240	8	...	...	.50	1.50	.04	.04	2.00	4-7	26-30	.50(t)	...
		HE	...	85	585	40	275	9	...	...	.20-.50	2.00	.04	.04	2.00	8-11	26-30	.50(t)	...
		HF	...	70	485	35	240	25	...	...	.20-.40	2.00	.04	.04	2.00	8-12	18-23	.50(t)	...
		HH	...	75	515	35	240	10	...	...	.20-.50	2.00	.04	.04	2.00	11-14	24-28	.50(t)	...
		HI	...	70	485	35	240	10	...	...	.20-.50	2.00	.04	.04	2.00	14-18	26-30	.50(t)	...
		HK	...	65	450	35	240	10	...	...	.20-.60	2.00	.04	.04	2.00	18-22	24-28	.50(t)	...
		HL	...	65	450	35	240	10	...	...	.20-.60	2.00	.04	.04	2.00	18-22	28-32	.50(t)	...
		HN	...	63	435	...	...	8	...	...	.20-.50	2.00	.04	.04	2.00	23-27	19-23	.50(t)	...
		HT	...	65	450	...	...	4	...	...	.35-.75	2.00	.04	.04	2.50	33-37	15-19	.50(t)	...
		HU	...	65	450	...	...	4	...	...	.35-.75	2.00	.04	.04	2.50	37-41	17-21	.50(t)	...
		HW	...	60	415	...	...	...	...	...	.35-.75	2.00	.04	.04	2.50	58-62	10-14	.50(t)	...
		HX	...	60	415	...	...	...	...	...	.35-.75	2.00	.04	.04	2.50	64-68	15-19	.50(t)	...
		HP	...	62.5	430	34	235	4.5	...	...	.35-.75	2.00	.04	.04	2.50	33-37	24-28	.50(t)	...
ASTM A-351-76		CF3	S	70	485	30	205	35.0	...	...	.03	1.50	.040	.040	2.00	8-12	17-21	...	...
		CF3A(u)	S	77	530	35	240	35.0	...	...	.03	1.50	.040	.040	2.00	8-12	17-21	...	...
		CF8	S	70	485	30	205	35.0	...	...	.08	1.50	.040	.040	2.00	8-11	18-21	...	...
		CF8A(u)	S	77	530	35	240	35.0	...	...	.08	1.50	.040	.040	2.00	8-11	18-21	...	...
		CF3M	S	70	485	30	205	30.0	...	...	.03	1.50	.040	.040	1.50	9-13	17-21	2-3	...
		CF3MA(u)	S	80	550	37	255	30.0	...	...	.03	1.50	.040	.040	1.50	9-13	17-21	2-3	...
		CF8M	S	70	485	30	205	30.0	...	...	.08	1.50	.040	.040	1.50	9-12	18-21	2-3	...
		CF8C	S	70	485	30	205	30.0	...	...	.08	1.50	.040	.040	2.00	9-12	18-21	...	Cb(v)
		CH8	S	65	450	28	195	30.0	...	...	.08	1.50	.040	.040	1.50	12-15	22-26	...	...
		CH10	S	70	485	30	205	30.0	...	...	.10	1.50	.040	.040	2.00	12-15	22-26	...	...
		CH20	S	70	485	30	205	30.0	...	...	.20	1.50	.040	.040	2.00	12-15	22-26	...	...
		CK20	S	65	450	28	195	30.0	...	...	.30	1.50	.040	.040	1.75	19-22	23-27	...	...
		HK30	As cast	65	450	35	240	10.0	...	...	.25-.35	1.50	.040	.040	1.75	19-22	23-27	...	...
		HK40	As cast	62	425	35	240	10.0	...	...	.35-.45	1.50	.040	.040	1.75	19-22	23-27	...	...
		HT30	As cast	65	450	28	195	15.0	...	...	.25-.35	2.00	.040	.040	2.50	33-37	13-17	.50	...
		CF10MC	SA	70	485	30	205	20.0	...	...	.10	1.50	.040	.040	1.50	13-16	15-18	1.75-2.25	Cb(w)
		CN7M	SA	62	425	25	170	35.0	...	M. perm. 1.70(z)	.07	1.50	.040	.040	1.50	27.5-30.5	19-22	2-3	Cu 3-4
		CD4MCu	Q 1900 F (1040 C)	100	690	70	485	16.0	...	Stress-rupture(z)	.04	1.00	.040	.040	1.00	4.75-6.00	24.5-26.5	1.75-2.25	Cu 2.75-3.25
ASTM A447-74		I	As cast(x)	80(y)	550(y)	...	...	9	...	M. perm. 1.70(z)	.20-.45	2.50	.05	.05	1.75	10-14(aa)	23-28	...	N .20
		II	As cast(x)	80(bb)	550(bb)	...	...	4	...	Stress-rupture(z)	.20-.45	2.50	.05	.05	1.75	10-14(aa)	23-28	...	N .20
				20(cc)	140(cc)	...	...	8(cc)											

High-alloy Cast Steels (continued)

Specification and heat treatment			Mechanical properties (minimum unless range is given)							Chemical composition (maximum percent unless range is given)							
Specification	Grade	Heat treatment	Tensile strength ksi	Tensile strength MPa	Yield strength ksi	Yield strength MPa	Elongation in 2", %	Reduction of area, %	Other tests: bend, impact hardness	C	Mn	P	S	Si	Ni	Cr	Mo	Other elements
ASTM A494-76	CZ-100	As cast	50	340	18	125	10.0	...	...	1.00	1.50	0.03	0.03	2.00	95 min	...	...	Fe 3.0, Cu 1.25
	M-35	As cast	65	450	30	205	25.0	...	...	.35	1.50	0.03	0.03	2.00	Remainder	...	...	Fe 3.5, Cu 26.0-33.0
	N-12M-1	S.A.	76	520	46	315	6.0	...	...	.12	1.00	0.040	0.030	1.00	Remainder	1.00	26.0-30.0	Fe 4.0-6.0, V .20-60, Co 2.50
	N-12M-2	S.A.	76	520	46	315	20.0	...	...	.07	1.00	0.040	0.030	1.00	Remainder	1.0	30.0-33.0	Fe 3.0
	CY-40	As cast, S.A.	70	480	28	195	30.0	...	...	.40	1.50	0.03	0.03	3.00	Remainder	14.0-17.0	16.0-18.0	Fe 11.0
	CW-12M-1	S.A.	72	500	46	315	4.0	...	...	.12	1.00	0.040	0.030	1.00	Remainder	15.50-17.50	17.0-20.0	Fe 4.5-7.5, V .20-40, Co 2.5, W 3.75-5.25
	CW-12M-2	S.A.	72	500	46	315	25.0	...	Unnotched Charpy ft.-lb.@°F J@°C 50-77 78-25	.07	1.00	0.040	0.030	1.00	Remainder	17.0-20.0	...	Fe 3.0
ASTM A560-74	50 Cr-50Ni	As cast(dd)	80	550	50	340	5	...	...	.10	.30	.02	.02	1.00	Balance	48-52	...	N .30, Fe 1.0, Ti .50, Al .25
	60 Cr-40Ni	As cast(dd)	110	760	85	590	...	...	10-77 14-25	.10	.30	.02	.02	1.00	Balance	58-62	...	N .30, Fe 1.0, Ti .50, Al .25
Military MIL-S 16993A	1	NT	90	621	65	448	18	30	...	.15	1.00	.05	.05	1.50	1.00 max	11.5-14.0	.50 max	...
	2	NT	90	621	65	448	18	30	...	.15	1.00	.05	.05	.50	.65-1.0	11.5-14.0	.50-.70	...
Military MIL-S 867 A	I	Q	70	483	28	193	35	...	...	.08(ee)	1.50	.05	.05	2.00	8-11	18-21	...	...
	II	Q	70	483	30	207	30	...	...	.08	1.50	.05	.05	2.00	9-12	18-21	...	Cb + Ta 1.1(ff)
	III	Q	70	483	30	207	30	...	...	.08	1.50	.05	.05	2.00	9-12	18-21	2-3	...

(a)Section size precludes the use of all grades and the producer should be consulted as to grades practically obtainable for a particular design required. Final selection shall be by mutual agreement between manufacturer and purchaser. (b)Unless otherwise specified, Grade A will be supplied. (c)Supplementary bend test if required by customer. (d)When adequate weldability is stipulated, the silicon content may have to be lowered, in which case the minimum required yield strength shall be 25 ksi (180 MPa). (e)Supplementary intergranular corrosion test if specified by customer. (f)Chemical analysis is not normally required for the elements phosphorus, sulfur, and molybdenum, but if they are present in amounts over those stated, they may be cause for rejection. (g)Grade CF-8C shall have a columbium content of not less than 8 times the carbon content and not more than 1.0 percent. If a columbium-plus-tantalum alloy in the approximate Cb:Ta ratio of 3:1 is used for stabilizing this grade, the total columbium-plus-tantalum content shall not be less than 9 times the carbon content and shall not exceed 1.1 percent. (h)For free machining properties, the composition of Grade CF-16F may contain suitable combinations of selenium, phosphorus, and molybdenum (Grade CF-16F) or of sulfur and molybdenum (Grade CF-16Fa) as follows. Selenium, phosphorus, and molybdenum: selenium, percent, 0.20-0.35; phosphorus, maximum percent, 0.17; molybdenum, maximum percent, 1.50. Sulfur and molybdenum: sulfur, percent, 0.20-40; molybdenum, percent 0.40-0.80. Other combinations of elements for free-machining properties may be agreed upon by the manufacturer and the purchaser. (i)For the more severe general corrosive conditions, and when so specified, the carbon content shall not exceed 0.10 percent. This low-carbon grade shall be designated as Grade CH-10. (j)Refer to original specification for additional heat treatment information. (k)For Grade CB-30 a copper content of 0.90 to 1.20 percent is optional. (l)For purposes of determining conformance with this specification, the observed or calculated value for carbon content shall be rounded to the nearest 0.01 percent in accordance with the rounding method of Recommended Practice E 29. (m)For low ferrite or nonmagnetic castings of this grade the following values shall apply: tensile strength, minimum 65 ksi (450 MPa); yield point minimum, 28 ksi (195 MPa). (n)Supplementary corrosion test if specified by the customer. (o)Grade CF-8C shall have a columbium content of not less than 8 times the carbon content and not more than 1.0 percent. If a columbium-plus-tantalum alloy in the approximate Cb:Ta ratio of 3:1 is used for stabilizing this grade, the total columbium-plus-tantalum content shall be not less than 9 times the carbon content and shall not exceed 1.1 percent. (p)For purposes of determining conformance with this specification, the observed or calculated value for carbon content shall be rounded to the nearest 0.01 percent in accordance with the rounding method of Recommended Practice E 29. (q)As agreed upon by the manufacturer and the purchaser so as to develop acceptable corrosion resistance. (r)When adequate weldability is stipulated, the silicon content may have to be lowered, in which case the minimum yield strength shall be 26 ksi (180 MPa). (s)Refer to original specification for additional heat treatment information. (t)Castings having a specified molybdenum range agreed upon by the manufacturer and the purchaser may also be furnished under these specifications. (u)The properties shown are obtained by adjusting the composition within the limits shown to obtain a ferrite-austenite ratio that will result in the higher ultimate and yield strengths indicated. Because of the thermal instability of Grades CF3A, CF3MA, and CF8A, they are not recommended for service at temperatures in excess of 800° F (425 °C). (v)Grade CF8C shall have a columbium content of not less than 8 times the carbon content but not over 1.00 percent. (w)Grade CF10MC shall have a columbium content of not less than 10 times the carbon content but not over 1.20 percent. (x)Heat treatment as agreed upon by manufacturer and purchaser. (y)Properties after aging. Additionally, the short time, high temperature tensile property requirements are to be agreed upon by manufacturer and purchaser. (z)Refer to original specification for details. Note that out of the 4 tests: tension after aging, magnetic permeability, stress rupture, and short time high-temperature, the purchaser shall specify no more than two tests! (aa)Commercial nickel usually carries a small amount of cobalt, and within the usual limits cobalt shall be counted as nickel. (bb)Properties after aging. (cc)Short time, high temperature tensile properties. (dd)Heat treatment as agreed upon by manufacturer and purchaser. (ee)If chromium is over 20 percent and nickel is over 10 percent, a maximum carbon content of .12 will be permitted. (ff)Columbium or columbium plus tantalum shall be not less than 10 times the carbon content and not more than 1.10 percent (tantalum shall not exceed .4 times the sum of the columbium and tantalum content) or titanium content shall be not less than 6 times the carbon content and not more than .75 percent.

Abbreviations

AA	Aluminum Association
ACI	Alloy Casting Institute
AISI	American Iron and Steel Institute
Ann	Anneal or annealed
AOD	Argon-oxygen-decarburization
A Rock	Hardness Rockwell A scale
ASTM	American Society for Testing and Materials
B	Flux density
Bal	Balance or remainder
BHN	Brinell hardness number
BOF	Basic oxygen furnace
Br	Remanence flux in gausses
Brin	Brinell hardness number
BS	British standard
CD	Cold drawn
CDA	Copper Development Association
CEL	Elastic limit in compression in pounds per square inch
CF	Cold finished
Ch	Charpy impact in foot pounds
cmf	Circular mil feet
Coef Exp	Coefficient of expansion per °C
Coef Res	Temperature coefficient of resistance per °C
CS	Crushing strength in pounds per square inch
CUS	Ultimate strength in compression in pounds per square inch
CYP	Yield point in compression in pounds per square inch
DCEN	Direct current, electrode negative
DCEP	Direct current, electrode positive
DPH	Depth penetration hardness (Vickers hardness)
DVM	Double vacuum melted
EFM	Electroflux melting
El	Elongation in percent
El Res	Electrical resistivity in microhms per centimeter cube (20 °C)
ESR	Electroslag melting
H	Magnetizing field strength in oersteds
Hc	Coercive force in gilberts per centimeter or oersteds
HSLA	High strength low alloy (steels)
Ht	Heat treated
Ht Tr	Heat treated
Iz	Izod impact value in foot pounds
kg/mm²	Kilograms per square millimeter
kg-fmm²	Kilograms force per square millimeter
ksi	Thousand pounds per square inch
ME	Modulus of elasticity in pounds per square inch
MP	Melting point
MPa	Megapascals (see chart in Appendix)
N/mm²	Newtons per square millimeter
N & T	Normalized and tempered
PM	Permament mold
ppm	Parts per million
psi	Pounds per square inch
Q & T	Quenched and tempered
RA	Reduction of area in percent
Ra or R_a	Hardness Rockwell A scale
Rb	Hardness Rockwell B scale
Rc	Hardness Rockwell C scale
SAE	Society of Automotive Engineers

SC	Sand cast
SS	Shear strength in pounds per square inch
STA	Solution treated and aged
TC	Total carbon content
Tr D	Transverse deflection in inches
Tr S	Transverse strength in pounds
TS	Tensile strength in pounds per square inch
tsi	Tons per square inch
UTS	Ultimate tensile strength
VAM	Vacuum arc melting
VDH	Vickers diamond hardness
VIM	Vacuum induction melting
W.Nr.	Werkstoff number (German standard)
YP	Yield point in tension
YS	Yield strength in tension

Sci Ref TA 483 .W64 1979
Woldman, Norman Emme, 1899-
   1969.
Woldman's Engineering alloys

JUN 5 1986